FEATURES AND BENEFITS

Algebra 2 ©2005

		See page(s):
Curriculum and Instructional Design	. . . presents a coherent curriculum that effectively organizes and integrates important mathematical ideas.	
	• Chapters are grouped by units to bring depth to algebra concepts.	iii
	• Lessons are divided into two related objectives to allow teachers flexibility in presenting the lesson.	96
Student Support	. . . is provided throughout the text to help all students succeed in algebra.	
	• Foldables™ Study Organizers help students actively organize key concepts and create their own review materials.	55
	• Key Concept and Concept Summary boxes help students identify main concepts.	75, 92
	• Study Tips in the margins help students understand new material.	12, 22
	• Homework Help in the margin of the exercise sets links homework exercises to corresponding examples within the lesson.	66
Reading / Writing In Mathematics	. . . strategies and activities are essential for student success in mathematics.	
	• Reading Math Study Tips help clarify mathematical terms.	82, 124
	• Practice in vocabulary usage in each lesson and at the end of each chapter builds reading and writing skills.	145, 276
	• Writing in Math exercises require students to summarize what they have learned in the lesson.	275
Daily Intervention	. . . opportunities are provided throughout the program.	
	• Prerequisite Skills at the beginning of each chapter and in each lesson assess student readiness.	285, 319
	• The Student Handbook contains review and practice of prerequisite skills.	813–827
	• Daily Intervention features provide suggestions for addressing various learning styles and helping students who are having difficulty.	65, 141
	• A variety of Online Study Tools are readily accessible to students.	1
Test Preparation and Assessment	. . . provides targeted practice for local, state, and national tests.	
	• Standardized Test Practice questions appear in each lesson.	122
	• Standardized Test Practice Examples help students learn how to approach test questions.	76
	• Two pages of Standardized Test Practice at the end of each chapter include multiple-choice, short-response/grid-in, and extended response questions.	106–107, 406–407
	• Preparing for Standardized Tests includes examples and practice to help students become better test takers.	877–892
	• Interactive Standardized Test Practice is available in the Online Study Tools.	1
Staff Development	. . . features are available to assist new teachers and those teaching outside of their primary subject area.	
	• Mathematical Connections and Background provides an overview of the mathematics in the chapter and links to prior knowledge and future topics.	4C, 4D
	• Building on Prior Knowledge links what students have previously learned to the content of the current lesson.	68
	• Tips for New Teachers provide helpful suggestions for classroom management, teaching techniques, and assessment.	62
Education Partnership	. . . strengthens the relevance of applications and projects.	
	• USA TODAY Snapshots® provide current topics and data in graphs, charts, and tables and enhance the unit WebQuest projects.	3, 135

"Sticky Notes" in Chapter 1 provide a "walk-through" of key features. pp. 4–65

GLENCOE MATHEMATICS

Algebra 2

Teacher Wraparound Edition

Teacher Handbook

Holliday
Cuevas
Moore-Harris
Carter
Marks
Casey
Day
Hayek

Glencoe
McGraw-Hill

New York, New York
Columbus, Ohio
Chicago, Illinois
Peoria, Illinois
Woodland Hills, California

McGraw Hill Glencoe

The McGraw·Hill Companies

Send all inquiries to:
Glencoe/McGraw-Hill
8787 Orion Place
Columbus, OH 43240

ISBN: 0-07-865609-5 (Student Edition)
ISBN: 0-07-865610-9 (Teacher Edition)

3 4 5 6 7 8 9 10 055/071 12 11 10 09 08 07 06 05

Contents in Brief

Unit 1 First Degree Equations and Inequalities 2

- **Chapter 1** Solving Equations and Inequalities 4
- **Chapter 2** Linear Relations and Functions 54
- **Chapter 3** Systems of Equations and Inequalities 108
- **Chapter 4** Matrices 152

Unit 2 Polynomial and Radical Equations and Inequalities 218

- **Chapter 5** Polynomials 220
- **Chapter 6** Quadratic Functions and Inequalities 284
- **Chapter 7** Polynomial Functions 344

Unit 3 Advanced Functions and Relations 408

- **Chapter 8** Conic Sections 410
- **Chapter 9** Rational Expressions and Equations 470
- **Chapter 10** Exponential and Logarithmic Relations 520

Unit 4 Discrete Mathematics 574

- **Chapter 11** Sequences and Series 576
- **Chapter 12** Probability and Statistics 630

Unit 5 Trigonometry 696

- **Chapter 13** Trigonometric Functions 698
- **Chapter 14** Trigonometric Graphs and Identities 760

Authors

Berchie Holliday, Ed.D.
Former Mathematics Teacher
Northwest Local
School District
Cincinnati, OH

Gilbert J. Cuevas, Ph.D.
Professor of Mathematics
Education
University of Miami
Miami, FL

Beatrice Moore-Harris
Educational Specialist
Bureau of Education
and Research
League City, TX

John A. Carter, Ph.D.
Director of Mathematics
Adlai E. Stevenson
High School
Lincolnshire, IL

Consulting Author

Carol Malloy, Ph.D.
Associate Professor, Curriculum Instruction,
Secondary Mathematics
The University of North Carolina at Chapel Hill
Chapel Hill, NC

Authors

Daniel Marks, Ed.D.
Associate Professor of Mathematics
Auburn University at Montgomery
Montgomery, AL

Ruth M. Casey
Mathematics Teacher Department Chair
Anderson County High School
Lawrenceburg, KY

Roger Day, Ph.D.
Associate Professor of Mathematics
Illinois State University
Normal, IL

Linda M. Hayek
Mathematics Teacher
Ralston Public Schools
Omaha, NE

Contributing Authors

USA TODAY
The USA TODAY Snapshots®, created by USA TODAY®, help students make the connection between real life and mathematics.

Dinah Zike
Educational Consultant
Dinah-Might Activities, Inc.
San Antonio, TX

Content Consultants

Each of the Content Consultants reviewed every chapter and gave suggestions for improving the effectiveness of the mathematics instruction.

Mathematics Consultants

Gunnar E. Carlsson, Ph.D.
Consulting Author
Professor of Mathematics
Stanford University
Stanford, CA

Ralph L. Cohen, Ph.D.
Consulting Author
Professor of Mathematics
Stanford University
Stanford, CA

Alan G. Foster
Former Mathematics Teacher & Department Chairperson
Addison Trail High School
Addison, IL

Les Winters
Instructor
California State University Northridge
Northridge, CA

William Collins
Director, The Sisyphus Math Learning Center
East Side Union High School District
San Jose, CA

Dora Swart
Mathematics Teacher
W.F. West High School
Chehalis, WA

David S. Daniels
Former Mathematics Chair
Longmeadow High School
Longmeadow, MA

Mary C. Enderson, Ph.D.
Associate Professor of Mathematics
Middle Tennessee State University
Murfreesboro, TN

Gerald A. Haber
Consultant, Mathematics Standards and Professional Development
New York, NY

C. Vincent Pané, Ed.D.
Associate Professor of Education/ Coordinator of Secondary & Special Subjects Education
Molloy College
Rockville Centre, NY

Reading Consultant

Lynn T. Havens
Director
Project CRISS
Kalispell, MT

ELL Consultant

Idania Dorta
Mathematics Educational Specialist
Miami-Dade County Public Schools
Miami, FL

Teacher Reviewers

Each Teacher Reviewer reviewed at least two chapters of the Student Edition, giving feedback and suggestions for improving the effectiveness of the mathematics instruction.

Yvonne Adonai
Assistant Principal, Mathematics
Middle College at Medgar Evers College
Brooklyn, NY

Ann Rushing Allred
Secondary Mathematics Coordinator
Bossier Parish Schools
Bossier City, LA

Thomas J. Altonjy
Mathematics Supervisor
Montville Township Public Schools
Montville, NJ

Susan J. Barr
Department Chair/Teacher
Dublin Coffman High School
Dublin, OH

Douglas W. Becker
Math Dept Chair/Senior Math Teacher
Gaylord High School
Gaylord, MI

Dr. Edward A. Brotak
Professor, Atmospheric Sciences
UNC Asheville
Asheville, NC

Teacher Reviewers

Sonya Smith Bryant
Mathematics Teacher
Booker T. Washington High School
Shreveport, LA

Judy Buchholtz
Math Department Chair/Teacher
Dublin Scioto High School
Dublin, OH

A. G. Chase
Mathematics Teacher
Evergreen High School
Vancouver, WA

Natalie Dillinger
Mathematics Teacher
Hurricane High School
Hurricane, WV

John M. Dunford, Jr.
Chairman Mathematics
Tuba City High School
Tuba City, AZ

Diana Flick
Mathematics Teacher
Harrisonburg High School
Harrisonburg, VA

Susan Hammer
Mathematics Department Head
Gaither High School
Tampa, FL

Deborah L. Hewitt
Mathematics Teacher
Chester High School
Chester, NY

Kristen L. Karbon
Mathematics Teacher
Troy High School
Troy, MI

William Leschensky
Former Mathematics Teacher
Glenbard South High School
College of DuPage
Glen Ellyn, IL

Patricia Lund
Mathematics Teacher
Divide County High School
Crosby, ND

Wallace J. Mack
Mathematics Department Chairperson
Ben Davis High School
Indianapolis, IN

T. E. Madre
Mathematics Department Chairperson
North Mecklenburg High School
Huntersville, NC

Marilyn Martau
Mathematics Teacher (Retired)
Lakewood High School
Lakewood, OH

Ron Millard
Mathematics Department Chair
Shawnee Mission South High School
Overland Park, KS

Rebecca D. Morrisey
Assistant Principal
Leavenworth High School
Leavenworth, KS

Constance D. Mosakowsky
Mathematics Teacher
Minnie Howard School
Alexandria, VA

Anne Newcomb
Mathematics Department Chairperson
Celina High School
Celina, OH

Barbara Nunn
Secondary Mathematics Curriculum Specialist
Broward County Schools
Ft. Lauderdale, FL

Shannon Collins Pan
Department of Mathematics
Waverly High School
Waverly, NY

Aletha T. Paskett
Mathematics Teacher
Indian Hills Middle School
Sandy, UT

Holly K. Plunkett
Mathematics Teacher
University High School
Morgantown, WV

Thomas M. Pond, Jr.
Mathematics Teacher
Matoaca High School
Chesterfield County Public Schools, VA

Debra K. Prowse
Mathematics Teacher
Beloit Memorial High School
Beloit, WI

B. J. Rasberry
Teacher
John T. Hoggard High School
Wilmington, NC

Harry Rattien
A.P. Supervisor (Math)
Townsend Harris High School at QC
Flushing, NY

Becky Reed
Teacher
John F. Kennedy High School
Mt. Angel, OR

Steve Sachs
Mathematics Department Chairperson
Lawrence North High School
Indianapolis, IN

Sue W. Sams
Mathematics Teacher/Department Chair
Providence High School
Charlotte, NC

Calvin Stuhmer
Mathematics Teacher
Sutton Public Schools
Sutton, NE

Ruth Stutzman
Math Department Chair & Teacher
Jefferson Forest High School
Forest, VA

Patricia Taepke
Mathematics Teacher and BTSA Trainer
South Hills High School
West Covina, CA

Christine Waddell
Mathematics Department Chair/Teacher
Albion Middle School
Sandy, UT

Gail Watson
Mathematics Teacher
Pineville High School
Pineville, LA

Linda E. Westbrook
Mathematics Department Chair
George Jenkins High School
Lakeland, FL

Cottina Woods
Lane Technical High School
Chicago, IL

Warren Zarrell
Mathematics Department Chairman
James Monroe High School
North Hills, CA

Teacher Advisory Board

Glencoe/McGraw-Hill wishes to thank the following teachers for their feedback on Glencoe *Algebra*. They were instrumental in providing valuable input toward the development of this program.

Mary Jo Ahler
Mathematics Teacher
Davis Drive Middle School
Apex, NC

David Armstrong
Mathematics Facilitator
Huntington Beach Union High School District
Huntington Beach, CA

Berta Guillen
Mathematics Department Chairperson
Barbara Goleman Sr. High School
Miami, FL

Bonnie Johnston
Academically Gifted Program Coordinator
Valley Springs Middle School
Arden, NC

JoAnn Lopykinski
Mathematics Teacher
Lincoln Way East High School
Frankfort, IL

David Lorkiewicz
Mathematics Teacher
Lockport High School
Lockport, IL

Norma Molina
Ninth Grade Success Initiative Campus Coordinator
Holmes High School
San Antonio, TX

Sarah Morrison
Mathematics Department Chairperson
Northwest Cabarrus High School
Concord, NC

Raylene Paustian
Mathematics Curriculum Coordinator
Clovis Unified School District
Clovis, CA

Tom Reardon
Mathematics Department Chairperson
Austintown Fitch High School
Youngstown, OH

Guy Roy
Mathematics Coordinator
Plymouth Public Schools
Plymouth, MA

Jenny Weir
Mathematics Department Chairperson
Felix Verela Sr. High School
Miami, FL

Teacher Handbook

Table of Contents

Dynamic Instructional Design T2
Time-Saving Teacher Resources T4
Reading and Writing T6
Intervention T8
Assessment T10
USA TODAY® T12
Internet Resources T13
Foldables™ T14
Meeting Student Needs T15
Staff Development T16
Program Validation T18
Pacing T20
NCTM Principles and Standards T22

Designed to be in more

The consistent instructional design of the lessons strengthens student learning—from the very beginning of each lesson where students see "What You'll Learn," to the very end of each lesson where they have a chance to prepare for the next lesson.

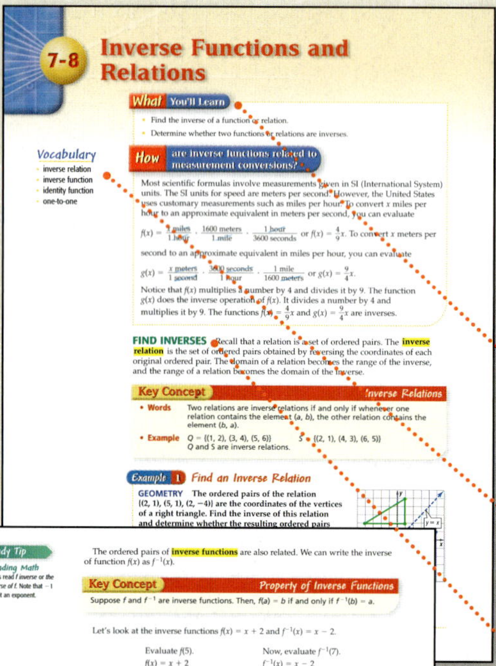

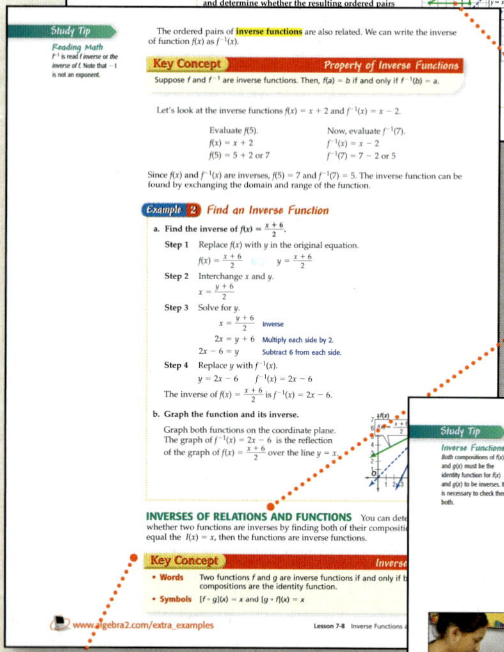

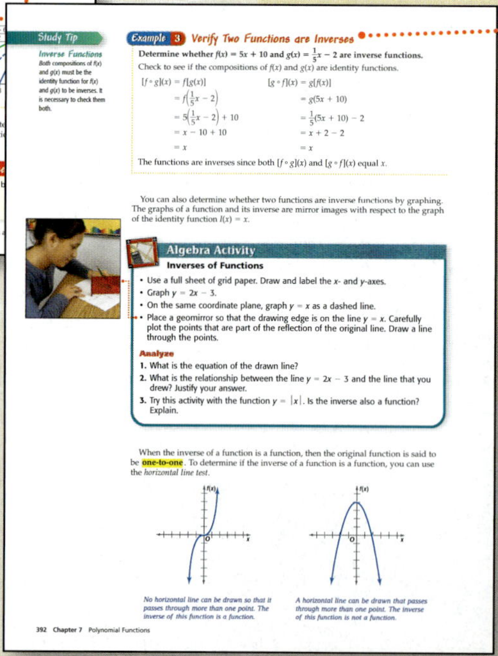

- **What You'll Learn** at the beginning of each lesson previews the topics to come, and engaging **Why** or **How** questions help students see why the concepts are relevant.
- ***Vocabulary*** terms are listed at the beginning of each lesson for easy reference, and they're also highlighted in yellow as they appear within the lesson.
- **Two-part lesson structure** gives you the flexibility to teach the two related lesson objectives together or separately. The Chapter Resource Masters contain a Study Guide and Intervention Master for each objective.

Completely worked-out examples with clear explanations are paralleled by the **Guided Practice** and **Practice and Apply** exercises that follow. Examples often include strategies for problem-solving and mastering standardized test questions.

Key Concept boxes use words, symbols, models, and examples to illustrate new rules, properties, and definitions, so students can build their reading skills as they build their math skills. **Concept Summary** boxes provide a concise overview of key topics.

effective,

ways than one.

Check for Understanding

You can use this portion of exercises in class to ensure that all students understand the concepts.

- *Concept Check* exercises give students opportunities to define, describe, and explain the mathematical concepts they've just learned.
- *Guided Practice* presents a representative sample of the exercises in the Practice and Apply section. A key is provided in the **Teacher Wraparound Edition** that correlates the exercises with appropriate examples.
- *Application* problems give students the opportunity to use the skills they have learned in a real-world setting.

Practice and Apply

- **Skill Exercises** correspond to the Guided Practice exercises and are structured so that students practice the same concepts whether they are assigned odd- or even-numbered problems. Homework Help is provided so students can refer to examples in the lesson as they complete the exercises.
- **Applications** give students frequent opportunities to apply concepts to both real-life and mathematical situations.
- **CRITICAL THINKING** exercises in each lesson require students to explain, make conjectures, and prove mathematical relationships.
- *Standardized Test Practice* questions provide students with ongoing opportunities to sharpen their test-taking skills.

Maintain Your Skills

- *Mixed Review* includes spiraled, cumulative exercises from the two previous lessons as well as earlier lessons.
- *Getting Ready for the Next Lesson* exercises give students the chance to preview prerequisite skills for the coming lesson. A reference is provided should students need additional help.

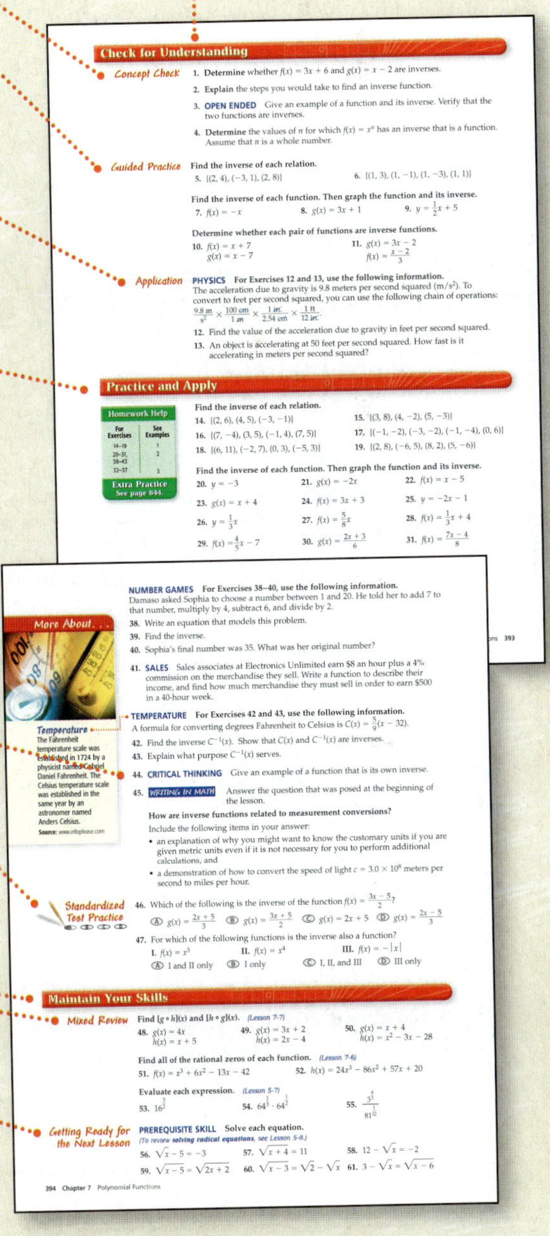

Accomplish more

Glencoe Algebra 2 provides so many resources for lesson planning and teaching that you can create a complete, customized course in Algebra 2 quickly . . . and easily.

This is where you start.

The **Teacher Wraparound Edition** is your key to all of the teaching resources in *Glencoe Algebra 2.* In addition to teaching suggestions, additional examples, and answers, the Teacher Wraparound Edition provides a guide for all of the print and software materials available for each lesson.

FAST FILE Chapter Resource Masters contain all of the core supplements you'll need to begin teaching a chapter of *Glencoe Algebra 2.* Each chapter booklet features convenient tabs for easy filing.

- **Vocabulary Builder** helps students locate and define key vocabulary words from the chapter.
- **Study Guide and Intervention** for each objective summarizes key concepts and provides practice.
- **Skills Practice** provides ample exercises to help students develop basic computational skills, lesson by lesson.
- **Practice** mimics the computational and verbal problems in each lesson at an average level.
- **Reading to Learn Mathematics** provides students with various reading strategies to master the mathematics presented in each lesson.
- **Enrichment** activities extend students' knowledge and widen their appreciation of how mathematics relates to the world around them.
- **Assessment** options for each chapter include six forms of chapter tests, assessment tasks, quizzes, mid-chapter test, cumulative review, and standardized test practice.

Reading and Writing

WebQuest and Project Resources include teacher notes and answers for the Internet WebQuest projects, as well as other long-term projects that can be used with *Glencoe Algebra 2.*

Reading and Writing in the Mathematics Classroom features suggestions and activities for including reading as an integral part of the mathematics curriculum, as well as differentiated approaches to teaching mathematics that promote English learning and inclusion.

FOLDABLES™ Study Organizer

Teaching Mathematics with Foldables™ offers guidelines for using Foldables interactive study organizers in your class. The booklet was written by Foldables creator Dinah Zike.

More information on options for reading and writing in Glencoe Algebra 2 is available on pages T6–T7.

Applications

Science and Mathematics Lab Manual includes lab activity masters and teaching suggestions for integrating science into the mathematics classroom.

School-to-Career Masters feature activities that show how mathematics relates to various careers.

Graphing Calculator and Spreadsheet Masters include activities to incorporate the TI-83 Plus calculator and spreadsheets into your Algebra 2 course.

Real-World Transparencies and Masters feature colorful transparencies with accompanying student worksheets to show how mathematics relates to real-world topics.

than you'd ever imagine

in less time than you'd ever believe.

Assessment and Intervention

5-Minute Check Transparencies with Standardized Test Practice include a transparency for each lesson that evaluates what students have learned in the previous lesson. Each transparency also includes a standardized test practice question.

Closing the Gap for Absent Students provides an easy-to-use summary of all the materials you have covered in the chapter in a format that can be posted or distributed to students who have missed class.

DAILY INTERVENTION **Guide to Daily Intervention** offers suggestions for daily assessment and tips on how to help students succeed.

Staff Development

Answer Key Transparencies provide answers to Student Edition exercises.

Lesson Planning Guide features a daily resource guide for planning your curriculum, as well as pacing for block scheduling.

Solutions Manual includes completely worked-out solutions for all exercises in the Student Edition.

Using the Internet in the Mathematics Classroom provides guidelines for using the Internet, as well as a guide to additional mathematics resources available on the Internet.

Teaching Algebra with Manipulatives features activities and teaching suggestions to help you present algebraic concepts with manipulatives and hands-on materials.

Technology Support for Teachers

Glencoe offers many timesaving software products to help you develop creative classroom presentations . . . fast.

TeacherWorks **All-in-One Lesson Planner and Resource Center** CD-ROM includes a lesson planner and interactive Teacher Edition, so you can customize lesson plans and reproduce classroom resources quickly and easily, from just about anywhere.

Answer Key Maker software allows you to customize answer keys for your assignments from the Student Edition exercises.

Interactive Chalkboard CD-ROM includes fully worked-out examples, the 5-Minute Check Transparencies, and Your Turn problems in a customizable Microsoft® PowerPoint® format.

And more . . . *Additional technology products and Internet resources for students, teachers, and parents are discussed on pages T6–T13 and T17.*

HELP your students become fluent

Glencoe Algebra 2 makes it easy for you to incorporate constructive reading and writing strategies into every class you teach.

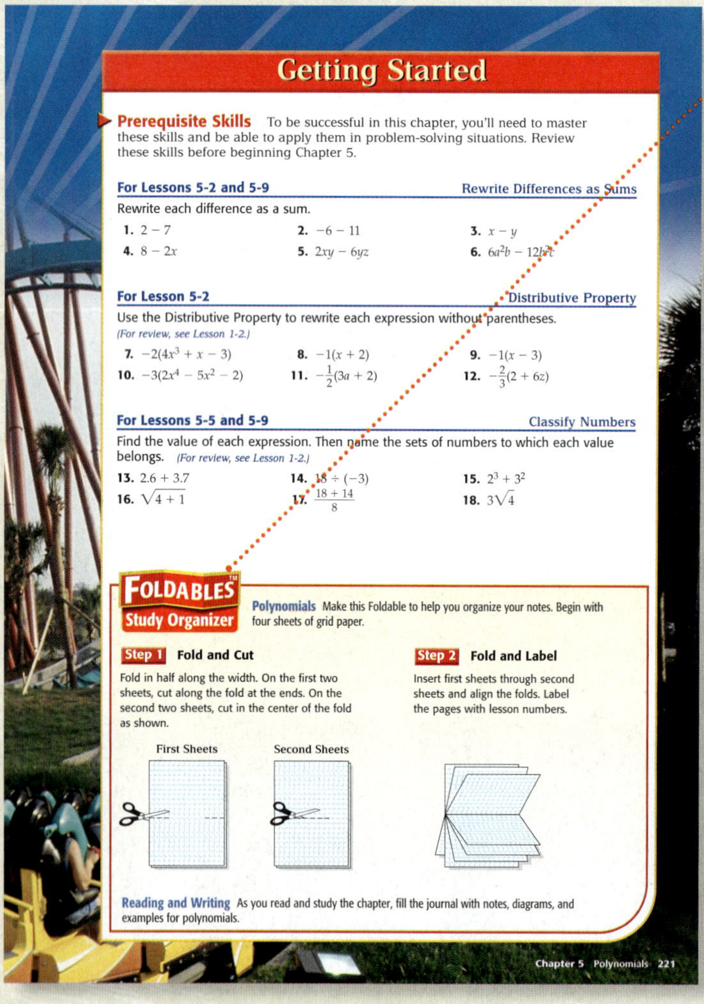
Getting Started

Prerequisite Skills To be successful in this chapter, you'll need to master these skills and be able to apply them in problem-solving situations. Review these skills before beginning Chapter 5.

For Lessons 5-2 and 5-9 — Rewrite Differences as Sums

Rewrite each difference as a sum.

1. $2 - 7$
2. $-6 - 11$
3. $x - y$
4. $8 - 2x$
5. $2xy - 6yz$
6. $6a^2b - 12b^2c$

For Lesson 5-2 — Distributive Property

Use the Distributive Property to rewrite each expression without parentheses. *(For review, see Lesson 1-2.)*

7. $-2(4x^3 + x - 3)$
8. $-1(x + 2)$
9. $-1(x - 3)$
10. $-3(2x^4 - 5x^2 - 2)$
11. $-\frac{1}{2}(3a + 2)$
12. $-\frac{2}{3}(2 + 6z)$

For Lessons 5-5 and 5-9 — Classify Numbers

Find the value of each expression. Then name the sets of numbers to which each value belongs. *(For review, see Lesson 1-2.)*

13. $2.6 + 3.7$
14. $18 \div (-3)$
15. $2^3 + 3^2$
16. $\sqrt{4 + 1}$
17. $\frac{18 + 14}{8}$
18. $3\sqrt{4}$

FOLDABLES™ Study Organizer

Polynomials Make this Foldable to help you organize your notes. Begin with four sheets of grid paper.

Step 1 Fold and Cut Fold in half along the width. On the first two sheets, cut along the fold at the ends. On the second two sheets, cut in the center of the fold as shown.

First Sheets — Second Sheets

Step 2 Fold and Label Insert first sheets through second sheets and align the folds. Label the pages with lesson numbers.

Reading and Writing As you read and study the chapter, fill the journal with notes, diagrams, and examples for polynomials.

Chapter 5 Polynomials 221

Student Edition

Foldables™ Study Organizers at the beginning of each chapter provide students with tools for organizing what they are reading and studying.

Reading Math Study Tips appear throughout each chapter, to help students learn and use the language of algebra.

Writing in Math questions in every lesson require students to use critical thinking skills to develop their answers.

Vocabulary terms are listed at the beginning of each lesson and highlighted when defined. The **Vocabulary and Concept Check** in each Study Guide and Review checks students' understanding of the key concepts of the chapter.

Key Concepts are illustrated using Words, Symbols, Models, and Examples, as appropriate. This approach improves reading comprehension by using multiple representations.

WebQuest Internet Projects are long-term projects that use problem-based learning to give students the chance to develop their research and creative writing skills.

in the Language of MATHEMATICS.

Teacher Wraparound Edition

Study Notebook suggestions provide motivational ideas to help students create study notebooks that are thorough and effective.

Concept Check questions require students to describe, write, and explain the mathematical concepts they have learned in each lesson.

Modeling, Speaking, and **Writing** in every lesson require students to summarize what they have learned by responding to open-ended prompts.

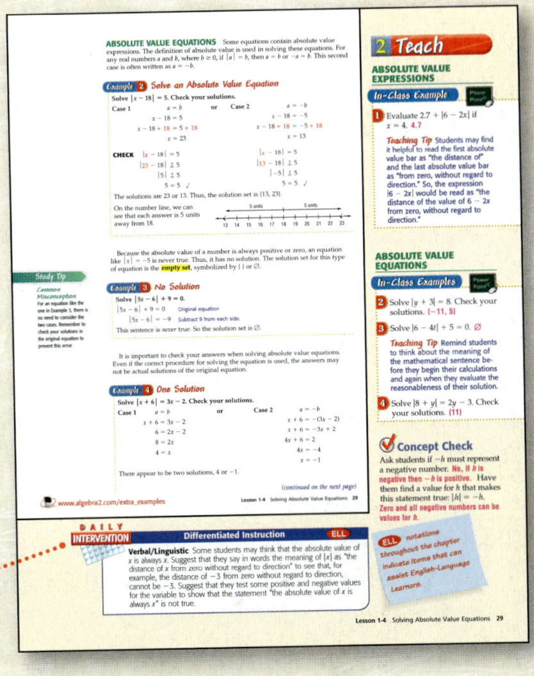

ELL Resources highlight features and activities that help English-Language Learners grasp content.

Differentiated Instruction features help students at all points on the learning spectrum develop their reading, writing, and comprehension skills.

Technology Support

StudentWorks™ Glencoe's backpack solution, includes the entire Student Edition, formatted like the hardbound book, so students can study from just about anywhere—no book required. Students can also print their own lesson worksheet pages and get instant access to interactive web resources.

www.algebra2.com/vocabulary_review is a Glencoe site that provides online study tools for reviewing the vocabulary of each chapter.

Vocabulary PuzzleMaker software creates crossword, jumble, and word search puzzles using vocabulary lists that you can customize.

Multimedia Applications: Virtual Activities CD-ROM provides in-depth interactive activities that help students explore the main concepts of each chapter in a real-world setting.

Additional Resources

Chapter Resource Masters

- Vocabulary Builder
- Reading to Learn Mathematics

Teaching Mathematics with Foldables™

Reading and Writing in the Mathematics Classroom

WebQuest and Project Resources

For more information on these products, see pp. T4–T5.

With these TOOLS, you'll always know

Whether you need daily intervention resources integrated right into the program, or supplemental materials for after school and summer school programs, *Glencoe Algebra 2* puts it all right at your fingertips!

Prerequisite Skills

Students often struggle in Algebra because they have not mastered the prerequisite skills needed to be successful. *Glencoe Algebra 2* provides several opportunities to check student skills and determine which students need additional review and practice.

- The **Prerequisite Skills** at the beginning of every chapter help students identify and practice the skills they'll need for each new concept.
- Additional **prerequisite skills practice** is provided at the end of each lesson and includes page references to help students get extra review whenever they need it. More prerequisite skill practice appears in the Student Handbook section at the back of the Student Edition.

The **Student Edition** contains additional problems to help students master each lesson before completing the chapter assessment.

- **Extra Practice**, located in the back of the Student Edition, provides additional, immediate practice with the concepts from each lesson.
- **Mixed Problem Solving**, also in the back of the Student Edition, includes numerous verbal problems to help students reinforce their problem-solving skills.

Daily Intervention Opportunities

Guide to Daily Intervention offers suggestions for using Glencoe materials to intercept students who are having difficulties and prescribe a system of reinforcement to promote student success.

The **Chapter Resource Masters** include several types of worksheets that can be used for daily intervention in each lesson. For a description of each worksheet, see page T4.

- **Study Guide and Intervention***
- **Skills Practice***
- **Practice***
- **Reading to Learn Mathematics**

** Each of these types of worksheets is available as a* ***consumable workbook****.*

who needs EXTRA HELP.

And you'll be able to *DELIVER* it.

Technology Resources for Intervention

In addition to print resources, Glencoe offers a variety of timesaving technology tools to help students build their math skills more effectively.

Alge2PASS: Tutorial Plus CD-ROM provides an interactive, self-paced tutorial for a complete Algebra 2 curriculum. The 28 lessons are correlated directly to *Glencoe Algebra 2*. Each lesson, or concept, includes a pretest, tutorial, guided practice, and posttest. Students' answers to the pretests automatically determine whether they need the tutorial for each concept, so students can take responsibility for their own learning—without taking teacher time for grading.

Online Study Tools include comprehensive review and intervention tools that are available anytime, anyplace simply by logging on to

www.algebra2.com.

Self-check quizzes are available for every lesson, and immediate feedback helps students check their progress and find specific pages and examples in the Student Edition whenever they need extra review. These Online Study Tools also include extra examples, chapter tests, standardized test practice, and vocabulary review.

ALEKS® is an online, intuitive, individualized tutor that students can take anywhere. This artificial intelligence-based system analyzes student answers and targets what the student is prepared to learn next. ALEKS is available by subscription only on the Internet.

www.k12.aleks.com

Additional Teacher Resources

The following materials are available to help you determine which students need intervention and allow you to develop strategies for giving students the help they need. For a description of each feature, see page T5.

- **5-Minute Check Transparencies with Standardized Test Practice**
- **Daily Intervention** features in the Teacher Wraparound Edition
- **Closing the Gap for Absent Students**

Give ASSESSMENT

Glencoe Algebra 2 gives you all the tools you need to prepare students for success–including Standardized Test Practice in each lesson and the powerful ExamView®Pro CD-ROM.

Student Edition

Every lesson contains two Standardized Test Practice questions, and every chapter contains a completely worked-out standardized test example as well as two full pages of Standardized Test Practice with Test-Taking Tips.

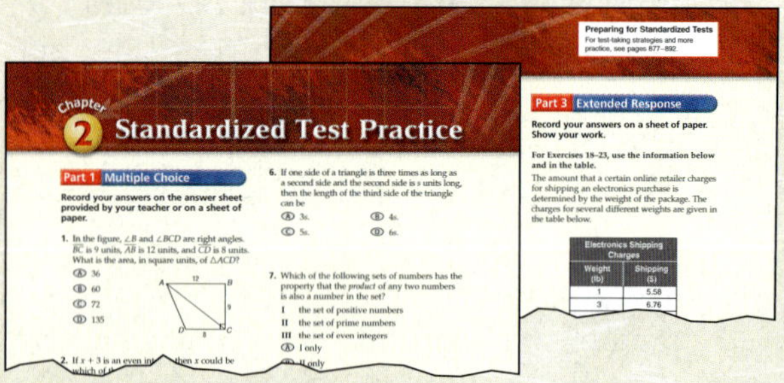

Preparing for Standardized Tests is designed to help your students become better test-takers. Included are examples and practice for the types of questions and concepts commonly seen on standardized tests.

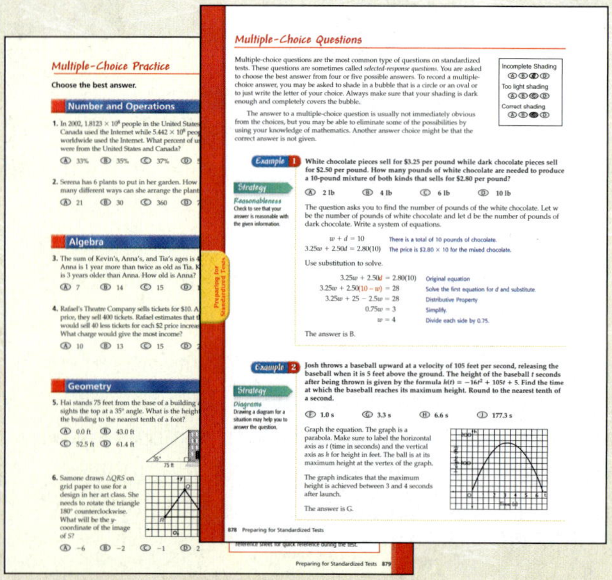

Chapter Study Guide and Review provides Vocabulary and Concept review—a Glencoe exclusive—and Lesson-by-Lesson Review, all at the point of use for students.

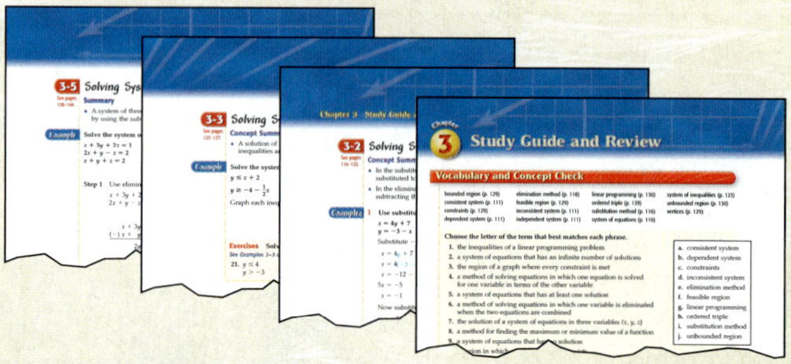

Practice Quizzes (2 per chapter) and a **Practice Test** for each chapter provide the variety of practice questions students need to succeed on tests.

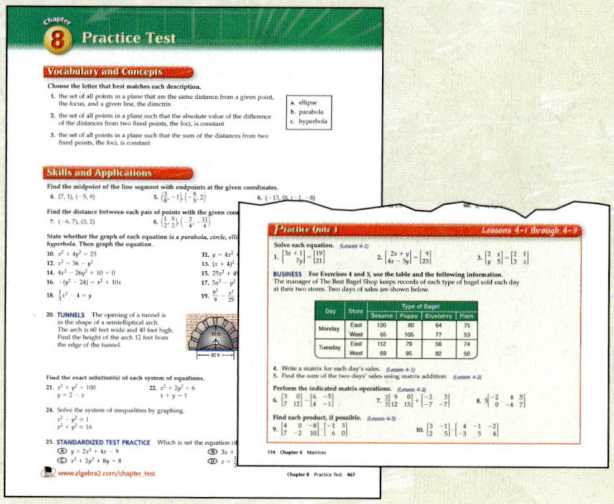

Teacher Wraparound Edition

An **Open-Ended Assessment** activity is provided in each lesson in the margin of the Teacher Wraparound Edition.

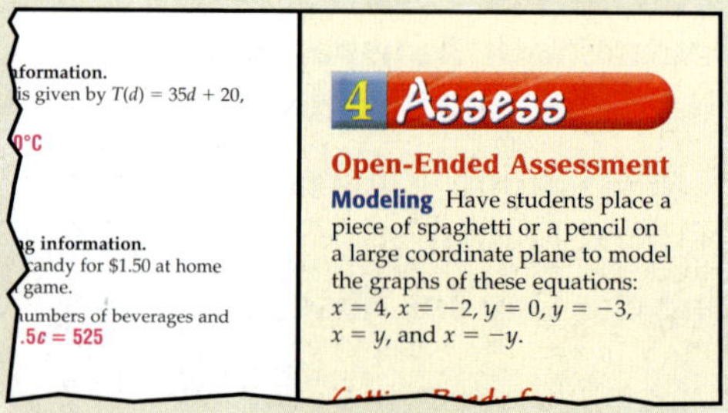

the extra attention it deserves, without the extra prep time.

Teacher Classroom Resources

5-Minute Check Transparencies with Standardized Test Practice provide full-size transparencies with questions covering the previous lesson or chapter. Standardized Test Practice Questions are also included.

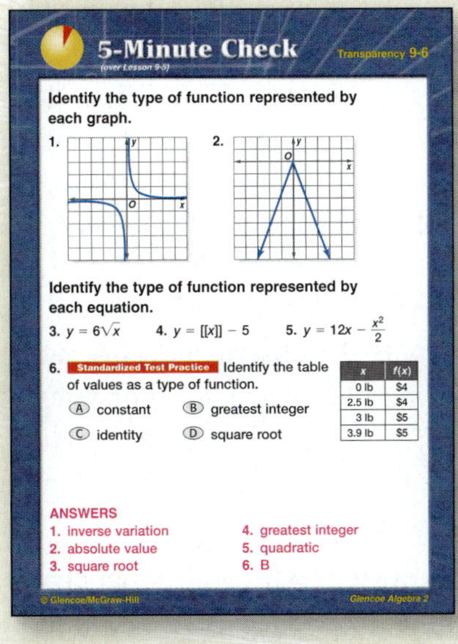

Technology Support

Use the networkable **ExamView® Pro** to:

- Create **multiple versions** of tests.
- Create **modified** tests for *Inclusion* students.
- **Edit** existing questions and **add** your own questions.
- Use built-in **state curriculum correlations** to create tests aligned with state standards.
- **Apply** art to your tests from a program bank of artwork.

MindJogger Videoquizzes present chapter-by-chapter review sessions in a game show format to make review more interesting and active to students . . . especially great for reluctant readers.

Online Study Tools

- Self-Check Quizzes
- Chapter Test Practice
- Vocabulary Review
- Standardized Test Practice

Assessment Options in the Chapter Resource Masters

These assessment resources are available for each chapter in *Glencoe Algebra 2*.

- 6 Chapter Tests
- Open-Ended Assessment with Scoring Rubric
- Vocabulary Test and Review **Glencoe Exclusive!**
- 4 Quizzes
- Mid-Chapter Test
- Cumulative Review
- 2-page Standardized Test Practice

Unit Tests, Semester Tests, and a **Final Test** are also available at point of use in the Chapter Resource Masters.

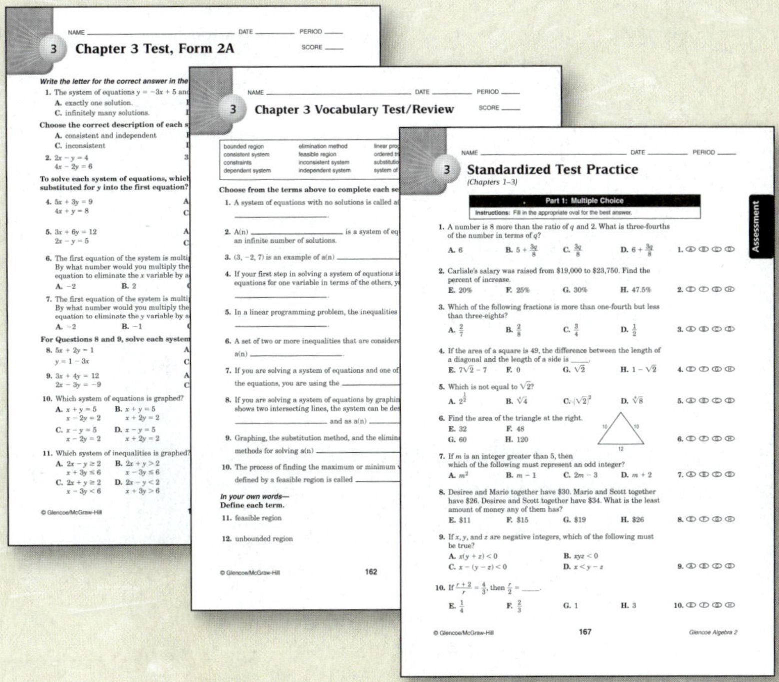

Introducing our new partner

USA TODAY® Education

USA TODAY Snapshots®

This is the same up-to-date data you know so well. But now, in an exclusive partnership with Glencoe/McGraw-Hill, USA TODAY® Education has brought its powerful, one-of-a-kind perspective and dynamic content to the pages of *Glencoe Algebra 2*. USA TODAY Snapshots® explode off the page to make Algebra come alive with current, relevant data.

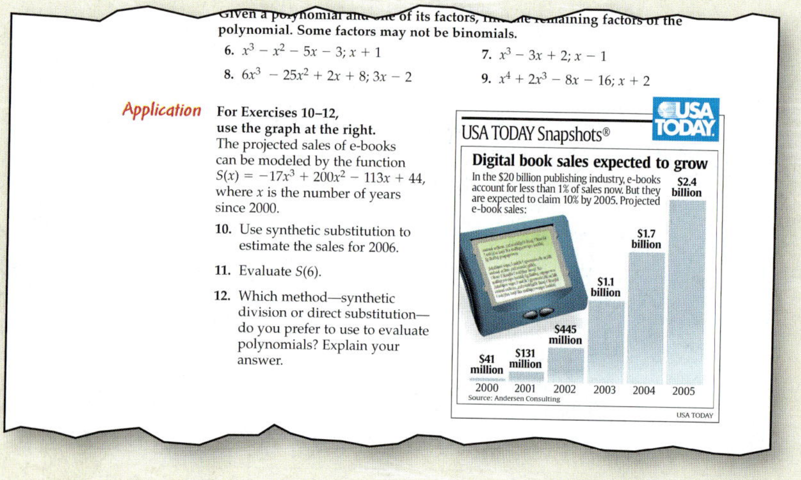

Given a polynomial and one of its factors, find the remaining factors of the polynomial. Some factors may not be binomials.

6. $x^3 - x^2 - 5x - 3; x + 1$
7. $x^3 - 3x + 2; x - 1$
8. $6x^3 - 25x^2 + 2x + 8; 3x - 2$
9. $x^4 + 2x^3 - 8x - 16; x + 2$

Application **For Exercises 10–12, use the graph at the right.** The projected sales of e-books can be modeled by the function $S(x) = -17x^3 + 200x^2 - 113x + 44$, where x is the number of years since 2000.

10. Use synthetic substitution to estimate the sales for 2006.
11. Evaluate $S(6)$.
12. Which method—synthetic division or direct substitution—do you prefer to use to evaluate polynomials? Explain your answer.

- www.algebra2.com/usa_today provides additional activities related to the topics presented in the USA TODAY Snapshots®.
- www.education.usatoday.com, USA TODAY® K–12 Education's Web site offers resources and interactive features connected to each day's newspaper. *Experience Today*, USA TODAY®'s daily lesson plan, is available on the site and delivered daily to subscribers. This plan provides instruction for integrating USA TODAY® graphics and key editorial features into your mathematics classroom.

Stay current with additional charts and graphs with USA TODAY®. Log on to www.education.usatoday.com, or call USA TODAY® at (800) 757-TEACH.

WebQuest: Online Projects

www.algebra2.com/webquest gives students the chance to work through a long-term project to enable them to develop their research, creative writing, and presentation skills.

- WebQuests often utilize USA TODAY Snapshots® or USA TODAY® articles.
- Special features in the Student Edition prompt students to complete each stage of their WebQuest.
- Parents can use the guided instruction to help students become familiar with the Internet in a safe, productive manner.

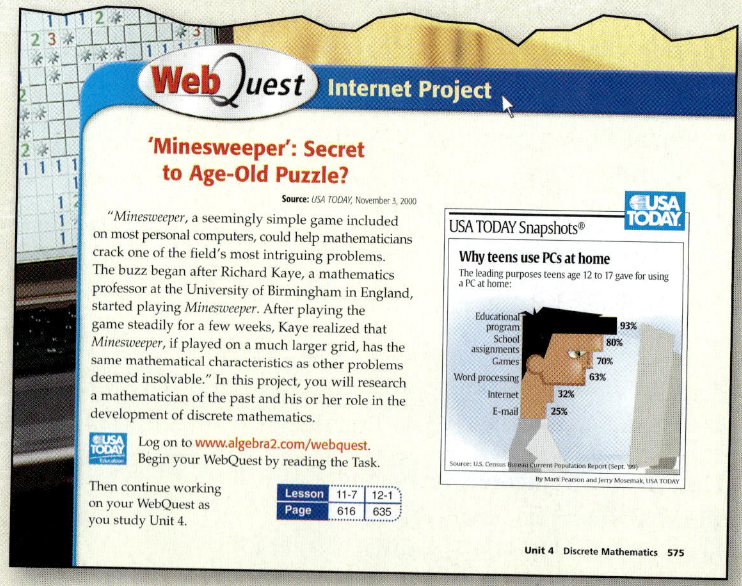

WebQuest Internet Project

'Minesweeper': Secret to Age-Old Puzzle?

Source: *USA TODAY*, November 3, 2000

"*Minesweeper*, a seemingly simple game included on most personal computers, could help mathematicians crack one of the field's most intriguing problems. The buzz began after Richard Kaye, a mathematics professor at the University of Birmingham in England, started playing *Minesweeper*. After playing the game steadily for a few weeks, Kaye realized that *Minesweeper*, if played on a much larger grid, has the same mathematical characteristics as other problems deemed insolvable." In this project, you will research a mathematician of the past and his or her role in the development of discrete mathematics.

Log on to www.algebra2.com/webquest. Begin your WebQuest by reading the Task.

Then continue working on your WebQuest as you study Unit 4.

Lesson	11-7	12-1
Page	616	635

Unit 4 Discrete Mathematics 575

The INTERNET: One TOOL. Endless possibilities.

Many of your students may already be familiar with the Internet, but may not have discovered the full potential of this powerful research tool. With *Glencoe Algebra 2*, your students can use the Internet to build their algebra skills. And you can access a wide variety of resources to help you plan classes, extend lessons, even meet professional development requirements.

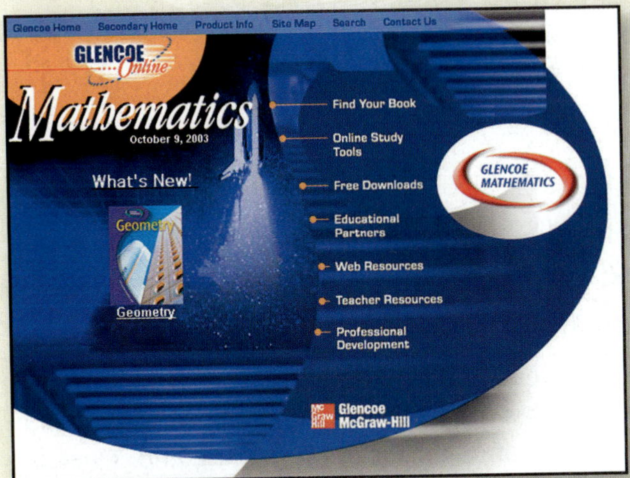

For Students

Online Study Tools, referenced on the Student Edition pages are keyed specifically to *Glencoe Algebra 2.*

- www.algebra2.com/extra_examples features additional fully worked-out examples.
- www.algebra2.com/self_check_quiz allows students to check their progress in each lesson.
- www.algebra2.com/vocabulary_review lets students check their vocabulary comprehension.
- www.algebra2.com/chapter_test provides additional practice in test taking.
- www.algebra2.com/standardized_test simulates questions that appear on standardized and proficiency tests.

Other Online Resources

- www.algebra2.com/webquest offers an online research project.
- www.algebra2.com/usa_today provides additional activities related to the topics presented in the USA TODAY Snapshots®.
- www.algebra2.com/data_update features links to updated statistical data presented in exercises.
- www.algebra2.com/careers offers information about career opportunities.
- www.algebra2.com/other_calculator_keystrokes provides keystroke instructions for various calculators to accompany graphing calculator activities and exercises in the Student Edition.

For Teachers

Powerful tools to make your job easier

- Classroom Games
- Problem of the Week Activities
- USA TODAY® K–12 Education daily lesson plans
- Sharing Ideas with Other Teachers
- Cool Math Links
- State and National Resources

Staff Development Sites

- NCTM links
- Teaching Today link
- McGraw-Hill Learning Network link
- Cooperative learning suggestions
- Using the Internet in the Mathematics Classroom

For Parents

Help parents get involved with their child's learning

- Involving Parents and Community in the Mathematics Classroom

DISCOVER how a simple sheet of paper can CHANGE the way your students THINK about math… forever.

Students love Foldables™ because they're fun. Teachers love them because they're effective.

Foldables are easy-to-make, three-dimensional interactive graphic organizers that students create out of simple sheets of paper. These unique hands-on tools for learning and reviewing were created exclusively for *Glencoe Algebra 2* by teaching specialist Dinah Zike.

Building Prereading Skills

At the beginning of each chapter, students construct one of a variety of Foldables. Each Foldable helps students create an interactive strategy for organizing what they read and observe. As they work through each chapter, students add more detail to their Foldable until they have created a comprehensive, interactive snapshot of the key concepts and vocabulary of the chapter.

Reading and Writing

Each Foldable helps students practice basic reading and writing skills, find and report main ideas, organize information, review key vocabulary terms, and more.

Review and Reinforcement

The completed Foldable is a comprehensive overview of the chapter concepts—perfect for preparing for chapter, unit, and even end-of-course tests.

Assessment

Foldables present an ideal opportunity to probe the depth of your students' understanding of chapter concepts. You'll get detailed feedback on what your students know and what misconceptions they may have.

Staff Development

Teaching Mathematics with Foldables™ equips teachers to extend the use of Foldables in their classrooms by exploring the different Foldable formats and providing suggestions for using them throughout the mathematics curriculum.

Give *students control,* so that they leave your classroom with *knowledge and power* over their *own learning.*

PROJECT CRISS[SM] Study Skill

Project CRISS[SM] (**CR**eating **I**ndependence through **S**tudent-owned **S**trategies) is a research-based staff development program created to help students better organize, understand, and retain course information. In short, students receiving the CRISS method of instruction will "LEARN HOW TO LEARN".

CRISS strategies are designed to develop thoughtful and independent readers and learners.

To enhance student learning, CRISS employs several concepts drawn from cognitive psychology.

- Students must be able to integrate new information with prior knowledge.
- Students need to be actively involved in their own learning by discussing, writing, and organizing information.
- Students must self-monitor to identify which strategies are the most effective for their own learning.

These behaviors need to be taught by content teachers to maximize student learning.

CReating Independence Through Student-Owned Strategies

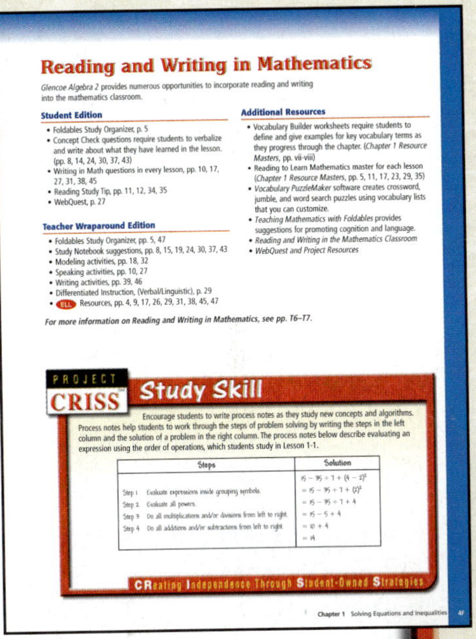

Reading and Writing in Mathematics

Glencoe Algebra 2 provides numerous opportunities to incorporate reading and writing into the mathematics classroom.

Student Edition

- Foldables Study Organizer, p. 5
- Concept Check questions require students to verbalize and write about what they have learned in the lesson. (pp. 8, 14, 24, 30, 37, 43)
- Writing in Math questions in every lesson, pp. 10, 17, 27, 31, 38, 45
- Reading Study Tip, pp. 11, 12, 34, 35
- WebQuest, p. 27

Teacher Wraparound Edition

- Foldables Study Organizer, pp. 5, 47
- Study Notebook suggestions, pp. 8, 15, 19, 24, 30, 37, 43
- Modeling activities, pp. 18, 32
- Speaking activities, pp. 10, 27
- Writing activities, pp. 39, 46
- Differentiated Instruction, (Verbal/Linguistic), p. 29
- ELL Resources, pp. 4, 9, 17, 26, 29, 31, 38, 45, 47

Additional Resources

- Vocabulary Builder worksheets require students to define and give examples for key vocabulary terms as they progress through the chapter. (*Chapter 1 Resource Masters*, pp. vii-viii)
- Reading to Learn Mathematics master for each lesson (*Chapter 1 Resource Masters*, pp. 5, 11, 17, 23, 29, 35)
- *Vocabulary PuzzleMaker* software creates crossword, jumble, and word search puzzles using vocabulary lists that you can customize.
- *Teaching Mathematics with Foldables* provides suggestions for promoting cognition and language.
- *Reading and Writing in the Mathematics Classroom*
- *WebQuest and Project Resources*

For more information on Reading and Writing in Mathematics, see pp. T6–T7.

PROJECT CRISS Study Skill

Encourage students to write process notes as they study new concepts and algorithms. Process notes help students to work through the steps of problem solving by writing the steps in the left column and the solution of a problem in the right column. The process notes below describe evaluating an expression using the order of operations, which students study in Lesson 1-1.

Steps	Solution
	$15 - 35 \div 7 + (4 - 2)^2$
Step 1 Evaluate expressions inside grouping symbols.	$= 15 - 35 \div 7 + (2)^2$
Step 2 Evaluate all powers.	$= 15 - 35 \div 7 + 4$
Step 3 Do all multiplications and/or divisions from left to right.	$= 15 - 5 + 4$
Step 4 Do all additions and/or subtractions from left to right.	$= 10 + 4$
	$= 14$

CReating Independence Through Student-Owned Strategies

Chapter 1 Solving Equations and Inequalities 4F

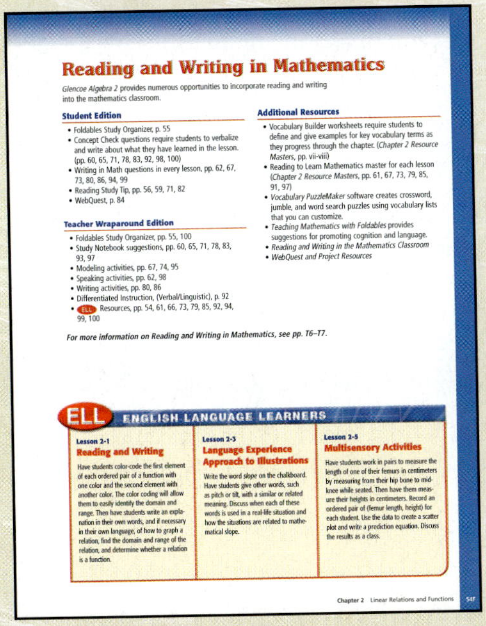

Reading and Writing in Mathematics

Glencoe Algebra 2 provides numerous opportunities to incorporate reading and writing into the mathematics classroom.

Student Edition

- Foldables Study Organizer, p. 55
- Concept Check questions require students to verbalize and write about what they have learned in the lesson. (pp. 60, 65, 71, 78, 83, 92, 98, 100)
- Writing in Math questions in every lesson, pp. 62, 67, 73, 80, 86, 94, 99
- Reading Study Tip, pp. 56, 59, 71, 82
- WebQuest, p. 84

Teacher Wraparound Edition

- Foldables Study Organizer, pp. 55, 100
- Study Notebook suggestions, pp. 60, 65, 71, 78, 83, 93, 97
- Modeling activities, pp. 67, 74, 95
- Speaking activities, pp. 62, 98
- Writing activities, pp. 80, 86
- Differentiated Instruction, (Verbal/Linguistic), p. 92
- ELL Resources, pp. 54, 61, 66, 73, 79, 85, 92, 94, 99, 100

Additional Resources

- Vocabulary Builder worksheets require students to define and give examples for key vocabulary terms as they progress through the chapter. (*Chapter 2 Resource Masters*, pp. vii-viii)
- Reading to Learn Mathematics master for each lesson (*Chapter 2 Resource Masters*, pp. 61, 67, 73, 79, 85, 91, 97)
- *Vocabulary PuzzleMaker* software creates crossword, jumble, and word search puzzles using vocabulary lists that you can customize.
- *Teaching Mathematics with Foldables* provides suggestions for promoting cognition and language.
- *Reading and Writing in the Mathematics Classroom*
- *WebQuest and Project Resources*

For more information on Reading and Writing in Mathematics, see pp. T6–T7.

ELL ENGLISH LANGUAGE LEARNERS

Lesson 2-1
Reading and Writing

Have students color-code the first element of each ordered pair of a function with one color and the second element with another color. The color coding will allow them to easily identify the domain and range. Then have students write an explanation in their own words, and if necessary in their own language, of how to graph a relation, find the domain and range of the relation, and determine whether a relation is a function.

Lesson 2-3
Language Experience Approach to Illustrations

Write the word *slope* on the chalkboard. Have students give other words, such as pitch or tilt, with a similar or related meaning. Discuss when each of these words is used in a real-life situation and how the situations are related to mathematical slope.

Lesson 2-5
Multisensory Activities

Have students work in pairs to measure the length of one of their femurs in centimeters by measuring from their hip bone to mid-knee while seated. Then have them measure their heights in centimeters. Record an ordered pair of (femur length, height) for each student. Use the data to create a scatter plot and write a prediction equation. Discuss the results as a class.

Chapter 2 Linear Relations and Functions 54F

Implementing CRISS Strategies

Project CRISS Study Skills were developed with leaders from Project CRISS to facilitate the teaching of each chapter of *Glencoe Algebra 2*. These strategies appear in the interleaf of the Teacher Wraparound Edition.

For more information on project CRISS[SM], visit www.projectcriss.com

ELL ENGLISH LANGUAGE LEARNERS

English language learners may need specialized help in overcoming a language barrier to learn mathematics. Hands-on activities, modeling, working in flexible groups, and vocabulary building activities are particularly helpful to ELL students. Suggested strategies appear in the interleaf of the Teacher Wraparound Edition.

It's Staff Development,

As professional development continues to take on greater importance for educators across the country, teachers are constantly looking for easy-to-use tools to help them stay abreast of current trends and issues. At Glencoe, we know how valuable your time is, so we've developed a variety of staff development tools to help you meet your district's requirements.

Teacher Wraparound Edition

Mathematical Connections and Background found at the beginning of each chapter gives you an overview of the mathematics skills required in each lesson. Information about prior knowledge as well as future connections lets you see the continuity of instruction.

Building on Prior Knowledge provides you with information that links what students have previously learned to the content of the lesson.

Tips for New Teachers offers helpful suggestions for such things as classroom management, assessment, teaching techniques, and more.

Teaching Tips can be found not only in the margins but also on the reduced student pages at point of use.

Teacher Classroom Resources

Glencoe Mathematics Staff Development Series is a series of publications that allows you to stay current with issues that affect your teaching effectiveness. The series is intended to help you implement new mathematics strategies and enhance your classroom performance.

Available in print

- *Using the Internet in the Mathematics Classroom*
- *Reading and Writing in the Mathematics Classroom*
- *Teaching Mathematics with Foldables™*
- *Teaching Algebra with Manipulatives*

Available online at www.math.glencoe.com

- *Graphing Calculators in the Mathematics Classroom*
- *Cooperative Learning in the Mathematics Classroom*
- *Alternative Assessment in the Mathematics Classroom*
- *Involving Parents and the Community in the Mathematics Classroom*

made convenient.

Technology Support

At www.math.glencoe.com, you'll find:

- a Staff Development site that addresses current issues in education.
- a Teacher Forum that allows teachers to discuss issues and ideas with colleagues.
- a State and National Resources site that links to math and math education resources, nationally and by state.

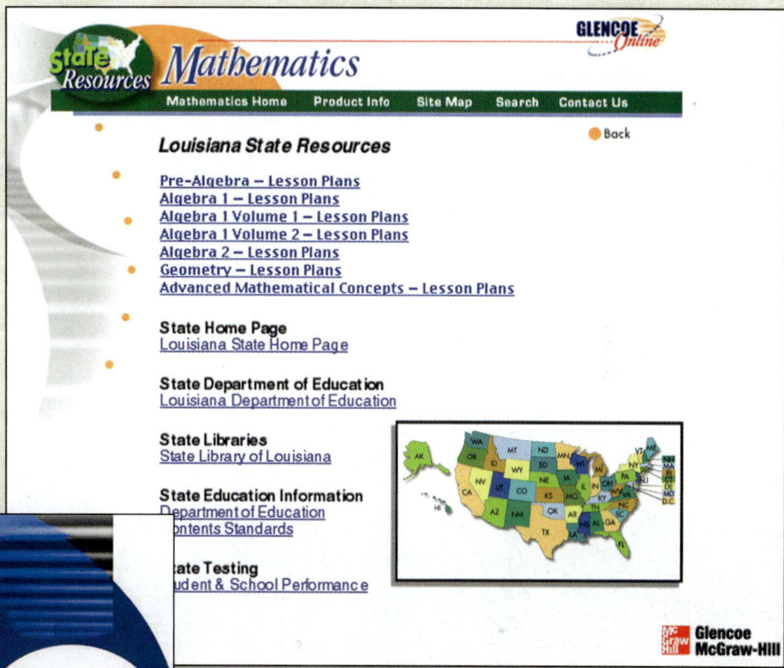

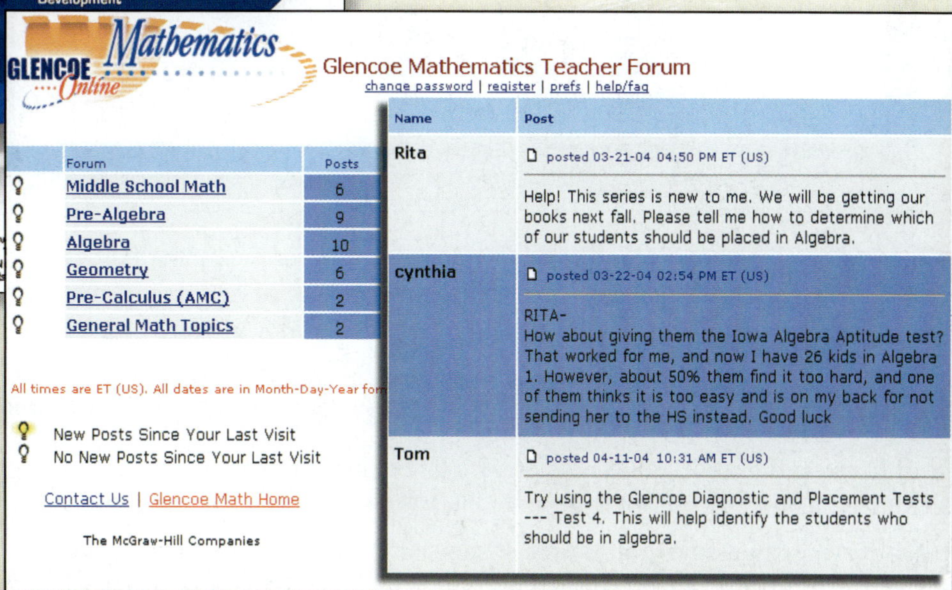

Glencoe Mathematics Programs— Research-Based and

Glencoe's mathematics programs are the product of ongoing classroom and educational research activities involving students, teachers, curriculum supervisors, administrators, parents, and college-level mathematics educators, mathematicians, and researchers.

SOUND

Prior to the publication of any Glencoe mathematics program, the following initial research is completed.

- Monitoring of national and state changes and trends such as graduation requirements, standardized test exams, and the latest NCTM and NAEP reports.
- Incorporating the most current and applicable educational research in which reported results show significant improvement on student learning and achievement.
- Analyzing returns from independently contracted mailing and telephone surveys.

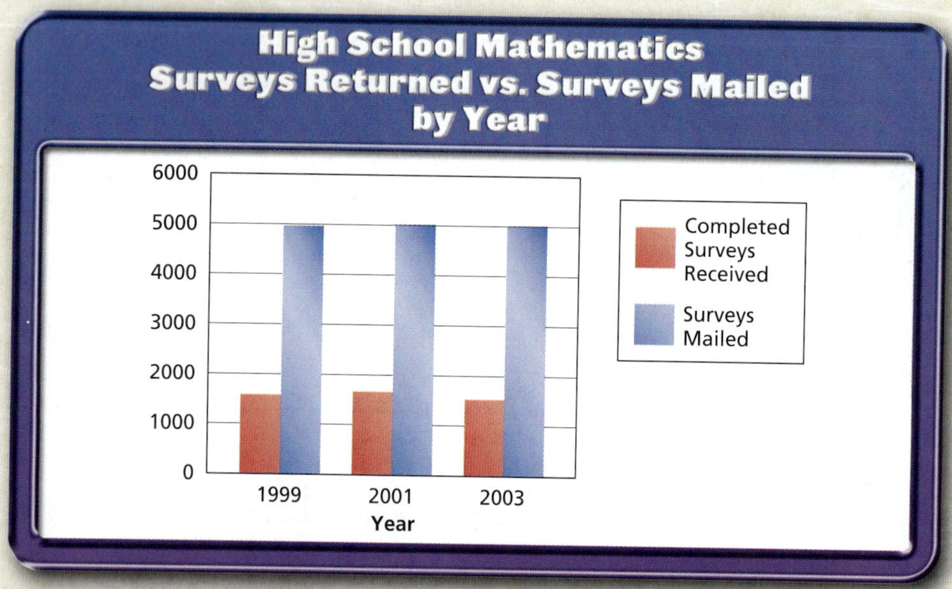

Source: High School Mathematics Longitudinal Surveys, 1999, 2001, 2003

- Reviewing all comments and correspondence on appropriate prior editions in terms of specific lessons. This helps Glencoe to build in staff development support, which makes the programs easy to implement from the first day of use.

PROVEN

Prior to the publication of *Glencoe Algebra 2*, extensive research was conducted using manuscript and pre-publication versions of the program.

- Nationwide discussion groups were conducted, which involved mathematics teachers, department chairpersons, supervisors, and educational learning specialists.

tested to ensure success.

- Face-to-face interviews were carried out with mathematics teachers.
- Reviewers and consultants reviewed *Glencoe Algebra 2* manuscripts for accuracy, content development, and thoroughness.
- Before the design of the Student Edition was completed, an independent research company was contracted to organize and conduct blind focus groups with high school algebra teachers in various cities. The teachers' reactions and comments were recorded and used for improvements.
- Follow-up interviews, observations, and surveys of users of Glencoe mathematics programs are continuously conducted and monitored.

PREFERRED

Glencoe's mathematics programs are currently used by millions of students and tens of thousands of teachers. The Glencoe author team—a combination of practicing classroom teachers, curriculum supervisors, college-level educators, and learning specialists—is a key reason why Glencoe mathematics programs continue to ensure success in the mathematics classroom. This proven mix of authors and consultants, along with the incorporation of national trends and research results, leads to a preferred rating among teachers and students.

Survey Results Responding *Very Satisfied* or *Satisfied* with Glencoe *High School Mathematics**

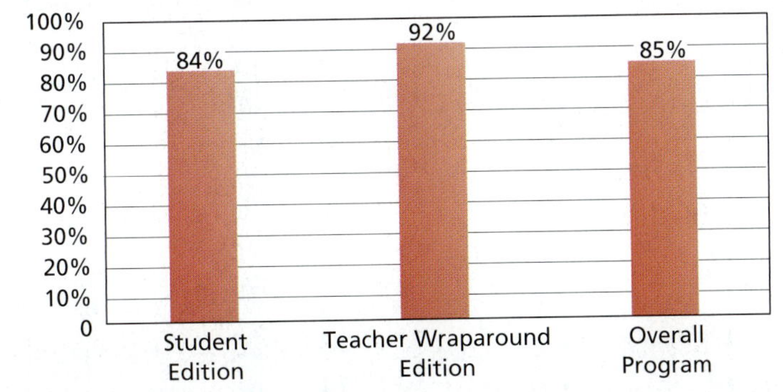

*Ratings for all other high school math programs were 83%, 85%, and 77%, respectively.

Source: High School Mathematics Longitudinal Survey, 2003

For more details of Glencoe's research, please contact us at www.math.glencoe.com.

Planning Your

Glencoe Algebra 2 and the accompanying support materials allow you to create an Algebra 2 course that meets the needs of each class of students. The charts below and at the top of the facing page offer suggestions for pacing for average and advanced levels with standard class periods and block schedule class periods.

Average Pacing is for those students who have completed Algebra 1 successfully but may need more review before studying Algebra 2 topics. You may want to use one of the six chapter tests provided in the *Chapter Resource Masters* as a pretest to determine how well your students are prepared for each chapter. If you find that they are well prepared, consider using the Study Guide and Review at the end of the chapter as a one-day lesson and proceed to the next chapter.

Advanced Pacing is for those students who have successfully completed an Algebra course and need little review before beginning Algebra 2. In advanced pacing, Chapters 1–3 may be considered as a review. You can use the Study Guide and Review at the end of each chapter to refresh students' memories of these topics and identify those with which they are less familiar. One of the six chapter tests provided in the Chapter Resource Masters can also be used as a pretest for each chapter.

Year-Long Schedule

45–50 minute periods

Grading Period	Average		Advanced	
	Chapter	Days	Chapter	Days
1	1	9	Review of 1, 2, and 3	11
	2	10	4	17
	3	10	5-1 to 5-7	10
	4-1 to 4-6	11		
2	4-7 to 4-8	6	5-8 to 5-9	6
	5	16	6	10
	6	11	7	15
	7-1 to 7-5	7	8-1 to 8-6	10
3	7-6 to 7-9	7	8-7	4
	8	12	9	13
	9	11	10	10
	10	10	11	13
4	11	13	12	13
	12	13	13	14
	13	14	14	14
optional	14	0		
Total		160	Total	160

The total number of days in each level of pacing is less than the typical 180-day school year and 90-day semester to allow for flexibility in planning due to testing, school cancellation, or shortened class periods. *A more detailed Suggested Pacing chart for one-year courses appears in the interleaf page A preceding each chapter in the Teacher Wraparound Edition.*

Modifying Average Pacing for Basic Students

For those students who are less prepared for Algebra 2, spend more time on Units 1 and 2 (Chapters 1-7). Unit 5 (Chapters 13-14) may be omitted.

Algebra Course

Block Scheduling pacing for average and advanced courses is given in the chart below. The pacing is designed for classes that meet once each day for one semester or every other day for one year.

Block Schedule

90-minute periods

<table>
<tr><th>Chapter</th><th>1</th><th>2</th><th>3</th><th>4</th><th>5</th><th>6</th><th>7</th><th>8</th><th>9</th><th>10</th><th>11</th><th>12</th><th>13</th><th>14</th><th>Total</th></tr>
<tr><td>Days (Average)</td><td>4.5</td><td>5</td><td>5</td><td>8</td><td>8</td><td>6</td><td>7</td><td>6</td><td>6</td><td>5</td><td>6</td><td>7</td><td>7.5</td><td>0</td><td>81</td></tr>
<tr><td>Days (Advanced)</td><td colspan="3">4.5</td><td>9</td><td>8</td><td>5</td><td>8</td><td>7</td><td>6</td><td>5</td><td>6</td><td>7</td><td>7</td><td>7.5</td><td>80</td></tr>
</table>

Two-year Pacing is for those students who want to take algebra 2, but find the abstract concepts difficult to grasp. This pacing allows students to cover the same material and work the same problems as students using the average or advanced pacing. Students will be able to spend more time on each concept and will have more time to complete hands-on labs and activities that help develop and internalize the abstract concepts presented in this course.

In year one, students cover the first seven chapters of the book. In year two, students begin by reviewing the first seven chapters and then continue with Chapters 8 through 13. If less time is needed for review of Chapters 1 through 7, more time will be available for covering the more difficult concepts and activities in the last five chapters of the book.

<table>
<tr><th colspan="2">Year 1 Schedule</th><th colspan="2">Year 2 Schedule</th></tr>
<tr><th>Chapter</th><th>Days</th><th>Chapter</th><th>Days</th></tr>
<tr><td>1</td><td>18</td><td>Review 1–7*</td><td>35</td></tr>
<tr><td>2</td><td>20</td><td>8</td><td>21</td></tr>
<tr><td>3</td><td>20</td><td>9</td><td>18</td></tr>
<tr><td>4</td><td>29</td><td>10</td><td>18</td></tr>
<tr><td>5</td><td>28</td><td>11</td><td>23</td></tr>
<tr><td>6</td><td>22</td><td>12</td><td>23</td></tr>
<tr><td>7</td><td>26</td><td>13</td><td>24</td></tr>
<tr><td>Total</td><td>163 Days</td><td>Total</td><td>162 Days</td></tr>
</table>

* Five days are allotted for each chapter in the review.

Implementing the NCTM *Principles and Standards*

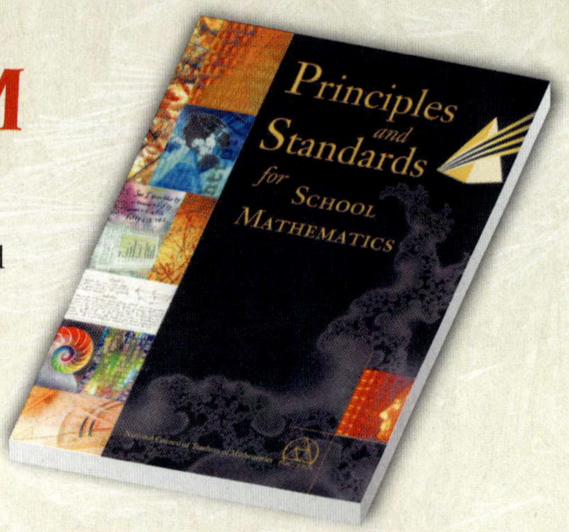

In 1989, the National Council of Teachers of Mathematics (NCTM) published their *Curriculum and Evaluation Standards for School Mathematics,* which gave mathematics teachers their first set of goals toward a national mathematics curriculum. Teachers and supervisors have embraced these Standards and developed state standards based on this framework. In 2000, the National Council of Teachers of Mathematics published a revision of these guidelines entitled *NCTM Principles and Standards for School Mathematics.*

NCTM Principles for School Mathematics	Glencoe Algebra 2
Equity *Excellence in mathematics education requires equity—high expectations and strong support for all students.*	Glencoe's product line encourages high achievement at every level. Numerous teacher support materials provide activities for **differentiated instruction,** promotion of **reading and writing, pacing** for individual levels of achievement, and **daily intervention.**
Curriculum *A curriculum is more than a collection of activities: it must be coherent, focused on important mathematics, and well articulated across the grades.*	Glencoe authors developed a philosophy and scope and sequence to ensure a continuum of mathematical learning that builds on **prior knowledge** and extends concepts toward more **advanced mathematical thinking.**
Teaching *Effective mathematics teaching requires understanding what students know and need to learn and then challenging and supporting them to learn it well.*	Glencoe offers a plethora of teacher support materials. A comprehensive *Teacher Wraparound Edition* provides **mathematical background,** teaching tips, resource management guidelines, and **tips for new teachers.**
Learning *Students must learn mathematics with understanding, actively building new knowledge from experience and prior knowledge.*	The *Teacher Wraparound Edition* includes instruction on building from prior knowledge with materials in each interleaf and in **Building On Prior Knowledge** features. **Find the Error** and **Unlocking Misconception** teaching tips help to evaluate how students are thinking and learning.
Assessment *Assessment should support the learning of important mathematics and furnish useful information to both teachers and students.*	The **Practice Quizzes** and the **Chapter Practice Test** provide ways for students to check their own progress. Online Study Tools, such as **Self-Check Quizzes,** offer a unique way for students with Internet access to monitor their progress. The assessment tools in the *Chapter Resource Masters* contain different levels and formats for tests, as well as intermediate opportunities for assessment.
Technology *Technology is essential in teaching and learning mathematics; it influences the mathematics that is taught and enhances students' learning.*	The *Student Edition* includes opportunities to utilize graphing calculators and spreadsheets in the exploration of Algebra 2 concepts. The *Teacher Wraparound Edition* offers teaching tips on using technology. *Graphing Calculator and Spreadsheet Masters* has additional activities. Glencoe's Web site is constantly updated to meet the needs of students and teachers in excelling in mathematics education.

NCTM Standards for School Mathematics

The Standards portion of the *NCTM Principles and Standards for School Mathematics* center upon ten areas of mathematics curriculum development. The number assigned to each standard is for easy reference and is not part of each standard's official title.

Instructional programs from prekindergarten through grade 12 should enable all students to:

1 Numbers and Operations

- Understand numbers, ways of representing numbers, relationships among numbers, and number systems
- Understand the meaning of operations and how they relate to each other
- Compute fluently and make reasonable estimates

Pages: 6-46, 56-99, 110-144, 154-208, 222-275, 294-299, 301-319, 346-358, 360-399, 492-498, 522-565, 578-617, 632-685, 709-751, 762-781, 786-797, 799-804

2 Algebra

- Understand patterns, relations, and functions
- Represent and analyze mathematical situations and structures using algebraic symbols
- Use mathematical models to represent and understand quantitative relationships
- Analyze change in various contexts

Pages: 6-10, 20-46, 56-99, 110-144, 154-158, 160-208, 222-275, 286-335, 346-399, 412-416, 419-431, 433-452, 455-460, 472-512, 522-565, 588-592, 594-604, 606-610, 612-621, 700-708, 762-804

3 Geometry

- Analyze characteristics and properties of two- and three-dimensional geometric shapes and develop mathematical arguments about geometric relationships
- Specify locations and describe spatial relationships using coordinate geometry and other representational systems
- Apply transformations and use symmetry to analyze mathematical situations
- Use visualization, spatial reasoning, and geometric modeling to solve problems

Pages: 19, 167-174, 239-244, 270-275, 286-293, 301-312, 329-335, 360-382, 390-394, 412-460, 522-540, 552-559, 611, 700-751, 769-776

4 Measurement

- Understand measurable attributes of objects and the units, systems, and processes of measurement
- Apply appropriate techniques, tools, and formulas to determine measurements

Pages: 6-10, 20-27, 63-74, 87-88, 182-188, 360-364, 371-382, 390-394, 432, 523-538, 541-551, 554-565, 611, 709-751, 769-776

5 Data Analysis and Probability

- Formulate questions that can be addressed with data and collect, organize, and display relevant data to answer them
- Select and use appropriate statistical methods to analyze data
- Develop and evaluate inferences and predictions that are based on data
- Understand and apply basic concepts of probability

Pages: 81-95, 154-158, 300, 359, 539-540, 632-686

Instructional programs from prekindergarten through grade 12 should enable all students to:

6 Problem Solving

- Build new mathematical knowledge through problem solving
- Solve problems that arise in mathematics and in other contexts
- Apply and adapt a variety of appropriate strategies to solve problems
- Monitor and reflect on the process of mathematical problem solving

Pages: 20-27, 33-46, 63-99, 110-127, 129-135, 138-144, 160-207, 222-267, 270-275, 286-319, 322-335, 346-399, 412-431, 433-452, 455-460, 472-512, 522-551, 554-565, 578-621, 632-685, 700-751, 762-781, 786-790, 799-804

7 Reasoning and Proof

- Recognize reasoning and proof as fundamental aspects of mathematics
- Make and investigate mathematical conjectures
- Develop and evaluate mathematical arguments and proofs
- Select and use various types of reasoning and methods of proof

Pages: 56-62, 68-74, 116-122, 138-144, 182-188, 202-207, 222-228, 233-238, 245-256, 270-275, 301-312, 322-328, 346-352, 371-399, 432, 453-454, 522, 531-538, 541-546, 554-559, 578-592, 594-604, 606-611, 618-621, 686, 762-768, 786-797, 799-804

8 Communication

- Organize and consolidate their mathematical thinking through communication
- Communicate their mathematical thinking coherently and clearly to peers, teachers, and others
- Analyze and evaluate the mathematical thinking and strategies of others
- Use the language of mathematics to express mathematical ideas precisely

Pages: 63-86, 89-99, 110-127, 129-135, 138-144, 154-158, 160-207, 222-267, 270-275, 286-335, 346-358, 360-399, 412-452, 455-460, 472-511, 522-551, 554-565, 578-592, 594-621, 632-680, 682-686, 701-751, 762-797, 799-804

9 Connections

- Recognize and use connections among mathematical ideas
- Understand how mathematical ideas build on one another to produce a coherent whole
- Recognize and apply mathematics in contexts outside of mathematics

Pages: 6-46, 56-99, 110-127, 129-135, 138-144, 154-158, 160-207, 222-267, 270-275, 286-299, 301-319, 322-335, 346-399, 412-416, 419-431, 433-452, 455-460, 472-490, 492-511, 523-538, 541-551, 554-565, 578-592, 594-610, 612-617, 632-686, 701-715, 717-751, 762-781, 786-797, 799-804

10 Representation

- Create and use representations to organize, record, and communicate mathematical ideas
- Select, apply, and translate among mathematical representations to solve problems
- Use representations to model and interpret physical, social, and mathematical phenomena

Pages: 19, 28-32, 40-46, 56-62, 68-99, 110-127, 129-144, 154-181, 189-208, 222-232, 239-244, 250-256, 263-275, 286-300, 306-312, 320-335, 346-377, 383-399, 412-431, 433-460, 472-490, 492-511, 522-530, 539-540, 583-592, 594-610, 612-617, 632-686, 701-715, 717-751, 762-781, 786-804

UNIT 1

First-Degree Equations and Inequalities

2

Chapter 1 Solving Equations and Inequalities

4

1-1 Expressions and Formulas 6
1-2 Properties of Real Numbers 11
Practice Quiz 1: Lessons 1-1 and 1-2 18
Algebra Activity: Investigating Polygons and Patterns 19
1-3 Solving Equations 20
1-4 Solving Absolute Value Equations 28
1-5 Solving Inequalities 33
Practice Quiz 2: Lessons 1-3 through 1-5 39
1-6 Solving Compound and Absolute Value Inequalities 40
Study Guide and Review 47
Practice Test 51
Standardized Test Practice 52

WebQuest Internet Project

- Introduction 3
- Follow-Ups 27, 84, 120, 192
- Culmination 207

Prerequisite Skills

- Getting Started 5
- Getting Ready for the Next Lesson 10, 18, 27, 32, 39

FOLDABLES™

Study Organizer 5

Reading and Writing Mathematics

- Reading Math Tips 11, 12, 34, 35
- Writing in Math 10, 17, 27, 31, 38, 45

Standardized Test Practice

- Multiple Choice 10, 17, 23, 24, 27, 31, 32, 39, 46, 51, 52
- Short Response/Grid In 46, 53
- Extended Response 53

USA TODAY Snapshots 17, 39

Lesson 1-4, page 31

Chapter 2 Linear Relations and Functions 54

2-1 Relations and Functions 56
2-2 Linear Equations 63
2-3 Slope 68
Practice Quiz 1: Lessons 2-1 through 2-3 74
2-4 Writing Linear Equations 75
2-5 Modeling Real-World Data: Using Scatter Plots 81
Graphing Calculator Investigation: Lines of Regression 87
2-6 Special Functions 89
Practice Quiz 2: Lessons 2-4 through 2-6 95
2-7 Graphing Inequalities 96
Study Guide and Review 100
Practice Test 105
Standardized Test Practice 106

Prerequisite Skills
- Getting Started 55
- Getting Ready for the Next Lesson 62, 67, 74, 80, 86, 95

FOLDABLES™ Study Organizer 55

Reading and Writing Mathematics
- Reading Tips 56, 59, 71, 82
- Writing in Math 62, 67, 73, 80, 86, 94, 99

Standardized Test Practice
- Multiple Choice 62, 67, 74, 76, 78, 80, 86, 95, 99, 105, 106
- Short Response/Grid In 107
- Extended Response 107

USA TODAY Snapshots 69, 84

Lesson 2-2, page 64

Chapter 3 Systems of Equations and Inequalities 108

3-1 Solving Systems of Equations by Graphing 110
3-2 Solving Systems of Equations Algebraically 116
Practice Quiz 1: Lessons 3-1 and 3-2 122
3-3 Solving Systems of Inequalities by Graphing 123
Graphing Calculator Investigation: Systems of Linear Inequalities 128
3-4 Linear Programming 129
Practice Quiz 2: Lessons 3-3 and 3-4 135
Algebra Activity: Graphing Equations in Three Variables 136
3-5 Solving Systems of Equations in Three Variables 138
Study Guide and Review 145
Practice Test 149
Standardized Test Practice 150

Prerequisite Skills

- Getting Started 109
- Getting Ready for the Next Lesson 115, 122, 127, 135

FOLDABLES™

Study Organizer 109

Reading and Writing Mathematics

- Reading Tips 124, 129
- Writing in Math 114, 121, 127, 134, 144

Standardized Test Practice

- Multiple Choice 115, 117, 120, 122, 127, 134, 144, 150
- Short Response/Grid In 149, 151
- Extended Response 127, 151

USA TODAY Snapshots 135

Lesson 3-4, page 131

Chapter 4 Matrices 152

4-1 Introduction to Matrices 154

Spreadsheet Investigation: Organizing Data 159

4-2 Operations with Matrices 160

4-3 Multiplying Matrices 167

Practice Quiz 1: Lessons 4-1 through 4-3 174

4-4 Transformations with Matrices 175

4-5 Determinants 182

4-6 Cramer's Rule 189

Practice Quiz 2: Lessons 4-4 through 4-6 194

4-7 Identity and Inverse Matrices 195

4-8 Using Matrices to Solve Systems of Equations 202

Graphing Calculator Investigation: Augmented Matrices 208

Study Guide and Review 209

Practice Test 215

Standardized Test Practice 216

Prerequisite Skills
- Getting Started 153
- Getting Ready for the Next Lesson 158, 166, 174, 181, 188, 194, 201

FOLDABLES™

Study Organizer 153

Reading and Writing Mathematics
- Reading Tips 154, 175, 182
- Writing in Math 158, 166, 173, 181, 187, 193, 200, 207

Standardized Test Practice
- Multiple Choice 158, 166, 173, 179, 181, 187, 194, 201, 207, 215, 216
- Short Response/Grid In 176, 194, 207, 217
- Extended Response 217

USA TODAY **Snapshots** 206

Lesson 4-6, page 193

UNIT 2

Polynomial and Radical Equations and Inequalities 218

Chapter 5 Polynomials 220

5-1 Monomials 222
5-2 Polynomials 229
5-3 Dividing Polynomials 233
Practice Quiz 1: Lessons 5-1 through 5-3 238
5-4 Factoring Polynomials 239
5-5 Roots of Real Numbers 245
5-6 Radical Expressions 250
Practice Quiz 2: Lessons 5-4 through 5-6 256
5-7 Rational Exponents 257
5-8 Radical Equations and Inequalities 263
Graphing Calculator Investigation: Solving Radical Equations and Inequalities by Graphing 268
5-9 Complex Numbers 270
Study Guide and Review 276
Practice Test 281
Standardized Test Practice 282

WebQuest Internet Project

- Introduction 219
- Follow-Ups 227, 326, 369
- Culmination 399

Lesson 5-7, page 259

Prerequisite Skills

- Getting Started 221
- Getting Ready for the Next Lesson 228, 232, 238, 244, 249, 256, 262, 267

FOLDABLES™ Study Organizer 221

Reading and Writing Mathematics

- Reading Tips 229, 246, 252, 270, 271, 273
- Writing in Math 227, 232, 238, 243, 249, 255, 262, 267, 275

Standardized Test Practice

- Multiple Choice 228, 232, 234, 236, 238, 244, 249, 255, 262, 267, 275, 281, 282
- Short Response/Grid In 283
- Extended Response 283

USA TODAY Snapshots 228

Chapter 6 Quadratic Functions and Inequalities 284

6-1 Graphing Quadratic Functions 286
6-2 Solving Quadratic Equations by Graphing....................294
Graphing Calculator Investigation: Modeling Real-World Data 300
6-3 Solving Quadratic Equations by Factoring 301
Practice Quiz 1: Lessons 6-1 through 6-3 305
6-4 Completing the Square 306
6-5 The Quadratic Formula and the Discriminant 313
Graphing Calculator Investigation: Families of Parabolas 320
6-6 Analyzing Graphs of Quadratic Functions 322
Practice Quiz 2: Lessons 6-4 through 6-6 328
6-7 Graphing and Solving Quadratic Inequalities....................329
Study Guide and Review 336
Practice Test 341
Standardized Test Practice 342

Prerequisite Skills
- Getting Started 285
- Getting Ready for the Next Lesson 293, 299, 305, 312, 319, 328

FOLDABLES™
Study Organizer 285

Reading and Writing Mathematics
- Reading Tips 294, 306, 313, 316, 323
- Writing in Math 292, 299, 305, 312, 319, 327, 334

Standardized Test Practice
- Multiple Choice 292, 293, 299, 302, 303, 305, 312, 319, 327, 335, 341, 342
- Short Response/Grid In 343
- Extended Response 343

USA TODAY **Snapshots** 328

Lesson 6-4, page 311

Chapter 7 Polynomial Functions 344

7-1 Polynomial Functions 346

7-2 Graphing Polynomial Functions 353

Graphing Calculator Investigation: Modeling Real-World Data 359

7-3 Solving Equations Using Quadratic Techniques 360

Practice Quiz 1: Lessons 7-1 through 7-3 364

7-4 The Remainder and Factor Theorems 365

7-5 Roots and Zeros 371

7-6 Rational Zero Theorem 378

Practice Quiz 2: Lessons 7-4 through 7-6 382

7-7 Operations on Functions 383

7-8 Inverse Functions and Relations 390

7-9 Square Root Functions and Inequalities 395

Study Guide and Review 400

Practice Test 405

Standardized Test Practice 406

Lesson 7-1, page 346

Prerequisite Skills
- Getting Started 345
- Getting Ready for the Next Lesson 352, 358, 364, 370, 377, 382, 389, 394

FOLDABLES™ Study Organizer 345

Reading and Writing Mathematics
- Reading Tips 354, 372, 384, 391
- Writing in Math 352, 357, 364, 370, 377, 382, 389, 394, 399

Standardized Test Practice
- Multiple Choice 352, 358, 364, 370, 377, 382, 389, 394, 399, 405, 406
- Short Response/Grid In 370, 374, 375, 407
- Extended Response 407

USA TODAY Snapshots 368

UNIT 3

Advanced Functions and Relations 408

Chapter 8 Conic Sections 410

8-1 Midpoint and Distance Formulas 412

Algebra Activity: Midpoint and Distance Formulas in Three Dimensions 417

8-2 Parabolas 419

8-3 Circles 426

Practice Quiz 1: Lessons 8-1 and 8-3 431

Algebra Activity: Investigating Ellipses 432

8-4 Ellipses 433

8-5 Hyperbolas 441

Practice Quiz 2: Lessons 8-4 and 8-5 448

8-6 Conic Sections 449

Algebra Activity: Conic Sections 453

8-7 Solving Quadratic Systems 455

Study Guide and Review 461

Practice Test 467

Standardized Test Practice 468

WebQuest Internet Project

- Introduction 409
- Follow-Ups 429, 502, 529
- Culmination 565

Prerequisite Skills

- Getting Started 411
- Getting Ready for the Next Lesson 416, 425, 431, 440, 448, 452

FOLDABLES™

Study Organizer 411

Reading and Writing Mathematics

- Reading Tips 442, 449
- Writing in Math 416, 425, 430, 439, 447, 452, 459

Standardized Test Practice

- Multiple Choice 413, 414, 416, 425, 431, 439, 440, 447, 452, 459, 467, 468
- Short Response/Grid In 469
- Extended Response 469

USA TODAY **Snapshots** 448

Lesson 8-4, page 435

Chapter 9 Rational Expressions and Equations 470

9-1 Multiplying and Dividing Rational Expressions......472

9-2 Adding and Subtracting Rational Expressions.........479

Practice Quiz 1: Lessons 9-1 and 9-2 484

9-3 Graphing Rational Functions 485

Graphing Calculator Investigation:
Graphing Rational Functions 491

9-4 Direct, Joint, and Inverse Variation 492

Practice Quiz 2: Lessons 9-3 and 9-4 498

9-5 Classes of Functions 499

9-6 Solving Rational Equations and Inequalities 505

Graphing Calculator Investigation:
Solving Rational Equations by Graphing 512

Study Guide and Review 513

Practice Test 517

Standardized Test Practice 518

Prerequisite Skills

- Getting Started 471
- Getting Ready for the Next Lesson 478, 484, 490, 498, 504

FOLDABLES™

Study Organizer 471

Reading and Writing Mathematics

- Writing in Math 477, 484, 490, 498, 503, 511

Standardized Test Practice

- Multiple Choice 473, 476, 478, 484, 490, 498, 503, 504, 511, 517, 518
- Short Response/Grid In 519
- Extended Response 519

USA TODAY

Snapshots 492

Lesson 9-5, page 503

Unit 3

Chapter 10 Exponential and Logarithmic Relations 520

Algebra Activity: Investigating Exponential Functions 522

10-1 Exponential Functions 523

10-2 Logarithms and Logarithmic Functions 531

Practice Quiz 1: Lessons 10-1 and 10-2 538

Graphing Calculator Investigation: Modeling Real-World Data: Curve Fitting 539

10-3 Properties of Logarithms 541

10-4 Common Logarithms 547

Graphing Calculator Investigation: Solving Exponential and Logarithmic Equations and Inequalities 552

10-5 Base *e* and Natural Logarithms 554

Practice Quiz 2: Lessons 10-3 through 10-5 559

10-6 Exponential Growth and Decay 560

Study Guide and Review 566

Practice Test 571

Standardized Test Practice 572

Lesson 10-6, page 564

Prerequisite Skills
- Getting Started 521
- Getting Ready for the Next Lesson 530, 538, 546, 551, 559

FOLDABLES Study Organizer 521

Reading and Writing Mathematics
- Writing in Math 530, 537, 546, 551, 559, 564

Standardized Test Practice
- Multiple Choice 530, 537, 538, 546, 551, 559, 562, 563, 564, 571, 572
- Short Response/Grid In 530, 546, 559, 564, 573
- Extended Response 573

USA TODAY Snapshots 535, 565

UNIT 4

Discrete Mathematics 574

Chapter 11 Sequences and Series 576

11-1 Arithmetic Sequences 578
11-2 Arithmetic Series 583
11-3 Geometric Sequences 588
Practice Quiz 1: Lessons 11-1 through 11-3 592
Graphing Calculator Investigation: Limits 593
11-4 Geometric Series 594
11-5 Infinite Geometric Series 599
Spreadsheet Investigation: Amortizing Loans 605
11-6 Recursion and Special Sequences 606
Algebra Activity: Fractals 611
11-7 The Binomial Theorem 612
Practice Quiz 2: Lessons 11-4 through 11-7 617
11-8 Proof and Mathematical Induction 618
Study Guide and Review 622
Practice Test 627
Standardized Test Practice 628

WebQuest Internet Project

- Introduction 575
- Follow-Ups 616, 635
- Culmination 685

Prerequisite Skills

- Getting Started 577
- Getting Ready for the Next Lesson 582, 587, 592, 598, 604, 610, 617

FOLDABLES™ Study Organizer 577

Reading and Writing Mathematics

- Reading Tips 606, 619
- Writing in Math 582, 587, 592, 598, 603, 610, 616, 621

Standardized Test Practice

- Multiple Choice 582, 587, 588, 591, 592, 598, 603, 610, 616, 621, 627, 628
- Short Response/Grid In 629
- Extended Response 629

USA TODAY Snapshots 604

Lesson 11-5, page 603

Chapter 12 Probability and Statistics 630

12-1 The Counting Principle 632
12-2 Permutations and Combinations 638
12-3 Probability 644
Practice Quiz 1: Lessons 12-1 through 12-3 650
12-4 Multiplying Probabilities 651
12-5 Adding Probabilities 658
12-6 Statistical Measures 664
Practice Quiz 2: Lessons 12-4 through 12-6 670
12-7 The Normal Distribution 671
12-8 Binomial Experiments 676
Algebra Activity: Simulations 681
12-9 Sampling and Error 682
Algebra Activity: Testing Hypotheses 686
Study Guide and Review 687
Practice Test 693
Standardized Test Practice 694

Lesson 12-3, page 648

Prerequisite Skills
- Getting Started 631
- Getting Ready for the Next Lesson 637, 643, 650, 657, 663, 670, 675, 680

FOLDABLES Study Organizer 631

Reading and Writing Mathematics
- Reading Tips 633, 638, 644, 646, 665, 669
- Writing in Math 636, 642, 649, 657, 662, 669, 675, 679, 685

Standardized Test Practice
- Multiple Choice 633, 634, 636, 642, 649, 657, 662, 669, 675, 680, 685, 693, 694
- Short Response/Grid In 680, 695
- Extended Response 695

USA TODAY Snapshots 658

UNIT 5

Trigonometry

696

Chapter 13 Trigonometric Functions

698

Spreadsheet Investigation: Special Right Triangles 700

13-1 Right Triangle Trigonometry 701

13-2 Angles and Angle Measure 709

Practice Quiz 1: Lessons 13-1 and 13-2 715

Algebra Activity: Investigating Regular Polygons Using Trigonometry 716

13-3 Trigonometric Functions of General Angles 717

13-4 Law of Sines 725

13-5 Law of Cosines 733

Practice Quiz 2: Lessons 13-3 through 13-5 738

13-6 Circular Functions 739

13-7 Inverse Trigonometric Functions 746

Study Guide and Review 752

Practice Test 757

Standardized Test Practice 758

WebQuest Internet Project

- Introduction 697
- Follow-Ups 708, 775
- Culmination 804

Prerequisite Skills

- Getting Started 699
- Getting Ready for the Next Lesson 708, 715, 724, 732, 738, 745

FOLDABLES™

Study Organizer 699

Reading and Writing Mathematics

- Reading Tips 701, 709, 711, 718, 740
- Writing in Math 708, 714, 724, 732, 737, 744, 751

Standardized Test Practice

- Multiple Choice 702, 706, 708, 714, 724, 732, 737, 738, 745, 751, 757, 758
- Short Response/Grid In 708, 724, 732, 745, 751, 759
- Extended Response 759

USA TODAY Snapshots 715

Lesson 13-6, page 744

Chapter 14 Trigonometric Graphs and Identities 760

14-1 Graphing Trigonometric Functions 762
14-2 Translations of Trigonometric Graphs 769
14-3 Trigonometric Identities 777
Practice Quiz 1: Lessons 14-1 through 14-3 781
14-4 Verifying Trigonometric Identities 782
14-5 Sum and Difference of Angles Formulas 786
14-6 Double-Angle and Half-Angle Formulas 791
Practice Quiz 2: Lessons 14-4 through 14-6 797
Graphing Calculator Investigation: Solving Trigonometric Equations 798
14-7 Solving Trigonometric Equations 799
Study Guide and Review 805
Practice Test 809
Standardized Test Practice 810

Lesson 14-7, page 803

Student Handbook

Skills

Prerequisite Skills 814
Extra Practice 828
Mixed Problem Solving 862
Preparing for Standardized Tests 876

Reference

English-Spanish Glossary R1
Selected Answers R20
Photo Credits R86
Index R87
Symbols and Formulas ....Inside Back Cover

Prerequisite Skills
- Getting Started 761
- Getting Ready for the Next Lesson 768, 776, 781, 785, 790, 797

FOLDABLES™ Study Organizer 761

Reading and Writing Mathematics
- Reading Tips 786, 788
- Writing in Math 768, 776, 781, 785, 790, 796, 804

Standardized Test Practice
- Multiple Choice 768, 776, 781, 783, 784, 785, 790, 796, 804, 809, 810
- Short Response/Grid In 811
- Extended Response 811

USA TODAY Snapshots 797

Student & Parent One-Stop Internet Resources

Need extra help or information? Log on to math.glencoe.com or any of the Web addresses below to learn more.

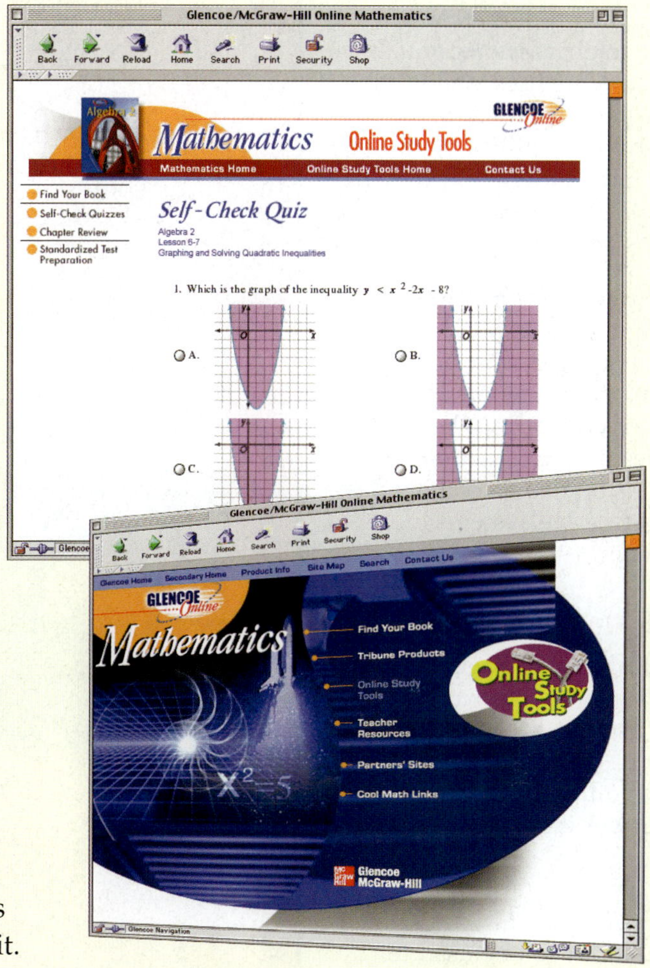

Online Study Tools

- www.algebra2.com/extra_examples shows you additional worked-out examples that mimic the ones in your book.
- www.algebra2.com/self_check_quiz provides you with a practice quiz for each lesson that grades itself.
- www.algebra2.com/vocabulary_review lets you check your understanding of the terms and definitions used in each chapter.
- www.algebra2.com/chapter_test allows you to take a self-checking test before the actual test.
- www.algebra2.com/standardized_test is another way to brush up on your standardized test-taking skills.

Research Options

- www.algebra2.com/webquest walks you step-by-step through a long-term project using the Web. One WebQuest for each unit is explored using the mathematics from that unit.
- www.algebra2.com/usa_today provides activities related to the concept of the lesson as well as up-to-date Snapshot data.
- www.algebra2.com/careers links you to additional information about interesting careers.
- www.algebra2.com/data_update links you to the most current data available for subjects such as basketball and family.

Calculator Help

- www.algebra2.com/other_calculator_keystrokes provides you with keystrokes other than the TI-83 Plus used in your textbook.

Get Started!

- To help you learn how to use your math book, use the Scavenger Hunt at www.algebra2.com

UNIT 1 Notes

Introduction

In this unit, students begin by applying the properties of real numbers to expressions, equalities, and inequalities, including absolute value inequalities and compound inequalities. Throughout the unit, students explore the relationship between linear equations and their graphs.

These explorations include modeling data with scatter plots and lines of regression, as well as linear programming and solving systems of equations. The unit concludes with instruction about operations on matrices and using matrices to solve systems of equations.

Assessment Options

Unit 1 Test Pages 237–238 of the *Chapter 4 Resource Masters* may be used as a test or review for Unit 1. This assessment contains both multiple-choice and short answer items.

ExamView® Pro

This CD-ROM can be used to create additional unit tests and review worksheets.

UNIT 1

First-Degree Equations and Inequalities

You can model and analyze real-world situations by using algebra. In this unit, you will solve and graph linear equations and inequalities and use matrices.

Chapter 1
Solving Equations and Inequalities

Chapter 2
Linear Relations and Functions

Chapter 3
Systems of Equations and Inequalities

Chapter 4
Matrices

Teaching Suggestions

Have students study the USA TODAY Snapshot®.

- Ask students to write an inequality using the data for two of the expenditure categories shown. **See students' work.**
- According to the data, what was the average cost per person for apparel in 1998? **$669.60**
- Point out to students that how they budget their money can affect their ability to buy a home. Their spending habits also affect what type of home they could afford.

Additional USA TODAY Snapshots® appearing in Unit 1:

Chapter 1	School shopping (p. 17)
	Just looking, thank you (p. 39)
Chapter 2	Cruises grow in popularity (p. 69)
	Cost of seeing the doctor (p. 84)
Chapter 3	Per-pupil spending is climbing (p. 135)
Chapter 4	Student-to-teacher ratios dropping (p. 206)

WebQuest Internet Project

Lessons in Home Buying, Selling

Source: *USA TODAY*, November 18, 1999

"'Buying a home,' says Housing and Urban Development Secretary Andrew Cuomo, 'is the most expensive, most complicated and most intimidating financial transaction most Americans ever make.'" In this project, you will be exploring how functions and equations relate to buying a home and your income.

Log on to www.algebra2.com/webquest. Begin your WebQuest by reading the Task.

Then continue working on your WebQuest as you study Unit 1.

Lesson	1-3	2-5	3-2	4-6
Page	27	84	120	192

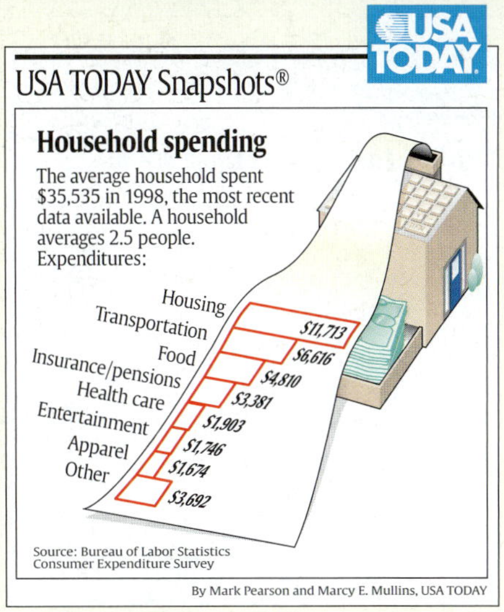

WebQuest Internet Project

Problem-Based Learning A WebQuest is an online project in which students do research on the Internet, gather data, and make presentations using word processing, graphing, page-making, or presentation software. In each chapter, students advance to the next step in their WebQuest. At the end of Chapter 4, the project culminates with a presentation of their findings.

Teaching suggestions and sample answers are available in the *WebQuest and Project Resources*.

Chapter 1

Solving Equations and Inequalities

Chapter Overview and Pacing

Year-long and two-year pacing: pages T20–T21.

LESSON OBJECTIVES	PACING (days) Regular Basic/Average	Regular Advanced	Block Basic/Average	Block Advanced
1-1 **Expressions and Formulas** *(pp. 6–10)* • Use the order of operations to evaluate expressions. • Use formulas.	1	optional	0.5	optional
1-2 **Properties of Real Numbers** *(pp. 11–19)* • Classify real numbers. • Use the properties of real numbers to evaluate expressions. ***Follow-Up:*** Investigating Polygons and Patterns	2 (with 1-2 Follow-Up)	optional	0.5	optional
1-3 **Solving Equations** *(pp. 20–27)* • Translate verbal expressions into algebraic expressions and equations, and vice versa. • Solve equations using the properties of equality.	1	optional	1 (with 1-2 Follow-Up)	optional
1-4 **Solving Absolute Value Equations** *(pp. 28–32)* • Evaluate expressions involving absolute values. • Solve absolute value equations.	1	optional	0.5	optional
1-5 **Solving Inequalities** *(pp. 33–39)* • Solve inequalities. • Solve real-world problems involving inequalities.	1	optional	0.5	optional
1-6 **Solving Compound and Absolute Value Inequalities** *(pp. 40–46)* • Solve compound inequalities. • Solve absolute value inequalities.	1	optional	0.5	optional
Study Guide and **Practice Test** *(pp. 47–51)* **Standardized Test Practice** *(pp. 52–53)*	1	2	0.5	1
Chapter Assessment	1	1	0.5	0.5
TOTAL	**9**	**3**	**4.5**	**1.5**

StudentWorks

An electronic version of this chapter is available on ***StudentWorks™****. This backpack solution CD-ROM allows students instant access to the Student Edition, lesson worksheet pages, and web resources.*

Timesaving Tools

TeacherWorks™

All-In-One Planner and Resource Center

See pages T5 and T21.

Chapter Resource Manager

Study Guide and Intervention	Practice (Skills and Average)	Reading to Learn Mathematics	Enrichment	Assessment	Applications*	5-Minute Check Transparencies	Interactive Chalkboard	Alge2PASS: Tutorial Plus (lessons)	Materials
1–2	3–4	5	6		SC 1, SM 91–96	1-1	1-1		graphing calculator, colored pencils
7–8	9–10	11	12	51		1-2	1-2		algebra tiles, index cards (*Follow:Up:* ruler or geometry software)
13–14	15–16	17	18	51, 53	GCS 27, SC 2	1-3	1-3		
19–20	21–22	23	24		GCS 28	1-4	1-4	1	
25–26	27–28	29	30	52		1-5	1-5	2	graphing calculator
31–32	33–34	35	36	52		1-6	1-6		masking tape
				38–50, 54–56					

Chapter 1 Resource Masters

**Key to Abbreviations:* GCS = Graphing Calculator and Speadsheet Masters,
SC = School-to-Career Masters,
SM = Science and Mathematics Lab Manual

Chapter 1

Mathematical Connections and Background

Continuity of Instruction

Prior Knowledge

Students have worked with linear equations in previous classes and they should be familiar, to some extent, with some of the properties of equality and inequality. Also, in earlier grades students have used number lines and have related inequalities to intervals on number lines.

This Chapter

Students review the real number system and the order of operations. They begin to study formulas, evaluating expressions, and additive and multiplicative inverses. They see how properties of equality and properties of the real number system can be used to solve equations, and they study other topics related to linear equations, linear inequalities, and absolute value.

Future Connections

Equations, inequalities, and absolute value expressions appear throughout all levels of mathematics. Solving equations and inequalities and justifying mathematical steps on the basis of properties is at the center of all mathematical analysis and presentation.

1-1 Expressions and Formulas

An algebraic expression usually contains at least one variable and may also contain numbers and operations. The order of operations is a mathematical convention for deciding which operations are performed before others in an algebraic expression. That order is: evaluate powers; multiply and divide from left to right; and add and subtract from left to right. There is one more part to the convention: any grouping symbol (parentheses, brackets, braces, fraction bar) takes first priority. To evaluate an expression means to replace each variable with its given value and then follow the order of operations to simplify. A formula is an equation in which one variable is set equal to an algebraic expression.

1-2 Properties of Real Numbers

The set N of natural numbers is $\{1, 2, 3, \ldots\}$; add zero and the result is the set W of whole numbers. The set Z of integers is $\{\ldots, -2, -1, 0, 1, 2, \ldots\}$ and the numbers in the set Q of rational numbers have the form $\frac{a}{b}$ where a and b are integers and $b \neq 0$. The rationals, along with the set I of irrational numbers, make up the set R of real numbers. There is a one-to-one correspondence between the real numbers and the points on a line in that each real number corresponds to exactly one point on a line and each point on a line corresponds to exactly one real number.

Properties of real numbers are used to justify the steps of solving equations and describing mathematical relationships. These include the commutative and associative properties of addition and the commutative and associative properties of multiplication. Another property, the distributive property, relates addition and multiplication. The real numbers include an identity element for the operation of addition, an identity element for the operation of multiplication, an additive inverse for every real number, and a multiplicative inverse for every real number except 0.

1-3 Solving Equations

A mathematic sentence with an equal sign between two algebraic or arithmetic expressions is called an *equation*. To solve an equation requires a series of equations, equivalent to the given equation, that result in a final equation that isolates the variable on one side. That final equation presents the solution to the original equation. However, solutions should always be substituted into the original equation to check for correctness.

The rules for writing equivalent equations are called Properties of Equality. We can write the equation

$a = a$; given $a = b$ then we can write $b = a$; given $a = b$ and $b = c$ then we can write $a = c$. A fourth rule is Substitution: if $a = b$, then we can write an equation replacing a with b or b with a. Also, if $a = b$ we can write $a + c = b + c$, we can write $a - c = b - c$, we can write $a \cdot c = b \cdot c$, and, if $c \neq 0$, we can write $\frac{a}{c} = \frac{b}{c}$.

1-4 Solving Absolute Value Equations

The absolute value of a number is its distance from zero. Described algebraically, the definition of absolute value is $|a| = a$ if $a \geq 0$ and $|a| = -a$ if $a < 0$. The absolute value symbols are a grouping symbol like parentheses or a fraction bar. For example, to evaluate $2 \cdot |15 - 31|$, first calculate inside the symbols. So, $2 \cdot |15 - 31| = 2 \cdot |-16| = 2 \cdot (16)$ or 32.

The equation $|a - 6| = 4$ can be interpreted as *the distance between a and 6 is 4 units*. The value $a - 6$ can be 4 or -4, so if $a - 6 = 4$, then $a = 10$. If $a - 6 = -4$, then $a = 2$. The solution is {2, 10}. "No solution" can be written as { } or ∅, the symbols for the empty set.

1-5 Solving Inequalities

An inequality is a mathematical sentence with one of the symbols $<$, $\leq$, $>$, or $\geq$ between two expressions. Solving an inequality means writing a series of equivalent inequalities, ending with one that isolates the variable. The rules for writing equivalent inequalities are called properties of inequality. (The properties hold for all inequalities, but are usually expressed initially in terms of $>$.) If $a > b$, then we can write $a + c > b + c$ and $a - c > b - c$. Also, if $a > b$ and $c > 0$, then we can write $ac > bc$ and $\frac{a}{c} > \frac{b}{c}$ or, if $c < 0$, we can write $ac < bc$ and $\frac{a}{c} < \frac{b}{c}$. In general, multiplying or dividing an inequality by a negative number *reverses* the order of the inequality. The Trichotomy Property states that for any two real numbers, either the values are equal or one value is greater than the other. In symbols, exactly one of these statements is true: $a < b$, $a = b$, or $a > b$.

When the solution to an inequality is graphed, an open circle indicates a value that is not included and a closed circle indicates a value that is included. Open circles are used with $<$ and $>$, and closed circles are used with $\leq$ and $\geq$. Solutions to inequalities are often written using set-builder notation, so a solution such as $x \geq 4$ would be written $\{x | x \geq 4\}$, read *the set of values x such that x is greater than or equal to 4.*

1-6 Solving Compound and Absolute Value Inequalities

There are important connections between compound inequalities and absolute value inequalities. An absolute value inequality using $<$ or $\leq$ is related to a compound inequality using the word *and*. For example, thinking of $|a| < 7$ as $|a - 0| < 7$, then the value of a is any number whose distance from 0 is less than 7 units.

Possible values for a

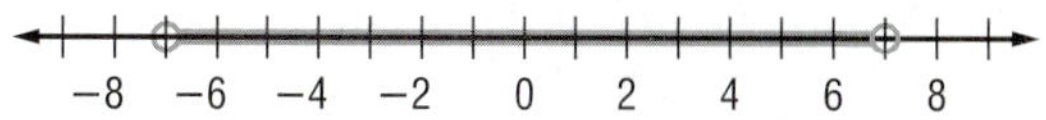

$|a| < 7$ means $-7 < a$ *and* $a < 7$ or $-7 < a < 7$

An absolute value inequality using $>$ or $\geq$ is related to a compound inequality using the word *or*. For example, thinking of $|b| > 5$ as $|b - 0| > 5$, then the value of b is any number whose distance from 0 is more than 5.

Possible values for b

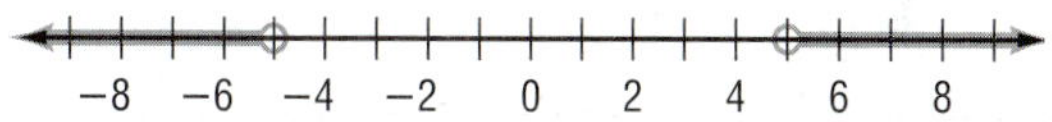

$|b| > 5$ means $b < -5$ *or* $b > 5$

To solve absolute value inequalities, use two patterns. One pattern is to rewrite $|A| < B$ as $-B < A$ *and* $A < B$ (or $-B < A < B$), so rewrite $|2x - 5| < 18$ as $-18 < 2x - 5$ *and* $2x - 5 < 18$. The solution is $-\frac{13}{2} < x < \frac{23}{2}$. The other pattern is to rewrite $|A| > B$ as $A < -B$ or $A >$ B, so rewrite the inequality $|3x + 1| > 15$ as $3x + 1 < -15$ *or* $3x + 1 > 15$. The solution is $x < -\frac{16}{3}$ *or* $x > \frac{14}{3}$.

Additional mathematical information and teaching notes are available at www.algebra2.com/key_concepts.

Chapter 1

DAILY INTERVENTION and Assessment

	Type	Student Edition	Teacher Resources	Technology/Internet
INTERVENTION	Ongoing	Prerequisite Skills, pp. 5, 10, 18, 27, 32, 39 Practice Quiz 1, p. 18 Practice Quiz 2, p. 39	5-Minute Check Transparencies Quizzes, *CRM* pp. 51–52 Mid-Chapter Test, *CRM* p. 53 Study Guide and Intervention, *CRM* pp. 1–2, 7–8, 13–14, 19–20, 25–26, 31–32	Alge2PASS: Tutorial Plus www.algebra2.com/self_check_quiz www.algebra2.com/extra_examples
	Mixed Review	pp. 18, 27, 32, 39, 46	Cumulative Review, *CRM* p. 54	
	Error Analysis	Find the Error, pp. 24, 43 Common Misconceptions, p. 12	Find the Error, *TWE* pp. 24, 44 Unlocking Misconceptions, *TWE* pp. 15, 18, 22 Tips for New Teachers, *TWE* pp. 10, 27	
ASSESSMENT	Standardized Test Practice	pp. 10, 17, 23, 24, 27, 31, 32, 39, 46, 51, 52–53	*TWE* p. 23 Standardized Test Practice, *CRM* pp. 55–56	Standardized Test Practice CD-ROM www.algebra2.com/standardized_test
	Open-Ended Assessment	Writing in Math, pp. 10, 17, 27, 31, 38, 45 Open Ended, pp. 8, 14, 24, 30, 37, 43	Modeling: *TWE* pp. 18, 32 Speaking: *TWE* pp. 10, 27 Writing: *TWE* pp. 39, 46 Open-Ended Assessment, *CRM* p. 49	
	Chapter Assessment	Study Guide, pp. 47–50 Practice Test, p. 51	Multiple-Choice Tests (Forms 1, 2A, 2B), *CRM* pp. 37–42 Free-Response Tests (Forms 2C, 2D, 3), *CRM* pp. 43–48 Vocabulary Test/Review, *CRM* p. 50	ExamView® Pro (see below) MindJogger Videoquizzes www.algebra2.com/vocabulary_review www.algebra2.com/chapter_test

Key to Abbreviations: TWE = Teacher Wraparound Edition; CRM = Chapter Resource Masters

Intervention Technology

Alge2PASS: Tutorial Plus CD-ROM offers a complete, self-paced algebra curriculum.

Algebra 2 Lesson	Alge2PASS Lesson
1-4	1 *Solving Multi-Operational Equations IV*
1-5	2 *Solving Inequalities*

ExamView® Pro

Use the networkable **ExamView® Pro** to:
- Create **multiple versions** of tests.
- Create **modified** tests for *Inclusion* students.
- **Edit** existing questions and **add** your own questions.
- Use built-in **state curriculum correlations** to create tests aligned with state standards.
- **Apply** art to your tests from a program bank of artwork.

For more information on Intervention and Assessment, see pp. T8–T11.

Reading and Writing in Mathematics

Glencoe Algebra 2 provides numerous opportunities to incorporate reading and writing into the mathematics classroom.

Student Edition

- Foldables Study Organizer, p. 5
- Concept Check questions require students to verbalize and write about what they have learned in the lesson. (pp. 8, 14, 24, 30, 37, 43)
- Writing in Math questions in every lesson, pp. 10, 17, 27, 31, 38, 45
- Reading Study Tip, pp. 11, 12, 34, 35
- WebQuest, p. 27

Teacher Wraparound Edition

- Foldables Study Organizer, pp. 5, 47
- Study Notebook suggestions, pp. 8, 15, 19, 24, 30, 37, 43
- Modeling activities, pp. 18, 32
- Speaking activities, pp. 10, 27
- Writing activities, pp. 39, 46
- Differentiated Instruction, (Verbal/Linguistic), p. 29
- ELL Resources, pp. 4, 9, 17, 26, 29, 31, 38, 45, 47

Additional Resources

- Vocabulary Builder worksheets require students to define and give examples for key vocabulary terms as they progress through the chapter. (*Chapter 1 Resource Masters,* pp. vii-viii)
- Reading to Learn Mathematics master for each lesson (*Chapter 1 Resource Masters*, pp. 5, 11, 17, 23, 29, 35)
- *Vocabulary PuzzleMaker* software creates crossword, jumble, and word search puzzles using vocabulary lists that you can customize.
- *Teaching Mathematics with Foldables* provides suggestions for promoting cognition and language.
- *Reading and Writing in the Mathematics Classroom*
- *WebQuest and Project Resources*

For more information on Reading and Writing in Mathematics, see pp. T6–T7.

PROJECT CRISS℠ Study Skill

Encourage students to write process notes as they study new concepts and algorithms. Process notes help students to work through the steps of problem solving by writing the steps in the left column and the solution of a problem in the right column. The process notes below describe evaluating an expression using the order of operations, which students study in Lesson 1-1.

Steps	Solution
	$15 - 35 \div 7 + (4 - 2)^2$
Step 1 Evaluate expressions inside grouping symbols.	$= 15 - 35 \div 7 + (2)^2$
Step 2 Evaluate all powers.	$= 15 - 35 \div 7 + 4$
Step 3 Do all multiplications and/or divisions from left to right.	$= 15 - 5 + 4$
Step 4 Do all additions and/or subtractions from left to right.	$= 10 + 4$
	$= 14$

CReating **I**ndependence Through **S**tudent-Owned **S**trategies

Chapter 1 Notes

What You'll Learn

Have students read over the list of objectives and make a list of any words with which they are not familiar.

Why It's Important

Point out to students that this is only one of many reasons why each objective is important. Others are provided in the introduction to each lesson.

The chart below correlates the objectives for each lesson to the NCTM Standards 2000. There is also space for you to reference your state and/or local objectives.

Lesson	NCTM Standards	Local Objectives
1-1	1, 2, 4, 8, 9	
1-2	1, 8, 9	
1-2 Follow-Up	1, 3, 9, 10	
1-3	1, 2, 4, 6, 8, 9	
1-4	1, 2, 8, 9, 10	
1-5	1, 2, 6, 8, 9	
1-6	1, 2, 6, 9, 10	

Key to NCTM Standards:

1=Number & Operations, 2=Algebra, 3=Geometry, 4=Measurement, 5=Data Analysis & Probability, 6=Problem Solving, 7=Reasoning & Proof, 8=Communication, 9=Connections, 10=Representation

Chapter 1 Solving Equations and Inequalities

What You'll Learn

- **Lesson 1-1** Simplify and evaluate algebraic expressions.
- **Lesson 1-2** Classify and use the properties of real numbers.
- **Lesson 1-3** Solve equations.
- **Lesson 1-4** Solve absolute value equations.
- **Lessons 1-5 and 1-6** Solve and graph inequalities.

Key Vocabulary

- order of operations (p. 6)
- algebraic expression (p. 7)
- Distributive Property (p. 12)
- equation (p. 20)
- absolute value (p. 28)

Why It's Important

Algebra allows you to write expressions, equations, and inequalities that hold true for most or all values of variables. Because of this, algebra is an important tool for describing relationships among quantities in the real world. For example, the angle at which you view fireworks and the time it takes you to hear the sound are related to the width of the fireworks burst. A change in one of the quantities will cause one or both of the other quantities to change. *In Lesson 1-1, you will use the formula that relates these quantities.*

Vocabulary Builder

ELL

The Key Vocabulary list introduces students to some of the main vocabulary terms included in this chapter. For a more thorough vocabulary list with pronunciations of new words, give students the Vocabulary Builder worksheets found on pages vii and viii of the *Chapter 1 Resource Masters*. Encourage them to complete the definition of each term as they progress through the chapter. You may suggest that they add these sheets to their study notebooks for future reference when studying for the Chapter 1 test.

Getting Started

Prerequisite Skills To be successful in this chapter, you'll need to master these skills and be able to apply them in problem-solving situations. Review these skills before beginning Chapter 1.

For Lessons 1-1 through 1-3 — Operations with Rational Numbers

Simplify.

1. $20 - 0.16$ **19.84**
2. $12.2 + (-8.45)$ **3.75**
3. $-3.01 - 14.5$ **-17.51**
4. $-1.8 + 17$ **15.2**
5. $\frac{1}{4} - \frac{2}{3}$ **$-\frac{5}{12}$**
6. $\frac{3}{5} + (-6)$ **$-5\frac{2}{5}$**
7. $-7\frac{1}{2} + 5\frac{1}{3}$ **$-2\frac{1}{6}$**
8. $-11\frac{5}{8} - \left(-4\frac{3}{7}\right)$ **$-7\frac{11}{56}$**
9. $(0.15)(3.2)$ **0.48**
10. $2 \div (-0.4)$ **-5**
11. $(-1.21) \div (-1.1)$ **1.1**
12. $(-9)(0.036)$ **-0.324**
13. $-4 \div \frac{3}{2}$ **$-2\frac{2}{3}$**
14. $\left(\frac{5}{4}\right)\left(-\frac{3}{10}\right)$ **$-\frac{3}{8}$**
15. $\left(-2\frac{3}{4}\right)\left(-3\frac{1}{5}\right)$ **$8\frac{4}{5}$**
16. $7\frac{1}{8} \div (-2)$ **$-3\frac{9}{16}$**

For Lesson 1-1 — Powers

Evaluate each power.

17. 2^3 **8**
18. 5^3 **125**
19. $(-7)^2$ **49**
20. $(-1)^3$ **-1**
21. $(-0.8)^2$ **0.64**
22. $-(1.2)^2$ **-1.44**
23. $\left(\frac{2}{3}\right)^2$ **$\frac{4}{9}$**
24. $\left(-\frac{4}{11}\right)^2$ **$\frac{16}{121}$**

For Lesson 1-5 — Compare Real Numbers

Identify each statement as *true* or *false*.

25. $-5 < -7$ **false**
26. $6 > -8$ **true**
27. $-2 \geq -2$ **true**
28. $-3 \geq -3.01$ **true**
29. $-9.02 < -9.2$ **false**
30. $\frac{1}{5} < \frac{1}{8}$ **false**
31. $\frac{2}{5} \geq \frac{16}{40}$ **true**
32. $\frac{3}{4} > 0.8$ **false**

FOLDABLES™ Study Organizer

Relations and Functions Make this Foldable to help you organize your notes. Begin with one sheet of notebook paper.

Step 1 Fold

Fold lengthwise to the holes.

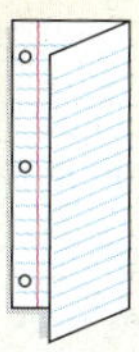

Step 2 Open and Label

Open and label the columns as shown.

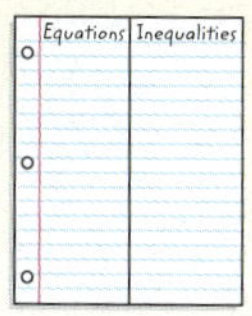

Reading and Writing As you read and study the chapter, write notes, examples, and graphs in each column.

Getting Started

This section provides a review of the basic concepts needed before beginning Chapter 1. Page references are included for additional student help.

Prerequisite Skills in the Getting Ready for the Next Lesson section at the end of each exercise set review a skill needed in the next lesson.

For Lesson	Prerequisite Skill
1-2	Evaluating Square Roots (p. 10)
1-3	Evaluating Expressions (p. 18)
1-4	Additive Inverses (p. 27)
1-5	Solving Equations (p. 32)
1-6	Solving Absolute Value Equations (p. 39)

Each chapter opens with Prerequisite Skills practice for lessons in the chapter. More Prerequisite Skill practice can be found at the end of each lesson.

FOLDABLES™ Study Organizer

For more information about Foldables, see *Teaching Mathematics with Foldables.*

Note-Taking and Charting Main Ideas Use this Foldable study guide for student notes about equations and inequalities. Note-taking is a skill that is based upon listening or reading for main ideas and then recording those ideas for future reference. In the columns of their Foldable, have students take notes about the processes and procedures that they learn. Encourage students to apply what they know and what they learn as they analyze and solve equations and inequalities.

Foldables™ are a unique way to enhance students' study skills. Encourage students to add to their Foldable as they work through the chapter, and use it to review for their chapter test.

1-1 Lesson Notes

1 Focus

5-Minute Check Transparency 1-1 Use as a quiz or review of prerequisite skills.

Mathematical Background notes are available for this lesson on p. 4C.

Building on Prior Knowledge

In previous courses, students have performed operations on integers and used the order of operations. In this lesson, they should realize that using formulas requires these skills.

How are formulas used by nurses?

Ask students:

- What are the units for the flow rate F? **drops per minute**
- Why is 12 hours multiplied by 60? **to convert the time from hours to minutes**
- **Medicine** What might happen if the flow rate is too fast or slow? **Too fast: the fluid might not be absorbed by the patient's body as expected; too slow: the medication might not be effective.**

1-1 Expressions and Formulas

What You'll Learn

- Use the order of operations to evaluate expressions.
- Use formulas.

Vocabulary

- order of operations
- variable
- algebraic expression
- formula

Lessons open with a question that is designed to engage students in the mathematics of the lesson. These opening problems should also help to answer the question "When am I ever going to use this?"

How are formulas used by nurses?

Nurses setting up intravenous or IV fluids must control the flow rate F, in drops per minute. They use the formula $F = \frac{V \times d}{t}$, where V is the volume of the solution in milliliters, d is the drop factor in drops per milliliter, and t is the time in minutes.

Suppose a doctor orders 1500 milliliters of IV saline to be given over 12 hours. Using a drop factor of 15 drops per milliliter, the expression $\frac{1500 \times 15}{12 \times 60}$ gives the correct flow rate for this patient's IV.

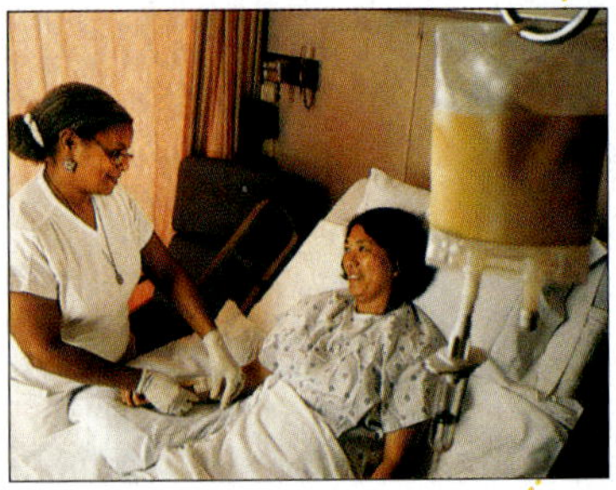

ORDER OF OPERATIONS A numerical expression such as $\frac{1500 \times 15}{12 \times 60}$ must have exactly one value. In order to find that value, you must follow the **order of operations**.

Key Concept *Order of Operations*

Step 1 Evaluate expressions inside grouping symbols, such as parentheses, ( ), brackets, [ ], braces, { }, and fraction bars, as in $\frac{5+7}{2}$.

Step 2 Evaluate all powers.

Step 3 Do all multiplications and/or divisions from left to right.

Step 4 Do all additions and/or subtractions from left to right.

Grouping symbols can be used to change or clarify the order of operations. When calculating the value of an expression, begin with the innermost set of grouping symbols.

Example 1 *Simplify an Expression*

Find the value of $[2(10 - 4)^2 + 3] \div 5$.

$[2(10 - 4)^2 + 3] \div 5 = [2(6)^2 + 3] \div 5$ First subtract 4 from 10.

$= [2(36) + 3] \div 5$ Then square 6.

$= (72 + 3) \div 5$ Multiply 36 by 2.

$= 75 \div 5$ Add 72 and 3.

$= 15$ Finally, divide 75 by 5.

The value is 15.

Resource Manager

Workbook and Reproducible Masters

Chapter 1 Resource Masters

- Study Guide and Intervention, pp. 1–2
- Skills Practice, p. 3
- Practice, p. 4
- Reading to Learn Mathematics, p. 5
- Enrichment, p. 6

School-to-Career Masters, p. 1

Science and Mathematics Lab Manual, pp. 91–96

Transparencies

5-Minute Check Transparency 1-1
Real-World Transparency 1
Answer Key Transparencies

Technology

Interactive Chalkboard

Scientific calculators follow the order of operations.

Graphing Calculator Investigation
Order of Operations

Think and Discuss 2, 4, 5. See margin.

1. Simplify $8 - 2 \times 4 + 5$ using a graphing calculator. 5
2. Describe the procedure the calculator used to get the answer.
3. Where should parentheses be inserted in $8 - 2 \times 4 + 5$ so that the expression has each of the following values?
   a. -10 around $4 + 5$ **b.** 29 around $8 - 2$ **c.** -5 around $2 \times 4 + 5$
4. Evaluate $18^2 \div (2 \times 3)$ using your calculator. Explain how the answer was calculated.
5. If you remove the parentheses in Exercise 4, would the solution remain the same? Explain.

Variables are symbols, usually letters, used to represent unknown quantities. Expressions that contain at least one variable are called **algebraic expressions**. You can evaluate an algebraic expression by replacing each variable with a number and then applying the order of operations.

Example 2 Evaluate an Expression

Evaluate $x^2 - y(x + y)$ if $x = 8$ and $y = 1.5$.

$x^2 - y(x + y) = 8^2 - 1.5(8 + 1.5)$	Replace x with 8 and y with 1.5.
$= 8^2 - 1.5(9.5)$	Add 8 and 1.5.
$= 64 - 1.5(9.5)$	Find 8^2.
$= 64 - 14.25$	Multiply 1.5 and 9.5.
$= 49.75$	Subtract 14.25 from 64.

The value is 49.75.

Study Tip

Common Misconception
A common error in this type of problem is to subtract before multiplying.

$64 - 1.5(9.5) \neq 62.5(9.5)$

Remember to follow the order of operations.

Example 3 Expression Containing a Fraction Bar

Evaluate $\frac{a^3 + 2bc}{c^2 - 5}$ if $a = 2$, $b = -4$, and $c = -3$.

The fraction bar acts as both an operation symbol, indicating division, and as a grouping symbol. Evaluate the expressions in the numerator and denominator separately before dividing.

$\frac{a^3 + 2bc}{c^2 - 5} = \frac{2^3 + 2(-4)(-3)}{(-3)^2 - 5}$	$a = 2$, $b = -4$, and $c = -3$
$= \frac{8 + (-8)(-3)}{9 - 5}$	Evaluate the numerator and the denominator separately.
$= \frac{8 + 24}{9 - 5}$	Multiply -8 by -3.
$= \frac{32}{4}$ or 8	Simplify the numerator and the denominator. Then divide.

The value is 8.

www.algebra2.com/extra_examples

2 Teach

ORDER OF OPERATIONS

In-Class Examples Power Point®

1. Find the value of $[384 - 3(7 - 2)^3] \div 3$. 3
2. Evaluate $s - t(s^2 - t)$ if $s = 2$ and $t = 3.4$. -0.04
3. Evaluate $\frac{8xy + z^3}{y^2 + 5}$ if $x = 5$, $y = -2$, and $z = -1$. -9

Teaching Tip Ask students what sign the cube of a negative number has. negative sign

Answers

Graphing Calculator Investigation

2. The calculator multiplies 2 by 4, subtracts the result from 8, and then adds 5.
4. 54; The calculator found the square of 18 and divided it by the product of 2 and 3.
5. No; you would square 18 and then divide it by 2. The result would then be multiplied by 3.

Interactive Chalkboard
PowerPoint® Presentations

This CD-ROM is a customizable Microsoft® PowerPoint® presentation that includes:

- Step-by-step, dynamic solutions of each In-Class Example from the Teacher Wraparound Edition
- Additional, Your Turn exercises for each example
- The 5-Minute Check Transparencies
- Hot links to Glencoe Online Study Tools

Graphing Calculator Investigation

Order of Operations To help find entry errors, have students work in pairs so one of them can watch as their partner performs the keystrokes to enter the expression. Sometimes it is necessary to use parentheses to obtain the correct answer with fractional expressions. For example, to evaluate $\frac{4(12)}{5(4)}$, you must enter 4 * 12/(5 * 4). Ask students why this is so.

FORMULAS

In-Class Example Power Point®

4 **GEOMETRY** Find the area of a trapezoid with base lengths of 13 meters and 25 meters and a height of 8 meters. **152 m^2**

3 Practice/Apply

Study Notebook

Have students—

- add the definitions/examples of the vocabulary terms to their Vocabulary Builder worksheets for Chapter 1.
- copy several of the formulas (for example, the area of a trapezoid), and include notes about when the formula is used.
- make a sketch of a trapezoid and label the variables used in the formula for its area.
- include any other item(s) that they find helpful in mastering the skills in this lesson.

About the Exercises...

Organization by Objective

- **Order of Operations:** 16–37
- **Formulas:** 38–54

Odd/Even Assignments

Exercises 16–49 are structured so that students practice the same concepts whether they are assigned odd or even problems.

Alert! Exercise 53 involves research on the Internet or other reference materials.

Assignment Guide

Basic: 17–33 odd, 37–47 odd, 53, 55–66

Average: 17–53 odd, 55–66

Advanced: 16–54 even, 55–58, (optional: 59–66)

Examples illustrate all of the concepts taught in the lesson and closely mirror the exercises in the Guided Practice and Practice and Apply sections.

FORMULAS A **formula** is a mathematical sentence that expresses the relationship between certain quantities. If you know the value of every variable in the formula except one, you can find the value of the remaining variable.

Example 4 Use a Formula

GEOMETRY **The formula for the area A of a trapezoid is $A = \frac{1}{2}h(b_1 + b_2)$, where h represents the height, and b_1 and b_2 represent the measures of the bases. Find the area of the trapezoid shown below.**

Substitute each value given into the formula. Then evaluate the expression using the order of operations.

$A = \frac{1}{2}h(b_1 + b_2)$	Area of a trapezoid
$= \frac{1}{2}(10)(16 + 52)$	Replace h with 10, b_1 with 16, and b_2 with 52.
$= \frac{1}{2}(10)(68)$	Add 16 and 52.
$= 5(68)$	Divide 10 by 2.
$= 340$	Multiply 5 by 68.

The area of the trapezoid is 340 square inches.

Check for Understanding

Concept Check

1. **Describe** how you would evaluate the expression $a + b[(c + d) \div e]$ given values for a, b, c, d, and e. **1. First, find the sum of c and d. Divide this sum by e. Multiply the quotient by b. Finally, add a.**
2. **OPEN ENDED** Give an example of an expression where subtraction is performed before division and the symbols ( ), [ ], or { } are not used. **2. Sample answer: $\frac{14 - 4}{5}$**
3. **Determine** which expression below represents the amount of change someone would receive from a \$50 bill if they purchased 2 children's tickets at \$4.25 each and 3 adult tickets at \$7 each at a movie theater. Explain.

   a. $50 - 2 \times 4.25 + 3 \times 7$ **b.** $50 - (2 \times 4.25 + 3 \times 7)$

   c. $(50 - 2 \times 4.25) + 3 \times 7$ **d.** $50 - (2 \times 4.25) + (3 \times 7)$

   b; See margin for explanation.

Guided Practice

GUIDED PRACTICE KEY	
Exercises	Examples
4–9	1, 3
10–12	2
13–15	4

Find the value of each expression.

4. $8(3 + 6)$ **72**
5. $10 - 8 \div 2$ **6**
6. $14 \cdot 2 - 5$ **23**
7. $[9 + 3(5 - 7)] \div 3$ **1**
8. $[6 - (12 - 8)^2] \div 5$ **−2**
9. $\frac{17(2 + 26)}{4}$ **119**

Evaluate each expression if $x = 4$, $y = -2$, and $z = 6$.

10. $z - x + y$ **0**
11. $x + (y - 1)^3$ **−23**
12. $x + [3(y + z) - y]$ **18**

DAILY INTERVENTION

Differentiated Instruction

Visual/Spatial Suggest that students first rewrite an expression they are to evaluate and then write the value for each variable on top of that variable before they start to evaluate the expression. Students may find it helpful to use colored pencils to color code the values for the different variables in an expression.

Application

BANKING **For Exercises 13–15, use the following information.**
Simple interest is calculated using the formula $I = prt$, where p represents the principal in dollars, r represents the annual interest rate, and t represents the time in years. Find the simple interest I given each of the following values.

13. $p = \$1800$, $r = 6\%$, $t = 4$ years **\$432**
14. $p = \$5000$, $r = 3.75\%$, $t = 10$ years **\$1875**
15. $p = \$31{,}000$, $r = 2\frac{1}{2}\%$, $t = 18$ months **\$1162.50**

★ indicates increased difficulty

Practice and Apply

Homework Help

For Exercises	See Examples
16–37	1, 3
38–50	2, 3
51–54	4

Extra Practice
See page 828.

Homework Help charts show students which examples to refer to if they need additional practice. Extra Practice for every lesson is provided on pages 828–861.

Find the value of each expression.

16. $18 + 6 \div 3$ **20**
17. $7 - 20 \div 5$ **3**
18. $3(8 + 3) - 4$ **29**
19. $(6 + 7)2 - 1$ **25**
20. $2(6^2 - 9)$ **54**
21. $-2(3^2 + 8)$ **−34**
22. $2 + 8(5) \div 2 - 3$ **19**
23. $4 + 64 \div (8 \times 4) \div 2$ **5**
24. $[38 - (8 - 3)] \div 3$ **11**
25. $10 - [5 + 9(4)]$ **−31**
26. $1 - \{30 \div [7 + 3(-4)]\}$ **7**
27. $12 + \{10 \div [11 - 3(2)]\}$ **14**
28. $\frac{1}{3}(4 - 7^2)$ **−15**
29. $\frac{1}{2}[9 + 5(-3)]$ **−3**
30. $\frac{16(9 - 22)}{4}$ **−52**
31. $\frac{45(4 + 32)}{10}$ **162**
32. $0.3(1.5 + 24) \div 0.5$ **15.3**
33. $1.6(0.7 + 3.3) \div 2.5$ **2.56**
★34. $\frac{1}{5} - \frac{20(81 \div 9)}{25}$ **−7**
★35. $\frac{12(52 \div 2^2)}{6} - \frac{2}{3}$ **$25\frac{1}{3}$**

36. **BICYCLING** The amount of pollutants saved by riding a bicycle rather than driving a car is calculated by adding the organic gases, carbon monoxide, and nitrous oxides emitted. To find the pounds of pollutants created by starting a typical car 10 times and driving it for 50 miles, find the value of the expression $\frac{(52.84 \times 10) + (5.955 \times 50)}{454}$. **about 1.8 lb**

37. **NURSING** Determine the IV flow rate for the patient described at the beginning of the lesson by finding the value of $\frac{1500 \times 15}{12 \times 60}$. **31.25 drops per min**

Evaluate each expression if $w = 6$, $x = 0.4$, $y = \frac{1}{2}$, and $z = -3$.

38. $w + x + z$ **3.4**
39. $w + 12 \div z$ **2**
40. $w(8 - y)$ **45**
41. $z(x + 1)$ **−4.2**
42. $w - 3x + y$ **5.3**
43. $5x + 2z$ **−4**
44. $z^4 - w$ **75**
45. $(5 - w)^2 + x$ **1.4**
46. $\frac{5wx}{z}$ **−4**
47. $\frac{2z - 15x}{3y}$ **−8**
★48. $(x - y)^2 - 2wz$ **36.01**
★49. $\frac{1}{y} + \frac{1}{w}$ **$2\frac{1}{6}$**

50. **GEOMETRY** The formula for the area A of a circle with diameter d is $A = \pi\left(\frac{d}{2}\right)^2$. Write an expression to represent the area of the circle. **$\pi\left(\frac{y+5}{2}\right)^2$**

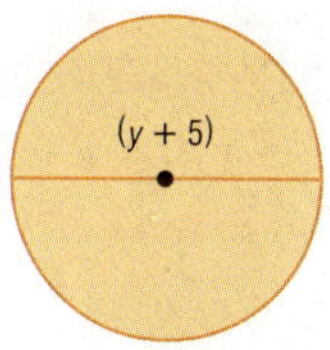

★51. Find the value of ab^n if $n = 3$, $a = 2000$, and $b = -\frac{1}{5}$. **−16**

www.algebra2.com/self_check_quiz

3. The sum of the cost of adult and children tickets should be subtracted from 50. Therefore parentheses need to be inserted around this sum to insure that this addition is done before subtraction.

Enrichment, p. 6

Significant Digits

All measurements are approximations. The **significant digits** of an approximate number are those which indicate the results of a measurement. For example, the mass of an object, measured to the nearest gram, is 210 grams. The measurement 210 g has 3 significant digits. The mass of the same object, measured to the nearest 100 g, is 200 g. The measurement 200 g has one significant digit.

1. Nonzero digits and zeros between significant digits are significant. For example, the measurement 9.071 m has 4 significant digits, 9, 0, 7, and 1.
2. Zeros at the end of a decimal fraction are significant. The measurement 0.050 mm has 2 significant digits, 5 and 0.
3. Underlined zeros in whole numbers are significant. The measurement 104,000 km has 5 significant digits, 1, 0, 4, 0, and 0.

In general, a computation involving multiplication or division of measurements *cannot* be more accurate than the least accurate measurement in the computation. Thus, the result of computation involving multiplication or division of measurements should be rounded to the number of significant digits in the least ...ent.

Study Guide and Intervention, p. 1 (shown) and p. 2

Order of Operations

Order of Operations	1. Simplify the expressions inside grouping symbols. 2. Evaluate all powers. 3. Do all multiplications and divisions from left to right. 4. Do all additions and subtractions from left to right.

Example 1 **Evaluate $[18 - (6 + 4)] \div 2$.**

$[18 - (6 + 4)] \div 2 = [18 - 10] \div 2$
$= 8 \div 2$
$= 4$

Example 2 **Evaluate $3x^2 + x(y - 5)$ if $x = 3$ and $y = 0.5$.**

Replace each variable with the given value.
$3x^2 + x(y - 5) = 3 \cdot (3)^2 + 3(0.5 - 5)$
$= 3 \cdot (9) + 3(-4.5)$
$= 27 - 13.5$
$= 13.5$

Exercises

Find the value of each expression.

1. $14 + (6 \div 2)$ **17**
2. $11 - (3 + 2)^2$ **−14**
3. $2 + (4 - 2)^3 - 6$ **4**
4. $9(3^2 + 6)$ **135**
5. $(5 + 2^3)^2 - 5^2$ **144**
6. $5^2 + \frac{1}{4} + 18 \div 2$ **34.25**
7. $\frac{16 + 2^3 \div 4}{1 - 2^2}$ **−6**
8. $(7 - 3^2)^2 + 6^2$ **40**
9. $20 \div 2^2 + 6$ **11**
10. $12 + 6 \div 3 - 2(4)$ **6**
11. $14 \div (8 - 20 \div 2)$ **−7**
12. $6(7) + 4 \div 4 - 5$ **38**
13. $8(4^2 \div 8 - 32)$ **−240**
14. $\frac{6 + 4 \div 2}{4 \div 6 - 1}$ **−24**
15. $\frac{6 + 9 \div 3 + 15}{8 - 2}$ **4**

Evaluate each expression if $a = 8.2$, $b = -3$, $c = 4$, and $d = -\frac{1}{2}$.

16. $\frac{ab}{d}$ **49.2**
17. $5(6c - 8b + 10d)$ **215**
18. $\frac{c^2 - 1}{b - d}$ **−6**
19. $ac - bd$ **31.3**
20. $(b - c)^2 + 4a$ **81.8**
21. $\frac{a}{d} + 6b - 5c$ **−54.4**
22. $3\left(\frac{c}{d}\right) - b$ **−21**
23. $cd + \frac{b}{d}$ **4**
24. $d(a + c)$ **−6.1**
25. $a + b \div c$ **7.45**
26. $b - c + 4 \div d$ **−15**
27. $\frac{a}{b + c} - d$ **8.7**

Skills Practice, p. 3 and Practice, p. 4 (shown)

Find the value of each expression.

1. $3(4 - 7) - 11$ **−20**
2. $4(12 - 4^2)$ **−16**
3. $1 + 2 - 3(4) \div 2$ **−3**
4. $12 - [20 - 2(6^2 \div 3 \times 2^2)]$ **88**
5. $20 \div (5 - 3) + 5^2(3)$ **85**
6. $(-2)^3 - (3)(8) + (5)(10)$ **18**
7. $18 - \{5 - [34 - (17 - 11)]\}$ **41**
8. $[4(5 - 3) - 2(4 - 8)] \div 16$ **1**
9. $\frac{1}{2}[6 - 4^2]$ **−5**
10. $\frac{1}{4}[-5 + 5(-3)]$ **−5**
11. $\frac{-8(13 - 37)}{6}$ **32**
12. $\frac{(-8)^2}{5 - 9} - (-1)^2 + 4(-9)$ **−53**

Evaluate each expression if $a = \frac{3}{4}$, $b = -8$, $c = -2$, $d = 3$, and $e = \frac{1}{3}$.

13. $ab^2 - d$ **45**
14. $(c + d)b$ **−8**
15. $\frac{ab}{c} + d^2$ **12**
16. $\frac{d(b - c)}{ac}$ **12**
17. $(b - de)e^2$ **−1**
18. $ac^3 - b^2de$ **−70**
19. $-b[a + (c - d)^2]$ **206**
20. $\frac{ac^4}{d} - \frac{c}{e^2}$ **22**
21. $9bc - \frac{1}{e}$ **141**
22. $2ab^2 - (d^3 - c)$ **67**

23. **TEMPERATURE** The formula $F = \frac{9}{5}C + 32$ gives the temperature in degrees Fahrenheit for a given temperature in degrees Celsius. What is the temperature in degrees Fahrenheit when the temperature is −40 degrees Celsius? **−40°F**

24. **PHYSICS** The formula $h = 120t - 16t^2$ gives the height h in feet of an object t seconds after it is shot upward from Earth's surface with an initial velocity of 120 feet per second. What will the height of the object be after 6 seconds? **144 ft**

25. **AGRICULTURE** Faith owns an organic apple orchard. From her experience the last few seasons, she has developed the formula $P = 20x - 0.01x^2 - 240$ to predict her profit P in dollars this season if her trees produce x bushels of apples. What is Faith's predicted profit this season if her orchard produces 300 bushels of apples? **\$4860**

Reading to Learn Mathematics, p. 5 — ELL

Pre-Activity **How are formulas used by nurses?**

Read the introduction to Lesson 1-1 at the top of page 6 in your textbook.

- Nurses use the formula $F = \frac{V \times d}{t}$ to control the flow rate for IVs. Name the quantity that each of the variables in this formula represents and the units in which each is measured.

  F represents the **flow rate** and is measured in **drops** per minute.

  V represents the **volume** of solution and is measured in **milliliters**.

  d represents the **drop factor** and is measured in **drops** per milliliter.

  t represents **time** and is measured in **minutes**.

- Write the expression that a nurse would use to calculate the flow rate of an IV if a doctor orders 1350 milliliters of IV saline to be given over 8 hours, with a drop factor of 20 drops per milliliter. Do not find the value of this expression. **$\frac{1350 \times 20}{8 \times 60}$**

Reading the Lesson

1. There is a customary order for grouping symbols. Brackets are used outside of parentheses. Braces are used outside of brackets. Identify the innermost expression(s) in each of the following expressions.
   a. $[(3 - 2^2) + 8] \div 4$ **$(3 - 2^2)$**
   b. $9 - [5(8 - 6) + 2(10 + 7)]$ **$(8 - 6)$ and $(10 + 7)$**
   c. $\{14 - [8 + (3 - 12)^2]\} \div (6^3 - 100)$ **$(3 - 12)$**
2. Read the following instructions. Then use grouping symbols to show how the instructions can be put in the form of a mathematical expression.
   Multiply the difference of 13 and 5 by the sum of 9 and 21. Add the result to 10. Then divide what you get by 2. **$[(13 - 5)(9 + 21) + 10] \div 2$**
3. Why is it important for everyone to use the same order of operations for evaluating expressions? **Sample answer: If everyone did not use the same order of operations, different people might get different answers.**

Helping You Remember

4. Think of a phrase or sentence to help you remember the order of operations. **Sample answer: Please excuse my dear Aunt Sally. (parentheses; exponents; multiplication and division; addition and subtraction)**

4 Assess

Open-Ended Assessment

Speaking Ask students to state various formulas they remember using in previous courses, and to explain what each variable represents (for example, $P = 2(\ell + w)$ to find the perimeter of a rectangle, where ℓ is the length and w is the width). Then have a volunteer suggest appropriate values for the variables in the formula. Ask the class as a whole to evaluate the given formula using the suggested values.

Tips for New Teachers

Intervention Students may be reluctant to take time to show all the steps they use when evaluating an expression, such as showing the substituted values before doing the computations. Help them see that these steps enable them to self-diagnose errors and to prevent calculation errors that might keep them from getting correct values.

Getting Ready for Lesson 1-2

PREREQUISITE SKILL Lesson 1-2 presents the properties of real numbers and the subsets of the real numbers, including irrationals. Remind students that the square root of a number is irrational if that number is not a perfect square. Exercises 59–66 should be used to determine your students' familiarity with evaluating square roots.

More About. . .

Fireworks

To estimate the width w in feet of a firework burst, use the formula $w = 20At$. In this formula, A is the estimated viewing angle of the firework display and t is the time in seconds from the instant you see the light until you hear the sound.

Source: www.efg2.com

New teachers, or teachers new to teaching mathematics, may especially appreciate the Tips for New Teachers.

52. **MEDICINE** Suppose a patient must take a blood pressure medication that is dispensed in 125-milligram tablets. The dosage is 15 milligrams per kilogram of body weight and is given every 8 hours. If the patient weighs 25 kilograms, how many tablets would be needed for a 30-day supply? Use the formula $n = [15b \div (125 \times 8)] \times 24d$, where n is the number of tablets, d is the number of days the supply should last, and b is the body weight of the patient in kilograms. **270**

53. **MONEY** In 1950, the average price of a car was about \$2000. This may sound inexpensive, but the average income in 1950 was much less than it is now. To compare dollar amounts over time, use the formula $V = \frac{A}{S}C$, where A is the old dollar amount, S is the starting year's Consumer Price Index (CPI), C is the converting year's CPI, and V is the current value of the old dollar amount. Buying a car for \$2000 in 1950 was like buying a car for how much money in 2000? **\$14,439.83**

Year	Average CPI
1950	42.1
1960	29.6
1970	38.8
1980	82.4
1990	130.7
2000	174.0

Source: U.S. Department of Labor

Online Research **Data Update** What is the current Consumer Price Index? Visit www.algebra2.com/data_update to learn more.

54. **FIREWORKS** Suppose you are about a mile from a fireworks display. You count 5 seconds between seeing the light and hearing the sound of the firework display. You estimate the viewing angle is about 4°. Using the information at the left, estimate the width of the firework display. **400 ft**

55. **CRITICAL THINKING** Write expressions having values from one to ten using exactly four 4s. You may use any combination of the operation symbols $+, -, \times, \div$, and/or grouping symbols, but no other numbers are allowed. An example of such an expression with a value of zero is $(4 + 4) - (4 + 4)$. **See margin.**

56. **WRITING IN MATH** Answer the question that was posed at the beginning of the lesson. **See margin.**

How are formulas used by nurses?

Include the following in your answer:
- an explanation of why a formula for the flow rate of an IV is more useful than a table of specific IV flow rates, and
- a description of the impact of using a formula, such as the one for IV flow rate, incorrectly.

Standardized Test Practice

57. Find the value of $1 + 3(5 - 17) \div 2 \times 6$. **C**

(A) -4 (B) 109
(C) -107 (D) -144

58. The following are the dimensions of four rectangles. Which rectangle has the same area as the triangle at the right? **D**

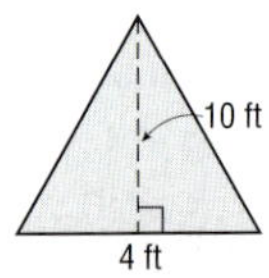

(A) 1.6 ft by 25 ft (B) 5 ft by 16 ft
(C) 3.5 ft by 4 ft (D) 0.4 ft by 50 ft

Maintain Your Skills

Getting Ready for the Next Lesson **PREREQUISITE SKILL** **Evaluate each expression.**

59. $\sqrt{9}$ **3** 60. $\sqrt{16}$ **4** 61. $\sqrt{100}$ **10** 62. $\sqrt{169}$ **13**

63. $-\sqrt{4}$ **−2** 64. $-\sqrt{25}$ **−5** 65. $\sqrt{\frac{4}{9}}$ **$\frac{2}{3}$** 66. $\sqrt{\frac{36}{49}}$ **$\frac{6}{7}$**

Answers

55. Sample answer:
$4 - 4 + 4 \div 4 = 1$
$4 \div 4 + 4 \div 4 = 2$
$(4 + 4 + 4) \div 4 = 3$
$4 \times (4 - 4) + 4 = 4$
$(4 \times 4 + 4) \div 4 = 5$
$(4 + 4) \div 4 + 4 = 6$
$44 \div 4 - 4 = 7$
$(4 + 4) \times (4 \div 4) = 8$
$4 + 4 + 4 \div 4 = 9$
$(44 - 4) \div 4 = 10$

56. Nurses use formulas to calculate a drug dosage given a supply dosage and a doctor's drug order. They also use formulas to calculate IV flow rates. Answers should include the following.
- A table of IV flow rates is limited to those situations listed, while a formula can be used to find any IV flow rate.
- If a formula used in a nursing setting is applied incorrectly, a patient could die.

1-2 Properties of Real Numbers

What You'll Learn

- Classify real numbers.
- Use the properties of real numbers to evaluate expressions.

Vocabulary

- real numbers
- rational numbers
- irrational numbers

How is the Distributive Property useful in calculating store savings?

Manufacturers often offer coupons to get consumers to try their products. Some grocery stores try to attract customers by doubling the value of manufacturers' coupons. You can use the Distributive Property to calculate these savings.

REAL NUMBERS All of the numbers that you use in everyday life are **real numbers**. Each real number corresponds to exactly one point on the number line, and every point on the number line represents exactly one real number.

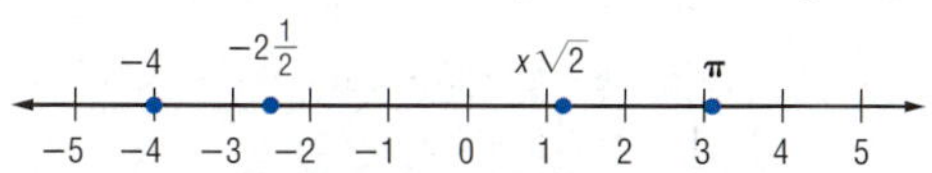

Real numbers can be classified as either **rational** or **irrational**.

Study Tip

Reading Math
A *ratio* is the comparison of two numbers by division.

Key Concept — Real Numbers

Rational Numbers

- **Words** A rational number can be expressed as a ratio $\frac{m}{n}$, where m and n are integers and n is not zero. The decimal form of a rational number is either a terminating or repeating decimal.
- **Examples** $\frac{1}{6}$, 1.9, 2.575757..., -3, $\sqrt{4}$, 0

Irrational Numbers

- **Words** A real number that is not rational is irrational. The decimal form of an irrational number neither terminates nor repeats.
- **Examples** $\sqrt{5}$, π, 0.010010001...

The sets of natural numbers, {1, 2, 3, 4, 5, ...}, whole numbers, {0, 1, 2, 3, 4, ...}, and integers, {..., −3, − 2, − 1, 0, 1, 2, ...} are all subsets of the rational numbers. The whole numbers are a subset of the rational numbers because every whole number n is equal to $\frac{n}{1}$.

1-2 Lesson Notes

1 Focus

5-Minute Check Transparency 1-2 Use as a quiz or review of Lesson 1-1.

Mathematical Background notes are available for this lesson on p. 4C.

Building on Prior Knowledge

In Lesson 1-1, students simplified and evaluated expressions. In this lesson, they broaden those skills to include using the real numbers and applying the commutative, associative, identity, inverse, and distributive properties of real numbers.

How is the Distributive Property useful in calculating store savings?

Ask students:

- In the list of Scanned Coupons and Bonus Coupons shown, what does 0.30 mean? **30¢**
- Why is there a negative sign after the decimal numbers? **The negative sign indicates that the amount is taken off or subtracted from the price.**

Resource Manager

Workbook and Reproducible Masters

Chapter 1 Resource Masters

- Study Guide and Intervention, pp. 7–8
- Skills Practice, p. 9
- Practice, p. 10
- Reading to Learn Mathematics, p. 11
- Enrichment, p. 12
- Assessment, p. 51

Teaching Algebra With Manipulatives Masters, p. 212

Transparencies

5-Minute Check Transparency 1-2
Answer Key Transparencies

Technology

Interactive Chalkboard

2 Teach

REAL NUMBERS

Teaching Tip Point out that a nonterminating decimal whose digits show a pattern but which has no repeating group of digits, such as the number 0.010010001… given in the Key Concepts examples on p. 11, is irrational. Another example is the number 1.232233222333….

In-Class Example Power Point®

1 Name the sets of numbers to which each number belongs.

a. $-\frac{2}{3}$ **Q, R**

b. 9.999… **Q, R**

c. $\sqrt{6}$ **I, R**

d. $\sqrt{100}$ **N, W, Z, Q, R**

e. −23.3 **Q, R**

Reading Tip Ask students whether *fraction* and *rational number* mean the same thing. (No; 4 is not a fraction but it is a rational number. *Fraction* refers to the form of a number: $\frac{8}{4}$ is in the form of a fraction but it is a whole number in value.)

PROPERTIES OF REAL NUMBERS

Reading Tip Help students remember the names of properties by connecting the term *commutative* with "commuting, or moving from one position to another," and by connecting the term *associative* with "the people you associate with, or your group."

Study Tips offer students helpful information about the topics they are studying.

Real Numbers (R)

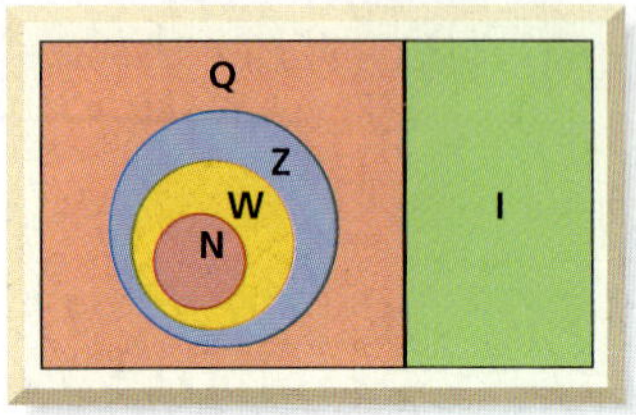

The Venn diagram shows the relationships among these sets of numbers.

R = reals Q = rationals
I = irrationals Z = integers
W = wholes N = naturals

The square root of any whole number is either a whole number or it is irrational. For example, $\sqrt{36}$ is a whole number, but $\sqrt{35}$, since it lies between 5 and 6, must be irrational.

Study Tip

Common Misconception
Do not assume that a number is irrational because it is expressed using the square root symbol. Find its value first.

Example 1 *Classify Numbers*

Name the sets of numbers to which each number belongs.

a. $\sqrt{16}$

$\sqrt{16} = 4$ naturals (N), wholes (W), integers (Z), rationals (Q), reals (R)

b. −185 integers (Z), rationals (Q), and reals (R)

c. $\sqrt{20}$ irrationals (I) and reals (R)

$\sqrt{20}$ lies between 4 and 5 so it is not a whole number.

d. $-\frac{7}{8}$ rationals (Q) and reals (R)

e. $0.\overline{45}$ rationals (Q) and reals (R)

The bar over the 45 indicates that those digits repeat forever.

PROPERTIES OF REAL NUMBERS The real number system is an example of a mathematical structure called a *field*. Some of the properties of a field are summarized in the table below.

Study Tip

Reading Math
$-a$ is read the opposite of a.

Key Concepts *Real Number Properties*

For any real numbers a, b, and c:

Property	Addition	Multiplication
Commutative	$a + b = b + a$	$a \cdot b = b \cdot a$
Associative	$(a + b) + c = a + (b + c)$	$(a \cdot b) \cdot c = a \cdot (b \cdot c)$
Identity	$a + 0 = a = 0 + a$	$a \cdot 1 = a = 1 \cdot a$
Inverse	$a + (-a) = 0 = (-a) + a$	If $a \neq 0$, then $a \cdot \frac{1}{a} = 1 = \frac{1}{a} \cdot a$.
Distributive	$a(b + c) = ab + ac$ and $(b + c)a = ba + ca$	

Example 2 Identify Properties of Real Numbers

Name the property illustrated by each equation.

a. $(5 + 7) + 8 = 8 + (5 + 7)$

Commutative Property of Addition

The Commutative Property says that the order in which you add does not change the sum.

b. $3(4x) = (3 \cdot 4)x$

Associative Property of Multiplication

The Associative Property says that the way you group three numbers when multiplying does not change the product.

Example 3 Additive and Multiplicative Inverses

Identify the additive inverse and multiplicative inverse for each number.

a. $-1\frac{3}{4}$

Since $-1\frac{3}{4} + \left(1\frac{3}{4}\right) = 0$, the additive inverse of $-1\frac{3}{4}$ is $1\frac{3}{4}$.

Since $-1\frac{3}{4} = -\frac{7}{4}$ and $\left(-\frac{7}{4}\right)\left(-\frac{4}{7}\right) = 1$, the multiplicative inverse of $-1\frac{3}{4}$ is $-\frac{4}{7}$.

b. 1.25

Since $1.25 + (-1.25) = 0$, the additive inverse of 1.25 is -1.25.

The multiplicative inverse of 1.25 is $\frac{1}{1.25}$ or 0.8.

CHECK Notice that $1.25 \times 0.8 = 1$. ✓

You can model the Distributive Property using algebra tiles.

Algebra Activity

Distributive Property

- A 1 tile is a square that is 1 unit wide and 1 unit long. Its area is 1 square unit. An x tile is a rectangle that is 1 unit wide and x units long. Its area is x square units.

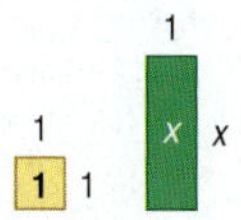

- To find the product $3(x + 1)$, model a rectangle with a width of 3 and a length of $x + 1$. Use your algebra tiles to mark off the dimensions on a product mat. Then make the rectangle with algebra tiles.
- The rectangle has 3 x tiles and 3 1 tiles. The area of the rectangle is $x + x + x + 1 + 1 + 1$ or $3x + 3$. Thus, $3(x + 1) = 3x + 3$.

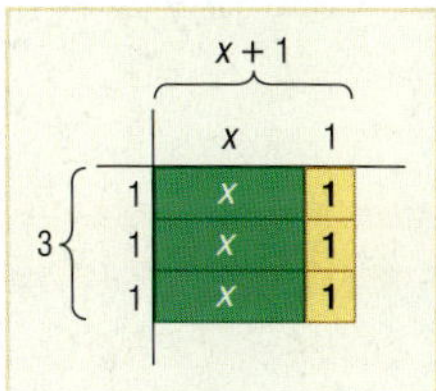

Model and Analyze

Tell whether each statement is *true* or *false*. Justify your answer with algebra tiles and a drawing. 1–4. See pp. 53A–53B for drawings.

1. $4(x + 2) = 4x + 2$ false
2. $3(2x + 4) = 6x + 7$ false
3. $2(3x + 5) = 6x + 10$ true
4. $(4x + 1)5 = 4x + 5$ false

www.algebra2.com/extra_examples

In-Class Examples

Power Point®

2 Name the property illustrated by each equation.

a. $(-8 + 8) + 15 = 0 + 15$ **Additive Inverse Property**

b. $5(8 - 6) = 5(8) - 5(6)$ **Distributive Property**

3 Identify the additive inverse and multiplicative inverse for each number.

a. -7 **additive: 7; multiplicative:** $-\frac{1}{7}$

b. $\sqrt{\frac{1}{9}}$ **additive:** $-\sqrt{\frac{1}{9}}$ **or** $-\frac{1}{3}$**; multiplicative: 3**

Teaching Tip Make sure students understand that additive inverses must have a sum of 0 and that multiplicative inverses must have a product of 1.

In-Class Examples, which are included for every example in the Student Edition, exactly parallel the examples in the text. Teaching Tips about the examples in the Student Edition are included where appropriate.

Concept Check

Real Number Properties Ask students to name some mathematical operations that are *not* commutative and to give examples supporting their choices. **subtraction and division;** $7 - 3 \neq 3 - 7$; $8 \div 2 \neq 2 \div 8$

Algebra Activity

Materials: algebra tiles, product mat

- Have students verify with their tiles that the length of an x tile is not a multiple of the side length of a 1 tile.
- Suggest that students can verify they have modeled an expression like $2(3x + 5)$ correctly if they read the expression as "2 rows of 3 x tiles and 5 1 tiles." If they arrange their models like the one shown in the book, the rows of tiles can be "read" from left to right just as when reading the text.

In-Class Examples

Power Point®

4 **POSTAGE** Audrey went to a post office and bought eight 34¢ stamps and eight 21¢ postcard stamps. How much did Audrey spend altogether on stamps? **8(0.34) + 8(0.21) or 8(0.34 + 0.21)**

5 Simplify $4(3a - b) + 2(b + 3a)$. $\mathbf{18a - 2b}$

Reading Tip Help students recall the Distributive Property by connecting the name to "distributing or handing out papers, one to each person." Point out that the factor outside of the parentheses acts as a multiplier for each term within the parentheses.

Answer

2. A rational number is the ratio of two integers. Since $\sqrt{3}$ is not an integer, $\frac{\sqrt{3}}{2}$ is not a rational number.

Daily Intervention notes help you help students when they need it most. Differentiated Instruction suggestions are keyed to eight commonly-accepted learning styles.

The Distributive Property is often used in real-world applications.

Example 4 Use the Distributive Property to Solve a Problem

FOOD SERVICE A restaurant adds a 20% tip to the bills of parties of 6 or more people. Suppose a server waits on five such tables. The bill without the tip for each party is listed in the table. How much did the server make in tips during this shift?

Party 1	Party 2	Party 3	Party 4	Party 5
\$185.45	\$205.20	\$195.05	\$245.80	\$262.00

There are two ways to find the total amount of tips received.

Method 1

Multiply each dollar amount by 20% or 0.2 and then add.

$T = 0.2(185.45) + 0.2(205.20) + 0.2(195.05) + 0.2(245.80) + 0.2(262)$

$= 37.09 + 41.04 + 39.01 + 49.16 + 52.40$

$= 218.70$

Method 2

Add the bills of all the parties and then multiply the total by 0.2.

$T = 0.2(185.45 + 205.20 + 195.05 + 245.80 + 262)$

$= 0.2(1093.50)$

$= 218.70$

The server made \$218.70 during this shift.

Notice that both methods result in the same answer.

More About. . .

Food Service

Leaving a "tip" began in 18th century English coffee houses and is believed to have originally stood for "To Insure Promptness." Today, the American Automobile Association suggests leaving a 15% tip.

Source: Market Facts, Inc.

The properties of real numbers can be used to simplify algebraic expressions.

Example 5 Simplify an Expression

Simplify $2(5m + n) + 3(2m - 4n)$.

$2(5m + n) + 3(2m - 4n)$

$= 2(5m) + 2(n) + 3(2m) - 3(4n)$	Distributive Property
$= 10m + 2n + 6m - 12n$	Multiply.
$= 10m + 6m + 2n - 12n$	Commutative Property (+)
$= (10 + 6)m + (2 - 12)n$	Distributive Property
$= 16m - 10n$	Simplify.

Check for Understanding

Concept Check

1. **OPEN ENDED** Give an example of each type of number. **Sample answers given.**
   - **a.** natural **2**
   - **b.** whole **5**
   - **c.** integer **−11**
   - **d.** rational **1.3**
   - **e.** irrational $\sqrt{2}$
   - **f.** real **−1.3**
2. **Explain** why $\frac{\sqrt{3}}{2}$ is *not* a rational number. **See margin.**
3. **Disprove** the following statement by giving a counterexample. A **counterexample** is a specific case that shows that a statement is false. Explain.
   Every real number has a multiplicative inverse.

3. 0; zero does not have a multiplicative inverse since $\frac{1}{0}$ is undefined.

DAILY INTERVENTION — Differentiated Instruction

Kinesthetic To model the Distributive Property, write $7(8 + 6)$ on the board. Then have a student distribute an index card with 7 on it to a student holding an index card with 8 written on it and also distribute an index card with 7 on it to a student holding an index card with 6 written on it. Ask each student holding 2 cards to name their product. Have the student who distributed the 7s find the sum of the products. Complete the equation on the board: $7(8 + 6) = 7(8) + 7(6)$.

Guided Practice

GUIDED PRACTICE KEY	
Exercises	Examples
4–6	1
7–9	2
10–12	3
13–16	5
17, 18	4

Name the sets of numbers to which each number belongs.

4. -4 **Z, Q, R**
5. 45 **N, W, Z, Q, R**
6. $6.\overline{23}$ **Q, R**

Name the property illustrated by each equation.

7. $\frac{2}{3} \cdot \frac{3}{2} = 1$ **Mult. Inv.**
8. $(a + 4) + 2 = a + (4 + 2)$ **Assoc. (+)**
9. $4x + 0 = 4x$ **Add. Iden.**

Identify the additive inverse and multiplicative inverse for each number.

10. -8 **$8, -\frac{1}{8}$**
11. $\frac{1}{3}$ **$-\frac{1}{3}, 3$**
12. 1.5 **$-1.5, \frac{2}{3}$**

Simplify each expression.

13. $3x + 4y - 5x$ **$-2x + 4y$**
14. $9p - 2n + 4p + 2n$ **$13p$**
15. $3(5c + 4d) + 6(d - 2c)$ **$3c + 18d$**
16. $\frac{1}{2}(16 - 4a) - \frac{3}{4}(12 + 20a)$ **$-17a - 1$**

Application

BAND BOOSTERS For Exercises 17 and 18, use the information below and in the table.

Ashley is selling chocolate bars for \$1.50 each to raise money for the band.

17. Write an expression to represent the total amount of money Ashley raised during this week.
18. Evaluate the expression from Exercise 17 by using the Distributive Property. **\$175.50**

17. 1.5(10 + 15 + 12 + 8 + 19 + 22 + 31) or 1.5(10) + 1.5(15) + 1.5(12) + 1.5(8) + 1.5(19) + 1.5(22) + 1.5(31)

Ashley's Sales for One Week

Day	Bars Sold
Monday	10
Tuesday	15
Wednesday	12
Thursday	8
Friday	19
Saturday	22
Sunday	31

★ indicates increased difficulty

Practice and Apply

Homework Help	
For Exercises	See Examples
19–27, 40–42, 59–62	1
28–39	2
43–48	3
63–65	4
49–58, 66–69	5

Extra Practice
See page 828.

Name the sets of numbers to which each number belongs. **19–26. See margin.**

19. 0
20. $-\frac{2}{9}$
21. $\sqrt{121}$
22. -4.55
23. $\sqrt{10}$
24. -31
25. $\frac{12}{2}$
★26. $\frac{3\pi}{2}$

★27. Name the sets of numbers to which all of the following numbers belong. Then arrange the numbers in order from least to greatest.
$2.\overline{49}, 2.4\overline{9}, 2.4, 2.49, 2.\overline{9}$ **Q, R; $2.4, 2.49, 2.\overline{49}, 2.4\overline{9}, 2.\overline{9}$**

Name the property illustrated by each equation. **31. Assoc. (+)**

28. $5a + (-5a) = 0$ **Add. Inv.**
29. $(3 \cdot 4) \cdot 25 = 3 \cdot (4 \cdot 25)$ **Assoc. (×)**
30. $-6xy + 0 = -6xy$ **Add. Iden.**
31. $[5 + (-2)] + (-4) = 5 + [-2 + (-4)]$
32. $(2 + 14) + 3 = 3 + (2 + 14)$ **Comm. (+)**
33. $\left(1\frac{2}{7}\right)\left(\frac{7}{9}\right) = 1$ **Mult. Inv.**
34. $2\sqrt{3} + 5\sqrt{3} = (2 + 5)\sqrt{3}$ **Dist.**
35. $ab = 1ab$ **Multi. Iden.**

NUMBER THEORY For Exercises 36–39, use the properties of real numbers to answer each question. **37. $-m$; Add. Inv.** **38. $\frac{1}{m}$; Multi. Inv.**

36. If $m + n = m$, what is the value of n? **0**
37. If $m + n = 0$, what is the value of n? What is n called with respect to m?
38. If $mn = 1$, what is the value of n? What is n called with respect to m?
39. If $mn = m$, what is the value of n? **1**

www.algebra2.com/self_check_quiz

DAILY INTERVENTION

Unlocking Misconceptions

Positive Root Remind students that $\sqrt{9}$ means only the positive root, if one exists, so $\sqrt{9} = 3$. To indicate both roots of the equation $x^2 = 9$, the mathematical notation is $x = \pm\sqrt{9}$ or $x = \pm 3$.

3 Practice/Apply

Study Notebook

Have students—

- add the definitions/examples of the vocabulary terms to their Vocabulary Builder worksheets for Chapter 1.
- copy the Venn diagram on p. 12, and add at least three examples for each set.
- copy the table of Real Number Properties and add examples that use whole numbers.
- include any other item(s) that they find helpful in mastering the skills in this lesson.

About the Exercises...

Organization by Objective

- **Real Numbers:** 19–27, 59–62
- **Properties of Real Numbers:** 28–58, 63–69

Exercises 19–26, 28–39, and 43–62 are structured so that students practice the same concepts whether they are assigned odd or even problems.

Assignment Guide

Basic: 19–25 odd, 29–39 odd, 40–42, 43–57 odd, 59, 61, 63–64, 65, 67, 69, 70–73, 78–86

Average: 19–39 odd, 40–42, 43–61 odd, 63–65, 67–73, 78–86 (optional: 74–77)

Advanced: 20–38 even, 40–42, 44–62 even, 66–82 (optional: 83–86)

All: Practice Quiz 1 (1–10)

Answers

19. W, Z, Q, R
20. Q, R
21. N, W, Z, Q, R
22. Q, R
23. I, R
24. Z, Q, R
25. N, W, Z, Q, R
26. I, R

Answers

65. $3\left(2\frac{1}{4}\right) + 2\left(1\frac{1}{8}\right)$

$= 3\left(2 + \frac{1}{4}\right) + 2\left(1 + \frac{1}{8}\right)$ Definition of a mixed number

$= 3(2) + 3\left(\frac{1}{4}\right) + 2(1) + 2\left(\frac{1}{8}\right)$ Distributive Property

$= 6 + \frac{3}{4} + 2 + \frac{1}{4}$ Multiply.

$= 6 + 2 + \frac{3}{4} + \frac{1}{4}$ Commutative Property of Addition

$= 8 + \frac{3}{4} + \frac{1}{4}$ Add.

$= 8 + \left(\frac{3}{4} + \frac{1}{4}\right)$ Associative Property of Addition

$= 8 + 1$ or 9 Add.

71. Answers should include the following.
- Instead of doubling each coupon value and then adding these values together, the Distributive Property could be applied allowing you to add the coupon values first and then double the sum.
- If a store had a 25% off sale on all merchandise, the Distributive Property could be used to calculate these savings. For example, the savings on a \$15 shirt, \$40 pair of jeans, and \$25 pair of slacks could be calculated as 0.25(15) + 0.25(40) + 0.25(25) or as 0.25(15 + 40 + 25) using the Distributive Property.

More About. . .

Math History

Pythagoras (572–497 B.C.), was a Greek philosopher whose followers came to be known as the Pythagoreans. It was their knowledge of what is called the Pythagorean Theorem that led to the first discovery of irrational numbers.

Source: *A History of Mathematics*

MATH HISTORY **For Exercises 40–42, use the following information.**
The Greek mathematician Pythagoras believed that all things could be described by numbers. By "number" he meant positive integers.

40. To what set of numbers was Pythagoras referring when he spoke of "numbers?" **natural numbers**

41. Use the formula $c = \sqrt{2s^2}$ to calculate the length of the hypotenuse c, or longest side, of this right triangle using s, the length of one leg. **$\sqrt{2}$ units**

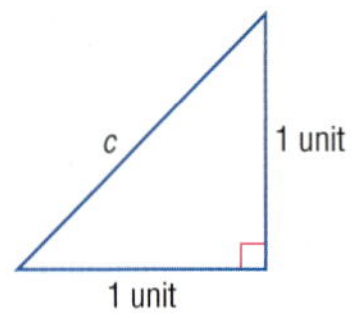

42. Explain why Pythagoras could not find a "number" to describe the value of c. **The square root of 2 is irrational and therefore cannot be described by a natural number.**

Name the additive inverse and multiplicative inverse for each number.

43. -10 **$10; -\frac{1}{10}$**
44. 2.5 **$-2.5; 0.4$**
45. -0.125 **$0.125; -8$**
46. $-\frac{5}{8}$ **$\frac{5}{8}; -\frac{8}{5}$**
47. $\frac{4}{3}$ **$-\frac{4}{3}, \frac{3}{4}$**
48. $-4\frac{3}{5}$ **$4\frac{3}{5}; -\frac{5}{23}$**

Simplify each expression. **55. $-3.4m + 1.8n$ 56. $4.4p - 2.9q$**

49. $7a + 3b - 4a - 5b$ **$3a - 2b$**
50. $3x + 5y + 7x - 3y$ **$10x + 2y$**
51. $3(15x - 9y) + 5(4y - x)$ **$40x - 7y$**
52. $2(10m - 7a) + 3(8a - 3m)$ **$11m + 10a$**
53. $8(r + 7t) - 4(13t + 5r)$ **$-12r + 4t$**
54. $4(14c - 10d) - 6(d + 4c)$ **$32c - 46d$**
55. $4(0.2m - 0.3n) - 6(0.7m - 0.5n)$
56. $7(0.2p + 0.3q) + 5(0.6p - q)$
57. $\frac{1}{4}(6 + 20y) - \frac{1}{2}(19 - 8y)$ **$-8 + 9y$**
★58. $\frac{1}{6}(3x + 5y) + \frac{2}{3}\left(\frac{3}{5}x - 6y\right)$ **$\frac{9}{10}x - \frac{19}{6}y$**

Determine whether each statement is *true* or *false*. If false, give a counterexample.

59. true
60. false; -3

59. Every whole number is an integer.
60. Every integer is a whole number.
61. Every real number is irrational. **false; 6**
62. Every integer is a rational number. **true**

WORK **For Exercises 63 and 64, use the information below and in the table.**
Andrea works as a hostess in a restaurant and is paid every two weeks.

63. If Andrea earns \$6.50 an hour, illustrate the Distributive Property by writing two expressions representing Andrea's pay last week.

63. 6.5(4.5 + 4.25 + 5.25 + 6.5 + 5) or 6.5(4.5) + 6.5(4.25) + (6.5)5.25 + 6.5(6.5) + 6.5(5)

64. Find the mean or average number of hours Andrea worked each day, to the nearest tenth of an hour. Then use this average to predict her pay for a two-week pay period. **3.6; \$327.60**

65. **BAKING** Mitena is making two types of cookies. The first recipe calls for $2\frac{1}{4}$ cups of flour, and the second calls for $1\frac{1}{8}$ cups of flour. If Mitena wants to make 3 batches of the first recipe and 2 batches of the second recipe, how many cups of flour will she need? Use the properties of real numbers to show how Mitena could compute this amount mentally. Justify each step. **See margin.**

Online Lesson Plans

USA TODAY Education's Online site offers resources and interactive features connected to each day's newspaper. *Experience TODAY*, USA TODAY's daily lesson plan, is available on the site and delivered daily to subscribers. This plan provides instruction for integrating USA TODAY graphics and key editorial features into your mathematics classroom. Log on to www.education.usatoday.com.

BASKETBALL For Exercises 66 and 67, use the diagram of an NCAA basketball court below.

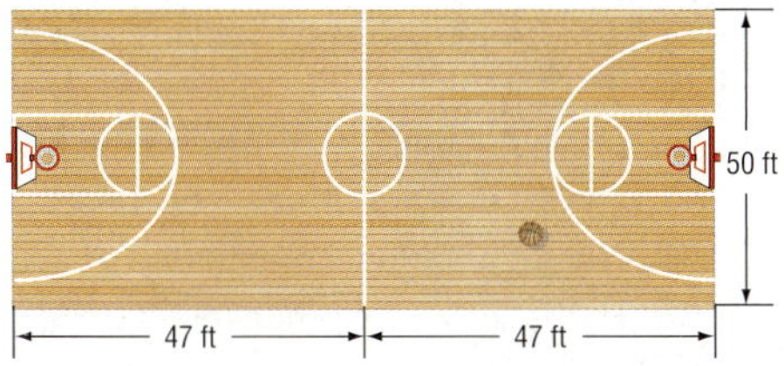

66. Illustrate the Distributive Property by writing two expressions for the area of the basketball court. **50(47 + 47); 50(47) + 50(47)**

67. Evaluate the expression from Exercise 66 using the Distributive Property. What is the area of an NCAA basketball court? **4700 ft²**

SCHOOL SHOPPING For Exercises 68 and 69, use the graph at the right.

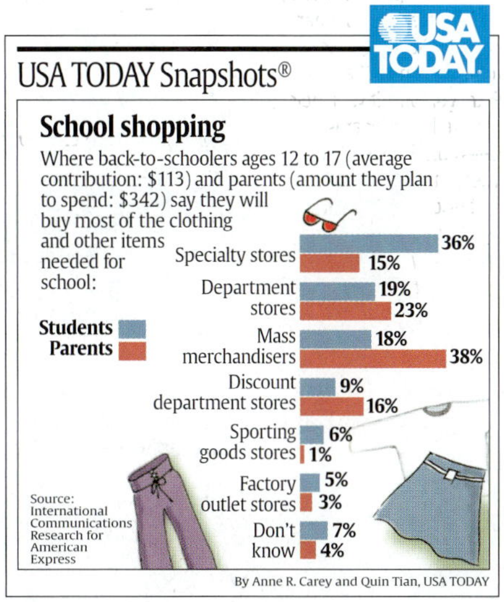

68. \$113(0.36 + 0.19); \$113(0.36) + \$113(0.19)

68. Illustrate the Distributive Property by writing two expressions to represent the amount that the average student spends shopping for school at specialty stores and department stores.

69. Evaluate the expression from Exercise 68 using the Distributive Property. **\$62.15**

70. Yes; $\frac{6+8}{2} = \frac{6}{2} + \frac{8}{2} = 7$; dividing by a number is the same as multiplying by its reciprocal.

70. **CRITICAL THINKING** Is the Distributive Property also true for division? In other words, does $\frac{b+c}{a} = \frac{b}{a} + \frac{c}{a}$, $a \neq 0$? If so, give an example and explain why it is true. If not true, give a counterexample.

Standardized Test Practice exercises were created to closely parallel those on actual state proficiency tests and college entrance exams.

71. **WRITING IN MATH** Answer the question that was posed at the beginning of the lesson. **See margin.**

How is the Distributive Property useful in calculating store savings?

Include the following in your answer:
- an explanation of how the Distributive Property could be used to calculate the coupon savings listed on a grocery receipt, and
- an example of how the Distributive Property could be used to calculate the savings from a clothing store sale where all items were discounted by the same percent.

Standardized Test Practice

72. If a and b are natural numbers, then which of the following must also be a natural number? **B**

I. $a - b$ II. ab III. $\frac{a}{b}$

(A) I only   (B) II only   (C) III only
(D) I and II only   (E) II and III only

73. If $x = 1.4$, find the value of $27(x + 1.2) - 26(x + 1.2)$. **C**

(A) 1   (B) −0.4   (C) 2.6   (D) 65

Study Guide and Intervention, p. 7 (shown) and p. 8

Real Numbers All real numbers can be classified as either rational or irrational. The set of rational numbers includes several subsets: natural numbers, whole numbers, and integers.

R	real numbers	{all rationals and irrationals}
Q	rational numbers	{all numbers that can be represented in the form $\frac{m}{n}$, where m and n are integers and n is not equal to 0}
I	irrational numbers	{all nonterminating, nonrepeating decimals}
N	natural numbers	{1, 2, 3, 4, 5, 6, 7, 8, 9, ...}
W	whole numbers	{0, 1, 2, 3, 4, 5, 6, 7, 8, ...}
Z	integers	{..., −3, −2, −1, 0, 1, 2, 3, ...}

Example **Name the sets of numbers to which each number belongs.**

a. $-\frac{11}{3}$ rationals (Q), reals (R)

b. $\sqrt{25}$
$\sqrt{25} = 5$ naturals (N), wholes (W), integers (Z), rationals (Q), reals (R)

Exercises

Name the sets of numbers to which each number belongs.

1. $\frac{6}{7}$ **Q, R**   2. $-\sqrt{81}$ **Z, Q, R**   3. 0 **W, Z, Q, R**   4. 192.0005 **Q, R**

5. 73 **N, W, Z, Q, R**   6. $34\frac{1}{2}$ **Q, R**   7. $\frac{\sqrt{36}}{9}$ **Q, R**   8. 26.1 **Q, R**

9. π **I, R**   10. $\frac{15}{3}$ **N, W, Z, Q, R**   11. $-4.\overline{17}$ **Q, R**

12. $\frac{\sqrt{25}}{5}$ **N, W, Z, Q, R**   13. −1 **Z, Q, R**   14. $\sqrt{42}$ **I, R**

15. −11.2 **Q, R**   16. $-\frac{8}{13}$ **Q, R**   17. $\frac{\sqrt{5}}{2}$ **I, R**

18. $33.\overline{3}$ **Q, R**   19. 894,000 **N, W, Z, Q, R**   20. −0.02 **Q, R**

Skills Practice, p. 9 and Practice, p. 10 (shown)

Name the sets of numbers to which each number belongs.

1. 6425 **N, W, Z, Q, R**   2. $\sqrt{7}$ **I, R**   3. 2π **I, R**   4. 0 **W, Z, Q, R**

5. $\sqrt{\frac{25}{36}}$ **Q, R**   6. $-\sqrt{16}$ **Z, Q, R**   7. −35 **Z, Q, R**   8. −31.8 **Q, R**

Name the property illustrated by each equation.

9. $5x \cdot (4y + 3x) = 5x \cdot (3x + 4y)$ **Comm. (+)**
10. $7x + (9x + 8) = (7x + 9x) + 8$ **Assoc. (+)**
11. $5(3x + y) = 5(3x + 1y)$ **Mult. Iden.**
12. $7n + 2n = (7 + 2)n$ **Distributive**
13. $3(2x)y = (3 \cdot 2)(xy)$ **Assoc. (×)**
14. $3x \cdot 2y = 3 \cdot 2 \cdot x \cdot y$ **Comm. (×)**
15. $(6 + -6)y = 0y$ **Add. Inv.**
16. $\frac{1}{4} \cdot 4y = 1y$ **Mult. Inv.**
17. $5(x + y) = 5x + 5y$ **Distributive**
18. $4n + 0 = 4n$ **Add. Iden.**

Name the additive inverse and multiplicative inverse for each number.

19. 0.4 **−0.4, 2.5**   20. −1.6 **1.6, −0.625**

21. $-\frac{11}{16}$ **$\frac{11}{16}, -\frac{16}{11}$**   22. $5\frac{5}{6}$ **$-5\frac{5}{6}, \frac{6}{35}$**

Simplify each expression.

23. $5x - 3y - 2x + 3y$ **3x**
24. $-11a - 13b + 7a - 3b$ **−4a − 16b**
25. $8x - 7y - (3 - 6y)$ **8x − y − 3**
26. $4c - 2c - (4c + 2c)$ **−4c**
27. $3(r - 10s) - 4(7s + 2r)$ **−5r − 58s**
28. $\frac{1}{5}(10a - 15) + \frac{1}{2}(8 + 4a)$ **4a + 1**
29. $2(4 - 2x + y) - 4(5 + x - y)$ **−12 − 8x + 6y**
30. $\frac{5}{6}\left(\frac{3}{5}x + 12y\right) - \frac{1}{4}(2x - 12y)$ **13y**

31. **TRAVEL** Olivia drives her car at 60 miles per hour for t hours. Ian drives his car at 50 miles per hour for $(t + 2)$ hours. Write a simplified expression for the sum of the distances traveled by the two cars. **(110t + 100) mi**

32. **NUMBER THEORY** Use the properties of real numbers to tell whether the following statement is true or false: If $a > b$, it follows that $a\left(\frac{1}{a}\right) > b\left(\frac{1}{b}\right)$. Explain your reasoning. **false; counterexample: $5\left(\frac{1}{5}\right) \not> 4\left(\frac{1}{4}\right)$**

Reading to Learn Mathematics, p. 11 — ELL

Pre-Activity **How is the Distributive Property useful in calculating store savings?**

Read the introduction to Lesson 1-2 at the top of page 11 in your textbook.

- Why are all of the amounts listed on the register slip at the top of page 11 followed by negative signs? **Sample answer: The amount of each coupon is subtracted from the total amount of purchases so that you save money by using coupons.**
- Describe two ways of calculating the amount of money you saved by using coupons if your register slip is the one shown on page 11. **Sample answer: Add all the individual coupon amounts or add the amounts for the scanned coupons and multiply the sum by 2.**

Reading the Lesson

1. Refer to the Key Concepts box on page 11. The numbers $2.\overline{57}$ and 0.010010001... both involve decimals that "go on forever." Explain why one of these numbers is rational and the other is irrational. **Sample answer: $2.\overline{57}$ = 2.5757... is a repeating decimal because there is a block of digits, 57, that repeats forever, so this number is rational. The number 0.010010001... is a non-repeating decimal because, although the digits follow a pattern, there is no block of digits that repeats. So this number is an irrational number.**

2. Write the Associative Property of Addition in symbols. Then illustrate this property by finding the sum 12 + 18 + 45 in two different ways. **$(a + b) + c = a + (b + c)$; Sample answer: (12 + 18) + 45 = 30 + 45 = 75; 12 + (18 + 45) = 12 + 63 = 75**

3. Consider the equations $(a \cdot b) \cdot c = a \cdot (b \cdot c)$ and $(a \cdot b) \cdot c = c \cdot (a \cdot b)$. One of the equations uses the Associative Property of Multiplication and one uses the Commutative Property of Multiplication. How can you tell which property is being used in each equation? **The first equation uses the Associative Property of Multiplication. The quantities *a*, *b*, and *c* are used in the same order, but they are grouped differently on the two sides of the equation. The second equation uses the quantities in different orders on the two sides of the equation. So the second equation uses the Commutative Property of Multiplication.**

Helping You Remember

4. How can the meanings of the words *commuter* and *association* help you to remember the difference between the commutative and associative properties? **Sample answer: A commuter is someone who travels back and forth to work or another place, and the commutative property says you can switch the *order* when two numbers that are being added or multiplied. An association is a group of people who are connected or united, and the associative property says that you can switch the *grouping* when three numbers are added or multiplied.**

Enrichment, p. 12

Properties of a Group

A set of numbers forms a group with respect to an operation if for that operation the set has (1) the Closure Property, (2) the Associative Property, (3) a member which is an identity, and (4) an inverse for each member of the set.

Example 1 **Does the set {0, 1, 2, 3, ...} form a group with respect to addition?**

Closure Property: For all numbers in the set, is $a + b$ in the set? 0 + 1 = 1, and 1 is in the set; 0 + 2 = 2, and 2 is in the set; and so on. The set has closure for addition.

Associative Property: For all numbers in the set, does $a + (b + c) = (a + b) + c$? 0 + (1 + 2) = (0 + 1) + 2; 1 + (2 + 3) = (1 + 2) + 3; and so on. The set is associative for addition.

Identity: Is there some number, i, in the set such that $i + a = a = a + i$ for all a? 0 + 1 = 1 = 1 + 0; 0 + 2 = 2 = 2 + 0; and so on. The identity for addition is 0.

Inverse: Does each number, a, have an inverse, a', such that $a' + a =$ ... $= i$? The integer inverse of 3 is −3 since ...

4 Assess

Open-Ended Assessment

Modeling Ask students to give examples of each of the properties (identity, inverse, commutative, associative, and distributive) and examples for each set of numbers (reals, rationals, irrationals, integers, wholes, and naturals).

Getting Ready for Lesson 1-3

PREREQUISITE SKILL Lesson 1-3 presents translating verbal expressions into algebraic expressions and using the properties of equality to solve equations. After solving an equation, the solution is checked in the original equation by evaluating the expression on each side after replacing the variable with its numerical value. Use Exercises 83–86 to determine your students' familiarity with evaluating expressions.

Assessment Options

Practice Quiz 1 The quiz provides students with a brief review of the concepts and skills in Lessons 1-1 and 1-2. Lesson numbers are given to the right of exercises or instruction lines so students can review concepts not yet mastered.

Quiz (Lessons 1-1 and 1-2) is available on p. 51 of the *Chapter 1 Resource Masters*.

Daily Intervention notes help you help students when they need it most. Unlocking Misconceptions suggestions help you analyze where students make common errors so you can point these trouble spots out to them.

Extending the Lesson

For Exercises 74–77, use the following information.
The product of any two whole numbers is always a whole number. So, the set of whole numbers is said to be *closed* under multiplication. This is an example of the **Closure Property**. State whether each statement is *true* or *false*. If false, give a counterexample. **75. False; $0 - 1 = -1$, which is not a whole number.**

74. The set of integers is closed under multiplication. **true**

75. The set of whole numbers is closed under subtraction.

76. The set of rational numbers is closed under addition. **true**

77. The set of whole numbers is closed under division.
False, $2 \div 3 = \frac{2}{3}$, which is not a whole number.

Maintain Your Skills

Mixed Review

Find the value of each expression. *(Lesson 1-1)*

78. $9(4 - 3)^5$ **9**

79. $5 + 9 \div 3(3) - 8$ **6**

Evaluate each expression if $a = -5$, $b = 0.25$, $c = \frac{1}{2}$, and $d = 4$. *(Lesson 1-1)*

80. $a + 2b - c$ **−5**

81. $b + 3(a + d)^3$ **−2.75**

82. **GEOMETRY** The formula for the surface area SA of a rectangular prism is $SA = 2\ell w + 2\ell h + 2wh$, where ℓ represents the length, w represents the width, and h represents the height. Find the surface area of the rectangular prism. *(Lesson 1-1)* **358 in^2**

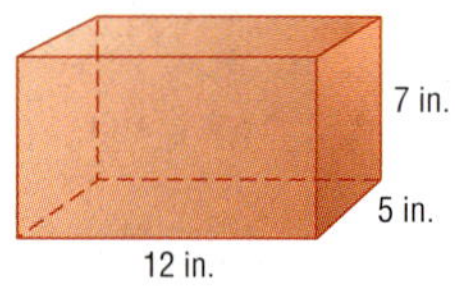

Getting Ready for the Next Lesson

PREREQUISITE SKILL **Evaluate each expression if $a = 2$, $b = -\frac{3}{4}$, and $c = 1.8$.**
*(To review **evaluating expressions**, see Lesson 1-1.)*

83. $8b - 5$ **−11**

84. $\frac{2}{5}b + 1$ **$\frac{7}{10}$**

85. $1.5c - 7$ **−4.3**

86. $-9(a - 6)$ **36**

Two Quizzes in each chapter review skills and concepts presented in previous lessons.

Practice Quiz 1 — Lessons 1-1 and 1-2

Find the value of each expression. *(Lesson 1-1)*

1. $18 - 12 \div 3$ **14**
2. $-4 + 5(7 - 2^3)$ **−9**
3. $\frac{18 + 3 \times 4}{13 - 8}$ **6**
4. Evaluate $a^3 + b(9 - c)$ if $a = -2$, $b = \frac{1}{3}$, and $c = -12$. *(Lesson 1-1)* **−1**
5. **ELECTRICITY** Find the amount of current I (in amperes) produced if the electromotive force E is 2.5 volts, the circuit resistance R is 1.05 ohms, and the resistance r within a battery is 0.2 ohm. Use the formula $I = \frac{E}{R + r}$. *(Lesson 1-1)* **2 amperes**

Name the sets of numbers to which each number belongs. *(Lesson 1-2)*

6. 3.5 **Q, R**
7. $\sqrt{100}$ **N, W, Z, Q, R**
8. Name the property illustrated by $bc + (-bc) = 0$. *(Lesson 1-2)* **Add. Inv.**
9. Name the additive inverse and multiplicative inverse of $\frac{6}{7}$. *(Lesson 1-2)* **$-\frac{6}{7}, \frac{7}{6}$**
10. Simplify $4(14x - 10y) - 6(x + 4y)$. *(Lesson 1-2)* **$50x - 64y$**

DAILY INTERVENTION

Unlocking Misconceptions

Associative or Commutative Students sometimes use inappropriate visual cues to name properties. For example, they may think that an expression can only have two terms to be an example of commutativity. Suggest that students look first at the change from one expression to the other and ask themselves if it is a change in grouping (associativity) or in position (commutativity).

Algebra Activity

A Follow-Up of Lesson 1-2

Investigating Polygons and Patterns

Collect the Data

Use a ruler or geometry drawing software to draw six large polygons with 3, 4, 5, 6, 7, and 8 sides. The polygons do not need to be regular. Convex polygons, ones whose diagonals lie in the interior, will be best for this activity.

1. Copy the table below and complete the column labeled *Diagonals* by drawing the diagonals for all six polygons and record your results.

Figure Name	Sides (n)	Diagonals	Diagonals From One Vertex
triangle	3	0	0
quadrilateral	4	2	1
pentagon	5	**5**	**2**
hexagon	6	**9**	**3**
heptagon	7	**14**	**4**
octagon	8	**20**	**5**

Analyze the Data

2. Describe the pattern shown by the number of diagonals in the table above. **See pp. 53A–53B.**

3. Complete the last column in the table above by recording the number of diagonals that can be drawn from one vertex of each polygon.

4. Write an expression in terms of n that relates the number of diagonals from one vertex to the number of sides for each polygon. **$n - 3$**

5. If a polygon has n sides, how many vertices does it have? **n**

6. How many vertices does one diagonal connect? **2**

Make a Conjecture

7. Write a formula in terms of n for the number of diagonals of a polygon of n sides. (*Hint:* Consider your answers to Exercises 2, 3, and 4.) **$[n(n - 3)] \div 2$**

8. Draw a polygon with 10 sides. Test your formula for the decagon. **See pp. 53A–53B.**

9. Explain how your formula relates to the number of vertices of the polygon and the number of diagonals that can be drawn from each vertex. **See pp. 53A–53B.**

Extend the Activity

10. Draw 3 noncollinear dots on your paper. Determine the number of lines that are needed to connect each dot to every other dot. Continue by drawing 4 dots, 5 dots, and so on and finding the number of lines to connect them. **See pp. 53A–53B.**

11. Copy and complete the table at the right. **See table.**

12. Use any method to find a formula that relates the number of dots, x, to the number of lines, y. **$y = [x(x - 1)] \div 2$ or $y = 0.5x^2 - 0.5x$**

13. Explain why the formula works. **See pp. 53A–53B.**

Dots (x)	Connection Lines (y)
3	3
4	**6**
5	**10**
6	**15**
7	**21**
8	**28**

Algebra Activities use manipulatives and models to help students learn key concepts. There are teacher notes for every Algebra Activity in the Student Edition.

Resource Manager

Teaching Algebra with Manipulatives
- p. 213 (student recording sheet)

Glencoe Mathematics Classroom Manipulative Kit
- ruler

Algebra Activity

A Follow-Up of Lesson 1-2

Getting Started

Objective Discover the relationship between the number of sides of a convex polygon and the total number of diagonals that can be drawn in the polygon.

Materials
ruler or geometry drawing software

Teach

- In Exercise 8, suggest to students that they draw a large decagon, draw all of its diagonals, and then carefully mark each diagonal as they count it.
- Guide students to recognize that each figure they create when connecting the dots in Exercises 10–13 is a polygon with all of its diagonals drawn. Relate this to the work in Exercises 1–9.

Assess

In Exercises 2–6, students should be able to see that there are consistent patterns in these relationships, and they should be able to make the generalizations that will form the parts of the formula. In Exercises 7–9, students should understand that the elements in the formula are not just arbitrary or mysterious, but are derived from the characteristics of the diagonals. They should also be able to apply the formula to a polygon with any number of sides.

Study Notebook

You may wish to have students summarize this activity and what they learned from it.

1-3 Lesson Notes

1 Focus

5-Minute Check Transparency 1-3 Use as a quiz or review of Lesson 1-2.

Mathematical Background notes are available for this lesson on p. 4C.

Building on Prior Knowledge

In Lesson 1-2, students evaluated expressions with real numbers. In this lesson, they apply this skill to writing expressions and solving equations.

How can you find the most effective level of intensity for your workout?

Ask students:

- How can the expression $6 \times P \div (220 - A)$ be written as a ratio? $\frac{6P}{220 - A}$
- To achieve a 100% intensity level, the numerator and denominator of the ratio you just found must be equal. At what 10-second pulse count would you achieve a 100% intensity level? **Answers will vary.**
- **Fitness** Find your 10-second pulse count P after running in place for 30 seconds. What is your level of intensity for this value of P? **Answers will vary.**

1-3 Solving Equations

What You'll Learn

- Translate verbal expressions into algebraic expressions and equations, and vice versa.
- Solve equations using the properties of equality.

Vocabulary

- open sentence
- equation
- solution

Vocabulary words are listed at the beginning of the lesson and are highlighted in yellow at point of use.

How can you find the most effective level of intensity for your workout?

When exercising, one goal is to find the best level of intensity as a percent of your maximum heart rate. To find the intensity level, multiply 6 and P, your 10-second pulse count. Then divide by the difference of 220 and your age A.

Multiply 6 and your pulse rate	and divide by	the difference of 220 and your age.
$6 \times P$	$\div$	$(220 - A)$

VERBAL EXPRESSIONS TO ALGEBRAIC EXPRESSIONS Verbal expressions can be translated into algebraic or mathematical expressions using the language of algebra. Any letter can be used as a variable to represent a number that is not known.

Example 1 Verbal to Algebraic Expression

Write an algebraic expression to represent each verbal expression.

a. 7 less than a number — $n - 7$

b. three times the square of a number — $3x^2$

c. the cube of a number increased by 4 times the same number — $p^3 + 4p$

d. twice the sum of a number and 5 — $2(y + 5)$

A mathematical sentence containing one or more variables is called an **open sentence**. A mathematical sentence stating that two mathematical expressions are equal is called an **equation**.

Example 2 Algebraic to Verbal Sentence

Write a verbal sentence to represent each equation.

a. $10 = 12 - 2$ — Ten is equal to 12 minus 2.

b. $n + (-8) = -9$ — The sum of a number and −8 is −9.

c. $\frac{n}{6} = n^2$ — A number divided by 6 is equal to that number squared.

Open sentences are neither true nor false until the variables have been replaced by numbers. Each replacement that results in a true sentence is called a **solution** of the open sentence.

Resource Manager

Workbook and Reproducible Masters

Chapter 1 Resource Masters

- Study Guide and Intervention, pp. 13–14
- Skills Practice, p. 15
- Practice, p. 16
- Reading to Learn Mathematics, p. 17
- Enrichment, p. 18
- Assessment, pp. 51, 53

Graphing Calculator and Spreadsheet Masters, p. 27

School-to-Career Masters, p. 2

Teaching Algebra With Manipulatives Masters, pp. 214–215

Transparencies

5-Minute Check Transparency 1-3

Answer Key Transparencies

Technology

Interactive Chalkboard

PROPERTIES OF EQUALITY To solve equations, we can use properties of equality. Some of these *equivalence relations* are listed in the table below.

Key Concept *Properties of Equality*

Property	Symbols	Examples
Reflexive	For any real number a, $a = a$.	$-7 + n = -7 + n$
Symmetric	For all real numbers a and b, if $a = b$, then $b = a$.	If $3 = 5x - 6$, then $5x - 6 = 3$.
Transitive	For all real numbers a, b, and c, if $a = b$ and $b = c$, then $a = c$.	If $2x + 1 = 7$ and $7 = 5x - 8$, then $2x + 1 = 5x - 8$.
Substitution	If $a = b$, then a may be replaced by b and b may be replaced by a.	If $(4 + 5)m = 18$, then $9m = 18$.

Study Tip

Properties of Equality
These properties are also known as *axioms of equality*.

Example 3 *Identify Properties of Equality*

Name the property illustrated by each statement.

a. **If $3m = 5n$ and $5n = 10p$, then $3m = 10p$.**

Transitive Property of Equality

b. **If $-11a + 2 = -3a$, then $-3a = -11a + 2$.**

Symmetric Property of Equality

Sometimes an equation can be solved by adding the same number to each side or by subtracting the same number from each side or by multiplying or dividing each side by the same number.

Key Concept *Properties of Equality*

Addition and Subtraction Properties of Equality

- **Symbols** For any real numbers a, b, and c, if $a = b$, then $a + c = b + c$ and $a - c = b - c$.
- **Examples** If $x - 4 = 5$, then $x - 4 + 4 = 5 + 4$. If $n + 3 = -11$, then $n + 3 - 3 = -11 - 3$.

Multiplication and Division Properties of Equality

- **Symbols** For any real numbers a, b, and c, if $a = b$, then $a \cdot c = b \cdot c$ and, if $c \neq 0$, $\frac{a}{c} = \frac{b}{c}$.
- **Examples** If $\frac{m}{4} = 6$, then $4 \cdot \frac{m}{4} = 4 \cdot 6$. If $-3y = 6$, then $\frac{-3y}{-3} = \frac{6}{-3}$.

Example 4 *Solve One-Step Equations*

Solve each equation. Check your solution.

a. $a + 4.39 = 76$

$a + 4.39 = 76$ Original equation

$a + 4.39 - 4.39 = 76 - 4.39$ Subtract 4.39 from each side.

$a = 71.61$ Simplify.

The solution is 71.61.

(continued on the next page)

www.algebra2.com/extra_examples

2 Teach

VERBAL EXPRESSIONS TO ALGEBRAIC EXPRESSIONS

In-Class Examples Power Point®

1 Write an algebraic expression to represent each verbal expression.

a. 3 more than a number $x + 3$

b. six times the cube of a number $6x^3$

c. the square of a number decreased by the product of 5 and the number $x^2 - 5x$

d. twice the difference of a number and 6 $2(x - 6)$

2 Write a verbal sentence to represent each equation.

a. $14 + 9 = 23$ **The sum of 14 and 9 is 23.**

b. $6 = -5 + x$ **Six is equal to −5 plus a number.**

c. $7y - 2 = 19$ **Seven times a number minus 2 is 19.**

PROPERTIES OF EQUALITY

In-Class Examples Power Point®

3 Name the property illustrated by each statement.

a. If $xy = 28$ and $x = 7$, then $7y = 28$. **Substitution Property of Equality**

b. $a - 2.03 = a - 2.03$ **Reflexive Property of Equality**

Reading Tip Help students remember the name of the Reflexive Property by relating $a = a$ to seeing your reflection in a mirror.

4 Solve each equation. Check your solution.

a. $s - 5.48 = 0.02$ **5.5**

b. $18 = \frac{1}{2}t$ **36**

The Resource Manager lists all of the resources available for the lesson, including workbooks, blackline masters, transparencies, and technology.

Teaching Tip Suggest that students ask themselves these questions: "What is being shown on the left side of the equation in In-Class Example 4a at the right?" **5.48 is subtracted from s.** "What is the opposite or inverse of subtracting 5.48?" **Adding 5.48.** "What must be done to both sides of the equation $s - 5.48 = 0.02$ to get the variable s alone on one side of the equation?" **Add 5.48 to both sides and simplify the resulting equation.**

In-Class Examples

5 Solve
$53 = 3(y - 2) - 2(3y - 1)$.
−19

6 **GEOMETRY** The area of a trapezoid is $A = \frac{1}{2}(b_1 + b_2)h$, where A is the area, b_1 is the length of one base, b_2 is the length of the other base, and h is the height of the trapezoid. Solve the formula for h.
$h = \frac{2A}{b_1 + b_2}$

CHECK

$a + 4.39 = 76$ Original equation

$71.61 + 4.39 \stackrel{?}{=} 76$ Substitute 71.61 for a.

$76 = 76$ ✓ Simplify.

Study Tip

Multiplication and Division Properties of Equality
Example 4b could also have been solved using the Division Property of Equality. Note that dividing each side of the equation by $-\frac{3}{5}$ is the same as multiplying each side by $-\frac{5}{3}$.

b. $-\frac{3}{5}d = 18$

$-\frac{3}{5}d = 18$ Original equation

$-\frac{5}{3}\left(-\frac{3}{5}\right)d = -\frac{5}{3}(18)$ Multiply each side by $-\frac{5}{3}$, the multiplicative inverse of $-\frac{3}{5}$.

$d = -30$ Simplify.

The solution is −30.

CHECK

$-\frac{3}{5}d = 18$ Original equation

$-\frac{3}{5}(-30) \stackrel{?}{=} 18$ Substitute −30 for d.

$18 = 18$ ✓ Simplify.

Sometimes you must apply more than one property to solve an equation.

Example 5 Solve a Multi-Step Equation

Solve $2(2x + 3) - 3(4x - 5) = 22$.

$2(2x + 3) - 3(4x - 5) = 22$ Original equation

$4x + 6 - 12x + 15 = 22$ Distributive and Substitution Properties

$-8x + 21 = 22$ Commutative, Distributive, and Substitution Properties

$-8x = 1$ Subtraction and Substitution Properties

$x = -\frac{1}{8}$ Division and Substitution Properties

The solution is $-\frac{1}{8}$.

You can use properties of equality to solve an equation or formula for a specified variable.

Example 6 Solve for a Variable

GEOMETRY **The surface area of a cone is $S = \pi r\ell + \pi r^2$, where S is the surface area, ℓ is the slant height of the cone, and r is the radius of the base. Solve the formula for ℓ.**

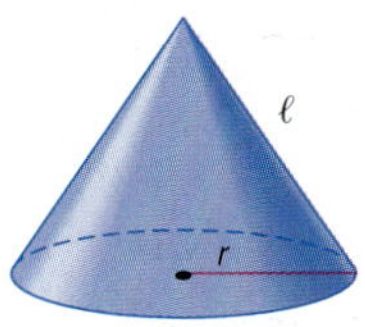

$S = \pi r\ell + \pi r^2$ Surface area formula

$S - \pi r^2 = \pi r\ell + \pi r^2 - \pi r^2$ Subtract πr^2 from each side.

$S - \pi r^2 = \pi r\ell$ Simplify.

$\frac{S - \pi r^2}{\pi r} = \frac{\pi r\ell}{\pi r}$ Divide each side by πr.

$\frac{S - \pi r^2}{\pi r} = \ell$ Simplify.

DAILY INTERVENTION — Unlocking Misconceptions

- **Solving Equations** Students may want to simplify, collect terms, and use the properties of equality to perform an operation on each side of an equation all in one or two steps. Help them see that it is more efficient to write down each step in the solution process than to have to solve the equation again because of a computational error.
- **Checking Solutions** Explain that checking solutions in order to discover possible errors is a vital procedure when you use math on the job.

Many standardized test questions can be solved by using properties of equality.

Standardized Test Practice

Example 7 Apply Properties of Equality

Multiple-Choice Test Item

If $3n - 8 = \frac{9}{5}$, what is the value of $3n - 3$?

Ⓐ $\frac{34}{5}$ Ⓑ $\frac{49}{15}$ Ⓒ $-\frac{16}{5}$ Ⓓ $-\frac{27}{5}$

Read the Test Item

You are asked to find the value of the expression $3n - 3$. Your first thought might be to find the value of n and then evaluate the expression using this value. Notice, however, that you are *not* required to find the value of n. Instead, you can use the Addition Property of Equality on the given equation to find the value of $3n - 3$.

Solve the Test Item

$3n - 8 = \frac{9}{5}$ Original equation

$3n - 8 + 5 = \frac{9}{5} + 5$ Add 5 to each side.

$3n - 3 = \frac{34}{5}$ $\frac{9}{5} + 5 = \frac{9}{5} + \frac{25}{5}$ or $\frac{34}{5}$

The answer is A.

Test-Taking Tip

If a problem seems to require lengthy calculations, look for a shortcut. There is probably a quicker way to solve it. Try using properties of equality.

To solve a word problem, it is often necessary to define a variable and write an equation. Then solve by applying the properties of equality.

Example 8 Write an Equation

HOME IMPROVEMENT **Josh and Pam have bought an older home that needs some repair. After budgeting a total of $1685 for home improvements, they started by spending $425 on small improvements. They would like to replace six interior doors next. What is the maximum amount they can afford to spend on each door?**

Explore Let c represent the cost to replace each door.

Plan Write and solve an equation to find the value of c.

The number of doors	times	the cost to replace each door	plus	previous expenses	equals	the total cost.
6	$\cdot$	c	+	425	=	1685

Solve

$6c + 425 = 1685$ Original equation

$6c + 425 - 425 = 1685 - 425$ Subtract 425 from each side.

$6c = 1260$ Simplify.

$\frac{6c}{6} = \frac{1260}{6}$ Divide each side by 6.

$c = 210$ Simplify.

They can afford to spend $210 on each door.

Examine The total cost to replace six doors at $210 each is 6(210) or $1260. Add the other expenses of $425 to that, and the total home improvement bill is 1260 + 425 or $1685. Thus, the answer is correct.

More About...

Home Improvement

Previously occupied homes account for approximately 85% of all U.S. home sales. Most homeowners remodel within 18 months of purchase. The top two remodeling projects are kitchens and baths.

Source: National Association of Remodeling Industry

In-Class Examples

Power Point®

7 If $4g + 5 = \frac{4}{9}$, what is the value of $4g - 2$? **B**

A $-\frac{41}{36}$ **B** $-\frac{59}{9}$

C $-\frac{41}{9}$ **D** $-\frac{67}{7}$

8 **HOME IMPROVEMENT** Carl wants to replace the five windows in the 2nd-story bedrooms of his house. His neighbor Will is a carpenter and he has agreed to help install them for $250. If Carl has budgeted $1000 for the total cost, what is the maximum amount he can spend on each window? **$150**

Teaching Tip Students, especially those with math anxiety, tend to omit the planning step. Encourage students to see that this step helps them find a way to write an equation, even if they only do the planning mentally.

Standardized Test Practice

Example 7 Point out to students that there are several ways to find the specified value. One alternate way would be to first solve the given equation for $3n$ and then subtract 3 from each side of that equation.

$3n - 8 = \frac{9}{5} \Rightarrow 3n = \frac{49}{5} \Rightarrow 3n - 3 = \frac{34}{5}$

Each chapter contains an example that gives students practice in solving problems on standardized tests. Standardized Test Practice suggestions give students additional methods for achieving success on standardized tests.

Lesson 1-3 Solving Equations 23

3 Practice/Apply

Study Notebook

Have students—

- add the definitions/examples of the vocabulary terms to their Vocabulary Builder worksheets for Chapter 1.
- add the properties of equality given in this lesson to their list of real number properties from Lesson 1-2.
- include the formula in Example 6 in the list of formulas they began in Lesson 1-2.
- use the content of Example 7 to start a list of test-taking tips that they can review as they prepare for standardized tests.
- include any other item(s) that they find helpful in mastering the skills in this lesson

DAILY INTERVENTION

FIND THE ERROR Encourage students to use correct mathematical language to state the error. For example, *Crystal needed to use the Distributive Property on the right side of the equation before subtracting.*

Answer

3. His method can be confirmed by solving the equation using an alternative method.

$$C = \frac{5}{9}(F - 32)$$

$$C = \frac{5}{9}F - \frac{5}{9}(32)$$

$$C + \frac{5}{9}(32) = \frac{5}{9}F$$

$$\frac{9}{5}\left[C + \frac{5}{9}(32)\right] = F$$

$$\frac{9}{5}C + 32 = F$$

Check for Understanding

Concept Check

1. Sample answer: $2x = -14$
2. Sometimes true; only when the expression you are dividing by does not equal zero.

Find the Error exercises help students identify and address common errors before they occur.

1. **OPEN ENDED** Write an equation whose solution is -7.
2. **Determine** whether the following statement is *sometimes*, *always*, or *never* true. Explain.

   Dividing each side of an equation by the same expression produces an equivalent equation.

3. **FIND THE ERROR** Crystal and Jamal are solving $C = \frac{5}{9}(F - 32)$ for F.

Crystal	Jamal
$C = \frac{5}{9}(F - 32)$	$C = \frac{5}{9}(F - 32)$
$C + 32 = \frac{5}{9}F$	$\frac{9}{5}C = F - 32$
$\frac{9}{5}(C + 32) = F$	$\frac{9}{5}C + 32 = F$

Who is correct? Explain your reasoning. **Jamal; see margin for explanation.**

Guided Practice

GUIDED PRACTICE KEY	
Exercises	Examples
4, 5	1
6, 7	2
8, 9	3
10–15	4, 5
16, 17	6
18	7

Write an algebraic expression to represent each verbal expression.

4. five increased by four times a number **$5 + 4n$**
5. twice a number decreased by the cube of the same number **$2n - n^3$**

Write a verbal expression to represent each equation. **6–7. Sample answers given.**

6. $9n - 3 = 6$ **9 times a number decreased by 3 is 6.**
7. $5 + 3x^2 = 2x$ **5 plus 3 times the square of a number is twice that number.**

Name the property illustrated by each statement.

8. $(3x + 2) - 5 = (3x + 2) - 5$ **Reflexive (=)**
9. If $4c = 15$, then $4c + 2 = 15 + 2$. **Addition (=)**

Solve each equation. Check your solution.

10. $y + 14 = -7$ **-21**
11. $7 + 3x = 49$ **14**
12. $-4(b + 7) = -12$ **-4**
13. $7q + q - 3q = -24$ **-4.8**
14. $1.8a - 5 = -2.3$ **1.5**
15. $-\frac{3}{4}n + 1 = -11$ **16**

Solve each equation or formula for the specified variable.

16. $4y - 2n = 9$, for y **$y = \frac{9 + 2n}{4}$**
17. $I = prt$, for p **$p = \frac{I}{rt}$**

Standardized Test Practice

18. If $4x + 7 = 18$, what is the value of $12x + 21$? **D**

    (A) 2.75 (B) 32 (C) 33 (D) 54

★ indicates increased difficulty

Practice and Apply

Homework Help	
For Exercises	See Examples
19–28	1
29–34	2
35–40	3
41–56	4, 5
57–62	6
63–74	7

Extra Practice See page 828.

Write an algebraic expression to represent each verbal expression.

19. the sum of 5 and three times a number **$5 + 3n$**
20. seven more than the product of a number and 10 **$10n + 7$**
21. four less than the square of a number **$n^2 - 4$**
22. the product of the cube of a number and -6 **$-6n^3$**
23. five times the sum of 9 and a number **$5(9 + n)$**
24. twice the sum of a number and 8 **$2(n + 8)$**
25. ★ the square of the quotient of a number and 4 **$\left(\frac{n}{4}\right)^2$**
26. ★ the cube of the difference of a number and 7 **$(n - 7)^3$**

DAILY INTERVENTION

Differentiated Instruction

Interpersonal Have students work in pairs to read, discuss, and plan a solution strategy for real-world problems such as the one given in Example 8. This interaction can help students identify individual difficulties with word problems and also to discover new strategies used by other students.

24 Chapter 1 Solving Equations and Inequalities

GEOMETRY **For Exercises 27 and 28, use the following information.**
The formula for the surface area of a cylinder with radius r and height h is π times twice the product of the radius and height plus twice the product of π and the square of the radius.

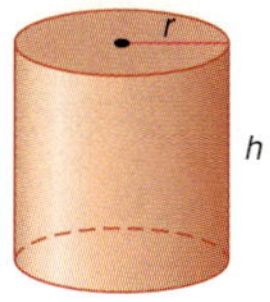

29–34. Sample answers are given.

27. Translate this verbal expression of the formula into an algebraic expression. $2\pi rh + 2\pi r^2$
28. Write an equivalent expression using the Distributive Property. $2\pi r(h + r)$

Write a verbal expression to represent each equation.

29. $x - 5 = 12$
30. $2n + 3 = -1$
31. $y^2 = 4y$
32. $3a^3 = a + 4$
33. $\frac{b}{4} = 2(b + 1)$
★ 34. $7 - \frac{1}{2}x = \frac{3}{x^2}$

29. 5 less than a number is 12.
30. Twice a number plus 3 is −1.
31. A number squared is equal to 4 times the number.
32. Three times the cube of a number is equal to the number plus 4.
33. A number divided by 4 is equal to twice the sum of that number and 1.
34. 7 minus half a number is equal to 3 divided by the square of the number.

Name the property illustrated by each statement.

35. If $[3(-2)]z = 24$, then $-6z = 24$. Substitution (=)
36. If $5 + b = 13$, then $b = 8$. Subtraction (=)
37. If $2x = 3d$ and $3d = -4$, then $2x = -4$. Transitive (=)
38. If $g - t = n$, then $g = n + t$. Addition (=)
39. If $14 = \frac{x}{2} + 11$, then $\frac{x}{2} + 11 = 14$. Symmetric (=)
★ 40. If $y - 2 = -8$, then $3(y - 2) = 3(-8)$. Multiplication (=)

Solve each equation. Check your solution.

41. $2p + 15 = 29$ 7
42. $14 - 3n = -10$ 8
43. $7a - 3a + 2a - a = 16$ 3.2
44. $x + 9x - 6x + 4x = 20$ 2.5
45. $\frac{1}{9} - \frac{2}{3}b = \frac{1}{18}$ $\frac{1}{12}$
46. $\frac{5}{8} + \frac{3}{4}x = \frac{1}{16}$ $-\frac{3}{4}$
47. $27 = -9(y + 5)$ -8
48. $-7(p + 8) = 21$ -11
49. $3f - 2 = 4f + 5$ -7
50. $3d + 7 = 6d + 5$ $\frac{2}{3}$
51. $4.3n + 1 = 7 - 1.7n$ 1
52. $1.7x - 8 = 2.7x + 4$ -12
★ 53. $3(2z + 25) - 2(z - 1) = 78$ $\frac{1}{4}$
★ 54. $4(k + 3) + 2 = 4.5(k + 1)$ 19
★ 55. $\frac{3}{11}a - 1 = \frac{7}{11}a + 9$ $-\frac{55}{2}$
★ 56. $\frac{2}{5}x + \frac{3}{7} = 1 - \frac{4}{7}x$ $\frac{10}{17}$

Solve each equation or formula for the specified variable.

57. $d = rt$, for r $\frac{d}{t} = r$
58. $x = \frac{-b}{2a}$, for a $a = \frac{-b}{2x}$
59. $V = \frac{1}{3}\pi r^2 h$, for h $\frac{3V}{\pi r^2} = h$
60. $A = \frac{1}{2}h(a + b)$, for b $\frac{2A}{h} - a = b$
★ 61. $\frac{a(b - 2)}{c - 3} = x$, for b $b = \frac{x(c - 3)}{a} + 2$
★ 62. $x = \frac{y}{y + 4}$, for y $\frac{4x}{1 - x} = y$

Define a variable, write an equation, and solve the problem.

63. **BOWLING** Jon and Morgan arrive at Sunnybrook Lanes with $16.75. Find the maximum number of games they can bowl if they each rent shoes. n = number of games; $2(1.50) + n(2.50) = 16.75$; 5

SUNNYBROOK LANES

Shoe Rental: *$1.50*

Games: *$2.50 each*

www.algebra2.com/self_check_quiz

About the Exercises…

Organization by Objective
- **Verbal Expressions to Algebraic Expressions:** 19–34
- **Properties of Equality:** 35–74

Exercises 19–26 and 29–70 are structured so that students practice the same concepts whether they are assigned odd or even problems.

Assignment Guide

Basic: 19–23 odd, 27–28, 29–39 odd, 41–51 odd, 57, 59, 63–69 odd, 75–89

Average: 19–25 odd, 27–28, 29–69 odd, 75–89

Advanced: 20–26 even, 30–70 even, 71–83 (optional: 84–89)

The Assignment Guides provide suggestions for exercises that are appropriate for basic, average, or advanced students. Many of the homework exercises are paired, so that students can do the odds one day and the evens the next day.

Study Guide and Intervention, p. 13 (shown) and p. 14

Verbal Expressions to Algebraic Expressions The chart suggests some ways to help you translate word expressions into algebraic expressions. Any letter can be used to represent a number that is not known.

Word Expression	Operation
and, plus, sum, increased by, more than	addition
minus, difference, decreased by, less than	subtraction
times, product, of (as in $\frac{1}{2}$ of a number)	multiplication
divided by, quotient	division

Example 1 **Write an algebraic expression to represent 18 less than the quotient of a number and 3.**

$\frac{n}{3} - 18$

Example 2 **Write a verbal sentence to represent $6(n - 2) = 14$.**

Six times the difference of a number and two is equal to 14.

Exercises

Write an algebraic expression to represent each verbal expression.

1. the sum of six times a number and 25 $6n + 25$
2. four times the sum of a number and 3 $4(n + 3)$
3. 7 less than fifteen times a number $15n - 7$
4. the difference of nine times a number and the quotient of 6 and the same number $9n - \frac{6}{n}$
5. the sum of 100 and four times a number $100 + 4n$
6. the product of 3 and the sum of 11 and a number $3(11 + n)$
7. four times the square of a number increased by five times the same number $4n^2 + 5n$
8. 23 more than the product of 7 and a number $7n + 23$

Write a verbal sentence to represent each equation. Sample answers are given.

9. $3n - 35 = 79$ The difference of three times a number and 35 is equal to 79.
10. $2(n^3 + 3n^2) = 4n$ Twice the sum of the cube of a number and three times the square of the number is equal to four times the number.
11. $\frac{5n}{n + 3} = n - 8$ The quotient of five times a number and the sum of the number and 3 is equal to the difference of the number and 8.

Skills Practice, p. 15 and Practice, p. 16 (shown)

Write an algebraic expression to represent each verbal expression.

1. 2 more than the quotient of a number and 5 $\frac{y}{5} + 2$
2. the sum of two consecutive integers $n + (n + 1)$
3. 5 times the sum of a number and 1 $5(m + 1)$
4. 1 less than twice the square of a number $2y^2 - 1$

Write a verbal expression to represent each equation. 5–8. Sample answers are given.

5. $5 - 2x = 4$ The difference of 5 and twice a number is 4.
6. $3y = 4y^3$ Three times a number is 4 times the cube of the number.
7. $3c = 2(c - 1)$ Three times a number is twice the difference of the number and 1.
8. $\frac{m}{5} = 3(2m + 1)$ The quotient of a number and 5 is 3 times the sum of twice the number and 1.

Name the property illustrated by each statement.

9. If $t - 13 = 52$, then $52 = t - 13$. Symmetric (=)
10. If $8(2q + 1) = 4$, then $2(2q + 1) = 1$. Division (=)
11. If $h + 12 = 22$, then $h = 10$. Subtraction (=)
12. If $4m = -15$, then $-12m = 45$. Multiplication (=)

Solve each equation. Check your solution.

13. $14 = 8 - 6r$ -1
14. $9 + 4n = -59$ -17
15. $\frac{3}{4} - \frac{1}{2}n = \frac{5}{8}$ $\frac{1}{4}$
16. $\frac{5}{6}s + \frac{3}{4} = \frac{11}{12}$ $\frac{1}{5}$
17. $-1.6r + 5 = -7.8$ 8
18. $6x - 5 = 7 - 9x$ $\frac{4}{5}$
19. $5(6 - 4v) = v + 21$ $\frac{3}{7}$
20. $6y - 5 = -3(2y + 1)$ $\frac{1}{6}$

Solve each equation or formula for the specified variable.

21. $E = mc^2$, for m $m = \frac{E}{c^2}$
22. $c = \frac{2d + 1}{3}$, for d $d = \frac{3c - 1}{2}$
23. $h = vt - gt^2$, for v $v = \frac{h + gt^2}{t}$
24. $E = \frac{1}{2}Iw^2 + U$, for I $I = \frac{2(E - U)}{w^2}$

Define a variable, write an equation, and solve the problem.

25. **GEOMETRY** The length of a rectangle is twice the width. Find the width if the perimeter is 60 centimeters. w = width; $2(2w) + 2w = 60$; 10 cm
26. **GOLF** Luis and three friends went golfing. Two of the friends rented clubs for \$6 each. The total cost of the rented clubs and the green fees for each person was \$76. What was the cost of the green fees for each person? g = green fees per person; $6(2) + 4g = 76$; \$16

Reading to Learn Mathematics, p. 17 ELL

Pre-Activity **How can you find the most effective level of intensity for your workout?**

Read the introduction to Lesson 1-3 at the top of page 20 in your textbook.

- To find your target heart rate, what two pieces of information must you supply? age (A) and desired intensity level (I)
- Write an equation that shows how to calculate your target heart rate. $P = \frac{(220 - A) \cdot I}{6}$ or $P = (220 - A) \cdot I \div 6$

Reading the Lesson

1. a. How are algebraic expressions and equations alike? Sample answer: Both contain variables, constants, and operation signs.
   b. How are algebraic expressions and equations different? Sample answer: Equations contain equal signs; expressions do not.
   c. How are algebraic expressions and equations related? Sample answer: An equation is a statement that says that two algebraic expressions are equal.

Read the following problem and then write an equation that you could use to solve it. Do not actually solve the equation. In your equation, let m be the number of miles driven.

2. When Louisa rented a moving truck, she agreed to pay \$28 per day plus \$0.42 per mile. If she kept the truck for 3 days and the rental charges (without tax) were \$153.72, how many miles did Louisa drive the truck? $3(28) + 0.42m = 153.72$

Helping You Remember

3. How can the words *reflection* and *symmetry* help you remember and distinguish between the reflexive and symmetric properties of equality? Think about how these words are used in everyday life or in geometry. Sample answer: When you look at your reflection, you are looking at yourself. The reflexive property says that every number is equal to itself. In geometry, symmetry with respect to a line means that the parts of a figure on the two sides of a line are identical. The symmetric property of equality allows you to interchange the two sides of an equation. The equal sign is like the line of symmetry.

For Exercises 64–70, define a variable, write an equation, and solve the problem.

64. **GEOMETRY** The perimeter of a regular octagon is 124 inches. Find the length of each side. s = length of a side; $8s = 124$; 15.5 in.

65. **CAR EXPENSES** Benito spent \$1837 to operate his car last year. Some of these expenses are listed below. Benito's only other expense was for gasoline. If he drove 7600 miles, what was the average cost of the gasoline per mile? x = cost of gasoline per mile; $972 + 114 + 105 + 7600x = 1837$; 8.5¢

Operating Expenses

Insurance: \$972

Registration: \$114

Maintenance: \$105

66. **SCHOOL** A school conference room can seat a maximum of 83 people. The principal and two counselors need to meet with the school's student athletes to discuss eligibility requirements. If each student must bring a parent with them, what is the maximum number of students that can attend each meeting? n = number of students that can attend each meeting; $2n + 3 = 83$; 40 students

67. **FAMILY** Chun-Wei's mother is 8 more than twice his age. His father is three years older than his mother is. If all three family members have lived 94 years, how old is each family member? a = Chun-Wei's age; $a + (2a + 8) + (2a + 8 + 3) = 94$; Chun-Wei: 15 yrs old, mother: 38 yrs old, father: 41 yrs old

68. **SCHOOL TRIP** The Parent Teacher Organization has raised \$1800 to help pay for a trip to an amusement park. They ask that there be one adult for every five students attending. Adult tickets cost \$45 and student tickets cost \$30. If the group wants to take 50 students, how much will each student need to pay so that adults agreeing to chaperone pay nothing? c = cost per student; $50(30 - c) + \frac{50}{5}(45) = 1800$; \$3

69. **BUSINESS** A trucking company is hired to deliver 125 lamps for \$12 each. The company agrees to pay \$45 for each lamp that is broken during transport. If the trucking company needs to receive a minimum payment of \$1365 for the shipment to cover their expenses, find the maximum number of lamps they can afford to break during the trip. n = number of lamps broken; $12(125) - 45n = 1365$; 3 lamps

70. **PACKAGING** Two designs for a soup can are shown at the right. If each can holds the same amount of soup, what is the height of can A? h = height of can A; $\pi(1.2^2)h = \pi(2^2)3$; $8\frac{1}{3}$ units

Can A   Can B

Career Choices

Industrial Design

Industrial designers use research on product use, marketing, materials, and production methods to create functional and appealing packaging designs.

Online Research

For information about a career as an industrial designer, visit: www.algebra2.com/careers

71. 15.1 mi/mo

RAILROADS **For Exercises 71–73, use the following information.**

The First Transcontinental Railroad was built by two companies. The Central Pacific began building eastward from Sacramento, California, while the Union Pacific built westward from Omaha, Nebraska. The two lines met at Promontory, Utah, in 1869, about 6 years after construction began.

71. The Central Pacific Company laid an average of 9.6 miles of track per month. Together the two companies laid a total of 1775 miles of track. Determine the average number of miles of track laid per month by the Union Pacific Company.
72. About how many miles of track did each company lay? See margin.
73. Why do you think the Union Pacific was able to lay track so much more quickly than the Central Pacific? See margin.

Enrichment, p. 18

Venn Diagrams

Relationships among sets can be shown using Venn diagrams. Study the diagrams below. The circles represent sets A and B, which are subsets of set S.

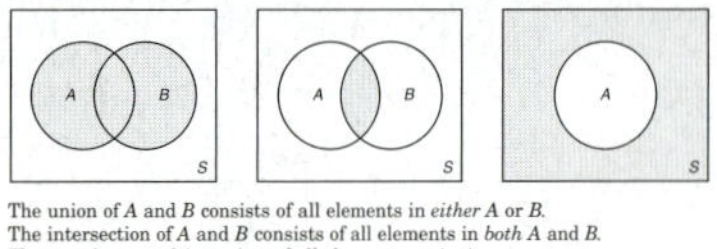

The union of A and B consists of all elements in *either* A or B.
The intersection of A and B consists of all elements in *both* A and B.
The complement of A consists of all elements *not* in A.

You can combine the operations of union, intersection, and finding the complement.

Example Shade the region $(A \cap B)'$.

Answers

72. Central: 690 mi.; Union: 1085 mi

73. The Central Pacific had to lay their track through the Rocky Mountains, while the Union Pacific mainly built track over flat prairie.

★ 74. **MONEY** Allison is saving money to buy a video game system. In the first week, her savings were \$8 less than $\frac{2}{5}$ the price of the system. In the second week, she saved 50 cents more than $\frac{1}{2}$ the price of the system. She was still \$37 short. Find the price of the system.
\$295

WebQuest You can write and solve equations to determine the monthly payment for a home. Visit www.algebra2.com/webquest to continue work on your WebQuest project.

75. **CRITICAL THINKING** Write a verbal expression to represent the algebraic expression $3(x - 5) + 4x(x + 1)$. **See margin.**

76. **WRITING IN MATH** Answer the question that was posed at the beginning of the lesson. **See pp. 53A–53B.**

How can you find the most effective level of intensity for your workout?

Include the following in your answer:

- an explanation of how to find the age of a person who is exercising at an 80% level of intensity I with a pulse count of 27, and
- a description of when it would be desirable to solve a formula like the one given for a specified variable.

Standardized Test Practice

77. If $-6x + 10 = 17$, then $3x - 5 =$ **B**
(A) $-\frac{7}{6}$. (B) $-\frac{17}{2}$. (C) 2. (D) $\frac{19}{3}$. (E) $\frac{5}{3}$.

78. In triangle PQR, $\overline{QS}$ and $\overline{SR}$ are angle bisectors and angle $P = 74°$. How many degrees are there in angle QSR? **D**
(A) 106 (B) 121 (C) 125
(D) 127 (E) 143

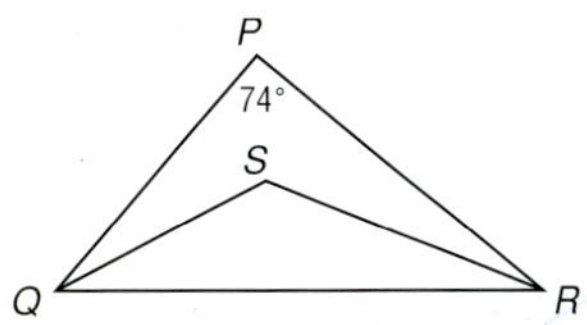

Maintain Your Skills

Mixed Review

Simplify each expression. *(Lesson 1-2)*

79. $2x + 9y + 4z - y - 8x$ **$-6x + 8y + 4z$**

80. $4(2a + 5b) - 3(4b - a)$ **$11a + 8b$**

Evaluate each expression if $a = 3$, $b = -2$, and $c = 1.2$. *(Lesson 1-1)*

81. $a - [b(a - c)]$ **6.6**

82. $c^2 - ab$ **7.44**

83. **GEOMETRY** The formula for the surface area S of a regular pyramid is $S = \frac{1}{2}P\ell + B$, where P is the perimeter of the base, ℓ is the slant height, and B is the area of the base. Find the surface area of the square-based pyramid shown at the right. *(Lesson 1-1)* **105 cm²**

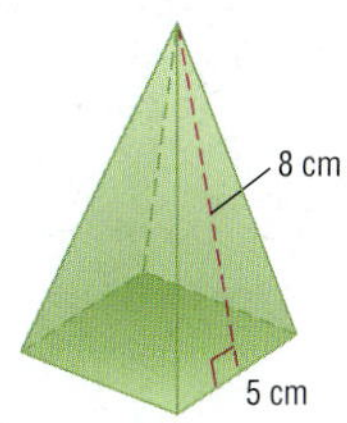

By having your students complete the Getting Ready exercises, you can target specific skills they will need for the next lesson.

Getting Ready for the Next Lesson

PREREQUISITE SKILL **Identify the additive inverse for each number or expression.** *(To review **additive inverses**, see Lesson 1-2.)*

84. 5 **−5**

85. −3 **3**

86. 2.5 **−2.5**

87. $\frac{1}{4}$ **$-\frac{1}{4}$**

88. $-3x$ **$3x$**

89. $5 - 6y$ **$-5 + 6y$**

Answer

75. the product of 3 and the difference of a number and 5 added to the product of four times the number and the sum of the number and 1

Assessment Options lists the quizzes and tests that are available in the Chapter Resource Masters.

4 Assess

Open-Ended Assessment

Speaking Have students discuss what difficulties they have with translating verbal problems into algebraic equations, including any anxieties that word problems may create. Ask students to share their strategies for overcoming these difficulties, using specific examples to illustrate their strategies.

Tips for New Teachers

Intervention Explain to students that they can solve verbal problems when they (1) face their anxiety about the words instead of avoiding the task, (2) ask questions about words they do not understand, and (3) take time to read, understand, and plan, using a sketch to help.

Getting Ready for Lesson 1-4

PREREQUISITE SKILL Lesson 1-4 presents solving equations that involve absolute value expressions. Solving equations often involves using additive inverses to isolate the variable on one side of an equation. Exercises 84–89 should be used to determine your students' familiarity with finding additive inverses.

Assessment Options

Quiz (Lesson 1–3) is available on p. 51 of the *Chapter 1 Resource Masters.*

Mid-Chapter Test (Lessons 1-1 through 1-3) is available on p. 53 of the *Chapter 1 Resource Masters.*

1-4 Lesson Notes

1 Focus

5-Minute Check Transparency 1-4 Use as a quiz or review of Lesson 1-3.

Mathematical Background notes are available for this lesson on p. 4D.

Building on Prior Knowledge

In Lesson 1-3, students wrote expressions and solved equations. In this lesson, they apply those skills to equations involving absolute values.

How can an absolute value equation describe the magnitude of an earthquake?

Ask students:

- In the absolute value equation $|E - 6.1| = 0.3$, what does the variable E represent? **the actual magnitude of the earthquake**
- What is the meaning of the number 0.3 in the equation? **the uncertainty of the estimated magnitude**
- What would the equation be for the magnitude of an earthquake estimated at 5.8 on the Richter scale? $|E - 5.8| = 0.3$

Key Concept boxes highlight definitions, formulas, and other important ideas. Multiple representations—words, symbols, examples, models—reach students of all learning styles

1-4 Solving Absolute Value Equations

What You'll Learn

- Evaluate expressions involving absolute values.
- Solve absolute value equations.

Vocabulary

- absolute value
- empty set

How can an absolute value equation describe the magnitude of an earthquake?

Seismologists use the Richter scale to express the magnitudes of earthquakes. This scale ranges from 1 to 10, 10 being the highest. The uncertainty in the estimate of a magnitude E is about plus or minus 0.3 unit. This means that an earthquake with a magnitude estimated at 6.1 on the Richter scale might actually have a magnitude as low as 5.8 or as high as 6.4. These extremes can be described by the absolute value equation $|E - 6.1| = 0.3$.

ABSOLUTE VALUE EXPRESSIONS The **absolute value** of a number is its distance from 0 on the number line. Since distance is nonnegative, the absolute value of a number is always nonnegative. The symbol $|x|$ is used to represent the absolute value of a number x.

Key Concept — *Absolute Value*

- **Words** For any real number a, if a is positive or zero, the absolute value of a is a. If a is negative, the absolute value of a is the opposite of a.
- **Symbols** For any real number a, $|a| = a$ if $a \geq 0$, and $|a| = -a$ if $a < 0$.
- **Model** $|-3| = 3$ and $|3| = 3$

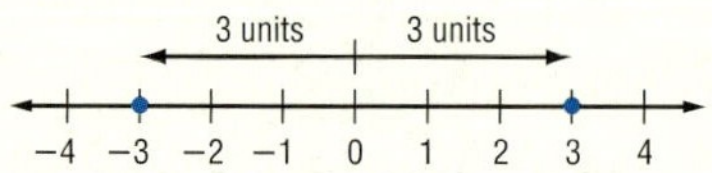

When evaluating expressions that contain absolute values, the absolute value bars act as a grouping symbol. Perform any operations inside the absolute value bars first.

Example 1 *Evaluate an Expression with Absolute Value*

Evaluate $1.4 + |5y - 7|$ if $y = -3$.

$1.4 + |5y - 7| = 1.4 + |5(-3) - 7|$	Replace y with -3.
$= 1.4 + |-15 - 7|$	Simplify $5(-3)$ first.
$= 1.4 + |-22|$	Subtract 7 from -15.
$= 1.4 + 22$	$|-22| = 22$
$= 23.4$	Add.

The value is 23.4.

Resource Manager

Workbook and Reproducible Masters

Chapter 1 Resource Masters

- Study Guide and Intervention, pp. 19–20
- Skills Practice, p. 21
- Practice, p. 22
- Reading to Learn Mathematics, p. 23
- Enrichment, p. 24

Graphing Calculator and Spreadsheet Masters, p. 28

Transparencies

5-Minute Check Transparency 1-4
Answer Key Transparencies

Technology

Alge2PASS: Tutorial Plus, Lesson 1
Interactive Chalkboard

ABSOLUTE VALUE EQUATIONS Some equations contain absolute value expressions. The definition of absolute value is used in solving these equations. For any real numbers a and b, where $b \geq 0$, if $|a| = b$, then $a = b$ or $-a = b$. This second case is often written as $a = -b$.

Example 2 Solve an Absolute Value Equation

Solve $|x - 18| = 5$. Check your solutions.

Case 1 $a = b$ **or** **Case 2** $a = -b$

$x - 18 = 5$ $\qquad$ $x - 18 = -5$

$x - 18 + 18 = 5 + 18$ $\qquad$ $x - 18 + 18 = -5 + 18$

$x = 23$ $\qquad$ $x = 13$

CHECK $|x - 18| = 5$ $\qquad$ $|x - 18| = 5$

$|23 - 18| \stackrel{?}{=} 5$ $\qquad$ $|13 - 18| \stackrel{?}{=} 5$

$|5| \stackrel{?}{=} 5$ $\qquad$ $|-5| \stackrel{?}{=} 5$

$5 = 5$ ✓ $\qquad$ $5 = 5$ ✓

The solutions are 23 or 13. Thus, the solution set is {13, 23}.

On the number line, we can see that each answer is 5 units away from 18.

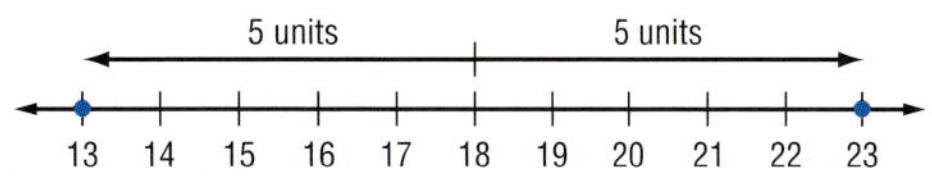

Because the absolute value of a number is always positive or zero, an equation like $|x| = -5$ is never true. Thus, it has no solution. The solution set for this type of equation is the **empty set**, symbolized by { } or ∅.

Study Tip

Common Misconception
For an equation like the one in Example 3, there is no need to consider the two cases. Remember to check your solutions in the original equation to prevent this error.

Example 3 No Solution

Solve $|5x - 6| + 9 = 0$.

$|5x - 6| + 9 = 0$ Original equation

$|5x - 6| = -9$ Subtract 9 from each side.

This sentence is *never* true. So the solution set is ∅.

It is important to check your answers when solving absolute value equations. Even if the correct procedure for solving the equation is used, the answers may not be actual solutions of the original equation.

Example 4 One Solution

Solve $|x + 6| = 3x - 2$. Check your solutions.

Case 1 $a = b$ **or** **Case 2** $a = -b$

$x + 6 = 3x - 2$ $\qquad$ $x + 6 = -(3x - 2)$

$6 = 2x - 2$ $\qquad$ $x + 6 = -3x + 2$

$8 = 2x$ $\qquad$ $4x + 6 = 2$

$4 = x$ $\qquad$ $4x = -4$

$x = -1$

There appear to be two solutions, 4 or −1.

(continued on the next page)

www.algebra2.com/extra_examples

2 Teach

ABSOLUTE VALUE EXPRESSIONS

In-Class Example Power Point®

1 Evaluate $2.7 + |6 - 2x|$ if $x = 4$. **4.7**

Teaching Tip Students may find it helpful to read the first absolute value bar as "the distance of" and the last absolute value bar as "from zero, without regard to direction." So, the expression $|6 - 2x|$ would be read as "the distance of the value of $6 - 2x$ from zero, without regard to direction."

ABSOLUTE VALUE EQUATIONS

In-Class Examples Power Point®

2 Solve $|y + 3| = 8$. Check your solutions. **{−11, 5}**

3 Solve $|6 - 4t| + 5 = 0$. **∅**

Teaching Tip Remind students to think about the meaning of the mathematical sentence before they begin their calculations and again when they evaluate the reasonableness of their solution.

4 Solve $|8 + y| = 2y - 3$. Check your solutions. **{11}**

Concept Check

Ask students if $-h$ must represent a negative number. **No, if h is negative then $-h$ is positive.** Have them find a value for h that makes this statement true: $|h| = -h$. **Zero and all negative numbers can be values for h.**

DAILY INTERVENTION

Differentiated Instruction ELL

Verbal/Linguistic Some students may think that the absolute value of x is always x. Suggest that they say in words the meaning of $|x|$ as "the distance of x from zero without regard to direction" to see that, for example, the distance of −3 from zero without regard to direction, cannot be −3. Suggest that they test some positive and negative values for the variable to show that the statement "the absolute value of x is always x" is not true.

ELL notations throughout the chapter indicate items that can assist English-Language Learners.

3 Practice/Apply

CHECK $|x + 6| = 3x - 2$ or $|x + 6| = 3x - 2$

$|4 + 6| \stackrel{?}{=} 3(4) - 2$ $\qquad$ $|-1 + 6| \stackrel{?}{=} 3(-1) - 2$

$|10| \stackrel{?}{=} 12 - 2$ $\qquad$ $|5| \stackrel{?}{=} -3 - 2$

$10 = 10$ ✓ $\qquad$ ~~$5 = -5$~~

Since $5 \neq -5$, the only solution is 4. Thus, the solution set is {4}.

Study Notebook

Have students—

- add the definitions/examples of the vocabulary terms to their Vocabulary Builder worksheets for Chapter 1.
- include any other item(s) that they find helpful in mastering the skills in this lesson.

About the Exercises…

Organization by Objective

- **Absolute Values Expressions:** 17–28
- **Absolute Value Equations:** 29–49

Odd/Even Assignments

Exercises 17–48 are structured so that students practice the same concepts whether they are assigned odd or even problems.

Assignment Guide

Basic: 17–25 odd, 29–43 odd, 47, 49, 50–54, 59–79

Average: 17–49 odd, 50–54, 59–79 (optional: 55–58)

Advanced: 18–48 even, 50–73 (optional: 74–79)

Check for Understanding

Concept Check

1. **Explain** why if the absolute value of a number is always nonnegative, $|a|$ can equal $-a$.

1. $|a| = -a$ when a is a negative number and the opposite of a negative number is positive.

2. **Write** an absolute value equation for each solution set graphed below.

a.

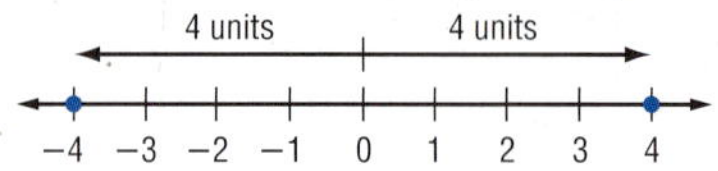

b.

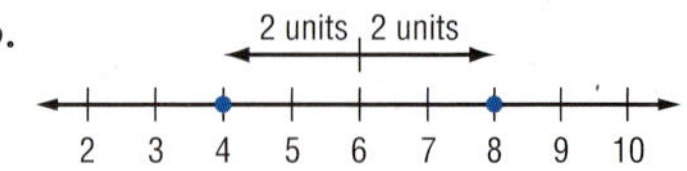

2a. $|x| = 4$

2b. $|x - 6| = 2$

3. **Determine** whether the following statement is *sometimes*, *always*, or *never* true. Explain. **See margin.**

For all real numbers a and b, $a \neq 0$, the equation $|ax + b| = 0$ will have one solution.

4. **OPEN ENDED** Write and evaluate an expression with absolute value. **Sample answer: $|4 - 6|$; 2**

Guided Practice

GUIDED PRACTICE KEY	
Exercises	**Examples**
5–7	1
8–13	2–4
14–16	2

Evaluate each expression if $a = -4$ and $b = 1.5$.

5. $|a + 12|$ **8**   **6.** $|-6b|$ **9**   **7.** $-|a + 21|$ **−17**

Solve each equation. Check your solutions.

8. $|x + 4| = 17$ **{−21, 13}**   **9.** $|b + 15| = 3$ **{−18, −12}**

10. $|a - 9| = 20$ **{−11, 29}**   **11.** $|y - 2| = 34$ **{−32, 36}**

12. $|2w + 3| + 6 = 2$ **∅**   **13.** $|c - 2| = 2c - 10$ **{8}**

Application

FOOD **For Exercises 14–16, use the following information.**
A meat thermometer is used to assure that a safe temperature has been reached to destroy bacteria. Most meat thermometers are accurate to within plus or minus 2°F. **Source:** U.S. Department of Agriculture

Check for Understanding exercises are intended to be completed in class. Concept Check exercises ensure that students understand the concepts in the lesson. The other exercises are representative of the exercises used for homework.

14. The ham you are baking needs to reach an internal temperature of 160°F. If the thermometer reads 160°F, write an equation to determine the least and greatest temperatures of the meat. **$|x - 160| = 2$**

15. Solve the equation you wrote in Exercise 14. **least: 158°F; greatest: 162°F**

16. To what temperature reading should you bake a ham to ensure that the minimum internal temperature is reached? Explain. **162°F; This would ensure a minimum internal temperature of 160°F.**

★ indicates increased difficulty

Practice and Apply

Evaluate each expression if $a = -5$, $b = 6$, and $c = 2.8$.

17. $|-3a|$ **15**   **18.** $|-4b|$ **24**   **19.** $|a + 5|$ **0**

20. $|2 - b|$ **4**   **21.** $|2b - 15|$ **3**   **22.** $|4a + 7|$ **13**

23. $-|18 - 5c|$ **−4**   **24.** $-|c - a|$ **−7.8**   **25.** $6 - |3c + 7|$ **−9.4**

26. $9 - |-2b + 8|$ **5**   ★**27.** $3|a - 10| + |2a|$ **55**   ★**28.** $|a - b| - |10c - a|$ **−22**

Answers

3. Always; since the opposite of 0 is still 0, this equation has only one case, $ax + b = 0$. The solution is $-\frac{b}{a}$.

52. Answers should include the following.

- This equation needs to show that the difference of the estimate E from the originally stated magnitude of 6.1 could be plus 0.3 or minus 0.3, as shown in the graph below. Instead of writing two equations, $E - 6.1 = 0.3$ and $E - 6.1 = -0.3$, absolute value symbols can be used to account for both possibilities, $|E - 6.1| = 0.3$.

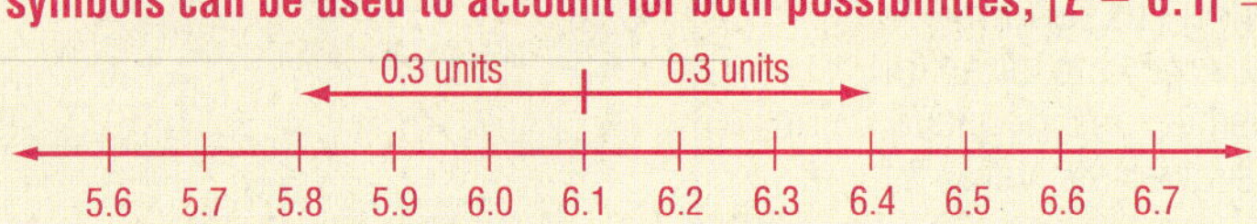

- Using an original magnitude of 5.9, the equation to represent the estimated extremes would be $|E - 5.9| = 0.3$.

Homework Help	
For Exercises	See Examples
17–28	1
29–49	2–4

Extra Practice
See page 829.

Solve each equation. Check your solutions.

29. $|x - 25| = 17$ {8, 42}
30. $|y + 9| = 21$ {12, −30}
31. $|a + 12| = 33$ {−45, 21}
32. $2|b + 4| = 48$ {−28, 20}
33. $8|w - 7| = 72$ {−2, 16}
34. $|3x + 5| = 11$ $\left\{2, -\frac{16}{3}\right\}$
35. $|2z - 3| = 0$ $\left\{\frac{3}{2}\right\}$
36. $|6c - 1| = -2$ ∅
37. $7|4x - 13| = 35$ $\left\{2, \frac{9}{2}\right\}$
38. $-3|2n + 5| = -9$ {−4, −1}
39. $-12|9x + 1| = 144$ ∅
40. $|5x + 9| + 6 = 1$ ∅
41. $|a - 3| - 14 = -6$ {−5, 11}
42. $3|p - 5| = 2p$ {3, 15}
43. $3|2a + 7| = 3a + 12$ $\left\{-\frac{11}{3}, -3\right\}$
44. $|3x - 7| - 5 = -3$ $\left\{3, \frac{5}{3}\right\}$
★ 45. $4|3t + 8| = 16t$ {8}
★ 46. $|15 + m| = -2m + 3$ {−4}

47. **COFFEE** Some say that to brew an excellent cup of coffee, you must have a brewing temperature of 200°F, plus or minus five degrees. Write and solve an equation describing the maximum and minimum brewing temperatures for an excellent cup of coffee. $|x - 200| = 5$; maximum: 205°F; minimum: 195°F

48. **MANUFACTURING** A machine is used to fill each of several bags with 16 ounces of sugar. After the bags are filled, another machine weighs them. If the bag weighs 0.3 ounce more or less than the desired weight, the bag is rejected. Write an equation to find the heaviest and lightest bag the machine will approve. $|x - 16| = 0.3$; heaviest: 16.3 oz, lightest: 15.7 oz

49. **METEOROLOGY** The troposphere is the layer of atmosphere closest to Earth. The average upper boundary of the layer is about 13 kilometers above Earth's surface. This height varies with latitude and with the seasons by as much as 5 kilometers. Write and solve an equation describing the maximum and minimum heights of the upper bound of the troposphere. $|x - 13| = 5$; maximum: 18 km, minimum: 8 km

More About. . .

Meteorology

The troposphere is characterized by the density of its air and an average vertical temperature change of 6°C per kilometer. All weather phenomena occur within the troposphere.

Source: NASA

CRITICAL THINKING For Exercises 50 and 51, determine whether each statement is *sometimes*, *always*, or *never* true. Explain your reasoning.

50. If a and b are real numbers, then $|a + b| = |a| + |b|$.

51. If a, b, and c are real numbers, then $c|a + b| = |ca + cb|$.
sometimes; true only if $c \geq 0$

52. **WRITING IN MATH** Answer the question that was posed at the beginning of the lesson. See margin.

How can an absolute value equation describe the magnitude of an earthquake?

Include the following in your answer:
- a verbal and graphical explanation of how $|E - 6.1| = 0.3$ describes the possible extremes in the variation of the earthquake's magnitude, and
- an equation to describe the extremes for a different magnitude.

50. sometimes; true only if $a \geq 0$ and $b \geq 0$ or if $a \leq 0$ and $b \leq 0$

Standardized Test Practice

53. Which of the graphs below represents the solution set for $|x - 3| - 4 = 0$? **B**

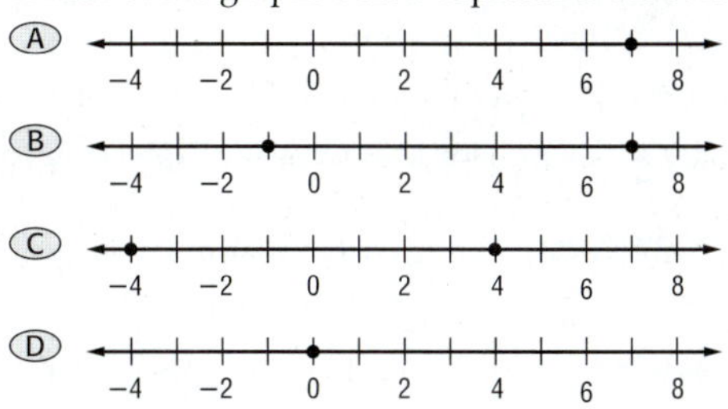

www.algebra2.com/self_check_quiz

There is a Study Guide and Intervention, Skills Practice, Practice, Reading to Learn Mathematics, and Enrichment Master for every lesson in the Student Edition. These masters can be found in the Chapter Resource Masters.

Study Guide and Intervention, p. 19 (shown) and p. 20

Absolute Value Expressions The **absolute value** of a number is the number of units it is from 0 on a number line. The symbol $|x|$ is used to represent the absolute value of a number x.

Absolute Value	
• Words	For any real number a, if a is positive or zero, the absolute value of a is a. If a is negative, the absolute value of a is the opposite of a.
• Symbols	For any real number a, $|a| = a$, if $a \geq 0$, and $|a| = -a$, if $a < 0$.

Example 1 Evaluate $|-4| - |-2x|$ if $x = 6$.

$$\begin{aligned}|-4| - |-2x| &= |-4| - |-2 \cdot 6| \\ &= |-4| - |-12| \\ &= 4 - 12 \\ &= -8\end{aligned}$$

Example 2 Evaluate $|2x - 3y|$ if $x = -4$ and $y = 3$.

$$\begin{aligned}|2x - 3y| &= |2(-4) - 3(3)| \\ &= |-8 - 9| \\ &= |-17| \\ &= 17\end{aligned}$$

Exercises

Evaluate each expression if $w = -4$, $x = 2$, $y = \frac{1}{2}$, and $z = -6$.

1. $|2x - 8|$ 4
2. $|6 + z| - |-7|$ −7
3. $5 + |w + z|$ 15
4. $|x + 5| - |2w|$ −1
5. $|x| - |y| - |z|$ $-4\frac{1}{2}$
6. $|7 - x| + |3x|$ 11
7. $|w - 4x|$ 12
8. $|wz| - |xy|$ 23
9. $|z| - 3|5yz|$ −39
10. $5|w| + 2|z - 2y|$ 34
11. $|z| - 4|2z + y|$ −40
12. $10 - |xw|$ 2
13. $|6y + z| + |yz|$ 6
14. $3|wx| + \frac{1}{4}|4x + 8y|$ 27
15. $7|yz| - 30$ −9
16. $14 - 2|w - xy|$ 4
17. $|2x - y| + 5y$ 6
18. $|xyz| + |wxz|$ 54
19. $z|z| + x|x|$ −32
20. $12 - |10x - 10y|$ −3
21. $\frac{1}{2}|5z + 8w|$ 31
22. $|yz - 4w| - w$ 17
23. $\frac{3}{4}|wz| + \frac{1}{2}|8y|$ 20
24. $xz - |xz|$ −24

Skills Practice, p. 21 and Practice, p. 22 (shown)

Evaluate each expression if $a = -1$, $b = -8$, $c = 5$, and $d = -1.4$.

1. $|6a|$ 6
2. $|2b + 4|$ 12
3. $-|10d + a|$ −15
4. $|17c| + |3b - 5|$ 114
5. $-6|10a - 12|$ −132
6. $|2b - 1| - |-8b + 5|$ −52
7. $|5a - 7| + |3c - 4|$ 23
8. $|1 - 7c| - |a|$ 33
9. $-3|0.5c + 2| - |-0.5b|$ −17.5
10. $|4d| + |5 - 2a|$ 12.6
11. $|a - b| + |b - a|$ 14
12. $|2 - 2d| - 3|b|$ −19.2

Solve each equation. Check your solutions.

13. $|n - 4| = 13$ {−9, 17}
14. $|x - 13| = 2$ {11, 15}
15. $|2y - 3| = 29$ {−13, 16}
16. $7|x + 3| = 42$ {−9, 3}
17. $|3u - 6| = 42$ {−12, 16}
18. $|5x - 4| = -6$ ∅
19. $-3|4x - 9| = 24$ ∅
20. $-6|5 - 2y| = -9$ $\left\{\frac{7}{4}, \frac{13}{4}\right\}$
21. $|8 + p| = 2p - 3$ {11}
22. $|4w - 1| = 5w + 37$ {−4}
23. $4|2y - 7| + 5 = 9$ {3, 4}
24. $-2|7 - 3y| - 6 = -14$ $\left\{1, \frac{11}{3}\right\}$
25. $2|4 - s| = -3s$ {−8}
26. $5 - 3|2 + 2w| = -7$ {−3, 1}
27. $5|2r + 3| - 5 = 0$ {−2, −1}
28. $3 - 5|2d - 3| = 4$ ∅

29. **WEATHER** A thermometer comes with a guarantee that the stated temperature differs from the actual temperature by no more than 1.5 degrees Fahrenheit. Write and solve an equation to find the minimum and maximum actual temperatures when the thermometer states that the temperature is 87.4 degrees Fahrenheit.
$|t - 87.4| = 1.5$; minimum: 85.9°F, maximum: 88.9°F

30. **OPINION POLLS** Public opinion polls reported in newspapers are usually given with a margin of error. For example, a poll with a margin of error of ±5% is considered accurate to within plus or minus 5% of the actual value. A poll with a stated margin of error of ±3% predicts that candidate Tonwe will receive 51% of an upcoming vote. Write and solve an equation describing the minimum and maximum percent of the vote that candidate Tonwe is expected to receive.
$|x - 51| = 3$; minimum: 48%, maximum: 54%

Reading to Learn Mathematics, p. 23 ELL

Pre-Activity How can an absolute value equation describe the magnitude of an earthquake?

Read the introduction to Lesson 1-4 at the top of page 28 in your textbook.

- What is a seismologist and what does magnitude of an earthquake mean? a scientist who studies earthquakes; a number from 1 to 10 that tells how strong an earthquake is
- Why is an absolute value equation rather than an equation without absolute value used to find the extremes in the actual magnitude of an earthquake in relation to its measured value on the Richter scale? Sample answer: The actual magnitude can vary from the measured magnitude by up to 0.3 unit in either direction, so an absolute value equation is needed.
- If the magnitude of an earthquake is estimated to be 6.9 on the Richter scale, it might actually have a magnitude as low as 6.6 or as high as 7.2.

Reading the Lesson

1. Explain how $-a$ could represent a positive number. Give an example. Sample answer: If a is negative, then $-a$ is positive. Example: If $a = -25$, then $-a = -(-25) = 25$.
2. Explain why the absolute value of a number can never be negative. Sample answer: The absolute value is the number of units it is from 0 on the number line. The number of units is never negative.
3. What does the sentence $b \geq 0$ mean? Sample answer: The number b is 0 or greater than 0.
4. What does the symbol ∅ mean as a solution set? Sample answer: If a solution set is ∅, then there are no solutions.

Helping You Remember

5. How can the number line model for absolute value that is shown on page 28 of your textbook help you remember that many absolute value equations have two solutions? Sample answer: The number line shows that for every positive number, there are two numbers that have that number as their absolute value.

Enrichment, p. 24

Considering All Cases in Absolute Value Equations

You have learned that absolute value equations with one set of absolute value symbols have two cases that must be considered. For example, $|x + 3| = 5$ must be broken into $x + 3 = 5$ or $-(x + 3) = 5$. For an equation with two sets of absolute value symbols, four cases must be considered.

Consider the problem $|x + 2| + 3 = |x + 6|$. First we must write the equations for the case where $x + 6 \geq 0$ and where $x + 6 < 0$. Here are the equations for these two cases:

$|x + 2| + 3 = x + 6$
$|x + 2| + 3 = -(x + 6)$

Each of these equations also has two cases. By writing the equations for both cases of each equation above, you end up with the following four equations:

$x + 2 + 3 = x + 6$ $\quad$ $x + 2 + 3 = -(x + 6)$
$-(x + 2) + 3 = x + 6$ $\quad$ $-x - 2 + 3 = -(x + 6)$

4 Assess

Open-Ended Assessment

Modeling Have students draw a number-line diagram like the one shown in Example 2 to model the equation $|x - 3| = 7$ and another number line to model the equation $|y| = 7$. You might suggest that students think of the equation $|y| = 7$ as $|y - 0| = 7$.

Each lesson ends with Open-Ended Assessment strategies for closing the lesson. These include writing, modeling, and speaking.

Getting Ready for Lesson 1-5

PREREQUISITE SKILL Lesson 1-5 presents solving inequalities using steps similar to those for solving equations. Exercises 74–79 should be used to determine your students' familiarity with solving equations.

54. Find the value of $-|-9| - |4| - 3|5 - 7|$. **A**

(A) −19 (B) −11 (C) −7 (D) 11

Extending the Lesson

For Exercises 55–58, consider the equation $|x + 1| + 2 = |x + 4|$.

55. To solve this equation, we must consider the case where $x + 4 \geq 0$ and the case where $x + 4 < 0$. Write the equations for each of these cases.

56. Notice that each equation you wrote in Exercise 55 has two cases. For each equation, write two other equations taking into consideration the case where $x + 1 \geq 0$ and the case where $x + 1 < 0$.

57. Solve each equation you wrote in Exercise 56. Then, check each solution in the original equation, $|x + 1| + 2 = |x + 4|$. What are the solution(s) to this absolute value equation? **{−1.5}**

58. **MAKE A CONJECTURE** For equations with one set of absolute value symbols, two cases must be considered. For an equation with two sets of absolute value symbols, four cases must be considered. How many cases must be considered for an equation containing three sets of absolute value symbols? **8**

55. $|x + 1| + 2 = x + 4$; $|x + 1| + 2 = -(x + 4)$

56. $x + 1 + 2 = x + 4$; $-x - 1 + 2 = x + 4$; $x + 1 + 2 = -x - 4$; $-x - 1 + 2 = -x - 4$

Maintain Your Skills

Mixed Review

Write an algebraic expression to represent each verbal expression. *(Lesson 1-3)*

59. twice the difference of a number and 11 **$2(n - 11)$**

60. the product of the square of a number and 5 **$5n^2$**

Solve each equation. Check your solution. *(Lesson 1-3)*

61. $3x + 6 = 22$ **$\frac{16}{3}$**

62. $7p - 4 = 3(4 + 5p)$ **−2**

63. $\frac{5}{7}y - 3 = \frac{3}{7}y + 1$ **14**

Name the property illustrated by each equation. *(Lesson 1-2)*

64. $(5 + 9) + 13 = 13 + (5 + 9)$ **Comm. (+)**

65. $m(4 - 3) = m \cdot 4 - m \cdot 3$ **Dist.**

66. $\left(\frac{1}{4}\right)4 = 1$ **Mult. Inv.**

67. $5x + 0 = 5x$ **Add. Iden.**

Determine whether each statement is *true* or *false*. If false, give a counterexample. *(Lesson 1-2)*

68. Every real number is a rational number. **false; $\sqrt{3}$**

69. Every natural number is an integer. **true**

70. Every irrational number is a real number. **true**

71. Every rational number is an integer. **false; 1.2**

GEOMETRY For Exercises 72 and 73, use the following information.

The formula for the area A of a triangle is $A = \frac{1}{2}bh$, where b is the measure of the base and h is the measure of the height. *(Lesson 1-1)*

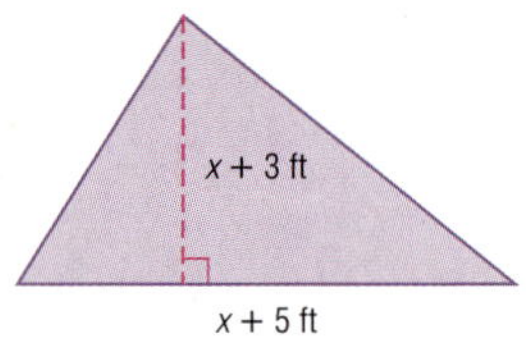

72. Write an expression to represent the area of the triangle above. **72. $\frac{1}{2}(x + 3)(x + 5)$**

73. Evaluate the expression you wrote in Exercise 72 for $x = 23$. **364 ft^2**

Getting Ready for the Next Lesson

PREREQUISITE SKILL Solve each equation. *(To review **solving equations**, see page 20.)*

74. $14y - 3 = 25$ **2**

75. $4.2x + 6.4 = 40$ **8**

76. $7w + 2 = 3w - 6$ **−2**

77. $2(a - 1) = 8a - 6$ **$\frac{2}{3}$**

78. $48 + 5y = 96 - 3y$ **6**

79. $\frac{2x + 3}{5} = \frac{3}{10}$ **$-\frac{3}{4}$**

1-5 Solving Inequalities

What You'll Learn

- Solve inequalities.
- Solve real-world problems involving inequalities.

How can inequalities be used to compare phone plans?

Kuni is trying to decide between two rate plans offered by a wireless phone company.

	Plan 1	Plan 2
Monthly Access Fee	\$35.00	\$55.00
Minutes Included	150	400
Additional Minutes	40¢	35¢

To compare these two rate plans, we can use inequalities. The monthly access fee for Plan 1 is less than the fee for Plan 2, $\$35 < \$55$. However, the additional minutes fee for Plan 1 is greater than that of Plan 2, $\$0.40 > \$0.35$.

Vocabulary
- set-builder notation
- interval notation

SOLVE INEQUALITIES For any two real numbers, a and b, exactly one of the following statements is true.

$$a < b \qquad a = b \qquad a > b$$

This is known as the **Trichotomy Property** or the *property of order*.

Adding the same number to, or subtracting the same number from, each side of an inequality does not change the truth of the inequality.

Study Tip

Properties of Inequality The properties of inequality are also known as *axioms* of inequality.

Key Concept — Properties of Inequality

Addition Property of Inequality

- **Words** For any real numbers, a, b, and c:
  If $a > b$, then $a + c > b + c$.
  If $a < b$, then $a + c < b + c$.
- **Example**
  $3 < 5$
  $3 + (-4) < 5 + (-4)$
  $-1 < 1$

Subtraction Property of Inequality

- **Words** For any real numbers, a, b, and c:
  If $a > b$, then $a - c > b - c$.
  If $a < b$, then $a - c < b - c$.
- **Example**
  $2 > -7$
  $2 - 8 > -7 - 8$
  $-6 > -15$

These properties are also true for $\leq$ and $\geq$.

These properties can be used to solve inequalities. The solution sets of inequalities in one variable can then be graphed on number lines. Use a circle with an arrow to the left for $<$ and an arrow to the right for $>$. Use a dot with an arrow to the left for $\leq$ and an arrow to the right for $\geq$.

1-5 Lesson Notes

1 Focus

5-Minute Check Transparency 1-5 Use as a quiz or review of Lesson 1-4.

Mathematical Background notes are available for this lesson on p. 4D.

Building on Prior Knowledge

In Lessons 1-3 and 1-4, students solved equations. In this lesson, students use similar steps to solve inequalities.

How can inequalities be used to compare phone plans?

Ask students:

- If Kuni knows that she will use no more than 150 minutes per month, which plan is best for her? **Plan 1**
- How much would she pay if she used 350 minutes under Plan 1? under Plan 2? **\$115; \$55**

Questions are provided at the beginning of each lesson to help you use the problem provided there to engage and inform students.

Resource Manager

Workbook and Reproducible Masters

Chapter 1 Resource Masters
- Study Guide and Intervention, pp. 25–26
- Skills Practice, p. 27
- Practice, p. 28
- Reading to Learn Mathematics, p. 29
- Enrichment, p. 30
- Assessment, p. 52

Transparencies

5-Minute Check Transparency 1-5
Answer Key Transparencies

Technology

Alge2PASS: Tutorial Plus, Lesson 2
Interactive Chalkboard
Multimedia Applications

A Four-step Teaching Plan shows you how to Focus, Teach, Practice/Apply, and Assess each lesson.

2 Teach

SOLVE INEQUALITIES

In-Class Example Power Point®

1 Solve $4y - 3 < 5y + 2$. Graph the solution set on a number line. $y > -5$

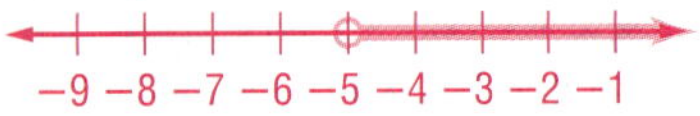

Teaching Tip Ask students what difference it makes when you use the Addition and Subtraction Properties of Inequality whether the inequality sign is <, >, ≤, or ≥. **There is no difference in the calculations but there is a difference in the direction and beginning of the graph of the solution set.**

Example 1 Solve an Inequality Using Addition or Subtraction

Solve $7x - 5 > 6x + 4$. Graph the solution set on a number line.

$7x - 5 > 6x + 4$	Original inequality
$7x - 5 + (-6x) > 6x + 4 + (-6x)$	Add $-6x$ to each side.
$x - 5 > 4$	Simplify.
$x - 5 + 5 > 4 + 5$	Add 5 to each side.
$x > 9$	Simplify.

Any real number greater than 9 is a solution of this inequality.

The graph of the solution set is shown at the right.

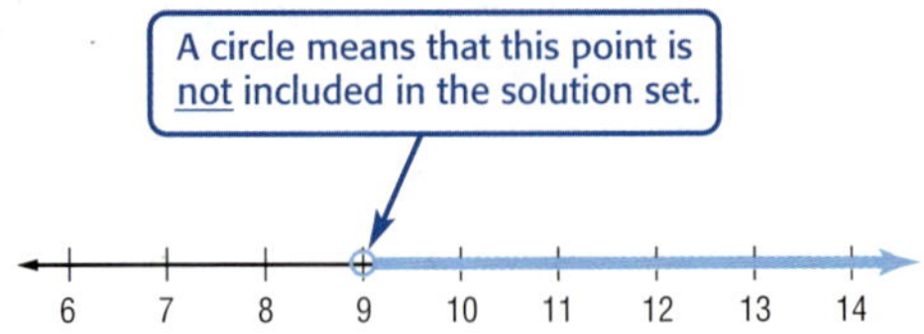

CHECK Substitute 9 for x in $7x - 5 > 6x + 4$. The two sides should be equal. Then substitute a number greater than 9. The inequality should be true.

Multiplying or dividing each side of an inequality by a positive number does not change the truth of the inequality. However, multiplying or dividing each side of an inequality by a *negative* number requires that the order of the inequality be *reversed*. For example, to reverse ≤, replace it with ≥.

Key Concept — *Properties of Inequality*

Multiplication Property of Inequality

• Words For any real numbers, a, b, and c, where

	Words	• Examples
c is positive:	if $a > b$, then $ac > bc$. if $a < b$, then $ac < bc$.	$-2 < 3$ $4(-2) < 4(3)$ $-8 < 12$
c is negative:	if $a > b$, then $ac < bc$. if $a < b$, then $ac > bc$.	$5 > -1$ $(-3)(5) < (-3)(-1)$ $-15 < 3$

Division Property of Inequality

• Words For any real numbers, a, b, and c, where

	Words	• Examples
c is positive:	if $a > b$, then $\frac{a}{c} > \frac{b}{c}$. if $a < b$, then $\frac{a}{c} < \frac{b}{c}$.	$-18 < -9$ $\frac{-18}{3} < \frac{-9}{3}$ $-6 < -3$
c is negative:	if $a > b$, then $\frac{a}{c} < \frac{b}{c}$. if $a < b$, then $\frac{a}{c} > \frac{b}{c}$.	$12 > 8$ $\frac{12}{-2} < \frac{8}{-2}$ $-6 < -4$

These properties are also true for ≤ and ≥.

Study Tip

Reading Math
$\{x \mid x > 9\}$ is read *the set of all numbers x such that x is greater than 9.*

The solution set of an inequality can be expressed by using **set-builder notation**. For example, the solution set in Example 1 can be expressed as $\{x \mid x > 9\}$.

Example 2 Solve an Inequality Using Multiplication or Division

Solve $-0.25y \geq 2$. Graph the solution set on a number line.

$-0.25y \geq 2$ — Original inequality

$\frac{-0.25y}{-0.25} \leq \frac{2}{-0.25}$ — Divide each side by -0.25, reversing the inequality symbol.

$y \leq -8$ — Simplify.

The solution set is $\{y \mid y \leq -8\}$.

The graph of the solution set is shown below.

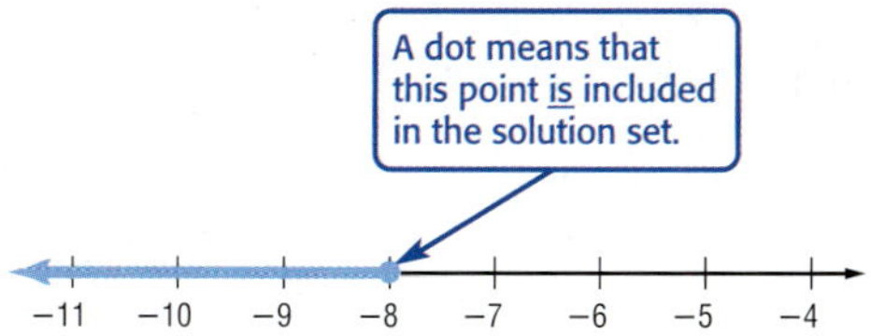

Study Tip

Reading Math
The symbol $+\infty$ is read *positive infinity*, and the symbol $-\infty$ is read *negative infinity*.

The solution set of an inequality can also be described by using **interval notation**. The infinity symbols $+\infty$ and $-\infty$ are used to indicate that a set is unbounded in the positive or negative direction, respectively. To indicate that an endpoint *is not* included in the set, a parenthesis, (or), is used.

$x < 2$ — interval notation $(-\infty, 2)$

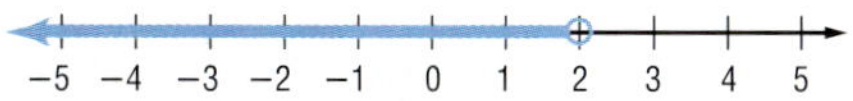

A bracket is used to indicate that the endpoint, -2, *is* included in the solution set below. Parentheses are always used with the symbols $+\infty$ and $-\infty$, because they do not include endpoints.

$x \geq -2$ — interval notation $[-2, +\infty)$

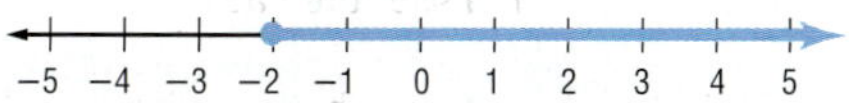

Study Tip

Solutions to Inequalities
When solving an inequality,
- if you arrive at a false statement, such as $3 > 5$, then the solution set for that inequality is the empty set, $\varnothing$.
- if you arrive at a true statement such as $3 > -1$, then the solution set for that inequality is the set of all real numbers.

Example 3 Solve a Multi-Step Inequality

Solve $-m \leq \frac{m+4}{9}$. Graph the solution set on a number line.

$-m \leq \frac{m+4}{9}$ — Original inequality

$-9m \leq m + 4$ — Multiply each side by 9.

$-10m \leq 4$ — Add $-m$ to each side.

$m \geq -\frac{4}{10}$ — Divide each side by -10, reversing the inequality symbol.

$m \geq -\frac{2}{5}$ — Simplify.

The solution set is $\left[-\frac{2}{5}, +\infty\right)$ and is graphed below.

www.algebra2.com/extra_examples

In-Class Examples

Power Point®

2 Solve $12 \geq -0.3p$. Graph the solution set on a number line. $\{p \mid p \geq -40\}$

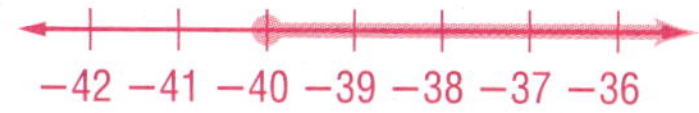

Teaching Tip Remind students that when solving an inequality, in order to keep each intermediate inequality equivalent to the original, they must show both the division by a negative number and the reversal of the inequality sign in the same step.

3 Solve $-x > \frac{x-7}{2}$. Graph the solution set on a number line. $\left(-\infty, \frac{7}{3}\right)$

Concept Check

Ask students to name three different ways to show the solution of an inequality. **four possible responses: as a graph on a number line, as an inequality, using set-builder notation, using interval notation**

DAILY INTERVENTION

Differentiated Instruction

Intrapersonal Have students discuss the differences between solving an equation and solving an inequality and then how the solution processes are the same.

REAL-WORLD PROBLEMS WITH INEQUALITIES

In-Class Example Power Point®

Teaching Tip To understand the situation given in Example 4, some students may find it helpful to make a sketch representing the elevator, the boxes, and the person.

4 **CONSUMER COSTS** Alida has at most $10.50 to spend at a convenience store. She buys a bag of potato chips and a can of soda for $1.55. If gasoline at this store costs $1.35 per gallon, how many gallons of gasoline can Alida buy for her car, to the nearest tenth of a gallon? **no more than 6.6 gal**

Answer

Graphing Calculator Investigation

1. The graph is of the line $y = 1$, for $x \geq -1$.

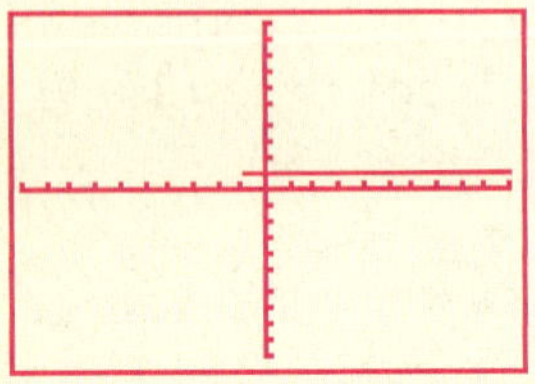

Answers (p. 37)

4. $(-\infty, 1.5)$	5. $\left(-\infty, \frac{5}{3}\right]$
6. $[3, +\infty)$	7. $(6, +\infty)$
8. $(-\infty, -7)$	9. $(15, +\infty)$
10. $(-\infty, -24]$	11. $(-\infty, +\infty)$

Every effort is ... show the An... exercises (1)... reduced St... page, or (2... margin of ... Wraparo... However, answers that do not fit in either of these places can be found in pages at the end of each chapter.

REAL-WORLD PROBLEMS WITH INEQUALITIES Inequalities can be used to solve many verbal and real-world problems.

Study Tip

Inequality Phrases

- $<$ is less than; is fewer than
- $>$ is greater than; is more than
- $\leq$ is at most; is no more than; is less than or equal to
- $\geq$ is at least; is no less than; is greater than or equal to

Example 4 Write an Inequality

DELIVERIES Craig is delivering boxes of paper to each floor of an office building. Each box weighs 64 pounds, and Craig weighs 160 pounds. If the maximum capacity of the elevator is 2000 pounds, how many boxes can Craig safely take on each elevator trip?

Explore Let b = the number of boxes Craig can safely take on each trip. A maximum capacity of 2000 pounds means that this weight must be less than or equal to 2000.

Plan The total weight of the boxes is $64b$. Craig's weight plus the total weight of the boxes must be less than or equal to 2000. Write an inequality.

Craig's weight	plus	the weight of the boxes	is less than or equal to	2000.
160	+	$64b$	$\leq$	2000

Solve

$160 + 64b \leq 2000$ Original inequality

$160 - 160 + 64b \leq 2000 - 160$ Subtract 160 from each side.

$64b \leq 1840$ Simplify.

$\frac{64b}{64} \leq \frac{1840}{64}$ Divide each side by 64.

$b \leq 28.75$ Simplify.

Examine Since he cannot take a fraction of a box, Craig can take no more than 28 boxes per trip and still meet the safety requirements of the elevator.

You can use a graphing calculator to find the solution set for an inequality.

Graphing Calculator Investigation

Solving Inequalities

The inequality symbols in the TEST menu on the TI-83 Plus are called *relational operators*. They compare values and return 1 if the test is true or 0 if the test is false.

You can use these relational operators to find the solution set of an inequality in one variable.

Think and Discuss 1. See margin.

1. Clear the Y= list. Enter $11x + 3 \geq 2x - 6$ as Y1. Put your calculator in DOT mode. Then, graph in the standard viewing window. Describe the graph.
2. Using the TRACE function, investigate the graph. What values of x are on the graph? What values of y are on the graph? **all real numbers; 0 and 1**
3. Based on your investigation, what inequality is graphed? $x \geq -1$
4. Solve $11x + 3 \geq 2x - 6$ algebraically. How does your solution compare to the inequality you wrote in Exercise 3? **The solutions are the same.**

Graphing Calculator Investigation

Solving Inequalities After students enter $11x + 3$, have them press [2nd] [MATH] 4 to insert the $\geq$ symbol before entering $2x - 6$. The values of x for which 0 is returned (where the inequality is false) are not visible on the screen because they overlay part of the x-axis. To help students realize this fact, have them use the Trace feature to travel from positive values of x to increasingly negative values of x along the graph shown in the window.

Check for Understanding

Concept Check

1. Dividing by a number is the same as multiplying by its inverse.

1. **Explain** why it is not necessary to state a division property for inequalities.
2. **Write** an inequality using the $>$ symbol whose solution set is graphed below. Sample answer: $-2n > -6$

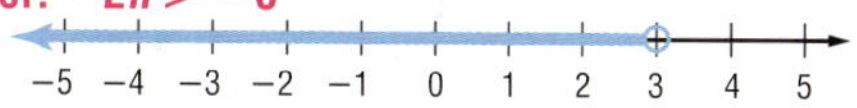

3. **OPEN ENDED** Write an inequality for which the solution set is the empty set. Sample answer: $x + 2 < x + 1$

Guided Practice

GUIDED PRACTICE KEY	
Exercises	Examples
4–11	1–3
12–14	4

4–11. See margin for interval notation. See pp. 53A–53B for graphs.

Solve each inequality. Describe the solution set using set-builder or interval notation. Then graph the solution set on a number line.

4. $a + 2 < 3.5$ $\{a \mid a < 1.5\}$
5. $5 \geq 3x$ $\left\{x \mid x \leq \frac{5}{3}\right\}$
6. $11 - c \leq 8$ $\{c \mid c \geq 3\}$
7. $4y + 7 > 31$ $\{y \mid y > 6\}$
8. $2w + 19 < 5$ $\{w \mid w < -7\}$
9. $-0.6p < -9$ $\{p \mid p > 15\}$
10. $\frac{n}{12} + 15 \leq 13$ $\{n \mid n \leq -24\}$
11. $\frac{5z + 2}{4} < \frac{5z}{4} + 2$ all real numbers

Define a variable and write an inequality for each problem. Then solve.

12. The product of 12 and a number is greater than 36. $12n > 36;\ n > 3$
13. Three less than twice a number is at most 5. $2n - 3 \leq 5;\ n \leq 4$

Application

14. **SCHOOL** The final grade for a class is calculated by taking 75% of the average test score and adding 25% of the score on the final exam. If all scores are out of 100 and a student has a 76 test average, what score does the student need to make on the final exam to have a final grade of at least 80? at least 92

★ indicates increased difficulty

Practice and Apply

Homework Help

For Exercises	See Examples
15–40	1–3
41–51	4

Extra Practice
See page 829.

15–38. See margin for interval notation. See pp. 53A–53B for graphs.

21. $\{k \mid k \geq -3.5\}$
23. $\{m \mid m > -4\}$
27. $\{n \mid n \geq 1.75\}$
28. $\left\{w \mid w > -\frac{1}{20}\right\}$
29. $\{x \mid x < -279\}$
30. $\{c \mid c > -18\}$
31. $\{d \mid d \geq -5\}$
32. $\{z \mid z > 2.6\}$
34. $\left\{a \mid a \geq \frac{5}{7}\right\}$

Solve each inequality. Describe the solution set using set-builder or interval notation. Then, graph the solution set on a number line.

15. $n + 4 \geq -7$ $\{n \mid n \geq -11\}$
16. $b - 3 \leq 15$ $\{b \mid b \leq 18\}$
17. $5x < 35$ $\{x \mid x < 7\}$
18. $\frac{d}{2} > -4$ $\{d \mid d > -8\}$
19. $\frac{g}{-3} \geq -9$ $\{g \mid g \leq 27\}$
20. $-8p \geq 24$ $\{p \mid p \leq -3\}$
21. $13 - 4k \leq 27$
22. ★ $14 > 7y - 21$ $\{y \mid y < 5\}$
23. $-27 < 8m + 5$
24. $6b + 11 \geq 15$ $\left\{b \mid b \geq \frac{2}{3}\right\}$
25. $2(4t + 9) \leq 18$ $\{t \mid t \leq 0\}$
26. $90 \geq 5(2r + 6)$ $\{r \mid r \leq 6\}$
27. $14 - 8n \leq 0$
28. $-4(5w - 8) < 33$
29. $0.02x + 5.58 < 0$
30. $1.5 - 0.25c < 6$
31. $6d + 3 \geq 5d - 2$
32. $9z + 2 > 4z + 15$
33. $2(g + 4) < 3g - 2(g - 5)$ $\{g \mid g < 2\}$
34. $3(a + 4) - 2(3a + 4) \leq 4a - 1$
35. $y < \frac{-y + 2}{9}$ $\left\{y \mid y < \frac{1}{5}\right\}$
36. $\frac{1 - 4p}{5} < 0.2$ $\{p \mid p > 0\}$
37. ★ $\frac{4x + 2}{6} < \frac{2x + 1}{3}$ $\varnothing$
38. $12\left(\frac{1}{4} - \frac{n}{3}\right) \leq -6n$ $\left\{n \mid n \leq -\frac{3}{2}\right\}$
39. **PART-TIME JOB** David earns \$5.60 an hour working at Box Office Videos. Each week, 25% of his total pay is deducted for taxes. If David wants his take-home pay to be at least \$105 a week, solve the inequality $5.6x - 0.25(5.6x) \geq 105$ to determine how many hours he must work. at least 25 h
40. **STATE FAIR** Juan's parents gave him \$35 to spend at the State Fair. He spends \$13.25 for food. If rides at the fair cost \$1.50 each, solve the inequality $1.5n + 13.25 \leq 35$ to determine how many rides he can afford. no more than 14 rides

www.algebra2.com/self_check_quiz

3 Practice/Apply

Study Notebook

Have students—

- add the definitions/examples of the vocabulary terms to their Vocabulary Builder worksheets for Chapter 1.
- add the properties of inequality given in this lesson to their list of real number properties.
- write several examples of both set-builder notation and interval notation.
- include any other item(s) that they find helpful in mastering the skills in this lesson.

About the Exercises...

Organization by Objective

- **Solve Inequalities:** 15–40
- **Real-World Problems with Inequalities:** 41–51

Exercises 15–46 are structured so that students practice the same concepts whether they are assigned odd or even problems.

Alert! Exercises 56–58 require a graphing calculator.

Assignment Guide

Basic: 15–35 odd, 39–43 odd, 47–49, 52–55, 59–72

Average: 15–47 odd, 48–49, 52–55, 59–72 (optional: 56–58)

Advanced: 16–46 even, 48–66 (optional: 67–72)

All: Practice Quiz 2 (1–5)

Answers

15. $[-11, +\infty)$
16. $(-\infty, 18]$
17. $(-\infty, 7)$
18. $(-8, +\infty)$
19. $(-\infty, 27]$
20. $(-\infty, -3]$
21. $[-3.5, +\infty)$
22. $(-\infty, 5)$
23. $(-4, +\infty)$
24. $\left[\frac{2}{3}, +\infty\right)$
25. $(-\infty, 0]$
26. $(-\infty, 6]$
27. $[1.75, +\infty)$
28. $\left(-\frac{1}{20}, +\infty\right)$
29. $(-\infty, -279)$
30. $(-18, +\infty)$
31. $[-5, +\infty)$
32. $(2.6, +\infty)$
33. $(-\infty, 2)$
34. $\left[\frac{5}{7}, +\infty\right)$
35. $\left(-\infty, \frac{1}{5}\right)$
36. $(0, +\infty)$
37. $\varnothing$
38. $\left(-\infty, -\frac{3}{2}\right]$

Study Guide and Intervention, p. 25 (shown) and p. 26

Solve Inequalities The following properties can be used to solve inequalities.

Addition and Subtraction Properties for Inequalities	Multiplication and Division Properties for Inequalities
For any real numbers a, b, and c: 1. If $a < b$, then $a + c < b + c$ and $a - c < b - c$. 2. If $a > b$, then $a + c > b + c$ and $a - c > b - c$.	For any real numbers a, b, and c, with $c \neq 0$: 1. If c is positive and $a < b$, then $ac < bc$ and $\frac{a}{c} < \frac{b}{c}$. 2. If c is positive and $a > b$, then $ac > bc$ and $\frac{a}{c} > \frac{b}{c}$. 3. If c is negative and $a < b$, then $ac > bc$ and $\frac{a}{c} > \frac{b}{c}$. 4. If c is negative and $a > b$, then $ac < bc$ and $\frac{a}{c} < \frac{b}{c}$.

These properties are also true for ≤ and ≥.

Example 1 Solve $2x + 4 > 36$. Then graph the solution set on a number line.

$2x + 4 - 4 > 36 - 4$
$2x > 32$
$x > 16$

The solution set is $\{x \mid x > 16\}$.

Example 2 Solve $17 - 3w \geq 35$. Then graph the solution set on a number line.

$17 - 3w \geq 35$
$17 - 3w - 17 \geq 35 - 17$
$-3w \geq 18$
$w \leq -6$

The solution set is $(-\infty, -6]$.

Exercises

Solve each inequality. Describe the solution set using set-builder or interval notation. Then graph the solution set on a number line.

1. $7(7a - 9) \leq 84$ $\{a \mid a \leq 3\}$ or $(-\infty, 3]$
2. $3(9z + 4) > 35z - 4$ $\{z \mid z < 2\}$ or $(-\infty, 2)$
3. $5(12 - 3n) < 165$ $\{n \mid n > -7\}$ or $(-7, +\infty)$
4. $18 - 4k < 2(k + 21)$ $\{k \mid k > -4\}$ or $(-4, +\infty)$
5. $4(b - 7) + 6 < 22$ $\{b \mid b < 11\}$ or $(-\infty, 11)$
6. $2 + 3(m + 5) \geq 4(m + 3)$ $\{m \mid m \leq 5\}$ or $(-\infty, 5]$
7. $4x - 2 > -7(4x - 2)$ $\left\{x \mid x > \frac{1}{2}\right\}$ or $\left(\frac{1}{2}, +\infty\right)$
8. $\frac{1}{3}(2y - 3) > y + 2$ $\{y \mid y < -9\}$ or $(-\infty, -9)$
9. $2.5d + 15 \leq 75$ $\{d \mid d \leq 24\}$ or $(-\infty, 24]$

Skills Practice, p. 27 and Practice, p. 28 (shown)

Solve each inequality. Describe the solution set using set-builder or interval notation. Then, graph the solution set on a number line.

1. $8x - 6 \geq 10$ $\{x \mid x \geq 2\}$ or $[2, \infty)$
2. $23 - 4u < 11$ $\{u \mid u > 3\}$ or $(3, \infty)$
3. $-16 - 8r \geq 0$ $\{r \mid r \leq -2\}$ or $(-\infty, -2]$
4. $14s < 9s + 5$ $\{s \mid s < 1\}$ or $(-\infty, 1)$
5. $9x - 11 > 6x - 9$ $\left\{x \mid x > \frac{2}{3}\right\}$ or $\left(\frac{2}{3}, \infty\right)$
6. $-3(4w - 1) > 18$ $\left\{w \mid w < -\frac{5}{4}\right\}$ or $\left(-\infty, -\frac{5}{4}\right)$
7. $1 - 8u \leq 3u - 10$ $\{u \mid u \geq 1\}$ or $[1, \infty)$
8. $17.5 < 19 - 2.5x$ $\{x \mid x < 0.6\}$ or $(-\infty, 0.6)$
9. $9(2r - 5) - 3 < 7r - 4$ $\{r \mid r < 4\}$ or $(-\infty, 4)$
10. $1 + 5(x - 8) \leq 2 - (x + 5)$ $\{x \mid x \leq 6\}$ or $(-\infty, 6]$
11. $\frac{4x - 3}{2} \geq -3.5$ $\{x \mid x \geq -1\}$ or $[-1, \infty)$
12. $q - 2(2 - q) \leq 0$ $\left\{q \mid q \leq \frac{4}{3}\right\}$ or $\left(-\infty, \frac{4}{3}\right]$
13. $-36 - 2(w + 77) > -4(2w + 52)$ $\{w \mid w > -3\}$ or $(-3, \infty)$
14. $4n - 5(n - 3) > 3(n + 1) - 4$ $\{n \mid n < 4\}$ or $(-\infty, 4)$

Define a variable and write an inequality for each problem. Then solve.

15. Twenty less than a number is more than twice the same number. $n - 20 > 2n$; $n < -20$
16. Four times the sum of twice a number and −3 is less than 5.5 times that same number. $4[2n + (-3)] < 5.5n$; $n < 4.8$
17. **HOTELS** The Lincoln's hotel room costs \$90 a night. An additional 10% tax is added. Hotel parking is \$12 per day. The Lincoln's expect to spend \$30 in tips during their stay. Solve the inequality $90x + 90(0.1)x + 12x + 30 \leq 600$ to find how many nights the Lincoln's can stay at the hotel without exceeding total hotel costs of \$600. 5 nights
18. **BANKING** Jan's account balance is \$3800. Of this, \$750 is for rent. Jan wants to keep a balance of at least \$500. Write and solve an inequality describing how much she can withdraw and still meet these conditions. $3800 - 750 - w \geq 500$; $w \leq \$2550$

Reading to Learn Mathematics, p. 29 ELL

Pre-Activity **How can inequalities be used to compare phone plans?**

Read the introduction to Lesson 1-5 at the top of page 33 in your textbook.

- Write an inequality comparing the number of minutes per month included in the two phone plans. $150 < 400$ or $400 > 150$
- Suppose that in one month you use 230 minutes of airtime on your wireless phone. Find your monthly cost with each plan.

  Plan 1: \$67 Plan 2: \$55

  Which plan should you choose? Plan 2

Reading the Lesson

1. There are several different ways to write or show inequalities. Write each of the following in interval notation.
   a. $\{x \mid x < -3\}$ $(-\infty, -3)$
   b. $\{x \mid x \geq 5\}$ $[5, +\infty)$
   c. (number line) $(-\infty, 2]$
   d. (number line) $(-1, +\infty)$
2. Show how you can write an inequality symbol followed by a number to describe each of the following situations.
   a. There are fewer than 600 students in the senior class. < 600
   b. A student may enroll in no more than six courses each semester. ≤ 6
   c. To participate in a concert, you must be willing to attend at least ten rehearsals. ≥ 10
   d. There is space for at most 165 students in the high school band. ≤ 165

Helping You Remember

3. One way to remember something is to explain it to another person. A common student error in solving inequalities is forgetting to reverse the inequality symbol when multiplying or dividing both sides of an inequality by a negative number. Suppose that your classmate is having trouble remembering this rule. How could you explain this rule to your classmate? Sample answer: Draw a number line. Plot two positive numbers, for example, 3 and 8. Then plot their additive inverses, −3 and −8. Write an inequality that compares the positive numbers and one that compares the negative numbers. Notice that $8 > 3$, but $-8 < -3$. The order changes when you multiply by −1.

More About. . .

Child Care

In 1995, 55% of children ages three to five were enrolled in center-based child care programs. Parents cared for 26% of children, relatives cared for 19% of children, and non-relatives cared for 17% of children.

Source: National Center for Education Statistics

Define a variable and write an inequality for each problem. Then solve.

41. The sum of a number and 8 is more than 2. $n + 8 > 2$; $n > -6$
42. The product of −4 and a number is at least 35. $-4n \geq 35$; $n \leq -8.75$
43. The difference of one half of a number and 7 is greater than or equal to 5.
44. One more than the product of −3 and a number is less than 16.
45. ★ Twice the sum of a number and 5 is no more than 3 times that same number increased by 11. $2(n + 5) \leq 3n + 11$; $n \geq -1$
46. ★ 9 less than a number is at most that same number divided by 2. $n - 9 \leq \frac{n}{2}$; $n \leq 18$
47. **CHILD CARE** By Ohio law, when children are napping, the number of children per child care staff member may be as many as twice the maximum listed at the right. Write and solve an inequality to determine how many staff members are required to be present in a room where 17 children are napping and the youngest child is 18 months old. $2(7m) \geq 17$; $m \geq \frac{17}{14}$; at least 2 child care staff members

Maximum Number of Children Per Child Care Staff Member
At least one child care staff member caring for:
Every 5 infants less than 12 months old (or 2 for every 12)
Every 6 infants who are at least 12 months olds, but less than 18 months old
Every 7 toddlers who are at least 18 months old, but less than 30 months old
Every 8 toddlers who are at least 30 months old, but less than 3 years old

Source: Ohio Department of Job and Family Services

CAR SALES For Exercises 48 and 49, use the following information.

Mrs. Lucas earns a salary of \$24,000 per year plus 1.5% commission on her sales. If the average price of a car she sells is \$30,500, about how many cars must she sell to make an annual income of at least \$40,000?

48. Write an inequality to describe this situation. $\$24{,}000 + 0.015(30{,}500n) \geq 40{,}000$
49. Solve the inequality and interpret the solution. $n \geq 34.97$; She must sell at least 35 cars.

TEST GRADES For Exercises 50 and 51, use the following information.

Ahmik's scores on the first four of five 100-point history tests were 85, 91, 89, and 94.

50. If a grade of at least 90 is an A, write an inequality to find the score Ahmik must receive on the fifth test to have an A test average. See margin.
51. Solve the inequality and interpret the solution. $s \geq 91$; Ahmik must score at least 91 on her next test to have an A test average.
52. **CRITICAL THINKING** Which of the following properties hold for inequalities? Explain your reasoning or give a counterexample.
    a. Reflexive b. Symmetric c. Transitive

    52a. It holds only for ≤ or ≥; $2 \not< 2$. 52b. $1 < 2$ but $2 \not< 1$

53. WRITING IN MATH Answer the question that was posed at the beginning of the lesson. See pp. 53A–53B.

    How can inequalities be used to compare phone plans?

    Include the following in your answer:
    - an inequality comparing the number of minutes offered by each plan, and
    - an explanation of how Kuni might determine when Plan 1 might be cheaper than Plan 2 if she typically uses more than 150 but less than 400 minutes.

43. $\frac{1}{2}n - 7 \geq 5$; $n \geq 24$

44. $-3n + 1 < 16$; $n > -5$

52c. For all real numbers a, b, and c, if $a < b$ and $b < c$ then $a < c$.

Enrichment, p. 30

Equivalence Relations

A relation R on a set A is an *equivalence relation* if it has the following properties.

Reflexive Property	For any element a of set A, a R a.
Symmetric Property	For all elements a and b of set A, if a R b, then b R a.
Transitive Property	For all elements a, b, and c of set A, if a R b and b R c, then a R c.

Equality on the set of all real numbers is reflexive, symmetric, and transitive. Therefore, it is an equivalence relation.

In each of the following, a relation and a set are given. Write *yes* if the relation is an equivalence relation on the given set. If it is not, tell which of the properties it fails to exhibit.

1. <, {all numbers} no; reflexive, symmetric

Answer

50. $\frac{85 + 91 + 89 + 94 + s}{5} \geq 90$

Standardized Test Practice

54. If $4 - 5n \geq -1$, then n could equal all of the following EXCEPT **D**

Ⓐ $-\frac{1}{5}$. Ⓑ $\frac{1}{5}$. Ⓒ 1. Ⓓ 2.

55. If $a < b$ and $c < 0$, which of the following are true? **D**

I. $ac > bc$ II. $a + c < b + c$ III. $a - c > b - c$

Ⓐ I only Ⓑ II only Ⓒ III only

Ⓓ I and II only Ⓔ I, II, and III

Graphing Calculator

Use a graphing calculator to solve each inequality.

56. $-5x - 8 < 7$ **$x > -3$**

57. $-4(6x - 3) \leq 60$ **$x \geq -2$**

58. $3(x + 3) \geq 2(x + 4)$ **$x \geq -1$**

Maintain Your Skills

Mixed Review

Solve each equation. Check your solutions. *(Lesson 1-4)*

59. $|x - 3| = 17$ **$\{-14, 20\}$**

60. $8|4x - 3| = 64$

61. $|x + 1| = x$ **$\varnothing$**

60. $\left\{-\frac{5}{4}, \frac{11}{4}\right\}$

62. **SHOPPING** On average, by how much did the number of people who just browse, but not necessarily buy, online increase each year from 1997 to 2003? Define a variable, write an equation, and solve the problem. *(Lesson 1-3)*

62. **b = online browsers each year; $6b + 19.2 = 106.6$; about 14.6 million browsers each year**

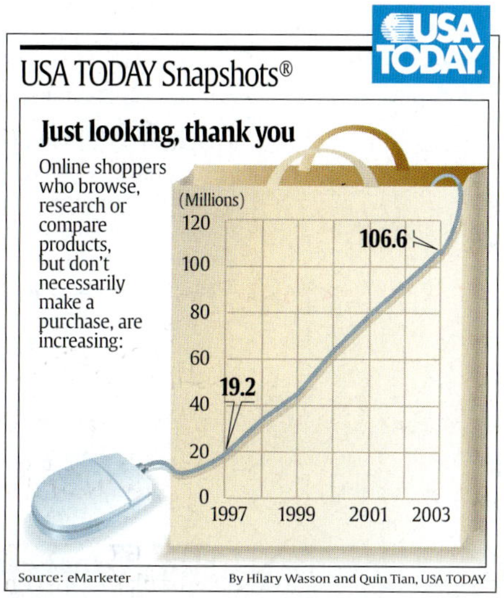

Name the sets of numbers to which each number belongs. *(Lesson 1-2)*

63. 31 64. $-4.\overline{2}$ **Q, R** 65. $\sqrt{7}$ **I, R**

63. **N, W, Z, Q, R**

66. **BABY-SITTING** Jenny baby-sat for $5\frac{1}{2}$ hours on Friday night and 8 hours on Saturday. She charges \$4.25 per hour. Use the Distributive Property to write two equivalent expressions that represent how much money Jenny earned. *(Lesson 1-2)* **4.25(5.5 + 8); 4.25(5.5) + 4.25(8)**

Getting Ready for the Next Lesson

PREREQUISITE SKILL **Solve each equation. Check your solutions.**

*(To review **solving absolute value equations**, see Lesson 1-4.)*

67. $|x| = 7$ **$\{-7, 7\}$**

68. $|x + 5| = 18$ **$\{13, -23\}$**

69. $|5y - 8| = 12$ **$\left\{4, -\frac{4}{5}\right\}$**

70. $|2x - 36| = 14$ **$\{11, 25\}$**

71. $2|w + 6| = 10$ **$\{-11, -1\}$**

72. $|x + 4| + 3 = 17$ **$\{-18, 10\}$**

Practice Quiz 2

Lessons 1-3 through 1-5

1. Solve $2d + 5 = 8d + 2$. Check your solution. *(Lesson 1-3)* **0.5**
2. Solve $s = \frac{1}{2}gt^2$ for g. *(Lesson 1-3)* **$\frac{2s}{t^2} = g$**
3. Evaluate $|x - 3y|$ if $x = -8$ and $y = 2$. *(Lesson 1-4)* **14**
4. Solve $3|3x + 2| = 51$. Check your solutions. *(Lesson 1-4)* **$\left\{-\frac{19}{3}, 5\right\}$**
5. Solve $2(m - 5) - 3(2m - 5) < 5m + 1$. Describe the solution set using set-builder or interval notation. Then graph the solution set on a number line. *(Lesson 1-5)* **See margin.**

4 Assess

Open-Ended Assessment

Writing Have students write their own list of tips for solving inequalities, including when to reverse the inequality sign and how to tell when the graph begins with a circle or with a dot.

Getting Ready for Lesson 1-6

PREREQUISITE SKILL Lesson 1-6 presents solving compound inequalities and absolute value inequalities. The procedure for solving absolute value inequalities are similar to those discussed for solving absolute value equations. Exercises 67–72 should be used to determine your students' familiarity with solving absolute value equations.

Assessment Options

Practice Quiz 2 The quiz provides students with a brief review of the concepts and skills in Lessons 1-3 through 1-5. Lesson numbers are given to the right of exercises or instruction lines so students can review concepts not yet mastered.

Quiz (Lessons 1-4 and 1-5) is available on p. 52 of the *Chapter 1 Resource Masters.*

Answer (Practice Quiz 2)

5. $\left\{m \mid m > \frac{4}{9}\right\}$ or $\left(\frac{4}{9}, +\infty\right)$

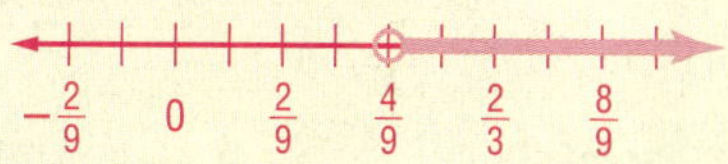

Online Lesson Plans

USA TODAY Education

USA TODAY Education's Online site offers resources and interactive features connected to each day's newspaper. *Experience TODAY*, USA TODAY's daily lesson plan, is available on the site and delivered daily to subscribers. This plan provides instruction for integrating USA TODAY graphics and key editorial features into your mathematics classroom. Log on to www.education.usatoday.com.

Glencoe's exclusive partnership with USA TODAY provides actual USA TODAY Snapshots® that illustrate mathematical concepts.

1-6 Lesson Notes

1 Focus

5-Minute Check Transparency 1-6 Use as a quiz or review of Lesson 1-5.

Mathematical Background notes are available for this lesson on p. 4D.

Building on Prior Knowledge

In Lesson 1-5 students solved inequalities, and in Lesson 1-4 they solved absolute value equations. In this lesson, they expand these skills to solving compound inequalities and absolute value inequalities.

How are compound inequalities used in medicine?

Ask students:

- If you are scheduled to have a glucose tolerance test at 10 A.M., at what hour should you begin fasting? **sometime between 6 P.M. and midnight**
- **Medicine** What does a glucose tolerance test measure? **how well the body processes sugar (glucose)**

1-6 Solving Compound and Absolute Value Inequalities

What You'll Learn

- Solve compound inequalities.
- Solve absolute value inequalities.

Vocabulary

- compound inequality
- intersection
- union

How are compound inequalities used in medicine?

One test used to determine whether a patient is diabetic and requires insulin is a glucose tolerance test. Patients start the test in a *fasting state*, meaning they have had no food or drink except water for at least 10 but no more than 16 hours. The acceptable number of hours h for fasting can be described by the following compound inequality.

$$h \geq 10 \text{ and } h \leq 16$$

COMPOUND INEQUALITIES A **compound inequality** consists of two inequalities joined by the word *and* or the word *or*. To solve a compound inequality, you must solve each part of the inequality. The graph of a compound inequality containing *and* is the **intersection** of the solution sets of the two inequalities.

Compound inequalities involving the word *and* are called *conjunctions.* Compound inequalities involving the word *or* are called *disjunctions.*

Key Concept — "And" Compound Inequalities

- **Words** A compound inequality containing the word *and* is true if and only if *both* inequalities are true.
- **Example** $x \geq -1$; $x < 2$; $x \geq -1$ and $x < 2$

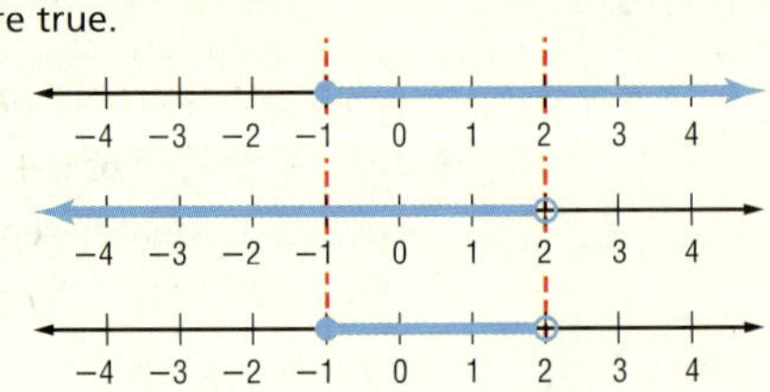

Another way of writing $x \geq -1$ and $x < 2$ is $-1 \leq x < 2$.
Both forms are read *x is greater than or equal to −1 and less than 2.*

Study Tip

Interval Notation

The compound inequality $-1 \leq x < 2$ can be written as $[-1, 2)$, indicating that the solution set is the set of all numbers between −1 and 2, including −1, but not including 2.

Example 1 Solve an "and" Compound Inequality

Solve $13 < 2x + 7 \leq 17$. Graph the solution set on a number line.

Method 1

Write the compound inequality using the word *and*. Then solve each inequality.

$13 < 2x + 7$	and	$2x + 7 \leq 17$
$6 < 2x$		$2x \leq 10$
$3 < x$		$x \leq 5$

$$3 < x \leq 5$$

Method 2

Solve both parts at the same time by subtracting 7 from each part. Then divide each part by 2.

$$13 < 2x + 7 \leq 17$$
$$6 < 2x \leq 10$$
$$3 < x \leq 5$$

Resource Manager

Workbook and Reproducible Masters

Chapter 1 Resource Masters

- Study Guide and Intervention, pp. 31–32
- Skills Practice, p. 33
- Practice, p. 34
- Reading to Learn Mathematics, p. 35
- Enrichment, p. 36
- Assessment, p. 52

Teaching Algebra With Manipulatives Masters, p. 216

Transparencies

5-Minute Check Transparency 1-6
Answer Key Transparencies

Technology

Interactive Chalkboard

Graph the solution set for each inequality and find their intersection.

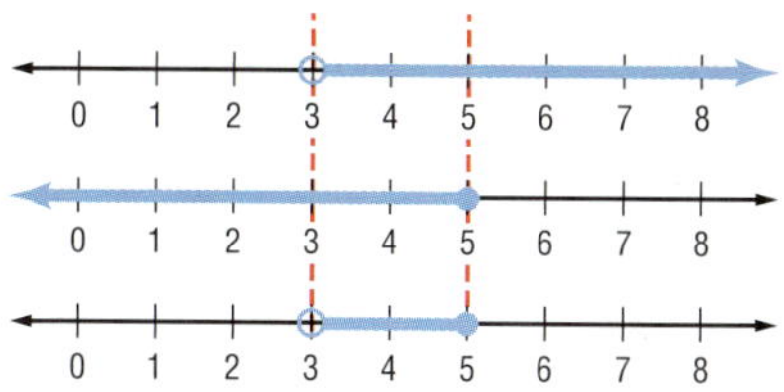

$x > 3$

$x \leq 5$

$3 < x \leq 5$

The solution set is $\{x \mid 3 < x \leq 5\}$.

The graph of a compound inequality containing *or* is the **union** of the solution sets of the two inequalities.

Key Concept — "Or" Compound Inequalities

- **Words** A compound inequality containing the word *or* is true if one or more of the inequalities is true.
- **Example** $x \leq 1$; $x > 4$; $x \leq 1$ or $x > 4$

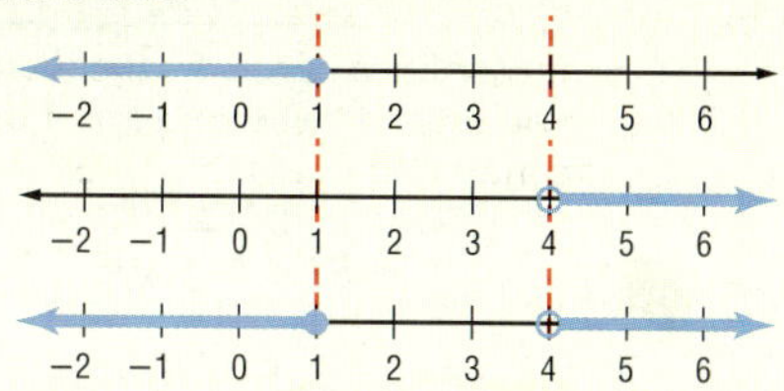

Example 2 Solve an "or" Compound Inequality

Solve $y - 2 > -3$ or $y + 4 \leq -3$. Graph the solution set on a number line.

Solve each inequality separately.

$y - 2 > -3$ or $y + 4 \leq -3$

$y > -1$ $\qquad$ $y \leq -7$

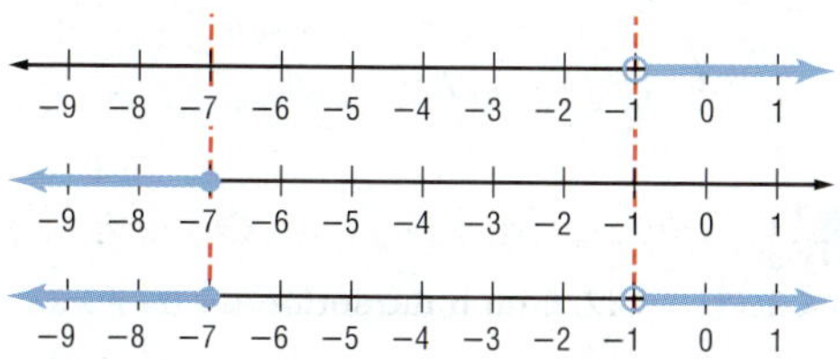

$y > -1$

$y \leq -7$

$y > -1$ or $y \leq -7$

The solution set is $\{y \mid y > -1 \text{ or } y \leq -7\}$.

Study Tip

Interval Notation

In interval notation, the symbol for the union of the two sets is $\cup$. The compound inequality $y > -1$ or $y \leq -7$ is written as $(-\infty, -7] \cup (-1, +\infty)$, indicating that all values less than and including -7 are part of the solution set. In addition, all values greater than -1, not including -1, are part of the solution set.

ABSOLUTE VALUE INEQUALITIES In Lesson 1-4, you learned that the absolute value of a number is its distance from 0 on the number line. You can use this definition to solve inequalities involving absolute value.

www.algebra2.com/extra_examples

2 Teach

COMPOUND INEQUALITIES

In-Class Examples Power Point®

1 Solve $10 \leq 3y - 2 < 19$. Graph the solution set on a number line. $\{y \mid 4 \leq y < 7\}$

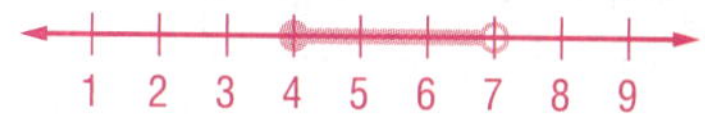

Teaching Tip Remind students that the word *and* used in Method 1 means the values for $2x + 7$ must meet *both* conditions. That is, a value must be both greater than 13 and less than or equal to 17.

2 Solve $x + 3 < 2$ or $-x \leq -4$. Graph the solution set on a number line. $\{x \mid x < -1 \text{ or } x \geq 4\}$

Reading Tip Students may make the mistake of wanting to associate *union* with the word *and* because union often indicates the joining of two or more things. As a memory device, point out that the word *or* begins with the letter *o* which is found in the word *union*, while *and* begins with the letter *a* which is not found in *union*.

Teacher to Teacher

Ron Millard — Shawnee Mission South H.S., Overland Park, KS

"To help make further work with absolute value more understandable, I teach my students to solve absolute value inequalities by using the definition of absolute value. Using this method, the statement $|3x - 12| \geq 6$ is rewritten as $3x - 12 \geq 6$ or $-(3x - 12) \geq 6$."

Teacher to Teacher features contain teaching suggestions from teachers who are creatively teaching Algebra in their classrooms.

ABSOLUTE VALUE INEQUALITIES

In-Class Examples

Power Point®

3 Solve $3 > |d|$. Graph the solution set on a number line.
$\{d \mid -3 < d < 3\}$

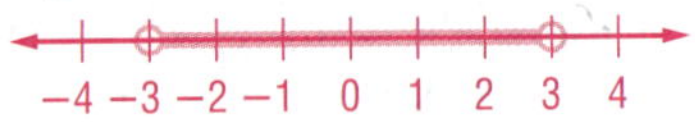

4 Solve $3 < |d|$. Graph the solution set on a number line.
$\{d \mid d < -3 \text{ or } d > 3\}$

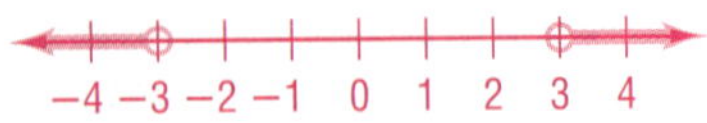

Reading Tip Make sure students understand the meaning of Examples 3 and 4 before they go on. Have them say the problem in words (for Example 3: "The distance of *a* from zero without regard to direction is less than 4.") and demonstrate where *a* can be located on a number line.

5 Solve $|2x - 2| \geq 4$. Graph the solution set on a number line.
$\{x \mid x \leq -1 \text{ or } x \geq 3\}$

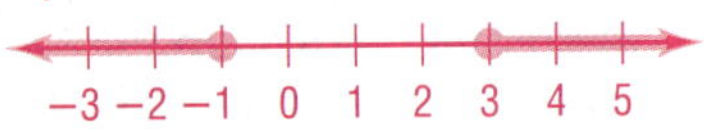

Study Tip

Absolute Value Inequalities
Because the absolute value of a number is never negative,
- the solution of an inequality like $|a| < -4$ is the empty set.
- the solution of an inequality like $|a| > -4$ is the set of all real numbers.

Example 3 Solve an Absolute Value Inequality (<)

Solve $|a| < 4$. Graph the solution set on a number line.

You can interpret $|a| < 4$ to mean that the distance between a and 0 on a number line is less than 4 units. To make $|a| < 4$ true, you must substitute numbers for a that are fewer than 4 units from 0.

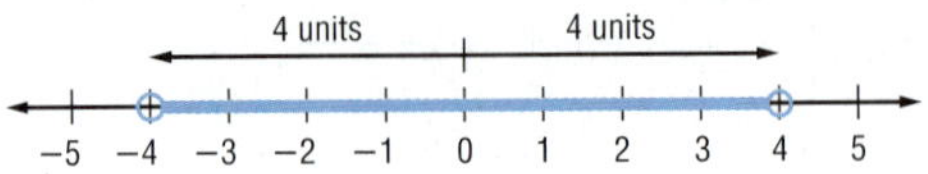

Notice that the graph of $|a| < 4$ is the same as the graph of $a > -4$ and $a < 4$.

All of the numbers between −4 and 4 are less than 4 units from 0. The solution set is $\{a \mid -4 < a < 4\}$.

Example 4 Solve an Absolute Value Inequality (>)

Solve $|a| > 4$. Graph the solution set on a number line.

You can interpret $|a| > 4$ to mean that the distance between a and 0 is greater than 4 units. To make $|a| > 4$ true, you must substitute values for a that are greater than 4 units from 0.

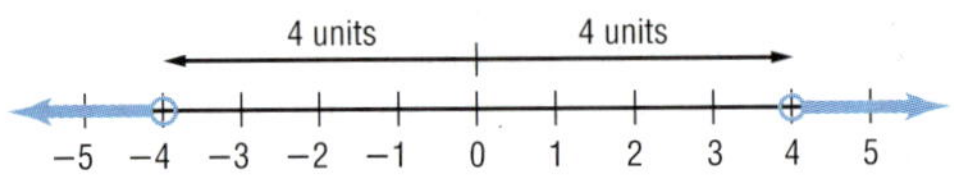

Notice that the graph of $|a| > 4$ is the same as the graph of $a > 4$ or $a < -4$.

All of the numbers *not* between −4 and 4 are greater than 4 units from 0. The solution set is $\{a \mid a > 4 \text{ or } a < -4\}$.

An absolute value inequality can be solved by rewriting it as a compound inequality.

Key Concept — Absolute Value Inequalities

- **Symbols** For all real numbers a and b, $b > 0$, the following statements are true.
  1. If $|a| < b$ then $-b < a < b$.
  2. If $|a| > b$ then $a > b$ or $a < -b$.
- **Examples** If $|2x + 1| < 5$, then $-5 < 2x + 1 < 5$.
  If $|2x + 1| > 5$, then $2x + 1 > 5$ or $2x + 1 < -5$.

These statements are also true for ≤ and ≥, respectively.

Example 5 Solve a Multi-Step Absolute Value Inequality

Solve $|3x - 12| \geq 6$. Graph the solution set on a number line.

$|3x - 12| \geq 6$ is equivalent to $3x - 12 \geq 6$ or $3x - 12 \leq -6$. Solve each inequality.

$3x - 12 \geq 6$	or	$3x - 12 \leq -6$
$3x \geq 18$		$3x \leq 6$
$x \geq 6$		$x \leq 2$

The solution set is $\{x \mid x \geq 6 \text{ or } x \leq 2\}$.

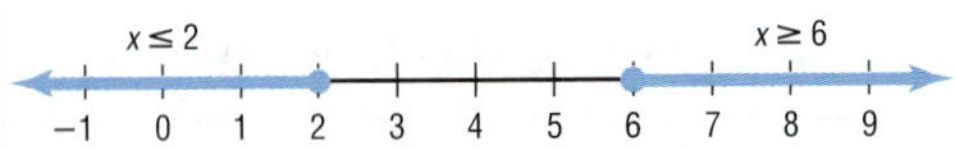

DAILY INTERVENTION — Differentiated Instruction

Kinesthetic Have students work in pairs to create a number line on the floor, perhaps using floor tiles and masking tape. Ask one partner to write or say an inequality such as $|x| < 5$ and then have the other partner walk from −5 to 5 on the number line to demonstrate the possible values for x.

Example 6 Write an Absolute Value Inequality

JOB HUNTING To prepare for a job interview, Megan researches the position's requirements and pay. She discovers that the average starting salary for the position is \$38,500, but her actual starting salary could differ from the average by as much as \$2450.

a. Write an absolute value inequality to describe this situation.

Let x = Megan's starting salary.

Her starting salary could differ from the average	by as much as	\$2450.
$\lvert 38{,}500 - x \rvert$	$\leq$	2450

b. Solve the inequality to find the range of Megan's starting salary.

Rewrite the absolute value inequality as a compound inequality. Then solve for x.

$$-2450 \leq 38{,}500 - x \leq 2450$$
$$-2450 - 38{,}500 \leq 38{,}500 - x - 38{,}500 \leq 2450 - 38{,}500$$
$$-40{,}950 \leq -x \leq -36{,}050$$
$$40{,}950 \geq x \geq 36{,}050$$

The solution set is $\{x \mid 36{,}050 \leq x \leq 40{,}950\}$. Thus, Megan's starting salary will fall between \$36,050 and \$40,950, inclusive.

More About. . .

Job Hunting When executives in a recent survey were asked to name one quality that impressed them the most about a candidate during a job interview, 32 percent said honesty and integrity.

Source: careerexplorer.net

In-Class Example

Power Point®

Teaching Tip Show students that $\lvert x - 38{,}500 \rvert \leq 2450$ will also work as the inequality for Example 6.

6 HOUSING According to a recent survey, the average monthly rent for a one-bedroom apartment in one city neighborhood is \$750. However, the actual rent for any given one-bedroom apartment might vary as much as \$250 from that average.

a. Write an absolute value inequality to describe this situation. $\lvert 750 - r \rvert \leq 250$

b. Solve the inequality to find the range of monthly rent. $\{r \mid 500 \leq r \leq 1000\}$; The actual rent falls between \$500 and \$1000.

Teaching Tip Suggest that students write some sample situations to help them understand problems that involve absolute value inequalities. In Example 6 for instance, students might ask themselves, "What are some possible salaries that fit this situation?"

Check for Understanding

Concept Check

1. **Write** a compound inequality to describe the following situation. *Buy a present that costs at least \$5 and at most \$15.* $5 \leq c \leq 15$
2. **OPEN ENDED** Write a compound inequality whose graph is the empty set. Sample answer: $x < -3$ and $x > 2$
3. **FIND THE ERROR** Sabrina and Isaac are solving $\lvert 3x + 7 \rvert > 2$.

Sabrina	Isaac
$\lvert 3x + 7 \rvert > 2$	$\lvert 3x + 7 \rvert > 2$
$3x + 7 > 2$ or $3x + 7 < -2$	$-2 < 3x + 7 < 2$
$3x > -5$ $\quad$ $3x < -9$	$-9 < 3x < -5$
$x > -\frac{5}{3}$ $\quad$ $x < -3$	$-3 < x < -\frac{5}{3}$

Who is correct? Explain your reasoning.

3. Sabrina; an absolute value inequality of the form $\lvert a \rvert > b$ should be rewritten as an *or* compound inequality, $a > b$ or $a < -b$.

Guided Practice

GUIDED PRACTICE KEY	
Exercises	**Examples**
4, 5, 6, 7	3–5
8–13	1–5
14	6

Write an absolute value inequality for each of the following. Then graph the solution set on a number line. 4–5. See margin for graphs.

4. all numbers between -8 and 8 $\lvert n \rvert < 8$
5. all numbers greater than 3 or less than -3 $\lvert n \rvert > 3$

Write an absolute value inequality for each graph.

6. 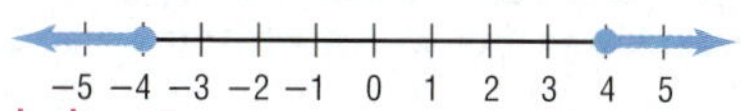
$\lvert n \rvert \geq 4$

7. 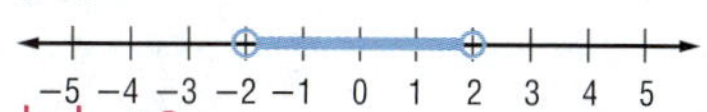
$\lvert n \rvert < 2$

3 Practice/Apply

Study Notebook

Have students—

- *complete the definitions/examples for the remaining terms on their Vocabulary Builder worksheets for Chapter 1.*
- *write a comparison between compound inequalities whose solutions involve the word "and," and compound inequalities whose solutions involve the word "or," including examples of both types.*
- *include any other item(s) that they find helpful in mastering the skills in this lesson.*

Study Notebook tips offer suggestions for helping your students keep notes they can use to study this chapter.

Answers

4. 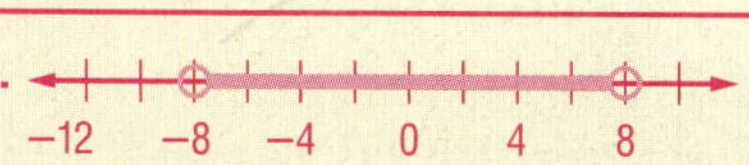

5. 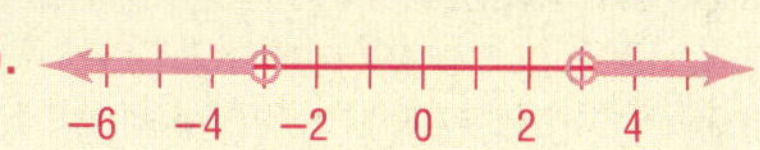

About the Exercises...

Organization by Objective

- **Compound Inequalities:** 27–32, 45–47, 49–52
- **Absolute Value Inequalities:** 15–26, 33–44, 48

Odd/Even Assignments

Exercises 15–44 are structured so that students practice the same concepts whether they are assigned odd or even problems.

Alert! Exercises 57–60 require a graphing calculator.

Assignment Guide

Basic: 15–23 odd, 27–39 odd, 45–47, 53–56, 61–75

Average: 15–45 odd, 46–47, 49–50, 53–56, 61–75 (optional: 57–60)

Advanced: 16–44 even, 48–75

DAILY INTERVENTION **FIND THE ERROR** Have students use a finger to cover up "$-2 <$" in the second line of Isaac's solution. Ask them to compare the remaining inequality to the original, emphasizing the direction of the inequality symbols. Stress that Isaac's symbol should point in the same direction as the original symbol.

8–13. See margin for graphs.

8. $\{y \mid y > 4$ or $y < -1\}$

Solve each inequality. Graph the solution set on a number line.

8. $y - 3 > 1$ or $y + 2 < 1$
9. $3 < d + 5 < 8$ $\{d \mid -2 < d < 3\}$
10. $|a| \geq 5$ $\{a \mid a \geq 5$ or $a \leq -5\}$
11. $|g + 4| \leq 9$ $\{g \mid -13 \leq g \leq 5\}$
12. $|4k - 8| < 20$ $\{k \mid -3 < k < 7\}$
13. $|w| \geq -2$ all real numbers

Application

14. **FLOORING** Deion estimates that he will need between 55 and 60 ceramic tiles to retile his kitchen floor. If each tile costs \$6.25, write and solve a compound inequality to determine what the cost c of the tile could be. $55 \leq \frac{c}{6.25} \leq 60$; $343.75 \leq c \leq 375$; between \$343.75 and \$375

★ indicates increased difficulty

Practice and Apply

Homework Help

For Exercises	See Examples
15–26, 33–44	3–5
27–32, 51, 52	1, 2
45–50	6

Extra Practice See page 829.

Write an absolute value inequality for each of the following. Then graph the solution set on a number line. 15–20. See margin for graphs.

15. all numbers greater than or equal to 5 or less than or equal to -5 $|n| \geq 5$
16. all numbers less than 7 and greater than -7 $|n| < 7$
17. all numbers between -4 and 4 $|n| < 4$
18. all numbers less than or equal to -6 or greater than or equal to 6 $|n| \geq 6$
19. all numbers greater than 8 or less than -8 $|n| > 8$
20. all number less than or equal to 1.2 and greater than or equal to -1.2 $|n| \leq 1.2$

Write an absolute value inequality for each graph.

27–44. See pp. 53A–53B for graphs.

27. $\{p \mid p \leq 2$ or $p \geq 8\}$

30. $\{c \mid c < -2$ or $c \geq 1\}$

32. all real numbers

21. 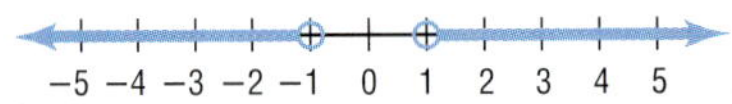
$|n| > 1$

22. 
$|n| \leq 5$

23. 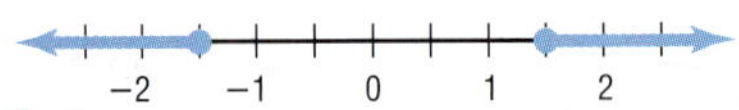
$|n| \geq 1.5$

24. 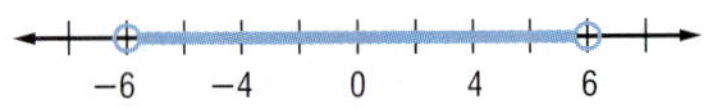
$|n| < 6$

★ 25. 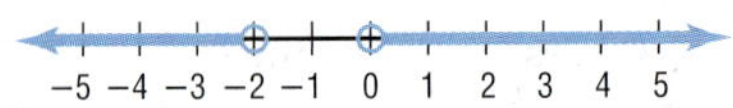
$|n + 1| > 1$

★ 26. 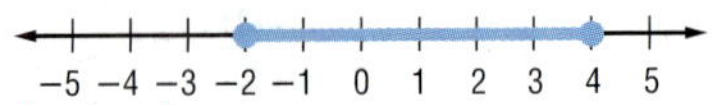
$|n - 1| \leq 3$

More About...

Betta Fish

Adult Male Size: 3 inches
Water pH: 6.8–7.4
Temperature: 75–86°F
Diet: omnivore, prefers live foods
Tank Level: top dweller
Difficulty of Care: easy to intermediate
Life Span: 2–3 years
Source: www.about.com

Solve each inequality. Graph the solution set on a number line.

27. $3p + 1 \leq 7$ or $2p - 9 \geq 7$
28. $9 < 3t + 6 < 15$ $\{t \mid 1 < t < 3\}$
29. $-11 < -4x + 5 < 13$ $\{x \mid -2 < x < 4\}$
30. $2c - 1 < -5$ or $3c + 2 \geq 5$
31. $-4 < 4f + 24 < 4$ $\{f \mid -7 < f < -5\}$
32. $a + 2 > -2$ or $a - 8 < 1$
33. $|g| \leq 9$ $\{g \mid -9 \leq g \leq 9\}$
34. $|2m| \geq 8$ $\{m \mid m \geq 4$ or $m \leq -4\}$
35. $|3k| < 0$ $\varnothing$
36. $|-5y| < 35$ $\{y \mid -7 < y < 7\}$
37. $|b - 4| > 6$ $\{b \mid b > 10$ or $b < -2\}$
38. $|6r - 3| < 21$ $\{r \mid -3 < r < 4\}$
39. $|3w + 2| \leq 5$ $\left\{w \,\middle|\, -\frac{7}{3} \leq w \leq 1\right\}$
40. $|7x| + 4 < 0$ $\varnothing$
41. ★ $|n| \geq n$ all real numbers
42. ★ $|n| \leq n$ $\{n \mid n \geq 0\}$
43. ★ $|2n - 7| \leq 0$ $\left\{n \,\middle|\, n = \frac{7}{2}\right\}$
44. ★ $|n - 3| < n$ $\{n \mid n > 1.5\}$
45. **BETTA FISH** A Siamese Fighting Fish, also known as a Betta fish, is one of the most recognized and colorful fish kept as a pet. Using the information at the left, write a compound inequality to describe the acceptable range of water pH levels for a male Betta. $6.8 < x < 7.4$

Answers

8. 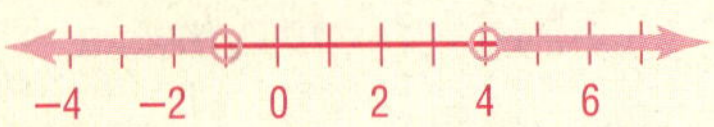

9. 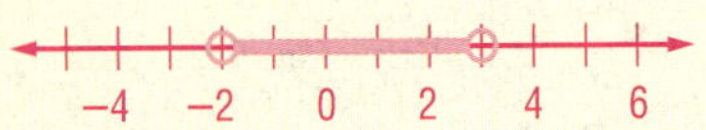

10. 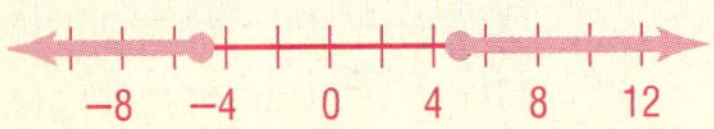

11. 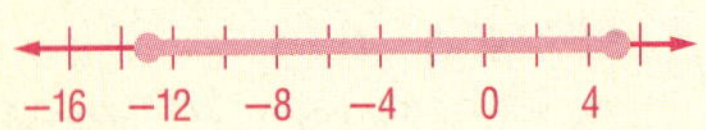

12. 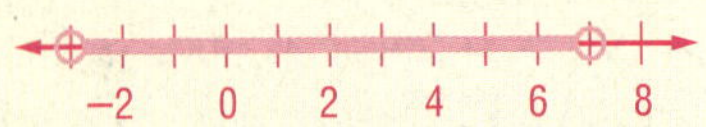

13. 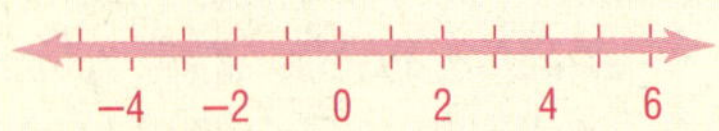

15. 

16. 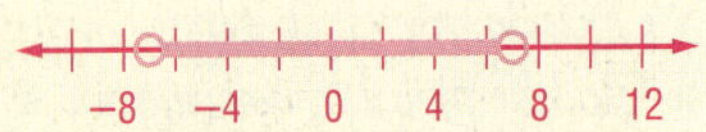

17. 

18. 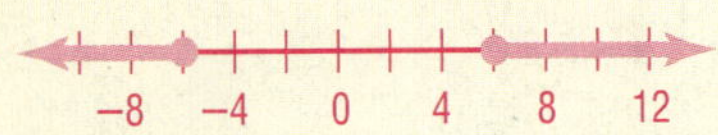

19. 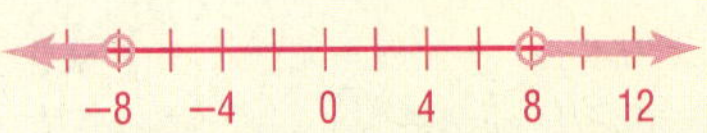

20. 

53a. 

53b. 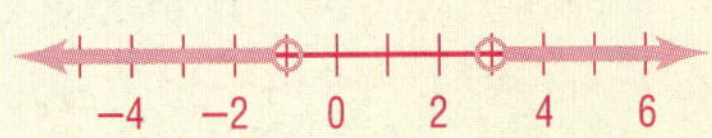

53c. 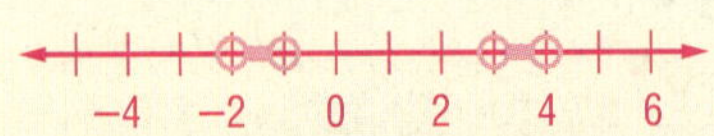

53d. $3 < |x + 2| \leq 8$ can be rewritten as $|x + 2| > 3$ and $|x + 2| \leq 8$. The solution of $|x + 2| > 3$ is $x > 1$ or $x < -5$. The solution of $|x + 2| \leq 8$ is $-10 \leq x \leq 6$. Therefore, the union of these two sets is ($x > 1$ or $x < -5$) and ($-10 \leq x \leq 6$). *(continued on the next page)*

SPEED LIMITS **For Exercises 46 and 47, use the following information.**
On some interstate highways, the maximum speed a car may drive is 65 miles per hour. A tractor-trailer may not drive more than 55 miles per hour. The minimum speed for all vehicles is 45 miles per hour.

46. Write an inequality to represent the allowable speed for a car on an interstate highway. **$45 \leq s \leq 65$**

47. Write an inequality to represent the speed at which a tractor-trailer may travel on an interstate highway. **$45 \leq s \leq 55$**

48. **HEALTH** *Hypothermia* and *hyperthermia* are similar words but have opposite meanings. Hypothermia is defined as a lowered body temperature. Hyperthermia means an extremely high body temperature. Both conditions are potentially dangerous and occur when a person's body temperature fluctuates by more than 8° from the normal body temperature of 98.6°F. Write and solve an absolute value inequality to describe body temperatures that are considered potentially dangerous. **$|b - 98.6| > 8$; $\{b \mid b > 106.6$ or $b < 90.6\}$**

MAIL **For Exercises 49 and 50, use the following information.**
The U.S. Postal Service defines an oversized package as one for which the length L of its longest side plus the distance D around its thickest part is more than 108 inches and less than or equal to 130 inches.

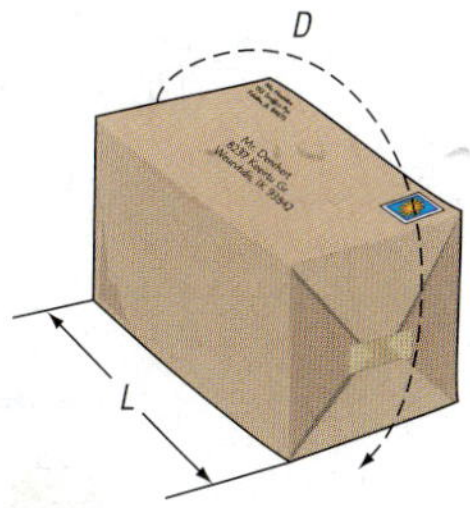

49. Write a compound inequality to describe this situation. **108 in. $< L + D \leq$ 130 in.**

50. If the distance around the thickest part of a package you want to mail is 24 inches, describe the range of lengths that would classify your package as oversized. **84 in. $< L \leq$ 106 in.**

GEOMETRY **For Exercises 51 and 52, use the following information.**
The *Triangle Inequality Theorem* states that the sum of the measures of any two sides of a triangle is greater than the measure of the third side.

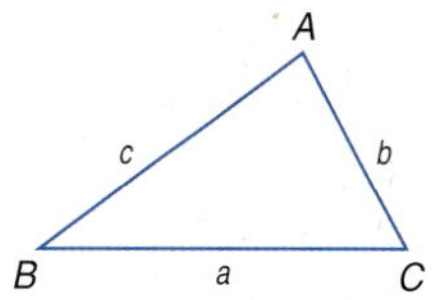

51. $a + b > c$, $a + c > b$, $b + c > a$

★51. Write three inequalities to express the relationships among the sides of $\triangle ABC$.

★52. Write a compound inequality to describe the range of possible measures for side c in terms of a and b. Assume that $a > b > c$. (*Hint*: Solve each inequality you wrote in Exercise 51 for c.) **$a - b < c < a + b$**

53. **CRITICAL THINKING** **Graph each set on a number line.** **a–d. See margin.**
   a. $-2 < x < 4$
   b. $x < -1$ or $x > 3$
   c. $(-2 < x < 4)$ and $(x < -1$ or $x > 3)$ (*Hint*: This is the intersection of the graphs in part **a** and part **b**.)
   d. Solve $3 < |x + 2| \leq 8$. Explain your reasoning and graph the solution set.

54. WRITING IN MATH Answer the question that was posed at the beginning of the lesson. **See pp. 53A–53B.**

How are compound inequalities used in medicine?

Include the following in your answer:
- an explanation as to when to use *and* and when to use *or* when writing a compound inequality,
- an alternative way to write $h \geq 10$ and $h \leq 16$, and
- an example of an acceptable number of hours for this fasting state and a graph to support your answer.

www.algebra2.com/self_check_quiz

The union of the graph of $x > 1$ or $x < -5$ and the graph of $-10 \leq x \leq 6$ is shown below. From this we can see that the solution can be rewritten as $(-10 \leq x < -5)$ or $(1 < x \leq 6)$.

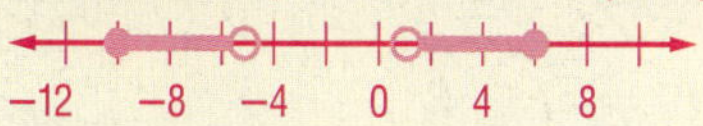

Enrichment, p. 36

Conjunctions and Disjunctions

An absolute value inequality may be solved as a compound sentence.

Example 1 **Solve $|2x| < 10$.**

$|2x| < 10$ means $2x < 10$ and $2x > -10$.

Solve each inequality. $x < 5$ and $x > -5$.

Every solution for $|2x| < 10$ is a replacement for x that makes both $x < 5$ and $x > -5$ true.

A compound sentence that combines two statements by the word *and* is a *conjunction*.

Example 2 **Solve $|3x - 7| \geq 11$.**

$|3x - 7| \geq 11$ means $3x - 7 \geq 11$ or $3x - 7 \leq -11$.

Solve each inequality. $3x \geq 18$ or $3x \leq -4$

Study Guide and Intervention, p. 31 (shown) and p. 32

Compound Inequalities A compound inequality consists of two inequalities joined by the word *and* or the word *or*. To solve a compound inequality, you must solve each part separately.

And Compound Inequalities	Example: $x > -4$ and $x < 3$	The graph is the intersection of solution sets of two inequalities.
Or Compound Inequalities	Example: $x \leq -3$ or $x > 1$	The graph is the union of solution sets of two inequalities.

Example 1 **Solve $-3 \leq 2x + 5 \leq 19$. Graph the solution set on a number line.**

$-3 \leq 2x + 5$ and $2x + 5 \leq 19$
$-8 \leq 2x$ $\quad 2x \leq 14$
$-4 \leq x$ $\quad x \leq 7$
$-4 \leq x \leq 7$

Example 2 **Solve $3y - 2 \geq 7$ or $2y - 1 \leq -9$. Graph the solution set on a number line.**

$3y - 2 \geq 7$ or $2y - 1 \leq -9$
$3y \geq 9$ or $2y \leq -8$
$y \geq 3$ or $y \leq -4$

Exercises

Solve each inequality. Graph the solution set on a number line.

1. $-10 < 3x + 2 \leq 14$ **$\{x \mid -4 < x \leq 4\}$**
2. $3a + 8 < 23$ or $\frac{1}{4}a - 6 > 7$ **$\{a \mid a < 5$ or $a > 52\}$**
3. $18 < 4x - 10 < 50$ **$\{x \mid 7 < x < 15\}$**
4. $5k + 2 < -13$ or $8k - 1 > 19$ **$\{k \mid k < -3$ or $k > 2.5\}$**
5. $100 \leq 5y - 45 \leq 225$ **$\{y \mid 29 \leq y \leq 54\}$**
6. $\frac{2}{3}b - 2 > 10$ or $\frac{3}{4}b + 5 < -4$ **$\{b \mid b < -12$ or $b > 18\}$**
7. $22 < 6w - 2 < 82$ **$\{w \mid 4 < w < 14\}$**
8. $4d - 1 > -9$ or $2d + 5 < 11$ **{all real numbers}**

Skills Practice, p. 33 and Practice, p. 34 (shown)

Write an absolute value inequality for each of the following. Then graph the solution set on a number line.

1. all numbers greater than 4 or less than -4 **$|n| > 4$**
2. all numbers between -1.5 and 1.5, including -1.5 and 1.5 **$|n| \leq 1.5$**

Write an absolute value inequality for each graph.

3. **$|n| \geq 10$** 4. **$|n| < \frac{4}{3}$**

Solve each inequality. Graph the solution set on a number line.

5. $-8 \leq 3y - 20 < 52$ **$\{y \mid 4 \leq y < 24\}$**
6. $3(5x - 2) < 24$ or $6x - 4 > 4 + 5x$ **$\{x \mid x < 2$ or $x > 8\}$**
7. $2x - 3 > 15$ or $3 - 7x < 17$ **$\{x \mid x > -2\}$**
8. $15 - 5x \leq 0$ and $5x + 6 \geq -14$ **$\{x \mid x \geq 3\}$**
9. $|2w| \geq 5$ **$\left\{w \mid w \leq -\frac{5}{2}\right.$ or $\left.w \geq \frac{5}{2}\right\}$**
10. $|y + 5| < 2$ **$\{x \mid -7 < x < -3\}$**
11. $|x - 8| \geq 3$ **$\{x \mid x \leq 5$ or $x \geq 11\}$**
12. $|2z - 2| \leq 3$ **$\left\{z \mid -\frac{1}{2} \leq z \leq \frac{5}{2}\right\}$**
13. $|2x + 2| - 7 \leq -5$ **$\{x \mid -2 \leq x \leq 0\}$**
14. $|x| > x - 1$ **all real numbers**
15. $|3b + 5| \leq -2$ **$\varnothing$**
16. $|3n - 2| - 2 < 1$ **$\left\{n \mid -\frac{1}{3} < n < \frac{5}{3}\right\}$**

17. **RAINFALL** In 90% of the last 30 years, the rainfall at Shell Beach has varied no more than 6.5 inches from its mean value of 24 inches. Write and solve an absolute value inequality to describe the rainfall in the other 10% of the last 30 years. **$|r - 24| > 6.5$; $\{r \mid r < 17.5$ or $r > 30.5\}$**

18. **MANUFACTURING** A company's guidelines call for each can of soup produced not to vary from its stated volume of 14.5 fluid ounces by more than 0.08 ounces. Write and solve an absolute value inequality to describe acceptable can volumes. **$|v - 14.5| \leq 0.08$; $\{v \mid 14.42 \leq v \leq 14.58\}$**

Reading to Learn Mathematics, p. 35

ELL

Pre-Activity **How are compound inequalities used in medicine?**

Read the introduction to Lesson 1-6 at the top of page 40 in your textbook.

- Five patients arrive at a medical laboratory at 11:30 A.M. for a glucose tolerance test. Each of them is asked when they last had something to eat or drink. Some of the patients are given the test and others are told that they must come back another day. Each of the patients is listed below with the times when they started to fast. (The P.M. times refer to the night before.) Which of the patients were accepted for the test?

Ora 5:00 A.M. Juanita 11:30 P.M.
Jason 1:30 A.M. Samir 5:00 P.M. **Jason and Juanita**

Reading the Lesson

1. a. Write a compound inequality that says, "x is greater than -3 *and* x is less than or equal to 4." **$-3 < x \leq 4$**
   b. Graph the inequality that you wrote in part a on a number line.
2. Use a compound inequality and set-builder notation to describe the following graph. **$\{x \mid x \leq -1$ or $x > 3\}$**
3. Write a statement equivalent to $|4x - 5| > 2$ that does not use the absolute value symbol. **$4x - 5 > 2$ or $4x - 5 < -2$**
4. Write a statement equivalent to $|3x + 7| < 8$ that does not use the absolute value symbol. **$-8 < 3x + 7 < 8$**

Helping You Remember

5. Many students have trouble knowing whether an absolute value inequality should be translated into an *and* or an *or* compound inequality. Describe a way to remember which of these applies to an absolute value inequality. Also describe how to recognize the difference from a number line graph. **Sample answer: If the absolute value quantity is followed by a < or ≤ symbol, the expression inside the absolute value bars must be *between* two numbers, so this becomes an *and* inequality. The number line graph will show a single interval *between* two numbers. If the absolute value quantity is followed by a > or ≥ symbol, it becomes an *or* inequality, and the graph will show two disconnected intervals with arrows going in opposite directions.**

4 Assess

Open-Ended Assessment

Writing Have students write a summary of the different kinds of inequalities they have seen in this chapter, with examples of each type and graphs of their solution sets.

Assessment Options

Quiz (Lesson 1-6) is available on p. 52 of the *Chapter 1 Resource Masters.*

Answers

57. 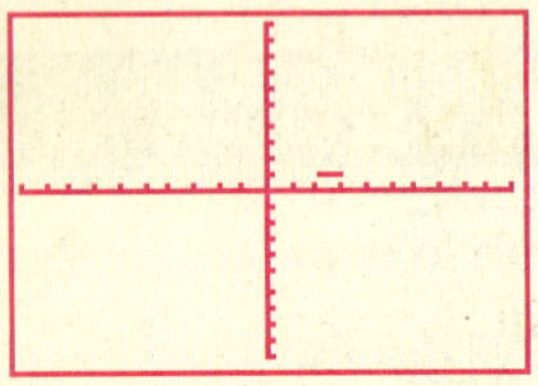

61. 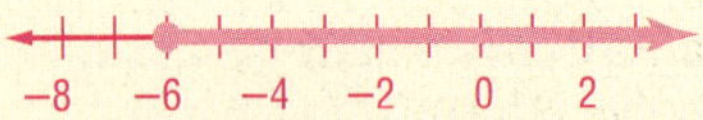

62. 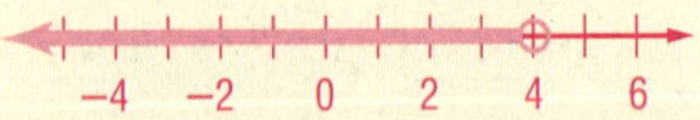

63. 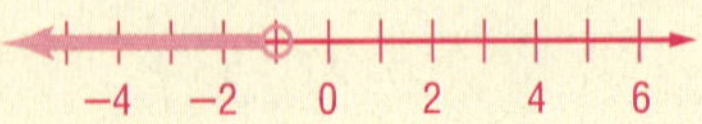

Standardized Test Practice

55. **SHORT RESPONSE** Solve $|2x + 11| > 1$ for x. **$x > -5$ or $x < -6$**

56. If $5 < a < 7 < b < 14$, then which of the following best defines $\frac{a}{b}$? **D**

   Ⓐ $\frac{5}{7} < \frac{a}{b} < \frac{1}{2}$ Ⓑ $\frac{5}{14} < \frac{a}{b} < \frac{1}{2}$

   Ⓒ $\frac{5}{7} < \frac{a}{b} < 1$ Ⓓ $\frac{5}{14} < \frac{a}{b} < 1$

Graphing Calculator

LOGIC MENU For Exercises 57–60, use the following information.
You can use the operators in the LOGIC menu on the TI-83 Plus to graph compound and absolute value inequalities. To display the LOGIC menu, press [2nd] [TEST] [▶].

57. Clear the Y= list. Enter $(5x + 2 > 12)$ and $(3x - 8 < 1)$ as Y1. With your calculator in DOT mode and using the standard viewing window, press [GRAPH]. Make a sketch of the graph displayed. **See margin for sketch.**

58. Using the TRACE function, investigate the graph. Based on your investigation, what inequality is graphed? **$2 < x < 3$**

59. Write the expression you would enter for Y1 to find the solution set of the compound inequality $5x + 2 \geq 3$ or $5x + 2 \leq -3$. Then use the graphing calculator to find the solution set.

59. $(5x + 2 \geq 3)$ or $(5x + 2 \leq -3)$; $\{x \mid x \geq 0.2$ or $x \leq -1\}$

60. A graphing calculator can also be used to solve absolute value inequalities. Write the expression you would enter for Y1 to find the solution set of the inequality $|2x - 6| > 10$. Then use the graphing calculator to find the solution set. (*Hint*: The absolute value operator is item 1 on the MATH NUM menu.)
abs$(2x - 6) > 10$; $\{x \mid x < -2$ or $x > 8\}$

Maintain Your Skills

Mixed Review

Solve each inequality. Describe the solution set using set builder or interval notation. Then graph the solution set on a number line. *(Lesson 1-5)*

61–63. See margin for graphs.

61. $2d + 15 \geq 3$ **$\{d \mid d \geq -6\}$ or $[-6, +\infty)$**
62. $7x + 11 > 9x + 3$ **$\{x \mid x < 4\}$ or $(-\infty, 4)$**
63. $3n + 4(n + 3) < 5(n + 2)$ **$\{n \mid n < -1\}$ or $(-\infty, -1)$**

64. **CONTESTS** To get a chance to win a car, you must guess the number of keys in a jar to within 5 of the actual number. Those who are within this range are given a key to try in the ignition of the car. Suppose there are 587 keys in the jar. Write and solve an equation to determine the highest and lowest guesses that will give contestants a chance to win the car. *(Lesson 1-4)* **$|x - 587| = 5$; highest: 592 keys, lowest: 582 keys**

Solve each equation. Check your solutions.

65. $5|x - 3| = 65$ **$\{-10, 16\}$**
66. $|2x + 7| = 15$ **$\{-11, 4\}$**
67. $|8c + 7| = -4$ **$\varnothing$**

Name the property illustrated by each statement. *(Lesson 1-3)*

68. If $3x = 10$, then $3x + 7 = 10 + 7$. **Addition (=)**
69. If $-5 = 4y - 8$, then $4y - 8 = -5$. **Symmetric (=)**
70. If $-2x - 5 = 9$ and $9 = 6x + 1$, then $-2x - 5 = 6x + 1$. **Transitive (=)**

Simplify each expression. *(Lesson 1-2)*

71. $6a - 2b - 3a + 9b$ **$3a + 7b$**
72. $-2(m - 4n) - 3(5n + 6)$ **$-2m - 7n - 18$**

Find the value of each expression. *(Lesson 1-1)*

73. $6(5 - 8) \div 9 + 4$ **2**
74. $(3 + 7)^2 - 16 \div 2$ **92**
75. $\frac{7(1 - 4)}{8 - 5}$ **−7**

Key concepts from the lesson, one or two examples, and several practice problems are included in the Lesson-by-Lesson Review.

46 Chapter 1 Solving Equations and Inequalities

Chapter 1 Study Guide and Review

Vocabulary and Concept Check

absolute value (p. 28)
Addition Property
  of Equality (p. 21)
  of Inequality (p. 33)
algebraic expression (p. 7)
Associative Property (p. 12)
Commutative Property (p. 12)
compound inequality (p. 40)
counterexample (p. 14)
Distributive Property (p. 12)
Division Property
  of Equality (p. 21)
  of Inequality (p. 34)
empty set (p. 29)
equation (p. 20)
formula (p. 8)
Identity Property (p. 12)
intersection (p. 40)
interval notation (p. 35)
Inverse Property (p. 12)
irrational numbers (p. 11)
Multiplication Property
  of Equality (p. 21)
  of Inequality (p. 34)
open sentence (p. 20)
order of operations (p. 6)
rational numbers (p. 11)
real numbers (p. 11)
Reflexive Property (p. 21)
set-builder notation (p. 34)
solution (p. 20)
Substitution Property (p. 21)
Subtraction Property
  of Equality (p. 21)
  of Inequality (p. 33)
Symmetric Property (p. 21)
Transitive Property (p. 21)
Trichotomy Property (p. 33)
union (p. 41)
variable (p. 7)

Choose the term from the list above that best matches each example.

1. $y > 3$ or $y < -2$ **compound inequality**
2. $0 + (-4b) = -4b$ **Iden. (+)**
3. $(m - 1)(-2) = -2(m - 1)$ **Comm. (×)**
4. $35x + 56 = 7(5x + 8)$ **Distributive**
5. $ab + 1 = ab + 1$ **Reflexive (=)**
6. If $2x = 3y - 4$, $3y - 4 = 7$, then $2x = 7$. **Trans. (=)**
7. $4(0.25) = 1$ **Multi. Inv.**
8. $2p + (4 + 9r) = (2p + 4) + 9r$ **Assoc. (+)**
9. $|5n|$ **absolute value**
10. $6y + 5z - 2(x + y)$ **algebraic expression**

Lesson-by-Lesson Review

1-1 Expressions and Formulas

See pages 6–10.

Concept Summary

- Order of Operations

Step 1 Simplify the expressions inside grouping symbols, such as parentheses, ( ), brackets, [ ], braces, { }, and fraction bars.

Step 2 Evaluate all powers.

Step 3 Do all multiplications and/or divisions from left to right.

Step 4 Do all additions and/or subtractions from left to right.

Example Evaluate $\frac{y^3}{3ab + 2}$ if $y = 4$, $a = -2$, and $b = -5$.

$$\frac{y^3}{3ab + 2} = \frac{4^3}{3(-2)(-5) + 2} \quad y = 4, a = -2, \text{ and } b = -5$$

$$= \frac{64}{3(10) + 2} \quad \text{Evaluate the numerator and denominator separately.}$$

$$= \frac{64}{32} \text{ or } 2$$

www.algebra2.com/vocabulary_review

FOLDABLES™ Study Organizer

For more information about Foldables, see *Teaching Mathematics with Foldables.*

Since this is your students' first use of the Foldables, you may want to show some good examples, and ask volunteers to name the main ideas and procedures that they included. Then have everyone add any information they may have overlooked.

Encourage students to refer to their Foldables while completing the Study Guide and Review and to use them in preparing for the Chapter Test.

Chapter 1 Study Guide and Review

Vocabulary and Concept Check

- This alphabetical list of vocabulary terms in Chapter 1 includes a page reference where each term was introduced.
- **Assessment** A vocabulary test/review for Chapter 1 is available on p. 50 of the *Chapter 1 Resource Masters.*

Lesson-by-Lesson Review

For each lesson,

- the main ideas are summarized,
- additional examples review concepts, and
- practice exercises are provided.

Vocabulary PuzzleMaker

ELL The Vocabulary PuzzleMaker software improves students' mathematics vocabulary using four puzzle formats—crossword, scramble, word search using a word list, and word search using clues. Students can work on a computer screen or from a printed handout.

MindJogger Videoquizzes

ELL MindJogger Videoquizzes provide an alternative review of concepts presented in this chapter. Students work in teams in a game show format to gain points for correct answers. The questions are presented in three rounds.

Round 1 Concepts (5 questions)
Round 2 Skills (4 questions)
Round 3 Problem Solving (4 questions)

Exercises **Find the value of each expression.** *See Example 1 on page 6.*

11. $10 + 16 \div 4 + 8$ **22** **12.** $[21 - (9 - 2)] \div 2$ **7** **13.** $\frac{14(8 - 15)}{2}$ **−49**

Evaluate each expression if $a = 12$, $b = 0.5$, $c = -3$, and $d = \frac{1}{3}$.
See Examples 2 and 3 on page 7.

14. $6b - 5c$ **18** **15.** $c^3 + ad$ **−23** **16.** $\frac{9c + ab}{c}$ **7** **17.** $a[b^2(b + a)]$ **37.5**

1-2 Properties of Real Numbers

See pages 11–18.

Concept Summary

- Real numbers (R) can be classified as rational (Q) or irrational (I).
- Rational numbers can be classified as natural numbers (N), whole numbers (W), and/or integers (Z).
- Use the properties of real numbers to simplify algebraic expressions.

Example **Simplify $4(2b + 6c) + 3b - c$.**

$4(2b + 6c) + 3b - c = 4(2b) + 4(6c) + 3b - c$	Distributive Property
$= 8b + 24c + 3b - c$	Multiply.
$= 8b + 3b + 24c - c$	Commutative Property (+)
$= (8 + 3)b + (24 - 1)c$	Distributive Property
$= 11b + 23c$	Add 3 to 8 and subtract 1 from 24.

Exercises **Name the sets of numbers to which each value belongs.**
See Example 1 on page 12.

18. $-\sqrt{9}$ **Z, Q, R** **19.** $1.\overline{6}$ **Q, R** **20.** $\frac{35}{7}$ **N, W, Z, Q, R** **21.** $\sqrt{18}$ **I, R**

Simplify each expression. *See Example 5 on page 14.*

22. $2m + 7n - 6m - 5n$ **$-4m + 2n$** **23.** $-5(a - 4b) + 4b$ **$-5a + 24b$** **24.** $2(5x + 4y) - 3(x + 8y)$ **$7x - 16y$**

1-3 Solving Equations

See pages 20–27.

Concept Summary

- Verbal expressions can be translated into algebraic expressions using the language of algebra, using variables to represent the unknown quantities.
- Use the properties of equality to solve equations.

Example **Solve $4(a + 5) - 2(a + 6) = 3$.**

$4(a + 5) - 2(a + 6) = 3$	Original equation
$4a + 20 - 2a - 12 = 3$	Distributive Property
$2a + 8 = 3$	Commutative, Distributive, and Substitution Properties
$2a = -5$	Subtraction Property (=)
$a = -2.5$	Division Property (=)

48 Chapter 1 Solving Equations and Inequalities

Exercises **Solve each equation. Check your solution.**
See Examples 4 and 5 on pages 21 and 22.

25. $x - 6 = -20$ **−14** **26.** $-\frac{2}{3}a = 14$ **−21** **27.** $7 + 5n = -58$ **−13**

28. $3w + 14 = 7w + 2$ **3** **29.** $5y + 4 = 2(y - 4)$ **−4** **30.** $\frac{n}{4} + \frac{n}{3} = \frac{1}{2}$ $\mathbf{\frac{6}{7}}$

Solve each equation or formula for the specified variable. *See Example 6 on page 22.*

31. $Ax + By = C$ for x $\mathbf{x = \frac{C - By}{A}}$ **32.** $\frac{a - 4b^2}{2c} = d$ for a $\mathbf{a = 2cd + 4b^2}$ **33.** $A = p + prt$ for p $\mathbf{p = \frac{A}{1 + rt}}$

1-4 Solving Absolute Value Equations

See pages 28–32.

Concept Summary

- For any real numbers a and b, where $b \geq 0$, if $|a| = b$, then $a = b$ or $a = -b$.

Example **Solve $|2x + 9| = 11$.**

Case 1 $a = b$ **or** **Case 2** $a = -b$

Case 1	Case 2
$2x + 9 = 11$	$2x + 9 = -11$
$2x = 2$	$2x = -20$
$x = 1$	$x = -10$

The solution set is $\{1, -10\}$. Check these solutions in the original equation.

Exercises **Solve each equation. Check your solutions.**
See Examples 1–4 on pages 28–30.

34. $|x + 11| = 42$ **{31, −53}** **35.** $3|x + 6| = 36$ **{6, −18}** **36.** $|4x - 5| = -25$ **∅**

37. $|x + 7| = 3x - 5$ **{6}** **38.** $|y - 5| - 2 = 10$ **{−7, 17}** **39.** $4|3x + 4| = 4x + 8$ $\mathbf{\left\{-\frac{3}{2}, -1\right\}}$

1-5 Solving Inequalities

See pages 33–39.

Concept Summary

- Adding the same number to, or subtracting the same number from, each side of an inequality does not change the truth of the inequality.
- When you multiply or divide each side of an inequality by a negative number, the direction of the inequality symbol must be *reversed*.

Example **Solve $5 - 4a > 8$. Graph the solution set on a number line.**

$5 - 4a > 8$ Original inequality

$-4a > 3$ Subtract 5 from each side.

$a < -\frac{3}{4}$ Divide each side by −4, reversing the inequality symbol.

The solution set is $\left\{a \mid a < -\frac{3}{4}\right\}$.

The graph of the solution set is shown at the right.

Chapter 1 For More ...
- Extra Practice, see pages 828–829.
- Mixed Problem Solving, see page 862.

Exercises **Solve each inequality. Describe the solution set using set builder or interval notation. Then graph the solution set on a number line.**
See Examples 1–3 on pages 34–35. **40–45. See margin.**

40. $-7w > 28$	**41.** $3x + 4 \geq 19$	**42.** $\frac{n}{12} + 5 \leq 7$
43. $3(6 - 5a) < 12a - 36$	**44.** $2 - 3z \geq 7(8 - 2z) + 12$	**45.** $8(2x - 1) > 11x - 17$

1-6 Solving Compound and Absolute Value Inequalities

See pages 40–46.

Concept Summary

- The graph of an *and* compound inequality is the intersection of the solution sets of the two inequalities.
- The graph of an *or* compound inequality is the union of the solution sets of the two inequalities.
- For all real numbers a and b, $b > 0$, the following statements are true.
  1. If $|a| < b$ then $-b < a < b$.
  2. If $|a| > b$ then $a > b$ or $a < -b$.

Examples **Solve each inequality. Graph the solution set on a number line.**

1 $-19 < 4d - 7 \leq 13$

$-19 < 4d - 7 \leq 13$ Original inequality

$-12 < 4d \leq 20$ Add 7 to each part.

$-3 < d \leq 5$ Divide each part by 4.

The solution set is $\{d \mid -3 < d \leq 5\}$.

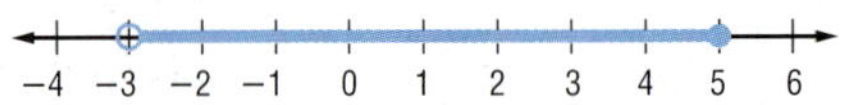

2 $|2x + 4| \geq 12$

$|2x + 4| \geq 12$ is equivalent to $2x + 4 \geq 12$ or $2x + 4 \leq -12$.

$2x + 4 \geq 12$ or $2x + 4 \leq -12$ Original inequality

$2x \geq 8$ $\quad 2x \leq -16$ Subtract 4 from each side.

$x \geq 4$ $\quad x \leq -8$ Divide each side by 2.

The solution set is $\{x \mid x \geq 4 \text{ or } x \leq -8\}$.

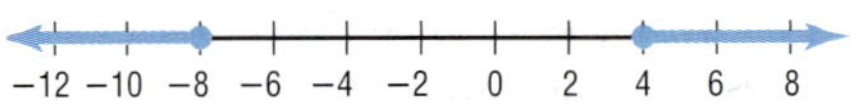

Exercises **Solve each inequality. Graph the solution set on a number line.**
See Examples 1–5 on pages 40–42. **46–51. See margin.**

46. $-1 < 3a + 2 < 14$	**47.** $-1 < 3(y - 2) \leq 9$	**48.** $|x| + 1 > 12$
49. $|2y - 9| \leq 27$	**50.** $|5n - 8| > -4$	**51.** $|3b + 11| > 1$

Answers

40. $\{w \mid w < -4\}$ or $(-\infty, -4)$

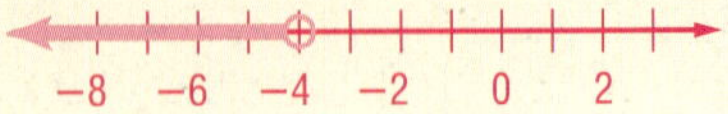

41. $\{x \mid x \geq 5\}$ or $[5, +\infty)$

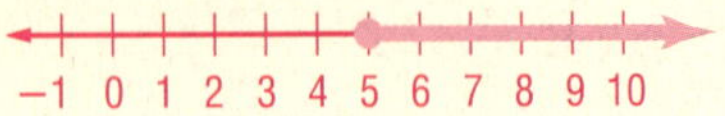

42. $\{n \mid n \leq 24\}$ or $(-\infty, 24]$

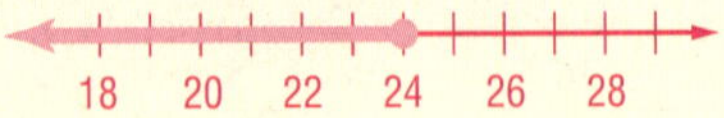

43. $\{a \mid a > 2\}$ or $(2, +\infty)$

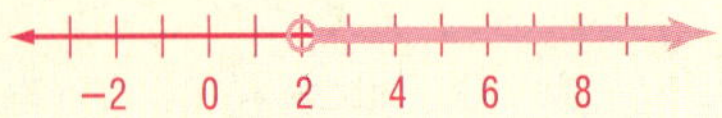

44. $\{z \mid z \geq 6\}$ or $[6, +\infty)$

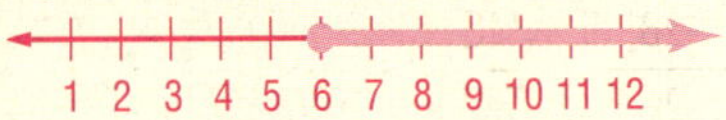

45. $\{x \mid x > -1.8\}$ or $(-1.8, +\infty)$

46. $\{a \mid -1 < a < 4\}$

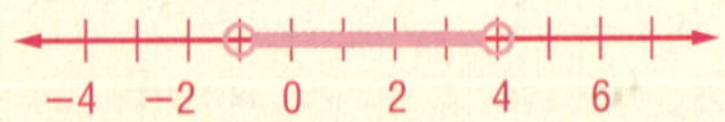

47. $\left\{y \,\middle|\, \frac{5}{3} < y \leq 5\right\}$

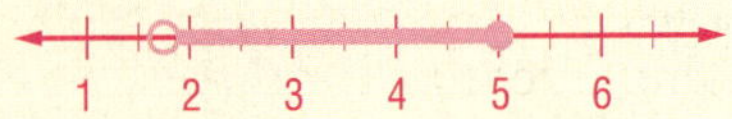

48. $\{x \mid x < -11 \text{ or } x > 11\}$

49. $\{y \mid -9 \leq y \leq 18\}$

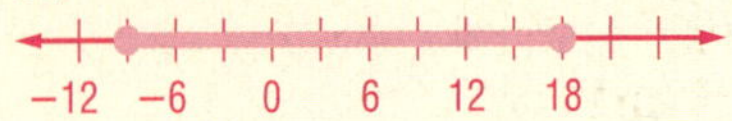

50. all real numbers

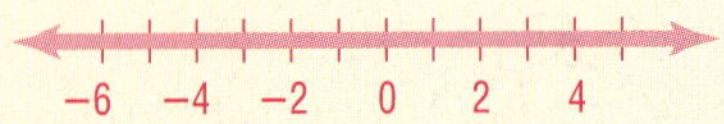

51. $\left\{b \,\middle|\, b < -4 \text{ or } b > -\frac{10}{3}\right\}$

Answers (p. 51)

25. $(-\infty, 3)$

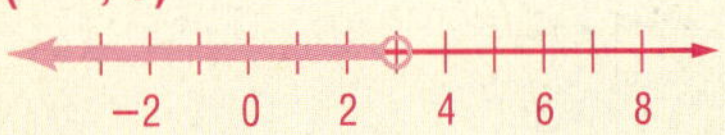

26. $[2, +\infty)$

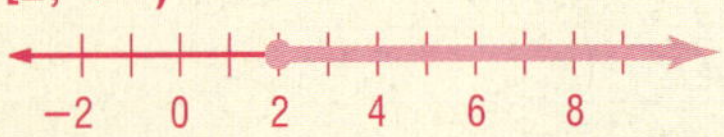

27. $(-\infty, 3)$

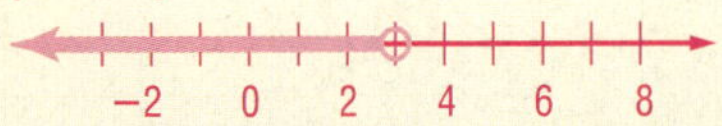

28. $[-13, 3]$

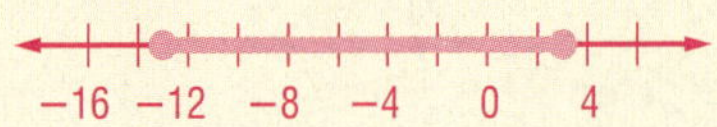

29. $(-1, 2]$

30. $\left\{y \,\middle|\, y < -\frac{4}{3} \text{ or } y > 2\right\}$

Chapter 1 Practice Test

Vocabulary and Concepts

Choose the term that best completes each sentence.

1. An algebraic *(equation, expression)* contains an equals sign.
2. *(Whole numbers, Rationals)* are a subset of the set of integers.
3. If $x + 3 = y$, then $y = x + 3$ is an example of the *(Transitive, Symmetric)* Property of Equality.

Skills and Applications

Find the value of each expression.

4. $[(3 + 6)^2 \div 3] \times 4$ **108**
5. $\frac{20 + 4 \times 3}{11 - 3}$ **4**
6. $0.5(2.3 + 25) \div 1.5$ **9.1**

Evaluate each expression if $a = -9$, $b = \frac{2}{3}$, $c = 8$, and $d = -6$.

7. $\frac{db + 4c}{a}$ $-\frac{28}{9}$
8. $\frac{a}{b^2} + c$ **−12.25**
9. $2b(4a + a^2)$ **60**

Name the sets of numbers to which each number belongs.

10. $\sqrt{17}$ **I, R**
11. 0.86 **Q, R**
12. $\sqrt{64}$ **N, W, Z, Q, R**

Name the property illustrated by each equation or statement. **14. Symm. (=)**

13. $(7 \cdot s) \cdot t = 7 \cdot (s \cdot t)$ **Assoc. (×)**
14. If $(r + s)t = rt + st$, then $rt + st = (r + s)t$.
15. $\left(3 \cdot \frac{1}{3}\right) \cdot 7 = \left(3 \cdot \frac{1}{3}\right) \cdot 7$ **Reflex. (=)**
16. $(6 - 2)a - 3b = 4a - 3b$ **Subst. (=)**
17. $(4 + x) + y = y + (4 + x)$ **Comm. (+)**
18. If $5(3) + 7 = 15 + 7$ and $15 + 7 = 22$, then $5(3) + 7 = 22$. **Trans. (=)**

Solve each equation. Check your solution(s). **21. all reals**

19. $5t - 3 = -2t + 10$ $\frac{13}{7}$
20. $2x - 7 - (x - 5) = 0$ **2**
21. $5m - (5 + 4m) = (3 + m) - 8$
22. $|8w + 2| + 2 = 0$ $\varnothing$
23. $12\left|\frac{1}{2}y + 3\right| = 6$ $\{-7, -5\}$
24. $2|2y - 6| + 4 = 8$ $\{2, 4\}$

Solve each inequality. Describe the solution set using set builder or interval notation. Then graph the solution set on a number line. **27. $\{x \mid x < 3\}$** **25–30. See margin for interval notation and graphs.**

25. $4 > b + 1$ $\{b \mid b < 3\}$
26. $3q + 7 \geq 13$ $\{q \mid q \geq 2\}$
27. $5(3x - 5) + x < 2(4x - 1) + 1$
28. $|5 + k| \leq 8$ $\{k \mid -13 \leq k \leq 3\}$
29. $-12 < 7d - 5 \leq 9$ $\{d \mid -1 < d \leq 2\}$
30. $|3y - 1| > 5$ **See margin.**

For Exercises 31 and 32, define a variable, write an equation or inequality, and solve the problem. **31. m = miles traveled; $19.50 + 0.18m = 33$; 75 mi**

31. **CAR RENTAL** Mrs. Denney is renting a car that gets 35 miles per gallon. The rental charge is \$19.50 a day plus 18¢ per mile. Her company will reimburse her for \$33 of this portion of her travel expenses. If Mrs. Denney rents the car for 1 day, find the maximum number of miles that will be paid for by her company.
32. **SCHOOL** To receive a B in his English class, Nick must have an average score of at least 80 on five tests. He scored 87, 89, 76, and 77 on his first four tests. What must he score on the last test to receive a B in the class? **32. s = score on last test; $\frac{s + 87 + 89 + 76 + 77}{5} \geq 80$; at least 71**
33. **STANDARDIZED TEST PRACTICE** If $\frac{a}{b} = 8$ and $ac - 5 = 11$, then $bc =$ **B**

    (A) 93. (B) 2. (C) $\frac{5}{8}$. (D) cannot be determined

www.algebra2.com/chapter_test

Portfolio Suggestion

Introduction Translating words into algebraic expressions involves reading the words, deciding what they mean mathematically, and using the correct notation to write the translation. One way to build the skills involved is to go in the opposite direction, translating algebraic expressions into words.

Ask Students Write an expression or equation and create a word problem about it. Exchange your problem with a partner and translate what you receive into an expression or equation. Place your problem in your portfolio.

Chapter 1 Practice Test

Assessment Options

Vocabulary Test A vocabulary test/review for Chapter 1 can be found on p. 50 of the *Chapter 1 Resource Masters*.

Chapter Tests There are six Chapter 1 Tests and an Open-Ended Assessment task available in the *Chapter 1 Resource Masters*.

Chapter 1 Tests			
Form	Type	Level	Pages
1	MC	basic	37–38
2A	MC	average	39–40
2B	MC	average	41–42
2C	FR	average	43–44
2D	FR	average	45–46
3	FR	advanced	47–48

MC = multiple-choice questions
FR = free-response questions

Open-Ended Assessment
Performance tasks for Chapter 1 can be found on p. 49 of the *Chapter 1 Resource Masters*. A sample scoring rubric for these tasks appears on p. A26.

ExamView® Pro

Use the networkable ExamView® Pro to:

- Create **multiple versions** of tests.
- Create **modified** tests for *Inclusion* students.
- **Edit** existing questions and **add** your own questions.
- Use built-in **state curriculum correlations** to create tests aligned with state standards.
- **Apply** art to your tests from a program bank of artwork.

Chapter 1 Standardized Test Practice

These two pages contain practice questions in the various formats that can be found on the most frequently given standardized tests.

A practice answer sheet for these two pages can be found on p. A1 of the *Chapter 1 Resource Masters*.

Standardized Test Practice Student Recording Sheet, p. A1

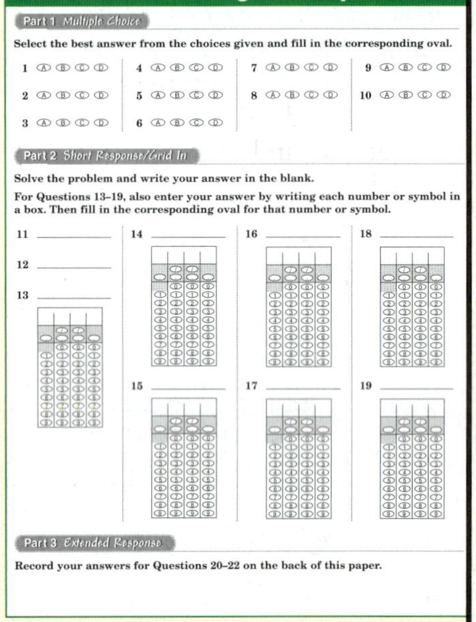
Part 1 Multiple Choice

Select the best answer from the choices given and fill in the corresponding oval.

Part 2 Short Response/Grid In

Solve the problem and write your answer in the blank.

For Questions 13–19, also enter your answer by writing each number or symbol in a box. Then fill in the corresponding oval for that number or symbol.

Part 3 Extended Response

Record your answers for Questions 20–22 on the back of this paper.

Additional Practice

See pp. 55–56 in the *Chapter 1 Resource Masters* for additional standardized test practice.

The items on the Standardized Test Practice pages were created to closely parallel those on SAT, ACT, TIMSS 12, NAEP 12, and state proficiency tests.

ExamView® Pro

Special banks of standardized test questions similar to those on the SAT, ACT, TIMSS 12, NAEP 12, and state proficiency tests can be found on this CD-ROM.

Chapter 1 Standardized Test Practice

Part 1 Multiple Choice

Record your answers on the answer sheet provided by your teacher or on a sheet of paper.

1. In the square at the right, what is the value of x? **B**

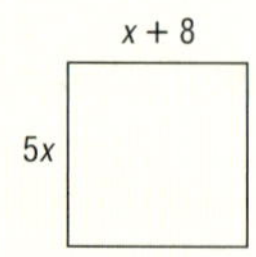

(A) 1 (B) 2 (C) 3 (D) 4

2. On a college math test, 18 students earned an A. This number is exactly 30% of the total number of students in the class. How many students are in the class? **D**

(A) 5 (B) 23 (C) 48 (D) 60

3. A student computed the average of her 7 test scores by adding the scores together and dividing this total by the number of tests. The average was 87. On her next test, she scored a 79. What is her new test average? **D**

(A) 83 (B) 84 (C) 85 (D) 86

4. If the perimeter of $\triangle PQR$ is 3 times the length of PQ, then $PR =$ ____. **D**

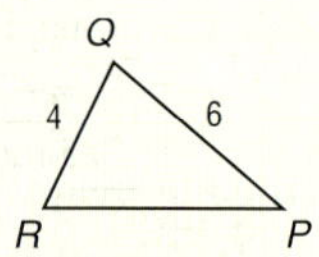

Note: Figure not drawn to scale.

(A) 4 (B) 6 (C) 7 (D) 8

5. If a different number is selected from each of the three sets shown below, what is the greatest sum these 3 numbers could have? **C**

R = {3, 6, 7}; S = {2, 4, 7}; T = {1, 3, 7}

(A) 13 (B) 14 (C) 17 (D) 21

6. A pitcher contains a ounces of orange juice. If b ounces of juice are poured from the pitcher into each of c glasses, which expression represents the amount of juice remaining in the pitcher? **C**

(A) $\frac{a}{b} + c$ (B) $ab - c$ (C) $a - bc$ (D) $\frac{a}{bc}$

7. The sum of three consecutive integers is 135. What is the greatest of the three integers? **D**

(A) 43 (B) 44 (C) 45 (D) 46

8. The ratio of girls to boys in a class is 5 to 4. If there are a total of 27 students in the class, how many are girls? **A**

(A) 15 (B) 12 (C) 9 (D) 5

9. For which of the following ordered pairs (x, y) is $x + y > 3$ and $x - y < -2$? **D**

(A) (0, 3) (B) (3, 4) (C) (5, 3) (D) (2, 5)

10. If the area of $\triangle ABD$ is 280, what is the area of the polygon $ABCD$? **B**

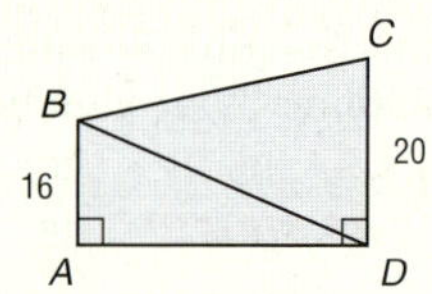

Note: Figure not drawn to scale.

(A) 560 (B) 630 (C) 700 (D) 840

Test-Taking Tip

Question 9

To solve equations or inequalities, you can replace the variables in the question with the values given in each answer choice. The answer choice that results in true statements is the correct answer choice.

Preparing for Standardized Tests
For test-taking strategies and more practice, see pages 877–892.

Part 2 Short Response/Grid In

Record your answers on the answer sheet provided by your teacher or on a sheet of paper.

11. In the triangle below, x and y are integers. If $25 < y < 30$, what is one possible value of x? **122, 124, 126, or 128**

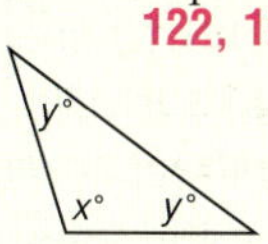

12. If n and p are each different positive integers and $n + p = 4$, what is one possible value of $3n + 4p$? **13 or 15**

13. In the figure at the right, what is the value of x? **55**

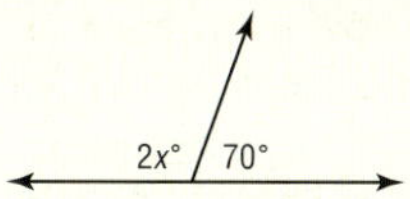

14. One half quart of lemonade concentrate is mixed with $1\frac{1}{2}$ quarts of water to make lemonade for 6 people. If you use the same proportions of concentrate and water, how many quarts of lemonade concentrate are needed to make lemonade for 21 people? **1.75 or 7/4**

15. If 25 percent of 300 is equal to 500 percent of t, then t is equal to what number? **15**

16. In the figure below, what is the area of the shaded square in square units? **13**

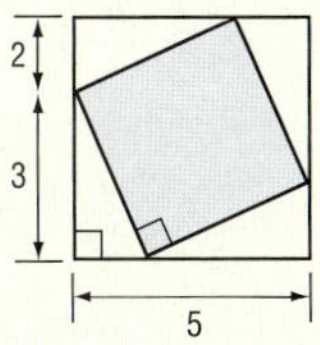

17. There are 140 students in the school band. One of these students will be selected at random to be the student representative. If the probability that a brass player is selected is $\frac{2}{5}$, how many brass players are in the band? **56**

www.algebra2.com/standardized_test

18. A shelf holds fewer than 50 cans. If all of the cans on this shelf were put into stacks of five cans each, no cans would remain. If the same cans were put into stacks of three cans each, one can would remain. What is the greatest number of cans that could be on the shelf? **40**

19. The area of a trapezoid is $\frac{1}{2}h(b_1 + b_2)$, where h is the altitude, and b_1 and b_2 are the lengths of the parallel bases. If a trapezoid has an altitude of 8 inches, an area of 56 square inches, and one base 4 inches long, what is the length of its other base in inches? **10**

Part 3 Extended Response

Record your answers on a sheet of paper. Show your work.

For Exercises 20–22, use the information below and in the table.

Amanda's hours at her summer job for one week are listed in the table below. She earns \$6 per hour.

Amanda's Work Hours	
Sunday	0
Monday	6
Tuesday	4
Wednesday	0
Thursday	2
Friday	6
Saturday	8

20. Write an expression for Amanda's total weekly earnings.
\$6(0 + 6 + 4 + 0 + 2 + 6 + 8)

21. Evaluate the expression from Exercise 20 by using the Distributive Property. **\$156**

22. Michael works with Amanda and also earns \$6 per hour. If Michael's earnings were \$192 this week, write and solve an equation to find how many more hours Michael worked than Amanda. **$6h + 156 = 192$; 6 h**

Evaluating Extended Response Questions

Extended Response questions are graded by using a multilevel rubric that guides you in assessing a student's knowledge of a particular concept.

Goal: Determine the dimensions of a rectangulaer garden for a given perimeter.

Sample Scoring Rubric: The following rubric is a sample scoring device. You may wish to add more detail to this sample to meet your individual scoring needs.

Score	Criteria
4	A correct solution that is supported by well-developed, accurate explanations
3	A generally correct solution, but may contain minor flaws in reasoning or computation
2	A partially correct interpretation and/or solution to the problem
1	A correct solution with no supporting evidence or explanation
0	An incorrect solution indicating no mathematical understanding of the concept or task, or no solution is given

Page 13, Algebra Activity

1. 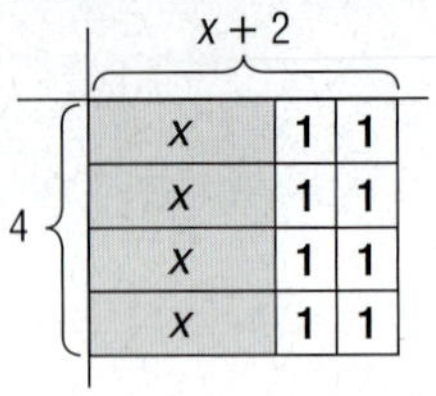

2. 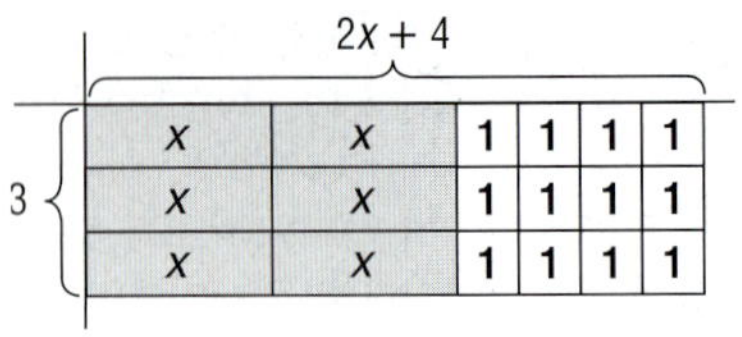

3. 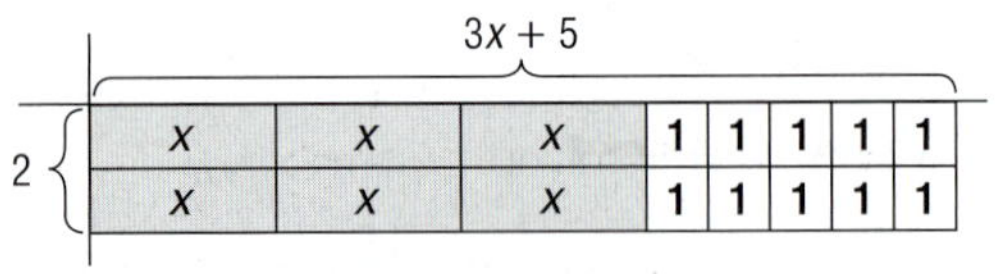

4. 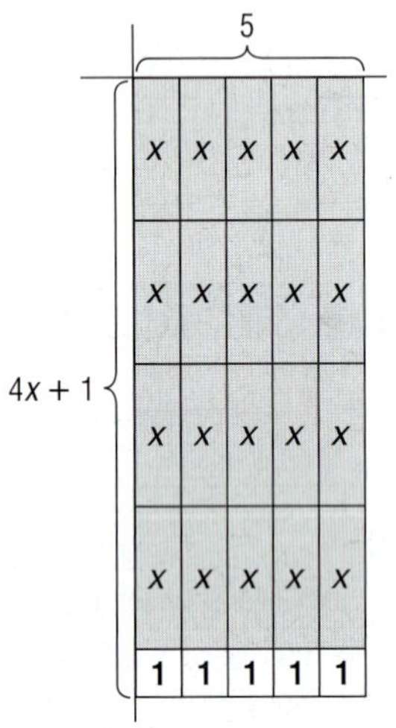

Page 19, Follow-Up of Lesson 1-2 Algebra Activity

2. Beginning with 2, you add the next consecutive integer to obtain the next number of diagonals: $0 + 2 = 2$, $2 + 3 = 5$, $5 + 4 = 9$, and so on.

8. See students' work; a decagon has 35 diagonals because $10(10 - 3) \div 2 = 35$.

9. A generic polygon has *n* sides and *n* vertices. From each vertex, $n - 3$ diagonals can be drawn. So $n(n - 3)$ is the number of diagonals except that this formula counts each diagonal twice since one diagonal connects 2 vertices. Therefore, you must divide the expression by 2. So the formula is $y = n(n - 3) \div 2$.

10. Sample answers:

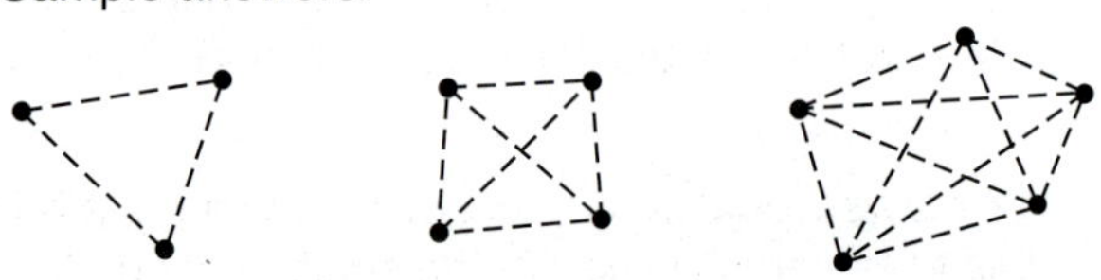

13. There are two possible answers. One answer is that the number of dots is *x*. From each dot $x - 1$ lines can be drawn to other dots, but then the lines are counted twice, so the formula is $y = x(x - 1) \div 2$. A second answer is that you can see that the number of lines needed to connect the dots is the number of diagonals for a polygon with that number of vertices or sides plus the number of sides. So, $y = [x(x - 3) \div 2] + x = (0.5x^2 - 1.5x) + x = 0.5x^2 - 0.5x$ or $y = 0.5x^2 - 0.5x$.

Page 27, Lesson 1-3

76. To find the most effective level of intensity for your workout, you need to use your age and 10-second pulse count. You must also be able to solve the formula given for *A*. Answers should include the following.
   - Substitute 0.80 for *I* and 27 for *P* in the formula $I = 6 \times P \div (220 - A)$ and solve for *A*. To solve this equation, divide the product of 6 and 28 by 0.8. Then subtract 220 and divide by -1. The result is 17.5. This means that this person is $17\frac{1}{2}$ years old.
   - To find the intensity level for different values of *A* and *P* would require solving a new equation but using the same steps as described above. Solving for *A* would mean that for future calculations of *A* you would only need to simplify an expression, $220 - \frac{6P}{I}$, rather than solve an equation.

Pages 37–38, Lesson 1-5

4. 

5. 

6. 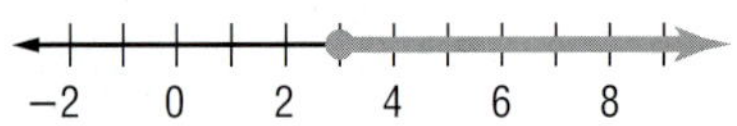

7. 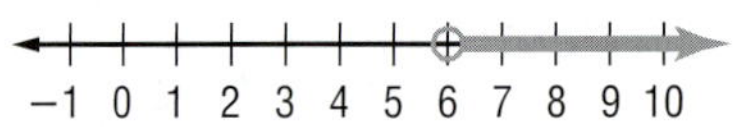

8. 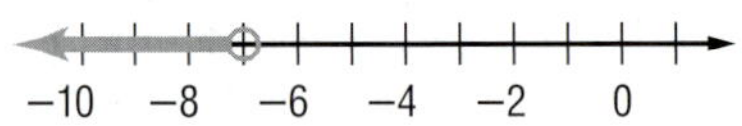

9. 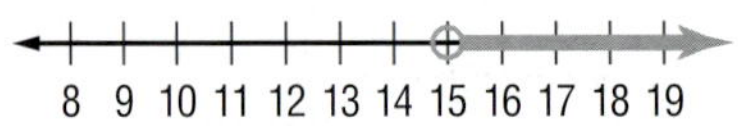

10. 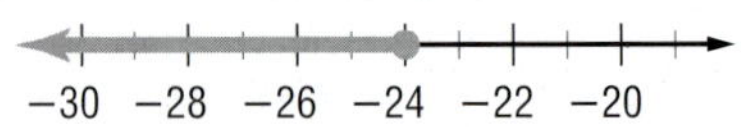

11. 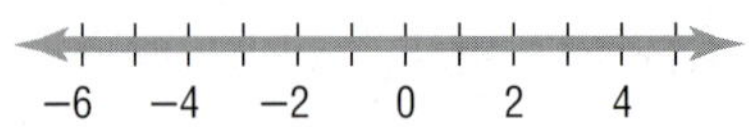

15. 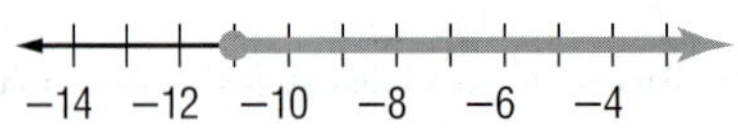

16. 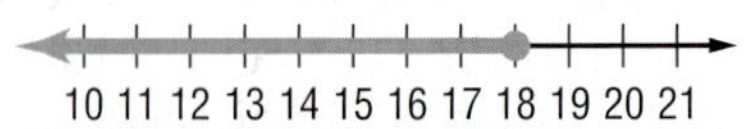

17. 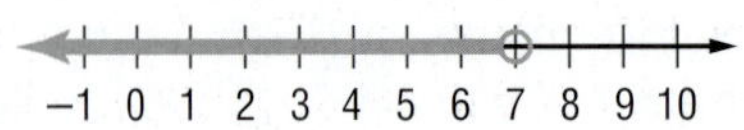

18. 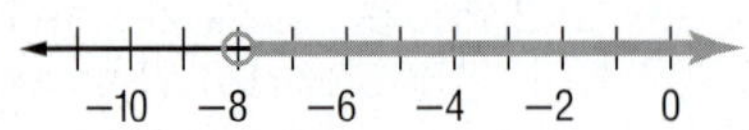

19. 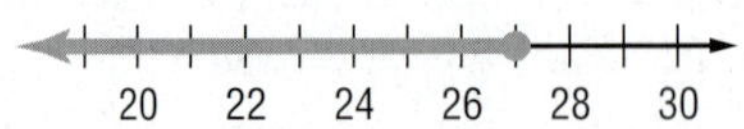

20. 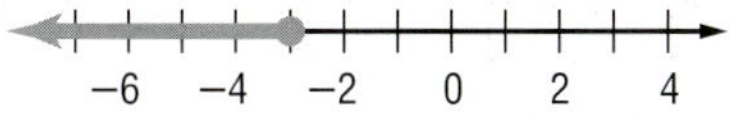

21. 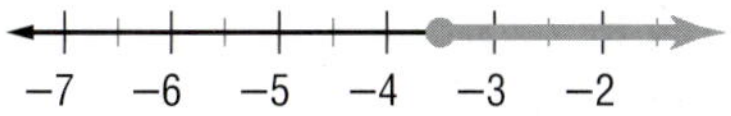

22. 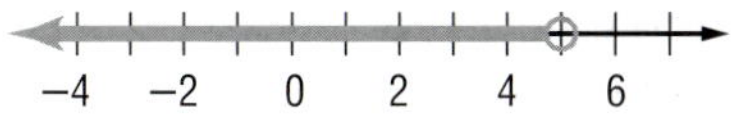

23. 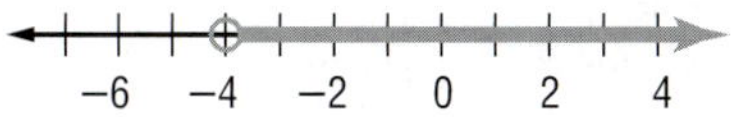

24. 

25. 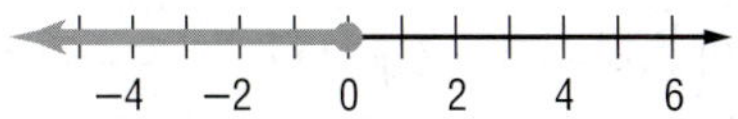

26. 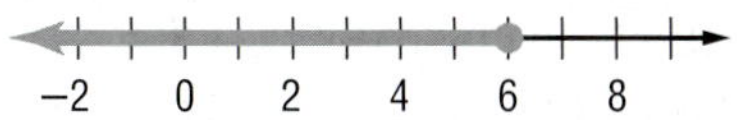

27. 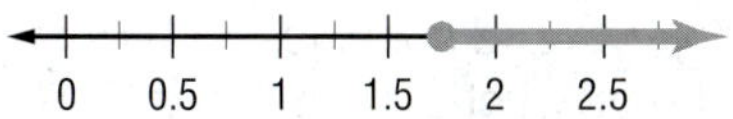

28. 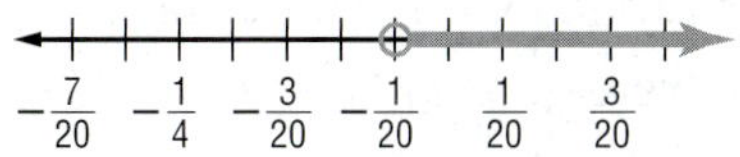

29. 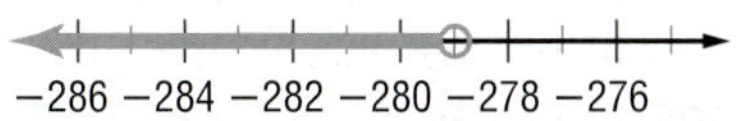

30. 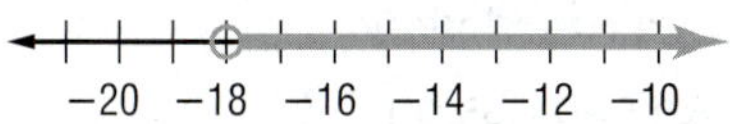

31. 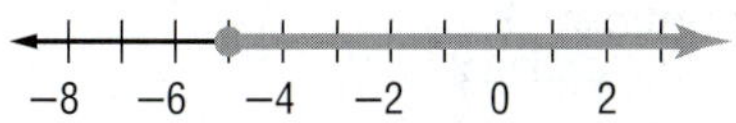

32. 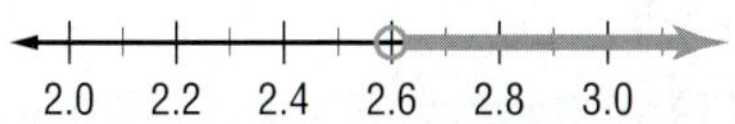

33. 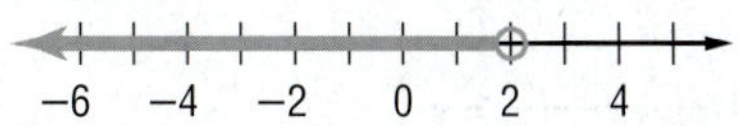

34. 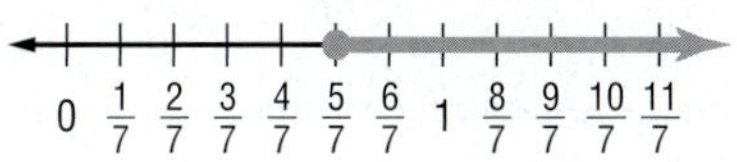

35. 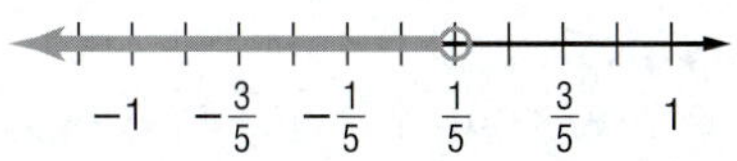

36. 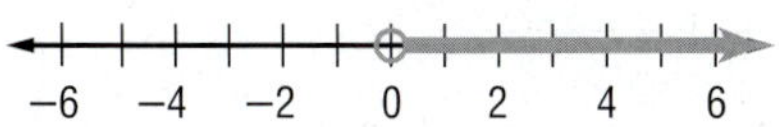

37. 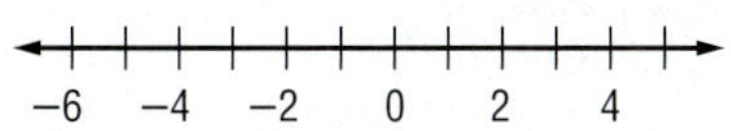

38. 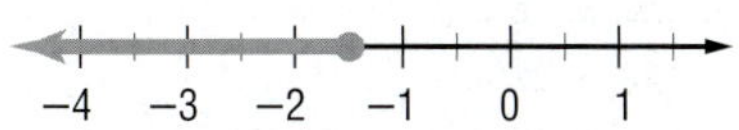

53. Answers should include the following.

- $150 < 400$
- Let n equal the number of minutes used. Write an expression representing the cost of Plan 1 and for Plan 2 for n minutes. The cost for Plan 1 would include a monthly access fee of \$35 plus 40¢ for each minute over 150 minutes or $35 + 0.4(n - 150)$. The cost for Plan 2 for 400 minutes or less would be \$55. To find where Plan 2 would cost less than Plan 1 solve $55 < 35 + 0.4(n - 150)$ for n. The solution set is $\{n | n > 200\}$, which means that for more than 200 minutes of calls, Plan 2 is cheaper.

Pages 44–45, Lesson 1-6

27. 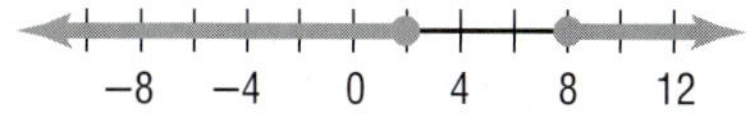

28. 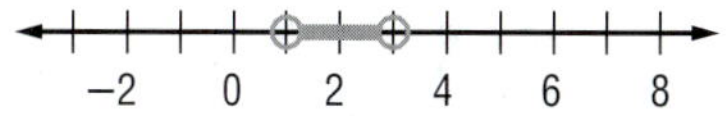

29. 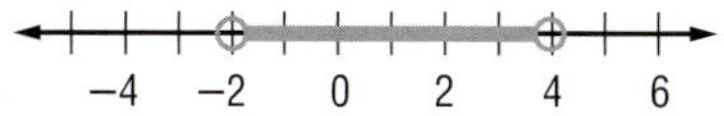

30. 

31. 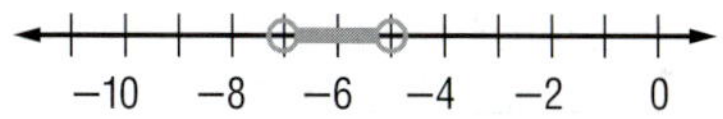

32. 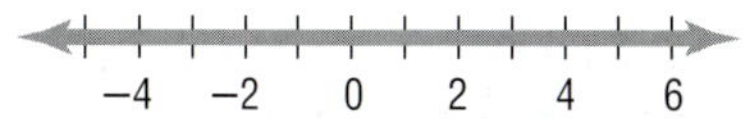

33. 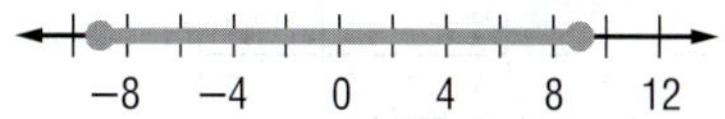

34. 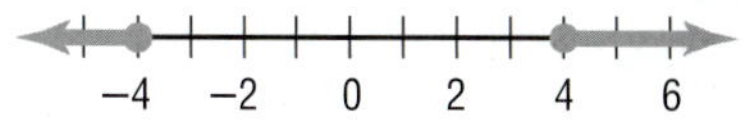

35. 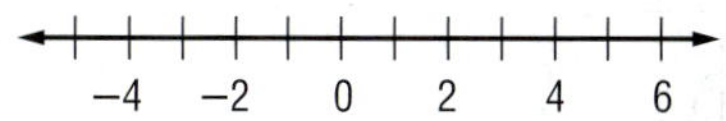

36. 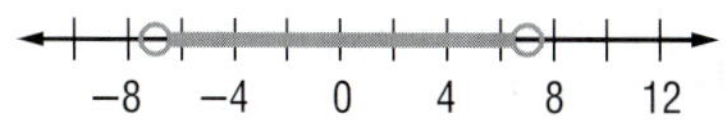

37. 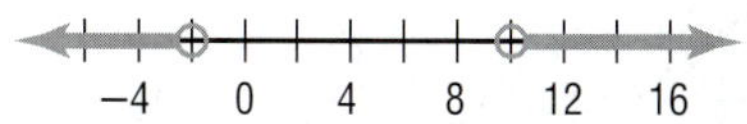

38. 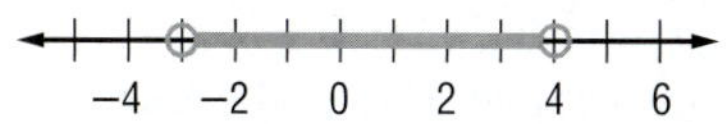

39. 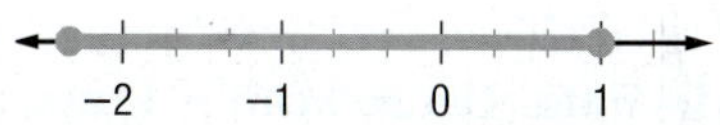

40. 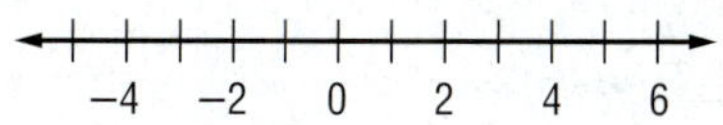

41. 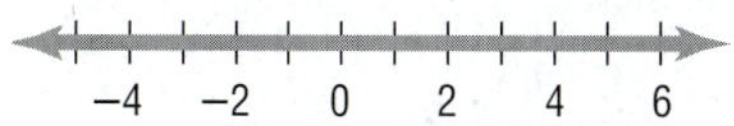

42. 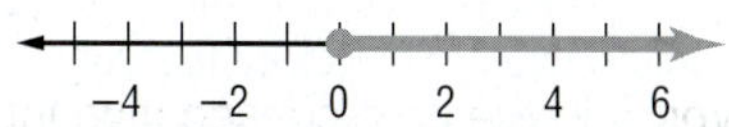

43. 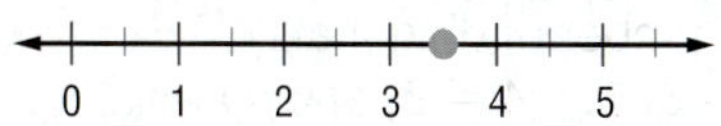

44. 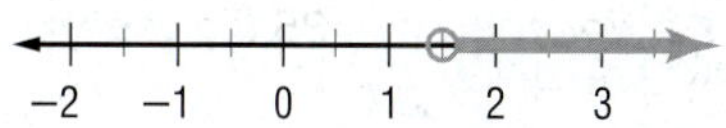

54. Compound inequalities can be used to describe the acceptable time frame for the fasting state before a glucose tolerance test is administered to a patient suspected of having diabetes. Answers should include the following.

- Use the word *and* when both inequalities must be satisfied. Use the word *or* when only one or the other of the inequalities must be satisfied.
- $10 \leq h \leq 16$
- 12 hours would be an acceptable fasting state for this test since it is part of the solution set of $10 \leq h \leq 16$, as indicated on the graph below.

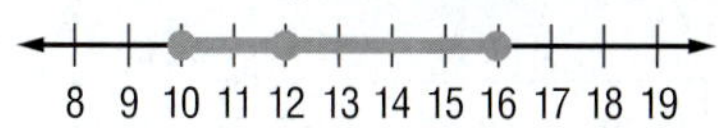

Chapter 2

Linear Relations and Functions

Chapter Overview and Pacing

Year-long and two-year pacing: pages T20–T21.

LESSON OBJECTIVES

LESSON OBJECTIVES	PACING (days) Regular		Block	
	Basic/ Average	Advanced	Basic/ Average	Advanced
2-1 Relations and Functions *(pp. 56–62)* • Analyze and graph relations. • Find functional values.	1	optional	0.5	optional
2-2 Linear Equations *(pp. 63–67)* • Identify linear equations and functions. • Write linear equations in standard form and graph them.	1	optional	0.5	optional
2-3 Slope *(pp. 68–74)* • Find and use the slope of a line. • Graph parallel and perpendicular lines.	1	optional	0.5	optional
2-4 Writing Linear Equations *(pp. 75–80)* • Write an equation of a line given the slope and a point on the line. • Write an equation of a line parallel or perpendicular to a given line.	1	optional	0.5	optional
2-5 Modeling Real-World Data: Using Scatter Plots *(pp. 81–88)* • Draw scatter plots. • Find and use prediction equations. ***Follow-Up:*** Lines of Regression	2 (with 2-5 Follow-Up)	optional	1	optional
2-6 Special Functions *(pp. 89–95)* • Identify and graph step, constant, and identity functions. • Identify and graph absolute value and piecewise functions.	1	optional	0.5	optional
2-7 Graphing Inequalities *(pp. 96–99)* • Graph linear inequalities. • Graph absolute value inequalities.	1	optional	0.5	optional
Study Guide and **Practice Test** *(pp. 100–105)* **Standardized Test Practice** *(pp. 106–107)*	1	2	0.5	0.5
Chapter Assessment	1	1	0.5	0.5
TOTAL	**10**	**3**	**5**	**1**

*An electronic version of this chapter is available on **StudentWorks**™. This backpack solution CD-ROM allows students instant access to the Student Edition, lesson worksheet pages, and web resources.*

Timesaving Tools
TeacherWorks™
All-In-One Planner and Resource Center
See pages T5 and T21.

Chapter Resource Manager

CHAPTER 2 RESOURCE MASTERS									
Study Guide and Intervention	Practice (Skills and Average)	Reading to Learn Mathematics	Enrichment	Assessment	Applications*	5-Minute Check Transparencies	Interactive Chalkboard	Alge2PASS: Tutorial Plus (lessons)	Materials
57–58	59–60	61	62			2-1	2-1		
63–64	65–66	67	68	113		2-2	2-2	3	spaghetti
69–70	71–72	73	74			2-3	2-3		graphing calculator, spaghetti
75–76	77–78	79	80	113, 115	GCS 30, SC 3	2-4	2-4		
81–82	83–84	85	86		SC 4 SM 97–102	2-5	2-5		tape measure, graph paper (*Follow-Up:* graphing calculator)
87–88	89–90	91	92	114	GCS 29	2-6	2-6		graphing calculator, toothpicks
93–94	95–96	97	98	114		2-7	2-7	4	
				99–112, 116–118					

**Key to Abbreviations:* GCS = Graphing Calculator and Speadsheet Masters,
SC = School-to-Career Masters,
SM = Science and Mathematics Lab Manual

Chapter 2

Mathematical Connections and Background

Continuity of Instruction

Prior Knowledge

In prior years students worked with coordinate systems, ordered pairs, and linear equations and functions. They manipulated and solved linear equations and inequalities algebraically. Also, they used graphs to represent two-variable data sets.

This Chapter

Students explore algebraic descriptions of linear functions, graphs of lines, and how to go back and forth between linear equations and graphed lines. They find the slope of a line containing two given points, relate the slope and *y*-intercept of a line to the values *m* and *b* in the slope-intercept form of an equation, and write equations for lines given two points or given one point and the slope. They find lines of fit for data and graph inequalities and special functions such as the greatest integer function and the absolute value function.

Future Connections

Students will continue to learn how algebraic expressions and the coordinate plane are related. At a simple level, they will learn how the "horizontal line test" identifies a one-to-one function. As activities become more complex, they will use graphs to explore quadratic and other non-linear equations and inequalities and use graphs to represent and solve systems of equations and inequalities.

2-1 Relations and Functions

This lesson begins an exploration of two important themes of algebra. One theme is how algebraic equations and the Cartesian coordinate system are related. The other theme is the relationships between equations that represent entire lines and numbers that represent points on or properties of that line.

For this lesson the central idea is ordered pairs. First, ordered pairs are explored as names for points in a coordinate plane. Second, several ordered pairs are used to describe how the set of the first elements (the domain) can be related to the set of the second elements (the range). Third, for equations that represent a line or a curve, ordered pairs are used to determine the graph that represents that equation in the coordinate plane.

In the lesson the relation between functions and relations is explored in two ways. Mappings of domain elements to range elements are used to identify functions that are one to one, functions that are not one to one, and relations that are not functions. In the coordinate plane, the vertical line test is used to distinguish a relation from a function. Relationships between ordered pairs and functions are explored in two ways. First, students are given an equation (for a curve or for a line) and make a table of ordered pairs for the equation. Second, students are given a function and a domain value, and evaluate the function to find the range value.

2-2 Linear Equations

In this lesson students deal with linear functions and intercepts. Linear functions and equations can be written in slope-intercept form, $f(x) = mx + b$ or $y = mx + b$, or in standard form, $Ax + By = C$. The graph of a linear function or equation is always a line.

2-3 Slope

Slope is a fundamental concept in algebra and higher mathematics. In this lesson, students calculate the slope of a line given two points on the line and explore the slopes of families or pairs of lines that are parallel and the slopes of pairs of lines that are perpendicular.

Students graph a line given two points or given one point and the slope. In the coordinate plane, students associate lines with slopes that are positive, negative, zero, or undefined.

2-4 Writing Linear Equations

This lesson focuses on the slope and y-intercept of a linear equation. In the slope-intercept form of a linear equation, $y = mx + b$, m represents the slope of the line and b is the y-intercept.

Students use two forms of a linear equation, the slope-intercept form and the point-slope form, to write an equation given two points, given a point and the slope, or given a point and the equation of a parallel or a perpendicular line.

2-5 Modeling Real-World Data: Using Scatter Plots

This lesson explores equations that approximate the relation between domain values and range values, extending the idea of using an algebraic equation to represent a set of points in a plane. Starting with a scatter plot of data, students mentally picture a line through the data. After selecting two points on that line, they calculate the slope and y-intercept of that line. The equation, called a line of fit or a prediction equation, may be used to calculate the value of one variable given a value of the other.

Activities in this lesson require three steps: given a set of ordered pairs, students identify a line that represents a set of ordered pairs; then they select two ordered pairs that lie on the line; and finally they calculate the slope and y-intercept for that line.

2-6 Special Functions

In this lesson, students explore special functions. The identity and constant functions are special linear functions. The graph of a step function is a series of line segments. An absolute value function has a V-shaped graph made up of portions of two lines. A piecewise function is a function written using two or more algebraic expressions.

2-7 Graphing Inequalities

In this lesson the graph of an equation is seen as the boundary between two regions of the coordinate plane. An inequality is a description of one of the two regions, and whether the boundary is part of that region depends on the inequality symbol that is used. Students explore how inequalities, including absolute value inequalities, are modeled by points in the coordinate plane, and vice versa.

Additional mathematical information and teaching notes are available at www.algebra2.com/key_concepts.

Chapter 2

DAILY INTERVENTION and Assessment

	Type	Student Edition	Teacher Resources	Technology/Internet
INTERVENTION	Ongoing	Prerequisite Skills, pp. 55, 62, 67, 74, 80, 86, 95 Practice Quiz 1, p. 74 Practice Quiz 2, p. 95	5-Minute Check Transparencies Quizzes, *CRM* pp. 113–114 Mid-Chapter Test, *CRM* p. 115 Study Guide and Intervention, *CRM* pp. 57–58, 63–64, 69–70, 75–76, 81–82, 87–88, 93–94	Alge2PASS: Tutorial Plus www.algebra2.com/self_check_quiz www.algebra2.com/extra_examples
	Mixed Review	pp. 62, 67, 74, 80, 86, 95, 99	Cumulative Review, *CRM* p. 116	
	Error Analysis	Find the Error, pp. 60, 71	Find the Error, *TWE* pp. 60, 71 Unlocking Misconceptions, *TWE* p. 58 Tips for New Teachers, *TWE* pp. 62, 74, 90	
ASSESSMENT	Standardized Test Practice	pp. 62, 67, 74, 76, 78, 80, 86, 95, 99, 105, 106–107	*TWE* p. 76 Standardized Test Practice, *CRM* pp. 117–118	Standardized Test Practice CD-ROM www.algebra2.com/standardized_test
	Open-Ended Assessment	Writing in Math, pp. 62, 67, 73, 80, 86, 94, 99 Open Ended, pp. 60, 65, 71, 78, 83, 92, 98	Modeling: *TWE* pp. 67, 74, 95 Speaking: *TWE* pp. 62, 98 Writing: *TWE* pp. 80, 86 Open-Ended Assessment, *CRM* p. 111	
	Chapter Assessment	Study Guide, pp. 100–104 Practice Test, p. 105	Multiple-Choice Tests (Forms 1, 2A, 2B), *CRM* pp. 99–104 Free-Response Tests (Forms 2C, 2D, 3), *CRM* pp. 105–110 Vocabulary Test/Review, *CRM* p. 112	ExamView® Pro (see below) MindJogger Videoquizzes www.algebra2.com/vocabulary_review www.algebra2.com/chapter_test

Key to Abbreviations: TWE = Teacher Wraparound Edition; CRM = Chapter Resource Masters

Intervention Technology

Alge2PASS: Tutorial Plus CD-ROM offers a complete, self-paced algebra curriculum.

Algebra 2 Lesson	Alge2PASS Lesson
2-2	3 *Graphing Linear Equations on the Coordinate Plane*
2-7	4 *Graphing Linear Inequalities on the Coordinate Plane*

ExamView® Pro

Use the networkable **ExamView® Pro** to:

- Create **multiple versions** of tests.
- Create **modified** tests for *Inclusion* students.
- **Edit** existing questions and **add** your own questions.
- Use built-in **state curriculum correlations** to create tests aligned with state standards.
- **Apply** art to your tests from a program bank of artwork.

For more information on Intervention and Assessment, see pp. T8–T11.

Reading and Writing in Mathematics

Glencoe Algebra 2 provides numerous opportunities to incorporate reading and writing into the mathematics classroom.

Student Edition

- Foldables Study Organizer, p. 55
- Concept Check questions require students to verbalize and write about what they have learned in the lesson. (pp. 60, 65, 71, 78, 83, 92, 98, 100)
- Writing in Math questions in every lesson, pp. 62, 67, 73, 80, 86, 94, 99
- Reading Study Tip, pp. 56, 59, 71, 82
- WebQuest, p. 84

Teacher Wraparound Edition

- Foldables Study Organizer, pp. 55, 100
- Study Notebook suggestions, pp. 60, 65, 71, 78, 83, 93, 97
- Modeling activities, pp. 67, 74, 95
- Speaking activities, pp. 62, 98
- Writing activities, pp. 80, 86
- Differentiated Instruction, (Verbal/Linguistic), p. 92
- ELL Resources, pp. 54, 61, 66, 73, 79, 85, 92, 94, 99, 100

Additional Resources

- Vocabulary Builder worksheets require students to define and give examples for key vocabulary terms as they progress through the chapter. (*Chapter 2 Resource Masters*, pp. vii-viii)
- Reading to Learn Mathematics master for each lesson (*Chapter 2 Resource Masters*, pp. 61, 67, 73, 79, 85, 91, 97)
- *Vocabulary PuzzleMaker* software creates crossword, jumble, and word search puzzles using vocabulary lists that you can customize.
- *Teaching Mathematics with Foldables* provides suggestions for promoting cognition and language.
- *Reading and Writing in the Mathematics Classroom*
- *WebQuest and Project Resources*

For more information on Reading and Writing in Mathematics, see pp. T6–T7.

ELL ENGLISH LANGUAGE LEARNERS

Lesson 2-1

Reading and Writing

Have students color-code the first element of each ordered pair of a function with one color and the second element with another color. The color coding will allow them to easily identify the domain and range. Then have students write an explanation in their own words, and if necessary in their own language, of how to graph a relation, find the domain and range of the relation, and determine whether a relation is a function.

Lesson 2-3

Language Experience Approach to Illustrations

Write the word *slope* on the chalkboard. Have students give other words, such as pitch or tilt, with a similar or related meaning. Discuss when each of these words is used in a real-life situation and how the situations are related to mathematical slope.

Lesson 2-5

Multisensory Activities

Have students work in pairs to measure the length of one of their femurs in centimeters by measuring from their hip bone to mid-knee while seated. Then have them measure their heights in centimeters. Record an ordered pair of (femur length, height) for each student. Use the data to create a scatter plot and write a prediction equation. Discuss the results as a class.

Chapter 2 Notes

What You'll Learn

Have students read over the list of objectives and make a list of any words with which they are not familiar.

Why It's Important

Point out to students that this is only one of many reasons why each objective is important. Others are provided in the introduction to each lesson.

Lesson	NCTM Standards	Local Objectives
2-1	1, 2, 7, 9, 10	
2-2	1, 2, 4, 6, 8, 9	
2-3	1, 2, 4, 6, 7, 8, 9, 10	
2-4	1, 2, 6, 8, 9, 10	
2-5	1, 2, 5, 6, 8, 9, 10	
2-5 Follow-Up	1, 2, 4, 5, 6, 9, 10	
2-6	1, 2, 5, 6, 8, 9, 10	
2-7	1, 2, 6, 8, 9, 10	

Key to NCTM Standards:

1=Number & Operations, 2=Algebra, 3=Geometry, 4=Measurement, 5=Data Analysis & Probability, 6=Problem Solving, 7=Reasoning & Proof, 8=Communication, 9=Connections, 10=Representation

Chapter 2 Linear Relations and Functions

What You'll Learn

- **Lesson 2-1** Analyze relations and functions.
- **Lessons 2-2 and 2-4** Identify, graph, and write linear equations.
- **Lesson 2-3** Find the slope of a line.
- **Lesson 2-5** Draw scatter plots and find prediction equations.
- **Lessons 2-6 and 2-7** Graph special functions, linear inequalities, and absolute value inequalities.

Key Vocabulary

- linear equation (p. 63)
- linear function (p. 63)
- slope (p. 68)
- slope-intercept form (p. 75)
- point-slope form (p. 76)

Why It's Important

Linear equations can be used to model relationships between many real-world quantities. One of the most common uses of a linear model is to make predictions.

Most hot springs are the result of groundwater passing through or near recently formed, hot, igneous rocks. Iceland, Yellowstone Park in the United States, and North Island of New Zealand are noted for their hot springs. *You will use a linear equation to find the temperature of underground rocks in Lesson 2-2.*

Vocabulary Builder

ELL

The Key Vocabulary list introduces students to some of the main vocabulary terms included in this chapter. For a more thorough vocabulary list with pronunciations of new words, give students the Vocabulary Builder worksheets found on pages vii and viii of the *Chapter 2 Resource Masters*. Encourage them to complete the definition of each term as they progress through the chapter. You may suggest that they add these sheets to their study notebooks for future reference when studying for the Chapter 2 test.

Getting Started

Prerequisite Skills To be successful in this chapter, you'll need to master these skills and be able to apply them in problem-solving situations. Review these skills before beginning Chapter 2.

For Lesson 2-1 — **Identify Points on a Coordinate Plane**

Write the ordered pair for each point.

1. A **(−3, 3)** **2.** B **(2, 3)**
3. C **(−3, −1)** **4.** D **(2, 0)**
5. E **(0, −4)** **6.** F **(3, −2)**

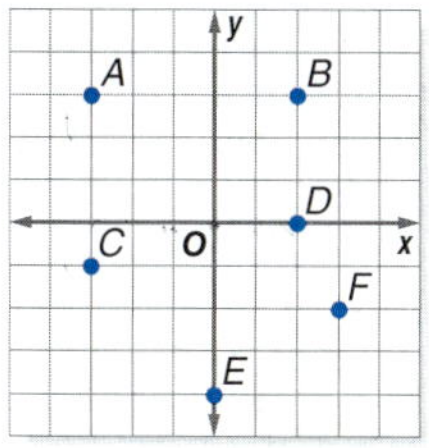

For Lesson 2-1 — **Evaluate Expressions**

Evaluate each expression if $a = -1$, $b = 3$, $c = -2$, and $d = 0$. *(For review, see Lesson 1-1.)*

7. $c + d$ **−2** **8.** $4c - b$ **−11** **9.** $a^2 - 5a + 3$ **9**
10. $2b^2 + b + 7$ **28** **11.** $\frac{a - b}{c - d}$ **2** **12.** $\frac{a + c}{b + c}$ **−3**

For Lesson 2-4 — **Simplify Expressions**

Simplify each expression. *(For review, see Lesson 1-2.)*

13. $x - (-1)$ $x + 1$ **14.** $x - (-5)$ $x + 5$ **15.** $2[x - (-3)]$ $2x + 6$
16. $4[x - (-2)]$ $4x + 8$ **17.** $\frac{1}{2}[x - (-4)]$ $\frac{1}{2}x + 2$ **18.** $\frac{1}{3}[x - (-6)]$ $\frac{1}{3}x + 2$

For Lessons 2-6 and 2-7 — **Evaluate Expressions with Absolute Value**

Evaluate each expression if $x = -3$, $y = 4$, and $z = -4.5$. *(For review, see Lesson 1-4.)*

19. $|x|$ **3** **20.** $|y|$ **4** **21.** $|5x|$ **15**
22. $-|2z|$ **−9** **23.** $5|y + z|$ **2.5** **24.** $-3|x + y| - |x + z|$ **−10.5**

FOLDABLES™ Study Organizer

Relations and Functions Make this Foldable to help you organize your notes. Begin with two sheets of grid paper.

Step 1 Fold

Fold in half along the width and staple along the fold.

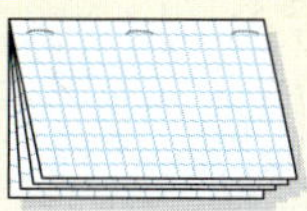

Step 2 Cut and Label

Cut the top three sheets and label as shown.

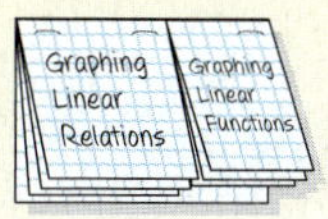

Reading and Writing As you read and study the chapter, write notes, examples, and graphs under the tabs.

Getting Started

This section provides a review of the basic concepts needed before beginning Chapter 2. Page references are included for additional student help.

Prerequisite Skills in the Getting Ready for the Next Lesson section at the end of each exercise set review a skill needed in the next lesson.

For Lesson	Prerequisite Skill
2-2	Solving Equations (p. 62)
2-4	Solving Equations (p. 74)
2-5	Finding a Median (p. 80)
2-6	Absolute Value (p. 86)
2-7	Inequalities (p. 95)

FOLDABLES™ Study Organizer

For more information about Foldables, see *Teaching Mathematics with Foldables.*

Organization of Data: Annotating As students read and work their way through the chapter, have them make annotations under the appropriate tabs of their Foldable. Explain to them that annotations are usually notes taken in the margins of books, which we own, to organize the text for review or studying. Annotations often include questions that arise, reader comments and reactions, short summaries, steps or data numbered by the reader, and key points highlighted or underlined.

2-1 Lesson Notes

1 Focus

5-Minute Check Transparency 2-1 Use as a quiz or review of Chapter 1.

Mathematical Background notes are available for this lesson on p. 54C.

Building on Prior Knowledge

In Chapter 1, students solved equations and inequalities. In this lesson, students relate equations to functions and relations, as well as to their graphs.

How do relations and functions apply to biology?

Ask students:

- What is the difference between average lifetime and maximum lifetime? **The average lifetime is a representative number of years for any animal of that type, while the maximum lifetime is the greatest age ever attained by an animal of that type.**
- Why can you be sure that the second number in the ordered pairs for this data is always greater than or equal to the first? **For each animal, the maximum age will always equal or exceed the average age.**

2-1 Relations and Functions

What You'll Learn

- Analyze and graph relations.
- Find functional values.

How do relations and functions apply to biology?

The table shows the average lifetime and maximum lifetime for some animals. The data can also be represented as **ordered pairs**. The ordered pairs for the data are (12, 28), (15, 30), (8, 20), (12, 20), and (20, 50). The first number in each ordered pair is the average lifetime, and the second number is the maximum lifetime.

(12, 28)
average lifetime ↑ ↑ maximum lifetime

Animal	Average Lifetime (years)	Maximum Lifetime (years)
Cat	12	28
Cow	15	30
Deer	8	20
Dog	12	20
Horse	20	50

Source: *The World Almanac*

Vocabulary

- ordered pair
- Cartesian coordinate plane
- quadrant
- relation
- domain
- range
- function
- mapping
- one-to-one function
- vertical line test
- independent variable
- dependent variable
- functional notation

GRAPH RELATIONS You can graph the ordered pairs above on a *coordinate system* with two axes. Remember that each point in the coordinate plane can be named by exactly one ordered pair and that every ordered pair names exactly one point in the coordinate plane.

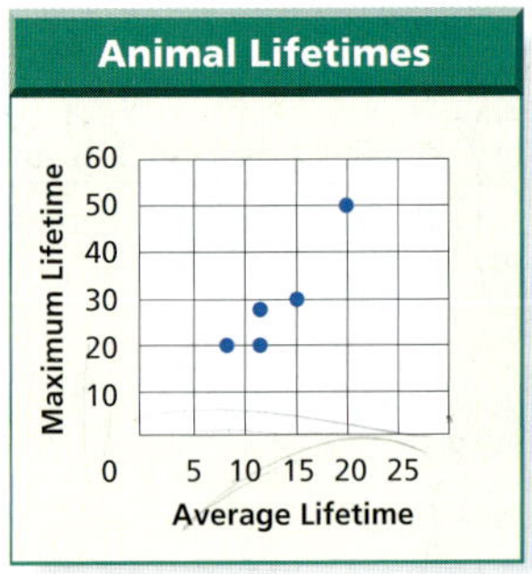

The graph of the animal lifetime data lies in only one part of the Cartesian coordinate plane—the part with all positive numbers. The **Cartesian coordinate plane** is composed of the *x-axis* (horizontal) and the *y-axis* (vertical), which meet at the *origin* (0, 0) and divide the plane into four **quadrants**. *The points on the two axes do not lie in any quadrant.*

In general, any ordered pair in the coordinate plane can be written in the form (x, y).

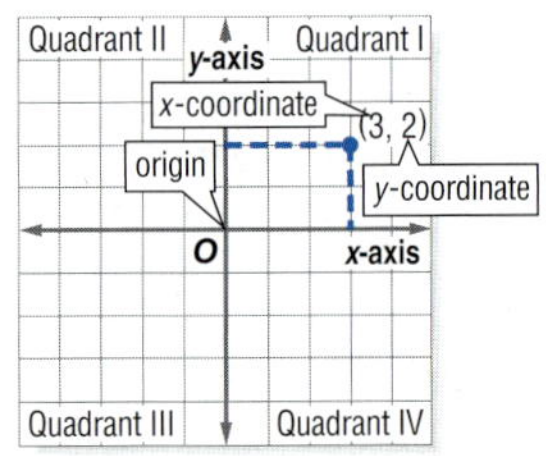

Assume that each square on a graph represents 1 unit unless otherwise labeled.

A **relation** is a set of ordered pairs, such as the one for the longevity of animals. The **domain** of a relation is the set of all first coordinates (x-coordinates) from the ordered pairs, and the **range** is the set of all second coordinates (y-coordinates) from the ordered pairs. The *graph* of a relation is the set of points in the coordinate plane corresponding to the ordered pairs in the relation.

Study Tip

Reading Math
An x-coordinate is sometimes called an *abscissa*, and a y-coordinate is sometimes called an *ordinate*.

Resource Manager

Workbook and Reproducible Masters

Chapter 2 Resource Masters

- Study Guide and Intervention, pp. 57–58
- Skills Practice, p. 59
- Practice, p. 60
- Reading to Learn Mathematics, p. 61
- Enrichment, p. 62

Transparencies

5-Minute Check Transparency 2-1
Answer Key Transparencies

Technology

Interactive Chalkboard

A **function** is a special type of relation in which each element of the domain is paired with *exactly one* element of the range. A **mapping** shows how each member of the domain is paired with each member of the range.

{(−3, 1), (0, 2), (2, 4)}

Domain Range

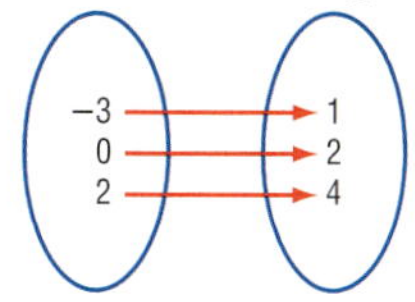

The first two relations shown below are functions. The third relation is not a function because the −3 in the domain is paired with both 0 and 6 in the range. A function like the first one below, where each element of the range is paired with exactly one element of the domain, is called a **one-to-one function**.

Concept Summary — *Functions*

{(−3, 1), (0, 2), (2, 4)}

Domain Range

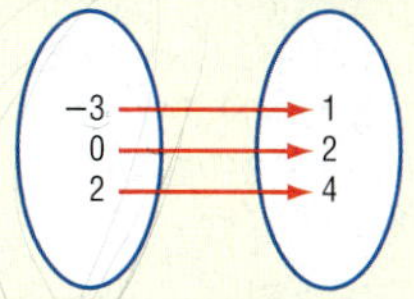

one-to-one function

{(−1, 5), (1, 3), (4, 5)}

Domain Range

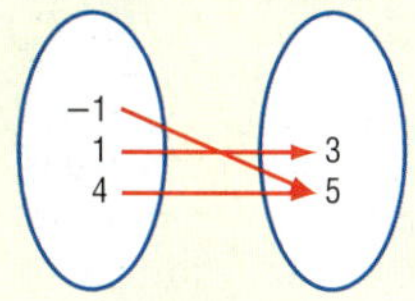

function, not one-to-one

{(5, 6), (−3, 0), (1, 1), (−3, 6)}

Domain Range

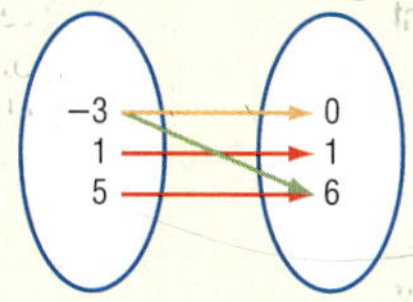

not a function

Example 1 *Domain and Range*

State the domain and range of the relation shown in the graph. Is the relation a function?

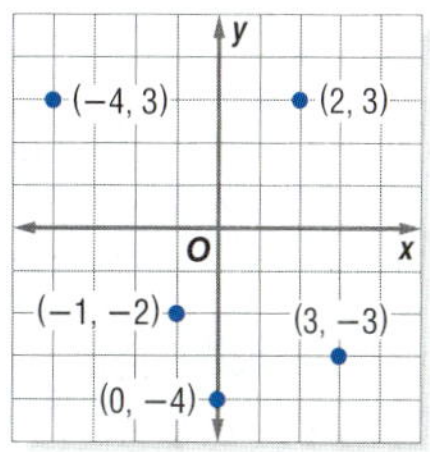

The relation is {(−4, 3), (−1, −2), (0, −4), (2, 3), (3, −3)}.
The domain is {−4, −1, 0, 2, 3}.
The range is {−4, −3, −2, 3}.

Each member of the domain is paired with exactly one member of the range, so this relation is a function.

You can use the **vertical line test** to determine whether a relation is a function.

Key Concept — *Vertical Line Test*

- **Words** If no vertical line intersects a graph in more than one point, the graph represents a function. If some vertical line intersects a graph in two or more points, the graph does not represent a function.
- **Models**

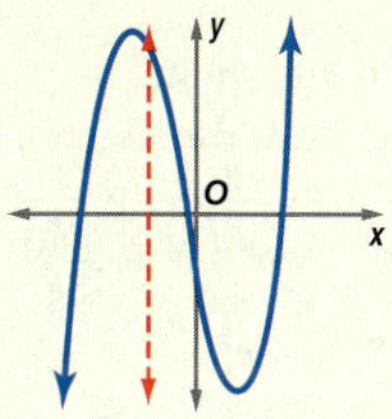

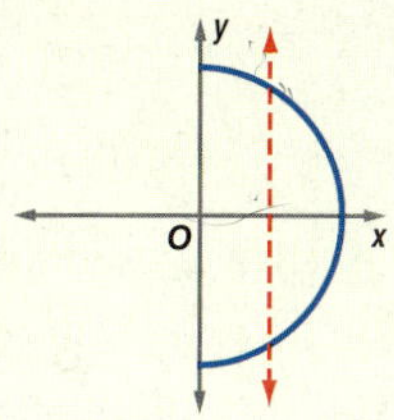

In Example 1, there is no vertical line that contains more than one of the points. Therefore, the relation is a function.

2 Teach

GRAPH RELATIONS

In-Class Example (PowerPoint®)

1 State the domain and range of the relation shown in the graph. Is the relation a function?

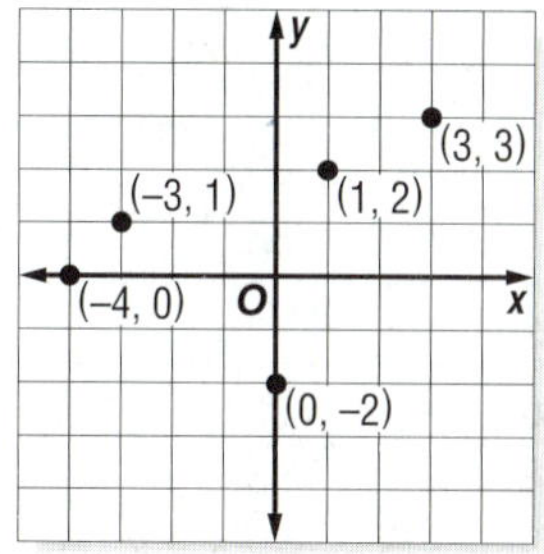

The domain is {−4, −3, 0, 1, 3}. The range is {−2, 0, 1, 2, 3}. Each member of the domain is paired with exactly one member of the range, so this relation is a function.

Teaching Tip Ask students to explain why the set of ordered pairs (9, 3), (9, −3), (4, 2), (4, −2) is not a function. **Sample answer: For at least one of the *x* values, there are two different *y* values.**

Interactive Chalkboard
PowerPoint® Presentations

This CD-ROM is a customizable Microsoft® PowerPoint® presentation that includes:

- Step-by-step, dynamic solutions of each In-Class Example from the Teacher Wraparound Edition
- Additional, Your Turn exercises for each example
- The 5-Minute Check Transparencies
- Hot links to Glencoe Online Study Tools

In-Class Example

2 **TRANSPORTATION** The table shows the average fuel efficiency in miles per gallon for light trucks for several years. Graph this information and determine whether it represents a function.

Year	Fuel Efficiency (mi/gal)
1995	20.5
1996	20.8
1997	20.6
1998	20.9
1999	20.5
2000	20.5
2001	20.4

Source: U.S. Environmental Protection Agency

Fuel Efficiency of Light Trucks

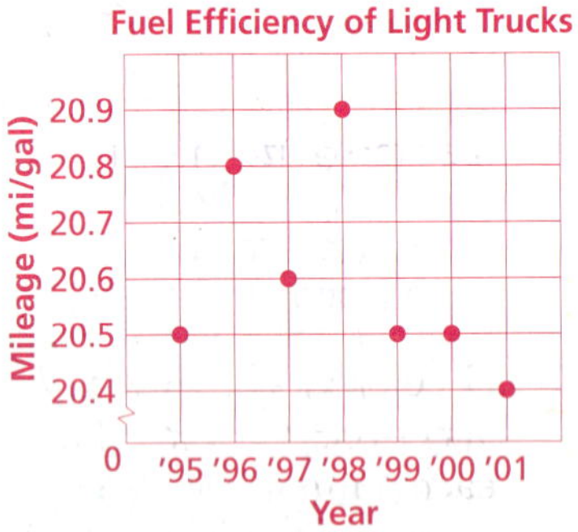

Yes, this relation is a function.

EQUATIONS OF FUNCTIONS AND RELATIONS

In-Class Example

3 a. Graph the relation represented by $y = 3x - 1$.

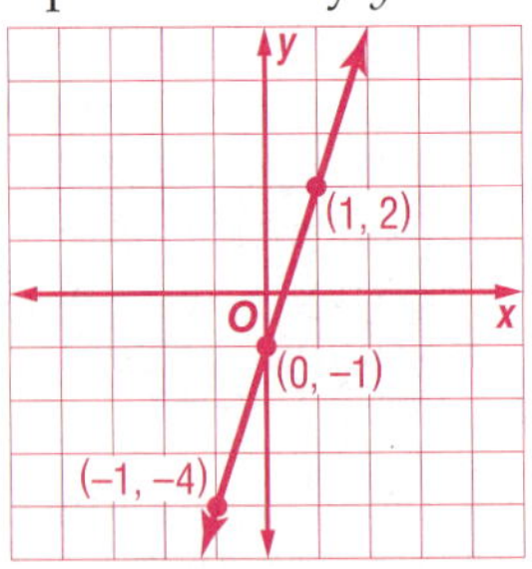

b. Find the domain and range. **The domain and range are both all real numbers.**

c. Determine whether the relation is a function. **Yes, the equation $y = 3x - 1$ represents a function.**

Study Tip

Vertical Line Test

You can use a pencil to represent a vertical line. Slowly move the pencil to the right across the graph to see if it intersects the graph at more than one point.

Example 2 Vertical Line Test

GEOGRAPHY **The table shows the population of the state of Indiana over the last several decades. Graph this information and determine whether it represents a function.**

Year	Population (millions)
1950	3.9
1960	4.7
1970	5.2
1980	5.5
1990	5.5
2000	6.1

Source: U.S. Census Bureau

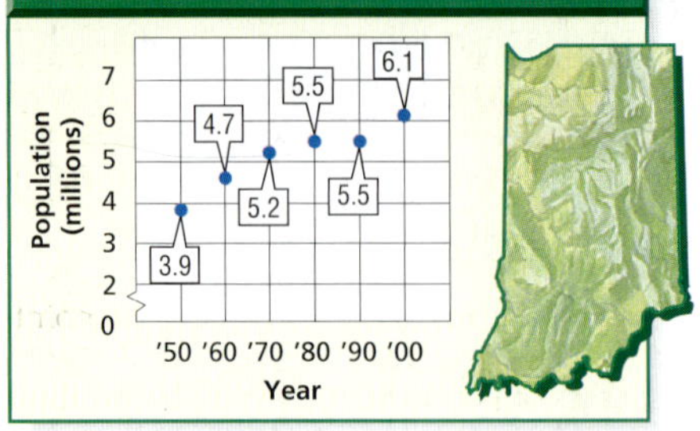

Use the vertical line test. Notice that no vertical line can be drawn that contains more than one of the data points. Therefore, this relation is a function. *Notice also that each year is paired with only one population value.*

EQUATIONS OF FUNCTIONS AND RELATIONS

Relations and functions can also be represented by equations. The solutions of an equation in x and y are the set of ordered pairs (x, y) that make the equation true.

Consider the equation $y = 2x - 6$. Since x can be any real number, the domain has an infinite number of elements. To determine whether an equation represents a function, it is often simplest to look at the graph of the relation.

Example 3 Graph Is a Line

a. Graph the relation represented by $y = 2x + 1$.

Make a table of values to find ordered pairs that satisfy the equation. Choose values for x and find the corresponding values for y. Then graph the ordered pairs.

x	y
–1	
0	
1	
2	

⟶

x	y
–1	–1
0	1
1	3
2	5

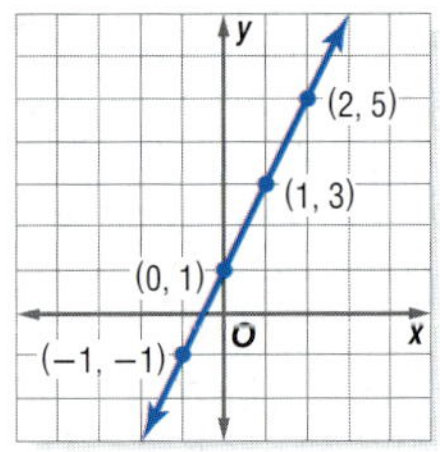

b. Find the domain and range.

Since x can be any real number, there is an infinite number of ordered pairs that can be graphed. All of them lie on the line shown. Notice that every real number is the x-coordinate of some point on the line. Also, every real number is the y-coordinate of some point on the line. So the domain and range are both all real numbers.

c. Determine whether the relation is a function.

This graph passes the vertical line test. For each x value, there is exactly one y value, so the equation $y = 2x + 1$ represents a function.

DAILY INTERVENTION — Unlocking Misconceptions

- **Relations and Functions** Some students may not realize that all functions are relations. Explain that a function is a special type of relation, analogous to how a square is a special type of rectangle.
- **Vertical Line Test** Make sure students understand that when two points on the graph of a relation are intersected by a vertical line, this means those two points have the same x value but different y values. That is, one domain value is paired with more than one range value.

Example 4 Graph Is a Curve

a. Graph the relation represented by $x = y^2 - 2$.

Make a table. In this case, it is easier to choose y values and then find the corresponding values for x. Then sketch the graph, connecting the points with a smooth curve.

x	y
	–2
	–1
	0
	1
	2

→

x	y
2	–2
–1	–1
–2	0
–1	1
2	2

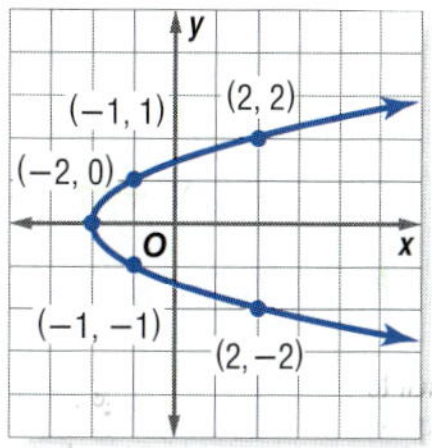

b. Find the domain and range.

Every real number is the y-coordinate of some point on the graph, so the range is all real numbers. But, only real numbers greater than or equal to -2 are x-coordinates of points on the graph. So the domain is $\{x \mid x \geq -2\}$.

c. Determine whether the relation is a function.

You can see from the table and the vertical line test that there are two y values for each x value except $x = -2$. Therefore, the equation $x = y^2 - 2$ does not represent a function.

Study Tip

Reading Math

Suppose you have a job that pays by the hour. Since your pay *depends* on the number of hours you work, you might say that your pay is a *function* of the number of hours you work.

When an equation represents a function, the variable, usually x, whose values make up the domain is called the **independent variable**. The other variable, usually y, is called the **dependent variable** because its values depend on x.

Equations that represent functions are often written in **functional notation**. The equation $y = 2x + 1$ can be written as $f(x) = 2x + 1$. The symbol $f(x)$ replaces the y and is read "f of x." The f is just the name of the function. It is not a variable that is multiplied by x. Suppose you want to find the value in the range that corresponds to the element 4 in the domain of the function. This is written as $f(4)$ and is read "f of 4." The value $f(4)$ is found by substituting 4 for each x in the equation. Therefore, $f(4) = 2(4) + 1$ or 9. *Letters other than f can be used to represent a function. For example, $g(x) = 2x + 1$.*

TEACHING TIP

Your pay is the dependent variable, and the number of hours you work is the independent variable.

Example 5 Evaluate a Function

Given $f(x) = x^2 + 2$ and $g(x) = 0.5x^2 - 5x + 3.5$, find each value.

a. $f(-3)$

$f(x) = x^2 + 2$ Original function

$f(-3) = (-3)^2 + 2$ Substitute.

$= 9 + 2$ or 11 Simplify.

b. $g(2.8)$

$g(x) = 0.5x^2 - 5x + 3.5$ Original function

$g(2.8) = 0.5(2.8)^2 - 5(2.8) + 3.5$ **Estimate:** $g(3) = 0.5(3)^2 - 5(3) + 3.5$ or -7

$= 3.92 - 14 + 3.5$ Multiply.

$= -6.58$ Compare with the estimate.

c. $f(3z)$

$f(x) = x^2 + 2$ Original function

$f(3z) = (3z)^2 + 2$ Substitute.

$= 9z^2 + 2$ $(ab)^2 = a^2b^2$

www.algebra2.com/extra_examples

In-Class Examples

Power Point®

4 **a.** Graph the relation represented by $x = y^2 + 1$.

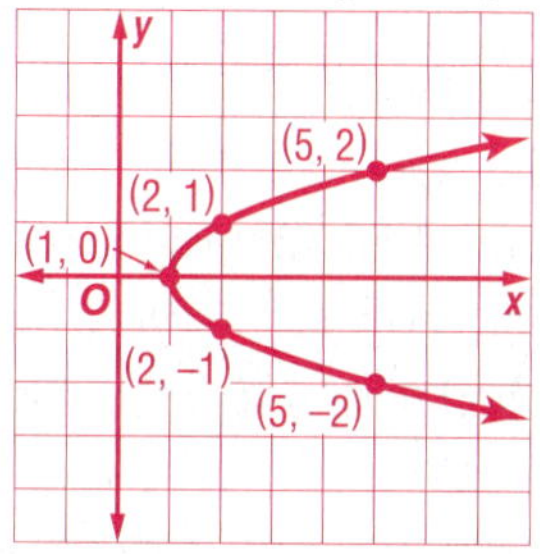

b. Find the domain and range. **The domain is $\{x \mid x \geq 1\}$ and the range is all real numbers.**

c. Determine whether the relation is a function. **No, the equation $x = y^2 + 1$ does not represent a function.**

Teaching Tip Some students may recognize this curve as a parabola. It is the same shape as the more familiar graph of the equation $y = x^2 + 1$.

5 Given $f(x) = x^3 - 3$ and $h(x) = 0.3x^2 - 3x - 2.7$, find each value.

a. $f(-2)$ **-11**

b. $h(1.6)$ **-6.732**

c. $f(2t)$ **$8t^3 - 3$**

DAILY INTERVENTION

Differentiated Instruction

Auditory/Musical Encourage students to relate to the mathematics in this lesson by asking those who are familiar with reading musical notation to explain to the class how graphing points on a coordinate plane compares to writing musical notes on a staff.

3 Practice/Apply

Study Notebook

Have students—
- add the definitions/examples of the vocabulary terms to their Vocabulary Builder worksheets for Chapter 2.
- draw their own diagrams, similar to those in the Concept Summary on p. 57.
- make a sketch illustrating how to use the vertical line test, both for a function and a non-function.
- include any other item(s) that they find helpful in mastering the skills in this lesson.

About the Exercises...

Organization by Objective
- **Graph Relations:** 17–28, 35–45, 55
- **Equations of Functions and Relations:** 29–34, 46–54, 56

Odd/Even Assignments

Exercises 17–34 are structured so that students practice the same concepts whether they are assigned odd or even problems.

Assignment Guide

Basic: 17–31 odd, 35–37, 47–53 odd, 55–58, 63–73

Average: 17–33 odd, 35–41, 47–53 odd, 55–58, 63–73 (optional: 59–62)

Advanced: 18–34 even, 42–45, 46–54 even, 55–69 (optional: 70–73)

DAILY INTERVENTION **FIND THE ERROR** Suggest that students rewrite the original function by substituting the expression (2*a*) for each variable *x* before they begin simplifying.

Check for Understanding

Concept Check

1. Sample answer: {(−4, 3), (−2, 3), (1, 5), (−2, 1)}
2. See pp. 107A–107H.

1. **OPEN ENDED** Write a relation of four ordered pairs that is *not* a function.
2. **Copy** the graph at the right. Then draw a vertical line that shows that the graph does not represent a function.

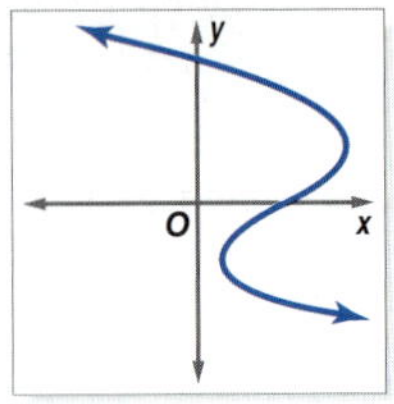

3. **FIND THE ERROR** Teisha and Molly are finding $g(2a)$ for the function $g(x) = x^2 + x - 1$.

Teisha	Molly
$g(2a) = 2(a^2 + a - 1)$ $= 2a^2 + 2a - 2$	$g(2a) = (2a)^2 + 2a - 1$ $= 4a^2 + 2a - 1$

Who is correct? Explain your reasoning. **See margin.**

Guided Practice

GUIDED PRACTICE KEY

Exercises	Examples
4–8	1, 2
9	3
10	4
11–14	2
15, 16	5

Determine whether each relation is a function. Write *yes* or *no*.

4. 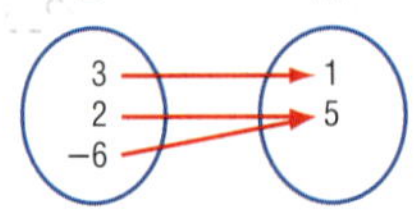

yes

5.

x	y
5	−2
10	−2
15	−2
20	−2

yes

6. 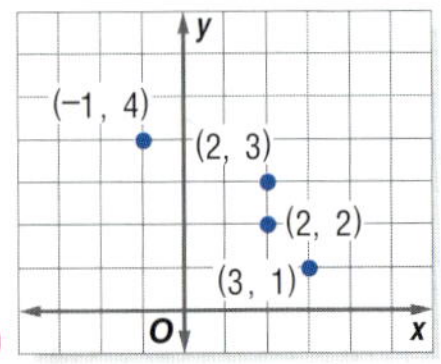

no

Graph each relation or equation and find the domain and range. Then determine whether the relation or equation is a function. **7–10. pp. 107A–107H.**

7. {(7, 8), (7, 5), (7, 2), (7, −1)}
8. {(6, 2.5), (3, 2.5), (4, 2.5)}
9. $y = -2x + 1$
10. $x = y^2$
11. Find $f(5)$ if $f(x) = x^2 - 3x$. **10**
12. Find $h(-2)$ if $h(x) = x^3 + 1$. **−7**

Application

13. D = {70, 72, 88}, R = {95, 97, 105, 114}
14–16. See margin.

WEATHER **For Exercises 13–16, use the table of record high temperatures (°F) for January and July.**

City	Jan.	July
Los Angeles	88	97
Sacramento	70	114
San Diego	88	95
San Francisco	72	105

Source: U.S. National Oceanic and Atmospheric Administration

13. Identify the domain and range. Assume that the January temperatures are the domain.
14. Write a relation of ordered pairs for the data.
15. Graph the relation.
16. Is this relation a function? Explain.

★ indicates increased difficulty

Practice and Apply

Homework Help

For Exercises	See Examples
17–28	1, 2
29–32	3
33, 34	4
35–45, 55	2
46–54, 56	5

Extra Practice See page 830.

Determine whether each relation is a function. Write *yes* or *no*.

17. 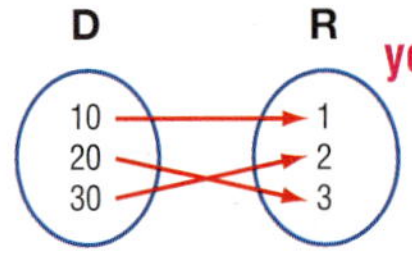

yes

18. 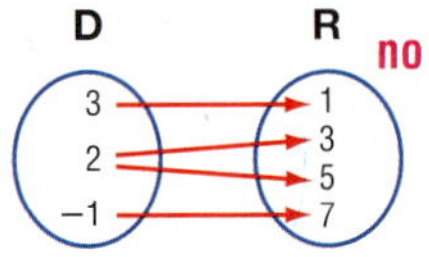

no

19.

x	y
0.5	−3
2	0.8
0.5	8

no

20.

x	y
2000	$4000
2001	$4300
2002	$4000
2003	$4500

yes

21. 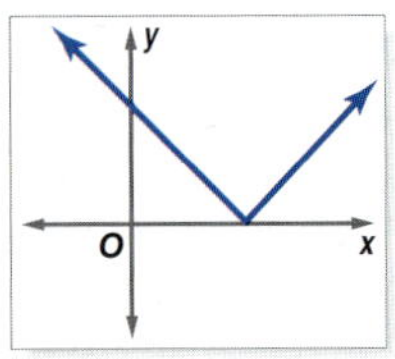

yes

22. 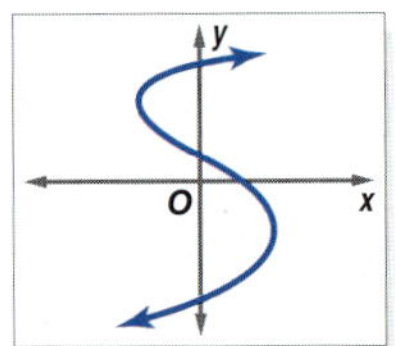

no

Answers

3. Molly; to find $g(2a)$, replace x with $2a$. Teisha found $2g(a)$, not $g(2a)$.
14. {(88, 97), (70, 114), (88, 95), (72, 105)}
15. See graph at right.
16. No; the domain value 88 is paired with two range values.

Record High Temperatures

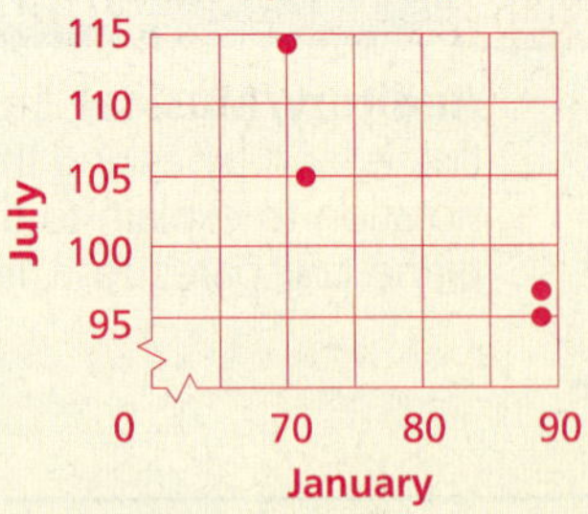

23. D = {−3, 1, 2}, R = {0, 1, 5}; yes
24. D = {3, 4, 6}, R = {5}; yes
25. D = {−2, 3}, R = {5, 7, 8}; no
26. D = {3, 4, 5, 6}, R = {3, 4, 5, 6}; yes
27. D = {−3.6, 0, 1.4, 2}, R = {−3, −1.1, 2, 8}; yes

Graph each relation or equation and find the domain and range. Then determine whether the relation or equation is a function. **23–34. See pp. 107A–107H for graphs.**

23. {(2, 1), (−3, 0), (1, 5)}
24. {(4, 5), (6, 5), (3, 5)}
25. {(−2, 5), (3, 7), (−2, 8)}
26. {(3, 4), (4, 3), (6, 5), (5, 6)}
27. {(0, −1.1), (2, −3), (1.4, 2), (−3.6, 8)}
28. {(−2.5, 1), (−1, −1), (0, 1), (−1, 1)}
29. $y = -5x$
30. $y = 3x$
31. $y = 3x - 4$
32. $y = 7x - 6$
★33. $y = x^2$ **D = all reals, R = {y | y ≥ 0}; yes**
★34. $x = 2y^2 - 3$ **D = {x | x ≥ −3}, R = all reals; no**

SPORTS For Exercises 35–37, use the table that shows the leading home run and runs batted in totals in the American League for 1996–2000.

Year	1996	1997	1998	1999	2000
HR	52	56	56	48	47
RBI	148	147	157	165	145

Source: *The World Almanac*

35. Make a graph of the data with home runs on the horizontal axis and runs batted in on the vertical axis. **See pp. 107A–107H.**
36. Identify the domain and range.
37. Does the graph represent a function? Explain your reasoning. **See margin.**

36. D = {47, 48, 52, 56}, R = {145, 147, 148, 157, 165}

More About. . .

Sports

The major league record for runs batted in (RBIs) is 191 by Hack Wilson.

Source: www.baseball-almanac.com

FINANCE For Exercises 38–41, use the table that shows a company's stock price in recent years. **38, 40. See margin.**

Year	Price
1997	\$39
1998	\$43
1999	\$48
2000	\$55
2001	\$61
2002	\$52

38. Write a relation to represent the data.
39. Graph the relation. **See pp. 107A–107H.**
40. Identify the domain and range.
41. Is the relation a function? Explain your reasoning. **Yes; each domain value is paired with only one range value.**

28. D = {−2.5, −1, 0}, R = {−1, 1}; no
29. D = all reals, R = all reals; yes
30. D = all reals, R = all reals, yes
31. D = all reals, R = all reals, yes
32. D = all reals, R = all reals; yes

GOVERNMENT For Exercises 42–45, use the table below that shows the number of members of the U.S. House of Representatives with 30 or more consecutive years of service in Congress from 1987 to 1999.

Year	1987	1989	1991	1993	1995	1997	1999
Representatives	12	13	11	12	9	6	3

Source: *Congressional Directory*

42. Write a relation to represent the data. **See margin.**
43. Graph the relation. **See pp.107A–107H.**
44. Identify the domain and range. **See pp. 107A–107H.**
45. Is the relation a function? If so, is it a one-to-one function? Explain. **Yes; no; see pp. 107A–107H for explanation.**

Find each value if $f(x) = 3x - 5$ and $g(x) = x^2 - x$.

46. $f(-3)$ **−14**
47. $g(3)$ **6**
48. $g\left(\frac{1}{3}\right)$ $-\frac{2}{9}$
49. $f\left(\frac{2}{3}\right)$ **−3**
50. $f(a)$ **3a − 5**
51. $g(5n)$ $25n^2 - 5n$

52. Find the value of $f(x) = -3x + 2$ when $x = 2$. **−4**
53. What is $g(4)$ if $g(x) = x^2 - 5$? **11**

www.algebra2.com/self_check_quiz

Answers

37. No; the domain value 56 is paired with two different range values.
38. {(1997, 39), (1998, 43), (1999, 48), (2000, 55), (2001, 61), (2002, 52)}
40. D = {1997, 1998, 1999, 2000, 2001, 2002}, R = {39, 43, 48, 52, 55, 61}
42. {(1987, 12), (1989, 13), (1991, 11), (1993, 12), (1995, 9), (1997, 6), (1999, 3)}

Enrichment, p. 62

Mappings

There are three special ways in which one set can be mapped to another. A set can be mapped *into* another set, *onto* another set, or can have a *one-to-one correspondence* with another set.

Into mapping	A mapping from set *A* to set *B* where every element of *A* is mapped to one or more elements of set *B*, but never to an element not in *B*.
Onto mapping	A mapping from set *A* to set *B* where each element of set *B* has at least one element of set *A* mapped to it.
One-to-one correspondence	A mapping from set *A* onto set *B* where each element of set *A* is mapped to exactly one element of set *B* and different elements of *A* are never mapped to the same element of *B*.

State whether each set is mapped into the second set, onto the second set, or has a one-to-one correspondence with the second set.

1. 2 → 7 2. 4, 12 → 0 3. a → 1 4. → 10

Study Guide and Intervention, p. 57 (shown) and p. 58

Graph Relations A **relation** can be represented as a set of ordered pairs or as an equation; the relation is then the set of all ordered pairs (x, y) that make the equation true. The **domain** of a relation is the set of all first coordinates of the ordered pairs, and the **range** is the set of all second coordinates.

A **function** is a relation in which each element of the domain is paired with exactly one element of the range. You can tell if a relation is a function by graphing, then using the **vertical line test**. If a vertical line intersects the graph at more than one point, the relation is not a function.

Example **Graph the equation $y = 2x - 3$ and find the domain and range. Does the equation represent a function?**

Make a table of values to find ordered pairs that satisfy the equation. Then graph the ordered pairs.

The domain and range are both all real numbers. The graph passes the vertical line test, so it is function.

x	y
−1	−5
0	−3
1	−1
2	1
3	3

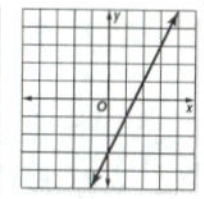

Exercises

Graph each relation or equation and find the domain and range. Then determine whether the relation or equation is a function.

1. {(1, 3), (−3, 5), (−2, 5), (2, 3)}
2. {(3, −4), (1, 0), (2, −2), (3, 2)}
3. {(0, 4), (−3, −2), (3, 2), (5, 1)}

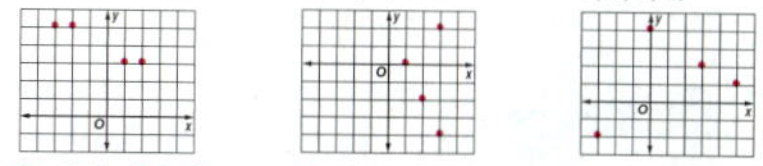

D = {−3, −2, 1, 2}, R = {3, 5}; yes
D = {1, 2, 3}, R = {−4, −2, 0, 2}; no
D = {−3, 0, 3, 5}, R = {−2, 1, 2, 4}; yes

4. $y = x^2 - 1$
5. $y = x - 4$
6. $y = 3x + 2$

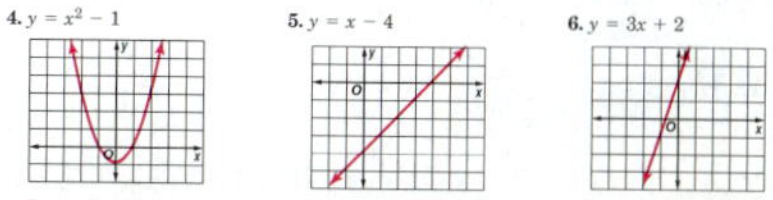

D = all reals, R = {y | y ≥ −1}; yes
D = all reals, R = all reals; yes
D = all reals, R = all reals; yes

Skills Practice, p. 59 and Practice, p. 60 (shown)

Determine whether each relation is a function. Write *yes* or *no*.

1. **no**
2. **yes**

3. **yes**

x	y
−3	0
−1	−1
0	0
2	−2
3	4

4. **no**

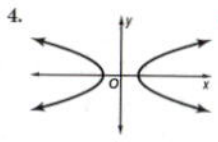

Graph each relation or equation and find the domain and range. Then determine whether the relation or equation is a function.

5. {(−4, −1), (4, 0), (0, 3), (2, 0)}
6. $y = 2x - 1$

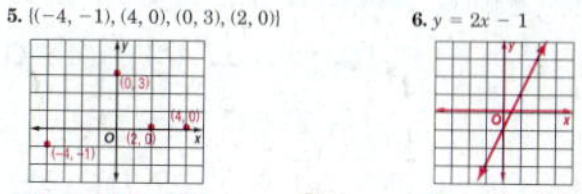

D = {−4, 0, 2, 4}, R = {−1, 0, 3}; yes
D = all reals, R = all reals; yes

Find each value if $f(x) = \frac{5}{x + 2}$ and $g(x) = -2x + 3$.

7. $f(3)$ **1**
8. $f(-4)$ $-\frac{5}{2}$
9. $g\left(\frac{1}{2}\right)$ **2**
10. $f(-2)$ **undefined**
11. $g(-6)$ **15**
12. $f(m - 2)$ $\frac{5}{m}$

13. **MUSIC** The ordered pairs (1, 16), (2, 16), (3, 32), (4, 32), and (5, 48) represent the cost of buying various numbers of CDs through a music club. Identify the domain and range of the relation. Is the relation a function? **D = {1, 2, 3, 4, 5}, R = {16, 32, 48}; yes**

14. **COMPUTING** If a computer can do one calculation in 0.0000000015 second, then the function $T(n) = 0.0000000015n$ gives the time required for the computer to do n

Reading to Learn Mathematics, p. 61 ELL

Pre-Activity **How do relations and functions apply to biology?**

Read the introduction to Lesson 2-1 at the top of page 56 in your textbook.

- Refer to the table. What does the ordered pair (8, 20) tell you? **For a deer, the average longevity is 8 years and the maximum longevity is 20 years.**
- Suppose that this table is extended to include more animals. Is it possible to have an ordered pair for the data in which the first number is larger than the second? **Sample answer: No, the maximum longevity must always be greater than the average longevity.**

Reading the Lesson

1. a. Explain the difference between a relation and a function. **Sample answer: A relation is any set of ordered pairs. A function is a special kind of relation in which each element of the domain is paired with exactly one element in the range.**
   b. Explain the difference between domain and range. **Sample answer: The domain of a relation is the set of all first coordinates of the ordered pairs. The range is the set of all second coordinates.**
2. a. Write the domain and range of the relation shown in the graph.

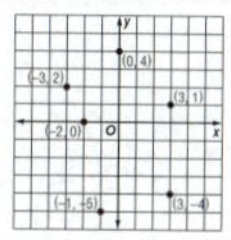

D: {−3, −2, −1, 0, 3}; R: {−5, −4, 0, 1, 2, 4}

   b. Is this relation a function? Explain. **Sample answer: No, it is not a function because one of the elements of the domain, 3, is paired with two elements of the range.**

Helping You Remember

3. Look up the words *dependent* and *independent* in a dictionary. How can the meaning of these words help you distinguish between independent and dependent variables in a function? **Sample answer: The variable whose values depend on, or are determined by, the values of the other variable is the dependent variable.**

4 Assess

Open-Ended Assessment

Speaking Ask students to discuss function notation, including what they may find confusing and how to read it out loud. Make sure they understand and can correctly explain the difference between f times x and f of x.

Tips for New Teachers

Intervention This lesson has a number of vocabulary words that may be new or challenging for some students. Make sure that all students are comfortable with the mathematical language of this lesson before they go on.

Getting Ready for Lesson 2-2

PREREQUISITE SKILL Lesson 2-2 presents identifying and graphing linear equations in two variables. The skills needed to write linear equations in standard form are similar to solving equations in one variable. Exercises 70–73 should be used to determine your students' familiarity with solving single-variable equations.

Answer

56. Relations and functions can be used to represent biological data. Answers should include the following.

- If the data are written as ordered pairs, then those ordered pairs are a relation.
- The maximum lifetime of an animal is not a function of its average lifetime.

54. **HOBBIES** Chaz has a collection of 15 CDs. After he gets a part-time job, he decides to buy 3 more CDs every time he goes to the music store. The function $C(t) = 15 + 3t$ counts the number of CDs, $C(t)$, he has after t trips to the music store. How many CDs will he have after he has been to the music store 8 times? **39**

55. **CRITICAL THINKING** If $f(3a - 1) = 12a - 7$, find $f(x)$. **$f(x) = 4x - 3$**

56. **WRITING IN MATH** Answer the question that was posed at the beginning of the lesson. **See margin.**

How do relations and functions apply to biology?

Include the following in your answer:

- an explanation of how a relation can be used to represent data, and
- a sentence that includes the words *average lifetime*, *maximum lifetime*, and *function*.

Standardized Test Practice

57. If $f(x) = 2x - 5$, then $f(0) =$ **B**

Ⓐ 0. Ⓑ -5. Ⓒ -3. Ⓓ $\frac{5}{2}$.

58. If $g(x) = x^2$, then $g(x + 1) =$ **C**

Ⓐ 1. Ⓑ $x^2 + 1$. Ⓒ $x^2 + 2x + 1$. Ⓓ $x^2 - x$.

Extending the Lesson

A function whose graph consists of disconnected points is called a *discrete function*. A function whose graph you can draw without lifting your pencil is called a *continuous function*. Determine whether each function is *discrete* or *continuous*.

59. 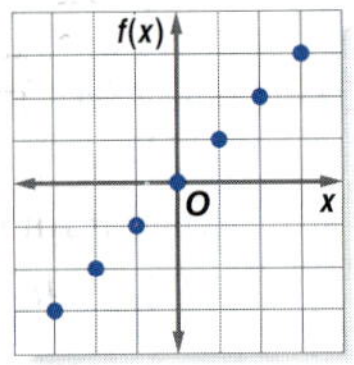

discrete

60. 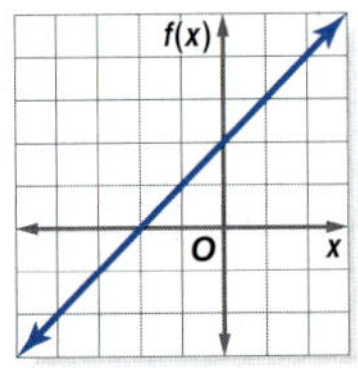

continuous

61. $\{(-3, 0), (-1, 1), (1, 3)\}$ **discrete**

62. $y = -x + 4$ **continuous**

Maintain Your Skills

Mixed Review

Solve each inequality. *(Lessons 1-5 and 1-6)* **63. $\{y \mid -8 < y < 6\}$ 64. $\{m \mid 4 < m < 6\}$**

63. $|y + 1| < 7$
64. $|5 - m| < 1$
65. $x - 5 < 0.1$ **$\{x \mid x < 5.1\}$**

SHOPPING For Exercises 66 and 67, use the following information.
Javier had $25.04 when he went to the mall. His friend Sally had $32.67. Javier wanted to buy a shirt for $27.89. *(Lesson 1-3)*

66. How much money did he have to borrow from Sally to buy the shirt? **$2.85**
67. How much money did that leave Sally? **$29.82**

Simplify each expression. *(Lessons 1-1 and 1-2)*

68. $3^2(2^2 - 1^2) + 4^2$ **43**
69. $3(5a + 6b) + 8(2a - b)$ **$31a + 10b$**

Getting Ready for the Next Lesson

PREREQUISITE SKILL Solve each equation. Check your solution.
*(To review **solving equations**, see Lesson 1-3.)*

70. $x + 3 = 2$ **-1**
71. $-4 + 2y = 0$ **2**
72. $0 = \frac{1}{2}x - 3$ **6**
73. $\frac{1}{3}x - 4 = 1$ **15**

2-2 Linear Equations

What You'll Learn

- Identify linear equations and functions.
- Write linear equations in standard form and graph them.

Vocabulary

- linear equation
- linear function
- standard form
- y-intercept
- x-intercept

How do linear equations relate to time spent studying?

Lolita has 4 hours after dinner to study and do homework. She has brought home math and chemistry. If she spends x hours on math and y hours on chemistry, a portion of the graph of the equation $x + y = 4$ can be used to relate how much time she spends on each.

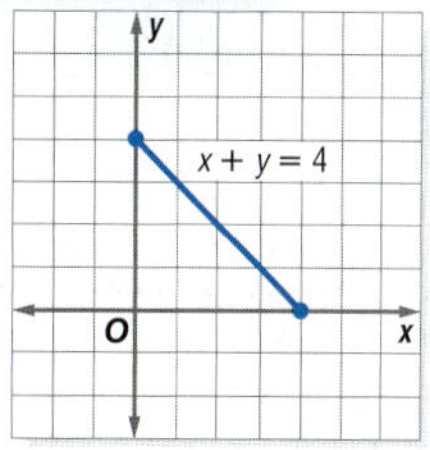

IDENTIFY LINEAR EQUATIONS AND FUNCTIONS An equation such as $x + y = 4$ is called a linear equation. A **linear equation** has no operations other than addition, subtraction, and multiplication of a variable by a constant. The variables may not be multiplied together or appear in a denominator. A linear equation does not contain variables with exponents other than 1. The graph of a linear equation is always a line.

Linear equations	Not linear equations
$5x - 3y = 7$	$7a + 4b^2 = -8$
$x = 9$	$y = \sqrt{x + 5}$
$6s = -3t - 15$	$x + xy = 1$
$y = \frac{1}{2}x$	$y = \frac{1}{x}$

TEACHING TIP
When variables other than x and y are used, the letter coming first in the alphabet usually represents the domain variable or horizontal coordinate.

A **linear function** is a function whose ordered pairs satisfy a linear equation. Any linear function can be written in the form $f(x) = mx + b$, where m and b are real numbers.

Example 1 Identify Linear Functions

State whether each function is a linear function. Explain.

a. $f(x) = 10 - 5x$ — This is a linear function because it can be written as $f(x) = -5x + 10$. $m = -5$, $b = 10$

b. $g(x) = x^4 - 5$ — This is not a linear function because x has an exponent other than 1.

c. $h(x, y) = 2xy$ — This is not a linear function because the two variables are multiplied together.

2-2 Lesson Notes

1 Focus

5-Minute Check Transparency 2-2 Use as a quiz or review of Lesson 2-1.

Mathematical Background notes are available for this lesson on p. 54C.

Building on Prior Knowledge

In Lesson 2-1, students graphed functions and equations by using a table of points. In this lesson, they generalize these skills to write equations in standard form before graphing them.

How do linear equations relate to time spent studying?

Ask students:

- What does a value of zero for x mean in this situation? **Lolita spends 0 hours studying math.**
- What does a negative value for x or for y mean in this situation? **No meaning; she cannot spend negative hours studying.**

Resource Manager

Workbook and Reproducible Masters

Chapter 2 Resource Masters

- Study Guide and Intervention, pp. 63–64
- Skills Practice, p. 65
- Practice, p. 66
- Reading to Learn Mathematics, p. 67
- Enrichment, p. 68
- Assessment, p. 113

Transparencies

5-Minute Check Transparency 2-2
Real-World Transparency 2
Answer Key Transparencies

Technology

Alge2PASS: Tutorial Plus, Lesson 3
Interactive Chalkboard

2 Teach

IDENTIFY LINEAR EQUATIONS AND FUNCTIONS

In-Class Examples

Power Point®

1 State whether each function is a linear function. Explain.

a. $g(x) = 2x - 5$ **yes; $m = 2$; $b = -5$**

b. $p(x) = x^3 + 2$ **No; x has an exponent other than 1.**

c. $t(x) = 4 + 7x$ **yes; $m = 7$; $b = 4$**

2 **METEOROLOGY** The linear function $f(C) = 1.8C + 32$ can be used to find the number of degrees Fahrenheit, f, that are equivalent to a given number of degrees Celsius, C.

a. On the Celsius scale, normal body temperature is 37°C. What is normal body temperature in degrees Fahrenheit? **98.6°F**

b. There are 100 Celsius degrees between the freezing and boiling points of water and 180 Fahrenheit degrees between these two points. How many Fahrenheit degrees equal 1 Celsius degree? **1.8°F = 1°C**

STANDARD FORM

In-Class Example

Power Point®

3 Write each equation in standard form. Identify A, B, and C.

a. $y = 3x - 9$ **$3x - y = 9$; $A = 3$, $B = -1$, $C = 9$**

b. $-\frac{2}{3}x = 2y - 1$ **$2x + 6y = 3$; $A = 2$, $B = 6$, $C = 3$**

c. $8x - 6y + 4 = 0$ **$4x - 3y = -2$; $A = 4$, $B = -3$, $C = -2$**

More About. . .

Military

To avoid decompression sickness, it is recommended that divers ascend no faster than 30 feet per minute.

Source: www.emedicine.com

Example 2 Evaluate a Linear Function

MILITARY **In August 2000, the Russian submarine *Kursk* sank to a depth of 350 feet in the Barents Sea. The linear function $P(d) = 62.5d + 2117$ can be used to find the pressure (lb/ft²) at a depth of d feet below the surface of the water.**

a. Find the pressure at a depth of 350 feet.

$P(d) = 62.5d + 2117$ Original function

$P(350) = 62.5(350) + 2117$ Substitute.

$= 23{,}992$ Simplify.

The pressure at a depth of 350 feet is about 24,000 lb/ft².

b. The term 2117 in the function represents the atmospheric pressure at the surface of the water. How many times as great is the pressure at a depth of 350 feet as the pressure at the surface?

Divide the pressure 350 feet below the surface by the pressure at the surface.

$\frac{23{,}992}{2117} \approx 11.33$ Use a calculator.

The pressure at that depth is more than 11 times as great as the pressure at the surface.

STANDARD FORM Any linear equation can be written in **standard form**, $Ax + By = C$, where A, B, and C are integers whose greatest common factor is 1.

> **Key Concept** — *Standard Form of a Linear Equation*
>
> The standard form of a linear equation is $Ax + By = C$, where $A \geq 0$, A and B are not both zero.

Example 3 Standard Form

Write each equation in standard form. Identify A, B, and C.

a. $y = -2x + 3$

$y = -2x + 3$ Original equation

$2x + y = 3$ Add $2x$ to each side.

So, $A = 2$, $B = 1$, and $C = 3$.

b. $-\frac{3}{5}x = 3y - 2$

$-\frac{3}{5}x = 3y - 2$ Original equation

$-\frac{3}{5}x - 3y = -2$ Subtract $3y$ from each side.

$3x + 15y = 10$ Multiply each side by -5 so that the coefficients are integers and $A \geq 0$.

So, $A = 3$, $B = 15$, and $C = 10$.

c. $3x - 6y - 9 = 0$

$3x - 6y - 9 = 0$ Original equation

$3x - 6y = 9$ Add 9 to each side.

$x - 2y = 3$ Divide each side by 3 so that the coefficients have a GCF of 1.

So, $A = 1$, $B = -2$, and $C = 3$.

Reading Tip Make sure that students understand the difference between the *x*- and *y*-intercepts. Some students may use the word *intersect* instead of the correct term *intercept*. Help them see that an intercept is the nonzero coordinate of the point where the graph intersects either axis.

Answer (page 65)

1. The function can be written as $f(x) = \frac{1}{2}x + 1$, so it is of the form $f(x) = mx + b$, where $m = \frac{1}{2}$ and $b = 1$.

Study Tip

Vertical and Horizontal Lines
An equation of the form $x = C$ represents a vertical line, which has only an x-intercept. $y = C$ represents a horizontal line, which has only a y-intercept.

TEACHING TIP
Some students may find it helpful to remember that if there is no y in the equation, the graph cannot cross the y-axis. Similarly, if there is no x in the equation, the graph cannot cross the x-axis.

In Lesson 2-1, you graphed an equation or function by making a table of values, graphing enough ordered pairs to see a pattern, and connecting the points with a line or smooth curve. Since two points determine a line, there are quicker ways to graph a linear equation or function. One way is to find the points at which the graph intersects each axis and connect them with a line. The y-coordinate of the point at which a graph crosses the y-axis is called the **y-intercept**. Likewise, the x-coordinate of the point at which it crosses the x-axis is the **x-intercept**.

Example 4 Use Intercepts to Graph a Line

Find the x-intercept and the y-intercept of the graph of $3x - 4y + 12 = 0$. Then graph the equation.

The x-intercept is the value of x when $y = 0$.

$3x - 4y + 12 = 0$	Original equation
$3x - 4(0) + 12 = 0$	Substitute 0 for y.
$3x = -12$	Subtract 12 from each side.
$x = -4$	Divide each side by 3.

The x-intercept is -4. The graph crosses the x-axis at $(-4, 0)$.

Likewise, the y-intercept is the value of y when $x = 0$.

$3x - 4y + 12 = 0$	Original equation
$3(0) - 4y + 12 = 0$	Substitute 0 for x.
$-4y = -12$	Subtract 12 from each side.
$y = 3$	Divide each side by -4.

The y-intercept is 3. The graph crosses the y-axis at $(0, 3)$.

Use these ordered pairs to graph the equation.

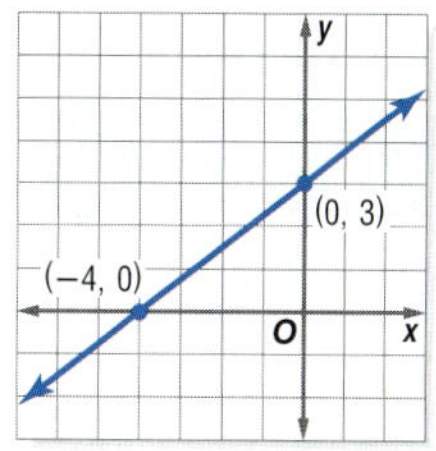

Check for Understanding

Concept Check

1. **Explain** why $f(x) = \frac{x+2}{2}$ is a linear function. **See margin.**
2. **Name** the x- and y-intercepts of the graph shown at the right. **5, −2**
3. **OPEN ENDED** Write an equation of a line with an x-intercept of 2. **Sample answer: $x + y = 2$**

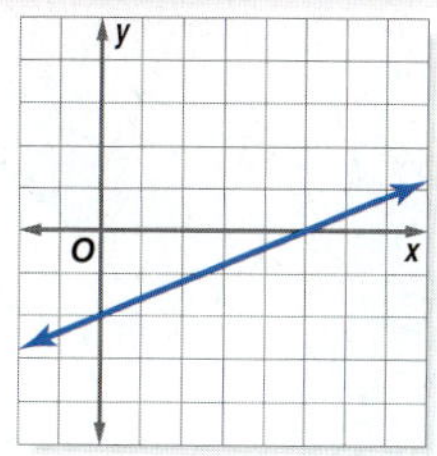

Guided Practice

GUIDED PRACTICE KEY

Exercises	Examples
4, 5	1
6–8	3
9–12	4
13, 14	2

State whether each equation or function is linear. Write *yes* or *no*. If no, explain your reasoning.

4. $x^2 + y^2 = 4$ **No, the variables have an exponent other than 1.**
5. $h(x) = 1.1 - 2x$ **yes**

Write each equation in standard form. Identify A, B, and C.

6. $y = 3x - 5$ **$3x - y = 5$; 3, −1, 5**
7. $4x = 10y + 6$ **$2x - 5y = 3$; 2, −5, 3**
8. $y = \frac{2}{3}x + 1$ **$2x - 3y = -3$; 2, −3, −3**

Find the x-intercept and the y-intercept of the graph of each equation. Then graph the equation. **9–12. See pp. 107A–107H for graphs.**

9. $y = -3x - 5$ **$-\frac{5}{3}$, −5**
10. $x - y - 2 = 0$ **2, −2**
11. $3x + 2y = 6$ **2, 3**
12. $4x + 8y = 12$ **3, $\frac{3}{2}$**

www.algebra2.com/extra_examples

DAILY INTERVENTION

Differentiated Instruction

Visual/Spatial Encourage students to relate the intercepts and the graph of the equation to the standard form of the equation, $Ax + By = C$. Point out that the ratio $-\frac{A}{B}$ is equivalent to the ratio of the graph's intercepts: $-\frac{y\text{-intercept}}{x\text{-intercept}}$. These ratios can be visualized on the graph by counting grid squares vertically and horizontally from one intercept to the other. Recognition of the ratios leads to the discussion of slope in Lesson 2-3.

In-Class Example Power Point®

4 Find the x-intercept and the y-intercept of the graph of $-2x + y - 4 = 0$. Then graph the equation. **x-intercept: −2; y-intercept: 4**

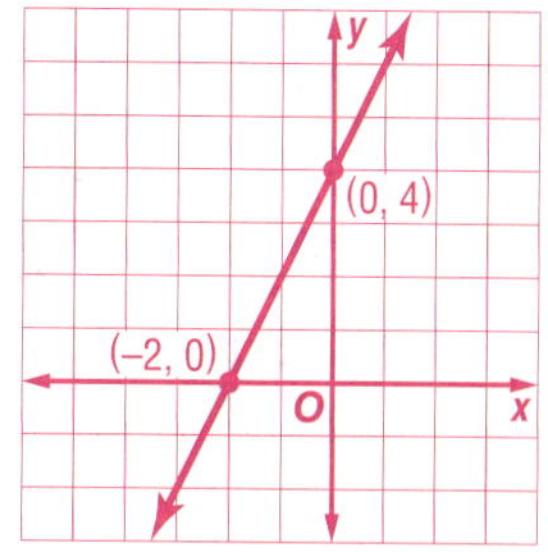

3 Practice/Apply

Study Notebook

Have students—
- add the definitions/examples of the vocabulary terms to their Vocabulary Builder worksheets for Chapter 2.
- include any other items(s) that students find helpful in mastering the skills in the lesson.

About the Exercises...

Organization by Objective
- **Identify Linear Equations and Functions:** 15–26
- **Standard Form:** 27–52

Odd/Even Assignments
Exercises 15–24 and 27–50 are structured so that students practice the same concepts whether they are assigned odd or even problems.

Assignment Guide

Basic: 15–23 odd, 25, 26, 27–35 odd, 39–47 odd, 51–55, 61–78

Average: 15–23 odd, 25, 26, 27–49 odd, 51–55, 61–78

Advanced: 16–24 even, 28–50 even, 51–70 (optional: 71–78)

Study Guide and Intervention, p. 63 (shown) and p. 64

Identify Linear Equations and Functions A **linear equation** has no operations other than addition, subtraction, and multiplication of a variable by a constant. The variables may not be multiplied together or appear in a denominator. A linear equation does not contain variables with exponents other than 1. The graph of a linear equation is a line.

A **linear function** is a function whose ordered pairs satisfy a linear equation. Any linear function can be written in the form $f(x) = mx + b$, where m and b are real numbers.

If an equation is linear, you need only two points that satisfy the equation in order to graph the equation. One way is to find the x-intercept and the y-intercept and connect these two points with a line.

Example 1 **Is $f(x) = 0.2 - \frac{x}{5}$ a linear function? Explain.**

Yes; it is a linear function because it can be written in the form $f(x) = -\frac{1}{5}x + 0.2$.

Example 2 **Is $2x + xy - 3y = 0$ a linear function? Explain.**

No; it is not a linear function because the variables x and y are multiplied together in the middle term.

Example 3 **Find the x-intercept and the y-intercept of the graph of $4x - 5y = 20$. Then graph the equation.**

The x-intercept is the value of x when $y = 0$.

$4x - 5y = 20$ Original equation
$4x - 5(0) = 20$ Substitute 0 for y.
$x = 5$ Simplify.

So the x-intercept is 5. Similarly, the y-intercept is -4.

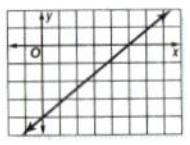

Exercises

State whether each equation or function is linear. Write *yes* or *no*. If no, explain.

1. $6y - x = 7$ **yes**
2. $9x = \frac{18}{y}$ **No; the variable y appears in the denominator.**
3. $f(x) = 2 - \frac{x}{11}$ **yes**

Find the x-intercept and the y-intercept of the graph of each equation. Then graph the equation.

4. $2x + 7y = 14$ **x-int: 7; y-int: 2**

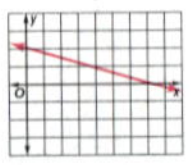

5. $5y - x = 10$ **x-int: −10; y-int: 2**

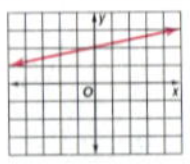

6. $2.5x - 5y + 7.5 = 0$ **x-int: −3; y-int: 1.5**

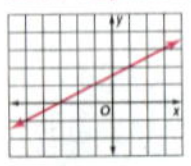

Skills Practice, p. 65 and Practice, p. 66 (shown)

State whether each equation or function is linear. Write *yes* or *no*. If no, explain your reasoning.

1. $h(x) = 23$ **yes**
2. $y = \frac{2}{3}x$ **yes**
3. $y = \frac{5}{x}$ **No; x is a denominator.**
4. $9 - 5xy = 2$ **No; x and y are multiplied.**

Write each equation in standard form. Identify A, B, and C.

5. $y = 7x - 5$ **$7x - y = 5$; 7, −1, 5**
6. $y = \frac{3}{8}x + 5$ **$3x - 8y = -40$; 3, −8, −40**
7. $3y - 5 = 0$ **$3y = 5$; 0, 3, 5**
8. $x = -\frac{2}{7}y + \frac{3}{4}$ **$28x + 8y = 21$; 28, 8, 21**

Find the x-intercept and the y-intercept of the graph of each equation. Then graph the equation.

9. $y = 2x + 4$ **−2, 4**

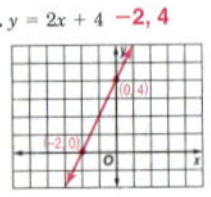

10. $2x + 7y = 14$ **7, 2**

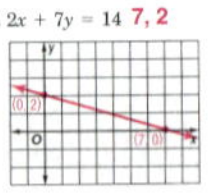

11. $y = -2x - 4$ **−2, −4**

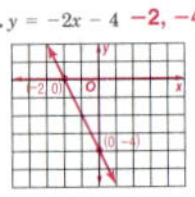

12. $6x + 2y = 6$ **1, 3**

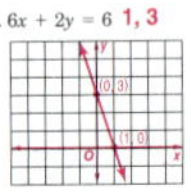

13. MEASURE The equation $y = 2.54x$ gives the length in centimeters corresponding to a length x in inches. What is the length in centimeters of a 1-foot ruler? **30.48 cm**

LONG DISTANCE **For Exercises 14 and 15, use the following information.**

For Meg's long-distance calling plan, the monthly cost C in dollars is given by the linear function $C(t) = 6 + 0.05t$, where t is the number of minutes talked.

14. What is the total cost of talking 8 hours? of talking 20 hours? **\$30; \$66**
15. What is the effective cost per minute (the total cost divided by the number of minutes talked) of talking 8 hours? of talking 20 hours? **\$0.0625; \$0.055**

Reading to Learn Mathematics, p. 67 ELL

Pre-Activity **How do linear equations relate to time spent studying?**

Read the introduction to Lesson 2-2 at the top of page 63 in your textbook.

- If Lolita spends $2\frac{1}{2}$ hours studying math, how many hours will she have to study chemistry? **$1\frac{1}{2}$ hours**
- Suppose that Lolita decides to stay up one hour later so that she now has 5 hours to study and do homework. Write a linear equation that describes this situation. **$x + y = 5$**

Reading the Lesson

1. Write *yes* or *no* to tell whether each linear equation is in standard form. If it is not, explain why it is not.
   a. $-x + 2y = 5$ **No; A is negative.**
   b. $9x - 12y = -5$ **yes**
   c. $5x - 7y = 3$ **yes**
   d. $2x - \frac{4}{7}y = 1$ **No; B is not an integer.**
   e. $0x + 0y = 0$ **No; A and B are both 0.**
   f. $2x + 4y = 8$ **No; The greatest common factor of 2, 4, and 8 is 2, not 1.**
2. How can you use the standard form of a linear equation to tell whether the graph is a horizontal line or a vertical line? **If $A = 0$, then the graph is a horizontal line. If $B = 0$, then the graph is a vertical line.**

Helping You Remember

3. One way to remember something is to explain it to another person. Suppose that you are studying this lesson with a friend who thinks that she should let $x = 0$ to find the x-intercept and let $y = 0$ to find the y-intercept. How would you explain to her how to remember the correct way to find intercepts of a line? **Sample answer: The x-intercept is the x-coordinate of a point on the x-axis. Every point on the x-axis has y-coordinate 0, so let $y = 0$ to find an x-intercept. The y-intercept is the y-coordinate of a point on the y-axis. Every point on the y-axis has x-coordinate 0, so let $x = 0$ to find a y-intercept.**

Application **ECONOMICS For Exercises 13 and 14, use the following information.**
On January 1, 1999, the euro became legal tender in 11 participating countries in Europe. Based on the exchange rate on March 22, 2001, the linear function $d(x) = 0.8881x$ could be used to convert x euros to U.S. dollars.

13. On that date, what was the value in U.S. dollars of 200 euros? **\$177.62**
14. On that date, what was the value in euros of 500 U.S. dollars? **563.00 euros**

Online Research **Data Update** How do the dollar and the euro compare today? Visit www.algebra2.com/data_update to convert among currencies.

★ indicates increased difficulty

Practice and Apply

Homework Help	
For Exercises	**See Examples**
15–24	1
25, 26	2
27–38	3
39–52	4
53–60	2, 4

Extra Practice See page 830.

State whether each equation or function is linear. Write *yes* or *no*. If no, explain your reasoning. **16–19, 21–22. See margin for explanations.**

15. $x + y = 5$ **yes**
16. $\frac{1}{x} + 3y = -5$ **no**
17. $x + \sqrt{y} = 4$ **no**
18. $h(x) = 2x^3 - 4x^2 + 5$ **no**
19. $g(x) = 10 + \frac{2}{x^2}$ **no**
20. $f(x) = 6x - 19$ **yes**
21. $f(x) = 7x^5 + x - 1$ **no**
22. $y = \sqrt{2x - 5}$ **no**
23. Which of the equations $x + 9y = 7$, $x^2 + 5y = 0$, and $y = 3x - 1$ is not linear? **23. $x^2 + 5y = 0$**
24. Which of the functions $f(x) = 2x + 4$, $g(x) = 7$, and $h(x) = x^3 - x^2 + 3x$ is not linear? **$h(x) = x^3 - x^2 + 3x$**

PHYSICS For Exercises 25 and 26, use the following information.
When a sound travels through water, the distance y in meters that the sound travels in x seconds is given by the equation $y = 1440x$.

25. How far does a sound travel underwater in 5 seconds? **7200 m**
26. In air, the equation is $y = 343x$. Does sound travel faster in air or water? Explain. **Sound travels only 1715 m in 5 seconds in air, so it travels faster underwater.**

Write each equation in standard form. Identify A, B, and C.

27. $y = -3x + 4$ **27. $3x + y = 4$; 3, 1, 4**
28. $y = 12x$ **28. $12x - y = 0$; 12, −1, 0**
29. $x = 4y - 5$ **29. $x - 4y = -5$; 1, −4, −5**
30. $x = 7y + 2$ **30. $x - 7y = 2$; 1, −7, 2**
31. $5y = 10x - 25$ **31. $2x - y = 5$; 2, −1, 5**
32. $4x = 8y - 12$ **32. $x - 2y = -3$; 1, −2, −3**
33. $\frac{1}{2}x + \frac{1}{2}y = 6$ **33. $x + y = 12$; 1, 1, 12**
34. $\frac{1}{3}x - \frac{1}{3}y = -2$ **34. $x - y = -6$; 1, −1, −6**
35. $0.5x = 3$ **$x = 6$; 1, 0, 6**
36. $0.25y = 10$ **$y = 40$; 0, 1, 40**
★37. $\frac{5}{6}x + \frac{1}{15}y = \frac{3}{10}$ **$25x + 2y = 9$; 25, 2, 9**
★38. $0.25x = 0.1 + 0.2y$ **$5x - 4y = 2$; 5, −4, 2**

Find the x-intercept and the y-intercept of the graph of each equation. Then graph the equation. **39–50. See pp. 107A–107H for graphs.**

39. $5x + 3y = 15$ **3, 5**
40. $2x - 6y = 12$ **6, −2**
41. $3x - 4y - 10 = 0$ **41. $\frac{10}{3}$, $-\frac{5}{2}$**
42. $2x + 5y - 10 = 0$ **5, 2**
43. $y = x$ **0, 0**
44. $y = 4x - 2$ **$\frac{1}{2}$, −2**
45. $y = -2$ **none, −2**
46. $y = 4$ **none, 4**
47. $x = 8$ **8, none**
48. $x = 1$ **1, none**
★49. $f(x) = 4x - 1$ **$\frac{1}{4}$, −1**
★50. $g(x) = 0.5x - 3$ **6, −3**

CRITICAL THINKING For Exercises 51 and 52, use $x + y = 0$, $x + y = 5$, and $x + y = -5$.

51. Graph the equations on a coordinate plane. Compare and contrast the graphs. **51. See margin for graph. The lines are parallel but have different y-intercepts.**
52. Write a linear equation whose graph is between the graphs of $x + y = 0$ and $x + y = 5$. **Sample answer: $x + y = 2$**

Enrichment, p. 68

Greatest Common Factor

Suppose we are given a linear equation $ax + by = c$ where a, b, and c are nonzero integers, and we want to know if there exist *integers* x and y that satisfy the equation. We could try guessing a few times, but this process would be time consuming for an equation such as $588x + 432y = 72$. By using the Euclidean Algorithm, we can determine not only if such integers x and y exist, but also find them. The following example shows how this algorithm works.

Example **Find integers x and y that satisfy $588x + 432y = 72$.**

Divide the greater of the two coefficients by the lesser to get a quotient and remainder. Then, repeat the process by dividing the divisor by the remainder until you get a remainder of 0. The process can be written as follows.

$588 = 432(1) + 156$ (1)
$432 = 156(2) + 120$ (2)
$156 = 120(1) + 36$ (3)
$120 = 36(3) + 12$ (4)

Answers

16. No; x appears in a denominator.
17. No; y is inside a square root.
18. No; x has exponents other than 1.
19. No; x appears in a denominator.
20. Yes; this is a linear function because it is written as $f(x) = 6x - 19$, $m = 6$, $b = -19$.
21. No; x has an exponent other than 1.
22. No; x is inside a square root.

More About. . .

Geology

Geothermal energy from hot springs is being used for electricity in California, Italy, and Iceland.

GEOLOGY **For Exercises 53–55, use the following information.**
Suppose the temperature T (°C) below Earth's surface is given by $T(d) = 35d + 20$, where d is the depth (km).

53. Find the temperature at a depth of 2 kilometers. **90°C**
54. Find the depth if the temperature is 160°C. **4 km**
55. Graph the linear function. **See margin.**

FUND-RAISING **For Exercises 56–59, use the following information.**
The Jackson Band Boosters sell beverages for $1.75 and candy for $1.50 at home games. Their goal is to have total sales of $525 for each game.

56. Write an equation that is a model for the different numbers of beverages and candy that can be sold to meet the goal. **$1.75b + 1.5c = 525$**
57. Graph the equation. **See margin.**
58. Does this equation represent a function? Explain.
59. If they sell 100 beverages and 200 pieces of candy, will the Band Boosters meet their goal? **no**

58. Yes; the graph passes the vertical line test.

60. **GEOMETRY** Find the area of the shaded region in the graph. (*Hint:* The area of a trapezoid is given by $A = \frac{1}{2}h(b_1 + b_2)$.) **$\frac{21}{2}$ units^2**

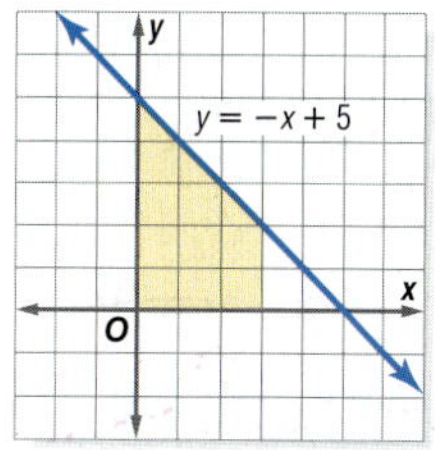

61. **WRITING IN MATH** Answer the question that was posed at the beginning of the lesson.
How do linear equations relate to time spent studying?
Include the following in your answer: **See margin.**
- why only the part of the graph in the first quadrant is shown, and
- an interpretation of the graph's intercepts in terms of the amount of time Lolita spends on each subject.

Standardized Test Practice

62. Which function is linear? **B**
(A) $f(x) = x^2$ (B) $g(x) = 2.7$ (C) $g(x) = \sqrt{x - 1}$ (D) $f(x) = \sqrt{9 - x^2}$

63. What is the y-intercept of the graph of $10 - x = 2y$? **B**
(A) 2 (B) 5 (C) 6 (D) 10

Maintain Your Skills

Mixed Review

State the domain and range of each relation. Then graph the relation and determine whether it is a function. *(Lesson 2-1)* **64–65. See pp. 107A–107H for graphs.**

64. {(−1, 5), (1, 3), (2, −4), (4, 3)}
65. {(0, 2), (1, 3), (2, −1), (1, 0)}

64. D = {−1, 1, 2, 4}, R = {−4, 3, 5}; yes
65. D = {0, 1, 2}, R = {−1, 0, 2, 3}; no

Solve each inequality. *(Lesson 1-6)*

66. $-2 < 3x + 1 < 7$ **$\{x \mid -1 < x < 2\}$**
67. $|x + 4| > 2$ **$\{x \mid x < -6 \text{ or } x > -2\}$**

68. **TAX** Including a 6% sales tax, a paperback book costs $8.43. What is the price before tax? *(Lesson 1-3)* **$7.95**

Simplify each expression. *(Lesson 1-1)*

69. $(9s - 4) - 3(2s - 6)$ **$3s + 14$**
70. $[19 - (8 - 1)] \div 3$ **4**

Getting Ready for the Next Lesson

BASIC SKILL **Find the reciprocal of each number.**

71. 3 **$\frac{1}{3}$**
72. −4 **$-\frac{1}{4}$**
73. $\frac{1}{2}$ **2**
74. $-\frac{2}{3}$ **$-\frac{3}{2}$**
75. $-\frac{1}{5}$ **−5**
76. $3\frac{3}{4}$ **$\frac{4}{15}$**
77. 2.5 **0.4**
78. −1.25 **−0.8**

www.algebra2.com/self_check_quiz

4 Assess

Open-Ended Assessment

Modeling Have students place a piece of spaghetti or a pencil on a large coordinate plane to model the graphs of these equations: $x = 4$, $x = -2$, $y = 0$, $y = -3$, $x = y$, and $x = -y$.

Getting Ready for Lesson 2-3

BASIC SKILL Lesson 2-3 presents the fact that perpendicular lines have slopes that are negative reciprocals. Exercises 71–78 should be used to determine your students' familiarity with finding reciprocals.

Assessment Options

Quiz (Lessons 2-1 and 2-2) is available on p. 113 of the *Chapter 2 Resource Masters.*

Answers

51.

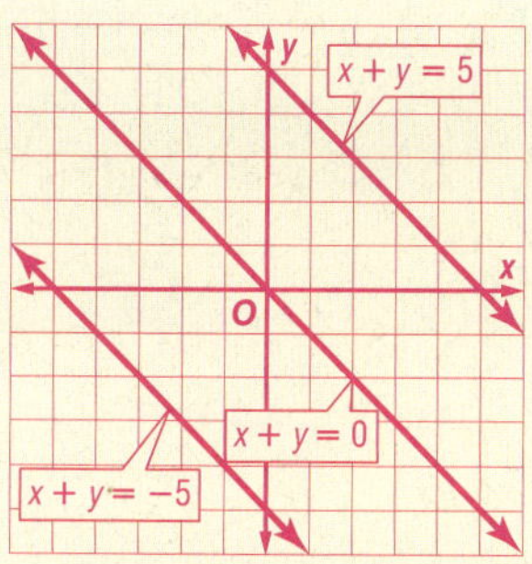

55.

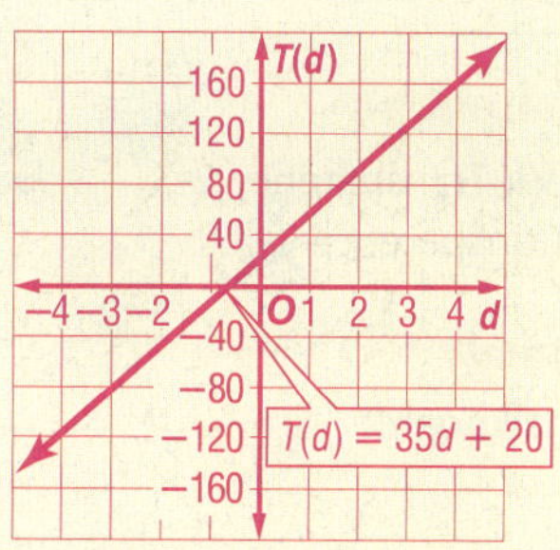

57.

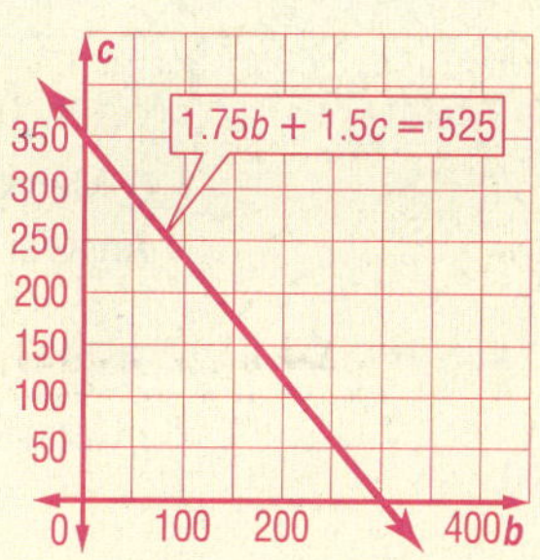

61. A linear equation can be used to relate the amounts of time that a student spends on each of two subjects if the total amount of time is fixed. Answers should include the following.

- **x and y must be nonnegative because Lolita cannot spend a negative amount of time studying a subject.**
- **The intercepts represent Lolita spending all of her time on one subject. The x-intercept represents her spending all of her time on math, and the y-intercept represents her spending all of her time on chemistry.**

2-3 Lesson Notes

1 Focus

5-Minute Check Transparency 2-3 Use as a quiz or review of Lesson 2-2.

Mathematical Background notes are available for this lesson on p. 54C.

Building on Prior Knowledge

In Lesson 2-2, students wrote linear equations in standard form and graphed them using intercepts. In this lesson, they apply this skill to finding the slope of a line and to graphing parallel and perpendicular lines.

How does slope apply to the steepness of roads?

Ask students:

- An engineer designed a road with a rise of 4 feet for each horizontal distance of 100 feet. What is the grade of this road? **4%**
- If a road has a grade of 3%, what is its rise for each horizontal distance of 50 feet? **1.5 ft**

2-3 Slope

What You'll Learn

- Find and use the slope of a line.
- Graph parallel and perpendicular lines.

How does slope apply to the steepness of roads?

The grade of a road is a percent that measures the steepness of the road. It is found by dividing the amount the road rises by the corresponding horizontal distance.

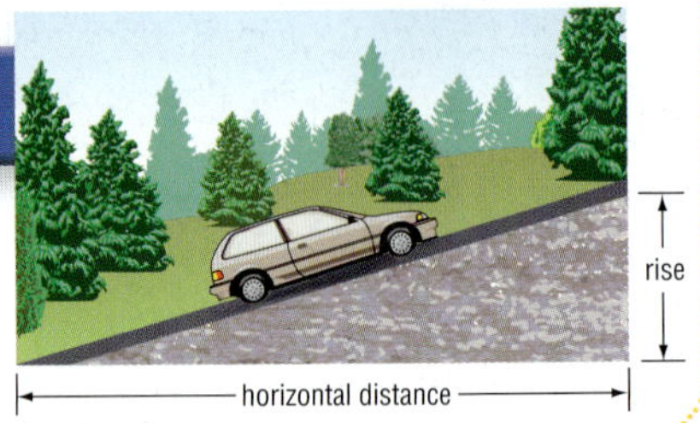

Vocabulary

- slope
- rate of change
- family of graphs
- parent graph
- oblique

SLOPE The **slope** of a line is the ratio of the change in y-coordinates to the corresponding change in x-coordinates. The slope measures how steep a line is. Suppose a line passes through points at (x_1, y_1) and (x_2, y_2).

$$\text{slope} = \frac{\text{change in } y\text{-coordinates}}{\text{change in } x\text{-coordinates}}$$

$$= \frac{y_2 - y_1}{x_2 - x_1}$$

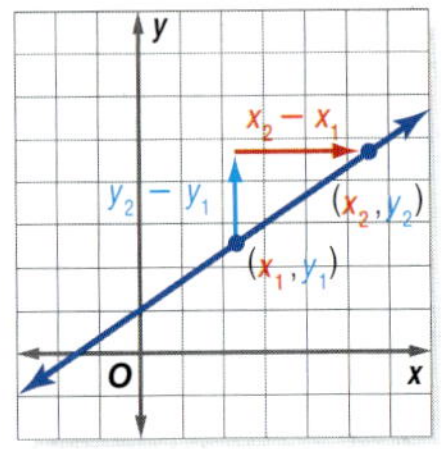

The slope of a line is the same, no matter what two points on the line are used.

Study Tip

Slope

The formula for slope is often remembered as *rise over run*, where the rise is the difference in y-coordinates and the run is the difference in x-coordinates.

Key Concept — *Slope of a Line*

- **Words** The slope of a line is the ratio of the change in y-coordinates to the change in x-coordinates.
- **Symbols** The slope m of the line passing through (x_1, y_1) and (x_2, y_2) is given by $m = \frac{y_2 - y_1}{x_2 - x_1}$, where $x_1 \neq x_2$.

Example 1 *Find Slope*

Find the slope of the line that passes through (−1, 4) and (1, −2). Then graph the line.

$m = \frac{y_2 - y_1}{x_2 - x_1}$ Slope formula

$= \frac{-2 - 4}{1 - (-1)}$ $(x_1, y_1) = (-1, 4), (x_2, y_2) = (1, -2)$

$= \frac{-6}{2}$ or -3 Simplify.

The slope of the line is −3.

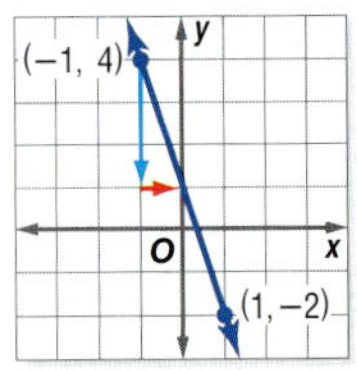

TEACHING TIP

Make sure students realize that y_2 must come from the same ordered pair as x_2.

Graph the two ordered pairs and draw the line. Use the slope to check your graph by selecting any point on the line. Then go down 3 units and right 1 unit or go up 3 units and left 1 unit. This point should also be on the line.

Resource Manager

Workbook and Reproducible Masters

Chapter 2 Resource Masters

- Study Guide and Intervention, pp. 69–70
- Skills Practice, p. 71
- Practice, p. 72
- Reading to Learn Mathematics, p. 73
- Enrichment, p. 74

Transparencies

5-Minute Check Transparency 2-3
Answer Key Transparencies

Technology

Interactive Chalkboard

Example 2 Use Slope to Graph a Line

Graph the line passing through $(-4, -3)$ with a slope of $\frac{2}{3}$.

Graph the ordered pair $(-4, -3)$. Then, according to the slope, go up 2 units and right 3 units. Plot the new point at $(-1, -1)$. *You can also go right 3 units and then up 2 units to plot the new point.*

Draw the line containing the points.

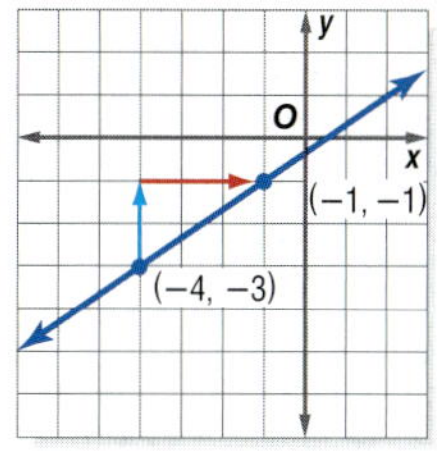

The slope of a line tells the direction in which it rises or falls.

Concept Summary

If the line rises to the right, then the slope is *positive*.	If the line is horizontal, then the slope is *zero*.	If the line falls to the right, then the slope is *negative*.	If the line is vertical, then the slope is *undefined*.
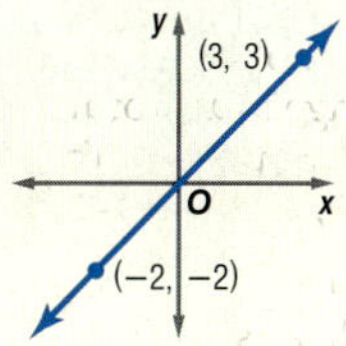	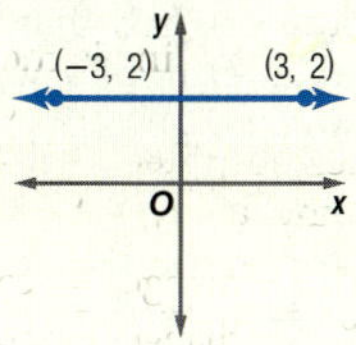	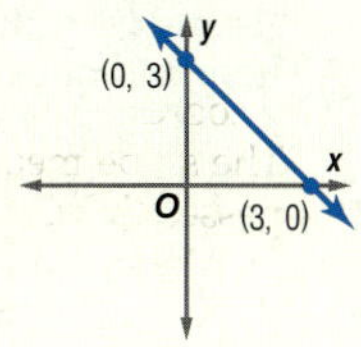	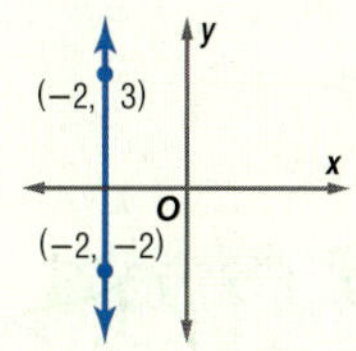
$m = \frac{3-(-2)}{3-(-2)}$ $= 1$	$m = \frac{2-2}{3-(-3)}$ $= 0$	$m = \frac{0-3}{3-0}$ $= -1$	$x_1 = x_2$, so m is undefined.

Slope is often referred to as **rate of change**. It measures how much a quantity changes, on average, relative to the change in another quantity, often time.

Example 3 Rate of Change

TRAVEL Refer to the graph at the right. Find the rate of change of the number of people taking cruises from 1985 to 2000.

$m = \frac{y_2 - y_1}{x_2 - x_1}$ — Slope formula

$= \frac{6.9 - 2.2}{2000 - 1985}$ — Substitute.

≈ 0.31 — Simplify.

Between 1985 and 2000, the number of people taking cruises increased at an average rate of about 0.31(1,000,000) or 310,000 people per year.

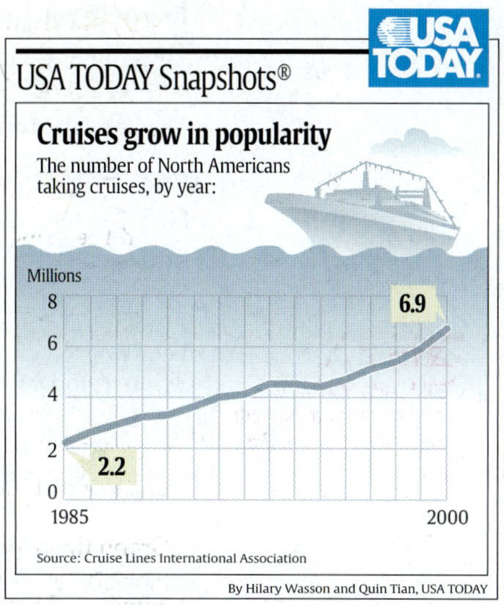

Log on for:
- Updated data
- More activities on rate of change

www.algebra2.com/usa_today

www.algebra2.com/extra_examples

2 Teach

SLOPE

In-Class Examples

Power Point®

1 Find the slope of the line that passes through (1, 3) and (−2, −3). Then graph the line. **2**

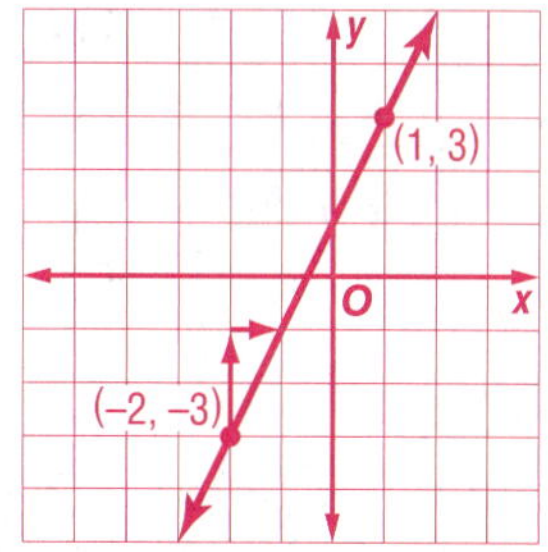

2 Graph the line passing through (1, −3) with a slope of $-\frac{3}{4}$.

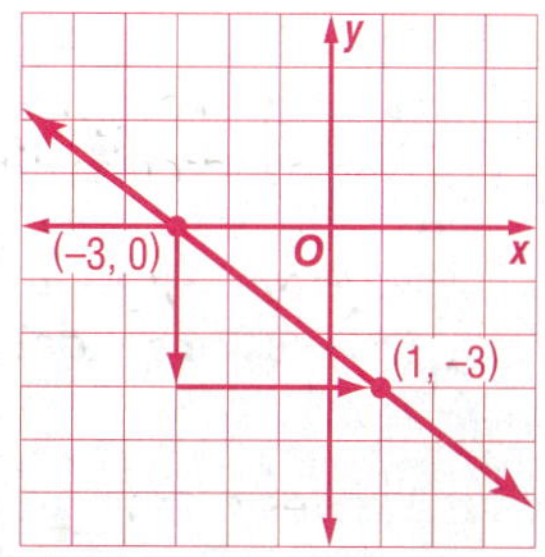

3 **COMMUNICATION** Refer to the graph below. Find the rate of change of the number of radio stations on the air in the United States from 1990 to 1998.

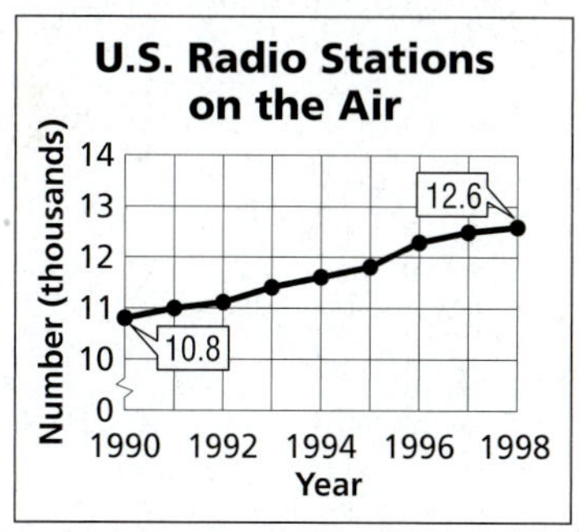

Source: *The New York Times Almanac*

Between 1990 and 1998, the number of radio stations on the air in the United States increased at an average rate of 0.225(1000) or 225 stations per year.

Teacher to Teacher

Judy Buchholtz — Dublin Scioto H.S., Dublin, OH

"A match it, graph it CBL-motion detector lab is a fun way for students to 'experience' slope."

PARALLEL AND PERPENDICULAR LINES

In-Class Example

Power Point®

4 Graph the line through $(1, -2)$ that is parallel to the line with equation $x - y = -2$.

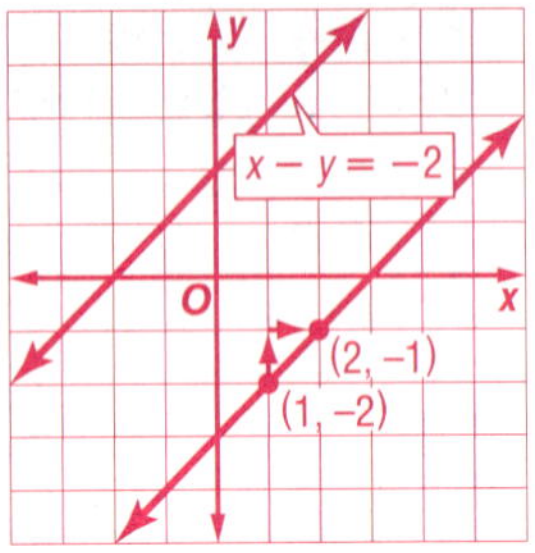

Answer

1. $y = 3x$; The graphs are parallel lines, but they have different y-intercepts.

PARALLEL AND PERPENDICULAR LINES A **family of graphs** is a group of graphs that displays one or more similar characteristics. The **parent graph** is the simplest of the graphs in a family. A graphing calculator can be used to graph several graphs in a family on the same screen.

Graphing Calculator Investigation

Lines with the Same Slope

The calculator screen shows the graphs of $y = 3x$, $y = 3x + 2$, $y = 3x - 2$, and $y = 3x + 5$.

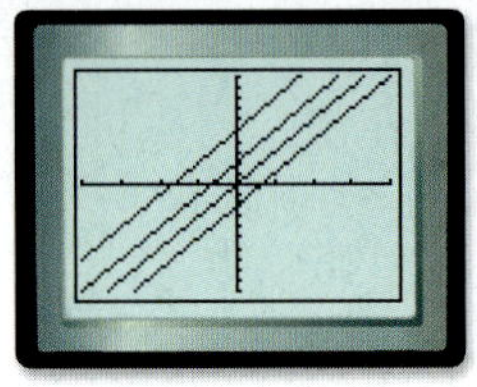

$[-4, 4]$ scl: 1 by $[-10, 10]$ scl: 1

Think and Discuss 1. See margin.

1. Identify the parent function and describe the family of graphs. What is similar about the graphs? What is different about the graphs?
2. Find the slope of each line. 3
3. Write another function that has the same characteristics as this family of graphs. Check by graphing. Sample answer: $y = 3x - 4$

In the Investigation, you saw that lines that have the same slope are parallel. These and other similar examples suggest the following rule.

Study Tip

Horizontal Lines

All horizontal lines are parallel because they all have a slope of 0.

Key Concept — Parallel Lines

- **Words** In a plane, nonvertical lines with the same slope are parallel. All vertical lines are parallel.
- **Model**

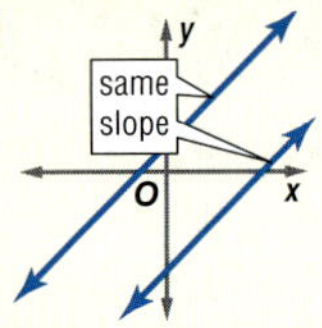

Example 4 Parallel Lines

Graph the line through $(-1, 3)$ that is parallel to the line with equation $x + 4y = -4$.

The x-intercept is -4, and the y-intercept is -1. Use the intercepts to graph $x + 4y = -4$.

The line falls 1 unit for every 4 units it moves to the right, so the slope is $-\frac{1}{4}$.

Now use the slope and the point at $(-1, 3)$ to graph the line parallel to the graph of $x + 4y = -4$.

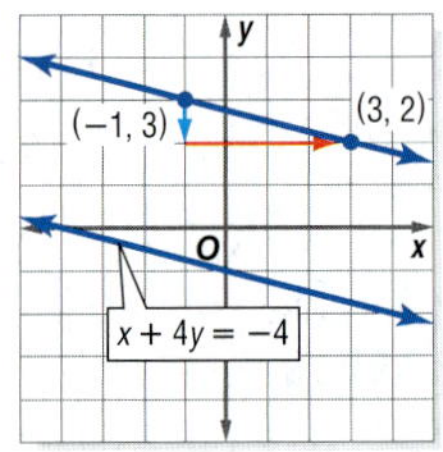

The figure at the right shows the graphs of two lines that are perpendicular. You know that parallel lines have the same slope. What is the relationship between the slopes of two perpendicular lines?

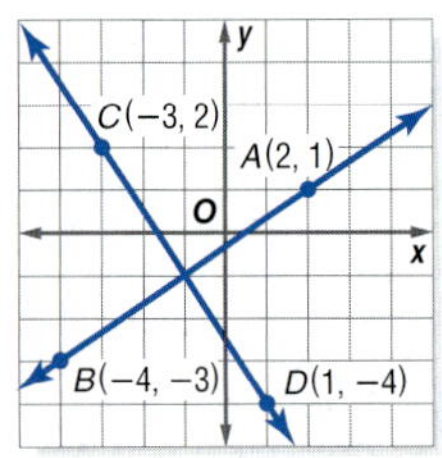

slope of line AB	slope of line CD
$\frac{-3-1}{-4-2} = \frac{-4}{-6}$ or $\frac{2}{3}$	$\frac{-4-2}{1-(-3)} = \frac{-6}{4}$ or $-\frac{3}{2}$

The slopes are opposite reciprocals of each other. This relationship is true in general. When you multiply the slopes of two perpendicular lines, the product is always -1.

Graphing Calculator Investigation

Lines with the Same Slope Point out that the simplest of the graphs in a family is often the one that passes through the origin, where the values of x and y are both zero. Suggest that students substitute 0 for x in each equation to find a point that will help them identify which graph goes with each equation.

Study Tip

Reading Math
An *oblique* line is a line that is neither horizontal nor vertical.

Key Concept — Perpendicular Lines

- **Words** In a plane, two oblique lines are perpendicular if and only if the product of their slopes is -1.
- **Symbols** Suppose m_1 and m_2 are the slopes of two oblique lines. Then the lines are perpendicular if and only if
$m_1 m_2 = -1$, or $m_1 = -\frac{1}{m_2}$.
- **Model**

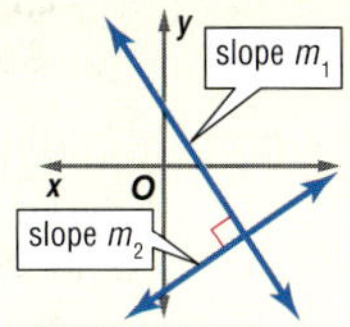

Any vertical line is perpendicular to any horizontal line.

Example 5 Perpendicular Line

Graph the line through $(-3, 1)$ that is perpendicular to the line with equation $2x + 5y = 10$.

The x-intercept is 5, and the y-intercept is 2. Use the intercepts to graph $2x + 5y = 10$.

The line falls 2 units for every 5 units it moves to the right, so the slope is $-\frac{2}{5}$. The slope of the perpendicular line is the opposite reciprocal of $-\frac{2}{5}$, or $\frac{5}{2}$.

Start at $(-3, 1)$ and go up 5 units and right 2 units. Use this point and $(-3, 1)$ to graph the line.

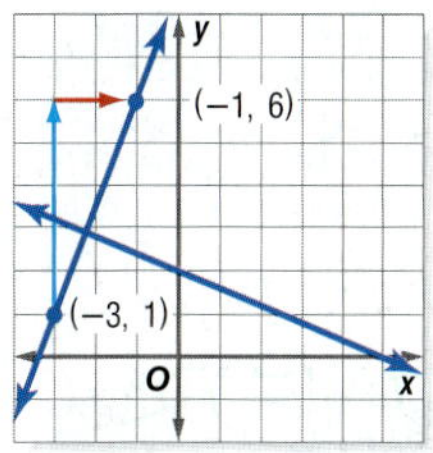

Check for Understanding

Concept Check

2. Sometimes; the slope of a vertical line is undefined.

1. **OPEN ENDED** Write an equation of a line with slope 0. **Sample answer: $y = 1$**
2. **Decide** whether the statement below is *sometimes*, *always*, or *never* true. Explain.
The slope of a line is a real number.
3. **FIND THE ERROR** Mark and Luisa are finding the slope of the line through (2, 4) and (−1, 5). Who is correct? Explain your reasoning. **See margin.**

Mark	Luisa
$m = \frac{5-4}{2-(-1)}$ or $\frac{1}{3}$	$m = \frac{4-5}{2-(-1)}$ or $-\frac{1}{3}$

Guided Practice

GUIDED PRACTICE KEY	
Exercises	Examples
4–6	1
7, 8	2
9	4
10, 11	5
12–14	3

Find the slope of the line that passes through each pair of points.

4. (1, 1), (3, 1) **0**
5. (−1, 0), (3, −2) $-\frac{1}{2}$
6. (3, 4), (1, 2) **1**

Graph the line passing through the given point with the given slope.

7. (2, −1), −3
8. (−3, −4), $\frac{3}{2}$ **7–8. See pp. 107A–107H.**

Graph the line that satisfies each set of conditions. **9–11. See pp. 107A–107H.**

9. passes through (0, 3), parallel to graph of $6y - 10x = 30$
10. passes through (4, −2), perpendicular to graph of $3x - 2y = 6$
11. passes through (−1, 5), perpendicular to graph of $5x - 3y - 3 = 0$

DAILY INTERVENTION — **Differentiated Instruction**

Naturalist Ask students to apply what they have learned about slope, grade, or steepness as they describe or sketch what they see around them in nature, such as the shapes of hills, branches of trees, and roof lines of houses.

In-Class Example

5 Graph the line through (2, 1) that is perpendicular to the line with equation $2x - 3y = 3$.

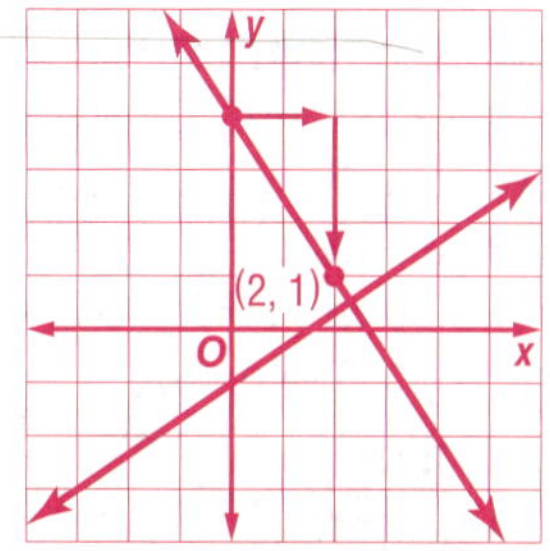

3 Practice/Apply

Study Notebook

Have students—
- add the definitions/examples of the vocabulary terms to their Vocabulary Builder worksheets for Chapter 2.
- draw sample graphs to compare and contrast lines with positive slope, negative slope, zero slope, and a slope that is undefined.
- include any other item(s) that they find helpful in mastering the skills in this lesson.

DAILY INTERVENTION

FIND THE ERROR Point out that when finding a slope, if you use $y_1 - y_2$ as the numerator, you must use $x_1 - x_2$ as the denominator. To find the slope, you can move from point *A* to point *B*, or from point *B* to point *A*, but you must move in a consistent direction for both the rise and the run.

Answer

3. Luisa; Mark did not subtract in a consistent manner when using the slope formula. If $y_2 = 5$ and $y_1 = 4$, then x_2 must be -1 and x_1 must be 2, not vice versa.

About the Exercises...

Organization by Objective

- **Slope:** 15–42
- **Parallel and Perpendicular Lines:** 43–52

Odd/Even Assignments

Exercises 15–36 and 43–50 are structured so that students practice the same concepts whether they are assigned odd or even problems.

Alert! Exercises 56 and 57 require a graphing calculator.

Assignment Guide

Basic: 15–21 odd, 31–35 odd, 37–39, 43–47 odd, 51–55, 58–75

Average: 15–35 odd, 37–39, 43–51 odd, 52–55, 58–75 (optional: 56, 57)

Advanced: 16–36 even, 40–42, 44–50 even, 52–69 (optional: 70–75)

All: Practice Quiz 1 (1–5)

Application **WEATHER** **For Exercises 12–14, use the table that shows the temperatures at different times on March 23, 2002.**

Time	8:00 A.M.	10:00 A.M.	12:00 P.M.	2:00 P.M.	4:00 P.M.
Temp (°F)	36	47	55	58	60

12. What was the average rate of change of the temperature from 8:00 A.M. to 10:00 A.M.? **5.5°/h**

13. What was the average rate of change of the temperature from 12:00 P.M. to 4:00 P.M.? **1.25°/h**

14. During what 2-hour period was the average rate of change of the temperature the least? **2:00 P.M.–4:00 P.M.**

★ indicates increased difficulty

Practice and Apply

For Exercises	See Examples
15–30	1
31–36	2
37–42	3
43–52	4, 5

Homework Help — **Extra Practice** See page 830.

Find the slope of the line that passes through each pair of points. **15.** $-\frac{5}{2}$

15. (6, 1), (8, −4)

16. (6, 8), (5, −5) **13**

17. (−6, −5), (4, 1) $\frac{3}{5}$

18. (2, −7), (4, 1) **4**

19. (7, 8), (1, 8) **0**

20. (−2, −3), (0, −5) **−1**

21. (2.5, 3), (1, −9) **8**

22. (4, −1.5), (4, 4.5) **undefined**

★**23.** $\left(\frac{1}{2}, -\frac{1}{3}\right), \left(\frac{1}{4}, \frac{2}{3}\right)$ **−4**

★**24.** $\left(\frac{1}{2}, \frac{2}{3}\right), \left(\frac{5}{6}, \frac{1}{4}\right)$ $-\frac{5}{4}$

★**25.** $(a, 2), (a, -2)$ **undefined**

★**26.** $(3, b), (-5, b)$ **0**

★**27.** Determine the value of r so that the line through $(6, r)$ and $(9, 2)$ has slope $\frac{1}{3}$. **1**

★**28.** Determine the value of r so that the line through $(5, r)$ and $(2, 3)$ has slope 2. **9**

ANCIENT CULTURES **Mayan Indians of Mexico and Central America built pyramids that were used as their temples. Ancient Egyptians built pyramids to use as tombs for the pharohs. Estimate the slope that a face of each pyramid makes with its base.**

★**29.**

The Pyramid of the Sun in Teotihuacán, Mexico, measures about 700 feet on each side of its square base and is about 210 feet high. **about 0.6**

★**30.**

The Great Pyramid in Egypt measures 756 feet on each side of its square base and was originally 481 feet high. **about 1.3**

Graph the line passing through the given point with the given slope.

31–36. See margin.

31. $(2, 6), m = \frac{2}{3}$

32. $(-3, -1), m = -\frac{1}{5}$

33. $(3, -4), m = 2$

34. $(1, 2), m = -3$

35. $(6, 2), m = 0$

36. $(-2, -3)$, undefined

Answers

31.

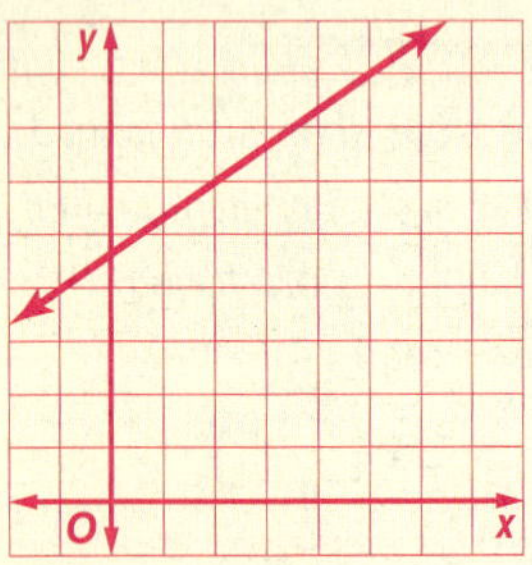

32.

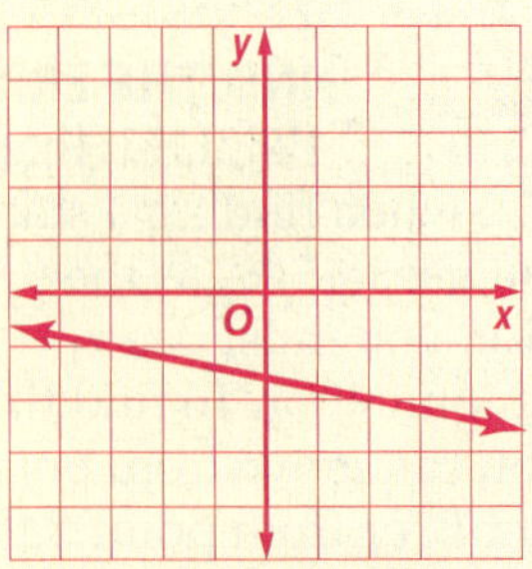

33.

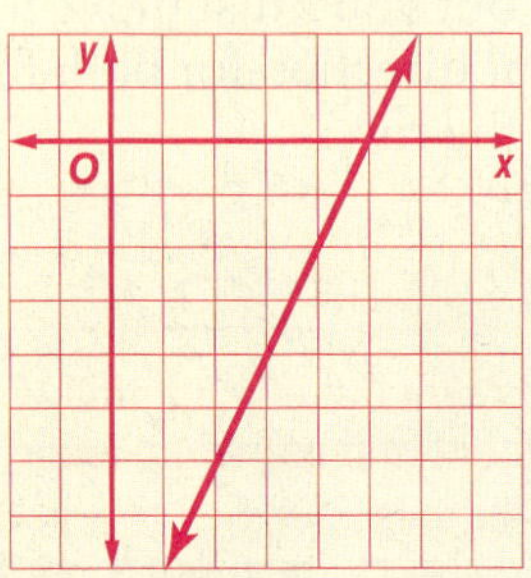

34.

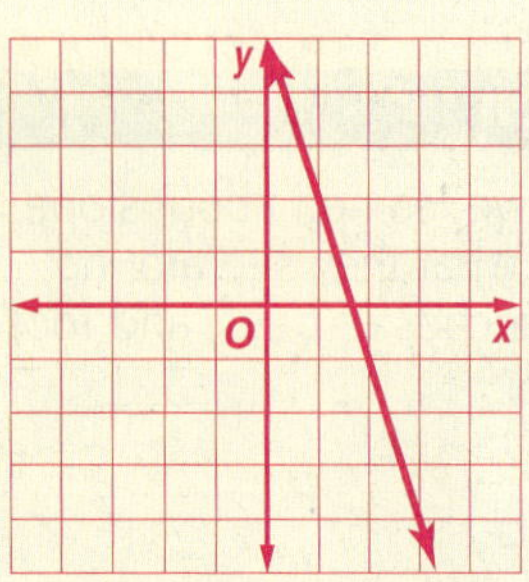

35.

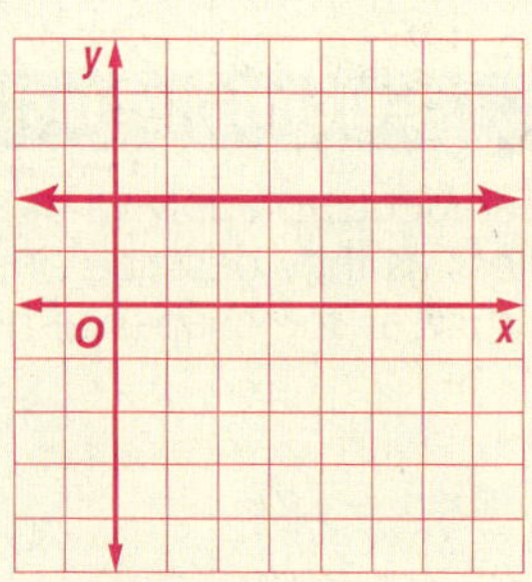

36.

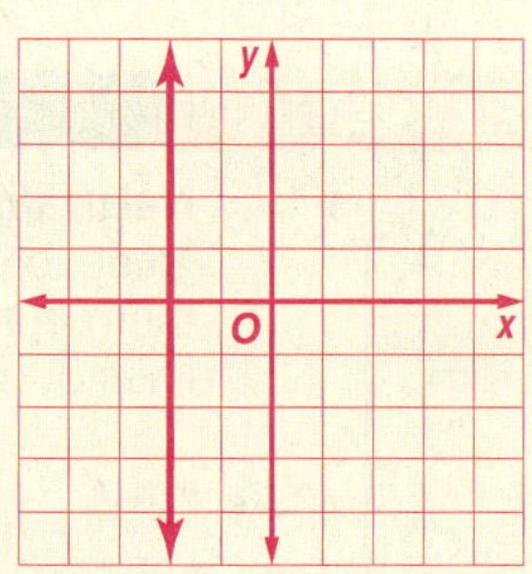

ENTERTAINMENT For Exercises 37–39, refer to the graph that shows the number of CDs and cassette tapes shipped by manufacturers to retailers in recent years.

CD and Tape Shipments

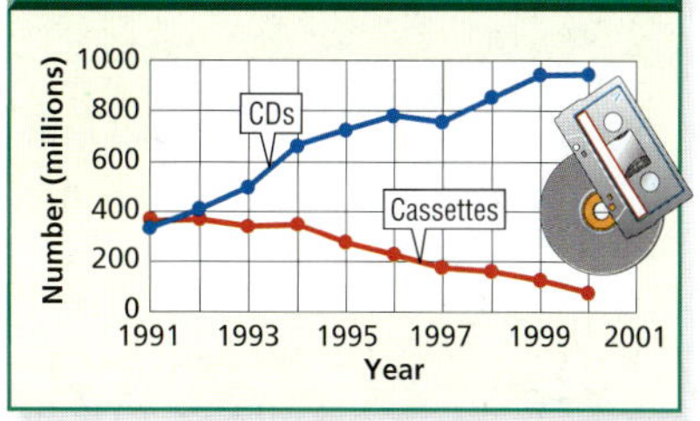

Source: Recording Industry Association of America

37. about 68 million per year

38. about −32 million per year

37. Find the average rate of change of the number of CDs shipped from 1991 to 2000.

38. Find the average rate of change of the number of cassette tapes shipped from 1991 to 2000.

39. Interpret the sign of your answer to Exercise 38. **The number of cassette tapes shipped has been decreasing.**

TRAVEL For Exercises 40–42, use the following information.

Mr. and Mrs. Wellman are taking their daughter to college. The table shows their distance from home after various amounts of time.

Time (h)	Distance (mi)
0	0
1	55
2	110
3	165
4	165
5	225

40. Find the average rate of change of their distance from home between 1 and 3 hours after leaving home. **55 mph**

41. Find the average rate of change of their distance from home between 0 and 5 hours after leaving home. **45 mph**

42. What is another word for *rate of change* in this situation? **speed or velocity**

Graph the line that satisfies each set of conditions. **43–50. See pp. 107A–107H.**

43. passes through $(-2, 2)$, parallel to a line whose slope is -1

44. passes through $(-4, 1)$, perpendicular to a line whose slope is $-\frac{3}{2}$

45. passes through $(3, 3)$, perpendicular to graph of $y = 3$

46. passes through $(2, -5)$, parallel to graph of $x = 4$

47. passes through $(2, -1)$, parallel to graph of $2x + 3y = 6$

48. passes through origin, parallel to graph of $x + y = 10$

★ 49. perpendicular to graph of $3x - 2y = 24$, intersects that graph at its x-intercept

★ 50. perpendicular to graph of $2x + 5y = 10$, intersects that graph at its y-intercept

51. **GEOMETRY** Determine whether quadrilateral $ABCD$ with vertices $A(-2, -1)$, $B(1, 1)$, $C(3, -2)$, and $D(0, -4)$ is a rectangle. Explain. **Yes; slopes show that adjacent sides are perpendicular.**

52. **CRITICAL THINKING** If the graph of the equation $ax + 3y = 9$ is perpendicular to the graph of the equation $3x + y = -4$, find the value of a. **−1**

53. **WRITING IN MATH** Answer the question that was posed at the beginning of the lesson. **See pp. 107A–107H.**

How does slope apply to the steepness of roads?

Include the following in your answer:

- a few sentences explaining the relationship between the grade of a road and the slope of a line, and
- a graph of $y = 0.08x$, which corresponds to a grade of 8%. (A road with a grade of 6% to 8% is considered to be fairly steep. The scales on your x- and y-axes should be the same.)

www.algebra2.com/self_check_quiz

Study Guide and Intervention, p. 69 (shown) and p. 70

Slope

Slope m of a Line: For points (x_1, y_1) and (x_2, y_2), where $x_1 \neq x_2$, $m = \frac{\text{change in } y}{\text{change in } x} = \frac{y_2 - y_1}{x_2 - x_1}$

Example 1 Determine the slope of the line that passes through $(2, -1)$ and $(-4, 5)$.

$m = \frac{y_2 - y_1}{x_2 - x_1}$ Slope formula

$= \frac{5 - (-1)}{-4 - 2}$ $(x_1, y_1) = (2, -1)$, $(x_2, y_2) = (-4, 5)$

$= \frac{6}{-6} = -1$ Simplify.

The slope of the line is -1.

Example 2 Graph the line passing through $(-1, -3)$ with a slope of $\frac{4}{5}$.

Graph the ordered pair $(-1, -3)$. Then, according to the slope, go up 4 units and right 5 units. Plot the new point (4, 1). Connect the points and draw the line.

Exercises

Find the slope of the line that passes through each pair of points.

1. (4, 7) and (6, 13) **3**
2. (6, 4) and (3, 4) **0**
3. (5, 1) and (7, −3) **−2**
4. (5, −3) and (−4, 3) $-\frac{2}{3}$
5. (5, 10) and (−1, −2) **2**
6. (−1, −4) and (−13, 2) $-\frac{1}{2}$
7. (7, −2) and (3, 3) $-\frac{5}{4}$
8. (−5, 9) and (5, 5) $-\frac{2}{5}$
9. (4, −2) and (−4, −8) $\frac{3}{4}$

Graph the line passing through the given point with the given slope.

10. slope $= -\frac{1}{3}$ passes through (0, 2)
11. slope = 2 passes through (1, 4)
12. slope = 0 passes through (−2, −5)
13. slope = 1 passes through (−4, 6)
14. slope $= -\frac{3}{4}$ passes through (−3, 0)
15. slope $= \frac{1}{5}$ passes through (0, 0)

Skills Practice, p. 71 and Practice, p. 72 (shown)

Find the slope of the line that passes through each pair of points.

1. (3, −8), (−5, 2) $-\frac{5}{4}$
2. (−10, −3), (7, 2) $\frac{5}{17}$
3. (−7, −6), (3, −6) **0**
4. (8, 2), (8, −1) **undefined**
5. (4, 3), (7, −2) $-\frac{5}{3}$
6. (−6, −3), (−8, 4) $-\frac{7}{2}$

Graph the line passing through the given point with the given slope.

7. (0, −3), $m = 3$
8. (2, 1), $m = -\frac{3}{4}$
9. (0, 2), $m = 0$
10. (2, −3), $m = \frac{4}{5}$

Graph the line that satisfies each set of conditions.

11. passes through (3, 0), perpendicular to a line whose slope is $\frac{3}{2}$
12. passes through (−3, −1), parallel to a line whose slope is −1

DEPRECIATION For Exercises 13–15, use the following information.

A machine that originally cost $15,600 has a value of $7500 at the end of 3 years. The same machine has a value of $2800 at the end of 8 years.

13. Find the average rate of change in value (depreciation) of the machine between its purchase and the end of 3 years. **−$2700 per year**
14. Find the average rate of change in value of the machine between the end of 3 years and the end of 8 years. **−$940 per year**
15. Interpret the sign of your answers. **It is negative because the value is decreasing.**

Reading to Learn Mathematics, p. 73 ELL

Pre-Activity How does slope apply to the steepness of roads?

Read the introduction to Lesson 2-3 at the top of page 68 in your textbook.

- What is the grade of a road that rises 40 feet over a horizontal distance of 1000 feet? **4%**
- What is the grade of a road that rises 525 meters over a horizontal distance of 10 kilometers? (1 kilometer = 1000 meters) **5.25%**

Reading the Lesson

1. Describe each type of slope and include a sketch.

Type of Slope	Description of Graph	Sketch
Positive	The line rises to the right.	
Zero	The line is horizontal.	
Negative	The line falls to the right.	
Undefined	The line is vertical.	

2. a. How are the slopes of two nonvertical parallel lines related? **They are equal.**
b. How are the slopes of two oblique perpendicular lines related? **Their product is −1.**

Helping You Remember

3. Look up the terms *grade*, *pitch*, *slant*, and *slope*. How can everyday meanings of these words help you remember the definition of slope? **Sample answer: All these words can be used when you describe how much a thing slants upward or downward. You can describe this numerically by comparing rise to run.**

Enrichment, p. 74

Aerial Surveyors and Area

Many land regions have irregular shapes. Aerial surveyors supply aerial mappers with lists of coordinates and elevations for the areas that need to be photographed from the air. These maps provide information about the horizontal and vertical features of the land.

Step 1 List the ordered pairs for the vertices in counterclockwise order, repeating the first ordered pair at the bottom of the list.

Step 2 Find D, the sum of the downward diagonal products (from left to right).
$D = (5 \cdot 5) + (2 \cdot 1) + (2 \cdot 3) + (6 \cdot 7)$
$= 25 + 2 + 6 + 42$ or 75

Step 3 Find U, the sum of the upward diagonal products (from left to right).

4 Assess

Open-Ended Assessment

Modeling Have students use a piece of spaghetti or a pencil on a coordinate plane to model lines with slopes of 1, −1, and 0, as well as a line with undefined slope. Then have them use a second piece of spaghetti or a second pencil to model a pair of parallel lines and then a pair of perpendicular lines.

Tips for New Teachers

Intervention If there is any doubt whether your students thoroughly understand the concept of slope, consider spending an extra day on this lesson. Use the Extra Practice on p. 830, the Study Guide and Intervention Masters, or the Practice Masters to reinforce this concept.

Getting Ready for Lesson 2-4

PREREQUISITE SKILL Lesson 2-4 presents writing the equation of a line given the slope and a point on the line. Exercises 70–75 should be used to determine your students' familiarity with solving equations for a given variable.

Assessment Options

Practice Quiz 1 The quiz provides students with a brief review of the concepts and skills in Lessons 2-1 through 2-3. Lesson numbers are given to the right of exercises or instruction lines so students can review concepts not yet mastered.

Standardized Test Practice

54. What is the slope of the line shown in the graph at the right? **D**

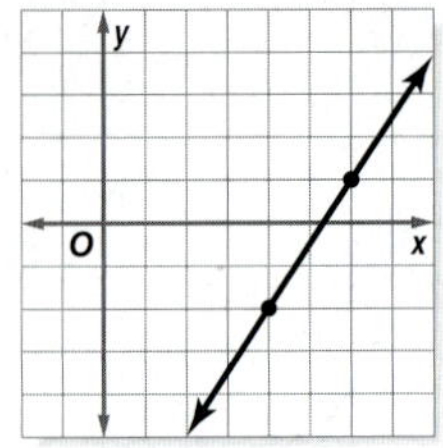

Ⓐ $-\frac{3}{2}$ Ⓑ $-\frac{2}{3}$ Ⓒ $\frac{2}{3}$ Ⓓ $\frac{3}{2}$

55. What is the slope of a line perpendicular to a line with slope $-\frac{1}{2}$? **D**

Ⓐ -2 Ⓑ $-\frac{1}{2}$ Ⓒ $\frac{1}{2}$ Ⓓ 2

Graphing Calculator

FAMILY OF GRAPHS Use a graphing calculator to investigate each family of graphs. Explain how changing the slope affects the graph of the line.

56. $y = 2x + 3, y = 4x + 3, y = 8x + 3, y = x + 3$ **56–57. See margin.**

57. $y = -3x + 1, y = -x + 1, y = -5x + 1, y = -7x + 1$

Maintain Your Skills

Mixed Review

Find the x-intercept and the y-intercept of the graph of each equation. Then graph the equation. *(Lesson 2-2)* **58–60. See pp. 107A–107H for graphs.**

58. $-2x + 5y = 20$ **−10, 4**
59. $4x - 3y + 8 = 0$ $\mathbf{-2, \frac{8}{3}}$
60. $y = 7x$ **0, 0**

Find each value if $f(x) = 3x - 4$. *(Lesson 2-1)*

61. $f(-1)$ **−7**
62. $f(3)$ **5**
63. $f\left(\frac{1}{2}\right)$ $\mathbf{-\frac{5}{2}}$
64. $f(a)$ **3a − 4**

Solve each inequality. *(Lessons 1-5 and 1-6)*

65. $5 < 2x + 7 < 13$ $\{x \mid -1 < x < 3\}$
66. $2z + 5 \geq 1475$ $\{z \mid z \geq 735\}$

67. **SCHOOL** A test has multiple-choice questions worth 4 points each and true-false questions worth 3 points each. Marco answers 14 multiple-choice questions correctly. How many true-false questions must he answer correctly to get at least 80 points total? *(Lesson 1-5)* **at least 8**

Simplify. *(Lessons 1-1 and 1-2)*

68. $\frac{1}{3}(15a + 9b) - \frac{1}{7}(28b - 84a)$ **17a − b**
69. $3 + (21 \div 7) \times 8 \div 4$ **9**

Getting Ready for the Next Lesson

PREREQUISITE SKILL Solve each equation for y.
*(To review **solving equations**, see Lesson 1-3.)* **70–75. See margin.**

70. $x + y = 9$
71. $4x + y = 2$
72. $-3x - y + 7 = 0$
73. $5x - 2y - 1 = 0$
74. $3x - 5y + 4 = 0$
75. $2x + 3y - 11 = 0$

Practice Quiz 1 — Lessons 2-1 through 2-3

1. State the domain and range of the relation $\{(2, 5), (-3, 2), (2, 1), (-7, 4), (0, -2)\}$. *(Lesson 2-1)* **D = {−7, −3, 0, 2}, R = {−2, 1, 2, 4, 5}**
2. Find the value of $f(15)$ if $f(x) = 100x - 5x^2$. *(Lesson 2-1)* **375**
3. Write $y = -6x + 4$ in standard form. *(Lesson 2-2)* **6x + y = 4**
4. Find the x-intercept and the y-intercept of the graph of $3x + 5y = 30$. Then graph the equation. *(Lesson 2-2)* **10, 6; See margin for graph.**
5. Graph the line that goes through $(4, -3)$ and is parallel to the line whose equation is $2x + 5y = 10$. *(Lesson 2-3)* **See margin.**

Answers

56. The graphs have the same y-intercept. As the slopes increase, the lines get steeper.

57. The graphs have the same y-intercept. As the slopes become more negative, the lines get steeper.

70. $y = 9 - x$
71. $y = -4x + 2$
72. $y = -3x + 7$
73. $y = \frac{5}{2}x - \frac{1}{2}$
74. $y = \frac{3}{5}x + \frac{4}{5}$
75. $y = -\frac{2}{3}x + \frac{11}{3}$

Answers (Practice Quiz 1)

4.

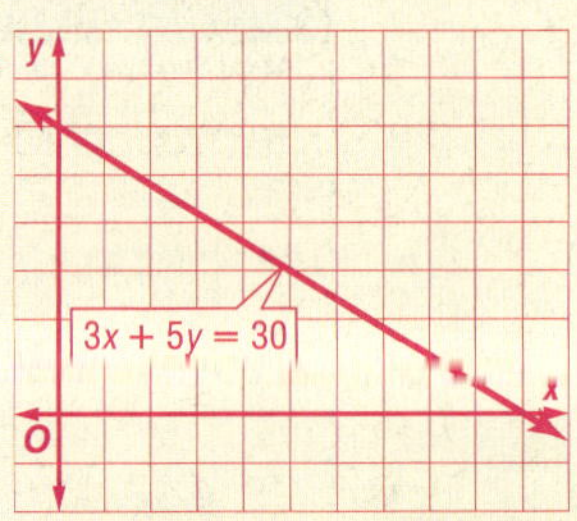

5.

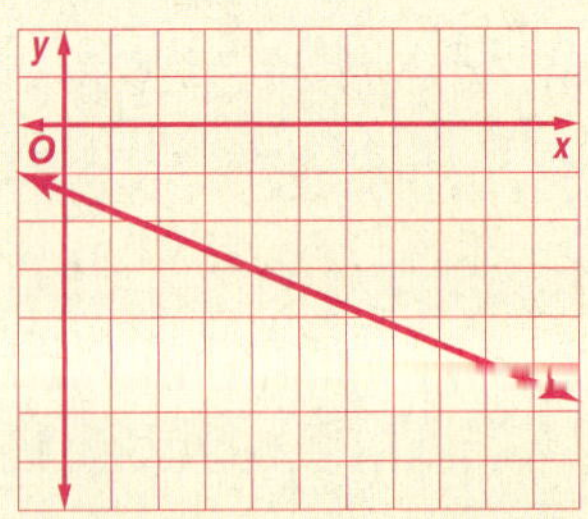

74 Chapter 2 Linear Relations and Functions

2-4 Writing Linear Equations

What You'll Learn

- Write an equation of a line given the slope and a point on the line.
- Write an equation of a line parallel or perpendicular to a given line.

Vocabulary

- slope-intercept form
- point-slope form

How do linear equations apply to business?

When a company manufactures a product, they must consider two types of cost. There is the *fixed cost*, which they must pay no matter how many of the product they produce, and there is *variable cost*, which depends on how many of the product they produce. In some cases, the total cost can be found using a linear equation such as $y = 5400 + 1.37x$.

FORMS OF EQUATIONS Consider the graph at the right. The line passes through $A(0, b)$ and $C(x, y)$. Notice that b is the y-intercept of $\overleftrightarrow{AC}$. You can use these two points to find the slope of $\overleftrightarrow{AC}$. Substitute the coordinates of points A and C into the slope formula.

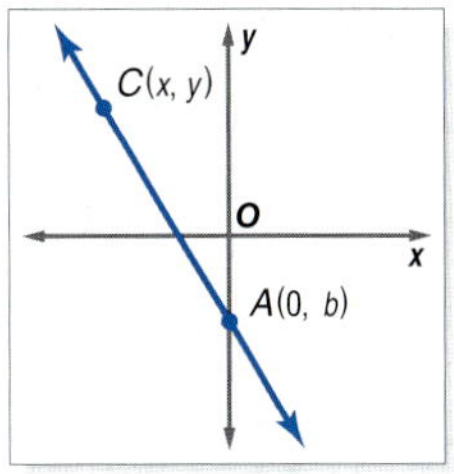

$$m = \frac{y_2 - y_1}{x_2 - x_1}$$ Slope formula

$$m = \frac{y - b}{x - 0}$$ $(x_1, y_1) = (0, b)$, $(x_2, y_2) = (x, y)$

$$m = \frac{y - b}{x}$$ Simplify.

Now solve the equation for y.

$mx = y - b$ Multiply each side by x.

$mx + b = y$ Add b to each side.

$y = mx + b$ Symmetric Property of Equality

When an equation is written in this form, it is in **slope-intercept form**.

Study Tip

Slope-intercept Form
The equation of a vertical line cannot be written in slope-intercept form because its slope is undefined.

Key Concept — *Slope-Intercept Form of a Linear Equation*

- **Words** The slope-intercept form of the equation of a line is $y = mx + b$, where m is the slope and b is the y-intercept.
- **Symbols** $y = mx + b$ (slope ↑ m, y-intercept ↑ b)
- **Model**

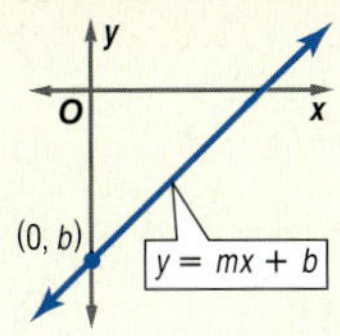

If you are given the slope and y-intercept of a line, you can find an equation of the line by substituting the values of m and b into the slope-intercept form. For example, if you know that the slope of a line is -3 and the y-intercept is 4, the equation of the line is $y = -3x + 4$, or, in standard form, $3x + y = 4$.

You can also use the slope-intercept form to find an equation of a line if you know the slope and the coordinates of any point on the line.

2-4 Lesson Notes

1 *Focus*

5-Minute Check Transparency 2-4 Use as a quiz or review of Lesson 2-3.

Mathematical Background notes are available for this lesson on p. 54D.

Building on Prior Knowledge

In Lesson 2-3, students studied slope and the graphs of parallel and perpendicular lines. In this lesson, students apply these ideas to writing the equations of lines.

How do linear equations apply to business?

Ask students:

- What are some examples of costs that might be included in the fixed cost? **utilities, salaries, rent**
- What are some examples of costs that might be included in the variable cost? **raw materials, packaging, shipping**

Resource Manager

Workbook and Reproducible Masters

Chapter 2 Resource Masters

- Study Guide and Intervention, pp. 75–76
- Skills Practice, p. 77
- Practice, p. 78
- Reading to Learn Mathematics, p. 79
- Enrichment, p. 80
- Assessment, pp. 113, 115

Graphing Calculator and Spreadsheet Masters, p. 30

School-to-Career Masters, p. 3

Transparencies

5-Minute Check Transparency 2-4

Answer Key Transparencies

Technology

Interactive Chalkboard

2 Teach

FORMS OF EQUATIONS

In-Class Examples

Power Point®

1 Write an equation in slope-intercept form for the line that has a slope of $-\frac{3}{5}$ and passes through (5, −2). $y = -\frac{3}{5}x + 1$

Teaching Tip Make sure that students understand that the letter m is always used for slope, and b for the y-intercept, in $y = mx + b$.

2 What is an equation of the line through (2, −3) and (−3, 7)? **D**

A $y = -2x - 1$

B $y = -\frac{1}{2}x + 1$

C $y = \frac{1}{2}x + 1$

D $y = -2x + 1$

Example 1 Write an Equation Given Slope and a Point

Write an equation in slope-intercept form for the line that has a slope of $-\frac{3}{2}$ and passes through (−4, 1).

Substitute for m, x, and y in the slope-intercept form.

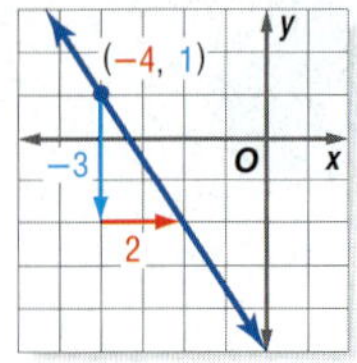

$y = mx + b$	Slope-intercept form
$1 = \left(-\frac{3}{2}\right)(-4) + b$	$(x, y) = (-4, 1)$, $m = -\frac{3}{2}$
$1 = 6 + b$	Simplify.
$-5 = b$	Subtract 6 from each side.

The y-intercept is −5. So, the equation in slope-intercept form is $y = -\frac{3}{2}x - 5$.

If you are given the coordinates of two points on a line, you can use the **point-slope form** to find an equation of the line that passes through them.

Key Concept — Point-Slope Form of a Linear Equation

- **Words** The point-slope form of the equation of a line is $y - y_1 = m(x - x_1)$, where (x_1, y_1) are the coordinates of a point on the line and m is the slope of the line.
- **Symbols** $y - y_1 = m(x - x_1)$ — slope: m; coordinates of point on line: (x_1, y_1)

Standardized Test Practice

Example 2 Write an Equation Given Two Points

Multiple-Choice Test Item

> What is an equation of the line through (−1, 4) and (−4, 5)?
>
> Ⓐ $y = -\frac{1}{3}x + \frac{11}{3}$ Ⓑ $y = \frac{1}{3}x + \frac{13}{3}$ Ⓒ $y = -\frac{1}{3}x + \frac{13}{3}$ Ⓓ $y = -3x + 1$

Read the Test Item

You are given the coordinates of two points on the line. Notice that the answer choices are in slope-intercept form.

Solve the Test Item

- First, find the slope of the line.

$m = \frac{y_2 - y_1}{x_2 - x_1}$	Slope formula
$= \frac{5 - 4}{-4 - (-1)}$	$(x_1, y_1) = (-1, 4)$, $(x_2, y_2) = (-4, 5)$
$= \frac{1}{-3}$ or $-\frac{1}{3}$	Simplify.

The slope is $-\frac{1}{3}$. That eliminates choices B and D.

- Then use the point-slope form to find an equation.

$y - y_1 = m(x - x_1)$	Point-slope form
$y - 4 = -\frac{1}{3}[x - (-1)]$	$m = -\frac{1}{3}$; you can use either point for (x_1, y_1).
$y - 4 = -\frac{1}{3}x - \frac{1}{3}$	Distributive Property
$y = -\frac{1}{3}x + \frac{11}{3}$	The answer is A.

Test-Taking Tip

To check your answer, substitute each ordered pair into your answer. Each should satisfy the equation.

Standardized Test Practice

Example 2 Point out that finding the slope eliminated two of the choices. Emphasize that eliminating some of the answer choices helps you to use your time efficiently when taking a timed test.

When changes in real-world situations occur at a linear rate, a linear equation can be used as a model for describing the situation.

Example 3 Write an Equation for a Real-World Situation

SALES As a salesperson, Eric Fu is paid a daily salary plus commission. When his sales are $1000, he makes $100. When his sales are $1400, he makes $120.

a. Write a linear equation to model this situation.

Let x be his sales and let y be the amount of money he makes. Use the points (1000, 100) and (1400, 120) to make a graph to represent the situation.

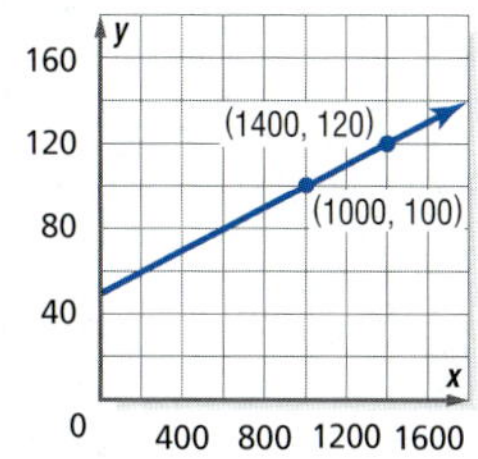

$m = \frac{y_2 - y_1}{x_2 - x_1}$ Slope formula

$= \frac{120 - 100}{1400 - 1000}$ $(x_1, y_1) = (1000, 100)$, $(x_2, y_2) = (1400, 120)$

$= 0.05$ Simplify.

Now use the slope and either of the given points with the point-slope form to write the equation.

$y - y_1 = m(x - x_1)$ Point-slope form

$y - 100 = 0.05(x - 1000)$ $m = 0.05$, $(x_1, y_1) = (1000, 100)$

$y - 100 = 0.05x - 50$ Distributive Property

$y = 0.05x + 50$ Add 100 to each side.

The slope-intercept form of the equation is $y = 0.05x + 50$.

b. What are Mr. Fu's daily salary and commission rate?

The y-intercept of the line is 50. The y-intercept represents the money Eric would make if he had no sales. In other words, $50 is his daily salary.

The slope of the line is 0.05. Since the slope is the coefficient of x, which is his sales, he makes 5% commission.

c. How much would he make in a day if Mr. Fu's sales were $2000?

Find the value of y when $x = 2000$.

$y = 0.05x + 50$ Use the equation you found in part **a**.

$= 0.05(2000) + 50$ Replace x with 2000.

$= 100 + 50$ or 150 Simplify.

Mr. Fu would make $150 if his sales were $2000.

Study Tip

Alternative Method
You could also find Mr. Fu's salary in part **c** by extending the graph. Then find the y value when x is 2000.

PARALLEL AND PERPENDICULAR LINES The slope-intercept and point-slope forms can be used to find equations of lines that are parallel or perpendicular to given lines.

Example 4 Write an Equation of a Perpendicular Line

Write an equation for the line that passes through $(-4, 3)$ and is perpendicular to the line whose equation is $y = -4x - 1$.

The slope of the given line is -4. Since the slopes of perpendicular lines are opposite reciprocals, the slope of the perpendicular line is $\frac{1}{4}$.

(continued on the next page)

www.algebra2.com/extra_examples

In-Class Example

PowerPoint®

3 **SALES** As a part-time salesperson, Jean Stock is paid a daily salary plus commission. When her sales are $100, she makes $58. When her sales are $300, she make $78.

a. Write a linear equation to model this situation. $y = 0.1x + 48$

b. What are Ms. Stock's daily salary and commission rate? **$48; 10%**

c. How much would Jean make in a day if her sales were $500? **$98**

Teaching Tip Point out that when the units differ on the two axes, you cannot estimate the slope of the graphed line by comparing it to the slope of $x = y$ as the 45-degree line with a slope of 1.

PARALLEL AND PERPENDICULAR LINES

In-Class Example

PowerPoint®

4 Write an equation for the line that passes through $(3, -2)$ and is perpendicular to the line whose equation is $y = -5x + 1$. $y = \frac{1}{5}x - \frac{13}{5}$

Lesson 2-4 Writing Linear Equations 77

3 Practice/Apply

Study Notebook

Have students—
- add the definitions/examples of the vocabulary terms to their Vocabulary Builder worksheets for Chapter 2.
- add the slope-intercept and point-slope forms of a linear equation to their notebooks, with examples of both types of equations.
- add items from Example 2 to their list of test-taking tips, which they can review as they prepare for standardized tests.
- include any other item(s) that they find helpful in mastering the skills in this lesson.

About the Exercises...

Organization by Objective
- **Forms of Equations:** 13–34
- **Parallel and Perpendicular Lines:** 35–38

Odd/Even Assignments
Exercises 13–40 are structured so that students practice the same concepts whether they are assigned odd or even problems.

Assignment Guide

Basic: 13, 15, 19–35 odd, 39, 41–43, 45, 49–54, 57–67

Average: 13–39 odd, 41–43, 45–54, 57–67 (optional: 55, 56)

Advanced: 14–40 even, 44, 46–63 (optional: 64–67)

Answer

3. Solve the equation for y to get $y = \frac{3}{5}x - \frac{2}{5}$. The slope of this line is $\frac{3}{5}$. The slope of a parallel line is the same.

TEACHING TIP
You could also use slope-intercept form.

Use the point-slope form and the ordered pair $(-4, 3)$ to write the equation.

$y - y_1 = m(x - x_1)$ Point-slope form

$y - 3 = \frac{1}{4}[x - (-4)]$ $(x_1, y_1) = (-4, 3)$, $m = \frac{1}{4}$

$y - 3 = \frac{1}{4}x + 1$ Distributive Property

$y = \frac{1}{4}x + 4$ Add 3 to each side.

An equation of the line is $y = \frac{1}{4}x + 4$.

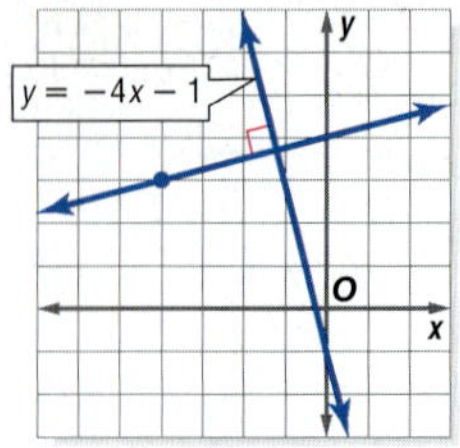

Check for Understanding

Concept Check

1. **OPEN ENDED** Write an equation of a line in slope-intercept form. **1. Sample answer: $y = 3x + 2$**
2. **Identify** the slope and y-intercept of the line with equation $y = 6x$. **6, 0**
3. **Explain** how to find the slope of a line parallel to the graph of $3x - 5y = 2$. **See margin.**

Guided Practice

GUIDED PRACTICE KEY	
Exercises	Examples
4–7	1
8, 9, 11, 12	2, 3
10	4

State the slope and y-intercept of the graph of each equation.

4. $y = 2x - 5$ **2, −5**
5. $3x + 2y - 10 = 0$ **$-\frac{3}{2}$, 5**

Write an equation in slope-intercept form for the line that satisfies each set of conditions. **6. $y = 0.5x + 1$ 7. $y = -\frac{3}{4}x + 2$ 8. $y = -\frac{5}{2}x + 16$**

6. slope 0.5, passes through (6, 4)
7. slope $-\frac{3}{4}$, passes through $\left(2, \frac{1}{2}\right)$
8. passes through (6, 1) and (8, −4)
9. passes through (−3, 5) and (2, 2) **9. $y = -\frac{3}{5}x + \frac{16}{5}$**
10. passes through (0, −2), perpendicular to the graph of $y = x - 2$ **$y = -x - 2$**
11. Write an equation in slope-intercept form for the graph at the right. **$y = \frac{5}{4}x + 7$**

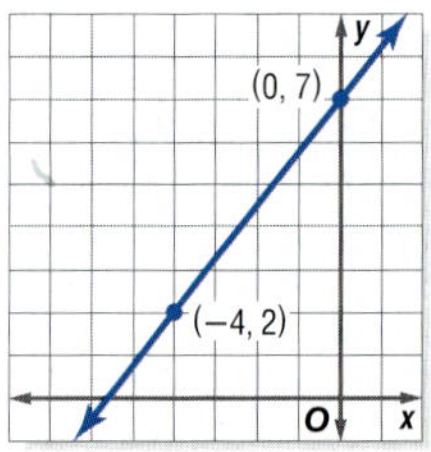

Standardized Test Practice

12. What is an equation of the line through (2, −4) and (−3, −1)? **B**
 (A) $y = -\frac{3}{5}x + \frac{26}{5}$ (B) $y = -\frac{3}{5}x - \frac{14}{5}$
 (C) $y = \frac{3}{5}x - \frac{26}{5}$ (D) $y = \frac{3}{5}x + \frac{14}{5}$

★ indicates increased difficulty

Practice and Apply

Homework Help

For Exercises	See Examples
13–18, 21–28	1
19, 20, 29–34, 39, 40	2, 3
35–38	4
41–52	1–3

Extra Practice
See page 831.

State the slope and y-intercept of the graph of each equation.

13. $y = -\frac{2}{3}x - 4$ **$-\frac{2}{3}$, −4**
14. $y = \frac{3}{4}x$ **$\frac{3}{4}$, 0**
15. $2x - 4y = 10$ **$\frac{1}{2}$, $-\frac{5}{2}$**
16. $3x + 5y - 30 = 0$ **16. $-\frac{3}{5}$, 6**
★17. $x = 7$ **undefined, none**
★18. $cx + y = d$ **$-c$, d**

Write an equation in slope-intercept form for each graph.

19. **$y = 0.8x$**

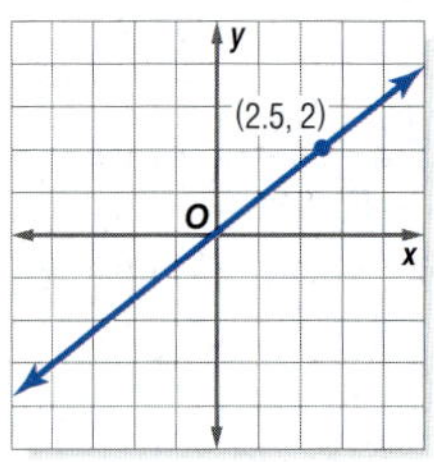

20. **$y = -\frac{5}{3}x + \frac{29}{3}$**

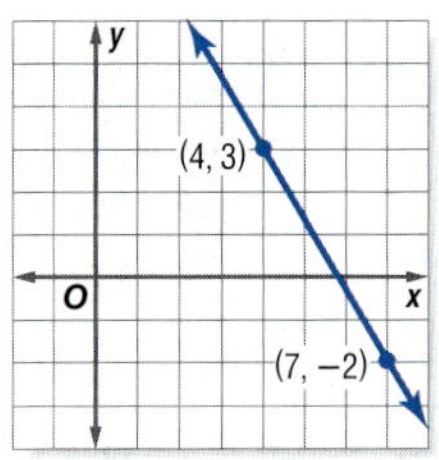

DAILY INTERVENTION

Differentiated Instruction

Intrapersonal Have students write a paragraph summarizing any problems they have with relating graphs to their equations. Ask them to include specific techniques they find useful to help them with this task.

Write an equation in slope-intercept form for each graph.

21. 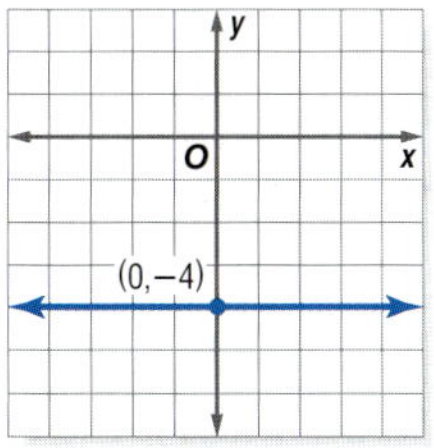

$y = -4$

22. 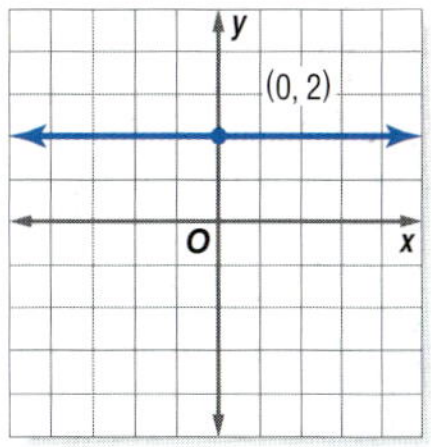

$y = 2$

Write an equation in slope-intercept form for the line that satisfies each set of conditions.

23. $y = 3x - 6$
24. $y = 0.25x + 4$
25. $y = -\frac{1}{2}x + \frac{7}{2}$
26. $y = \frac{3}{2}x + \frac{17}{2}$
27. $y = -0.5x - 2$
28. $y = 4x$
29. $y = -\frac{4}{5}x + \frac{17}{5}$
30. no slope-intercept form for $x = 7$
31. $y = 0$
32. $y = \frac{3}{2}x$
34. $y = \frac{3}{4}x - \frac{1}{4}$
37. $y = -\frac{1}{15}x - \frac{23}{5}$

23. slope 3, passes through $(0, -6)$
24. slope 0.25, passes through $(0, 4)$
25. slope $-\frac{1}{2}$, passes through $(1, 3)$
26. slope $\frac{3}{2}$, passes through $(-5, 1)$
27. slope -0.5, passes through $(2, -3)$
28. slope 4, passes through the origin
29. passes through $(-2, 5)$ and $(3, 1)$
30. passes through $(7, 1)$ and $(7, 8)$
31. passes through $(-4, 0)$ and $(3, 0)$
32. passes through $(-2, -3)$ and $(0, 0)$
33. x-intercept -4, y-intercept 4 $y = x + 4$
34. x-intercept $\frac{1}{3}$, y-intercept $-\frac{1}{4}$
35. passes through $(4, 6)$, parallel to the graph of $y = \frac{2}{3}x + 5$ $y = \frac{2}{3}x + \frac{10}{3}$
36. passes through $(2, -5)$, perpendicular to the graph of $y = \frac{1}{4}x + 7$ $y = -4x + 3$

★37. passes through $(6, -5)$, perpendicular to the line whose equation is $3x - \frac{1}{5}y = 3$

★38. passes through $(-3, -1)$, parallel to the line that passes through $(3, 3)$ and $(0, 6)$ $y = -x - 4$

39. Write an equation in slope-intercept form of the line that passes through the points indicated in the table. $y = 3x - 2$

x	y
−1	−5
1	1
3	7

40. Write an equation in slope-intercept form of the line that passes through $(-2, 10)$, $(2, 2)$, and $(4, -2)$. $y = -2x + 6$

GEOMETRY For Exercises 41–43, use the equation $d = 180(c - 2)$ that gives the total number of degrees d in any convex polygon with c sides.

41. Write this equation in slope-intercept form. $d = 180c - 360$
42. Identify the slope and d-intercept. 180, −360
43. Find the number of degrees in a pentagon. 540°

44. **ECOLOGY** A park ranger at Blendon Woods estimates there are 6000 deer in the park. She also estimates that the population will increase by 75 deer each year thereafter. Write an equation that represents how many deer will be in the park in x years. $y = 75x + 6000$

45. **BUSINESS** Refer to the signs below. At what distance do the two stores charge the same amount for a balloon arrangement? 10 mi

www.algebra2.com/self_check_quiz

Enrichment, p. 80

Two-Intercept Form of a Linear Equation

You are already familiar with the slope-intercept form of a linear equation, $y = mx + b$. Linear equations can also be written in the form $\frac{x}{a} + \frac{y}{b} = 1$ with x-intercept a and y-intercept b. This is called two-intercept form.

Example 1 **Draw the graph of $\frac{x}{-3} + \frac{y}{6} = 1$.**

The graph crosses the x-axis at -3 and the y-axis at 6. Graph $(-3, 0)$ and $(0, 6)$, then draw a straight line through them.

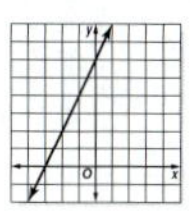

Example 2 **Write $3x + 4y = 12$ in two-intercept form.**

$\frac{3x}{12} + \frac{4y}{12} = \frac{12}{12}$ Divide by 12 to obtain 1 on the right side.

Study Guide and Intervention, p. 75 (shown) and p. 76

Forms of Equations

Slope-Intercept Form of a Linear Equation	$y = mx + b$, where m is the slope and b is the y-intercept
Point-Slope Form of a Linear Equation	$y - y_1 = m(x - x_1)$, where (x_1, y_1) are the coordinates of a point on the line and m is the slope of the line

Example 1 **Write an equation in slope-intercept form for the line that has slope −2 and passes through the point (3, 7).**

Substitute for m, x, and y in the slope-intercept form.

$y = mx + b$ Slope-intercept form
$7 = (-2)(3) + b$ $(x, y) = (3, 7)$, $m = -2$
$7 = -6 + b$ Simplify.
$13 = b$ Add 6 to both sides.

The y-intercept is 13. The equation in slope-intercept form is $y = -2x + 13$.

Example 2 **Write an equation in slope-intercept form for the line that has slope $\frac{1}{3}$ and x-intercept 5.**

$y = mx + b$ Slope-intercept form
$0 = \left(\frac{1}{3}\right)(5) + b$ $(x, y) = (5, 0)$, $m = \frac{1}{3}$
$0 = \frac{5}{3} + b$ Simplify.
$-\frac{5}{3} = b$ Subtract $\frac{5}{3}$ from both sides.

The y-intercept is $-\frac{5}{3}$. The slope-intercept form is $y = \frac{1}{3}x - \frac{5}{3}$.

Exercises

Write an equation in slope-intercept form for the line that satisfies each set of conditions.

1. slope -2, passes through $(-4, 6)$ $y = -2x - 2$
2. slope $\frac{3}{2}$, y-intercept 4 $y = \frac{3}{2}x + 4$
3. slope 1, passes through $(2, 5)$ $y = x + 3$
4. slope $-\frac{13}{5}$, passes through $(5, -7)$ $y = -\frac{13}{5}x + 6$

Write an equation in slope-intercept form for each graph.

5. 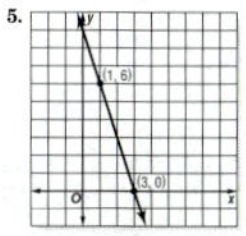

$y = -3x + 9$

6. 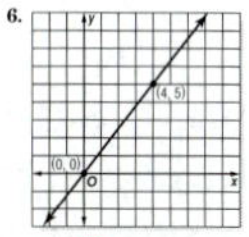

$y = \frac{5}{4}x$

7. 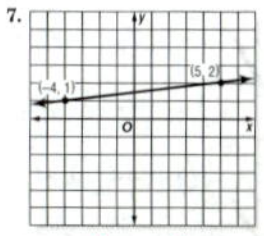

$y = \frac{1}{9}x + 1\frac{4}{9}$

Skills Practice, p. 77 and Practice, p. 78 (shown)

State the slope and y-intercept of the graph of each equation.

1. $y = 8x + 12$ 8, 12
2. $y = 0.25x - 1$ 0.25, −1
3. $y = -\frac{3}{5}x$ $-\frac{3}{5}$, 0
4. $3y = 7$ 0, $\frac{7}{3}$
5. $3x = -15 + 5y$ $\frac{3}{5}$, 3
6. $2x - 3y = 10$ $\frac{2}{3}$, $-\frac{10}{3}$

Write an equation in slope-intercept form for each graph.

7. 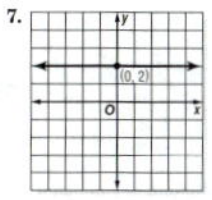

$y = 2$

8. 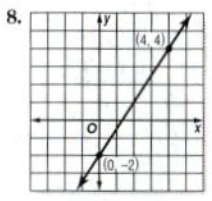

$y = \frac{3}{2}x - 2$

9. 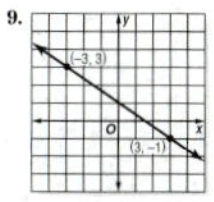

$y = -\frac{2}{3}x + 1$

Write an equation in slope-intercept form for the line that satisfies each set of conditions.

10. slope -5, passes through $(-3, -8)$ $y = -5x - 23$
11. slope $\frac{4}{5}$, passes through $(10, -3)$ $y = \frac{4}{5}x - 11$
12. slope 0, passes through $(0, -10)$ $y = -10$
13. slope $-\frac{2}{3}$, passes through $(6, -8)$ $y = -\frac{2}{3}x - 4$
14. passes through $(3, 11)$ and $(-6, 5)$ $y = \frac{2}{3}x + 9$
15. passes through $(7, -2)$ and $(3, -1)$ $y = -\frac{1}{4}x - \frac{1}{4}$
16. x-intercept 3, y-intercept 2 $y = -\frac{2}{3}x + 2$
17. x-intercept -5, y-intercept 7 $y = \frac{7}{5}x + 7$
18. passes through $(-8, -7)$, perpendicular to the graph of $y = 4x - 3$ $y = -\frac{1}{4}x - 9$
19. **RESERVOIRS** The surface of Grand Lake is at an elevation of 648 feet. During the current drought, the water level is dropping at a rate of 3 inches per day. If this trend continues, write an equation that gives the elevation in feet of the surface of Grand Lake after x days. $y = -0.25x + 648$
20. **BUSINESS** Tony Marconi's company manufactures CD-ROM drives. The company will make \$150,000 profit if it manufactures 100,000 drives, and \$1,750,000 profit if it manufactures 500,000 drives. The relationship between the number of drives manufactured and the profit is linear. Write an equation that gives the profit P when n drives are manufactured. $P = 4n - 250,000$

Reading to Learn Mathematics, p. 79 ELL

Pre-Activity **How do linear equations apply to business?**

Read the introduction to Lesson 2-4 at the top of page 75 in your textbook.

- If the total cost of producing a product is given by the equation $y = 5400 + 1.37x$, what is the fixed cost? What is the variable cost (for each item produced)? \$5400; \$1.37
- Write a linear equation that describes the following situation: A company that manufactures computers has a fixed cost of \$228,750 and a variable cost of \$852 to produce each computer. $y = 228,750 + 852x$

Reading the Lesson

1. a. Write the slope-intercept form of the equation of a line. Then explain the meaning of each of the variables in the equation. $y = mx + b$; m is the slope and b is the y-intercept. The variables x and y are the coordinates of any point on the line.

   b. Write the point-slope form of the equation of a line. Then explain the meaning of each of the variables in the equation. $y - y_1 = m(x - x_1)$; m is the slope. x and y are the coordinates of any point on the line. x_1 and y_1 are the coordinates of one specific point on the line.

2. Suppose that your algebra teacher asks you to write the point-slope form of the equation of the line through the points $(-6, 7)$ and $(-3, -2)$. You write $y + 2 = -3(x + 3)$ and your classmate writes $y - 7 = -3(x + 6)$. Which of you is correct? Explain. You are both correct. Either point may be used as (x_1, y_1) in the point-slope form. You used $(-3, -2)$, and your classmate used $(-6, 7)$.

3. You are asked to write an equation of two lines that pass through $(3, -5)$, one of them parallel to and one of them perpendicular to the line whose equation is $y = -3x + 4$. The first step in finding these equations is to find their slopes. What is the slope of the parallel line? What is the slope of the perpendicular line? -3; $\frac{1}{3}$

Helping You Remember

4. Many students have trouble remembering the point-slope form for a linear equation. How can you use the definition of slope to remember this form? Sample answer: Write the definition of slope: $m = \frac{y_2 - y_1}{x_2 - x_1}$. Multiply both sides of this equation by $x_2 - x_1$. Drop the subscripts in y_2 and x_2. This gives the point-slope form of the equation of a line.

4 Assess

Open-Ended Assessment

Writing Have students write their own summary of how to relate graphs to their equations, including slope, intercepts, and parallel and perpendicular lines.

Getting Ready for Lesson 2-5

PREREQUISITE SKILL Lesson 2-5 presents modeling real-world data using scatter plots. Exercises 64–67 should be used to determine your students' familiarity with finding the median of a set of numbers.

Assessment Options

Quiz (Lessons 2-3 and 2-4) is available on p. 113 of the *Chapter 2 Resource Masters*.

Mid-Chapter Test (Lessons 2-1 through 2-4) is available on p. 115 of the *Chapter 2 Resource Masters*.

Answers

46.

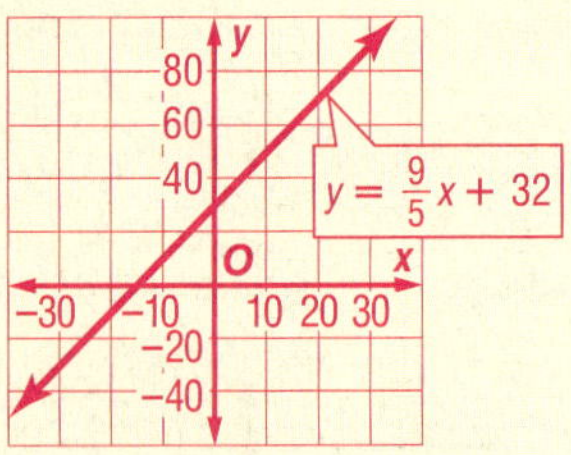

52. A linear equation can sometimes be used to relate a company's cost to the number they produce of a product. Answers should include the following.

- The y-intercept, 5400, is the cost the company must pay if they produce 0 units, so it is the fixed cost. The slope, 1.37, means that it costs $1.37 to produce each unit. The variable cost is $1.37x$.
- $6770

More About. . .

Science Ice forms at a temperature of 0°C, which corresponds to a temperature of 32°F. A temperature of 100°C corresponds to a temperature of 212°F.

SCIENCE For Exercises 46–48, use the information on temperatures at the left.

46. Write and graph the linear equation that gives the number y of degrees Fahrenheit in terms of the number x of degrees Celsius. $y = \frac{9}{5}x + 32$; See margin for graph.

47. What temperature corresponds to 20°C? 68°F

48. What temperature is the same on both scales? −40°

TELEPHONES For Exercises 49 and 50, use the following information.
Namid is examining the calling card portion of his phone bill. A 4-minute call at the night rate cost $2.65. A 10-minute call at the night rate cost $4.75.

49. Write a linear equation to model this situation. $y = 0.35x + 1.25$

50. How much would it cost to talk for half an hour at the night rate? $11.75

51. CRITICAL THINKING Given $\triangle ABC$ with vertices $A(-6, -8)$, $B(6, 4)$, and $C(-6, 10)$, write an equation of the line containing the altitude from A. (*Hint:* The altitude from A is a segment that is perpendicular to $\overline{BC}$.) $y = 2x + 4$

52. WRITING IN MATH Answer the question that was posed at the beginning of the lesson.

How do linear equations apply to business?

Include the following in your answer: See margin.

- the *fixed cost* and the *variable cost* in the equation $y = 5400 + 1.37x$, where y is the cost for a company to produce x units of its product, and
- the cost for the company to produce 1000 units of its product.

Standardized Test Practice

53. Find an equation of the line through (0, −3) and (4, 1). C

Ⓐ $y = -x + 3$ Ⓑ $y = -x - 3$ Ⓒ $y = x - 3$ Ⓓ $y = -x + 3$

54. Choose the equation of the line through $\left(\frac{1}{2}, -\frac{3}{2}\right)$ and $\left(-\frac{1}{2}, \frac{1}{2}\right)$. A

Ⓐ $y = -2x - \frac{1}{2}$ Ⓑ $y = -3x$ Ⓒ $y = 2x - \frac{5}{2}$ Ⓓ $y = \frac{1}{2}x + 1$

Extending the Lesson

For Exercises 55 and 56, use the following information.
The form $\frac{x}{a} + \frac{y}{b} = 1$ is known as the **intercept form** of the equation of a line because a is the x-intercept and b is the y-intercept.

55. Write the equation $2x - y - 5 = 0$ in intercept form. $\frac{x}{\frac{5}{2}} - \frac{y}{5} = 1$

56. Identify the x- and y-intercepts of the graph of $2x - y - 5 = 0$. $\frac{5}{2}$, −5

Maintain Your Skills

Mixed Review Find the slope of the line that passes through each pair of points. *(Lesson 2-3)*

57. (7, 2), (5, 6) −2 **58.** (1, −3), (3, 3) 3 **59.** (−5, 0), (4, 0) 0

60. INTERNET A Webmaster estimates that the time (seconds) required to connect to the server when n people are connecting is given by $t(n) = 0.005n + 0.3$. Estimate the time required to connect when 50 people are connecting. *(Lesson 2-2)* 0.55 s

Solve each inequality. *(Lessons 1-5 and 1-6)*

61. $|x - 2| \leq -99$ ∅ **62.** $-4x + 7 \leq 31$ $\{x \mid x \geq -6\}$ **63.** $2(r - 4) + 5 \geq 9$ $\{r \mid r \geq 6\}$

Getting Ready for the Next Lesson

PREREQUISITE SKILL Find the median of each set of numbers.
*(To review **finding a median**, see pages 822 and 823.)*

64. {3, 2, 1, 3, 4, 8, 4} 3 **65.** {9, 3, 7, 5, 6, 3, 7, 9} 6.5

66. {138, 235, 976, 230, 412, 466} 323.5 **67.** {2.5, 7.8, 5.5, 2.3, 6.2, 7.8} 5.85

80 Chapter 2 Linear Relations and Functions

2-5 Modeling Real-World Data: Using Scatter Plots

What You'll Learn

- Draw scatter plots.
- Find and use prediction equations.

Vocabulary

- bivariate data
- scatter plot
- line of fit
- prediction equation

How can a linear equation model the number of Calories you burn exercising?

The table shows the number of Calories burned per hour by a 140-pound person running at various speeds. A linear function can be used to model these data.

Speed (mph)	Calories
5	508
6	636
7	731
8	858

SCATTER PLOTS Data with two variables, such as speed and Calories, is called **bivariate data**. A set of bivariate data graphed as ordered pairs in a coordinate plane is called a **scatter plot**. A scatter plot can show whether there is a relationship between the data.

Example 1 Draw a Scatter Plot

HOUSING The table below shows the median selling price of new, privately-owned, one-family houses for some recent years. Make a scatter plot of the data.

Year	1990	1992	1994	1996	1998	2000
Price ($1000)	122.9	121.5	130.0	140.0	152.5	169.0

Source: U.S. Census Bureau and U.S. Department of Housing and Urban Development

Graph the data as ordered pairs, with the number of years since 1990 on the horizontal axis and the price on the vertical axis.

Study Tip

Choosing the Independent Variable

Letting x be the number of years since the first year in the data set sometimes simplifies the calculations involved in finding a function to model the data.

PREDICTION EQUATIONS Except for (0, 122.9), the data in Example 1 appear to lie nearly on a straight line. When you find a line that closely approximates a set of data, you are finding a **line of fit** for the data. An equation of such a line is often called a **prediction equation** because it can be used to predict one of the variables given the other variable.

2-5 Lesson Notes

1 Focus

5-Minute Check Transparency 2-5 Use as a quiz or review of Lesson 2-4.

Mathematical Background notes are available for this lesson on p. 54D.

Building on Prior Knowledge

In Lesson 2-4, students wrote linear equations based on information provided about their graphs. In this lesson, students apply those skills to modeling real-world data with scatter plots and writing their prediction equations.

How can a linear equation model the number of Calories you burn exercising?

Ask students:

- Will a person that weighs less than 140 pounds burn more or less Calories than shown in the table at the given speeds? **less**
- What is a reasonable estimate of the number of Calories a 140-pound person burns running at a speed of 4 miles per hour? **about 400 Calories**

Resource Manager

Workbook and Reproducible Masters

Chapter 2 Resource Masters

- Study Guide and Intervention, pp. 81–82
- Skills Practice, p. 83
- Practice, p. 84
- Reading to Learn Mathematics, p. 85
- Enrichment, p. 86

School-to-Career Masters, p. 4

Science and Mathematics Lab Manual, pp. 97–102

Teaching Algebra With Manipulatives Masters, p. 218

Transparencies

5-Minute Check Transparency 2-5
Answer Key Transparencies

Technology

Interactive Chalkboard
Multimedia Applications

2 Teach

SCATTER PLOTS

In-Class Example Power Point®

1 **EDUCATION** The table below shows the approximate percent of students who sent applications to two colleges in various years since 1985. Make a scatter plot of the data.

Years Since 1985	0	3	6	9	12	15
Percent	20	18	15	15	14	13

Source: *U.S. News & World Report*

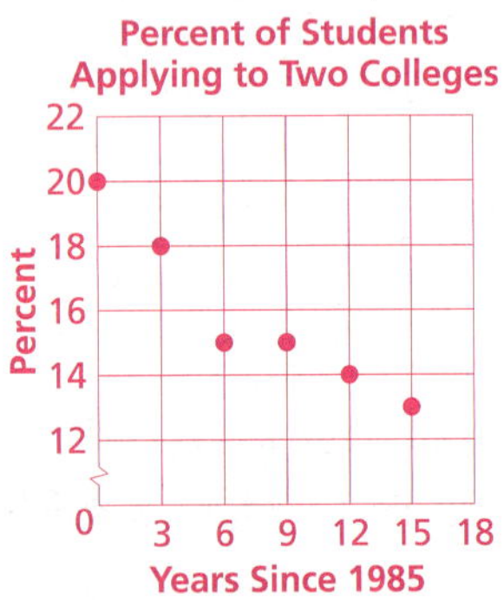

PREDICTION EQUATIONS

In-Class Example Power Point®

2 **EDUCATION** Refer to the data in In-Class Example 1.

a. Draw a line of fit for the data. How well does the line fit the data?

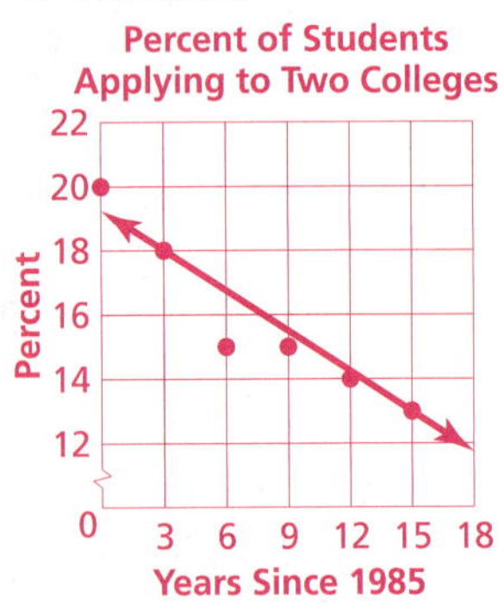

Except for (6, 15), the line fits the data fairly well.

(continued on the next page)

To find a line of fit and a prediction equation for a set of data, select two points that appear to represent the data well. This is a matter of personal judgment, so your line and prediction equation may be different from someone else's.

Study Tip

Reading Math

A data point that does not appear to belong with the rest of the set is called an **outlier**.

Example 2 Find and Use a Prediction Equation

HOUSING **Refer to the data in Example 1.**

a. Draw a line of fit for the data. How well does the line fit the data?

Ignore the point (0, 122.9) since it would not be close to a line that represents the rest of the data points. The points (4, 130.0) and (8, 152.5) appear to represent the data well. Draw a line through these two points. Except for (0, 122.9), this line fits the data very well.

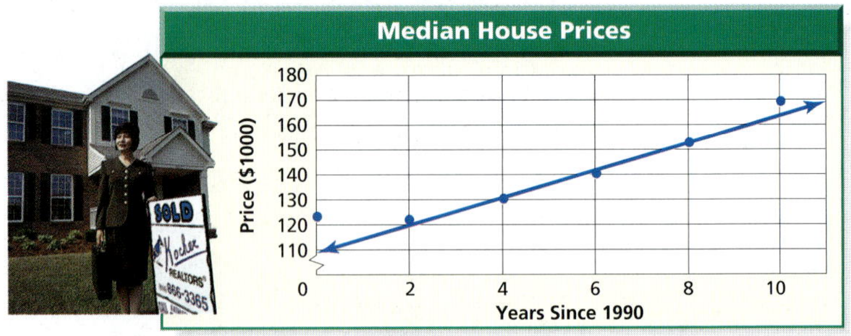

b. Find a prediction equation. What do the slope and y-intercept indicate?

Find an equation of the line through (4, 130.0) and (8, 152.5). Begin by finding the slope.

$m = \frac{y_2 - y_1}{x_2 - x_1}$ Slope formula

$= \frac{152.5 - 130.0}{8 - 4}$ Substitute.

≈ 5.63 Simplify.

$y - y_1 = m(x - x_1)$ Point-slope form

$y - 130.0 = 5.63(x - 4)$ $m = 5.63$, $(x_1, y_1) = (4, 130.0)$

$y - 130.0 = 5.63x - 22.52$ Distributive Property

$y = 5.63x + 107.48$ Add 130.0 to each side.

One prediction equation is $y = 5.63x + 107.48$. The slope indicates that the median price is increasing at a rate of about $5630 per year. The y-intercept indicates that, according to the trend of the rest of the data, the median price in 1990 should have been about $107,480.

Study Tip

Reading Math

When you are predicting for an x value greater than any in the data set, the process is known as **extrapolation**. When you are predicting for an x value between the least and greatest in the data set, the process is known as **interpolation**.

c. Predict the median price in 2010.

The year 2010 is 20 years after 1990, so use the prediction equation to find the value of y when $x = 20$.

$y = 5.63x + 107.48$ Prediction equation

$= 5.63(20) + 107.48$ $x = 20$

$= 220.08$ Simplify.

The model predicts that the median price in 2010 will be about $220,000.

d. How accurate is the prediction?

Except for the outlier, the line fits the data very well, so the predicted value should be fairly accurate.

TEACHING TIP

On pages 87 and 88, students will learn about a number that measures how well a line fits a set of data.

DAILY INTERVENTION

Differentiated Instruction

Naturalist Ask students how scatter plots and prediction equations might be used to relate local insect and animal populations to food supplies, temperature, and rainfall. Have interested students devise a plan for conducting the research necessary for developing such a prediction equation.

82 Chapter 2 Linear Relations and Functions

Algebra Activity

Head versus Height

Collect the Data

- Collect data from several of your classmates. Use a tape measure to measure the circumference of each person's head and his or her height. Record the data as ordered pairs of the form (height, circumference).

Analyze the Data 1–5. See students' work.

1. Graph the data in a scatter plot.
2. Choose two ordered pairs and write a prediction equation.
3. Explain the meaning of the slope in the prediction equation.

Make a Conjecture

4. Predict the head circumference of a person who is 66 inches tall.
5. Predict the height of an individual whose head circumference is 18 inches.

Study Tip

Outliers
If your scatter plot includes points that are far from the others on the graph, check your data before deciding it is an outlier. You may have made a graphing or recording mistake.

b. Find a prediction equation. What do the slope and y-intercept indicate? Using (3, 18) and (15, 13): $y = -0.42x + 19.25$; The slope indicates that the percent is falling at about 0.4% each year. The y-intercept indicates the percent in 1985 should have been about 19%.

c. Predict the percent in 2010. about 9%

d. How accurate is the prediction? The fit is only approximate, so the prediction may not be very accurate.

Check for Understanding

Concept Check

1. **Choose** the scatter plot with data that could best be modeled by a linear function. d

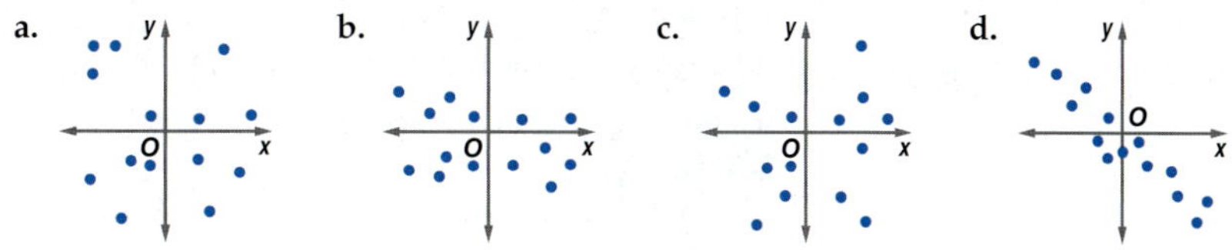

2. **Identify** the domain and range of the relation in the graph at the right. **Predict** the value of y when $x = 5$.

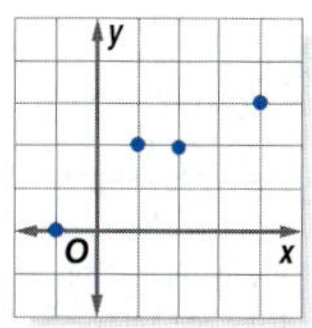

2. $D = \{-1, 1, 2, 4\}$, $R = \{0, 2, 3\}$; Sample answer using $(-1, 0)$ and $(2, 2)$: 4

3. **OPEN ENDED** Write a different prediction equation for the data in Examples 1 and 2 on pages 81 and 82.
Sample answer using (4, 130.0) and (6, 140.0): $y = 5x + 110$

Guided Practice

GUIDED PRACTICE KEY	
Exercises	Examples
4, 5	1, 2

Complete parts a–c for each set of data in Exercises 4 and 5.

a. **Draw a scatter plot.**
b. **Use two ordered pairs to write a prediction equation.**
c. **Use your prediction equation to predict the missing value.**

4. **SCIENCE** Whether you are climbing a mountain or flying in an airplane, the higher you go, the colder the air gets. The table shows the temperature in the atmosphere at various altitudes. 4–5. See pp. 107A–107H.

Altitude (ft)	0	1000	2000	3000	4000	5000
Temp (°C)	15.0	13.0	11.0	9.1	7.1	?

Source: NASA

5. **TELEVISION** As more channels have been added, cable television has become attractive to more viewers. The table shows the number of U.S. households with cable service in some recent years.

Year	1990	1992	1994	1996	1998	2010
Households (millions)	55	57	59	65	67	?

Source: Nielsen Media Research

www.algebra2.com/extra_examples

3 Practice/Apply

Study Notebook

Have students—

- add the definitions/examples of the vocabulary terms to their Vocabulary Builder worksheets for Chapter 2.
- write a summary of the procedure for placing a line of fit on a scatter plot and how to find a prediction equation using this line.
- include any other item(s) that they find helpful in mastering the skills in this lesson.

Algebra Activity

Materials: tape measure, grid paper

- You can use string and a ruler as an alternative way to measure the circumference of a person's head as well as their height.
- Measurements can be made in inches or in centimeters. You may want to have half the class use one system and the other half the alternate system.

About the Exercises...

Organization by Objective

- **Scatter Plots:** 6a, 7a, 8a, 9a, 15
- **Prediction Equations:** 6b, 6c, 7b, 7c, 8b, 8c, 9b, 9c, 10–14, 16–21

Odd/Even Assignments

Exercises 6–9 are structured so that students practice the same concepts whether they are assigned odd or even problems.

Alert! Exercise 21 involves research on the Internet or other reference materials.

Assignment Guide

Basic: 7, 10–12, 19–24, 31–42

Average: 7, 9, 10–12, 15–24, 31–42 (optional: 25–30)

Advanced: 6, 8, 13–37 (optional: 38–42)

Answers

6a. **Lives Saved by Minimum Drinking Age**

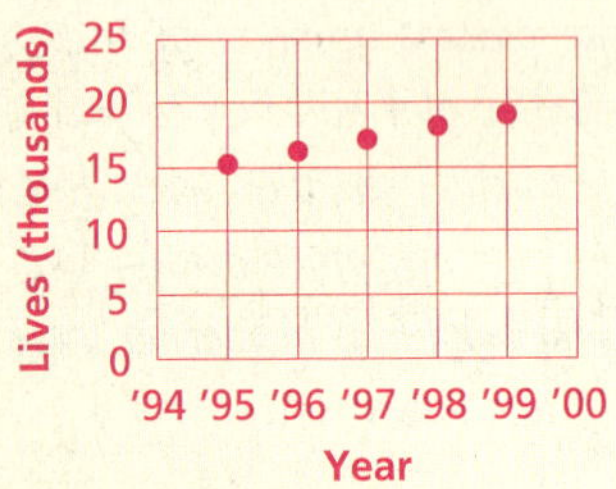

7a. **2000–2001 Detroit Red Wings**

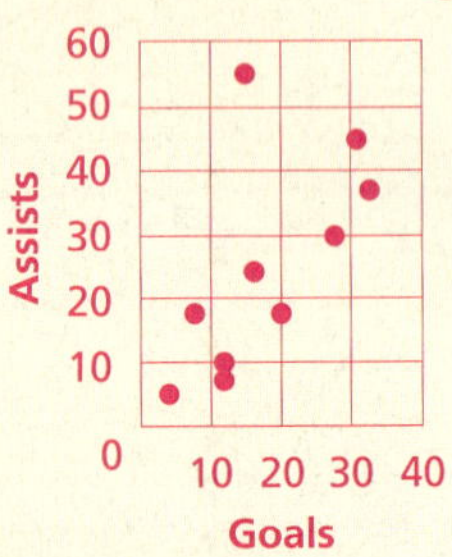

8a. **Bottled Water Consumption**

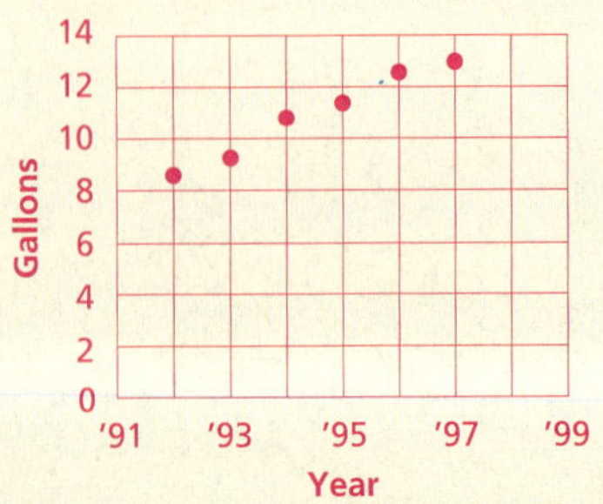

★ indicates increased difficulty

Practice and Apply

Homework Help

For Exercises	See Examples
6–21	1, 2

Extra Practice See page 831.

Complete parts a–c for each set of data in Exercises 6–9.

a. **Draw a scatter plot.**

b. **Use two ordered pairs to write a prediction equation.**

c. **Use your prediction equation to predict the missing value.**

6. **SAFETY** All states and the District of Columbia have enacted laws setting 21 as the minimum drinking age. The table shows the estimated cumulative number of lives these laws have saved by reducing traffic fatalities.

Year	1995	1996	1997	1998	1999	2010
Lives (1000s)	15.7	16.5	17.4	18.2	19.1	?

Source: National Highway Traffic Safety Administration

7. **HOCKEY** Each time a hockey player scores a goal, up to two teammates may be credited with assists. The table shows the number of goals and assists for some of the members of the Detroit Red Wings in the 2000–2001 NHL season.

Goals	31	15	32	27	16	20	8	4	12	12	?
Assists	45	56	37	30	24	18	17	5	10	7	15

Source: www.detroitredwings.com

8. **HEALTH** Bottled water has become very popular. The table shows the number of gallons of bottled water consumed per person in some recent years.

Year	1992	1993	1994	1995	1996	1997	2010
Gallons	8.2	9.4	10.7	11.6	12.5	13.1	?

Source: U.S. Department of Agriculture

★ 9. **THEATER** Broadway, in New York City, is the center of American theater. The table shows the total revenue of all Broadway plays for some recent seasons.

Season	'95–'96	'96–'97	'97–'98	'98–'99	'99–'00	'09–'10
Revenue ($ millions)	436	499	558	588	603	?

Source: The League of American Theatres and Producers, Inc.

6a. See margin.
6b. Sample answer using (1996, 16.5) and (1998, 18.2): $y = 0.85x - 1680.1$
6c. Sample answer: 28,400
7a. See margin.
7b. Sample answer using (4,5) and (32,37): $y = 1.14x + 0.44$
7c. Sample answer: about 13

WebQuest

A scatter plot of loan payments can help you analyze home loans. Visit www.algebra2.com/webquest to continue work on your WebQuest project.

8a. See margin.
8b. Sample answer using (1993, 9.4) and (1996, 12.5): $y = 1.03x - 2043.39$
8c. Sample answer: about 26.9 gal
9a. See pp. 107A–107H.
9b. Sample answer using (1, 499) and (3, 588): $y = 44.5x + 454.5$, where x is the number of seasons since 1995–1996
9c. Sample answer: about $1078 million or $1.1 billion

MEDICINE **For Exercises 10–12, use the graph that shows how much Americans spent on doctors' visits in some recent years.**

10. Write a prediction equation from the data for 1990, 1995, and 2000.

11. Use your equation to predict the amount for 2005.

12. Compare your prediction to the one given in the graph. **The value predicted by the equation is somewhat lower than the one given in the graph.**

10. Sample answer using (1990, 563) and (1995, 739): $y = 35.2x - 69,485$
11. Sample answer: $1091

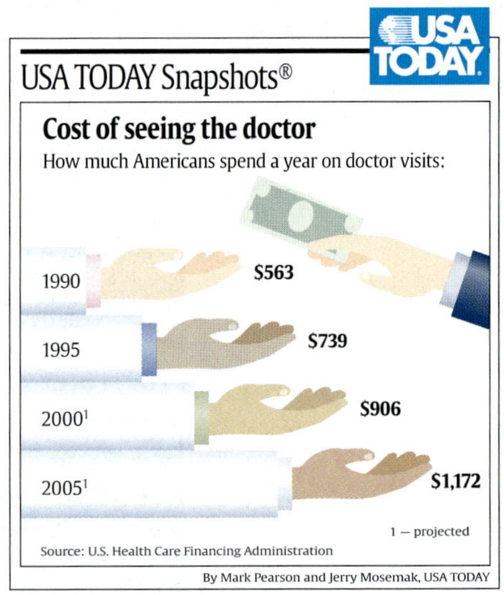

Online Lesson Plans

USA TODAY Education's Online site offers resources and interactive features connected to each day's newspaper. *Experience TODAY*, USA TODAY's daily lesson plan, is available on the site and delivered daily to subscribers. This plan provides instruction for integrating USA TODAY graphics and key editorial features into your mathematics classroom. Log on to www.education.usatoday.com.

Career Choices

Finance

A financial analyst can advise people about how to invest their money and plan for retirement.

Online Research For information about a career as a financial analyst, visit: www.algebra2.com/careers

FINANCE **For Exercises 13 and 14, use the following information.**
Della has $1000 that she wants to invest in the stock market. She is considering buying stock in either Company 1 or Company 2. The values of the stocks at the ends of the last four months are shown in the tables below.

Company 1	
Month	Share Price ($)
Aug.	25.13
Sept.	22.94
Oct.	24.19
Nov.	22.56

Company 2	
Month	Share Price ($)
Aug.	31.25
Sept.	32.38
Oct.	32.06
Nov.	32.44

★13. Based only on these data, which stock should Della buy? Explain.

★14. Do you think investment decisions should be based on this type of reasoning? If not, what other factors should be considered?

13–14. See pp. 107A–107H.

GEOGRAPHY **For Exercises 15–18, use the table below that shows the elevation and average precipitation for selected cities.**

City	Elevation (feet)	Average Precip. (inches)	City	Elevation (feet)	Average Precip. (inches)
Rome, Italy	79	33	London, England	203	30
Algiers, Algeria	82	27	Paris, France	213	26
Istanbul, Turkey	108	27	Bucharest, Romania	298	23
Montreal, Canada	118	37	Budapest, Hungary	456	20
Stockholm, Sweden	171	21	Toronto, Canada	567	31
Berlin, Germany	190	23			

Source: World Meteorological Association

★15. Draw a scatter plot with elevation as the independent variable. See pp. 107A–107H.

★16. Write a prediction equation.

16. Sample answer using (213, 26) and (298, 23): $y = -0.04x + 34.52$

★17. Predict the average annual precipitation for Dublin, Ireland, which has an elevation of 279 feet. Sample answer: about 23 in.

★18. Compare your prediction to the actual value of 29 inches. See margin.

CRITICAL THINKING **For Exercises 19 and 20, use the table that shows the percent of people ages 25 and over with a high school diploma over the last few decades.**

High School Graduates

Year	Percent
1970	52.3
1975	62.5
1980	66.5
1985	73.9
1990	77.6
1995	81.7
1999	83.4

Source: U.S. Census Bureau

19. Use a prediction equation to predict the percent in 2010.

19. Sample answer using (1975, 62.5) and (1995, 81.7): 96.1%

20. Do you think your prediction is accurate? Explain. See margin.

21. **RESEARCH** Use the Internet or other resource to look up the population of your community or state in several past years. Use a prediction equation to predict the population in some future year. See students' work.

www.algebra2.com/self_check_quiz

Answers

18. Sample answer: The predicted value differs from the actual value by more than 20%, possibly because no line fits the data very well.

20. Sample answer: The predicted percent is almost certainly too high. Since the percent cannot exceed 100%, it cannot continue to increase indefinitely at a linear rate.

Enrichment, p. 86

Median-Fit Lines

A **median-fit line** is a particular type of line of fit. Follow the steps below to find the equation of the median-fit line for the data.

Approximate Percentage of Violent Crimes Committed by Juveniles That Victims Reported to Law Enforcement

Year	1980	1982	1984	1986	1988	1990	1992	1994	1996
Offenders	36	35	33	32	31	30	29	29	30

Source: U.S. Bureau of Justice Statistics

1. Divide the data into three approximately equal groups. There should always be the same number of points in the first and third groups. In this case, there will be three data points in each group.

Group 1		Group 2		Group 3	
Year	Offenders	Year	Offenders	Year	Offenders

Study Guide and Intervention, p. 81 (shown) and p. 82

Scatter Plots When a set of data points is graphed as ordered pairs in a coordinate plane, the graph is called a **scatter plot**. A scatter plot can be used to determine if there is a relationship among the data.

Example **BASEBALL The table below shows the number of home runs and runs batted in for various baseball players who won the Most Valuable Player Award during the 1990s. Make a scatter plot of the data.**

Home Runs	Runs Batted In
33	114
39	116
40	130
28	61
41	128
47	144

Source: *New York Times Almanac*

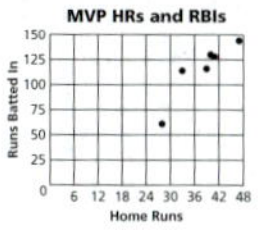

Exercises

Make a scatter plot for the data in each table below.

1. **FUEL EFFICIENCY** The table below shows the average fuel efficiency in miles per gallon of new cars manufactured during the years listed.

Year	Fuel Efficiency (mpg)
1960	15.5
1970	14.1
1980	22.6
1990	26.9

Source: *New York Times Almanac*

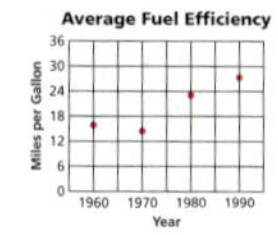

2. **CONGRESS** The table below shows the number of women serving in the United States Congress during the years 1987–1999.

Congressional Session	Number of Women
100	25
101	31
102	33
103	55
104	58
105	62

Source: *Wall Street Journal Almanac*

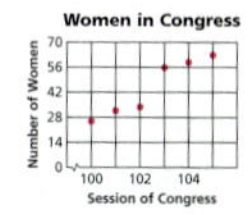

Skills Practice, p. 83 and Practice, p. 84 (shown)

For Exercises 1–3, complete parts a–c for each set of data.
a. Draw a scatter plot.
b. Use two ordered pairs to write a prediction equation.
c. Use your prediction equation to predict the missing value.

1. **FUEL ECONOMY** The table gives the approximate weights in tons and estimates for overall fuel economy in miles per gallon for several cars.

Weight (tons)	1.3	1.4	1.5	1.8	2	2.1	2.4
Miles per Gallon	29	24	23	21	?	17	15

1b. Sample answer using (1.4, 24) and (2.4, 15): $y = -9x + 36.6$

1c. Sample answer: 18.6 mi/gal

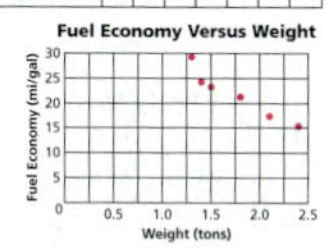

2. **ALTITUDE** In most cases, temperature decreases with increasing altitude. As Anchara drives into the mountains, her car thermometer registers the temperatures (°F) shown in the table at the given altitudes (feet).

Altitude (ft)	7500	8200	8600	9200	9700	10,400	12,000
Temperature (°F)	61	58	56	53	50	46	?

2b. Sample answer using (7500, 61) and (9700, 50): $y = -0.005x + 98.5$

2c. Sample answer: 38.5°F

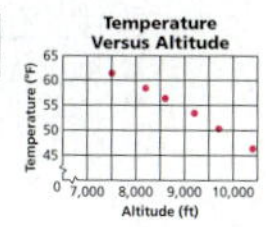

3. **HEALTH** Alton has a treadmill that uses the time on the treadmill and the speed of walking or running to estimate the number of Calories he burns during a workout. The table gives workout times and Calories burned for several workouts.

Time (min)	18	24	30	40	42	48	52	60
Calories Burned	260	280	320	380	400	440	475	?

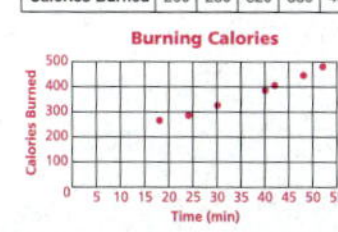

3b. Sample answer using (24, 280) and (48, 440): $y = 6.67x + 119.92$

3c. Sample answer: about 520 calories

Reading to Learn Mathematics, p. 85 ELL

Pre-Activity **How can a linear equation model the number of Calories you burn exercising?**

Read the introduction to Lesson 2-5 at the top of page 81 in your textbook.

- If a woman runs 5.5 miles per hour, about how many Calories will she burn in an hour? Sample answer: 572 Calories
- If a man runs 7.5 miles per hour, about how many Calories will he burn in half an hour? Sample answer: 397 Calories

Reading the Lesson

1. Suppose that a set of data can be modeled by a linear equation. Explain the difference between a scatter plot of the data and a graph of the linear equation that models that data.
Sample answer: The scatter plot is a discrete graph. It is made up just of the individual points that represent the data points. The linear equation has a continuous graph that is the line that best fits the data points.

2. Suppose that tuition at a state college was $3500 per year in 1995 and has been increasing at a rate of $225 per year.
   a. Write a prediction equation that expresses this information.
   $y = 3500 + 225x$
   b. Explain the meaning of each variable in your prediction equation.
   x represents the number of year since 1995 and y represents the tuition in that year.

3. Use this model to predict the tuition at this college in 2007. $6200

Helping You Remember

4. Look up the word *scatter* in a dictionary. How can its definition help you to remember the meaning of the difference between a scatter plot and the graph of a linear equation?
Sample answer: To *scatter* means to break up and go in many directions. The points on a scatter plot are broken up. In a scatter plot, the points are scattered or broken up. In the graph of a linear equation, the points are connected to form a continuous line.

4 Assess

Open-Ended Assessment

Writing Ask students to write instructions that could be used to teach a friend how to read a graph such as the one shown in Example 2 on p. 82. The importance of reading titles, captions, and text, as well as the scales should be included as part of the instructions.

Getting Ready for Lesson 2-6

PREREQUISITE SKILL Lesson 2-6 presents the graphing of special functions, including absolute value functions. Exercises 38–42 should be used to determine your students' familiarity with finding absolute values.

Answer

22. Data can be used to write a linear equation that approximates the number of Calories burned per hour in terms of the speed that a person runs. Answers should include the following.

- 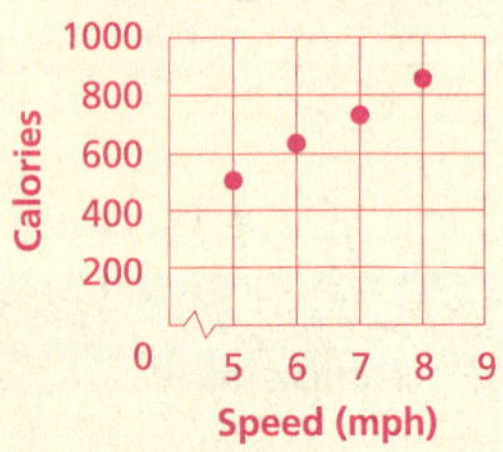

  Sample answer using (5, 508) and (8, 858): $y = 116.67x - 75.35$
- about 975 calories; Sample answer: The predicted value differs from the actual value by only about 2%.

---

22. **WRITING IN MATH** Answer the question that was posed at the beginning of the lesson.

How can a linear equation model the number of Calories you burn exercising?

Include the following in your answer: **See margin.**

- a scatter plot and a prediction equation for the data, and
- a prediction of the number of Calories burned in an hour by a 140-pound person running at 9 miles per hour, with a comparison of your predicted value with the actual value of 953.

Standardized Test Practice

23. Which line best fits the data in the graph at the right? **D**

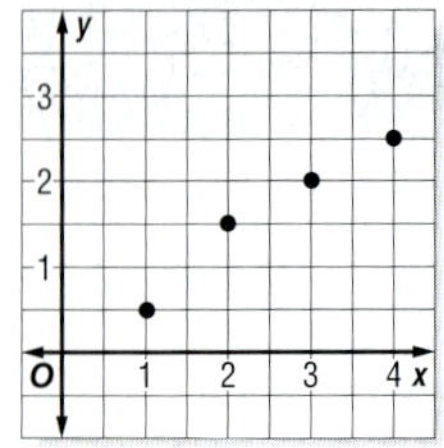

Ⓐ $y = x$ Ⓑ $y = -0.5x + 4$

Ⓒ $y = -0.5x - 4$ Ⓓ $y = 0.5 + 0.5x$

24. A prediction equation for a set of data is $y = 0.63x + 4.51$. For which x value is the predicted y value 6.4? **A**

Ⓐ 3 Ⓑ 4.5

Ⓒ 6 Ⓓ 8.54

Extending the Lesson

For Exercises 25–30, use the following information.

A **median-fit line** is a particular type of line of fit. Follow the steps below to find the equation of the median-fit line for the data.

Federal and State Prisoners (per 100,000 U.S. citizens)								
Year	1986	1988	1990	1992	1994	1996	1998	1999
Prisoners	217	247	297	332	389	427	461	476

Source: U.S. Bureau of Justice Statistics

25. 1988, 1993, 1998; 247, 360.5, 461

25. Divide the data into three approximately equal groups. There should always be the same number of points in the first and third groups. Find x_1, x_2, and x_3, the medians of the x values in Groups 1, 2, and 3, respectively. Find y_1, y_2, and y_3, the medians of the y values in Groups 1, 2, and 3, respectively.

26. Find an equation of the line through (x_1, y_1) and (x_3, y_3). $y = 21.4x - 42{,}296.2$

27. Find Y, the y-coordinate of the point on the line in Exercise 26 with an x-coordinate of x_2. **354**

28. about (1993, 356.17)

28. The median-fit line is parallel to the line in Exercise 26, but is one-third closer to (x_2, y_2). This means it passes through $\left(x_2, \frac{2}{3}Y + \frac{1}{3}y_2\right)$. Find this ordered pair.

29. Write an equation of the median-fit line. $y = 21.4x - 42{,}294.03$

30. Predict the number of prisoners per 100,000 citizens in 2005 and 2010. **about 613, about 720**

Maintain Your Skills

Mixed Review

Write an equation in slope-intercept form that satisfies each set of conditions. *(Lesson 2-4)* **31.** $y = 4x + 6$

31. slope 4, passes through (0, 6)

32. passes through (5, −3) and (−2, 0)

32. $y = -\frac{3}{7}x - \frac{6}{7}$

Find each value if $g(x) = -\frac{4x}{3} + 7$. *(Lesson 2-1)*

33. $g(3)$ **3**
34. $g(0)$ **7**
35. $g(-2)$ $\frac{29}{3}$
36. $g(-4)$ $\frac{37}{3}$
37. Solve $|x + 4| > 3$. *(Lesson 1-6)* $\{x \mid x < -7 \text{ or } x > -1\}$

Getting Ready for the Next Lesson

PREREQUISITE SKILL **Find each absolute value.** *(To review **absolute value**, see Lesson 1-4.)*

38. $|-3|$ **3**
39. $|11|$ **11**
40. $|0|$ **0**
41. $\left|-\frac{2}{3}\right|$ $\frac{2}{3}$
42. $|-1.5|$ **1.5**

Teacher to Teacher

Susan Nelson Spring H.S., Spring, TX

"I have my students take measurements of their height and arm span and record them. We enter the entire class' data into a graphing calculator and find the linear regression. Then we use the regression equation to make predictions."

86 Chapter 2 Linear Relations and Functions

Lines of Regression

You can use a TI-83 Plus graphing calculator to find a line that best fits a set of data. This line is called a **regression line** or **line of best fit**. You can also use the calculator to draw scatter plots and make predictions.

INCOME The table shows the median income of U.S. families for the period 1970–1998.

Year	1970	1980	1985	1990	1995	1998
Income ($)	9867	21,023	27,735	35,353	40,611	46,737

Source: U.S. Census Bureau

Find and graph a regression equation. Then predict the median income in 2010.

Step 1 *Find a regression equation.*

- Enter the years in L1 and the incomes in L2.

  KEYSTROKES: STAT ENTER 1970 ENTER

- Find the regression equation by selecting LinReg(ax+b) on the STAT CALC menu.

  KEYSTROKES: STAT ▶ 4 ENTER

The regression equation is about $y = 1304.19x - 2{,}560{,}335.07$.

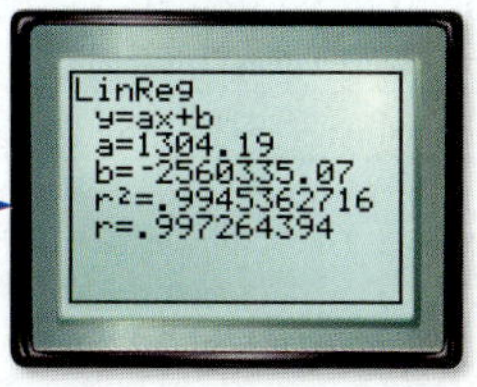

The slope indicates that family incomes were increasing at a rate of about $1300 per year.

The number r is called the **linear correlation coefficient**. The closer the value of r is to 1 or -1, the closer the data points are to the line.

If the values of r^2 and r are not displayed, use DiagnosticOn from the CATALOG menu.

Step 2 *Graph the regression equation.*

- Use STAT PLOT to graph a scatter plot.

  KEYSTROKES: 2nd [STAT PLOT] ENTER ENTER

- Select the scatter plot, L1 as the Xlist, and L2 as the Ylist.
- Copy the equation to the Y= list and graph.

  KEYSTROKES: Y= VARS 5 ▶ ▶ 1 GRAPH

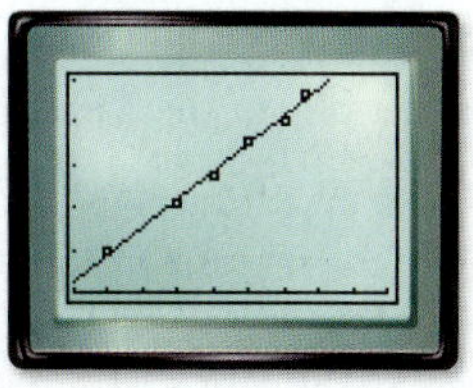

[1965, 2010] scl: 5 by [0, 50,000] scl: 10,000

Notice that the regression line does not pass through any of the data points, but comes close to all of them. The line fits the data very well.

Step 3 *Predict using the regression equation.*

- Find y when $x = 2010$. Use value on the CALC menu.

  KEYSTROKES: 2nd CALC 1 2010 ENTER

According to the regression equation, the median family income in 2010 will be about $61,087.

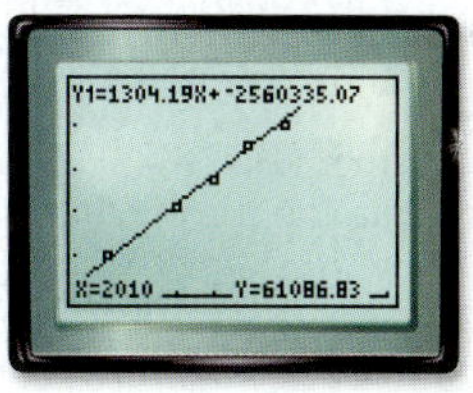

www.algebra2.com/other_calculator_keystrokes

Graphing Calculator Investigation

A Follow-Up of Lesson 2-5

Getting Started

Know Your Calculator The TI-83 Plus graphing calculator has a linear regression function, LinReg($ax + b$), that uses a least-squares fit method to determine the values for a and b. This involves calculus and finding the distance from each point to the line of best fit.

Correlation Coefficient With DiagnosticOn, the calculator also displays values for r^2 and r. The closer the value of $|r|$ is to 1, the better the equation fits the data.

Teach

- Make sure students have cleared the L_1 and L_2 lists before they enter the new data.
- After students have entered data, have them work in pairs to verify that the data display shows the correct numbers before proceeding.
- Have students complete Exercises 1–15.

Exercises

GOVERNMENT **For Exercises 1–3, use the table below that shows the population and the number of representatives in Congress for selected states.**

State	CA	NY	TX	FL	NC	IN	AL
Population (millions)	29.8	18.0	17.0	12.9	6.6	5.5	4.0
Representatives	52	31	30	23	12	10	7

Source: *The World Almanac*

1. Make a scatter plot of the data. **See margin.**
2. Find a regression equation for the data. **$y = 1.73x + 0.39$**
3. Predict the number of representatives for Oregon, which has a population of about 2.8 million. **5**

BASEBALL **For Exercises 4–6, use the table at the right that shows the total attendance for minor league baseball in some recent years.**

Year	Attendance (millions)
1985	18.4
1990	25.2
1995	33.1
2000	37.6

Source: National Association of Professional Baseball Leagues

4. Make a scatter plot of the data. **See margin.**
5. Find a regression equation for the data. **$y = 1.31x - 2581.6$**
6. Predict the attendance in 2010. **51,500,000**

TRANSPORTATION **For Exercises 7–11, use the table below that shows the retail sales of motor vehicles in the United States for the period 1992–1999.**

Motor Vehicle Sales								
Year	1992	1993	1994	1995	1996	1997	1998	1999
Vehicles (thousands)	13,118	14,199	15,413	15,118	15,456	15,498	15,963	17,414

Source: American Automobile Manufacturers Association

9. about 470,000 vehicles more per year

7. Make a scatter plot of the data. **See margin.**
8. Find a regression equation for the data. **$y = 470.06x - 922{,}731.40$**
9. According to the regression equation, what was the average rate of change of vehicle sales during the period?
10. Predict the sales in 2010. **about 22,089,000**
11. How accurate do you think your prediction is? Explain. **See margin.**

RECREATION **For Exercises 12–15, use the table at the right that shows the amount of money spent on skin diving and scuba equipment in some recent years.** **14. about $440,000,000**

Skin Diving and Scuba Equipment	
Year	Sales ($ millions)
1993	315
1994	322
1995	328
1996	340
1997	332
1998	345
1999	363

Source: National Sporting Goods Association

12. Find a regression equation for the data. **$y = 6.93x - 13{,}494.43$**
13. Delete the outlier (1997, 332) from the data set. Then find a new regression equation for the data. **$y = 7.36x - 14{,}354.33$**
14. Use the new regression equation to predict the sales in 2010.
15. Compare the new correlation coefficient to the old value and state whether the regression line fits the data better. **See margin.**

Assess

Refer students to the calculator display shown in Step 1 on p. 87, and ask the following questions.

- What is the regression equation for the income data? **$y = 1304.19x - 2{,}560{,}335.07$**
- What does the value of r tell you about the regression equation? **The value of r is very close to 1, so the data points are very close to the graph of that equation.**

Answers

1. 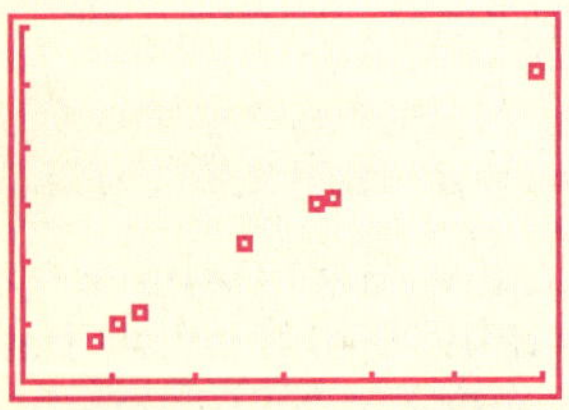

[0, 30] scl: 5 by [0, 60] scl: 10

4. 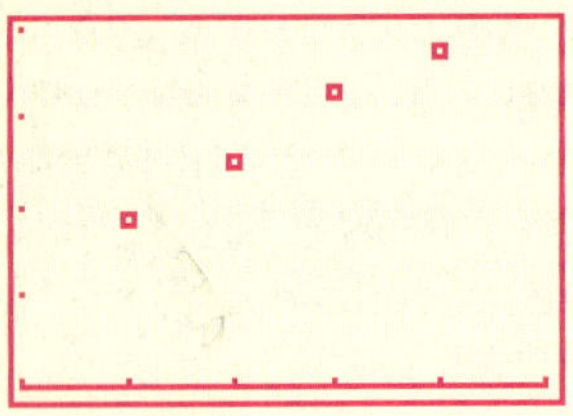

[1980, 2005] scl: 5 by [0, 40] scl: 10

7. 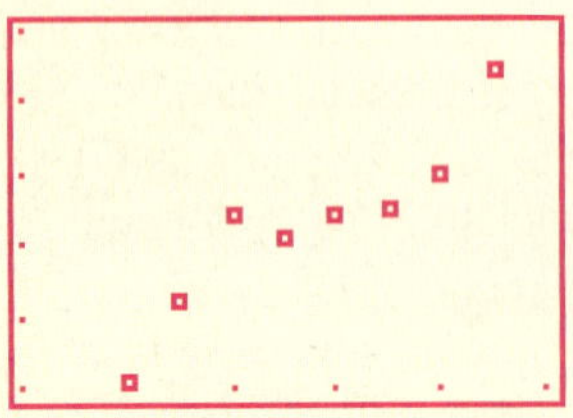

[1990, 2000] scl: 2 by [13,000, 18,000] scl: 1000

11. **The prediction may not be accurate because different parts of the data could be represented by lines with different slopes. The sales could drop, as they did in 1995, or they could level out, as they did in 1996 and 1997.**

15. **The correlation coefficient, 0.9761660092, is closer to 1. The new regression line fits the data better.**

2-6 Special Functions

What You'll Learn

- Identify and graph step, constant, and identity functions.
- Identify and graph absolute value and piecewise functions.

How do step functions apply to postage rates?

The cost of the postage to mail a letter is a function of the weight of the letter. But the function is not linear. It is a special function called a **step function**.

Vocabulary

- step function
- greatest integer function
- constant function
- identity function
- absolute value function
- piecewise function

STEP FUNCTIONS, CONSTANT FUNCTIONS, AND THE IDENTITY FUNCTION The graph of a step function is not linear. It consists of line segments or rays. The **greatest integer function**, written $f(x) = [\![x]\!]$, is an example of a step function. The symbol $[\![x]\!]$ means *the greatest integer less than or equal to x*. For example, $[\![7.3]\!] = 7$ and $[\![-1.5]\!] = -2$ because $-1 > -1.5$. Study the table and graph below.

Study Tip

Greatest Integer Function

Notice that the domain of this step function is all real numbers and the range is all integers.

$f(x) = [\![x]\!]$	
x	$f(x)$
$-3 \le x < -2$	-3
$-2 \le x < -1$	-2
$-1 \le x < 0$	-1
$0 \le x < 1$	0
$1 \le x < 2$	1
$2 \le x < 3$	2
$3 \le x < 4$	3

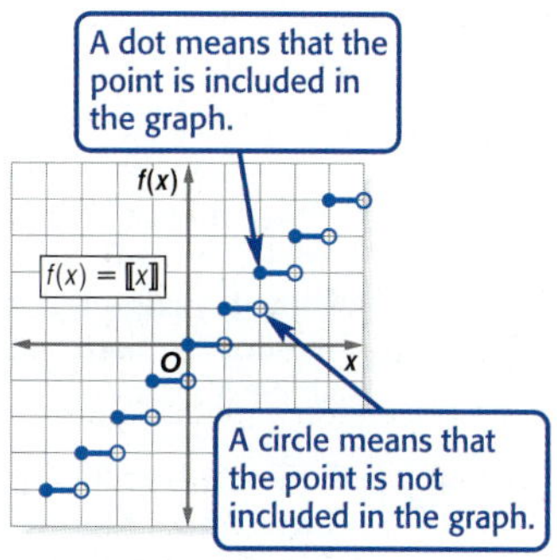

Example 1 *Step Function*

BUSINESS Labor costs at the Fix-It Auto Repair Shop are $60 per hour or any fraction thereof. Draw a graph that represents this situation.

Explore The total labor charge must be a multiple of $60, so the graph will be the graph of a step function.

Plan If the time spent on labor is greater than 0 hours, but less than or equal to 1 hour, then the labor cost is $60. If the time is greater than 1 hour but less than or equal to 2 hours, then the labor cost is $120, and so on.

Solve Use the pattern of times and costs to make a table, where x is the number of hours of labor and $C(x)$ is the total labor cost. Then draw the graph.

(continued on the next page)

2-6 Lesson Notes

1 Focus

5-Minute Check Transparency 2-6 Use as a quiz or review of Lesson 2-5.

Mathematical Background notes are available for this lesson on p. 54D.

Building on Prior Knowledge

In Lesson 2-5, students drew scatter plots. In this lesson they draw graphs of several special functions: step, constant, identity, absolute value, and piecewise.

How do step functions apply to postage rates?

Ask students:

- What is the cost of mailing a letter when the weight is 0.9 ounces? **$0.34** when it is 1.1 ounces? **$0.55** What is the ratio of the change in price over the change in weight from 0.9 ounce to 1.1 ounces? **1.05** from 1.8 ounces to 2.0 ounces? **0**
- How can you tell that this is not a linear function? **Sample answer: The slope between some pairs of points is not the same as the slope between other pairs of points.**

Resource Manager

Workbook and Reproducible Masters

Chapter 2 Resource Masters

- Study Guide and Intervention, pp. 87–88
- Skills Practice, p. 89
- Practice, p. 90
- Reading to Learn Mathematics, p. 91
- Enrichment, p. 92
- Assessment, p. 114

Graphing Calculator and Spreadsheet Masters, p. 29

Teaching Algebra With Manipulatives Masters, p. 219

Transparencies

5-Minute Check Transparency 2-6
Answer Key Transparencies

Technology

Interactive Chalkboard

2 Teach

STEP FUNCTIONS, CONSTANT FUNCTIONS, AND THE IDENTITY FUNCTION

In-Class Examples

Power Point®

1 **PSYCHOLOGY** One psychologist charges for counseling sessions at the rate of \$85 per hour or any fraction thereof. Draw a graph that represents this situation.

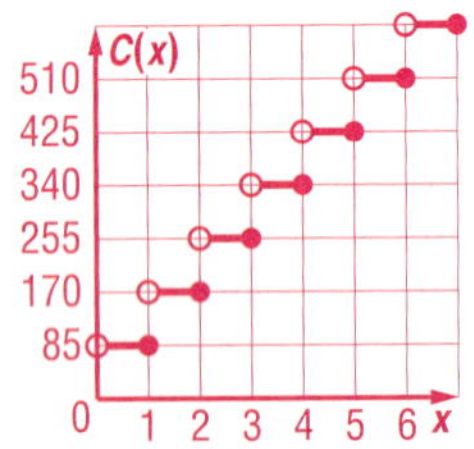

Teaching Tip Help clear up confusions about step functions by asking students to choose several sample times, in hours and minutes, and find the associated costs. Lead them to see that two different times (x values) can have the same cost (C value).

2 Graph $g(x) = -3$.

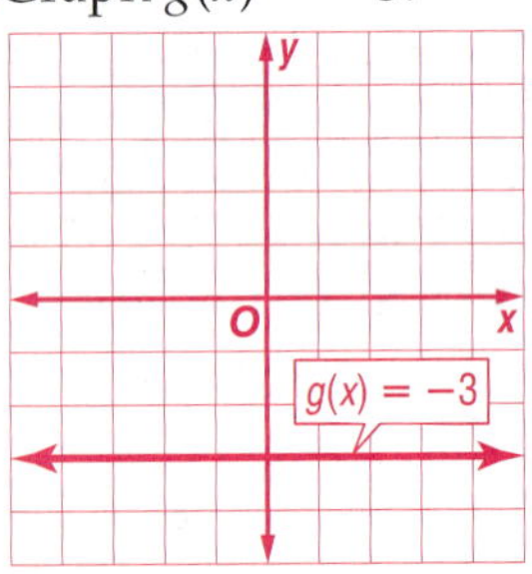

Tips for New Teachers

Intervention Students who are having trouble with the absolute value graphs may see more clearly what is happening if they draw their own graph of the parent function $y = |x|$.

x	$C(x)$
$0 < x \le 1$	\$60
$1 < x \le 2$	\$120
$2 < x \le 3$	\$180
$3 < x \le 4$	\$240
$4 < x \le 5$	\$300

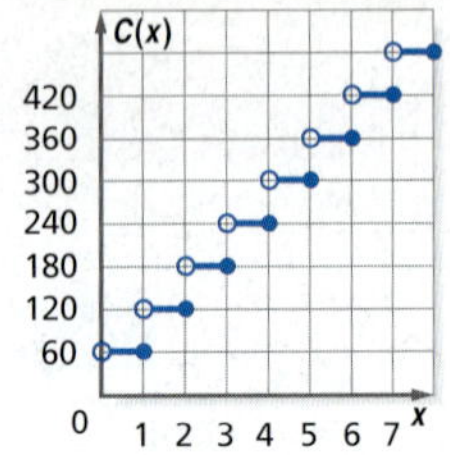

Examine Since the shop rounds any fraction of an hour up to the next whole number, each segment on the graph has a circle at the left endpoint and a dot at the right endpoint.

You learned in Lesson 2-4 that the slope-intercept form of a linear function is $y = mx + b$, or in functional notation, $f(x) = mx + b$. When $m = 0$, the value of the function is $f(x) = b$ for every x value. So, $f(x) = b$ is called a **constant function**.
The function $f(x) = 0$ is called the zero function.

Example 2 Constant Function

Graph $f(x) = 3$.

For every value of x, $f(x) = 3$. The graph is a horizontal line.

$f(x) = 3$	
x	$f(x)$
−2	3
−0.5	3
0	3
$\frac{1}{3}$	3

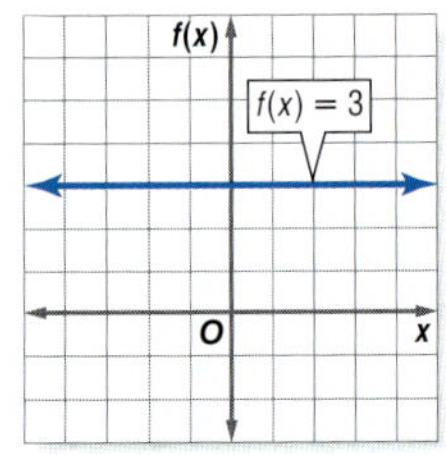

Another special case of slope-intercept form is $m = 1, b = 0$. This is the function $f(x) = x$. The graph is the line through the origin with slope 1.

Since the function does not change the input value, $f(x) = x$ is called the **identity function**.

$f(x) = x$	
x	$f(x)$
−2	−2
−0.5	−0.5
0	0
$\frac{1}{3}$	$\frac{1}{3}$

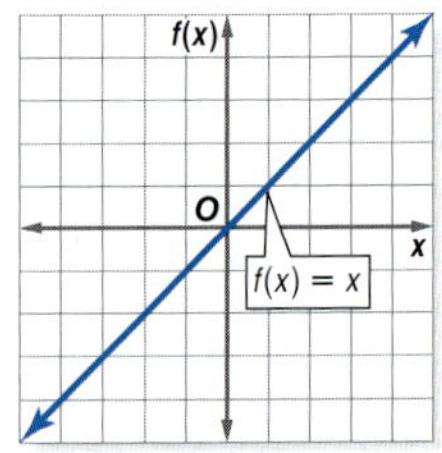

Study Tip

Absolute Value Function
Notice that the domain is all real numbers and the range is all nonnegative real numbers.

ABSOLUTE VALUE AND PIECEWISE FUNCTIONS Another special function is the **absolute value function**, $f(x) = |x|$.

$f(x) = \|x\|$	
x	$f(x)$
−3	3
−2	2
−1	1
0	0
1	1
2	2
3	3

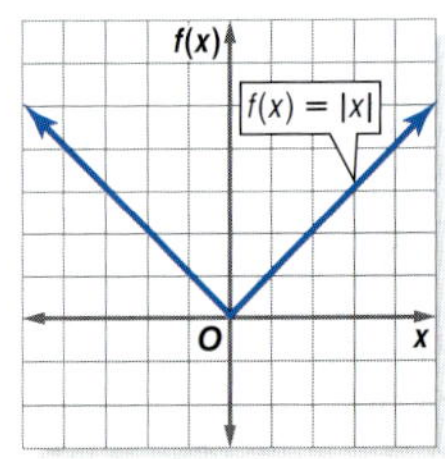

90 Chapter 2 Linear Relations and Functions

The absolute value function can be written as $f(x) = \begin{cases} -x \text{ if } x < 0 \\ x \text{ if } x \geq 0 \end{cases}$. A function that is written using two or more expressions is called a **piecewise function**.

Recall that a family of graphs is a group of graphs that displays one or more similar characteristics. The parent graph of most absolute value functions is $y = |x|$.

Study Tip

Look Back
To review **families of graphs**, see Lesson 2-3.

Example 3 Absolute Value Functions

Graph $f(x) = |x| + 1$ and $g(x) = |x| - 2$ on the same coordinate plane. Determine the similarities and differences in the two graphs.

Find several ordered pairs for each function.

x	|x| + 1
−2	3
−1	2
0	1
1	2
2	3

x	|x| − 2
−2	0
−1	−1
0	−2
1	−1
2	0

Graph the points and connect them.

- The domain of each function is all real numbers.
- The range of $f(x) = |x| + 1$ is $\{y \mid y \geq 1\}$.
  The range of $g(x) = |x| - 2$ is $\{y \mid y \geq -2\}$.
- The graphs have the same shape, but different y-intercepts.
- The graph of $g(x) = |x| - 2$ is the graph of $f(x) = |x| + 1$ translated down 3 units.

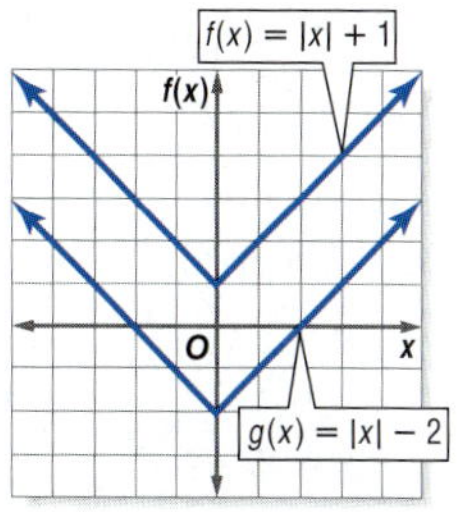

You can also use a graphing calculator to investigate families of absolute value graphs.

TEACHING TIP
abs(is located on the MATH NUM menu.

Graphing Calculator Investigation

Families of Absolute Value Graphs

The calculator screen shows the graphs of $y = |x|$, $y = 2|x|$, $y = 3|x|$, and $y = 5|x|$.

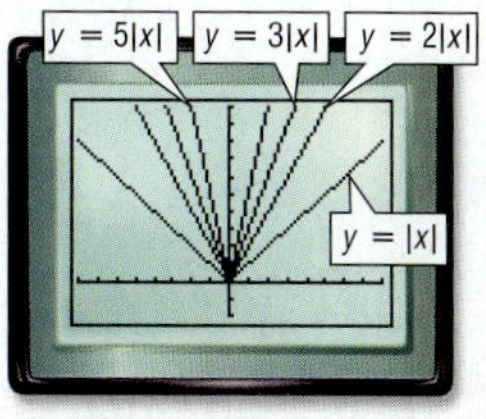

[−8, 8] scl: 1 by [−2, 10] scl: 1

Think and Discuss

1. What do these graphs have in common?
2. Describe how the graph of $y = a|x|$ changes as a increases. Assume $a > 0$.
3. Write an absolute value function whose graph is between the graphs of $y = 2|x|$ and $y = 3|x|$. **Sample answer: $y = 2.5|x|$**
4. Graph $y = |x|$ and $y = -|x|$ on the same screen. Then graph $y = 2|x|$ and $y = -2|x|$ on the same screen. What is true in each case?
5. In general, what is true about the graph of $y = a|x|$ when $a < 0$?

1. All of the graphs have a corner point at the origin.
2. The graph becomes narrower.
4. The graphs are reflections of each other about the x-axis.
5. The graph opens downward.

www.algebra2.com/extra_examples

ABSOLUTE VALUE AND PIECEWISE FUNCTIONS

In-Class Example

Power Point®

Teaching Tip Lead students to see that the y-intercept, b, of the graph of an absolute value function of the form $y = |x| + b$ indicates how the parent graph of $y = |x|$ is translated.

3 Graph $f(x) = |x - 3|$ and $g(x) = |x + 2|$ on the same coordinate plane. Determine the similarities and differences in the two graphs.

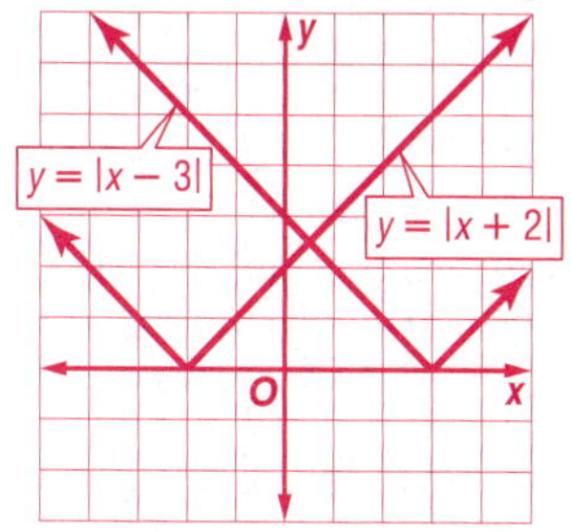

The domain of both graphs is all real numbers. The range of $(x) = |x - 3|$ is $\{y|y \geq 0\}$. The range of $(x) = |x + 2|$ is $\{y|y \geq 0\}$. The graphs have the same shape, but different x-intercepts. The graph of $(x) = |x + 2|$ is the graph of $(x) = |x - 3|$ translated 5 units to the left.

Graphing Calculator Investigation

Families of Absolute Value Graphs It is important that students arrive at the conclusion in Exercise 2 that the graph of the function narrows as the coefficient a increases. As an extension, ask students to compare the graph of $y = |x|$ with the graph of $y = |x| + 2$ and with $y = 2|x|$. Make sure they see the difference between adding 2 and having 2 as a coefficient.

In-Class Examples

Power Point®

4 Graph $f(x) = \begin{cases} x - 1 \text{ if } x \leq 3 \\ -1 \text{ if } x > 3 \end{cases}$. Identify the domain and range.

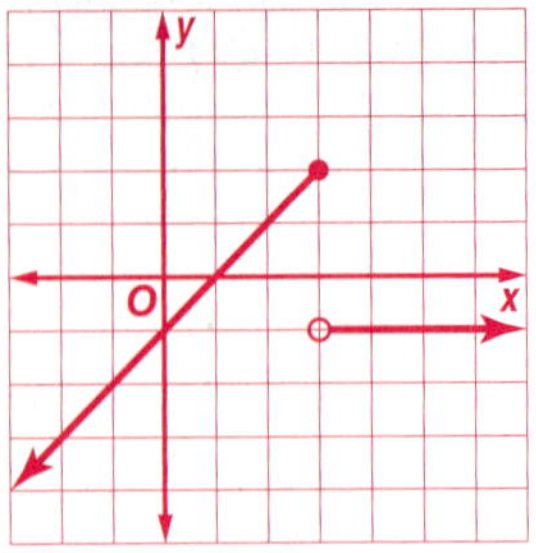

The domain is all real numbers. The range is $\{y | y \leq 2\}$.

5 Determine whether each graph represents a step function, a constant function, an absolute value function, or a piecewise function.

a.

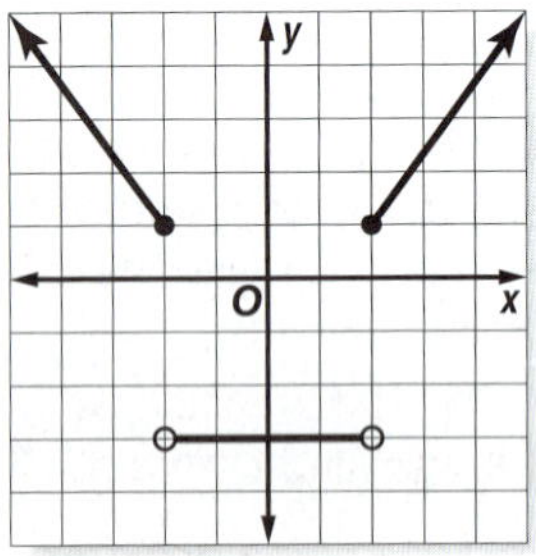

piecewise function

b.

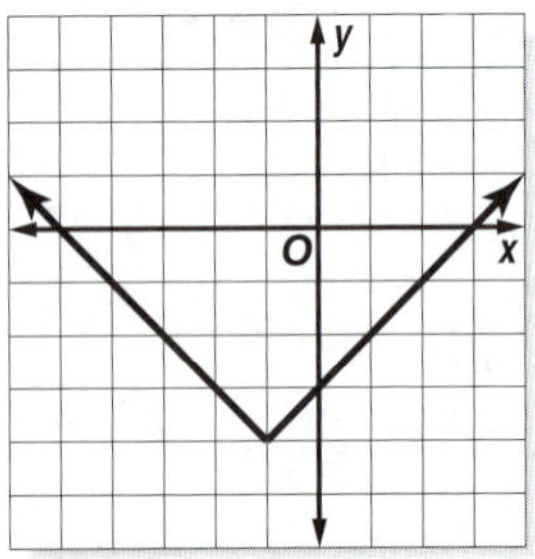

absolute value function

Study Tip

Graphs of Piecewise Functions

The graphs of each part of a piecewise function may or may not connect. A graph may stop at a given x value and then begin again at a different y value for the same x value.

To graph other piecewise functions, examine the inequalities in the definition of the function to determine how much of each piece to include.

Example 4 Piecewise Function

Graph $f(x) = \begin{cases} x - 4 \text{ if } x < 2 \\ 1 \text{ if } x \geq 2 \end{cases}$. Identify the domain and range.

Step 1 Graph the linear function $f(x) = x - 4$ for $x < 2$. Since 2 does not satisfy this inequality, stop with an open circle at $(2, -2)$.

Step 2 Graph the constant function $f(x) = 1$ for $x \geq 2$. Since 2 does satisfy this inequality, begin with a closed circle at $(2, 1)$ and draw a horizontal ray to the right.

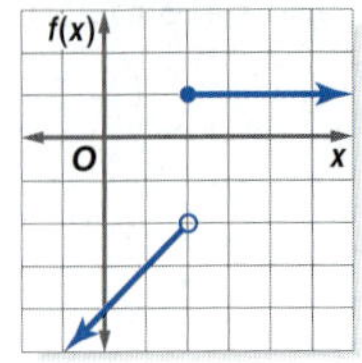

The function is defined for all values of x, so the domain is all real numbers. The values that are y-coordinates of points on the graph are 1 and all real numbers less than -2, so the range is $\{y | y < -2 \text{ or } y = 1\}$.

Concept Summary — Special Functions

Step Function	Constant Function	Absolute Value Function	Piecewise Function
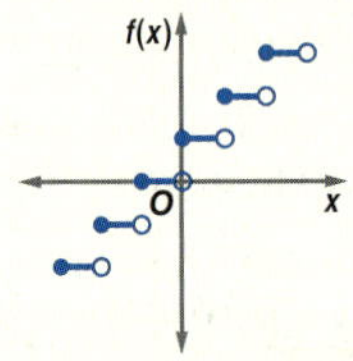	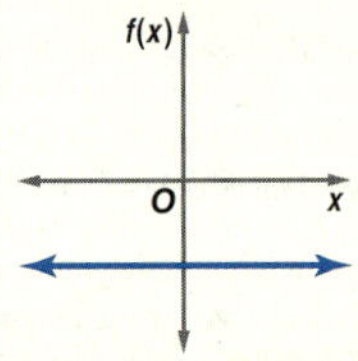	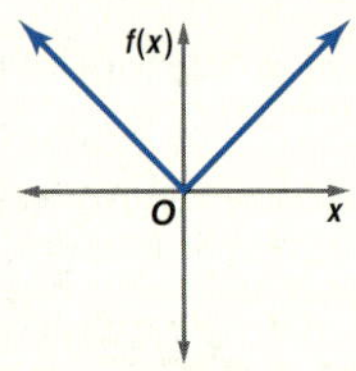	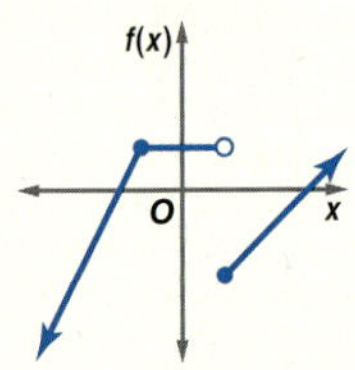
horizontal segments and rays	horizontal line	V-shape	different rays, segments, and curves

Example 5 Identify Functions

Determine whether each graph represents a step function, a constant function, an absolute value function, or a piecewise function.

a.

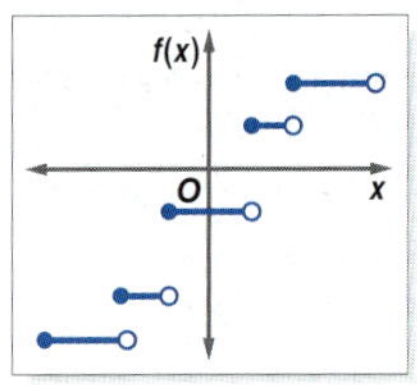

The graph has multiple horizontal segments. It represents a step function.

b.

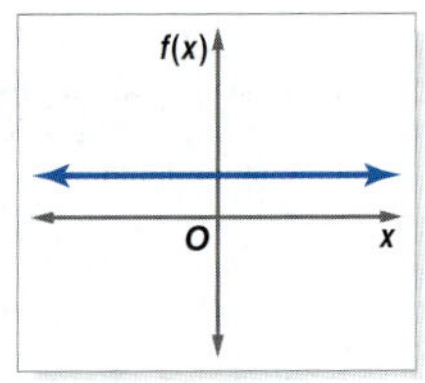

The graph is a horizontal line. It represents a constant function.

Check for Understanding

Concept Check

1. **Find a counterexample** to the statement *To find the greatest integer function of x when x is not an integer, round x to the nearest integer.* **Sample answer: [[1.9]] = 1**

2. **Evaluate** $g(4.3)$ if $g(x) = [\![x - 5]\!]$. **−1**

3. **OPEN ENDED** Write a function involving absolute value for which $f(-2) = 3$.

3. Sample answer: $f(x) = |x - 1|$

DAILY INTERVENTION — **Differentiated Instruction** **ELL**

Verbal/Linguistic Have students explain why step functions, constant functions, and piecewise functions are so named.

Guided Practice

GUIDED PRACTICE KEY

Exercises	Examples
4, 5	5
6–11	1–4
12–14	1

Identify each function as S for step, C for constant, A for absolute value, or P for piecewise.

4. 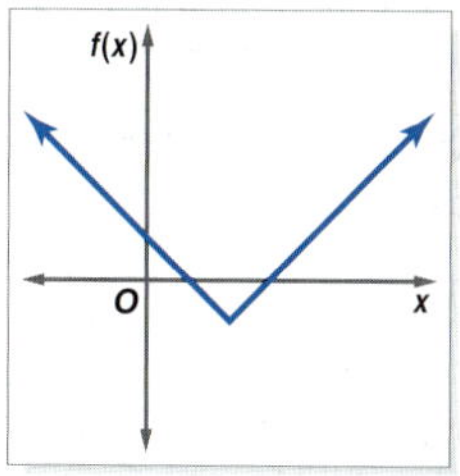

A

5. 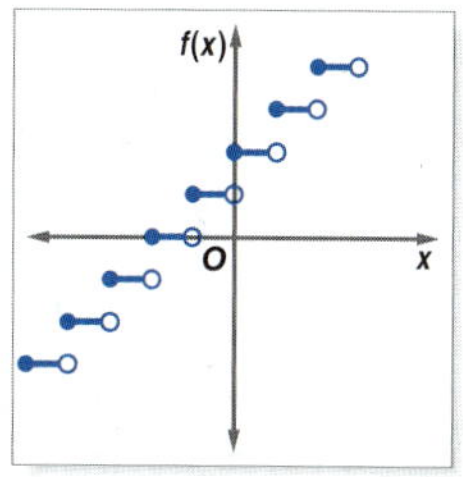

S

Graph each function. Identify the domain and range. **6–11. See pp. 107A–107H for graphs.**

6. $f(x) = -[\![x]\!]$
7. $g(x) = [\![2x]\!]$
8. $h(x) = |x - 4|$
9. $f(x) = |3x - 2|$
10. $g(x) = \begin{cases} -1 \text{ if } x < 0 \\ -x + 2 \text{ if } x \geq 0 \end{cases}$ **D = all reals, R = {y|y ≤ 2}**
11. $h(x) = \begin{cases} x + 3 \text{ if } x \leq -1 \\ 2x \text{ if } x > -1 \end{cases}$ **D = all reals, R = all reals**

6. D = all reals, R = all integers
7. D = all reals, R = all integers
8. D = all reals, R = all nonnegative reals
9. D = all reals, R = all nonnegative reals

Application

PARKING For Exercises 12–14, use the following information.
A downtown parking lot charges \$2 for the first hour and \$1 for each additional hour or part of an hour.

12. What type of special function models this situation? **step function**
13. Draw a graph of a function that represents this situation. **See margin.**
14. Use the graph to find the cost of parking there for $4\frac{1}{2}$ hours. **\$6**

★ indicates increased difficulty

Practice and Apply

Homework Help

For Exercises	See Examples
15–20	5
21–29	1
30–37, 45–47, 49	3
38–41, 44, 48	2, 4
42, 43	1, 3

Extra Practice
See page 831.

Identify each function as S for step, C for constant, A for absolute value, or P for piecewise.

15. 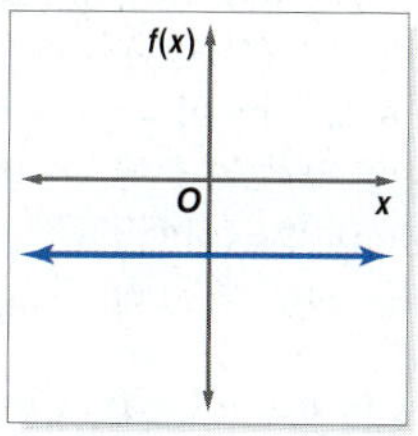

C

16. 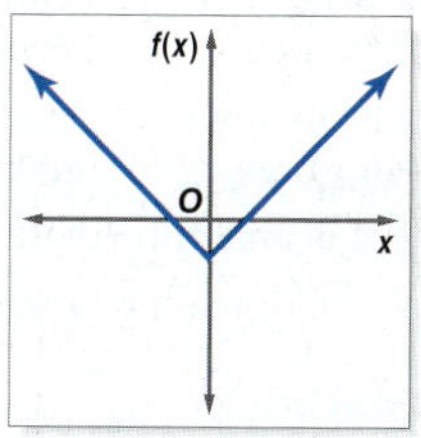

A

17. 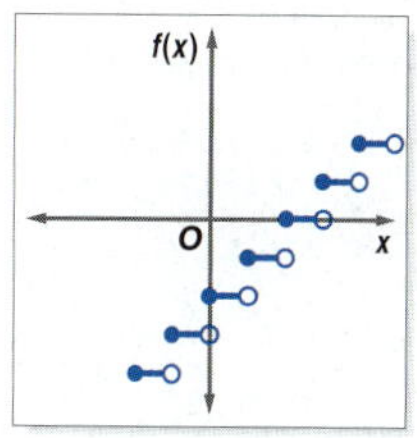

S

18. 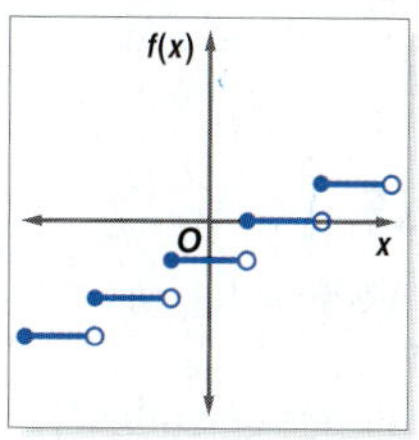

S

19. 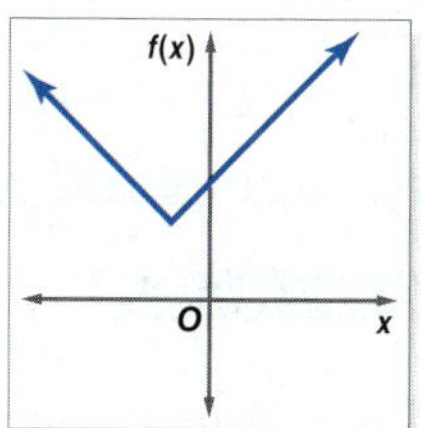

A

20. 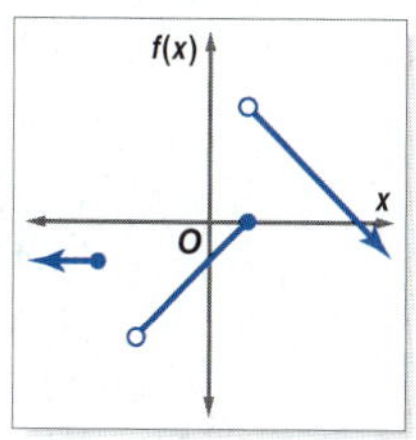

P

21. **TRANSPORTATION** Bluffton High School chartered buses so the student body could attend the girls' basketball state tournament games. Each bus held a maximum of 60 students. Draw a graph of a step function that shows the relationship between the number of students x who went to the game and the number of buses y that were needed. **See margin.**

www.algebra2.com/self_check_quiz

3 Practice/Apply

Study Notebook

Have students—
- add the definitions/examples of the vocabulary terms to their Vocabulary Builder worksheets for Chapter 2.
- draw their own graphs to compare and contrast step, constant, absolute value, and piecewise functions.
- Include any other items(s) that they find helpful in mastering the skills in the lesson.

About the Exercises...

Organization by Objective
- **Step Functions, Constant Functions, and the Identity Function:** 24–29
- **Absolute Value and Piecewise Functions:** 30–41

Odd/Even Assignments
Exercises 15–20 and 24–43 are structured so that students practice the same concepts whether they are assigned odd or even problems.

Assignment Guide

Basic: 15–21 odd, 22, 23, 25–41 odd, 49–65

Average: 15–21 odd, 22, 23, 25–43 odd, 49–65

Advanced: 16–20 even, 22, 23, 24–44 even, 45–59 (optional: 60–65)

All: Practice Quiz 2 (1–5)

Answers

13. 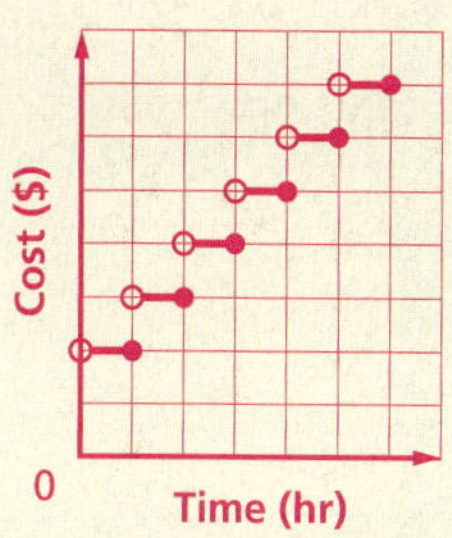

21. 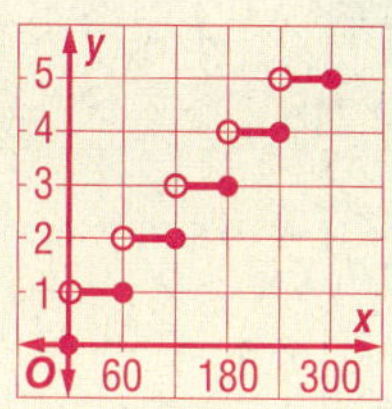

Study Guide and Intervention, p. 87 (shown) and p. 88

Step Functions, Constant Functions, and the Identity Function The chart below lists some special functions you should be familiar with.

Function	Written as	Graph
Constant	$f(x) = c$	horizontal line
Identity	$f(x) = x$	line through the origin with slope 1
Greatest Integer Function	$f(x) = [\![x]\!]$	one-unit horizontal segments, with right endpoints missing, arranged like steps

The greatest integer function is an example of a **step function**, a function with a graph that consists of horizontal segments.

Example **Identify each function as a constant function, the identity function, or a step function.**

a. 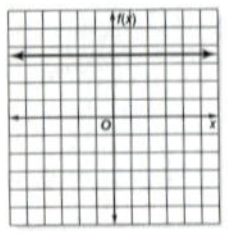

a constant function

b. 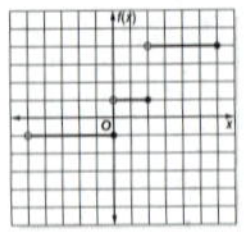

a step function

Exercises

Identify each function as a constant function, the identity function, a greatest integer function, or a step function.

1. 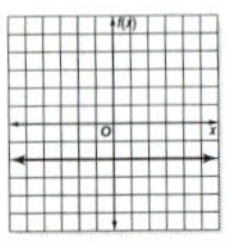
a constant function

2. 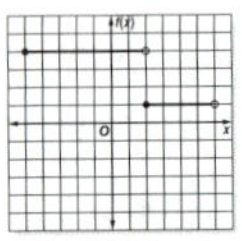
a step function

3. 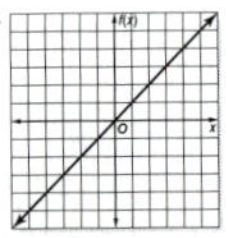
the identity function

Skills Practice, p. 89 and Practice, p. 90 (shown)

Graph each function. Identify the domain and range.

1. $f(x) = [\![0.5x]\!]$

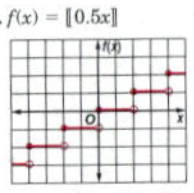

D = all reals, R = all integers

2. $f(x) = [\![x]\!] - 2$

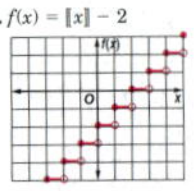

D = all reals, R = all integers

3. $g(x) = -2|x|$

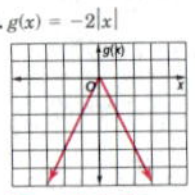

D = all reals, R = nonpositive reals

4. $f(x) = |x + 1|$

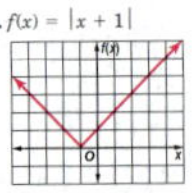

D = all reals, R = nonnegative reals

5. $f(x) = \begin{cases} x + 2 \text{ if } x \leq -2 \\ 3x \text{ if } x > -2 \end{cases}$

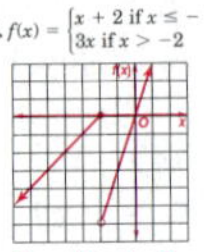

D = all reals, R = all reals

6. $h(x) = \begin{cases} 4 - x \text{ if } x > 0 \\ -2x - 2 \text{ if } x < 0 \end{cases}$

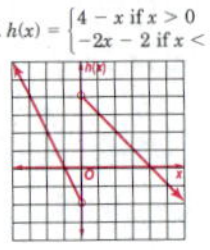

D = all nonzero reals, R = all reals

7. **BUSINESS** *A Stitch in Time* charges \$40 per hour or any fraction thereof for labor. Draw a graph of the step function that represents this situation.

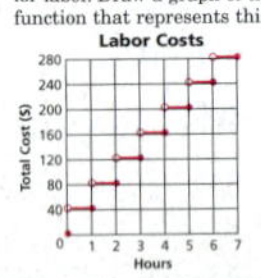

8. **BUSINESS** A wholesaler charges a store \$3.00 per pound for less than 20 pounds of candy and \$2.50 per pound for 20 or more pounds. Draw a graph of the function that represents this situation.

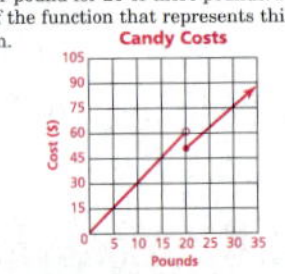

Reading to Learn Mathematics, p. 91 ELL

Pre-Activity **How do step functions apply to postage rates?**

Read the introduction to Lesson 2-6 at the top of page 89 in your textbook.

- What is the cost of mailing a letter that weighs 0.5 ounce? \$0.34 or 34 cents
- Give three different weights of letters that would each cost 55 cents to mail. Answers will vary. Sample answer: 1.1 ounces, 1.9 ounces, 2.0 ounces

Reading the Lesson

1. Find the value of each expression.

a. $|-3| =$ 3 $[\![-3]\!] =$ −3

b. $|6.2| =$ 6.2 $[\![6.2]\!] =$ 6

c. $|-4.01| =$ 4.01 $[\![-4.01]\!] =$ −5

2. Tell how the name of each kind of function can help you remember what the graph looks like.

a. constant function Sample answer: Something is constant if it does not change. The *y*-values of a constant function do not change, so the graph is a horizontal line.

b. absolute value function Sample answer: The absolute value of a number tells you how far it is from 0 on the number line. It makes no difference whether you go to the left or right so long as you go the same distance each time.

c. step function Sample answer: A step function's graph looks like steps that go up or down.

d. identity function Sample answer: The *x*- and *y*-values are always identically the same for any point on the graph. So the graph is a line through the origin that has slope 1.

Helping You Remember

3. Many students find the greatest integer function confusing. Explain how you can use a number line to find the value of this function for any real number. Answers will vary. Sample answer: Draw a number line that shows the integers. To find the value of the greatest integer function for any real number, place that number on the number line. If it is an integer, the value of the function is the number itself. If not, move to the integer directly to the left of the number you chose. This integer will give the value you need.

TELEPHONE RATES For Exercises 22 and 23, use the following information.
Sarah has a long-distance telephone plan where she pays 10¢ for each minute or part of a minute that she talks, regardless of the time of day.

22. Graph a step function that represents this situation. See pp. 107A–107H.

23. Sarah made a call to her brother that lasted 9 minutes and 40 seconds. How much did the call cost? **\$1.00**

Graph each function. Identify the domain and range. **24–43. See pp. 107A–107H.**

24. $f(x) = [\![x + 3]\!]$
25. $g(x) = [\![x - 2]\!]$
26. $f(x) = 2[\![x]\!]$
27. $h(x) = -3[\![x]\!]$
28. $g(x) = [\![x]\!] + 3$
29. $f(x) = [\![x]\!] - 1$
30. $f(x) = |2x|$
31. $h(x) = |-x|$
32. $g(x) = |x| + 3$
33. $g(x) = |x| - 4$
34. $h(x) = |x + 3|$
35. $f(x) = |x + 2|$
36. $f(x) = \left|x - \frac{1}{4}\right|$
37. $f(x) = \left|x + \frac{1}{2}\right|$
38. $f(x) = \begin{cases} -x \text{ if } x \leq 3 \\ 2 \text{ if } x > 3 \end{cases}$
39. $h(x) = \begin{cases} -1 \text{ if } x < -2 \\ 1 \text{ if } x > 2 \end{cases}$
40. $f(x) = \begin{cases} x \text{ if } x < -3 \\ 2 \text{ if } -3 \leq x < 1 \\ -2x + 2 \text{ if } x \geq 1 \end{cases}$
41. $g(x) = \begin{cases} -1 \text{ if } x \leq -2 \\ x \text{ if } -2 < x < 2 \\ -x + 1 \text{ if } x \geq 2 \end{cases}$

★42. $f(x) = [\![\,|x|\,]\!]$
★43. $g(x) = |[\![x]\!]|$

★44. Write the function shown in the graph.

$$f(x) = \begin{cases} 2 \text{ if } x < -1 \\ 2x \text{ if } -1 \leq x \leq 1 \\ -x \text{ if } x > 1 \end{cases}$$

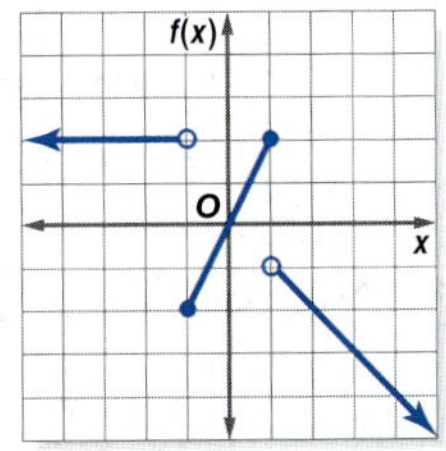

More About. . .

Nutrition

Good sources of vitamin C include citrus fruits and juices, cantaloupe, broccoli, brussels sprouts, potatoes, sweet potatoes, tomatoes, and cabbage.

Source: *The World Almanac*

NUTRITION For Exercises 45–47, use the following information.
The recommended dietary allowance for vitamin C is 2 micrograms per day.

45. Write an absolute value function for the difference between the number of micrograms of vitamin C you ate today x and the recommended amount.

45. $f(x) = |x - 2|$

46. What is an appropriate domain for the function? $\{x \mid x \geq 0\}$

47. Use the domain to graph the function. See pp. 107A–107H.

48. **INSURANCE** According to the terms of Lavon's insurance plan, he must pay the first \$300 of his annual medical expenses. The insurance company pays 80% of the rest of his medical expenses. Write a function for how much the insurance company pays if x represents Lavon's annual medical expenses. See margin.

49. **CRITICAL THINKING** Graph $|x| + |y| = 3$. See pp. 107A–107H.

50. *WRITING IN MATH* Answer the question that was posed at the beginning of the lesson.

How do step functions apply to postage rates?

Include the following in your answer: See pp. 107A–107H.

- an explanation of why a step function is the best model for this situation, while your gas mileage as a function of time as you drive to the post office cannot be modeled with a step function, and
- a graph of a function that represents the cost of a first-class letter.

Enrichment, p. 92

Greatest Integer Functions

Use the greatest integer function $[\![x]\!]$ to explore some unusual graphs. It will be helpful to make a chart of values for each functions and to use a colored pen or pencil.

Graph each function.

1. $y = 2x - [\![x]\!]$

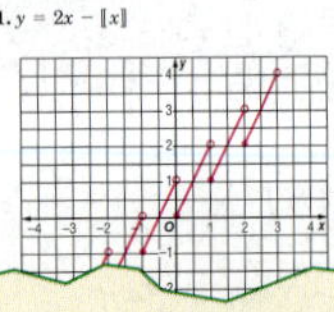

2. $y = \dfrac{[\![x]\!]}{[\![x]\!]}$

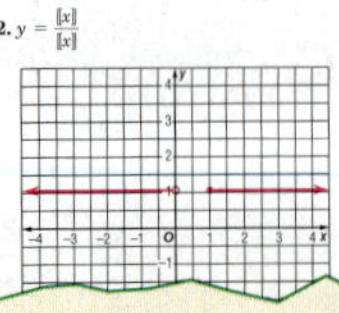

Answer

48. $f(x) = \begin{cases} 0 \text{ if } 0 \leq x \leq 300 \\ 0.8(x - 300) \text{ if } x > 300 \end{cases}$

Standardized Test Practice

51. For which function does $f\left(-\frac{1}{2}\right) \neq -1$? **B**

(A) $f(x) = 2x$ (B) $f(x) = |-2x|$ (C) $f(x) = [\![x]\!]$ (D) $f(x) = [\![2x]\!]$

52. For which function is the range $\{y \mid y \leq 0\}$? **D**

(A) $f(x) = -x$ (B) $f(x) = [\![x]\!]$ (C) $f(x) = |x|$ (D) $f(x) = -|x|$

Maintain Your Skills

Mixed Review

HEALTH **For Exercises 53–55, use the table that shows the life expectancy for people born in various years.** *(Lesson 2-5)*

Year	1950	1960	1970	1980	1990	1997
Expectancy	68.2	69.7	70.8	73.7	75.4	76.5

Source: National Center for Health Statistics

53. Draw a scatter plot in which x is the number of years since 1950. **See margin.**

54. Find a prediction equation.

55. Predict the life expectancy of a person born in 2010. **Sample answer: 78.7 yr**

54. Sample answer using (10, 69.7) and (47, 76.5): $y = 0.18x + 67.9$

Write an equation in slope-intercept form that satisfies each set of conditions. *(Lesson 2-4)*

56. slope 3, passes through $(-2, 4)$ **$y = 3x + 10$**

57. passes through $(0, -2)$ and $(4, 2)$ **$y = x - 2$**

Solve each inequality. Graph the solution set. *(Lesson 1-5)*

58. $3x - 5 \geq 4$ **$\{x \mid x \geq 3\}$**

59. $28 - 6y < 23$ **$\left\{y \mid y > \frac{5}{6}\right\}$**

58–59. See margin for graphs.

Getting Ready for the Next Lesson

PREREQUISITE SKILL **Determine whether (0, 0) satisfies each inequality. Write** *yes* **or** *no*. *(To review* ***inequalities****, see Lesson 1-5.)*

60. $y < 2x + 3$ **yes**

61. $y \geq -x + 1$ **no**

62. $y \leq \frac{3}{4}x - 5$ **no**

63. $2x + 6y + 3 > 0$ **yes**

64. $y > |x|$ **no**

65. $|x| + y \leq 3$ **yes**

Practice Quiz 2 — *Lessons 2-4 through 2-6*

1. Write an equation in slope-intercept form of the line with slope $-\frac{2}{3}$ that passes through $(-2, 5)$. *(Lesson 2-4)* **$y = -\frac{2}{3}x + \frac{11}{3}$**

BASKETBALL **For Exercises 2–4, use the following information.**
On August 26, 2000, the Houston Comets beat the New York Liberty to win their fourth straight WNBA championship. The table shows the heights and weights of the Comets who played in that final game. *(Lesson 2-5)*

Height (in.)	74	71	76	70	66	74	72
Weight (lb)	178	147	195	150	138	190	?

Source: WNBA

2. Draw a scatter plot. **See margin.**

3. Use two ordered pairs to write a prediction equation. **Sample answer using (66, 138) and (74, 178): $y = 5x - 192$**

4. Use your prediction equation to predict the missing value. **Sample answer: 168 lb**

5. Graph $f(x) = |x - 1|$. Identify the domain and range. *(Lesson 2-6)* **See margin.**

4 Assess

Open-Ended Assessment

Modeling Have students draw a large coordinate plane on a sheet of paper. Then have them use toothpicks (or other similar objects) to model the general shapes of step, constant, and absolute value functions. Students should identify each type of graph as they model it.

Getting Ready for Lesson 2-7

PREREQUISITE SKILL Lesson 2-7 presents graphing inequalities. It is sometimes necessary to solve an inequality for y in order to determine whether to shade above or below the boundary. Exercises 60–65 should be used to determine your students' familiarity with inequalities.

Assessment Options

Practice Quiz 2 The quiz provides students with a brief review of the concepts and skills in Lessons 2-4 through 2-6. Lesson numbers are given to the right of exercises or instruction lines so students can review concepts not yet mastered.

Quiz (Lessons 2-5 and 2-6) is available on p. 114 of the *Chapter 2 Resource Masters*.

Answers (Practice Quiz 2)

2.

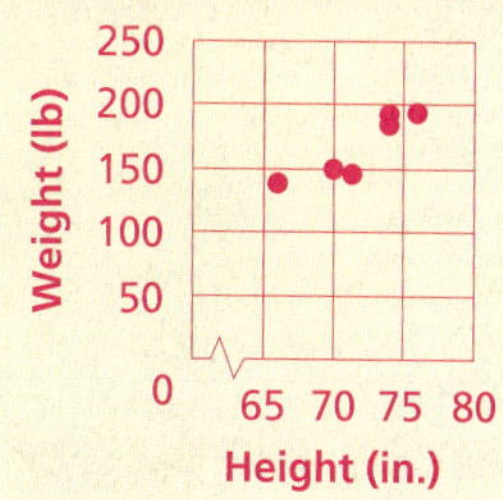

5.

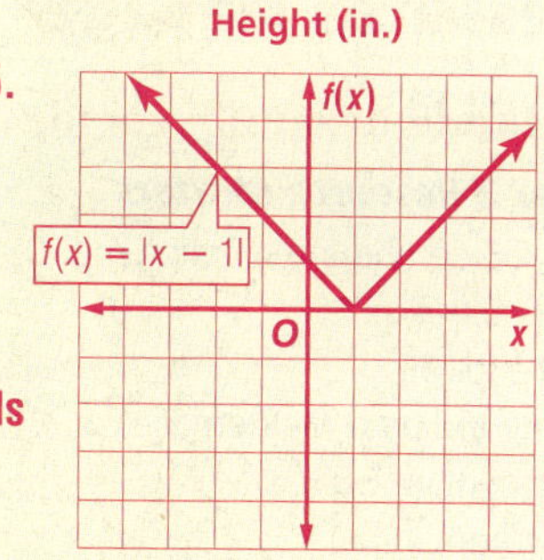

D = all reals
R = nonnegative reals

Answers

53.

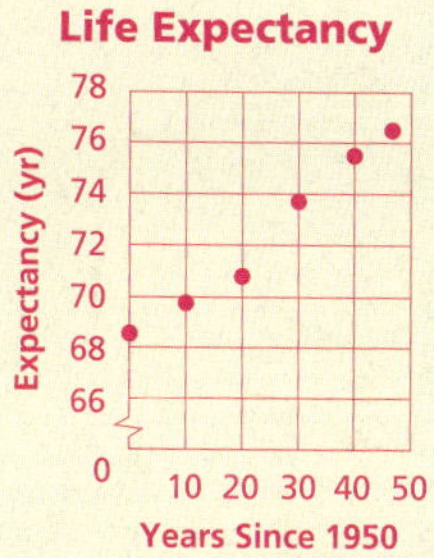

58.

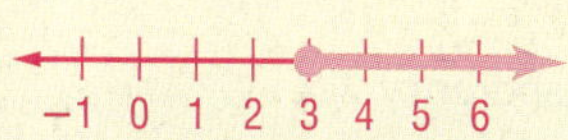

59.

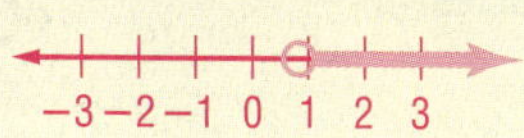

2-7 Lesson Notes

1 Focus

5-Minute Check Transparency 2-7 Use as a quiz or review of Lesson 2-6.

Mathematical Background notes are available for this lesson on p. 54D.

How do inequalities apply to fantasy football?

Ask students:

- What is the meaning of "receiving yards?" **the number of yards that a team advances down the field when the receiver of a pass catches the ball**

2 Teach

GRAPH LINEAR INEQUALITIES

In-Class Example Power Point®

1 Graph $x - 2y < 4$.

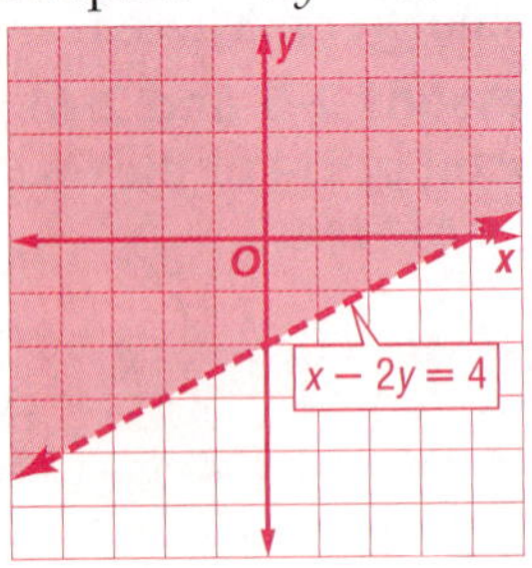

2-7 Graphing Inequalities

What You'll Learn

- Graph linear inequalities.
- Graph absolute value inequalities.

Vocabulary

- boundary

How do inequalities apply to fantasy football?

Dana has Vikings receiver Randy Moss as a player on his online fantasy football team. Dana gets 5 points per receiving yard that Moss gets and 100 points per touchdown that Moss scores. He considers 1000 points or more to be a good game. Dana can use a linear inequality to check whether certain combinations of yardage and touchdowns, such as those in the table, result in 1000 points or more.

	Yards	TDs
Game 1	168	3
Game 2	144	2
Game 3	136	1

GRAPH LINEAR INEQUALITIES A linear inequality resembles a linear equation, but with an inequality symbol instead of an equals symbol. For example, $y \leq 2x + 1$ is a linear inequality and $y = 2x + 1$ is the related linear equation.

The graph of the inequality $y \leq 2x + 1$ is the shaded region. Every point in the shaded region satisfies the inequality. The graph of $y = 2x + 1$ is the **boundary** of the region. It is drawn as a solid line to show that points on the line satisfy the inequality. If the inequality symbol were $<$ or $>$, then points on the boundary would not satisfy the inequality, so the boundary would be drawn as a dashed line.

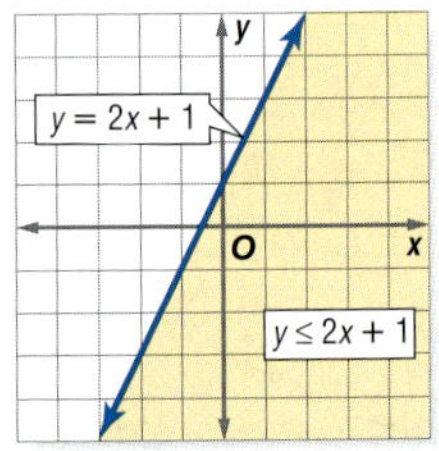

You can graph an inequality by following these steps.

Step 1 Determine whether the boundary should be solid or dashed. Graph the boundary.

Step 2 Choose a point not on the boundary and test it in the inequality.

Step 3 If a true inequality results, shade the region containing your test point. If a false inequality results, shade the other region.

Example 1 *Dashed Boundary*

Graph $2x + 3y > 6$.

The boundary is the graph of $2x + 3y = 6$. Since the inequality symbol is $>$, the boundary will be dashed. Use the slope-intercept form, $y = -\frac{2}{3}x + 2$.

Now test the point (0, 0). *The point (0, 0) is usually a good point to test because it results in easy calculations.*

$2x + 3y > 6$ Original inequality

$2(0) + 3(0) > 6$ $(x, y) = (0, 0)$

$0 > 6$ false

Shade the region that does *not* contain (0, 0).

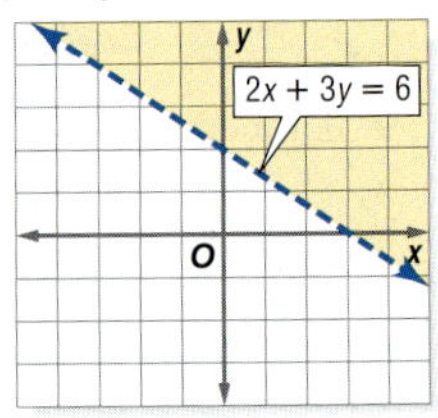

TEACHING TIP

Students may wish to test a point in the shaded region as a check of their work.

Resource Manager

Workbook and Reproducible Masters

Chapter 2 Resource Masters

- Study Guide and Intervention, pp. 93–94
- Skills Practice, p. 95
- Practice, p. 96
- Reading to Learn Mathematics, p. 97
- Enrichment, p. 98
- Assessment, p. 114

Transparencies

5-Minute Check Transparency 2-7
Answer Key Transparencies

Technology

Alge2PASS: Tutorial Plus, Lesson 4
Interactive Chalkboard

Inequalities can sometimes be used to model real-world situations.

Example 2 Solid Boundary

BUSINESS **A mail-order company is hiring temporary employees to help in their packing and shipping departments during their peak season.**

a. Write an inequality to describe the number of employees that can be assigned to each department if the company has 20 temporary employees available.

Let p be the number of employees assigned to packing and let s be the number assigned to shipping. Since the company can assign *at most* 20 employees total to the two departments, use a $\leq$ symbol.

The number of employees for packing	and	the number of employees for shipping	is at most	twenty.
p	$+$	s	$\leq$	20

b. Graph the inequality.

Since the inequality symbol is $\leq$, the graph of the related linear equation $p + s = 20$ is solid. This is the boundary of the inequality.

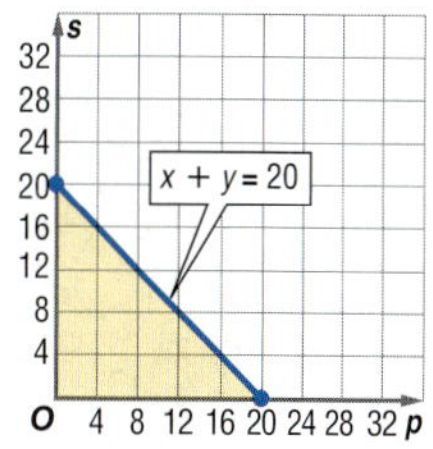

Test (0, 0).

$p + s \leq 20$ Original inequality

$0 + 0 \leq 20$ $(p, s) = (0, 0)$

$0 \leq 20$ true

Shade the region that contains (0, 0). *Since the variables cannot be negative, shade only the part in the first quadrant.*

c. Can the company assign 8 employees to packing and 10 employees to shipping?

The point (8, 10) is in the shaded region, so it satisfies the inequality. The company can assign 8 employees to packing and 10 to shipping.

Study Tip

Look Back

To review **translating verbal expressions to inequalities,** see Lesson 1-5.

GRAPH ABSOLUTE VALUE INEQUALITIES Graphing absolute value inequalities is similar to graphing linear inequalities. The inequality symbol determines whether the boundary is solid or dashed, and you can test a point to determine which region to shade.

Example 3 Absolute Value Inequality

Graph $y < |x| + 1$.

Since the inequality symbol is $<$, the graph of the related equation $y = |x| + 1$ is dashed. Graph the equation.

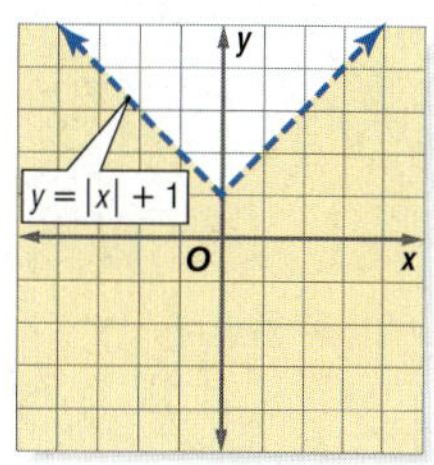

Test (0, 0).

$y < |x| + 1$ Original inequality

$0 < |0| + 1$ $(x, y) = (0, 0)$

$0 < 0 + 1$ $|0| = 0$

$0 < 1$ true

Shade the region that includes (0, 0).

www.algebra2.com/extra_examples

In-Class Example Power Point®

2 EDUCATION The SAT has two parts. One tutoring company advertises that it specializes in helping students who have a combined score on the SAT that is 900 or less.

a. Write an inequality to describe the combined scores of students who are prospective tutoring clients. $x + y \leq 900$

b. Graph the inequality.

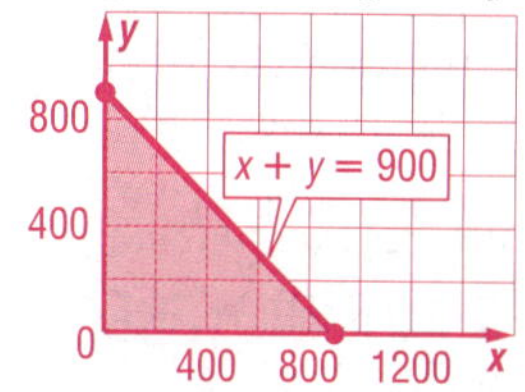

c. Does a student with a verbal score of 480 and a math score of 410 fit the tutoring company's guidelines? **yes**

GRAPH ABSOLUTE VALUE INEQUALITIES

In-Class Example Power Point®

3 Graph $y \geq |x| - 2$.

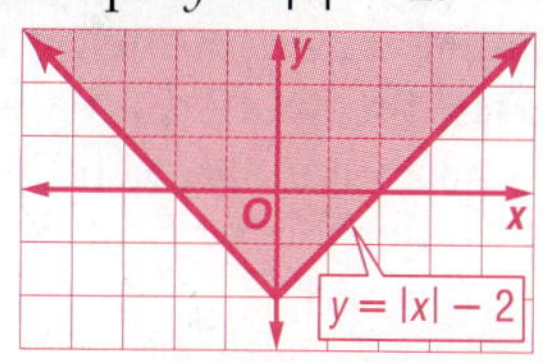

3 Practice/Apply

Study Notebook

Have students—

- *complete the definitions/examples for the remaining terms on their Vocabulary Builder worksheets for Chapter 2.*
- *include any other items(s) that they find helpful in mastering the skills in this lesson.*

DAILY INTERVENTION **Differentiated Instruction**

Interpersonal Have students work in pairs as they graph the inequalities, so they can check and correct each other's procedures and results. When graphing, one partner can work to determine the location of the boundary while the other partner determines whether the shaded region is above or below the boundary.

About the Exercises...

Organization by Objective

- **Graph Linear Inequalities:** 13–24, 33–39
- **Graph Absolute Value Inequalities:** 25–30

Odd/Even Assignments

Exercises 13–32 are structured so that students practice the same concepts whether they are assigned odd or even problems.

Alert! Exercises 44–47 require a graphing calculator.

Assignment Guide

Basic: 13–27 odd, 31, 33, 34, 40–43, 48–56

Average: 13–31 odd, 33–37, 40–43, 48–56 (optional: 44–47)

Advanced: 14–32 even, 38–56

4 Assess

Open-Ended Assessment

Speaking Have students explain how to tell just from looking at an inequality with y alone on the left whether the shaded area will be below or above the boundary, as well as whether the boundary is solid or dashed.

Assessment Options

Quiz (Lesson 2-7) is available on p. 114 of the *Chapter 2 Resource Masters.*

Answers

2. Substitute the coordinates of a point not on the boundary into the inequality. If the inequality is satisfied, shade the region containing the point. If the inequality is not satisfied, shade the region that does not contain the point.

44.

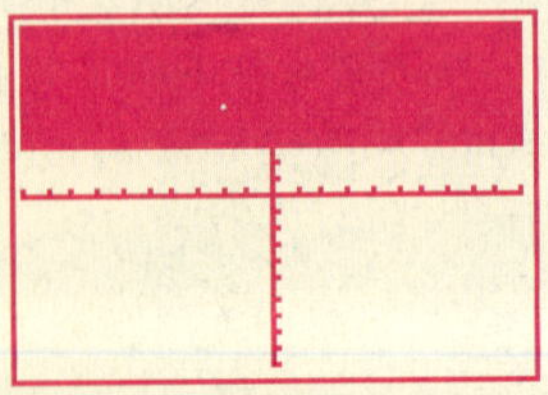

[−10, 10] scl: 1 by [−10, 10] scl: 1

45.

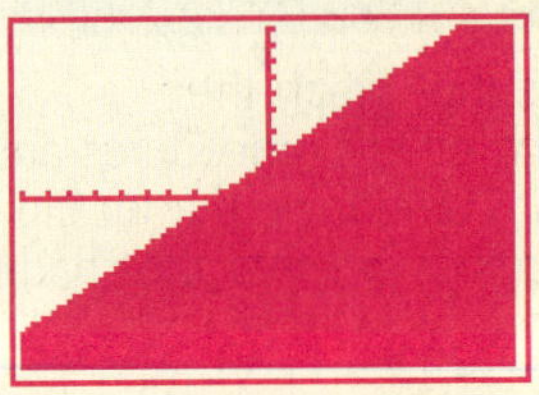

[−10, 10] scl: 1 by [−10, 10] scl: 1

46.

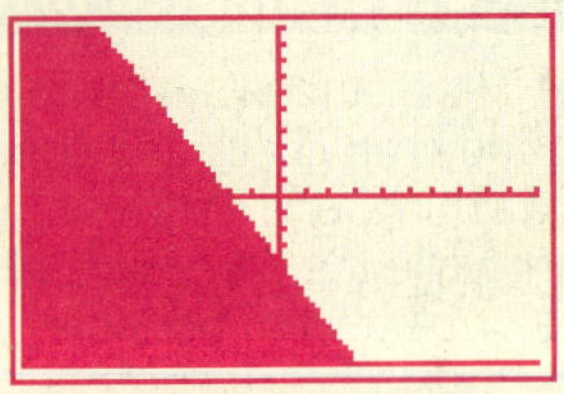

[−10, 10] scl: 1 by [−10, 10] scl: 1

47.

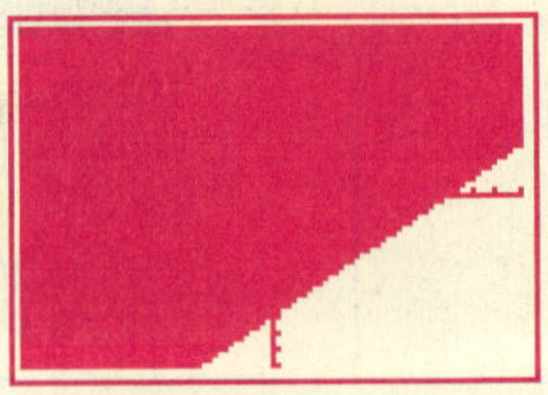

[−10, 10] scl: 1 by [−10, 10] scl: 1

Check for Understanding

Concept Check

1. **Write** an inequality for the graph at the right. 1. $y \leq -3x + 4$

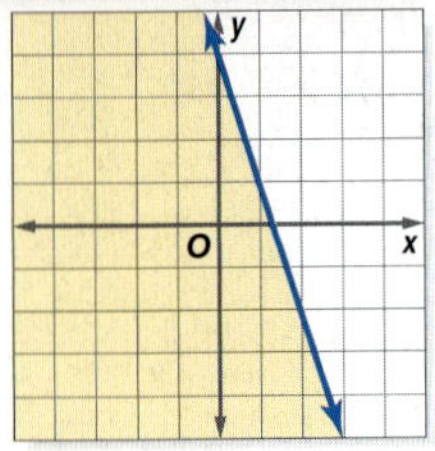

2. **Explain** how to determine which region to shade when graphing an inequality. See margin.
3. **OPEN ENDED** Write an absolute value inequality for which the boundary is solid and the solution is the region above the graph of the related equation. Sample answer: $y \geq |x|$

Guided Practice

Graph each inequality. 4–9. See pp. 107A–107H.

4. $y < 2$
5. $y > 2x - 3$
6. $x - y \geq 0$
7. $x - 2y \leq 5$
8. $y > |2x|$
9. $y \leq 3|x| - 1$

GUIDED PRACTICE KEY	
Exercises	**Examples**
4–7	1, 2
8, 9	3
10–12	2

Application

SHOPPING For Exercises 10–12, use the following information.

Gwen wants to buy some cassettes that cost \$10 each and some CDs that cost \$13 each. She has \$40 to spend.

10. Write an inequality to represent the situation, where c is the number of cassettes she buys and d is the number of CDs. $10c + 13d \leq 40$
11. Graph the inequality. See pp. 107A–107H.
12. Can she buy 3 cassettes and 2 CDs? Explain. No; (3, 2) is not in the shaded region.

★ indicates increased difficulty

Practice and Apply

Homework Help	
For Exercises	**See Examples**
13–24, 31, 32	1, 2
25–30, 41	3
33–40	2

Extra Practice See page 832.

Graph each inequality. 13–30. See pp. 107A–107H.

13. $x + y > -5$
14. $3 \geq x - 3y$
15. $y > 6x - 2$
16. $x - 5 \leq y$
17. $y \geq -4x + 3$
18. $y - 2 < 3x$
19. $y \geq 1$
20. $y + 1 < 4$
21. $4x - 5y - 10 \leq 0$
22. $x - 6y + 3 > 0$
23. $y > \frac{1}{3}x + 5$
24. $y \geq \frac{1}{2}x - 5$
25. $y \leq |x|$
26. $y > |4x|$
27. $y + |x| < 3$
28. $y \geq |x - 1| - 2$
29. ★ $|x + y| > 1$
30. ★ $|x| \leq |y|$

31. Graph all the points on the coordinate plane to the left of the graph of $x = -2$. Write an inequality to describe these points. $x < -2$
32. Graph all the points on the coordinate plane below the graph of $y = 3x - 5$. Write an inequality to describe these points. $y < 3x - 5$

31–32. See pp. 107A–107H for graphs.

SCHOOL For Exercises 33 and 34, use the following information.

Rosa's professor says that the midterm exam will count for 40% of each student's grade and the final exam will count for 60%. A score of at least 90 is required for an A.

33. The inequality $0.4x + 0.6y \geq 90$ represents this situation, where x is the midterm score and y is the final exam score. Graph this inequality. See pp. 107A–107H.
34. If Rosa scores 85 on the midterm and 95 on the final, will she get an A? yes

DRAMA For Exercises 35–37, use the following information.

Tickets for the Prestonville High School Drama Club's spring play cost \$4 for adults and \$3 for students. In order to cover expenses, at least \$2000 worth of tickets must be sold.

35. Write an inequality that describes this situation. $4a + 3s \geq 2000$
36. Graph the inequality. See pp. 107A–107H.
37. If 180 adult and 465 student tickets are sold, will the club cover its expenses? yes

More About. . .

Finance

A *dividend* is a payment from a company to an investor. It is a way to make money on a stock without selling it.

FINANCE For Exercises 38–40, use the following information.

Carl Talbert estimates that he will need to earn at least $9000 per year combined in dividend income from the two stocks he owns to supplement his retirement plan.

Company	Dividend per Share
Able Rentals	$1.20
Best Bikes	$1.80

38. Write and graph an inequality for this situation. **See pp. 107A–107H.**

39. Will he make enough from 3000 shares of each company? **yes**

40. **CRITICAL THINKING** Graph $|y| < x$. **See pp. 107A–107H.**

41. **WRITING IN MATH** Answer the question that was posed at the beginning of the lesson.

How do inequalities apply to fantasy football?

Include the following in your answer: **See pp. 107A–107H.**

- an inequality, and an explanation of how you obtained it, to represent a good game for Randy Moss in Dana's fantasy football league,
- a graph of your inequality (remember that the number of touchdowns cannot be negative, but receiving yardage can be), and
- which of the games with statistics in the table qualify as good games.

Standardized Test Practice

42. Which could be the inequality for the graph? **A**

Ⓐ $y < 3x + 2$ Ⓑ $y \leq 3x + 2$
Ⓒ $y > 3x + 2$ Ⓓ $y \geq 3x + 2$

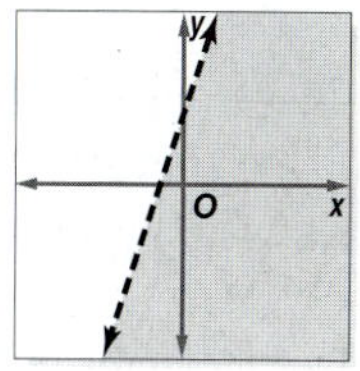

43. Which point satisfies $y > 5|x| - 3$? **B**

Ⓐ (2, 2) Ⓑ (−1, 3)
Ⓒ (3, 7) Ⓓ (−2, 4)

Graphing Calculator

SHADE(COMMAND You can graph inequalities with a graphing calculator by using the Shade(command located in the DRAW menu. You must enter two functions.

- The first function defines the lower boundary of the region to be shaded.
- The second function defines the upper boundary of the region.
- If the inequality is "$y \leq$," use the Ymin window value as the lower boundary.
- If the inequality is "$y \geq$," use the Ymax window value as the upper boundary.

Graph each inequality. **44–47. See margin.**

44. $y \geq 3$ 45. $y \leq x + 2$ 46. $y \leq -2x - 4$ 47. $x - 7 \leq y$

Maintain Your Skills

Mixed Review

48–50. See pp. 107A–107H for graphs.

48. D = all reals, R = all integers

49. D = all reals, R = {y | y ≥ −1}

50. D = all reals, R = all nonnegative reals

52. Sample answer using (4, 6000) and (6, 8000): y = 1000x + 2000

Graph each function. Identify the domain and range. *(Lesson 2-6)*

48. $f(x) = [\![x]\!] - 4$ 49. $g(x) = |x| - 1$ 50. $h(x) = |x - 3|$

SALES For Exercises 51–53, use the table that shows the years of experience for eight sales representatives and their sales during a given period of time. *(Lesson 2-5)*

Years	6	5	3	1	4	3	6	2
Sales ($)	9000	6000	4000	3000	6000	5000	8000	2000

51. Draw a scatter plot. **See margin.**

52. Find a prediction equation.

53. Predict the sales for a representative with 8 years of experience. **Sample answer: $10,000**

Solve each equation. Check your solution. *(Lesson 1-3)*

54. $4x - 9 = 23$ **8** 55. $11 - 2y = 5$ **3** 56. $2z - 3 = -6z + 1$ $\frac{1}{2}$

www.algebra2.com/self_check_quiz

51.

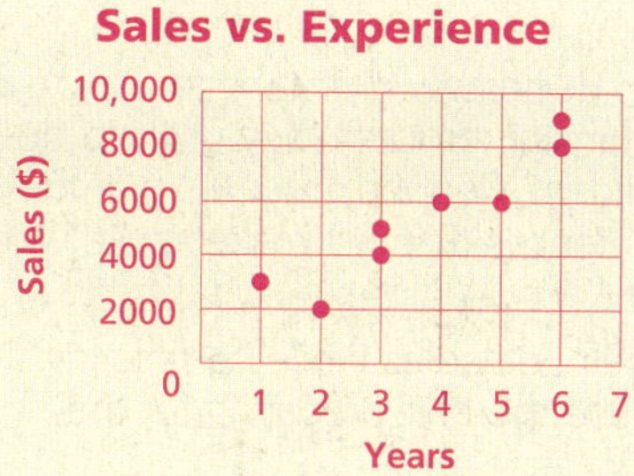

Study Guide and Intervention, p. 93 (shown) and p. 94

Graph Linear Inequalities. A **linear inequality**, like $y \geq 2x - 1$, resembles a linear equation, but with an inequality sign instead of an equals sign. The graph of the related linear equation separates the coordinate plane into two half-planes. The line is the boundary of each half-plane.

To graph a linear inequality, follow these steps.

1. Graph the boundary, that is, the related linear equation. If the inequality symbol is ≤ or ≥, the boundary is solid. If the inequality symbol is < or >, the boundary is dashed.
2. Choose a point not on the boundary and test it in the inequality. (0, 0) is a good point to choose if the boundary does not pass through the origin.
3. If a true inequality results, shade the half-plane containing your test point. If a false inequality results, shade the other half-plane.

Example **Graph $x + 2y \geq 4$.**

The boundary is the graph of $x + 2y = 4$.

Use the slope-intercept form, $y = -\frac{1}{2}x + 2$, to graph the boundary line.

The boundary line should be solid.

Now test the point (0, 0).

$0 + 2(0) \overset{?}{\geq} 4$ $(x, y) = (0, 0)$

$0 \geq 4$ false

Shade the region that does *not* contain (0, 0).

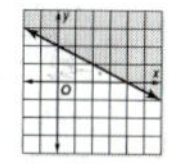

Exercises

Graph each inequality.

1. $y < 3x + 1$ 2. $y \geq x - 5$ 3. $4x + y \leq -1$

4. $y < \frac{x}{2} - 4$ 5. $x + y > 6$ 6. $0.5x - 0.25y < 1.5$

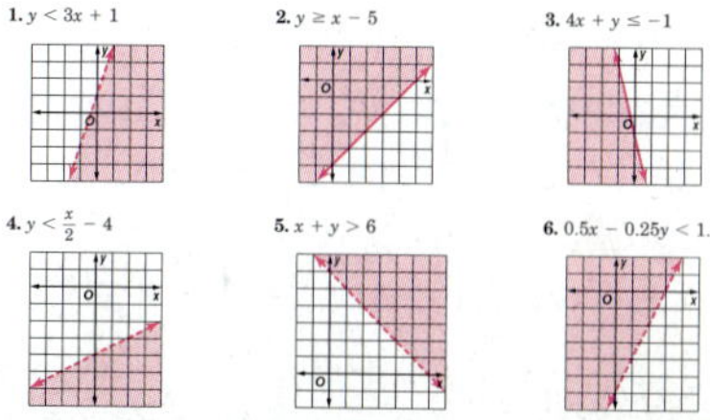

Skills Practice, p. 95 and Practice, p. 96 (shown)

Graph each inequality.

1. $y \leq -3$ 2. $x > 2$ 3. $x + y \leq -4$

4. $y < -3x + 5$ 5. $y < \frac{1}{2}x + 3$ 6. $y - 1 \geq -x$

7. $x - 3y \leq 6$ 8. $y > |x| - 1$ 9. $y > -3|x + 1| - 2$

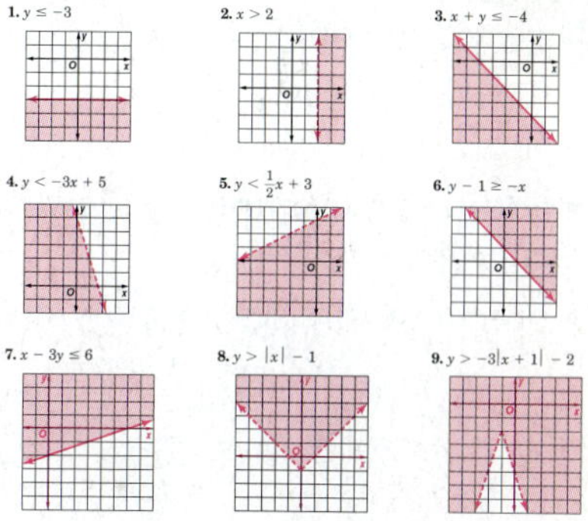

COMPUTERS **For Exercises 10–12, use the following information.**

A school system is buying new computers. They will buy desktop computers costing $1000 per unit, and notebook computers costing $1200 per unit. The total cost of the computers cannot exceed $80,000.

10. Write an inequality that describes this situation. **1000d + 1200n ≤ 80,000**
11. Graph the inequality.
12. If the school wants to buy 50 of the desktop computers and 25 of the notebook computers, will they have enough money? **yes**

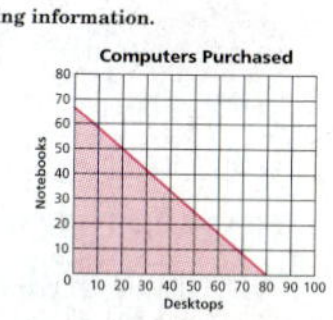

Reading to Learn Mathematics, p. 97 ELL

Pre-Activity **How do inequalities apply to fantasy football?**

Read the introduction to Lesson 2-7 at the top of page 96 in your textbook.

- Which of the combinations of yards and touchdowns listed would Dana consider a good game? **The first one: 168 yards and 3 touchdowns**
- Suppose that in one of the games Dana plays, Moss gets 157 receiving yards. What is the smallest number of touchdowns he must get in order for Dana to consider this a good game? **3**

Reading the Lesson

1. When graphing a linear inequality in two variables, how do you know whether to make the boundary a solid line or a dashed line? **If the symbol is ≥ or ≤, the line is solid. If the symbol is > or <, the line is dashed.**
2. How do you know which side of the boundary to shade? **Sample answer: If the test point gives a true inequality, shade the region containing the test point. If the test point gives a false inequality, shade the region *not* containing the test point.**
3. Match each inequality with its graph.

a. $y > 2x - 3$ **iii** b. $y < -2x + 3$ **iv** c. $y \geq 2x - 3$ **ii** d. $y \geq -2x + 3$ **i**

i. ii. iii. iv.

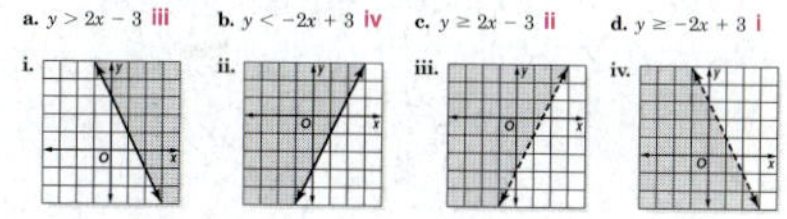

Helping You Remember

4. Describe some ways in which graphing an inequality in one variable on a number line is similar to graphing an inequality in two variables in a coordinate plane. How can what you know about graphing inequalities on a number line help you to graph inequalities in a coordinate plane? **Sample answer: A boundary on a coordinate graph is similar to an endpoint on a number line graph. A dashed line is similar to a circle on a number line: both are open and mean *not included*; they represent the symbols > and <. A solid line is similar to a dot on a number line: both are closed and mean *included*; they represent the symbols ≥ and ≤.**

Enrichment, p. 98

Algebraic Proof

The following paragraph states a result you might be asked to prove in a mathematics course. Parts of the paragraph are numbered.

01 Let n be a positive integer.

02 Also, let $n_1 = s(n_1)$ be the sum of the squares of the digits in n.

03 Then $n_2 = s(n_1)$ is the sum of the squares of the digits of n_1, and $n_3 = s(n_2)$ is the sum of the squares of the digits of n_2.

04 In general, $n_k = s(n_{k-1})$ is the sum of the squares of the digits of n_{k-1}.

05 Consider the sequence: $n, n_1, n_2, n_3, \cdots, n_k, \cdots$.

06 In this sequence either all the terms from some k on have the value 1,

07 or some term, say n_j, has the value 4, so that the eight terms 4, 16, 37, 58, 89, 145, 42, and 20 keep repeating from that point on.

Use the paragraph to answer these questions.

Chapter 2 Study Guide and Review

Vocabulary and Concept Check

- This alphabetical list of vocabulary terms in Chapter 2 includes a page reference where each term was introduced.
- **Assessment** A vocabulary test/review for Chapter 2 is available on p. 112 of the *Chapter 2 Resource Masters*.

Lesson-by-Lesson Review

For each lesson,

- the main ideas are summarized,
- additional examples review concepts, and
- practice exercises are provided.

Vocabulary PuzzleMaker

ELL The Vocabulary PuzzleMaker software improves students' mathematics vocabulary using four puzzle formats—crossword, scramble, word search using a word list, and word search using clues. Students can work on a computer screen or from a printed handout.

MindJogger Videoquizzes

ELL MindJogger Videoquizzes provide an alternative review of concepts presented in this chapter. Students work in teams in a game show format to gain points for correct answers. The questions are presented in three rounds.

Round 1 Concepts (5 questions)
Round 2 Skills (4 questions)
Round 3 Problem Solving (4 questions)

Chapter 2 Study Guide and Review

Vocabulary and Concept Check

absolute value function (p. 90)	independent variable (p. 59)	range (p. 56)
bivariate data (p. 81)	linear equation (p. 63)	rate of change (p. 69)
boundary (p. 96)	linear function (p. 63)	relation (p. 56)
Cartesian coordinate plane (p. 56)	line of fit (p. 81)	scatter plot (p. 81)
constant function (p. 90)	mapping (p. 57)	slope (p. 68)
dependent variable (p. 59)	one-to-one function (p. 57)	slope-intercept form (p. 75)
domain (p. 56)	ordered pair (p. 56)	standard form (p. 64)
family of graphs (p. 70)	parent graph (p. 70)	step function (p. 89)
function (p. 57)	piecewise function (p. 91)	vertical line test (p. 57)
functional notation (p. 59)	point-slope form (p. 76)	*x*-intercept (p. 65)
greatest integer function (p. 89)	prediction equation (p. 81)	*y*-intercept (p. 65)
identity function (p. 90)	quadrant (p. 56)	

Choose the correct term to complete each sentence.

1. The (*constant*, *identity*) function is a linear function described by $f(x) = x$.
2. The graph of the (*absolute value*, *greatest integer*) function forms a V-shape.
3. The (*slope-intercept*, *standard*) form of the equation of a line is $Ax + By = C$, where A and B are not both zero.
4. Two lines in the same plane having the same slope are (*parallel*, *perpendicular*).
5. The (*domain*, *range*) is the set of all x-coordinates of the ordered pairs of a relation.
6. The set of all y-coordinates of the ordered pairs of a relation is the (*domain*, *range*).
7. The ratio of the change in y-coordinates to the corresponding change in x-coordinates is called the (*slope*, *y-intercept*) of a line.
8. The (*line of fit*, *vertical line test*) can be used to determine if a relation is a function.

Lesson-by-Lesson Review

2-1 Relations and Functions

See pages 56–62.

Concept Summary

- A relation is a set of ordered pairs. The domain is the set of all x-coordinates, and the range is the set of all y-coordinates.
- A function is a relation where each member of the domain is paired with exactly one member of the range.

Example **Graph the relation $\{(-3, 1), (0, 2), (2, 5)\}$ and find the domain and range. Then determine whether the relation is a function.**

The domain is $\{-3, 0, 2\}$, and the range is $\{1, 2, 5\}$.

Graph the ordered pairs. Since each x value is paired with exactly one y value, the relation is a function.

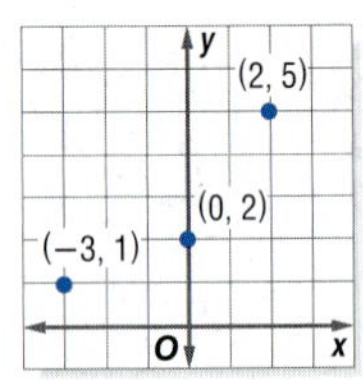

www.algebra2.com/vocabulary_review

FOLDABLES™ Study Organizer

For more information about Foldables, see *Teaching Mathematics with Foldables.*

Some students may need help in deciding which tabs to use. Ask student volunteers to share the thinking process they used to decide where various material should go. Remind students that their notations should be complete sentences that will make sense when they are reviewed weeks later.

Encourage students to refer to their Foldables while completing the Study Guide and Review and to use them in preparing for the Chapter Test.

Exercises Graph each relation or equation and find the domain and range. Then determine whether the relation or equation is a function.
See Examples 1 and 2 on pages 57 and 58. **9–12. See margin for graphs.**

9. $\{(6, 3), (2, 1), (-2, 3)\}$
10. $\{(-5, 2), (2, 4), (1, 1), (-5, -2)\}$
11. $y = 0.5x$ **D = all reals, R = all reals; yes**
12. $y = 2x + 1$ **D = all reals, R = all reals; yes**

Find each value if $f(x) = 5x - 9$. *See Example 5 on page 59.*

13. $f(6)$ **21**
14. $f(-2)$ **−19**
15. $f(y)$ **5y − 9**
16. $f(-2v)$ **−10v − 9**

9. D = {−2, 2, 6}, R = {1, 3}; yes 10. D = {−5, 1 2}, R = {−2, 1, 2, 4}; no

2-2 Linear Equations

See pages 63–67.

Concept Summary

- A linear equation is an equation whose graph is a line. A linear function can be written in the form $f(x) = mx + b$.
- The standard form of a linear equation is $Ax + By = C$.

Example **Write $2x - 6 = y + 8$ in standard form. Identify A, B, and C.**

$2x - 6 = y + 8$ Original equation

$2x - y - 6 = 8$ Subtract y from each side.

$2x - y = 14$ Add 6 to each side.

The standard form is $2x - y = 14$. So, $A = 2$, $B = -1$, and $C = 14$.

Exercises State whether each equation or function is linear. Write *yes* or *no*. If no, explain your reasoning. *See Example 1 on page 63.*

17. $3x^2 - y = 6$ **No; x has an exponent other than 1.**
18. $2x + y = 11$ **yes**
19. $h(x) = \sqrt{2x + 1}$ **No; x is inside a square root.**

Write each equation in standard form. Identify A, B, and C. *See Example 3 on page 64.*

20. $y = 7x + 15$ **$7x - y = -15$; 7, −1, −15**
21. $0.5x = -0.2y - 0.4$ **$5x + 2y = -4$; 5, 2, −4**
22. $\frac{2}{3}x - \frac{3}{4}y = 6$ **$8x - 9y = 72$; 8, −9, 72**

Find the x-intercept and the y-intercept of the graph of each equation. Then graph the equation. *See Example 4 on page 65.* **23–25. See margin for graphs.**

23. $-\frac{1}{5}y = x + 4$ **−4, −20**
24. $6x = -12y + 48$ **8, 4**
25. $y - x = -9$ **9, −9**

2-3 Slope

See pages 68–74.

Concept Summary

- The slope of a line is the ratio of the change in y-coordinates to the corresponding change in x-coordinates.

$$m = \frac{y_2 - y_1}{x_2 - x_1}$$

- Lines with the same slope are parallel. Lines with slopes that are opposite reciprocals are perpendicular.

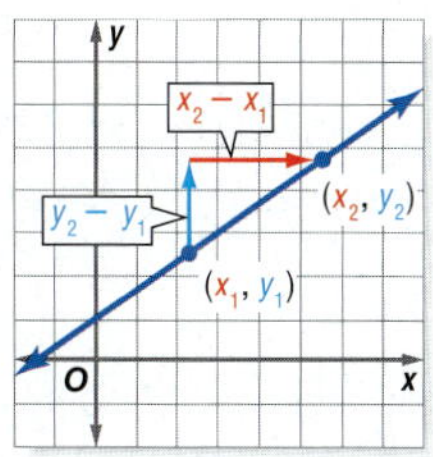

Answers

11.
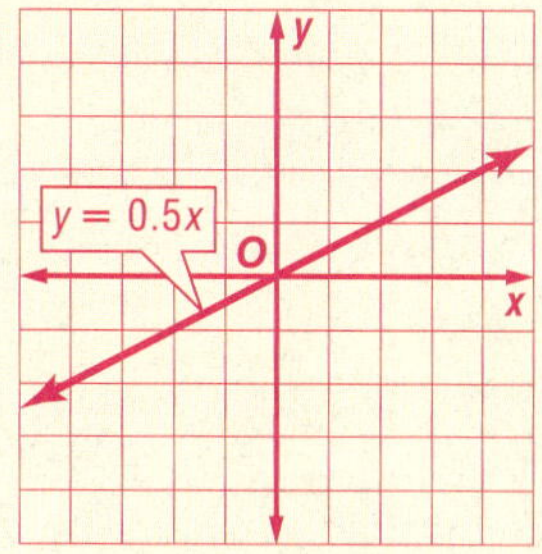

12.
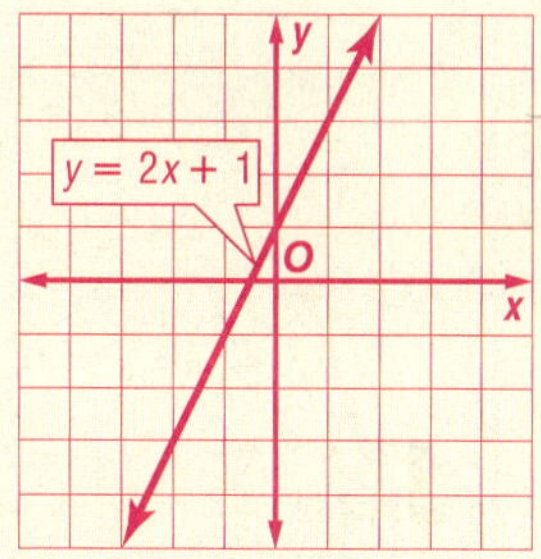

23.
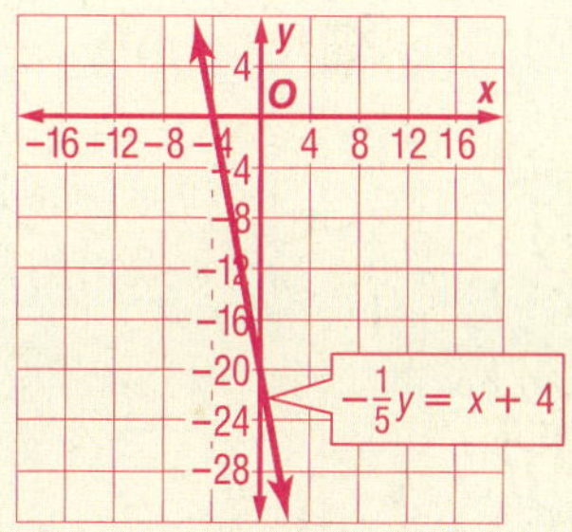

24.
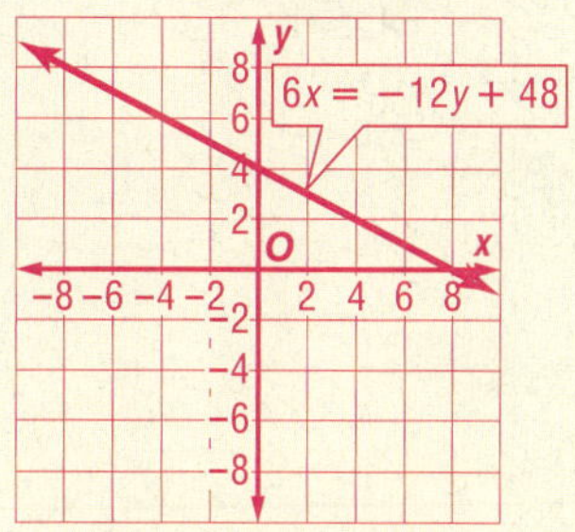

25.
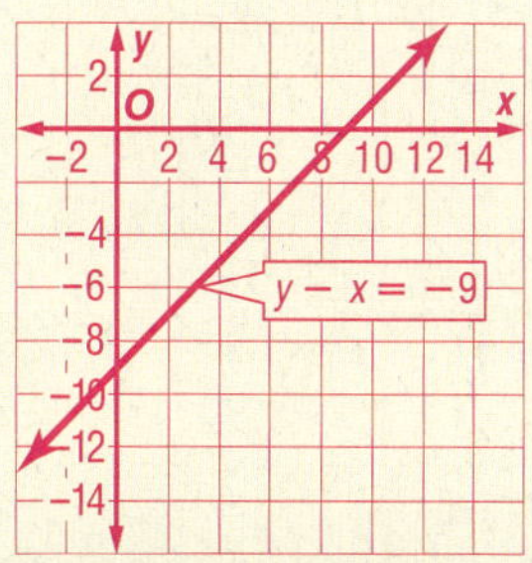

9.
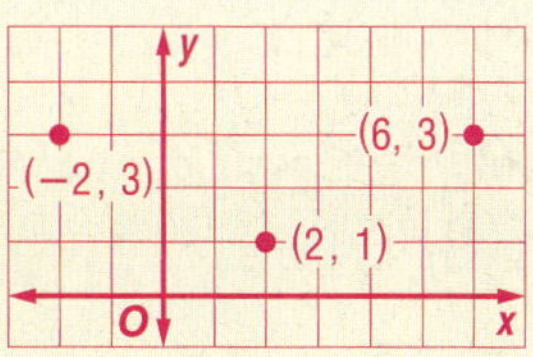

10.
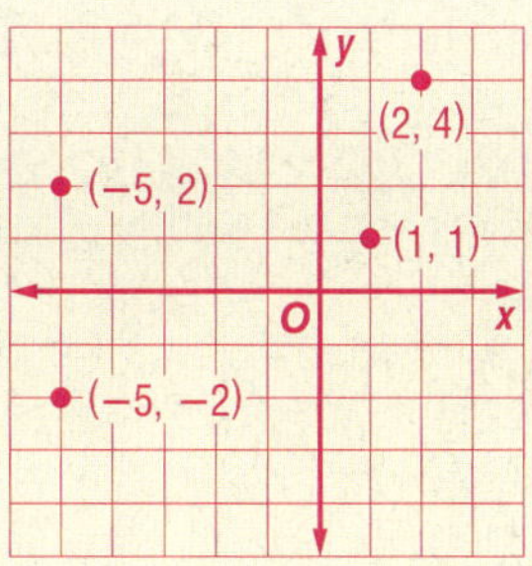

Example Find the slope of the line that passes through $(-5, 3)$ and $(7, 9)$.

$m = \frac{y_2 - y_1}{x_2 - x_1}$ Slope formula

$= \frac{9 - 3}{7 - (-5)}$ $(x_1, y_1) = (-5, 3), (x_2, y_2) = (7, 9)$

$= \frac{6}{12}$ or $\frac{1}{2}$ Simplify.

Exercises Find the slope of the line that passes through each pair of points.
See Example 1 on page 68.

26. $(-6, -3), (6, 7)$ $\frac{5}{6}$
27. $(5.5, -5.5), (11, -7)$ $-\frac{3}{11}$
28. $(-3, 24), (10, -41)$ -5

Graph the line passing through the given point with the given slope.
See Example 2 on page 69. **29–31. See margin.**

29. $(0, 1), m = 2$
30. $(3, -2), m = \frac{5}{2}$
31. $(-5, 2), m = -\frac{1}{4}$

Graph the line that satisfies each set of conditions.
See Examples 4 and 5 on pages 70 and 71. **32–35. See margin.**

32. passes through $(2, 0)$, parallel to a line whose slope is 3
33. passes through $(-1, -2)$, perpendicular to a line whose slope is $\frac{1}{2}$
34. passes through $(4, 1)$, perpendicular to graph of $2x + 3y = 1$
35. passes through $(-2, 2)$, parallel to graph of $-2x + y = 4$

2-4 Writing Linear Equations

See pages 75–80.

Concept Summary

- Slope-Intercept Form: $y = mx + b$
- Point-Slope Form: $y - y_1 = m(x - x_1)$

Example Write an equation in slope-intercept form for the line through $(4, 5)$ that is parallel to the line through $(-1, -3)$ and $(2, -1)$.

First, find the slope of the given line.

$m = \frac{y_2 - y_1}{x_2 - x_1}$ Slope formula

$= \frac{-1 - (-3)}{2 - (-1)}$ $(x_1, y_1) = (-1, -3), (x_2, y_2) = (2, -1)$

$= \frac{2}{3}$ Simplify.

The parallel line will also have slope $\frac{2}{3}$.

$y - y_1 = m(x - x_1)$ Point-slope form

$y - 5 = \frac{2}{3}(x - 4)$ $(x_1, y_1) = (4, 5), m = \frac{2}{3}$

$y = \frac{2}{3}x + \frac{7}{3}$ Slope-intercept form

Exercises Write an equation in slope-intercept form for the line that satisfies each set of conditions. *See Examples 1, 2, and 4 on pages 76–78.*

36. slope $\frac{3}{4}$, passes through $(-6, 9)$ $y = \frac{3}{4}x + \frac{27}{2}$
37. passes through $(3, -8)$ and $(-3, 2)$ $y = -\frac{5}{3}x - 3$
38. passes through $(-1, 2)$, parallel to the graph of $x - 3y = 14$ $y = \frac{1}{3}x + \frac{7}{3}$
39. passes through $(3, 2)$, perpendicular to the graph of $4x - 3y = 12$ $y = -\frac{3}{4}x + \frac{17}{4}$

Answers

29.
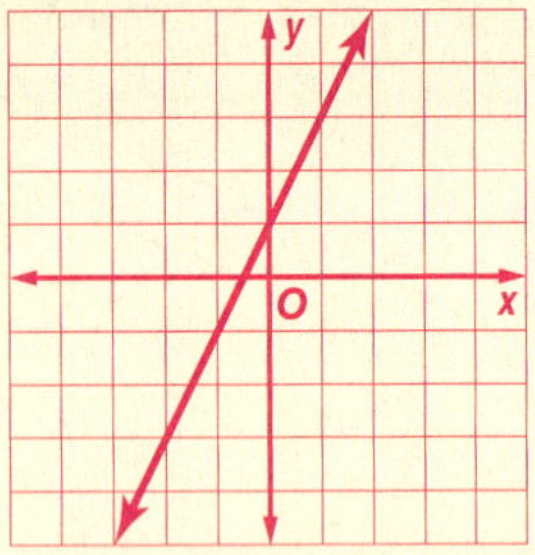

30.
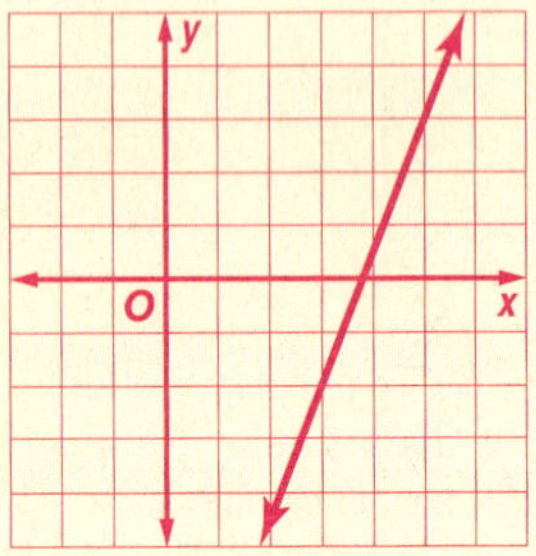

31.
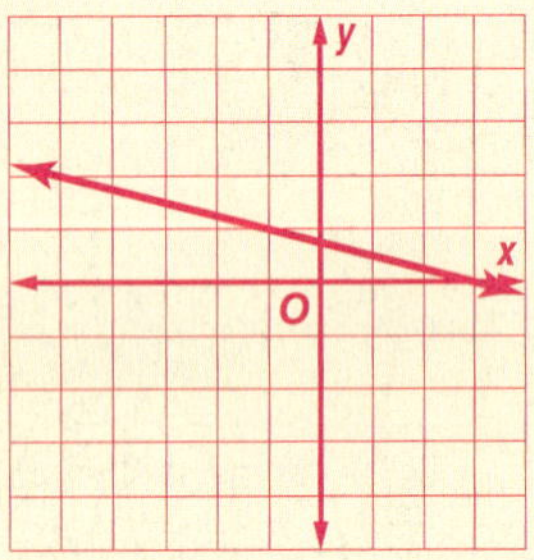

32.
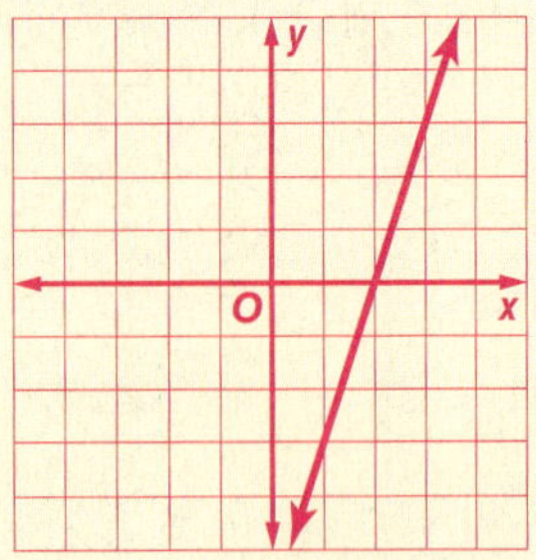

33.
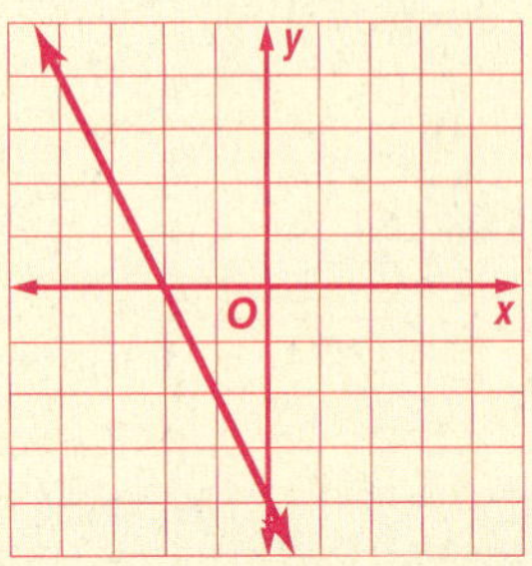

34.
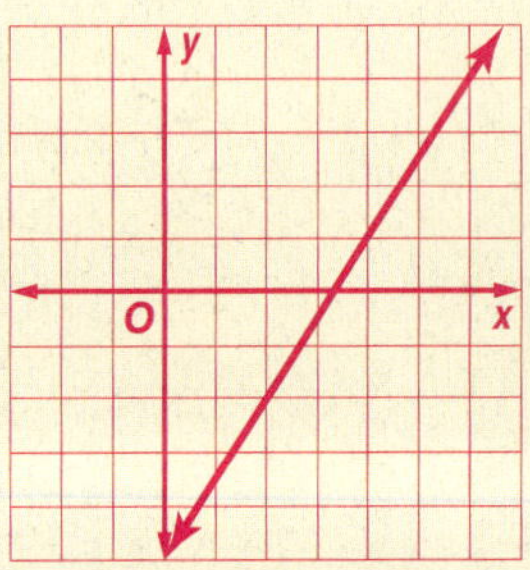

35.
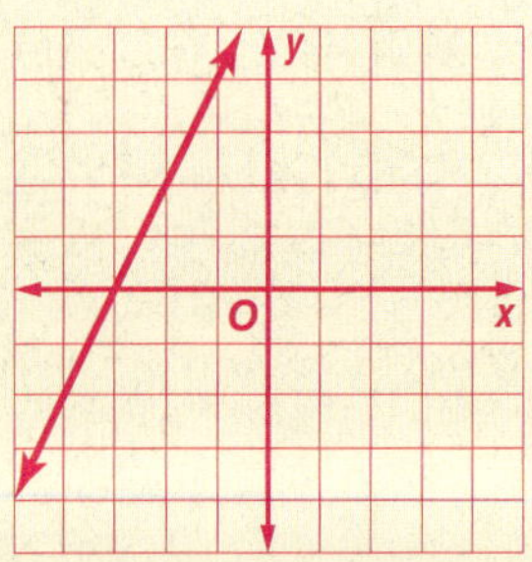

102 Chapter 2 Linear Relations and Functions

2-5 Modeling Real-World Data: Using Scatter Plots

See pages 81–86.

Concept Summary

- A scatter plot is a graph of ordered pairs of data.
- A prediction equation can be used to predict one of the variables given the other variable.

Example **WEEKLY PAY** **The table below shows the median weekly earnings for American workers for the period 1985–1999. Predict the median weekly earnings for 2010.**

Year	1985	1990	1995	1999	2010
Earnings ($)	343	412	479	549	?

Source: U.S. Bureau of Labor Statistics

A scatter plot suggests that any two points could be used to find a prediction equation. Use (1985, 343) and (1990, 412).

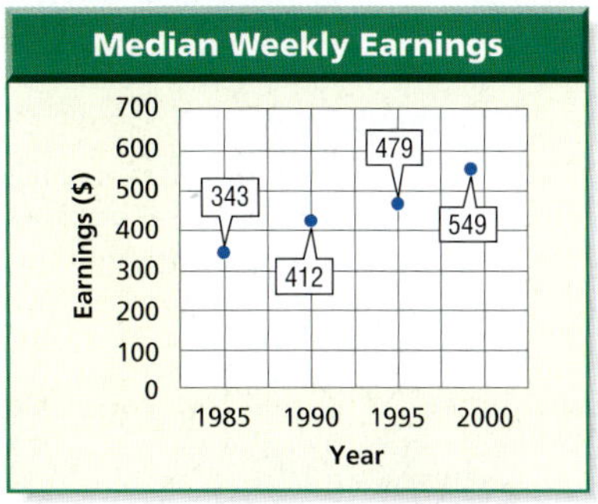

Source: U.S. Bureau of Labor Statistics

$m = \frac{y_2 - y_1}{x_2 - x_1}$ Slope formula

$= \frac{412 - 343}{1990 - 1985}$ $(x_1, y_1) = (1985, 343)$, $(x_2, y_2) = (1990, 412)$

$= \frac{69}{5}$ or 13.8 Simplify.

$y - y_1 = m(x - x_1)$ Point-slope form

$y - 343 = 13.8(x - 1985)$ Substitute.

$y = 13.8x - 27{,}050$ Add 343 to each side.

To predict the earnings for 2010, substitute 2010 for x.

$y = 13.8(2010) - 27{,}050$ $x = 2010$

$= 688$ Simplify.

The model predicts median weekly earnings of $688 in 2010.

Exercises **For Exercises 40–42, use the table that shows the number of people below the poverty level for the period 1980–1998.** *See Examples 1 and 2 on pages 81 and 82.*

40. Draw a scatter plot. **See margin.**

41. Use two ordered pairs to write a prediction equation.

42. Use your prediction equation to predict the number for 2010. **Sample answer: 42.2 million**

41. Sample answer using (1980, 29.3) and (1990, 33.6): $y = 0.43x - 822.1$

Year	People (millions)
1980	29.3
1985	33.1
1990	33.6
1995	36.4
1998	34.5
2010	?

Source: U.S. Census Bureau

Answer

40. **People Below Poverty Level**

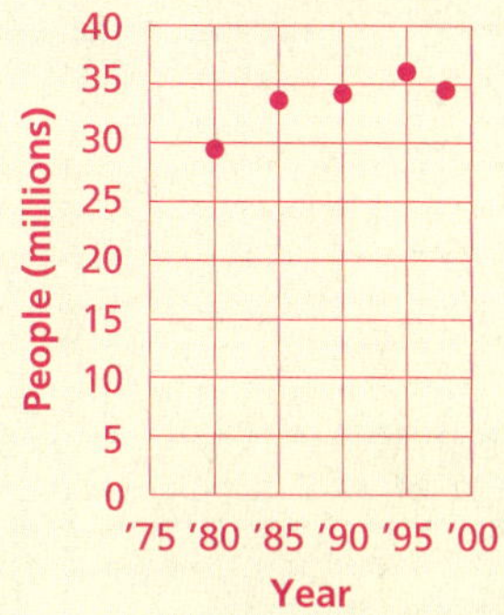

Chapter 2 Study Guide and Review 103

Chapter 2 For More ...
• Extra Practice, see pages 830–832.
• Mixed Problem Solving, see page 863.

Answers

49.

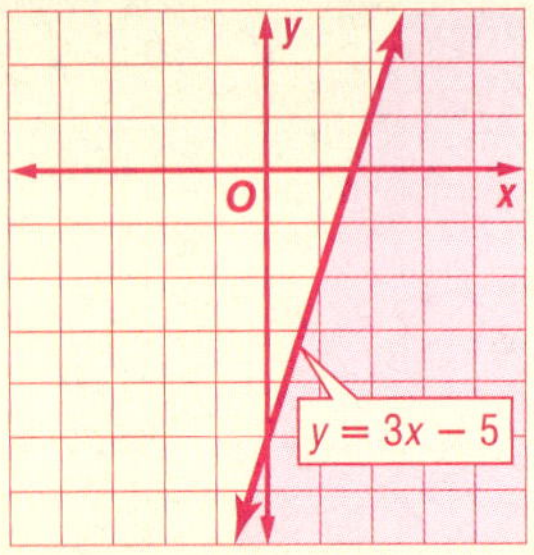

50.

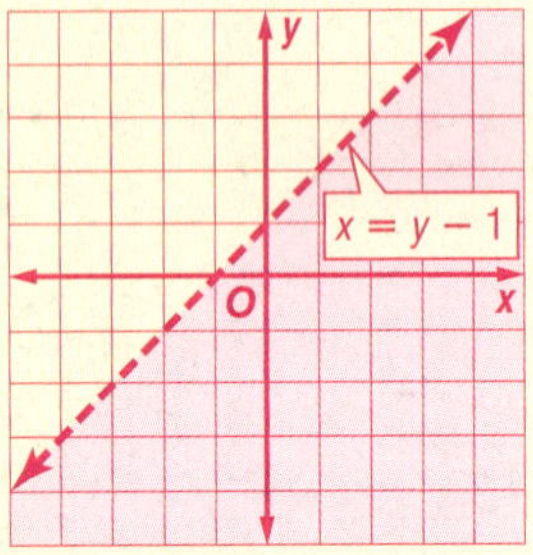

51.

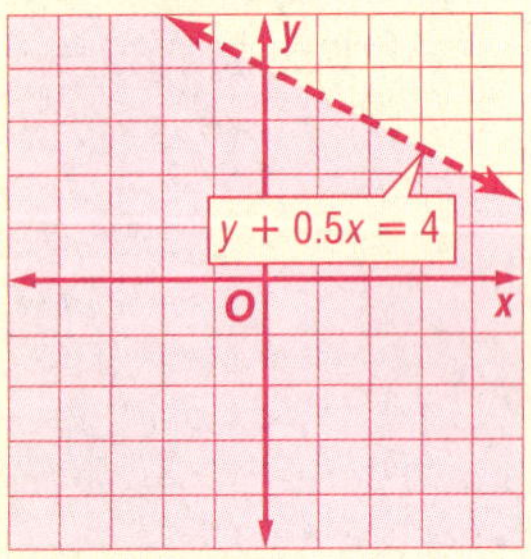

52.

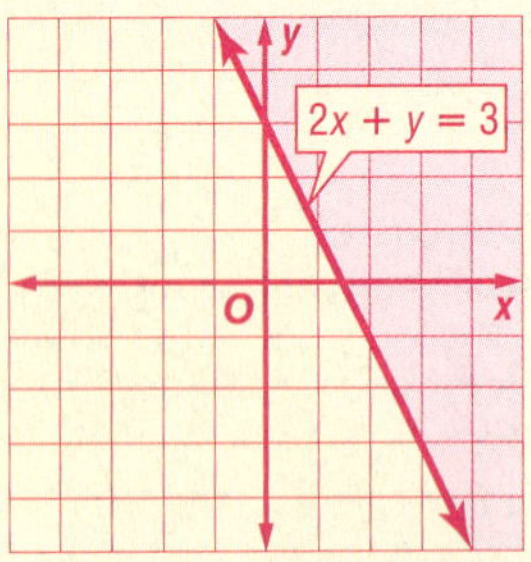

53.

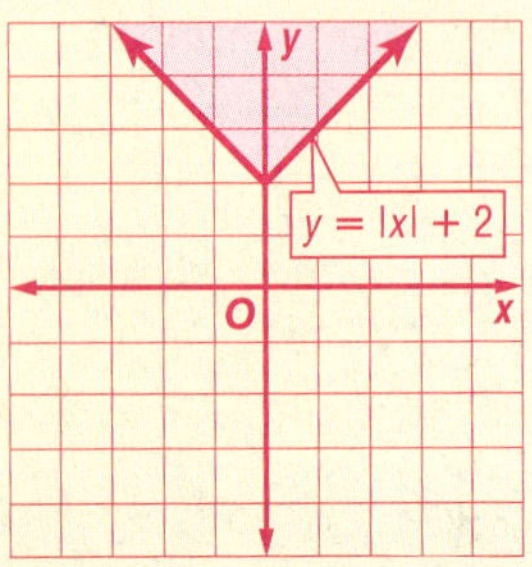

54.

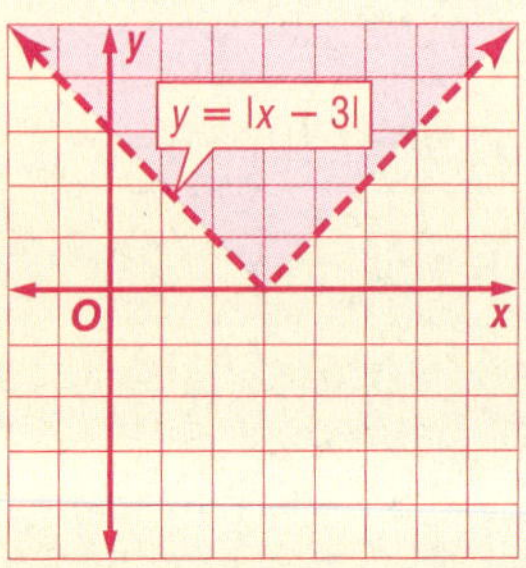

2-6 Special Functions

See pages 89–95.

Concept Summary

Greatest Integer

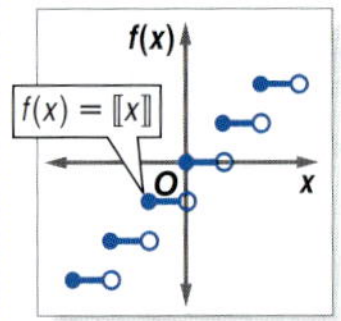

Constant

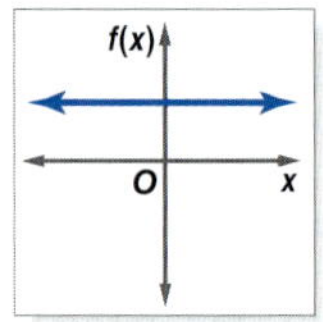

Absolute Value

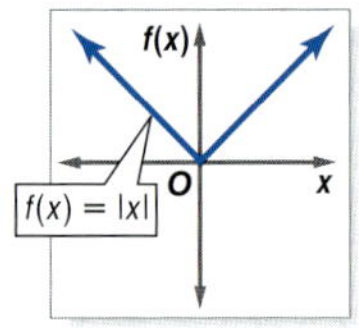

Piecewise

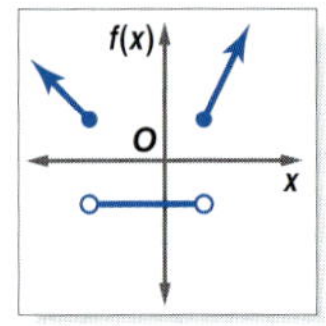

Example **Graph the function $f(x) = 3|x| - 2$. Identify the domain and range.**

The domain is all real numbers. The range is all real numbers greater than or equal to -2.

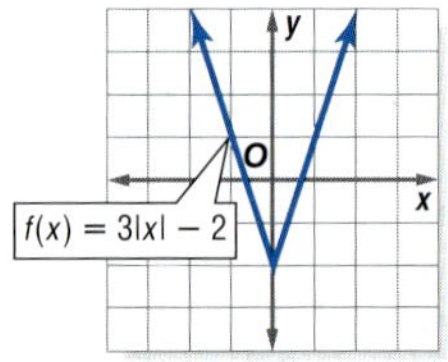

Exercises **Graph each function. Identify the domain and range.**
See Examples 1–3 on pages 89–91. **43–48. See pp. 107A–107H.**

43. $f(x) = [\![x]\!] - 2$
44. $h(x) = [\![2x - 1]\!]$
45. $g(x) = |x| + 4$
46. $h(x) = |x - 1| - 7$
47. $f(x) = \begin{cases} 2 \text{ if } x < -1 \\ -x - 1 \text{ if } x \geq -1 \end{cases}$
48. $g(x) = \begin{cases} -2x - 3 \text{ if } x < 1 \\ x - 4 \text{ if } x > 1 \end{cases}$

2-7 Graphing Inequalities

See pages 96–99.

Concept Summary

You can graph an inequality by following these steps.

Step 1 Determine whether the boundary is solid or dashed. Graph the boundary.

Step 2 Choose a point not on the boundary and test it in the inequality.

Step 3 If a true inequality results, shade the region containing your test point. If a false inequality results, shade the other region.

Example **Graph $x + 4y \leq 4$.**

Since the inequality symbol is $\leq$, the graph of the boundary should be solid. Graph the equation. Test (0, 0).

$x + 4y \leq 4$ Original inequality

$0 + 4(0) \leq 4$ $(x, y) = (0, 0)$

$0 \leq 4$ Shade the region that contains (0, 0).

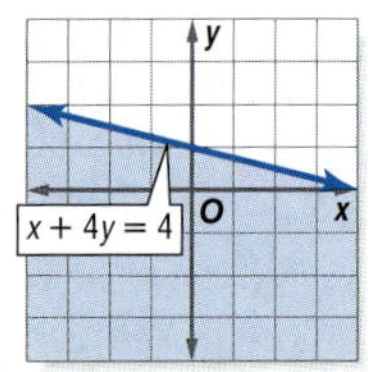

Exercises **Graph each inequality.** *See Examples 1–3 on pages 96 and 97.*

49. $y \leq 3x - 5$
50. $x > y - 1$
51. $y + 0.5x < 4$
52. $2x + y \geq 3$
53. $y \geq |x| + 2$
54. $y > |x - 3|$

49–54. See margin.

Chapter 2 Practice Test

Vocabulary and Concepts

Choose the correct term to complete each sentence.

1. The variable whose values make up the domain of a function is called the (**independent**, *dependent*) variable.
2. To find the (**x-intercept**, y-intercept) of the graph of a linear equation, let $y = 0$.
3. An equation of the form ($Ax + By = C$, **$y = mx + b$**) is in slope-intercept form.

Skills and Applications

Graph each relation and find the domain and range. Then determine whether the relation is a function. **4–5. See pp. 107A–107H for graphs.**

4. $\{(-4, -8), (-2, 2), (0, 5), (2, 3), (4, -9)\}$ **D = {−4, −2, 0, 2, 4}, R = {−9, −8, 2, 3, 5}; yes**
5. $y = 3x - 3$ **D = all reals, R = all reals; yes**

Find each value.

6. $f(3)$ if $f(x) = 7 - x^2$ **−2**
7. $f(0)$ if $f(x) = x - 3x^2$ **0**

Graph each equation or inequality. **8–19. See pp. 107A–107H.**

8. $y = \frac{3}{5}x - 4$
9. $4x - y = 2$
10. $x = -4$
11. $y = 2x - 5$
12. $f(x) = 3x - 1$
13. $f(x) = [\![3x]\!] + 3$
14. $g(x) = |x + 2|$
15. $h(x) = \begin{cases} x + 2 \text{ if } x < -2 \\ 2x - 1 \text{ if } x \geq -2 \end{cases}$
16. $y \leq 10$
17. $x > 6$
18. $-2x + 5 \leq 3y$
19. $y < 4|x - 1|$

Find the slope of the line that passes through each pair of points.

20. $(8, -4), (6, 1)$ **$-\frac{5}{2}$**
21. $(-2, 5), (4, 5)$ **0**
22. $(5, 7), (4, -6)$ **13**

Graph the line passing through the given point with the given slope. **23–25. See pp. 107A–107H.**

23. $(1, -3), 2$
24. $(-2, 2), -\frac{1}{3}$
25. $(3, -2)$, undefined

Write an equation in slope-intercept form for the line that satisfies each set of conditions.

26. slope -5, y-intercept 11 **$y = -5x + 11$**
27. x-intercept 9, y-intercept -4 **$y = \frac{4}{9}x - 4$**
28. passes through $(-6, 15)$, parallel to the graph of $2x + 3y = 1$ **$y = -\frac{2}{3}x + 11$**
29. passes through $(5, 2)$, perpendicular to the graph of $x + 3y = 7$ **$y = 3x - 13$**

RECREATION For Exercises 30–32, use the table that shows the amount Americans spent on recreation in recent years.

Year	1995	1996	1997	1998
Amount ($ billions)	401.6	429.6	457.8	494.7

Source: U.S. Bureau of Economic Analysis

30. Draw a scatter plot, where x represents the number of years since 1995. **See margin.**
31. Write a prediction equation. **Sample answer using (0, 401.6) and (1, 429.6): $y = 28x + 401.6$**
32. Predict the amount that will be spent on recreation in 2010. **Sample answer: $821.6 billion**

33. **STANDARDIZED TEST PRACTICE** What is the slope of a line parallel to $y - 2 = 4(x + 1)$? **D**

Ⓐ -4 Ⓑ $-\frac{1}{4}$ Ⓒ $\frac{1}{4}$ Ⓓ 4

www.algebra2.com/chapter_test

Portfolio Suggestion

Introduction In this chapter, you have graphed many different kinds of functions. The appearances of these graphs were also very different from one another.

Ask Students Select one kind of graph that you found difficult to master and explain why you felt this to be the case. Suggest ways that this topic might be presented in a different way to help other students who have the same difficulty.

Chapter 2 Practice Test

Assessment Options

Vocabulary Test A vocabulary test/review for Chapter 2 can be found on p. 112 of the *Chapter 2 Resource Masters*.

Chapter Tests There are six Chapter 2 Tests and an Open-Ended Assessment task available in the *Chapter 2 Resource Masters*.

Chapter 2 Tests			
Form	**Type**	**Level**	**Pages**
1	MC	basic	99–100
2A	MC	average	101–102
2B	MC	average	103–104
2C	FR	average	105–106
2D	FR	average	107–108
3	FR	advanced	109–110

MC = multiple-choice questions
FR = free-response questions

Open-Ended Assessment Performance tasks for Chapter 2 can be found on p. 111 of the *Chapter 2 Resource Masters*. A sample scoring rubric for these tasks appears on p. A28.

ExamView® Pro

Use the networkable ExamView® Pro to:

- Create **multiple versions** of tests.
- Create **modified** tests for *Inclusion* students.
- **Edit** existing questions and **add** your own questions.

Answer

30. **Money Spent on Recreation**

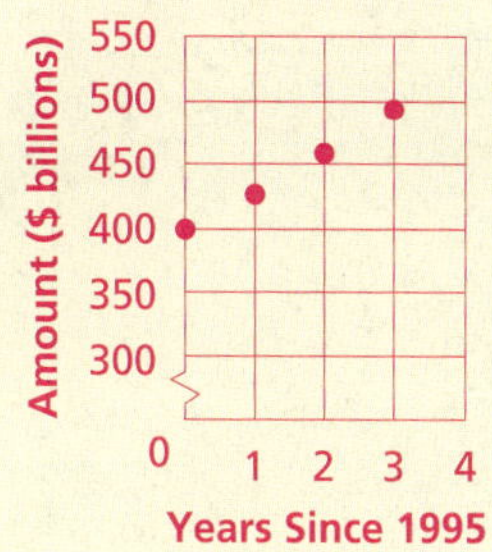

Chapter 2 Standardized Test Practice

These two pages contain practice questions in the various formats that can be found on the most frequently given standardized tests.

A practice answer sheet for these two pages can be found on p. A1 of the *Chapter 2 Resource Masters*.

Standardized Test Practice
Student Recording Sheet, p. A1

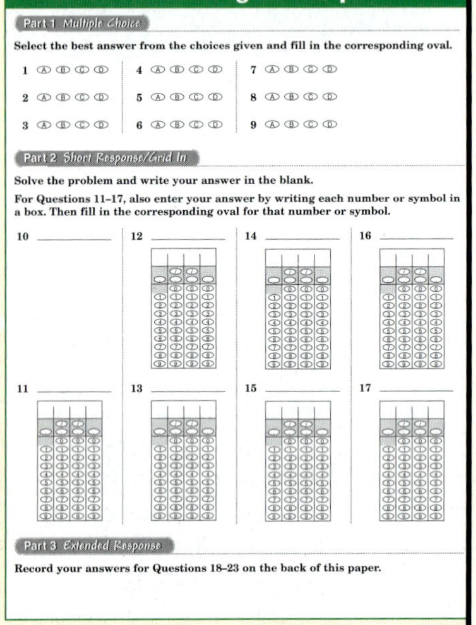
Part 1 Multiple Choice

Select the best answer from the choices given and fill in the corresponding oval.

1 Ⓐ Ⓑ Ⓒ Ⓓ 4 Ⓐ Ⓑ Ⓒ Ⓓ 7 Ⓐ Ⓑ Ⓒ Ⓓ
2 Ⓐ Ⓑ Ⓒ Ⓓ 5 Ⓐ Ⓑ Ⓒ Ⓓ 8 Ⓐ Ⓑ Ⓒ Ⓓ
3 Ⓐ Ⓑ Ⓒ Ⓓ 6 Ⓐ Ⓑ Ⓒ Ⓓ 9 Ⓐ Ⓑ Ⓒ Ⓓ

Part 2 Short Response/Grid In

Solve the problem and write your answer in the blank.

For Questions 11–17, also enter your answer by writing each number or symbol in a box. Then fill in the corresponding oval for that number or symbol.

10 ____ 12 ____ 14 ____ 16 ____
11 ____ 13 ____ 15 ____ 17 ____

Part 3 Extended Response

Record your answers for Questions 18–23 on the back of this paper.

Additional Practice

See pp. 117–118 in the *Chapter 2 Resource Masters* for additional standardized test practice.

Chapter 2 Standardized Test Practice

Part 1 Multiple Choice

Record your answers on the answer sheet provided by your teacher or on a sheet of paper.

1. In the figure, $\angle B$ and $\angle BCD$ are right angles. $\overline{BC}$ is 9 units, $\overline{AB}$ is 12 units, and $\overline{CD}$ is 8 units. What is the area, in square units, of $\triangle ACD$? **A**

Ⓐ 36
Ⓑ 60
Ⓒ 72
Ⓓ 135

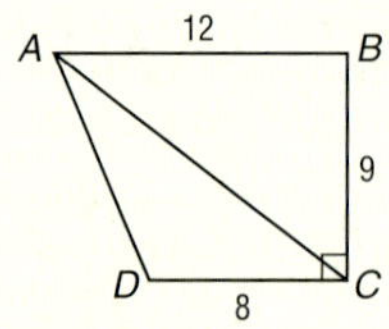

2. If $x + 3$ is an even integer, then x could be which of the following? **B**

Ⓐ -2 Ⓑ -1
Ⓒ 0 Ⓓ 2

3. What is the slope of the line that contains the points (15, 7) and (6, 4)? **B**

Ⓐ $\frac{1}{4}$ Ⓑ $\frac{1}{3}$
Ⓒ $\frac{3}{8}$ Ⓓ $\frac{2}{3}$

4. In 2000, Matt had a collection of 30 music CDs. Since then he has given away 2 CDs, purchased 6 new CDs, and traded 3 of his CDs to Kashan for 4 of Kashan's CDs. Since 2000, what has been the percent of increase in the number of CDs in Matt's collection? **D**

Ⓐ $3\frac{1}{3}\%$ Ⓑ 10%
Ⓒ $14\frac{2}{7}\%$ Ⓓ $16\frac{2}{3}\%$

5. If the product of (2 + 3), (3 + 4), and (4 + 5) is equal to three times the sum of 40 and x, then $x =$ ______. **B**

Ⓐ 43 Ⓑ 65
Ⓒ 105 Ⓓ 195

6. If one side of a triangle is three times as long as a second side and the second side is s units long, then the length of the third side of the triangle can be **A**

Ⓐ $3s$. Ⓑ $4s$.
Ⓒ $5s$. Ⓓ $6s$.

7. Which of the following sets of numbers has the property that the *product* of any two numbers is also a number in the set? **D**

I the set of positive numbers
II the set of prime numbers
III the set of even integers

Ⓐ I only
Ⓑ II only
Ⓒ III only
Ⓓ I and III only

8. If $\frac{3 + x}{7 + x} = \frac{3}{7} + \frac{3}{7}$, then $x =$ ______. **D**

Ⓐ $\frac{3}{7}$ Ⓑ 3
Ⓒ 7 Ⓓ 21

9. The average (arithmetic mean) of r, s, x, and y is 8, and the average of x and y is 4. What is the average of r and s? **D**

Ⓐ 4 Ⓑ 6
Ⓒ 8 Ⓓ 12

Test-Taking Tip

Questions 1–9
On multiple-choice questions, try to compute the answer first. Then compare your answer to the given answer choices. If you don't find your answer among the choices, check your calculations.

ExamView® Pro

Special banks of standardized test questions similar to those on the SAT, ACT, TIMSS 12, NAEP 12, and state proficiency tests can be found on this CD-ROM.

Preparing for Standardized Tests
For test-taking strategies and more practice, see pages 877–892.

Part 2 Short Response/Grid In

Record your answers on the answer sheet provided by your teacher or on a sheet of paper.

10. If n is a prime integer such that $2n > 19 \geq \frac{7}{8}n$ what is one possible value of n? **11, 13, 17, or 19**

11. If $\overline{AC}$ is 2 units, what is the value of t? **1/4 or .25**

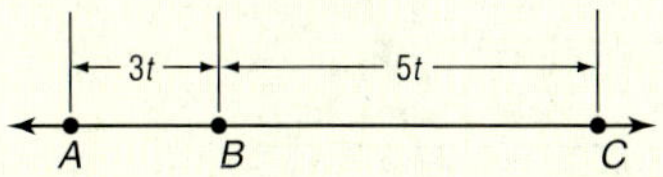

12. If $0.85x = 8.5$, what is the value of $\frac{1}{x}$? $\frac{1}{10}$ **or .1**

13. In $\triangle ABC$, what is the value of $w + x + y + z$? **240**

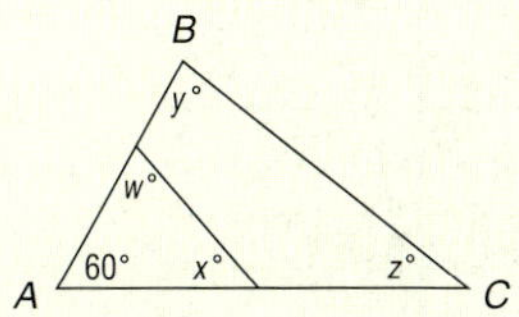

14. In an election, a total of 4000 votes were cast for three candidates, *A*, *B*, and *C*. Candidate *C* received 800 votes. If candidate *B* received more votes than candidate *C* and candidate *A* received more votes than candidate *B*, what is the least number of votes that candidate *A* could have received? **1601**

15. If the points $P(-2, 3)$, $Q(2, 5)$, and $R(2, 3)$ are vertices of a triangle, what is the area of the triangle? **4**

16. How many of the first one hundred positive integers contain the digit 7? **19**

17. A triangle has a base of length 17, and the other two sides are equal in length. If the lengths of the sides of the triangle are integers, what is the shortest possible length of a side? **9**

www.algebra2.com/standardized_test

Part 3 Extended Response

Record your answers on a sheet of paper. Show your work.

For Exercises 18–23, use the information below and in the table.

The amount that a certain online retailer charges for shipping an electronics purchase is determined by the weight of the package. The charges for several different weights are given in the table below.

Electronics Shipping Charges	
Weight (lb)	Shipping ($)
1	5.58
3	6.76
4	7.35
7	9.12
10	10.89
13	12.66
15	13.84

18. Write a relation to represent the data. Use weight as the independent variable and the shipping charge as the dependent variable. **{(1, 5.58), (3, 6.76), (4, 7.35), (7, 9.12), (10, 10.89), (13, 12.66), (15, 13.84)}**

19. Graph the relation on a coordinate plane. **See margin.**

20. Is the relation a function? Explain your reasoning. **Yes, each value of the domain is paired with one value of the range.**

21. Find the rate of change of the shipping charge per pound. **$0.59**

22. Write an equation that could be used to find the shipping charge y for a package that weighs x pounds. $y = 0.59x + 4.99$

23. Find the shipping charge for a package that weighs 19 pounds. **$16.20**

Evaluating Extended Response Questions

Extended Response questions are graded by using a multilevel rubric that guides you in assessing a student's knowledge of a particular concept.

Goal: Determine the dimensions of a rectangular garden for a given perimeter.

Sample Scoring Rubric: The following rubric is a sample scoring device. You may wish to add more detail to this sample to meet your individual scoring needs.

Score	Criteria
4	A correct solution that is supported by well-developed, accurate explanations
3	A generally correct solution, but may contain minor flaws in reasoning or computation
2	A partially correct interpretation and/or solution to the problem
1	A correct solution with no supporting evidence or explanation
0	An incorrect solution indicating no mathematical understanding of the concept or task, or no solution is given

Answer

19.

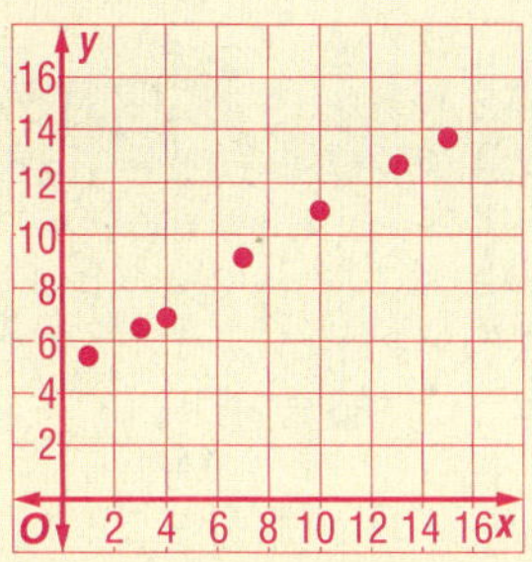

Pages 60–61, Lesson 2-1

2. Sample answer:

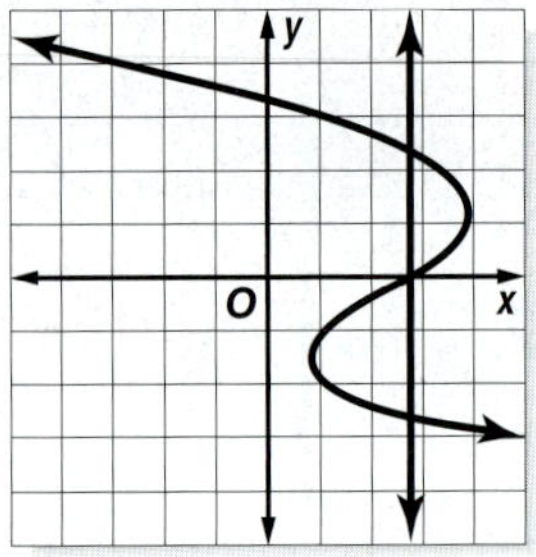

7. D = {7},
R = {−1, 2, 5, 8}, no

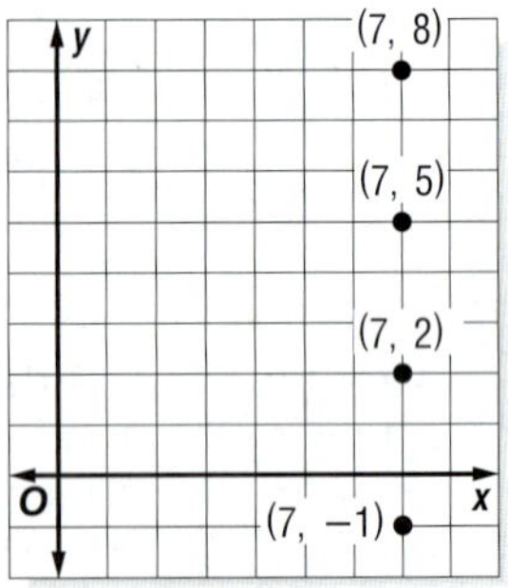

8. D = {3, 4, 6},
R = {2.5}, yes

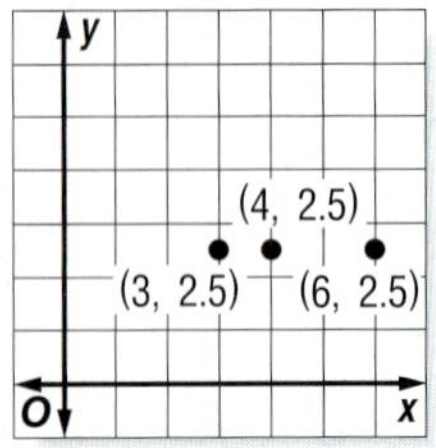

9. D = all reals,
R = all reals, yes

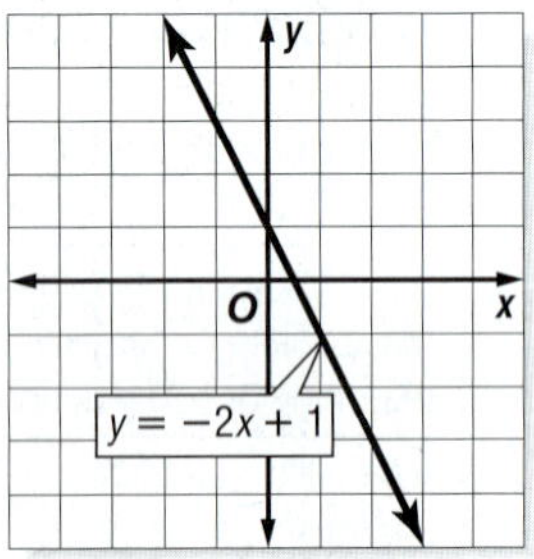

10. D = $\{x|x \geq 0\}$,
R = all reals, no

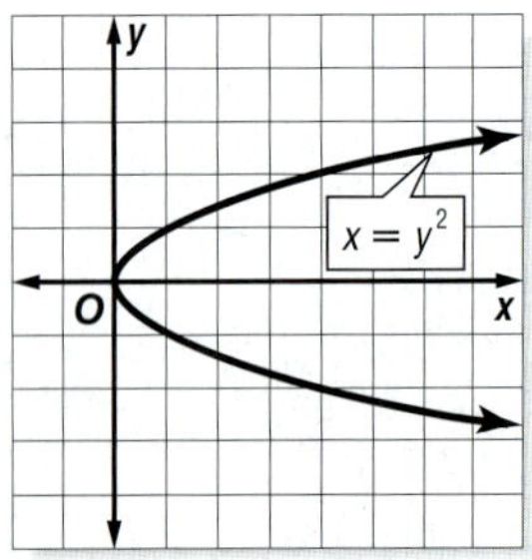

23.

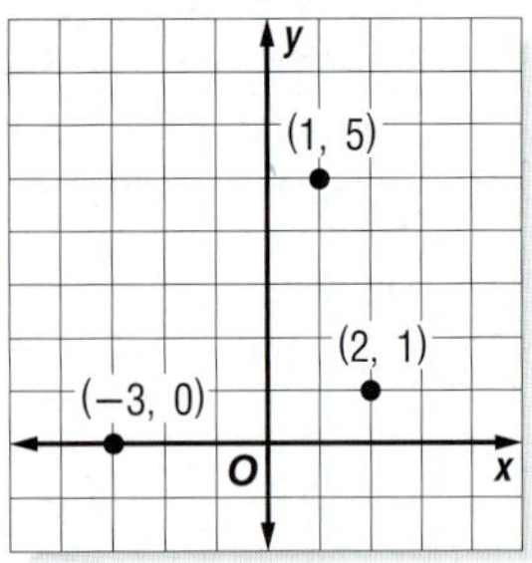

24.

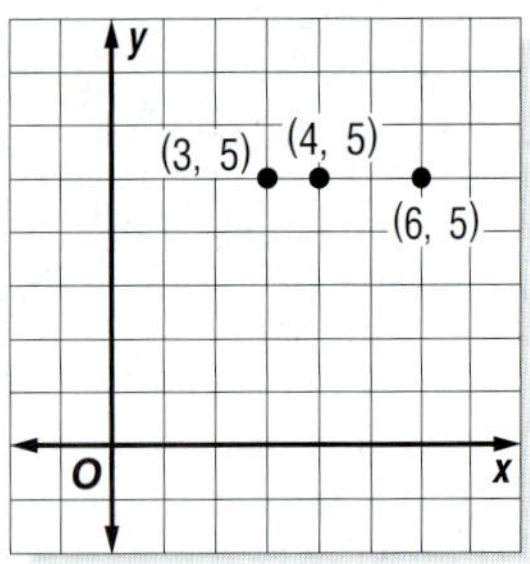

25.

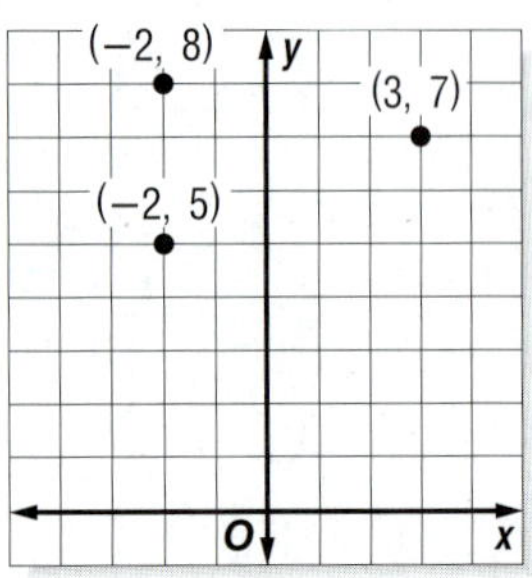

26.

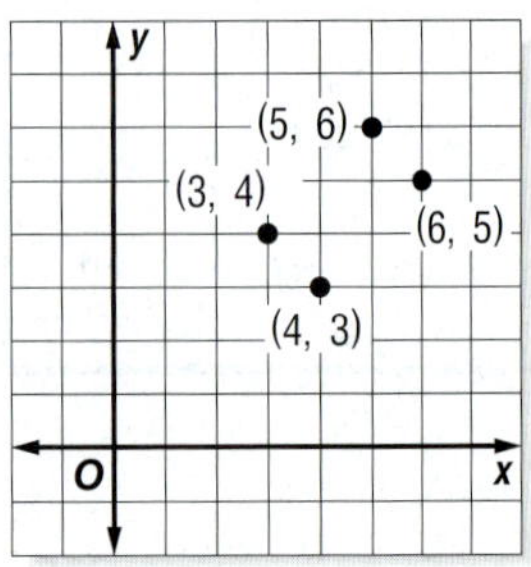

27.

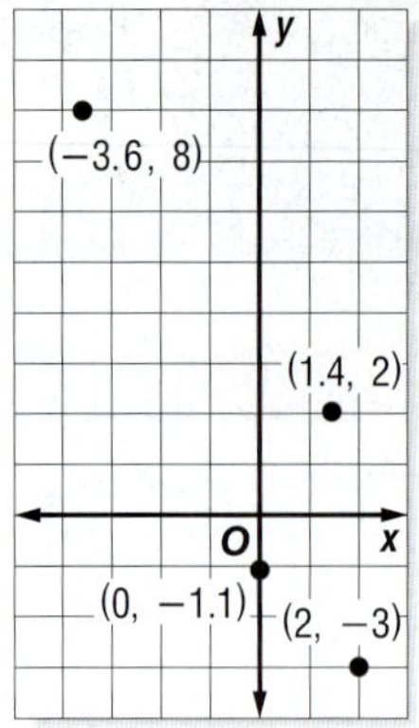

28.

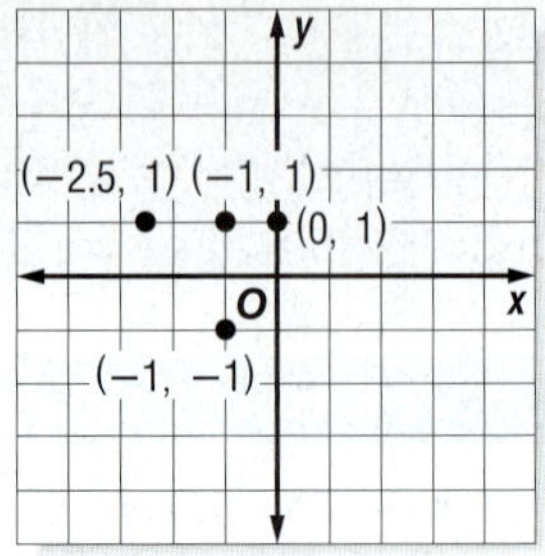

29.

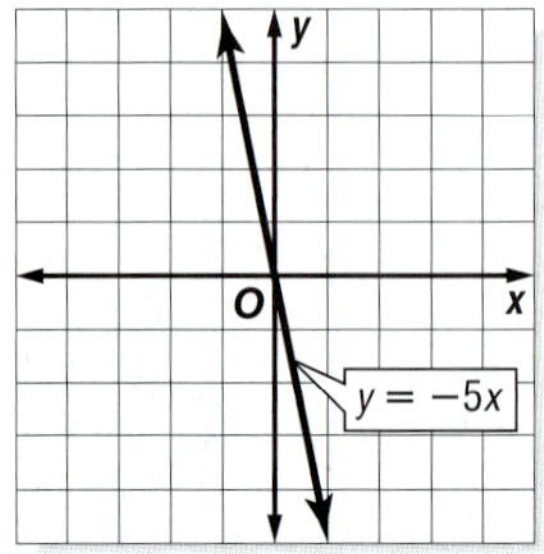

30.

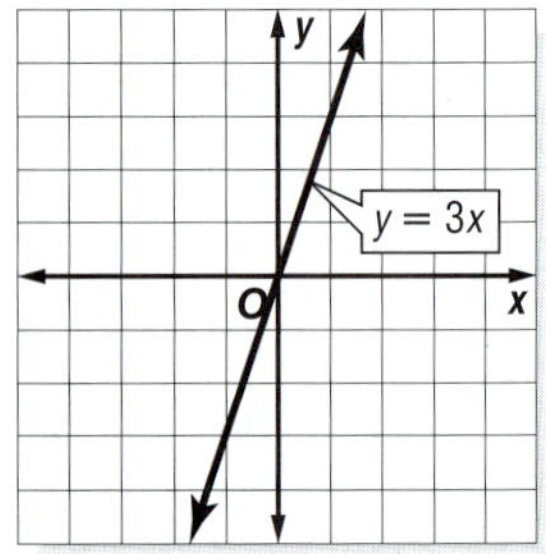

31.

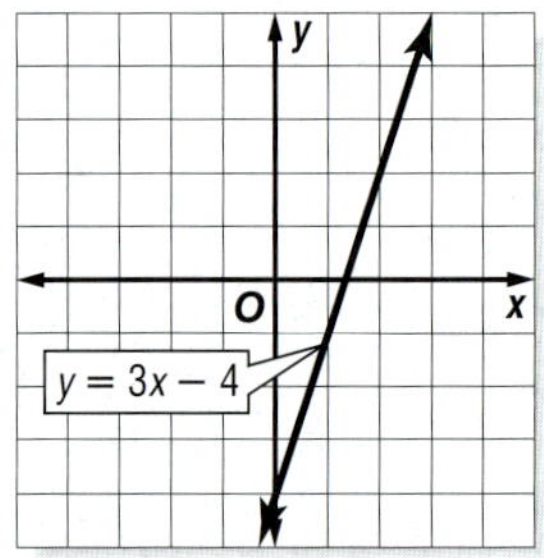

32.

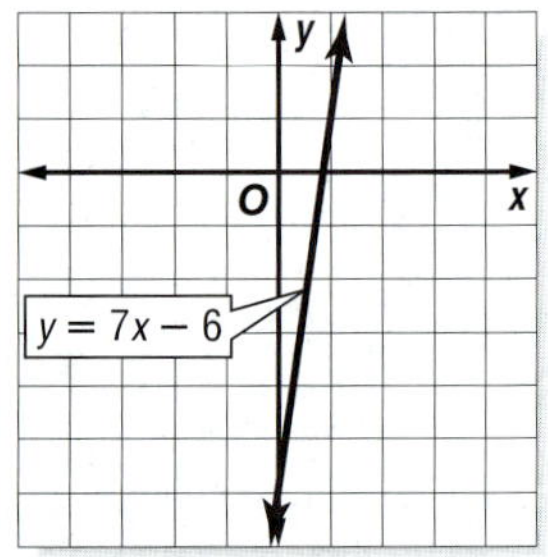

33.

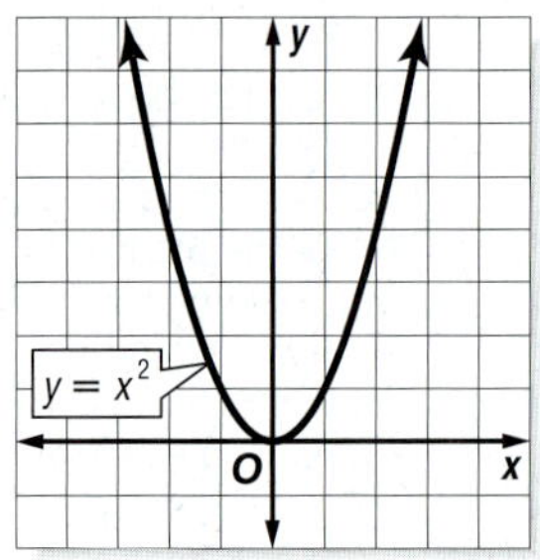

34.

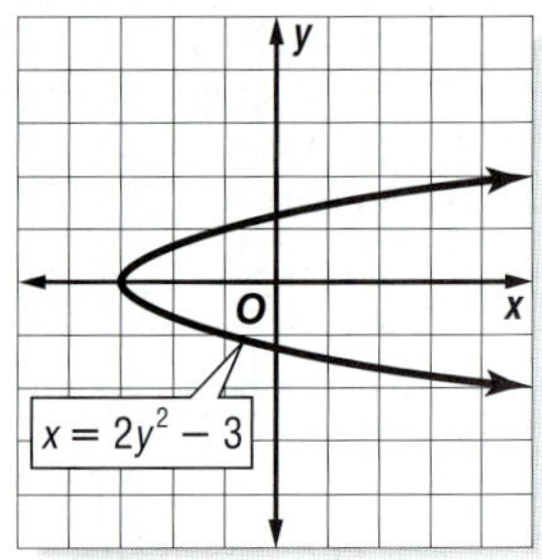

35. American League Leaders

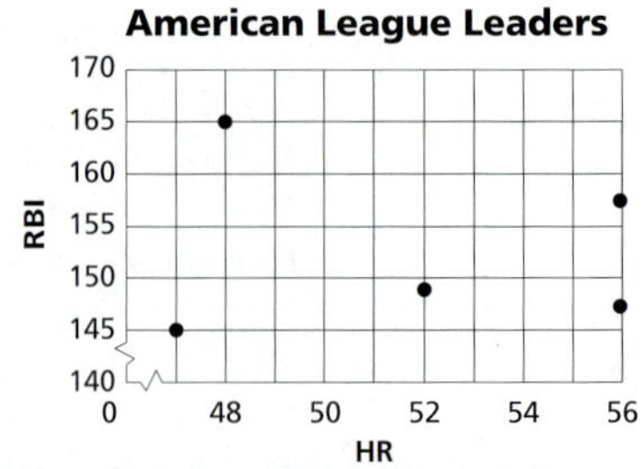

39. Stock Price

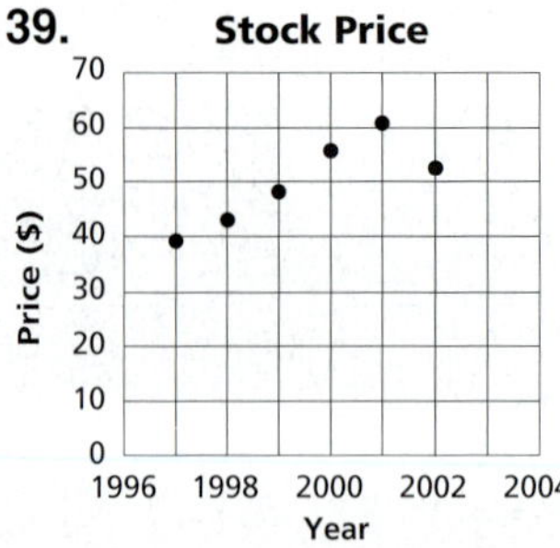

43. 30+ Years of Service

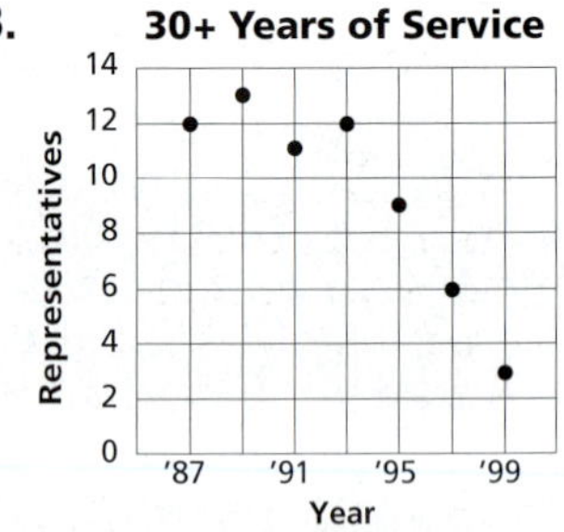

44. D = {1987, 1989, 1991, 1993, 1995, 1997, 1999}, R = {3, 6, 9, 11, 12, 13}

45. Each domain value is paired with only one range value so the relation is a function, but the range value 12 is paired with two domain values so the function is not one-to-one.

Pages 65–67, Lesson 2-2

9.

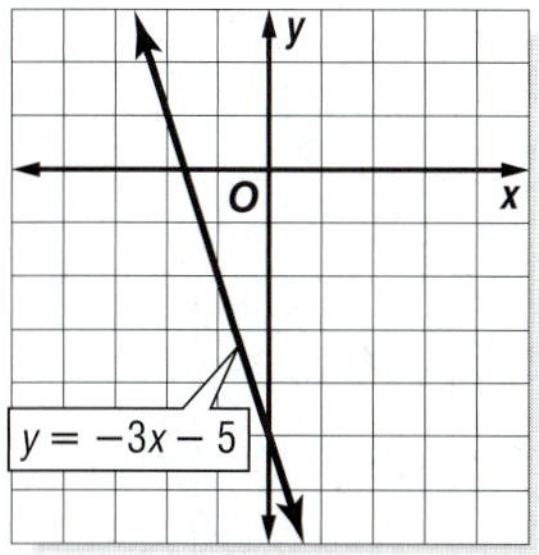

10.

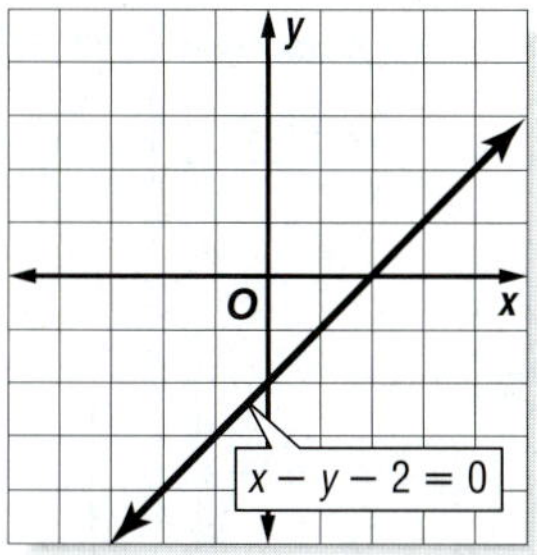

11.

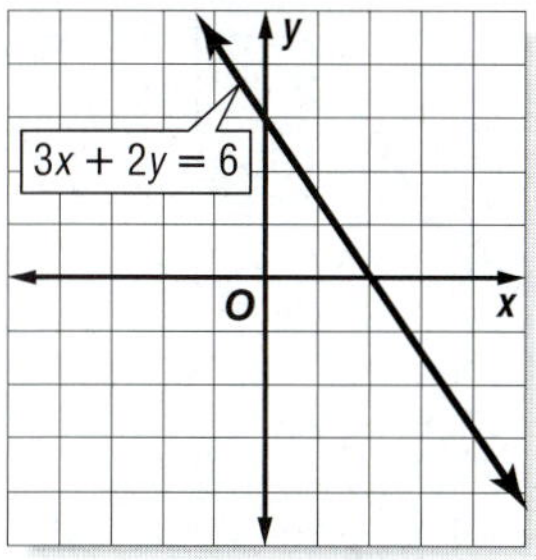

12.

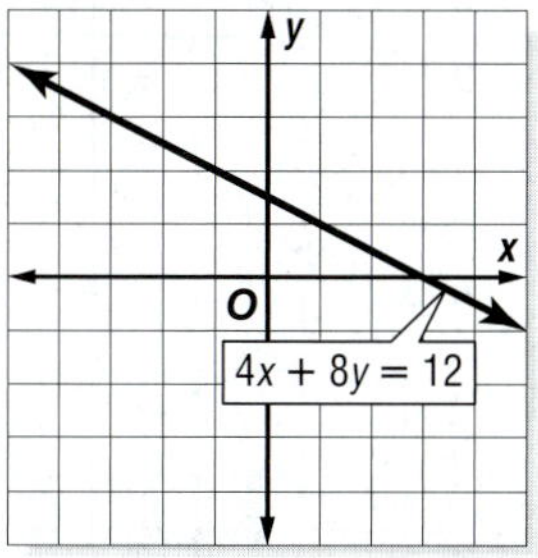

39.

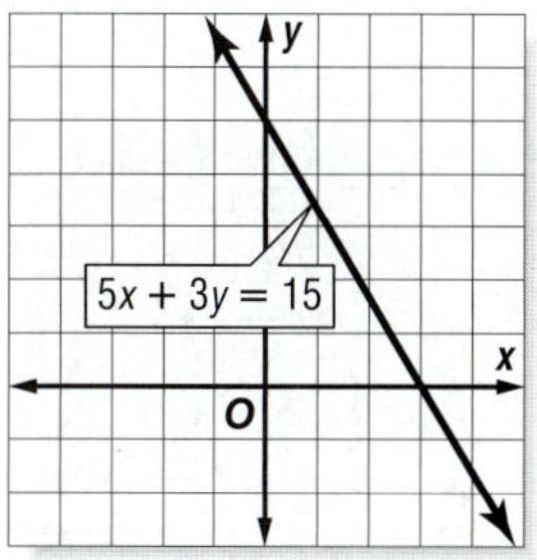

40.

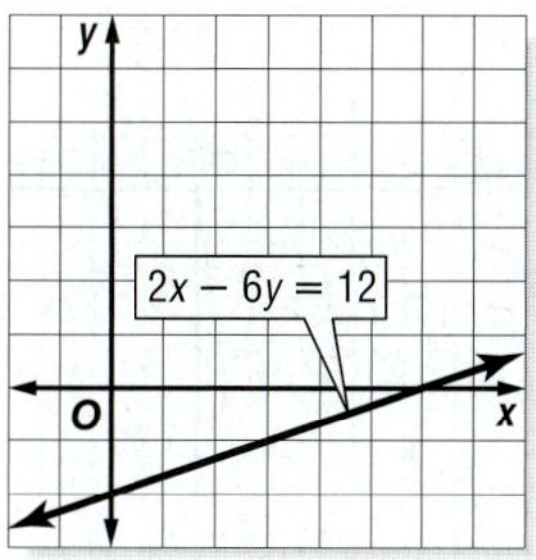

41.

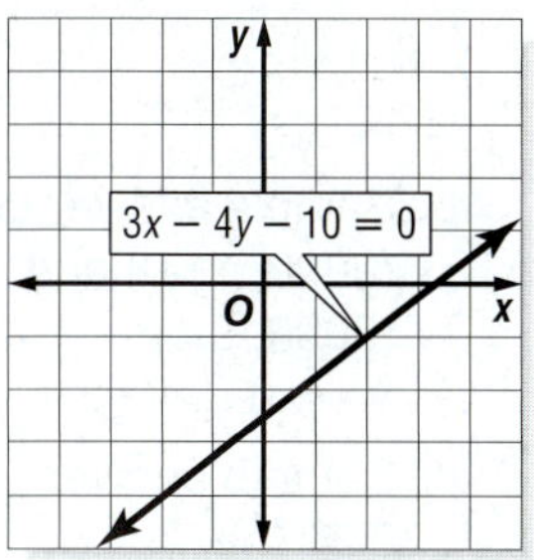

42.

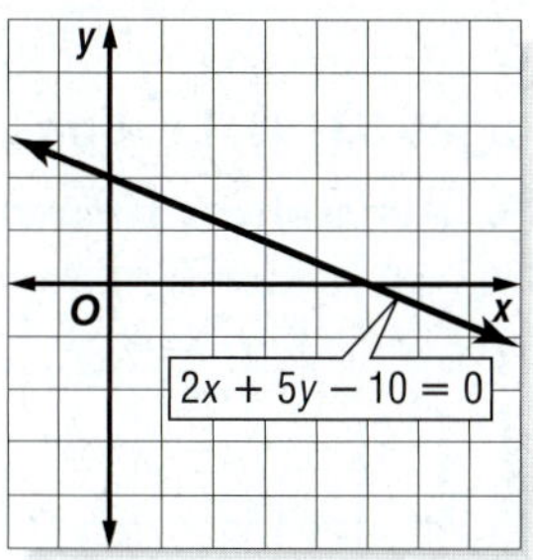

43.

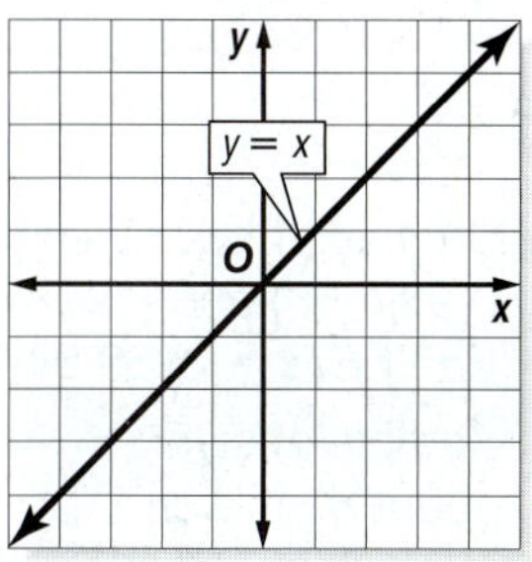

44.

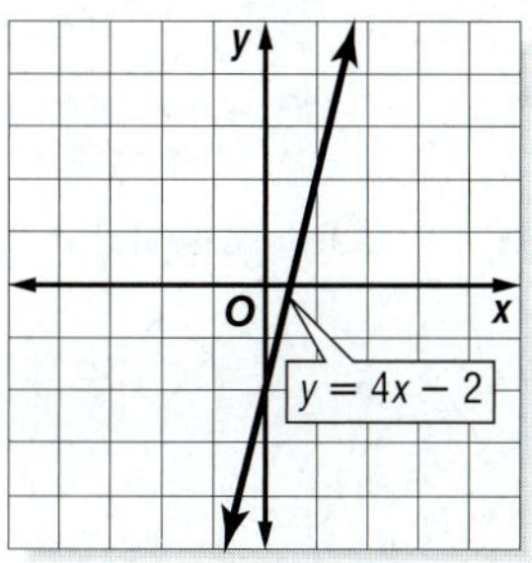

45.

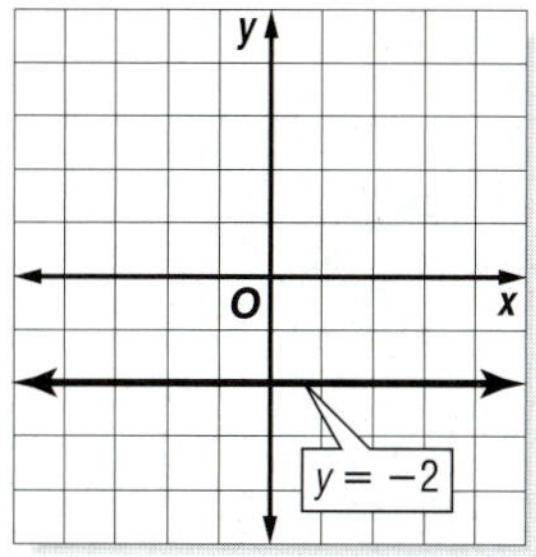

46.

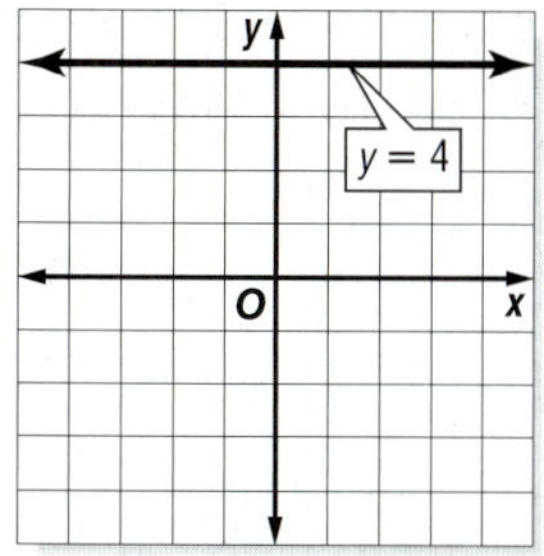

47.

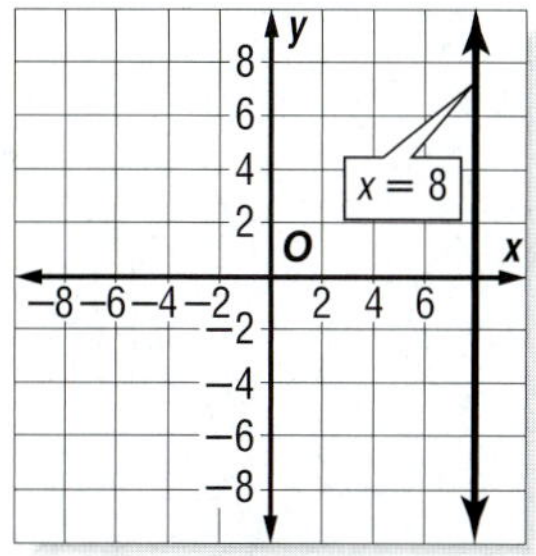

48.

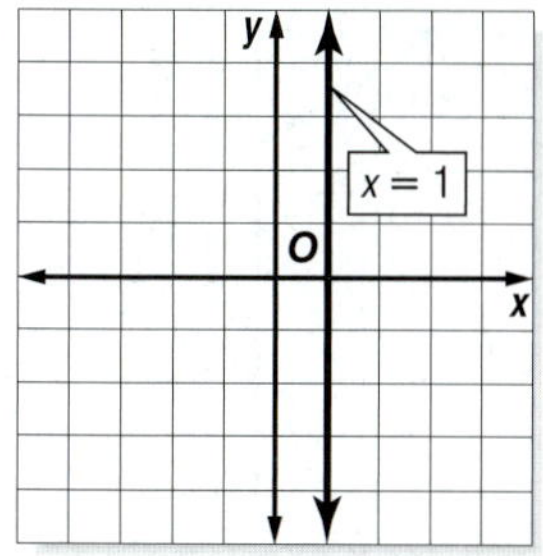

49.

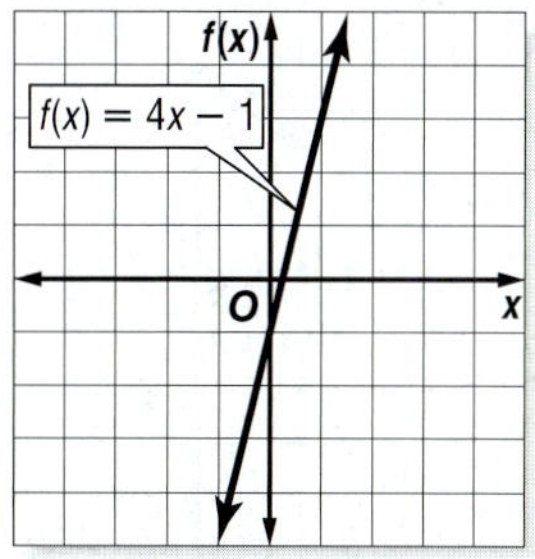

50.

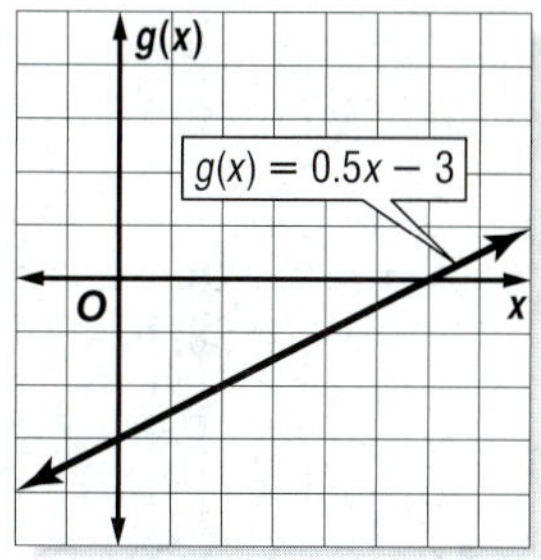

64.

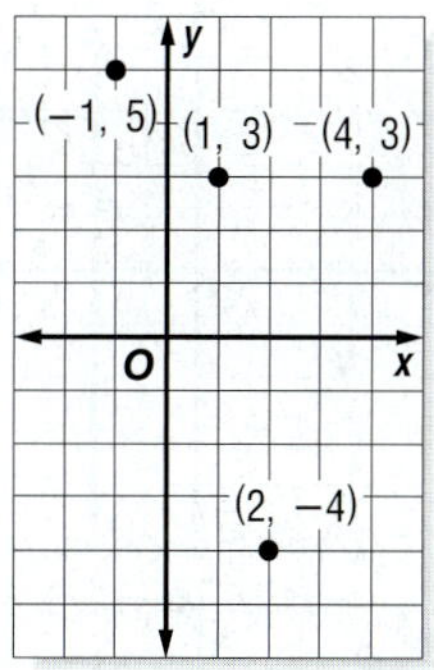

65.

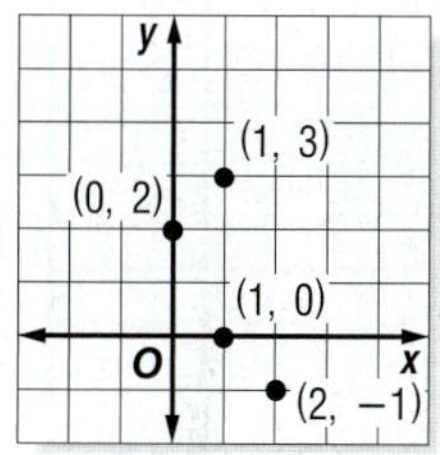

Pages 71–74, Lesson 2-3

7.

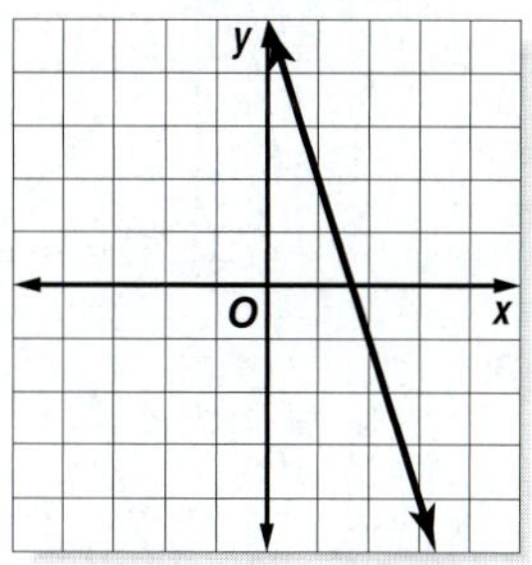

8.

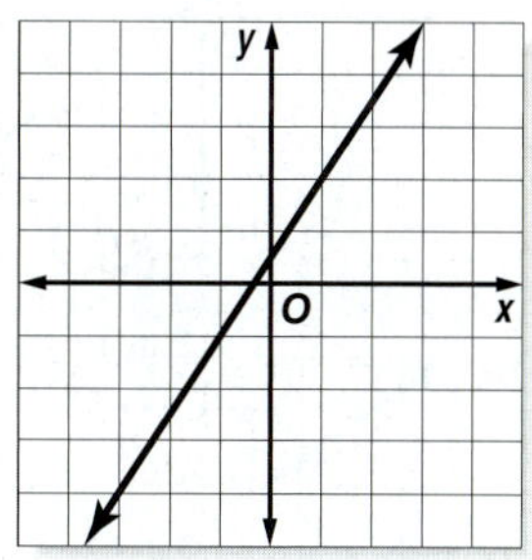

9.

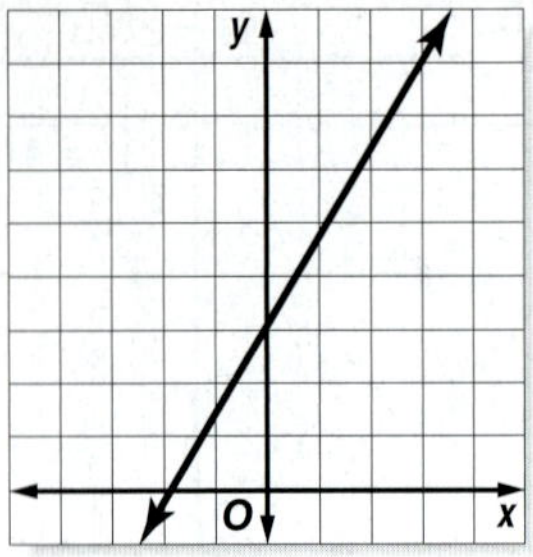

10.

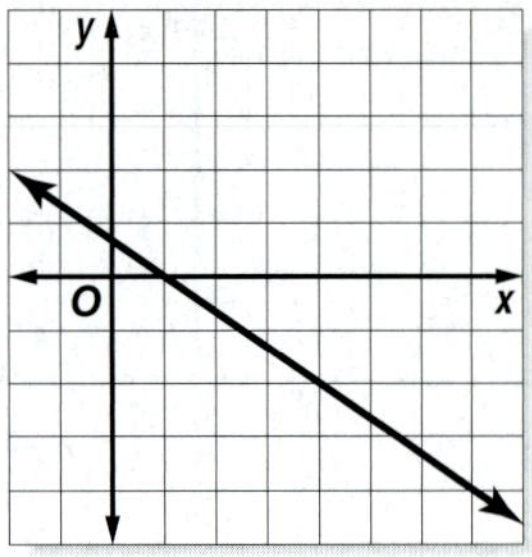

11.

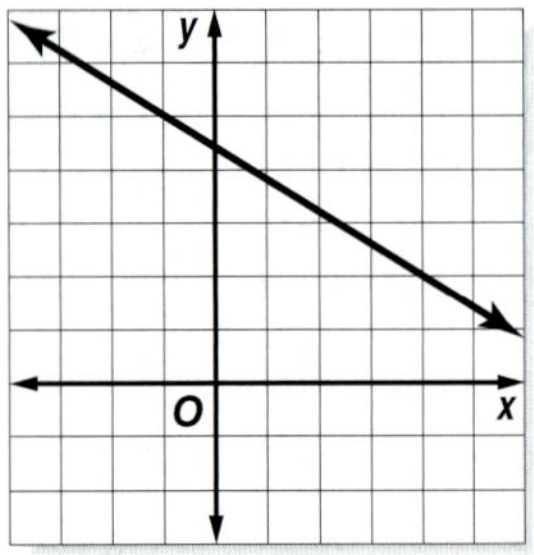

43.

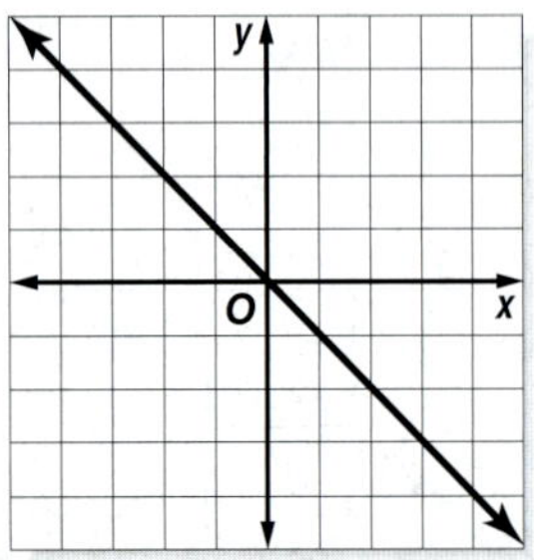

44.

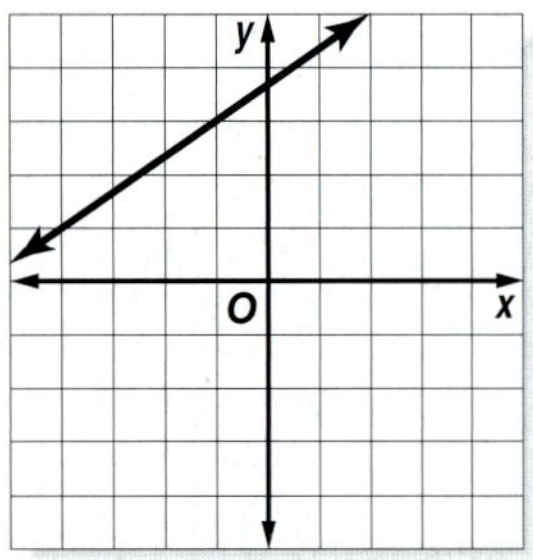

45.

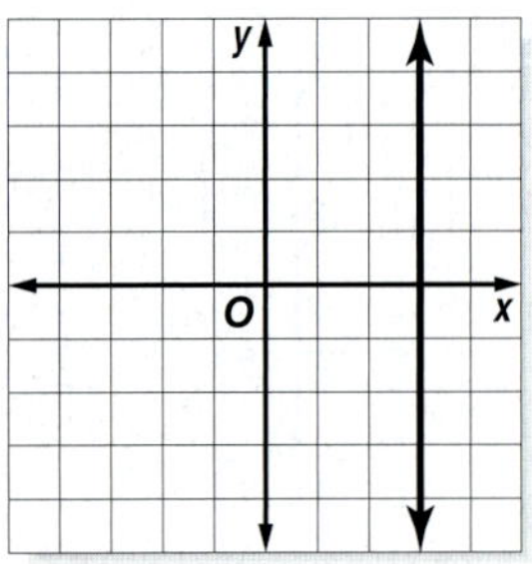

46.

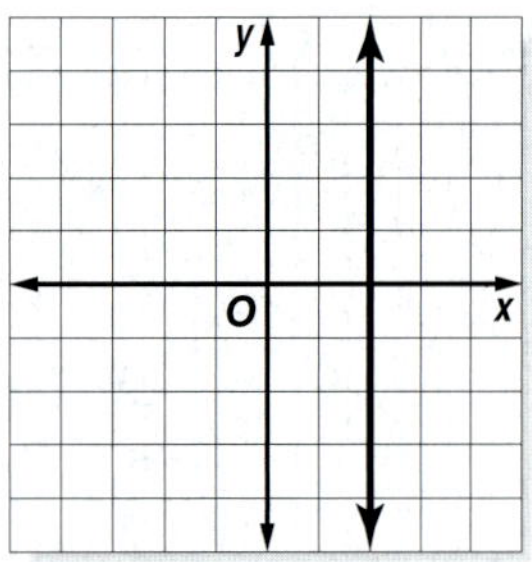

47.

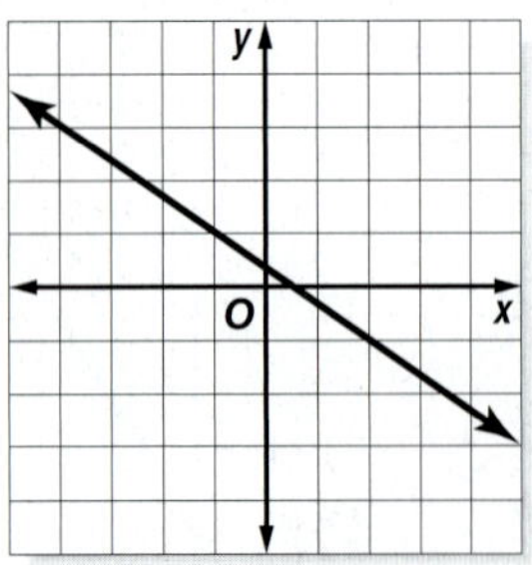

48.

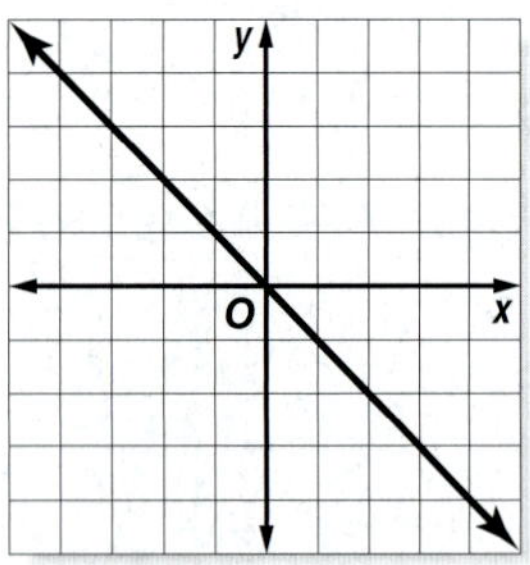

49.

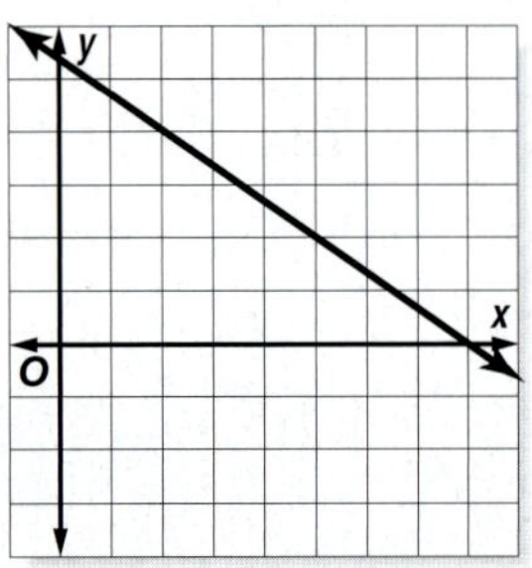

50.

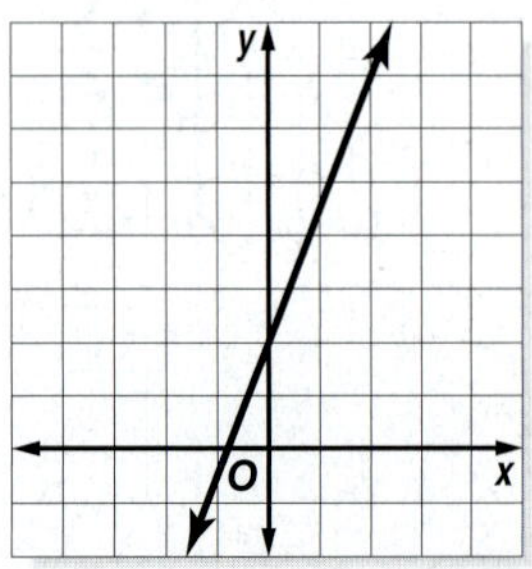

53. The grade or steepness of a road can be interpreted mathematically as a slope. Answers should include the following.

- Think of the diagram at the beginning of the lesson as being in a coordinate plane. Then the rise is a change in *y*-coordinates and the horizontal distance is a change in *x*-coordinates. Thus, the grade is a slope expressed as a percent.
- 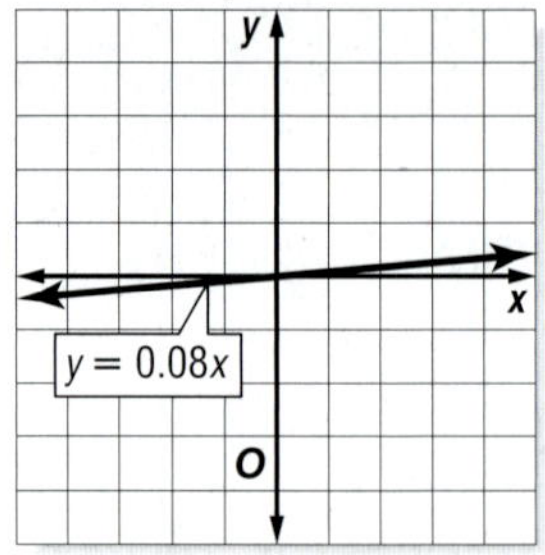

58.

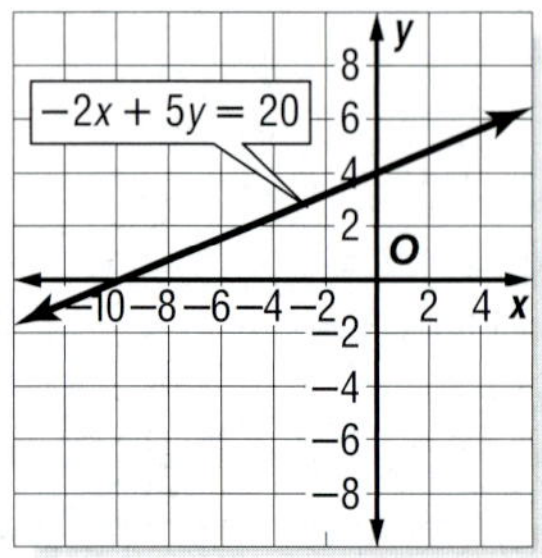

59.

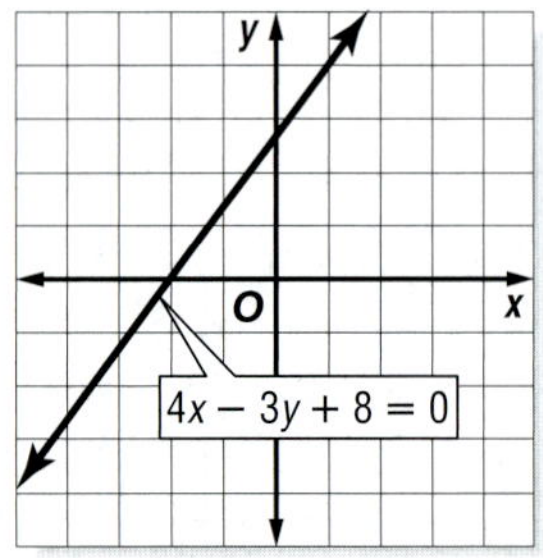

60.

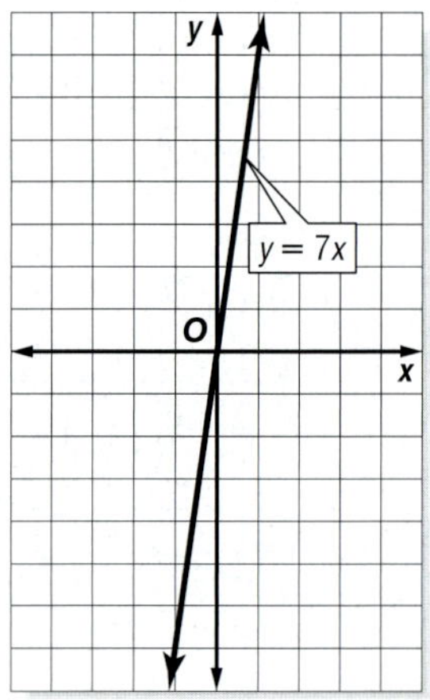

Pages 83–85, Lesson 2-5

4a. **Atmospheric Temperature**

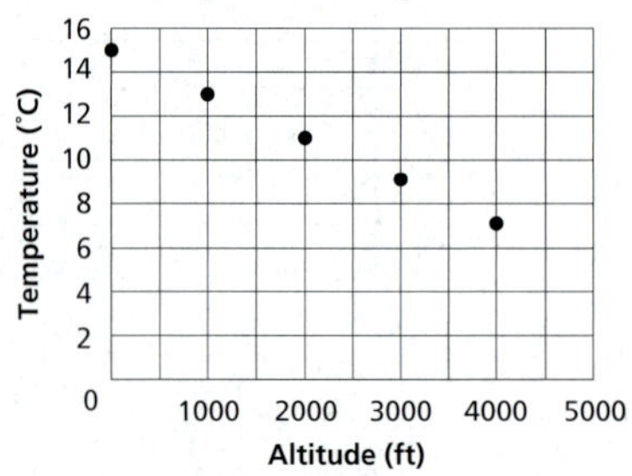

4b. Sample answer using (2000, 11.0) and (3000, 9.1): $y = -0.0019x + 14.8$

4c. Sample answer: 5.3°C

5a. **Cable Television**

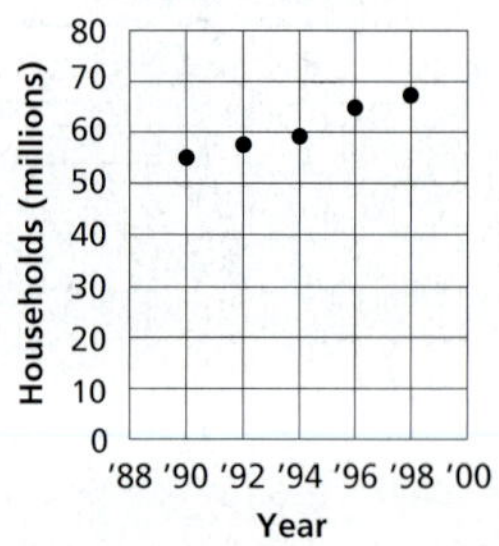

5b. Sample answer using (1992, 57) and (1998, 67): $y = 1.67x - 3269.64$

5c. Sample answer: about 87 million

9a. **Broadway Play Revenue**

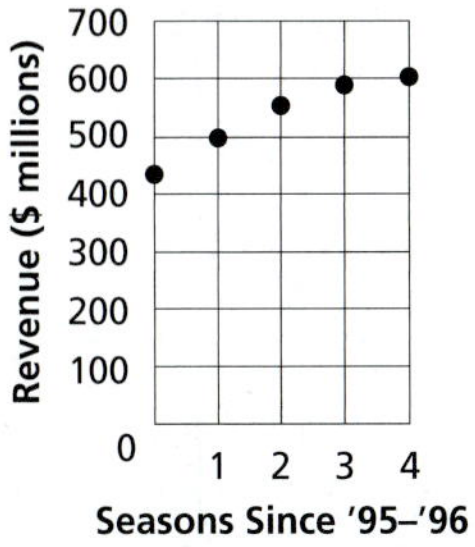

13. Sample answer: Using the data for August and November, a prediction equation for Company 1 is $y = -0.86x + 25.13$, where x is the number of months since August. The negative slope suggests that the value of Company 1's stock is going down. Using the data for October and November, a prediction equation for Company 2 is $y = 0.38x + 31.3$, where x is the number of months since August. The positive slope suggests that the value of Company 2's stock is going up. Since the value of Company 1's stock appears to be going down, and the value of Company 2's stock appears to be going up, Della should buy Company 2.

14. No. Past performance is no guarantee of the future performance of a stock. Other factors that should be considered include the companies' earnings data and how much debt they have.

15. **World Cities**

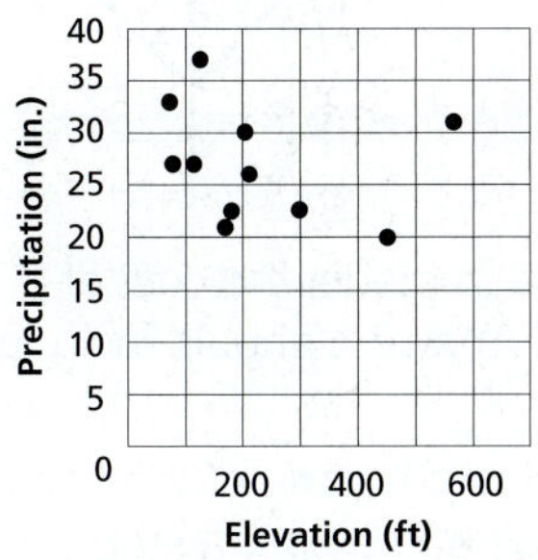

Pages 93–94, Lesson 2-6

6.

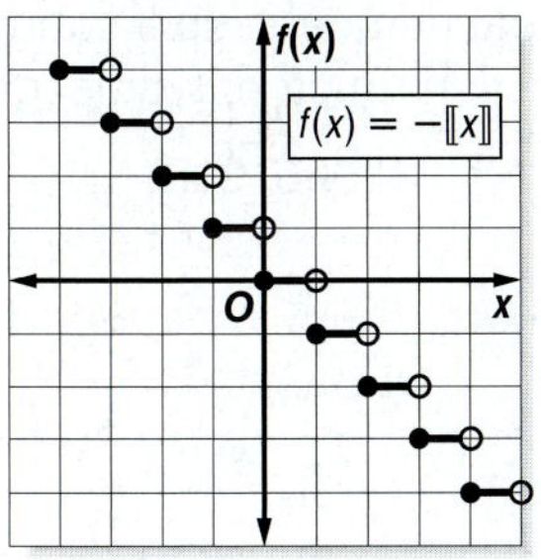

7.

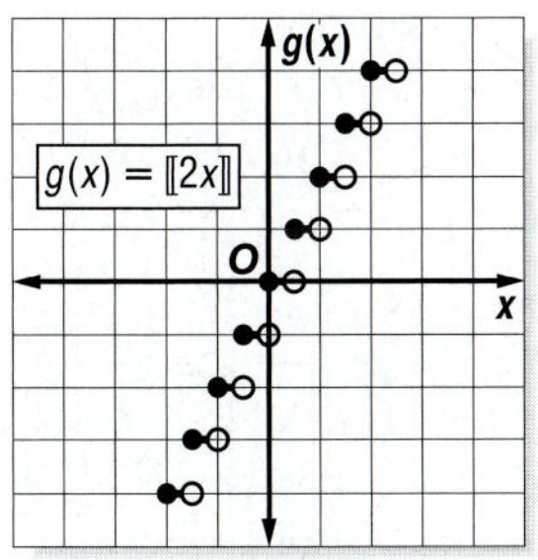

8.

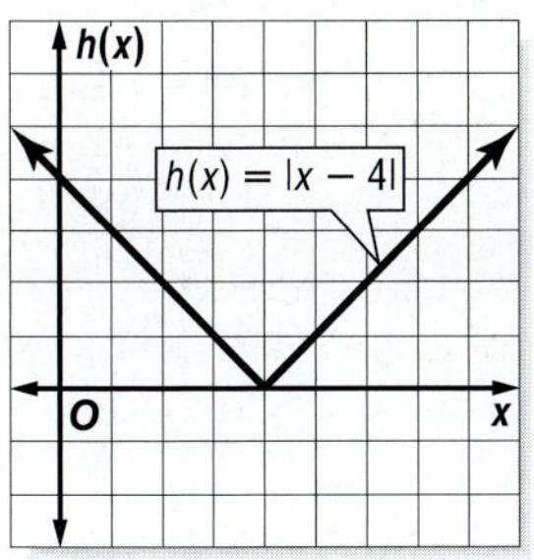

9.

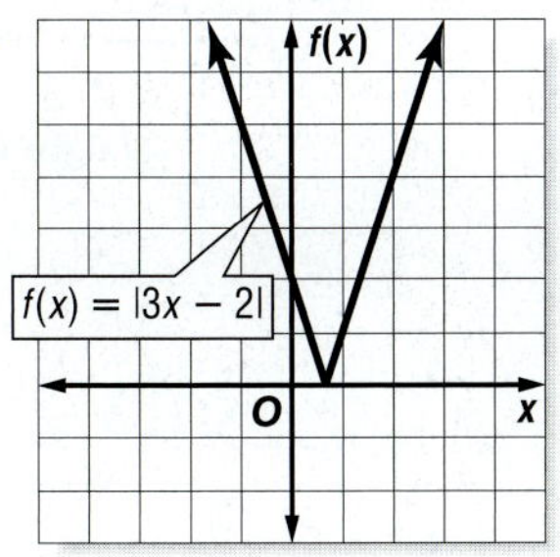

10.

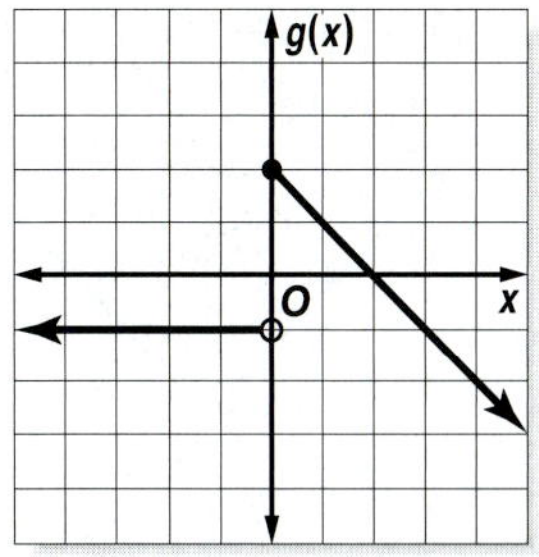

11.

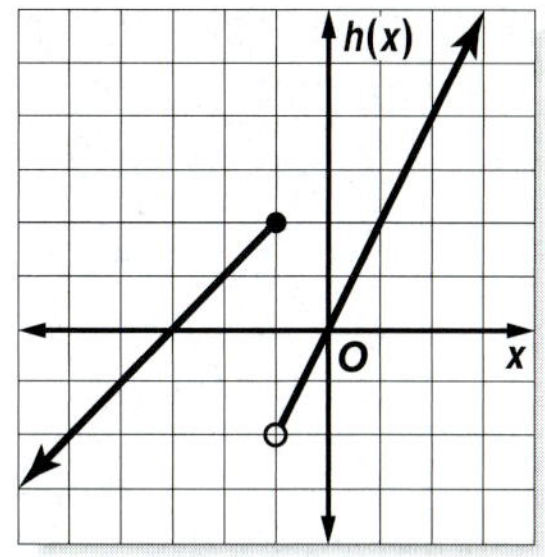

22.

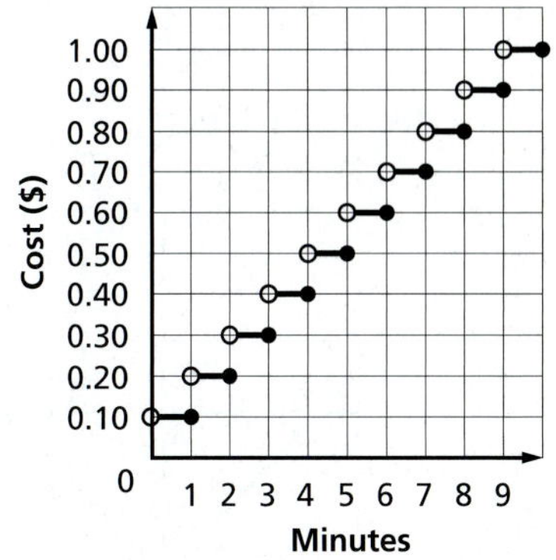

24. D = all reals, R = all integers

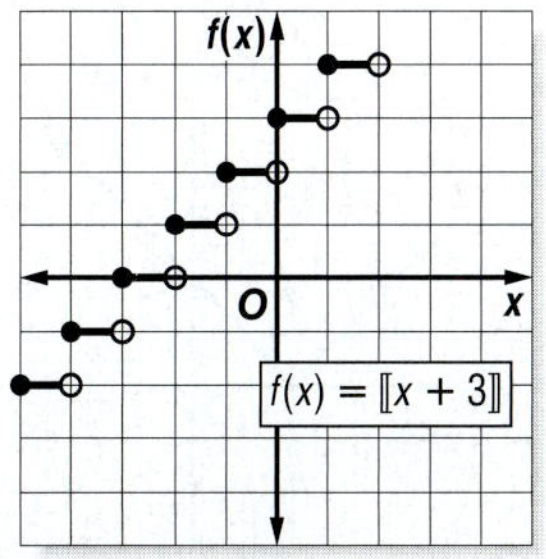

25. D = all reals, R = all integers

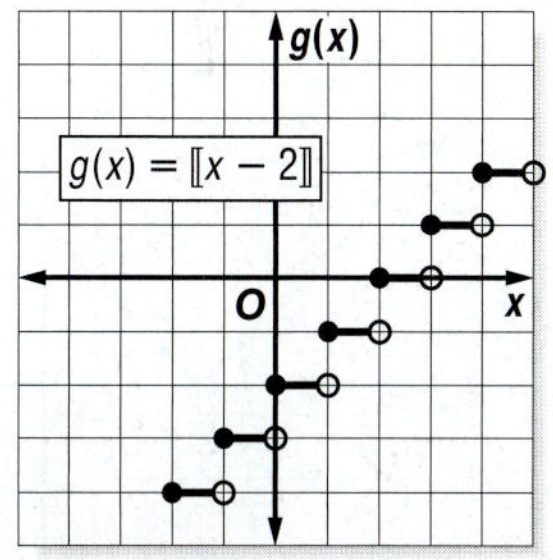

26. D = all reals, R = all even integers

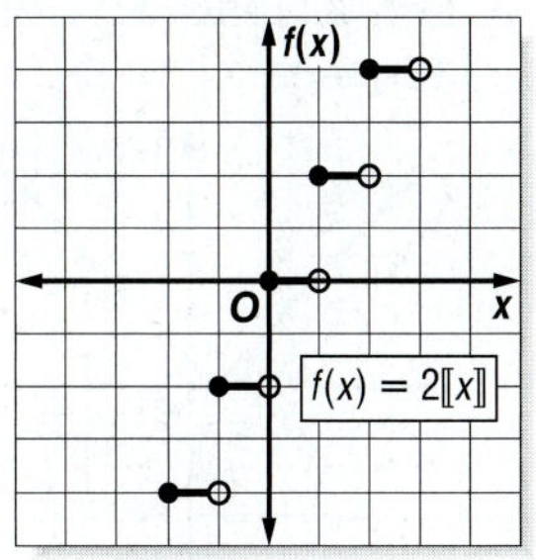

27. D = all reals, R = {3a | a is an integer}

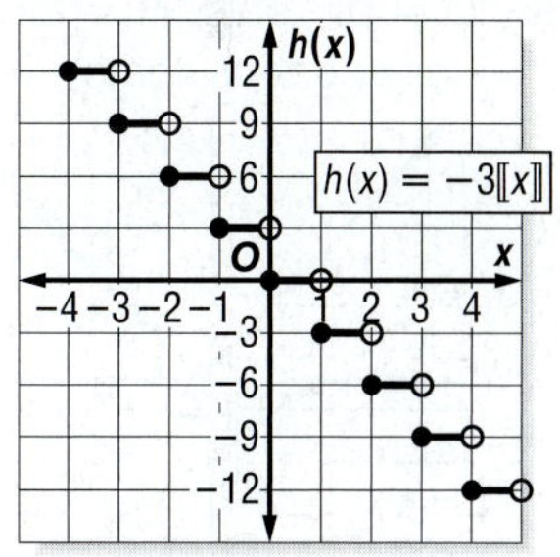

28. D = all reals, R = all integers

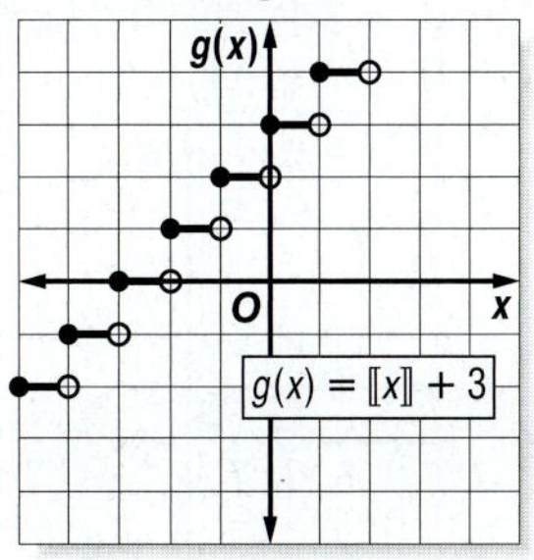

29. D = all reals, R = all integers

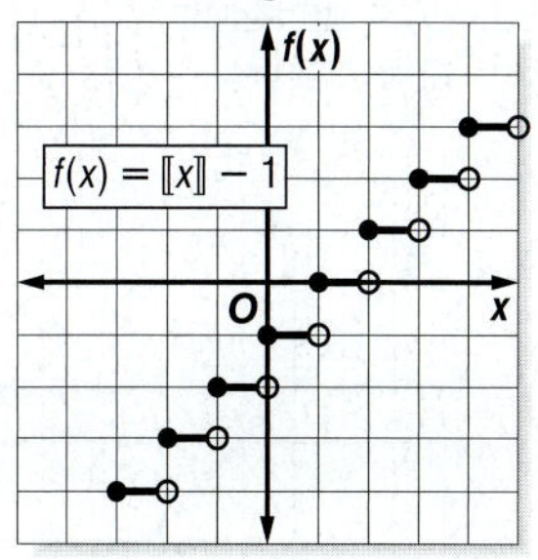

30. D = all reals, R = all nonnegative reals

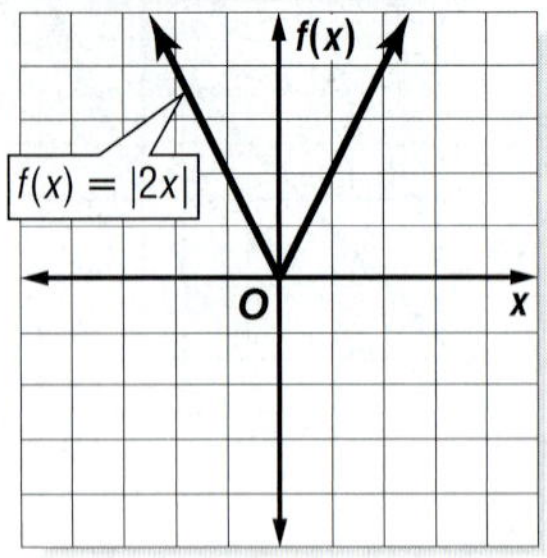

31. D = all reals, R = all nonnegative reals

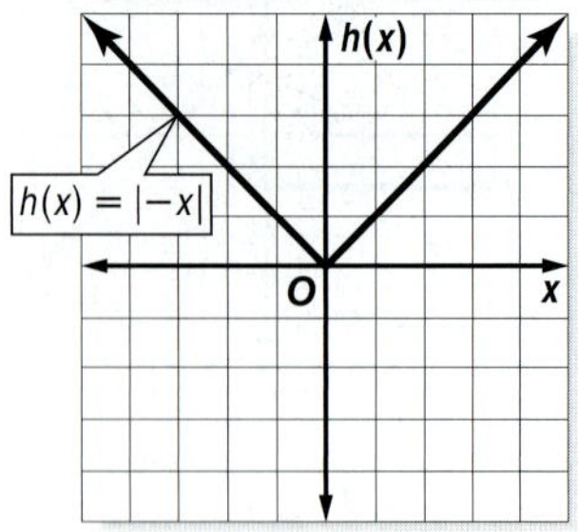

32. D = all reals, R = $\{y \mid y \geq 3\}$

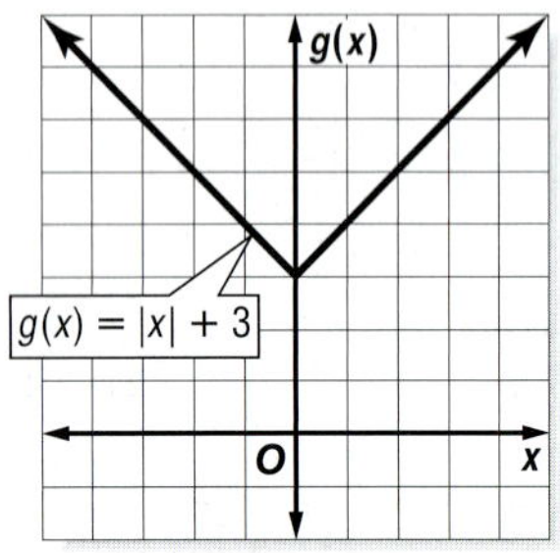

33. D = all reals, R = $\{y \mid y \geq -4\}$

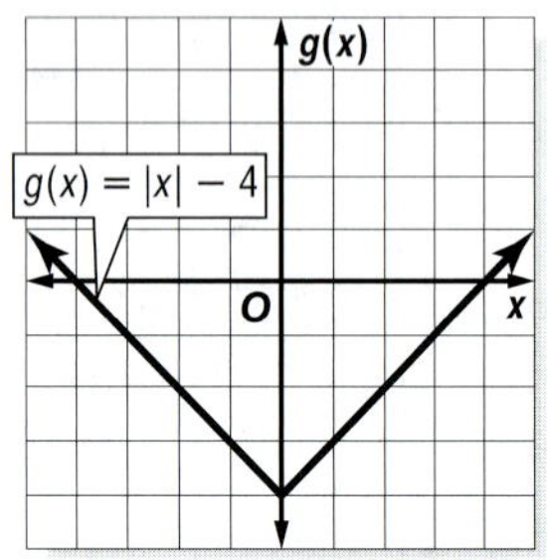

34. D = all reals, R = all nonnegative reals

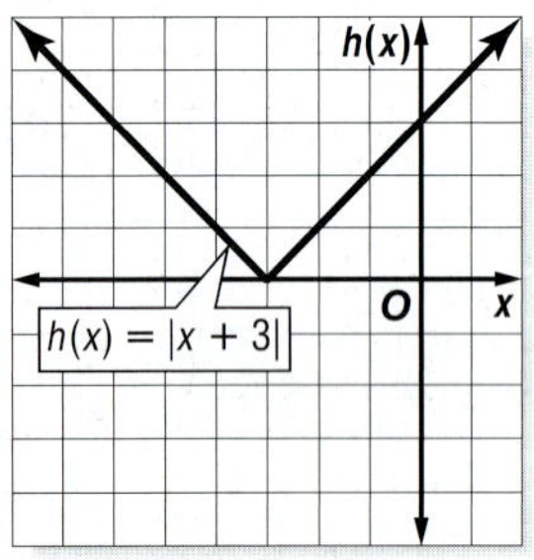

35. D = all reals, R = all nonnegative reals

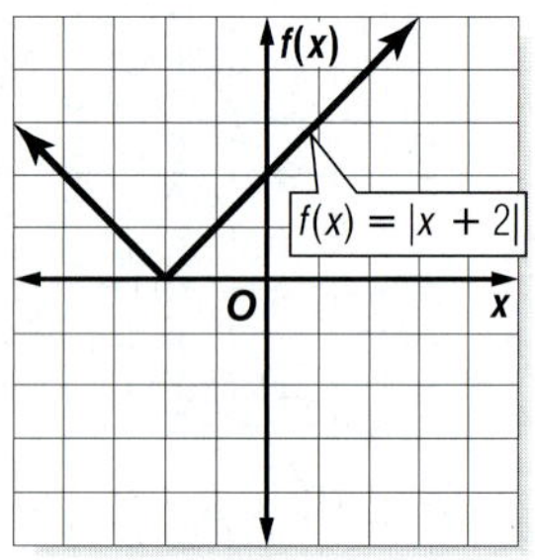

36. D = all reals, R = all nonnegative reals

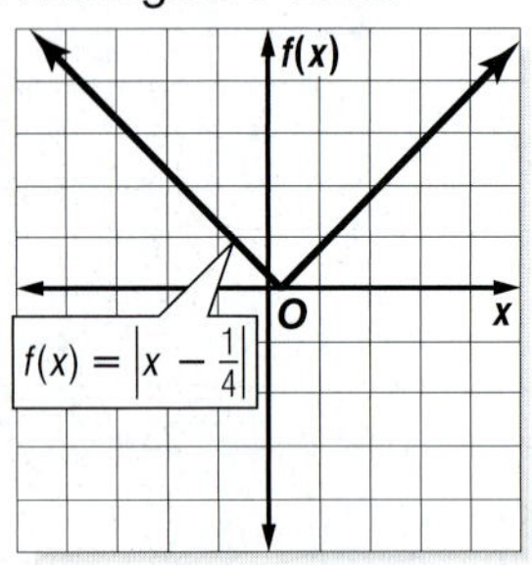

37. D = all reals, R = all nonnegative reals

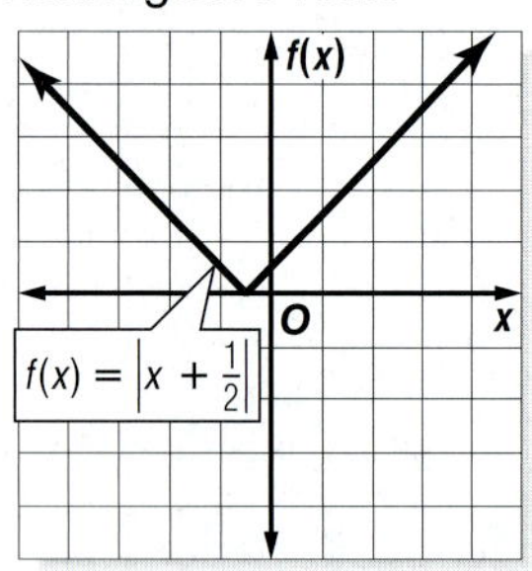

38. D = all reals, R = $\{y \mid y \geq -3\}$

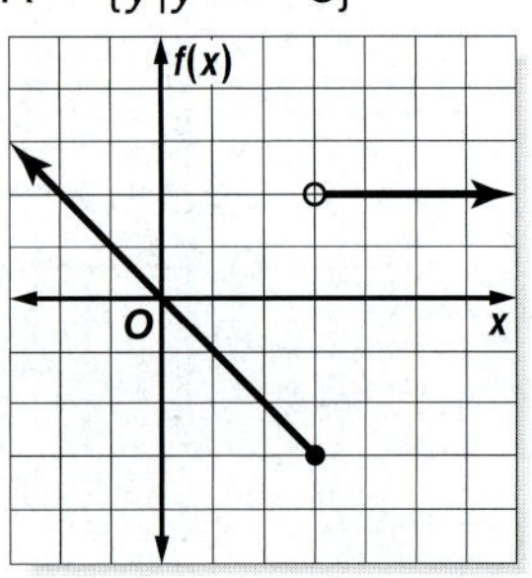

39. D = $\{x \mid x < -2 \text{ or } x > 2\}$, R = $\{-1, 1\}$

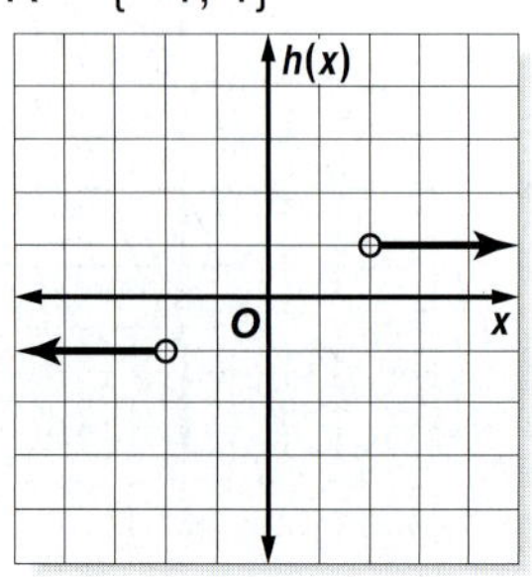

40. D = all reals, R = $\{y \mid y \leq 0 \text{ or } y = 2\}$

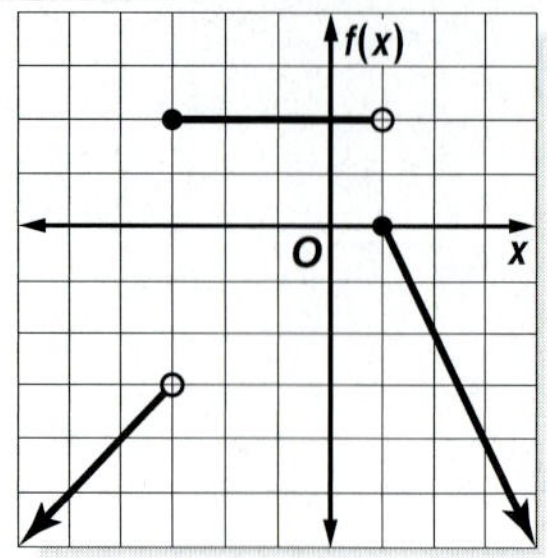

41. D = all reals, R = $\{y \mid y < 2\}$

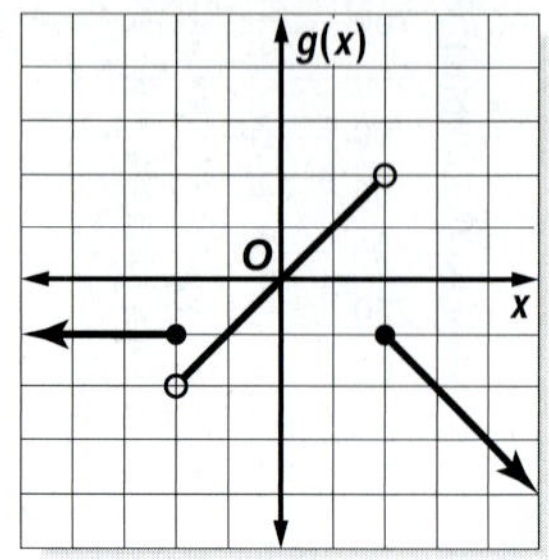

42. D = all reals, R = all whole numbers

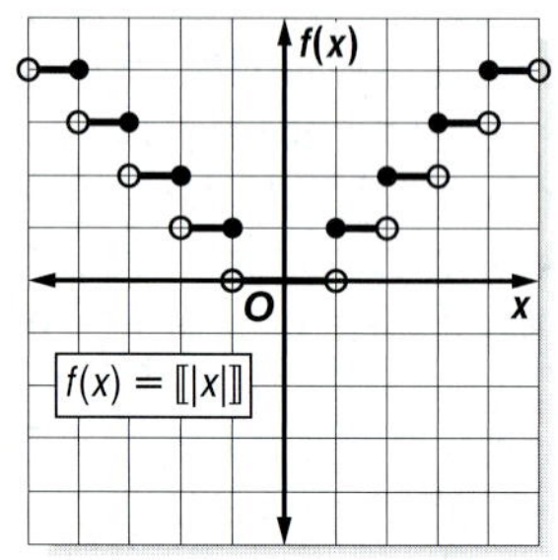

43. D = all reals, R = all whole numbers

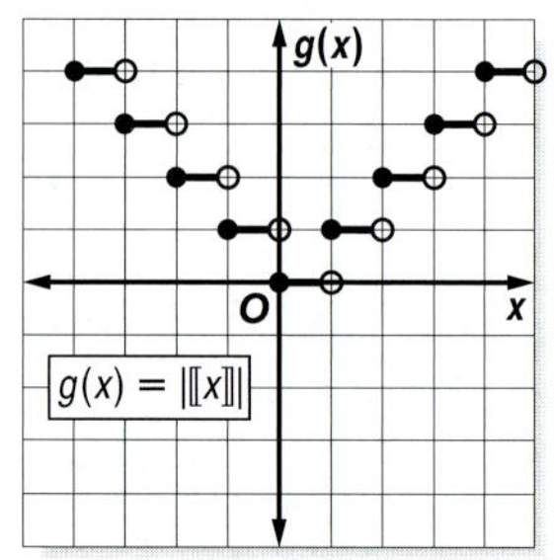

47.

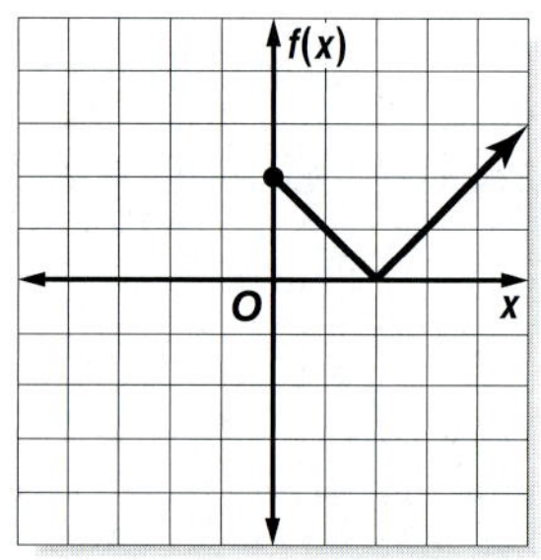

49.

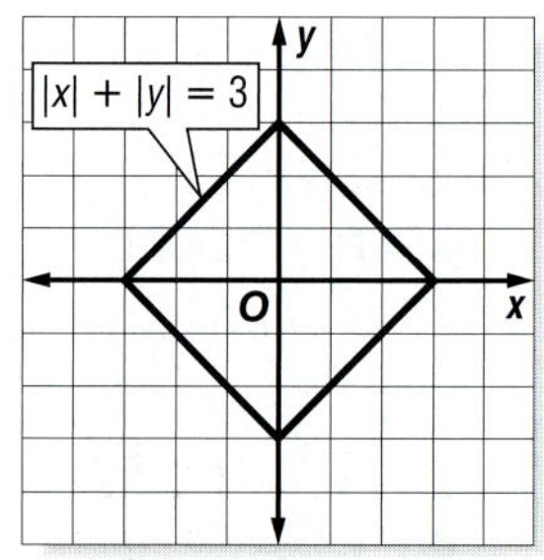

50. A step function can be used to model the cost of a letter in terms of its weight. Answers should include the following.

- Since the cost of a letter must be one of the values \$0.34, \$0.55, \$0.76, \$0.97, and so on, a step function is the best model for the cost of mailing a letter. The gas mileage of a car can be any real number in an interval of real numbers, so it cannot be modeled by a step function. In other words, gas mileage is a continuous function of time.
- 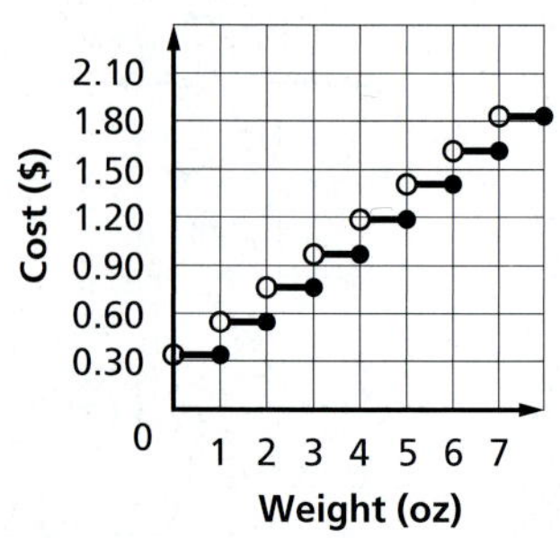

Pages 98–99, Lesson 2-7

4.
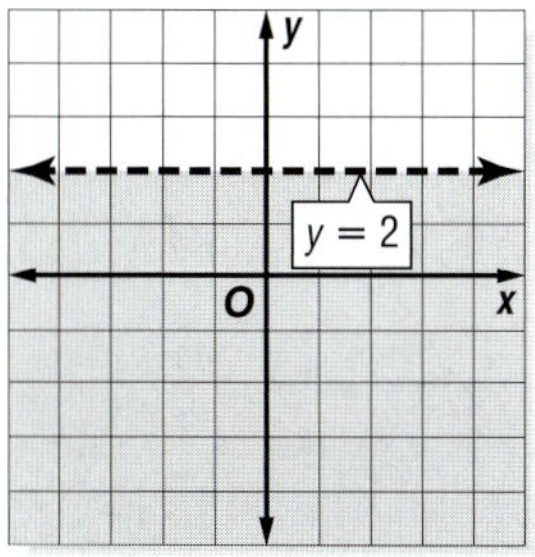

5.
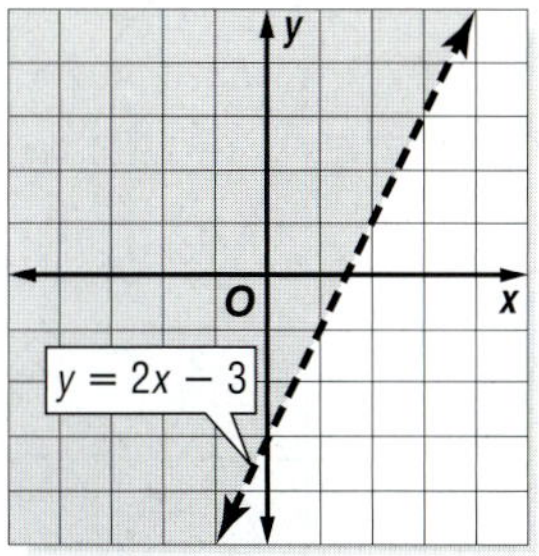

6.
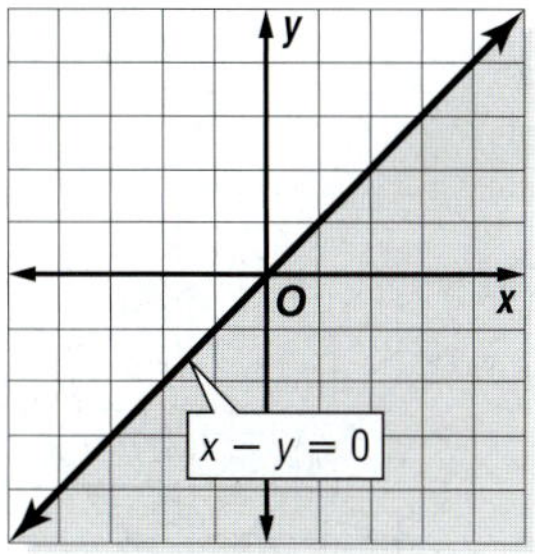

7.
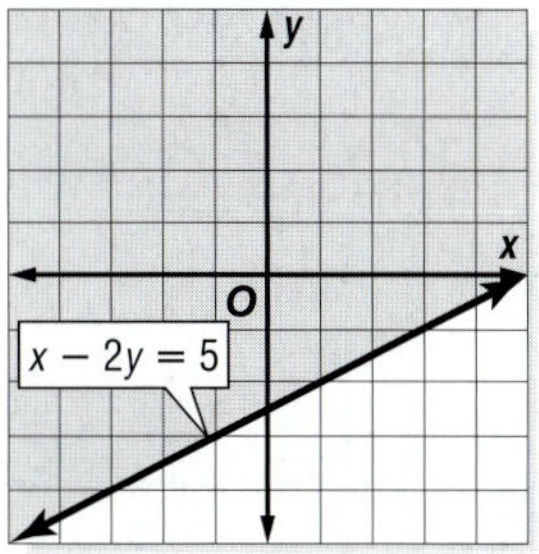

8.
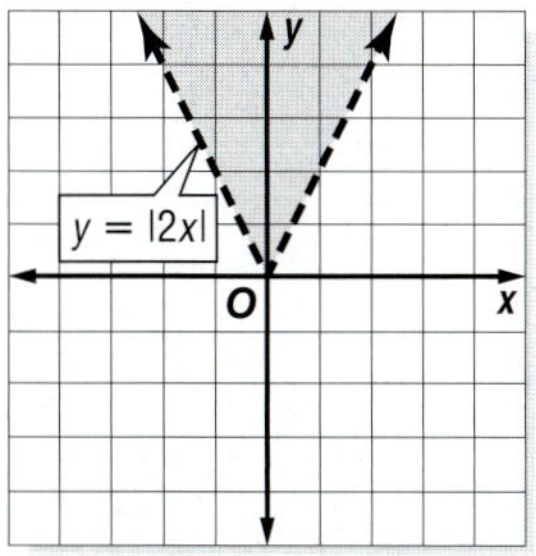

9.
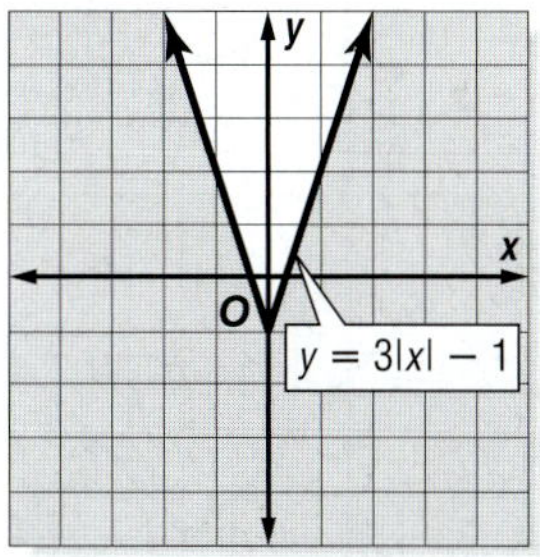

11.
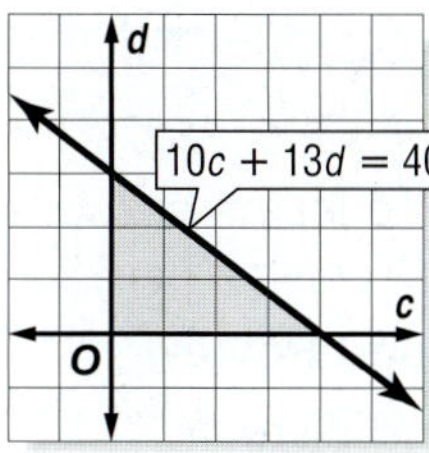

13.
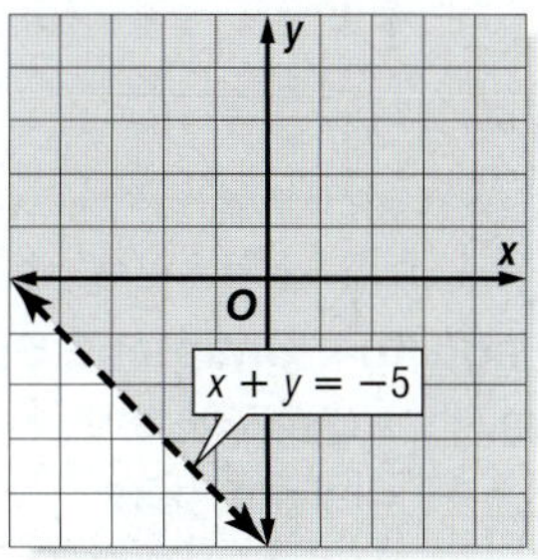

14.
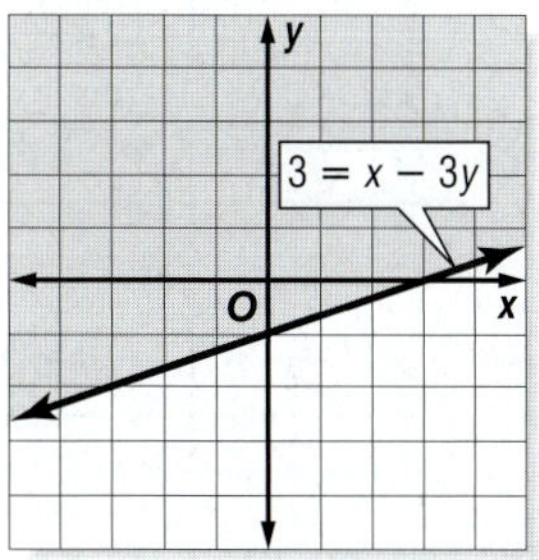

15.
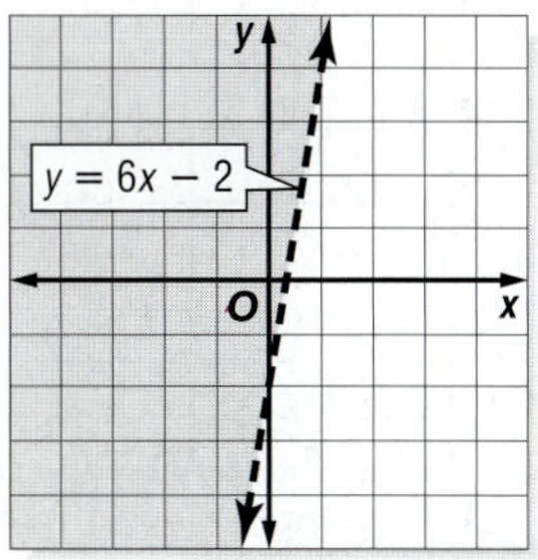

16.
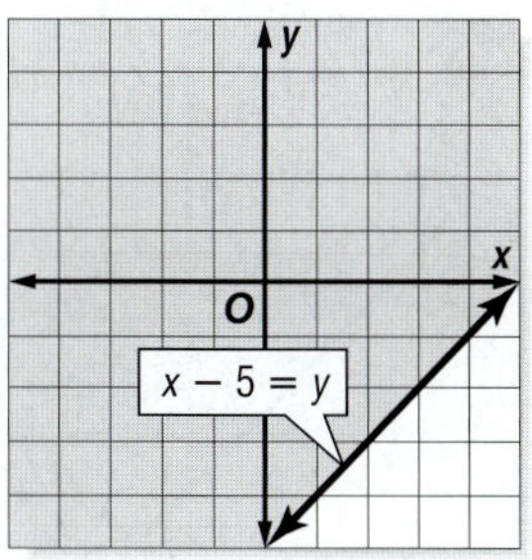

17.
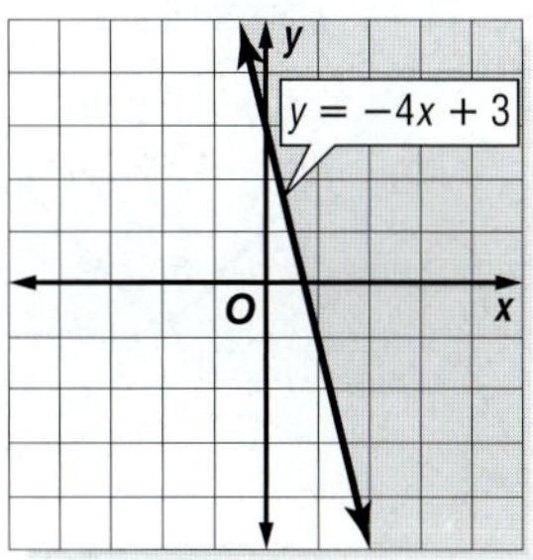

18.
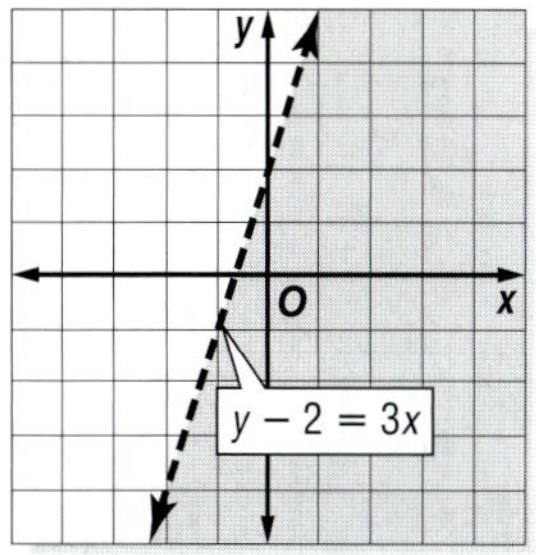

19.
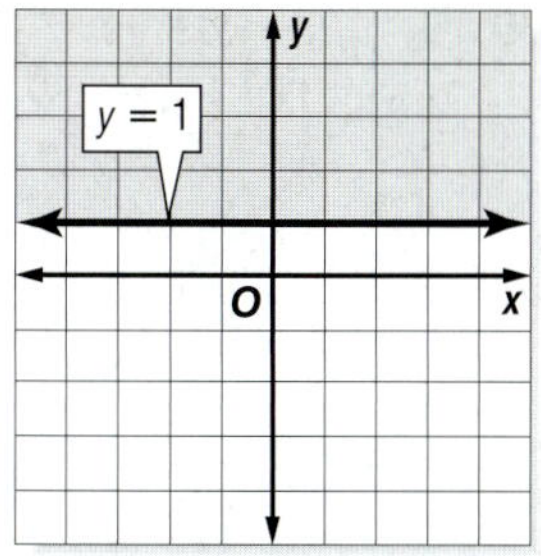

20.
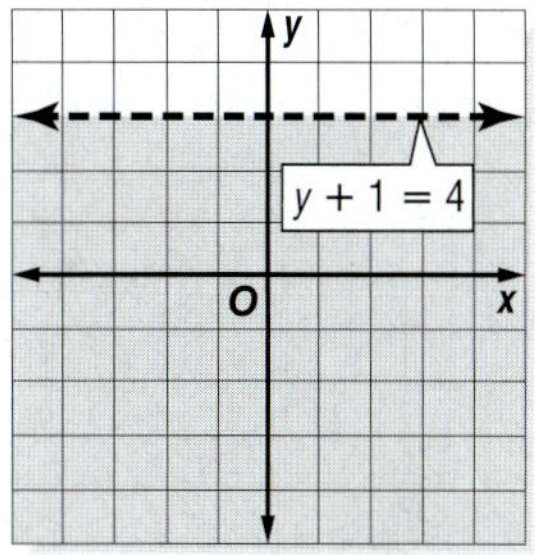

21.
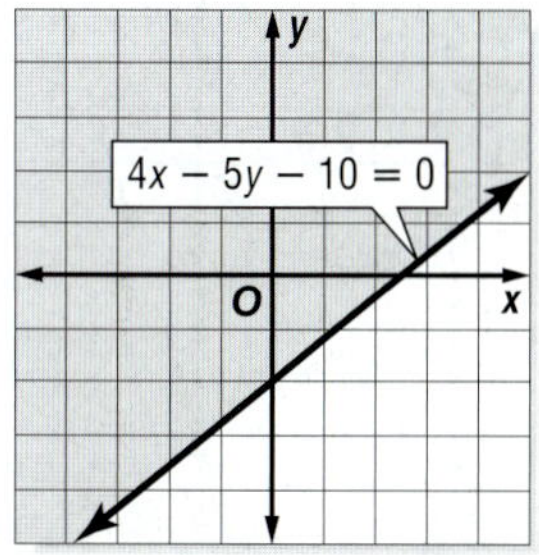

22.
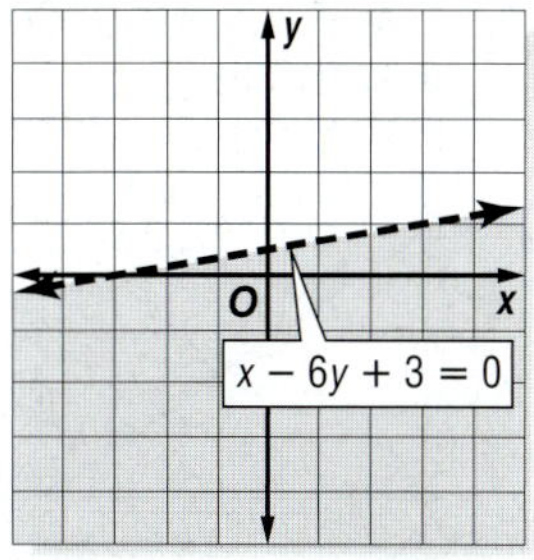

23.
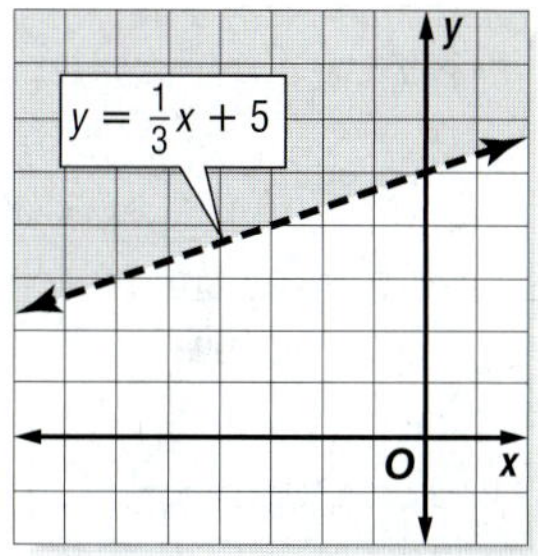

24.
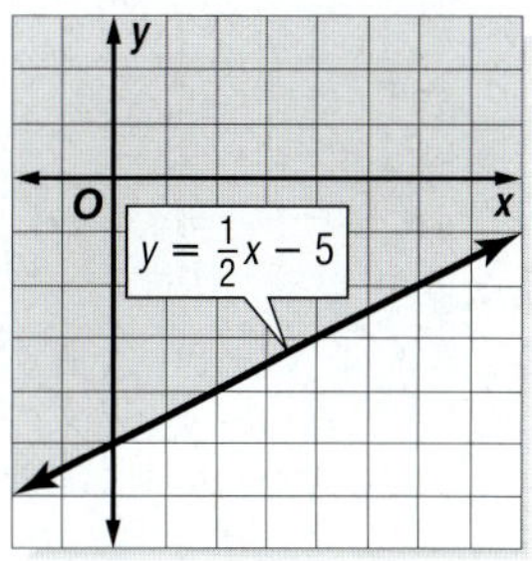

25.
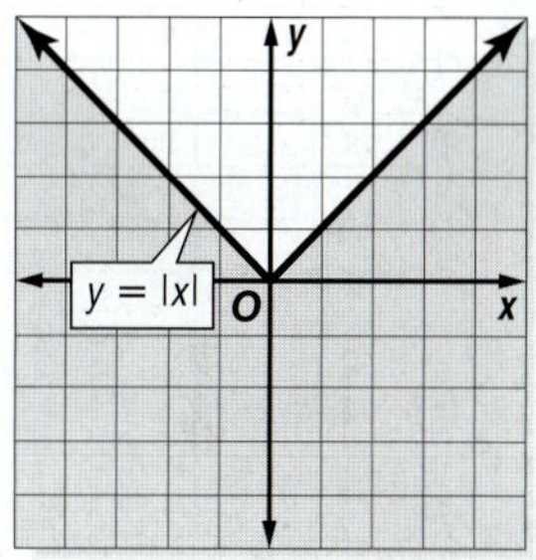

26.
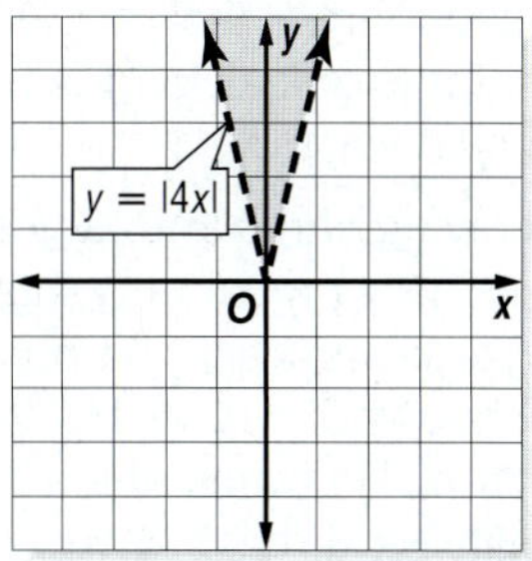

27.
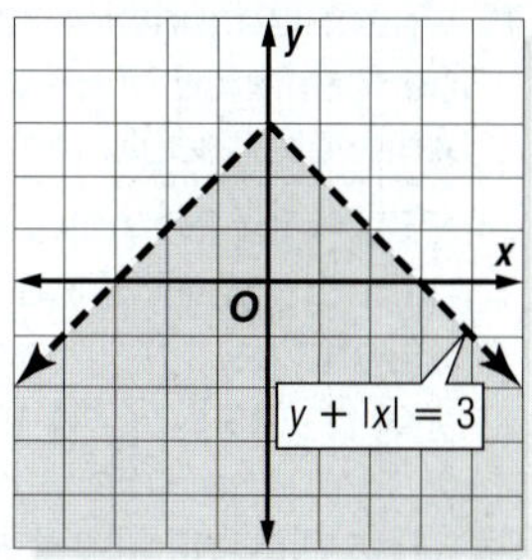

28.
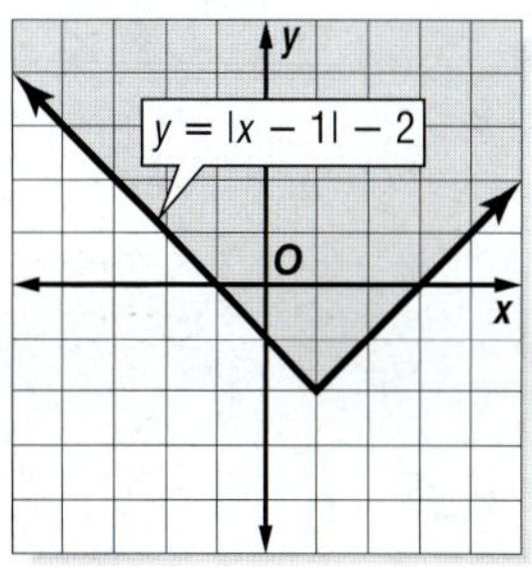

29.
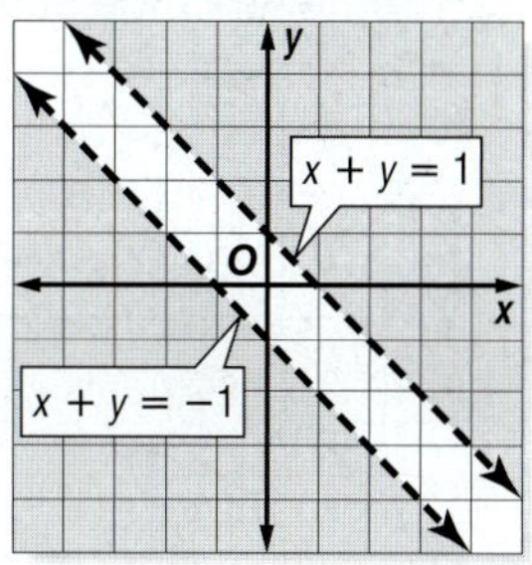

30.

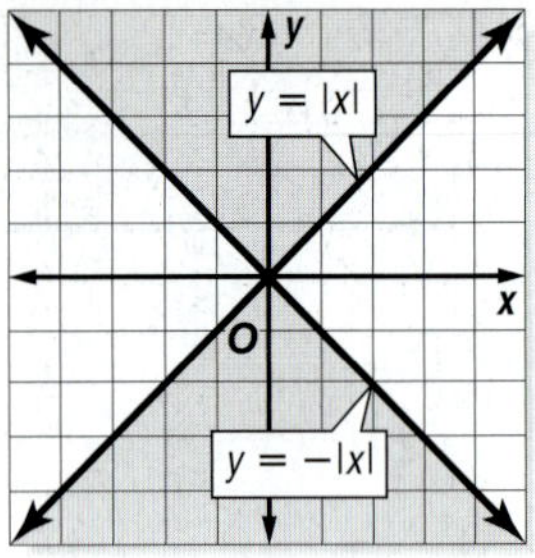

31.

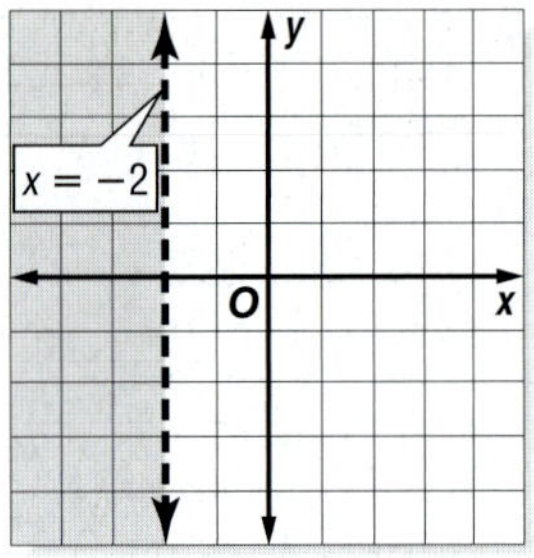

32.

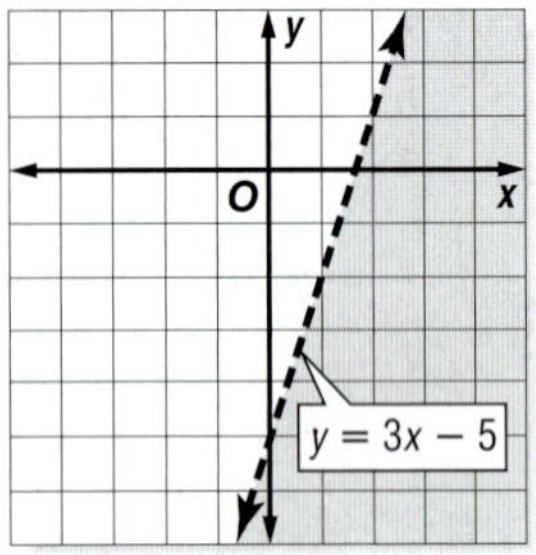

33.

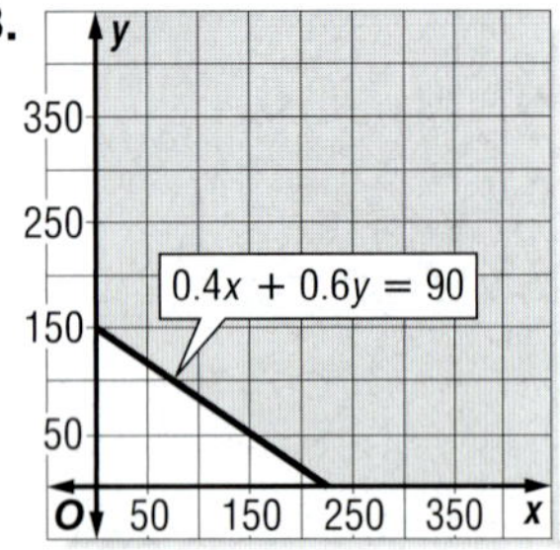

36.

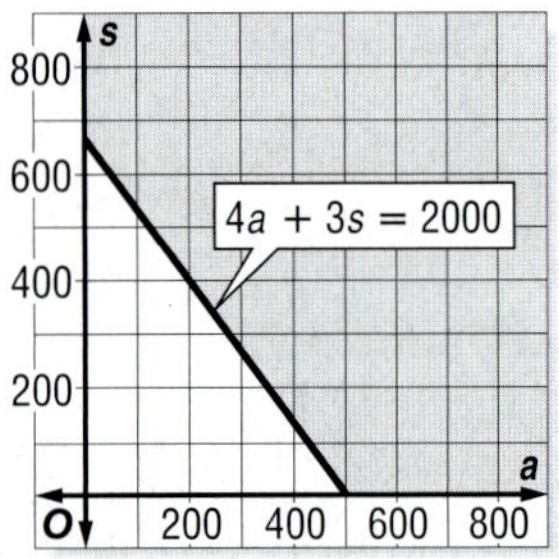

38. $1.2a + 1.8b \geq 9000$

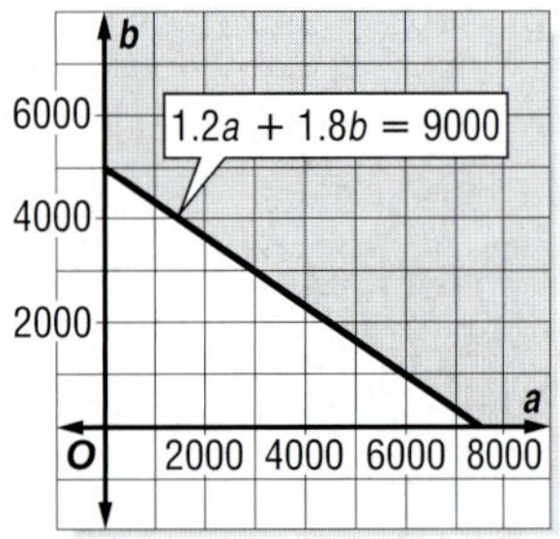

40.

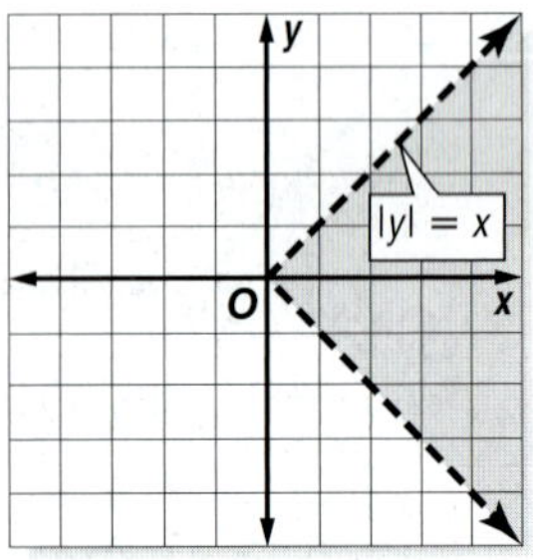

41. Linear inequalities can be used to track the performance of players in fantasy football leagues. Answers should include the following.

- Let x be the number of receiving yards and let y be the number of touchdowns. The number of points Dana gets from receiving yards is $5x$ and the number of points he gets from touchdowns is $100y$. His total number of points is $5x + 100y$. He wants at least 1000 points, so the inequality $5x + 100y \geq 1000$ represents the situation.

- 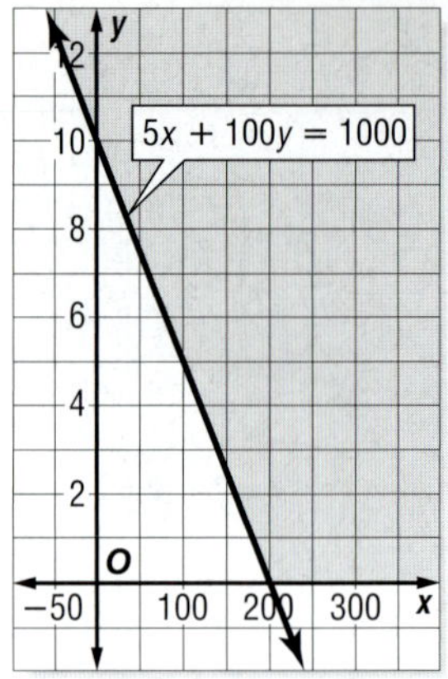

- the first one

48.

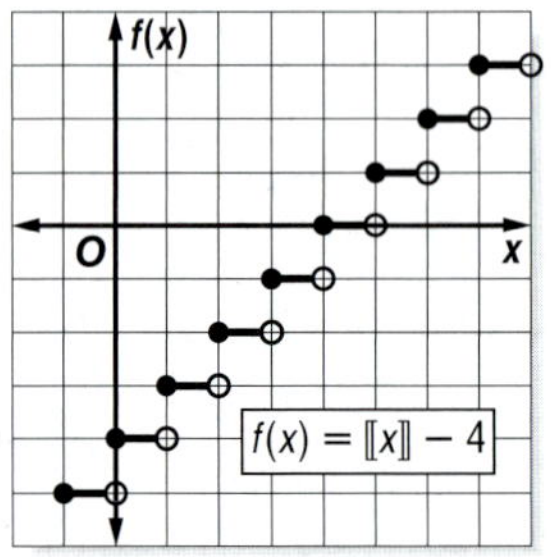

49.

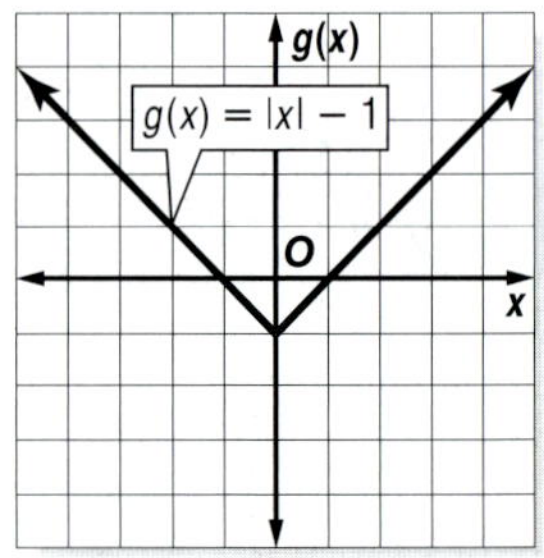

50.

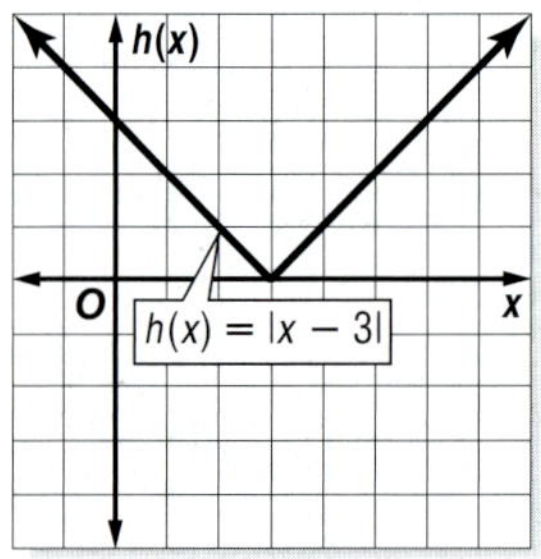

Page 104, Chapter 2 Study Guide and Review

43. D = all reals,
R = all integers

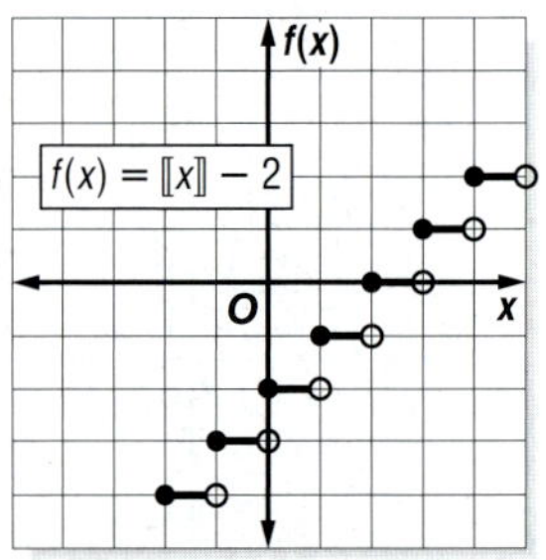

44. D = all reals,
R = all integers

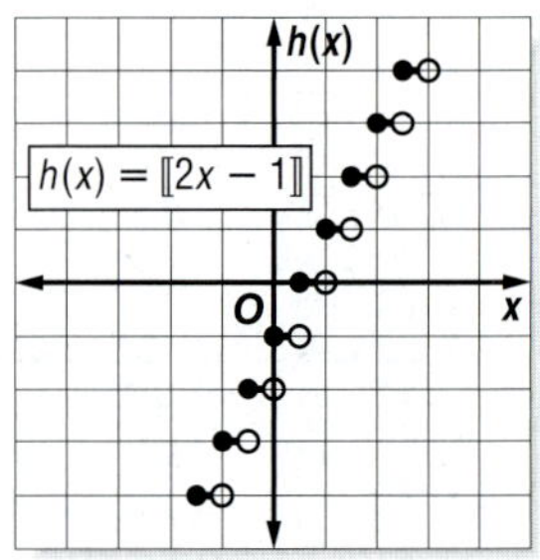

45. D = all reals,
R = $\{y \mid y \geq 4\}$

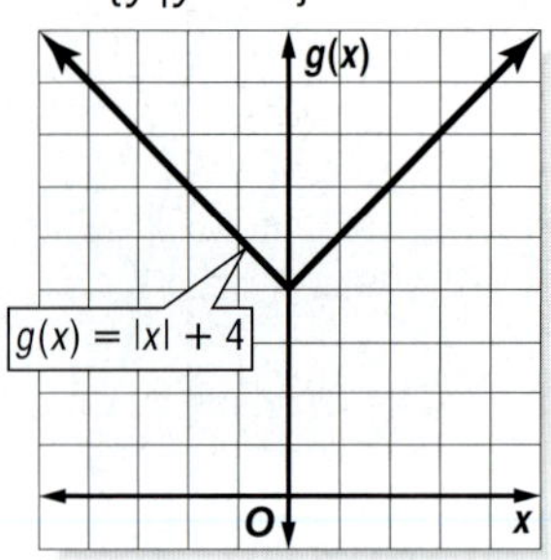

46. D = all reals,
R = $\{y \mid y \geq -7\}$

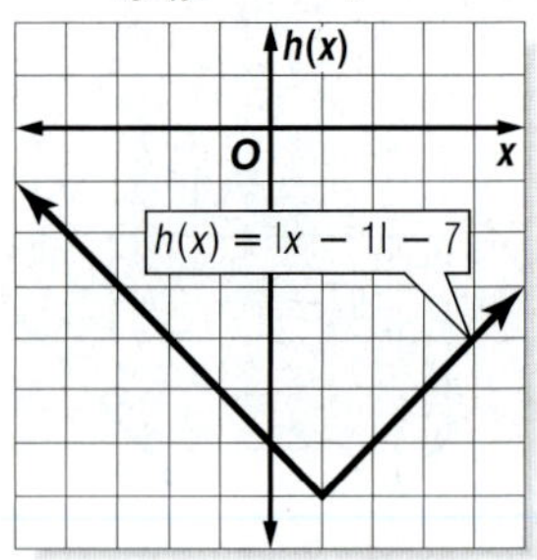

47. D = all reals, R = $\{y \mid y \leq 0 \text{ or } y = 2\}$

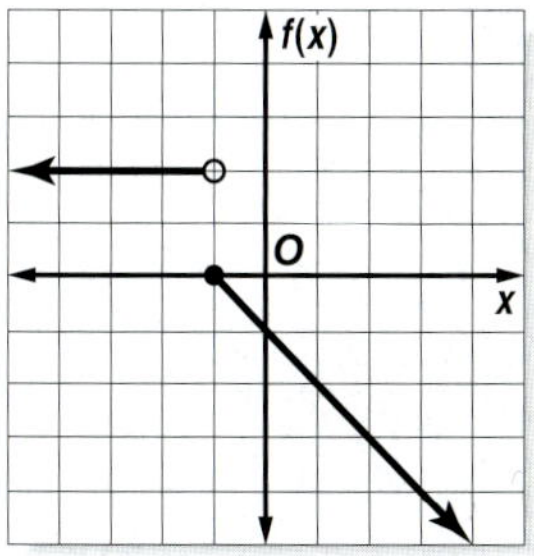

48. D = $\{x \mid x \neq 1\}$, R = $\{y \mid y > -5\}$

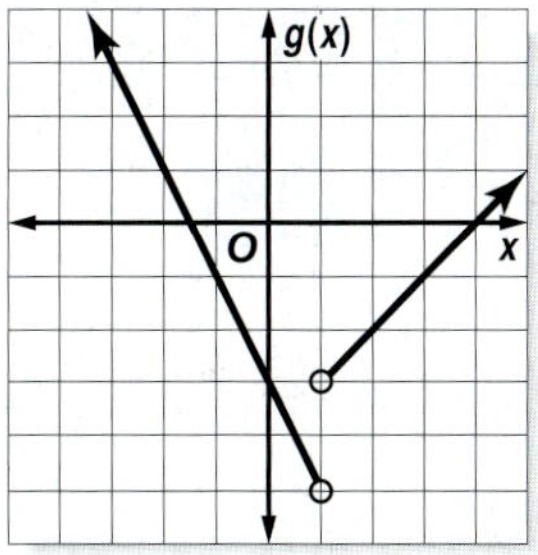

Page 105, Chapter 2 Practice Test

4.

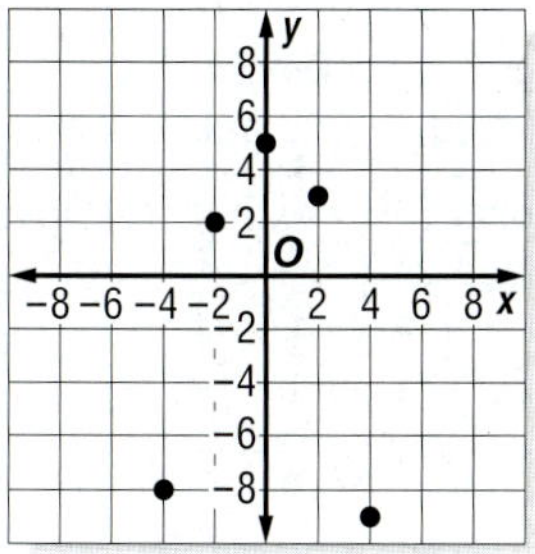

5.

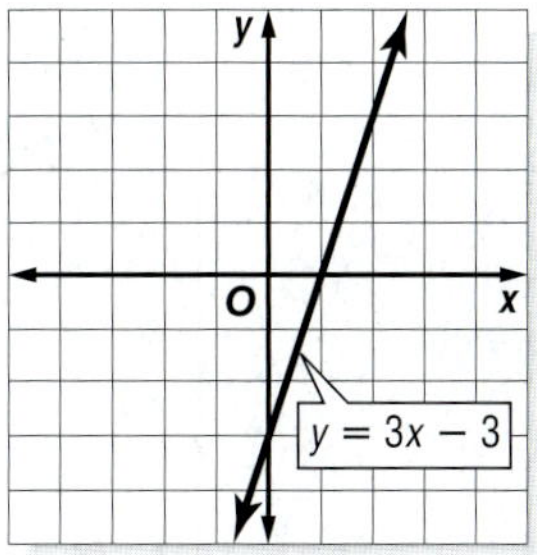

8.

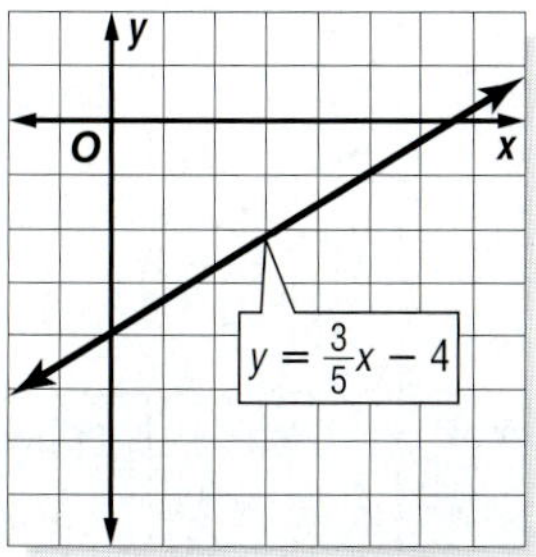

9.

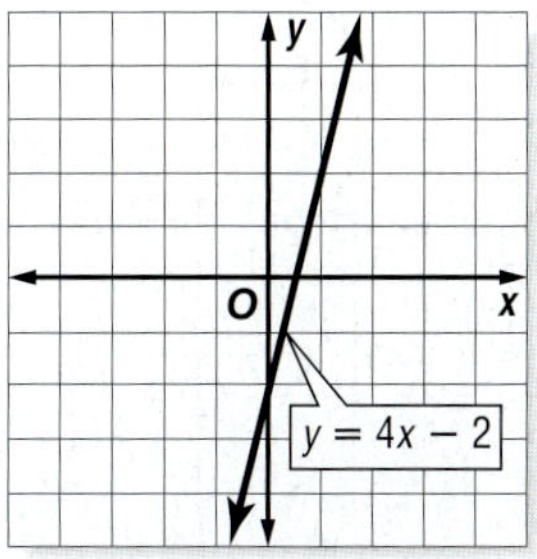

10.

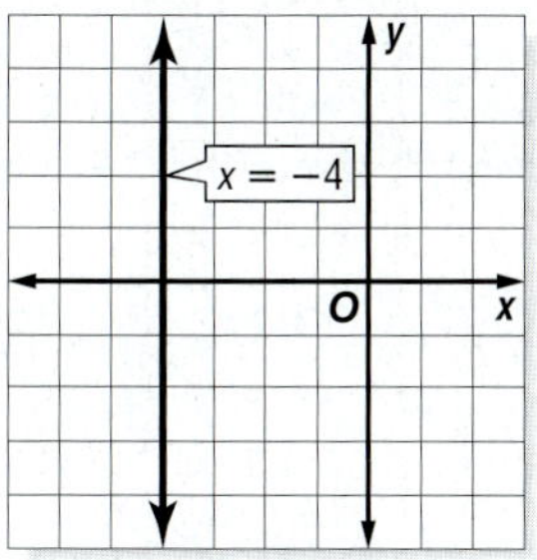

11.

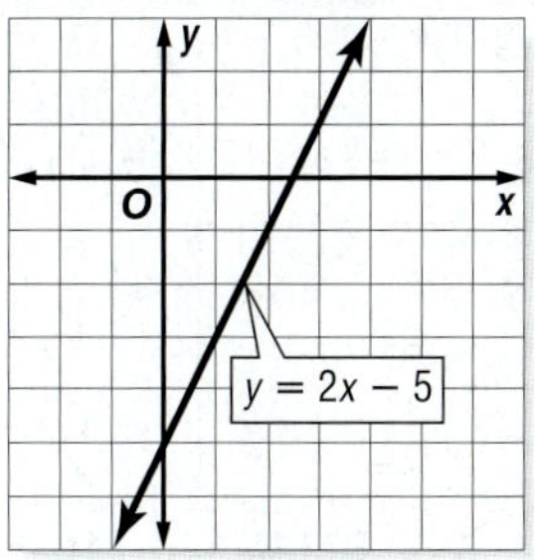

12.

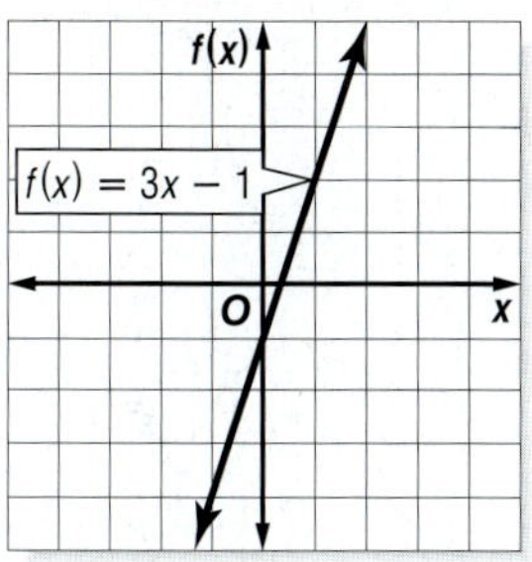

13.

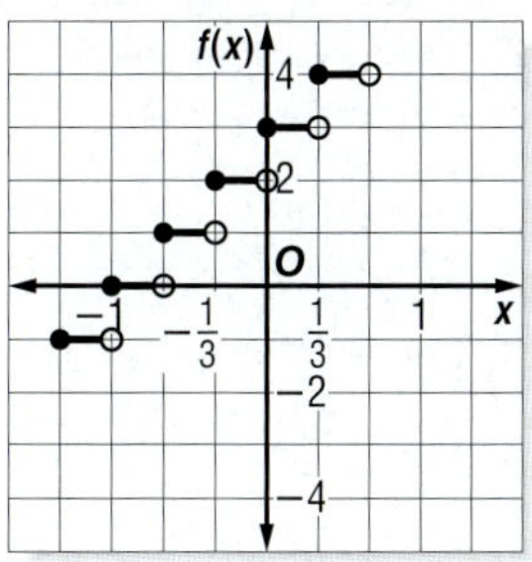

14.

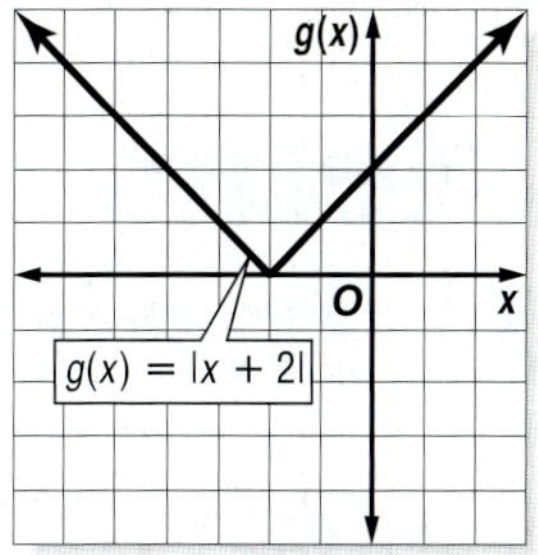

15.

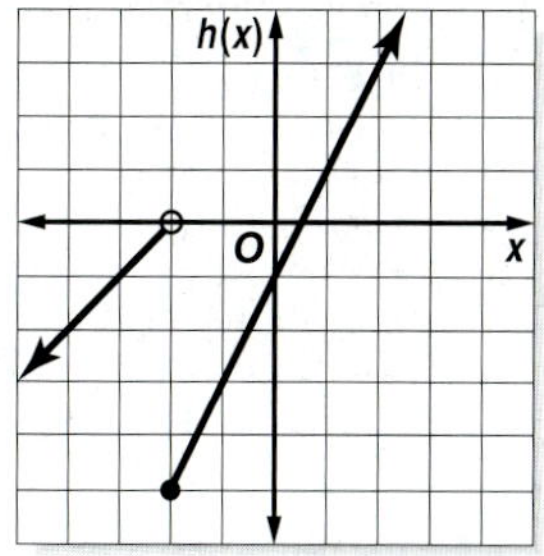

16.

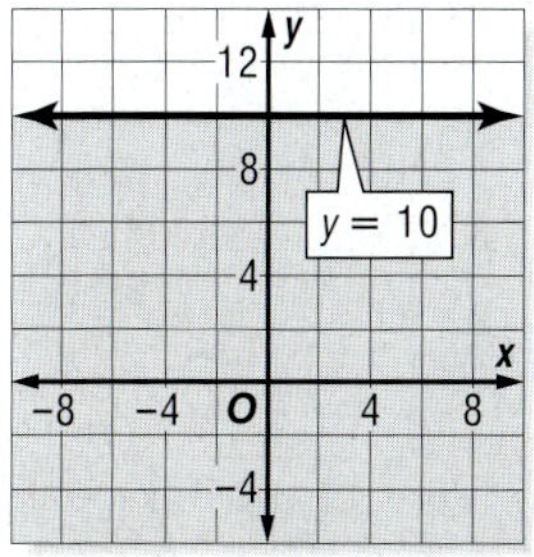

17.

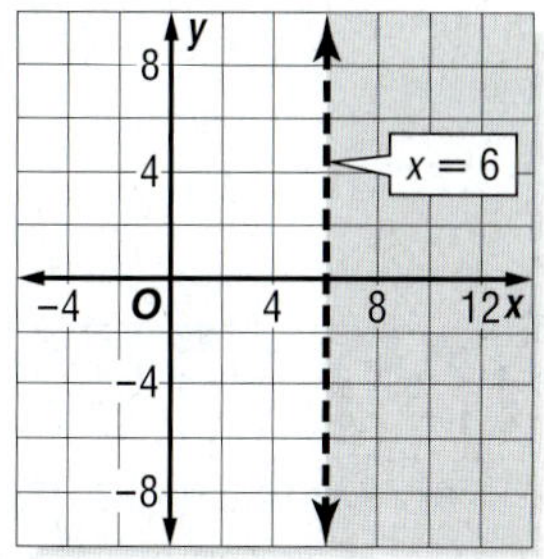

18.

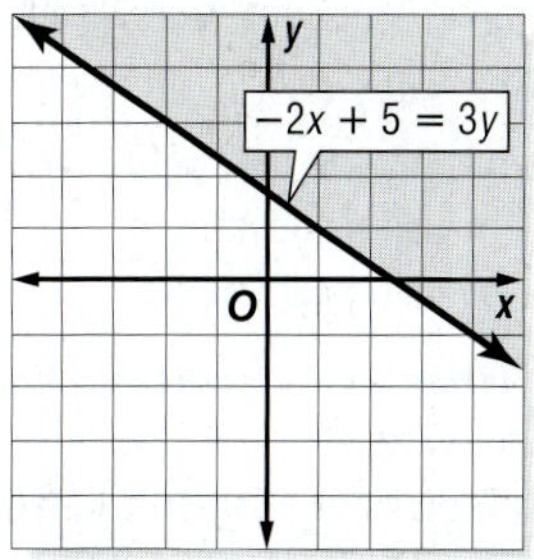

19.

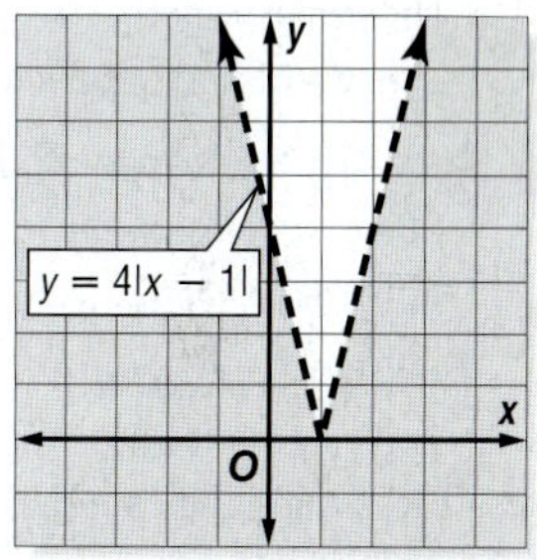

23.

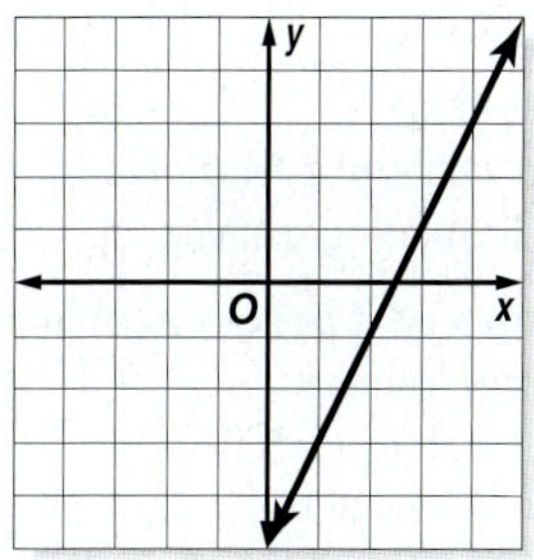

24.

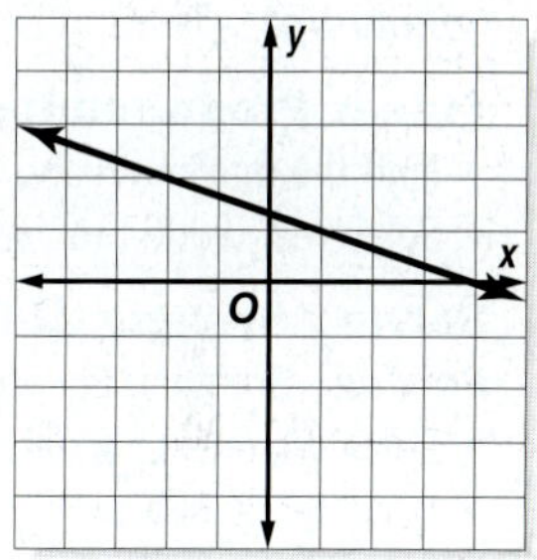

25.

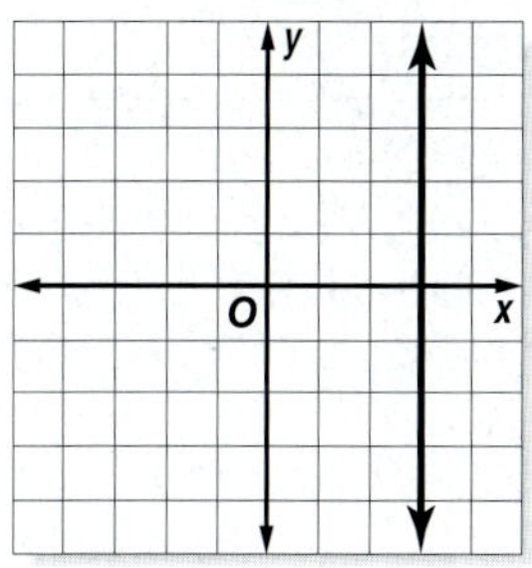

Chapter 3

Systems of Equations and Inequalities

Chapter Overview and Pacing

Year-long and two-year pacing: pages T20–T21.

LESSON OBJECTIVES	PACING (days) Regular		PACING (days) Block	
	Basic/ Average	Advanced	Basic/ Average	Advanced
3-1 Solving Systems of Equations by Graphing *(pp. 110–115)* • Solve systems of linear equations by graphing. • Determine whether a system of linear equations is consistent and independent, consistent and dependent, or inconsistent.	1	optional	0.5	optional
3-2 Solving Systems of Equations Algebraically *(pp. 116–122)* • Solve systems of linear equations by using substitution. • Solve systems of linear equations by using elimination.	1	optional	0.5	optional
3-3 Solving Systems of Inequalities by Graphing *(pp. 123–128)* • Solve systems of inequalities by graphing. • Determine the coordinates of the vertices of a region formed by the graph of a system of inequalities. ***Follow-Up:*** Systems of Linear Inequalities	2 (with 3-3 Follow-Up)	optional	1	optional
3-4 Linear Programming *(pp. 129–135)* • Find the maximum and minimum values of a function over a region. • Solve real-world problems using linear programming.	1	optional	0.5	optional
3-5 Solving Systems of Equations in Three Variables *(pp. 136–144)* ***Preview:*** Graphing Equations in Three Variables • Solve systems of linear equations in three variables. • Solve real-world problems using systems of linear equations in three variables.	3 (with 3-5 Preview)	2 (with 3-5 Preview)	1.5 (with 3-5 Preview)	1 (with 3-5 Preview)
Study Guide and **Practice Test** *(pp. 145–149)* **Standardized Test Practice** *(pp. 150–151)*	1	2	0.5	0.5
Chapter Assessment	1	1	0.5	0.5
TOTAL	**10**	**5**	**5**	**2**

*An electronic version of this chapter is available on **StudentWorks**™. This backpack solution CD-ROM allows students instant access to the Student Edition, lesson worksheet pages, and web resources.*

Timesaving Tools
TeacherWorks™
All-In-One Planner and Resource Center
See pages T5 and T21.

Chapter Resource Manager

Study Guide and Intervention	Practice (Skills and Average)	Reading to Learn Mathematics	Enrichment	Assessment	Applications*	5-Minute Check Transparencies	Interactive Chalkboard	Alge2PASS: Tutorial Plus (lessons)	Materials
Chapter 3 Resource Masters									
119–120	121–122	123	124		GCS 32	3-1	3-1		
125–126	127–128	129	130	163	SC 5, SM 63–66	3-2	3-2	5	
131–132	133–134	135	136	163, 165	SC 6	3-3	3-3		(*Follow-Up:* graphing calculator)
137–138	139–140	141	142	164	GCS 31	3-4	3-4		colored pencils
143–144	145–146	147	148	164		3-5	3-5		(*Preview:* isometric dot paper, two cardboard boxes)
				149–162, 166–168					

**Key to Abbreviations:* GCS = Graphing Calculator and Speadsheet Masters,
SC = School-to-Career Masters,
SM = Science and Mathematics Lab Manual

Chapter 3

Mathematical Connections and Background

Continuity of Instruction

Prior Knowledge

Students have worked with two-variable linear equations, representing an equation as a line on a graph, testing whether an ordered pair satisfies the equation, and solving for one variable in terms of the other. Also, students have manipulated one- and two-variable linear inequalities using properties of inequalities.

This Chapter

Students use graphical methods to represent two equations as lines in a single plane, identify points common to the two lines, and explore when the two lines have points in common. Also, they develop algebraic methods to solve systems of two equations in two variables and extend those skills to solve systems of three equations. Students represent linear inequalities as regions of the plane, find regions that represent solutions, and solve linear programming problems.

Future Connections

Students will relate equations and graphs for circles, parabolas, and hyperbolas. They will develop algebraic techniques for using matrices and their inverses to represent and solve systems of equations. They will also explore systems that involve nonlinear equations and/or nonlinear inequalities.

3-1 Solving Systems of Equations by Graphing

This lesson extends the relationship between linear equations and graphed lines by dealing with systems of two linear equations. One main idea of the lesson is that the intersection of the two lines represents a point that satisfies both equations, so the coordinates of that point are the solution to the system. In the lesson, students draw the graph for each equation, find the coordinates of the intersection, and confirm that those coordinates satisfy both equations. A second idea in the lesson uses the fact that, for any two lines in a plane, the lines must be intersecting, parallel, or coincident. If the lines intersect, then the system has exactly one solution and is called *consistent and independent*. If the lines are parallel, then the system has no solution and is called *inconsistent*. If the lines coincide, then the system has infinitely many solutions and is called *consistent and dependent*.

3-2 Solving Systems of Equations Algebraically

Students explore three algebraic properties as the basis for solving systems of two linear equations (all three properties were discussed in Lesson 1-3). One property is substitution. Students solve one of two equations for one of the two variables and then substitute the resulting expression into the other equation. That gives them a single equation in one variable which they solve and then use that solution in either of the two-variable equations to solve for the other variable.

A second property is addition. Given two equations in standard form such that one variable has coefficients that are additive inverses, students add the two equations. The resulting equation is in one variable. Students solve that equation and then solve for the other variable.

A third property is multiplication. Given a system of two linear equations in standard form, students can select two numbers so that, after multiplying each equation by one of the numbers, the result is another system of two equations in which one variable has coefficients that are additive inverses. This new system can then be solved by addition.

The method of solution based on addition or multiplication-and-addition is called *solving by elimination*.

3-3 Solving Systems of Inequalities by Graphing

This lesson combines two skills, modeling inequalities as regions of the coordinate plane and solving systems of linear inequalities. The region modeled by an inequality has a boundary line. Rewriting the inequality as an equality describes that boundary line algebraically. For a system of two inequalities, students identify the region that is common to the two inequalities. Also, given a system with at least three inequalities, students identify the region bounded by the system. Then, using pairs of boundary-line equations as a system of equations, they find the coordinates of the vertices of the region.

Two cases are explored for a system of two inequalities where the boundary lines are parallel. In one case, the region common to the two inequalities is between the two parallel boundary lines. In the other case, there is no region common to the two inequalities, so the solution to the system is ∅.

3-4 Linear Programming

Solving a linear programming problem adds two steps to the skills of the previous lesson, evaluating a given function using the vertex coordinates and then comparing the results to identify the point that maximizes or minimizes the function. In a linear programming problem, each of the inequalities in the system is called a *constraint*. The region that represents the system's solution is called the *feasible region,* and the intersections of pairs of boundary lines are referred to as the *vertices of the feasible region*. The function to be evaluated usually describes net income or cost. A bounded feasible region is one whose outline is a polygon, and an unbounded feasible region is one whose boundary is not closed. Students are given the fact that for a bounded region, the maximum and minimum values of the income or cost function always occur at a vertex. For an unbounded feasible region, the income or cost function may have no maximum or minimum value.

3-5 Solving Systems of Equations in Three Variables

This lesson stresses graphical models to explore four possibilities for the solution of a system as the intersection of three planes. (1) The three planes can intersect in a point, and the ordered triple for that point is the single solution to the system. (2) The three planes can intersect in a line, and the system has an infinite number of solutions, any point on that line. (3) Two equations represent one plane, which intersects the third plane; the solution is any point on the line of intersection. (4) The three planes have no point in common, either because the planes are parallel or the lines of intersection of pairs of planes are parallel, and the solution is the empty set. (Another possibility, not mentioned, is that all three equations represent the same plane and the solution is any point on that plane.

The lesson stresses algebraic manipulation for solving a system of three equations in three variables; selecting two of the equations and using elimination to produce a system of two equations in two variables; solving that two-equation, two-variable system and finding the values of the two variables; substituting the values of those two variables into any of the three-variable equations and finding the value of the third variable.

During the algebraic process of solving for the variables, finding always-false statements such as $0 = 1$ means that at least two of the planes are parallel; that system can have no solution. Finding an always-true statement such as $0 = 0$ means at least two of the planes coincide. If the third plane intersects or coincides with that plane, then the number of solutions is infinite; if the third plane is parallel to the first plane, then the system has no solution.

Additional mathematical information and teaching notes are available at www.algebra2.com/key_concepts.

Chapter 3

DAILY INTERVENTION and Assessment

	Type	Student Edition	Teacher Resources	Technology/Internet
INTERVENTION	Ongoing	Prerequisite Skills, pp. 109, 115, 122, 127, 135 Practice Quiz 1, p. 122 Practice Quiz 2, p. 135	5-Minute Check Transparencies Quizzes, *CRM* pp. 163–164 Mid-Chapter Test, *CRM* p. 165 Study Guide and Intervention, *CRM* pp. 119–120, 125–126, 131–132, 137–138, 143–144	Alge2PASS: Tutorial Plus www.algebra2.com/self_check_quiz www.algebra2.com/extra_examples
	Mixed Review	pp. 115, 122, 127, 135, 144	Cumulative Review, *CRM* p. 166	
	Error Analysis	Find the Error, pp. 119, 142 Common Misconceptions, pp. 118, 130	Find the Error, *TWE* pp. 119, 142 Unlocking Misconceptions, *TWE* p. 140 Tips for New Teachers, *TWE* p. 144	
	Standardized Test Practice	pp. 115, 117, 120, 122, 127, 134, 144, 149, 150–151	*TWE* p. 117 Standardized Test Practice, *CRM* pp. 167–168	Standardized Test Practice CD-ROM www.algebra2.com/standardized_test
ASSESSMENT	Open-Ended Assessment	Writing in Math, pp. 114, 121, 127, 134, 144 Open Ended, pp. 112, 119, 125, 132, 142	Modeling: *TWE* pp. 115, 127 Speaking: *TWE* pp. 122, 144 Writing: *TWE* p. 135 Open-Ended Assessment, *CRM* p. 161	
	Chapter Assessment	Study Guide, pp. 145–148 Practice Test, p. 149	Multiple-Choice Tests (Forms 1, 2A, 2B), *CRM* pp. 149–154 Free-Response Tests (Forms 2C, 2D, 3), *CRM* pp. 155–160 Vocabulary Test/Review, *CRM* p. 162	ExamView® Pro (see below) MindJogger Videoquizzes www.algebra2.com/vocabulary_review www.algebra2.com/chapter_test

Key to Abbreviations: TWE = Teacher Wraparound Edition; CRM = Chapter Resource Masters

Intervention Technology

Alge2PASS: Tutorial Plus CD-ROM offers a complete, self-paced algebra curriculum.

Algebra 2 Lesson	Alge2PASS Lesson
3-2	5 *Solving a System of Simultaneous Equations*

ExamView® Pro

Use the networkable **ExamView® Pro** to:

- Create **multiple versions** of tests.
- Create **modified** tests for *Inclusion* students.
- **Edit** existing questions and **add** your own questions.
- Use built-in **state curriculum correlations** to create tests aligned with state standards.
- **Apply** art to your tests from a program bank of artwork.

For more information on Intervention and Assessment, see pp. T8–T11.

Reading and Writing in Mathematics

Glencoe Algebra 2 provides numerous opportunities to incorporate reading and writing into the mathematics classroom.

Student Edition

- Foldables Study Organizer, p. 109
- Concept Check questions require students to verbalize and write about what they have learned in the lesson. (pp. 112, 119, 125, 132, 142, 145)
- Writing in Math questions in every lesson, pp. 114, 121, 127, 134, 144
- Reading Study Tip, pp. 124, 129
- WebQuest, p. 120

Teacher Wraparound Edition

- Foldables Study Organizer, pp. 109, 145
- Study Notebook suggestions, pp. 113, 119, 125, 132, 136, 142
- Modeling activities, pp. 115. 127
- Speaking activities, pp. 122, 144
- Writing activities, p. 135
- Differentiated Instruction, (Verbal/Linguistic), pp. 125, 141
- ELL Resources, pp. 108, 114, 121, 125, 126, 134, 141, 143, 145

Additional Resources

- Vocabulary Builder worksheets require students to define and give examples for key vocabulary terms as they progress through the chapter. (*Chapter 3 Resource Masters*, pp. vii-viii)
- Reading to Learn Mathematics master for each lesson (*Chapter 3 Resource Masters*, pp. 123, 129, 135, 141, 147)
- *Vocabulary PuzzleMaker* software creates crossword, jumble, and word search puzzles using vocabulary lists that you can customize.
- *Teaching Mathematics with Foldables* provides suggestions for promoting cognition and language.
- *Reading and Writing in the Mathematics Classroom*
- *WebQuest and Project Resources*

For more information on Reading and Writing in Mathematics, see pp. T6–T7.

PROJECT CRISS[SM]

Study Skill

Students may find it helpful to organize information they are learning in a frame. A sample content frame showing major concepts about systems of linear equations is shown below. Students can complete this sample and/or develop their own frames to explain other concepts from Chapter 3.

Term	Sample Graph	Description	Definition (own words)
Consistent system	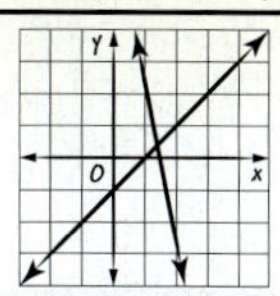	intersecting	a system of equations with one or more solutions
Inconsistent	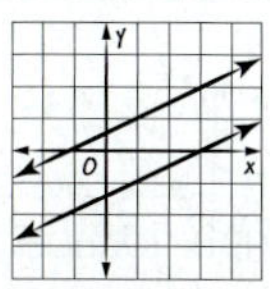	parallel	a system of equations with no solution

CReating **I**ndependence Through **S**tudent-Owned **S**trategies

Chapter 3 Notes

What You'll Learn

Have students read over the list of objectives and make a list of any words with which they are not familiar.

Why It's Important

Point out to students that this is only one of many reasons why each objective is important. Others are provided in the introduction to each lesson.

Lesson	NCTM Standards	Local Objectives
3-1	1, 2, 6, 8, 9, 10	
3-2	1, 2, 6, 7, 8, 9, 10	
3-3	1, 2, 6, 8, 9, 10	
3-3 Follow-Up	1, 2	
3-4	1, 2, 6, 8, 9, 10	
3-5 Preview	1, 2, 10	
3-5	1, 2, 6, 7, 8, 9, 10	

Key to NCTM Standards:

1=Number & Operations, 2=Algebra, 3=Geometry, 4=Measurement, 5=Data Analysis & Probability, 6=Problem Solving, 7=Reasoning & Proof, 8=Communication, 9=Connections, 10=Representation

Chapter 3 Systems of Equations and Inequalities

What You'll Learn

- **Lessons 3-1, 3-2, and 3-5** Solve systems of linear equations in two or three variables.
- **Lesson 3-3** Solve systems of inequalities.
- **Lesson 3-4** Use linear programming to find maximum and minimum values of functions.

Key Vocabulary

- system of equations (p. 110)
- substitution method (p. 116)
- elimination method (p. 118)
- linear programming (p. 130)
- ordered triple (p. 136)

Why It's Important

Systems of linear equations and inequalities can be used to model real-world situations in which many conditions must be met. For example, hurricanes are classified using inequalities that involve wind speed and storm surge. Weather satellites provide images of hurricanes, which are rated on a scale of 1 to 5. *You will learn how to classify the strength of a hurricane in Lesson 3-3.*

Vocabulary Builder

ELL

The Key Vocabulary list introduces students to some of the main vocabulary terms included in this chapter. For a more thorough vocabulary list with pronunciations of new words, give students the Vocabulary Builder worksheets found on pages vii and viii of the *Chapter 3 Resource Masters*. Encourage them to complete the definition of each term as they progress through the chapter. You may suggest that they add these sheets to their study notebooks for future reference when studying for the Chapter 3 test.

Getting Started

Prerequisite Skills To be successful in this chapter, you'll need to master these skills and be able to apply them in problem-solving situations. Review these skills before beginning Chapter 3.

For Lesson 3-1 — Graph Linear Equations

Graph each equation. *(For review, see Lesson 2-2.)* **1–6. See pp. 151A–151F.**

1. $2y = x$ **2.** $y = x - 4$ **3.** $y = 2x - 3$

4. $x + 3y = 6$ **5.** $2x + 3y = -12$ **6.** $4y - 5x = 10$

For Lesson 3-2 — Solve for a Specified Variable

Solve each equation for y. *(For review, see Lesson 1-3.)*

7. $2x + y = 0$ $y = -2x$ **8.** $x - y = -4$ $y = x + 4$ **9.** $6x + 2y = 12$ $y = 6 - 3x$

10. $8 - 4y = 5x$ $y = 2 - \frac{5}{4}x$ **11.** $\frac{1}{2}y + 3x = 1$ $y = 2 - 6x$ **12.** $\frac{1}{3}x - 2y = 8$ $y = \frac{1}{6}x - 4$

For Lessons 3-3 and 3-4 — Graph Inequalities

Graph each inequality. *(For review, see Lesson 2-7.)* **13–18. See pp. 151A–151F.**

13. $y \geq -2$ **14.** $x + y \leq 0$ **15.** $y < 2x - 2$

16. $x + 4y < 3$ **17.** $2x - y \geq 6$ **18.** $3x - 4y < 10$

For Lesson 3-5 — Evaluate Expressions

Evaluate each expression if $x = -3$, $y = 1$, and $z = 2$. *(For review, see Lesson 1-1.)*

19. $3x + 2y - z$ -9 **20.** $3y - 8z$ -13 **21.** $x - 5y + 4z$ 0

22. $2x + 9y + 4z$ 11 **23.** $2x - 6y - 5z$ -22 **24.** $7x - 3y + 2z$ -20

FOLDABLES™ Study Organizer

Systems of Linear Equations and Inequalities Make this Foldable to organize your notes. Begin with one sheet of 11" × 17" paper and four sheets of grid paper.

Step 1 Fold and Cut

Fold the short sides of the 11" × 17" paper to meet in the middle. Cut each tab in half as shown.

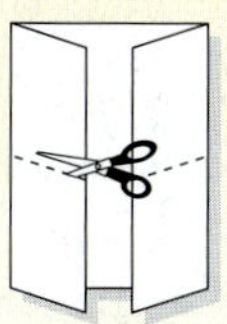

Step 2 Staple and Label

Insert 2 folded half-sheets of grid paper in each tab. Staple at edges. Label each tab as shown.

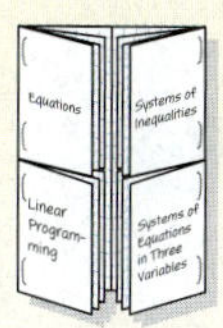

Reading and Writing As you read and study the chapter, fill the tabs with notes, diagrams, and examples for each topic.

Getting Started

This section provides a review of the basic concepts needed before beginning Chapter 3. Page references are included for additional student help.

Prerequisite Skills in the Getting Ready for the Next Lesson section at the end of each exercise set review a skill needed in the next lesson.

For Lesson	Prerequisite Skill
3-2	Simplifying Expressions (p. 115)
3-3	Inequalities (p. 122)
3-4	Functions (p. 127)
3-5	Evaluating Expressions (p. 135)

FOLDABLES™ Study Organizer

For more information about Foldables, see *Teaching Mathematics with Foldables.*

Paraphrasing or Summarizing After students make their Foldable, have them label the tabs to correspond to the lessons in this chapter, combining Lessons 3-1 and 3-2 on the first tab. Students use their Foldable to take notes, define terms, record concepts, and write examples. At the end of each lesson, ask students to paraphrase or write a summary of the main ideas and supporting details presented in the lesson. Summaries are useful for condensing data and realizing what is important.

3-1 Lesson Notes

1 Focus

5-Minute Check Transparency 3-1 Use as a quiz or review of Chapter 2.

Mathematical Background notes are available for this lesson on p. 108C.

Building on Prior Knowledge

In Chapter 2, students solved and graphed equations. In this lesson, students use similar procedures to solve and graph systems of equations.

How can a system of equations be used to predict sales?

Ask students:

- Are in-store sales or online sales growing at a faster rate? **online sales**
- How can you tell which type of sales are growing at a faster rate? **The slope, 7.5, of the online sales graph is greater than the other slope of 4.2, which means the line for online sales is steeper.**

3-1 Solving Systems of Equations by Graphing

What You'll Learn

- Solve systems of linear equations by graphing.
- Determine whether a system of linear equations is consistent and independent, consistent and dependent, or inconsistent.

Vocabulary

- system of equations
- consistent
- inconsistent
- independent
- dependent

How can a system of equations be used to predict sales?

Since 1999, the growth of in-store sales for Custom Creations by Cathy can be modeled by $y = 4.2x + 29$. The growth of her online sales can be modeled by $y = 7.5x + 9.3$. In these equations, x represents the number of years since 1999, and y represents the amount of sales in thousands of dollars.

The equations $y = 4.2x + 29$ and $y = 7.5x + 9.3$ are called a system of equations.

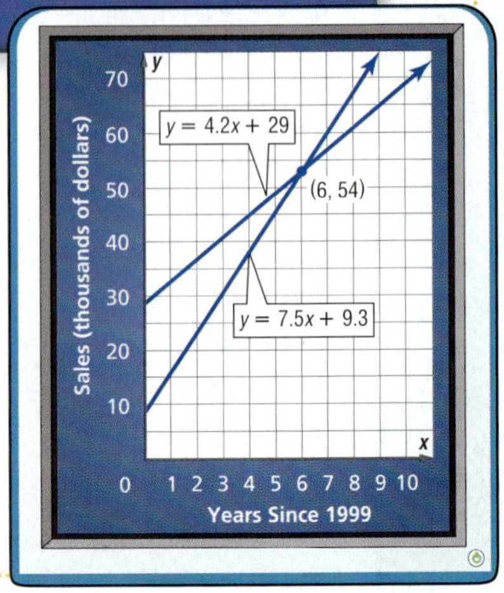

GRAPH SYSTEMS OF EQUATIONS A **system of equations** is two or more equations with the same variables. To solve a system of equations, find the ordered pair that satisfies all of the equations. One way to do this is to graph the equations on the same coordinate plane. The point of intersection represents the solution.

Example 1 Solve by Graphing

Solve the system of equations by graphing.

$2x + y = 5$

$x - y = 1$

Write each equation in slope-intercept form.

$2x + y = 5 \rightarrow y = -2x + 5$

$x - y = 1 \rightarrow y = x - 1$

The graphs appear to intersect at (2, 1).

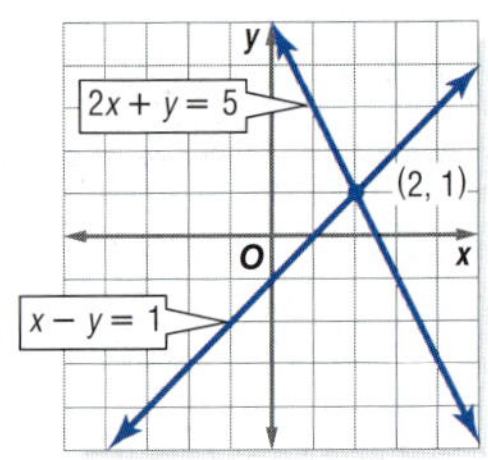

CHECK Substitute the coordinates into each equation.

$2x + y = 5$	$x - y = 1$	Original equations
$2(2) + 1 \stackrel{?}{=} 5$	$2 - 1 \stackrel{?}{=} 1$	Replace x with 2 and y with 1.
$5 = 5$ ✓	$1 = 1$ ✓	Simplify.

The solution of the system is (2, 1).

Study Tip

Checking Solutions
When using a graph to find a solution, always check the ordered pair in *both* original equations.

Systems of equations are used in businesses to determine the *break-even point*. The break-even point is the point at which the income equals the cost.

Resource Manager

Workbook and Reproducible Masters

Chapter 3 Resource Masters
- Study Guide and Intervention, pp. 119–120
- Skills Practice, p. 121
- Practice, p. 122
- Reading to Learn Mathematics, p. 123
- Enrichment, p. 124

Graphing Calculator and Spreadsheet Masters, p. 32

Transparencies

5-Minute Check Transparency 3-1
Answer Key Transparencies

Technology

Interactive Chalkboard
Multimedia Applications

Example 2 Break-Even Point Analysis

MUSIC Travis and his band are planning to record their first CD. The initial start-up cost is $1500, and each CD will cost $4 to produce. They plan to sell their CDs for $10 each. How many CDs must the band sell before they make a profit?

Let x = the number of CDs, and let y = the number of dollars.

Cost of x CDs	is	cost per CD	plus	startup cost.
y	$=$	$4x$	$+$	1500

Income from x CDs	is	price per CD	times	number of CDs.
y	$=$	10	$\cdot$	x

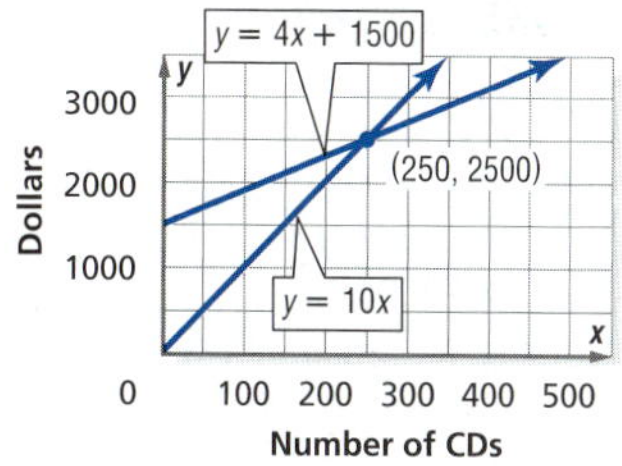

The graphs intersect at (250, 2500). This is the break-even point. If the band sells fewer than 250 CDs, they will lose money. If the band sells more than 250 CDs, they will make a profit.

More About. . .

Music

Compact discs (CDs) store music digitally. The recorded sound is converted to a series of 1s and 0s. This coded pattern can then be read by an infrared laser in a CD player.

CLASSIFY SYSTEMS OF EQUATIONS Graphs of systems of linear equations may be intersecting lines, parallel lines, or the same line. A system of equations is **consistent** if it has at least one solution and **inconsistent** if it has no solutions. A consistent system is **independent** if it has exactly one solution or **dependent** if it has an infinite number of solutions.

Example 3 Intersecting Lines

Graph the system of equations and describe it as *consistent and independent, consistent and dependent,* or *inconsistent.*

$$x + \frac{1}{2}y = 5$$
$$3y - 2x = 6$$

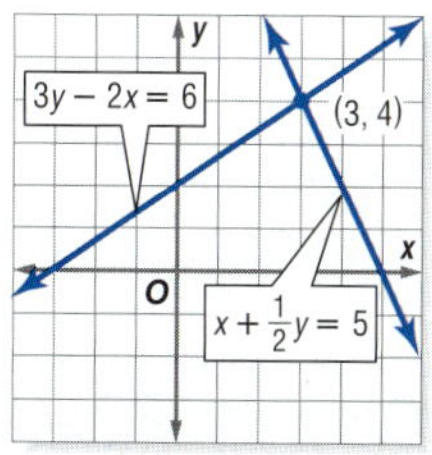

Write each equation in slope-intercept form.

$x + \frac{1}{2}y = 5 \rightarrow y = -2x + 10$

$3y - 2x = 6 \rightarrow y = \frac{2}{3}x + 2$

The graphs intersect at (3, 4). Since there is one solution, this system is *consistent and independent*.

Example 4 Same Line

Graph the system of equations and describe it as *consistent and independent, consistent and dependent,* or *inconsistent.*

$$9x - 6y = 24$$
$$6x - 4y = 16$$

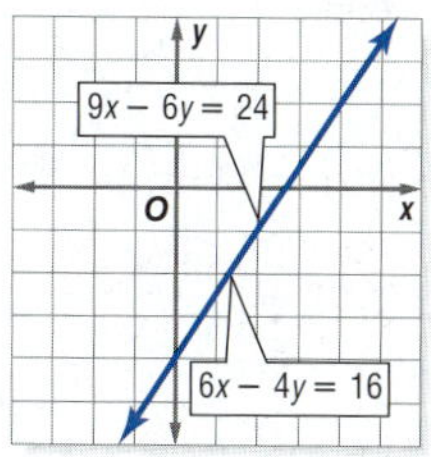

$9x - 6y = 24 \rightarrow y = \frac{3}{2}x - 4$

$6x - 4y = 16 \rightarrow y = \frac{3}{2}x - 4$

Since the equations are equivalent, their graphs are the same line. Any ordered pair representing a point on that line will satisfy both equations.

So, there are infinitely many solutions to this system. This system is *consistent and dependent*.

www.algebra2.com/extra_examples

2 Teach

GRAPH SYSTEMS OF EQUATIONS

In-Class Examples Power Point®

1 Solve the system of equations by graphing.
$x - 2y = 0$
$x + y = 6$ **(4, 2)**

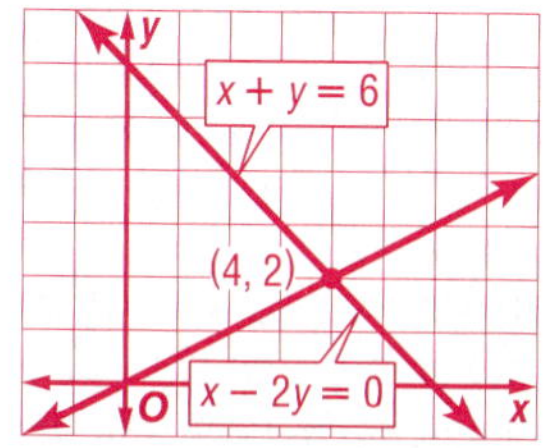

2 A service club is selling copies of their holiday cookbook to raise funds for a project. The printer's set-up charge is $200, and each book costs $2 to print. The cookbooks will sell for $6 each. How many cookbooks must the members sell before they make a profit? **more than 50 cookbooks**

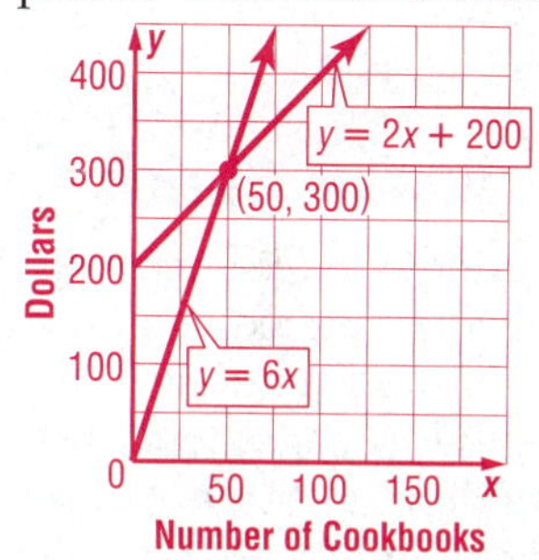

CLASSIFY SYSTEMS OF EQUATIONS

In-Class Example Power Point®

3 Graph the system of equations and describe it as *consistent and independent, consistent and dependent,* or *inconsistent.*
$x - y = 5$
$x + 2y = -4$ **consistent and independent**

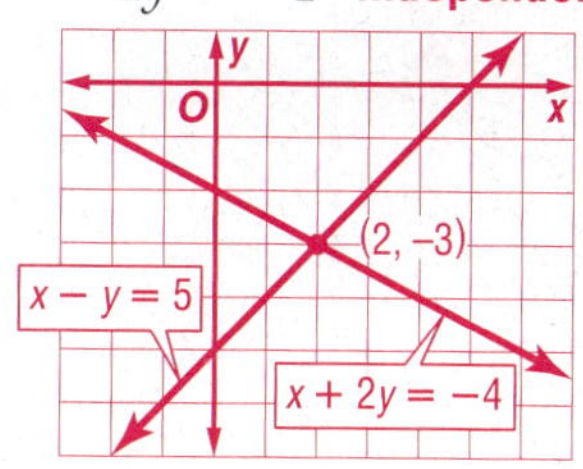

DAILY INTERVENTION — Differentiated Instruction

Interpersonal Have students work in pairs to write three systems of equations, one that is *consistent and independent*, one that is *consistent and dependent*, and another that is *inconsistent*. To simplify the activity somewhat, you might require that each system include the equation $2x + y = 1$. Pairs can exchange their systems with another pair of students to have their work checked.

In-Class Examples

Power Point®

4 Graph the system of equations and describe it as *consistent and independent*, *consistent and dependent*, or *inconsistent*.

$9x - 6y = -6$ **consistent and**
$6x - 4y = -4$ **dependent**

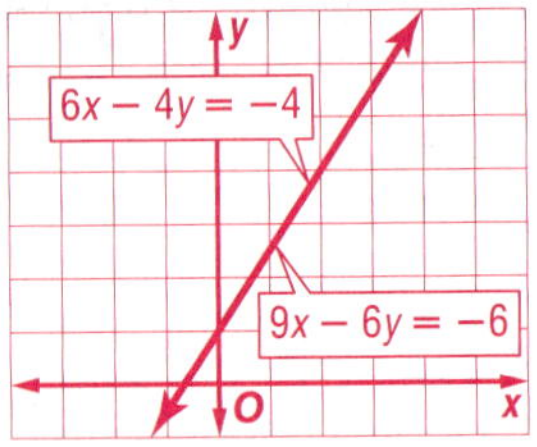

5 Graph the system of equations and describe it as *consistent and independent*, *consistent and dependent*, or *inconsistent*.

$15x - 6y = 0$
$5x - 2y = 10$ **inconsistent**

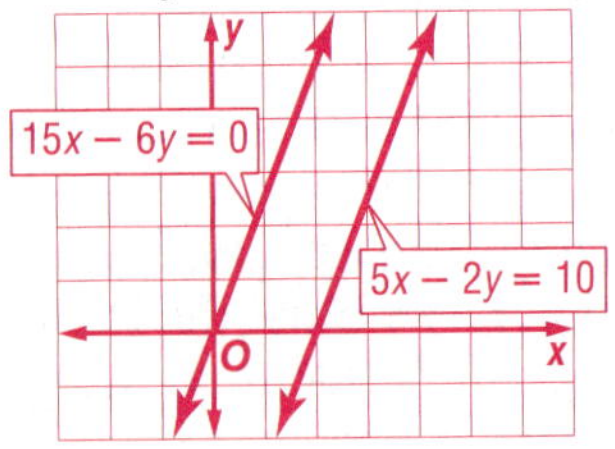

Concept Check

Ask students to describe the slopes and y-intercepts of the graphs of the equations in a system that is consistent and independent, a system that is consistent and dependent, and a system that is inconsistent. **different slopes, different y-intercepts; same slope, same y-intercept; same slope, different y-intercepts**

Teaching Tip Remind students that the graphs of equations with the same slope and different intercepts are, by definition, parallel.

Study Tip

Parallel Lines
Notice from their equations that the lines have the same slope and different y-intercepts.

Example 5 Parallel Lines

Graph the system of equations and describe it as *consistent and independent*, *consistent and dependent*, or *inconsistent*.

$\mathbf{3x + 4y = 12}$
$\mathbf{6x + 8y = -16}$

$3x + 4y = 12 \rightarrow y = -\frac{3}{4}x + 3$

$6x + 8y = -16 \rightarrow y = -\frac{3}{4}x - 2$

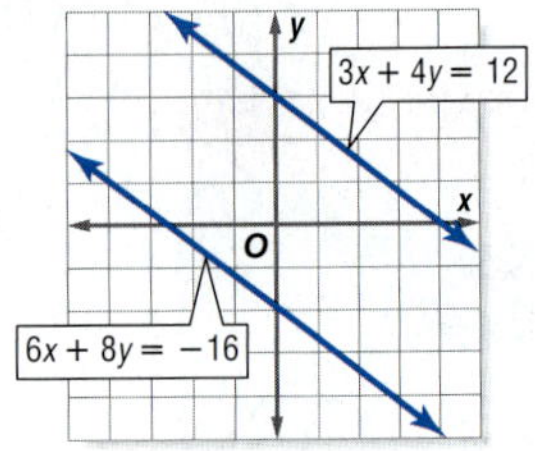

The lines do not intersect. Their graphs are parallel lines. So, there are no solutions that satisfy both equations. This system is *inconsistent*.

The relationship between the graph of a system of equations and the number of its solutions is summarized below.

Concept Summary — *Systems of Equations*

consistent and independent	consistent and dependent	inconsistent
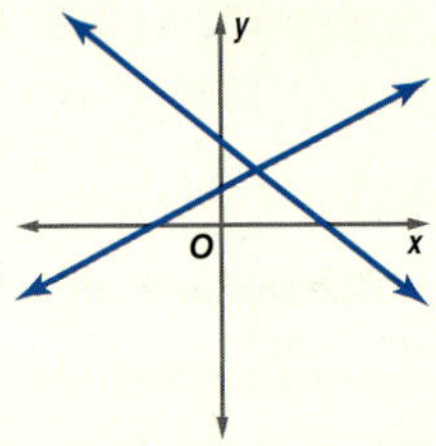	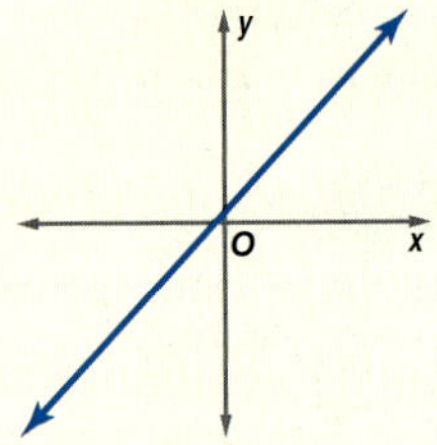	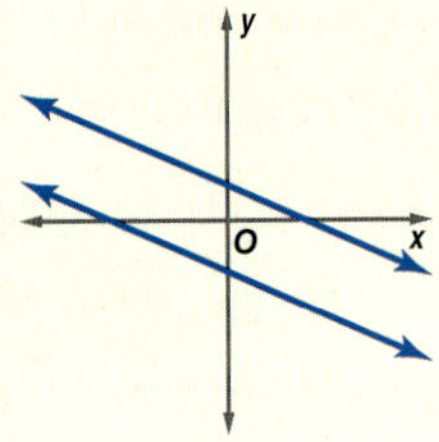
intersecting lines; one solution	same line; infinitely many solutions	parallel lines; no solution

Check for Understanding

Concept Check

1. **Explain** why a system of linear equations cannot have exactly two solutions. **1. Two lines cannot intersect in exactly two points.**
2. **OPEN ENDED** Give an example of a system of equations that is consistent and independent. **Sample answer: $x + y = 4$, $x - y = 2$**
3. **Explain** why it is important to check a solution found by graphing in both of the original equations. **A graph is used to estimate the solution. To determine that the point lies on both lines, you must check that it satisfies both equations.**

Guided Practice

GUIDED PRACTICE KEY	
Exercises	Examples
4–6	1
7–9	3–5
10–12	2

Solve each system of equations by graphing. **4–6. See margin for graphs.**

4. $y = 2x + 9$, $y = -x + 3$ **(–2, 5)**
5. $3x + 2y = 10$, $2x + 3y = 10$ **(2, 2)**
6. $4x - 2y = 22$, $6x + 9y = -3$ **(4, –3)**

Graph each system of equations and describe it as *consistent and independent*, *consistent and dependent*, or *inconsistent*. **7–9. See margin for graphs.**

7. $y = 6 - x$, $y = x + 4$ **cons. and ind.**
8. $x + 2y = 2$, $2x + 4y = 8$ **incon.**
9. $x - 2y = 8$, $\frac{1}{2}x - y = 4$ **cons. and dep.**

Answers

4.
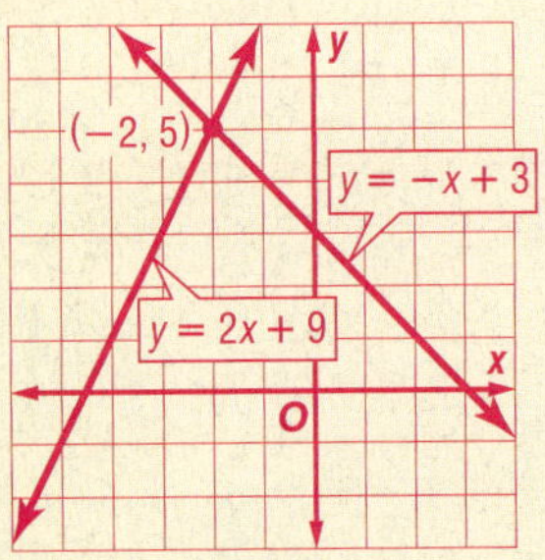

5.
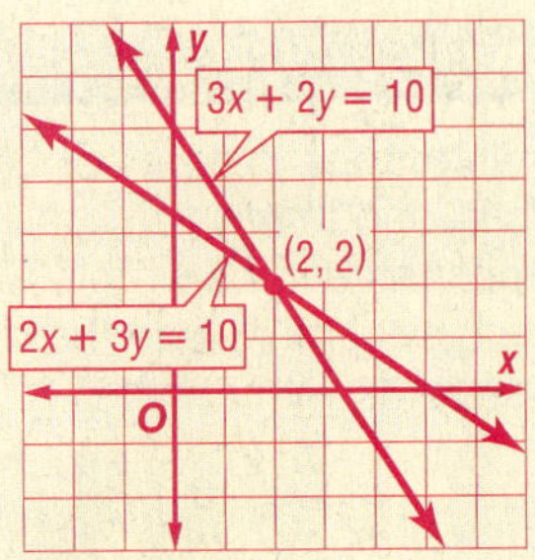

6.
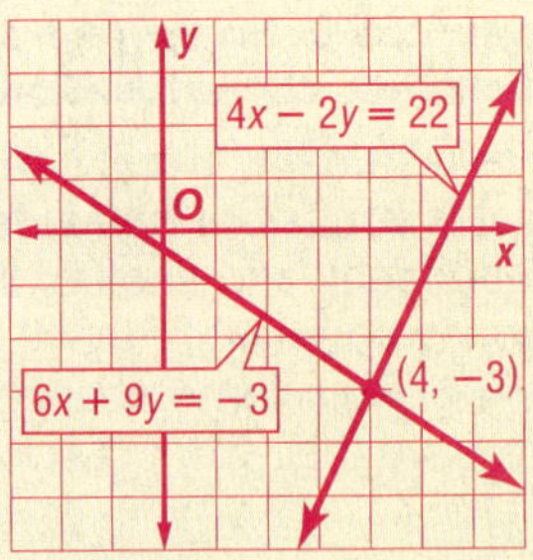

7.
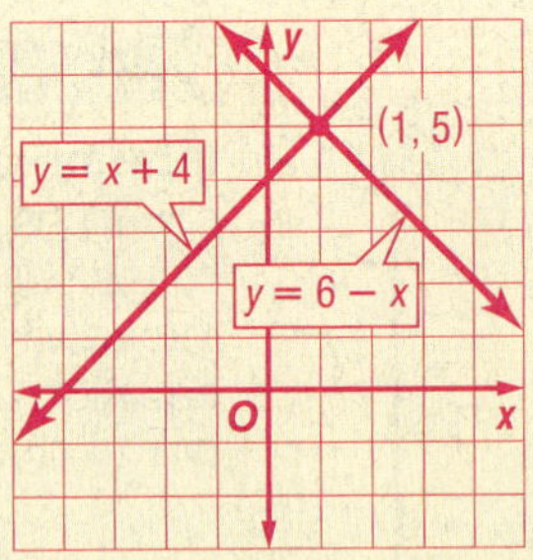

3 Practice/Apply

Study Notebook

Have students—

- add the definitions/ examples of the vocabulary terms to their Vocabulary Builder worksheets for Chapter 3.
- include any other items(s) that they find helpful in mastering the skills in this lesson.

About the Exercises...

Organization by Objective

- **Graph Systems of Equations:** 13–24, 37–47
- **Classify Systems of Equations:** 25–36, 48

Odd/Even Assignments

Exercises 13–38 are structured so that students practice the same concepts whether they are assigned odd or even problems.

Alert! Exercises 52–57 require a graphing calculator.

Assignment Guide

Basic: 13–21 odd, 25–37 odd, 39–41, 48–51, 58–79

Average: 13–37 odd, 39–44, 48–51, 58–79 (optional: 52–57)

Advanced: 14–38 even, 39–73 (optional: 74–79)

Application

PHOTOS For Exercises 10–12, use the graphic at the right.

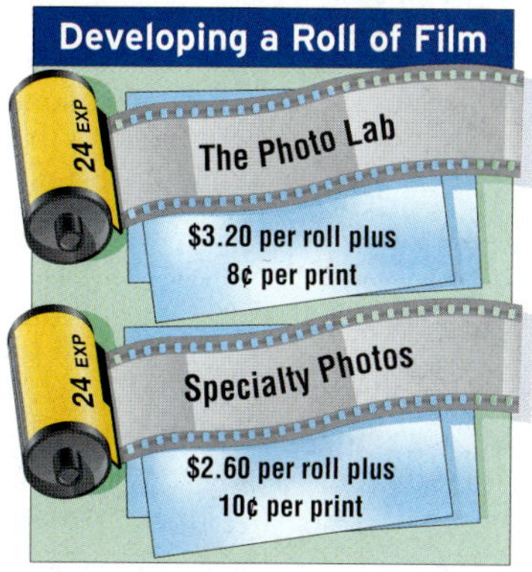

10. Write equations that represent the cost of developing a roll of film at each lab.

11. Under what conditions is the cost to develop a roll of film the same for either store?

12. When is it best to use The Photo Lab and when is it best to use Specialty Photos? **You should use Specialty Photos if you are developing less than 30 prints, and you should use The Photo Lab if you are developing more than 30 prints.**

10. $y = 0.08x + 3.2$, $y = 0.10x + 2.6$

11. The cost is \$5.60 for both stores to develop 30 prints.

★ indicates increased difficulty

Practice and Apply

Homework Help	
For Exercises	**See Examples**
13–24, 37, 38	1
25–36	3–5
39–47	2

Extra Practice
See page 832.

Solve each system of equations by graphing. **13–24. See pp. 151A–151F for graphs.**

13. $y = 2x - 4$, $y = -3x + 1$ **(1, −2)**

14. $y = 3x - 8$, $y = x - 8$ **(0, −8)**

15. $x + 2y = 6$, $2x + y = 9$ **(4, 1)**

16. $2x + 3y = 12$, $2x - y = 4$ **(3, 2)**

17. $3x - 7y = -6$, $x + 2y = 11$ **(5, 3)**

18. $5x - 11 = 4y$, $7x - 1 = 8y$ **(7, 6)**

19. $2x + 3y = 7$, $2x - 3y = 7$ **(3.5, 0)**

20. $8x - 3y = -3$, $4x - 2y = -4$ **(1.5, 5)**

21. $\frac{1}{4}x + 2y = 5$, $2x - y = 6$ **(4, 2)**

22. $\frac{2}{3}x + y = -3$, $y - \frac{1}{3}x = 6$ **(−9, 3)**

★**23.** $\frac{1}{2}x - y = 0$, $\frac{1}{4}x + \frac{1}{2}y = -2$ **(−4, −2)**

★**24.** $\frac{4}{3}x + \frac{1}{5}y = 3$, $\frac{2}{3}x - \frac{3}{5}y = 5$ **(3, −5)**

Graph each system of equations and describe it as *consistent and independent*, *consistent and dependent*, or *inconsistent*. **25–36. See pp. 151A–151F for graphs.**

25. $y = x + 4$, $y = x - 4$ **incon.**

26. $y = x + 3$, $y = 2x + 6$ **cons. and ind.**

27. $x + y = 4$, $-4x + y = 9$ **cons. and ind.**

28. $3x + y = 3$, $6x + 2y = 6$ **cons. and dep.**

29. $y - x = 5$, $2y - 2x = 8$ **incon.**

30. $4x - 2y = 6$, $6x - 3y = 9$ **cons. and dep.**

31. $2y = x$, $8y = 2x + 1$ **cons. and ind.**

32. $2y = 5 - x$, $6y = 7 - 3x$ **incon.**

33. $0.8x - 1.5y = -10$, $1.2x + 2.5y = 4$ **cons. and ind.**

34. $1.6y = 0.4x + 1$, $0.4y = 0.1x + 0.25$ **cons. and dep.**

35. $3y - x = -2$, $y - \frac{1}{3}x = 2$ **incon.**

36. $2y - 4x = 3$, $\frac{4}{3}x - y = -2$ **cons. and ind.**

37. GEOMETRY The sides of an angle are parts of two lines whose equations are $2y + 3x = -7$ and $3y - 2x = 9$. The angle's vertex is the point where the two sides meet. Find the coordinates of the vertex of the angle. **(−3, 1)**

38. GEOMETRY The graphs of $y - 2x = 1$, $4x + y = 7$, and $2y - x = -4$ contain the sides of a triangle. Find the coordinates of the vertices of the triangle. **See margin for graph; (1, 3), (2, −1),(−2, −3).**

TRAVEL For Exercises 39–41, use the following information.

Adam and his family are planning to rent a midsize car for a one-day trip. In the Standard Rental Plan, they can rent a car for \$52 per day plus 23 cents per mile. In the Deluxe Rental Plan, they can rent a car for \$80 per day with unlimited mileage.

39. For each plan, write an equation that represents the cost of renting a car.

40. Graph the equations. Estimate the break-even point of the rental costs.

41. If Adam's family plans to drive 150 miles, which plan should they choose?

39. $y = 52 + 0.23x$, $y = 80$

40. See margin for graph; (120, 80).

41. Deluxe Plan

www.algebra2.com/self_check_quiz

Answers

8.

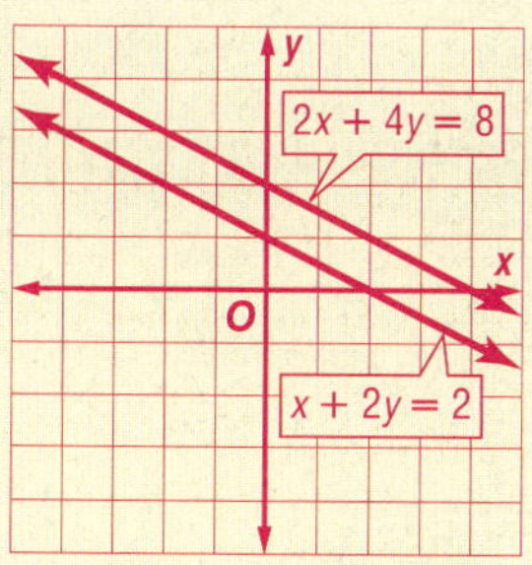

9.

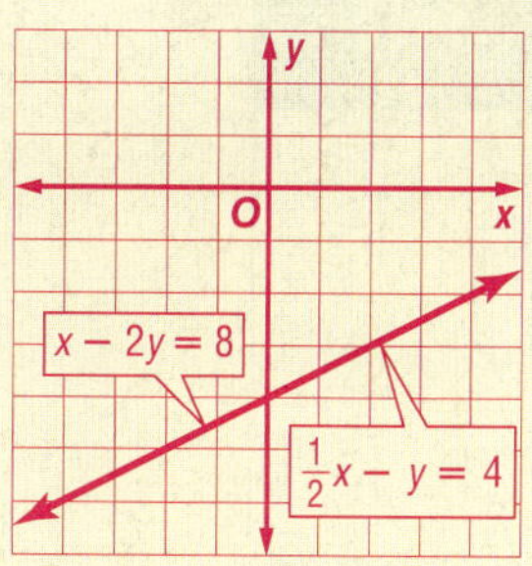

38.

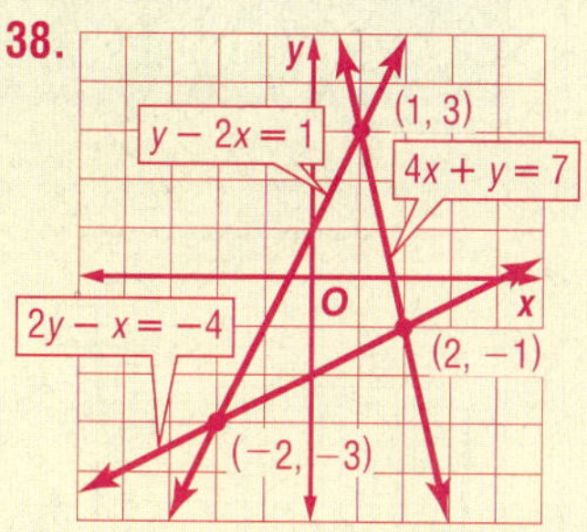

40.

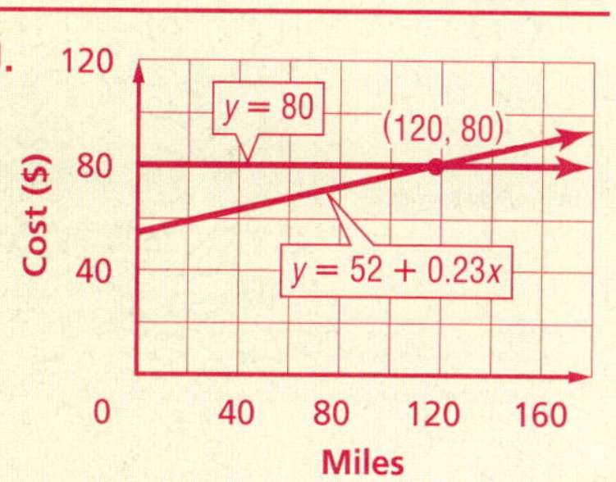

Study Guide and Intervention, p. 119 (shown) and p. 120

Graph Systems of Equations A system of equations is a set of two or more equations containing the same variables. You can solve a system of linear equations by graphing the equations on the same coordinate plane. If the lines intersect, the solution is that intersection point.

Example **Solve the system of equations by graphing.** $x - 2y = 4$, $x + y = -2$

Write each equation in slope-intercept form.

$x - 2y = 4 \rightarrow y = \frac{x}{2} - 2$

$x + y = -2 \rightarrow y = -x - 2$

The graphs appear to intersect at (0, −2).

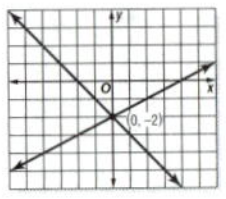

CHECK Substitute the coordinates into each equation.

$x - 2y = 4$ | $x + y = -2$
$0 - 2(-2) \stackrel{?}{=} 4$ | $0 + (-2) \stackrel{?}{=} -2$
$4 = 4$ ✓ | $-2 = -2$ ✓

The solution of the system is (0, −2).

Exercises

Solve each system of equations by graphing.

1. $y = -\frac{x}{3} + 1$, $y = \frac{x}{2} - 4$ **(6, −1)**
2. $y = 2x - 2$, $y = -x + 4$ **(2, 2)**
3. $y = -\frac{x}{2} + 3$, $y = \frac{x}{4}$ **(4, 1)**

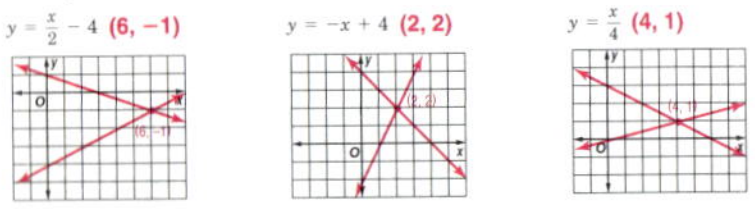

4. $3x - y = 0$, $x - y = -2$ **(1, 3)**
5. $2x + \frac{y}{3} = -7$, $\frac{x}{2} + y = 1$ **(−4, 3)**
6. $\frac{x}{2} - y = 2$, $2x - y = -1$ **(−2, −3)**

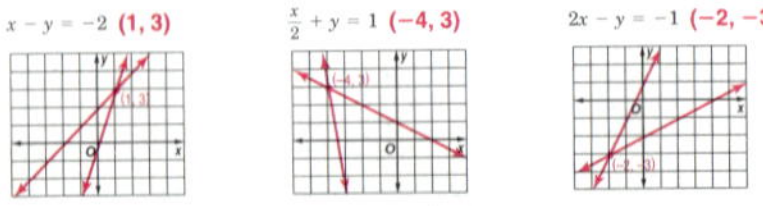

Skills Practice, p. 121 and Practice, p. 122 (shown)

Solve each system of equations by graphing.

1. $x - 2y = 0$, $y = 2x - 3$ **(2, 1)**
2. $x + 2y = 4$, $2x - 3y = 1$ **(2, 1)**
3. $2x + y = 3$, $y = \frac{1}{2}x - \frac{9}{2}$ **(3, −3)**

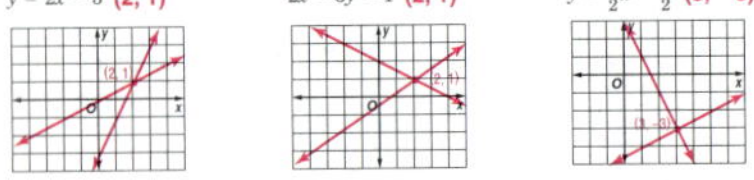

4. $y - x = 3$, $y = 1$ **(−2, 1)**
5. $2x - y = 6$, $x + 2y = -2$ **(2, −2)**
6. $5x - y = 4$, $-2x + 6y = 4$ **(1, 1)**

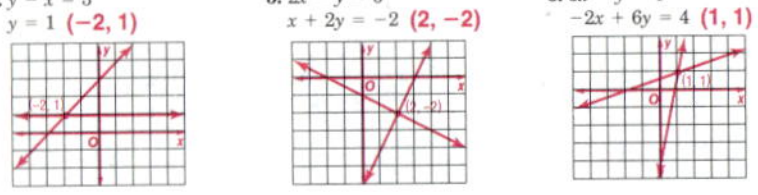

Graph each system of equations and describe it as *consistent and independent*, *consistent and dependent*, or *inconsistent*.

7. $2x - y = 4$, $x - y = 2$ **consistent and indep.**
8. $y = -x - 2$, $x + y = -4$ **inconsistent**
9. $2y - 8 = x$, $y = \frac{1}{2}x + 4$ **consistent and dep.**

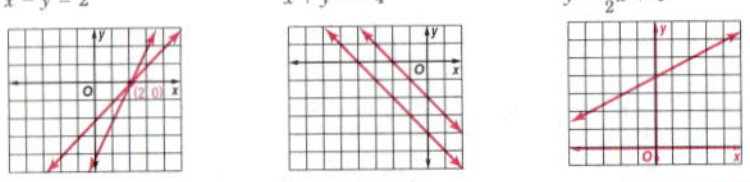

SOFTWARE **For Exercises 10–12, use the following information.**

Location Mapping needs new software. Software A costs $13,000 plus $500 per additional site license. Software B costs $2500 plus $1200 per additional site license.

10. Write two equations that represent the cost of each software. **A: $y = 13{,}000 + 500x$, B: $y = 2500 + 1200x$**

11. Graph the equations. Estimate the break-even point of the software costs. **15 additional licenses**

12. If Location Mapping plans to buy 10 additional site licenses, which software will cost less? **B**

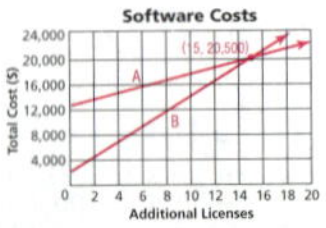

Reading to Learn Mathematics, p. 123 ELL

Pre-Activity **How can a system of equations be used to predict sales?**

Read the introduction to Lesson 3-1 at the top of page 110 in your textbook.

- Which are growing faster, in-store sales or online sales? **online sales**
- In what year will the in-store and online sales be the same? **2005**

Reading the Lesson

1. The Study Tip on page 110 of your textbook says that when you solve a system of equations by graphing and find a point of intersection of the two lines, you must always check the ordered pair in *both* of the original equations. Why is it not good enough to check the ordered pair in just one of the equations? **Sample answer: To be a solution of the system, the ordered pair must make both of the equations true.**

2. Under each system graphed below, write all of the following words that apply: *consistent*, *inconsistent*, *dependent*, and *independent*.

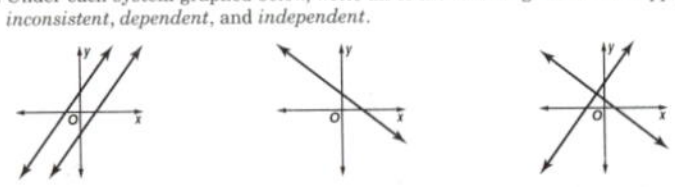

inconsistent | **consistent; dependent** | **consistent; independent**

Helping You Remember

3. Look up the words *consistent* and *inconsistent* in a dictionary. How can the meaning of these words help you distinguish between consistent and inconsistent systems of equations? **Sample answer: One meaning of *consistent* is "being in agreement," or "compatible," while one meaning of *inconsistent* is "not being in agreement" or "incompatible." When a system is consistent, the equations are compatible because both can be true at the same time (for the same values of x and y). When a system is inconsistent, the equations are incompatible because they can never be true at the same time.**

ECONOMICS **For Exercises 42–44, use the graph below that shows the supply and demand curves for a new multivitamin.**

In Economics, the point at which the supply equals the demand is the *equilibrium price*. If the supply of a product is greater than the demand, there is a surplus and prices fall. If the supply is less than the demand, there is a shortage and prices rise.

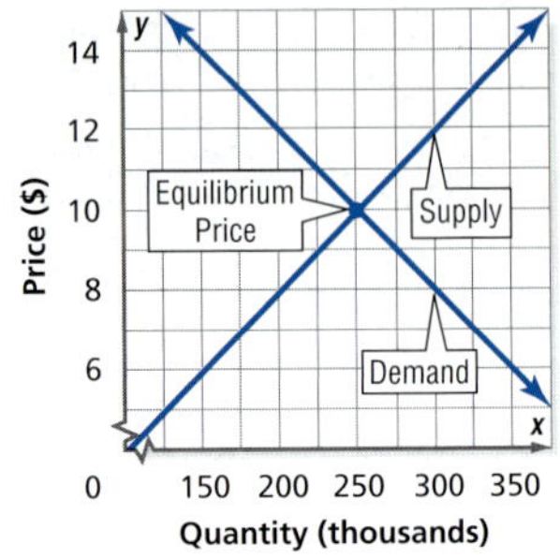

42. If the price for vitamins is $8 a bottle, what is the supply of the product and what is the demand? Will prices tend to rise or fall?

42. Supply, 200,000; demand, 300,000; prices will tend to rise.

43. If the price for vitamins is $12 a bottle, what is the supply of the product and what is the demand? Will prices tend to rise or fall?

43. Supply, 300,000; demand, 200,000; prices will tend to fall.

44. At what quantity will the prices stabilize? What is the equilibrium price for this product? **250,000; $10**

More About. . .

Population

In the United States there is approximately one birth every 8 seconds and one death every 14 seconds.

Source: U.S. Census Bureau

POPULATION **For Exercises 45–47, use the graphic that shows 2000 state populations.**

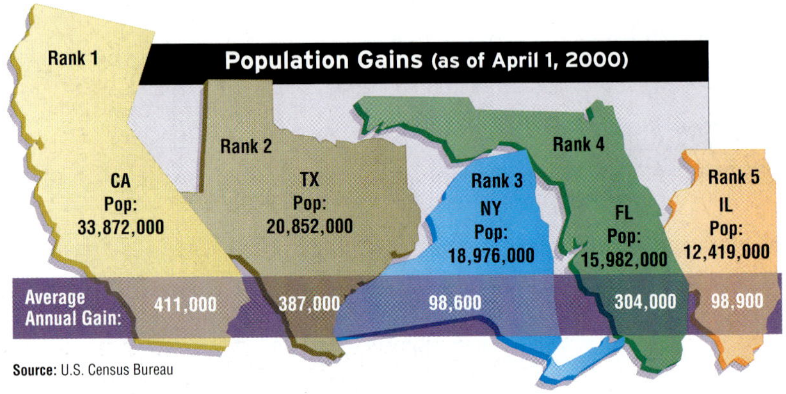

Source: U.S. Census Bureau

★ 45. Write equations that represent the populations of Florida and New York x years after 2000. Assume that both states continue to gain the same number of residents every year. Let y equal the population in thousands.

45. $y = 304x + 15{,}982$, $y = 98.6x + 18{,}976$

★ 46. Graph both equations for the years 2000 to 2020. Estimate when the populations of both states will be equal. **See margin for graph; 2015.**

47. Do you think Florida will overtake New York as the third most populous state by 2010? by 2020? Explain your reasoning.

47. Sample answer: FL will probably be ranked third by 2020. The graphs intersect in the year 2015, so NY will still have a higher population in 2010, but FL will have a higher population in 2020.

48. **CRITICAL THINKING** State the conditions for which the system below is: (a) consistent and dependent, (b) consistent and independent, (c) inconsistent.

$ax + by = c$
$dx + ey = f$

a. $\frac{a}{b} = \frac{d}{e'}\ \frac{c}{b} = \frac{f}{e}$ b. $\frac{a}{b} \neq \frac{d}{e}$ c. $\frac{a}{b} = \frac{d}{e'}\ \frac{c}{b} \neq \frac{f}{e}$

49. WRITING IN MATH Answer the question that was posed at the beginning of the lesson. **See margin.**

How can a system of equations be used to predict sales?

Include the following in your answer:

- an explanation of the real-world meaning of the solution of the system of equations in the application at the beginning of the lesson, and
- a description of what a business owner would learn if the system of equations representing the in-store and online sales is inconsistent.

Enrichment, p. 124

Investments

The following graph shows the value of two different investments over time. Line A represents an initial investment of $30,000 with a bank paying passbook savings interest. Line B represents an initial investment of $5,000 in a profitable mutual fund with dividends reinvested and capital gains accepted in shares. By deriving the linear equation $y = mx + b$ for A and B, you can predict the value of these investments for years to come.

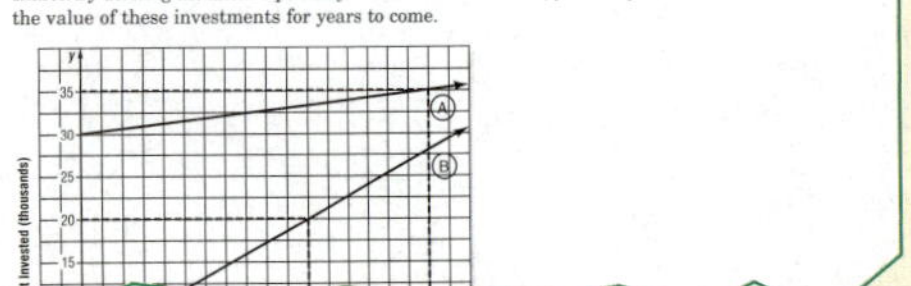

46.

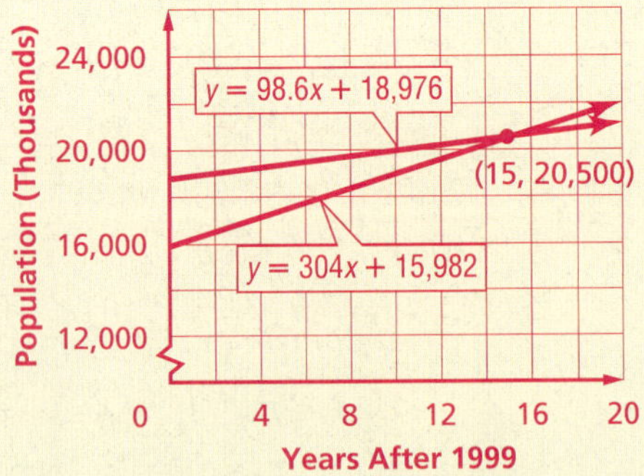

Standardized Test Practice

50. What are the coordinates (x, y) at which the graphs of $2x + 3y = 12$ and $2x - y = 4$ intersect? **A**

Ⓐ (3, 2) Ⓑ (2, 3) Ⓒ (1, −2) Ⓓ (−3, 6)

51. Which equation has the same graph as $4x + 8y = 12$? **C**

Ⓐ $x + y = 3$ Ⓑ $2x + y = 3$ Ⓒ $x + 2y = 3$ Ⓓ $2x + 2y = 6$

Graphing Calculator

INTERSECT FEATURE **To use a TI-83 Plus to solve a system of equations, graph both equations on the same screen. Then, select intersect, which is option 5 under the CALC menu, to find the coordinates of the point of intersection. Solve each system of equations to the nearest hundredth.** **54. (4, 3.42)**

52. $y = 0.125x - 3.005$
$y = -2.58$ **(3.40, −2.58)**

53. $3.6x - 2y = 4$
$-2.7x + y = 3$ **(−5.56, −12)**

54. $y = 0.18x + 2.7$
$y = -0.42x + 5.1$

55. $1.6x + 3.2y = 8$
$1.2x + 2.4y = 4$
no solution

56. $y - \frac{1}{4}x = 6$
$2y + \frac{1}{2}x = 3$
(−9, 3.75)

57. $\frac{1}{2}y - 5x = 8$
$\frac{1}{3}y - 8x = -7$
(2.64, 42.43)

Maintain Your Skills

Mixed Review

Graph each inequality. *(Lesson 2-7)* **58–60. See margin.**

58. $y \geq 5 + 3x$
59. $2x + y > -4$
60. $2y - 1 \leq x$

Identify each function as S for step, C for constant, A for absolute value, or P for piecewise. *(Lesson 2-6)*

61. 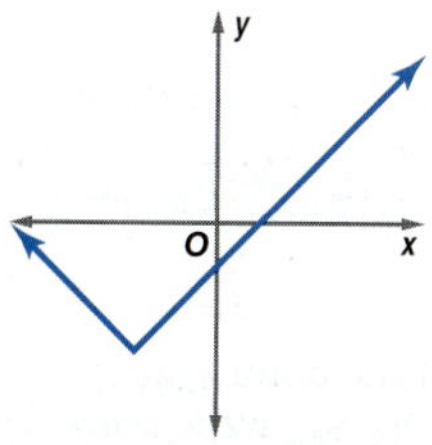
A

62. 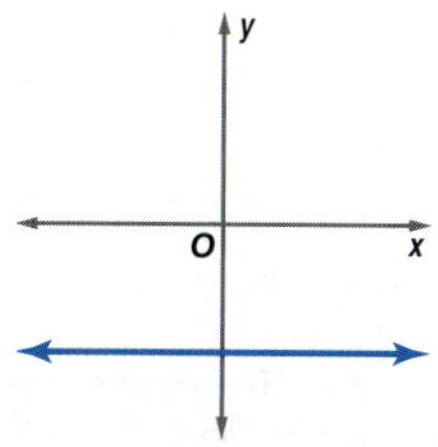
C

63. 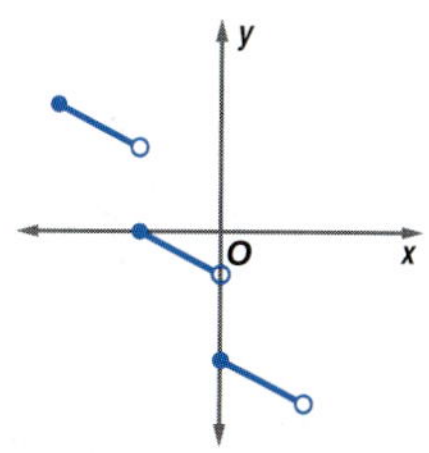
P

Solve each equation. Check your solutions. *(Lesson 1-4)*

64. $|x| - 5 = 8$ **{−13, 13}**
65. $|w + 3| = 12$ **{−15, 9}**
66. $|6a - 4| = -2$ **∅**
67. $3|2t - 1| = 15$ **{−2, 3}**
68. $|4r + 3| - 7 = 10$ **$\left\{-5, \frac{7}{2}\right\}$**
69. $|k + 7| = 3k - 11$ **{9}**

Write an algebraic expression to represent each verbal expression. *(Lesson 1-3)*

70. the sum of 8 and 2 times a number **$8 + 2n$**
71. six less than the square of a number **$x^2 - 6$**
72. four times the sum of a number and 5 **$4(a + 5)$**
73. the quotient of a number and 3 increased by 1 **$\frac{z}{3} + 1$**

Getting Ready for the Next Lesson

PREREQUISITE SKILL **Simplify each expression.**
*(To review **simplifying expressions**, see Lesson 1-2.)*

74. $(3x + 5) - (2x + 3)$ **$x + 2$**
75. $(3y - 11) + (6y + 12)$ **$9y + 1$**
76. $(5x - y) + (-8x + 7y)$ **$-3x + 6y$**
77. $6(2x + 3y - 1)$ **$12x + 18y - 6$**
78. $5(4x + 2y - x + 2)$ **$15x + 10y + 10$**
79. $3(x + 4y) - 2(x + 4y)$ **$x + 4y$**

Answer

49. You can use a system of equations to track sales and make predictions about future growth based on past performance and trends in the graphs. Answers should include the following.

- The coordinates (6, 54) represent that 6 years after 1999 both the in-store sales and online sales will be $54,000.
- The in-store sales and the online sales will never be equal and in-store sales will continue to be higher than online sales.

4 Assess

Open-Ended Assessment

Modeling Have students draw and label graphs showing each of the three kinds of systems, similar to the graphs in the Concept Summary following Example 5 on p. 112, but using the line given by $2x + 3y = 4$ in each system.

Getting Ready for Lesson 3-2

PREREQUISITE SKILL Lesson 3-2 presents solving systems of equations. The process of solving equations algebraically involves simplifying algebraic expressions. Exercises 74–79 should be used to determine your students' familiarity with simplifying expressions.

Answers

58. 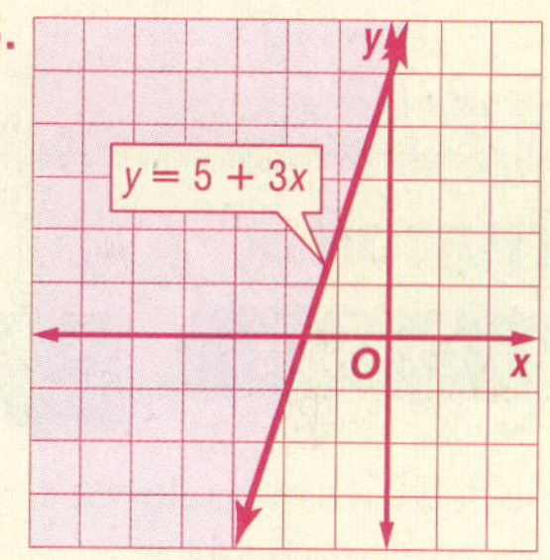

59. 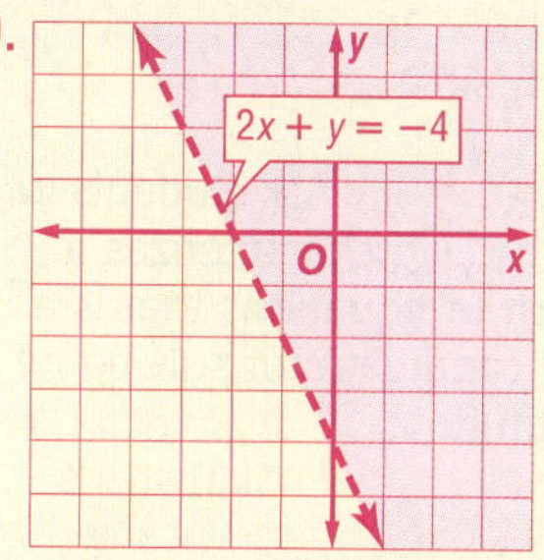

60. 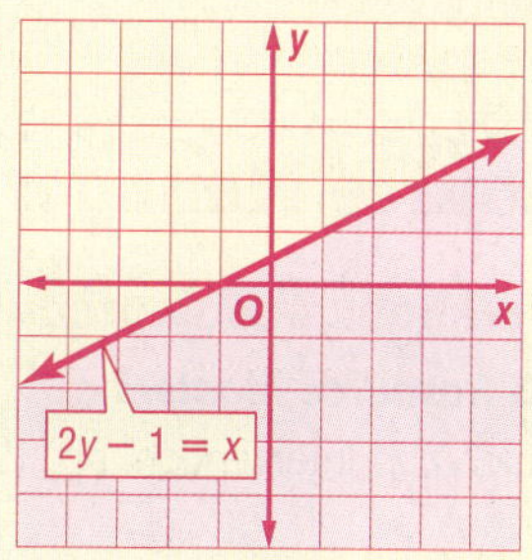

3-2 Lesson Notes

1 Focus

5-Minute Check Transparency 3-2 Use as a quiz or review of Lesson 3-1.

Mathematical Background notes are available for this lesson on p. 108C.

How can systems of equations be used to make consumer decisions?

Ask students:

- What is represented by the point of intersection of the two graphs? **The x-coordinate of the point of intersection is the per minute rate she pays for all her long-distance calls and the y-coordinate is the monthly fee she pays.**

2 Teach

SUBSTITUTION

In-Class Example Power Point®

1 Use substitution to solve the system of equations.

$x + 4y = 26$

$x - 5y = -10$ **(10, 4)**

Teaching Tip Ask students what they believe characterizes a system of equations that is a good candidate for solving by substitution. **Sample answer: when one of the equations is easily solved for one of the variables**

3-2 Solving Systems of Equations Algebraically

What You'll Learn

- Solve systems of linear equations by using substitution.
- Solve systems of linear equations by using elimination.

Vocabulary

- substitution method
- elimination method

How can systems of equations be used to make consumer decisions?

In January, Yolanda's long-distance bill was \$5.50 for 25 minutes of calls. The bill was \$6.54 in February, when Yolanda made 38 minutes of calls. What are the rate per minute and flat fee the company charges?

Let x equal the rate per minute, and let y equal the monthly fee.

January bill: $25x + y = 5.5$
February bill: $38x + y = 6.54$

It is difficult to determine the exact coordinates of the point where the lines intersect from the graph. For systems of equations like this one, it may be easier to solve the system by using algebraic methods.

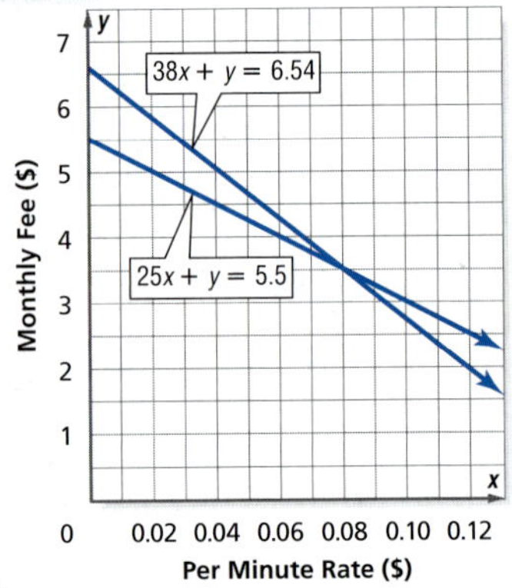

SUBSTITUTION One algebraic method is the **substitution method**. Using this method, one equation is solved for one variable in terms of the other. Then, this expression is substituted for the variable in the other equation.

Study Tip

Coefficient of 1
It is easier to solve for the variable that has a coefficient of 1.

Example 1 Solve by Using Substitution

Use substitution to solve the system of equations.

$\mathbf{x + 2y = 8}$

$\mathbf{\frac{1}{2}x - y = 18}$

Solve the first equation for x in terms of y.

$x + 2y = 8$ First equation

$x = 8 - 2y$ Subtract $2y$ from each side.

Substitute $8 - 2y$ for x in the second equation and solve for y.

$\frac{1}{2}x - y = 18$ Second equation

$\frac{1}{2}(8 - 2y) - y = 18$ Substitute $8 - 2y$ for x.

$4 - y - y = 18$ Distributive Property

$-2y = 14$ Subtract 4 from each side.

$y = -7$ Divide each side by -2.

Now, substitute the value for y in either original equation and solve for x.

$x + 2y = 8$ First equation

$x + 2(-7) = 8$ Replace y with -7.

$x - 14 = 8$ Simplify.

$x = 22$ The solution of the system is $(22, -7)$.

TEACHING TIP

Have the students substitute the value for y in the other original equation. Have them choose the equation that is easier to solve.

Resource Manager

Workbook and Reproducible Masters

Chapter 3 Resource Masters

- Study Guide and Intervention, pp. 125–126
- Skills Practice, p. 127
- Practice, p. 128
- Reading to Learn Mathematics, p. 129
- Enrichment, p. 130
- Assessment, p. 163

School-to-Career Masters, p. 5

Science and Mathematics Lab Manual, pp. 63–66

Teaching Algebra With Manipulatives Masters, p. 222

Transparencies

5-Minute Check Transparency 3-2
Answer Key Transparencies

Technology

Alge2PASS: Tutorial Plus, Lesson 5
Interactive Chalkboard

Standardized Test Practice

Example 2 Solve by Substitution

Extended Response Test Item

> Matthew stopped for gasoline twice on a long car trip. The price of gasoline at the first station where he stopped was $1.46 per gallon. At the second station, the price was $1.55 per gallon. Matthew bought a total of 36.1 gallons of gasoline and spent $54.29. Write and solve a system of equations to find the number of gallons of gasoline Matthew bought at each gas station. Show your work.

Read the Test Item

You are asked to find the number of gallons of gasoline that Matthew bought at each gas station.

Solve the Test Item

Step 1 Define variables and write the system of equations. Let x represent the number of gallons bought at the first station and y represent the number of gallons bought at the second station.

$x + y = 36.1$ *The total number of gallons was 36.1.*

$1.46x + 1.55y = 54.29$ *The total price was $54.29.*

Step 2 Solve one of the equations for one of the variables in terms of the other. Since the coefficient of y is 1, it makes sense to solve the first equation for y in terms of x.

$x + y = 36.1$ *First equation*

$y = 36.1 - x$ *Subtract x from each side.*

Step 3 Substitute $36.1 - x$ for y in the second equation.

$1.46x + 1.55y = 54.29$ *Second equation*

$1.46x + 1.55(36.1 - x) = 54.29$ *Substitute 36.1 − x for y.*

$1.46x + 55.955 - 1.55x = 54.29$ *Distributive Property*

$-0.09x = -1.665$ *Simplify.*

$x = 18.5$ *Divide each side by −0.09.*

Step 4 Now replace x with 18.5 in either equation to find the value of y.

$x + y = 36.1$ *First equation*

$18.5 + y = 36.1$ *Substitute 18.5 for x.*

$y = 17.6$ *Subtract 18.5 from each side.*

Step 5 Check the solution.

$x + y = 36.1$ *Original equation* $1.46x + 1.55y = 54.29$

$18.5 + 17.6 \stackrel{?}{=} 36.1$ *Substitute.* $1.46(18.5) + 1.55(17.6) \stackrel{?}{=} 54.29$

$36.1 = 36.1$ ✓ *Simplify.* $24.01 + 27.28 = 54.29$ ✓

Step 6 State the solution.

Matthew bought 18.5 gallons at the first station and 17.6 gallons at the second station.

Test-Taking Tip

Substituting 18.5 for x in either of the original equations would allow you to find the value of y. But the first equation is simpler and therefore easier to solve.

www.algebra2.com/extra_examples

In-Class Example

Power Point®

2 Lancaster Woodworkers Furniture Store builds two types of wooden outdoor chairs. A rocking chair sells for $265 and an Adirondack chair with footstool sells for $320. The books show that last month, the business earned $13,930 for the 48 outdoor chairs sold. How many of each type of chair were sold? Show your work. **26 rocking chairs, 22 Adirondack chairs**

Standardized Test Practice

Example 2 Recommend that when students find a solution, they check their answer against the facts presented in the original problem.

Interactive Chalkboard

PowerPoint® Presentations

This CD-ROM is a customizable Microsoft® PowerPoint® presentation that includes:

- Step-by-step, dynamic solutions of each In-Class Example from the Teacher Wraparound Edition
- Additional, Your Turn exercises for each example
- The 5-Minute Check Transparencies
- Hot links to Glencoe Online Study Tools

ELIMINATION

In-Class Examples

Power Point®

3 Use the elimination method to solve the system of equations.
$x + 2y = 10$
$x + y = 6$ **(2, 4)**

4 Use the elimination method to solve the system of equations.
$2x + 3y = 12$
$5x - 2y = 11$ **(3, 2)**

Teaching Tip Remind students to take time to plan their solution strategy before they start their calculations. Stress that a little forethought may show that it is easier to eliminate one of the variables rather than the other.

Concept Check

Ask students what they might look for to help them plan their strategy for solving a system of equations by elimination. **Sample answer: Look to see if the coefficients for one of the variables are already the same or if they are opposites. If so, that variable can be eliminated by subtracting or adding the given equations.**

ELIMINATION Another algebraic method is the **elimination method**. Using this method, you eliminate one of the variables by adding or subtracting the equations. When you add two true equations, the result is a new equation that is also true.

Example 3 Solve by Using Elimination

Use the elimination method to solve the system of equations.

$\mathbf{4a + 2b = 15}$
$\mathbf{2a + 2b = 7}$

In each equation, the coefficient of b is 2. If one equation is subtracted from the other, the variable b will be eliminated.

$$\begin{aligned} 4a + 2b &= 15 \\ (-)\ 2a + 2b &= 7 \\ \hline 2a \qquad &= 8 \quad \text{Subtract the equations.} \\ a &= 4 \quad \text{Divide each side by 2.} \end{aligned}$$

Now find b by substituting 4 for a in either original equation.

$$\begin{aligned} 2a + 2b &= 7 && \text{Second equation} \\ 2(4) + 2b &= 7 && \text{Replace } a \text{ with 4.} \\ 8 + 2b &= 7 && \text{Multiply.} \\ 2b &= -1 && \text{Subtract 8 from each side.} \\ b &= -\frac{1}{2} && \text{Divide each side by 2.} \end{aligned}$$

The solution is $\left(4, -\frac{1}{2}\right)$.

Study Tip

Common Misconception
You may find it confusing to subtract equations. It may be helpful to multiply the second equation by −1 and then add the equations.

Sometimes, adding or subtracting the two equations will not eliminate either variable. You may use multiplication to write an equivalent equation so that one of the variables has the same or opposite coefficient in both equations. When you multiply an equation by a nonzero number, the new equation has the same set of solutions.

Example 4 Multiply, Then Use Elimination

Use the elimination method to solve the system of equations.

$\mathbf{3x - 7y = -14}$
$\mathbf{5x + 2y = 45}$

Multiply the first equation by 2 and the second equation by 7. Then add the equations to eliminate the y variable.

$3x - 7y = -14$ Multiply by 2. → $6x - 14y = -28$

$5x + 2y = 45$ Multiply by 7. → $(+)\ 35x + 14y = 315$

$$\begin{aligned} 41x \qquad &= 287 \quad \text{Add the equations.} \\ x &= 7 \quad \text{Divide each side by 41.} \end{aligned}$$

Replace x with 7 and solve for y.

$$\begin{aligned} 3x - 7y &= -14 && \text{First equation} \\ 3(7) - 7y &= -14 && \text{Replace } x \text{ with 7.} \\ 21 - 7y &= -14 && \text{Multiply.} \\ -7y &= -35 && \text{Subtract 21 from each side.} \\ y &= 5 && \text{Divide each side by } -7. \end{aligned}$$

The solution is (7, 5).

Study Tip

Alternative Method
You could also multiply the first equation by 5 and the second equation by 3. Then subtract to eliminate the x variable.

Teacher to Teacher

Vickie McGlohon DH Conley H.S., Greenville, NC

"I have students make up 10 word problems that can be solved using a system of two or three equations. All problems must be related to a theme, such as sports. The problems must be illustrated and bound for presentation."

118 Chapter 3 Systems of Equations and Inequalities

If you add or subtract two equations in a system and the result is an equation that is never true, then the system is inconsistent and it has no solution. If the result when you add or subtract two equations in a system is an equation that is always true, then the system is dependent and it has infinitely many solutions.

Example 5 Inconsistent System

Use the elimination method to solve the system of equations.

$\mathbf{8x + 2y = 17}$
$\mathbf{-4x - y = 9}$

Use multiplication to eliminate x.

$8x + 2y = 17 \qquad\qquad\qquad 8x + 2y = 17$
$-4x - y = 9 \quad$ Multiply by 2. $\quad -8x - 2y = 18$
$\qquad\qquad\qquad\qquad\qquad\qquad 0 = 35 \quad$ Add the equations.

Since there are no values of x and y that will make the equation $0 = 35$ true, there are no solutions for this system of equations.

Check for Understanding

Concept Check

1. **OPEN ENDED** Give an example of a system of equations that is more easily solved by substitution and a system that is more easily solved by elimination.
2. **Make a conjecture** about the solution of a system of equations if the result of subtracting one equation from the other is $0 = 0$.
3. **FIND THE ERROR** Juanita and Vincent are solving the system $2x - y = 6$ and $2x + y = 10$. Who is correct? Explain your reasoning.

Juanita

$2x - y = 6$
$(-)2x + y = 10$
$0 = -4$

The statement $0 = -4$ is never true, so there is no solution.

Vincent

$2x - y = 6 \qquad 2x - y = 6$
$(+)2x + y = 10 \qquad 2(4) - y = 6$
$4x = 16 \qquad 8 - y = 6$
$x = 4 \qquad y = 2$

The solution is $(4, 2)$.

1. See students' work; one equation should have a variable with a coefficient of 1.
2. There are infinitely many solutions.
3. Vincent; Juanita subtracted the two equations incorrectly; $-y - y = -2y$, not 0.

Guided Practice

GUIDED PRACTICE KEY	
Exercises	Examples
4, 5	1, 2
6	3
7	4
8–11	1–5
12	2

Solve each system of equations by using substitution.

4. $y = 3x - 4$; $y = 4 + x$ **(4, 8)**
5. $4c + 2d = 10$; $c + 3d = 10$ **(1, 3)**

Solve each system of equations by using elimination.

6. $2r - 3s = 11$; $2r + 2s = 6$ **(4, −1)**
7. $2p + 4q = 18$; $3p - 6q = 3$ **(5, 2)**

Solve each system of equations by using either substitution or elimination.

8. $a - b = 2$; $-2a + 3b = 3$ **(9, 7)**
9. $5m + n = 10$; $4m + n = 4$ **(6, −20)**
10. $3g - 2h = -1$; $8h = 5 + 12g$ **no solution**
11. $\frac{1}{4}x + y = \frac{7}{2}$; $x - \frac{1}{2}y = 2$ $\left(3\frac{1}{3}, 2\frac{2}{3}\right)$

In-Class Example Power Point®

5 Use the elimination method to solve the system of equations.
$-3x + 5y = 12$
$6x - 10y = -21$
There are no solutions for this system of equations.

3 Practice/Apply

Study Notebook

Have students—
- add the definitions/examples of the vocabulary terms to their Vocabulary Builder worksheets for Chapter 3.
- add the tip for Example 2 to their list of test-taking tips that they can use to review before taking a standardized test.
- include any other item(s) that they find helpful in mastering the skills in this lesson.

DAILY INTERVENTION **FIND THE ERROR** Stress the use of the Distributive Property when subtracting the left sides of the equations using Juanita's method. Have students rewrite this subtraction as $2x - y - (2x + y)$ and then apply the property to obtain the expression $2x - y - 2x - y$, which equals $-2y$ and not 0.

DAILY INTERVENTION **Differentiated Instruction**

Logical Have students summarize the various algebraic methods for solving a system of equations using if-then statements and examples. Sample: "If one of the equations has a variable with a coefficient of 1 (such as $x + 3y = 9$ or $5x - y = 13$), consider the substitution method."

About the Exercises...

Organization by Objective
- **Substitution:** 13–18, 25–36
- **Elimination:** 19–36

Odd/Even Assignments
Exercises 13–38 are structured so that students practice the same concepts whether they are assigned odd or even problems.

Assignment Guide

Basic: 13–33 odd, 37, 39–41, 43, 44, 48–70

Average: 13–37 odd, 41, 43–70

Advanced: 14–38 even, 42, 43–66 (optional: 67–70)

All: Practice Quiz 1 (1–5)

Standardized Test Practice

12. A wholesaler is offering two different packages of custom printed T-shirts and sweatshirts to groups for fund-raisers. One package contains 10 dozen T-shirts and 14 dozen sweatshirts for $1544. The other package contains 18 dozen T-shirts and 8 dozen sweatshirts for $1472. Write and solve a system of linear equations to find the cost, in dollars, of a dozen T-shirts. **C**

   (A) $76
   (B) $58
   (C) $48
   (D) $36

★ indicates increased difficulty

Practice and Apply

Homework Help

For Exercises	See Examples
13–18	1, 2
19–24	3, 4
25–49	1–5

Extra Practice
See page 832.

Solve each system of equations by using substitution.

13. $2j - 3k = 3$, $j + k = 14$ **(9, 5)**
14. $2r + s = 11$, $6r - 2s = -2$ **(2, 7)**
15. $5a - b = 17$, $3a + 2b = 5$ **(3, −2)**
16. $-w - z = -2$, $4w + 5z = 16$ **(−6, 8)**
17. $6c + 3d = 12$, $2c = 8 - d$ **no solution**
18. $2x + 4y = 6$, $7x = 4 + 3y$ **(1, 1)**

Solve each system of equations by using elimination.

19. $u + v = 7$, $2u + v = 11$ **(4, 3)**
20. $m - n = -9$, $7m + 2n = 9$ **(−1, 8)**
21. $3p - 5q = 6$, $2p - 4q = 4$ **(2, 0)**
22. $4x - 5y = 17$, $3x + 4y = 5$ **(3, −1)**
23. $2c + 6d = 14$, $\frac{1}{2}c - 3d = 8$ **(10, −1)**
24. $3s + 2t = -3$, $s + \frac{1}{3}t = -4$ **(−7, 9)**

Solve each system of equations by using either substitution or elimination.

25. $r + 4s = -8$, $3r + 2s = 6$ **(4, −3)**
26. $10m - 9n = 15$, $5m - 4n = 10$ **(6, 5)**
27. $3c - 7d = -3$, $2c + 6d = -34$ **(−8, −3)**
28. $6g - 8h = 50$, $4g + 6h = 22$ **(7, −1)**
29. $2p = 7 + q$, $6p - 3q = 24$ **no solution**
30. $3x = -31 + 2y$, $5x + 6y = 23$ **(−5, 8)**
31. $3u + 5v = 6$, $2u - 4v = -7$ **$\left(-\frac{1}{2}, \frac{3}{2}\right)$**
32. $3a - 2b = -3$, $3a + b = 3$ **$\left(\frac{1}{3}, 2\right)$**
33. $s + 3t = 27$, $\frac{1}{2}s + 2t = 19$ **(−6, 11)**
★34. $f = 6 - 2g$, $\frac{1}{6}f + \frac{1}{3}g = 1$ **infinitely many**
★35. $0.25x + 1.75y = 1.25$, $0.5x + 2.5y = 2$ **(1.5, 0.5)**
★36. $0.4m + 1.8n = 8$, $1.2m + 3.4n = 16$ **(2, 4)**

WebQuest
A system of equations can be used to compare home loan options. Visit www.algebra2.com/webquest to continue work on your WebQuest project.

37. Three times one number added to five times another number is 54. The second number is two less than the first. Find the numbers. **8, 6**

★38. The average of two numbers is 7. Find the numbers if three times one of the numbers is one half the other number. **2, 12**

SKIING **For Exercises 39 and 40, use the following information.**
All 28 members in Crestview High School's Ski Club went on a one-day ski trip. Members can rent skis for $16.00 per day or snowboards for $19.00 per day. The club paid a total of $478 for rental equipment.

39. Write a system of equations that represents the number of members who rented the two types of equipment. **$x + y = 28$, $16x + 19y = 478$**

40. How many members rented skis and how many rented snowboards? **18 members rented skis and 10 members rented snowboards.**

Answer

51. You can use a system of equations to find the monthly fee and rate per minute charged during the months of January and February. Answers should include the following.
    - The coordinates of the point of intersection are (0.08, 3.5).
    - Currently, Yolanda is paying a monthly fee of $3.50 and an additional 8¢ per minute. If she graphs $y = 0.08x + 3.5$ (to represent what she is paying currently) and $y = 0.10x + 3$ (to represent the other long-distance plan) and finds the intersection, she can identify which plan would be better for a person with her level of usage.

41. **HOUSING** Campus Rentals rents 2- and 3-bedroom apartments for \$700 and \$900 per month, respectively. Last month they had six vacant apartments and reported \$4600 in lost rent. How many of each type of apartment were vacant? **4 2-bedroom, 2 3-bedroom**

42. **GEOMETRY** Find the coordinates of the vertices of the parallelogram whose sides are contained in the lines whose equations are $2x + y = -12$, $2x - y = -8$, $2x - y - 4 = 0$, and $4x + 2y = 24$. **(−5, −2), (4, 4), (−2, −8), (1, 10)**

INVENTORY **For Exercises 43 and 44, use the following information.**
Heung-Soo is responsible for checking a shipment of technology equipment that contains laser printers that cost \$700 each and color monitors that cost \$200 each. He counts 30 boxes on the loading dock. The invoice states that the order totals \$15,000. **43.** $x + y = 30$, $700x + 200y = 15{,}000$

43. Write a system of two equations that represents the number of each item.

44. How many laser printers and how many color monitors were delivered? **18 printers, 12 monitors**

TEACHING **For Exercises 45–47, use the following information.**
Mr. Talbot is writing a test for his science classes. The test will have true/false questions worth 2 points each and multiple-choice questions worth 4 points each for a total of 100 points. He wants to have twice as many multiple-choice questions as true/false.

45. Write a system of equations that represents the number of each type of question. $2x + 4y = 100$, $y = 2x$

46. How many true/false questions and multiple-choice questions will be on the test? **10 true/false, 20 multiple-choice**

47. If most of his students can answer true/false questions within 1 minute and multiple-choice questions within $1\frac{1}{2}$ minutes, will they have enough time to finish the test in 45 minutes? **Yes; they should finish the test within 40 minutes.**

EXERCISE **For Exercises 48 and 49, use the following information.**
Megan exercises every morning for 40 minutes. She does a combination of step aerobics, which burns about 11 Calories per minute, and stretching, which burns about 4 Calories per minute. Her goal is to burn 335 Calories during her routine.

48. Write a system of equations that represents Megan's morning workout.

49. How long should she participate in each activity in order to burn 335 Calories? **25 min of step aerobics, 15 min of stretching**

50. **CRITICAL THINKING** Solve the system of equations. **(4, 6)**

$$\frac{1}{x} + \frac{3}{y} = \frac{3}{4}$$
$$\frac{3}{x} - \frac{2}{y} = \frac{5}{12}$$

$\left(\textit{Hint}: \text{Let } m = \frac{1}{x} \text{ and } n = \frac{1}{y}.\right)$

51. **WRITING IN MATH** Answer the question that was posed at the beginning of the lesson. **See margin.**

How can a system of equations be used to make consumer decisions?

Include the following in your answer:
- a solution of the system of equations in the application at the beginning of the lesson, and
- an explanation of how Yolanda can use a graph to decide whether she should change to a long-distance plan that charges \$0.10 per minute and a flat fee of \$3.00 per month.

Career Choices

Teacher

Besides the time they spend in a classroom, teachers spend additional time preparing lessons, grading papers, and assessing students' progress.

Online Research For information about a career as a teacher, visit: www.algebra2.com/careers

48. $a + s = 40$, $11a + 4s = 335$

www.algebra2.com/self_check_quiz

Study Guide and Intervention, p. 125 (shown) and p. 126

Substitution To solve a system of linear equations by **substitution**, first solve for one variable in terms of the other in one of the equations. Then substitute this expression into the other equation and simplify.

Example **Use substitution to solve the system of equations.** $2x - y = 9$, $x + 3y = -6$

Solve the first equation for y in terms of x.

$2x - y = 9$ First equation
$-y = -2x + 9$ Subtract $2x$ from both sides.
$y = 2x - 9$ Multiply both sides by -1.

Substitute the expression $2x - 9$ for y into the second equation and solve for x.

$x + 3y = -6$ Second equation
$x + 3(2x - 9) = -6$ Substitute $2x - 9$ for y.
$x + 6x - 27 = -6$ Distributive Property
$7x - 27 = -6$ Simplify.
$7x = 21$ Add 27 to each side.
$x = 3$ Divide each side by 7.

Now, substitute the value 3 for x in either original equation and solve for y.

$2x - y = 9$ First equation
$2(3) - y = 9$ Replace x with 3.
$6 - y = 9$ Simplify.
$-y = 3$ Subtract 6 from each side.
$y = -3$ Multiply each side by -1.

The solution of the system is $(3, -3)$.

Exercises

Solve each system of linear equations by using substitution.

1. $3x + y = 7$, $4x + 2y = 16$ **(−1, 10)**
2. $2x + y = 5$, $3x - 3y = 3$ **(2, 1)**
3. $2x + 3y = -3$, $x + 2y = 2$ **(−12, 7)**
4. $2x - y = 7$, $6x - 3y = 14$ **no solution**
5. $4x - 3y = 4$, $2x + y = -8$ **(−2, −4)**
6. $5x + y = 6$, $3 - x = 0$ **(3, −9)**
7. $x + 8y = -2$, $x - 3y = 20$ **(14, −2)**
8. $2x - y = -4$, $4x + y = 1$ $\left(-\frac{1}{2}, 3\right)$
9. $x - y = -2$, $2x - 3y = 2$ **(−8, −6)**
10. $x - 4y = 4$, $2x + 12y = 13$ $\left(5, \frac{1}{4}\right)$
11. $x + 3y = 2$, $4x + 12y = 8$ **infinitely many**
12. $2x + 2y = 4$, $x - 2y = 0$ $\left(\frac{4}{3}, \frac{2}{3}\right)$

Skills Practice, p. 127 and Practice, p. 128 (shown)

Solve each system of equations by using substitution.

1. $2x + y = 4$, $3x + 2y = 1$ **(7, −10)**
2. $x - 3y = 9$, $x + 2y = -1$ **(3, −2)**
3. $g + 3h = 8$, $\frac{1}{3}g + h = 9$ **no solution**
4. $2a - 4b = 6$, $-a + 2b = -3$ **infinitely many**
5. $2m + n = 6$, $5m + 6n = 1$ **(5, −4)**
6. $4x - 3y = -6$, $-x - 2y = 7$ **(−3, −2)**
7. $u - 2v = \frac{1}{2}$, $-u + 2v = 5$ **no solution**
8. $x - 3y = 16$, $4x - y = 9$ **(1, −5)**
9. $w + 3z = 1$, $3w - 5z = -4$ $\left(-\frac{1}{2}, \frac{1}{2}\right)$

Solve each system of equations by using elimination.

10. $2r + s = 5$, $3r - s = 20$ **(5, −5)**
11. $2m - n = -1$, $3m + 2n = 30$ **(4, 9)**
12. $6x + 3y = 6$, $8x + 5y = 12$ **(−1, 4)**
13. $3j - k = 10$, $4j - k = 16$ **(6, 8)**
14. $2x - y = -4$, $-4x + 2y = 6$ **no solution**
15. $2g + h = 6$, $3g - 2h = 16$ **(4, −2)**
16. $2t + 4v = 6$, $-t - 2v = -3$ **infinitely many**
17. $3x - 2y = 12$, $2x + \frac{2}{3}y = 14$ **(6, 3)**
18. $\frac{1}{2}x + 3y = 11$, $8x - 5y = 17$ **(4, 3)**

Solve each system of equations by using either substitution or elimination.

19. $8x + 3y = -5$, $10x + 6y = -13$ $\left(\frac{1}{2}, -3\right)$
20. $8q - 15r = -40$, $4q + 2r = 56$ **(10, 8)**
21. $3x - 4y = 12$, $\frac{1}{3}x - \frac{4}{9}y = \frac{4}{3}$ **infinitely many**
22. $4b - 2d = 5$, $-2b + d = 1$ **no solution**
23. $s + 3y = 4$, $s = 1$ **(1, 1)**
24. $4m - 2p = 0$, $-3m + 9p = 5$ $\left(\frac{1}{3}, \frac{2}{3}\right)$
25. $5g + 4k = 10$, $-3g - 5k = 7$ **(6, −5)**
26. $0.5x + 2y = 5$, $x - 2y = -8$ **(−2, 3)**
27. $h - z = 3$, $-3h + 3z = 6$ **no solution**

SPORTS **For Exercises 28 and 29, use the following information.**
Last year the volleyball team paid \$5 per pair for socks and \$17 per pair for shorts on a total purchase of \$315. This year they spent \$342 to buy the same number of pairs of socks and shorts because the socks now cost \$6 a pair and the shorts cost \$18.

28. Write a system of two equations that represents the number of pairs of socks and shorts bought each year. $5x + 17y = 315$, $6x + 18y = 342$

29. How many pairs of socks and shorts did the team buy each year? **socks: 12, shorts: 15**

Reading to Learn Mathematics, p. 129 ELL

Pre-Activity **How can systems of equations be used to make consumer decisions?**

Read the introduction to Lesson 3-2 at the top of page 116 in your textbook.

- How many more minutes of long distance time did Yolanda use in February than in January? **13 minutes**
- How much more were the February charges than the January charges? **\$1.04**
- Using your answers for the questions above, how can you find the rate per minute? **Find \$1.04 ÷ 13.**

Reading the Lesson

1. Suppose that you are asked to solve the system of equations at the right by the substitution method. $4x - 5y = 7$, $3x + y = -9$

   The first step is to solve one of the equations for one variable in terms of the other. To make your work as easy as possible, which equation would you solve for which variable? Explain.
   Sample answer: Solve the second equation for *y* because in that equation the variable *y* has a coefficient of 1.

2. Suppose that you are asked to solve the system of equations at the right by the elimination method. $2x + 3y = -2$, $7x - y = 39$

   To make your work as easy as possible, which variable would you eliminate? Describe how you would do this.
   Sample answer: Eliminate the variable *y*; multiply the second equation by 3 and then add the result to the first equation.

Helping You Remember

3. The substitution method and elimination method for solving systems both have several steps, and it may be difficult to remember them. You may be able to remember them more easily if you notice what the methods have in common. What step is the same in both methods?
   Sample answer: After finding the value of one of the variables, you find the value of the other variable by substituting the value you have found in one of the original equations.

Enrichment, p. 130

Using Coordinates

From one observation point, the line of sight to a downed plane is given by $y = x - 1$. This equation describes the distance from the observation point to the plane in a straight line. From another observation point, the line of sight is given by $x + 3y = 21$. What are the coordinates of the point at which the crash occurred?

Solve the system of equations $\begin{cases} y = x - 1 \\ x + 3y = 21 \end{cases}$.

$x + 3y = 21$
$x + 3(x - 1) = 21$ Substitute $x - 1$ for y.
$x + 3x - 3 = 21$
$4x = 24$
$x = 6$

$x + 3y = 21$
$6 + 3y = 21$ Substitute 6 for x.
$3y = 15$

4 Assess

Open-Ended Assessment

Speaking Have students look at a specific system of equations and explain which method they would use, how they decided on that method, and what steps that method will involve when they are actually solving the system.

Getting Ready for Lesson 3-3

PREREQUISITE SKILL Lesson 3-3 presents the solving of systems of inequalities by graphing. Determining whether or not an ordered pair satisfies a system of inequalities in two variables is useful when checking the graphing of a system of inequalities. Exercises 67–70 should be used to determine your students' familiarity with inequalities.

Assessment Options

Practice Quiz 1 The quiz provides students with a brief review of the concepts and skills in Lessons 3-1 and 3-2. Lesson numbers are given to the right of exercises or instruction lines so students can review concepts not yet mastered.

Quiz (Lessons 3-1 and 3-2) is available on p. 163 of the *Chapter 3 Resource Masters*.

Answers (Practice Quiz 1)

1.

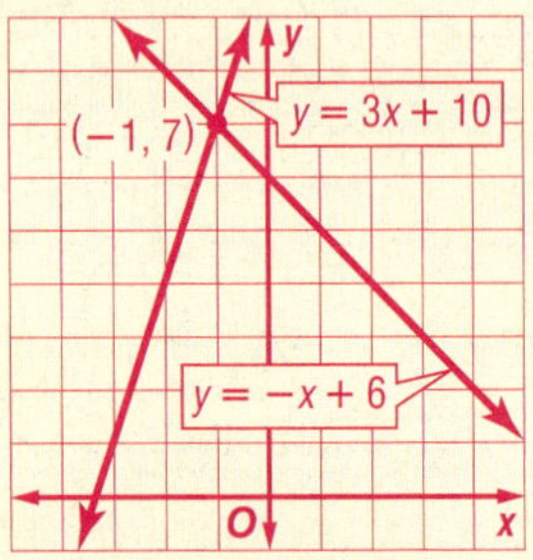

2.

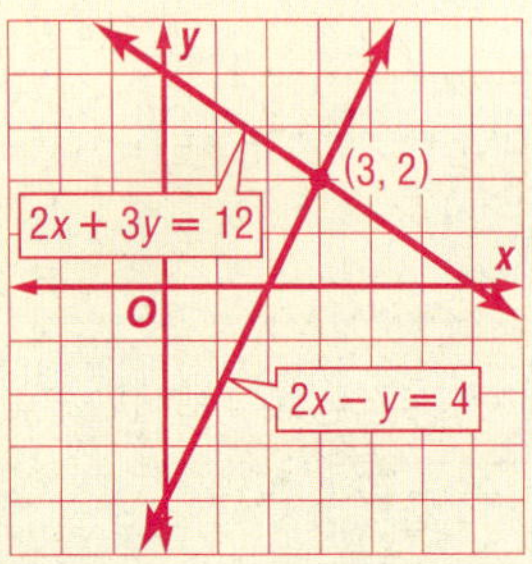

Standardized Test Practice

52. If $x = y + z$ and $x + y = 6$ and $x = 10$, then $z =$ **C**
Ⓐ 4. Ⓑ 8. Ⓒ 14. Ⓓ 16.

53. If the perimeter of the square shown at the right is 48 units, find the value of x. **A**
Ⓐ 3 Ⓑ 4
Ⓒ 6 Ⓓ 8

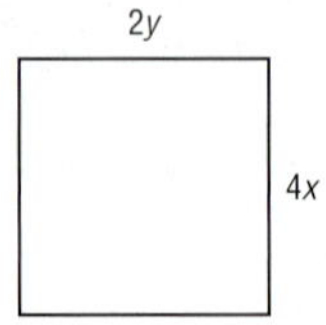

Maintain Your Skills

Mixed Review

Graph each system of equations and describe it as *consistent and independent*, *consistent and dependent*, or *inconsistent*. *(Lesson 3-1)*

54–56. See margin for graphs.

54. $y = x + 2$
$y = x - 1$ **incon.**

55. $4y - 2x = 4$
$y - \frac{1}{2}x = 1$ **cons. and dep.**

56. $3x + y = 1$
$y = 2x - 4$ **cons. and ind.**

Graph each inequality. *(Lesson 2-7)* **57–59. See pp. 151A–151F.**

57. $x + y \leq 3$

58. $5y - 4x < -20$

59. $3x + 9y \geq -15$

Write each equation in standard form. Identify A, B, and C. *(Lesson 2-2)*

60. $y = 7x + 4$

61. $x = y$

62. $3x = 2 - 5y$

63. $6x = 3y - 9$

64. $y = \frac{1}{2}x - 3$

65. $\frac{2}{3}y - 6 = 1 - x$

60. $7x - y = -4$; 7, −1, −4
61. $x - y = 0$; 1, −1, 0
62. $3x + 5y = 2$; 3, 5, 2
63. $2x - y = -3$; 2, −1, −3
64. $x - 2y = 6$; 1, −2, 6
65. $3x + 2y = 21$; 3, 2, 21

66. **ELECTRICITY** Use the formula $I = \frac{E}{R + r}$ to find the amount of current I (in amperes) produced if the electromotive force E is 1.5 volts, the circuit resistance R is 2.35 ohms, and the resistance r within a battery is 0.15 ohms. *(Lesson 1-1)* **0.6 ampere**

Getting Ready for the Next Lesson

PREREQUISITE SKILL Determine whether the given point satisfies each inequality. *(To review **inequalities**, see Lesson 2-7.)*

67. $3x + 2y \leq 10$; (2, −1) **yes**

68. $4x - 2y > 6$; (3, 3) **no**

69. $7x + 4y \geq -15$; (−4, 2) **no**

70. $7y + 6x < 50$; (−5, 5) **yes**

Practice Quiz 1 — Lessons 3-1 and 3-2

Solve each system of equations by graphing. *(Lesson 3-1)* **1–2. See margin for graphs.**

1. $y = 3x + 10$
$y = -x + 6$ **(−1, 7)**

2. $2x + 3y = 12$
$2x - y = 4$ **(3, 2)**

Solve each system of equations by using either substitution or elimination. *(Lesson 3-2)*

3. $y = x + 5$
$x + y = 9$ **(2, 7)**

4. $2x + 6y = 2$
$3x + 2y = 10$ **(4, −1)**

5. **AIRPORTS** According to the Airports Council International, the busiest airport in the world is Atlanta's Hartsfield International Airport, and the second busiest is Chicago's O'Hare Airport. Together they handled 150.5 million passengers in the first six months of 1999. If Hartsfield handled 5.5 million more passengers than O'Hare, how many were handled by each airport? *(Lesson 3-2)* **Hartsfield, 78 million; O'Hare, 72.5 million**

Answers

54.

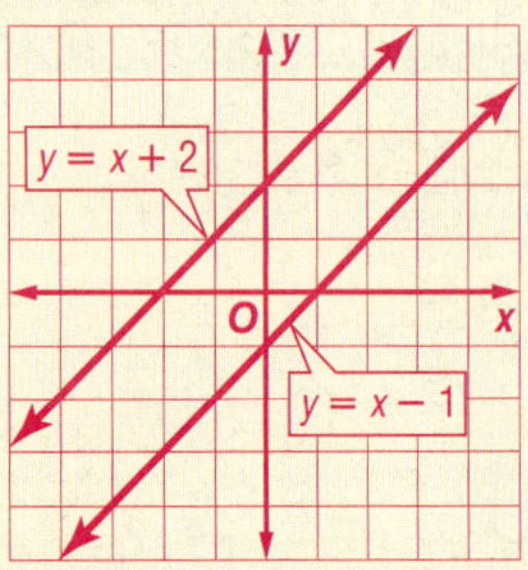

55.

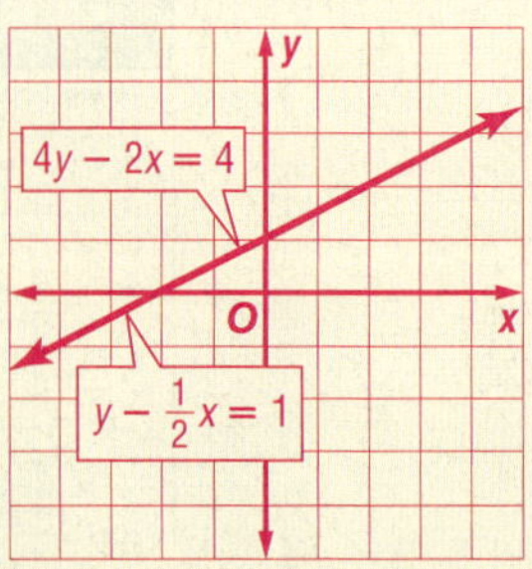

56.

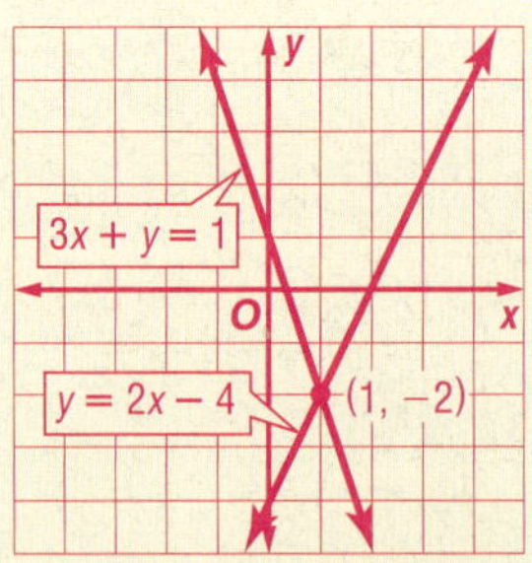

122 Chapter 3 Systems of Equations and Inequalities

3-3 Solving Systems of Inequalities by Graphing

What You'll Learn

- Solve systems of inequalities by graphing.
- Determine the coordinates of the vertices of a region formed by the graph of a system of inequalities.

Vocabulary

- system of inequalities

How can you determine whether your blood pressure is in a normal range?

During one heartbeat, blood pressure reaches a maximum pressure (systolic) and a minimum pressure (diastolic), which are measured in millimeters of mercury (mm Hg). Blood pressure is expressed as the maximum pressure over the minimum pressure—for example, 120/80. Normal blood pressure for people under 40 ranges from 100 to 140 mm Hg for the maximum and from 60 to 90 mm Hg for the minimum. This information can be represented by a system of inequalities.

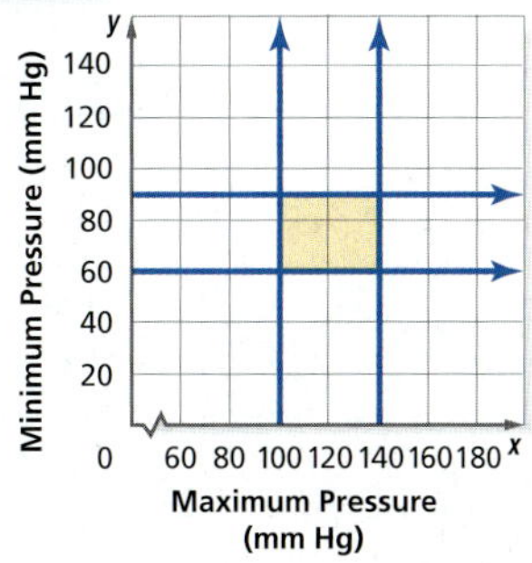

GRAPH SYSTEMS OF INEQUALITIES To solve a **system of inequalities**, we need to find the ordered pairs that satisfy all of the inequalities in the system. One way to solve a system of inequalities is to graph the inequalities on the same coordinate plane. The solution set is represented by the intersection of the graph.

Example 1 Intersecting Regions

Solve each system of inequalities by graphing.

a. $y > -2x + 4$
$y \leq x - 2$

solution of $y > -2x + 4 \rightarrow$ Regions 1 and 2
solution of $y \leq x - 2 \quad \rightarrow$ Regions 2 and 3

The intersection of these regions is Region 2, which is the solution of the system of inequalities. Notice that the solution is a region containing an infinite number of ordered pairs.

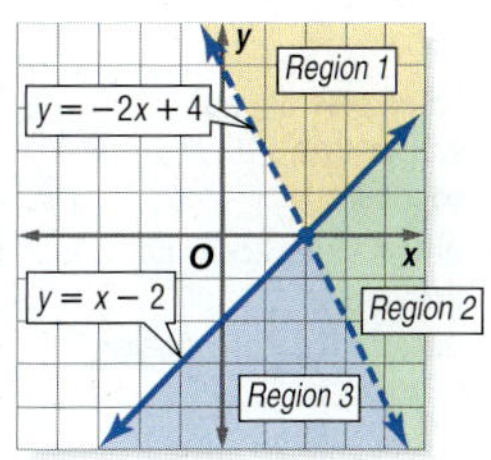

The green area represents where the yellow area of one graph overlaps the blue area of the other.

b. $y > x + 1$
$|y| \leq 3$

The inequality $|y| \leq 3$ can be written as $y \leq 3$ and $y \geq -3$.

Graph all of the inequalities on the same coordinate plane and shade the region or regions that are common to all.

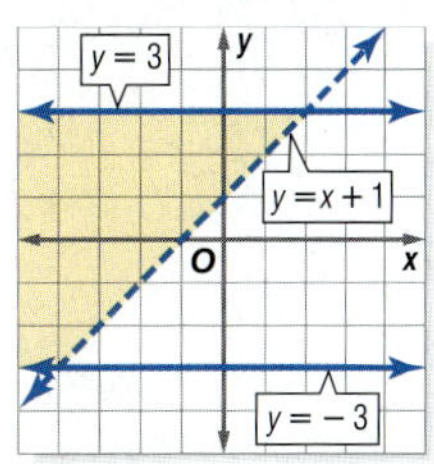

Study Tip

Look Back
To review **graphing inequalities**, see Lesson 2-7.

3-3 Lesson Notes

1 Focus

5-Minute Check Transparency 3-3 Use as a quiz or review of Lesson 3-2.

Mathematical Background notes are available for this lesson on p. 108D.

How can you determine whether your blood pressure is in a normal range?

Ask students:

- What do the letters Hg represent? **Hg is the symbol for mercury.**
- Do you think the blood pressure ranges for people over 40 are higher or lower than those for people under 40? **higher**

2 Teach

GRAPH SYSTEMS OF INEQUALITIES

In-Class Example Power Point®

1 Solve each system of inequalities by graphing.

a. $y \geq 2x - 3$
$y < -x + 2$

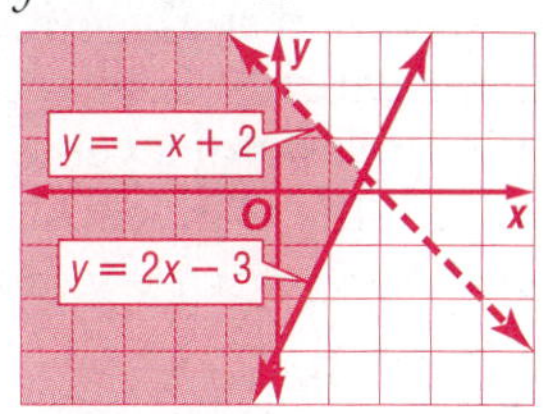

(continued on the next page)

Resource Manager

Workbook and Reproducible Masters

Chapter 3 Resource Masters
- Study Guide and Intervention, pp. 131–132
- Skills Practice, p. 133
- Practice, p. 134
- Reading to Learn Mathematics, p. 135
- Enrichment, p. 136
- Assessment, pp. 163, 165

School-to-Career Masters, p. 6
Teaching Algebra With Manipulatives Masters, pp. 223–224

Transparencies
5-Minute Check Transparency 3-3
Answer Key Transparencies

Technology
Interactive Chalkboard

In-Class Examples

Power Point®

1 *(continued)*

b. $y \le -x + 1$
$|x + 1| < 3$

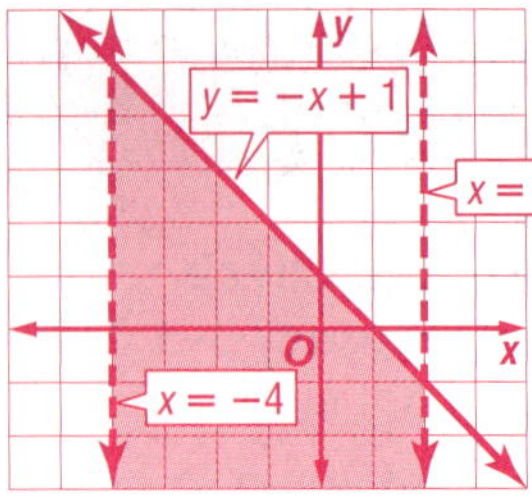

2 Solve the system of inequalities by graphing.

$y \ge -\frac{3}{4}x + 1$

$y \le -\frac{3}{4}x - 2$ ∅

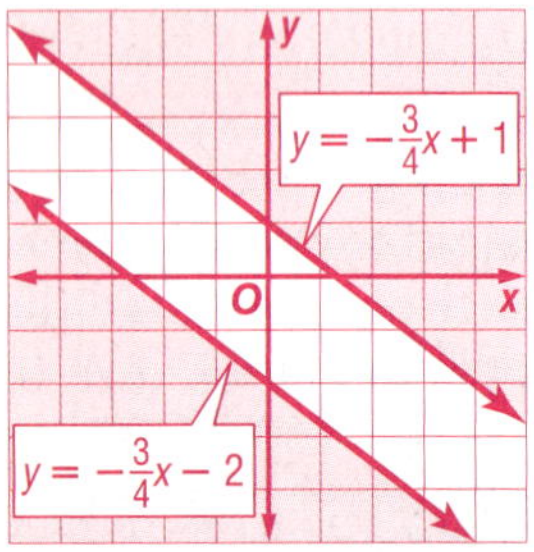

3 **MEDICINE** Medical professionals recommend that patients have a cholesterol level below 200 milligrams per deciliter (mg/dL) of blood and a triglyceride level below 150 mg/dL. Write and graph a system of inequalities that represents the range of cholesterol levels and triglyceride levels for patients. Let *c* represent the cholesterol levels and *t* represent the triglyceride levels. **Source:** American Heart Association

$0 \le c < 200$, $0 \le t < 150$

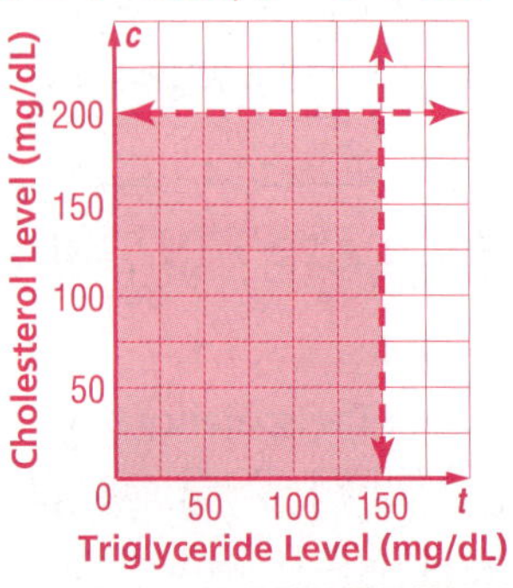

Study Tip

Reading Math
The empty set is also called the *null set*. It can be represented as ∅ or { }.

It is possible that two regions do *not* intersect. In such cases, we say the solution is the empty set ∅ and no solution exists.

Example 2 Separate Regions

Solve the system of inequalities by graphing.

$y > \frac{1}{2}x + 1$

$y < \frac{1}{2}x - 3$

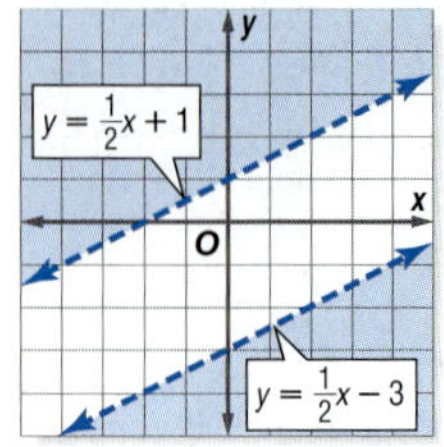

Graph both inequalities. The graphs do not overlap, so the solutions have no points in common. The solution set is ∅.

Example 3 Write and Use a System of Inequalities

SPACE EXPLORATION **In 1959, NASA wanted astronauts who were at least 5 feet 4 inches, but no more than 5 feet 11 inches tall, because of limited space inside the Mercury capsule. They were to be between 21 and 40 years of age. Write and graph a system of inequalities that represents the range of heights and ages for qualifying astronauts.**

Let *h* represent the height of an astronaut in inches. The acceptable heights are at least 5 feet 4 inches (or 64 inches) and no more than 5 feet 11 inches (or 71 inches). We can write two inequalities.

$64 \le h$ and $h \le 71$

Let *a* represent the age of an astronaut. The acceptable ages can also be written as two inequalities.

$a > 21$ and $a < 40$

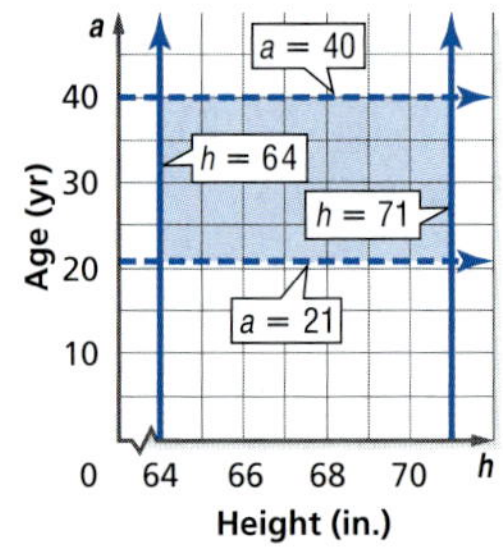

Graph all of the inequalities. Any ordered pair in the intersection of the graphs is a solution of the system.

More About. . .

Space Exploration

Today the basic physical qualifications for an astronaut are:

- blood pressure no greater than 140 over 90,
- distant visual acuity no greater than 20/100 uncorrected, correctable to 20/20, and
- height from 60 inches to 76 inches.

Source: NASA

FIND VERTICES OF A POLYGONAL REGION Sometimes, the graph of a system of inequalities forms a polygonal region. You can find the vertices of the region by determining the coordinates of the points at which the boundary lines intersect.

Example 4 Find Vertices

Find the coordinates of the vertices of the figure formed by $x + y \ge -1$, $x - y \le 6$, and $12y + x \le 32$.

Graph each inequality. The intersection of the graphs forms a triangle.

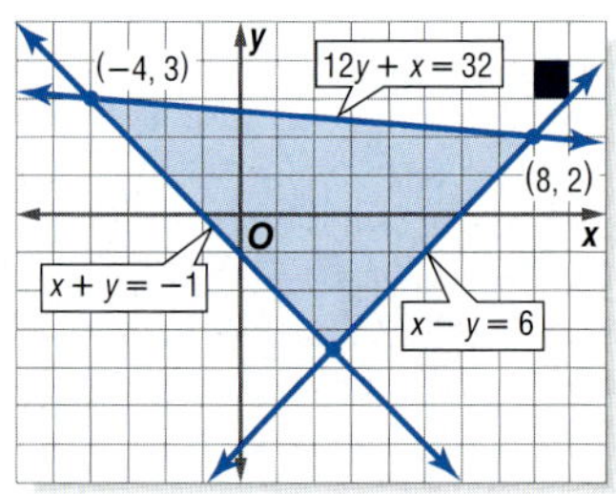

The coordinates (−4, 3) and (8, 2) can be determined from the graph. To find the coordinates of the third vertex, solve the system of equations $x + y = -1$ and $x - y = 6$.

124 Chapter 3 Systems of Equations and Inequalities

Add the equations to eliminate y.

$$\begin{array}{rrl} & x + y = & -1 \\ (+) & x - y = & \ \ 6 \\ \hline & 2x \quad = & \ \ 5 \end{array} \quad \text{Add the equations.}$$

$$x = \frac{5}{2} \quad \text{Divide each side by 2.}$$

Now find y by substituting $\frac{5}{2}$ for x in the first equation.

$x + y = -1$ First equation

$\frac{5}{2} + y = -1$ Replace x with $\frac{5}{2}$.

$y = -\frac{7}{2}$ Subtract $\frac{5}{2}$ from each side.

The vertices of the triangle are at $(-4, 3)$, $(8, 2)$, and $\left(\frac{5}{2}, -\frac{7}{2}\right)$.

Check for Understanding

Concept Check

1. Sample answer: $y > x + 3$, $y < x - 2$

1. **OPEN ENDED** Write a system of inequalities that has no solution.
2. **Tell** whether the following statement is *true* or *false*. If false, give a counterexample. *A system of two linear inequalities has either no points or infinitely many points in its solution.* **true**
3. **State** which region is the solution of the following systems of inequalities.

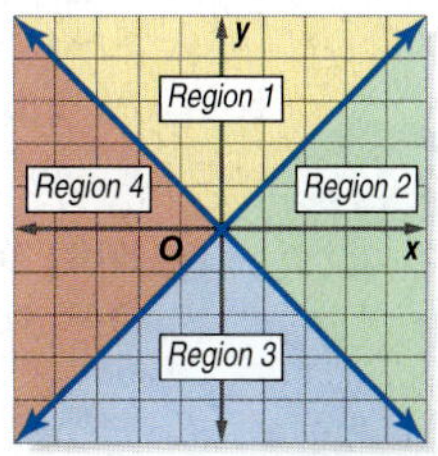

   a. $y \geq x$, $y \leq -x$ **4**
   b. $y \leq x$, $y \geq -x$ **2**
   c. $y \geq x$, $y \geq -x$ **1**
   d. $y \leq x$, $y \leq -x$ **3**

Guided Practice

GUIDED PRACTICE KEY	
Exercises	Examples
4–7	1, 2
8, 9	4
10, 11	3

Solve each system of inequalities by graphing. **4–7. See pp. 151A–151F.**

4. $x \leq 4$, $y > 2$
5. $y \geq x - 2$, $y \leq -2x + 4$
6. $|x - 1| \leq 2$, $x + y > 2$
7. $x \leq 1$, $y < 2x + 1$, $x + 2y \geq -3$

Find the coordinates of the vertices of the figure formed by each system of inequalities.

8. $y \leq x$, $y \geq -3$, $3y + 5x \leq 16$ **(−3, −3), (2, 2), (5, −3)**
9. $y \geq x - 3$, $y \leq x + 7$, $x + y \leq 11$, $x + y \geq -1$ **(−4, 3), (1, −2), (2, 9), (7, 4)**

Application

11. Sample answer: 3 pkgs. of bagels, 4 pkgs. of muffins; 4 pkgs. of bagels, 4 pkgs. of muffins; 3 pkgs. of bagels, 5 pkgs. of muffins

SHOPPING **For Exercises 10 and 11, use the following information.**
Willis has been sent to the grocery store to purchase bagels and muffins for the members of the track team. He can spend at most \$28. A package of bagels costs \$2.50 and contains 6 bagels. A package of muffins costs \$3.50 and contains 8 muffins. He needs to buy at least 12 bagels and 24 muffins.

10. Graph the region that shows how many packages of each item he can purchase. **See pp. 151A–151F.**
11. Give an example of three different purchases he can make.

FIND VERTICES OF A POLYGONAL REGION

In-Class Example Power Point®

4 Find the coordinates of the vertices of the figure formed by $2x - y \geq -1$, $x + y \leq 4$, and $x + 4y \geq 4$.

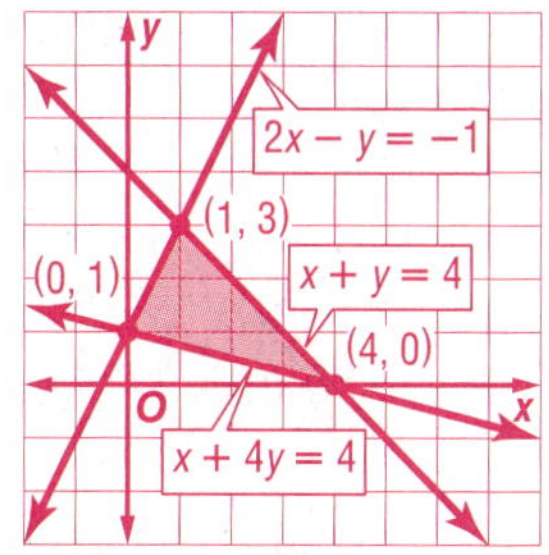

3 Practice/Apply

Study Notebook

Have students—

- add the definitions/examples of the vocabulary terms to their Vocabulary Builder worksheets for Chapter 3.
- include any other item(s) that they find helpful in mastering the skills in this lesson.

About the Exercises...

Organization by Objective

- **Graph Systems of Inequalities:** 12–23, 32–37
- **Find Vertices of a Polygonal Region:** 24–29

Odd/Even Assignments

Exercises 12–31 are structured so that students practice the same concepts whether they are assigned odd or even problems.

Assignment Guide

Basic: 13–27 odd, 33, 34, 38–54

Average: 13–31 odd, 33, 34, 38–54

Advanced: 12–32 even, 33–48 (optional: 49–54)

DAILY INTERVENTION

Differentiated Instruction **ELL**

Verbal/Linguistic Have students write a list of tips to help someone draw the graphs of systems of inequalities and find the vertices easily and efficiently.

Study Guide and Intervention, p. 131 (shown) and p. 132

Graph Systems of Inequalities To solve a system of inequalities, graph the inequalities in the same coordinate plane. The solution set is represented by the intersection of the graphs.

Example **Solve the system of inequalities by graphing.**

$y \le 2x - 1$ and $y > \frac{x}{3} + 2$

The solution of $y \le 2x - 1$ is Regions 1 and 2.

The solution of $y > \frac{x}{3} + 2$ is Regions 1 and 3.

The intersection of these regions is Region 1, which is the solution set of the system of inequalities.

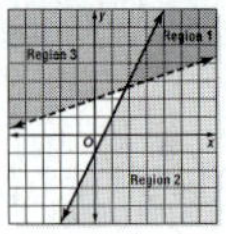

Exercises

Solve each system of inequalities by graphing.

1. $x - y \le 2$, $x + 2y \ge 1$
2. $3x - 2y \le -1$, $x + 4y \ge -12$
3. $|y| \le 1$, $x > 2$

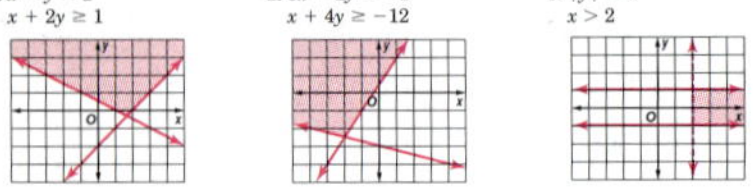

4. $y \ge \frac{x}{2} - 3$, $y < 2x$
5. $y < \frac{x}{3} + 2$, $y < -2x + 1$
6. $y \ge -\frac{x}{4} + 1$, $y < 3x - 1$

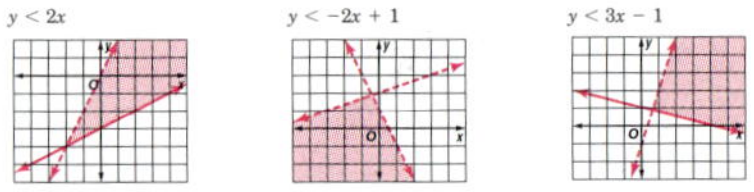

7. $x + y \ge 4$, $2x - y > 2$
8. $x + 3y < 3$, $x - 2y \ge 4$
9. $x - 2y > 6$, $x + 4y < -4$

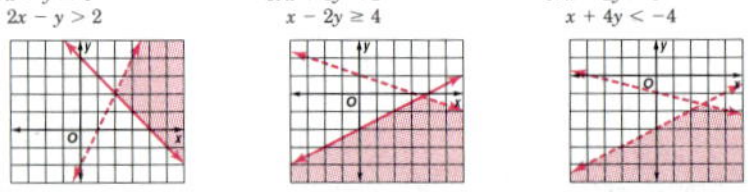

Skills Practice, p. 133 and Practice, p. 134 (shown)

Solve each system of inequalities by graphing.

1. $y + 1 < -x$, $y \ge 1$
2. $x > -2$, $2y \ge 3x + 6$
3. $y \le 2x - 3$, $y \le -\frac{1}{2}x + 2$

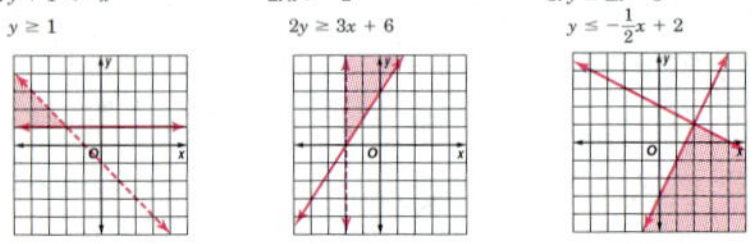

4. $x + y > -2$, $3x - y \ge -2$
5. $|y| \le 1$, $y < x - 1$
6. $3y > 4x$, $2x - 3y > -6$

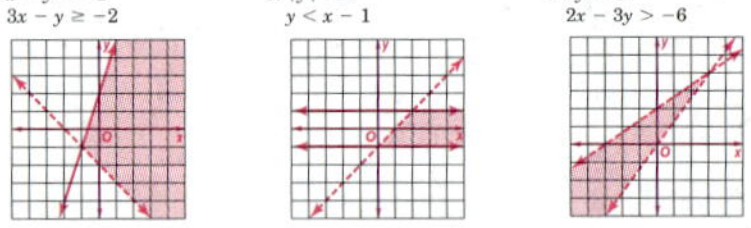

Find the coordinates of the vertices of the figure formed by each system of inequalities.

7. $y \ge 1 - x$, $y \le x - 1$, $x \le 3$ **(1, 0), (3, 2), (3, −2)**
8. $x - y \le 2$, $x + y \le 2$, $x \ge -2$ **(−2, 4), (−2, −4), (2, 0)**
9. $y \ge 2x - 2$, $2x + 3y \ge 6$, $y < 4$ **(−3, 4), $\left(\frac{3}{2}, 1\right)$, (3, 4)**

DRAMA **For Exercises 10 and 11, use the following information.**

The drama club is selling tickets to its play. An adult ticket costs \$15 and a student ticket costs \$11. The auditorium will seat 300 ticket-holders. The drama club wants to collect at least \$3630 from ticket sales.

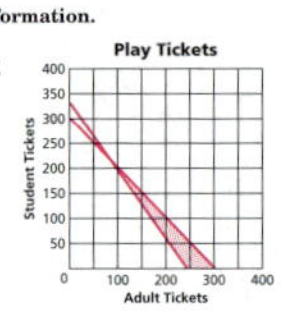

10. Write and graph a system of four inequalities that describe how many of each type of ticket the club must sell to meets its goal. **$x \ge 0$, $y \ge 0$, $x + y \le 300$, $15x + 11y \ge 3630$**

11. List three different combinations of tickets sold that satisfy the inequalities. **Sample answer: 250 adult and 50 student, 200 adult and 100 student, 145 adult and 148 student**

Reading to Learn Mathematics, p. 135 ELL

Pre-Activity **How can you determine whether your blood pressure is in a normal range?**

Read the introduction to Lesson 3-3 at the top of page 123 in your textbook.

Satish is 37 years old. He has a blood pressure reading of 135/99. Is his blood pressure within the normal range? Explain.

Sample answer: No; his systolic pressure is normal, but his diastolic pressure is too high. It should be between 60 and 90.

Reading the Lesson

1. Without actually drawing the graph, describe the boundary lines for the system of inequalities shown at the right. $|x| < 3$, $|y| \le 5$
Two dashed vertical lines ($x = 3$ and $x = -3$) and two solid horizontal lines ($y = -5$ and $y = 5$)

2. Think about how the graph would look for the system given above. What will be the shape of the shaded region? (It is not necessary to draw the graph. See if you can imagine it without drawing anything. If this is difficult to do, make a rough sketch to help you answer the question.)
a rectangle

3. Which system of inequalities matches the graph shown at the right? **B**

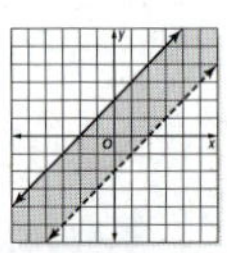

A. $x - y \le -2$, $x - y > 2$
B. $x - y \ge -2$, $x - y < 2$
C. $x + y \le -2$, $x + y > 2$
D. $x - y > -2$, $x - y \le 2$

Helping You Remember

4. To graph a system of inequalities, you must graph two or more boundary lines. When you graph each of these lines, how can the inequality symbols help you remember whether to use a dashed or solid line?
Use a dashed line if the inequality symbol is > or <, because these symbols do not include equality and the dashed line reminds you that the line itself is not included in the graph. Use a solid line if the symbol is ≥ or ≤, because these symbols include equality and tell you that the line itself is included in the graph.

★ indicates increased difficulty

Practice and Apply

Homework Help

For Exercises	See Examples
12–23	1, 2
24–31	4
32–37	3

Extra Practice See page 833.

Solve each system of inequalities by graphing. **12–23. See pp. 151A–151F.**

12. $x \ge 2$; $y > 3$
13. $x \le -1$; $y \ge -4$
14. $y < 2 - x$; $y > x + 4$
15. $y > x - 3$; $|y| \le 2$
16. $3x + 2y \ge 6$; $4x - y \ge 2$
17. $4x - 3y < 7$; $2y - x < -6$
18. $y < 2x - 3$; $y \le \frac{1}{2}x + 1$
19. $3y \le 2x - 8$; $y \ge \frac{2}{3}x - 1$
20. $|x| \le 3$; $|y| > 1$
21. $|x + 1| \le 3$; $x + 3y \ge 6$
22. $y \ge 2x + 1$; $y \le 2x - 2$; $3x + y \ge 9$
23. $x - 3y > 2$; $2x - y < 4$; $2x + 4y \ge -7$

Find the coordinates of the vertices of the figure formed by each system of inequalities.

24. $y \ge 0$; $x \ge 0$; $x + 2y \le 8$ **(0, 0), (0, 4), (8, 0)**
25. $y \ge -4$; $y \le 2x + 2$; $2x + y \le 6$ **(−3, −4), (5, −4), (1, 4)**
26. $x \le 3$; $-x + 3y \le 12$; $4x + 3y \ge 12$ **(0, 4), (3, 0), (3, 5)**
27. $x + y \le 9$; $x - 2y \le 12$; $y \le 2x + 3$ **(−6, −9), (2, 7), (10, −1)**
★28. $y \ge -3$; $x \le 6$; $y \ge x - 2$; $2y \le x + 5$ **(−11, −3), (−1, −3), (6, 4), $\left(6, 5\frac{1}{2}\right)$**
★29. $y \ge x - 5$; $y \le 2x + 11$; $x + 2y \le 12$; $x + 2y \ge 2$ **(−4, 3), (−2, 7), (4, −1), $\left(7\frac{1}{3}, 2\frac{1}{3}\right)$**

★30. Find the area of the region defined by the system of inequalities $y + x \le 3$, $y - x \le 3$, and $y \ge -1$. **16 units²**

★31. Find the area of the region defined by the system of inequalities $x \ge -3$, $y + x \le 8$, and $y - x \ge -2$. **64 units²**

32. **PART-TIME JOBS** Bryan Clark makes \$10 an hour cutting grass and \$12 an hour for raking leaves. He cannot work more than 15 hours per week. Graph two inequalities that Bryan can use to determine how many hours he needs to work at each job if he wants to earn at least \$120 per week. **See pp. 151A–151F.**

Career Choices

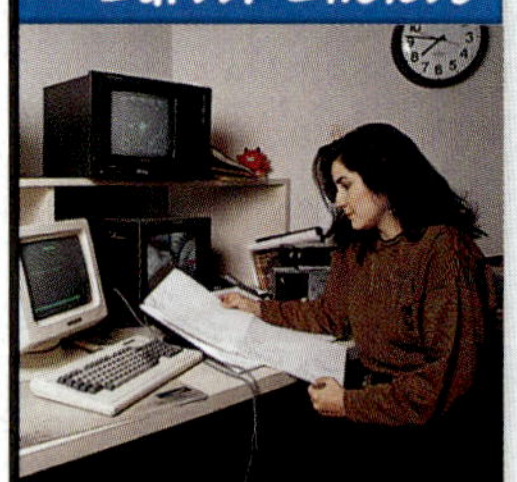

Atmospheric Scientist

The best known use of atmospheric science is for weather forecasting. However, weather information is also studied for air-pollution control, agriculture, and transportation.

Online Research For information about a career as an atmospheric scientist, visit: www.algebra2.com/careers

HURRICANES **For Exercises 33 and 34, use the following information.**
Hurricanes are divided into categories according to their wind speed and storm surge.

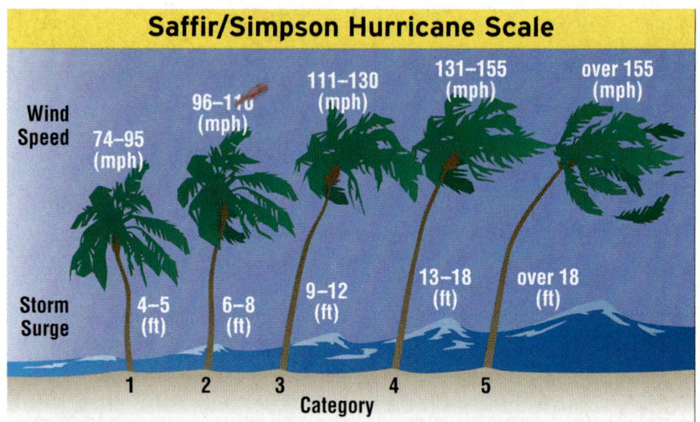

33. Write and graph the system of inequalities that represents the range of wind speeds s and storm surges h for a category 3 hurricane. **See pp. 151A–151F.**

34. On September 16, 1999, Hurricane Floyd hit the United States with winds of 140 mph. Classify Hurricane Floyd, and identify the heights of its storm surges. **category 4; 13–18 ft**

Enrichment, p. 136

Tracing Strategy

Try to trace over each of the figures below without tracing the same segment twice.

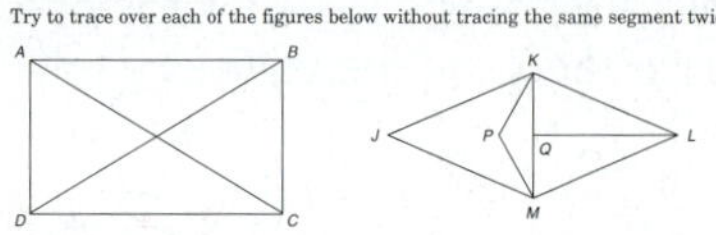

The figure at the left cannot be traced, but the one at the right can. The rule is that a figure is traceable if it has no more than two points where an odd number of segments meet. The figure at the left has three segments meeting at each of the four corners. However, the figure at the right has only two points, L and Q, where an odd number of segments meet.

Determine if each figure can be traced without tracing the same segment

BAKING For Exercises 35–37, use the recipes at the right.

The Merry Bakers are baking pumpkin bread and Swedish soda bread for this week's specials. They have 24 cups of flour and at most 26 teaspoons of baking powder.

35. See pp. 151A–151F.

35. Graph the inequalities that represent how many loaves of each type of bread the bakers can make.

36. Sample answer: 2 pumpkin, 8 soda; 4 pumpkin, 6 soda; 8 pumpkin, 4 soda

36. List three different combinations of breads they can make.

★ 37. Which combination uses all of the available flour and baking soda? **6 pumpkin, 8 soda**

38. **CRITICAL THINKING** Find the area of the region defined by $|x| + |y| \leq 5$ and $|x| + |y| \geq 2$. **42 units²**

39. **WRITING IN MATH** Answer the question that was posed at the beginning of the lesson. **See margin.**

How can you determine whether your blood pressure is in a normal range?

Include the following in your answer:
- an explanation of how to use the graph, and
- a description of the regions that indicate high blood pressure, both systolic and diastolic.

Standardized Test Practice

40. Choose the system of inequalities whose solution is represented by the graph. **B**

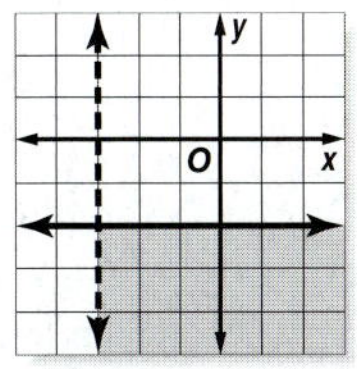

Ⓐ $y < -2$, $x < -3$ Ⓑ $y \leq -2$, $x > -3$

Ⓒ $x \leq -2$, $y > -3$ Ⓓ $x < -3$, $y < -3$

41. **EXTENDED RESPONSE** Create a system of inequalities for which the graph will be a square with its interior located in the first quadrant. **Sample answer: $y \leq 6$, $y \geq 2$, $x \leq 5$, $x \geq 1$**

Maintain Your Skills

Mixed Review **Solve each system of equations by using either substitution or elimination.** *(Lesson 3-2)*

42. $4x - y = -20$, $x + 2y = 13$ **(−3, 8)**

43. $3x - 4y = -2$, $5x + 2y = 40$ **(6, 5)**

44. $4x + 5y = 7$, $3x - 2y = 34$ **(8, −5)**

Solve each system of equations by graphing. *(Lesson 3-1)*

45–47. See margin for graphs.

45. $y = 2x + 1$, $y = -\frac{1}{2}x - 4$ **(−2, −3)**

46. $2x + y = -3$, $6x + 3y = -9$ **infinitely many**

47. $2x - y = 6$, $-x + 8y = 12$ **(4, 2)**

48. Write an equation in slope-intercept form of the line that passes through (−4, 4) and (6, 9). *(Lesson 2-4)* $y = \frac{1}{2}x + 6$

Getting Ready for the Next Lesson **PREREQUISITE SKILL** **Find each value if $f(x) = 4x + 3$ and $g(x) = 5x - 7$.** *(To review **functions**, see Lesson 2-1.)*

49. $f(-2)$ **−5**

50. $g(-1)$ **−12**

51. $g(3)$ **8**

52. $f(6)$ **27**

53. $f(0.5)$ **5**

54. $g(-0.25)$ **−8.25**

www.algebra2.com/self_check_quiz

4 Assess

Open-Ended Assessment

Modeling Have students create their own example of a system of three inequalities that bound a region, writing the inequalities and graphing them, and giving the coordinates of the vertices of the region.

Getting Ready for Lesson 3-4

PREREQUISITE SKILL Lesson 3-4 presents finding minimum and maximum values of a function. Linear programming, the topic of Lesson 3-4, utilizes these extreme values of a function. Exercises 49–54 should be used to determine your students' familiarity with evaluating functions.

Assessment Options

Quiz (Lesson 3-3) is available on p. 163 of the *Chapter 3 Resource Masters*.

Mid-Chapter Test (Lessons 3-1 through 3-3) is available on p. 165 of the *Chapter 3 Resource Masters*.

Answers

46.

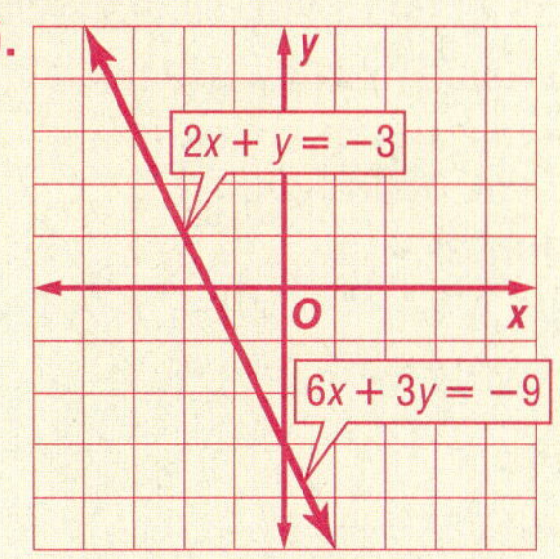

39. The range for normal blood pressure satisfies four inequalities that can be graphed to find their intersection. Answers should include the following.
- Graph the blood pressure as an ordered pair; if the point lies in the shaded region, it is in the normal range.
- High systolic pressure is represented by the region to the right of $x = 140$ and high diastolic pressure is represented by the region above $y = 90$.

45.

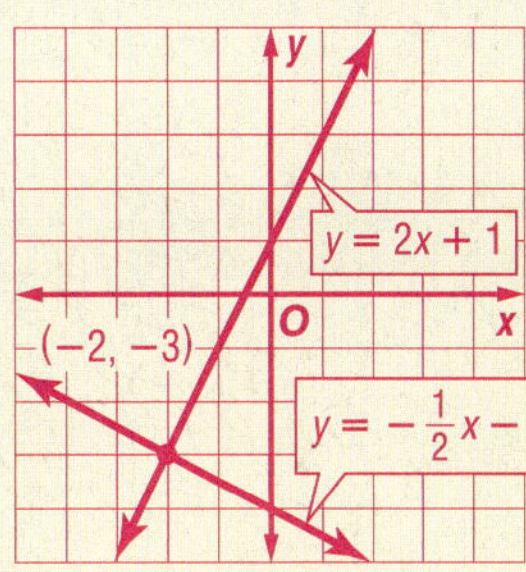

47.

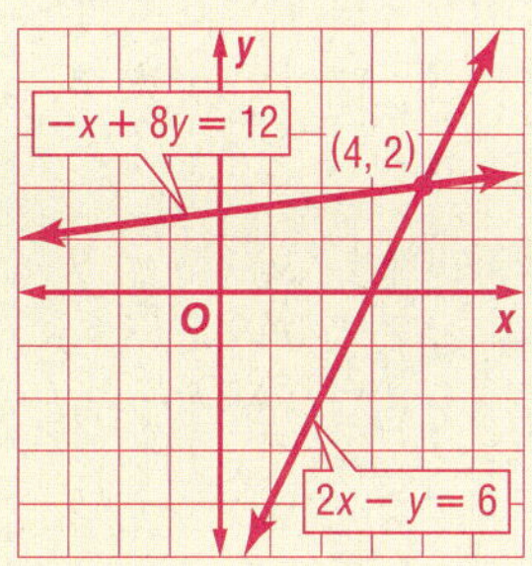

Graphing Calculator Investigation

A Follow-Up of Lesson 3-3

Getting Started

Error Source One possible source of an error message when entering the expression in Step 1 is that the student used the subtract key [−] rather than the negative key [(−)].

Teach

- Explain to students that when they sketch the graph from their calculator window on paper, they need to draw the axes, the intercepts, and the points of intersection, as well as the lines.
- Have students complete Exercises 1–8.

Assess

Ask students:

- How many points are in the solution set of a system of inequalities such as those in this activity? **infinitely many**
- When is the point of intersection of the two boundary lines included in the solution set? **The point of intersection is in the solution set when both boundary lines on the graph are solid and when the symbol in both inequalities is either $\geq$ or $\leq$.**

Graphing Calculator Investigation

A Follow-Up of Lesson 3-3

Systems of Linear Inequalities

You can graph systems of linear inequalities with a TI-83 Plus calculator using the Y= menu. You can choose different graphing styles to shade above or below a line.

Example **Graph the system of inequalities in the standard viewing window.**

$y \geq -2x + 3$
$y \leq x + 5$

Step 1

- Enter $-2x + 3$ as Y1. Since y is greater than $-2x + 3$, shade above the line.

  KEYSTROKES: −2 [X,T,θ,n] [+] 3

- Use the left arrow key to move your cursor as far left as possible. Highlight the graph style icon. Press [ENTER] until the shade above icon, ◥, appears.

Step 2

- Enter $x + 5$ as Y2. Since y is less than $x + 5$, shade below the line.

  KEYSTROKES: [X,T,θ,n] [+] 5

- Use the arrow and [ENTER] keys to choose the shade below icon, ◣.

Step 3

- Display the graphs by pressing [GRAPH].

  Notice the shading pattern above the line $y = -2x + 3$ and the shading pattern below the line $y = x + 5$. The intersection of the graphs is the region where the patterns overlap. This region includes all the points that satisfy the system $y \geq -2x + 3$ and $y \leq x + 5$.

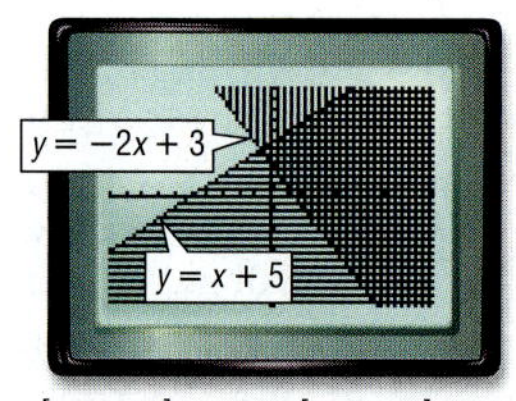

[−10, 10] scl: 1 by [−10, 10] scl: 1

Exercises **1–8. See pp. 151A–151F.**

Solve each system of inequalities. Sketch each graph on a sheet of paper.

1. $y \geq 4$
$y \leq -x$

2. $y \geq -2x$
$y \leq -3$

3. $y \geq 1 - x$
$y \leq x + 5$

4. $y \geq x + 2$
$y \leq -2x - 1$

5. $3y \geq 6x - 15$
$2y \leq -x + 3$

6. $y + 3x \geq 6$
$y - 2x \leq 9$

7. $6y + 4x \geq 12$
$5y - 3x \leq -10$

8. $\frac{1}{4}y - x \geq -2$
$\frac{1}{3}y + 2x \leq 4$

www.algebra2.com/other_calculator_keystrokes

3-4 Linear Programming

What You'll Learn

- Find the maximum and minimum values of a function over a region.
- Solve real-world problems using linear programming.

Vocabulary

- constraints
- feasible region
- bounded
- vertices
- unbounded
- linear programming

How is linear programming used in scheduling work?

One of the primary tasks of the U.S. Coast Guard is to maintain the buoys that ships use to navigate. The ships that service buoys are called buoy tenders.

Suppose a certain buoy tender can carry up to 8 replacement buoys. Their crew can check and repair a buoy in 1 hour. It takes the crew $2\frac{1}{2}$ hours to replace a buoy.

MAXIMUM AND MINIMUM VALUES The buoy tender captain can use a system of inequalities to represent the limitations of time and the number of replacement buoys on the ship. If these inequalities are graphed, all of the points in the intersection are the combinations of repairs and replacements that the buoy tender can schedule.

The inequalities are called the **constraints**. The intersection of the graphs is called the **feasible region**. When the graph of a system of constraints is a polygonal region like the one graphed at the right, we say that the region is **bounded**.

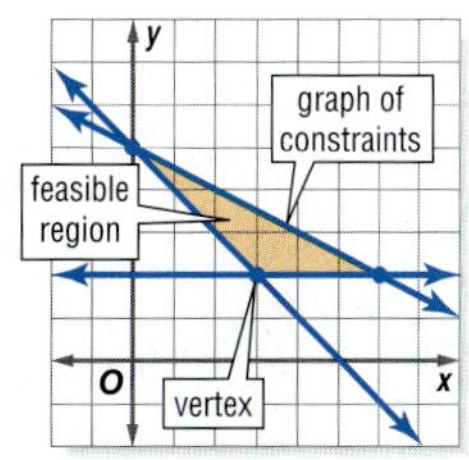

Sometimes it is necessary to find the maximum or minimum values that a linear function has for the points in a feasible region. For example, the buoy tender captain wishes to maximize the total number of buoys serviced. The maximum or minimum value of a related function *always* occurs at one of the **vertices** of the feasible region.

Study Tip

Reading Math
The notation $f(x, y)$ is used to represent a function with two variables x and y. It is read *f of x and y*.

Example 1 Bounded Region

Graph the following system of inequalities. Name the coordinates of the vertices of the feasible region. Find the maximum and minimum values of the function $f(x, y) = 3x + y$ for this region.

$x \geq 1$
$y \geq 0$
$2x + y \leq 6$

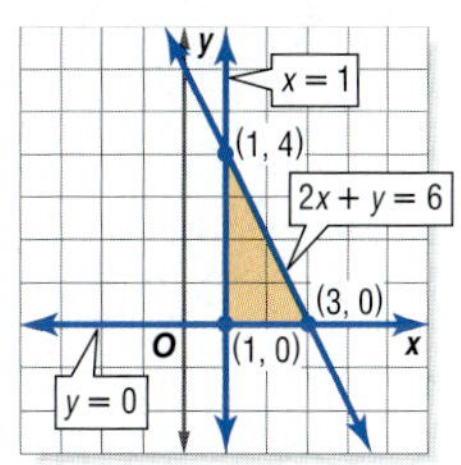

Step 1 Find the vertices of the region. Graph the inequalities.

The polygon formed is a triangle with vertices at (1, 4), (3, 0), and (1, 0).

(continued on the next page)

3-4 Lesson Notes

1 Focus

5-Minute Check Transparency 3-4 Use as a quiz or review of Lesson 3-3.

Mathematical Background notes are available for this lesson on p. 108D.

Building on Prior Knowledge

In Lessons 3-2 and 3-3, students graphed systems of equations and inequalities. In this lesson, students will use these graphing skills to determine a feasible region.

How is linear programming used in scheduling work?

Ask students:

- If none of the buoys need to be replaced, how many buoys can be checked and repaired in 24 hours? **24 buoys**
- If the crew of a buoy tender uses all 8 of their replacement buoys in one 24-hour period, how much time did they spend making replacements? **20 h**

Resource Manager

Workbook and Reproducible Masters

Chapter 3 Resource Masters

- Study Guide and Intervention, pp. 137–138
- Skills Practice, p. 139
- Practice, p. 140
- Reading to Learn Mathematics, p. 141
- Enrichment, p. 142
- Assessment, p. 164

Graphing Calculator and Spreadsheet Masters, p. 31

Transparencies

5-Minute Check Transparency 3-4
Real-World Transparency 3
Answer Key Transparencies

Technology

Interactive Chalkboard

2 Teach

MAXIMUM AND MINIMUM VALUES

In-Class Examples

Power Point®

1 Graph the following system of inequalities. Name the coordinates of the vertices of the feasible region. Find the maximum and minimum values of the function $f(x, y) = 3x - 2y$ for this region.

$x \leq 5$
$y \leq 4$
$x + y \geq 2$

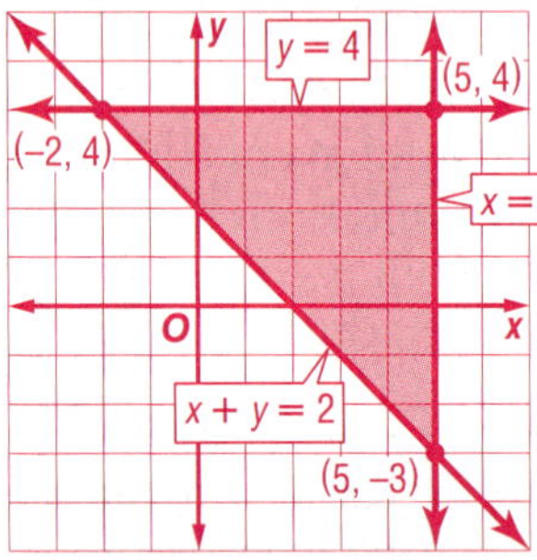

The maximum value is 21 at (5, −3). The minimum value is −14 at (−2, 4).

2 Graph the following system of inequalities. Name the coordinates of the vertices of the feasible region. Find the maximum and minimum values of the function $f(x, y) = 2x + 3y$ for this region.

$-x + 2y \leq 2$
$x - 2y \leq 4$
$x + y \geq -2$

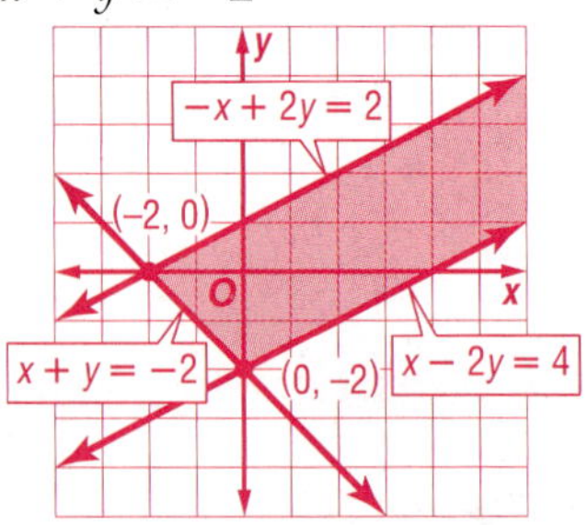

There is no maximum value. The minimum value is −6 at (0, −2).

Step 2 Use a table to find the maximum and minimum values of $f(x, y)$. Substitute the coordinates of the vertices into the function.

(x, y)	$3x + y$	$f(x, y)$	
(1, 4)	3(1) + 4	7	
(3, 0)	3(3) + 0	9	← maximum
(1, 0)	3(1) + 0	3	← minimum

The maximum value is 9 at (3, 0). The minimum value is 3 at (1, 0).

Sometimes a system of inequalities forms a region that is open. In this case, the region is said to be **unbounded**.

Example 2 Unbounded Region

Graph the following system of inequalities. Name the coordinates of the vertices of the feasible region. Find the maximum and minimum values of the function $f(x, y) = 5x + 4y$ for this region.

$2x + y \geq 3$
$3y - x \leq 9$
$2x + y \leq 10$

Graph the system of inequalities. There are only two points of intersection, (0, 3) and (3, 4).

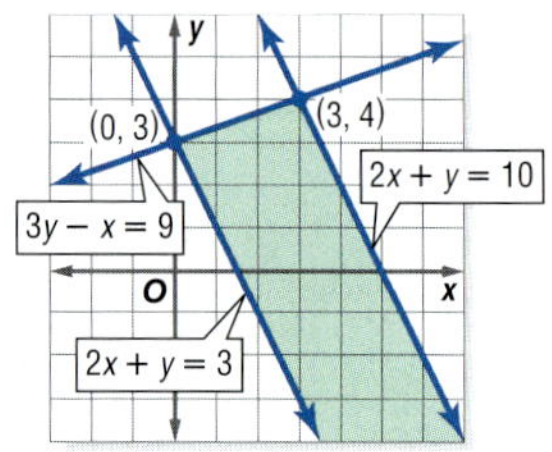

(x, y)	$5x + 4y$	$f(x, y)$
(0, 3)	5(0) + 4(3)	12
(3, 4)	5(3) + 4(4)	31

The maximum is 31 at (3, 4).

Although $f(0, 3)$ is 12, it is not the minimum value since there are other points in the solution that produce lesser values. For example, $f(3, -2) = 7$ and $f(20, -35) = -40$. It appears that because the region is unbounded, $f(x, y)$ has no minimum value.

Study Tip

Common Misconception

Always test a point contained in the feasible region when the graph is unbounded. Do not assume that there is no minimum value if the feasible region is unbounded below the line, or that there is no maximum value if the feasible region is unbounded above the line.

REAL-WORLD PROBLEMS The process of finding maximum or minimum values of a function for a region defined by inequalities is called **linear programming**. The steps used to solve a problem using linear programming are listed below.

Key Concept *Linear Programming Procedure*

Step 1 Define the variables.
Step 2 Write a system of inequalities.
Step 3 Graph the system of inequalities.
Step 4 Find the coordinates of the vertices of the feasible region.
Step 5 Write a function to be maximized or minimized.
Step 6 Substitute the coordinates of the vertices into the function.
Step 7 Select the greatest or least result. Answer the problem.

Linear programming can be used to solve many types of real-world problems. These problems have certain restrictions placed on the variables, and some function of the variable must be maximized or minimized.

Teaching Tip Help students see that the constraints and feasible region represent a way to model the facts in a complex situation. Within those constraints, the function $f(x, y)$ represents the relationship for which you need to find a maximum or minimum value.

Example 3 Linear Programming

VETERINARY MEDICINE **As a receptionist for a veterinarian, one of Dolores Alvarez's tasks is to schedule appointments. She allots 20 minutes for a routine office visit and 40 minutes for a surgery. The veterinarian cannot do more than 6 surgeries per day. The office has 7 hours available for appointments. If an office visit costs $55 and most surgeries cost $125, find a combination of office visits and surgeries that will maximize the income the veterinarian practice receives per day.**

Step 1 Define the variables.

v = the number of office visits
s = the number of surgeries

Step 2 Write a system of inequalities.

Since the number of appointments cannot be negative, v and s must be nonnegative numbers.

$v \geq 0$ and $s \geq 0$

An office visit is 20 minutes, and a surgery is 40 minutes. There are 7 hours available for appointments.

$20v + 40s \leq 420$ 7 hours = 420 minutes

The veterinarian cannot do more than 6 surgeries per day.

$s \leq 6$

Step 3 Graph the system of inequalities.

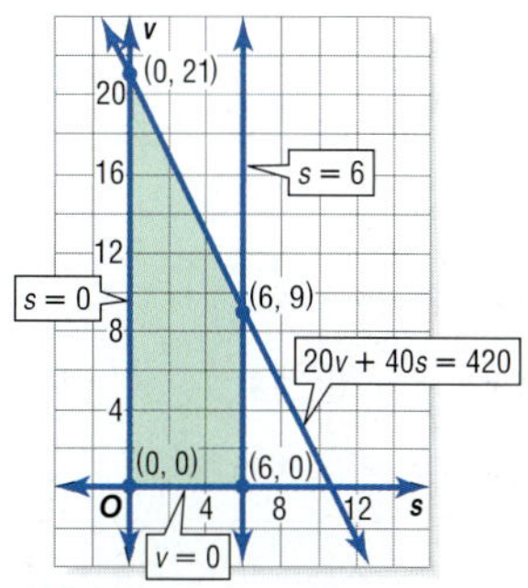

Step 4 Find the coordinates of the vertices of the feasible region.

From the graph, the vertices of the feasible region are at (0, 0), (6, 0), (6, 9), and (0, 21). If the vertices could not be read from the graph easily, we could also solve a system of equations using the boundaries of the inequalities.

Step 5 Write a function to be maximized or minimized.

The function that describes the income is $f(s, v) = 125s + 55v$. We wish to find the maximum value for this function.

Step 6 Substitute the coordinates of the vertices into the function.

(s, v)	125s + 55v	f(s, v)
(0, 0)	125(0) + 55(0)	0
(6, 0)	125(6) + 55(0)	750
(6, 9)	125(6) + 55(9)	1245
(0, 21)	125(0) + 55(21)	1155

Step 7 Select the greatest or least result. Answer the problem.

The maximum value of the function is 1245 at (6, 9). This means that the maximum income is $1245 when Dolores schedules 6 surgeries and 9 office visits.

More About...

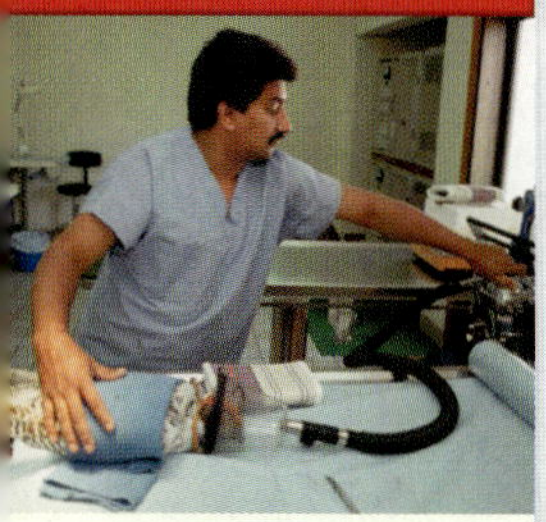

Veterinary Medicine

Surgeries are usually performed in the morning so that the animal can recover throughout the day while there is plenty of staff to monitor its progress.

Source: www.vetmedicine.miningco.com

www.algebra2.com/extra_examples

REAL-WORLD PROBLEMS

In-Class Example

Power Point®

3 **LANDSCAPING** A landscaping company has crews who mow lawns and prune shrubbery. The company schedules 1 hour for mowing jobs and 3 hours for pruning jobs. Each crew is scheduled for no more than 2 pruning jobs per day. Each crew's schedule is set up for a maximum of 9 hours per day. On the average, the charge for mowing a lawn is $40 and the charge for pruning shrubbery is $120. Find a combination of mowing lawns and pruning shrubs that will maximize the income the company receives per day for one of its crews. **either 3 mowings and 2 prunings, or 9 mowings and no prunings**

DAILY INTERVENTION

Differentiated Instruction

Visual/Spatial Have students use different colored pencils to shade the different regions of a graph defined by the inequalities in a linear programming problem. This should help students clarify the relationship between the various regions in these graphs.

Check for Understanding

Concept Check

1. **Determine** whether the following statement is *always*, *sometimes*, or *never* true. *A feasible region has a minimum and a maximum value.* **sometimes**

2. **OPEN ENDED** Give an example of a system of inequalities that forms a bounded region. **Sample answer: $y \geq -x$, $y \geq x - 5$, $y \leq 0$**

Guided Practice

Graph each system of inequalities. Name the coordinates of the vertices of the feasible region. Find the maximum and minimum values of the given function for this region. **3–8. See margin.**

GUIDED PRACTICE KEY	
Exercises	Examples
3, 5–8	1
4	2
9–14	3

3. $y \geq 2$
$x \geq 1$
$x + 2y \leq 9$
$f(x, y) = 2x - 3y$

4. $x \geq -3$
$y \leq 1$
$3x + y \leq 6$
$f(x, y) = 5x - 2y$

5. $y \leq 2x + 1$
$1 \leq y \leq 3$
$x + 2y \leq 12$
$f(x, y) = 3x + y$

6. $y \geq -x + 2$
$2 \leq x \leq 7$
$y \leq \frac{1}{2}x + 5$
$f(x, y) = 8x + 3y$

7. $x + 2y \leq 6$
$2x - y \leq 7$
$x \geq -2, y \geq -3$
$f(x, y) = x - y$

8. $x - 3y \geq -7$
$5x + y \leq 13$
$x + 6y \geq -9$
$3x - 2y \geq -7$
$f(x, y) = x - y$

Application

MANUFACTURING **For Exercises 9–14, use the following information.**
The Future Homemakers Club is making canvas tote bags and leather tote bags for a money making project. Both types of tote bags will be lined with canvas and have leather handles. For the canvas tote bags, they need 4 yards of canvas and 1 yard of leather. For the leather tote bags, they need 3 yards of leather and 2 yards of canvas. Their advisor has purchased 56 yards of leather and 104 yards of canvas.

9. Let c represent the number of canvas tote bags and let ℓ represent the number of leather tote bags. Write a system of inequalities to represent the number of tote bags that can be produced. **$c \geq 0$, $\ell \geq 0$, $c + 3\ell \leq 56$, $4c + 2\ell \leq 104$**
10. Draw the graph showing the feasible region. **See margin.**
11. List the coordinates of the vertices of the feasible region.
12. If the club plans to sell the canvas bags at a profit of \$20 each and the leather bags at a profit of \$35 each, write a function for the total profit on the bags.
13. Determine the number of canvas and leather bags that they need to make for a maximum profit. **20 canvas tote bags and 12 leather tote bags**
14. What is the maximum profit? **\$820**

11. (0, 0), (26, 0), (20, 12), $\left(0, 18\frac{2}{3}\right)$

12. $f(c, \ell) = 20c + 35\ell$

★ indicates increased difficulty

Practice and Apply

Homework Help	
For Exercises	See Examples
15–29	1, 2
31–36, 38–42	3

Extra Practice
See page 833.

Graph each system of inequalities. Name the coordinates of the vertices of the feasible region. Find the maximum and minimum values of the given function for this region. **15–20. See pp. 151A–151F.**

15. $y \geq 1$
$x \leq 6$
$y \leq 2x + 1$
$f(x, y) = x + y$

16. $y \geq -4$
$x \leq 3$
$y \leq 3x - 4$
$f(x, y) = x - y$

17. $y \geq 2$
$1 \leq x \leq 5$
$y \leq x + 3$
$f(x, y) = 3x - 2y$

18. $y \geq 1$
$2 \leq x \leq 4$
$x - 2y \geq -4$
$f(x, y) = 3y + x$

19. $y \leq x + 2$
$y \leq 11 - 2x$
$2x + y \geq -7$
$f(x, y) = 4x - 3y$

20. $y \leq x + 6$
$y + 2x \geq 6$
$2 \leq x \leq 6$
$f(x, y) = -x + 3y$

3 Practice/Apply

Study Notebook

Have students—
- add the definitions/examples of the vocabulary terms to their Vocabulary Builder worksheets for Chapter 3.
- include any other item(s) that they find helpful in mastering the skills in this lesson.

About the Exercises...

Organization by Objective
- **Maximum and Minimum Values:** 15–30
- **Real-World Problems:** 31–42

Odd/Even Assignments
Exercises 15–29 are structured so that students practice the same concepts whether they are assigned odd or even problems.

Alert! Exercise 37 involves research on the Internet or other reference materials.

Assignment Guide

Basic: 15–27 odd, 30–37, 43–62
Average: 15–29 odd, 30–41, 43–62
Advanced: 16–28 even, 30–56, (optional: 57–62)
All: Practice Quiz 2 (1–5)

Answers

3.
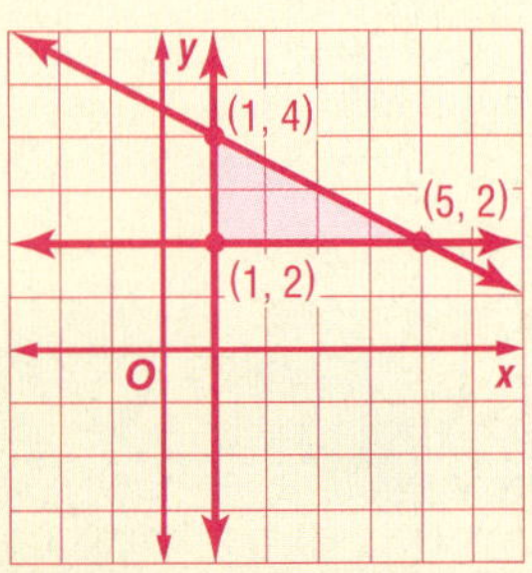

vertices: (1, 2), (1, 4), (5, 2);
max: $f(5, 2) = 4$,
min: $f(1, 4) = -10$

4.
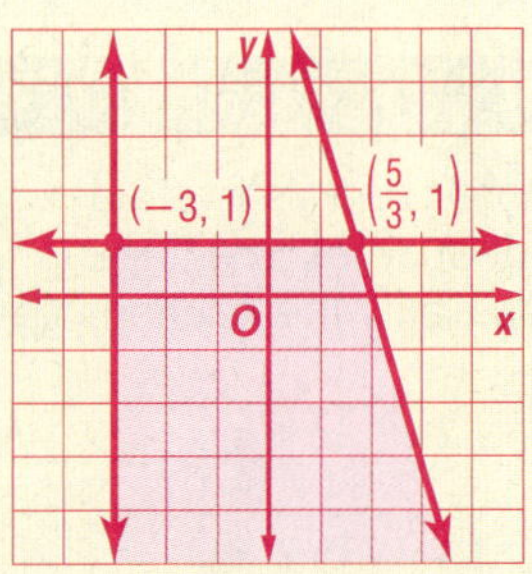

vertices: $(-3, 1)$, $\left(\frac{5}{3}, 1\right)$;
no maximum;
min: $f(-3, 1) = -17$

132 Chapter 3 Systems of Equations and Inequalities

21–29. See pp. 151A–151F.

21. $x + y \le 3$
$x + 2y \le 4$
$x \ge 0, y \ge 0$
$f(x, y) = 3y - 4x$

22. $y \le 7 - x$
$3x - 2y \le 6$
$x \ge 0, y \ge 0$
$f(x, y) = 5x - 2y$

23. $y \ge x - 3$
$y \le 6 - 2x$
$2x + y \ge -3$
$f(x, y) = 3x + 4y$

24. $x + y \ge 4$
$3x - 2y \le 12$
$x - 4y \ge -16$
$f(x, y) = x - 2y$

25. $x + y \ge 2$
$4y \le x + 8$
$y \ge 2x - 5$
$f(x, y) = 4x + 3y$

26. $2x + 2y \ge 4$
$2y \ge 3x - 6$
$4y \le x + 8$
$f(x, y) = 3y + x$

27. $2x + 3y \ge 6$
$3x - 2y \ge -4$
$5x + y \ge 15$
$f(x, y) = x + 3y$

★ **28.** $x \ge 0$
$y \ge 0$
$x + 2y \le 6$
$2y - x \le 2$
$x + y \le 5$
$f(x, y) = 3x - 5y$

★ **29.** $x \ge 2$
$y \ge 1$
$x - 2y \ge -4$
$x + y \le 8$
$2x - y \le 7$
$f(x, y) = x - 4y$

30. **CRITICAL THINKING** The vertices of a feasible region are $A(1, 2)$, $B(5, 2)$, and $C(1, 4)$. Write a function that satisfies each condition. **Sample answers given.**

a. A is the maximum and B is the minimum. $f(x, y) = -2x - y$

b. C is the maximum and B is the minimum. $f(x, y) = 3y - 2x$

c. B is the maximum and A is the minimum. $f(x, y) = x + y$

d. A is the maximum and C is the minimum. $f(x, y) = -x - 3y$

e. B and C are both maxima and A is the minimum. $f(x, y) = x + 2y$

PRODUCTION **For Exercises 31–36, use the following information.**
There are a total of 85 workers' hours available per day for production at a calculator manufacturer. There are 40 workers' hours available for encasement and quality control each day. The table below shows the number of hours needed in each department for two different types of calculators.

Calculator Production Time

Calculator Type	Production Time	Encasement and Quality Control
graphing calculator	$1\frac{1}{2}$ hours	2 hours
computer-algebra systems (CAS)	1 hour	$\frac{1}{2}$ hour

31. Let g represent the number of graphing calculators and let c represent the number of CAS calculators. Write a system of inequalities to represent the number of calculators that can be produced. $g \ge 0$, $c \ge 0$, $1.5g + c \le 85$, $2g + 0.5c \le 40$

32. Draw the graph showing the feasible region. **See margin.**

33. (0, 0), (0, 20), (80, 0)

33. List the coordinates of the vertices of the feasible region.

34. If the profit on a graphing calculator is \$50 and the profit on a CAS calculator is \$65, write a function for the total profit on the calculators. $f(g, c) = 50g + 65c$

35. Determine the number of each type of calculator that is needed to make a maximum profit. **0 graphing calculators, 80 CAS calculators**

36. What is the maximum profit? **\$5200**

37. **RESEARCH** Use the Internet or other reference to find an industry that uses linear programming. Describe the restrictions or constraints of the problem and explain how linear programming is used to solve the problem. **See students' work.**

www.algebra2.com/self_check_quiz

Answers

7.

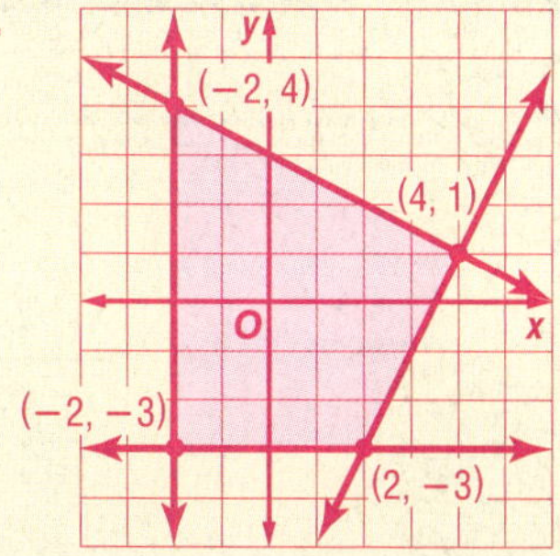

vertices: (−2, 4), (−2, −3), (2, −3), (4, 1); max: $f(2, -3) = 5$; min: $f(-2, 4) = -6$

8.

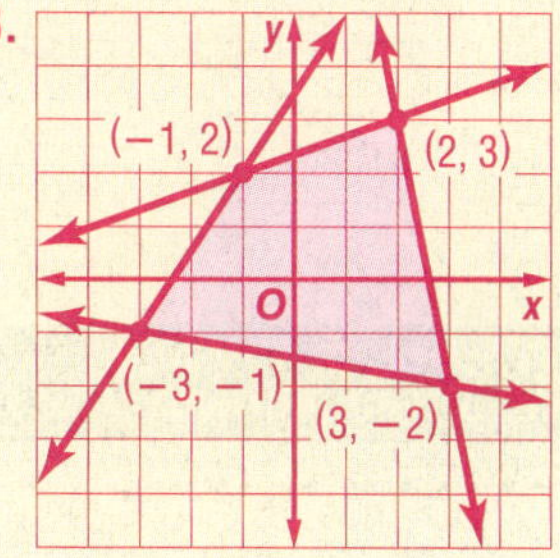

vertices: (−3, −1), (−1, 2), (2, 3), (3, −2); max: $f(3, -2) = 5$, min: $f(-1, 2) = -3$

10.

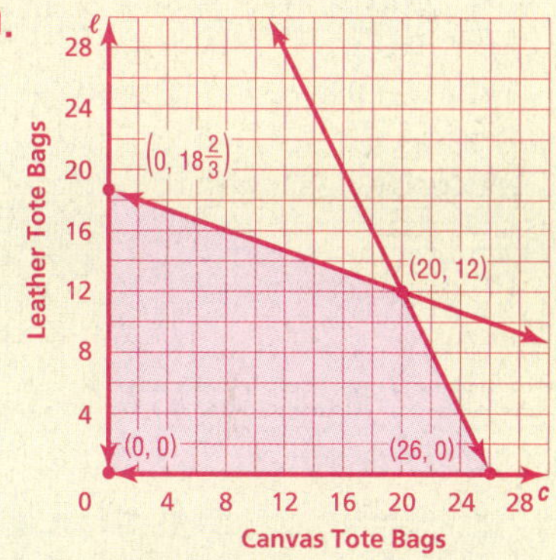

32.

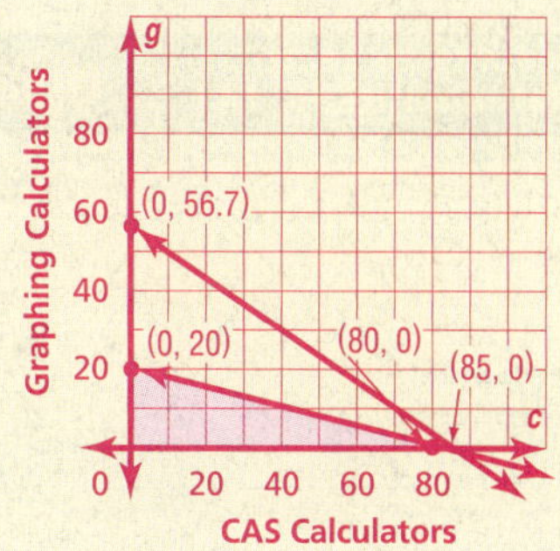

5.

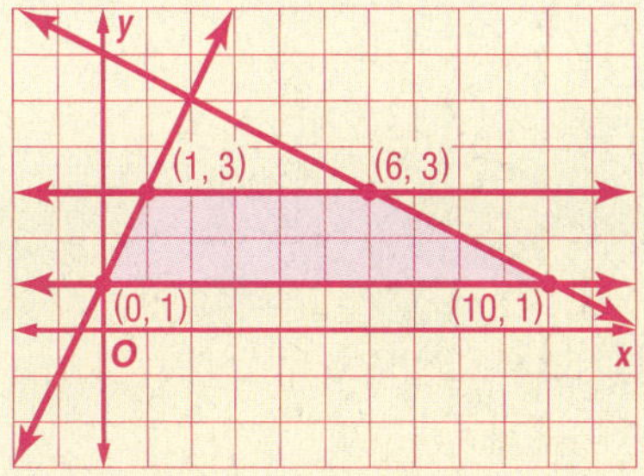

vertices: (0, 1), (1, 3), (6, 3), (10, 1); max: $f(10, 1) = 31$, min: $f(0, 1) = 1$

6.

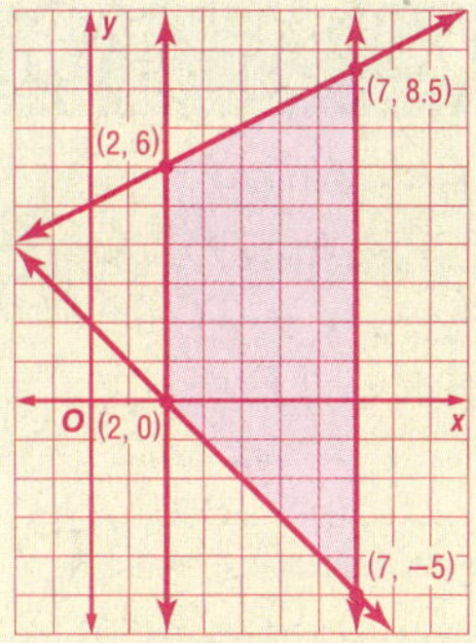

vertices: (2, 0), (2, 6), (7, 8.5), (7, −5); max: $f(7, 8.5) = 81.5$, min: $f(2, 0) = 16$

Study Guide and Intervention, p. 137 (shown) and p. 138

Maximum and Minimum Values When a system of linear inequalities produces a bounded polygonal region, the *maximum* or *minimum* value of a related function will occur at a vertex of the region.

Example **Graph the system of inequalities. Name the coordinates of the vertices of the feasible region. Find the maximum and minimum values of the function $f(x, y) = 3x + 2y$ for this polygonal region.**

$y \le 4$
$y \le -x + 6$
$y \ge \frac{1}{2}x - \frac{3}{2}$
$y \le 6x + 4$

First find the vertices of the bounded region. Graph the inequalities.

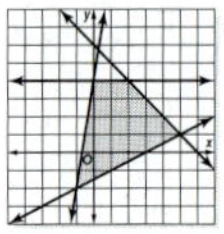

The polygon formed is a quadrilateral with vertices at (0, 4), (2, 4), (5, 1), and (−1, −2). Use the table to find the maximum and minimum values of $f(x, y) = 3x + 2y$.

(x, y)	$3x + 2y$	$f(x, y)$
(0, 4)	3(0) + 2(4)	8
(2, 4)	3(2) + 2(4)	14
(5, 1)	3(5) + 2(1)	17
(−1, −2)	3(−1) + 2(−2)	−7

The maximum value is 17 at (5, 1). The minimum value is −7 at (−1, −2).

Exercises

Graph each system of inequalities. Name the coordinates of the vertices of the feasible region. Find the maximum and minimum values of the given function for this region.

1. $y \ge 2$
$1 \le x \le 5$
$y \le x + 3$
$f(x, y) = 3x - 2y$

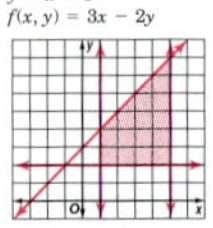

vertices: (1, 2), (1, 4), (5, 8), (5, 2); max: 11; min; −5

2. $y \ge -2$
$y \ge 2x - 4$
$x - 2y \ge -1$
$f(x, y) = 4x - y$

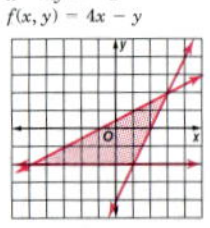

vertices: (−5, −2), (3, 2), (1, −2); max: 10; min: −18

3. $x + y \ge 2$
$4y \le x + 8$
$y \ge 2x - 5$
$f(x, y) = 4x + 3y$

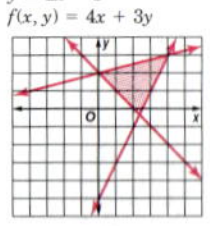

vertices (0, 2), (4, 3), $\left(\frac{7}{3}, -\frac{1}{3}\right)$; max: 25; min: 6

Skills Practice, p. 139 and Practice, p. 140 (shown)

Graph each system of inequalities. Name the coordinates of the vertices of the feasible region. Find the maximum and minimum values of the given function for this region.

1. $2x - 4 \le y$
$-2x - 4 \le y$
$y \le 2$
$f(x, y) = -2x + y$

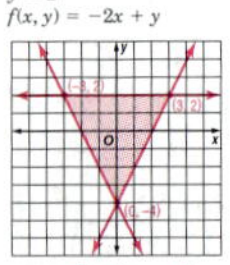

max.: 8, min.: −4

2. $3x - y \le 7$
$2x - y \ge 3$
$y \ge x - 3$
$f(x, y) = x - 4y$

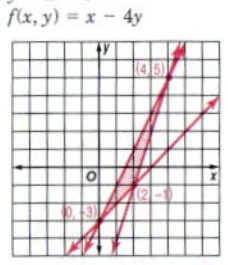

max.: 12, min.: −16

3. $x \ge 0$
$y \ge 0$
$y \le 6$
$y \le -3x + 15$
$f(x, y) = 3x + y$

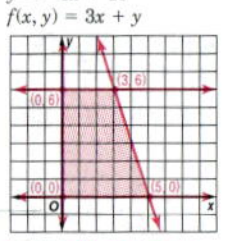

max.: 15, min.: 0

4. $x \le 0$
$y \le 0$
$4x + y \ge -7$
$f(x, y) = -x - 4y$

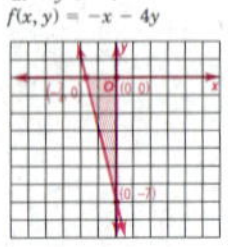

max.: 28, min.: 0

5. $y \le 3x + 6$
$4y + 3x \le 3$
$x \ge -2$
$f(x, y) = -x + 3y$

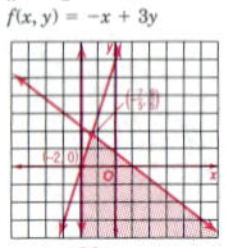

max.: $\frac{34}{5}$, no min.

6. $2x + 3y \ge 6$
$2x - y \le 2$
$x \ge 0$
$y \ge 0$
$f(x, y) = x + 4y + 3$

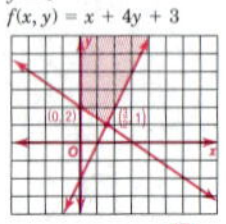

no max., min.: $\frac{17}{2}$

PRODUCTION For Exercises 7–9, use the following information.

A glass blower can form 8 simple vases or 2 elaborate vases in an hour. In a work shift of no more than 8 hours, the worker must form at least 40 vases.

7. Let s represent the hours forming simple vases and e the hours forming elaborate vases. Write a system of inequalities involving the time spent on each type of vase. $s \ge 0, e \ge 0, s + e \le 8, 8s + 2e \ge 40$

8. If the glass blower makes a profit of \$30 per hour worked on the simple vases and \$35 per hour worked on the elaborate vases, write a function for the total profit on the vases. $f(s, e) = 30s + 35e$

9. Find the number of hours the worker should spend on each type of vase to maximize profit. What is that profit? 4 h on each; \$260

Reading to Learn Mathematics, p. 141 ELL

Pre-Activity **How is linear programming used in scheduling work?**

Read the introduction to Lesson 3-4 at the top of page 129 in your textbook.

Name two or more facts that indicate that you will need to use inequalities to model this situation.

Sample answer: The buoy tender can carry up to 8 new buoys. There seems to be a limit of 24 hours on the time the crew has at sea. The crew will want to repair or replace the maximum number of buoys possible.

Reading the Lesson

1. Complete each sentence.

   a. When you find the feasible region for a linear programming problem, you are solving a system of linear __inequalities__ called __constraints__. The points in the feasible region are __solutions__ of the system.

   b. The corner points of a polygonal region are the __vertices__ of the feasible region.

2. A polygonal region always takes up only a limited part of the coordinate plane. One way to think of this is to imagine a circle or rectangle that the region would fit inside. In the case of a polygonal region, you can always find a circle or rectangle that is large enough to contain all the points of the polygonal region. What word is used to describe a region that can be enclosed in this way? What word is used to describe a region that is too large to be enclosed in this way? **bounded; unbounded**

3. How do you find the corner points of the polygonal region in a linear programming problem? **You solve a system of two linear equations.**

4. What are some everyday meanings of the word *feasible* that remind you of the mathematical meaning of the term *feasible region*?
Sample answer: possible or achievable

Helping You Remember

5. Look up the word *constraint* in a dictionary. If more than one definition is given, choose the one that seems closest to the idea of a *constraint* in a linear programming problem. How can this definition help you to remember the meaning of *constraint* as it is used in this lesson? **Sample answer: A *constraint* is a restriction or limitation. The constraints in a linear programming problem are restrictions on the variables that translate into inequality statements.**

FARMING **For Exercises 38–41, use the following information.**
Dean Stadler has 20 days in which to plant corn and soybeans. The corn can be planted at a rate of 250 acres per day and the soybeans at a rate of 200 acres per day. He has 4500 acres available for planting these two crops.

38. Let c represent the number of acres of corn and let s represent the number of acres of soybeans. Write a system of inequalities to represent the possible ways Mr. Stadler can plant the available acres.

38. $c \ge 0$, $s \ge 0$, $c + s \le 4500$, $4c + 5s \le 20{,}000$

39. Draw the graph showing the feasible region and list the coordinates of the vertices of the feasible region. **See pp. 151A–151F.**

40. If the profit on corn is \$26 per acre and the profit on soybeans is \$30 per acre, how much of each should Mr. Stadler plant? What is the maximum profit?

40. 2500 acres corn, 2000 acres soybeans; \$125,000

41. How much of each should Mr. Stadler plant if the profit on corn is \$29 per acre and the profit on soybeans is \$24 per acre? What is the maximum profit?
4500 acres corn, 0 acres soybeans; \$130,500

★ 42. **PACKAGING** The Cookie Factory's best selling items are chocolate chip cookies and peanut butter cookies. They want to sell both types of cookies together in combination packages. The different-sized packages will contain between 6 and 12 cookies, inclusively. At least three of each type of cookie should be in each package. The cost of making a chocolate chip cookie is 19¢, and the selling price is 44¢ each. The cost of making a peanut butter cookie is 13¢, and the selling price is 39¢. How many of each type of cookie should be in each package to maximize the profit? **3 chocolate chip, 9 peanut butter**

43. *WRITING IN MATH* Answer the question that was posed at the beginning of the lesson. **See pp. 151A–151F.**

How is linear programming used in scheduling work?

Include the following in your answer:

- a system of inequalities that represents the constraints that are used to schedule buoy repair and replacement,
- an explanation of the linear function that the buoy tender captain would wish to maximize, and
- a demonstration of how to solve the linear programming problem to find the maximum number of buoys the buoy tender could service in 24 hours at sea.

Standardized Test Practice

44. A feasible region has vertices at (0, 0), (4, 0), (5, 5), and (0, 8). Find the maximum and minimum of the function $f(x, y) = x + 3y$ over this region. **A**

   (A) maximum: $f(0, 8) = 24$
   minimum: $f(0, 0) = 0$

   (B) minimum: $f(0, 0) = 0$
   maximum: $f(5, 5) = 20$

   (C) maximum: $f(5, 5) = 20$
   minimum: $f(0, 8) = 8$

   (D) minimum: $f(4, 0) = 4$
   maximum: $f(0, 0) = 0$

45. What is the area of square $ABCD$? **C**

   (A) 25 units^2
   (B) $4\sqrt{29}$ units^2
   (C) 29 units^2
   (D) $25 + \sqrt{2}$ units^2

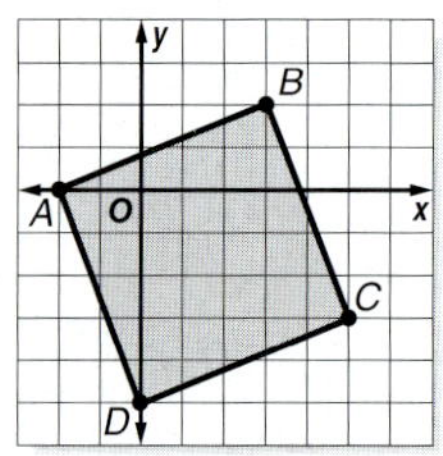

Enrichment, p. 142

Computer Circuits and Logic

Computers operate according to the laws of logic. The circuits of a computer can be described using logic.

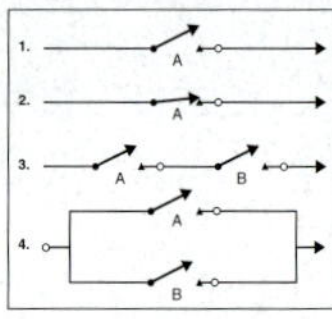

1. With switch A open, no current flows. The value 0 is assigned to an open switch.
2. With switch A closed, current flows. The value 1 is assigned to a closed switch.
3. With switches A and B open, no current flows. This circuit can be described by the conjunction, A · B.
4. In this circuit, current flows if either A or B is closed. This circuit can be described by the disjunction, A + B.

A	B	A + …

Truth tables are used to describe the flow of current in a circuit. The table …

Maintain Your Skills

Mixed Review Solve each system of inequalities by graphing. *(Lesson 3-3)* **46–47. See margin.**

46. $2y + x \geq 4$
$y \geq x - 4$

47. $3x - 2y \leq -6$
$y \leq \frac{3}{2}x - 1$

Solve each system of equations by using either substitution or elimination. *(Lesson 3-2)*

48. $4x + 5y = 20$
$5x + 4y = 7$ **(−5, 8)**

49. $6x + y = 15$
$x - 4y = -10$ **(2, 3)**

50. $3x + 8y = 23$
$5x - y = 24$ **(5, 1)**

SCHOOLS For Exercises 51 and 52, use the graph at the right. *(Lesson 1-3)*

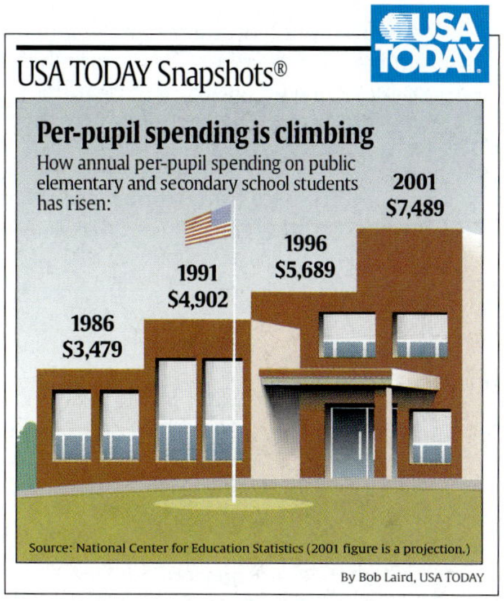

51. c = average cost each year; $15c + 3479 = 7489$

51. Define a variable and write an equation that can be used to determine on average how much the annual per-pupil spending has increased from 1986 to 2001.

52. Solve the problem. **about $267 per year**

Name the property illustrated by each equation. *(Lesson 1-2)*

53. $4n + (-4n) = 0$ **Add. Inv.**

54. $(2 \cdot 5) \cdot 6 = 2 \cdot (5 \cdot 6)$ **Assoc. (×)**

55. $\left(-\frac{3}{2}\right)\left(-\frac{2}{3}\right) = 1$ **Mult. Inv.**

56. $6(x + 9) = 6x + 6(9)$ **Distributive**

Getting Ready for the Next Lesson **PREREQUISITE SKILL** Evaluate each expression if $x = -2$, $y = 6$, and $z = 5$.
*(To review **evaluating expressions**, see Lesson 1-1.)*

57. $x + y + z$ **9**

58. $2x - y + 3z$ **5**

59. $-x + 4y - 2z$ **16**

60. $5x + 2y - z$ **−3**

61. $3x - y + 4z$ **8**

62. $-2x - 3y + 2z$ **−4**

Practice Quiz 2 — *Lessons 3-3 and 3-4*

Solve each system of inequalities by graphing. *(Lesson 3-3)* **1–3. See pp. 151A–151F.**

1. $y - x > 0$
$y + x < 4$

2. $y \geq 3x - 4$
$y \leq x + 3$

3. $x + 3y \geq 15$
$4x + y \leq 16$

Graph each system of inequalities. Name the coordinates of the vertices of the feasible region. Find the maximum and minimum values of the given function for this region. *(Lesson 3-4)* **4–5. See pp. 151A–151F for graphs.**

4. $x \geq 0$
$y \geq 0$
$y \leq 2x + 4$
$3x + y \leq 9$
$f(x, y) = 2x + y$
vertices: (0, 0), (0, 4), (1, 6), (3, 0); max: $f(1, 6) = 8$, min: $f(0, 0) = 0$

5. $x \leq 5$
$y \geq -3x$
$2y \leq x + 7$
$y \geq x - 4$
$f(x, y) = 4x - 3y$
vertices: (1, −3), (−1, 3), (5, 6), (5, 1); max: $f(5, 1) = 17$, min: $f(-1, 3) = -13$

Online Lesson Plans

USA TODAY Education's Online site offers resources and interactive features connected to each day's newspaper. *Experience TODAY*, USA TODAY's daily lesson plan, is available on the site and delivered daily to subscribers. This plan provides instruction for integrating USA TODAY graphics and key editorial features into your mathematics classroom. Log on to www.education.usatoday.com.

4 Assess

Open-Ended Assessment

Writing Have students write a summary of the steps they use to approach, graph, and solve a linear programming problem.

Getting Ready for Lesson 3-5

PREREQUISITE SKILL Lesson 3-5 presents systems of equations in three variables. Verifying the solution to such a system involves evaluating the three equations for the solution. Exercises 57–62 should be used to determine your students' familiarity with evaluating expressions involving three variables.

Assessment Options

Practice Quiz 2 The quiz provides students with a brief review of the concepts and skills in Lessons 3-3 and 3-4. Lesson numbers are given to the right of exercises or instruction lines so students can review concepts not yet mastered.

Quiz (Lesson 3-4) is available on p. 164 of the *Chapter 3 Resource Masters*.

Answers

46.
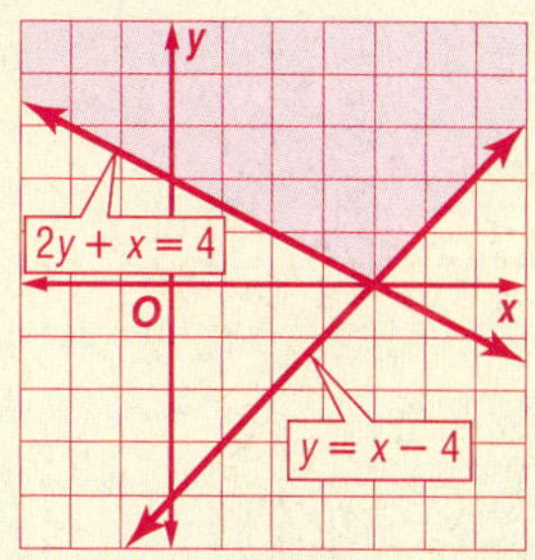

47.
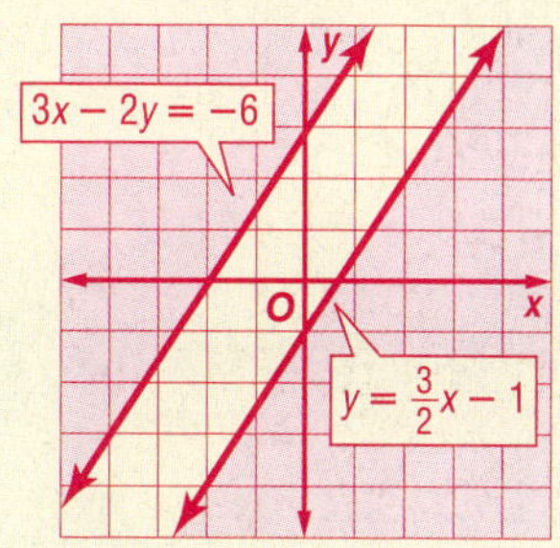

∅

Algebra Activity

A Preview of Lesson 3-5

Graphing Equations in Three Variables

To graph an equation in three variables, it is necessary to add a third dimension to our coordinate system. The graph of an equation of the form $Ax + By + Cz = D$, where A, B, C, and D can not all be equal to zero is a plane.

When graphing in three-dimensional space, begin with the xy-coordinate plane in a horizontal position. Then draw the z-axis as a vertical line passing through the origin. There are now three coordinate planes: the xy-plane, the xz-plane, and the yz-plane. These planes intersect at right angles and divide space into eight regions, called **octants**.

A point in space (three dimensions) has three coordinates and is represented by an **ordered triple** (x, y, z).

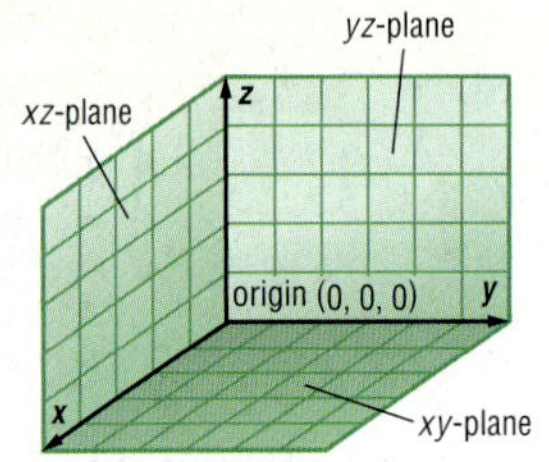

The first octant contains the points in space for which all three coordinates are positive.

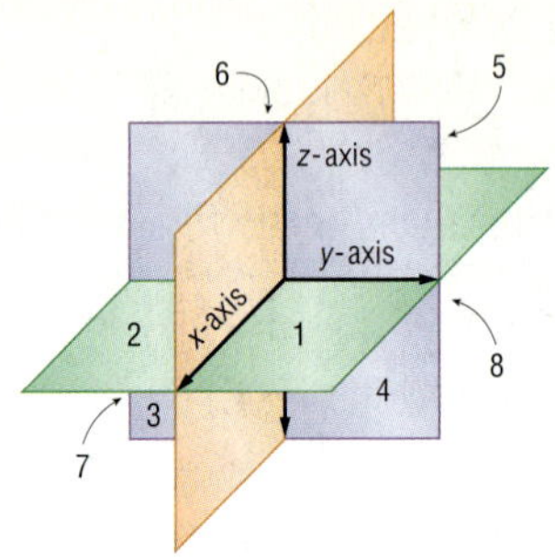

The octants are numbered as shown.

Activity 1

Use isometric dot paper to graph (3, 4, 2) on a three-dimensional coordinate system. Name the octant in which it lies.

Draw the x-, y-, and z-axes as shown.

Begin by finding the point (3, 4, 0) in the xy-plane.

The z-coordinate is 2, so move the point up two units parallel to the z-axis.

The point lies in octant 1.

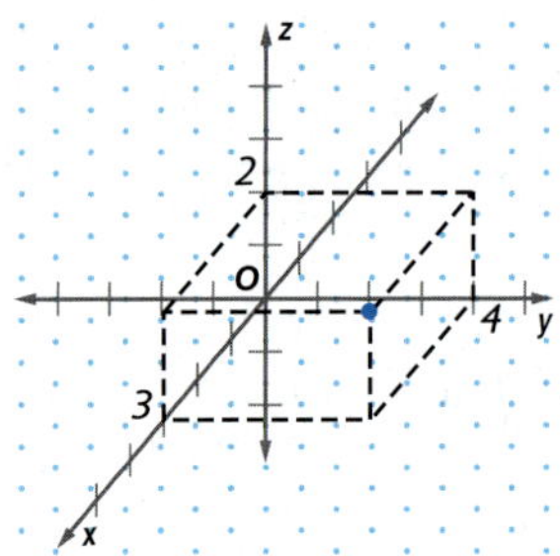

To graph a linear equation in three variables, first find the intercepts of the graph. Connect the intercepts on each axis. This forms a portion of a plane that lies in a single octant.

Algebra Activity

A Preview of Lesson 3-5

Getting Started

Objective Discover how to graph equations in three variables.

Materials
isometric dot paper
ruler

Teach

- To help students visualize this activity, consider cutting all four bottom corners from each of two cardboard boxes and taping them together to model the three-dimensional graphing system. The axes, planes, and octants can then be labeled to complete the model.
- Help students realize that the ordered triple (x, y, z) is similar to the familiar ordered pair (x, y) in a coordinate plane.
- Have students complete Exercises 1–14.

Answers (p. 137)

4.

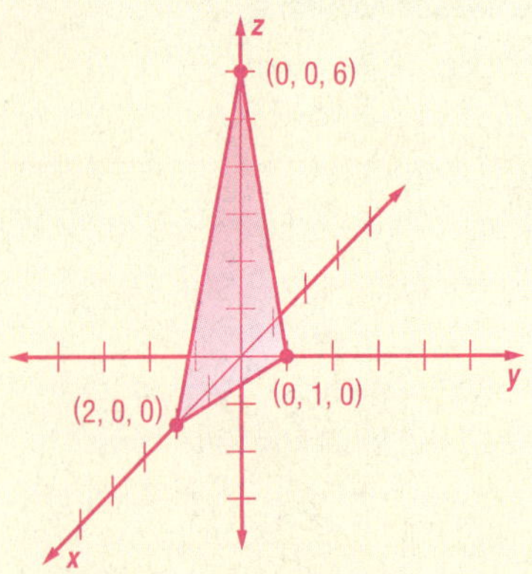

5.

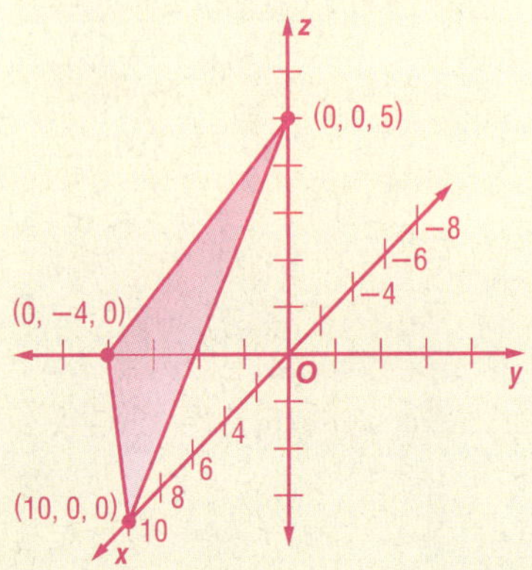

Resource Manager

Teaching Algebra with Manipulatives
- p. 19 (master for isometric dot paper)
- p. 225 (student recording sheet)

Glencoe Mathematics Classroom Manipulative Kit
- isometric dot grid stamp
- ink pad
- ruler

Activity 2

Graph $2x + 3y + 4z = 12$.

Begin by finding the x-, y-, and z-intercepts.

x-intercept	y-intercept	z-intercept
Let $y = 0$ and $z = 0$.	Let $x = 0$ and $z = 0$.	Let $x = 0$ and $y = 0$.
$2x = 12$	$3y = 12$	$4z = 12$
$x = 6$	$y = 4$	$z = 3$

To sketch the plane, graph the intercepts, which have coordinates (6, 0, 0), (0, 4, 0), and (0, 0, 3). Then connect the points. Remember this is only a portion of the plane that extends indefinitely.

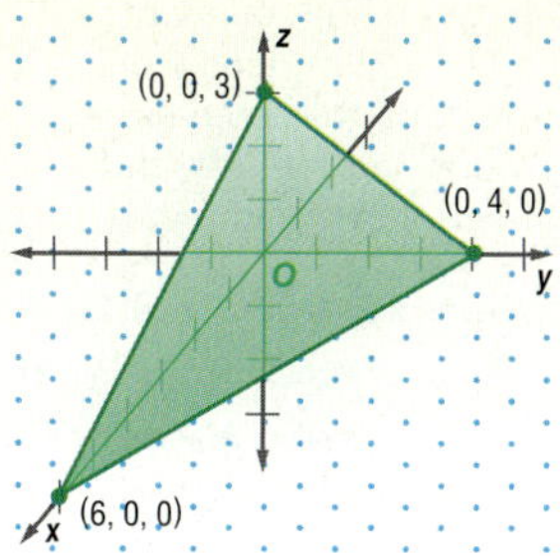

Model and Analyze ★ indicates increased difficulty

Graph each ordered triple on a three-dimensional coordinate system. Name the octant in which each point lies. 1–3. See pp. 151A–151F for graphs.

1. (5, 3, 6) 1 **2.** (−2, 4, 3) 5 **3.** (1, −5, 7) 2

Graph each equation. Name the coordinates for the x-, y-, and z-intercepts.

4. $3x + 6y + z = 6$ **5.** $2x - 5y + 4z = 20$ **6.** $x + 3y - 6z = 3$

7. $-3x + 5y + 10z = 15$ **8.** $6x + 9z = 18$ **9.** $4x - 6y = 24$

4–9. See margin for graphs.

4. (2, 0, 0), (0, 1, 0), (0, 0, 6)
5. (10, 0, 0), (0, −4, 0), (0, 0, 5)
6. (3, 0, 0), (0, 1, 0), (0, 0, −0.5)
7. (−5, 0, 0), (0, 3, 0), (0, 0, 1.5)
8. (3, 0, 0), none, (0, 0, 2)
9. (6, 0, 0), (0, −4, 0), none

★ **Write an equation of the plane given its x-, y-, and z-intercepts, respectively.**

10. 8, −3, 6 ★ **11.** 10, 4, −5 ★ **12.** $\frac{1}{2}$, 4, −12

$3x - 8y + 4z = 24$ $2x + 5y - 4z = 20$ $24x + 3y - z = 12$

10–12. Sample answers are given.

13. Describe the values of x, y, and z as either positive or negative for each octant. See margin.

14. Consider the graph $x = -3$ in one, two, and three dimensions. a–e. See pp. 151A–151F.

a. Graph the equation on a number line.

b. Graph the equation on a coordinate plane.

c. Graph the equation in a three-dimensional coordinate axis.

d. Describe and compare the graphs in parts **a**, **b**, and **c**.

e. **Make a conjecture** about the graph of $x > -3$ in one, two, and three dimensions.

Assess

In Exercises 1–12, students should be able to—

- graph an ordered triple in a three-dimensional coordinate system.
- graph equations and find the three intercepts.
- write an equation of a plane given its three intercepts.

Study Notebook

You may wish to have students summarize this activity and what they learned from it.

Answers

9.

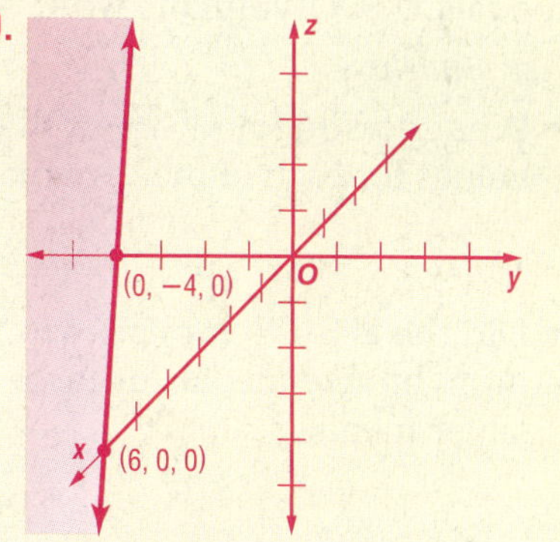

13. 1: x is positive, y is positive, and z is positive.
2: x is positive, y is negative, and z is positive.
3: x is positive, y is negative, and z is negative.
4: x is positive, y is positive, and z is negative.
5: x is negative, y is positive, and z is positive.
6: x is negative, y is negative, and z is positive.
7: x is negative, y is negative, and z is negative.
8: x is negative, y is positive, and z is negative.

6.

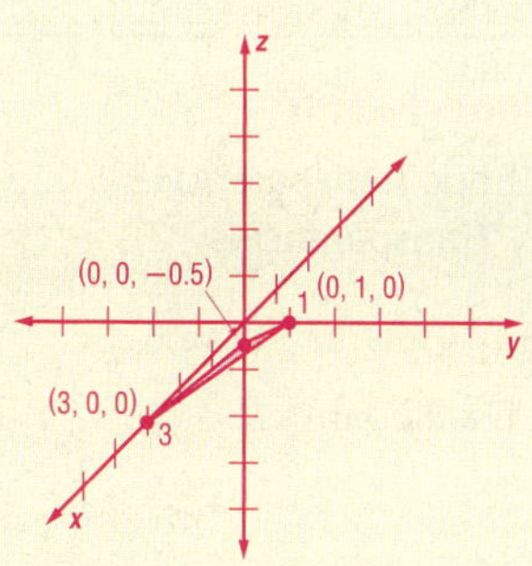

7.

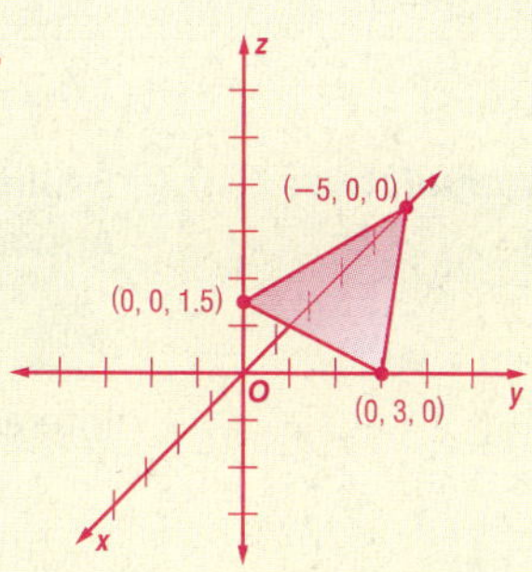

8.

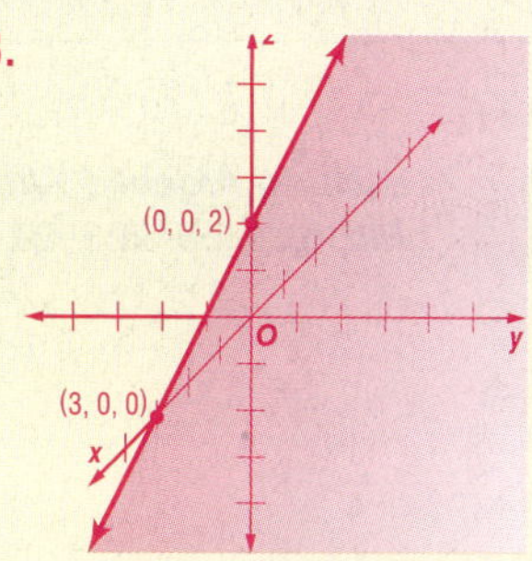

Algebra Activity Graphing Equations in Three Variables **137**

3-5 Lesson Notes

3-5 Solving Systems of Equations in Three Variables

1 Focus

5-Minute Check Transparency 3-5 Use as a quiz or review of Lesson 3-4.

Mathematical Background notes are available for this lesson on p. 108D.

Building on Prior Knowledge

In Lesson 3-2, students solved systems of two equations in two variables using algebra. In this lesson, they use many of the same skills as they solve systems of three equations in three variables.

How can you determine the number and type of medals U.S. Olympians won?

Ask students:

- What do you know about the values for each of the variables? **They are greater than or equal to zero.**
- Did the U.S. Olympians win more bronze medals or more silver medals? **bronze**

What You'll Learn

- Solve systems of linear equations in three variables.
- Solve real-world problems using systems of linear equations in three variables.

How can you determine the number and type of medals U.S. Olympians won?

At the 2000 Summer Olympics in Sydney, Australia, the United States won 97 medals. They won 6 more gold medals than bronze and 8 fewer silver medals than bronze.

You can write and solve a system of three linear equations to determine how many of each type of medal the U.S. Olympians won. Let g represent the number of gold medals, let s represent the number of silver medals, and let b represent the number of bronze medals.

$g + s + b = 97$	The U.S. won a total of 97 medals.
$g = b + 6$	They won 6 more gold medals than bronze.
$s = b - 8$	They won 8 fewer silver medals than bronze.

SYSTEMS IN THREE VARIABLES The system of equations above has three variables. The graph of an equation in three variables, all to the first power, is a plane. The solution of a system of three equations in three variables can have one solution, infinitely many solutions, or no solution.

Key Concept — *System of Equations in Three Variables*

One Solution

- planes intersect in one point

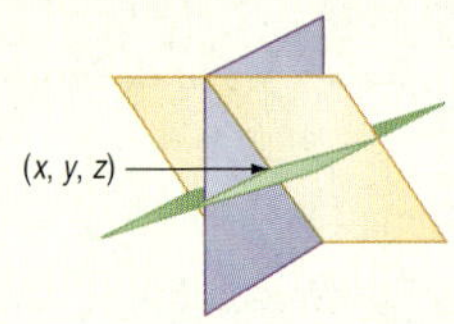

Infinite Solutions

- planes intersect in a line
- planes intersect in the same plane

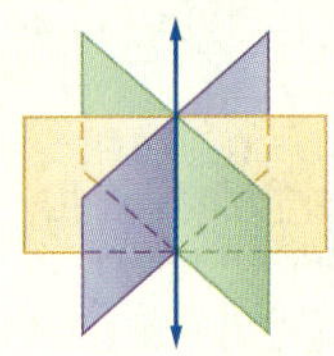

No Solution

- planes have no point in common

Resource Manager

Workbook and Reproducible Masters

Chapter 3 Resource Masters

- Study Guide and Intervention, pp. 143–144
- Skills Practice, p. 145
- Practice, p. 146
- Reading to Learn Mathematics, p. 147
- Enrichment, p. 148
- Assessment, p. 164

Teaching Algebra With Manipulatives Masters, pp. 226–227

Transparencies

5-Minute Check Transparency 3-5
Answer Key Transparencies

Technology

Interactive Chalkboard

Solving systems of equations in three variables is similar to solving systems of equations in two variables. Use the strategies of substitution and elimination. The solution of a system of equations in three variables x, y, and z is called an **ordered triple** and is written as (x, y, z).

Example 1 One Solution

Solve the system of equations.

$$x + 2y + z = 10$$
$$2x - y + 3z = -5$$
$$2x - 3y - 5z = 27$$

Study Tip

Elimination

Remember that you can eliminate any of the three variables.

Step 1 Use elimination to make a system of two equations in two variables.

$$\begin{aligned} x + 2y + z &= 10 \\ 2x - y + 3z &= -5 \end{aligned} \quad \text{Multiply by 2.} \quad \begin{aligned} 2x + 4y + 2z &= 20 \\ (-)\ 2x - y + 3z &= -5 \\ \hline 5y - z &= 25 \end{aligned} \quad \text{Subtract to eliminate } x.$$

$$\begin{aligned} 2x - y + 3z &= -5 && \text{Second equation} \\ (-)\ 2x - 3y - 5z &= 27 && \text{Third equation} \\ \hline 2y + 8z &= -32 && \text{Subtract to eliminate } x. \end{aligned}$$

Notice that the x terms in each equation have been eliminated. The result is two equations with the same two variables y and z.

Step 2 Solve the system of two equations.

$$\begin{aligned} 5y - z &= 25 \\ 2y + 8z &= -32 \end{aligned} \quad \text{Multiply by 8.} \quad \begin{aligned} 40y - 8z &= 200 \\ (+)\ 2y + 8z &= -32 \\ \hline 42y &= 168 && \text{Add to eliminate } z. \\ y &= 4 && \text{Divide by 42.} \end{aligned}$$

Substitute 4 for y in one of the equations with two variables and solve for z.

$$\begin{aligned} 5y - z &= 25 && \text{Equation with two variables} \\ 5(4) - z &= 25 && \text{Replace } y \text{ with 4.} \\ 20 - z &= 25 && \text{Multiply.} \\ z &= -5 && \text{Simplify.} \end{aligned}$$

The result is $y = 4$ and $z = -5$.

Step 3 Substitute 4 for y and -5 for z in one of the original equations with three variables.

$$\begin{aligned} x + 2y + z &= 10 && \text{Original equation with three variables} \\ x + 2(4) + (-5) &= 10 && \text{Replace } y \text{ with 4 and } z \text{ with } -5. \\ x + 8 - 5 &= 10 && \text{Multiply.} \\ x &= 7 && \text{Simplify.} \end{aligned}$$

The solution is $(7, 4, -5)$. You can check this solution in the other two original equations.

www.algebra2.com/extra_examples

2 Teach

SYSTEMS IN THREE VARIABLES

In-Class Example Power Point®

1 Solve the system of equations.
$5x + 3y + 2z = 2$
$2x + y - z = 5$
$x + 4y + 2z = 16$ **$(-2, 6, -3)$**

Teaching Tip Suggest that when solving systems of equations in three variables students should take a few moments before beginning their calculations to examine the equations and make a plan for solving the system. Also, remind students that the three values of the solution represent a point in space that is contained in all three planes, the x-plane, the y-plane, and the z-plane discussed in the Algebra Activity presented on pp. 136–137.

In-Class Examples

Power Point®

2 Solve the system of equations.

$2x + y - 3z = 5$
$x + 2y - 4z = 7$
$6x + 3y - 9z = 15$

There are an infinite number of solutions.

Teaching Tip Ask students to sketch a drawing of three planes whose equations form a system with an infinite number of solutions.

3 Solve the system of equations.

$3x - y - 2z = 4$
$6x + 4y + 8z = 11$
$9x + 6y + 12z = -3$

There is no solution of this system.

Teaching Tip Ask students to sketch a drawing of three planes whose equations form a system with no solutions.

Example 2 Infinite Solutions

Solve the system of equations.

$4x - 6y + 4z = 12$
$6x - 9y + 6z = 18$
$5x - 8y + 10z = 20$

Eliminate x in the first two equations.

$4x - 6y + 4z = 12$ Multiply by 3. $\quad 12x - 18y + 12z = 36$

$6x - 9y + 6z = 18$ Multiply by -2. $\quad (+)\ -12x + 18y - 12z = -36$ Add the equations.

$0 = 0$

The equation $0 = 0$ is always true. This indicates that the first two equations represent the same plane. Check to see if this plane intersects the third plane.

$4x - 6y + 4z = 12$ Multiply by 5. $\quad 20x - 30y + 20z = 60$

$5x - 8y + 10z = 20$ Multiply by -2. $\quad (+)\ -10x + 16y - 20z = -40$ Add the equations.

$10x - 14y = 20$

$5x - 7y = 10$ Divide by the GCF, 2.

The planes intersect in the line. So, there are an infinite number of solutions.

Example 3 No Solution

Solve the system of equations.

$6a + 12b - 8c = 24$
$9a + 18b - 12c = 30$
$4a + 8b - 7c = 26$

Eliminate a in the first two equations.

$6a + 12b - 8c = 24$ Multiply by 3. $\quad 18a + 36b - 24c = 72$

$9a + 18b - 12c = 30$ Multiply by 2. $\quad (-)\ 18a + 36b - 24c = 60$ Subtract the equations.

$0 = 12$

The equation $0 = 12$ is never true. So, there is no solution of this system.

More About...

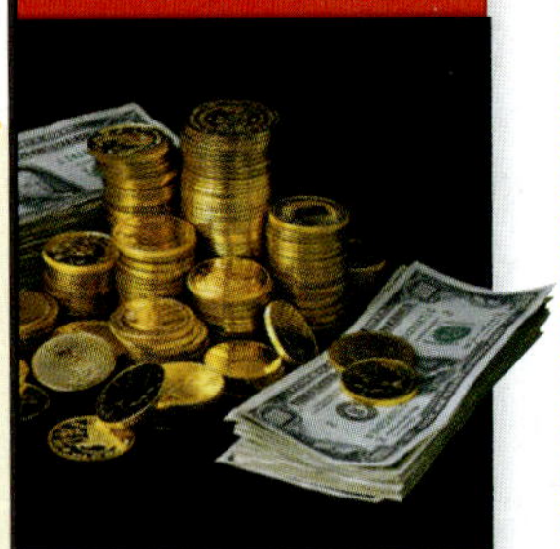

Investments

A certificate of deposit (CD) is a way to invest your money with a bank. The bank generally pays higher interest rates on CDs than savings accounts. However, you must invest your money for a specific time period, and there are penalties for early withdrawal.

REAL-WORLD PROBLEMS When solving problems involving three variables, use the four-step plan to help organize the information.

Example 4 Write and Solve a System of Equations

INVESTMENTS **Andrew Chang has $15,000 that he wants to invest in certificates of deposit (CDs). For tax purposes, he wants his total interest per year to be $800. He wants to put $1000 more in a 2-year CD than in a 1-year CD and invest the rest in a 3-year CD. How much should Mr. Chang invest in each type of CD?**

Number of Years	1	2	3
Rate	3.4%	5.0%	6.0%

Explore Read the problem and define the variables.

a = the amount of money invested in a 1-year certificate
b = the amount of money in a 2-year certificate
c = the amount of money in a 3-year certificate

DAILY INTERVENTION — Unlocking Misconceptions

Systems in Three Variables Some students may think that *any* ordered triple will be a solution to a system in three variables where there are an infinite number of solutions (as in Example 2). Explain that only the infinite set of ordered triples that names points on the line of intersection contains solutions to the system of equations.

140 Chapter 3 Systems of Equations and Inequalities

Plan Mr. Chang has \$15,000 to invest.

$a + b + c = 15{,}000$

The interest he earns should be \$800. The interest equals the rate times the amount invested.

$0.034a + 0.05b + 0.06c = 800$

There is \$1000 more in the 2-year certificate than in the 1-year certificate.

$b = a + 1000$

Solve Substitute $b = a + 1000$ in each of the first two equations.

$a + (a + 1000) + c = 15{,}000$ Replace b with $(a + 1000)$.

$2a + 1000 + c = 15{,}000$ Simplify.

$2a + c = 14{,}000$ Subtract 1000 from each side.

$0.034a + 0.05(a + 1000) + 0.06c = 800$ Replace b with $(a + 1000)$.

$0.034a + 0.05a + 50 + 0.06c = 800$ Distributive Property

$0.084a + 0.06c = 750$ Simplify.

Now solve the system of two equations in two variables.

$2a + c = 14{,}000$ Multiply by 0.06. $0.12a + 0.06c = 840$

$0.084a + 0.06c = 750$ $(-)\ 0.084a + 0.06c = 750$

$0.036a = 90$

$a = 2500$

Substitute 2500 for a in one of the original equations.

$b = a + 1000$ Third equation

$= 2500 + 1000$ $a = 2500$

$= 3500$ Add.

Substitute 2500 for a and 3500 for b in one of the original equations.

$a + b + c = 15{,}000$ First equation

$2500 + 3500 + c = 15{,}000$ $a = 2500, b = 3500$

$6000 + c = 15{,}000$ Add.

$c = 9000$ Subtract 6000 from each side.

So, Mr. Chang should invest \$2500 in a 1-year certificate, \$3500 in a 2-year certificate, and \$9000 in a 3-year certificate.

Examine Check to see if all the criteria are met.

The total investment is \$15,000.

$2500 + 3500 + 9000 = 15{,}000$ ✓

The interest earned will be \$800.

$0.034(2500) + 0.05(3500) + 0.06(9000) = 800$

$85 + 175 + 540 = 800$ ✓

There is \$1000 more in the 2-year certificate than the 1-year certificate.

$3500 = 2500 + 1000$ ✓

REAL-WORLD PROBLEMS

In-Class Example Power Point®

4 **SPORTS** There are 49,000 seats in a sports stadium. Tickets for the seats in the upper level sell for \$25, the ones in the middle level cost \$30, and the ones in the bottom level are \$35 each. The number of seats in the middle and bottom levels together equals the number of seats in the upper level. When all of the seats are sold for an event, the total revenue is \$1,419,500. How many seats are there in each level? **lower level: 14,400 seats; middle level: 10,100 seats; upper level: 24,500 seats**

DAILY INTERVENTION **Differentiated Instruction** ELL

Verbal/Linguistic Encourage students to explain to a partner their explorations and plans for solving problems using systems of three equations in three variables. Suggest that the listening partner take notes about their partner's strategies, asking questions as needed for clarification.

3 Practice/Apply

Study Notebook

Have students—
- complete the definitions/examples for the remaining terms on their Vocabulary Builder worksheets for Chapter 3.
- draw their own diagrams, similar to those on p. 138, showing the various possibilities for the intersection of three planes.
- include any other item(s) that they find helpful in mastering the skills in this lesson.

About the Exercises...

Organization by Objective
- **Systems in Three Variables:** 12–25
- **Real-World Problems:** 26–30

Odd/Even Assignments
Exercises 12–25 are structured so that students practice the same concepts whether they are assigned odd or even problems.

Assignment Guide

Basic: 13–19 odd, 25, 27, 28, 31–44

Average: 13–25 odd, 27–44

Advanced: 12–26 even, 27–44

DAILY INTERVENTION **FIND THE ERROR** Make sure students realize their work is not complete when they have determined two equations represent the same plane. Stress that they must determine that this plane intersects the plane represented by the third equation.

Check for Understanding

Concept Check

1. **Explain** how you can use the methods of solving a system of two equations in two variables to solve a system of three equations in three variables.

1. You can use elimination or substitution to eliminate one of the variables. Then you can solve two equations in two variables.

2. **FIND THE ERROR** Melissa is solving the system of equations $r + 2s + t = 3$, $2r + 4s + 2t = 6$, and $3r + 6s + 3t = 12$.

$r + 2s + t = 3 \rightarrow \quad 2r + 4s + 2t = 6$

$2r + 4s + 2t = 6 \rightarrow (-)2r + 4s + 2t = 6$

$0 = 0$

The second equation is a multiple of the first, so they are the same plane. There are infinitely many solutions.

Is she correct? Explain your reasoning.

2. No; the first two equations do represent the same plane, however they do not intersect the third plane, so there is no solution of this system.

3. **OPEN ENDED** Give an example of a system of three equations in three variables that has $(-3, 5, 2)$ as a solution. Show that the ordered triple satisfies all three equations. **See margin.**

Guided Practice

GUIDED PRACTICE KEY	
Exercises	Examples
4–9	1–3
10, 11	4

Solve each system of equations.

4. $x + 2y = 12$
$3y - 4z = 25$
$x + 6y + z = 20$
(6, 3, −4)

5. $9a + 7b = -30$
$8b + 5c = 11$
$-3a + 10c = 73$
(−1, −3, 7)

6. $r - 3s + t = 4$
$3r - 6s + 9t = 5$
$4r - 9s + 10t = 9$
infinitely many

7. $2r + 3s - 4t = 20$
$4r - s + 5t = 13$
$3r + 2s + 4t = 15$
(5, 2, −1)

8. $2x - y + z = 1$
$x + 2y - 4z = 3$
$4x + 3y - 7z = -8$
no solution

9. $x + y + z = 12$
$6x - 2y - z = 16$
$3x + 4y + 2z = 28$
(4, 0, 8)

Application

COOKING **For Exercises 10 and 11, use the following information.**
Jambalaya is a Cajun dish made from chicken, sausage, and rice. Simone is making a large pot of jambalaya for a party. Chicken costs \$6 per pound, sausage costs \$3 per pound, and rice costs \$1 per pound. She spends \$42 on $13\frac{1}{2}$ pounds of food. She buys twice as much rice as sausage.

10. Write a system of three equations that represents how much food Simone purchased. **$6c + 3s + r = 42$, $c + s + r = 13\frac{1}{2}$, $r = 2s$**

11. How much chicken, sausage, and rice will she use in her dish?
$4\frac{1}{2}$ lb chicken, 3 lb sausage, 6 lb rice

★ indicates increased difficulty

Practice and Apply

Homework Help	
For Exercises	See Examples
12–23	1–3
24–30	4

Extra Practice See page 833.

Solve each system of equations.

12. $2x - y = 2$
$3z = 21$
$4x + z = 19$ **(3, 4, 7)**

13. $-4a = 8$
$5a + 2c = 0$
$7b + 3c = 22$ **(−2, 1, 5)**

14. $5x + 2y = 4$ **(2, −3, 6)**
$3x + 4y + 2z = 6$
$7x + 3y + 4z = 29$

15. $8x - 6z = 38$
$2x - 5y + 3z = 5$
$x + 10y - 4z = 8$
(4, 0, −1)

16. $4a + 2b - 6c = 2$
$6a + 3b - 9c = 3$
$8a + 4b - 12c = 6$
no solution

17. $2r + s + t = 14$
$-r - 3s + 2t = -2$
$4r - 6s + 3t = -5$
(1, 5, 7)

142 Chapter 3 Systems of Equations and Inequalities

Answer

3. Sample answer: $x + y + z = 4$, $2x - y + z = -9$, $x + 2y - z = 5$; $-3 + 5 + 2 = 4$, $2(-3) - 5 + 2 = -9$, $-3 + 2(5) - 2 = 5$

18. (1, 2, −1)
19. infinitely many
20. $\left(\frac{1}{2}, \frac{3}{2}, \frac{9}{2}\right)$
21. $\left(\frac{1}{3}, -\frac{1}{2}, \frac{1}{4}\right)$
22. (8, 3, −6)
23. (−5, 9, 4)

18. $3x + y + z = 4$
$2x + 2y + 3z = 3$
$x + 3y + 2z = 5$

19. $4a - 2b + 8c = 30$
$a + 2b - 7c = -12$
$2a - b + 4c = 15$

★ **20.** $2r + s + t = 7$
$r + 2s + t = 8$
$r + s + 2t = 11$

★ **21.** $6x + 2y + 4z = 2$
$3x + 4y - 8z = -3$
$-3x - 6y + 12z = 5$

★ **22.** $r + s + t = 5$
$2r - 7s - 3t = 13$
$\frac{1}{2}r - \frac{1}{3}s + \frac{2}{3}t = -1$

★ **23.** $2a - b + 3c = -7$
$4a + 5b + c = 29$
$a - \frac{2b}{3} + \frac{c}{4} = -10$

24. The sum of three numbers is 20. The second number is 4 times the first, and the sum of the first and third is 8. Find the numbers. **3, 12, 5**

25. The sum of three numbers is 12. The first number is twice the sum of the second and third. The third number is 5 less than the first. Find the numbers. **8, 1, 3**

26. TRAVEL Jonathan and members of his Spanish Club are going to Costa Rica over spring break. Before his trip, he purchases 10 travelers checks in denominations of \$20, \$50, and \$100, totaling \$370. He has twice as many \$20 checks as \$50 checks. How many of each type of denomination of travelers checks does he have? **1-\$100, 3-\$50, and 6-\$20 checks**

DINING **For Exercises 27 and 28, use the following information.**
Maka loves the lunch combinations at Rosita's Mexican Restaurant. Today however, she wants a different combination than the ones listed on the menu.

Lunch Combo Meals

1. Two Tacos, One Burrito \$6.55
2. One Enchilada, One Taco, One Burrito \$7.10
3. Two Enchiladas, Two Tacos \$8.90

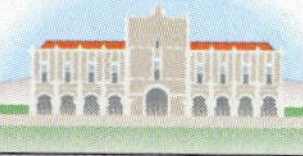

27. enchilada, \$2.50; taco, \$1.95; burrito, \$2.65

27. Assume that the price of a combo meal is the same price as purchasing each item separately. Find the price for an enchilada, a taco, and a burrito.

28. If Maka wants 2 burritos and 1 enchilada, how much should she plan to spend? **\$7.80**

More About. . .

Basketball

In 2001, Katie Smith was ranked 1st in the WNBA for points per game, three-point field goals, and minutes per game. She was also ranked 5th for free-throw percentage.

Source: www.wnba.com

BASKETBALL **For Exercises 29 and 30, use the following information.**
In the 2000–2001 season, Minnesota's Katie Smith was ranked first in the WNBA for total points and three-point goals made. She scored 646 points making 355 shots, including 3-point field goals, 2-point field goals, and 1-point free throws. She made 27 more 2-point field goals than 3-point field goals.

29. Write a system of three equations that represents the number of goals Katie Smith made. $x + y + z = 355$, $x + 2y + 3z = 646$, $y = z + 27$

30. Find the number of each type of goal she made. **88 3-point goals, 115 2-point goals, 152 1-point free throws**

Online Research **Data Update** What are the current rankings for the WNBA? Visit www.algebra2.com/data_update to learn more.

31. CRITICAL THINKING The general form of an equation for a parabola is $y = ax^2 + bx + c$, where (x, y) is a point on the parabola. Determine the values of a, b, c for the parabola at the right. Write the general form of the equation.
$a = \frac{3}{2}, b = 0, c = 3; y = \frac{3}{2}x^2 + 0x + 3$ or $y = \frac{3}{2}x^2 + 3$

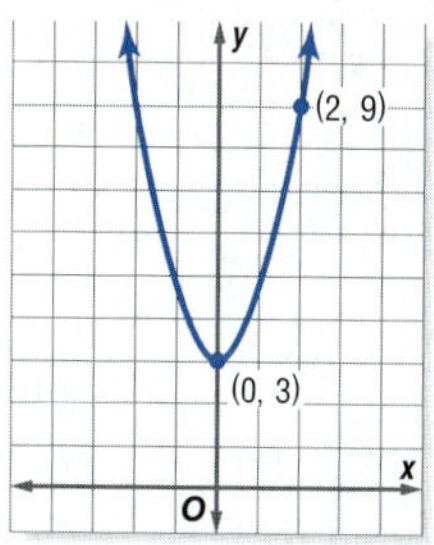

www.algebra2.com/self_check_quiz

Study Guide and Intervention, p. 143 (shown) and p. 144

Systems in Three Variables Use the methods used for solving systems of linear equations in two variables to solve systems of equations in three variables. A system of three equations in three variables can have a unique solution, infinitely many solutions, or no solution. A solution is an **ordered triple**.

Example **Solve this system of equations.** $3x + y - z = -6$; $2x - y + 2z = 8$; $4x + y - 3z = -21$

Step 1 Use elimination to make a system of two equations in two variables.

$3x + y - z = -6$ First equation
$(+)\ 2x - y + 2z = 8$ Second equation
$5x + z = 2$ Add to eliminate y.

$2x - y + 2z = 8$ Second equation
$(+)\ 4x + y - 3z = -21$ Third equation
$6x - z = -13$ Add to eliminate y.

Step 2 Solve the system of two equations.

$5x + z = 2$
$(+)\ 6x - z = -13$
$11x = -11$ Add to eliminate z.
$x = -1$ Divide both sides by 11.

Substitute −1 for x in one of the equations with two variables and solve for z.

$5x + z = 2$ Equation with two variables
$5(-1) + z = 2$ Replace x with −1.
$-5 + z = 2$ Multiply.
$z = 7$ Add 5 to both sides.

The result so far is $x = -1$ and $z = 7$.

Step 3 Substitute −1 for x and 7 for z in one of the original equations with three variables.

$3x + y - z = -6$ Original equation with three variables
$3(-1) + y - 7 = -6$ Replace x with −1 and z with 7.
$-3 + y - 7 = -6$ Multiply.
$y = 4$ Simplify.

The solution is (−1, 4, 7).

Exercises

Solve each system of equations.

1. $2x + 3y - z = 0$; $x - 2y - 4z = 14$; $3x + y - 8z = 17$ **(4, −3, −1)**
2. $2x - y + 4z = 11$; $x + 2y - 6z = -11$; $3x - 2y - 10z = 11$ $\left(2, -5, \frac{1}{2}\right)$
3. $x - 2y + z = 8$; $2x + y - z = 0$; $3x - 6y + 3z = 24$ **infinitely many solutions**
4. $3x - y - z = 5$; $3x + 2y - z = 11$; $6x - 3y + 2z = -12$ $\left(\frac{2}{3}, 2, -5\right)$
5. $2x - 4y - z = 10$; $4x - 8y - 2z = 16$; $3x + y + z = 12$ **no solution**
6. $x - 6y + 4z = 2$; $2x + 4y - 8z = 16$; $x - 2y = 5$ $\left(6, \frac{1}{2}, -\frac{1}{4}\right)$

Skills Practice, p. 145 and Practice, p. 146 (shown)

Solve each system of equations.

1. $2x - y + 2z = 15$; $-x + y + z = 3$; $3x - y + 2z = 18$ **(3, 1, 5)**
2. $x - 4y + 3z = -27$; $2x + 2y - 3z = 22$; $4z = -16$ **(1, 4, −4)**
3. $a + b = 3$; $-b + c = 3$; $a + 2c = 10$ **(2, 1, 4)**
4. $3m - 2n + 4p = 15$; $m - n + p = 3$; $m + 4n - 5p = 0$ **(3, 3, 3)**
5. $2g + 3h - 8j = 10$; $g - 4h = 1$; $-2g - 3h + 8j = 5$ **no solution**
6. $2x + y - z = -8$; $4x - y + 2z = -3$; $-3x + y + 2z = 5$ **(−2, −3, 1)**
7. $2x - 5y + z = 5$; $3x + 2y - z = 17$; $4x - 3y + 2z = 17$ **(5, 1, 0)**
8. $2x + 3y + 4z = 2$; $5x - 2y + 3z = 0$; $x - 5y - 2z = -4$ **(2, 2, −2)**
9. $p + 4r = -7$; $p - 3q = -8$; $q + r = 1$ **(1, 3, −2)**
10. $4x + 4y - 2z = 8$; $3x - 5y + 3z = 0$; $2x + 2y - z = 4$ **infinitely many**
11. $d + 3e + f = 0$; $-d + 2e + f = -1$; $4d + e - f = 1$ **(1, −1, 2)**
12. $4x + y + 5z = -9$; $x - 4y - 2z = -2$; $2x + 3y - 2z = 21$ **(2, 3, −4)**
13. $5x + 9y + z = 20$; $2x - y - z = -21$; $5x + 2y + 2z = -21$ **(−7, 6, 1)**
14. $2x + y - 3z = -3$; $3x + 2y + 4z = 5$; $-6x - 3y + 9z = 9$ **infinitely many**
15. $3x + 3y + z = 10$; $5x + 2y + 2z = 7$; $3x - 2y + 3z = -9$ **(1, 3, −2)**
16. $2u + v + w = 2$; $-3u + 2v + 3w = 7$; $-u - v + 2w = 7$ **(0, −1, 3)**
17. $x + 5y - 3z = -18$; $3x - 2y + 5z = 22$; $-2x - 3y + 8z = 28$ **(1, −2, 3)**
18. $x - 2y + z = -1$; $-x + 2y - z = 6$; $-4y + 2z = 1$ **no solution**
19. $2x - 2y - 4z = -2$; $3x - 3y - 6z = -3$; $-2x + 3y + z = 7$ **(4, 5, 0)**
20. $x - y + 9z = -27$; $2x - 4y - z = -1$; $3x + 6y - 3z = 27$ **(2, 2, −3)**
21. $2x - 5y - 3z = 7$; $-4x + 10y + 2z = 6$; $6x - 15y - z = -19$ **(1, 2, −5)**

22. The sum of three numbers is 6. The third number is the sum of the first and second numbers. The first number is one more than the third number. Find the numbers. **4, −1, 3**

23. The sum of three numbers is −4. The second number decreased by the third is equal to the first. The sum of the first and second numbers is −5. Find the numbers. **−3, −2, 1**

24. **SPORTS** Alexandria High School scored 37 points in a football game. Six points are awarded for each touchdown. After each touchdown, the team can earn one point for the extra kick or two points for a 2-point conversion. The team scored one fewer 2-point conversions than extra kicks. The team scored 10 times during the game. How many touchdowns were made during the game? **5**

Reading to Learn Mathematics, p. 147 ELL

Pre-Activity **How can you determine the number and type of medals U.S. Olympians won?**

Read the introduction to Lesson 3-5 at the top of page 138 in your textbook.

At the 1996 Summer Olympics in Atlanta, Georgia, the United States won 101 medals. The U.S. team won 12 more gold medals than silver and 7 fewer bronze medals than silver. Using the same variables as those in the introduction, write a system of equations that describes the medals won for the 1996 Summer Olympics.
$g + s + b = 101; g = s + 12; b = s - 7$

Reading the Lesson

1. The planes for the equations in a system of three linear equations in three variables determine the number of solutions. Match each graph description below with the description of the number of solutions of the system. (Some of the items on the right may be used more than once, and not all possible types of graphs are listed.)

a. three parallel planes **II**
b. three planes that intersect in a line **III**
c. three planes that intersect in one point **I**
d. one plane that represents all three equations **III**

I. one solution
II. no solutions
III. infinite solutions

2. Suppose that three classmates, Monique, Josh, and Lilly, are studying for a quiz on this lesson. They work together on solving a system of equations in three variables, x, y, and z, following the algebraic method shown in your textbook. They first find that $z = 3$, then that $y = -2$, and finally that $x = -1$. The students agree on these values, but disagree on how to write the solution. Here are their answers:

Monique: (3, −2, −1) Josh: (−2, −1, 3) Lilly: (−1, −2, 3)

a. How do you think each student decided on the order of the numbers in the ordered triple? **Sample answer: Monique arranged the values in the order in which she found them. Josh arranged them from smallest to largest. Lilly arranged them in alphabetical order of the variables.**

b. Which student is correct? **Lilly**

Helping You Remember

3. How can you remember that obtaining the equation $0 = 0$ indicates a system with infinitely many solutions, while obtaining an equation such as $0 = 8$ indicates a system with no solutions? **$0 = 0$ is always true, while $0 = 8$ is never true.**

Enrichment, p. 148

Billiards

The figure at the right shows a billiard table. The object is to use a cue stick to strike the ball at point C so that the ball will hit the sides (or cushions) of the table at least once before hitting the ball located at point A. In playing the game, you need to locate point P.

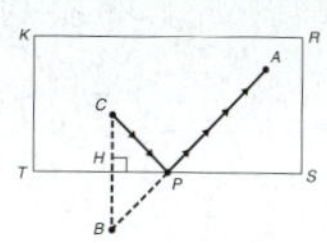

Step 1 Find point B so that $\overline{BC} \perp \overline{ST}$ and $\overline{BH} \cong \overline{CH}$. B is called the reflected image of C in $\overline{ST}$.

Step 2 Draw $\overline{AB}$.

Step 3 $\overline{AB}$ intersects $\overline{ST}$ at the desired point P.

For each billiards problem, the cue ball at point C must strike the indicated cushion(s) and then strike the ball at point A. Draw and label the correct path

32. **WRITING IN MATH** Answer the question that was posed at the beginning of the lesson. **See margin.**

How can you determine the number and type of medals U.S. Olympians won?

Include the following in your answer:

- a demonstration of how to find the number of each type of medal won by the U.S. Olympians, and
- a description of another situation where you can use a system of three equations in three variables to solve a problem.

Standardized Test Practice

33. If $a + b = 16$, $a - c = 4$, and $b - c = -4$, which statements are true? **D**

I. $b + c = 12$

II. $a - b = 8$

III. $a + c = 20$

Ⓐ I only　　Ⓑ II only

Ⓒ I and II only　　Ⓓ I, II, and III

34. If $x + y = 1$, $y + z = 10$, and $x + z = 3$, what is $x + y + z$? **A**

Ⓐ 7　　Ⓑ 8　　Ⓒ 13　　Ⓓ 14

Maintain Your Skills

Mixed Review

35. **PAPER** Wood pulp can be converted to either notebook paper or newsprint. The Canyon Pulp and Paper Mill can produce at most 200 units of paper a day. Regular customers require at least 10 units of notebook paper and 80 units of newspaper daily. If the profit on a unit of notebook paper is \$500 and the profit on a unit of newsprint is \$350, how many units of each type of paper should the mill produce each day to maximize profits? *(Lesson 3-4)* **120 units of notebook paper and 80 units of newsprint**

Solve each system of inequalities by graphing. *(Lesson 3-3)* **36–38. See margin.**

36. $y \le x + 2$
$y \ge 7 - 2x$

37. $4y - 2x > 4$
$3x + y > 3$

38. $3x + y \ge 1$
$2y - x \le -4$

STAMPS **For Exercises 39 and 40, use the following information.**
The table shows the price for first-class stamps since the U.S. Postal Service was created on July 1, 1971. *(Lesson 2-5)*

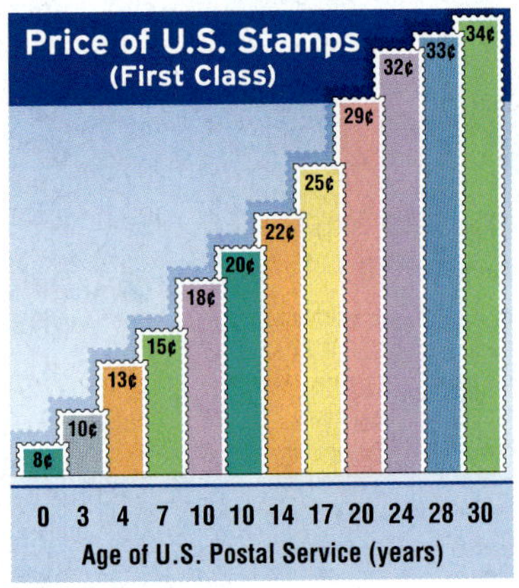

39. Write a prediction equation for this relationship.

39. Sample answer using (7, 15) and (14, 22): $y = x + 8$

40. Predict the price for a first-class stamp issued in the year 2010. **about 47¢**

Simplify each expression. *(Lesson 1-2)*

41. $5x + 2y - 4x + y$ **$x + 3y$**

42. $(4z + 1) - (6z - 7)$ **$-2z + 8$**

43. $(8s - 5t) + (9t + s)$ **$9s + 4t$**

44. $4(6a + 5b) - 2(3a + 2b)$ **$18a + 16b$**

4 Assess

Open-Ended Assessment

Speaking Ask students to state the various ways that three planes can intersect, explaining what will occur when solving the system of equations in three variables in each situation.

Tips for New Teachers

Intervention Some students have difficulty visualizing situations in three dimensions. Encourage them to use paper, pencils, and other objects to help them model the situations they encounter in this lesson. If you or some of your students have access to computer software that draws models in three dimensions, encourage them to use this tool to demonstrate what happens visually with these systems.

Assessment Options

Quiz (Lesson 3-5) is available on p. 164 of the *Chapter 3 Resource Masters*.

Answers

32. You can write a system of three equations in three variables to find the number of each type of medal. Answers should include the following.

- **You can substitute $b + 6$ for g and $b - 8$ for s in the equation $g + s + b = 97$. This equation is now in terms of b. Once you find b, you can substitute again to find g and s. The U.S. Olympians won 39 gold medals, 25 silver medals, and 33 bronze medals.**
- **Another situation involving three variables is winning times of the first, second, and third place finishers of a race.**

36.

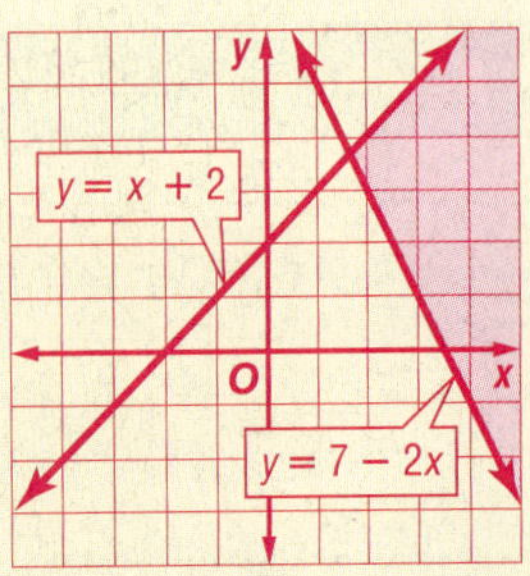

37.

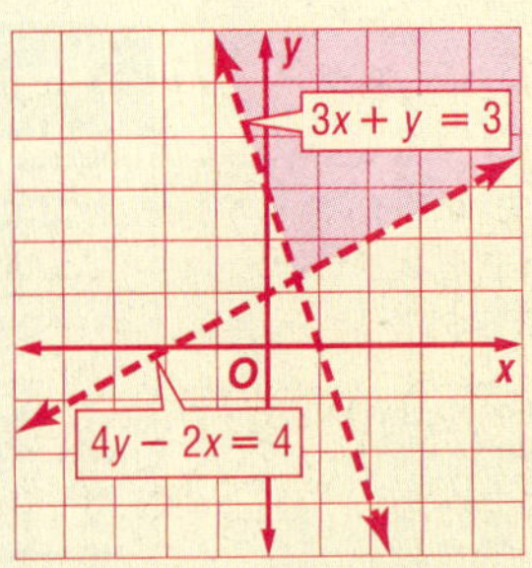

38.

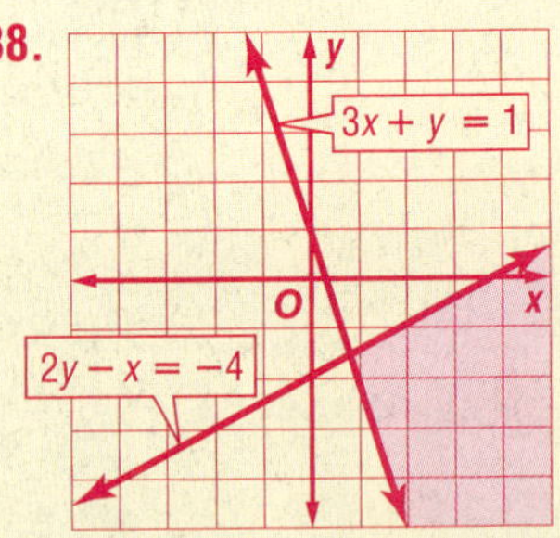

Chapter 3 Study Guide and Review

Vocabulary and Concept Check

bounded region (p. 129)	elimination method (p. 118)	linear programming (p. 130)	system of inequalities (p. 123)
consistent system (p. 111)	feasible region (p. 129)	ordered triple (p. 139)	unbounded region (p. 130)
constraints (p. 129)	inconsistent system (p. 111)	substitution method (p. 116)	vertices (p. 129)
dependent system (p. 111)	independent system (p. 111)	system of equations (p. 110)	

Choose the letter of the term that best matches each phrase.

1. the inequalities of a linear programming problem **c**
2. a system of equations that has an infinite number of solutions **b**
3. the region of a graph where every constraint is met **f**
4. a method of solving equations in which one equation is solved for one variable in terms of the other variable **i**
5. a system of equations that has at least one solution **a**
6. a method of solving equations in which one variable is eliminated when the two equations are combined **e**
7. the solution of a system of equations in three variables (x, y, z) **h**
8. a method for finding the maximum or minimum value of a function **g**
9. a system of equations that has no solution **d**
10. a region in which no maximum value exists **j**

a. consistent system
b. dependent system
c. constraints
d. inconsistent system
e. elimination method
f. feasible region
g. linear programming
h. ordered triple
i. substitution method
j. unbounded region

Lesson-by-Lesson Review

3-1 Solving Systems of Equations by Graphing

See pages 110–115.

Concept Summary

- The solution of a system of equations can be found by graphing the two lines and determining if they intersect and at what point they intersect.

Example **Solve the system of equations by graphing.**

$x + y = 3$
$3x - y = 1$

Graph both equations on the same coordinate plane.

The solution of the system is (1, 2).

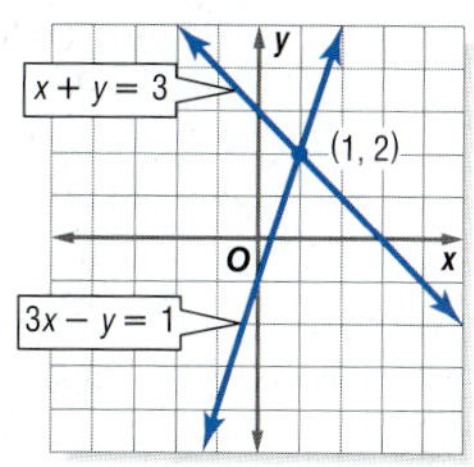

Exercises **Solve each system of equations by graphing.** *See Example 1 on page 110.*

11–14. See pp. 151A–151F for graphs.

11. $3x + 2y = 12$
$x - 2y = 4$ **(4, 0)**
12. $8x - 10y = 7$
$4x - 5y = 7$ **no solution**
13. $y - 2x = 8$
$y = \frac{1}{2}x - 4$ **(−8, −8)**
14. $20y + 13x = 10$
$0.65x + y = 0.5$ **infinitely many**

www.algebra2.com/vocabulary_review

FOLDABLES™ Study Organizer

For more information about Foldables, see *Teaching Mathematics with Foldables.*

Have students compare the length of their summaries for each lesson. Discuss what might make a summary too long (excessive detail) and what might make it too short (entries without enough detail). Have students work in small groups to compare their summaries for each lesson.

Encourage students to refer to their Foldables while completing the Study Guide and Review and to use them in preparing for the Chapter Test.

Chapter 3 Study Guide and Review

Vocabulary and Concept Check

- This alphabetical list of vocabulary terms in Chapter 3 includes a page reference where each term was introduced.
- **Assessment** A vocabulary test/review for Chapter 3 is available on p. 162 of the *Chapter 3 Resource Masters.*

Lesson-by-Lesson Review

For each lesson,

- the main ideas are summarized,
- additional examples review concepts, and
- practice exercises are provided.

Vocabulary PuzzleMaker

ELL The Vocabulary PuzzleMaker software improves students' mathematics vocabulary using four puzzle formats—crossword, scramble, word search using a word list, and word search using clues. Students can work on a computer screen or from a printed handout.

MindJogger Videoquizzes

ELL MindJogger Videoquizzes provide an alternative review of concepts presented in this chapter. Students work in teams in a game show format to gain points for correct answers. The questions are presented in three rounds.

Round 1 Concepts (5 questions)
Round 2 Skills (4 questions)
Round 3 Problem Solving (4 questions)

3-2 Solving Systems of Equations Algebraically

See pages 116–122.

Concept Summary

- In the substitution method, one equation is solved for a variable and substituted to find the value of another variable.
- In the elimination method, one variable is eliminated by adding or subtracting the equations.

Examples

1 **Use substitution to solve the system of equations.**

$x = 4y + 7$
$y = -3 - x$

Substitute $-3 - x$ for y in the first equation.

$x = 4y + 7$	First equation
$x = 4(-3 - x) + 7$	Substitute $-3 - x$ for y.
$x = -12 - 4x + 7$	Distributive Property
$5x = -5$	Add $4x$ to each side.
$x = -1$	Divide each side by 5.

Now substitute the value for x in either original equation.

$y = -3 - x$	Second equation
$y = -3 - (-1)$ or -2	The solution is $(-1, -2)$.

2 **Use the elimination method to solve the system of equations.**

$3x - 2y = 8$
$-x + y = 9$

Multiply the second equation by 2. Then add the equations to eliminate the y variable.

$3x - 2y = 8$
$-x + y = 9$

Multiply by 2.

$$\begin{aligned} 3x - 2y &= 8 \\ (+)\ -2x + 2y &= 18 \\ \hline x \qquad &= 26 \end{aligned}$$ Add the equations.

Replace x with 26 and solve for y.

$3x - 2y = 8$	Original equation.
$3(26) - 2y = 8$	Replace x with 26.
$78 - 2y = 8$	Multiply.
$-2y = -70$	Subtract 78 from each side.
$y = 35$	The solution is (26, 35).

Exercises **Solve each system of equations by using either substitution or elimination.** *See Examples 1–4 on pages 116–119.*

15. $x + y = 5$, $2x - y = 4$ **(3, 2)**

16. $2x - 3y = 9$, $4x + 2y = -22$ **(−3, −5)**

17. $7y - 2x = 10$, $-3y + x = -3$ **(9, 4)**

18. $-2x - 6y = 0$, $3x + 11y = 4$ **(−6, 2)**

19. $3x - 5y = -13$, $4x + 2y = 0$ **(−1, 2)**

20. $x + y = 4$, $x - y = 8.5$ **(6.25, −2.25)**

3-3 Solving Systems of Inequalities by Graphing

See pages 123–127.

Concept Summary

- A solution of a system of inequalities is found by graphing the inequalities and determining the intersection of the graphs.

Example **Solve the system of inequalities by graphing.**

$y \le x + 2$

$y \ge -4 - \frac{1}{2}x$

Graph each inequality and shade the intersection.

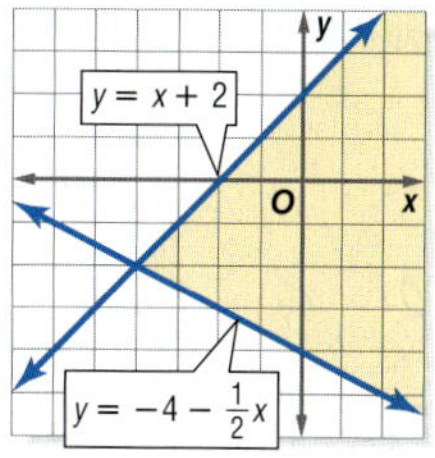

Exercises **Solve each system of inequalities by graphing.**
See Examples 1–3 on pages 123–124. **21–24. See margin.**

21. $y \le 4$
$y > -3$

22. $|y| > 3$
$x \le 1$

23. $y < x + 1$
$x > 5$

24. $y \le x + 4$
$2y \ge x - 3$

3-4 Linear Programming

See pages 129–135.

Concept Summary

- The maximum and minimum values of a function are determined by linear programming techniques.

Example **The available parking area of a parking lot is 600 square meters. A car requires 6 square meters of space, and a bus requires 30 square meters of space. The attendant can handle no more than 60 vehicles. If a car is charged \$3 to park and a bus is charged \$8, how many of each should the attendant accept to maximize income?**

Let c = the number of cars and b = the number of buses.

$c \ge 0$, $b \ge 0$, $6c + 30b \le 600$, and $c + b \le 60$

Graph the inequalities. The vertices of the feasible region are (0, 0), (0, 20), (50, 10), and (60, 0).

The profit function is $f(c, b) = 3c + 8b$. The maximum value of \$230 occurs at (50, 10). So the attendant should accept 50 cars and 10 buses.

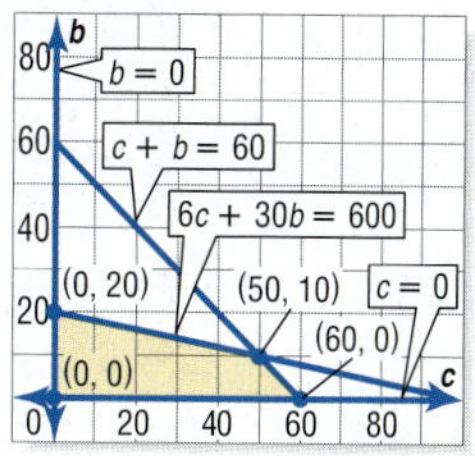

Exercise *See Example 3 on page 131.*

25. MANUFACTURING A toy manufacturer is introducing two new dolls, My First Baby and My Real Baby. In one hour, the company can produce 8 First Babies or 20 Real Babies. Because of demand, the company produces at least twice as many First Babies as Real Babies. The company spends no more than 48 hours per week making these two dolls. The profit on each First Baby is \$3.00, and the profit on each Real Baby is \$7.50. Find the number and type of dolls that should be produced to maximize profit. **160 My Real Babies, 320 My First Babies**

Answers

21.

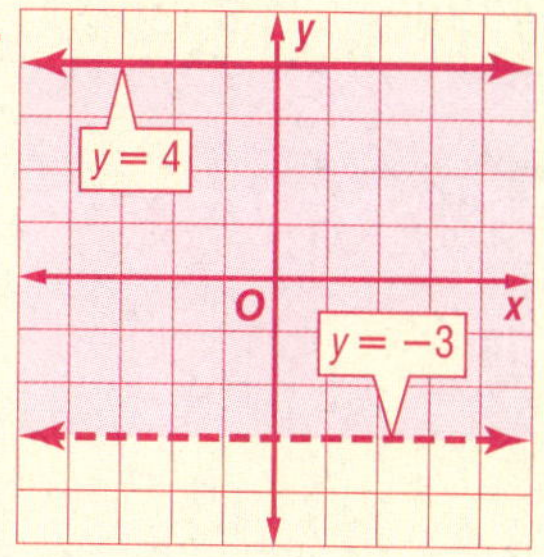

22.

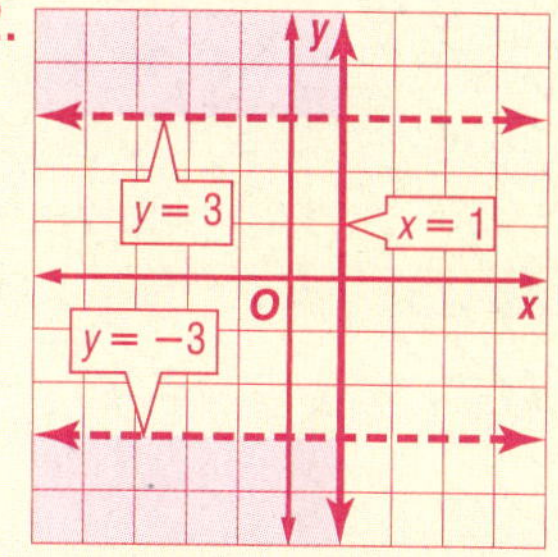

23.

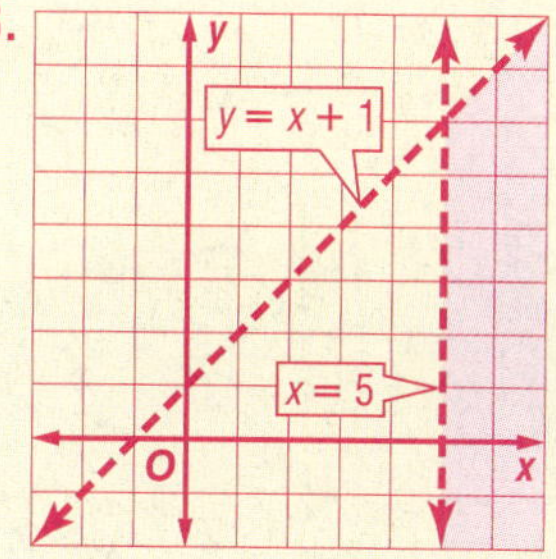

24.

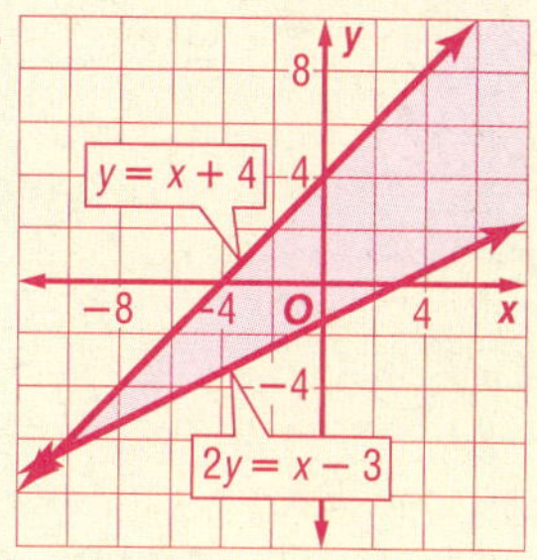

Chapter 3 **For More ...**

• Extra Practice, see pages 832–833.
• Mixed Problem Solving, see page 864.

3-5 Solving Systems of Equations in Three Variables

See pages 138–144.

Summary

- A system of three equations in three variables can be solved algebraically by using the substitution method or the elimination method.

Example **Solve the system of equations.**

$$x + 3y + 2z = 1$$
$$2x + y - z = 2$$
$$x + y + z = 2$$

Step 1 Use elimination to make a system of two equations in two variables.

$x + 3y + 2z = 1$, $2x + y - z = 2$ → Multiply by 2.

$$\begin{aligned} 2x + 6y + 4z &= 2 && \text{First equation} \\ (-)\ 2x + y - z &= 2 && \text{Second equation} \\ \hline 5y + 5z &= 0 && \text{Subtract.} \end{aligned}$$

$$\begin{aligned} x + 3y + 2z &= 1 && \text{First equation} \\ (-)\ x + y + z &= 2 && \text{Third equation} \\ \hline 2y + z &= -1 && \text{Subtract to eliminate } x. \end{aligned}$$

Step 2 Solve the system of two equations.

$5y + 5z = 0$, $2y + z = -1$ → Multiply by 5.

$$\begin{aligned} 5y + 5z &= 0 \\ (-)\ 10y + 5z &= -5 \\ \hline -5y &= 5 && \text{Subtract to eliminate } z. \\ y &= -1 && \text{Divide by } -5. \end{aligned}$$

Substitute -1 for y in one of the equations with two variables and solve for z.

$$\begin{aligned} 5y + 5z &= 0 && \text{Equation with two variables} \\ 5(-1) + 5z &= 0 && \text{Replace } y \text{ with } -1. \\ 5z &= 5 && \text{Add 5 to each side.} \\ z &= 1 && \text{Divide each side by 5.} \end{aligned}$$

Step 3 Substitute -1 for y and 1 for z in one of the equations with three variables.

$$\begin{aligned} 2x + y - z &= 2 && \text{Original equation with three variables} \\ 2x + (-1) - 1 &= 2 && \text{Replace } y \text{ with } -1 \text{ and } z \text{ with 1.} \\ 2x &= 4 && \text{Add 2 to each side.} \\ x &= 2 && \text{Divide each side by 2.} \end{aligned}$$

The solution is $(2, -1, 1)$.

Exercises **Solve each system of equations.** *See Examples 2–4 on pages 140–141.*

26. $x + 4y - z = 6$
$3x + 2y + 3z = 16$
$2x - y + z = 3$ **(1, 2, 3)**

27. $2a + b - c = 5$
$a - b + 3c = 9$
$3a - 6c = 6$ **(4, −2, 1)**

28. $e + f = 4$
$2d + 4e - f = -3$
$3e = -3$ **(3, −1, 5)**

Answers (p. 149)

10.

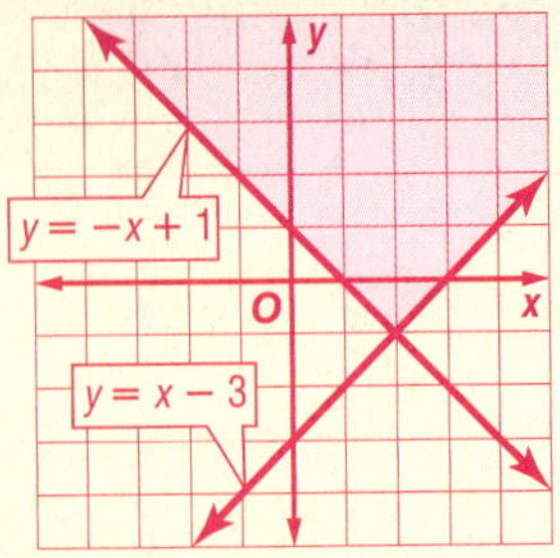

11.

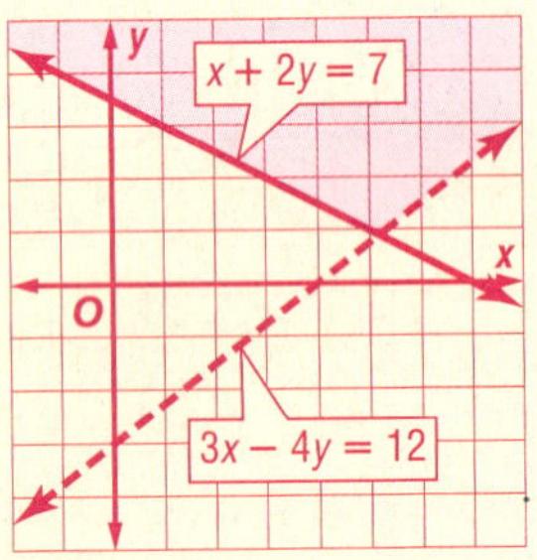

12.

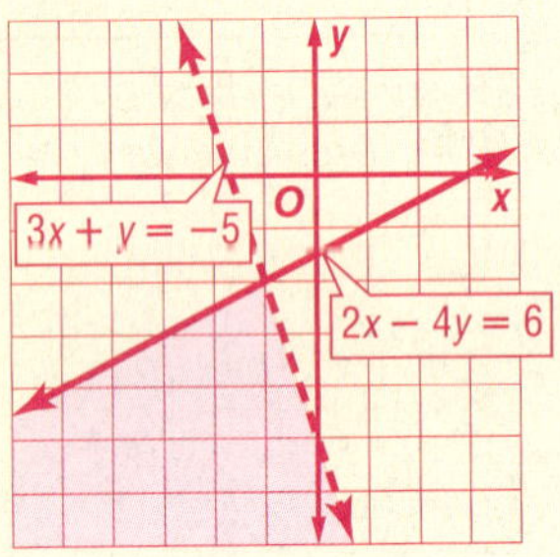

13.

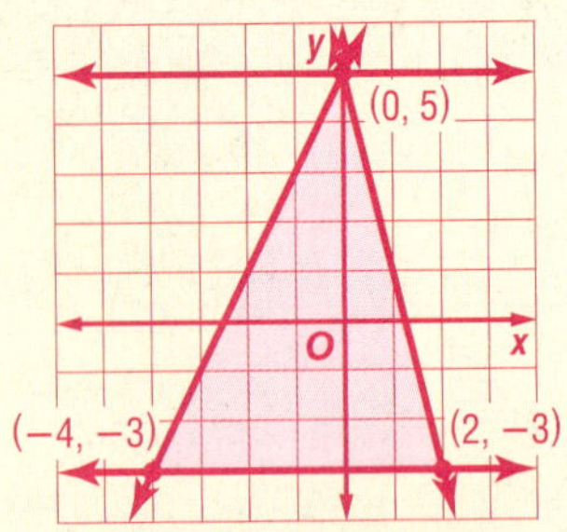

vertices: (−4, −3), (0, 5), (2, −3); max: f(2, −3) = 17, min: f(0, 5) = −15

14.

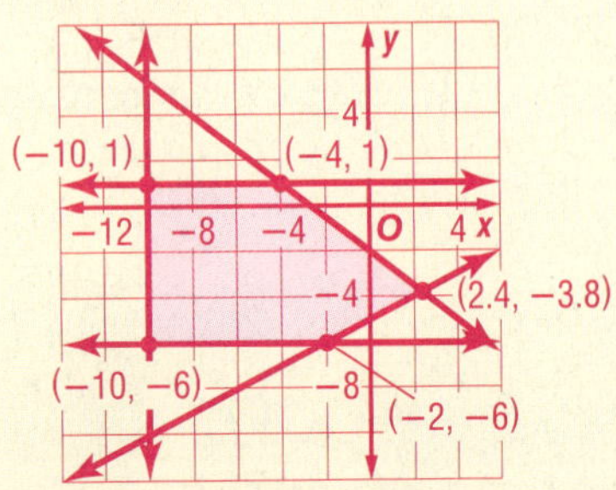

vertices: (−10, 1), (−4, 1), (2.4, −3.8), (−2, −6), (−10, −6); max: f(2.4, −3.8) = 1, min: f(−10, −6) = −26

148 Chapter 3 Systems of Equations and Inequalities

Chapter 3 Practice Test

Vocabulary and Concepts

Choose the word or term that best completes each statement or phrase.

1. Finding the maximum and minimum value of a linear function subject to constraints is called (*linear*, *polygonal*) programming.
2. The process of adding or subtracting equations to remove a variable and simplify solving the system of equations is called (*substitution*, *elimination*).
3. If a system of three equations in three variables has one solution, the graphs of the equations intersect in a (*point*, *plane*).

Skills and Applications

Solve each system of equations by graphing, substitution, or elimination.

4. $-4x + y = -5$, $2x + y = 7$ **(2, 3)**
5. $x + y = -8$, $-3x + 2y = 9$ **(−5, −3)**
6. $3x + 2y = 18$, $y = 6x - 6$ **(2, 6)**
7. $-6x + 3y = 33$, $-4x + y = 16$ **(−2.5, 6)**
8. $-7x + 6y = 42$, $3x + 4y = 28$ **(0, 7)**
9. $2y = 5x - 1$, $x + y = -1$ **$\left(-\frac{1}{7}, -\frac{6}{7}\right)$**

Solve each system of inequalities by graphing. **10–12. See margin.**

10. $y \geq x - 3$, $y \geq -x + 1$
11. $x + 2y \geq 7$, $3x - 4y < 12$
12. $3x + y < -5$, $2x - 4y \geq 6$

Graph each system of inequalities. Name the coordinates of the vertices of the feasible region. Find the maximum and the minimum values of the given function. **13–14. See margin.**

13. $5 \geq y \geq -3$, $4x + y \leq 5$, $-2x + y \leq 5$, $f(x, y) = 4x - 3y$
14. $x \geq -10$, $1 \geq y \geq -6$, $3x + 4y \leq -8$, $2y \geq x - 10$, $f(x, y) = 2x + y$

MANUFACTURING For Exercises 15 and 16, use the following information.
A sporting goods manufacturer makes a \$5 profit on soccer balls and a \$4 profit on volleyballs. Cutting requires 2 hours to make 75 soccer balls and 3 hours to make 60 volleyballs. Sewing needs 3 hours to make 75 soccer balls and 2 hours to make 60 volleyballs. Cutting has 500 hours available, and Sewing has 450 hours available.

15. How many soccer balls and volleyballs should be made to maximize the profit? **15. 11,250 soccer balls, 0 volleyballs**
16. What is the maximum profit the company can make from these two products? **\$56,250**

Solve each system of equations.

17. $x + y + z = -1$, $2x + 4y + z = 1$, $x + 2y - 3z = -3$ **(−4, 2, 1)**
18. $x + z = 7$, $2y - z = -3$, $-x - 3y + 2z = 11$ **(−2, 3, 9)**
19. **SHOPPING** Carla bought 3 shirts, 4 pairs of pants, and 2 pairs of shoes for a total of \$149.79. Beth bought 5 shirts, 3 pairs of pants, and 3 pairs of shoes totaling \$183.19. Kayla bought 6 shirts, 5 pairs of pants, and a pair of shoes for \$181.14. Assume that all of the shirts were the same price, all of the pants were the same price, and all of the shoes were the same price. What was the price of each item? **shirt, \$12.95; pants, \$15.99; shoes, \$23.49**
20. **STANDARDIZED TEST PRACTICE** Find the point at which the graphs of $2x + 3y = 7$ and $3x - 4y = 2$ intersect. **(2, 1)**

www.algebra2.com/chapter_test

Portfolio Suggestion

Introduction Your portfolio represents the mathematics you have done in this course. It shows you, your family, and your teacher a sampling of what you have learned and accomplished.

Ask Students Select one of the assignments from this chapter that you found especially challenging and place it in your portfolio. Write a short paragraph explaining why you found the assignment challenging and discuss how you were able to complete the assignment.

Chapter 3 Practice Test

Assessment Options

Vocabulary Test A vocabulary test/review for Chapter 3 can be found on p. 162 of the *Chapter 3 Resource Masters*.

Chapter Tests There are six Chapter 3 Tests and an Open-Ended Assessment task available in the *Chapter 3 Resource Masters*.

Chapter 3 Tests			
Form	**Type**	**Level**	**Pages**
1	MC	basic	149–150
2A	MC	average	151–152
2B	MC	average	153–154
2C	FR	average	155–156
2D	FR	average	157–158
3	FR	advanced	159–160

MC = multiple-choice questions
FR = free-response questions

Open-Ended Assessment
Performance tasks for Chapter 3 can be found on p. 161 of the *Chapter 3 Resource Masters*. A sample scoring rubric for these tasks appears on p. A22.

ExamView® Pro

Use the networkable **ExamView® Pro** to:

- Create **multiple versions** of tests.
- Create **modified** tests for *Inclusion* students.
- **Edit** existing questions and **add** your own questions.
- Use built-in **state curriculum correlations** to create tests aligned with state standards.
- **Apply** art to your tests from a program bank of artwork.

Chapter 3 Standardized Test Practice

These two pages contain practice questions in the various formats that can be found on the most frequently given standardized tests.

A practice answer sheet for these two pages can be found on p. A1 of the *Chapter 3 Resource Masters*.

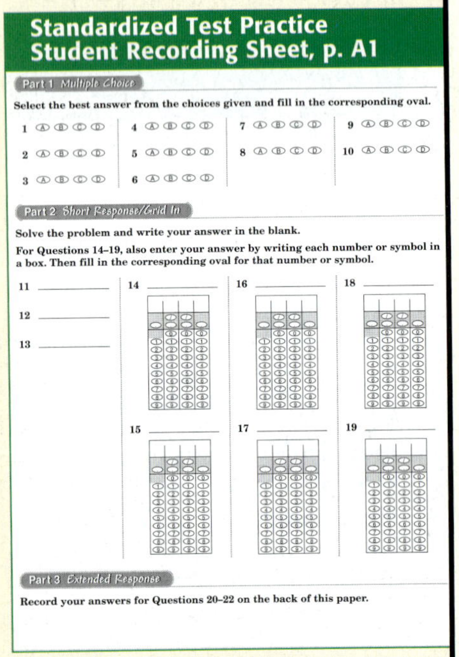
Standardized Test Practice Student Recording Sheet, p. A1

Part 1 Multiple Choice

Select the best answer from the choices given and fill in the corresponding oval.

1 Ⓐ Ⓑ Ⓒ Ⓓ 4 Ⓐ Ⓑ Ⓒ Ⓓ 7 Ⓐ Ⓑ Ⓒ Ⓓ 9 Ⓐ Ⓑ Ⓒ Ⓓ
2 Ⓐ Ⓑ Ⓒ Ⓓ 5 Ⓐ Ⓑ Ⓒ Ⓓ 8 Ⓐ Ⓑ Ⓒ Ⓓ 10 Ⓐ Ⓑ Ⓒ Ⓓ
3 Ⓐ Ⓑ Ⓒ Ⓓ 6 Ⓐ Ⓑ Ⓒ Ⓓ

Part 2 Short Response/Grid In

Solve the problem and write your answer in the blank.

For Questions 14–19, also enter your answer by writing each number or symbol in a box. Then fill in the corresponding oval for that number or symbol.

11 ____ 12 ____ 13 ____ 14 ____ 15 ____ 16 ____ 17 ____ 18 ____ 19 ____

Part 3 Extended Response

Record your answers for Questions 20–22 on the back of this paper.

Additional Practice

See pp. 167–168 in the *Chapter 3 Resource Masters* for additional standardized test practice.

Chapter 3 Standardized Test Practice

Part 1 Multiple Choice

Record your answers on the answer sheet provided by your teacher or on a sheet of paper.

1. What is the slope of any line parallel to the graph of $6x + 5y = 9$? **B**

Ⓐ -6 Ⓑ $-\frac{6}{5}$

Ⓒ $\frac{2}{3}$ Ⓓ 6

2. In the figure, $\triangle MOQ$ is similar to $\triangle NOP$. What is the length of $\overline{MQ}$? **D**

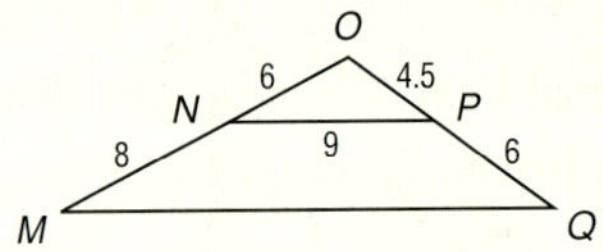

Ⓐ 12 Ⓑ 12.5

Ⓒ 19 Ⓓ 21

3. If $3x - y = -3$ and $x + 5y = 15$, what is the value of y? **D**

Ⓐ -3 Ⓑ 0

Ⓒ 1 Ⓓ 3

4. When 3 times x is increased by 4, the result is less than 16. Which of the following is a graph of the real numbers x that satisfy this relationship? **D**

Ⓐ 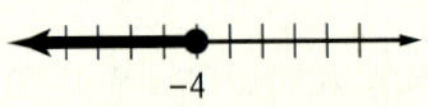

Ⓑ 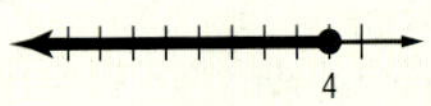

Ⓒ 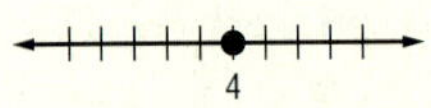

Ⓓ 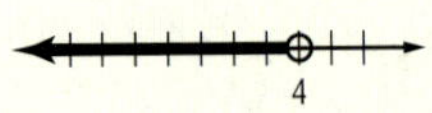

5. What is the area of the square $ABCD$? **C**

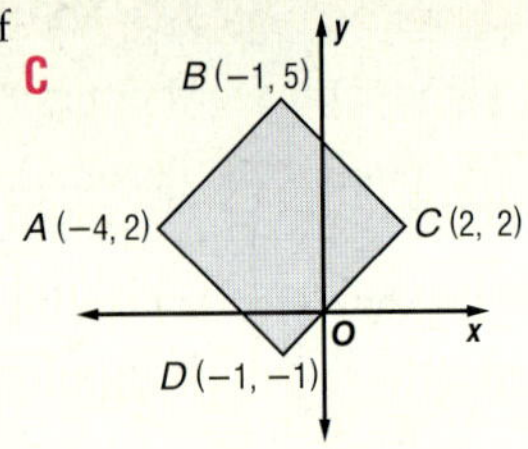

Ⓐ 27 units^2

Ⓑ $9\sqrt{2}$ units^2

Ⓒ 18 units^2

Ⓓ $12\sqrt{2}$ units^2

6. Twenty-seven white cubes of the same size are put together to form a larger cube. The larger cube is painted red. How many of the smaller cubes have exactly one red face ? **B**

Ⓐ 4 Ⓑ 6

Ⓒ 9 Ⓓ 12

7. Find the value of $|-4| \cdot |3|$. **D**

Ⓐ -12 Ⓑ -1

Ⓒ 7 Ⓓ 12

8. If two sides of a triangle measure 30 and 60, which of the following *cannot* be the measure of the third side? **A**

Ⓐ 30 Ⓑ 31

Ⓒ 40 Ⓓ 60

9. Marcus tried to compute the average of his 8 test scores. He mistakenly divided the correct total S of his scores by 7. The result was 12 more than what it should have been. Which equation would determine the value of S? **D**

Ⓐ $8S - 12 = 7S$ Ⓑ $\frac{S}{7} = \frac{S + 12}{18}$

Ⓒ $\frac{S}{7} + 12 = \frac{S}{8}$ Ⓓ $\frac{S}{7} - 12 = \frac{S}{8}$

10. If $x = -2$, then $15 - 3(x + 1) =$ **C**

Ⓐ 6. Ⓑ 12.

Ⓒ 18. Ⓓ 21.

ExamView® Pro

Special banks of standardized test questions similar to those on the SAT, ACT, TIMSS 12, NAEP 12, and state proficiency tests can be found on this CD-ROM.

Preparing for Standardized Tests
For test-taking strategies and more practice, see pages 877–892.

Part 2 Short Response/Grid In

Record your answers on the answer sheet provided by your teacher or on a sheet of paper.

11. Six of the 13 members of a club are boys, and the rest are girls. What is the ratio of girls to boys in the club? **7:6**

12. The integer k is greater than 50 and less than 100. When k is divided by 3, the remainder is 1. When k is divided by 8, the remainder is 2. What is one possible value of k? **58 or 82**

13. The area of the base of the rectangular box shown below is 35 square units. The area of one of the faces is 56 square units. Each of the dimensions a, b, and c is an integer greater than 1. What is the volume of the rectangular box? **280 cubic units**

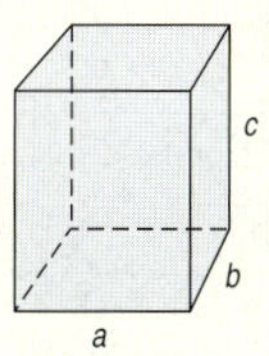

14. Four lines on a plane intersect in one point, forming 8 equal angles that are nonoverlapping. What is the measure, in degrees, of one of these angles? **45**

15. What is the greatest of five consecutive integers if the sum of these integers equals 135? **29**

Test-Taking Tip

Questions 8, 12, 16, and 17
If the question involves a geometric object but does not include a figure, draw one. A diagram can help you see relationships among the given values that will help you answer the question.

www.algebra2.com/standardized_test

16. If the perimeter of a rectangle is 12 times the width of the rectangle, then the length of the rectangle is how many times the width? **5**

17. Points A, B, C, and D lie in consecutive order on a line. If $AC = \frac{4}{3}AB$ and $BD = 6BC$, then what is $\frac{AB}{CD}$? **3/5 or .6**

18. The average (arithmetic mean) of the test scores of Mrs. Hilgart's first period class of x Algebra 2 students is 74. The average of the test scores of Mrs. Hilgart's fifth period class of y Algebra 2 students is 88. When the scores of both classes are combined, the average is 76. What is the value of $\frac{x}{y}$? **6**

19. A box contains 19 cups of rice. At most, how many servings can Alicia make from this box of rice if each serving is at least $\frac{3}{4}$ cup? **25**

Part 3 Extended Response

Record your answers on a sheet of paper. Show your work.

For Exercises 20–22, use the information below.

Christine had one dress and three sweaters cleaned at the dry cleaner and the charge was \$19.50. The next week, she had two dresses and two sweaters cleaned for a total charge of \$23.00.

20. Let d represent the price of cleaning a dress and s represent the price of cleaning a sweater. Write a system of linear equations to represent the prices of cleaning each item.
$d + 3s = 19.5$; $2d + 2s = 23$

21. Solve the system of equations using substitution or elimination. Explain your choice of method. **See margin.**

22. What will the charge be if Christine takes two dresses and four sweaters to be cleaned? **\$31**

Answer

20. (7.5, 4); Sample answer: Substitution; the coefficient of d is 1 in the first equation.

Evaluating Extended-Response Assessment Questions

Extended-Response Assessment questions are graded by using a multilevel rubric that guides you in assessing a student's knowledge of a particular concept.

Goal: Determine the dimensions of a rectangular garden for a given perimeter.

Sample Scoring Rubric: The following rubric is a sample scoring device. You may wish to add more detail to this sample to meet your individual scoring needs.

Score	Criteria
4	A correct solution that is supported by well-developed, accurate explanations
3	A generally correct solution, but may contain minor flaws in reasoning or computation
2	A partially correct interpretation and/or solution to the problem
1	A correct solution with no supporting evidence or explanation
0	An incorrect solution indicating no mathematical understanding of the concept or task, or no solution is given

Page 109, Chapter 3 Getting Started

1.
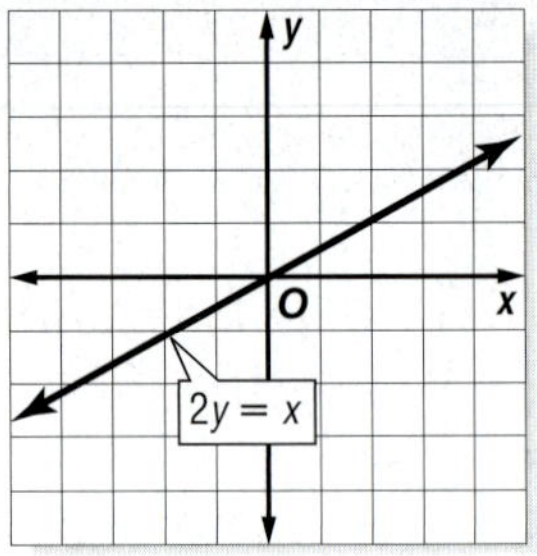

2.
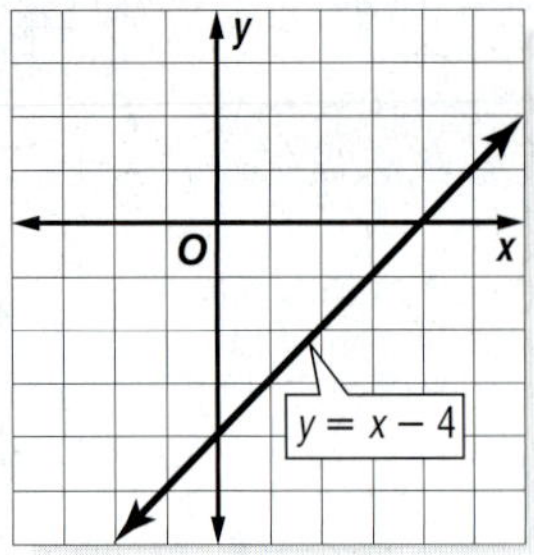

3.
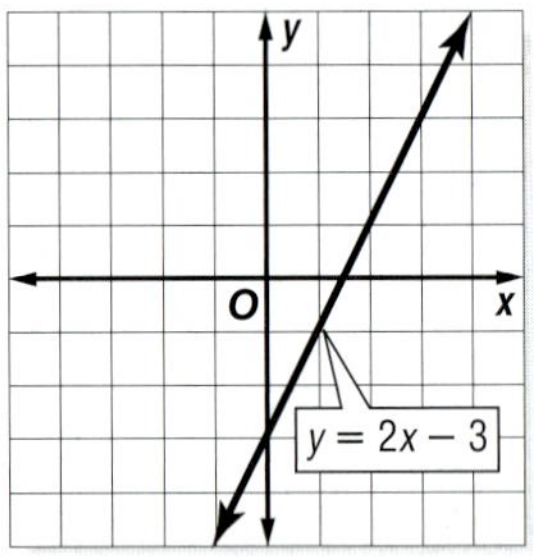

4.
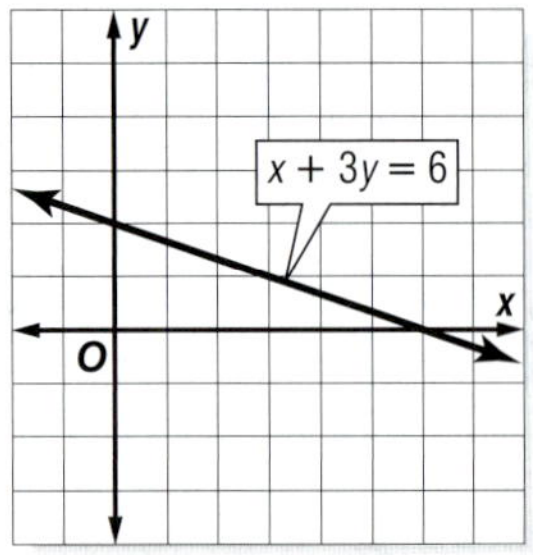

5.
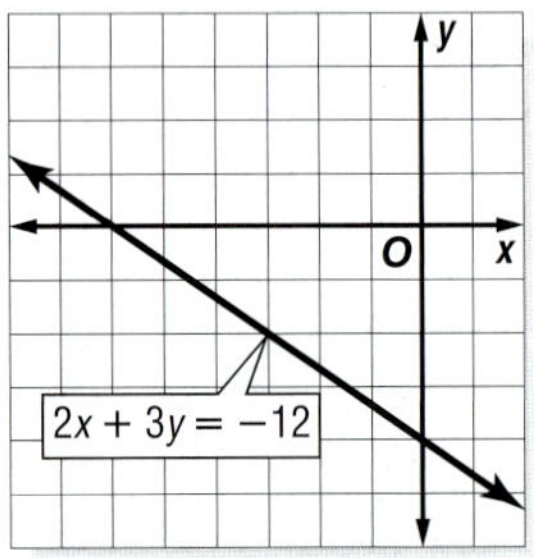

6.
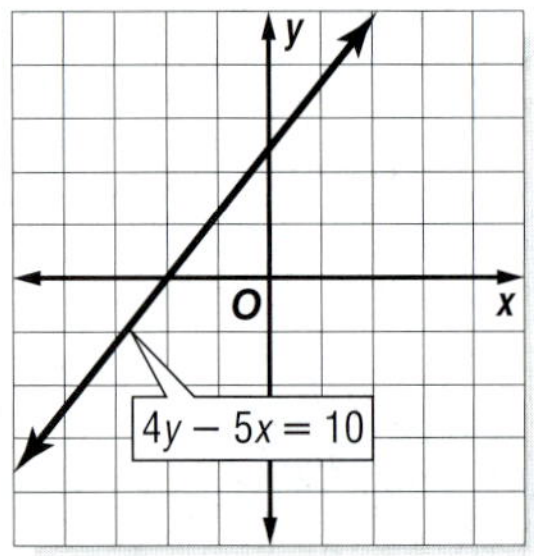

13.
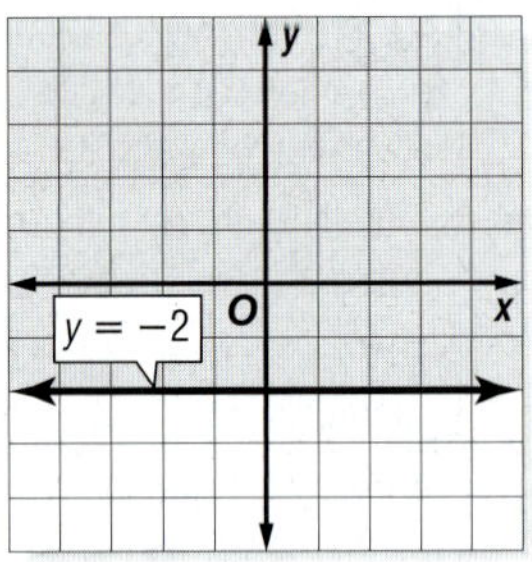

14.
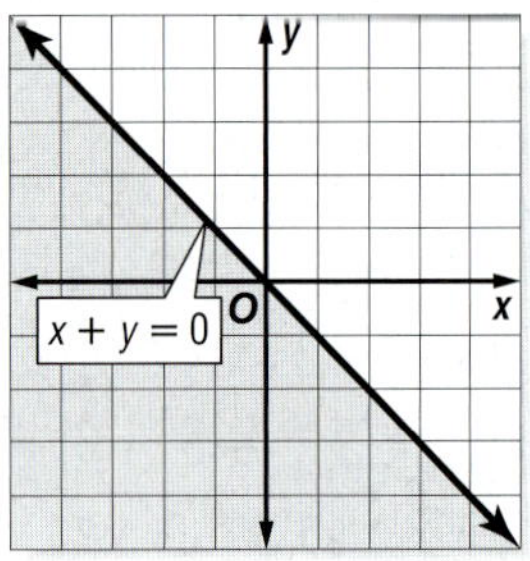

15.
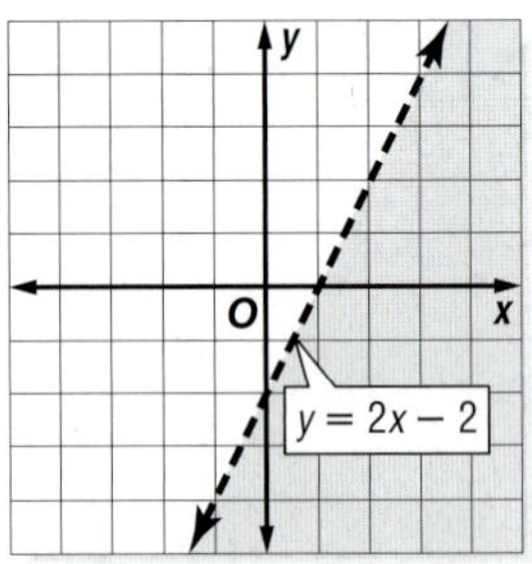

16.
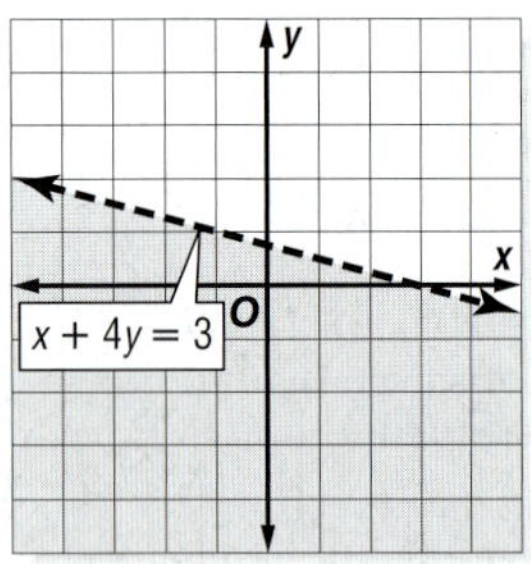

17.
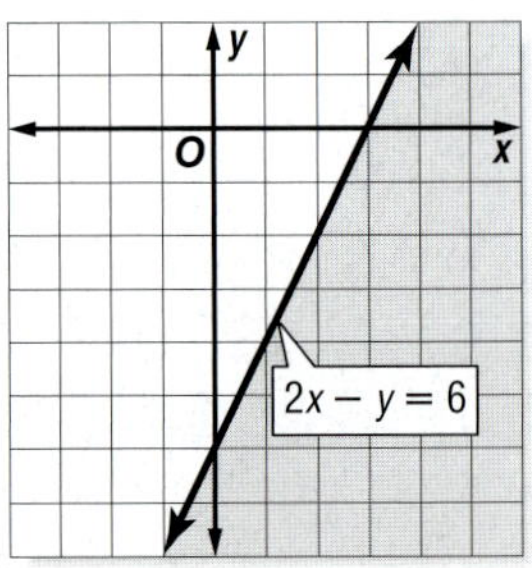

18.
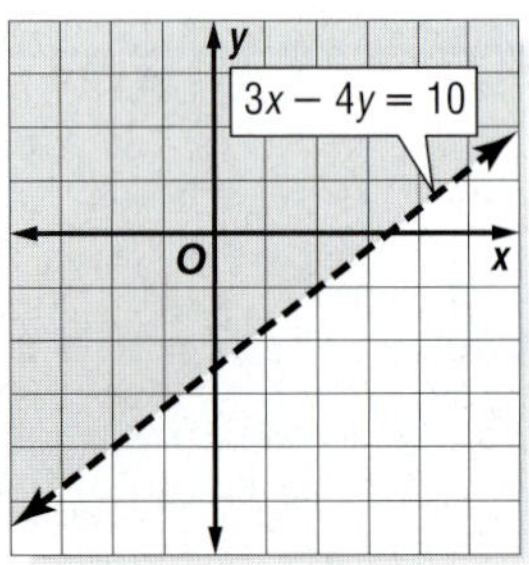

Page 113, Lesson 3-1

13.
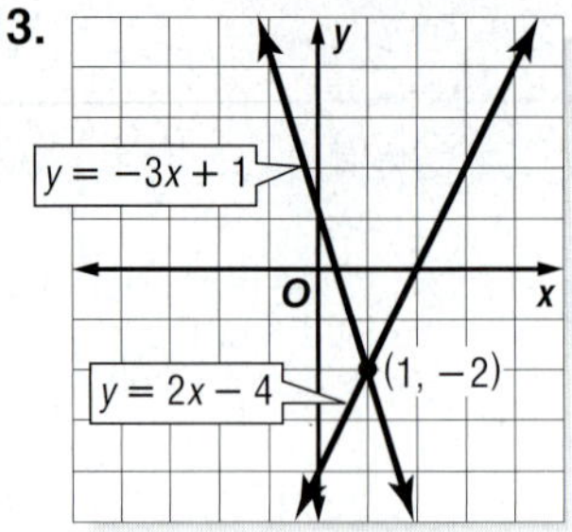

14.
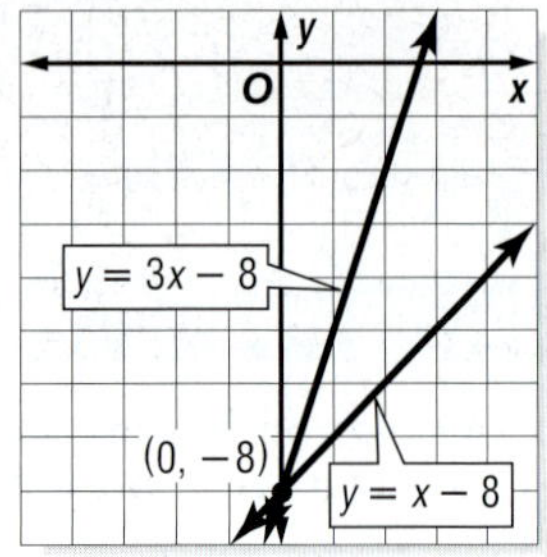

15.
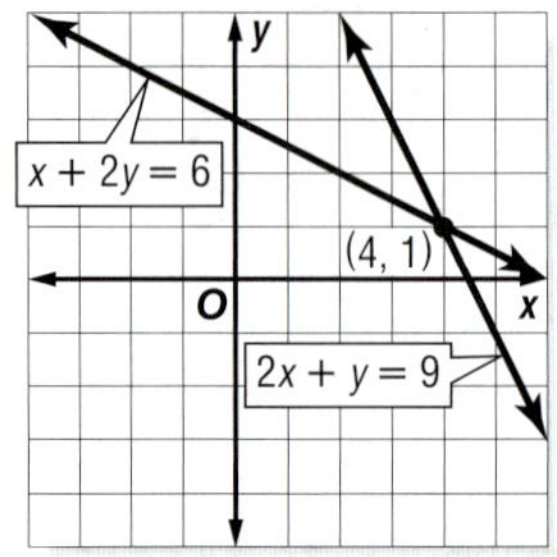

16.
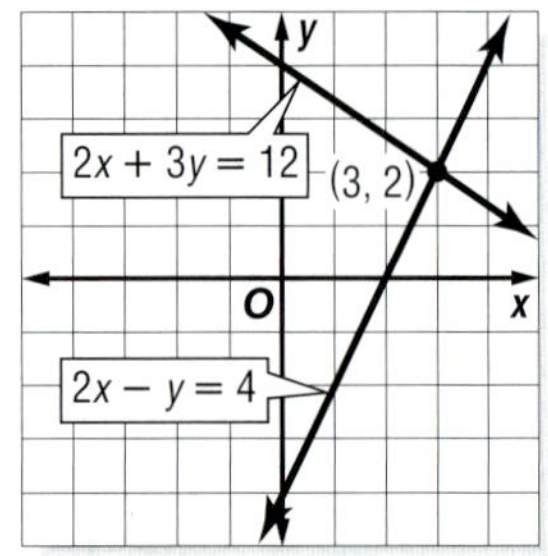

17.
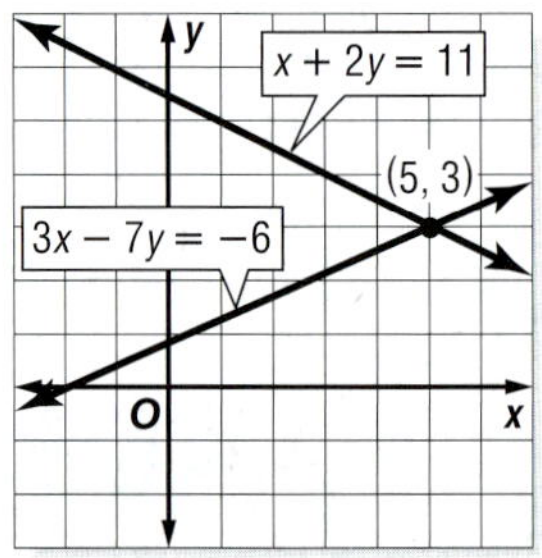

18.
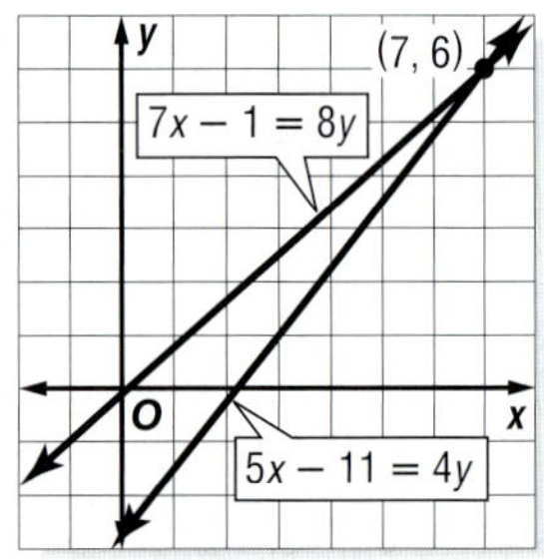

19.
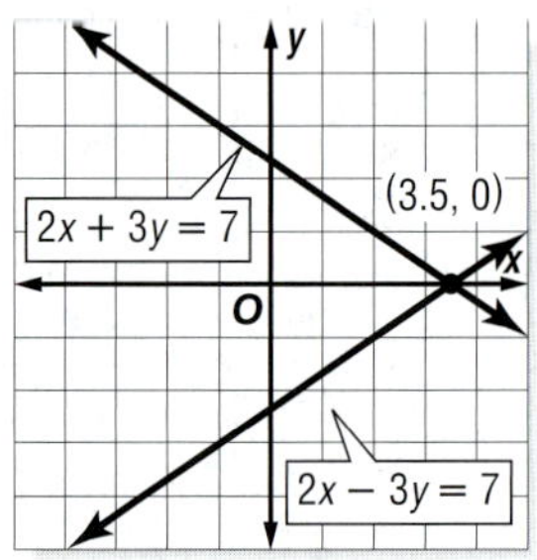

20.
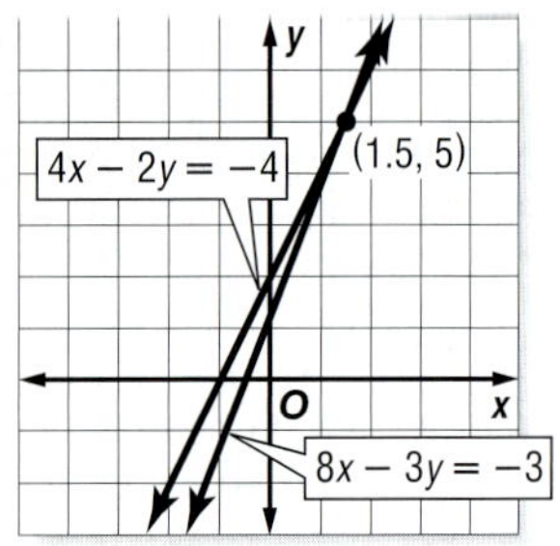

21.
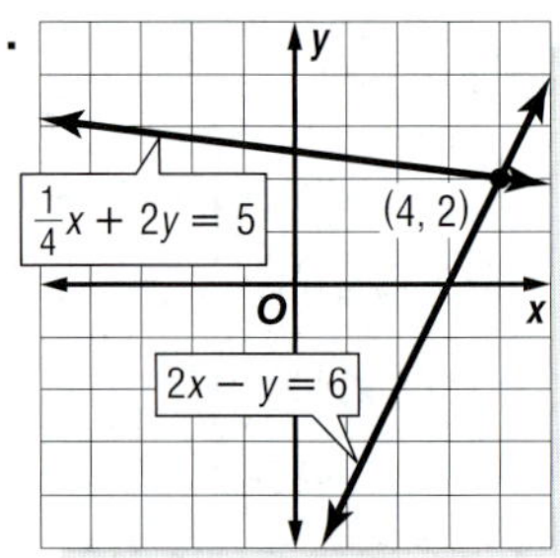

22.
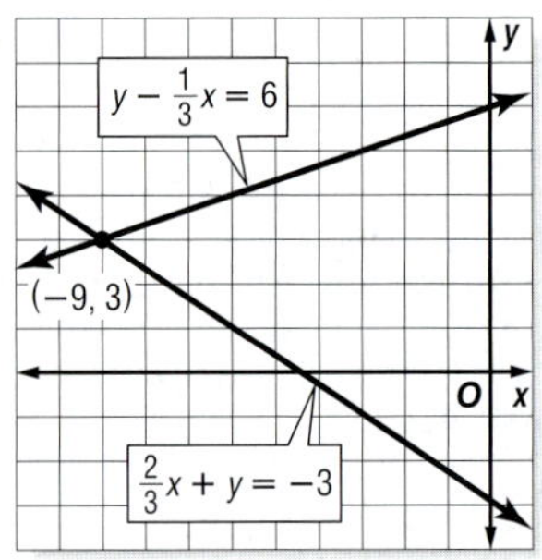

23.
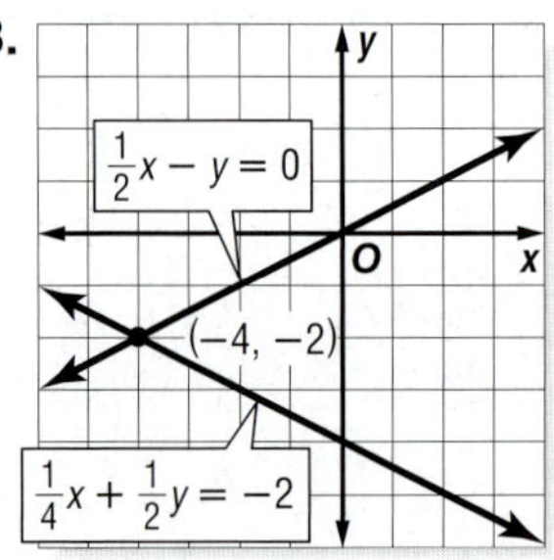

24.
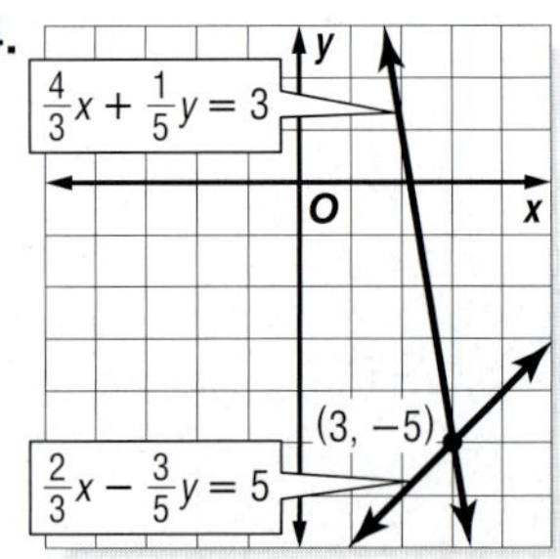

25.
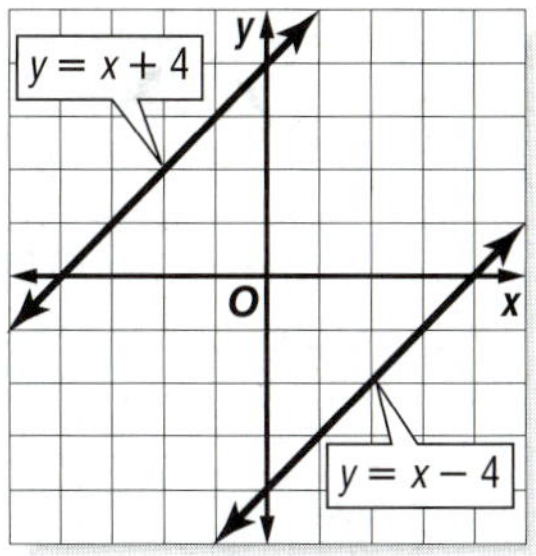

26.
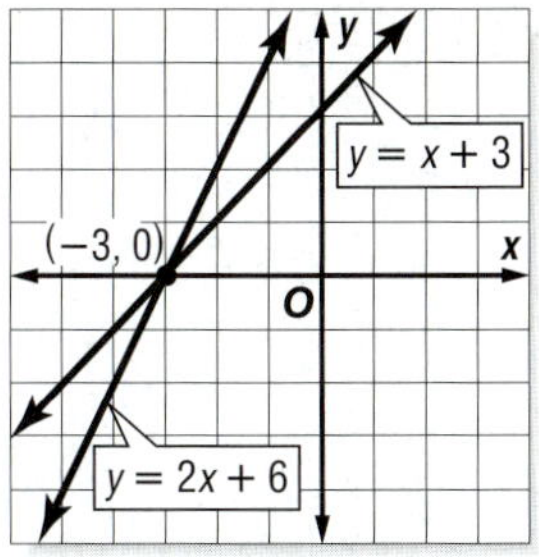

27.
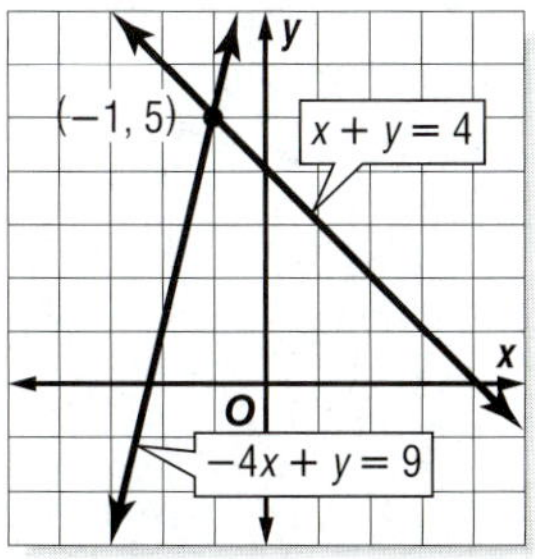

28.
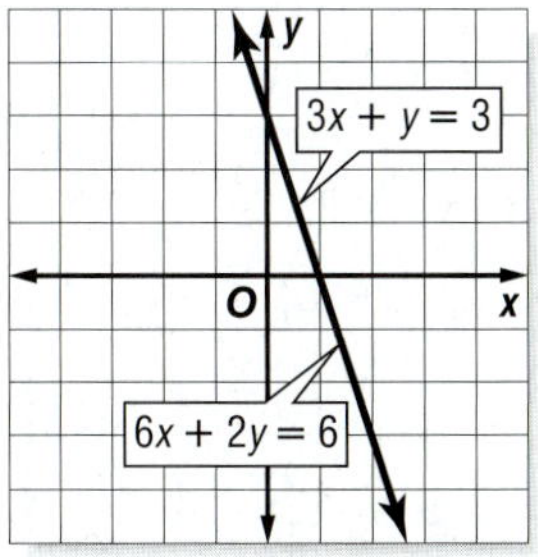

29.
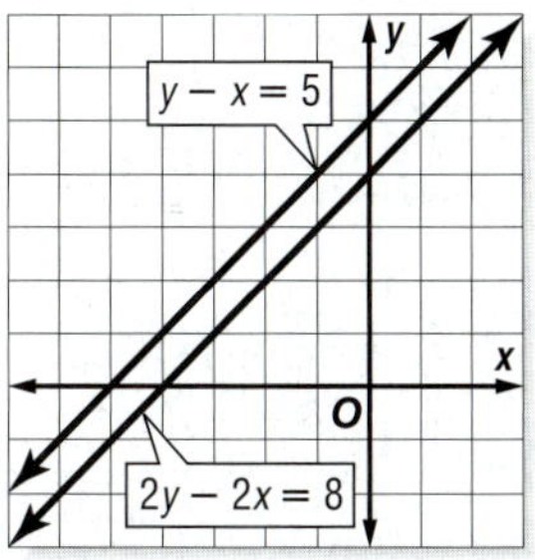

30.
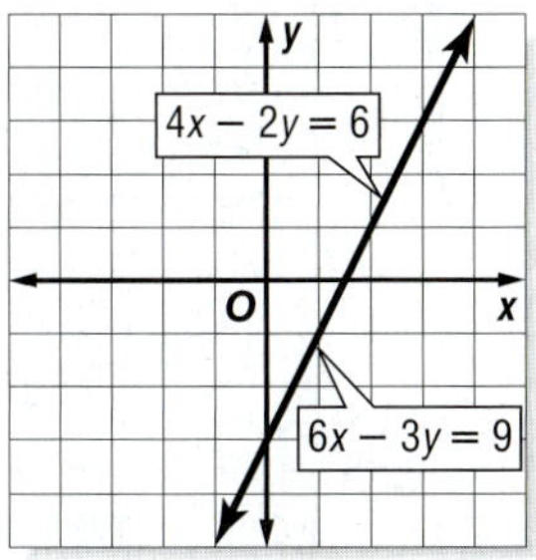

31.
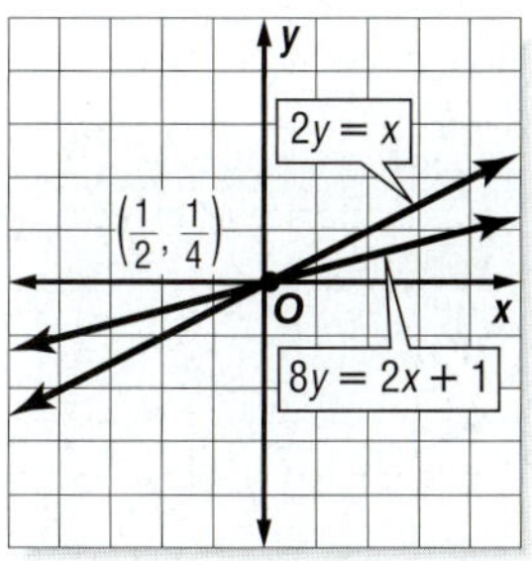

32.
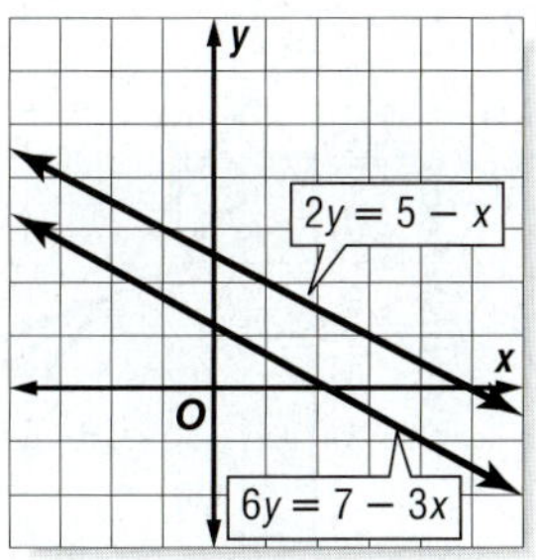

33.
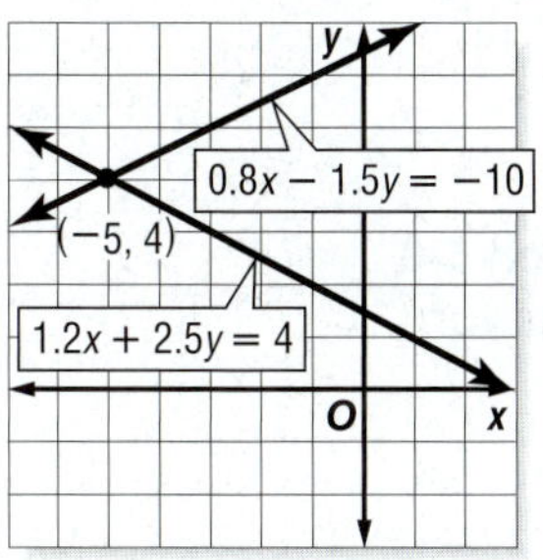

34.
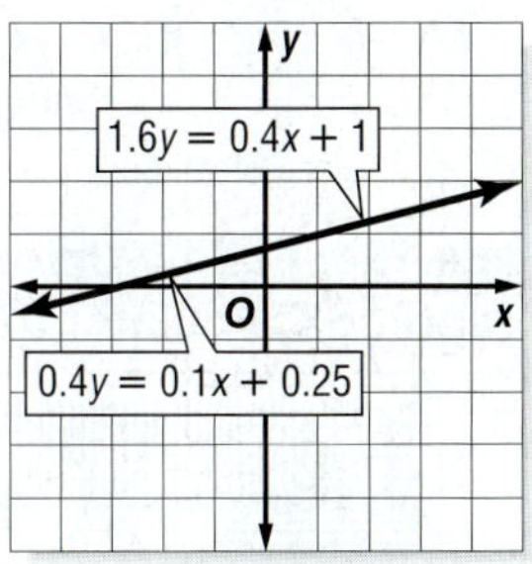

35.
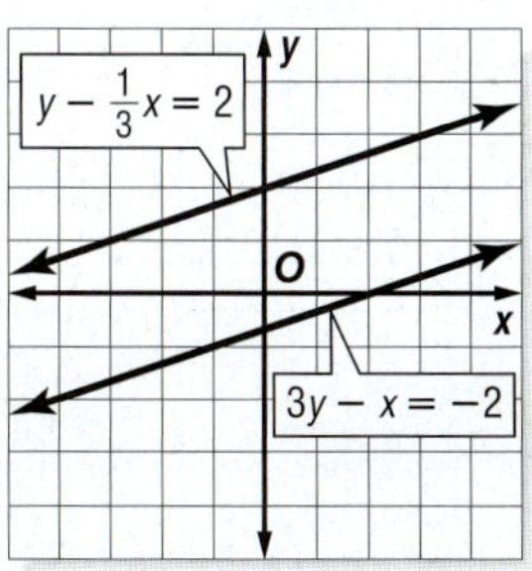

36.
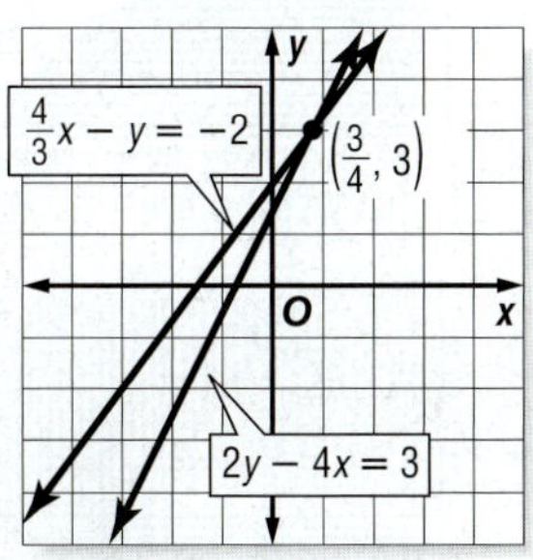

Page 122, Lesson 3-2

57.
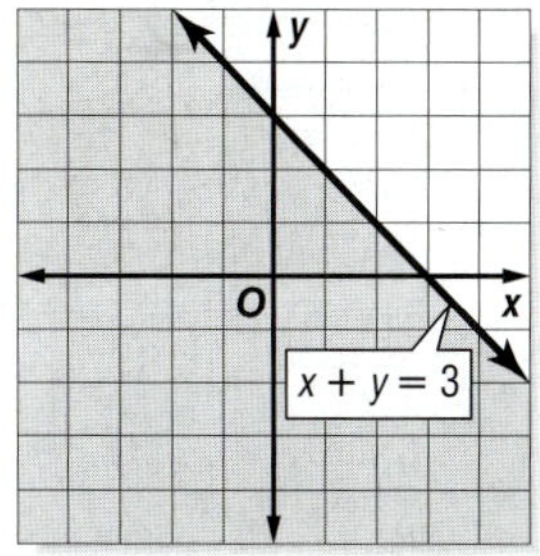

58.
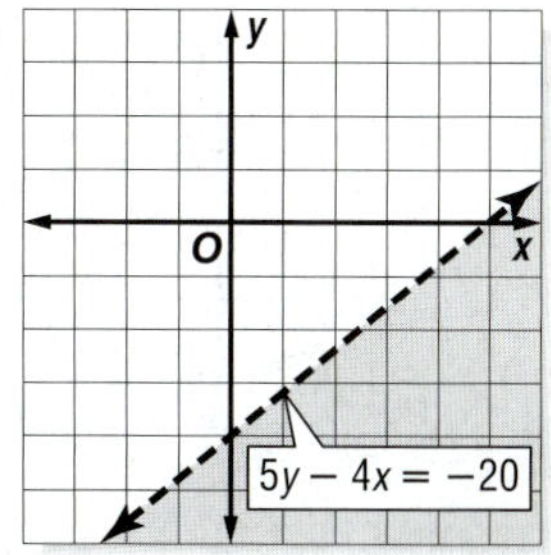

59.
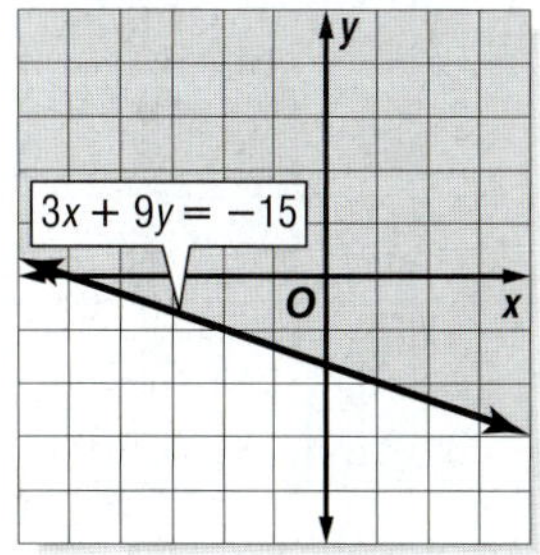

Pages 125–127, Lesson 3-3

4.
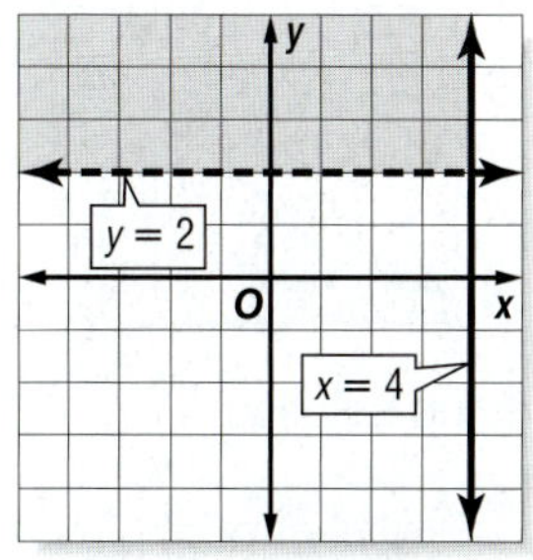

5.
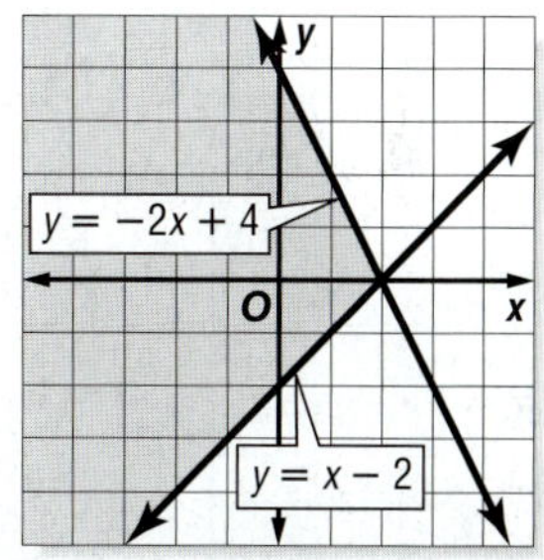

6.
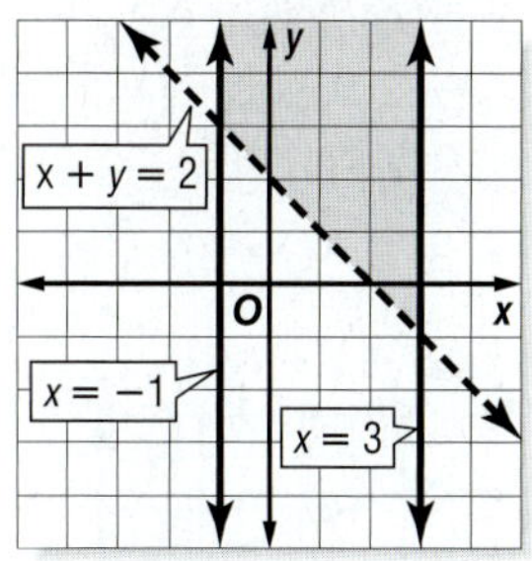

7.
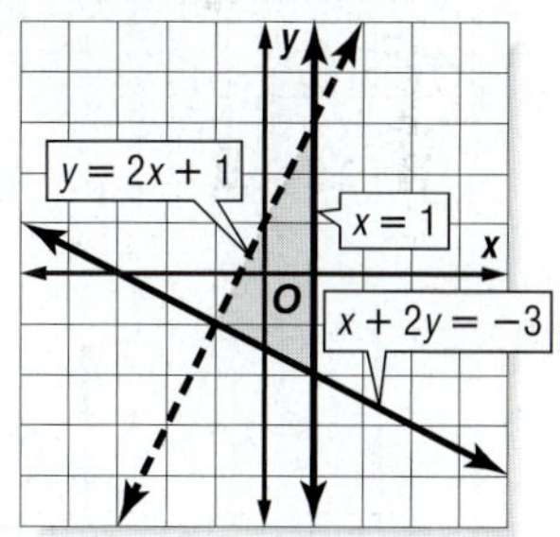

10.
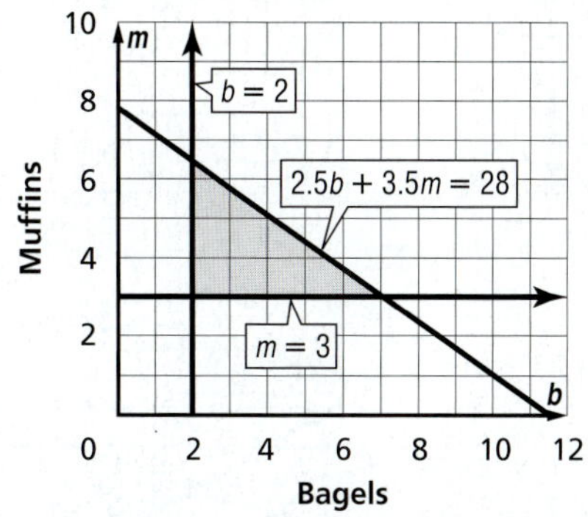

12.

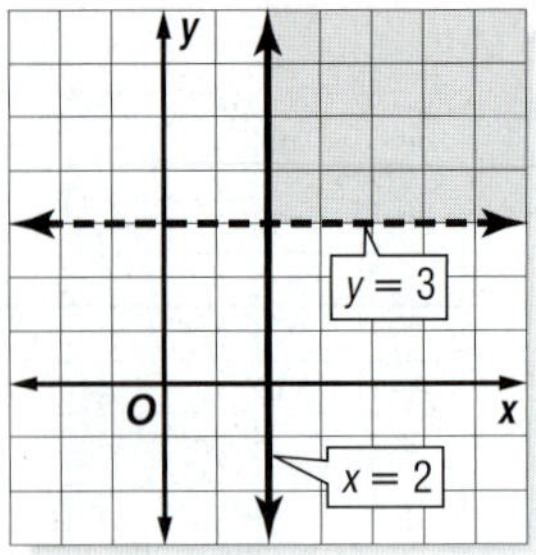

13.

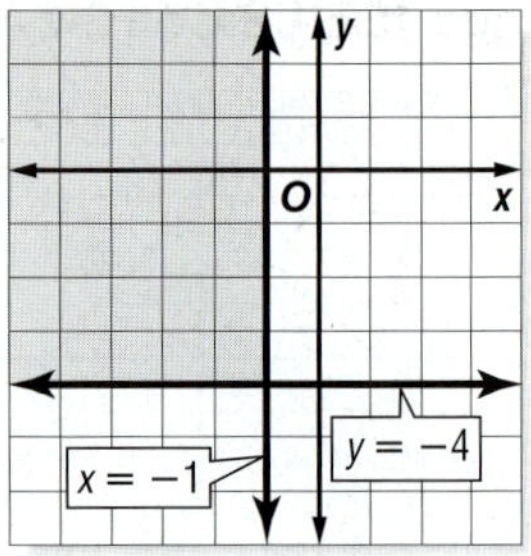

14.

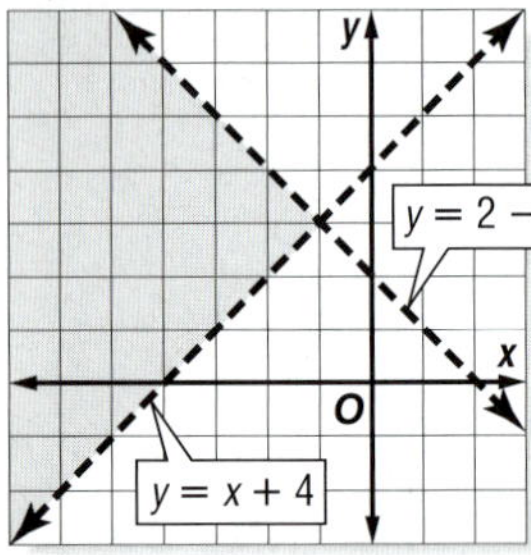

15.

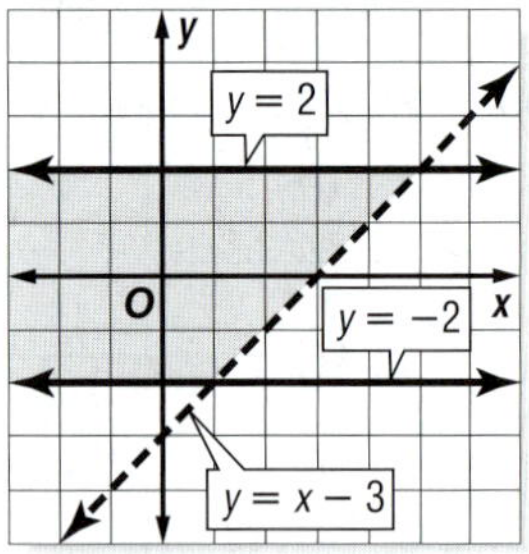

16.

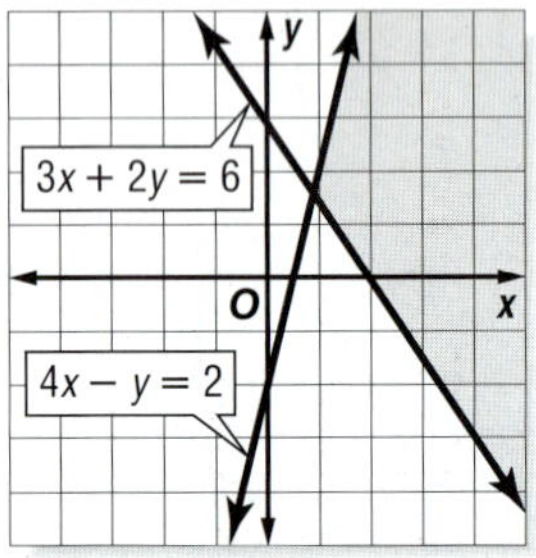

17.

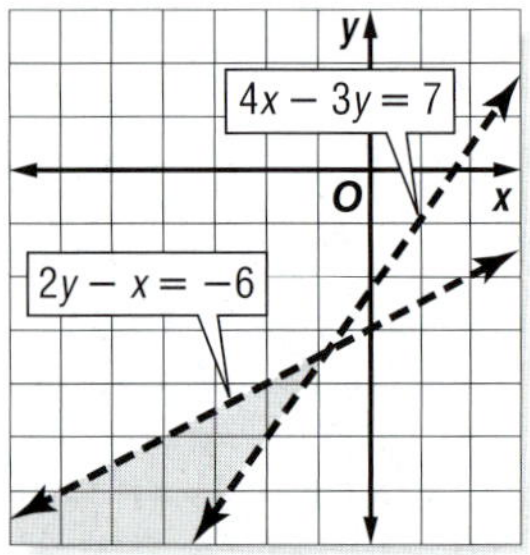

18.

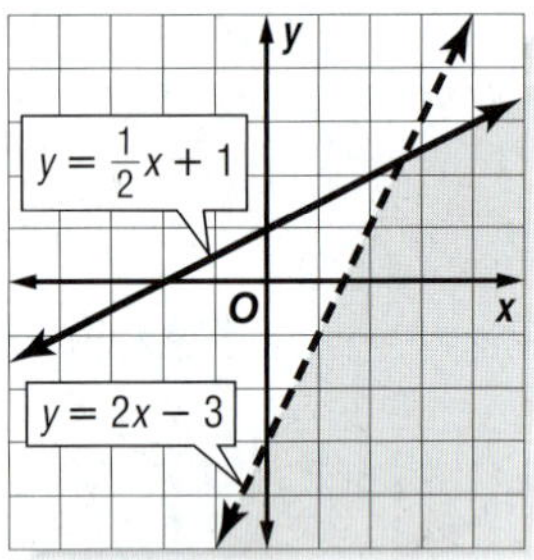

19. no solution

20.

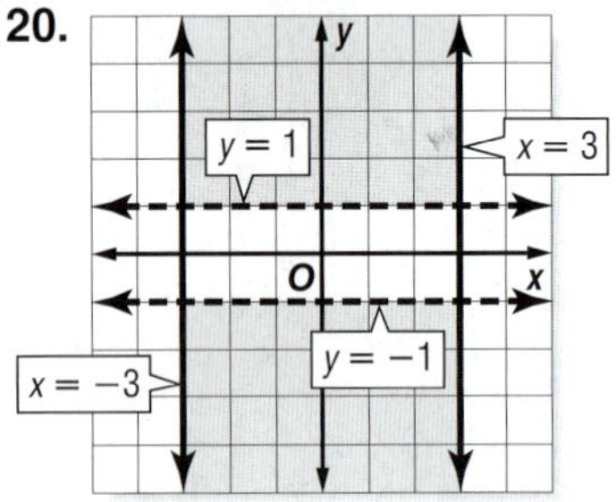

21.

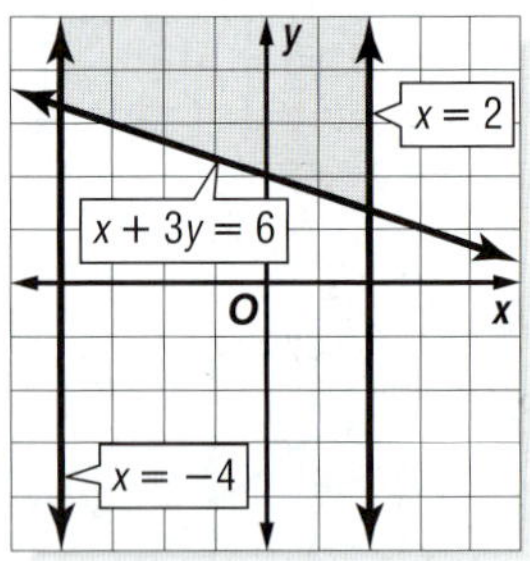

22. no solution

23.

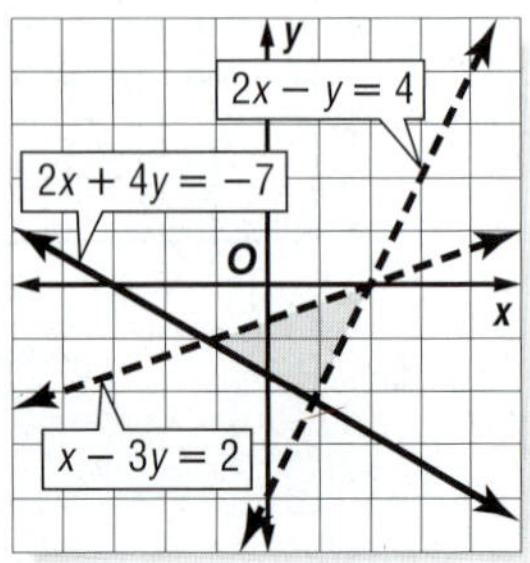

32.

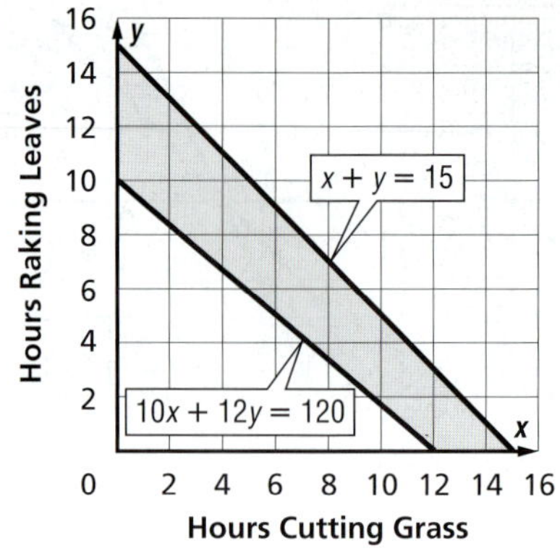

33. $s \geq 111$, $s \leq 130$, $h \geq 9$, $h \leq 12$

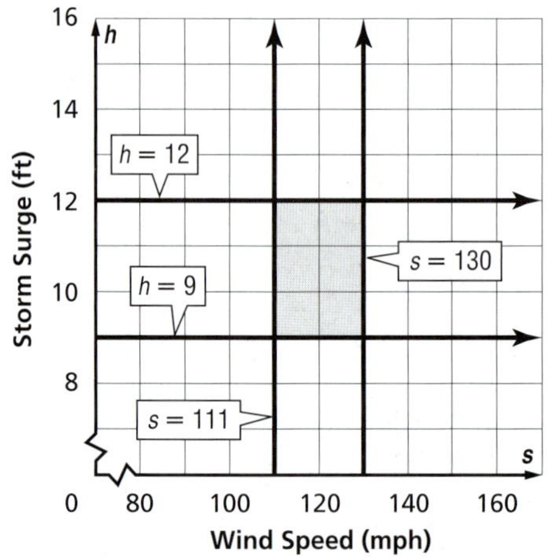

35.

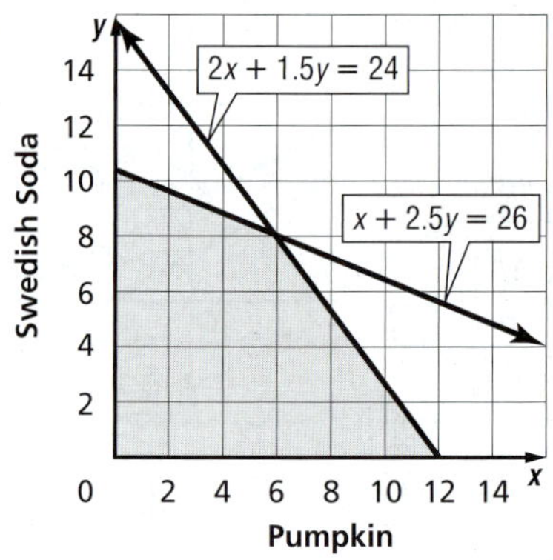

Page 128, Follow-Up of Lesson 3-3 Graphing Calculator Investigation

1.

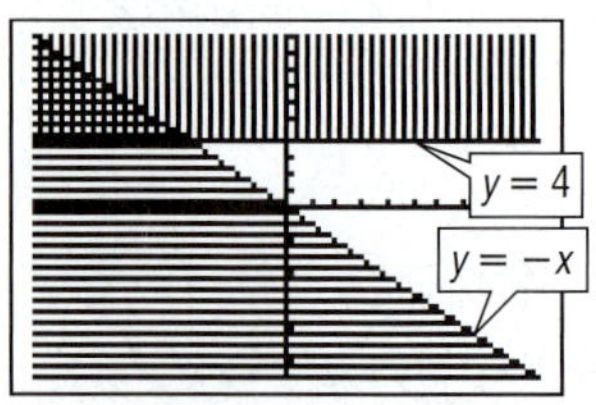

[−10, 10] scl: 1 by [−10, 10] scl: 1

2.

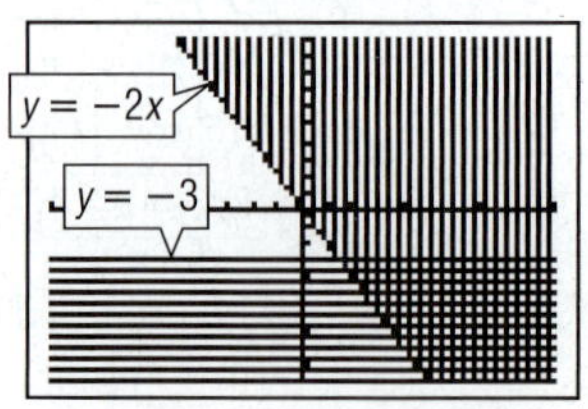

[−10, 10] scl: 1 by [−10, 10] scl: 1

3.

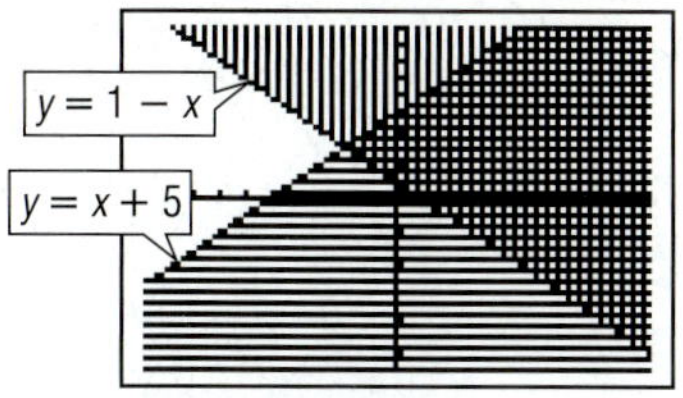

[−10, 10] scl: 1 by [−10, 10] scl: 1

4.

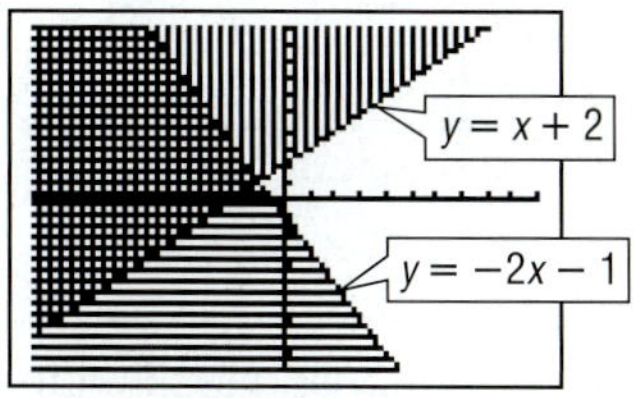

[−10, 10] scl: 1 by [−10, 10] scl: 1

5.

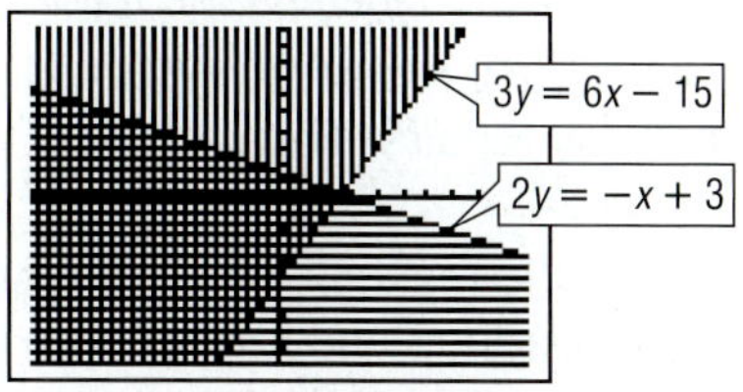

[−10, 10] scl: 1 by [−10, 10] scl: 1

6.

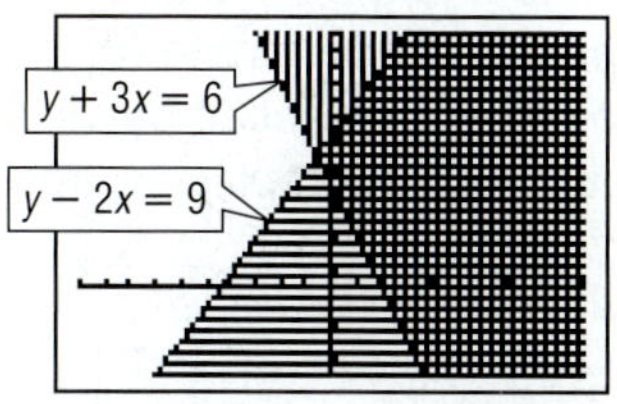

[−10, 10] scl: 1 by [−5, 15] scl: 1

7.

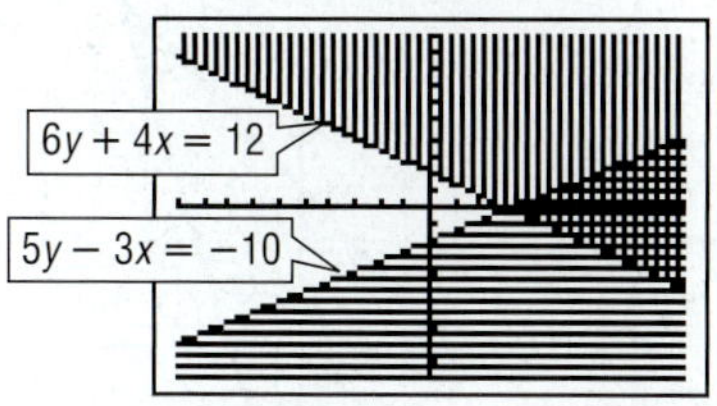

[−10, 10] scl: 1 by [−10, 10] scl: 1

8.

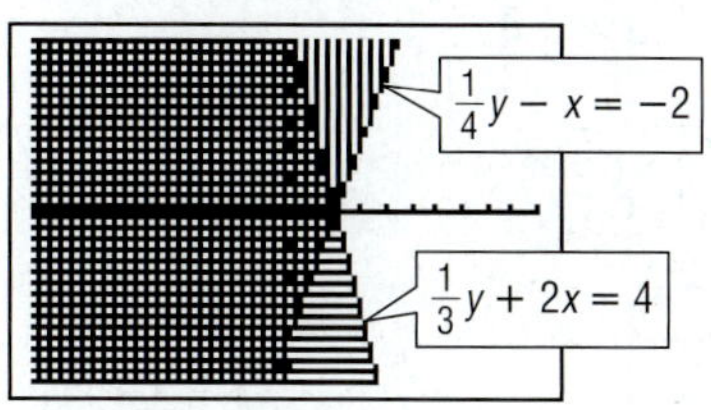

[−10, 10] scl: 1 by [−10, 10] scl: 1

Pages 132–134, Lesson 3-4

15.

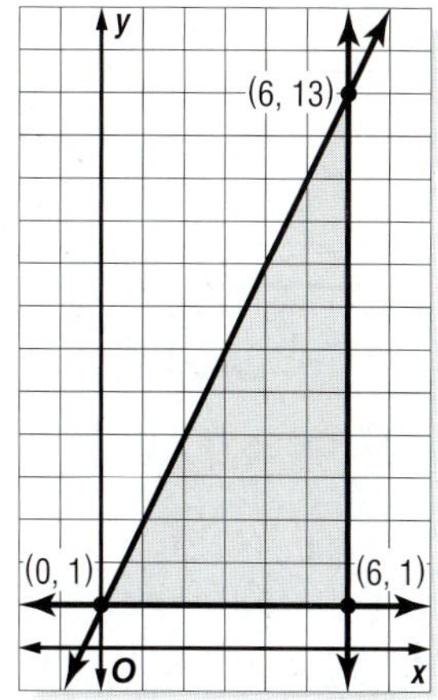

vertices: (0, 1), (6, 1), (6, 13);
max: $f(6, 13) = 19$;
min: $f(0, 1) = 1$

16.

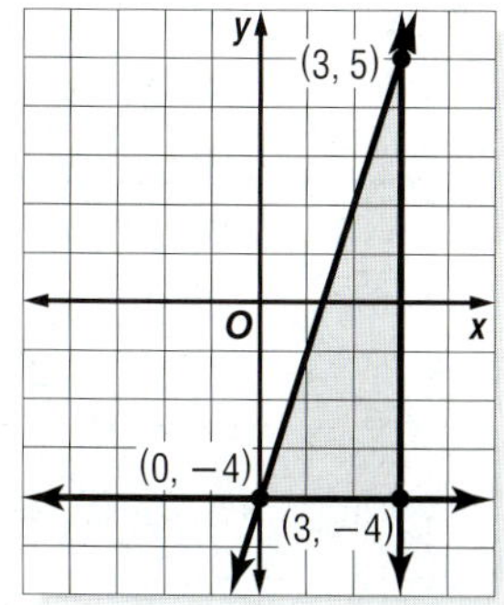

vertices: (0, −4), (3, 5), (3, −4);
max: $f(3, -4) = 7$;
min: $f(3, 5) = -2$

17.

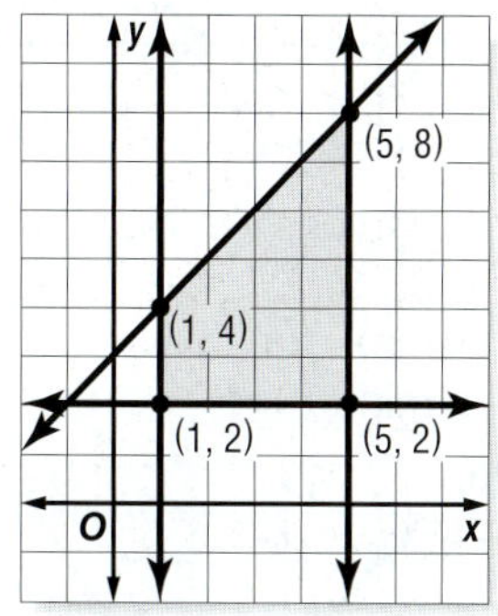

vertices: (1, 4), (5, 8), (5, 2), (1, 2);
max: $f(5, 2) = 11$;
min: $f(1, 4) = -5$

18.

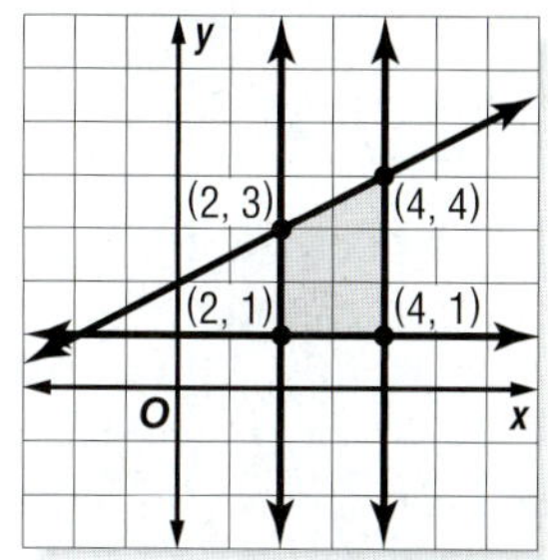

vertices: (2, 1), (2, 3), (4, 4), (4, 1);
max: $f(4, 4) = 16$;
min: $f(2, 1) = 5$

19.

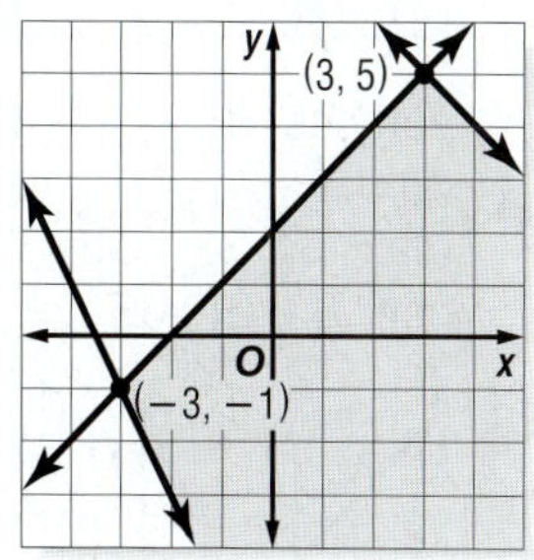

vertices: (−3, −1), (3, 5);
no maximum;
min: $f(-3, -1) = -9$

20.

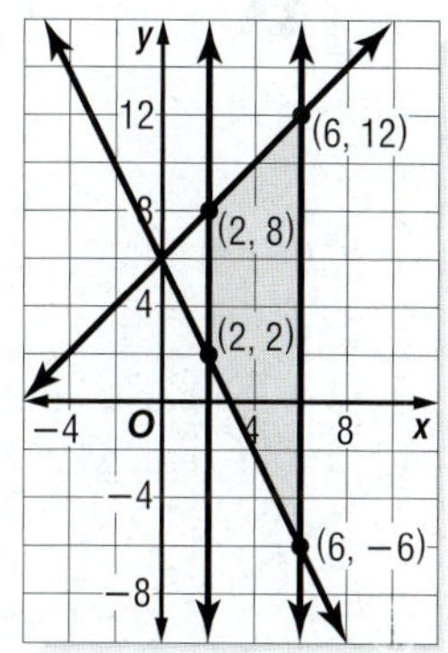

vertices: (2, 2), (2, 8), (6, 12), (6, −6);
max: $f(6, 12) = 30$;
min: $f(6, -6) = -24$

21.

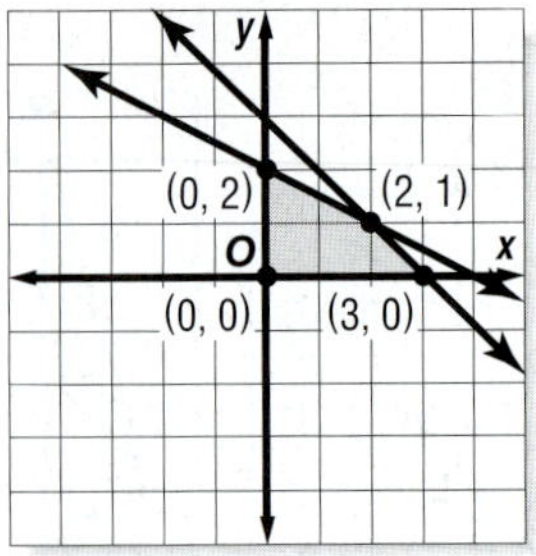

vertices: (0, 0), (0, 2), (2, 1), (3, 0);
max: $f(0, 2) = 6$;
min: $f(3, 0) = -12$

22.

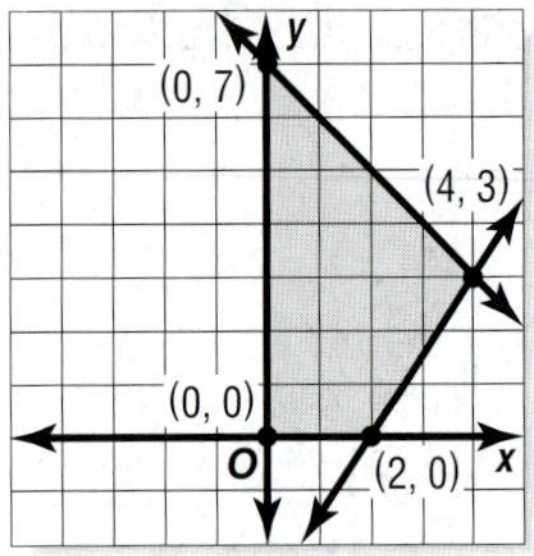

vertices: (0, 0), (0, 7), (4, 3), (2, 0);
max: $f(4, 3) = 14$;
min: $f(0, 7) = -14$

23.

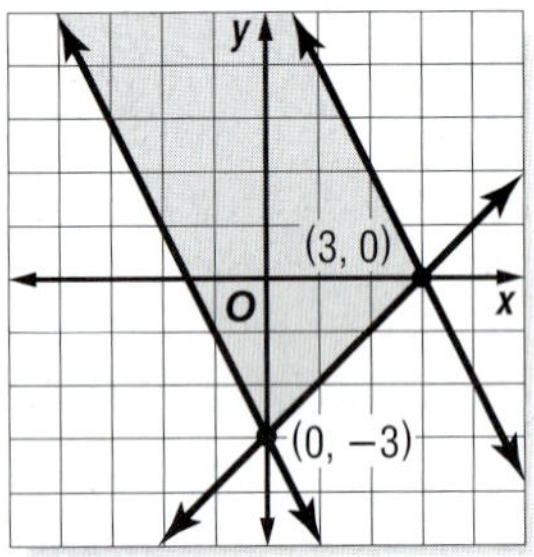

vertices: (3, 0), (0, −3);
no maximum;
min: $f(0, -3) = -12$

24.

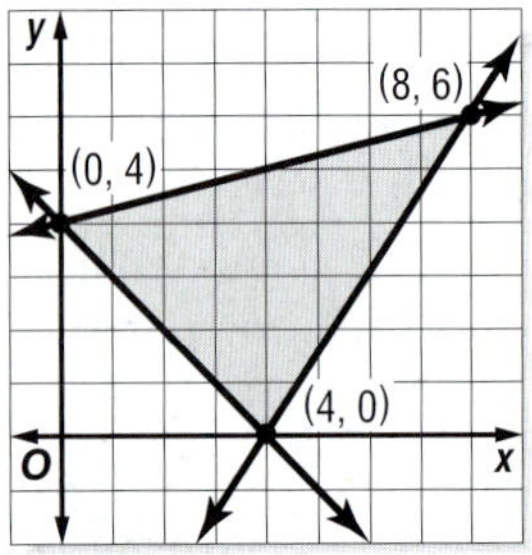

vertices: (0, 4), (4, 0), (8, 6);
max: $f(4, 0) = 4$;
min: $f(0, 4) = -8$

25.

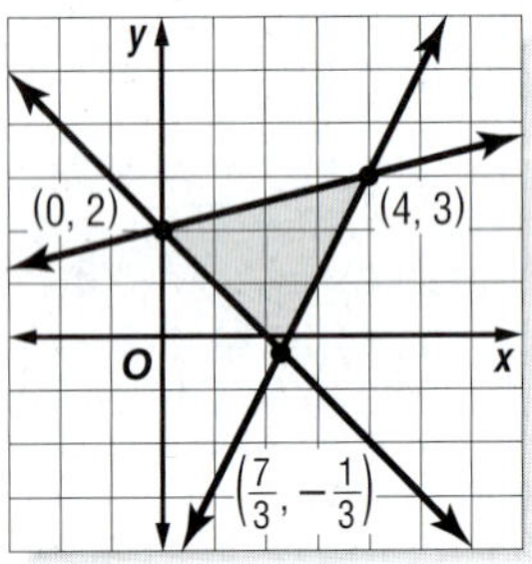

vertices: (0, 2), (4, 3), $\left(\frac{7}{3}, -\frac{1}{3}\right)$;
max: $f(4, 3) = 25$;
min: $f(0, 2) = 6$

26.

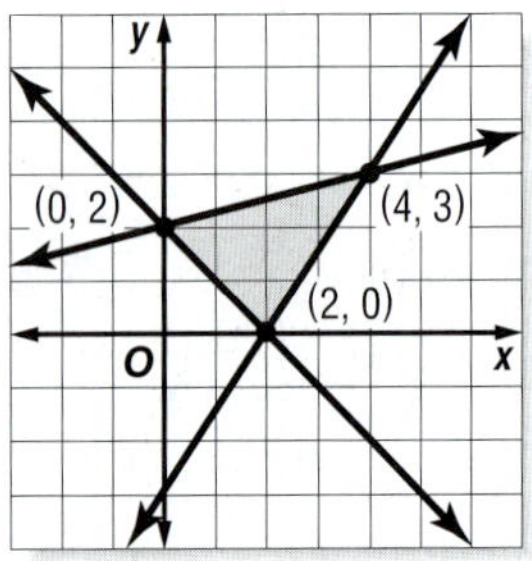

vertices: (0, 2), (4, 3), (2, 0);
max: $f(4, 3) = 13$;
min: $f(2, 0) = 2$

27.

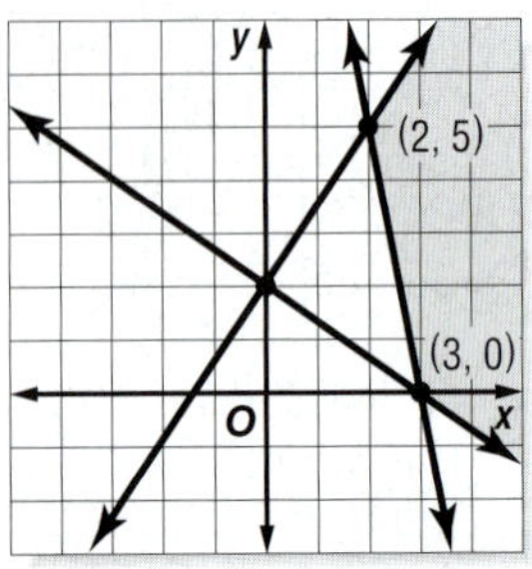

vertices: (2, 5), (3, 0);
no maximum;
no minimum

28.

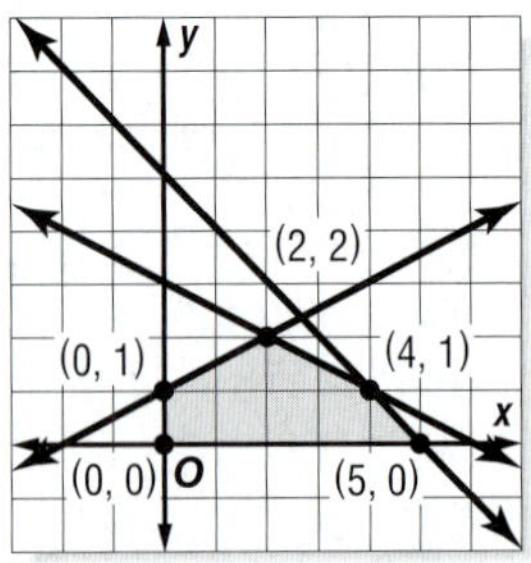

vertices: (0, 0), (0, 1), (2, 2), (4, 1), (5, 0);
max: $f(5, 0) = 15$;
min: $f(0, 1) = -5$

29.

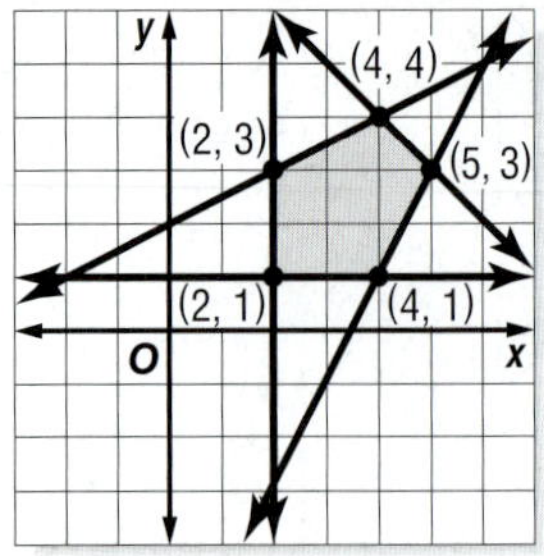

vertices: (2, 1), (2, 3), (4, 1), (4, 4), (5, 3);
max: $f(4, 1) = 0$;
min: $f(4, 4) = -12$

39. (0, 0), (0, 4000), (2500, 2000), (4500, 0)

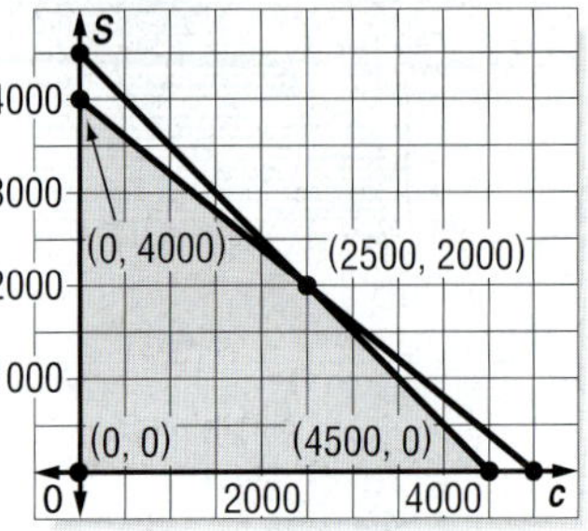

43. There are many variables in scheduling tasks. Linear programming can help make sure that all the requirements are met. Answers should include the following.

- Let x = the number of buoy replacements and let y = the number of buoy repairs. Then, $x \geq 0$, $y \geq 0$, $x \leq 8$ and $2.5x + y \leq 24$.
- The captain would want to maximize the number of buoys that a crew could repair and replace so $f(x, y) = x + y$.
- Graph the inequalities and find the vertices of the intersection of the graphs. The coordinate (0, 24) maximizes the function. So the crew can service the maximum number of buoys if they replace 0 and repair 24 buoys.

Page 135, Practice Quiz 2

1.

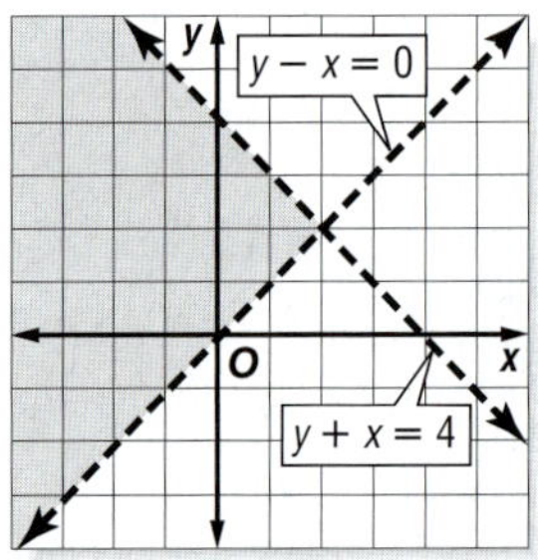

2.

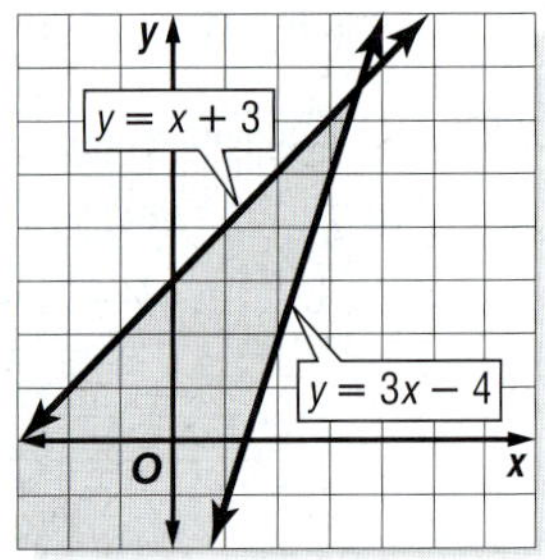

3.

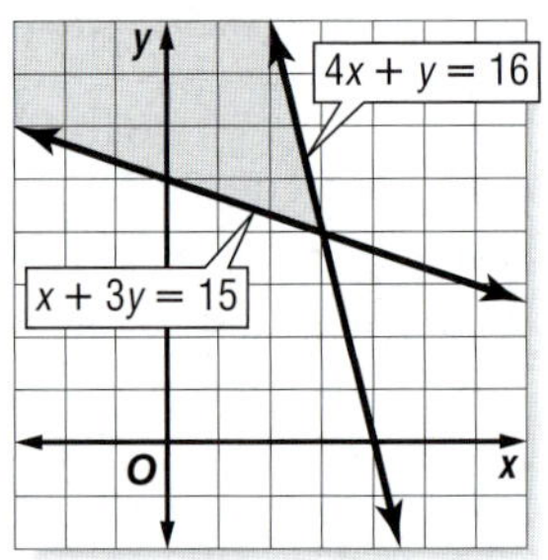

4.

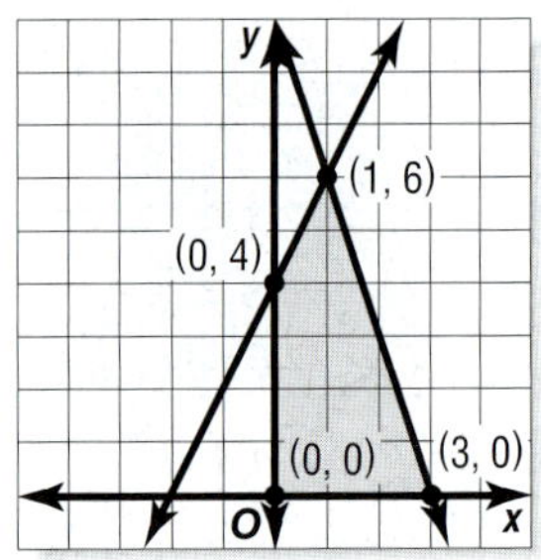

5.

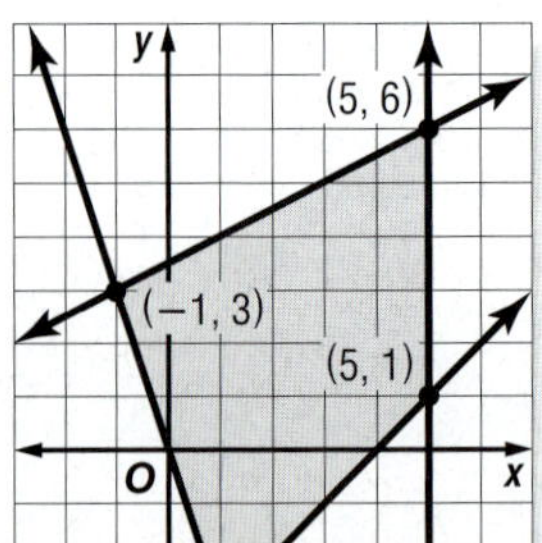

Page 137, Preview of Lesson 3-5
Algebra Activity

1.

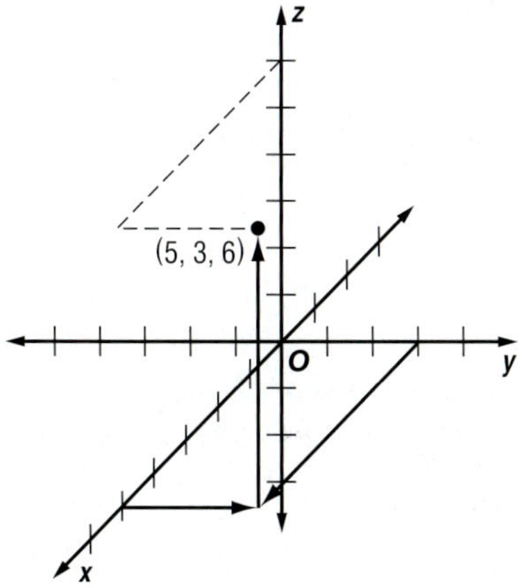

2.

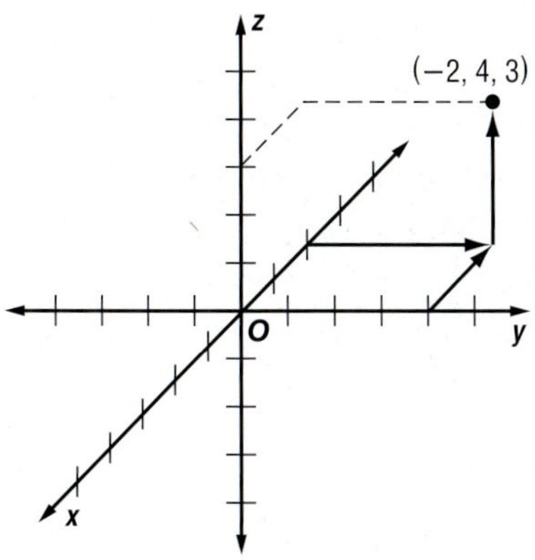

3.

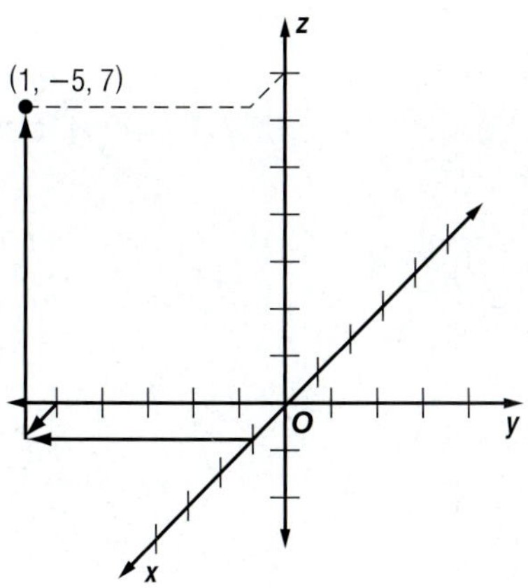

14a.

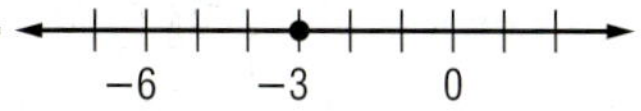

14b.

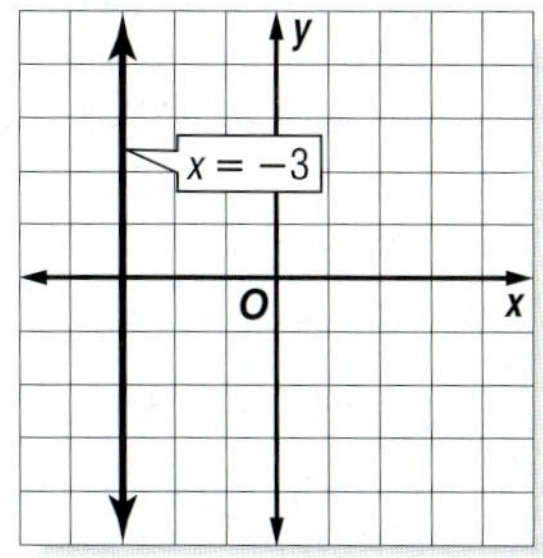

14c.

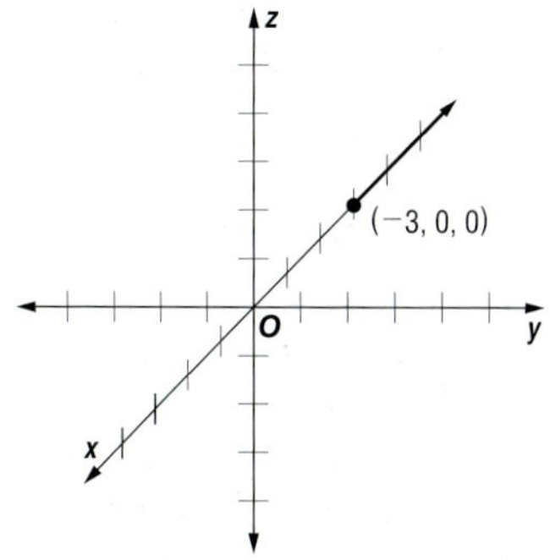

14d. One is a point (one-dimensional), one is a line (two-dimensional), and one is a plane (three-dimensional).

14e. The graph of $x > -3$ in one dimension includes all of the numbers that lie to the right of the point $x = -3$ on a number line. The graph of $x > -3$ in two dimensions is a half-plane and includes all of the ordered pairs that lie to the right of the line $x = -3$. The graph of $x > -3$ in three dimensions includes all of the space that lies in front of the plane $x = -3$.

Page 145, Chapter 3 Study Guide and Review

11.

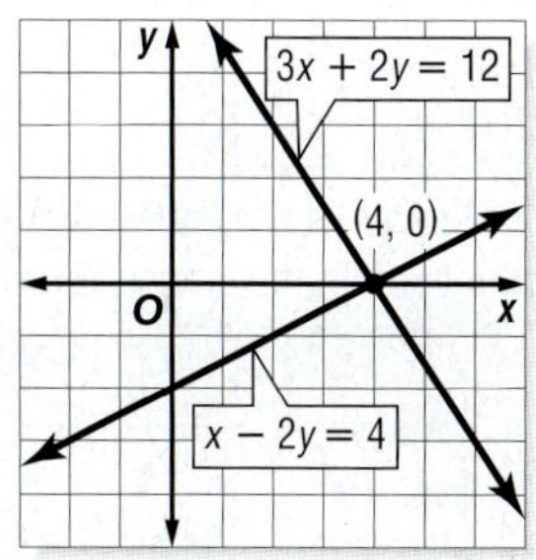

12.

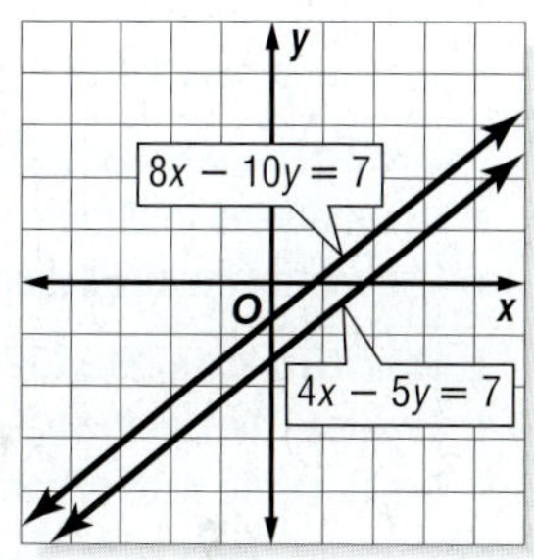

13.

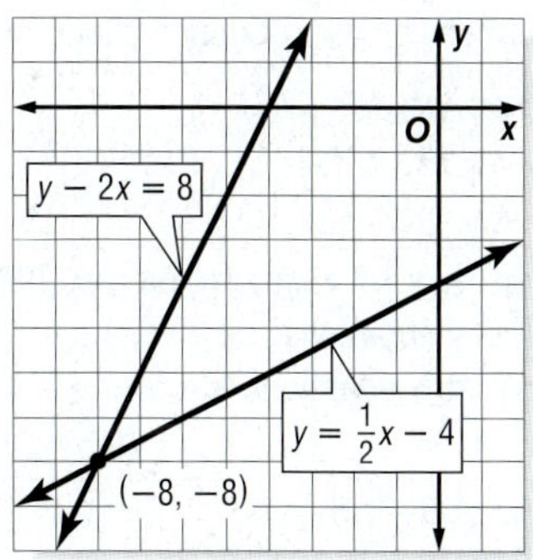

14.

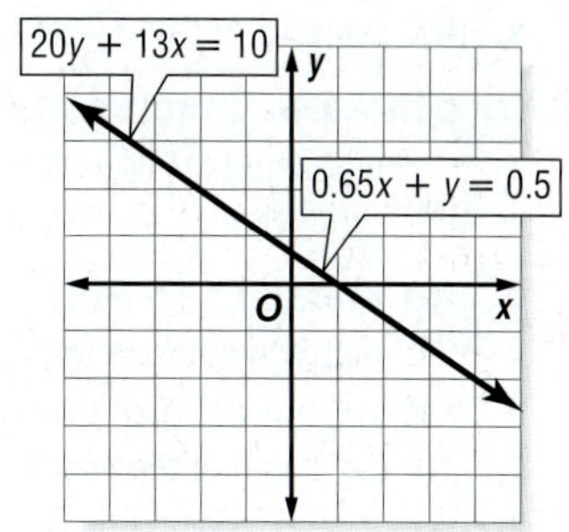

Chapter 4

Matrices

Chapter Overview and Pacing

Year-long and two-year pacing: pages T20–T21.

LESSON OBJECTIVES

LESSON OBJECTIVES	PACING (days)			
	Regular		Block	
	Basic/ Average	Advanced	Basic/ Average	Advanced
4-1 Introduction to Matrices *(pp. 154–159)* • Organize data in matrices. • Solve equations involving matrices. ***Follow-Up:*** Organizing Data	1	1	0.5	0.5
4-2 Operations with Matrices *(pp. 160–166)* • Add and subtract matrices. • Multiply by a matrix scalar.	2 (with 4-1 Follow-Up)	2 (with 4-1 Follow-Up)	1 (with 4-1 Follow-Up)	1.5 (with 4-1 Follow-Up)
4-3 Multiplying Matrices *(pp. 167–174)* • Multiply matrices. • Use the properties of matrix multiplication.	2	2	1	1
4-4 Transformations with Matrices *(pp. 175–181)* • Use matrices to determine the coordinates of a translated or dilated figure. • Use matrix multiplication to find the coordinates of a reflected or rotated figure.	2	2	1	1
4-5 Determinants *(pp. 182–188)* • Evaluate the determinant of a 2×2 matrix. • Evaluate the determinant of a 3×3 matrix.	2	2	1	1
4-6 Cramer's Rule *(pp. 189–194)* • Solve systems of two linear equations by using Cramer's Rule. • Solve systems of three linear equations by using Cramer's Rule.	2	2	1	1
4-7 Identity and Inverse Matrices *(pp. 195–201)* • Determine whether two matrices are inverses. • Find the inverse of a 2×2 matrix.	2	2	1	1
4-8 Using Matrices to Solve Systems of Equations *(pp. 202–208)* • Write matrix equations for systems of equations. • Solve systems of equations using matrix equations. ***Follow-Up:*** Augmented Matrices	2	2 (with 4-8 Follow-Up)	0.5	1
Study Guide and **Practice Test** *(pp. 209–215)* **Standardized Test Practice** *(pp. 216–217)*	1	1	0.5	0.5
Chapter Assessment	1	1	0.5	0.5
TOTAL	**17**	**17**	**8**	**9**

*An electronic version of this chapter is available on **StudentWorks™**. This backpack solution CD-ROM allows students instant access to the Student Edition, lesson worksheet pages, and web resources.*

Timesaving Tools
TeacherWorks™
All-In-One Planner and Resource Center
See pages T5 and T21.

Chapter Resource Manager

Study Guide and Intervention	Practice (Skills and Average)	Reading to Learn Mathematics	Enrichment	Assessment	Applications*	5-Minute Check Transparencies	Interactive Chalkboard	Alge2PASS: Tutorial Plus (lessons)	Materials
Chapter 4 Resource Masters									
169–170	171–172	173	174			4-1	4-1		(*Follow-Up:* spreadsheet software)
175–176	177–178	179	180	231		4-2	4-2	6	graphing calculator
181–182	183–184	185	186			4-3	4-3		
187–188	189–190	191	192	231, 233	GCS 33	4-4	4-4		colored pencils
193–194	195–196	197	198		GCS 34, SC 7	4-5	4-5		posterboard, colored markers
199–200	201–202	203	204	232		4-6	4-6	7	
205–206	207–208	209	210			4-7	4-7		
211–212	213–214	215	216	232	SC 8	4-8	4-8		graphing calculator (*Follow-Up:* graphing calculator)
				217–230, 234–236					

**Key to Abbreviations:* GCS = Graphing Calculator and Speadsheet Masters,
SC = School-to-Career Masters,
SM = Science and Mathematics Lab Manual

Chapter 4

Mathematical Connections and Background

Continuity of Instruction

Prior Knowledge

The main idea of this chapter, matrices, introduces a new notation in which the position of a number within that notation is important. Students have seen this idea, to focus on the position of a number, for exponents and subscripts. An important topic of the chapter, solving systems of equations, has been studied in Chapter 3.

This Chapter

Students explore matrices as a notation in which they have to attend to the position of a number as well as to its magnitude. They learn when matrices can be added, subtracted, or multiplied, and learn how to find the sums, differences, products, and scalar products of matrices. They use matrices to represent transformations and to represent and solve systems of equations.

Future Connections

Attending to the position of a number within a notation occurs frequently in mathematics; students will see such notation when they read and write symbols for permutations and combinations. Also, in later courses they will derive and justify the matrix properties used in this chapter and will explore other algebraic properties of matrices.

4-1 Introduction to Matrices

Matrices are introduced as a way to organize data. In a matrix students have to attend to both the *magnitude* and the *position* of a number within a notation. Students explore equal matrices. They identify corresponding elements in the two matrices and write statements equating the corresponding elements. Then they apply previously-developed skills to solve equations or a system of equations.

4-2 Operations with Matrices

Students continue to explore the positions of elements in matrices and also look at some general properties of matrices. Students find the sum or difference of two matrices by checking that the matrices have the same dimensions and then calculating the sums or differences of corresponding elements. Students find the product of a scalar and a matrix by multiplying every element in the matrix by that scalar. For addition, subtraction, and scalar multiplication, the resulting matrix and the given matrices (or matrix) have identical dimensions.

For general properties of matrices, students look at examples that illustrate the commutative and associative properties of matrix addition.

4-3 Multiplying Matrices

Matrix multiplication extends two ideas of working with matrices. One idea deals with identifying the dimensions of a matrix. Students explore this idea by determining when matrix products are defined and, if so, stating the dimensions of the product matrix. The second idea deals with combining elements in two matrices to find an element in the resulting matrix. Unlike matrix addition or subtraction, which combined single elements of two matrices, matrix multiplication combines all the elements of a row in one matrix with all the elements of a column in the other matrix. Like matrix addition and subtraction, the result of a combination is a single element.

Students also explore matrix multiplication to show that there is an Associative Property of Matrix Multiplication and a Distributive Property of Matrix Multiplication over Matrix Addition. However, there cannot be a Commutative Property of Matrix Multiplication.

4-4 Transformations with Matrices

Students practice matrix addition, scalar multiplication, and matrix multiplication by using matrices to describe transformations. To describe a translation, the coordinates of the n vertices of the object are written as the n columns of a 2-by-n matrix. The translation is written as a 2-by-n matrix, each column the same. The sum of the two matrices gives the coordinates of the result of the translation. To describe a dilation (change in size) of a figure by a factor of n, the matrix of the coordinates for the figure is multiplied by the scalar n.

Reflections and rotations of figures are described using matrix multiplication. Each reflection and each rotation has a unique 2-by-2 matrix. When the matrix of the coordinates for a figure is multiplied by such a matrix, the resulting matrix gives the coordinates of the reflected or rotated figure.

4-5 Determinants

Students learn how to find the determinant of a 2-by-2 and a 3-by-3 matrix. This lesson introduces just a single application of a determinant, finding the area of a triangle, but other applications are seen in other lessons in this chapter. The determinant of a matrix is a single number or expression; for a 2-by-2 matrix it is the difference of two 2-factor terms, the products of the two diagonals. For a 3-by-3 matrix, the determinant is the sum or difference of six 3-factor terms.

To find the area of a triangle, students write a matrix where each row has the x-coordinate and y-coordinate of a vertex, and the number 1. The area of the triangle is the absolute value of one-half of the determinant for the matrix.

4-6 Cramer's Rule

An important algebraic application of determinants, called Cramer's Rule, is to solve systems of equations. When using Cramer's Rule to solve a system of equations, each equation is first written in standard form. Then the value of each variable in the solution is given by a fraction whose numerator and denominator are both determinants. The denominator is always the determinant of the coefficients of the variables. The numerators begin with the same determinant, but the column of coefficients for that particular variable is replaced with the column of constants. Since the determinant of the coefficients appears in each denominator, so the system has no solution if the value of that determinant is zero.

4-7 Identity and Inverse Matrices

This lesson returns to a focus on the positions of elements within a matrix and uses matrices to explore two algebraic ideas. One idea is that an element I of a set is a multiplicative identity for that set if, for any element A in the set, $A \cdot I = I \cdot A = A$; that is, multiplying by the multiplicative identity element results in no change. The other idea is that two elements are multiplicative inverses if their product is the multiplicative identity element.

For matrices, the multiplicative identity element is a square matrix with 1's down the left-to-right diagonal and 0's in every other position. Students verify that this matrix serves as a multiplicative identity. The multiplicative inverse for a 2-by-2 matrix is given as a product of two factors. One factor is a unit fraction whose denominator is the determinant of the given matrix. The other factor is the given matrix, but down the left-to-right diagonal the elements are switched and along the other diagonal the elements are multiplied by –1. The lesson points out that since the determinant appears in a denominator, a matrix must have a non-zero determinant in order to have an inverse.

4-8 Using Matrices to Solve Systems of Equations

In this culminating lesson of the chapter, students see how matrices can provide an efficient way to represent a system of equations and how to use matrix inverses and matrix multiplication as an efficient way to solve the system. To represent a system of equations, they write a single matrix equation. The left-hand side shows the matrix of coefficients multiplied by a column matrix of the variables. The right-hand side is a column matrix of the constants. To solve the system, students first find the inverse of the matrix of coefficients. When they multiply both sides of the original matrix equation by that inverse, the result gives the column matrix of variables equal to another column matrix. The value of each variable can be read directly from that equation.

Additional mathematical information and teaching notes are available at www.algebra2.com/key_concepts.

Chapter 4

DAILY INTERVENTION and Assessment

	Type	Student Edition	Teacher Resources	Technology/Internet
INTERVENTION	Ongoing	Prerequisite Skills, pp. 153, 158, 166, 174, 181, 188, 194, 201 Practice Quiz 1, p. 174 Practice Quiz 2, p. 194	5-Minute Check Transparencies Quizzes, *CRM* pp. 231–232 Mid-Chapter Test, *CRM* p. 233 Study Guide and Intervention, *CRM* pp. 169–170, 175–176, 181–182, 187–188, 193–194, 199–200, 205–206, 211–212	Alge2PASS: Tutorial Plus www.algebra2.com/self_check_quiz www.algebra2.com/extra_examples
	Mixed Review	pp. 158, 166, 174, 181, 188, 194, 201, 207	Cumulative Review, *CRM* p. 234	
	Error Analysis	Find the Error, pp. 185, 205	Find the Error, *TWE* pp. 185, 205 Unlocking Misconceptions, *TWE* pp. 161, 183 Tips for New Teachers, *TWE* pp. 159, 166, 167	
	Standardized Test Practice	pp. 158, 166, 173, 176, 179, 181, 187, 194, 201, 207, 215, 216–217	*TWE* p. 176 Standardized Test Practice, *CRM* pp. 235–236	Standardized Test Practice CD-ROM www.algebra2.com/standardized_test
ASSESSMENT	Open-Ended Assessment	Writing in Math, pp. 158, 166, 173, 181, 187, 193, 200, 207 Open Ended, pp. 156, 163, 171, 178, 185, 192, 198, 205	Modeling: *TWE* pp. 174, 188 Speaking: *TWE* pp. 166, 194, 207 Writing: *TWE* pp. 158, 181, 201 Open-Ended Assessment, *CRM* p. 229	
	Chapter Assessment	Study Guide, pp. 209–214 Practice Test, p. 215	Multiple-Choice Tests (Forms 1, 2A, 2B), *CRM* pp. 217–222 Free-Response Tests (Forms 2C, 2D, 3), *CRM* pp. 223–228 Vocabulary Test/Review, *CRM* p. 230	ExamView® Pro (see below) MindJogger Videoquizzes www.algebra2.com/vocabulary_review www.algebra2.com/chapter_test

Key to Abbreviations: TWE = Teacher Wraparound Edition; CRM = Chapter Resource Masters

Intervention Technology

Alge2PASS: Tutorial Plus CD-ROM offers a complete, self-paced algebra curriculum.

Algebra 2 Lesson	Alge2PASS Lesson	
4-2	6	*Adding and Subtracting Matrices and Scalar Multiplication*
4-6	7	*Multiplying Matrices*

ExamView® Pro

Use the networkable **ExamView® Pro** to:

- Create **multiple versions** of tests.
- Create **modified** tests for *Inclusion* students.
- **Edit** existing questions and **add** your own questions.
- Use built-in **state curriculum correlations** to create tests aligned with state standards.
- **Apply** art to your tests from a program bank of artwork.

For more information on Intervention and Assessment, see pp. T8–T11.

Reading and Writing in Mathematics

Glencoe Algebra 2 provides numerous opportunities to incorporate reading and writing into the mathematics classroom.

Student Edition

- Foldables Study Organizer, p. 153
- Concept Check questions require students to verbalize and write about what they have learned in the lesson. (pp. 156, 163, 171, 178, 185, 192, 198, 205, 209)
- Writing in Math questions in every lesson, pp. 158, 166, 173, 181, 187, 193, 200, 207
- Reading Study Tip, pp. 154, 175, 182
- WebQuest, pp. 192, 207

Teacher Wraparound Edition

- Foldables Study Organizer, pp. 153, 209
- Study Notebook suggestions, pp. 156, 163, 171, 178, 185, 192, 198, 205
- Modeling activities, pp. 174, 188
- Speaking activities, pp. 166, 194, 207
- Writing activities, pp. 158, 181, 201
- Differentiated Instruction, (Verbal/Linguistic), p. 162
- ELL Resources, pp. 152, 157, 162, 165, 173, 180, 187, 193, 200, 206, 209

Additional Resources

- Vocabulary Builder worksheets require students to define and give examples for key vocabulary terms as they progress through the chapter. (*Chapter 4 Resource Masters,* pp. vii-viii)
- Reading to Learn Mathematics master for each lesson (*Chapter 4 Resource Masters*, pp. 173, 179, 185, 191, 197, 203, 209, 215)
- *Vocabulary PuzzleMaker* software creates crossword, jumble, and word search puzzles using vocabulary lists that you can customize.
- *Teaching Mathematics with Foldables* provides suggestions for promoting cognition and language.
- *Reading and Writing in the Mathematics Classroom*
- *WebQuest and Project Resources*

For more information on Reading and Writing in Mathematics, see pp. T6–T7.

ELL ENGLISH LANGUAGE LEARNERS

Lesson 4-1

Using Applications

Provide pairs of students with data in paragraphs or in tables. Have students organize the data into one or more matrices. Discuss how the students chose what data to record in each row and column of the matrix.

Lesson 4-4

Using Manipulatives

Have students draw a triangle on grid paper. Then cut out the triangle and place it on a coordinate plane made out of the same-sized grid paper. Outline the triangle. Then maintain the orientation of the triangle and slide it to a new position on the coordinate plane and outline again. Write matrices to represent the coordinates of each triangle. Ask students to compare the matrices.

Lesson 4-8

Flexible Groups

Have students work in groups to write and solve systems of equations using matrices. Ask students to compare this method of solving equations with the methods of elimination and substitution that they studied in Chapter 3.

Chapter

4 Notes

What You'll Learn

Have students read over the list of objectives and make a list of any words with which they are not familiar.

Why It's Important

Point out to students that this is only one of many reasons why each objective is important. Others are provided in the introduction to each lesson.

Lesson	NCTM Standards	Local Objectives
4-1	1, 2, 5, 8, 9, 10	
4-1 Follow-Up	1, 5, 10	
4-2	1, 2, 6, 8, 9, 10	
4-3	1, 2, 3, 6, 8, 9, 10	
4-4	1, 2, 6, 8, 9, 10	
4-5	1, 2, 4, 6, 7, 8, 9	
4-6	1, 2, 6, 8, 9, 10	
4-7	1, 2, 6, 8, 9, 10	
4-8	1, 2, 6, 7, 8, 9, 10	
4-8 Follow-Up	1, 2, 10	

Key to NCTM Standards:

1=Number & Operations, 2=Algebra, 3=Geometry, 4=Measurement, 5=Data Analysis & Probability, 6=Problem Solving, 7=Reasoning & Proof, 8=Communication, 9=Connections, 10=Representation

Chapter

4 Matrices

What You'll Learn

- **Lesson 4-1** Organize data in matrices.
- **Lessons 4-2, 4-3, and 4-5** Perform operations with matrices and determinants.
- **Lesson 4-4** Transform figures on a coordinate plane.
- **Lessons 4-6 and 4-8** Use matrices to solve systems of equations.
- **Lesson 4-7** Find the inverse of a matrix.

Key Vocabulary

- matrix (p. 154)
- determinant (p.182)
- expansion by minors (p. 183)
- Cramer's Rule (p. 189)
- matrix equation (p. 202)

Why It's Important

Data are often organized into matrices. For example, the National Federation of State High School Associations uses matrices to record student participation in sports by category for males and females. To find the total participation of both groups in each sport, you can add the two matrices. *You will learn how to add matrices in Lesson 4-2.*

Vocabulary Builder ELL

The Key Vocabulary list introduces students to some of the main vocabulary terms included in this chapter. For a more thorough vocabulary list with pronunciations of new words, give students the Vocabulary Builder worksheets found on pages vii and viii of the *Chapter 4 Resource Masters*. Encourage them to complete the definition of each term as they progress through the chapter. You may suggest that they add these sheets to their study notebooks for future reference when studying for the Chapter 4 test.

Getting Started

Prerequisite Skills To be successful in this chapter, you'll need to master these skills and be able to apply them in problem-solving situations. Review these skills before beginning Chapter 4.

For Lesson 4-1 — Solve Equations

Solve each equation. *(For review, see Lesson 1-3.)*

1. $3x = 18$ **6**
2. $2a - 3 = -11$ **−4**
3. $4t - 5 = 14$ $\mathbf{4\frac{3}{4}}$
4. $\frac{1}{3}y + 5 = 9$ **12**
5. $3k + 5 = 2k - 8$ **−13**
6. $5m - 6 = 7m - 8$ **1**

For Lessons 4-2 and 4-7 — Additive and Multiplicative Inverses

Name the additive inverse and the multiplicative inverse for each number.
(For review, see Lesson 1-2.)

7. 3 $-3; \frac{1}{3}$
8. −11 $11; -\frac{1}{11}$
9. 8 $-8; \frac{1}{8}$
10. −0.5 **0.5; −2**
11. 1.25 **−1.25; 0.8**
12. $\frac{5}{9}$ $-\frac{5}{9}; \frac{9}{5}$
13. $-\frac{8}{3}$ $\frac{8}{3}; -\frac{3}{8}$
14. $-1\frac{1}{5}$ $\frac{6}{5}; -\frac{5}{6}$

For Lesson 4-4 — Graph Ordered Pairs

Graph each set of ordered pairs on a coordinate plane. *(For review, see Lesson 2-1.)*

15. $\{(0, 0), (1, 3), (-2, 4)\}$
16. $\{(-1, 5), (2, -3), (4, 0)\}$
17. $\{(-3, -3), (-1, 2), (1, -3), (3, -6)\}$
18. $\{(-2, 5), (1, 3), (4, -2), (4, 7)\}$

15–18. See pp. 217A–217B.

For Lessons 4-6 and 4-8 — Solve Systems of Equations

Solve each system of equations by using either substitution or elimination.
(For review, see Lesson 3-2.)

19. $x = y + 5$, $3x + y = 19$ **(6, 1)**
20. $3x - 2y = 1$, $4x + 2y = 20$ **(3, 4)**
21. $5x + 3y = 25$, $4x + 7y = -3$ **(8, −5)**
22. $y = x - 7$, $2x - 8y = 2$ **(9, 2)**
23. $5x - 3y = 16$, $x - 3y = 8$ **(2, −2)**
24. $9x + 4y = 17$, $3x - 2y = 29$ **(5, −7)**

FOLDABLES™ Study Organizer

Matrices Make this Foldable to help you organize your notes. Begin with one sheet of notebook paper.

Step 1 Fold and Cut
Fold lengthwise to the holes. Cut eight tabs in the top sheet.

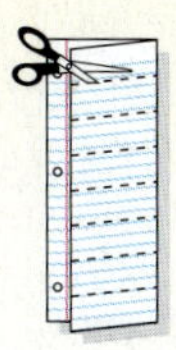

Step 2 Label
Label each tab with a lesson number and title.

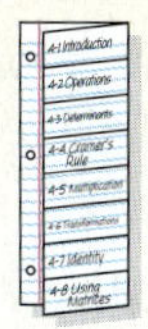

Reading and Writing As you read and study the chapter, write notes and examples under the tabs.

Getting Started

This section provides a review of the basic concepts needed before beginning Chapter 4. Page references are included for additional student help.

Prerequisite Skills in the Getting Ready for the Next Lesson section at the end of each exercise set review a skill needed in the next lesson.

For Lesson	Prerequisite Skill
4-2	Evaluating Expressions (p. 158)
4-3	Properties of Equality (p. 166)
4-4	Graphing Ordered Pairs (p. 174)
4-6	Solving Systems of Equations (p. 188)
4-7	Multiplying Matrices (p. 194)
4-8	Solving Multi-Step Equations (p. 201)

FOLDABLES™ Study Organizer

For more information about Foldables, see *Teaching Mathematics with Foldables.*

Organizing Data and Descriptive Writing After students make their Foldable, have them use their Foldable to take notes, define terms, record concepts, and write examples about matrices. At the end of each lesson, ask students to use their notes to write a descriptive paragraph sharing their learning experiences with matrices. For example, a student might descriptively write about how they use Cramer's Rule to solve systems of two or three linear equations.

4-1 Lesson Notes

1 Focus

5-Minute Check Transparency 4-1 Use as a quiz or review of Chapter 3.

Mathematical Background notes are available for this lesson on p. 152C.

Building on Prior Knowledge

Students have used tables and arrays of data in a number of contexts. In previous courses, they may have been introduced to the concept of a matrix.

How are matrices used to make decisions?

Ask students:

- How many rows of numbers appear in the SUV matrix? 4
- How many columns of numbers are there in the SUV matrix? 5

4-1 Introduction to Matrices

What You'll Learn

- Organize data in matrices.
- Solve equations involving matrices.

How are matrices used to make decisions?

Sabrina wants to buy a sports-utility vehicle (SUV). There are many types of SUVs in many prices and styles. So, Sabrina makes a list of the qualities for different models and organizes the information in a matrix.

	Base Price	Horse-power	Towing Capacity (lb)	Cargo Space (ft^3)	Fuel Economy (mpg)
Large SUV	\$32,450	285	12,000	46	17
Standard SUV	\$29,115	275	8700	16	17.5
Mid-Size SUV	\$27,975	190	5700	34	20
Compact SUV	\$18,180	127	3000	15	26.5

Source: *Car and Driver Buyer's Guide*

When the information is organized in a matrix, it is easy to compare the features of each vehicle.

Vocabulary

- matrix
- element
- dimension
- row matrix
- column matrix
- square matrix
- zero matrix
- equal matrices

ORGANIZE DATA A **matrix** is a rectangular array of variables or constants in horizontal rows and vertical columns, usually enclosed in brackets.

Study Tip

Reading Math
The plural of *matrix* is *matrices*. Matrices are sometimes called *ordered arrays*.

Example 1 Organize Data in a Matrix

Sharon wants to install cable television in her new apartment. There are two cable companies in the area whose prices are listed below. Use a matrix to organize the information. When is each company's service less expensive?

Metro Cable	
Basic Service (26 channels)	\$11.95
Standard Service (53 channels)	\$30.75
Premium Channels (in addition to Standard Service)	
• One Premium	\$10.00
• Two Premiums	\$19.00
• Three Premiums	\$25.00

Cable City	
Basic Service (26 channels)	\$9.95
Standard Service (53 channels)	\$31.95
Premium Channels (in addition to Standard Service)	
• One Premium	\$8.95
• Two Premiums	\$16.95
• Three Premiums	\$22.95

Organize the costs into labeled columns and rows.

	Basic	Standard	Standard Plus One Premium	Standard Plus Two Premiums	Standard Plus Three Premiums
Metro Cable	11.95	30.75	40.75	49.75	55.75
Cable City	9.95	31.95	40.90	48.90	54.90

Metro Cable has the best price for standard service and standard plus one premium channel. Cable City has the best price for the other categories.

Resource Manager

Workbook and Reproducible Masters

Chapter 4 Resource Masters
- Study Guide and Intervention, pp. 169–170
- Skills Practice, p. 171
- Practice, p. 172
- Reading to Learn Mathematics, p. 173
- Enrichment, p. 174

Teaching Algebra With Manipulatives Masters, pp. 229–230

Transparencies

5-Minute Check Transparency 4-1
Real-World Transparency 4
Answer Key Transparencies

Technology

Interactive Chalkboard

Study Tip

Element

The elements of a matrix can be represented using double subscript notation. The element a_{ij} is the element in row i column j.

In a matrix, numbers or data are organized so that each position in the matrix has a purpose. Each value in the matrix is called an **element**. A matrix is usually named using an uppercase letter.

$$A = \begin{bmatrix} 2 & 6 & 1 \\ 7 & 1 & 5 \\ 9 & 3 & 0 \\ 12 & 15 & 26 \end{bmatrix} \text{ 4 rows, 3 columns}$$

The element 15 is in row 4, column 2.

A matrix can be described by its **dimensions**. A matrix with m rows and n columns is an $m \times n$ matrix (read "m by n"). Matrix A above is a 4×3 matrix since it has 4 rows and 3 columns.

Example 2 Dimensions of a Matrix

State the dimensions of matrix B if $B = \begin{bmatrix} 1 & -3 \\ -5 & 18 \\ 0 & -2 \end{bmatrix}$.

$$B = \begin{bmatrix} 1 & -3 \\ -5 & 18 \\ 0 & -2 \end{bmatrix} \text{ 3 rows, 2 columns}$$

Since matrix B has 3 rows and 2 columns, the dimensions of matrix B are 3×2.

Certain matrices have special names. A matrix that has only one row is called a **row matrix**, while a matrix that has only one column is called a **column matrix**. A matrix that has the same number of rows and columns is called a **square matrix**. Another special type of matrix is the **zero matrix**, in which every element is 0. The zero matrix can have any dimension.

EQUATIONS INVOLVING MATRICES Two matrices are considered **equal matrices** if they have the same dimensions and if each element of one matrix is equal to the corresponding element of the other matrix.

$\begin{bmatrix} 6 & 3 \\ 0 & 9 \\ 1 & 3 \end{bmatrix} \neq \begin{bmatrix} 6 & 0 & 1 \\ 3 & 9 & 3 \end{bmatrix}$ The matrices have different dimensions. They are not equal.

$\begin{bmatrix} 1 & 2 \\ 8 & 5 \end{bmatrix} \neq \begin{bmatrix} 1 & 8 \\ 2 & 5 \end{bmatrix}$ Corresponding elements are not equal. The matrices are not equal.

$\begin{bmatrix} 5 & 6 & 0 \\ 0 & 7 & 2 \\ 3 & 1 & 4 \end{bmatrix} = \begin{bmatrix} 5 & 6 & 0 \\ 0 & 7 & 2 \\ 3 & 1 & 4 \end{bmatrix}$ The matrices have the same dimensions and the corresponding elements are equal. The matrices are equal.

The definition of equal matrices can be used to find values when elements of equal matrices are algebraic expressions.

Example 3 Solve an Equation Involving Matrices

Solve $\begin{bmatrix} y \\ 3x \end{bmatrix} = \begin{bmatrix} 6 - 2x \\ 31 + 4y \end{bmatrix}$ for x and y.

Since the matrices are equal, the corresponding elements are equal. When you write the sentences to show this equality, two linear equations are formed.

$y = 6 - 2x$

$3x = 31 + 4y$

(continued on the next page)

www.algebra2.com/extra_examples

2 Teach

ORGANIZE DATA

In-Class Examples Power Point®

1 **COLLEGE** Kaitlin wants to attend one of three Iowa universities next year. She has gathered information about tuition (T), room and board (R/B), and enrollment (E) for the universities. Use a matrix to organize the information. Which university's total cost is lowest?

Iowa State University: T–\$3132, R/B–\$4432, E–26,845

University of Iowa: T–\$3204, R/B–\$4597, E–28,311

University of Northern Iowa: T–\$3130, R/B–\$4149, E–14,106

Source: *The World Almanac 2002*

	T	R/B	E
ISU	3132	4432	26845
UI	3204	4597	28311
UNI	3130	4149	14106

The University of Northern Iowa has the lowest total cost.

2 State the dimensions of matrix G if $G = \begin{bmatrix} 2 & -1 & 0 & 3 \\ 1 & 5 & -3 & -1 \end{bmatrix}$. **2 × 4**

Teaching Tip Stress that matrix dimensions are always given as "rows by columns."

EQUATIONS INVOLVING MATRICES

In-Class Example Power Point®

3 Solve $\begin{bmatrix} y \\ 3 \end{bmatrix} = \begin{bmatrix} 3x - 2 \\ 2y + x \end{bmatrix}$ for x and y. **(1, 1)**

Teaching Tip Caution students to set equal those elements of the matrices that are in corresponding positions.

DAILY INTERVENTION **Differentiated Instruction**

Kinesthetic To help students associate the terms *row* and *column* with the correct parts of a matrix, use this activity. Instruct students to say either "row" or "column" when you point to them. When a student says "row," tell students to stretch out their arms parallel to the floor. When a student calls out "column," everyone is to stretch both of their arms overhead to form a column. Relate these movements to the side-to-side aspect of matrix rows and the up-and-down aspect of matrix columns.

3 Practice/Apply

Study Notebook

Have students—
- add the definitions/examples of the vocabulary terms in this lesson to their Vocabulary Builder worksheets for Chapter 4.
- add labeled examples of various kinds of matrices to their notebook.
- include any other item(s) that they find helpful in mastering the skills in this lesson.

About the Exercises...

Organization by Objective
Organize Data: 10–15, 26–33
Equations Involving Matrices: 16–25

Odd/Even Assignments
Exercises 10–25 are structured so that students practice the same concepts whether they are assigned odd or even problems.

Assignment Guide

Basic: 11–23 odd, 26, 27, 32–57
Average: 11–25 odd, 26–29, 31–57
Advanced: 10–24 even, 26–49 (optional: 50–57)

Answers

1. The matrices must have the same dimensions and each element of one matrix must be equal to the corresponding element of the other matrix.
2. Sample answers: row matrix, $[1 \;\; 2 \;\; 3]$, 1×3; column matrix, $\begin{bmatrix} 1 \\ 2 \end{bmatrix}$, 2×1; square matrix, $\begin{bmatrix} 1 & 2 \\ 3 & 4 \end{bmatrix}$, 2×2; zero matrix, $\begin{bmatrix} 0 & 0 \\ 0 & 0 \end{bmatrix}$, 2×2

This system can be solved using substitution.

$3x = 31 + 4y$	Second equation
$3x = 31 + 4(6 - 2x)$	Substitute $6 - 2x$ for y.
$3x = 31 + 24 - 8x$	Distributive Property
$11x = 55$	Add $8x$ to each side.
$x = 5$	Divide each side by 11.

To find the value for y, substitute 5 for x in either equation.

$y = 6 - 2x$	First equation
$y = 6 - 2(5)$	Substitute 5 for x.
$y = -4$	Simplify.

The solution is $(5, -4)$.

Check for Understanding

Concept Check

1–2. See margin.

1. **Describe** the conditions that must be met in order for two matrices to be considered equal.
2. **OPEN ENDED** Give examples of a row matrix, a column matrix, a square matrix, and a zero matrix. State the dimensions of each matrix.
3. **Explain** what is meant by corresponding elements. **Corresponding elements are elements in the same row and column positions.**

Guided Practice

GUIDED PRACTICE KEY

Exercises	Examples
4, 5	2
6, 7	3
8, 9	1

State the dimensions of each matrix.

4. $[3 \;\; 4 \;\; 5 \;\; 6 \;\; 7]$ **1×5**

5. $\begin{bmatrix} 10 & -6 & 18 & 0 \\ -7 & 5 & 2 & 4 \\ 3 & 11 & 9 & 7 \end{bmatrix}$ **3×4**

Solve each equation.

6. $\begin{bmatrix} x + 4 \\ 2y \end{bmatrix} = \begin{bmatrix} 9 \\ 12 \end{bmatrix}$ **(5, 6)**

7. $[9 \;\; 13] = [x + 2y \;\; 4x + 1]$ **(3, 3)**

Application

WEATHER **For Exercises 8 and 9, use the table that shows a five-day forecast indicating high (H) and low (L) temperatures.**

Fri	Sat	Sun	Mon	Tue
H 88	H 88	H 90	H 86	H 85
L 54	L 54	L 56	L 53	L 52

8. Organize the temperatures in a matrix. **See margin.**
9. What are the dimensions of the matrix? **2×5**

★ indicates increased difficulty

Practice and Apply

Homework Help

For Exercises	See Examples
10–15	2
16–25	3
26–31	1

Extra Practice
See page 834.

State the dimensions of each matrix.

10. $\begin{bmatrix} 6 & -1 & 5 \\ -2 & 3 & -4 \end{bmatrix}$ **2×3**

11. $\begin{bmatrix} 7 \\ 8 \\ 9 \end{bmatrix}$ **3×1**

12. $\begin{bmatrix} 0 & 0 & 8 \\ 6 & 2 & 4 \\ 1 & 3 & 6 \\ 5 & 9 & 2 \end{bmatrix}$ **4×3**

13. $\begin{bmatrix} -3 & 17 & -22 \\ 9 & 31 & 16 \\ 20 & -15 & 4 \end{bmatrix}$ **3×3**

14. $\begin{bmatrix} 17 & -2 & 8 & -9 & 6 \\ 5 & 11 & 20 & -1 & 4 \end{bmatrix}$ **2×5**

15. $\begin{bmatrix} 16 & 8 \\ 10 & 5 \\ 0 & 0 \end{bmatrix}$ **3×2**

8.

	Fri	Sat	Sun	Mon	Tue
High	88	88	90	86	85
Low	54	54	56	53	52

26.

	Evening	Matinee	Twilight
Adult	7.50	5.50	3.75
Child	4.50	4.50	3.75
Senior	5.50	5.50	3.75

Solve each equation.

16. $[2x \quad 3 \quad 3z] = [5 \quad 3y \quad 9]$ **(2.5, 1, 3)**

17. $[4x \quad 3y] = [12 \quad -1]$ **$\left(3, -\frac{1}{3}\right)$**

18. $\begin{bmatrix} 4x \\ 5 \end{bmatrix} = \begin{bmatrix} 15 + x \\ 2y - 1 \end{bmatrix}$ **(5, 3)**

19. $\begin{bmatrix} 4x - 3 & 3y \\ 7 & 13 \end{bmatrix} = \begin{bmatrix} 9 & -15 \\ 7 & 2z + 1 \end{bmatrix}$ **(3, −5, 6)**

20. $\begin{bmatrix} x + 3y \\ 3x + y \end{bmatrix} = \begin{bmatrix} -13 \\ 1 \end{bmatrix}$ **(2, −5)**

21. $\begin{bmatrix} 2x + y \\ x - 3y \end{bmatrix} = \begin{bmatrix} 5 \\ 13 \end{bmatrix}$ **(4, −3)**

22. $\begin{bmatrix} 2x \\ 2x + 3y \end{bmatrix} = \begin{bmatrix} y \\ 12 \end{bmatrix}$ **(1.5, 3)**

23. $\begin{bmatrix} 4x \\ y - 1 \end{bmatrix} = \begin{bmatrix} 11 + 3y \\ x \end{bmatrix}$ **(14, 15)**

★ 24. $\begin{bmatrix} x^2 + 1 & 5 - y \\ x + y & y - 4 \end{bmatrix} = \begin{bmatrix} 5 & x \\ 5 & 3 \end{bmatrix}$ **(−2, 7)**

★ 25. $\begin{bmatrix} 3x - 5 & x + y \\ 12 & 9z \end{bmatrix} = \begin{bmatrix} 10 & 8 \\ 12 & 3x + y \end{bmatrix}$ **(5, 3, 2)**

More About. . .

Movies

Adjusting for inflation, *Cleopatra* (1963) is the most expensive movie ever made. Its $44 million budget is equivalent to $306,867,120 today.

Source: *The Guinness Book of Records*

MOVIES For Exercises 26 and 27, use the advertisement shown at the right.

26. Write a matrix for the prices of movie tickets for adults, children, and seniors. **See margin.**

27. What are the dimensions of the matrix? **3 × 3**

DINING OUT For Exercises 28 and 29, use the following information.

A newspaper rated several restaurants by cost, level of service, atmosphere, and location using a scale of ★ being low and ★★★★ being high.

Catalina Grill: cost ★★, service ★, atmosphere ★, location ★

Oyster Club: cost ★★★, service ★★, atmosphere ★, location ★★

Casa di Pasta: cost ★★★★, service ★★★, atmosphere ★★★, location ★★★

Mason's Steakhouse: cost ★★, service ★★★★, atmosphere ★★★★, location ★★★

28. Write a 4 × 4 matrix to organize this information. **28–29. See pp. 217A–217B.**

29. Which restaurant would you select based on this information and why?

HOTELS For Exercises 30 and 31, use the costs for an overnight stay at a hotel that are given below.

Single Room: $60 weekday; $79 weekend

Double Room: $70 weekday; $89 weekend

Suite: $75 weekday; $95 weekend

30. Write a 3 × 2 matrix that represents the cost of each room. **See margin.**

★ 31. Write a 2 × 3 matrix that represents the cost of each room. **See margin.**

CRITICAL THINKING For Exercises 32 and 33, use the matrix at the right.

$$\begin{bmatrix} 1 & 3 & 6 & 10 & 15 & \dots \\ 2 & 5 & 9 & 14 & 20 & \dots \\ 4 & 8 & 13 & 19 & 26 & \dots \\ 7 & 12 & 18 & 25 & 33 & \dots \\ 11 & 17 & 24 & 32 & 41 & \dots \\ 16 & 23 & 31 & 40 & 50 & \dots \\ \vdots & \vdots & \vdots & \vdots & \vdots & \vdots \end{bmatrix}$$

32. Study the pattern of numbers. Complete the matrix for column 6 and row 7.

33. In which row and column will 100 occur? **row 6, column 9**

32.

$$\begin{bmatrix} 1 & 3 & 6 & 10 & 15 & 21 \\ 2 & 5 & 9 & 14 & 20 & 27 \\ 4 & 8 & 13 & 19 & 26 & 34 \\ 7 & 12 & 18 & 25 & 33 & 42 \\ 11 & 17 & 24 & 32 & 41 & 51 \\ 16 & 23 & 31 & 40 & 50 & 61 \\ 22 & 30 & 39 & 49 & 60 & 72 \end{bmatrix}$$

www.algebra2.com/self_check_quiz

Answers

30.

	Weekday	Weekend
Single	60	79
Double	70	89
Suite	75	95

31.

	Single	Double	Suite
Weekday	60	70	75
Weekend	79	89	95

Enrichment, p. 174

Tessellations

A **tessellation** is an arangement of polygons covering a plane without any gaps or overlapping. One example of a tessellation is a honeycomb. Three congruent regular hexagons meet at each vertex, and there is no wasted space between cells. This tessellation is called a regular tessellation since it is formed by congruent regular polygons.

A **semi-regular tessellation** is a tessellation formed by two or more regular polygons such that the number of sides of the polygons meeting at each vertex is the same.

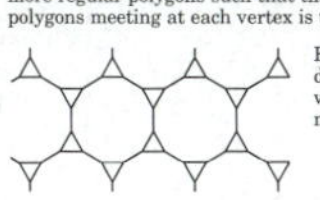

For example, the tessellation at the left has two regular dodecagons and one equilateral triangle meeting at each vertex. We can name this tessellation a 3-12-12 for the number of sides of each polygon that meet at one vertex.

Study Guide and Intervention, p. 169 (shown) and p. 170

Organize Data

Matrix	a rectangular array of variables or constants in horizontal rows and vertical columns, usually enclosed in brackets.

A matrix can be described by its **dimensions**. A matrix with m rows and n columns is an $m \times n$ matrix.

Example 1 **Owls' eggs incubate for 30 days and their fledgling period is also 30 days. Swifts' eggs incubate for 20 days and their fledgling period is 44 days. Pigeon eggs incubate for 15 days, and their fledgling period is 17 days. Eggs of the king penguin incubate for 53 days, and the fledgling time for a king penguin is 360 days. Write a 2 × 4 matrix to organize this information.** Source: *The Cambridge Factfinder*

	Owl	Swift	Pigeon	King Penguin
Incubation	30	20	15	53
Fledgling	30	44	17	360

Example 2 **What are the dimensions of matrix A if $A = \begin{bmatrix} 13 & 10 & -3 & 45 \\ 2 & 8 & 15 & 80 \end{bmatrix}$?**

Since matrix A has 2 rows and 4 columns, the dimensions of A are 2 × 4.

Exercises

State the dimensions of each matrix.

1. $\begin{bmatrix} 15 & 5 & 27 & -4 \\ 23 & 6 & 0 & 5 \\ 14 & 70 & 24 & -3 \\ 63 & 3 & 42 & 90 \end{bmatrix}$ **4 × 4**

2. $[16 \quad 12 \quad 0]$ **1 × 3**

3. $\begin{bmatrix} 71 & 44 \\ 39 & 27 \\ 45 & 16 \\ 92 & 53 \\ 78 & 65 \end{bmatrix}$ **5 × 2**

4. A travel agent provides for potential travelers the normal high temperatures for the months of January, April, July, and October for various cities. In Boston these figures are 36°, 56°, 82°, and 63°. In Dallas they are 54°, 76°, 97°, and 79°. In Los Angeles they are 68°, 72°, 84°, and 79°. In Seattle they are 46°, 58°, 74°, and 60°, and in St. Louis they are 38°, 67°, 89°, and 69°. Organize this information in a 4 × 5 matrix. Source: *The New York Times Almanac*

	Boston	Dallas	Los Angeles	Seattle	St. Louis
January	36	54	68	46	38
April	56	76	72	58	67
July	82	97	84	74	89
October	63	79	79	60	69

Skills Practice, p. 171 and Practice, p. 172 (shown)

State the dimensions of each matrix.

1. $[-3 \quad -3 \quad 7]$ **1 × 3**

2. $\begin{bmatrix} 5 & 8 & -1 \\ -2 & 1 & 8 \end{bmatrix}$ **2 × 3**

3. $\begin{bmatrix} -2 & 2 & -2 & 3 \\ 5 & 16 & 0 & 0 \\ 4 & 7 & -1 & 4 \end{bmatrix}$ **3 × 4**

Solve each equation.

4. $[4x \quad 42] = [24 \quad 6y]$ **(6, 7)**

5. $[-2x \quad 22 \quad -3z] = [6x \quad -2y \quad 45]$ **(0, −11, −15)**

6. $\begin{bmatrix} 6x \\ 2y + 3 \end{bmatrix} = \begin{bmatrix} -36 \\ 17 \end{bmatrix}$ **(−6, 7)**

7. $\begin{bmatrix} 7x - 8 \\ 8y - 3 \end{bmatrix} = \begin{bmatrix} 20 \\ 2y + 3 \end{bmatrix}$ **(4, 1)**

8. $\begin{bmatrix} -4x - 3 \\ 6y \end{bmatrix} = \begin{bmatrix} -3x \\ -2y + 16 \end{bmatrix}$ **(−3, 2)**

9. $\begin{bmatrix} 6x - 12 \\ -3y + 6 \end{bmatrix} = \begin{bmatrix} -3x - 21 \\ 8y - 5 \end{bmatrix}$ **(−1, 1)**

10. $\begin{bmatrix} -5 & 3x + 1 \\ 2y - 1 & 3z - 2 \end{bmatrix} = \begin{bmatrix} -5 & x - 1 \\ 3y & 5z - 4 \end{bmatrix}$ **(−1, −1, 1)**

11. $\begin{bmatrix} 3x \\ y + 4 \end{bmatrix} = \begin{bmatrix} y + 8 \\ 17 \end{bmatrix}$ **(7, 13)**

12. $\begin{bmatrix} 5x + 8y \\ 3x - 11 \end{bmatrix} = \begin{bmatrix} -1 \\ y \end{bmatrix}$ **(3, −2)**

13. $\begin{bmatrix} 2x + y \\ 3x + 2y \end{bmatrix} = \begin{bmatrix} 0 \\ -2 \end{bmatrix}$ **(2, −4)**

14. **TICKET PRICES** The table at the right gives ticket prices for a concert. Write a 2 × 3 matrix that represents the cost of a ticket.

	Child	Student	Adult
Cost Purchased in Advance	$6	$12	$18
Cost Purchased at the Door	$8	$15	$22

$\begin{bmatrix} 6 & 12 & 18 \\ 8 & 15 & 22 \end{bmatrix}$

CONSTRUCTION For Exercises 15 and 16, use the following information.

During each of the last three weeks, a road-building crew has used three truck-loads of gravel. The table at the right shows the amount of gravel in each load.

Week 1	Week 2	Week 3
Load 1 40 tons	Load 1 40 tons	Load 1 32 tons
Load 2 32 tons	Load 2 40 tons	Load 2 24 tons
Load 3 24 tons	Load 3 32 tons	Load 3 24 tons

15. Write a matrix for the amount of gravel in each load.

$\begin{bmatrix} 40 & 32 & 24 \\ 40 & 40 & 32 \\ 32 & 24 & 24 \end{bmatrix}$ or $\begin{bmatrix} 40 & 40 & 32 \\ 32 & 40 & 24 \\ 24 & 32 & 24 \end{bmatrix}$

16. What are the dimensions of the matrix? **3 × 3**

Reading to Learn Mathematics, p. 173 ELL

Pre-Activity **How are matrices used to make decisions?**

Read the introduction to Lesson 4-1 at the top of page 154 in your textbook.

What is the base price of a Mid-Size SUV?
$27,975

Reading the Lesson

1. Give the dimensions of each matrix.

a. $\begin{bmatrix} 3 & 2 & 5 \\ -1 & 0 & 6 \end{bmatrix}$ **2 × 3**

b. $[1 \quad 4 \quad 0 \quad -8 \quad 2]$ **1 × 5**

2. Identify each matrix with as many of the following descriptions that apply: *row matrix, column matrix, square matrix, zero matrix.*

a. $[6 \quad 5 \quad 4 \quad 3]$ **row matrix**

b. $\begin{bmatrix} 0 \\ 0 \\ 0 \end{bmatrix}$ **column matrix; zero matrix**

c. $[0]$ **row matrix; column matrix; square matrix; zero matrix**

3. Write a system of equations that you could use to solve the following matrix equation for x, y, and z. (Do not actually solve the system.)

$\begin{bmatrix} 3x \\ x + y \\ y - z \end{bmatrix} = \begin{bmatrix} -9 \\ 5 \\ 6 \end{bmatrix}$ **$3x = -9, x + y = 5, y - z = 6$**

Helping You Remember

4. Some students have trouble remembering which number comes first in writing the dimensions of a matrix. Think of an easy way to remember this.

Sample answer: Read the matrix from top to bottom, then from left to right. Reading down gives the number of rows, which is written first in the dimensions of the matrix. Reading across gives the number of columns, which is written second.

4 Assess

Open-Ended Assessment

Writing Have students write a matrix equation that is equivalent to the following system of equations.

$2x - y = -4$

$3x + y = -1$

$\begin{bmatrix} 2x - y \\ 3x + y \end{bmatrix} = \begin{bmatrix} -4 \\ -1 \end{bmatrix}$

Getting Ready for Lesson 4-2

PREREQUISITE SKILL Lesson 4-2 presents the addition and subtraction of matrices, and the multiplication of a matrix by a scalar. These operations involve evaluating various expressions with real numbers. Exercises 50–57 should be used to determine your students' familiarity with evaluating expressions.

Answers

34. Matrices are used to organize information so it can be read and compared more easily. Answers should include the following.
    - If you want the least expensive vehicle, the compact SUV has the best price; the large SUV has the most horsepower, towing capacity, and cargo space, and the standard SUV has the best fuel economy.
    - Sample answer: Matrices are used to report stock prices in the newspaper.

40.

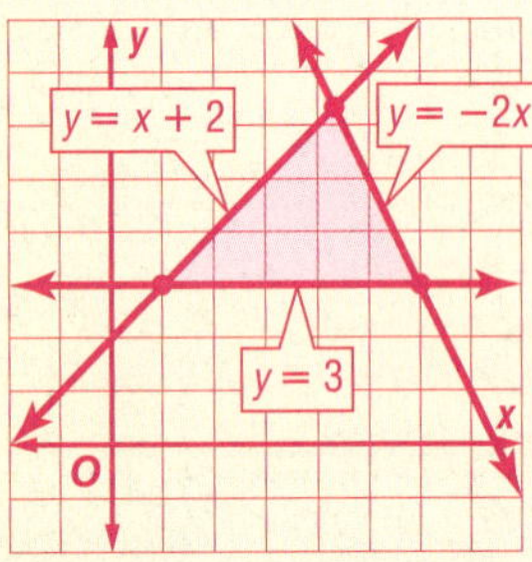

34. WRITING IN MATH Answer the question that was posed at the beginning of the lesson. **See margin.**

How are matrices used to make decisions?

Include the following in your answer:
- the circumstances under which each vehicle best fits a person's needs, and
- an example of how matrices are used in newspapers or magazines.

Standardized Test Practice

35. In matrix $A = \begin{bmatrix} 1 & 5 & -2 \\ -4 & 0 & 6 \\ 3 & 7 & 8 \end{bmatrix}$, element 3 is in which row and column? **B**

(A) row 1, column 3 (B) row 3, column 1
(C) row 1, column 1 (D) row 3, column 3

36. What is the value of y if $\begin{bmatrix} 3x \\ y + 5 \end{bmatrix} = \begin{bmatrix} 9 + y \\ x \end{bmatrix}$? **C**

(A) 2 (B) 4 (C) −3 (D) −1

Maintain Your Skills

Mixed Review

Solve each system of equations. *(Lesson 3-5)*

37. $3x - 3y = 6$ **(7, 5, 4)**
$-6y = -30$
$5z - 2x = 6$

38. $3a + 2b = 27$ **(7, 3, −9)**
$5a - 7b + c = 5$
$-2a + 10b + 5c = -29$

39. $3r - 15s + 4t = -57$
$9r + 45s - t = 26$
$-6r + 10s + 3t = -19$

Graph each system of inequalities. Name the coordinates of the vertices of the feasible region. Find the maximum and minimum values of the given function for this region. *(Lesson 3-4)* **40–42. See margin for graphs.**

40. $y \geq 3$
$y \leq x + 2$
$y \leq -2x + 15$
$f(x, y) = 2x + 3y$

41. $y \geq \frac{1}{3}x$
$y \geq -5x + 16$
$y \leq -x + 10$
$f(x, y) = 5x - y$

42. $y \geq \frac{1}{2}x$
$y \geq -x + 3$
$y \leq -\frac{3}{2}x + 12$
$f(x, y) = 3y - x$

39. $\left(-\frac{4}{3}, \frac{3}{5}, -11\right)$

40. vertices: (1, 3), (6, 3), $\left(\frac{13}{3}, \frac{19}{3}\right)$; max: $f\left(\frac{13}{3}, \frac{19}{3}\right) = \frac{83}{8}$, min: $f(1, 3) = 11$

41. vertices: (3, 1), $\left(\frac{15}{2}, \frac{5}{2}\right)$, $\left(\frac{3}{2}, \frac{17}{2}\right)$; max: $f\left(\frac{15}{2}, \frac{5}{2}\right) = 35$, min: $f\left(\frac{3}{2}, \frac{17}{2}\right) = -1$

42. vertices: (2, 1), (6, 3); no maximum, min: $f(2, 1) = 1$

BUSINESS For Exercises 43–45, use the following information.
The parking garage at Burrough's Department Store charges \$1.50 for each hour or fraction of an hour for parking. *(Lesson 2-6)*

43. Graph the function. **See pp. 217A–217B.**

44. What type of function represents this situation? **step function**

45. Jada went shopping at Burrough's Department Store yesterday. She left her car in the parking garage for two hours and twenty-seven minutes. How much did Jada pay for parking? **\$4.50**

Find each value if $f(x) = x^2 - 3x + 2$. *(Lesson 2-1)*

46. $f(3)$ **2** 47. $f(0)$ **2** 48. $f(2)$ **0** 49. $f(-3)$ **20**

Getting Ready for the Next Lesson

PREREQUISITE SKILL **Find the value of each expression.**
*(To review **evaluating expressions**, see Lesson 1-2.)*

50. $8 + (-5)$ **3** 51. $-2 - 8$ **−10** 52. $3.5 + 2.7$ **6.2** 53. $6(-3)$ **−18**

54. $\frac{1}{2}(34)$ **17** 55. $6(4) + 3(-9)$ **−3** 56. $-5(3 - 18)$ **75** 57. $14\left(\frac{1}{4}\right) - 12\left(\frac{1}{6}\right)$ $\frac{3}{2}$

41.

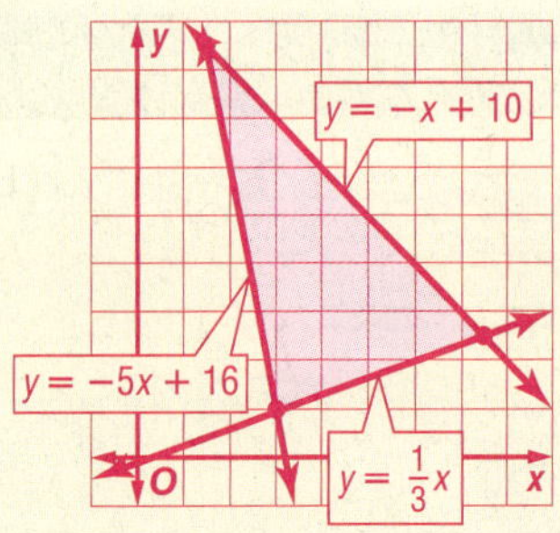

42.

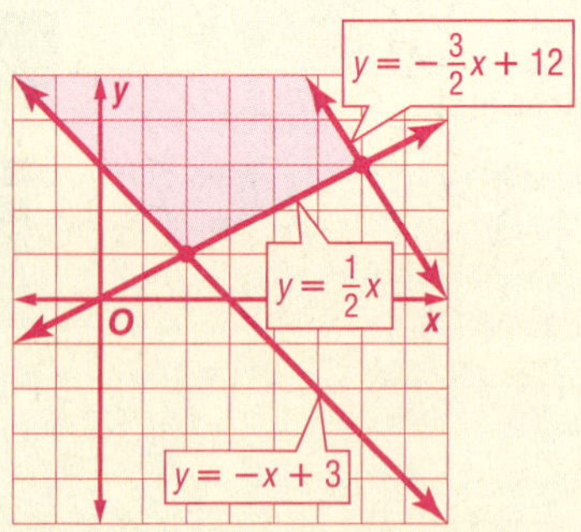

158 Chapter 4 Matrices

Organizing Data

You can use a computer spreadsheet to organize and display data. Then you can use the data to create graphs or perform calculations.

Example

Enter the data on the Atlantic Coast Conference Men's Basketball scoring into a spreadsheet.

Atlantic Coast Conference 2000–2001 Men's Basketball			
Team	Free Throws	2-Point Field Goals	3-Point Field Goals
Clemson	456	549	248
Duke	697	810	407
Florida State	453	594	148
Georgia Tech	457	516	260
Maryland	622	915	205
North Carolina	532	756	189
North Carolina State	507	562	170
Virginia	556	648	204
Wake Forest	443	661	177

Source: Atlantic Coast Conference

Use Column A for the team names, Column B for the numbers of free throws, Column C for the numbers of 2-point field goals, and Column D for the numbers of 3-point field goals.

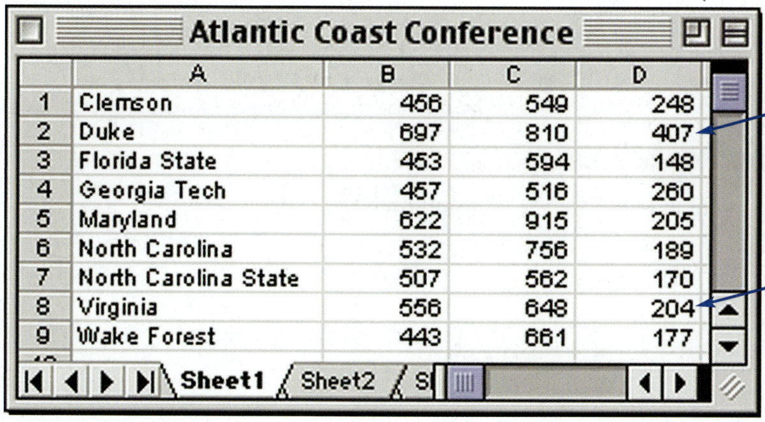

Each **row** contains data for a different team. Row 2 represents Duke.

Each **cell** of the spreadsheet contains one piece of data. Cell 8D contains the value 204, representing the number of 3-point field goals made by Virginia.

Model and Analyze **1. See pp. 217A–217B.**

1. Enter the data about sports-utility vehicles on page 154 into a spreadsheet.
2. Compare and contrast how data are organized in a spreadsheet and how they are organized in a matrix. **See margin.**

Answer

2. **Both use rows and columns. In a spreadsheet, the rows are designated by numbers and the columns are designated by letters. In a matrix, both rows and columns are designated by numbers.**

Spreadsheet Investigation

A Follow-Up of Lesson 4-1

Getting Started

You may want to have students format the cells before they start to enter the data. Use the Format Cells command. You can use the General or Text format for all the data unless number operations are going to be performed with the contents of the cells.

Teach

- Have students practice their skills in using a spreadsheet by entering the data in the example.
- Clear up any confusion that students may have about how to move from cell to cell in a spreadsheet.
- Have students complete Exercises 1 and 2.

Tips for New Teachers

Intervention Many sources of confusion can be avoided when students work with a partner on technology investigations. Assign partners so that a student with more knowledge of spreadsheets is paired with one who has less experience with them.

Assess

Ask students about the structure of a given spreadsheet. For example, ask them which cells contain data values and which contain labels to identify the data.

4-2 Lesson Notes

1 Focus

5-Minute Check Transparency 4-2 Use as a quiz or review of Lesson 4-1.

Mathematical Background notes are available for this lesson on p. 152C.

How can matrices be used to calculate daily dietary needs?

Ask students:

- Suppose a 3 × 3 matrix is created for the breakfast data with the days as the rows and the Calories, protein, and fat contents as the columns. What is the row and column location for the entry for 17? **row 2, column 3**
- Suppose a matrix for the dinner data is created with the days as the rows. What is the meaning of the entry that will be located in row 3, column 2? **The entry 29 means that there are 29 grams of protein in the dinner meal on day 3.**

4-2 Operations with Matrices

What You'll Learn

- Add and subtract matrices.
- Multiply by a matrix scalar.

Vocabulary

- scalar
- scalar multiplication

How can matrices be used to calculate daily dietary needs?

In her job as a hospital dietician, Celeste designs weekly menus for her patients and tracks various nutrients for each daily diet. The table shows the Calories, protein, and fat in a patient's meals over a three-day period.

Day	Breakfast			Lunch			Dinner		
	Calories	Protein (g)	Fat (g)	Calories	Protein (g)	Fat (g)	Calories	Protein (g)	Fat (g)
1	566	18	7	785	22	19	1257	40	26
2	482	12	17	622	23	20	987	32	45
3	530	10	11	710	26	12	1380	29	38

These data can be organized in three matrices representing breakfast, lunch, and dinner. The daily totals can then be found by adding the three matrices.

ADD AND SUBTRACT MATRICES

Matrices can be added if and only if they have the same dimensions.

Key Concept *Addition of Matrices*

- **Words** If A and B are two $m \times n$ matrices, then $A + B$ is an $m \times n$ matrix in which each element is the sum of the corresponding elements of A and B.
- **Symbols** $\begin{bmatrix} a & b & c \\ d & e & f \\ g & h & i \end{bmatrix} + \begin{bmatrix} j & k & l \\ m & n & o \\ p & q & r \end{bmatrix} = \begin{bmatrix} a+j & b+k & c+l \\ d+m & e+n & f+o \\ g+p & h+q & i+r \end{bmatrix}$

Example 1 *Add Matrices*

a. Find $A + B$ if $A = \begin{bmatrix} 4 & -6 \\ 2 & 3 \end{bmatrix}$ and $B = \begin{bmatrix} -3 & 7 \\ 5 & -9 \end{bmatrix}$.

$A + B = \begin{bmatrix} 4 & -6 \\ 2 & 3 \end{bmatrix} + \begin{bmatrix} -3 & 7 \\ 5 & -9 \end{bmatrix}$ Definition of matrix addition

$= \begin{bmatrix} 4 + (-3) & -6 + 7 \\ 2 + 5 & 3 + (-9) \end{bmatrix}$ Add corresponding elements.

$= \begin{bmatrix} 1 & 1 \\ 7 & -6 \end{bmatrix}$ Simplify.

b. Find $A + B$ if $A = \begin{bmatrix} 3 & -7 & 4 \\ 12 & 5 & 0 \end{bmatrix}$ and $B = \begin{bmatrix} 2 & 9 \\ 4 & -6 \end{bmatrix}$.

Since the dimensions of A are 2×3 and the dimensions of B are 2×2, you cannot add these matrices.

Resource Manager

Workbook and Reproducible Masters

Chapter 4 Resource Masters

- Study Guide and Intervention, pp. 175–176
- Skills Practice, p. 177
- Practice, p. 178
- Reading to Learn Mathematics, p. 179
- Enrichment, p. 180
- Assessment, p. 231

Transparencies

5-Minute Check Transparency 4-2
Answer Key Transparencies

Technology

Alge2PASS: Tutorial Plus, Lesson 6
Interactive Chalkboard

You can subtract matrices in a similar manner.

Key Concept — *Subtraction of Matrices*

- **Words** If A and B are two $m \times n$ matrices, then $A - B$ is an $m \times n$ matrix in which each element is the difference of the corresponding elements of A and B.
- **Symbols** $\begin{bmatrix} a & b & c \\ d & e & f \\ g & h & i \end{bmatrix} - \begin{bmatrix} j & k & l \\ m & n & o \\ p & q & r \end{bmatrix} = \begin{bmatrix} a-j & b-k & c-l \\ d-m & e-n & f-o \\ g-p & h-q & i-r \end{bmatrix}$

Example 2 Subtract Matrices

Find $A - B$ if $A = \begin{bmatrix} 9 & 2 \\ -4 & 7 \end{bmatrix}$ and $B = \begin{bmatrix} 3 & 6 \\ 8 & -2 \end{bmatrix}$.

$A - B = \begin{bmatrix} 9 & 2 \\ -4 & 7 \end{bmatrix} - \begin{bmatrix} 3 & 6 \\ 8 & -2 \end{bmatrix}$ Definition of matrix subtraction

$= \begin{bmatrix} 9-3 & 2-6 \\ -4-8 & 7-(-2) \end{bmatrix}$ Subtract corresponding elements.

$= \begin{bmatrix} 6 & -4 \\ -12 & 9 \end{bmatrix}$ Simplify.

More About. . .

Animals

The rarest animal in the world today is a giant tortoise that lives in the Galapagos Islands. "Lonesome George" is the only remaining representative of his species *(Geochelone elephantopus abingdoni)*. With virtually no hope of discovering another specimen, this species is now effectively extinct.

Source: www.ecoworld.com

Example 3 Use Matrices to Model Real-World Data

ANIMALS The table below shows the number of endangered and threatened species in the United States and in the world. How many more endangered and threatened species are there on the world list than on the U.S. list?

Endangered and Threatened Species

Type of Animal	United States		World	
	Endangered	Threatened	Endangered	Threatened
Mammals	61	8	309	24
Birds	74	15	252	21
Reptiles	14	22	79	36
Amphibians	9	8	17	9
Fish	69	42	80	42

Source: Fish and Wildlife Service, U.S. Department of Interior

The data in the table can be organized in two matrices. Find the difference of the matrix that represents species in the world and the matrix that represents species in the U.S.

World **U.S.**

$\begin{bmatrix} 309 & 24 \\ 252 & 21 \\ 79 & 36 \\ 17 & 9 \\ 80 & 42 \end{bmatrix} - \begin{bmatrix} 61 & 8 \\ 74 & 15 \\ 14 & 22 \\ 9 & 8 \\ 69 & 42 \end{bmatrix} = \begin{bmatrix} 309-61 & 24-8 \\ 252-74 & 21-15 \\ 79-14 & 36-22 \\ 17-9 & 9-8 \\ 80-69 & 42-42 \end{bmatrix}$ Subtract corresponding elements.

(continued on the next page)

www.algebra2.com/extra_examples

2 Teach

ADD AND SUBTRACT MATRICES

In-Class Examples Power Point®

1 **a.** Find $A + B$ if $A = \begin{bmatrix} 6 & 4 \\ -1 & 0 \end{bmatrix}$ and $B = \begin{bmatrix} -3 & 1 \\ 0 & 3 \end{bmatrix}$. $\begin{bmatrix} 3 & 5 \\ -1 & 3 \end{bmatrix}$

b. Find $A + B$ if $A = \begin{bmatrix} 4 & -2 & 0 \\ 1 & 5 & -1 \end{bmatrix}$ and $B = \begin{bmatrix} -6 & 7 \\ -9 & 3 \end{bmatrix}$. **Since their dimensions are different, these matrices cannot be added.**

2 Find $A - B$ if $A = \begin{bmatrix} 3 & 2 \\ -1 & 0 \end{bmatrix}$ and $B = \begin{bmatrix} -2 & 1 \\ 0 & -1 \end{bmatrix}$. $\begin{bmatrix} 5 & 1 \\ -1 & 1 \end{bmatrix}$

Teaching Tip Remind students that subtraction can be rewritten as "adding the opposite."

3 The table below shows the total number of households in the United States and the number of U.S. households that had a computer (PC) in given years. Use matrices to find the number of U.S. households without a PC.

Year	U.S. Households	
	Total Households (millions)	Households with PCs (millions)
1995	97.7	33.2
1996	98.9	38.7
1997	100.0	44.0
1998	101.0	47.8
1999	101.7	51.9

Source: *The New York Times Almanac, 2000*

64.5 million households in 1995, 60.2 million households in 1996, 56.0 million households in 1997, 53.2 million households in 1998, 49.8 million households in 1999

DAILY INTERVENTION

Unlocking Misconceptions

In order to help students see why matrices must have the same dimensions if they are to be added or subtracted, suggest that they try to add a 2×3 matrix to a 3×2 matrix.

SCALAR MULTIPLICATION

Building on Prior Knowledge

Point out that scalar multiplication of matrices is similar to using the Distributive Property to multiply the expression $3(x + y)$.

In-Class Example Power Point®

4 If $A = \begin{bmatrix} 2 & 1 \\ -1 & 3 \\ 0 & 5 \end{bmatrix}$, find $2A$.

$\begin{bmatrix} 4 & 2 \\ -2 & 6 \\ 0 & 10 \end{bmatrix}$

Interactive Chalkboard PowerPoint® Presentations

This CD-ROM is a customizable Microsoft® PowerPoint® presentation that includes:

- Step-by-step, dynamic solutions of each In-Class Example from the Teacher Wraparound Edition
- Additional, Your Turn exercises for each example
- The 5-Minute Check Transparencies
- Hot links to Glencoe Online Study Tools

$$= \begin{bmatrix} 248 & 16 \\ 178 & 6 \\ 65 & 14 \\ 8 & 1 \\ 11 & 0 \end{bmatrix} \begin{matrix} \text{mammals} \\ \text{birds} \\ \text{reptiles} \\ \text{amphibians} \\ \text{fish} \end{matrix}$$

(columns: Endangered, Threatened)

The first column represents the difference in the number of endangered species on the world and U.S. lists. There are 248 mammals, 178 birds, 65 reptiles, 8 amphibians, and 11 fish species in this category.

The second column represents the difference in the number of threatened species on the world and U.S. lists. There are 16 mammals, 6 birds, 14 reptiles, 1 amphibian, and no fish in this category.

SCALAR MULTIPLICATION You can multiply any matrix by a constant called a **scalar**. This operation is called **scalar multiplication**.

Key Concept *Scalar Multiplication*

- **Words** The product of a scalar k and an $m \times n$ matrix is an $m \times n$ matrix in which each element equals k times the corresponding elements of the original matrix.
- **Symbols** $k\begin{bmatrix} a & b & c \\ d & e & f \end{bmatrix} = \begin{bmatrix} ka & kb & kc \\ kd & ke & kf \end{bmatrix}$

Example 4 *Multiply a Matrix by a Scalar*

If $A = \begin{bmatrix} 2 & 8 & -3 \\ 5 & -9 & 2 \end{bmatrix}$, find $3A$.

$3A = 3\begin{bmatrix} 2 & 8 & -3 \\ 5 & -9 & 2 \end{bmatrix}$ Substitution.

$= \begin{bmatrix} 3(2) & 3(8) & 3(-3) \\ 3(5) & 3(-9) & 3(2) \end{bmatrix}$ Multiply each element by 3.

$= \begin{bmatrix} 6 & 24 & -9 \\ 15 & -27 & 6 \end{bmatrix}$ Simplify.

Many properties of real numbers also hold true for matrices.

Study Tip

Additive Identity

The matrix $\begin{bmatrix} 0 & 0 \\ 0 & 0 \end{bmatrix}$ is called a *zero matrix*. It is the *additive identity matrix* for any 2×2 matrix. How is this similar to the additive identity for real numbers?

Concept Summary *Properties of Matrix Operations*

For any matrices A, B, and C with the same dimensions and any scalar c, the following properties are true.

Commutative Property of Addition	$A + B = B + A$
Associative Property of Addition	$(A + B) + C = A + (B + C)$
Distributive Property	$c(A + B) = cA + cB$

162 Chapter 4 Matrices

DAILY INTERVENTION **Differentiated Instruction** ELL

Verbal/Linguistic Students may find it helpful to talk softly, or even silently, to themselves as they work with matrices. For example, they might recite the words "row by column" to remind themselves how to write the dimensions of a matrix. When multiplying by a scalar, students may find that it helps to say, for example, "5 times 1 is 5 and 5 times negative 3 is negative 15." In this way, students use more than one of their senses to check their calculations.

Example 5 Combination of Matrix Operations

If $A = \begin{bmatrix} 7 & 3 \\ -4 & -1 \end{bmatrix}$ and $B = \begin{bmatrix} 9 & 6 \\ 3 & 10 \end{bmatrix}$, find $5A - 2B$.

Perform the scalar multiplication first. Then subtract the matrices.

$5A - 2B = 5\begin{bmatrix} 7 & 3 \\ -4 & -1 \end{bmatrix} - 2\begin{bmatrix} 9 & 6 \\ 3 & 10 \end{bmatrix}$ Substitution

$= \begin{bmatrix} 5(7) & 5(3) \\ 5(-4) & 5(-1) \end{bmatrix} - \begin{bmatrix} 2(9) & 2(6) \\ 2(3) & 2(10) \end{bmatrix}$ Multiply each element in the first matrix by 5 and multiply each element in the second matrix by 2.

$= \begin{bmatrix} 35 & 15 \\ -20 & -5 \end{bmatrix} - \begin{bmatrix} 18 & 12 \\ 6 & 20 \end{bmatrix}$ Simplify.

$= \begin{bmatrix} 35 - 18 & 15 - 12 \\ -20 - 6 & -5 - 20 \end{bmatrix}$ Subtract corresponding elements.

$= \begin{bmatrix} 17 & 3 \\ -26 & -25 \end{bmatrix}$ Simplify.

Study Tip

Matrix Operations
The order of operations for matrices is similar to that of real numbers. Perform scalar multiplication before matrix addition and subtraction.

Graphing Calculator Investigation

Matrix Operations

Most graphing calculators can perform operations with matrices. On the TI-83 Plus, 2nd [MATRX] accesses the matrix menu. Choose EDIT to define a matrix. Enter the dimensions of the matrix A using the ▶ key. Then enter each element by pressing ENTER after each entry. To display and use the matrix in calculations, choose the matrix under NAMES from the [MATRX] menu.

Think and Discuss

1. Enter $A = \begin{bmatrix} 3 & -2 \\ 5 & 4 \end{bmatrix}$ with a graphing calculator. Does the calculator enter elements row by row or column by column? **row by row**
2. Notice that there are two numbers in the bottom left corner of the screen. What do these numbers represent?
3. Clear the screen. Find the matrix $18A$. $\begin{bmatrix} 54 & -36 \\ 90 & 72 \end{bmatrix}$
4. Enter $B = \begin{bmatrix} 1 & 9 & -3 \\ 8 & 6 & -5 \end{bmatrix}$. Find $A + B$. What is the result and why? **There is an error on the screen because the dimensions are not equal.**

2. the row and column of the element being entered

Check for Understanding

Concept Check

1. **Describe** the conditions under which matrices can be added or subtracted.
2. **OPEN ENDED** Give an example of two matrices whose sum is a zero matrix.
3. **Write** a matrix that, when added to a 3×2 matrix, increases each element in the matrix by 4. **See margin.**

1. They must have the same dimensions.
2. Sample answer: [−3 1], [3 −1]

Guided Practice

Perform the indicated matrix operations. If the matrix does not exist, write *impossible*. **5–7. See margin.**

4. $[5 \quad 8 \quad -4] + [12 \quad 5]$ **impossible**

5. $\begin{bmatrix} 3 & 7 \\ -2 & 1 \end{bmatrix} - \begin{bmatrix} 2 & -3 \\ 5 & -4 \end{bmatrix}$

6. $3\begin{bmatrix} 6 & -1 & 5 & 2 \\ 7 & 3 & -2 & 8 \end{bmatrix}$

7. $4\begin{bmatrix} 2 & 7 \\ -3 & 6 \end{bmatrix} + 5\begin{bmatrix} -6 & -4 \\ 3 & 0 \end{bmatrix}$

GUIDED PRACTICE KEY	
Exercises	Examples
4–10	1, 2, 4, 5
11–13	3

In-Class Example

Power Point®

5 If $A = \begin{bmatrix} 2 & 3 \\ -1 & 0 \end{bmatrix}$ and $B = \begin{bmatrix} -2 & 1 \\ 0 & -1 \end{bmatrix}$, find $4A - 3B$. $\begin{bmatrix} 14 & 9 \\ -4 & 3 \end{bmatrix}$

3 Practice/Apply

Study Notebook

Have students—

- add the definitions/examples of the vocabulary terms in this lesson to their Vocabulary Builder worksheets for Chapter 4.
- write the general formulas for matrix addition, matrix subtraction, and scalar multiplication, along with an example of the use of each formula.
- copy the properties of matrix operations, including the study tips shown on pp. 162–163, into their notebooks.
- include any other item(s) that they find helpful in mastering the skills in this lesson.

Answers

3. $\begin{bmatrix} 4 & 4 \\ 4 & 4 \\ 4 & 4 \end{bmatrix}$

5. $\begin{bmatrix} 1 & 10 \\ -7 & 5 \end{bmatrix}$

6. $\begin{bmatrix} 18 & -3 & 15 & 6 \\ 21 & 9 & -6 & 24 \end{bmatrix}$

7. $\begin{bmatrix} -22 & 8 \\ 3 & 24 \end{bmatrix}$

Graphing Calculator Investigation

If students are using a TI-83 and not the TI-83 Plus, point out that the MATRX key is not a **2nd** function. Also, point out that the row and column designation of the element being entered appears at the bottom of the screen, and that the matrix elements are entered row by row.

Lesson 4-2 Operations with Matrices 163

About the Exercises...

Organization by Objective

- **Add and Subtract Matrices:** 14–16, 20, 24, 25, 36–39
- **Scalar Multiplication:** 17–19, 21–23, 26–29

Odd/Even Assignments

Exercises 14–29 are structured so that students practice the same concepts whether they are assigned odd or even problems.

Alert! Exercise 35 involves research on the Internet or other reference materials.

Assignment Guide

Basic: 15–19 odd, 25–29 odd, 30–35, 40–62

Average: 15–29 odd, 30–35, 40–62

Advanced: 14–28 even, 33–58 (optional: 59–62)

Use matrices *A*, *B*, and *C* to find the following.

$A = \begin{bmatrix} 2 & 3 \\ 5 & 6 \end{bmatrix}$ $B = \begin{bmatrix} -1 & 7 \\ 0 & -4 \end{bmatrix}$ $C = \begin{bmatrix} 9 & -4 \\ -6 & 5 \end{bmatrix}$

8. $A + B + C$ $\begin{bmatrix} 10 & 6 \\ -1 & 7 \end{bmatrix}$
9. $3B - 2C$ $\begin{bmatrix} -21 & 29 \\ 12 & -22 \end{bmatrix}$
10. $4A + 2B - C$ $\begin{bmatrix} -3 & 30 \\ 26 & 11 \end{bmatrix}$

Application

SPORTS For Exercises 11–13, use the table below that shows high school participation in various sports.

Sport	Males		Females	
	Schools	Participants	Schools	Participants
Basketball	16,763	549,499	16,439	456,873
Track and Field	14,620	477,960	14,545	405,163
Baseball/Softball	14,486	455,305	12,679	340,480
Soccer	9041	321,416	7931	257,586
Swimming and Diving	5234	83,411	5450	133,235

Source: National Federation of State High School Associations

11. Write two matrices that represent these data for males and females. **See margin.**
12. Find the total number of students that participate in each individual sport expressed as a matrix.
13. Could you add the two matrices to find the total number of schools that offer a particular sport? Why or why not? **No; many schools offer the same sport for males and females, so those schools would be counted twice.**

12. $\begin{bmatrix} 1{,}006{,}372 \\ 883{,}123 \\ 795{,}785 \\ 579{,}002 \\ 216{,}646 \end{bmatrix}$

★ indicates increased difficulty

Practice and Apply

Homework Help

For Exercises	See Examples
14–29	1, 2, 4, 5
30–39	3

Extra Practice See page 834.

Perform the indicated matrix operations. If the matrix does not exist, write *impossible*.

14. $\begin{bmatrix} 4 \\ 1 \\ -3 \end{bmatrix} + \begin{bmatrix} 6 \\ -5 \\ 8 \end{bmatrix}$ $\begin{bmatrix} 10 \\ -4 \\ 5 \end{bmatrix}$
15. $\begin{bmatrix} -5 & 7 \\ 6 & 8 \end{bmatrix} - \begin{bmatrix} 4 & 0 & -2 \\ 9 & 0 & 1 \end{bmatrix}$ **impossible**
16. $\begin{bmatrix} 12 & 0 & 8 \\ 9 & 15 & -11 \end{bmatrix} - \begin{bmatrix} -3 & 0 & 4 \\ 9 & 2 & -6 \end{bmatrix}$
17. $-2\begin{bmatrix} 2 & -4 & 1 \\ -3 & 5 & 8 \\ 7 & 6 & -2 \end{bmatrix}$
18. $5[0 \quad -1 \quad 7 \quad 2] + 3[5 \quad -8 \quad 10 \quad -4]$ $[15 \quad -29 \quad 65 \quad -2]$
19. $5\begin{bmatrix} 1 \\ -1 \\ -3 \end{bmatrix} + 6\begin{bmatrix} -4 \\ 3 \\ 5 \end{bmatrix} - 2\begin{bmatrix} -3 \\ 8 \\ -4 \end{bmatrix}$ $\begin{bmatrix} -13 \\ -3 \\ 23 \end{bmatrix}$
20. ★ $\begin{bmatrix} 1.35 & 5.80 \\ 1.24 & 14.32 \\ 6.10 & 35.26 \end{bmatrix} + \begin{bmatrix} 0.45 & 3.28 \\ 1.94 & 16.72 \\ 4.31 & 21.30 \end{bmatrix}$
21. ★ $8\begin{bmatrix} 0.25 & 0.5 \\ 0.75 & 1.5 \end{bmatrix} - 2\begin{bmatrix} 0.25 & 0.5 \\ 0.75 & 1.5 \end{bmatrix}$ $\begin{bmatrix} 1.5 & 3 \\ 4.5 & 9 \end{bmatrix}$
22. ★ $\frac{1}{2}\begin{bmatrix} 4 & 6 \\ 3 & 0 \end{bmatrix} - \frac{2}{3}\begin{bmatrix} 9 & 27 \\ 0 & 3 \end{bmatrix}$ $\begin{bmatrix} -4 & -15 \\ \frac{3}{2} & -2 \end{bmatrix}$
23. ★ $5\begin{bmatrix} \frac{1}{2} & 0 & 1 \\ 2 & \frac{1}{3} & -1 \end{bmatrix} + 4\begin{bmatrix} -2 & \frac{3}{4} & 1 \\ \frac{1}{6} & 0 & \frac{5}{8} \end{bmatrix}$

16. $\begin{bmatrix} 15 & 0 & 4 \\ 0 & 13 & -5 \end{bmatrix}$

17. $\begin{bmatrix} -4 & 8 & -2 \\ 6 & -10 & -16 \\ -14 & -12 & 4 \end{bmatrix}$

20. $\begin{bmatrix} 1.8 & 9.08 \\ 3.18 & 31.04 \\ 10.41 & 56.56 \end{bmatrix}$

23. $\begin{bmatrix} -5\frac{1}{2} & 3 & 9 \\ 10\frac{2}{3} & 1\frac{2}{3} & -2\frac{1}{2} \end{bmatrix}$

Use matrices *A*, *B*, *C*, and *D* to find the following. **24–29. See margin.**

$A = \begin{bmatrix} 5 & 7 \\ -1 & 6 \\ 3 & -9 \end{bmatrix}$ $B = \begin{bmatrix} 8 & 3 \\ 5 & 1 \\ 4 & 4 \end{bmatrix}$ $C = \begin{bmatrix} 0 & 4 \\ -2 & 5 \\ 7 & -1 \end{bmatrix}$ $D = \begin{bmatrix} 6 & 2 \\ 9 & 0 \\ -3 & 0 \end{bmatrix}$

24. $A + B$
25. $D - B$
26. $4C$
27. $6B - 2A$
28. $3C - 4A + B$
29. $C + \frac{1}{3}D$

Answers

11. Males = $\begin{bmatrix} 16{,}763 & 549{,}499 \\ 14{,}620 & 477{,}960 \\ 14{,}486 & 455{,}305 \\ 9041 & 321{,}416 \\ 5234 & 83{,}411 \end{bmatrix}$, Females = $\begin{bmatrix} 16{,}439 & 456{,}873 \\ 14{,}545 & 405{,}163 \\ 12{,}679 & 340{,}480 \\ 7931 & 257{,}586 \\ 5450 & 133{,}235 \end{bmatrix}$

24. $\begin{bmatrix} 13 & 10 \\ 4 & 7 \\ 7 & -5 \end{bmatrix}$

25. $\begin{bmatrix} -2 & -1 \\ 4 & -1 \\ -7 & -4 \end{bmatrix}$

26. $\begin{bmatrix} 0 & 16 \\ -8 & 20 \\ 28 & -4 \end{bmatrix}$

27. $\begin{bmatrix} 38 & 4 \\ 32 & -6 \\ 18 & 42 \end{bmatrix}$

28. $\begin{bmatrix} -12 & -13 \\ 3 & -8 \\ 13 & 37 \end{bmatrix}$

29. $\begin{bmatrix} 2 & 4\frac{2}{3} \\ 1 & 5 \\ 6 & -1 \end{bmatrix}$

36. Residents: $\begin{array}{c} \\ \text{Before 6} \\ \text{After 6} \end{array} \begin{array}{c} \text{Child} \quad \text{Adult} \\ \begin{bmatrix} 3.00 & 4.50 \\ 2.00 & 3.50 \end{bmatrix} \end{array}$, Nonresidents: $\begin{array}{c} \\ \text{Before 6} \\ \text{After 6} \end{array} \begin{array}{c} \text{Child} \quad \text{Adult} \\ \begin{bmatrix} 4.50 & 6.75 \\ 3.00 & 5.25 \end{bmatrix} \end{array}$

37. $\begin{bmatrix} 1.50 & 2.25 \\ 1.00 & 1.75 \end{bmatrix}$

38. Before 6:00: $\begin{array}{c} \\ \text{Residents} \\ \text{Nonresidents} \end{array} \begin{array}{c} \text{Child} \quad \text{Adult} \\ \begin{bmatrix} 3.00 & 4.50 \\ 4.50 & 6.75 \end{bmatrix} \end{array}$, After 6:00: $\begin{array}{c} \\ \text{Residents} \\ \text{Nonresidents} \end{array} \begin{array}{c} \text{Child} \quad \text{Adult} \\ \begin{bmatrix} 2.00 & 3.50 \\ 3.00 & 5.25 \end{bmatrix} \end{array}$

39. $\begin{bmatrix} 1.00 & 1.00 \\ 1.50 & 1.50 \end{bmatrix}$

BUSINESS **For Exercises 30–32, use the following information.**
The Cookie Cutter Bakery records each type of cookie sold at three of their branch stores. Two days of sales are shown in the spreadsheets below.

30. Write a matrix for each day's sales.

31. Find the sum of the two days' sales expressed as a matrix.

32. Find the difference in cookie sales from Friday to Saturday expressed as a matrix.

	A	B	C	D	E
1	Friday	chocolate chip	peanut butter	sugar	cut-out
2	Store 1	120	97	64	75
3	Store 2	80	59	36	60
4	Store 3	72	84	29	48

	A	B	C	D	E
1	Saturday	chocolate chip	peanut butter	sugar	cut-out
2	Store 1	112	87	56	74
3	Store 2	84	65	39	70
4	Store 3	88	98	43	60

30. Friday: $\begin{bmatrix} 120 & 97 & 64 & 75 \\ 80 & 59 & 36 & 60 \\ 72 & 84 & 29 & 48 \end{bmatrix}$

Saturday: $\begin{bmatrix} 112 & 87 & 56 & 74 \\ 84 & 65 & 39 & 70 \\ 88 & 98 & 43 & 60 \end{bmatrix}$

31. $\begin{bmatrix} 232 & 184 & 120 & 149 \\ 164 & 124 & 75 & 130 \\ 160 & 182 & 72 & 108 \end{bmatrix}$

32. $\begin{bmatrix} -8 & -10 & -8 & -1 \\ 4 & 6 & 3 & 10 \\ 16 & 14 & 14 & 12 \end{bmatrix}$

33. $\begin{bmatrix} 245 \\ 228 \\ 319 \\ 227 \\ 117 \end{bmatrix}$ 34. $\begin{bmatrix} 15 \\ 41 \\ 35 \\ 27 \\ 51 \end{bmatrix}$

WEATHER **For Exercises 33–35, use the table that shows the total number of deaths due to severe weather.**

Year	Lightning	Tornadoes	Floods	Hurricanes
1996	52	25	131	37
1997	42	67	118	1
1998	44	130	136	9
1999	46	94	68	19
2000	51	29	37	0

Source: National Oceanic & Atmospheric Administration

33. Find the total number of deaths due to severe weather for each year expressed as a column matrix.

34. Write a matrix that represents how many more people died as a result of lightning than hurricanes for each year.

35. What type of severe weather accounted for the most deaths each year?
1996, floods; 1997, floods; 1998, floods; 1999, tornadoes; 2000, lightning

More About. . .

Weather

Flash floods and floods are the number 1 weather-related killer recorded in the U.S. each year. The large majority of deaths due to flash flooding are a result of people driving through flooded areas.

Source: National Oceanic & Atmospheric Administration

Online Research **Data Update** What are the current weather statistics? Visit www.algebra2.com/data_update to learn more.

RECREATION **For Exercises 36–39, use the following price list for one-day admissions to the community pool.** **36–39. See margin.**

36. Write a matrix that represents the cost of admission for residents and a matrix that represents the cost of admission for nonresidents.

37. Find the matrix that represents the additional cost for nonresidents.

38. Write a matrix that represents the cost of admission before 6:00 P.M. and a matrix that represents the cost of admission after 6:00 P.M.

39. Find a matrix that represents the difference in cost if a child or adult goes to the pool after 6:00 P.M.

Daily Admission Fees

Residents		
Time of day	Child	Adult
Before 6:00 P.M.	$3.00	$4.50
After 6:00 P.M.	$2.00	$3.50
Nonresidents		
Time of day	Child	Adult
Before 6:00 P.M.	$4.50	$6.75
After 6:00 P.M.	$3.00	$5.25

www.algebra2.com/self_check_quiz

Study Guide and Intervention, p. 175 (shown) and p. 176

Add and Subtract Matrices

Addition of Matrices	$\begin{bmatrix} a & b & c \\ d & e & f \\ g & h & i \end{bmatrix} + \begin{bmatrix} j & k & l \\ m & n & o \\ p & q & r \end{bmatrix} = \begin{bmatrix} a+j & b+k & c+l \\ d+m & e+n & f+o \\ g+p & h+q & i+r \end{bmatrix}$
Subtraction of Matrices	$\begin{bmatrix} a & b & c \\ d & e & f \\ g & h & i \end{bmatrix} - \begin{bmatrix} j & k & l \\ m & n & o \\ p & q & r \end{bmatrix} = \begin{bmatrix} a-j & b-k & c-l \\ d-m & e-n & f-o \\ g-p & h-q & i-r \end{bmatrix}$

Example 1 Find $A + B$ if $A = \begin{bmatrix} 6 & -7 \\ 2 & -12 \end{bmatrix}$ and $B = \begin{bmatrix} 4 & 2 \\ -5 & -6 \end{bmatrix}$.

$$A + B = \begin{bmatrix} 6 & -7 \\ 2 & -12 \end{bmatrix} + \begin{bmatrix} 4 & 2 \\ -5 & -6 \end{bmatrix} = \begin{bmatrix} 6+4 & -7+2 \\ 2+(-5) & -12+(-6) \end{bmatrix} = \begin{bmatrix} 10 & -5 \\ -3 & -18 \end{bmatrix}$$

Example 2 Find $A - B$ if $A = \begin{bmatrix} -2 & 8 \\ 3 & -4 \\ 10 & 7 \end{bmatrix}$ and $B = \begin{bmatrix} 4 & -3 \\ -2 & 1 \\ -6 & 8 \end{bmatrix}$.

$$A - B = \begin{bmatrix} -2 & 8 \\ 3 & -4 \\ 10 & 7 \end{bmatrix} - \begin{bmatrix} 4 & -3 \\ -2 & 1 \\ -6 & 8 \end{bmatrix} = \begin{bmatrix} -2-4 & 8-(-3) \\ 3-(-2) & -4-1 \\ 10-(-6) & 7-8 \end{bmatrix} = \begin{bmatrix} -6 & 11 \\ 5 & -5 \\ 16 & -1 \end{bmatrix}$$

Exercises

Perform the indicated operations. If the matrix does not exist, write *impossible*.

1. $\begin{bmatrix} 8 & 7 \\ -10 & -6 \end{bmatrix} - \begin{bmatrix} -4 & 3 \\ 2 & -12 \end{bmatrix}$ $\begin{bmatrix} 12 & 4 \\ -12 & 6 \end{bmatrix}$

2. $\begin{bmatrix} 6 & -5 & 9 \\ -3 & 4 & 5 \end{bmatrix} + \begin{bmatrix} -4 & 3 & 2 \\ 6 & 9 & -4 \end{bmatrix}$ $\begin{bmatrix} 2 & -2 & 11 \\ 3 & 13 & 1 \end{bmatrix}$

3. $\begin{bmatrix} 6 \\ -3 \\ 2 \end{bmatrix} + [-6 \quad 3 \quad -2]$ impossible

4. $\begin{bmatrix} 5 & -2 \\ -4 & 6 \\ 7 & 9 \end{bmatrix} + \begin{bmatrix} -11 & 6 \\ 2 & -5 \\ 4 & -7 \end{bmatrix}$ $\begin{bmatrix} -6 & 4 \\ -2 & 1 \\ 11 & 2 \end{bmatrix}$

5. $\begin{bmatrix} 8 & 0 & -6 \\ 4 & 5 & -11 \\ -7 & 3 & 4 \end{bmatrix} - \begin{bmatrix} -2 & 1 & 7 \\ 3 & -4 & 3 \\ -8 & 5 & 6 \end{bmatrix}$ $\begin{bmatrix} 10 & -1 & -13 \\ 1 & 9 & -14 \\ 1 & -2 & -2 \end{bmatrix}$

6. $\begin{bmatrix} \frac{3}{4} & \frac{2}{5} \\ -\frac{1}{2} & \frac{4}{3} \end{bmatrix} - \begin{bmatrix} \frac{1}{2} & \frac{2}{3} \\ \frac{2}{3} & -\frac{1}{2} \end{bmatrix}$ $\begin{bmatrix} \frac{1}{4} & -\frac{4}{15} \\ -\frac{7}{6} & \frac{11}{6} \end{bmatrix}$

Skills Practice, p. 177 and Practice, p. 178 (shown)

Perform the indicated matrix operations. If the matrix does not exist, write *impossible*.

1. $\begin{bmatrix} 2 & -1 \\ 3 & 7 \\ 14 & -9 \end{bmatrix} + \begin{bmatrix} -6 & 9 \\ 7 & -11 \\ -8 & 17 \end{bmatrix}$ $\begin{bmatrix} -4 & 8 \\ 10 & -4 \\ 6 & 8 \end{bmatrix}$

2. $\begin{bmatrix} 4 \\ -71 \\ 18 \end{bmatrix} - \begin{bmatrix} -67 \\ 45 \\ -24 \end{bmatrix}$ $\begin{bmatrix} 71 \\ -116 \\ 42 \end{bmatrix}$

3. $-3\begin{bmatrix} -1 & 0 \\ 17 & -11 \end{bmatrix} + 4\begin{bmatrix} -3 & 16 \\ -21 & 12 \end{bmatrix}$ $\begin{bmatrix} -9 & 64 \\ -135 & 81 \end{bmatrix}$

4. $7\begin{bmatrix} 2 & -1 & 8 \\ 4 & 7 & 9 \end{bmatrix} - 2\begin{bmatrix} -1 & 4 & -3 \\ 7 & 2 & -6 \end{bmatrix}$ $\begin{bmatrix} 16 & -15 & 62 \\ 14 & 45 & 75 \end{bmatrix}$

5. $-2\begin{bmatrix} 1 \\ 2 \end{bmatrix} + 4\begin{bmatrix} 0 \\ 5 \end{bmatrix} - \begin{bmatrix} 10 \\ 18 \end{bmatrix}$ $\begin{bmatrix} -12 \\ -2 \end{bmatrix}$

6. $\frac{3}{4}\begin{bmatrix} 8 & 12 \\ -16 & 20 \end{bmatrix} + \frac{2}{3}\begin{bmatrix} 27 & -9 \\ 54 & -18 \end{bmatrix}$ $\begin{bmatrix} 24 & 3 \\ 24 & 3 \end{bmatrix}$

Use $A = \begin{bmatrix} 4 & -1 & 0 \\ -3 & 6 & 2 \end{bmatrix}$, $B = \begin{bmatrix} -2 & 4 & 5 \\ 1 & 0 & -9 \end{bmatrix}$, and $C = \begin{bmatrix} 10 & -8 & 6 \\ -6 & -4 & 20 \end{bmatrix}$ to find the following.

7. $A - B$ $\begin{bmatrix} 6 & -5 & -5 \\ -4 & 6 & 11 \end{bmatrix}$

8. $A - C$ $\begin{bmatrix} -6 & 7 & -6 \\ 3 & 10 & -18 \end{bmatrix}$

9. $-3B$ $\begin{bmatrix} 6 & -12 & -15 \\ -3 & 0 & 27 \end{bmatrix}$

10. $4B - A$ $\begin{bmatrix} -12 & 17 & 20 \\ 7 & -6 & -38 \end{bmatrix}$

11. $-2B - 3C$ $\begin{bmatrix} -26 & 16 & -28 \\ 16 & 12 & -42 \end{bmatrix}$

12. $A + 0.5C$ $\begin{bmatrix} 9 & -5 & 3 \\ -6 & 4 & 12 \end{bmatrix}$

ECONOMICS **For Exercises 13 and 14, use the table that shows loans by an economic development board to women and men starting new businesses.**

	Women		Men	
	Businesses	Loan Amount ($)	Businesses	Loan Amount ($)
1999	27	$567,000	36	$864,000
2000	41	$902,000	32	$672,000
2001	35	$777,000	28	$562,000

13. Write two matrices that represent the number of new businesses and loan amounts, one for women and one for men. $\begin{bmatrix} 27 & 567{,}000 \\ 41 & 902{,}000 \\ 35 & 777{,}000 \end{bmatrix}$, $\begin{bmatrix} 36 & 864{,}000 \\ 32 & 672{,}000 \\ 28 & 562{,}000 \end{bmatrix}$

14. Find the sum of the numbers of new businesses and loan amounts for both men and women over the three-year period expressed as a matrix. $\begin{bmatrix} 63 & 1{,}431{,}000 \\ 73 & 1{,}574{,}000 \\ 63 & 1{,}339{,}000 \end{bmatrix}$

15. **PET NUTRITION** Use the table that gives nutritional information for two types of dog food. Find the difference in the percent of protein, fat, and fiber between Mix B and Mix A expressed as a matrix. $[2 \quad -4 \quad 3]$

	% Protein	% Fat	% Fiber
Mix A	22	12	5
Mix B	24	8	8

Reading to Learn Mathematics, p. 179 ELL

Pre-Activity **How can matrices be used to calculate daily dietary needs?**

Read the introduction to Lesson 4-2 at the top of page 160 in your textbook.

- Write a sum that represents the total number of Calories in the patient's diet for Day 2. (Do not actually calculate the sum.) 482 + 622 + 987
- Write the sum that represents the total fat content in the patient's diet for Day 3. (Do not actually calculate the sum.) 11 + 12 + 38

Reading the Lesson

1. For each pair of matrices, give the dimensions of the indicated sum, difference, or scalar product. If the indicated sum, difference, or scalar product does not exist, write *impossible*.

$A = \begin{bmatrix} 3 & 5 & 6 \\ -2 & 8 & 1 \end{bmatrix}$ $B = \begin{bmatrix} -4 & 0 \\ 0 & -5 \end{bmatrix}$

$C = \begin{bmatrix} 5 & 10 \\ -3 & 6 \\ 4 & 12 \end{bmatrix}$ $D = \begin{bmatrix} -3 & 6 & 0 \\ -8 & 4 & 0 \end{bmatrix}$

$A + D$: 2×3 $C + D$: impossible $5B$: 2×2

$-4C$: 3×2 $2D - 3A$: 2×3

2. Suppose that M, N, and P are nonzero 2×4 matrices and k is a negative real number. Indicate whether each of the following statements is *true* or *false*.

a. $M + (N + P) = M + (P + N)$ true
b. $M - N = N - M$ false
c. $M - (N - P) = (M - N) - P$ false
d. $k(M - N) = kM - kN$ true

Helping You Remember

3. The mathematical term *scalar* may be unfamiliar, but its meaning is related to the word *scale* as used in a *scale of miles* on a map. How can this usage of the word *scale* help you remember the meaning of *scalar*?

Sample answer: A scale of miles tells you how to multiply the distances you measure on a map by a certain number to get the actual distance between two locations. This multiplier is often called a *scale* factor. A *scalar* represents the same idea: It is a real number by which a matrix can be multiplied to change all the elements of the matrix by a uniform *scale* factor.

Enrichment, p. 180

Sundaram's Sieve

The properties and patterns of prime numbers have fascinated many mathematicians. In 1934, a young East Indian student named Sundaram constructed the following matrix.

4	7	10	13	16	19	22	25	. . .
7	12	17	22	27	32	37	42	. . .
10	17	24	31	38	45	52	59	. . .
13	22	31	40	49	58	67	76	. . .
16	27	38	49	60	71	82	93	. . .
.	.	.	.	.	.	.	.	. . .

A surprising property of this matrix is that it can be used to determine whether or not some numbers are prime.

Complete these problems to discover this property.

1. The first row and the first column are created by using an arithmetic pattern. What is the common difference used in the pattern? 3

Lesson 4-2 Operations with Matrices 165

4 Assess

Open-Ended Assessment

Speaking Have students discuss and explain how adding and subtracting matrices, and multiplying a matrix by a scalar, are similar to these operations with numbers.

Tips for New Teachers

Intervention Adding and subtracting matrices offers numerous chances for students to make errors. They might make a mistake in choosing which elements to combine when adding or subtracting the entries. They may not write the result in the correct position in the answer matrix. Ask students to develop tips for keeping track of where they are in the matrix. Students may find it helpful to circle the entries they are combining, to use color coding, or to point with their fingers to the two entries to be added or subtracted as they do the calculation.

Assessment Options

Quiz (Lessons 4-1 and 4-2) is available on p. 231 of the *Chapter 4 Resource Masters.*

Getting Ready for Lesson 4-3

PREREQUISITE SKILL Lesson 4-3 presents matrix multiplication and the properties pertaining to it. Students will use their familiarity with the properties of real numbers to compare them to matrix properties. Exercises 59–62 should be used to determine your students' familiarity with the properties of equality.

40. $2\begin{bmatrix} 0.5 & 0.75 & 3 \\ 1 & 4 & 0.1 \end{bmatrix} = \begin{bmatrix} 1 & 1.5 & 6 \\ 2 & 8 & 0.2 \end{bmatrix}$

40. **CRITICAL THINKING** Determine values for each variable if $d = 1$, $e = 4d$, $z + d = e$, $f = \frac{x}{5}$, $ay = 1.5$, $x = \frac{d}{2}$, and $y = x + \frac{x}{2}$.

$$a\begin{bmatrix} x & y & z \\ d & e & f \end{bmatrix} = \begin{bmatrix} ax & ay & az \\ ad & ae & af \end{bmatrix}$$

41. **WRITING IN MATH** Answer the question that was posed at the beginning of the lesson. **See pp. 217A–217B.**

How can matrices be used to calculate daily dietary needs?

Include the following in your answer:
- three matrices that represent breakfast, lunch, and dinner over the three-day period, and
- a matrix that represents the total Calories, protein, and fat consumed each day.

Standardized Test Practice

42. Which matrix equals $\begin{bmatrix} 5 & -2 \\ -3 & 7 \end{bmatrix} - \begin{bmatrix} 3 & 4 \\ -5 & 6 \end{bmatrix}$? **D**

Ⓐ $\begin{bmatrix} 2 & 2 \\ -8 & 1 \end{bmatrix}$ Ⓑ $\begin{bmatrix} 8 & -6 \\ -8 & 1 \end{bmatrix}$ Ⓒ $\begin{bmatrix} 2 & 2 \\ 2 & 1 \end{bmatrix}$ Ⓓ $\begin{bmatrix} 2 & -6 \\ 2 & 1 \end{bmatrix}$

43. Solve for x and y in the matrix equation $\begin{bmatrix} x \\ 7 \end{bmatrix} + \begin{bmatrix} 3y \\ -x \end{bmatrix} = \begin{bmatrix} 16 \\ 12 \end{bmatrix}$. **A**

Ⓐ $(-5, 7)$ Ⓑ $(7, 5)$ Ⓒ $(7, 3)$ Ⓓ $(5, 7)$

Maintain Your Skills

Mixed Review **State the dimensions of each matrix.** *(Lesson 4-1)*

44. $\begin{bmatrix} 1 & 0 \\ 0 & 1 \end{bmatrix}$ **2 × 2**

45. $[2 \quad 0 \quad 3 \quad 0]$ **1 × 4**

46. $\begin{bmatrix} 5 & 1 & -6 & 2 \\ -38 & 5 & 7 & 3 \end{bmatrix}$ **2 × 4**

47. $\begin{bmatrix} 7 & -3 & 5 \\ 0 & 2 & -9 \\ 6 & 5 & 1 \end{bmatrix}$ **3 × 3**

48. $\begin{bmatrix} 8 & 6 \\ 5 & 2 \\ -4 & -1 \end{bmatrix}$ **3 × 2**

49. $\begin{bmatrix} 7 & 5 & 0 \\ -8 & 3 & 8 \\ 9 & -1 & 15 \\ 4 & 2 & 11 \end{bmatrix}$ **4 × 3**

Solve each system of equations. *(Lesson 3-5)*

50. $2a + b = 2$; $5a = 15$; $a + b + c = -1$ **(3, −4, 0)**

51. $r + s + t = 15$; $r + t = 12$; $s + t = 10$ **(5, 3, 7)**

52. $6x - 2y - 3z = -10$; $-6x + y + 9z = 3$; $8x - 3y = -16$

52. $\left(\frac{1}{4}, 6, -\frac{1}{6}\right)$

Solve each system by using substitution or elimination. *(Lesson 3-2)*

53. $2s + 7t = 39$; $5s - t = 5$ **(2, 5)**

54. $3p + 6q = -3$; $2p - 3q = -9$ **(−3, 1)**

55. $a + 5b = 1$; $7a - 2b = 44$ **(6, −1)**

SCRAPBOOKS For Exercises 56–58, use the following information.
Ian has $6.00, and he wants to buy paper for his scrapbook. A sheet of printed paper costs 30¢, and a sheet of solid color paper costs 15¢. *(Lesson 2-7)*

56. Write an inequality that describes this situation. **$0.30p + 0.15s \le 6$**

57. Graph the inequality. **See margin.**

58. Does Ian have enough money to buy 14 pieces of each type of paper? **No, it would cost $6.30.**

Getting Ready for the Next Lesson **PREREQUISITE SKILL Name the property illustrated by each equation.**
(To review the ***properties of equality****, see Lesson 1-2.)*

59. $\frac{7}{9} \cdot \frac{9}{7} = 1$ **Mult. Inverse**

60. $7 + (w + 5) = (7 + w) + 5$ **Assoc. (+)**

61. $3(x + 12) = 3x + 3(12)$ **Dist.**

62. $6(9a) = 9a(6)$ **Comm. (×)**

Answer

57.

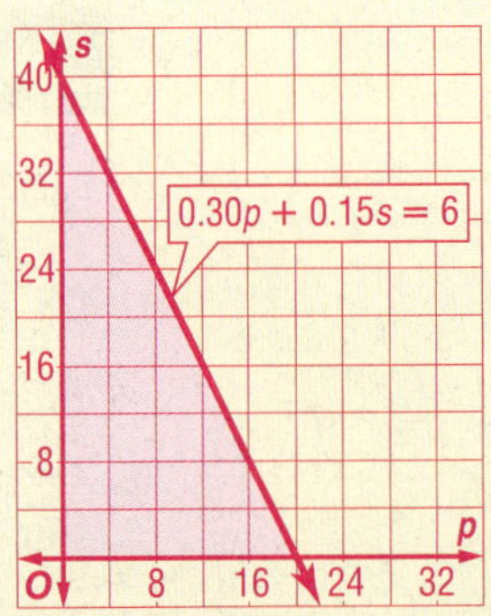

4-3 Multiplying Matrices

What You'll Learn

- Multiply matrices.
- Use the properties of matrix multiplication.

How can matrices be used in sports statistics?

Professional football teams track many statistics throughout the season to help evaluate their performance. The table shows the scoring summary of the Oakland Raiders for the 2000 season. The team's record can be summarized in the record matrix R. The values for each type of score can be organized in the point values matrix P.

Oakland Raiders Regular Season Scoring	
Type	Number
Touchdown	58
Extra Point	56
Field Goal	23
2-Point Conversion	1
Safety	2

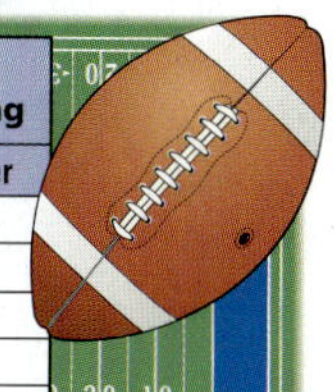

Source: National Football League

Record

$$R = \begin{bmatrix} 58 \\ 56 \\ 23 \\ 1 \\ 2 \end{bmatrix} \begin{matrix} \text{touchdown} \\ \text{extra point} \\ \text{field goal} \\ \text{2-point conversion} \\ \text{safety} \end{matrix}$$

Point Values

touchdown, extra point, field goal, 2-point conversion, safety

$$P = [6 \quad 1 \quad 3 \quad 2 \quad 2]$$

You can use matrix multiplication to find the total points scored.

MULTIPLY MATRICES You can multiply two matrices if and only if the number of columns in the first matrix is equal to the number of rows in the second matrix. When you multiply two matrices $A_{m \times n}$ and $B_{n \times r}$, the resulting matrix AB is an $m \times r$ matrix.

outer dimensions = dimensions of AB

$$\begin{matrix} A & \cdot & B & = & AB \\ 2 \times 3 & & 3 \times 4 & & 2 \times 4 \end{matrix}$$

inner dimensions are equal

Example 1 Dimensions of Matrix Products

Determine whether each matrix product is defined. If so, state the dimensions of the product.

a. $A_{2 \times 5}$ **and** $B_{5 \times 4}$

$$\begin{matrix} A & \cdot & B & = & AB \\ 2 \times 5 & & 5 \times 4 & & 2 \times 4 \end{matrix}$$

The inner dimensions are equal so the matrix product is defined. The dimensions of the product are 2×4.

b. $A_{1 \times 3}$ **and** $B_{4 \times 3}$

$$\begin{matrix} A & \cdot & B \\ 1 \times 3 & & 4 \times 3 \end{matrix}$$

The inner dimensions are not equal, so the matrix product is not defined.

4-3 Lesson Notes

1 Focus

5-Minute Check Transparency 4-3 Use as a quiz or review of Lesson 4-2.

Mathematical Background notes are available for this lesson on p. 152C.

How can matrices be used in sports statistics?

Ask students:

- The table shows that the Raiders scored 58 touchdowns. A touchdown is worth 6 points. How many points were scored by touchdowns? **348 points**
- A field goal is worth 3 points. How many points did the Raiders score by field goals? **69 points**

Tips for New Teachers

Ask a volunteer to explain the various football plays for the benefit of those who may not be familiar with them.

Resource Manager

Workbook and Reproducible Masters

Chapter 4 Resource Masters

- Study Guide and Intervention, pp. 181–182
- Skills Practice, p. 183
- Practice, p. 184
- Reading to Learn Mathematics, p. 185
- Enrichment, p. 186

Teaching Algebra With Manipulatives Masters, p. 231

Transparencies

5-Minute Check Transparency 4-3

Answer Key Transparencies

Technology

Interactive Chalkboard

2 Teach

MULTIPLY MATRICES

In-Class Examples

Power Point®

1 Determine whether each matrix product is defined. If so, state the dimensions of the product.

a. $A_{3 \times 4}$ and $B_{4 \times 2}$ **The matrix product is defined. The dimensions are 3 × 2.**

b. $A_{3 \times 2}$ and $B_{4 \times 3}$ **The matrix product is not defined.**

2 Find RS if $R = \begin{bmatrix} 3 & 2 \\ -1 & 0 \end{bmatrix}$ and $S = \begin{bmatrix} -2 & 1 \\ 0 & -1 \end{bmatrix}$. $\begin{bmatrix} -6 & 1 \\ 2 & -1 \end{bmatrix}$

The product of two matrices is found by multiplying columns and rows. The entry in the first row and first column of AB is found by multiplying corresponding elements in the first row of A and the first column of B and then adding.

Key Concept — *Multiply Matrices*

- **Words** The element a_{ij} of AB is the sum of the products of the corresponding elements in row i of A and column j of B.
- **Symbols** $\begin{bmatrix} a_1 & b_1 \\ a_2 & b_2 \end{bmatrix} \cdot \begin{bmatrix} x_1 & y_1 \\ x_2 & y_2 \end{bmatrix} = \begin{bmatrix} a_1x_1 + b_1x_2 & a_1y_1 + b_1y_2 \\ a_2x_1 + b_2x_2 & a_2y_1 + b_2y_2 \end{bmatrix}$

Example 2 Multiply Square Matrices

Find RS if $R = \begin{bmatrix} 2 & -1 \\ 3 & 4 \end{bmatrix}$ and $S = \begin{bmatrix} 3 & -9 \\ 5 & 7 \end{bmatrix}$.

$$RS = \begin{bmatrix} 2 & -1 \\ 3 & 4 \end{bmatrix} \cdot \begin{bmatrix} 3 & -9 \\ 5 & 7 \end{bmatrix}$$

Step 1 Multiply the numbers in the first row of R by the numbers in the first column of S, add the products, and put the result in the first row, first column of RS.

$$\begin{bmatrix} 2 & -1 \\ 3 & 4 \end{bmatrix} \cdot \begin{bmatrix} 3 & -9 \\ 5 & 7 \end{bmatrix} = \begin{bmatrix} 2(3) + (-1)(5) & \\ & \end{bmatrix}$$

Step 2 Multiply the numbers in the first row of R by the numbers in the second column of S, add the products, and put the result in the first row, second column of RS.

$$\begin{bmatrix} 2 & -1 \\ 3 & 4 \end{bmatrix} \cdot \begin{bmatrix} 3 & -9 \\ 5 & 7 \end{bmatrix} = \begin{bmatrix} 2(3) + (-1)(5) & 2(-9) + (-1)(7) \\ & \end{bmatrix}$$

Step 3 Multiply the numbers in the second row of R by the numbers in the first column of S, add the products, and put the result in the second row, first column of RS.

$$\begin{bmatrix} 2 & -1 \\ 3 & 4 \end{bmatrix} \cdot \begin{bmatrix} 3 & -9 \\ 5 & 7 \end{bmatrix} = \begin{bmatrix} 2(3) + (-1)(5) & 2(-9) + (-1)(7) \\ 3(3) + 4(5) & \end{bmatrix}$$

Step 4 Multiply the numbers in the second row of R by the numbers in the second column of S, add the products, and put the result in the second row, second column of RS.

$$\begin{bmatrix} 2 & -1 \\ 3 & 4 \end{bmatrix} \cdot \begin{bmatrix} 3 & -9 \\ 5 & 7 \end{bmatrix} = \begin{bmatrix} 2(3) + (-1)(5) & 2(-9) + (-1)(7) \\ 3(3) + 4(5) & 3(-9) + 4(7) \end{bmatrix}$$

Step 5 Simplify the product matrix.

$$\begin{bmatrix} 2(3) + (-1)(5) & 2(-9) + (-1)(7) \\ 3(3) + 4(5) & 3(-9) + 4(7) \end{bmatrix} = \begin{bmatrix} 1 & -25 \\ 29 & 1 \end{bmatrix}$$

So, $RS = \begin{bmatrix} 1 & -25 \\ 29 & 1 \end{bmatrix}$.

Study Tip

Multiplying Matrices
To avoid any miscalculations, find the product of the matrices in order as shown in Example 2. It may also help to cover rows or columns not being multiplied as you find elements of the product matrix.

When solving real-world problems, make sure to multiply the matrices in the order for which the product is defined.

Example 3 Multiply Matrices with Different Dimensions

TRACK AND FIELD **In a four-team track meet, 5 points were awarded for each first-place finish, 3 points for each second, and 1 point for each third. Find the total number of points for each school. Which school won the meet?**

School	First Place	Second Place	Third Place
Jefferson	8	4	5
London	6	3	7
Springfield	5	7	3
Madison	7	5	4

Explore The final scores can be found by multiplying the track results for each school by the points awarded for each first-, second-, and third-place finish.

Plan Write the results of the races and the points awarded in matrix form. Set up the matrices so that the number of rows in the points matrix equals the number of columns in the results matrix.

Results **Points**

$$R = \begin{bmatrix} 8 & 4 & 5 \\ 6 & 3 & 7 \\ 5 & 7 & 3 \\ 7 & 5 & 4 \end{bmatrix} \qquad P = \begin{bmatrix} 5 \\ 3 \\ 1 \end{bmatrix}$$

Solve Multiply the matrices.

$$RP = \begin{bmatrix} 8 & 4 & 5 \\ 6 & 3 & 7 \\ 5 & 7 & 3 \\ 7 & 5 & 4 \end{bmatrix} \cdot \begin{bmatrix} 5 \\ 3 \\ 1 \end{bmatrix}$$ Write an equation.

$$= \begin{bmatrix} 8(5) + 4(3) + 5(1) \\ 6(5) + 3(3) + 7(1) \\ 5(5) + 7(3) + 3(1) \\ 7(5) + 5(3) + 4(1) \end{bmatrix}$$ Multiply columns by rows.

$$= \begin{bmatrix} 57 \\ 46 \\ 49 \\ 54 \end{bmatrix}$$ Simplify.

The labels for the product matrix are shown below.

Total Points

Jefferson	57
London	46
Springfield	49
Madison	54

Jefferson won the track meet with a total of 57 points.

Examine R is a 4×3 matrix and P is a 3×1 matrix; so their product should be a 4×1 matrix. *Why?*

MULTIPLICATIVE PROPERTIES Recall that the same properties for real numbers also held true for matrix addition. However, some of these properties do *not* always hold true for matrix multiplication.

www.algebra2.com/extra_examples

More About. . .

Track and Field

Running and hurdling contests make up the track events. Jumping and throwing contests make up the field events. More than 950,000 high school students participate in track and field competitions each year.

Source: www.encarta.msn.com

TEACHING TIP

Ask the students if they could multiply the number of points times the results. Ask them to explain their answer.

In-Class Example

Power Point®

3 **CHESS** Three teams competed in the final round of the Chess Club's championships. For each win, a team was awarded 3 points and for each draw a team received 1 point.

Team	Wins	Draws
Blue	5	4
Red	6	3
Green	4	5

Find the total number of points for each team. Which team won the tournament? **The Blue Team had 19 points, the Red Team had 21 points, and the Green Team had 17 points. So, the Red Team won the tournament.**

Lesson 4-3 Multiplying Matrices 169

MULTIPLICATIVE PROPERTIES

In-Class Examples

Power Point®

4 Find each product if

$K = \begin{bmatrix} -3 & 2 & 2 \\ -1 & -2 & 0 \end{bmatrix}$ and

$L = \begin{bmatrix} 1 & -2 \\ 4 & 3 \\ 0 & -1 \end{bmatrix}$.

a. KL $\begin{bmatrix} 5 & 10 \\ -9 & -4 \end{bmatrix}$

b. LK $\begin{bmatrix} -1 & 6 & 2 \\ -15 & 2 & 8 \\ 1 & 2 & 0 \end{bmatrix}$

5 Find each product if

$A = \begin{bmatrix} -1 & 2 \\ 0 & 1 \end{bmatrix}$, $B = \begin{bmatrix} 1 & 0 \\ 3 & -2 \end{bmatrix}$,

and $C = \begin{bmatrix} -3 & 1 \\ -1 & 0 \end{bmatrix}$.

a. $A(B + C)$ $\begin{bmatrix} 6 & -5 \\ 2 & -2 \end{bmatrix}$

b. $AB + AC$ $\begin{bmatrix} 6 & -5 \\ 2 & -2 \end{bmatrix}$

Building on Prior Knowledge

Ask students to compare the process in Example 5 to the procedures they have used previously when applying the Distributive Property to real numbers and algebraic expressions.

Example 4 Commutative Property

Find each product if $P = \begin{bmatrix} 8 & -7 \\ -2 & 4 \\ 0 & 3 \end{bmatrix}$ and $Q = \begin{bmatrix} 9 & -3 & 2 \\ 6 & -1 & -5 \end{bmatrix}$.

a. PQ

$$PQ = \begin{bmatrix} 8 & -7 \\ -2 & 4 \\ 0 & 3 \end{bmatrix} \cdot \begin{bmatrix} 9 & -3 & 2 \\ 6 & -1 & -5 \end{bmatrix} \quad \text{Substitution}$$

$$= \begin{bmatrix} 72 - 42 & -24 + 7 & 16 + 35 \\ -18 + 24 & 6 - 4 & -4 - 20 \\ 0 + 18 & 0 - 3 & 0 - 15 \end{bmatrix} \quad \text{Multiply columns by rows.}$$

$$= \begin{bmatrix} 30 & -17 & 51 \\ 6 & 2 & -24 \\ 18 & -3 & -15 \end{bmatrix} \quad \text{Simplify.}$$

b. QP

$$QP = \begin{bmatrix} 9 & -3 & 2 \\ 6 & -1 & -5 \end{bmatrix} \cdot \begin{bmatrix} 8 & -7 \\ -2 & 4 \\ 0 & 3 \end{bmatrix} \quad \text{Substitution}$$

$$= \begin{bmatrix} 72 + 6 + 0 & -63 - 12 + 6 \\ 48 + 2 + 0 & -42 - 4 - 15 \end{bmatrix} \quad \text{Multiply columns by rows.}$$

$$= \begin{bmatrix} 78 & -69 \\ 50 & -61 \end{bmatrix} \quad \text{Simplify.}$$

In Example 4, notice that $PQ \neq QP$ because $\begin{bmatrix} 30 & -17 & 51 \\ 6 & 2 & -24 \\ 18 & -3 & -15 \end{bmatrix} \neq \begin{bmatrix} 78 & -69 \\ 50 & -61 \end{bmatrix}$. This demonstrates that the Commutative Property of Multiplication does not hold for matrix multiplication. The order in which you multiply matrices is very important.

Example 5 Distributive Property

Find each product if $A = \begin{bmatrix} 3 & 2 \\ -1 & 4 \end{bmatrix}$, $B = \begin{bmatrix} -2 & 5 \\ 6 & 7 \end{bmatrix}$, and $C = \begin{bmatrix} 1 & 1 \\ -5 & 3 \end{bmatrix}$.

a. $A(B + C)$

$$A(B + C) = \begin{bmatrix} 3 & 2 \\ -1 & 4 \end{bmatrix} \cdot \left(\begin{bmatrix} -2 & 5 \\ 6 & 7 \end{bmatrix} + \begin{bmatrix} 1 & 1 \\ -5 & 3 \end{bmatrix} \right) \quad \text{Substitution}$$

$$= \begin{bmatrix} 3 & 2 \\ -1 & 4 \end{bmatrix} \cdot \begin{bmatrix} -1 & 6 \\ 1 & 10 \end{bmatrix} \quad \text{Add corresponding elements.}$$

$$= \begin{bmatrix} 3(-1) + 2(1) & 3(6) + 2(10) \\ -1(-1) + 4(1) & -1(6) + 4(10) \end{bmatrix} \text{ or } \begin{bmatrix} -1 & 38 \\ 5 & 34 \end{bmatrix} \quad \text{Multiply columns by rows.}$$

b. $AB + AC$

$$AB + AC = \begin{bmatrix} 3 & 2 \\ -1 & 4 \end{bmatrix} \cdot \begin{bmatrix} -2 & 5 \\ 6 & 7 \end{bmatrix} + \begin{bmatrix} 3 & 2 \\ -1 & 4 \end{bmatrix} \cdot \begin{bmatrix} 1 & 1 \\ -5 & 3 \end{bmatrix} \quad \text{Substitution}$$

$$= \begin{bmatrix} 3(-2) + 2(6) & 3(5) + 2(7) \\ -1(-2) + 4(6) & -1(5) + 4(7) \end{bmatrix} + \begin{bmatrix} 3(1) + 2(-5) & 3(1) + 2(3) \\ -1(1) + 4(-5) & -1(1) + 4(3) \end{bmatrix}$$

$$= \begin{bmatrix} 6 & 29 \\ 26 & 23 \end{bmatrix} + \begin{bmatrix} -7 & 9 \\ -21 & 11 \end{bmatrix} \quad \text{Simplify.}$$

$$= \begin{bmatrix} -1 & 38 \\ 5 & 34 \end{bmatrix} \quad \text{Add corresponding elements.}$$

DAILY INTERVENTION — Differentiated Instruction

Auditory/Musical Ask students to write their own verse or rap to explain the steps in multiplying a 2 × 3 matrix and a 3 × 2 matrix. Invite some students to share their efforts with the class.

170 Chapter 4 Matrices

Notice that in Example 5, $A(B + C) = AB + AC$. This and other examples suggest that the Distributive Property is true for matrix multiplication. Some properties of matrix multiplication are shown below.

Concept Summary — *Properties of Matrix Multiplication*

For any matrices A, B, and C for which the matrix product is defined, and any scalar c, the following properties are true.

Associative Property of Matrix Multiplication	$(AB)C = A(BC)$
Associative Property of Scalar Multiplication	$c(AB) = (cA)B = A(cB)$
Left Distributive Property	$C(A + B) = CA + CB$
Right Distributive Property	$(A + B)C = AC + BC$

To show that a property is true for all cases, you must show it is true for the general case. To show that a property is *not* true for all cases, you only need to find a counterexample.

Check for Understanding

Concept Check

1. Sample answer: $\begin{bmatrix} 1 & 2 \\ 3 & 4 \\ 5 & 6 \end{bmatrix} \cdot \begin{bmatrix} 7 & 8 \\ 9 & 10 \end{bmatrix}$

1. **OPEN ENDED** Give an example of two matrices whose product is a 3×2 matrix.
2. **Determine** whether the following statement is *always*, *sometimes*, or *never* true. Explain your reasoning. **Never; the inner dimensions will never be equal.**
   For any matrix $A_{m \times n}$ for $m \neq n$, A^2 is defined.
3. **Explain** why, in most cases, $(A + B)C \neq CA + CB$. **See margin.**

Guided Practice

GUIDED PRACTICE KEY	
Exercises	Examples
4, 5	1
6–9	2
10	4, 5
11, 12	3

Determine whether each matrix product is defined. If so, state the dimensions of the product.

4. $A_{3 \times 5} \cdot B_{5 \times 2}$ **3 × 2**
5. $X_{2 \times 3} \cdot Y_{2 \times 3}$ **undefined**

Find each product, if possible.

6. $\begin{bmatrix} 3 & -5 \end{bmatrix} \cdot \begin{bmatrix} 3 & 5 \\ -2 & 0 \end{bmatrix}$ **$\begin{bmatrix} 19 & 15 \end{bmatrix}$**
7. $\begin{bmatrix} 5 \\ 8 \end{bmatrix} \cdot \begin{bmatrix} 3 & -1 & 4 \end{bmatrix}$ **$\begin{bmatrix} 15 & -5 & 20 \\ 24 & -8 & 32 \end{bmatrix}$**
8. $\begin{bmatrix} 5 & -2 & -1 \\ 8 & 0 & 3 \end{bmatrix} \cdot \begin{bmatrix} -4 & 2 \\ 1 & 0 \end{bmatrix}$ **not possible**
9. $\begin{bmatrix} 4 & -1 \\ 3 & 5 \end{bmatrix} \cdot \begin{bmatrix} 7 \\ 4 \end{bmatrix}$ **$\begin{bmatrix} 24 \\ 41 \end{bmatrix}$**
10. Use $A = \begin{bmatrix} 2 & -1 \\ 3 & 5 \end{bmatrix}$, $B = \begin{bmatrix} -4 & 1 \\ 8 & 0 \end{bmatrix}$, and $C = \begin{bmatrix} 3 & 2 \\ -1 & 2 \end{bmatrix}$ to determine whether $A(BC) = (AB)C$ is true for the given matrices. **See margin.**

Application

SPORTS For Exercises 11 and 12, use the table below that shows the number of kids registered for baseball and softball.

The Westfall Youth Baseball and Softball League charges the following registration fees: ages 7–8, $45; ages 9–10, $55; and ages 11–14, $65.

Team Members		
Age	Baseball	Softball
7–8	350	280
9–10	320	165
11–14	180	120

11. Write a matrix for the registration fees and a matrix for the number of players.
   11. $\begin{bmatrix} 45 & 55 & 65 \end{bmatrix}$, $\begin{bmatrix} 350 & 280 \\ 320 & 165 \\ 180 & 120 \end{bmatrix}$
12. Find the total amount of money the League received from baseball and softball registrations. **$74,525**

3 Practice/Apply

Study Notebook

Have students—

- add the definitions/examples of the vocabulary terms in this lesson to their Vocabulary Builder worksheets for Chapter 4.
- copy the general formula for multiplying matrices into their notebooks.
- write their own examples for testing whether two matrices can be multiplied and for multiplying two matrices.
- copy the properties of matrix multiplication.
- include any other item(s) that they find helpful in mastering the skills in this lesson.

Answers

3. The Right Distributive Property says that $(A + B)C = AC + BC$, but $AC + BC \neq CA + CB$ since the Commutative Property does not hold for matrix multiplication in most cases.

10. yes;
$A(BC)$
$= \begin{bmatrix} 2 & -1 \\ 3 & 5 \end{bmatrix} \cdot \left(\begin{bmatrix} -4 & 1 \\ 8 & 0 \end{bmatrix} \cdot \begin{bmatrix} 3 & 2 \\ -1 & 2 \end{bmatrix} \right)$
$= \begin{bmatrix} 2 & -1 \\ 3 & 5 \end{bmatrix} \cdot \begin{bmatrix} -13 & -6 \\ 24 & 16 \end{bmatrix}$
$= \begin{bmatrix} -50 & -28 \\ 81 & 62 \end{bmatrix}$
$(AB)C$
$= \left(\begin{bmatrix} 2 & -1 \\ 3 & 5 \end{bmatrix} \cdot \begin{bmatrix} -4 & 1 \\ 8 & 0 \end{bmatrix} \right) \cdot \begin{bmatrix} 3 & 2 \\ -1 & 2 \end{bmatrix}$
$= \begin{bmatrix} -16 & 2 \\ 28 & 3 \end{bmatrix} \cdot \begin{bmatrix} 3 & 2 \\ -1 & 2 \end{bmatrix}$
$= \begin{bmatrix} -50 & -28 \\ 81 & 62 \end{bmatrix}$

About the Exercises...

Organization by Objective

- **Multiply Matrices:** 13–26, 31–34, 36–42
- **Multiplicative Properties:** 27–30, 35

Odd/Even Assignments

Exercises 13–30 are structured so that students practice the same concepts whether they are assigned odd or even problems.

Assignment Guide

Basic: 13–29 odd, 31–39, 43–60

Average: 13–29 odd, 31–39, 43–60

Advanced: 14–30 even, 35–56 (optional: 57–60)

All: Practice Quiz 1 (1–10)

Practice and Apply

Homework Help

For Exercises	See Examples
13–18	1
19–26	2
27–30	4,5
31–41	3

Extra Practice
See page 834.

Determine whether each matrix product is defined. If so, state the dimensions of the product. 15. undefined

13. $A_{4 \times 3} \cdot B_{3 \times 2}$ 4×2
14. $X_{2 \times 2} \cdot Y_{2 \times 2}$ 2×2
15. $P_{1 \times 3} \cdot Q_{4 \times 1}$
16. $R_{1 \times 4} \cdot S_{4 \times 5}$ 1×5
17. $M_{4 \times 3} \cdot N_{4 \times 3}$ undefined
18. $A_{3 \times 1} \cdot B_{1 \times 5}$ 3×5

Find each product, if possible.

19. $\begin{bmatrix} 2 & -1 \end{bmatrix} \cdot \begin{bmatrix} 5 \\ 4 \end{bmatrix}$ $[6]$

20. $\begin{bmatrix} 3 & -2 \\ 5 & 1 \end{bmatrix} \cdot \begin{bmatrix} 4 & 1 \\ 2 & 7 \end{bmatrix}$ $\begin{bmatrix} 8 & -11 \\ 22 & 12 \end{bmatrix}$

21. $\begin{bmatrix} 4 & -1 & 6 \\ 1 & 5 & -8 \end{bmatrix} \cdot \begin{bmatrix} 1 & 3 \\ 9 & -6 \end{bmatrix}$ not possible

22. $\begin{bmatrix} 4 & -2 & -7 \\ 6 & 3 & 5 \end{bmatrix} \cdot \begin{bmatrix} -2 \\ 5 \\ 3 \end{bmatrix}$ $\begin{bmatrix} -39 \\ 18 \end{bmatrix}$

23. $\begin{bmatrix} 2 & -1 \\ 3 & 4 \end{bmatrix} \cdot \begin{bmatrix} 3 & -9 & -2 \\ 5 & 7 & -6 \end{bmatrix}$

23. $\begin{bmatrix} 1 & -25 & 2 \\ 29 & 1 & -30 \end{bmatrix}$

24. $\begin{bmatrix} 7 & 3 \\ 0 & 2 \\ 5 & 5 \end{bmatrix} \cdot \begin{bmatrix} -2 & 1 & 4 \\ 3 & -5 & 2 \\ 4 & 3 & 1 \end{bmatrix}$ not possible

25. $\begin{bmatrix} 4 & 0 \\ -3 & 7 \\ -5 & 9 \end{bmatrix} \cdot \begin{bmatrix} 6 & 4 \\ -2 & 1 \end{bmatrix}$ $\begin{bmatrix} 24 & 16 \\ -32 & -5 \\ -48 & -11 \end{bmatrix}$

26. $\begin{bmatrix} 0 & 8 \\ 3 & 1 \\ -1 & 5 \end{bmatrix} \cdot \begin{bmatrix} 3 & 1 & -2 \\ 0 & 8 & -5 \end{bmatrix}$

26. $\begin{bmatrix} 0 & 64 & -40 \\ 9 & 11 & -11 \\ -3 & 39 & -23 \end{bmatrix}$

Use $A = \begin{bmatrix} 1 & -2 \\ 4 & 3 \end{bmatrix}$, $B = \begin{bmatrix} -5 & 2 \\ 4 & 3 \end{bmatrix}$, $C = \begin{bmatrix} 5 & 1 \\ 2 & -4 \end{bmatrix}$, **and scalar $c = 3$ to determine whether the following equations are true for the given matrices.**

27–30. See pp. 217A–217B for work.

27. $AC + BC = (A + B)C$ yes
28. $c(AB) = A(cB)$ yes
29. $C(A + B) = AC + BC$ no
30. $ABC = CBA$ no

PRODUCE **For Exercises 31–34, use the table and the following information.**

Carmen Fox owns three fruit farms on which he grows apples, peaches, and apricots. He sells apples for $22 a case, peaches for $25 a case, and apricots for $18 a case.

Number of Cases in Stock of Each Type of Fruit

Farm	Apples	Peaches	Apricots
1	290	165	210
2	175	240	190
3	110	75	0

31. Write an inventory matrix for the number of cases for each type of fruit for each farm.
32. Write a cost matrix for the price per case for each type of fruit.
33. Find the total income of the three fruit farms expressed as a matrix.
34. What is the total income from all three fruit farms combined? $31,850

31. $\begin{bmatrix} 290 & 165 & 210 \\ 175 & 240 & 190 \\ 110 & 75 & 0 \end{bmatrix}$

32. $\begin{bmatrix} 22 \\ 25 \\ 18 \end{bmatrix}$

33. $\begin{bmatrix} 14{,}285 \\ 13{,}270 \\ 4295 \end{bmatrix}$

35. any two matrices $\begin{bmatrix} a & b \\ c & d \end{bmatrix}$ and $\begin{bmatrix} e & f \\ g & h \end{bmatrix}$ where $bg = cf$, $a = d$, and $e = h$

35. **CRITICAL THINKING** Give an example of two matrices A and B whose product is commutative so that $AB = BA$.

Answer

44. Sports statistics are often listed in columns and matrices. In this case, you can find the total number of points scored by multiplying the point matrix, which doesn't change, by the record matrix, which changes for each season. Answers should include the following.

- $P \cdot R = [479]$
- Basketball and wrestling use different point values in scoring.

172 Chapter 4 Matrices

FUND-RAISING **For Exercises 36–39, use the table and the information below.** Lawrence High School sold wrapping paper and boxed cards for their fund-raising event. The school receives \$1.00 for each roll of wrapping paper sold and \$0.50 for each box of cards sold.

Total Amounts for Each Class		
Class	Wrapping Paper	Cards
Freshmen	72	49
Sophomores	68	63
Juniors	90	56
Seniors	86	62

36. $\begin{bmatrix} 72 & 49 \\ 68 & 63 \\ 90 & 56 \\ 86 & 62 \end{bmatrix}, \begin{bmatrix} 1.00 \\ 0.50 \end{bmatrix}$

37. $\begin{bmatrix} 96.50 \\ 99.50 \\ 118 \\ 117 \end{bmatrix}$

36. Write a matrix that represents the amounts sold for each class and a matrix that represents the amount of money the school earns for each item sold.

37. Write a matrix that shows how much each class earned.

38. Which class earned the most money? **Juniors**

39. What is the total amount of money the school made from the fund-raiser? **\$431**

FINANCE **For Exercises 40–42, use the table below that shows the purchase price and selling price of stock for three companies.**

For a class project, Taini "bought" shares of stock in three companies. She bought 150 shares of a utility company, 100 shares of a computer company, and 200 shares of a food company. At the end of the project she "sold" all of her stock.

Company	Purchase Price (per share)	Selling Price (per share)
Utility	\$54.00	\$55.20
Computer	\$48.00	\$58.60
Food	\$60.00	\$61.10

40. Organize the data in two matrices and use matrix multiplication to find the total amount she spent for the stock. **\$24,900**

41. Write two matrices and use matrix multiplication to find the total amount she received for selling the stock. **\$26,360**

42. Use matrix operations to find how much money Taini "made" or "lost." **\$1460**

43. $a = 1$, $b = 0$, $c = 0$, $d = 1$; the original matrix

43. **CRITICAL THINKING** Find the values of a, b, c, and d to make the statement $\begin{bmatrix} 3 & 5 \\ -1 & 7 \end{bmatrix} \cdot \begin{bmatrix} a & b \\ c & d \end{bmatrix} = \begin{bmatrix} 3 & 5 \\ -1 & 7 \end{bmatrix}$ true. If the matrix $\begin{bmatrix} a & b \\ c & d \end{bmatrix}$ is multiplied by any other matrix containing two columns, what do you think the result would be?

44. **WRITING IN MATH** Answer the question that was posed at the beginning of the lesson. **See margin.**

How can matrices be used in sports statistics?

Include the following in your answer:
- a matrix that represents the total points scored in the 2000 season, and
- an example of another sport where different point values are used in scoring.

Standardized Test Practice

45. If C is a 5×1 matrix and D is a 3×5 matrix, what are the dimensions of DC? **B**

Ⓐ 5×5 Ⓑ 3×1 Ⓒ 1×3 Ⓓ DC is not defined.

46. What is the product of $[5 \quad -2 \quad 3]$ and $\begin{bmatrix} 1 & -2 \\ 0 & 3 \\ 2 & 5 \end{bmatrix}$? **A**

Ⓐ $[11 \quad -1]$ Ⓑ $\begin{bmatrix} 11 \\ -1 \end{bmatrix}$ Ⓒ $\begin{bmatrix} 5 & -10 \\ 0 & -6 \\ 6 & -15 \end{bmatrix}$ Ⓓ undefined

www.algebra2.com/self_check_quiz

Enrichment, p. 186

Properties of Matrices

Computing with matrices is different from computing with real numbers. Stated below are some properties of the real number system. Are these also true for matrices? In the problems on this page, you will investigate this question.

For all real numbers a and b, $ab = 0$ if and only if $a = 0$ or $b = 0$.

Multiplication is commutative. For all real numbers a and b, $ab = ba$.

Multiplication is associative. For all real numbers a, b, and c, $a(bc) = (ab)c$.

Use the matrices A, B, and C for the problems. Write whether each statement is true. Assume that a 2-by-2 matrix is the 0 matrix if and only if all of its elements are zero.

$A = \begin{bmatrix} 3 & 1 \\ 1 & 3 \end{bmatrix}$ $B = \begin{bmatrix} 1 & -3 \\ -1 & 3 \end{bmatrix}$ $C = \begin{bmatrix} 3 & 6 \\ 1 & 2 \end{bmatrix}$

1. $AB = 0$ **no**
2. $AC = 0$ **no**
3. $BC = 0$ **yes**

Study Guide and Intervention, p. 181 (shown) and p. 182

Multiply Matrices You can multiply two matrices if and only if the number of columns in the first matrix is equal to the number of rows in the second matrix.

Multiplication of Matrices	$\begin{bmatrix} a_1 & b_1 \\ a_2 & b_2 \end{bmatrix} \cdot \begin{bmatrix} x_1 & y_1 \\ x_2 & y_2 \end{bmatrix} = \begin{bmatrix} a_1x_1 + b_1x_2 & a_1y_1 + b_1y_2 \\ a_2x_1 + b_2x_2 & a_2y_1 + b_2y_2 \end{bmatrix}$

Example **Find AB if $A = \begin{bmatrix} -4 & 3 \\ 2 & -2 \\ 1 & 7 \end{bmatrix}$ and $B = \begin{bmatrix} 5 & -2 \\ -1 & 3 \end{bmatrix}$.**

$AB = \begin{bmatrix} -4 & 3 \\ 2 & -2 \\ 1 & 7 \end{bmatrix} \cdot \begin{bmatrix} 5 & -2 \\ -1 & 3 \end{bmatrix}$ Substitution

$= \begin{bmatrix} -4(5) + 3(-1) & -4(-2) + 3(3) \\ 2(5) + (-2)(-1) & 2(-2) + (-2)(3) \\ 1(5) + 7(-1) & 1(-2) + 7(3) \end{bmatrix}$ Multiply columns by rows.

$= \begin{bmatrix} -23 & 17 \\ 12 & -10 \\ -2 & 19 \end{bmatrix}$ Simplify.

Exercises

Find each product, if possible.

1. $\begin{bmatrix} 4 & 1 \\ -2 & 3 \end{bmatrix} \cdot \begin{bmatrix} 3 & 0 \\ 0 & 3 \end{bmatrix}$ $\begin{bmatrix} 12 & 3 \\ -6 & 9 \end{bmatrix}$
2. $\begin{bmatrix} -1 & 0 \\ 3 & 7 \end{bmatrix} \cdot \begin{bmatrix} 3 & 2 \\ -1 & 4 \end{bmatrix}$ $\begin{bmatrix} -3 & -2 \\ 2 & 34 \end{bmatrix}$
3. $\begin{bmatrix} 3 & -1 \\ 2 & 4 \end{bmatrix} \cdot \begin{bmatrix} 3 & -1 \\ 2 & 4 \end{bmatrix}$ $\begin{bmatrix} 7 & -7 \\ 14 & 14 \end{bmatrix}$
4. $\begin{bmatrix} -3 & 1 \\ 5 & -2 \end{bmatrix} \cdot \begin{bmatrix} 4 & 0 & -2 \\ -3 & 1 & 1 \end{bmatrix}$ $\begin{bmatrix} -15 & 1 & 7 \\ 26 & -2 & -12 \end{bmatrix}$
5. $\begin{bmatrix} 3 & -2 \\ 0 & 4 \\ -5 & 1 \end{bmatrix} \cdot \begin{bmatrix} 1 & 2 \\ 2 & 1 \end{bmatrix}$ $\begin{bmatrix} -1 & 4 \\ 8 & 4 \\ -3 & -9 \end{bmatrix}$
6. $\begin{bmatrix} 5 & -2 \\ 2 & -3 \end{bmatrix} \cdot \begin{bmatrix} 4 & -1 \\ -2 & 5 \end{bmatrix}$ $\begin{bmatrix} 24 & -15 \\ 14 & -17 \end{bmatrix}$
7. $\begin{bmatrix} 6 & 10 \\ -4 & 3 \\ -2 & 7 \end{bmatrix} \cdot [0 \quad 4 \quad -3]$ not possible
8. $\begin{bmatrix} 7 & -2 \\ 5 & -4 \end{bmatrix} \cdot \begin{bmatrix} 1 & -3 \\ -2 & 0 \end{bmatrix}$ $\begin{bmatrix} 11 & -21 \\ 13 & -15 \end{bmatrix}$
9. $\begin{bmatrix} 2 & 0 & -3 \\ 1 & 4 & -2 \\ -1 & 3 & 1 \end{bmatrix} \cdot \begin{bmatrix} 2 & -2 \\ 3 & 1 \\ -2 & 4 \end{bmatrix}$ $\begin{bmatrix} 10 & -16 \\ 18 & -6 \\ 5 & 9 \end{bmatrix}$

Skills Practice, p. 183 and Practice, p. 184 (shown)

Determine whether each matrix product is defined. If so, state the dimensions of the product.

1. $A_{7 \times 4} \cdot B_{4 \times 3}$ 7×3
2. $A_{3 \times 5} \cdot M_{5 \times 8}$ 3×8
3. $M_{2 \times 1} \cdot A_{1 \times 6}$ 2×6
4. $M_{3 \times 2} \cdot A_{3 \times 2}$ undefined
5. $P_{1 \times 9} \cdot Q_{9 \times 1}$ 1×1
6. $P_{9 \times 1} \cdot Q_{1 \times 9}$ 9×9

Find each product, if possible.

7. $\begin{bmatrix} 2 & 4 \\ 3 & -1 \end{bmatrix} \cdot \begin{bmatrix} 3 & -2 & 7 \\ 6 & 0 & -5 \end{bmatrix}$ $\begin{bmatrix} 30 & -4 & -6 \\ 3 & -6 & 26 \end{bmatrix}$
8. $\begin{bmatrix} 2 & 4 \\ 7 & -1 \end{bmatrix} \cdot \begin{bmatrix} -3 & 0 \\ 2 & 5 \end{bmatrix}$ $\begin{bmatrix} 2 & 20 \\ -23 & -5 \end{bmatrix}$
9. $\begin{bmatrix} -3 & 0 \\ 2 & 5 \end{bmatrix} \cdot \begin{bmatrix} 2 & 4 \\ 7 & -1 \end{bmatrix}$ $\begin{bmatrix} -6 & -12 \\ 39 & 3 \end{bmatrix}$
10. $\begin{bmatrix} 3 & -2 & 7 \\ 6 & 0 & -5 \end{bmatrix} \cdot \begin{bmatrix} 3 & -2 & 7 \\ 6 & 0 & -5 \end{bmatrix}$ not possible
11. $[4 \quad 0 \quad 2] \cdot \begin{bmatrix} 1 \\ 3 \\ -1 \end{bmatrix}$ $[2]$
12. $\begin{bmatrix} 1 \\ 3 \\ -1 \end{bmatrix} \cdot [4 \quad 0 \quad 2]$ $\begin{bmatrix} 4 & 0 & 2 \\ 12 & 0 & 6 \\ -4 & 0 & -2 \end{bmatrix}$
13. $\begin{bmatrix} -6 & 2 \\ 3 & -1 \end{bmatrix} \cdot \begin{bmatrix} 5 & 0 \\ 0 & 5 \end{bmatrix}$ $\begin{bmatrix} -30 & 10 \\ 15 & -5 \end{bmatrix}$
14. $[-15 \quad -9] \cdot \begin{bmatrix} 6 & 11 \\ 23 & -10 \end{bmatrix}$ $[-297 \quad -75]$

Use $A = \begin{bmatrix} 1 & 3 \\ 3 & 1 \end{bmatrix}$, $B = \begin{bmatrix} 4 & 0 \\ -2 & -1 \end{bmatrix}$, $C = \begin{bmatrix} -1 & 0 \\ 0 & -1 \end{bmatrix}$, and scalar $c = 3$ to determine whether the following equations are true for the given matrices.

15. $AC = CA$ yes
16. $A(B + C) = BA + CA$ no
17. $(AB)c = c(AB)$ yes
18. $(A + C)B = B(A + C)$ no

RENTALS **For Exercises 19–21, use the following information.**

For their one-week vacation, the Montoyas can rent a 2-bedroom condominium for \$1796, a 3-bedroom condominium for \$2165, or a 4-bedroom condominium for \$2538. The table shows the number of units in each of three complexes.

	2-Bedroom	3-Bedroom	4-Bedroom
Sun Haven	36	24	22
Surfside	29	32	42
Seabreeze	18	22	18

19. Write a matrix that represents the number of each type of unit available at each complex and a matrix that represents the weekly charge for each type of unit. $\begin{bmatrix} 36 & 24 & 22 \\ 29 & 32 & 42 \\ 18 & 22 & 18 \end{bmatrix}, \begin{bmatrix} \$1796 \\ \$2165 \\ \$2538 \end{bmatrix}$

20. If all of the units in the three complexes are rented for the week at the rates given the Montoyas, express the income of each of the three complexes as a matrix. $\begin{bmatrix} 172{,}452 \\ 227{,}960 \\ 125{,}642 \end{bmatrix}$

21. What is the total income of all three complexes for the week? \$526,054

Reading to Learn Mathematics, p. 185

ELL

Pre-Activity **How can matrices be used in sports statistics?**

Read the introduction to Lesson 4-3 at the top of page 167 in your textbook.

Write a sum that shows the total points scored by the Oakland Raiders during the 2000 season. (The sum will include multiplications. Do not actually calculate this sum.)

$6 \cdot 58 + 1 \cdot 56 + 3 \cdot 23 + 2 \cdot 1 + 2 \cdot 2$

Reading the Lesson

1. Determine whether each indicated matrix product is defined. If so, state the dimensions of the product. If not, write *undefined*.

a. $M_{3 \times 2}$ and $N_{2 \times 3}$ MN: 3×3 NM: 2×2

b. $M_{1 \times 2}$ and $N_{1 \times 2}$ MN: undefined NM: undefined

c. $M_{4 \times 1}$ and $N_{1 \times 4}$ MN: 4×4 NM: 1×1

d. $M_{3 \times 4}$ and $N_{4 \times 4}$ MN: 3×4 NM: undefined

2. The regional sales manager for a chain of computer stores wants to compare the revenue from sales of one model of notebook computer and one model of printer for three stores in his area. The notebook computer sells for \$1850 and the printer for \$175. The number of computers and printers sold at the three stores during September are shown in the following table.

Store	Computers	Printers
A	128	101
B	205	166
C	97	73

Write a matrix product that the manager could use to find the total revenue for computers and printers for each of the three stores. (Do not calculate the product.)

$\begin{bmatrix} 128 & 101 \\ 205 & 166 \\ 97 & 73 \end{bmatrix} \cdot \begin{bmatrix} 1850 \\ 175 \end{bmatrix}$

Helping You Remember

3. Many students find the procedure of matrix multiplication confusing at first because it is unfamiliar. Think of an easy way to use the letters R and C to remember how to multiply matrices and what the dimensions of the product will be. **Sample answer: Just remember RC for "row, column." Multiply each *row* of the first matrix by each *column* of the second matrix. The dimensions of the product are the number of rows of the first matrix and the number of *columns* of the second matrix.**

Lesson 4-3 Multiplying Matrices 173

4 Assess

Open-Ended Assessment

Modeling Have students use the class seating arrangement to form matrices, with each student's desk as an element. Have students make 24 large cards showing the values for the elements of two matrices. Form two matrices $A_{4 \times 3}$ and $B_{3 \times 4}$ with student seats and give each student sitting in these "matrices" a card to show the element in that position. Have students model the matrix multiplication $AB = C$. Begin by drawing a blank 4×4 matrix on the board. For each element of matrix C, have students walk out the products, compute the sums, and then write the results in the correct locations of the matrix on the board.

Assessment Options

Practice Quiz 1 The quiz provides students with a brief review of the concepts and skills in Lessons 4-1 through 4-3. Lesson numbers are given to the right of exercises or instruction lines so students can review concepts not yet mastered.

Getting Ready for Lesson 4-4

PREREQUISITE SKILL Lesson 4-4 shows how to use a matrix to find the coordinates of a figure that has been transformed by translation, dilation, reflection, or rotation. Students will use their familiarity with graphing ordered pairs to graph transformed figures. Exercises 57–60 should be used to determine your students' familiarity with graphing ordered pairs on a coordinate plane.

Answer (Practice Quiz 1)

4. $\begin{bmatrix} 120 & 80 & 64 & 75 \\ 65 & 105 & 77 & 53 \end{bmatrix}$, $\begin{bmatrix} 112 & 79 & 56 & 74 \\ 69 & 95 & 82 & 50 \end{bmatrix}$

Maintain Your Skills

Mixed Review

Perform the indicated matrix operations. If the matrix does not exist, write *impossible*. *(Lesson 4-2)*

47. $3\begin{bmatrix} 4 & -2 \\ -1 & 7 \end{bmatrix}$ $\begin{bmatrix} 12 & -6 \\ -3 & 21 \end{bmatrix}$

48. $[3 \quad 5 \quad 9] + \begin{bmatrix} 5 \\ 2 \\ 6 \end{bmatrix}$ impossible

49. $2\begin{bmatrix} 6 & 3 \\ -8 & -2 \end{bmatrix} - 4\begin{bmatrix} 8 & 1 \\ 3 & -4 \end{bmatrix}$ $\begin{bmatrix} -20 & 2 \\ -28 & 12 \end{bmatrix}$

Solve each equation. *(Lesson 4-1)*

50. $\begin{bmatrix} 3x + 2 \\ 15 \end{bmatrix} = \begin{bmatrix} 23 \\ -4y - 1 \end{bmatrix}$ (7, −4)

51. $\begin{bmatrix} x + 3y \\ 2x - y \end{bmatrix} = \begin{bmatrix} -22 \\ 19 \end{bmatrix}$ (5, −9)

52. $\begin{bmatrix} x + 3z \\ -2x + y - z \\ 5y - 7z \end{bmatrix} = \begin{bmatrix} -19 \\ -2 \\ 24 \end{bmatrix}$ (2, −5, −7)

53. **CAMERA SUPPLIES** Mrs. Franklin is planning a family vacation. She bought 8 rolls of film and 2 camera batteries for \$23. The next day, her daughter went back and bought 6 more rolls of film and 2 batteries for her camera. This bill was \$18. What is the price of a roll of film and a camera battery? *(Lesson 3-2)* **\$2.50; \$1.50**

Find the x-intercept and the y-intercept of the graph of each equation. Then graph the equation. *(Lesson 2-2)* **54–56. See margin for graphs.**

54. $y = 3 - 2x$ $\frac{3}{2}$; 3

55. $x - \frac{1}{2}y = 8$ 8; −16

56. $5x - 2y = 10$ 2; −5

Getting Ready for the Next Lesson

PREREQUISITE SKILL Graph each set of ordered pairs on a coordinate plane.
*(To review **graphing ordered pairs**, see Lesson 2-1.)* **57–60. See pp. 217A–217B.**

57. {(2, 4), (−1, 3), (0, −2)}

58. {(−3, 5), (−2, −4), (3, −2)}

59. {(−1, 2), (2, 4), (3, −3), (4, −1)}

60. {(−3, 3), (1, 3), (4, 2), (−1, −5)}

Practice Quiz 1 — Lessons 4-1 through 4-3

Solve each equation. *(Lesson 4-1)*

1. $\begin{bmatrix} 3x + 1 \\ 7y \end{bmatrix} = \begin{bmatrix} 19 \\ 21 \end{bmatrix}$ (6, 3)

2. $\begin{bmatrix} 2x + y \\ 4x - 3y \end{bmatrix} = \begin{bmatrix} 9 \\ 23 \end{bmatrix}$ (5, −1)

3. $\begin{bmatrix} 2 & x \\ y & 5 \end{bmatrix} = \begin{bmatrix} 2 & 1 \\ 3 & z \end{bmatrix}$ (1, 3, 5)

BUSINESS For Exercises 4 and 5, use the table and the following information.
The manager of The Best Bagel Shop keeps records of each type of bagel sold each day at their two stores. Two days of sales are shown below.

Day	Store	Type of Bagel			
		Sesame	Poppy	Blueberry	Plain
Monday	East	120	80	64	75
	West	65	105	77	53
Tuesday	East	112	79	56	74
	West	69	95	82	50

4. Write a matrix for each day's sales. *(Lesson 4-1)* **See margin.**
5. Find the sum of the two days' sales using matrix addition. *(Lesson 4-2)* $\begin{bmatrix} 232 & 159 & 120 & 149 \\ 134 & 200 & 159 & 103 \end{bmatrix}$

Perform the indicated matrix operations. *(Lesson 4-2)*

6. $\begin{bmatrix} 3 & 0 \\ 7 & 12 \end{bmatrix} - \begin{bmatrix} 6 & -5 \\ 4 & -1 \end{bmatrix}$ $\begin{bmatrix} -3 & 5 \\ 3 & 13 \end{bmatrix}$

7. $\frac{2}{3}\begin{bmatrix} 9 & 0 \\ 12 & 15 \end{bmatrix} + \begin{bmatrix} -2 & 3 \\ -7 & -7 \end{bmatrix}$ $\begin{bmatrix} 4 & 3 \\ 1 & 3 \end{bmatrix}$

8. $5\begin{bmatrix} -2 & 4 & 5 \\ 0 & -4 & 7 \end{bmatrix}$ $\begin{bmatrix} -10 & 20 & 25 \\ 0 & -20 & 35 \end{bmatrix}$

Find each product, if possible. *(Lesson 4-3)*

9. $\begin{bmatrix} 4 & 0 & -8 \\ 7 & -2 & 10 \end{bmatrix} \cdot \begin{bmatrix} -1 & 3 \\ 6 & 0 \end{bmatrix}$ not possible

10. $\begin{bmatrix} 3 & -1 \\ 2 & 5 \end{bmatrix} \cdot \begin{bmatrix} 4 & -1 & -2 \\ -3 & 5 & 4 \end{bmatrix}$ $\begin{bmatrix} 15 & -8 & -10 \\ -7 & 23 & 16 \end{bmatrix}$

Answers

54.
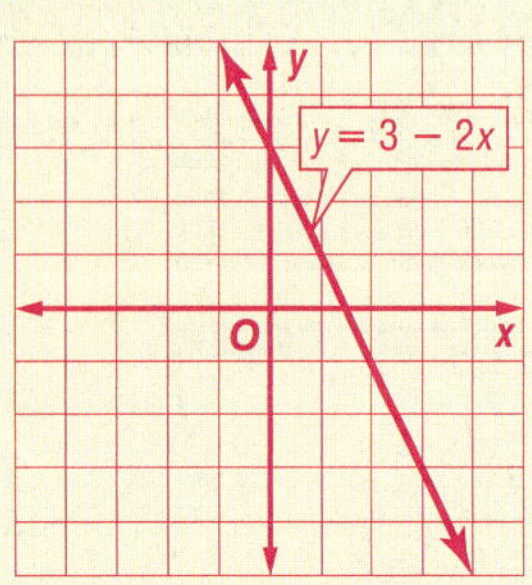

55.
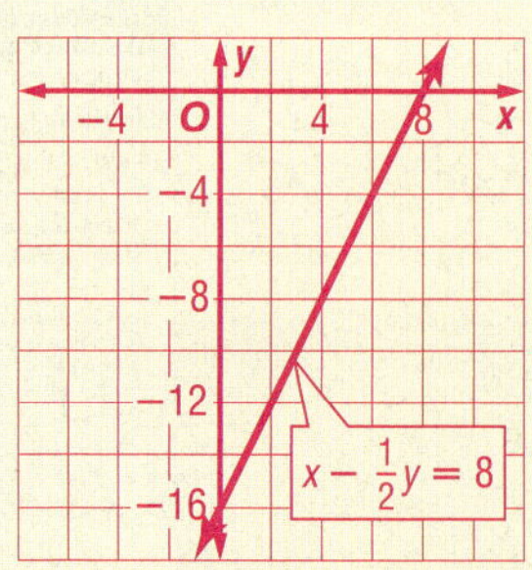

56.
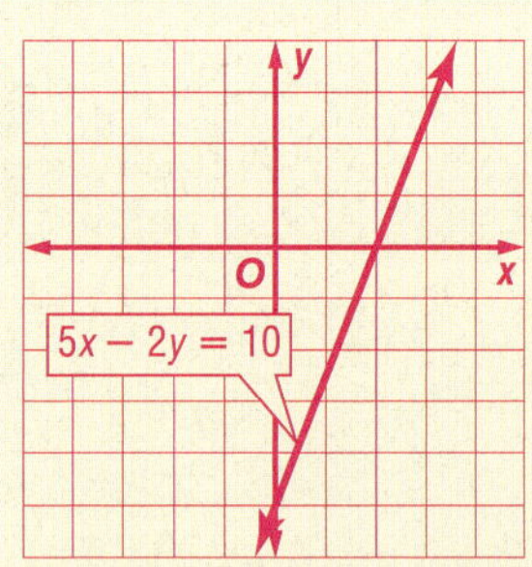

4-4 Transformations with Matrices

What You'll Learn

- Use matrices to determine the coordinates of a translated or dilated figure.
- Use matrix multiplication to find the coordinates of a reflected or rotated figure.

Vocabulary

- vertex matrix
- transformation
- preimage
- image
- isometry
- translation
- dilation
- reflection
- rotation

How are transformations used in computer animation?

Computer animation creates the illusion of motion by using a succession of computer-generated still images. Computer animation is used to create movie special effects and to simulate images that would be impossible to show otherwise.

Complex geometric figures can be broken into simple triangles and then moved to other parts of the screen.

TRANSLATIONS AND DILATIONS Points on a coordinate plane can be represented by matrices. The ordered pair (x, y) can be represented by the column matrix $\begin{bmatrix} x \\ y \end{bmatrix}$. Likewise, polygons can be represented by placing all of the column matrices of the coordinates of the vertices into one matrix, called a **vertex matrix**.

Triangle ABC with vertices $A(3, 2)$, $B(4, -2)$, and $C(2, -1)$ can be represented by the following vertex matrix.

$$\begin{matrix} & A & B & C \\ \triangle ABC = & \begin{bmatrix} 3 & 4 & 2 \\ 2 & -2 & -1 \end{bmatrix} & \begin{matrix} \leftarrow x\text{-coordinates} \\ \leftarrow y\text{-coordinates} \end{matrix} \end{matrix}$$

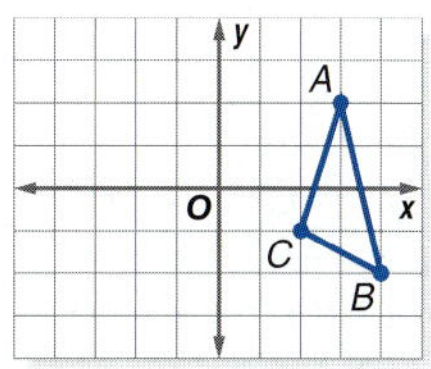

Matrices can be used to perform transformations. **Transformations** are functions that map points of a **preimage** onto its **image**. If the image and preimage are congruent figures, the transformation is an **isometry**.

One type of isometry is a translation. A **translation** occurs when a figure is moved from one location to another without changing its size, shape, or orientation. You can use matrix addition and a *translation matrix* to find the coordinates of a translated figure.

Study Tip

Reading Math
A matrix containing coordinates of a geometric figure is also called a *coordinate matrix*.

Example 1 Translate a Figure

Find the coordinates of the vertices of the image of quadrilateral $QUAD$ with $Q(2, 3)$, $U(5, 2)$, $A(4, -2)$, and $D(1, -1)$, if it is moved 4 units to the left and 2 units up. Then graph $QUAD$ and its image $Q'U'A'D'$.

Write the vertex matrix for quadrilateral $QUAD$. $\begin{bmatrix} 2 & 5 & 4 & 1 \\ 3 & 2 & -2 & -1 \end{bmatrix}$

To translate the quadrilateral 4 units to the left, add -4 to each x-coordinate. To translate the figure 2 units up, add 2 to each y-coordinate. This can be done by adding the translation matrix $\begin{bmatrix} -4 & -4 & -4 & -4 \\ 2 & 2 & 2 & 2 \end{bmatrix}$ to the vertex matrix of $QUAD$.

(continued on the next page)

4-4 Lesson Notes

1 Focus

5-Minute Check Transparency 4-4 Use as a quiz or review of Lesson 4-3.

Mathematical Background notes are available for this lesson on p. 152D.

How are transformations used in computer animation?

Ask students:

- What does it mean to "simulate" an image? **Sample answer: to use a series of drawings that together create the illusion of real life**
- How would you break a geometric figure (a polygon) into simple triangles? **Draw all the diagonals from one vertex.**

Resource Manager

Workbook and Reproducible Masters

Chapter 4 Resource Masters

- Study Guide and Intervention, pp. 187–188
- Skills Practice, p. 189
- Practice, p. 190
- Reading to Learn Mathematics, p. 191
- Enrichment, p. 192
- Assessment, pp. 231, 233

Graphing Calculator and Spreadsheet Masters, p. 33

Transparencies

5-Minute Check Transparency 4-4
Answer Key Transparencies

Technology

Interactive Chalkboard

2 Teach

TRANSLATIONS AND DILATIONS

In-Class Examples Power Point®

1 Find the coordinates of the vertices of the image of quadrilateral $ABCD$ with $A(-5, -1)$, $B(-2, -1)$, $C(-1, -4)$, $D(-3, -5)$, if it is moved 3 units to the right and 4 units up. Then graph $ABCD$ and its image $A'B'C'D'$. **$A'(-2, 3)$, $B'(1, 3)$, $C'(2, 0)$, $D'(0, -1)$**

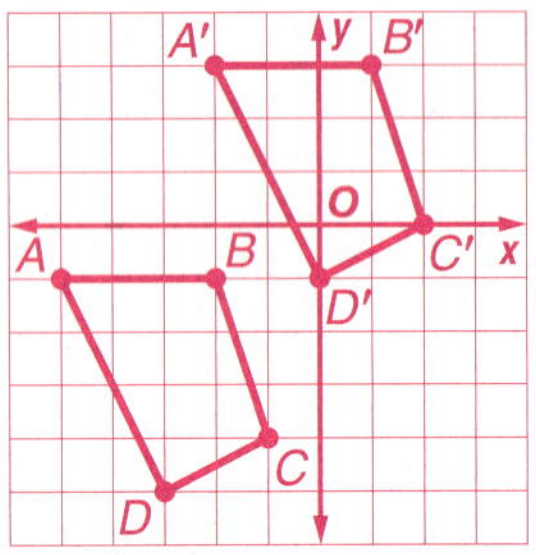

2 Rectangle $E'F'G'H'$ is the result of a translation of rectangle $EFGH$. A table of the vertices of each rectangle is shown. Find the coordinates of F and G'.

Rectangle *EFGH*	Rectangle *E'F'G'H'*
$E(-2, 2)$	$E'(-5, 0)$
F	$F'(1, 0)$
$G(4, -2)$	G'
$H(-2, -2)$	$H'(-5, -4)$

The coordinates of F are $(4, 2)$ and the coordinates of G' are $(1, -4)$.

Vertex Matrix of *QUAD* + Translation Matrix = Vertex Matrix of *Q'U'A'D'*

$$\begin{bmatrix} 2 & 5 & 4 & 1 \\ 3 & 2 & -2 & -1 \end{bmatrix} + \begin{bmatrix} -4 & -4 & -4 & -4 \\ 2 & 2 & 2 & 2 \end{bmatrix} = \begin{bmatrix} -2 & 1 & 0 & -3 \\ 5 & 4 & 0 & 1 \end{bmatrix}$$

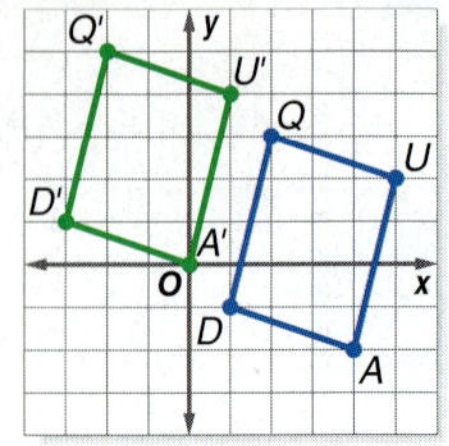

The coordinates of $Q'U'A'D'$ are $Q'(-2, 5)$, $U'(1, 4)$, $A'(0, 0)$, and $D'(-3, 1)$. Graph the preimage and the image. The two quadrilaterals have the same size and shape.

Standardized Test Practice

Example 2 Find a Translation Matrix

Short-Response Test Item

Rectangle $A'B'C'D'$ is the result of a translation of rectangle $ABCD$. A table of the vertices of each rectangle is shown. Find the coordinates of A and D'.

Rectangle *ABCD*	Rectangle *A'B'C'D'*
A	$A'(-1, 1)$
$B(1, 5)$	$B'(4, 1)$
$C(1, -2)$	$C'(4, -6)$
$D(-4, -2)$	D'

Read the Test Item

- You are given the coordinates of the preimage and image of points B and C. Use this information to find the translation matrix. Then you can use the translation matrix to find the coordinates of A and D'.

Solve the Test Item

- Write a matrix equation. Let (a, b) represent the coordinates of A and let (c, d) represent the coordinates of D'.

$$\begin{bmatrix} a & 1 & 1 & -4 \\ b & 5 & -2 & -2 \end{bmatrix} + \begin{bmatrix} x & x & x & x \\ y & y & y & y \end{bmatrix} = \begin{bmatrix} -1 & 4 & 4 & c \\ 1 & 1 & -6 & d \end{bmatrix}$$

$$\begin{bmatrix} a+x & 1+x & 1+x & -4+x \\ b+y & 5+y & -2+y & -2+y \end{bmatrix} = \begin{bmatrix} -1 & 4 & 4 & c \\ 1 & 1 & -6 & d \end{bmatrix}$$

- Since these two matrices are equal, corresponding elements are equal.

Solve an equation for x.

$1 + x = 4$
$x = 3$

Solve an equation for y.

$5 + y = 1$
$y = -4$

- Use the values for x and y to find the values for $A(a, b)$ and $D'(c, d)$.

$a + x = -1$
$a + 3 = -1$
$a = -4$

$b + y = 1$
$b + (-4) = 1$
$b = 5$

$-4 + x = c$
$-4 + 3 = c$
$-1 = c$

$-2 + y = d$
$-2 + (-4) = d$
$-6 = d$

So the coordinates of A are $(-4, 5)$, and the coordinates for D' are $(-1, -6)$.

Test-Taking Tip

Sometimes you need to solve for unknown value(s) before you can solve for the value(s) requested in the question.

When a geometric figure is enlarged or reduced, the transformation is called a **dilation**. In a dilation, all linear measures of the image change in the same ratio. For example, if the length of each side of a figure doubles, then the perimeter doubles, and vice versa. You can use scalar multiplication to perform dilations.

Standardized Test Practice

Example 2 Test items involving figures in coordinate planes can often be solved quickly by sketching the figure. After plotting points B, C, and D, students can use their knowledge of rectangles to quickly determine the location of point A. Students can then plot point B' or point C', determine the direction of the translation, and then apply it to point D to find the coordinates of point D'.

Example 3 Dilation

$\triangle JKL$ has vertices $J(-2, -3)$, $K(-5, 4)$, and $L(3, 2)$. Dilate $\triangle JKL$ so that its perimeter is one-half the original perimeter. What are the coordinates of the vertices of $\triangle J'K'L'$?

If the perimeter of a figure is one-half the original perimeter, then the lengths of the sides of the figure will be one-half the measure of the original lengths. Multiply the vertex matrix by the scale factor of $\frac{1}{2}$.

$$\frac{1}{2}\begin{bmatrix} -2 & -5 & 3 \\ -3 & 4 & 2 \end{bmatrix} = \begin{bmatrix} -1 & -\frac{5}{2} & \frac{3}{2} \\ -\frac{3}{2} & 2 & 1 \end{bmatrix}$$

The coordinates of the vertices of $\triangle J'K'L'$ are $J'\left(-1, -\frac{3}{2}\right)$, $K'\left(-\frac{5}{2}, 2\right)$, and $L'\left(\frac{3}{2}, 1\right)$.

Graph $\triangle JKL$ and $\triangle J'K'L'$. The triangles are not congruent. The image has sides that are half the length of those of the original figure.

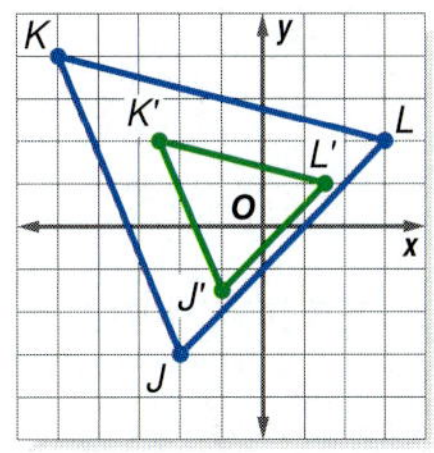

REFLECTIONS AND ROTATIONS In addition to translations, reflections and rotations are also isometries. A **reflection** occurs when every point of a figure is mapped to a corresponding image across a line of symmetry using a *reflection matrix*. The matrices used for three common reflections are shown below.

Concept Summary *Reflection Matrices*

For a reflection over the:	*x*-axis	*y*-axis	line $y = x$
Multiply the vertex matrix on the left by:	$\begin{bmatrix} 1 & 0 \\ 0 & -1 \end{bmatrix}$	$\begin{bmatrix} -1 & 0 \\ 0 & 1 \end{bmatrix}$	$\begin{bmatrix} 0 & 1 \\ 1 & 0 \end{bmatrix}$

Example 4 Reflection

Find the coordinates of the vertices of the image of pentagon $QRSTU$ with $Q(1, 3)$, $R(3, 2)$, $S(3, -1)$, $T(1, -2)$, and $U(-1, 1)$ after a reflection across the y-axis.

Write the ordered pairs as a vertex matrix. Then multiply the vertex matrix by the reflection matrix for the y-axis.

$$\begin{bmatrix} -1 & 0 \\ 0 & 1 \end{bmatrix} \cdot \begin{bmatrix} 1 & 3 & 3 & 1 & -1 \\ 3 & 2 & -1 & -2 & 1 \end{bmatrix} = \begin{bmatrix} -1 & -3 & -3 & -1 & 1 \\ 3 & 2 & -1 & -2 & 1 \end{bmatrix}$$

The coordinates of the vertices of $Q'R'S'T'U'$ are $Q'(-1, 3)$, $R'(-3, 2)$, $S'(-3, -1)$, $T'(-1, -2)$, and $U'(1, 1)$. Notice that the preimage and image are congruent. Both figures have the same size and shape.

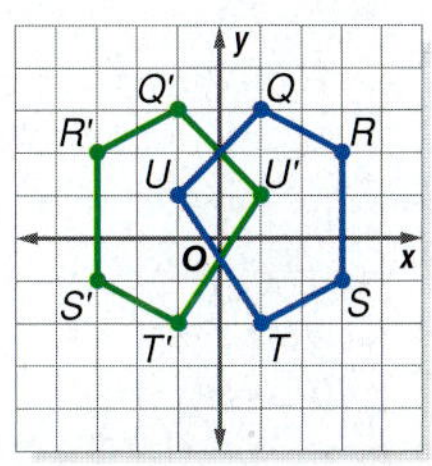

www.algebra2.com/extra_examples

In-Class Example Power Point®

3 $\triangle XYZ$ has vertices $X(1, 2)$, $Y(3, -1)$, and $Z(-1, -2)$. Dilate $\triangle XYZ$ so that its perimeter is twice the original perimeter. What are the coordinates of the vertices of $\triangle X'Y'Z'$? **$X'(2, 4)$, $Y'(6, -2)$, $Z'(-2, -4)$**

REFLECTIONS AND ROTATIONS

In-Class Example Power Point®

4 Find the coordinates of the vertices of the image of pentagon $PENTA$ with $P(-3, 1)$, $E(0, -1)$, $N(-1, -3)$, $T(-3, -4)$, and $A(-4, -1)$ after a reflection across the x-axis. **$P'(-3, -1)$, $E'(0, 1)$, $N'(-1, 3)$, $T'(-3, 4)$, $A'(-4, 1)$**

Teaching Tip Ask students to describe the pattern of coordinates for corresponding points after a reflection over either axis. **When a figure is reflected over the x-axis, the sign of the y-coordinate for each point on the figure changes. When a figure is reflected over the y-axis, the sign of the x-coordinate for each point on the figure changes.** Then ask them to describe the pattern after a reflection over the line $y = x$. **When a figure is reflected over the line $y = x$, the x- and y-coordinates of each point are transposed.**

DAILY INTERVENTION **Differentiated Instruction**

Visual/Spatial When students are drawing the figures and their images in a coordinate plane in this lesson, have them use different colored pencils for the original figure and its image.

In-Class Example

Power Point®

5 Find the coordinates of the vertices of the image of $\triangle DEF$ with $D(4, 3)$, $E(1, 1)$, and $F(2, 5)$ after it is rotated 90° counterclockwise about the origin. **$D'(-3, 4)$, $E'(-1, 1)$, and $F'(-5, 2)$**

3 Practice/Apply

Study Notebook

Have students—
- add the definitions/examples of the vocabulary terms in this lesson to their Vocabulary Builder worksheets for Chapter 4.
- write their own examples for translating, dilating, reflecting, and rotating polygons.
- add the reflection matrices for reflections over the x-axis, the y-axis, and the line y = x, and the rotation matrices for counterclockwise rotations of 90°, 180°, and 270° about the origin.
- include any other item(s) that they find helpful in mastering the skills in this lesson.

A **rotation** occurs when a figure is moved around a center point, usually the origin. To determine the vertices of a figure's image by rotation, multiply its vertex matrix by a *rotation matrix*. Commonly used rotation matrices are summarized below.

Concept Summary — *Rotation Matrices*

For a counterclockwise rotation about the origin of:	90°	180°	270°
Multiply the vertex matrix on the left by:	$\begin{bmatrix} 0 & -1 \\ 1 & 0 \end{bmatrix}$	$\begin{bmatrix} -1 & 0 \\ 0 & -1 \end{bmatrix}$	$\begin{bmatrix} 0 & 1 \\ -1 & 0 \end{bmatrix}$

Example 5 *Rotation*

Find the coordinates of the vertices of the image of $\triangle ABC$ with $A(4, 3)$, $B(2, 1)$, and $C(1, 5)$ after it is rotated 90° counterclockwise about the origin.

Write the ordered pairs in a vertex matrix. Then multiply the vertex matrix by the rotation matrix.

$$\begin{bmatrix} 0 & -1 \\ 1 & 0 \end{bmatrix} \cdot \begin{bmatrix} 4 & 2 & 1 \\ 3 & 1 & 5 \end{bmatrix} = \begin{bmatrix} -3 & -1 & -5 \\ 4 & 2 & 1 \end{bmatrix}$$

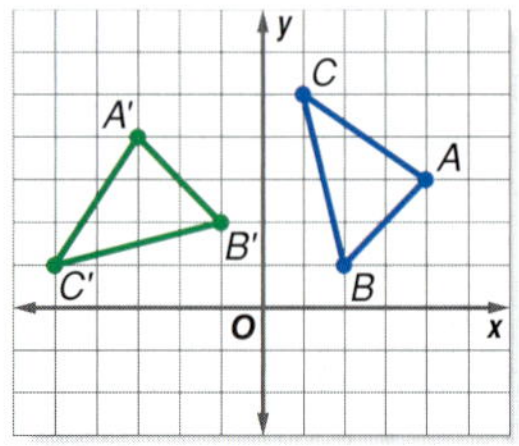

The coordinates of the vertices of $\triangle A'B'C'$ are $A'(-3, 4)$, $B'(-1, 2)$, and $C'(-5, 1)$. The image is congruent to the preimage.

Check for Understanding

Concept Check

1. **Compare and contrast** the size and shape of the preimage and image for each type of transformation. Tell which transformations are isometries. **See margin.**

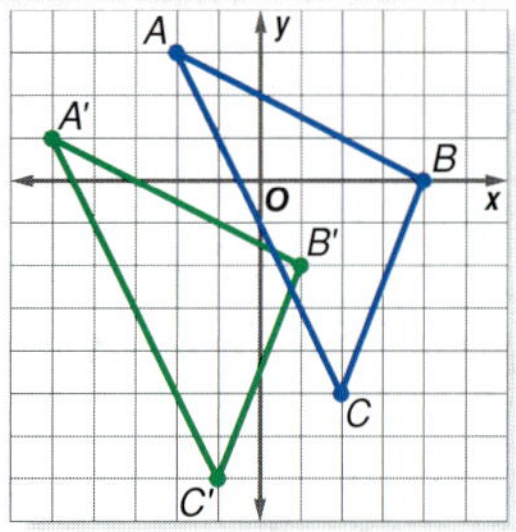

2. **Write** the translation matrix for $\triangle ABC$ and its image $\triangle A'B'C'$ shown at the right.
3. **OPEN ENDED** Write a translation matrix that moves $\triangle DEF$ up and left on the coordinate plane. **Sample answer: $\begin{bmatrix} -4 & -4 & -4 \\ 1 & 1 & 1 \end{bmatrix}$**

2. $\begin{bmatrix} -3 & -3 & -3 \\ -2 & -2 & -2 \end{bmatrix}$

7. $\begin{bmatrix} 0 & 5 & 5 & 0 \\ 4 & 4 & 0 & 0 \end{bmatrix}$

Guided Practice

Triangle ABC with vertices $A(1, 4)$, $B(2, -5)$, and $C(-6, -6)$ is translated 3 units right and 1 unit down.

4. Write the translation matrix. **$\begin{bmatrix} 3 & 3 & 3 \\ -1 & -1 & -1 \end{bmatrix}$**
5. Find the coordinates of $\triangle A'B'C'$. **$A'(4, 3)$, $B'(5, -6)$, $C'(-3, -7)$**
6. Graph the preimage and the image. **See margin.**

For Exercises 7–10, use the rectangle at the right.

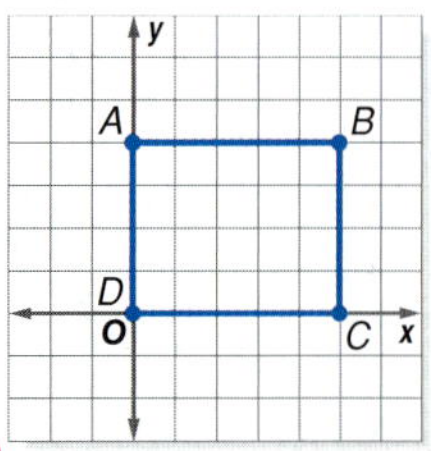

7. Write the coordinates in a vertex matrix.
8. Find the coordinates of the image after a dilation by a scale factor of 3.
9. Find the coordinates of the image after a reflection over the x-axis.
10. Find the coordinates of the image after a rotation of 180°. **$A'(0, -4)$, $B'(-5, -4)$, $C'(-5, 0)$, $D'(0, 0)$**

GUIDED PRACTICE KEY

Exercises	Examples
4–6	1
7, 8	3
9	4
10	5
11	2

8. $A'(0, 12)$, $B'(15, 12)$, $C'(15, 0)$, $D'(0, 0)$

9. $A'(0, -4)$, $B'(5, -4)$, $C'(5, 0)$, $D'(0, 0)$

Answers

1.

Transformation	Size	Shape	Isometry
reflection	same	same	yes
rotation	same	same	yes
translation	same	same	yes
dilation	changes	same	no

6.

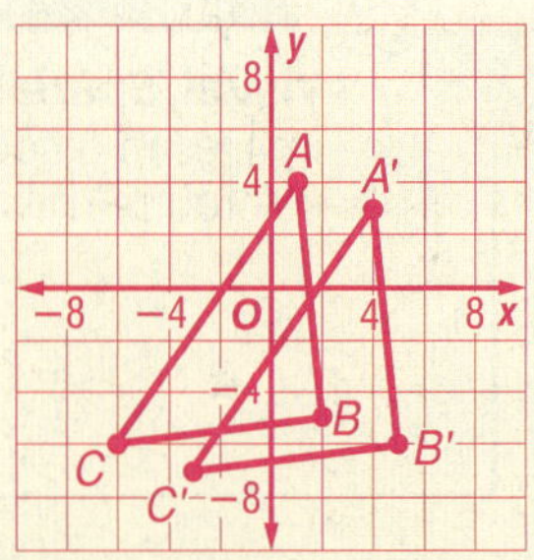

Standardized Test Practice

11. A point is translated from *B* to *C* as shown at the right. If a point at (−4, 3) is translated in the same way, what will be its new coordinates? **B**

Ⓐ (3, 4) Ⓑ (1, 1)
Ⓒ (−7, 8) Ⓓ (1, 6)

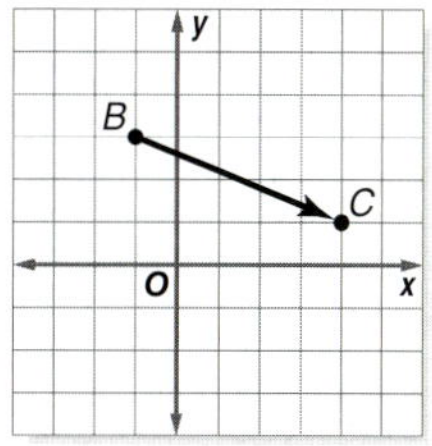

★ indicates increased difficulty

Practice and Apply

Homework Help

For Exercises	See Examples
12–14, 35, 36	1
15–17, 33, 34	3
18–20	4
21–23, 25, 37	5
24	2
26–32, 38–41	1–5

Extra Practice
See page 835.

For Exercises 12–14, use the following information.
Triangle *DEF* with vertices *D*(1, 4), *E*(2, −5), and *F*(−6, −6) is translated 4 units left and 2 units up.

12. Write the translation matrix. $\begin{bmatrix} -4 & -4 & -4 \\ 2 & 2 & 2 \end{bmatrix}$
13. Find the coordinates of $\triangle D'E'F'$. **D′(−3, 6), E′(−2, −3), F′(−10, −4)**
14. Graph the preimage and the image. **See margin.**

For Exercises 15–17, use the following information.
The vertices of $\triangle ABC$ are *A*(0, 2), *B*(1.5, −1.5), and *C*(−2.5, 0). The triangle is dilated so that its perimeter is three times the original perimeter.

15. Write the coordinates for $\triangle ABC$ in a vertex matrix.
16. Find the coordinates of the image $\triangle A'B'C'$. **A′(0, 6), B′(4.5, −4.5), C′(−7.5, 0)**
17. Graph $\triangle ABC$ and $\triangle A'B'C'$. **See margin.**

15. $\begin{bmatrix} 0 & 1.5 & -2.5 \\ 2 & -1.5 & 0 \end{bmatrix}$

For Exercises 18–20, use the following information.
The vertices of $\triangle XYZ$ are *X*(1, −1), *Y*(2, −4), and *Z*(7, −1). The triangle is reflected over the line $y = x$.

18. Write the coordinates of $\triangle XYZ$ in a vertex matrix. $\begin{bmatrix} 1 & 2 & 7 \\ -1 & -4 & -1 \end{bmatrix}$
19. Find the coordinates of $\triangle X'Y'Z'$. **X′(−1, 1), Y′(−4, 2), Z′(−1, 7)**
20. Graph $\triangle XYZ$ and $\triangle X'Y'Z'$. **See margin.**

For Exercises 21–23, use the following information.
Parallelogram *DEFG* with *D*(2, 4), *E*(5, 4), *F*(4, 1), and *G*(1, 1) is rotated 270° counterclockwise about the origin.

21. Write the coordinates of the parallelogram in a vertex matrix. $\begin{bmatrix} 2 & 5 & 4 & 1 \\ 4 & 4 & 1 & 1 \end{bmatrix}$
22. Find the coordinates of parallelogram *D′E′F′G′*.
23. Graph the preimage and the image. **See margin.**

22. **D′(4, −2), E′(4, −5), F′(1, −4), G′(1, −1)**

★ 24. Triangle *DEF* with vertices *D*(−2, 2), *E*(3, 5), and *F*(5, −2) is translated so that *D′* is at (1, −5). Find the coordinates of *E′* and *F′*. **E′(6, −2), F′(8, −9)**

★ 25. A triangle is rotated 90° counterclockwise about the origin. The coordinates of the vertices are *J′*(−3, −5), *K′*(−2, 7), and *L′*(1, 4). What were the coordinates of the triangle in its original position? **J(−5, 3), K(7, 2), L(4, −1)**

For Exercises 26–28, use quadrilateral *QRST* shown at the right.

26. Write the vertex matrix. Multiply the vertex matrix by −1.
27. Graph the preimage and image. **See right.**
28. What type of transformation does the graph represent? **180° rotation**

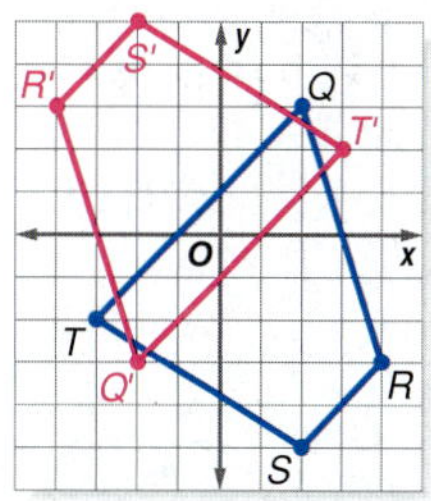

26. $\begin{bmatrix} 2 & 4 & 2 & -3 \\ 3 & -3 & -5 & -2 \end{bmatrix} \cdot (-1) = \begin{bmatrix} -2 & -4 & -2 & 3 \\ -3 & 3 & 5 & 2 \end{bmatrix}$

www.algebra2.com/self_check_quiz

About the Exercises...

Organization by Objective

- **Translations and Dilations:** 12–17, 24, 33–36
- **Reflections and Rotations:** 18–23, 25–32, 37

Assignment Guide

Basic: 12–23, 26–28, 33–36, 42–64

Average: 12–23, 25–32, 35–37, 42–64

Advanced: 12–24, 26–32, 38–58 (optional: 59–64)

Answers

14.

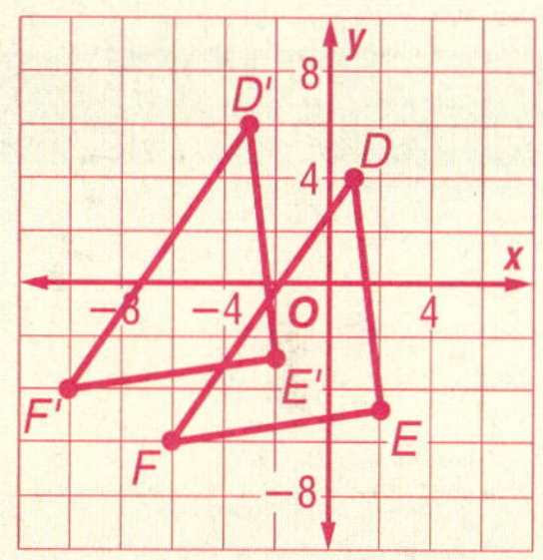

17.

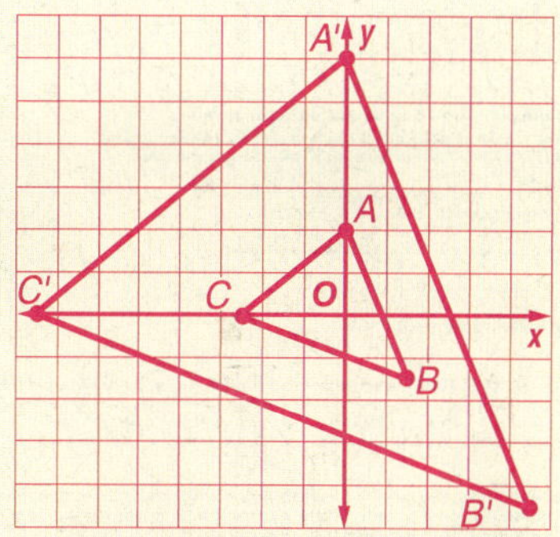

20.

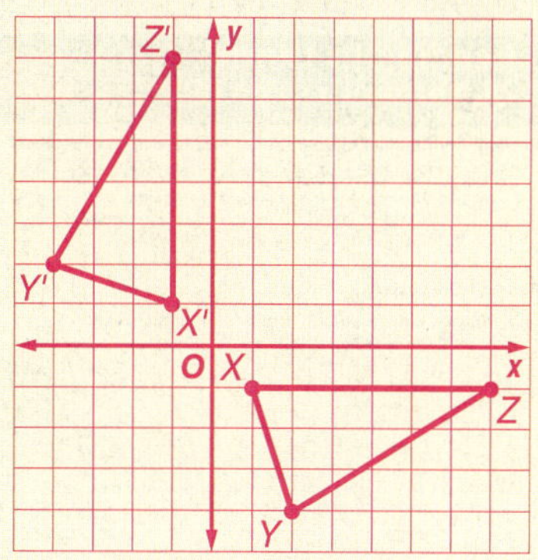

23.

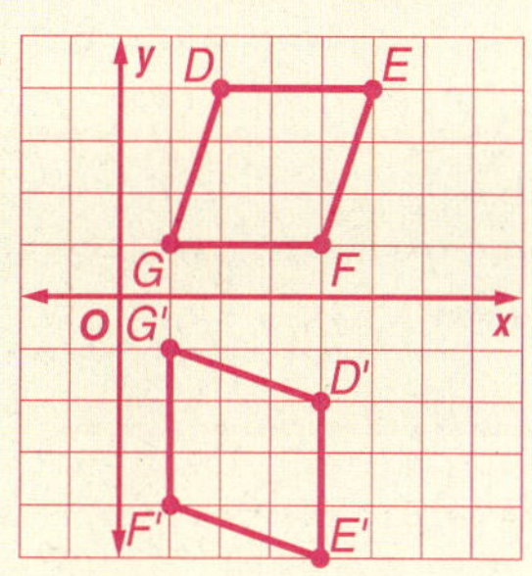

Study Guide and Intervention, p. 187 (shown) and p. 188

Translations and Dilations Matrices that represent coordinates of points on a plane are useful in describing transformations.

Translation	a transformation that moves a figure from one location to another on the coordinate plane

You can use matrix addition and a translation matrix to find the coordinates of the translated figure.

Dilation	a transformation in which a figure is enlarged or reduced

You can use scalar multiplication to perform dilations.

Example **Find the coordinates of the vertices of the image of $\triangle ABC$ with vertices $A(-5, 4)$, $B(-1, 5)$, and $C(-3, -1)$ if it is moved 6 units to the right and 4 units down. Then graph $\triangle ABC$ and its image $\triangle A'B'C'$.**

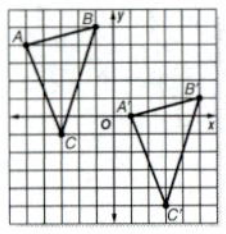

Write the vertex matrix for $\triangle ABC$. $\begin{bmatrix} -5 & -1 & -3 \\ 4 & 5 & -1 \end{bmatrix}$

Add the translation matrix $\begin{bmatrix} 6 & 6 & 6 \\ -4 & -4 & -4 \end{bmatrix}$ to the vertex matrix of $\triangle ABC$.

$\begin{bmatrix} -5 & -1 & -3 \\ 4 & 5 & -1 \end{bmatrix} + \begin{bmatrix} 6 & 6 & 6 \\ -4 & -4 & -4 \end{bmatrix} = \begin{bmatrix} 1 & 5 & 3 \\ 0 & 1 & -5 \end{bmatrix}$

The coordinates of the vertices of $\triangle A'B'C'$ are $A'(1, 0)$, $B'(5, 1)$, and $C'(3, -5)$.

Exercises

For Exercises 1 and 2 use the following information. Quadrilateral *QUAD* with vertices $Q(-1, -3)$, $U(0, 0)$, $A(5, -1)$, and $D(2, -5)$ is translated 3 units to the left and 2 units up.

1. Write the translation matrix. $\begin{bmatrix} -3 & -3 & -3 & -3 \\ 2 & 2 & 2 & 2 \end{bmatrix}$

2. Find the coordinates of the vertices of $Q'U'A'D'$.
$Q'(-4, -1)$, U', $(-3, 2)$, $A'(2, 1)$, $D'(-1, -3)$

For Exercises 3–5, use the following information. The vertices of $\triangle ABC$ are $A(4, -2)$, $B(2, 8)$, and $C(8, 2)$. The triangle is dilated so that its perimeter is one-fourth the original perimeter.

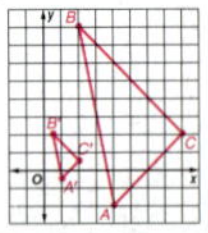

3. Write the coordinates of the vertices of $\triangle ABC$ in a vertex matrix. $\begin{bmatrix} 4 & 2 & 8 \\ -2 & 8 & 2 \end{bmatrix}$

4. Find the coordinates of the vertices of image $\triangle A'B'C'$.
$A'\left(1, -\frac{1}{2}\right)$, $B'\left(\frac{1}{2}, 2\right)$, $C'\left(2, \frac{1}{2}\right)$

5. Graph the preimage and the image.

Skills Practice, p. 189 and Practice, p. 190 (shown)

For Exercises 1–3, use the following information.

Quadrilateral *WXYZ* with vertices $W(-3, 2)$, $X(-2, 4)$, $Y(4, 1)$, and $Z(3, 0)$ is translated 1 unit left and 3 units down.

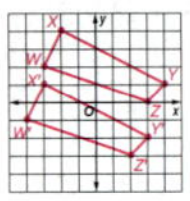

1. Write the translation matrix. $\begin{bmatrix} -1 & -1 & -1 & -1 \\ -3 & -3 & -3 & -3 \end{bmatrix}$

2. Find the coordinates of quadrilateral $W'X'Y'Z'$.
$W'(-4, -1)$, $X'(-3, 1)$, $Y'(3, -2)$, $Z'(2, -3)$

3. Graph the preimage and the image.

For Exercises 4–6, use the following information.

The vertices of $\triangle RST$ are $R(6, 2)$, $S(3, -3)$, and $T(-2, 5)$. The triangle is dilated so that its perimeter is one half the original perimeter.

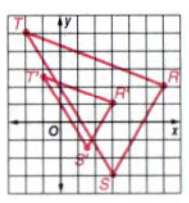

4. Write the coordinates of $\triangle RST$ in a vertex matrix. $\begin{bmatrix} 6 & 3 & -2 \\ 2 & -3 & 5 \end{bmatrix}$

5. Find the coordinates of the image $\triangle R'S'T'$.
$R'(3, 1)$, $S'(1.5, -1.5)$, $T'(-1, 2.5)$

6. Graph $\triangle RST$ and $\triangle R'S'T'$.

For Exercises 7–10, use the following information.

The vertices of quadrilateral *ABCD* are $A(-3, 2)$, $B(0, 3)$, $C(4, -4)$, and $D(-2, -2)$. The quadrilateral is reflected over the y-axis.

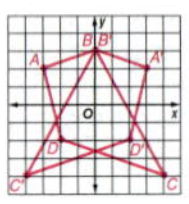

7. Write the coordinates of *ABCD* in a vertex matrix. $\begin{bmatrix} -3 & 0 & 4 & -2 \\ 2 & 3 & -4 & -2 \end{bmatrix}$

8. Write the reflection matrix for this situation. $\begin{bmatrix} -1 & 0 \\ 0 & 1 \end{bmatrix}$

9. Find the coordinates of $A'B'C'D'$.
$A'(3, 2)$, $B'(0, 3)$, $C'(-4, -4)$, $D'(2, -2)$

10. Graph *ABCD* and $A'B'C'D'$.

11. **ARCHITECTURE** Using architectural design software, the Bradleys plot their kitchen plans on a grid with each unit representing 1 foot. They place the corners of an island at (2, 8), (8, 11), (3, 5), and (9, 8). If the Bradleys wish to move the island 1.5 feet to the right and 2 feet down, what will the new coordinates of its corners be?
(3.5, 6), (9.5, 9), (4.5, 3), and (10.5, 6)

12. **BUSINESS** The design of a business logo calls for locating the vertices of a triangle at (1.5, 5), (4, 1), and (1, 0) on a grid. If design changes require rotating the triangle 90° counterclockwise, what will the new coordinates of the vertices be?
(−5, 1.5), (−1, 4), and (0, 1)

Reading to Learn Mathematics, p. 191 ELL

Pre-Activity **How are transformations used in computer animation?**

Read the introduction to Lesson 4-4 at the top of page 175 in your textbook.

Describe how you can change the orientation of a figure without changing its size or shape.

Flip (or reflect) the figure over a line.

Reading the Lesson

1. a. Write the vertex matrix for the quadrilateral *ABCD* shown in the graph at the right.
$\begin{bmatrix} -4 & 2 & 1 & -2 \\ 1 & 3 & -4 & -3 \end{bmatrix}$

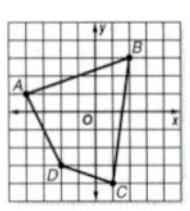

b. Write the vertex matrix that represents the position of the quadrilateral $A'B'C'D'$ that results when quadrilateral *ABCD* is translated 3 units to the right and 2 units down.
$\begin{bmatrix} -1 & 5 & 4 & 1 \\ -1 & 1 & -6 & -5 \end{bmatrix}$

2. Describe the transformation that corresponds to each of the following matrices.

a. $\begin{bmatrix} -1 & 0 \\ 0 & -1 \end{bmatrix}$ counterclockwise rotation about the origin of 180°

b. $\begin{bmatrix} 3 & 3 & 3 \\ -4 & -4 & -4 \end{bmatrix}$ translation 4 units down and 3 units to the right

c. $\begin{bmatrix} -1 & 0 \\ 0 & 1 \end{bmatrix}$ reflection over the y-axis

d. $\begin{bmatrix} 0 & 1 \\ 1 & 0 \end{bmatrix}$ reflection over the line $y = x$

Helping You Remember

3. Describe a way to remember which of the reflection matrices corresponds to reflection over the x-axis.
Sample answer: The only elements used in the reflection matrices are 0, 1, and −1. For such a 2 × 2 matrix M to have the property that $M \cdot \begin{bmatrix} x \\ y \end{bmatrix} = \begin{bmatrix} x \\ -y \end{bmatrix}$, the elements in the top row must be 1 and 0 (in that order), and elements in the bottom row must be 0 and −1 (in that order).

29. $\begin{bmatrix} 4 & -4 & -4 & 4 \\ -4 & -4 & 4 & 4 \end{bmatrix}$

30. $\begin{bmatrix} 4 & -4 & -4 & 4 \\ -4 & -4 & 4 & 4 \end{bmatrix}$

31. $\begin{bmatrix} 4 & 4 & -4 & -4 \\ -4 & 4 & 4 & -4 \end{bmatrix}$

For Exercises 29–32, use rectangle *ABCD* with vertices $A(-4, 4)$, $B(4, 4)$, $C(4, -4)$, and $D(-4, -4)$.

★ 29. Find the coordinates of the image in matrix form after a reflection over the x-axis followed by a reflection over the y-axis.

30. Find the coordinates of the image after a 180° rotation about the origin.

31. Find the coordinates of the image after a reflection over the line $y = x$.

32. What do you observe about these three matrices? Explain.

LANDSCAPING **For Exercises 33 and 34, use the following information.**
A garden design is plotted on a coordinate grid. The original plan shows a fountain with vertices at (−2, −2), (−6, −2), (−8, −5), and (−4, −5). Changes to the plan now require that the fountain's perimeter be three-fourths that of the original.

33. Determine the new coordinates for the fountain.

34. The center of the fountain was at (−5, −3.5). What will be the coordinate of the center after the changes in the plan have been made? **(−3.75, −2.625)**

More About. . .

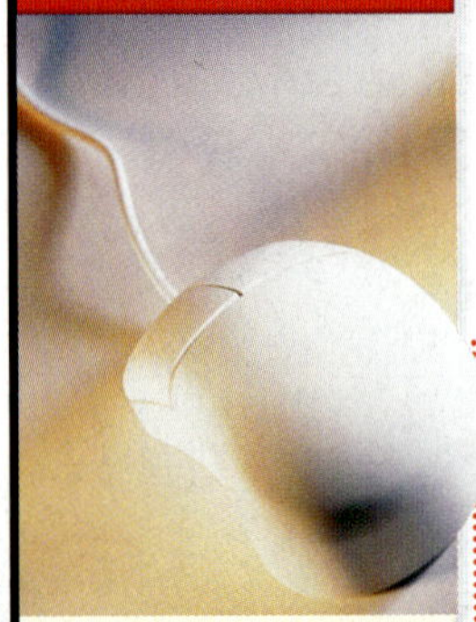

Technology

Douglas Engelbart invented the "X-Y position indicator for a display system" in 1964. He nicknamed this invention "the mouse" because a tail came out the end.

Source: www.about.com

TECHNOLOGY **For Exercises 35 and 36, use the following information.**
As you move the mouse for your computer, a corresponding arrow is translated on the screen. Suppose the position of the cursor on the screen is given in inches with the origin at the bottom left-hand corner of the screen. **35. See margin.**

35. You want to move your cursor 3 inches to the right and 4 inches up. Write a translation matrix that can be used to move the cursor to the new position.

36. If the cursor is currently at (3.5, 2.25), what are the coordinates of the position after the translation? **(6.5, 6.25)**

37. **GYMNASTICS** The drawing at the right shows four positions of a man performing the giant swing in the high bar event. Suppose this drawing is placed on a coordinate grid with the hand grips at $H(0, 0)$ and the toe of the figure in the upper right corner at $T(7, 8)$. Find the coordinates of the toes of the other three figures, if each successive figure has been rotated 90° counterclockwise about the origin.
(−8, 7), (−7, −8), and (8, −7)

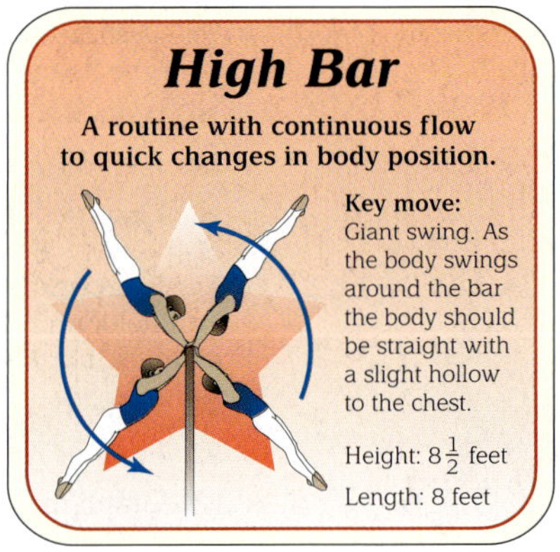

32. The figures in Exercise 29 and Exercise 30 have the same coordinates, but the figure in Exercise 31 has different coordinates.

33. (−1.5, −1.5), (−4.5, −1.5), (−6, −3.75), (−3, −3.75)

38. The object is reflected over the x-axis, then translated 6 units to the right.

39. Multiply the coordinates by $\begin{bmatrix} 1 & 0 \\ 0 & -1 \end{bmatrix}$, then add the result to $\begin{bmatrix} 6 \\ 0 \end{bmatrix}$.

FOOTPRINTS **For Exercises 38–41, use the following information.**
The combination of a reflection and a translation is called a *glide reflection*. An example is a set of footprints.

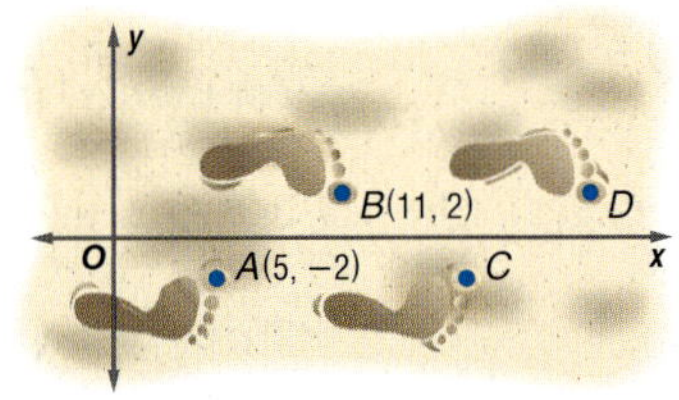

★ 38. Describe the reflection and transformation combination shown at the right.

★ 39. Write two matrix operations that can be used to find the coordinates of point C.

★ 40. Does it matter which operation you do first? Explain. **See margin.**

★ 41. What are the coordinates of the next two footprints? **(17, −2), (23, 2)**

Enrichment, p. 192

Communications Networks

The diagram at the right represents a communications network linking five computer remote stations. The arrows indicate the direction in which signals can be transmitted and received by each computer. We can generate a matrix to describe this network.

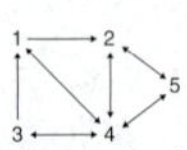

$$A = \underset{\text{computer } i}{\text{from}} \overset{\text{to computer } j}{\begin{bmatrix} 0 & 1 & 0 & 1 & 0 \\ 0 & 0 & 0 & 1 & 1 \\ 1 & 0 & 0 & 1 & 0 \\ 1 & 1 & 1 & 0 & 1 \\ 0 & 1 & 0 & 1 & 0 \end{bmatrix}}$$

The entry in position a_{ij} represents the number of ways to send a message from computer i to computer j directly. Compare the entries of matrix A to the diagram to verify the entries. For example, there is one way to send a message from computer 3 to computer 4, so $A_{3,4} = 1$. A computer cannot send a message to itself, so $A_{1,1} = 0$.

Matrix A is a communications network for direct communication. Suppose you want to send a message from one computer to another using exactly one other computer as a relay point. It can be shown that the entries of matrix A^2 represent the number of ways to send a message from one point to another by going through a third station. For example, a message may be sent from station 1 to station 5 by going through station 2 or station 4 on the way. Therefore, $A^2_{1,5} = 2$.

Answer

35. $\begin{bmatrix} 3 \\ 4 \end{bmatrix}$

42. There is no single matrix to achieve this. However, you could reflect the object over the y-axis and then translate it 2(3) or 6 units to the right.

42. **CRITICAL THINKING** Do you think a matrix exists that would represent a reflection over the line $x = 3$? If so, make a conjecture and verify it.

43. **WRITING IN MATH** Answer the question that was posed at the beginning of the lesson. See pp. 217A–217B.

How are transformations used in computer animation?

Include the following in your answer:

- an example of how a figure with 5 points (coordinates) could be written in a matrix and multiplied by a rotation matrix, and
- a description of the motion that is a result of repeated dilations with a scale factor of one-fourth.

Standardized Test Practice

44. Which matrix represents a reflection over the y-axis followed by a reflection over the x-axis? **B**

(A) $\begin{bmatrix} 0 & -1 \\ -1 & 0 \end{bmatrix}$ (B) $\begin{bmatrix} -1 & 0 \\ 0 & -1 \end{bmatrix}$ (C) $\begin{bmatrix} -1 & -1 \\ -1 & -1 \end{bmatrix}$ (D) none of these

45. Triangle ABC has vertices with coordinates $A(-4, 2)$, $B(-4, -3)$, and $C(3, -2)$. After a dilation, triangle $A'B'C'$ has coordinates $A'(-12, 6)$, $B'(-12, -9)$, and $C'(9, -6)$. How many times as great is the perimeter of $A'B'C'$ as ABC? **A**

(A) 3 (B) 6 (C) 12 (D) $\frac{1}{3}$

Maintain Your Skills

Mixed Review

Determine whether each matrix product is defined. If so, state the dimensions of the product. *(Lesson 4-3)*

46. $A_{2 \times 3} \cdot B_{3 \times 2}$ **2 × 2**
47. $A_{4 \times 1} \cdot B_{2 \times 1}$ **undefined**
48. $A_{2 \times 5} \cdot B_{5 \times 5}$ **2 × 5**

Perform the indicated matrix operations. If the matrix does not exist, write *impossible*. *(Lesson 4-2)*

49. $2\begin{bmatrix} 4 & 9 & -8 \\ 6 & -11 & -2 \\ 12 & -10 & 3 \end{bmatrix} + 3\begin{bmatrix} 1 & 2 & 3 \\ 2 & 3 & 4 \\ 3 & 4 & 5 \end{bmatrix}$

50. $4\begin{bmatrix} 3 & 4 & -7 \\ 6 & -9 & -2 \\ -3 & 1 & 3 \end{bmatrix} - \begin{bmatrix} -8 & 6 & -4 \\ -7 & 10 & 1 \\ -2 & 1 & 5 \end{bmatrix}$

49. $\begin{bmatrix} 11 & 24 & -7 \\ 18 & -13 & 8 \\ 33 & -8 & 21 \end{bmatrix}$

50. $\begin{bmatrix} 20 & 10 & -24 \\ 31 & -46 & -9 \\ -10 & 3 & 7 \end{bmatrix}$

Graph each relation or equation and find the domain and range. Then determine whether the relation or equation is a function. *(Lesson 2-1)*

51. (3, 5), (4, 6), (5, −4)
52. $x = -5y + 2$
53. $x = y^2$

51–53. See margin for graphs.

51. D = {3, 4, 5}, R = {−4, 5, 6}; yes
52. D = {all real numbers}, R = {all real numbers}; yes
53. D = $\{x \mid x \geq 0\}$, R = {all real numbers}; no

Write an absolute value inequality for each graph. *(Lesson 1-6)*

54. 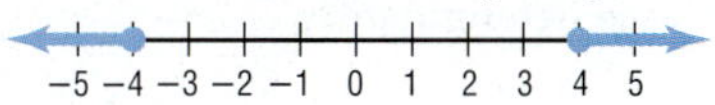

55. 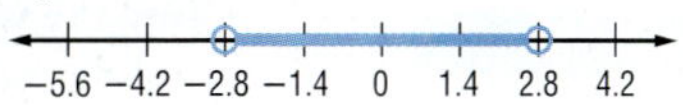

56. 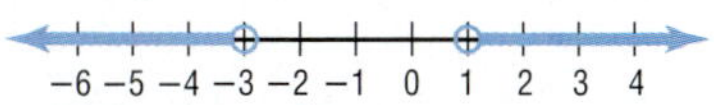

$|x + 1| > 2$

57. 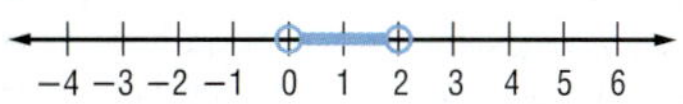

$|x - 1| < 1$

54. $|x| \geq 4$
55. $|x| < 2.8$

58. **BUSINESS** Reliable Rentals rents cars for \$12.95 per day plus 15¢ per mile. Luis Romero works for a company that limits expenses for car rentals to \$90 per day. What is the maximum number of miles that Mr. Romero can drive each day? *(Lesson 1-5)* **$513\frac{2}{3}$ mi**

Getting Ready for the Next Lesson

BASIC SKILL Use cross products to solve each proportion.

59. $\frac{x}{8} = \frac{3}{4}$ **6**
60. $\frac{4}{20} = \frac{1}{m}$ **5**
61. $\frac{2}{3} = \frac{a}{42}$ **28**
62. $\frac{5}{6} = \frac{k}{4}$ **$\frac{10}{3}$**
63. $\frac{2}{y} = \frac{8}{9}$ **$\frac{9}{4}$**
64. $\frac{x}{5} = \frac{x+1}{8}$ **$\frac{5}{3}$**

4 Assess

Open-Ended Assessment

Writing Have students write steps or procedures that they can use to translate, dilate, reflect, and rotate a polygon using matrices as described in this lesson.

Assessment Options

Quiz (Lessons 4-3 and 4-4) is available on p. 231 of the *Chapter 4 Resource Masters.*

Mid-Chapter Test (Lessons 4-1 through 4-4) is available on p. 233 of the *Chapter 4 Resource Masters.*

Getting Ready for Lesson 4-5

BASIC SKILL Lesson 4-5 shows how to evaluate the determinant of a 2 × 2 matrix and the determinant of a 3 × 3 matrix. Students will use their familiarity with finding cross products as they compute the determinant. Exercises 59–64 should be used to determine your students' familiarity with using cross products.

Answers

40. No; since the translation does not change the y-coordinate, it does not matter whether you do the translation or the reflection over the x-axis first. However, if the translation did change the y-coordinate, then order would be important.

51. 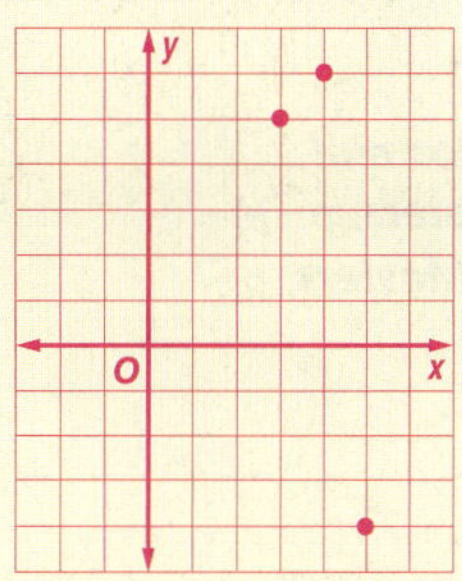

52. 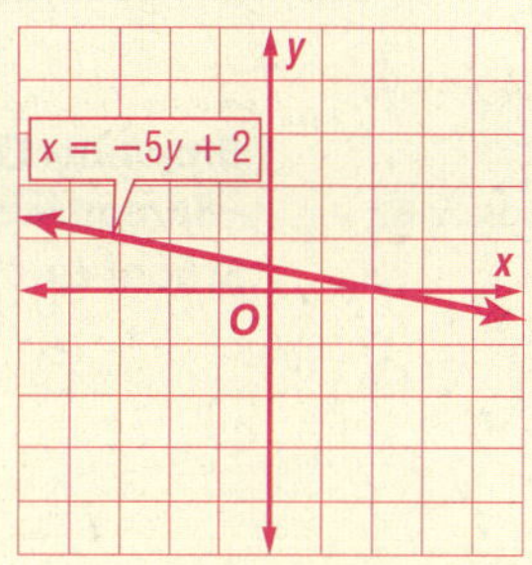

53. 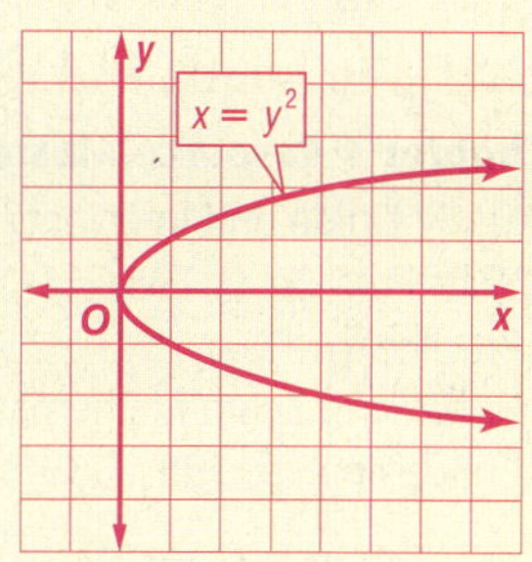

4-5 Lesson Notes

1 Focus

5-Minute Check Transparency 4-5 Use as a quiz or review of Lesson 4-4.

Mathematical Background notes are available for this lesson on p. 152D.

How are determinants used to find areas of polygons?

Ask students:

- What method do you already know for finding the area of a triangle? **the formula $A = \frac{1}{2}bh$**
- Why would it be difficult to use this method in this situation? **The height of the triangle would be difficult to find.**

4-5 Determinants

What You'll Learn

- Evaluate the determinant of a 2×2 matrix.
- Evaluate the determinant of a 3×3 matrix.

Vocabulary

- determinant
- second-order determinant
- third-order determinant
- expansion by minors
- minor

How are determinants used to find areas of polygons?

The "Bermuda Triangle" is an area located off the southeastern Atlantic coast of the United States that is noted for a high incidence of unexplained losses of ships, small boats, and aircraft. You can estimate the area of this triangular region by finding the determinant of the matrix that contains the coordinates of the vertices of the triangle.

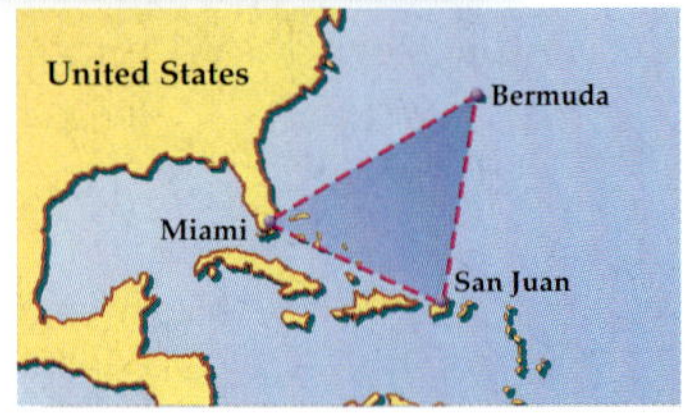

DETERMINANTS OF 2 × 2 MATRICES Every square matrix has a number associated with it called its determinant. A **determinant** is a square array of numbers or variables enclosed between two parallel lines. For example, the determinant of $\begin{bmatrix} 3 & -1 \\ 2 & 5 \end{bmatrix}$ can be represented by $\begin{vmatrix} 3 & -1 \\ 2 & 5 \end{vmatrix}$ or $\det \begin{bmatrix} 3 & -1 \\ 2 & 5 \end{bmatrix}$. The determinant of a 2×2 matrix is called a **second-order determinant**.

Key Concept *Second-Order Determinant*

- **Words** The value of a second-order determinant is found by calculating the difference of the products of the two diagonals.
- **Symbols** $\begin{vmatrix} a & b \\ c & d \end{vmatrix} = ad - bc$

Study Tip

Reading Math
The term *determinant* is often used to mean the *value* of the determinant.

Example 1 *Second-Order Determinant*

Find the value of each determinant.

a. $\begin{vmatrix} -2 & 5 \\ 6 & 8 \end{vmatrix}$

$\begin{vmatrix} -2 & 5 \\ 6 & 8 \end{vmatrix} = (-2)(8) - 5(6)$ Definition of determinant

$= -16 - 30$ Multiply.

$= -46$ Simplify.

b. $\begin{vmatrix} 7 & 4 \\ -3 & 2 \end{vmatrix}$

$\begin{vmatrix} 7 & 4 \\ -3 & 2 \end{vmatrix} = (7)(2) - 4(-3)$ Definition of determinant

$= 14 - (-12)$ Multiply.

$= 26$ Simplify.

Resource Manager

Workbook and Reproducible Masters

Chapter 4 Resource Masters
- Study Guide and Intervention, pp. 193–194
- Skills Practice, p. 195
- Practice, p. 196
- Reading to Learn Mathematics, p. 197
- Enrichment, p. 198

Graphing Calculator and Spreadsheet Masters, p. 34
School-to-Career Masters, p. 7

Transparencies
5-Minute Check Transparency 4-5
Answer Key Transparencies

Technology
Interactive Chalkboard

DETERMINANTS OF 3 × 3 MATRICES Determinants of 3 × 3 matrices are called **third-order determinants**. One method of evaluating third-order determinants is **expansion by minors**. The **minor** of an element is the determinant formed when the row and column containing that element are deleted.

$\begin{vmatrix} a_1 & b_1 & c_1 \\ a_2 & b_2 & c_2 \\ a_3 & b_3 & c_3 \end{vmatrix}$ The minor of a_1 is $\begin{vmatrix} b_2 & c_2 \\ b_3 & c_3 \end{vmatrix}$.

$\begin{vmatrix} a_1 & b_1 & c_1 \\ a_2 & b_2 & c_2 \\ a_3 & b_3 & c_3 \end{vmatrix}$ The minor of b_1 is $\begin{vmatrix} a_2 & c_2 \\ a_3 & c_3 \end{vmatrix}$.

$\begin{vmatrix} a_1 & b_1 & c_1 \\ a_2 & b_2 & c_2 \\ a_3 & b_3 & c_3 \end{vmatrix}$ The minor of c_1 is $\begin{vmatrix} a_2 & b_2 \\ a_3 & b_3 \end{vmatrix}$.

To use expansion by minors with third-order determinants, each member of one row is multiplied by its minor and its *position sign*, and the results are added together. The position signs alternate between positive and negative, beginning with a positive sign in the first row, first column.

$$\begin{bmatrix} + & - & + \\ - & + & - \\ + & - & + \end{bmatrix}$$

Key Concept *Third-Order Determinant*

$$\begin{vmatrix} a & b & c \\ d & e & f \\ g & h & i \end{vmatrix} = a\begin{vmatrix} e & f \\ h & i \end{vmatrix} - b\begin{vmatrix} d & f \\ g & i \end{vmatrix} + c\begin{vmatrix} d & e \\ g & h \end{vmatrix}$$

The definition of third-order determinants shows an expansion using the elements in the first row of the determinant. However, any row can be used.

Example 2 Expansion by Minors

Evaluate $\begin{vmatrix} 2 & 7 & -3 \\ -1 & 5 & -4 \\ 6 & 9 & 0 \end{vmatrix}$ using expansion by minors.

Decide which row of elements to use for the expansion. For this example, we will use the first row.

$$\begin{vmatrix} 2 & 7 & -3 \\ -1 & 5 & -4 \\ 6 & 9 & 0 \end{vmatrix} = 2\begin{vmatrix} 5 & -4 \\ 9 & 0 \end{vmatrix} - 7\begin{vmatrix} -1 & -4 \\ 6 & 0 \end{vmatrix} + (-3)\begin{vmatrix} -1 & 5 \\ 6 & 9 \end{vmatrix}$$ Expansion by minors

$= 2(0 - (-36)) - 7(0 - (-24)) - 3(-9 - 30)$ Evaluate 2 × 2 determinants.

$= 2(36) - 7(24) - 3(-39)$

$= 72 - 168 + 117$ Multiply.

$= 21$ Simplify.

Another method for evaluating a third-order determinant is by using diagonals.

Step 1 Begin by writing the first two columns on the right side of the determinant.

$$\begin{vmatrix} a & b & c \\ d & e & f \\ g & h & i \end{vmatrix} \Rightarrow \begin{vmatrix} a & b & c \\ d & e & f \\ g & h & i \end{vmatrix}\begin{matrix} a & b \\ d & e \\ g & h \end{matrix}$$

(continued on the next page)

www.algebra2.com/extra_examples

2 Teach

DETERMINANTS OF 2 × 2 MATRICES

In-Class Example Power Point®

1 Find the value of each determinant.

a. $\begin{vmatrix} 6 & 4 \\ -1 & 0 \end{vmatrix}$ 4

b. $\begin{vmatrix} -6 & 7 \\ -9 & 3 \end{vmatrix}$ 45

DETERMINANTS OF 3 × 3 MATRICES

In-Class Example Power Point®

Teaching Tip When using expansion by minors for the first time, urge students to write down each step in the procedure. Then have them compare their work with a classmate to find any errors in either their calculations or their use of the procedure.

2 Evaluate $\begin{vmatrix} 1 & 0 & -1 \\ 2 & -1 & 3 \\ 4 & -2 & -3 \end{vmatrix}$ using expansion by minors. 9

DAILY INTERVENTION Unlocking Misconceptions

Not All Matrices Have a Determinant Discuss with students what the dimensions of a matrix must be in order for it to have a determinant. Stress that not every matrix has a determinant. Compare these restrictions to those relating to whether or not two matrices can be multiplied.

In-Class Example

3 Evaluate $\begin{vmatrix} 3 & -2 & -1 \\ 2 & -1 & 0 \\ 1 & 2 & -3 \end{vmatrix}$ using diagonals. **−8**

Step 2 Next, draw diagonals from each element of the top row of the determinant downward to the right. Find the product of the elements on each diagonal.

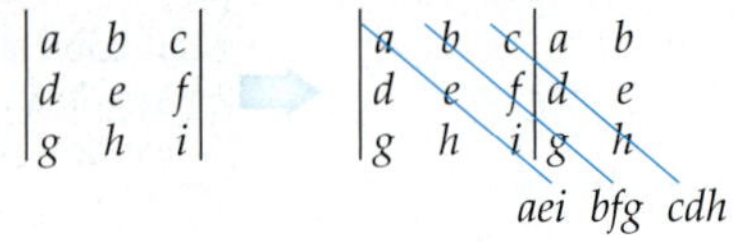

Then, draw diagonals from the elements in the third row of the determinant upward to the right. Find the product of the elements on each diagonal.

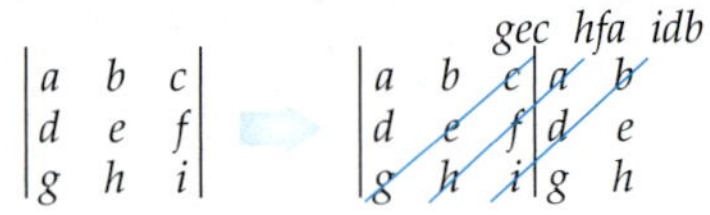

Step 3 To find the value of the determinant, add the products of the first set of diagonals and then subtract the products of the second set of diagonals. The sum is $aei + bfg + cdh - gec - hfa - idb$.

Example 3 Use Diagonals

Evaluate $\begin{vmatrix} -1 & 3 & -3 \\ 4 & -2 & -1 \\ 0 & -5 & 2 \end{vmatrix}$ using diagonals.

Step 1 Rewrite the first two columns to the right of the determinant.

$$\begin{vmatrix} -1 & 3 & -3 \\ 4 & -2 & -1 \\ 0 & -5 & 2 \end{vmatrix} \begin{matrix} -1 & 3 \\ 4 & -2 \\ 0 & -5 \end{matrix}$$

Step 2 Find the products of the elements of the diagonals.

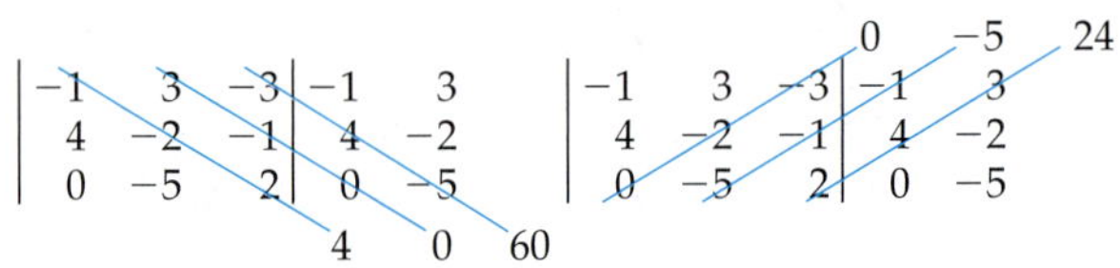

Step 3 Add the bottom products and subtract the top products.
$4 + 0 + 60 - 0 - (-5) - 24 = 45$

The value of the determinant is 45.

One very useful application of determinants is finding the areas of polygons. The formula below shows how determinants can be used to find the area of a triangle using the coordinates of the vertices.

Study Tip

Area Formula
Notice that it is necessary to use the absolute value of A to guarantee a nonnegative value for the area.

Key Concept — Area of a Triangle

The area of a triangle having vertices at (a, b), (c, d), and (e, f) is $|A|$, where

$$A = \frac{1}{2}\begin{vmatrix} a & b & 1 \\ c & d & 1 \\ e & f & 1 \end{vmatrix}.$$

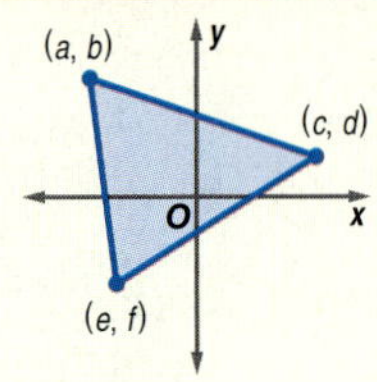

DAILY INTERVENTION — Differentiated Instruction

Interpersonal Organize students into small groups so that those for whom the kind of detailed work involved in finding determinants is easier can work with those who are having trouble carrying through the process to get the correct answer. Ask students to act as coaches to those who are having difficulties, helping them find the sources of their difficulties.

Example 4 Area of a Triangle

GEOMETRY Find the area of a triangle whose vertices are located at (−1, 6), (2, 4), and (0, 0).

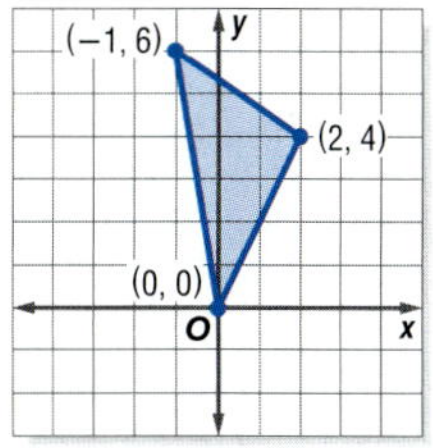

Assign values to a, b, c, d, e, and f and substitute them into the Area Formula. Then evaluate.

$$A = \frac{1}{2}\begin{vmatrix} a & b & 1 \\ c & d & 1 \\ e & f & 1 \end{vmatrix} \quad \text{Area Formula}$$

$$= \frac{1}{2}\begin{vmatrix} -1 & 6 & 1 \\ 2 & 4 & 1 \\ 0 & 0 & 1 \end{vmatrix} \quad (a, b) = (-1, 6), (c, d) = (2, 4), (e, f) = (0, 0)$$

$$= \frac{1}{2}\left[-1\begin{vmatrix} 4 & 1 \\ 0 & 1 \end{vmatrix} - 6\begin{vmatrix} 2 & 1 \\ 0 & 1 \end{vmatrix} + 1\begin{vmatrix} 2 & 4 \\ 0 & 0 \end{vmatrix}\right] \quad \text{Expansion by minors}$$

$$= \frac{1}{2}[-1(4 - 0) - 6(2 - 0) + 1(0 - 0)] \quad \text{Evaluate } 2 \times 2 \text{ determinants.}$$

$$= \frac{1}{2}[-4 - 12 + 0] \quad \text{Multiply.}$$

$$= \frac{1}{2}[-16] \text{ or } -8 \quad \text{Simplify.}$$

Remember that the area of a triangle is the absolute value of A. Thus, the area is $|-8|$ or 8 square units.

In-Class Example

Power Point®

4 Find the area of a triangle whose vertices are located at (0, −1), (−2, −6), and (3, −2). **8.5 units^2**

3 Practice/Apply

Study Notebook

Have students—

- add the definitions/examples of the vocabulary terms in this lesson to their Vocabulary Builder worksheets for Chapter 4.
- write the formula for evaluating a second-order determinant.
- include their own examples for both methods of finding the determinant of a 3 × 3 matrix.
- write the determinant formula for the area of a triangle.
- include any other item(s) that they find helpful in mastering the skills in this lesson.

Check for Understanding

Concept Check

1. **OPEN ENDED** Write a matrix whose determinant is zero.
2. **FIND THE ERROR** Khalid and Erica are finding the determinant of $\begin{bmatrix} 8 & 3 \\ -5 & 2 \end{bmatrix}$.

Khalid

$$\begin{vmatrix} 8 & 3 \\ -5 & 2 \end{vmatrix} = 16 - (-15) = 31$$

Erica

$$\begin{vmatrix} 8 & 3 \\ -5 & 2 \end{vmatrix} = 16 - 15 = 1$$

Who is correct? Explain your reasoning.

3. **Explain** why $\begin{bmatrix} 2 & 1 & 7 \\ 3 & -5 & 0 \end{bmatrix}$ does not have a determinant.
4. **Find a counterexample** to disprove the following statement. *Two different matrices can never have the same determinant.*
5. **Describe** how to find the minor of 6 in $\begin{bmatrix} 5 & 11 & 7 \\ -1 & 3 & 8 \\ 6 & 0 & -2 \end{bmatrix}$. **See margin.**
6. **Show** that the value of $\begin{vmatrix} -2 & 3 & 5 \\ 0 & -1 & 4 \\ 9 & 7 & 2 \end{vmatrix}$ is the same whether you use expansion by minors or diagonals. **See margin.**

1. Sample answer: $\begin{bmatrix} 2 & 1 \\ 8 & 4 \end{bmatrix}$

2. Khalid; the value of the determinant is the difference of the products of the diagonals.

3. It is not a square matrix.

4. Sample answer: $\begin{bmatrix} 3 & 1 \\ 6 & 5 \end{bmatrix}, \begin{bmatrix} 4 & 3 \\ 1 & 3 \end{bmatrix}$

Guided Practice Find the value of each determinant.

7. $\begin{vmatrix} 7 & 8 \\ 3 & -2 \end{vmatrix}$ **−38**
8. $\begin{vmatrix} -3 & -6 \\ 4 & 8 \end{vmatrix}$ **0**
9. $\begin{vmatrix} 0 & 8 \\ 5 & 9 \end{vmatrix}$ **−40**

DAILY INTERVENTION

FIND THE ERROR To avoid the mistake made by Erica, suggest that students begin finding a determinant by identifying the values of a, b, c, and d in the formula for a second-order determinant. Stress that the value of b is negative in this problem.

Answers

5. Cross out the column and row that contains 6. The minor is the remaining 2 × 2 matrix.

6. $\begin{vmatrix} -2 & 3 & 5 \\ 0 & -1 & 4 \\ 9 & 7 & 2 \end{vmatrix} = -2\begin{vmatrix} -1 & 4 \\ 7 & 2 \end{vmatrix} - 3\begin{vmatrix} 0 & 4 \\ 9 & 2 \end{vmatrix} + 5\begin{vmatrix} 0 & -1 \\ 9 & 7 \end{vmatrix}$

$= -2(-2 - 28) - 3(0 - 36) + 5(0 - (-9))$

$= -2(-30) - 3(-36) + 5(9)$

$= 60 + 108 + 45$

$= 213$

$$\left|\begin{matrix} -2 & 3 & 5 \\ 0 & -1 & 4 \\ 9 & 7 & 2 \end{matrix}\right|\begin{matrix} -2 & 3 \\ 0 & -1 \\ 9 & 7 \end{matrix}$$

4 108 0 (products down); −45 −56 0 (products up)

$4 + 108 + 0 - (-45) - (-56) - 0 = 213$

About the Exercises...

Organization by Objective

- Determinants of 2 × 2 Matrices: 15–26, 39
- Determinants of 3 × 3 Matrices: 27–38, 40–45

Odd/Even Assignments

Exercises 15–46 are structured so that students practice the same concepts whether they are assigned odd or even problems.

Alert! Exercises 50–55 require a graphing calculator.

Assignment Guide

Basic: 15–23 odd, 27–35 odd, 45–49, 56–75

Average: 15–43 odd, 45–49, 56–75 (optional: 50–55)

Advanced: 16–44 even, 45–69 (optional: 70–75)

Answers

46. Multiply each member in the top row by its minor and position sign. In this case the minor is a 3 × 3 matrix. Evaluate the 3 × 3 matrix using expansion by minors again.

47. If you know the coordinates of the vertices of a triangle, you can use a determinant to find the area. This is convenient since you don't need to know any additional information such as the measure of the angles. Answers should include the following.
    - You could place a coordinate grid over a map of the Bermuda Triangle with one vertex at the origin. By using the scale of the map, you could determine coordinates to represent the other two vertices and use a determinant to estimate the area.
    - The determinant method is advantageous since you don't need to physically measure the lengths of each side or the measure of the angles between the vertices.

GUIDED PRACTICE KEY	
Exercises	Examples
7–9	1
10, 11	2
12, 13	3
14	4

Evaluate each determinant using expansion by minors.

10. $\begin{vmatrix} 0 & -4 & 0 \\ 3 & -2 & 5 \\ 2 & -1 & 1 \end{vmatrix}$ **−28**

11. $\begin{vmatrix} 2 & 3 & 4 \\ 6 & 5 & 7 \\ 1 & 2 & 8 \end{vmatrix}$ **−43**

Evaluate each determinant using diagonals.

12. $\begin{vmatrix} 1 & 6 & 4 \\ -2 & 3 & 1 \\ 1 & 6 & 4 \end{vmatrix}$ **0**

13. $\begin{vmatrix} -1 & 4 & 0 \\ 3 & -2 & -5 \\ -3 & -1 & 2 \end{vmatrix}$ **45**

Application

14. **GEOMETRY** Find the area of the triangle shown at the right. **26 units^2**

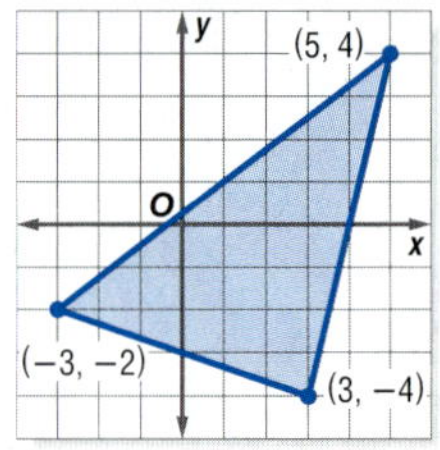

★ indicates increased difficulty

Practice and Apply

Homework Help	
For Exercises	See Examples
15–26, 39	1
27–32, 40	2
33–38	3
41–44	4

Extra Practice
See page 835.

Find the value of each determinant.

15. $\begin{vmatrix} 10 & 6 \\ 5 & 5 \end{vmatrix}$ **20**

16. $\begin{vmatrix} 8 & 5 \\ 6 & 1 \end{vmatrix}$ **−22**

17. $\begin{vmatrix} -7 & 3 \\ -9 & 7 \end{vmatrix}$ **−22**

18. $\begin{vmatrix} -2 & 4 \\ 3 & -6 \end{vmatrix}$ **0**

19. $\begin{vmatrix} 2 & -7 \\ -5 & 3 \end{vmatrix}$ **−29**

20. $\begin{vmatrix} -6 & -2 \\ 8 & 5 \end{vmatrix}$ **−14**

21. $\begin{vmatrix} -9 & 0 \\ -12 & -7 \end{vmatrix}$ **63**

22. $\begin{vmatrix} 6 & 14 \\ -3 & -8 \end{vmatrix}$ **−6**

23. $\begin{vmatrix} 15 & 11 \\ 23 & 19 \end{vmatrix}$ **32**

24. $\begin{vmatrix} 21 & 43 \\ 16 & 31 \end{vmatrix}$ **−37**

★25. $\begin{vmatrix} 7 & 5.2 \\ -4 & 1.6 \end{vmatrix}$ **32**

★26. $\begin{vmatrix} -3.2 & -5.8 \\ 4.1 & 3.9 \end{vmatrix}$ **11.3**

Evaluate each determinant using expansion by minors.

27. $\begin{vmatrix} 3 & 1 & 2 \\ 0 & 6 & 4 \\ 2 & 5 & 1 \end{vmatrix}$ **−58**

28. $\begin{vmatrix} 7 & 3 & -4 \\ -2 & 9 & 6 \\ 0 & 0 & 0 \end{vmatrix}$ **0**

29. $\begin{vmatrix} -2 & 7 & -2 \\ 4 & 5 & 2 \\ 1 & 0 & -1 \end{vmatrix}$ **62**

30. $\begin{vmatrix} -3 & 0 & 6 \\ 6 & 5 & -2 \\ 1 & 4 & 2 \end{vmatrix}$ **60**

31. $\begin{vmatrix} 1 & 5 & -4 \\ -7 & 3 & 2 \\ 6 & 3 & -1 \end{vmatrix}$ **172**

32. $\begin{vmatrix} 3 & 7 & 6 \\ -1 & 6 & 2 \\ 8 & -3 & -5 \end{vmatrix}$ **−265**

Evaluate each determinant using diagonals.

33. $\begin{vmatrix} 1 & 1 & 1 \\ 3 & 9 & 5 \\ 8 & 7 & 4 \end{vmatrix}$ **−22**

34. $\begin{vmatrix} 1 & 5 & 2 \\ -6 & -7 & 8 \\ 5 & 9 & -3 \end{vmatrix}$ **21**

35. $\begin{vmatrix} 8 & -9 & 0 \\ 1 & 5 & 4 \\ 6 & -2 & 3 \end{vmatrix}$ **−5**

36. $\begin{vmatrix} 4 & 10 & 7 \\ 3 & 3 & 1 \\ 0 & 5 & 2 \end{vmatrix}$ **49**

★37. $\begin{vmatrix} 2 & -3 & 4 \\ -2 & 1 & 5 \\ 5 & 3 & -2 \end{vmatrix}$ **−141**

★38. $\begin{vmatrix} 4 & -2 & 3 \\ -2 & 3 & 4 \\ 3 & 4 & 2 \end{vmatrix}$ **−123**

★39. Solve for x if $\det \begin{bmatrix} 2 & x \\ 5 & -3 \end{bmatrix} = 24$. **−6**

★40. Solve $\det \begin{bmatrix} 4 & x & -2 \\ -x & -3 & 1 \\ -6 & 2 & 3 \end{bmatrix} = -3$ for x. **$\frac{5}{3}$, −1**

186 Chapter 4 Matrices

★ 41. **GEOMETRY** Find the area of the polygon shown at the right. **14.5 units^2**

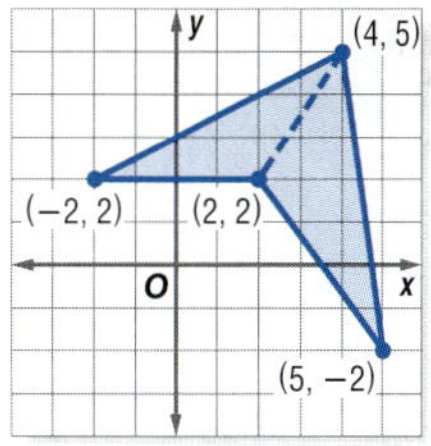

★ 42. **GEOMETRY** Find the value of x such that the area of a triangle whose vertices have coordinates (6, 5), (8, 2), and (x, 11) is 15 square units. **−8 or 12**

Career Choices

Archaeologist

Archaeologists attempt to reconstruct past ways of life by examining preserved bones, the ruins of buildings, and artifacts such as tools, pottery, and jewelry.

Source: www.encarta.msn.com

Online Research For information about a career as an archaeologist, visit: www.algebra2.com/careers

43. ★ **ARCHAEOLOGY** During an archaeological dig, a coordinate grid is laid over the site to identify the location of artifacts as they are excavated. During a dig, three corners of a rectangular building have been partially unearthed at (−1, 6), (4, 5), and (−3, −4). If each square on the grid measures one square foot, estimate the area of the floor of the building. **about 26 ft^2**

★ 44. **GEOGRAPHY** Mr. Cardona is a regional sales manager for a company in Florida. Tampa, Orlando, and Ocala outline his region. If a coordinate grid in which 1 unit = 10 miles is placed over the map of Florida with Tampa at the origin, the coordinates of the three cities are (0, 0), (7, 5), and (2.5, 10). Use a determinant to estimate the area of his sales territory. **2875 mi^2**

45. **CRITICAL THINKING** Find a third-order determinant in which no element is 0, but for which the determinant is 0. **Sample answer:** $\begin{vmatrix} 1 & 1 & 1 \\ 1 & 1 & 1 \\ 1 & 1 & 1 \end{vmatrix}$

46. **CRITICAL THINKING** Make a conjecture about how you could find the determinant of a 4 × 4 matrix using the expansion by minors method. Use a diagram in your explanation. **See margin.**

47. **WRITING IN MATH** Answer the question that was posed at the beginning of the lesson. **See margin.**

How are determinants used to find areas of polygons?

Include the following in your answer:

- an explanation of how you could use a coordinate grid to estimate the area of the Bermuda Triangle, and
- some advantages of using this method in this situation.

Standardized Test Practice

48. Find the value of det A if $A = \begin{bmatrix} 0 & 3 & -2 \\ -4 & 0 & 1 \\ 3 & 2 & 0 \end{bmatrix}$. **C**

Ⓐ 0 Ⓑ 12 Ⓒ 25 Ⓓ 36

49. Find the area of triangle ABC. **C**

Ⓐ 10 units^2
Ⓑ 12 units^2
Ⓒ 14 units^2
Ⓓ 16 units^2
Ⓔ none of these

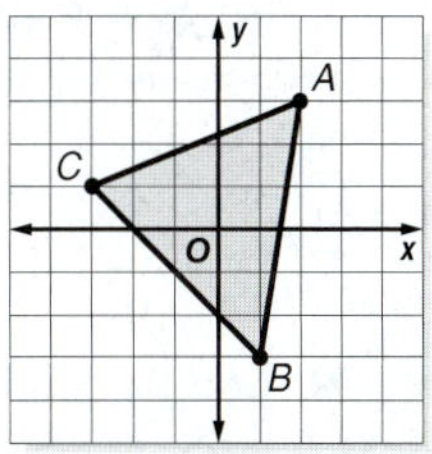

www.algebra2.com/self_check_quiz

Study Guide and Intervention, p. 193 (shown) and p. 194

Determinants of 2 × 2 Matrices

Second-Order Determinant	For the matrix $\begin{bmatrix} a & b \\ c & d \end{bmatrix}$, the determinant is $\begin{vmatrix} a & b \\ c & d \end{vmatrix} = ad - bc$.

Example Find the value of each determinant.

a. $\begin{vmatrix} 6 & 3 \\ -8 & 5 \end{vmatrix}$

$\begin{vmatrix} 6 & 3 \\ -8 & 5 \end{vmatrix} = 6(5) - 3(-8)$
$= 30 - (-24)$ or 54

b. $\begin{vmatrix} 11 & -5 \\ 9 & 3 \end{vmatrix}$

$\begin{vmatrix} 11 & -5 \\ 9 & 3 \end{vmatrix} = 11(3) - (-5)(9)$
$= 33 - (-45)$ or 12

Exercises

Find the value of each determinant.

1. $\begin{vmatrix} 6 & -2 \\ 5 & 7 \end{vmatrix}$ **52**
2. $\begin{vmatrix} -8 & 3 \\ -2 & 1 \end{vmatrix}$ **−2**
3. $\begin{vmatrix} 3 & 9 \\ 4 & 6 \end{vmatrix}$ **−18**
4. $\begin{vmatrix} 5 & 12 \\ -7 & -4 \end{vmatrix}$ **64**
5. $\begin{vmatrix} -6 & -3 \\ -4 & -1 \end{vmatrix}$ **−6**
6. $\begin{vmatrix} 4 & 7 \\ 5 & 9 \end{vmatrix}$ **1**
7. $\begin{vmatrix} 14 & 8 \\ 9 & -3 \end{vmatrix}$ **−114**
8. $\begin{vmatrix} 15 & 12 \\ 23 & 28 \end{vmatrix}$ **144**
9. $\begin{vmatrix} -8 & 35 \\ 5 & 20 \end{vmatrix}$ **−335**
10. $\begin{vmatrix} 10 & 16 \\ 22 & 40 \end{vmatrix}$ **48**
11. $\begin{vmatrix} 24 & -8 \\ 7 & -3 \end{vmatrix}$ **−16**
12. $\begin{vmatrix} 13 & 62 \\ -4 & 19 \end{vmatrix}$ **495**
13. $\begin{vmatrix} 0.2 & 8 \\ -1.5 & 15 \end{vmatrix}$ **15**
14. $\begin{vmatrix} 8.6 & 0.5 \\ 14 & 5 \end{vmatrix}$ **36**
15. $\begin{vmatrix} 20 & 110 \\ 0.1 & 1.4 \end{vmatrix}$ **17**
16. $\begin{vmatrix} 4.8 & 2.1 \\ 3.4 & 5.3 \end{vmatrix}$ **18.3**
17. $\begin{vmatrix} \frac{2}{3} & -\frac{1}{2} \\ \frac{1}{6} & \frac{1}{5} \end{vmatrix}$ $\mathbf{\frac{13}{60}}$
18. $\begin{vmatrix} 6.8 & 15 \\ -0.2 & 5 \end{vmatrix}$ **37**

Skills Practice, p. 195 and Practice, p. 196 (shown)

Find the value of each determinant.

1. $\begin{vmatrix} 1 & 6 \\ 2 & 7 \end{vmatrix}$ **−5**
2. $\begin{vmatrix} 9 & 6 \\ 3 & 2 \end{vmatrix}$ **0**
3. $\begin{vmatrix} 4 & 1 \\ -2 & -5 \end{vmatrix}$ **−18**
4. $\begin{vmatrix} -14 & -3 \\ 2 & -2 \end{vmatrix}$ **34**
5. $\begin{vmatrix} 4 & -3 \\ -12 & 4 \end{vmatrix}$ **−20**
6. $\begin{vmatrix} 2 & -5 \\ 5 & -11 \end{vmatrix}$ **3**
7. $\begin{vmatrix} 4 & 0 \\ -2 & 9 \end{vmatrix}$ **36**
8. $\begin{vmatrix} 3 & -4 \\ 7 & 9 \end{vmatrix}$ **55**
9. $\begin{vmatrix} -1 & -11 \\ 10 & -2 \end{vmatrix}$ **112**
10. $\begin{vmatrix} 3 & -4 \\ 3.75 & 5 \end{vmatrix}$ **30**
11. $\begin{vmatrix} 2 & -1 \\ 3 & -9.5 \end{vmatrix}$ **−16**
12. $\begin{vmatrix} 0.5 & -0.7 \\ 0.4 & -0.3 \end{vmatrix}$ **0.13**

Evaluate each determinant using expansion by minors.

13. $\begin{vmatrix} -2 & 3 & 1 \\ 0 & 4 & -3 \\ 2 & 5 & -1 \end{vmatrix}$ **−48**
14. $\begin{vmatrix} 2 & -4 & 1 \\ 3 & 0 & 9 \\ -1 & 5 & 7 \end{vmatrix}$ **45**
15. $\begin{vmatrix} 2 & 1 & 1 \\ 1 & -1 & -2 \\ 1 & 1 & -1 \end{vmatrix}$ **7**
16. $\begin{vmatrix} 0 & -4 & 0 \\ 2 & -1 & 1 \\ 3 & -2 & 5 \end{vmatrix}$ **28**
17. $\begin{vmatrix} 2 & 7 & -6 \\ 8 & 4 & 0 \\ 1 & -1 & 3 \end{vmatrix}$ **−72**
18. $\begin{vmatrix} -12 & 0 & 3 \\ 7 & 5 & -1 \\ 4 & 2 & -6 \end{vmatrix}$ **318**

Evaluate each determinant using diagonals.

19. $\begin{vmatrix} -4 & 3 & -1 \\ 2 & 1 & -2 \\ 4 & 1 & -4 \end{vmatrix}$ **10**
20. $\begin{vmatrix} 2 & 2 & 3 \\ 1 & -1 & 1 \\ 3 & 1 & 1 \end{vmatrix}$ **12**
21. $\begin{vmatrix} 1 & -4 & -1 \\ 1 & -6 & -2 \\ 2 & 3 & 1 \end{vmatrix}$ **5**
22. $\begin{vmatrix} 1 & 2 & -4 \\ 1 & 4 & -6 \\ 2 & 3 & 3 \end{vmatrix}$ **20**
23. $\begin{vmatrix} 2 & -1 & -2 \\ 4 & 0 & -2 \\ 0 & 3 & 2 \end{vmatrix}$ **−4**
24. $\begin{vmatrix} 2 & 1 & 3 \\ 1 & 8 & 0 \\ 0 & 5 & -1 \end{vmatrix}$ **0**

25. **GEOMETRY** Find the area of a triangle whose vertices have coordinates (3, 5), (6, −5), and (−4, 10). **27.5 units^2**

26. **LAND MANAGEMENT** A fish and wildlife management organization uses a GIS (geographic information system) to store and analyze data for the parcels of land it manages. All of the parcels are mapped on a grid in which 1 unit represents 1 acre. If the coordinates of the corners of a parcel are (−8, 10), (6, 17), and (2, −4), how many acres is the parcel? **133 acres**

Reading to Learn Mathematics, p. 197 ELL

Pre-Activity **How are determinants used to find areas of polygons?**

Read the introduction to Lesson 4-5 at the top of page 182 in your textbook.

In this lesson, you will learn how to find the area of a triangle if you know the coordinates of its vertices using determinants. Describe a method you already know for finding the area of the Bermuda Triangle. **Sample answer: Use the map. Choose any side of the triangle as the base, and measure this side with a ruler. Multiply this length by the scale factor for the map. Next, draw a segment from the opposite vertex perpendicular to the base. Measure this segment, and multiply its length by the scale for the map. Finally, find the area by using the formula $A = \frac{1}{2}bh$.**

Reading the Lesson

1. Indicate whether each of the following statements is *true* or *false*.
   a. Every matrix has a determinant. **false**
   b. If both rows of a 2 × 2 matrix are identical, the determinant of the matrix will be 0. **true**
   c. Every element of a 3 × 3 matrix has a minor. **true**
   d. In order to evaluate a third-order determinant by expansion by minors it is necessary to find the minor of every element of the matrix. **false**
   e. If you evaluate a third-order determinant by expansion about the second row, the position signs you will use are − + −. **true**
2. Suppose that triangle RST has vertices $R(-2, 5)$, $S(4, 1)$, and $T(0, 6)$.
   a. Write a determinant that could be used in finding the area of triangle RST. $\begin{vmatrix} -2 & 5 & 1 \\ 4 & 1 & 1 \\ 0 & 6 & 1 \end{vmatrix}$
   b. Explain how you would use the determinant you wrote in part **a** to find the area of the triangle. **Sample answer: Evaluate the determinant and multiply the result by $\frac{1}{2}$. Then take the absolute value to make sure the final answer is positive.**

Helping You Remember

3. A good way to remember a complicated procedure is to break it down into steps. Write a list of steps for evaluating a third-order determinant using expansion by minors. **Sample answer: 1. Choose a row of the matrix. 2. Find the position signs for the row you have chosen. 3. Find the minor of each element in that row. 4. Multiply each element by its position sign and by its minor. 5. Add the results.**

Enrichment, p. 198

Properties of Determinants

The following properties often help when evaluating determinants.

- If all the elements of a row (or column) are zero, the value of the determinant is zero.
  $\begin{vmatrix} a & b \\ 0 & 0 \end{vmatrix} = 0$ **$(a \cdot 0) - (0 \cdot b) = 0$**
- Multiplying all the elements of a row (or column) by a constant is equivalent to multiplying the value of the determinant by the constant.
  $3\begin{vmatrix} 4 & -1 \\ 5 & 3 \end{vmatrix} = \begin{vmatrix} 12 & -3 \\ 5 & 3 \end{vmatrix}$ **$3[4(3) - 5(-1)] = 3[12 + 5] = 51$; $12(3) - 5(-3) = 51$**
- If two rows (or columns) have equal corresponding elements, the value of the determinant is zero.

4 Assess

Open-Ended Assessment

Modeling Have students make posters that show the two processes for finding the determinant of a 3 × 3 matrix, using colored markers to clearly identify the minors or the diagonals in each procedure.

Getting Ready for Lesson 4-6

PREREQUISITE SKILL Lesson 4-6 presents Cramer's Rule for solving systems of two linear equations. Students will use their familiarity with previous methods for solving systems of equations as they learn this new method. Exercises 70–75 should be used to determine your students' familiarity with solving systems of two equations in two variables.

Answer

58.

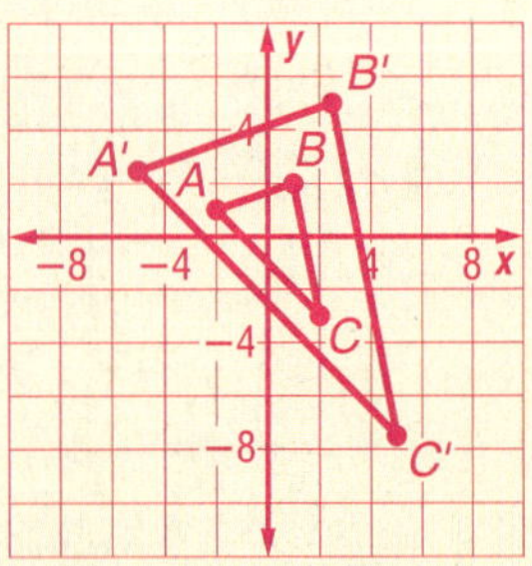

Graphing Calculator

MATRX FUNCTION You can use a TI-83 Plus to find determinants of square matrices using the MATRX functions. Enter the matrix under the NAMES menu. Then use the arrow keys to highlight the MATH menu. Choose det(, which is option 1, to calculate the determinant.

Use a graphing calculator to find the value of each determinant.

50. $\begin{vmatrix} 3 & -6.5 \\ 8 & 3.75 \end{vmatrix}$ **63.25**

51. $\begin{vmatrix} 1.3 & 7.2 \\ 6.1 & 5.4 \end{vmatrix}$ **−36.9**

52. $\begin{vmatrix} 6.1 & 4.8 \\ 9.7 & 3.5 \end{vmatrix}$ **−25.21**

53. $\begin{vmatrix} 8 & 6 & -5 \\ 10 & -7 & 3 \\ 9 & 14 & -6 \end{vmatrix}$ **−493**

54. $\begin{vmatrix} 10 & 20 & 30 \\ 40 & 50 & 60 \\ 70 & 80 & 90 \end{vmatrix}$ **0**

55. $\begin{vmatrix} 10 & 12 & 4 \\ -3 & 18 & -9 \\ 16 & -2 & -1 \end{vmatrix}$ **−3252**

Maintain Your Skills

Mixed Review

For Exercises 56–58, use the following information.
The vertices of $\triangle ABC$ are $A(-2, 1)$, $B(1, 2)$ and $C(2, -3)$. The triangle is dilated so that its perimeter is $2\frac{1}{2}$ times the original perimeter. *(Lesson 4-4)*

56. Write the coordinates of $\triangle ABC$ in a vertex matrix. $\begin{bmatrix} -2 & 1 & 2 \\ 1 & 2 & -3 \end{bmatrix}$

57. Find the coordinates of $\triangle A'B'C'$. **$A'(-5, 2.5)$, $B'(2.5, 5)$, $C'(5, -7.5)$**

58. Graph $\triangle ABC$ and $\triangle A'B'C'$. **See margin.**

Find each product, if possible. *(Lesson 4-3)*

59. $[5 \quad 2] \cdot \begin{bmatrix} -2 \\ 3 \end{bmatrix}$ **[−4]**

60. $\begin{bmatrix} 2 & 4 \\ -2 & 3 \end{bmatrix} \cdot \begin{bmatrix} 3 & 9 \\ -1 & 2 \end{bmatrix}$ $\begin{bmatrix} 2 & 26 \\ -9 & -12 \end{bmatrix}$

61. $\begin{bmatrix} 5 \\ 7 \end{bmatrix} \cdot \begin{bmatrix} 1 & 6 \\ -4 & 2 \end{bmatrix}$ **undefined**

62. $\begin{bmatrix} 7 & 4 \\ -1 & 2 \\ -3 & 5 \end{bmatrix} \cdot [3 \quad 5]$ **undefined**

63. $[4 \quad 2 \quad 0] \cdot \begin{bmatrix} 3 & -2 \\ 1 & 0 \\ 5 & 6 \end{bmatrix}$ **[14 −8]**

64. $\begin{bmatrix} 7 & -5 & 4 \\ 6 & 1 & 3 \end{bmatrix} \cdot \begin{bmatrix} -1 & 3 \\ -2 & -8 \\ 1 & 2 \end{bmatrix}$ $\begin{bmatrix} 7 & 69 \\ -5 & 16 \end{bmatrix}$

65. **RUNNING** The length of a marathon was determined by the first marathon in the 1908 Olympic Games in London, England. The race began at Windsor Castle and ended in front of the royal box at London's Olympic Stadium, which was a distance of 26 miles 385 yards. Determine how many feet the marathon covers using the formula $f(m, y) = 5280m + 3y$, where m is the number of miles and y is the number of yards. *(Lesson 3-4)* **138,435 ft**

Write an equation in slope-intercept form for the line that satisfies each set of conditions. *(Lesson 2-4)* **66. $y = x - 2$ 67. $y = -\frac{4}{3}x$**

66. slope 1 passes through (5, 3)

67. slope $-\frac{4}{3}$ passes through (6, −8)

68. passes through (3, 7) and (−2, −3) **$y = 2x + 1$**

69. passes through (0, 5) and (10, 10) **$y = \frac{1}{2}x + 5$**

Getting Ready for the Next Lesson

PREREQUISITE SKILL **Solve each system of equations.**
*(To review **solving systems of equations**, see Lesson 3-2.)*

70. $x + y = -3$ **(0, −3)**
$3x + 4y = -12$

71. $x + y = 10$ **(1, 9)**
$2x + y = 11$

72. $2x + y = 5$ **(2, 1)**
$4x + y = 9$

73. $3x + 5y = 2$ **(−1, 1)**
$2x - y = -3$

74. $6x + 2y = 22$ **(2, 5)**
$3x + 7y = 41$

75. $3x - 2y = -2$ **(4, 7)**
$4x + 7y = 65$

4-6 Cramer's Rule

What You'll Learn

- Solve systems of two linear equations by using Cramer's Rule.
- Solve systems of three linear equations by using Cramer's Rule.

Vocabulary

- Cramer's Rule

How is Cramer's Rule used to solve systems of equations?

Two sides of a triangle are contained in lines whose equations are $1.4x + 3.8y = 3.4$ and $2.5x - 1.7y = -10.9$. To find the coordinates of the vertex of the triangle between these two sides, you must solve the system of equations. However, solving this system by using substitution or elimination would require many calculations. Another method for solving systems of equations is Cramer's Rule.

SYSTEMS OF TWO LINEAR EQUATIONS **Cramer's Rule** uses determinants to solve systems of equations. Consider the following system.

$$ax + by = e \quad \text{a, b, c, d, e, and f represent constants, *not* variables.}$$

$$cx + dy = f$$

Study Tip

Look Back To review **solving systems of equations**, see Lesson 3-2.

Solve for x by using elimination.

$$adx + bdy = de \quad \text{Multiply the first equation by } d.$$

$$(-)\ bcx + bdy = bf \quad \text{Multiply the second equation by } b.$$

$$adx - bcx = de - bf \quad \text{Subtract.}$$

$$(ad - bc)x = de - bf \quad \text{Factor.}$$

$$x = \frac{de - bf}{ad - bc} \quad \text{Notice that } ad - bc \text{ must not be zero.}$$

Solving for y in the same way produces the following expression.

$$y = \frac{af - ce}{ad - bc}$$

So the solution of the system of equations $ax + by = e$ and $cx + dy = f$ is $\left(\frac{de - bf}{ad - bc}, \frac{af - ce}{ad - bc}\right)$.

Notice that the denominators for each expression are the same. It can be written using a determinant. The numerators can also be written as determinants.

$$ad - bc = \begin{vmatrix} a & b \\ c & d \end{vmatrix} \qquad de - bf = \begin{vmatrix} e & b \\ f & d \end{vmatrix} \qquad af - ce = \begin{vmatrix} a & e \\ c & f \end{vmatrix}$$

Key Concept — *Cramer's Rule for Two Variables*

The solution of the system of linear equations

$$ax + by = e$$
$$cx + dy = f$$

is (x, y), where $x = \dfrac{\begin{vmatrix} e & b \\ f & d \end{vmatrix}}{\begin{vmatrix} a & b \\ c & d \end{vmatrix}}$, $y = \dfrac{\begin{vmatrix} a & e \\ c & f \end{vmatrix}}{\begin{vmatrix} a & b \\ c & d \end{vmatrix}}$, and $\begin{vmatrix} a & b \\ c & d \end{vmatrix} \neq 0$.

4-6 Lesson Notes

1 Focus

5-Minute Check Transparency 4-6 Use as a quiz or review of Lesson 4-5.

Mathematical Background notes are available for this lesson on p. 152D.

Building on Prior Knowledge

Students will use their knowledge of determinants from Lesson 4-5 when using Cramer's Rule to solve systems of equations in this lesson.

How is Cramer's rule used to solve systems of equations?

Ask students:

- Why does the text say that the sides of the triangle "are contained in" the two lines whose equations are given? **because lines extend forever while the sides of a triangle are line segments**
- What makes solving the system of these two equations more difficult using substitution or elimination? **The coefficients in the equations are decimals rather than integers.**

Resource Manager

Workbook and Reproducible Masters

Chapter 4 Resource Masters

- Study Guide and Intervention, pp. 199–200
- Skills Practice, p. 201
- Practice, p. 202
- Reading to Learn Mathematics, p. 203
- Enrichment, p. 204
- Assessment, p. 232

Transparencies

5-Minute Check Transparency 4-6
Answer Key Transparencies

Technology

Alge2PASS: Tutorial Plus, Lesson 7
Interactive Chalkboard
Multimedia Applications

2 Teach

SYSTEMS OF TWO LINEAR EQUATIONS

In-Class Examples

Power Point®

1 Use Cramer's Rule to solve the system of equations.
$5x + 4y = 28$
$3x - 2y = 8$ **(4, 2)**

2 **VOTING** In a vote for the school colors of a new high school, blue-and-gold received 440 votes from 10th and 11th graders and red-and-black received 210 votes from 10th and 11th graders. In the 10th grade, blue-and-gold received 72% of the votes, and red-and-black received 28% of the votes. In the 11th grade, blue-and-gold received 64% of the votes and red-and-black received 36%.

a. Write a system of equations that represents the total number of votes cast for each set of colors in these two grades. Let t represent the total number of 10th grade votes and let e represent the total number of 11th grade votes. **$0.72t + 0.64e = 440$** **$0.28t + 0.36e = 210$**

b. Find the total number of votes cast in the 10th and in the 11th grades. **There were 300 votes cast by 10th graders and 350 votes cast by 11th graders.**

Example 1 System of Two Equations

Use Cramer's Rule to solve the system of equations.

$5x + 7y = 13$
$2x - 5y = 13$

$$x = \frac{\begin{vmatrix} e & b \\ f & d \end{vmatrix}}{\begin{vmatrix} a & b \\ c & d \end{vmatrix}} \qquad \text{Cramer's Rule} \qquad y = \frac{\begin{vmatrix} a & e \\ c & f \end{vmatrix}}{\begin{vmatrix} a & b \\ c & d \end{vmatrix}}$$

$$= \frac{\begin{vmatrix} 13 & 7 \\ 13 & -5 \end{vmatrix}}{\begin{vmatrix} 5 & 7 \\ 2 & -5 \end{vmatrix}} \qquad a = 5, b = 7, c = 2, d = -5, e = 13, \text{ and } f = 13 \qquad = \frac{\begin{vmatrix} 5 & 13 \\ 2 & 13 \end{vmatrix}}{\begin{vmatrix} 5 & 7 \\ 2 & -5 \end{vmatrix}}$$

$$= \frac{13(-5) - 13(7)}{5(-5) - 2(7)} \qquad \text{Evaluate each determinant.} \qquad = \frac{5(13) - 2(13)}{5(-5) - 2(7)}$$

$$= \frac{-156}{-39} \text{ or } 4 \qquad \text{Simplify.} \qquad = \frac{39}{-39} \text{ or } -1$$

The solution is $(4, -1)$.

Cramer's Rule is especially useful when the coefficients are large or involve fractions or decimals.

Example 2 Use Cramer's Rule

ELECTIONS **In the 2000 presidential election, George W. Bush received about 8,400,000 votes in California and Texas while Al Gore received about 8,300,000 votes in those states. The graph shows the percent of the popular vote that each candidate received in those states.**

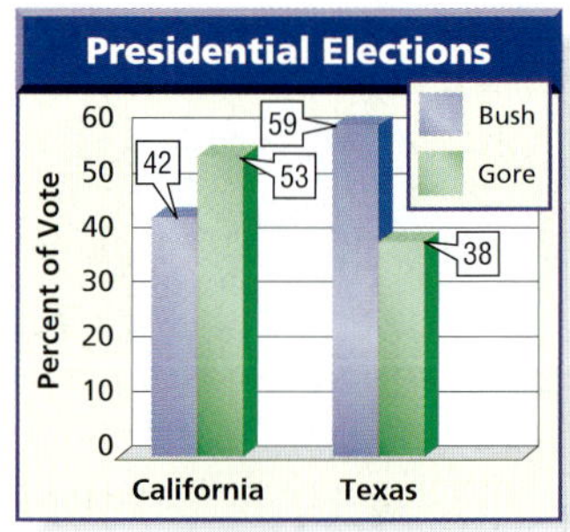

Source: States' Elections Offices

a. Write a system of equations that represents the total number of votes cast for each candidate in these two states.

Let x represent the total number of votes in California.

Let y represent the total number of votes in Texas.

$0.42x + 0.59y = 8{,}400{,}000$ Votes for Bush

$0.53x + 0.38y = 8{,}300{,}000$ Votes for Gore

b. Find the total number of popular votes cast in California and in Texas.

$$x = \frac{\begin{vmatrix} e & b \\ f & d \end{vmatrix}}{\begin{vmatrix} a & b \\ c & d \end{vmatrix}} \qquad \text{Cramer's Rule} \qquad y = \frac{\begin{vmatrix} a & e \\ c & f \end{vmatrix}}{\begin{vmatrix} a & b \\ c & d \end{vmatrix}}$$

$$= \frac{\begin{vmatrix} 8{,}400{,}000 & 0.59 \\ 8{,}300{,}000 & 0.38 \end{vmatrix}}{\begin{vmatrix} 0.42 & 0.59 \\ 0.53 & 0.38 \end{vmatrix}} \qquad = \frac{\begin{vmatrix} 0.42 & 8{,}400{,}000 \\ 0.53 & 8{,}300{,}000 \end{vmatrix}}{\begin{vmatrix} 0.42 & 0.59 \\ 0.53 & 0.38 \end{vmatrix}}$$

More About. . .

Elections

In 1936, Franklin D. Roosevelt received a record 523 electoral college votes to Alfred M. Landon's 8 votes. This is the largest electoral college majority.

Source: *The Guinness Book of Records*

$$x = \frac{8{,}400{,}000(0.38) - 8{,}300{,}000(0.59)}{0.42(0.38) - 0.53(0.59)} \qquad y = \frac{0.42(8{,}300{,}000) - 0.53(8{,}400{,}000)}{0.42(0.38) - 0.53(0.59)}$$

$$= \frac{-1{,}705{,}000}{-0.1531} \qquad = \frac{-966{,}000}{-0.1531}$$

$$\approx 11{,}136{,}512.08 \qquad \approx 6{,}309{,}601.57$$

The solution of the system is about (11,136,512.08, 6,309,601.57).

So, there were about 11,100,000 popular votes cast in California and about 6,300,000 popular votes cast in Texas.

SYSTEMS OF THREE LINEAR EQUATIONS

You can also use Cramer's Rule to solve a system of three equations in three variables.

TEACHING TIP Point out that the numerator for x is found by replacing the coefficients of x with the constant terms of the system. Similarly, the numerators for y and z are found by replacing the coefficients of y and z, respectively, with the constant terms of the system.

Key Concept — Cramer's Rule for Three Variables

The solution of the system whose equations are

$$ax + by + cz = j$$
$$dx + ey + fz = k$$
$$gx + hy + iz = \ell$$

is (x, y, z), where $x = \dfrac{\begin{vmatrix} j & b & c \\ k & e & f \\ \ell & h & i \end{vmatrix}}{\begin{vmatrix} a & b & c \\ d & e & f \\ g & h & i \end{vmatrix}}$, $y = \dfrac{\begin{vmatrix} a & j & c \\ d & k & f \\ g & \ell & i \end{vmatrix}}{\begin{vmatrix} a & b & c \\ d & e & f \\ g & h & i \end{vmatrix}}$, $z = \dfrac{\begin{vmatrix} a & b & j \\ d & e & k \\ g & h & \ell \end{vmatrix}}{\begin{vmatrix} a & b & c \\ d & e & f \\ g & h & i \end{vmatrix}}$, and $\begin{vmatrix} a & b & c \\ d & e & f \\ g & h & i \end{vmatrix} \neq 0$.

Example 3 System of Three Equations

Use Cramer's Rule to solve the system of equations.

$$3x + y + z = -1$$
$$-6x + 5y + 3z = -9$$
$$9x - 2y - z = 5$$

$$x = \frac{\begin{vmatrix} j & b & c \\ k & e & f \\ \ell & h & i \end{vmatrix}}{\begin{vmatrix} a & b & c \\ d & e & f \\ g & h & i \end{vmatrix}} \qquad y = \frac{\begin{vmatrix} a & j & c \\ d & k & f \\ g & \ell & i \end{vmatrix}}{\begin{vmatrix} a & b & c \\ d & e & f \\ g & h & i \end{vmatrix}} \qquad z = \frac{\begin{vmatrix} a & b & j \\ d & e & k \\ g & h & \ell \end{vmatrix}}{\begin{vmatrix} a & b & c \\ d & e & f \\ g & h & i \end{vmatrix}}$$

$$= \frac{\begin{vmatrix} -1 & 1 & 1 \\ -9 & 5 & 3 \\ 5 & -2 & -1 \end{vmatrix}}{\begin{vmatrix} 3 & 1 & 1 \\ -6 & 5 & 3 \\ 9 & -2 & -1 \end{vmatrix}} \qquad = \frac{\begin{vmatrix} 3 & -1 & 1 \\ -6 & -9 & 3 \\ 9 & 5 & -1 \end{vmatrix}}{\begin{vmatrix} 3 & 1 & 1 \\ -6 & 5 & 3 \\ 9 & -2 & -1 \end{vmatrix}} \qquad = \frac{\begin{vmatrix} 3 & 1 & -1 \\ -6 & 5 & -9 \\ 9 & -2 & 5 \end{vmatrix}}{\begin{vmatrix} 3 & 1 & 1 \\ -6 & 5 & 3 \\ 9 & -2 & -1 \end{vmatrix}}$$

Use a calculator to evaluate each determinant.

$x = \frac{-2}{-9}$ or $\frac{2}{9}$ $\qquad$ $y = \frac{12}{-9}$ or $-\frac{4}{3}$ $\qquad$ $z = \frac{3}{-9}$ or $-\frac{1}{3}$

The solution is $\left(\frac{2}{9}, -\frac{4}{3}, -\frac{1}{3}\right)$.

www.algebra2.com/extra_examples

DAILY INTERVENTION — Differentiated Instruction

Intrapersonal Ask students to write a brief reflection on their reaction to the various methods they have learned involving the manipulation of matrices. Ask them to comment on what aspects they found efficient and helpful, and what aspects they found to be difficult or confusing.

SYSTEMS OF THREE LINEAR EQUATIONS

In-Class Example

Power Point®

3 Use Cramer's Rule to solve the system of equations.

$$2x - 3y + z = 5$$
$$x + 2y + z = -1$$
$$x - 3y + 2z = 1 \quad \left(\frac{9}{4}, -\frac{3}{4}, -\frac{7}{4}\right)$$

Teaching Tip Ask students to compare this method for speed and ease of use to other methods they could use.

3 Practice/Apply

Study Notebook

Have students—

- *add the definitions/examples of the vocabulary terms in this lesson to their Vocabulary Builder worksheets for Chapter 4.*
- *write Cramer's Rule for two variables and for three variables, including examples of how to use both rules.*
- *include any other item(s) that they find helpful in mastering the skills in this lesson.*

About the Exercises...

Organization by Objective

- **Systems of Two Linear Equations:** 12–25, 32–35
- **Systems of Three Linear Equations:** 26–31, 36, 37

Odd/Even Assignments

Exercises 12–31 are structured so that students practice the same concepts whether they are assigned odd or even problems.

Assignment Guide

Basic: 13–19 odd, 25, 27, 29, 32, 33, 38–54

Average: 13–31 odd, 32–35, 38–54

Advanced: 12–30 even, 34–51 (optional: 52–54)

All: Practice Quiz 2 (1–10)

Check for Understanding

Concept Check

1. **Describe** the condition that must be met in order to use Cramer's Rule. **The determinant of the coefficient matrix cannot be zero.**

2. **OPEN ENDED** Write a system of equations that *cannot* be solved using Cramer's Rule. **Sample answer: $2x + y = 5$ and $6x + 3y = 8$**

3. **Write** a system of equations whose solution is $x = \dfrac{\begin{vmatrix} -6 & 5 \\ 30 & -2 \end{vmatrix}}{\begin{vmatrix} 3 & 5 \\ 4 & -2 \end{vmatrix}}$, $y = \dfrac{\begin{vmatrix} 3 & -6 \\ 4 & 30 \end{vmatrix}}{\begin{vmatrix} 3 & 5 \\ 4 & -2 \end{vmatrix}}$. **$3x + 5y = -6$, $4x - 2y = 30$**

Guided Practice

GUIDED PRACTICE KEY	
Exercises	Examples
4–6	1
7–9	3
10, 11	2

Use Cramer's Rule to solve each system of equations.

4. $x - 4y = 1$; $2x + 3y = 13$ **(5, 1)**

5. $0.2a = 0.3b$; $0.4a - 0.2b = 0.2$ **(0.75, 0.5)**

6. $\frac{1}{2}r - \frac{2}{3}s = 2\frac{1}{3}$; $\frac{3}{5}r + \frac{4}{5}s = -10$ **(−6, −8)**

7. $2x - y + 3z = 5$; $3x + 2y - 5z = 4$; $x - 4y + 11z = 3$ **no solution**

8. $a + 9b - 2c = 2$; $-a - 3b + 4c = 1$; $2a + 3b - 6c = -5$ **8. $\left(-5, \frac{2}{3}, -\frac{1}{2}\right)$**

9. $r + 4s + 3t = 10$; $2r - 2s + t = 15$; $r + 2s - 3t = -1$ **9. $\left(6, -\frac{1}{2}, 2\right)$**

Application

INVESTING **For Exercises 10 and 11, use the following information.**
Jarrod Wright has $4000 he would like to invest. He could put it in a savings account paying 6.5% interest annually, or in a certificate of deposit with an annual rate of 8%. He wants his interest for the year to be $297.50, because earning more than this would put him into a higher tax bracket.

10. Write a system of equations, in which the unknowns s and d stand for the amounts of money Jarrod should deposit in the savings account and the certificate of deposit, respectively. **$s + d = 4000$, $0.065s + 0.08d = 297.50$**

11. How much should he put in a savings account, and how much should he put in the certificate of deposit? **savings account, $1500; certificate of deposit, $2500**

★ indicates increased difficulty

Practice and Apply

Homework Help	
For Exercises	See Examples
12–25	1
26–31	3
32–37	2

Extra Practice See page 835.

Use Cramer's Rule to solve each system of equations.

12. $5x + 2y = 8$; $2x - 3y = 7$ **(2, −1)**

13. $2m + 7n = 4$; $m - 2n = -20$ **(−12, 4)**

14. $2r - s = 1$; $3r + 2s = 19$ **(3, 5)**

15. $3a + 5b = 33$; $5a + 7b = 51$ **(6, 3)**

16. $2m - 4n = -1$; $3n - 4m = -5$ **(2.3, 1.4)**

17. $4x + 3y = 6$; $8x - y = -9$ **(−0.75, 3)**

18. $0.5r - s = -1$; $0.75r + 0.5s = -0.25$ **(−0.75, 0.625)**

19. $1.5m - 0.7n = 0.5$; $2.2m - 0.6n = -7.4$ **(−8.5625, −19.0625)**

★**20.** $3x - 2y = 4$; $\frac{1}{2}x - \frac{2}{3}y = 1$ **$\left(\frac{2}{3}, -1\right)$**

★**21.** $2a + 3b = -16$; $\frac{3}{4}a - \frac{7}{8}b = 10$ **(4, −8)**

★**22.** $\frac{1}{3}r + \frac{2}{5}s = 5$; $\frac{2}{3}r - \frac{1}{2}s = -3$ **(3, 10)**

★**23.** $\frac{3}{4}x + \frac{1}{2}y = \frac{11}{12}$; $\frac{1}{2}x - \frac{1}{4}y = \frac{1}{8}$ **$\left(\frac{2}{3}, \frac{5}{6}\right)$**

WebQuest

You can use Cramer's Rule to compare home loans. Visit www.algebra2.com/webquest to continue work on your WebQuest project.

24. **GEOMETRY** The two sides of an angle are contained in lines whose equations are $4x + y = -4$ and $2x - 3y = -9$. Find the coordinates of the vertex of the angle. **(−1.5, 2)**

25. **GEOMETRY** Two sides of a parallelogram are contained in the lines whose equations are $2.3x + 1.2y = 2.1$ and $4.1x - 0.5y = 14.3$. Find the coordinates of a vertex of the parallelogram. **(3, −4)**

Answers

38. If the determinant is zero, there is no unique solution to the system. There is either no solution or there are infinitely many solutions. Sample answer: $2x + y = 4$ and $4x + 2y = 8$ has a det = 0; there are infinitely many solutions of this system. $2x + y = 4$ and $4x + 2y = 10$ has a det = 0; there are no solutions of this system.

39. Cramer's Rule is a formula for the variables x and y where (x, y) is a solution for a system of equations. Answers should include the following.

- Cramer's Rule uses determinants composed of the coefficients and constants in a system of linear equations to solve the system.
- Cramer's Rule is convenient when coefficients are large or involve fractions or decimals. Finding the value of the determinant is sometimes easier than trying to find a greatest common factor if you are solving by using elimination or substituting complicated numbers.

Use Cramer's Rule to solve each system of equations.

26. $x + y + z = 6$, $2x + y - 4z = -15$, $5x - 3y + z = -10$ **(−1, 3, 4)**

27. $a - 2b + c = 7$, $6a + 2b - 2c = 4$, $4a + 6b + 4c = 14$ **(2, −1, 3)**

28. $r - 2s - 5t = -1$, $r + 2s - 2t = 5$, $4r + s + t = -1$ **$\left(-\frac{11}{19}, \frac{39}{19}, -\frac{14}{19}\right)$**

29. $3a + c = 23$, $4a + 7b - 2c = -22$, $8a - b - c = 34$ **$\left(\frac{141}{29}, -\frac{102}{29}, \frac{244}{29}\right)$**

★ 30. $4x + 2y - 3z = -32$, $-x - 3y + z = 54$, $2y + 8z = 78$ **(11, −17, 14)**

★ 31. $2r + 25s = 40$, $10r + 12s + 6t = -2$, $36r - 25s + 50t = -10$ **$\left(-\frac{155}{28}, \frac{143}{70}, \frac{673}{140}\right)$**

GAMES For Exercises 32 and 33, use the following information.

Marcus purchased a game card to play virtual games at the arcade. His favorite games are the race car simulator, which costs 7 points for each play, and the snowboard simulator, which costs 5 points for each play. Marcus came with enough money to buy a 50-point card, and he has time to play 8 games.

32. Write a system of equations. **$r + s = 8$, $7r + 5s = 50$**

33. Solve the system using Cramer's Rule to find the number of times Marcus can play race car simulator and snowboard simulator. **race car, 5 plays; snowboard, 3 plays**

INTERIOR DESIGN For Exercises 34 and 35, use the following information.

An interior designer is preparing invoices for two of her clients. She has ordered silk dupioni and cotton damask fabric for both of them.

Client	Fabric	Yards	Total Cost
Harada	silk cotton	8 13	\$604.79
Martina	silk cotton	$5\frac{1}{2}$ 14	\$542.30

34. Write a system of two equations using the information given. **$8s + 13c = 604.79$, $5\frac{1}{2}s + 14c = 542.30$**

35. Find the price per yard of each fabric. **silk, \$34.99; cotton, \$24.99**

Career Choices

Interior Designer

Interior designers work closely with architects and clients to develop a design that is not only aesthetic, but also functional, within budget, and meets all building codes.

Source: www.uwec.edu

Online Research

For information about a career as an interior designer, visit: www.algebra2.com/careers

PRICING For Exercises 36 and 37, use the following information.

The Harvest Nut Company sells made-to-order trail mixes. Santito's favorite mix contains peanuts, raisins, and carob-coated pretzels. Peanuts sell for \$3.20 per pound, raisins are \$2.40 per pound, and the carob-coated pretzels are \$4.00 per pound. Santito chooses to have twice as many pounds of pretzels as raisins, wants 5 pounds of mix, and can afford \$16.80.

36. Write a system of three equations using the information given.

37. How many pounds of peanuts, raisins, and carob-coated pretzels can Santito buy? **peanuts, 2 lb; raisins, 1 lb; pretzels, 2 lb**

36. $p + r + c = 5$, $2r - p = 0$, $3.2p + 2.4r + 4c = 16.8$

38. **CRITICAL THINKING** In Cramer's Rule, if the value of the determinant is zero, what must be true of the graph of the system of equations represented by the determinant? Give examples to support your answer. **See margin.**

39. **WRITING IN MATH** Answer the question that was posed at the beginning of the lesson. **See margin.**

How is Cramer's Rule used to solve systems of equations?

Include the following in your answer.

- an explanation of how Cramer's rule uses determinants, and
- a situation where Cramer's rule would be easier to solve a system of equations than substitution or elimination and why.

www.algebra2.com/self_check_quiz

Enrichment, p. 204

Fourth-Order Determinants

To find the value of a 4 × 4 determinant, use a method called **expansion by minors**.

First write the expansion. Use the first row of the determinant. Remember that the signs of the terms alternate.

$$\begin{vmatrix} 6 & -3 & 2 & 7 \\ 0 & 4 & 3 & 5 \\ 0 & 2 & 1 & -4 \\ 6 & 0 & -2 & 0 \end{vmatrix} = 6\begin{vmatrix} 4 & 3 & 5 \\ 2 & 1 & -4 \\ 0 & -2 & 0 \end{vmatrix} - (-3)\begin{vmatrix} 0 & 3 & 5 \\ 0 & 1 & -4 \\ 6 & -2 & 0 \end{vmatrix} + 2\begin{vmatrix} 0 & 4 & 5 \\ 0 & 2 & -4 \\ 6 & 0 & 0 \end{vmatrix} - 7\begin{vmatrix} 0 & 4 & 3 \\ 0 & 2 & 1 \\ 6 & 0 & -2 \end{vmatrix}$$

Then evaluate each 3 × 3 determinant. Use any row.

$$\begin{vmatrix} 4 & 3 & 5 \\ 2 & 1 & -4 \\ 0 & -2 & 0 \end{vmatrix} = -(-2)\begin{vmatrix} 4 & 5 \\ 2 & -4 \end{vmatrix}$$

$$= 2(-16 - 10)$$

$$\begin{vmatrix} 0 & 3 & 5 \\ 0 & 1 & -4 \\ 6 & -2 & 0 \end{vmatrix} = -3\begin{vmatrix} 0 & -4 \\ 6 & 0 \end{vmatrix} + 5\begin{vmatrix} 0 & 1 \\ 6 & -2 \end{vmatrix}$$

$$= -3(24) + 5(-6)$$

Study Guide and Intervention, p. 199 (shown) and p. 200

Systems of Two Linear Equations Determinants provide a way for solving systems of equations.

Cramer's Rule for Two-Variable Systems	The solution of the linear system of equations $ax + by = e$, $cx + dy = f$ is (x, y) where $x = \frac{\begin{vmatrix} e & b \\ f & d \end{vmatrix}}{\begin{vmatrix} a & b \\ c & d \end{vmatrix}}$, $y = \frac{\begin{vmatrix} a & e \\ c & f \end{vmatrix}}{\begin{vmatrix} a & b \\ c & d \end{vmatrix}}$, and $\begin{vmatrix} a & b \\ c & d \end{vmatrix} \neq 0$.

Example Use Cramer's Rule to solve the system of equations. $5x - 10y = 8$, $10x + 25y = -2$

$x = \frac{\begin{vmatrix} e & b \\ f & d \end{vmatrix}}{\begin{vmatrix} a & b \\ c & d \end{vmatrix}}$ Cramer's Rule $y = \frac{\begin{vmatrix} a & e \\ c & f \end{vmatrix}}{\begin{vmatrix} a & b \\ c & d \end{vmatrix}}$

$= \frac{\begin{vmatrix} 8 & -10 \\ -2 & 25 \end{vmatrix}}{\begin{vmatrix} 5 & -10 \\ 10 & 25 \end{vmatrix}}$ $a = 5, b = -10, c = 10, d = 25, e = 8, f = -2$ $= \frac{\begin{vmatrix} 5 & 8 \\ 10 & -2 \end{vmatrix}}{\begin{vmatrix} 5 & -10 \\ 10 & 25 \end{vmatrix}}$

$= \frac{8(25) - (-2)(-10)}{5(25) - (-10)(10)}$ Evaluate each determinant. $= \frac{5(-2) - 8(10)}{5(25) - (-10)(10)}$

$= \frac{180}{225}$ or $\frac{4}{5}$ Simplify. $= -\frac{90}{225}$ or $-\frac{2}{5}$

The solution is $\left(\frac{4}{5}, -\frac{2}{5}\right)$.

Exercises

Use Cramer's Rule to solve each system of equations.

1. $3x - 2y = 7$, $2x + 7y = 38$ **(5, 4)**
2. $x - 4y = 17$, $3x - y = 29$ **(9, −2)**
3. $2x - y = -2$, $4x - y = 4$ **(3, 8)**
4. $2x - y = 1$, $5x + 2y = -29$ **(−3, −7)**
5. $4x + 2y = 1$, $5x - 4y = 24$ **$\left(2, -\frac{7}{2}\right)$**
6. $6x - 3y = -3$, $2x + y = 21$ **(5, 11)**
7. $2x + 7y = 16$, $x - 2y = 30$ **(22, −4)**
8. $2x - 3y = -2$, $3x - 4y = 9$ **(35, 24)**
9. $\frac{x}{3} + \frac{y}{5} = 2$, $\frac{x}{4} - \frac{y}{6} = -8$ **(−12, 30)**
10. $6x - 9y = -1$, $3x + 18y = 12$ **$\left(\frac{2}{3}, \frac{5}{9}\right)$**
11. $3x - 12y = -14$, $9x + 6y = -7$ **$\left(-\frac{4}{3}, \frac{5}{6}\right)$**
12. $8x + 2y = \frac{3}{7}$, $5x - 4y = -\frac{27}{7}$ **$\left(-\frac{1}{7}, \frac{11}{14}\right)$**

Skills Practice, p. 201 and Practice, p. 202 (shown)

Use Cramer's Rule to solve each system of equations.

1. $2x + y = 0$, $3x + 2y = -2$ **(2, −4)**
2. $5c + 9d = 19$, $2c - d = -20$ **(−7, 6)**
3. $2x + 3y = 5$, $3x - 2y = 1$ **(1, 1)**
4. $20m - 3n = 28$, $2m + 3n = 16$ **(2, 4)**
5. $x - 3y = 6$, $3x + y = -22$ **(−6, −4)**
6. $5x - 6y = -45$, $9x + 8y = 13$ **(−3, 5)**
7. $-2e + f = 4$, $-3e + 5f = -15$ **(−5, −6)**
8. $2x - y = -1$, $2x - 4y = 8$ **(−2, −3)**
9. $8a + 3b = 24$, $2a + b = 4$ **(6, −8)**
10. $-3x + 15y = 45$, $-2x + 7y = 18$ **(5, 4)**
11. $3u - 5v = 11$, $6u + 7v = -12$ **$\left(\frac{1}{3}, -2\right)$**
12. $-6g + h = -10$, $-3g - 4h = 4$ **$\left(\frac{4}{3}, -2\right)$**
13. $x - 3y = 8$, $x - 0.5y = 3$ **(2, −2)**
14. $0.2x - 0.5y = -1$, $0.6x - 3y = -9$ **(5, 4)**
15. $0.3d - 0.6g = 1.8$, $0.2d + 0.3g = 0.5$ **(4, −1)**

16. **GEOMETRY** The two sides of an angle are contained in the lines whose equations are $x - \frac{4}{3}y = 6$ and $2x + y = 1$. Find the coordinates of the vertex of the angle. **(2, −3)**

17. **GEOMETRY** Two sides of a parallelogram are contained in the lines whose equations are $0.2x - 0.5y = 1$ and $0.02x - 0.3y = -0.9$. Find the coordinates of a vertex of the parallelogram. **(15, 4)**

Use Cramer's Rule to solve each system of equations.

18. $x + 3y + 3z = 4$, $-x + 2y + z = -1$, $4x + y - 2z = -1$ **(1, −1, 2)**
19. $-5a + b - 4c = 7$, $-3a + 2b - c = 0$, $2a + 3b - c = 17$ **(3, 2, −5)**
20. $2x + y - 3z = -5$, $5x + 2y - 2z = 8$, $3x - 3y + 5z = 17$ **(2, 3, 4)**
21. $2c + 3d - e = 17$, $4c + d + 5e = -9$, $c + 2d - e = 12$ **(2, 3, −4)**
22. $2j + k - 3m = -3$, $3j + 2k + 4m = 5$, $-4j - k + 2m = 4$ **(−1, 2, 1)**
23. $3x - 2y + 5z = 3$, $2x + 2y - 4z = 3$, $-5x + 10y + 7z = -3$ **(1.2, 0.3, 0)**

24. **LANDSCAPING** A memorial garden being planted in front of a municipal library will contain three circular beds that are tangent to each other. A landscape architect has prepared a sketch of the design for the garden using CAD (computer-aided drafting) software, as shown at the right. The centers of the three circular beds are represented by points *A*, *B*, and *C*. The distance from *A* to *B* is 15 feet, the distance from *B* to *C* is 13 feet, and the distance from *A* to *C* is 16 feet. What is the radius of each of the circular beds? **circle *A*: 9 ft, circle *B*: 6 ft, circle *C*: 7 ft**

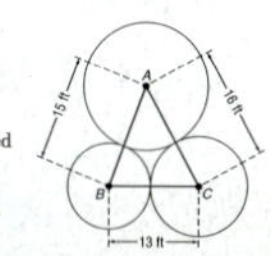

Reading to Learn Mathematics, p. 203 ELL

Pre-Activity How is Cramer's Rule used to solve systems of equations?

Read the introduction to Lesson 4-6 at the top of page 189 in your textbook.

A triangle is bounded by the x-axis, the line $y = \frac{1}{2}x$, and the line $y = -2x + 10$. Write three systems of equations that you could use to find the three vertices of the triangle. (Do not actually find the vertices.)

$x = 0, y = \frac{1}{2}x$; $x = 0, y = -2x + 10$; $y = \frac{1}{2}x, y = -2x + 10$

Reading the Lesson

1. Suppose that you are asked to solve the following system of equations by Cramer's Rule.

$3x + 2y = 7$
$2x - 3y = 22$

Without actually evaluating any determinants, indicate which of the following ratios of determinants gives the correct value for x. **B**

A. $\frac{\begin{vmatrix} 3 & 2 \\ 2 & -3 \end{vmatrix}}{\begin{vmatrix} 7 & 2 \\ 22 & -3 \end{vmatrix}}$ B. $\frac{\begin{vmatrix} 7 & 2 \\ 22 & -3 \end{vmatrix}}{\begin{vmatrix} 3 & 2 \\ 2 & -3 \end{vmatrix}}$ C. $\frac{\begin{vmatrix} 3 & 7 \\ 2 & 22 \end{vmatrix}}{\begin{vmatrix} 3 & 2 \\ 2 & -3 \end{vmatrix}}$

2. In your textbook, the statements of Cramer's Rule for two variables and three variables specify that the determinant formed from the coefficients of the variables cannot be 0. If the determinant is zero, what do you know about the system and its solutions?

The system could be a dependent system and have infinitely many solutions, or it could be an inconsistent system and have no solutions.

Helping You Remember

3. Some students have trouble remembering how to arrange the determinants that are used in solving a system of two linear equations by Cramer's Rule. What is a good way to remember this?

Sample answer: Let D be the determinant of the coefficients. Let D_x be the determinant formed by replacing the first column of D with the constants from the right-hand side of the system, and let D_y be the determinant formed by replacing the second column of D with the constants. Then $x = \frac{D_x}{D}$ and $y = \frac{D_y}{D}$.

4 Assess

Open-Ended Assessment

Speaking Have students share tips and techniques for remembering the correct order of the calculations they must perform in the multi-step processes discussed in this lesson.

Getting Ready for Lesson 4-7

PREREQUISITE SKILL Lesson 4-7 shows how to determine if two matrices are inverses and how to find the inverse of a 2 × 2 matrix. Students will use their familiarity with multiplying matrices as they work with inverses. Exercises 52–54 should be used to determine your students' familiarity with multiplying matrices.

Assessment Options

Practice Quiz 2 The quiz provides students with a brief review of the concepts and skills in Lessons 4-4 through 4-6. Lesson numbers are given to the right of exercises or instruction lines so students can review concepts not yet mastered.

Quiz (Lessons 4-5 and 4-6) is available on p. 232 of the *Chapter 4 Resource Masters*.

Standardized Test Practice

40. Use Cramer's Rule to solve the system of equations $3x + 8y = 28$ and $5x - 7y = -55$. **B**

Ⓐ (3, 5) Ⓑ (−4, 5) Ⓒ (4, 2) Ⓓ (−5, 4)

41. **SHORT RESPONSE** Find the measures of $\angle ABC$ and $\angle CBD$. **111°; 69°**

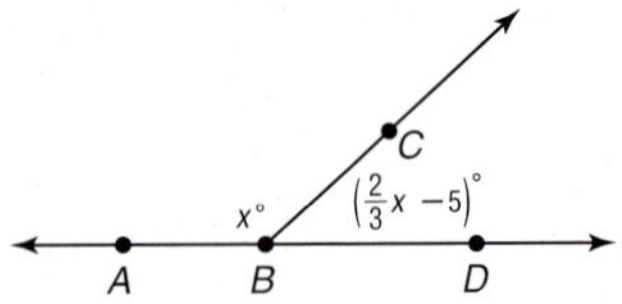

Maintain Your Skills

Mixed Review

Find the value of each determinant. *(Lesson 4-5)*

42. $\begin{vmatrix} 3 & 2 \\ -2 & 4 \end{vmatrix}$ **16**   43. $\begin{vmatrix} 8 & 6 \\ 4 & 8 \end{vmatrix}$ **40**   44. $\begin{vmatrix} -5 & 2 \\ 4 & 9 \end{vmatrix}$ **−53**

For Exercises 45–47, use the following information.
Triangle ABC with vertices $A(0, 2)$, $B(-3, -1)$, and $C(-2, -4)$ is translated 1 unit right and 3 units up. *(Lesson 4-4)*

45. Write the translation matrix. $\begin{bmatrix} 1 & 1 & 1 \\ 3 & 3 & 3 \end{bmatrix}$

46. Find the coordinates of $\triangle A'B'C'$. **$A'(1, 5)$, $B'(-2, 2)$, $C'(-1, -1)$**

47. Graph the preimage and the image. **See margin.**

48–50. See margin for graphs.

Solve each system of equations by graphing. *(Lesson 3-1)*

48. $y = 3x + 5$, $y = -2x - 5$ **(−2, −1)**   49. $x + y = 7$, $\frac{1}{2}x - y = -1$ **(4, 3)**   50. $x - 2y = 10$, $2x - 4y = 12$ **no solution**

51. **BUSINESS** The Friendly Fix-It Company charges a base fee of \$35 for any in-home repair. In addition, the technician charges \$10 per hour. Write an equation for the cost c of an in-home repair of h hours. *(Lesson 1-3)* **$c = 10h + 35$**

Getting Ready for the Next Lesson

PREREQUISITE SKILL Find each product, if possible.
*(To review **multiplying matrices**, see Lesson 4-3.)*

52. $[2 \quad 5] \cdot \begin{bmatrix} 3 & 1 \\ -2 & 6 \end{bmatrix}$ **$[-4 \quad 32]$**   53. $\begin{bmatrix} 0 & 9 \\ 5 & 7 \end{bmatrix} \cdot \begin{bmatrix} 2 & -6 \\ 8 & 1 \end{bmatrix}$ $\begin{bmatrix} 72 & 9 \\ 66 & -23 \end{bmatrix}$ 54. $\begin{bmatrix} 5 & -4 \\ 8 & 3 \end{bmatrix} \cdot \begin{bmatrix} 5 \\ 1 \end{bmatrix}$ $\begin{bmatrix} 21 \\ 43 \end{bmatrix}$

Practice Quiz 2 — *Lessons 4-4 through 4-6*

For Exercises 1–3, reflect square $ABCD$ with vertices $A(1, 2)$, $B(4, -1)$, $C(1, -4)$, and $D(-2, -1)$ over the y-axis. *(Lesson 4-4)*

1. Write the coordinates in a vertex matrix. $\begin{bmatrix} 1 & 4 & 1 & -2 \\ 2 & -1 & -4 & -1 \end{bmatrix}$
2. Find the coordinates of $A'B'C'D'$. **$A'(-1, 2)$, $B'(-4, -1)$, $C'(-1, -4)$, $D'(2, -1)$**
3. Graph $ABCD$ and $A'B'C'D'$. **See pp. 217A–217B.**

Find the value of each determinant. *(Lesson 4-5)*

4. $\begin{vmatrix} 3 & -2 \\ 5 & 4 \end{vmatrix}$ **22**   5. $\begin{vmatrix} -8 & 3 \\ 6 & 5 \end{vmatrix}$ **−58**   6. $\begin{vmatrix} 1 & 3 & -2 \\ 7 & 0 & 4 \\ -3 & 5 & -1 \end{vmatrix}$ **−105**   7. $\begin{vmatrix} 3 & 4 & 4 \\ 2 & 1 & 5 \\ 0 & -8 & 6 \end{vmatrix}$ **26**

Use Cramer's Rule to solve each system of equations. *(Lesson 4-6)*

8. $3x - 2y = 7$, $4x - y = 6$ **(1, −2)**   9. $7r + 5s = 3$, $3r - 2s = 22$ **(4, −5)**   10. $3a - 5b + 2c = -5$, $4a + b + 3c = 9$, $2a - c = 1$ **(1, 2, 1)**

Answers

47.

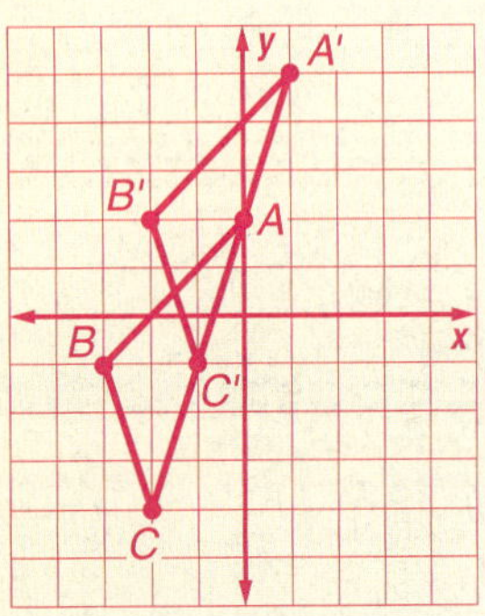

48.

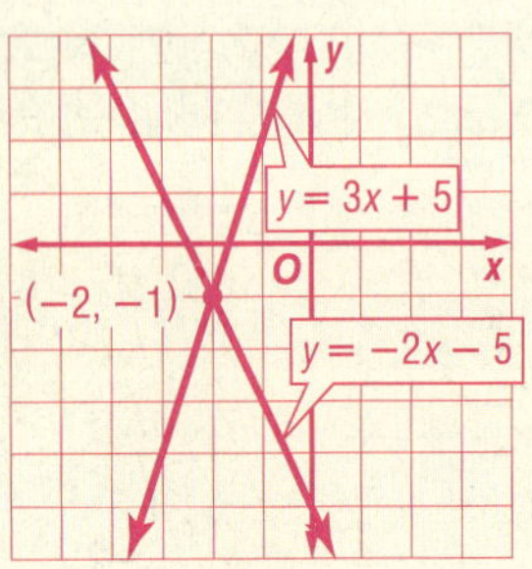

49.

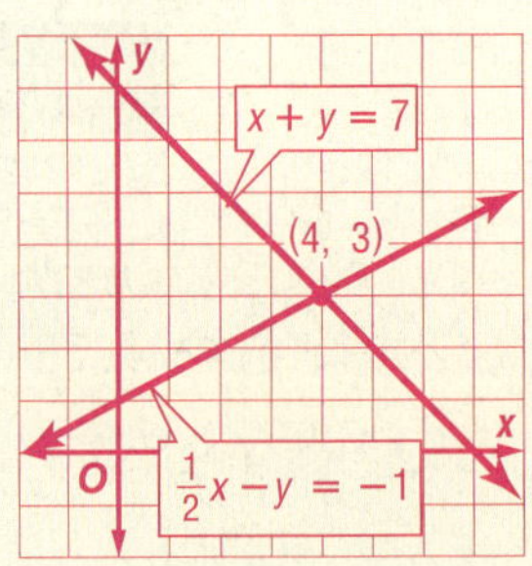

50.

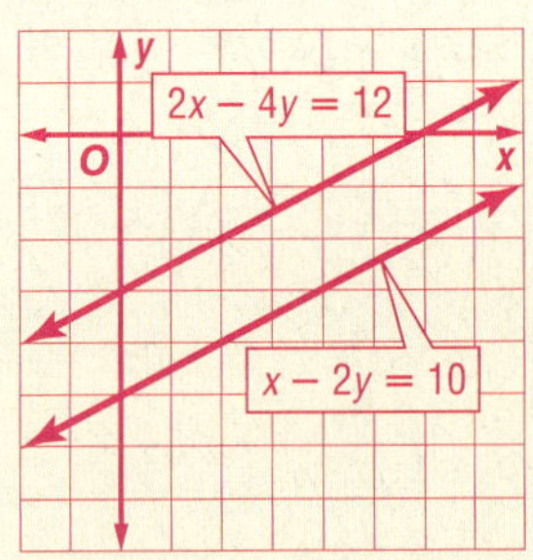

4-7 Identity and Inverse Matrices

What You'll Learn

- Determine whether two matrices are inverses.
- Find the inverse of a 2×2 matrix.

How are inverse matrices used in cryptography?

With the rise of Internet shopping, ensuring the privacy of the user's personal information has become an important priority. Companies protect their computers by using codes. Cryptography is a method of preparing coded messages that can only be deciphered by using the "key" to the message.

The following technique is a simplified version of how cryptography works.

- First, assign a number to each letter of the alphabet.
- Convert your message into a matrix and multiply it by the coding matrix. The message is now unreadable to anyone who does not have the key to the code.
- To decode the message, the recipient of the coded message would multiply by the opposite, or inverse, of the coding matrix.

Code

_ 0	A 1	B 2	C 3	D 4	E 5	F 6	G 7	H 8
I 9	J 10	K 11	L 12	M 13	N 14	O 15	P 16	Q 17
R 18	S 19	T 20	U 21	V 22	W 23	X 24	Y 25	Z 26

Vocabulary

- identity matrix
- inverse

IDENTITY AND INVERSE MATRICES Recall that in real numbers, two numbers are inverses if their product is the identity, 1. Similarly, for matrices, the **identity matrix** is a square matrix that, when multiplied by another matrix, equals that same matrix. If A is any $n \times n$ matrix and I is the $n \times n$ identity matrix, then $A \cdot I = A$ and $I \cdot A = A$.

2 × 2 Identity Matrix

$$\begin{bmatrix} 1 & 0 \\ 0 & 1 \end{bmatrix}$$

3 × 3 Identity Matrix

$$\begin{bmatrix} 1 & 0 & 0 \\ 0 & 1 & 0 \\ 0 & 0 & 1 \end{bmatrix}$$

Key Concept — Identity Matrix for Multiplication

- **Words** The identity matrix for multiplication I is a square matrix with 1 for every element of the main diagonal, from upper left to lower right, and 0 in all other positions. For any square matrix A of the same dimension as I, $A \cdot I = I \cdot A = A$.
- **Symbols** If $A = \begin{bmatrix} a & b \\ c & d \end{bmatrix}$, then $I = \begin{bmatrix} 1 & 0 \\ 0 & 1 \end{bmatrix}$ such that

$$\begin{bmatrix} a & b \\ c & d \end{bmatrix} \cdot \begin{bmatrix} 1 & 0 \\ 0 & 1 \end{bmatrix} = \begin{bmatrix} 1 & 0 \\ 0 & 1 \end{bmatrix} \cdot \begin{bmatrix} a & b \\ c & d \end{bmatrix} = \begin{bmatrix} a & b \\ c & d \end{bmatrix}.$$

Two $n \times n$ matrices are **inverses** of each other if their product is the identity matrix. If matrix A has an inverse symbolized by A^{-1}, then $A \cdot A^{-1} = A^{-1} \cdot A = I$.

4-7 Lesson Notes

1 Focus

5-Minute Check Transparency 4-7 Use as a quiz or review of Lesson 4-6.

Mathematical Background notes are available for this lesson on p. 152D.

How are inverse matrices used in cryptography?

Ask students:

- Using just the simple code given in the table, what word is given by the numbers 13, 1, 20, 8? MATH
- The decoding process mentions the "inverse of the coding matrix." What other inverses have you seen in mathematics? multiplicative inverses (or reciprocals), and additive inverses (or opposites)

Resource Manager

Workbook and Reproducible Masters

Chapter 4 Resource Masters

- Study Guide and Intervention, pp. 205–206
- Skills Practice, p. 207
- Practice, p. 208
- Reading to Learn Mathematics, p. 209
- Enrichment, p. 210

Transparencies

5-Minute Check Transparency 4-7

Answer Key Transparencies

Technology

Interactive Chalkboard

2 Teach

IDENTITY AND INVERSE MATRICES

In-Class Example

Power Point®

1 Determine whether each pair of matrices are inverses.

a. $X = \begin{bmatrix} 3 & -2 \\ -1 & 1 \end{bmatrix}$ and $Y = \begin{bmatrix} 1 & 2 \\ 1 & 3 \end{bmatrix}$ **yes**

b. $P = \begin{bmatrix} 3 & -1 \\ 4 & -2 \end{bmatrix}$ and $Q = \begin{bmatrix} 1 & -3 \\ 2 & 4 \end{bmatrix}$ **no**

Example 1 Verify Inverse Matrices

Determine whether each pair of matrices are inverses.

a. $X = \begin{bmatrix} 2 & 2 \\ -1 & 4 \end{bmatrix}$ **and** $Y = \begin{bmatrix} \frac{1}{2} & \frac{1}{2} \\ -1 & \frac{1}{4} \end{bmatrix}$

Check to see if $X \cdot Y = I$.

$X \cdot Y = \begin{bmatrix} 2 & 2 \\ -1 & 4 \end{bmatrix} \cdot \begin{bmatrix} \frac{1}{2} & \frac{1}{2} \\ -1 & \frac{1}{4} \end{bmatrix}$ — Write an equation.

$= \begin{bmatrix} 1 - 2 & 1 + \frac{1}{2} \\ -\frac{1}{2} + (-4) & -\frac{1}{2} + 1 \end{bmatrix}$ or $\begin{bmatrix} -1 & 1\frac{1}{2} \\ -4\frac{1}{2} & \frac{1}{2} \end{bmatrix}$ — Matrix multiplication

Since $X \cdot Y \neq I$, they are *not* inverses.

b. $P = \begin{bmatrix} 3 & 4 \\ 1 & 2 \end{bmatrix}$ **and** $Q = \begin{bmatrix} 1 & -2 \\ -\frac{1}{2} & \frac{3}{2} \end{bmatrix}$

Find $P \cdot Q$.

$P \cdot Q = \begin{bmatrix} 3 & 4 \\ 1 & 2 \end{bmatrix} \cdot \begin{bmatrix} 1 & -2 \\ -\frac{1}{2} & \frac{3}{2} \end{bmatrix}$ — Write an equation.

$= \begin{bmatrix} 3 - 2 & -6 + 6 \\ 1 - 1 & -2 + 3 \end{bmatrix}$ or $\begin{bmatrix} 1 & 0 \\ 0 & 1 \end{bmatrix}$ — Matrix multiplication

Now find $Q \cdot P$.

$Q \cdot P = \begin{bmatrix} 1 & -2 \\ -\frac{1}{2} & \frac{3}{2} \end{bmatrix} \cdot \begin{bmatrix} 3 & 4 \\ 1 & 2 \end{bmatrix}$ — Write an equation.

$= \begin{bmatrix} 3 - 2 & 4 - 4 \\ -\frac{3}{2} + \frac{3}{2} & -2 + 3 \end{bmatrix}$ or $\begin{bmatrix} 1 & 0 \\ 0 & 1 \end{bmatrix}$ — Matrix multiplication

Since $P \cdot Q = Q \cdot P = I$, P and Q are inverses.

Study Tip

Verifying Inverses
Since multiplication of matrices is not commutative, it is necessary to check the products in both orders.

FIND INVERSE MATRICES Some matrices do not have an inverse. You can determine whether a matrix has an inverse by using the determinant.

Key Concept — *Inverse of a 2 × 2 Matrix*

The inverse of matrix $A = \begin{bmatrix} a & b \\ c & d \end{bmatrix}$ is $A^{-1} = \frac{1}{ad - bc}\begin{bmatrix} d & -b \\ -c & a \end{bmatrix}$, where $ad - bc \neq 0$.

Notice that $ad - bc$ is the value of det A. Therefore, if the value of the determinant of a matrix is 0, the matrix cannot have an inverse.

Teacher to Teacher

Beatrice Moore-Harris — Educational Specialist, League City, TX

"Have students make a three-column table. List the terms dilation, dimension, element, image, minor, reflection, rotation, transformation, and translation in one column. Have them provide a real-world definition, example, and/or illustration of each term in another column and the mathematical definition in a third column. Finally, ask students to compare and contrast how the terms are used outside and inside of the math classroom. Are any definitions the same?"

Example 2 Find the Inverse of a Matrix

Find the inverse of each matrix, if it exists.

a. $R = \begin{bmatrix} -4 & -3 \\ 8 & 6 \end{bmatrix}$

Find the value of the determinant.

$$\begin{vmatrix} -4 & -3 \\ 8 & 6 \end{vmatrix} = -24 - (-24) = 0$$

Since the determinant equals 0, R^{-1} does not exist.

b. $P = \begin{bmatrix} 3 & 1 \\ 5 & 2 \end{bmatrix}$

Find the value of the determinant.

$$\begin{vmatrix} 3 & 1 \\ 5 & 2 \end{vmatrix} = 6 - 5 \text{ or } 1$$

Since the determinant does not equal 0, P^{-1} exists.

$$P^{-1} = \frac{1}{ad - bc}\begin{bmatrix} d & -b \\ -c & a \end{bmatrix} \quad \text{Definition of inverse}$$

$$= \frac{1}{3(2) - 1(5)}\begin{bmatrix} 2 & -1 \\ -5 & 3 \end{bmatrix} \quad a = 3, b = 1, c = 5, d = 2$$

$$= 1\begin{bmatrix} 2 & -1 \\ -5 & 3 \end{bmatrix} \text{ or } \begin{bmatrix} 2 & -1 \\ -5 & 3 \end{bmatrix} \quad \text{Simplify.}$$

CHECK $\begin{bmatrix} 2 & -1 \\ -5 & 3 \end{bmatrix} \cdot \begin{bmatrix} 3 & 1 \\ 5 & 2 \end{bmatrix} = \begin{bmatrix} 6 - 5 & 2 - 2 \\ -15 + 15 & -5 + 6 \end{bmatrix} = \begin{bmatrix} 1 & 0 \\ 0 & 1 \end{bmatrix}$ ✓

Matrices can be used to code messages by placing the message in a $2 \times n$ matrix.

Example 3 Use Inverses to Solve a Problem

a. CRYPTOGRAPHY Use the table at the beginning of the lesson to assign a number to each letter in the message GO_TONIGHT. Then code the message with the matrix $A = \begin{bmatrix} 2 & 1 \\ 4 & 3 \end{bmatrix}$.

Convert the message to numbers using the table.

G O _ T O N I G H T
7 | 15 | 0 | 20 | 15 | 14 | 9 | 7 | 8 | 20

Write the message in matrix form. Then multiply the message matrix B by the coding matrix A.

$$BA = \begin{bmatrix} 7 & 15 \\ 0 & 20 \\ 15 & 14 \\ 9 & 7 \\ 8 & 20 \end{bmatrix} \cdot \begin{bmatrix} 2 & 1 \\ 4 & 3 \end{bmatrix} \quad \text{Write an equation.}$$

$$= \begin{bmatrix} 14 + 60 & 7 + 45 \\ 0 + 80 & 0 + 60 \\ 30 + 56 & 15 + 42 \\ 18 + 28 & 9 + 21 \\ 16 + 80 & 8 + 60 \end{bmatrix} \quad \text{Matrix multiplication}$$

(continued on the next page)

More About...

Cryptography

The Enigma was a German coding machine used in World War II. Its code was considered to be unbreakable. However, the code was eventually solved by a group of Polish mathematicians.

Source: www.bletchleypark.org.uk

www.algebra2.com/extra_examples

FIND INVERSE MATRICES

In-Class Examples

Power Point®

2 Find the inverse of each matrix, if it exists.

a. $S = \begin{bmatrix} -1 & 0 \\ 8 & -2 \end{bmatrix}$ $\begin{bmatrix} -1 & 0 \\ -4 & -\frac{1}{2} \end{bmatrix}$

b. $T = \begin{bmatrix} -4 & 6 \\ -2 & 3 \end{bmatrix}$ **No inverse exists.**

3

a. Use the table at the beginning of the lesson to assign a number to each letter in the message ALWAYS_SMILE. Then code the message with the matrix $A = \begin{bmatrix} 1 & 2 \\ 1 & 3 \end{bmatrix}$.

The coded message is 13 | 38 | 24 | 49 | 44 | 107 | 19 | 57 | 22 | 53 | 17 | 39.

b. Use the inverse matrix A^{-1} to decode the message in In-Class Example 3a.

1 | 12 | 23 | 1 | 25 | 19 | 0 | 19 | 13 | 9 | 12 | 5; ALWAYS_SMILE

Teaching Tip Remind students to add a zero at the end of the message when it contains an odd number of letters.

Lesson 4-7 Identity and Inverse Matrices 197

3 Practice/Apply

Study Notebook

Have students—
- add the definitions/examples of the vocabulary terms in this lesson to their Vocabulary Builder worksheets for Chapter 4.
- write the 2 × 2 identity matrix, and the 3 × 3 identity matrix.
- write the formula for the inverse of a 2 × 2 matrix, including an example showing how to find the inverse of such a matrix.
- include any other item(s) that they find helpful in mastering the skills in this lesson.

Answers

1. $\begin{bmatrix} 1 & 0 & 0 & 0 \\ 0 & 1 & 0 & 0 \\ 0 & 0 & 1 & 0 \\ 0 & 0 & 0 & 1 \end{bmatrix}$

2. Exchange the values for *a* and *d* in the first diagonal in the matrix. Multiply the values for *b* and *c* by −1 in the second diagonal in the matrix. Find the determinant of the original matrix. Multiply the negative reciprocal of the determinant by the matrix with the above mentioned changes.

3. Sample answer: $\begin{bmatrix} 3 & 3 \\ 3 & 3 \end{bmatrix}$

Study Tip

Messages
If there is an odd number of letters to be coded, add a 0 at the end of the message.

$$= \begin{bmatrix} 74 & 52 \\ 80 & 60 \\ 86 & 57 \\ 46 & 30 \\ 96 & 68 \end{bmatrix} \quad \text{Simplify.}$$

The coded message is 74 | 52 | 80 | 60 | 86 | 57 | 46 | 30 | 96 | 68.

b. Use the inverse matrix A^{-1} to decode the message in Example 3a.

First find the inverse matrix of $A = \begin{bmatrix} 2 & 1 \\ 4 & 3 \end{bmatrix}$.

$$A^{-1} = \frac{1}{ad - bc}\begin{bmatrix} d & -b \\ -c & a \end{bmatrix} \quad \text{Definition of inverse}$$

$$= \frac{1}{2(3) - (1)(4)}\begin{bmatrix} 3 & -1 \\ -4 & 2 \end{bmatrix} \quad a = 2, b = 1, c = 4, d = 3$$

$$= \frac{1}{2}\begin{bmatrix} 3 & -1 \\ -4 & 2 \end{bmatrix} \text{ or } \begin{bmatrix} \frac{3}{2} & -\frac{1}{2} \\ -2 & 1 \end{bmatrix} \quad \text{Simplify.}$$

Next, decode the message by multiplying the coded matrix C by A^{-1}.

$$CA^{-1} = \begin{bmatrix} 74 & 52 \\ 80 & 60 \\ 86 & 57 \\ 46 & 30 \\ 96 & 68 \end{bmatrix} \cdot \begin{bmatrix} \frac{3}{2} & -\frac{1}{2} \\ -2 & 1 \end{bmatrix}$$

$$= \begin{bmatrix} 111 - 104 & -37 + 52 \\ 120 - 120 & -40 + 60 \\ 129 - 114 & -43 + 57 \\ 69 - 60 & -23 + 30 \\ 144 - 136 & -48 + 68 \end{bmatrix}$$

$$= \begin{bmatrix} 7 & 15 \\ 0 & 20 \\ 15 & 14 \\ 9 & 7 \\ 8 & 20 \end{bmatrix}$$

Use the table again to convert the numbers to letters. You can now read the message.

7 | 15 | 0 | 20 | 15 | 14 | 9 | 7 | 8 | 20

G O _ T O N I G H T

Check for Understanding

Concept Check

1. **Write** the 4 × 4 identity matrix. **1–3. See margin.**
2. **Explain** how to find the inverse of a 2 × 2 matrix.
3. **OPEN ENDED** Create a square matrix that does not have an inverse.

DAILY INTERVENTION — Differentiated Instruction

Logical Ask students to write a comparison of the inverse of a matrix to the multiplicative and additive inverses of a number, discussing what is the same about these inverses, and what is different.

Guided Practice

GUIDED PRACTICE KEY

Exercises	Examples
4, 5	1
6–8	2
9	3

Determine whether each pair of matrices are inverses.

4. $A = \begin{bmatrix} 2 & -1 \\ 1 & -3 \end{bmatrix}, B = \begin{bmatrix} \frac{1}{2} & 0 \\ 0 & -\frac{1}{3} \end{bmatrix}$ **no**

5. $X = \begin{bmatrix} 3 & 1 \\ 5 & 2 \end{bmatrix}, Y = \begin{bmatrix} 2 & -1 \\ -5 & 3 \end{bmatrix}$ **yes**

Find the inverse of each matrix, if it exists.

6. $\begin{bmatrix} 8 & -5 \\ -3 & 2 \end{bmatrix}$ $\begin{bmatrix} 2 & 5 \\ 3 & 8 \end{bmatrix}$

7. $\begin{bmatrix} 4 & -8 \\ -1 & 2 \end{bmatrix}$ **no inverse exists**

8. $\begin{bmatrix} -5 & 1 \\ 7 & 4 \end{bmatrix}$ $-\frac{1}{27}\begin{bmatrix} 4 & -1 \\ -7 & -5 \end{bmatrix}$

Application

9. CRYPTOGRAPHY Select a headline from a newspaper or the title of a magazine article and code it using your own coding matrix. Give your message and the coding matrix to a friend to decode. (*Hint*: Use a coding matrix whose determinant is 1 and that has all positive elements.) **See students' work.**

★ indicates increased difficulty

Practice and Apply

Homework Help

For Exercises	See Examples
10–19, 32, 33	1
20–31	2
34–41	3

Extra Practice
See page 836.

Determine whether each pair of matrices are inverses.

10. $P = \begin{bmatrix} 0 & 1 \\ 1 & 1 \end{bmatrix}, Q = \begin{bmatrix} -1 & 1 \\ 1 & 0 \end{bmatrix}$ **yes**

11. $R = \begin{bmatrix} 2 & 2 \\ 3 & 4 \end{bmatrix}, S = \begin{bmatrix} 2 & -1 \\ -\frac{3}{2} & 1 \end{bmatrix}$ **yes**

12. $A = \begin{bmatrix} 6 & 2 \\ 5 & 2 \end{bmatrix}, B = \begin{bmatrix} 1 & 1 \\ -\frac{5}{2} & -3 \end{bmatrix}$ **no**

13. $X = \begin{bmatrix} \frac{1}{3} & -\frac{2}{3} \\ \frac{2}{3} & -\frac{1}{3} \end{bmatrix}, Y = \begin{bmatrix} 1 & 2 \\ 2 & 1 \end{bmatrix}$ **no**

14. $C = \begin{bmatrix} 1 & 5 \\ 1 & -2 \end{bmatrix}, D = \begin{bmatrix} \frac{2}{7} & \frac{5}{7} \\ \frac{1}{7} & -\frac{1}{7} \end{bmatrix}$ **yes**

★ **15.** $J = \begin{bmatrix} 1 & 2 & 3 \\ 2 & 3 & 1 \\ 1 & 1 & 2 \end{bmatrix}, K = \begin{bmatrix} -\frac{5}{4} & \frac{1}{4} & \frac{7}{4} \\ \frac{3}{4} & \frac{1}{4} & -\frac{5}{4} \\ \frac{1}{4} & -\frac{1}{4} & \frac{1}{4} \end{bmatrix}$ **yes**

Determine whether each statement is *true* or *false*.

16. Only square matrices have multiplicative identities. **true**

17. Only square matrices have multiplicative inverses. **true**

18. Some square matrices do not have multiplicative inverses. **true**

19. Some square matrices do not have multiplicative identities. **false**

Find the inverse of each matrix, if it exists.

20. $\begin{bmatrix} 5 & 0 \\ 0 & 1 \end{bmatrix}$ $\frac{1}{5}\begin{bmatrix} 1 & 0 \\ 0 & 5 \end{bmatrix}$

21. $\begin{bmatrix} 6 & 3 \\ 8 & 4 \end{bmatrix}$ **no inverse exists**

22. $\begin{bmatrix} 1 & 2 \\ 2 & 1 \end{bmatrix}$ $-\frac{1}{3}\begin{bmatrix} 1 & -2 \\ -2 & 1 \end{bmatrix}$

23. $\begin{bmatrix} 3 & 1 \\ -4 & 1 \end{bmatrix}$ $\frac{1}{7}\begin{bmatrix} 1 & -1 \\ 4 & 3 \end{bmatrix}$

24. $\begin{bmatrix} -3 & -2 \\ 6 & 4 \end{bmatrix}$ **no inverse exists**

25. $\begin{bmatrix} -3 & 7 \\ 2 & -6 \end{bmatrix}$

26. $\begin{bmatrix} 4 & -3 \\ 2 & 7 \end{bmatrix}$ $\frac{1}{34}\begin{bmatrix} 7 & 3 \\ -2 & 4 \end{bmatrix}$

27. $\begin{bmatrix} -2 & 0 \\ 5 & 6 \end{bmatrix}$ $-\frac{1}{12}\begin{bmatrix} 6 & 0 \\ -5 & -2 \end{bmatrix}$

28. $\begin{bmatrix} -4 & 6 \\ 6 & -9 \end{bmatrix}$ **no inverse exists**

29. $\begin{bmatrix} 2 & -5 \\ 6 & 1 \end{bmatrix}$ $\frac{1}{32}\begin{bmatrix} 1 & 5 \\ -6 & 2 \end{bmatrix}$

★ **30.** $\begin{bmatrix} \frac{1}{2} & -\frac{3}{4} \\ \frac{1}{6} & \frac{1}{4} \end{bmatrix}$

★ **31.** $\begin{bmatrix} \frac{3}{10} & \frac{5}{8} \\ \frac{1}{5} & \frac{3}{4} \end{bmatrix}$

25. $\frac{1}{4}\begin{bmatrix} -6 & -7 \\ -2 & -3 \end{bmatrix}$

30. $4\begin{bmatrix} \frac{1}{4} & \frac{3}{4} \\ -\frac{1}{6} & \frac{1}{2} \end{bmatrix}$

31. $10\begin{bmatrix} \frac{3}{4} & -\frac{5}{8} \\ -\frac{1}{5} & \frac{3}{10} \end{bmatrix}$

www.algebra2.com/self_check_quiz

About the Exercises...

Organization by Objective

- **Identity and Inverse Matrices:** 10–17, 32, 33
- **Find Inverse Matrices:** 18–31, 34–41

Odd/Even Assignments

Exercises 10–33 are structured so that students practice the same concepts whether they are assigned odd or even problems.

Alert! Exercise 42 involves research on the Internet or other reference materials. Exercises 47–52 require a graphing calculator.

Assignment Guide

Basic: 11, 13, 17–29 odd, 33, 43–46, 53–77

Average: 11–33 odd, 34–38, 43–46, 53–77 (optional: 47–52)

Advanced: 10–32 even, 34–71 (optional: 72–77)

Lesson 4-7 Identity and Inverse Matrices 199

Study Guide and Intervention, p. 205 (shown) and p. 206

Identity and Inverse Matrices The identity matrix for matrix multiplication is a square matrix with 1s for every element of the main diagonal and zeros elsewhere.

Identity Matrix for Multiplication	If A is an $n \times n$ matrix and I is the identity matrix, then $A \cdot I = A$ and $I \cdot A = A$.

If an $n \times n$ matrix A has an inverse A^{-1}, then $A \cdot A^{-1} = A^{-1} \cdot A = I$.

Example Determine whether $X = \begin{bmatrix} 7 & 4 \\ 10 & 6 \end{bmatrix}$ and $Y = \begin{bmatrix} 3 & -2 \\ -5 & \frac{7}{2} \end{bmatrix}$ are inverse matrices.

Find $X \cdot Y$.

$$X \cdot Y = \begin{bmatrix} 7 & 4 \\ 10 & 6 \end{bmatrix} \cdot \begin{bmatrix} 3 & -2 \\ -5 & \frac{7}{2} \end{bmatrix} = \begin{bmatrix} 21-20 & -14+14 \\ 30-30 & -20+21 \end{bmatrix} \text{ or } \begin{bmatrix} 1 & 0 \\ 0 & 1 \end{bmatrix}$$

Find $Y \cdot X$.

$$Y \cdot X = \begin{bmatrix} 3 & -2 \\ -5 & \frac{7}{2} \end{bmatrix} \cdot \begin{bmatrix} 7 & 4 \\ 10 & 6 \end{bmatrix} = \begin{bmatrix} 21-20 & 12-12 \\ -35+35 & -20+21 \end{bmatrix} \text{ or } \begin{bmatrix} 1 & 0 \\ 0 & 1 \end{bmatrix}$$

Since $X \cdot Y = Y \cdot X = I$, X and Y are inverse matrices.

Exercises

Determine whether each pair of matrices are inverses.

1. $\begin{bmatrix} 4 & 5 \\ 3 & 4 \end{bmatrix}$ and $\begin{bmatrix} 4 & -5 \\ -3 & 4 \end{bmatrix}$ **yes**
2. $\begin{bmatrix} 3 & 2 \\ 5 & 4 \end{bmatrix}$ and $\begin{bmatrix} 2 & -1 \\ -\frac{5}{2} & \frac{3}{2} \end{bmatrix}$ **yes**
3. $\begin{bmatrix} 2 & 3 \\ 5 & -1 \end{bmatrix}$ and $\begin{bmatrix} 2 & 3 \\ -1 & -2 \end{bmatrix}$ **no**
4. $\begin{bmatrix} 8 & 11 \\ 3 & 4 \end{bmatrix}$ and $\begin{bmatrix} -4 & 11 \\ 3 & -8 \end{bmatrix}$ **yes**
5. $\begin{bmatrix} 4 & -1 \\ 5 & 3 \end{bmatrix}$ and $\begin{bmatrix} 1 & 2 \\ 3 & 8 \end{bmatrix}$ **no**
6. $\begin{bmatrix} 5 & 2 \\ 11 & 4 \end{bmatrix}$ and $\begin{bmatrix} -2 & 1 \\ \frac{11}{2} & -\frac{5}{2} \end{bmatrix}$ **yes**
7. $\begin{bmatrix} 4 & 2 \\ 6 & -2 \end{bmatrix}$ and $\begin{bmatrix} -\frac{1}{5} & -\frac{1}{10} \\ \frac{3}{10} & \frac{1}{10} \end{bmatrix}$ **no**
8. $\begin{bmatrix} 5 & 8 \\ 4 & 6 \end{bmatrix}$ and $\begin{bmatrix} -3 & 4 \\ 2 & -\frac{5}{2} \end{bmatrix}$ **yes**
9. $\begin{bmatrix} 3 & 7 \\ 2 & 4 \end{bmatrix}$ and $\begin{bmatrix} \frac{7}{2} & -\frac{3}{2} \\ 1 & -2 \end{bmatrix}$ **no**
10. $\begin{bmatrix} 3 & 2 \\ 4 & -6 \end{bmatrix}$ and $\begin{bmatrix} 3 & 2 \\ -4 & -3 \end{bmatrix}$ **no**
11. $\begin{bmatrix} 7 & 2 \\ 17 & 5 \end{bmatrix}$ and $\begin{bmatrix} 5 & -2 \\ -17 & 7 \end{bmatrix}$ **yes**
12. $\begin{bmatrix} 4 & 3 \\ 7 & 5 \end{bmatrix}$ and $\begin{bmatrix} -5 & 3 \\ 7 & -4 \end{bmatrix}$ **yes**

Skills Practice, p. 207 and Practice, p. 208 (shown)

Determine whether each pair of matrices are inverses.

1. $M = \begin{bmatrix} 2 & 1 \\ 3 & 2 \end{bmatrix}, N = \begin{bmatrix} -2 & 1 \\ 3 & -2 \end{bmatrix}$ **no**
2. $X = \begin{bmatrix} -3 & 2 \\ 5 & -3 \end{bmatrix}, Y = \begin{bmatrix} 3 & 2 \\ 5 & 3 \end{bmatrix}$ **yes**
3. $A = \begin{bmatrix} 3 & 1 \\ -4 & 2 \end{bmatrix}, B = \begin{bmatrix} \frac{1}{5} & -\frac{1}{10} \\ \frac{2}{5} & \frac{3}{10} \end{bmatrix}$ **yes**
4. $P = \begin{bmatrix} 6 & -2 \\ -2 & 3 \end{bmatrix}, Q = \begin{bmatrix} \frac{3}{14} & \frac{1}{7} \\ \frac{1}{7} & \frac{3}{7} \end{bmatrix}$ **yes**

Determine whether each statement is *true* or *false*.

5. All square matrices have multiplicative inverses. **false**
6. All square matrices have multiplicative identities. **true**

Find the inverse of each matrix, if it exists.

7. $\begin{bmatrix} 4 & 5 \\ -4 & -3 \end{bmatrix}$ $\frac{1}{8}\begin{bmatrix} -3 & -5 \\ 4 & 4 \end{bmatrix}$
8. $\begin{bmatrix} 2 & 0 \\ 3 & 5 \end{bmatrix}$ $\frac{1}{10}\begin{bmatrix} 5 & 0 \\ -3 & 2 \end{bmatrix}$
9. $\begin{bmatrix} -1 & 3 \\ 4 & -7 \end{bmatrix}$ $-\frac{1}{5}\begin{bmatrix} -7 & -3 \\ -4 & -1 \end{bmatrix}$
10. $\begin{bmatrix} 2 & 5 \\ -1 & 3 \end{bmatrix}$ $\frac{1}{11}\begin{bmatrix} 3 & -5 \\ 1 & 2 \end{bmatrix}$
11. $\begin{bmatrix} 2 & -5 \\ 3 & 1 \end{bmatrix}$ $\frac{1}{17}\begin{bmatrix} 1 & 5 \\ -3 & 2 \end{bmatrix}$
12. $\begin{bmatrix} 4 & 6 \\ 6 & 9 \end{bmatrix}$ **no inverse exists**

GEOMETRY For Exercises 13–16, use the figure at the right.

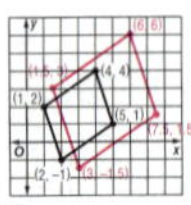

13. Write the vertex matrix A for the rectangle. $\begin{bmatrix} 1 & 4 & 5 & 2 \\ 2 & 4 & 1 & -1 \end{bmatrix}$
14. Use matrix multiplication to find BA if $B = \begin{bmatrix} 1.5 & 0 \\ 0 & 1.5 \end{bmatrix}$. $\begin{bmatrix} 1.5 & 6 & 7.5 & 3 \\ 3 & 6 & 1.5 & -1.5 \end{bmatrix}$
15. Graph the vertices of the transformed triangle on the previous graph. Describe the transformation. **dilation by a scale factor of 1.5**
16. Make a conjecture about what transformation B^{-1} describes on a coordinate plane. **dilation by a scale factor of $\frac{2}{3}$**
17. CODES Use the alphabet table below and the inverse of coding matrix $C = \begin{bmatrix} 1 & 2 \\ 2 & 1 \end{bmatrix}$ to decode this message:
19|14|11|13|11|22|55|65|57|60|2|1|52|47|33|51|56|55. **CHECK_YOUR_ANSWERS**

CODE						
A 1	B 2	C 3	D 4	E 5	F 6	G 7
H 8	I 9	J 10	K 11	L 12	M 13	N 14
O 15	P 16	Q 17	R 18	S 19	T 20	U 21
V 22	W 23	X 24	Y 25	Z 26	– 0	

Reading to Learn Mathematics, p. 209 ELL

Pre-Activity **How are inverse matrices used in cryptography?**

Read the introduction to Lesson 4-7 at the top of page 195 in your textbook.

Refer to the code table given in the introduction to this lesson. Suppose that you receive a message coded by this system as follows:

16 12 5 1 19 5   2 5   13 25   6 18 9 5 14 4.

Decode the message. **Please be my friend.**

Reading the Lesson

1. Indicate whether each of the following statements is *true* or *false*.
   a. Every element of an identity matrix is 1. **false**
   b. There is a 3×2 identity matrix. **false**
   c. Two matrices are inverses of each other if their product is the identity matrix. **true**
   d. If M is a matrix, M^{-1} represents the reciprocal of M. **false**
   e. No 3×2 matrix has an inverse. **true**
   f. Every square matrix has an inverse. **false**
   g. If the two columns of a 2×2 matrix are identical, the matrix does not have an inverse. **true**
2. Explain how to find the inverse of a 2×2 matrix. Do not use any special mathematical symbols in your explanation.
   Sample answer: First find the determinant of the matrix. If it is zero, then the matrix has no inverse. If the determinant is not zero, form a new matrix as follows. Interchange the top left and bottom right elements. Change the signs but not the positions of the other two elements. Multiply the resulting matrix by the reciprocal of the determinant of the original matrix. The resulting matrix is the inverse of the original matrix.

Helping You Remember

3. One way to remember something is to explain it to another person. Suppose that you are studying with a classmate who is having trouble remembering how to find the inverse of a 2×2 matrix. He remembers how to move elements and change signs in the matrix, but thinks that he should multiply by the determinant of the original matrix. How can you help him remember that he must multiply by the *reciprocal* of this determinant?
   Sample answer: If the determinant of the matrix is 0, its reciprocal is undefined. This agrees with the fact that if the determinant of a matrix is 0, the matrix does not have an inverse.

32. Compare the matrix used to reflect a figure over the x-axis to the matrix used to reflect a figure over the y-axis.
    a. Are they inverses? **no**
    b. Does your answer make sense based on the geometry? Use a drawing to support your answer. **See margin.**

33. The matrix used to rotate a figure 270° counterclockwise about the origin is $\begin{bmatrix} 0 & 1 \\ -1 & 0 \end{bmatrix}$. Compare this matrix with the matrix used to rotate a figure 90° counterclockwise about the origin.
    a. Are they inverses? **yes**
    b. Does your answer make sense based on the geometry? Use a drawing to support your answer. **See margin.**

GEOMETRY For Exercises 34–38, use the figure below.

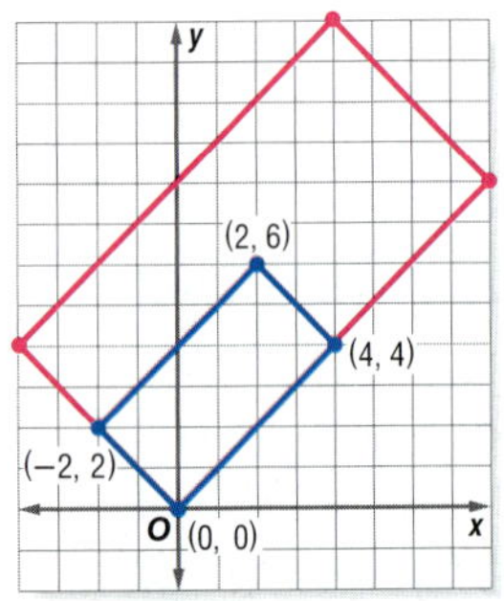

34. Write the vertex matrix A for the rectangle.
    34. $\begin{bmatrix} 0 & -2 & 2 & 4 \\ 0 & 2 & 6 & 4 \end{bmatrix}$
35. Use matrix multiplication to find BA if $B = \begin{bmatrix} 2 & 0 \\ 0 & 2 \end{bmatrix}$. $\begin{bmatrix} 0 & -4 & 4 & 8 \\ 0 & 4 & 12 & 8 \end{bmatrix}$
36. Graph the vertices of the transformed rectangle. Describe the transformation.
    36. dilation by a scale factor of 2
37. Make a conjecture about what transformation B^{-1} describes on a coordinate plane.
    37. dilation by a scale factor of $\frac{1}{2}$
38. Test your conjecture. Find B^{-1} and multiply it by the result of BA. Make a drawing to verify your conjecture.
    38. $B^{-1} = \begin{bmatrix} \frac{1}{2} & 0 \\ 0 & \frac{1}{2} \end{bmatrix}$; **the graph of the inverse transformation is the original figure.**

CRYPTOGRAPHY For Exercises 39–41, use the alphabet table below.

Your friend has sent you a series of messages that were coded with the coding matrix $C = \begin{bmatrix} 2 & 1 \\ 1 & 1 \end{bmatrix}$. Use the inverse of matrix C to decode each message.

Code		
A 26	J 17	S 8
B 25	K 16	T 7
C 24	L 15	U 6
D 23	M 14	V 5
E 22	N 13	W 4
F 21	O 12	X 3
G 20	P 11	Y 2
H 19	Q 10	Z 1
I 18	R 9	_ 0

39. 50|36|51|29|18|18|26|13|33|26|44|22|48|33|59|34|61|35|4|2
    39. MEET_IN_THE_LIBRARY
40. 59|33|8|8|39|21|7|7|56|37|25|16|4|2 **AT_SIX_THIRTY**
41. 59|34|49|31|40|20|16|14|21|15|25|25|36|24|32|16 **BRING_YOUR_BOOK**

42. **RESEARCH** Use the Internet or other reference to find examples of codes used throughout history. Explain how messages were coded. **See students' work.**

43. **CRITICAL THINKING** For which values of a, b, c, and d will $A = \begin{bmatrix} a & b \\ c & d \end{bmatrix} = A^{-1}$? $a = \pm 1, d = \pm 1, b = c = 0$

44. **WRITING IN MATH** Answer the question that was posed at the beginning of the lesson. **See margin.**

    How are inverse matrices used in cryptography?

    Include the following in your answer:
    - an explanation of why the inverse matrix works in decoding a message, and
    - a description of the conditions you must consider when writing a message in matrix form.

Enrichment, p. 210

Permutation Matrices

A permutation matrix is a square matrix in which each row and each column has one entry that is 1. All the other entries are 0. Find the inverse of a permutation matrix interchanging the rows and columns. For example, row 1 is interchanged with column 1, row 2 is interchanged with column 2.

$$P = \begin{bmatrix} 0 & 0 & 1 & 0 \\ 0 & 1 & 0 & 0 \\ 0 & 0 & 0 & 1 \\ 1 & 0 & 0 & 0 \end{bmatrix} \quad P^{-1} = \begin{bmatrix} 0 & 0 & 0 & 1 \\ 0 & 1 & 0 & 0 \\ 1 & 0 & 0 & 0 \\ 0 & 0 & 1 & 0 \end{bmatrix}$$

P is a 4×4 permutation matrix. P^{-1} is the inverse of P.

Solve each problem.

1. There is just one 2×2 permutation matrix that is not also an identity matrix. Write this matrix. $\begin{bmatrix} 0 & 1 \\ 1 & 0 \end{bmatrix}$
2. Find the inverse of the matrix you wrote in Exercise 1. What do you notice? $\begin{bmatrix} 0 & 1 \\ 1 & 0 \end{bmatrix}$ **The two matrices are the same.**
3. Show th[illegible] two matrices in Exercises 1 and 2 are inverses.

Standardized Test Practice

45. What is the inverse of $\begin{bmatrix} 4 & 1 \\ 10 & 2 \end{bmatrix}$? **A**

Ⓐ $\begin{bmatrix} -1 & \frac{1}{2} \\ 5 & -2 \end{bmatrix}$ Ⓑ $\begin{bmatrix} 2 & -1 \\ -10 & 4 \end{bmatrix}$ Ⓒ $\begin{bmatrix} 1 & 5 \\ \frac{1}{2} & 2 \end{bmatrix}$ Ⓓ $\begin{bmatrix} -2 & \frac{1}{2} \\ 5 & -1 \end{bmatrix}$

46. Which matrix does *not* have an inverse? **D**

Ⓐ $\begin{bmatrix} 5 & 3 \\ 2 & 4 \end{bmatrix}$ Ⓑ $\begin{bmatrix} 1 & 2 \\ 2 & 1 \end{bmatrix}$ Ⓒ $\begin{bmatrix} -3 & -3 \\ 6 & -6 \end{bmatrix}$ Ⓓ $\begin{bmatrix} -10 & -5 \\ 8 & 4 \end{bmatrix}$

Graphing Calculator

INVERSE FUNCTION The x^{-1} key on a TI-83 Plus is used to find the inverse of a matrix. If you get a SINGULAR MATRIX error on the screen, then the matrix has no inverse.

Use a graphing calculator to find the inverse of each matrix.

47. $\begin{bmatrix} -11 & 9 \\ 6 & -5 \end{bmatrix}$ **$\begin{bmatrix} -5 & -9 \\ -6 & -11 \end{bmatrix}$** 48. $\begin{bmatrix} 12 & 4 \\ 15 & 5 \end{bmatrix}$ **no inverse exists** 49. $\begin{bmatrix} 2 & -1 \\ 1 & -3 \end{bmatrix}$

50. $\begin{bmatrix} 25 & -4 \\ -35 & 6 \end{bmatrix}$ 51. $\begin{bmatrix} 2 & 5 & 2 \\ 1 & 4 & 1 \\ 6 & 3 & 3 \end{bmatrix}$ 52. $\begin{bmatrix} 3 & 1 & 2 \\ -2 & 0 & 4 \\ 3 & 5 & 2 \end{bmatrix}$

49. $\begin{bmatrix} \frac{3}{5} & -\frac{1}{5} \\ \frac{1}{5} & -\frac{2}{5} \end{bmatrix}$

50. $\begin{bmatrix} \frac{3}{5} & \frac{2}{5} \\ 3\frac{1}{2} & 2\frac{1}{2} \end{bmatrix}$

Maintain Your Skills

Mixed Review

51. $\begin{bmatrix} -1 & 1 & \frac{1}{3} \\ -\frac{1}{3} & \frac{2}{3} & 0 \\ \frac{7}{3} & -\frac{8}{3} & -\frac{1}{3} \end{bmatrix}$

52. $\begin{bmatrix} \frac{5}{16} & -\frac{1}{8} & -\frac{1}{16} \\ -\frac{1}{4} & 0 & \frac{1}{4} \\ \frac{5}{32} & \frac{3}{16} & -\frac{1}{32} \end{bmatrix}$

Use Cramer's Rule to solve each system of equations. *(Lesson 4-6)*

53. $3x + 2y = -2$, $x - 3y = 14$ **(2, −4)**

54. $2x + 5y = 35$, $7x - 4y = -28$ **(0, 7)**

55. $4x - 3z = -23$, $-2x - 5y + z = -9$, $y - z = 3$ **(−5, 4, 1)**

Evaluate each determinant by using diagonals or expansion by minors. *(Lesson 4-5)*

56. $\begin{vmatrix} 2 & 8 & -6 \\ 4 & 5 & 2 \\ -3 & -6 & -1 \end{vmatrix}$ **52** 57. $\begin{vmatrix} -3 & -3 & 1 \\ -9 & -2 & 3 \\ 5 & -2 & -1 \end{vmatrix}$ **−14** 58. $\begin{vmatrix} 5 & -7 & 3 \\ -1 & 2 & -9 \\ 5 & -7 & 3 \end{vmatrix}$ **0**

Find the slope of the line that passes through each pair of points. *(Lesson 2-3)*

59. (2, 5), (6, 9) **1** 60. (1, 0), (−2, 9) **−3** 61. (−5, 4), (−3, −6) **−5**

62. (−2, 2), (−5, 1) **$\frac{1}{3}$** 63. (0, 3), (−2, −2) **$\frac{5}{2}$** 64. (−8, 9), (0, 6) **$-\frac{3}{8}$**

65. **OCEANOGRAPHY** The deepest point in any ocean, the bottom of the Mariana Trench in the Pacific Ocean, is 6.8 miles below sea level. Water pressure in the ocean is represented by the function $f(x) = 1.15x$, where x is the depth in miles and $f(x)$ is the pressure in tons per square inch. Find the water pressure at the deepest point in the Mariana Trench. *(Lesson 2-1)* **7.82 tons/in²**

Evaluate each expression. *(Lesson 1-1)*

66. $3(2^3 + 1)$ **27** 67. $7 - 5 \div 2 + 1$ **$5\frac{1}{2}$** 68. $\frac{9 - 4 \cdot 3}{6}$ **$-\frac{1}{2}$**

69. $[40 - (7 + 9)] \div 8$ **3** 70. $[(-2 + 8)6 + 1]8$ **296** 71. $(4 - 1)(8 + 2)^2$ **300**

Getting Ready for the Next Lesson

PREREQUISITE SKILL Solve each equation.

*(To review **solving multi-step equations**, see Lesson 1-3.)*

72. $3k + 8 = 5$ **−1** 73. $12 = -5h + 2$ **−2** 74. $7z - 4 = 5z + 8$ **6**

75. $\frac{x}{2} + 5 = 7$ **4** 76. $\frac{3 + n}{6} = -4$ **−27** 77. $6 = \frac{s - 8}{-7}$ **−34**

44. A matrix can be used to code a message. The key to the message is the inverse of the matrix. Answers should include the following.

- The inverse matrix undoes the work of the matrix. So if you multiply a numeric message by a matrix it changes the message. When you multiply the changed message by the inverse matrix, the result is the original numeric message.
- You must consider the dimensions of the coding matrix so that you can write the numeric message in a matrix with dimensions that can be multiplied by the coding matrix.

4 Assess

Open-Ended Assessment

Writing Have students write a one-sentence message and assign a number to each letter using the chart on p. 195. Then have them write a 2 × 2 matrix of their own that has an inverse, and use it to code their message. Students should exchange their coded message with a classmate along with their coding matrix. The recipient should then find the inverse of the coding matrix they receive, use it to decode the numbers, and read the original message.

Getting Ready for Lesson 4-8

PREREQUISITE SKILL Lesson 4-8 shows how to use matrices to solve systems of equations. Students will use their familiarity with solving linear equations as they solve systems using matrices. Exercises 72–77 should be used to determine your students' familiarity with solving multi-step equations.

Answers

32b. Sample answer:

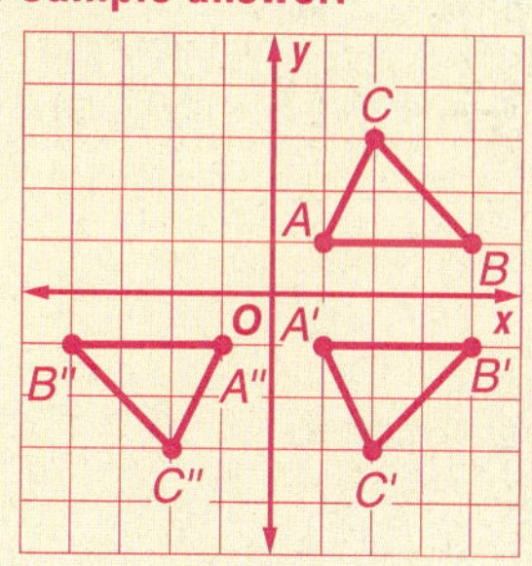

33b. Sample answer:

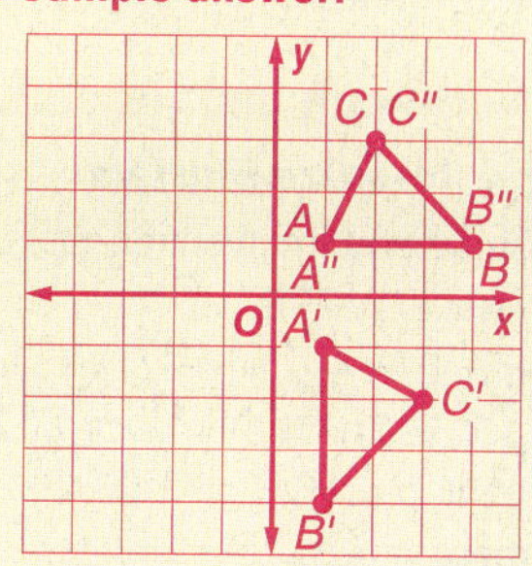

4-8 Lesson Notes

1 Focus

5-Minute Check Transparency 4-8 Use as a quiz or review of Lesson 4-7.

Building on Prior Knowledge

To solve a system of equations using a matrix equation, students will use the matrix multiplication skills they acquired in Lesson 4-3.

Mathematical Background notes are available for this lesson on p. 152D.

How can matrices be used in population ecology?

Ask students:

- Which of the two species of birds needs more pounds of food per nesting pair during its nesting season? **Species A**
- Which of the two species needs more territory? **Species A**

4-8 Using Matrices to Solve Systems of Equations

What You'll Learn

- Write matrix equations for systems of equations.
- Solve systems of equations using matrix equations.

Vocabulary

- matrix equation

How can matrices be used in population ecology?

Population ecology is the study of a species or a group of species that inhabits the same area. A biologist is studying two species of birds that compete for food and territory. He estimates that a particular region with an area of 14.25 acres (approximately 69,000 square yards) can supply 20,000 pounds of food for the birds during their nesting season.

Species A needs 140 pounds of food and has a territory of 500 square yards per nesting pair. Species B needs 120 pounds of food and has a territory of 400 square yards per nesting pair. The biologist can use this information to find the number of birds of each species that the area can support.

WRITE MATRIX EQUATIONS The situation above can be represented using a system of equations that can be solved using matrices. Consider the system of equations below. You can write this system with matrices by using the left and right sides of the equations.

$$\begin{aligned} 5x + 7y &= 11 \\ 3x + 8y &= 18 \end{aligned} \rightarrow \begin{bmatrix} 5x + 7y \\ 3x + 8y \end{bmatrix} = \begin{bmatrix} 11 \\ 18 \end{bmatrix}$$

Write the matrix on the left as the product of the coefficients and the variables.

$$\underset{\text{coefficient matrix}}{\underbrace{\begin{bmatrix} 5 & 7 \\ 3 & 8 \end{bmatrix}}^{A}} \cdot \underset{\text{variable matrix}}{\underbrace{\begin{bmatrix} x \\ y \end{bmatrix}}^{X}} = \underset{\text{constant matrix}}{\underbrace{\begin{bmatrix} 11 \\ 18 \end{bmatrix}}^{B}}$$

The system of equations is now expressed as a **matrix equation**.

Example 1 Two-Variable Matrix Equation

Write a matrix equation for the system of equations.
$5x - 6y = -47$
$3x + 2y = -17$

Determine the coefficient, variable, and constant matrices.

$$\begin{aligned} 5x - 6y &= -47 \\ 3x + 2y &= -17 \end{aligned} \rightarrow \begin{bmatrix} 5 & -6 \\ 3 & 2 \end{bmatrix} \quad \begin{bmatrix} x \\ y \end{bmatrix} \quad \begin{bmatrix} -47 \\ -17 \end{bmatrix}$$

Write the matrix equation.

$$\overset{A}{\begin{bmatrix} 5 & -6 \\ 3 & 2 \end{bmatrix}} \cdot \overset{X}{\begin{bmatrix} x \\ y \end{bmatrix}} = \overset{B}{\begin{bmatrix} -47 \\ -17 \end{bmatrix}}$$

Resource Manager

Workbook and Reproducible Masters

Chapter 4 Resource Masters

- Study Guide and Intervention, pp. 211–212
- Skills Practice, p. 213
- Practice, p. 214
- Reading to Learn Mathematics, p. 215
- Enrichment, p. 216
- Assessment, p. 232

School-to-Career Masters, p. 8

Transparencies

5-Minute Check Transparency 4-8
Answer Key Transparencies

Technology

Interactive Chalkboard

You can use a matrix equation to determine the weight of an atom of an element.

Example 2 Solve a Problem Using a Matrix Equation

CHEMISTRY The molecular formula for glucose is $C_6H_{12}O_6$, which represents that a molecule of glucose has 6 carbon (C) atoms, 12 hydrogen (H) atoms, and 6 oxygen (O) atoms. One molecule of glucose weighs 180 atomic mass units (amu), and one oxygen atom weighs 16 atomic mass units. The formulas and weights for glucose and another sugar, sucrose, are listed below.

Sugar	Formula	Atomic Weight (amu)
glucose	$C_6H_{12}O_6$	180
sucrose	$C_{12}H_{22}O_{11}$	342

More About. . .

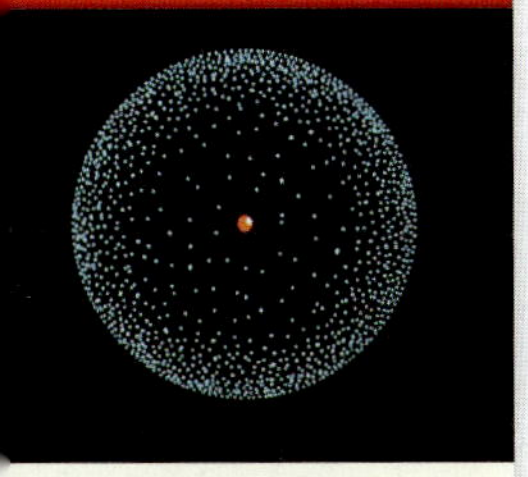

Chemistry

Atomic mass units are relative units of weight because they were compared to the weight of a hydrogen atom. So a molecule of nitrogen, whose weight is 14.0 amu, weighs 14 times as much as a hydrogen atom.

Source: www.sizes.com

a. Write a system of equations that represents the weight of each atom.

Let c represent the weight of a carbon atom.
Let h represent the weight of a hydrogen atom.

Write an equation for the weight of each sugar. The subscript represents how many atoms of each element are in the molecule.

Glucose:	$6c + 12h + 6(16) = 180$	Equation for glucose
	$6c + 12h + 96 = 180$	Simplify.
	$6c + 12h = 84$	Subtract 96 from each side.
Sucrose:	$12c + 22h + 11(16) = 342$	Equation for sucrose
	$12c + 22h + 176 = 342$	Simplify.
	$12c + 22h = 166$	Subtract 176 from each side.

b. Write a matrix equation for the system of equations.

Determine the coefficient, variable, and constant matrices.

$$\begin{array}{l} 6c + 12h = 84 \\ 12c + 22h = 166 \end{array} \rightarrow \begin{bmatrix} 6 & 12 \\ 12 & 22 \end{bmatrix} \quad \begin{bmatrix} c \\ h \end{bmatrix} \quad \begin{bmatrix} 84 \\ 166 \end{bmatrix}$$

Write the matrix equation.

$$\begin{matrix} A & \cdot & X & = & B \end{matrix}$$

$$\begin{bmatrix} 6 & 12 \\ 12 & 22 \end{bmatrix} \cdot \begin{bmatrix} c \\ h \end{bmatrix} = \begin{bmatrix} 84 \\ 166 \end{bmatrix}$$ You will solve this matrix equation in Exercise 11.

SOLVE SYSTEMS OF EQUATIONS

You can solve a system of linear equations by solving a matrix equation. A matrix equation in the form $AX = B$, where A is a coefficient matrix, X is a variable matrix, and B is a constant matrix, can be solved in a similar manner as a linear equation of the form $ax = b$.

$ax = b$	Write the equation.	$AX = B$
$\left(\frac{1}{a}\right)ax = \left(\frac{1}{a}\right)b$	Multiply each side by the inverse of the coefficient, if it exists.	$A^{-1}AX = A^{-1}B$
$1x = \left(\frac{1}{a}\right)b$	$\left(\frac{1}{a}\right)a = 1, A^{-1}A = I$	$IX = A^{-1}B$
$x = \left(\frac{1}{a}\right)b$	$1x = x, IX = X$	$X = A^{-1}B$

Notice that the solution of the matrix equation is the product of the inverse of the coefficient matrix and the constant matrix.

www.algebra2.com/extra_examples

2 Teach

WRITE MATRIX EQUATIONS

In-Class Examples

Power Point®

1 Write a matrix equation for the system of equations.
$x + 3y = 3$
$x + 2y = 7$

$$\begin{bmatrix} 1 & 3 \\ 1 & 2 \end{bmatrix} \cdot \begin{bmatrix} x \\ y \end{bmatrix} = \begin{bmatrix} 3 \\ 7 \end{bmatrix}$$

2 **FABRICS** The table below shows the composition of three types of fabrics and the cost per yard of each type.

Type	Wool	Silk	Cotton	Cost
R	10%	20%	70%	\$7
S	20%	30%	50%	\$8
T	20%	50%	30%	\$10

a. Write a system of equations that represents the total cost for each of the three fabric components.
$0.1w + 0.2s + 0.7c = 7$
$0.2w + 0.3s + 0.5c = 8$
$0.2w + 0.5s + 0.3c = 10$

b. Write a matrix equation for the system of equations.

$$\begin{bmatrix} 0.1 & 0.2 & 0.7 \\ 0.2 & 0.3 & 0.5 \\ 0.2 & 0.5 & 0.3 \end{bmatrix} \cdot \begin{bmatrix} w \\ s \\ c \end{bmatrix} = \begin{bmatrix} 7 \\ 8 \\ 10 \end{bmatrix}$$

Lesson 4-8 Using Matrices to Solve Systems of Equations 203

SOLVE SYSTEMS OF EQUATIONS

In-Class Examples

Power Point®

3 Use a matrix equation to solve the system of equations.
$5x + 3y = 13$
$4x + 7y = -8$ **(5, −4)**

4 Use a matrix equation to solve the system of equations.
$10x + 5y = 15$
$6x + 3y = -6$
There is no unique solution of this system.

Example 3 Solve a System of Equations

Use a matrix equation to solve the system of equations.
$6x + 2y = 11$
$3x - 8y = 1$

The matrix equation is $\begin{bmatrix} 6 & 2 \\ 3 & -8 \end{bmatrix} \cdot \begin{bmatrix} x \\ y \end{bmatrix} = \begin{bmatrix} 11 \\ 1 \end{bmatrix}$, when $A = \begin{bmatrix} 6 & 2 \\ 3 & -8 \end{bmatrix}$, $X = \begin{bmatrix} x \\ y \end{bmatrix}$, and $B = \begin{bmatrix} 11 \\ 1 \end{bmatrix}$.

Step 1 Find the inverse of the coefficient matrix.

$$A^{-1} = \frac{1}{-48 - 6}\begin{bmatrix} -8 & -2 \\ -3 & 6 \end{bmatrix} \text{ or } -\frac{1}{54}\begin{bmatrix} -8 & -2 \\ -3 & 6 \end{bmatrix}$$

Step 2 Multiply each side of the matrix equation by the inverse matrix.

$$-\frac{1}{54}\begin{bmatrix} -8 & -2 \\ -3 & 6 \end{bmatrix} \cdot \begin{bmatrix} 6 & 2 \\ 3 & -8 \end{bmatrix} \cdot \begin{bmatrix} x \\ y \end{bmatrix} = -\frac{1}{54}\begin{bmatrix} -8 & -2 \\ -3 & 6 \end{bmatrix} \cdot \begin{bmatrix} 11 \\ 1 \end{bmatrix} \quad \text{Multiply each side by } A^{-1}$$

$$\begin{bmatrix} 1 & 0 \\ 0 & 1 \end{bmatrix} \cdot \begin{bmatrix} x \\ y \end{bmatrix} = -\frac{1}{54}\begin{bmatrix} -90 \\ -27 \end{bmatrix} \quad \text{Multiply matrices.}$$

$$\begin{bmatrix} x \\ y \end{bmatrix} = \begin{bmatrix} \frac{5}{3} \\ \frac{1}{2} \end{bmatrix} \quad \begin{bmatrix} 1 & 0 \\ 0 & 1 \end{bmatrix} = I$$

The solution is $\left(\frac{5}{3}, \frac{1}{2}\right)$. Check this solution in the original equation.

Study Tip

Identity Matrix
The identity matrix on the left verifies that the inverse matrix has been calculated correctly.

Example 4 System of Equations with No Solution

Use a matrix equation to solve the system of equations.
$6a - 9b = -18$
$8a - 12b = 24$

The matrix equation is $\begin{bmatrix} 6 & -9 \\ 8 & -12 \end{bmatrix} \cdot \begin{bmatrix} a \\ b \end{bmatrix} = \begin{bmatrix} -18 \\ 24 \end{bmatrix}$, when $A = \begin{bmatrix} 6 & -9 \\ 8 & -12 \end{bmatrix}$, $X = \begin{bmatrix} a \\ b \end{bmatrix}$, and $B = \begin{bmatrix} -18 \\ 24 \end{bmatrix}$.

Find the inverse of the coefficient matrix.

$$A^{-1} = \frac{1}{-72 + 72}\begin{bmatrix} -12 & 9 \\ -8 & 6 \end{bmatrix}$$

The determinant of the coefficient matrix $\begin{bmatrix} 6 & -9 \\ 8 & -12 \end{bmatrix}$ is 0, so A^{-1} does not exist.

There is no unique solution of this system.

Graph the system of equations. Since the lines are parallel, this system has no solution. Therefore, the system is inconsistent.

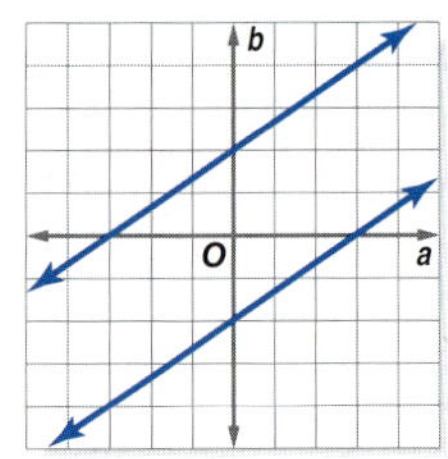

Study Tip

Look Back
To review **inconsistent systems of equations**, see Lesson 3-1.

To solve a system of equations with three variables, you can use the 3 × 3 identity matrix. However, finding the inverse of a 3 × 3 matrix may be tedious. Graphing calculators and computer programs offer fast and accurate methods for performing the necessary calculations.

DAILY INTERVENTION | **Differentiated Instruction**

Logical Ask students to list those matrix operations that they know how to perform using their graphing calculator. Ask those students who are adept with the calculator to explain each series of keystrokes and their purpose to a classmate who is experiencing difficulty.

Graphing Calculator Investigation

Systems of Three Equations in Three Variables

You can use a graphing calculator and a matrix equation to solve systems of equations. Consider the system of equations below.

$3x - 2y + z = 0$
$2x + 3y - z = 17$
$5x - y + 4z = -7$

Think and Discuss

1. Write a matrix equation for the system of equations.
2. Enter the coefficient matrix as matrix *A* and the constant matrix as matrix *B* in the graphing calculator. Find the product of A^{-1} and *B*. Recall that the $\boxed{x^{-1}}$ key is used to find A^{-1}.
3. How is the result related to the solution? **(3, 2, –5)**

1. $\begin{bmatrix} 3 & -2 & 1 \\ 2 & 3 & -1 \\ 5 & -1 & 4 \end{bmatrix} \cdot \begin{bmatrix} x \\ y \\ z \end{bmatrix} = \begin{bmatrix} 0 \\ 17 \\ -7 \end{bmatrix}$

2. $\begin{bmatrix} 3 \\ 2 \\ -5 \end{bmatrix}$

Check for Understanding

Concept Check

1. Write the matrix equation $\begin{bmatrix} 2 & -3 \\ 1 & 4 \end{bmatrix} \cdot \begin{bmatrix} r \\ s \end{bmatrix} = \begin{bmatrix} 4 \\ -2 \end{bmatrix}$ as a system of linear equations.
2. **OPEN ENDED** Write a system of equations that does not have a unique solution. **Sample answer: $x + 3y = 8$ and $2x + 6y = 16$**
3. **FIND THE ERROR** Tommy and Laura are solving a system of equations. They find that $A^{-1} = \begin{bmatrix} 3 & -2 \\ -7 & 5 \end{bmatrix}$, $B = \begin{bmatrix} -7 \\ -9 \end{bmatrix}$, and $X = \begin{bmatrix} x \\ y \end{bmatrix}$.

Tommy

$\begin{bmatrix} x \\ y \end{bmatrix} = \begin{bmatrix} 3 & -2 \\ -7 & 5 \end{bmatrix} \cdot \begin{bmatrix} -7 \\ -9 \end{bmatrix}$

$\begin{bmatrix} x \\ y \end{bmatrix} = \begin{bmatrix} -3 \\ 4 \end{bmatrix}$

Laura

$\begin{bmatrix} x \\ y \end{bmatrix} = \begin{bmatrix} -7 \\ -9 \end{bmatrix} \cdot \begin{bmatrix} 3 & -2 \\ -7 & 5 \end{bmatrix}$

$\begin{bmatrix} x \\ y \end{bmatrix} = \begin{bmatrix} 42 \\ 31 \end{bmatrix}$

Who is correct? Explain your reasoning.

1. $2r - 3s = 4$, $r + 4s = -2$

3. Tommy; a 2×1 matrix cannot be multiplied by a 2×2 matrix.

Guided Practice

GUIDED PRACTICE KEY	
Exercises	Examples
4–6	1
7–10	3, 4
11	2

Write a matrix equation for each system of equations. **4–6. See pp. 217A–217B.**

4. $x - y = -3$
$x + 3y = 5$

5. $2g + 3h = 8$
$-4g - 7h = -5$

6. $3a - 5b + 2c = 9$
$4a + 7b + c = 3$
$2a - c = 12$

Solve each matrix equation or system of equations by using inverse matrices.

7. $\begin{bmatrix} 3 & 1 \\ 4 & -2 \end{bmatrix} \cdot \begin{bmatrix} x \\ y \end{bmatrix} = \begin{bmatrix} 13 \\ 24 \end{bmatrix}$ **(5, –2)**

8. $\begin{bmatrix} 8 & -1 \\ 2 & 3 \end{bmatrix} \cdot \begin{bmatrix} a \\ b \end{bmatrix} = \begin{bmatrix} 16 \\ -9 \end{bmatrix}$ **(1.5, –4)**

9. $5x - 3y = -30$ **(–3, 5)**
$8x + 5y = 1$

10. $5s + 4t = 12$ **(1, 1.75)**
$4s - 3t = -1.25$

Application

11. **CHEMISTRY** Refer to Example 2 on page 203. Solve the system of equations to find the weight of a carbon, hydrogen, and oxygen atom. **$h = 1$, $c = 12$**

3 Practice/Apply

Study Notebook

Have students—

- complete the definitions/examples for the remaining terms on their Vocabulary Builder worksheets for Chapter 4.
- show the conversion of a system of equations to a matrix equation.
- write an explanation of the procedure for solving a matrix equation.
- include any other item(s) that they find helpful in mastering the skills in this lesson.

DAILY INTERVENTION

FIND THE ERROR Emphasize that multiplication of the constant matrix *B* by the inverse matrix A^{-1} is "multiplication on the left." That is, $X = A^{-1}B$ and not $X = BA^{-1}$.

About the Exercises...

Organization by Objective

- **Write Matrix Equations:** 12–19
- **Solve Systems of Equations:** 20–34

Odd/Even Assignments

Exercises 12–33 are structured so that students practice the same concepts whether they are assigned odd or even problems.

Alert! Exercises 39–41 require a graphing calculator.

Assignment Guide

Basic: 13–33 odd, 35–38, 42–51

Average: 13–33 odd, 35–38, 42–51 (optional: 39–41)

Advanced: 12–34 even, 35–51

Graphing Calculator Investigation

On the TI-83 calculator, to find A^{-1} once you have entered the elements of matrix *A* and have a clear screen, students should press MATRX 1 (to choose Matrix *A*, the first in the list), and then the $\boxed{x^{-1}}$ key followed by ENTER. Also, stress that matrix *B* must be multiplied *on the left* by A^{-1}.

Practice and Apply

Homework Help

For Exercises	See Examples
12–19	1
20–31	3, 4
32–34	2

Extra Practice
See page 836.

Write a matrix equation for each system of equations. 12–19. See margin.

12. $3x - y = 0$
$x + 2y = -21$

13. $4x - 7y = 2$
$3x + 5y = 9$

14. $5a - 6b = -47$
$3a + 2b = -17$

15. $3m - 7n = -43$
$6m + 5n = -10$

16. $2a + 3b - 5c = 1$
$7a + 3c = 7$
$3a - 6b + c = -5$

17. $3x - 5y + 2z = 9$
$x - 7y + 3z = 11$
$4x - 3z = -1$

18. $x - y = 8$
$-2x - 5y - 6z = -27$
$9x + 10y - z = 54$

19. $3r - 5s + 6t = 21$
$11r - 12s + 16t = 15$
$-5r + 8s - 3t = -7$

Solve each matrix equation or system of equations by using inverse matrices.

20. $\begin{bmatrix} 7 & -3 \\ 2 & 5 \end{bmatrix} \cdot \begin{bmatrix} m \\ n \end{bmatrix} = \begin{bmatrix} 41 \\ 0 \end{bmatrix}$ **(5, −2)**

21. $\begin{bmatrix} 3 & 1 \\ 2 & -1 \end{bmatrix} \cdot \begin{bmatrix} a \\ b \end{bmatrix} = \begin{bmatrix} 13 \\ 2 \end{bmatrix}$ **(3, 4)**

22. $\begin{bmatrix} 4 & -3 \\ 5 & 2 \end{bmatrix} \cdot \begin{bmatrix} a \\ b \end{bmatrix} = \begin{bmatrix} -17 \\ -4 \end{bmatrix}$ **(−2, 3)**

23. $\begin{bmatrix} 7 & 1 \\ 3 & -8 \end{bmatrix} \cdot \begin{bmatrix} x \\ y \end{bmatrix} = \begin{bmatrix} 43 \\ 10 \end{bmatrix}$ **(6, 1)**

24. $\begin{bmatrix} 2 & -9 \\ 6 & 5 \end{bmatrix} \cdot \begin{bmatrix} c \\ d \end{bmatrix} = \begin{bmatrix} 28 \\ -12 \end{bmatrix}$ $\left(\frac{1}{2}, -3\right)$

25. $\begin{bmatrix} 6 & 5 \\ 3 & 2 \end{bmatrix} \cdot \begin{bmatrix} a \\ b \end{bmatrix} = \begin{bmatrix} 18 \\ 7 \end{bmatrix}$ $\left(-\frac{1}{3}, 4\right)$

26. $6r + s = 9$ **(2, −3)**
$3r = -2s$

27. $5a + 9b = -28$ **(−2, −2)**
$2a - b = -2$

28. $p - 2q = 1$ **(7, 3)**
$p + 5q = 22$

29. $4m - 7n = -63$ **(0, 9)**
$3m + 2n = 18$

30. $x + 2y = 8$
$3x + 2y = 6$ $\left(-1, \frac{9}{2}\right)$

31. $4x - 3y = 5$
$2x + 9y = 6$ $\left(\frac{3}{2}, \frac{1}{3}\right)$

32. **PILOT TRAINING** Hai-Ling is training for his pilot's license. Flight instruction costs \$105 per hour, and the simulator costs \$45 per hour. The school requires students to spend 4 more hours in airplane training than in the simulator. If Hai-Ling can afford to spend \$3870 on training, how many hours can he spend training in an airplane and in a simulator?
27 h of flight instruction and 23 h in the simulator

33. **SCHOOLS** The graphic shows that student-to-teacher ratios are dropping in both public and private schools. If these rates of change remain constant, predict when the student-to-teacher ratios for private and public schools will be the same. **2010**

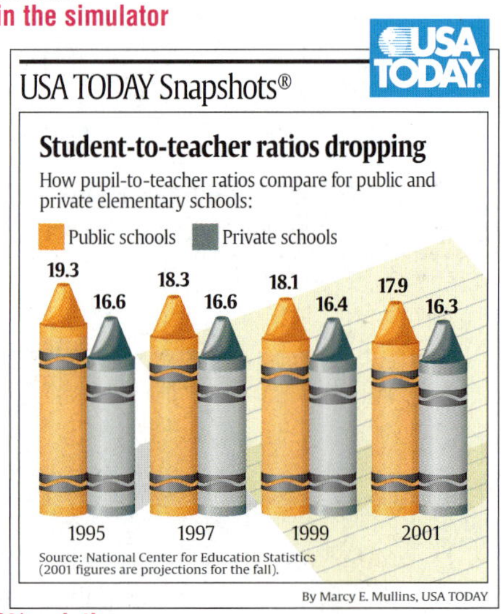

34. **CHEMISTRY** Cara is preparing an acid solution. She needs 200 milliliters of 48% concentration solution. Cara has 60% and 40% concentration solutions in her lab. How many milliliters of 40% acid solution should be mixed with 60% acid solution to make the required amount of 48% acid solution? **80 mL of the 60% solution and 120 mL of the 40% solution**

Study Guide and Intervention, p. 211 (shown) and p. 212

Write Matrix Equations A **matrix equation** for a system of equations consists of the product of the coefficient and variable matrices on the left and the constant matrix on the right of the equals sign.

Example **Write a matrix equation for each system of equations.**

a. $3x - 7y = 12$
$x + 5y = -8$
Determine the coefficient, variable, and constant matrices.
$\begin{bmatrix} 3 & -7 \\ 1 & 5 \end{bmatrix} \cdot \begin{bmatrix} x \\ y \end{bmatrix} = \begin{bmatrix} 12 \\ -8 \end{bmatrix}$

b. $2x - y + 3z = -7$
$x + 3y - 4z = 15$
$7x + 2y + z = -28$
$\begin{bmatrix} 2 & -1 & 3 \\ 1 & 3 & -4 \\ 7 & 2 & 1 \end{bmatrix} \cdot \begin{bmatrix} x \\ y \\ z \end{bmatrix} = \begin{bmatrix} -7 \\ 15 \\ -28 \end{bmatrix}$

Exercises

Write a matrix equation for each system of equations.

1. $2x + y = 8$, $5x - 3y = -12$ $\begin{bmatrix} 2 & 1 \\ 5 & -3 \end{bmatrix} \cdot \begin{bmatrix} x \\ y \end{bmatrix} = \begin{bmatrix} 8 \\ -12 \end{bmatrix}$

2. $4x - 3y = 18$, $x + 2y = 12$ $\begin{bmatrix} 4 & -3 \\ 1 & 2 \end{bmatrix} \cdot \begin{bmatrix} x \\ y \end{bmatrix} = \begin{bmatrix} 18 \\ 12 \end{bmatrix}$

3. $7x - 2y = 15$, $3x + y = -10$ $\begin{bmatrix} 7 & -2 \\ 3 & 1 \end{bmatrix} \cdot \begin{bmatrix} x \\ y \end{bmatrix} = \begin{bmatrix} 15 \\ -10 \end{bmatrix}$

4. $4x - 6y = 20$, $3x + y + 8 = 0$ $\begin{bmatrix} 4 & -6 \\ 3 & 1 \end{bmatrix} \cdot \begin{bmatrix} x \\ y \end{bmatrix} = \begin{bmatrix} 20 \\ -8 \end{bmatrix}$

5. $5x + 2y = 18$, $x = -4y + 25$ $\begin{bmatrix} 5 & 2 \\ 1 & 4 \end{bmatrix} \cdot \begin{bmatrix} x \\ y \end{bmatrix} = \begin{bmatrix} 18 \\ 25 \end{bmatrix}$

6. $3x - y = 24$, $3y = 80 - 2x$ $\begin{bmatrix} 3 & -1 \\ 2 & 3 \end{bmatrix} \cdot \begin{bmatrix} x \\ y \end{bmatrix} = \begin{bmatrix} 24 \\ 80 \end{bmatrix}$

7. $2x + y + 7z = 12$, $5x - y + 3z = 15$, $x + 2y - 6z = 25$ $\begin{bmatrix} 2 & 1 & 7 \\ 5 & -1 & 3 \\ 1 & 2 & -6 \end{bmatrix} \cdot \begin{bmatrix} x \\ y \\ z \end{bmatrix} = \begin{bmatrix} 12 \\ 15 \\ 25 \end{bmatrix}$

8. $5x - y + 7z = 32$, $x + 3y - 2z = -18$, $2x + 4y - 3z = 12$ $\begin{bmatrix} 5 & -1 & 7 \\ 1 & 3 & -2 \\ 2 & 4 & -3 \end{bmatrix} \cdot \begin{bmatrix} x \\ y \\ z \end{bmatrix} = \begin{bmatrix} 32 \\ -18 \\ 12 \end{bmatrix}$

9. $4x - 3y - z = -100$, $2x + y - 3z = -64$, $5x + 3y - 2z = 8$ $\begin{bmatrix} 4 & -3 & -1 \\ 2 & 1 & -3 \\ 5 & 3 & -2 \end{bmatrix} \cdot \begin{bmatrix} x \\ y \\ z \end{bmatrix} = \begin{bmatrix} -100 \\ -64 \\ 8 \end{bmatrix}$

10. $x - 3y + 7z = 27$, $2x + y - 5z = 48$, $4x - 2y + 3z = 72$ $\begin{bmatrix} 1 & -3 & 7 \\ 2 & 1 & -5 \\ 4 & -2 & 3 \end{bmatrix} \cdot \begin{bmatrix} x \\ y \\ z \end{bmatrix} = \begin{bmatrix} 27 \\ 48 \\ 72 \end{bmatrix}$

11. $2x + 3y - 9z = -108$, $x + 5z = 40 + 2y$, $3x + 5y = 89 + 4z$ $\begin{bmatrix} 2 & 3 & -9 \\ 1 & -2 & 5 \\ 3 & 5 & -4 \end{bmatrix} \cdot \begin{bmatrix} x \\ y \\ z \end{bmatrix} = \begin{bmatrix} -108 \\ 40 \\ 89 \end{bmatrix}$

12. $z = 45 - 3x + 2y$, $2x + 3y - z = 60$, $x = 4y - 2z + 120$ $\begin{bmatrix} 3 & -2 & 1 \\ 2 & 3 & -1 \\ 1 & -4 & 2 \end{bmatrix} \cdot \begin{bmatrix} x \\ y \\ z \end{bmatrix} = \begin{bmatrix} 45 \\ 60 \\ 120 \end{bmatrix}$

Skills Practice, p. 213 and Practice, p. 214 (shown)

Write a matrix equation for each system of equations.

1. $-3x + 2y = 9$, $5x - 3y = -13$ $\begin{bmatrix} -3 & 2 \\ 5 & -3 \end{bmatrix} \cdot \begin{bmatrix} x \\ y \end{bmatrix} = \begin{bmatrix} 9 \\ -13 \end{bmatrix}$

2. $6x - 2y = -2$, $3x + 3y = 10$ $\begin{bmatrix} 6 & -2 \\ 3 & 3 \end{bmatrix} \cdot \begin{bmatrix} x \\ y \end{bmatrix} = \begin{bmatrix} -2 \\ 10 \end{bmatrix}$

3. $2a + b = 0$, $3a + 2b = -2$ $\begin{bmatrix} 2 & 1 \\ 3 & 2 \end{bmatrix} \cdot \begin{bmatrix} a \\ b \end{bmatrix} = \begin{bmatrix} 0 \\ -2 \end{bmatrix}$

4. $r + 5s = 10$, $2r - 3s = 7$ $\begin{bmatrix} 1 & 5 \\ 2 & -3 \end{bmatrix} \cdot \begin{bmatrix} r \\ s \end{bmatrix} = \begin{bmatrix} 10 \\ 7 \end{bmatrix}$

5. $3x - 2y + 5z = 3$, $x + y - 4z = 2$, $-2x + 2y + 7z = -5$ $\begin{bmatrix} 3 & -2 & 5 \\ 1 & 1 & -4 \\ -2 & 2 & 7 \end{bmatrix} \cdot \begin{bmatrix} x \\ y \\ z \end{bmatrix} = \begin{bmatrix} 3 \\ 2 \\ -5 \end{bmatrix}$

6. $2m + n - 3p = -5$, $5m + 2n - 2p = 8$, $3m - 3n + 5p = 17$ $\begin{bmatrix} 2 & 1 & -3 \\ 5 & 2 & -2 \\ 3 & -3 & 5 \end{bmatrix} \cdot \begin{bmatrix} m \\ n \\ p \end{bmatrix} = \begin{bmatrix} -5 \\ 8 \\ 17 \end{bmatrix}$

Solve each matrix equation or system of equations by using inverse matrices.

7. $\begin{bmatrix} 2 & 1 \\ 3 & 2 \end{bmatrix} \cdot \begin{bmatrix} g \\ h \end{bmatrix} = \begin{bmatrix} 0 \\ -2 \end{bmatrix}$ **(2, −4)**

8. $\begin{bmatrix} -2 & 3 \\ 1 & 5 \end{bmatrix} \cdot \begin{bmatrix} x \\ y \end{bmatrix} = \begin{bmatrix} -7 \\ 10 \end{bmatrix}$ **(5, 1)**

9. $\begin{bmatrix} -1 & -3 \\ 3 & 4 \end{bmatrix} \cdot \begin{bmatrix} a \\ b \end{bmatrix} = \begin{bmatrix} 12 \\ -11 \end{bmatrix}$ **(3, −5)**

10. $\begin{bmatrix} -5 & 3 \\ 6 & 4 \end{bmatrix} \cdot \begin{bmatrix} c \\ d \end{bmatrix} = \begin{bmatrix} 16 \\ 34 \end{bmatrix}$ **(1, 7)**

11. $\begin{bmatrix} -4 & 2 \\ 7 & 4 \end{bmatrix} \cdot \begin{bmatrix} r \\ s \end{bmatrix} = \begin{bmatrix} 17 \\ -26 \end{bmatrix}$ $\left(-4, \frac{1}{2}\right)$

12. $\begin{bmatrix} 8 & 3 \\ 12 & 6 \end{bmatrix} \cdot \begin{bmatrix} y \\ z \end{bmatrix} = \begin{bmatrix} -1 \\ -1 \end{bmatrix}$ $\left(-\frac{1}{4}, \frac{1}{3}\right)$

13. $2x + 3y = 5$, $3x - 2y = 1$ **(1, 1)**

14. $8d + 9f = 13$, $-6d + 5f = -45$ **(5, −3)**

15. $5m + 9n = 19$, $2m - n = -20$ **(−7, 6)**

16. $-4j + 9k = -8$, $6j + 12k = -5$ $\left(\frac{1}{2}, -\frac{2}{3}\right)$

17. **AIRLINE TICKETS** Last Monday at 7:30 A.M., an airline flew 89 passengers on a commuter flight from Boston to New York. Some of the passengers paid \$120 for their tickets and the rest paid \$230 for their tickets. The total cost of all of the tickets was \$14,200. How many passengers bought \$120 tickets? How many bought \$230 tickets? **57; 32**

18. **NUTRITION** A single dose of a dietary supplement contains 0.2 gram of calcium and 0.2 gram of vitamin C. A single dose of a second dietary supplement contains 0.1 gram of calcium and 0.4 gram of vitamin C. If a person wants to take 0.6 gram of calcium and 1.2 grams of vitamin C, how many doses of each supplement should she take? **2 doses of the first supplement and 2 doses of the second supplement**

Reading to Learn Mathematics, p. 215 ELL

Pre-Activity **How are inverse matrices used in population ecology?**

Read the introduction to Lesson 4-8 at the top of page 202 in your textbook.

Write a 2 × 2 matrix that summarizes the information given in the introduction about the food and territory requirements for the two species.

$\begin{bmatrix} 140 & 500 \\ 120 & 400 \end{bmatrix}$

Reading the Lesson

1. a. Write a matrix equation for the following system of equations.
$3x + 5y = 10$
$2x - 4y = -7$
$\begin{bmatrix} 3 & 5 \\ 2 & -4 \end{bmatrix} \cdot \begin{bmatrix} x \\ y \end{bmatrix} = \begin{bmatrix} 10 \\ -7 \end{bmatrix}$

b. Explain how to use the matrix equation you wrote above to solve the system. Use as few mathematical symbols in your explanation as you can. Do not actually solve the system.
Sample answer: Find the inverse of the 2 × 2 matrix of coefficients. Multiply this inverse by the 2 × 1 matrix of constants, with the 2 × 2 matrix on the left. The product will be a 2 × 1 matrix. The number in the first row will be the value of *x*, and the number in the second row will be the value of *y*.

2. Write a system of equations that corresponds to the following matrix equation.
$\begin{bmatrix} 3 & 2 & -4 \\ 2 & -1 & 0 \\ 0 & 5 & 6 \end{bmatrix} \cdot \begin{bmatrix} x \\ y \\ z \end{bmatrix} = \begin{bmatrix} -2 \\ 6 \\ -4 \end{bmatrix}$ $3x + 2y - 4z = -2$, $2x - y = 6$, $5y + 6z = -4$

Helping You Remember

3. Some students have trouble remembering how to set up a matrix equation to solve a system of linear equations. What is an easy way to remember the order in which to write the three matrices that make up the equation?
Sample answer: Just remember "CVC" for "coefficients, variables, constants." The variable matrix is on the left side of the equals sign, just as the variables are in the system of linear equations. The constant matrix is on the right side of the equals sign, just as the constants are in the system of linear equations.

Enrichment, p. 216

Matrix Multiplication

A furniture manufacturer makes upholstered chairs and wood tables. Matrix *A* shows the number of hours spent on each item by three different workers. One day the factory receives an order for 10 chairs and 3 tables. This is shown in matrix *B*.

	woodworker (hours)	finsher (hours)	upholsterer (hours)
chair	4	2	12
table	18	15	0

$= A$

	chair	table
number ordered	10	3

$= B$

$[10 \;\; 3]\begin{bmatrix} 4 & 2 & 12 \\ 18 & 15 & 0 \end{bmatrix} = [10(4) + 3(18) \;\; 10(2) + 3(15) \;\; 10(12) + 3(0)] = [94 \;\; 65 \;\; 120]$

The product of the two matrices shows the number of hours needed for each type of worker to complete the order: 94 hours for woodworking, 65 hours for finishing, and 120 hours for upholstering.

To find the total labor cost, multiply by a matrix that shows the hourly rate for each worker: \$15 for woodworking, \$9 for finishing, and \$12 for upholstering.

[15] ... 20] = [94(15 ... 120(1

206 Chapter 4 Matrices

35. The solution set is the empty set or infinite solutions.

35. **CRITICAL THINKING** Describe the solution set of a system of equations if the coefficient matrix does not have an inverse.

36. **WRITING IN MATH** Answer the question that was posed at the beginning of the lesson. See pp. 217A–217B.

How can matrices be used in population ecology?

Include the following in your answer:

- a system of equations that can be used to find the number of each species the region can support, and
- a solution of the problem using matrices.

Standardized Test Practice

37. Solve the system of equations $6a + 8b = 5$ and $10a - 12b = 2$. **D**

Ⓐ $\left(\frac{3}{4}, \frac{1}{2}\right)$ Ⓑ $\left(\frac{1}{2}, \frac{3}{4}\right)$ Ⓒ $\left(\frac{1}{2}, -\frac{1}{2}\right)$ Ⓓ $\left(\frac{1}{2}, \frac{1}{4}\right)$

38. **SHORT RESPONSE** The Yogurt Shoppe sells cones in three sizes: small \$0.89; medium, \$1.19; and large, \$1.39. One day Scott sold 52 cones. He sold seven more medium cones than small cones. If he sold \$58.98 in cones, how many of each size did he sell? **17 small, 24 medium, 11 large**

Graphing Calculator

INVERSE MATRICES **Use a graphing calculator to solve each system of equations using inverse matrices.**

39. $2a - b + 4c = 6$
$a + 5b - 2c = -6$
$3a - 2b + 6c = 8$
(−6, 2, 5)

40. $3x - 5y + 2z = 22$
$2x + 3y - z = -9$
$4x + 3y + 3z = 1$
(1, −3, 2)

41. $2q + r + s = 2$
$-q - r + 2s = 7$
$-3q + 2r + 3s = 7$
(0, −1, 3)

Maintain Your Skills

Mixed Review

Find the inverse of each matrix, if it exists. *(Lesson 4-7)*

42. $\begin{bmatrix} 4 & 4 \\ 2 & 3 \end{bmatrix}$

43. $\begin{bmatrix} 9 & 5 \\ 7 & 4 \end{bmatrix}$ $\begin{bmatrix} 4 & -5 \\ -7 & 9 \end{bmatrix}$

44. $\begin{bmatrix} -3 & -6 \\ 5 & 10 \end{bmatrix}$ **no inverse exists**

42. $\begin{bmatrix} \frac{3}{4} & -1 \\ -\frac{1}{2} & 1 \end{bmatrix}$

46. (4.27, −5.11)

Use Cramer's Rule to solve each system of equations. *(Lesson 4-6)*

45. $6x + 7y = 10$
$3x - 4y = 20$
(4, −2)

46. $6a + 7b = -10.15$
$9.2a - 6b = 69.944$

47. $\frac{x}{2} - \frac{2y}{3} = 2\frac{1}{3}$
$3x + 4y = -50$
(−6, −8)

48. **ECOLOGY** If you recycle a $3\frac{1}{2}$-foot stack of newspapers, one less 20-foot loblolly pine tree will be needed for paper. Use a prediction equation to determine how many feet of loblolly pine trees will *not* be needed for paper if you recycle a pile of newspapers 20 feet tall. *(Lesson 2-5)* **about 114.3 ft**

Solve each equation. Check your solutions. *(Lesson 1-4)*

49. $|x - 3| = 7$ **{−4, 10}**

50. $-4|d + 2| = -12$ **{−5, 1}**

51. $5|k - 4| = k + 8$ **{2, 7}**

WebQuest Internet Project

Lessons in Home Buying, Selling

It is time to complete your project. Use the information and data you have gathered about home buying and selling to prepare a portfolio or Web page. Be sure to include your tables, graphs, and calculations. You may also wish to include additional data, information, or pictures.

www.algebra2.com/webquest

www.algebra2.com/self_check_quiz

4 Assess

Open-Ended Assessment

Speaking Have students describe how matrices can be used to solve systems of two equations in two variables.

Assessment Options

Quiz (Lessons 4-7 and 4-8) is available on p. 232 of the *Chapter 4 Resource Masters*.

Answers

12. $\begin{bmatrix} 3 & -1 \\ 1 & 2 \end{bmatrix} \cdot \begin{bmatrix} x \\ y \end{bmatrix} = \begin{bmatrix} 0 \\ -21 \end{bmatrix}$

13. $\begin{bmatrix} 4 & -7 \\ 3 & 5 \end{bmatrix} \cdot \begin{bmatrix} x \\ y \end{bmatrix} = \begin{bmatrix} 2 \\ 9 \end{bmatrix}$

14. $\begin{bmatrix} 5 & -6 \\ 3 & 2 \end{bmatrix} \cdot \begin{bmatrix} a \\ b \end{bmatrix} = \begin{bmatrix} -47 \\ -17 \end{bmatrix}$

15. $\begin{bmatrix} 3 & -7 \\ 6 & 5 \end{bmatrix} \cdot \begin{bmatrix} m \\ n \end{bmatrix} = \begin{bmatrix} -43 \\ -10 \end{bmatrix}$

16. $\begin{bmatrix} 2 & 3 & -5 \\ 7 & 0 & 3 \\ 3 & -6 & 1 \end{bmatrix} \cdot \begin{bmatrix} a \\ b \\ c \end{bmatrix} = \begin{bmatrix} 1 \\ 7 \\ -5 \end{bmatrix}$

17. $\begin{bmatrix} 3 & -5 & 2 \\ 1 & -7 & 3 \\ 4 & 0 & -3 \end{bmatrix} \cdot \begin{bmatrix} x \\ y \\ z \end{bmatrix} = \begin{bmatrix} 9 \\ 11 \\ -1 \end{bmatrix}$

18. $\begin{bmatrix} 1 & -1 & 0 \\ -2 & -5 & -6 \\ 9 & 10 & -1 \end{bmatrix} \cdot \begin{bmatrix} x \\ y \\ z \end{bmatrix} = \begin{bmatrix} 8 \\ -27 \\ 54 \end{bmatrix}$

19. $\begin{bmatrix} 3 & -5 & 6 \\ 11 & -12 & 16 \\ -5 & 8 & -3 \end{bmatrix} \cdot \begin{bmatrix} r \\ s \\ t \end{bmatrix} = \begin{bmatrix} 21 \\ 15 \\ -7 \end{bmatrix}$

Online Lesson Plans

USA TODAY Education's Online site offers resources and interactive features connected to each day's newspaper. *Experience TODAY*, USA TODAY's daily lesson plan, is available on the site and delivered daily to subscribers. This plan provides instruction for integrating USA TODAY graphics and key editorial features into your mathematics classroom. Log on to www.education.usatoday.com.

Augmented Matrices

Using a TI-83 Plus, you can solve a system of linear equations using the MATRX function. An **augmented matrix** contains the coefficient matrix with an extra column containing the constant terms.

The reduced row echelon function of a graphing calculator reduces the augmented matrix so that the solution of the system of equations can be easily determined.

Write an augmented matrix for the following system of equations. Then solve the system by using the reduced row echelon form on the graphing calculator.

$3x + y + 3z = 2$
$2x + y + 2z = 1$
$4x + 2y + 5z = 5$

Step 1 Write the augmented matrix and enter it into a calculator.

The augmented matrix $B = \left[\begin{array}{ccc|c} 3 & 1 & 3 & 2 \\ 2 & 1 & 2 & 1 \\ 4 & 2 & 5 & 5 \end{array}\right]$.

Begin by entering the matrix.

KEYSTROKES: *Review matrices on page 163.*

Step 2 Find the reduced row echelon form (rref) using the graphing calculator.

KEYSTROKES: [2nd] [MATRX] [▶] [ALPHA] [B] [2nd] [MATRX] 2 [)] [ENTER]

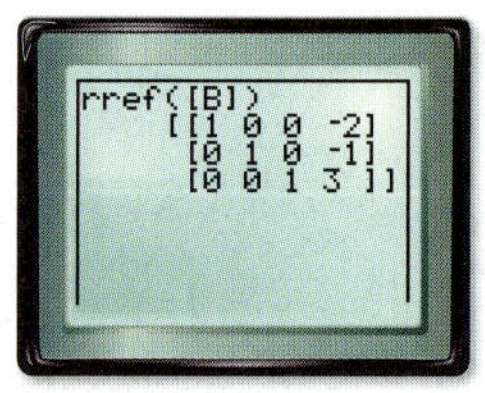

Study the reduced echelon matrix. The first three columns are the same as a 3 × 3 identity matrix. The first row represents $x = -2$, the second row represents $y = -1$, and the third row represents $z = 3$. The solution is $(-2, -1, 3)$.

Exercises

Write an augmented matrix for each system of equations. Then solve with a graphing calculator. **1–6. See margin for matrices.**

1. $x - 3y = 5$
   $2x + y = 1$ **(1.14, −1.29)**
2. $15x + 11y = 36$
   $4x - 3y = -26$ **(−2, 6)**
3. $2x + y = 5$
   $2x - 3y = 1$ **(2, 1)**
4. $3x - y = 0$
   $2x - 3y = 1$ **(−0.14, −0.43)**
5. $3x - 2y + z = -2$
   $x - y + 3z = 5$
   $-x + y + z = -1$ **(−7, −9, 1)**
6. $x - y + z = 2$
   $x - z = 1$
   $y + 2z = 0$ **(1.25, −0.5, 0.25)**

www.algebra2.com/other_calculator_keystrokes

Graphing Calculator Investigation

A Follow-Up of Lesson 4-8

Getting Started

Know Your Calculator The TI-83 Plus allows you to perform row operations on matrices. These row operations are items C through F on the [MATRX] Math menu. Successive use of row operations allows you to transform a matrix to reduced row echelon form. The rref(function performs all the steps at once, thereby saving a great deal of time.

Teach

- Point out that if one of the variables is absent from an equation in a system of equations, then its coefficient is zero. In Exercise 6, students may find it helpful to rewrite the equation showing 0 as the coefficient of the missing variables in the second and third equations in order to determine the correct augmented matrix.
- Have students complete Exercises 1–6.

Assess

Ask students which method they prefer for solving systems of two equations in two variables, the graphing calculator method shown in this investigation or the method presented in Lesson 4-8. Have them choose a preferred method for solving systems of three equations in three variables. Have them explain their choices.

Answers

1. $A = \left[\begin{array}{cc|c} 1 & -3 & 5 \\ 2 & 1 & 1 \end{array}\right]$

2. $A = \left[\begin{array}{cc|c} 15 & 11 & 36 \\ 4 & -3 & -26 \end{array}\right]$

3. $A = \left[\begin{array}{cc|c} 2 & 1 & 5 \\ 2 & -3 & 1 \end{array}\right]$

4. $A = \left[\begin{array}{cc|c} 3 & -1 & 0 \\ 2 & -3 & 1 \end{array}\right]$

5. $A = \left[\begin{array}{ccc|c} 3 & -2 & 1 & -2 \\ 1 & -1 & 3 & 5 \\ -1 & 1 & 1 & -1 \end{array}\right]$

6. $A = \left[\begin{array}{ccc|c} 1 & -1 & 1 & 2 \\ 1 & 0 & -1 & 1 \\ 0 & 1 & 2 & 0 \end{array}\right]$

Chapter 4 Study Guide and Review

Vocabulary and Concept Check

column matrix (p. 155)
Cramer's Rule (p. 189)
determinant (p. 182)
dilation (p. 176)
dimension (p. 155)
element (p. 155)
equal matrices (p. 155)
expansion by minors (p. 183)
identity matrix (p. 195)
image (p. 175)
inverse (p. 195)
isometry (p. 175)
matrix (p. 154)
matrix equation (p. 202)
minor (p. 183)
preimage (p. 175)
reflection (p. 177)
rotation (p. 178)
row matrix (p. 155)
scalar (p. 162)
scalar multiplication (p. 162)
second-order determinant (p. 182)
square matrix (p. 155)
third-order determinant (p. 183)
transformation (p. 175)
translation (p. 175)
vertex matrix (p. 175)
zero matrix (p. 155)

Choose the correct term to complete each sentence. **3. Scalar multiplication**

determinant
dilation
dimensions
equal matrices
identity matrix
isometry
matrix equation
rotation
scalar multiplication
translation

1. The matrix $\begin{bmatrix}1 & 0 & 0\\0 & 1 & 0\\0 & 0 & 1\end{bmatrix}$ is a(n) ______ for multiplication. **identity matrix**
2. When an image and a preimage are congruent, then the transformation is called a(n) ______. **isometry**
3. ______ is the process of multiplying a matrix by a constant.
4. A(n) **rotation** is when a figure is moved around a center point.
5. The ______ of $\begin{bmatrix}-1 & 4\\2 & -3\end{bmatrix}$ is −5. **determinant**
6. A(n) ______ is the product of the coefficient matrix and the variable matrix equal to the constant matrix. **matrix equation**
7. The ______ of a matrix tell how many rows and columns are in the matrix. **dimensions**
8. A(n) ______ occurs when a figure is moved from one location to another on the coordinate plane. **translation**
9. The matrices $\begin{bmatrix}3x\\x+2y\end{bmatrix}$ and $\begin{bmatrix}y\\7\end{bmatrix}$ are ______ if $x = 1$ and $y = 3$. **equal matrices**
10. A(n) **dilation** is when a geometric figure is enlarged or reduced.

Lesson-by-Lesson Review

4-1 Introduction to Matrices

See pages 154–158.

Concept Summary

- A matrix is a rectangular array of variables or constants in horizontal rows and vertical columns.
- Equal matrices have the same dimensions and corresponding elements equal.

Example Solve $\begin{bmatrix}2x\\y\end{bmatrix} = \begin{bmatrix}32+6y\\7-x\end{bmatrix}$ for x and y.

Since the matrices are equal, corresponding elements are equal. You can write two linear equations.

$2x = 32 + 6y$
$y = 7 - x$

(continued on the next page)

www.algebra2.com/vocabulary_review

Chapter 4 Study Guide and Review

Vocabulary and Concept Check

- This alphabetical list of vocabulary terms in Chapter 4 includes a page reference where each term was introduced.
- **Assessment** A vocabulary test/review for Chapter 4 is available on p. 230 of the *Chapter 4 Resource Masters*.

Lesson-by-Lesson Review

For each lesson,

- the main ideas are summarized,
- additional examples review concepts, and
- practice exercises are provided.

Vocabulary PuzzleMaker

ELL The Vocabulary PuzzleMaker software improves students' mathematics vocabulary using four puzzle formats—crossword, scramble, word search using a word list, and word search using clues. Students can work on a computer screen or from a printed handout.

MindJogger Videoquizzes

ELL MindJogger Videoquizzes provide an alternative review of concepts presented in this chapter. Students work in teams in a game show format to gain points for correct answers. The questions are presented in three rounds.

Round 1 Concepts (5 questions)
Round 2 Skills (4 questions)
Round 3 Problem Solving (4 questions)

FOLDABLES™ Study Organizer

For more information about Foldables, see *Teaching Mathematics with Foldables.*

Ask students to check over their notes and descriptions about matrices to see if they wish to add any further information, either about the definitions of operations with matrices or about ways that matrices are applied to systems of equations.

Encourage students to refer to their Foldables while completing the Study Guide and Review and to use them in preparing for the Chapter Test.

Solve the system of equations.

$2x = 32 + 6y$ First equation

$2x = 32 + 6(7 - x)$ Substitute $7 - x$ for y.

$2x = 32 + 42 - 6x$ Distributive Property

$8x = 74$ Add $6x$ to each side.

$x = 9.25$ Divide each side by 8.

The solution is (9.25, −2.25).

To find the value for y, substitute 9.25 for x in either equation.

$y = 7 - x$ Second equation

$= 7 - 9.25$ Substitute 9.25 for x.

$= -2.25$ Simplify.

Exercises **Solve each equation.** *See Example 3 on pages 155 and 156.*

11. $\begin{bmatrix} 2y - x \\ x \end{bmatrix} = \begin{bmatrix} 3 \\ 4y - 1 \end{bmatrix}$ **(−5, −1)**

12. $\begin{bmatrix} 7x \\ x + y \end{bmatrix} = \begin{bmatrix} 5 + 2y \\ 11 \end{bmatrix}$ **(3, 8)**

13. $\begin{bmatrix} 3x + y \\ x - 3y \end{bmatrix} = \begin{bmatrix} -3 \\ -1 \end{bmatrix}$ **(−1, 0)**

14. $\begin{bmatrix} 2x - y \\ 6x - y \end{bmatrix} = \begin{bmatrix} 2 \\ 22 \end{bmatrix}$ **(5, 8)**

4-2 Operations with Matrices

See pages 160–166.

Concept Summary

- Matrices can be added or subtracted if they have the same dimensions. Add or subtract corresponding elements.
- To multiply a matrix by a scalar k, multiply each element in the matrix by k.

Examples

1 Find $A - B$ if $A = \begin{bmatrix} 3 & 8 \\ -5 & 2 \end{bmatrix}$ and $B = \begin{bmatrix} -4 & 6 \\ 1 & 9 \end{bmatrix}$.

$A - B = \begin{bmatrix} 3 & 8 \\ -5 & 2 \end{bmatrix} - \begin{bmatrix} -4 & 6 \\ 1 & 9 \end{bmatrix}$ Definition of matrix subtraction

$= \begin{bmatrix} 3 - (-4) & 8 - 6 \\ -5 - 1 & 2 - 9 \end{bmatrix}$ Subtract corresponding elements.

$= \begin{bmatrix} 7 & 2 \\ -6 & -7 \end{bmatrix}$ Simplify.

2 If $X = \begin{bmatrix} 3 & 2 & -1 \\ 4 & -6 & 0 \end{bmatrix}$, find $4X$.

$4X = 4\begin{bmatrix} 3 & 2 & -1 \\ 4 & -6 & 0 \end{bmatrix}$

$= \begin{bmatrix} 4(3) & 4(2) & 4(-1) \\ 4(4) & 4(-6) & 4(0) \end{bmatrix}$ or $\begin{bmatrix} 12 & 8 & -4 \\ 16 & -24 & 0 \end{bmatrix}$ Multiply each element by 4.

Exercises **Perform the indicated matrix operations. If the matrix does not exist, write *impossible*.** *See Examples 1, 2, and 4 on pages 160–162.*

15. $\begin{bmatrix} -4 & 3 \\ -5 & 2 \end{bmatrix} + \begin{bmatrix} 1 & -3 \\ 3 & -8 \end{bmatrix}$ $\begin{bmatrix} -3 & 0 \\ -2 & -6 \end{bmatrix}$

16. $[0.2 \quad 1.3 \quad -0.4] - [2 \quad 1.7 \quad 2.6]$ $[-1.8 \quad -0.4 \quad -3]$

17. $\begin{bmatrix} 1 & -5 \\ -2 & 3 \end{bmatrix} + \frac{3}{4}\begin{bmatrix} 0 & 4 \\ -16 & 8 \end{bmatrix}$ $\begin{bmatrix} 1 & -2 \\ -14 & 9 \end{bmatrix}$

18. $\begin{bmatrix} 1 & 0 & -3 \\ 4 & -5 & 2 \end{bmatrix} - 2\begin{bmatrix} -2 & 3 & 5 \\ -3 & -1 & 2 \end{bmatrix}$ $\begin{bmatrix} 5 & -6 & -13 \\ 10 & -3 & -2 \end{bmatrix}$

210 Chapter 4 Matrices

4-3 Multiplying Matrices

See pages 167–174.

Concept Summary

- Two matrices can be multiplied if and only if the number of columns in the first matrix is equal to the number of rows in the second matrix.

Example **Find XY if $X = [6 \quad 4 \quad 1]$ and $Y = \begin{bmatrix} 2 & 5 \\ -3 & 0 \\ -1 & 3 \end{bmatrix}$.**

$XY = [6 \quad 4 \quad 1] \cdot \begin{bmatrix} 2 & 5 \\ -3 & 0 \\ -1 & 3 \end{bmatrix}$ Write an equation.

$= [6(2) + 4(-3) + 1(-1) \quad 6(5) + 4(0) + 1(3)]$ Multiply columns by rows.

$= [-1 \quad 33]$ Simplify.

Exercises **Find each product, if possible.** *See Example 2 on page 168.*

19. $[2 \quad 7] \cdot \begin{bmatrix} 5 \\ -4 \end{bmatrix}$ $[-18]$

20. $\begin{bmatrix} 8 & -3 \\ 6 & 1 \end{bmatrix} \cdot \begin{bmatrix} 2 & -3 \\ 1 & -5 \end{bmatrix}$ $\begin{bmatrix} 13 & -9 \\ 13 & -23 \end{bmatrix}$

21. $\begin{bmatrix} 3 & 4 \\ 1 & 0 \\ 2 & -5 \end{bmatrix} \cdot \begin{bmatrix} -2 & 4 & 5 \\ 3 & 0 & -1 \\ 1 & 0 & -1 \end{bmatrix}$ not possible

22. $\begin{bmatrix} 3 & 0 & -1 \\ 4 & -2 & 3 \end{bmatrix} \cdot \begin{bmatrix} 7 & 1 \\ 6 & -3 \\ 2 & 1 \end{bmatrix}$ $\begin{bmatrix} 19 & 2 \\ 22 & 13 \end{bmatrix}$

4-4 Transformations with Matrices

See pages 175–181.

Concept Summary

- Use matrix addition and a translation matrix to find the coordinates of a translated figure.
- Use scalar multiplication to perform dilations.
- To reflect a figure, multiply the vertex matrix on the left by a reflection matrix.

reflection over x-axis: $\begin{bmatrix} 1 & 0 \\ 0 & -1 \end{bmatrix}$ reflection over y-axis: $\begin{bmatrix} -1 & 0 \\ 0 & 1 \end{bmatrix}$ reflection over line $y = x$: $\begin{bmatrix} 0 & 1 \\ 1 & 0 \end{bmatrix}$

- To rotate a figure counterclockwise about the origin, multiply the vertex matrix on the left by a rotation matrix.

90° rotation: $\begin{bmatrix} 0 & -1 \\ 1 & 0 \end{bmatrix}$ 180° rotation: $\begin{bmatrix} -1 & 0 \\ 0 & -1 \end{bmatrix}$ 270° rotation: $\begin{bmatrix} 0 & 1 \\ -1 & 0 \end{bmatrix}$

Example **Find the coordinates of the vertices of the image of $\triangle PQR$ with $P(4, 2)$, $Q(6, 5)$, and $R(0, 5)$ after it is rotated 90° counterclockwise about the origin.**

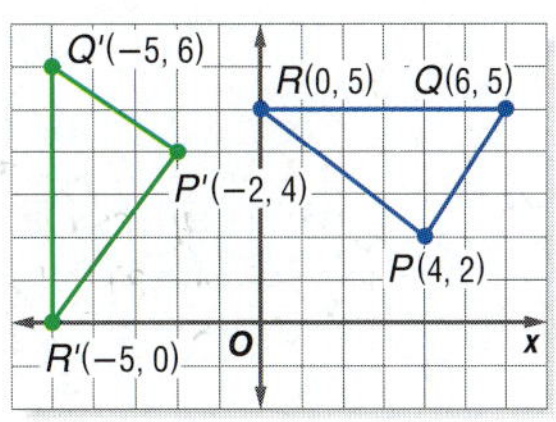

Write the ordered pairs in a vertex matrix. Then multiply the vertex matrix by the rotation matrix.

$$\begin{bmatrix} 0 & -1 \\ 1 & 0 \end{bmatrix} \cdot \begin{bmatrix} 4 & 6 & 0 \\ 2 & 5 & 5 \end{bmatrix} = \begin{bmatrix} -2 & -5 & -5 \\ 4 & 6 & 0 \end{bmatrix}$$

The coordinates of the vertices of $\triangle P'Q'R'$ are $P'(-2, 4)$, $Q'(-5, 6)$, and $R'(-5, 0)$.

Chapter 4 Study Guide and Review 211

Exercises For Exercises 23–26, use the figure at the right.
See Examples 1–5 on pages 175–178.

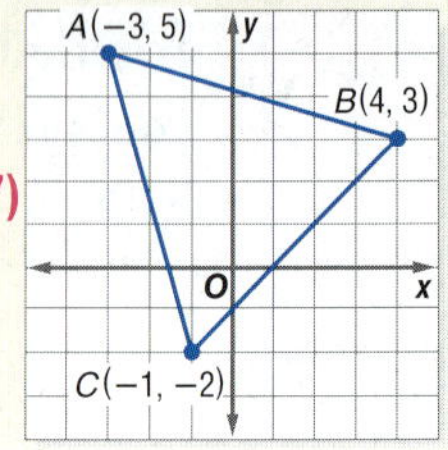

23. Find the coordinates of the image after a translation 4 units right and 5 units down. **A′(1, 0), B′(8, −2), C′(3, −7)**
24. Find the coordinates of the image of the figure after a dilation by a scale factor of 2.
25. Find the coordinates of the image after a reflection over the y-axis. **A′(3, 5), B′(−4, 3), C′(1, −2)**
26. Find the coordinates of the image of the figure after a rotation of 180°. **A′(3, −5), B′(−4, −3), C′(1, 2)**

24. A′(−6, 10), B′(8, 6), C′(−2, −4)

4-5 Determinants

See pages 182–188.

Concept Summary

- Determinant of a 2 × 2 matrix: $\begin{vmatrix} a & b \\ c & d \end{vmatrix} = ad - bc$
- Determinant of a 3 × 3 matrix: $\begin{vmatrix} a & b & c \\ d & e & f \\ g & h & i \end{vmatrix} = a\begin{vmatrix} e & f \\ h & i \end{vmatrix} - b\begin{vmatrix} d & f \\ g & i \end{vmatrix} + c\begin{vmatrix} d & e \\ g & h \end{vmatrix}$
- Area of a triangle with vertices at (a, b), (c, d), and (e, f):

  $|A|$ where $A = \frac{1}{2}\begin{vmatrix} a & b & 1 \\ c & d & 1 \\ e & f & 1 \end{vmatrix}$

Examples

1 **Find the value of $\begin{vmatrix} 3 & 6 \\ -4 & 2 \end{vmatrix}$.**

$\begin{vmatrix} 3 & 6 \\ -4 & 2 \end{vmatrix} = 3(2) - (-4)(6)$ Definition of determinant

$= 6 - (-24)$ or 30 Simplify.

2 **Evaluate $\begin{vmatrix} 3 & 1 & 5 \\ 1 & -2 & 1 \\ 0 & -1 & 2 \end{vmatrix}$ using expansion by minors.**

$\begin{vmatrix} 3 & 1 & 5 \\ 1 & -2 & 1 \\ 0 & -1 & 2 \end{vmatrix} = 3\begin{vmatrix} -2 & 1 \\ -1 & 2 \end{vmatrix} - 1\begin{vmatrix} 1 & 1 \\ 0 & 2 \end{vmatrix} + 5\begin{vmatrix} 1 & -2 \\ 0 & -1 \end{vmatrix}$ Expansion by minors

$= 3(-4 - (-1)) - 1(2 - 0) + 5(-1 - 0)$ Evaluate 2 × 2 determinants.

$= -9 - 2 - 5$ or -16 Simplify.

Exercises Find the value of each determinant. *See Examples 1–3 on pages 182–184.*

27. $\begin{vmatrix} 4 & 11 \\ -7 & 8 \end{vmatrix}$ **109**
28. $\begin{vmatrix} 6 & -7 \\ 5 & 3 \end{vmatrix}$ **53**
29. $\begin{vmatrix} 12 & 8 \\ 9 & 6 \end{vmatrix}$ **0**
30. $\begin{vmatrix} 2 & -3 & 1 \\ 0 & 7 & 8 \\ 2 & 1 & 3 \end{vmatrix}$ **−36**
31. $\begin{vmatrix} 7 & -4 & 5 \\ 1 & 3 & -6 \\ 5 & -1 & -2 \end{vmatrix}$ **−52**
32. $\begin{vmatrix} 6 & 3 & -2 \\ -4 & 2 & 5 \\ -3 & -1 & 0 \end{vmatrix}$ **−35**

212 Chapter 4 Matrices

4-6 Cramer's Rule

See pages 189–194.

Concept Summary

- Cramer's Rule for two variables:
  The solution of the system of equations $ax + by = e$ and $cx + dy = f$ is (x, y), where $x = \frac{\begin{vmatrix} e & b \\ f & d \end{vmatrix}}{\begin{vmatrix} a & b \\ c & d \end{vmatrix}}$, $y = \frac{\begin{vmatrix} a & e \\ c & f \end{vmatrix}}{\begin{vmatrix} a & b \\ c & d \end{vmatrix}}$, and $\begin{vmatrix} a & b \\ c & d \end{vmatrix} \neq 0$.

- Cramer's Rule for three variables:
  The solution of the system whose equations are $ax + by + cz = j$, $dx + ey + fz = k$, $gx + hy + iz = \ell$ is (x, y, z), where

$$x = \frac{\begin{vmatrix} j & b & c \\ k & e & f \\ \ell & h & i \end{vmatrix}}{\begin{vmatrix} a & b & c \\ d & e & f \\ g & h & i \end{vmatrix}}, \; y = \frac{\begin{vmatrix} a & j & c \\ d & k & f \\ g & \ell & i \end{vmatrix}}{\begin{vmatrix} a & b & c \\ d & e & f \\ g & h & i \end{vmatrix}}, \; z = \frac{\begin{vmatrix} a & b & j \\ d & e & k \\ g & h & \ell \end{vmatrix}}{\begin{vmatrix} a & b & c \\ d & e & f \\ g & h & i \end{vmatrix}}, \text{ and } \begin{vmatrix} a & b & c \\ d & e & f \\ g & h & i \end{vmatrix} \neq 0.$$

Example **Use Cramer's Rule to solve each system of equations $5a - 3b = 7$ and $3a + 9b = -3$.**

$$a = \frac{\begin{vmatrix} 7 & -3 \\ -3 & 9 \end{vmatrix}}{\begin{vmatrix} 5 & -3 \\ 3 & 9 \end{vmatrix}} \quad \text{Cramer's Rule} \qquad b = \frac{\begin{vmatrix} 5 & 7 \\ 3 & -3 \end{vmatrix}}{\begin{vmatrix} 5 & -3 \\ 3 & 9 \end{vmatrix}}$$

$$= \frac{63 - 9}{45 + 9} \quad \text{Evaluate each determinant.} \qquad = \frac{-15 - 21}{45 + 9}$$

$$= \frac{54}{54} \text{ or } 1 \quad \text{Simplify.} \qquad = \frac{-36}{54} \text{ or } -\frac{2}{3}$$

The solution is $\left(1, -\frac{2}{3}\right)$.

Exercises **Use Cramer's Rule to solve each system of equations.**
See Examples 1 and 3 on pages 190 and 191.

33. $9a - b = 1$
$3a + 2b = 12$ **$\left(\frac{2}{3}, 5\right)$**

34. $x + 5y = 14$ **(4, 2)**
$-2x + 6y = 4$

35. $3x + 4y = -15$ **(−1, −3)**
$2x - 7y = 19$

36. $8a + 5b = 2$
$-6a - 4b = -1$ **$\left(\frac{3}{2}, -2\right)$**

37. $6x - 7z = 13$ **(1, 2, −1)**
$8y + 2z = 14$
$7x + z = 6$

38. $2a - b - 3c = -20$
$4a + 2b + c = 6$
$2a + b - c = -6$ **$\left(-\frac{1}{2}, 1, 6\right)$**

4-7 Identity and Inverse Matrices

See pages 195–201.

Concept Summary

- An identity matrix is a square matrix with ones on the diagonal and zeros in the other positions.
- Two matrices are inverses of each other if their product is the identity matrix.
- The inverse of matrix $A = \begin{bmatrix} a & b \\ c & d \end{bmatrix}$ is $A^{-1} \frac{1}{ad - bc} \begin{bmatrix} d & -b \\ -c & a \end{bmatrix}$, where $ad - bc \neq 0$.

Chapter 4 Study Guide and Review 213

Example Find the inverse of $S = \begin{bmatrix} 3 & -4 \\ 2 & 1 \end{bmatrix}$.

Find the value of the determinant.

$\begin{vmatrix} 3 & -4 \\ 2 & 1 \end{vmatrix} = 3 - (-8) \text{ or } 11$

Use the formula for the inverse matrix.

$S^{-1} = \frac{1}{11}\begin{bmatrix} 1 & 4 \\ -2 & 3 \end{bmatrix}$

Exercises **Find the inverse of each matrix, if it exists.** *See Example 2 on page 197.*

39. $\begin{bmatrix} 3 & 2 \\ 4 & -2 \end{bmatrix}$ $-\frac{1}{14}\begin{bmatrix} -2 & -2 \\ -4 & 3 \end{bmatrix}$

40. $\begin{bmatrix} 8 & 6 \\ 9 & 7 \end{bmatrix}$ $\frac{1}{2}\begin{bmatrix} 7 & -6 \\ -9 & 8 \end{bmatrix}$

41. $\begin{bmatrix} 2 & 4 \\ -3 & 6 \end{bmatrix}$ $\frac{1}{24}\begin{bmatrix} 6 & -4 \\ 3 & 2 \end{bmatrix}$

42. $\begin{bmatrix} 6 & -2 \\ 3 & -1 \end{bmatrix}$ no inverse exists

43. $\begin{bmatrix} 0 & 2 \\ 5 & -4 \end{bmatrix}$ $-\frac{1}{10}\begin{bmatrix} -4 & -2 \\ -5 & 0 \end{bmatrix}$

44. $\begin{bmatrix} 6 & -1 & 0 \\ 5 & 8 & -2 \end{bmatrix}$ no inverse exists

4-8 Using Matrices to Solve Systems of Equations

See pages 202–207.

Concept Summary

- A system of equations can be written as a matrix equation in the form $A \cdot X = B$.

  $\begin{matrix} 2x + 3y = 12 \\ x - 4y = 6 \end{matrix} \rightarrow \begin{bmatrix} 2 & 3 \\ 1 & -4 \end{bmatrix} \cdot \begin{bmatrix} x \\ y \end{bmatrix} = \begin{bmatrix} 12 \\ 6 \end{bmatrix}$

- To solve a matrix equation, find the inverse of the coefficient matrix. Then multiply each side by the inverse matrix, so $X = A^{-1}B$.

Example Solve $\begin{bmatrix} 4 & 8 \\ 2 & -3 \end{bmatrix} \cdot \begin{bmatrix} x \\ y \end{bmatrix} = \begin{bmatrix} 12 \\ 13 \end{bmatrix}$.

Step 1 Find the inverse of the coefficient matrix.

$A^{-1} = \frac{1}{-12 - 16}\begin{bmatrix} -3 & -8 \\ -2 & 4 \end{bmatrix} \text{ or } -\frac{1}{28}\begin{bmatrix} -3 & -8 \\ -2 & 4 \end{bmatrix}$

Step 2 Multiply each side by the inverse matrix.

$$-\frac{1}{28}\begin{bmatrix} -3 & -8 \\ -2 & 4 \end{bmatrix} \cdot \begin{bmatrix} 4 & 8 \\ 2 & -3 \end{bmatrix} \cdot \begin{bmatrix} x \\ y \end{bmatrix} = -\frac{1}{28}\begin{bmatrix} -3 & -8 \\ -2 & 4 \end{bmatrix} \cdot \begin{bmatrix} 12 \\ 13 \end{bmatrix}$$

$$\begin{bmatrix} 1 & 0 \\ 0 & 1 \end{bmatrix} \cdot \begin{bmatrix} x \\ y \end{bmatrix} = -\frac{1}{28}\begin{bmatrix} -140 \\ 28 \end{bmatrix}$$

$$\begin{bmatrix} x \\ y \end{bmatrix} = \begin{bmatrix} 5 \\ -1 \end{bmatrix}$$

The solution is $(5, -1)$.

Exercises **Solve each matrix equation or system of equations by using inverse matrices.** *See Example 3 on page 204.*

45. $\begin{bmatrix} 5 & -2 \\ 1 & 3 \end{bmatrix} \cdot \begin{bmatrix} x \\ y \end{bmatrix} = \begin{bmatrix} 16 \\ 10 \end{bmatrix}$ (4, 2)

46. $\begin{bmatrix} 4 & 1 \\ 3 & -2 \end{bmatrix} \cdot \begin{bmatrix} a \\ b \end{bmatrix} = \begin{bmatrix} 9 \\ 4 \end{bmatrix}$ (2, 1)

47. $3x + 8 = -y$
    $4x - 2y = -14$ (−3, 1)

48. $3x - 5y = -13$
    $4x + 3y = 2$ (−1, 2)

214 Chapter 4 Matrices

Chapter 4 Practice Test

Vocabulary and Concepts

Choose the letter that best matches each description.

1. $\begin{vmatrix} a & b \\ c & d \end{vmatrix} = ad - bc$ **b**
2. $\begin{bmatrix} a & b \\ c & d \end{bmatrix} \cdot \begin{bmatrix} x \\ y \end{bmatrix} = \begin{bmatrix} e \\ f \end{bmatrix}$ **c**
3. $\dfrac{1}{ad - bc}\begin{bmatrix} d & -b \\ -c & a \end{bmatrix}$ **a**

- **a.** inverse of $\begin{bmatrix} a & b \\ c & d \end{bmatrix}$
- **b.** determinant of $\begin{bmatrix} a & b \\ c & d \end{bmatrix}$
- **c.** matrix equation for $ax + by = e$ and $cx + dy = f$

Skills and Applications

Solve each equation.

4. $\begin{bmatrix} 3x + 1 \\ 2y \end{bmatrix} = \begin{bmatrix} 10 \\ 4 + y \end{bmatrix}$ **(3, 4)**

5. $\begin{bmatrix} 2x & y + 1 \\ 13 & -2 \end{bmatrix} = \begin{bmatrix} -16 & -7 \\ 13 & z - 8 \end{bmatrix}$ **(−8, −8, 6)**

Perform the indicated matrix operations. If the matrix does not exist, write *impossible*.

6. $\begin{bmatrix} 2 & -4 & 1 \\ 3 & 8 & -2 \end{bmatrix} - 2\begin{bmatrix} 1 & 2 & -4 \\ -2 & 3 & 7 \end{bmatrix}$ $\begin{bmatrix} 0 & -8 & 9 \\ 7 & 2 & -16 \end{bmatrix}$

7. $\begin{bmatrix} 1 & 6 & 7 \\ 1 & -3 & -4 \end{bmatrix} \cdot \begin{bmatrix} -4 & 3 \\ -1 & -2 \\ 2 & 5 \end{bmatrix}$ $\begin{bmatrix} 4 & 26 \\ -9 & -11 \end{bmatrix}$

Find the value of each determinant.

8. $\begin{vmatrix} -1 & 4 \\ -6 & 3 \end{vmatrix}$ **21**

9. $\begin{vmatrix} 5 & -3 & 2 \\ -6 & 1 & 3 \\ -1 & 4 & -7 \end{vmatrix}$ **−6**

Find the inverse of each matrix, if it exists.

10. $\begin{bmatrix} -2 & 5 \\ 3 & 1 \end{bmatrix}$ $-\dfrac{1}{17}\begin{bmatrix} 1 & -5 \\ -3 & -2 \end{bmatrix}$

11. $\begin{bmatrix} -6 & -3 \\ 8 & 4 \end{bmatrix}$ **no inverse exists**

12. $\begin{bmatrix} 5 & -2 \\ 6 & 3 \end{bmatrix}$ $\dfrac{1}{27}\begin{bmatrix} 3 & 2 \\ -6 & 5 \end{bmatrix}$

Solve each matrix equation or system of equations by using inverse matrices.

13. $\begin{bmatrix} 1 & 8 \\ 2 & -6 \end{bmatrix} \cdot \begin{bmatrix} x \\ y \end{bmatrix} = \begin{bmatrix} -3 \\ -17 \end{bmatrix}$ $\left(-7, \dfrac{1}{2}\right)$

14. $\begin{bmatrix} 5 & 7 \\ -9 & 3 \end{bmatrix} \cdot \begin{bmatrix} m \\ n \end{bmatrix} = \begin{bmatrix} 41 \\ -105 \end{bmatrix}$ **(11, −2)**

15. $5a + 2b = -49$
$2a + 9b = 5$ **(−11, 3)**

For Exercises 16–18, use $\triangle ABC$ whose vertices have coordinates $A(6, 3)$, $B(1, 5)$, and $C(-1, 4)$.

16. Use the determinant to find the area of $\triangle ABC$. **4.5 units^2**
17. Translate $\triangle ABC$ so that the coordinates of B' are (3, 1). What are the coordinates of A' and C'? **$A'(8, -1)$, $C'(1, 0)$**
18. Find the coordinates of the vertices of a similar triangle whose perimeter is five times that of $\triangle ABC$. **$A'(30, 15)$, $B'(5, 25)$, $C'(-5, 20)$**

19. **RETAIL SALES** Brittany is preparing boxes of assorted chocolates. Chocolate-covered peanuts cost \$7 per pound. Chocolate-covered caramels cost \$6.50 per pound. The boxes of assorted candies contain five more pounds of peanut candies than caramel candies. If the total amount sold was \$575, how many pounds of each candy were needed to make the boxes? **40 lb caramel, 45 lb peanut**

20. **STANDARDIZED TEST PRACTICE** If $\begin{bmatrix} 43 & z \\ 7x - 2 & 2x + 3 \end{bmatrix} = \begin{bmatrix} z + 3 & 2m + 5 \\ y & 37 \end{bmatrix}$, then $y =$ **B**

Ⓐ 120. Ⓑ 117. Ⓒ 22. Ⓓ not enough information

www.algebra2.com/chapter_test

Portfolio Suggestion

Introduction The Associative, Commutative, and Distributive Properties are familiar to students. These properties can also be used, with a few differences, in operations with matrices.

Ask Students Using matrix addition, subtraction, and multiplication, determine to what extent these three properties apply to matrix operations. Write a convincing argument for each property.

Chapter 4 Practice Test

Assessment Options

Vocabulary Test A vocabulary test/review for Chapter 4 can be found on p. 230 of the *Chapter 4 Resource Masters*.

Chapter Tests There are six Chapter 4 Tests and an Open-Ended Assessment task available in the *Chapter 4 Resource Masters*.

Chapter 4 Tests

Form	Type	Level	Pages
1	MC	basic	217–218
2A	MC	average	219–220
2B	MC	average	221–222
2C	FR	average	223–224
2D	FR	average	225–226
3	FR	advanced	227–228

MC = multiple-choice questions
FR = free-response questions

Open-Ended Assessment Performance tasks for Chapter 4 can be found on p. 229 of the *Chapter 4 Resource Masters*. A sample scoring rubric for these tasks appears on p. A31.

Unit 1 Test A unit test/review can be found on pp. 237–238 of the *Chapter 4 Resource Masters*.

ExamView® Pro

Use the networkable ExamView® Pro to:

- Create **multiple versions** of tests.
- Create **modified** tests for *Inclusion* students.
- **Edit** existing questions and **add** your own questions.
- Use built-in **state curriculum correlations** to create tests aligned with state standards.
- **Apply** art to your tests from a program bank of artwork.

Chapter 4 Standardized Test Practice

These two pages contain practice questions in the various formats that can be found on the most frequently given standardized tests.

A practice answer sheet for these two pages can be found on p. A1 of the *Chapter 4 Resource Masters*.

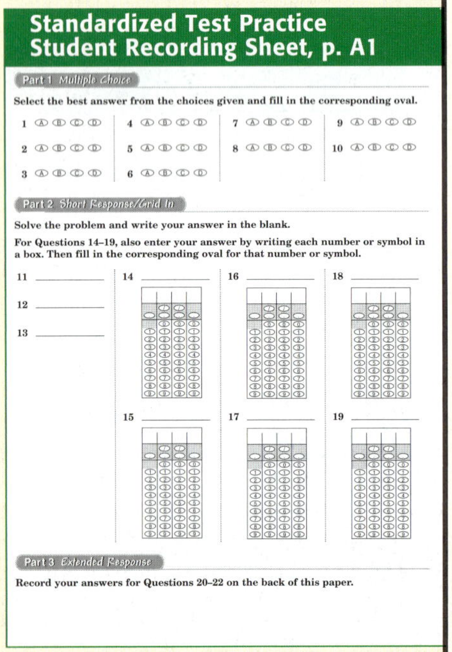

Standardized Test Practice
Student Recording Sheet, p. A1

Part 1 Multiple Choice

Select the best answer from the choices given and fill in the corresponding oval.

1 Ⓐ Ⓑ Ⓒ Ⓓ 4 Ⓐ Ⓑ Ⓒ Ⓓ 7 Ⓐ Ⓑ Ⓒ Ⓓ 9 Ⓐ Ⓑ Ⓒ Ⓓ

2 Ⓐ Ⓑ Ⓒ Ⓓ 5 Ⓐ Ⓑ Ⓒ Ⓓ 8 Ⓐ Ⓑ Ⓒ Ⓓ 10 Ⓐ Ⓑ Ⓒ Ⓓ

3 Ⓐ Ⓑ Ⓒ Ⓓ 6 Ⓐ Ⓑ Ⓒ Ⓓ

Part 2 Short Response/Grid In

Solve the problem and write your answer in the blank.

For Questions 14–19, also enter your answer by writing each number or symbol in a box. Then fill in the corresponding oval for that number or symbol.

11 ______ 12 ______ 13 ______ 14 ______ 15 ______ 16 ______ 17 ______ 18 ______ 19 ______

Part 3 Extended Response

Record your answers for Questions 20–22 on the back of this paper.

Teaching Tip In Questions 4 and 6, students may want to write down the basic formulas for the circumference and area of a circle before they begin their calculations.

Additional Practice

See pp. 235–236 in the *Chapter 4 Resource Masters* for additional standardized test practice.

Chapter 4 Standardized Test Practice

Part 1 Multiple Choice

Record your answers on the answer sheet provided by your teacher or on a sheet of paper.

1. If the average (arithmetic mean) of ten numbers is 18 and the average of six of these numbers is 12, what is the average of the other four numbers? **C**

Ⓐ 15 Ⓑ 18
Ⓒ 27 Ⓓ 28

2. A car travels 65 miles per hour for 2 hours. A truck travels 60 miles per hour for 1.5 hours. What is the difference between the number of miles traveled by the car and the number of miles traveled by the truck? **B**

Ⓐ 31.25 Ⓑ 40
Ⓒ 70 Ⓓ 220

3. In the figure, $a =$ **B**

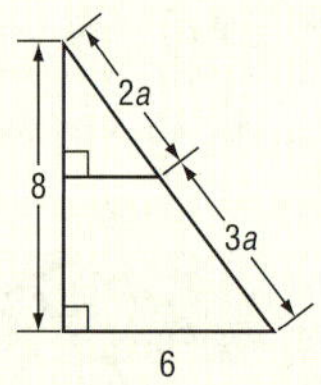

Ⓐ 1.
Ⓑ 2.
Ⓒ 3.
Ⓓ 4.

4. If the circumference of a circle is $\frac{4\pi}{3}$, then what is half of its area? **A**

Ⓐ $\frac{2\pi}{9}$ Ⓑ $\frac{4\pi}{9}$
Ⓒ $\frac{8\pi}{9}$ Ⓓ $\frac{2\pi^2}{9}$

5. A line is represented by the equation $x = 6$. What is the slope of the line? **D**

Ⓐ 0 Ⓑ $\frac{5}{6}$
Ⓒ 6 Ⓓ undefined

6. In the figure, $ABCD$ is a square inscribed in the circle centered at O. If $\overline{OB}$ is 10 units long, how many units long is minor arc BC? **B**

Ⓐ $\frac{5}{2}\pi$ units
Ⓑ 5π units
Ⓒ 10π units
Ⓓ 20π units

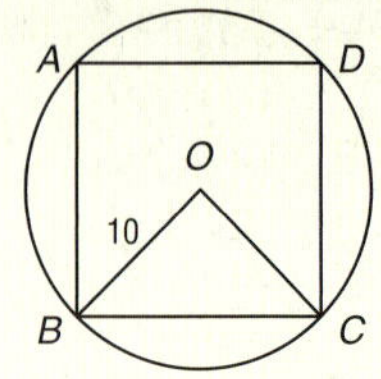

7. If $3 < x < 5 < y < 10$, then which of the following best defines $\frac{x}{y}$? **A**

Ⓐ $\frac{3}{10} < \frac{x}{y} < 1$
Ⓑ $\frac{3}{10} < \frac{x}{y} < \frac{1}{2}$
Ⓒ $\frac{3}{5} < \frac{x}{y} < \frac{1}{2}$
Ⓓ $\frac{3}{5} < \frac{x}{y} < 1$

8. If $x + 3y = 12$ and $\frac{2}{3}x - y = 5$, then $x =$ **C**

Ⓐ 1. Ⓑ 8.
Ⓒ 9. Ⓓ 13.5.

9. At what point do the two lines with the equations $7x - 3y = 13$ and $y = 2x - 3$ intersect? **C**

Ⓐ $(-4, -11)$ Ⓑ $(4, 11)$
Ⓒ $(4, 5)$ Ⓓ $(5, 4)$

10. If $N = \begin{bmatrix} -1 & 0 \\ 5 & -2 \end{bmatrix}$ and $M = \begin{bmatrix} -1 & 0 \\ 5 & 2 \end{bmatrix}$, find $N - M$. **C**

Ⓐ $\begin{bmatrix} 0 & 0 \\ 0 & 0 \end{bmatrix}$ Ⓑ $\begin{bmatrix} 1 & 0 \\ 0 & 1 \end{bmatrix}$
Ⓒ $\begin{bmatrix} 0 & 0 \\ 0 & -4 \end{bmatrix}$ Ⓓ $\begin{bmatrix} -2 & 0 \\ 0 & -4 \end{bmatrix}$

ExamView® Pro

Special banks of standardized test questions similar to those on the SAT, ACT, TIMSS 12, NAEP 12, and state proficiency tests can be found on this CD-ROM.

216 Chapter 4 Matrices

Preparing for Standardized Tests
For test-taking strategies and more practice, see pages 877–892.

Part 2 Short Response/Grid In

Record your answers on the answer sheet provided by your teacher or on a sheet of paper.

11. A computer manufacturer reduced the price of its Model X computer by 3%. If the new price of the Model X computer is \$2489, then how much did the computer cost, in dollars, before its price was reduced? (Round to the nearest dollar.) **\$2566**

12. In square *PQRS*, $PQ = 4$, $PU = UQ$, and $PT = TS$. What is the area of the shaded region? **6 units²**

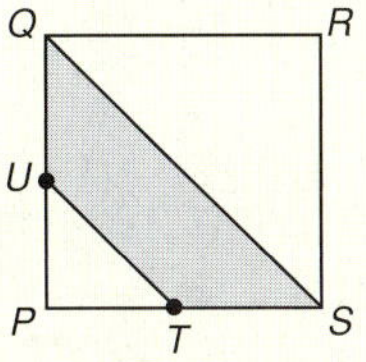

13. Write an equation of a line that passes through the origin and is parallel to the line with equation $3x - y = 5$. **$3x - y = 0$**

14. A rectangular solid has two faces the same size and shape as Figure 1 and four faces the same size and shape as Figure 2. What is the volume of the solid in cubic units? **54**

Figure 1 Figure 2

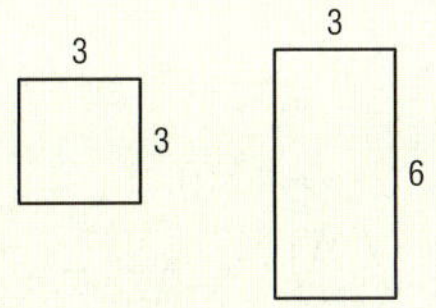

15. If the average (arithmetic mean) of three different positive integers is 60, what is the greatest possible value of one of the integers? **177**

Test-Taking Tip

Questions 15 and 16
Watch for the phrases "greatest possible" or "least possible." Think logically about the conditions that make an expression greatest or least. Notice what types of numbers are used—positive, even, prime, integers.

www.algebra2.com/standardized_test

16. The perimeter of a triangle is 15. The lengths of the sides are integers. If the length of one side is 6, what is the shortest possible length of another side of the triangle? **2**

17. In this sequence below, each term after the first term is $\frac{1}{4}$ of the term preceding it. What is the sixth term of this sequence? **5/16**

320, 80, 20, …

18. If the sum of two numbers is 5 and their difference is 2, what is their product? **5.25**

19. What positive value of k would make the lines below parallel in the coordinate plane? **6**

$$9x + ky = 16$$
$$kx + 4y = 11$$

Part 3 Extended Response

Record your answers on a sheet of paper. Show your work.

For Exercises 20–22, use the information below.

The Colonial High School Yearbook Staff is selling yearbooks and chrome picture frames engraved with the year. The number of yearbooks and frames sold to members of each grade is shown in the table.

Sales for Each Class		
Grade	Yearbooks	Frames
9th	423	256
10th	464	278
11th	546	344
12th	575	497

20. Find the difference in the sales of yearbooks and frames made to the 10th and 11th grade classes. **82; 66**

21. Find the total numbers of yearbooks and frames sold. **2008, 1375**

22. A yearbook costs \$48, and a frame costs \$18. Find the total sales of books and frames sold to each class. **yearbooks: \$20,304, \$22,272, \$26,208, \$27,600; frames: \$4608, \$5004, \$6192, \$8946**

Evaluating Extended-Response Assessment Questions

Extended-Response Assessment questions are graded by using a multilevel rubric that guides you in assessing a student's knowledge of a particular concept.

Goal: Determine the dimensions of a rectangular garden for a given perimeter.

Sample Scoring Rubric: The following rubric is a sample scoring device. You may wish to add more detail to this sample to meet your individual scoring needs.

Score	Criteria
4	A correct solution that is supported by well-developed, accurate explanations
3	A generally correct solution, but may contain minor flaws in reasoning or computation
2	A partially correct interpretation and/or solution to the problem
1	A correct solution with no supporting evidence or explanation
0	An incorrect solution indicating no mathematical understanding of the concept or task, or no solution is given

Chapter 4 Standardized Test Practice 217

Page 153, Chapter 4 Getting Started

15.

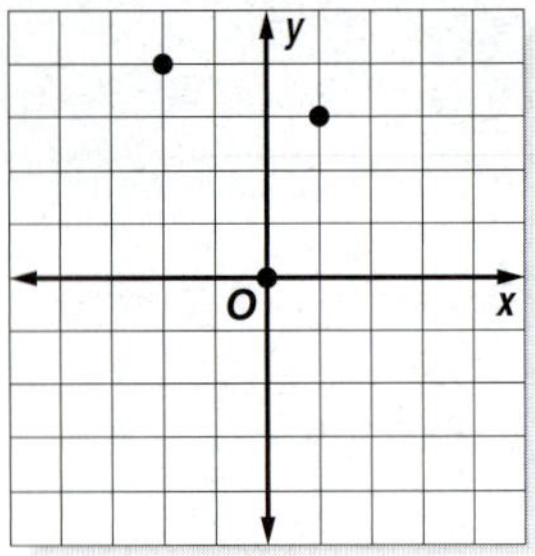

16.

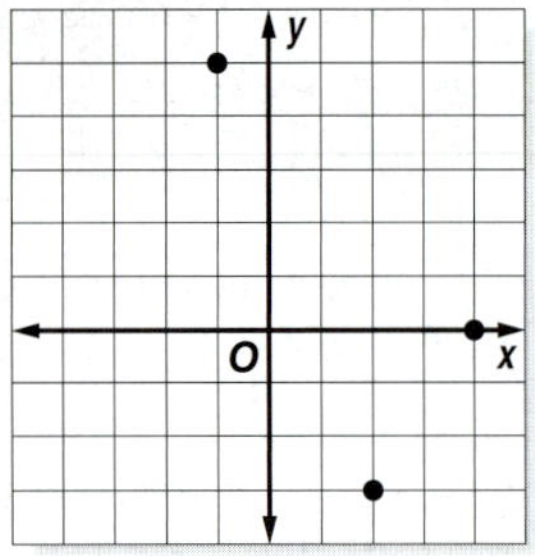

17.

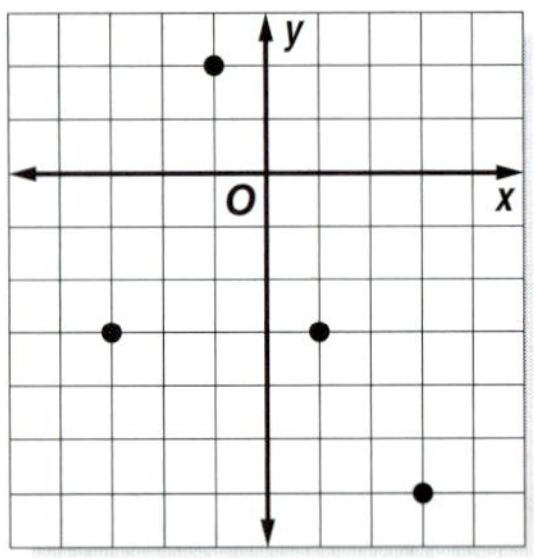

18.

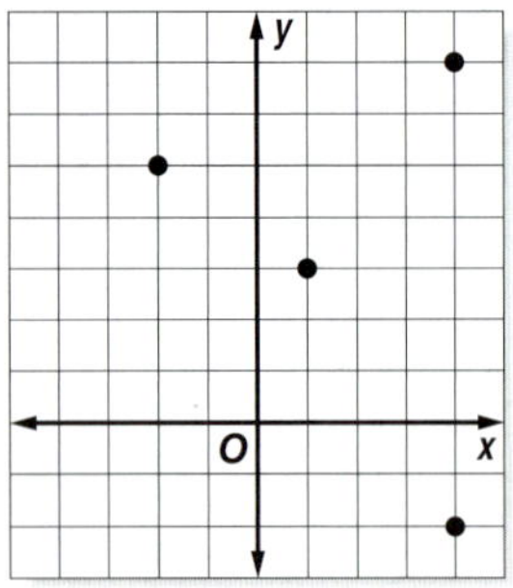

Pages 157–158, Lesson 4-1

28.

	Cost	Service	Atmosphere	Location
Catalina Grill	★★	★	★	★
Oyster Club	★★★	★★	★	★★
Casa di Pasta	★★★★	★★★	★★★	★★★
Mason's Steakhouse	★★	★★★★	★★★★	★★★

29. Sample answer: Mason's Steakhouse; it was given the highest rating possible for service and atmosphere, location was given one of the highest ratings, and it is moderately priced.

43.

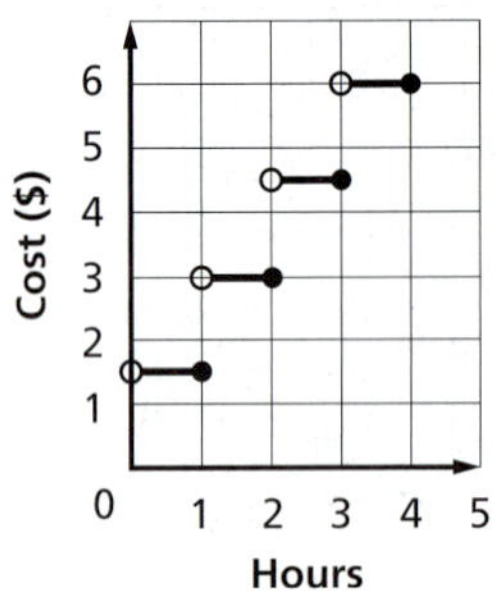

Page 159, Follow-Up of Lesson 4-1 Spreadsheet Investigation

1.

	A	B	C	D	E	F
1		Base Price	Horse-power	Towing Capacity (lb)	Cargo Capacity (ft^3)	Fuel Economy (mpg)
2	Large	\$32,450	285	12,000	46	17
3	Standard	\$29,115	275	8700	16	17.5
4	Mid-Size	\$27,975	190	5700	34	20
5	Compact	\$18,180	127	3000	15	26.5

Page 166, Lesson 4-2

41. You can use matrices to track dietary requirements and add them to find the total each day or each week. Answers should include the following.

- Breakfast $= \begin{bmatrix} 566 & 18 & 7 \\ 482 & 12 & 17 \\ 530 & 10 & 11 \end{bmatrix}$, Lunch $= \begin{bmatrix} 785 & 22 & 19 \\ 622 & 23 & 20 \\ 710 & 26 & 12 \end{bmatrix}$,

  Dinner $= \begin{bmatrix} 1257 & 40 & 26 \\ 987 & 32 & 45 \\ 1380 & 29 & 38 \end{bmatrix}$

- Add the three matrices: $\begin{bmatrix} 2608 & 80 & 52 \\ 2091 & 67 & 82 \\ 2620 & 65 & 61 \end{bmatrix}$.

Pages 172–174, Lesson 4-3

27.
$$AC + BC = \begin{bmatrix} 1 & -2 \\ 4 & 3 \end{bmatrix} \cdot \begin{bmatrix} 5 & 1 \\ 2 & -4 \end{bmatrix} + \begin{bmatrix} -5 & 2 \\ 4 & 3 \end{bmatrix} \cdot \begin{bmatrix} 5 & 1 \\ 2 & -4 \end{bmatrix}$$
$$= \begin{bmatrix} 1 & 9 \\ 26 & -8 \end{bmatrix} + \begin{bmatrix} -21 & -13 \\ 26 & -8 \end{bmatrix}$$
$$= \begin{bmatrix} -20 & -4 \\ 52 & -16 \end{bmatrix}$$

$$(A + B)C = \left(\begin{bmatrix} 1 & -2 \\ 4 & 3 \end{bmatrix} + \begin{bmatrix} -5 & 2 \\ 4 & 3 \end{bmatrix}\right) \cdot \begin{bmatrix} 5 & 1 \\ 2 & -4 \end{bmatrix}$$
$$= \begin{bmatrix} -4 & 0 \\ 8 & 6 \end{bmatrix} \cdot \begin{bmatrix} 5 & 1 \\ 2 & -4 \end{bmatrix}$$
$$= \begin{bmatrix} -20 & -4 \\ 52 & -16 \end{bmatrix}$$

28.
$$c(AB) = 3\left(\begin{bmatrix} 1 & -2 \\ 4 & 3 \end{bmatrix} + \begin{bmatrix} -5 & 2 \\ 4 & 3 \end{bmatrix}\right)$$
$$= 3\begin{bmatrix} -13 & -4 \\ -8 & 17 \end{bmatrix} = \begin{bmatrix} -39 & -12 \\ -24 & 51 \end{bmatrix}$$

$$A(cB) = \begin{bmatrix} 1 & -2 \\ 4 & 3 \end{bmatrix} \cdot \left(3\begin{bmatrix} -5 & 2 \\ 4 & 3 \end{bmatrix}\right)$$
$$= \begin{bmatrix} 1 & -2 \\ 4 & 3 \end{bmatrix} \cdot \begin{bmatrix} -15 & 6 \\ 12 & 9 \end{bmatrix} = \begin{bmatrix} -39 & -12 \\ -24 & 51 \end{bmatrix}$$

29.
$$C(A + B) = \begin{bmatrix} 5 & 1 \\ 2 & -4 \end{bmatrix} \cdot \left(\begin{bmatrix} 1 & -2 \\ 4 & 3 \end{bmatrix} + \begin{bmatrix} -5 & 2 \\ 4 & 3 \end{bmatrix}\right)$$
$$= \begin{bmatrix} 5 & 1 \\ 2 & -4 \end{bmatrix} \cdot \begin{bmatrix} -4 & 0 \\ 8 & 6 \end{bmatrix}$$
$$= \begin{bmatrix} -12 & 6 \\ -40 & -24 \end{bmatrix}$$

$$AC + BC = \begin{bmatrix} 1 & -2 \\ 4 & 3 \end{bmatrix} \cdot \begin{bmatrix} 5 & 1 \\ 2 & -4 \end{bmatrix} + \begin{bmatrix} -5 & 2 \\ 4 & 3 \end{bmatrix} \cdot \begin{bmatrix} 5 & 1 \\ 2 & -4 \end{bmatrix}$$
$$= \begin{bmatrix} 1 & 9 \\ 26 & -8 \end{bmatrix} + \begin{bmatrix} -21 & -13 \\ 26 & -8 \end{bmatrix}$$
$$= \begin{bmatrix} -20 & -4 \\ 52 & -16 \end{bmatrix}$$

30. $ABC = \begin{bmatrix} 1 & -2 \\ 4 & 3 \end{bmatrix} \cdot \begin{bmatrix} -5 & 2 \\ 4 & 3 \end{bmatrix} \cdot \begin{bmatrix} 5 & 1 \\ 2 & -4 \end{bmatrix}$

$= \begin{bmatrix} -13 & -4 \\ -8 & 17 \end{bmatrix} \cdot \begin{bmatrix} 5 & 1 \\ 2 & -4 \end{bmatrix}$

$= \begin{bmatrix} -73 & 3 \\ -6 & -76 \end{bmatrix}$

$CBA = \begin{bmatrix} 5 & 1 \\ 2 & -4 \end{bmatrix} \cdot \begin{bmatrix} -5 & 2 \\ 4 & 3 \end{bmatrix} \cdot \begin{bmatrix} 1 & -2 \\ 4 & 3 \end{bmatrix}$

$= \begin{bmatrix} -21 & 13 \\ -26 & -8 \end{bmatrix} \cdot \begin{bmatrix} 1 & -2 \\ 4 & 3 \end{bmatrix}$

$= \begin{bmatrix} 31 & 81 \\ -58 & 28 \end{bmatrix}$

57.

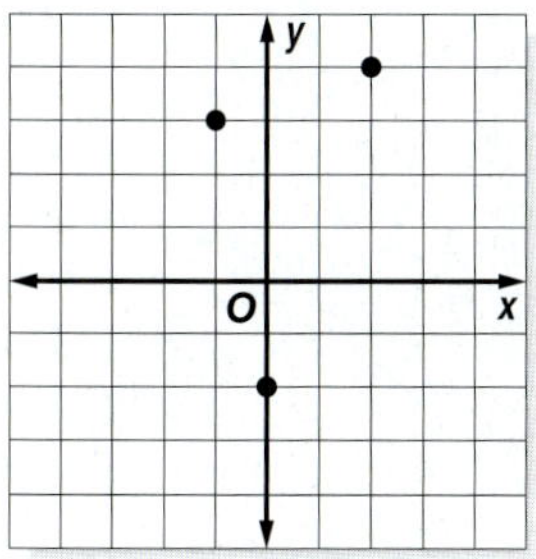

58.

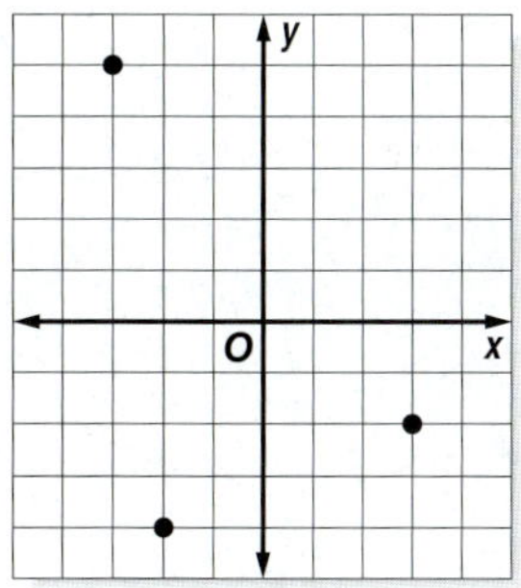

59.

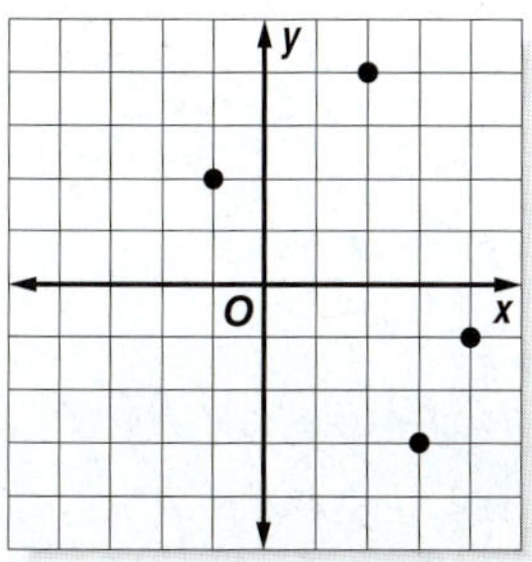

60.

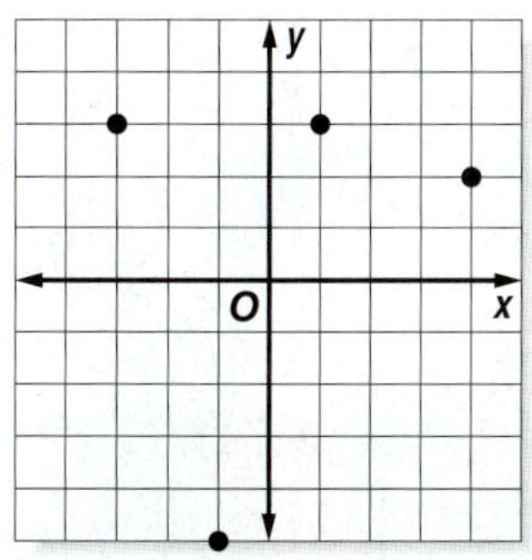

Page 181, Lesson 4-4

43. Transformations are used in computer graphics to create special effects. You can simulate the movement of an object, like in space, which you wouldn't be able to recreate otherwise. Answers should include the following.

- A figure with points (a, b), (c, d), (e, f), (g, h), and (i, j) could be written in a 2×5 matrix $\begin{bmatrix} a & c & e & g & i \\ b & d & f & h & j \end{bmatrix}$ and multiplied on the left by the 2×2 rotation matrix.
- The object would get smaller and appear to be moving away from you.

Page 194, Practice Quiz 2

3.

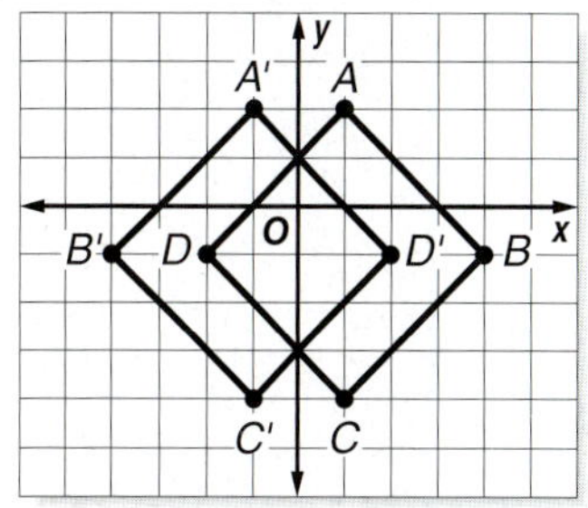

Pages 205–207, Lesson 4-8

4. $\begin{bmatrix} 1 & -1 \\ 1 & 3 \end{bmatrix} \cdot \begin{bmatrix} x \\ y \end{bmatrix} = \begin{bmatrix} -3 \\ 5 \end{bmatrix}$

5. $\begin{bmatrix} 2 & 3 \\ -4 & -7 \end{bmatrix} \cdot \begin{bmatrix} g \\ h \end{bmatrix} = \begin{bmatrix} 8 \\ -5 \end{bmatrix}$

6. $\begin{bmatrix} 3 & -5 & 2 \\ 4 & 7 & 1 \\ 2 & 0 & -1 \end{bmatrix} \cdot \begin{bmatrix} a \\ b \\ c \end{bmatrix} = \begin{bmatrix} 9 \\ 3 \\ 12 \end{bmatrix}$

36. The food and territory that two species of birds require form a system of equations. Any independent system of equations can be solved using a matrix equation. Answers should include the following.

- Let a represent the number of nesting pairs of Species A and let b represent the number of nesting pairs of Species B. Then, $140a + 120b = 20{,}000$ and $500a + 400b = 69{,}000$.
- $\begin{bmatrix} a \\ b \end{bmatrix} = -\frac{1}{4000} \begin{bmatrix} 400 & -120 \\ -500 & 140 \end{bmatrix} \cdot \begin{bmatrix} 20{,}000 \\ 69{,}000 \end{bmatrix}$; $a = 70$ and $b = 85$, so the area can support 70 pairs of Species A and 85 pairs of Species B.

UNIT 2 **Notes**

Introduction

In this unit, students extend their knowledge of first-degree equations and their graphs to radical equations and inequalities. Then they graph quadratic functions and solve quadratic equations and inequalities by various methods, including completing the square and using the Quadratic Formula.

The unit concludes with methods for evaluating polynomial functions, including the Remainder and Factor Theorems. Students graph polynomial functions and investigate their roots and zeros. Finally, they study the composition of two functions, and then find the inverse of a function.

Assessment Options

Unit 2 Test Pages 449–450 of the *Chapter 7 Resource Masters* may be used as a test or review for Unit 2. This assessment contains both multiple-choice and short answer items.

ExamView® Pro

This CD-ROM can be used to create additional unit tests and review worksheets.

UNIT 2

Polynomial and Radical Equations and Inequalities

Equations that model real-world data allow you to make predictions about the future. In this unit, you will learn about nonlinear equations, including polynomial and radical equations, and inequalities.

Chapter 5
Polynomials

Chapter 6
Quadratic Functions and Inequalities

Chapter 7
Polynomial Functions

218 Unit 2 Polynomial and Radical Equations and Inequalities

WebQuest Internet Project

Population Explosion

The United Nations estimated that the world's population reached 6 billion in 1999. The population had doubled in about 40 years and gained 1 billion people in just 12 years. Assuming middle-range birth and death trends, world population is expected to exceed 9 billion by 2050, with most of the increase in countries that are less economically developed. In this project, you will use quadratic and polynomial mathematical models that will help you to project future populations.

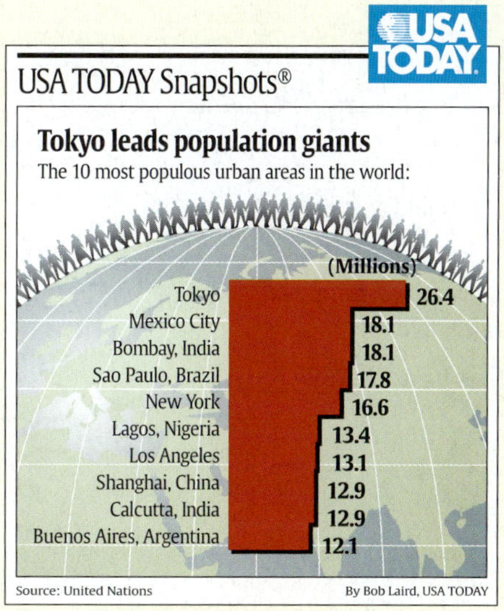

USA TODAY Education

Log on to www.algebra2.com/webquest. Begin your WebQuest by reading the Task.

Then continue working on your WebQuest as you study Unit 2.

Lesson	5-1	6-6	7-4
Page	227	326	369

USA TODAY Education Teaching Suggestions

Have students study the USA TODAY Snapshot®.

- Have students make conjectures about why a quadratic or polynomial model may be better than a linear one for modeling population data.
- According to the given data, what urban area has a population nearly as great as that of New York and Los Angeles combined? **Tokyo**

Additional USA TODAY Snapshots® appearing in Unit 2:

Chapter 5 Hanging on to the old buggy (p. 228)
Chapter 6 More Americans study abroad (p. 328)
Chapter 7 Digital book sales expected to grow (p. 368)

WebQuest Internet Project

Problem-Based Learning A WebQuest is an online project in which students do research on the Internet, gather data, and make presentations using word processing, graphing, page-making, or presentation software. In each chapter, students advance to the next step in their WebQuest. At the end of Chapter 7, the project culminates with a presentation of their findings.

Teaching suggestions and sample answers are available in the *WebQuest and Project Resources.*

Chapter 5

Polynomials
Chapter Overview and Pacing

Year-long and two-year pacing: pages T20–T21.

LESSON OBJECTIVES

	PACING (days)			
	Regular		Block	
	Basic/ Average	Advanced	Basic/ Average	Advanced
5-1 Monomials *(pp. 222–228)* • Multiply and divide monomials. • Use expressions written in scientific notation.	1	1	0.5	0.5
5-2 Polynomials *(pp. 229–232)* • Add and subtract polynomials. • Multiply polynomials.	1	1	0.5	0.5
5-3 Dividing Polynomials *(pp. 233–238)* • Divide polynomials using long division. • Divide polynomials using synthetic division.	1	1	0.5	0.5
5-4 Factoring Polynomials *(pp. 239–244)* • Factor polynomials. • Simplify polynomial quotients by factoring.	2	2	1	1
5-5 Roots of Real Numbers *(pp. 245–249)* • Simplify radicals. • Use a calculator to approximate radicals.	1	1	0.5	0.5
5-6 Radical Expressions *(pp. 250–256)* • Simplify radical expressions. • Add, subtract, multiply, and divide radical expressions.	2	2	1	1
5-7 Rational Exponents *(pp. 257–262)* • Write expressions with rational exponents in radical form, and vice versa. • Simplify expressions in exponential or radical form.	2	2	1	1
5-8 Radical Equations and Inequalities *(pp. 263–269)* • Solve equations containing radicals. • Solve inequalities containing radicals. ***Follow-Up:*** Solving Radical Equations and Inequalities by Graphing	2	2 (with 4-8 Follow-Up)	1	1
5-9 Complex Numbers *(pp. 270–275)* • Add and subtract complex numbers. • Multiply and divide complex numbers.	2	2	1	1
Study Guide and **Practice Test** *(pp. 276–281)* **Standardized Test Practice** *(pp. 282–283)*	1	1	0.5	0.5
Chapter Assessment	1	1	0.5	0.5
TOTAL	**16**	**16**	**8**	**8**

*An electronic version of this chapter is available on **StudentWorks**™. This backpack solution CD-ROM allows students instant access to the Student Edition, lesson worksheet pages, and web resources.*

Timesaving Tools

TeacherWorks™

All-In-One Planner and Resource Center

See pages T5 and T21.

Chapter Resource Manager

Chapter 5 Resource Masters									
Study Guide and Intervention	Practice (Skills and Average)	Reading to Learn Mathematics	Enrichment	Assessment	Applications*	5-Minute Check Transparencies	Interactive Chalkboard	Alge2PASS: Tutorial Plus (lessons)	Materials
239–240	241–242	243	244		SM 109–114	5-1	5-1		
245–246	247–248	249	250		SC 9	5-2	5-2		algebra tiles
251–252	253–254	255	256	307		5-3	5-3		
257–258	259–260	261	262		GCS 35	5-4	5-4	8	algebra tiles, graphing calculator
263–264	265–266	267	268	307, 309		5-5	5-5		index cards, string, small weights
269–270	271–272	273	274			5-6	5-6		
275–276	277–278	279	280	308	GCS 36	5-7	5-7		
281–282	283–284	285	286			5-8	5-8	9	(*Follow-Up:* graphing calculator)
287–288	289–290	291	292	308	SC 10	5-9	5-9		grid paper
				293–306, 310–312					

**Key to Abbreviations:* GCS = Graphing Calculator and Speadsheet Masters,
SC = School-to-Career Masters,
SM = Science and Mathematics Lab Manual

Chapter 5

Mathematical Connections and Background

Continuity of Instruction

Prior Knowledge

Students are familiar with the coefficients, variables, and positive integer exponents that make up monomials, and they have worked with radicals as square roots. They are familiar with the basic arithmetic operations (addition, subtraction, multiplication, division) that will be applied in new situations in this chapter.

This Chapter

Students learn how to apply the basic arithmetic operations to polynomials, radical expressions, and complex numbers. They explore factoring polynomials and solving radical equations and inequalities. They learn how to write equivalent statements by using the Distributive Property, properties of exponents, or properties of radicals. They solve equations and inequalities involving polynomials, radicals, or complex numbers.

Future Connections

As early as the next two chapters, students will factor polynomials to solve quadratic equations, use complex numbers to express the solution to equations or inequalities, and relate complex numbers and roots of polynomials. In their exploration of complex numbers, they will greatly expand on this chapter's brief introduction to the complex coordinate plane.

5-1 Monomials

Throughout this chapter students are introduced to new symbols and new notation. Each time, the new symbols are related to each other and to familiar ideas. In this lesson the new symbol is a negative exponent. The familiar idea is simplifying an expression, and to simplify a monomial means to write an equivalent expression without negative exponents and without parentheses. The rules for writing equivalent expressions include the definition of negative exponents and properties of exponents.

The lesson introduces the terms *coefficient* and *degree* when discussing monomials. It also explores writing and operating on numbers written in scientific notation, and using dimensional analysis to compute with units of measure.

5-2 Polynomials

In this lesson students find the sum or difference of several monomials, which is called a *polynomial*. The familiar ideas in this lesson are the operations of addition, subtraction, and multiplication as applied to polynomials. To add polynomials means to rewrite an indicated sum of polynomials as a sum of terms and then, by combining like terms, to rewrite that sum as a single polynomial. To subtract polynomials means to use the Distributive Property to rewrite the subtraction as a sum of terms, and then to combine like terms. To multiply a polynomial by a monomial means to use the Distributive Property to rewrite the product as a single polynomial.

To multiply two binomials, the lesson shows how two applications of the Distributive Property result in finding the products of the First, Outer, Inner, and Last pairs of terms of the binomials. The two-time application of the Distributive Property is called the FOIL method.

5-3 Dividing Polynomials

This lesson explores polynomial division; with the previous lesson, the four basic arithmetic operations are interpreted for polynomials. Dividing by a monomial uses the Distributive Property. Dividing a polynomial by a binomial (or by any polynomial) uses a process and format analogous to the long division algorithm for whole numbers. The student follows the four steps of "divide, multiply, subtract, bring down"; the steps are repeated until there are no more terms in the dividend.

The lesson also introduces an abbreviated form, called synthetic division, that records and manipulates just the coefficients of the polynomial terms. The divisor must be a binomial of degree 1, the terms of the dividend must be in descending order, using zeros to represent any missing terms, and the polynomial quotient must be written so that the leading coefficient of the divisor is 1.

5-4 Factoring Polynomials

In factoring, a polynomial of degree two or more is rewritten as a product of polynomials each having a lesser degree. Taking out a common factor uses the Distributive Property; factoring by grouping uses two applications of the Distributive Property.

Binomials written in the form of the difference of two squares or as the sum or difference of two cubes can be rewritten as a product of two factors, and a perfect square trinomial can be rewritten as the square of a binomial. Some other trinomials can be factored as the product of two binomials.

Students reduce quotients of polynomials by removing common factors in the numerator and denominator. A record is kept of values of variables that would imply division by zero.

5-5 Roots of Real Numbers

This lesson begins with the familiar idea of a square root: a is a square root of b if $a^2 = b$. Then two new ideas are introduced. One idea is to use the same kind of definition to introduce the nth root of a number: a is an nth root of b if $a^n = b$. The second new idea is to introduce the symbols for principal roots. The principal square root is always a nonnegative number. The value of a principal nth root depends on the sign of the radicand and whether its index is even or odd. The lesson explores how to use absolute value symbols to simplify nth roots.

5-6 Radical Expressions

This lesson explains that "simplifying," as it pertains to radical expressions, takes into account the index of the radical and the form of the radicand. The lesson also presents some rules for writing equivalent radical expressions, and students apply the rules as they add, subtract, multiply, and divide radicals.

5-7 Rational Exponents

The new symbol introduced in this lesson is a fraction used as an exponent. The rules for writing equivalent expressions for rational exponents include properties that describe how to translate between radical form and exponential form. Those rules include dealing with rational exponents that are unit fractions, either positive or negative. The rules for dealing with a base raised to a fractional exponent require that the denominator of the fraction is a positive integer and take into account the sign of the radicand and whether its index is even or odd.

5-8 Radical Equations and Inequalities

This lesson deals with the familiar skills of solving equations and inequalities; the new concept is that the equations contain a variable inside a radical. No new properties are needed to solve these equations or inequalities; equivalent equations or inequalities are written until the variable is isolated on one side. At least once in the solution, both sides of the equation or inequality are raised to a power in order to remove a radical symbol. A most-important idea in the lesson is that sometimes this process of raising both sides to a power does not produce an equivalent statement. For example, it is clear that $\sqrt{x} = -5$ has no real number solution. Squaring both sides results in $x = 25$; the two statements are not equivalent and $x = 25$ is not a solution to the original equation. Raising both sides to a power can introduce an extraneous solution, which is an apparent solution that will not satisfy the original equation or inequality.

5-9 Complex Numbers

This lesson introduces not simply a new symbol but a new set of numbers that are not part of the real number system. The new symbol is i, and a new rule is that an expression such as $\sqrt{-5}$ can be rewritten as the equivalent expression $i\sqrt{5}$. The complex number $a + bi$ can be treated as if it is a binomial, and operations on complex numbers follow the properties for adding, subtracting, multiplying, and dividing binomials, with one exception. That exception is to replace i^2 with -1 whenever i^2 appears in an expression.

Additional mathematical information and teaching notes are available at www.algebra2.com/key_concepts.

Chapter 5

DAILY INTERVENTION and Assessment

INTERVENTION

ASSESSMENT

Type	Student Edition	Teacher Resources	Technology/Internet
Ongoing	Prerequisite Skills, pp. 221, 228, 232, 238, 244, 249, 256, 262, 267 Practice Quiz 1, p. 238 Practice Quiz 2, p. 256	5-Minute Check Transparencies Quizzes, *CRM* pp. 307–308 Mid-Chapter Test, *CRM* p. 309 Study Guide and Intervention, *CRM* pp. 239–240, 245–246, 251–252, 257–258, 263–264, 269–270, 275–276, 281–282, 287–288	Alge2PASS: Tutorial Plus www.algebra2.com/self_check_quiz www.algebra2.com/extra_examples
Mixed Review	pp. 228, 232, 238, 244, 249, 256, 262, 267, 275	Cumulative Review, *CRM* p. 310	
Error Analysis	Find the Error, pp. 226, 236	Find the Error, *TWE* pp. 226, 236 Unlocking Misconceptions, *TWE* pp. 223, 235, 244, 246, 253, 258 Tips for New Teachers, *TWE* pp. 228, 238, 244, 246, 256, 262, 267, 275	
Standardized Test Practice	pp. 228, 232, 234, 236, 238, 244, 249, 255, 262, 267, 275, 281, 282–283	*TWE* p. 234 Standardized Test Practice, *CRM* pp. 311–312	Standardized Test Practice CD-ROM www.algebra2.com/standardized_test
Open-Ended Assessment	Writing in Math, pp. 227, 232, 238, 243, 249, 255, 262, 267, 275 Open Ended, pp. 226, 231, 236, 242, 247, 254, 260, 265, 273	Modeling: *TWE* pp. 244, 249 Speaking: *TWE* pp. 228, 256, 262, 275 Writing: *TWE* pp. 232, 238, 267 Open-Ended Assessment, *CRM* p. 305	
Chapter Assessment	Study Guide, pp. 276–280 Practice Test, p. 281	Multiple-Choice Tests (Forms 1, 2A, 2B), *CRM* pp. 293–298 Free-Response Tests (Forms 2C, 2D, 3), *CRM* pp. 299–304 Vocabulary Test/Review, *CRM* p. 306	ExamView® Pro (see below) MindJogger Videoquizzes www.algebra2.com/vocabulary_review www.algebra2.com/chapter_test

Key to Abbreviations: TWE = Teacher Wraparound Edition; CRM = Chapter Resource Masters

Intervention Technology

Alge2PASS: Tutorial Plus CD-ROM offers a complete, self-paced algebra curriculum.

Algebra 2 Lesson	Alge2PASS Lesson
5-4	**8** *Factoring Expressions II*
5-8	**9** *Solving Radical Equations*

ExamView® Pro

Use the networkable **ExamView® Pro** to:

- Create **multiple versions** of tests.
- Create **modified** tests for *Inclusion* students.
- **Edit** existing questions and **add** your own questions.
- Use built-in **state curriculum correlations** to create tests aligned with state standards.
- **Apply** art to your tests from a program bank of artwork.

For more information on Intervention and Assessment, see pp. T8–T11.

Reading and Writing in Mathematics

Glencoe Algebra 2 provides numerous opportunities to incorporate reading and writing into the mathematics classroom.

Student Edition

- Foldables Study Organizer, p. 221
- Concept Check questions require students to verbalize and write about what they have learned in the lesson. (pp. 226, 231, 236, 242, 247, 254, 260, 265, 273, 280)
- Writing in Math questions in every lesson, pp. 227, 232, 238, 243, 249, 255, 262, 267, 275
- Reading Study Tip, pp. 229, 246, 252, 270, 271, 273
- WebQuest, p. 227

Teacher Wraparound Edition

- Foldables Study Organizer, pp. 221, 276
- Study Notebook suggestions, pp. 226, 230, 236, 242, 247, 254, 260, 265, 273
- Modeling activities, pp. 244, 249
- Speaking activities, pp. 228, 256, 262, 275
- Writing activities, pp. 232, 238, 267
- Differentiated Instruction, (Verbal/Linguistic), p. 271
- ELL Resources, pp. 220, 227, 231, 237, 243, 248, 255, 261, 266, 271, 274, 276

Additional Resources

- Vocabulary Builder worksheets require students to define and give examples for key vocabulary terms as they progress through the chapter. (*Chapter 5 Resource Masters,* pp. vii-viii)
- Reading to Learn Mathematics master for each lesson (*Chapter 5 Resource Masters*, pp. 243, 249, 255, 261, 267, 273, 279, 285, 291)
- *Vocabulary PuzzleMaker* software creates crossword, jumble, and word search puzzles using vocabulary lists that you can customize.
- *Teaching Mathematics with Foldables* provides suggestions for promoting cognition and language.
- *Reading and Writing in the Mathematics Classroom*
- *WebQuest and Project Resources*

For more information on Reading and Writing in Mathematics, see pp. T6–T7.

PROJECT CRISS℠ Study Skill

Students may find a concept of definition map helpful in visualizing the components of a definition. Often students benefit from making a concept of definition map as a cooperative group activity. A sample concept of definition map on polynomials is shown at the right. You may wish to have students use this sample to develop their own maps to organize other definitions they will study in Chapter 5. Students can use large sheets of paper to extend their maps as they read by adding other details, properties, and explanations.

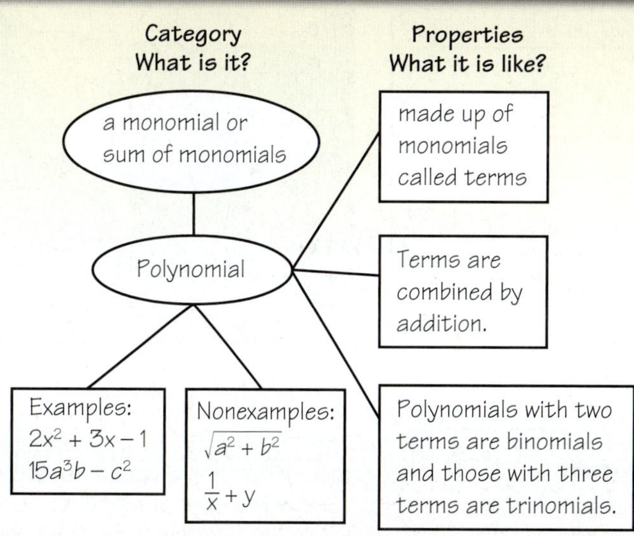

CReating **I**ndependence Through **S**tudent-Owned **S**trategies

Chapter 5 Notes

What You'll Learn

Have students read over the list of objectives and make a list of any words with which they are not familiar.

Why It's Important

Point out to students that this is only one of many reasons why each objective is important. Others are provided in the introduction to each lesson.

Lesson	NCTM Standards	Local Objectives
5-1	1, 2, 6, 7, 8, 9, 10	
5-2	1, 2, 6, 8, 9, 10	
5-3	1, 2, 6, 7, 8, 9	
5-4	1, 2, 3, 6, 8, 9, 10	
5-5	1, 2, 6, 7, 8, 9	
5-6	1, 2, 6, 7, 8, 9, 10	
5-7	1, 2, 6, 8, 9	
5-8	1, 2, 6, 8, 9, 10	
5-8 Follow-Up	1, 2, 10	
5-9	1, 2, 3, 6, 7, 8, 9, 10	

Key to NCTM Standards:

1=Number & Operations, 2=Algebra, 3=Geometry, 4=Measurement, 5=Data Analysis & Probability, 6=Problem Solving, 7=Reasoning & Proof, 8=Communication, 9=Connections, 10=Representation

Chapter 5 Polynomials

What You'll Learn

- **Lessons 5-1 through 5-4** Add, subtract, multiply, divide, and factor polynomials.
- **Lessons 5-5 through 5-8** Simplify and solve equations involving roots, radicals, and rational exponents.
- **Lesson 5-9** Perform operations with complex numbers.

Key Vocabulary

- scientific notation (p. 225)
- polynomial (p. 229)
- FOIL method (p. 230)
- synthetic division (p. 234)
- complex number (p. 271)

Why It's Important

Many formulas involve polynomials and/or square roots. For example, equations involving speeds or velocities of objects are often written with square roots. You can use such an equation to find the velocity of a roller coaster. *You will use an equation relating the velocity of a roller coaster and the height of a hill in Lesson 5-6.*

Vocabulary Builder

ELL

The Key Vocabulary list introduces students to some of the main vocabulary terms included in this chapter. For a more thorough vocabulary list with pronunciations of new words, give students the Vocabulary Builder worksheets found on pages vii and viii of the *Chapter 5 Resource Masters*. Encourage them to complete the definition of each term as they progress through the chapter. You may suggest that they add these sheets to their study notebooks for future reference when studying for the Chapter 5 test.

Getting Started

Prerequisite Skills To be successful in this chapter, you'll need to master these skills and be able to apply them in problem-solving situations. Review these skills before beginning Chapter 5.

For Lessons 5-2 and 5-9 — Rewrite Differences as Sums

Rewrite each difference as a sum.

1. $2 - 7$ **$2 + (-7)$**
2. $-6 - 11$ **$-6 + (-11)$**
3. $x - y$ **$x + (-y)$**
4. $8 - 2x$ **$8 + (-2x)$**
5. $2xy - 6yz$ **$2xy + (-6yz)$**
6. $6a^2b - 12b^2c$ **$6a^2b + (-12b^2c)$**

For Lesson 5-2 — Distributive Property

Use the Distributive Property to rewrite each expression without parentheses. *(For review, see Lesson 1-2.)* **7. $-8x^3 - 2x + 6$**

7. $-2(4x^3 + x - 3)$
8. $-1(x + 2)$ **$-x - 2$**
9. $-1(x - 3)$ **$-x + 3$**
10. $-3(2x^4 - 5x^2 - 2)$ **$-6x^4 + 15x^2 + 6$**
11. $-\frac{1}{2}(3a + 2)$ **$-\frac{3}{2}a - 1$**
12. $-\frac{2}{3}(2 + 6z)$ **$-\frac{4}{3} - 4z$**

For Lessons 5-5 and 5-9 — Classify Numbers

Find the value of each expression. Then name the sets of numbers to which each value belongs. *(For review, see Lesson 1-2.)* **13–18. See margin.**

13. $2.6 + 3.7$
14. $18 \div (-3)$
15. $2^3 + 3^2$
16. $\sqrt{4 + 1}$
17. $\frac{18 + 14}{8}$
18. $3\sqrt{4}$

FOLDABLES™ Study Organizer

Polynomials Make this Foldable to help you organize your notes. Begin with four sheets of grid paper.

Step 1 Fold and Cut

Fold in half along the width. On the first two sheets, cut along the fold at the ends. On the second two sheets, cut in the center of the fold as shown.

First Sheets

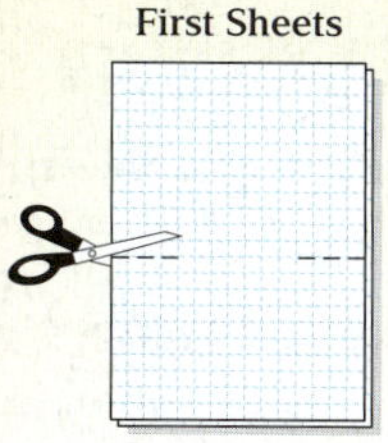

Second Sheets

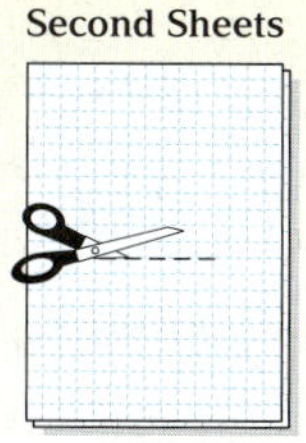

Step 2 Fold and Label

Insert first sheets through second sheets and align the folds. Label the pages with lesson numbers.

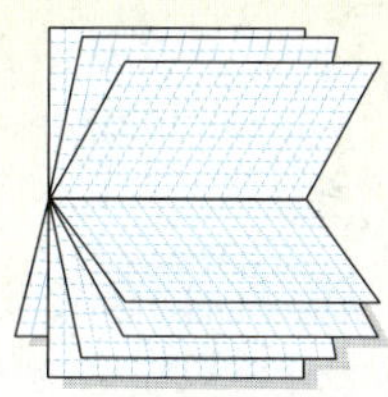

Reading and Writing As you read and study the chapter, fill the journal with notes, diagrams, and examples for polynomials.

Getting Started

This section provides a review of the basic concepts needed before beginning Chapter 5. Page references are included for additional student help.

Prerequisite Skills in the Getting Ready for the Next Lesson section at the end of each exercise set review a skill needed in the next lesson.

For Lesson	Prerequisite Skill
5-2	Distributive Property (p. 228)
5-3	Properties of Exponents (p. 232)
5-5	Rational and Irrational Numbers (p. 244)
5-6	Multiplying Binomials (p. 249)
5-8	Multiplying Radicals (p. 262)
5-9	Binomials (p. 267)

Answers

13. 6.3; reals, rationals
14. -6; reals, rationals, integers
15. 17; reals, rationals, integers, whole numbers, natural numbers
16. $\sqrt{5}$; reals, irrationals
17. 4; reals, rationals, integers, whole numbers, natural numbers
18. 6; reals, rationals, integers, whole numbers, natural numbers

FOLDABLES™ Study Organizer

For more information about Foldables, see *Teaching Mathematics with Foldables.*

Organization of Data and Journal Writing When labeling the pages for the lessons, combine Lessons 5-1 and 5-2 on the same page and Lessons 5-8 and 5-9 on the same page. Use extra pages for vocabulary lists and applications. Writer's journals can also be used to record the direction and progress of learning, to describe positive and negative experiences during learning, to write about personal associations and experiences while learning, and to list examples of ways in which new knowledge has or will be used in their daily life.

5-1 Lesson Notes

1 Focus

5-Minute Check Transparency 5-1 Use as a quiz or a review of Chapter 4.

Mathematical Background notes are available for this lesson on p. 220C.

Why is scientific notation useful in economics?

Ask students:

- What are the powers of ten? **…, 10^{-3}, 10^{-2}, 10^{-1}, 10^{0}, 10^{1}, 10^{2}, 10^{3}, …**
- What are some other fields that use scientific notation for very large or very small numbers? **astronomy, biology, computer science**

2 Teach

MONOMIALS

In-Class Example Power Point®

Teaching Tip Help students think carefully about the meaning of exponents by asking them to read this expression aloud correctly. If students read x^3 as "x three," instead of correctly saying "x cubed" or "x to the third (power)," they are apt to confuse x^3 with $3x$.

1 Simplify $(-2a^3b)(-5ab^4)$. **$10a^4b^5$**

5-1 Monomials

What You'll Learn

- Multiply and divide monomials.
- Use expressions written in scientific notation.

Why is scientific notation useful in economics?

Economists often deal with very large numbers. For example, the table shows the U.S. public debt for several years in the last century. Such numbers, written in standard notation, are difficult to work with because they contain so many digits. Scientific notation uses powers of ten to make very large or very small numbers more manageable.

Source: U.S. Department of the Treasury

Vocabulary

- monomial
- constant
- coefficient
- degree
- power
- simplify
- standard notation
- scientific notation
- dimensional analysis

MONOMIALS A **monomial** is an expression that is a number, a variable, or the product of a number and one or more variables. Monomials cannot contain variables in denominators, variables with exponents that are negative, or variables under radicals.

Monomials	Not Monomials
$5b, -w, 23, x^2, \frac{1}{3}x^3y^4$	$\frac{1}{n^4}, \sqrt[3]{x}, x + 8, a^{-1}$

Constants are monomials that contain no variables, like 23 or -1. The numerical factor of a monomial is the **coefficient** of the variable(s). For example, the coefficient of m in $-6m$ is -6. The **degree** of a monomial is the sum of the exponents of its variables. For example, the degree of $12g^7h^4$ is 7 + 4 or 11. The degree of a constant is 0.

A **power** is an expression of the form x^n. The word *power* is also used to refer to the exponent itself. Negative exponents are a way of expressing the multiplicative inverse of a number. For example, $\frac{1}{x^2}$ can be written as x^{-2}. Note that an expression such as x^{-2} is not a monomial. *Why?*

Key Concept *Negative Exponents*

- **Words** For any real number $a \neq 0$ and any integer n, $a^{-n} = \frac{1}{a^n}$ and $\frac{1}{a^{-n}} = a^n$.
- **Examples** $2^{-3} = \frac{1}{2^3}$ and $\frac{1}{b^{-8}} = b^8$

To **simplify** an expression containing powers means to rewrite the expression without parentheses or negative exponents.

Example 1 *Simplify Expressions with Multiplication*

Simplify $(3x^3y^2)(-4x^2y^4)$.

$$(3x^3y^2)(-4x^2y^4) = (3 \cdot x \cdot x \cdot x \cdot y \cdot y)(-4 \cdot x \cdot x \cdot y \cdot y \cdot y \cdot y) \quad \text{Definition of exponents}$$
$$= 3(-4) \cdot x \cdot x \cdot x \cdot x \cdot x \cdot y \cdot y \cdot y \cdot y \cdot y \cdot y \quad \text{Commutative Property}$$
$$= -12x^5y^6 \quad \text{Definition of exponents}$$

Resource Manager

Workbook and Reproducible Masters

Chapter 5 Resource Masters

- Study Guide and Intervention, pp. 239–240
- Skills Practice, p. 241
- Practice, p. 242
- Reading to Learn Mathematics, p. 243
- Enrichment, p. 244

Science and Mathematics Lab Manual, pp. 109–114

Transparencies

5-Minute Check Transparency 5-1
Answer Key Transparencies

Technology

Interactive Chalkboard

Example 1 suggests the following property of exponents.

Key Concept *Product of Powers*

- **Words** For any real number a and integers m and n, $a^m \cdot a^n = a^{m+n}$.
- **Examples** $4^2 \cdot 4^9 = 4^{11}$ and $b^3 \cdot b^5 = b^8$

To multiply powers of the same variable, add the exponents. Knowing this, it seems reasonable to expect that when dividing powers, you would subtract exponents. Consider $\frac{x^9}{x^5}$.

$$\frac{x^9}{x^5} = \frac{\overset{1}{\cancel{x}} \cdot \overset{1}{\cancel{x}} \cdot \overset{1}{\cancel{x}} \cdot \overset{1}{\cancel{x}} \cdot \overset{1}{\cancel{x}} \cdot x \cdot x \cdot x \cdot x}{\underset{1}{\cancel{x}} \cdot \underset{1}{\cancel{x}} \cdot \underset{1}{\cancel{x}} \cdot \underset{1}{\cancel{x}} \cdot \underset{1}{\cancel{x}}}$$ Remember that $x \neq 0$.

$= x \cdot x \cdot x \cdot x$ Simplify.

$= x^4$ Definition of exponents

It appears that our conjecture is true. To divide powers of the same base, you subtract exponents.

Key Concept *Quotient of Powers*

- **Words** For any real number $a \neq 0$, and integers m and n, $\frac{a^m}{a^n} = a^{m-n}$.
- **Examples** $\frac{5^3}{5} = 5^{3-1}$ or 5^2 and $\frac{x^7}{x^3} = x^{7-3}$ or x^4

Example 2 Simplify Expressions with Division

Simplify $\frac{p^3}{p^8}$. Assume that $p \neq 0$.

$\frac{p^3}{p^8} = p^{3-8}$ Subtract exponents.

$= p^{-5}$ or $\frac{1}{p^5}$ Remember that a simplified expression cannot contain negative exponents.

CHECK $$\frac{p^3}{p^8} = \frac{\overset{1}{\cancel{p}} \cdot \overset{1}{\cancel{p}} \cdot \overset{1}{\cancel{p}}}{\underset{1}{\cancel{p}} \cdot \underset{1}{\cancel{p}} \cdot \underset{1}{\cancel{p}} \cdot p \cdot p \cdot p \cdot p \cdot p}$$ Definition of exponents

$= \frac{1}{p^5}$ Simplify.

You can use the Quotient of Powers property and the definition of exponents to simplify $\frac{y^4}{y^4}$, if $y \neq 0$.

Method 1

$\frac{y^4}{y^4} = y^{4-4}$ Quotient of Powers

$= y^0$ Subtract.

Method 2

$$\frac{y^4}{y^4} = \frac{\overset{1}{\cancel{y}} \cdot \overset{1}{\cancel{y}} \cdot \overset{1}{\cancel{y}} \cdot \overset{1}{\cancel{y}}}{\underset{1}{\cancel{y}} \cdot \underset{1}{\cancel{y}} \cdot \underset{1}{\cancel{y}} \cdot \underset{1}{\cancel{y}}}$$ Definition of exponents

$= 1$ Divide.

In order to make the results of these two methods consistent, we define $y^0 = 1$, where $y \neq 0$. In other words, any nonzero number raised to the zero power is equal to 1. *Notice that 0^0 is undefined.*

In-Class Example Power Point®

2 Simplify $\frac{s^2}{s^{10}}$. Assume that $s \neq 0$. $\frac{1}{s^8}$

Teaching Tip When discussing In-Class Example 2, if any students get the incorrect answer $\frac{1}{s^5}$, lead them to understand that they divided the exponents instead of subtracting them as a method for dividing the two expressions.

Interactive Chalkboard
PowerPoint® Presentations

This CD-ROM is a customizable Microsoft® PowerPoint® presentation that includes:

- Step-by-step, dynamic solutions of each In-Class Example from the Teacher Wraparound Edition
- Additional, Your Turn exercises for each example
- The 5-Minute Check Transparencies
- Hot links to Glencoe Online Study Tools

DAILY INTERVENTION **Unlocking Misconceptions**

- **Correcting Errors** Encourage students to analyze the error or errors they made when they get an incorrect answer. Stress that students should use errors as an opportunity to clarify their thinking.
- **Using Definitions** Suggest that students return to the basic definitions for exponents rather than just memorizing rules. For example, they can derive the rule for multiplying quantities such as $x^2 \cdot x^3$ by rewriting the problem as $x \cdot x \cdot x \cdot x \cdot x$.

In-Class Examples

3 Simplify each expression.

a. $(b^2)^4$ b^8

b. $(-3c^2d^5)^3$ $-27c^6d^{15}$

c. $\left(\frac{-2a}{b^2}\right)^5$ $-\frac{32a^5}{b^{10}}$

d. $\left(\frac{x}{3}\right)^{-4}$ $\frac{81}{x^4}$

Teaching Tip Encourage students to begin each step of a simplification by naming the operation they are about to perform, such as finding the power of a power.

4 Simplify $\left(\frac{-3a^{5y}}{a^{6y}b^4}\right)^5$. $-\frac{243}{a^{5y}b^{20}}$

Concept Check

Monomials Have students write their own summary of the properties of exponents, such as "to multiply expressions with exponents, you add the exponents; to divide, you subtract the exponents" and so on.

The properties we have presented can be used to verify the properties of powers that are listed below.

Key Concept — Properties of Powers

- **Words** Suppose a and b are real numbers and m and n are integers. Then the following properties hold.

Words	Examples
Power of a Power: $(a^m)^n = a^{mn}$	$(a^2)^3 = a^6$
Power of a Product: $(ab)^m = a^mb^m$	$(xy)^2 = x^2y^2$
Power of a Quotient: $\left(\frac{a}{b}\right)^n = \frac{a^n}{b^n}$, $b \neq 0$ and	$\left(\frac{a}{b}\right)^3 = \frac{a^3}{b^3}$
$\left(\frac{a}{b}\right)^{-n} = \left(\frac{b}{a}\right)^n$ or $\frac{b^n}{a^n}$, $a \neq 0$, $b \neq 0$	$\left(\frac{x}{y}\right)^{-4} = \frac{y^4}{x^4}$

Example 3 Simplify Expressions with Powers

Simplify each expression.

a. $(a^3)^6$

$(a^3)^6 = a^{3(6)}$ Power of a power

$= a^{18}$

b. $(-2p^3s^2)^5$

$(-2p^3s^2)^5 = (-2)^5 \cdot (p^3)^5 \cdot (s^2)^5$

$= -32p^{15}s^{10}$ Power of a power

c. $\left(\frac{-3x}{y}\right)^4$

$\left(\frac{-3x}{y}\right)^4 = \frac{(-3x)^4}{y^4}$ Power of a quotient

$= \frac{(-3)^4x^4}{y^4}$ Power of a product

$= \frac{81x^4}{y^4}$ $(-3)^4 = 81$

d. $\left(\frac{a}{4}\right)^{-3}$

$\left(\frac{a}{4}\right)^{-3} = \left(\frac{4}{a}\right)^3$ Negative exponent

$= \frac{4^3}{a^3}$ Power of a quotient

$= \frac{64}{a^3}$ $4^3 = 64$

Study Tip

Simplified Expressions

A monomial expression is in simplified form when:

- there are no powers of powers,
- each base appears exactly once,
- all fractions are in simplest form, and
- there are no negative exponents.

With complicated expressions, you often have a choice of which way to start simplifying.

Example 4 Simplify Expressions Using Several Properties

Simplify $\left(\frac{-2x^{3n}}{x^{2n}y^3}\right)^4$.

Method 1

Raise the numerator and denominator to the fourth power before simplifying.

$$\left(\frac{-2x^{3n}}{x^{2n}y^3}\right)^4 = \frac{(-2x^{3n})^4}{(x^{2n}y^3)^4}$$

$$= \frac{(-2)^4(x^{3n})^4}{(x^{2n})^4(y^3)^4}$$

$$= \frac{16x^{12n}}{x^{8n}y^{12}}$$

$$= \frac{16x^{12n-8n}}{y^{12}}$$

$$= \frac{16x^{4n}}{y^{12}}$$

Method 2

Simplify the fraction before raising to the fourth power.

$$\left(\frac{-2x^{3n}}{x^{2n}y^3}\right)^4 = \left(\frac{-2x^{3n-2n}}{y^3}\right)^4$$

$$= \left(\frac{-2x^n}{y^3}\right)^4$$

$$= \frac{16x^{4n}}{y^{12}}$$

224 Chapter 5 Polynomials

SCIENTIFIC NOTATION The form that you usually write numbers in is **standard notation**. A number is in **scientific notation** when it is in the form $a \times 10^n$, where $1 \le a < 10$ and n is an integer. Scientific notation is used to express very large or very small numbers.

Study Tip

Graphing Calculators

To solve scientific notation problems on a graphing calculator, use the EE function. Enter 6.38×10^6 as 6.38 [2nd] [EE] 6.

Example 5 Express Numbers in Scientific Notation

Express each number in scientific notation.

a. 6,380,000

$6{,}380{,}000 = 6.38 \times 1{,}000{,}000$ — $1 \le 6.38 < 10$

$= 6.38 \times 10^6$ — Write 1,000,000 as a power of 10.

b. 0.000047

$0.000047 = 4.7 \times 0.00001$ — $1 \le 4.7 < 10$

$= 4.7 \times \frac{1}{10^5}$ — $0.00001 = \frac{1}{100{,}000}$ or $\frac{1}{10^5}$

$= 4.7 \times 10^{-5}$ — Use a negative exponent.

You can use properties of powers to multiply and divide numbers in scientific notation.

Example 6 Multiply Numbers in Scientific Notation

Evaluate. Express the result in scientific notation.

a. $(4 \times 10^5)(2 \times 10^7)$

$(4 \times 10^5)(2 \times 10^7) = (4 \cdot 2) \times (10^5 \cdot 10^7)$ — Associative and Commutative Properties

$= 8 \times 10^{12}$ — $4 \cdot 2 = 8$, $10^5 \cdot 10^7 = 10^{5+7}$ or 10^{12}

b. $(2.7 \times 10^{-2})(3 \times 10^6)$

$(2.7 \times 10^{-2})(3 \times 10^6) = (2.7 \cdot 3) \times (10^{-2} \cdot 10^6)$ — Associative and Commutative Properties

$= 8.1 \times 10^4$ — $2.7 \cdot 3 = 8.1$, $10^{-2} \cdot 10^6 = 10^{-2+6}$ or 10^4

Real-world problems often involve units of measure. Performing operations with units is known as **dimensional analysis**.

More About. . .

Astronomy

Light travels at a speed of about 3.00×10^8 m/s. The distance that light travels in a year is called a *light-year*.

Source: www.britannica.com

Example 7 Divide Numbers in Scientific Notation

ASTRONOMY **After the Sun, the next-closest star to Earth is Alpha Centauri C, which is about 4×10^{16} meters away. How long does it take light from Alpha Centauri C to reach Earth? Use the information at the left.**

Begin with the formula $d = rt$, where d is distance, r is rate, and t is time.

$t = \frac{d}{r}$ — Solve the formula for time.

$= \frac{4 \times 10^{16} \text{ m}}{3.00 \times 10^8 \text{ m/s}}$ — ← Distance from Alpha Centauri C to Earth

$= \frac{4}{3.00} \cdot \frac{10^{16}}{10^8 \text{ 1/s}}$ — **Estimate:** The result should be slightly greater than $\frac{10^{16}}{10^8}$ or 10^8.

$\approx 1.33 \times 10^8 \text{ s}$ — $\frac{4}{3.00} \approx 1.33$, $\frac{10^{16}}{10^8} = 10^{16-8}$ or 10^8

It takes about 1.33×10^8 seconds or 4.2 years for light from Alpha Centauri C to reach Earth.

SCIENTIFIC NOTATION

In-Class Examples Power Point®

5 Express each number in scientific notation.

a. 4,560,000 **4.56×10^6**

b. 0.000092 **9.2×10^{-5}**

6 Evaluate. Express the result in scientific notation.

a. $(5 \times 10^3)(7 \times 10^8)$ **3.5×10^{12}**

b. $(1.8 \times 10^{-4})(4 \times 10^7)$ **7.2×10^3**

7 **BIOLOGY** There are about 5×10^6 red blood cells in one milliliter of blood. A certain blood sample contains 8.32×10^6 red blood cells. About how many milliliters of blood are in the sample? **about 1.66 mL**

DAILY INTERVENTION

Differentiated Instruction

Interpersonal Have students discuss with a partner or in a small group the methods for multiplying and dividing monomial expressions with exponents, and also numbers written in scientific notation. Ask them to work together to develop a list of common errors for such problems, and to suggest ways to correct and avoid these errors.

Check for Understanding

Concept Check

1. **OPEN ENDED** Write an example that illustrates a property of powers. Then use multiplication or division to explain why it is true. **See margin.**

2. Determine whether $x^y \cdot x^z = x^{yz}$ is *sometimes*, *always*, or *never* true. Explain.

2. Sometimes; in general $x^y \cdot x^z = x^{y+z}$, so $x^y \cdot x^z = x^{yz}$ when $y + z = yz$, such as when $y = 2$ and $z = 2$.

3. **FIND THE ERROR** Alejandra and Kyle both simplified $\frac{2a^2b}{(-2ab^3)^{-2}}$.

Alejandra	Kyle
$\frac{2a^2b}{(-2ab^3)^{-2}} = (2a^2b)(-2ab^3)^2$	$\frac{2a^2b}{(-2ab^3)^{-2}} = \frac{2a^2b}{(-2)^{-2}a(b^3)^{-2}}$
$= (2a^2b)(-2)^2a^2(b^3)^2$	$= \frac{2a^2b}{4ab^{-6}}$
$= (2a^2b)2^2a^2b^6$	$= \frac{2a^2bb^6}{4a}$
$= 8a^4b^7$	$= \frac{ab^7}{2}$

Who is correct? Explain your reasoning.

3. Alejandra; when Kyle used the Power of a Product property in his first step, he forgot to put an exponent of -2 on a. Also, in his second step, $(-2)^{-2}$ should be $\frac{1}{4}$, not 4.

Guided Practice

GUIDED PRACTICE KEY

Exercises	Examples
4–9	1–3
10–12	4
13, 14	5
15	6
16, 17	7

Simplify. Assume that no variable equals 0.

4. $x^2 \cdot x^8$ $\mathbf{x^{10}}$
5. $(2b)^4$ $\mathbf{16b^4}$
6. $(n^3)^3(n^{-3})^3$ **1**
7. $\frac{30y^4}{-5y^2}$ $\mathbf{-6y^2}$
8. $\frac{-2a^3b^6}{18a^2b^2}$ $\mathbf{-\frac{ab^4}{9}}$
9. $\frac{81p^6q^5}{(3p^2q)^2}$ $\mathbf{9p^2q^3}$
10. $\left(\frac{1}{w^4z^2}\right)^3$ $\mathbf{\frac{1}{w^{12}z^6}}$
11. $\left(\frac{cd}{3}\right)^{-2}$ $\mathbf{\frac{9}{c^2d^2}}$
12. $\left(\frac{-6x^6}{3x^3}\right)^{-2}$ $\mathbf{\frac{1}{4x^6}}$

Express each number in scientific notation.

13. 421,000 $\mathbf{4.21 \times 10^5}$
14. 0.000862 $\mathbf{8.62 \times 10^{-4}}$

Evaluate. Express the result in scientific notation.

15. $(3.42 \times 10^8)(1.1 \times 10^{-5})$ $\mathbf{3.762 \times 10^3}$
16. $\frac{8 \times 10^{-1}}{16 \times 10^{-2}}$ $\mathbf{5 \times 10^0}$

Application

17. **ASTRONOMY** Refer to Example 7 on page 225. The average distance from Earth to the Moon is about 3.84×10^8 meters. How long would it take a radio signal traveling at the speed of light to cover that distance? **about 1.28 s**

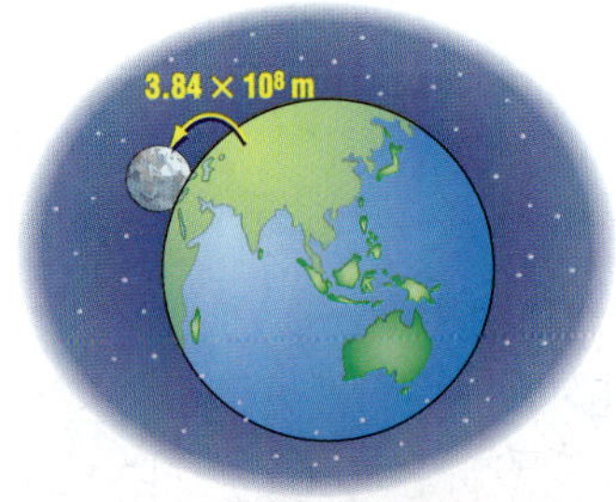

★ indicates increased difficulty

Practice and Apply

Simplify. Assume that no variable equals 0.

18. $a^2 \cdot a^6$ $\mathbf{a^8}$
19. $b^{-3} \cdot b^7$ $\mathbf{b^4}$
20. $(n^4)^4$ $\mathbf{n^{16}}$
21. $(z^2)^5$ $\mathbf{z^{10}}$
22. $(2x)^4$ $\mathbf{16x^4}$
23. $(-2c)^3$ $\mathbf{-8c^3}$
24. $\frac{a^2n^6}{an^5}$ $\mathbf{an}$
25. $\frac{-y^5z^7}{y^2z^5}$ $\mathbf{-y^3z^2}$
26. $(7x^3y^{-5})(4xy^3)$ $\mathbf{\frac{28x^4}{y^2}}$
27. $(-3b^3c)(7b^2c^2)$ $\mathbf{-21b^5c^3}$
28. $(a^3b^3)(ab)^{-2}$ $\mathbf{ab}$
29. $(-2r^2s)^3(3rs^2)$ $\mathbf{-24r^7s^5}$
30. $2x^2(6y^3)(2x^2y)$ $\mathbf{24x^4y^4}$
31. $3a(5a^2b)(6ab^3)$ $\mathbf{90a^4b^4}$
32. $\frac{-5x^3y^3z^4}{20x^3y^7z^4}$ $\mathbf{-\frac{1}{4y^4}}$

3 Practice/Apply

Study Notebook

Have students—
- add the definitions/examples of the vocabulary terms to their Vocabulary Builder worksheets for Chapter 5.
- add the information about the meaning of dimensional analysis to their notebook.
- include any other item(s) that they find helpful in mastering the skills in this lesson.

About the Exercises...

Organization by Objective
- **Monomials:** 18–43
- **Scientific Notation:** 44–60

Odd/Even Assignments
Exercises 18–55 are structured so that students practice the same concepts whether they are assigned odd or even problems.

Alert! Exercise 59 involves research on the Internet or other reference materials.

Assignment Guide
Basic: 19–39 odd, 45–57 odd, 59–84

Average: 19–57 odd, 59–84

Advanced: 18–58 even, 60–78 (optional: 79–84)

DAILY INTERVENTION

FIND THE ERROR

Suggest that students use two steps to simplify expressions such as $\frac{1}{(-2)^{-2}}$ by first rewriting with the reciprocal and then squaring.

Answers

1. Sample answer: $(2x^2)^3 = 8x^6$ since $(2x^2)^3 = (2x^2) \cdot (2x^2) \cdot (2x^2) = 2x^2 \cdot 2x^2 \cdot 2x^2 = 2x \cdot x \cdot 2x \cdot x \cdot 2x \cdot x = 8x^6$

62. $(ab)^m = \underbrace{ab \cdot ab \cdot \ldots \cdot ab}_{m \text{ factors}} = \underbrace{a \cdot a \cdot \ldots \cdot a}_{m \text{ factors}} \cdot \underbrace{b \cdot b \cdot \ldots \cdot b}_{m \text{ factors}} = a^mb^m$

63. Economics often involves large amounts of money. Answers should include the following.
- The national debt in 2000 was five trillion, six hundred seventy-four billion, two hundred million or 5.6742×10^{12} dollars. The population was two hundred eighty-one million or 2.81×10^8.
- Divide the national debt by the population: $\frac{5.6742 \times 10^{12}}{2.81 \times 10^8} \approx \$2.0193 \times 10^4$ or about \$20,193 per person.

Homework Help

For Exercises	See Examples
18–35, 60	1–3
36–39	4
40–43	1, 2
44–49, 56, 57	5
50–55, 58, 59	6, 7

Extra Practice
See page 836.

33. $\frac{3a^5b^3c^3}{9a^3b^7c}$ $\frac{a^2c^2}{3b^4}$

34. $\frac{2c^3d(3c^2d^5)}{30c^4d^2}$ $\frac{cd^4}{5}$

35. $\frac{-12m^4n^8(m^3n^2)}{36m^3n}$ $-\frac{m^4n^9}{3}$

36. $\left(\frac{8a^3b^2}{16a^2b^3}\right)^4$ $\frac{a^4}{16b^4}$

37. $\left(\frac{6x^2y^4}{3x^4y^3}\right)^3$ $\frac{8y^3}{x^6}$

38. $\left(\frac{x}{y^{-1}}\right)^{-2}$ $\frac{1}{x^2y^2}$

39. $\left(\frac{v}{w^{-2}}\right)^{-3}$ $\frac{1}{v^3w^6}$

★ 40. $\frac{30a^{-2}b^{-6}}{60a^{-6}b^{-8}}$ $\frac{a^4b^2}{2}$

★ 41. $\frac{12x^{-3}y^{-2}z^{-8}}{30x^{-6}y^{-4}z^{-1}}$ $\frac{2x^3y^2}{5z^7}$

★ 42. If $2^{r+5} = 2^{2r-1}$, what is the value of r? **6**

★ 43. What value of r makes $y^{28} = y^{3r} \cdot y^7$ true? **7**

Express each number in scientific notation.

44. 462.3 4.623×10^2
45. 43,200 4.32×10^4
46. 0.0001843 1.843×10^{-4}
47. 0.006810 6.81×10^{-3}
48. 502,020,000 5.0202×10^8
49. 675,400,000 6.754×10^8

Evaluate. Express the result in scientific notation.

50. $(4.15 \times 10^3)(3.0 \times 10^6)$ 1.245×10^{10}

51. $(3.01 \times 10^{-2})(2 \times 10^{-3})$ 6.02×10^{-5}

52. $\frac{6.3 \times 10^5}{1.4 \times 10^3}$ 4.5×10^2

53. $\frac{9.3 \times 10^7}{1.5 \times 10^{-3}}$ 6.2×10^{10}

54. $(6.5 \times 10^4)^2$ 4.225×10^9

55. $(4.1 \times 10^{-4})^2$ 1.681×10^{-7}

56. **POPULATION** The population of Earth is about 6,080,000,000. Write this number in scientific notation. 6.08×10^9

57. **BIOLOGY** Use the diagram at the right to write the diameter of a typical flu virus in scientific notation. 2×10^{-7} m

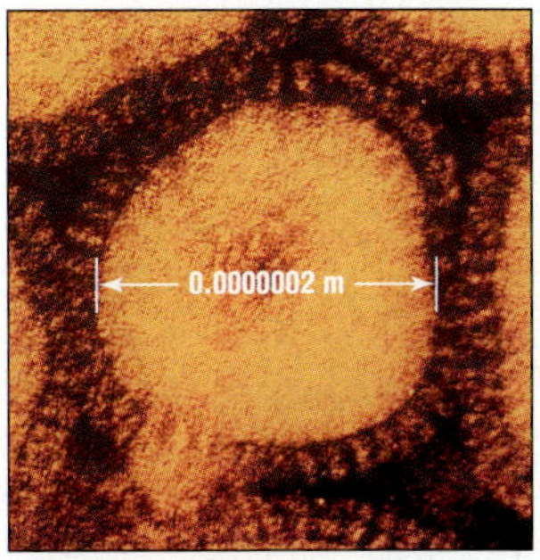

58. **CHEMISTRY** One gram of water contains about 3.34×10^{22} molecules. About how many molecules are in 500 grams of water? 1.67×10^{25}

59. **RESEARCH** Use the Internet or other source to find the masses of Earth and the Sun. About how many times as large as Earth is the Sun? **about 330,000 times**

60. **CRITICAL THINKING** Determine which is greater, 100^{10} or 10^{100}. Explain. $100^{10} = (10^2)^{10}$ or 10^{20}, and $10^{100} > 10^{20}$, so $10^{100} > 100^{10}$.

CRITICAL THINKING For Exercises 61 and 62, use the following proof of the Power of a Power Property. **61. Definition of an exponent**

$$a^m a^n = \underbrace{a \cdot a \cdot \ldots \cdot a}_{m \text{ factors}} \cdot \underbrace{a \cdot a \cdot \ldots \cdot a}_{n \text{ factors}}$$
$$= \underbrace{a \cdot a \cdot \ldots \cdot a}_{m+n \text{ factors}}$$
$$= a^{m+n}$$

61. What definition or property allows you to make each step of the proof?

62. Prove the Power of a Product Property, $(ab)^m = a^m b^m$. **See margin.**

63. **WRITING IN MATH** Answer the question that was posed at the beginning of the lesson. **See margin.**

Why is scientific notation useful in economics?

Include the following in your answer:
- the 2000 national debt of \$5,674,200,000,000 and the U.S. population of 281,000,000, both written in words and in scientific notation, and
- an explanation of how to find the amount of debt per person, with the result written in scientific notation and in standard notation.

WebQuest

A scatter plot of populations will help you make a model for the data. Visit www.algebra2.com/webquest to continue work on your WebQuest project.

www.algebra2.com/self_check_quiz

Study Guide and Intervention, p. 239 (shown) and p. 240

Monomials A **monomial** is a number, a variable, or the product of a number and one or more variables. Constants are monomials that contain no variables.

Negative Exponent	$a^{-n} = \frac{1}{a^n}$ and $\frac{1}{a^{-n}} = a^n$ for any real number $a \neq 0$ and any integer n.

When you **simplify an expression**, you rewrite it without parentheses or negative exponents. The following properties are useful when simplifying expressions.

Product of Powers	$a^m \cdot a^n = a^{m+n}$ for any real number a and integers m and n.
Quotient of Powers	$\frac{a^m}{a^n} = a^{m-n}$ for any real number $a \neq 0$ and integers m and n.
Properties of Powers	For a, b real numbers and m, n integers: $(a^m)^n = a^{mn}$; $(ab)^m = a^mb^m$; $\left(\frac{a}{b}\right)^n = \frac{a^n}{b^n}, b \neq 0$; $\left(\frac{a}{b}\right)^{-n} = \left(\frac{b}{a}\right)^n$ or $\frac{b^n}{a^n}, a \neq 0, b \neq 0$

Example Simplify. Assume that no variable equals 0.

a. $(3m^4n^{-2})(-5mn)^2$

$(3m^4n^{-2})(-5mn)^2 = 3m^4n^{-2} \cdot 25m^2n^2 = 75m^4m^2n^{-2}n^2 = 75m^{4+2}n^{-2+2} = 75m^6$

b. $\frac{(-m^4)^3}{(2m^2)^{-2}}$

$\frac{(-m^4)^3}{(2m^2)^{-2}} = \frac{-m^{12}}{\frac{1}{4m^4}} = -m^{12} \cdot 4m^4 = -4m^{16}$

Exercises

Simplify. Assume that no variable equals 0.

1. $c^{12} \cdot c^{-4} \cdot c^6$ c^{14}
2. $\frac{b^8}{b^2}$ b^6
3. $(a^4)^5$ a^{20}
4. $\frac{x^{-2}y}{x^4y^{-1}}$ $\frac{y^2}{x^6}$
5. $\left(\frac{a^2b}{a^{-3}b^2}\right)^{-1}$ $\frac{b}{a^5}$
6. $\left(\frac{x^2y}{xy^3}\right)^2$ $\frac{x^2}{y^4}$
7. $\frac{1}{5}(-5a^2b^3)^2(abc)^2$ $5a^6b^8c^2$
8. $m^7 \cdot m^8$ m^{15}
9. $\frac{8m^3n^2}{4mn^3}$ $\frac{2m^2}{n}$
10. $\frac{2^3c^4t^2}{2^2c^4t^2}$ 2
11. $4j(2j^{-2}k^2)(3j^3k^{-7})$ $\frac{24j^2}{k^5}$
12. $\frac{2mn^2(3m^2n)^2}{12m^3n^4}$ $\frac{3}{2}m^2$

Skills Practice, p. 241 and Practice, p. 242 (shown)

Simplify. Assume that no variable equals 0.

1. $n^5 \cdot n^2$ n^7
2. $y^7 \cdot y^3 \cdot y^2$ y^{12}
3. $t^9 \cdot t^{-8}$ t
4. $x^{-4} \cdot x^{-4} \cdot x^4$ $\frac{1}{x^4}$
5. $(2f^4)^6$ $64f^{24}$
6. $(-2b^{-2}c^3)^3$ $-\frac{8c^9}{b^6}$
7. $(4d^2t^5v^{-4})(-5dt^{-3}v^{-1})$ $-\frac{20d^3t^2}{v^5}$
8. $8u(2z)^3$ $64uz^3$
9. $\frac{12m^8y^6}{-9my^4}$ $-\frac{4m^7y^2}{3}$
10. $\frac{-6s^5x^3}{18sx^7}$ $-\frac{s^4}{3x^4}$
11. $\frac{-27x^3(-x^7)}{16x^4}$ $\frac{27x^6}{16}$
12. $\left(\frac{2}{3r^2s^3z^6}\right)^2$ $\frac{4}{9r^4s^6z^{12}}$
13. $-(4w^{-3}z^{-5})(8w)^2$ $-\frac{256}{wz^5}$
14. $(m^4n^6)^4(m^3n^2p^5)^6$ $m^{34}n^{36}p^{30}$
15. $\left(\frac{3}{2}d^2f^4\right)^4\left(-\frac{4}{3}d^5f\right)^3$ $-12d^{23}f^{19}$
16. $\left(\frac{2x^3y^2}{-x^2y^5}\right)^{-2}$ $\frac{y^6}{4x^2}$
17. $\frac{(3x^{-2}y^3)(5xy^{-8})}{(x^{-3})^4y^{-2}}$ $\frac{15x^{11}}{y^3}$
18. $\frac{-20(m^2v)(-v)^3}{5(-v)^2(-m^4)}$ $-\frac{4v^2}{m^2}$

Express each number in scientific notation.

19. 896,000 8.96×10^5
20. 0.000056 5.6×10^{-5}
21. 433.7×10^8 4.337×10^{10}

Evaluate. Express the result in scientific notation.

22. $(4.8 \times 10^2)(6.9 \times 10^4)$ 3.312×10^7
23. $(3.7 \times 10^9)(8.7 \times 10^2)$ 3.219×10^{12}
24. $\frac{2.7 \times 10^6}{9 \times 10^{10}}$ 3×10^{-5}

25. **COMPUTING** The term *bit*, short for *binary digit*, was first used in 1946 by John Tukey. A single bit holds a zero or a one. Some computers use 32-bit numbers, or strings of 32 consecutive bits, to identify each address in their memories. Each 32-bit number corresponds to a number in our base-ten system. The largest 32-bit number is nearly 4,295,000,000. Write this number in scientific notation. 4.295×10^9

26. **LIGHT** When light passes through water, its velocity is reduced by 25%. If the speed of light in a vacuum is 1.86×10^5 miles per second, at what velocity does it travel through water? Write your answer in scientific notation. 1.395×10^5 mi/s

27. **TREES** Deciduous and coniferous trees are hard to distinguish in a black-and-white photo. But because deciduous trees reflect infrared energy better than coniferous trees, the two types of trees are more distinguishable in an infrared photo. If an infrared wavelength measures about 8×10^{-7} meters and a blue wavelength measures about 4.5×10^{-7} meters, about how many times longer is the infrared wavelength than the blue wavelength? **about 1.8 times**

Reading to Learn Mathematics, p. 243 ELL

Pre-Activity **Why is scientific notation useful in economics?**

Read the introduction to Lesson 5-1 at the top of page 222 in your textbook.

Your textbook gives the U.S. public debt as an example from economics that involves large numbers that are difficult to work with when written in standard notation. Give an example from science that involves very large numbers and one that involves very small numbers. **Sample answer: distances between Earth and the stars, sizes of molecules and atoms**

Reading the Lesson

1. Tell whether each expression is a monomial or not a monomial. If it is a monomial, tell whether it is a constant or not a constant.
   a. $3x^2$ **monomial; not a constant**
   b. $y^2 + 5y - 6$ **not a monomial**
   c. -73 **monomial; constant**
   d. $\frac{1}{z}$ **not a monomial**

2. Complete the following definitions of a negative exponent and a zero exponent.

   For any real number $a \neq 0$ and any integer n, $a^{-n} = \underline{\frac{1}{a^n}}$.

   For any real number $a \neq 0$, $a^0 = \underline{1}$.

3. Name the property or properties of exponents that you would use to simplify each expression. (Do not actually simplify.)
   a. $\frac{m^8}{m^3}$ **quotient of powers**
   b. $y^6 \cdot y^9$ **product of powers**
   c. $(3r^2s)^4$ **power of a product and power of a power**

Helping You Remember

4. When writing a number in scientific notation, some students have trouble remembering when to use positive exponents and when to use negative ones. What is an easy way to remember this? **Sample answer: Use a positive exponent if the number is 10 or greater. Use a negative number if the number is less than 1.**

Enrichment, p. 244

Properties of Exponents

The rules about powers and exponents are usually given with letters such as m, n, and k to represent exponents. For example, one rule states that $a^m \cdot a^n = a^{m+n}$.

In practice, such exponents are handled as algebraic expressions and the rules of algebra apply.

Example 1 Simplify $2a^2(a^{n+1} + a^{4n})$.

$2a^2(a^{n+1} + a^{4n}) = 2a^2 \cdot a^{n+1} + 2a^2 \cdot a^{4n}$ Use the Distributive Law.
$= 2a^{2+n+1} + 2a^{2+4n}$ Recall $a^m \cdot a^n = a^{m+n}$.
$= 2a^{n+3} + 2a^{2+4n}$ Simplify the exponent $2 + n + 1$ as $n + 3$.

It is important always to collect *like* terms only.

Example 2 Simplify $(a^n + b^m)^2$.

$(a^n + b^m)^2 = (a^n + b^m)(a^n + b^m)$

4 Assess

Open-Ended Assessment

Speaking Have students explain in informal language how to tell whether a number is written in scientific notation. Then have them explain how to simplify monomial expressions involving negative exponents.

Tips for New Teachers

Intervention Simplifying expressions with exponents is a skill that is needed frequently in algebra. Take time to clear up any misconceptions at this point and to help students develop an understanding of the properties so that they remember the procedures correctly for later use.

Getting Ready for Lesson 5-2

PREREQUISITE SKILL Lesson 5-2 presents multiplying polynomials. This multiplication involves the use of the Distributive Property. Exercises 79–84 should be used to determine your students' familiarity with the Distributive Property.

Answer

74. **Median Age of Vehicles**

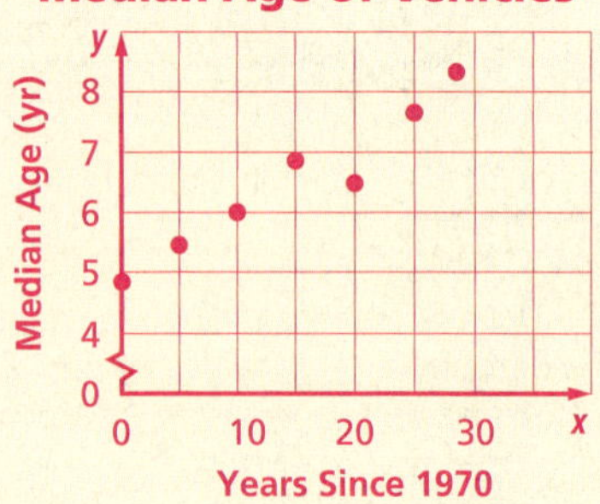

Standardized Test Practice

64. Simplify $\frac{(2x^2)^3}{12x^4}$. **D**

(A) $\frac{x}{2}$ (B) $\frac{2x}{3}$ (C) $\frac{1}{2x^2}$ (D) $\frac{2x^2}{3}$

65. $7.3 \times 10^5 = ?$ **B**

(A) 73,000 (B) 730,000 (C) 7,300,000 (D) 73,000,000

Maintain Your Skills

Mixed Review

Solve each system of equations by using inverse matrices. *(Lesson 4-8)*

66. $2x + 3y = 8$, $x - 2y = -3$ **(1, 2)**

67. $x + 4y = 9$, $3x + 2y = -3$ **(−3, 3)**

Find the inverse of each matrix, if it exists. *(Lesson 4-7)*

68. $\begin{bmatrix} 2 & 5 \\ -1 & -2 \end{bmatrix}$ $\begin{bmatrix} -2 & -5 \\ 1 & 2 \end{bmatrix}$

69. $\begin{bmatrix} 4 & 3 \\ 2 & 1 \end{bmatrix}$ $\begin{bmatrix} -\frac{1}{2} & \frac{3}{2} \\ 1 & -2 \end{bmatrix}$

Evaluate each determinant. *(Lesson 4-3)*

70. $\begin{vmatrix} 3 & 0 \\ 2 & -2 \end{vmatrix}$ **−6**

71. $\begin{vmatrix} 1 & 0 & -3 \\ 2 & -1 & 4 \\ -3 & 0 & 2 \end{vmatrix}$ **7**

Solve each system of equations. *(Lesson 3-5)*

72. $x + y = 5$, $x + y + z = 4$, $2x - y + 2z = -1$ **(2, 3, −1)**

73. $a + b + c = 6$, $2a - b + 3c = 16$, $a + 3b - 2c = -6$ **(2, 0, 4)**

TRANSPORTATION For Exercises 74–76, refer to the graph at the right. *(Lesson 2-5)*

74. Make a scatter plot of the data, where the horizontal axis is the number of years since 1970. **74. See margin.**

75. Write a prediction equation. **75. Sample answer using (0, 4.9) and (28, 8.3): $y = 0.12x + 4.9$**

76. Predict the median age of vehicles on the road in 2010. **Sample answer: 9.7 yr**

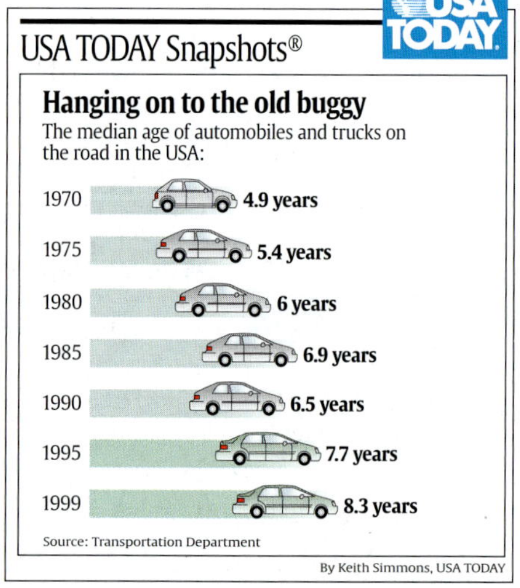

Solve each equation. *(Lesson 1-3)*

77. $2x + 11 = 25$ **7**

78. $-12 - 5x = 3$ **−3**

Getting Ready for the Next Lesson

Use the Distributive Property to find each product.
(To review the ***Distributive Property****, see Lesson 1-2.)*

79. $2(x + y)$ **$2x + 2y$**

80. $3(x - z)$ **$3x - 3z$**

81. $4(x + 2)$ **$4x + 8$**

82. $-2(3x - 5)$ **$-6x + 10$**

83. $-5(x - 2y)$ **$-5x + 10y$**

84. $-3(-y + 5)$ **$3y - 15$**

Online Lesson Plans

USA TODAY Education's Online site offers resources and interactive features connected to each day's newspaper. *Experience TODAY*, USA TODAY's daily lesson plan, is available on the site and delivered daily to subscribers. This plan provides instruction for integrating USA TODAY graphics and key editorial features into your mathematics classroom. Log on to www.education.usatoday.com.

5-2 Polynomials

What You'll Learn

- Add and subtract polynomials.
- Multiply polynomials.

How can polynomials be applied to financial situations?

Shenequa wants to attend Purdue University in Indiana, where the out-of-state tuition is \$8820. Suppose the tuition increases at a rate of 4% per year. You can use polynomials to represent the increasing tuition costs.

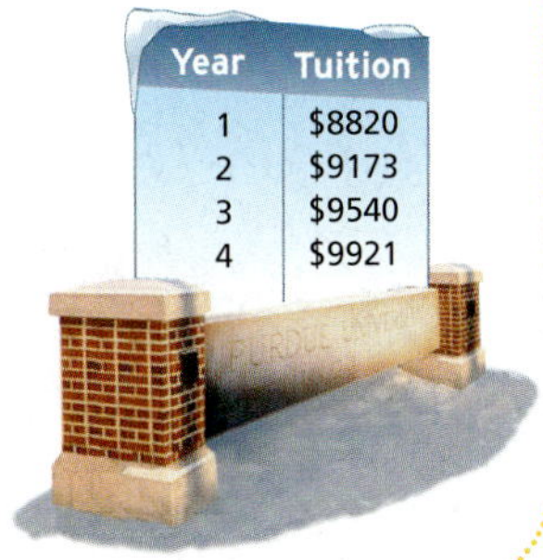

Year	Tuition
1	\$8820
2	\$9173
3	\$9540
4	\$9921

Vocabulary

- polynomial
- terms
- like terms
- trinomial
- binomial
- FOIL method

ADD AND SUBTRACT POLYNOMIALS If r represents the rate of increase of tuition, then the tuition for the second year will be $8820(1 + r)$. For the third year, it will be $8820(1 + r)^2$, or $8820r^2 + 17{,}640r + 8820$ in expanded form. This expression is called a polynomial. A **polynomial** is a monomial or a sum of monomials.

The monomials that make up a polynomial are called the **terms** of the polynomial. In a polynomial such as $x^2 + 2x + x + 1$, the two monomials $2x$ and x can be combined because they are **like terms**. The result is $x^2 + 3x + 1$. The polynomial $x^2 + 3x + 1$ is a **trinomial** because it has three unlike terms. A polynomial such as $xy + z^3$ is a **binomial** because it has two unlike terms. The *degree* of a polynomial is the degree of the monomial with the greatest degree. For example, the degree of $x^2 + 3x + 1$ is 2, and the degree of $xy + z^3$ is 3.

Study Tip

Reading Math
The prefix *bi-* means *two*, and the prefix *tri-* means *three*.

Example 1 Degree of a Polynomial

Determine whether each expression is a polynomial. If it is a polynomial, state the degree of the polynomial.

a. $\frac{1}{6}x^3y^5 - 9x^4$

This expression is a polynomial because each term is a monomial.
The degree of the first term is 3 + 5 or 8, and the degree of the second term is 4. The degree of the polynomial is 8.

b. $x + \sqrt{x} + 5$

This expression is not a polynomial because $\sqrt{x}$ is not a monomial.

To *simplify* a polynomial means to perform the operations indicated and combine like terms.

Example 2 Subtract and Simplify

Simplify $(3x^2 - 2x + 3) - (x^2 + 4x - 2)$.

$(3x^2 - 2x + 3) - (x^2 + 4x - 2) = 3x^2 - 2x + 3 - x^2 - 4x + 2$ Distribute the −1.

$= (3x^2 - x^2) + (-2x - 4x) + (3 + 2)$ Group like terms.

$= 2x^2 - 6x + 5$ Combine like terms.

www.algebra2.com/extra_examples

5-2 Lesson Notes

1 Focus

5-Minute Check Transparency 5-2 Use as a quiz or review of Lesson 5-1.

Mathematical Background notes are available for this lesson on p. 220C.

How can polynomials be applied to financial situations?

Ask students:

- What is meant by "tuition increases at a rate of 4% per year?" **Each year the tuition is 4% higher than it was the year before.**
- Will the amount of the tuition increase be the same each year? **no**

2 Teach

ADD AND SUBTRACT POLYNOMIALS

In-Class Examples Power Point®

1 Determine whether each expression is a polynomial. If it is a polynomial, state the degree of the polynomial.

a. $c^4 - 4\sqrt{c} + 18$ **no**

b. $-16p^5 + \frac{3}{4}p^2q^7$ **yes, 9**

2 Simplify $(2a^3 + 5a - 7) - (a^3 - 3a + 2)$. $a^3 + 8a - 9$

Resource Manager

Workbook and Reproducible Masters

Chapter 5 Resource Masters
- Study Guide and Intervention, pp. 245–246
- Skills Practice, p. 247
- Practice, p. 248
- Reading to Learn Mathematics, p. 249
- Enrichment, p. 250

School-to-Career Masters, p. 9

Teaching Algebra With Manipulatives Masters, p. 234

Transparencies

5-Minute Check Transparency 5-2
Answer Key Transparencies

Technology

Interactive Chalkboard

MULTIPLY POLYNOMIALS

In-Class Examples Power Point®

3 Find $-y(4y^2 + 2y - 3)$.
$-4y^3 - 2y^2 + 3y$

4 Find $(2p + 3)(4p + 1)$.
$8p^2 + 14p + 3$

5 Find $(a^2 + 3a - 4)(a + 2)$.
$a^3 + 5a^2 + 2a - 8$

3 Practice/Apply

Study Notebook

Have students—

- *add the definitions/examples of the vocabulary terms to their Vocabulary Builder worksheets for Chapter 5.*
- *include any other item(s) that they find helpful in mastering the skills in this lesson.*

About the Exercises...

Organization by Objective

- **Add and Subtract Polynomials:** 16–27
- **Multiply Polynomials:** 28–33, 37–50

Odd/Even Assignments

Exercises 16–33 and 37–50 are structured so that students practice the same concepts whether they are assigned odd or even problems.

Assignment Guide

Basic: 17–33 odd, 35, 36, 37–45 odd, 51, 53–69

Average: 17–33 odd, 35, 36, 37–53 odd, 54–69

Advanced: 16–34 even, 35, 36, 38–52 even, 54–65 (optional: 66–69)

MULTIPLY POLYNOMIALS You can use the Distributive Property to multiply polynomials.

Example 3 Multiply and Simplify

Find $2x(7x^2 - 3x + 5)$.

$2x(7x^2 - 3x + 5) = 2x(7x^2) + 2x(-3x) + 2x(5)$ Distributive Property

$= 14x^3 - 6x^2 + 10x$ Multiply the monomials.

You can use algebra tiles to model the product of two binomials.

Algebra Activity

Multiplying Binomials

Use algebra tiles to find the product of $x + 5$ and $x + 2$.

- Draw a 90° angle on your paper.
- Use an x tile and a 1 tile to mark off a length equal to $x + 5$ along the top.
- Use the tiles to mark off a length equal to $x + 2$ along the side.
- Draw lines to show the grid formed.
- Fill in the lines with the appropriate tiles to show the area product. The model shows the polynomial $x^2 + 7x + 10$.

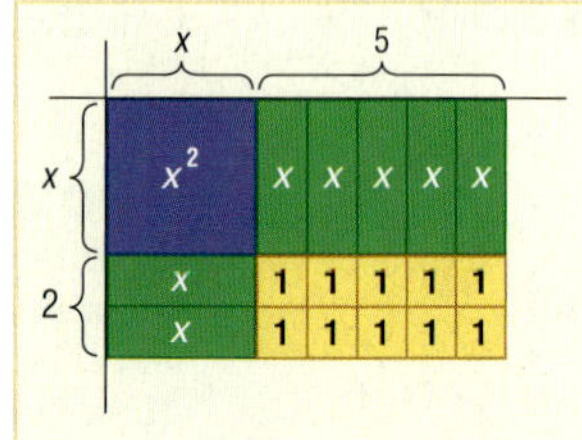

The area of the rectangle is the product of its length and width. So, $(x + 5)(x + 2) = x^2 + 7x + 10$.

The **FOIL method** uses the Distributive Property to multiply binomials.

Key Concept *FOIL Method for Multiplying Binomials*

The product of two binomials is the sum of the products of **F** the *first* terms, **O** the *outer* terms, **I** the *inner* terms, and **L** the *last* terms.

Study Tip

Vertical Method
You may also want to use the vertical method to multiply polynomials.

$$\begin{array}{r} 3y + 2 \\ (\times)\ 5y + 4 \\ \hline 12y + 8 \\ 15y^2 + 10y \\ \hline 15y^2 + 22y + 8 \end{array}$$

Example 4 Multiply Two Binomials

Find $(3y + 2)(5y + 4)$.

$(3y + 2)(5y + 4) = \underbrace{3y \cdot 5y}_{\text{First terms}} + \underbrace{3y \cdot 4}_{\text{Outer terms}} + \underbrace{2 \cdot 5y}_{\text{Inner terms}} + \underbrace{2 \cdot 4}_{\text{Last terms}}$

$= 15y^2 + 22y + 8$ Multiply monomials and add like terms.

Example 5 Multiply Polynomials

Find $(n^2 + 6n - 2)(n + 4)$.

$(n^2 + 6n - 2)(n + 4)$

$= n^2(n + 4) + 6n(n + 4) + (-2)(n + 4)$ Distributive Property

$= n^2 \cdot n + n^2 \cdot 4 + 6n \cdot n + 6n \cdot 4 + (-2) \cdot n + (-2) \cdot 4$ Distributive Property

$= n^3 + 4n^2 + 6n^2 + 24n - 2n - 8$ Multiply monomials.

$= n^3 + 10n^2 + 22n - 8$ Combine like terms.

Answers

22. $4x^2 + 3x - 7$
23. $-3y - 3y^2$
24. $r^2 - r + 6$
25. $10m^2 + 5m - 15$
26. $4x^2 - 3xy - 6y^2$

Algebra Activity

Materials: protractor, ruler/straightedge, algebra tiles

- Remind students that the length of an x tile is *not* a multiple of the length of a side of a unit tile.
- Point out to students that the width of an x tile is exactly one unit (the same as the length of a side of a unit tile).

230 Chapter 5 Polynomials

Check for Understanding

Concept Check

1. Sample answer: $x^5 + x^4 + x^3$

1. **OPEN ENDED** Write a polynomial of degree 5 that has three terms.
2. **Identify** the degree of the polynomial $2x^3 - x^2 + 3x^4 - 7$. **4**
3. **Model** $3x(x + 2)$ using algebra tiles. **See pp. 283A–283B.**

Guided Practice

GUIDED PRACTICE KEY	
Exercises	Examples
4–6	1
7, 8	2
9, 10	3
11–14	4
15	5

Determine whether each expression is a polynomial. If it is a polynomial, state the degree of the polynomial.

4. $2a + 5b$ **yes, 1**
5. $\frac{1}{3}x^3 - 9y$ **yes, 3**
6. $\frac{mw^2 - 3}{nz^3 + 1}$ **no**

Simplify. **8. $-3x^2 - 7x + 8$ 10. $10p^3q^2 - 6p^5q^3 + 8p^3q^5$**

7. $(2a + 3b) + (8a - 5b)$ **$10a - 2b$**
8. $(x^2 - 4x + 3) - (4x^2 + 3x - 5)$
9. $2x(3y + 9)$ **$6xy + 18x$**
10. $2p^2q(5pq - 3p^3q^2 + 4pq^4)$
11. $(y - 10)(y + 7)$ **$y^2 - 3y - 70$**
12. $(x + 6)(x + 3)$ **$x^2 + 9x + 18$**
13. $(2z - 1)(2z + 1)$ **$4z^2 - 1$**
14. $(2m - 3n)^2$ **$4m^2 - 12mn + 9n^2$**

Application

15. **GEOMETRY** Find the area of the triangle. **$7.5x^2 + 12.5x$ ft^2**

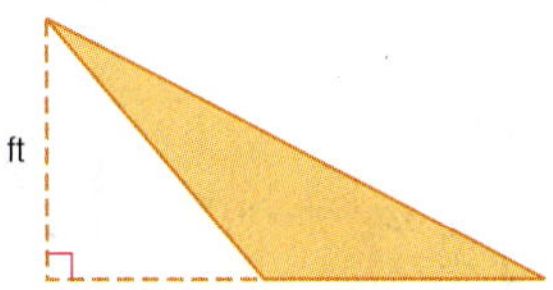

★ indicates increased difficulty

Practice and Apply

Homework Help	
For Exercises	See Examples
16–21	1
22–27, 35, 36, 51	2
28–33, 47, 48	3
34	2, 3
37–46, 52, 53	4
49, 50, 54	5

Extra Practice
See page 837.

Determine whether each expression is a polynomial. If it is a polynomial, state the degree of the polynomial.

16. $3z^2 - 5z + 11$ **yes, 2**
17. $x^3 - 9$ **yes, 3**
18. $\frac{6xy}{z} - \frac{3c}{d}$ **no**
19. $\sqrt{m - 5}$ **no**
20. $5x^2y^4 + x\sqrt{3}$ **yes, 6**
21. $\frac{4}{3}y^2 + \frac{5}{6}y^7$ **yes, 7**

Simplify. **22–33. See margin.**

22. $(3x^2 - x + 2) + (x^2 + 4x - 9)$
23. $(5y + 3y^2) + (-8y - 6y^2)$
24. $(9r^2 + 6r + 16) - (8r^2 + 7r + 10)$
25. $(7m^2 + 5m - 9) + (3m^2 - 6)$
26. $(4x^2 - 3y^2 + 5xy) - (8xy + 3y^2)$
27. $(10x^2 - 3xy + 4y^2) - (3x^2 + 5xy)$
28. $4b(cb - zd)$
29. $4a(3a^2 + b)$
30. $-5ab^2(-3a^2b + 6a^3b - 3a^4b^4)$
31. $2xy(3xy^3 - 4xy + 2y^4)$
32. $\frac{3}{4}x^2(8x + 12y - 16xy^2)$
33. $\frac{1}{2}a^3(4a - 6b + 8ab^4)$

34. **PERSONAL FINANCE** Toshiro has $850 to invest. He can invest in a savings account that has an annual interest rate of 3.7%, and he can invest in a money market account that pays about 5.5% per year. Write a polynomial to represent the amount of interest he will earn in 1 year if he invests x dollars in the savings account and the rest in the money market account. **$46.75 - 0.018x$**

E-SALES **For Exercises 35 and 36, use the following information.**
A small online retailer estimates that the cost, in dollars, associated with selling x units of a particular product is given by the expression $0.001x^2 + 5x + 500$. The revenue from selling x units is given by $10x$. **35. $-0.001x^2 + 5x - 500$**

35. Write a polynomial to represent the profit generated by the product.
36. Find the profit from sales of 1850 units. **$5327.50**

www.algebra2.com/self_check_quiz

Answers

27. $7x^2 - 8xy + 4y^2$
28. $4b^2c - 4bdz$
29. $12a^3 + 4ab$
30. $15a^3b^3 - 30a^4b^3 + 15a^5b^6$
31. $6x^2y^4 - 8x^2y^2 + 4xy^5$
32. $6x^3 + 9x^2y - 12x^3y^2$
33. $2a^4 - 3a^3b + 4a^4b^4$

Enrichment, p. 250

Polynomials with Fractional Coefficients

Polynomials may have fractional coefficients as long as there are no variables in the denominators. Computing with fractional coefficients is performed in the same way as computing with whole-number coefficients.

Simpliply. Write all coefficients as fractions.

1. $\left(\frac{3}{5}m - \frac{2}{7}p - \frac{1}{3}n\right) - \left(\frac{7}{3}p - \frac{5}{2}m - \frac{3}{4}n\right)$ **$\frac{31}{10}m + \frac{5}{12}n - \frac{55}{21}p$**
2. $\left(\frac{3}{2}x - \frac{4}{3}y - \frac{5}{4}z\right) + \left(-\frac{1}{4}x + y + \frac{2}{5}z\right) + \left(-\frac{7}{8}x - \frac{6}{7}y + \frac{1}{2}z\right)$ **$\frac{3}{8}x - \frac{25}{21}y - \frac{7}{20}z$**
3. $\left(\frac{1}{2}a^2 - \frac{1}{3}ab + \frac{1}{4}b^2\right) + \left(\frac{5}{6}a^2 + \frac{2}{3}ab - \frac{3}{4}b^2\right)$ **$\frac{4}{3}a^2 + \frac{1}{3}ab - \frac{1}{2}b^2$**

Study Guide and Intervention, p. 245 (shown) and p. 246

Add and Subtract Polynomials

Polynomial	a monomial or a sum of monomials
Like Terms	terms that have the same variable(s) raised to the same power(s)

To add or subtract polynomials, perform the indicated operations and combine like terms.

Example 1 **Simplify $-6rs + 18r^2 - 5s^2 - 14r^2 + 8rs - 6s^2$.**

$-6rs + 18r^2 - 5s^2 - 14r^2 + 8rs - 6s^2$
$= (18r^2 - 14r^2) + (-6rs + 8rs) + (-5s^2 - 6s^2)$ Group like terms.
$= 4r^2 + 2rs - 11s^2$ Combine like terms.

Example 2 **Simplify $4xy^2 + 12xy - 7x^2y - (20xy + 5xy^2 - 8x^2y)$.**

$4xy^2 + 12xy - 7x^2y - (20xy + 5xy^2 - 8x^2y)$
$= 4xy^2 + 12xy - 7x^2y - 20xy - 5xy^2 + 8x^2y$ Distribute the minus sign.
$= (-7x^2y + 8x^2y) + (4xy^2 - 5xy^2) + (12xy - 20xy)$ Group like terms.
$= x^2y - xy^2 - 8xy$ Combine like terms.

Exercises

Simplify.

1. $(6x^2 - 3x + 2) - (4x^2 + x - 3)$ **$2x^2 - 4x + 5$**
2. $(7y^2 + 12xy - 5x^2) + (6xy - 4y^2 - 3x^2)$ **$3y^2 + 18xy - 8x^2$**
3. $(-4m^2 - 6m) - (6m + 4m^2)$ **$-8m^2 - 12m$**
4. $27x^2 - 5y^2 + 12y^2 - 14x^2$ **$13x^2 + 7y^2$**
5. $(18p^2 + 11pq - 6q^2) - (15p^2 - 3pq + 4q^2)$ **$3p^2 + 14pq - 10q^2$**
6. $17j^2 - 12k^2 + 3j^2 - 15j^2 + 14k^2$ **$5j^2 + 2k^2$**
7. $(8m^2 - 7n^2) - (n^2 - 12m^2)$ **$20m^2 - 8n^2$**
8. $14bc + 6b - 4c + 8b - 8c + 8bc$ **$14b + 22bc - 12c$**
9. $6r^2s + 11rs^2 + 3r^2s - 7rs^2 + 15r^2s - 9rs^2$ **$24r^2s - 5rs^2$**
10. $-9xy + 11x^2 - 14y^2 - (6y^2 - 5xy - 3x^2)$ **$14x^2 - 4xy - 20y^2$**
11. $(12xy - 8x + 3y) + (15x - 7y - 8xy)$ **$7x + 4xy - 4y$**
12. $10.8b^2 - 5.7b + 7.2 - (2.9b^2 - 4.6b - 3.1)$ **$7.9b^2 - 1.1b + 10.3$**
13. $(3bc - 9b^2 - 6c^2) + (4c^2 - b^2 + 5bc)$ **$-10b^2 + 8bc - 2c^2$**
14. $11x^2 + 4y^2 + 6xy + 3y^2 - 5xy - 10x^2$ **$x^2 + xy + 7y^2$**
15. $\frac{1}{4}x^2 - \frac{3}{8}xy + \frac{1}{2}y^2 - \frac{1}{2}xy + \frac{1}{4}y^2 - \frac{3}{8}x^2$ **$-\frac{1}{8}x^2 - \frac{7}{8}xy + \frac{3}{4}y^2$**
16. $24p^3 - 15p^2 + 3p - 15p^3 + 13p^2 - 7p$ **$9p^3 - 2p^2 - 4p$**

Skills Practice, p. 247 and Practice, p. 248 (shown)

Determine whether each expression is a polynomial. If it is a polynomial, state the degree of the polynomial.

1. $5x^3 + 2xy^4 + 6xy$ **yes; 5**
2. $-\frac{4}{3}ac - a^5d^3$ **yes; 8**
3. $\frac{12m^8n^9}{(m - n)^2}$ **no**
4. $25x^3z - x\sqrt{78}$ **yes; 4**
5. $6c^{-2} + c - 1$ **no**
6. $\frac{5}{r} + \frac{6}{s}$ **no**

Simplify.

7. $(3n^2 + 1) + (8n^2 - 8)$ **$11n^2 - 7$**
8. $(6w - 11w^2) - (4 + 7w^2)$ **$-18w^2 + 6w - 4$**
9. $(-6n - 13n^2) + (-3n + 9n^2)$ **$-9n - 4n^2$**
10. $(8x^2 - 3x) - (4x^2 + 5x - 3)$ **$4x^2 - 8x + 3$**
11. $(5m^2 - 2mp - 6p^2) - (-3m^2 + 5mp + p^2)$ **$8m^2 - 7mp - 7p^2$**
12. $(2x^2 - xy + y^2) + (-3x^2 + 4xy + 3y^2)$ **$-x^2 + 3xy + 4y^2$**
13. $(5t - 7) + (2t^2 + 3t + 12)$ **$2t^2 + 8t + 5$**
14. $(u - 4) - (6 + 3u^2 - 4u)$ **$-3u^2 + 5u - 10$**
15. $-9(y^2 - 7w)$ **$-9y^2 + 63w$**
16. $-9r^4y^2(-3ry^7 + 2r^3y^4 - 8r^{10})$ **$27r^5y^9 - 18r^7y^6 + 72r^{14}y^2$**
17. $-6a^2w(a^3w - aw^4)$ **$-6a^5w^2 + 6a^3w^5$**
18. $5a^2w^3(a^2w^6 - 3a^4w^2 + 9aw^6)$ **$5a^4w^9 - 15a^6w^5 + 45a^3w^9$**
19. $2x^2(x^2 + xy - 2y^2)$ **$2x^4 + 2x^3y - 4x^2y^2$**
20. $-\frac{3}{5}ab^3d^2(-5ab^2d^5 - 5ab)$ **$3a^2b^5d^7 + 3a^2b^4d^2$**
21. $(v^2 - 6)(v^2 + 4)$ **$v^4 - 2v^2 - 24$**
22. $(7a + 9y)(2a - y)$ **$14a^2 + 11ay - 9y^2$**
23. $(y - 8)^2$ **$y^2 - 16y + 64$**
24. $(x^2 + 5y)^2$ **$x^4 + 10x^2y + 25y^2$**
25. $(5x + 4w)(5x - 4w)$ **$25x^2 - 16w^2$**
26. $(2n^4 - 3)(2n^4 + 3)$ **$4n^8 - 9$**
27. $(w + 2s)(w^2 - 2ws + 4s^2)$ **$w^3 + 8s^3$**
28. $(x + y)(x^2 - 3xy + 2y^2)$ **$x^3 - 2x^2y - xy^2 + 2y^3$**
29. **BANKING** Terry invests $1500 in two mutual funds. The first year, one fund grows 3.8% and the other grows 6%. Write a polynomial to represent the amount Terry's $1500 grows to in that year if x represents the amount he invested in the fund with the lesser growth rate. **$-0.022x + 1590$**
30. **GEOMETRY** The area of the base of a rectangular box measures $2x^2 + 4x - 3$ square units. The height of the box measures x units. Find a polynomial expression for the volume of the box. **$2x^3 + 4x^2 - 3x$ units^3**

Reading to Learn Mathematics, p. 249 ELL

Pre-Activity **How can polynomials be applied to financial situations?**

Read the introduction to Lesson 5-2 at the top of page 229 in your textbook.

Suppose that Shenequa decides to enroll in a five-year engineering program rather than a four-year program. Using the model given in your textbook, how could she estimate the tuition for the fifth year of her program? (Do not actually calculate, but describe the calculation that would be necessary.) **Multiply $15,604 by 1.04.**

Reading the Lesson

1. State whether the terms in each of the following pairs are *like terms* or *unlike terms*.
   a. $3x^2, 3y^2$ **unlike terms**
   b. $-m^4, 5m^4$ **like terms**
   c. $8r^3, 8s^3$ **unlike terms**
   d. $-6, 6$ **like terms**
2. State whether each of the following expressions is a *monomial*, *binomial*, *trinomial*, or *not a polynomial*. If the expression is a polynomial, give its degree.
   a. $4r^4 - 2r + 1$ **trinomial; degree 4**
   b. $\sqrt{3x}$ **not a polynomial**
   c. $5x + 4y$ **binomial; degree 1**
   d. $2ab + 4ab^2 - 6ab^3$ **trinomial; degree 4**
3. a. What is the FOIL method used for in algebra? **to multiply binomials**
   b. The FOIL method is an application of what property of real numbers? **Distributive Property**
   c. In the FOIL method, what do the letters F, O, I, and L mean? **first, outer, inner, last**
   d. Suppose you want to use the FOIL method to multiply $(2x + 3)(4x + 1)$. Show the terms you would multiply, but do not actually multiply them.
   F **(2x)(4x)**
   O **(2x)(1)**
   I **(3)(4x)**
   L **(3)(1)**

Helping You Remember

4. You can remember the difference between *monomials*, *binomials*, and *trinomials* by thinking of common English words that begin with the same prefixes. Give two words unrelated to mathematics that start with *mono-*, two that begin with *bi-*, and two that begin with *tri-*. **Sample answer: monotonous, monogram; bicycle, bifocal; tricycle, tripod**

4 Assess

Open-Ended Assessment

Writing Have students write an explanation, including an example, showing why the FOIL method is a valid alternative to applying the Distributive Property when multiplying two binomials.

Getting Ready for Lesson 5-3

PREREQUISITE SKILL Lesson 5-3 presents dividing polynomials. Dividing polynomials requires the use of the properties of exponents. Exercises 66–69 should be used to determine your students' familiarity with the properties of exponents.

Answers

63.

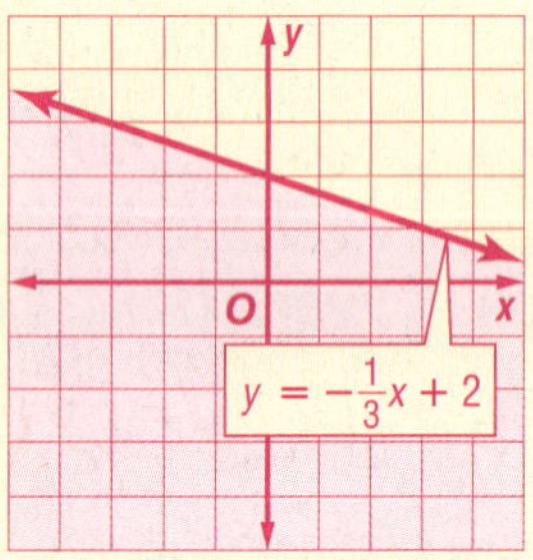

64.

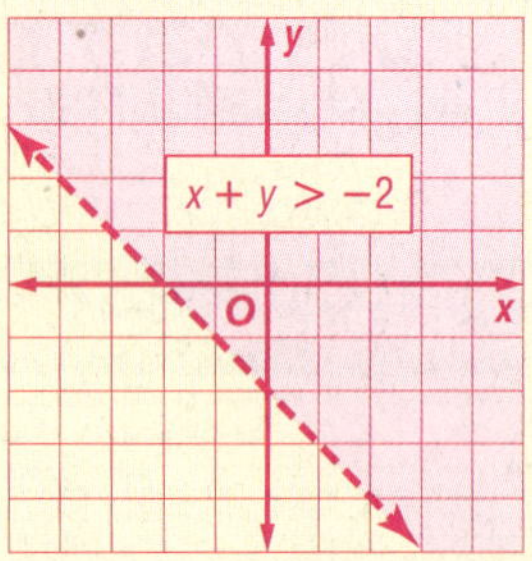

65.

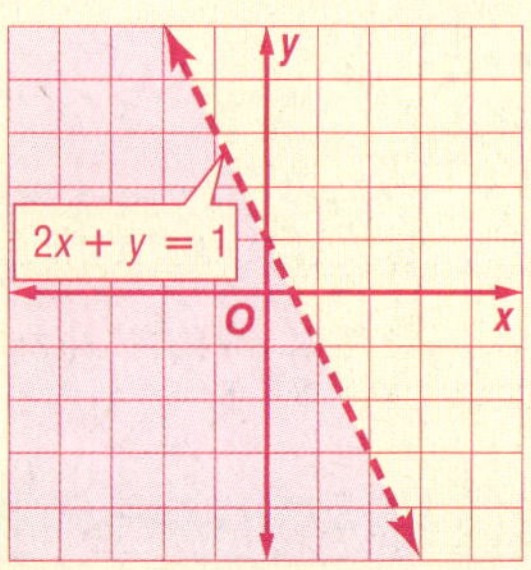

48. $xy^3 + y + \frac{1}{x}$

51. $9c^2 - 12cd + 7d^2$

More About. . .

Genetics

The possible genes of parents and offspring can be summarized in a *Punnett square*, such as the one above.

Source: *Biology: The Dynamics of Life*

Simplify.

37. $(p + 6)(p - 4)$ $p^2 + 2p - 24$

38. $(a + 6)(a + 3)$ $a^2 + 9a + 18$

39. $(b + 5)(b - 5)$ $b^2 - 25$

40. $(6 - z)(6 + z)$ $36 - z^2$

41. $(3x + 8)(2x + 6)$ $6x^2 + 34x + 48$

42. $(4y - 6)(2y + 7)$ $8y^2 + 16y - 42$

43. $(a^3 - b)(a^3 + b)$ $a^6 - b^2$

44. $(m^2 - 5)(2m^2 + 3)$ $2m^4 - 7m^2 - 15$

45. $(x - 3y)^2$ $x^2 - 6xy + 9y^2$

46. $(1 + 4c)^2$ $1 + 8c + 16c^2$

★ 47. $d^{-3}(d^5 - 2d^3 + d^{-1})$ $d^2 - 2 + \frac{1}{d^4}$

★ 48. $x^{-3}y^2(yx^4 + y^{-1}x^3 + y^{-2}x^2)$

★ 49. $(3b - c)^3$ $27b^3 - 27b^2c + 9bc^2 - c^3$

★ 50. $(x^2 + xy + y^2)(x - y)$ $x^3 - y^3$

51. Simplify $(c^2 - 6cd - 2d^2) + (7c^2 - cd + 8d^2) - (-c^2 + 5cd - d^2)$.

52. Find the product of $6x - 5$ and $-3x + 2$. $-18x^2 + 27x - 10$

53. **GENETICS** Suppose R and W represent two genes that a plant can inherit from its parents. The terms of the expansion of $(R + W)^2$ represent the possible pairings of the genes in the offspring. Write $(R + W)^2$ as a polynomial. $R^2 + 2RW + W^2$

54. **CRITICAL THINKING** What is the degree of the product of a polynomial of degree 8 and a polynomial of degree 6? Include an example in support of your answer. **14; Sample answer:** $(x^8 + 1)(x^6 + 1) = x^{14} + x^8 + x^6 + 1$

55. **WRITING IN MATH** Answer the question that was posed at the beginning of the lesson. See pp. 283A–283B.

How can polynomials be applied to financial situations?

Include the following in your answer:

- an explanation of how a polynomial can be applied to a situation with a fixed percent rate of increase,
- two expressions in terms of r for the tuition in the fourth year, and
- an explanation of how to use one of the expressions and the 4% rate of increase to estimate Shenequa's tuition in the fourth year, and a comparison of the value you found to the value given in the table.

Standardized Test Practice

56. Which polynomial has degree 3? **D**

(A) $x^3 + x^2 - 2x^4$	(B) $-2x^2 - 3x + 4$
(C) $x^2 + x + 12^3$	(D) $1 + x + x^3$

57. $(x + y) - (y + z) - (x + z) = ?$ **B**

(A) $2x + 2y + 2z$	(B) $-2z$
(C) $2y$	(D) $x - y - z$

Maintain Your Skills

Mixed Review

Simplify. Assume that no variable equals 0. *(Lesson 5-1)*

58. $(-4d^2)^3$ $-64d^6$

59. $5rt^2(2rt)^2$ $20r^3t^4$

60. $\frac{x^2yz^4}{xy^3z^2}$ $\frac{xz^2}{y^2}$

61. $\left(\frac{3ab^2}{6a^2b}\right)^2$ $\frac{b^2}{4a^2}$

62. Solve the system $4x - y = 0$, $2x + 3y = 14$ by using inverse matrices. *(Lesson 4-8)* **(1, 4)**

Graph each inequality. *(Lesson 2-7)* **63–65. See margin.**

63. $y \leq -\frac{1}{3}x + 2$

64. $x + y > -2$

65. $2x + y < 1$

Getting Ready for the Next Lesson

PREREQUISITE SKILL **Simplify. Assume that no variable equals 0.**
*(To review **properties of exponents**, see Lesson 5-1.)*

66. $\frac{x^3}{x}$ x^2

67. $\frac{4y^5}{2y^2}$ $2y^3$

68. $\frac{x^2y^3}{xy}$ xy^2

69. $\frac{9a^3b}{3ab}$ $3a^2$

DAILY INTERVENTION

Differentiated Instruction

Logical Have students demonstrate how to use algebra tiles to multiply two binomials that contain at least one negative coefficient.

232 Chapter 5 Polynomials

5-3 Dividing Polynomials

What You'll Learn

- Divide polynomials using long division.
- Divide polynomials using synthetic division.

Vocabulary

- synthetic division

How can you use division of polynomials in manufacturing?

A machinist needed $32x^2 + x$ square inches of metal to make a square pipe $8x$ inches long. In figuring the area needed, she allowed a fixed amount of metal for overlap of the seam. If the width of the finished pipe will be x inches, how wide is the seam? You can use a quotient of polynomials to help find the answer.

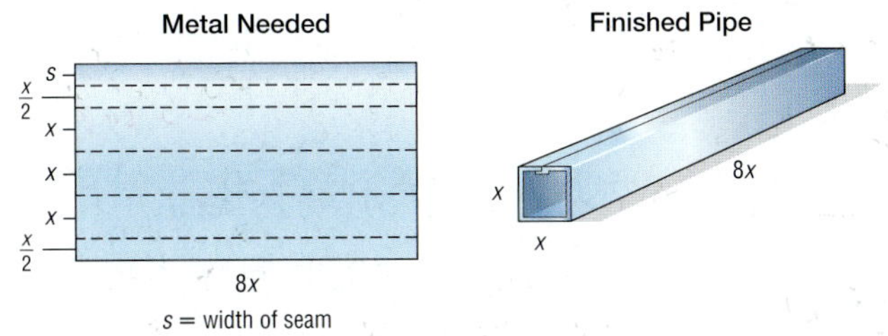

USE LONG DIVISION In Lesson 5-1, you learned to divide monomials. You can divide a polynomial by a monomial by using those same skills.

Example 1 Divide a Polynomial by a Monomial

Simplify $\frac{4x^3y^2 + 8xy^2 - 12x^2y^3}{4xy}$.

$$\frac{4x^3y^2 + 8xy^2 - 12x^2y^3}{4xy} = \frac{4x^3y^2}{4xy} + \frac{8xy^2}{4xy} - \frac{12x^2y^3}{4xy} \quad \text{Sum of quotients}$$

$$= \frac{4}{4} \cdot x^{3-1}y^{2-1} + \frac{8}{4} \cdot x^{1-1}y^{2-1} - \frac{12}{4} \cdot x^{2-1}y^{3-1} \quad \text{Divide.}$$

$$= x^2y + 2y - 3xy^2 \quad x^{1-1} = x^0 \text{ or } 1$$

You can use a process similar to long division to divide a polynomial by a polynomial with more than one term. The process is known as the *division algorithm*. When doing the division, remember that you can only add or subtract like terms.

Example 2 Division Algorithm

Use long division to find $(z^2 + 2z - 24) \div (z - 4)$.

$$\begin{array}{r} z \\ z-4\overline{)z^2+2z-24} \\ \underline{(-)z^2-4z} \\ 6z-24 \end{array} \qquad \begin{array}{l} \\ \\ z(z-4) = z^2 - 4z \\ 2z - (-4z) = 6z \end{array}$$

$$\begin{array}{r} z+6 \\ z-4\overline{)z^2+2z-24} \\ \underline{(-)z^2-4z} \\ 6z-24 \\ \underline{(-)6z-24} \\ 0 \end{array}$$

The quotient is $z + 6$. The remainder is 0.

5-3 Lesson Notes

1 Focus

5-Minute Check Transparency 5-3 Use as a quiz or review of Lesson 5-2.

Mathematical Background notes are available for this lesson on p. 220C.

How can you use division of polynomials in manufacturing?

Ask students:

- What does the expression $\frac{x}{2}$ shown in the figure represent? **one half of the side length of the pipe opening**
- What happens to the width of the pipe opening as the length of the pipe increases? **The width of the pipe opening, x, increases also.**

2 Teach

USE LONG DIVISION

In-Class Example PowerPoint®

1 Simplify $\frac{5a^2b - 15ab^3 + 10a^3b^4}{5ab}$. $a - 3b^2 + 2a^2b^3$

Resource Manager

Workbook and Reproducible Masters

Chapter 5 Resource Masters

- Study Guide and Intervention, pp. 251–252
- Skills Practice, p. 253
- Practice, p. 254
- Reading to Learn Mathematics, p. 255
- Enrichment, p. 256
- Assessment, p. 307

Transparencies

5-Minute Check Transparency 5-3
Answer Key Transparencies

Technology

Interactive Chalkboard

In-Class Examples

Power Point®

2 Use long division to find $(x^2 - 2x - 15) \div (x - 5)$. $x + 3$

Teaching Tip If students are having difficulty with the use of the division algorithm, review the algorithm as it is used for long division of numbers.

3 Which expression is equal to $(a^2 - 5a + 3)(2 - a)^{-1}$? **D**

A $a + 3$

B $-a + 3 + \frac{3}{2 - a}$

C $-a - 3 + \frac{3}{2 - a}$

D $-a + 3 - \frac{3}{2 - a}$

USE SYNTHETIC DIVISION

In-Class Example

Power Point®

Teaching Tip When discussing Example 4, stress that the divisor must be of the form $x - r$ in order to use synthetic division.

4 Use synthetic division to find $(x^3 - 4x^2 + 6x - 4) \div (x - 2)$. $x^2 - 2x + 2$

Teaching Tip Ask students to discuss whether they would rather use long division or synthetic division, giving a reason for their choice.

Standardized Test Practice

Example 3 Point out that the denominator $5 - t$ is rewritten as $-t + 5$ before starting the division in order to have both numerator and denominator written in descending order of the variable. Point out that the first step in the long division eliminates choice A. Students could also quickly eliminate choice B by multiplying $-t - 8$ by $-t + 5$ and noting that the product is not $t^2 + 3t - 9$.

Just as with the division of whole numbers, the division of two polynomials may result in a quotient with a remainder. Remember that $9 \div 4 = 2 + \text{R}1$ and is often written as $2\frac{1}{4}$. The result of a division of polynomials with a remainder can be written in a similar manner.

Standardized Test Practice

Example 3 Quotient with Remainder

Multiple-Choice Test Item

Which expression is equal to $(t^2 + 3t - 9)(5 - t)^{-1}$?

Ⓐ $t + 8 - \frac{31}{5 - t}$ Ⓑ $-t - 8$

Ⓒ $-t - 8 + \frac{31}{5 - t}$ Ⓓ $-t - 8 - \frac{31}{5 - t}$

Read the Test Item

Since the second factor has an exponent of -1, this is a division problem.

$(t^2 + 3t - 9)(5 - t)^{-1} = \frac{t^2 + 3t - 9}{5 - t}$

Test-Taking Tip

You may be able to eliminate some of the answer choices by substituting the same value for t in the original expression and the answer choices and evaluating.

Solve the Test Item

$$\begin{array}{r|l} \begin{array}{r} -t - 8 \\ -t + 5\overline{)t^2 + 3t - 9} \\ (-)t^2 - 5t \\ 8t - 9 \\ (-)8t - 40 \\ 31 \end{array} & \begin{array}{l} \\ \text{For ease in dividing, rewrite } 5 - t \text{ as } -t + 5. \\ -t(-t + 5) = t^2 - 5t \\ 3t - (-5t) = 8t \\ -8(-t + 5) = 8t - 40 \\ \text{Subtract. } -9 - (-40) = 31 \end{array} \end{array}$$

The quotient is $-t - 8$, and the remainder is 31. Therefore, $(t^2 + 3t - 9)(5 - t)^{-1} = -t - 8 + \frac{31}{5 - t}$. The answer is C.

USE SYNTHETIC DIVISION **Synthetic division** is a simpler process for dividing a polynomial by a binomial. Suppose you want to divide $5x^3 - 13x^2 + 10x - 8$ by $x - 2$ using long division. Compare the coefficients in this division with those in Example 4.

$$\begin{array}{r} 5x^2 - 3x + 4 \\ x - 2\overline{)5x^3 - 13x^2 + 10x - 8} \\ (-)5x^3 - 10x^2 \qquad\qquad \\ -3x^2 + 10x \qquad \\ (-)-3x^2 + 6x \qquad \\ 4x - 8 \\ (-)4x - 8 \\ 0 \end{array}$$

Example 4 Synthetic Division

Use synthetic division to find $(5x^3 - 13x^2 + 10x - 8) \div (x - 2)$.

Step 1 Write the terms of the dividend so that the degrees of the terms are in descending order. Then write just the coefficients as shown at the right.

$$\begin{array}{cccc} 5x^3 & -13x^2 & +10x & -8 \\ \downarrow & \downarrow & \downarrow & \downarrow \\ 5 & -13 & 10 & -8 \end{array}$$

Step 2 Write the constant r of the divisor $x - r$ to the left. In this case, $r = 2$. Bring the first coefficient, 5, down as shown.

$$\begin{array}{c|cccc} 2 & 5 & -13 & 10 & -8 \\ & & & & \\ \hline & 5 & & & \end{array}$$

Teacher to Teacher

Christine Waddell — Albion M.S., Sandy, UT

"To help students better understand the division algorithm for polynomials, I first work through a long division problem with large whole numbers, such as 3248 ÷ 24, step by step. Then I work through Example 2 and point out the similarities in each process."

Step 3 Multiply the first coefficient by r: $2 \cdot 5 = 10$. Write the product under the second coefficient. Then add the product and the second coefficient: $-13 + 10 = -3$.

$$\begin{array}{r|rrrr} 2 & 5 & -13 & 10 & -8 \\ & & 10 & & \\ \hline & 5 & -3 & & \end{array}$$

Step 4 Multiply the sum, -3, by r: $2(-3) = -6$. Write the product under the next coefficient and add: $10 + (-6) = 4$.

$$\begin{array}{r|rrrr} 2 & 5 & -13 & 10 & -8 \\ & & 10 & -6 & \\ \hline & 5 & -3 & 4 & \end{array}$$

Step 5 Multiply the sum, 4, by r: $2 \cdot 4 = 8$. Write the product under the next coefficient and add: $-8 + 8 = 0$. The remainder is 0.

$$\begin{array}{r|rrrr} 2 & 5 & -13 & 10 & -8 \\ & & 10 & -6 & 8 \\ \hline & 5 & -3 & 4 & 0 \end{array}$$

The numbers along the bottom row are the coefficients of the quotient. Start with the power of x that is one less than the degree of the dividend. Thus, the quotient is $5x^2 - 3x + 4$.

To use synthetic division, the divisor must be of the form $x - r$. If the coefficient of x in a divisor is not 1, you can rewrite the division expression so that you can use synthetic division.

Example 5 Divisor with First Coefficient Other than 1

Use synthetic division to find $(8x^4 - 4x^2 + x + 4) \div (2x + 1)$.

Use division to rewrite the divisor so it has a first coefficient of 1.

$\frac{8x^4 - 4x^2 + x + 4}{2x + 1} = \frac{(8x^4 - 4x^2 + x + 4) \div 2}{(2x + 1) \div 2}$ Divide numerator and denominator by 2.

$= \frac{4x^4 - 2x^2 + \frac{1}{2}x + 2}{x + \frac{1}{2}}$ Simplify the numerator and denominator.

Since the numerator does not have an x^3-term, use a coefficient of 0 for x^3.

$x - r = x + \frac{1}{2}$, so $r = -\frac{1}{2}$.

$$\begin{array}{r|rrrrr} -\frac{1}{2} & 4 & 0 & -2 & \frac{1}{2} & 2 \\ & & -2 & 1 & \frac{1}{2} & -\frac{1}{2} \\ \hline & 4 & -2 & -1 & 1 & \frac{3}{2} \end{array}$$

The result is $4x^3 - 2x^2 - x + 1 + \frac{\frac{3}{2}}{x + \frac{1}{2}}$. Now simplify the fraction.

$\frac{\frac{3}{2}}{x + \frac{1}{2}} = \frac{3}{2} \div \left(x + \frac{1}{2}\right)$ Rewrite as a division expression.

$= \frac{3}{2} \div \frac{2x + 1}{2}$ $\quad x + \frac{1}{2} = \frac{2x}{2} + \frac{1}{2} = \frac{2x + 1}{2}$

$= \frac{3}{2} \cdot \frac{2}{2x + 1}$ Multiply by the reciprocal.

$= \frac{3}{2x + 1}$ Multiply.

The solution is $4x^3 - 2x^2 - x + 1 + \frac{3}{2x + 1}$.

(continued on the next page)

www.algebra2.com/extra_examples

In-Class Example PowerPoint®

5 Use synthetic division to find $(4y^4 - 5y^2 + 2y + 4) \div (2y - 1)$. $2y^3 + y^2 - 2y + \frac{4}{2y - 1}$

Teaching Tip Remind students to include a coefficient of 0 for any missing terms of the variable in the dividend.

DAILY INTERVENTION **Unlocking Misconceptions**

- **Subtracting** Have students analyze any errors they make when using long division. Verify that they are using the signs correctly.
- **Remainders** Have students do a simple numeric division example, such as $8 \div 5$, to help them remember how to write the remainder as part of the quotient.

3 Practice/Apply

Study Notebook

Have students—
- add the definitions/examples of the vocabulary terms to their Vocabulary Builder worksheets for Chapter 5.
- add the Test-Taking Tip on p. 234 to their list of tips which they can review as they prepare for standardized tests.
- include any other item(s) that they find helpful in mastering the skills in this lesson.

About the Exercises...

Organization by Objective
- **Use Long Division:** 15–20, 45–48
- **Use Synthetic Division:** 21–44, 49, 50

Odd/Even Assignments
Exercises 15–50 are structured so that students practice the same concepts whether they are assigned odd or even problems.

Assignment Guide

Basic: 15–43 odd, 49, 51, 53, 54, 58–74

Average: 15–51 odd, 53, 54, 58–74

Advanced: 16–52 even, 55–68 (optional: 69–74)

All: Practice Quiz 1 (1–10)

DAILY INTERVENTION **FIND THE ERROR** Suggest that students recheck their calculations immediately whenever they begin to get large numbers as the coefficients of their quotient. While nothing forbids large coefficients, this is sometimes the first indication that they have made an error in their calculations.

CHECK Divide using long division.

$$\begin{array}{r} 4x^3 - 2x^2 - x + 1 \\ 2x + 1\overline{)8x^4 + 0x^3 - 4x^2 + x + 4} \\ \underline{(-)8x^4 + 4x^3} \qquad\qquad\qquad \\ -4x^3 - 4x^2 \qquad\qquad \\ \underline{(-)-4x^3 - 2x^2} \qquad\qquad \\ -2x^2 + x \qquad \\ \underline{(-)-2x^2 - x} \qquad \\ 2x + 4 \\ \underline{(-)2x + 1} \\ 3 \end{array}$$

The result is $4x^3 - 2x^2 - x + 1 + \frac{3}{2x + 1}$. ✓

Check for Understanding

Concept Check

1. **OPEN ENDED** Write a quotient of two polynomials such that the remainder is 5. **Sample answer: $(x^2 + x + 5) \div (x + 1)$**
2. **Explain** why synthetic division cannot be used to simplify $\frac{x^3 - 3x + 1}{x^2 + 1}$.
3. **FIND THE ERROR** Shelly and Jorge are dividing $x^3 - 2x^2 + x - 3$ by $x - 4$.

Shelly

4	1	-2	1	-3
		4	-24	100
	1	-6	25	-103

Jorge

4	1	-2	1	-3
		4	8	36
	1	2	9	33

Who is correct? Explain your reasoning.

2. The divisor contains an x^2 term.

3. Jorge; Shelly is subtracting in the columns instead of adding.

10. $x^2 + 11x - 34 + \frac{60}{x + 2}$

11. $b^3 + b - 1$

Guided Practice

GUIDED PRACTICE KEY	
Exercises	**Examples**
4, 5	1
6–10	2, 4
11, 14	3
12, 13	5

Simplify. **7. $3a^3 - 9a^2 + 7a - 6$** **8. $z^4 + 2z^3 + 4z^2 + 5z + 10$**

4. $\frac{6xy^2 - 3xy + 2x^2y}{xy}$ **$6y - 3 + 2x$**
5. $(5ab^2 - 4ab + 7a^2b)(ab)^{-1}$ **$5b - 4 + 7a$**
6. $(x^2 - 10x - 24) \div (x + 2)$ **$x - 12$**
7. $(3a^4 - 6a^3 - 2a^2 + a - 6) \div (a + 1)$
8. $(z^5 - 3z^2 - 20) \div (z - 2)$
9. $(x^3 + y^3) \div (x + y)$ **$x^2 - xy + y^2$**
10. $\frac{x^3 + 13x^2 - 12x - 8}{x + 2}$
11. $(b^4 - 2b^3 + b^2 - 3b + 2)(b - 2)^{-1}$
12. $(12y^2 + 36y + 15) \div (6y + 3)$ **$2y + 5$**
13. $\frac{9b^2 + 9b - 10}{3b - 2}$ **$3b + 5$**

Standardized Test Practice

14. Which expression is equal to $(x^2 - 4x + 6)(x - 3)^{-1}$? **B**

Ⓐ $x - 1$ Ⓑ $x - 1 + \frac{3}{x - 3}$

Ⓒ $x - 1 - \frac{3}{x - 3}$ Ⓓ $-x + 1 - \frac{3}{x - 3}$

★ **indicates increased difficulty**

Practice and Apply

Simplify.

15. $\frac{9a^3b^2 - 18a^2b^3}{3a^2b}$ **$3ab - 6b^2$**
16. $\frac{5xy^2 - 6y^3 + 3x^2y^3}{xy}$ **$5y - \frac{6y^2}{x} + 3xy^2$**
17. $(28c^3d - 42cd^2 + 56cd^3) \div (14cd)$
18. $(12mn^3 + 9m^2n^2 - 15m^2n) \div (3mn)$
19. $(2y^3z + 4y^2z^2 - 8y^4z^5)(yz)^{-1}$ **$2y^2 + 4yz - 8y^3z^4$**
20. $(a^3b^2 - a^2b + 2a)(-ab)^{-1}$ **$-a^2b + a - \frac{2}{b}$**

17. $2c^2 - 3d + 4d^2$

18. $4n^2 + 3mn - 5m$

DAILY INTERVENTION **Differentiated Instruction**

Interpersonal To help discover confusions and catch careless errors, have students work in pairs as they do division problems. One person should write the solution steps, explaining each step out loud while the other person watches, listens, and checks the work. The students should then exchange roles and repeat the activity.

Homework Help

For Exercises	See Examples
15–20, 51	1
21–34, 49, 50, 52–54	2, 4
35–38	3, 4
39–48	2, 3, 5

Extra Practice
See page 837.

21–48. See pp. 283A–283B.

21. $(b^3 + 8b^2 - 20b) \div (b - 2)$
22. $(x^2 - 12x - 45) \div (x + 3)$
23. $(n^3 + 2n^2 - 5n + 12) \div (n + 4)$
24. $(2c^3 - 3c^2 + 3c - 4) \div (c - 2)$
25. $(x^4 - 3x^3 + x^2 - 5) \div (x + 2)$
26. $(6w^5 - 18w^2 - 120) \div (w - 2)$
27. $(x^3 - 4x^2) \div (x - 4)$
28. $(x^3 - 27) \div (x - 3)$
29. $\dfrac{y^3 + 3y^2 - 5y - 4}{y + 4}$
30. $\dfrac{m^3 + 3m^2 - 7m - 21}{m + 3}$
31. $\dfrac{a^4 - 5a^3 - 13a^2 + 10}{a + 1}$
32. $\dfrac{2m^4 - 5m^3 - 10m + 8}{m - 3}$
33. $\dfrac{x^5 - 7x^3 + x + 1}{x + 3}$
34. $\dfrac{3c^5 + 5c^4 + c + 5}{c + 2}$
35. $(g^2 + 8g + 15)(g + 3)^{-1}$
36. $(2b^3 + b^2 - 2b + 3)(b + 1)^{-1}$
37. $(t^5 - 3t^2 - 20)(t - 2)^{-1}$
38. $(y^5 + 32)(y + 2)^{-1}$
39. $(6t^3 + 5t^2 + 9) \div (2t + 3)$
40. $(2h^3 - 5h^2 + 22h) \div (2h + 3)$
41. $\dfrac{9d^3 + 5d - 8}{3d - 2}$
42. $\dfrac{4x^3 + 5x^2 - 3x - 1}{4x + 1}$
43. $\dfrac{2x^4 + 3x^3 - 2x^2 - 3x - 6}{2x + 3}$
44. $\dfrac{6x^4 + 5x^3 + x^2 - 3x + 1}{3x + 1}$

★ 45. $\dfrac{x^3 - 3x^2 + x - 3}{x^2 + 1}$

★ 46. $\dfrac{x^4 + x^2 - 3x + 5}{x^2 + 2}$

★ 47. $\dfrac{x^3 + 3x^2 + 3x + 2}{x^2 + x + 1}$

★ 48. $\dfrac{x^3 - 4x^2 + 5x - 6}{x^2 - x + 2}$

49. What is $x^3 - 2x^2 + 4x - 3$ divided by $x - 1$? $x^2 - x + 3$
50. Divide $2y^3 + y^2 - 5y + 2$ by $y + 2$. $2y^2 - 3y + 1$
51. **BUSINESS** A company estimates that it costs $0.03x^2 + 4x + 1000$ dollars to produce x units of a product. Find an expression for the average cost per unit.
52. **ENTERTAINMENT** A magician gives these instructions to a volunteer.
    - Choose a number and multiply it by 3.
    - Then add the sum of your number and 8 to the product you found.
    - Now divide by the sum of your number and 2.

    What number will the volunteer always have at the end? Explain.
    4; See margin for explanation.

MEDICINE **For Exercises 53 and 54, use the following information.**
The number of students at a large high school who will catch the flu during an outbreak can be estimated by $n = \dfrac{170t^2}{t^2 + 1}$, where t is the number of weeks from the beginning of the epidemic and n is the number of ill people.

53. Perform the division indicated by $\dfrac{170t^2}{t^2 + 1}$. $170 - \dfrac{170}{t^2 + 1}$
54. Use the formula to estimate how many people will become ill during the first week. 85 people

PHYSICS **For Exercises 55–57, suppose an object moves in a straight line so that after t seconds, it is $t^3 + t^2 + 6t$ feet from its starting point.** 55. $x^3 + x^2 + 6x - 24$ ft

55. Find the distance the object travels between the times $t = 2$ and $t = x$.
56. How much time elapses between $t = 2$ and $t = x$? $x - 2$ s
57. Find a simplified expression for the average speed of the object between times $t = 2$ and $t = x$. $x^2 + 3x + 12$ ft/s
58. **CRITICAL THINKING** Suppose the result of dividing one polynomial by another is $r^2 - 6r + 9 - \dfrac{1}{r - 3}$. What two polynomials might have been divided?

Career Choices

Cost Analyst

Cost analysts study and write reports about the factors involved in the cost of production.

Online Research For information about a career in cost analysis, visit: www.algebra2.com/careers

51. $\$0.03x + 4 + \dfrac{1000}{x}$

58. Sample answer: $r^3 - 9r^2 + 27r - 28$ and $r - 3$

www.algebra2.com/self_check_quiz

Answer

52. Let x be the number. Multiplying by 3 results in $3x$. The sum of the number, 8, and the result of the multiplication is $x + 8 + 3x$ or $4x + 8$. Dividing by the sum of the number and 2 gives $\dfrac{4x + 8}{x + 2}$ or 4. The end result is always 4.

Study Guide and Intervention, p. 251 (shown) and p. 252

Use Long Division To divide a polynomial by a monomial, use the properties of powers from Lesson 5-1.

To divide a polynomial by a polynomial, use a long division pattern. Remember that only like terms can be added or subtracted.

Example 1 **Simplify $\dfrac{12p^3t^2r - 21p^2qtr^2 - 9p^3tr}{3p^2tr}$.**

$$\frac{12p^3t^2r - 21p^2qtr^2 - 9p^3tr}{3p^2tr} = \frac{12p^3t^2r}{3p^2tr} - \frac{21p^2qtr^2}{3p^2tr} - \frac{9p^3tr}{3p^2tr}$$
$$= \frac{12}{3}p^{3-2}t^{2-1}r^{1-1} - \frac{21}{3}p^{2-2}qt^{1-1}r^{2-1} - \frac{9}{3}p^{3-2}t^{1-1}r^{1-1}$$
$$= 4pt - 7qr - 3p$$

Example 2 **Use long division to find $(x^3 - 8x^2 + 4x - 9) \div (x - 4)$.**

```
          x² − 4x − 12
x − 4)x³ − 8x² + 4x − 9
   (−)x³ − 4x²
          −4x² + 4x
       (−)−4x² + 16x
                −12x − 9
             (−)−12x + 48
                     −57
```

The quotient is $x^2 - 4x - 12$, and the remainder is -57.
Therefore $\dfrac{x^3 - 8x^2 + 4x - 9}{x - 4} = x^2 - 4x - 12 - \dfrac{57}{x - 4}$.

Exercises

Simplify.

1. $\dfrac{18a^3 + 30a^2}{3a}$ $6a^2 + 10a$
2. $\dfrac{24mn^6 - 40m^2n^3}{4m^2n^3}$ $\dfrac{6n^3}{m} - 10$
3. $\dfrac{60a^2b^3 - 48b^4 + 84a^5b^2}{12ab^2}$ $5ab - \dfrac{4b^2}{a} + 7a^4$
4. $(2x^2 - 5x - 3) \div (x - 3)$ $2x + 1$
5. $(m^2 - 3m - 7) \div (m + 2)$ $m - 5 + \dfrac{3}{m + 2}$
6. $(p^3 - 6) \div (p - 1)$ $p^2 + p + 1 - \dfrac{5}{p - 1}$
7. $(t^3 - 6t^2 + 1) \div (t + 2)$ $t^2 - 8t + 16 - \dfrac{31}{t + 2}$
8. $(x^5 - 1) \div (x - 1)$ $x^4 + x^3 + x^2 + x + 1$
9. $(2x^3 - 5x^2 + 4x - 4) \div (x - 2)$ $2x^2 - x + 2$

Skills Practice, p. 253 and Practice, p. 254 (shown)

Simplify.

1. $\dfrac{15r^{10} - 5r^8 + 40r^2}{5r^4}$ $3r^6 - r^4 + \dfrac{8}{r^2}$
2. $\dfrac{6k^2m - 12k^3m^2 + 9m^3}{2km^2}$ $\dfrac{3k}{m} - 6k^2 + \dfrac{9m}{2k}$
3. $(-30x^3y + 12x^2y^2 - 18x^2y) \div (-6x^2y)$ $5x - 2y + 3$
4. $(-6w^3z^4 - 3w^2z^5 + 4w + 5z) \div (2w^2z)$ $-3wz^3 - \dfrac{3z^4}{2} + \dfrac{2}{wz} + \dfrac{5}{2w^2}$
5. $(4a^3 - 8a^2 + a^2)(4a)^{-1}$ $a^2 - 2a + \dfrac{a}{4}$
6. $(28d^3k^2 + d^2k^2 - 4dk^2)(4dk^2)^{-1}$ $7d^2 + \dfrac{d}{4} - 1$
7. $\dfrac{f^2 + 7f + 10}{f + 2}$ $f + 5$
8. $\dfrac{2x^2 + 3x - 14}{x - 2}$ $2x + 7$
9. $(a^3 - 64) \div (a - 4)$ $a^2 + 4a + 16$
10. $(b^3 + 27) \div (b + 3)$ $b^2 - 3b + 9$
11. $\dfrac{2x^3 + 6x + 152}{x + 4}$ $2x^2 - 8x + 38$
12. $\dfrac{2x^3 + 4x - 6}{x + 3}$ $2x^2 - 6x + 22 - \dfrac{72}{x + 3}$
13. $(3w^3 + 7w^2 - 4w + 3) \div (w + 3)$ $3w^2 - 2w + 2 - \dfrac{3}{w + 3}$
14. $(6y^4 + 15y^3 - 28y - 6) \div (y + 2)$ $6y^3 + 3y^2 - 6y - 16 + \dfrac{26}{y + 2}$
15. $(x^4 - 3x^3 - 11x^2 + 3x + 10) \div (x - 5)$ $x^3 + 2x^2 - x - 2$
16. $(3m^5 + m - 1) \div (m + 1)$ $3m^4 - 3m^3 + 3m^2 - 3m + 4 - \dfrac{5}{m + 1}$
17. $(x^4 - 3x^3 + 5x - 6)(x + 2)^{-1}$ $x^3 - 5x^2 + 10x - 15 + \dfrac{24}{x + 2}$
18. $(6y^2 - 5y - 15)(2y + 3)^{-1}$ $3y - 7 + \dfrac{6}{2y + 3}$
19. $\dfrac{4x^2 - 2x + 6}{2x - 3}$ $2x + 2 + \dfrac{12}{2x - 3}$
20. $\dfrac{6x^2 - x - 7}{3x + 1}$ $2x - 1 - \dfrac{6}{3x + 1}$
21. $(2r^3 + 5r^2 - 2r - 15) \div (2r - 3)$ $r^2 + 4r + 5$
22. $(6t^3 + 5t^2 - 2t + 1) \div (3t + 1)$ $2t^2 + t - 1 + \dfrac{2}{3t + 1}$
23. $\dfrac{4p^4 - 17p^2 + 14p - 3}{2p - 3}$ $2p^3 + 3p^2 - 4p + 1$
24. $\dfrac{2h^4 - h^3 + h^2 + h - 3}{h^2 - 1}$ $2h^2 - h + 3$
25. **GEOMETRY** The area of a rectangle is $2x^2 - 11x + 15$ square feet. The length of the rectangle is $2x - 5$ feet. What is the width of the rectangle? $x - 3$ ft
26. **GEOMETRY** The area of a triangle is $15x^4 + 3x^3 + 4x^2 - x - 3$ square meters. The length of the base of the triangle is $6x^2 - 2$ meters. What is the height of the triangle? $5x^2 + x + 3$ m

Reading to Learn Mathematics, p. 255 ELL

Pre-Activity **How can you use division of polynomials in manufacturing?**

Read the introduction to Lesson 5-3 at the top of page 233 in your textbook.

Using the division symbol (÷), write the division problem that you would use to answer the question asked in the introduction. (Do not actually divide.) $(32x^2 + x) \div (8x)$

Reading the Lesson

1. a. Explain in words how to divide a polynomial by a monomial. Divide each term of the polynomial by the monomial.
   b. If you divide a trinomial by a monomial and get a polynomial, what kind of polynomial will the quotient be? trinomial
2. Look at the following division example that uses the division algorithm for polynomials.

```
         2x + 4
x − 4)2x² − 4x + 7
      2x² − 8x
            4x + 7
            4x − 16
                 23
```

   Which of the following is the correct way to write the quotient? C
   A. $2x + 4$ B. $x - 4$ C. $2x + 4 + \dfrac{23}{x - 4}$ D. $\dfrac{23}{x - 4}$
3. If you use synthetic division to divide $x^3 + 3x^2 - 5x - 8$ by $x - 2$, the division will look like this:

```
2| 1   3   −5   −8
       2   10   10
   1   5    5 |  2
```

   Which of the following is the answer for this division problem? B
   A. $x^2 + 5x + 5$ B. $x^2 + 5x + 5 + \dfrac{2}{x - 2}$
   C. $x^3 + 5x^2 + 5x + \dfrac{2}{x - 2}$ D. $x^3 + 5x^2 + 5x + 2$

Helping You Remember

4. When you translate the numbers in the last row of a synthetic division into the quotient and remainder, what is an easy way to remember which exponents to use in writing the terms of the quotient? Sample answer: Start with the power that is one less than the degree of the dividend. Decrease the power by one for each term after the first. The final number will be the remainder. Drop any term that is represented by a 0.

Enrichment, p. 256

Oblique Asymptotes

The graph of $y = ax + b$, where $a \neq 0$, is called an oblique asymptote of $y = f(x)$ if the graph of f comes closer and closer to the line as $x \to \infty$ or $x \to -\infty$. ∞ is the mathematical symbol for **infinity**, which means *endless*.

For $f(x) = 3x + 4 + \frac{2}{x}$, $y = 3x + 4$ is an oblique asymptote because $f(x) - 3x - 4 = \frac{2}{x}$, and $\frac{2}{x} \to 0$ as $x \to \infty$ or $-\infty$. In other words, as $|x|$ increases, the value of $\frac{2}{x}$ gets smaller and smaller approaching 0.

Example **Find the oblique asymptote for $f(x) = \dfrac{x^2 + 8x + 15}{x + 2}$.**

```
−2| 1   8    15     Use synthetic division.
       −2   −12
    1   6 |   3
```

$y = \dfrac{x^2 + 8x + 15}{x + 2} = x + 6 + \dfrac{3}{x + 2}$

4 Assess

Open-Ended Assessment

Writing Have students write their own list of tips for how to do division problems, describing the techniques they use to help avoid making errors.

Tips for New Teachers

Intervention Some students may have trouble keeping their concentration throughout the sequence of steps required in long division. Encourage them to compare intermediate results with a partner, so that they can ask questions and catch errors before completing the entire problem.

Getting Ready for Lesson 5-4

BASIC SKILL Lesson 5-4 presents factoring polynomials. This requires a knowledge of the greatest common factor. Exercises 69–74 should be used to determine your students' familiarity with the greatest common factor of a set of numbers.

Assessment Options

Practice Quiz 1 The quiz provides students with a brief review of the concepts and skills in Lessons 5-1 through 5-3. Lesson numbers are given to the right of exercises or instruction lines so students can review concepts not yet mastered.

Quiz (Lessons 5-1 through 5-3) is available on p. 307 of the *Chapter 5 Resource Masters*.

---

59. **WRITING IN MATH** Answer the question that was posed at the beginning of the lesson. **See pp. 283A–283B.**

How can you use division of polynomials in manufacturing?

Include the following in your answer:
- the dimensions of the piece of metal that the machinist needs,
- the formula from geometry that applies to this situation, and
- an explanation of how to use division of polynomials to find the width s of the seam.

Standardized Test Practice

60. An office employs x women and 3 men. What is the ratio of the total number of employees to the number of women? **A**

Ⓐ $1 + \frac{3}{x}$ Ⓑ $\frac{x}{x+3}$ Ⓒ $\frac{3}{x}$ Ⓓ $\frac{x}{3}$

61. If $a + b = c$ and $a = b$, then all of the following are true EXCEPT **D**

Ⓐ $a - c = b - c$. Ⓑ $a - b = 0$.
Ⓒ $2a + 2b = 2c$. Ⓓ $c - b = 2a$.

Maintain Your Skills

Mixed Review

Simplify. *(Lesson 5-2)* **62.** $-x^2 - 4x + 14$ **63.** $y^4z^4 - y^3z^3 + 3y^2z$

62. $(2x^2 - 3x + 5) - (3x^2 + x - 9)$
63. $y^2z(y^2z^3 - yz^2 + 3)$
64. $(y + 5)(y - 3)$ **$y^2 + 2y - 15$**
65. $(a - b)^2$ **$a^2 - 2ab + b^2$**

66. **ASTRONOMY** Earth is an average of 1.5×10^{11} meters from the Sun. Light travels at 3×10^8 meters per second. About how long does it take sunlight to reach Earth? *(Lesson 5-1)* **5×10^2 s or 8 min 20 s**

Write an equation in slope-intercept form for each graph. *(Lesson 2-4)*

67. 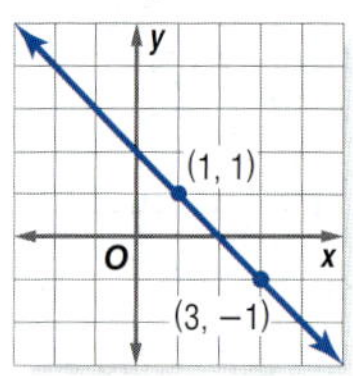

$y = -x + 2$

68. 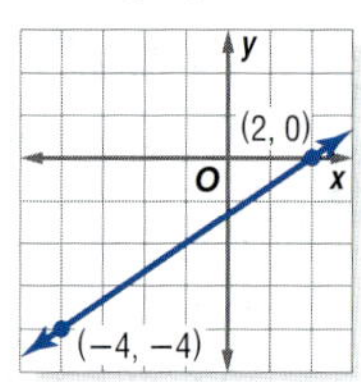

$y = \frac{2}{3}x - \frac{4}{3}$

Getting Ready for the Next Lesson

BASIC SKILL **Find the greatest common factor of each set of numbers.**

69. 18, 27 **9**
70. 24, 84 **12**
71. 16, 28 **4**
72. 12, 27, 48 **3**
73. 12, 30, 54 **6**
74. 15, 30, 65 **5**

Practice Quiz 1 — *Lessons 5-1 through 5-3*

Express each number in scientific notation. *(Lesson 5-1)*

1. 653,000,000 **6.53×10^8**
2. 0.0072 **7.2×10^{-3}**

Simplify. *(Lessons 5-1 and 5-2)*

3. $(-3x^2y)^3(2x)^2$ **$-108x^8y^3$**
4. $\frac{a^6b^{-2}c}{a^3b^2c^4}$ **$\frac{a^3}{b^4c^3}$**
5. $\left(\frac{x^2z}{xz^4}\right)^2$ **$\frac{x^2}{z^6}$**
6. $(9x + 2y) - (7x - 3y)$ **$2x + 5y$**
7. $(t + 2)(3t - 4)$ **$3t^2 + 2t - 8$**
8. $(n + 2)(n^2 - 3n + 1)$ **$n^3 - n^2 - 5n + 2$**

Simplify. *(Lesson 5-3)* **9.** $m^2 - 3 - \frac{19}{m-4}$

9. $(m^3 - 4m^2 - 3m - 7) \div (m - 4)$
10. $\frac{2d^3 - d^2 - 9d + 9}{2d - 3}$ **$d^2 + d - 3$**

5-4 Factoring Polynomials

What You'll Learn

- Factor polynomials.
- Simplify polynomial quotients by factoring.

How does factoring apply to geometry?

Suppose the expression $4x^2 + 10x - 6$ represents the area of a rectangle. Factoring can be used to find possible dimensions of the rectangle.

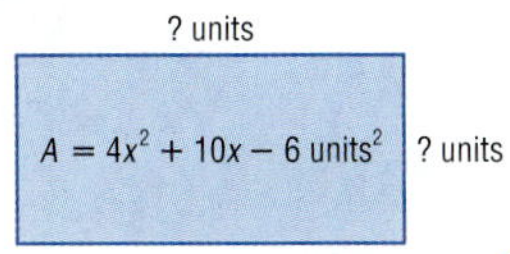

FACTOR POLYNOMIALS Whole numbers are factored using prime numbers. For example, $100 = 2 \cdot 2 \cdot 5 \cdot 5$. Many polynomials can also be factored. Their factors, however, are other polynomials. Polynomials that cannot be factored are called *prime*.

The table below summarizes the most common factoring techniques used with polynomials.

Concept Summary — *Factoring Techniques*

Number of Terms	Factoring Technique	General Case
any number	Greatest Common Factor (GCF)	$a^3b^2 + 2a^2b - 4ab^2 = ab(a^2b + 2a - 4b)$
two	Difference of Two Squares Sum of Two Cubes Difference of Two Cubes	$a^2 - b^2 = (a + b)(a - b)$ $a^3 + b^3 = (a + b)(a^2 - ab + b^2)$ $a^3 - b^3 = (a - b)(a^2 + ab + b^2)$
three	Perfect Square Trinomials	$a^2 + 2ab + b^2 = (a + b)^2$ $a^2 - 2ab + b^2 = (a - b)^2$
	General Trinomials	$acx^2 + (ad + bc)x + bd = (ax + b)(cx + d)$
four or more	Grouping	$ax + bx + ay + by = x(a + b) + y(a + b) = (a + b)(x + y)$

Whenever you factor a polynomial, always look for a common factor first. Then determine whether the resulting polynomial factor can be factored again using one or more of the methods listed in the table above.

Example 1 GCF

Factor $6x^2y^2 - 2xy^2 + 6x^3y$.

$$6x^2y^2 - 2xy^2 + 6x^3y = (2 \cdot 3 \cdot x \cdot x \cdot y \cdot y) - (2 \cdot x \cdot y \cdot y) + (2 \cdot 3 \cdot x \cdot x \cdot x \cdot y)$$
$$= (2xy \cdot 3xy) - (2xy \cdot y) + (2xy \cdot 3x^2)$$
$$= 2xy(3xy - y + 3x^2)$$

The GCF is $2xy$. The remaining polynomial cannot be factored using the methods above.

Check this result by finding the product.

A GCF is also used in grouping to factor a polynomial of four or more terms.

5-4 Lesson Notes

1 Focus

5-Minute Check Transparency 5-4 Use as a quiz or review of Lesson 5-3.

Mathematical Background notes are available for this lesson on p. 220D.

Building on Prior Knowledge

In this lesson, students will need to recall how to find the area of a rectangle, and they will also need to remember the set of prime numbers.

How does factoring apply to geometry?

Ask students:

- How do Examples 1 and 2 in Lesson 5-3 relate to factoring, the topic of this lesson? **The quotient and the divisor are factors of the dividend.**

2 Teach

FACTOR POLYNOMIALS

In-Class Example (Power Point®)

1 Factor $10a^3b^2 + 15a^2b - 5ab^3$. $5ab(2a^2b + 3a - b^2)$

Resource Manager

Workbook and Reproducible Masters

Chapter 5 Resource Masters
- Study Guide and Intervention, pp. 257–258
- Skills Practice, p. 259
- Practice, p. 260
- Reading to Learn Mathematics, p. 261
- Enrichment, p. 262

Graphing Calculator and Spreadsheet Masters, p. 35

Teaching Algebra With Manipulatives Masters, p. 235

Transparencies

5-Minute Check Transparency 5-4
Answer Key Transparencies

Technology

Alge2PASS: Tutorial Plus, Lesson 8
Interactive Chalkboard

In-Class Example

Power Point®

2 Factor $x^3 + 5x^2 - 2x - 10$. $(x + 5)(x^2 - 2)$

Teaching Tip Point out to students that it is often difficult to recognize that grouping can be used to factor a polynomial. Stress that this technique should only be considered when trying to factor a polynomial with four terms.

Example 2 Grouping

Factor $a^3 - 4a^2 + 3a - 12$.

$$\begin{aligned} a^3 - 4a^2 + 3a - 12 &= (a^3 - 4a^2) + (3a - 12) && \text{Group to find a GCF.} \\ &= a^2(a - 4) + 3(a - 4) && \text{Factor the GCF of each binomial.} \\ &= (a - 4)(a^2 + 3) && \text{Distributive Property} \end{aligned}$$

You can use algebra tiles to model factoring a polynomial.

Algebra Activity

Factoring Trinomials

Use algebra tiles to factor $2x^2 + 7x + 3$.

Model and Analyze

- Use algebra tiles to model $2x^2 + 7x + 3$.
- To find the product that resulted in this polynomial, arrange the tiles to form a rectangle.

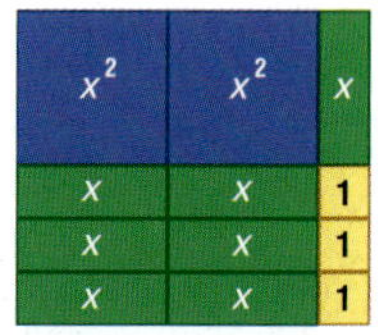

- Notice that the total area can be expressed as the sum of the areas of two smaller rectangles.

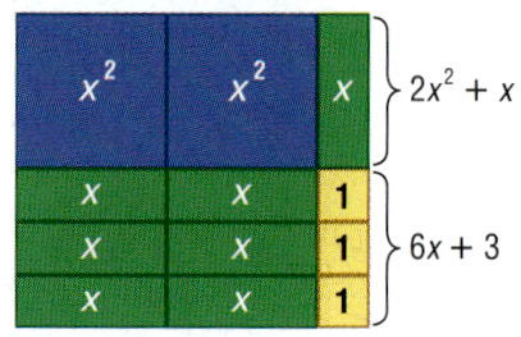

Use these expressions to rewrite the trinomial. Then factor.

$$\begin{aligned} 2x^2 + 7x + 3 &= (2x^2 + x) + (6x + 3) && \text{total area = sum of areas of smaller rectangles} \\ &= x(2x + 1) + 3(2x + 1) && \text{Factor out each GCF.} \\ &= (2x + 1)(x + 3) && \text{Distributive Property} \end{aligned}$$

Make a Conjecture

Study the factorization of $2x^2 + 7x + 3$ above.

1. What are the coefficients of the two x terms in $(2x^2 + x) + (6x + 3)$? Find their sum and their product. 1 and 6; 7; 6
2. Compare the sum you found in Exercise 1 to the coefficient of the x term in $2x^2 + 7x + 3$. They are the same.
3. Find the product of the coefficient of the x^2 term and the constant term in $2x^2 + 7x + 3$. How does it compare to the product in Exercise 1?
4. Make a conjecture about how to factor $3x^2 + 7x + 2$.

Study Tip

Algebra Tiles
When modeling a polynomial with algebra tiles, it is easiest to arrange the x^2 tiles first, then the x tiles and finally the 1 tiles to form a rectangle.

3. 6; It is the same.

4. Find two numbers with a product of $3 \cdot 2$ or 6 and a sum of 7. Use those numbers to rewrite the trinomial. Then factor.

The FOIL method can help you factor a polynomial into the product of two binomials. Study the following example.

$$\begin{aligned} (ax + b)(cx + d) &= \overbrace{ax \cdot cx}^{F} + \overbrace{ax \cdot d}^{O} + \overbrace{b \cdot cx}^{I} + \overbrace{b \cdot d}^{L} \\ &= acx^2 + (ad + bc)x + bd \end{aligned}$$

Notice that the product of the coefficient of x^2 and the constant term is $abcd$. The product of the two terms in the coefficient of x is also $abcd$.

Algebra Activity

Materials: algebra tiles

- Inform students that the two factors of the trinomial can be read directly from the completed array of tiles in another way. Point out that the length of the array is $2x + 1$, and the height is $x + 3$. The product of the length and width gives the area, $2x^2 + 7x + 3$, of the array.
- You might wish to have students experiment to see if there is another way to form a rectangle with the tiles.

Example 3 Two or Three Terms

Factor each polynomial.

a. $5x^2 - 13x + 6$

To find the coefficients of the x-terms, you must find two numbers whose product is $5 \cdot 6$ or 30, and whose sum is -13. The two coefficients must be -10 and -3 since $(-10)(-3) = 30$ and $-10 + (-3) = -13$.

Rewrite the expression using $-10x$ and $-3x$ in place of $-13x$ and factor by grouping.

$5x^2 - 13x + 6 = 5x^2 - 10x - 3x + 6$ Substitute $-10x - 3x$ for $-13x$.

$= (5x^2 - 10x) + (-3x + 6)$ Associative Property

$= 5x(x - 2) - 3(x - 2)$ Factor out the GCF of each group.

$= (5x - 3)(x - 2)$ Distributive Property

b. $3xy^2 - 48x$

$3xy^2 - 48x = 3x(y^2 - 16)$ Factor out the GCF.

$= 3x(y + 4)(y - 4)$ $y^2 - 16$ is the difference of two squares.

c. $c^3d^3 + 27$

$c^3d^3 = (cd)^3$ and $27 = 3^3$. Thus, this is the sum of two cubes.

$c^3d^3 + 27 = (cd + 3)[(cd)^2 - 3(cd) + 3^2]$ Sum of two cubes formula with $a = cd$ and $b = 3$

$= (cd + 3)(c^2d^2 - 3cd + 9)$ Simplify.

d. $m^6 - n^6$

This polynomial could be considered the difference of two squares or the difference of two cubes. The difference of two squares should always be done before the difference of two cubes. This will make the next step of the factorization easier.

$m^6 - n^6 = (m^3 + n^3)(m^3 - n^3)$ Difference of two squares

$= (m + n)(m^2 - mn + n^2)(m - n)(m^2 + mn + n^2)$ Sum and difference of two cubes

You can use a graphing calculator to check that the factored form of a polynomial is correct.

Graphing Calculator Investigation

Factoring Polynomials

Is the factored form of $2x^2 - 11x - 21$ equal to $(2x - 7)(x + 3)$? You can find out by graphing $y = 2x^2 - 11x - 21$ and $y = (2x - 7)(x + 3)$. If the two graphs coincide, the factored form is probably correct.

- Enter $y = 2x^2 - 11x - 21$ and $y = (2x - 7)(x + 3)$ on the Y= screen.
- Graph the functions. Since two different graphs appear, $2x^2 - 11x - 21 \neq (2x - 7)(x + 3)$.

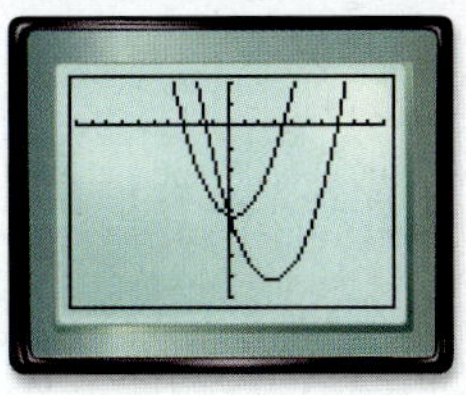

[−10, 10] scl: 1 by [−40, 10] scl: 5

Think and Discuss

1. Determine if $x^2 + 5x - 6 = (x - 3)(x - 2)$ is a true statement. If not, write the correct factorization. **no; $(x + 6)(x - 1)$**
2. Does this method guarantee a way to check the factored form of a polynomial? Why or why not?

2. No; in some cases, the graphs might be so close in shape that they seem to coincide but do not.

www.algebra2.com/extra_examples

In-Class Example

Power Point®

3 Factor each polynomial.

a. $3y^2 - 2y - 5$ **$(3y - 5)(y + 1)$**

b. $5mp^2 - 45m$ **$5m(p + 3)(p - 3)$**

c. $x^3y^3 + 8$ **$(xy + 2)(x^2y^2 - 2xy + 4)$**

d. $64x^6 - y^6$ **$(2x - y)(4x^2 + 2xy + y^2)(2x + y)(4x^2 - 2xy + y^2)$**

Teaching Tip Emphasize the importance of checking each factor to make sure it is prime before deciding that the final group of factors has been found.

Concept Check

Ask students to write an ordered list describing what they will check for as they factor a polynomial. **Sample answer: First look for any common factors of the terms; if there is a common factor, find the GCF. After factoring out the GCF, look for the difference of two squares, a perfect square trinomial, and so on. Use the factoring techniques listed on p. 239 to factor the expression further. Examine the resulting factored form to see if there are any factors that are not prime. If so, continue the process. If not, the factoring is complete.**

Graphing Calculator Investigation

Factoring Polynomials So that students see what happens when a polynomial and a correct factorization are graphed, have students graph the functions $y = x^2 - 81$ and $y = (x - 9)(x + 9)$ in the same screen. It looks like only one graph appears on the screen because both graphs are the same.

SIMPLIFY QUOTIENTS

In-Class Example

Power Point®

4 Simplify $\frac{a^2 - a - 6}{a^2 + 7a + 10}$.

$\frac{a - 3}{a + 5}$, if $a \neq -5, -2$

3 Practice/Apply

Study Notebook

Have students—

- add the definitions/examples of the vocabulary terms to their Vocabulary Builder worksheets for Chapter 5.
- add a list of factoring techniques to their notebook, including the factoring of the special cases listed in the Concept Summary on p. 239 and the FOIL method described on p. 240.
- include any other item(s) that they find helpful in mastering the skills in this lesson.

About the Exercises...

Organization by Objective

- **Factor Polynomials:** 15–45
- **Simplify Quotients:** 46–51

Odd/Even Assignments

Exercises 15–44 and 46–51 are structured so that students practice the same concepts whether they are assigned odd or even problems.

Assignment Guide

Basic: 15–37 odd, 43–49 odd, 55–58, 63–81

Average: 15–51 odd, 55–58, 63–81 (optional: 59–62)

Advanced: 16–50 even, 52–75 (optional: 76–81)

SIMPLIFY QUOTIENTS In Lesson 5-3, you learned to simplify the quotient of two polynomials by using long division or synthetic division. Some quotients can be simplified using factoring.

Example 4 Quotient of Two Trinomials

Simplify $\frac{x^2 + 2x - 3}{x^2 + 7x + 12}$.

$\frac{x^2 + 2x - 3}{x^2 + 7x + 12} = \frac{\overset{1}{\cancel{(x+3)}}(x - 1)}{(x + 4)\underset{1}{\cancel{(x+3)}}}$ Factor the numerator and denominator.

$= \frac{x - 1}{x + 4}$ Divide. Assume $x \neq -3, -4$.

Therefore, $\frac{x^2 + 2x - 3}{x^2 + 7x + 12} = \frac{x - 1}{x + 4}$, if $x \neq -3, -4$.

Check for Understanding

Concept Check

1. **OPEN ENDED** Write an example of a perfect square trinomial.
2. **Find a counterexample** to the statement $a^2 + b^2 = (a + b)^2$.
3. **Decide** whether the statement $\frac{x - 2}{x^2 + x - 6} = \frac{1}{x + 3}$ is *sometimes, always,* or *never* true. **sometimes**

1. Sample answer: $x^2 + 2x + 1$

2. Sample answer: If $a = 1$ and $b = 1$, then $a^2 + b^2 = 2$ but $(a + b)^2 = 4$.

Guided Practice

GUIDED PRACTICE KEY	
Exercises	Examples
4–5	1
6	2
7–11, 14	3
12–13	4

Factor completely. If the polynomial is not factorable, write *prime*.

4. $-12x^2 - 6x$ $-6x(2x + 1)$
5. $a^2 + 5a + ab$ $a(a + 5 + b)$
6. $21 - 7y + 3x - xy$ $(x + 7)(3 - y)$
7. $y^2 - 6y + 8$ $(y - 2)(y - 4)$
8. $z^2 - 4z - 12$ $(z - 6)(z + 2)$
9. $3b^2 - 48$ $3(b - 4)(b + 4)$
10. $16w^2 - 169$ $(4w + 13)(4w - 13)$
11. $h^3 + 8000$ $(h + 20)(h^2 - 20h + 400)$

Simplify. Assume that no denominator is equal to 0.

12. $\frac{x^2 - 2x - 8}{x^2 - 5x - 14}$ $\frac{x - 4}{x - 7}$
13. $\frac{2y^2 + 8y}{y^2 - 16}$ $\frac{2y}{y - 4}$

Application

14. **GEOMETRY** Find the width of rectangle *ABCD* if its area is $3x^2 + 9xy + 6y^2$ square centimeters. $x + y$ cm

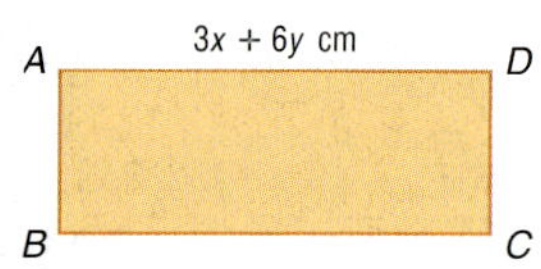

★ indicates increased difficulty

Practice and Apply

Factor completely. If the polynomial is not factorable, write *prime*.

15. $2xy^3 - 10x$ $2x(y^3 - 5)$
16. $6a^2b^2 + 18ab^3$ $6ab^2(a + 3b)$
17. $12cd^3 - 8c^2d^2 + 10c^5d^3$
18. $3a^2bx + 15cx^2y + 25ad^3y$ prime
19. $8yz - 6z - 12y + 9$ $(2z - 3)(4y - 3)$
20. $3ax - 15a + x - 5$ $(3a + 1)(x - 5)$
21. $x^2 + 7x + 6$ $(x + 1)(x + 6)$
22. $y^2 - 5y + 4$ $(y - 1)(y - 4)$
23. $2a^2 + 3a + 1$ $(2a + 1)(a + 1)$
24. $2b^2 + 13b - 7$ $(2b - 1)(b + 7)$
25. $6c^2 + 13c + 6$ $(2c + 3)(3c + 2)$
26. $12m^2 - m - 6$ $(3m + 2)(4m - 3)$
27. $3n^2 + 21n - 24$ $3(n + 8)(n - 1)$
28. $3z^2 + 24z + 45$ $3(z + 3)(z + 5)$

17. $2cd^2(6d - 4c + 5c^4d)$

DAILY INTERVENTION

Differentiated Instruction

Auditory/Musical Ask students to create songs or raps to help them remember the factoring techniques for the difference of two squares, for the sum or difference of two cubes, or for one of the two perfect square trinomial types.

242 Chapter 5 Polynomials

Homework Help	
For Exercises	See Examples
15–18	1
19, 20	2
21–38, 43–45, 55	3
39–42	2, 3
46–54	4

Extra Practice
See page 837.

29. $x^2 + 12x + 36$ $(x + 6)^2$
30. $x^2 - 6x + 9$ $(x - 3)^2$
31. $16a^2 + 25b^2$ prime
32. $3m^2 - 3n^2$ $3(m + n)(m - n)$
33. $y^4 - z^2$ $(y^2 + z)(y^2 - z)$
34. $3x^2 - 27y^2$ $3(x + 3y)(x - 3y)$
35. $z^3 + 125$ $(z + 5)(z^2 - 5z + 25)$
36. $t^3 - 8$ $(t - 2)(t^2 + 2t + 4)$
37. $p^4 - 1$ $(p^2 + 1)(p + 1)(p - 1)$
38. $x^4 - 81$ $(x^2 + 9)(x + 3)(x - 3)$

★ 39. $7ac^2 + 2bc^2 - 7ad^2 - 2bd^2$ $(7a + 2b)(c + d)(c - d)$

★ 40. $8x^2 + 8xy + 8xz + 3x + 3y + 3z$ $(8x + 3)(x + y + z)$

★ 41. $5a^2x + 4aby + 3acz - 5abx - 4b^2y - 3bcz$ $(a - b)(5ax + 4by + 3cz)$

★ 42. $3a^3 + 2a^2 - 5a + 9a^2b + 6ab - 15b$ $(a + 3b)(3a + 5)(a - 1)$

43. Find the factorization of $3x^2 + x - 2$. $(3x - 2)(x + 1)$

44. What are the factors of $2y^2 + 9y + 4$? $(2y + 1)(y + 4)$

45. **LANDSCAPING** A boardwalk that is x feet wide is built around a rectangular pond. The combined area of the pond and the boardwalk is $4x^2 + 140x + 1200$ square feet. What are the dimensions of the pond? 30 ft by 40 ft

Simplify. Assume that no denominator is equal to 0.

46. $\dfrac{x^2 + 4x + 3}{x^2 - x - 12}$ $\dfrac{x + 1}{x - 4}$
47. $\dfrac{x^2 + 4x - 5}{x^2 - 7x + 6}$ $\dfrac{x + 5}{x - 6}$
48. $\dfrac{x^2 - 25}{x^2 + 3x - 10}$ $\dfrac{x - 5}{x - 2}$
49. $\dfrac{x^2 - 6x + 8}{x^3 - 8}$ $\dfrac{x - 4}{x^2 + 2x + 4}$

★ 50. $\dfrac{x^2}{(x^2 - x)(x - 1)^{-1}}$ x

★ 51. $\dfrac{x + 1}{(x^2 + 3x + 2)(x + 2)^{-2}}$ $x + 2$

BUILDINGS **For Exercises 52 and 53, use the following information.**
When an object is dropped from a tall building, the distance it falls between 1 second after it is dropped and x seconds after it is dropped is $16x^2 - 16$ feet.

52. How much time elapses between 1 second after it is dropped and x seconds after it is dropped? $x - 1$ s

53. What is the average speed of the object during that time period? $16x + 16$ ft/s

54. **GEOMETRY** The length of one leg of a right triangle is $x - 6$ centimeters, and the area is $\frac{1}{2}x^2 - 7x + 24$ square centimeters. What is the length of the other leg? $x - 8$ cm

55. **CRITICAL THINKING** Factor $64p^{2n} + 16p^n + 1$. $(8p^n + 1)^2$

56. **WRITING IN MATH** Answer the question that was posed at the beginning of the lesson. See pp. 283A–283B.

How does factoring apply to geometry?

Include the following in your answer:
- an explanation of how to use factoring to find possible dimensions for the rectangle described at the beginning of the lesson, and
- why your dimensions are not the only ones possible, even if you assume that the dimensions are binomials with integer coefficients.

More About. . .

Buildings

The tallest buildings in the world are the Petronas Towers in Kuala Lumpur, Malaysia. Each is 1483 feet tall.

Source: www.worldstallest.com

www.algebra2.com/self_check_quiz

Study Guide and Intervention, p. 257 (shown) and p. 258

Factor Polynomials

Techniques for Factoring Polynomials	
	For any number of terms, check for: greatest common factor
	For two terms, check for: Difference of two squares $a^2 - b^2 = (a + b)(a - b)$; Sum of two cubes $a^3 + b^3 = (a + b)(a^2 - ab + b^2)$; Difference of two cubes $a^3 - b^3 = (a - b)(a^2 + ab + b^2)$
	For three terms, check for: Perfect square trinomials $a^2 + 2ab + b^2 = (a + b)^2$, $a^2 - 2ab + b^2 = (a - b)^2$; General trinomials $acx^2 + (ad + bc)x + bd = (ax + b)(cx + d)$
	For four terms, check for: Grouping $ax + bx + ay + by = x(a + b) + y(a + b) = (a + b)(x + y)$

Example **Factor $24x^2 - 42x - 45$.**

First factor out the GCF to get $24x^2 - 42x - 45 = 3(8x^2 - 14x - 15)$. To find the coefficients of the x terms, you must find two numbers whose product is $8 \cdot (-15) = -120$ and whose sum is -14. The two coefficients must be -20 and 6. Rewrite the expression using $-20x$ and $6x$ and factor by grouping.

$8x^2 - 14x - 15 = 8x^2 - 20x + 6x - 15$ Group to find a GCF.
$= 4x(2x - 5) + 3(2x - 5)$ Factor the GCF of each binomial.
$= (4x + 3)(2x - 5)$ Distributive Property

Thus, $24x^2 - 42x - 45 = 3(4x + 3)(2x - 5)$.

Exercises

Factor completely. If the polynomial is not factorable, write *prime*.

1. $14x^2y^2 + 42xy^3$ $14xy^2(x + 3y)$
2. $6mn + 18m - n - 3$ $(6m - 1)(n + 3)$
3. $2x^2 + 18x + 16$ $2(x + 8)(x + 1)$
4. $x^4 - 1$ $(x^2 + 1)(x + 1)(x - 1)$
5. $35x^3y^4 - 60x^4y$ $5x^3y(7y^3 - 12x)$
6. $2r^3 + 250$ $2(r + 5)(r^2 - 5r + 25)$
7. $100m^8 - 9$ $(10m^4 - 3)(10m^4 + 3)$
8. $x^2 + x + 1$ prime
9. $c^4 + c^3 - c^2 - c$ $c(c + 1)^2(c - 1)$

Skills Practice, p. 259 and Practice, p. 260 (shown)

Factor completely. If the polynomial is not factorable, write *prime*.

1. $15a^2b - 10ab^2$ $5ab(3a - 2b)$
2. $3st^2 - 9s^3t + 6s^2t^2$ $3st(t - 3s^2 + 2st)$
3. $3x^3y^2 - 2x^2y + 5xy$ $xy(3x^2y - 2x + 5)$
4. $2x^3y - x^2y + 5xy^2 + xy^3$ $xy(2x^2 - x + 5y + y^2)$
5. $21 - 7t + 3r - rt$ $(7 + r)(3 - t)$
6. $x^2 - xy + 2x - 2y$ $(x + 2)(x - y)$
7. $y^2 + 20y + 96$ $(y + 8)(y + 12)$
8. $4ab + 2a + 6b + 3$ $(2a + 3)(2b + 1)$
9. $6n^2 - 11n - 2$ $(6n + 1)(n - 2)$
10. $6x^2 + 7x - 3$ $(3x - 1)(2x + 3)$
11. $x^2 - 8x - 8$ prime
12. $6p^2 - 17p - 45$ $(2p - 9)(3p + 5)$
13. $r^3 + 3r^2 - 54r$ $r(r + 9)(r - 6)$
14. $8a^2 + 2a - 6$ $2(4a - 3)(a + 1)$
15. $c^2 - 49$ $(c - 7)(c + 7)$
16. $x^3 + 8$ $(x + 2)(x^2 - 2x + 4)$
17. $16r^2 - 169$ $(4r + 13)(4r - 13)$
18. $b^4 - 81$ $(b^2 + 9)(b + 3)(b - 3)$
19. $8m^3 - 25$ prime
20. $2t^3 + 32t^2 + 128t$ $2t(t + 8)^2$
21. $5y^5 + 135y^2$ $5y^2(y + 3)(y^2 - 3y + 9)$
22. $81x^4 - 16$ $(9x^2 + 4)(3x + 2)(3x - 2)$

Simplify. Assume that no denominator is equal to 0.

23. $\dfrac{x^2 - 16}{x^2 + x - 20}$ $\dfrac{x + 4}{x + 5}$
24. $\dfrac{x^2 - 16x + 64}{x^2 + x - 72}$ $\dfrac{x - 8}{x + 9}$
25. $\dfrac{3x^2 - 27}{x^3 - 27}$ $\dfrac{3(x + 3)}{x^2 + 3x + 9}$

26. **DESIGN** Bobbi Jo is using a software package to create a drawing of a cross section of a brace as shown at the right. Write a simplified, factored expression that represents the area of the cross section of the brace. $x(20.2 - x)$ cm^2

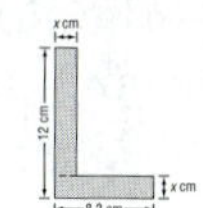

27. **COMBUSTION ENGINES** In an internal combustion engine, the up and down motion of the pistons is converted into the rotary motion of the crankshaft, which drives the flywheel. Let r_1 represent the radius of the flywheel at the right and let r_2 represent the radius of the crankshaft passing through it. If the formula for the area of a circle is $A = \pi r^2$, write a simplified, factored expression for the area of the cross section of the flywheel outside the crankshaft. $\pi(r_1 - r_2)(r_1 + r_2)$

Reading to Learn Mathematics, p. 261 ELL

Pre-Activity **How does factoring apply to geometry?**

Read the introduction to Lesson 5-4 at the top of page 239 in your textbook.

If a trinomial that represents the area of a rectangle is factored into two binomials, what might the two binomials represent? the length and width of the rectangle

Reading the Lesson

1. Name three types of binomials that it is always possible to factor. difference of two squares, sum of two cubes, difference of two cubes

2. Name a type of trinomial that it is always possible to factor. perfect square trinomial

3. Complete: Since $x^2 + y^2$ cannot be factored, it is an example of a ___prime___ polynomial.

4. On an algebra quiz, Marlene needed to factor $2x^2 - 4x - 70$. She wrote the following answer: $(x + 5)(2x - 14)$. When she got her quiz back, Marlene found that she did not get full credit for her answer. She thought she should have gotten full credit because she checked her work by multiplication and showed that $(x + 5)(2x - 14) = 2x^2 - 4x - 70$.

a. If you were Marlene's teacher, how would you explain to her that her answer was not entirely correct? Sample answer: When you are asked to factor a polynomial, you must factor it *completely*. The factorization was not complete, because $2x - 14$ can be factored further as $2(x - 7)$.

b. What advice could Marlene's teacher give her to avoid making the same kind of error in factoring in the future? Sample answer: Always look for a common factor first. If there is a common factor, factor it out first, and then see if you can factor further.

Helping You Remember

5. Some students have trouble remembering the correct signs in the formulas for the sum and difference of two cubes. What is an easy way to remember the correct signs? Sample answer: In the binomial factor, the operation sign is the same as in the expression that is being factored. In the trinomial factor, the operation sign before the middle term is the opposite of the sign in the expression that is being factored. The sign before the last term is always a plus.

Enrichment, p. 262

Using Patterns to Factor

Study the patterns below for factoring the sum and the difference of cubes.

$a^3 + b^3 = (a + b)(a^2 - ab + b^2)$
$a^3 - b^3 = (a - b)(a^2 + ab + b^2)$

This pattern can be extended to other odd powers. Study these examples.

Example 1 **Factor $a^5 + b^5$.**

Extend the first pattern to obtain $a^5 + b^5 = (a + b)(a^4 - a^3b + a^2b^2 - ab^3 + b^4)$.

Check: $(a + b)(a^4 - a^3b + a^2b^2 - ab^3 + b^4) = a^5 - a^4b + a^3b^2 - a^2b^3 + ab^4 + a^4b - a^3b^2 + a^2b^3 - ab^4 + b^5 = a^5 + b^5$

Example 2 **Factor $a^5 - b^5$.**

Extend the second pattern to obtain $a^5 - b^5 = (a - b)(a^4 + a^3b + a^2b^2 + ab^3 + b^4)$.

Check: $(a - b)(a^4 + a^3b + a^2b^2 + ab^3 + b^4) = a^5 + a^4b + a^3b^2 + a^2b^3 + ab^4$

4 Assess

Open-Ended Assessment

Modeling Have students use algebra tiles to model and determine the factored form of the polynomial $3x^2 + 11x + 6$. $(3x + 2)(x + 3)$

Tips for New Teachers

Intervention Point out that factoring involving $a^2 - b^2$, $a^3 - b^3$, $a^3 + b^3$, $(a - b)^2$, and $(a + b)^2$ occurs so frequently in algebra that students should memorize these forms. Then students can easily recognize and use them whenever the need arises.

Getting Ready for Lesson 5-5

PREREQUISITE SKILL Lesson 5-5 discusses radicals. Many radicals are irrational numbers whose approximate value can be found using a calculator. Exercises 76–81 should be used to determine your students' familiarity with rational and irrational numbers.

Standardized Test Practice

57. Which of the following is the factorization of $2x - 15 + x^2$? **B**

   (A) $(x - 3)(x - 5)$ (B) $(x - 3)(x + 5)$
   (C) $(x + 3)(x - 5)$ (D) $(x + 3)(x + 5)$

58. Which is not a factor of $x^3 - x^2 - 2x$? **C**

   (A) x (B) $x + 1$ (C) $x - 1$ (D) $x - 2$

Graphing Calculator

CHECK FACTORING Use a graphing calculator to determine if each polynomial is factored correctly. Write *yes* or *no*. If the polynomial is not factored correctly, find the correct factorization. **60. no; $(x + 2)(x^2 - 2x + 4)$**

59. $3x^2 + 5x + 2 \stackrel{?}{=} (3x + 2)(x + 1)$ **yes**
60. $x^3 + 8 \stackrel{?}{=} (x + 2)(x^2 - x + 4)$
61. $2x^2 - 5x - 3 \stackrel{?}{=} (x - 1)(2x + 3)$ **no; $(2x + 1)(x - 3)$**
62. $3x^2 - 48 \stackrel{?}{=} 3(x + 4)(x - 4)$ **yes**

Maintain Your Skills

Mixed Review

Simplify. *(Lesson 5-3)*

63. $(t^3 - 3t + 2) \div (t + 2)$ **$t^2 - 2t + 1$**
64. $(y^2 + 4y + 3)(y + 1)^{-1}$ **$y + 3$**
65. $\dfrac{x^3 - 3x^2 + 2x - 6}{x - 3}$ **$x^2 + 2$**
66. $\dfrac{3x^4 + x^3 - 8x^2 + 10x - 3}{3x - 2}$ **$x^3 + x^2 - 2x + 2 + \dfrac{1}{3x - 2}$**

Simplify. *(Lesson 5-2)*

67. $(3x^2 - 2xy + y^2) + (x^2 + 5xy - 4y^2)$ **$4x^2 + 3xy - 3y^2$**
68. $(2x + 4)(7x - 1)$ **$14x^2 + 26x - 4$**

Perform the indicated operations, if possible. *(Lesson 4-5)*

69. $[3 \quad -1] \cdot \begin{bmatrix} 0 \\ 2 \end{bmatrix}$ **$[-2]$**
70. $\begin{bmatrix} 1 & -4 \\ 2 & 2 \end{bmatrix} \cdot \begin{bmatrix} 0 & 3 \\ 9 & -1 \end{bmatrix}$ **$\begin{bmatrix} -36 & 7 \\ 18 & 4 \end{bmatrix}$**

71. **PHOTOGRAPHY** The perimeter of a rectangular picture is 86 inches. Twice the width exceeds the length by 2 inches. What are the dimensions of the picture? *(Lesson 3-2)* **15 in. by 28 in.**

Determine whether each relation is a function. Write *yes* or *no*. *(Lesson 2-1)*

72. 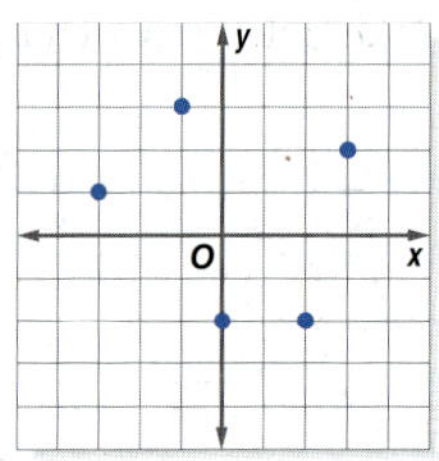

yes

73. 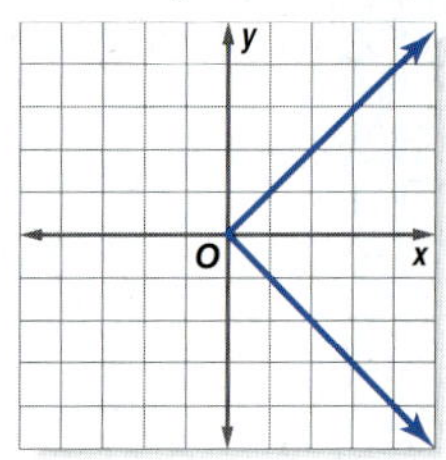

no

State the property illustrated by each equation. *(Lesson 1-2)*

74. $(3 + 8)5 = 3(5) + 8(5)$ **Distributive**
75. $1 + (7 + 4) = (1 + 7) + 4$ **Associative (+)**

Getting Ready for the Next Lesson

PREREQUISITE SKILL Determine whether each number is *rational* or *irrational*. *(To review **rational and irrational numbers**, see Lesson 1-2.)* **80. irrational**

76. 4.63 **rational**
77. π **irrational**
78. $\frac{16}{3}$ **rational**
79. 8.333... **rational**
80. 7.323223222...
81. $9.7\overline{1}$ **rational**

DAILY INTERVENTION — **Unlocking Misconceptions**

- **Difference of Two Squares** Many people think that the expressions $a^2 - b^2$ and $(a - b)^2$ are the same. Have students choose values for a and b, such as $a = 5$ and $b = 3$, to see that this is not true.
- **Sum of Two Squares** Students may need to be convinced that $a^2 + b^2$ cannot be factored after seeing that $a^3 + b^3$ can be factored. Have them substitute values for a and b to test possible factored forms, such as $(a + b)(a + b)$, to verify they do not equal $a^2 + b^2$.

5-5 Roots of Real Numbers

What You'll Learn

- Simplify radicals.
- Use a calculator to approximate radicals.

Vocabulary

- square root
- *n*th root
- principal root

How do square roots apply to oceanography?

The speed s in knots of a wave can be estimated using the formula $s = 1.34\sqrt{\ell}$, where ℓ is the length of the wave in feet. This is an example of an equation that contains a square root.

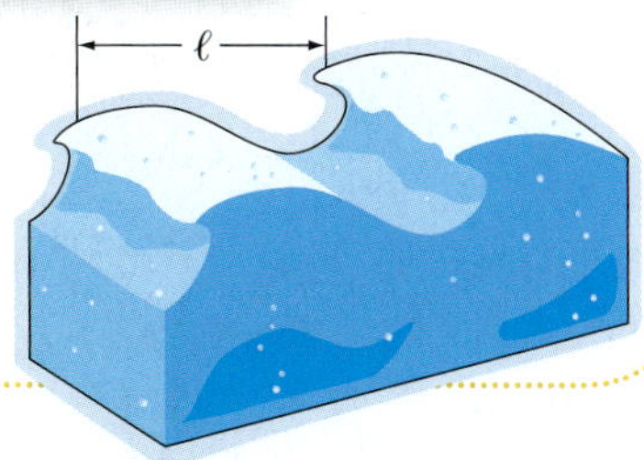

SIMPLIFY RADICALS Finding the square root of a number and squaring a number are inverse operations. To find the **square root** of a number n, you must find a number whose square is n. For example, 7 is a square root of 49 since $7^2 = 49$. Since $(-7)^2 = 49$, -7 is also a square root of 49.

Key Concept — *Definition of Square Root*

- **Words** For any real numbers a and b, if $a^2 = b$, then a is a square root of b.
- **Example** Since $5^2 = 25$, 5 is a square root of 25.

Since finding the square root of a number and squaring a number are inverse operations, it makes sense that the inverse of raising a number to the nth power is finding the **nth root** of a number. The table below shows the relationship between raising a number to a power and taking that root of a number.

Powers	Factors	Roots
$a^3 = 125$	$5 \cdot 5 \cdot 5 = 125$	5 is a cube root of 125.
$a^4 = 81$	$3 \cdot 3 \cdot 3 \cdot 3 = 81$	3 is a fourth root of 81.
$a^5 = 32$	$2 \cdot 2 \cdot 2 \cdot 2 \cdot 2 = 32$	2 is a fifth root of 32.
$a^n = b$	$\underbrace{a \cdot a \cdot a \cdot a \cdot \ldots \cdot a}_{n \text{ factors of } a} = b$	a is an nth root of b.

This pattern suggests the following formal definition of an nth root.

Key Concept — *Definition of nth Root*

- **Words** For any real numbers a and b, and any positive integer n, if $a^n = b$, then a is an nth root of b.
- **Example** Since $2^5 = 32$, 2 is a fifth root of 32.

5-5 Lesson Notes

1 Focus

5-Minute Check Transparency 5-5 Use as a quiz or review of Lesson 5-4.

Mathematical Background notes are available for this lesson on p. 220D.

How do square roots apply to oceanography?

Ask students:

- One *knot* means one nautical mile per hour and one nautical mile is about 6076 feet. One mile on land (called a statute mile) is 5280 feet. Which is faster, 1 knot or 1 statute mile per hour? **1 knot**
- As the length of a wave (represented by ℓ in the diagram) increases, does the speed of the wave increase or decrease? **increases**

Resource Manager

Workbook and Reproducible Masters

Chapter 5 Resource Masters

- Study Guide and Intervention, pp. 263–264
- Skills Practice, p. 265
- Practice, p. 266
- Reading to Learn Mathematics, p. 267
- Enrichment, p. 268
- Assessment, pp. 307, 309

Teaching Algebra With Manipulatives Masters, pp. 236–237

Transparencies

5-Minute Check Transparency 5-5
Answer Key Transparencies

Technology

Interactive Chalkboard

2 Teach

SIMPLIFY RADICALS

In-Class Example

Power Point®

1 Simplify.

a. $\pm\sqrt{16x^6}$ $\pm 4x^3$

b. $-\sqrt{(q^3+5)^4}$ $-(q^3+5)^2$

c. $\sqrt[5]{243a^{10}b^{15}}$ $3a^2b^3$

d. $\sqrt{-4}$ not a real number

Teaching Tip Be sure students understand that since $3^2 = 9$ and $(-3)^2 = 9$, then the equation $x^2 = 9$ has two roots, 3 and -3. However, the value of the expression $\sqrt{9}$ is 3 only. To indicate *both* square roots and not just the principal root, the expression must be given as $\pm\sqrt{9}$.

Teaching Tip When discussing the information following Example 1, offer this alternative. Another way to simplify radicals that involve only numbers and no variables, is to simplifying the expression under the radical sign first. For example, $\sqrt{(-5)^2}$ could be rewritten by *first* simplifying under the radical sign to get $\sqrt{25}$, and then taking the principal root to get 5. Similarly, $\sqrt{(-2)^6}$ simplifies to $\sqrt{64}$, whose principal square root is 8.

Tips for New Teachers

Reading Tip Make sure that students understand what it means to say that the radical sign $\sqrt{\ }$ designates the principal root.

Study Tip

Reading Math
$\sqrt[n]{50}$ is read *the nth root of 50.*

The symbol $\sqrt[n]{\ }$ indicates an *n*th root.

index → $\sqrt[n]{50}$ ← radical sign, radicand

Some numbers have more than one real *n*th root. For example, 36 has two square roots, 6 and -6. When there is more than one real root, the nonnegative root is called the **principal root**. When no index is given, as in $\sqrt{36}$, the radical sign indicates the principal square root. The symbol $\sqrt[n]{b}$ stands for the principal *n*th root of *b*. If *n* is odd and *b* is negative, there will be no nonnegative root. In this case, the principal root is negative.

$\sqrt{16} = 4$ — $\sqrt{16}$ indicates the principal square root of 16.

$-\sqrt{16} = -4$ — $-\sqrt{16}$ indicates the opposite of the principal square root of 16.

$\pm\sqrt{16} = \pm 4$ — $\pm\sqrt{16}$ indicates both square roots of 16. $\pm$ means positive or negative.

$\sqrt[3]{-125} = -5$ — $\sqrt[3]{-125}$ indicates the principal cube root of -125.

$-\sqrt[4]{81} = -3$ — $-\sqrt[4]{81}$ indicates the opposite of the principal fourth root of 81.

Concept Summary — *Real nth roots of b, $\sqrt[n]{b}$, or $-\sqrt[n]{b}$*

n	$\sqrt[n]{b}$ if $b > 0$	$\sqrt[n]{b}$ if $b < 0$	$b = 0$
even	one positive root, one negative root $\pm\sqrt[4]{625} = \pm 5$	no real roots $\sqrt{-4}$ is not a real number.	one real root, 0 $\sqrt[n]{0} = 0$
odd	one positive root, no negative roots $\sqrt[3]{8} = 2$	no positive roots, one negative root $\sqrt[5]{-32} = -2$	

Example 1 Find Roots

Simplify.

a. $\pm\sqrt{25x^4}$

$\pm\sqrt{25x^4} = \pm\sqrt{(5x^2)^2}$
$= \pm 5x^2$

The square roots of $25x^4$ are $\pm 5x^2$.

b. $-\sqrt{(y^2+2)^8}$

$-\sqrt{(y^2+2)^8} = -\sqrt{[(y^2+2)^4]^2}$
$= -(y^2+2)^4$

The opposite of the principal square root of $(y^2+2)^8$ is $-(y^2+2)^4$.

c. $\sqrt[5]{32x^{15}y^{20}}$

$\sqrt[5]{32x^{15}y^{20}} = \sqrt[5]{(2x^3y^4)^5}$
$= 2x^3y^4$

The principal fifth root of $32x^{15}y^{20}$ is $2x^3y^4$.

d. $\sqrt{-9}$

$\sqrt{-9} = \sqrt[2]{-9}$ ← *n* is even. ← *b* is negative.

Thus, $\sqrt{-9}$ is not a real number.

When you find the *n*th root of an even power and the result is an odd power, you must take the absolute value of the result to ensure that the answer is nonnegative.

$$\sqrt{(-5)^2} = |-5| \text{ or } 5 \qquad \sqrt{(-2)^6} = |(-2)^3| \text{ or } 8$$

If the result is an even power or you find the *n*th root of an odd power, there is no need to take the absolute value. *Why?*

DAILY INTERVENTION

Unlocking Misconceptions

- **Variables** Some students tend to think that x must represent a positive number and $-x$ must represent a negative number. Reading $-x$ as "the opposite of x" should help them understand that $-x$ is 5 if $x = -5$.
- **Square Roots of Negative Numbers** Explain that -9 has no square root that is a real number. That is, no real number can be squared to give -9. However, inform students that $\sqrt{-9}$ *does* represent a number, called an *imaginary number*. Lesson 5-9 discusses such numbers.

Example 2 Simplify Using Absolute Value

Simplify.

a. $\sqrt[8]{x^8}$

Note that x is an eighth root of x^8. The index is even, so the principal root is nonnegative. Since x could be negative, you must take the absolute value of x to identify the principal root.

$\sqrt[8]{x^8} = |x|$

b. $\sqrt[4]{81(a+1)^{12}}$

$\sqrt[4]{81(a+1)^{12}} = \sqrt[4]{[3(a+1)^3]^4}$

Since the index 4 is even and the exponent 3 is odd, you must use the absolute value of $(a+1)^3$.

$\sqrt[4]{81(a+1)^{12}} = 3|(a+1)^3|$

In-Class Example

Power Point®

2 Simplify.

a. $\sqrt[6]{t^6}$ $|t|$

b. $\sqrt[5]{243(x+2)^{15}}$ $3(x+2)^3$

APPROXIMATE RADICALS WITH A CALCULATOR

Recall that real numbers that cannot be expressed as terminating or repeating decimals are *irrational numbers*. $\sqrt{2}$ and $\sqrt{3}$ are examples of irrational numbers. Decimal approximations for irrational numbers are often used in applications.

Example 3 Approximate a Square Root

PHYSICS **The time T in seconds that it takes a pendulum to make a complete swing back and forth is given by the formula $T = 2\pi\sqrt{\frac{L}{g}}$, where L is the length of the pendulum in feet and g is the acceleration due to gravity, 32 feet per second squared. Find the value of T for a 3-foot-long pendulum.**

Explore You are given the values of L and g and must find the value of T. Since the units on g are feet per second squared, the units on the time T should be seconds.

Plan Substitute the values for L and g into the formula. Use a calculator to evaluate.

Solve

$T = 2\pi\sqrt{\frac{L}{g}}$ Original formula

$= 2\pi\sqrt{\frac{3}{32}}$ $L = 3, g = 32$

≈ 1.92 Use a calculator.

It takes the pendulum about 1.92 seconds to make a complete swing.

Examine The closest square to $\frac{3}{32}$ is $\frac{1}{9}$, and π is approximately 3, so the answer should be close to $2(3)\sqrt{\frac{1}{9}} = 2(3)\left(\frac{1}{3}\right)$ or 2. The answer is reasonable.

Study Tip

Graphing Calculators

To find a root of index greater than 2, first type the index. Then select $\sqrt[x]{\ }$ from the MATH MATH menu. Finally, enter the radicand.

APPROXIMATE RADICALS WITH A CALCULATOR

In-Class Example

Power Point®

3 **PHYSICS** Use the formula given in Example 3. Find the value of T for a 1.5-foot-long pendulum. **about 1.36 s**

Teaching Tip Help students see how time and the length of a pendulum are related by having them experiment using weights with different lengths of string.

3 Practice/Apply

Check for Understanding

Concept Check

1. **OPEN ENDED** Write a number whose principal square root and cube root are both integers. **Sample answer: 64**
2. **Explain** why it is not always necessary to take the absolute value of a result to indicate the principal root. **See margin.**
3. **Determine** whether the statement $\sqrt[4]{(-x)^4} = x$ is *sometimes*, *always*, or *never* true. Explain your reasoning. **Sometimes; it is true when $x > 0$.**

Study Notebook

Have students—

- add the definitions/examples of the vocabulary terms to their Vocabulary Builder worksheets for Chapter 5.
- keep a list of study tips for the graphing calculator, including the one in this lesson.
- include any other item(s) that they find helpful in mastering the skills in this lesson.

www.algebra2.com/extra_examples

Answer

2. If all of the powers in the result of an even root have even exponents, the result is nonnegative without taking absolute value.

DAILY INTERVENTION — Differentiated Instruction

Visual/Spatial Have students create various rectangles using index cards or cardboard, or using masking tape on the classroom floor. Have them use the formula $d = \sqrt{\ell^2 + w^2}$ to find the length of the diagonal of each rectangle. After creating several rectangles, have students experiment with using the diagonal measures of two rectangles to create another rectangle whose length and width are irrational numbers. Students should then find the length of the diagonal of this new rectangle.

Study Guide and Intervention, p. 263 (shown) and p. 264

Simplify Radicals

Square Root	For any real numbers a and b, if $a^2 = b$, then a is a square root of b.
nth Root	For any real numbers a and b, and any positive integer n, if $a^n = b$, then a is an nth root of b.
Real nth Roots of b, $\sqrt[n]{b}$, $-\sqrt[n]{b}$	1. If n is even and $b > 0$, then b has one positive root and one negative root. 2. If n is odd and $b > 0$, then b has one positive root. 3. If n is even and $b < 0$, then b has no real roots. 4. If n is odd and $b < 0$, then b has one negative root.

Example 1 Simplify $\sqrt{49z^8}$.

$\sqrt{49z^8} = \sqrt{(7z^4)^2} = 7z^4$

z^4 must be positive, so there is no need to take the absolute value.

Example 2 Simplify $-\sqrt[3]{(2a-1)^6}$

$-\sqrt[3]{(2a-1)^6} = -\sqrt[3]{[(2a-1)^2]^3} = (2a-1)^2$

Exercises

Simplify.

1. $\sqrt{81}$ **9**
2. $\sqrt[3]{-343}$ **−7**
3. $\sqrt{144p^6}$ **$12|p^3|$**
4. $\pm\sqrt{4a^{10}}$ **$\pm 2a^5$**
5. $\sqrt[5]{243p^{10}}$ **$3p^2$**
6. $-\sqrt[3]{m^6n^9}$ **$-m^2n^3$**
7. $\sqrt[3]{-b^{12}}$ **$-b^4$**
8. $\sqrt{16a^{10}b^8}$ **$4|a^5|b^4$**
9. $\sqrt{121x^6}$ **$11|x^3|$**
10. $\sqrt{(4k)^4}$ **$16k^2$**
11. $\pm\sqrt{169r^4}$ **$\pm 13r^2$**
12. $-\sqrt[3]{-27p^6}$ **$3p^2$**
13. $-\sqrt{625y^2z^4}$ **$-25|y|z^2$**
14. $\sqrt{36q^{34}}$ **$6|q^{17}|$**
15. $\sqrt{100x^2y^4z^6}$ **$10|x|y^2|z^3|$**
16. $\sqrt[3]{-0.027}$ **−0.3**
17. $-\sqrt{-0.36}$ **not a real number**
18. $\sqrt{0.64p^{10}}$ **$0.8|p^5|$**
19. $\sqrt[4]{(2x)^8}$ **$4x^2$**
20. $\sqrt{(11y^2)^4}$ **$121y^4$**
21. $\sqrt[3]{(5a^2b)^6}$ **$25a^4b^2$**
22. $\sqrt{(3x-1)^2}$ **$|3x-1|$**
23. $\sqrt[3]{(m-5)^6}$ **$(m-5)^2$**
24. $\sqrt{36x^2 - 12x + 1}$ **$|6x-1|$**

Skills Practice, p. 265 and Practice, p. 266 (shown)

Use a calculator to approximate each value to three decimal places.

1. $\sqrt{7.8}$ **2.793**
2. $-\sqrt{89}$ **−9.434**
3. $\sqrt[3]{25}$ **2.924**
4. $\sqrt[3]{-4}$ **−1.587**
5. $\sqrt[4]{1.1}$ **1.024**
6. $\sqrt[5]{-0.1}$ **−0.631**
7. $\sqrt[6]{5555}$ **4.208**
8. $\sqrt[4]{(0.94)^2}$ **0.970**

Simplify.

9. $\sqrt{0.81}$ **0.9**
10. $-\sqrt{324}$ **−18**
11. $-\sqrt[4]{256}$ **−4**
12. $\sqrt[6]{64}$ **2**
13. $\sqrt[3]{-64}$ **−4**
14. $\sqrt[3]{0.512}$ **0.8**
15. $\sqrt[5]{-243}$ **−3**
16. $-\sqrt[4]{1296}$ **−6**
17. $\sqrt[5]{\frac{-1024}{243}}$ **$-\frac{4}{3}$**
18. $\sqrt[5]{243x^{10}}$ **$3x^2$**
19. $\sqrt{(14a)^2}$ **$14|a|$**
20. $\sqrt{-(14a)^2}$ **not a real number**
21. $\sqrt{49m^2t^8}$ **$7|m|t^4$**
22. $\sqrt{\frac{16m^2}{25}}$ **$\frac{4|m|}{5}$**
23. $\sqrt[3]{-64r^6w^{15}}$ **$-4r^2w^5$**
24. $\sqrt{(2x)^8}$ **$16x^4$**
25. $-\sqrt[4]{625s^8}$ **$-5s^2$**
26. $\sqrt[3]{216p^3q^9}$ **$6pq^3$**
27. $\sqrt{676x^4y^6}$ **$26x^2|y^3|$**
28. $\sqrt[3]{-27x^9y^{12}}$ **$-3x^3y^4$**
29. $-\sqrt{144m^8n^6}$ **$-12m^4|n^3|$**
30. $\sqrt[5]{-32x^5y^{10}}$ **$-2xy^2$**
31. $\sqrt[6]{(m+4)^6}$ **$|m+4|$**
32. $\sqrt[3]{(2x+1)^3}$ **$2x+1$**
33. $-\sqrt{49a^{10}b^{16}}$ **$-7|a^5|b^8$**
34. $\sqrt[4]{(x-5)^8}$ **$(x-5)^2$**
35. $\sqrt[3]{343d^6}$ **$7d^2$**
36. $\sqrt{x^2 + 10x + 25}$ **$|x+5|$**

37. **RADIANT TEMPERATURE** Thermal sensors measure an object's *radiant* temperature, which is the amount of energy radiated by the object. The *internal* temperature of an object is called its *kinetic* temperature. The formula $T_r = T_k\sqrt[4]{e}$ relates an object's radiant temperature T_r to its kinetic temperature T_k. The variable e in the formula is a measure of how well the object radiates energy. If an object's kinetic temperature is 30°C and $e = 0.94$, what is the object's radiant temperature to the nearest tenth of a degree? **29.5°C**

38. **HERO'S FORMULA** Salvatore is buying fertilizer for his triangular garden. He knows the lengths of all three sides, so he is using Hero's formula to find the area. Hero's formula states that the area of a triangle is $\sqrt{s(s-a)(s-b)(s-c)}$, where a, b, and c are the lengths of the sides of the triangle and s is half the perimeter of the triangle. If the lengths of the sides of Salvatore's garden are 15 feet, 17 feet, and 20 feet, what is the area of the garden? Round your answer to the nearest whole number. **124 ft²**

Reading to Learn Mathematics, p. 267 ELL

Pre-Activity **How do square roots apply to oceanography?**

Read the introduction to Lesson 5-5 at the top of page 245 in your textbook.

Suppose the length of a wave is 5 feet. Explain how you would estimate the speed of the wave to the nearest tenth of a knot using a calculator. (Do not actually calculate the speed.) **Sample answer: Using a calculator, find the positive square root of 5. Multiply this number by 1.34. Then round the answer to the nearest tenth.**

Reading the Lesson

1. For each radical below, identify the radicand and the index.
   a. $\sqrt[3]{23}$ radicand: **23** index: **3**
   b. $\sqrt{15x^2}$ radicand: **$15x^2$** index: **2**
   c. $\sqrt[5]{-343}$ radicand: **−343** index: **5**

2. Complete the following table. (Do not actually find any of the indicated roots.)

Number	Number of Positive Square Roots	Number of Negative Square Roots	Number of Positive Cube Roots	Number of Negative Cube Roots
27	1	1	1	0
−16	0	0	0	1

3. State whether each of the following is *true* or *false*.
   a. A negative number has no real fourth roots. **true**
   b. $\pm\sqrt{121}$ represents both square roots of 121. **true**
   c. When you take the fifth root of x^5, you must take the absolute value of x to identify the principal fifth root. **false**

Helping You Remember

4. What is an easy way to remember that a negative number has no real square roots but has one real cube root? **Sample answer: The square of a positive or negative number is positive, so there is no real number whose square is negative. However, the cube of a negative number is negative, so a negative number has one real cube root, which is a negative number.**

248 Chapter 5 Polynomials

Guided Practice

GUIDED PRACTICE KEY	
Exercises	Examples
4–6	3
7–14	1, 2
15	3

Use a calculator to approximate each value to three decimal places.

4. $\sqrt{77}$ **8.775**
5. $-\sqrt[3]{19}$ **−2.668**
6. $\sqrt[4]{48}$ **2.632**

Simplify. **10. not a real number 13. $6|a|b^2$ 14. $|4x + 3y|$**

7. $\sqrt[3]{64}$ **4**
8. $\sqrt{(-2)^2}$ **2**
9. $\sqrt[5]{-243}$ **−3**
10. $\sqrt[4]{-4096}$
11. $\sqrt[3]{x^3}$ **x**
12. $\sqrt[4]{y^4}$ **$|y|$**
13. $\sqrt{36a^2b^4}$
14. $\sqrt{(4x + 3y)^2}$

Application

15. **OPTICS** The distance D in miles from an observer to the horizon over flat land or water can be estimated using the formula $D = 1.23\sqrt{h}$, where h is the height in feet of the point of observation. How far is the horizon for a person whose eyes are 6 feet above the ground? **about 3.01 mi**

★ indicates increased difficulty

Practice and Apply

Homework Help	
For Exercises	See Examples
16–27, 60–62	3
28–59	1, 2

Extra Practice See page 838.

Use a calculator to approximate each value to three decimal places.

16. $\sqrt{129}$ **11.358**
17. $-\sqrt{147}$ **−12.124**
18. $\sqrt{0.87}$ **0.933**
19. $\sqrt{4.27}$ **2.066**
20. $\sqrt[3]{59}$ **3.893**
21. $\sqrt[3]{-480}$ **−7.830**
22. $\sqrt[4]{602}$ **4.953**
23. $\sqrt[5]{891}$ **3.890**
24. $\sqrt[6]{4123}$ **4.004**
25. $\sqrt[7]{46{,}815}$ **4.647**
26. $\sqrt[6]{(723)^3}$ **26.889**
27. $\sqrt[4]{(3500)^2}$ **59.161**

Simplify.

30. not a real number

28. $\sqrt{225}$ **15**
29. $\pm\sqrt{169}$ **±13**
30. $\sqrt{-(-7)^2}$
31. $\sqrt{(-18)^2}$ **18**
32. $\sqrt[3]{-27}$ **−3**
33. $\sqrt[7]{-128}$ **−2**
34. $\sqrt{\frac{1}{16}}$ **$\frac{1}{4}$**
35. $\sqrt[3]{\frac{1}{125}}$ **$\frac{1}{5}$**
36. $\sqrt{0.25}$ **0.5**
37. $\sqrt[3]{-0.064}$ **−0.4**
38. $\sqrt[4]{z^8}$ **z^2**
39. $-\sqrt[6]{x^6}$ **$-|x|$**
40. $\sqrt{49m^6}$ **$7|m^3|$**
41. $\sqrt{64a^8}$ **$8a^4$**
42. $\sqrt[3]{27r^3}$ **$3r$**
43. $\sqrt[3]{-c^6}$ **$-c^2$**
44. $\sqrt{(5g)^4}$ **$25g^2$**
45. $\sqrt[3]{(2z)^6}$ **$4z^2$**
46. $\sqrt{25x^4y^6}$ **$5x^2|y^3|$**
47. $\sqrt{36x^4z^4}$ **$6x^2z^2$**
48. $\sqrt{169x^8y^4}$ **$13x^4y^2$**
49. $\sqrt{9p^{12}q^6}$ **$3p^6|q^3|$**
50. $\sqrt[3]{8a^3b^3}$ **$2ab$**
51. $\sqrt[3]{-27c^9d^{12}}$ **$-3c^3d^4$**
52. $\sqrt{(4x-y)^2}$ **$|4x-y|$**
53. $\sqrt[3]{(p+q)^3}$ **$p+q$**
★54. $-\sqrt{x^2 + 4x + 4}$ **54. $-|x+2|$**
★55. $\sqrt{z^2 + 8z + 16}$ **$|z+4|$**
★56. $\sqrt{4a^2 + 4a + 1}$ **$|2a+1|$**
★57. $\sqrt{-9x^2 - 12x - 4}$ **not a real number**

58. Find the principal fifth root of 32. **2**

59. What is the third root of −125? **−5**

60. **SPORTS** Refer to the drawing at the right. How far does the catcher have to throw a ball from home plate to second base? **about 127.28 ft**

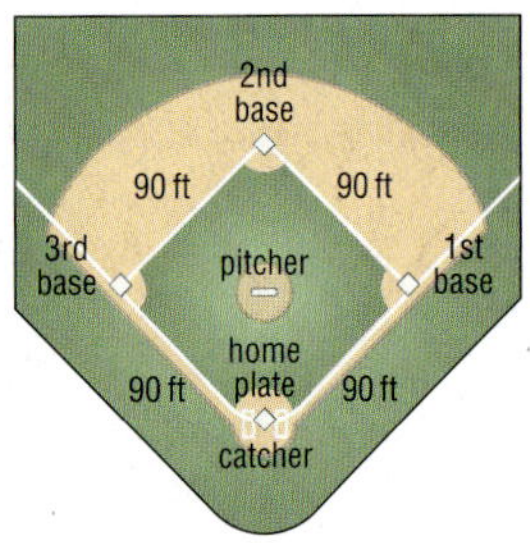

61. **FISH** The relationship between the length and mass of Pacific halibut can be approximated by the equation $L = 0.46\sqrt[3]{M}$, where L is the length in meters and M is the mass in kilograms. Use this equation to predict the length of a 25-kilogram Pacific halibut. **about 1.35 m**

Enrichment, p. 268

Approximating Square Roots

Consider the following expansion.

$$\left(a + \frac{b}{2a}\right)^2 = a^2 + \frac{2ab}{2a} + \frac{b^2}{4a^2}$$
$$= a^2 + b + \frac{b^2}{4a^2}$$

Think what happens if a is very great in comparison to b. The term $\frac{b^2}{4a^2}$ is very small and can be disregarded in an approximation.

$$\left(a + \frac{b}{2a}\right)^2 \approx a^2 + b$$
$$a + \frac{b}{2a} \approx \sqrt{a^2 + b}$$

Suppose a number can be expressed as $a^2 + b$, $a > b$. Then an approximate value of the square root is $a + \frac{b}{2a}$. You should also see that $a - \frac{b}{2a} \approx \sqrt{a^2 - b}$.

More About. . .

Space Science

The escape velocity for the Moon is about 2400 m/s. For the Sun, it is about 618,000 m/s.

Source: NASA

62. **SPACE SCIENCE** The velocity v required for an object to escape the gravity of a planet or other body is given by the formula $v = \sqrt{\frac{2GM}{R}}$, where M is the mass of the body, R is the radius of the body, and G is Newton's gravitational constant. Use $M = 5.98 \times 10^{24}$ kg, $R = 6.37 \times 10^6$ m, and $G = 6.67 \times 10^{-11}\ \text{N} \cdot \text{m}^2/\text{kg}^2$ to find the escape velocity for Earth. **about 11,200 m/s**

63. **CRITICAL THINKING** Under what conditions does $\sqrt{x^2 + y^2} = x + y$? **$x = 0$ and $y \geq 0$, or $y = 0$ and $x \geq 0$**

64. **WRITING IN MATH** Answer the question that was posed at the beginning of the lesson. **See margin.**

How do square roots apply to oceanography?

Include the following in your answer:
- the values of s for $\ell = 2, 5,$ and 10 feet, and
- an observation of what happens to the value of s as the value of ℓ increases.

Standardized Test Practice

65. Which of the following is closest to $\sqrt{7.32}$? **B**

(A) 2.6 (B) 2.7 (C) 2.8 (D) 2.9

66. In the figure, $\triangle ABC$ is an equilateral triangle with sides 9 units long. What is the length of $\overline{BD}$ in units? **D**

(A) 3 (B) 9
(C) $9\sqrt{2}$ (D) 18

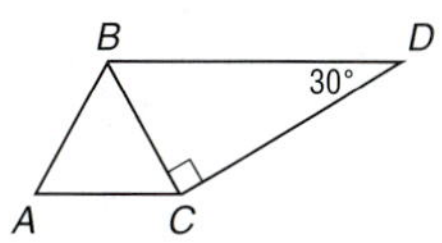

Maintain Your Skills

Mixed Review

Factor completely. If the polynomial is not factorable, write *prime*. *(Lesson 5-4)*

67. $7xy^3 - 14x^2y^5 + 28x^3y^2$ **67. $7xy^2(y - 2xy^3 + 4x^2)$**
68. $ab - 5a + 3b - 15$ **$(a + 3)(b - 5)$**
69. $2x^2 + 15x + 25$ **$(2x + 5)(x + 5)$**
70. $c^3 - 216$ **$(c - 6)(c^2 + 6c + 36)$**

Simplify. *(Lesson 5-3)*

71. $(4x^3 - 7x^2 + 3x - 2) \div (x - 2)$ **71. $4x^2 + x + 5 + \frac{8}{x - 2}$**
72. $\frac{x^4 + 4x^3 - 4x^2 + 5x}{x + 5}$ **$x^3 - x^2 + x$**

73. **TRAVEL** The matrix at the right shows the costs of airline flights between some cities. Write a matrix that shows the costs of two tickets for these flights. *(Lesson 4-2)* **$\begin{bmatrix} 810 & 2320 \\ 1418 & 2504 \end{bmatrix}$**

	New York	LA
Atlanta	405	1160
Chicago	709	1252

Solve each system of equations by using either substitution or elimination. *(Lesson 3-2)*

74. $a + 4b = 6$; $3a + 2b = -2$ **$(-2, 2)$**
75. $10x - y = 13$; $3x - 4y = 15$ **$(1, -3)$**
76. $3c - 7d = -1$; $2c - 6d = -6$ **$(9, 4)$**

Getting Ready for the Next Lesson

PREREQUISITE SKILL **Find each product.**
*(To review **multiplying binomials**, see Lesson 5-2.)*

77. $(x + 3)(x + 8)$ **$x^2 + 11x + 24$**
78. $(y - 2)(y + 5)$ **$y^2 + 3y - 10$**
79. $(a + 2)(a - 9)$ **$a^2 - 7a - 18$**
80. $(a + b)(a + 2b)$ **$a^2 + 3ab + 2b^2$**
81. $(x - 3y)(x + 3y)$ **$x^2 - 9y^2$**
82. $(2w + z)(3w - 5z)$ **$6w^2 - 7wz - 5z^2$**

www.algebra2.com/self_check_quiz

Answer

64. The speed and length of a wave are related by an expression containing a square root. Answers should include the following.
- about 1.90 knots, about 3.00 knots, and 4.24 knots
- As the value of ℓ increases, the value of s increases.

About the Exercises...

Organization by Objective
- **Simplify Radicals:** 28–59
- **Approximate Radicals with a Calculator:** 16–27

Odd/Even Assignments
Exercises 16–59 are structured so that students practice the same concepts whether they are assigned odd or even problems.

Assignment Guide

Basic: 17–53 odd, 59, 61, 63–82
Average: 17–61 odd, 63–82
Advanced: 16–62 even, 63–76 (optional: 77–82)

4 Assess

Open-Ended Assessment

Modeling Have students use the hypotenuse of a right triangle with both legs one unit long to demonstrate and explain that $\sqrt{2}$ is a number on the real number line.

Getting Ready for Lesson 5-6

PREREQUISITE SKILL Lesson 5-6 presents operations with radical expressions. In Example 5 on p. 253, they will encounter the multiplication of two binomials involving radical expressions. Exercises 77–82 should be used to determine your students' familiarity with multiplying binomials.

Assessment Options

Quiz (Lessons 5-4 and 5-5) is available on p. 307 of the *Chapter 5 Resource Masters.*

Mid-Chapter Test (Lessons 5-1 through 5-5) is available on p. 309 of the *Chapter 5 Resource Masters.*

5-6 Lesson Notes

1 Focus

5-Minute Check Transparency 5-6 Use as a quiz or a review of Lesson 5-5.

Mathematical Background notes are available for this lesson on p. 220D.

Building on Prior Knowledge

In Lesson 5-5, students simplified radicals. In this lesson, students build on the skills they learned in that lesson to simplify and combine radical expressions.

How do radical expressions apply to falling objects?

Ask students:

- If the value of d in the formula doubles, will the value of t also double? **no**
- Is the relationship between time t and distance d for a falling object *linear* or *nonlinear*? **nonlinear**

5-6 Radical Expressions

What You'll Learn

- Simplify radical expressions.
- Add, subtract, multiply, and divide radical expressions.

Vocabulary

- rationalizing the denominator
- like radical expressions
- conjugates

How do radical expressions apply to falling objects?

The amount of time t in seconds that it takes for an object to drop d feet is given by $t = \sqrt{\frac{2d}{g}}$, where $g = 32$ ft/s^2 is the acceleration due to gravity. In this lesson, you will learn how to simplify radical expressions like $\sqrt{\frac{2d}{g}}$.

SIMPLIFY RADICAL EXPRESSIONS You can use the Commutative Property and the definition of square root to find an equivalent expression for a product of radicals such as $\sqrt{3} \cdot \sqrt{5}$. Begin by squaring the product.

$$(\sqrt{3} \cdot \sqrt{5})^2 = \sqrt{3} \cdot \sqrt{5} \cdot \sqrt{3} \cdot \sqrt{5}$$
$$= \sqrt{3} \cdot \sqrt{3} \cdot \sqrt{5} \cdot \sqrt{5} \quad \text{Commutative Property of Multiplication}$$
$$= 3 \cdot 5 \text{ or } 15 \quad \text{Definition of square root}$$

Since $\sqrt{3} \cdot \sqrt{5} > 0$ and $(\sqrt{3} \cdot \sqrt{5})^2 = 15$, $\sqrt{3} \cdot \sqrt{5}$ is the principal square root of 15. That is, $\sqrt{3} \cdot \sqrt{5} = \sqrt{15}$. This illustrates the following property of radicals.

Key Concept — Product Property of Radicals

For any real numbers a and b and any integer $n > 1$,

1. if n is even and a and b are both nonnegative, then $\sqrt[n]{ab} = \sqrt[n]{a} \cdot \sqrt[n]{b}$, and
2. if n is odd, then $\sqrt[n]{ab} = \sqrt[n]{a} \cdot \sqrt[n]{b}$.

Follow these steps to simplify a square root.

Step 1 Factor the radicand into as many squares as possible.

Step 2 Use the Product Property to isolate the perfect squares.

Step 3 Simplify each radical.

Example 1 Square Root of a Product

Simplify $\sqrt{16p^8q^7}$.

$$\sqrt{16p^8q^7} = \sqrt{4^2 \cdot (p^4)^2 \cdot (q^3)^2 \cdot q} \quad \text{Factor into squares where possible.}$$
$$= \sqrt{4^2} \cdot \sqrt{(p^4)^2} \cdot \sqrt{(q^3)^2} \cdot \sqrt{q} \quad \text{Product Property of Radicals}$$
$$= 4p^4|q^3|\sqrt{q} \quad \text{Simplify.}$$

However, for $\sqrt{16p^8q^7}$ to be defined, $16p^8q^7$ must be nonnegative. If that is true, q must be nonnegative, since it is raised to an odd power. Thus, the absolute value is unnecessary, and $\sqrt{16p^8q^7} = 4p^4q^3\sqrt{q}$.

Resource Manager

Workbook and Reproducible Masters

Chapter 5 Resource Masters

- Study Guide and Intervention, pp. 269–270
- Skills Practice, p. 271
- Practice, p. 272
- Reading to Learn Mathematics, p. 273
- Enrichment, p. 274

Teaching Algebra With Manipulatives Masters, p. 238

Transparencies

5-Minute Check Transparency 5-6
Real-World Transparency 5
Answer Key Transparencies

Technology

Interactive Chalkboard

Look at a radical that involves division to see if there is a quotient property for radicals that is similar to the Product Property. Consider $\frac{49}{9}$. The radicand is a perfect square, so $\sqrt{\frac{49}{9}} = \sqrt{\left(\frac{7}{3}\right)^2}$ or $\frac{7}{3}$. Notice that $\frac{7}{3} = \frac{\sqrt{49}}{\sqrt{9}}$. This suggests the following property.

Key Concept — *Quotient Property of Radicals*

- **Words** For any real numbers a and $b \neq 0$, and any integer $n > 1$, $\sqrt[n]{\frac{a}{b}} = \frac{\sqrt[n]{a}}{\sqrt[n]{b}}$, if all roots are defined.
- **Example** $\frac{\sqrt{27}}{\sqrt{3}} = \sqrt{9}$ or 3

You can use the properties of radicals to write expressions in simplified form.

Concept Summary — *Simplifying Radical Expressions*

A radical expression is in simplified form when the following conditions are met.

- The index n is as small as possible.
- The radicand contains no factors (other than 1) that are nth powers of an integer or polynomial.
- The radicand contains no fractions.
- No radicals appear in a denominator.

Study Tip

Rationalizing the Denominator

You may want to think of rationalizing the denominator as making the denominator a rational number.

To eliminate radicals from a denominator or fractions from a radicand, you can use a process called **rationalizing the denominator**. To rationalize a denominator, multiply the numerator and denominator by a quantity so that the radicand has an exact root. Study the examples below.

Example 2 Simplify Quotients

Simplify each expression.

a. $\sqrt{\frac{x^4}{y^5}}$

$$\sqrt{\frac{x^4}{y^5}} = \frac{\sqrt{x^4}}{\sqrt{y^5}} \quad \text{Quotient Property}$$

$$= \frac{\sqrt{(x^2)^2}}{\sqrt{(y^2)^2 \cdot y}} \quad \text{Factor into squares.}$$

$$= \frac{\sqrt{(x^2)^2}}{\sqrt{(y^2)^2} \cdot \sqrt{y}} \quad \text{Product Property}$$

$$= \frac{x^2}{y^2\sqrt{y}} \quad \sqrt{(x^2)^2} = x^2$$

$$= \frac{x^2}{y^2\sqrt{y}} \cdot \frac{\sqrt{y}}{\sqrt{y}} \quad \text{Rationalize the denominator.}$$

$$= \frac{x^2\sqrt{y}}{y^3} \quad \sqrt{y} \cdot \sqrt{y} = y$$

b. $\sqrt[5]{\frac{5}{4a}}$

$$\sqrt[5]{\frac{5}{4a}} = \frac{\sqrt[5]{5}}{\sqrt[5]{4a}} \quad \text{Quotient Property}$$

$$= \frac{\sqrt[5]{5}}{\sqrt[5]{4a}} \cdot \frac{\sqrt[5]{8a^4}}{\sqrt[5]{8a^4}} \quad \text{Rationalize the denominator.}$$

$$= \frac{\sqrt[5]{5 \cdot 8a^4}}{\sqrt[5]{4a \cdot 8a^4}} \quad \text{Product Property}$$

$$= \frac{\sqrt[5]{40a^4}}{\sqrt[5]{32a^5}} \quad \text{Multiply.}$$

$$= \frac{\sqrt[5]{40a^4}}{2a} \quad \sqrt[5]{32a^5} = 2a$$

www.algebra2.com/extra_examples

2 Teach

SIMPLIFY RADICAL EXPRESSIONS

In-Class Examples Power Point®

Teaching Tip When discussing the Product Property of Radicals, stress the fact that a and b must both be nonnegative if n is even. This means that $\sqrt{-2}$ times $\sqrt{-8}$ may *not* be written as $\sqrt{16}$. This condition is necessary because $\sqrt{-2}$ and $\sqrt{-8}$ are *not* real numbers.

1 Simplify $\sqrt{25a^4b^9}$. $5a^2b^4\sqrt{b}$

2 Simplify each expression.

a. $\sqrt{\frac{y^8}{x^7}}$ $\frac{y^4\sqrt{x}}{x^4}$

b. $\sqrt[3]{\frac{2}{9x}}$ $\frac{\sqrt[3]{6x^2}}{3x}$

Teaching Tip Urge students to verify that each of their final answers is in simplified form by testing it against the four conditions listed in the Concept Summary for Simplifying Radical Expressions.

DAILY INTERVENTION — **Differentiated Instruction**

Intrapersonal Have students think about irrational numbers, radicals, and the rules for operations with radicals. Ask them to write down what puzzles them most about these concepts, including a list of the definitions and operations about which they feel some confusion. Invite students to share these concerns with you so that they can be cleared up.

Lesson 5-6 Radical Expressions 251

OPERATIONS WITH RADICALS

In-Class Example Power Point®

3 Simplify $5\sqrt[3]{100a^2} \cdot \sqrt[3]{10a}$. **$50a$**

Teaching Tip After discussing the information presented in the Algebra Activity, make sure students also understand that $\sqrt{a} + \sqrt{b}$ is not equivalent to $\sqrt{a + b}$. Suggest they use the values $a = 16$ and $b = 9$ to verify this fact.

OPERATIONS WITH RADICALS You can use the Product and Quotient Properties to multiply and divide some radicals, respectively.

Example 3 Multiply Radicals

Simplify $6\sqrt[3]{9n^2} \cdot 3\sqrt[3]{24n}$.

$6\sqrt[3]{9n^2} \cdot 3\sqrt[3]{24n} = 6 \cdot 3 \cdot \sqrt[3]{9n^2 \cdot 24n}$	Product Property of Radicals
$= 18 \cdot \sqrt[3]{2^3 \cdot 3^3 \cdot n^3}$	Factor into cubes where possible.
$= 18 \cdot \sqrt[3]{2^3} \cdot \sqrt[3]{3^3} \cdot \sqrt[3]{n^3}$	Product Property of Radicals
$= 18 \cdot 2 \cdot 3 \cdot n$ or $108n$	Multiply.

Can you add radicals in the same way that you multiply them? In other words, if $\sqrt{a} \cdot \sqrt{a} = \sqrt{a \cdot a}$, does $\sqrt{a} + \sqrt{a} = \sqrt{a + a}$?

Algebra Activity

Adding Radicals

You can use dot paper to show the sum of two like radicals, such as $\sqrt{2} + \sqrt{2}$.

Model and Analyze

Step 1 First, find a segment of length $\sqrt{2}$ units by using the Pythagorean Theorem with the dot paper.

$a^2 + b^2 = c^2$
$1^2 + 1^2 = c^2$
$2 = c^2$

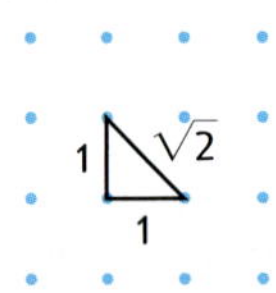

Step 2 Extend the segment to twice its length to represent $\sqrt{2} + \sqrt{2}$.

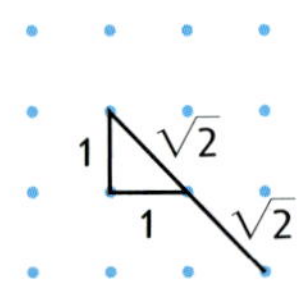

Make a Conjecture

1. Is $\sqrt{2} + \sqrt{2} = \sqrt{2 + 2}$ or 2? Justify your answer using the geometric models above.
2. Use this method to model other irrational numbers. Do these models support your conjecture? **See students' work.**

1. No; $\sqrt{2} + \sqrt{2}$ units is the length of the hypotenuse of an isosceles right triangle whose legs have length 2 units. Therefore, $\sqrt{2} + \sqrt{2} > 2$.

In the activity, you discovered that you cannot add radicals in the same manner as you multiply them. You add radicals in the same manner as adding monomials. That is, you can add only the like terms or like radicals.

Two radical expressions are called **like radical expressions** if both the indices and the radicands are alike. Some examples of like and unlike radical expressions are given below.

Study Tip

Reading Math
Indices is the plural of *index*.

$\sqrt{3}$ and $\sqrt[3]{3}$ are not like expressions.	Different indices
$\sqrt[4]{5x}$ and $\sqrt[4]{5}$ are not like expressions.	Different radicands
$2\sqrt[4]{3a}$ and $5\sqrt[4]{3a}$ are like expressions.	Radicands are $3a$; indices are 4.

Algebra Activity

Materials: rectangular dot paper, ruler/straightedge

- Ask students what leg lengths they could use on a right triangle to find a line whose length is $\sqrt{5}$. **lengths of 1 and 2 units**
- Point out that there are other instances where you can perform a multiplication but not an addition. For example, you can multiply fractions by multiplying the numerators and denominators separately, but you do not add fractions this way.

Example 4 Add and Subtract Radicals

Simplify $2\sqrt{12} - 3\sqrt{27} + 2\sqrt{48}$.

$2\sqrt{12} - 3\sqrt{27} + 2\sqrt{48}$

$= 2\sqrt{2^2 \cdot 3} - 3\sqrt{3^2 \cdot 3} + 2\sqrt{2^2 \cdot 2^2 \cdot 3}$ — Factor using squares.

$= 2\sqrt{2^2} \cdot \sqrt{3} - 3\sqrt{3^2} \cdot \sqrt{3} + 2\sqrt{2^2} \cdot \sqrt{2^2} \cdot \sqrt{3}$ — Product Property

$= 2 \cdot 2 \cdot \sqrt{3} - 3 \cdot 3 \cdot \sqrt{3} + 2 \cdot 2 \cdot 2 \cdot \sqrt{3}$ — $\sqrt{2^2} = 2$, $\sqrt{3^2} = 3$

$= 4\sqrt{3} - 9\sqrt{3} + 8\sqrt{3}$ — Multiply.

$= 3\sqrt{3}$ — Combine like radicals.

Just as you can add and subtract radicals like monomials, you can multiply radicals using the FOIL method as you do when multiplying binomials.

Example 5 Multiply Radicals

Simplify each expression.

a. $(3\sqrt{5} - 2\sqrt{3})(2 + \sqrt{3})$

F O I L

$(3\sqrt{5} - 2\sqrt{3})(2 + \sqrt{3}) = 3\sqrt{5} \cdot 2 + 3\sqrt{5} \cdot \sqrt{3} - 2\sqrt{3} \cdot 2 - 2\sqrt{3} \cdot \sqrt{3}$

$= 6\sqrt{5} + 3\sqrt{5 \cdot 3} - 4\sqrt{3} - 2\sqrt{3^2}$ — Product Property

$= 6\sqrt{5} + 3\sqrt{15} - 4\sqrt{3} - 6$ — $2\sqrt{3^2} = 2 \cdot 3$ or 6

b. $(5\sqrt{3} - 6)(5\sqrt{3} + 6)$

$(5\sqrt{3} - 6)(5\sqrt{3} + 6) = 5\sqrt{3} \cdot 5\sqrt{3} + 5\sqrt{3} \cdot 6 - 6 \cdot 5\sqrt{3} - 6 \cdot 6$ — FOIL

$= 25\sqrt{3^2} + 30\sqrt{3} - 30\sqrt{3} - 36$ — Multiply.

$= 75 - 36$ — $25\sqrt{3^2} = 25 \cdot 3$ or 75

$= 39$ — Subtract.

Binomials like those in Example 5b, of the form $a\sqrt{b} + c\sqrt{d}$ and $a\sqrt{b} - c\sqrt{d}$ where a, b, c, and d are rational numbers, are called **conjugates** of each other. The product of conjugates is always a rational number. You can use conjugates to rationalize denominators.

Example 6 Use a Conjugate to Rationalize a Denominator

Simplify $\frac{1 - \sqrt{3}}{5 + \sqrt{3}}$.

$\frac{1 - \sqrt{3}}{5 + \sqrt{3}} = \frac{(1 - \sqrt{3})(5 - \sqrt{3})}{(5 + \sqrt{3})(5 - \sqrt{3})}$ — Multiply by $\frac{5 - \sqrt{3}}{5 - \sqrt{3}}$ because $5 - \sqrt{3}$ is the conjugate of $5 + \sqrt{3}$.

$= \frac{1 \cdot 5 - 1 \cdot \sqrt{3} - \sqrt{3} \cdot 5 + (\sqrt{3})^2}{5^2 - (\sqrt{3})^2}$ — FOIL; Difference of squares

$= \frac{5 - \sqrt{3} - 5\sqrt{3} + 3}{25 - 3}$ — Multiply.

$= \frac{8 - 6\sqrt{3}}{22}$ — Combine like terms.

$= \frac{4 - 3\sqrt{3}}{11}$ — Divide numerator and denominator by 2.

In-Class Examples

Power Point®

4 Simplify $3\sqrt{45} - 5\sqrt{80} + 4\sqrt{20}$. $-3\sqrt{5}$

5 Simplify each expression.

a. $(2\sqrt{3} + 3\sqrt{5})(3 - \sqrt{3})$ $6\sqrt{3} - 6 + 9\sqrt{5} - 3\sqrt{15}$

b. $(4\sqrt{2} + 7)(4\sqrt{2} - 7)$ -17

6 Simplify $\frac{2 + \sqrt{3}}{4 - \sqrt{3}}$. $\frac{11 + 6\sqrt{3}}{13}$

DAILY INTERVENTION — Unlocking Misconceptions

Radical Expressions When presented with a radical expression such as $11 + 6\sqrt{3}$, some students may persist in trying to add the 11 and the 6. Help them understand why this cannot be done by comparing this radical expression $11 + 6\sqrt{3}$ to the expression $11 + 6x$. Stress that the radical $6\sqrt{3}$ is a multiplication expression just like $6x$. Remind students that the order of operations requires that multiplication be performed before addition. Students may find it helpful to rewrite $11 + 6\sqrt{3}$, as $11 + 6 \cdot \sqrt{3}$.

3 Practice/Apply

Study Notebook

Have students—

- add the definitions/examples of the vocabulary terms to their Vocabulary Builder worksheets for Chapter 5.
- add the information from the Key Concept and Concept Summary features to their notebook.
- include any other item(s) that they find helpful in mastering the skills in this lesson.

About the Exercises...

Organization by Objective

- **Simplify Radical Expressions:** 15–30
- **Operations with Radicals:** 31–48

Odd/Even Assignments

Exercises 15–48 are structured so that students practice the same concepts whether they are assigned odd or even problems.

Assignment Guide

Basic: 15–45 odd, 49–51, 55–82

Average: 15–49 odd, 50–53, 55–82

Advanced: 16–48 even, 50–74 (optional: 75–82)

All: Practice Quiz 2 (1–10)

Check for Understanding

Concept Check

1. **Determine** whether the statement $\frac{1}{\sqrt[n]{a}} = \sqrt[n]{a}$ is *sometimes*, *always*, or *never* true. Explain. **1. Sometimes; $\frac{1}{\sqrt[n]{a}} = \sqrt[n]{a}$ only when $a = 1$.**
2. **OPEN ENDED** **Write** a sum of three radicals that contains two like terms.
3. **Explain** why the product of two conjugates is always a rational number. **2–3. See margin.**

Guided Practice

GUIDED PRACTICE KEY	
Exercises	**Examples**
4–5, 14	1
6	2
7–9	3
10, 11	4
12	5
13	6

Simplify.

4. $5\sqrt{63}$ **$15\sqrt{7}$**
5. $\sqrt[4]{16x^5y^4}$ **$2x|y|\sqrt[4]{x}$**
6. $\sqrt{\frac{7}{8y}}$ **$\frac{\sqrt{14y}}{4y}$**
7. $(-2\sqrt{15})(4\sqrt{21})$ **$-24\sqrt{35}$**
8. $\frac{\sqrt[3]{625}}{\sqrt[3]{25}}$ **$\sqrt[3]{25}$**
9. $\sqrt{2ab^2} \cdot \sqrt{6a^3b^2}$ **$2a^2b^2\sqrt{3}$**
10. $\sqrt{3} - 2\sqrt[4]{3} + 4\sqrt{3} + 5\sqrt[4]{3}$ **$5\sqrt{3} + 3\sqrt[4]{3}$**
11. $3\sqrt[3]{128} + 5\sqrt[3]{16}$ **$22\sqrt[3]{2}$**
12. $(3 - \sqrt{5})(1 + \sqrt{3})$ **$3 + 3\sqrt{3} - \sqrt{5} - \sqrt{15}$**
13. $\frac{1 + \sqrt{5}}{3 - \sqrt{5}}$ **$2 + \sqrt{5}$**

Application

14. **LAW ENFORCEMENT** A police accident investigator can use the formula $s = 2\sqrt{5\ell}$ to estimate the speed s of a car in miles per hour based on the length ℓ in feet of the skid marks it left. How fast was a car traveling that left skid marks 120 feet long? **about 49 mph**

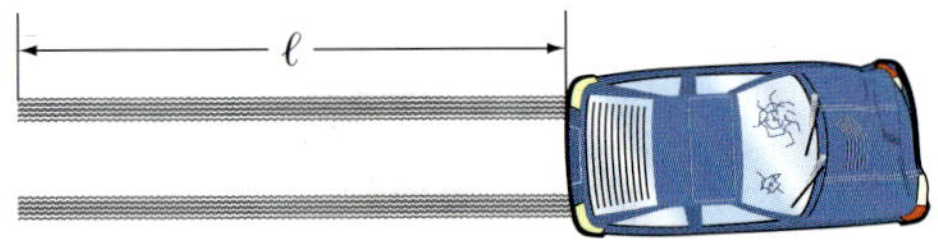

★ indicates increased difficulty

Practice and Apply

Homework Help	
For Exercises	**See Examples**
15–26	1
27–30	2
31–34	3
35–38	4
39–42	5
43–48	6

Extra Practice See page 838.

Simplify. **21. $3|x|y\sqrt{2y}$ 22. $2ab^2\sqrt{10a}$ 23. $6y^2z\sqrt[3]{7}$ 24. $4mn\sqrt[3]{3mn^2}$**

15. $\sqrt{243}$ **$9\sqrt{3}$**
16. $\sqrt{72}$ **$6\sqrt{2}$**
17. $\sqrt[3]{54}$ **$3\sqrt[3]{2}$**
18. $\sqrt[4]{96}$ **$2\sqrt[4]{6}$**
19. $\sqrt{50x^4}$ **$5x^2\sqrt{2}$**
20. $\sqrt[3]{16y^3}$ **$2y\sqrt[3]{2}$**
21. $\sqrt{18x^2y^3}$
22. $\sqrt{40a^3b^4}$
23. $3\sqrt[3]{56y^6z^3}$
24. $2\sqrt[3]{24m^4n^5}$
25. $\sqrt[4]{\frac{1}{81}c^5d^4}$ **25. $\frac{1}{3}c|d|\sqrt[4]{c}$**
26. $\sqrt[5]{\frac{1}{32}w^6z^7}$ **26. $\frac{1}{2}wz\sqrt[5]{wz^2}$**
27. $\sqrt[3]{\frac{3}{4}}$ **$\frac{\sqrt[3]{6}}{2}$**
28. $\sqrt[4]{\frac{2}{3}}$ **$\frac{\sqrt[4]{54}}{3}$**
29. $\sqrt{\frac{a^4}{b^3}}$ **$\frac{a^2\sqrt{b}}{b^2}$**
30. $\sqrt{\frac{4r^8}{t^9}}$ **$\frac{2r^4\sqrt{t}}{t^5}$**
31. $(3\sqrt{12})(2\sqrt{21})$ **$36\sqrt{7}$**
32. $(-3\sqrt{24})(5\sqrt{20})$ **$-60\sqrt{30}$**
33. What is $\sqrt{39}$ divided by $\sqrt{26}$? **$\frac{\sqrt{6}}{2}$**
34. Divide $\sqrt{14}$ by $\sqrt{35}$. **$\frac{\sqrt{10}}{5}$**

Simplify. **37. $7\sqrt{3} - 2\sqrt{2}$ 38. $4\sqrt{5} + 23\sqrt{6}$ 39. $25 - 5\sqrt{2} + 5\sqrt{6} - 2\sqrt{3}$**

35. $\sqrt{12} + \sqrt{48} - \sqrt{27}$ **$3\sqrt{3}$**
36. $\sqrt{98} - \sqrt{72} + \sqrt{32}$ **$5\sqrt{2}$**
37. $\sqrt{3} + \sqrt{72} - \sqrt{128} + \sqrt{108}$
38. $5\sqrt{20} + \sqrt{24} - \sqrt{180} + 7\sqrt{54}$
39. $(5 + \sqrt{6})(5 - \sqrt{2})$
40. $(3 + \sqrt{7})(2 + \sqrt{6})$ **40. $6 + 3\sqrt{6} + 2\sqrt{7} + \sqrt{42}$**
41. $(\sqrt{11} - \sqrt{2})^2$ **$13 - 2\sqrt{22}$**
42. $(\sqrt{3} - \sqrt{5})^2$ **$8 - 2\sqrt{15}$**
43. $\frac{7}{4 - \sqrt{3}}$ **$\frac{28 + 7\sqrt{3}}{13}$**
44. $\frac{\sqrt{6}}{5 + \sqrt{3}}$ **$\frac{5\sqrt{6} - 3\sqrt{2}}{22}$**
45. $\frac{-2 - \sqrt{3}}{1 + \sqrt{3}}$ **$\frac{-1 - \sqrt{3}}{2}$**
46. $\frac{2 + \sqrt{2}}{5 - \sqrt{2}}$ **$\frac{12 + 7\sqrt{2}}{23}$**
47. ★ $\frac{x + 1}{\sqrt{x^2 - 1}}$ **$\frac{\sqrt{x^2 - 1}}{x - 1}$**
48. ★ $\frac{x - 1}{\sqrt{x} - 1}$ **$\sqrt{x} + 1$**

Answers

2. Sample answer: $\sqrt{2} + \sqrt{3} + \sqrt{2}$
3. The product of two conjugates yields a difference of two squares. Each square produces a rational number and the difference of two rational numbers is a rational number.

50. The square root of a difference is not the difference of the square roots.

56. The formula for the time it takes an object to fall a certain distance can be written in various forms involving radicals. Answers should include the following.
    - By the Quotient Property of Radicals, $t = \frac{\sqrt{2d}}{\sqrt{g}}$. Multiply by $\frac{\sqrt{g}}{\sqrt{g}}$ to rationalize the denominator. The result is $\frac{\sqrt{2dg}}{g}$.
    - about 1.12 s

49. **GEOMETRY** Find the perimeter and area of the rectangle. **$6 + 16\sqrt{2}$ yd, $24 + 6\sqrt{2}$ yd²**

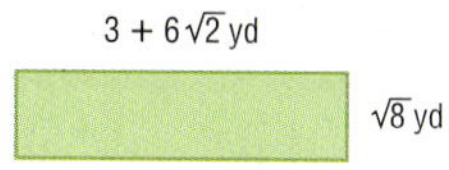

More About. . .

Amusement Parks

Attendance at the top 50 theme parks in North America increased to 175.1 million in 2000.

Source: *Amusement Business*

AMUSEMENT PARKS **For Exercises 50 and 51, use the following information.** The velocity v in feet per second of a roller coaster at the bottom of a hill is related to the vertical drop h in feet and the velocity v_0 in feet per second of the coaster at the top of the hill by the formula $v_0 = \sqrt{v^2 - 64h}$.

50. Explain why $v_0 = v - 8\sqrt{h}$ is not equivalent to the given formula. **See margin.**

51. What velocity must a coaster have at the top of a 225-foot hill to achieve a velocity of 120 feet per second at the bottom? **0 ft/s**

Online Research **Data Update** What are the values of v and h for some of the world's highest and fastest roller coasters? Visit www.algebra2.com/data_update to learn more.

SPORTS **For Exercises 52 and 53, use the following information.** A ball that is hit or thrown horizontally with a velocity of v meters per second will travel a distance of d meters before hitting the ground, where $d = v\sqrt{\frac{h}{4.9}}$ and h is the height in meters from which the ball is hit or thrown.

52. Use the properties of radicals to rewrite the formula. **$d = v\frac{\sqrt{4.9h}}{4.9}$**

53. How far will a ball that is hit horizontally with a velocity of 45 meters per second at a height of 0.8 meter above the ground travel before hitting the ground? **about 18.18 m**

54. **AUTOMOTIVE ENGINEERING** An automotive engineer is trying to design a safer car. The maximum force a road can exert on the tires of the car being redesigned is 2000 pounds. What is the maximum velocity v in ft/s at which this car can safely round a turn of radius 320 feet? Use the formula $v = \sqrt{\frac{F_c r}{100}}$, where F_c is the force the road exerts on the car and r is the radius of the turn. **80 ft/s or about 55 mph**

55. **CRITICAL THINKING** Under what conditions is the equation $\sqrt{x^3y^2} = xy\sqrt{x}$ true? **x and y are nonnegative.**

56. **WRITING IN MATH** Answer the question that was posed at the beginning of the lesson. **See margin.**

How do radical expressions apply to falling objects?

Include the following in your answer:
- an explanation of how you can use the properties in this lesson to rewrite the formula $t = \sqrt{\frac{2d}{g}}$, and
- the amount of time a 5-foot tall student has to get out of the way after a balloon is dropped from a window 25 feet above.

Standardized Test Practice

57. The expression $\sqrt{180}$ is equivalent to which of the following? **B**

Ⓐ $5\sqrt{6}$ Ⓑ $6\sqrt{5}$ Ⓒ $3\sqrt{10}$ Ⓓ $36\sqrt{5}$

58. Which of the following is *not* a length of a side of the triangle? **D**

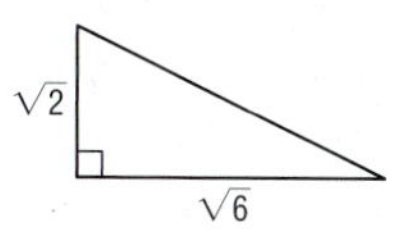

Ⓐ $\sqrt{8}$ Ⓑ $2\sqrt{2}$

Ⓒ $\sqrt{4+2}$ Ⓓ $\sqrt{4} + \sqrt{2}$

www.algebra2.com/self_check_quiz

Study Guide and Intervention, p. 269 (shown) and p. 270

Simplify Radical Expressions

Product Property of Radicals	For any real numbers a and b, and any integer $n > 1$: 1. if n is even and a and b are both nonnegative, then $\sqrt[n]{ab} = \sqrt[n]{a} \cdot \sqrt[n]{b}$. 2. if n is odd, then $\sqrt[n]{ab} = \sqrt[n]{a} \cdot \sqrt[n]{b}$.

To simplify a square root, follow these steps:
1. Factor the radicand into as many squares as possible.
2. Use the Product Property to isolate the perfect squares.
3. Simplify each radical.

Quotient Property of Radicals	For any real numbers a and $b \neq 0$, and any integer $n > 1$, $\sqrt[n]{\frac{a}{b}} = \frac{\sqrt[n]{a}}{\sqrt[n]{b}}$, if all roots are defined.

To eliminate radicals from a denominator or fractions from a radicand, multiply the numerator and denominator by a quantity so that the radicand has an exact root.

Example 1 **Simplify $\sqrt[3]{-16a^5b^7}$.**

$$\sqrt[3]{-16a^5b^7} = \sqrt[3]{(-2)^3 \cdot 2 \cdot a^3 \cdot a^2 \cdot (b^2)^3 \cdot b} = -2ab^2\sqrt[3]{2a^2b}$$

Example 2 **Simplify $\sqrt{\frac{8x^3}{45y^5}}$.**

$\sqrt{\frac{8x^3}{45y^5}} = \frac{\sqrt{8x^3}}{\sqrt{45y^5}}$ Quotient Property

$= \frac{\sqrt{(2x)^2 \cdot 2x}}{\sqrt{(3y^2)^2 \cdot 5y}}$ Factor into squares.

$= \frac{\sqrt{(2x)^2} \cdot \sqrt{2x}}{\sqrt{(3y^2)^2} \cdot \sqrt{5y}}$ Product Property

$= \frac{2|x|\sqrt{2x}}{3y^2\sqrt{5y}}$ Simplify.

$= \frac{2|x|\sqrt{2x}}{3y^2\sqrt{5y}} \cdot \frac{\sqrt{5y}}{\sqrt{5y}}$ Rationalize the denominator.

$= \frac{2|x|\sqrt{10xy}}{15y^3}$ Simplify.

Exercises

Simplify.

1. $5\sqrt{54}$ **$15\sqrt{6}$**
2. $\sqrt[4]{32a^9b^{20}}$ **$2a^2|b^5|\sqrt[4]{2a}$**
3. $\sqrt{75x^4y^7}$ **$5x^2y^3\sqrt{5y}$**
4. $\sqrt{\frac{36}{125}}$ **$\frac{6\sqrt{5}}{25}$**
5. $\sqrt{\frac{a^6b^3}{98}}$ **$\frac{|a^3|b\sqrt{2b}}{14}$**
6. $\sqrt[3]{\frac{p^5q^3}{40}}$ **$\frac{pq\sqrt[3]{5p^2}}{10}$**

Skills Practice, p. 271 and Practice, p. 272 (shown)

Simplify.

1. $\sqrt{540}$ **$6\sqrt{15}$**
2. $\sqrt[3]{-432}$ **$-6\sqrt[3]{2}$**
3. $\sqrt[3]{128}$ **$4\sqrt[3]{2}$**
4. $-\sqrt[4]{405}$ **$-3\sqrt[4]{5}$**
5. $\sqrt[3]{-5000}$ **$-10\sqrt[3]{5}$**
6. $\sqrt[5]{-1215}$ **$-3\sqrt[5]{5}$**
7. $\sqrt[3]{125t^6w^2}$ **$5t^2\sqrt[3]{w^2}$**
8. $\sqrt[4]{48v^8z^{13}}$ **$2v^2z^3\sqrt[4]{3z}$**
9. $\sqrt[3]{8g^3k^8}$ **$2gk^2\sqrt[3]{k^2}$**
10. $\sqrt{45x^3y^8}$ **$3xy^4\sqrt{5x}$**
11. $\sqrt{\frac{11}{9}}$ **$\frac{\sqrt{11}}{3}$**
12. $\sqrt[3]{\frac{216}{24}}$ **$\sqrt[3]{9}$**
13. $\sqrt{\frac{1}{128}c^4d^7}$ **$\frac{1}{16}c^2d^3\sqrt{2d}$**
14. $\sqrt{\frac{9a^5}{64b^4}}$ **$\frac{3a^2\sqrt{a}}{8b^2}$**
15. $\sqrt[4]{\frac{8}{9a^3}}$ **$\frac{\sqrt[4]{72a}}{3a}$**
16. $(3\sqrt{15})(-4\sqrt{45})$ **$-180\sqrt{3}$**
17. $(2\sqrt{24})(7\sqrt{18})$ **$168\sqrt{3}$**
18. $\sqrt{810} + \sqrt{240} - \sqrt{250}$ **$4\sqrt{10} + 4\sqrt{15}$**
19. $6\sqrt{20} + 8\sqrt{5} - 5\sqrt{45}$ **$5\sqrt{5}$**
20. $8\sqrt{48} - 6\sqrt{75} + 7\sqrt{80}$ **$2\sqrt{3} + 28\sqrt{5}$**
21. $(3\sqrt{2} + 2\sqrt{3})^2$ **$30 + 12\sqrt{6}$**
22. $(3 - \sqrt{7})^2$ **$16 - 6\sqrt{7}$**
23. $(\sqrt{5} - \sqrt{6})(\sqrt{5} + \sqrt{2})$ **$5 + \sqrt{10} - \sqrt{30} - 2\sqrt{3}$**
24. $(\sqrt{2} + \sqrt{10})(\sqrt{2} - \sqrt{10})$ **-8**
25. $(1 + \sqrt{6})(5 - \sqrt{7})$ **$5 - \sqrt{7} + 5\sqrt{6} - \sqrt{42}$**
26. $(\sqrt{3} + 4\sqrt{7})^2$ **$115 + 8\sqrt{21}$**
27. $(\sqrt{108} - 6\sqrt{3})^2$ **0**
28. $\frac{\sqrt{3}}{\sqrt{5} - 2}$ **$\sqrt{15} + 2\sqrt{3}$**
29. $\frac{6}{\sqrt{2} - 1}$ **$6\sqrt{2} + 6$**
30. $\frac{5 + \sqrt{3}}{4 + \sqrt{3}}$ **$\frac{17 - \sqrt{3}}{13}$**
31. $\frac{3 + \sqrt{2}}{2 - \sqrt{2}}$ **$\frac{8 + 5\sqrt{2}}{2}$**
32. $\frac{3 + \sqrt{6}}{5 - \sqrt{24}}$ **$27 + 11\sqrt{6}$**
33. $\frac{3 + \sqrt{x}}{2 - \sqrt{x}}$ **$\frac{6 + 5\sqrt{x} + x}{4 - x}$**

34. **BRAKING** The formula $s = 2\sqrt{5\ell}$ estimates the speed s in miles per hour of a car when it leaves skid marks ℓ feet long. Use the formula to write a simplified expression for s if $\ell = 85$. Then evaluate s to the nearest mile per hour. **$10\sqrt{17}$; 41 mi/h**

35. **PYTHAGOREAN THEOREM** The measures of the legs of a right triangle can be represented by the expressions $6x^2y$ and $9x^2y$. Use the Pythagorean Theorem to find a simplified expression for the measure of the hypotenuse. **$3x^2|y|\sqrt{13}$**

Reading to Learn Mathematics, p. 273 ELL

Pre-Activity **How do radical expressions apply to falling objects?**

Read the introduction to Lesson 5-6 at the top of page 250 in your textbook.

Describe how you could use the formula given in your textbook and a calculator to find the time, to the nearest tenth of a second, that it would take for the water balloons to drop 22 feet. (Do not actually calculate the time.) **Sample answer: Multiply 22 by 2 (giving 44) and divide by 32. Use the calculator to find the square root of the result. Round this square root to the nearest tenth.**

Reading the Lesson

1. Complete the conditions that must be met for a radical expression to be in simplified form.
   - The **index** n is as **small** as possible.
   - The **radicand** contains no **factors** (other than 1) that are nth powers of a(n) **integer** or polynomial.
   - The radicand contains no **fractions**.
   - No **radicals** appear in the **denominator**.

2. a. What are conjugates of radical expressions used for? **to rationalize binomial denominators**

   b. How would you use a conjugate to simplify the radical expression $\frac{1 + \sqrt{2}}{3 - \sqrt{2}}$? **Multiply numerator and denominator by $3 + \sqrt{2}$.**

   c. In order to simplify the radical expression in part b, two multiplications are necessary. The multiplication in the numerator would be done by the **FOIL** method, and the multiplication in the denominator would be done by finding the **difference** of **two** **squares**.

Helping You Remember

3. One way to remember something is to explain it to another person. When rationalizing the denominator in the expression $\frac{1}{\sqrt[3]{2}}$, many students think they should multiply numerator and denominator by $\frac{\sqrt[3]{2}}{\sqrt[3]{2}}$. How would you explain to a classmate why this is incorrect and what he should do instead. **Sample answer: Because you are working with cube roots, not square roots, you need to make the radicand in the denominator a perfect cube, not a perfect square. Multiply numerator and denominator by $\frac{\sqrt[3]{4}}{\sqrt[3]{4}}$ to make the denominator $\sqrt[3]{8}$, which equals 2.**

Enrichment, p. 274

Special Products with Radicals

Notice that $(\sqrt{3})(\sqrt{3}) = 3$, or $(\sqrt{3})^2 = 3$.

In general, $(\sqrt{x})^2 = x$ when $x \geq 0$.

Also, notice that $(\sqrt{9})(\sqrt{4}) = \sqrt{36}$.

In general, $(\sqrt{x})(\sqrt{y}) = \sqrt{xy}$ when x and y are not negative.

You can use these ideas to find the special products below.

$(\sqrt{a} + \sqrt{b})(\sqrt{a} - \sqrt{b}) = (\sqrt{a})^2 - (\sqrt{b})^2 = a - b$

$(\sqrt{a} + \sqrt{b})^2 = (\sqrt{a})^2 + 2\sqrt{ab} + (\sqrt{b})^2 = a + 2\sqrt{ab} + b$

$(\sqrt{a} - \sqrt{b})^2 = (\sqrt{a})^2 - 2\sqrt{ab} + (\sqrt{b})^2 = a - 2\sqrt{ab} + b$

Example 1 **Find the product: $(\sqrt{2} + \sqrt{5})(\sqrt{2} - \sqrt{5})$.**

$(\sqrt{2} + \sqrt{5})(\sqrt{2} - \sqrt{5}) = (\sqrt{2})^2 - (\sqrt{5})^2 = 2 - 5 = -3$

Example 2 **Evaluate $(\sqrt{2} + \ldots$**

Lesson 5-6 Radical Expressions 255

4 Assess

Open-Ended Assessment

Speaking Ask students to describe how combining radicals is the same as combining expressions with variables, and how it differs from working with variables.

Tips for New Teachers

Intervention Students will need to simplify expressions involving radicals in much of their further work in algebra. Take time to help students uncover and correct their misconceptions by analyzing the errors they make.

Getting Ready for Lesson 5-7

BASIC SKILL Lesson 5-7 presents working with rational exponents. This often involves adding, subtracting, or multiplying fractions. Exercises 75–82 should be used to determine your students' familiarity with rational numbers.

Assessment Options

Practice Quiz 2 The quiz provides students with a brief review of the concepts and skills in Lessons 5-4 through 5-6. Lesson numbers are given to the right of exercises or instruction lines so students can review concepts not yet mastered.

Maintain Your Skills

Mixed Review

Simplify. *(Lesson 5-5)*

59. $\sqrt{144z^8}$ **$12z^4$** 60. $\sqrt[3]{216a^3b^9}$ **$6ab^3$** 61. $\sqrt{(y+2)^2}$ **$|y+2|$**

Simplify. Assume that no denominator is equal to 0. *(Lesson 5-4)*

62. $\frac{x^2+5x-14}{x^2-6x+8}$ **$\frac{x+7}{x-4}$** 63. $\frac{x^2-3x-4}{x^2-16}$ **$\frac{x+1}{x+4}$**

Perform the indicated operations. *(Lesson 4-2)*

64. $\begin{bmatrix}3 & -4\\ 2 & 8\\ 0 & 1\end{bmatrix}+\begin{bmatrix}-5 & 0\\ 7 & 7\\ 3 & -6\end{bmatrix}$ **$\begin{bmatrix}-2 & -4\\ 9 & 15\\ 3 & -5\end{bmatrix}$** 65. $\begin{bmatrix}3 & 3\\ 0 & -2\end{bmatrix}-\begin{bmatrix}2 & -1\\ 5 & 2\end{bmatrix}$ **$\begin{bmatrix}1 & 4\\ -5 & -4\end{bmatrix}$**

66. Find the maximum and minimum values of the function $f(x, y) = 2x + 3y$ for the region with vertices at (2, 4), (−1, 3), (−3, −3), and (2, −5). *(Lesson 3-4)* **16, −15**

67. State whether the system of equations shown at the right is *consistent and independent*, *consistent and dependent*, or *inconsistent*. *(Lesson 3-1)*
consistent and independent

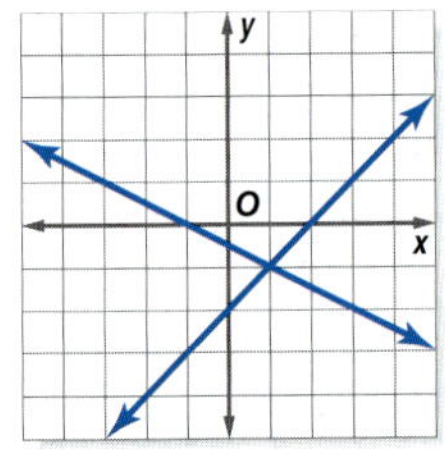

68. **BUSINESS** The amount that a mail-order company charges for shipping and handling is given by the function $c(x) = 3 + 0.15x$, where x is the weight in pounds. Find the charge for an 8-pound order. *(Lesson 2-2)* **$4.20**

Solve. *(Lessons 1-3, 1-4, and 1-5)*

69. $2x + 7 = -3$ **−5** 70. $-5x + 6 = -4$ **2**
71. $|x - 1| = 3$ **−2, 4** 72. $|3x + 2| = 5$ **$-\frac{7}{3}$, 1**
73. $2x - 4 > 8$ **$\{x \mid x > 6\}$** 74. $-x - 3 \le 4$ **$\{x \mid x \ge -7\}$**

Getting Ready for the Next Lesson

BASIC SKILL **Evaluate each expression.**

75. $2\left(\frac{1}{8}\right)$ **$\frac{1}{4}$** 76. $3\left(\frac{1}{6}\right)$ **$\frac{1}{2}$** 77. $\frac{1}{2}+\frac{1}{3}$ **$\frac{5}{6}$** 78. $\frac{1}{3}+\frac{3}{4}$ **$\frac{13}{12}$**

79. $\frac{1}{8}+\frac{5}{12}$ **$\frac{13}{24}$** 80. $\frac{5}{6}-\frac{1}{5}$ **$\frac{19}{30}$** 81. $\frac{5}{8}-\frac{1}{4}$ **$\frac{3}{8}$** 82. $\frac{1}{4}-\frac{2}{3}$ **$-\frac{5}{12}$**

Practice Quiz 2 — Lessons 5-4 through 5-6

Factor completely. If the polynomial is not factorable, write prime. *(Lesson 5-4)*

1. $3x^3y + x^2y^2 + x^2y$ **$x^2y(3x + y + 1)$** 2. $3x^2 - 2x - 2$ **prime**
3. $ax^2 + 6ax + 9a$ **$a(x + 3)^2$** 4. $8r^3 - 64s^6$ **$8(r - 2s^2)(r^2 + 2rs^2 + 4s^4)$**

Simplify. *(Lessons 5-5 and 5-6)*

5. $\sqrt{36x^2y^6}$ **$6|x||y^3|$** 6. $\sqrt[3]{-64a^6b^9}$ **$-4a^2b^3$** 7. $\sqrt{4n^2+12n+9}$ **$|2n+3|$**

8. $\sqrt{\frac{x^4}{y^3}}$ **$\frac{x^2\sqrt{y}}{y^2}$** 9. $(3+\sqrt{7})(2-\sqrt{7})$ **$-1-\sqrt{7}$** 10. $\frac{5+\sqrt{2}}{2+\sqrt{2}}$ **$\frac{8-3\sqrt{2}}{2}$**

256 Chapter 5 Polynomials

5-7 Rational Exponents

What You'll Learn

- Write expressions with rational exponents in radical form, and vice versa.
- Simplify expressions in exponential or radical form.

How do rational exponents apply to astronomy?

Astronomers refer to the space around a planet where the planet's gravity is stronger than the Sun's as the *sphere of influence* of the planet. The radius r of the sphere of influence is given by the formula $r = D\left(\frac{M_p}{M_S}\right)^{\frac{2}{5}}$, where M_p is the mass of the planet, M_S is the mass of the Sun, and D is the distance between the planet and the Sun.

RATIONAL EXPONENTS AND RADICALS You know that squaring a number and taking the square root of a number are inverse operations. But how would you evaluate an expression that contains a fractional exponent such as the one above? You can investigate such an expression by assuming that fractional exponents behave as integral exponents.

$$\left(b^{\frac{1}{2}}\right)^2 = b^{\frac{1}{2}} \cdot b^{\frac{1}{2}}$$ Write the square as multiplication.

$$= b^{\frac{1}{2} + \frac{1}{2}}$$ Add the exponents.

$$= b^1 \text{ or } b$$ Simplify.

Thus, $b^{\frac{1}{2}}$ is a number whose square equals b. So it makes sense to define $b^{\frac{1}{2}} = \sqrt{b}$.

Key Concept — $b^{\frac{1}{n}}$

- **Words** For any real number b and for any positive integer n, $b^{\frac{1}{n}} = \sqrt[n]{b}$, except when $b < 0$ and n is even.
- **Example** $8^{\frac{1}{3}} = \sqrt[3]{8}$ or 2

Example 1 Radical Form

Write each expression in radical form.

a. $a^{\frac{1}{4}}$

$a^{\frac{1}{4}} = \sqrt[4]{a}$ Definition of $b^{\frac{1}{n}}$

b. $x^{\frac{1}{5}}$

$x^{\frac{1}{5}} = \sqrt[5]{x}$ Definition of $b^{\frac{1}{n}}$

5-7 Lesson Notes

1 Focus

5-Minute Check Transparency 5-7 Use as a quiz or review of Lesson 5-6.

Mathematical Background notes are available for this lesson on p. 220D.

How do rational exponents apply to astronomy?

Ask students:

- Is the p in M_p an exponent? Is it a variable? **No, it is neither an exponent nor a variable; it is a subscript.**
- Would you expect the radius of the sphere of influence for one of the larger planets in our solar system to be greater than the radius of the sphere of influence for Earth? Use the formula to justify your answer. **Yes; for the planets larger than Earth, the value of M_p would be greater than the value of M_p for Earth while the value of M_S is the same. So the value of the ratio $\frac{M_p}{M_S}$ is greater for the larger planets.**

Resource Manager

Workbook and Reproducible Masters

Chapter 5 Resource Masters
- Study Guide and Intervention, pp. 275–276
- Skills Practice, p. 277
- Practice, p. 278
- Reading to Learn Mathematics, p. 279
- Enrichment, p. 280
- Assessment, p. 308

Graphing Calculator and Spreadsheet Masters, p. 36

Transparencies

5-Minute Check Transparency 5-7
Answer Key Transparencies

Technology

Interactive Chalkboard

2 Teach

RATIONAL EXPONENTS AND RADICALS

In-Class Examples

Power Point®

1 Write each expression in radical form.

a. $a^{\frac{1}{6}}$ $\sqrt[6]{a}$

b. $m^{\frac{1}{2}}$ $\sqrt{m}$

2 Write each radical using rational exponents.

a. $\sqrt[5]{b}$ $b^{\frac{1}{5}}$

b. $\sqrt{w}$ $w^{\frac{1}{2}}$

3 Evaluate each expression.

a. $49^{-\frac{1}{2}}$ $\frac{1}{7}$

b. $32^{\frac{2}{5}}$ 4

Teaching Tip If students are having difficulty remembering which part of the fractional exponent is the index, suggest that they recall the basic definition $b^{\frac{1}{2}} = \sqrt{b}$.

Example 2 Exponential Form

Write each radical using rational exponents.

a. $\sqrt[3]{y}$

$\sqrt[3]{y} = y^{\frac{1}{3}}$ Definition of $b^{\frac{1}{n}}$

b. $\sqrt[8]{c}$

$\sqrt[8]{c} = c^{\frac{1}{8}}$ Definition of $b^{\frac{1}{n}}$

Many expressions with fractional exponents can be evaluated using the definition of $b^{\frac{1}{n}}$ or the properties of powers.

Study Tip

Negative Base
Suppose the base of a monomial is negative such as $(-9)^2$ or $(-9)^3$. The expression is undefined if the exponent is even because there is no number that, when multiplied an even number of times, results in a negative number. However, the expression is defined for an odd exponent.

Example 3 Evaluate Expressions with Rational Exponents

Evaluate each expression.

a. $16^{-\frac{1}{4}}$

Method 1

$16^{-\frac{1}{4}} = \frac{1}{16^{\frac{1}{4}}}$ $b^{-n} = \frac{1}{b^n}$

$= \frac{1}{\sqrt[4]{16}}$ $16^{\frac{1}{4}} = \sqrt[4]{16}$

$= \frac{1}{\sqrt[4]{2^4}}$ $16 = 2^4$

$= \frac{1}{2}$ Simplify.

Method 2

$16^{-\frac{1}{4}} = (2^4)^{-\frac{1}{4}}$ $16 = 2^4$

$= 2^{4\left(-\frac{1}{4}\right)}$ Power of a Power

$= 2^{-1}$ Multiply exponents.

$= \frac{1}{2}$ $2^{-1} = \frac{1}{2^1}$

b. $243^{\frac{3}{5}}$

Method 1

$243^{\frac{3}{5}} = 243^{3\left(\frac{1}{5}\right)}$ Factor.

$= (243^3)^{\frac{1}{5}}$ Power of a Power

$= \sqrt[5]{243^3}$ $b^{\frac{1}{5}} = \sqrt[5]{b}$

$= \sqrt[5]{(3^5)^3}$ $243 = 3^5$

$= \sqrt[5]{3^5 \cdot 3^5 \cdot 3^5}$ Expand the cube.

$= 3 \cdot 3 \cdot 3$ or 27 Find the fifth root.

Method 2

$243^{\frac{3}{5}} = (3^5)^{\frac{3}{5}}$ $243 = 3^5$

$= 3^{5\left(\frac{3}{5}\right)}$ Power of a Power

$= 3^3$ Multiply exponents.

$= 27$ $3^3 = 3 \cdot 3 \cdot 3$

In Example 3b, Method 1 uses a combination of the definition of $b^{\frac{1}{n}}$ and the properties of powers. This example suggests the following general definition of rational exponents.

Key Concept — Rational Exponents

- **Words** For any nonzero real number b, and any integers m and n, with $n > 1$, $b^{\frac{m}{n}} = \sqrt[n]{b^m} = \left(\sqrt[n]{b}\right)^m$, except when $b < 0$ and n is even.
- **Example** $8^{\frac{2}{3}} = \sqrt[3]{8^2} = \left(\sqrt[3]{8}\right)^2$ or 4

In general, we define $b^{\frac{m}{n}}$ as $\left(b^{\frac{1}{n}}\right)^m$ or $(b^m)^{\frac{1}{n}}$. Now apply the definition of $b^{\frac{1}{n}}$ to $\left(b^{\frac{1}{n}}\right)^m$ and $(b^m)^{\frac{1}{n}}$.

$\left(b^{\frac{1}{n}}\right)^m = \left(\sqrt[n]{b}\right)^m$ $(b^m)^{\frac{1}{n}} = \sqrt[n]{b^m}$

DAILY INTERVENTION — Unlocking Misconceptions

- **Exponents** Students may be confused because they are not perceiving and reading the exponent in a way that distinguishes it from a coefficient or multiplier. Ask them to practice reading the exponent correctly, for example, reading x^3 as "x to the third power" or as "x cubed.".
- **Radicals** Ask students to practice reading radical expressions correctly, for example, reading $\sqrt{y^3}$ as "the square root of y cubed."

More About. . .

Weight Lifting

With origins in both the ancient Egyptian and Greek societies, weightlifting was among the sports on the program of the first Modern Olympic Games, in 1896, in Athens, Greece.

Source: International Weightlifting Association

Example 4 Rational Exponent with Numerator Other Than 1

WEIGHT LIFTING **The formula $M = 512 - 146{,}230B^{-\frac{8}{5}}$ can be used to estimate the maximum total mass that a weight lifter of mass B kilograms can lift in two lifts, the snatch and the clean and jerk, combined.**

a. According to the formula, what is the maximum amount that 2000 Olympic champion Xugang Zhan of China can lift if he weighs 72 kilograms?

$M = 512 - 146{,}230B^{-\frac{8}{5}}$ Original formula

$= 512 - 146{,}230(72)^{-\frac{8}{5}}$ $B = 72$

≈ 356 kg Use a calculator.

The formula predicts that he can lift at most 356 kilograms.

b. Xugang Zhan's winning total in the 2000 Olympics was 367.50 kg. Compare this to the value predicted by the formula.

The formula prediction is close to the actual weight, but slightly lower.

SIMPLIFY EXPRESSIONS All of the properties of powers you learned in Lesson 5-1 apply to rational exponents. When simplifying expressions containing rational exponents, leave the exponent in rational form rather than writing the expression as a radical. To simplify such an expression, you must write the expression with all positive exponents. Furthermore, any exponents in the denominator of a fraction must be positive *integers*. So, it may be necessary to rationalize a denominator.

Example 5 Simplify Expressions with Rational Exponents

Simplify each expression.

a. $x^{\frac{1}{5}} \cdot x^{\frac{7}{5}}$

$x^{\frac{1}{5}} \cdot x^{\frac{7}{5}} = x^{\frac{1}{5} + \frac{7}{5}}$ Multiply powers.

$= x^{\frac{8}{5}}$ Add exponents.

b. $y^{-\frac{3}{4}}$

$y^{-\frac{3}{4}} = \frac{1}{y^{\frac{3}{4}}}$ $b^{-n} = \frac{1}{b^n}$

$= \frac{1}{y^{\frac{3}{4}}} \cdot \frac{y^{\frac{1}{4}}}{y^{\frac{1}{4}}}$ Why use $\frac{y^{\frac{1}{4}}}{y^{\frac{1}{4}}}$?

$= \frac{y^{\frac{1}{4}}}{y^{\frac{4}{4}}}$ $y^{\frac{3}{4}} \cdot y^{\frac{1}{4}} = y^{\frac{3}{4} + \frac{1}{4}}$

$= \frac{y^{\frac{1}{4}}}{y}$ $y^{\frac{4}{4}} = y^1$ or y

TEACHING TIP

Tell students that if they are simplifying an expression that was originally written with radicals, they should write the answer with radicals. If the expression was originally written with rational exponents, they should write the answer with rational exponents.

When simplifying a radical expression, always use the smallest index possible. Using rational exponents makes this process easier, but the answer should be written in radical form.

www.algebra2.com/extra_examples

In-Class Example Power Point®

Teaching Tip Be sure students notice that the fractional exponent is negative in the formula for the maximum total mass M.

4 **WEIGHT LIFTING** Use the formula given in Example 4.

a. U.S. weightlifter Oscar Chaplin III competed in the same weight class as Xugang Zhan, finishing in 7th place. According to the formula, what is the maximum that Chaplin can lift if he weighs 77 kilograms? **Source:** cnnsi.com **The formula predicts that he can lift at most 372 kilograms.**

b. Oscar Chaplin's total in the 2000 Olympics was 335 kg. Compare this to the value predicted by the formula. **The formula prediction is somewhat higher than his actual total.**

SIMPLIFY EXPRESSIONS

In-Class Example Power Point®

5 Simplify each expression.

a. $y^{\frac{1}{7}} \cdot y^{\frac{4}{7}}$ $y^{\frac{5}{7}}$

b. $x^{-\frac{2}{3}}$ $\frac{x^{\frac{1}{3}}}{x}$

DAILY INTERVENTION — Differentiated Instruction

Auditory/Musical Have students name and demonstrate on a keyboard, guitar, or other instrument, the sounds of the notes described in Exercises 67 and 68. Other students can work to associate the sound of the note with the number of vibrations per second given by the formula.

In-Class Example

Power Point®

6 Simplify each expression.

a. $\dfrac{\sqrt[6]{16}}{\sqrt[3]{2}}$ $2^{\frac{1}{3}}$ or $\sqrt[3]{2}$

b. $\sqrt[6]{4x^4}$ $\sqrt[3]{2x^2}$

c. $\dfrac{y^{\frac{1}{2}}+1}{y^{\frac{1}{2}}-1}$ $\dfrac{y+2y^{\frac{1}{2}}+1}{y-1}$

3 Practice/Apply

Study Notebook

Have students—

- add the definitions/examples of the vocabulary terms to their Vocabulary Builder worksheets for Chapter 5.
- add the information listed in the Concept Summary below Example 6 to their notebook.
- include any other item(s) that they find helpful in mastering the skills in this lesson.

About the Exercises...

Organization by Objective

- Rational Exponents and Radicals: 21–40
- Simplify Expressions: 41–64

Odd/Even Assignments

Exercises 21–66 are structured so that students practice the same concepts whether they are assigned odd or even problems.

Assignment Guide

Basic: 21–61 odd, 65, 69–84

Average: 21–65 odd, 69–84

Advanced: 22–66 even, 67, 68, 70–80 (optional: 81–84)

Example 6 Simplify Radical Expressions

Simplify each expression.

a. $\dfrac{\sqrt[8]{81}}{\sqrt[6]{3}}$

$\dfrac{\sqrt[8]{81}}{\sqrt[6]{3}} = \dfrac{81^{\frac{1}{8}}}{3^{\frac{1}{6}}}$ Rational exponents

$= \dfrac{(3^4)^{\frac{1}{8}}}{3^{\frac{1}{6}}}$ $81 = 3^4$

$= \dfrac{3^{\frac{1}{2}}}{3^{\frac{1}{6}}}$ Power of a Power

$= 3^{\frac{1}{2}-\frac{1}{6}}$ Quotient of Powers

$= 3^{\frac{1}{3}}$ or $\sqrt[3]{3}$ Simplify.

b. $\sqrt[4]{9z^2}$

$\sqrt[4]{9z^2} = (9z^2)^{\frac{1}{4}}$ Rational exponents

$= (3^2 \cdot z^2)^{\frac{1}{4}}$ $9 = 3^2$

$= 3^{2\left(\frac{1}{4}\right)} \cdot z^{2\left(\frac{1}{4}\right)}$ Power of a Power

$= 3^{\frac{1}{2}} \cdot z^{\frac{1}{2}}$ Multiply.

$= \sqrt{3} \cdot \sqrt{z}$ $3^{\frac{1}{2}} = \sqrt{3}, z^{\frac{1}{2}} = \sqrt{z}$

$= \sqrt{3z}$ Simplify.

c. $\dfrac{m^{\frac{1}{2}}-1}{m^{\frac{1}{2}}+1}$

$\dfrac{m^{\frac{1}{2}}-1}{m^{\frac{1}{2}}+1} = \dfrac{m^{\frac{1}{2}}-1}{m^{\frac{1}{2}}+1} \cdot \dfrac{m^{\frac{1}{2}}-1}{m^{\frac{1}{2}}-1}$ $m^{\frac{1}{2}}-1$ is the conjugate of $m^{\frac{1}{2}}+1$.

$= \dfrac{m-2m^{\frac{1}{2}}+1}{m-1}$ Multiply.

Concept Summary — Expressions with Rational Exponents

An expression with rational exponents is simplified when all of the following conditions are met.

- It has no negative exponents.
- It has no fractional exponents in the denominator.
- It is not a complex fraction.
- The index of any remaining radical is the least number possible.

Check for Understanding

Concept Check

1. **OPEN ENDED** Determine a value of b for which $b^{\frac{1}{6}}$ is an integer.
2. Explain why $(-16)^{\frac{1}{2}}$ is not a real number. See margin.
3. Explain why $\sqrt[n]{b^m} = \left(\sqrt[n]{b}\right)^m$. See margin.

1. Sample answer: 64

Answers

2. In radical form, the expression would be $\sqrt{-16}$, which is not a real number because the index is even and the radicand is negative.

3. In exponential form $\sqrt[n]{b^m}$ is equal to $(b^m)^{\frac{1}{n}}$. By the Power of a Power Property, $(b^m)^{\frac{1}{n}} = b^{\frac{m}{n}}$. But, $b^{\frac{m}{n}}$ is also equal to $\left(b^{\frac{1}{n}}\right)^m$ by the Power of a Power Property. This last expression is equal to $\left(\sqrt[n]{b}\right)^m$. Thus, $\sqrt[n]{b^m} = \left(\sqrt[n]{b}\right)^m$.

Guided Practice

GUIDED PRACTICE KEY	
Exercises	Examples
4, 5	1
6, 7	2
8–11	3
12–17	5
18, 19	6
20	4

Write each expression in radical form.

4. $7^{\frac{1}{3}}$ $\sqrt[3]{7}$
5. $x^{\frac{2}{3}}$ $\sqrt[3]{x^2}$ or $(\sqrt[3]{x})^2$

Write each radical using rational exponents.

6. $\sqrt[4]{26}$ $26^{\frac{1}{4}}$
7. $\sqrt[3]{6x^5y^7}$ $6^{\frac{1}{3}}x^{\frac{5}{3}}y^{\frac{7}{3}}$

Evaluate each expression.

8. $125^{\frac{1}{3}}$ 5
9. $81^{-\frac{1}{4}}$ $\frac{1}{3}$
10. $27^{\frac{2}{3}}$ 9
11. $\frac{54}{9^{\frac{3}{2}}}$ 2

Simplify each expression.

12. $a^{\frac{2}{3}} \cdot a^{\frac{1}{4}}$ $a^{\frac{11}{12}}$
13. $\frac{x^{\frac{5}{6}}}{x^{\frac{1}{6}}}$ $x^{\frac{2}{3}}$
14. $\frac{1}{2z^{\frac{1}{3}}}$ $\frac{z^{\frac{2}{3}}}{2z}$
15. $\frac{a^2}{b^{\frac{1}{3}}} \cdot \frac{b}{a^{\frac{1}{2}}}$ $a^{\frac{3}{2}}b^{\frac{2}{3}}$
16. $(mn^2)^{-\frac{1}{3}}$ $\frac{m^{\frac{2}{3}}n^{\frac{1}{3}}}{mn}$
17. $z(x - 2y)^{-\frac{1}{2}}$
18. $\sqrt[6]{27x^3}$ $\sqrt{3x}$
19. $\frac{\sqrt[4]{27}}{\sqrt[4]{3}}$ $\sqrt{3}$

17. $\frac{z(x-2y)^{\frac{1}{2}}}{x-2y}$

Application

20. **ECONOMICS** When inflation causes the price of an item to increase, the new cost C and the original cost c are related by the formula $C = c(1 + r)^n$, where r is the rate of inflation per year as a decimal and n is the number of years. What would be the price of a \$4.99 item after six months of 5% inflation? **\$5.11**

★ indicates increased difficulty

Practice and Apply

Homework Help	
For Exercises	See Examples
21–24	1
25–28	2
29–40	3
41–52, 64–66	5
53–63	6

Extra Practice
See page 838.

Write each expression in radical form.

21. $6^{\frac{1}{5}}$ $\sqrt[5]{6}$
22. $4^{\frac{1}{3}}$ $\sqrt[3]{4}$
23. $c^{\frac{2}{5}}$ $\sqrt[5]{c^2}$ or $(\sqrt[5]{c})^2$
24. $(x^2)^{\frac{4}{3}}$ $x^2\sqrt[3]{x^2}$

Write each radical using rational exponents.

25. $\sqrt{23}$ $23^{\frac{1}{2}}$
26. $\sqrt[3]{62}$ $62^{\frac{1}{3}}$
27. $\sqrt[4]{16z^2}$ $2z^{\frac{1}{2}}$
28. $\sqrt[3]{5x^2y}$ $5^{\frac{1}{3}}x^{\frac{2}{3}}y^{\frac{1}{3}}$

Evaluate each expression.

29. $16^{\frac{1}{4}}$ 2
30. $216^{\frac{1}{3}}$ 6
31. $25^{-\frac{1}{2}}$ $\frac{1}{5}$
32. $81^{-\frac{3}{4}}$ $\frac{1}{27}$
33. $(-27)^{-\frac{2}{3}}$ $\frac{1}{9}$
34. $(-32)^{-\frac{3}{5}}$ $-\frac{1}{8}$
35. $81^{-\frac{1}{2}} \cdot 81^{\frac{3}{2}}$ 81
36. $8^{\frac{3}{2}} \cdot 8^{\frac{5}{2}}$ 4096
37. $\left(\frac{8}{27}\right)^{\frac{1}{3}}$ $\frac{2}{3}$
38. $\left(\frac{1}{243}\right)^{-\frac{3}{5}}$ 27
39. $\frac{16^{\frac{1}{2}}}{9^{\frac{1}{2}}}$ $\frac{4}{3}$
40. $\frac{8^{\frac{1}{3}}}{64^{\frac{1}{3}}}$ $\frac{1}{2}$

Simplify each expression.

41. $y^{\frac{5}{3}} \cdot y^{\frac{7}{3}}$ y^4
42. $x^{\frac{3}{4}} \cdot x^{\frac{9}{4}}$ x^3
43. $\left(b^{\frac{1}{3}}\right)^{\frac{3}{5}}$ $b^{\frac{1}{5}}$
44. $\left(a^{-\frac{2}{3}}\right)^{-\frac{1}{6}}$ $a^{\frac{1}{9}}$
45. $w^{-\frac{4}{5}}$ $\frac{w^{\frac{1}{5}}}{w}$
46. $x^{-\frac{1}{6}}$ $\frac{x^{\frac{5}{6}}}{x}$
47. $\frac{t^{\frac{3}{4}}}{t^{\frac{1}{2}}}$ $t^{\frac{1}{4}}$
48. $\frac{r^{\frac{2}{3}}}{r^{\frac{1}{6}}}$ $r^{\frac{1}{2}}$
49. $\frac{a^{-\frac{1}{2}}}{6a^{\frac{1}{3}} \cdot a^{-\frac{1}{4}}}$ $\frac{a^{\frac{5}{12}}}{6a}$
50. $\frac{2c^{\frac{1}{8}}}{c^{-\frac{1}{16}} \cdot c^{\frac{1}{4}}}$ $\frac{2c^{\frac{15}{16}}}{c}$
51. $\frac{y^{\frac{3}{2}}}{y^{\frac{1}{2}} + 2}$ $\frac{y^2 - 2y^{\frac{3}{2}}}{y - 4}$
52. $\frac{x^{\frac{1}{2}} + 2}{x^{\frac{1}{2}} - 1}$ $\frac{x + 3x^{\frac{1}{2}} + 2}{x - 1}$
53. $\sqrt[4]{25}$ $\sqrt{5}$
54. $\sqrt[6]{27}$ $\sqrt{3}$
55. $\sqrt{17} \cdot \sqrt[3]{17^2}$ $17\sqrt[6]{17}$
56. $\sqrt[3]{5} \cdot \sqrt{5^3}$ $5\sqrt[6]{5^5}$
57. $\sqrt[8]{25x^4y^4}$ $\sqrt[4]{5x^2y^2}$
58. $\sqrt[6]{81a^4b^8}$ $b\sqrt[3]{9a^2b}$
59. $\frac{xy}{\sqrt{z}}$ $\frac{xy\sqrt{z}}{z}$
60. $\frac{ab}{\sqrt[3]{c}}$ $\frac{ab\sqrt[3]{c^2}}{c}$
61. $\sqrt{\sqrt{8}}$ $\sqrt[4]{2}$
62. $\sqrt{\sqrt[3]{36}}$ $\sqrt[3]{6}$

★ 63. $\frac{8^{\frac{1}{6}} - 9^{\frac{1}{4}}}{\sqrt{3} + \sqrt{2}}$ $2\sqrt{6} - 5$

★ 64. $\frac{x^{\frac{5}{3}} - x^{\frac{1}{3}}z^{\frac{4}{3}}}{x^{\frac{2}{3}} + z^{\frac{2}{3}}}$ $x - x^{\frac{1}{3}}z^{\frac{2}{3}}$

www.algebra2.com/self_check_quiz

Study Guide and Intervention, p. 275 (shown) and p. 276

Rational Exponents and Radicals

Definition of $b^{\frac{1}{n}}$	For any real number b and any positive integer n, $b^{\frac{1}{n}} = \sqrt[n]{b}$, except when $b < 0$ and n is even.
Definition of $b^{\frac{m}{n}}$	For any nonzero real number b, and any integers m and n, with $n > 1$, $b^{\frac{m}{n}} = \sqrt[n]{b^m} = (\sqrt[n]{b})^m$, except when $b < 0$ and n is even.

Example 1 Write $28^{\frac{1}{2}}$ in radical form.

Notice that $28 > 0$.

$28^{\frac{1}{2}} = \sqrt{28}$
$= \sqrt{2^2 \cdot 7}$
$= \sqrt{2^2} \cdot \sqrt{7}$
$= 2\sqrt{7}$

Example 2 Evaluate $\left(\frac{-8}{-125}\right)^{\frac{1}{3}}$.

Notice that $-8 < 0$, $-125 < 0$, and 3 is odd.

$\left(\frac{-8}{-125}\right)^{\frac{1}{3}} = \frac{\sqrt[3]{-8}}{\sqrt[3]{-125}}$
$= \frac{-2}{-5}$
$= \frac{2}{5}$

Exercises

Write each expression in radical form.

1. $11^{\frac{1}{7}}$ $\sqrt[7]{11}$
2. $15^{\frac{1}{3}}$ $\sqrt[3]{15}$
3. $300^{\frac{3}{2}}$ $\sqrt{300^3}$

Write each radical using rational exponents.

4. $\sqrt{47}$ $47^{\frac{1}{2}}$
5. $\sqrt[3]{3a^5b^2}$ $3^{\frac{1}{3}}a^{\frac{5}{3}}b^{\frac{2}{3}}$
6. $\sqrt[4]{162p^5}$ $3 \cdot 2^{\frac{1}{4}} \cdot p^{\frac{5}{4}}$

Evaluate each expression.

7. $-27^{\frac{2}{3}}$ 9
8. $\frac{5^{-\frac{1}{2}}}{2\sqrt{5}}$ $\frac{1}{10}$
9. $(0.0004)^{\frac{1}{2}}$ 0.02
10. $8^{\frac{2}{3}} \cdot 4^{\frac{3}{2}}$ 32
11. $\frac{144^{-\frac{1}{2}}}{27^{-\frac{1}{3}}}$ $\frac{1}{4}$
12. $\frac{16^{-\frac{1}{2}}}{(0.25)^{\frac{1}{2}}}$ $\frac{1}{2}$

Skills Practice, p. 277 and Practice, p. 278 (shown)

Write each expression in radical form.

1. $5^{\frac{1}{3}}$ $\sqrt[3]{5}$
2. $6^{\frac{2}{5}}$ $\sqrt[5]{6^2}$ or $(\sqrt[5]{6})^2$
3. $m^{\frac{4}{7}}$ $\sqrt[7]{m^4}$ or $(\sqrt[7]{m})^4$
4. $(n^3)^{\frac{2}{5}}$ $n\sqrt[5]{n}$

Write each radical using rational exponents.

5. $\sqrt{79}$ $79^{\frac{1}{2}}$
6. $\sqrt[4]{153}$ $153^{\frac{1}{4}}$
7. $\sqrt[3]{27m^6n^4}$ $3m^2n^{\frac{4}{3}}$
8. $5\sqrt{2a^{10}b}$ $5 \cdot 2^{\frac{1}{2}}|a^5|b^{\frac{1}{2}}$

Evaluate each expression.

9. $81^{\frac{1}{4}}$ 3
10. $1024^{-\frac{1}{5}}$ $\frac{1}{4}$
11. $8^{-\frac{5}{3}}$ $\frac{1}{32}$
12. $-256^{-\frac{3}{4}}$ $-\frac{1}{64}$
13. $(-64)^{-\frac{2}{3}}$ $\frac{1}{16}$
14. $27^{\frac{1}{3}} \cdot 27^{\frac{4}{3}}$ 243
15. $\left(\frac{125}{216}\right)^{\frac{2}{3}}$ $\frac{25}{36}$
16. $\frac{64^{\frac{2}{3}}}{343^{\frac{2}{3}}}$ $\frac{16}{49}$
17. $\left(25^{\frac{1}{2}}\right)\left(-64^{-\frac{1}{3}}\right)$ $-\frac{5}{4}$

Simplify each expression.

18. $g^{\frac{4}{7}} \cdot g^{\frac{3}{7}}$ g
19. $s^{\frac{3}{4}} \cdot s^{\frac{13}{4}}$ s^4
20. $\left(u^{-\frac{1}{3}}\right)^{-\frac{4}{5}}$ $u^{\frac{4}{15}}$
21. $y^{-\frac{1}{2}}$ $\frac{y^{\frac{1}{2}}}{y}$
22. $b^{-\frac{3}{5}}$ $\frac{b^{\frac{2}{5}}}{b}$
23. $\frac{q^{\frac{3}{5}}}{q^{\frac{2}{5}}}$ $q^{\frac{1}{5}}$
24. $\frac{t^{\frac{2}{3}}}{5t^{\frac{1}{2}} \cdot t^{-\frac{3}{4}}}$ $\frac{t^{\frac{11}{12}}}{5}$
25. $\frac{2z^{\frac{1}{2}}}{z^{\frac{1}{2}} - 1}$ $\frac{2z + 2z^{\frac{1}{2}}}{z - 1}$
26. $\sqrt[10]{8^5}$ $2\sqrt{2}$
27. $\sqrt{12} \cdot \sqrt[5]{12^3}$ $12\sqrt[10]{12}$
28. $\sqrt[4]{6} \cdot 3\sqrt[4]{6}$ $3\sqrt{6}$
29. $\frac{a}{\sqrt{3b}}$ $\frac{a\sqrt{3b}}{3b}$

30. **ELECTRICITY** The amount of current in amperes I that an appliance uses can be calculated using the formula $I = \left(\frac{P}{R}\right)^{\frac{1}{2}}$, where P is the power in watts and R is the resistance in ohms. How much current does an appliance use if P = 500 watts and R = 10 ohms? Round your answer to the nearest tenth. **7.1 amps**

31. **BUSINESS** A company that produces DVDs uses the formula $C = 88n^{\frac{1}{3}} + 330$ to calculate the cost C in dollars of producing n DVDs per day. What is the company's cost to produce 150 DVDs per day? Round your answer to the nearest dollar. **\$798**

Reading to Learn Mathematics, p. 279 ELL

Pre-Activity **How do rational exponents apply to astronomy?**

Read the introduction to Lesson 5-7 at the top of page 257 in your textbook. The formula in the introduction contains the exponent $\frac{2}{5}$. What do you think it might mean to raise a number to the $\frac{2}{5}$ power?
Sample answer: Take the fifth root of the number and square it.

Reading the Lesson

1. Complete the following definitions of rational exponents.
   - For any real number b and for any positive integer n, $b^{\frac{1}{n}}$ = ___$\sqrt[n]{b}$___ except when b ___< 0___ and n is ___even___.
   - For any nonzero real number b, and any integers m and n, with n ___> 1___, $b^{\frac{m}{n}}$ = ___$\sqrt[n]{b^m}$___ = ___$(\sqrt[n]{b})^m$___, except when b ___< 0___ and n is ___even___.
2. Complete the conditions that must be met in order for an expression with rational exponents to be simplified.
   - It has no ___negative___ exponents.
   - It has no ___fractional___ exponents in the ___denominator___.
   - It is not a ___complex___ fraction.
   - The ___index___ of any remaining ___radical___ is the ___least___ number possible.
3. Margarita and Pierre were working together on their algebra homework. One exercise asked them to evaluate the expression $27^{\frac{4}{3}}$. Margarita thought that they should raise 27 to the fourth power first and then take the cube root of the result. Pierre thought that they should take the cube root of 27 first and then raise the result to the fourth power. Whose method is correct? **Both methods are correct.**

Helping You Remember

4. Some students have trouble remembering which part of the fraction in a rational exponent gives the power and which part gives the root. How can your knowledge of integer exponents help you to keep this straight? **Sample answer: An integer exponent can be written as a rational exponent. For example, $2^3 = 2^{\frac{3}{1}}$. You know that this means that 2 is raised to the third power, so the numerator must give the power, and, therefore, the denominator must give the root.**

Enrichment, p. 280

Lesser-Known Geometric Formulas

Many geometric formulas involve radical expressions.

Make a drawing to illustrate each of the formulas given on this page. Then evaluate the formula for the given value of the variable. Round answers to the nearest hundredth.

1. The area of an isosceles triangle. Two sides have length a; the other side has length c. Find A when a = 6 and c = 7.
   $A = \frac{c}{4}\sqrt{4a^2 - c^2}$
   A ≈ 17.06

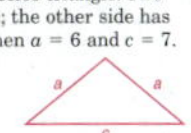

2. The area of an equilateral triangle with a side of length a. Find A when a = 8.
   $A = \frac{a^2}{4}\sqrt{3}$
   A ≈ 27.71

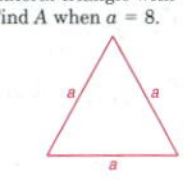

3. The area of a regular pentagon with a side of length a. Find A when a = 4.
4. The area of a regular hexagon with a side of length a. Find A when a = 9.

4 Assess

Open-Ended Assessment

Speaking Have students write two expressions with rational exponents, one that is in simplified form and another that is not. Ask them to explain the difference between them, using the four conditions listed in the Concept Summary on p. 260.

Tips for New Teachers

Intervention In order to help students see why the exception "except when $b < 0$ and n is even" is necessary when defining rational exponents, ask them to choose values for b and n that violate these constraints, and see what results when applying the definition.

Getting Ready for Lesson 5-8

PREREQUISITE SKILL Lesson 5-8 presents solving equations and inequalities that contain radicals. Solving such equations and inequalities involves finding the power of an expression involving a radical. Exercises 81–84 should be used to determine your students' familiarity with multiplying radicals.

Assessment Options

Quiz (Lessons 5-6 and 5-7) is available on p. 308 of the *Chapter 5 Resource Masters*.

65. Find the simplified form of $32^{\frac{1}{2}} + 3^{\frac{1}{2}} - 8^{\frac{1}{2}}$. $2^{\frac{3}{2}} + 3^{\frac{1}{2}}$

66. What is the simplified form of $81^{\frac{1}{3}} - 24^{\frac{1}{3}} + 3^{\frac{1}{3}}$? $2 \cdot 3^{\frac{1}{3}}$

More About. . .

Music

The first piano was made in about 1709 by Bartolomeo Cristofori, a maker of harpsichords in Florence, Italy.

Source: www.infoplease.com

MUSIC **For Exercises 67 and 68, use the following information.**
On a piano, the frequency of the A note above middle C should be set at 440 vibrations per second. The frequency f_n of a note that is n notes above that A should be $f_n = 440 \cdot 2^{\frac{n}{12}}$.

67. At what frequency should a piano tuner set the A that is one octave, or 12 notes, above the A above middle C? **880 vibrations per second**

68. Middle C is nine notes below the A that has a frequency of 440 vibrations per second. What is the frequency of middle C? **about 262 vibrations per second**

69. **BIOLOGY** Suppose a culture has 100 bacteria to begin with and the number of bacteria doubles every 2 hours. Then the number N of bacteria after t hours is given by $N = 100 \cdot 2^{\frac{t}{2}}$. How many bacteria will be present after 3 and a half hours? **about 336**

70. **CRITICAL THINKING** Explain how to solve $9^x = 3^{x + \frac{1}{2}}$ for x. **See margin.**

71. **WRITING IN MATH** Answer the question that was posed at the beginning of the lesson. **See pp. 283A–283B.**

How do rational exponents apply to astronomy?

Include the following in your answer:
- an explanation of how to write the formula $r = D\left(\frac{M_P}{M_S}\right)^{\frac{2}{5}}$ in radical form and simplify it, and
- an explanation of what happens to the value of r as the value of D increases assuming that M_p and M_S are constant.

Standardized Test Practice

72. Which is the value of $4^{\frac{1}{2}} + \left(\frac{1}{2}\right)^4$? **C**

(A) 1 (B) 2 (C) $2\frac{1}{16}$ (D) $2\frac{1}{2}$

73. If $4x + 2y = 5$ and $x - y = 1$, then what is the value of $3x + 3y$? **C**

(A) 1 (B) 2 (C) 4 (D) 6

Maintain Your Skills

Mixed Review **Simplify.** *(Lessons 5-5 and 5-6)*

74. $\sqrt{4x^3y^2}$ $2x|y|\sqrt{x}$

75. $(2\sqrt{6})(3\sqrt{12})$ $36\sqrt{2}$

76. $\sqrt{32} + \sqrt{18} - \sqrt{50}$ $2\sqrt{2}$

77. $\sqrt[4]{(-8)^4}$ 8

78. $4\sqrt{(x-5)^2}$ $4|x-5|$

79. $\sqrt{\frac{9}{36}x^4}$ $\frac{1}{2}x^2$

80. **BIOLOGY** Humans blink their eyes about once every 5 seconds. How many times do humans blink their eyes in two hours? *(Lesson 1-1)* **1440**

Getting Ready for the Next Lesson **PREREQUISITE SKILL** **Find each power.** *(To review **multiplying radicals**, see Lesson 5-6.)*

81. $(\sqrt{x-2})^2$ $x - 2$

82. $(\sqrt[3]{2x-3})^3$ $2x - 3$

83. $(\sqrt{x} + 1)^2$ $x + 2\sqrt{x} + 1$

84. $(2\sqrt{x} - 3)^2$ $4x - 12\sqrt{x} + 9$

Answer

70. Rewrite the equation so that the bases are the same on each side.

$$9^x = 3^{x + \frac{1}{2}}$$
$$(3^2)^x = 3^{x + \frac{1}{2}}$$
$$3^{2x} = 3^{x + \frac{1}{2}}$$

Since the bases are the same and this is an equation, the exponents must be equal. Solve $2x = x + \frac{1}{2}$. The result is $x = \frac{1}{2}$.

5-8 Radical Equations and Inequalities

What You'll Learn

- Solve equations containing radicals.
- Solve inequalities containing radicals.

How do radical equations apply to manufacturing?

Computer chips are made from the element silicon, which is found in sand. Suppose a company that manufactures computer chips uses the formula $C = 10n^{\frac{2}{3}} + 1500$ to estimate the cost C in dollars of producing n chips. This equation can be rewritten as a radical equation.

Vocabulary

- radical equation
- extraneous solution
- radical inequality

SOLVE RADICAL EQUATIONS Equations with radicals that have variables in the radicands are called **radical equations**. To solve this type of equation, raise each side of the equation to a power equal to the index of the radical to eliminate the radical.

Example 1 Solve a Radical Equation

Solve $\sqrt{x+1} + 2 = 4$.

$\sqrt{x+1} + 2 = 4$	Original equation
$\sqrt{x+1} = 2$	Subtract 2 from each side to isolate the radical.
$(\sqrt{x+1})^2 = 2^2$	Square each side to eliminate the radical.
$x + 1 = 4$	Find the squares.
$x = 3$	Subtract 1 from each side.

CHECK

$\sqrt{x+1} + 2 = 4$	Original equation
$\sqrt{3+1} + 2 \stackrel{?}{=} 4$	Replace x with 3.
$4 = 4$ ✓	Simplify.

The solution checks. The solution is 3.

When you solve a radical equation, it is very important that you check your solution. Sometimes you will obtain a number that does not satisfy the original equation. Such a number is called an **extraneous solution**. You can use a graphing calculator to predict the number of solutions of an equation or to determine whether the solution you obtain is reasonable.

Example 2 Extraneous Solution

Solve $\sqrt{x-15} = 3 - \sqrt{x}$.

$\sqrt{x-15} = 3 - \sqrt{x}$	Original equation
$(\sqrt{x-15})^2 = (3 - \sqrt{x})^2$	Square each side.
$x - 15 = 9 - 6\sqrt{x} + x$	Find the squares.
$-24 = -6\sqrt{x}$	Isolate the radical.
$4 = \sqrt{x}$	Divide each side by -6.
$4^2 = (\sqrt{x})^2$	Square each side again.
$16 = x$	Evaluate the squares.

(continued on the next page)

5-8 Lesson Notes

1 Focus

5-Minute Check Transparency 5-8 Use as a quiz or review of Lesson 5-7.

Mathematical Background notes are available for this lesson on p. 220D.

How do radical equations apply to manufacturing?

Ask students:

- Why can the equation be rewritten as a radical equation? **because the variable n has a rational exponent**
- **Manufacturing** There are production costs associated with manufactured goods that occur even before the first item is made. That is, there is still a cost even if no items have been produced yet. How much are these costs for the production of this company's computer chips? **$1500**

Resource Manager

Workbook and Reproducible Masters

Chapter 5 Resource Masters

- Study Guide and Intervention, pp. 281–282
- Skills Practice, p. 283
- Practice, p. 284
- Reading to Learn Mathematics, p. 285
- Enrichment, p. 286

Transparencies

5-Minute Check Transparency 5-8
Answer Key Transparencies

Technology

Alge2PASS: Tutorial Plus, Lesson 9
Interactive Chalkboard

2 Teach

SOLVE RADICAL EQUATIONS

In-Class Examples Power Point®

1 Solve $\sqrt{y-2} - 1 = 5$. **38**

2 Solve $\sqrt{x-12} = 2 - \sqrt{x}$. **no solution**

Teaching Tip Remind students that the square root sign in an equation means the principal root.

3 Solve $(3y + 1)^{\frac{1}{3}} + 5 = 0$. **−42**

Teaching Tip Have a discussion with students about which operations may introduce extraneous solutions when solving a radical equation.

CHECK

$$\sqrt{x-15} = 3 - \sqrt{x}$$
$$\sqrt{16-15} \stackrel{?}{=} 3 - \sqrt{16}$$
$$\sqrt{1} \stackrel{?}{=} 3 - 4$$
$$1 \neq -1$$

The solution does not check, so the equation has no real solution.

The graphing calculator screen shows the graphs of $y = \sqrt{x-15}$ and $y = 3 - \sqrt{x}$. The graphs do not intersect, which confirms that there is no solution.

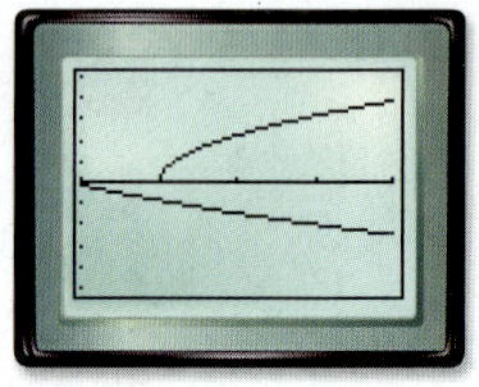

[10, 30] scl: 5 by [−5, 5] scl: 1

You can apply the same methods used in solving square root equations to solving equations with roots of any index. Remember that to undo a square root, you square the expression. To undo an nth root, you must raise the expression to the nth power.

Study Tip

Alternative Method

To solve a radical equation, you can substitute a variable for the radical expression. In Example 3, let $A = 5n - 1$.

$$3A^{\frac{1}{3}} - 2 = 0$$
$$3A^{\frac{1}{3}} = 2$$
$$A^{\frac{1}{3}} = \frac{2}{3}$$
$$A = \frac{8}{27}$$
$$5n - 1 = \frac{8}{27}$$
$$n = \frac{7}{27}$$

Example 3 Cube Root Equation

Solve $3(5n - 1)^{\frac{1}{3}} - 2 = 0$.

In order to remove the $\frac{1}{3}$ power, or cube root, you must first isolate it and then raise each side of the equation to the third power.

$3(5n-1)^{\frac{1}{3}} - 2 = 0$	Original equation
$3(5n-1)^{\frac{1}{3}} = 2$	Add 2 to each side.
$(5n-1)^{\frac{1}{3}} = \frac{2}{3}$	Divide each side by 3.
$\left[(5n-1)^{\frac{1}{3}}\right]^3 = \left(\frac{2}{3}\right)^3$	Cube each side.
$5n - 1 = \frac{8}{27}$	Evaluate the cubes.
$5n = \frac{35}{27}$	Add 1 to each side.
$n = \frac{7}{27}$	Divide each side by 5.

CHECK

$3(5n-1)^{\frac{1}{3}} - 2 = 0$	Original equation
$3\left(5 \cdot \frac{7}{27} - 1\right)^{\frac{1}{3}} - 2 \stackrel{?}{=} 0$	Replace n with $\frac{7}{27}$.
$3\left(\frac{8}{27}\right)^{\frac{1}{3}} - 2 \stackrel{?}{=} 0$	Simplify.
$3\left(\frac{2}{3}\right) - 2 \stackrel{?}{=} 0$	The cube root of $\frac{8}{27}$ is $\frac{2}{3}$.
$0 = 0 \checkmark$	Subtract.

The solution is $\frac{7}{27}$.

SOLVE RADICAL INEQUALITIES You can use what you know about radical equations to help solve radical inequalities. A **radical inequality** is an inequality that has a variable in a radicand.

264 Chapter 5 Polynomials

Example 4 Radical Inequality

Solve $2 + \sqrt{4x - 4} \leq 6$.

Since the radicand of a square root must be greater than or equal to zero, first solve $4x - 4 \geq 0$ to identify the values of x for which the left side of the given inequality is defined.

$$4x - 4 \geq 0$$
$$4x \geq 4$$
$$x \geq 1$$

Now solve $2 + \sqrt{4x - 4} \leq 6$.

$2 + \sqrt{4x - 4} \leq 6$ Original inequality

$\sqrt{4x - 4} \leq 4$ Isolate the radical.

$4x - 4 \leq 16$ Eliminate the radical.

$4x \leq 20$ Add 4 to each side.

$x \leq 5$ Divide each side by 4.

It appears that $1 \leq x \leq 5$. You can test some x values to confirm the solution. Let $f(x) = 2 + \sqrt{4x - 4}$. Use three test values: one less than 1, one between 1 and 5, and one greater than 5. Organize the test values in a table.

$x = 0$	$x = 2$	$x = 7$
$f(0) = 2 + \sqrt{4(0) - 4}$ $= 2 + \sqrt{-4}$ Since $\sqrt{-4}$ is not a real number, the inequality is not satisfied.	$f(2) = 2 + \sqrt{4(2) - 4}$ $= 4$ Since $4 \leq 6$, the inequality is satisfied.	$f(7) = 2 + \sqrt{4(7) - 4}$ ≈ 6.90 Since $6.90 \not\leq 6$, the inequality is not satisfied.

The solution checks. Only values in the interval $1 \leq x \leq 5$ satisfy the inequality. You can summarize the solution with a number line.

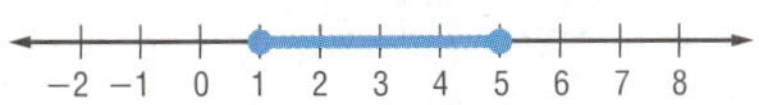

Study Tip

Check Your Solution
You may also want to use a graphing calculator to check. Graph each side of the original inequality and examine the intersection.

Concept Summary — Solving Radical Inequalities

To solve radical inequalities, complete the following steps.

Step 1 If the index of the root is even, identify the values of the variable for which the radicand is nonnegative.

Step 2 Solve the inequality algebraically.

Step 3 Test values to check your solution.

Check for Understanding

Concept Check

1. **Explain** why you do not have to square each side to solve $2x + 1 = \sqrt{3}$. Then solve the equation. **See margin.**
2. **Show** how to solve $x - 6\sqrt{x} + 9 = 0$ by factoring. Name the properties of equality that you use. **See margin.**
3. **OPEN ENDED** Write an equation containing two radicals for which 1 is a solution. **Sample answer: $\sqrt{x} + \sqrt{x + 3} = 3$**

www.algebra2.com/extra_examples

SOLVE RADICAL INEQUALITIES

In-Class Example Power Point®

4 Solve $\sqrt{3x - 6} + 4 \leq 7$. $2 \leq x \leq 5$

Teaching Tip Emphasize the importance of checking key test values in the appropriate ranges.

3 Practice/Apply

Study Notebook

Have students—
- add the definitions/examples of the vocabulary terms to their Vocabulary Builder worksheets for Chapter 5.
- write a list of the steps for solving radical equations and copy the list of steps for solving radical inequalities given in the Concept Summary on p. 265.
- include any other item(s) that they find helpful in mastering the skills in this lesson.

Answers

1. Since x is not under the radical, the equation is a linear equation, not a radical equation. The solution is $\frac{\sqrt{3} - 1}{2}$.
2. The trinomial is a perfect square in terms of $\sqrt{x}$. $x - 6\sqrt{x} + 9 = (\sqrt{x} - 3)^2$, so the equation can be written as $(\sqrt{x} - 3)^2 = 0$. Take the square root of each side to get $\sqrt{x} - 3 = 0$. Use the Addition Property of Equality to add 3 to each side, then square each side to get $x = 9$.

DAILY INTERVENTION — **Differentiated Instruction**

Logical Have students compare solving radical equations and inequalities to solving other types of equations and inequalities. Have them write or give a short presentation about the similarities and differences between the procedures used in the solution processes.

Guided Practice

GUIDED PRACTICE KEY	
Exercises	Examples
4–9, 12	1–3
10, 11	4

Solve each equation or inequality.

4. $\sqrt{4x+1} = 3$ **2**
5. $4 - (7 - y)^{\frac{1}{2}} = 0$ **−9**
6. $1 + \sqrt{x+2} = 0$ **no solution**
7. $\sqrt{z-6} - 3 = 0$ **15**
8. $\frac{1}{6}(12a)^{\frac{1}{3}} = 1$ **18**
9. $\sqrt[3]{x-4} = 3$ **31**
10. $\sqrt{2x+3} - 4 \le 5$ $-\frac{3}{2} \le x \le 39$
11. $\sqrt{b+12} - \sqrt{b} > 2$ $0 \le b < 4$

Application

12. **GEOMETRY** The surface area S of a cone can be found by using $S = \pi r\sqrt{r^2 + h^2}$, where r is the radius of the base and h is the height of the cone. Find the height of the cone. **about 13.42 cm**

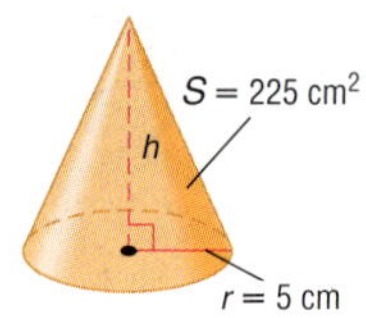

★ indicates increased difficulty

Practice and Apply

Homework Help	
For Exercises	See Examples
13–24, 29–32, 37–42	1–3
25–28, 33–36	4

Extra Practice
See page 839.

Solve each equation or inequality.

13. $\sqrt{x} = 4$ **16**
14. $\sqrt{y} - 7 = 0$ **49**
15. $a^{\frac{1}{2}} + 9 = 0$ **no solution**
16. $2 + 4z^{\frac{1}{2}} = 0$ **no solution**
17. $\sqrt[3]{c-1} = 2$ **9**
18. $\sqrt[3]{5m+2} = 3$ **5**
19. $7 + \sqrt{4x+8} = 9$ **−1**
20. $5 + \sqrt{4y-5} = 12$ $\frac{27}{2}$
21. $(6n-5)^{\frac{1}{3}} + 3 = -2$ **−20**
22. $(5x+7)^{\frac{1}{5}} + 3 = 5$ **5**
23. $\sqrt{x-5} = \sqrt{2x-4}$ **no solution**
24. $\sqrt{2t-7} = \sqrt{t+2}$ **9**
25. $1 + \sqrt{7x-3} > 3$ $x > 1$
26. $\sqrt{3x+6} + 2 \le 5$ $-2 \le x \le 1$
27. $-2 + \sqrt{9-5x} \ge 6$ $x \le -11$
28. $6 - \sqrt{2y+1} < 3$ $y > 4$

★29. $\sqrt{x-6} - \sqrt{x} = 3$ **no solution**
★30. $\sqrt{y+21} - 1 = \sqrt{y+12}$ **4**
★31. $\sqrt{b+1} = \sqrt{b+6} - 1$ **3**
★32. $\sqrt{4z+1} = 3 + \sqrt{4z-2}$ **no solution**
★33. $\sqrt{2} - \sqrt{x+6} \le -\sqrt{x}$ $0 \le x \le 2$
★34. $\sqrt{a+9} - \sqrt{a} > \sqrt{3}$ $0 \le a < 3$
★35. $\sqrt{b-5} - \sqrt{b+7} \le 4$ $b \ge 5$
★36. $\sqrt{c+5} + \sqrt{c+10} > 2.5$ $c > -\frac{79}{16}$

37. What is the solution of $2 - \sqrt{x+6} = -1$? **3**
38. Solve $\sqrt{2x+4} - 4 = 2$. **16**

39. **CONSTRUCTION** The minimum depth d in inches of a beam required to support a load of s pounds is given by the formula $d = \sqrt{\frac{s\ell}{576w}}$, where ℓ is the length of the beam in feet and w is the width in feet. Find the load that can be supported by a board that is 25 feet long, 2 feet wide, and 5 inches deep. **1152 lb**

40. **AEROSPACE ENGINEERING** The radius r of the orbit of a satellite is given by $r = \sqrt[3]{\frac{GMt^2}{4\pi^2}}$, where G is the universal gravitational constant, M is the mass of the central object, and t is the time it takes the satellite to complete one orbit. Solve this formula for t. $t = \sqrt{\frac{4\pi^2 r^3}{GM}}$

Study Guide and Intervention, p. 281 (shown) and p. 282

Solve Radical Equations The following steps are used in solving equations that have variables in the radicand. Some algebraic procedures may be needed before you use these steps.

Step 1	Isolate the radical on one side of the equation.
Step 2	To eliminate the radical, raise each side of the equation to a power equal to the index of the radical.
Step 3	Solve the resulting equation.
Step 4	Check your solution in the original equation to make sure that you have not obtained any extraneous roots.

Example 1 **Solve $2\sqrt{4x+8} - 4 = 8$.**

$2\sqrt{4x+8} - 4 = 8$ Original equation
$2\sqrt{4x+8} = 12$ Add 4 to each side.
$\sqrt{4x+8} = 6$ Isolate the radical.
$4x + 8 = 36$ Square each side.
$4x = 28$ Subtract 8 from each side.
$x = 7$ Divide each side by 4.

Check
$2\sqrt{4(7)+8} - 4 \stackrel{?}{=} 8$
$2\sqrt{36} - 4 \stackrel{?}{=} 8$
$2(6) - 4 \stackrel{?}{=} 8$
$8 = 8$
The solution $x = 7$ checks.

Example 2 **Solve $\sqrt{3x+1} = \sqrt{5x} - 1$.**

$\sqrt{3x+1} = \sqrt{5x} - 1$ Original equation
$3x + 1 = 5x - 2\sqrt{5x} + 1$ Square each side.
$2\sqrt{5x} = 2x$ Simplify.
$\sqrt{5x} = x$ Isolate the radical.
$5x = x^2$ Square each side.
$x^2 - 5x = 0$ Subtract 5x from each side.
$x(x-5) = 0$ Factor.
$x = 0$ or $x = 5$

Check
$\sqrt{3(0)+1} = 1$, but $\sqrt{5(0)} - 1 = -1$, so 0 is not a solution.
$\sqrt{3(5)+1} = 4$, and $\sqrt{5(5)} - 1 = 4$, so the solution is $x = 5$.

Exercises

Solve each equation.

1. $3 + 2x\sqrt{3} = 5$ $\frac{\sqrt{3}}{3}$
2. $2\sqrt{3x+4} + 1 = 15$ **15**
3. $8 + \sqrt{x+1} = 2$ **no solution**
4. $\sqrt{5-x} - 4 = 6$ **−95**
5. $12 + \sqrt{2x-1} = 4$ **no solution**
6. $\sqrt{12-x} = 0$ **12**
7. $\sqrt{21} - \sqrt{5x-4} = 0$ **5**
8. $10 - \sqrt{2x} = 5$ **12.5**
9. $\sqrt{x^2+7x} = \sqrt{7x-9}$ **no solution**
10. $4\sqrt[3]{2x+11} - 2 = 10$ **8**
11. $2\sqrt{x+11} = \sqrt{x+2} + \sqrt{3x-6}$ **14**
12. $\sqrt{9x-11} = x+1$ **3, 4**

Skills Practice, p. 283 and Practice, p. 284 (shown)

Solve each equation or inequality.

1. $\sqrt{x} = 8$ **64**
2. $4 - \sqrt{x} = 3$ **1**
3. $\sqrt{2p} + 3 = 10$ $\frac{49}{2}$
4. $4\sqrt{3h} - 2 = 0$ $\frac{1}{12}$
5. $c^{\frac{1}{2}} + 6 = 9$ **9**
6. $18 + 7h^{\frac{1}{2}} = 12$ **no solution**
7. $\sqrt[3]{d+2} = 7$ **341**
8. $\sqrt[5]{w-7} = 1$ **8**
9. $6 + \sqrt[3]{q-4} = 9$ **31**
10. $\sqrt[4]{y-9} + 4 = 0$ **no solution**
11. $\sqrt{2m-6} - 16 = 0$ **131**
12. $\sqrt[3]{4m+1} - 2 = 2$ $\frac{63}{4}$
13. $\sqrt{8n-5} - 1 = 2$ $\frac{7}{4}$
14. $\sqrt{1-4t} - 8 = -6$ $-\frac{3}{4}$
15. $\sqrt{2t-5} - 3 = 3$ $\frac{41}{2}$
16. $(7v-2)^{\frac{1}{4}} + 12 = 7$ **no solution**
17. $(3g+1)^{\frac{1}{2}} - 6 = 4$ **33**
18. $(6u-5)^{\frac{1}{3}} + 2 = -3$ **−20**
19. $\sqrt{2d-5} = \sqrt{d-1}$ **4**
20. $\sqrt{4r-6} = \sqrt{r}$ **2**
21. $\sqrt{6x-4} = \sqrt{2x+10}$ $\frac{7}{2}$
22. $\sqrt{2x+5} = \sqrt{2x+1}$ **no solution**
23. $3\sqrt{a} \ge 12$ $a \ge 16$
24. $\sqrt{z+5} + 4 \le 13$ $-5 \le z \le 76$
25. $8 + \sqrt{2q} \le 5$ **no solution**
26. $\sqrt{2a-3} < 5$ $\frac{3}{2} < a < 14$
27. $9 - \sqrt{c+4} \le 6$ $c \ge 5$
28. $\sqrt[3]{x-1} < -2$ $x < -7$

29. **STATISTICS** Statisticians use the formula $\sigma = \sqrt{v}$ to calculate a standard deviation σ, where v is the variance of a data set. Find the variance when the standard deviation is 15. **225**

30. **GRAVITATION** Helena drops a ball from 25 feet above a lake. The formula $t = \frac{1}{4}\sqrt{25-h}$ describes the time t in seconds that the ball is h feet above the water. How many feet above the water will the ball be after 1 second? **9 ft**

Reading to Learn Mathematics, p. 285 ELL

Pre-Activity **How do radical equations apply to manufacturing?**

Read the introduction to Lesson 5-8 at the top of page 263 in your textbook.

Explain how you would use the formula in your textbook to find the cost of producing 125,000 computer chips. (Describe the steps of the calculation in the order in which you would perform them, but do not actually do the calculation.) **Sample answer: Raise 125,000 to the $\frac{2}{3}$ power by taking the cube root of 125,000 and squaring the result (or raise 125,000 to the $\frac{2}{3}$ power by entering 125,000 ^ (2/3) on a calculator). Multiply the number you get by 10 and then add 1500.**

Reading the Lesson

1. a. What is an *extraneous solution* of a radical equation? **Sample answer: a number that satisfies an equation obtained by raising both sides of the original equation to a higher power but does not satisfy the original equation**

   b. Describe two ways you can check the proposed solutions of a radical equation in order to determine whether any of them are extraneous solutions. **Sample answer: One way is to check each proposed solution by substituting it into the original equation. Another way is to use a graphing calculator to graph both sides of the original equation. See where the graphs intersect. This can help you identify solutions that may be extraneous.**

2. Complete the steps that should be followed in order to solve a radical inequality.

   Step 1 If the ___**index**___ of the root is ___**even**___, identify the values of the variable for which the ___**radicand**___ is ___**nonnegative**___.

   Step 2 Solve the ___**inequality**___ algebraically.

   Step 3 Test ___**values**___ to check your solution.

Helping You Remember

3. One way to remember something is to explain it to another person. Suppose that your friend Leora thinks that she does not need to check her solutions to radical equations by substitution because she knows she is very careful and seldom makes mistakes in her work. How can you explain to her that she should nevertheless check every proposed solution in the original equation? **Sample answer: Squaring both sides of an equation can produce an equation that is not equivalent to the original one. For example, the only solution of $x = 5$ is 5, but the squared equation $x^2 = 25$ has two solutions, 5 and −5.**

Enrichment, p. 286

Truth Tables

In mathematics, the basic operations are addition, subtraction, multiplication, division, finding a root, and raising to a power. In logic, the basic operations are the following: *not* (~), *and* (∧), *or* (∨), and *implies* (→).

If P and Q are statements, then $\sim P$ means not P; $\sim Q$ means not Q; $P \wedge Q$ means P and Q; $P \vee Q$ means P or Q; and $P \rightarrow Q$ means P implies Q. The operations are defined by truth tables. On the left below is the truth table for the statement $\sim P$. Notice that there are two possible conditions for P, true (T) or false (F). If P is true, $\sim P$ is false; if P is false, $\sim P$ is true. Also shown are the truth tables for $P \wedge Q$, $P \vee Q$, and $P \rightarrow Q$.

P	~P
T	F
F	T

P	Q	P ∧ Q
T	T	T
T	F	F
F	T	F
F	F	F

P	Q	P ∨ Q
T	T	T
T	F	T
F	T	T
F	F	F

P	Q	P → Q
T	T	T
T	F	F
F	T	T
F	F	T

You can use this information to find out under what conditions a complex

266 Chapter 5 Polynomials

More About. . .

Health

A ponderal index p is a measure of a person's body based on height h in meters and mass m in kilograms. One such formula is $p = \frac{\sqrt[3]{m}}{h}$.

Source: *A Dictionary of Food and Nutrition*

41. **PHYSICS** When an object is dropped from the top of a 50-foot tall building, the object will be h feet above the ground after t seconds, where $\frac{\sqrt{50 - h}}{4} = t$. How far above the ground will the object be after 1 second? **34 ft**

42. **HEALTH** **Use the information about health at the left.** A 70-kilogram person who is 1.8 meters tall has a ponderal index of about 2.29. How much weight could such a person gain and still have an index of at most 2.5? **21.125 kg**

43. **CRITICAL THINKING** Explain how you know that $\sqrt{x + 2} + \sqrt{2x - 3} = -1$ has no real solution without trying to solve it. **See margin.**

44. **WRITING IN MATH** Answer the question that was posed at the beginning of the lesson. **See pp. 283A–283B.**

    How do radical equations apply to manufacturing?

    Include the following in your answer:
    - the equation $C = 10n^{\frac{2}{3}} + 1500$ rewritten as a radical equation, and
    - a step-by-step explanation of how to determine the maximum number of chips the company could make for $10,000.

Standardized Test Practice

45. If $\sqrt{x + 5} + 1 = 4$, what is the value of x? **D**

    (A) -4 (B) 0 (C) 2 (D) 4

46. Side $\overline{AC}$ of triangle ABC contains which of the following points? **C**

    (A) (3, 4) (B) (3, 5) (C) (4, 3)

    (D) (4, 5) (E) (4, 6)

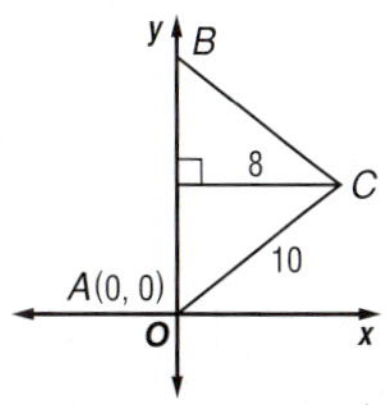

Maintain Your Skills

Mixed Review **Write each radical using rational exponents.** *(Lesson 5-7)*

47. $\sqrt[7]{5^3}$ $\mathbf{5^{\frac{3}{7}}}$
48. $\sqrt{x + 7}$ $\mathbf{(x + 7)^{\frac{1}{2}}}$
49. $(\sqrt[3]{x^2 + 1})^2$ $\mathbf{(x^2 + 1)^{\frac{2}{3}}}$

Simplify. *(Lesson 5-6)*

50. $\sqrt{72x^6y^3}$ $\mathbf{6|x^3|y\sqrt{2y}}$
51. $\frac{1}{\sqrt[3]{10}}$ $\mathbf{\frac{\sqrt[3]{100}}{10}}$
52. $(5 - \sqrt{3})^2$ $\mathbf{28 - 10\sqrt{3}}$

53. **BUSINESS** A dry cleaner ordered 7 drums of two different types of cleaning fluid. One type cost \$30 per drum, and the other type cost \$20 per drum. The total cost was \$160. How much of each type of fluid did the company order? Write a system of equations and solve by graphing. *(Lesson 3-1)*
    $x + y = 7$, $30x + 20y = 160$; See margin for graph; (2, 5).

Getting Ready for the Next Lesson **PREREQUISITE SKILL** **Simplify each expression.**
*(To review **binomials**, see Lesson 5-2.)*

54. $(5 + 2x) + (-1 - x)$ $\mathbf{4 + x}$
55. $(-3 - 2y) + (4 + y)$ $\mathbf{1 - y}$
56. $(4 + x) - (2 - 3x)$ $\mathbf{2 + 4x}$
57. $(-7 - 3x) - (4 - 3x)$ $\mathbf{-11}$
58. $(1 + z)(4 + 2z)$ $\mathbf{4 + 6z + 2z^2}$
59. $(-3 - 4x)(1 + 2x)$ $\mathbf{-3 - 10x - 8x^2}$

www.algebra2.com/self_check_quiz

About the Exercises...

Organization by Objective
- **Solve Radical Equations:** 13–24, 29–32, 37–42
- **Solve Radical Inequalities:** 25–28, 33–36

Odd/Even Assignments
Exercises 13–38 are structured so that students practice the same concepts whether they are assigned odd or even problems.

Assignment Guide

Basic: 13–27 odd, 37–41 odd, 43–59

Average: 13–41 odd, 43–59

Advanced: 14–42 even, 43–53 (optional: 54–59)

4 Assess

Open-Ended Assessment

Writing Have students write a list of examples showing how to solve each of the different types of radical equations and inequalities discussed in this lesson.

Tips for New Teachers

Intervention Make sure that students know the constraints on the values of the variables in a radical equation so that the solutions are real numbers.

Getting Ready for Lesson 5-9

PREREQUISITE SKILL Lesson 5-9 presents calculating with complex numbers, often written in the form of binomials. Exercises 54–59 should be used to determine your students' familiarity with adding, subtracting, and multiplying binomials.

Answers

43. Since $\sqrt{x + 2} \geq 0$ and $\sqrt{2x - 3} \geq 0$, the left side of the equation is nonnegative. Therefore, the left side of the equation cannot equal -1. Thus, the equation has no solution.

53.

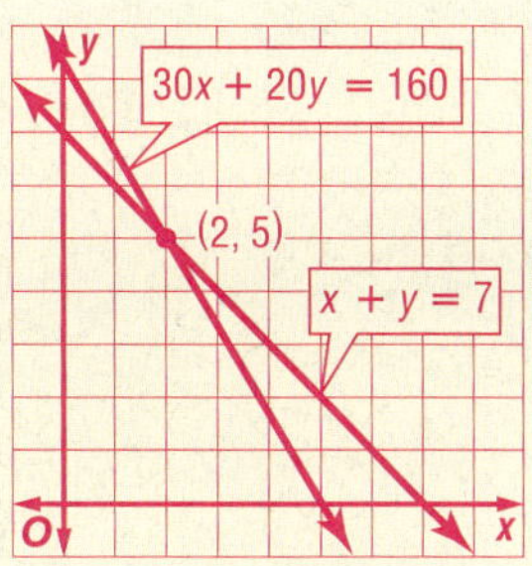

Graphing Calculator Investigation

A Follow-Up of Lesson 5-8

Getting Started

Know Your Calculator The TI-83 Plus automatically supplies a left parenthesis after each radical sign. When functions are entered on the Y= list, it is important to supply right parentheses as needed to ensure correct graphs and correct numerical results.

Displaying Tables In Step 2 on p. 268, students should check to be sure that the AUTO option has been selected on each of the last two lines of the TABLE SETUP screen.

Exact Solutions The approximate zero displayed on the screen shown in Step 4 on p. 268 appears to be a repeating decimal. If you go to the home screen immediately after Step 4 and use the keystrokes [X,T,θ,n] [MATH] 1 [ENTER], the calculator will display a fraction for the exact solution, $\frac{49}{36}$. The calculator can be used to verify that this is indeed the exact solution of the equation.

Teach

- After reading the sentence at the top of p. 269, have students solve the radical equation on p. 268 again treating each side as a separate function. Point out that the right side will simply be graphed as the function $y = 3$.
- After completing the discussion of the procedure on p. 269 for solving a radical inequality, have students solve it again by first subtracting $2\sqrt{x}$ from both sides and then graphing the function $y = \sqrt{x+2} + 1 - 2\sqrt{x}$. Point out that the portion of the graph below the x-axis shows the solution.
- Have students complete Exercises 1–10.

Graphing Calculator

A Follow-Up of Lesson 5-8

Solving Radical Equations and Inequalities by Graphing

You can use a TI-83 Plus to solve radical equations and inequalities. One way to do this is by rewriting the equation or inequality so that one side is 0 and then using the zero feature on the calculator.

Solve $\sqrt{x} + \sqrt{x+2} = 3$.

Step 1 *Rewrite the equation.*

- Subtract 3 from each side of the equation to obtain $\sqrt{x} + \sqrt{x+2} - 3 = 0$.
- Enter the function $y = \sqrt{x} + \sqrt{x+2} - 3$ in the Y= list.

KEYSTROKES: *Review entering a function on page 128.*

Step 2 *Use a table.*

- You can use the TABLE function to locate intervals where the solution(s) lie. First, enter the starting value and the interval for the table.

KEYSTROKES: [2nd] [TBLSET] 0 [ENTER] 1 [ENTER]

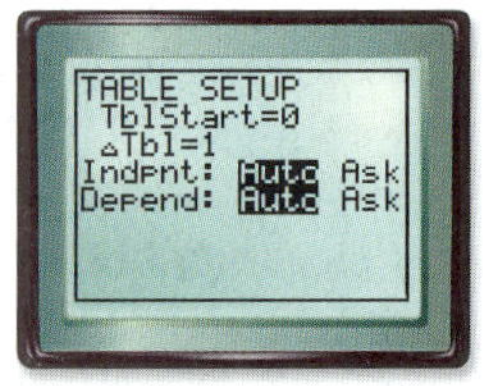

Step 3 *Estimate the solution.*

- Complete the table and estimate the solution(s).

KEYSTROKES: [2nd] [TABLE]

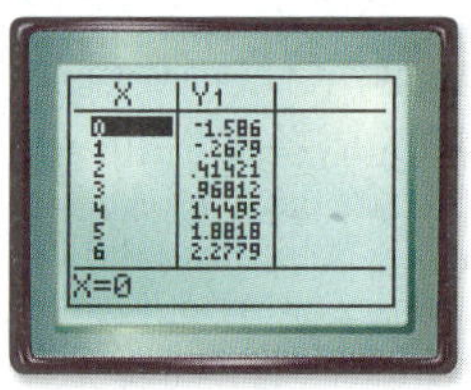

Since the function changes sign from negative to positive between $x = 1$ and $x = 2$, there is a solution between 1 and 2.

Step 4 *Use the zero feature.*

- Graph, then select zero from the CALC menu.

KEYSTROKES: [GRAPH] [2nd] [CALC] 2

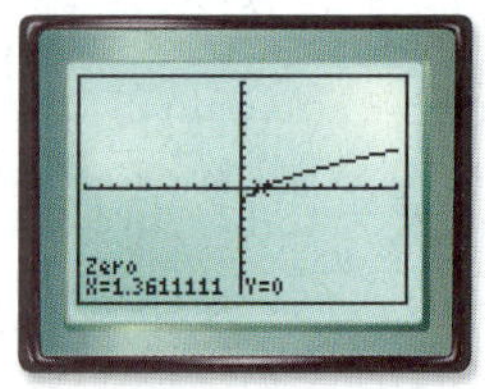

[−10, 10] scl: 1 by [−10, 10] scl: 1

Place the cursor to the left of the zero and press [ENTER] for the Left Bound. Then place the cursor to the right of the zero and press [ENTER] for the Right Bound. Press [ENTER] to solve.

The solution is about 1.36. This agrees with the estimate made by using the TABLE.

www.algebra2.com/other_calculator_keystrokes

268 Chapter 5 Polynomials

Investigation

Instead of rewriting an equation or inequality so that one side is 0, you can also treat each side of the equation or inequality as a separate function and graph both.

Solve $2\sqrt{x} > \sqrt{x+2} + 1$.

Step 1 *Graph each side of the inequality.*

- In the Y= list, enter $y_1 = 2\sqrt{x}$ and $y_2 = \sqrt{x+2} + 1$. Then press GRAPH.

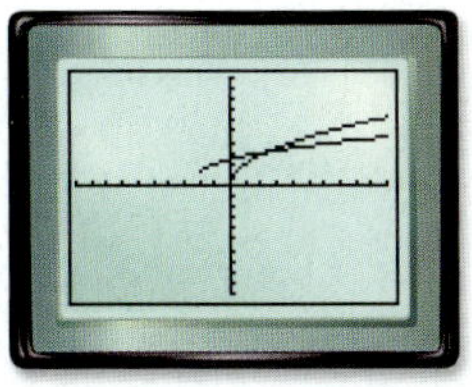

[−10, 10] scl: 1 by [−10, 10] scl: 1

Step 2 *Use the trace feature.*

- Press . You can use ▲ or ▼ to switch the cursor between the two curves.

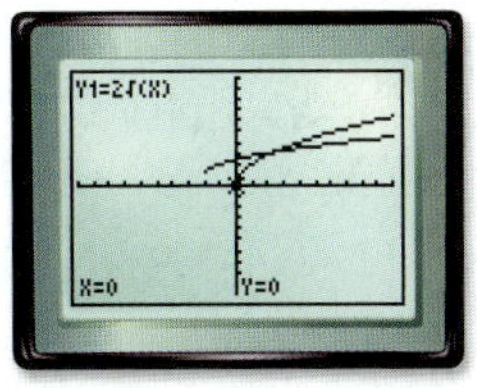

[−10, 10] scl: 1 by [−10, 10] scl: 1

The calculator screen above shows that, for points to the left of where the curves cross, $Y_1 < Y_2$ or $2\sqrt{x} < \sqrt{x+2} + 1$. To solve the original inequality, you must find points for which $Y_1 > Y_2$. These are the points to the right of where the curves cross.

Step 3 *Use the intersect feature.*

- You can use the INTERSECT feature on the CALC menu to approximate the x-coordinate of the point at which the curves cross.

KEYSTROKES: 2nd [CALC] 5

- Press ENTER for each of First curve?, Second curve?, and Guess?.

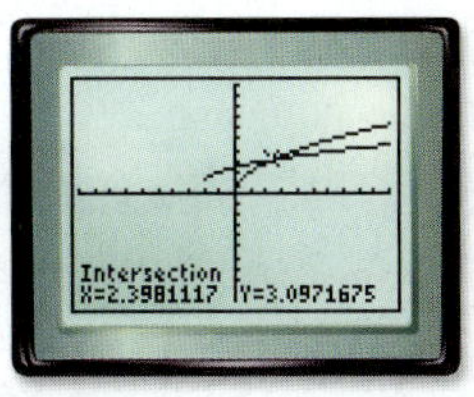

[−10, 10] scl: 1 by [−10, 10] scl: 1

The calculator screen shows that the x-coordinate of the point at which the curves cross is about 2.40. Therefore, the solution of the inequality is about $x > 2.40$. *Use the symbol > instead of ≥ in the solution because the symbol in the original inequality is >.*

Exercises **4. about 3.89 5. about 2.52 8. about $0 \leq x < 1$ 9. about $1 \leq x < 4.52$**

Solve each equation or inequality.

1. $\sqrt{x+4} = 3$ **5**

2. $\sqrt{3x-5} = 1$ **2**

3. $\sqrt{x+5} = \sqrt{3x+4}$ **0.5**

4. $\sqrt{x+3} + \sqrt{x-2} = 4$

5. $\sqrt{3x-7} = \sqrt{2x-2} - 1$

6. $\sqrt{x+8} - 1 = \sqrt{x+2}$ **4.25**

7. $\sqrt{x-3} \geq 2$ **$x \geq 7$**

8. $\sqrt{x+3} > 2\sqrt{x}$

9. $\sqrt{x} + \sqrt{x-1} < 4$

10. Explain how you could apply the technique in the first example to solving an inequality. **See margin.**

Assess

- When examining the table of values for the equation on p. 268, why do you know a solution lies between the two x values where the graphed function changes signs? **Sample answer: Since the original equation was solved so that one side was zero, the function representing the other side has a value of 0 when the value of y is 0. That must occur for a value of x somewhere between two values of x whose corresponding values of y have different signs (because 0 lies between the positive numbers and the negative numbers).**

Answer

10. Rewrite the inequality so that one side is 0. Then graph the other side and find the x values for which the graph is above or below the x-axis, according to the inequality symbol. Use the zero feature to approximate the x-coordinate of the point at which the graph crosses the x-axis.

Graphing Calculator Investigation Solving Radical Equations and Inequalities by Graphing 269

5-9 Lesson Notes

1 Focus

5-Minute Check Transparency 5-9 Use as a quiz or a review of Lesson 5-8.

Mathematical Background notes are available for this lesson on p. 220D.

How do complex numbers apply to polynomial equations?

Ask students:

- If by definition $i^2 = -1$, then what do you think is the value of i^4? Justify your answer. **Since $i^4 = (i^2)^2$, replacing i^2 with -1 gives $i^4 = (-1)^2$ or 1.**

2 Teach

ADD AND SUBTRACT COMPLEX NUMBERS

In-Class Examples Power Point®

1 Simplify.

a. $\sqrt{-28}$ **$2i\sqrt{7}$**

b. $\sqrt{-32y^3}$ **$4i|y|\sqrt{2y}$**

2 Simplify.

a. $-3i \cdot 2i$ **6**

b. $\sqrt{-12} \cdot \sqrt{-2}$ **$-2\sqrt{6}$**

3 Simplify i^{35}. **$-i$**

5-9 Complex Numbers

What You'll Learn

- Add and subtract complex numbers.
- Multiply and divide complex numbers.

How do complex numbers apply to polynomial equations?

Consider the equation $2x^2 + 2 = 0$. If you solve this equation for x^2, the result is $x^2 = -1$. Since there is no real number whose square is -1, the equation has no real solutions. French mathematician René Descartes (1596–1650) proposed that a number i be defined such that $i^2 = -1$.

Vocabulary

- imaginary unit
- pure imaginary number
- complex number
- absolute value
- complex conjugates

ADD AND SUBTRACT COMPLEX NUMBERS Since i is defined to have the property that $i^2 = -1$, the number i is the principal square root of -1; that is, $i = \sqrt{-1}$. i is called the **imaginary unit**. Numbers of the form $3i$, $-5i$, and $i\sqrt{2}$ are called **pure imaginary numbers**. Pure imaginary numbers are square roots of negative real numbers. For any positive real number b, $\sqrt{-b^2} = \sqrt{b^2} \cdot \sqrt{-1}$ or bi.

Example 1 Square Roots of Negative Numbers

Simplify.

a. $\sqrt{-18}$

$$\sqrt{-18} = \sqrt{-1 \cdot 3^2 \cdot 2} = \sqrt{-1} \cdot \sqrt{3^2} \cdot \sqrt{2} = i \cdot 3 \cdot \sqrt{2} \text{ or } 3i\sqrt{2}$$

b. $\sqrt{-125x^5}$

$$\sqrt{-125x^5} = \sqrt{-1 \cdot 5^2 \cdot x^4 \cdot 5x} = \sqrt{-1} \cdot \sqrt{5^2} \cdot \sqrt{x^4} \cdot \sqrt{5x} = i \cdot 5 \cdot x^2 \cdot \sqrt{5x} \text{ or } 5ix^2\sqrt{5x}$$

Study Tip

Reading Math
i is usually written before radical symbols to make it clear that it is not under the radical.

The Commutative and Associative Properties of Multiplication hold true for pure imaginary numbers.

Example 2 Multiply Pure Imaginary Numbers

Simplify.

a. $-2i \cdot 7i$

$-2i \cdot 7i = -14i^2$
$= -14(-1)$ $i^2 = -1$
$= 14$

b. $\sqrt{-10} \cdot \sqrt{-15}$

$\sqrt{-10} \cdot \sqrt{-15} = i\sqrt{10} \cdot i\sqrt{15}$
$= i^2\sqrt{150}$
$= -1 \cdot \sqrt{25} \cdot \sqrt{6}$
$= -5\sqrt{6}$

TEACHING TIP
Point out that when multiplying radicals with negative radicands, students should first take the roots, then multiply. Otherwise, their answers may be off by a factor of -1.

Example 3 Simplify a Power of i

Simplify i^{45}.

$i^{45} = i \cdot i^{44}$ Multiplying powers
$= i \cdot (i^2)^{22}$ Power of a Power
$= i \cdot (-1)^{22}$ $i^2 = -1$
$= i \cdot 1$ or i $(-1)^{22} = 1$

Resource Manager

Workbook and Reproducible Masters

Chapter 5 Resource Masters
- Study Guide and Intervention, pp. 287–288
- Skills Practice, p. 289
- Practice, p. 290
- Reading to Learn Mathematics, p. 291
- Enrichment, p. 292
- Assessment, p. 308

School-to-Career Masters, p. 10
Teaching Algebra With Manipulatives Masters, pp. 239, 240

Transparencies
5-Minute Check Transparency 5-9
Answer Key Transparencies

Technology
Interactive Chalkboard
Multimedia Applications

The solutions of some equations involve pure imaginary numbers.

Example 4 Equation with Imaginary Solutions

Solve $3x^2 + 48 = 0$.

$3x^2 + 48 = 0$ Original equation

$3x^2 = -48$ Subtract 48 from each side.

$x^2 = -16$ Divide each side by 3.

$x = \pm\sqrt{-16}$ Take the square root of each side.

$x = \pm 4i$ $\sqrt{-16} = \sqrt{16} \cdot \sqrt{-1}$

Study Tip

Quadratic Solutions

Quadratic equations always have complex solutions. If the discriminant is:

- negative, there are two imaginary roots,
- zero, there are two equal real roots, or
- positive, there are two unequal real roots.

Consider an expression such as $5 + 2i$. Since 5 is a real number and $2i$ is a pure imaginary number, the terms are not like terms and cannot be combined. This type of expression is called a **complex number**.

Key Concept — Complex Numbers

- **Words** A complex number is any number that can be written in the form $a + bi$, where a and b are real numbers and i is the imaginary unit. a is called the real part, and b is called the imaginary part.
- **Examples** $7 + 4i$ and $2 - 6i = 2 + (-6)i$

The Venn diagram at the right shows the complex number system.

- If $b = 0$, the complex number is a real number.
- If $b \neq 0$, the complex number is imaginary.
- If $a = 0$, the complex number is a pure imaginary number.

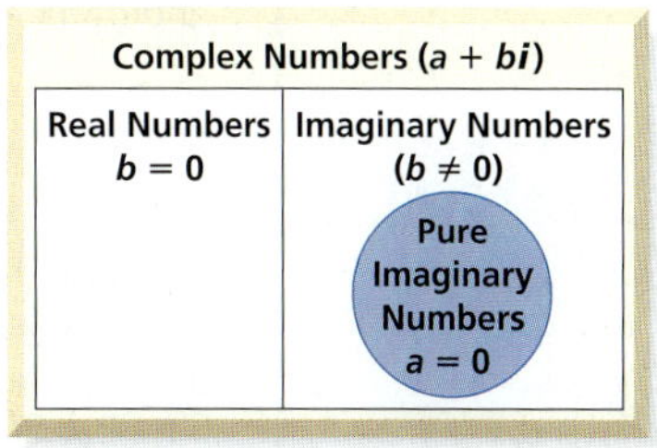

Two complex numbers are equal if and only if their real parts are equal and their imaginary parts are equal. That is, $a + bi = c + di$ if and only if $a = c$ and $b = d$.

Example 5 Equate Complex Numbers

Find the values of x and y that make the equation $2x - 3 + (y - 4)i = 3 + 2i$ true.

Set the real parts equal to each other and the imaginary parts equal to each other.

$2x - 3 = 3$ Real parts

$2x = 6$ Add 3 to each side.

$x = 3$ Divide each side by 2.

$y - 4 = 2$ Imaginary parts

$y = 6$ Add 4 to each side.

Study Tip

Reading Math

The form $a + bi$ is sometimes called the **standard form** of a complex number.

www.algebra2.com/extra_examples

In-Class Examples

Power Point®

4 Solve $5y^2 + 20 = 0$. $\pm 2i$

Teaching Tip Make sure students understand that when they take the square root of both sides of an equation, they must use the ± symbol in front of the radical sign.

5 Find the values of x and y that make the equation $2x + yi = -14 - 3i$ true. $x = -7, y = -3$

Teaching Tip Emphasize that two complex numbers are equal if and only if their real parts are equal and their imaginary parts are equal.

DAILY INTERVENTION — Differentiated Instruction ELL

Verbal/Linguistic Have students write poems about the imaginary number i and the repeating values of its powers, perhaps including wordplay with the terms *real* and *imaginary*. The content of the poems should be helpful for remembering the mathematical characteristics of i.

In-Class Example

Power Point®

6 Simplify.

a. $(3 + 5i) + (2 - 4i)$ **$5 + i$**

b. $(4 - 6i) - (3 - 7i)$ **$1 + i$**

Answers

Algebra Activity

1.

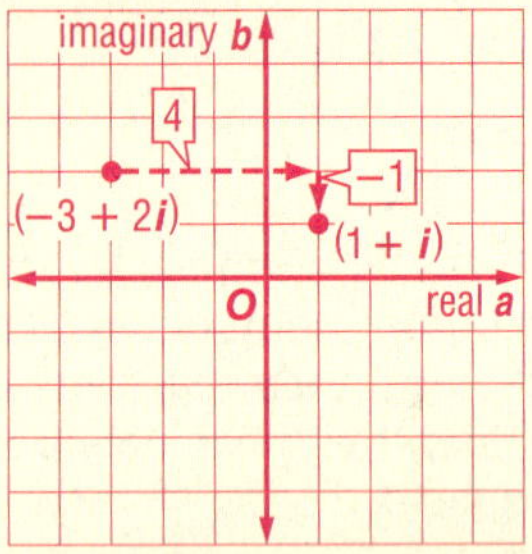

2. Rewrite the difference as a sum, $(-3 + 2i) - (4 - i) = (-3 + 2i) + (-4 + i)$. Then apply the method discussed in this activity.

To add or subtract complex numbers, combine like terms. That is, combine the real parts and combine the imaginary parts.

Example 6 Add and Subtract Complex Numbers

Simplify.

a. $(6 - 4i) + (1 + 3i)$

$(6 - 4i) + (1 + 3i) = (6 + 1) + (-4 + 3)i$ Commutative and Associative Properties

$= 7 - i$ Simplify.

b. $(3 - 2i) - (5 - 4i)$

$(3 - 2i) - (5 - 4i) = (3 - 5) + [-2 - (-4)]i$ Commutative and Associative Properties

$= -2 + 2i$ Simplify.

You can model the addition and subtraction of complex numbers geometrically.

Algebra Activity

Adding Complex Numbers

You can model the addition of complex numbers on a coordinate plane. The horizontal axis represents the real part a of the complex number, and the vertical axis represents the imaginary part b of the complex number.

Use a coordinate plane to find $(4 + 2i) + (-2 + 3i)$.

- Create a coordinate plane and label the axes appropriately.
- Graph $4 + 2i$ by drawing a segment from the origin to (4, 2) on the coordinate plane.
- Represent the addition of $-2 + 3i$ by moving 2 units to the left and 3 units up from (4, 2).
- You end at the point (2, 5), which represents the complex number $2 + 5i$.
So, $(4 + 2i) + (-2 + 3i) = 2 + 5i$.

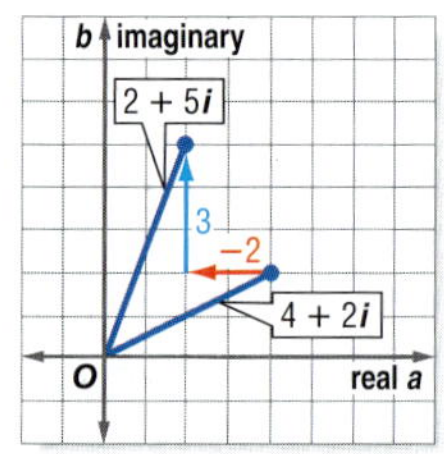

Model and Analyze

1. Model $(-3 + 2i) + (4 - i)$ on a coordinate plane. **See margin.**
2. Describe how you could model the difference $(-3 + 2i) - (4 - i)$ on a coordinate plane. **See margin.**
3. The **absolute value** of a complex number is the distance from the origin to the point representing that complex number in a coordinate plane. Refer to the graph above. Find the absolute value of $2 + 5i$. **$\sqrt{29}$**
4. Find an expression for the absolute value of $a + bi$. **$\sqrt{a^2 + b^2}$**

MULTIPLY AND DIVIDE COMPLEX NUMBERS Complex numbers are used with electricity. In a circuit with alternating current, the voltage, current, and impedance, or hindrance to current, can be represented by complex numbers.

Algebra Activity

Materials: grid paper, ruler/straightedge

- The horizontal axis is often called a real number line. What might be a corresponding name for the vertical axis? **an imaginary number line**
- Where do real numbers lie on this coordinate plane? **on the horizontal axis** Where do pure imaginary numbers lie? **on the vertical axis**
- Where do complex numbers for which neither a nor b is 0 lie on this coordinate plane? **on the regions of the plane other than the axes**

272 Chapter 5 Polynomials

You can use the FOIL method to multiply complex numbers.

Study Tip

Reading Math
Electrical engineers use j as the imaginary unit to avoid confusion with the I for current.

Example 7 Multiply Complex Numbers

ELECTRICITY **In an AC circuit, the voltage E, current I, and impedance Z are related by the formula $E = I \cdot Z$. Find the voltage in a circuit with current $1 + 3j$ amps and impedance $7 - 5j$ ohms.**

$E = I \cdot Z$	Electricity formula
$= (1 + 3j) \cdot (7 - 5j)$	$I = 1 + 3j, Z = 7 - 5j$
$= 1(7) + 1(-5j) + (3j)7 + 3j(-5j)$	FOIL
$= 7 - 5j + 21j - 15j^2$	Multiply.
$= 7 + 16j - 15(-1)$	$j^2 = -1$
$= 22 + 16j$	Add.

The voltage is $22 + 16j$ volts.

Two complex numbers of the form $a + bi$ and $a - bi$ are called **complex conjugates**. The product of complex conjugates is always a real number. For example, $(2 + 3i)(2 - 3i) = 4 - 6i + 6i + 9$ or 13. You can use this fact to simplify the quotient of two complex numbers.

Example 8 Divide Complex Numbers

Simplify.

a. $\frac{3i}{2 + 4i}$

$\frac{3i}{2+4i} = \frac{3i}{2+4i} \cdot \frac{2-4i}{2-4i}$	$2 + 4i$ and $2 - 4i$ are conjugates.
$= \frac{6i - 12i^2}{4 - 16i^2}$	Multiply.
$= \frac{6i + 12}{20}$	$i^2 = -1$
$= \frac{3}{5} + \frac{3}{10}i$	Standard form

b. $\frac{5 + i}{2i}$

$\frac{5+i}{2i} = \frac{5+i}{2i} \cdot \frac{i}{i}$	Why multiply by $\frac{i}{i}$ intead of $\frac{-2i}{-2i}$?
$= \frac{5i + i^2}{2i^2}$	Multiply.
$= \frac{5i - 1}{-2}$	$i^2 = -1$
$= \frac{1}{2} - \frac{5}{2}i$	Standard form

MULTIPLY AND DIVIDE COMPLEX NUMBERS

In-Class Examples Power Point®

7 **ELECTRICITY** In an AC circuit, the voltage E, current I, and impedence Z are related by the formula $E = I \cdot Z$. Find the voltage in a circuit with current $1 + 4j$ amps and impedence $3 - 6j$ ohms. $27 + 6j$

8 Simplify.

a. $\frac{5i}{3 + 2i}$ $\frac{10}{13} + \frac{15}{13}i$

b. $\frac{4 - i}{5i}$ $-\frac{1}{5} - \frac{4}{5}i$

3 Practice/Apply

Study Notebook

Have students—

- complete the definitions/examples for the remaining terms on their Vocabulary Builder worksheets for Chapter 5.
- write a list of the first four powers of i: $i^1 = \sqrt{-1}$, $i^2 = -1$, $i^3 = -i$, $i^4 = 1$.
- include any other item(s) that they find helpful in mastering the skills in this lesson.

Check for Understanding

Concept Check

1. **Determine** if each statement is *true* or *false*. If false, find a counterexample.
   a. Every real number is a complex number. **true**
   b. Every imaginary number is a complex number. **true**
2. **Decide** which of the properties of a field and the properties of equality that the set of complex numbers satisfies. **all of them**
3. **OPEN ENDED** Write two complex numbers whose product is 10. **Sample answer: $1 + 3i$ and $1 - 3i$**

Study Tip

Look Back
Refer to Chapter 1 to review the properties of fields and the properties of equality.

Guided Practice

Simplify.

4. $\sqrt{-36}$ $6i$
5. $\sqrt{-50x^2y^2}$ $5i|xy|\sqrt{2}$
6. $(6i)(-2i)$ 12
7. $5\sqrt{-24} \cdot 3\sqrt{-18}$ $-180\sqrt{3}$
8. i^{29} i
9. $(8 + 6i) - (2 + 3i)$ $6 + 3i$
10. $(3 - 5i)(4 + 6i)$ $42 - 2i$
11. $\frac{3 + i}{1 + 4i}$ $\frac{7}{17} - \frac{11}{17}i$

GUIDED PRACTICE KEY	
Exercises	Examples
4, 5	1
6, 7	2
8	3
9	6
10, 11	7, 8

Lesson 5-9 Complex Numbers 273

Study Guide and Intervention, p. 287 (shown) and p. 288

Add and Subtract Complex Numbers

Complex Number	A complex number is any number that can be written in the form $a + bi$, where a and b are real numbers and i is the imaginary unit ($i^2 = -1$). a is called the real part, and b is called the imaginary part.
Addition and Subtraction of Complex Numbers	Combine like terms. $(a + bi) + (c + di) = (a + c) + (b + d)i$ $(a + bi) - (c + di) = (a - c) + (b - d)i$

Example 1 **Simplify $(6 + i) + (4 - 5i)$.**

$(6 + i) + (4 - 5i)$
$= (6 + 4) + (1 - 5)i$
$= 10 - 4i$

Example 2 **Simplify $(8 + 3i) - (6 - 2i)$.**

$(8 + 3i) - (6 - 2i)$
$= (8 - 6) + [3 - (-2)]i$
$= 2 + 5i$

To solve a quadratic equation that does not have real solutions, you can use the fact that $i^2 = -1$ to find complex solutions.

Example 3 **Solve $2x^2 + 24 = 0$.**

$2x^2 + 24 = 0$ Original equation
$2x^2 = -24$ Subtract 24 from each side.
$x^2 = -12$ Divide each side by 2.
$x = \pm\sqrt{-12}$ Take the square root of each side.
$x = \pm 2i\sqrt{3}$ $\sqrt{-12} = \sqrt{4} \cdot \sqrt{-1} \cdot \sqrt{3}$

Exercises

Simplify.

1. $(-4 + 2i) + (6 - 3i)$ **$2 - i$**
2. $(5 - i) - (3 - 2i)$ **$2 + i$**
3. $(6 - 3i) + (4 - 2i)$ **$10 - 5i$**
4. $(-11 + 4i) - (1 - 5i)$ **$-12 + 9i$**
5. $(8 + 4i) + (8 - 4i)$ **16**
6. $(5 + 2i) - (-6 - 3i)$ **$11 + 5i$**
7. $(12 - 5i) - (4 + 3i)$ **$8 - 8i$**
8. $(9 + 2i) + (-2 + 5i)$ **$7 + 7i$**
9. $(15 - 12i) + (11 - 13i)$ **$26 - 25i$**
10. i^4 **1**
11. i^6 **-1**
12. i^{15} **$-i$**

Solve each equation.

13. $5x^2 + 45 = 0$ **$\pm 3i$**
14. $4x^2 + 24 = 0$ **$\pm i\sqrt{6}$**
15. $-9x^2 = 9$ **$\pm i$**

Skills Practice, p. 289 and Practice, p. 290 (shown)

Simplify.

1. $\sqrt{-49}$ **$7i$**
2. $6\sqrt{-12}$ **$12i\sqrt{3}$**
3. $\sqrt{-121s^8}$ **$11s^4i$**
4. $\sqrt{-36a^3b^4}$ **$6|a|b^2i\sqrt{a}$**
5. $\sqrt{-8} \cdot \sqrt{-32}$ **-16**
6. $\sqrt{-15} \cdot \sqrt{-25}$ **$-5\sqrt{15}$**
7. $(-3i)(4i)(-5i)$ **$-60i$**
8. $(7i)^2(6i)$ **$-294i$**
9. i^{42} **-1**
10. i^{55} **$-i$**
11. i^{89} **i**
12. $(5 - 2i) + (-13 - 8i)$ **$-8 - 10i$**
13. $(7 - 6i) + (9 + 11i)$ **$16 + 5i$**
14. $(-12 + 48i) + (15 + 21i)$ **$3 + 69i$**
15. $(10 + 15i) - (48 - 30i)$ **$-38 + 45i$**
16. $(28 - 4i) - (10 - 30i)$ **$18 + 26i$**
17. $(6 - 4i)(6 + 4i)$ **52**
18. $(8 - 11i)(8 - 11i)$ **$-57 - 176i$**
19. $(4 + 3i)(2 - 5i)$ **$23 - 14i$**
20. $(7 + 2i)(9 - 6i)$ **$75 - 24i$**
21. $\frac{6 + 5i}{-2i}$ **$\frac{-5 + 6i}{2}$**
22. $\frac{2}{7 - 8i}$ **$\frac{14 + 16i}{113}$**
23. $\frac{3 - i}{2 - i}$ **$\frac{7 + i}{5}$**
24. $\frac{2 - 4i}{1 + 3i}$ **$-1 - i$**

Solve each equation.

25. $5n^2 + 35 = 0$ **$\pm i\sqrt{7}$**
26. $2m^2 + 10 = 0$ **$\pm i\sqrt{5}$**
27. $4m^2 + 76 = 0$ **$\pm i\sqrt{19}$**
28. $-2m^2 - 6 = 0$ **$\pm i\sqrt{3}$**
29. $-5m^2 - 65 = 0$ **$\pm i\sqrt{13}$**
30. $\frac{3}{4}x^2 + 12 = 0$ **$\pm 4i$**

Find the values of m and n that make each equation true.

31. $15 - 28i = 3m + 4ni$ **5, -7**
32. $(6 - m) + 3ni = -12 + 27i$ **18, 9**
33. $(3m + 4) + (3 - n)i = 16 - 3i$ **4, 6**
34. $(7 + n) + (4m - 10)i = 3 - 6i$ **1, -4**

35. **ELECTRICITY** The impedance in one part of a series circuit is $1 + 3j$ ohms and the impedance in another part of the circuit is $7 - 5j$ ohms. Add these complex numbers to find the total impedance in the circuit. **$8 - 2j$ ohms**

36. **ELECTRICITY** Using the formula $E = IZ$, find the voltage E in a circuit when the current I is $3 - j$ amps and the impedance Z is $3 + 2j$ ohms. **$11 + 3j$ volts**

Reading to Learn Mathematics, p. 291 ELL

Pre-Activity **How do complex numbers apply to polynomial equations?**

Read the introduction to Lesson 5-9 at the top of page 270 in your textbook.

Suppose the number i is defined such that $i^2 = -1$. Complete each equation.

$2i^2 =$ **-2** $(2i)^2 =$ **-4** $i^4 =$ **1**

Reading the Lesson

1. Complete each statement.
   a. The form $a + bi$ is called the **standard form** of a complex number.
   b. In the complex number $4 + 5i$, the real part is **4** and the imaginary part is **5**. This is an example of a complex number that is also a(n) **imaginary** number.
   c. In the complex number 3, the real part is **3** and the imaginary part is **0**. This is example of complex number that is also a(n) **real** number.
   d. In the complex number $7i$, the real part is **0** and the imaginary part is **7**. This is an example of a complex number that is also a(n) **pure imaginary** number.
2. Give the complex conjugate of each number.
   a. $3 + 7i$ **$3 - 7i$**
   b. $2 - i$ **$2 + i$**
3. Why are complex conjugates used in dividing complex numbers? **The product of complex conjugates is always a real number.**
4. Explain how you would use complex conjugates to find $(3 + 7i) \div (2 - i)$. **Write the division in fraction form. Then multiply numerator and denominator by $2 + i$.**

Helping You Remember

5. How can you use what you know about simplifying an expression such as $\frac{1 + \sqrt{3}}{2 - \sqrt{5}}$ to help you remember how to simplify fractions with imaginary numbers in the denominator? **Sample answer: In both cases, you can multiply the numerator and denominator by the conjugate of the denominator.**

Solve each equation.

12. $2x^2 + 18 = 0$ **$\pm 3i$**
13. $4x^2 + 32 = 0$ **$\pm 2i\sqrt{2}$**
14. $-5x^2 - 25 = 0$ **$\pm i\sqrt{5}$**

GUIDED PRACTICE KEY	
Exercises	Examples
12–14	4
15, 16	5

Find the values of m and n that make each equation true.

15. $2m + (3n + 1)i = 6 - 8i$ **3, -3**
16. $(2n - 5) + (-m - 2)i = 3 - 7i$ **5, 4**

Application

17. **ELECTRICITY** The current in one part of a series circuit is $4 - j$ amps. The current in another part of the circuit is $6 + 4j$ amps. Add these complex numbers to find the total current in the circuit. **$10 + 3j$ amps**

★ indicates increased difficulty

Practice and Apply

Homework Help	
For Exercises	See Examples
18–21	1
22–25	2
26–29	3
30–33, 46, 47	6
34–37, 42, 43	7
38–41, 44, 45	8
48–55	4
56–61	5

Extra Practice
See page 839.

Simplify. **22. $-13\sqrt{2}$ 30. $9 + 2i$ 33. $4 - 5i$ 35. $6 - 7i$**

18. $\sqrt{-144}$ **$12i$**
19. $\sqrt{-81}$ **$9i$**
20. $\sqrt{-64x^4}$ **$8x^2i$**
21. $\sqrt{-100a^4b^2}$ **$10a^2|b|i$**
22. $\sqrt{-13} \cdot \sqrt{-26}$
23. $\sqrt{-6} \cdot \sqrt{-24}$ **-12**
24. $(-2i)(-6i)(4i)$ **$-48i$**
25. $3i(-5i)^2$ **$-75i$**
26. i^{13} **i**
27. i^{24} **1**
28. i^{38} **-1**
29. i^{63} **$-i$**
30. $(5 - 2i) + (4 + 4i)$
31. $(3 - 5i) + (3 + 5i)$ **6**
32. $(3 - 4i) - (1 - 4i)$ **2**
33. $(7 - 4i) - (3 + i)$
34. $(3 + 4i)(3 - 4i)$ **25**
35. $(1 - 4i)(2 + i)$
36. $(6 - 2i)(1 + i)$ **$8 + 4i$**
37. $(-3 - i)(2 - 2i)$ **$-8 + 4i$**
38. $\frac{4i}{3 + i}$ **$\frac{2}{5} + \frac{6}{5}i$**
39. $\frac{4}{5 + 3i}$ **$\frac{10}{17} - \frac{6}{17}i$**
40. $\frac{10 + i}{4 - i}$ **$\frac{39}{17} + \frac{14}{17}i$**
41. $\frac{2 - i}{3 - 4i}$ **$\frac{2}{5} + \frac{1}{5}i$**

★42. $(-5 + 2i)(6 - i)(4 + 3i)$ **$-163 - 16i$**

★43. $(2 + i)(1 + 2i)(3 - 4i)$ **$20 + 15i$**

★44. $\frac{5 - i\sqrt{3}}{5 + i\sqrt{3}}$ **$\frac{11}{14} - \frac{5\sqrt{3}}{14}i$**

★45. $\frac{1 - i\sqrt{2}}{1 + i\sqrt{2}}$ **$-\frac{1}{3} - \frac{2\sqrt{2}}{3}i$**

46. Find the sum of $ix^2 - (2 + 3i)x + 2$ and $4x^2 + (5 + 2i)x - 4i$. **46. $(i + 4)x^2 + (3 - i)x + 2 - 4i$**

47. Simplify $[(3 + i)x^2 - ix + 4 + i] - [(-2 + 3i)x^2 + (1 - 2i)x - 3]$. **$(5 - 2i)x^2 + (-1 + i)x + 7 + i$**

Solve each equation.

48. $5x^2 + 5 = 0$ **$\pm i$**
49. $4x^2 + 64 = 0$ **$\pm 4i$**
50. $2x^2 + 12 = 0$ **$\pm i\sqrt{6}$**
51. $6x^2 + 72 = 0$ **$\pm 2i\sqrt{3}$**
52. $-3x^2 - 9 = 0$ **$\pm i\sqrt{3}$**
53. $-2x^2 - 80 = 0$ **$\pm 2i\sqrt{10}$**
54. $\frac{2}{3}x^2 + 30 = 0$ **$\pm 3i\sqrt{5}$**
55. $\frac{4}{5}x^2 + 1 = 0$ **$\pm\frac{\sqrt{5}i}{2}$**

Find the values of m and n that make each equation true. **61. $\frac{67}{11}, \frac{19}{11}$**

56. $8 + 15i = 2m + 3ni$ **4, 5**
57. $(m + 1) + 3ni = 5 - 9i$ **4, -3**
58. $(2m + 5) + (1 - n)i = -2 + 4i$ **$-\frac{7}{2}, -3$**
59. $(4 + n) + (3m - 7)i = 8 - 2i$ **$\frac{5}{3}, 4$**

★60. $(m + 2n) + (2m - n)i = 5 + 5i$ **3, 1**

★61. $(2m - 3n)i + (m + 4n) = 13 + 7i$

62. **ELECTRICITY** The impedance in one part of a series circuit is $3 + 4j$ ohms, and the impedance in another part of the circuit is $2 - 6j$. Add these complex numbers to find the total impedance in the circuit. **$5 - 2j$ ohms**

ELECTRICAL ENGINEERING For Exercises 63 and 64, use the formula $E = I \cdot Z$.

63. The current in a circuit is $2 + 5j$ amps, and the impedance is $4 - j$ ohms. What is the voltage? **$13 + 18j$ volts**

64. The voltage in a circuit is $14 - 8j$ volts, and the impedance is $2 - 3j$ ohms. What is the current? **$4 + 2j$ amps**

Career Choices

Electrical Engineering

The chips and circuits in computers are designed by electrical engineers.

Online Research

To learn more about electrical engineering, visit: www.algebra2.com/careers

Enrichment, p. 292

Conjugates and Absolute Value

When studying complex numbers, it is often convenient to represent a complex number by a single variable. For example, we might let $z = x + yi$. We denote the conjugate of z by $\bar{z}$. Thus, $\bar{z} = x - yi$.

We can define the absolute value of a complex number as follows.

$|z| = |x + yi| = \sqrt{x^2 + y^2}$

There are many important relationships involving conjugates and absolute values of complex numbers.

Example 1 **Show $|z|^2 = z\bar{z}$ for any complex number z.**

Let $z = x + yi$. Then,
$z = (x + yi)(x - yi)$
$= x^2 + y^2$
$= \sqrt{(x^2 + y^2)^2}$

65. **CRITICAL THINKING** Show that the order relation "<" does not make sense for the set of complex numbers. (*Hint*: Consider the two cases $i > 0$ and $i < 0$. In each case, multiply each side by i.) **See pp. 283A–283B.**

66. **WRITING IN MATH** Answer the question that was posed at the beginning of the lesson. **See margin.**

How do complex numbers apply to polynomial equations?

Include the following in your answer:

- how the a and c must be related if the equation $ax^2 + c = 0$ has complex solutions, and
- the solutions of the equation $2x^2 + 2 = 0$.

Standardized Test Practice

67. If $i^2 = -1$, then what is the value of i^{71}? **C**

Ⓐ -1 Ⓑ 0 Ⓒ $-i$ Ⓓ i

68. The area of the square is 16 square units. What is the area of the circle? **C**

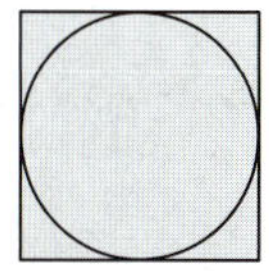

Ⓐ 2π units^2 Ⓑ 12 units^2

Ⓒ 4π units^2 Ⓓ 16π units^2

Extending the Lesson

PATTERN OF POWERS OF i **69. $-1, -i, 1, i, -1, -i, 1, i, -1$**

69. Find the simplified forms of i^6, i^7, i^8, i^9, i^{10}, i^{11}, i^{12}, i^{13}, and i^{14}.

70. Explain how to use the exponent to determine the simplified form of any power of i. **See margin.**

Maintain Your Skills

Mixed Review

Solve each equation. *(Lesson 5-8)*

71. $\sqrt{2x+1} = 5$ **12**
72. $\sqrt[3]{x-3} + 1 = 3$ **11**
73. $\sqrt{x+5} + \sqrt{x} = 5$ **4**

Simplify each expression. *(Lesson 5-7)*

74. $x^{-\frac{1}{5}} \cdot x^{\frac{2}{3}}$ $\mathbf{x^{\frac{7}{15}}}$
75. $\left(y^{-\frac{1}{2}}\right)^{-\frac{2}{3}}$ $\mathbf{y^{\frac{1}{3}}}$
76. $a^{-\frac{3}{4}}$ $\mathbf{\frac{a^{\frac{1}{4}}}{a}}$

For Exercises 77–80, triangle ABC is reflected over the x-axis. *(Lesson 4-6)*

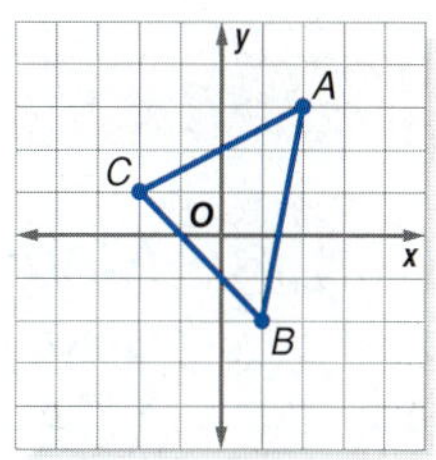

77. Write a vertex matrix for the triangle.
78. Write the reflection matrix.
79. Write the vertex matrix for $\triangle A'B'C'$.
80. Graph $\triangle A'B'C'$. **See pp. 283A–283B.**

77. $\begin{bmatrix} 2 & 1 & -2 \\ 3 & -2 & 1 \end{bmatrix}$

78. $\begin{bmatrix} 1 & 0 \\ 0 & -1 \end{bmatrix}$

79. $\begin{bmatrix} 2 & 1 & -2 \\ -3 & 2 & -1 \end{bmatrix}$

81. **FURNITURE** A new sofa, love seat, and coffee table cost \$2050. The sofa costs twice as much as the love seat. The sofa and the coffee table together cost \$1450. How much does each piece of furniture cost? *(Lesson 3-5)* **sofa: \$1200, love seat: \$600, coffee table: \$250**

Graph each system of inequalities. *(Lesson 3-3)* **82–83. See pp. 283A–283B.**

82. $y < x + 1$
$y > -2x - 2$

83. $x + y \geq 1$
$x - 2y \leq 4$

Find the slope of the line that passes through each pair of points. *(Lesson 2-3)*

84. $(-2, 1), (8, 2)$ $\mathbf{\frac{1}{10}}$
85. $(4, -3), (5, -3)$ **0**

www.algebra2.com/self_check_quiz

Answers

66. Some polynomial equations have complex solutions. Answers should include the following.

- a and c must have the same sign.
- $\pm i$

70. Examine the remainder when the exponent is divided by 4. If the remainder is 0, the result is 1. If the remainder is 1, the result is i. If the remainder is 2, the result is -1. And if the remainder is 3, the result is $-i$.

About the Exercises...

Organization by Objective

- **Add and Subtract Complex Numbers:** 18–33, 46–61
- **Multiply and Divide Complex Numbers:** 34–45

Odd/Even Assignments

Exercises 18–61 are structured so that students practice the same concepts whether they are assigned odd or even problems.

Assignment Guide

Basic: 19–41 odd, 47–59 odd, 63, 65–68, 71–85

Average: 19–63 odd, 65–68, 71–85 (optional: 69, 70)

Advanced: 18–64 even, 65–85

4 Assess

Open-Ended Assessment

Speaking Have students discuss the meaning and "reality" of imaginary numbers, including their graphical representation and their usefulness in electrical engineering.

Tips for New Teachers

Intervention Suggest that students who are confused by imaginary numbers think of i as a very special kind of variable that most of the time can be treated similar to the variable x.

Assessment Options

Quiz (Lessons 5-8 and 5-9) is available on p. 308 of the *Chapter 5 Resource Masters*.

Chapter 5 Study Guide and Review

Vocabulary and Concept Check

- This alphabetical list of vocabulary terms in Chapter 5 includes a page reference where each term was introduced.
- **Assessment** A vocabulary test/review for Chapter 5 is available on p. 306 of the *Chapter 5 Resource Masters*.

Lesson-by-Lesson Review

For each lesson,

- the main ideas are summarized,
- additional examples review concepts, and
- practice exercises are provided.

Vocabulary PuzzleMaker

ELL The Vocabulary PuzzleMaker software improves students' mathematics vocabulary using four puzzle formats—crossword, scramble, word search using a word list, and word search using clues. Students can work on a computer screen or from a printed handout.

MindJogger Videoquizzes

ELL MindJogger Videoquizzes provide an alternative review of concepts presented in this chapter. Students work in teams in a game show format to gain points for correct answers. The questions are presented in three rounds.

Round 1 Concepts (5 questions)
Round 2 Skills (4 questions)
Round 3 Problem Solving (4 questions)

Chapter 5 Study Guide and Review

Vocabulary and Concept Check

absolute value (p. 272)	dimensional analysis (p. 225)	polynomial (p. 229)	scientific notation (p. 225)
binomial (p. 229)	extraneous solution (p. 263)	power (p. 222)	simplify (p. 222)
coefficient (p. 222)	FOIL method (p. 230)	principal root (p. 246)	square root (p. 245)
complex conjugates (p. 273)	imaginary unit (p. 270)	pure imaginary number (p. 270)	standard notation (p. 225)
complex number (p. 271)	like radical expressions (p. 252)	radical equation (p. 263)	synthetic division (p. 234)
conjugates (p. 253)	like terms (p. 229)	radical inequality (p. 264)	terms (p. 229)
constant (p. 222)	monomial (p. 222)	rationalizing the denominator (p. 251)	trinomial (p. 229)
degree (p. 222)	*n*th root (p. 245)		

Choose a word or term from the list above that best completes each statement or phrase.

1. A number is expressed in _______ when it is in the form $a \times 10^n$, where $1 \leq a < 10$ and n is an integer. **scientific notation**
2. A shortcut method known as _______ is used to divide polynomials by binomials. **synthetic division**
3. The _______ is used to multiply two binomials. **FOIL method**
4. A(n) _______ is an expression that is a number, a variable, or the product of a number and one or more variables. **monomial**
5. A solution of a transformed equation that is not a solution of the original equation is a(n) _______. **extraneous solution**
6. _______ are imaginary numbers of the form $a + bi$ and $a - bi$. **Complex conjugates**
7. For any number a and b, if $a^2 = b$, then a is a(n) _______ of b. **square root**
8. A polynomial with three terms is known as a(n) _______. **trinomial**
9. When a number has more than one real root, the _______ is the nonnegative root. **principal root**
10. i is called the _______. **imaginary unit**

Lesson-by-Lesson Review

5-1 Monomials

See pages 222–228.

Concept Summary

- The properties of powers for real numbers a and b and integers m and n are as follows.

$a^{-n} = \frac{1}{a^n}, a \neq 0$ $\qquad (a^m)^n = a^{mn}$

$a^m \cdot a^n = a^{m+n}$ $\qquad (ab)^m = a^m b^m$

$\frac{a^m}{a^n} = a^{m-n}, a \neq 0$ $\qquad \left(\frac{a}{b}\right)^n = \frac{a^n}{b^n}, b \neq 0$

- Use scientific notation to represent very large or very small numbers.

Examples

1 **Simplify $(3x^4y^6)(-8x^3y)$.**

$(3x^4y^6)(-8x^3y) = (3)(-8)x^{4+3}y^{6+1}$ Commutative Property and products of powers

$= -24x^7y^7$ Simplify.

www.algebra2.com/vocabulary_review

FOLDABLES™ Study Organizer

For more information about Foldables, see *Teaching Mathematics with Foldables.*

Ask students to review their Foldable and make sure that their notes, diagrams, and examples are complete. Since journal entries are personal, remind students that these journals are shared only with their consent. Ask if anyone would like to describe one of their journal entries, perhaps something they had difficulty with but later cleared up by asking questions.

Encourage students to refer to their Foldables while completing the Study Guide and Review and to use them in preparing for the Chapter Test.

2 Express each number in scientific notation.

a. 31,000

$31{,}000 = 3.1 \times 10{,}000$

$= 3.1 \times 10^4$ $\quad 10{,}000 = 10^4$

b. 0.007

$0.007 = 7 \times 0.001$

$= 7 \times 10^{-3}$ $\quad 0.001 = \frac{1}{1000}$ or $\frac{1}{10^3}$

Exercises Simplify. Assume that no variable equals 0.
See Examples 1–4 on pages 222–224.

11. $f^{-7} \cdot f^4$ $\frac{1}{f^3}$
12. $(3x^2)^3$ $27x^6$
13. $(2y)(4xy^3)$ $8xy^4$
14. $\left(\frac{3}{5}c^2f\right)\left(\frac{4}{3}cd\right)^2$ $\frac{16}{15}c^4d^2f$

Evaluate. Express the result in scientific notation. *See Examples 5–7 on page 225.*

15. $(2000)(85{,}000)$ 1.7×10^8
16. $(0.0014)^2$ 1.96×10^{-6}
17. $\frac{5{,}400{,}000}{6000}$ 9×10^2

5-2 Polynomials

See pages 229–232.

Concept Summary

- Add or subtract polynomials by combining like terms.
- Multiply polynomials by using the Distributive Property.
- Multiply binomials by using the FOIL method.

Examples

1 Simplify $(5x^2 + 4x) - (3x^2 + 6x - 7)$.

$5x^2 + 4x - (3x^2 + 6x - 7)$
$= 5x^2 + 4x - 3x^2 - 6x + 7$
$= (5x^2 - 3x^2) + (4x - 6x) + 7$
$= 2x^2 - 2x + 7$

2 Find $(9k + 4)(7k - 6)$.

$(9k + 4)(7k - 6)$
$= (9k)(7k) + (9k)(-6) + (4)(7k) + (4)(-6)$
$= 63k^2 - 54k + 28k - 24$
$= 63k^2 - 26k - 24$

18. $-3c + 1$ 19. $4x^2 + 22x - 34$ 20. $-18m^3n - 78m^3 + 30m^2n$

Exercises Simplify. *See Examples 2–5 on pages 229 and 230.*

18. $(4c - 5) - (c + 11) + (-6c + 17)$
19. $(11x^2 + 13x - 15) - (7x^2 - 9x + 19)$
20. $-6m^2(3mn + 13m - 5n)$
21. $x^{-8}y^{10}(x^{11}y^{-9} + x^{10}y^{-6})$ $x^3y + x^2y^4$
22. $(d - 5)(d + 3)$ $d^2 - 2d - 15$
23. $(2a^2 + 6)^2$ $4a^4 + 24a^2 + 36$
24. $(2b - 3c)^3$ $8b^3 - 36b^2c + 54bc^2 - 27c^3$

5-3 Dividing Polynomials

See pages 233–238.

Concept Summary

- Use the division algorithm or synthetic division to divide polynomials.

Example Use synthetic division to find $(4x^4 - x^3 - 19x^2 + 11x - 2) \div (x - 2)$.

2	4	−1	−19	11	−2
		8	14	−10	2
	4	7	−5	1	0

0 → The quotient is $4x^3 + 7x^2 - 5x + 1$.

26. $10x^3 - 5x^2 + 9x - 9$

Exercises Simplify. *See Examples 1–5 on pages 233–235.* 25. $2x^3 + x - \frac{3}{x - 3}$

25. $(2x^4 - 6x^3 + x^2 - 3x - 3) \div (x - 3)$
26. $(10x^4 + 5x^3 + 4x^2 - 9) \div (x + 1)$
27. $(x^2 - 5x + 4) \div (x - 1)$ $x - 4$
28. $(5x^4 + 18x^3 + 10x^2 + 3x) \div (x^2 + 3x)$ $5x^2 + 3x + 1$

Chapter 5 Study Guide and Review 277

5-4 Factoring Polynomials

See pages 239–244.

Concept Summary

- You can factor polynomials using the GCF, grouping, or formulas involving squares and cubes.

Examples

1 Factor $4x^3 - 6x^2 + 10x - 15$.

$4x^3 - 6x^2 + 10x - 15 = (4x^3 - 6x^2) + (10x - 15)$ Group to find the GCF.

$= 2x^2(2x - 3) + 5(2x - 3)$ Factor the GCF of each binomial.

$= (2x^2 + 5)(2x - 3)$ Distributive Property

2 Factor $3m^2 + m - 4$.

Find two numbers whose product is 3(−4) or 12, and whose sum is 1. The two numbers must be 4 and −3 because 4(−3) = −12 and 4 + (−3) = 1.

$3m^2 + m - 4 = 3m^2 + 4m - 3m - 4$

$= (3m^2 + 4m) - (3m + 4)$

$= m(3m + 4) + (-1)(3m + 4)$

$= (3m + 4)(m - 1)$

Exercises **Factor completely. If the polynomial is not factorable, write *prime*.**
See Examples 1–3 on pages 239 and 241. **30. $2(5a^2 - 1)(a - 2)$** **31. $(5w^2 + 3)(w - 4)$**

29. $200x^2 - 50$ **$50(2x + 1)(2x - 1)$**
30. $10a^3 - 20a^2 - 2a + 4$
31. $5w^3 - 20w^2 + 3w - 12$
32. $x^4 - 7x^3 + 12x^2$ **$x^2(x - 3)(x - 4)$**
33. $s^3 + 512$ **$(s + 8)(s^2 - 8s + 64)$**
34. $x^2 - 7x + 5$ **prime**

5-5 Roots of Real Numbers

See pages 245–249.

Concept Summary

Real *n*th roots of *b*, $\sqrt[n]{b}$, or $-\sqrt[n]{b}$			
n	$\sqrt[n]{b}$ if $b > 0$	$\sqrt[n]{b}$ if $b < 0$	$\sqrt[n]{b}$ if $b = 0$
even	one positive root one negative root	no real roots	one real root, 0
odd	one positive root no negative roots	no positive roots one negative root	one real root, 0

Examples

1 Simplify $\sqrt{81x^6}$.

$\sqrt{81x^6} = \sqrt{(9x^3)^2}$ $81x^6 = (9x^3)^2$

$= 9|x^3|$ Use absolute value.

2 Simplify $\sqrt[7]{2187x^{14}y^{35}}$.

$\sqrt[7]{2187x^{14}y^{35}} = \sqrt[7]{(3x^2y^5)^7}$ $2187x^{14}y^{35} = (3x^2y^5)^7$

$= 3x^2y^5$ Evaluate.

Exercises **Simplify.** *See Examples 1 and 2 on pages 246 and 247.*

35. $\pm\sqrt{256}$ **±16**
36. $\sqrt[3]{-216}$ **−6**
37. $\sqrt{(-8)^2}$ **8**
38. $\sqrt[5]{c^5d^{15}}$ ***cd^3***
39. $\sqrt{(x^4 - 3)^2}$ **$|x^4 - 3|$**
40. $\sqrt[3]{(512 + x^2)^3}$ **$512 + x^2$**
41. $\sqrt[4]{16m^8}$ **$2m^2$**
42. $\sqrt{a^2 - 10a + 25}$ **$|a - 5|$**

278 Chapter 5 Polynomials

5-6 Radical Expressions

See pages 250–256.

Concept Summary

For any real numbers a and b and any integer $n > 1$,

- Product Property: $\sqrt[n]{ab} = \sqrt[n]{a} \cdot \sqrt[n]{b}$
- Quotient Property: $\sqrt[n]{\frac{a}{b}} = \frac{\sqrt[n]{a}}{\sqrt[n]{b}}$

Example Simplify $6\sqrt[5]{32m^3} \cdot 5\sqrt[5]{1024m^2}$.

$6\sqrt[5]{32m^3} \cdot 5\sqrt[5]{1024m^2} = 6 \cdot 5\sqrt[5]{(32m^3 \cdot 1024m^2)}$	Product Property of Radicals
$= 30\sqrt[5]{2^5 \cdot 4^5 \cdot m^5}$	Factor into exponents of 5 if possible.
$= 30\sqrt[5]{2^5} \cdot \sqrt[5]{4^5} \cdot \sqrt[5]{m^5}$	Product Property of Radicals
$= 30 \cdot 2 \cdot 4 \cdot m$ or $240m$	Write the fifth roots.

Exercises Simplify. *See Examples 1–6 on pages 250–253.* **47. $20 + 8\sqrt{6}$**

43. $\sqrt[6]{128}$ **$2\sqrt[6]{2}$**
44. $\sqrt{5} + \sqrt{20}$ **$3\sqrt{5}$**
45. $5\sqrt{12} - 3\sqrt{75}$ **$-5\sqrt{3}$**
46. $6\sqrt[5]{11} - 8\sqrt[5]{11}$ **$-2\sqrt[5]{11}$**
47. $(\sqrt{8} + \sqrt{12})^2$
48. $\sqrt{8} \cdot \sqrt{15} \cdot \sqrt{21}$ **$6\sqrt{70}$**
49. $\frac{\sqrt{243}}{\sqrt{3}}$ **9**
50. $\frac{1}{3 + \sqrt{5}}$ **$\frac{3 - \sqrt{5}}{4}$**
51. $\frac{\sqrt{10}}{4 + \sqrt{2}}$ **$\frac{2\sqrt{10} - \sqrt{5}}{7}$**

5-7 Radical Exponents

See pages 257–262.

Concept Summary

- For any nonzero real number b, and any integers m and n, with $n > 1$, $b^{\frac{m}{n}} = \sqrt[n]{b^m} = \left(\sqrt[n]{b}\right)^m$

Examples

1 Write $32^{\frac{4}{5}} \cdot 32^{\frac{2}{5}}$ in radical form.

$32^{\frac{4}{5}} \cdot 32^{\frac{2}{5}} = 32^{\frac{4}{5} + \frac{2}{5}}$	Product of powers
$= 32^{\frac{6}{5}}$	Add.
$= (2^5)^{\frac{6}{5}}$	$32 = 2^5$
$= 2^6$ or 64	Power of a power

2 Simplify $\frac{3x}{\sqrt[3]{z}}$.

$\frac{3x}{\sqrt[3]{z}} = \frac{3x}{z^{\frac{1}{3}}}$	Rational exponents
$= \frac{3x}{z^{\frac{1}{3}}} \cdot \frac{z^{\frac{2}{3}}}{z^{\frac{2}{3}}}$	Rationalize the denominator.
$= \frac{3xz^{\frac{2}{3}}}{z}$ or $\frac{3x\sqrt[3]{z^2}}{z}$	Rewrite in radical form.

Exercises Evaluate. *See Examples 3 and 5 on pages 258 and 259.*

52. $27^{-\frac{2}{3}}$ **$\frac{1}{9}$**
53. $9^{\frac{1}{3}} \cdot 9^{\frac{5}{3}}$ **81**
54. $\left(\frac{8}{27}\right)^{-\frac{2}{3}}$ **$\frac{9}{4}$**

Simplify. *See Example 5 on page 259.*

55. $\frac{1}{y^{\frac{2}{5}}}$ **$\frac{y^{\frac{3}{5}}}{y}$**
56. $\frac{xy}{\sqrt[3]{z}}$ **$\frac{xyz^{\frac{2}{3}}}{z}$**
57. $\frac{3x + 4x^2}{x^{-\frac{2}{3}}}$ **$3x^{\frac{5}{3}} + 4x^{\frac{8}{3}}$**

Chapter 5 Study Guide and Review 279

Chapter 5 For More ...
- Extra Practice, see pages 836–839.
- Mixed Problem Solving, see page 866.

5-8 Radical Equations and Inequalities

See pages 263–267.

Concept Summary

- To solve a radical equation, isolate the radical. Then raise each side of the equation to a power equal to the index of the radical.

Example **Solve $\sqrt{3x-8} + 1 = 3$.**

$\sqrt{3x-8} + 1 = 3$	Original equation
$\sqrt{3x-8} = 2$	Subtract 1 from each side.
$(\sqrt{3x-8})^2 = 2^2$	Square each side.
$3x - 8 = 4$	Evaluate the squares.
$x = 4$	Solve for *x*.

Exercises **Solve each equation.** *See Examples 1–3 on pages 263 and 264.*

58. $\sqrt{x} = 6$ **36**
59. $y^{\frac{1}{3}} - 7 = 0$ **343**
60. $(x-2)^{\frac{3}{2}} = -8$ **no solution**
61. $\sqrt{x+5} - 3 = 0$ **4**
62. $\sqrt{3t-5} - 3 = 4$ **18**
63. $\sqrt{2x-1} = 3$ **5**
64. $\sqrt[4]{2x-1} = 2$ **8.5**
65. $\sqrt{y+5} = \sqrt{2y-3}$ **8**
66. $\sqrt{y+1} + \sqrt{y-4} = 5$ **8**

5-9 Complex Numbers

See pages 270–275.

Concept Summary

- $i^2 = -1$ and $i = \sqrt{-1}$
- Complex conjugates can be used to simplify quotients of complex numbers.

Examples

1 **Simplify $(15 - 2i) + (-11 + 5i)$.**

$(15 - 2i) + (-11 + 5i) = [15 + (-11)] + (-2 + 5)i$	Group the real and imaginary parts.
$= 4 + 3i$	Add.

2 **Simplify $\frac{7i}{2+3i}$.**

$\frac{7i}{2+3i} = \frac{7i}{2+3i} \cdot \frac{2-3i}{2-3i}$	$2 + 3i$ and $2 - 3i$ are conjugates.
$= \frac{14i - 21i^2}{4 - 9i^2}$	Multiply.
$= \frac{21 + 14i}{13}$ or $\frac{21}{13} + \frac{14}{13}i$	$i^2 = -1$

Exercises **Simplify.** *See Examples 1–3 and 6–8 on pages 270, 272, and 273.* **68. $10 - 10i$**

67. $\sqrt{-64m^{12}}$ **$8m^6 i$**
68. $(7 - 4i) - (-3 + 6i)$
69. $-6\sqrt{-9} \cdot 2\sqrt{-4}$ **72**
70. i^6 **−1**
71. $(3 + 4i)(5 - 2i)$ **$23 + 14i$**
72. $(\sqrt{6} + i)(\sqrt{6} - i)$ **7**
73. $\frac{1+i}{1-i}$ **i**
74. $\frac{4-3i}{1+2i}$ **$-\frac{2}{5} - \frac{11}{5}i$**
75. $\frac{3-9i}{4+2i}$ **$\frac{-3-21i}{10}$**

280 Chapter 5 Polynomials

Chapter 5

Practice Test

Vocabulary and Concepts

Choose the term that best describes the shaded part of each trinomial.

1. $\boxed{2}x^2 - 3x + 4$ **c**
2. $4x^{\boxed{2}} - 6x - 3$ **a**
3. $9x^2 + 2x + \boxed{7}$ **b**

a. degree
b. constant term
c. coefficient

Skills and Applications

Simplify.

4. $(5b)^4(6c)^2$ **$22{,}500b^4c^2$**
5. $(13x - 1)(x + 3)$ **$13x^2 + 38x - 3$**
6. $(2h - 6)^3$ **$8h^3 - 72h^2 + 216h - 216$**

Evaluate. Express the result in scientific notation.

7. $(3.16 \times 10^3)(24 \times 10^2)$ **7.584×10^6**
8. $\dfrac{7{,}200{,}000 \cdot 0.0011}{0.018}$ **4.4×10^5**

Simplify.

9. $(x^4 - x^3 - 10x^2 + 4x + 24) \div (x - 2)$ **$x^3 + x^2 - 8x - 12$**
10. $(2x^3 + 9x^2 - 2x + 7) \div (x + 2)$ **$2x^2 + 5x - 12 + \dfrac{31}{x + 2}$**

Factor completely. If the polynomial is not factorable, write *prime*.

11. $x^2 - 14x + 45$ **$(x - 5)(x - 9)$**
12. $2r^2 + 3pr - 2p^2$ **$(2r - p)(r + 2p)$**
13. $x^2 + 2\sqrt{3}x + 3$ **$(x + \sqrt{3})^2$**

Simplify.

14. $\sqrt{175}$ **$5\sqrt{7}$**
15. $(5 + \sqrt{3})(7 - 2\sqrt{3})$ **$29 - 3\sqrt{3}$**
16. $3\sqrt{6} + 5\sqrt{54}$ **$18\sqrt{6}$**
17. $\dfrac{9}{5 - \sqrt{3}}$ **$\dfrac{45 + 9\sqrt{3}}{22}$**
18. $\left(9^{\frac{1}{2}} \cdot 9^{\frac{2}{3}}\right)^{\frac{1}{6}}$ **$3^{\frac{7}{18}}$**
19. $11^{\frac{1}{2}} \cdot 11^{\frac{7}{3}} \cdot 11^{\frac{1}{6}}$ **1331**
20. $\sqrt[6]{256s^{11}t^{18}}$ **$2s|t^3|\sqrt[6]{4s^5}$**
21. $v^{-\frac{7}{11}}$ **$\dfrac{v^{\frac{4}{11}}}{v}$**
22. $\dfrac{b^{\frac{1}{2}}}{b^{\frac{3}{2}} - b^{\frac{1}{2}}}$ **$\dfrac{1}{b - 1}$**

Solve each equation.

23. $\sqrt{b + 15} = \sqrt{3b + 1}$ **7**
24. $\sqrt{2x} = \sqrt{x - 4}$ **no solution**
25. $\sqrt[4]{y + 2} + 9 = 14$ **623**
26. $\sqrt[3]{2w - 1} + 11 = 18$ **172**
27. $\sqrt{4x + 28} = \sqrt{6x + 38}$ **−5**
28. $1 + \sqrt{x + 5} = \sqrt{x + 12}$ **4**

Simplify.

29. $(5 - 2i) - (8 - 11i)$ **$-3 + 9i$**
30. $(14 - 5i)^2$ **$171 - 140i$**

31. **SKYDIVING** The approximate time t in seconds that it takes an object to fall a distance of d feet is given by $t = \sqrt{\dfrac{d}{16}}$. Suppose a parachutist falls 11 seconds before the parachute opens. How far does the parachutist fall during this time period? **1936 ft**

32. **GEOMETRY** The area of a triangle with sides of length a, b, and c is given by $\sqrt{s(s - a)(s - b)(s - c)}$, where $s = \frac{1}{2}(a + b + c)$. If the lengths of the sides of a triangle are 6, 9, and 12 feet, what is the area of the triangle expressed in radical form? **$\dfrac{27\sqrt{15}}{4}$ ft²**

33. **STANDARDIZED TEST PRACTICE** $2 + \left(x + \dfrac{1}{x}\right)^2 =$ **D**

Ⓐ 2   Ⓑ 4   Ⓒ $x^2 + \dfrac{1}{x^2}$ Ⓓ $x^2 + \dfrac{1}{x^2} + 4$

www.algebra2.com/chapter_test

Chapter 5 Practice Test

Assessment Options

Vocabulary Test A vocabulary test/review for Chapter 5 can be found on p. 306 of the *Chapter 5 Resource Masters*.

Chapter Tests There are six Chapter 5 Tests and an Open-Ended Assessment task available in the *Chapter 5 Resource Masters*.

Chapter 5 Tests			
Form	**Type**	**Level**	**Pages**
1	MC	basic	293–294
2A	MC	average	295–296
2B	MC	average	297–298
2C	FR	average	299–300
2D	FR	average	301–302
3	FR	advanced	303–304

MC = multiple-choice questions
FR = free-response questions

Open-Ended Assessment
Performance tasks for Chapter 5 can be found on p. 305 of the *Chapter 5 Resource Masters*. A sample scoring rubric for these tasks appears on p. A34.

ExamView® Pro

Use the networkable ExamView® Pro to:

- Create **multiple versions** of tests.
- Create **modified** tests for *Inclusion* students.
- **Edit** existing questions and **add** your own questions.
- Use built-in **state curriculum correlations** to create tests aligned with state standards.
- **Apply** art to your tests from a program bank of artwork.

Portfolio Suggestion

Introduction In this chapter, you have divided and simplified monomials, polynomials, radical expressions, and complex numbers, often using procedures that involved a series of steps.

Ask Students Write a description for your portfolio comparing these various division problems. Identify which type of division problems was most challenging for you and explain why you think this is true. Be sure to include several examples of your work from this chapter.

Chapter 5 Standardized Test Practice

These two pages contain practice questions in the various formats that can be found on the most frequently given standardized tests.

A practice answer sheet for these two pages can be found on p. A1 of the *Chapter 5 Resource Masters*.

Standardized Test Practice Student Recording Sheet, p. A1

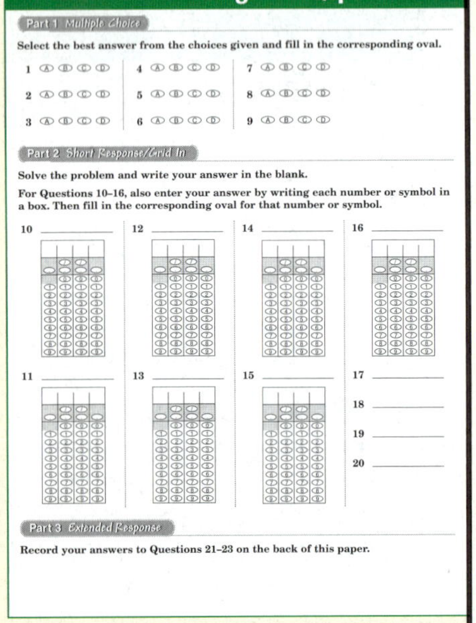
Part 1 Multiple Choice

Select the best answer from the choices given and fill in the corresponding oval.

1 Ⓐ Ⓑ Ⓒ Ⓓ 4 Ⓐ Ⓑ Ⓒ Ⓓ 7 Ⓐ Ⓑ Ⓒ Ⓓ

2 Ⓐ Ⓑ Ⓒ Ⓓ 5 Ⓐ Ⓑ Ⓒ Ⓓ 8 Ⓐ Ⓑ Ⓒ Ⓓ

3 Ⓐ Ⓑ Ⓒ Ⓓ 6 Ⓐ Ⓑ Ⓒ Ⓓ 9 Ⓐ Ⓑ Ⓒ Ⓓ

Part 2 Short Response/Grid In

Solve the problem and write your answer in the blank.

For Questions 10–16, also enter your answer by writing each number or symbol in a box. Then fill in the corresponding oval for that number or symbol.

10 ____ 12 ____ 14 ____ 16 ____

11 ____ 13 ____ 15 ____ 17 ____

18 ____

19 ____

20 ____

Part 3 Extended Response

Record your answers to Questions 21–23 on the back of this paper.

Additional Practice

See pp. 311–312 in the *Chapter 5 Resource Masters* for additional standardized test practice.

Chapter 5 Standardized Test Practice

Part 1 Multiple Choice

Record your answers on the answer sheet provided by your teacher or on a sheet of paper.

1. If $x^3 = 30$ and x is a real number, then x lies between which two consecutive integers? **B**

Ⓐ 2 and 3

Ⓑ 3 and 4

Ⓒ 4 and 5

Ⓓ 5 and 6

2. If $12x + 7y = 19$ and $4x - y = 3$, then what is the value of $8x + 8y$? **C**

Ⓐ 2

Ⓑ 8

Ⓒ 16

Ⓓ 22

3. For all positive integers n,

$\boxed{n} = n - 1$, if n is even and

$\boxed{n} = \frac{1}{2}(n + 1)$, if n is odd.

What is $\boxed{8} \times \boxed{13}$? **B**

Ⓐ 42

Ⓑ 49

Ⓒ 56

Ⓓ 82

4. Let $x \circledast y = xy - y$ for all integers x and y. If $x \circledast y = 0$ and $y \neq 0$, what must x equal? **D**

Ⓐ −2

Ⓑ −1

Ⓒ 0

Ⓓ 1

5. The sum of a number and its square is three times the number. What is the number? **D**

Ⓐ 0 only

Ⓑ −2 only

Ⓒ 2 only

Ⓓ 0 or 2

6. In rectangle $ABCD$, $\overline{AD}$ is 8 units long. What is the length of $\overline{AB}$ in units? **C**

Ⓐ 4

Ⓑ 8

Ⓒ $8\sqrt{3}$

Ⓓ 16

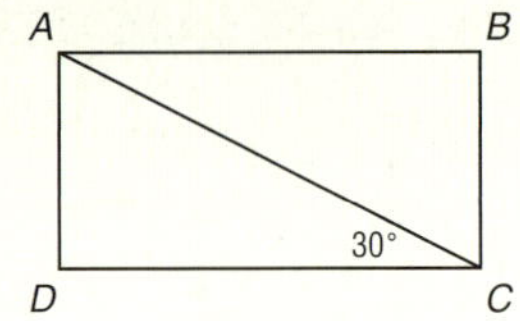

7. The sum of two positive consecutive integers is s. In terms of s, what is the value of the greater integer? **D**

Ⓐ $\frac{s}{2} - 1$ Ⓑ $\frac{s-1}{2}$

Ⓒ $\frac{s}{2}$ Ⓓ $\frac{s+1}{2}$

8. Latha, Renee, and Cindy scored a total of 30 goals for their soccer team this season. Latha scored three times as many goals as Renee. The combined number of goals scored by Latha and Cindy is four times the number scored by Renee. How many goals did Latha score? **C**

Ⓐ 5 Ⓑ 6

Ⓒ 18 Ⓓ 20

9. If $s = t + 1$ and $t \geq 1$, then which of the following must be equal to $s^2 - t^2$? **D**

Ⓐ $(s - t)^2$ Ⓑ $t^2 - 1$

Ⓒ $s^2 - 1$ Ⓓ $s + t$

Test-Taking Tip

Question 9

If you simplify an expression and do not find your answer among the given answer choices, follow these steps. First, check your answer. Then, compare your answer with each of the given answer choices to determine whether it is equivalent to any of the answer choices.

ExamView® Pro

Special banks of standardized test questions similar to those on the SAT, ACT, TIMSS 12, NAEP 12, and state proficiency tests can be found on this CD-ROM.

Preparing for Standardized Tests
For test-taking strategies and more practice, see pages 877–892.

Part 2 Short Response/Grid In

Record your answers on the answer sheet provided by your teacher or on a sheet of paper.

10. Let a ✿ $b = a + \frac{1}{b}$, where $b \neq 0$. What is the value of 3 ✿ 4? **3.25 or 13/4**

11. If $3x^2 = 27$, what is the value of $3x^4$? **243**

12. In the figure, if $x = 25$ and $z = 50$, what is the value of y? **105**

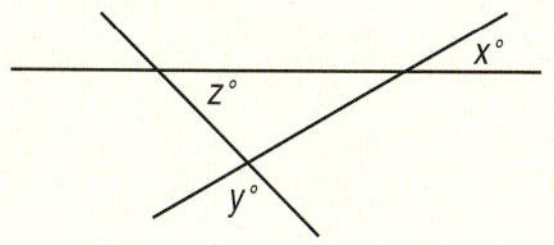

13. For all positive integers n, let $\boxed{n}$ equal the greatest prime number that is a divisor of n. What does $\frac{\boxed{70}}{\boxed{27}}$ equal? **7/3**

14. If $3x + 2y = 36$ and $\frac{5y}{3x} = 5$, then $x =$ __?__. **4**

15. In the figure, a square with side of length $2\sqrt{2}$ is inscribed in a circle. If the area of the circle is $k\pi$, what is the exact value of k? **4**

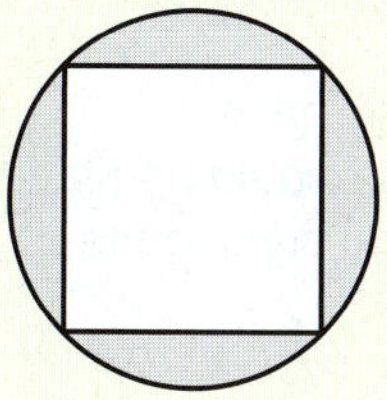

16. For all nonnegative numbers n, let $\boxed{n}$ be defined by $\boxed{n} = \frac{\sqrt{n}}{2}$. If $\boxed{n} = 4$, what is the value of n? **64**

17. For the numbers a, b, and c, the average (arithmetic mean) is twice the median. If $a = 0$, and $a < b < c$, what is the value of $\frac{c}{b}$? **5**

18. Write the numbers $\{\sqrt[3]{64}, \sqrt{18}, 4.1, 16^{\frac{1}{4}}\}$ in order from least to greatest. Show or describe your method. $\{16^{\frac{1}{4}}, \sqrt[3]{64}, 4.1, \sqrt{18}\}$

19. Divide $x^3 - 7x^2 + 18x - 18$ by $x - 3$. Show each step of the process. $x^2 - 4x + 6$

20. Simplify $\frac{x^2 - 2x - 8}{2x^2 - 9x + 4}$. Assume that the denominator is not equal to zero. $\frac{x + 2}{2x - 1}$

Part 3 Extended Response

Record your answers on a sheet of paper. Show your work.

For Exercises 21–23, use the information below.

The period of a pendulum is the time it takes for the pendulum to make one complete swing back and forth. The formula $T = 2\pi\sqrt{\frac{L}{32}}$ gives the period T in seconds for a pendulum L feet long.

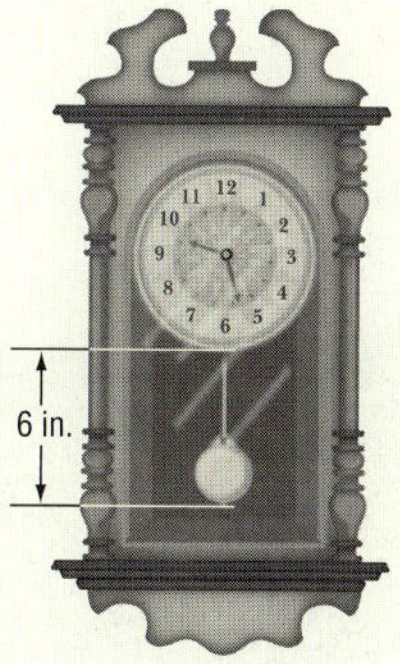

21. What is the period of the pendulum in the wall clock shown? Round to the nearest hundredth of a second. **0.79 s**

22. Solve the formula for the length of the pendulum L in terms of the time T. Show each step of the process. $L = 32\left(\frac{T}{2\pi}\right)^2$

23. If you are building a grandfather clock and you want the pendulum to have a period of 2 seconds, how long should you make the pendulum? Round to the nearest tenth of a foot. **3.2 ft**

www.algebra2.com/standardized_test

Evaluating Extended Response Questions

Extended Response questions are graded by using a multilevel rubric that guides you in assessing a student's knowledge of a particular concept.

Goal: Determine the dimensions of a rectangular garden for a given perimeter.

Sample Scoring Rubric: The following rubric is a sample scoring device. You may wish to add more detail to this sample to meet your individual scoring needs.

Score	Criteria
4	A correct solution that is supported by well-developed, accurate explanations
3	A generally correct solution, but may contain minor flaws in reasoning or computation
2	A partially correct interpretation and/or solution to the problem
1	A correct solution with no supporting evidence or explanation
0	An incorrect solution indicating no mathematical understanding of the concept or task, or no solution is given

Chapter 5 Standardized Test Practice 283

Pages 231–232, Lesson 5-2

3.

	x	x	x
x	x^2	x^2	x^2
2	x	x	x
	x	x	x

55. The expression for how much an amount of money will grow to is a polynomial in terms of the interest rate. Answers should include the following.

- If an amount A grows by r percent for n years, the amount will be $A(1 + r)^n$ after n years. When this expression is expanded, a polynomial results.
- $8820(1 + r)^3$, $8820r^3 + 26{,}460r^2 + 26{,}460r + 8820$
- Evaluate one of the expressions when $r = 0.04$. For example, $8820(1 + r)^3 = 8820(1.04)^3$ or \$9921.30 to the nearest cent. The value given in the table is \$9921 rounded to the nearest dollar.

Pages 237–238, Lesson 5-3

21. $b^2 + 10b$

22. $x - 15$

23. $n^2 - 2n + 3$

24. $2c^2 + c + 5 + \frac{6}{c - 2}$

25. $x^3 - 5x^2 + 11x - 22 + \frac{39}{x + 2}$

26. $6w^4 + 12w^3 + 24w^2 + 30w + 60$

27. x^2

28. $x^2 + 3x + 9$

29. $y^2 - y - 1$

30. $m^2 - 7$

31. $a^3 - 6a^2 - 7a + 7 + \frac{3}{a + 1}$

32. $2m^3 + m^2 + 3m - 1 + \frac{5}{m - 3}$

33. $x^4 - 3x^3 + 2x^2 - 6x + 19 - \frac{56}{x + 3}$

34. $3c^4 - c^3 + 2c^2 - 4c + 9 - \frac{13}{c + 2}$

35. $g + 5$

36. $2b^2 - b - 1 + \frac{4}{b + 1}$

37. $t^4 + 2t^3 + 4t^2 + 5t + 10$

38. $y^4 - 2y^3 + 4y^2 - 8y + 16$

39. $3t^2 - 2t + 3$

40. $h^2 - 4h + 17 - \frac{51}{2h + 3}$

41. $3d^2 + 2d + 3 - \frac{2}{3d - 2}$

42. $x^2 + x - 1$

43. $x^3 - x - \frac{6}{2x + 3}$

44. $2x^3 + x^2 - 1 + \frac{2}{3x + 1}$

45. $x - 3$

46. $x^2 - 1 + \frac{-3x + 7}{x^2 + 2}$

47. $x + 2$

48. $x - 3$

59. Division of polynomials can be used to solve for unknown quantities in geometric formulas that apply to manufacturing situations. Answers should include the following.

- $8x$ in. by $4x + s$ in.
- The area of a rectangle is equal to the length times the width. That is, $A = \ell w$.
- Substitute $32x^2 + x$ for A, $8x$ for ℓ, and $4x + s$ for w. Solving for s involves dividing $32x^2 + x$ by $8x$.

$$A = \ell w$$
$$32x^2 + x = 8x(4x + s)$$
$$\frac{32x^2 + x}{8x} = 4x + s$$
$$4x + \frac{1}{8} = 4x + s$$
$$\frac{1}{8} = s$$

The seam is $\frac{1}{8}$ inch.

Page 243, Lesson 5-4

56. Factoring can be used to find possible dimensions of a geometric figure, given the area. Answers should include the following.

- Since the area of the rectangle is the product of its length and its width, the length and width are factors of the area. One set of possible dimensions is $4x - 2$ by $x + 3$.
- The complete factorization of the area is $2(2x - 1)(x + 3)$, so the factor of 2 could be placed with either $2x - 1$ or $x + 3$ when assigning the dimensions.

Page 262, Lesson 5-7

71. The equation that determines the size of the region around a planet where the planet's gravity is stronger than the Sun's can be written in terms of a fractional exponent. Answers should include the following.

- The radical form of the equation is $r = D\sqrt[5]{\left(\frac{M_p}{M_S}\right)^2}$ or $r = D\sqrt[5]{\frac{M_p^2}{M_S^2}}$. Multiply the fraction under the radical by $\frac{M_S^3}{M_S^3}$.

$$r = D\sqrt[5]{\frac{M_p^2}{M_S^2} \cdot \frac{M_p^3}{M_S^3}}$$

$$= D\sqrt[5]{\frac{M_p^2 M_S^3}{M_S^5}}$$

$$= D\frac{\sqrt[5]{M_p^2 M_S^3}}{\sqrt[5]{M_S^5}}$$

$$= \frac{D\sqrt[5]{M_p^2 M_S^3}}{M_S}$$

The simplified radical form is $\frac{D\sqrt[5]{M_p^2 M_S^3}}{M_S}$.

- If M_p and M_S are constant, then r increases as D increases because r is a linear function of D with positive slope.

Page 267, Lesson 5-8

44. If a company's cost and number of units manufactured are related by an equation involving radicals or rational exponents, then the production level associated with a given cost can be found by solving a radical equation. Answers should include the following.

- $C = 10\sqrt[3]{n^2} + 1500$
-

$10{,}000 = 10n^{\frac{2}{3}} + 1500$	$C = 10{,}000$
$8500 = 10n^{\frac{2}{3}}$	Subtract 1500 from each side.
$850 = n^{\frac{2}{3}}$	Divide each side by 10.
$850^{\frac{3}{2}} = n$	Raise each side to the $\frac{3}{2}$ power.
$24{,}781.55 \approx n$	Use a calculator.

Round down so that the cost does not exceed \$10,000. The company can make at most 24,781 chips.

Page 275, Lesson 5-9

65. Case 1: $i > 0$

Multiply each side by i to get $i^2 > 0 \cdot i$ or $-1 > 0$. This is a contradiction.

Case 2: $i < 0$

Since you are assuming i is negative in this case, you must change the inequality symbol when you multiply each side by i. The result is again $i^2 > 0 \cdot i$ or $-1 > 0$, a contradiction.

Since both possible cases result in contradictions, the order relation "$<$" cannot be applied to the complex numbers.

80.

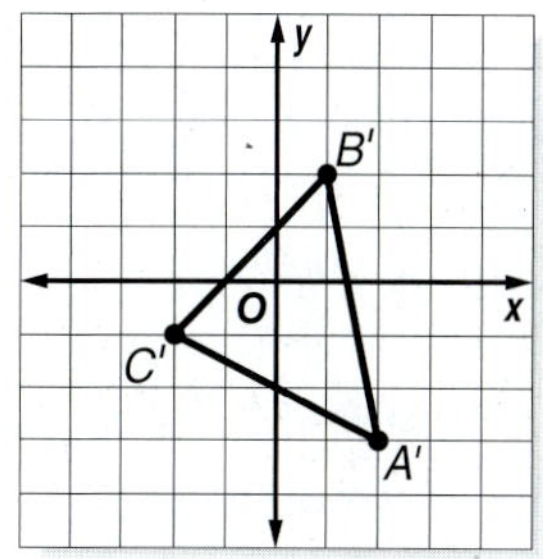

82.

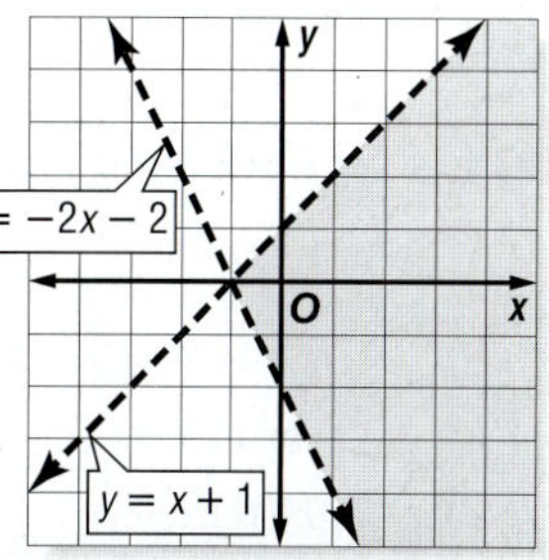

83.

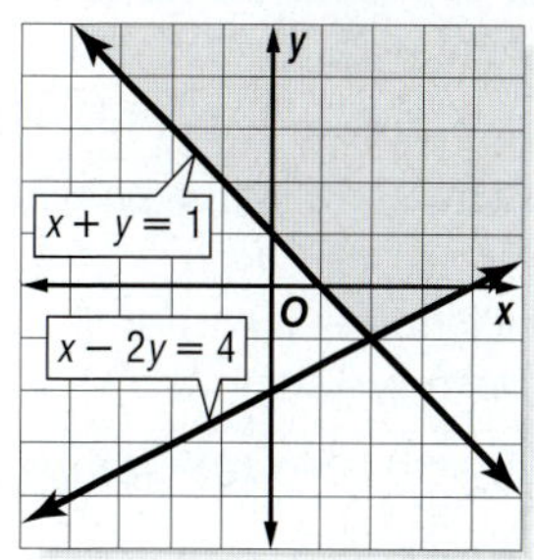

Chapter 6

Quadratic Functions and Inequalities

Chapter Overview and Pacing

Year-long and two-year pacing: pages T20–T21.

LESSON OBJECTIVES

LESSON OBJECTIVES	PACING (days) Regular		Block	
	Basic/ Average	Advanced	Basic/ Average	Advanced
6-1 Graphing Quadratic Functions *(pp. 286–293)* • Graph quadratic functions. • Find and interpret the maximum and minimum values of a quadratic function.	1	1	0.5	0.5
6-2 Solving Quadratic Equations by Graphing *(pp. 294–300)* • Solve quadratic equations by graphing. • Estimate solutions of quadratic equations by graphing. ***Follow-Up:*** Modeling Real-World Data	1	2 (with 6-2 Follow-Up)	0.5	1
6-3 Solving Quadratic Equations by Factoring *(pp. 301–305)* • Solve quadratic equations by factoring. • Write a quadratic equation with given roots.	1	1	0.5	0.5
6-4 Completing the Square *(pp. 306–312)* • Solve quadratic equations by using the Square Root Property. • Solve quadratic equations by completing the square.	2	1	1	0.5
6-5 The Quadratic Formula and the Discriminant *(pp. 313–319)* • Solve quadratic equations by using the Quadratic Formula. • Use the discriminant to determine the number and type of roots of a quadratic equation.	1	1	0.5	0.5
6-6 Analyzing Graphs of Quadratic Functions *(pp. 320–328)* ***Preview:*** Families of Parabolas • Analyze quadratic functions of the form $y = a(x - h)^2 + k$. • Write a quadratic function in the form $y = a(x - h)^2 + k$.	2 (with 6-6 Preview)	1	1.5 (with 6-6 Preview)	0.5
6-7 Graphing and Solving Quadratic Inequalities *(pp. 329–335)* • Graph quadratic inequalities in two variables. • Solve quadratic inequalities in one variable.	1	1	0.5	0.5
Study Guide and **Practice Test** *(pp. 336–341)* **Standardized Test Practice** *(pp. 342–343)*	1	1	0.5	0.5
Chapter Assessment	1	1	0.5	0.5
TOTAL	**11**	**10**	**6**	**5**

StudentWorks

*An electronic version of this chapter is available on **StudentWorks**™. This backpack solution CD-ROM allows students instant access to the Student Edition, lesson worksheet pages, and web resources.*

Timesaving Tools
TeacherWorks™
All-In-One Planner and Resource Center
See pages T5 and T21.

Chapter Resource Manager

CHAPTER 6 RESOURCE MASTERS: Study Guide and Intervention	Practice (Skills and Average)	Reading to Learn Mathematics	Enrichment	Assessment	Applications*	5-Minute Check Transparencies	Interactive Chalkboard	Alge2PASS: Tutorial Plus (lessons)	Materials
313–314	315–316	317	318			6-1	6-1	10	
319–320	321–322	323	324	369	SC 11	6-2	6-2		(*Follow-Up:* graphing calculator)
325–326	327–328	329	330			6-3	6-3		grid paper
331–332	333–334	335	336	369, 371		6-4	6-4		algebra tiles
337–338	339–340	341	342		GCS 38	6-5	6-5	11, 12	posterboard
343–344	345–346	347	348	370		6-6	6-6		(*Preview:* graphing calculator) graphing calculator, index cards
349–350	351–352	353	354	370	GCS 37, SC 12	6-7	6-7		
				355–368, 372–374					

**Key to Abbreviations:* GCS = Graphing Calculator and Speadsheet Masters,
SC = School-to-Career Masters,
SM = Science and Mathematics Lab Manual

Chapter 6

Mathematical Connections and Background

Continuity of Instruction

Prior Knowledge

In the previous chapter students explored how to factor quadratic expressions, how to work with radicals, and how to perform operations on complex numbers. They have also solved linear equations and inequalities, and they are familiar with using formulas.

This Chapter

Students solve quadratic equations by graphing, by factoring, by completing the square, and by using the Quadratic Formula. They explore how values of a quadratic equation are reflected in the parabola that represents it, and use equations and graphs to explore quadratic equations that have 0, 1, or 2 roots. They relate the value of the discriminant to the number of roots and to whether the roots are rational, irrational, or complex.

Future Connections

Students will continue to explore roots and zeros of equations, examining higher-order polynomial equations in Chapter 7. Students will learn to recognize other types of equations that can be solved using the Quadratic Formula. They will explore systems of quadratic inequalities in Chapter 8.

6-1 Graphing Quadratic Functions

When graphing a quadratic function, use all the available information to produce the most accurate possible graph. This includes a table of values containing the vertex and the y-intercept. Be sure the points on the graph are connected with a smooth curve and that the graph has a U-shape and not a V-shape at the vertex of the graph. Unlike the letter U, however, the graph should become progressively wider and have arrows indicating that the graph continues to infinity.

6-2 Solving Quadratic Equations by Graphing

It is important to distinguish between finding solutions or roots of an equation and finding zeros of its related function. An equation of the form $ax^2 + bx + c = 0$ has a related function, $f(x) = ax^2 + bx + c$. The zeros of $f(x)$ are the x-coordinates of the points where the graph crosses the x-axis. These x values are the solutions of the related quadratic equation. Without the use of a graphing calculator, this method of solving quadratic equations will usually provide only an estimate of solutions. Solutions that appear to be integers should be verified by substituting them into the original equation.

6-3 Solving Quadratic Equations by Factoring

When solving a quadratic equation by factoring, it is important to review factoring techniques. These include the techniques for factoring general trinomials, perfect square trinomials, and a difference of squares. Also remember to look for a greatest common factor (GCF) that might be factored out or the possibility of factoring by grouping. Before factoring, the equation should be rewritten so that one side of the equation is 0. If the GCF of the terms of the polynomial being factored is a variable or the product of a number and a variable, such as $3x$, one solution to the equation is 0.

6-4 Complete the Square

Completing the square is a technique most often used to solve quadratic equations that are not factorable. To use this technique it is desirable to rewrite the equation so that it is equal to a constant. Then divide the coefficient of the linear term by 2 and square the result. Add this value to both sides of the equation. One side of the equation will now be a perfect square trinomial that can be rewritten as the square of a binomial. To help isolate the variable on one side of the equation, take the square root of both sides, remembering that the square root of the constant on one side of the equation will result in two values, one positive and one negative, represented by a $\pm$ sign. Finally, isolate the variable using the Addition, Subtraction, Multiplication, and/or Division Properties of Equality. Simplify any solution that involves the symbol $\pm$ by simplifying two separate expressions, one using the plus sign and the other using the minus sign. When using the technique of completing the square, be sure that the coefficient of the quadratic term is 1. If it is not 1, divide both sides of the equation by that coefficient. You can then solve the equation by completing the square.

6-5 The Quadratic Formula and the Discriminant

While the technique of completing the square can be used to solve any quadratic equation, implementing this technique can lead to operations involving unwieldy fractions. The Quadratic Formula can also be used to solve any quadratic equation. Using this formula, the variable is isolated in the very first step and the other steps involve simplifying the solutions by simplifying radicals and fractions. The discriminant is simply the portion of the Quadratic Formula that appears underneath the radical, $b^2 - 4ac$. This value alone will determine the number and type of roots (solutions) of the equation. This is because the square root of this value could result in a rational number, such as 6, an irrational number, such as $\sqrt{2}$, the value 0, or an imaginary number, such as $3\boldsymbol{i}$. By finding and examining just the value of the discriminant, you can tell very quickly what type of solutions a quadratic equation will have. This can serve as a check when solving the equation.

6-6 Analyzing Graphs of Quadratic Functions

To write a quadratic equation in the form $y = a(x - h)^2 + k$, called *vertex form*, it is important to remember that an equation is a statement of equality. When an equation is rewritten in a different form, this equality must be maintained. In previous lessons, if a value was added to one side of an equation, it was also added to the other side of the equation in order to maintain equality. Another way to maintain equality is to add a value to one side and then subtract that same value from that side. For example, if an addition of 5 is shown on the right side of an equation, a subtraction of 5 would also be shown on the right side. So in essence this would be an overall addition of $5 - 5$ or 0, which does not affect the equality of the statement. For example, $y = 3x$ is equivalent to the statement $y = 3x + 5 - 5$. Be especially careful when adding a value inside a set of parentheses, since the use of parentheses often involves multiplication. For example $y = 3(x)$ is not equivalent to $y = 3(x + 1) - 1$, but instead to $y = 3(x + 1) - 3(1)$.

6-7 Graphing and Solving Quadratic Inequalities

One way of solving a linear inequality not discussed in Chapter 1 is to first solve its related linear equation and then test values on either side of this value in the original inequality. For example, to solve $3x + 2 > -4$, you would solve the equation $3x + 2 = -4$ and find that $x = -2$. Testing a value less than -2 and a value greater than -2 in the inequality reveals that the solution to the inequality is the set of values greater than -2. The approach to solving a quadratic inequality algebraically is similar. The difference lies in the fact that many quadratic inequalities have not one but two solutions. This means that your number line is divided in three possible solution sets. Testing a value from each interval on the number line reveals which solution set or sets are correct. The solution set of a quadratic inequality will often be a compound inequality, so you will want to review this topic from Chapter 1.

Additional mathematical information and teaching notes are available at www.algebra2.com/key_concepts.

Chapter 6

DAILY INTERVENTION and Assessment

	Type	Student Edition	Teacher Resources	Technology/Internet
INTERVENTION	Ongoing	Prerequisite Skills, pp. 285, 293, 299, 305, 312, 319, 328 Practice Quiz 1, p. 305 Practice Quiz 2, p. 328	5-Minute Check Transparencies Quizzes, *CRM* pp. 369–370 Mid-Chapter Test, *CRM* p. 371 Study Guide and Intervention, *CRM* pp. 313–314, 319–320, 325–326, 331–332, 337–338, 343–344, 349–350	Alge2PASS: Tutorial Plus www.algebra2.com/self_check_quiz www.algebra2.com/extra_examples
	Mixed Review	pp. 293, 299, 305, 312, 319, 328, 335	Cumulative Review, *CRM* p. 372	
	Error Analysis	Find the Error, pp. 303, 310, 325 Common Misconceptions, pp. 289, 308	Find the Error, *TWE* pp. 303, 310, 325 Unlocking Misconceptions, *TWE* pp. 288, 295 Tips for New Teachers, *TWE* pp. 288, 305, 312, 323	
ASSESSMENT	Standardized Test Practice	pp. 292, 293, 299, 302, 303, 305, 312, 319, 327, 335, 341, 342–343	*TWE* p. 302 Standardized Test Practice, *CRM* pp. 373–374	Standardized Test Practice CD-ROM www.algebra2.com/standardized_test
	Open-Ended Assessment	Writing in Math, pp. 292, 299, 305, 312, 319, 327, 334 Open Ended, pp. 290, 297, 303, 317, 325, 332	Modeling: *TWE* pp. 299, 319 Speaking: *TWE* pp. 293, 305, 335 Writing: *TWE* pp. 312, 328 Open-Ended Assessment, *CRM* p. 367	
	Chapter Assessment	Study Guide, pp. 336–340 Practice Test, p. 341	Multiple-Choice Tests (Forms 1, 2A, 2B), *CRM* pp. 355–360 Free-Response Tests (Forms 2C, 2D, 3), *CRM* pp. 361–366 Vocabulary Test/Review, *CRM* p. 368	ExamView® Pro (see below) MindJogger Videoquizzes www.algebra2.com/vocabulary_review www.algebra2.com/chapter_test

Key to Abbreviations: TWE = Teacher Wraparound Edition; CRM = Chapter Resource Masters

Intervention Technology

Alge2PASS: Tutorial Plus CD-ROM offers a complete, self-paced algebra curriculum.

Algebra 2 Lesson	Alge2PASS Lesson
6-1	**10** *Graphing Quadratic Equations*
6-5	**11** *Solving Quadratic Equations Using the Quadratic Formula*
6-5	**12** *Solving Word Problems Using Quadratic Equations*

ExamView® Pro

Use the networkable **ExamView® Pro** to:

- Create **multiple versions** of tests.
- Create **modified** tests for *Inclusion* students.
- **Edit** existing questions and **add** your own questions.
- Use built-in **state curriculum correlations** to create tests aligned with state standards.
- **Apply** art to your tests from a program bank of artwork.

For more information on Intervention and Assessment, see pp. T8–T11.

Reading and Writing in Mathematics

Glencoe Algebra 2 provides numerous opportunities to incorporate reading and writing into the mathematics classroom.

Student Edition

- Foldables Study Organizer, p. 285
- Concept Check questions require students to verbalize and write about what they have learned in the lesson. (pp. 290, 297, 303, 310, 317, 325, 332, 336)
- Writing in Math questions in every lesson, pp. 292, 299, 305, 312, 319, 327, 334
- Reading Study Tip, pp. 306, 313, 316
- WebQuest, p. 328

Teacher Wraparound Edition

- Foldables Study Organizer, pp. 285, 336
- Study Notebook suggestions, pp. 290, 297, 303, 310, 317, 325, 332
- Modeling activities, pp. 299, 319
- Speaking activities, pp. 293, 305, 335
- Writing activities, pp. 312, 328
- Differentiated Instruction, (Verbal/Linguistic), p. 296
- ELL Resources, pp. 284, 292, 296, 298, 304, 311, 318, 327, 334, 336

Additional Resources

- Vocabulary Builder worksheets require students to define and give examples for key vocabulary terms as they progress through the chapter. (*Chapter 6 Resource Masters*, pp. vii-viii)
- Reading to Learn Mathematics master for each lesson (*Chapter 6 Resource Masters*, pp. 317, 323, 329, 335, 341, 347, 353)
- *Vocabulary PuzzleMaker* software creates crossword, jumble, and word search puzzles using vocabulary lists that you can customize.
- *Teaching Mathematics with Foldables* provides suggestions for promoting cognition and language.
- *Reading and Writing in the Mathematics Classroom*
- *WebQuest and Project Resources*

For more information on Reading and Writing in Mathematics, see pp. T6–T7.

ELL ENGLISH LANGUAGE LEARNERS

Lesson 6-4

Flexible Groups

Have pairs of students draw flowcharts to illustrate the process of completing the square. The charts can be created by hand or by using computer software if it is available. Make sure that students illustrate completing the square for an expression in which the coefficient of the x^2 term is not 1. Have students test their chart by completing Exercises 6 and 7 using the steps they have shown.

Lesson 6-5

Language Experience Approach to Illustrations

Have students research the root words that form the words *quadratic* and *discriminant.* Both words have Latin origins: *quadrus—square; discrimen—distinction.*

Discuss how the root words relate to the mathematical meanings of quadratic and discriminant. Ask bilingual students to tell how they think the words in their language may be related to these root words.

Lesson 6-6

Multisensory Activities

Provide students with grid paper and tracing or tissue paper. Have them place a piece of tracing paper over the grid paper and graph $y = x^2$. Then have students slide the tracing paper in order to translate the graph. Discuss the equation of the new graph. Then reflect the graph over the x-axis. How is the equation affected?

Chapter 6 Notes

What You'll Learn

Have students read over the list of objectives and make a list of any words with which they are not familiar.

Why It's Important

Point out to students that this is only one of many reasons why each objective is important. Others are provided in the introduction to each lesson.

Lesson	NCTM Standards	Local Objectives
6-1	2, 3, 6, 8, 9, 10	
6-2	1, 2, 6, 8, 9, 10	
6-2 Follow-Up	2, 5, 6, 8, 10	
6-3	1, 2, 3, 6, 7, 8, 9	
6-4	1, 2, 3, 6, 7, 8, 9, 10	
6-5	1, 2, 6, 8, 9	
6-6 Preview	2, 8, 10	
6-6	2, 6, 7, 8, 9, 10	
6-7	2, 3, 6, 8, 9, 10	

Key to NCTM Standards:

1=Number & Operations, 2=Algebra, 3=Geometry, 4=Measurement, 5=Data Analysis & Probability, 6=Problem Solving, 7=Reasoning & Proof, 8=Communication, 9=Connections, 10=Representation

Chapter 6 Quadratic Functions and Inequalities

What You'll Learn

- **Lesson 6-1** Graph quadratic functions.
- **Lessons 6-2 through 6-5** Solve quadratic equations.
- **Lesson 6-3** Write quadratic equations and functions.
- **Lesson 6-6** Analyze graphs of quadratic functions.
- **Lesson 6-7** Graph and solve quadratic inequalities.

Key Vocabulary

- root (p. 294)
- zero (p. 294)
- completing the square (p. 307)
- Quadratic Formula (p. 313)
- discriminant (p. 316)

Why It's Important

Quadratic functions can be used to model real-world phenomena like the motion of a falling object. They can also be used to model the shape of architectural structures such as the supporting cables of a suspension bridge. *You will learn to calculate the value of the discriminant of a quadratic equation in order to describe the position of the supporting cables of the Golden Gate Bridge in Lesson 6-5.*

Vocabulary Builder

ELL

The Key Vocabulary list introduces students to some of the main vocabulary terms included in this chapter. For a more thorough vocabulary list with pronunciations of new words, give students the Vocabulary Builder worksheets found on pages vii and viii of the *Chapter 6 Resource Masters*. Encourage them to complete the definition of each term as they progress through the chapter. You may suggest that they add these sheets to their study notebooks for future reference when studying for the Chapter 6 test.

Getting Started

Prerequisite Skills To be successful in this chapter, you'll need to master these skills and be able to apply them in problem-solving situations. Review these skills before beginning Chapter 6.

For Lessons 6-1 and 6-2 — Graph Functions

Graph each equation by making a table of values. *(For review, see Lesson 2-1.)*

1. $y = 2x + 3$ **2.** $y = -x - 5$ **3.** $y = x^2 + 4$ **4.** $y = -x^2 - 2x + 1$

1–4. See pp. 343A–343F.

For Lessons 6-1, 6-2, and 6-5 — Multiply Polynomials

Find each product. *(For review, see Lesson 5-2.)*

5. $(x - 4)(7x + 12)$ $7x^2 - 16x - 48$

6. $(x + 5)^2$ $x^2 + 10x + 25$

7. $(3x - 1)^2$ $9x^2 - 6x + 1$

8. $(3x - 4)(2x - 9)$ $6x^2 - 35x + 36$

For Lessons 6-3 and 6-4 — Factor Polynomials

Factor completely. If the polynomial is not factorable, write *prime*. *(For review, see Lesson 5-4.)*

9. $x^2 + 11x + 30$ **10.** $x^2 - 13x + 36$ **11.** $x^2 - x - 56$ **12.** $x^2 - 5x - 14$

13. $x^2 + x + 2$ prime

14. $x^2 + 10x + 25$ $(x + 5)^2$

15. $x^2 - 22x + 121$ $(x - 11)^2$

16. $x^2 - 9$ $(x + 3)(x - 3)$

For Lessons 6-4 and 6-5 — Simplify Radical Expressions

Simplify. *(For review, see Lessons 5-6 and 5-9.)*

17. $\sqrt{225}$ 15 **18.** $\sqrt{48}$ $4\sqrt{3}$ **19.** $\sqrt{180}$ $6\sqrt{5}$ **20.** $\sqrt{68}$ $2\sqrt{17}$

21. $\sqrt{-25}$ $5i$ **22.** $\sqrt{-32}$ $4i\sqrt{2}$ **23.** $\sqrt{-270}$ $3i\sqrt{30}$ **24.** $\sqrt{-15}$ $i\sqrt{15}$

9. $(x + 6)(x + 5)$ 10. $(x - 4)(x - 9)$ 11. $(x - 8)(x + 7)$ 12. $(x + 2)(x - 7)$

FOLDABLES™ Study Organizer

Quadratic Functions and Inequalities Make this Foldable to help you organize your notes. Begin with one sheet of 11" by 17" paper.

Step 1 Fold and Cut

Fold in half lengthwise. Then fold in fourths crosswise. Cut along the middle fold from the edge to the last crease as shown.

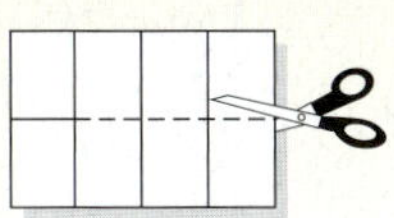

Step 2 Refold and Label

Refold along the lengthwise fold and staple the uncut section at the top. Label each section with a lesson number and close to form a booklet.

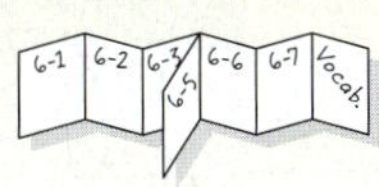

Reading and Writing As you read and study the chapter, fill the journal with notes, diagrams, and examples for each lesson.

Getting Started

This section provides a review of the basic concepts needed before beginning Chapter 6. Page references are included for additional student help.

Prerequisite Skills in the Getting Ready for the Next Lesson section at the end of each exercise set review a skill needed in the next lesson.

For Lesson	Prerequisite Skill
6-2	Evaluating Functions (p. 293)
6-3	Factoring Trinomials (p. 299)
6-4	Simplifying Radicals (p. 305)
6-5	Evaluating Expressions (p. 312)
6-6	Perfect Square Trinomials (p. 319)
6-7	Inequalities (p. 328)

FOLDABLES™ Study Organizer

For more information about Foldables, see *Teaching Mathematics with Foldables.*

Sequencing Information and Progression of Knowledge After students make their Foldable, have them label a section for each lesson in Chapter 6 and a section for vocabulary. As students progress through the lessons, have them summarize key concepts and note the order in which they are presented. Ask students to write about why the concepts were presented in this sequence. If they cannot see the logic in the sequence, have them reorder the key concepts and justify their reasoning.

Chapter 6 Quadratic Functions and Inequalities 285

6-1 Lesson Notes

1 Focus

5-Minute Check Transparency 6-1 Use as a quiz or a review of Chapter 5.

Mathematical Background notes are available for this lesson on p. 284C.

Building on Prior Knowledge

In Chapter 5, students wrote and solved various equations and inequalities. In this lesson, they will relate quadratic equations to their graphs.

How can income from a rock concert be maximized?

Ask students:

- How is the income represented in the given function? **by $P(x)$**
- What is significant about the value of $P(x)$ when $x = 40$ (the ticket price of \$40)? **The value of $P(x)$ is greatest when $x = 40$, or the income is at its maximum value when $x = 40$.**

6-1 Graphing Quadratic Functions

What You'll Learn

- Graph quadratic functions.
- Find and interpret the maximum and minimum values of a quadratic function.

Vocabulary

- quadratic function
- quadratic term
- linear term
- constant term
- parabola
- axis of symmetry
- vertex
- maximum value
- minimum value

How can income from a rock concert be maximized?

Rock music managers handle publicity and other business issues for the artists they manage. One group's manager has found that based on past concerts, the predicted income for a performance is $P(x) = -50x^2 + 4000x - 7500$, where x is the price per ticket in dollars.

The graph of this quadratic function is shown at the right. At first the income increases as the price per ticket increases, but as the price continues to increase, the income declines.

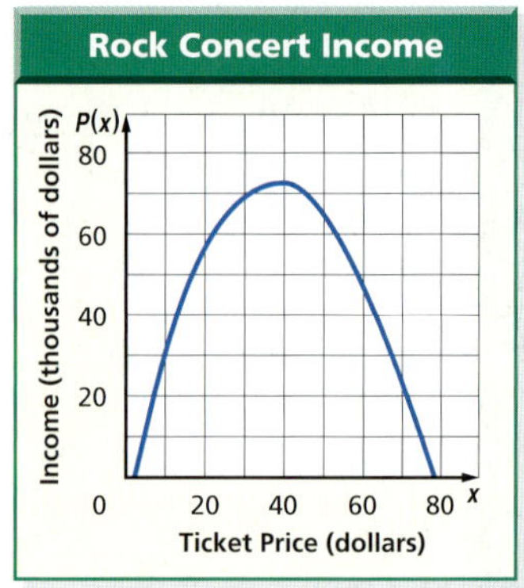

GRAPH QUADRATIC FUNCTIONS A **quadratic function** is described by an equation of the following form.

$$f(x) = ax^2 + bx + c, \text{ where } a \neq 0$$

linear term (bx); quadratic term (ax^2); constant term (c)

The graph of any quadratic function is called a **parabola**. One way to graph a quadratic function is to graph ordered pairs that satisfy the function.

Example 1 Graph a Quadratic Function

Graph $f(x) = 2x^2 - 8x + 9$ by making a table of values.

First, choose integer values for x. Then, evaluate the function for each x value. Graph the resulting coordinate pairs and connect the points with a smooth curve.

x	$2x^2 - 8x + 9$	$f(x)$	$(x, f(x))$
0	$2(0)^2 - 8(0) + 9$	9	(0, 9)
1	$2(1)^2 - 8(1) + 9$	3	(1, 3)
2	$2(2)^2 - 8(2) + 9$	1	(2, 1)
3	$2(3)^2 - 8(3) + 9$	3	(3, 3)
4	$2(4)^2 - 8(4) + 9$	9	(4, 9)

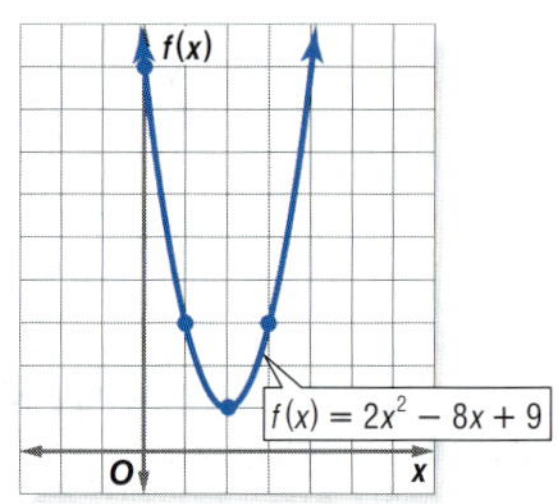

Resource Manager

Workbook and Reproducible Masters

Chapter 6 Resource Masters

- Study Guide and Intervention, pp. 313–314
- Skills Practice, p. 315
- Practice, p. 316
- Reading to Learn Mathematics, p. 317
- Enrichment, p. 318

Teaching Algebra With Manipulatives Masters, p. 243

Transparencies

5-Minute Check Transparency 6-1
Answer Key Transparencies

Technology

Alge2PASS: Tutorial Plus, Lesson 10
Interactive Chalkboard
Multimedia Applications

TEACHING TIP

Tell students they will derive the equation for the axis of symmetry in Lesson 6-6, Exercise 53, after they have learned about a technique called completing the square.

All parabolas have an **axis of symmetry**. If you were to fold a parabola along its axis of symmetry, the portions of the parabola on either side of this line would match.

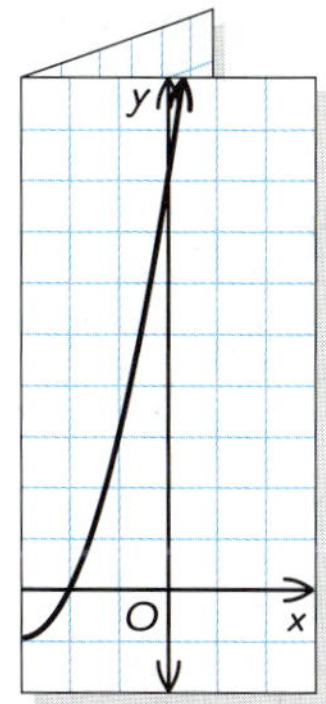

The point at which the axis of symmetry intersects a parabola is called the **vertex**. The y-intercept of a quadratic function, the equation of the axis of symmetry, and the x-coordinate of the vertex are related to the equation of the function as shown below.

Key Concept — *Graph of a Quadratic Function*

- **Words** Consider the graph of $y = ax^2 + bx + c$, where $a \neq 0$.
  - The y-intercept is $a(0)^2 + b(0) + c$ or c.
  - The equation of the axis of symmetry is $x = -\frac{b}{2a}$.
  - The x-coordinate of the vertex is $-\frac{b}{2a}$.
- **Model**

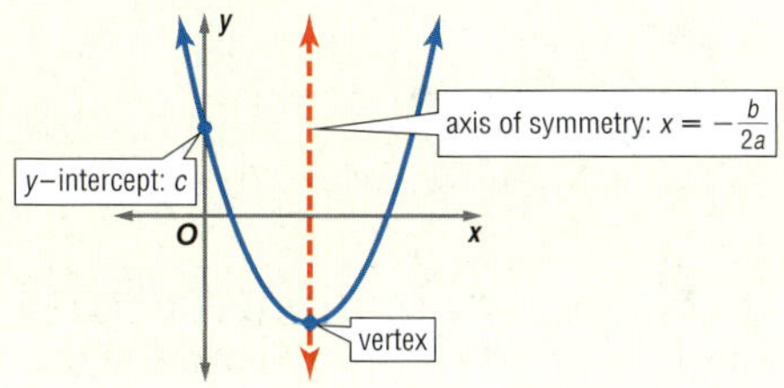

Knowing the location of the axis of symmetry, y-intercept, and vertex can help you graph a quadratic function.

Example 2 Axis of Symmetry, y-Intercept, and Vertex

Consider the quadratic function $f(x) = x^2 + 9 + 8x$.

a. Find the y-intercept, the equation of the axis of symmetry, and the x-coordinate of the vertex.

Begin by rearranging the terms of the function so that the quadratic term is first, the linear term is second, and the constant term is last. Then identify a, b, and c.

$$f(x) = ax^2 + bx + c$$

$$f(x) = x^2 + 9 + 8x \rightarrow f(x) = 1x^2 + 8x + 9$$

So, $a = 1$, $b = 8$, and $c = 9$.

The y-intercept is 9. You can find the equation of the axis of symmetry using a and b.

$x = -\frac{b}{2a}$ Equation of the axis of symmetry

$x = -\frac{8}{2(1)}$ $a = 1, b = 8$

$x = -4$ Simplify.

The equation of the axis of symmetry is $x = -4$. Therefore, the x-coordinate of the vertex is -4.

www.algebra2.com/extra_examples

2 Teach

GRAPH QUADRATIC FUNCTIONS

In-Class Examples

Power Point®

1 Graph $f(x) = x^2 + 3x - 1$ by making a table of values.

x	−3	−2	−1	0	1
$f(x)$	−1	−3	−3	−1	3

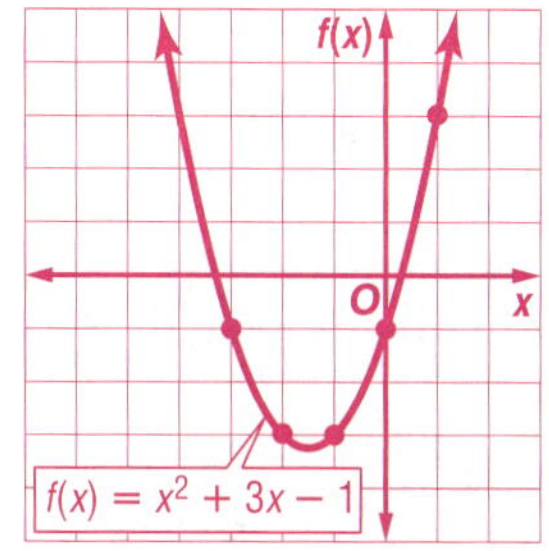

Teaching Tip Make sure students understand that the graph shows values for all the points that satisfy the function, even when the x value is not an integer. For example, the vertex is $(-1.5, -3.25)$.

2 Consider the quadratic function $f(x) = 2 - 4x + x^2$.

a. Find the y-intercept, the equation of the axis of symmetry, and the x-coordinate of the vertex. **2; $x = 2$; 2**

b. Make a table of values that includes the vertex.

x	0	1	2	3	4
$f(x)$	2	−1	−2	−1	2

c. Use this information to graph the function.

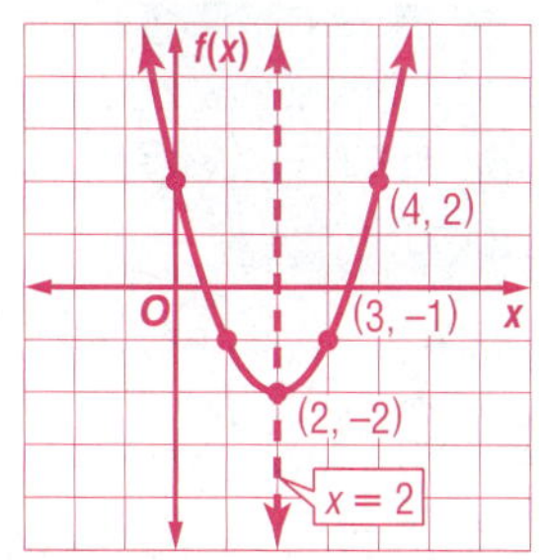

Lesson 6-1 Graphing Quadratic Functions 287

Tips for New Teachers

Intervention When discussing Example 2, make sure students realize that $f(x)$ and y can be used interchangeably, and also that the maximum or minimum value of the function is given by the y-coordinate of the vertex of the parabola.

MAXIMUM AND MINIMUM VALUES

In-Class Example Power Point®

3 Consider the function $f(x) = -x^2 + 2x + 3$.

a. Determine whether the function has a maximum or a minimum value. **maximum**

b. State the maximum or minimum value of the function. **4**

Teaching Tip Remind students to use the equation for the axis of symmetry to determine the x-coordinate of the vertex and then to find the y-coordinate of the vertex.

Interactive Chalkboard
PowerPoint® Presentations

This CD-ROM is a customizable Microsoft® PowerPoint® presentation that includes:

- Step-by-step, dynamic solutions of each In-Class Example from the Teacher Wraparound Edition
- Additional, Your Turn exercises for each example
- The 5-Minute Check Transparencies
- Hot links to Glencoe Online Study Tools

Study Tip

Symmetry
Sometimes it is convenient to use symmetry to help find other points on the graph of a parabola. Each point on a parabola has a mirror image located the same distance from the axis of symmetry on the other side of the parabola.

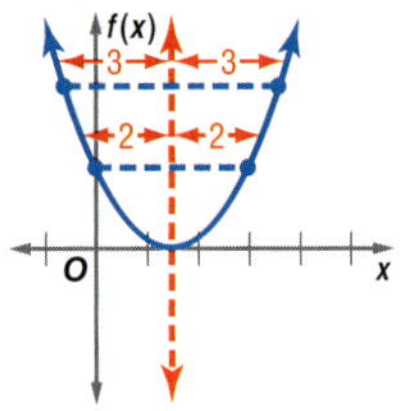

b. Make a table of values that includes the vertex.

Choose some values for x that are less than -4 and some that are greater than -4. This ensures that points on each side of the axis of symmetry are graphed.

x	$x^2 + 8x + 9$	$f(x)$	$(x, f(x))$
-6	$(-6)^2 + 8(-6) + 9$	-3	$(-6, -3)$
-5	$(-5)^2 + 8(-5) + 9$	-6	$(-5, -6)$
-4	$(-4)^2 + 8(-4) + 9$	-7	$(-4, -7)$ ← Vertex
-3	$(-3)^2 + 8(-3) + 9$	-6	$(-3, -6)$
-2	$(-2)^2 + 8(-2) + 9$	-3	$(-2, -3)$

c. Use this information to graph the function.

Graph the vertex and y-intercept. Then graph the points from your table connecting them and the y-intercept with a smooth curve. As a check, draw the axis of symmetry, $x = -4$, as a dashed line. The graph of the function should be symmetrical about this line.

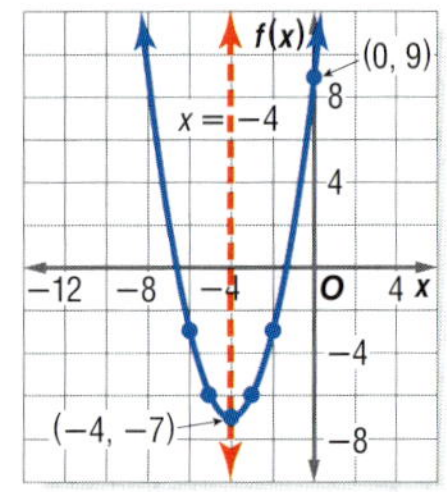

MAXIMUM AND MINIMUM VALUES The y-coordinate of the vertex of a quadratic function is the **maximum value** or **minimum value** obtained by the function.

Key Concept *Maximum and Minimum Value*

- **Words** The graph of $f(x) = ax^2 + bx + c$, where $a \neq 0$,
  - opens up and has a minimum value when $a > 0$, and
  - opens down and has a maximum value when $a < 0$.
- **Models** a is positive. a is negative.

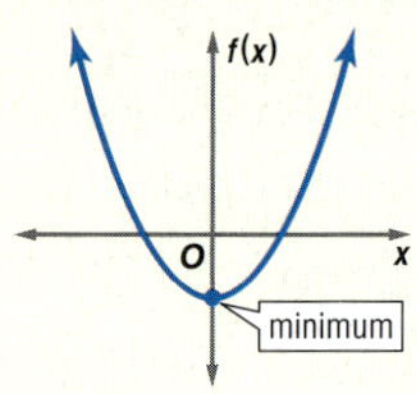

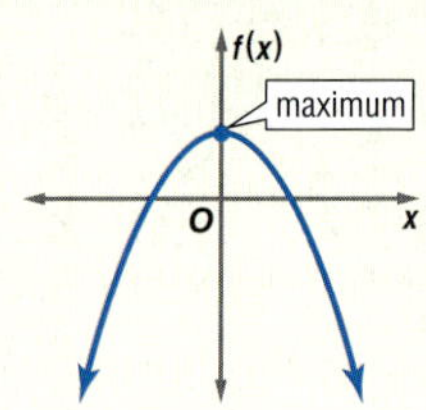

Example 3 *Maximum or Minimum Value*

Consider the function $f(x) = x^2 - 4x + 9$.

a. Determine whether the function has a maximum or a minimum value.

For this function, $a = 1$, $b = -4$, and $c = 9$. Since $a > 0$, the graph opens up and the function has a minimum value.

DAILY INTERVENTION **Unlocking Misconceptions**

Minimum and Maximum Values Make sure students understand that a parabola which opens upward is the graph of a function with a minimum value and that a parabola which opens downward is the graph of a function with a maximum value. Compare these parabolas to valleys (where the altitude of the valley floor is a minimum) and hills (where the peak of the hill is the maximum altitude).

Study Tip

Common Misconception

The terms minimum *point* and minimum *value* are not interchangeable. The minimum point on the graph of a quadratic function is the set of coordinates that describe the location of the vertex. The minimum value of a function is the y-coordinate of the minimum point. It is the smallest value obtained when $f(x)$ is evaluated for all values of x.

b. State the maximum or minimum value of the function.

The minimum value of the function is the y-coordinate of the vertex.

The x-coordinate of the vertex is $-\frac{-4}{2(1)}$ or 2.

Find the y-coordinate of the vertex by evaluating the function for $x = 2$.

$f(x) = x^2 - 4x + 9$ Original function

$f(2) = (2)^2 - 4(2) + 9$ or 5 $x = 2$

Therefore, the minimum value of the function is 5.

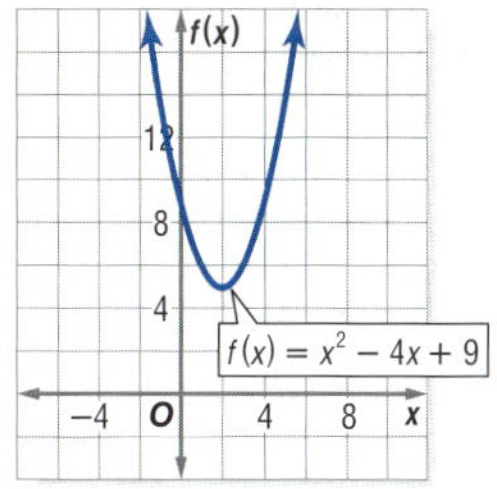

When quadratic functions are used to model real-world situations, their maximum or minimum values can have real-world meaning.

Example 4 Find a Maximum Value

FUND-RAISING **Four hundred people came to last year's winter play at Sunnybrook High School. The ticket price was \$5. This year, the Drama Club is hoping to earn enough money to take a trip to a Broadway play. They estimate that for each \$0.50 increase in the price, 10 fewer people will attend their play.**

a. How much should the tickets cost in order to maximize the income from this year's play?

Words The income is the number of tickets multiplied by the price per ticket.

Variables Let x = the number of \$0.50 price increases.
Then $5 + 0.50x$ = the price per ticket and
$400 - 10x$ = the number of tickets sold.

Let $I(x)$ = income as a function of x.

	The income	is	the number of tickets	multiplied by	the price per ticket.
Equation	$I(x)$	$=$	$(400 - 10x)$	$\cdot$	$(5 + 0.50x)$

$= 400(5) + 400(0.50x) - 10x(5) - 10x(0.50x)$

$= 2000 + 200x - 50x - 5x^2$ Multiply.

$= 2000 + 150x - 5x^2$ Simplify.

$= -5x^2 + 150x + 2000$ Rewrite in $ax^2 + bx + c$ form.

$I(x)$ is a quadratic function with $a = -5$, $b = 150$, and $c = 2000$. Since $a < 0$, the function has a maximum value at the vertex of the graph. Use the formula to find the x-coordinate of the vertex.

$x\text{-coordinate of the vertex} = -\frac{b}{2a}$ Formula for the x-coordinate of the vertex

$= -\frac{150}{2(-5)}$ $a = -5, b = 150$

$= 15$ Simplify.

This means the Drama Club should make 15 price increases of \$0.50 to maximize their income. Thus, the ticket price should be $5 + 0.50(15)$ or \$12.50.

(continued on the next page)

In-Class Example Power Point®

4 **ECONOMICS** A souvenir shop sells about 200 coffee mugs each month for \$6 each. The shop owner estimates that for each \$0.50 increase in the price, he will sell about 10 fewer coffee mugs per month.

a. How much should the owner charge for each mug in order to maximize the monthly income from their sales? **\$8**

b. What is the maximum monthly income the owner can expect to make from these items? **\$1280**

DAILY INTERVENTION

Differentiated Instruction

Auditory/Musical Ask students to suggest the kinds of musical sounds they might associate with a parabola that has a maximum value, and have them contrast this to a sound that might be associated with a parabola that has a minimum value. For example, an orchestra piece that rises to a crescendo and then gradually returns to the previous volume might be seen as being related to a parabola with a maximum value.

3 Practice/Apply

Study Notebook

Have students—
- add the definitions/examples of the vocabulary terms to their Vocabulary Builder worksheets for Chapter 6.
- add the information in the Key Concept box on p. 287 about the graph of a quadratic function to their notebook.
- make labeled sketches similar to those on p. 288 illustrating the maximum and minimum values of the graphs of quadratic functions.
- include any other item(s) that they find helpful in mastering the skills in this lesson.

About the Exercises...

Organization by Objective
- **Graph Quadratic Functions:** 14–31, 44
- **Maximum and Minimum Values:** 32–43, 45–53

Odd/Even Assignments

Exercises 14–31 and 32–43 are structured so that students practice the same concepts whether they are assigned odd or even problems.

Alert! Exercises 58–63 require a graphing calculator.

Assignment Guide

Basic: 15–27 odd, 33–43 odd, 44–47, 54–57, 64–78

Average: 15–43 odd, 46–50, 53–57, 64–78 (optional: 58–63)

Advanced: 14–42 even, 46–74 (optional: 75–78)

b. What is the maximum income the Drama Club can expect to make?

To determine maximum income, find the maximum value of the function by evaluating $I(x)$ for $x = 15$.

$$I(x) = -5x^2 + 150x + 2000 \quad \text{Income function}$$
$$I(15) = -5(15)^2 + 150(15) + 2000 \quad x = 15$$
$$= 3125 \quad \text{Use a calculator.}$$

Thus, the maximum income the Drama Club can expect is $3125.

CHECK Graph this function on a graphing calculator, and use the CALC menu to confirm this solution.

KEYSTROKES: [2nd] [CALC] 4 0 [ENTER] 25 [ENTER] [ENTER]

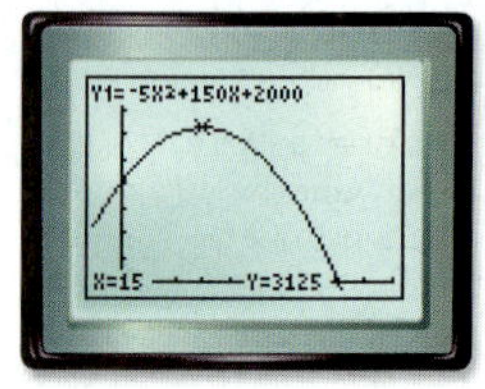

[−5, 50] scl: 5 by [−100, 4000] scl: 500

At the bottom of the display are the coordinates of the maximum point on the graph of $y = -5x^2 + 150x + 2000$. The y value of these coordinates is the maximum value of the function, or 3125. ✓

Check for Understanding

Concept Check

1. **OPEN ENDED** Give an example of a quadratic function. Identify its quadratic term, linear term, and constant term. **1. Sample answer: $f(x) = 3x^2 + 5x - 6$; $3x^2$, $5x$, -6**
2. **Identify** the vertex and the equation of the axis of symmetry for each function graphed below. **a. (2, 1); $x = 2$ b. (−3, −2); $x = -3$**

a. 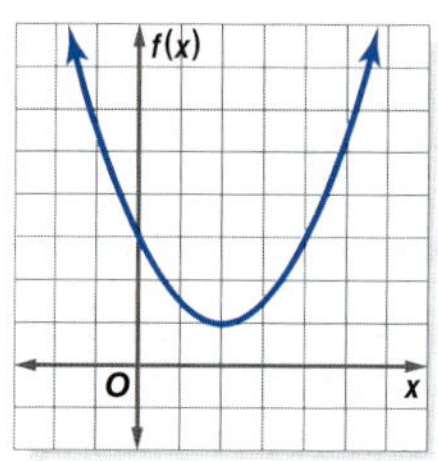

b. 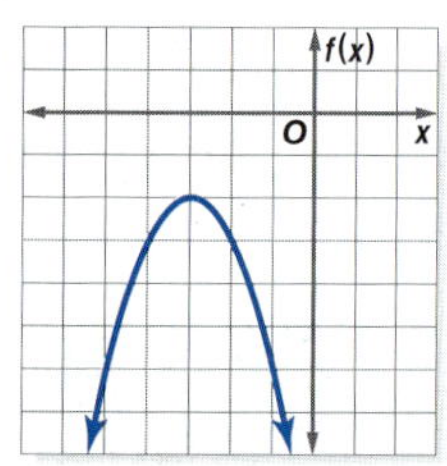

3. **State** whether the graph of each quadratic function opens *up* or *down*. Then state whether the function has a *maximum* or *minimum* value.
   - a. $f(x) = 3x^2 + 4x - 5$ **up; min.**
   - b. $f(x) = -2x^2 + 9$ **down; max.**
   - c. $f(x) = -5x^2 - 8x + 2$ **down; max.**
   - d. $f(x) = 6x^2 - 5x$ **up; min.**

Guided Practice

GUIDED PRACTICE KEY	
Exercises	Examples
4–9	1, 2
10–12	3
13	4

Complete parts a–c for each quadratic function.

a. **Find the y-intercept, the equation of the axis of symmetry, and the x-coordinate of the vertex.**

b. **Make a table of values that includes the vertex.**

c. **Use this information to graph the function.** **4–9. See margin.**

4. $f(x) = -4x^2$
5. $f(x) = x^2 + 2x$
6. $f(x) = -x^2 + 4x - 1$
7. $f(x) = x^2 + 8x + 3$
8. $f(x) = 2x^2 - 4x + 1$
9. $f(x) = 3x^2 + 10x$

Answers

4a. 0; $x = 0$; 0

4b.

x	$f(x)$
−1	−4
0	0
1	−4

4c. 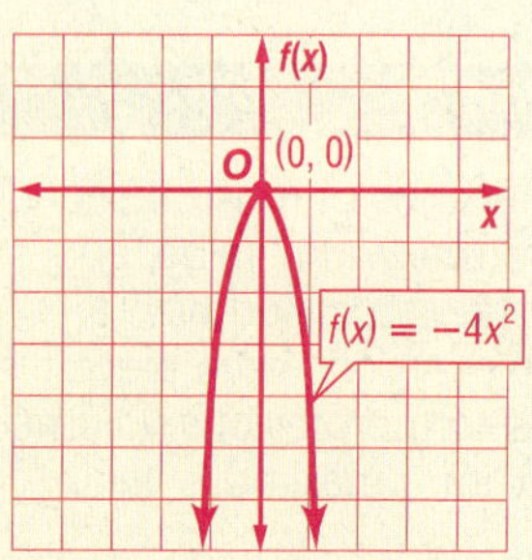

5a. 0; $x = -1$; −1

5b.

x	$f(x)$
−3	3
−2	0
−1	−1
0	0
1	3

5c. 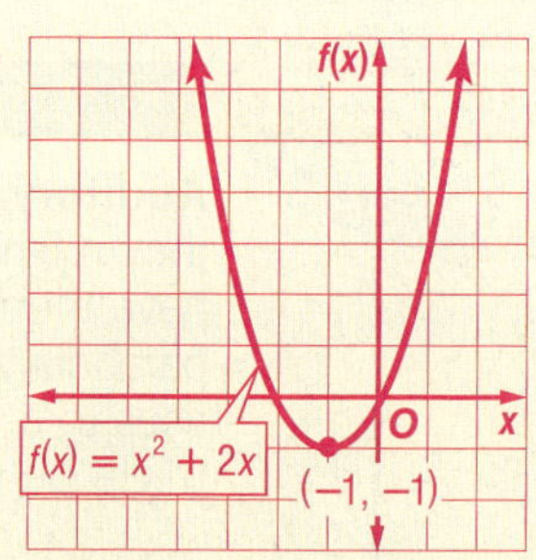

Determine whether each function has a maximum or a minimum value. Then find the maximum or minimum value of each function.

10. $f(x) = -x^2 + 7$ **max.; 7**

11. $f(x) = x^2 - x - 6$ **min.; $-\frac{25}{4}$**

12. $f(x) = 4x^2 + 12x + 9$ **min.; 0**

Application

13. **NEWSPAPERS** Due to increased production costs, the Daily News must increase its subscription rate. According to a recent survey, the number of subscriptions will decrease by about 1250 for each 25¢ increase in the subscription rate. What weekly subscription rate will maximize the newspaper's income from subscriptions? **$8.75**

★ indicates increased difficulty

Practice and Apply

Homework Help

For Exercises	See Examples
14–19	1
20–31	2
32–43, 54	3
44–53	4

Extra Practice
See page 839.

Complete parts a–c for each quadratic function.

a. Find the y-intercept, the equation of the axis of symmetry, and the x-coordinate of the vertex.

b. Make a table of values that includes the vertex.

c. Use this information to graph the function. **14–31. See pp. 343A–343F.**

14. $f(x) = 2x^2$

15. $f(x) = -5x^2$

16. $f(x) = x^2 + 4$

17. $f(x) = x^2 - 9$

18. $f(x) = 2x^2 - 4$

19. $f(x) = 3x^2 + 1$

20. $f(x) = x^2 - 4x + 4$

21. $f(x) = x^2 - 9x + 9$

22. $f(x) = x^2 - 4x - 5$

23. $f(x) = x^2 + 12x + 36$

24. $f(x) = 3x^2 + 6x - 1$

25. $f(x) = -2x^2 + 8x - 3$

26. $f(x) = -3x^2 - 4x$

27. $f(x) = 2x^2 + 5x$

★28. $f(x) = 0.5x^2 - 1$

★29. $f(x) = -0.25x^2 - 3x$

★30. $f(x) = \frac{1}{2}x^2 + 3x + \frac{9}{2}$

★31. $f(x) = x^2 - \frac{2}{3}x - \frac{8}{9}$

Determine whether each function has a maximum or a minimum value. Then find the maximum or minimum value of each function.

32. $f(x) = 3x^2$ **min.; 0**

33. $f(x) = -x^2 - 9$ **max.; −9**

34. $f(x) = x^2 - 8x + 2$ **min.; −14**

35. $f(x) = x^2 + 6x - 2$ **min.; −11**

36. $f(x) = 4x - x^2 + 1$ **max.; 5**

37. $f(x) = 3 - x^2 - 6x$ **max.; 12**

38. $f(x) = 2x + 2x^2 + 5$ **min.; $\frac{9}{2}$**

39. $f(x) = x - 2x^2 - 1$ **max.; $-\frac{7}{8}$**

40. $f(x) = -7 - 3x^2 + 12x$ **max.; 5**

41. $f(x) = -20x + 5x^2 + 9$ **min.; −11**

42. $f(x) = -\frac{1}{2}x^2 - 2x + 3$ **max.; 5**

43. $f(x) = \frac{3}{4}x^2 - 5x - 2$ **min.; $-10\frac{1}{3}$**

More About . . .

Architecture

The Exchange House in London, England, is supported by two interior and two exterior steel arches. V-shaped braces add stability to the structure.

Source: Council on Tall Buildings and Urban Habitat

ARCHITECTURE **For Exercises 44 and 45, use the following information.**
The shape of each arch supporting the Exchange House can be modeled by $h(x) = -0.025x^2 + 2x$, where $h(x)$ represents the height of the arch and x represents the horizontal distance from one end of the base in meters.

44. Write the equation of the axis of symmetry, and find the coordinates of the vertex of the graph of $h(x)$. **$x = 40$; (40, 40)**

45. According to this model, what is the maximum height of the arch? **40 m**

www.algebra2.com/self_check_quiz

Answers

8a. 1; $x = 1$; 1

8b.

x	$f(x)$
−1	7
0	1
1	−1
2	1
3	7

8c.

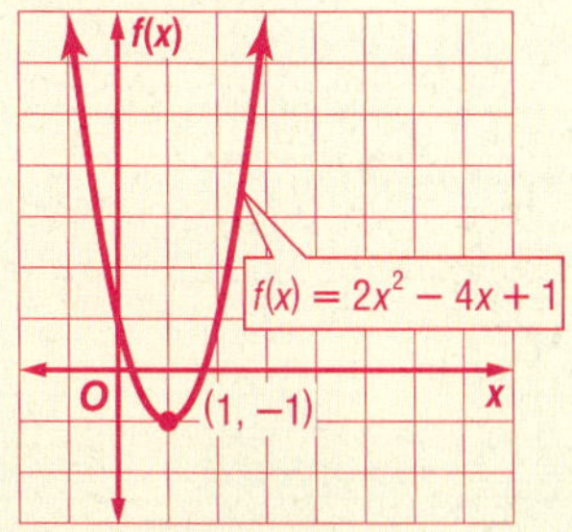

9a. 0; $x = -\frac{5}{3}$; $-\frac{5}{3}$

9b.

x	$f(x)$
−3	−3
−2	−8
$-\frac{5}{3}$	$-\frac{25}{3}$
−1	−7
0	0

9c.

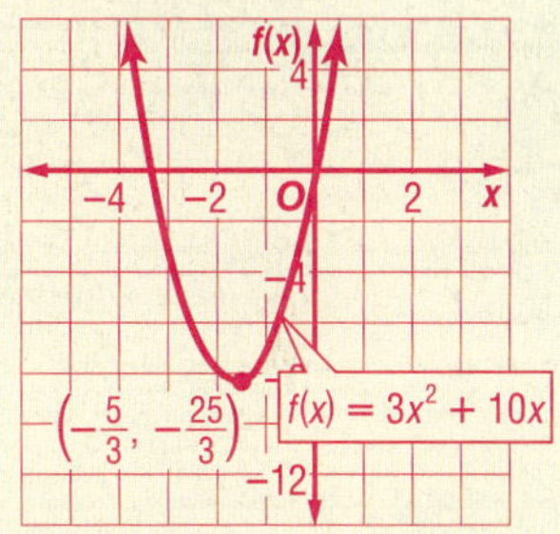

6a. −1; $x = 2$; 2

6b.

x	$f(x)$
0	−1
1	2
2	3
3	2
4	−1

6c.

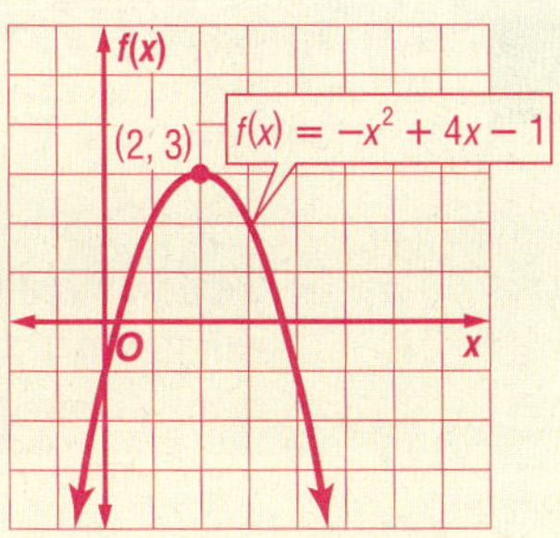

7a. 3; $x = -4$; −4

7b.

x	$f(x)$
−6	−9
−5	−12
−4	−13
−3	−12
−2	−9

7c.

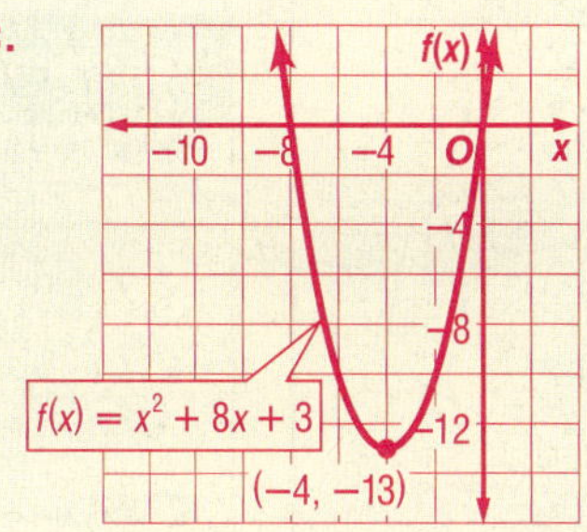

Study Guide and Intervention, p. 313 (shown) and p. 314

Graph Quadratic Functions

Quadratic Function	A function defined by an equation of the form $f(x) = ax^2 + bx + c$, where $a \neq 0$
Graph of a Quadratic Function	A **parabola** with these characteristics: y intercept: c; axis of symmetry: $x = \frac{-b}{2a}$; x-coordinate of vertex: $\frac{-b}{2a}$

Example **Find the y-intercept, the equation of the axis of symmetry, and the x-coordinate of the vertex for the graph of $f(x) = x^2 - 3x + 5$. Use this information to graph the function.**

$a = 1$, $b = -3$, and $c = 5$, so the y-intercept is 5. The equation of the axis of symmetry is $x = \frac{-(-3)}{2(1)}$ or $\frac{3}{2}$. The x-coordinate of the vertex is $\frac{3}{2}$.

Next make a table of values for x near $\frac{3}{2}$.

x	$x^2 - 3x + 5$	$f(x)$	$(x, f(x))$
0	$0^2 - 3(0) + 5$	5	(0, 5)
1	$1^2 - 3(1) + 5$	3	(1, 3)
$\frac{3}{2}$	$\left(\frac{3}{2}\right)^2 - 3\left(\frac{3}{2}\right) + 5$	$\frac{11}{4}$	$\left(\frac{3}{2}, \frac{11}{4}\right)$
2	$2^2 - 3(2) + 5$	3	(2, 3)
3	$3^2 - 3(3) + 5$	5	(3, 5)

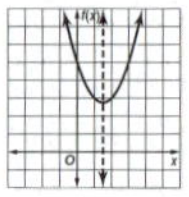

Exercises

For Exercises 1–3, complete parts a–c for each quadratic function.

a. Find the y-intercept, the equation of the axis of symmetry, and the x-coordinate of the vertex.

b. Make a table of values that includes the vertex.

c. Use this information to graph the function.

1. $f(x) = x^2 + 6x + 8$
8, $x = -3$, -3

x	−3	−2	−1	−4
$f(x)$	−1	0	3	0

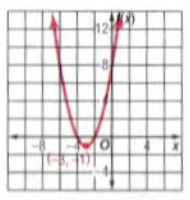

2. $f(x) = -x^2 - 2x + 2$
2, $x = -1$, -1

x	−1	0	−2	1
$f(x)$	3	2	2	−1

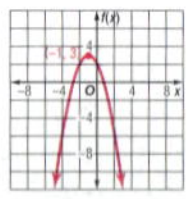

3. $f(x) = 2x^2 - 4x + 3$
3, $x = 1$, 1

x	1	0	2	3
$f(x)$	1	3	3	9

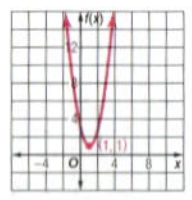

Skills Practice, p. 315 and Practice, p. 316 (shown)

Complete parts a–c for each quadratic function.

a. Find the y-intercept, the equation of the axis of symmetry, and the x-coordinate of the vertex.

b. Make a table of values that includes the vertex.

c. Use this information to graph the function.

1. $f(x) = x^2 - 8x + 15$
15; $x = 4$; 4

x	0	2	4	6	8
$f(x)$	15	3	−1	3	15

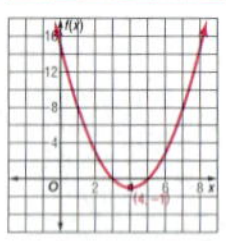

2. $f(x) = -x^2 - 4x + 12$
12; $x = -2$; -2

x	−6	−4	−2	0	2
$f(x)$	0	12	16	12	0

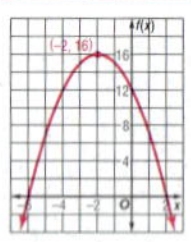

3. $f(x) = 2x^2 - 2x + 1$
1; $x = 0.5$; 0.5

x	−1	0	0.5	1	2
$f(x)$	5	1	0.5	1	5

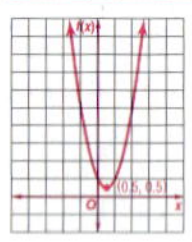

Determine whether each function has a maximum or a minimum value. Then find the maximum or minimum value of each function.

4. $f(x) = x^2 + 2x - 8$ **min.; −9**
5. $f(x) = x^2 - 6x + 14$ **min.; 5**
6. $v(x) = -x^2 + 14x - 57$ **max.; −8**
7. $f(x) = 2x^2 + 4x - 6$ **min.; −8**
8. $f(x) = -x^2 + 4x - 1$ **max.; 3**
9. $f(x) = -\frac{2}{3}x^2 + 8x - 24$ **max.; 0**

10. **GRAVITATION** From 4 feet above a swimming pool, Susan throws a ball upward with a velocity of 32 feet per second. The height $h(t)$ of the ball t seconds after Susan throws it is given by $h(t) = -16t^2 + 32t + 4$. Find the maximum height reached by the ball and the time that this height is reached. **20 ft; 1 s**

11. **HEALTH CLUBS** Last year, the SportsTime Athletic Club charged \$20 to participate in an aerobics class. Seventy people attended the classes. The club wants to increase the class price this year. They expect to lose one customer for each \$1 increase in the price.
   a. What price should the club charge to maximize the income from the aerobics classes? **\$45**
   b. What is the maximum income the SportsTime Athletic Club can expect to make? **\$2025**

Reading to Learn Mathematics, p. 317 ELL

Pre-Activity **How can income from a rock concert be maximized?**

Read the introduction to Lesson 6-1 at the top of page 286 in your textbook.

- Based on the graph in your textbook, for what ticket price is the income the greatest? **\$40**
- Use the graph to estimate the maximum income. **about \$72,000**

Reading the Lesson

1. a. For the quadratic function $f(x) = 2x^2 + 5x + 3$, $2x^2$ is the **quadratic** term, $5x$ is the **linear** term, and 3 is the **constant** term.
   b. For the quadratic function $f(x) = -4 + x - 3x^2$, $a =$ **−3**, $b =$ **1**, and $c =$ **−4**.
2. Consider the quadratic function $f(x) = ax^2 + bx + c$, where $a \neq 0$.
   a. The graph of this function is a **parabola**.
   b. The y-intercept is **c**.
   c. The axis of symmetry is the line **$x = -\frac{b}{2a}$**.
   d. If $a > 0$, then the graph opens **upward** and the function has a **minimum** value.
   e. If $a < 0$, then the graph opens **downward** and the function has a **maximum** value.
3. Refer to the graph at the right as you complete the following sentences.

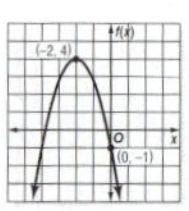

   a. The curve is called a **parabola**.
   b. The line $x = -2$ is called the **axis of symmetry**.
   c. The point $(-2, 4)$ is called the **vertex**.
   d. Because the graph contains the point $(0, -1)$, -1 is the **y-intercept**.

Helping You Remember

4. How can you remember the way to use the x^2 term of a quadratic function to tell whether the function has a maximum or a minimum value? **Sample answer: Remember that the graph of $f(x) = x^2$ (with $a > 0$) is a U-shaped curve that opens up and has a minimum. The graph of $g(x) = -x^2$ (with $a < 0$) is just the opposite. It opens down and has a maximum.**

PHYSICS For Exercises 46 and 47, use the following information.

An object is fired straight up from the top of a 200-foot tower at a velocity of 80 feet per second. The height $h(t)$ of the object t seconds after firing is given by $h(t) = -16t^2 + 80t + 200$.

46. Find the maximum height reached by the object and the time that the height is reached. **300 ft, 2.5 s**

47. Interpret the meaning of the y-intercept in the context of this problem. **The y-intercept is the initial height of the object.**

CONSTRUCTION For Exercises 48–50, use the following information. **49. 60 ft by 30 ft**

Steve has 120 feet of fence to make a rectangular kennel for his dogs. He will use his house as one side.

48. Write an algebraic expression for the kennel's length. **$120 - 2x$**

49. What dimensions produce a kennel with the greatest area?

50. Find the maximum area of the kennel. **1800 ft²**

More About . . .

Tourism

Known as the Hostess City of the South, Savannah, Georgia, is a popular tourist destination. One of the first planned cities in the Americas, Savannah's Historic District is based on a grid-like pattern of streets and alleys surrounding open spaces called squares.

Source: savannah-online.com

TOURISM For Exercises 51 and 52, use the following information.

A tour bus in the historic district of Savannah, Georgia, serves 300 customers a day. The charge is \$8 per person. The owner estimates that the company would lose 20 passengers a day for each \$1 fare increase.

51. What charge would give the most income for the company? **\$11.50**

52. If the company raised their fare to this price, how much daily income should they expect to bring in? **\$2645**

★ 53. **GEOMETRY** A rectangle is inscribed in an isosceles triangle as shown. Find the dimensions of the inscribed rectangle with maximum area. (*Hint:* Use similar triangles.) **5 in. by 4 in.**

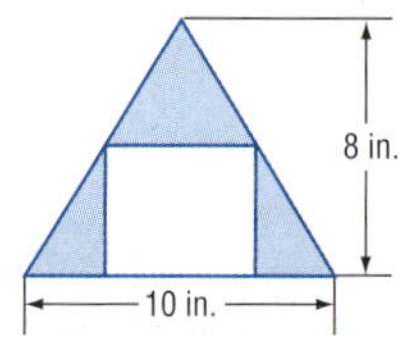

54. **CRITICAL THINKING** Write an expression for the minimum value of a function of the form $y = ax^2 + c$, where $a > 0$. Explain your reasoning. Then use this function to find the minimum value of $y = 8.6x^2 - 12.5$. **c; See margin for explanation; −12.5.**

55. **WRITING IN MATH** Answer the question that was posed at the beginning of the lesson. **See margin.**

How can income from a rock concert be maximized?

Include the following in your answer:
- an explanation of why income increases and then declines as the ticket price increases, and
- an explanation of how to algebraically and graphically determine what ticket price should be charged to achieve maximum income.

Standardized Test Practice

56. The graph of which of the following equations is symmetrical about the y-axis? **C**

Ⓐ $y = x^2 + 3x - 1$ Ⓑ $y = -x^2 + x$
Ⓒ $y = 6x^2 + 9$ Ⓓ $y = 3x^2 - 3x + 1$

292 Chapter 6 Quadratic Functions and Inequalities

Enrichment, p. 318

Finding the Axis of Symmetry of a Parabola

As you know, if $f(x) = ax^2 + bx + c$ is a quadratic function, the values of x that make $f(x)$ equal to zero are $\frac{-b + \sqrt{b^2 - 4ac}}{2a}$ and $\frac{-b - \sqrt{b^2 - 4ac}}{2a}$.

The average of these two number values is $-\frac{b}{2a}$. The function $f(x)$ has its maximum or minimum value when $x = -\frac{b}{2a}$. Since the axis of symmetry of the graph of $f(x)$ passes through the point where the maximum or minimum occurs, the axis of symmetry has the equation $x = -\frac{b}{2a}$.

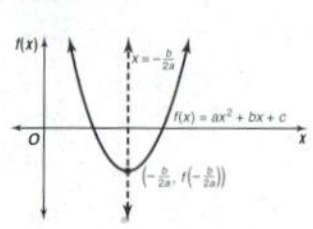

Example **Find the vertex and axis of symmetry for $f(x) = 5x^2 + 10x - 7$.**

Answer

54. The x-coordinate of the vertex of $y = ax^2 + c$ is $-\frac{0}{2a}$ or 0, so the y-coordinate of the vertex, the minimum of the function, is $a(0)^2 + c$ or c.

57. Which of the following tables represents a quadratic relationship between the two variables x and y? **C**

Ⓐ

x	1	2	3	4	5
y	3	3	3	3	3

Ⓑ

x	1	2	3	4	5
y	5	4	3	2	1

Ⓒ

x	1	2	3	4	5
y	6	3	2	3	6

Ⓓ

x	1	2	3	4	5
y	−4	−3	−4	−3	−4

Graphing Calculator

MAXIMA AND MINIMA You can use the MINIMUM or MAXIMUM feature on a graphing calculator to find the minimum or maximum value of a quadratic function. This involves defining an interval that includes the vertex of the parabola. A lower bound is an x value left of the vertex, and an upper bound is an x value right of the vertex.

Step 1 Graph the function so that the vertex of the parabola is visible.

Step 2 Select 3:minimum or 4:maximum from the CALC menu.

Step 3 Using the arrow keys, locate a left bound and press [ENTER].

Step 4 Locate a right bound and press [ENTER] twice. The cursor appears on the maximum or minimum point of the function. The maximum or minimum value is the y-coordinate of that point.

Find the maximum or minimum value of each quadratic function to the nearest hundredth.

58. $f(x) = 3x^2 - 7x + 2$ **−2.08**

59. $f(x) = -5x^2 + 8x$ **3.20**

60. $f(x) = 2x^2 - 3x + 2$ **0.88**

61. $f(x) = -6x^2 + 9x$ **3.38**

62. $f(x) = 7x^2 + 4x + 1$ **0.43**

63. $f(x) = -4x^2 + 5x$ **1.56**

Maintain Your Skills

Mixed Review

Simplify. *(Lesson 5-9)*

64. i^{14} **−1**

65. $(4 - 3i) - (5 - 6i)$ **−1 + 3*i***

66. $(7 + 2i)(1 - i)$ **9 − 5*i***

Solve each equation. *(Lesson 5-8)*

67. $5 - \sqrt{b + 2} = 0$ **23**

68. $\sqrt[3]{x + 5} + 6 = 4$ **−13**

69. $\sqrt{n + 12} - \sqrt{n} = 2$ **4**

Perform the indicated operations. *(Lesson 4-2)*

70. $[4 \quad 1 \quad -3] + [6 \quad -5 \quad 8]$ $[10 \quad -4 \quad 5]$

71. $[2 \quad -5 \quad 7] - [-3 \quad 8 \quad -1]$ $[5 \quad -13 \quad 8]$

72. $4\begin{bmatrix} -7 & 5 & -11 \\ 2 & -4 & 9 \end{bmatrix}$ $\begin{bmatrix} -28 & 20 & -44 \\ 8 & -16 & 36 \end{bmatrix}$

73. $-2\begin{bmatrix} -3 & 0 & 12 \\ -7 & \frac{1}{3} & 4 \end{bmatrix}$ $\begin{bmatrix} 6 & 0 & -24 \\ 14 & -\frac{2}{3} & -8 \end{bmatrix}$

74. Graph the system of equations $y = -3x$ and $y - x = 4$. State the solution. Is the system of equations *consistent* and *independent*, *consistent* and *dependent*, or *inconsistent*? *(Lesson 3-1)* **See margin for graph; (−1, 3); consistent and independent.**

Getting Ready for the Next Lesson

PREREQUISITE SKILL **Evaluate each function for the given value.**
*(To review **evaluating functions**, see Lesson 2-1.)*

75. $f(x) = x^2 + 2x - 3, x = 2$ **5**

76. $f(x) = -x^2 - 4x + 5, x = -3$ **8**

77. $f(x) = 3x^2 + 7x, x = -2$ **−2**

78. $f(x) = \frac{2}{3}x^2 + 2x - 1, x = -3$ **−1**

4 Assess

Open-Ended Assessment

Speaking Ask students to explain how to tell by examining a quadratic function whether its graph will have a maximum or minimum value. Then ask them to give an example of what such a value might mean in a real-world problem.

Getting Ready for Lesson 6-2

PREREQUISITE SKILL Lesson 6-2 presents solving quadratic equations by graphing. Finding points on the graph of the function involves evaluating quadratic functions. Exercises 75–78 should be used to determine your students' familiarity with evaluating functions.

Answer

74.

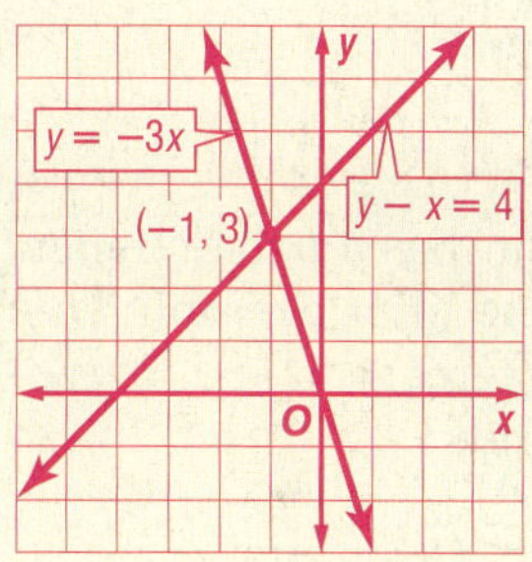

Answer

55. If a quadratic function can be used to model ticket price versus profit, then by finding the x-coordinate of the vertex of the parabola you can determine the price per ticket that should be charged to achieve maximum profit. Answers should include the following.

- If the price of a ticket is too low, then you won't make enough money to cover your costs, but if the ticket price is too high fewer people will buy them.
- You can locate the vertex of the parabola on the graph of the function. It occurs when $x = 40$. Algebraically, this is found by calculating $x = -\frac{b}{2a}$ which, for this case, is $x = \frac{-4000}{2(-50)}$ or 40. Thus the ticket price should be set at $40 each to achieve maximum profit.

Lesson 6-1 Graphing Quadratic Functions 293

6-2 Lesson Notes

1 Focus

5-Minute Check Transparency 6-2 Use as a quiz or a review of Lesson 6-1.

Mathematical Background notes are available for this lesson on p. 284C.

How does a quadratic function model a free-fall ride?

Ask students:

- The acceleration of a free-falling object due to Earth's gravity is -32 ft/sec^2. It is given as a negative value because the acceleration is downward, toward Earth's surface. How is this fact represented in the height function? **The coefficient -16 is the one half of the acceleration due to gravity in a downward direction.**
- How far has a person fallen 1 second after beginning a free fall? after 2 seconds? after 3 seconds? **16 ft; 64 ft; 144 ft**

6-2 Solving Quadratic Equations by Graphing

What You'll Learn

- Solve quadratic equations by graphing.
- Estimate solutions of quadratic equations by graphing.

How does a quadratic function model a free-fall ride?

As you speed to the top of a free-fall ride, you are pressed against your seat so that you feel like you're being pushed downward. Then as you free-fall, you fall at the same rate as your seat. Without the force of your seat pressing on you, you *feel* weightless. The height above the ground (in feet) of an object in free-fall can be determined by the quadratic function $h(t) = -16t^2 + h_0$, where t is the time in seconds and the initial height is h_0 feet.

Vocabulary

- quadratic equation
- root
- zero

SOLVE QUADRATIC EQUATIONS When a quadratic function is set equal to a value, the result is a quadratic equation. A **quadratic equation** can be written in the form $ax^2 + bx + c = 0$, where $a \neq 0$.

The solutions of a quadratic equation are called the **roots** of the equation. One method for finding the roots of a quadratic equation is to find the **zeros** of the related quadratic function. The zeros of the function are the x-intercepts of its graph. These are the solutions of the related equation because $f(x) = 0$ at those points. The zeros of the function graphed at the right are 1 and 3.

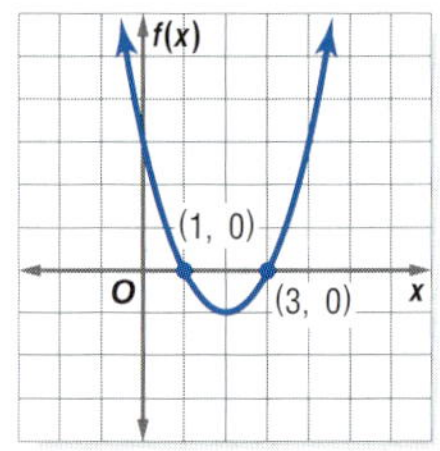

Study Tip

Reading Math

In general, equations have roots, functions have zeros, and graphs of functions have x-intercepts.

Example 1 Two Real Solutions

Solve $x^2 + 6x + 8 = 0$ by graphing.

Graph the related quadratic function $f(x) = x^2 + 6x + 8$. The equation of the axis of symmetry is $x = -\frac{6}{2(1)}$ or -3. Make a table using x values around -3. Then, graph each point.

x	-5	-4	-3	-2	-1
$f(x)$	3	0	-1	0	3

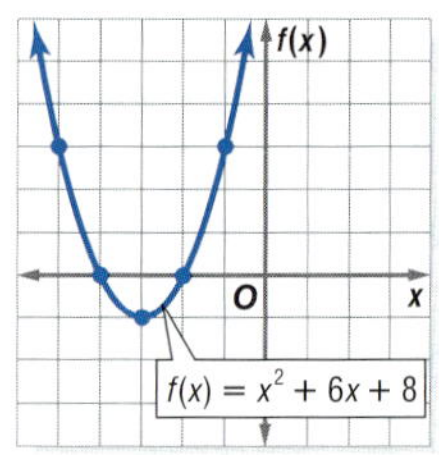

From the table and the graph, we can see that the zeros of the function are -4 and -2. Therefore, the solutions of the equation are -4 and -2.

CHECK Check the solutions by substituting each solution into the equation to see if it is satisfied.

$$x^2 + 6x + 8 = 0 \qquad\qquad x^2 + 6x + 8 = 0$$
$$(-4)^2 + 6(-4) + 8 \stackrel{?}{=} 0 \qquad\qquad (-2)^2 + 6(-2) + 8 \stackrel{?}{=} 0$$
$$0 = 0\ \checkmark \qquad\qquad 0 = 0\ \checkmark$$

The graph of the related function in Example 1 had two zeros; therefore, the quadratic equation had two real solutions. This is one of the three possible outcomes when solving a quadratic equation.

Resource Manager

Workbook and Reproducible Masters

Chapter 6 Resource Masters

- Study Guide and Intervention, pp. 319–320
- Skills Practice, p. 321
- Practice, p. 322
- Reading to Learn Mathematics, p. 323
- Enrichment, p. 324
- Assessment, p. 369

School-to-Career Masters, p. 11

Transparencies

5-Minute Check Transparency 6-2
Answer Key Transparencies

Technology

Interactive Chalkboard

Key Concept — Solutions of a Quadratic Equation

- **Words** A quadratic equation can have one real solution, two real solutions, or no real solution.
- **Models** One Real Solution — Two Real Solutions — No Real Solution

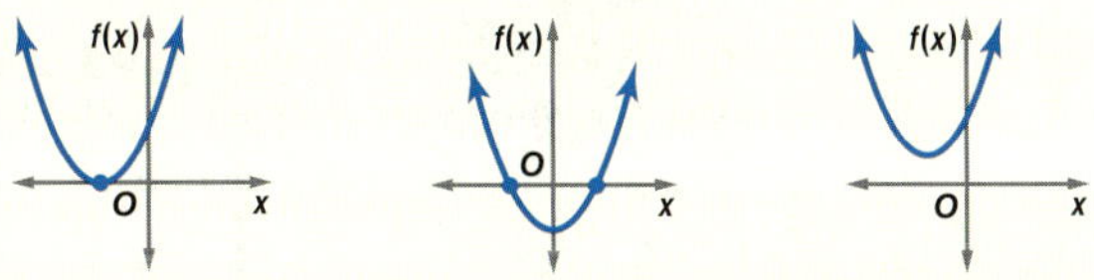

Study Tip

One Real Solution
When a quadratic equation has one real solution, it really has two solutions that are the same number.

Example 2 One Real Solution

Solve $8x - x^2 = 16$ by graphing.

Write the equation in $ax^2 + bx + c = 0$ form.

$8x - x^2 = 16 \rightarrow -x^2 + 8x - 16 = 0$ Subtract 16 from each side.

Graph the related quadratic function $f(x) = -x^2 + 8x - 16$.

x	2	3	4	5	6
f(x)	−4	−1	0	−1	−4

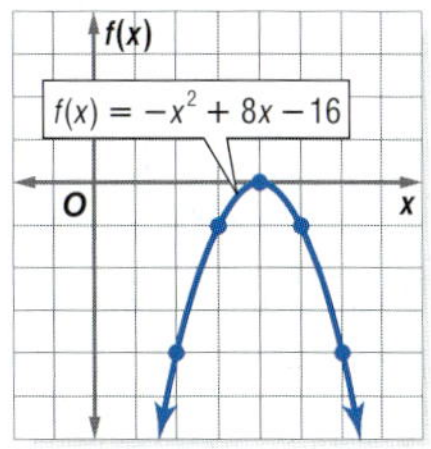

Notice that the graph has only one x-intercept, 4. Thus, the equation's only solution is 4.

Example 3 No Real Solution

NUMBER THEORY **Find two real numbers whose sum is 6 and whose product is 10 or show that no such numbers exist.**

Explore Let x = one of the numbers. Then $6 - x$ = the other number.

Plan Since the product of the two numbers is 10, you know that $x(6 - x) = 10$.

$x(6 - x) = 10$ Original equation

$6x - x^2 = 10$ Distributive Property

$-x^2 + 6x - 10 = 0$ Subtract 10 from each side.

Solve You can solve $-x^2 + 6x - 10 = 0$ by graphing the related function $f(x) = -x^2 + 6x - 10$.

x	1	2	3	4	5
f(x)	−5	−2	−1	−2	−5

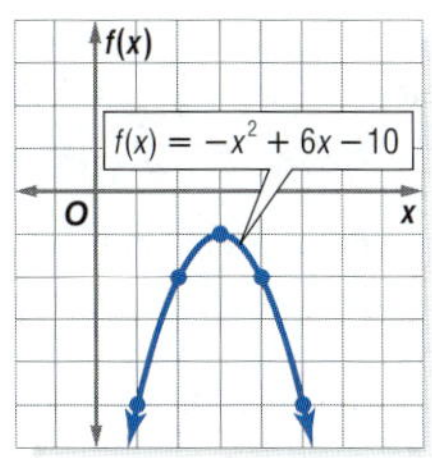

Notice that the graph has no x-intercepts. This means that the original equation has no real solution. Thus, it is *not* possible for two numbers to have a sum of 6 and a product of 10.

Examine Try finding the product of several pairs of numbers whose sum is 6, Is the product of each pair less than 10 as the graph suggests?

www.algebra2.com/extra_examples

2 Teach

SOLVE QUADRATIC EQUATIONS

In-Class Examples Power Point®

1 Solve $x^2 - 3x - 4 = 0$ by graphing. **−1 and 4**

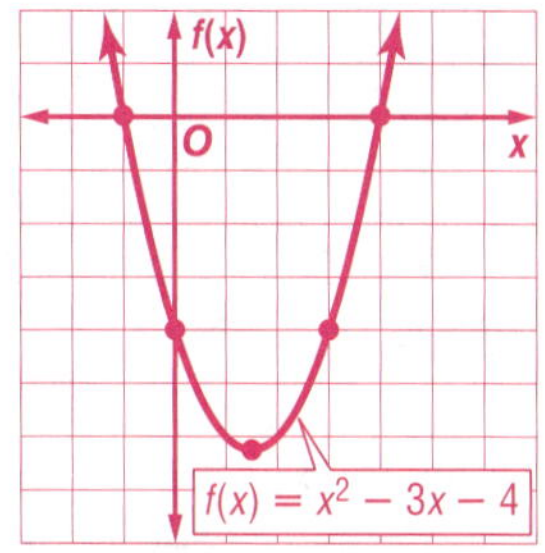

2 Solve $x^2 - 4x = -4$ by graphing. **2**

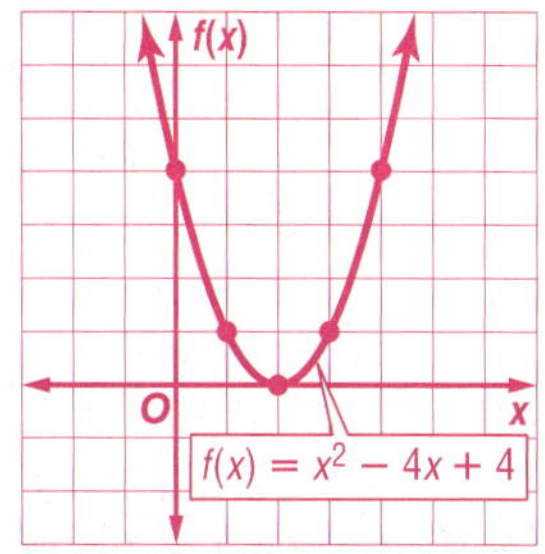

Teaching Tip In Example 3, inform students that while this equation does not have any real solutions, it does have a solution in the set of complex numbers, the topic of Lesson 5-9. Such equations will be discussed again in Lessons 6-4 and 6-5.

3 **NUMBER THEORY** Find two real numbers whose sum is 4 and whose product is 5 or show that no such numbers exist.

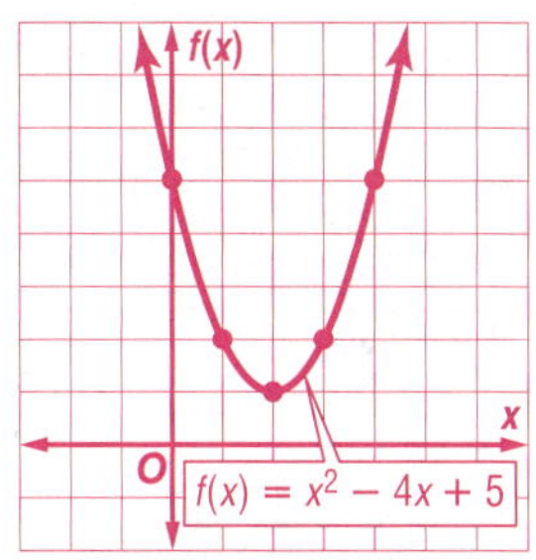

The graph of the related function does not intersect the x-axis. Therefore, no such numbers exist.

DAILY INTERVENTION — Unlocking Misconceptions

Equations and Functions Some students may notice that the equation derived in Example 2, $-x^2 + 8x - 16 = 0$, is equivalent to the equation $x^2 - 8x + 16 = 0$. Either equation can be produced from the other by multiplying each side by −1. These two equations have the same solution, 4. However, stress that the related functions, $f(x) = -x^2 + 8x - 16$ and $f(x) = x^2 - 8x + 16$ are *not* equivalent. This can be seen by looking at their graphs, which open in opposite directions.

ESTIMATE SOLUTIONS

In-Class Examples

Power Point®

4 Solve $x^2 - 6x + 3 = 0$ by graphing. If exact roots cannot be found, state the consecutive integers between which the roots are located. **One solution is between 0 and 1, and the other is between 5 and 6.**

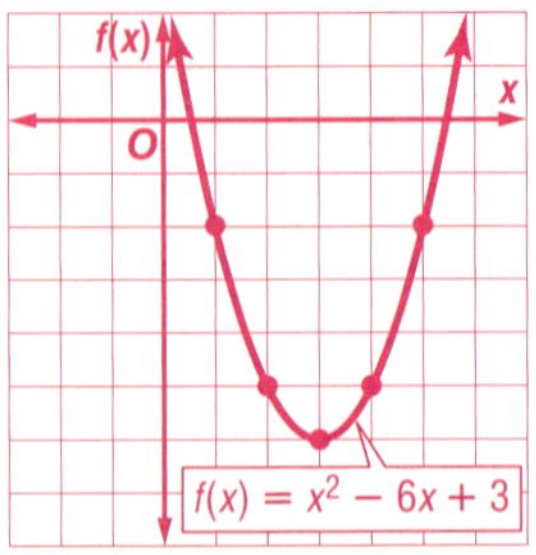

5 **ROYAL GORGE BRIDGE** The highest bridge in the United States is the Royal Gorge Bridge in Colorado. The deck of the bridge is 1053 feet above the river below. Suppose a marble is dropped over the railing from a height of 3 feet above the bridge deck. How long will it take the marble to reach the surface of the water, assuming there is no air resistance? Use the formula $h(t) = -16t^2 + h_0$, where t is the time in seconds and h_0 is the initial height above the water in feet. **about 8 s**

ESTIMATE SOLUTIONS Often exact roots cannot be found by graphing. In this case, you can estimate solutions by stating the consecutive integers between which the roots are located.

Example 4 Estimate Roots

Solve $-x^2 + 4x - 1 = 0$ by graphing. If exact roots cannot be found, state the consecutive integers between which the roots are located.

The equation of the axis of symmetry of the related function is $x = -\frac{4}{2(-1)}$ or 2.

x	0	1	2	3	4
f(x)	−1	2	3	2	−1

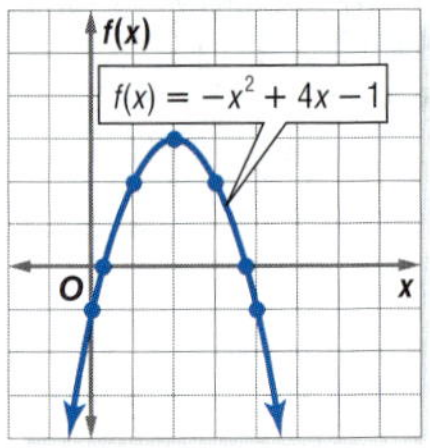

The x-intercepts of the graph are between 0 and 1 and between 3 and 4. So, one solution is between 0 and 1, and the other is between 3 and 4.

> **Study Tip**
>
> *Location of Roots*
>
> Notice in the table of values that the value of the function changes from negative to positive between the x values of 0 and 1, and 3 and 4.

For many applications, an exact answer is not required, and approximate solutions are adequate. Another way to estimate the solutions of a quadratic equation is by using a graphing calculator.

Example 5 Write and Solve an Equation

EXTREME SPORTS **On March 12, 1999, Adrian Nicholas broke the world record for the longest human flight. He flew 10 miles from his drop point in 4 minutes 55 seconds using a specially designed, aerodynamic suit. Using the information at the right and ignoring air resistance, how long would Mr. Nicholas have been in free-fall had he not used this special suit? Use the formula $h(t) = -16t^2 + h_0$, where the time t is in seconds and the initial height h_0 is in feet.**

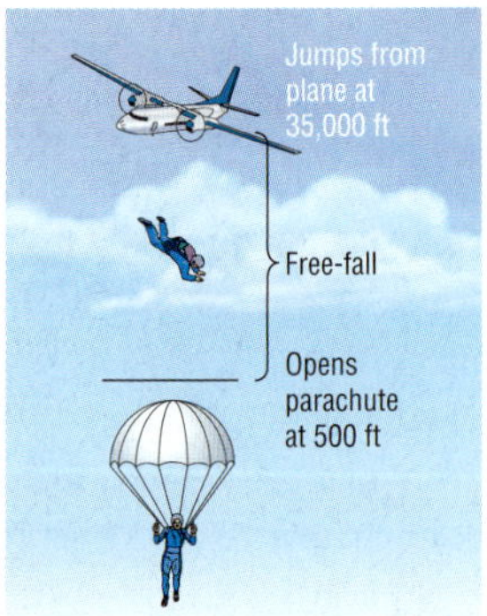

We need to find t when $h_0 = 35{,}000$ and $h(t) = 500$. Solve $500 = -16t^2 + 35{,}000$.

$500 = -16t^2 + 35{,}000$ Original equation

$0 = -16t^2 + 34{,}500$ Subtract 500 from each side.

Graph the related function $y = -16t^2 + 34{,}500$ using a graphing calculator. Adjust your window so that the x-intercepts of the graph are visible.

Use the ZERO feature, [2nd] [CALC], to find the positive zero of the function, since time cannot be negative. Use the arrow keys to locate a left bound for the zero and press [ENTER].

Then, locate a right bound and press [ENTER] twice. The positive zero of the function is approximately 46.4. Mr. Nicholas would have been in free-fall for about 46 seconds.

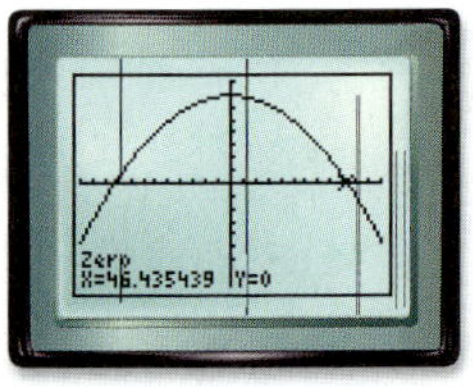

[−60, 60] scl: 5 by [−40000, 40000] scl: 5000

DAILY INTERVENTION **Differentiated Instruction** **ELL**

Verbal/Linguistic Have students discuss with a partner or in a small group the methods for multiplying and dividing monomial expressions with exponents, and also numbers written in scientific notation. Ask them to work together to develop a list of common errors for such problems, and to suggest ways to correct and avoid these errors.

Check for Understanding

Concept Check

1. **Define** each term and explain how they are related. **See margin.**
   a. solution b. root c. zero of a function d. x-intercept

2. **OPEN ENDED** Give an example of a quadratic function and state its related quadratic equation. **Sample answer: $f(x) = 3x^2 + 2x - 1$; $3x^2 + 2x - 1 = 0$**

3. **Explain** how you can estimate the solutions of a quadratic equation by examining the graph of its related function. **See margin.**

Guided Practice

GUIDED PRACTICE KEY	
Exercises	Examples
4–6	1–3
7–12	1–4
13	3

Use the related graph of each equation to determine its solutions.

4. $x^2 + 3x - 4 = 0$ **−4, 1**
5. $2x^2 + 2x - 4 = 0$ **−2, 1**
6. $x^2 + 8x + 16 = 0$ **−4**

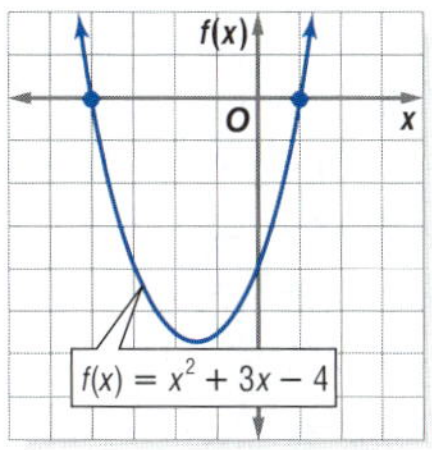

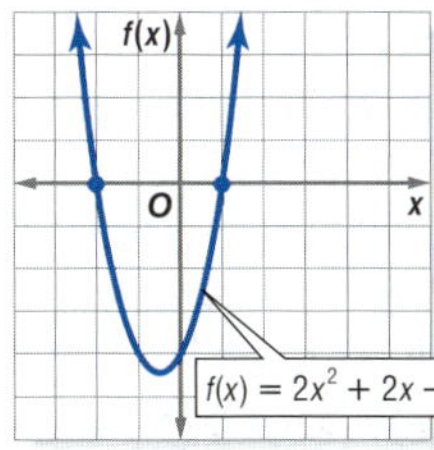

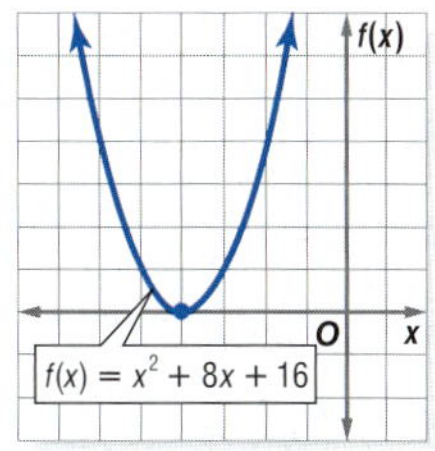

Solve each equation by graphing. If exact roots cannot be found, state the consecutive integers between which the roots are located.

7. $-x^2 - 7x = 0$ **−7, 0**
8. $x^2 - 2x - 24 = 0$ **−4, 6**
9. $x^2 + 3x = 28$ **−7, 4**
10. $25 + x^2 + 10x = 0$ **−5**
11. $4x^2 - 7x - 15 = 0$ **between −2 and −1; 3**
12. $2x^2 - 2x - 3 = 0$

12. between −1 and 0; between 1 and 2

Application

13. **NUMBER THEORY** Use a quadratic equation to find two real numbers whose sum is 5 and whose product is −14, or show that no such numbers exist. **−2, 7**

★ indicates increased difficulty

Practice and Apply

Homework Help	
For Exercises	See Examples
14–19	1–3
20–37	1–4
38–41	3
42–46	5

Extra Practice See page 840.

Use the related graph of each equation to determine its solutions.

14. $x^2 - 6x = 0$ **0, 6**
15. $x^2 - 6x + 9 = 0$ **3**
16. $-2x^2 - x + 6 = 0$ **$-2, 1\frac{1}{2}$**

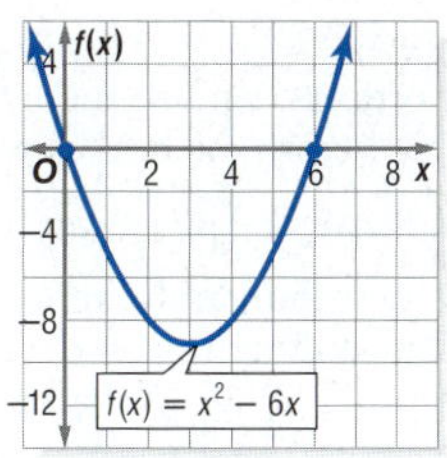

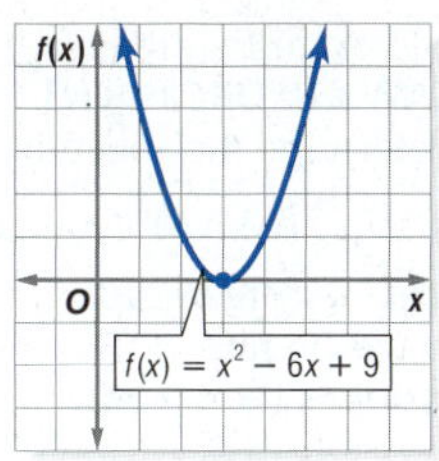

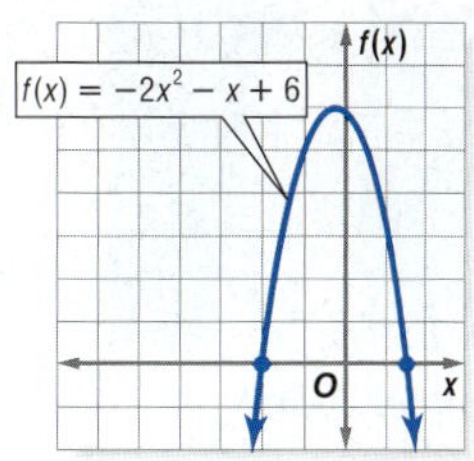

17. $-0.5x^2 = 0$ **0**
18. $2x^2 - 5x - 3 = 0$ **$-\frac{1}{2}, 3$**
19. $-3x^2 - 1 = 0$ **no real solutions**

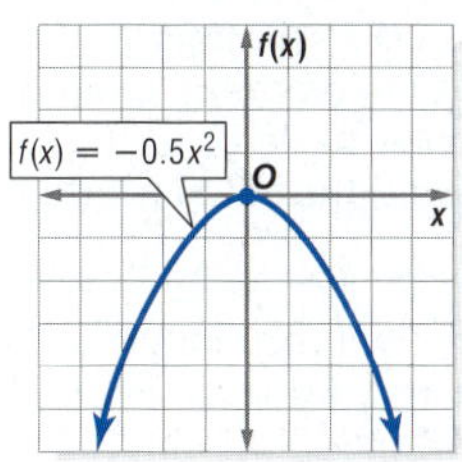

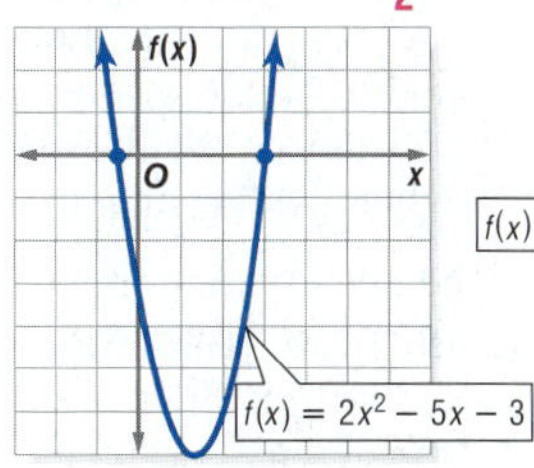

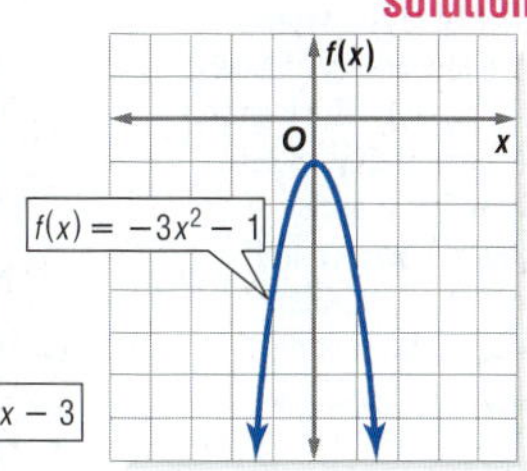

www.algebra2.com/self_check_quiz

3 Practice/Apply

Study Notebook

Have students—
- add the definitions/examples of the vocabulary terms to their Vocabulary Builder worksheets for Chapter 6.
- add the graphic representations of the three possible types of real solutions of a quadratic equation shown in the Key Concept box on p. 295.
- include any other item(s) that they find helpful in mastering the skills in this lesson.

About the Exercises…

Organization by Objective
- **Solve Quadratic Equations:** 14–21, 24–31, 36–41
- **Estimate Solutions:** 22, 23, 32–35, 42–46

Odd/Even Assignments
Exercises 14–41 are structured so that students practice the same concepts whether they are assigned odd or even problems.

Alert! Exercises 51–56 require a graphing calculator.

Assignment Guide

Basic: 15–45 odd, 47–50, 57–72

Average: 15–45 odd, 47–50, 57–72 (optional: 51–56)

Advanced: 14–46 even, 47–66 (optional: 67–72)

Answers

1a. The solution is the value that satisfies an equation.

1b. A root is a solution of an equation.

1c. A zero is the x value of a function that makes the function equal to 0.

1d. An x-intercept is the point at which a graph crosses the x-axis. The solutions, or roots, of a quadratic equation are the zeros of the related quadratic function. You can find the zeros of a quadratic function by finding the x-intercepts of its graph.

3. The x-intercepts of the related function are the solutions to the equation. You can estimate the solutions by stating the consecutive integers between which the x-intercepts are located.

Study Guide and Intervention, p. 319 (shown) and p. 320

Solve Quadratic Equations

Quadratic Equation	A quadratic equation has the form $ax^2 + bx + c = 0$, where $a \neq 0$.
Roots of a Quadratic Equation	solution(s) of the equation, or the zero(s) of the related quadratic function

The zeros of a quadratic function are the x-intercepts of its graph. Therefore, finding the x-intercepts is one way of solving the related quadratic equation.

Example **Solve $x^2 + x - 6 = 0$ by graphing.**

Graph the related function $f(x) = x^2 + x - 6$.

The x-coordinate of the vertex is $\frac{-b}{2a} = -\frac{1}{2}$, and the equation of the axis of symmetry is $x = -\frac{1}{2}$.

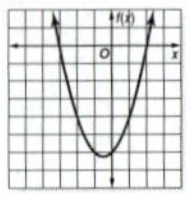

Make a table of values using x-values around $-\frac{1}{2}$.

x	-1	$-\frac{1}{2}$	0	1	2
$f(x)$	-6	$-6\frac{1}{4}$	-6	-4	0

From the table and the graph, we can see that the zeros of the function are 2 and -3.

Exercises

Solve each equation by graphing.

1. $x^2 + 2x - 8 = 0$ **2, −4**
2. $x^2 - 4x - 5 = 0$ **5, −1**
3. $x^2 - 5x + 4 = 0$ **1, 4**

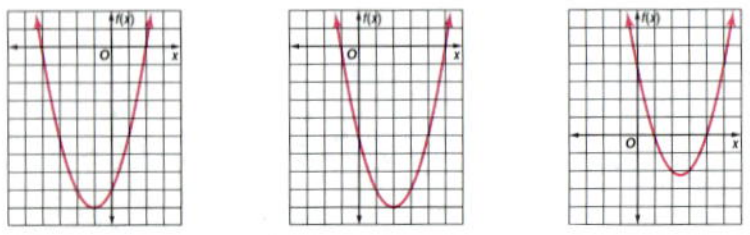

4. $x^2 - 10x + 21 = 0$
5. $x^2 + 4x + 6 = 0$
6. $4x^2 + 4x + 1 = 0$

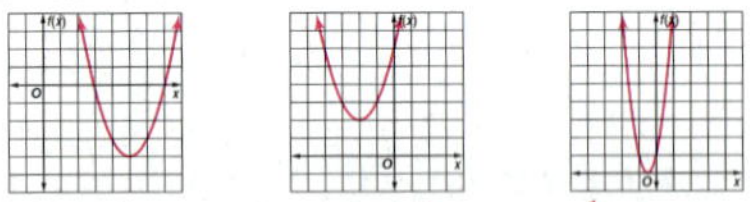

3, 7 **no real solutions** $-\frac{1}{2}$

Skills Practice, p. 321 and Practice, p. 322 (shown)

Use the related graph of each equation to determine its solutions.

1. $-3x^2 + 3 = 0$
2. $3x^2 + x + 3 = 0$
3. $x^2 - 3x + 2 = 0$

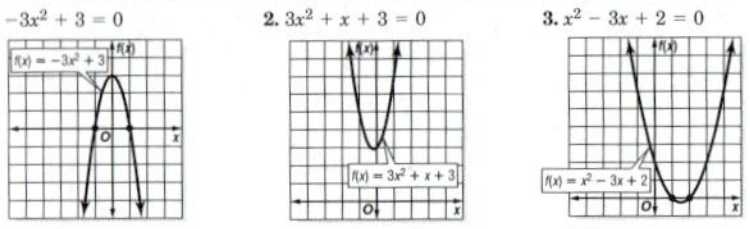

−1, 1 **no real solutions** **1, 2**

Solve each equation by graphing. If exact roots cannot be found, state the consecutive integers between which the roots are located.

4. $-2x^2 - 6x + 5 = 0$ **between 0 and 1; between −4 and −3**
5. $x^2 + 10x + 24 = 0$ **−6, −4**
6. $2x^2 - x - 6 = 0$ **between −2 and −1, 2**

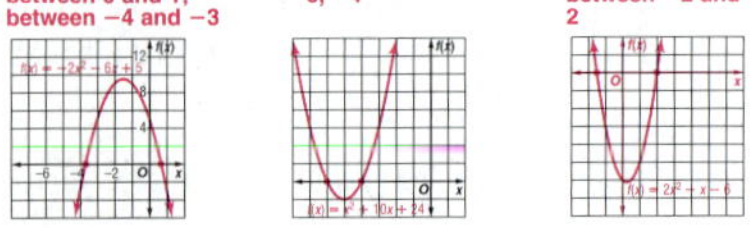

Use a quadratic equation to find two real numbers that satisfy each situation, or show that no such numbers exist.

7. Their sum is 1, and their product is −6. **$-x^2 + x + 6 = 0$; 3, −2**
8. Their sum is 5, and their product is 8. **$-x^2 + 5x - 8 = 0$; no such real numbers exist**

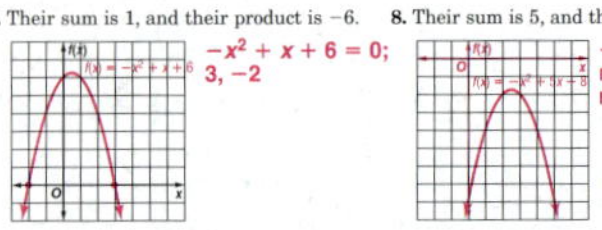

For Exercises 9 and 10, use the formula $h(t) = v_0t - 16t^2$, where $h(t)$ is the height of an object in feet, v_0 is the object's initial velocity in feet per second, and t is the time in seconds.

9. **BASEBALL** Marta throws a baseball with an initial upward velocity of 60 feet per second. Ignoring Marta's height, how long after she releases the ball will it hit the ground? **3.75 s**
10. **VOLCANOES** A volcanic eruption blasts a boulder upward with an initial velocity of 240 feet per second. How long will it take the boulder to hit the ground if it lands at the same elevation from which it was ejected? **15 s**

Reading to Learn Mathematics, p. 323 ELL

Pre-Activity **How does a quadratic function model a free-fall ride?**

Read the introduction to Lesson 6-2 at the top of page 294 in your textbook.

Write a quadratic function that describes the height of a ball t seconds after it is dropped from a height of 125 feet. **$h(t) = -16t^2 + 125$**

Reading the Lesson

1. The graph of the quadratic function $f(x) = -x^2 + x + 6$ is shown at the right. Use the graph to find the solutions of the quadratic equation $-x^2 + x + 6 = 0$. **−2 and 3**

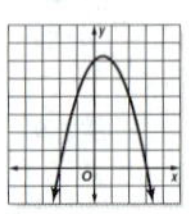

2. Sketch a graph to illustrate each situation.
   a. A parabola that opens downward and represents a quadratic function with two real zeros, both of which are negative numbers.
   b. A parabola that opens upward and represents a quadratic function with exactly one real zero. The zero is a positive number.
   c. A parabola that opens downward and represents a quadratic function with no real zeros.

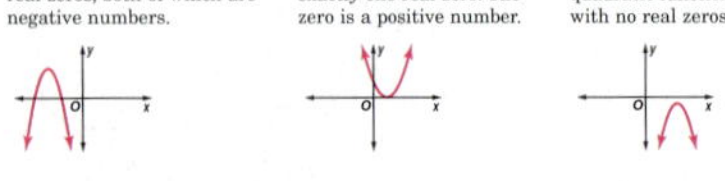

Helping You Remember

3. Think of a memory aid that can help you recall what is meant by the *zeros* of a quadratic function.

Sample answer: The basic facts about a subject are sometimes called the ABCs. In the case of zeros, the ABCs are the XYZs, because the zeros are the *x*-values that make the *y*-values equal to *z*ero.

22. between −5 and −4; between 0 and 1
23. between −1 and 0; between 2 and 3
32. between −4 and −3; between 0 and 1
33. between 0 and 1; between 3 and 4
34. between −1 and 0; between 2 and 3
35. between −3 and −2; between 2 and 3

Solve each equation by graphing. If exact roots cannot be found, state the consecutive integers between which the roots are located.

20. $x^2 - 3x = 0$ **0, 3**
21. $-x^2 + 4x = 0$ **0, 4**
22. $x^2 + 4x - 4 = 0$
23. $x^2 - 2x - 1 = 0$
24. $-x^2 + x = -20$ **−4, 5**
25. $x^2 - 9x = -18$ **3, 6**
26. $14x + x^2 + 49 = 0$ **−7**
27. $-12x + x^2 = -36$ **6**
28. $2x^2 - 3x = 9$ **$-1\frac{1}{2}$, 3**
29. $4x^2 - 8x = 5$ **$-\frac{1}{2}$, $2\frac{1}{2}$**
30. $2x^2 = -5x + 12$ **−4, $1\frac{1}{2}$**
31. $2x^2 = x + 15$ **$-2\frac{1}{2}$, 3**
32. $x^2 + 3x - 2 = 0$
33. $x^2 - 4x + 2 = 0$
34. $-2x^2 + 3x + 3 = 0$
35. $0.5x^2 - 3 = 0$
36. $x^2 + 2x + 5 = 0$ **no real solutions**
37. $-x^2 + 4x - 6 = 0$ **no real solutions**

NUMBER THEORY Use a quadratic equation to find two real numbers that satisfy each situation, or show that no such numbers exist.

38. Their sum is −17, and their product is 72. **−8, −9**
39. Their sum is 7, and their product is 14. **See pp. 343A–343F.**
40. Their sum is −9, and their product is 24. **See pp. 343A–343F.**
41. Their sum is 12, and their product is −28. **−2, 14**

For Exercises 42–44, use the formula $h(t) = v_0t - 16t^2$ where $h(t)$ is the height of an object in feet, v_0 is the object's initial velocity in feet per second, and t is the time in seconds.

42. **ARCHERY** An arrow is shot upward with a velocity of 64 feet per second. Ignoring the height of the archer, how long after the arrow is released does it hit the ground? **4 s**

43. **TENNIS** A tennis ball is hit upward with a velocity of 48 feet per second. Ignoring the height of the tennis player, how long does it take for the ball to fall to the ground? **3 s**

44. **BOATING** A boat in distress launches a flare straight up with a velocity of 190 feet per second. Ignoring the height of the boat, how many seconds will it take for the flare to hit the water? **about 12 s**

More About. . .

Empire State Building

Located on the 86th floor, 1050 feet (320 meters) above the streets of New York City, the Observatory offers panoramic views from within a glass-enclosed pavilion and from the surrounding open-air promenade.

Source: www.esbnyc.com

45. **LAW ENFORCEMENT** Police officers can use the length of skid marks to help determine the speed of a vehicle before the brakes were applied. If the skid marks are on dry concrete, the formula $\frac{s^2}{24} = d$ can be used. In the formula, s represents the speed in miles per hour, and d represents the length of the skid marks in feet. If the length of the skid marks on dry concrete are 50 feet, how fast was the car traveling? **about 35 mph**

46. **EMPIRE STATE BUILDING** Suppose you could conduct an experiment by dropping a small object from the Observatory of the Empire State Building. How long would it take for the object to reach the ground, assuming there is no air resistance? Use the information at the left and the formula $h(t) = -16t^2 + h_0$, where t is the time in seconds and the initial height h_0 is in feet. **about 8 s**

47. **CRITICAL THINKING** A quadratic function has values $f(-4) = -11$, $f(-2) = 9$, and $f(0) = 5$. Between which two x values must $f(x)$ have a zero? Explain your reasoning. **−4 and −2; See margin for explanation.**

Enrichment, p. 324

Graphing Absolute Value Equations

You can solve absolute value equations in much the same way you solved quadratic equations. Graph the related absolute value function for each equation using a graphing calculator. Then use the ZERO feature in the CALC menu to find its real solutions, if any. Recall that solutions are points where the graph intersects the x-axis.

For each equation, make a sketch of the related graph and find the solutions rounded to the nearest hundredth.

1. $|x + 5| = 0$ **−5**
2. $|4x - 3| + 5 = 0$ **No solutions**
3. $|x - 7| = 0$ **7**

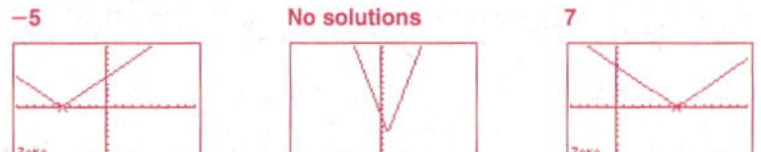

Answer

47. The value of the function changes from negative to positive, therefore the value of the function is zero between these two numbers.

48. **WRITING IN MATH** Answer the question that was posed at the beginning of the lesson. **See pp. 343A–343F.**

How does a quadratic function model a free-fall ride?

Include the following in your answer:

- a graph showing the height at any given time of a free-fall ride that lifts riders to a height of 185 feet, and
- an explanation of how to use this graph to estimate how long the riders would be in free-fall if the ride were allowed to hit the ground before stopping.

Standardized Test Practice

49. If one of the roots of the equation $x^2 + kx - 12 = 0$ is 4, what is the value of k? **A**

Ⓐ −1 Ⓑ 0 Ⓒ 1 Ⓓ 3

50. For what value of x does $f(x) = x^2 + 5x + 6$ reach its minimum value? **B**

Ⓐ −3 Ⓑ $-\frac{5}{2}$ Ⓒ −2 Ⓓ −5

Extending the Lesson

SOLVE ABSOLUTE VALUE EQUATIONS BY GRAPHING Similar to quadratic equations, you can solve absolute value equations by graphing. Graph the related absolute value function for each equation using a graphing calculator. Then use the ZERO feature, [2nd] [CALC], to find its real solutions, if any, rounded to the nearest hundredth.

51. $|x + 1| = 0$ **−1**
52. $|x| - 3 = 0$ **±3**
53. $|x - 4| - 1 = 0$ **3, 5**
54. $-|x + 4| + 5 = 0$ **−9, 1**
55. $2|3x| - 8 = 0$ **±1.33**
56. $|2x - 3| + 1 = 0$ **no real solutions**

Maintain Your Skills

Mixed Review

Find the y-intercept, the equation of the axis of symmetry, and the x-coordinate of the vertex for each quadratic function. Then graph the function by making a table of values. *(Lesson 6-1)* **57–59. See margin for graphs.**

57. $f(x) = x^2 - 6x + 4$ **4; $x = 3$; 3**
58. $f(x) = -4x^2 + 8x - 1$ **−1; $x = 1$; 1**
59. $f(x) = \frac{1}{4}x^2 + 3x + 4$ **4; $x = -6$; −6**

Simplify. *(Lesson 5-9)*

60. $\frac{2i}{3 + i}$ $\frac{1}{5} + \frac{3}{5}i$
61. $\frac{4}{5 - i}$ $\frac{10}{13} + \frac{2}{13}i$
62. $\frac{1 + i}{3 - 2i}$ $\frac{1}{13} + \frac{5}{13}i$

Evaluate the determinant of each matrix. *(Lesson 4-3)*

63. $\begin{bmatrix} 6 & 4 \\ -3 & 2 \end{bmatrix}$ **24**
64. $\begin{bmatrix} 2 & -1 & -6 \\ 5 & 0 & 3 \\ -3 & 2 & 11 \end{bmatrix}$ **−8**
65. $\begin{bmatrix} 6 & 5 & -2 \\ -3 & 0 & 6 \\ 1 & 4 & 2 \end{bmatrix}$ **−60**

66. **COMMUNITY SERVICE** A drug awareness program is being presented at a theater that seats 300 people. Proceeds will be donated to a local drug information center. If every two adults must bring at least one student, what is the maximum amount of money that can be raised? *(Lesson 3-4)* **$500**

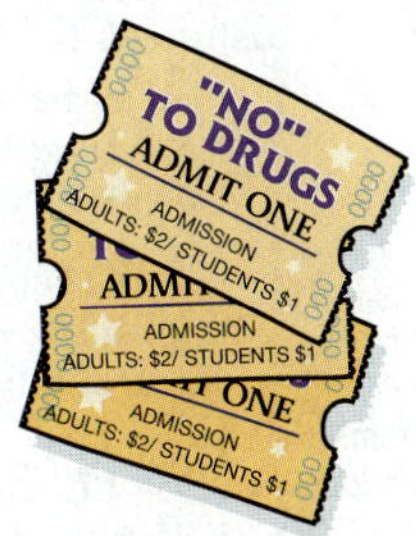

68. $(x - 10)(x + 10)$
69. $(x - 7)(x - 4)$

Getting Ready for the Next Lesson

PREREQUISITE SKILL **Factor completely.**
*(To review **factoring trinomials**, see Lesson 5-4.)*

71. $(3x + 2)(x + 2)$
72. $2(3x + 2)(x - 3)$

67. $x^2 + 5x$ **$x(x + 5)$**
68. $x^2 - 100$
69. $x^2 - 11x + 28$
70. $x^2 - 18x + 81$ **$(x - 9)^2$**
71. $3x^2 + 8x + 4$
72. $6x^2 - 14x - 12$

4 Assess

Open-Ended Assessment

Modeling Have students draw parabolas in various positions and label them to show how many real roots they have and approximately where those roots occur.

Getting Ready for Lesson 6-3

PREREQUISITE SKILL Lesson 6-3 presents solving quadratic equations by factoring. Frequently this involves factoring a trinomial expression on one side of an equation. Exercises 67–72 should be used to determine your students' familiarity with factoring trinomials.

Assessment Options

Quiz (Lessons 6-1 and 6-2) is available on p. 369 of the *Chapter 6 Resource Masters*.

Answers

57.
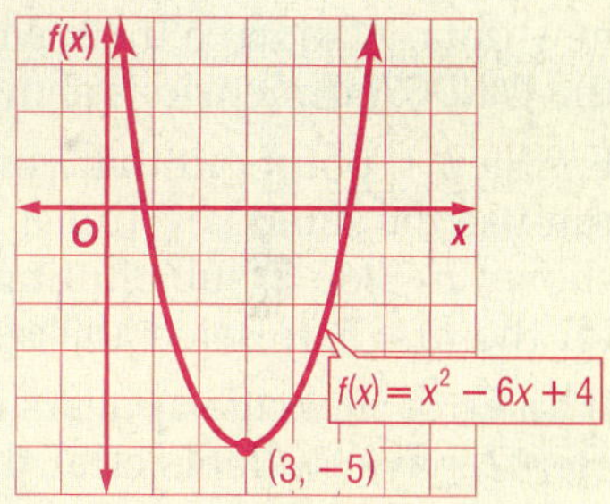

58.
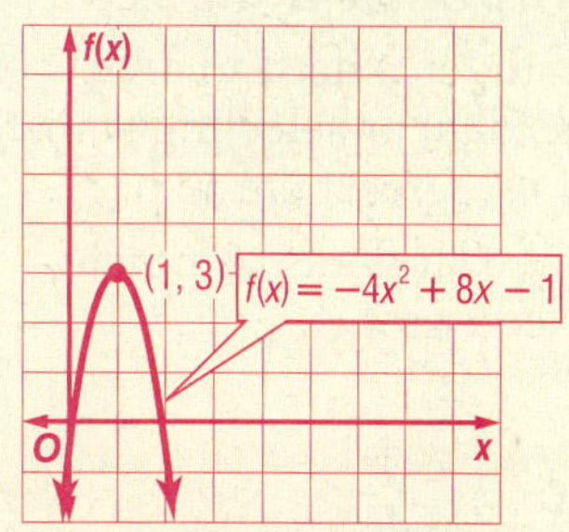

59.
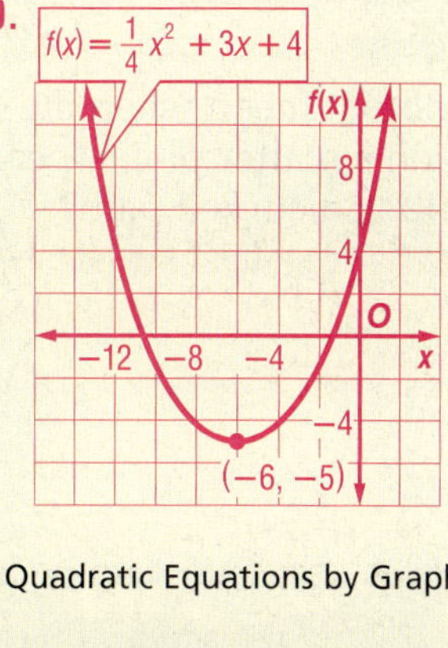

Graphing Calculator Investigation

A Follow-Up of Lesson 6-2

Getting Started

Know Your Calculator When students use the procedure in Step 2 to copy the regression equation from Step 1 to the Y= list, the coefficients will have several more digits than the coefficients displayed on the home screen. The coefficients on the home screen are rounded versions of those in the Y= list.

Scientific Notation In Step 1, the value of the coefficient a is displayed as 2.1035215E−4. Point out that this is how the calculator displays the scientific notation 2.1035215×10^{-4}.

Teach

- Make sure students have cleared the L1 and L2 lists before entering new data. Also have them enter the WINDOW dimensions shown.
- For Step 1, point out that you can use the same keystrokes shown in Step 2, substituting 4 for the first 5, to select LinReg.
- If an error message appears in Step 2, have students clear the Y= list before trying Step 2 again.
- If students need to review entering data or selecting statistical plots, refer them to p. 87.
- Have students complete Exercises 1–4.

Assess

Ask students:

- What does it mean when the points on a scatter plot appear to lie along a curved path? **The equation that best models the situation may be quadratic, and is probably not linear.**

Graphing Calculator Investigation

A Follow-Up of Lesson 6-2

Modeling Real-World Data

You can use a TI-83 Plus to model data points whose curve of best fit is quadratic.

FALLING WATER **Water is allowed to drain from a hole made in a 2-liter bottle. The table shows the level of the water y measured in centimeters from the bottom of the bottle after x seconds. Find and graph a linear regression equation and a quadratic regression equation. Determine which equation is a better fit for the data.**

Time (s)	0	20	40	60	80	100	120	140	160	180	200	220
Water level (cm)	42.6	40.7	38.9	37.2	35.8	34.3	33.3	32.3	31.5	30.8	30.4	30.1

Step 1 *Find a linear regression equation.*

- Enter the times in L1 and the water levels in L2. Then find a linear regression equation.

  KEYSTROKES: *Review lists and finding a linear regression equation on page 87.*

- Graph a scatter plot and the regression equation.

  KEYSTROKES: *Review graphing a regression equation on page 87.*

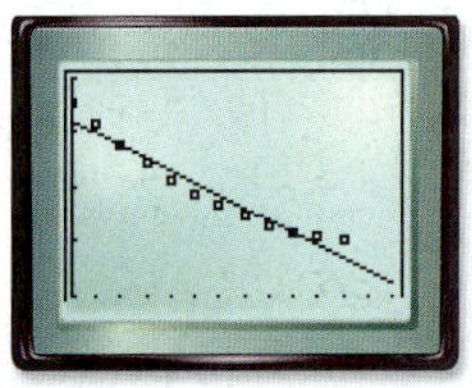

[0, 260] scl: 20 by [25, 45] scl: 5

Step 2 *Find a quadratic regression equation.*

- Find the quadratic regression equation. Then copy the equation to the Y= list and graph.

  KEYSTROKES: STAT ▶ 5 ENTER Y= VARS 5 ▶ ▶ ENTER GRAPH

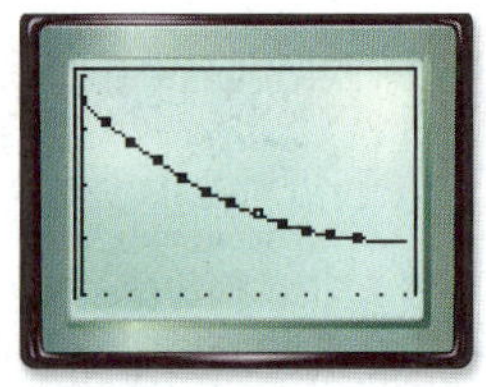

[0, 260] scl: 20 by [25, 45] scl: 5

The graph of the linear regression equation appears to pass through just two data points. However, the graph of the quadratic regression equation fits the data very well.

Exercises 1–4. See margin.

For Exercises 1–4, use the graph of the braking distances for dry pavement.

1. Find and graph a linear regression equation and a quadratic regression equation for the data. Determine which equation is a better fit for the data.
2. Use the CALC menu with each regression equation to estimate the braking distance at speeds of 100 and 150 miles per hour.
3. How do the estimates found in Exercise 2 compare?
4. How might choosing a regression equation that does not fit the data well affect predictions made by using the equation?

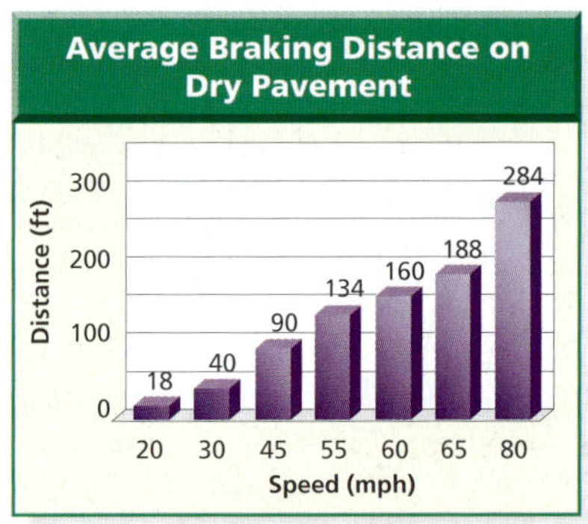

Source: Missouri Department of Revenue

www.algebra2.com/other_calculator_keystrokes

Answers

1. See pp. 343A–343F.
2. linear: (100, 345), (150, 562); quadratic: (100, 440), (150, 990)
3. The quadratic estimates are much greater.
4. Sample answer: Choosing a model that does not fit the data well may cause inaccurate predictions when the data are very large or small.

6-3 Solving Quadratic Equations by Factoring

What You'll Learn

- Solve quadratic equations by factoring.
- Write a quadratic equation with given roots.

How is the Zero Product Property used in geometry?

The length of a rectangle is 5 inches more than its width, and the area of the rectangle is 24 square inches. To find the dimensions of the rectangle you need to solve the equation $x(x + 5) = 24$ or $x^2 + 5x = 24$.

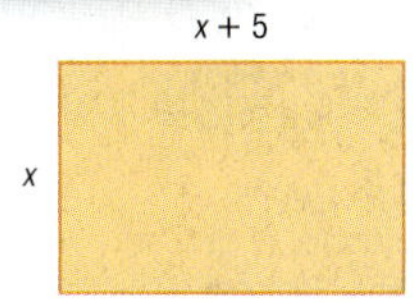

SOLVE EQUATIONS BY FACTORING In the last lesson, you learned to solve a quadratic equation like the one above by graphing. Another way to solve this equation is by factoring. Consider the following products.

$$7(0) = 0 \qquad 0(-2) = 0$$
$$(6 - 6)(0) = 0 \qquad -4(-5 + 5) = 0$$

Notice that in each case, *at least one* of the factors is zero. These examples illustrate the **Zero Product Property**.

Key Concept *Zero Product Property*

- **Words** For any real numbers a and b, if $ab = 0$, then either $a = 0$, $b = 0$, or both a and b equal zero.
- **Example** If $(x + 5)(x - 7) = 0$, then $x + 5 = 0$ and/or $x - 7 = 0$.

Example 1 *Two Roots*

Solve each equation by factoring.

a. $x^2 = 6x$

$x^2 = 6x$	Original equation
$x^2 - 6x = 0$	Subtract $6x$ from each side.
$x(x - 6) = 0$	Factor the binomial.
$x = 0$ or $x - 6 = 0$	Zero Product Property
$x = 6$	Solve the second equation.

The solution set is $\{0, 6\}$.

CHECK Substitute 0 and 6 for x in the original equation.

$$x^2 = 6x \qquad x^2 = 6x$$
$$(0)^2 \stackrel{?}{=} 6(0) \qquad (6)^2 \stackrel{?}{=} 6(6)$$
$$0 = 0 \checkmark \qquad 36 = 36 \checkmark$$

6-3 Lesson Notes

1 Focus

5-Minute Check Transparency 6-3 Use as a quiz or review of Lesson 6-2.

Mathematical Background notes are available for this lesson on p. 284C.

Building on Prior Knowledge

In Lesson 6-2, students solved quadratic equations by graphing. In this lesson, they use factoring as a method for finding the roots of a quadratic equation.

How is the Zero Product Property used in geometry?

Ask students:

- What is the product of the length and the width of the rectangle? **24 in²**
- What is the difference between the length and the width of the rectangle? **5 in.**

Resource Manager

Workbook and Reproducible Masters

Chapter 6 Resource Masters

- Study Guide and Intervention, pp. 325–326
- Skills Practice, p. 327
- Practice, p. 328
- Reading to Learn Mathematics, p. 329
- Enrichment, p. 330

Transparencies

5-Minute Check Transparency 6-3
Real-World Transparency 6
Answer Key Transparencies

Technology

Interactive Chalkboard

2 Teach

SOLVE EQUATIONS BY FACTORING

In-Class Examples

Power Point®

Teaching Tip In Example 1a, some students may suggest solving the equation by dividing both sides by x. Point out that this cannot be done because the value of x could be zero, and division by zero is undefined.

1 Solve each equation by factoring.

a. $x^2 = -4x$ $\{0, -4\}$

b. $3x^2 = 5x + 2$ $\left\{-\frac{1}{3}, 2\right\}$

2 Solve $x^2 - 6x = -9$ by factoring. $\{3\}$

Teaching Tip Point out that the term *repeated root* is sometimes used as a substitute for the term *double root*.

3 What is the positive solution of the equation $2x^2 - 8x - 42 = 0$? **D**

A -3 **B** 5

C 6 **D** 7

Teaching Tip Ask students why dividing each side of the equation in this example results in an equivalent equation, without the possibility of losing a root. (The right side of the equation is 0, not $f(x)$ or y, and dividing by 2 means that you can be sure that you are not dividing by zero.)

WRITE QUADRATIC EQUATIONS

In-Class Example

Power Point®

4 Write a quadratic equation with $-\frac{2}{3}$ and 6 as its roots. Write the equation in the form $ax^2 + bx + c = 0$, where a, b, and c are integers. **Sample answer: $3x^2 - 16x - 12 = 0$**

---

b. $2x^2 + 7x = 15$

$2x^2 + 7x = 15$	Original equation
$2x^2 + 7x - 15 = 0$	Subtract 15 from each side.
$(2x - 3)(x + 5) = 0$	Factor the trinomial.
$2x - 3 = 0$ or $x + 5 = 0$	Zero Product Property
$2x = 3$ $\quad x = -5$	Solve each equation.
$x = \frac{3}{2}$	

The solution set is $\left\{-5, \frac{3}{2}\right\}$. Check each solution.

Study Tip

Double Roots The application of the Zero Product Property produced two identical equations, $x - 8 = 0$, both of which have a root of 8. For this reason, 8 is called the *double root* of the equation.

Example 2 Double Root

Solve $x^2 - 16x + 64 = 0$ by factoring.

$x^2 - 16x + 64 = 0$	Original equation
$(x - 8)(x - 8) = 0$	Factor.
$x - 8 = 0$ or $x - 8 = 0$	Zero Product Property
$x = 8$ $\quad x = 8$	Solve each equation.

The solution set is $\{8\}$.

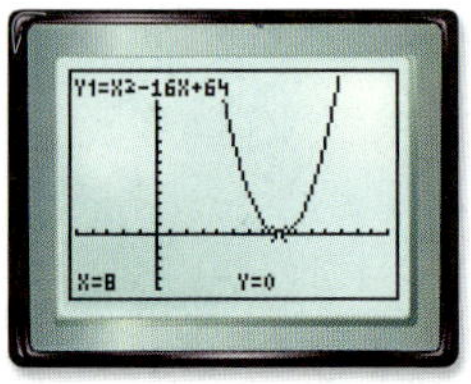

CHECK The graph of the related function, $f(x) = x^2 - 16x + 64$, intersects the x-axis only once. Since the zero of the function is 8, the solution of the related equation is 8.

Standardized Test Practice

Example 3 Greatest Common Factor

Multiple-Choice Test Item

What is the positive solution of the equation $3x^2 - 3x - 60 = 0$?

Ⓐ -4 Ⓑ 2 Ⓒ 5 Ⓓ 10

Read the Test Item

You are asked to find the *positive* solution of the given quadratic equation. This implies that the equation also has a solution that is not positive. Since a quadratic equation can either have one, two, or no solutions, we should expect to find two solutions to this equation.

Solve the Test Item

Solve this equation by factoring. But before trying to factor $3x^2 - 3x - 60$ into two binomials, look for a greatest common factor. Notice that each term is divisible by 3.

$3x^2 - 3x - 60 = 0$	Original equation
$3(x^2 - x - 20) = 0$	Factor.
$x^2 - x - 20 = 0$	Divide each side by 3.
$(x + 4)(x - 5) = 0$	Factor.
$x + 4 = 0$ or $x - 5 = 0$	Zero Product Property
$x = -4$ $\quad x = 5$	Solve each equation.

Both solutions, -4 and 5, are listed among the answer choices. Since the question asked for the positive solution, the answer is C.

Test-Taking Tip

Because the problem asked for a *positive* solution, choice A could have been eliminated even before the expression was factored.

Standardized Test Practice

Example 3 Point out to students that by reading the question carefully and noting exactly what is asked for (the *positive* solution), they can quickly eliminate answer choice A because it is negative. Choice A is an attractive (though incorrect) choice because it is indeed a solution of the equation, just not the positive one.

302 Chapter 6 Quadratic Functions and Inequalities

WRITE QUADRATIC EQUATIONS You have seen that a quadratic equation of the form $(x - p)(x - q) = 0$ has roots p and q. You can use this pattern to find a quadratic equation for a given pair of roots.

Study Tip

Writing an Equation

The pattern $(x - p)(x - q) = 0$ produces one equation with roots p and q. In fact, there are an infinite number of equations that have these same roots.

Example 4 Write an Equation Given Roots

Write a quadratic equation with $\frac{1}{2}$ and -5 as its roots. Write the equation in the form $ax^2 + bx + c = 0$, where a, b, and c are integers.

$(x - p)(x - q) = 0$ Write the pattern.

$\left(x - \frac{1}{2}\right)[x - (-5)] = 0$ Replace p with $\frac{1}{2}$ and q with -5.

$\left(x - \frac{1}{2}\right)(x + 5) = 0$ Simplify.

$x^2 + \frac{9}{2}x - \frac{5}{2} = 0$ Use FOIL.

$2x^2 + 9x - 5 = 0$ Multiply each side by 2 so that b and c are integers.

A quadratic equation with roots $\frac{1}{2}$ and -5 and integral coefficients is $2x^2 + 9x - 5 = 0$. You can check this result by graphing the related function.

Check for Understanding

Concept Check

1. Sample answer: If the product of two factors is zero, then at least one of the factors must be zero.
2. Sample answer: roots 6 and −5; $x^2 - x - 30 = 0$
3. Kristin; the Zero Product Property applies only when one side of the equation is 0.

1. **Write** the meaning of the Zero Product Property.
2. **OPEN ENDED** Choose two integers. Then, write an equation with those roots in the form $ax^2 + bx + c = 0$, where a, b, and c are integers.
3. **FIND THE ERROR** Lina and Kristin are solving $x^2 + 2x = 8$.

Lina	Kristin
$x^2 + 2x = 8$	$x^2 + 2x = 8$
$x(x + 2) = 8$	$x^2 + 2x - 8 = 0$
$x = 8$ or $x + 2 = 8$	$(x + 4)(x - 2) = 0$
$x = 6$	$x + 4 = 0$ or $x - 2 = 0$
	$x = -4$ $x = 2$

Who is correct? Explain your reasoning.

Guided Practice

GUIDED PRACTICE KEY	
Exercises	Examples
4–9	1, 2
10–12	4
13	3

Solve each equation by factoring.

4. $x^2 - 11x = 0$ **{0, 11}**
5. $x^2 + 6x - 16 = 0$ **{−8, 2}**
6. $x^2 = 49$ **{−7, 7}**
7. $x^2 + 9 = 6x$ **{3}**
8. $4x^2 - 13x = 12$ $\left\{-\frac{3}{4}, 4\right\}$
9. $5x^2 - 5x - 60 = 0$ **{−3, 4}**

Write a quadratic equation with the given roots. Write the equation in the form $ax^2 + bx + c = 0$, where a, b, and c are integers.

10. $-4, 7$ $x^2 - 3x - 28 = 0$
11. $\frac{1}{2}, \frac{4}{3}$ $6x^2 - 11x + 4 = 0$
12. $-\frac{3}{5}, -\frac{1}{3}$ $15x^2 + 14x + 3 = 0$

Standardized Test Practice

13. Which of the following is the sum of the solutions of $x^2 - 2x - 8 = 0$? **D**

Ⓐ −6 Ⓑ −4 Ⓒ −2 Ⓓ 2

www.algebra2.com/extra_examples

3 Practice/Apply

Study Notebook

Have students—

- add the definitions/examples of the vocabulary terms to their Vocabulary Builder worksheets for Chapter 6.
- include any other item(s) that they find helpful in mastering the skills in this lesson.

About the Exercises...

Organization by Objective

- **Solve Equations by Factoring:** 14–33, 42–26
- **Write Quadratic Equations:** 34–41

Odd/Even Assignments

Exercises 14–41 are structured so that students practice the same concepts whether they are assigned odd or even problems.

Assignment Guide

Basic: 15–31 odd, 35–43 odd, 47–68

Average: 15–43 odd, 47–68

Advanced: 14–44 even, 45–62 (optional: 63–68)

All: Practice Quiz 1 (1–5)

DAILY INTERVENTION **FIND THE ERROR**

To show that neither factor on the left side of Lina's second equation needs to be 8, ask students to name several pairs of numbers whose product is 8. Possible pairs are 2 and 4, 0.5 and 16, and −1 and −8.

DAILY INTERVENTION **Differentiated Instruction**

Visual/Spatial Provide each student with a sheet of grid paper. Have students begin by first drawing a coordinate grid with two points on the x-axis plotted as the roots of a quadratic equation. Then ask students to draw several different parabolas that might be the graphs of different equations having those two points as their solutions. Point out that this demonstrates that the steps demonstrated in Example 4 yield just *one* of the possible equations having the given roots.

★ indicates increased difficulty

Practice and Apply

Homework Help

For Exercises	See Examples
14–33, 42–46	1, 2
34–41	4
51–52	3

Extra Practice
See page 840.

Solve each equation by factoring.

14. $x^2 + 5x - 24 = 0$ **{−8, 3}**
15. $x^2 - 3x - 28 = 0$ **{−4, 7}**
16. $x^2 = 25$ **{−5, 5}**
17. $x^2 = 81$ **{−9, 9}**
18. $x^2 + 3x = 18$ **{−6, 3}**
19. $x^2 - 4x = 21$ **{−3, 7}**
20. $3x^2 = 5x$ $\left\{0, \frac{5}{3}\right\}$
21. $4x^2 = -3x$ $\left\{0, -\frac{3}{4}\right\}$
22. $x^2 + 36 = 12x$ **{6}**
23. $x^2 + 64 = 16x$ **{8}**
24. $4x^2 + 7x = 2$ $\left\{-2, \frac{1}{4}\right\}$
25. $4x^2 - 17x = -4$ $\left\{\frac{1}{4}, 4\right\}$
26. $4x^2 + 8x = -3$
27. $6x^2 + 6 = -13x$
28. $9x^2 + 30x = -16$ $\left\{-\frac{8}{3}, -\frac{2}{3}\right\}$
29. $16x^2 - 48x = -27$ $\left\{\frac{3}{4}, \frac{9}{4}\right\}$
30. $-2x^2 + 12x - 16 = 0$ **{2, 4}**
31. $-3x^2 - 6x + 9 = 0$ **{−3, 1}**

26. $\left\{-\frac{1}{2}, -\frac{3}{2}\right\}$

27. $\left\{-\frac{2}{3}, -\frac{3}{2}\right\}$

★ 32. Find the roots of $x(x + 6)(x - 5) = 0$. **0, −6, 5**

★ 33. Solve $x^3 = 9x$ by factoring. **0, −3, 3**

Write a quadratic equation with the given roots. Write the equation in the form $ax^2 + bx + c = 0$, where a, b, and c are integers.

34. 4, 5
35. −2, 7
36. 4, −5
37. −6, −8
38. $\frac{1}{2}, 3$ $2x^2 - 7x + 3 = 0$
39. $\frac{1}{3}, 5$ $3x^2 - 16x + 5 = 0$
40. $-\frac{2}{3}, \frac{3}{4}$ $12x^2 - x - 6 = 0$
41. $-\frac{3}{2}, -\frac{4}{5}$

34. $x^2 - 9x + 20 = 0$
35. $x^2 - 5x - 14 = 0$
36. $x^2 + x - 20 = 0$
37. $x^2 + 14x + 48 = 0$
41. $10x^2 + 23x + 12 = 0$

42. **DIVING** To avoid hitting any rocks below, a cliff diver jumps up and out. The equation $h = -16t^2 + 4t + 26$ describes her height h in feet t seconds after jumping. Find the time at which she returns to a height of 26 feet. $\frac{1}{4}$ **s**

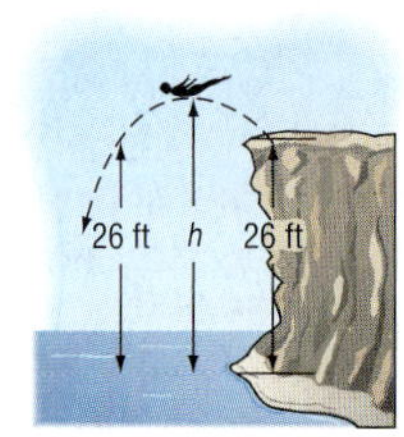

43. **NUMBER THEORY** Find two consecutive even integers whose product is 224. **14, 16 or −14, −16**

44. **PHOTOGRAPHY** A rectangular photograph is 8 centimeters wide and 12 centimeters long. The photograph is enlarged by increasing the length and width by an equal amount in order to double its area. What are the dimensions of the new photograph? **12 cm by 16 cm**

FORESTRY **For Exercises 45 and 46, use the following information.**
Lumber companies need to be able to estimate the number of board feet that a given log will yield. One of the most commonly used formulas for estimating board feet is the *Doyle Log Rule*, $B = \frac{L}{16}(D^2 - 8D + 16)$, where B is the number of board feet, D is the diameter in inches, and L is the length of the log in feet.

45. Rewrite Doyle's formula for logs that are 16 feet long. $B = D^2 - 8D + 16$

★ 46. Find the root(s) of the quadratic equation you wrote in Exercise 45. What do the root(s) tell you about the kinds of logs for which Doyle's rule makes sense? **See margin.**

47. **CRITICAL THINKING** For a quadratic equation of the form $(x - p)(x - q) = 0$, show that the axis of symmetry of the related quadratic function is located halfway between the x-intercepts p and q. **See margin.**

CRITICAL THINKING **Find a value of k that makes each statement true.**

48. −3 is a root of $2x^2 + kx - 21 = 0$. **−1**
49. $\frac{1}{2}$ is a root of $2x^2 + 11x = -k$. **−6**

More About. . .

Forestry

A board foot is a measure of lumber volume. One piece of lumber 1 foot long by 1 foot wide by 1 inch thick measures one board foot.

Source: www.wood-worker.com

Study Guide and Intervention, p. 325 (shown) and p. 326

Solve Equations by Factoring When you use factoring to solve a quadratic equation, you use the following property.

Zero Product Property	For any real numbers a and b, if $ab = 0$, then either $a = 0$ or $b = 0$, or both a and $b = 0$.

Example **Solve each equation by factoring.**

a. $3x^2 = 15x$

$3x^2 = 15x$ Original equation
$3x^2 - 15x = 0$ Subtract 15x from both sides.
$3x(x - 5) = 0$ Factor the binomial.
$3x = 0$ or $x - 5 = 0$ Zero Product Property
$x = 0$ or $x = 5$ Solve each equation.
The solution set is {0, 5}.

b. $4x^2 - 5x = 21$

$4x^2 - 5x = 21$ Original equation
$4x^2 - 5x - 21 = 0$ Subtract 21 from both sides.
$(4x + 7)(x - 3) = 0$ Factor the trinomial.
$4x + 7 = 0$ or $x - 3 = 0$ Zero Product Property
$x = -\frac{7}{4}$ or $x = 3$ Solve each equation.
The solution set is $\left\{-\frac{7}{4}, 3\right\}$.

Exercises

Solve each equation by factoring.

1. $6x^2 - 2x = 0$ $\left\{0, \frac{1}{3}\right\}$
2. $x^2 = 7x$ {0, 7}
3. $20x^2 = -25x$ $\left\{0, -\frac{5}{4}\right\}$
4. $6x^2 = 7x$ $\left\{0, \frac{7}{6}\right\}$
5. $6x^2 - 27x = 0$ $\left\{0, \frac{9}{2}\right\}$
6. $12x^2 - 8x = 0$ $\left\{0, \frac{2}{3}\right\}$
7. $x^2 + x - 30 = 0$ {5, −6}
8. $2x^2 - x - 3 = 0$ $\left\{\frac{3}{2}, -1\right\}$
9. $x^2 + 14x + 33 = 0$ {−11, −3}
10. $4x^2 + 27x - 7 = 0$ $\left\{\frac{1}{4}, -7\right\}$
11. $3x^2 + 29x - 10 = 0$ $\left\{-10, \frac{1}{3}\right\}$
12. $6x^2 - 5x - 4 = 0$ $\left\{-\frac{1}{2}, \frac{4}{3}\right\}$
13. $12x^2 - 8x + 1 = 0$ $\left\{\frac{1}{6}, \frac{1}{2}\right\}$
14. $5x^2 + 28x - 12 = 0$ $\left\{\frac{2}{5}, -6\right\}$
15. $2x^2 - 250x + 5000 = 0$ {100, 25}
16. $2x^2 - 11x - 40 = 0$ $\left\{8, -\frac{5}{2}\right\}$
17. $2x^2 + 21x - 11 = 0$ $\left\{-11, \frac{1}{2}\right\}$
18. $3x^2 + 2x - 21 = 0$ $\left\{\frac{7}{3}, -3\right\}$
19. $8x^2 - 14x + 3 = 0$ $\left\{\frac{3}{2}, \frac{1}{4}\right\}$
20. $6x^2 + 11x - 2 = 0$ $\left\{-2, \frac{1}{6}\right\}$
21. $5x^2 + 17x - 12 = 0$ $\left\{\frac{3}{5}, -4\right\}$
22. $12x^2 + 25x + 12 = 0$ $\left\{-\frac{4}{3}, -\frac{3}{4}\right\}$
23. $12x^2 + 18x + 6 = 0$ $\left\{-\frac{1}{2}, -1\right\}$
24. $7x^2 - 36x + 5 = 0$ $\left\{\frac{1}{7}, 5\right\}$

Skills Practice, p. 327 and Practice, p. 328 (shown)

Solve each equation by factoring.

1. $x^2 - 4x - 12 = 0$ {6, −2}
2. $x^2 - 16x + 64 = 0$ {8}
3. $x^2 - 20x + 100 = 0$ {10}
4. $x^2 - 6x + 8 = 0$ {2, 4}
5. $x^2 + 3x + 2 = 0$ {−2, −1}
6. $x^2 - 9x + 14 = 0$ {2, 7}
7. $x^2 - 4x = 0$ {0, 4}
8. $7x^2 = 4x$ $\left\{0, \frac{4}{7}\right\}$
9. $x^2 + 25 = 10x$ {5}
10. $10x^2 = 9x$ $\left\{0, \frac{9}{10}\right\}$
11. $x^2 = 2x + 99$ {−9, 11}
12. $x^2 + 12x = -36$ {−6}
13. $5x^2 - 35x + 60 = 0$ {3, 4}
14. $36x^2 = 25$ $\left\{\frac{5}{6}, -\frac{5}{6}\right\}$
15. $2x^2 - 8x - 90 = 0$ {9, −5}
16. $3x^2 + 2x - 1 = 0$ $\left\{\frac{1}{3}, -1\right\}$
17. $6x^2 = 9x$ $\left\{0, \frac{3}{2}\right\}$
18. $3x^2 + 24x + 45 = 0$ {−5, −3}
19. $15x^2 + 19x + 6 = 0$ $\left\{-\frac{3}{5}, -\frac{2}{3}\right\}$
20. $3x^2 - 8x = -4$ $\left\{2, \frac{2}{3}\right\}$
21. $6x^2 = 5x + 6$ $\left\{\frac{3}{2}, -\frac{2}{3}\right\}$

Write a quadratic equation with the given roots. Write the equation in the form $ax^2 + bx + c = 0$, where a, b, and c are integers.

22. 7, 2 $x^2 - 9x + 14 = 0$
23. 0, 3 $x^2 - 3x = 0$
24. −5, 8 $x^2 - 3x - 40 = 0$
25. −7, −8 $x^2 + 15x + 56 = 0$
26. −6, −3 $x^2 + 9x + 18 = 0$
27. 3, −4 $x^2 + x - 12 = 0$
28. $1, \frac{1}{2}$ $2x^2 - 3x + 1 = 0$
29. $\frac{1}{3}, 2$ $3x^2 - 7x + 2 = 0$
30. $0, -\frac{7}{2}$ $2x^2 + 7x = 0$
31. $\frac{1}{3}, -3$ $3x^2 + 8x - 3 = 0$
32. $4, \frac{1}{3}$ $3x^2 - 13x + 4 = 0$
33. $-\frac{2}{3}, -\frac{4}{5}$ $15x^2 + 22x + 8 = 0$

34. **NUMBER THEORY** Find two consecutive even positive integers whose product is 624. **24, 26**
35. **NUMBER THEORY** Find two consecutive odd positive integers whose product is 323. **17, 19**
36. **GEOMETRY** The length of a rectangle is 2 feet more than its width. Find the dimensions of the rectangle if its area is 63 square feet. **7 ft by 9 ft**
37. **PHOTOGRAPHY** The length and width of a 6-inch by 8-inch photograph are reduced by the same amount to make a new photograph whose area is half that of the original. By how many inches will the dimensions of the photograph have to be reduced? **2 in.**

Reading to Learn Mathematics, p. 329 ELL

Pre-Activity **How is the Zero Product Property used in geometry?**

Read the introduction to Lesson 6-3 at the top of page 301 in your textbook.

What does the expression $x(x + 5)$ mean in this situation?

It represents the area of the rectangle, since the area is the product of the width and length.

Reading the Lesson

1. The solution of a quadratic equation by factoring is shown below. Give the reason for each step of the solution.

$x^2 - 10x = -21$ Original equation
$x^2 - 10x + 21 = 0$ **Add 21 to each side.**
$(x - 3)(x - 7) = 0$ **Factor the trinomial.**
$x - 3 = 0$ or $x - 7 = 0$ **Zero Product Property**
$x = 3$ $x = 7$ **Solve each equation.**

The solution set is **{3, 7}**.

2. On an algebra quiz, students were asked to write a quadratic equation with −7 and 5 as its roots. The work that three students in the class wrote on their papers is shown below.

Marla	Rosa	Larry
$(x - 7)(x + 5) = 0$	$(x + 7)(x - 5) = 0$	$(x + 7)(x - 5) = 0$
$x^2 - 2x - 35 = 0$	$x^2 + 2x - 35 = 0$	$x^2 - 2x - 35 = 0$

Who is correct? **Rosa**

Explain the errors in the other two students' work.

Sample answer: Marla used the wrong factors. Larry used the correct factors but multiplied them incorrectly.

Helping You Remember

3. A good way to remember a concept is to represent it in more than one way. Describe an algebraic way and a graphical way to recognize a quadratic equation that has a double root.

Sample answer: Algebraic: Write the equation in the standard form $ax^2 + bx + c = 0$ and examine the trinomial. If it is a perfect square trinomial, the quadratic function has a double root. Graphical: Graph the related quadratic function. If the parabola has exactly one x-intercept, then the equation has a double root.

Enrichment, p. 330

Euler's Formula for Prime Numbers

Many mathematicians have searched for a formula that would generate prime numbers. One such formula was proposed by Euler and uses a quadratic polynomial, $x^2 + x + 41$.

Find the values of $x^2 + x + 41$ for the given values of x. State whether each value of the polynomial is or is not a prime number.

1. $x = 0$ **41, prime**
2. $x = 1$ **43, prime**
3. $x = 2$ **47, prime**
4. $x = 3$ **53, prime**
5. $x = 4$ **61, prime**
6. $x = 5$ **71, prime**

Answer

46. 4; The logs must have a diameter greater than 4 in. for the rule to produce positive board feet values.

304 Chapter 6 Quadratic Functions and Inequalities

50. WRITING IN MATH Answer the question that was posed at the beginning of the lesson. **See pp. 343A–343F.**

How is the Zero Product Property used in geometry?

Include the following in your answer:

- an explanation of how to find the dimensions of the rectangle using the Zero Product Property, and
- why the equation $x(x + 5) = 24$ is not solved by using $x = 24$ and $x + 5 = 24$.

Standardized Test Practice

51. Which quadratic equation has roots $\frac{1}{2}$ and $\frac{1}{3}$? **D**

(A) $5x^2 - 5x - 2 = 0$ (B) $5x^2 - 5x + 1 = 0$

(C) $6x^2 + 5x - 1 = 0$ (D) $6x^2 - 5x + 1 = 0$

52. If the roots of a quadratic equation are 6 and -3, what is the equation of the axis of symmetry? **B**

(A) $x = 1$ (B) $x = \frac{3}{2}$ (C) $x = \frac{1}{2}$ (D) $x = -2$

Maintain Your Skills

Mixed Review **Solve each equation by graphing. If exact roots cannot be found, state the consecutive integers between which the roots are located.** *(Lesson 6-2)*

53. $f(x) = -x^2 - 4x + 5$ 54. $f(x) = 4x^2 + 4x + 1$ 55. $f(x) = 3x^2 - 10x - 4$

56. Determine whether $f(x) = 3x^2 - 12x - 7$ has a maximum or a minimum value. Then find the maximum or minimum value. *(Lesson 6-1)* **min.; −19**

Simplify. *(Lesson 5-6)*

57. $\sqrt{3}(\sqrt{6} - 2)$ 58. $\sqrt{108} - \sqrt{48} + (\sqrt{3})^3$ 59. $(5 + \sqrt{8})^2$ **$33 + 20\sqrt{2}$**

Solve each system of equations. *(Lesson 3-2)*

60. $4a - 3b = -4$, $3a - 2b = -4$ **(−4, −4)** 61. $2r + s = 1$, $r - s = 8$ **(3, −5)** 62. $3x - 2y = -3$, $3x + y = 3$ **$\left(\frac{1}{3}, 2\right)$**

53. −5, 1
54. $-\frac{1}{2}$
55. between −1 and 0; between 3 and 4
57. $3\sqrt{2} - 2\sqrt{3}$
58. $5\sqrt{3}$

Getting Ready for the Next Lesson **PREREQUISITE SKILL Simplify.** *(To review **simplifying radicals**, see Lesson 5-5.)*

63. $\sqrt{8}$ **$2\sqrt{2}$** 64. $\sqrt{20}$ **$2\sqrt{5}$** 65. $\sqrt{27}$ **$3\sqrt{3}$**

66. $\sqrt{-50}$ **$5i\sqrt{2}$** 67. $\sqrt{-12}$ **$2i\sqrt{3}$** 68. $\sqrt{-48}$ **$4i\sqrt{3}$**

Practice Quiz 1 — *Lessons 6-1 through 6-3*

1. Find the y-intercept, the equation of the axis of symmetry, and the x-coordinate of the vertex for $f(x) = 3x^2 - 12x + 4$. Then graph the function by making a table of values. *(Lesson 6-1)* **4, x = 2; 2; See margin for graph.**
2. Determine whether $f(x) = 3 - x^2 + 5x$ has a maximum or minimum value. Then find this maximum or minimum value. *(Lesson 6-1)* **max.; $\frac{37}{4}$ or $9\frac{1}{4}$**
3. Solve $2x^2 - 11x + 12 = 0$ by graphing. If exact roots cannot be found, state the consecutive integers between which the roots are located. *(Lesson 6-2)* **$1\frac{1}{2}$, 4**
4. Solve $2x^2 - 5x - 3 = 0$ by factoring. *(Lesson 6-3)* **$\left\{3, -\frac{1}{2}\right\}$**
5. Write a quadratic equation with roots -4 and $\frac{1}{3}$. Write the equation in the form $ax^2 + bx + c = 0$, where a, b, and c are integers. *(Lesson 6-3)* **$3x^2 + 11x - 4 = 0$**

www.algebra2.com/self_check_quiz

4 Assess

Open-Ended Assessment

Speaking Ask students to give a verbal explanation of the Zero Product Property. They should discuss why it is true and how it is used in finding the roots of a quadratic equation, demonstrating the technique using an example.

Tips for New Teachers

Intervention Suggest that students who have difficulty understanding the Zero Product Property try to find two nonzero numbers whose product is zero. Students should quickly determine that at least one of the numbers must be zero in order for their product to be zero.

Getting Ready for Lesson 6-4

PREREQUISITE SKILL Lesson 6-4 presents solving quadratic equations by completing the square. The process involves evaluating radicals. Exercises 63–68 should be used to determine your students' familiarity with simplifying radicals.

Assessment Options

Practice Quiz 1 The quiz provides students with a brief review of the concepts and skills in Lessons 6-1 through 6-3. Lesson numbers are given to the right of exercises or instruction lines so students can review concepts not yet mastered.

Answer (Practice Quiz 1)

1.

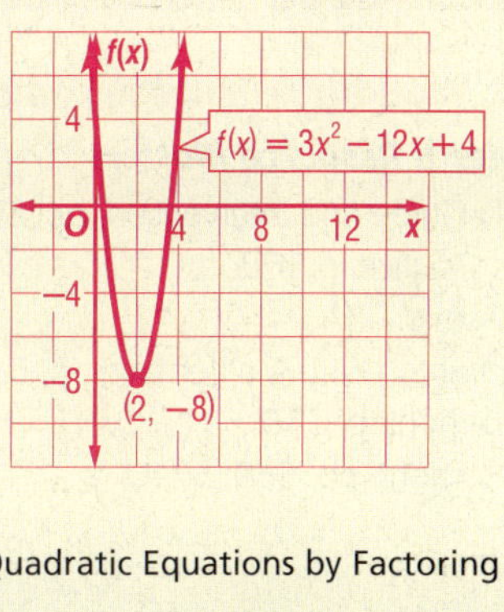

47. $y = (x - p)(x - q)$
$y = x^2 - px - qx + pq$
$y = x^2 - (p + q)x + pq$
$a = 1, b = -(p + q), c = +pq$
axis of symmetry: $x = -\frac{b}{2a}$
$x = -\frac{-(p + q)}{2(1)}$
$x = \frac{p + q}{2}$

The axis of symmetry is the average of the x-intercepts. Therefore the axis of symmetry is located halfway between the x-intercepts.

6-4 Lesson Notes

1 Focus

5-Minute Check Transparency 6-4 Use as a quiz or review of Lesson 6-3.

Mathematical Background notes are available for this lesson on p. 284C.

Building on Prior Knowledge

In Lesson 6-3, students solved quadratic equations by factoring. In this lesson they use two other methods to solve equations: the Square Root Property and completing the square.

How can you find the time it takes an accelerating race car to reach the finish line?

Ask students:

- The number 121 is a perfect square. What is the square root of 121? **11**
- How does the number 11 relate to the coefficient of the x term? **The coefficient of x is twice 11.**

6-4 Completing the Square

What You'll Learn

- Solve quadratic equations by using the Square Root Property.
- Solve quadratic equations by completing the square.

Vocabulary

- completing the square

How can you find the time it takes an accelerating race car to reach the finish line?

Under a yellow caution flag, race car drivers slow to a speed of 60 miles per hour. When the green flag is waved, the drivers can increase their speed.

Suppose the driver of one car is 500 feet from the finish line. If the driver accelerates at a constant rate of 8 feet per second squared, the equation $t^2 + 22t + 121 = 246$ represents the time t it takes the driver to reach this line. To solve this equation, you can use the Square Root Property.

SQUARE ROOT PROPERTY You have solved equations like $x^2 - 25 = 0$ by factoring. You can also use the **Square Root Property** to solve such an equation. This method is useful with equations like the one above that describes the race car's speed. In this case, the quadratic equation contains a perfect square trinomial set equal to a constant.

Study Tip

Reading Math
$\pm\sqrt{n}$ is read plus or minus the square root of n.

Key Concept — *Square Root Property*

For any real number n, if $x^2 = n$, then $x = \pm\sqrt{n}$.

TEACHING TIP

Have students solve Example 1 by factoring and compare the results. Point out that Example 2 cannot be solved by factoring.

Example 1 *Equation with Rational Roots*

Solve $x^2 + 10x + 25 = 49$ by using the Square Root Property.

$x^2 + 10x + 25 = 49$	Original equation
$(x + 5)^2 = 49$	Factor the perfect square trinomial.
$x + 5 = \pm\sqrt{49}$	Square Root Property
$x + 5 = \pm 7$	$\sqrt{49} = 7$
$x = -5 \pm 7$	Add -5 to each side.
$x = -5 + 7$ or $x = -5 - 7$	Write as two equations.
$x = 2$ $\quad$ $x = -12$	Solve each equation.

The solution set is $\{2, -12\}$. You can check this result by using factoring to solve the original equation.

Roots that are irrational numbers may be written as exact answers in radical form or as *approximate* answers in decimal form when a calculator is used.

Resource Manager

Workbook and Reproducible Masters

Chapter 6 Resource Masters
- Study Guide and Intervention, pp. 331–332
- Skills Practice, p. 333
- Practice, p. 334
- Reading to Learn Mathematics, p. 335
- Enrichment, p. 336
- Assessment, pp. 369, 371

Teaching Algebra With Manipulatives Masters, pp. 244, 245–246

Transparencies

5-Minute Check Transparency 6-4
Answer Key Transparencies

Technology

Interactive Chalkboard

Example 2 Equation with Irrational Roots

Solve $x^2 - 6x + 9 = 32$ by using the Square Root Property.

$x^2 - 6x + 9 = 32$	Original equation
$(x - 3)^2 = 32$	Factor the perfect square trinomial.
$x - 3 = \pm\sqrt{32}$	Square Root Property
$x = 3 \pm 4\sqrt{2}$	Add 3 to each side; $\sqrt{32} = 4\sqrt{2}$
$x = 3 + 4\sqrt{2}$ or $x = 3 - 4\sqrt{2}$	Write as two equations.
$x \approx 8.7$ $\quad x \approx -2.7$	Use a calculator.

The exact solutions of this equation are $3 - 4\sqrt{2}$ and $3 + 4\sqrt{2}$. The approximate solutions are -2.7 and 8.7. Check these results by finding and graphing the related quadratic function.

$x^2 - 6x + 9 = 32$	Original equation
$x^2 - 6x - 23 = 0$	Subtract 32 from each side.
$y = x^2 - 6x - 23$	Related quadratic function

CHECK Use the ZERO function of a graphing calculator. The approximate zeros of the related function are -2.7 and 8.7.

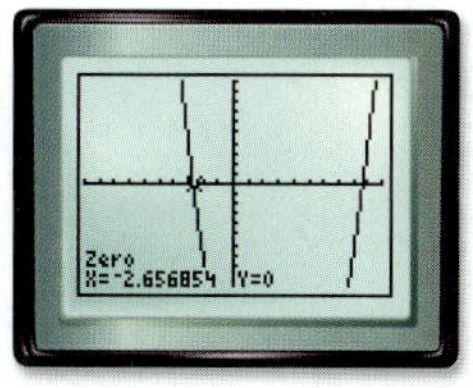

COMPLETE THE SQUARE The Square Root Property can only be used to solve quadratic equations when the side containing the quadratic expression is a perfect square. However, few quadratic expressions are perfect squares. To make a quadratic expression a perfect square, a method called **completing the square** may be used.

In a perfect square trinomial, there is a relationship between the coefficient of the linear term and the constant term. Consider the pattern for squaring a sum.

$$(x + 7)^2 = x^2 + 2(7)x + 7^2 \quad \text{Square of a sum pattern}$$
$$= x^2 + 14x + 49 \quad \text{Simplify.}$$
$$\left(\frac{14}{2}\right)^2 \longrightarrow 7^2 \quad \text{Notice that 49 is } 7^2 \text{ and 7 is one-half of 14.}$$

You can use this pattern of coefficients to complete the square of a quadratic expression.

Key Concept — Completing the Square

- **Words** To complete the square for any quadratic expression of the form $x^2 + bx$, follow the steps below.
  - **Step 1** Find one half of b, the coefficient of x.
  - **Step 2** Square the result in Step 1.
  - **Step 3** Add the result of Step 2 to $x^2 + bx$.
- **Symbols** $x^2 + bx + \left(\frac{b}{2}\right)^2 = \left(x + \frac{b}{2}\right)^2$

www.algebra2.com/extra_examples

2 Teach

SQUARE ROOT PROPERTY

In-Class Examples PowerPoint®

Teaching Tip In Example 1, point out that both constants in the equation, 25 and 49, are perfect squares.

1. Solve $x^2 + 14x + 49 = 64$ by using the Square Root Property. $\{-15, 1\}$

Teaching Tip Point out that the constant on the right side of the equation given in Example 2, is not a perfect square. Stress that this occurrence means the roots will be irrational numbers involving radicals. Also emphasize the use of the $\pm$ symbol in the step where the Square Root Property is utilized.

2. Solve $x^2 - 10x + 25 = 12$ by using the Square Root Property. $\{5 \pm 2\sqrt{3}\}$

COMPLETE THE SQUARE

Teaching Tip When discussing the steps for completing the square, emphasize that the coefficient of the quadratic term must be 1.

In-Class Examples

Power Point®

3 Find the value of c that makes $x^2 + 16x + c$ a perfect square. Then write the trinomial as a perfect square. **64; $(x + 8)^2$**

4 Solve $x^2 + 4x - 12 = 0$ by completing the square. **$\{-6, 2\}$**

Teaching Tip Remind students to think carefully about the difference between multiplying a quantity by a factor of 2 and squaring a quantity.

Example 3 Complete the Square

Find the value of c that makes $x^2 + 12x + c$ a perfect square. Then write the trinomial as a perfect square.

Step 1 Find one half of 12. $\frac{12}{2} = 6$

Step 2 Square the result of Step 1. $6^2 = 36$

Step 3 Add the result of Step 2 to $x^2 + 12x$. $x^2 + 12x + 36$

The trinomial $x^2 + 12x + 36$ can be written as $(x + 6)^2$.

You can solve any quadratic equation by completing the square. Because you are solving an equation, add the value you use to complete the square to each side.

Algebra Activity

Completing the Square

Use algebra tiles to complete the square for the equation $x^2 + 2x - 3 = 0$.

Step 1 Represent $x^2 + 2x - 3 = 0$ on an equation mat.

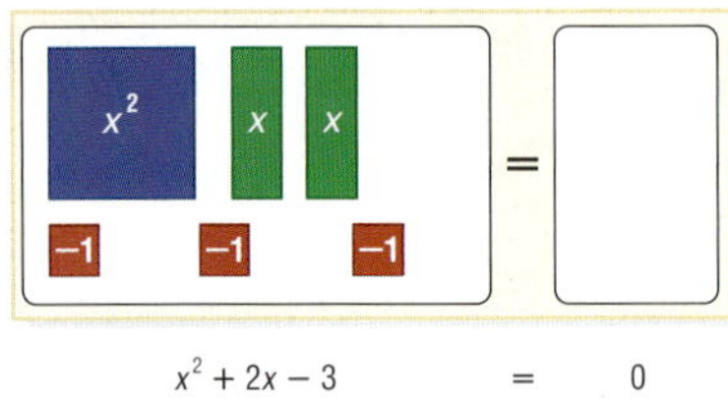

Step 2 Add 3 to each side of the mat. Remove the zero pairs.

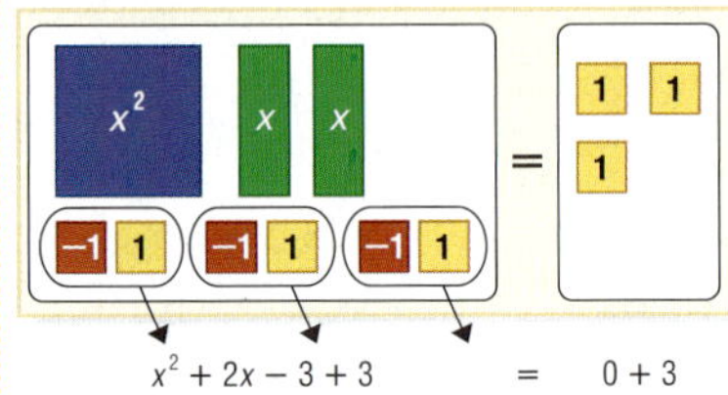

Step 3 Begin to arrange the x^2 and x tiles into a square.

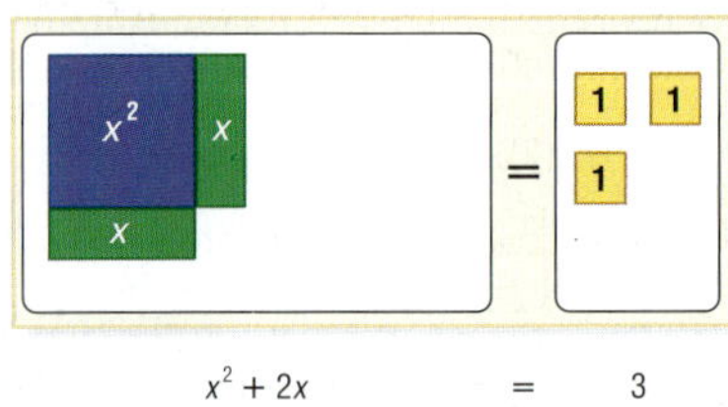

Step 4 To complete the square, add 1 yellow 1 tile to each side. The completed equation is $x^2 + 2x + 1 = 4$ or $(x + 1)^2 = 4$.

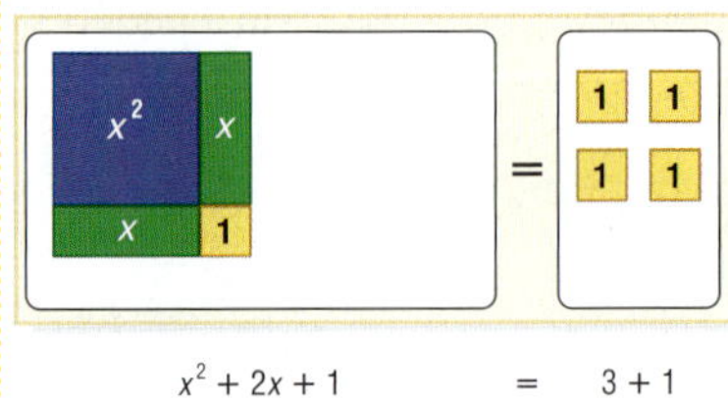

Model

Use algebra tiles to complete the square for each equation.

1. $x^2 + 2x - 4 = 0$ **$(x + 1)^2 = 5$**

2. $x^2 + 4x + 1 = 0$ **$(x + 2)^2 = 3$**

3. $x^2 - 6x = -5$ **$(x - 3)^2 = 4$**

4. $x^2 - 2x = -1$ **$(x - 1)^2 = 0$**

Study Tip

Common Misconception
When solving equations by completing the square, don't forget to add $\left(\frac{b}{2}\right)^2$ to *each* side of the equation.

Example 4 Solve an Equation by Completing the Square

Solve $x^2 + 8x - 20 = 0$ by completing the square.

$x^2 + 8x - 20 = 0$	Notice that $x^2 + 8x - 20$ is not a perfect square.
$x^2 + 8x = 20$	Rewrite so the left side is of the form $x^2 + bx$.
$x^2 + 8x + 16 = 20 + 16$	Since $\left(\frac{8}{2}\right)^2 = 16$, add 16 to each side.
$(x + 4)^2 = 36$	Write the left side as a perfect square by factoring.

Algebra Activity

Materials: algebra tiles, equation mat

- Ask students why the choice was made to add 3 unit tiles to each side of the equation mat in Step 2. **Sample answer: In order to simplify the work arranging the tiles into a square in Step 3.**
- Remind students that an x tile is x units long and 1 unit wide. Stress that the width is the same as the length of each side of a unit tile.

$x + 4 = \pm 6$ Square Root Property

$x = -4 \pm 6$ Add −4 to each side.

$x = -4 + 6$ or $x = -4 - 6$ Write as two equations.

$x = 2$ $x = -10$ The solution set is $\{-10, 2\}$.

You can check this result by using factoring to solve the original equation.

When the coefficient of the quadratic term is not 1, you must first divide the equation by that coefficient before completing the square.

Example 5 Equation with $a \neq 1$

Solve $2x^2 - 5x + 3 = 0$ by completing the square.

$2x^2 - 5x + 3 = 0$ Notice that $2x^2 - 5x + 3$ is not a perfect square.

$x^2 - \frac{5}{2}x + \frac{3}{2} = 0$ Divide by the coefficient of quadratic term, 2.

$x^2 - \frac{5}{2}x = -\frac{3}{2}$ Subtract $\frac{3}{2}$ from each side.

$x^2 - \frac{5}{2}x + \frac{25}{16} = -\frac{3}{2} + \frac{25}{16}$ Since $\left(-\frac{5}{2} \div 2\right)^2 = \frac{25}{16}$, add $\frac{25}{16}$ to each side.

$\left(x - \frac{5}{4}\right)^2 = \frac{1}{16}$ Write the left side as a perfect square by factoring. Simplify the right side.

$x - \frac{5}{4} = \pm\frac{1}{4}$ Square Root Property

$x = \frac{5}{4} \pm \frac{1}{4}$ Add $\frac{5}{4}$ to each side.

$x = \frac{5}{4} + \frac{1}{4}$ or $x = \frac{5}{4} - \frac{1}{4}$ Write as two equations.

$x = \frac{3}{2}$ $x = 1$ The solution set is $\left\{1, \frac{3}{2}\right\}$.

Not all solutions of quadratic equations are real numbers. In some cases, the solutions are complex numbers of the form $a + bi$, where $b \neq 0$.

Example 6 Equation with Complex Solutions

Solve $x^2 + 4x + 11 = 0$ by completing the square.

$x^2 + 4x + 11 = 0$ Notice that $x^2 + 4x + 11$ is not a perfect square.

$x^2 + 4x = -11$ Rewrite so the left side is of the form $x^2 + bx$.

$x^2 + 4x + 4 = -11 + 4$ Since $\left(\frac{4}{2}\right)^2 = 4$, add 4 to each side.

$(x + 2)^2 = -7$ Write the left side as a perfect square by factoring.

$x + 2 = \pm\sqrt{-7}$ Square Root Property

$x + 2 = \pm i\sqrt{7}$ $\sqrt{-1} = i$

$x = -2 \pm i\sqrt{7}$ Subtract 2 from each side.

The solution set is $\{-2 + i\sqrt{7}, -2 - i\sqrt{7}\}$. Notice that these are imaginary solutions.

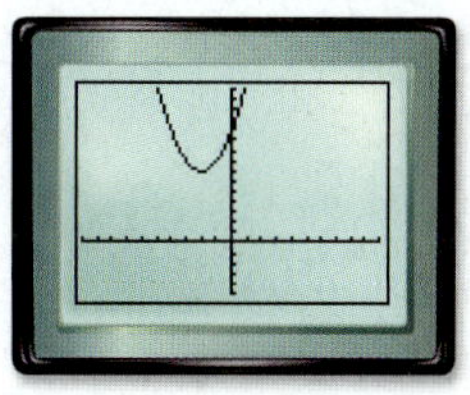

CHECK A graph of the related function shows that the equation has no real solutions since the graph has no x-intercepts. Imaginary solutions must be checked algebraically by substituting them in the original equation.

In-Class Examples

Power Point®

Teaching Tip Some students may notice that the left side of the equation in Example 5 can be factored into the product of two binomials: $(2x - 3)(x - 1)$. Then the Zero Product Property can be used to obtain the same solutions. This is a good time to point out that more than one method of solution is often possible when solving a quadratic equation.

5 Solve $3x^2 - 2x - 1 = 0$ by completing the square. $\left\{-\frac{1}{3}, 1\right\}$

6 Solve $x^2 + 2x + 3 = 0$ by completing the square. $\{-1 \pm i\sqrt{2}\}$

Teaching Tip Ask students to name three possible ways to check the solutions to an equation that they have solved by completing the square. **graphing the related function, factoring the equation, substituting the solutions into the original equation**

DAILY INTERVENTION — Differentiated Instruction

Kinesthetic Have students work with algebra tiles to help them write five equations that can be solved by completing the square. Provide each student with one x^2 tile, several x tiles, and several unit tiles. Have students begin by creating a square arrangement of their tiles and then work backwards through the steps shown in the Algebra Activity on p. 308 to find a quadratic equation. After students have written their five equations, ask them to trade their equations with another student and then use their algebra tiles to find the solutions of the equations they receive.

3 Practice/Apply

Study Notebook

Have students—

- add the definitions/examples of the vocabulary terms to their Vocabulary Builder worksheets for Chapter 6.
- include any other item(s) that they find helpful in mastering the skills in this lesson.

About the Exercises...

Organization by Objective

- **Square Root Property:** 14–23
- **Complete the Square:** 24–51, 53

Odd/Even Assignments

Exercises 14–47 are structured so that students practice the same concepts whether they are assigned odd or even problems.

Alert! Exercise 51 involves research on the Internet or other reference materials.

Assignment Guide

Basic: 15–19 odd, 23–47 odd, 52, 54–72

Average: 15–47 odd, 52–72

Advanced: 14–48 even, 49–52, 54–68 (optional: 69–72)

DAILY INTERVENTION

FIND THE ERROR Point out that, while it is possible to complete the square when the coefficient of the x^2 term is something other than 1, it is much easier to first divide each side by the coefficient and students will also be less likely to make an error like the one made by Rashid shown here.

Check for Understanding

Concept Check

1. **Explain** what it means to *complete the square.*
2. **Determine** whether the value of c that makes $ax^2 + bx + c$ a perfect square trinomial is *sometimes, always,* or *never* negative. Explain your reasoning.
3. **FIND THE ERROR** Rashid and Tia are solving $2x^2 - 8x + 10 = 0$ by completing the square. Who is correct? Explain your reasoning. **Tia; see margin for explanation.**

Rashid	Tia
$2x^2 - 8x + 10 = 0$	$2x^2 - 8x + 10 = 0$
$2x^2 - 8x = -10$	$x^2 - 4x = 0 - 5$
$2x^2 - 8x + 16 = -10 + 16$	$x^2 - 4x + 4 = -5 + 4$
$(x - 4)^2 = 6$	$(x - 2)^2 = -1$
$x - 4 = \pm\sqrt{6}$	$x - 2 = \pm i$
$x = 4 \pm \sqrt{6}$	$x = 2 \pm i$

1. Completing the square allows you to rewrite one side of a quadratic equation in the form of a perfect square. Once in this form, the equation is solved by using the Square Root Property.

2. Never; see margin for explanation.

Guided Practice

GUIDED PRACTICE KEY

Exercises	Examples
4, 5, 12, 13	1, 2
6, 7	3
8–11	4–6

Solve each equation by using the Square Root Property.

4. $x^2 + 14x + 49 = 9$ **$\{-10, -4\}$**
5. $9x^2 - 24x + 16 = 2$ **$\left\{\frac{4 \pm \sqrt{2}}{3}\right\}$**

Find the value of c that makes each trinomial a perfect square. Then write the trinomial as a perfect square.

6. $x^2 - 12x + c$ **$36; (x - 6)^2$**
7. $x^2 - 3x + c$ **$\frac{9}{4}; \left(x - \frac{3}{2}\right)^2$**

Solve each equation by completing the square.

8. $x^2 + 3x - 18 = 0$ **$\{-6, 3\}$**
9. $x^2 - 8x + 11 = 0$ **$\{4 \pm \sqrt{5}\}$**
10. $x^2 + 2x + 6 = 0$ **$\{-1 \pm i\sqrt{5}\}$**
11. $2x^2 - 3x - 3 = 0$ **$\left\{\frac{3 \pm \sqrt{33}}{4}\right\}$**

Application

ASTRONOMY **For Exercises 12 and 13, use the following information.**

The height h of an object t seconds after it is dropped is given by $h = -\frac{1}{2}gt^2 + h_0$, where h_0 is the initial height and g is the acceleration due to gravity. The acceleration due to gravity near Earth's surface is 9.8 m/s^2, while on Jupiter it is 23.1 m/s^2. Suppose an object is dropped from an initial height of 100 meters from the surface of each planet.

12. On which planet should the object reach the ground first? **Jupiter**
13. Find the time it takes for the object to reach the ground on each planet to the nearest tenth of a second. **Earth: 4.5 s, Jupiter: 2.9 s**

18. $\left\{\frac{7 \pm \sqrt{5}}{2}\right\}$

19. $\left\{\frac{-5 \pm \sqrt{11}}{3}\right\}$

★ **indicates increased difficulty**

Practice and Apply

Homework Help

For Exercises	See Examples
14–23, 48	1, 2
24–31	3
32–47, 49–50, 53	4–6

Extra Practice
See page 840.

Solve each equation by using the Square Root Property.

14. $x^2 + 4x + 4 = 25$ **$\{3, -7\}$**
15. $x^2 - 10x + 25 = 49$ **$\{-2, 12\}$**
16. $x^2 + 8x + 16 = 7$ **$\{-4 \pm \sqrt{7}\}$**
17. $x^2 - 6x + 9 = 8$ **$\{3 \pm 2\sqrt{2}\}$**
18. $4x^2 - 28x + 49 = 5$
19. $9x^2 + 30x + 25 = 11$

★ 20. $x^2 + x + \frac{1}{4} = \frac{9}{16}$ **$\left\{-\frac{5}{4}, \frac{1}{4}\right\}$**

★ 21. $x^2 + 1.4x + 0.49 = 0.81$ **$\{-1.6, 0.2\}$**

22. **MOVIE SCREENS** The area A in square feet of a projected picture on a movie screen is given by $A = 0.16d^2$, where d is the distance from the projector to the screen in feet. At what distance will the projected picture have an area of 100 square feet? **25 ft**

Answers

2. The value of c that makes $ax^2 + bx + c$ a perfect square trinomial is the square of $\frac{b}{2}$ and the square of a number can never be negative.
3. Before completing the square, you must first check to see that the coefficient of the quadratic term is 1. If it is not, you must first divide the equation by that coefficient.

23. **ENGINEERING** In an engineering test, a rocket sled is propelled into a target. The sled's distance d in meters from the target is given by the formula $d = -1.5t^2 + 120$, where t is the number of seconds after rocket ignition. How many seconds have passed since rocket ignition when the sled is 10 meters from the target? **about 8.56 s**

More About. . .

Engineering

Reverse ballistic testing—accelerating a target on a sled to impact a stationary test item at the end of the track—was pioneered at the Sandia National Laboratories' Rocket Sled Track Facility in Albuquerque, New Mexico. This facility provides a 10,000-foot track for testing items at very high speeds.

Source: www.sandia.gov

Find the value of c that makes each trinomial a perfect square. Then write the trinomial as a perfect square.

24. $x^2 + 16x + c$ **$64; (x + 8)^2$**
25. $x^2 - 18x + c$ **$81; (x - 9)^2$**
26. $x^2 - 15x + c$ **$\frac{225}{4}; \left(x - \frac{15}{2}\right)^2$**
27. $x^2 + 7x + c$ **$\frac{49}{4}; \left(x + \frac{7}{2}\right)^2$**
28. $x^2 + 0.6x + c$ **$0.09; (x + 0.3)^2$**
29. $x^2 - 2.4x + c$ **$1.44; (x - 1.2)^2$**
30. $x^2 - \frac{8}{3}x + c$ **$\frac{16}{9}; \left(x - \frac{4}{3}\right)^2$**
31. $x^2 + \frac{5}{2}x + c$ **$\frac{25}{16}; \left(x + \frac{5}{4}\right)^2$**

Solve each equation by completing the square.

32. $x^2 - 8x + 15 = 0$ **$\{3, 5\}$**
33. $x^2 + 2x - 120 = 0$ **$\{-12, 10\}$**
34. $x^2 + 2x - 6 = 0$ **$\{-1 \pm \sqrt{7}\}$**
35. $x^2 - 4x + 1 = 0$ **$\{2 \pm \sqrt{3}\}$**
36. $x^2 - 4x + 5 = 0$ **$\{2 \pm i\}$**
37. $x^2 + 6x + 13 = 0$ **$\{-3 \pm 2i\}$**
38. $2x^2 + 3x - 5 = 0$ **$\left\{-\frac{5}{2}, 1\right\}$**
39. $2x^2 - 3x + 1 = 0$ **$\left\{\frac{1}{2}, 1\right\}$**
40. $3x^2 - 5x + 1 = 0$
41. $3x^2 - 4x - 2 = 0$
42. $2x^2 - 7x + 12 = 0$
43. $3x^2 + 5x + 4 = 0$
44. $x^2 + 1.4x = 1.2$ **$\{-2, 0.6\}$**
45. $x^2 - 4.7x = -2.8$ **$\{0.7, 4\}$**
46. $x^2 - \frac{2}{3}x - \frac{26}{9} = 0$ **$\left\{\frac{1}{3} \pm \sqrt{3}\right\}$**
47. $x^2 - \frac{3}{2}x - \frac{23}{16} = 0$ **$\left\{\frac{3}{4} \pm \sqrt{2}\right\}$**

40. $\left\{\frac{5 \pm \sqrt{13}}{6}\right\}$

41. $\left\{\frac{2 \pm \sqrt{10}}{3}\right\}$

42. $\left\{\frac{7 \pm i\sqrt{47}}{4}\right\}$

43. $\left\{\frac{-5 \pm i\sqrt{23}}{6}\right\}$

48. **FRAMING** A picture has a square frame that is 2 inches wide. The area of the picture is one-third of the total area of the picture and frame. What are the dimensions of the picture to the nearest quarter of an inch? **$5\frac{1}{2}$ in. by $5\frac{1}{2}$ in.**

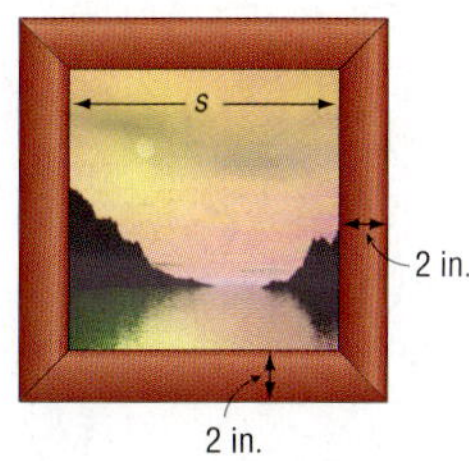

GOLDEN RECTANGLE **For Exercises 49–51, use the following information.**

A *golden rectangle* is one that can be divided into a square and a second rectangle that is geometrically similar to the original rectangle. The ratio of the length of the longer side to the shorter side of a golden rectangle is called the *golden ratio*.

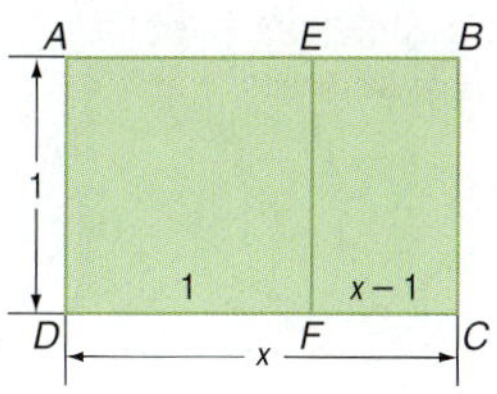

49. Find the ratio of the length of the longer side to the length of the shorter side for rectangle $ABCD$ and for rectangle $EBCF$.

50. Find the exact value of the golden ratio by setting the two ratios in Exercise 49 equal and solving for x. (*Hint*: The golden ratio is a positive value.)

51. **RESEARCH** Use the Internet or other reference to find examples of the golden rectangle in architecture. What applications does the reciprocal of the golden ratio have in music? **See margin.**

49. $\frac{x}{1}, \frac{1}{x-1}$

50. $\frac{1 + \sqrt{5}}{2}$

52. **CRITICAL THINKING** Find all values of n such that $x^2 + bx + \left(\frac{b}{2}\right)^2 = n$ has

a. one real root. **$n = 0$**
b. two real roots. **$n > 0$**
c. two imaginary roots. **$n < 0$**

www.algebra2.com/self_check_quiz

51. Sample answers: The golden rectangle is found in much of ancient Greek architecture, such as the Parthenon, as well as in modern architecture, such as in the windows of the United Nations building. Many songs have their climax at a point occurring 61.8% of the way through the piece, with 0.618 being about the reciprocal of the golden ratio. The reciprocal of the golden ratio is also used in the design of some violins.

Study Guide and Intervention, p. 331 (shown) and p. 332

Square Root Property Use the following property to solve a quadratic equation that is in the form "perfect square trinomial = constant."

Square Root Property	For any real number x if $x^2 = n$, then $x = \pm n$.

Example Solve each equation by using the Square Root Property.

a. $x^2 - 8x + 16 = 25$

$x^2 - 8x + 16 = 25$
$(x - 4)^2 = 25$
$x - 4 = \sqrt{25}$ or $x - 4 = -\sqrt{25}$
$x = 5 + 4 = 9$ or $x = -5 + 4 = -1$
The solution set is $\{9, -1\}$.

b. $4x^2 - 20x + 25 = 32$

$4x^2 - 20x + 25 = 32$
$(2x - 5)^2 = 32$
$2x - 5 = \sqrt{32}$ or $2x - 5 = -\sqrt{32}$
$2x - 5 = 4\sqrt{2}$ or $2x - 5 = -4\sqrt{2}$
$x = \frac{5 \pm 4\sqrt{2}}{2}$
The solution set is $\left\{\frac{5 \pm 4\sqrt{2}}{2}\right\}$.

Exercises

Solve each equation by using the Square Root Property.

1. $x^2 - 18x + 81 = 49$ **$\{2, 16\}$**
2. $x^2 + 20x + 100 = 64$ **$\{-2, -18\}$**
3. $4x^2 + 4x + 1 = 16$ **$\left\{\frac{3}{2}, -\frac{5}{2}\right\}$**
4. $36x^2 + 12x + 1 = 18$ **$\left\{\frac{-1 \pm 3\sqrt{2}}{6}\right\}$**
5. $9x^2 - 12x + 4 = 4$ **$\left\{0, \frac{4}{3}\right\}$**
6. $25x^2 + 40x + 16 = 28$ **$\left\{\frac{-4 \pm 2\sqrt{7}}{5}\right\}$**
7. $4x^2 - 28x + 49 = 64$ **$\left\{\frac{15}{2}, -\frac{1}{2}\right\}$**
8. $16x^2 + 24x + 9 = 81$ **$\left\{\frac{3}{2}, -3\right\}$**
9. $100x^2 - 60x + 9 = 121$ **$\{-0.8, 1.4\}$**
10. $25x^2 + 20x + 4 = 75$ **$\left\{\frac{-2 \pm 5\sqrt{3}}{5}\right\}$**
11. $36x^2 + 48x + 16 = 12$ **$\left\{\frac{-2 \pm \sqrt{3}}{3}\right\}$**
12. $25x^2 - 30x + 9 = 96$ **$\left\{\frac{3 \pm 4\sqrt{6}}{5}\right\}$**

Skills Practice, p. 333 and Practice, p. 334 (shown)

Solve each equation by using the Square Root Property.

1. $x^2 + 8x + 16 = 1$ **$-5, -3$**
2. $x^2 + 6x + 9 = 1$ **$-4, -2$**
3. $x^2 + 10x + 25 = 16$ **$-9, -1$**
4. $x^2 - 14x + 49 = 9$ **$4, 10$**
5. $4x^2 + 12x + 9 = 4$ **$-\frac{1}{2}, -\frac{5}{2}$**
6. $x^2 - 8x + 16 = 8$ **$4 \pm 2\sqrt{2}$**
7. $x^2 - 6x + 9 = 5$ **$3 \pm \sqrt{5}$**
8. $x^2 - 2x + 1 = 2$ **$1 \pm \sqrt{2}$**
9. $9x^2 - 6x + 1 = 2$ **$\frac{1 \pm \sqrt{2}}{3}$**

Find the value of c that makes each trinomial a perfect square. Then write the trinomial as a perfect square.

10. $x^2 + 12x + c$ **$36; (x + 6)^2$**
11. $x^2 - 20x + c$ **$100; (x - 10)^2$**
12. $x^2 + 11x + c$ **$\frac{121}{4}; \left(x + \frac{11}{2}\right)^2$**
13. $x^2 + 0.8x + c$ **$0.16; (x + 0.4)^2$**
14. $x^2 - 2.2x + c$ **$1.21; (x - 1.1)^2$**
15. $x^2 - 0.36x + c$ **$0.0324; (x - 0.18)^2$**
16. $x^2 + \frac{5}{6}x + c$ **$\frac{25}{144}; \left(x + \frac{5}{12}\right)^2$**
17. $x^2 - \frac{1}{4}x + c$ **$\frac{1}{64}; \left(x - \frac{1}{8}\right)^2$**
18. $x^2 - \frac{5}{3}x + c$ **$\frac{25}{36}; \left(x - \frac{5}{6}\right)^2$**

Solve each equation by completing the square.

19. $x^2 + 6x + 8 = 0$ **$-4, -2$**
20. $3x^2 + x - 2 = 0$ **$\frac{2}{3}, -1$**
21. $3x^2 - 5x + 2 = 0$ **$1, \frac{2}{3}$**
22. $x^2 + 18 = 9x$ **$6, 3$**
23. $x^2 - 14x + 19 = 0$ **$7 \pm \sqrt{30}$**
24. $x^2 + 16x - 7 = 0$ **$-8 \pm \sqrt{71}$**
25. $2x^2 + 8x - 3 = 0$ **$\frac{-4 \pm \sqrt{22}}{2}$**
26. $x^2 + x - 5 = 0$ **$\frac{-1 \pm \sqrt{21}}{2}$**
27. $2x^2 - 10x + 5 = 0$ **$\frac{5 \pm \sqrt{15}}{2}$**
28. $x^2 + 3x + 6 = 0$ **$\frac{-3 \pm i\sqrt{15}}{2}$**
29. $2x^2 + 5x + 6 = 0$ **$\frac{-5 \pm i\sqrt{23}}{4}$**
30. $7x^2 + 6x + 2 = 0$ **$\frac{-3 \pm i\sqrt{5}}{7}$**

31. **GEOMETRY** When the dimensions of a cube are reduced by 4 inches on each side, the surface area of the new cube is 864 square inches. What were the dimensions of the original cube? **16 in. by 16 in. by 16 in.**

32. **INVESTMENTS** The amount of money A in an account in which P dollars is invested for 2 years is given by the formula $A = P(1 + r)^2$, where r is the interest rate compounded annually. If an investment of \$800 in the account grows to \$882 in two years, at what interest rate was it invested? **5%**

Reading to Learn Mathematics, p. 335 ELL

Pre-Activity **How can you find the time it takes an accelerating race car to reach the finish line?**

Read the introduction to Lesson 6-4 at the top of page 306 in your textbook.

Explain what it means to say that the driver accelerates at a constant rate of 8 feet per second square.

If the driver is traveling at a certain speed at a particular moment, then one second later, the driver is traveling 8 feet per second faster.

Reading the Lesson

1. Give the reason for each step in the following solution of an equation by using the Square Root Property.

$x^2 - 12x + 36 = 81$	Original equation
$(x - 6)^2 = 81$	**Factor the perfect square trinomial.**
$x - 6 = \pm\sqrt{81}$	**Square Root Property**
$x - 6 = \pm 9$	**81 = 9**
$x - 6 = 9$ or $x - 6 = -9$	**Rewrite as two equations.**
$x = 15$ $x = -3$	**Solve each equation.**

2. Explain how to find the constant that must be added to make a binomial into a perfect square trinomial.
Sample answer: Find half of the coefficient of the linear term and square it.

3. a. What is the first step in solving the equation $3x^2 + 6x = 5$ by completing the square? **Divide the equation by 3.**
b. What is the first step in solving the equation $x^2 + 5x - 12 = 0$ by completing the square? **Add 12 to each side.**

Helping You Remember

4. How can you use the rules for squaring a binomial to help you remember the procedure for changing a binomial into a perfect square trinomial?
One of the rules for squaring a binomial is $(x + y)^2 = x^2 + 2xy + y^2$. In completing the square, you are starting with $x^2 + bx$ and need to find y^2. This shows you that $b = 2y$, so $y = \frac{b}{2}$. That is why you must take half of the coefficient and square it to get the constant that must be added to complete the square.

Enrichment, p. 336

The Golden Quadratic Equations

A **golden rectangle** has the property that its length can be written as $a + b$, where a is the width of the rectangle and $\frac{a+b}{a} = \frac{a}{b}$. Any golden rectangle can be divided into a square and a smaller golden rectangle, as shown.

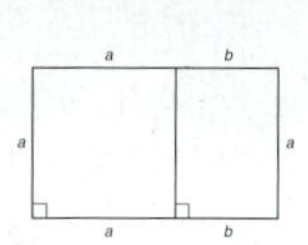

The proportion used to define golden rectangles can be used to derive two quadratic equations. These are sometimes called *golden quadratic equations*.

Solve each problem.

1. In the proportion for the golden rectangle, let a equal 1. Write the resulting quadratic equation and solve for b.
$b^2 + b - 1 = 0$
$b = \frac{-1 + \sqrt{5}}{2}$

4 Assess

Open-Ended Assessment

Writing Have students write a summary of the various techniques for solving quadratic equations using the Square Root Property. Students should provide written examples of each technique.

Tips for New Teachers

Intervention Suggest students write a summary of the various methods that can be used to solve quadratic equations. Ask them which method they prefer to use, and why they like that method the best.

Getting Ready for Lesson 6-5

PREREQUISITE SKILL Lesson 6-5 presents the Quadratic Formula. The first step in evaluating the formula is to evaluate the expression under the radical sign. Use Exercises 69–72 to determine your students' familiarity with evaluating expressions.

Assessment Options

Quiz (Lessons 6-3 and 6-4) is available on p. 369 of the *Chapter 6 Resource Masters.*

Mid-Chapter Test (Lessons 6-1 through 6-4) is available on p. 371 of the *Chapter 6 Resource Masters.*

★ 53. **KENNEL** A kennel owner has 164 feet of fencing with which to enclose a rectangular region. He wants to subdivide this region into three smaller rectangles of equal length, as shown. If the total area to be enclosed is 576 square feet, find the dimensions of the entire enclosed region. (*Hint*: Write an expression for ℓ in terms of w.) **18 ft by 32 ft or 64 ft by 9 ft**

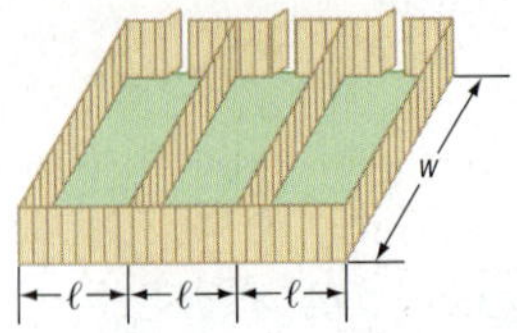

54. **WRITING IN MATH** Answer the question that was posed at the beginning of the lesson. **See margin.**

How can you find the time it takes an accelerating race car to reach the finish line?

Include the following in your answer:

- an explanation of why $t^2 + 22t + 121 = 246$ cannot be solved by factoring, and
- a description of the steps you would take to solve the equation $t^2 + 22t + 121 = 246$.

Standardized Test Practice

55. What is the absolute value of the product of the two solutions for x in $x^2 - 2x - 2 = 0$? **D**

Ⓐ −1 Ⓑ 0 Ⓒ 1 Ⓓ 2

56. For which value of c will the roots of $x^2 + 4x + c = 0$ be real and equal? **D**

Ⓐ 1 Ⓑ 2 Ⓒ 3 Ⓓ 4 Ⓔ 5

Maintain Your Skills

Mixed Review Write a quadratic equation with the given root(s). Write the equation in the form $ax^2 + bx + c = 0$, where a, b, and c are integers. *(Lesson 6-3)*

57. 2, 1 **$x^2 - 3x + 2 = 0$**

58. −3, 9 **$x^2 - 6x - 27 = 0$**

59. $6, \frac{1}{3}$ **$3x^2 - 19x + 6 = 0$**

60. $-\frac{1}{3}, -\frac{3}{4}$ **$12x^2 + 13x + 3 = 0$**

Solve each equation by graphing. If exact roots cannot be found, state the consecutive integers between which the roots are located. *(Lesson 6-2)*

61. $3x^2 = 4 - 8x$ **61. between −4 and −3; between 0 and 1**

62. $x^2 + 48 = 14x$ **6, 8**

63. $2x^2 + 11x = -12$ **$-4, -1\frac{1}{2}$**

64. Write *the seventh root of 5 cubed* using exponents. *(Lesson 5-7)* **$5^{\frac{3}{7}}$**

Solve each system of equations by using inverse matrices. *(Lesson 4-8)*

65. $5x + 3y = -5$
$7x + 5y = -11$ **(2, −5)**

66. $6x + 5y = 8$
$3x - y = 7$ **$\left(\frac{43}{21}, -\frac{6}{7}\right)$**

CHEMISTRY For Exercises 67 and 68, use the following information.
For hydrogen to be a liquid, its temperature must be within 2°C of −257°C. *(Lesson 1-4)*

67. Write an equation to determine the greatest and least temperatures for this substance. **$|x - (-257)| = 2$**

68. Solve the equation. **greatest: −255°C; least: −259°C**

Getting Ready for the Next Lesson **PREREQUISITE SKILL** Evaluate $b^2 - 4ac$ for the given values of a, b, and c.
*(To review **evaluating expressions**, see Lesson 1-1.)*

69. $a = 1, b = 7, c = 3$ **37**

70. $a = 1, b = 2, c = 5$ **−16**

71. $a = 2, b = -9, c = -5$ **121**

72. $a = 4, b = -12, c = 9$ **0**

Answer

54. To find the distance traveled by the accelerating race car in the given situation, you must solve the equation $t^2 + 22t + 121 = 246$ or $t^2 + 22t - 125 = 0$. Answers should include the following.

- Since the expression $t^2 + 22t - 125$ is prime, the solutions of $t^2 + 22t + 121 = 246$ cannot be obtained by factoring.
- Rewrite $t^2 + 22t + 121$ as $(t + 11)^2$. Solve $(t + 11)^2 = 246$ by applying the Square Root Property. Then, subtract 11 from each side. Using a calculator, the two solutions are about 4.7 or −26.7. Since time cannot be negative, the driver takes about 4.7 seconds to reach the finish line.

6-5 The Quadratic Formula and the Discriminant

What You'll Learn

- Solve quadratic equations by using the Quadratic Formula.
- Use the discriminant to determine the number and type of roots of a quadratic equation.

Vocabulary

- Quadratic Formula
- discriminant

How is blood pressure related to age?

As people age, their arteries lose their elasticity, which causes blood pressure to increase. For healthy women, average systolic blood pressure is estimated by $P = 0.01A^2 + 0.05A + 107$, where P is the average blood pressure in millimeters of mercury (mm Hg) and A is the person's age. For healthy men, average systolic blood pressure is estimated by $P = 0.006A^2 - 0.02A + 120$.

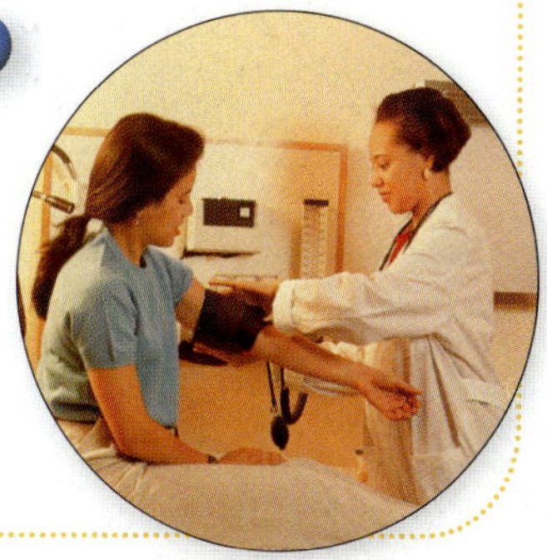

QUADRATIC FORMULA You have seen that exact solutions to some quadratic equations can be found by graphing, by factoring, or by using the Square Root Property. While completing the square can be used to solve any quadratic equation, the process can be tedious if the equation contains fractions or decimals. Fortunately, a formula exists that can be used to solve any quadratic equation of the form $ax^2 + bx + c = 0$. This formula can be derived by solving the general form of a quadratic equation.

$ax^2 + bx + c = 0$	General quadratic equation
$x^2 + \frac{b}{a}x + \frac{c}{a} = 0$	Divide each side by a.
$x^2 + \frac{b}{a}x = -\frac{c}{a}$	Subtract $\frac{c}{a}$ from each side.
$x^2 + \frac{b}{a}x + \frac{b^2}{4a^2} = -\frac{c}{a} + \frac{b^2}{4a^2}$	Complete the square.
$\left(x + \frac{b}{2a}\right)^2 = \frac{b^2 - 4ac}{4a^2}$	Factor the left side. Simplify the right side.
$x + \frac{b}{2a} = \pm\frac{\sqrt{b^2 - 4ac}}{2a}$	Square Root Property
$x = -\frac{b}{2a} \pm \frac{\sqrt{b^2 - 4ac}}{2a}$	Subtract $\frac{b}{2a}$ from each side.
$x = \frac{-b \pm \sqrt{b^2 - 4ac}}{2a}$	Simplify.

This equation is known as the **Quadratic Formula**.

Study Tip

Reading Math
The Quadratic Formula is read *x equals the opposite of b, plus or minus the square root of b squared minus 4ac, all divided by 2a.*

Key Concept — *Quadratic Formula*

The solutions of a quadratic equation of the form $ax^2 + bx + c = 0$, where $a \neq 0$, are given by the following formula.

$$x = \frac{-b \pm \sqrt{b^2 - 4ac}}{2a}$$

6-5 Lesson Notes

1 Focus

5-Minute Check Transparency 6-5 Use as a quiz or review of Lesson 6-4.

Mathematical Background notes are available for this lesson on p. 284D.

Building on Prior Knowledge

In Lesson 6-4, students solved quadratic equations by completing the square. In this lesson, students generalize this procedure as they complete the square for the general quadratic equation to derive the Quadratic Formula.

How is blood pressure related to age?

Ask students:

- As the value of A increases in these equations, what happens to the value of P? **It increases.**
- Which way do the parabolas open that are the graphs of these equations? **upward**

Resource Manager

Workbook and Reproducible Masters

Chapter 6 Resource Masters
- Study Guide and Intervention, pp. 337–338
- Skills Practice, p. 339
- Practice, p. 340
- Reading to Learn Mathematics, p. 341
- Enrichment, p. 342

Graphing Calculator and Spreadsheet Masters, p. 38

Transparencies

5-Minute Check Transparency 6-5
Answer Key Transparencies

Technology

Alge2PASS: Tutorial Plus, Lessons 11, 12
Interactive Chalkboard

2 Teach

QUADRATIC FORMULA

In-Class Examples

Power Point®

1 Solve $x^2 - 8x = 33$ by using the Quadratic Formula. **−3, 11**

Teaching Tip Encourage students to write down the values of *a*, *b*, and *c* from the standard form of the quadratic equation before they begin substituting into the formula.

2 Solve $x^2 - 34x + 289 = 0$ by using the Quadratic Formula. **17**

Example 1 Two Rational Roots

Solve $x^2 - 12x = 28$ by using the Quadratic Formula.

First, write the equation in the form $ax^2 + bx + c = 0$ and identify a, b, and c.

$$x^2 - 12x = 28 \longrightarrow \begin{matrix} ax^2 + bx + c = 0 \\ \downarrow \quad \downarrow \quad \downarrow \\ 1x^2 - 12x - 28 = 0 \end{matrix}$$

Then, substitute these values into the Quadratic Formula.

$x = \dfrac{-b \pm \sqrt{b^2 - 4ac}}{2a}$ Quadratic Formula

$x = \dfrac{-(-12) \pm \sqrt{(-12)^2 - 4(1)(-28)}}{2(1)}$ Replace *a* with 1, *b* with −12, and *c* with −28.

$x = \dfrac{12 \pm \sqrt{144 + 112}}{2}$ Simplify.

$x = \dfrac{12 \pm \sqrt{256}}{2}$ Simplify.

$x = \dfrac{12 \pm 16}{2}$ $\sqrt{256} = 16$

$x = \dfrac{12 + 16}{2}$ or $x = \dfrac{12 - 16}{2}$ Write as two equations.

$= 14$ $\quad = -2$ Simplify.

The solutions are −2 and 14. Check by substituting each of these values into the original equation.

Study Tip

Quadratic Formula
Although factoring may be an easier method to solve the equations in Examples 1 and 2, the Quadratic Formula can be used to solve any quadratic equation.

When the value of the radicand in the Quadratic Formula is 0, the quadratic equation has exactly one rational root.

Example 2 One Rational Root

Solve $x^2 + 22x + 121 = 0$ by using the Quadratic Formula.

Identify a, b, and c. Then, substitute these values into the Quadratic Formula.

$x = \dfrac{-b \pm \sqrt{b^2 - 4ac}}{2a}$ Quadratic Formula

$x = \dfrac{-(22) \pm \sqrt{(22)^2 - 4(1)(121)}}{2(1)}$ Replace *a* with 1, *b* with 22, and *c* with 121.

$x = \dfrac{-22 \pm \sqrt{0}}{2}$ Simplify.

$x = \dfrac{-22}{2}$ or −11 $\sqrt{0} = 0$

The solution is −11.

CHECK A graph of the related function shows that there is one solution at $x = -11$.

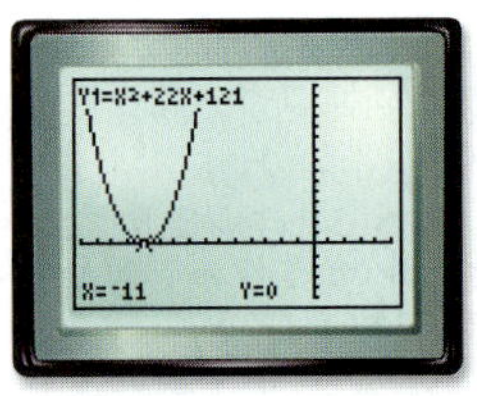

[−15, 5] scl: 1 by [−5, 15] scl: 1

Teacher to Teacher

Lori Haldorson & Cathy Hokkanen — Blaine H.S., Blaine, MN

"To help students memorize the Quadratic Formula, we sing it to the tune of 'Pop Goes the Weasel'."

You can express irrational roots exactly by writing them in radical form.

Example 3 Irrational Roots

Solve $2x^2 + 4x - 5 = 0$ by using the Quadratic Formula.

$x = \frac{-b \pm \sqrt{b^2 - 4ac}}{2a}$ Quadratic Formula

$x = \frac{-(4) \pm \sqrt{(4)^2 - 4(2)(-5)}}{2(2)}$ Replace a with 2, b with 4, and c with -5.

$x = \frac{-4 \pm \sqrt{56}}{4}$ Simplify.

$x = \frac{-4 \pm 2\sqrt{14}}{4}$ or $\frac{-2 \pm \sqrt{14}}{2}$ $\sqrt{56} = \sqrt{4 \cdot 14}$ or $2\sqrt{14}$

The exact solutions are $\frac{-2 - \sqrt{14}}{2}$ and $\frac{-2 + \sqrt{14}}{2}$. The approximate solutions are -2.9 and 0.9.

CHECK Check these results by graphing the related quadratic function, $y = 2x^2 + 4x - 5$. Using the ZERO function of a graphing calculator, the approximate zeros of the related function are -2.9 and 0.9.

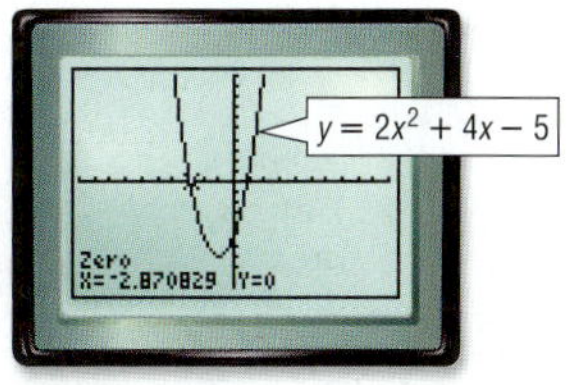

[–10, 10] scl: 1 by [–10, 10] scl: 1

When using the Quadratic Formula, if the radical contains a negative value, the solutions will be complex. Complex solutions always appear in conjugate pairs.

Example 4 Complex Roots

Solve $x^2 - 4x = -13$ by using the Quadratic Formula.

$x = \frac{-b \pm \sqrt{b^2 - 4ac}}{2a}$ Quadratic Formula

$x = \frac{-(-4) \pm \sqrt{(-4)^2 - 4(1)(13)}}{2(1)}$ Replace a with 1, b with -4, and c with 13.

$x = \frac{4 \pm \sqrt{-36}}{2}$ Simplify.

$x = \frac{4 \pm 6i}{2}$ $\sqrt{-36} = \sqrt{36(-1)}$ or $6i$

$x = 2 \pm 3i$ Simplify.

The solutions are the complex numbers $2 + 3i$ and $2 - 3i$.

A graph of the related function shows that the solutions are complex, but it cannot help you find them.

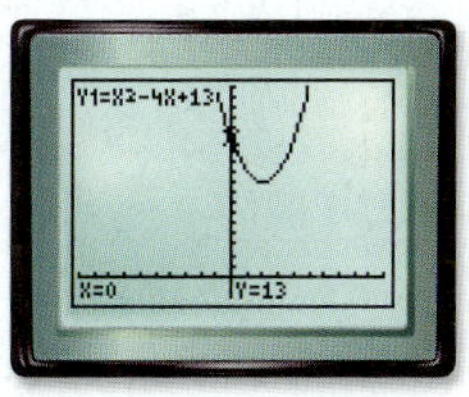

[–15, 5] scl: 1 by [–2, 18] scl: 1

Study Tip

Using the Quadratic Formula
Remember that to correctly identify a, b, and c for use in the Quadratic Formula, the equation must be written in the form $ax^2 + bx + c = 0$.

www.algebra2.com/extra_examples

In-Class Examples

Power Point®

Teaching Tip Some students may have commented that the equations in Examples 1 and 2 could have been solved by factoring. Before beginning Example 3, stress that many quadratic equations cannot easily be solved by factoring. State that the quadratic equation presented in Example 3 is such an equation. Emphasize that the Quadratic Formula provides a way to find the roots for *any* quadratic equation.

3 Solve $x^2 - 6x + 2 = 0$ by using the Quadratic Formula. **$3 \pm \sqrt{7}$, or approximately 0.4 and 5.6**

Teaching Tip Remind students that conjugate pairs are two complex numbers of the form $a + bi$ and $a - bi$.

4 Solve $x^2 + 13 = 6x$ by using the Quadratic Formula. **$3 \pm 2i$**

Concept Check

Real Roots or Imaginary Roots Ask students to look back at the first four examples in this lesson and see how they might predict whether the roots will be real or imaginary. **If the values of a and c have the same sign and 4 times their product is greater than the square of the value of b, then the roots will be imaginary.**

ROOTS AND THE DISCRIMINANT

In-Class Example Power Point®

5 Find the value of the discriminant for each quadratic equation. Then describe the number and type of roots for the equation.

a. $x^2 + 6x + 9 = 0$
0; one rational root

b. $x^2 + 3x + 5 = 0$
−11; two complex roots

c. $x^2 + 8x - 4 = 0$
80; two irrational roots

d. $x^2 - 11x + 10 = 0$
81; two rational roots

CHECK To check complex solutions, you must substitute them into the original equation. The check for $2 + 3i$ is shown below.

$x^2 - 4x = -13$	Original equation
$(2 + 3i)^2 - 4(2 + 3i) \stackrel{?}{=} -13$	$x = 2 + 3i$
$4 + 12i + 9i^2 - 8 - 12i \stackrel{?}{=} -13$	Sum of a square; Distributive Property
$-4 + 9i^2 \stackrel{?}{=} -13$	Simplify.
$-4 - 9 = -13$ ✓	$i^2 = -1$

Study Tip

Reading Math
Remember that the solutions of an equation are called *roots*.

ROOTS AND THE DISCRIMINANT

In Examples 1, 2, 3, and 4, observe the relationship between the value of the expression under the radical and the roots of the quadratic equation. The expression $b^2 - 4ac$ is called the **discriminant**.

$$x = \frac{-b \pm \sqrt{b^2 - 4ac}}{2a} \quad \longleftarrow \text{discriminant}$$

The value of the discriminant can be used to determine the number and type of roots of a quadratic equation.

Key Concept — *Discriminant*

Consider $ax^2 + bx + c = 0$.

Value of Discriminant	Type and Number of Roots	Example of Graph of Related Function
$b^2 - 4ac > 0$; $b^2 - 4ac$ is a perfect square.	2 real, rational roots	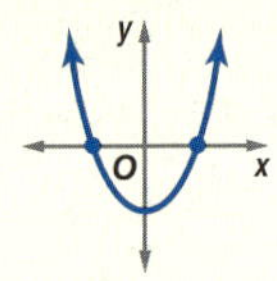
$b^2 - 4ac > 0$; $b^2 - 4ac$ is *not* a perfect square.	2 real, irrational roots	
$b^2 - 4ac = 0$	1 real, rational root	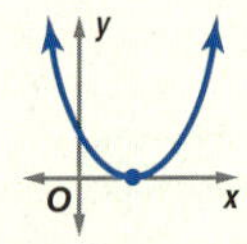
$b^2 - 4ac < 0$	2 complex roots	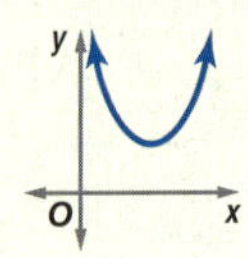

Study Tip

Using the Discriminant
The discriminant can help you check the solutions of a quadratic equation. Your solutions must match in number and in type to those determined by the discriminant.

Example 5 *Describe Roots*

Find the value of the discriminant for each quadratic equation. Then describe the number and type of roots for the equation.

a. $9x^2 - 12x + 4 = 0$

$a = 9, b = -12, c = 4$

$b^2 - 4ac = (-12)^2 - 4(9)(4)$
$= 144 - 144$
$= 0$

The discriminant is 0, so there is one rational root.

b. $2x^2 + 16x + 33 = 0$

$a = 2, b = 16, c = 33$

$b^2 - 4ac = (16)^2 - 4(2)(33)$
$= 256 - 264$
$= -8$

The discriminant is negative, so there are two complex roots.

DAILY INTERVENTION — **Differentiated Instruction**

Logical Have students use their classification skills to create a classroom poster listing the four different types of roots that can result when solving a quadratic equation. Each listing should include a sample equation that results in that type of roots and an explanation of how the value of the discriminant is indicative of the root type. Graphs like those shown on p. 316 can be added to the poster.

c. $-5x^2 + 8x - 1 = 0$

$a = -5, b = 8, c = -1$

$$b^2 - 4ac = (8)^2 - 4(-5)(-1)$$
$$= 64 - 20$$
$$= 44$$

The discriminant is 44, which is not a perfect square. Therefore, there are two irrational roots.

d. $-7x + 15x^2 - 4 = 0$

$a = 15, b = -7, c = -4$

$$b^2 - 4ac = (-7)^2 - 4(15)(-4)$$
$$= 49 + 240$$
$$= 289 \text{ or } 17^2$$

The discriminant is 289, which is a perfect square. Therefore, there are two rational roots.

You have studied a variety of methods for solving quadratic equations. The table below summarizes these methods.

Concept Summary — *Solving Quadratic Equations*

Method	Can be Used	When to Use
Graphing	sometimes	Use only if an exact answer is not required. Best used to check the reasonableness of solutions found algebraically.
Factoring	sometimes	Use if the constant term is 0 or if the factors are easily determined. **Example** $x^2 - 3x = 0$
Square Root Property	sometimes	Use for equations in which a perfect square is equal to a constant. **Example** $(x + 13)^2 = 9$
Completing the Square	always	Useful for equations of the form $x^2 + bx + c = 0$, where b is even. **Example** $x^2 + 14x - 9 = 0$
Quadratic Formula	always	Useful when other methods fail or are too tedious. **Example** $3.4x^2 - 2.5x + 7.9 = 0$

Check for Understanding

Concept Check

1. **OPEN ENDED** Sketch the graph of a quadratic equation whose discriminant is
   a. positive. b. negative. c. zero. **a–c. See margin.**

2. **Explain** why the roots of a quadratic equation are complex if the value of the discriminant is less than 0.

2. The square root of a negative number is a complex number.

3. **Describe** the relationship that must exist between a, b, and c in the equation $ax^2 + bx + c = 0$ in order for the equation to have exactly one solution.
   $b^2 - 4ac$ must equal 0.

Guided Practice

GUIDED PRACTICE KEY	
Exercises	Examples
4–7	1–5
8–11	1–4
12, 13	1–4

Complete parts a–c for each quadratic equation.

a. Find the value of the discriminant. **4–7. See margin.**

b. Describe the number and type of roots.

c. Find the exact solutions by using the Quadratic Formula.

4. $8x^2 + 18x - 5 = 0$
5. $2x^2 - 4x + 1 = 0$
6. $4x^2 + 4x + 1 = 0$
7. $x^2 + 3x + 8 = 5$

3 Practice/Apply

Study Notebook

Have students—
- add the definitions/examples of the vocabulary terms to their Vocabulary Builder worksheets for Chapter 6.
- copy the information provided in the Concept Summary on p. 317 into their notebook.
- include any other item(s) that they find helpful in mastering the skills in this lesson.

About the Exercises...

Organization by Objective
- **Quadratic Formula:** 14–39, 42–44
- **Roots and the Discriminant:** 14–27, 40, 41

Odd/Even Assignments

Exercises 14–39 are structured so that students practice the same concepts whether they are assigned odd or even problems.

Assignment Guide

Basic: 15–25 odd, 29–39 odd, 40, 41, 45–66

Average: 15–39 odd, 40–43, 45–66

Advanced: 14–38 even, 42–60 (optional: 61–66)

Answers

4a. 484 4b. 2 rational

4c. $\frac{1}{4}, -\frac{5}{2}$

5a. 8 5b. 2 irrational

5c. $\frac{2 \pm \sqrt{2}}{2}$

6a. 0 6b. one rational

6c. $-\frac{1}{2}$

7a. −3 7b. 2 complex

7c. $\frac{-3 \pm i\sqrt{3}}{2}$

1a. Sample answer:

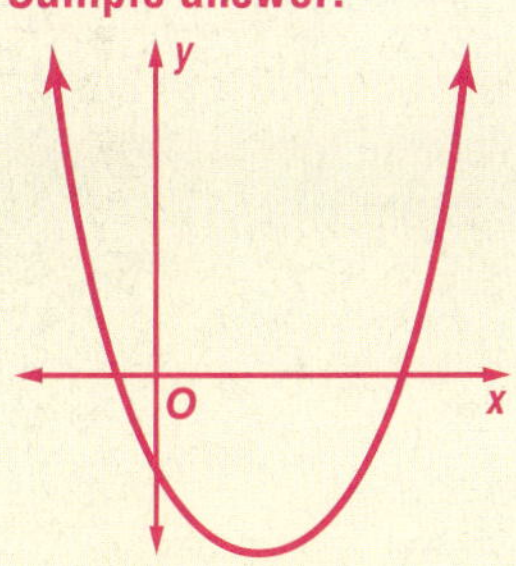

1b. Sample answer:

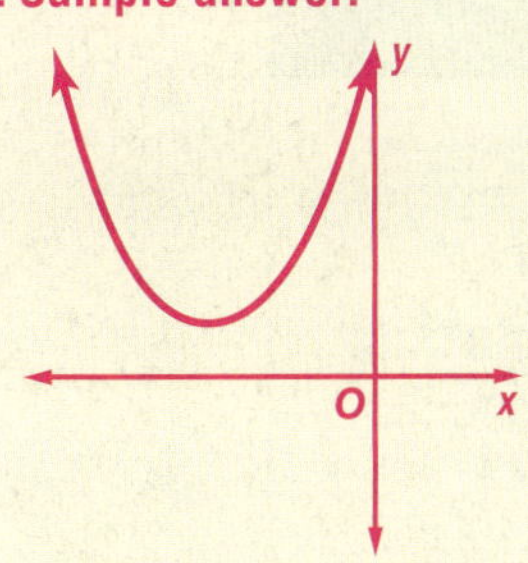

1c. Sample answer:

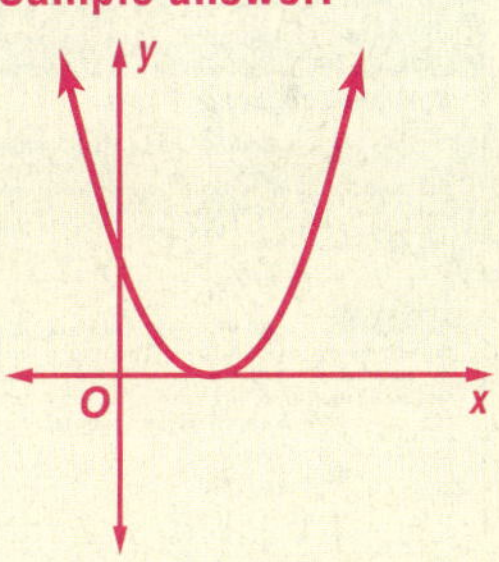

Lesson 6-5 The Quadratic Formula and the Discriminant 317

Study Guide and Intervention, p. 337 (shown) and p. 338

Quadratic Formula The **Quadratic Formula** can be used to solve *any* quadratic equation once it is written in the form $ax^2 + bx + c = 0$.

Quadratic Formula	The solutions of $ax^2 + bx + c = 0$, with $a \neq 0$, are given by $x = \frac{-b \pm \sqrt{b^2 - 4ac}}{2a}$.

Example **Solve $x^2 - 5x = 14$ by using the Quadratic Formula.**

Rewrite the equation as $x^2 - 5x - 14 = 0$.

$x = \frac{-b \pm \sqrt{b^2 - 4ac}}{2a}$ Quadratic Formula

$= \frac{-(-5) \pm \sqrt{(-5)^2 - 4(1)(-14)}}{2(1)}$ Replace *a* with 1, *b* with −5, and *c* with −14.

$= \frac{5 \pm \sqrt{81}}{2}$ Simplify.

$= \frac{5 \pm 9}{2}$

$= 7$ or -2

The solutions are −2 and 7.

Exercises

Solve each equation by using the Quadratic Formula.

1. $x^2 + 2x - 35 = 0$ **5, −7**
2. $x^2 + 10x + 24 = 0$ **−4, −6**
3. $x^2 - 11x + 24 = 0$ **3, 8**
4. $4x^2 + 19x - 5 = 0$ $\frac{1}{4}, -5$
5. $14x^2 + 9x + 1 = 0$ $-\frac{1}{2}, -\frac{1}{7}$
6. $2x^2 - x - 15 = 0$ $3, -\frac{5}{2}$
7. $3x^2 + 5x = 2$ $-2, \frac{1}{3}$
8. $2y^2 + y - 15 = 0$ $\frac{5}{2}, -3$
9. $3x^2 - 16x + 16 = 0$ $4, \frac{4}{3}$
10. $8x^2 + 6x - 9 = 0$ $-\frac{3}{2}, \frac{3}{4}$
11. $r^2 - \frac{3r}{5} + \frac{2}{25} = 0$ $\frac{2}{5}, \frac{1}{5}$
12. $x^2 - 10x - 50 = 0$ $5 \pm 5\sqrt{3}$
13. $x^2 + 6x - 23 = 0$ $-3 \pm 4\sqrt{2}$
14. $4x^2 - 12x - 63 = 0$ $\frac{3 \pm 6\sqrt{2}}{2}$
15. $x^2 - 6x + 21 = 0$ $3 \pm 2i\sqrt{3}$

Skills Practice, p. 339 and Practice, p. 340 (shown)

Complete parts a–c for each quadratic equation.

a. Find the value of the discriminant.

b. Describe the number and type of roots.

c. Find the exact solutions by using the Quadratic Formula.

1. $x^2 - 16x + 64 = 0$ **0; 1 rational; 8**
2. $x^2 = 3x$ **9; 2 rational; 0, 3**
3. $9x^2 - 24x + 16 = 0$ **0; 1 rational;** $\frac{4}{3}$
4. $x^2 - 3x = 40$ **169; 2 rational; −5, 8**
5. $3x^2 + 9x - 2 = 0$ **105; 2 irrational;** $\frac{-9 \pm \sqrt{105}}{6}$
6. $2x^2 + 7x = 0$ **49; 2 rational;** $0, -\frac{7}{2}$
7. $5x^2 - 2x + 4 = 0$ **−76; 2 complex;** $\frac{1 \pm i\sqrt{19}}{5}$
8. $12x^2 - x - 6 = 0$ **289; 2 rational;** $\frac{3}{4}, -\frac{2}{3}$
9. $7x^2 + 6x + 2 = 0$ **−20; 2 complex;** $\frac{-3 \pm i\sqrt{5}}{7}$
10. $12x^2 + 2x - 4 = 0$ **196; 2 rational;** $\frac{1}{2}, -\frac{2}{3}$
11. $6x^2 - 2x - 1 = 0$ **28; 2 irrational;** $\frac{1 \pm \sqrt{7}}{6}$
12. $x^2 + 3x + 6 = 0$ **−15; 2 complex;** $\frac{-3 \pm i\sqrt{15}}{2}$
13. $4x^2 - 3x^2 - 6 = 0$ **105; 2 irrational;** $\frac{3 \pm \sqrt{105}}{8}$
14. $16x^2 - 8x + 1 = 0$ **0; 1 rational;** $\frac{1}{4}$
15. $2x^2 - 5x - 6 = 0$ **73; 2 irrational;** $\frac{5 \pm \sqrt{73}}{4}$

Solve each equation by using the method of your choice. Find exact solutions.

16. $7x^2 - 5x = 0$ $0, \frac{5}{7}$
17. $4x^2 - 9 = 0$ $\pm\frac{3}{2}$
18. $3x^2 + 8x = 3$ $\frac{1}{3}, -3$
19. $x^2 - 21 = 4x$ **−3, 7**
20. $3x^2 - 13x + 4 = 0$ $\frac{1}{3}, 4$
21. $15x^2 + 22x = -8$ $-\frac{2}{3}, -\frac{4}{5}$
22. $x^2 - 6x + 3 = 0$ $3 \pm \sqrt{6}$
23. $x^2 - 14x + 53 = 0$ $7 \pm 2i$
24. $3x^2 = -54$ $\pm 3i\sqrt{2}$
25. $25x^2 - 20x - 6 = 0$ $\frac{2 \pm \sqrt{10}}{5}$
26. $4x^2 - 4x + 17 = 0$ $\frac{1 \pm 4i}{2}$
27. $8x - 1 = 4x^2$ $\frac{2 \pm \sqrt{3}}{2}$
28. $x^2 = 4x - 15$ $2 \pm i\sqrt{11}$
29. $4x^2 - 12x + 7 = 0$ $\frac{3 \pm \sqrt{2}}{2}$

30. **GRAVITATION** The height $h(t)$ in feet of an object t seconds after it is propelled straight up from the ground with an initial velocity of 60 feet per second is modeled by the equation $h(t) = -16t^2 + 60t$. At what times will the object be at a height of 56 feet? **1.75 s, 2 s**

31. **STOPPING DISTANCE** The formula $d = 0.05s^2 + 1.1s$ estimates the minimum stopping distance d in feet for a car traveling s miles per hour. If a car stops in 200 feet, what is the

Reading to Learn Mathematics, p. 341 ELL

Pre-Activity **How is blood pressure related to age?**

Read the introduction to Lesson 6-5 at the top of page 313 in your textbook.

Describe how you would calculate your normal blood pressure using one of the formulas in your textbook.

Sample answer: Substitute your age for *A* in the appropriate formula (for females or males) and evaluate the expression.

Reading the Lesson

1. a. Write the Quadratic Formula. $x = \frac{-b \pm \sqrt{b^2 - 4ac}}{2a}$

   b. Identify the values of a, b, and c that you would use to solve $2x^2 - 5x = -7$, but do not actually solve the equation.

   $a =$ **2** $b =$ **−5** $c =$ **7**

2. Suppose that you are solving four quadratic equations with rational coefficients and have found the value of the discriminant for each equation. In each case, give the number of roots and describe the type of roots that the equation will have.

Value of Discriminant	Number of Roots	Type of Roots
64	2	real, rational
−8	2	complex
21	2	real, irrational
0	1	real, rational

Helping You Remember

3. How can looking at the Quadratic Formula help you remember the relationships between the value of the discriminant and the number of roots of a quadratic equation and whether the roots are real or complex?

Sample answer: The discriminant is the expression under the radical in the Quadratic Formula. Look at the Quadratic Formula and consider what happens when you take the principal square root of $b^2 - 4ac$ and apply ± in front of the result. If $b^2 - 4ac$ is positive, its principal square root will be a positive number and applying ± will give two different real solutions, which may be rational or irrational. If $b^2 - 4ac = 0$, its principal square root is 0, so applying ± in the Quadratic Formula will only lead to one solution, which will be rational (assuming *a*, *b*, and *c* are integers). If $b^2 - 4ac$ is negative, since the square roots of negative numbers are not real numbers, you will get two complex roots, corresponding to the + and − in the ± symbol.

Solve each equation using the method of your choice. Find exact solutions.

8. $x^2 + 8x = 0$ **0, −8**
9. $x^2 + 5x + 6 = 0$ **−3, −2**
10. $x^2 - 2x - 2 = 0$ $1 \pm \sqrt{3}$
11. $4x^2 + 20x + 25 = -2$ $\frac{-5 \pm i\sqrt{2}}{2}$

Application

PHYSICS **For Exercises 12 and 13, use the following information.**

The height $h(t)$ in feet of an object t seconds after it is propelled straight up from the ground with an initial velocity of 85 feet per second is modeled by $h(t) = -16t^2 + 85t$.

12. at about 0.7 s and again at about 4.6 s

12. When will the object be at a height of 50 feet?
13. Will the object ever reach a height of 120 feet? Explain your reasoning. **No; see margin for explanation.**

★ **indicates increased difficulty**

Practice and Apply

Homework Help

For Exercises	See Examples
14–27	1-5
28–39, 42–44	1-4
40–41	5

Extra Practice
See page 841.

Complete parts a–c for each quadratic equation.

a. Find the value of the discriminant. **14–27. See margin.**

b. Describe the number and type of roots.

c. Find the exact solutions by using the Quadratic Formula.

14. $x^2 + 3x - 3 = 0$
15. $x^2 - 16x + 4 = 0$
16. $x^2 - 2x + 5 = 0$
17. $x^2 - x + 6 = 0$
18. $-12x^2 + 5x + 2 = 0$
19. $-3x^2 - 5x + 2 = 0$
20. $x^2 + 4x + 3 = 4$
21. $2x - 5 = -x^2$
22. $9x^2 - 6x - 4 = -5$
23. $25 + 4x^2 = -20x$
24. $4x^2 + 7 = 9x$
25. $3x + 6 = -6x^2$

★ 26. $\frac{3}{4}x^2 - \frac{1}{3}x - 1 = 0$

★ 27. $0.4x^2 + x - 0.3 = 0$

28–39. See pp. 343A–343F.

Solve each equation by using the method of your choice. Find exact solutions.

28. $x^2 - 30x - 64 = 0$
29. $7x^2 + 3 = 0$
30. $x^2 - 4x + 7 = 0$
31. $2x^2 + 6x - 3 = 0$
32. $4x^2 - 8 = 0$
33. $4x^2 + 81 = 36x$
34. $-4(x + 3)^2 = 28$
35. $3x^2 - 10x = 7$
36. $x^2 + 9 = 8x$
37. $10x^2 + 3x = 0$
38. $2x^2 - 12x + 7 = 5$
39. $21 = (x - 2)^2 + 5$

More About. . .

Bridges

The Golden Gate, located in San Francisco, California, is the tallest bridge in the world, with its towers extending 746 feet above the water and the floor of the bridge extending 220 feet above water.

Source: www.goldengatebridge.org

BRIDGES **For Exercises 40 and 41, use the following information.**

The supporting cables of the Golden Gate Bridge approximate the shape of a parabola. The parabola can be modeled by the quadratic function $y = 0.00012x^2 + 6$, where x represents the distance from the axis of symmetry and y represents the height of the cables. The related quadratic equation is $0.00012x^2 + 6 = 0$.

40. Calculate the value of the discriminant. **≈ −0.00288**
41. What does the discriminant tell you about the supporting cables of the Golden Gate Bridge? **See pp. 343A–343F.**

FOOTBALL **For Exercises 42 and 43, use the following information.**

The average NFL salary $A(t)$ (in thousands of dollars) from 1975 to 2000 can be estimated using the function $A(t) = 2.3t^2 - 12.4t + 73.7$, where t is the number of years since 1975. **42. D: $0 \le t \le 25$, R: $73.7 \le A(t) \le 1201.2$**

42. Determine a domain and range for which this function makes sense.
43. According to this model, in what year did the average salary first exceed 1 million dollars? **1998**

Online Research **Data Update** What is the current average NFL salary? How does this average compare with the average given by the function used in Exercises 42 and 43? Visit www.algebra2.com/data_update to learn more.

Enrichment, p. 342

Sum and Product of Roots

Sometimes you may know the roots of a quadratic equation without knowing the equation itself. Using your knowledge of factoring to solve an equation, you can work backward to find the quadratic equation. The rule for finding the sum and product of roots is as follows:

Sum and Product of Roots	If the roots of $ax^2 + bx + c = 0$, with $a \neq 0$, are s_1 and s_2, then $s_1 + s_2 = -\frac{b}{a}$ and $s_1 \cdot s_2 = \frac{c}{a}$.

Example **A road with an initial gradient, or slope, of 3% can be represented by the formula $y = ax^2 + 0.03x + c$, where y is the elevation and x is the distance along the curve. Suppose the elevation of the road is 1105 feet at points 200 feet and 1000 feet along the curve. You can find the equation of the transition curve. Equations of transition curves are used by civil engineers to design smooth and safe roads.**

The roots are $x = 3$ and $x = -8$.

$3 + (-8) = -5$ Add the roots.

$3(-8) = -24$ Multiply the roots.

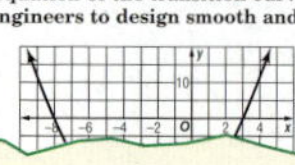

− 24 = 0

Answer

13. The discriminant of $-16t^2 + 85t = 120$ is −455, indicating that the equation has no real solutions.

44. **HIGHWAY SAFETY** Highway safety engineers can use the formula $d = 0.05s^2 + 1.1s$ to estimate the minimum stopping distance d in feet for a vehicle traveling s miles per hour. If a car is able to stop after 125 feet, what is the fastest it could have been traveling when the driver first applied the brakes? **about 40.2 mph**

45. **CRITICAL THINKING** Find all values of k such that $x^2 - kx + 9 = 0$ has

a. one real root. $k = \pm 6$ b. two real roots. $k < -6$ or $k > 6$ c. no real roots. $-6 < k < 6$

46. **WRITING IN MATH** Answer the question that was posed at the beginning of the lesson. **See pp. 343A–343F.**

How is blood pressure related to age?

Include the following in your answer:
- an expression giving the average systolic blood pressure for a person of your age, and
- an example showing how you could determine A in either formula given a specific value of P.

Standardized Test Practice

47. If $2x^2 - 5x - 9 = 0$, then x could equal which of the following? **D**

(A) −1.12 (B) 1.54 (C) 2.63 (D) 3.71

48. Which best describes the nature of the roots of the equation $x^2 - 3x + 4 = 0$? **C**

(A) real and equal (B) real and unequal
(C) complex (D) real and complex

Maintain Your Skills

Mixed Review

Solve each equation by using the Square Root Property. *(Lesson 6-4)*

49. $x^2 + 18x + 81 = 25$ $-14, -4$
50. $x^2 - 8x + 16 = 7$ $4 \pm \sqrt{7}$
51. $4x^2 - 4x + 1 = 8$ $\frac{1 \pm 2\sqrt{2}}{2}$

Solve each equation by factoring. *(Lesson 6-3)*

52. $4x^2 + 8x = 0$ $-2, 0$
53. $x^2 - 5x = 14$ $-2, 7$
54. $3x^2 + 10 = 17x$ $\frac{2}{3}, 5$

Simplify. *(Lesson 5-5)*

55. $\sqrt{a^8b^{20}}$ a^4b^{10}
56. $\sqrt{100p^{12}q^2}$ $10p^6|q|$
57. $\sqrt[3]{64b^6c^6}$ $4b^2c^2$

58. **ANIMALS** The fastest-recorded physical action of any living thing is the wing beat of the common midge. This tiny insect normally beats its wings at a rate of 133,000 times per minute. At this rate, how many times would the midge beat its wings in an hour? Write your answer in scientific notation. *(Lesson 5-1)* 7.98×10^6

Solve each system of inequalities. *(Lesson 3-3)* **59–60. See pp. 343A–343F.**

59. $x + y \le 9$
$x - y \le 3$
$y - x \ge 4$

60. $x \ge 1$
$y \le -1$
$y \le x$

Getting Ready for the Next Lesson

PREREQUISITE SKILL **State whether each trinomial is a perfect square. If it is, factor it.** *(To review **perfect square trinomials**, see Lesson 5-4.)*

61. $x^2 - 5x - 10$ **no**
62. $x^2 - 14x + 49$ **yes;** $(x - 7)^2$
63. $4x^2 + 12x + 9$ **yes;** $(2x + 3)^2$
64. $25x^2 + 20x + 4$ **yes;** $(5x + 2)^2$
65. $9x^2 - 12x + 16$ **no**
66. $36x^2 - 60x + 25$ **yes;** $(6x - 5)^2$

www.algebra2.com/self_check_quiz

4 Assess

Open-Ended Assessment

Modeling Ask students to sketch graphs of parabolas that illustrate each of the four types of roots for quadratic equations. Have them label each sketch with the type of value the discriminant of the corresponding quadratic equation would have.

Getting Ready for Lesson 6-6

PREREQUISITE SKILL Lesson 6-6 presents the analysis of the graphs of quadratic functions. To graph a quadratic function, it is helpful if the function is written in vertex form, which often requires students to complete the square. Recognition of perfect square trinomials is an important part of completing the square. Exercises 61–66 should be used to determine your students' familiarity with perfect square trinomials.

Answers

14a. 21
14b. 2 irrational
14c. $\frac{-3 \pm \sqrt{21}}{2}$
15a. 240
15b. 2 irrational
15c. $8 \pm 2\sqrt{15}$
16a. −16
16b. 2 complex
16c. $1 \pm 2i$
17a. −23
17b. 2 complex
17c. $\frac{1 \pm i\sqrt{23}}{2}$
18a. 121
18b. 2 rational
18c. $-\frac{1}{4}, \frac{2}{3}$
19a. 49
19b. 2 rational
19c. $-2, \frac{1}{3}$
20a. 20
20b. 2 irrational
20c. $-2 \pm \sqrt{5}$
21a. 24
21b. 2 irrational
21c. $-1 \pm \sqrt{6}$
22a. 0
22b. one rational
22c. $\frac{1}{3}$
23a. 0
23b. one rational
23c. $-\frac{5}{2}$
24a. −31
24b. 2 complex
24c. $\frac{9 \pm i\sqrt{31}}{8}$
25a. −135
25b. 2 complex
25c. $\frac{-1 \pm i\sqrt{15}}{4}$
26a. $\frac{28}{9}$
26b. 2 irrational
26c. $\frac{2 \pm 4\sqrt{7}}{9}$
27a. 1.48
27b. 2 irrational
27c. $\frac{-1 \pm 2\sqrt{0.37}}{0.8}$

Graphing Calculator Investigation

A Preview of Lesson 6-6

Getting Started

Know Your Calculator Students can use the calculator to confirm the location of the vertex of each parabola. A good way to do this is to change the window settings for the x-axis to $[-9.4, 9.4]$. Then use the Trace feature and symmetry properties of parabolas to check that the graph is symmetric with respect to the vertical line through the point that appears to be the vertex.

Teach

- Ask students to describe the three constants (a, h, and k) in the general form of a quadratic equation $y = a(x - h)^2 + k$. **Sample answer: a: coefficient of the squared quantity involving the variable x; h: value subtracted from x in the quantity being squared and then multiplied by a; k: value added at the end**
- Before discussing the examples, have students make a conjecture about the effect of the value of each of the constants a, h, and k on the graph of the parabola.
- After completing the discussion of Example 3, have students compare the conjectures they made at the beginning of the investigation to the knowledge they gained during the discussions.
- Have students complete Exercises 1–15.

Graphing Calculator

A Preview of Lesson 6-6

Families of Parabolas

The general form of a quadratic equation is $y = a(x - h)^2 + k$. Changing the values of a, h, and k results in a different parabola in the family of quadratic functions. The parent graph of the family of parabolas is the graph of $y = x^2$.

You can use a TI-83 Plus graphing calculator to analyze the effects that result from changing each of the parameters a, h, and k.

Example 1

Graph each set of equations on the same screen in the standard viewing window. Describe any similarities and differences among the graphs.

$$y = x^2,\ y = x^2 + 3,\ y = x^2 - 5$$

The graphs have the same shape, and all open up. The vertex of each graph is on the y-axis. However, the graphs have different vertical positions.

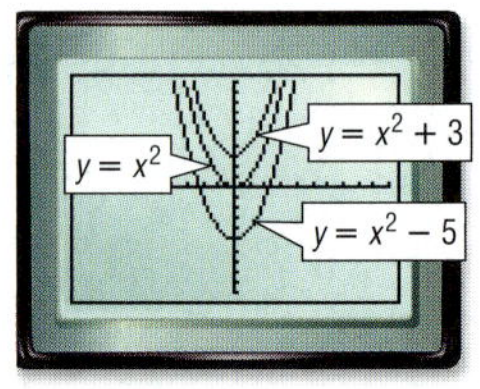

Example 1 shows how changing the value of k in the equation $y = a(x - h)^2 + k$ *translates* the parabola along the y-axis. If $k > 0$, the parabola is translated k units up, and if $k < 0$, it is translated k units down.

How do you think changing the value of h will affect the graph of $y = (x - h)^2$ as compared to the graph of $y = x^2$?

Example 2

Graph each set of equations on the same screen in the standard viewing window. Describe any similarities and differences among the graphs.

$$y = x^2,\ y = (x + 3)^2,\ y = (x - 5)^2$$

These three graphs all open up and have the same shape. The vertex of each graph is on the x-axis. However, the graphs have different horizontal positions.

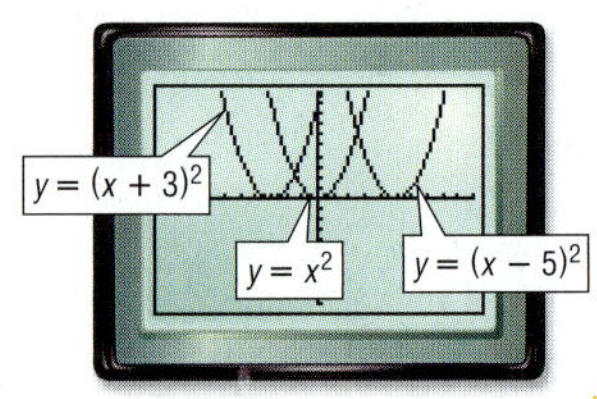

Example 2 shows how changing the value of h in the equation $y = a(x - h)^2 + k$ *translates* the graph horizontally. If $h > 0$, the graph translates to the right h units. If $h < 0$, the graph translates to the left h units.

www.algebra2.com/other_calculator_keystrokes

Investigation

How does the value a affect the graph of $y = ax^2$?

Example 3

Graph each set of equations on the same screen in the standard viewing window. Describe any similarities and differences among the graphs.

a. $y = x^2$, $y = -x^2$

The graphs have the same vertex and the same shape. However, the graph of $y = x^2$ opens up and the graph of $y = -x^2$ opens down.

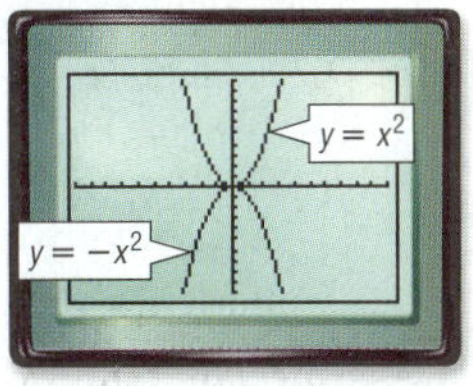

b. $y = x^2$, $y = 4x^2$, $y = \frac{1}{4}x^2$

The graphs have the same vertex, (0, 0), but each has a different shape. The graph of $y = 4x^2$ is narrower than the graph of $y = x^2$. The graph of $y = \frac{1}{4}x^2$ is wider than the graph of $y = x^2$.

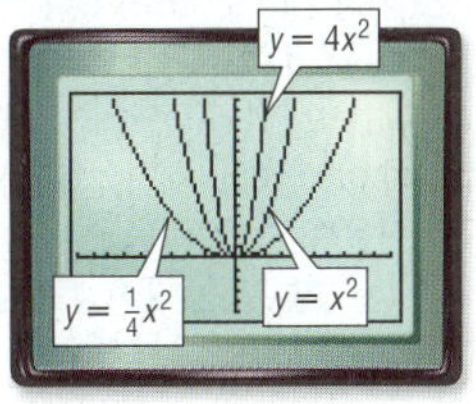

[−10, 10] scl: 1 by [−5, 15] scl: 1

Changing the value of a in the equation $y = a(x - h)^2 + k$ can affect the direction of the opening and the shape of the graph. If $a > 0$, the graph opens up, and if $a < 0$, the graph opens down or is *reflected* over the x-axis. If $|a| > 1$, the graph is narrower than the graph of $y = x^2$. If $|a| < 1$, the graph is wider than the graph of $y = x^2$. Thus, a change in the absolute value of a results in a *dilation* of the graph of $y = x^2$.

Exercises **1–3. See margin.**

Consider $y = a(x - h)^2 + k$.

1. How does changing the value of h affect the graph? Give an example.

2. How does changing the value of k affect the graph? Give an example.

3. How does using $-a$ instead of a affect the graph? Give an example.

Examine each pair of equations and predict the similarities and differences in their graphs. Use a graphing calculator to confirm your predictions. Write a sentence or two comparing the two graphs. **4–15. See pp. 343A–343F.**

4. $y = x^2$, $y = x^2 + 2.5$

5. $y = -x^2$, $y = x^2 - 9$

6. $y = x^2$, $y = 3x^2$

7. $y = x^2$, $y = -6x^2$

8. $y = x^2$, $y = (x + 3)^2$

9. $y = -\frac{1}{3}x^2$, $y = -\frac{1}{3}x^2 + 2$

10. $y = x^2$, $y = (x - 7)^2$

11. $y = x^2$, $y = 3(x + 4)^2 - 7$

12. $y = x^2$, $y = -\frac{1}{4}x^2 + 1$

13. $y = (x + 3)^2 - 2$, $y = (x + 3)^2 + 5$

14. $y = 3(x + 2)^2 - 1$, $y = 6(x + 2)^2 - 1$

15. $y = 4(x - 2)^2 - 3$, $y = \frac{1}{4}(x - 2)^2 - 1$

Assess

Ask students:

- In the general form of a quadratic equation, which constant would you change to move the graph left or right? ***h***
- Which constant would you change to move the graph up or down? ***k***
- Which constant would you change to make the graph wider or narrower? ***a***

Answers

1. Changing the value of h moves the graph to the left and the right. If $h > 0$, the graph translates to the right, and if $h < 0$, it translates to the left. In $y = x^2$, the vertex is at (0, 0) and in $y = (x - 2)^2$, the vertex is at (2, 0). The graph has been translated to the right.
2. Changing the value of k moves the graph up and down. If $k > 0$, the graph translates upward, and if $k < 0$, it translates downward. In $y = x^2$, the vertex is at (0, 0) and in $y = x^2 - 3$, the vertex is at (0, −3). The graph has been translated downward.
3. Using $-a$ instead of a reflects the graph over the x-axis. The graph of $y = x^2$ opens upward, while the graph of $y = -x^2$ opens downward.

Graphing Calculator Investigation Families of Parabolas 321

6-6 Lesson Notes

1 Focus

5-Minute Check Transparency 6-6 Use as a quiz or review of Lesson 6-5.

Mathematical Background notes are available for this lesson on p. 284D.

How can the graph of $y = x^2$ be used to graph any quadratic function?

Ask students:

- For the function $y = x^2$, what value of x makes y equal 0? **0** What value of x makes y equal 0 if the function is $y = (x - 3)^2$? **3**
- Compare the graph of $y = x^2 + 2$ with the graph of $y = (x + 2)^2$. What difference does adding the 2 within the parentheses make? **Sample answer: Adding the 2 inside the parentheses moves the graph 2 units to the left rather than 2 units up when compared to the graph of $y = x^2$.**

6-6 Analyzing Graphs of Quadratic Functions

What You'll Learn

- Analyze quadratic functions of the form $y = a(x - h)^2 + k$.
- Write a quadratic function in the form $y = a(x - h)^2 + k$.

Vocabulary

- vertex form

How can the graph of $y = x^2$ be used to graph any quadratic function?

A *family of graphs* is a group of graphs that displays one or more similar characteristics. The graph of $y = x^2$ is called the *parent graph* of the family of quadratic functions.

Study the graphs of $y = x^2$, $y = x^2 + 2$, and $y = (x - 3)^2$. Notice that adding a constant to x^2 moves the graph up. Subtracting a constant from x before squaring it moves the graph to the right.

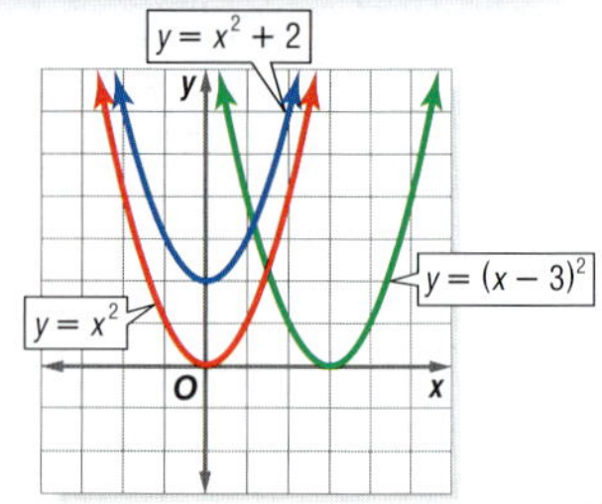

ANALYZE QUADRATIC FUNCTIONS Each function above can be written in the form $y = (x - h)^2 + k$, where (h, k) is the vertex of the parabola and $x = h$ is its axis of symmetry. This is often referred to as the **vertex form** of a quadratic function.

Equation	Vertex	Axis of Symmetry
$y = x^2$ or $y = (x - 0)^2 + 0$	(0, 0)	$x = 0$
$y = x^2 + 2$ or $y = (x - 0)^2 + 2$	(0, 2)	$x = 0$
$y = (x - 3)^2$ or $y = (x - 3)^2 + 0$	(3, 0)	$x = 3$

Recall that a translation slides a figure on the coordinate plane without changing its shape or size. As the values of h and k change, the graph of $y = a(x - h)^2 + k$ is the graph of $y = x^2$ translated:

- $|h|$ units *left* if h is negative or $|h|$ units *right* if h is positive, and
- $|k|$ units *up* if k is positive or $|k|$ units *down* if k is negative.

Example 1 Graph a Quadratic Function in Vertex Form

Analyze $y = (x + 2)^2 + 1$. Then draw its graph.

This function can be rewritten as $y = [x - (-2)]^2 + 1$. Then $h = -2$ and $k = 1$.

The vertex is at (h, k) or $(-2, 1)$, and the axis of symmetry is $x = -2$. The graph has the same shape as the graph of $y = x^2$, but is translated 2 units left and 1 unit up.

Now use this information to draw the graph.

Step 1 Plot the vertex, $(-2, 1)$.

Step 2 Draw the axis of symmetry, $x = -2$.

Step 3 Find and plot two points on one side of the axis of symmetry, such as $(-1, 2)$ and $(0, 5)$.

Step 4 Use symmetry to complete the graph.

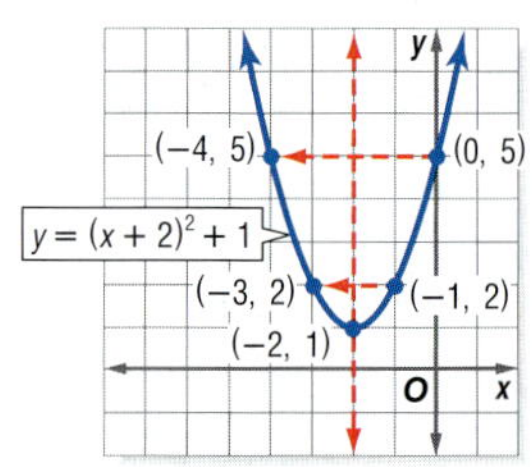

Resource Manager

Workbook and Reproducible Masters

Chapter 6 Resource Masters

- Study Guide and Intervention, pp. 343–344
- Skills Practice, p. 345
- Practice, p. 346
- Reading to Learn Mathematics, p. 347
- Enrichment, p. 348
- Assessment, p. 370

Teaching Algebra With Manipulatives Masters, pp. 247–248

Transparencies

5-Minute Check Transparency 6-6
Answer Key Transparencies

Technology

Interactive Chalkboard

How does the value of a in the general form $y = a(x - h)^2 + k$ affect a parabola? Compare the graphs of the following functions to the parent function, $y = x^2$.

a. $y = 2x^2$ **b.** $y = \frac{1}{2}x^2$

c. $y = -2x^2$ **d.** $y = -\frac{1}{2}x^2$

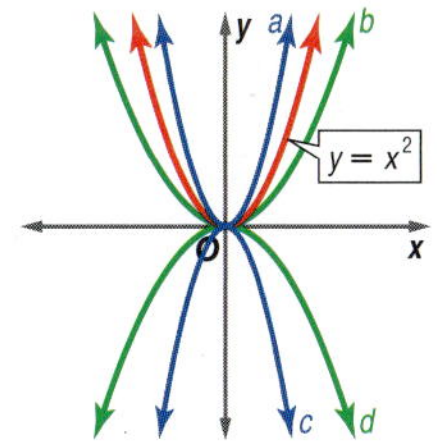

All of the graphs have the vertex (0, 0) and axis of symmetry $x = 0$.

Notice that the graphs of $y = 2x^2$ and $y = \frac{1}{2}x^2$ are *dilations* of the graph of $y = x^2$. The graph of $y = 2x^2$ is narrower than the graph of $y = x^2$, while the graph of $y = \frac{1}{2}x^2$ is wider. The graphs of $y = -2x^2$ and $y = 2x^2$ are *reflections* of each other over the x-axis, as are the graphs of $y = -\frac{1}{2}x^2$ and $y = \frac{1}{2}x^2$.

Changing the value of a in the equation $y = a(x - h)^2 + k$ can affect the direction of the opening and the shape of the graph.

- If $a > 0$, the graph opens up.
- If $a < 0$, the graph opens down.
- If $|a| > 1$, the graph is narrower than the graph of $y = x^2$.
- If $|a| < 1$, the graph is wider than the graph of $y = x^2$.

Study Tip

Reading Math

$|a| < 1$ means that a is a real number between 0 and 1, such as $\frac{2}{5}$, or a real number between -1 and 0, such as $-\frac{\sqrt{2}}{2}$.

Concept Summary — *Quadratic Functions in Vertex Form*

The vertex form of a quadratic function is $y = a(x - h)^2 + k$.

h and k

Vertex and Axis of Symmetry

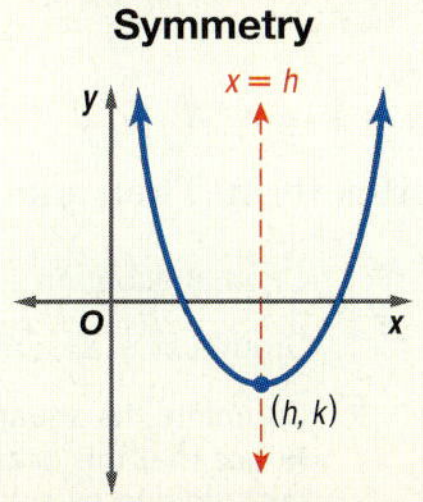

k

Vertical Translation

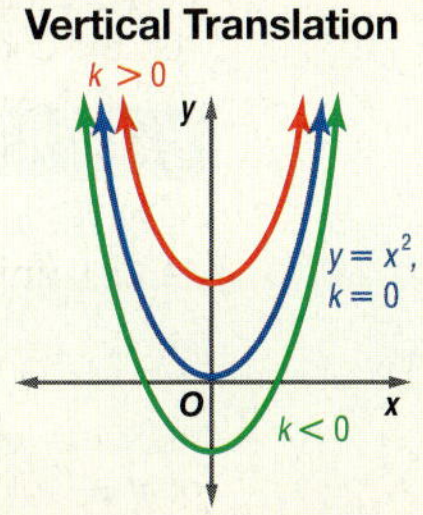

h

Horizontal Translation

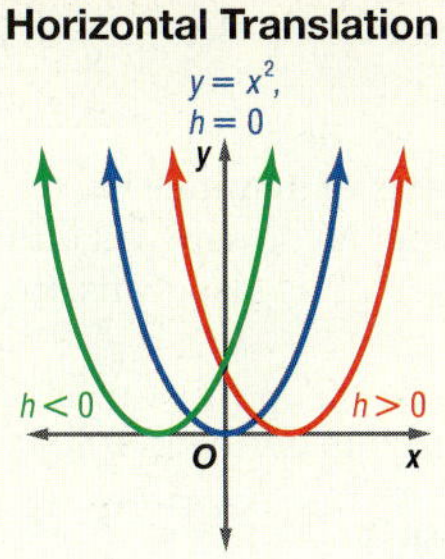

a

Direction of Opening and Shape of Parabola

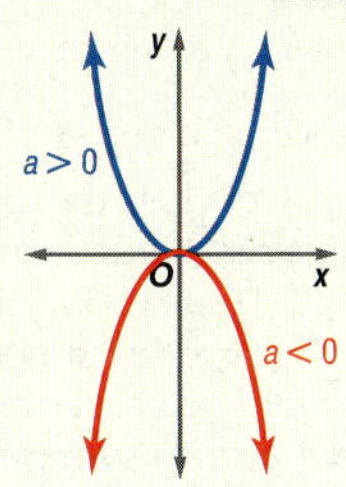

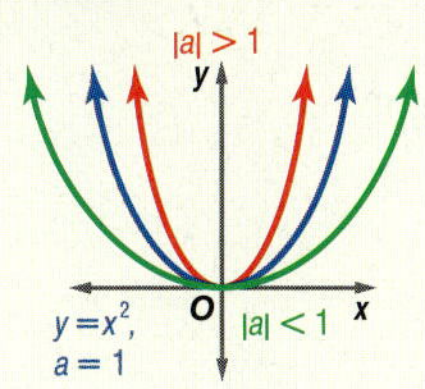

www.algebra2.com/extra_examples

2 Teach

ANALYZE QUADRATIC FUNCTIONS

In-Class Example PowerPoint®

1 Analyze $y = (x - 3)^2 + 2$. Then draw its graph. **The vertex of the graph is at (3, 2) and the axis of symmetry is $x = 3$. The graph has the same shape as the graph of $y = x^2$, but is translated 3 units right and 2 units up.**

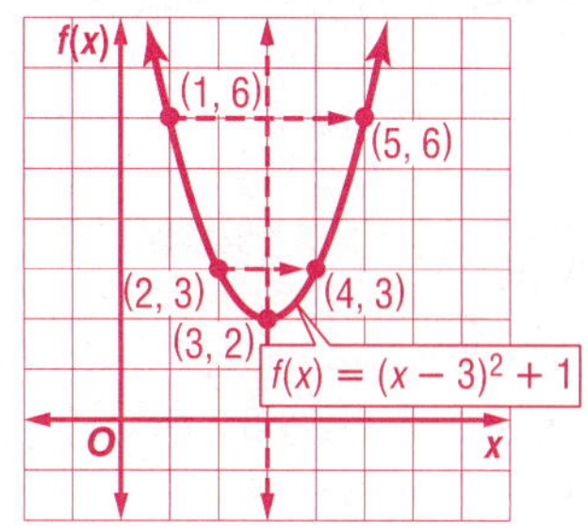

Teaching Tip To help students remember that as $|a|$ increases, the graph gets narrower and not wider, discuss the fact that a greater multiplier for the quantity $(x - h)^2$ will make the corresponding y value greater as well. Point out that greater values of y result in a steeper (and thus narrower) graph.

Tips for New Teachers

Intervention Encourage students to ask questions about any aspects they may find confusing that are covered in the Concept Summary chart on this page. Ask them to write and use their own summary on an index card. Explain to students that a thorough understanding of these concepts will save them time since this knowledge will enable them to sketch approximate graphs quickly.

WRITE QUADRATIC FUNCTIONS IN VERTEX FORM

In-Class Examples Power Point®

2 Write $y = x^2 + 2x + 4$ in vertex form. Then analyze the function. $y = (x + 1)^2 + 3$; vertex: $(-1, 3)$; axis of symmetry: $x = -1$; opens up; The graph has the same shape as the graph of $y = x^2$, but it is translated 1 unit left and 3 units up.

3 Write $y = -2x^2 - 4x + 2$ in vertex form. Then analyze and graph the function. $y = -2(x + 1)^2 + 4$; vertex: $(-1, 4)$; axis of symmetry: $x = -1$; opens down; The graph is narrower than the graph of $y = x^2$, and it is translated 1 unit left and 4 units up.

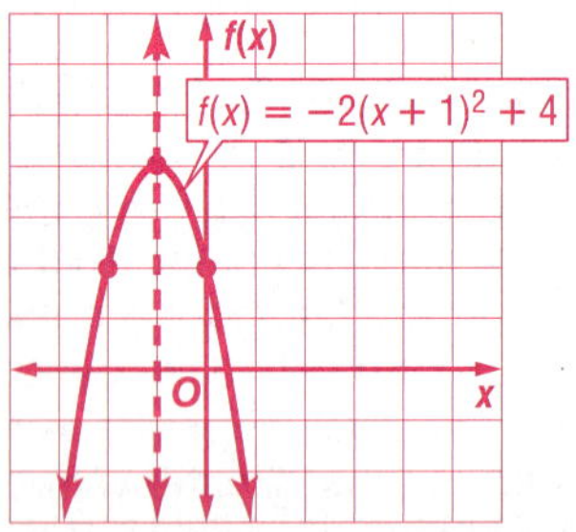

WRITE QUADRATIC FUNCTIONS IN VERTEX FORM Given a function of the form $y = ax^2 + bx + c$, you can complete the square to write the function in vertex form.

Example 2 Write $y = x^2 + bx + c$ in Vertex Form

Write $y = x^2 + 8x - 5$ in vertex form. Then analyze the function.

$y = x^2 + 8x - 5$	Notice that $x^2 + 8x - 5$ is not a perfect square.
$y = (x^2 + 8x + 16) - 5 - 16$	Complete the square by adding $\left(\frac{8}{2}\right)^2$ or 16. Balance this addition by subtracting 16.
$y = (x + 4)^2 - 21$	Write $x^2 + 8x + 16$ as a perfect square.

This function can be rewritten as $y = [x - (-4)]^2 + (-21)$. Written in this way, you can see that $h = -4$ and $k = -21$.

The vertex is at $(-4, -21)$, and the axis of symmetry is $x = -4$. Since $a = 1$, the graph opens up and has the same shape as the graph of $y = x^2$, but it is translated 4 units left and 21 units down.

CHECK You can check the vertex and axis of symmetry using the formula $x = -\frac{b}{2a}$. In the original equation, $a = 1$ and $b = 8$, so the axis of symmetry is $x = -\frac{8}{2(1)}$ or -4. Thus, the x-coordinate of the vertex is -4, and the y-coordinate of the vertex is $y = (-4)^2 + 8(-4) - 5$ or -21.

Study Tip

Check
As an additional check, graph the function in Example 2 to verify the location of its vertex and axis of symmetry.

When writing a quadratic function in which the coefficient of the quadratic term is not 1 in vertex form, the first step is to factor out that coefficient from the quadratic and linear terms. Then you can complete the square and write in vertex form.

Example 3 Write $y = ax^2 + bx + c$ in Vertex Form, $a \neq 1$

Write $y = -3x^2 + 6x - 1$ in vertex form. Then analyze and graph the function.

$y = -3x^2 + 6x - 1$	Original equation
$y = -3(x^2 - 2x) - 1$	Group $ax^2 + bx$ and factor, dividing by a.
$y = -3(x - 2x + 1) - 1 - (-3)(1)$	Complete the square by adding 1 inside the parentheses. Notice that this is an overall addition of $-3(1)$. Balance this addition by subtracting $-3(1)$.
$y = -3(x - 1)^2 + 2$	Write $x^2 - 2x + 1$ as a perfect square.

The vertex form of this function is $y = -3(x - 1)^2 + 2$. So, $h = 1$ and $k = 2$.

The vertex is at $(1, 2)$, and the axis of symmetry is $x = 1$. Since $a = -3$, the graph opens downward and is narrower than the graph of $y = x^2$. It is also translated 1 unit right and 2 units up.

Now graph the function. Two points on the graph to the right of $x = 1$ are $(1.5, 1.25)$ and $(2, -1)$. Use symmetry to complete the graph.

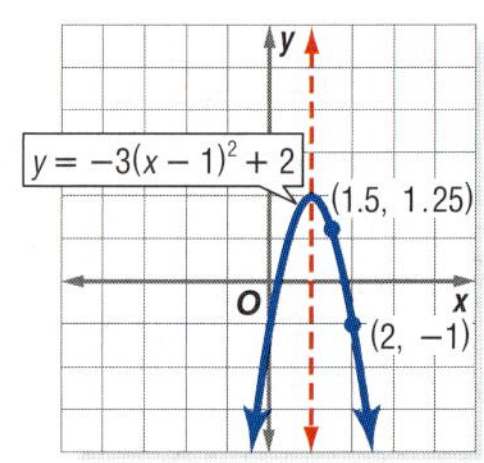

DAILY INTERVENTION **Differentiated Instruction**

Naturalist Have students observe or research some natural events that can be modeled by parabolas, such as the fountain's water stream discussed in Exercise 14 on p. 326. Students should report their observations and findings to the class. If students are able to determine a quadratic function that models the event, they should present the function and explain how the characteristics of the equation can be used to analyze its graph.

If the vertex and one other point on the graph of a parabola are known, you can write the equation of the parabola in vertex form.

Example 4 Write an Equation Given Points

Write an equation for the parabola whose vertex is at $(-1, 4)$ and passes through $(2, 1)$.

The vertex of the parabola is at $(-1, 4)$, so $h = -1$ and $k = 4$. Since $(2, 1)$ is a point on the graph of the parabola, let $x = 2$ and $y = 1$. Substitute these values into the vertex form of the equation and solve for a.

$y = a(x - h)^2 + k$	Vertex form
$1 = a[2 - (-1)]^2 + 4$	Substitute 1 for y, 2 for x, -1 for h, and 4 for k.
$1 = a(9) + 4$	Simplify.
$-3 = 9a$	Subtract 4 from each side.
$-\frac{1}{3} = a$	Divide each side by 9.

The equation of the parabola in vertex form is $y = -\frac{1}{3}(x + 1)^2 + 4$.

CHECK A graph of $y = -\frac{1}{3}(x + 1)^2 + 4$ verifies that the parabola passes through the point at $(2, 1)$.

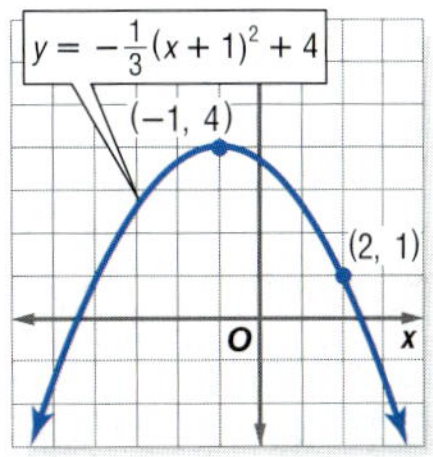

Check for Understanding

Concept Check

1d. $y = 2(x - 2)^2 + 3$
1e. Sample answer: $y = 4(x + 1)^2 + 3$
1f. Sample answer: $y = (x + 1)^2 + 3$
3. Sample answer: $y = 2(x - 2)^2 - 1$

1. **Write** a quadratic equation that transforms the graph of $y = 2(x + 1)^2 + 3$ so that it is:
   a. 2 units up. $y = 2(x + 1)^2 + 5$
   b. 3 units down. $y = 2(x + 1)^2$
   c. 2 units to the left. $y = 2(x + 3)^2 + 3$
   d. 3 units to the right.
   e. narrower.
   f. wider.
   g. opening in the opposite direction. $y = -2(x + 1)^2 + 3$

2. **Explain** how you can find an equation of a parabola using its vertex and one other point on its graph. See margin.

3. **OPEN ENDED** Write the equation of a parabola with a vertex of $(2, -1)$.

4. Jenny; when completing the square is used to write a quadratic function in vertex form, the quantity added is then subtracted from the same side of the equation to maintain equality.

4. **FIND THE ERROR** Jenny and Ruben are writing $y = x^2 - 2x + 5$ in vertex form.

Jenny	Ruben
$y = x^2 - 2x + 5$	$y = x^2 - 2x + 5$
$y = (x^2 - 2x + 1) + 5 - 1$	$y = (x^2 - 2x + 1) + 5 + 1$
$y = (x - 1)^2 + 4$	$y = (x - 1)^2 + 6$

Who is correct? Explain your reasoning.

Guided Practice

Write each quadratic function in vertex form, if not already in that form. Then identify the vertex, axis of symmetry, and direction of opening. 5–7. See margin.

5. $y = 5(x + 3)^2 - 1$
6. $y = x^2 + 8x - 3$
7. $y = -3x^2 - 18x + 11$

In-Class Example Power Point®

4 Write an equation for the parabola whose vertex is at $(1, 2)$ and passes through $(3, 4)$. $y = \frac{1}{2}(x - 1)^2 + 2$

3 Practice/Apply

Study Notebook

Have students—
- add the definitions/examples of the vocabulary terms to their Vocabulary Builder worksheets for Chapter 6.
- write a summary in their own words of everything you can tell about the graph of a parabola when the function is written in vertex form.
- include any other item(s) that they find helpful in mastering the skills in this lesson.

DAILY INTERVENTION **FIND THE ERROR**

Students may find that it will help them avoid errors of this type if they specifically write the addition and subtraction in a separate step before placing parentheses around the perfect square trinomial, as in $y = x^2 - 2x + 1 - 1 + 5$.

Answers

2. Substitute the x-coordinate of the vertex for h and the y-coordinate of the vertex for k in the equation $y = a(x - h)^2 + k$. Then substitute the x-coordinate of the other point for x and the y-coordinate for y into this equation and solve for a. Replace a with this value in the equation you wrote with h and k.
5. $(-3, -1)$; $x = -3$; up
6. $y = (x + 4)^2 - 19$, $(-4, -19)$; $x = -4$; up
7. $y = -3(x + 3)^2 + 38$; $(-3, 38)$; $x = -3$; down

Lesson 6-6 Analyzing Graphs of Quadratic Functions 325

About the Exercises...

Organization by Objective

- **Analyze Quadratic Functions:** 15, 16, 27–30, 37, 38, 47, 51, 52
- **Write Quadratic Functions in Vertex Form:** 19–26, 31–36, 39–46, 48–50

Odd/Even Assignments

Exercises 15–46 are structured so that students practice the same concepts whether they are assigned odd or even problems.

Assignment Guide

Basic: 15–33 odd, 39–45 odd, 53–71

Average: 15–47 odd, 48–50, 53–71

Advanced: 16–46 even, 48–67 (optional: 68–71)

All: Practice Quiz 2 (1–10)

GUIDED PRACTICE KEY	
Exercises	Examples
5–7	2
8–10	1, 3
11–14	4

Graph each function. **8–10. See margin.**

8. $y = 3(x + 3)^2$ 9. $y = \frac{1}{3}(x - 1)^2 + 3$ 10. $y = -2x^2 + 16x - 31$

Write an equation for the parabola with the given vertex that passes through the given point.

11. vertex: (2, 0) point: (1, 4) $y = 4(x - 2)^2$
12. vertex: (−3, 6) point: (−5, 2) $y = -(x + 3)^2 + 6$
13. vertex: (−2, −3) point: (−4, −5) $y = -\frac{1}{2}(x + 2)^2 - 3$

Application

14. **FOUNTAINS** The height of a fountain's water stream can be modeled by a quadratic function. Suppose the water from a jet reaches a maximum height of 8 feet at a distance 1 foot away from the jet. If the water lands 3 feet away from the jet, find a quadratic function that models the height $h(d)$ of the water at any given distance d feet from the jet. $h(d) = -2d^2 + 4d + 6$

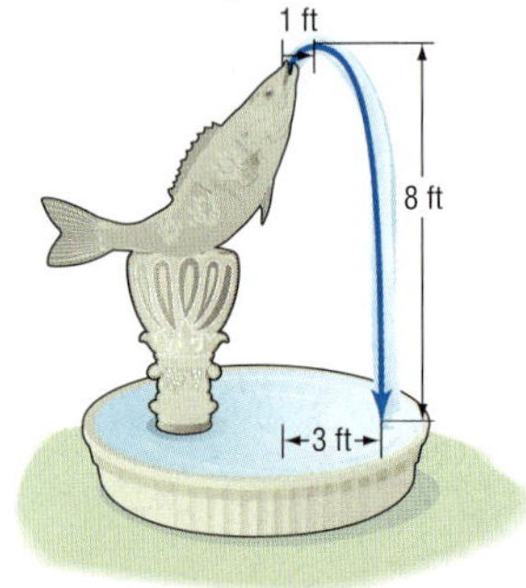

★ indicates increased difficulty

Practice and Apply

Homework Help	
For Exercises	See Examples
15–26	2
27–38, 47, 48, 50–52	1, 3
39–46, 49	4

Extra Practice See page 841.

Write each quadratic function in vertex form, if not already in that form. Then identify the vertex, axis of symmetry, and direction of opening. **19–26. See margin.**

15. $y = -2(x + 3)^2$ **(−3, 0); $x = -3$; down**
16. $y = \frac{1}{3}(x - 1)^2 + 2$ **(1, 2); $x = 1$; up**
17. $y = 5x^2 - 6$ **(0, −6); $x = 0$; up**
18. $y = -8x^2 + 3$ **(0, 3); $x = 0$; down**
19. $y = -x^2 - 4x + 8$
20. $y = x^2 - 6x + 1$
21. $y = -3x^2 + 12x$
22. $y = 4x^2 + 24x$
23. $y = 4x^2 + 8x - 3$
24. $y = -2x^2 + 20x - 35$
25. $y = 3x^2 + 3x - 1$
26. $y = 4x^2 - 12x - 11$

Graph each function. **27–36. See pp. 343A–343F.**

27. $y = 4(x + 3)^2 + 1$
28. $y = -(x - 5)^2 - 3$
29. $y = \frac{1}{4}(x - 2)^2 + 4$
30. $y = \frac{1}{2}(x - 3)^2 - 5$
31. $y = x^2 + 6x + 2$
32. $y = x^2 - 8x + 18$
33. $y = -4x^2 + 16x - 11$
34. $y = -5x^2 - 40x - 80$

★ 35. $y = -\frac{1}{2}x^2 + 5x - \frac{27}{2}$

★ 36. $y = \frac{1}{3}x^2 - 4x + 15$

★ 37. Write one sentence that compares the graphs of $y = 0.2(x + 3)^2 + 1$ and $y = 0.4(x + 3)^2 + 1$. **Sample answer: the graph of $y = 0.4(x + 3)^2 + 1$ is narrower than the graph of $y = 0.2(x + 3)^2 + 1$.**

★ 38. Compare the graphs of $y = 2(x - 5)^2 + 4$ and $y = 2(x - 4)^2 - 1$.

38. Sample answer: the graphs have the same shape, but the graph of $y = 2(x - 4)^2 + 1$ is 1 unit to the left and 5 units below the graph of $y = 2(x - 5)^2 - 4$.

Write an equation for the parabola with the given vertex that passes through the given point.

39. vertex: (6, 1) point: (5, 10) $y = 9(x - 6)^2 + 1$
40. vertex: (−4, 3) point: (−3, 6) $y = 3(x + 4)^2 + 3$
41. vertex: (3, 0) point: (6, −6) $y = -\frac{2}{3}(x - 3)^2$
42. vertex: (5, 4) point: (3, −8) $y = -3(x - 5)^2 + 4$
43. vertex: (0, 5) point: (3, 8) $y = \frac{1}{3}x^2 + 5$
44. vertex: (−3, −2) point: (−1, 8) $y = \frac{5}{2}(x + 3)^2 - 2$

WebQuest

You can use a quadratic function to model the world population. Visit www.algebra2.com/webquest to continue work on your WebQuest project.

Answers

8.

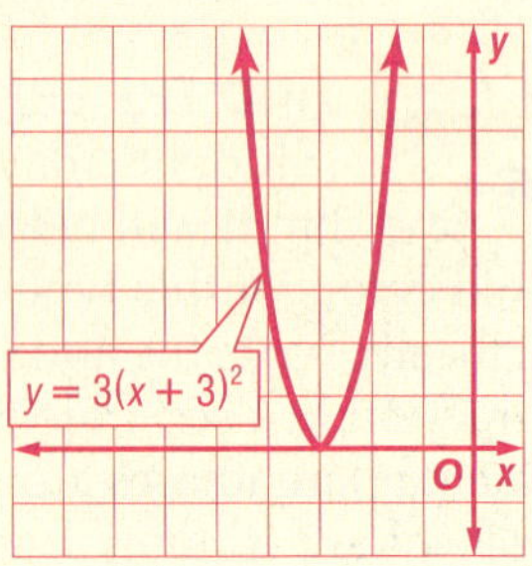

9.

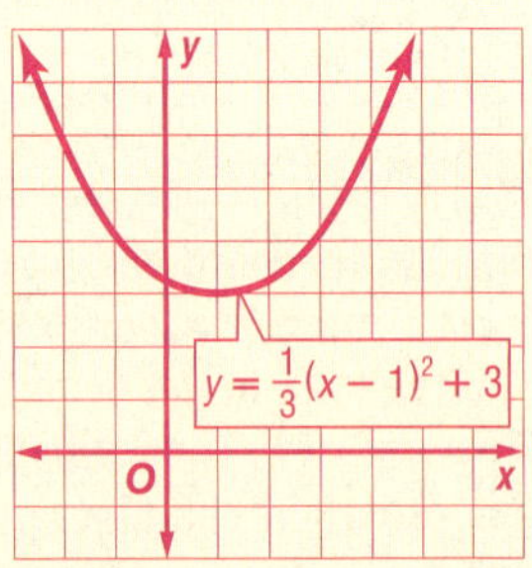

10.

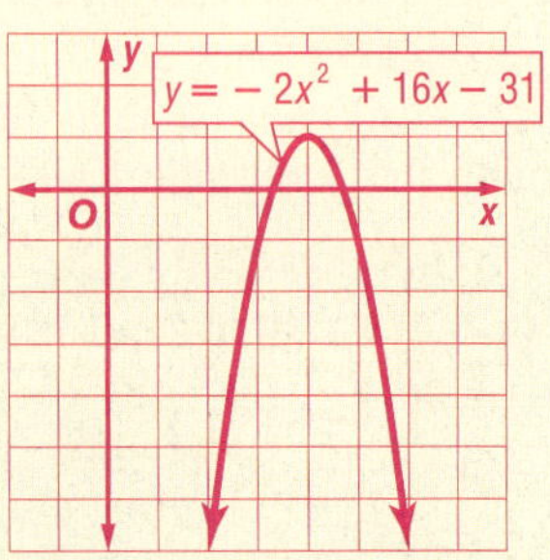

19. $y = -(x + 2)^2 + 12$; $(-2, 12)$; $x = -2$; down
20. $y = (x - 3)^2 - 8$; $(3, -8)$; $x = 3$; up
21. $y = -3(x - 2)^2 + 12$; $(2, 12)$; $x = 2$; down
22. $y = 4(x + 3)^2 - 36$; $(-3, -36)$; $x = -3$; up
23. $y = 4(x + 1)^2 - 7$; $(-1, -7)$; $x = -1$; up
24. $y = -2(x - 5)^2 + 15$; $(5, 15)$; $x = 5$; down
25. $y = 3\left(x + \frac{1}{2}\right)^2 - \frac{7}{4}$; $\left(-\frac{1}{2}, -\frac{7}{4}\right)$; $x = -\frac{1}{2}$; up
26. $y = 4\left(x - \frac{3}{2}\right)^2 - 20$; $\left(\frac{3}{2}, -20\right)$; $x = \frac{3}{2}$; up

More About. . .

Aerospace

The KC135A has the nickname "Vomit Comet." It starts its ascent at 24,000 feet. As it approaches maximum height, the engines are stopped, and the aircraft is allowed to free-fall at a determined angle. Zero gravity is achieved for 25 seconds as the plane reaches the top of its flight and begins its descent.

Source: NASA

45. Write an equation for a parabola whose vertex is at the origin and passes through $(2, -8)$. **$y = -2x^2$**

46. Write an equation for a parabola with vertex at $(-3, -4)$ and y-intercept 8. **$y = \frac{4}{3}(x + 3)^2 - 4$**

47. **AEROSPACE** NASA's KC135A aircraft flies in parabolic arcs to simulate the weightlessness experienced by astronauts in space. The height h of the aircraft (in feet) t seconds after it begins its parabolic flight can be modeled by the equation $h(t) = -9.09(t - 32.5)^2 + 34{,}000$. What is the maximum height of the aircraft during this maneuver and when does it occur? **34,000 feet; 32.5 s after the aircraft begins its parabolic flight**

DIVING For Exercises 48–50, use the following information.

The distance of a diver above the water $d(t)$ (in feet) t seconds after diving off a platform is modeled by the equation $d(t) = -16t^2 + 8t + 30$.

48. Find the time it will take for the diver to hit the water. **about 1.6 s**

49. Write an equation that models the diver's distance above the water if the platform were 20 feet higher. **$d(t) = -16t^2 + 8t + 50$**

50. Find the time it would take for the diver to hit the water from this new height. **about 2.0 s**

LAWN CARE For Exercises 51 and 52, use the following information.

The path of water from a sprinkler can be modeled by a quadratic function. The three functions below model paths for three different angles of the water.

Angle A: $y = -0.28(x - 3.09)^2 + 3.27$

Angle B: $y = -0.14(x - 3.57)^2 + 2.39$

Angle C: $y = -0.09(x - 3.22)^2 + 1.53$

51. Which sprinkler angle will send water the highest? Explain your reasoning.

52. Which sprinkler angle will send water the farthest? Explain your reasoning.

51. Angle A; the graph of the equation for angle A is higher than the other two since 3.27 is greater than 2.39 or 1.53.

52. Angle B; the vertex of the equation for angle B is farther to the right than the other two since 3.57 is greater than 3.09 or 3.22.

53. **CRITICAL THINKING** Given $y = ax^2 + bx + c$ with $a \neq 0$, derive the equation for the axis of symmetry by completing the square and rewriting the equation in the form $y = a(x - h)^2 + k$. **See pp. 343A–343F.**

54. **WRITING IN MATH** Answer the question that was posed at the beginning of the lesson. **See pp. 343A–343F.**

How can the graph $y = x^2$ be used to graph any quadratic function?

Include the following in your answer:

- a description of the effects produced by changing a, h, and k in the equation $y = a(x - h)^2 + k$, and
- a comparison of the graph of $y = x^2$ and the graph of $y = a(x - h)^2 + k$ using values of your own choosing for a, h, and k.

Standardized Test Practice

55. If $f(x) = x^2 - 5x$ and $f(n) = -4$, then which of the following could be n? **D**

   (A) -5 (B) -4 (C) -1 (D) 1

56. The vertex of the graph of $y = 2(x - 6)^2 + 3$ is located at which of the following points? **B**

   (A) $(2, 3)$ (B) $(6, 3)$ (C) $(6, -3)$ (D) $(-2, 3)$

www.algebra2.com/self_check_quiz

Study Guide and Intervention, p. 343 (shown) and p. 344

Analyze Quadratic Functions

Vertex Form of a Quadratic Function	The graph of $y = a(x - h)^2 + k$ has the following characteristics: • Vertex: (h, k) • Axis of symmetry: $x = h$ • Opens up if $a > 0$ • Opens down if $a < 0$ • Narrower than the graph of $y = x^2$ if $\lvert a \rvert > 1$ • Wider than the graph of $y = x^2$ if $\lvert a \rvert < 1$

Example Identify the vertex, axis of symmetry, and direction of opening of each graph.

a. $y = 2(x + 4)^2 - 11$

The vertex is at (h, k) or $(-4, -11)$, and the axis of symmetry is $x = -4$. The graph opens up, and is narrower than the graph of $y = x^2$.

a. $y = -\frac{1}{4}(x - 2)^2 + 10$

The vertex is at (h, k) or $(2, 10)$, and the axis of symmetry is $x = 2$. The graph opens down, and is wider than the graph of $y = x^2$.

Exercises

Each quadratic function is given in vertex form. Identify the vertex, axis of symmetry, and direction of opening of the graph.

1. $y = (x - 2)^2 + 16$ **(2, 16); x = 2; up**
2. $y = 4(x + 3)^2 - 7$ **(−3, −7); x = −3; up**
3. $y = \frac{1}{2}(x - 5)^2 + 3$ **(5, 3); x = 5; up**
4. $y = -7(x + 1)^2 - 9$ **(−1, −9); x = −1; down**
5. $y = \frac{1}{5}(x - 4)^2 - 12$ **(4, −12); x = 4; up**
6. $y = 6(x + 6)^2 + 6$ **(−6, 6); x = −6; up**
7. $y = \frac{2}{5}(x - 9)^2 + 12$ **(9, 12); x = 9; up**
8. $y = 8(x - 3)^2 - 2$ **(3, −2); x = 3; up**
9. $y = -3(x - 1)^2 - 2$ **(1, −2); x = 1; down**
10. $y = -\frac{5}{2}(x + 5)^2 + 12$ **(−5, 12); x = −5; down**
11. $y = \frac{4}{3}(x - 7)^2 + 22$ **(7, 22); x = 7; up**
12. $y = 16(x - 4)^2 + 1$ **(4, 1); x = 4; up**
13. $y = 3(x - 1.2)^2 + 2.7$ **(1.2, 2.7); x = 1.2; up**
14. $y = -0.4(x - 0.6)^2 - 0.2$ **(0.6, −0.2); x = 0.6; down**
15. $y = 1.2(x + 0.8)^2 + 6.5$ **(−0.8, 6.5); x = −0.8; up**

Skills Practice, p. 345 and Practice, p. 346 (shown)

Write each quadratic function in vertex form, if not already in that form. Then identify the vertex, axis of symmetry, and direction of opening.

1. $y = -6(x + 2)^2 - 1$ **$y = -6(x + 2)^2 - 1$; (−2, −1); x = −2; down**
2. $y = 2x^2 + 2$ **$y = 2(x + 0)^2 + 2$; (0, 2); x = 0; up**
3. $y = -4x^2 + 8x$ **$y = -4(x - 1)^2 + 4$; (1, 4); x = 1; down**
4. $y = x^2 + 10x + 20$ **$y = (x + 5)^2 - 5$; (−5, −5); x = −5; up**
5. $y = 2x^2 + 12x + 18$ **$y = 2(x + 3)^2$; (−3, 0); x = −3; up**
6. $y = 3x^2 - 6x + 5$ **$y = 3(x - 1)^2 + 2$; (1, 2); x = 1; up**
7. $y = -2x^2 - 16x - 32$ **$y = -2(x + 4)^2$; (−4, 0); x = −4; down**
8. $y = -3x^2 + 18x - 21$ **$y = -3(x - 3)^2 + 6$; (3, 6); x = 3; down**
9. $y = 2x^2 + 16x + 29$ **$y = 2(x + 4)^2 - 3$; (−4, −3); x = −4; up**

Graph each function.

10. $y = (x + 3)^2 - 1$
11. $y = -x^2 + 6x - 5$
12. $y = 2x^2 - 2x + 1$

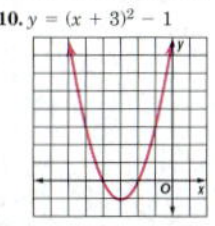

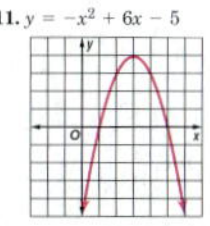

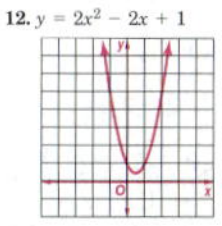

Write an equation for the parabola with the given vertex that passes through the given point.

13. vertex: (1, 3) point: (−2, −15) **$y = -2(x - 1)^2 + 3$**
14. vertex: (−3, 0) point: (3, 18) **$y = \frac{1}{2}(x + 3)^2$**
15. vertex: (10, −4) point: (5, 6) **$y = \frac{2}{5}(x - 10)^2 - 4$**

16. Write an equation for a parabola with vertex at (4, 4) and x-intercept 6. **$y = -(x - 4)^2 + 4$**

17. Write an equation for a parabola with vertex at (−3, −1) and y-intercept 2. **$y = \frac{1}{3}(x + 3)^2 - 1$**

18. **BASEBALL** The height h of a baseball t seconds after being hit is given by $h(t) = -16t^2 + 80t + 3$. What is the maximum height that the baseball reaches, and when does this occur? **103 ft; 2.5 s**

19. **SCULPTURE** A modern sculpture in a park contains a parabolic arc that starts at the ground and reaches a maximum height of 10 feet after a horizontal distance of 4 feet. Write a quadratic function in vertex form that describes the shape of the outside of the arc, where y is the height of a point on the arc and x is its horizontal distance from the left-hand starting point of the arc. **$y = -\frac{5}{8}(x - 4)^2 + 10$**

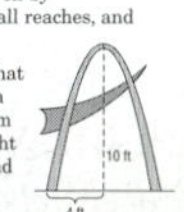

Reading to Learn Mathematics, p. 347 ELL

Pre-Activity How can the graph of $y = x^2$ be used to graph any quadratic function?

Read the introduction to Lesson 6-6 at the top of page 322 in your textbook.

- What does adding a positive number to x^2 do to the graph of $y = x^2$? **It moves the graph up.**
- What does subtracting a positive number to x before squaring do to the graph of $y = x^2$? **It moves the graph to the right.**

Reading the Lesson

1. Complete the following information about the graph of $y = a(x - h)^2 + k$.
   a. What are the coordinates of the vertex? **(h, k)**
   b. What is the equation of the axis of symmetry? **x = h**
   c. In which direction does the graph open if $a > 0$? If $a < 0$? **up; down**
   d. What do you know about the graph if $|a| < 1$? **It is wider than the graph of $y = x^2$.**
   If $|a| > 1$? **It is narrower than the graph of $y = x^2$.**

2. Match each graph with the description of the constants in the equation in vertex form.
   a. $a > 0, h > 0, k < 0$ **iii**
   b. $a < 0, h < 0, k < 0$ **iv**
   c. $a < 0, h < 0, k > 0$ **ii**
   d. $a > 0, h = 0, k < 0$ **i**

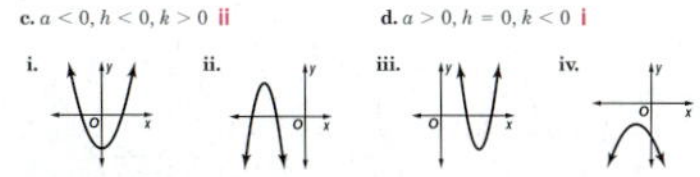

Helping You Remember

3. When graphing quadratic functions such as $y = (x + 4)^2$ and $y = (x - 5)^2$, many students have trouble remembering which represents a translation of the graph of $y = x^2$ to the left and which represents a translation to the right. What is an easy way to remember this?
Sample answer: In functions like $y = (x + 4)^2$, the plus sign puts the graph "ahead" so that the vertex comes "sooner" than the origin and the translation is to the left. In functions like $y = (x - 5)^2$, the minus puts the graph "behind" so that the vertex comes "later" than the origin and the translation is to the right.

Enrichment, p. 348

Patterns with Differences and Sums of Squares

Some whole numbers can be written as the difference of two squares and some cannot. Formulas can be developed to describe the sets of numbers algebraically.

If possible, write each number as the difference of two squares. Look for patterns.

1. 0 **$0^2 - 0^2$**
2. 1 **$1^2 - 0^2$**
3. 2 **cannot**
4. 3 **$2^2 - 1^2$**
5. 4 **$2^2 - 0^2$**
6. 5 **$3^2 - 2^2$**
7. 6 **cannot**
8. 7 **$4^2 - 3^2$**
9. 8 **$3^2 - 1^2$**
10. 9 **$3^2 - 0^2$**
11. 10 **cannot**
12. 11 **$6^2 - 5^2$**
13. 12 **$4^2 - 2^2$**
14. 13 **$7^2 - 6^2$**
15. 14 **cannot**
16. 15 **$4^2 - 1^2$**

Even numbers can be written as $2n$, where n is one of the numbers 0, 1, 2, 3, and so on. Odd numbers can be written $2n + 1$. Use these ... these proble...

Maintain Your Skills

Mixed Review

Find the value of the discriminant for each quadratic equation. Then describe the number and type of roots for the equation. *(Lesson 6-5)*

57. $3x^2 - 6x + 2 = 0$ **12; 2 irrational**
58. $4x^2 + 7x = 11$ **225; 2 rational**
59. $2x^2 - 5x + 6 = 0$ **−23; 2 complex**

Solve each equation by completing the square. *(Lesson 6-4)*

60. $x^2 + 10x + 17 = 0$ $\{-5 \pm 2\sqrt{2}\}$
61. $x^2 - 6x + 18 = 0$ $\{3 \pm 3i\}$
62. $4x^2 + 8x = 9$ $\left\{\frac{-2 \pm \sqrt{13}}{2}\right\}$

Find each quotient. *(Lesson 5-3)* **63–66. See margin.**

63. $(2t^3 - 2t - 3) \div (t - 1)$
64. $(t^3 - 3t + 2) \div (t + 2)$
65. $(n^4 - 8n^3 + 54n + 105) \div (n - 5)$
66. $(y^4 + 3y^3 + y - 1) \div (y + 3)$

67. **EDUCATION** The graph shows the number of U.S. students in study-abroad programs. *(Lesson 2-5)*
   a. Write a prediction equation from the data given.
   b. Use your equation to predict the number of students in these programs in 2005. **Sample answer: 161,167**

67a. Sample answer using (1994, 76,302) and (1997, 99,448): $y = 7715x - 15{,}307{,}408$

Getting Ready for the Next Lesson

PREREQUISITE SKILL **Determine whether the given value satisfies the inequality.** *(To review **inequalities**, see Lesson 1-6.)*

68. $-2x^2 + 3 < 0; x = 5$ **yes**
69. $4x^2 + 2x - 3 \geq 0; x = -1$ **no**
70. $4x^2 - 4x + 1 \leq 10; x = 2$ **yes**
71. $6x^2 + 3x > 8; x = 0$ **no**

Practice Quiz 2 — *Lessons 6-4 through 6-6*

Solve each equation by completing the square. *(Lesson 6-4)*

1. $x^2 + 14x + 37 = 0$ $\{-7 \pm 2\sqrt{3}\}$
2. $2x^2 - 2x + 5 = 0$ $\left\{\frac{1 \pm 3i}{2}\right\}$

Find the value of the discriminant for each quadratic equation. Then describe the number and type of roots for the equation. *(Lesson 6-5)*

3. $5x^2 - 3x + 1 = 0$ **−11; 2 complex**
4. $3x^2 + 4x - 7 = 0$ **100; 2 rational**

Solve each equation by using the Quadratic Formula. *(Lesson 6-5)*

5. $x^2 + 9x - 11 = 0$ $\left\{\frac{-9 \pm 5\sqrt{5}}{2}\right\}$
6. $-3x^2 + 4x = 4$ $\left\{\frac{2 \pm 2i\sqrt{2}}{3}\right\}$
7. Write an equation for a parabola with vertex at (2, −5) that passes through (−1, 1). *(Lesson 6-6)* $y = \frac{2}{3}(x - 2)^2 - 5$

Write each equation in vertex form. Then identify the vertex, axis of symmetry, and direction of opening. *(Lesson 6-6)* **8–10. See margin.**

8. $y = x^2 + 8x + 18$
9. $y = -x^2 + 12x - 36$
10. $y = 2x^2 + 12x + 13$

4 Assess

Open-Ended Assessment

Writing Give students the equation of a specific parabola and ask them to put it in vertex form, analyze it, and sketch the graph. Instruct them to show all the steps in their procedures, with notes and explanations for each step, similar to the notes in the right column of the Examples.

Getting Ready for Lesson 6-7

PREREQUISITE SKILL Lesson 6-7 presents quadratic inequalities. Use Exercises 68–71 to determine your students' familiarity with inequalities.

Assessment Options

Practice Quiz 2 The quiz provides students with a brief review of the concepts and skills in Lessons 6-4 through 6-6. Lesson numbers are given to the right of exercises or instruction lines so students can review concepts not yet mastered.

Quiz (Lessons 6-5 and 6-6) is available on p. 370 of the *Chapter 6 Resource Masters.*

Answers

63. $2t^2 + 2t - \frac{3}{t - 1}$
64. $t^2 - 2t + 1$
65. $n^3 - 3n^2 - 15n - 21$
66. $y^3 + 1 - \frac{4}{y + 3}$

Answers (Practice Quiz 2)

8. $y = (x + 4)^2 + 2$; $(-4, 2)$, $x = -4$; up
9. $y = -(x - 6)^2$; $(6, 0)$, $x = 6$; down
10. $y = 2(x + 3)^2 - 5$; $(-3, -5)$, $x = -3$; up

Online Lesson Plans

USA TODAY Education's Online site offers resources and interactive features connected to each day's newspaper. *Experience TODAY*, USA TODAY's daily lesson plan, is available on the site and delivered daily to subscribers. This plan provides instruction for integrating USA TODAY graphics and key editorial features into your mathematics classroom. Log on to www.education.usatoday.com.

6-7 Graphing and Solving Quadratic Inequalities

What You'll Learn

- Graph quadratic inequalities in two variables.
- Solve quadratic inequalities in one variable.

How can you find the time a trampolinist spends above a certain height?

Trampolining was first featured as an Olympic sport at the 2000 Olympics in Sydney, Australia. Suppose the height $h(t)$ in feet of a trampolinist above the ground during one bounce is modeled by the quadratic function $h(t) = -16t^2 + 42t + 3.75$. We can solve a quadratic inequality to determine how long this performer is more than a certain distance above the ground.

Vocabulary

- quadratic inequality

Study Tip

Look Back

For review of **graphing linear inequalities,** see Lesson 2-7.

TEACHING TIP

Remind students that (0, 0) is a good point to use as a test point.

GRAPH QUADRATIC INEQUALITIES

You can graph **quadratic inequalities** in two variables using the same techniques you used to graph linear inequalities in two variables.

Step 1 Graph the related quadratic equation, $y = ax^2 + bx + c$. Decide if the parabola should be solid or dashed.

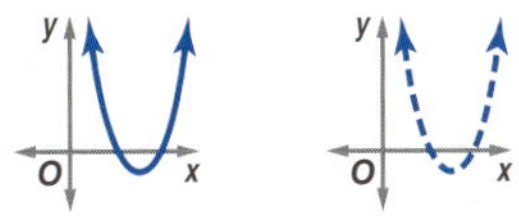

$\leq$ or $\geq$ $\qquad$ $<$ or $>$

Step 2 Test a point (x_1, y_1) inside the parabola. Check to see if this point is a solution of the inequality.

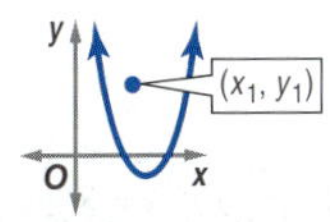

$$y_1 \overset{?}{\geq} a(x_1)^2 + b(x_1) + c$$

Step 3 If (x_1, y_1) is a solution, shade the region *inside* the parabola. If (x_1, y_1) is *not* a solution, shade the region *outside* the parabola.

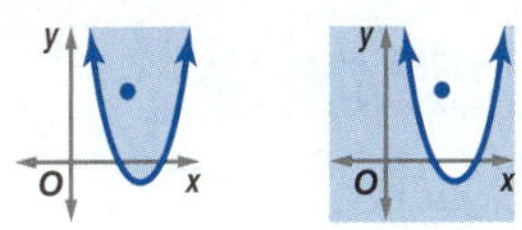

(x_1, y_1) is a solution. $\qquad$ (x_1, y_1) is not a solution.

Example 1 Graph a Quadratic Inequality

Graph $y > -x^2 - 6x - 7$.

Step 1 Graph the related quadratic equation, $y = -x^2 - 6x - 7$.

Since the inequality symbol is $>$, the parabola should be dashed.

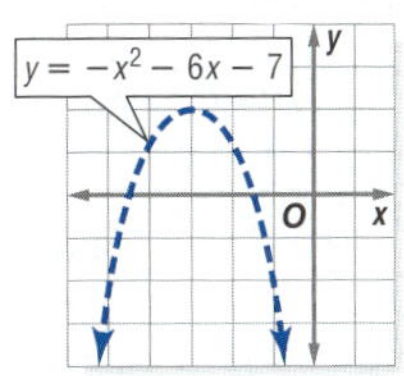

(continued on the next page)

6-7 Lesson Notes

1 Focus

5-Minute Check Transparency 6-7 Use as a quiz or review of Lesson 6-6.

Mathematical Background notes are available for this lesson on p. 284D.

Building on Prior Knowledge

In Lesson 6-6, students analyzed and graphed equations. In this lesson, students use the same techniques to graph and solve inequalities.

How can you find the time a trampolinist spends above a certain height?

Ask students:

- What is a trampoline and how does a trampolinist use it? **Ask a student who is familiar with this sport to explain it to those who may not have seen it.**
- Which way does the parabola for the given quadratic equation open? **downward**

Resource Manager

Workbook and Reproducible Masters

Chapter 6 Resource Masters

- Study Guide and Intervention, pp. 349–350
- Skills Practice, p. 351
- Practice, p. 352
- Reading to Learn Mathematics, p. 353
- Enrichment, p. 354
- Assessment, p. 370

Graphing Calculator and Spreadsheet Masters, p. 37

School-to-Career Masters, p. 12

Transparencies

5-Minute Check Transparency 6-7

Answer Key Transparencies

Technology

Interactive Chalkboard

2 Teach

GRAPH QUADRATIC INEQUALITIES

In-Class Example (Power Point®)

1 Graph $y > x^2 - 3x + 2$.

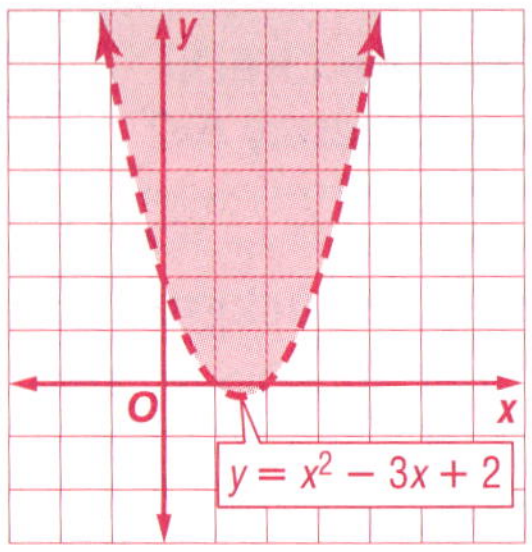

Teaching Tip For In-Class Example 1, encourage students to write the calculations as in Step 2 of Example 1 when testing a point inside the parabola. Emphasize that they must write a question mark over each inequality sign after substituting the coordinates of the point for the variables in the inequality.

SOLVE QUADRATIC INEQUALITIES

In-Class Example (Power Point®)

2 Solve $x^2 - 4x + 3 > 0$ by graphing.

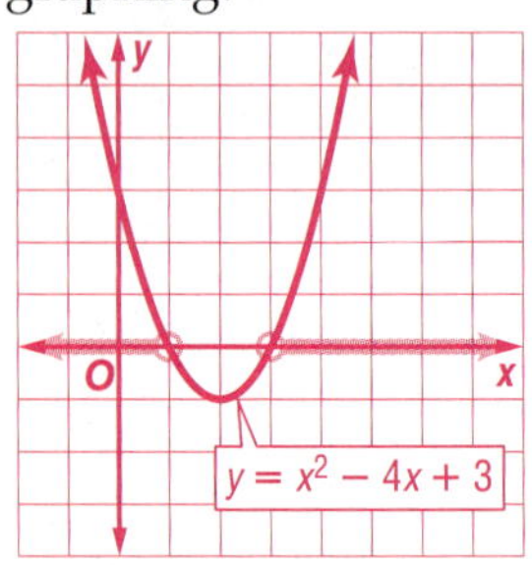

$\{x \mid x < 1 \text{ or } x > 3\}$

Teaching Tip Suggest that students try solving first by factoring, but if that does not quickly yield a solution, then use the Quadratic Formula.

Step 2 Test a point inside the parabola, such as $(-3, 0)$.

$y > -x^2 - 6x - 7$

$0 \overset{?}{>} -(-3)^2 - 6(-3) - 7$

$0 \overset{?}{>} -9 + 18 - 7$

$0 \overset{?}{>} 2$ ✗

So, $(-3, 0)$ is *not* a solution of the inequality.

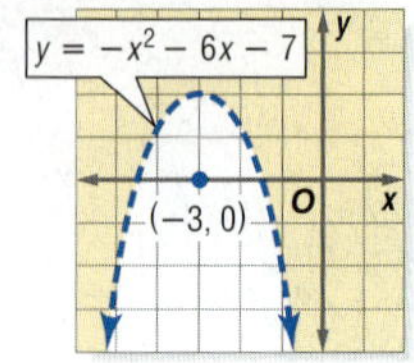

Step 3 Shade the region outside the parabola.

SOLVE QUADRATIC INEQUALITIES To solve a quadratic inequality in one variable, you can use the graph of the related quadratic function.

To solve $ax^2 + bx + c < 0$, graph $y = ax^2 + bx + c$. Identify the x values for which the graph lies *below* the x-axis.

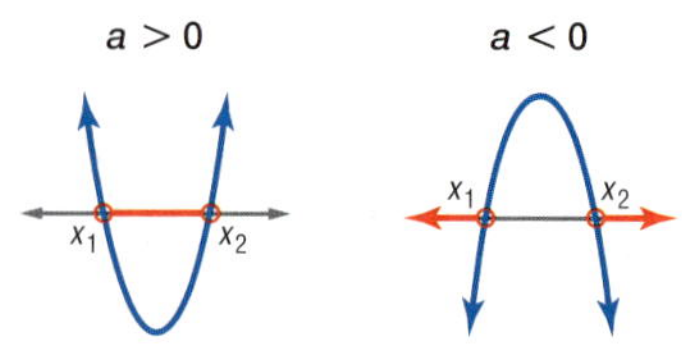

$\{x \mid x_1 < x < x_2\}$

For $\leq$, include the x-intercepts in the solution.

To solve $ax^2 + bx + c > 0$, graph $y = ax^2 + bx + c$. Identify the x values for which the graph lies *above* the x-axis.

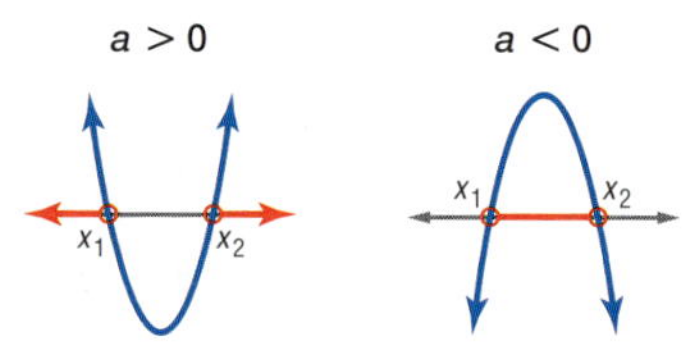

$\{x \mid x < x_1 \text{ or } x > x_2\}$

For $\geq$, include the x-intercepts in the solution.

> **Study Tip**
>
> *Solving Quadratic Inequalities by Graphing*
>
> A precise graph of the related quadratic function is not necessary since the zeros of the function were found algebraically.

Example 2 Solve $ax^2 + bx + c > 0$

Solve $x^2 + 2x - 3 > 0$ by graphing.

The solution consists of the x values for which the graph of the related quadratic function lies *above* the x-axis. Begin by finding the roots of the related equation.

$x^2 + 2x - 3 = 0$	Related equation
$(x + 3)(x - 1) = 0$	Factor.
$x + 3 = 0$ or $x - 1 = 0$	Zero Product Property
$x = -3$ $\quad$ $x = 1$	Solve each equation.

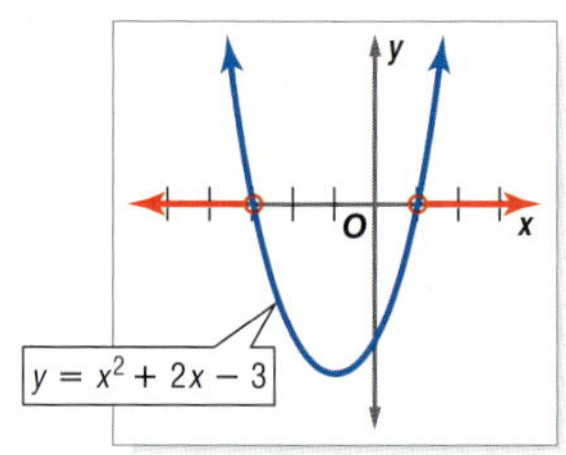

Sketch the graph of a parabola that has x-intercepts at -3 and 1. The graph should open up since $a > 0$.

The graph lies above the x-axis to the left of $x = -3$ and to the right of $x = 1$. Therefore, the solution set is $\{x \mid x < -3 \text{ or } x > 1\}$.

330 Chapter 6 Quadratic Functions and Inequalities

Example 3 Solve $ax^2 + bx + c \le 0$

Solve $0 \ge 3x^2 - 7x - 1$ by graphing.

This inequality can be rewritten as $3x^2 - 7x - 1 \le 0$. The solution consists of the x values for which the graph of the related quadratic function lies *on and below* the x-axis. Begin by finding the roots of the related equation.

$3x^2 - 7x - 1 = 0$ — Related equation

$x = \frac{-b \pm \sqrt{b^2 - 4ac}}{2a}$ — Use the Quadratic Formula.

$x = \frac{-(-7) \pm \sqrt{(-7)^2 - 4(3)(-1)}}{2(3)}$ — Replace a with 3, b with -7, and c with -1.

$x = \frac{7 + \sqrt{61}}{6}$ or $x = \frac{7 - \sqrt{61}}{6}$ — Simplify and write as two equations.

$x \approx 2.47$ $\quad x \approx -0.14$ — Simplify.

Sketch the graph of a parabola that has x-intercepts of 2.47 and -0.14. The graph should open up since $a > 0$.

The graph lies on and below the x-axis at $x = -0.14$ and $x = 2.47$ and between these two values. Therefore, the solution set of the inequality is approximately $\{x \mid -0.14 \le x \le 2.47\}$.

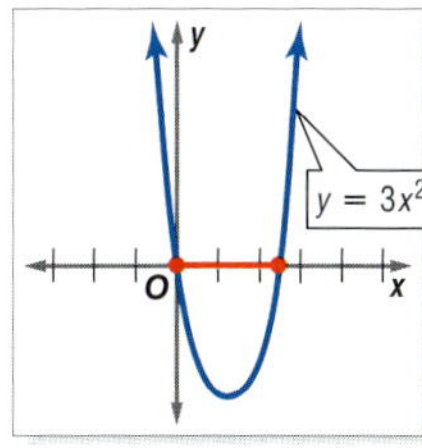

CHECK Test one value of x less than -0.14, one between -0.14 and 2.47, and one greater than 2.47 in the original inequality.

Test $x = -1$.	Test $x = 0$.	Test $x = 3$.
$0 \ge 3x^2 - 7x - 1$	$0 \ge 3x^2 - 7x - 1$	$0 \ge 3x^2 - 7x - 1$
$0 \overset{?}{\ge} 3(-1)^2 - 7(-1) - 1$	$0 \overset{?}{\ge} 3(0)^2 - 7(0) - 1$	$0 \overset{?}{\ge} 3(3)^2 - 7(3) - 1$
$0 \ge 9$ X	$0 \ge -1$ ✓	$0 \ge 5$ X

Real-world problems that involve vertical motion can often be solved by using a quadratic inequality.

Example 4 Write an Inequality

FOOTBALL The height of a punted football can be modeled by the function $H(x) = -4.9x^2 + 20x + 1$, where the height $H(x)$ is given in meters and the time x is in seconds. At what time in its flight is the ball within 5 meters of the ground?

The function $H(x)$ describes the height of the football. Therefore, you want to find the values of x for which $H(x) \le 5$.

$H(x) \le 5$ — Original inequality

$-4.9x^2 + 20x + 1 \le 5$ — $H(x) = -4.9x^2 + 20x + 1$

$-4.9x^2 + 20x - 4 \le 0$ — Subtract 5 from each side.

Graph the related function $y = -4.9x^2 + 20x - 4$ using a graphing calculator. The zeros of the function are about 0.21 and 3.87, and the graph lies below the x-axis when $x < 0.21$ or $x > 3.87$.

Thus, the ball is within 5 meters of the ground for the first 0.21 second of its flight and again after 3.87 seconds until the ball hits the ground at 4.13 seconds.

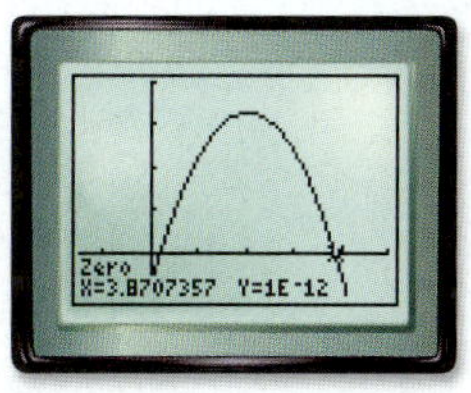

[−1.5, 5] scl: 1 by [−5, 20] scl: 5

More About...

Football A long hang time allows the kicking team time to provide good coverage on a punt return. The suggested hang time for high school and college punters is 4.5–4.6 seconds.

Source: www.takeaknee.com

www.algebra2.com/extra_examples

In-Class Examples

Power Point®

3 Solve $0 \le -2x^2 - 6x + 1$ by graphing.

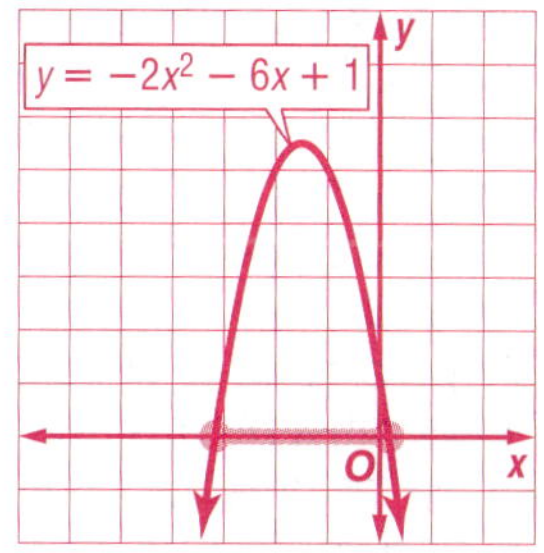

$\{x \mid -3.16 \le x \le 0.16\}$

Teaching Tip When discussing Example 4, be aware that some students may not be familiar with all of the aspects of the game of football. Ask students who are familiar with the terms in this example and the margin note to explain them to the class.

4 **SPORTS** The height of a ball above the ground after it is thrown upwards at 40 feet per second can be modeled by the function $h(x) = 40x - 16x^2$, where the height $h(x)$ is given in feet and the time x is in seconds. At what time in its flight is the ball within 15 feet of the ground? **The ball is within 15 feet of the ground for the first 0.46 second of its flight and again after 2.04 seconds until the ball hits the ground at 2.5 seconds.**

DAILY INTERVENTION — Differentiated Instruction

Intrapersonal Have students think about how the graph of a quadratic inequality helps them understand what the inequality means. Ask them to explore which is more meaningful to them (and therefore easier for them to grasp), the quadratic inequality itself or the graph of the inequality. Ask them to give an explanation of their choice.

Lesson 6-7 Graphing and Solving Quadratic Inequalities 331

In-Class Example

Teaching Tip An alternative way to solve the inequality in Example 5 is to first subtract 6 from both sides of the inequality, obtaining $x^2 + x - 6 > 0$. After factoring the left side as $(x + 3)(x - 2)$, point out that for the product $(x + 3)(x - 2)$ to be greater than 0, either both binomials must be positive or both must be negative. This fact can be used to test the three intervals of the number line.

5 Solve $x^2 + x \leq 2$ algebraically. $\{x \mid -2 \leq x \leq 1\}$

3 Practice/Apply

Study Notebook

Have students—
- complete the definitions/examples for the remaining terms on their Vocabulary Builder worksheets for Chapter 6.
- write the basic steps shown in Example 1 for solving a quadratic inequality by graphing.
- include any other item(s) that they find helpful in mastering the skills in this lesson.

You can also solve quadratic inequalities algebraically.

Study Tip

Solving Quadratic Inequalities Algebraically
As with linear inequalities, the solution set of a quadratic inequality can be all real numbers or the empty set, ∅. The solution is all real numbers when all three test points satisfy the inequality. It is the empty set when none of the tests points satisfy the inequality.

Example 5 Solve a Quadratic Inequality

Solve $x^2 + x > 6$ algebraically.

First solve the related quadratic equation $x^2 + x = 6$.

$x^2 + x = 6$	Related quadratic equation
$x^2 + x - 6 = 0$	Subtract 6 from each side.
$(x + 3)(x - 2) = 0$	Factor.
$x + 3 = 0$ or $x - 2 = 0$	Zero Product Property
$x = -3$ $\quad$ $x = 2$	Solve each equation.

Plot –3 and 2 on a number line. Use circles since these values are not solutions of the original inequality. Notice that the number line is now separated into three intervals.

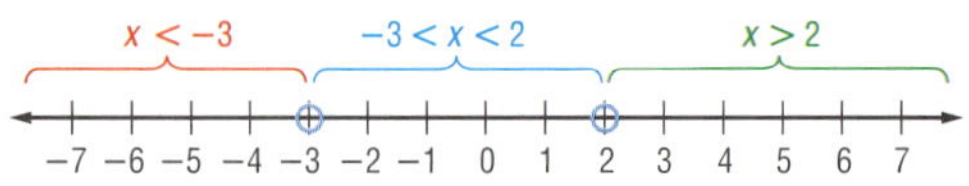

Test a value in each interval to see if it satisfies the original inequality.

$x < -3$	$-3 < x < 2$	$x > 2$
Test $x = -4$.	Test $x = 0$.	Test $x = 4$.
$x^2 + x > 6$	$x^2 + x > 6$	$x^2 + x > 6$
$(-4)^2 + (-4) \overset{?}{>} 6$	$0^2 + 0 \overset{?}{>} 6$	$4^2 + 4 \overset{?}{>} 6$
$12 > 6$ ✓	$0 > 6$ ✗	$20 > 6$ ✓

The solution set is $\{x \mid x < -3 \text{ or } x > 2\}$. This is shown on the number line below.

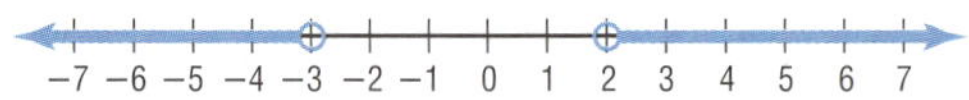

Check for Understanding

Concept Check

1. **Determine** which inequality, $y \geq (x - 3)^2 - 1$ or $y \leq (x - 3)^2 - 1$, describes the graph at the right. $y \geq (x - 3)^2 - 1$

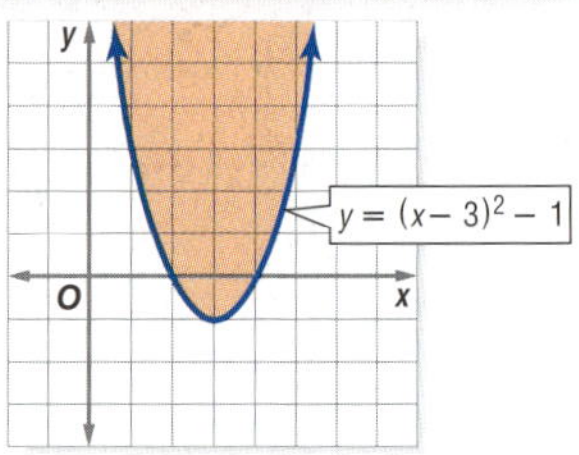

2. **OPEN ENDED** List three points you might test to find the solution of $(x + 3)(x - 5) < 0$.
Sample answer: one number less than –3, one number between –3 and 5, and one number greater than 5

3. **Examine** the graph of $y = x^2 - 4x - 5$ at the right. **a. $x = -1, 5$ b. $x \leq -1$ or $x \geq 5$**
   a. What are the solutions of $0 = x^2 - 4x - 5$?
   b. What are the solutions of $x^2 - 4x - 5 \geq 0$?
   c. What are the solutions of $x^2 - 4x - 5 \leq 0$? $-1 \leq x \leq 5$

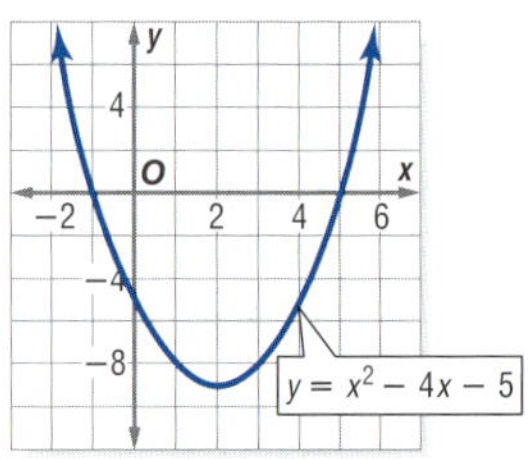

332 Chapter 6 Quadratic Functions and Inequalities

Guided Practice

GUIDED PRACTICE KEY	
Exercises	Examples
4–7	1
8	2, 3
9–12	2, 3, 5
13	4

Graph each inequality. **4–7. See margin.**

4. $y \geq x^2 - 10x + 25$
5. $y < x^2 - 16$
6. $y > -2x^2 - 4x + 3$
7. $y \leq -x^2 + 5x + 6$

8. Use the graph of the related function of $-x^2 + 6x - 5 < 0$, which is shown at the right, to write the solutions of the inequality. **$x < 1$ or $x > 5$**

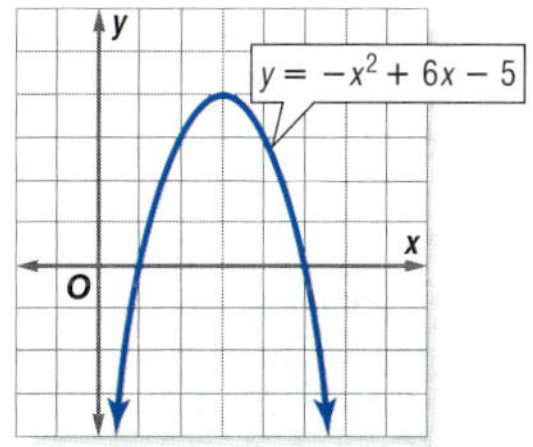

Solve each inequality algebraically.

9. $x^2 - 6x - 7 < 0$ **$\{x \mid -1 < x < 7\}$**
10. $x^2 - x - 12 > 0$ **$\{x \mid x < -3 \text{ or } x > 4\}$**
11. $x^2 < 10x - 25$ **$\varnothing$**
12. $x^2 \leq 3$ **$\{x \mid -\sqrt{3} \leq x \leq \sqrt{3}\}$**

Application

13. **BASEBALL** A baseball player hits a high pop-up with an initial upward velocity of 30 meters per second, 1.4 meters above the ground. The height $h(t)$ of the ball in meters t seconds after being hit is modeled by $h(t) = -4.9t^2 + 30t + 1.4$. How long does a player on the opposing team have to catch the ball if he catches it 1.7 meters above the ground? **about 6.1 s**

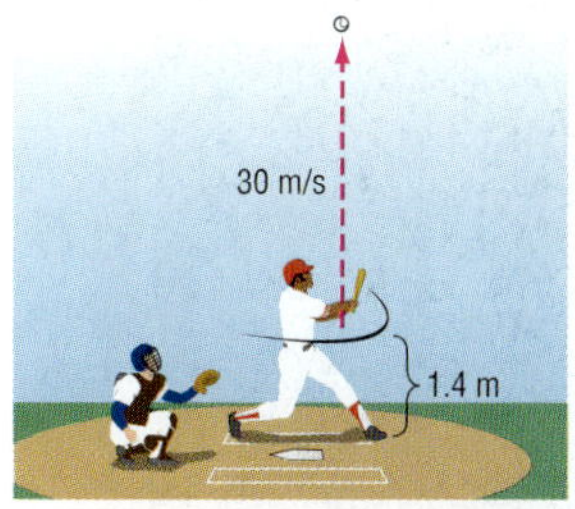

★ indicates increased difficulty

Practice and Apply

Homework Help	
For Exercises	See Examples
14–25	1
26–29	2, 3
30–42	2, 3, 5
43–48	4

Extra Practice
See page 841.

Graph each inequality. **14–25. See pp. 343A–343F.**

14. $y \geq x^2 + 3x - 18$
15. $y < -x^2 + 7x + 8$
16. $y \leq x^2 + 4x + 4$
17. $y \leq x^2 + 4x$
18. $y > x^2 - 36$
19. $y > x^2 + 6x + 5$
20. $y \leq -x^2 - 3x + 10$
21. $y \geq -x^2 - 7x + 10$
22. $y > -x^2 + 10x - 23$
23. $y < -x^2 + 13x - 36$
24. $y < 2x^2 + 3x - 5$
25. $y \geq 2x^2 + x - 3$

Use the graph of its related function to write the solutions of each inequality.

26. $-x^2 + 10x - 25 \geq 0$ **5**

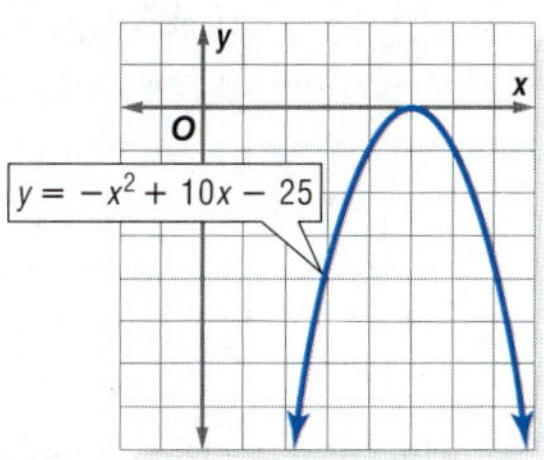

27. $x^2 - 4x - 12 \leq 0$ **$-2 \leq x \leq 6$**

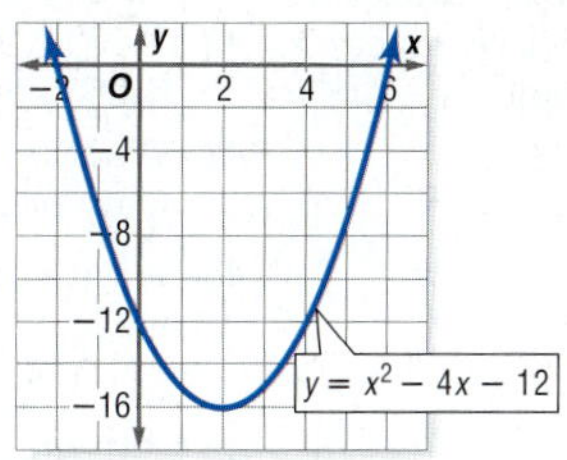

28. $x^2 - 9 > 0$ **$x < -3$ or $x > 3$**

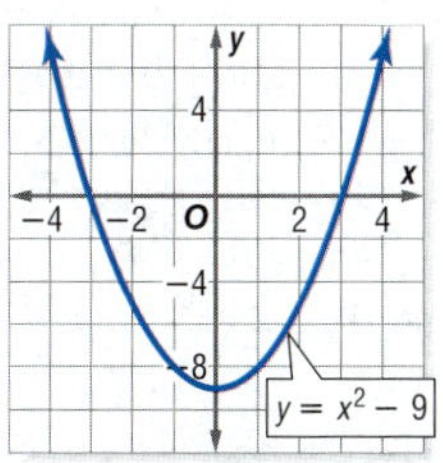

29. $-x^2 - 10x - 21 < 0$ **$x < -7$ or $x > -3$**

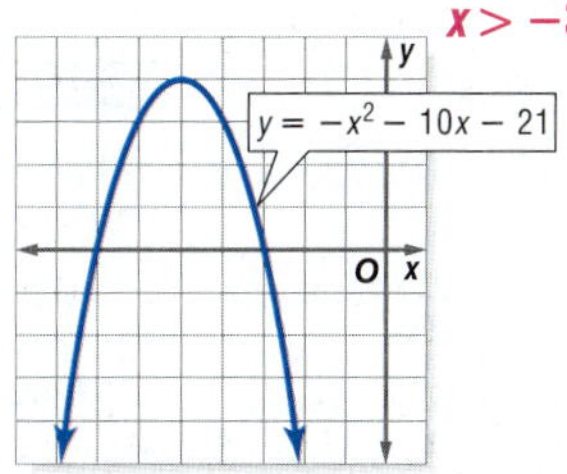

www.algebra2.com/self_check_quiz

About the Exercises...

Organization by Objective
- **Graph Quadratic Inequalities:** 14–25
- **Solve Quadratic Inequalities:** 26–48

Odd/Even Assignments

Exercises 14–41 are structured so that students practice the same concepts whether they are assigned odd or even problems.

Alert! Exercises 53–58 require a graphing calculator.

Assignment Guide

Basic: 15–41 odd, 45, 49–52, 59–71

Average: 15–45 odd, 49–52, 59–71 (optional: 53–58)

Advanced: 14–44 even, 46–71

Answers

4.

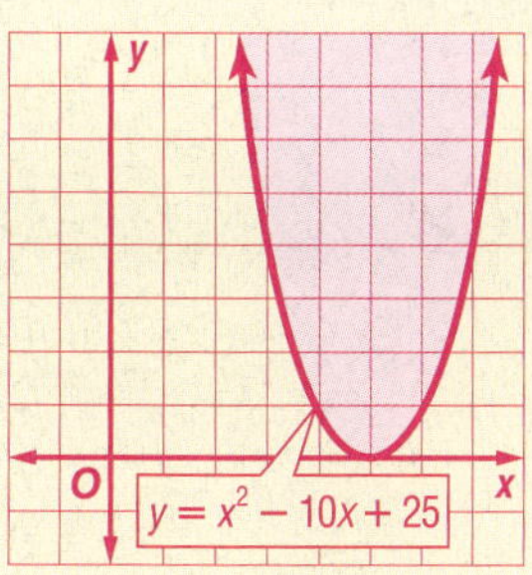

5.

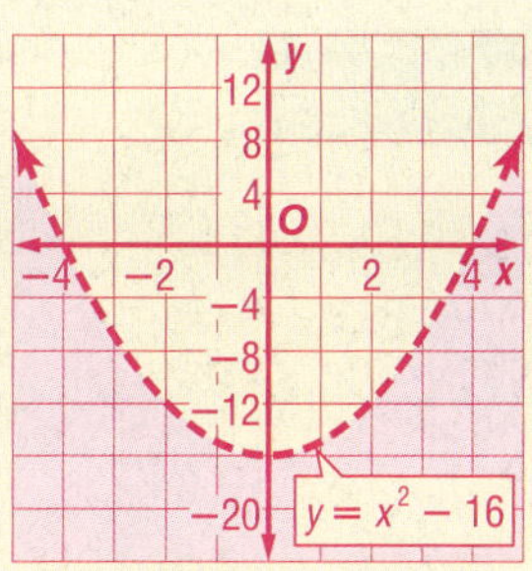

6.

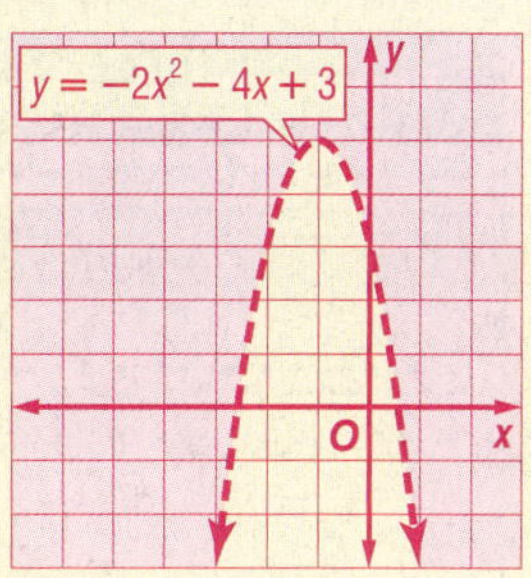

7.

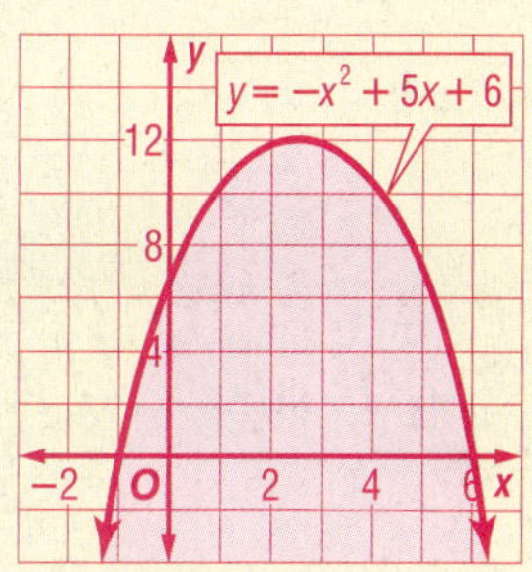

Solve each inequality algebraically.

30. $x^2 - 3x - 18 > 0$ $\{x \mid x < -3 \text{ or } x > 6\}$
31. $x^2 + 3x - 28 < 0$ $\{x \mid -7 < x < 4\}$
32. $x^2 - 4x \leq 5$ $\{x \mid -1 \leq x \leq 5\}$
33. $x^2 + 2x \geq 24$ $\{x \mid x \leq -6 \text{ or } x \geq 4\}$
34. $-x^2 - x + 12 \geq 0$ $\{x \mid -4 \leq x \leq 3\}$
35. $-x^2 - 6x + 7 \leq 0$ $\{x \mid x \leq -7 \text{ or } x \geq 1\}$
36. $9x^2 - 6x + 1 \leq 0$ $\left\{x \mid x = \frac{1}{3}\right\}$
37. $4x^2 + 20x + 25 \geq 0$ all reals
38. $x^2 + 12x < -36$ $\varnothing$
39. $-x^2 + 14x - 49 \geq 0$ $\{x \mid x = 7\}$
40. $18x - x^2 \leq 81$ all reals
41. $16x^2 + 9 < 24x$ $\varnothing$

★42. Solve $(x - 1)(x + 4)(x - 3) > 0$. $\{x \mid -4 < x < 1 \text{ or } x > 3\}$

43. **LANDSCAPING** Kinu wants to plant a garden and surround it with decorative stones. She has enough stones to enclose a rectangular garden with a perimeter of 68 feet, but she wants the garden to cover no more than 240 square feet. What could the width of her garden be? **0 to 10 ft or 24 to 34 ft**

44. **BUSINESS** A mall owner has determined that the relationship between monthly rent charged for store space r (in dollars per square foot) and monthly profit $P(r)$ (in thousands of dollars) can be approximated by the function $P(r) = -8.1r^2 + 46.9r - 38.2$. Solve each quadratic equation or inequality. Explain what each answer tells about the relationship between monthly rent and profit for this mall. **a–d. See margin.**
   a. $-8.1r^2 + 46.9r - 38.2 = 0$
   b. $-8.1r^2 + 46.9r - 38.2 > 0$
   c. $-8.1r^2 + 46.9r - 38.2 > 10$
   d. $-8.1r^2 + 46.9r - 38.2 < 10$

45. **GEOMETRY** A rectangle is 6 centimeters longer than it is wide. Find the possible dimensions if the area of the rectangle is more than 216 square centimeters. **The width should be greater than 12 cm and the length shoud be greater than 18 cm.**

FUND-RAISING **For Exercises 46–48, use the following information.**
The girls' softball team is sponsoring a fund-raising trip to see a professional baseball game. They charter a 60-passenger bus for \$525. In order to make a profit, they will charge \$15 per person if all seats on the bus are sold, but for each empty seat, they will increase the price by \$1.50 per person.

★46. Write a quadratic function giving the softball team's profit $P(n)$ from this fund-raiser as a function of the number of passengers n.

★47. What is the minimum number of passengers needed in order for the softball team not to lose money? **6**

★48. What is the maximum profit the team can make with this fund-raiser, and how many passengers will it take to achieve this maximum? **\$1312.50; 35 passengers**

49. **CRITICAL THINKING** Graph the intersection of the graphs of $y \leq -x^2 + 4$ and $y \geq x^2 - 4$. **See margin.**

50. **WRITING IN MATH** Answer the question that was posed at the beginning of the lesson. **See margin.**

How can you find the time a trampolinist spends above a certain height?

Include the following in your answer:
- a quadratic inequality that describes the time the performer spends more than 10 feet above the ground, and
- two approaches to solving this quadratic inequality.

Career Choices

Landscape Architect

Landscape architects design outdoor spaces so that they are not only functional, but beautiful and compatible with the natural environment.

Online Research
For information about a career as a landscape architect, visit: www.algebra2.com/careers

46. $P(n) = n[15 + 1.5(60 - n)] - 525$ or $-1.5n^2 + 105n - 525$

Study Guide and Intervention, p. 349 (shown) and p. 350

Graph Quadratic Inequalities To graph a quadratic inequality in two variables, use the following steps:

1. Graph the related quadratic equation, $y = ax^2 + bx + c$. Use a dashed line for < or >; use a solid line for ≤ or ≥.
2. Test a point inside the parabola. If it satisfies the inequality, shade the region inside the parabola; otherwise, shade the region outside the parabola.

Example **Graph the inequality $y > x^2 + 6x + 7$.**

First graph the equation $y = x^2 + 6x + 7$. By completing the square, you get the vertex form of the equation $y = (x + 3)^2 - 2$, so the vertex is $(-3, -2)$. Make a table of values around $x = -3$, and graph. Since the inequality includes >, use a dashed line. Test the point $(-3, 0)$, which is inside the parabola. Since $(-3)^2 + 6(-3) + 7 = -2$, and $0 > -2$, $(-3, 0)$ satisfies the inequality. Therefore, shade the region inside the parabola.

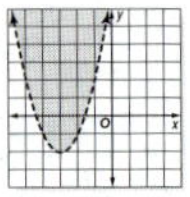

Exercises

Graph each inequality.

1. $y > x^2 - 8x + 17$
2. $y \leq x^2 + 6x + 4$
3. $y \geq x^2 + 2x + 2$
4. $y < -x^2 + 4x - 6$
5. $y \geq 2x^2 + 4x$
6. $y > -2x^2 - 4x + 2$

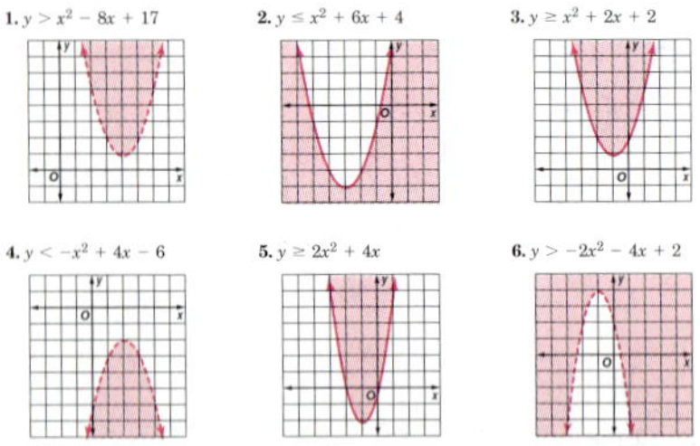

Skills Practice, p. 351 and Practice, p. 352 (shown)

Graph each inequality.

1. $y \leq x^2 + 4$
2. $y > x^2 + 6x + 6$
3. $y < 2x^2 - 4x - 2$

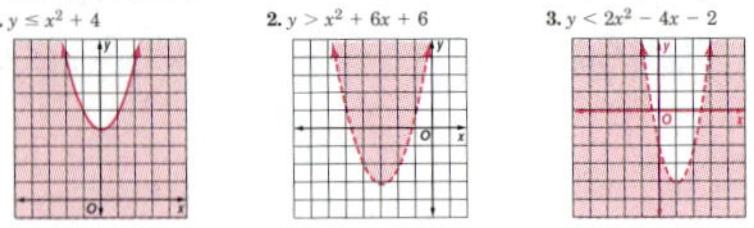

Use the graph of its related function to write the solutions of each inequality.

4. $x^2 - 8x > 0$ $x < 0$ or $x > 8$
5. $-x^2 - 2x + 3 \geq 0$ $-3 \leq x \leq 1$
6. $x^2 - 9x + 14 \leq 0$ $2 \leq x \leq 7$

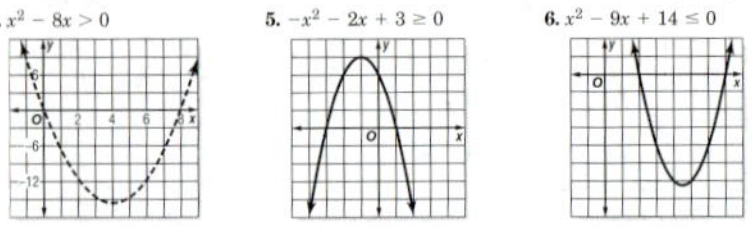

Solve each inequality algebraically.

7. $x^2 - x - 20 > 0$ $\{x \mid x < -4 \text{ or } x > 5\}$
8. $x^2 - 10x + 16 < 0$ $\{x \mid 2 < x < 8\}$
9. $x^2 + 4x + 5 \leq 0$ $\varnothing$
10. $x^2 + 14x + 49 \geq 0$ all reals
11. $x^2 - 5x > 14$ $\{x \mid x < -2 \text{ or } x > 7\}$
12. $-x^2 - 15 \geq 8x$ $\{x \mid -5 \leq x \leq -3\}$
13. $-x^2 + 5x - 7 \leq 0$ all reals
14. $9x^2 + 36x + 36 \leq 0$ $\{x \mid x = -2\}$
15. $9x \leq 12x^2$ $\left\{x \mid x \leq 0 \text{ or } x \geq \frac{3}{4}\right\}$
16. $4x^2 + 4x + 1 > 0$ $\left\{x \mid x \neq -\frac{1}{2}\right\}$
17. $5x^2 + 10 \geq 27x$ $\left\{x \mid x \leq \frac{2}{5} \text{ or } x \geq 5\right\}$
18. $9x^2 + 31x + 12 \leq 0$ $\left\{x \mid -3 \leq x \leq -\frac{4}{9}\right\}$

19. **FENCING** Vanessa has 180 feet of fencing that she intends to use to build a rectangular play area for her dog. She wants the play area to enclose at least 1800 square feet. What are the possible widths of the play area? 30 ft to 60 ft

20. **BUSINESS** A bicycle maker sold 300 bicycles last year at a profit of \$300 each. The maker wants to increase the profit margin this year, but predicts that each \$20 increase in profit will reduce the number of bicycles sold by 10. How many \$20 increases in profit can the maker add in and expect to make a total profit of at least \$100,000? from 5 to 10

Reading to Learn Mathematics, p. 353 ELL

Pre-Activity **How can you find the time a trampolinist spends above a certain height?**

Read the introduction to Lesson 6-7 at the top of page 329 in your textbook.

- How far above the ground is the trampoline surface? 3.75 feet
- Using the quadratic function given in the introduction, write a quadratic inequality that describes the times at which the trampolinist is more than 20 feet above the ground. $-16t^2 + 42t + 3.75 > 20$

Reading the Lesson

1. Answer the following questions about how you would graph the inequality $y \geq x^2 + x - 6$.
   a. What is the related quadratic equation? $y = x^2 + x - 6$
   b. Should the parabola be solid or dashed? How do you know? solid; The inequality symbol is ≥.
   c. The point (0, 2) is inside the parabola. To use this as a test point, substitute __0__ for x and __2__ for y in the quadratic inequality.
   d. Is the statement $2 \geq 0^2 + 0 - 6$ true or false? true
   e. Should the region inside or outside the parabola be shaded? inside

2. The graph of $y = -x^2 + 4x$ is shown at the right. Match each of the following related inequalities with its solution set.

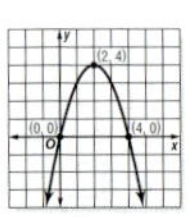

   a. $-x^2 + 4x > 0$ ii
   b. $-x^2 + 4x \leq 0$ iii
   c. $-x^2 + 4x \geq 0$ iv
   d. $-x^2 + 4x < 0$ i

   i. $\{x \mid x < 0 \text{ or } x > 4\}$
   ii. $\{x \mid 0 < x < 4\}$
   iii. $\{x \mid x \leq 0 \text{ or } x \geq 4\}$
   iv. $\{x \mid 0 \leq x \leq 4\}$

Helping You Remember

3. A quadratic inequality in two variables may have the form $y > ax^2 + bx + c$, $y < ax^2 + bx + c$, $y \geq ax^2 + bx + c$, or $y \leq ax^2 + bx + c$. Describe a way to remember which region to shade by looking at the inequality symbol and without using a test point. Sample answer: If the symbol is > or ≥, shade the region above the parabola. If the symbol is < or ≤, shade the region below the parabola.

Enrichment, p. 354

Graphing Absolute Value Inequalities

You can solve absolute value inequalities by graphing in much the same manner you graphed quadratic inequalities. Graph the related absolute function for each inequality by using a graphing calculator. For > and ≥, identify the x-values, if any, for which the graph lies *below* the x-axis. For < and ≤, identify the x values, if any, for which the graph lies *above* the x-axis.

For each inequality, make a sketch of the related graph and find the solutions rounded to the nearest hundredth.

1. $|x - 3| > 0$ $x > 3$ or $x < 3$
2. $|x| - 6 < 0$ $-6 < x < 6$
3. $-|x + 4| + 8 < 0$ $-12 < x < 4$

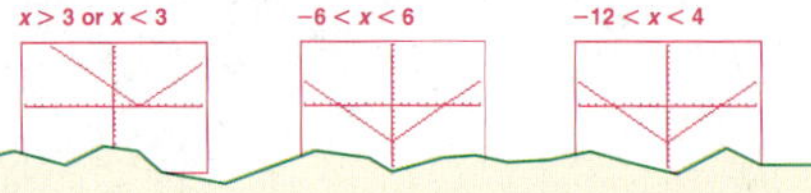

Standardized Test Practice

51. Which is a reasonable estimate of the area under the curve from $x = 0$ to $x = 18$? **C**

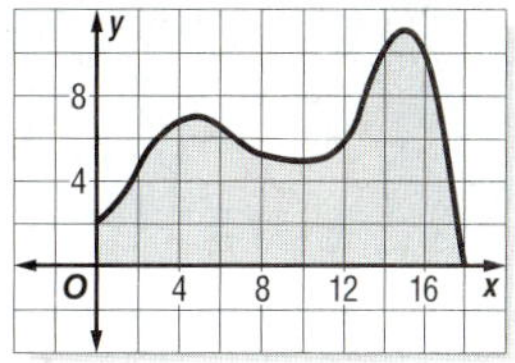

Ⓐ 29 square units
Ⓑ 58 square units
Ⓒ 116 square units
Ⓓ 232 square units

52. If $(x + 1)(x - 2)$ is positive, then **A**

Ⓐ $x < -1$ or $x > 2$.
Ⓑ $x > -1$ or $x < 2$.
Ⓒ $-1 < x < 2$.
Ⓓ $-2 < x < 1$.

Extending the Lesson

SOLVE ABSOLUTE VALUE INEQUALITIES BY GRAPHING Similar to quadratic inequalities, you can solve absolute value inequalities by graphing.

Graph the related absolute value function for each inequality using a graphing calculator. For > and ≥, identify the *x* values, if any, for which the graph lies *below* the *x*-axis. For < and ≤, identify the *x* values, if any, for which the graph lies *above* the *x*-axis.

53. $|x - 2| > 0$ **$\{x \mid \text{all reals}, x \neq 2\}$**

54. $|x| - 7 < 0$ **$\{x \mid -7 < x < 7\}$**

55. $-|x + 3| + 6 < 0$ **$\{x \mid x < -9 \text{ or } x > 3\}$**

56. $2|x + 3| - 1 \geq 0$ **$\{x \mid x \leq -3.5 \text{ or } x \geq -2.5\}$**

57. $|5x + 4| - 2 \leq 0$ **$\{x \mid -1.2 \leq x \leq -0.4\}$**

58. $|4x - 1| + 3 < 0$ **no real solutions**

Maintain Your Skills

Mixed Review

Write each equation in vertex form. Then identify the vertex, axis of symmetry, and direction of opening. *(Lesson 6-6)*

59. $y = x^2 - 2x + 9$

60. $y = -2x^2 + 16x - 32$

61. $y = \frac{1}{2}x^2 + 6x + 18$

59. $y = (x - 1)^2 + 8$; $(1, 8)$, $x = 1$; up
60. $y = -2(x - 4)^2$; $(4, 0)$, $x = 4$; down
61. $y = \frac{1}{2}(x + 6)^2$; $(-6, 0)$, $x = -6$; up
63. $\frac{-5 \pm i\sqrt{3}}{2}$

Solve each equation using the method of your choice. Find exact solutions. *(Lesson 6-5)*

62. $x^2 + 12x + 32 = 0$ **$-4, -8$**

63. $x^2 + 7 = -5x$

64. $3x^2 + 6x - 2 = 3$ **$\frac{-3 \pm 2\sqrt{6}}{3}$**

Simplify. *(Lesson 5-2)* **65. $4a^2b^2 + 2a^2b + 4ab^2 + 12a - 7b$**

65. $(2a^2b - 3ab^2 + 5a - 6b) + (4a^2b^2 + 7ab^2 - b + 7a)$

66. $(x^3 - 3x^2y + 4xy^2 + y^3) - (7x^3 + x^2y - 9xy^2 + y^3)$ **$-6x^3 - 4x^2y + 13xy^2$**

67. $x^{-3}y^2(x^4y + x^3y^{-1} + x^2y^{-2})$ **$xy^3 + y + \frac{1}{x}$**

68. $(5a - 3)(1 - 3a)$ **$-15a^2 + 14a - 3$**

Find each product, if possible. *(Lesson 4-3)*

69. $\begin{bmatrix} -6 & 3 \\ 4 & 7 \end{bmatrix} \cdot \begin{bmatrix} 2 & -5 \\ -3 & 6 \end{bmatrix}$ **$\begin{bmatrix} -21 & 48 \\ -13 & 22 \end{bmatrix}$**

70. $\begin{bmatrix} 2 & -6 & 3 \end{bmatrix} \cdot \begin{bmatrix} 3 & -3 \\ 9 & 0 \\ -2 & 4 \end{bmatrix}$ **$\begin{bmatrix} -54 & 6 \end{bmatrix}$**

71. **LAW ENFORCEMENT** Thirty-four states classify drivers having at least a 0.1 blood alcohol content (BAC) as intoxicated. An infrared device measures a person's BAC through an analysis of his or her breath. A certain detector measures BAC to within 0.002. If a person's actual blood alcohol content is 0.08, write and solve an absolute value equation to describe the range of BACs that might register on this device. *(Lesson 1-6)* **$|x - 0.08| \leq 0.002$; $0.078 \leq x \leq 0.082$**

4 Assess

Open-Ended Assessment

Speaking Have students explain how to test points in the coordinate plane in order to determine which region represents the solution to a quadratic inequality. Also ask them to explain how to analyze the graph of a quadratic equation in order to determine the solution set for a quadratic inequality.

Assessment Options

Quiz (Lesson 6-7) is available on p. 370 of the *Chapter 6 Resource Masters*.

Answers

49.

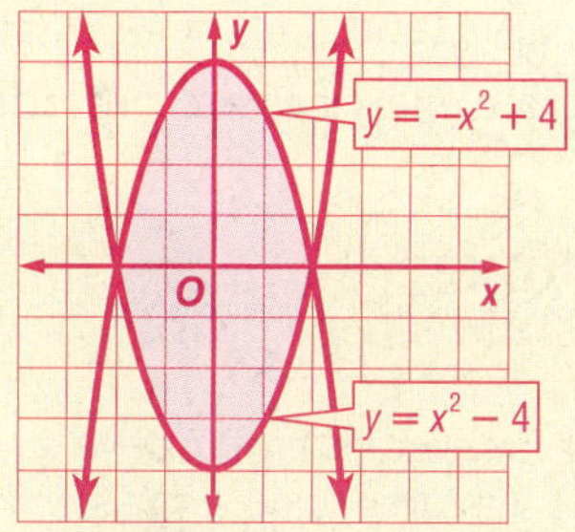

50. Answers should include the following.

- **$-16t^2 + 42t + 3.75 > 10$**
- **One method of solving this inequality is to graph the related quadratic function $h(t) = -16t^2 + 42t + 3.75 - 10$. The interval(s) at which the graph is above the *x*-axis represents the times when the trampolinist is above 10 feet. A second method of solving this inequality would be find the roots of the related quadratic equation $-16t^2 + 42t + 3.75 - 10 = 0$ and then test points in the three intervals determined by these roots to see if they satisfy the inequality. The interval(s) at which the inequality is satisfied represent the times when the trampolinist is above 10 feet.**

Answers

44a. 0.98, 4.81; The owner will break even if he charges \$0.98 or \$4.81 per square foot.

44b. $0.98 < r < 4.81$; The owner will make a profit if the rent is between \$0.98 and \$4.81.

44c. $1.34 < r < 4.45$; If rent is set between \$1.34 and \$4.45 per sq ft, the profit will be greater than \$10,000.

44d. $r < 1.34$ or $r > 4.45$; If rent is set between \$0 and \$1.34 or above \$4.45 per sq ft, the profit will be less than \$10,000.

Chapter 6 Study Guide and Review

Vocabulary and Concept Check

- This alphabetical list of vocabulary terms in Chapter 6 includes a page reference where each term was introduced.
- **Assessment** A vocabulary test/review for Chapter 6 is available on p. 368 of the *Chapter 6 Resource Masters*.

Lesson-by-Lesson Review

For each lesson,

- the main ideas are summarized,
- additional examples review concepts, and
- practice exercises are provided.

Vocabulary PuzzleMaker

ELL The Vocabulary PuzzleMaker software improves students' mathematics vocabulary using four puzzle formats—crossword, scramble, word search using a word list, and word search using clues. Students can work on a computer screen or from a printed handout.

MindJogger Videoquizzes

ELL MindJogger Videoquizzes provide an alternative review of concepts presented in this chapter. Students work in teams in a game show format to gain points for correct answers. The questions are presented in three rounds.

Round 1 Concepts (5 questions)
Round 2 Skills (4 questions)
Round 3 Problem Solving (4 questions)

Chapter 6 Study Guide and Review

Vocabulary and Concept Check

axis of symmetry (p. 287)	parabola (p. 286)	Square Root Property (p. 306)
completing the square (p. 307)	quadratic equation (p. 294)	vertex (p. 287)
constant term (p. 286)	Quadratic Formula (p. 313)	vertex form (p. 322)
discriminant (p. 316)	quadratic function (p. 286)	Zero Product Property (p. 301)
linear term (p. 286)	quadratic inequality (p. 329)	zeros (p. 294)
maximum value (p. 288)	quadratic term (p. 286)	
minimum value (p. 288)	roots (p. 294)	

Choose the letter of the term that best matches each phrase.

1. the graph of any quadratic function **f**
2. process used to create a perfect square trinomial **b**
3. the line passing through the vertex of a parabola and dividing the parabola into two mirror images **a**
4. a function described by an equation of the form $f(x) = ax^2 + bx + c$, where $a \neq 0$ **h**
5. the solutions of an equation **i**
6. $y = a(x - h)^2 + k$ **j**
7. in the Quadratic Formula, the expression under the radical sign, $b^2 - 4ac$ **c**
8. $x = \dfrac{-b \pm \sqrt{b^2 - 4ac}}{2a}$ **g**

a. axis of symmetry
b. completing the square
c. discriminant
d. constant term
e. linear term
f. parabola
g. Quadratic Formula
h. quadratic function
i. roots
j. vertex form

Lesson-by-Lesson Review

6-1 Graphing Quadratic Functions

See pages 286–293.

Concept Summary

The graph of $y = ax^2 + bx + c, a \neq 0$,

- opens up, and the function has a minimum value when $a > 0$, and
- opens down, and the function has a maximum value when $a < 0$.

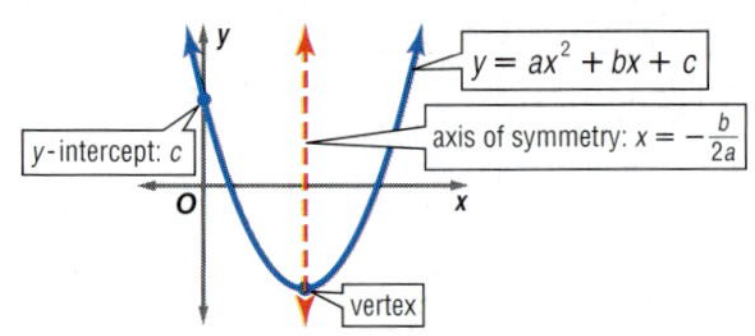

Example **Find the maximum or minimum value of $f(x) = -x^2 + 4x - 12$.**

Since $a < 0$, the graph opens down and the function has a maximum value. The maximum value of the function is the y-coordinate of the vertex. The x-coordinate of the vertex is $x = -\frac{4}{2(-1)}$ or 2. Find the y-coordinate by evaluating the function for $x = 2$.

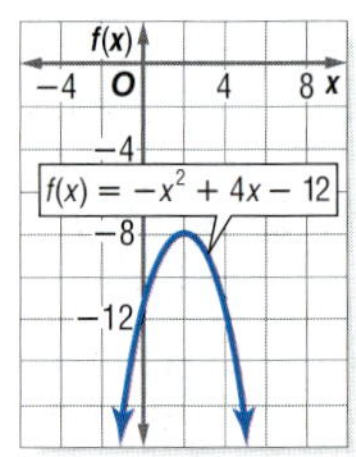

$f(x) = -x^2 + 4x - 12$ Original function

$f(2) = -(2)^2 + 4(2) - 12$ or -8 Replace x with 2.

Therefore, the maximum value of the function is -8.

www.algebra2.com/vocabulary_review

FOLDABLES™ Study Organizer

For more information about Foldables, see *Teaching Mathematics with Foldables.*

Discuss with students how they might recognize a key concept that needs to be included in the Foldable. Ask them to include a transition sentence or two in their notes that relates one topic to the next. Suggest that they use this discussion as they review their Foldable to add, delete, or reorganize material in order to make it more useful to them.

Encourage students to refer to their Foldables while completing the Study Guide and Review and to use them in preparing for the Chapter Test.

Exercises Complete parts a–c for each quadratic function.

a. Find the y-intercept, the equation of the axis of symmetry, and the x-coordinate of the vertex.

b. Make a table of values that includes the vertex.

c. Use this information to graph the function. *(See Example 2 on pages 287 and 288.)*

9. $f(x) = x^2 + 6x + 20$
10. $f(x) = x^2 - 2x - 15$
11. $f(x) = x^2 - 8x + 7$
12. $f(x) = -2x^2 + 12x - 9$
13. $f(x) = -x^2 - 4x - 3$
14. $f(x) = 3x^2 + 9x + 6$

9–14. See margin.

Determine whether each function has a maximum or a minimum value. Then find the maximum or minimum value of each function.

(See Example 3 on pages 288 and 289.)

15. $f(x) = 4x^2 - 3x - 5$ **min.; $-\frac{89}{16}$**
16. $f(x) = -3x^2 + 2x - 2$ **max.; $-\frac{5}{3}$**
17. $f(x) = -2x^2 + 7$ **max.; 7**

6-2 Solving Quadratic Equations by Graphing

See pages 294–299.

Concept Summary

- The solutions, or roots, of a quadratic equation are the zeros of the related quadratic function. You can find the zeros of a quadratic function by finding the x-intercepts of its graph.
- A quadratic equation can have one real solution, two real solutions, or no real solution.

One Real Solution

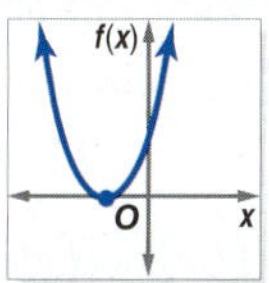

Two Real Solutions

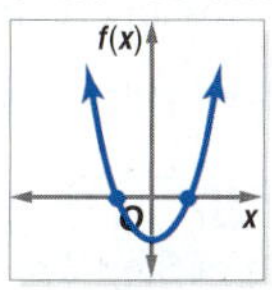

No Real Solution

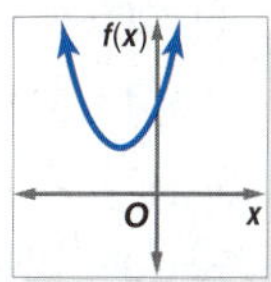

Example Solve $2x^2 - 5x + 2 = 0$ by graphing.

The equation of the axis of symmetry is $x = -\frac{-5}{2(2)}$ or $x = \frac{5}{4}$.

x	0	$\frac{1}{2}$	$\frac{5}{4}$	2	$\frac{5}{2}$
$f(x)$	2	0	$-\frac{9}{8}$	0	2

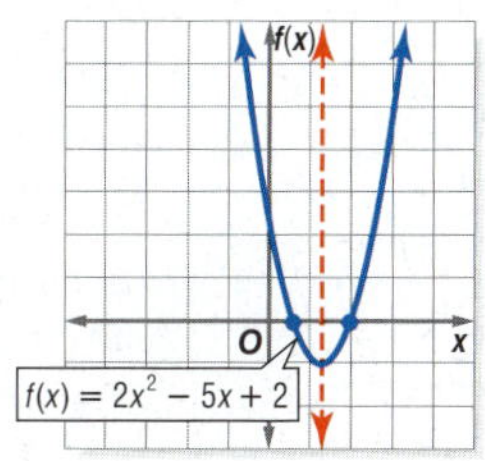

The zeros of the related function are $\frac{1}{2}$ and 2. Therefore, the solutions of the equation are $\frac{1}{2}$ and 2.

Exercises Solve each equation by graphing. If exact roots cannot be found, state the consecutive integers between which the roots are located.

(See Examples 1–3 on pages 294 and 295.) **19. 2, −5**

18. $x^2 - 36 = 0$ **6, −6**
19. $-x^2 - 3x + 10 = 0$
20. $2x^2 + x - 3 = 0$ **1, $-\frac{3}{2}$**
21. $-x^2 - 40x - 80 = 0$
22. $-3x^2 - 6x - 2 = 0$
23. $\frac{1}{5}(x + 3)^2 - 5 = 0$ **2, −8**

21. between −3 and −2; between −38 and −37

22. between −2 and −1; between −1 and 0

Answers

9a. 20; $x = -3$; −3

9b.

x	$f(x)$
−5	15
−4	12
−3	11
−2	12
−1	15

9c.

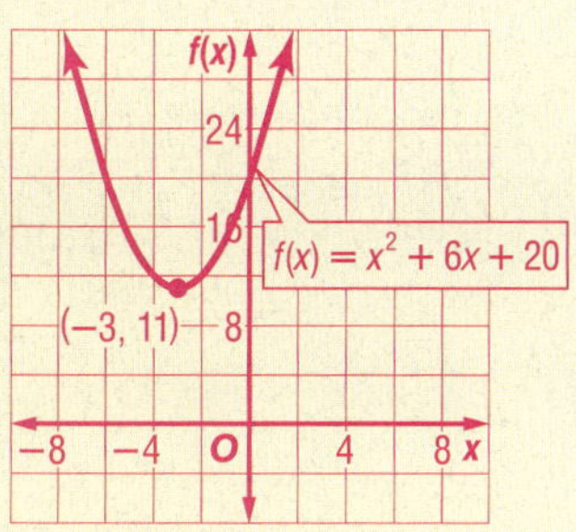

10a. −15; $x = 1$; 1

10b.

x	$f(x)$
−1	−12
0	−15
1	−16
2	−15
3	−12

10c.

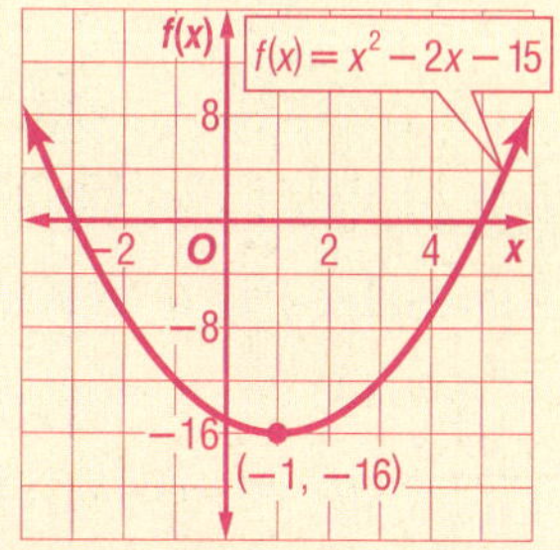

11a. 7; $x = 4$; 4

11b.

x	$f(x)$
2	−5
3	−8
4	−9
5	−8
6	−5

11c.

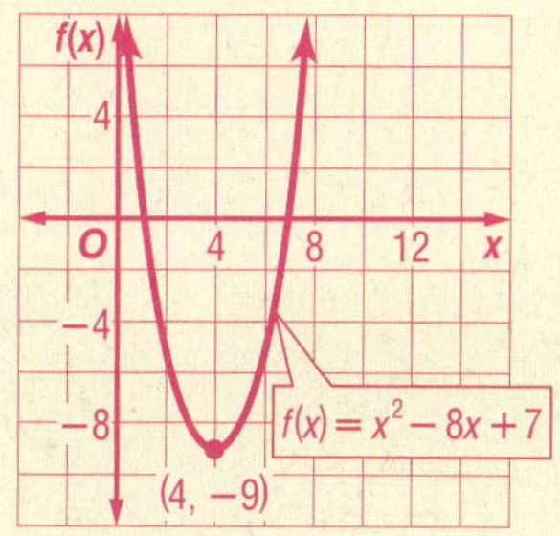

12a. −9; $x = 3$; 3

12b.

x	$f(x)$
1	1
2	7
3	9
4	7
5	1

12c.

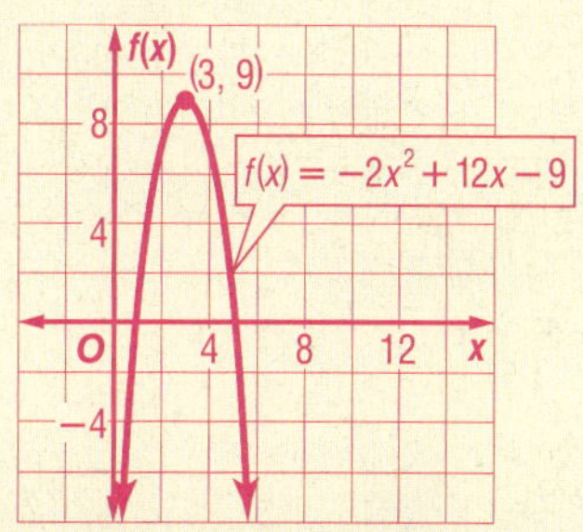

(continued on the next page)

Answers

13a. -3; $x = -2$; -2

13b.

x	$f(x)$
-4	-3
-3	0
-2	1
-1	0
0	-3

13c.

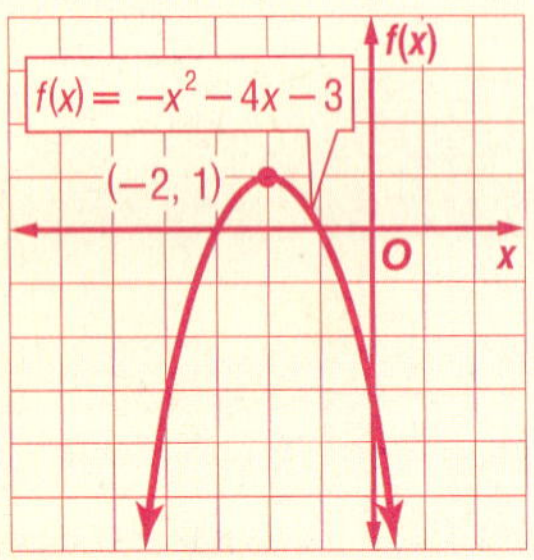

14a. 6; $x = -\frac{3}{2}$; $-\frac{3}{2}$

14b.

x	$f(x)$
-3	6
-2	0
$-\frac{3}{2}$	$-\frac{3}{4}$
-1	0
0	6

14c.

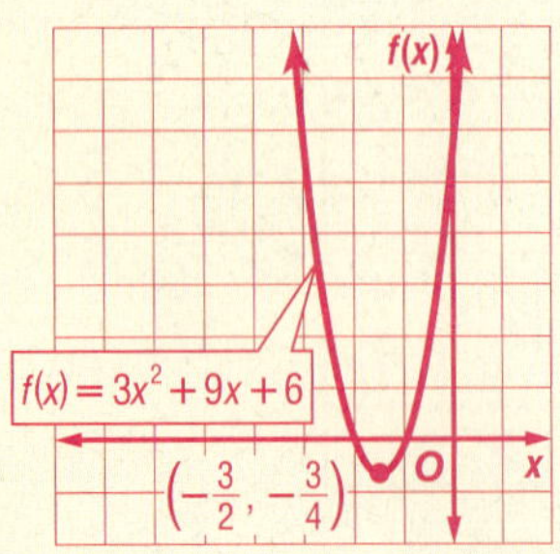

34. $\frac{121}{4}$; $\left(x - \frac{11}{2}\right)^2$

36. $\left\{-\frac{3}{2}, 5\right\}$

37. $\{3 \pm 2\sqrt{5}\}$

38. $\left\{\frac{5 \pm i\sqrt{7}}{4}\right\}$

6-3 Solving Quadratic Equations by Factoring

See pages 301–305.

Concept Summary

- Zero Product Property: For any real numbers a and b, if $ab = 0$, then either $a = 0$, $b = 0$, or both a and $b = 0$.

Example **Solve $x^2 + 9x + 20 = 0$ by factoring.**

$x^2 + 9x + 20 = 0$		Original equation
$(x + 4)(x + 5) = 0$		Factor the trinomial.
$x + 4 = 0$ or	$x + 5 = 0$	Zero Product Property
$x = -4$	$x = -5$	The solution set is $\{-5, -4\}$.

Exercises **Solve each equation by factoring.** *(See Examples 1–3 on pages 301 and 302.)*

24. $x^2 - 4x - 32 = 0$
25. $3x^2 + 6x + 3 = 0$ $\{-1\}$
26. $5y^2 = 80$ $\{-4, 4\}$
27. $2c^2 + 18c - 44 = 0$
28. $25x^2 - 30x = -9$ $\left\{\frac{3}{5}\right\}$
29. $6x^2 + 7x = 3$ $\left\{\frac{1}{3}, -\frac{3}{2}\right\}$

24. $\{-4, 8\}$ 27. $\{-11, 2\}$

Write a quadratic equation with the given root(s). Write the equation in the form $ax^2 + bx + c$, where a, b, and c are integers. *(See Example 4 on page 303.)*

30. $-4, -25$ $x^2 + 29x + 100 = 0$
31. $10, -7$ $x^2 - 3x - 70 = 0$
32. $\frac{1}{3}, 2$ $3x^2 - 7x + 2 = 0$

6-4 Completing the Square

See pages 306–312.

Concept Summary

- To complete the square for any quadratic expression $x^2 + bx$:

Step 1 Find one half of b, the coefficient of x.

Step 2 Square the result in Step 1.

Step 3 Add the result of Step 2 to $x^2 + bx$. $x^2 + bx + \left(\frac{b}{2}\right)^2 = \left(x + \frac{b}{2}\right)^2$

Example **Solve $x^2 + 10x - 39 = 0$ by completing the square.**

$x^2 + 10x - 39 = 0$		Notice that $x^2 + 10x - 39 = 0$ is not a perfect square.
$x^2 + 10x = 39$		Rewrite so the left side is of the form $x^2 + bx$.
$x^2 + 10x + 25 = 39 + 25$		Since $\left(\frac{10}{2}\right)^2 = 25$, add 25 to each side.
$(x + 5)^2 = 64$		Write the left side as a perfect square by factoring.
$x + 5 = \pm 8$		Square Root Property
$x + 5 = 8$ or	$x + 5 = -8$	Rewrite as two equations.
$x = 3$	$x = -13$	The solution set is $\{-13, 3\}$.

Exercises **Find the value of c that makes each trinomial a perfect square. Then write the trinomial as a perfect square.** *(See Example 3 on page 307.)*

33. $x^2 + 34x + c$ 289; $(x + 17)^2$
34. $x^2 - 11x + c$
35. $x^2 + \frac{7}{2}x + c$ $\frac{49}{16}$; $\left(x + \frac{7}{4}\right)^2$

Solve each equation by completing the square. *(See Examples 4–6 on pages 308 and 309.)*

36. $2x^2 - 7x - 15 = 0$
37. $2n^2 - 12n - 22 = 0$
38. $2x^2 - 5x + 7 = 3$

6-5 The Quadratic Formula and the Discriminant

See pages 313–319.

Concept Summary

- Quadratic Formula: $x = \frac{-b \pm \sqrt{b^2 - 4ac}}{2a}$ where $a \neq 0$

Solve $x^2 - 5x - 66 = 0$ by using the Quadratic Formula.

$x = \frac{-b \pm \sqrt{b^2 - 4ac}}{2a}$ — Quadratic Formula

$= \frac{-(-5) \pm \sqrt{(-5)^2 - 4(1)(-66)}}{2(1)}$ — Replace a with 1, b with -5, and c with -66.

$= \frac{5 \pm 17}{2}$ — Simplify.

$x = \frac{5 + 17}{2}$ or $x = \frac{5 - 17}{2}$ — Write as two equations.

$= 11$ $\quad = -6$ — The solution set is $\{11, -6\}$.

Exercises Complete parts a–c for each quadratic equation.

a. Find the value of the discriminant.

b. Describe the number and type of roots.

c. Find the exact solutions by using the Quadratic Formula.

(See Examples 1–4 on pages 314–316.) **39–41. See margin.**

39. $x^2 + 2x + 7 = 0$ 40. $-2x^2 + 12x - 5 = 0$ 41. $3x^2 + 7x - 2 = 0$

6-6 Analyzing Graphs of Quadratic Functions

See pages 322–328.

Concept Summary

- As the values of h and k change, the graph of $y = (x - h)^2 + k$ is the graph of $y = x^2$ translated
  - $|h|$ units left if h is negative or $|h|$ units right if h is positive.
  - $|k|$ units up if k is positive or $|k|$ units down if k is negative.
- Consider the equation $y = a(x - h)^2 + k$.
  - If $a > 0$, the graph opens up; if $a < 0$ the graph opens down.
  - If $|a| > 1$, the graph is narrower than the graph of $y = x^2$.
  - If $|a| < 1$, the graph is wider than the graph of $y = x^2$.

Example **Write the quadratic function $y = 3x^2 + 42x + 142$ in vertex form. Then identify the vertex, axis of symmetry, and direction of opening.**

$y = 3x^2 + 42x + 142$ — Original equation

$y = 3(x^2 + 14x) + 142$ — Group $ax^2 + bx$ and factor, dividing by a.

$y = 3(x^2 + 14x + 49) + 142 - 3(49)$ — Complete the square by adding $3\left(\frac{14}{2}\right)^2$. Balance this with a subtraction of 3(49).

$y = 3(x + 7)^2 - 5$ — Write $x^2 + 14x + 49$ as a perfect square.

So, $a = 3$, $h = -7$, and $k = -5$. The vertex is at $(-7, -5)$, and the axis of symmetry is $x = -7$. Since a is positive, the graph opens up.

Answers

39a. -24

39b. 2 complex

39c. $-1 \pm i\sqrt{6}$

40a. 104

40b. 2 irrational

40c. $3 \pm \frac{\sqrt{26}}{2}$

41a. 73

41b. 2 irrational

41c. $\frac{-7 \pm \sqrt{73}}{6}$

Chapter 6 For More ...
- Extra Practice, see pages 839–841.
- Mixed Problem Solving, see page 867.

Exercises Write each equation in vertex form, if not already in that form. Then identify the vertex, axis of symmetry, and direction of opening. *(See Examples 1 and 3 on pages 322 and 324.)* **42–44. See margin.**

42. $y = -6(x + 2)^2 + 3$ **43.** $y = 5x^2 + 35x + 58$ **44.** $y = -\frac{1}{3}x^2 + 8x$

Graph each function. *(See Examples 1–3 on pages 322 and 324.)* **45–47. See margin.**

45. $y = (x - 2)^2 - 2$ **46.** $y = 2x^2 + 8x + 10$ **47.** $y = -9x^2 - 18x - 6$

Write an equation for the parabola with the given vertex that passes through the given point. *(See Example 4 on page 325.)*

48. vertex: (4, 1)
point: (2, 13)
$y = 3(x - 4)^2 + 1$

49. vertex: (−2, 3)
point: (−6, 11)
$y = \frac{1}{2}(x + 2)^2 + 3$

50. vertex: (−3, −5)
point: (0, −14)
$y = -(x + 3)^2 - 5$

6-7 Graphing and Solving Quadratic Inequalities

See pages 329–335.

Concept Summary

- Graph quadratic inequalities in two variables as follows.

Step 1 Graph the related quadratic equation, $y = ax^2 + bx + c$. Decide if the parabola should be solid or dashed.

Step 2 Test a point (x_1, y_1) inside the parabola. Check to see if this point is a solution of the inequality.

Step 3 If (x_1, y_1) *is* a solution, shade the region *inside* the parabola. If (x_1, y_1) is *not* a solution, shade the region *outside* the parabola.

- To solve a quadratic inequality in one variable, graph the related quadratic function. Identify the x values for which the graph lies *below* the x-axis for $<$ and $\leq$. Identify the x values for which the graph lies *above* the x-axis for $>$ and $\geq$.

Example **Solve $x^2 + 3x - 10 < 0$ by graphing.**

Find the roots of the related equation.

$0 = x^2 + 3x - 10$		Related equation
$0 = (x + 5)(x - 2)$		Factor.
$x + 5 = 0$ or	$x - 2 = 0$	Zero Product Property
$x = -5$	$x = 2$	Solve each equation.

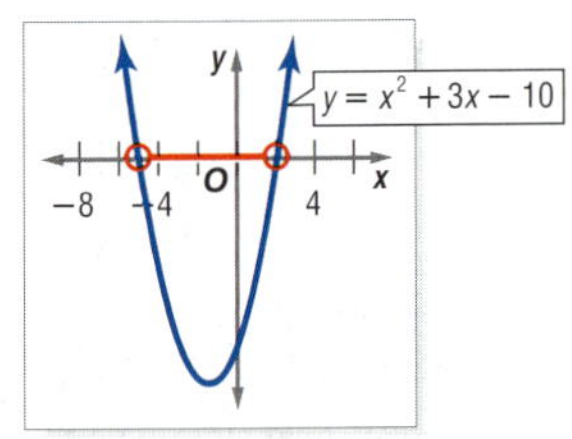

Sketch the graph of the parabola that has x-intercepts at −5 and 2. The graph should open up since $a > 0$. The graph lies below the x-axis between $x = -5$ and $x = 2$. Therefore, the solution set is $\{x \mid -5 < x < 2\}$.

Exercises Graph each inequality. *(See Example 1 on pages 329 and 330.)* **51–53. See margin.**

51. $y > x^2 - 5x + 15$ **52.** $y \leq 4x^2 - 36x + 17$ **53.** $y \geq -x^2 + 7x - 11$

Solve each inequality. *(See Examples 2, 3, and 5 on pages 330–332.)* **54–59. See pp. 343A–343F.**

54. $6x^2 + 5x > 4$ **55.** $8x + x^2 \geq -16$ **56.** $2x^2 + 5x < 12$

57. $2x^2 - 5x > 3$ **58.** $4x^2 - 9 \leq -4x$ **59.** $3x^2 - 5 > 6x$

Answers

42. (−2, 3); $x = -2$; down

43. $y = 5\left(x + \frac{7}{2}\right)^2 - \frac{13}{4}$; $\left(-\frac{7}{2}, -\frac{13}{4}\right)$; $x = -\frac{7}{2}$; up

44. $y = -\frac{1}{3}(x - 12)^2 + 48$; (12, 48); $x = 12$; down

45.
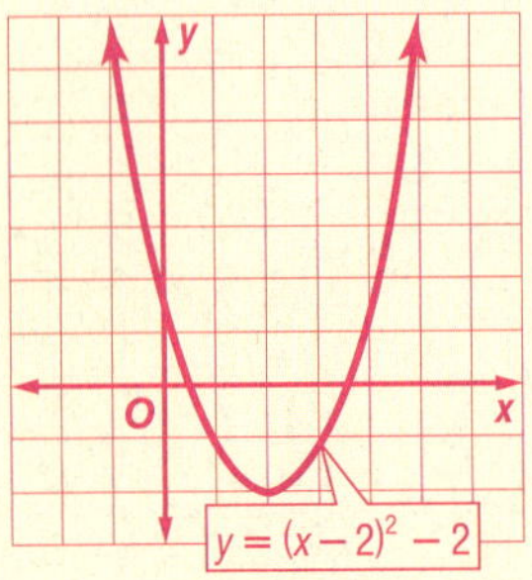

46.
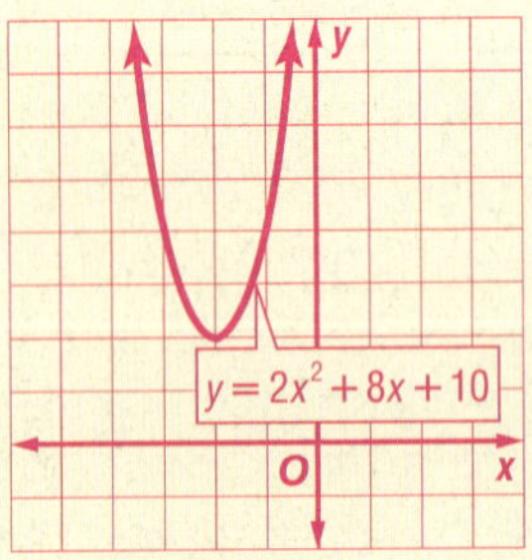

47.
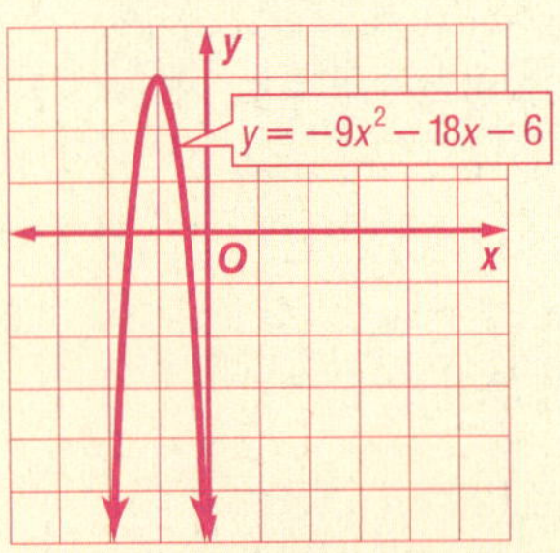

51.
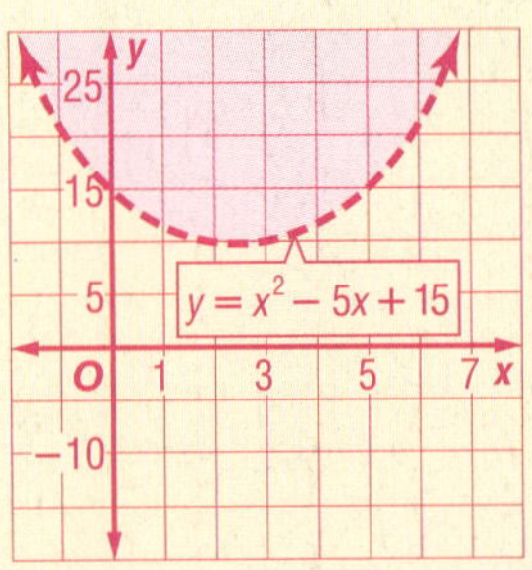

52.
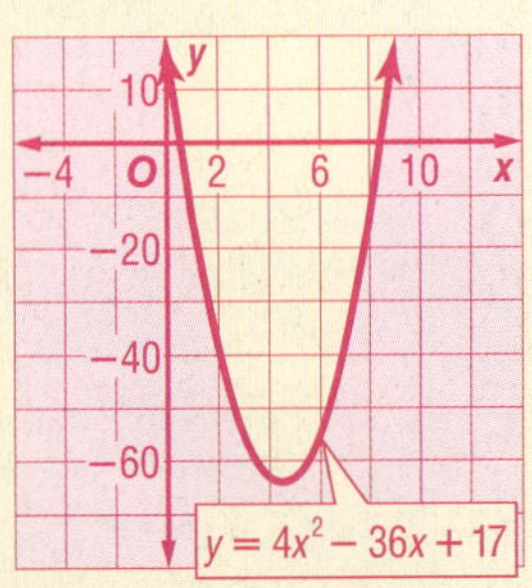

53.
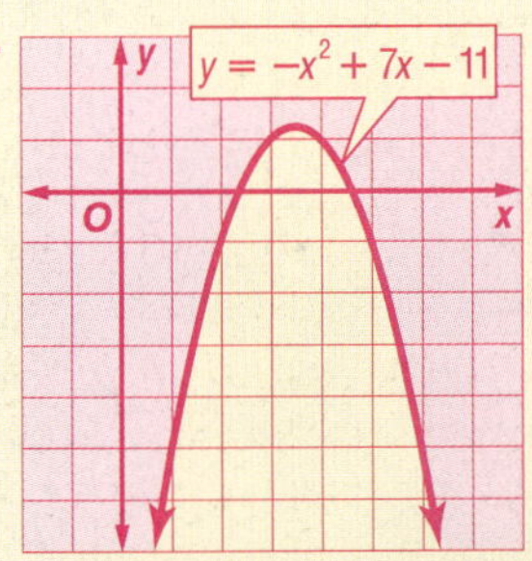

340 Chapter 6 Quadratic Functions and Inequalities

Chapter 6 Practice Test

Vocabulary and Concepts

Choose the word or term that best completes each statement.

1. The y-coordinate of the vertex of the graph of $y = ax^2 + bx + c$ is the (*maximum*, *minimum*) value obtained by the function when a is positive.
2. (*The Square Root Property*, *Completing the square*) can be used to solve any quadratic equation.

Skills and Applications

Complete parts a–c for each quadratic function.

a. **Find the y-intercept, the equation of the axis of symmetry, and the x-coordinate of the vertex.**

b. **Make a table of values that includes the vertex.**

c. **Use this information to graph the function.** **3–5. See pp. 343A–343F.**

3. $f(x) = x^2 - 2x + 5$
4. $f(x) = -3x^2 + 8x$
5. $f(x) = -2x^2 - 7x - 1$

Determine whether each function has a maximum or a minimum value. Then find the maximum or minimum value of each function.

6. $f(x) = x^2 + 6x + 9$ **min.; 0**
7. $f(x) = 3x^2 - 12x - 24$ **min.; −36**
8. $f(x) = -x^2 + 4x$ **max.; 4**
9. Write a quadratic equation with roots −4 and 5. Write the equation in the form $ax^2 + bx + c = 0$, where a, b, and c are integers. **$x^2 - x - 20 = 0$**

Solve each equation using the method of your choice. Find exact solutions. **10–18. See margin.**

10. $x^2 + x - 42 = 0$
11. $-1.6x^2 - 3.2x + 18 = 0$
12. $15x^2 + 16x - 7 = 0$
13. $x^2 + 8x - 48 = 0$
14. $x^2 + 12x + 11 = 0$
15. $x^2 - 9x - \frac{19}{4} = 0$
16. $3x^2 + 7x - 31 = 0$
17. $10x^2 + 3x = 1$
18. $-11x^2 - 174x + 221 = 0$
19. **BALLOONING** At a hot-air balloon festival, you throw a weighted marker straight down from an altitude of 250 feet toward a bull's eye below. The initial velocity of the marker when it leaves your hand is 28 feet per second. Find how long it will take the marker to hit the target by solving the equation $-16t^2 - 28t + 250 = 0$. **about 3.17 s**

Write each equation in vertex form, if not already in that form. Then identify the vertex, axis of symmetry, and direction of opening. **20–22. See pp. 343A–343F.**

20. $y = (x + 2)^2 - 3$
21. $y = x^2 + 10x + 27$
22. $y = -9x^2 + 54x - 8$

Graph each inequality. **23–25. See pp. 343A–343F.**

23. $y \leq x^2 + 6x - 7$
24. $y > -2x^2 + 9$
25. $y \geq -\frac{1}{2}x^2 - 3x + 1$

Solve each inequality. **26. $\{x \mid -7 < x < 5\}$ 27–28. See pp. 343A–343F.**

26. $(x - 5)(x + 7) < 0$
27. $3x^2 \geq 16$
28. $-5x^2 + x + 2 < 0$
29. **PETS** A rectangular turtle pen is 6 feet long by 4 feet wide. The pen is enlarged by increasing the length and width by an equal amount in order to double its area. What are the dimensions of the new pen? **8 ft by 6 ft**
30. **STANDARDIZED TEST PRACTICE** Which of the following is the sum of both solutions of the equation $x^2 + 8x - 48 = 0$? **B**

(A) −16 (B) −8 (C) −4 (D) 12

www.algebra2.com/chapter_test

Chapter 6 Practice Test

Assessment Options

Vocabulary Test A vocabulary test/review for Chapter 6 can be found on p. 368 of the *Chapter 6 Resource Masters*.

Chapter Tests There are six Chapter 6 Tests and an Open-Ended Assessment task available in the *Chapter 6 Resource Masters*.

Chapter 6 Tests			
Form	Type	Level	Pages
1	MC	basic	355–356
2A	MC	average	357–358
2B	MC	average	359–360
2C	FR	average	361–362
2D	FR	average	363–364
3	FR	advanced	365–366

MC = multiple-choice questions
FR = free-response questions

Open-Ended Assessment
Performance tasks for Chapter 6 can be found on p. 367 of the *Chapter 6 Resource Masters*. A sample scoring rubric for these tasks appears on p. A28.

ExamView® Pro

Use the networkable **ExamView® Pro** to:

- Create **multiple versions** of tests.
- Create **modified** tests for *Inclusion* students.
- **Edit** existing questions and **add** your own questions.

Portfolio Suggestion

Introduction In this chapter quadratic equations have been graphed and solved using many different methods, often following a process that involved numerous steps.

Ask Students Select an item from this chapter that shows your best work, including a graph, and place it in your portfolio. Explain why you believe it to be your best work and how you came to choose this particular piece of work.

Answers

10. −7, 6
11. $-\frac{9}{2}, \frac{5}{2}$
12. $-\frac{7}{5}, \frac{1}{3}$
13. −12, 4
14. −11, −1
15. $-\frac{1}{2}, \frac{19}{2}$
16. $\frac{-7 \pm \sqrt{421}}{6}$
17. $-\frac{1}{2}, \frac{1}{5}$
18. $-17, \frac{13}{11}$

Chapter 6 Standardized Test Practice

These two pages contain practice questions in the various formats that can be found on the most frequently given standardized tests.

A practice answer sheet for these two pages can be found on p. A1 of the *Chapter 6 Resource Masters*.

Standardized Test Practice Student Recording Sheet, p. A1

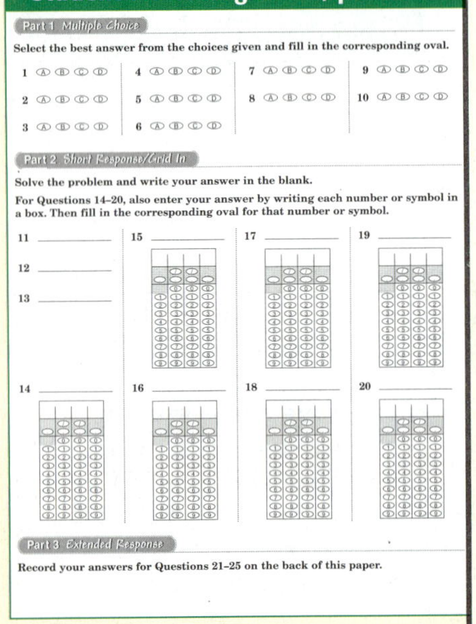
Part 1 Multiple Choice

Select the best answer from the choices given and fill in the corresponding oval.

1 Ⓐ Ⓑ Ⓒ Ⓓ 4 Ⓐ Ⓑ Ⓒ Ⓓ 7 Ⓐ Ⓑ Ⓒ Ⓓ 9 Ⓐ Ⓑ Ⓒ Ⓓ

2 Ⓐ Ⓑ Ⓒ Ⓓ 5 Ⓐ Ⓑ Ⓒ Ⓓ 8 Ⓐ Ⓑ Ⓒ Ⓓ 10 Ⓐ Ⓑ Ⓒ Ⓓ

3 Ⓐ Ⓑ Ⓒ Ⓓ 6 Ⓐ Ⓑ Ⓒ Ⓓ

Part 2 Short Response/Grid In

Solve the problem and write your answer in the blank.

For Questions 14–20, also enter your answer by writing each number or symbol in a box. Then fill in the corresponding oval for that number or symbol.

11 ____ 15 ____ 17 ____ 19 ____

12 ____

13 ____

14 ____ 16 ____ 18 ____ 20 ____

Part 3 Extended Response

Record your answers for Questions 21–25 on the back of this paper.

Additional Practice

See pp. 373–374 in the *Chapter 6 Resource Masters* for additional standardized test practice.

Chapter 6 Standardized Test Practice

Part 1 Multiple Choice

Record your answers on the answer sheet provided by your teacher or on a sheet of paper.

1. In a class of 30 students, half are girls and 24 ride the bus to school. If 4 of the girls do not ride the bus to school, how many boys in this class ride the bus to school? **C**

Ⓐ 2 Ⓑ 11

Ⓒ 13 Ⓓ 15

2. In the figure below, the measures of $\angle m + \angle n + \angle p =$ __?__ **D**

Ⓐ 90 Ⓑ 180

Ⓒ 270 Ⓓ 360

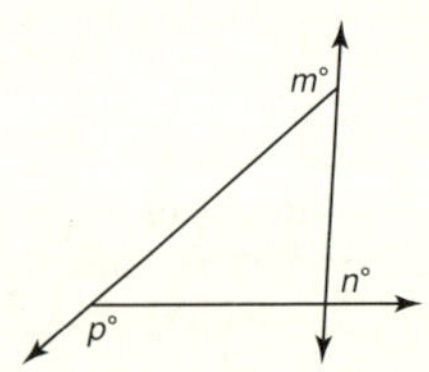

3. Of the points (−4, −2), (1, −3), (−1, 3), (3, 1), and (−2, 1), which three lie on the same side of the line $y - x = 0$? **C**

Ⓐ (−4, −2), (1, −3), (−2, 1)

Ⓑ (−4, −2), (1, −3), (3, 1)

Ⓒ (−4, −2), (−1, 3), (−2, 1)

Ⓓ (1, −3), (−1, 3), (3, 1)

4. If k is an integer, then which of the following must also be integers? **B**

I. $\frac{5k+5}{5k}$ II. $\frac{5k+5}{k+1}$ III. $\frac{5k^2+k}{5k}$

Ⓐ I only Ⓑ II only

Ⓒ I and II Ⓓ II and III

5. Which of the following is a factor of $x^2 - 7x - 8$? **D**

Ⓐ $x + 2$ Ⓑ $x - 1$

Ⓒ $x - 4$ Ⓓ $x - 8$

6. If $x > 0$, then $\frac{\sqrt{16x^2 + 64x + 64}}{x + 2} =$ __?__ **B**

Ⓐ 2 Ⓑ 4

Ⓒ 8 Ⓓ 16

7. If x and p are both greater than zero and $4x^2p^2 + xp - 33 = 0$, then what is the value of p in terms of x? **D**

Ⓐ $-\frac{3}{x}$ Ⓑ $-\frac{11}{4x}$

Ⓒ $\frac{3}{4x}$ Ⓓ $\frac{11}{4x}$

8. For all positive integers n, $\langle n \rangle = 3\sqrt{n}$. Which of the following equals 12? **C**

Ⓐ $\langle 4 \rangle$ Ⓑ $\langle 8 \rangle$

Ⓒ $\langle 16 \rangle$ Ⓓ $\langle 32 \rangle$

9. Which number is the sum of both solutions of the equation $x^2 - 3x - 18 = 0$? **C**

Ⓐ −6 Ⓑ −3

Ⓒ 3 Ⓓ 6

10. One of the roots of the polynomial $6x^2 + kx + 20 = 0$ is $-\frac{5}{2}$. What is the value of k? **C**

Ⓐ −23 Ⓑ $-\frac{4}{3}$

Ⓒ 23 Ⓓ 7

Test-Taking Tip

Questions 8, 11, 13, 16, 21, and 27

Be sure to use the information that describes the variables in any standardized test item. For example, if an item says that $x > 0$, check to be sure that your solution for x is not a negative number.

ExamView® Pro

Special banks of standardized test questions similar to those on the SAT, ACT, TIMSS 12, NAEP 12, and state proficiency tests can be found on this CD-ROM.

342 Chapter 6 Quadratic Functions and Inequalities

Preparing for Standardized Tests
For test-taking strategies and more practice, see pages 877–892.

Part 2 Short Response/Grid In

Record your answers on the answer sheet provided by your teacher or on a sheet of paper.

11. If n is a three-digit number that can be expressed as the product of three consecutive *even* integers, what is one possible value of n? **192, 480, or 960**

12. If x and y are *different* positive integers and $x + y = 6$, what is one possible value of $3x + 5y$? **20, 22, 26, or 28**

13. If a circle of radius 12 inches has its radius decreased by 6 inches, by what percent is its area decreased? **75%**

14. What is the least positive integer k for which $12k$ is the cube of an integer? **18**

15. If $AB = BC$ in the figure, what is the y-coordinate of point B? **7**

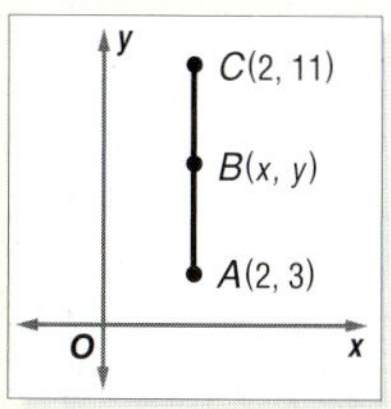

16. In the figure, if O is the center of the circle, what is the value of x? **35**

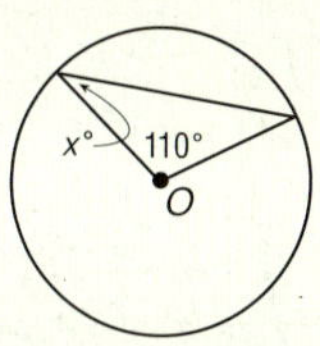

17. Let $a \blacklozenge b$ be defined as the sum of all integers greater than a and less than b. For example, $6 \blacklozenge 10 = 7 + 8 + 9$ or 24. What is the value of $(75 \blacklozenge 90) - (76 \blacklozenge 89)$? **165**

18. If $x^2 - y^2 = 42$ and $x + y = 6$, what is the value of $x - y$? **7**

19. By what amount does the sum of the roots exceed the product of the roots of the equation $(x - 7)(x + 3) = 0$? **25**

www.algebra2.com/standardized_test

20. If $x^2 = 36$ and $y^2 = 9$, what is the greatest possible value of $(x - y)^2$? **81**

Part 3 Extended Response

Record your answers on a sheet of paper. Show your work.

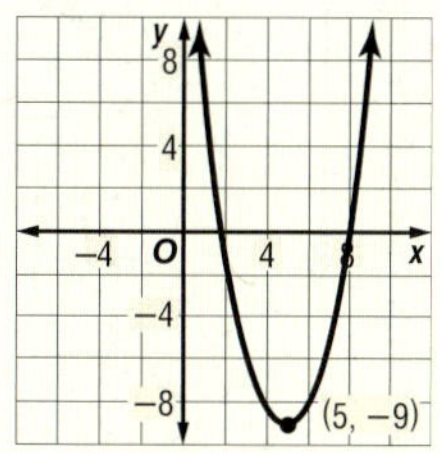

For Exercises 21–23, use the graph below.

21. What are the solutions of $f(x) = 0$? Explain how you found the solutions. **2, 8; The solutions are the x-intercepts of the graph.**

22. Does the function have a minimum value or a maximum value? Find the minimum or maximum value of the function. **minimum value; −9**

23. Write the function whose graph is shown. Explain your method. **See margin.**

For Exercises 24 and 25, use the information below.

Scott launches a model rocket from ground level. The rocket's height h in meters is given by the equation $h = -4.9t^2 + 56t$, where t is the time in seconds after the launch.

24. What is the maximum height the rocket will reach? Round to the nearest tenth of a meter. Show each step and explain your method. **See margin.**

25. How long after it is launched will the rocket reach its maximum height? Round to the nearest tenth of a second. **5.7 s**

Evaluating Extended Response Questions

Extended Response questions are graded by using a multilevel rubric that guides you in assessing a student's knowledge of a particular concept.

Goal: Determine the dimensions of a rectangular garden for a given perimeter.

Sample Scoring Rubric: The following rubric is a sample scoring device. You may wish to add more detail to this sample to meet your individual scoring needs.

Score	Criteria
4	A correct solution that is supported by well-developed, accurate explanations
3	A generally correct solution, but may contain minor flaws in reasoning or computation
2	A partially correct interpretation and/or solution to the problem
1	A correct solution with no supporting evidence or explanation
0	An incorrect solution indicating no mathematical understanding of the concept or task, or no solution is given

Answers

23. $f(x) = x^2 - 10x + 16$; Sample answer: Substitute the coordinates of the vertex $(5, -9)$ and a point $(2, 0)$ into $y = a(x - h)^2 + k$ and solve for a. Then write the equation.

24. 160 m; Sample answer: the vertex form of the equation is $h = -4.9\left(t - \frac{40}{7}\right)^2 - 160$, so the vertex is $\left(\frac{40}{7}, 160\right)$.

Page 284, Chapter 6 Getting Started

1.

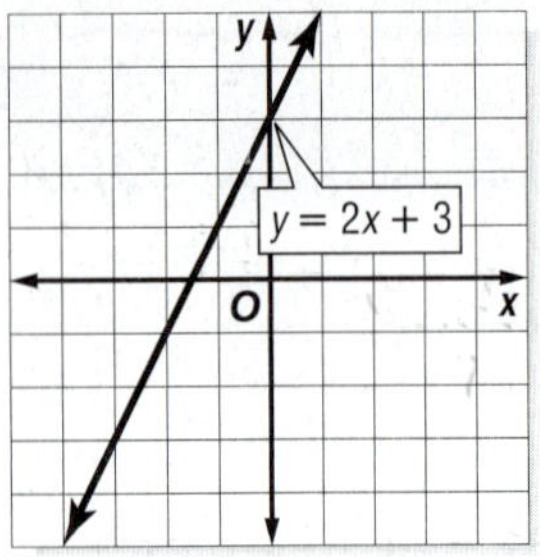

2.

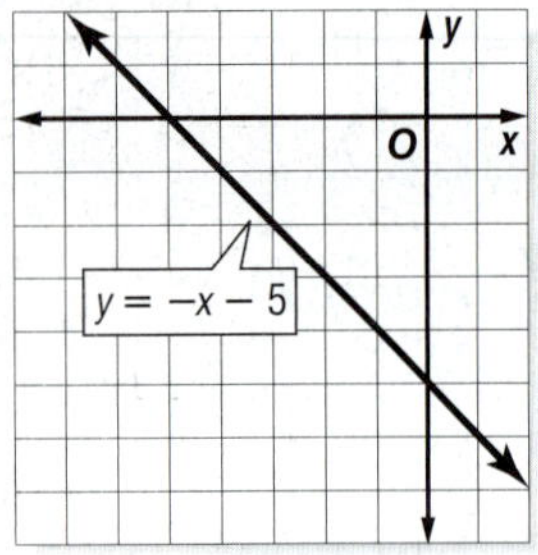

3.

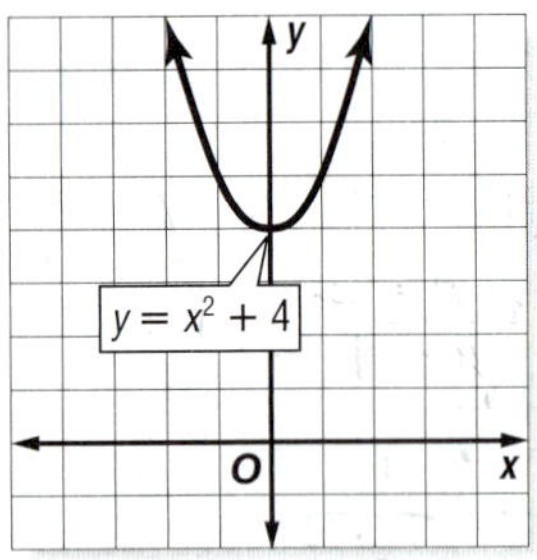

4.

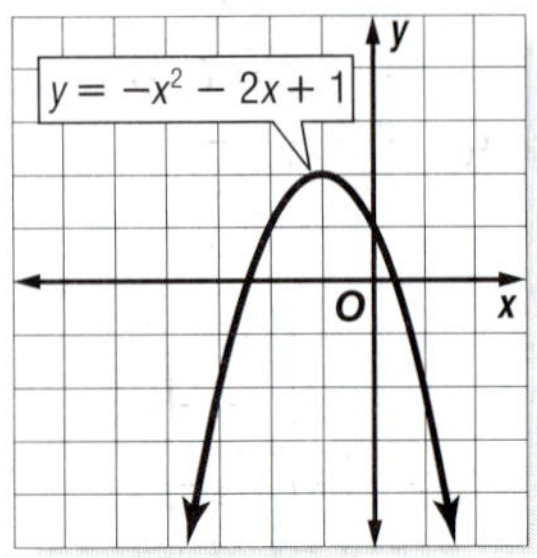

Page 291, Lesson 6-1

14a. 0; $x = 0$; 0

14b.

x	$f(x)$
−2	8
−1	2
0	0
1	2
2	8

14c.

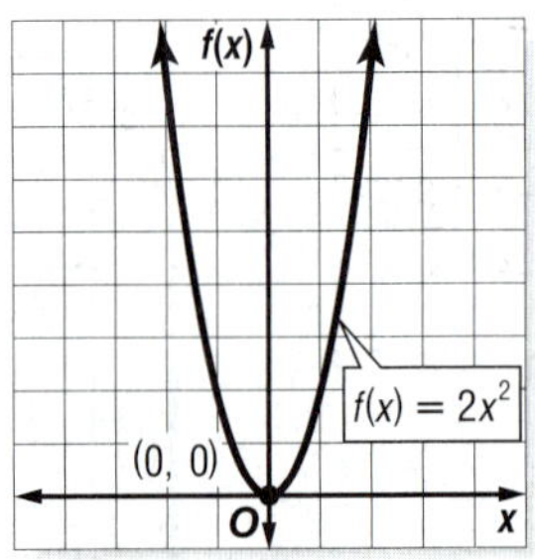

15a. 0; $x = 0$; 0

15b.

x	$f(x)$
−2	−20
−1	−5
0	0
1	−5
2	−20

15c.

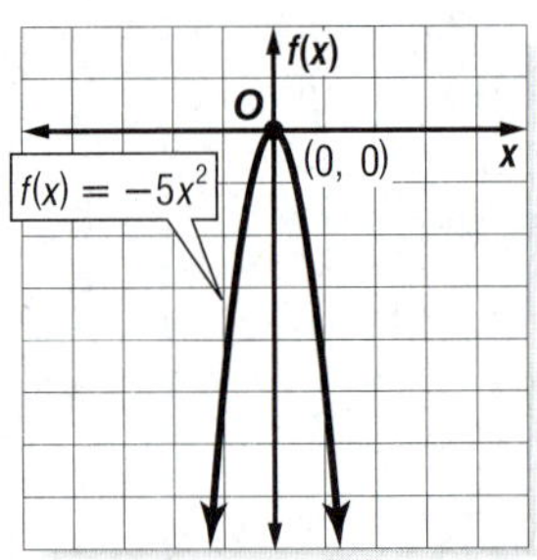

16a. 4; $x = 0$; 0

16b.

x	$f(x)$
−2	8
−1	5
0	4
1	5
2	8

16c.

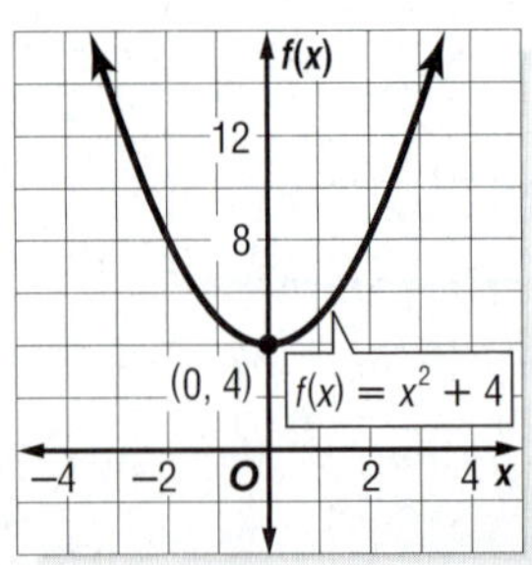

17a. −9; $x = 0$; 0

17b.

x	$f(x)$
−2	−5
−1	−8
0	−9
1	−8
2	−5

17c.

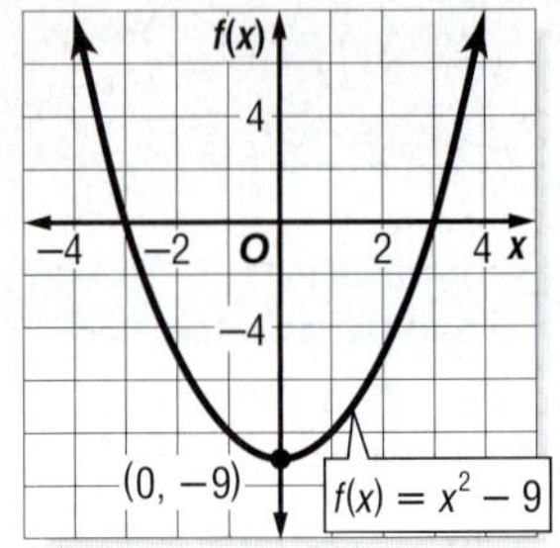

18a. −4; $x = 0$; 0

18b.

x	$f(x)$
−2	4
−1	−2
0	−4
1	−2
2	4

18c.

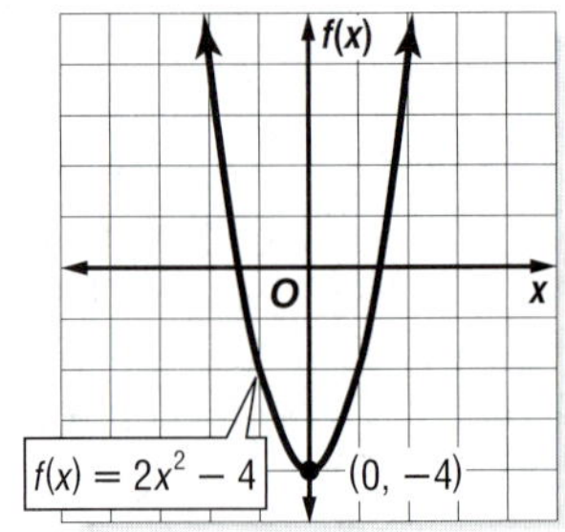

19a. 1; $x = 0$; 0

19b.

x	$f(x)$
−2	13
−1	4
0	1
1	4
2	13

19c.

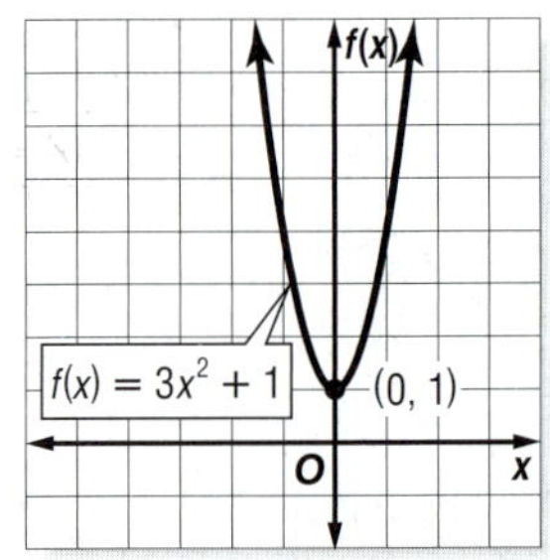

20a. 4; $x = 2$; 2

20b.

x	$f(x)$
0	4
1	1
2	0
3	1
4	4

20c.

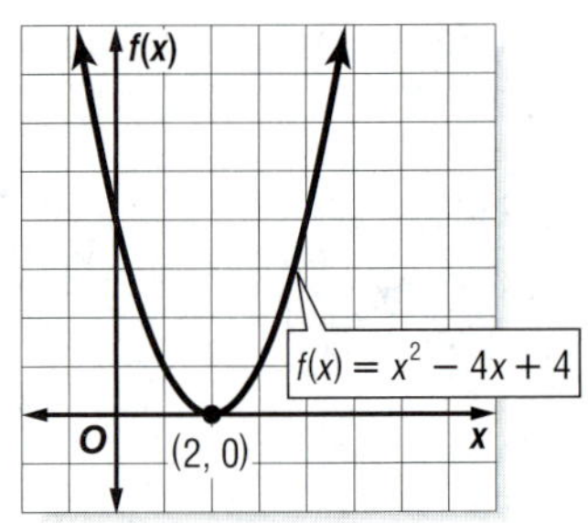

21a. 9; $x = 4.5$; 4.5

21b.

x	$f(x)$
3	−9
4	−11
4.5	−11.25
5	−11
6	−9

21c.

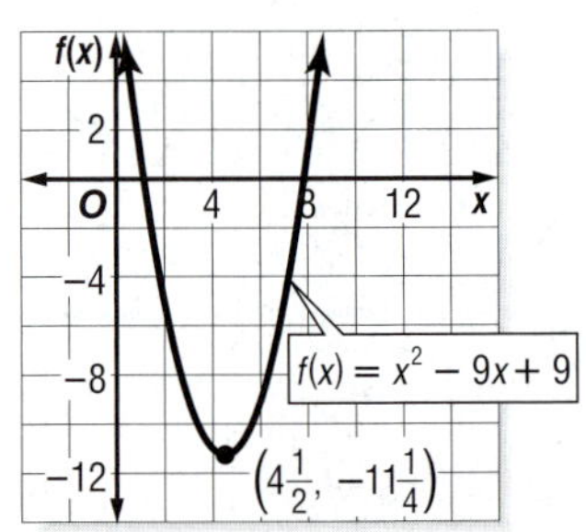

22a. -5; $x = 2$; 2

22b.

x	$f(x)$
0	-5
1	-8
2	-9
3	-8
4	-5

22c.

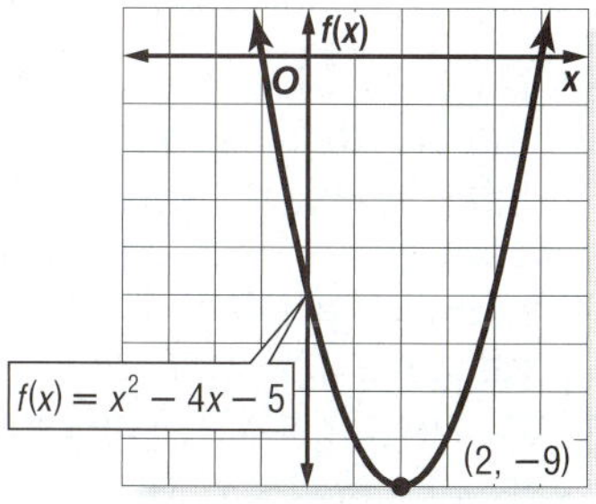

23a. 36; $x = -6$; -6

23b.

x	$f(x)$
-8	4
-7	1
-6	0
-5	1
-4	4

23c.

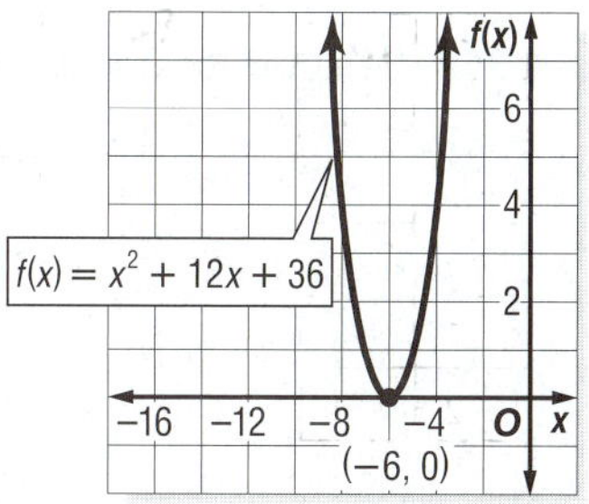

24a. -1; $x = -1$; -1

24b.

x	$f(x)$
-3	8
-2	-1
-1	-4
0	-1
1	8

24c.

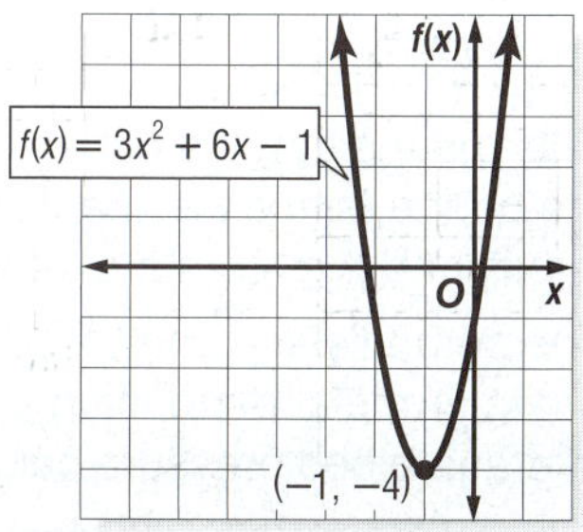

25a. -3; $x = 2$, 2

25b.

x	$f(x)$
0	-3
1	3
2	5
3	3
4	-3

25c.

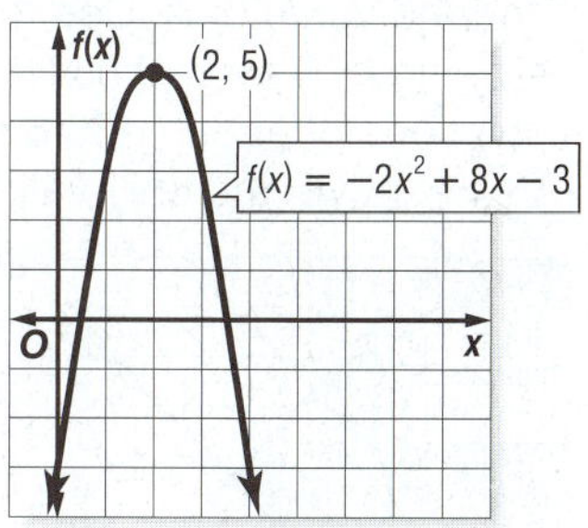

26a. 0; $x = -\frac{2}{3}$, $-\frac{2}{3}$

26b.

x	$f(x)$
-2	-4
-1	1
$-\frac{2}{3}$	$\frac{4}{3}$
0	0
1	-7

26c.

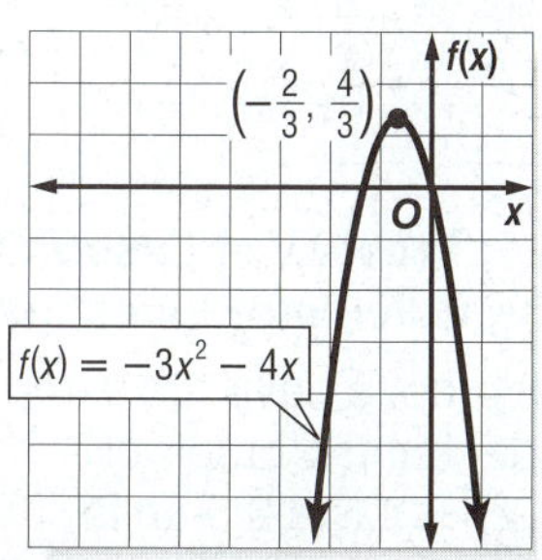

27a. 0; $x = -\frac{5}{4}$; $-\frac{5}{4}$

27b.

x	$f(x)$
-3	3
-2	-2
$-\frac{5}{4}$	$-\frac{25}{8}$
-1	-3
0	0

27c.

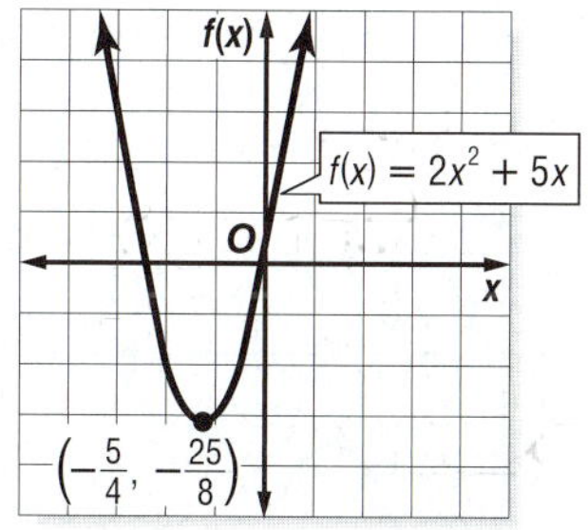

28a. -1; $x = 0$; 0

28b.

x	$f(x)$
-2	1
-1	$-\frac{1}{2}$
0	-1
1	$-\frac{1}{2}$
2	1

28c.

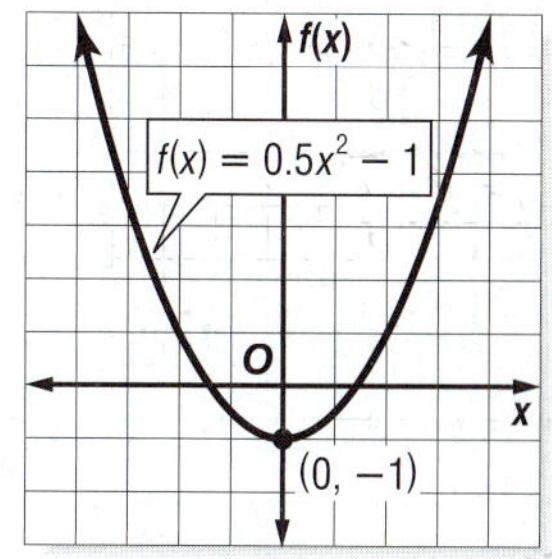

29a. 0; $x = -6$; -6

29b.

x	$f(x)$
-8	8
-7	8.75
-6	9
-5	8.75
-4	8

29c.

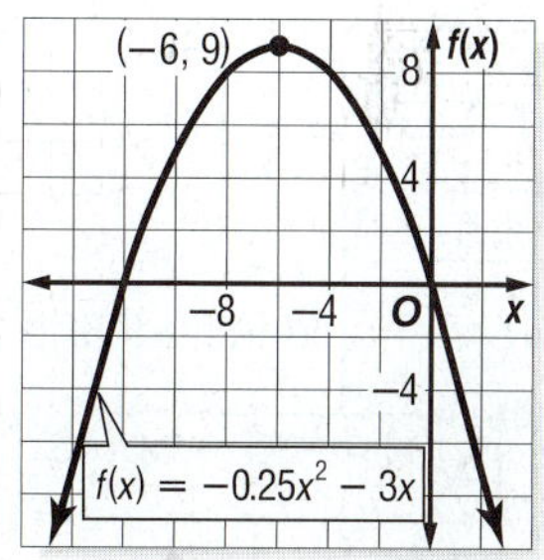

30a. $\frac{9}{2}$; $x = -3$, -3

30b.

x	$f(x)$
-5	2
-4	0.5
-3	0
-2	0.5
-1	2

30c.

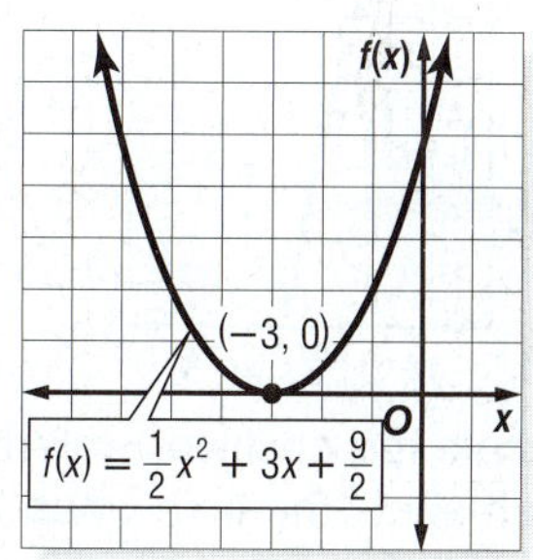

31a. $-\frac{8}{9}$; $x = \frac{1}{3}$; $\frac{1}{3}$

31b.

x	$f(x)$
-1	$\frac{7}{9}$
0	$-\frac{8}{9}$
$\frac{1}{3}$	-1
1	$-\frac{5}{9}$
2	$1\frac{7}{9}$

31c.

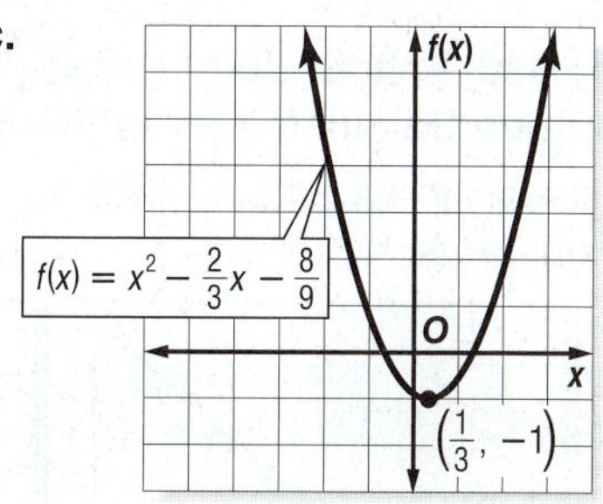

Pages 298–299, Lesson 6-2

39. Let x be the first number.
Then, $7 - x$ is the other number.

$$x(7 - x) = 14$$
$$-x^2 + 7x - 14 = 0$$

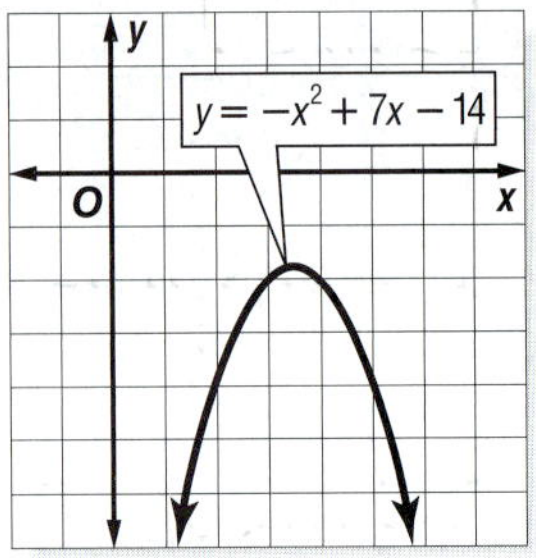

Since the graph of the related function does not intersect the x-axis, this equation has no real solutions. Therefore no such numbers exist.

40. Let x be the first number.
Then, $-9 - x$ is the other number.

$$x(-9 - x) = 24$$
$$-x^2 - 9x - 24 = 0$$

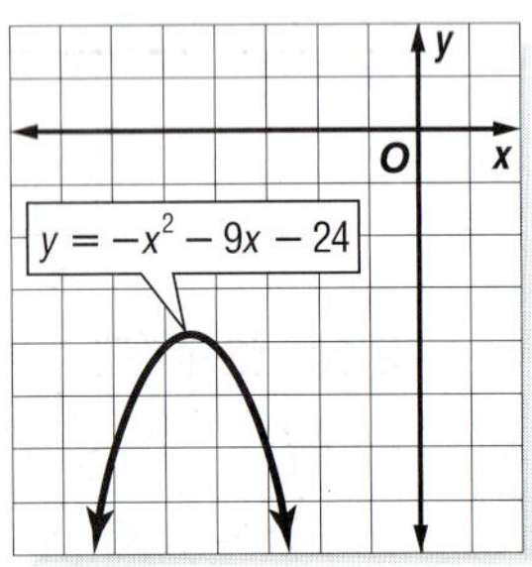

Since the graph of the related function does not intersect the x-axis, this equation has no real solutions. Therefore no such numbers exist.

48. Answers should include the following.

- 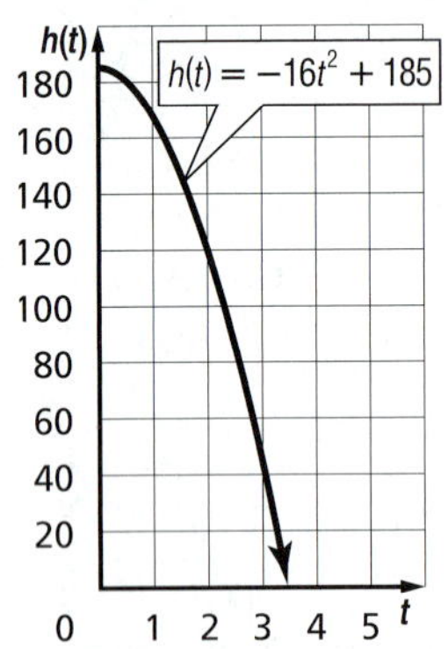

- Locate the positive x-intercept at about 3.4. This represents the time when the height of the ride is 0. Thus, if the ride were allowed to fall to the ground, it would take about 3.4 seconds.

Page 300, Follow-Up of Lesson 6-2 Graphing Calculator Investigation

1. linear: $y = 4.343x - 89.669$;
quadratic: $y = 0.044x^2 - 0.003x + 0.218$

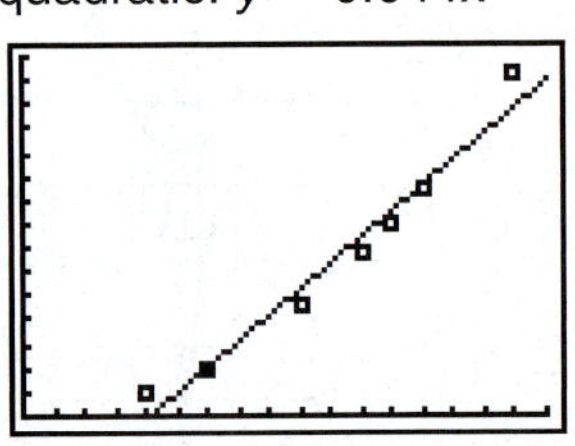

[0, 85] scl: 5 by [0, 300] scl: 20

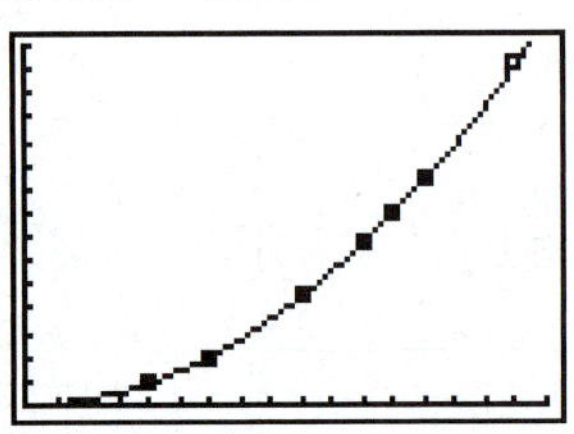

[0, 85] scl: 5 by [0, 300] scl: 20

The quadratic equation fits the data better.

Page 305, Lesson 6-3

50. Answers should include the following.

- Subtract 24 from each side of $x^2 + 5x = 24$ so that the equation becomes $x^2 + 5x - 24 = 0$. Factor the left side as $(x - 3)(x + 8)$. Set each factor equal to zero. Solve each equation for x. The solutions to the equation are 3 and -8. Since length cannot be negative, the width of the rectangle is 3 inches, and the length is $3 + 5$ or 8 inches.
- To use the Zero Product Property, one side of the equation must equal zero.

Pages 318–319, Lesson 6-5

28. $-2, 32$

29. $\pm i\frac{\sqrt{21}}{7}$

30. $2 \pm i\sqrt{3}$

31. $\frac{-3 \pm \sqrt{15}}{2}$

32. $\pm\sqrt{2}$

33. $\frac{9}{2}$

34. $-3 \pm i\sqrt{7}$

35. $\frac{5 \pm \sqrt{46}}{3}$

36. $4 \pm \sqrt{7}$

37. $0, -\frac{3}{10}$

38. $3 \pm 2\sqrt{2}$

39. $-2, 6$

41. This means that the cables do not touch the floor of the bridge, since the graph does not intersect the x-axis and the roots are imaginary.

46. The person's age can be substituted for A in the appropriate formula, depending upon their gender, and their average blood pressure calculated. See student's work.

- If a woman's blood pressure is given to be 118, then solve the equation $118 = 0.01A^2 + 0.05A + 107$ to find the value of A. Use the Quadratic Formula, substituting 0.01 for a, 0.05 for b, and -11 for c. This gives solutions of about -35.8 or 30.8. Since age cannot be negative, the only valid solution for A is 30.8.

59.

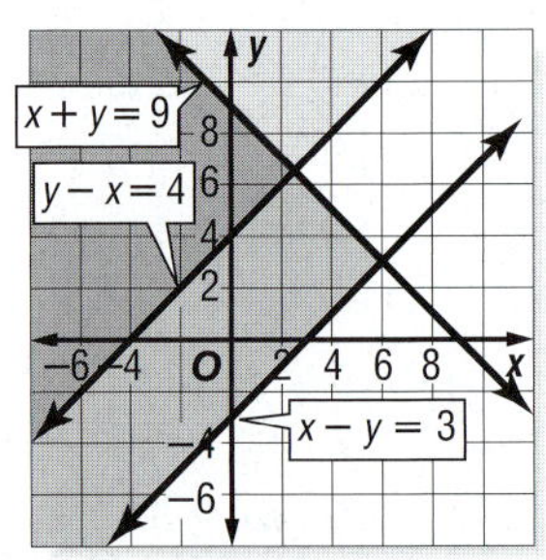

60.

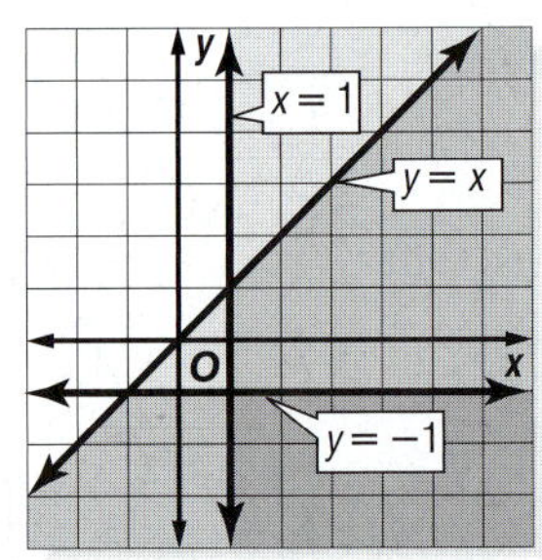

Page 321, Preview of Lesson 6-6 Graphing Calculator Investigation

4. Both graphs have the same shape, but the graph of $y = x^2 + 2.5$ is 2.5 units above the graph of $y = x^2$.

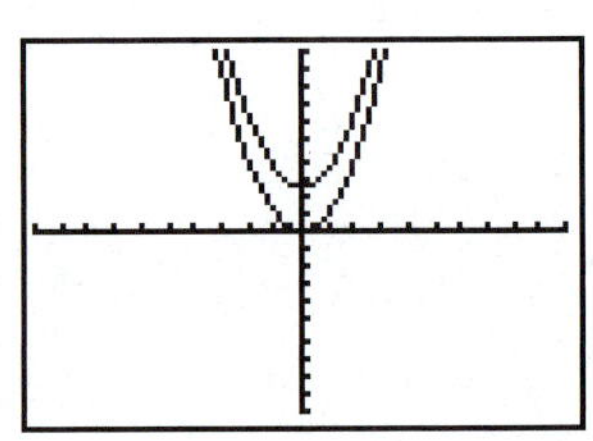

5. Both graphs have the same shape, but the graph of $y = -x^2$ opens downward while the graph of $y = x^2 - 9$ opens upward and is 9 units lower than the graph of $y = x^2$.

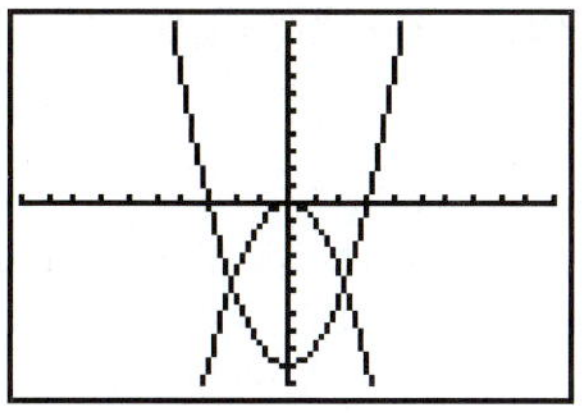

6. The graph of $y = 3x^2$ is narrower than the graph of $y = x^2$.

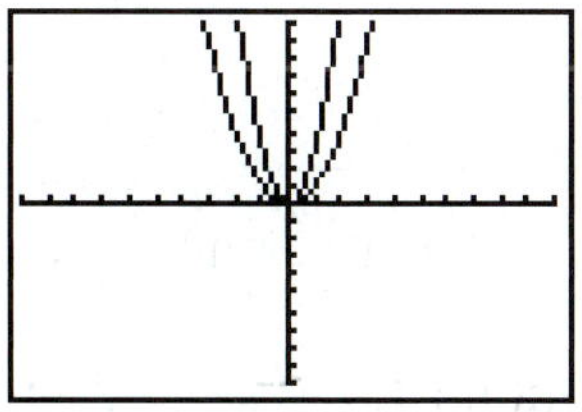

7. The graph of $y = -6x^2$ opens downward and is narrower than the graph of $y = x^2$.

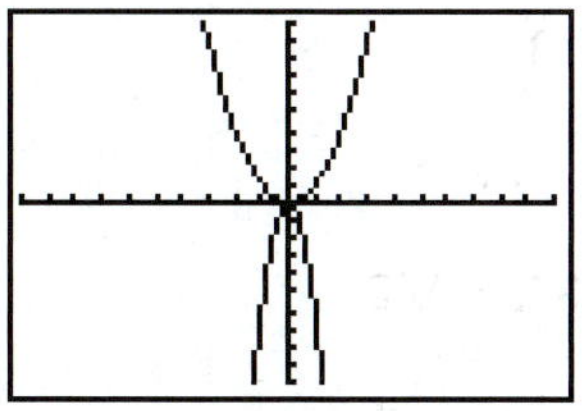

8. The graphs have the same shape, but the graph of $y = (x + 3)^2$ is 3 units to the left of the graph of $y = x^2$.

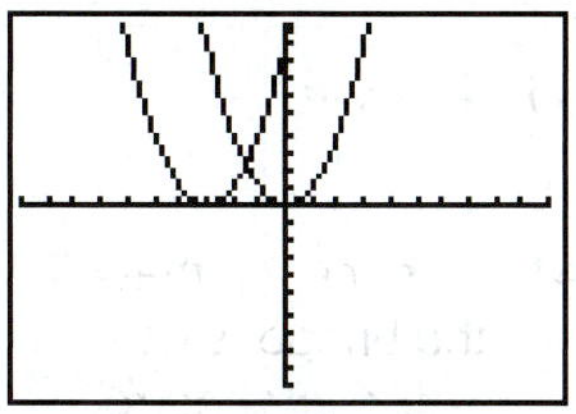

9. The graphs have the same shape and open downward, but the graph of $y = -\frac{1}{3}x^2 + 2$ is 2 units above the graph of $y = -\frac{1}{3}x^2$.

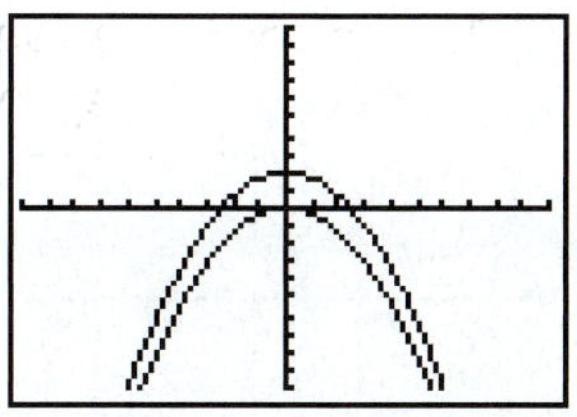

10. The graphs have the same shape, but the graph of $y = (x - 7)^2$ is 7 units to the right of the graph of $y = x^2$.

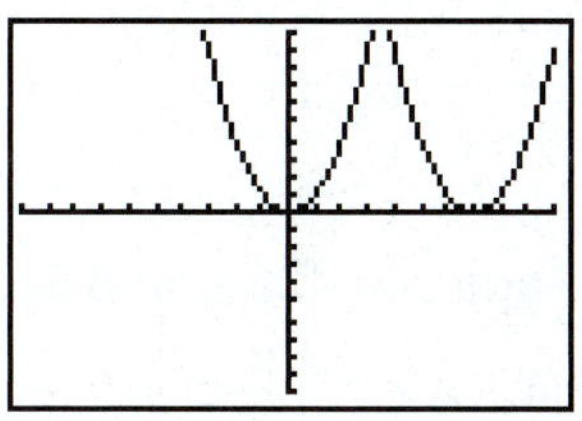

11. The graph of $y = 3(x + 4)^2 - 7$ is 4 units to the left, 7 units below, and narrower than the graph of $y = x^2$.

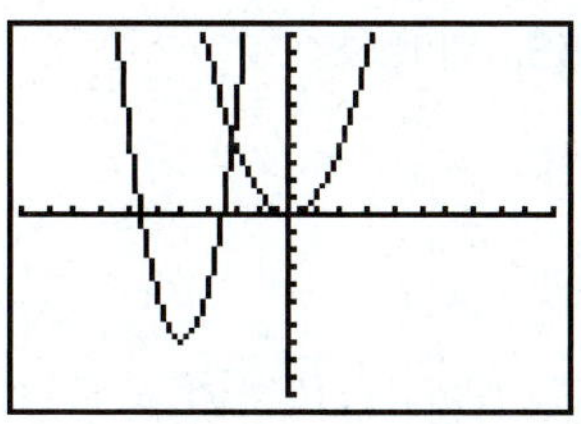

12. The graph of $y = -\frac{1}{4}x^2 + 1$ opens downward, is wider than and 1 unit above the graph of $y = -\frac{1}{4}x^2$.

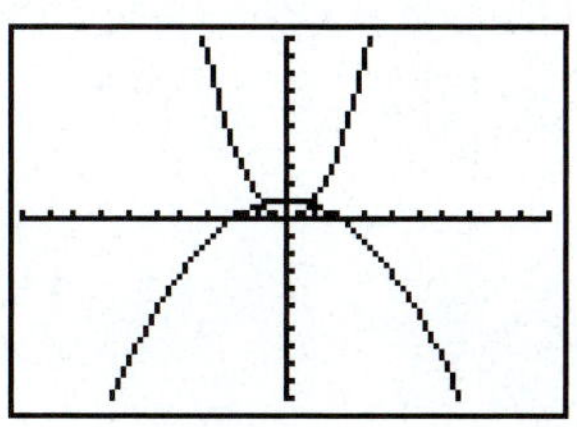

13. The graphs have the same shape, but the graph of $y = (x + 3)^2 + 5$ is 7 units above the graph of $y = (x + 3)^2 - 2$.

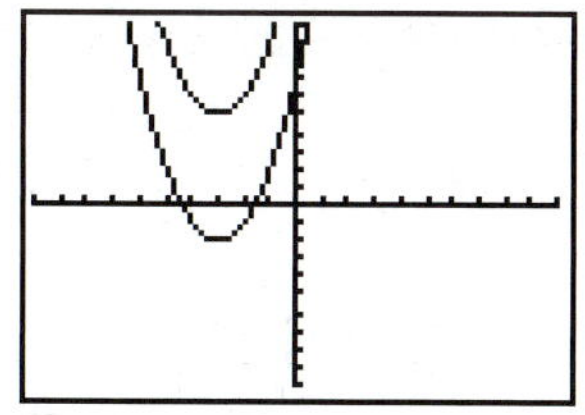

14. The graph of $y = 6(x + 2)^2 - 1$ is narrower than the graph of $y = 3(x + 2)^2 - 1$.

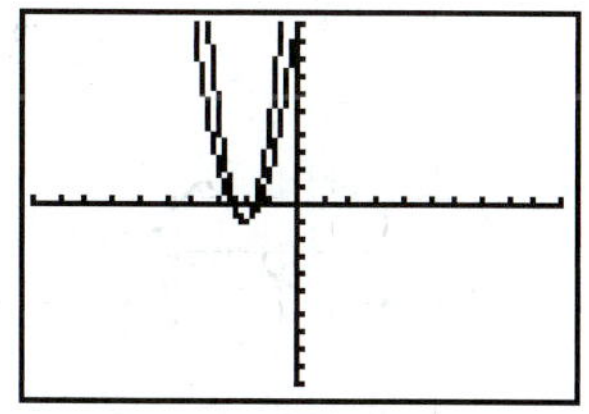

15. The graph of $y = \frac{1}{4}(x - 2)^2 - 1$ is wider than the graph of $y = 4(x - 2)^2 - 3$, and its vertex is 2 units above the vertex of $y = 4(x - 2)^2 - 3$.

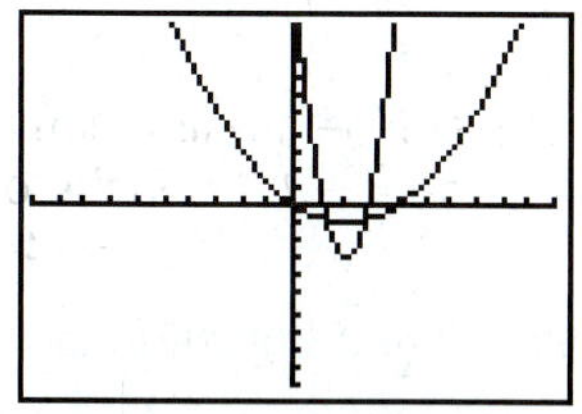

Pages 326–327, Lesson 6-6

27.

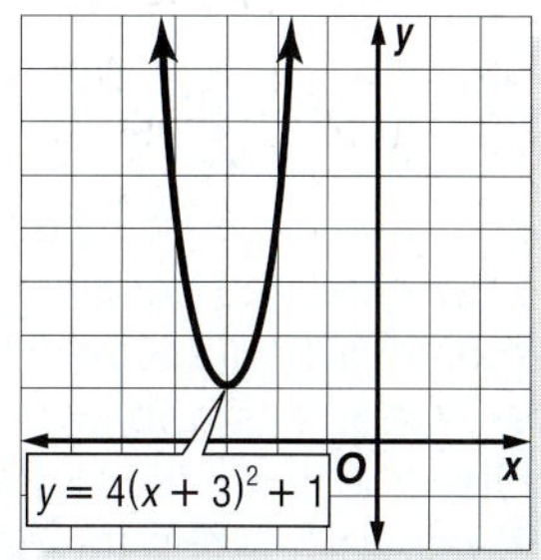

28.

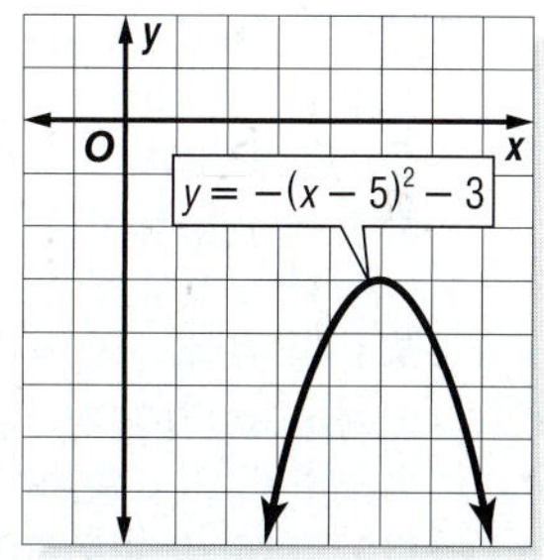

29.

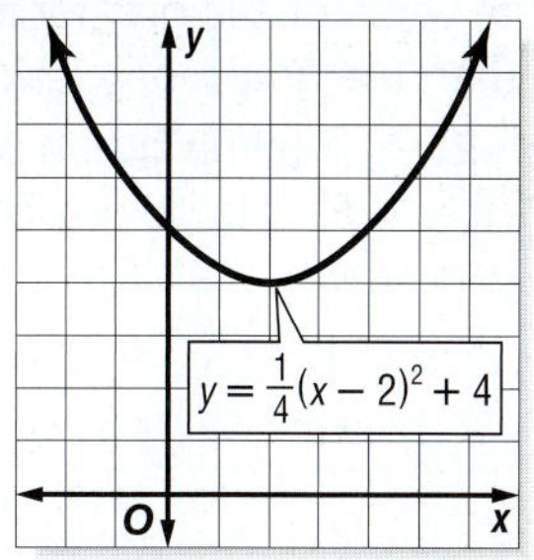

30.

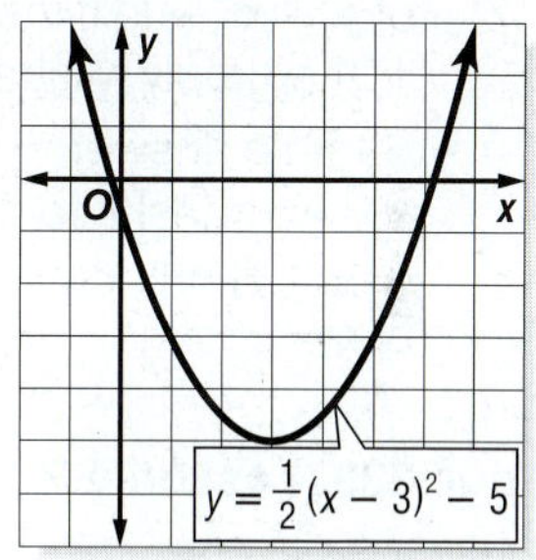

31.

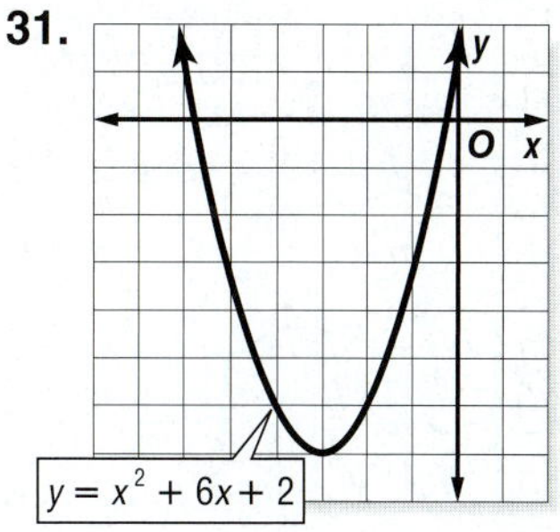

32.

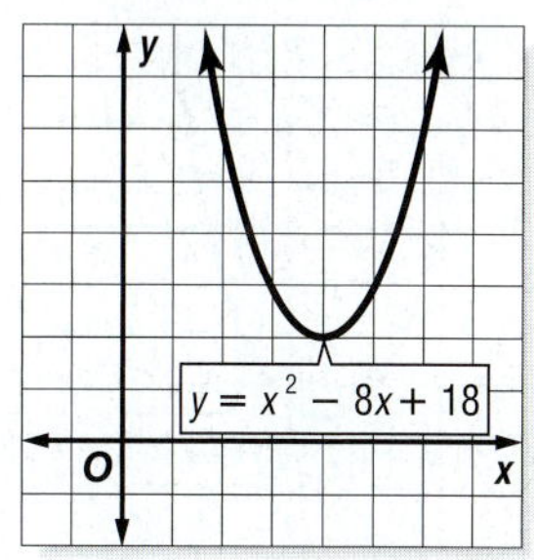

33.

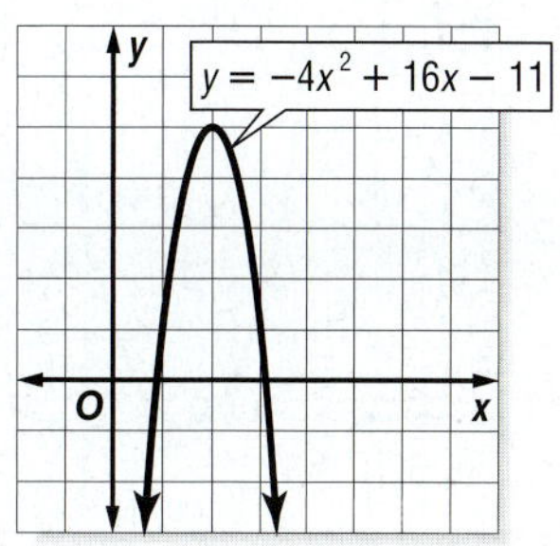

34.

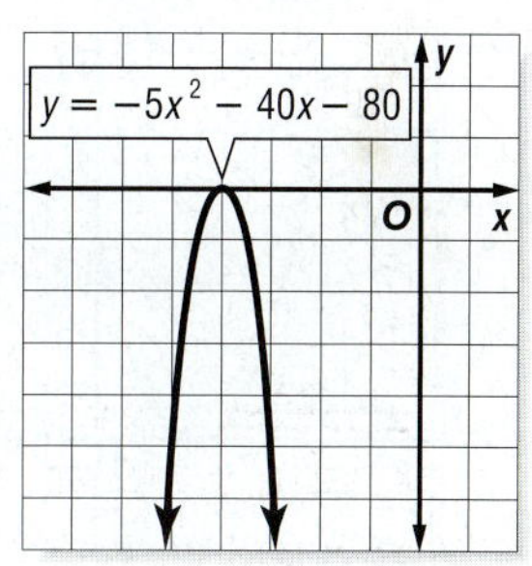

35.

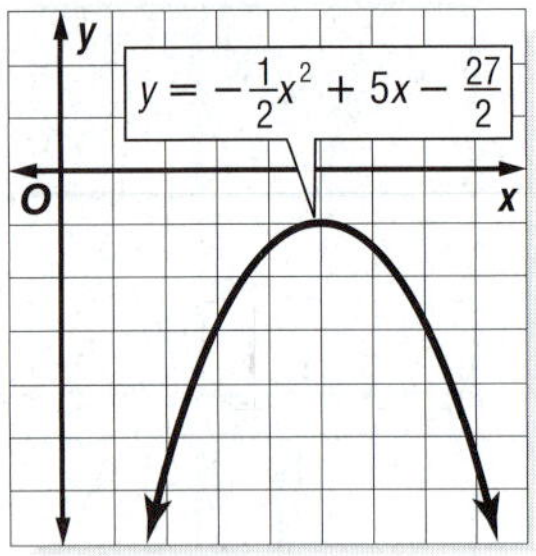

36.

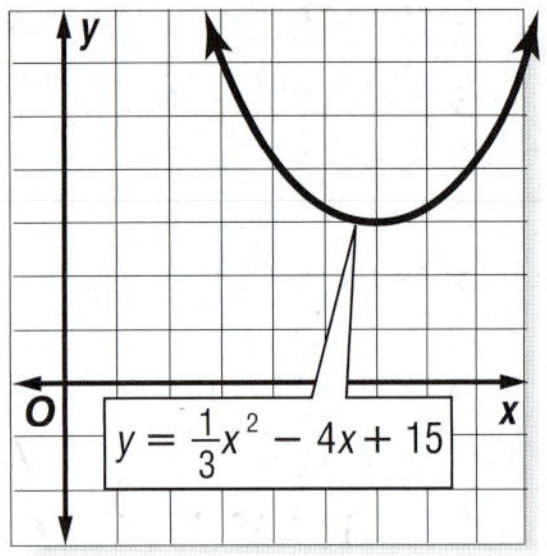

53. $y = ax^2 + bx + c$

$y = a\left(x^2 + \frac{b}{a}x\right) + c$

$y = a\left[x^2 + \frac{b}{a}x + \left(\frac{b}{2a}\right)^2\right] + c - a\left(\frac{b}{2a}\right)^2$

$y = a\left(x + \frac{b}{2a}\right)^2 + c - \frac{b^2}{4a}$

The axis of symmetry is $x = h$ or $-\frac{b}{2a}$.

54. All quadratic equations are a transformation of the parent graph $y = x^2$. By identifying these transformations when a quadratic function is written in vertex form, you can redraw the graph of $y = x^2$. Answers should include the following.

- In the equation $y = a(x - h)^2 + k$, h translated the graph of $y = x^2$ h units to the right when h is positive and h units to the left when h is negative. The graph of $y = x^2$ is translated k units up when k is positive and k units down when k is negative. When a is positive, the graph opens upward and when a is negative, the graph opens downward. If the absolute value of a is less than 1, the graph will be narrower than the graph of $y = x^2$, and if the absolute value of a is greater than 1, the graph will be wider than the graph of $y = x^2$.
- Sample answer: $y = 2(x + 2)^2 - 3$ is the graph of $y = x^2$ translated 2 units left and 3 units down. The graph opens upward, but is narrower that the graph of $y = x^2$.

Page 333, Lesson 6-7

14.

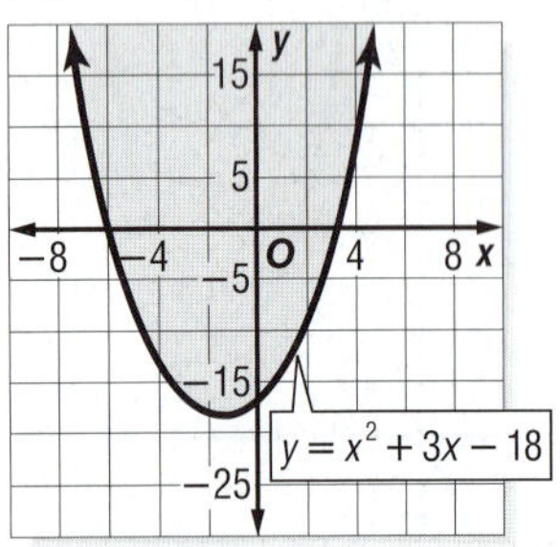

15.

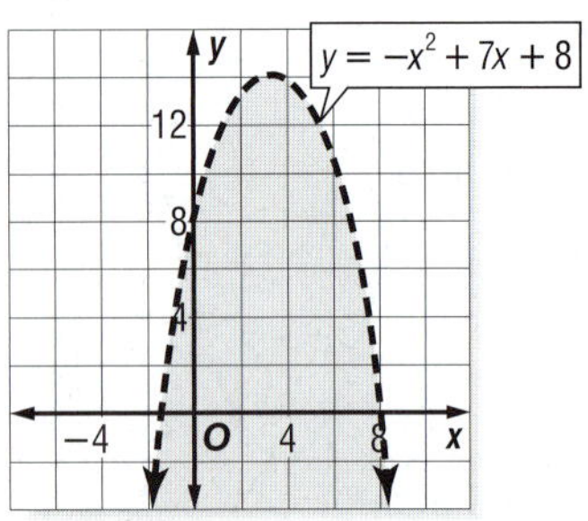

16.

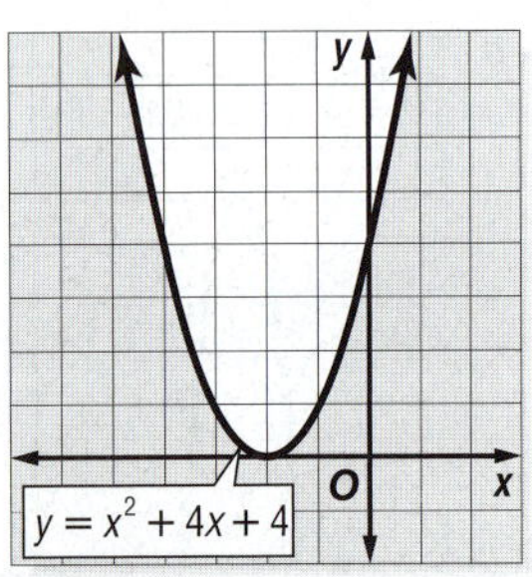

17.

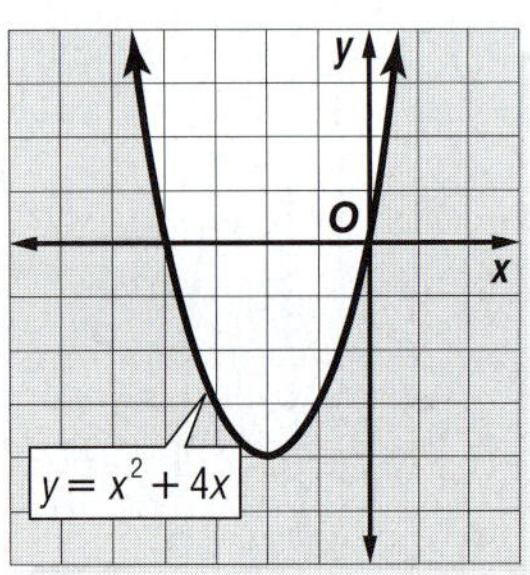

18.

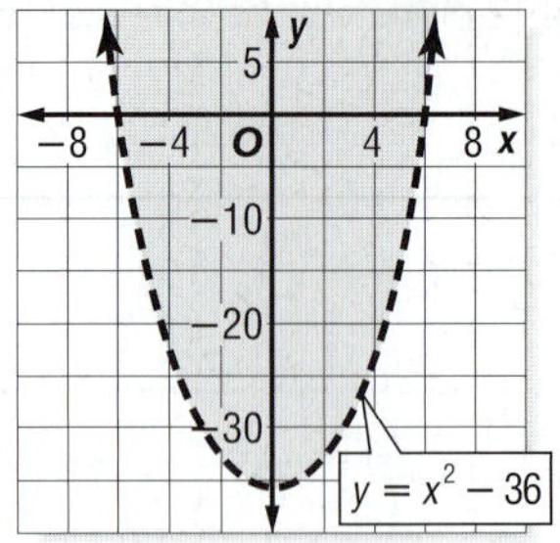

19.

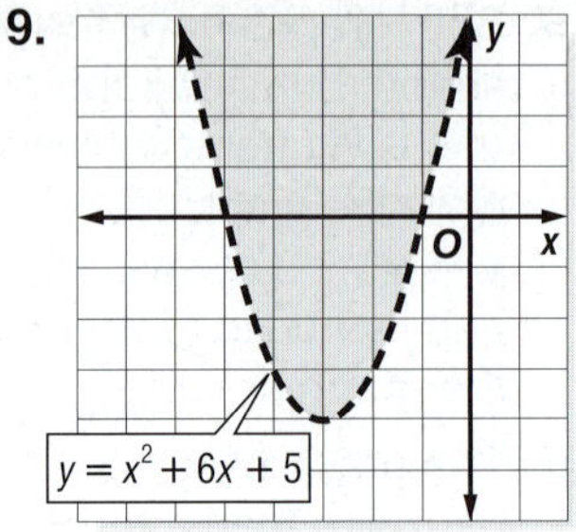

20.

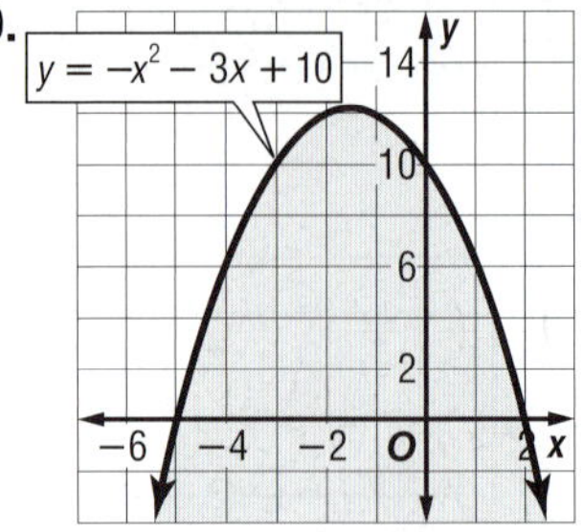

21.

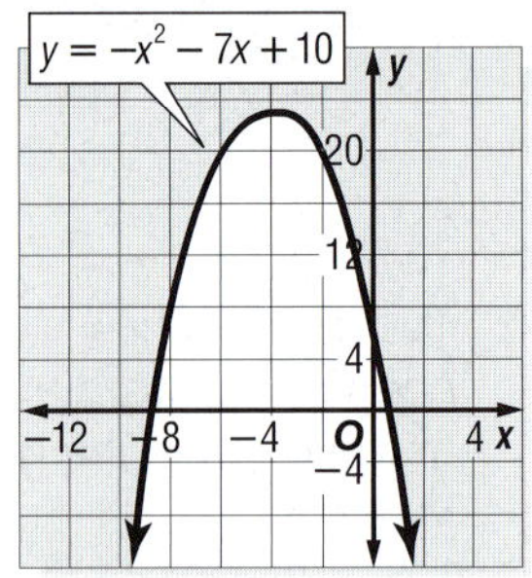

22.

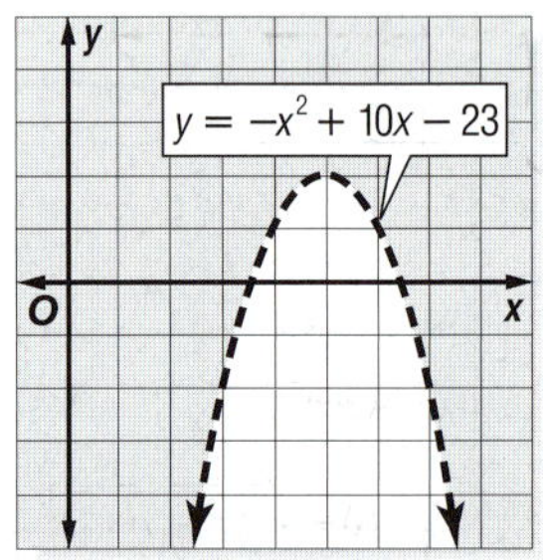

23.

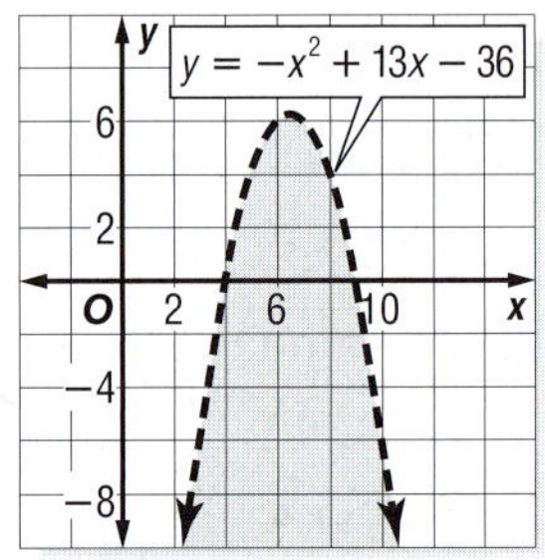

24.

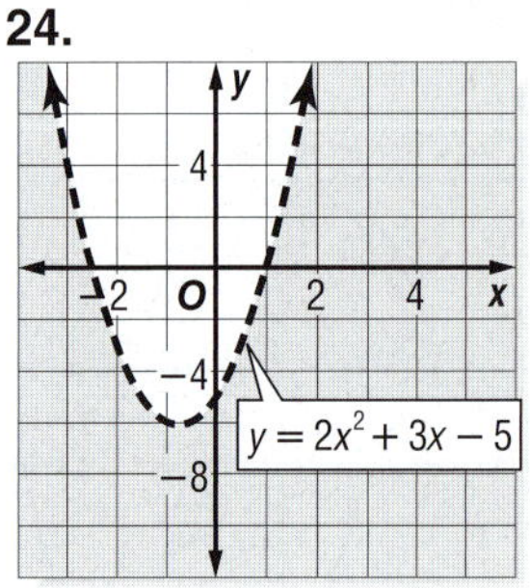

25.

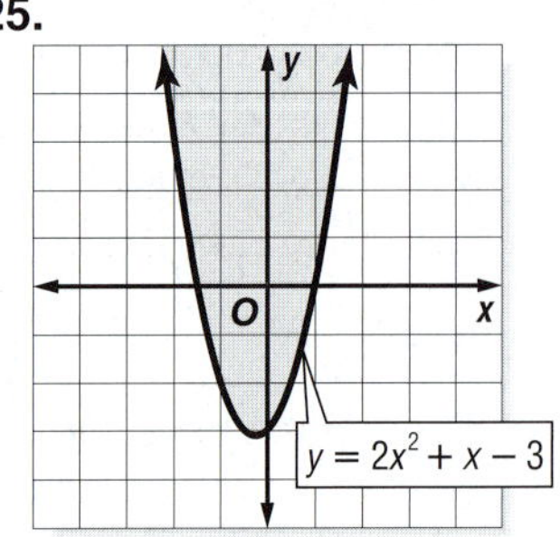

Page 340, Chapter 6 Study Guide and Review

54. $\left\{x \middle| x < -\frac{4}{3} \text{ or } x > \frac{1}{2}\right\}$

55. all reals

56. $\left\{x \middle| -4 < x < \frac{3}{2}\right\}$

57. $\left\{x \middle| x < -\frac{1}{2} \text{ or } x > 3\right\}$

58. $\left\{x \middle| \frac{-1 - \sqrt{10}}{2} \le x \le \frac{-1 + \sqrt{10}}{2}\right\}$

59. $\left\{x \middle| x < \frac{3 - 2\sqrt{6}}{3} \text{ or } x > \frac{3 + 2\sqrt{6}}{3}\right\}$

Page 341, Chapter 6 Practice Test

3a. (0, 5); $x = 1$; 1

3b.

x	$f(x)$
−1	8
0	5
1	4
2	5
3	8

3c.

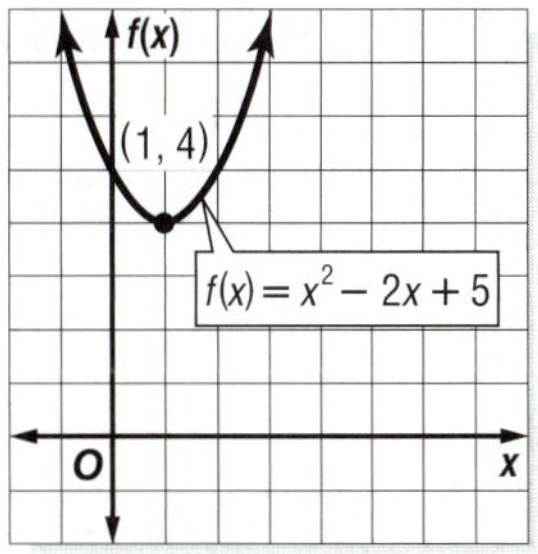

4a. (0, 0); $x = \frac{4}{3}$; $\frac{4}{3}$

4b.

x	$f(x)$
0	0
1	5
$\frac{4}{3}$	$\frac{16}{3}$
2	4
3	−3

4c.

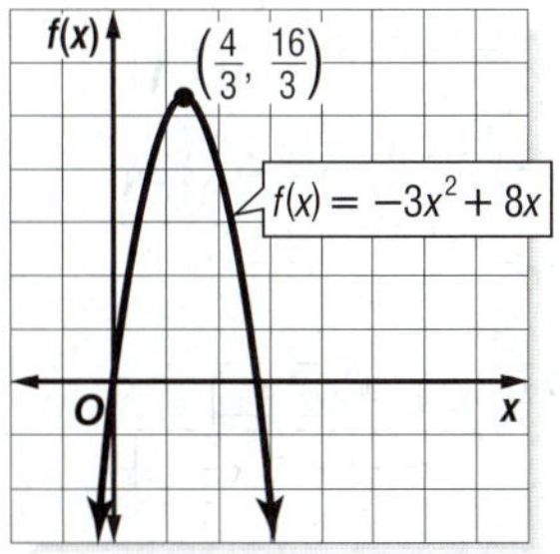

5a. (0, −1); $x = -\frac{7}{4}$; $-\frac{7}{4}$

5b.

x	$f(x)$
−3	2
−2	5
$-\frac{7}{4}$	$\frac{41}{8}$
−1	4
0	−1

5c.

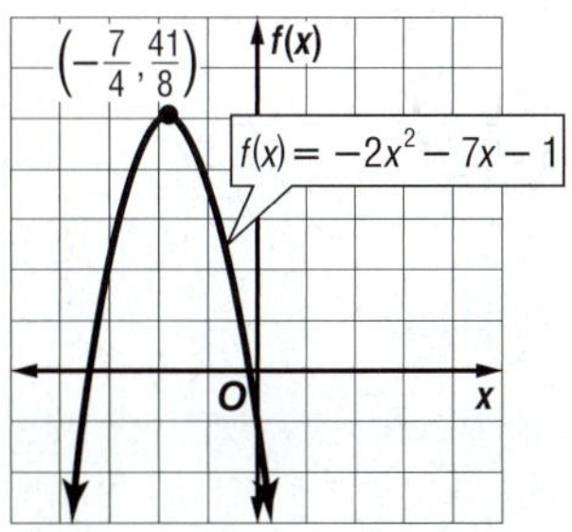

20. (−2, −3); $x = -2$; up

21. $y = (x + 5)^2 + 2$; (−5, 2); $x = -5$; up

22. $y = -9(x - 3)^2 + 73$; (3, 73); $x = 3$; down

23.

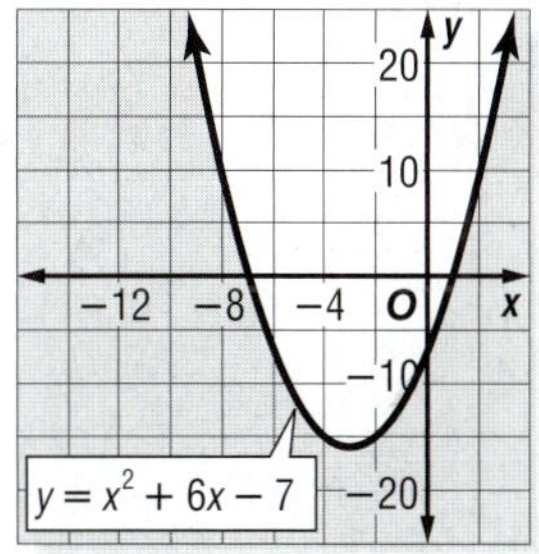

24.

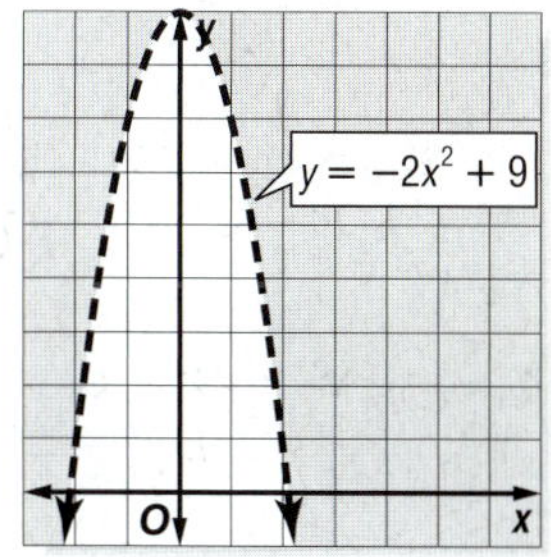

25.

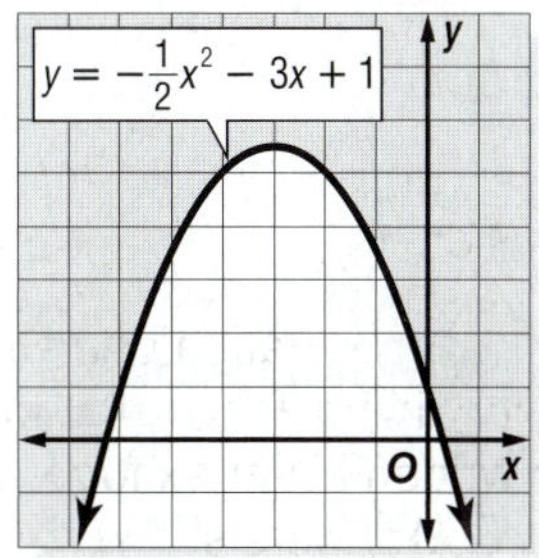

27. $\left\{x \middle| x \le -\frac{4\sqrt{3}}{3} \text{ or } x \ge \frac{4\sqrt{3}}{3}\right\}$

28. $\left\{x \middle| x < \frac{1 - \sqrt{41}}{10} \text{ or } x > \frac{1 + \sqrt{41}}{10}\right\}$

Chapter 7 Polynomial Functions

Chapter Overview and Pacing

Year-long and two-year pacing: pages T20–T21.

LESSON OBJECTIVES	PACING (days)			
	Regular		Block	
	Basic/ Average	Advanced	Basic/ Average	Advanced
7-1 Polynomial Functions *(pp. 346–352)* • Evaluate polynomial functions. • Identify general shapes of graphs of polynomial functions.	1	1	0.5	0.5
7-2 Graphing Polynomial Functions *(pp. 353–359)* • Graph polynomial functions and locate their real zeros. • Find the maxima and minima of polynomial functions. ***Follow-Up:*** Modeling Real-World Data	1	2 (with 7-2 Follow-Up)	0.5	1.5 (with 7-2 Follow-Up)
7-3 Solving Equations Using Quadratic Techniques *(pp. 360–364)* • Write expressions in quadratic form. • Use quadratic techniques to solve equations.	2	2	1	1
7-4 The Remainder and Factor Theorems *(pp. 365–370)* • Evaluate functions using synthetic substitution. • Determine whether a binomial is a factor of a polynomial by using synthetic substitution.	2	2	1	1
7-5 Roots and Zeros *(pp. 371–377)* • Determine the number and type of roots for a polynomial equation. • Find the zeros of a polynomial function.	1	1	0.5	0.5
7-6 Rational Zero Theorem *(pp. 378–382)* • Identify the possible rational zeros of a polynomial function. • Find all the rational zeros of a polynomial function.	2	2	1	1
7-7 Operations on Functions *(pp. 383–389)* • Find the sum, difference, product, and quotient of functions. • Find the composition of functions.	1	1	0.5	0.5
7-8 Inverse Functions and Relations *(pp. 390–394)* • Find the inverse of a function or relation. • Determine whether two functions or relations are inverses.	1	1	0.5	0.5
7-9 Square Root Functions and Inequalities *(pp. 395–399)* • Graph and analyze square root functions. • Graph square root inequalities.	1	1	0.5	0.5
Study Guide and **Practice Test** *(pp. 400–405)* **Standardized Test Practice** *(pp. 406–407)*	1	1	0.5	0.5
Chapter Assessment	1	1	0.5	0.5
TOTAL	**14**	**15**	**7**	**8**

*An electronic version of this chapter is available on **StudentWorks**™. This backpack solution CD-ROM allows students instant access to the Student Edition, lesson worksheet pages, and web resources.*

Timesaving Tools
TeacherWorks™
All-In-One Planner and Resource Center
See pages T5 and T21.

Chapter Resource Manager

Study Guide and Intervention	Practice (Skills and Average)	Reading to Learn Mathematics	Enrichment	Assessment	Applications*	5-Minute Check Transparencies	Interactive Chalkboard	Alge2PASS: Tutorial Plus (lessons)	Materials
375–376	377–378	379	380			7-1	7-1		graphing calculator, grid paper, string
381–382	383–384	385	386		SM 71–76	7-2	7-2		graphing calculator (*Follow-Up:* graphing calculator)
387–388	389–390	391	392	443		7-3	7-3	13	colored pencils
393–394	395–396	397	398			7-4	7-4		
399–400	401–402	403	404	443, 445		7-5	7-5	14	slips of paper
405–406	407–408	409	410		GCS 39	7-6	7-6		
411–412	413–414	415	416	444	GCS 40, SC 13	7-7	7-7		
417–418	419–420	421	422		SC 14	7-8	7-8		grid paper, string, spaghetti
423–424	425–426	427	428	444		7-9	7-9		graphing calculator
				429–442, 446–448					

Chapter 7 Resource Masters

Key to Abbreviations: GCS = Graphing Calculator and Speadsheet Masters,
SC = School-to-Career Masters,
SM = Science and Mathematics Lab Manual

Chapter 7

Mathematical Connections and Background

Continuity of Instruction

Prior Knowledge

Students have evaluated polynomials and terms of polynomials, graphed quadratic functions, and solved quadratic equations and inequalities. They explored synthetic division and the equivalence between zeros of functions and roots of equations. Also, they worked with function notation and manipulated and evaluated functions.

This Chapter

Students look at algebraic relationships among algebraic entities. They use the leading term of a polynomial function to describe and identify the end behavior of the function. They see how synthetic division of polynomials helps them explore the number of roots of a polynomial function and then find those roots. They perform operations on functions, and examine restrictions on domains and ranges necessary for combining functions.

Future Connections

Students will return to polynomials in later mathematics and other courses. They will continue their study of complex numbers, and extend it to study the complex plane. They will frequently use functions, combinations of functions, and inverses of functions in future mathematics topics.

7-1 Polynomial Functions

While this chapter uses some graphs to illustrate ideas, the primary focus of the chapter is on algebraic relationships between algebraic entities. In this lesson, the algebraic entities are the terms of a polynomial function. The main algebraic relationship in the lesson is how the degree and coefficient of one of those entities, the leading term of a polynomial function, determines the end behavior of the function.

The lesson begins by evaluating polynomial functions for various values of the variable, and uses the resulting sets of ordered pairs to illustrate the functions. Students use those illustrations to discuss when a graph may or must cross the x-axis. The illustrations also lead to the lesson's primarily idea of end behavior: for even-degree polynomial functions the left and right extremes of the graph are both positive or both negative, while for odd-degree polynomial functions one extreme is positive and the other extreme is negative.

7-2 Graphing Polynomial Functions

In this lesson, the primary algebraic entities are sets of x values, each set representing an interval on the x-axis. These entities are used to explore two types of changes in the values of a polynomial function. One type of change, a change between positive and negative function values, indicates that the graph of the function is crossing the x-axis, which is where the function has a zero. The other type of change, a change between increasing values and decreasing values for that x-axis interval, is called a turning point of the function. A turning point is a relative maximum or a relative minimum of the function depending on whether the switch is from increasing values to decreasing values, or vice versa. The lesson also uses the end-behavior analysis of the previous lesson and the two types of changes to help sketch graphs of polynomial functions.

7-3 Solving Equations Using Quadratic Techniques

In this lesson, the algebraic entities are polynomial expressions that can be rewritten as quadratic trinomials or binomials; for example, $a^4 - 3a^2 + 2$ can be rewritten as $(a^2)^2 - 3(a^2) + 2$. Students rewrite polynomial expressions whose exponents are integers or rational numbers, and then factor or use the Quadratic Formula to solve related quadratic equations.

7-4 The Remainder and Factor Theorems

In this lesson, the algebraic entity is a statement of polynomial division, $\frac{f(x)}{x-a} = q(x) + \frac{k}{x-a}$, where the right side of the equation is a polynomial plus a fraction whose numerator is a constant and whose denominator is the divisor. The lesson observes two properties of polynomials. One property focuses on the remainder. Multiplying both sides by the divisor $(x - a)$ and then finding the value of the polynomial at a gives the result $f(a) = q(a)(a - a) + k$. This simplifies to $f(a) = k$. In words, if a polynomial is divided by $(x - a)$, then the remainder k is equal to the value of the polynomial at a, or $f(a) = k$. This property is called the Remainder Theorem. The other property focuses on $(x - a)$, a binomial that divides a polynomial $f(x)$ evenly. In such a case, $(x - a)$ is a factor of $f(x)$ and the remainder must be zero; since $f(a)$ is the value of the remainder, $f(a) = 0$. This property is called the Factor Theorem. Students use these properties, along with the result that the degree of $q(x)$ is one less than the degree of $f(x)$, to find all the factors of third-degree polynomials.

7-5 Roots and Zeros

This lesson focuses on the equivalent algebraic entities of roots of a polynomial equation, zeros of functions, factors of polynomials, and x-intercepts of graphs. The lesson presents two properties. One property, called the Fundamental Theorem of Algebra, states the existence of at least one complex root for a polynomial equation. Based on this property, students learn that the degree of a polynomial equation is the same as the number of its complex roots. The other property, called Descartes' Rule of Signs, lets students calculate the number of positive and negative zeros of a polynomial equation. Then, given a polynomial equation, students use the two properties to find out how many roots they are looking for, how many will be real, and how many (as complex conjugates) will be complex. With that information, they use methods from the previous lessons to find the roots of the polynomial equation.

7-6 Rational Zero Theorem

In this lesson, the algebraic entities are the leading term a_0x_n and constant term a_n of a polynomial function with integer coefficients. The algebraic property is that if a reduced fraction is a zero of a function, then the fraction's numerator and denominator must be factors of a_0 and a_n, respectively. Students use this property in a two-step process for finding all the rational zeros of a polynomial function. First, they find all possible factors p of a_0, all possible factors q of a_n, and list (and reduce) all possible fractions $\frac{p}{q}$. Second, they test whether particular values are zeros. Once they find zeros, they can also use the Remainder Theorem to find the depressed polynomial for that zero.

7-7 Operations on Functions

In this lesson (and the next), the algebraic entities are functions themselves. The algebraic relationship is to look at the result of combining functions. Four combinations interpret the arithmetic operations addition, subtraction, multiplication, and division for functions. Another type of combination of functions is the composition $f \circ g$, where $[f \circ g](x) = f[g(x)]$. Students also identify relationships between and restrictions on ranges and domains for combining functions.

7-8 Inverse Functions and Relations

In this lesson, the algebraic entities are functions and the relationship explored is that of inverse functions. Given a function $f(x)$, students find another function $g(x)$ by switching the variables x and y and solving the resulting equation for y. Then they test whether $f(x)$ and $g(x)$ are inverses by checking that each of the two compositions $[f \circ g](x)$ and $[g \circ f](x)$ has the value x. The lesson uses the notation $f^{-1}(x)$ for the inverse function of $f(x)$ and the notation $I(x)$ for the identify function, and introduces the term *one-to-one* for a function that passes the horizontal line test.

7-9 Square Root Functions and Inequalities

In this lesson, the algebraic entity is a square root function, so called because the function contains a variable inside a square root symbol. The lesson explores two main ideas. One idea is that while the inverse of a quadratic function is not a function, you can restrict the domain of the inverse so that the result is a function. The other idea is that to graph a square root inequality, first you graph the related square root equation, forming two regions. Then you check points to see which region is represented by the inequality, and you use the inequality symbol to decide whether or not the boundary line is part of the solution region.

Additional mathematical information and teaching notes are available at www.algebra2.com/key_concepts.

Chapter 7

DAILY INTERVENTION and Assessment

	Type	Student Edition	Teacher Resources	Technology/Internet
INTERVENTION	Ongoing	Prerequisite Skills, pp. 345, 352, 358, 364, 370, 377, 382, 389, 394 Practice Quiz 1, p. 364 Practice Quiz 2, p. 382	5-Minute Check Transparencies Quizzes, *CRM* pp. 443–444 Mid-Chapter Test, *CRM* p. 445 Study Guide and Intervention, *CRM* pp. 375–376, 381–382, 387–388, 393–394, 399–400, 405–406, 411–412, 417–418, 423–424	Alge2PASS: Tutorial Plus www.algebra2.com/self_check_quiz www.algebra2.com/extra_examples
	Mixed Review	pp. 352, 358, 364, 370, 377, 382, 389, 394, 399	Cumulative Review, *CRM* p. 446	
	Error Analysis	Find the Error, pp. 380, 386	Find the Error, *TWE* pp. 380, 386 Unlocking Misconceptions, *TWE* pp. 354, 361, 375 Tips for New Teachers, *TWE* p. 384	
	Standardized Test Practice	pp. 352, 358, 364, 370, 374, 375, 377, 382, 389, 394, 399, 405, 406–407	*TWE* p. 374 Standardized Test Practice, *CRM* pp. 447–448	Standardized Test Practice CD-ROM www.algebra2.com/standardized_test
ASSESSMENT	Open-Ended Assessment	Writing in Math, pp. 352, 357, 364, 370, 377, 382, 389, 394, 399 Open Ended, pp. 350, 356, 362, 368, 375, 380, 382, 386, 393, 397	Modeling: *TWE* pp. 352, 389, 394 Speaking: *TWE* pp. 364, 370, 382 Writing: *TWE* pp. 358, 377, 399 Open-Ended Assessment, *CRM* p. 441	
	Chapter Assessment	Study Guide, pp. 400–404 Practice Test, p. 405	Multiple-Choice Tests (Forms 1, 2A, 2B), *CRM* pp. 429–434 Free-Response Tests (Forms 2C, 2D, 3), *CRM* pp. 435–440 Vocabulary Test/Review, *CRM* p. 442	ExamView® Pro (see below) MindJogger Videoquizzes www.algebra2.com/vocabulary_review www.algebra2.com/chapter_test

Key to Abbreviations: TWE = Teacher Wraparound Edition; CRM = Chapter Resource Masters

Intervention Technology

Alge2PASS: Tutorial Plus CD-ROM offers a complete, self-paced algebra curriculum.

Algebra 2 Lesson	Alge2PASS Lesson
7-3	**13** *Graphing Polynomial Functions*
7-5	**14** *Finding Roots and Zeros*

ExamView® Pro

Use the networkable **ExamView® Pro** to:

- Create **multiple versions** of tests.
- Create **modified** tests for *Inclusion* students.
- **Edit** existing questions and **add** your own questions.
- Use built-in **state curriculum correlations** to create tests aligned with state standards.
- **Apply** art to your tests from a program bank of artwork.

For more information on Intervention and Assessment, see pp. T8–T11.

Reading and Writing in Mathematics

Glencoe Algebra 2 provides numerous opportunities to incorporate reading and writing into the mathematics classroom.

Student Edition

- Foldables Study Organizer, p. 345
- Concept Check questions require students to verbalize and write about what they have learned in the lesson. (pp. 350, 356, 362, 368, 375, 380, 386, 393, 397, 400)
- Writing in Math questions in every lesson, pp. 352, 357, 364, 370, 377, 382, 389, 394, 399
- Reading Study Tip, pp. 354, 372, 384, 391
- WebQuest, p. 399

Teacher Wraparound Edition

- Foldables Study Organizer, pp. 345, 400
- Study Notebook suggestions, pp. 350, 356, 362, 368, 375, 380, 386, 392, 397
- Modeling activities, pp. 352, 389, 394
- Speaking activities, pp. 364, 370, 382
- Writing activities, pp. 358, 377, 399
- Differentiated Instruction, (Verbal/Linguistic), p. 356
- ELL Resources, pp. 344, 351, 356, 357, 363, 369, 376, 381, 388, 393, 398, 400

Additional Resources

- Vocabulary Builder worksheets require students to define and give examples for key vocabulary terms as they progress through the chapter. (*Chapter 7 Resource Masters,* pp. vii-viii)
- Reading to Learn Mathematics master for each lesson (*Chapter 7 Resource Masters*, pp. 379, 385, 391, 397, 403, 409, 415, 421, 427)
- *Vocabulary PuzzleMaker* software creates crossword, jumble, and word search puzzles using vocabulary lists that you can customize.
- *Teaching Mathematics with Foldables* provides suggestions for promoting cognition and language.
- *Reading and Writing in the Mathematics Classroom*
- *WebQuest and Project Resources*

For more information on Reading and Writing in Mathematics, see pp. T6–T7.

PROJECT CRISS[SM]

Study Skill

As you begin studying a new lesson or chapter, it may be helpful to students to complete a KWL frame. In a KWL frame, students record what they **K**now, what they **W**ant to know, and what they **L**earned. The completed sample frame below shows information that a student recorded before and after studying Lesson 7-1 on polynomial functions.

What I Know	What I Want To Know	What I Learned
• A polynomial is a sum of monomials. • A function is a relation in which each element of the domain is paired with exactly one element of the range.	• How can you predict the shape of the graph of a polynomial function by looking at the equation?	• The degree and the leading coefficient of the polynomial determine the graph's end behavior. • The graph of an odd function always crosses the x-axis at least once.

CReating **I**ndependence Through **S**tudent-Owned **S**trategies

Chapter 7 Notes

What You'll Learn

Have students read over the list of objectives and make a list of any words with which they are not familiar.

Why It's Important

Point out to students that this is only one of many reasons why each objective is important. Others are provided in the introduction to each lesson.

Lesson	NCTM Standards	Local Objectives
7-1	1, 2, 6, 7, 8, 9, 10	
7-2	1, 2, 6, 8, 9, 10	
7-2 Follow-Up	2, 5, 6, 9, 10	
7-3	1, 2, 3, 4, 6, 8, 9, 10	
7-4	1, 2, 3, 6, 8, 9, 10	
7-5	1, 2, 3, 4, 6, 7, 8, 9, 10	
7-6	1, 2, 3, 4, 6, 7, 8, 9	
7-7	1, 2, 6, 7, 8, 9, 10	
7-8	1, 2, 3, 4, 6, 7, 8, 9, 10	
7-9	1, 2, 6, 7, 8, 9, 10	

Key to NCTM Standards:

1=Number & Operations, 2=Algebra, 3=Geometry, 4=Measurement, 5=Data Analysis & Probability, 6=Problem Solving, 7=Reasoning & Proof, 8=Communication, 9=Connections, 10=Representation

Chapter 7 Polynomial Functions

What You'll Learn

- **Lessons 7-1 and 7-3** Evaluate polynomial functions and solve polynomial equations.
- **Lessons 7-2 and 7-9** Graph polynomial and square root functions.
- **Lessons 7-4, 7-5, and 7-6** Find factors and zeros of polynomial functions.
- **Lesson 7-7** Find the composition of functions.
- **Lesson 7-8** Determine the inverses of functions or relations.

Key Vocabulary

- polynomial function (p. 347)
- synthetic substitution (p. 365)
- Fundamental Theorem of Algebra (p. 371)
- composition of functions (p. 384)
- inverse function (p. 391)

Why It's Important

According to the Fundamental Theorem of Algebra, every polynomial equation has at least one root. Sometimes the roots have real-world meaning. Many real-world situations that cannot be modeled using a linear function can be approximated using a polynomial function. *You will learn how the power generated by a windmill can be modeled by a polynomial function in Lesson 7-1.*

Vocabulary Builder

ELL

The Key Vocabulary list introduces students to some of the main vocabulary terms included in this chapter. For a more thorough vocabulary list with pronunciations of new words, give students the Vocabulary Builder worksheets found on pages vii and viii of the *Chapter 7 Resource Masters*. Encourage them to complete the definition of each term as they progress through the chapter. You may suggest that they add these sheets to their study notebooks for future reference when studying for the Chapter 7 test.

Getting Started

Prerequisite Skills To be successful in this chapter, you'll need to master these skills and be able to apply them in problem-solving situations. Review these skills before beginning Chapter 7.

For Lesson 7-2 **3. between −5 and −4, between 0 and 1** **Solve Equations by Graphing**

Use the related graph of each equation to determine its roots. If exact roots cannot be found, state the consecutive integers between which the roots are located.
(For review, see Lesson 6-2.) **1. between 0 and 1, between 4 and 5** **2. between −2 and −1, 1**

1. $x^2 - 5x + 2 = 0$ **2.** $3x^2 + x - 4 = 0$ **3.** $\frac{2}{3}x^2 + 3x - 1 = 0$

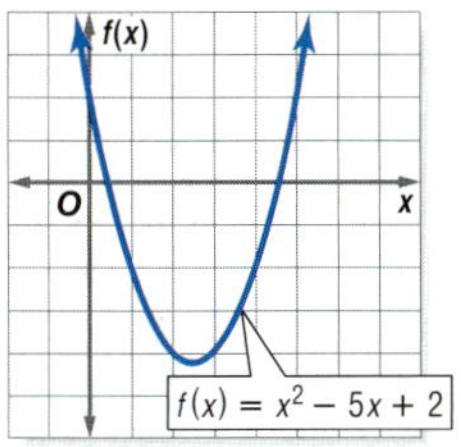

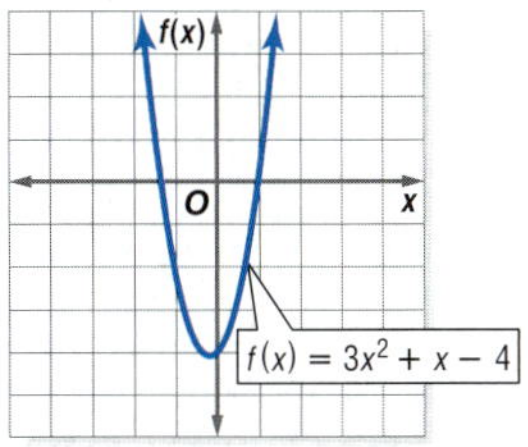

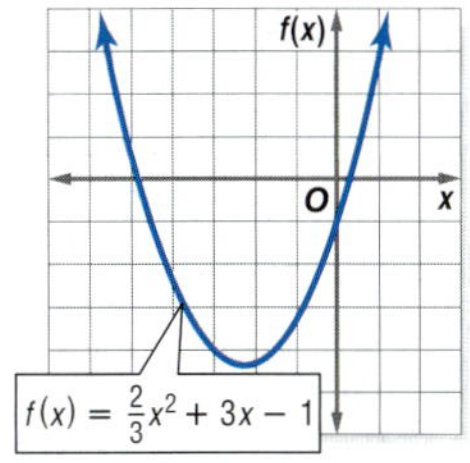

For Lesson 7-3 **Quadratic Formula**

Solve each equation. *(For review, see Lesson 6-5.)*

4. $x^2 - 17x + 60 = 0$ **5, 12** **5.** $14x^2 + 23x + 3 = 0$ $-\frac{3}{2}, -\frac{1}{7}$ **6.** $2x^2 + 5x + 1 = 0$ $\frac{-5 \pm \sqrt{17}}{4}$

For Lessons 7-4 through 7-6 **Synthetic Division**

Simplify each expression using synthetic division. *(For review, see Lesson 5-3.)*

7. $(3x^2 - 14x - 24) \div (x - 6)$ $3x + 4$ **8.** $(a^2 - 2a - 30) \div (a + 7)$ $a - 9 + \frac{33}{a + 7}$

For Lessons 7-1 and 7-7 **Evaluating Functions**

Find each value if $f(x) = 4x - 7$ and $g(x) = 2x^2 - 3x + 1$. *(For review, see Lesson 2-1.)*

9. $f(-3)$ -19 **10.** $g(2a)$ $8a^2 - 6a + 1$ **11.** $f(4b^2) + g(b)$ $18b^2 - 3b - 6$

FOLDABLES™ Study Organizer

Polynomial Functions Make this Foldable to help you organize your notes. Begin with five sheets of plain $8\frac{1}{2}$" by 11" paper.

Step 1 Stack and Fold

Stack sheets of paper with edges $\frac{3}{4}$-inch apart. Fold up the the bottom edges to create equal tabs.

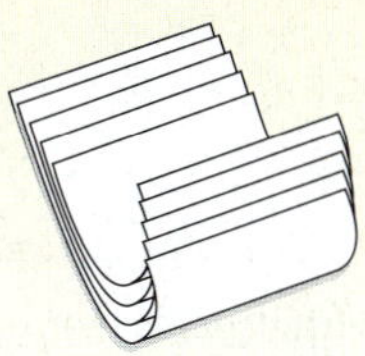

Step 2 Staple and Label

Staple along the fold. Label the tabs with lesson numbers.

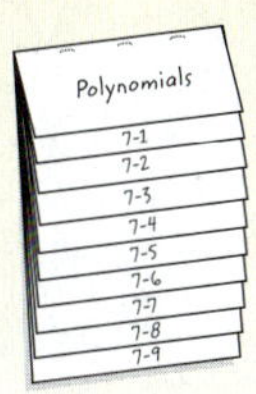

Reading and Writing As you read and study the chapter, use each page to write notes and examples.

Getting Started

This section provides a review of the basic concepts needed before beginning Chapter 7. Page references are included for additional student help.

Prerequisite Skills in the Getting Ready for the Next Lesson section at the end of each exercise set review a skill needed in the next lesson.

For Lesson	Prerequisite Skill
7-2	Graphing Quadratic Functions (p. 352)
7-3	Factoring Polynomials (p. 358)
7-4	Dividing Polynomials (p. 364)
7-5	Quadratic Formula (p. 370)
7-7	Operations with Polynomials (p. 382)
7-8	Solving Equations for a Variable (p. 389)
7-9	Solving Radical Equations (p. 394)

FOLDABLES™ Study Organizer

Making Generalizations while Previewing and Reviewing Data Have students label each tab of their Foldable to correspond to a lesson in Chapter 7. Before reading each lesson, have students preview it and write generalizations about what they think they will learn. As they read and study each lesson, students use their Foldable to take notes, define terms, record concepts, and write examples. After each lesson, ask them to compare what they thought they would learn with what they did learn as they review their notes.

For more information about Foldables, see *Teaching Mathematics with Foldables.*

7-1 Lesson Notes

1 Focus

5-Minute Check Transparency 7-1 Use as a quiz or review of Chapter 6.

Mathematical Background notes are available for this lesson on p. 344C.

Where are polynomial functions found in nature?

Ask students:

- How could you confirm that this model works for $r = 1$ and $r = 2$? **Substitute 1 and 2 for r in the formula and compare the results with the number of hexagons you count in the photograph.**
- Using the formula, what is the value of $f(8)$? **169**
- Why might it be useful for a beekeeper to know the approximate number of hexagons in a honeycomb? **Sample answer: The amount of honey stored in a honeycomb could be predicted by the number of hexagons.**

7-1 Polynomial Functions

What You'll Learn

- Evaluate polynomial functions.
- Identify general shapes of graphs of polynomial functions.

Vocabulary

- polynomial in one variable
- degree of a polynomial
- leading coefficients
- polynomial function
- end behavior

Where are polynomial functions found in nature?

If you look at a cross section of a honeycomb, you see a pattern of hexagons. This pattern has one hexagon surrounded by six more hexagons. Surrounding these is a third ring of 12 hexagons, and so on. The total number of hexagons in a honeycomb can be modeled by the function $f(r) = 3r^2 - 3r + 1$, where r is the number of rings and $f(r)$ is the number of hexagons.

POLYNOMIAL FUNCTIONS Recall that a polynomial is a monomial or a sum of monomials. The expression $3r^2 - 3r + 1$ is a **polynomial in one variable** since it only contains one variable, r.

Key Concept — Polynomial in One Variable

- **Words** A polynomial of degree n in one variable x is an expression of the form $a_0x^n + a_1x^{n-1} + \ldots + a_{n-2}x^2 + a_{n-1}x + a_n$, where the coefficients $a_0, a_1, a_2, \ldots, a_n$, represent real numbers, a_0 is not zero, and n represents a nonnegative integer.
- **Examples** $3x^5 + 2x^4 - 5x^3 + x^2 + 1$
  $n = 5$, $a_0 = 3$, $a_1 = 2$, $a_2 = -5$, $a_3 = 1$, $a_4 = 0$, and $a_5 = 1$

The **degree of a polynomial** in one variable is the greatest exponent of its variable. The **leading coefficient** is the coefficient of the term with the highest degree.

Polynomial	Expression	Degree	Leading Coefficient
Constant	9	0	9
Linear	$x - 2$	1	1
Quadratic	$3x^2 + 4x - 5$	2	3
Cubic	$4x^3 - 6$	3	4
General	$a_0x^n + a_1x^{n-1} + \ldots + a_{n-2}x^2 + a_{n-1}x + a_n$	n	a_0

Example 1 Find Degrees and Leading Coefficients

State the degree and leading coefficient of each polynomial in one variable. If it is not a polynomial in one variable, explain why.

a. $7x^4 + 5x^2 + x - 9$

This is a polynomial in one variable.

The degree is 4, and the leading coefficient is 7.

Resource Manager

Workbook and Reproducible Masters

Chapter 7 Resource Masters

- Study Guide and Intervention, pp. 375–376
- Skills Practice, p. 377
- Practice, p. 378
- Reading to Learn Mathematics, p. 379
- Enrichment, p. 380

Transparencies

5-Minute Check Transparency 7-1
Answer Key Transparencies

Technology

Interactive Chalkboard

b. $8x^2 + 3xy - 2y^2$

This is not a polynomial in one variable. It contains two variables, x and y.

c. $7x^6 - 4x^3 + \frac{1}{x}$

This is not a polynomial. The term $\frac{1}{x}$ cannot be written in the form x^n, where n is a nonnegative integer.

d. $\frac{1}{2}x^2 + 2x^3 - x^5$

Rewrite the expression so the powers of x are in decreasing order.

$$-x^5 + 2x^3 + \frac{1}{2}x^2$$

This is a polynomial in one variable with degree of 5 and leading coefficient of -1.

Study Tip

Power Function

A common type of function is a **power function**, which has an equation in the form $f(x) = ax^b$, where a and b are real numbers. When b is a positive integer, $f(x) = ax^b$ is a polynomial function.

A polynomial equation used to represent a function is called a **polynomial function**. For example, the equation $f(x) = 4x^2 - 5x + 2$ is a quadratic polynomial function, and the equation $p(x) = 2x^3 + 4x^2 - 5x + 7$ is a cubic polynomial function. Other polynomial functions can be defined by the following general rule.

Key Concept — Definition of a Polynomial Function

- **Words** A polynomial function of degree n can be described by an equation of the form $P(x) = a_0x^n + a_1x^{n-1} + \ldots + a_{n-2}x^2 + a_{n-1}x + a_n$, where the coefficients $a_0, a_1, a_2, \ldots, a_n$, represent real numbers, a_0 is not zero, and n represents a nonnegative integer.
- **Examples** $f(x) = 4x^2 - 3x + 2$
  $n = 2, a_0 = 4, a_1 = -3, a_2 = 2$

If you know an element in the domain of any polynomial function, you can find the corresponding value in the range. Recall that $f(3)$ can be found by evaluating the function for $x = 3$.

Example 2 Evaluate a Polynomial Function

NATURE **Refer to the application at the beginning of the lesson.**

a. Show that the polynomial function $f(r) = 3r^2 - 3r + 1$ gives the total number of hexagons when $r = 1, 2,$ and 3.

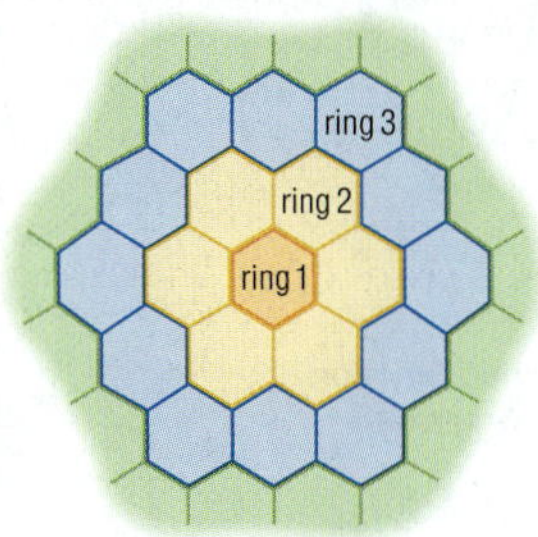

Rings of a Honeycomb

Find the values of $f(1)$, $f(2)$, and $f(3)$.

$f(r) = 3r^2 - 3r + 1$	$f(r) = 3r^2 - 3r + 1$	$f(r) = 3r^2 - 3r + 1$
$f(1) = 3(1)^2 - 3(1) + 1$	$f(2) = 3(2)^2 - 3(2) + 1$	$f(3) = 3(3)^2 - 3(3) + 1$
$= 3 - 3 + 1$ or 1	$= 12 - 6 + 1$ or 7	$= 27 - 9 + 1$ or 19

From the information given, you know the number of hexagons in the first ring is 1, the number of hexagons in the second ring is 6, and the number of hexagons in the third ring is 12. So, the total number of hexagons with one ring is 1, two rings is 6 + 1 or 7, and three rings is 12 + 6 + 1 or 19. These match the functional values for $r = 1, 2,$ and 3, respectively.

b. Find the total number of hexagons in a honeycomb with 12 rings.

$f(r) = 3r^2 - 3r + 1$ Original function

$f(12) = 3(12)^2 - 3(12) + 1$ Replace r with 12.

$= 432 - 36 + 1$ or 397 Simplify.

www.algebra2.com/extra_examples

2 Teach

POLYNOMIAL FUNCTIONS

In-Class Examples Power Point®

Teaching Tip Stress that the leading coefficient is not always the coefficient of the first term of a polynomial.

1 State the degree and leading coefficient of each polynomial in one variable. If it is not a polynomial in one variable, explain why.

a. $7z^3 - 4z^2 + z$ **degree 3, leading coefficient 7**

b. $6a^3 - 4a^2 + ab^2$ **This is not a polynomial in one variable. It contains two variables, a and b.**

c. $3c^2 + 4c - 2c^{-1}$ **This is not a polynomial. The term $-2c^{-1}$ is not of the form a_nc^n, where n is a nonnegative integer.**

d. $9y - 3y^2 + y^4$ **degree 4, leading coefficient 1**

2 **NATURE** Refer to the application at the beginning of the lesson. A sketch of the arrangement of hexagons shows a fourth ring of 18 hexagons, a fifth ring of 24 hexagons, and a sixth ring of 30 hexagons.

a. Show that the polynomial function $f(r) = 3r^2 - 3r + 1$ gives the total number of hexagons when $r = 4, 5,$ and 6. **$f(4) = 48 - 12 + 1$, or 37; $f(5) = 75 - 15 + 1$, or 61; $f(6) = 108 - 18 + 1$, or 91; The total number of hexagons for four rings is 19 + 18 or 37, five rings is 37 + 24 or 61, and six rings is 61 + 30 or 91. These match the functional values for $r = 4, 5,$ and 6, respectively.**

b. Find the total number of hexagons in a honeycomb with 20 rings. **1141**

In-Class Example

Power Point®

3 **a.** Find $p(y^3)$ if $p(x) = 2x^4 - x^3 + 3x$. $2y^{12} - y^9 + 3y^3$

b. Find $b(2x - 1) - 3b(x)$ if $b(m) = 2m^2 + m - 1$. $2x^2 - 9x + 3$

GRAPHS OF POLYNOMIAL FUNCTIONS

Teaching Tip Since the graphs on this page show the *maximum* number of times each type of graph may intersect the x-axis, some students may ask about the minimum number of times each graph type may intersect the x-axis. Have students work in pairs using the given graphs to discuss this issue. Lead students to see that for functions of degree 1 the minimum is 1 (the same as the maximum), for functions of degree 2 the minimum is 0, for functions of degree 3 the minimum is 1, for functions of degree 4 the minimum is 0, and for functions of degree 5 the minimum is 1. Some students may notice the pattern for functions with odd and even degrees.

Interactive Chalkboard

PowerPoint® Presentations

This CD-ROM is a customizable Microsoft® PowerPoint® presentation that includes:

- Step-by-step, dynamic solutions of each In-Class Example from the Teacher Wraparound Edition
- Additional, Your Turn exercises for each example
- The 5-Minute Check Transparencies
- Hot links to Glencoe Online Study Tools

You can also evaluate functions for variables and algebraic expressions.

Example 3 Functional Values of Variables

a. Find $p(a^2)$ if $p(x) = x^3 + 4x^2 - 5x$.

$p(x) = x^3 + 4x^2 - 5x$	Original function
$p(a^2) = (a^2)^3 + 4(a^2)^2 - 5(a^2)$	Replace x with a^2.
$= a^6 + 4a^4 - 5a^2$	Property of powers

b. Find $q(a + 1) - 2q(a)$ if $q(x) = x^2 + 3x + 4$.

To evaluate $q(a + 1)$, replace x in $q(x)$ with $a + 1$.

$q(x) = x^2 + 3x + 4$	Original function
$q(a + 1) = (a + 1)^2 + 3(a + 1) + 4$	Replace x with $a + 1$.
$= a^2 + 2a + 1 + 3a + 3 + 4$	Evaluate $(a + 1)^2$ and $3(a + 1)$.
$= a^2 + 5a + 8$	Simplify.

To evaluate $2q(a)$, replace x with a in $q(x)$, then multiply the expression by 2.

$q(x) = x^2 + 3x + 4$	Original function
$2q(a) = 2(a^2 + 3a + 4)$	Replace x with a.
$= 2a^2 + 6a + 8$	Distributive Property

Now evaluate $q(a + 1) - 2q(a)$.

$q(a + 1) - 2q(a) = a^2 + 5a + 8 - (2a^2 + 6a + 8)$	Replace $q(a + 1)$ and $2q(a)$ with evaluated expressions.
$= a^2 + 5a + 8 - 2a^2 - 6a - 8$	
$= -a^2 - a$	Simplify.

GRAPHS OF POLYNOMIAL FUNCTIONS The general shapes of the graphs of several polynomial functions are shown below. These graphs show the *maximum* number of times the graph of each type of polynomial may intersect the x-axis. Recall that the x-coordinate of the point at which the graph intersects the x-axis is called a *zero* of a function. How does the degree compare to the maximum number of real zeros?

Constant function
Degree 0

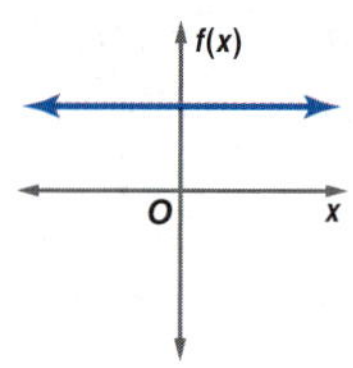

Linear function
Degree 1

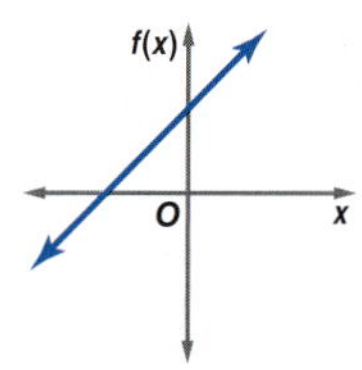

Quadratic function
Degree 2

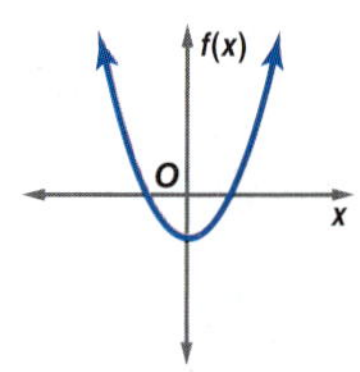

Cubic function
Degree 3

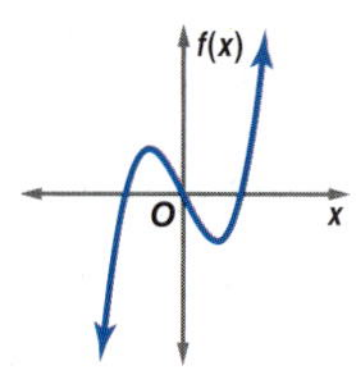

Quartic function
Degree 4

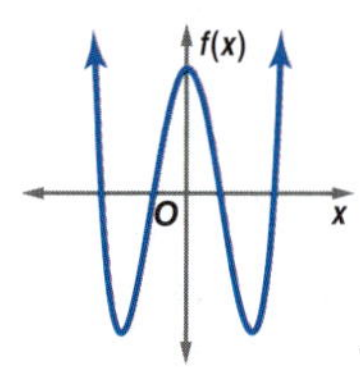

Quintic function
Degree 5

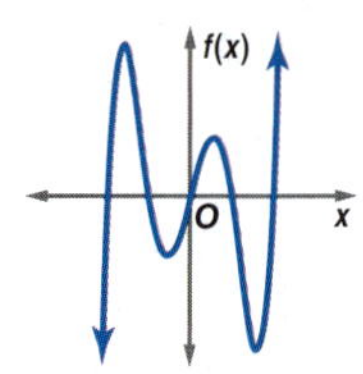

Notice the shapes of the graphs for even-degree polynomial functions and odd-degree polynomial functions. The degree and leading coefficient of a polynomial function determine the graph's end behavior.

The **end behavior** is the behavior of the graph as x approaches positive infinity $(+\infty)$ or negative infinity $(-\infty)$. This is represented as $x \to +\infty$ and $x \to -\infty$, respectively. $x \to +\infty$ is read *x approaches positive infinity.*

Concept Summary — *End Behavior of a Polynomial Function*

Degree: even
Leading Coefficient: positive
End Behavior:

$f(x) \to +\infty$ as $x \to -\infty$ $f(x) \to +\infty$ as $x \to +\infty$

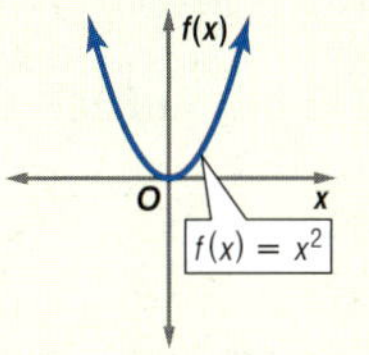

Degree: odd
Leading Coefficient: positive
End Behavior:

$f(x) \to +\infty$ as $x \to +\infty$

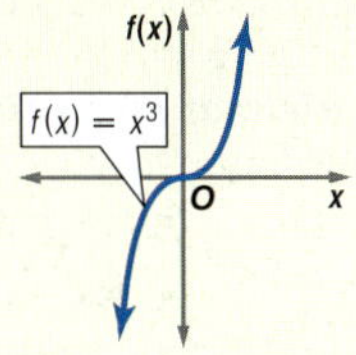

$f(x) \to -\infty$ as $x \to -\infty$

Degree: even
Leading Coefficient: negative
End Behavior:

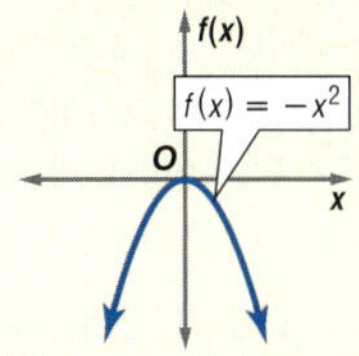

$f(x) \to -\infty$ as $x \to -\infty$ $f(x) \to -\infty$ as $x \to +\infty$

Degree: odd
Leading Coefficient: negative
End Behavior:

$f(x) \to +\infty$ as $x \to -\infty$

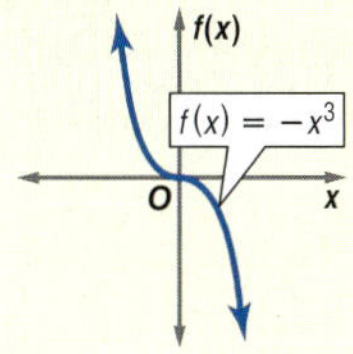

$f(x) \to -\infty$ as $x \to +\infty$

The graph of an even-degree function may or may not intersect the x-axis, depending on its location in the coordinate plane. If it intersects the x-axis in two places, the function has two real zeros. If it does not intersect the x-axis, the roots of the related equation are imaginary and cannot be determined from the graph. If the graph is tangent to the x-axis, as shown above, there are two zeros that are the same number. The graph of an odd-degree function always crosses the x-axis at least once, and thus the function always has at least one real zero.

Study Tip

Number of Zeros
The number of zeros of an odd-degree function may be less than the maximum by a multiple of 2. For example, the graph of a quintic function may only cross the x-axis 3 times.

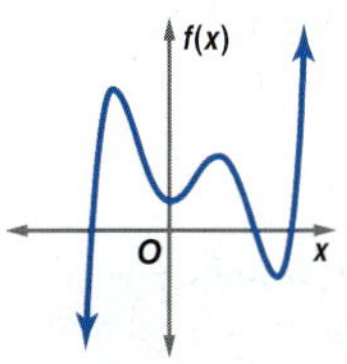

The same is true for an even-degree function. One exception is when the graph of $f(x)$ touches the x-axis.

Example 4 *Graphs of Polynomial Functions*

For each graph,

- **describe the end behavior,**
- **determine whether it represents an odd-degree or an even-degree polynomial function, and**
- **state the number of real zeros.**

a.

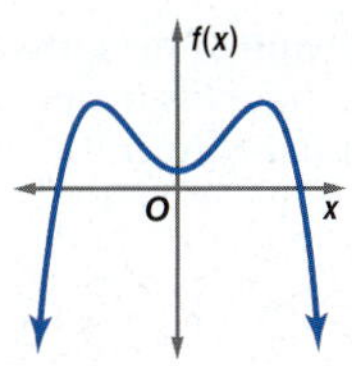

b.

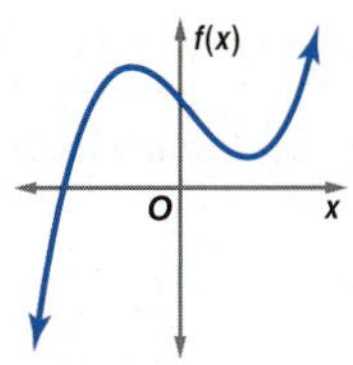

c.

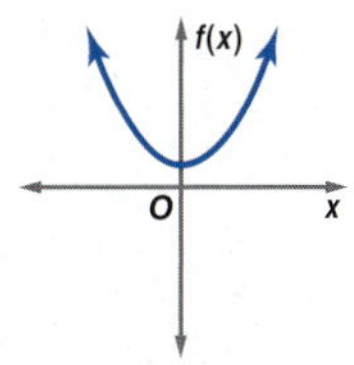

a. • $f(x) \to -\infty$ as $x \to +\infty$. $f(x) \to -\infty$ as $x \to -\infty$.
- It is an even-degree polynomial function.
- The graph intersects the x-axis at two points, so the function has two real zeros.

b. • $f(x) \to +\infty$ as $x \to +\infty$. $f(x) \to -\infty$ as $x \to -\infty$.
- It is an odd-degree polynomial function.
- The graph has one real zero.

c. • $f(x) \to +\infty$ as $x \to +\infty$. $f(x) \to +\infty$ as $x \to -\infty$.
- It is an even-degree polynomial function.
- This graph does not intersect the x-axis, so the function has no real zeros.

In-Class Example

PowerPoint®

4 For each graph,
- describe the end behavior,
- determine whether it represents an odd-degree or an even-degree polynomial function, and
- state the number of real zeros.

a.

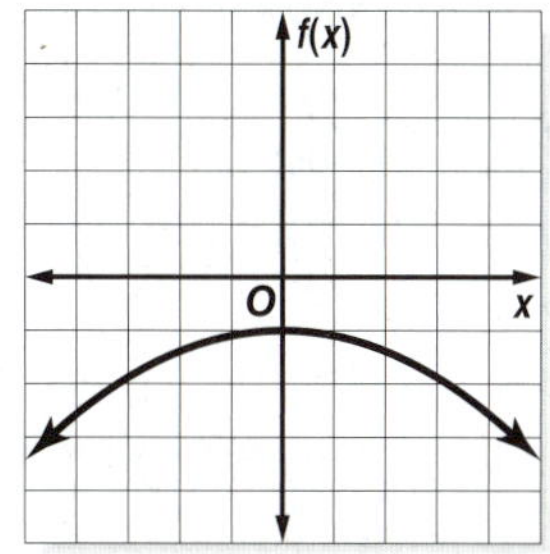

- **$f(x) \to -\infty$ as $x \to +\infty$. $f(x) \to -\infty$ as $x \to -\infty$.**
- **It is an even-degree polynomial function.**
- **The graph does not intersect the x-axis, so the function has no real zeros.**

b.

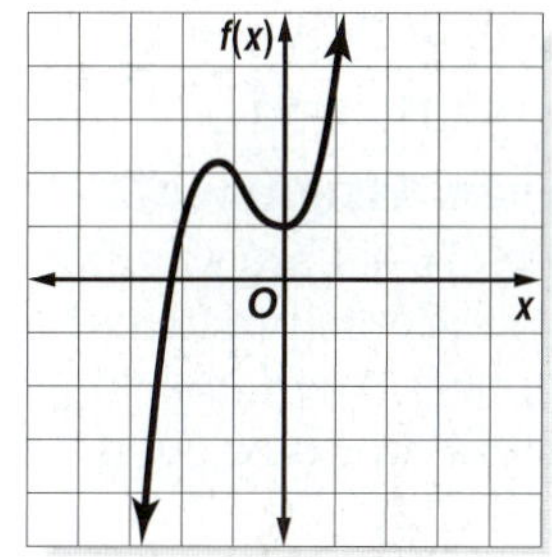

- **$f(x) \to +\infty$ as $x \to +\infty$. $f(x) \to -\infty$ as $x \to -\infty$.**
- **It is an odd-degree polynomial function.**
- **The graph intersects the x-axis at one point, so the function has one real zero.**

c.

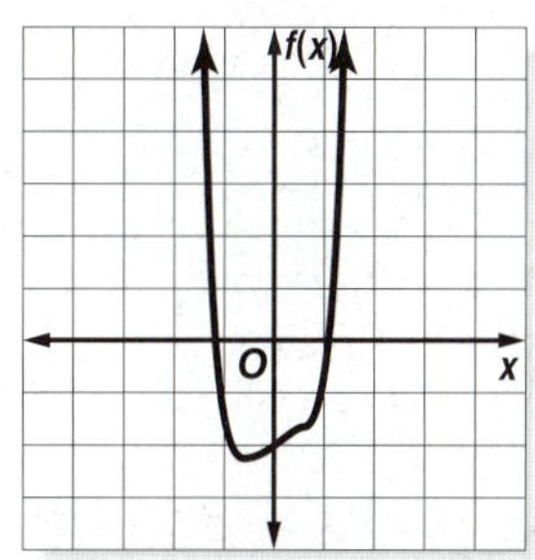

- **$f(x) \to +\infty$ as $x \to +\infty$. $f(x) \to +\infty$ as $x \to -\infty$.**
- **It is an even-degree polynomial function.**
- **The function has two real zeros.**

DAILY INTERVENTION — Differentiated Instruction

Interpersonal Arrange students in groups of 3 or 4, providing each group with a graphing calculator. Have each student write a polynomial function. As a group, have students state whether each function is an odd-degree or an even-degree polynomial function before predicting the end behavior and the number of zeros of the function. Then have students check their predictions by graphing each function. Challenge students to find at least one polynomial function that crosses the x-axis 3 or 4 times.

3 Practice/Apply

Study Notebook

Have students—
- add the definitions/examples of the vocabulary terms to their Vocabulary Builder worksheets for Chapter 7.
- copy the information in the Concept Summary about end behavior shown on p. 349.
- include any other item(s) that they find helpful in mastering the skills in this lesson.

About the Exercises...

Organization by Objective
- **Polynomial Functions:** 16–38
- **Graphs of Polynomial Functions:** 39–44

Odd/Even Assignments
Exercises 16–44 are structured so that students practice the same concepts whether they are assigned odd or even problems.

Assignment Guide

Basic: 17–45 odd, 49–52, 56–70

Average: 17–45 odd, 46–52, 56–70

Advanced: 16–44 even, 46–67 (optional: 68–70)

Answers

2. Sample answer: Even-degree polynomial functions with positive leading coefficients have graphs in which $f(x) \to +\infty$ as $x \to +\infty$ and as $x \to -\infty$. Odd-degree polynomial functions with positive leading coefficients have graphs in which $f(x) \to +\infty$ as $x \to +\infty$ and $f(x) \to -\infty$ as $x \to -\infty$.

Check for Understanding

Concept Check

1. **Explain** why a constant polynomial such as $f(x) = 4$ has degree 0 and a linear polynomial such as $f(x) = x + 5$ has degree 1. $4 = 4x^0$; $x = x^1$
2. **Describe** the characteristics of the graphs of odd-degree and even-degree polynomial functions whose leading coefficients are positive. See margin.
3. **OPEN ENDED** Sketch the graph of an odd-degree polynomial function with a negative leading coefficient and three real roots. See margin.
4. **Tell** whether the following statement is *always, sometimes* or *never* true. Explain.
A polynomial function that has four real roots is a fourth-degree polynomial.

4. Sometimes; a polynomial function with 4 real roots may be a sixth-degree polynomial function with 2 imaginary roots. A polynomial function that has 4 real roots is *at least* a fourth-degree polynomial.

Guided Practice

GUIDED PRACTICE KEY	
Exercises	Examples
5, 6	1
7, 8, 15	2
9–11	3
12–14	4

State the degree and leading coefficient of each polynomial in one variable. If it is not a polynomial in one variable, explain why.

5. $5x^6 - 8x^2$ 6; 5
6. $2b + 4b^3 - 3b^5 - 7$ 5; −3

Find $p(3)$ and $p(-1)$ for each function.

7. $p(x) = -x^3 + x^2 - x$ −21; 3
8. $p(x) = x^4 - 3x^3 + 2x^2 - 5x + 1$ 4; 12

If $p(x) = 2x^3 + 6x - 12$ and $q(x) = 5x^2 + 4$, find each value. 11. $6a^3 - 5a^2 + 8a - 45$

9. $p(a^3)$ $2a^9 + 6a^3 - 12$
10. $5[q(2a)]$ $100a^2 + 20$
11. $3p(a) - q(a + 1)$

For each graph,
a. describe the end behavior,
b. determine whether it represents an odd-degree or an even-degree polynomial function, and
c. state the number of real zeros.

12. 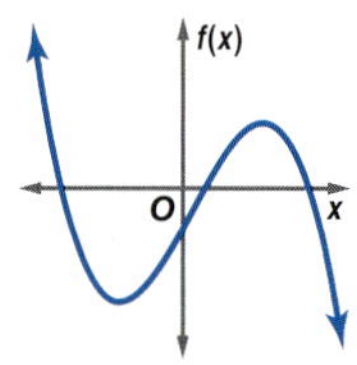

13. 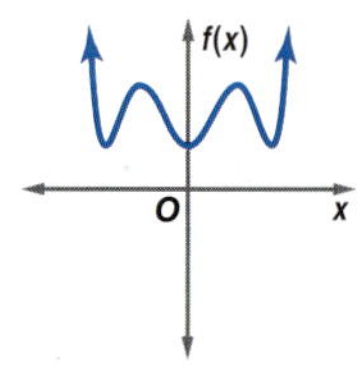

14. 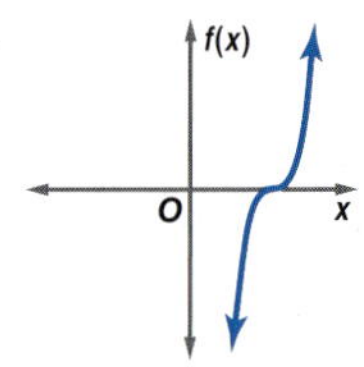

12. a. $f(x) \to -\infty$ as $x \to +\infty$, $f(x) \to +\infty$ as $x \to -\infty$; b. odd; c. 3
13. a. $f(x) \to +\infty$ as $x \to +\infty$, $f(x) \to +\infty$ as $x \to -\infty$; b. even; c. 0
14. a. $f(x) \to +\infty$ as $x \to +\infty$, $f(x) \to -\infty$ as $x \to -\infty$; b. odd; c. 1

Application

15. **BIOLOGY** The intensity of light emitted by a firefly can be determined by $L(t) = 10 + 0.3t + 0.4t^2 - 0.01t^3$, where t is temperature in degrees Celsius and $L(t)$ is light intensity in lumens. If the temperature is 30°C, find the light intensity. 109 lumens

★ indicates increased difficulty

Practice and Apply

Homework Help	
For Exercises	See Examples
16–21	1
22–29, 45	2
30–38	3
39–44, 46–48	4

Extra Practice
See page 842.

State the degree and leading coefficient of each polynomial in one variable. If it is not a polynomial in one variable, explain why.

16. $7 - x$ 1; −1
17. $(a + 1)(a^2 - 4)$ 3; 1
18. $a^2 + 2ab + b^2$ See margin.
19. $6x^4 + 3x^2 + 4x - 8$ 4; 6
20. $7 + 3x^2 - 5x^3 + 6x^2 - 2x$ 3; −5
21. $c^2 + c - \frac{1}{c}$ See margin.

Find $p(4)$ and $p(-2)$ for each function.

22. $p(x) = 2 - x$ −2; 4
23. $p(x) = x^2 - 3x + 8$ 12; 18
24. $p(x) = 2x^3 - x^2 + 5x - 7$ 125; −37
25. $p(x) = x^5 - x^2$ 1008; −36
26. $p(x) = x^4 - 7x^3 + 8x - 6$ −166; 50
27. $p(x) = 7x^2 - 9x + 10$ 86; 56
28. $p(x) = \frac{1}{2}x^4 - 2x^2 + 4$ 100; 4
29. $p(x) = \frac{1}{8}x^3 - \frac{1}{4}x^2 - \frac{1}{2}x + 5$ 7; 4

3. Sample answer:

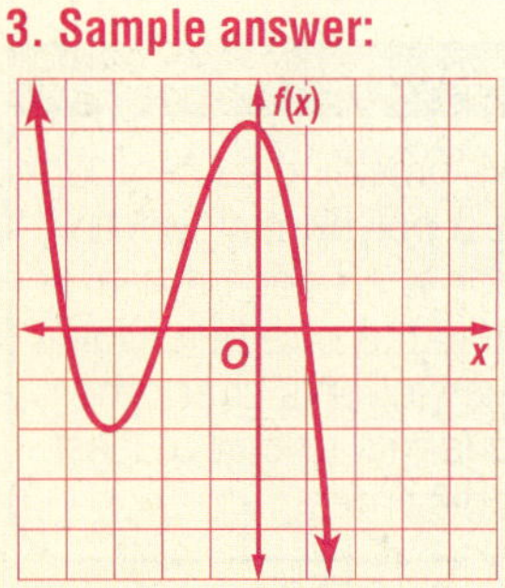

18. No, the polynomial contains two variables, a and b.
21. No, this is not a polynomial because the term $\frac{1}{c}$ cannot be written in the form x^n, where n is a nonnegative integer.

39a. $f(x) \to +\infty$ as $x \to +\infty$, $f(x) \to -\infty$ as $x \to -\infty$; b. odd; c. 3
40a. $f(x) \to +\infty$ as $x \to +\infty$, $f(x) \to +\infty$ as $x \to -\infty$; b. even; c. 4
41a. $f(x) \to -\infty$ as $x \to +\infty$, $f(x) \to -\infty$ as $x \to -\infty$; b. even; c. 0
42a. $f(x) \to +\infty$ as $x \to +\infty$, $f(x) \to -\infty$ as $x \to -\infty$; b. odd; c. 5
43a. $f(x) \to +\infty$ as $x \to +\infty$, $f(x) \to -\infty$ as $x \to -\infty$; b. odd; c. 1
44a. $f(x) \to -\infty$ as $x \to +\infty$, $f(x) \to -\infty$ as $x \to -\infty$; b. even; c. 2

350 Chapter 7 Polynomial Functions

32. $3a^4 - 2a^2 + 5$
34. $x^3 + 3x^2 + 4x + 3$
35. $3x^4 + 16x^2 + 26$

If $p(x) = 3x^2 - 2x + 5$ and $r(x) = x^3 + x + 1$, find each value.

30. $r(3a)$ $27a^3 + 3a + 1$
31. $4p(a)$ $12a^2 - 8a + 20$
32. $p(a^2)$
33. $p(2a^3)$ $12a^6 - 4a^3 + 5$
34. $r(x + 1)$
35. $p(x^2 + 3)$
36. $2[p(x + 4)]$ $6x^2 + 44x + 90$
37. $r(x + 1) - r(x^2)$ $-x^6 + x^3 + 2x^2 + 4x + 2$
38. $3[p(x^2 - 1)] + 4p(x)$ $9x^4 - 12x^2 - 8x + 50$

For each graph,

a. describe the end behavior,

b. determine whether it represents an odd-degree or an even-degree polynomial function, and

c. state the number of real zeros. **39–44. See margin.**

39.
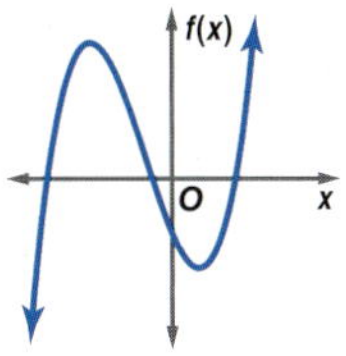

40.
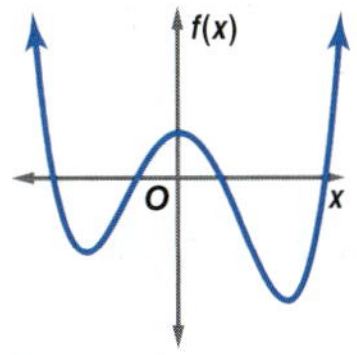

41.
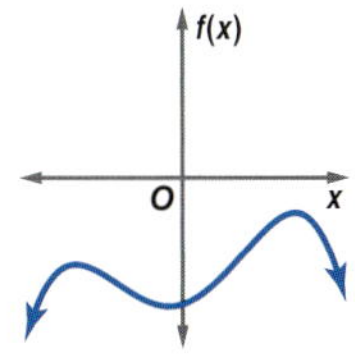

42.
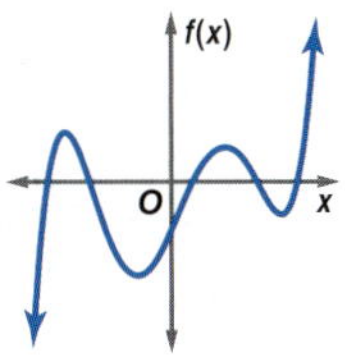

43.
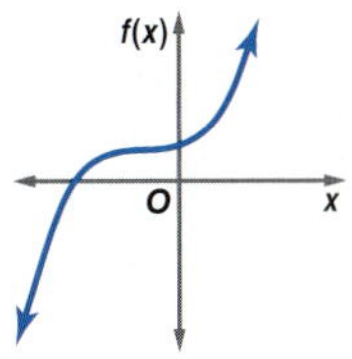

44.
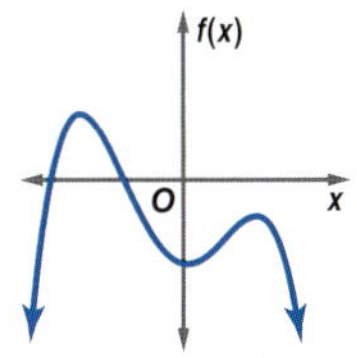

More About. . .

Theater

In 1997, *Cats* surpassed *A Chorus Line* as the longest-running Broadway show.

Source: www.newsherald.com

45. **ENERGY** The power generated by a windmill is a function of the speed of the wind. The approximate power is given by the function $P(s) = \frac{s^3}{1000}$, where s represents the speed of the wind in kilometers per hour. Find the units of power $P(s)$ generated by a windmill when the wind speed is 18 kilometers per hour. **5.832 units**

THEATER For Exercises 46–48, use the graph that models the attendance to Broadway plays (in millions) from 1970–2000.

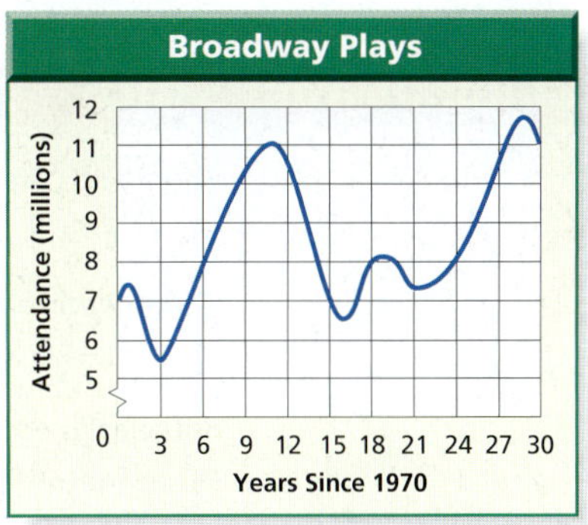

46. Is the graph an odd-degree or even-degree function? **even**
47. Discuss the end behavior of the graph.
48. Do you think attendance at Broadway plays will increase or decrease after 2000? Explain your reasoning.

47. $f(x) \to -\infty$ as $x \to +\infty$; $f(x) \to -\infty$ as $x \to -\infty$

48. Sample answer: Decrease; the graph appears to be turning at $x = 30$ indicating a relative maximum at that point. So attendance will decrease after 2000.

CRITICAL THINKING For Exercises 49–52, use the following information.
The graph of the polynomial function $f(x) = ax(x - 4)(x + 1)$ goes through the point at (5, 15).

49. Find the value of a. $\frac{1}{2}$
50. For what value(s) of x will $f(x) = 0$? $-1, 0, 4$
51. Rewrite the function as a cubic function. $f(x) = \frac{1}{2}x^3 - \frac{3}{2}x^2 - 2x$
52. Sketch the graph of the function. **See margin.**

www.algebra2.com/self_check_quiz

52.
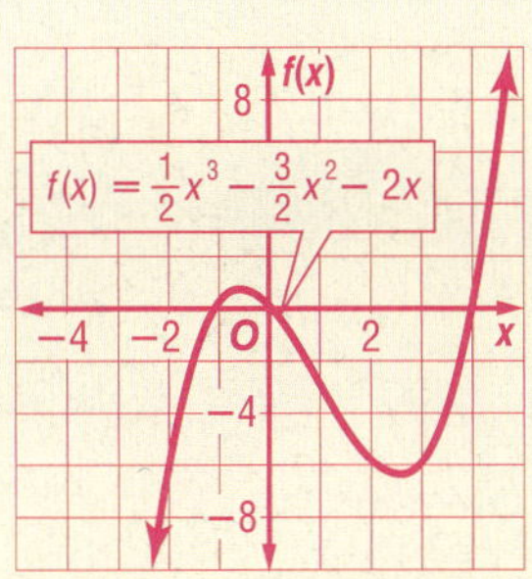

Study Guide and Intervention, p. 375 (shown) and p. 376

Polynomial Functions

Polynomial in One Variable	A polynomial of degree n in one variable x is an expression of the form $a_0x^n + a_1x^{n-1} + \ldots + a_{n-2}x^2 + a_{n-1}x + a_n$, where the coefficients $a_0, a_1, a_2, \ldots, a_n$ represent real numbers, a_0 is not zero, and n represents a nonnegative integer.

The **degree of a polynomial** in one variable is the greatest exponent of its variable. The **leading coefficient** is the coefficient of the term with the highest degree.

Polynomial Function	A polynomial function of degree n can be described by an equation of the form $P(x) = a_0x^n + a_1x^{n-1} + \ldots + a_{n-2}x^2 + a_{n-1}x + a_n$, where the coefficients $a_0, a_1, a_2, \ldots, a_n$ represent real numbers, a_0 is not zero, and n represents a nonnegative integer.

Example 1 **What are the degree and leading coefficient of $3x^2 - 2x^4 - 7 + x^3$?**

Rewrite the expression so the powers of x are in decreasing order.
$-2x^4 + x^3 + 3x^2 - 7$
This is a polynomial in one variable. The degree is 4, and the leading coefficient is −2.

Example 2 **Find $f(-5)$ if $f(x) = x^3 + 2x^2 - 10x + 20$.**

$f(x) = x^3 + 2x^2 - 10x + 20$ Original function
$f(-5) = (-5)^3 + 2(-5)^2 - 10(-5) + 20$ Replace x with −5.
$= -125 + 50 + 50 + 20$ Evaluate.
$= -5$ Simplify.

Example 3 **Find $g(a^2 - 1)$ if $g(x) = x^2 + 3x - 4$.**

$g(x) = x^2 + 3x - 4$ Original function
$g(a^2 - 1) = (a^2 - 1)^2 + 3(a^2 - 1) - 4$ Replace x with $a^2 - 1$.
$= a^4 - 2a^2 + 1 + 3a^2 - 3 - 4$ Evaluate.
$= a^4 + a^2 - 6$ Simplify.

Exercises

State the degree and leading coefficient of each polynomial in one variable. If it is not a polynomial in one variable, explain why.

1. $3x^4 + 6x^3 - x^2 + 12$ **4; 3**
2. $100 - 5x^3 + 10x^7$ **7; 10**
3. $4x^6 + 6x^4 + 8x^8 - 10x^2 + 20$ **8; 8**
4. $4x^2 - 3xy + 16y^2$ **not a polynomial in one variable; contains two variables**
5. $8x^3 - 9x^5 + 4x^2 - 36$ **5; −9**
6. $\frac{x^2}{18} - \frac{x^6}{25} + \frac{x^3}{36} - \frac{1}{72}$ **6; $-\frac{1}{25}$**

Find $f(2)$ and $f(-5)$ for each function.

7. $f(x) = x^2 - 9$ **−5; 16**
8. $f(x) = 4x^3 - 3x^2 + 2x - 1$ **23; −586**
9. $f(x) = 9x^3 - 4x^2 + 5x + 7$ **73; −1243**

Skills Practice, p. 377 and Practice, p. 378 (shown)

State the degree and leading coefficient of each polynomial in one variable. If it is not a polynomial in one variable, explain why.

1. $(3x^2 + 1)(2x^2 - 9)$ **4; 6**
2. $\frac{1}{5}a^3 - \frac{3}{5}a^2 + \frac{4}{5}a$ **3; $\frac{1}{5}$**
3. $\frac{2}{m^2} + 3m - 12$ **Not a polynomial; $\frac{2}{m^2}$ cannot be written in the form m^n for a nonnegative integer n.**
4. $27 + 3xy^3 - 12x^2y^2 - 10y$ **No, this polynomial contains two variables, x and y.**

Find $p(-2)$ and $p(3)$ for each function.

5. $p(x) = x^3 - x^5$ **24; −216**
6. $p(x) = -7x^2 + 5x + 9$ **−29; −39**
7. $p(x) = -x^5 + 4x^3$ **0; −135**
8. $p(x) = 3x^3 - x^2 + 2x - 5$ **−37; 73**
9. $p(x) = x^4 + \frac{1}{2}x^3 - \frac{1}{2}x$ **13; 93**
10. $p(x) = \frac{1}{3}x^3 + \frac{2}{3}x^2 + 3x$ **−6; 24**

If $p(x) = 3x^2 - 4$ and $r(x) = 2x^2 - 5x + 1$, find each value.

11. $p(8a)$ **$192a^2 - 4$**
12. $r(a^2)$ **$2a^4 - 5a^2 + 1$**
13. $-5r(2a)$ **$-40a^2 + 50a - 5$**
14. $r(x + 2)$ **$2x^2 + 3x - 1$**
15. $p(x^2 - 1)$ **$3x^4 - 6x^2 - 1$**
16. $5[p(x + 2)]$ **$15x^2 + 60x + 40$**

For each graph,
a. describe the end behavior,
b. determine whether it represents an odd-degree or an even-degree polynomial function, and
c. state the number of real zeroes.

17.
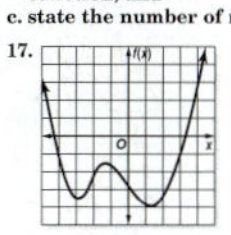
$f(x) \to +\infty$ as $x \to +\infty$, $f(x) \to +\infty$ as $x \to -\infty$; even; 2

18.
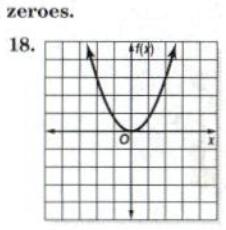
$f(x) \to +\infty$ as $x \to +\infty$, $f(x) \to +\infty$ as $x \to -\infty$; even; 1

19.
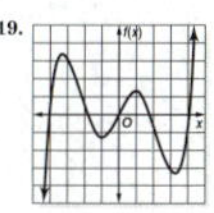
$f(x) \to +\infty$ as $x \to +\infty$, $f(x) \to -\infty$ as $x \to -\infty$; odd; 5

20. **WIND CHILL** The function $C(s) = 0.013s^2 - s - 7$ estimates the wind chill temperature $C(s)$ at 0°F for wind speeds s from 5 to 30 miles per hour. Estimate the wind chill temperature at 0°F if the wind speed is 20 miles per hour. **about −22°F**

Reading to Learn Mathematics, p. 379 ELL

Pre-Activity **Where are polynomial functions found in nature?**

Read the introduction to Lesson 7-1 at the top of page 346 in your textbook.

- In the honeycomb cross section shown in your textbook, there is 1 hexagon in the center, 6 hexagons in the second ring, and 12 hexagons in the third ring. How many hexagons will there be in the fourth, fifth, and sixth rings? **18; 24; 30**
- There is 1 hexagon in a honeycomb with 1 ring. There are 7 hexagons in a honeycomb with 2 rings. How many hexagons are there in honeycombs with 3 rings, 4 rings, 5 rings, and 6 rings? **19; 37; 61; 91**

Reading the Lesson

1. Give the degree and leading coefficient of each polynomial in one variable.

	degree	leading coefficient
a. $10x^3 + 3x^2 - x + 7$	3	10
b. $7y^2 - 2y^5 + y - 4y^3$	5	−2
c. 100	0	100

2. Match each description of a polynomial function from the list on the left with the corresponding end behavior from the list on the right.

a. even degree, negative leading coefficient **iii**
b. odd degree, positive leading coefficient **iv**
c. odd degree, negative leading coefficient **ii**
d. even degree, positive leading coefficient **i**

i. $f(x) \to +\infty$ as $x \to +\infty$; $f(x) \to +\infty$ as $x \to -\infty$
ii. $f(x) \to -\infty$ as $x \to +\infty$; $f(x) \to +\infty$ as $x \to -\infty$
iii. $f(x) \to -\infty$ as $x \to +\infty$; $f(x) \to -\infty$ as $x \to -\infty$
iv. $f(x) \to +\infty$ as $x \to +\infty$; $f(x) \to -\infty$ as $x \to -\infty$

Helping You Remember

3. What is an easy way to remember the difference between the end behavior of the graphs of even-degree and odd-degree polynomial functions?

Sample answer: Both ends of the graph of an even-degree function eventually keep going in the same direction. For odd-degree functions, the two ends eventually head in opposite directions, one upward, the other downward.

Enrichment, p. 380

Approximation by Means of Polynomials

Many scientific experiments produce pairs of numbers $[x, f(x)]$ that can be related by a formula. If the pairs form a function, you can fit a polynomial to the pairs in exactly one way. Consider the pairs given by the following table.

x	1	2	4	7
$f(x)$	6	11	39	−54

We will assume the polynomial is of degree three. Substitute the given values into this expression.

$f(x) = A + B(x - x_0) + C(x - x_0)(x - x_1) + D(x - x_0)(x - x_1)(x - x_2)$

You will get the system of equations shown below. You can solve this system and use the values for A, B, C, and D to find the desired polynomial.

$6 = A$
$11 = A + B(2 - 1) = A + B$

4 Assess

Open-Ended Assessment

Modeling Provide students with grid paper and a length of string. Describe the end behavior and number of real zeros of the graph of a function and have students use their string to model a possible graph that exhibits these characteristics.

Getting Ready for Lesson 7-2

PREREQUISITE SKILL In Lesson 7-2, students will graph polynomial functions by making a table of values. It is important that students know how to make tables of values and how to use them to graph equations. Use Exercises 68–70 to determine your students' familiarity with graphing quadratic functions by making a table of values.

Answers

54.

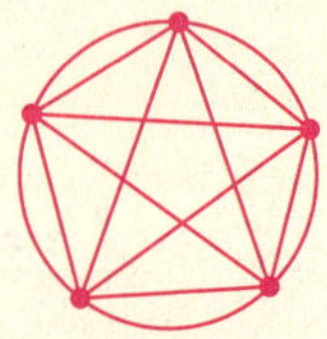

56. Many relationships in nature can be modeled by polynomial functions; for example, the pattern in a honeycomb or the rings in a tree trunk. Answers should include the following.
   - You can use the equation to find the number of hexagons in a honeycomb with 10 rings and the number of hexagons in a honeycomb with 9 rings. The difference is the number of hexagons in the tenth ring.
   - Other examples of patterns found in nature include pinecones, pineapples, and flower petals.

PATTERNS For Exercises 53–55, use the diagrams below that show the maximum number of regions formed by connecting points on a circle.

1 point, 1 region | 2 points, 2 regions | 3 points, 4 regions | 4 points, 8 regions

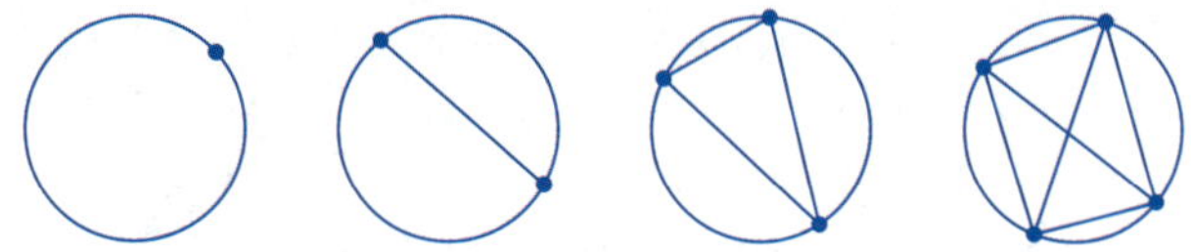

53. The maximum number of regions formed by connecting n points of a circle can be described by the function $f(n) = \frac{1}{24}(n^4 - 6n^3 + 23n^2 - 18n + 24)$. What is the degree of this polynomial function? **4**

★ 54. Find the maximum number of regions formed by connecting 5 points of a circle. Draw a diagram to verify your solution. **16 regions; See margin for diagram.**

★ 55. How many points would you have to connect to form 99 regions? **8 points**

56. **WRITING IN MATH** Answer the question that was posed at the beginning of the lesson. **See margin.**

Where are polynomial functions found in nature?

Include the following in your answer:
- an explanation of how you could use the equation to find the number of hexagons in the tenth ring, and
- any other examples of patterns found in nature that might be modeled by a polynomial equation.

Standardized Test Practice

57. The figure at the right shows the graph of the polynomial function $f(x)$. Which of the following could be the degree of $f(x)$? **C**

(A) 2 (B) 3 (C) 4 (D) 5

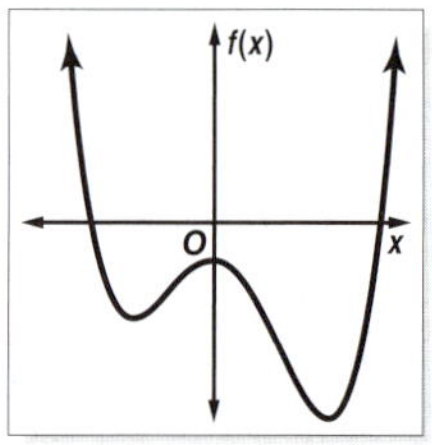

58. If $\frac{1}{2}x^2 - 6x + 2 = 0$, then x could equal which of the following? **C**

(A) −1.84 (B) −0.81 (C) 0.34 (D) 2.37

Maintain Your Skills

Mixed Review

59. $\{x \mid 2 < x < 6\}$
60. $\{x \mid x \leq -9$ or $x \geq 7\}$

Solve each inequality algebraically. *(Lesson 6-7)*

59. $x^2 - 8x + 12 < 0$
60. $x^2 + 2x - 86 \geq -23$
61. $15x^2 + 3x - 12 \leq 0$ $\left\{x \mid -1 \leq x \leq \frac{4}{5}\right\}$

Graph each function. *(Lesson 6-6)* **62–64. See margin.**

62. $y = -2(x - 2)^2 + 3$
63. $y = \frac{1}{3}(x + 5)^2 - 1$
64. $y = \frac{1}{2}x^2 + x + \frac{3}{2}$

Solve each equation by completing the square. *(Lesson 6-4)*

65. $x^2 - 8x - 2 = 0$ $\{4 \pm 3\sqrt{2}\}$
66. $x^2 + \frac{1}{3}x - \frac{35}{36} = 0$ $\left\{-\frac{7}{6}, \frac{5}{6}\right\}$

67. $23{,}450(1 + p)$; $23{,}450(1 + p)^3$

67. **BUSINESS** Becca is writing a computer program to find the salaries of her employees after their annual raise. The percent of increase is represented by p. Marty's salary is \$23,450 now. Write a polynomial to represent Marty's salary after one year and another to represent Marty's salary after three years. Assume that the rate of increase will be the same for each of the three years. *(Lesson 5-2)*

Getting Ready for the Next Lesson

68–70. See pp. 407A–407H.

PREREQUISITE SKILL Graph each equation by making a table of values.
(To review ***graphing quadratic functions****, see Lesson 6-1.)*

68. $y = x^2 + 4$
69. $y = -x^2 + 6x - 5$
70. $y = \frac{1}{2}x^2 + 2x - 6$

62.

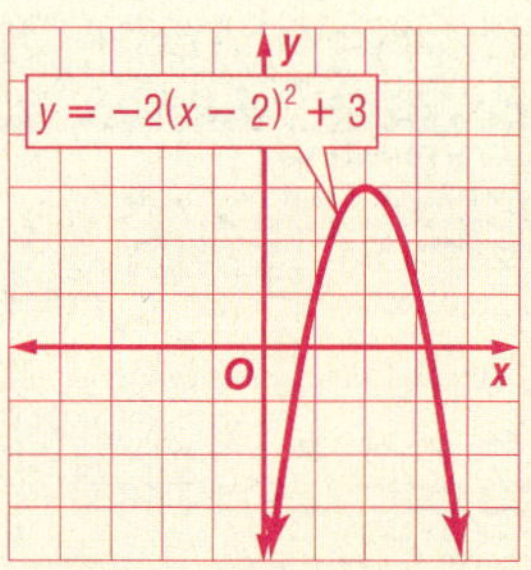

63.

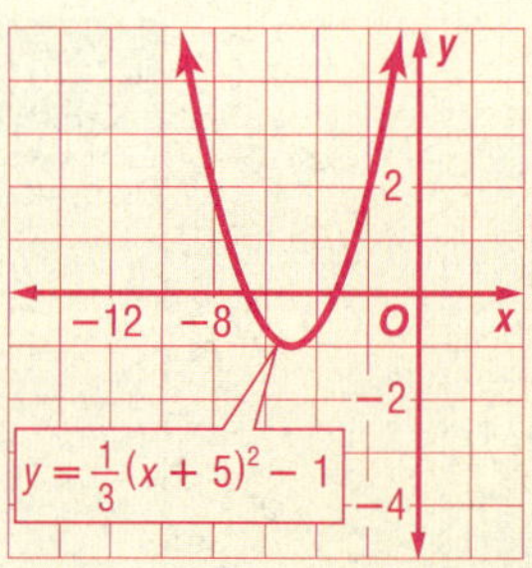

64.

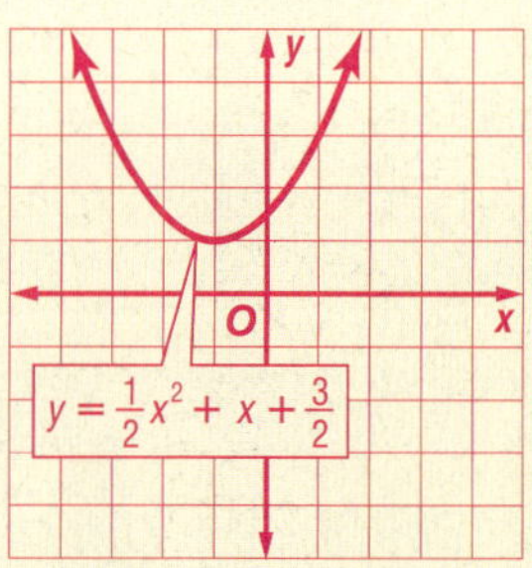

7-2 Graphing Polynomial Functions

What You'll Learn

- Graph polynomial functions and locate their real zeros.
- Find the maxima and minima of polynomial functions.

Vocabulary
- Location Principle
- relative maximum
- relative minimum

How can graphs of polynomial functions show trends in data?

The percent of the United States population that was foreign-born since 1900 can be modeled by $P(t) = 0.00006t^3 - 0.007t^2 + 0.05t + 14$, where $t = 0$ in 1900. Notice that the graph is decreasing from $t = 5$ to $t = 75$ and then it begins to increase. The points at $t = 5$ and $t = 75$ are turning points in the graph.

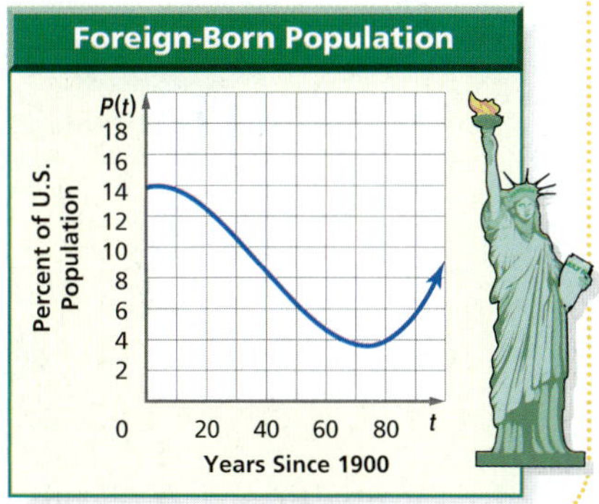

GRAPH POLYNOMIAL FUNCTIONS To graph a polynomial function, make a table of values to find several points and then connect them to make a smooth curve. Knowing the end behavior of the graph will assist you in completing the sketch of the graph.

Example 1 Graph a Polynomial Function

Graph $f(x) = x^4 + x^3 - 4x^2 - 4x$ by making a table of values.

x	$f(x)$
−2.5	≈ 8.4
−2.0	0.0
−1.5	≈ −1.3
−1.0	0.0
−0.5	≈ 0.9

x	$f(x)$
0.0	0.0
0.5	≈ −2.8
1.0	−6.0
1.5	≈ −6.6
2.0	0.0

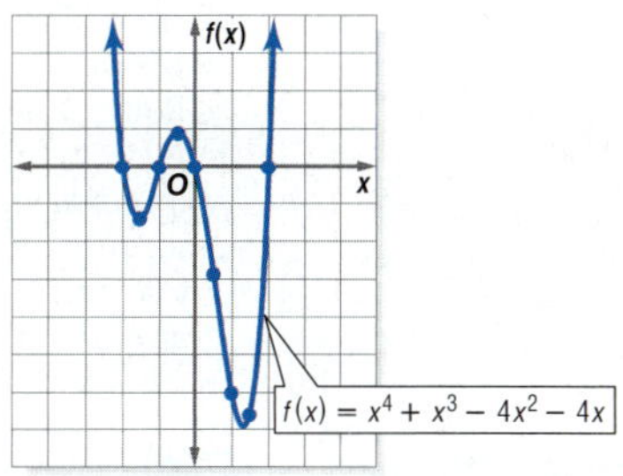

This is an even-degree polynomial with a positive leading coefficient, so $f(x) \to +\infty$ as $x \to +\infty$, and $f(x) \to +\infty$ as $x \to -\infty$. Notice that the graph intersects the x-axis at four points, indicating there are four real zeros of this function.

Study Tip

Graphing Polynomial Functions
To graph polynomial functions it will often be necessary to include x values that are not integers.

In Example 1, the zeros occur at integral values that can be seen in the table used to plot the function. Notice that the values of the function before and after each zero are different in sign. In general, the graph of a polynomial function will cross the x-axis somewhere between pairs of x values at which the corresponding $f(x)$ values change signs. Since zeros of the function are located at x-intercepts, there is a zero between each pair of these x values. This property for locating zeros is called the **Location Principle**.

7-2 Lesson Notes

1 Focus

5-Minute Check Transparency 7-2 Use as a quiz or review of Lesson 7-1.

Mathematical Background notes are available for this lesson on p. 344C.

Building on Prior Knowledge

In Chapter 6, students learned to graph quadratic functions. Those same skills will be used in this lesson to graph polynomial functions.

How can graphs of polynomial functions show trends in data?

Ask students:

- When the graph is sloping downward to the right, what does that tell you about the population it represents? **The percent of the U.S. population that is foreign-born is decreasing during that span of time.**
- If the United States government banned any further immigration, what would happen to the graph? **It would gradually approach the horizontal axis.**
- Why would the graph not immediately reach the horizontal axis, where $P(t) = 0$? **All of the current foreign-born residents of the U.S. may still be part of the population.**

Resource Manager

Workbook and Reproducible Masters

Chapter 7 Resource Masters
- Study Guide and Intervention, pp. 381–382
- Skills Practice, p. 383
- Practice, p. 384
- Reading to Learn Mathematics, p. 385
- Enrichment, p. 386

Science and Mathematics Lab Manual, pp. 71–76

Transparencies

5-Minute Check Transparency 7-2
Answer Key Transparencies

Technology

Interactive Chalkboard

2 Teach

GRAPH POLYNOMIAL FUNCTIONS

In-Class Examples

Power Point®

1 Graph $f(x) = -x^3 - 4x^2 + 5$ by making a table of values.

x	$f(x)$
−4	5
−3	−4
−2	−3
−1	2
0	5
1	0
2	−19

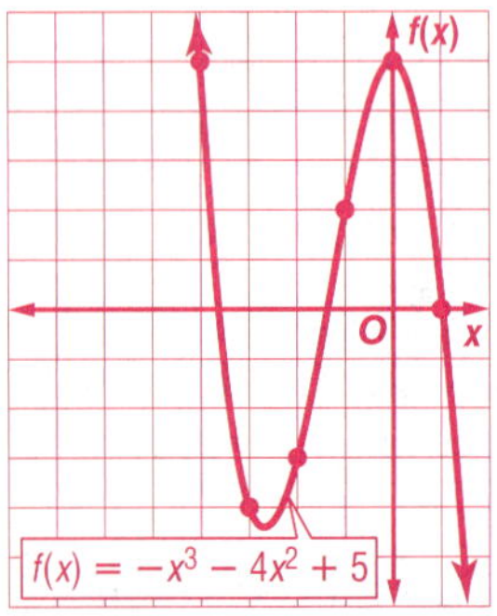

2 Determine consecutive values of x between which each real zero of the function $f(x) = x^4 - x^3 - 4x^2 + 1$ is located. Then draw the graph.

x	$f(x)$	
−2	9	} change in signs
−1	−1	} change in signs
0	1	} change in signs
1	−3	
2	−7	} change in signs
3	19	

There are zeros between $x = -2$ and $x = -1$, between $x = -1$ and $x = 0$, between $x = 0$ and $x = 1$, and between $x = 2$ and $x = 3$.

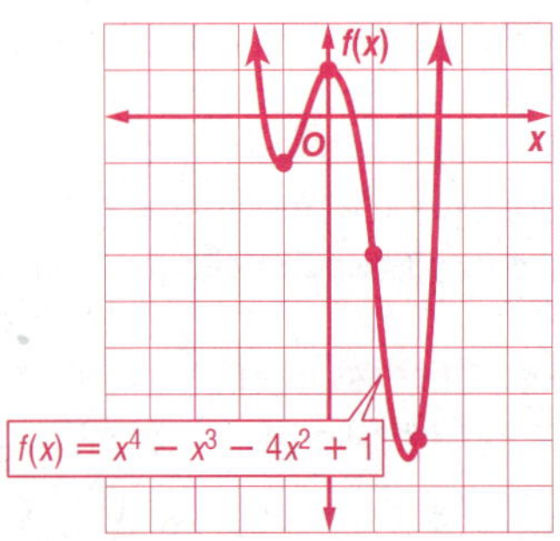

Key Concept — Location Principle

- **Words** Suppose $y = f(x)$ represents a polynomial function and a and b are two numbers such that $f(a) < 0$ and $f(b) > 0$. Then the function has at least one real zero between a and b.
- **Model**

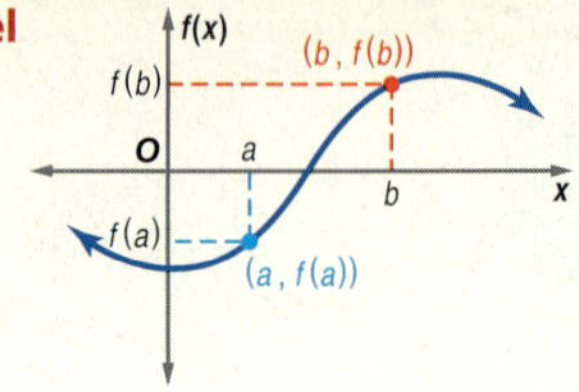

Example 2 Locate Zeros of a Function

Determine consecutive values of x between which each real zero of the function $f(x) = x^3 - 5x^2 + 3x + 2$ is located. Then draw the graph.

Make a table of values. Since $f(x)$ is a third-degree polynomial function, it will have either 1, 2, or 3 real zeros. Look at the values of $f(x)$ to locate the zeros. Then use the points to sketch a graph of the function.

x	$f(x)$	
−2	−32	
−1	−7	} change in signs
0	2	
1	1	} change in signs
2	−4	
3	−7	
4	−2	} change in signs
5	17	

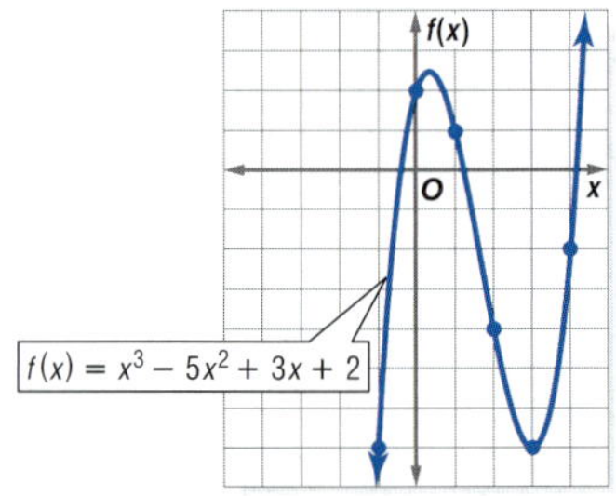

The changes in sign indicate that there are zeros between $x = -1$ and $x = 0$, between $x = 1$ and $x = 2$, and between $x = 4$ and $x = 5$.

MAXIMUM AND MINIMUM POINTS The graph at the right shows the shape of a general third-degree polynomial function.

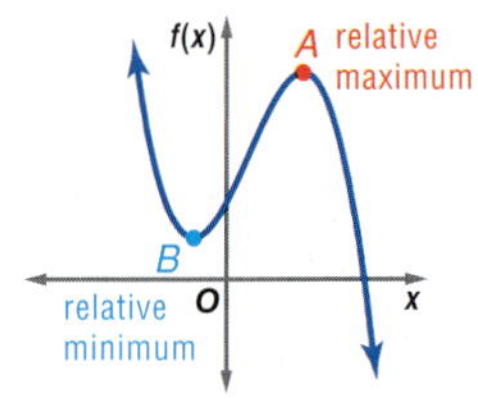

Point A on the graph is a **relative maximum** of the cubic function since no other nearby points have a greater y-coordinate. Likewise, point B is a **relative minimum** since no other nearby points have a lesser y-coordinate. These points are often referred to as *turning points*. The graph of a polynomial function of degree n has at most $n - 1$ turning points.

Study Tip

Reading Math
The plurals of maximum and minimum are *maxima* and *minima*.

Example 3 Maximum and Minimum Points

Graph $f(x) = x^3 - 3x^2 + 5$. Estimate the x-coordinates at which the relative maxima and relative minima occur.

Make a table of values and graph the equation.

x	$f(x)$	
−2	−15	} zero between $x = -2$ and $x = -1$
−1	1	
0	5	← indicates a relative maximum
1	3	
2	1	← indicates a relative minimum
3	5	

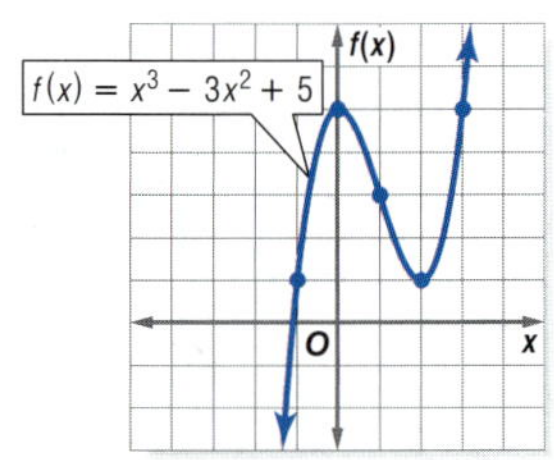

DAILY INTERVENTION — Unlocking Misconceptions

Modeling Real-World Data Students may incorrectly assume that functions exactly describe every member of a set of real-world data. Stress that a function is just a model of the data, and often it is only a reasonable model for a limited domain of values. Make sure students understand that the function is just an approximation of the real-world data and does not completely describe the data.

Look at the table of values and the graph.

- The values of $f(x)$ change signs between $x = -2$ and $x = -1$, indicating a zero of the function.
- The value of $f(x)$ at $x = 0$ is greater than the surrounding points, so it is a relative maximum.
- The value of $f(x)$ at $x = 2$ is less than the surrounding points, so it is a relative minimum.

The graph of a polynomial function can reveal trends in real-world data.

Example 4 Graph a Polynomial Model

ENERGY **The average fuel (in gallons) consumed by individual vehicles in the United States from 1960 to 2000 is modeled by the cubic equation $F(t) = 0.025t^3 - 1.5t^2 + 18.25t + 654$, where t is the number of years since 1960.**

a. Graph the equation.

Make a table of values for the years 1960–2000. Plot the points and connect with a smooth curve. Finding and plotting the points for every fifth year gives a good approximation of the graph.

t	$F(t)$
0	654
5	710.88
10	711.5
15	674.63
20	619
25	563.38
30	526.5
35	527.13
40	584

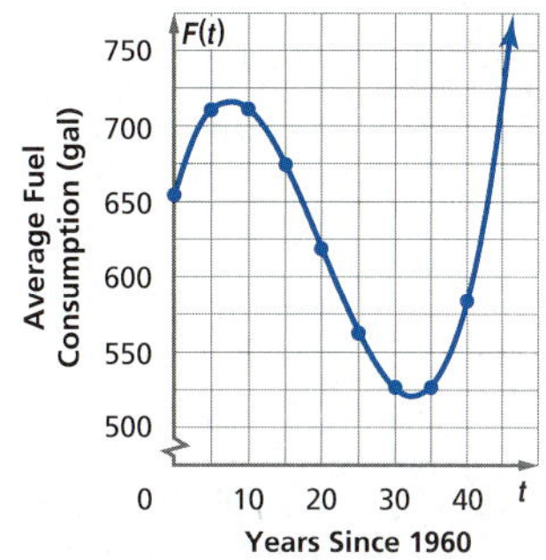

b. Describe the turning points of the graph and its end behavior.

There is a relative maximum between 1965 and 1970 and a relative minimum between 1990 and 1995. For the end behavior, as t increases, $F(t)$ increases.

c. What trends in fuel consumption does the graph suggest?

Average fuel consumption hit a maximum point around 1970 and then started to decline until 1990. Since 1990, fuel consumption has risen and continues to rise.

More About. . .

Energy

Gasoline and diesel fuels are the most familiar transportation fuels in this country, but other energy sources are available, including ethanol, a grain alcohol that can be produced from corn or other crops.

Source: U.S. Environmental Protection Agency

A graphing calculator can be helpful in finding the relative maximum and relative minimum of a function.

Graphing Calculator Investigation

Maximum and Minimum Points

You can use a TI-83 Plus to find the coordinates of relative maxima and relative minima. Enter the polynomial function in the Y= list and graph the function. Make sure that all the turning points are visible in the viewing window. Find the coordinates of the minimum and maximum points, respectively.

KEYSTROKES: *Refer to page 293 to review finding maxima and minima.*

(continued on the next page)

www.algebra2.com/extra_examples

MAXIMUM AND MINIMUM POINTS

In-Class Examples

Power Point®

3 Graph $f(x) = x^3 - 4x^2 + 5$. Estimate the x-coordinates at which the relative maxima and relative minima occur.

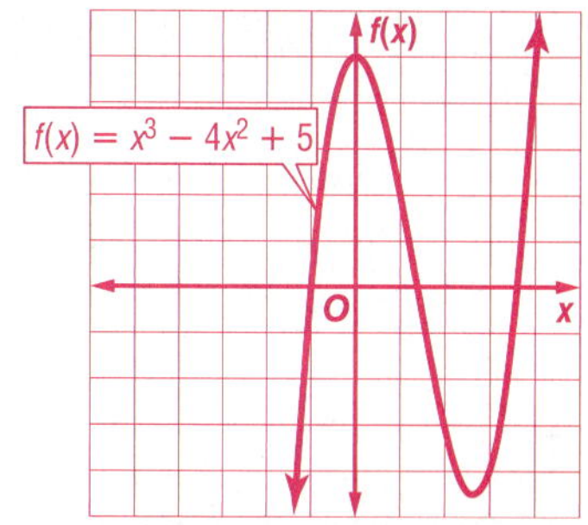

- **The value of $f(x)$ at $x = 0$ is greater than the surrounding points, so it is a relative maximum.**
- **The value of $f(x)$ at $x \approx 3$ is less than the surrounding points, so it is a relative minimum.**

4 **HEALTH** The weight w, in pounds, of a patient during a 7-week illness is modeled by the cubic equation $w(n) = 0.1n^3 - 0.6n^2 + 110$, where n is the number of weeks since the patient became ill.

a. Graph the equation.

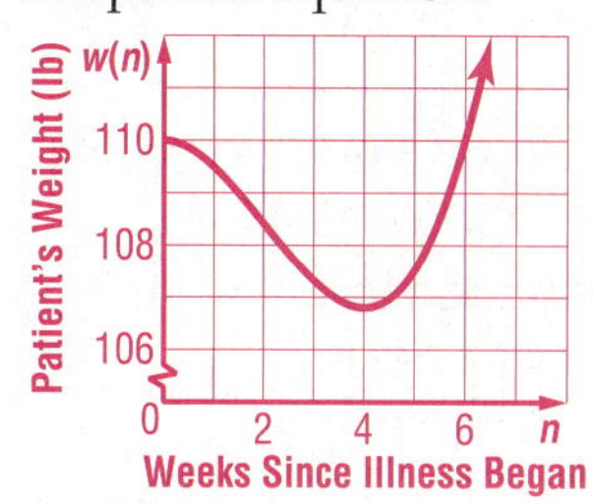

b. Describe the turning points of the graph and its end behavior. **There is a relative minimum point at week 4. For the end behavior, $w(n)$ increases as n increases.**

c. What trends in the patient's weight does the graph suggest? **The patient lost weight for each of 4 weeks after becoming ill. After 4 weeks, the patient gained weight and continues to gain weight.**

Graphing Calculator Investigation

Maximum and Minimum Points Remind students of the procedure for finding relative minima and maxima using the calculator. First, press 2nd [CALC] and select either 3 or 4, depending on whether you are finding a minimum or maximum. Then set the left bound. Use the arrow buttons to move the cursor well to the left of the point you suspect is the minimum or maximum, and press ENTER.

Move the cursor well to the right of the suspect point. Press ENTER twice to display the coordinates of the relative maximum/minimum.

3 Practice/Apply

Study Notebook

Have students—
- add the definitions/examples of the vocabulary terms to their Vocabulary Builder worksheets for Chapter 7.
- include any other item(s) that they find helpful in mastering the skills in this lesson.

About the Exercises...

Organization by Objective
- **Graph Polynomial Functions:** 13–26
- **Maximum and Minimum Points:** 13–26

Odd/Even Assignments
Exercises 13–26 are structured so that students practice the same concepts whether they are assigned odd or even problems.

Alert! Exercise 30 involves research on the Internet or other reference materials.

Assignment Guide

Basic: 13–25 odd, 27–32, 36–42, 47–66

Average: 13–25 odd, 27–42, 47–66 (optional: 43–46)

Advanced: 14–26 even, 27–60 (optional: 61–66)

Answers

1. There must be at least one real zero between two points on a graph when one of the points lies below the x-axis and the other point lies above the x-axis.

3.
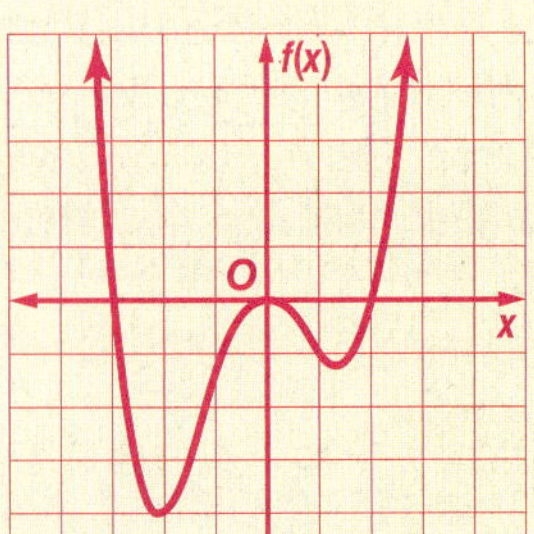

Think and Discuss 1–3. See pp. 407A–407H.

1. Graph $f(x) = x^3 - 3x^2 + 4$. Estimate the x-coordinates of the relative maximum and relative minimum points from the graph.
2. Use the maximum and minimum options from the CALC menu to find the exact coordinates of these points. You will need to use the arrow keys to select points to the left and to the right of the point.
3. Graph $f(x) = \frac{1}{2}x^4 - 4x^3 + 7x^2 - 8$. How many relative maximum and relative minimum points does the graph contain? What are the coordinates?

Check for Understanding

Concept Check

1. **Explain** the Location Principle in your own words. See margin.
2. **State** the number of turning points of the graph of a fifth-degree polynomial if it has five distinct real zeros. 4
3. **OPEN ENDED** Sketch a graph of a function that has one relative maximum point and two relative minimum points. See margin.

7. between −2 and −1, between −1 and 0, between 0 and 1, and between 1 and 2

Guided Practice

GUIDED PRACTICE KEY	
Exercises	Examples
4, 5	1
6, 7	2
8, 9	3
10–12	4

Graph each polynomial function by making a table of values. 4–5. See pp. 407A–407H.

4. $f(x) = x^3 - x^2 - 4x + 4$
5. $f(x) = x^4 - 7x^2 + x + 5$

Determine consecutive values of x between which each real zero of each function is located. Then draw the graph. 6–7. See pp. 407A–407H for graphs.

6. $f(x) = x^3 - x^2 + 1$ between −1 and 0
7. $f(x) = x^4 - 4x^2 + 2$

Graph each polynomial function. Estimate the x-coordinates at which the relative maxima and relative minima occur. 8–9. See pp. 407A–407H for graphs.

8. $f(x) = x^3 + 2x^2 - 3x - 5$
9. $f(x) = x^4 - 8x^2 + 10$

8. Sample answer: rel. max. at $x = -2$, rel. min. at $x = 0.5$

9. Sample answer: rel. max. at $x = 0$, rel. min. at $x = -2$ and at $x = 2$

Application

CABLE TV **For Exercises 10–12, use the following information.**
The number of cable TV systems after 1985 can be modeled by the function $C(t) = -43.2t^2 + 1343t + 790$, where t represents the number of years since 1985.

10. Graph this equation for the years 1985 to 2005. 10–12. See pp. 407A–407H.
11. Describe the turning points of the graph and its end behavior.
12. What trends in cable TV subscriptions does the graph suggest?

Practice and Apply

Homework Help

For Exercises	See Examples
13–26	1, 2, 3
27–35	4

Extra Practice
See page 842.

For Exercises 13–26, complete each of the following.

a. **Graph each function by making a table of values.**
b. **Determine consecutive values of x between which each real zero is located.**
c. **Estimate the x-coordinates at which the relative maxima and relative minima occur.** 13–26. See pp. 407A–407H.

13. $f(x) = -x^3 - 4x^2$
14. $f(x) = x^3 - 2x^2 + 6$
15. $f(x) = x^3 - 3x^2 + 2$
16. $f(x) = x^3 + 5x^2 - 9$
17. $f(x) = -3x^3 + 20x^2 - 36x + 16$
18. $f(x) = x^3 - 4x^2 + 2x - 1$
19. $f(x) = x^4 - 8$
20. $f(x) = x^4 - 10x^2 + 9$
21. $f(x) = -x^4 + 5x^2 - 2x - 1$
22. $f(x) = -x^4 + x^3 + 8x^2 - 3$
23. $f(x) = x^4 - 9x^3 + 25x^2 - 24x + 6$
24. $f(x) = 2x^4 - 4x^3 - 2x^2 + 3x - 5$
25. $f(x) = x^5 + 4x^4 - x^3 - 9x^2 + 3$
26. $f(x) = x^5 - 6x^4 + 4x^3 + 17x^2 - 5x - 6$

DAILY INTERVENTION — **Differentiated Instruction** ELL

Verbal/Linguistic Have the class work in groups of 3 or 4 students. Instruct students to take turns explaining how to make a table of values for a polynomial function, how to plot several points to begin a graph of the function, how to locate the zeros of the function, and how to estimate the x-coordinates at which the relative maxima and relative minima of the function occur.

27. highest: 1982; lowest: 2000

28. Rel. max. between 1980 and 1985 and between 1990 and 1995, rel. min. between 1975 and 1980 and between 1985 and 1990; as the number of years increases, the percent of the labor force that is unemployed decreases.

30. Sample answer: increase, based on the past fluctuations of the graph

EMPLOYMENT For Exercises 27–30, use the graph that models the unemployment rates from 1975–2000.

27. In what year was the unemployment rate the highest? the lowest?

28. Describe the turning points and end behavior of the graph.

29. If this graph was modeled by a polynomial equation, what is the least degree the equation could have? **5**

30. Do you expect the unemployment rate to increase or decrease from 2001 to 2005? Explain your reasoning.

Online Research Data Update What is the current unemployment rate? Visit www.algebra2.com/data_update to learn more.

More About. . .

Child Devolpment

As children develop, their sleeping needs change. Infants sleep about 16–18 hours a day. Toddlers usually sleep 10–12 hours at night and take one or two daytime naps. School-age children need 9–11 hours of sleep, and teens need at least 9 hours of sleep.

Source: www.kidshealth.org

CHILD DEVELOPMENT For Exercises 31 and 32, use the following information.

The average height (in inches) for boys ages 1 to 20 can be modeled by the equation $B(x) = -0.001x^4 + 0.04x^3 - 0.56x^2 + 5.5x + 25$, where x is the age (in years). The average height for girls ages 1 to 20 is modeled by the equation $G(x) = -0.0002x^4 + 0.006x^3 - 0.14x^2 + 3.7x + 26$.

31. Graph both equations by making a table of values. Use $x = \{0, 2, 4, 6, 8, 10, 12, 14, 16, 18, 20\}$ as the domain. Round values to the nearest inch. **See pp. 407A–407H.**

32. Compare the graphs. What do the graphs suggest about the growth rate for both boys and girls? **See margin.**

PHYSIOLOGY For Exercises 33–35, use the following information.

During a regular respiratory cycle, the volume of air in liters in the human lungs can be described by the function $V(t) = 0.173t + 0.152t^2 - 0.035t^3$, where t is the time in seconds.

33. Estimate the real zeros of the function by graphing. **about 5.3 s**

34. About how long does a regular respiratory cycle last? **5 s**

35. Estimate the time in seconds from the beginning of this respiratory cycle for the lungs to fill to their maximum volume of air. **about 3.4 s**

CRITICAL THINKING For Exercises 36–39, sketch a graph of each polynomial.

36. even-degree polynomial function with one relative maximum and two relative minima **36–39. See pp. 407A–407H for sample graphs.**

37. odd-degree polynomial function with one relative maximum and one relative minimum; the leading coefficient is negative

38. even-degree polynomial function with four relative maxima and three relative minima

39. odd-degree polynomial function with three relative maxima and three relative minima; the leftmost points are negative

40. **WRITING IN MATH** Answer the question that was posed at the beginning of the lesson. **See pp. 407A–407H.**

How can graphs of polynomial functions show trends in data?

Include the following in your answer:

- a description of the types of data that are best modeled by polynomial equations rather than linear equations, and
- an explanation of how you would determine when the percent of foreign-born citizens was at its highest and when the percent was at its lowest since 1900.

www.algebra2.com/self_check_quiz

Study Guide and Intervention, p. 381 (shown) and p. 382

Graph Polynomial Functions

Location Principle	Suppose $y = f(x)$ represents a polynomial function and a and b are two numbers such that $f(a) < 0$ and $f(b) > 0$. Then the function has at least one real zero between a and b.

Example **Determine the values of x between which each real zero of the function $f(x) = 2x^4 - x^3 - 5$ is located. Then draw the graph.**

Make a table of values. Look at the values of $f(x)$ to locate the zeros. Then use the points to sketch a graph of the function.

x	$f(x)$
−2	35
−1	−2
0	−5
1	−4
2	19

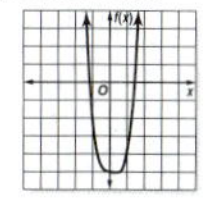

The changes in sign indicate that there are zeros between $x = -2$ and $x = -1$ and between $x = 1$ and $x = 2$.

Exercises

Graph each function by making a table of values. Determine the values of x at which or between which each real zero is located.

1. $f(x) = x^3 - 2x^2 + 1$

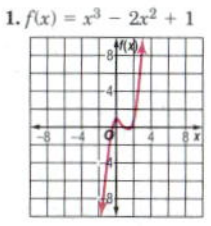

between 0 and −1; at 1; between 1 and 2

2. $f(x) = x^4 + 2x^3 - 5$

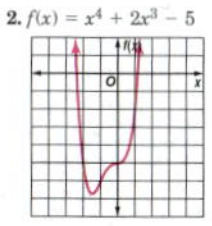

between −2 and −3; between 1 and 2

3. $f(x) = -x^4 + 2x^2 - 1$

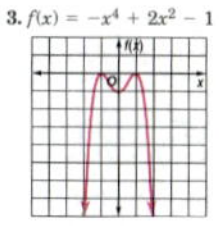

at ±1

4. $f(x) = x^3 - 3x^2 + 4$

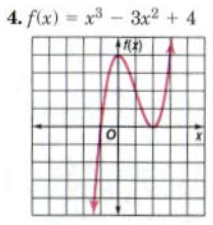

at −1, 2

5. $f(x) = 3x^3 + 2x - 1$

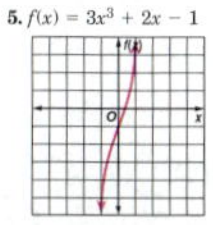

between 0 and 1

6. $f(x) = x^4 - 3x^3 + 1$

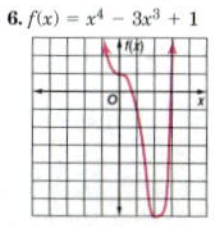

between 0 and 1; between 2 and 3

Skills Practice, p. 383 and Practice, p. 384 (shown)

Complete each of the following.
a. Graph each function by making a table of values.
b. Determine consecutive values of x between which each real zero is located.
c. Estimate the x-coordinates at which the relative and relative minima occur.

1. $f(x) = -x^3 + 3x^2 - 3$

x	$f(x)$
−2	17
−1	1
0	−3
1	−1
2	1
3	−3
4	−19

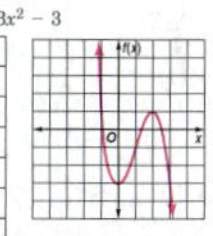

zeros between −1 and 0, 1 and 2, and 2 and 3; rel. max. at $x = 2$, rel. min. at $x = 0$

2. $f(x) = x^3 - 1.5x^2 - 6x + 1$

x	$f(x)$
−2	−1
−1	4.5
0	1
1	−5.5
2	−9
3	−3.5
4	17

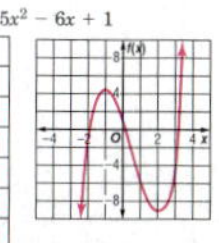

zeros between −2 and −1, 0 and 1, and 3 and 4; rel. max. at $x = -1$, rel. min. at $x = 2$

3. $f(x) = 0.75x^4 + x^3 - 3x^2 + 4$

x	$f(x)$
−3	10.75
−2	−4
−1	0.75
0	4
1	2.75
2	12

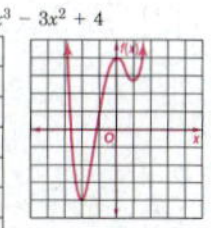

zeros between −3 and −2, and −2 and −1; rel. max. at $x = 0$, rel. min. at $x = -2$ and $x = 1$

4. $f(x) = x^4 + 4x^3 + 6x^2 + 4x - 3$

x	$f(x)$
−3	12
−2	−3
−1	−4
0	−3
1	12
2	77

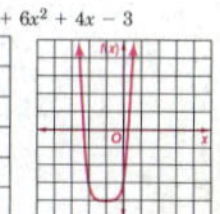

zeros between −3 and −2, and 0 and 1; rel. min. at $x = -1$

PRICES For Exercises 5 and 6, use the following information.

The Consumer Price Index (CPI) gives the relative price for a fixed set of goods and services. The CPI from September, 2000 to July, 2001 is shown in the graph.

Source: U. S. Bureau of Labor Statistics

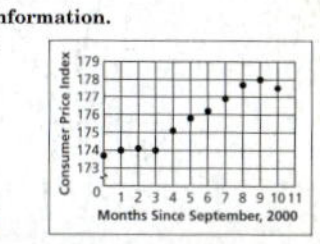

5. Describe the turning points of the graph. **rel max. in Nov. and June; rel. min in Dec.**

6. If the graph were modeled by a polynomial equation, what is the least degree the equation could have? **4**

7. **LABOR** A town's jobless rate can be modeled by (1, 3.3), (2, 4.9), (3, 5.3), (4, 6.4), (5, 4.5), (6, 5.6), (7, 2.5), (8, 2.7). How many turning points would the graph of a polynomial function through these points have? Describe them. **4: 2 rel. max. and 2 rel. min.**

Reading to Learn Mathematics, p. 385

ELL

Pre-Activity **How can graphs of polynomial functions show trends in data?**

Read the introduction to Lesson 7-2 at the top of page 353 in your textbook.

Three points on the graph shown in your textbook are (0, 14), (70, 3.78), and (100, 9). Give the real-world meaning of the coordinates of these points. **Sample answer: In 1900, 14% of the U. S. population was foreign born. In 1970, 3.78% of the population was foreign born. In 2000, 9% of the population was foreign born.**

Reading the Lesson

1. Suppose that $f(x)$ is a third-degree polynomial function and that c and d are real numbers, with $d > c$. Indicate whether each statement is *true* or *false*. (Remember that *true* means *always* true.)
   a. If $f(c) > 0$ and $f(d) < 0$, there is exactly one real zero between c and d. **false**
   b. If $f(c) = f(d) \neq 0$, there are no real zeros between c and d. **false**
   c. If $f(c) < 0$ and $f(d) > 0$, there is at least one real zero between c and d. **true**

2. Match each graph with its description.
   a. third-degree polynomial with one relative maximum and one relative minimum; leading coefficient negative **iii**
   b. fourth-degree polynomial with two relative minima and one relative maximum **i**
   c. third-degree polynomial with one relative maximum and one relative minimum; leading coefficient positive **iv**
   d. fourth-degree polynomial with two relative maxima and one relative minimum **ii**

i. 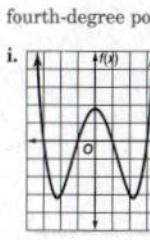
ii. 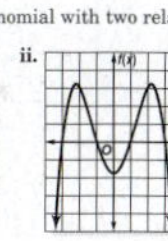
iii. 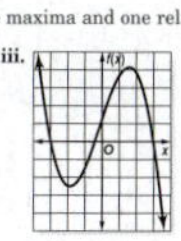
iv. 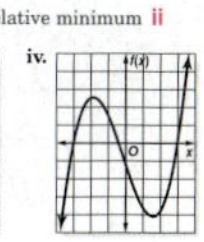

Helping You Remember

3. The origins of words can help you to remember their meaning and to distinguish between similar words. Look up *maximum* and *minimum* in a dictionary and describe their origins (original language and meaning). **Sample answer: *Maximum* comes from the Latin word *maximus*, meaning *greatest*. *Minimum* comes from the Latin word *minimus*, meaning *least*.**

Answer

32. The growth rate for both boys and girls increases steadily until age 18 and then begins to level off, with boys averaging a height of 71 in. and girls a height of 60 in.

Enrichment, p. 386

Golden Rectangles

Use a straightedge, a compass, and the instructions below to construct a golden rectangle.

1. Construct square $ABCD$ with sides of 2 centimeters.

2. Construct the midpoint of $\overline{AB}$. Call the midpoint M.

3. Using M as the center, set your compass opening at MC. Construct an arc with center M that intersects $\overline{AB}$. Call the point of intersection P.

4. Construct a line through P that is perpendicular to $\overline{AB}$.

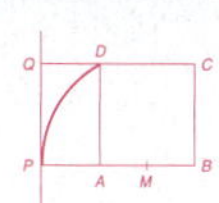

4 Assess

Open-Ended Assessment

Writing Have students write a paragraph describing how to find the turning points of the graph of a polynomial function.

Getting Ready for Lesson 7-3

PREREQUISITE SKILL Lesson 7-3 presents solving equations using quadratic techniques. Students will factor polynomials to find the solutions of polynomial equations. Use Exercises 61–66 to determine your students' familiarity with factoring polynomials.

Answers

53.

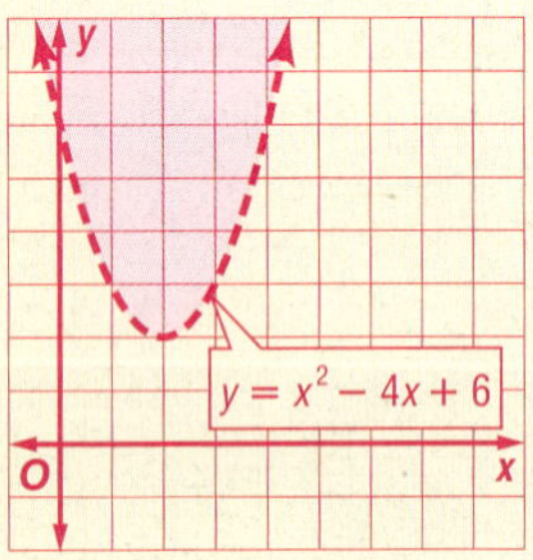

54.

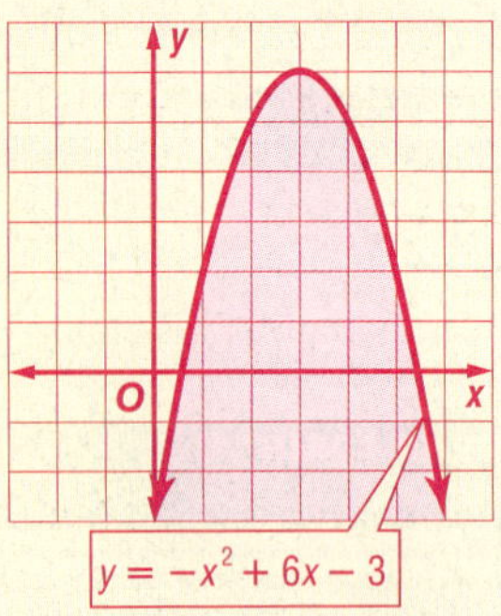

55.

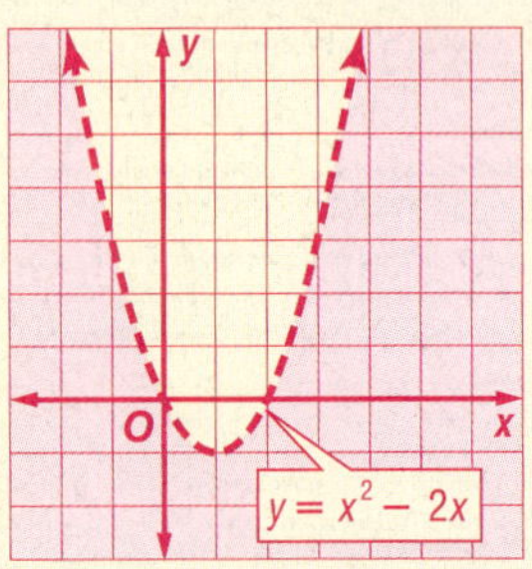

Standardized Test Practice

41. Which of the following could be the graph of $f(x) = x^3 + x^2 - 3x$? **D**

(A)

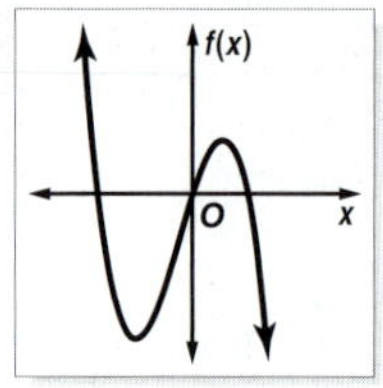

(B)

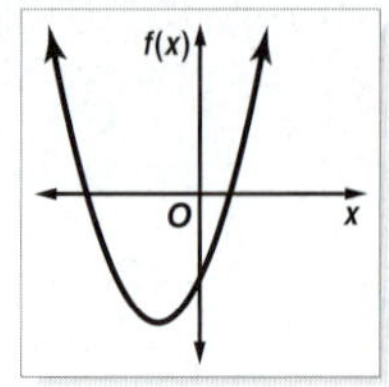

(C)

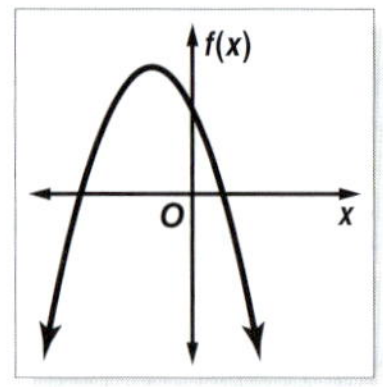

(D)

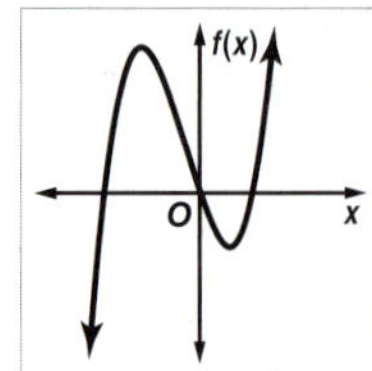

42. The function $f(x) = x^2 - 4x + 3$ has a relative minimum located at which of the following x values? **B**

(A) −2 (B) 2 (C) 3 (D) 4

Graphing Calculator

Use a graphing calculator to estimate the x-coordinates at which the maxima and minima of each function occur. Round to the nearest hundredth.

43. $f(x) = x^3 + x^2 - 7x - 3$ **−1.90; 1.23**
44. $f(x) = -x^3 + 6x^2 - 6x - 5$ **3.41; 0.59**
45. $f(x) = -x^4 + 3x^2 - 8$ **0; −1.22, 1.22**
46. $f(x) = 3x^4 - 7x^3 + 4x - 5$ **0.52; −0.39, 1.62**

Maintain Your Skills

Mixed Review

If $p(x) = 2x^2 - 5x + 4$ and $r(x) = 3x^3 - x^2 - 2$, find each value. *(Lesson 7-1)*

47. $r(2a)$ **$24a^3 - 4a^2 - 2$**
48. $5p(c)$ **$10c^2 - 25c + 20$**
49. $p(2a^2)$ **$8a^4 - 10a^2 + 4$**
50. $r(x - 1)$ **$3x^3 - 10x^2 + 11x - 6$**
51. $p(x^2 + 4)$ **$2x^4 + 11x^2 + 16$**
52. $2[p(x^2 + 1)] - 3r(x - 1)$ **$4x^4 - 9x^3 + 28x^2 - 33x + 20$**

Graph each inequality. *(Lesson 6-7)* **53–55. See margin.**

53. $y > x^2 - 4x + 6$
54. $y \leq -x^2 + 6x - 3$
55. $y < x^2 - 2x$

Solve each matrix equation or system of equations by using inverse matrices. *(Lesson 4-8)*

56. $\begin{bmatrix} 3 & 6 \\ 2 & -1 \end{bmatrix} \cdot \begin{bmatrix} a \\ b \end{bmatrix} = \begin{bmatrix} -3 \\ 18 \end{bmatrix}$ **(7, −4)**
57. $\begin{bmatrix} 5 & -7 \\ -3 & 4 \end{bmatrix} \cdot \begin{bmatrix} m \\ n \end{bmatrix} = \begin{bmatrix} -1 \\ 1 \end{bmatrix}$ **(−3, −2)**
58. $3j + 2k = 8$, $j - 7k = 18$ **(4, −2)**
59. $5y + 2z = 11$, $10y - 4z = -2$ **(1, 3)**

60. **SPORTS** Bob and Minya want to build a ramp that they can use while rollerblading. If they want the ramp to have a slope of $\frac{1}{4}$, how tall should they make the ramp? *(Lesson 2-3)* **1 ft**

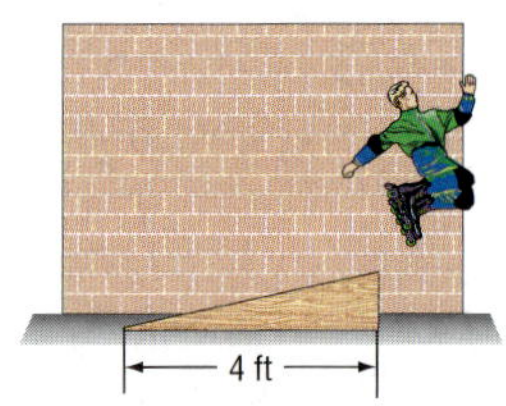

61. **$(x + 5)(x - 6)$**
62. **$(2b - 1)(b - 4)$**
63. **$(3a + 1)(2a + 5)$**
64. **$(2m + 3)(2m - 3)$**

Getting Ready for the Next Lesson

PREREQUISITE SKILL **Factor each polynomial.**
*(To review **factoring polynomials**, see Lesson 5-4.)*

61. $x^2 - x - 30$
62. $2b^2 - 9b + 4$
63. $6a^2 + 17a + 5$
64. $4m^2 - 9$
65. $t^3 - 27$ **$(t - 3)(t^2 + 3t + 9)$**
66. $r^4 - 1$ **$(r^2 + 1)(r + 1)(r - 1)$**

358 Chapter 7 Polynomial Functions

Graphing Calculator Investigation

A Follow-Up of Lesson 7-2

Modeling Real-World Data

You can use a TI-83 Plus to model data whose curve of best fit is a polynomial function.

Example

The table shows the distance a seismic wave can travel based on its distance from an earthquake's epicenter. Draw a scatter plot and a curve of best fit that relates distance to travel time. Then determine approximately how far from the epicenter the wave will be felt 8.5 minutes after the earthquake occurs.

Source: University of Arizona

Travel Time (min)	1	2	5	7	10	12	13
Distance (km)	400	800	2500	3900	6250	8400	10,000

Step 1 Enter the travel times in L1 and the distances in L2.

KEYSTROKES: *Refer to page 87 to review how to enter lists.*

Step 2 Graph the scatter plot.

KEYSTROKES: *Refer to page 87 to review how to graph a scatter plot.*

Step 3 Compute and graph the equation for the curve of best fit. A quartic curve is the best fit for these data.

KEYSTROKES: STAT ▶ 7 2nd [L1] , 2nd [L2] ENTER Y= VARS 5 ▶ ▶ 1 GRAPH

Step 4 Use the [CALC] feature to find the value of the function for $x = 8.5$.

KEYSTROKES: *Refer to page 87 to review how to find function values.*

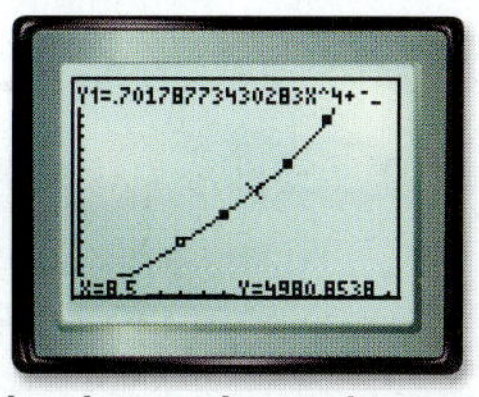

[0, 15] scl: 1 by [0, 10000] scl: 500

After 8.5 minutes, you would expect the wave to be felt approximately 5000 kilometers away.

Exercises 1. See pp. 407A–407H.

Use the table that shows how many minutes out of each eight-hour work day are used to pay one day's worth of taxes.

Year	Minutes
1940	83
1950	117
1960	130
1970	141
1980	145
1990	145
2000	160

Source: Tax Foundation

1. Draw a scatter plot of the data. Then graph several curves of best fit that relate the number of minutes to the year. Try LinReg, QuadReg, and CubicReg.
2. Write the equation for the curve that best fits the data. **See margin.**
3. Based on this equation, how many minutes should you expect to work each day in the year 2010 to pay one day's taxes? **about 184 min**

www.algebra2.com/other_calculator_keystrokes

Answer

2. $(9.4444444 \times 10^{-4})x^3 - 0.1057143x^2 + 4.21031746x - 83.1904762$; The equation is a good fit because $r^2 \approx 0.994$.

Graphing Calculator Investigation

A Follow-Up of Lesson 7-2

Getting Started

Knowing Your Calculator
Students should clear lists L1 and L2 before entering the data from the table in Step 1. This is a more reliable approach than simply "overwriting" old data with new data.

Teach

- Step 3 asserts that a quartic curve will fit the data best. An easy way to verify this is to find quadratic and cubic regression equations and then copy them to the Y= list along with the quartic regression equation. Change the window settings for the x-axis to [0, 18.8]. Turn off the scatter plot, and graph all three regression equations on the same screen. Then use the Trace feature to go to each x value in the first row of the table. While at each x value, use the up/down arrow keys to move between the curves, comparing the y value for each regression curve with the y value given in the table.
- Have students complete Exercises 1–3.

Assess

In **Exercises 1 and 2**, make sure students have graphed several curves before choosing the one they feel best fits the data. Students should be able to justify their final choice.

In **Exercise 3**, students' answers may vary slightly depending on their best-fit curve.

7-3 Lesson Notes

1 Focus

5-Minute Check Transparency 7-3 Use as a quiz or review of Lesson 7-2.

Mathematical Background notes are available for this lesson on p. 344C.

Building on Prior Knowledge

In Chapter 6, students learned to solve quadratic equations. Students should recognize that the same techniques are used in this lesson to solve higher-degree polynomial equations that can be written using quadratic form.

How can solving polynomial equations help you to find dimensions?

Ask students:

- How does this metal sheet become a box? **by folding up the sides after the squares have been removed from the corners**
- Why are squares, and not rectangles, cut from the corners? **Squares are used so that the sides of the box have the same height. If rectangles are cut from the corners, there are two different heights for the sides and a box cannot be formed.**
- What are the length, width, and height of the box in terms of x? **length: $50 - 2x$; width: $32 - 2x$; height: x**

7-3 Solving Equations Using Quadratic Techniques

What You'll Learn

- Write expressions in quadratic form.
- Use quadratic techniques to solve equations.

Vocabulary

- quadratic form

How can solving polynomial equations help you to find dimensions?

The Taylor Manufacturing Company makes open metal boxes of various sizes. Each sheet of metal is 50 inches long and 32 inches wide. To make a box, a square is cut from each corner. The volume of the box depends on the side length x of the cut squares. It is given by $V(x) = 4x^3 - 164x^2 + 1600x$. You can solve a polynomial equation to find the dimensions of the square to cut for a box with specific volume.

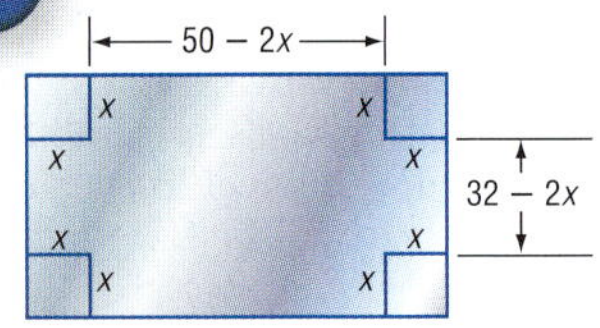

QUADRATIC FORM In some cases, you can rewrite a polynomial in x in the form $au^2 + bu + c$. For example, by letting $u = x^2$ the expression $x^4 - 16x^2 + 60$ can be written as $(x^2)^2 - 16(x^2) + 60$ or $u^2 - 16u + 60$. This new, but equivalent, expression is said to be in **quadratic form**.

TEACHING TIP
This method of substituting u for x^2 is called *u substitution.*

Key Concept — *Quadratic Form*

An expression that is quadratic in form can be written as $au^2 + bu + c$ for any numbers a, b, and c, $a \neq 0$, where u is some expression in x. The expression $au^2 + bu + c$ is called the quadratic form of the original expression.

Example 1 *Write Expressions in Quadratic Form*

Write each expression in quadratic form, if possible.

a. $x^4 + 13x^2 + 36$

$x^4 + 13x^2 + 36 = (x^2)^2 + 13(x^2) + 36$ $\quad (x^2)^2 = x^4$

b. $16x^6 - 625$

$16x^6 - 625 = (4x^3)^2 - 625$ $\quad (x^3)^2 = x^6$

c. $12x^8 - x^2 + 10$

This cannot be written in quadratic form since $x^8 \neq (x^2)^2$.

d. $x - 9x^{\frac{1}{2}} + 8$

$x - 9x^{\frac{1}{2}} + 8 = \left(x^{\frac{1}{2}}\right)^2 - 9\left(x^{\frac{1}{2}}\right) + 8$ $\quad x^1 = \left(x^{\frac{1}{2}}\right)^2$

SOLVE EQUATIONS USING QUADRATIC FORM In Chapter 6, you learned to solve quadratic equations by using the Zero Product Property and the Quadratic Formula. You can extend these techniques to solve higher-degree polynomial equations that can be written using quadratic form or have an expression that contains a quadratic factor.

Resource Manager

Workbook and Reproducible Masters

Chapter 7 Resource Masters

- Study Guide and Intervention, pp. 387–388
- Skills Practice, p. 389
- Practice, p. 390
- Reading to Learn Mathematics, p. 391
- Enrichment, p. 392
- Assessment, p. 443

Transparencies

5-Minute Check Transparency 7-3
Answer Key Transparencies

Technology

Alge2PASS: Tutorial Plus, Lesson 13
Interactive Chalkboard

Example 2 Solve Polynomial Equations

Solve each equation.

a. $x^4 - 13x^2 + 36 = 0$

$x^4 - 13x^2 + 36 = 0$ Original equation

$(x^2)^2 - 13(x^2) + 36 = 0$ Write the expression on the left in quadratic form.

$(x^2 - 9)(x^2 - 4) = 0$ Factor the trinomial.

$(x - 3)(x + 3)(x - 2)(x + 2) = 0$ Factor each difference of squares.

Use the Zero Product Property.

$x - 3 = 0$ or $x + 3 = 0$ or $x - 2 = 0$ or $x + 2 = 0$

$x = 3$ $\quad x = -3$ $\quad x = 2$ $\quad x = -2$

The solutions are -3, -2, 2, and 3.

CHECK The graph of $f(x) = x^4 - 13x^2 + 36$ shows that the graph intersects the x-axis at -3, -2, 2, and 3. ✓

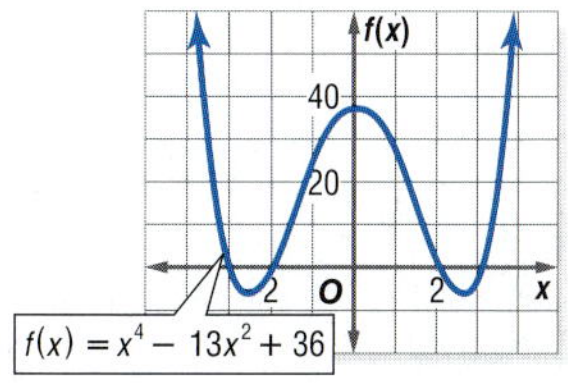

b. $x^3 + 343 = 0$

$x^3 + 343 = 0$ Original equation

$(x)^3 + 7^3 = 0$ This is the sum of two cubes.

$(x + 7)[x^2 - x(7) + 7^2] = 0$ Sum of two cubes formula with $a = x$ and $b = 7$

$(x + 7)(x^2 - 7x + 49) = 0$ Simplify.

$x + 7 = 0$ or $x^2 - 7x + 49 = 0$ Zero Product Property

The solution of the first equation is -7. The second equation can be solved by using the Quadratic Formula.

$x = \frac{-b \pm \sqrt{b^2 - 4ac}}{2a}$ Quadratic Formula

$= \frac{-(-7) \pm \sqrt{(-7)^2 - 4(1)(49)}}{2(1)}$ Replace a with 1, b with -7, and c with 49.

$= \frac{7 \pm \sqrt{-147}}{2}$ Simplify.

$= \frac{7 \pm i\sqrt{147}}{2}$ or $\frac{7 \pm 7i\sqrt{3}}{2}$ $\quad \sqrt{147} \times \sqrt{-1} = 7i\sqrt{3}$

Thus, the solutions of the original equation are -7, $\frac{7 + 7i\sqrt{3}}{2}$, and $\frac{7 - 7i\sqrt{3}}{2}$.

Some equations involving rational exponents can be solved by using a quadratic technique.

Example 3 Solve Equations with Rational Exponents

Solve $x^{\frac{2}{3}} - 6x^{\frac{1}{3}} + 5 = 0$.

$x^{\frac{2}{3}} - 6x^{\frac{1}{3}} + 5 = 0$ Original equation

$\left(x^{\frac{1}{3}}\right)^2 - 6\left(x^{\frac{1}{3}}\right) + 5 = 0$ Write the expression on the left in quadratic form.

(continued on the next page)

Study Tip

Look Back

To review the formula for factoring the **sum of two cubes**, see Lesson 5-4.

Study Tip

Substitution

To avoid confusion, you can substitute another variable for the expression in parentheses.

For example, $\left(x^{\frac{1}{3}}\right)^2 - 6\left(x^{\frac{1}{3}}\right) + 5 = 0$ could be written as $u^2 - 6u + 5 = 0$. Then, once you have solved the equation for u, substitute $x^{\frac{1}{3}}$ for u and solve for x.

www.algebra2.com/extra_examples

2 Teach

QUADRATIC FORM

In-Class Example

Power Point®

1 Write each expression in quadratic form, if possible.

a. $2x^6 + x^3 + 9$ $2(x^3)^2 + (x^3) + 9$

b. $7x^{10} + 6$ $7(x^5)^2 + 6$

c. $x^4 + 2x^3 - 1$ This cannot be written in quadratic form since $x^4 \neq (x^3)^2$.

d. $x^{\frac{2}{3}} + 2x^{\frac{1}{3}} - 4$ $\left(x^{\frac{1}{3}}\right)^2 + 2\left(x^{\frac{1}{3}}\right) - 4$

SOLVE EQUATIONS USING QUADRATIC FORM

In-Class Examples

Power Point®

2 Solve each equation.

a. $x^4 - 29x^2 + 100 = 0$ $-5, -2, 2, 5$

b. $x^3 + 216 = 0$ $-6, 3 + 3i\sqrt{3}, 3 - 3i\sqrt{3}$

3 Solve $x^{\frac{1}{2}} - x^{\frac{1}{4}} - 6 = 0$. 81

DAILY INTERVENTION

Unlocking Misconceptions

Quadratic Form In Example 1 on p. 360, students may mistakenly conclude that variables must have even powers in order for the expression to be written in quadratic form. Draw students' attention to Example 1d. Clarify that the relationship between the powers of two terms is what indicates whether an expression can be written in quadratic form. In Example 1d, the power of the x term is twice the power of the $x^{\frac{1}{2}}$ term, so the expression can be rewritten in quadratic form.

Lesson 7-3 Solving Equations Using Quadratic Techniques 361

In-Class Example

4 Solve $x + \sqrt{x} = 12$. **9**

3 Practice/Apply

Study Notebook

Have students—

- add the definitions/examples of the vocabulary terms to their Vocabulary Builder worksheets for Chapter 7.
- include any other item(s) that they find helpful in mastering the skills in this lesson.

About the Exercises...

Organization by Objective

- **Quadratic Form:** 11–16
- **Solve Equations Using Quadratic Form:** 17–30

Odd/Even Assignments

Exercises 11–30 are structured so that students practice the same concepts whether they are assigned odd or even problems.

Assignment Guide

Basic: 11–27 odd, 32–34, 37–52

Average: 11–31 odd, 32–34, 37–52

Advanced: 12–30 even, 32–48 (optional: 49–52)

All: Practice Quiz 1 (1–5)

Answers

1. Sample answer: $16x^4 - 12x^2 = 0$; $4[4(x^2)^2 - 3x^2] = 0$
2. The solutions of a polynomial equation are the points at which the graph intersects the x-axis.
3. Factor out an x and write the equation in quadratic form so you have $x[(x^2)^2 - 2(x^2) + 1] = 0$. Factor the trinomial and solve for x using the Zero Product Property. The solutions are −1, 0, and 1.

$\left(x^{\frac{1}{3}} - 1\right)\left(x^{\frac{1}{3}} - 5\right) = 0$ Factor the trinomial.

$x^{\frac{1}{3}} - 1 = 0$ or $x^{\frac{1}{3}} - 5 = 0$ Zero Product Property

$x^{\frac{1}{3}} = 1$ $x^{\frac{1}{3}} = 5$ Isolate x on one side of the equation.

$\left(x^{\frac{1}{3}}\right)^3 = 1^3$ $\left(x^{\frac{1}{3}}\right)^3 = 5^3$ Cube each side.

$x = 1$ $x = 125$ Simplify.

CHECK Substitute each value into the original equation.

$$x^{\frac{2}{3}} - 6x^{\frac{1}{3}} + 5 = 0 \qquad x^{\frac{2}{3}} - 6x^{\frac{1}{3}} + 5 = 0$$

$$1^{\frac{2}{3}} - 6(1)^{\frac{1}{3}} + 5 \stackrel{?}{=} 0 \qquad 125^{\frac{2}{3}} - 6(125)^{\frac{1}{3}} + 5 \stackrel{?}{=} 0$$

$$1 - 6 + 5 \stackrel{?}{=} 0 \qquad 25 - 30 + 5 \stackrel{?}{=} 0$$

$$0 = 0 \ \checkmark \qquad 0 = 0 \ \checkmark$$

The solutions are 1 and 125.

To use a quadratic technique, rewrite the equation so one side is equal to zero.

Example 4 Solve Radical Equations

Solve $x - 6\sqrt{x} = 7$.

$x - 6\sqrt{x} = 7$ Original equation

$x - 6\sqrt{x} - 7 = 0$ Rewrite so that one side is zero.

$(\sqrt{x})^2 - 6(\sqrt{x}) - 7 = 0$ Write the expression on the left in quadratic form.

You can use the Quadratic Formula to solve this equation.

$\sqrt{x} = \dfrac{-b \pm \sqrt{b^2 - 4ac}}{2a}$ Quadratic Formula

$\sqrt{x} = \dfrac{-(-6) \pm \sqrt{(-6)^2 - 4(1)(-7)}}{2(1)}$ Replace a with 1, b with −6, and c with −7.

$\sqrt{x} = \dfrac{6 \pm 8}{2}$ Simplify.

$\sqrt{x} = \dfrac{6 + 8}{2}$ or $\sqrt{x} = \dfrac{6 - 8}{2}$ Write as two equations.

$\sqrt{x} = 7$ $\sqrt{x} = -1$ Simplify.

$x = 49$

Since the principal square root of a number cannot be negative, the equation $\sqrt{x} = -1$ has no solution. Thus, the only solution of the original equation is 49.

Study Tip

Look Back
To review **principal roots**, see Lesson 5-5.

Check for Understanding

Concept Check

1. **OPEN ENDED** Give an example of an equation that is not quadratic but can be written in quadratic form. Then write it in quadratic form. **1–3. See margin.**
2. **Explain** how the graph of the related polynomial function can help you verify the solution to a polynomial equation.
3. **Describe** how to solve $x^5 - 2x^3 + x = 0$.

DAILY INTERVENTION

Differentiated Instruction

Visual/Spatial Have students repeat Example 3, using the substitution method discussed in the Study Tip at the bottom of p. 361. Instruct students to write the given equation in pencil and then use a colored pencil to write the statement "Let $u = x^{\frac{1}{3}}$." Have students continue solving the problem using the colored pencil to help them remember they are working with a substituted variable. After solving for u, when students substitute $x^{\frac{1}{3}}$ for u they should resume using their regular pencil.

Guided Practice

GUIDED PRACTICE KEY	
Exercises	Examples
4, 5	1
6–9	2
10	3

Write each expression in quadratic form, if possible.

4. $5y^4 + 7y^3 - 8$ **not possible**
5. $84n^4 - 62n^2$ **$84(n^2)^2 - 62(n^2)$**

Solve each equation. **8. $6, -3 + 3i\sqrt{3}, -3 - 3i\sqrt{3}$**

6. $x^3 + 9x^2 + 20x = 0$ **$0, -5, -4$**
7. $x^4 - 17x^2 + 16 = 0$ **$-4, -1, 4, 1$**
8. $x^3 - 216 = 0$
9. $x - 16x^{\frac{1}{2}} = -64$ **64**

Application

10. **POOL** The Shelby University swimming pool is in the shape of a rectangular prism and has a volume of 28,000 cubic feet. The dimensions of the pool are x feet deep by $7x - 6$ feet wide by $9x - 2$ feet long. How deep is the pool? **8 ft**

★ indicates increased difficulty **11. $2(x^2)^2 + 6(x^2) - 10$ 12. not possible 13. $11(n^3)^2 + 44(n^3)$ 15. not possible**

Practice and Apply

Homework Help	
For Exercises	See Examples
11–16	1
17–28	2–4
29–36	2

Extra Practice
See page 842.

Write each expression in quadratic form, if possible. **14. $b[7(b^2)^2 - 4(b^2) + 2)]$**

11. $2x^4 + 6x^2 - 10$
12. $a^8 + 10a^2 - 16$
13. $11n^6 + 44n^3$
14. $7b^5 - 4b^3 + 2b$
15. $7x^{\frac{1}{9}} - 3x^{\frac{1}{3}} + 4$
16. $6x^{\frac{2}{5}} - 4x^{\frac{1}{5}} - 16$ **$6\left(x^{\frac{1}{5}}\right)^2 - 4\left(x^{\frac{1}{5}}\right) - 16 = 0$**

Solve each equation. **17–28. See pp. 407A–407H.**

17. $m^4 + 7m^3 + 12m^2 = 0$
18. $a^5 + 6a^4 + 5a^3 = 0$
19. $b^4 = 9$
20. $t^5 - 256t = 0$
21. $d^4 + 32 = 12d^2$
22. $x^4 + 18 = 11x^2$
23. $x^3 + 729 = 0$
24. $y^3 - 512 = 0$
25. $x^{\frac{1}{2}} - 8x^{\frac{1}{4}} + 15 = 0$
26. $p^{\frac{2}{3}} + 11p^{\frac{1}{3}} + 28 = 0$
27. $y - 19\sqrt{y} = -60$
28. $z = 8\sqrt{z} + 240$

★29. $s^3 + 4s^2 - s - 4 = 0$ **$1, -1, -4$**

★30. $h^3 - 8h^2 + 3h - 24 = 0$ **$8, i\sqrt{3}, -i\sqrt{3}$**

31. **GEOMETRY** The width of a rectangular prism is w centimeters. The height is 2 centimeters less than the width. The length is 4 centimeters more than the width. If the volume of the prism is 8 times the measure of the length, find the dimensions of the prism. **$w = 4$ cm, $\ell = 8$ cm, $h = 2$ cm**

Career Choices

Designer

Designers combine practical knowledge with artistic ability to turn abstract ideas into formal designs. Designers usually specialize in a particular area, such as clothing, or home interiors.

Online Research
For information about a career as a designer, visit: www.algebra2.com/careers

DESIGN For Exercises 32–34, use the following information.

Jill is designing a picture frame for an art project. She plans to have a square piece of glass in the center and surround it with a decorated ceramic frame, which will also be a square. The dimensions of the glass and frame are shown in the diagram at the right. Jill determines that she needs 27 square inches of material for the frame.

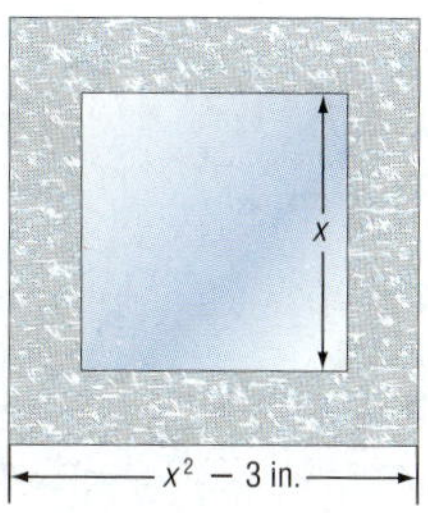

32. Write a polynomial equation that models the area of the frame. **$x^4 - 7x^2 + 9 = 27$**
33. What are the dimensions of the glass piece? **3 in. × 3 in.**
34. What are the dimensions of the frame? **6 in. × 6 in.**

PACKAGING For Exercises 35 and 36, use the following information.

A computer manufacturer needs to change the dimensions of its foam packaging for a new model of computer. The width of the original piece is three times the height, and the length is equal to the height squared. The volume of the new piece can be represented by the equation $V(h) = 3h^4 + 11h^3 + 18h^2 + 44h + 24$, where h is the height of the original piece. **35. $h^2 + 4, 3h + 2, h + 3$**

★35. Factor the equation for the volume of the new piece to determine three expressions that represent the height, length, and width of the new piece.

★36. How much did each dimension of the packaging increase for the new foam piece? **The height increased by 3, the width increased by 2, and the length increased by 4.**

www.algebra2.com/self_check_quiz

Study Guide and Intervention, p. 387 (shown) and p. 388

Quadratic Form Certain polynomial expressions in x can be written in the quadratic form $au^2 + bu + c$ for any numbers a, b, and c, $a \neq 0$, where u is an expression in x.

Example **Write each polynomial in quadratic form, if possible.**

a. $3a^6 - 9a^3 + 12$
Let $u = a^3$.
$3a^6 - 9a^3 + 12 = 3(a^3)^2 - 9(a^3) + 12$

b. $101b - 49\sqrt{b} + 42$
Let $u = \sqrt{b}$.
$101b - 49\sqrt{b} + 42 = 101(\sqrt{b})^2 - 49(\sqrt{b}) + 42$

c. $24a^5 + 12a^3 + 18$
This expression cannot be written in quadratic form, since $a^5 \neq (a^3)^2$.

Exercises

Write each polynomial in quadratic form, if possible.

1. $x^4 + 6x^2 - 8$ **$(x^2)^2 + 6(x^2) - 8$**
2. $4p^4 + 6p^2 + 8$ **$4(p^2)^2 + 6(p^2) + 8$**
3. $x^8 + 2x^4 + 1$ **$(x^4)^2 + 2(x^4) + 1$**
4. $x^{\frac{1}{8}} + 2x^{\frac{1}{16}} + 1$ **$\left(x^{\frac{1}{16}}\right)^2 + 2\left(x^{\frac{1}{16}}\right) + 1$**
5. $6x^4 + 3x^3 + 18$ **not possible**
6. $12x^4 + 10x^2 - 4$ **$12(x^2)^2 + 10(x^2) - 4$**
7. $24x^8 + x^4 + 4$ **$24(x^4)^2 + x^4 + 4$**
8. $18x^6 - 2x^3 + 12$ **$18(x^3)^2 - 2(x^3) + 12$**
9. $100x^4 - 9x^2 - 15$ **$100(x^2)^2 - 9(x^2) - 15$**
10. $25x^8 + 36x^6 - 49$ **not possible**
11. $48x^6 - 32x^3 + 20$ **$48(x^3)^2 - 32(x^3) + 20$**
12. $63x^8 + 5x^4 - 29$ **$63(x^4)^2 + 5(x^4) - 29$**
13. $32x^{10} + 14x^5 - 143$ **$32(x^5)^2 + 14(x^5) - 143$**
14. $50x^3 - 15x\sqrt{x} - 18$ **$50\left(x^{\frac{3}{2}}\right)^2 - 15\left(x^{\frac{3}{2}}\right) - 18$**
15. $60x^6 - 7x^3 + 3$ **$60(x^3)^2 - 7(x^3) + 3$**
16. $10x^{10} - 7x^5 - 7$ **$10(x^5)^2 - 7(x^5) - 7$**

Skills Practice, p. 389 and Practice, p. 390 (shown)

Write each expression in quadratic form, if possible.

1. $10b^4 + 3b^2 - 11$ **$10(b^2)^2 + 3(b^2) - 11$**
2. $-5x^8 + x^2 + 6$ **not possible**
3. $28d^6 + 25d^3$ **$28(d^3)^2 + 25(d^3)$**
4. $4s^8 + 4s^4 + 7$ **$4(s^4)^2 + 4(s^4) + 7$**
5. $500x^4 - x^2$ **$500(x^2)^2 - x^2$**
6. $8b^5 - 8b^3 - 1$ **not possible**
7. $32w^5 - 56w^3 + 8w$ **$8w[4(w^2)^2 - 7(w^2) + 1]$**
8. $e^{\frac{2}{3}} + 7e^{\frac{1}{3}} - 10$ **$\left(e^{\frac{1}{3}}\right)^2 + 7\left(e^{\frac{1}{3}}\right) - 10$**
9. $x^{\frac{1}{5}} + 29x^{\frac{1}{10}} + 2$ **$\left(x^{\frac{1}{10}}\right)^2 + 29\left(x^{\frac{1}{10}}\right) + 2$**

Solve each equation.

10. $y^4 - 7y^3 - 18y^2 = 0$ **$-2, 0, 9$**
11. $s^5 + 4s^4 - 32s^3 = 0$ **$-8, 0, 4$**
12. $m^4 - 625 = 0$ **$-5, 5, -5i, 5i$**
13. $n^4 - 49n^2 = 0$ **$0, -7, 7$**
14. $x^4 - 50x^2 + 49 = 0$ **$-1, 1, -7, 7$**
15. $t^4 - 21t^2 + 80 = 0$ **$-4, 4, \sqrt{5}, -\sqrt{5}$**
16. $4r^6 - 9r^4 = 0$ **$0, \frac{3}{2}, -\frac{3}{2}$**
17. $x^4 - 24 = -2x^2$ **$-2, 2, -i\sqrt{6}, i\sqrt{6}$**
18. $d^4 = 16d^2 - 48$ **$-2, 2, -2\sqrt{3}, 2\sqrt{3}$**
19. $t^3 - 343 = 0$ **$7, \frac{-7 - 7i\sqrt{3}}{2}, \frac{-7 + 7i\sqrt{3}}{2}$**
20. $x^{\frac{1}{2}} - 5x^{\frac{1}{4}} + 6 = 0$ **16, 81**
21. $x^{\frac{4}{3}} - 29x^{\frac{2}{3}} + 100 = 0$ **8, 125**
22. $y^3 - 28y^{\frac{3}{2}} + 27 = 0$ **1, 9**
23. $n - 10\sqrt{n} + 25 = 0$ **25**
24. $w - 12\sqrt{w} + 27 = 0$ **9, 81**
25. $x - 2\sqrt{x} - 80 = 0$ **100**

26. **PHYSICS** A proton in a magnetic field follows a path on a coordinate grid modeled by the function $f(x) = x^4 - 2x^2 - 15$. What are the x-coordinates of the points on the grid where the proton crosses the x-axis? **$-\sqrt{5}, \sqrt{5}$**

27. **SURVEYING** Vista county is setting aside a large parcel of land to preserve it as open space. The county has hired Meghan's surveying firm to survey the parcel, which is in the shape of a right triangle. The longer leg of the triangle measures 5 miles less than the square of the shorter leg, and the hypotenuse of the triangle measures 13 miles less than twice the square of the shorter leg. The length of each boundary is a whole number. Find the length of each boundary. **3 mi, 4 mi, 5 mi**

Reading to Learn Mathematics, p. 391 ELL

Pre-Activity **How can solving polynomial equations help you to find dimensions?**

Read the introduction to Lesson 7-3 at the top of page 360 in your textbook.

Explain how the formula given for the volume of the box can be obtained from the dimensions shown in the figure.

Sample answer: The volume of a rectangular box is given by the formula $V = \ell wh$. Substitute $50 - 2x$ for ℓ, $32 - 2x$ for w, and x for h to get $V(x) = (50 - 2x)(32 - 2x)(x) = 4x^3 - 164x^2 + 1600x$.

Reading the Lesson

1. Which of the following expressions can be written in quadratic form? **b, c, d, f, g, h, i**
   a. $x^3 + 6x^2 + 9$
   b. $x^4 - 7x^2 + 6$
   c. $m^6 + 4m^3 + 4$
   d. $y - 2y^{\frac{1}{2}} - 15$
   e. $x^5 + x^3 + 1$
   f. $r^4 + 6 - r^8$
   g. $p^{\frac{1}{4}} + 8p^{\frac{1}{2}} + 12$
   h. $r^{\frac{1}{3}} + 2r^{\frac{1}{6}} - 3$
   i. $5\sqrt{z} + 2z - 3$

2. Match each expression from the list on the left with its factorization from the list on the right.
   a. $x^4 - 3x^2 - 40$ **vi**
   b. $x^4 - 10x^2 + 25$ **v**
   c. $x^6 - 9$ **i**
   d. $x - 9$ **ii**
   e. $x^6 + 1$ **iv**
   f. $x + 6\sqrt{x} + 9$ **iii**

   i. $(x^3 + 3)(x^3 - 3)$
   ii. $(\sqrt{x} + 3)(\sqrt{x} - 3)$
   iii. $(\sqrt{x} + 3)^2$
   iv. $(x^2 + 1)(x^4 - x^2 + 1)$
   v. $(x^2 - 5)^2$
   vi. $(x^2 + 5)(x^2 - 8)$

Helping You Remember

3. What is an easy way to tell whether a trinomial in one variable containing one constant term can be written in quadratic form?

Sample answer: Look at the two terms that are not constants and compare the exponents on the variable. If one of the exponents is twice the other, the trinomial can be written in quadratic form.

Enrichment, p. 392

Odd and Even Polynomial Functions

Functions whose graphs are symmetric with respect to the origin are called *odd* functions. If $f(-x) = -f(x)$ for all x in the domain of $f(x)$, then $f(x)$ is odd.

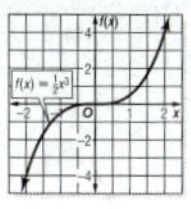

Functions whose graphs are symmetric with respect to the y-axis are called *even* functions. If $f(-x) = f(x)$ for all x in the domain of $f(x)$, then $f(x)$ is even.

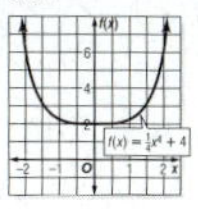

Example **Determine whether $f(x) = x^3 - 3x$ is odd, even, or neither.**

$f(x) = x^3 - 3x$

4 Assess

Open-Ended Assessment

Speaking Have students explain how the equation in Example 4 on p. 362 can be solved by first substituting a for $\sqrt{x}$. If students have difficulty getting started, ask them how they could express x in terms of a, given that $a = \sqrt{x}$.

Getting Ready for Lesson 7-4

PREREQUISITE SKILL Lesson 7-4 introduces students to the Remainder and Factor Theorems. Students will use division to find the factors of polynomials. Use Exercises 49–52 to determine your students' familiarity with dividing polynomials by a binomial.

Assessment Options

Practice Quiz 1 The quiz provides students with a brief review of the concepts and skills in Lessons 7-1 through 7-3. Lesson numbers are given to the right of the exercises or instruction lines so students can review concepts not yet mastered.

Quiz (Lessons 7-1 through 7-3) is available on p. 443 of the *Chapter 7 Resource Masters*.

Answers

48.

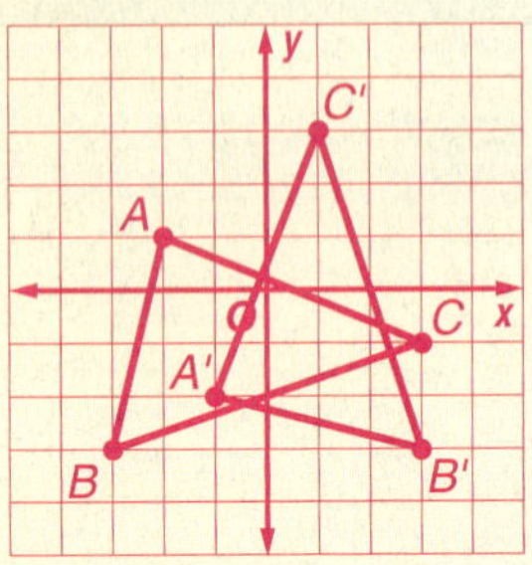

49. $x^2 + 5x - 4$

50. $4x^2 - 16x + 27 - \frac{64}{x + 2}$

51. $x^3 - 6x - 20 - \frac{54}{x - 3}$

52. $x^3 + 2x^2 - 10x + 15 - \frac{21}{x + 1}$

---

37. Write the equation in quadratic form, $u^2 - 9n + 8 = 0$, where $u = |a - 3|$. Then factor and use the Zero Product Property to solve for a; 11, 4, 2, and −5.

37. **CRITICAL THINKING** Explain how you would solve $|a - 3|^2 - 9|a - 3| = -8$. Then solve the equation.

38. **WRITING IN MATH** Answer the question that was posed at the beginning of the lesson. See pp. 407A–407H.

How can solving polynomial equations help you to find dimensions?

Include the following items in your answer:
- an explanation of how you could determine the dimensions of the cut square if the desired volume was 3600 cubic inches, and
- an explanation of why there can be more than one square that can be cut to produce the same volume.

Standardized Test Practice

39. Which of the following is a solution of $x^4 - 2x^2 - 3 = 0$? **D**

Ⓐ $\sqrt[4]{2}$ Ⓑ 1   Ⓒ −3   Ⓓ $\sqrt{3}$

40. **EXTENDED RESPONSE** Solve $18x + 9\sqrt{2x} - 4 = 0$ by first rewriting it in quadratic form. Show your work. $\frac{1}{18}$

Maintain Your Skills

Mixed Review

41–42. See pp. 407A–407H.

Graph each function by making a table of values. *(Lesson 7-2)*

41. $f(x) = x^3 - 4x^2 + x + 5$

42. $f(x) = x^4 - 6x^3 + 10x^2 - x - 3$

Find $p(7)$ and $p(-3)$ for each function. *(Lesson 7-1)*

43. $p(x) = x^2 - 5x + 3$ **17; 27**

44. $p(x) = x^3 - 11x - 4$ **262; 2**

45. $p(x) = \frac{2}{3}x^4 - 3x^3$ $\frac{1715}{3}$; **135**

For Exercises 46–48, use the following information.

Triangle ABC with vertices $A(-2, 1)$, $B(-3, -3)$, and $C(3, -1)$ is rotated 90° counterclockwise about the origin. *(Lesson 4-4)*

46. Write the coordinates of the triangle in a vertex matrix. $\begin{bmatrix} -2 & -3 & 3 \\ 1 & -3 & -1 \end{bmatrix}$

47. Find the coordinates of $\triangle A'B'C'$. **$A'(-1, -2)$, $B'(3, -3)$, $C'(1, 3)$**

48. Graph the preimage and the image. **See margin.**

Getting Ready for the Next Lesson

PREREQUISITE SKILL **Find each quotient.**

*(To review **dividing polynomials**, see Lesson 5-3.)* **49–52. See margin.**

49. $(x^3 + 4x^2 - 9x + 4) \div (x - 1)$

50. $(4x^3 - 8x^2 - 5x - 10) \div (x + 2)$

51. $(x^4 - 9x^2 - 2x + 6) \div (x - 3)$

52. $(x^4 + 3x^3 - 8x^2 + 5x - 6) \div (x + 1)$

Practice Quiz 1

Lessons 7-1 through 7-3

1. If $p(x) = 2x^3 - x$, find $p(a - 1)$. *(Lesson 7-1)* $2a^3 - 6a^2 + 5a - 1$

2. Describe the end behavior of the graph at the right. Then determine whether it represents an odd-degree or an even-degree polynomial function and state the number of real zeros. *(Lesson 7-1)* **See margin.**

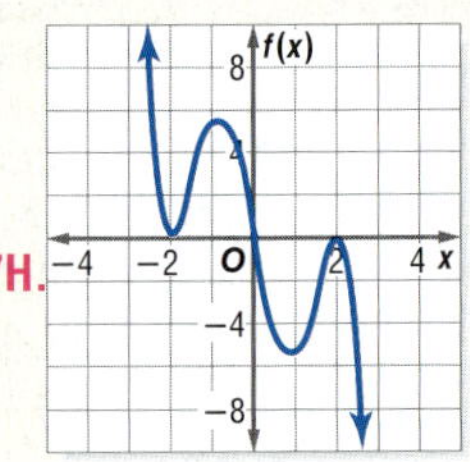

3. Graph $y = x^3 + 2x^2 - 4x - 6$. Estimate the x-coordinates at which the relative maxima and relative minima occur. *(Lesson 7-2)* **See pp. 407A–407H.**

4. Write the expression $18x^{\frac{1}{3}} + 36x^{\frac{2}{3}} + 5$ in quadratic form. *(Lesson 7-3)* **See margin.**

5. Solve $a^4 = 6a^2 + 27$. *(Lesson 7-3)* $-3, 3, -i\sqrt{3}, i\sqrt{3}$

Answers (Practice Quiz 1)

2. $f(x) \to -\infty$ as $x \to +\infty$, $f(x) \to +\infty$ as $x \to -\infty$; odd; 3

4. $\left(6x^{\frac{1}{3}}\right)^2 + 3\left(6x^{\frac{1}{3}}\right) + 5$ or $36\left(\sqrt[3]{x}\right)^2 + 18\left(\sqrt[3]{x}\right) + 5$

7-4 The Remainder and Factor Theorems

What You'll Learn

- Evaluate functions using synthetic substitution.
- Determine whether a binomial is a factor of a polynomial by using synthetic substitution.

Vocabulary

- synthetic substitution
- depressed polynomial

How can you use the Remainder Theorem to evaluate polynomials?

The number of international travelers to the United States since 1986 can be modeled by the equation $T(x) = 0.02x^3 - 0.6x^2 + 6x + 25.9$, where x is the number of years since 1986 and $T(x)$ is the number of travelers in millions. To estimate the number of travelers in 2006, you can evaluate the function for $x = 20$, or you can use synthetic substitution.

Study Tip

Look Back

To review **dividing polynomials** and **synthetic division**, see Lesson 5-3.

SYNTHETIC SUBSTITUTION Synthetic division is a shorthand method of long division. It can also be used to find the value of a function. Consider the polynomial function $f(a) = 4a^2 - 3a + 6$. Divide the polynomial by $a - 2$.

Method 1 Long Division

$$\begin{array}{r} 4a \;+\; 5 \\ a - 2\overline{)4a^2 - 3a + 6} \\ \underline{4a^2 - 8a} \\ 5a + \;\;6 \\ \underline{5a - 10} \\ 16 \end{array}$$

Method 2 Synthetic Division

$$\begin{array}{r|rrr} 2 & 4 & -3 & 6 \\ & & 8 & 10 \\ \hline & 4 & 5 & |\,16 \end{array}$$

Compare the remainder of 16 to $f(2)$.

$$\begin{aligned} f(2) &= 4(2)^2 - 3(2) + 6 && \text{Replace } a \text{ with 2.} \\ &= 16 - 6 + 6 && \text{Multiply.} \\ &= 16 && \text{Simplify.} \end{aligned}$$

Notice that the value of $f(2)$ is the same as the remainder when the polynomial is divided by $a - 2$. This illustrates the **Remainder Theorem**.

Key Concept *Remainder Theorem*

If a polynomial $f(x)$ is divided by $x - a$, the remainder is the constant $f(a)$, and

$$\underbrace{f(x)}_{\text{Dividend}} \;\underbrace{=}_{\text{equals}}\; \underbrace{q(x)}_{\text{quotient}} \;\underbrace{\cdot}_{\text{times}}\; \underbrace{(x - a)}_{\text{divisor}} \;\underbrace{+}_{\text{plus}}\; \underbrace{f(a),}_{\text{remainder.}}$$

where $q(x)$ is a polynomial with degree one less than the degree of $f(x)$.

When synthetic division is used to evaluate a function, it is called **synthetic substitution**. It is a convenient way of finding the value of a function, especially when the degree of the polynomial is greater than 2.

7-4 Lesson Notes

1 Focus

5-Minute Check Transparency 7-4 Use as a quiz or review of Lesson 7-3.

Mathematical Background notes are available for this lesson on p. 344D.

How can you use the Remainder Theorem to evaluate polynomials?

Ask students:

- What would be the value of x if you wanted to use the function to estimate the number of travelers in 1987? **1**
- Would you expect the actual number of travelers in 2006 to exactly match the number predicted by the function? Explain. **No. Sample answer: The equation is a model based on data available at this time.**
- Do you think the model is more accurate for the years immediately following 1986 or for years in the future? Explain. **Sample answer: The model closely matches actual data for the years immediately following 1986. For years in the future, the model is less likely to be as accurate.**

Resource Manager

Workbook and Reproducible Masters

Chapter 7 Resource Masters

- Study Guide and Intervention, pp. 393–394
- Skills Practice, p. 395
- Practice, p. 396
- Reading to Learn Mathematics, p. 397
- Enrichment, p. 398

Teaching Algebra With Manipulatives Masters, p. 252

Transparencies

5-Minute Check Transparency 7-4
Answer Key Transparencies

Technology

Interactive Chalkboard

Building on Prior Knowledge

In Lesson 5-3, students learned synthetic division. In this lesson, students will use synthetic division to evaluate a function and to find factors of polynomials.

2 Teach

SYNTHETIC SUBSTITUTION

In-Class Example Power Point®

Teaching Tip Most of this lesson relies heavily on synthetic substitution. Before finishing your discussion of Example 1, be certain that students understand the method. Ask a student volunteer to demonstrate the steps of the synthetic substitution shown in Method 1 of Example 1, and invite students to discuss any problems they have with the technique before moving on.

1 If $f(x) = 3x^4 - 2x^3 + x^2 - 2$, find $f(4)$. **654**

Example 1 Synthetic Substitution

If $f(x) = 2x^4 - 5x^2 + 8x - 7$, find $f(6)$.

Method 1 Synthetic Substitution

By the Remainder Theorem, $f(6)$ should be the remainder when you divide the polynomial by $x - 6$.

6	2	0	−5	8	−7
		12	72	402	2460
	2	12	67	410	2453

Notice that there is no x^3 term. A zero is placed in this position as a placeholder.

The remainder is 2453. Thus, by using synthetic substitution, $f(6) = 2453$.

Method 2 Direct Substitution

Replace x with 6.

$f(x) = 2x^4 - 5x^2 + 8x - 7$ — Original function

$f(6) = 2(6)^4 - 5(6)^2 + 8(6) - 7$ — Replace x with 6.

$= 2592 - 180 + 48 - 7$ or 2453 — Simplify.

By using direct substitution, $f(6) = 2453$.

FACTORS OF POLYNOMIALS Divide $f(x) = x^4 + x^3 - 17x^2 - 20x + 32$ by $x - 4$.

4	1	1	−17	−20	32
		4	20	12	−32
	1	5	3	−8	0

Study Tip

Depressed Polynomial
A *depressed polynomial* has a degree that is one less than the original polynomial.

The quotient of $f(x)$ and $x - 4$ is $x^3 + 5x^2 + 3x - 8$. When you divide a polynomial by one of its binomial factors, the quotient is called a **depressed polynomial**. From the results of the division and by using the Remainder Theorem, we can make the following statement.

$$\underbrace{x^4 + x^3 - 17x^2 - 20x + 32}_{\text{Dividend}} \underbrace{=}_{\text{equals}} \underbrace{(x^3 + 5x^2 + 3x - 8)}_{\text{quotient}} \underbrace{\cdot}_{\text{times}} \underbrace{(x - 4)}_{\text{divisor}} \underbrace{+}_{\text{plus}} \underbrace{0}_{\text{remainder.}}$$

Since the remainder is 0, $f(4) = 0$. This means that $x - 4$ is a factor of $x^4 + x^3 - 17x^2 - 20x + 32$. This illustrates the **Factor Theorem**, which is a special case of the Remainder Theorem.

Key Concept *Factor Theorem*

The binomial $x - a$ is a factor of the polynomial $f(x)$ if and only if $f(a) = 0$.

Suppose you wanted to find the factors of $x^3 - 3x^2 - 6x + 8$. One approach is to graph the related function, $f(x) = x^3 - 3x^2 - 6x + 8$. From the graph at the right, you can see that the graph of $f(x)$ crosses the x-axis at −2, 1, and 4. These are the zeros of the function. Using these zeros and the Zero Product Property, we can express the polynomial in factored form.

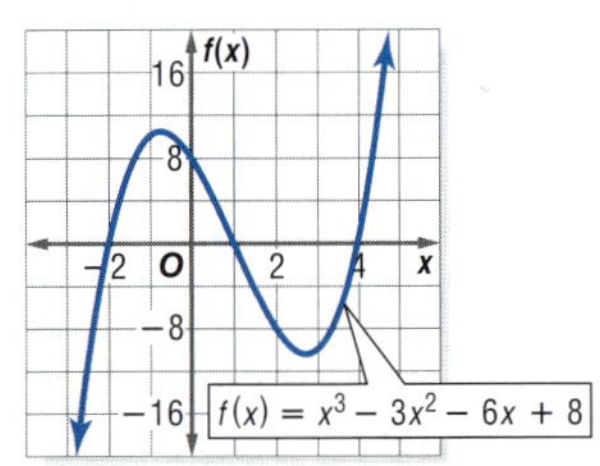

366 Chapter 7 Polynomial Functions

$$f(x) = [x - (-2)](x - 1)(x - 4)$$
$$= (x + 2)(x - 1)(x - 4)$$

This method of factoring a polynomial has its limitations. Most polynomial functions are not easily graphed and once graphed, the exact zeros are often difficult to determine.

The Factor Theorem can help you find all factors of a polynomial.

Example 2 Use the Factor Theorem

Show that $x + 3$ is a factor of $x^3 + 6x^2 - x - 30$. Then find the remaining factors of the polynomial.

The binomial $x + 3$ is a factor of the polynomial if -3 is a zero of the related polynomial function. Use the Factor Theorem and synthetic division.

−3	1	6	−1	−30
		−3	−9	30
	1	3	−10	0

Since the remainder is 0, $x + 3$ is a factor of the polynomial. The polynomial $x^3 + 6x^2 - x - 30$ can be factored as $(x + 3)(x^2 + 3x - 10)$. The polynomial $x^2 + 3x - 10$ is the depressed polynomial. Check to see if this polynomial can be factored.

$x^2 + 3x - 10 = (x - 2)(x + 5)$ Factor the trinomial.

So, $x^3 + 6x^2 - x - 30 = (x + 3)(x - 2)(x + 5)$.

CHECK You can see that the graph of the related function $f(x) = x^3 + 6x^2 - x - 30$ crosses the x-axis at -5, -3, and 2. Thus, $f(x) = [x - (-5)][x - (-3)](x - 2)$. ✓

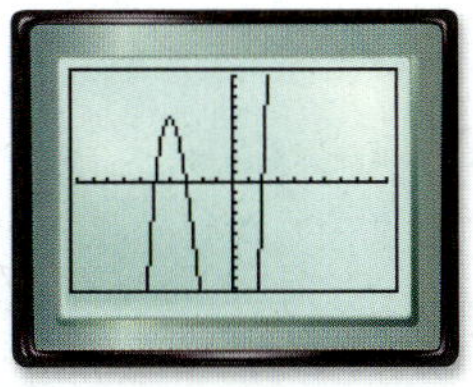

Study Tip

Factoring
The factors of a polynomial do not have to be binomials. For example, the factors of $x^3 + x^2 - x + 15$ are $x + 3$ and $x^2 - 2x + 5$.

Example 3 Find All Factors of a Polynomial

GEOMETRY The volume of the rectangular prism is given by $V(x) = x^3 + 3x^2 - 36x + 32$. Find the missing measures.

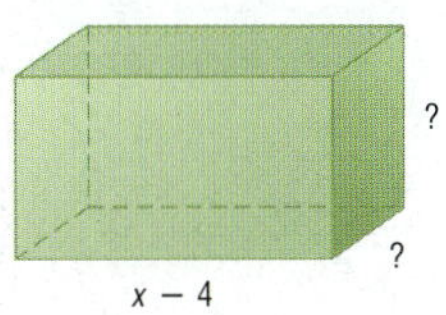

The volume of a rectangular prism is $\ell \times w \times h$.

You know that one measure is $x - 4$, so $x - 4$ is a factor of $V(x)$.

4	1	3	−36	32
		4	28	−32
	1	7	−8	0

The quotient is $x^2 + 7x - 8$. Use this to factor $V(x)$.

$V(x) = x^3 + 3x^2 - 36x + 32$ Volume function

$= (x - 4)(x^2 + 7x - 8)$ Factor.

$= (x - 4)(x + 8)(x - 1)$ Factor the trinomial $x^2 + 7x - 8$.

So, the missing measures of the prism are $x + 8$ and $x - 1$.

www.algebra2.com/extra_examples

FACTORS OF POLYNOMIALS

Teaching Tip Remind students that not all polynomials can be factored. Emphasize that the factors indicate where the graph of the function crosses the x-axis. If the graph of a polynomial function has no x-intercepts, then the polynomial cannot be factored. Students can graph the function $f(x) = x^4 - x^3 - x^2 + 2$ to see an example of a polynomial that cannot be factored.

In-Class Examples

Power Point®

2 Show that $x - 3$ is a factor of $x^3 + 4x^2 - 15x - 18$. Then find the remaining factors of the polynomial.

3	1	4	−15	−18
		3	21	18
	1	7	6	0

So, $x^3 + 4x^2 - 15x - 18 = (x - 3)(x^2 + 7x + 6)$. Since $x^2 + 7x + 6 = (x + 1)(x + 6)$, $x^3 + 4x^2 - 15x - 18 = (x - 3)(x + 1)(x + 6)$.

Teaching Tip Point out that the Factor Theorem does not say anything about which numbers to try. Techniques for identifying potential factors will be introduced later in the chapter.

3 **GEOMETRY** The volume of the rectangular prism is given by $V(x) = x^3 + 7x^2 + 2x - 40$. Find the missing measures.

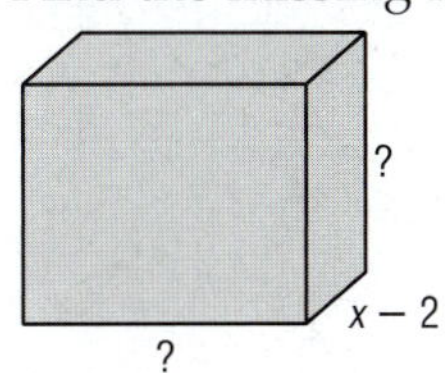

The missing measures of the rectangular prism are $x + 4$ and $x + 5$.

DAILY INTERVENTION

Differentiated Instruction

Intrapersonal Have students describe two or three things about this lesson that they found difficult to understand. Then have them address each item by writing an explanation that will help them review the material later or that they can refer to if they become confused in later lessons. Suggest that students add these notes to their Study Notebooks.

3 Practice/Apply

Study Notebook

Have students—

- add the definitions/examples of the vocabulary terms to their Vocabulary Builder worksheets for Chapter 7.
- write a note explaining how to determine whether a given binomial is a factor of a given polynomial.
- include any other item(s) that they find helpful in mastering the skills in this lesson.

About the Exercises...

Organization by Objective

- **Synthetic Substitution:** 13–20
- **Factors of Polynomials:** 21–30

Odd/Even Assignments

Exercises 13–30 are structured so that students practice the same concepts whether they are assigned odd or even problems.

Assignment Guide

Basic: 13–31 odd, 37, 38, 41, 46–62

Average: 13–35 odd, 37–41, 46–62

Advanced: 14–36 even, 39–59 (optional: 60–62)

Answers

25. $x + 3, x - \frac{1}{2}$ or $2x - 1$

26. $x - 1, x + \frac{4}{3}$ or $3x + 4$

27. $x + 7, x - 4$

28. $x - 1, x + 6$

29. $x - 1, x^2 + 2x + 3$

30. $2x - 3, 2x + 3, 4x^2 + 9$

Check for Understanding

Concept Check

1. **OPEN ENDED** Give an example of a polynomial function that has a remainder of 5 when divided by $x - 4$. **Sample answer: $f(x) = x^2 - 2x - 3$**

2. **State** the degree of the depressed polynomial that is the result of dividing $x^5 + 3x^4 - 16x - 48$ by one of its first-degree binomial factors. **4**

3. **Write** the dividend, divisor, quotient, and remainder represented by the synthetic division at the right.

−2	1	0	6	32
		−2	4	−20
	1	−2	10	12

3. dividend: $x^3 + 6x + 32$; divisor: $x + 2$; quotient: $x^2 - 2x + 10$; remainder: 12

Guided Practice

GUIDED PRACTICE KEY	
Exercises	Examples
4, 5	1
6–9	2
10	3

Use synthetic substitution to find $f(3)$ and $f(-4)$ for each function.

4. $f(x) = x^3 - 2x^2 - x + 1$ **7, −91**

5. $f(x) = 5x^4 - 6x^2 + 2$ **353, 1186**

Given a polynomial and one of its factors, find the remaining factors of the polynomial. Some factors may not be binomials.

6. $x^3 - x^2 - 5x - 3; x + 1$ **$x + 1, x - 3$**

7. $x^3 - 3x + 2; x - 1$ **$x - 1, x + 2$**

8. $6x^3 - 25x^2 + 2x + 8; 3x - 2$ **$2x + 1, x - 4$**

9. $x^4 + 2x^3 - 8x - 16; x + 2$ **$x - 2, x^2 + 2x + 4$**

Application

For Exercises 10–12, use the graph at the right.

The projected sales of e-books can be modeled by the function $S(x) = -17x^3 + 200x^2 - 113x + 44$, where x is the number of years since 2000. **10. $2.894 billion**

10. Use synthetic substitution to estimate the sales for 2006 in billions of dollars.

11. Evaluate $S(6)$. **$2.894 billion**

12. Which method—synthetic division or direct substitution—do you prefer to use to evaluate polynomials? Explain your answer. **Sample answer: Direct substitution, because it can be done quickly with a calculator.**

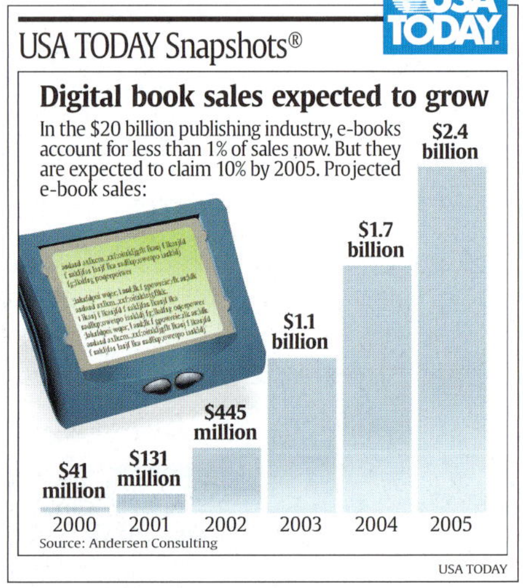

★ **indicates increased difficulty**

Practice and Apply

Homework Help	
For Exercises	See Examples
13–20	1
21–36	2
37–44	3

Extra Practice See page 843.

Use synthetic substitution to find $g(3)$ and $g(-4)$ for each function. **18. 267, 680**

13. $g(x) = x^2 - 8x + 6$ **−9, 54**

14. $g(x) = x^3 + 2x^2 - 3x + 1$ **37, −19**

15. $g(x) = x^3 - 5x + 2$ **14, −42**

16. $g(x) = x^4 - 6x - 8$ **55, 272**

17. $g(x) = 2x^3 - 8x^2 - 2x + 5$ **−19, −243**

18. $g(x) = 3x^4 + x^3 - 2x^2 + x + 12$

19. $g(x) = x^5 + 8x^3 + 2x - 15$ **450, −1559**

20. $g(x) = x^6 - 4x^4 + 3x^2 - 10$ **422, 3110**

Given a polynomial and one of its factors, find the remaining factors of the polynomial. Some factors may not be binomials. **24. $x - 3, x - 1$**

21. $x^3 + 2x^2 - x - 2; x - 1$ **$x + 1, x + 2$**

22. $x^3 - x^2 - 10x - 8; x + 1$ **$x - 4, x + 2$**

23. $x^3 + x^2 - 16x - 16; x + 4$ **$x - 4, x + 1$**

24. $x^3 - 6x^2 + 11x - 6; x - 2$

Online Lesson Plans

USA TODAY Education's Online site offers resources and interactive features connected to each day's newspaper. *Experience TODAY*, USA TODAY's daily lesson plan, is available on the site and delivered daily to subscribers. This plan provides instruction for integrating USA TODAY graphics and key editorial features into your mathematics classroom. Log on to www.education.usatoday.com.

368 Chapter 7 Polynomial Functions

25–30. See margin.

25. $2x^3 - 5x^2 - 28x + 15;\ x - 5$
26. $3x^3 + 10x^2 - x - 12;\ x + 3$
27. $2x^3 + 7x^2 - 53x - 28;\ 2x + 1$
28. $2x^3 + 17x^2 + 23x - 42;\ 2x + 7$
29. $x^4 + 2x^3 + 2x^2 - 2x - 3;\ x + 1$
30. $16x^5 - 32x^4 - 81x + 162;\ x - 2$

WebQuest

Changes in world population can be modeled by a polynomial function. Visit www.algebra2.com/webquest to continue work on your WebQuest project.

31. Use the graph of the polynomial function at the right to determine at least one binomial factor of the polynomial. Then find all the factors of the polynomial. **$x - 2$, $x + 2$, $x^2 + 1$**

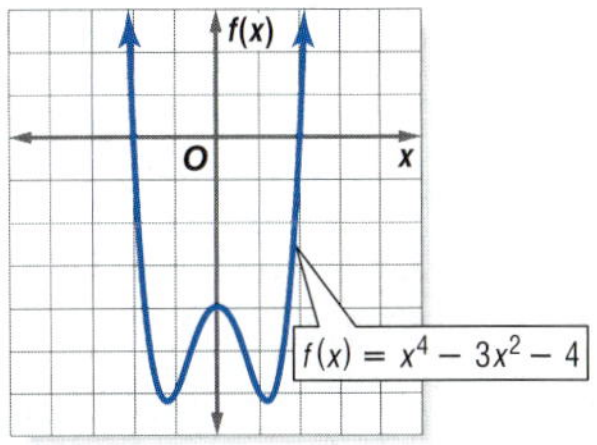

32. Use synthetic substitution to show that $x - 8$ is a factor of $x^3 - 4x^2 - 29x - 24$. Then find any remaining factors. **See pp. 407A–407H; $(x + 3)(x + 1)$.**

Find values of k so that each remainder is 3.

★33. $(x^2 - x + k) \div (x - 1)$ **3**
★34. $(x^2 + kx - 17) \div (x - 2)$ **8**
★35. $(x^2 + 5x + 7) \div (x + k)$ **1, 4**
★36. $(x^3 + 4x^2 + x + k) \div (x + 2)$ **−3**

ENGINEERING For Exercises 37 and 38, use the following information.
When a certain type of plastic is cut into sections, the length of each section determines its strength. The function $f(x) = x^4 - 14x^3 + 69x^2 - 140x + 100$ can describe the relative strength of a section of length x feet. Sections of plastic x feet long, where $f(x) = 0$, are extremely weak. After testing the plastic, engineers discovered that sections 5 feet long were extremely weak.

37. Show that $x - 5$ is a factor of the polynomial function. **See pp. 407A–407H.**
38. Are there other lengths of plastic that are extremely weak? Explain your reasoning. **Yes, 2-ft lengths; the binomial $x - 2$ is a factor of the polynomial since $f(2) = 0$.**

ARCHITECTURE For Exercises 39 and 40, use the following information.
Elevators traveling from one floor to the next do not travel at a constant speed. Suppose the speed of an elevator in feet per second is given by the function $f(t) = -0.5t^4 + 4t^3 - 12t^2 + 16t$, where t is the time in seconds.

39. Find the speed of the elevator at 1, 2, and 3 seconds. **7.5 ft/s, 8 ft/s, 7.5 ft/s**
40. It takes 4 seconds for the elevator to go from one floor to the next. Use synthetic substitution to find $f(4)$. Explain what this means. **0; The elevator is stopped.**

41. **CRITICAL THINKING** Consider the polynomial $f(x) = ax^4 + bx^3 + cx^2 + dx + e$, where $a + b + c + d + e = 0$. Show that this polynomial is divisible by $x - 1$. **See margin.**

PERSONAL FINANCE For Exercises 42–45, use the following information.
Zach has purchased some home theater equipment for $2000, which he is financing through the store. He plans to pay $340 per month and wants to have the balance paid off after six months. The formula $B(x) = 2000x^6 - 340(x^5 + x^4 + x^3 + x^2 + x + 1)$ represents his balance after six months if x represents 1 plus the monthly interest rate (expressed as a decimal).

42. Find his balance after 6 months if the annual interest rate is 12%. (*Hint*: The monthly interest rate is the annual rate divided by 12, so $x = 1.01$.) **$31.36**
43. Find his balance after 6 months if the annual interest rate is 9.6%. **$16.70**
44. How would the formula change if Zach wanted to pay the balance in five months? **$B(x) = 2000x^5 - 340(x^4 + x^3 + x^2 + x + 1)$**
45. Suppose he finances his purchase at 10.8% and plans to pay $410 every month. Will his balance be paid in full after five months? **No, he will still owe $4.40.**

www.algebra2.com/self_check_quiz

Study Guide and Intervention, p. 393 (shown) and p. 394

Synthetic Substitution

Remainder Theorem	The remainder, when you divide the polynomial $f(x)$ by $(x - a)$, is the constant $f(a)$. $f(x) = q(x) \cdot (x - a) + f(a)$, where $q(x)$ is a polynomial with degree one less than the degree of $f(x)$.

Example 1 **If $f(x) = 3x^4 + 2x^3 - 5x^2 + x - 2$, find $f(-2)$.**

Method 1 Synthetic Substitution
By the Remainder Theorem, $f(-2)$ should be the remainder when you divide the polynomial by $x + 2$.

−2	3	2	−5	1	−2
		−6	8	−6	10
	3	−4	3	−5	8

The remainder is 8, so $f(-2) = 8$.

Method 2 Direct Substitution
Replace x with −2.
$f(x) = 3x^4 + 2x^3 - 5x^2 + x - 2$
$f(-2) = 3(-2)^4 + 2(-2)^3 - 5(-2)^2 + (-2) - 2$
$= 48 - 16 - 20 - 2 - 2$ or 8
So $f(-2) = 8$.

Example 2 **If $f(x) = 5x^3 + 2x - 1$, find $f(3)$.**
Again, by the Remainder Theorem, $f(3)$ should be the remainder when you divide the polynomial by $x - 3$.

3	5	0	2	−1
		15	45	141
	5	15	47	140

The remainder is 140, so $f(3) = 140$.

Exercises

Use synthetic substitution to find $f(-5)$ and $f\left(\frac{1}{2}\right)$ for each function.

1. $f(x) = -3x^2 + 5x - 1$ **$-101; \frac{3}{4}$**
2. $f(x) = 4x^2 + 6x - 7$ **63; −3**
3. $f(x) = -x^3 + 3x^2 - 5$ **$195; -\frac{35}{8}$**
4. $f(x) = x^4 + 11x^2 - 1$ **$899; \frac{29}{16}$**

Use synthetic substitution to find $f(4)$ and $f(-3)$ for each function.

5. $f(x) = 2x^3 + x^2 - 5x + 3$ **127; −27**
6. $f(x) = 3x^3 - 4x + 2$ **178; −67**
7. $f(x) = 5x^3 - 4x^2 + 2$ **258; −169**
8. $f(x) = 2x^4 - 4x^3 + 3x^2 + x - 6$ **302; 288**
9. $f(x) = 5x^4 + 3x^3 - 4x^2 - 2x + 4$ **1404; 298**
10. $f(x) = 3x^4 - 2x^3 - x^2 + 2x - 5$ **627; 277**
11. $f(x) = 2x^4 - 4x^3 - x^2 - 6x + 3$ **219; 282**
12. $f(x) = 4x^4 - 4x^3 + 3x^2 - 2x - 3$ **805; 462**

Skills Practice, p. 395 and Practice, p. 396 (shown)

Use synthetic substitution to find $f(-3)$ and $f(4)$ for each function.

1. $f(x) = x^2 + 2x + 3$ **6, 27**
2. $f(x) = x^2 - 5x + 10$ **34, 6**
3. $f(x) = x^2 - 5x - 4$ **20, −8**
4. $f(x) = x^3 - x^2 - 2x + 3$ **−27, 43**
5. $f(x) = x^3 + 2x^2 + 5$ **−4, 101**
6. $f(x) = x^3 - 6x^2 + 2x$ **−87, −24**
7. $f(x) = x^3 - 2x^2 - 2x + 8$ **−31, 32**
8. $f(x) = x^3 - x^2 + 4x - 4$ **−52, 60**
9. $f(x) = x^3 + 3x^2 + 2x - 50$ **−56, 70**
10. $f(x) = x^4 + x^3 - 3x^2 - x + 12$ **42, 280**
11. $f(x) = x^4 - 2x^2 - x + 7$ **73, 227**
12. $f(x) = 2x^4 - 3x^3 + 4x^2 - 2x + 1$ **286, 377**
13. $f(x) = 2x^4 - x^3 + 2x^2 - 26$ **181, 454**
14. $f(x) = 3x^4 - 4x^3 + 3x^2 - 5x - 3$ **390, 537**
15. $f(x) = x^5 + 7x^3 - 4x - 10$ **−430, 1446**
16. $f(x) = x^6 + 2x^5 - x^4 + x^3 - 9x^2 + 20$ **74, 5828**

Given a polynomial and one of its factors, find the remaining factors of the polynomial. Some factors may not be binomials.

17. $x^3 + 3x^2 - 6x - 8;\ x - 2$ **$x + 1, x + 4$**
18. $x^3 + 7x^2 + 7x - 15;\ x - 1$ **$x + 3, x + 5$**
19. $x^3 - 9x^2 + 27x - 27;\ x - 3$ **$x - 3, x - 3$**
20. $x^3 - x^2 - 8x + 12;\ x + 3$ **$x - 2, x - 2$**
21. $x^3 + 5x^2 - 2x - 24;\ x - 2$ **$x + 3, x + 4$**
22. $x^3 - x^2 - 14x + 24;\ x + 4$ **$x - 3, x - 2$**
23. $3x^3 - 4x^2 - 17x + 6;\ x + 2$ **$x - 3, 3x - 1$**
24. $4x^3 - 12x^2 - x + 3;\ x - 3$ **$2x - 1, 2x + 1$**
25. $18x^3 + 9x^2 - 2x - 1;\ 2x + 1$ **$3x + 1, 3x - 1$**
26. $6x^3 + 5x^2 - 3x - 2;\ 3x - 2$ **$2x + 1, x + 1$**
27. $x^5 + x^4 - 5x^3 - 5x^2 + 4x + 4;\ x + 1$ **$x - 1, x + 1, x - 2, x + 2$**
28. $x^5 - 2x^4 + 4x^3 - 8x^2 - 5x + 10;\ x - 2$ **$x - 1, x + 1, x^2 + 5$**

29. **POPULATION** The projected population in thousands for a city over the next several years can be estimated by the function $P(x) = x^3 + 2x^2 - 8x + 520$, where x is the number of years since 2000. Use synthetic substitution to estimate the population for 2005. **655,000**

30. **VOLUME** The volume of water in a rectangular swimming pool can be modeled by the polynomial $2x^3 - 9x^2 + 7x + 6$. If the depth of the pool is given by the polynomial $2x + 1$, what polynomials express the length and width of the pool? **$x - 3$ and $x - 2$**

Reading to Learn Mathematics, p. 397 — ELL

Pre-Activity **How can you use the Remainder Theorem to evaluate polynomials?**

Read the introduction to Lesson 7-4 at the top of page 365 in your textbook.

Show how you would use the model in the introduction to estimate the number of international travelers (in millions) to the United States in the year 2000. (Show how you would substitute numbers, but do not actually calculate the result.)
Sample answer: $0.02(14)^3 - 0.6(14)^2 + 6(14) + 25.9$

Reading the Lesson

1. Consider the following synthetic division.

1	3	2	−6	4
		3	5	−1
	3	5	−1	3

a. Using the division symbol ÷, write the division problem that is represented by this synthetic division. (Do not include the answer.) **$(3x^3 + 2x^2 - 6x + 4) \div (x - 1)$**

b. Identify each of the following for this division.
dividend **$3x^3 + 2x^2 - 6x + 4$** divisor **$x - 1$**
quotient **$3x^3 + 5x - 1$** remainder **3**

c. If $f(x) = 3x^3 + 2x^2 - 6x + 4$, what is $f(1)$? **3**

2. Consider the following synthetic division.

−3	1	0	0	27
		−3	9	−27
	1	−3	9	0

a. This division shows that **$x + 3$** is a factor of **$x^3 + 27$**.

b. The division shows that **−3** is a zero of the polynomial function $f(x) =$ **$x^3 + 27$**.

c. The division shows that the point **(−3, 0)** is on the graph of the polynomial function $f(x) =$ **$x^3 + 27$**.

Helping You Remember

3. Think of a mnemonic for remembering the sentence, "Dividend equals quotient times divisor plus remainder."
Sample answer: Definitely every quiet teacher deserves proper rewards.

Answer

41. **By the Remainder Theorem, the remainder when $f(x)$ is divided by $x - 1$ is equivalent to $f(1)$, or $a + b + c + d + e$. Since $a + b + c + d + e = 0$, the remainder when $f(x)$ is divided by $x - 1$ is 0. Therefore, $x - 1$ is a factor of $f(x)$.**

Enrichment, p. 398

Using Maximum Values

Many times maximum solutions are needed for different situations. For instance, what is the area of the largest rectangular field that can be enclosed with 2000 feet of fencing?

Let x and y denote the length and width of the field, respectively.

Perimeter: $2x + 2y = 2000 \rightarrow y = 1000 - x$
Area: $A = xy = x(1000 - x) = -x^2 + 1000x$

This problem is equivalent to finding the highest point on the graph of $A(x) = -x^2 + 1000x$ shown on the right.

Complete the square for $-x^2 + 1000x$.

$A = -(x^2 - 1000x + 500^2) + 500^2$

4 Assess

Open-Ended Assessment

Speaking Ask students to offer a verbal comparison of the use of synthetic substitution and the use of direct substitution to determine factors of a polynomial.

Getting Ready for Lesson 7-5

PREREQUISITE SKILL In Lesson 7-5, students will find the zeros of polynomial functions by using the Quadratic Formula. Use Exercises 60–62 to determine your students' familiarity with using the Quadratic Formula.

Answers

46. Using the Remainder Theorem you can evaluate a polynomial for a value of a by dividing the polynomial by $x - a$ using synthetic division. Answers should include the following.
   - It is easier to use the Remainder Theorem when you have polynomials of degree 2 and lower or when you have access to a calculator.
   - The estimated number of international traveler to the U.S. in 2006 is 65.9 million.

52.

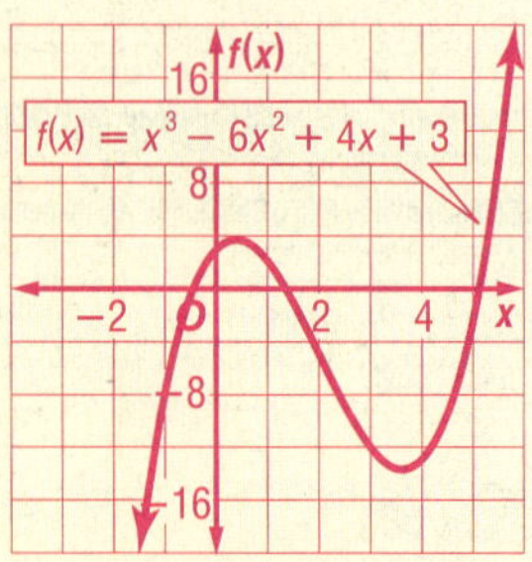

53.

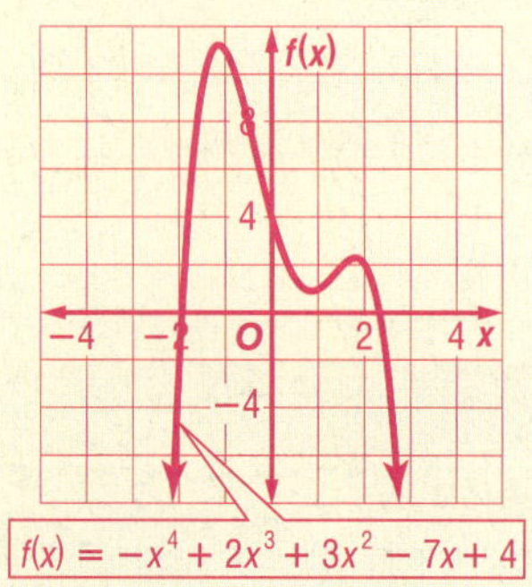

---

46. **WRITING IN MATH** Answer the question that was posed at the beginning of the lesson. **See margin.**

How can you use the Remainder Theorem to evaluate polynomials?

Include the following items in your answer:
- an explanation of when it is easier to use the Remainder Theorem to evaluate a polynomial rather than substitution, and
- evaluate the expression for the number of international travelers to the U.S. for $x = 20$.

Standardized Test Practice

47. Determine the zeros of the function $f(x) = x^2 + 7x + 12$ by factoring. **D**

Ⓐ 7, 12 Ⓑ 3, 4 Ⓒ −5, 5 Ⓓ −4, −3

48. **SHORT RESPONSE** Using the graph of the polynomial function at the right, find all the factors of the polynomial $x^5 + x^4 - 3x^3 - 3x^2 - 4x - 4$.
$x - 2$, $x + 2$, $x + 1$, $x^2 + 1$

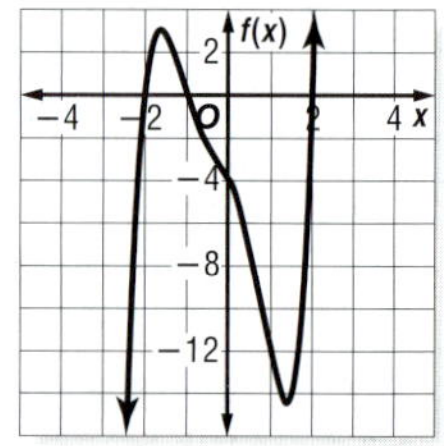

Maintain Your Skills

Mixed Review **Write each expression in quadratic form, if possible.** *(Lesson 7-3)*

49. $x^4 - 8x^2 + 4$ **$(x^2)^2 - 8(x^2) + 4$**
50. $9d^6 + 5d^3 - 2$ **$9(d^3)^2 + 5(d^3) - 2$**
51. $r^4 - 5r^3 + 18r$ **not possible**

Graph each polynomial function. Estimate the x-coordinates at which the relative maxima and relative minima occur. *(Lesson 7-2)* **52–53. See margin for graphs.**

52. $f(x) = x^3 - 6x^2 + 4x + 3$
53. $f(x) = -x^4 + 2x^3 + 3x^2 - 7x + 4$

52. Sample answer: rel. max. at $x = 0.5$, rel. min. at $x = 3.5$

53. Sample answer: rel. max. at $x = -1$ and $x = 1.5$, rel. min. at $x = 1$

54. **PHYSICS** A model airplane is fixed on a string so that it flies around in a circle. The formula $F_c = m\left(\frac{4\pi^2 r}{T^2}\right)$ describes the force required to keep the airplane going in a circle, where m represents the mass of the plane, r represents the radius of the circle, and T represents the time for a revolution. Solve this formula for T. Write in simplest radical form. *(Lesson 5-8)* $T = \frac{2\pi\sqrt{mrF_c}}{F_c}$

Solve each matrix equation. *(Lesson 4-1)*

55. $\begin{bmatrix} 7x \\ 12 \end{bmatrix} = \begin{bmatrix} 28 \\ -6y \end{bmatrix}$ **(4, −2)**

56. $\begin{bmatrix} 5a + 2b \\ a - 7b \end{bmatrix} = \begin{bmatrix} -17 \\ 4 \end{bmatrix}$ **(−3, −1)**

Identify each function as S for step, C for constant, A for absolute value, or P for piecewise. *(Lesson 2-6)*

57. **A**

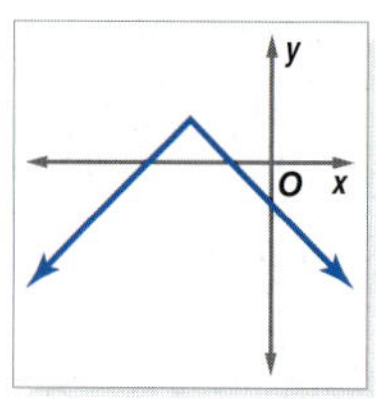

58. **C**

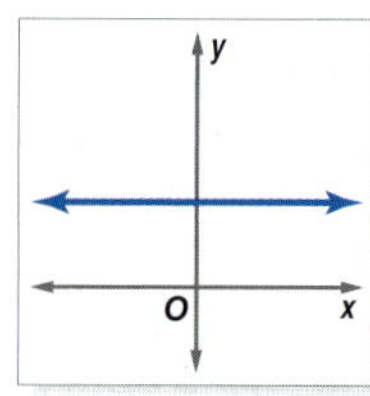

59. **S**

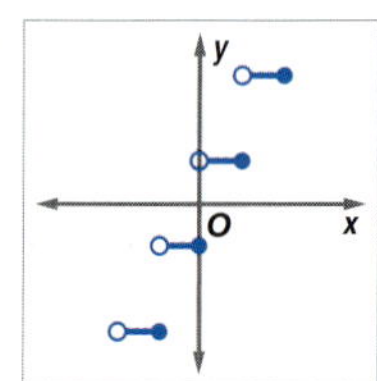

Getting Ready for the Next Lesson **PREREQUISITE SKILL Find the exact solutions of each equation by using the Quadratic Formula.** *(For review of the **Quadratic Formula**, see Lesson 6-5.)*

60. $x^2 + 7x + 8 = 0$ $\frac{-7 \pm \sqrt{17}}{2}$
61. $3x^2 - 9x + 2 = 0$ $\frac{9 \pm \sqrt{57}}{6}$
62. $2x^2 + 3x + 2 = 0$ $\frac{-3 \pm i\sqrt{7}}{4}$

7-5 Roots and Zeros

What You'll Learn

- Determine the number and type of roots for a polynomial equation.
- Find the zeros of a polynomial function.

How can the roots of an equation be used in pharmacology?

When doctors prescribe medication, they give patients instructions as to how much to take and how often it should be taken. The amount of medication in your body varies with time.

Suppose the equation $M(t) = 0.5t^4 + 3.5t^3 - 100t^2 + 350t$ models the number of milligrams of a certain medication in the bloodstream t hours after it has been taken. The doctor can use the roots of this equation to determine how often the patient should take the medication to maintain a certain concentration in the body.

TYPES OF ROOTS You have already learned that a zero of a function $f(x)$ is any value c such that $f(c) = 0$. When the function is graphed, the real zeros of the function are the x-intercepts of the graph.

Concept Summary — *Zeros, Factors, and Roots*

Let $f(x) = a_nx^n + \ldots + a_1x + a_0$ be a polynomial function. Then

- c is a zero of the polynomial function $f(x)$,
- $x - c$ is a factor of the polynomial $f(x)$, and
- c is a root or solution of the polynomial equation $f(x) = 0$.

In addition, if c is a real number, then $(c, 0)$ is an intercept of the graph of $f(x)$.

When you solve a polynomial equation with degree greater than zero, it may have one or more real roots, or no real roots (the roots are imaginary numbers). Since real numbers and imaginary numbers both belong to the set of complex numbers, all polynomial equations with degree greater than zero will have at least one root in the set of complex numbers. This is the **Fundamental Theorem of Algebra**.

Study Tip

Look Back
For review of **complex numbers**, see Lesson 5-9.

Key Concept — *Fundamental Theorem of Algebra*

Every polynomial equation with degree greater than zero has at least one root in the set of complex numbers.

Example 1 Determine Number and Type of Roots

Solve each equation. State the number and type of roots.

a. $x + 3 = 0$

$x + 3 = 0$ Original equation

$x = -3$ Subtract 3 from each side.

This equation has exactly one real root, -3.

7-5 Lesson Notes

1 Focus

5-Minute Check Transparency 7-5 Use as a quiz or review of Lesson 7-4.

Mathematical Background notes are available for this lesson on p. 344D.

Building on Prior Knowledge

In Chapter 6, students learned several methods for finding the roots of quadratic equations. In this lesson, students will incorporate those techniques into finding roots of polynomial equations.

How can the roots of an equation be used in pharmacology?

Ask students:

- Would the given equation be valid for any value of t? Explain. **No; the value of t must be positive because the number of hours cannot be negative. Also, once the number of milligrams, $M(t)$, reaches zero as the value of t increases, any greater values of t will be meaningless also.**
- What would a root of this equation tell the doctor? **There is no more medication in the patient's bloodstream.**

Resource Manager

Workbook and Reproducible Masters

Chapter 7 Resource Masters

- Study Guide and Intervention, pp. 399–400
- Skills Practice, p. 401
- Practice, p. 402
- Reading to Learn Mathematics, p. 403
- Enrichment, p. 404
- Assessment, pp. 443, 445

Transparencies

5-Minute Check Transparency 7-5
Answer Key Transparencies

Technology

Alge2PASS: Tutorial Plus, Lesson 14
Interactive Chalkboard
Multimedia Applications

2 Teach

TYPES OF ROOTS

In-Class Example

Power Point®

1 Solve each equation. State the number and type of roots.

a. $a - 10 = 0$ **This equation has exactly one real root, 10.**

b. $x^2 + 2x - 48 = 0$ **This equation has two real roots, 6 and −8.**

c. $3a^3 + 18a = 0$ **This equation has one real root, 0, and two imaginary roots $i\sqrt{6}$ and $-i\sqrt{6}$.**

d. $y^4 - 16 = 0$ **This equation has two real roots, 2 and −2, and two imaginary roots, $2i$ and $-2i$.**

Study Tip

Reading Math
In addition to double roots, equations can have triple or quadruple roots. In general, these roots are referred to as *repeated roots.*

b. $x^2 - 8x + 16 = 0$

$$\begin{aligned} x^2 - 8x + 16 &= 0 && \text{Original equation} \\ (x - 4)^2 &= 0 && \text{Factor the left side as a perfect square trinomial.} \\ x &= 4 && \text{Solve for } x \text{ using the Square Root Property.} \end{aligned}$$

Since $x - 4$ is twice a factor of $x^2 - 8x + 16$, 4 is a double root. So this equation has two real roots, 4 and 4.

c. $x^3 + 2x = 0$

$$\begin{aligned} x^3 + 2x &= 0 && \text{Original equation} \\ x(x^2 + 2) &= 0 && \text{Factor out the GCF.} \end{aligned}$$

Use the Zero Product Property.

$$\begin{aligned} x = 0 \quad \text{or} \quad x^2 + 2 &= 0 \\ x^2 &= -2 && \text{Subtract two from each side.} \\ x &= \pm\sqrt{-2} \text{ or } \pm i\sqrt{2} && \text{Square Root Property} \end{aligned}$$

This equation has one real root, 0, and two imaginary roots, $i\sqrt{2}$ and $-i\sqrt{2}$.

d. $x^4 - 1 = 0$

$$\begin{aligned} x^4 - 1 &= 0 \\ (x^2 + 1)(x^2 - 1) &= 0 \\ (x^2 + 1)(x + 1)(x - 1) &= 0 \end{aligned}$$

$$x^2 + 1 = 0 \quad \text{or} \quad x + 1 = 0 \quad \text{or} \quad x - 1 = 0$$
$$x^2 = -1 \qquad x = -1 \qquad x = 1$$
$$x = \pm\sqrt{-1} \text{ or } \pm i$$

This equation has two real roots, 1 and −1, and two imaginary roots, i and $-i$.

Compare the degree of each equation and the number of roots of each equation in Example 1. The following corollary of the Fundamental Theorem of Algebra is an even more powerful tool for problem solving.

Key Concept — *Corollary*

A polynomial equation of the form $P(x) = 0$ of degree n with complex coefficients has exactly n roots in the set of complex numbers.

Similarly, a polynomial function of nth degree has exactly n zeros.

More About. . .

Descartes

René Descartes (1596–1650) was a French mathematician and philosopher. One of his best-known quotations comes from his *Discourse on Method*: "I think, therefore I am."

Source: *A History of Mathematics*

French mathematician René Descartes made more discoveries about zeros of polynomial functions. His rule of signs is given below.

Key Concept — *Descartes' Rule of Signs*

If $P(x)$ is a polynomial with real coefficients whose terms are arranged in descending powers of the variable,

- the number of positive real zeros of $y = P(x)$ is the same as the number of changes in sign of the coefficients of the terms, or is less than this by an even number, and
- the number of negative real zeros of $y = P(x)$ is the same as the number of changes in sign of the coefficients of the terms of $P(-x)$, or is less than this number by an even number.

Teacher to Teacher

Warren Zarrell — James Monroe H.S., North Hills, CA

"My students have difficulty finding $p(-x)$, as found in Example 2. I tell them to change the sign of every odd degree term in the polynomial."

372 Chapter 7 Polynomial Functions

Example 2 Find Numbers of Positive and Negative Zeros

State the possible number of positive real zeros, negative real zeros, and imaginary zeros of $p(x) = x^5 - 6x^4 - 3x^3 + 7x^2 - 8x + 1$.

Since $p(x)$ has degree 5, it has five zeros. However, some of them may be imaginary. Use Descartes' Rule of Signs to determine the number and type of real zeros. Count the number of changes in sign for the coefficients of $p(x)$.

$$p(x) = x^5 \; - \; 6x^4 \; - \; 3x^3 \; + \; 7x^2 \; - \; 8x \; + \; 1$$

yes (+ to −), no (− to −), yes (− to +), yes (+ to −), yes (− to +)

Since there are 4 sign changes, there are 4, 2, or 0 positive real zeros.

Find $p(-x)$ and count the number of changes in signs for its coefficients.

$$p(x) = (-x)^5 - 6(-x)^4 - 3(-x)^3 + 7(-x)^2 - 8(-x) + 1$$
$$= -x^5 - 6x^4 + 3x^3 + 7x^2 + 8x + 1$$

no (− to −), yes (− to +), no (+ to +), no (+ to +), no (+ to +)

Since there is 1 sign change, there is exactly 1 negative real zero.

Thus, the function $p(x)$ has either 4, 2, or 0 positive real zeros and exactly 1 negative real zero. Make a chart of the possible combinations of real and imaginary zeros.

Number of Positive Real Zeros	Number of Negative Real Zeros	Number of Imaginary Zeros	Total Number of Zeros
4	1	0	4 + 1 + 0 = 5
2	1	2	2 + 1 + 2 = 5
0	1	4	0 + 1 + 4 = 5

Study Tip

Zero at the Origin
Recall that the number 0 has no sign. Therefore, if 0 is a zero of a function, the sum of the number of positive real zeros, negative real zeros, and imaginary zeros is reduced by how many times 0 is a zero of the function.

FIND ZEROS We can find all of the zeros of a function using some of the strategies you have already learned.

Example 3 Use Synthetic Substitution to Find Zeros

Find all of the zeros of $f(x) = x^3 - 4x^2 + 6x - 4$.

Since $f(x)$ has degree 3, the function has three zeros. To determine the possible number and type of real zeros, examine the number of sign changes for $f(x)$ and $f(-x)$.

$f(x) = x^3 - 4x^2 + 6x - 4$ — yes, yes, yes

$f(-x) = -x^3 - 4x^2 - 6x - 4$ — no, no, no

Since there are 3 sign changes for the coefficients of $f(x)$, the function has 3 or 1 positive real zeros. Since there are no sign changes for the coefficient of $f(-x)$, $f(x)$ has no negative real zeros. Thus, $f(x)$ has either 3 real zeros, or 1 real zero and 2 imaginary zeros.

To find these zeros, first list some possibilities and then eliminate those that are not zeros. Since none of the zeros are negative and $f(0)$ is -4, begin by evaluating $f(x)$ for positive integral values from 1 to 4. You can use a shortened form of synthetic substitution to find $f(a)$ for several values of a.

(continued on the next page)

www.algebra2.com/extra_examples

In-Class Example Power Point®

Teaching Tip Point out to students the method for determining the number of imaginary zeros for a polynomial function. In Example 2, the polynomial has degree 5, so it has a maximum of 5 real zeros. You find the numbers of positive and negative real zeros, and subtract the sum of these two numbers from 5 to find the number of imaginary zeros. Remind students that imaginary zeros come in conjugate pairs, so the number of imaginary zeros must be an even number.

2 State the possible number of positive real zeros, negative real zeros, and imaginary zeros of $p(x) = -x^6 + 4x^3 - 2x^2 - x - 1$. **The function has either 2 or 0 positive real zeros, 2 or 0 negative real zeros, and 6, 4, or 2 imaginary zeros.**

FIND ZEROS

In-Class Example Power Point®

3 Find all of the zeros of $f(x) = x^3 - x^2 + 2x + 4$. **The function has one real zero at $x = -1$, and two imaginary zeros at $x = 1 + i\sqrt{3}$ and $x = 1 - i\sqrt{3}$.**

DAILY INTERVENTION — Differentiated Instruction

Kinesthetic As you work Example 3 and Guided Practice Exercises 8–11 in class, provide each student with approximately 20 slips of paper. As students begin the process of finding the zeros of each polynomial function, have them first determine from the degree of the polynomial the number of zeros they need to find. Students should then count off slips of paper, one for each zero. As students work the problem and find the zeros, they should record the information about each zero (positive, negative, imaginary) on one of the slips of paper.

In-Class Example

4 Write a polynomial function of least degree with integral coefficients whose zeros include 4 and $4 - i$.
$f(x) = x^3 - 12x^2 + 49x - 68$ is a polynomial function of least degree with integral coefficients whose zeros are 4, $4 - i$, and $4 + i$.

Study Tip

Finding Zeros
While direct substitution could be used to find each real zero of a polynomial, using synthetic substitution provides you with a depressed polynomial that can be used to find any imaginary zeros.

x	1	−4	6	−4
1	1	−3	3	−1
2	1	−2	2	**0**
3	1	−1	3	5
4	1	0	6	20

Each row in the table shows the coefficients of the depressed polynomial and the remainder.

From the table, we can see that one zero occurs at $x = 2$. Since the depressed polynomial of this zero, $x^2 - 2x + 2$, is quadratic, use the Quadratic Formula to find the roots of the related quadratic equation, $x^2 - 2x + 2 = 0$.

$$x = \frac{-b \pm \sqrt{b^2 - 4ac}}{2a}$$ Quadratic Formula

$$= \frac{-(-2) \pm \sqrt{(-2)^2 - 4(1)(2)}}{2(1)}$$ Replace a with 1, b with −2, and c with 2.

$$= \frac{2 \pm \sqrt{-4}}{2}$$ Simplify.

$$= \frac{2 \pm 2i}{2}$$ $\sqrt{4} \times \sqrt{-1} = 2i$

$$= 1 \pm i$$ Simplify.

Thus, the function has one real zero at $x = 2$ and two imaginary zeros at $x = 1 + i$ and $x = 1 - i$. The graph of the function verifies that there is only one real zero.

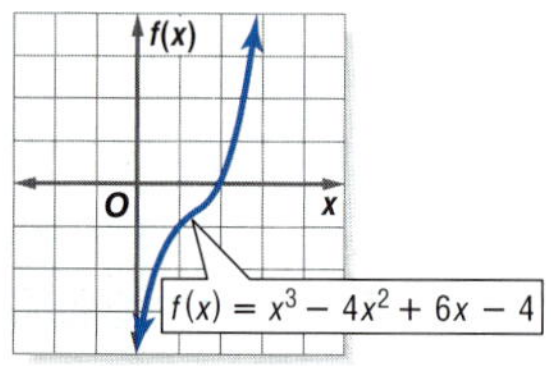

In Chapter 6, you learned that solutions of a quadratic equation that contains imaginary numbers come in pairs. This applies to the zeros of polynomial functions as well. For any polynomial function, if an imaginary number is a zero of that function, its conjugate is also a zero. This is called the **Complex Conjugates Theorem**.

Key Concept — *Complex Conjugates Theorem*

Suppose a and b are real numbers with $b \neq 0$. If $a + bi$ is a zero of a polynomial function with real coefficients, then $a - bi$ is also a zero of the function.

Standardized Test Practice

Example 4 Use Zeros to Write a Polynomial Function

Short-Response Test Item

Write a polynomial function of least degree with integral coefficients whose zeros include 3 and $2 - i$.

Read the Test Item

- If $2 - i$ is a zero, then $2 + i$ is also a zero according to the Complex Conjugates Theorem. So, $x - 3$, $x - (2 - i)$, and $x - (2 + i)$ are factors of the polynomial function.

Solve the Test Item

- Write the polynomial function as a product of its factors.
  $f(x) = (x - 3)[x - (2 - i)][x - (2 + i)]$

Standardized Test Practice

Example 4 Ask students to indicate with a show of hands how many of them have made mistakes in mathematics exercises because they could not read their own handwriting. Stress that throughout this course, students must work using neat and careful handwriting. It is extremely easy to misread coefficients and exponents, or misread i as the number one. In addition, when answering short-response items on standardized tests, students must be aware that if their handwriting is illegible or difficult to read then their answers will not be graded or they may be penalized.

- Multiply the factors to find the polynomial function.

$$f(x) = (x - 3)[x - (2 - i)][x - (2 + i)]$$ Write an equation.
$$= (x - 3)[(x - 2) + i][(x - 2) - i]$$ Regroup terms.
$$= (x - 3)[(x - 2)^2 - i^2]$$ Rewrite as the difference of two squares.
$$= (x - 3)[x^2 - 4x + 4 - (-1)]$$ Square $x - 2$ and replace i^2 with -1.
$$= (x - 3)(x^2 - 4x + 5)$$ Simplify.
$$= x^3 - 4x^2 + 5x - 3x^2 + 12x - 15$$ Multiply using the Distributive Property.
$$= x^3 - 7x^2 + 17x - 15$$ Combine like terms.

$f(x) = x^3 - 7x^2 + 17x - 15$ is a polynomial function of least degree with integral coefficients whose zeros are 3, $2 - i$, and $2 + i$.

1. Sample answer: $p(x) = x^3 - 6x^2 + x + 1$; $p(x)$ has either 2 or 0 positive real zeros, 1 negative real zero, and 2 or 0 imaginary zeros.

Check for Understanding

Concept Check

1. **OPEN ENDED** Write a polynomial function $p(x)$ whose coefficients have two sign changes. Then describe the nature of its zeros.
2. **Explain** why an odd-degree function must always have at least one real root. See margin.
3. **State** the least degree a polynomial equation with real coefficients can have if it has roots at $x = 5 + i$, $x = 3 - 2i$, and a double root at $x = 0$. **6**

Guided Practice

GUIDED PRACTICE KEY	
Exercises	Examples
4, 5	1
6, 7	2
8–11	3
12	4

Solve each equation. State the number and type of roots. 5. -7, 0, and 3; 3 real

4. $x^2 + 4 = 0$ $2i$, $-2i$; 2 imaginary
5. $x^3 + 4x^2 - 21x = 0$

State the possible number of positive real zeros, negative real zeros, and imaginary zeros of each function.

6. $f(x) = 5x^3 + 8x^2 - 4x + 3$ 2 or 0; 1; 2 or 0
7. $r(x) = x^5 - x^3 - x + 1$ 2 or 0; 1; 2 or 4

Find all of the zeros of each function.

8. $p(x) = x^3 + 2x^2 - 3x + 20$
9. $f(x) = x^3 - 4x^2 + 6x - 4$
10. $v(x) = x^3 - 3x^2 + 4x - 12$ $2i$, $-2i$, 3
11. $f(x) = x^3 - 3x^2 + 9x + 13$ $2 + 3i$, $2 - 3i$, -1

8. -4, $1 + 2i$, $1 - 2i$
9. 2, $1 + i$, $1 - i$

Standardized Test Practice

12. **SHORT RESPONSE** Write a polynomial function of least degree with integral coefficients whose zeros include 2 and $4i$. $f(x) = x^3 - 2x^2 + 16x - 32$

★ indicates increased difficulty 15. 0, $3i$, $-3i$; 1 real, 2 imaginary 16. 3, -3, $3i$, $-3i$; 2 real, 2 imaginary

Practice and Apply

Homework Help	
For Exercises	See Examples
13–18	1
19–24, 41	2
25–34, 44–48	3
35–40, 42, 43	4

Extra Practice
See page 843.

Solve each equation. State the number and type of roots.

13. $3x + 8 = 0$ $-\frac{8}{3}$; 1 real
14. $2x^2 - 5x + 12 = 0$ $\frac{5 \pm i\sqrt{71}}{4}$; 2 imaginary
15. $x^3 + 9x = 0$
16. $x^4 - 81 = 0$
17. $x^4 - 16 = 0$ 2, -2, $2i$, and $-2i$; 2 real, 2 imaginary
18. $x^5 - 8x^3 + 16x = 0$ -2, -2, 0, 2, and 2, 5 real

State the possible number of positive real zeros, negative real zeros, and imaginary zeros of each function. 19–24. See margin.

19. $f(x) = x^3 - 6x^2 + 1$
20. $g(x) = 5x^3 + 8x^2 - 4x + 3$
21. $h(x) = 4x^3 - 6x^2 + 8x - 5$
22. $q(x) = x^4 + 5x^3 + 2x^2 - 7x - 9$
23. $p(x) = x^5 - 6x^4 - 3x^3 + 7x^2 - 8x + 1$
24. $f(x) = x^{10} - x^8 + x^6 - x^4 + x^2 - 1$

www.algebra2.com/self_check_quiz

3 Practice/Apply

Study Notebook

Have students—
- add the definitions/examples of the vocabulary terms to their Vocabulary Builder worksheets for Chapter 7.
- summarize what they know so far about identifying types of zeros and finding some of them.
- include any other item(s) that they find helpful in mastering the skills in this lesson.

About the Exercises...

Organization by Objective
- **Types of Roots:** 13–24
- **Find Zeros:** 25–40

Odd/Even Assignments
Exercises 13–40 are structured so that students practice the same concepts whether they are assigned odd or even problems.

Assignment Guide

Basic: 13–41 odd, 49–70

Average: 13–41 odd, 42–45, 49–70

Advanced: 14–40 even, 44–66 (optional: 67–70)

DAILY INTERVENTION

Unlocking Misconceptions

Finding Zeros Students may incorrectly assume that they now know how to find all zeros. However, in this lesson they are using the guess-and-check technique to test possible zeros. In later lessons, students will learn further techniques for finding the zeros of polynomial functions.

Answers

2. An odd-degree function approaches positive infinity in one direction and negative infinity in the other direction, so the graph must cross the x-axis at least once, giving it at least one real root.
19. 2 or 0; 1; 2 or 0
20. 2 or 0; 1; 2 or 0
21. 3 or 1; 0; 2 or 0
22. 1; 3 or 1; 2 or 0
23. 4, 2, or 0; 1; 4, 2, or 0
24. 5, 3, or 1; 5, 3, or 1; 0, 2, 4, 6, or 8

25. $-2, -2+3i, -2i-3i$

26. $4, 1+i, 1-i$

27. $2i, -2i, \frac{i}{2}, -\frac{i}{2}$

28. $5i, -5i, 7$

29. $-\frac{3}{2}, 1+4i, 1-4i$

30. $\frac{1}{2}, 4+5i, 4-5i$

31. $4-i, 4+i, -3$

32. $3-i, 3+i, 4, -1$

Find all of the zeros of each function.

25. $g(x) = x^3 + 6x^2 + 21x + 26$
26. $h(x) = x^3 - 6x^2 + 10x - 8$
27. $h(x) = 4x^4 + 17x^2 + 4$
28. $f(x) = x^3 - 7x^2 + 25x - 175$
29. $g(x) = 2x^3 - x^2 + 28x + 51$
30. $q(x) = 2x^3 - 17x^2 + 90x - 41$
31. $f(x) = x^3 - 5x^2 - 7x + 51$
32. $p(x) = x^4 - 9x^3 + 24x^2 - 6x - 40$
33. $r(x) = x^4 - 6x^3 + 12x^2 + 6x - 13$ **$3-2i, 3+2i, -1, 1$**
34. $h(x) = x^4 - 15x^3 + 70x^2 - 70x - 156$ **$5-i, 5+i, -1, 6$**

Write a polynomial function of least degree with integral coefficients that has the given zeros. **35–40. See margin.**

35. $-4, 1, 5$
36. $-2, 2, 4, 6$
37. $4i, 3, -3$
38. $2i, 3i, 1$
39. $9, 1+2i$
40. $6, 2+2i$

41. Sketch the graph of a polynomial function that has the indicated number and type of zeros. **See pp. 407A–407H.**
   a. 3 real, 2 imaginary
   b. 4 real
   c. 2 imaginary

SCULPTING **For Exercises 42 and 43, use the following information.**
Antonio is preparing to make an ice sculpture. He has a block of ice that he wants to reduce in size by shaving off the same amount from the length, width, and height. He wants to reduce the volume of the ice block to 24 cubic feet. **42. $(3-x)(4-x)(5-x) = 24$**

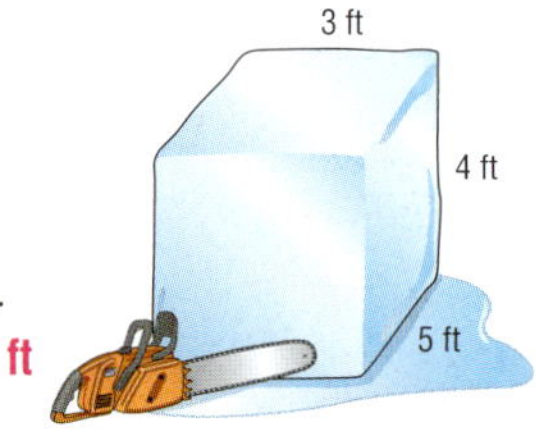

42. Write a polynomial equation to model this situation.
43. How much should he take from each dimension? **1 ft**

SPACE EXPLORATION **For Exercises 44 and 45, use the following information.**
The space shuttle has an external tank for the fuel that the main engines need for the launch. This tank is shaped like a capsule, a cylinder with a hemispherical dome at either end. The cylindrical part of the tank has an approximate volume of 336π cubic meters and a height of 17 meters more than the radius of the tank. (*Hint:* $V(r) = \pi r^2 h$)

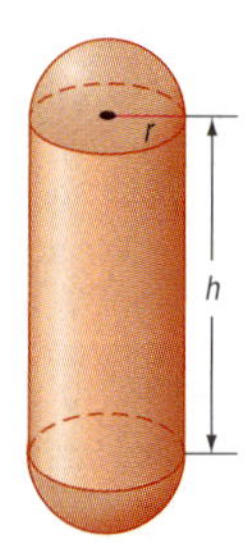

44. Write an equation that represents the volume of the cylinder. **$V(r) = \pi r^3 + 17\pi r^2$**
45. What are the dimensions of the tank? **radius = 4 m, height = 21 m**

MEDICINE **For Exercises 46–48, use the following information.**
Doctors can measure cardiac output in patients at high risk for a heart attack by monitoring the concentration of dye injected into a vein near the heart. A normal heart's dye concentration is given by $d(x) = -0.006x^4 - 0.15x^3 - 0.05x^2 + 1.8x$, where x is the time in seconds.

★ 46. How many positive real zeros, negative real zeros, and imaginary zeros exist for this function? (*Hint:* Notice that 0, which is neither positive nor negative, is a zero of this function since $d(0) = 0$.) **1; 2 or 0; 2 or 0**

★ 47. Approximate all real zeros to the nearest tenth by graphing the function using a graphing calculator. **See margin for graph; −24.1, −4.0, 0, and 3.1.**

★ 48. What is the meaning of the roots in this problem? **Nonnegative roots represent times when there is no concentration of dye registering on the monitor.**

49. **CRITICAL THINKING** Find a counterexample to disprove the following statement.
The polynomial function of least degree with integral coefficients with zeros at $x = 4$, $x = -1$, and $x = 3$, is unique.

49. Sample answer: $f(x) = x^3 - 6x^2 + 5x + 12$ and $g(x) = 2x^3 - 12x^2 + 10x + 24$; each have zeros at $x = 4$, $x = -1$, and $x = 3$.

More About. . .

Space Exploration
A space shuttle is a reusable vehicle, launched like a rocket, which can put people and equipment in orbit around Earth. The first space shuttle was launched in 1981.
Source: kidsastronomy.about.com

Study Guide and Intervention, p. 399 (shown) and p. 400

Types of Roots The following statements are equivalent for any polynomial function $f(x)$.
- c is a zero of the polynomial function $f(x)$.
- $(x - c)$ is a factor of the polynomial $f(x)$.
- c is a root or solution of the polynomial equation $f(x) = 0$.

If c is real, then $(c, 0)$ is an intercept of the graph of $f(x)$.

Fundamental Theorem of Algebra	Every polynomial equation with degree greater than zero has at least one root in the set of complex numbers.
Corollary to the Fundamental Theorem of Algebras	A polynomial equation of the form $P(x) = 0$ of degree n with complex coefficients has exactly n roots in the set of complex numbers.
Descartes' Rule of Signs	If $P(x)$ is a polynomial with real coefficients whose terms are arranged in descending powers of the variable, • the number of positive real zeros of $y = P(x)$ is the same as the number of changes in sign of the coefficients of the terms, or is less than this by an even number, and • the number of negative real zeros of $y = P(x)$ is the same as the number of changes in sign of the coefficients of the terms of $P(-x)$, or is less than this number by an even number.

Example 1 **Solve the equation $6x^3 + 3x = 0$ and state the number and type of roots.**

$6x^3 + 3x = 0$
$3x(2x^2 + 1) = 0$
Use the Zero Product Property.
$3x = 0$ or $2x^2 + 1 = 0$
$x = 0$ or $2x^2 = -1$
$x = \pm\frac{i\sqrt{2}}{2}$

The equation has one real root, 0, and two imaginary roots, $\pm\frac{i\sqrt{2}}{2}$.

Example 2 **State the number of positive real zeros, negative real zeros, and imaginary zeros for $p(x) = 4x^4 - 3x^3 + x^2 + 2x - 5$.**

Since $p(x)$ has degree 4, it has 4 zeros.
Use Descartes' Rule of Signs to determine the number and type of real zeros. Since there are three sign changes, there are 3 or 1 positive real zeros.
Find $p(-x)$ and count the number of changes in sign for its coefficients.
$p(-x) = 4(-x)^4 - 3(-x)^3 + (-x)^2 + 2(-x) - 5$
$= 4x^4 + 3x^3 + x^2 - 2x - 5$
Since there is one sign change, there is exactly 1 negative real zero.

Exercises

Solve each equation and state the number and type of roots.

1. $x^2 + 4x - 21 = 0$ **3, −7; 2 real**
2. $2x^3 - 50x = 0$ **0, ±5; 3 real**
3. $12x^3 + 100x = 0$ **$0, \pm\frac{5i\sqrt{3}}{3}$; 1 real, 2 imaginary**

State the number of positive real zeros, negative real zeros, and imaginary zeros for each function.

4. $f(x) = 3x^3 + x^2 - 8x - 12$ **1; 2 or 0; 0 or 2**
5. $f(x) = 2x^4 - x^3 - 3x + 7$ **2 or 0; 0; 2 or 4**
6. $f(x) = 3x^5 - x^4 - x^3 + 6x^2 - 5$ **3 or 1; 2 or 0; 0, 2, or 4**

Skills Practice, p. 401 and Practice, p. 402 (shown)

Solve each equation. State the number and type of roots.

1. $-9x - 15 = 0$ **$-\frac{5}{3}$; 1 real**
2. $x^4 - 5x^2 + 4 = 0$ **−1, 1, −2, 2; 4 real**
3. $x^5 = 81x$ **0, −3, 3, −3i, 3i; 3 real, 2 imaginary**
4. $x^3 + x^2 - 3x - 3 = 0$ **$-1, -\sqrt{3}, \sqrt{3}$; 3 real**
5. $x^3 + 6x + 20 = 0$ **$-2, 1 \pm 3i$; 1 real, 2 imaginary**
6. $x^4 - x^3 - x^2 - x - 2 = 0$ **2, −1, −i, i; 2 real, 2 imaginary**

State the possible number of positive real zeros, negative real zeros, and imaginary zeros of each function.

7. $f(x) = 4x^3 - 2x^2 + x + 3$ **2 or 0; 1; 2 or 0**
8. $p(x) = 2x^4 - 2x^3 + 2x^2 - x - 1$ **3 or 1; 1; 2 or 0**
9. $q(x) = 3x^4 + x^3 - 3x^2 + 7x + 5$ **2 or 0; 2 or 0; 4, 2, or 0**
10. $h(x) = 7x^4 + 3x^3 - 2x^2 - x + 1$ **2 or 0; 2 or 0; 4, 2, or 0**

Find all the zeros of each function.

11. $h(x) = 2x^3 + 3x^2 - 65x + 84$ **$-7, \frac{3}{2}, 4$**
12. $p(x) = x^3 - 3x^2 + 9x - 7$ **$1, 1 + i\sqrt{6}, 1 - i\sqrt{6}$**
13. $h(x) = x^3 - 7x^2 + 17x - 15$ **$3, 2 + i, 2 - i$**
14. $q(x) = x^4 + 50x^2 + 49$ **$-i, i, -7i, 7i$**
15. $g(x) = x^4 + 4x^3 - 3x^2 - 14x - 8$ **−1, −1, 2, −4**
16. $f(x) = x^4 - 6x^3 + 6x^2 + 24x - 40$ **$-2, 2, 3 - i, 3 + i$**

Write a polynomial function of least degree with integral coefficients that has the given zeros.

17. $-5, 3i$ **$f(x) = x^3 + 5x^2 + 9x + 45$**
18. $-2, 3 + i$ **$f(x) = x^3 - 4x^2 - 2x + 20$**
19. $-1, 4, 3i$ **$f(x) = x^4 - 3x^3 + 5x^2 - 27x - 36$**
20. $2, 5, 1 + i$ **$f(x) = x^4 - 9x^3 + 26x^2 - 34x + 20$**

21. **CRAFTS** Stephan has a set of plans to build a wooden box. He wants to reduce the volume of the box to 105 cubic inches. He would like to reduce the length of each dimension in the plan by the same amount. The plans call for the box to be 10 inches by 8 inches by 6 inches. Write and solve a polynomial equation to find out how much Stephen should take from each dimension. **$(10 - x)(8 - x)(6 - x) = 105$; 3 in.**

Reading to Learn Mathematics, p. 403 ELL

Pre-Activity **How can the roots of an equation be used in pharmacology?**

Read the introduction to Lesson 7-5 at the top of page 371 in your textbook.

Using the model given in the introduction, write a polynomial equation with 0 on one side that can be solved to find the time or times at which there is 100 milligrams of medication in a patient's bloodstream.
$0.5t^4 + 3.5t^3 - 100t^2 + 350t - 100 = 0$

Reading the Lesson

1. Indicate whether each statement is *true* or *false*.
   a. Every polynomial equation of degree greater than one has at least one root in the set of real numbers. **false**
   b. If c is a root of the polynomial equation $f(x) = 0$, then $(x - c)$ is a factor of the polynomial $f(x)$. **true**
   c. If $(x + c)$ is a factor of the polynomial $f(x)$, then c is a zero of the polynomial function f. **false**
   d. A polynomial function f of degree n has exactly $(n - 1)$ complex zeros. **false**
2. Let $f(x) = x^6 - 2x^5 + 3x^4 - 4x^3 + 5x^2 + 6x - 7$.
   a. What are the possible numbers of positive real zeros of f? **5, 3, or 1**
   b. Write $f(-x)$ in simplified form (with no parentheses).
   $x^6 + 2x^5 + 3x^4 + 4x^3 + 5x^2 - 6x - 7$
   What are the possible numbers of negative real zeros of f? **1**
   c. Complete the following chart to show the possible combinations of positive real zeros, negative real zeros, and imaginary zeros of the polynomial function f.

Number of Positive Real Zeros	Number of Negative Real Zeros	Number of Imaginary Zeros	Total Number of Zeros
5	1	0	6
3	1	2	6
1	1	4	6

Helping You Remember

3. It is easier to remember mathematical concepts and results if you relate them to each other. How can the Complex Conjugates Theorem help you remember the part of Descartes' Rule of Signs that says, "or is less than this number by an even number." **Sample answer: For a polynomial function in which the polynomial has real coefficients, imaginary zeros come in conjugate pairs. Therefore, there must be an even number of imaginary zeros. For each pair of imaginary zeros, the number of positive or negative zeros decreases by 2.**

376 Chapter 7 Polynomial Functions

Enrichment, p. 404

The Bisection Method for Approximating Real Zeros

The **bisection method** can be used to approximate zeros of polynomial functions like $f(x) = x^3 + x^2 - 3x - 3$.

Since $f(1) = -4$ and $f(2) = 3$, there is at least one real zero between 1 and 2. The midpoint of this interval is $\frac{1+2}{2} = 1.5$. Since $f(1.5) = -1.875$, the zero is between 1.5 and 2. The midpoint of this interval is $\frac{1.5+2}{2} = 1.75$. Since $f(1.75)$ is about 0.172, the zero is between 1.5 and 1.75. The midpoint of this interval is $\frac{1.5+1.75}{2} = 1.625$ and $f(1.625)$ is about -0.94. The zero is between 1.625 and 1.75. The midpoint of this interval is $\frac{1.625+1.75}{2} = 1.6875$. Since $f(1.6875)$ is about -0.41, the zero is between 1.6875 and 1.75. Therefore, the zero is 1.7 to the nearest tenth.

The diagram below summarizes the results obtained by the bisection method.

sign of $f(x)$: − − − − + +

Answers

35. $f(x) = x^3 - 2x^2 - 19x + 20$
36. $f(x) = x^4 - 10x^3 + 20x^2 + 40x - 96$
37. $f(x) = x^4 + 7x^2 - 144$
38. $f(x) = x^5 - x^4 + 13x^3 - 13x^2 + 36x - 36$
39. $f(x) = x^3 - 11x^2 + 23x - 45$
40. $f(x) = x^3 - 10x^2 + 32x - 48$

50. **CRITICAL THINKING** If a sixth-degree polynomial equation has exactly five distinct real roots, what can be said of one of its roots? Draw a graph of this situation. **One root is a double root; see margin for sample graph.**

51. **WRITING IN MATH** Answer the question that was posed at the beginning of the lesson. **See pp. 407A–407H.**

How can the roots of an equation be used in pharmacology?

Include the following items in your answer:
- an explanation of what the roots of this equation represent, and
- an explanation of what the roots of this equation reveal about how often a patient should take this medication.

Standardized Test Practice

52. The equation $x^4 - 1 = 0$ has exactly ___?___ complex root(s). **A**

(A) 4 (B) 0 (C) 2 (D) 1

53. How many negative real zeros does $f(x) = x^5 - 2x^4 - 4x^3 + 4x^2 - 5x + 6$ have? **C**

(A) 3 (B) 2 (C) 1 (D) 0

Maintain Your Skills

Mixed Review

Use synthetic substitution to find $f(-3)$ and $f(4)$ for each function. *(Lesson 7-4)*

54. $f(x) = x^3 - 5x^2 + 16x - 7$ **−127, 41**

55. $f(x) = x^4 + 11x^3 - 3x^2 + 2x - 5$ **−254, 915**

56. **RETAIL** The store Bunches of Boxes and Bags assembles boxes for mailing. The store manager found that the volume of a box made from a rectangular piece of cardboard with a square of length x inches cut from each corner is $4x^3 - 168x^2 + 1728x$ cubic inches. If the piece of cardboard is 48 inches long, what is the width? *(Lesson 7-3)* **36 in.**

Determine whether each function has a maximum or a minimum value. Then find the maximum or minimum value of each function. *(Lesson 6-1)*

57. $f(x) = x^2 - 8x + 3$ **min.; −13**

58. $f(x) = -3x^2 - 18x + 5$ **max.; 32**

59. $f(x) = -7 + 4x^2$ **min.; −7**

Factor completely. If the polynomial is not factorable, write *prime*. *(Lesson 5-4)*

60. $15a^2b^2 - 5ab^2c^2$ **$5ab^2(3a - c^2)$**

61. $12p^2 - 64p + 45$ **$(6p - 5)(2p - 9)$**

62. $4y^3 + 24y^2 + 36y$ **$4y(y + 3)^2$**

Use matrices A, B, C, and D to find the following. *(Lesson 4-2)*

$$A = \begin{bmatrix} -4 & 4 \\ 2 & -3 \\ 1 & 5 \end{bmatrix} \quad B = \begin{bmatrix} 7 & 0 \\ 4 & 1 \\ 6 & -2 \end{bmatrix} \quad C = \begin{bmatrix} -4 & -5 \\ -3 & 1 \\ 2 & 3 \end{bmatrix} \quad D = \begin{bmatrix} 1 & -2 \\ 1 & -1 \\ -3 & 4 \end{bmatrix}$$

63. $A + D$ 64. $B - C$ 65. $3B - 2A$

63. $\begin{bmatrix} -3 & 2 \\ 3 & -4 \\ -2 & 9 \end{bmatrix}$

64. $\begin{bmatrix} 11 & 5 \\ 7 & 0 \\ 4 & -5 \end{bmatrix}$

65. $\begin{bmatrix} 29 & -8 \\ 8 & 9 \\ 16 & -16 \end{bmatrix}$

66. Write an inequality for the graph at the right. *(Lesson 2-7)*

$y \geq -\frac{2}{3}x - 1$

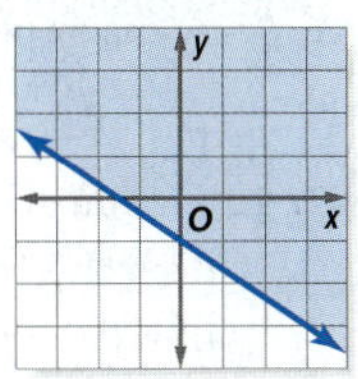

Getting Ready for the Next Lesson

67–70. See margin.

BASIC SKILL **Find all values of $\pm\frac{a}{b}$ given each replacement set.**

67. $a = \{1, 5\}; b = \{1, 2\}$

68. $a = \{1, 2\}; b = \{1, 2, 7, 14\}$

69. $a = \{1, 3\}; b = \{1, 3, 9\}$

70. $a = \{1, 2, 4\}; b = \{1, 2, 4, 8, 16\}$

4 Assess

Open-Ended Assessment

Writing Have students summarize the method for determining the number of real zeros for a polynomial function, and how to determine how many of them are positive, negative, and imaginary.

Getting Ready for Lesson 7-6

BASIC SKILL In Lesson 7-6, students will be introduced to the Rational Zero Theorem. They will factor integral coefficients in order to list every possible rational zero of a polynomial. Use Exercises 67–70 to determine your students' familiarity with finding all possible rational values of an expression $\pm\frac{a}{b}$, given all the possible values of a and b.

Assessment Options

Quiz (Lessons 7-4 and 7-5) is available on p. 443 of the *Chapter 7 Resource Masters*.

Mid-Chapter Test (Lessons 7-1 through 7-5) is available on p. 445 of the *Chapter 7 Resource Masters*.

Answers

47.

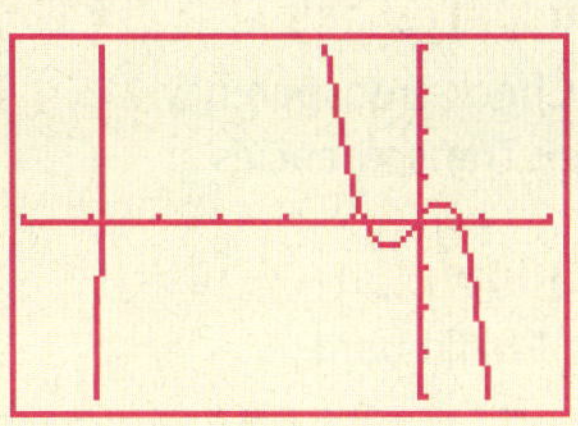

[−30, 10] scl: 5 by [−20, 20] scl: 5

50. Sample graph:

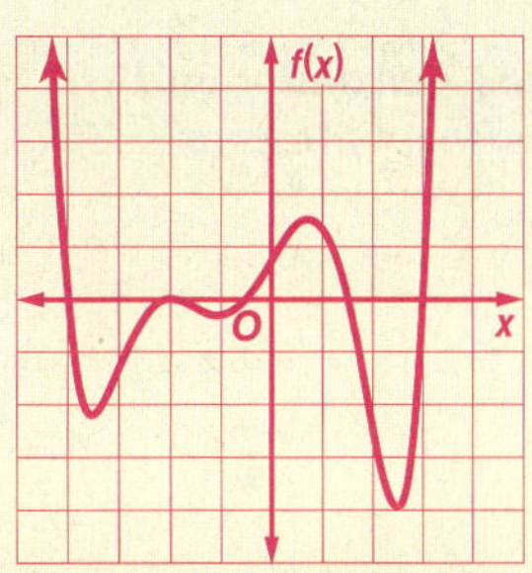

67. $\pm\frac{1}{2}, \pm1, \pm\frac{5}{2}, \pm5$

68. $\pm\frac{1}{14}, \pm\frac{1}{7}, \pm\frac{2}{7}, \pm\frac{1}{2}, \pm1, \pm2$

69. $\pm\frac{1}{9}, \pm\frac{1}{3}, \pm1, \pm3$

70. $\pm\frac{1}{16}, \pm\frac{1}{8}, \pm\frac{1}{4}, \pm\frac{1}{2}, \pm1, \pm2, \pm4$

Lesson 7-5 Roots and Zeros 377

7-6 Lesson Notes

1 Focus

5-Minute Check Transparency 7-6 Use as a quiz or review of Lesson 7-5.

Mathematical Background notes are available for this lesson on p. 344D.

How can the Rational Zero Theorem solve problems involving large numbers?

Ask students:

- Why does the polynomial equation contain an expression in which the variable h appears three times? **The actual width and length of the overhead compartment are not known; only their relationships to the height of the compartment are known.**
- How do you know that $h + 8$ is the length? **The length is 8 inches longer than the height, which is represented by the variable h.**
- Could the height be 5 inches? Explain. **No. If the height were 5 inches, then the width would be 0 inches.**

7-6 Rational Zero Theorem

What You'll Learn

- Identify the possible rational zeros of a polynomial function.
- Find all the rational zeros of a polynomial function.

How can the Rational Zero Theorem solve problems involving large numbers?

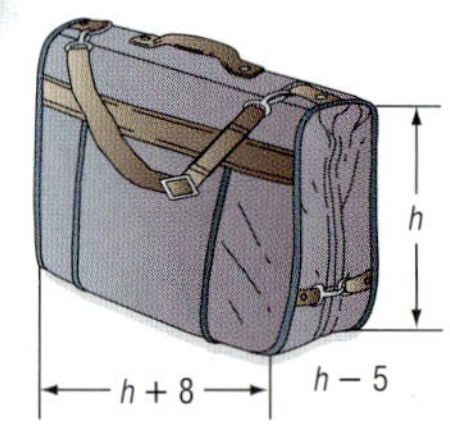

On an airplane, carry-on baggage must fit into the overhead compartment above the passenger's seat. The length of the compartment is 8 inches longer than the height, and the width is 5 inches shorter than the height. The volume of the compartment is 2772 cubic inches. You can solve the polynomial equation $h(h + 8)(h - 5) = 2772$, where h is the height, $h + 8$ is the length, and $h - 5$ is the width, to find the dimensions of the overhead compartment in which your luggage must fit.

IDENTIFY RATIONAL ZEROS Usually it is not practical to test all possible zeros of a polynomial function using only synthetic substitution. The **Rational Zero Theorem** can help you choose some possible zeros to test.

Key Concept — *Rational Zero Theorem*

- **Words** Let $f(x) = a_0x^n + a_1x^{n-1} + \ldots + a_{n-2}x^2 + a_{n-1}x + a_n$ represent a polynomial function with integral coefficients. If $\frac{p}{q}$ is a rational number in simplest form and is a zero of $y = f(x)$, then p is a factor of a_n and q is a factor of a_0.
- **Example** Let $f(x) = 2x^3 + 3x^2 - 17x + 12$. If $\frac{3}{2}$ is a zero of $f(x)$, then 3 is a factor of 12 and 2 is a factor of 2.

In addition, if the coefficient of the x term with the highest degree is 1, we have the following corollary.

Key Concept — *Corollary (Integral Zero Theorem)*

If the coefficients of a polynomial function are integers such that $a_0 = 1$ and $a_n \neq 0$, any rational zeros of the function must be factors of a_n.

Example 1 *Identify Possible Zeros*

List all of the possible rational zeros of each function.

a. $f(x) = 2x^3 - 11x^2 + 12x + 9$

If $\frac{p}{q}$ is a rational zero, then p is a factor of 9 and q is a factor of 2. The possible values of p are ±1, ±3, and ±9. The possible values for q are ±1 and ±2. So, $\frac{p}{q} = \pm1, \pm3, \pm9, \pm\frac{1}{2}, \pm\frac{3}{2}$, and $\pm\frac{9}{2}$.

Resource Manager

Workbook and Reproducible Masters

Chapter 7 Resource Masters
- Study Guide and Intervention, pp. 405–406
- Skills Practice, p. 407
- Practice, p. 408
- Reading to Learn Mathematics, p. 409
- Enrichment, p. 410

Graphing Calculator and Spreadsheet Masters, p. 39

Transparencies

5-Minute Check Transparency 7-6
Answer Key Transparencies

Technology

Interactive Chalkboard

b. $f(x) = x^3 - 9x^2 - x + 105$

Since the coefficient of x^3 is 1, the possible rational zeros must be a factor of the constant term 105. So, the possible rational zeros are the integers ±1, ±3, ±5, ±7, ±15, ±21, ±35, and ±105.

FIND RATIONAL ZEROS Once you have written the possible rational zeros, you can test each number using synthetic substitution.

Example 2 Use the Rational Zero Theorem

GEOMETRY The volume of a rectangular solid is 675 cubic centimeters. The width is 4 centimeters less than the height, and the length is 6 centimeters more than the height. Find the dimensions of the solid.

Let x = the height, $x - 4$ = the width, and $x + 6$ = the length.

Write an equation for the volume.

$x(x - 4)(x + 6) = 675$ Formula for volume

$x^3 + 2x^2 - 24x = 675$ Multiply.

$x^3 + 2x^2 - 24x - 675 = 0$ Subtract 675.

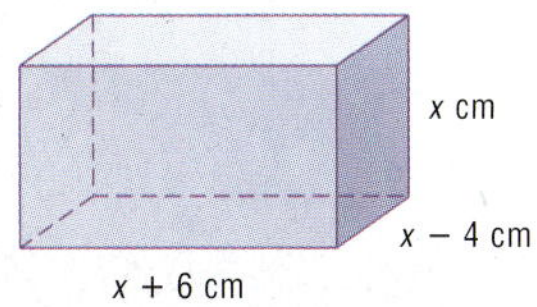

The leading coefficient is 1, so the possible integer zeros are factors of 675, ±1, ±3, ±5, ±9, ±15, ±25, ±27, ±45, ±75, ±135, ±225, and ±675. Since length can only be positive, we only need to check positive zeros. From Descartes' Rule of Signs, we also know there is only one positive real zero. Make a table and test possible real zeros.

p	1	2	−24	−675
1	1	3	−21	−696
3	1	5	−9	−702
5	1	7	11	−620
9	1	11	75	0

One zero is 9. Since there is only one positive real zero, we do not have to test the other numbers. The other dimensions are 9 − 4 or 5 centimeters and 9 + 6 or 15 centimeters.

CHECK Verify that the dimensions are correct. $5 \times 9 \times 15 = 675$ ✓

You usually do not need to test all of the possible zeros. Once you find a zero, you can try to factor the depressed polynomial to find any other zeros.

Example 3 Find All Zeros

Find all of the zeros of $f(x) = 2x^4 - 13x^3 + 23x^2 - 52x + 60$.

From the corollary to the Fundamental Theorem of Algebra, we know there are exactly 4 complex roots. According to Descartes' Rule of Signs, there are 4, 2, or 0 positive real roots and 0 negative real roots. The possible rational zeros are ±1, ±2, ±3, ±4, ±5, ±6, ±10, ±12, ±15, ±20, ±30, ±60, $\pm\frac{1}{2}$, $\pm\frac{3}{2}$, $\pm\frac{5}{2}$, and $\pm\frac{15}{2}$. Make a table and test some possible rational zeros.

$\frac{p}{q}$	2	−13	23	−52	60
1	2	−11	12	−40	20
2	2	−9	5	−42	−24
3	2	−7	2	−46	−78
5	2	−3	8	−12	0

(continued on the next page)

www.algebra2.com/extra_examples

Study Tip

Descartes' Rule of Signs
Examine the signs of the coefficients in the equation, + + − −. There is one change of sign, so there is only one positive real zero.

2 Teach

IDENTIFY RATIONAL ZEROS

In-Class Example Power Point®

Teaching Tip While discussing Example 1, point out that ±1 will always be possible rational zeros. Also make sure students clearly understand that these are just *possible* zeros. Until each potential zero has been tested by synthetic substitution, it should not be referred to as a zero.

1 List all of the possible rational zeros of each function.

a. $f(x) = 3x^4 - x^3 + 4$
$\pm1, \pm2, \pm4, \pm\frac{1}{3}, \pm\frac{2}{3}, \pm\frac{4}{3}$

b. $f(x) = x^4 + 7x^3 - 15$
$\pm1, \pm3, \pm5, \pm15$

FIND RATIONAL ZEROS

In-Class Examples Power Point®

2 **GEOMETRY** The volume of a rectangular solid is 1120 cubic feet. The width is 2 feet less than the height, and the length is 4 feet more than the height. Find the dimensions of the solid.

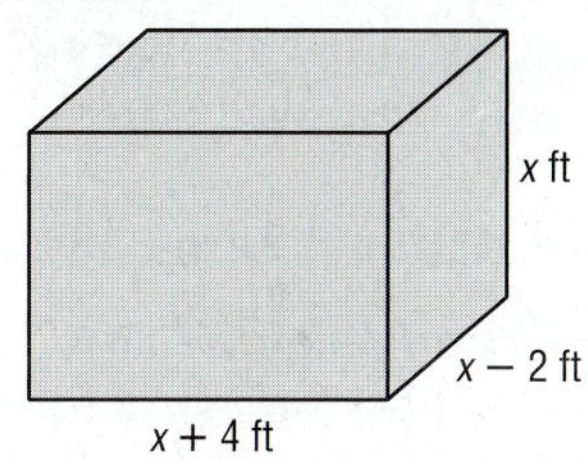

length: 14 ft, width: 8 ft, height: 10 ft

3 Find all of the zeros of $f(x) = x^4 + x^3 - 19x^2 + 11x + 30$.
−5, −1, 2, 3

DAILY INTERVENTION **Differentiated Instruction**

Logical Organize the students in groups of four or five. Have the students in each group split the work shown in Example 3 into four or five steps, depending on the size of their group. Each student then prepares and gives an explanation to the group of their part of the Example. In particular, students should explain any mathematical processes, what the result of their step is to be, and how the result relates to the next step in the process.

3 Practice/Apply

Study Notebook

Have students—

- add the definitions/examples of the vocabulary terms to their Vocabulary Builder worksheets for Chapter 7.
- write notes summarizing the Rational Zero Theorem and the Integral Zero Theorem.
- include any other item(s) that they find helpful in mastering the skills in this lesson.

DAILY INTERVENTION **FIND THE ERROR** Suggest students begin by assuming that *both* Lauren and Luis are incorrect. While students may spot Lauren's mistake quickly by looking at the possible integral zeros she has listed, ask students not to conclude that Luis must be correct without actually checking the possibilities he listed.

About the Exercises...

Organization by Objective

- **Identify Rational Zeros:** 12–17
- **Find Rational Zeros:** 18–33

Odd/Even Assignments

Exercises 12–33 are structured so that students practice the same concepts whether they are assigned odd or even problems.

Assignment Guide

Basic: 13–33 odd, 37, 38, 42–61

Average: 13–33 odd, 34–38, 42–61

Advanced: 12–32 even, 34–36, 39–55 (optional: 56–61)

All: Practice Quiz 2 (1–5)

Since $f(5) = 0$, you know that $x = 5$ is a zero. The depressed polynomial is $2x^3 - 3x^2 + 8x - 12$.

Factor $2x^3 - 3x^2 + 8x - 12$.

$2x^3 - 3x^2 + 8x - 12 = 0$	Write the depressed polynomial.
$2x^3 + 8x - 3x^2 - 12 = 0$	Regroup terms.
$2x(x^2 + 4) - 3(x^2 + 4) = 0$	Factor by grouping.
$(x^2 + 4)(2x - 3) = 0$	Distributive Property
$x^2 + 4 = 0$ or $2x - 3 = 0$	Zero Product Property
$x^2 = -4$ $\quad 2x = 3$	
$x = \pm 2i$ $\quad x = \frac{3}{2}$	

There is another real zero at $x = \frac{3}{2}$ and two imaginary zeros at $x = 2i$ and $x = -2i$.

The zeros of this function are $5, \frac{3}{2}, 2i$ and $-2i$.

Check for Understanding

Concept Check

1. **Explain** why it is useful to use the Rational Zero Theorem when finding the zeros of a polynomial function. **Sample answer: You limit the number of possible solutions.**
2. **OPEN ENDED** Write a polynomial function that has possible rational zeros of $\pm 1, \pm 3, \pm\frac{1}{2}, \pm\frac{3}{2}$. **Sample answer: $2x^2 - 8x + 3$**
3. **FIND THE ERROR** Lauren and Luis are listing the possible rational zeros of $f(x) = 4x^5 + 4x^4 - 3x^3 + 2x^2 - 5x + 6$.

Lauren	Luis
$\pm 1, \pm\frac{1}{2}, \pm\frac{1}{3}, \pm\frac{1}{6}, \pm 2, \pm\frac{2}{3}, \pm 4, \pm\frac{4}{3}$	$\pm 1, \pm\frac{1}{2}, \pm\frac{1}{4}, \pm 2, \pm 3, \pm\frac{3}{2}, \pm\frac{3}{4}, \pm 6$

Who is correct? Explain your reasoning. **Luis; Lauren found numbers in the form $\frac{q}{p}$, not $\frac{p}{q}$ as Luis did according to the Rational Zero Theorem.**

Guided Practice

GUIDED PRACTICE KEY	
Exercises	Examples
4, 5	1
6–9, 11	2
10	3

List all of the possible rational zeros of each function.

4. $p(x) = x^4 - 10$ **$\pm 1, \pm 2, \pm 5, \pm 10$**
5. $d(x) = 6x^3 + 6x^2 - 15x - 2$ **$\pm 1, \pm 2, \pm\frac{1}{2}, \pm\frac{1}{3}, \pm\frac{1}{6}, \pm\frac{2}{3}$**

Find all of the rational zeros of each function.

6. $p(x) = x^3 - 5x^2 - 22x + 56$ **$-4, 2, 7$**
7. $f(x) = x^3 - x^2 - 34x - 56$ **$-2, -4, 7$**
8. $t(x) = x^4 - 13x^2 + 36$ **$2, -2, 3, -3$**
9. $f(x) = 2x^3 - 7x^2 - 8x + 28$ **$-2, 2, \frac{7}{2}$**
10. Find all of the zeros of $f(x) = 6x^3 + 5x^2 - 9x + 2$. **$\frac{2}{3}, \frac{-3 \pm \sqrt{17}}{4}$**

Application

11. **GEOMETRY** The volume of the rectangular solid is 1430 cubic centimeters. Find the dimensions of the solid. **10 cm × 11 cm × 13 cm**

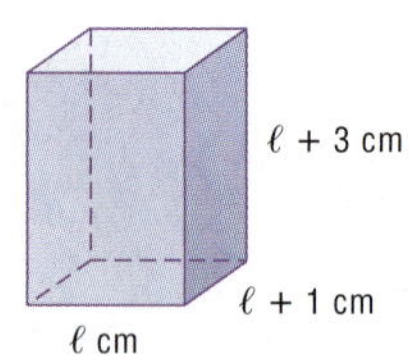

380 Chapter 7 Polynomial Functions

Practice and Apply

Homework Help

For Exercises	See Examples
12–17	1
18–29, 34–41	2
30–33	3

Extra Practice
See page 843.

14. $\pm1, \pm3, \pm5, \pm15, \pm\frac{1}{3}, \pm\frac{5}{3}$
15. $\pm1, \pm2, \pm3, \pm6, \pm9, \pm18$
16. $\pm1, \pm\frac{1}{3}, \pm3$
17. $\pm1, \pm\frac{1}{3}, \pm\frac{1}{9}, \pm3, \pm9, \pm27$

List all of the possible rational zeros of each function.

12. $f(x) = x^3 + 6x + 2$ **$\pm1, \pm2$**
13. $h(x) = x^3 + 8x + 6$ **$\pm1, \pm2, \pm3, \pm6$**
14. $f(x) = 3x^4 + 15$
15. $n(x) = x^5 + 6x^3 - 12x + 18$
16. $p(x) = 3x^3 - 5x^2 - 11x + 3$
17. $h(x) = 9x^6 - 5x^3 + 27$

Find all of the rational zeros of each function. **18. $-6, -5, 10$**

18. $f(x) = x^3 + x^2 - 80x - 300$
19. $p(x) = x^3 - 3x - 2$ **$-1, -1, 2$**
20. $h(x) = x^4 + x^2 - 2$ **$1, -1$**
21. $g(x) = x^4 - 3x^3 - 53x^2 - 9x$ **$0, 9$**
22. $f(x) = 2x^5 - x^4 - 2x + 1$ **$\frac{1}{2}, -1, 1$**
23. $f(x) = x^5 - 6x^3 + 8x$ **$0, 2, -2$**
24. $g(x) = x^4 - 3x^3 + x^2 - 3x$ **$0, 3$**
25. $p(x) = x^4 + 10x^3 + 33x^2 + 38x + 8$
26. $p(x) = x^3 + 3x^2 - 25x + 21$ **$-7, 1, 3$**
27. $h(x) = 6x^3 + 11x^2 - 3x - 2$
28. $h(x) = 10x^3 - 17x^2 - 7x + 2$
29. $g(x) = 48x^4 - 52x^3 + 13x - 3$

25. $-2, -4$ 27. $\frac{1}{2}, -\frac{1}{3}, -2$ 28. $-\frac{1}{2}, \frac{1}{5}, 2$ 29. $-\frac{1}{2}, \frac{1}{3}, \frac{1}{2}, \frac{3}{4}$

Find all of the zeros of each function. **30–33. See margin.**

30. $p(x) = 6x^4 + 22x^3 + 11x^2 - 38x - 40$
31. $g(x) = 5x^4 - 29x^3 + 55x^2 - 28x$
32. $h(x) = 9x^5 - 94x^3 + 27x^2 + 40x - 12$
33. $p(x) = x^5 - 2x^4 - 12x^3 - 12x^2 - 13x - 10$

FOOD For Exercises 34–36, use the following information. **35. $2, -3 \pm i\sqrt{3}$; 2**
Terri's Ice Cream Parlor makes gourmet ice cream cones. The volume of each cone is 8π cubic inches. The height is 4 inches more than the radius of the cone's opening.

34. Write a polynomial equation that represents the volume of an ice cream cone. Use the formula for the volume of a cone, $V = \frac{1}{3}\pi r^2 h$. **$V = \frac{1}{3}\pi r^3 + \frac{4}{3}\pi r^2$**
35. What are the possible values of r? Which of these values are reasonable?
36. Find the dimensions of the cone. **$r = 2$ in., $h = 6$ in.**

More About. . .

Food
The largest ice cream sundae, weighing 24.91 tons, was made in Edmonton, Alberta, in July 1988.
Source: *The Guinness Book of Records.*

AUTOMOBILES For Exercises 37 and 38, use the following information.
The length of the cargo space in a sport-utility vehicle is 4 inches greater than the height of the space. The width is sixteen inches less than twice the height. The cargo space has a total volume of 55,296 cubic inches.

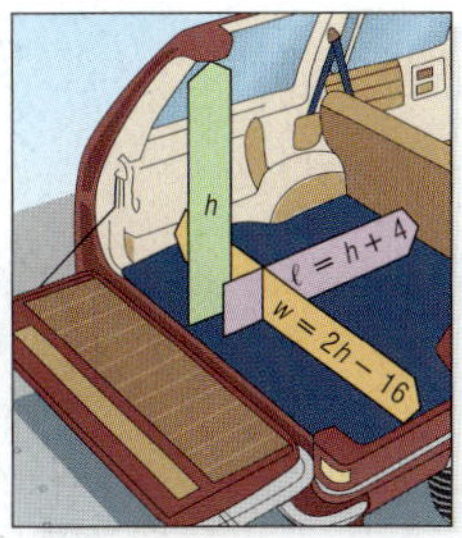

37. Write a polynomial function that represents the volume of the cargo space.
38. Find the dimensions of the cargo space.

37. $V = 2h^3 - 8h^2 - 64h$ 38. $\ell = 36$ in., $w = 48$ in., $h = 32$ in.

AMUSEMENT PARKS For Exercises 39–41, use the following information.
An amusement park owner wants to add a new wilderness water ride that includes a mountain that is shaped roughly like a pyramid. Before building the new attraction, engineers must build and test a scale model. **39. $V = \frac{1}{3}\ell^3 - 3\ell^2$**

39. If the height of the scale model is 9 inches less than its length and its base is a square, write a polynomial function that describes the volume of the model in terms of its length. Use the formula for the volume of a pyramid, $V = \frac{1}{3}Bh$.
40. If the volume of the model is 6300 cubic inches, write an equation for the situation.
41. What are the dimensions of the scale model? **$\ell = 30$ in., $w = 30$ in., $h = 21$ in.**

40. $6300 = \frac{1}{3}\ell^3 - 3\ell^2$

42. **CRITICAL THINKING** Suppose k and $2k$ are zeros of $f(x) = x^3 + 4x^2 + 9kx - 90$. Find k and all three zeros of $f(x)$. **$k = -3$; $-3, -6, 5$**

www.algebra2.com/self_check_quiz

Answers

30. $-2, \frac{4}{3}, \frac{-3 \pm i}{2}$
31. $\frac{4}{5}, 0, \frac{5 \pm i\sqrt{3}}{2}$
32. $3, \frac{2}{3}, -\frac{2}{3}, \frac{-3 \pm \sqrt{13}}{2}$
33. $-1, -2, 5, i, -i$

Enrichment, p. 410

Infinite Continued Fractions

Some infinite expressions are actually equal to real numbers! The infinite continued fraction at the right is one example.

$$x = 1 + \cfrac{1}{1 + \cfrac{1}{1 + \cfrac{1}{1 + \cfrac{1}{1 + \dots}}}}$$

If you use x to stand for the infinite fraction, then the entire denominator of the first fraction on the right is also equal to x. This observation leads to the following equation:

$$x = 1 + \frac{1}{x}$$

Write a decimal for each continued fraction.

1. $1 + \frac{1}{1}$ **2**
2. $1 + \cfrac{1}{1 + \frac{1}{1}}$ **1.5**
3. $1 + \cfrac{1}{1 + \cfrac{1}{1 + \frac{1}{1}}}$ **$1.66\overline{6}$**

Study Guide and Intervention, p. 405 (shown) and p. 406

Identify Rational Zeros

Rational Zero Theorem	Let $f(x) = a_0x^n + a_1x^{n-1} + \dots + a_{n-2}x^2 + a_{n-1}x + a^n$ represent a polynomial function with integral coefficients. If $\frac{p}{q}$ is a rational number in simplest form and is a zero of $y = f(x)$, then p is a factor of a_n and q is a factor of a_0.
Corollary (Integral Zero Theorem)	If the coefficients of a polynomial are integers such that $a_0 = 1$ and $a_n \neq 0$, any rational zeros of the function must be factors of a_n.

Example **List all of the possible rational zeros of each function.**

a. $f(x) = 3x^4 - 2x^2 + 6x - 10$
If $\frac{p}{q}$ is a rational root, then p is a factor of -10 and q is a factor of 3. The possible values for p are $\pm1, \pm2, \pm5$, and ±10. The possible values for q are ±1 and ±3. So all of the possible rational zeros are $\frac{p}{q} = \pm1, \pm2, \pm5, \pm10, \pm\frac{1}{3}, \pm\frac{2}{3}, \pm\frac{5}{3}$, and $\pm\frac{10}{3}$.

b. $q(x) = x^3 - 10x^2 + 14x - 36$
Since the coefficient of x^3 is 1, the possible rational zeros must be the factors of the constant term -36. So the possible rational zeros are $\pm1, \pm2, \pm3, \pm4, \pm6, \pm9, \pm12, \pm18$, and ±36.

Exercises

List all of the possible rational zeros of each function.

1. $f(x) = x^3 + 3x^2 - x + 8$ **$\pm1, \pm2, \pm4, \pm8$**
2. $g(x) = x^5 - 7x^4 + 3x^2 + x - 20$ **$\pm1, \pm2, \pm4, \pm5, \pm10, \pm20$**
3. $h(x) = x^4 - 7x^3 - 4x^2 + x - 49$ **$\pm1, \pm7, \pm49$**
4. $p(x) = 2x^4 - 5x^3 + 8x^2 + 3x - 5$ **$\pm1, \pm5, \pm\frac{1}{2}, \pm\frac{5}{2}$**
5. $q(x) = 3x^4 - 5x^3 + 10x + 12$ **$\pm1, \pm2, \pm3, \pm4, \pm6, \pm12, \pm\frac{1}{3}, \pm\frac{2}{3}, \pm\frac{4}{3}$**
6. $r(x) = 4x^5 - 2x + 18$ **$\pm1, \pm2, \pm3, \pm6, \pm9, \pm18, \pm\frac{1}{2}, \pm\frac{3}{2}, \pm\frac{9}{2}, \pm\frac{1}{4}, \pm\frac{3}{4}, \pm\frac{9}{4}$**
7. $f(x) = x^7 - 6x^5 - 3x^4 + x^3 + 4x^2 - 120$ **$\pm1, \pm2, \pm3, \pm4, \pm5, \pm6, \pm8, \pm10, \pm12, \pm15, \pm20, \pm24, \pm30, \pm40, \pm60, \pm120$**
8. $g(x) = 5x^6 - 3x^4 + 5x^3 + 2x^2 - 15$ **$\pm1, \pm3, \pm5, \pm15, \pm\frac{1}{5}, \pm\frac{3}{5}$**
9. $h(x) = 6x^5 - 3x^4 + 12x^3 + 18x^2 - 9x + 21$ **$\pm1, \pm3, \pm7, \pm21, \pm\frac{1}{2}, \pm\frac{3}{2}, \pm\frac{7}{2}, \pm\frac{21}{2}, \pm\frac{1}{3}, \pm\frac{7}{3}, \pm\frac{1}{6}, \pm\frac{7}{6}$**
10. $p(x) = 2x^7 - 3x^6 + 11x^5 - 20x^2 + 11$ **$\pm1, \pm11, \pm\frac{1}{2}, \pm\frac{11}{2}$**

Skills Practice, p. 407 and Practice, p. 408 (shown)

List all of the possible rational zeros of each function.

1. $h(x) = x^3 - 5x^2 + 2x + 12$ **$\pm1, \pm2, \pm3, \pm4, \pm6, \pm12$**
2. $s(x) = x^4 - 8x^3 + 7x - 14$ **$\pm1, \pm2, \pm7, \pm14$**
3. $f(x) = 3x^5 - 5x^2 + x + 6$ **$\pm\frac{1}{3}, \pm\frac{2}{3}, \pm1, \pm2, \pm3, \pm6$**
4. $p(x) = 3x^2 + x + 7$ **$\pm\frac{1}{3}, \pm\frac{7}{3}, \pm1, \pm7$**
5. $g(x) = 5x^3 + x^2 - x + 8$ **$\pm\frac{1}{5}, \pm\frac{2}{5}, \pm\frac{4}{5}, \pm\frac{8}{5}, \pm1, \pm2, \pm4, \pm8$**
6. $q(x) = 6x^5 + x^3 - 3$ **$\pm\frac{1}{6}, \pm\frac{1}{3}, \pm\frac{1}{2}, \pm\frac{3}{2}, \pm1, \pm3$**

Find all of the rational zeros of each function.

7. $q(x) = x^3 + 3x^2 - 6x - 8 = 0$ **$-4, -1, 2$**
8. $v(x) = x^3 - 9x^2 + 27x - 27$ **3**
9. $c(x) = x^3 - x^2 - 8x + 12$ **$-3, 2$**
10. $f(x) = x^4 - 49x^2$ **$0, -7, 7$**
11. $h(x) = x^3 - 7x^2 + 17x - 15$ **3**
12. $b(x) = x^3 + 6x + 20$ **-2**
13. $f(x) = x^3 - 6x^2 + 4x - 24$ **6**
14. $g(x) = 2x^3 + 3x^2 - 4x - 4$ **-2**
15. $h(x) = 2x^3 - 7x^2 - 21x + 54 = 0$ **$-3, 2, \frac{9}{2}$**
16. $z(x) = x^4 - 3x^3 + 5x^2 - 27x - 36$ **$-1, 4$**
17. $d(x) = x^4 + x^3 + 16$ **no rational zeros**
18. $n(x) = x^4 - 2x^3 - 3$ **-1**
19. $p(x) = 2x^4 - 7x^3 + 4x^2 + 7x - 6$ **$-1, 1, \frac{3}{2}, 2$**
20. $q(x) = 6x^4 + 29x^3 + 40x^2 + 7x - 12$ **$-\frac{3}{2}, -\frac{4}{3}$**

Find all of the zeros of each function.

21. $f(x) = 2x^4 + 7x^3 - 2x^2 - 19x - 12$ **$-1, -3, \frac{1 + \sqrt{33}}{4}, \frac{1 - \sqrt{33}}{4}$**
22. $q(x) = x^4 - 4x^3 + x^2 + 16x - 20$ **$-2, 2, 2 + i, 2 - i$**
23. $h(x) = x^6 - 8x^3$ **$0, 2, -1 + i\sqrt{3}, -1 - i\sqrt{3}$**
24. $g(x) = x^6 - 1$ **$-1, 1, \frac{-1 + i\sqrt{3}}{2}, \frac{-1 - i\sqrt{3}}{2}, \frac{1 + i\sqrt{3}}{2}, \frac{1 - i\sqrt{3}}{2}$**
25. **TRAVEL** The height of a box that Joan is shipping is 3 inches less than the width of the box. The length is 2 inches more than twice the width. The volume of the box is 1540 in^3. What are the dimensions of the box? **22 in. by 10 in. by 7 in.**
26. **GEOMETRY** The height of a square pyramid is 3 meters shorter than the side of its base. If the volume of the pyramid is 432 m^3, how tall is it? Use the formula $V = \frac{1}{3}Bh$. **9 m**

Reading to Learn Mathematics, p. 409 ELL

Pre-Activity **How can the Rational Zero Theorem solve problems involving large numbers?**

Read the introduction to Lesson 7-6 at the top of page 378 in your textbook.

Rewrite the polynomial equation $w(w + 8)(w - 5) = 2772$ in the form $f(x) = 0$, where $f(x)$ is a polynomial written in descending powers of x. **$w^3 + 3w^2 - 40w - 2772 = 0$**

Reading the Lesson

1. For each of the following polynomial functions, list all the possible values of p, all the possible values of q, and all the possible rational zeros $\frac{p}{q}$.
   a. $f(x) = x^3 - 2x^2 - 11x + 12$
   possible values of p: **$\pm1, \pm2, \pm3, \pm4, \pm6, \pm12$**
   possible values of q: **±1**
   possible values of $\frac{p}{q}$: **$\pm1, \pm2, \pm3, \pm4, \pm6, \pm12$**
   b. $f(x) = 2x^4 + 9x^3 - 23x^2 - 81x + 45$
   possible values of p: **$\pm1, \pm3, \pm5, \pm9, \pm15, \pm45$**
   possible values of q: **$\pm1, \pm2$**
   possible values of $\frac{p}{q}$: **$\pm1, \pm3, \pm5, \pm9, \pm15, \pm45, \pm\frac{1}{2}, \pm\frac{3}{2}, \pm\frac{5}{2}, \pm\frac{9}{2}, \pm\frac{15}{2}, \pm\frac{45}{2}$**
2. Explain in your own words how Descartes' Rule of Signs, the Rational Zero Theorem, and synthetic division can be used together to find all of the rational zeros of a polynomial function with integer coefficients.
   Sample answer: Use Descartes' Rule to find the possible numbers of positive and negative real zeros. Use the Rational Zero Theorem to list all possible rational zeros. Use synthetic division to test which of the numbers on the list of possible rational zeros are actually zeros of the polynomial function. (Descartes' Rule may help you to limit the possibilities.)

Helping You Remember

3. Some students have trouble remembering which numbers go in the numerators and which go in the denominators when forming a list of possible rational zeros of a polynomial function. How can you use the linear polynomial equation $ax + b = 0$, where a and b are nonzero integers, to remember this?
   Sample answer: The solution of the equation is $-\frac{b}{a}$. The numerator b is a factor of the constant term in $ax + b$. The denominator a is a factor of the leading coefficient in $ax + b$.

4 Assess

Open-Ended Assessment

Speaking Have students explain how to find all the possible rational zeros of a polynomial function. Ask them to demonstrate the technique using a polynomial function of degree 3 or higher while explaining the process.

Getting Ready for Lesson 7-7

PREREQUISITE SKILL Students will perform arithmetic operations on functions and find the composition of functions in Lesson 7-7. These skills will rely on students' ability to correctly perform operations on polynomials. Use Exercises 56–61 to determine your students' familiarity with performing operations with polynomials.

Assessment Options

Practice Quiz 2 The quiz provides students with a brief review of the concepts and skills in Lessons 7-4 through 7-6. Lesson numbers are given to the right of the exercises or instruction lines so students can review concepts not yet mastered.

Answer

43. The Rational Zero Theorem helps factor large numbers by eliminating some possible zeros because it is not practical to test all of them using synthetic substitution. Answers should include the following.
   - The polynomial equation that represents the volume of the compartment is $V = w^3 + 3w^2 - 40w$.
   - Reasonable measures of the width of the compartment are, in inches, 1, 2, 3, 4, 6, 7, 9, 12, 14, 18, 21, 22, 28, 33, 36, 42, 44, 63, 66, 77, and 84. The solution shows that $w = 14$ in., $\ell = 22$ in., and $d = 9$ in.

43. **WRITING IN MATH** Answer the question that was posed at the beginning of the lesson. See margin.

How can the Rational Zero Theorem solve problems involving large numbers?

Include the following items in your answer:
- the polynomial equation that represents the volume of the compartment, and
- a list of all reasonable measures of the width of the compartment, assuming that the width is a whole number.

Standardized Test Practice Ⓐ Ⓑ Ⓒ Ⓓ

44. Using the Rational Zero Theorem, determine which of the following is a zero of the function $f(x) = 12x^5 - 5x^3 + 2x - 9$. **D**

Ⓐ -6 Ⓑ $\frac{3}{8}$ Ⓒ $-\frac{2}{3}$ Ⓓ 1

45. **OPEN ENDED** Write a polynomial with $-5, -2, 1, 3$, and 4 as roots.
Sample answer: $x^5 - x^4 - 27x^3 + 41x^2 + 106x - 120$

Maintain Your Skills

Mixed Review **Given a function and one of its zeros, find all of the zeros of the function.** *(Lesson 7-5)*

46. $g(x) = x^3 + 4x^2 - 27x - 90; -3$ — $-6, -3, 5$
47. $h(x) = x^3 - 11x + 20; 2 + i$ — $-4, 2 + i, 2 - i$
48. $f(x) = x^3 + 5x^2 + 9x + 45; -5$ — $-5, 3i, -3i$
49. $g(x) = x^3 - 3x^2 - 41x + 203; -7$ — $-7, 5 + 2i, 5 - 2i$

Given a polynomial and one of its factors, find the remaining factors of the polynomial. Some factors may not be binomials. *(Lesson 7-4)*

50. $20x^3 - 29x^2 - 25x + 6; x - 2$ — $4x + 3, 5x - 1$
51. $3x^4 - 21x^3 + 38x^2 - 14x + 24; x - 3$ — $x - 4, 3x^2 + 2$

Simplify. *(Lesson 5-5)*

52. $\sqrt{245}$ — $7\sqrt{5}$
53. $\pm\sqrt{18x^3y^2}$ — $\pm 3xy\sqrt{2x}$
54. $\sqrt{16x^2 - 40x + 25}$ — $|4x - 5|$

55. **GEOMETRY** The perimeter of a right triangle is 24 centimeters. Three times the length of the longer leg minus two times the length of the shorter leg exceeds the hypotenuse by 2 centimeters. What are the lengths of all three sides? *(Lesson 3-5)* 6 cm, 8 cm, 10 cm

Getting Ready for the Next Lesson **PREREQUISITE SKILL** **Simplify.**
*(To review **operations with polynomials**, see Lessons 5-2 and 5-3.)*

56. $(x^2 - 7) + (x^3 + 3x^2 + 1)$
57. $(8x^2 - 3x) - (4x^2 + 5x - 3)$
58. $(x + 2)(x^2 + 3x - 5)$
59. $(x^3 + 3x^2 - 3x + 1)(x - 5)^2$
60. $(x^2 - 2x - 30) \div (x + 7)$
61. $(x^3 + 2x^2 - 3x + 1) \div (x + 1)$

56. $x^3 + 4x^2 - 6$
57. $4x^2 - 8x + 3$
58. $x^3 + 5x^2 + x - 10$
59. $x^5 - 7x^4 - 8x^3 + 106x^2 - 85x + 25$
60. $x - 9 + \frac{33}{x + 7}$
61. $x^2 + x - 4 + \frac{5}{x + 1}$

Practice Quiz 2 — *Lessons 7-4 through 7-6*

Use synthetic substitution to find $f(-2)$ and $f(3)$ for each function. *(Lesson 7-4)*

1. $f(x) = 7x^5 - 25x^4 + 17x^3 - 32x^2 + 10x - 22$ $-930, -145$
2. $f(x) = 3x^4 - 12x^3 - 21x^2 + 30x$ $0, -180$
3. Write the polynomial equation of degree 4 with leading coefficient 1 that has roots at $-2, -1, 3$, and 4. *(Lesson 7-5)* $x^4 - 4x^3 - 7x^2 + 22x + 24 = 0$

Find all of the rational zeros of each function. *(Lesson 7-6)*

4. $f(x) = 5x^3 - 29x^2 + 55x - 28$ $\frac{4}{5}$
5. $g(x) = 4x^3 + 16x^2 - x - 24$ $-\frac{3}{2}$

7-7 Operations on Functions

What You'll Learn

- Find the sum, difference, product, and quotient of functions.
- Find the composition of functions.

Vocabulary

- composition of functions

Why is it important to combine functions in business?

Carol Coffmon owns a garden store where she sells birdhouses. The revenue from the sale of the birdhouses is given by $r(x) = 125x$. The function for the cost of making the birdhouses is given by $c(x) = 65x + 5400$. Her profit p is the revenue minus the cost or $p = r - c$. So the profit function $p(x)$ can be defined as $p(x) = (r - c)(x)$. If you have two functions, you can form a new function by performing arithmetic operations on them.

ARITHMETIC OPERATIONS Let $f(x)$ and $g(x)$ be any two functions. You can add, subtract, multiply, and divide functions according to the following rules.

Key Concept — *Operations with Function*

Operation	Definition	Examples if $f(x) = x + 2$, $g(x) = 3x$
Sum	$(f + g)(x) = f(x) + g(x)$	$(x + 2) + 3x = 4x + 2$
Difference	$(f - g)(x) = f(x) - g(x)$	$(x + 2) - 3x = -2x + 2$
Product	$(f \cdot g)(x) = f(x) \cdot g(x)$	$(x + 2)3x = 3x^2 + 6x$
Quotient	$\left(\frac{f}{g}\right)(x) = \frac{f(x)}{g(x)}, g(x) \neq 0$	$\frac{x + 2}{3x}$

Example 1 Add and Subtract Functions

Given $f(x) = x^2 - 3x + 1$ and $g(x) = 4x + 5$, find each function.

a. $(f + g)(x)$

$$(f + g)(x) = f(x) + g(x) \quad \text{Addition of functions}$$
$$= (x^2 - 3x + 1) + (4x + 5) \quad f(x) = x^2 - 3x + 1 \text{ and } g(x) = 4x + 5$$
$$= x^2 + x + 6 \quad \text{Simplify.}$$

b. $(f - g)(x)$

$$(f - g)(x) = f(x) - g(x) \quad \text{Subtraction of functions}$$
$$= (x^2 - 3x + 1) - (4x + 5) \quad f(x) = x^2 - 3x + 1 \text{ and } g(x) = 4x + 5$$
$$= x^2 - 7x - 4 \quad \text{Simplify.}$$

Notice that the functions f and g have the same domain of all real numbers. The functions $f + g$ and $f - g$ also have domains that include all real numbers. For each new function, the domain consists of the intersection of the domains of $f(x)$ and $g(x)$. The domain of the quotient function is further restricted by excluded values that make the denominator equal to zero.

7-7 Lesson Notes

1 Focus

5-Minute Check Transparency 7-7 Use as a quiz or review of Lesson 7-6.

Mathematical Background notes are available for this lesson on p. 344D.

Why is it important to combine functions in business?

Ask students:

- What is the difference between revenue and profit? **Revenue is the amount of money generated by the sales of the birdhouses. Profit is the amount that revenue exceeds the cost (or expenses) of producing the birdhouses.**
- Is profit always a positive value? Explain. **No; profit is negative when expenses exceed revenue, in which case it is called *loss*.**

2 Teach

ARITHMETIC OPERATIONS

In-Class Example Power Point®

1 Given $f(x) = 3x^2 + 7x$ and $g(x) = 2x^2 - x - 1$, find each function.

a. $(f + g)(x)$ **$5x^2 + 6x - 1$**

b. $(f - g)(x)$ **$x^2 + 8x + 1$**

Resource Manager

Workbook and Reproducible Masters

Chapter 7 Resource Masters
- Study Guide and Intervention, pp. 411–412
- Skills Practice, p. 413
- Practice, p. 414
- Reading to Learn Mathematics, p. 415
- Enrichment, p. 416
- Assessment, p. 444

Graphing Calculator and Spreadsheet Masters, p. 40
School-to-Career Masters, p. 13
Teaching Algebra With Manipulatives Masters, pp. 253–255

Transparencies

5-Minute Check Transparency 7-7
Real-World Transparency 7
Answer Key Transparencies

Technology

Interactive Chalkboard

In-Class Example

Power Point®

2 Given $f(x) = 3x^2 - 2x + 1$ and $g(x) = x - 4$, find each function.

a. $(f \cdot g)(x)$ $\mathbf{3x^3 - 14x^2 + 9x - 4}$

b. $\left(\frac{f}{g}\right)(x)$ $\mathbf{\frac{3x^2 - 2x + 1}{x - 4}, x \neq 4}$

COMPOSITION OF FUNCTIONS

Tips for New Teachers

Intervention Some students may read $f \circ g$ as the word *fog*. Listen for students making this verbal error. Stress that students must learn to read this correctly because the correct wording will help them understand the meaning. Lead students to understand the similarity in meaning between $f(x)$ (read "*f* of *x*") and $f \circ g$ (read "*f* of *g* of *x*"). Reinforce this understanding by showing how $f \circ g$ can also be written as $f[g(x)]$. You can also relate $f[g(x)]$ to an expression containing nested parentheses, such as $(1 + (3 \cdot 5(4)))$ in which the expressions in parentheses are evaluated from the innermost parentheses to the outermost.

Example 2 Multiply and Divide Functions

Given $f(x) = x^2 + 5x - 1$ and $g(x) = 3x - 2$, find each function.

a. $(f \cdot g)(x)$

$(f \cdot g)(x) = f(x) \cdot g(x)$	Product of functions
$= (x^2 + 5x - 1)(3x - 2)$	$f(x) = x^2 + 5x - 1$ and $g(x) = 3x - 2$
$= x^2(3x - 2) + 5x(3x - 2) - 1(3x - 2)$	Distributive Property
$= 3x^3 - 2x^2 + 15x^2 - 10x - 3x + 2$	Distributive Property
$= 3x^3 + 13x^2 - 13x + 2$	Simplify.

b. $\left(\frac{f}{g}\right)(x)$

$\left(\frac{f}{g}\right)(x) = \frac{f(x)}{g(x)}$	Division of functions
$= \frac{x^2 + 5x - 1}{3x - 2}, x \neq \frac{2}{3}$	$f(x) = x^2 + 5x - 1$ and $g(x) = 3x - 2$

Because $x = \frac{2}{3}$ makes $3x - 2 = 0$, $\frac{2}{3}$ is excluded from the domain of $\left(\frac{f}{g}\right)(x)$.

Study Tip

Reading Math
$[f \circ g](x)$ and $f[g(x)]$ are both read *f of g of x*.

COMPOSITION OF FUNCTIONS Functions can also be combined using **composition of functions**. In a composition, a function is performed, and then a second function is performed on the result of the first function. The composition of *f* and *g* is denoted by $f \circ g$.

Key Concept — *Composition of Functions*

Suppose *f* and *g* are functions such that the range of *g* is a subset of the domain of *f*. Then the composite function $f \circ g$ can be described by the equation

$$[f \circ g](x) = f[g(x)].$$

The composition of functions can be shown by mappings. Suppose $f = \{(3, 4), (2, 3), (-5, 0)\}$ and $g = \{(3, -5), (4, 3), (0, 2)\}$. The composition of these functions is shown below.

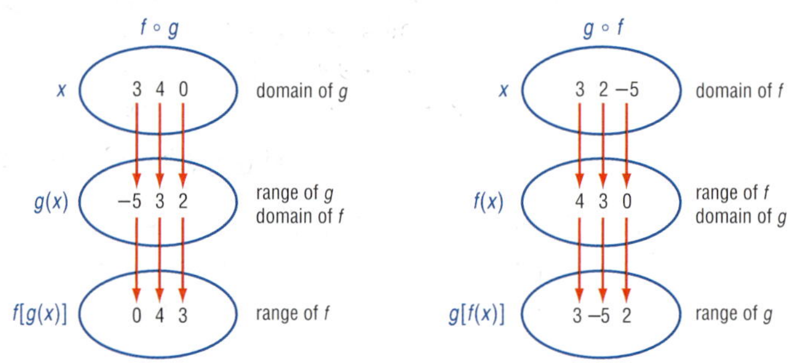

$f \circ g = \{(3, 0), (4, 4), (0, 3)\}$ $\quad$ $g \circ f = \{(3, 3), (2, -5), (-5, 2)\}$

The composition of two functions may not exist. Given two functions *f* and *g*, $[f \circ g](x)$ is defined only if the range of $g(x)$ is a subset of the domain of $f(x)$. Similarly, $[g \circ f](x)$ is defined only if the range of $f(x)$ is a subset of the domain of $g(x)$.

384 Chapter 7 Polynomial Functions

Example 3 Evaluate Composition of Relations

If $f(x) = \{(7, 8), (5, 3), (9, 8), (11, 4)\}$ and $g(x) = \{(5, 7), (3, 5), (7, 9), (9, 11)\}$, find $f \circ g$ and $g \circ f$.

To find $f \circ g$, evaluate $g(x)$ first. Then use the range of g as the domain of f and evaluate $f(x)$.

$f[g(5)] = f(7)$ or 8 $\quad g(5) = 7$
$f[g(3)] = f(5)$ or 3 $\quad g(3) = 5$
$f[g(7)] = f(9)$ or 8 $\quad g(7) = 9$
$f[g(9)] = f(11)$ or 4 $\quad g(9) = 11$

$f \circ g = \{(5, 8), (3, 3), (7, 8), (9, 4)\}$

To find $g \circ f$, evaluate $f(x)$ first. Then use the range of f as the domain of g and evaluate $g(x)$.

$g[f(7)] = g(8)$ $\quad$ $g(8)$ is undefined.
$g[f(5)] = g(3)$ or 5 $\quad f(5) = 3$
$g[f(9)] = g(8)$ $\quad$ $g(8)$ is undefined.
$g[f(11)] = g(4)$ $\quad$ $g(4)$ is undefined.

Since 8 and 4 are not in the domain of g, $g \circ f$ is undefined for $x = 7$, $x = 9$, and $x = 11$. However, $g[f(5)] = 5$ so $g \circ f = \{(5, 5)\}$.

Notice that in most instances $f \circ g \neq g \circ f$. Therefore, the order in which you compose two functions is very important.

Example 4 Simplify Composition of Functions

a. Find $[f \circ g](x)$ and $[g \circ f](x)$ for $f(x) = x + 3$ and $g(x) = x^2 + x - 1$.

$[f \circ g](x) = f[g(x)]$ $\quad$ Composition of functions
$= f(x^2 + x - 1)$ $\quad$ Replace $g(x)$ with $x^2 + x - 1$.
$= (x^2 + x - 1) + 3$ $\quad$ Substitute $x^2 + x - 1$ for x in $f(x)$.
$= x^2 + x + 2$ $\quad$ Simplify.

$[g \circ f](x) = g[f(x)]$ $\quad$ Composition of functions
$= g(x + 3)$ $\quad$ Replace $f(x)$ with $x + 3$.
$= (x + 3)^2 + (x + 3) - 1$ $\quad$ Substitute $x + 3$ for x in $g(x)$.
$= x^2 + 6x + 9 + x + 3 - 1$ $\quad$ Evaluate $(x + 3)^2$.
$= x^2 + 7x + 11$ $\quad$ Simplify.

So, $[f \circ g](x) = x^2 + x + 2$ and $[g \circ f](x) = x^2 + 7x + 11$.

b. Evaluate $[f \circ g](x)$ and $[g \circ f](x)$ for $x = 2$.

$[f \circ g](x) = x^2 + x + 2$ $\quad$ Function from part a
$[f \circ g](2) = (2)^2 + 2 + 2$ $\quad$ Replace x with 2.
$= 8$ $\quad$ Simplify.

$[g \circ f](x) = x^2 + 7x + 11$ $\quad$ Function from part a
$[g \circ f](2) = (2)^2 + 7(2) + 11$ $\quad$ Replace x with 2.
$= 29$ $\quad$ Simplify.

So, $[f \circ g](2) = 8$ and $[g \circ f](2) = 29$.

www.algebra2.com/extra_examples

In-Class Examples

Power Point®

3 If $f(x) = \{(2, 6), (9, 4), (7, 7), (0, -1)\}$ and $g(x) = \{(7, 0), (-1, 7), (4, 9), (8, 2)\}$, find $f \circ g$ and $g \circ f$. **$f \circ g = \{(7, -1), (-1, 7), (4, 4), (8, 6)\}$; $g \circ f = \{(9, 9), (7, 0), (0, 7)\}$**

4

a. Find $[f \circ g](x)$ and $[g \circ f](x)$ for $f(x) = 3x^2 - x + 4$ and $g(x) = 2x - 1$. **$[f \circ g](x) = 12x^2 - 14x + 8$; $[g \circ f](x) = 6x^2 - 2x + 7$**

b. Evaluate $[f \circ g](x)$ and $[g \circ f](x)$ for $x = -2$. **$[f \circ g](-2) = 84$; $[g \circ f](-2) = 35$**

DAILY INTERVENTION — Differentiated Instruction

Naturalist Invite students to think of events in nature whose occurrence students can relate to as being similar to the order in which the composition of functions must be carried out. Students interested in science might think of the stages of metamorphosis in insects, during which a larva changes to a pupa and finally to an adult. Each stage is dependent on the one before and the order of the stages is fixed.

In-Class Example

Power Point®

5 **TAXES** Tracie Long has \$100 deducted from every paycheck for retirement. She can have this deduction taken before state taxes are applied, which reduces her taxable income. Her state income tax rate is 4%. If Tracie earns \$1500 every pay period, find the difference in her net income if she has the retirement deduction taken before or after state taxes. **Her net pay is \$4 more by having her retirement deduction taken before state taxes.**

3 Practice/Apply

Study Notebook

Have students—

- *add the definitions/examples of the vocabulary terms to their Vocabulary Builder worksheets for Chapter 7.*
- *include any other item(s) that they find helpful in mastering the skills in this lesson.*

DAILY INTERVENTION

FIND THE ERROR Point out that the order in which the functions are applied is related to the proximity of the function name to the variable x. Since f is closer to x here, it is the first function to be evaluated. Invite students to suggest other ways they can avoid confusing $f \circ g$ with $g \circ f$.

Example 5 Use Composition of Functions

TAXES **Tyrone Davis has \$180 deducted from every paycheck for retirement. He can have these deductions taken before taxes are applied, which reduces his taxable income. His federal income tax rate is 18%. If Tyrone earns \$2200 every pay period, find the difference in his net income if he has the retirement deduction taken before taxes or after taxes.**

Study Tip

Combining Functions
By combining functions, you can make the evaluation of the functions more efficient.

Explore Let x = Tyrone's income per paycheck, $r(x)$ = his income after the deduction for retirement, and $t(x)$ = his income after the deduction for federal income tax.

Plan Write equations for $r(x)$ and $t(x)$.
\$180 is deducted from every paycheck for retirement: $r(x) = x - 180$.

Tyrone's tax rate is 18%: $t(x) = x - 0.18x$.

Solve If Tyrone has his retirement deducted *before* taxes, then his net income is represented by $[t \circ r](2200)$.

$$\begin{aligned}[t \circ r](2200) &= t(2200 - 180) && \text{Replace } x \text{ with 2200 in } r(x) = x - 180.\\ &= t(2020)\\ &= 2020 - 0.18(2020) && \text{Replace } x \text{ with 2020 in } t(x) = x - 0.18x.\\ &= 1656.40\end{aligned}$$

If Tyrone has his retirement deducted *after* taxes, then his net income is represented by $[r \circ t](2200)$.

$$\begin{aligned}[r \circ t](2200) &= r[2200 - 0.18(2200)] && \text{Replace } x \text{ with 2200 in } t(x) = x - 0.18x.\\ &= r(1804)\\ &= 1804 - 180 && \text{Replace } x \text{ with 1804 in } r(x) = x - 180.\\ &= 1624\end{aligned}$$

$[t \circ r](2200) = 1656.40$ and $[r \circ t](2200) = 1624$. The difference is \$1656.40 − \$1624 or \$32.40. So, his net pay is \$32.40 more by having his retirement deducted before taxes.

Examine The answer makes sense. Since the taxes are being applied to a smaller amount, less taxes will be deducted from his paycheck.

Check for Understanding

Concept Check

1. **Determine** whether the following statement is *always*, *sometimes*, or *never* true. Support your answer with an example.
Given two functions f and g, $f \circ g = g \circ f$.
2. **OPEN ENDED** Write a set of ordered pairs for functions f and g, given that $f \circ g = \{(4, 3), (-1, 9), (-2, 7)\}$.
3. **FIND THE ERROR** Danette and Marquan are finding $[g \circ f](3)$ for $f(x) = x^2 + 4x + 5$ and $g(x) = x - 7$. Who is correct? Explain your reasoning. **See margin.**

Danette

$$\begin{aligned}[g \circ f](3) &= g[(3)^2 + 4(3) + 5]\\ &= g(26)\\ &= 26 - 7\\ &= 19\end{aligned}$$

Marquan

$$\begin{aligned}[g \circ f](3) &= f(3 - 7)\\ &= f(-4)\\ &= (-4)^2 + 4(-4) + 5\\ &= 5\end{aligned}$$

1. Sometimes; sample answer: If $f(x) = x - 2$, $g(x) = x + 8$, then $f \circ g = x + 6$ and $g \circ f = x + 6$.

2. Sample answer: $g(x) = \{(-2, 1), (-1, 2), (4, 3)\}$, $f(x) = \{(1, 7), (2, 9), (3, 3)\}$

Answer

3. Danette; $[g \circ f](x) = g[f(x)]$ means to evaluate the f function first and then the g function. Marquan evaluated the functions in the wrong order.

386 Chapter 7 Polynomial Functions

Guided Practice

GUIDED PRACTICE KEY

Exercises	Examples
4–7	1, 2
8, 9	3
10–14	4
15, 16	5

8–9. See students' work for justification.

Find $(f+g)(x)$, $(f-g)(x)$, $(f \cdot g)(x)$, and $\left(\frac{f}{g}\right)(x)$ for each $f(x)$ and $g(x)$. Justify each step. **4–5. See margin. See Students' work for justification.**

4. $f(x) = 3x + 4$
   $g(x) = 5 + x$
5. $f(x) = x^2 + 3$
   $g(x) = x - 4$

For each set of ordered pairs, find $f \circ g$ and $g \circ f$, if they exist.

6. $f = \{(-1, 9), (4, 7)\}$
   $g = \{(-5, 4), (7, 12), (4, -1)\}$
   $\{(-5, 7), (4, 9)\}$; $\{(4, 12)\}$
7. $f = \{(0, -7), (1, 2), (2, -1)\}$
   $g = \{(-1, 10), (2, 0)\}$
   $\{(2, -7)\}$; $\{(1, 0), (2, 10)\}$

Find $[g \circ h](x)$ and $[h \circ g](x)$. Justify each step.

8. $g(x) = 2x$
   $h(x) = 3x - 4$ **$6x - 8$; $6x - 4$**
9. $g(x) = x + 5$ **$x^2 + 11$; $x^2 + 10x + 31$**
   $h(x) = x^2 + 6$

If $f(x) = 3x$, $g(x) = x + 7$, and $h(x) = x^2$, find each value.

10. $f[g(3)]$ **30**
11. $g[h(-2)]$ **11**
12. $h[h(1)]$ **1**

Application

14. $32.49; price of CD when 25% discount is taken and then the coupon is subtracted

15. $33.74; price of CD when coupon is subtracted and then 25% discount is taken

SHOPPING For Exercises 13–16, use the following information.
Mai-Lin is shopping for computer software. She finds a CD-ROM program that costs $49.99, but is on sale at a 25% discount. She also has a $5 coupon for the product.

13. Express the price of the CD after the discount and the price of the CD after the coupon using function notation. Let x represent the price of the CD, $p(x)$ represent the price after the 25% discount, and $c(x)$ represent the price after the coupon. **$p(x) = \frac{3}{4}x$; $c(x) = x - 5$**
14. Find $c[p(x)]$ and explain what this value represents.
15. Find $p[c(x)]$ and explain what this value represents.
16. Which method results in the lower sale price? Explain your reasoning. **See margin.**

Practice and Apply

Homework Help

For Exercises	See Examples
17–22	1, 2
23–28	3
29–46	4
47–55	5

Extra Practice
See page 844.

23. $\{(1, -3), (-3, 1), (2, 1)\}$; $\{(1, 0), (0, 1)\}$

24. $\{(2, 4), (4, 4)\}$; $\{(1, 5), (3, 3), (5, 3)\}$

25. $\{(0, 0), (8, 3), (3, 3)\}$; $\{(3, 6), (4, 4), (6, 6), (7, 8)\}$

26. $\{(4, 5), (2, 5), (6, 12), (8, 12)\}$; does not exist

34. $2x^2 - 5x + 9$; $2x^2 - x + 5$

Find $(f+g)(x)$, $(f-g)(x)$, $(f \cdot g)(x)$, and $\left(\frac{f}{g}\right)(x)$ for each $f(x)$ and $g(x)$. Justify each step. **17–22. See margin. See students' work for justification.**

17. $f(x) = x + 9$
    $g(x) = x - 9$
18. $f(x) = 2x - 3$
    $g(x) = 4x + 9$
19. $f(x) = 2x^2$
    $g(x) = 8 - x$
20. $f(x) = x^2 + 6x + 9$
    $g(x) = 2x + 6$
21. $f(x) = x^2 - 1$
    $g(x) = \frac{x}{x+1}$
22. $f(x) = x^2 - x - 6$
    $g(x) = \frac{x-3}{x+2}$

For each set of ordered pairs, find $f \circ g$ and $g \circ f$ if they exist.

23. $f = \{(1, 1), (0, -3)\}$
    $g = \{(1, 0), (-3, 1), (2, 1)\}$
24. $f = \{(1, 2), (3, 4), (5, 4)\}$
    $g = \{(2, 5), (4, 3)\}$
25. $f = \{(3, 8), (4, 0), (6, 3), (7, -1)\}$
    $g = \{(0, 4), (8, 6), (3, 6), (-1, 8)\}$
26. $f = \{(4, 5), (6, 5), (8, 12), (10, 12)\}$
    $g = \{4, 6), (2, 4), (6, 8), (8, 10)\}$
27. $f = \{(2, 5), (3, 9), (-4, 1)\}$
    $g = \{(5, -4), (8, 3), (2, -2)\}$
    $\{(5, 1), (8, 9)\}$; $\{(2, -4)\}$
28. $f = \{(7, 0), (-5, 3), (8, 3), (-9, 2)\}$
    $g = \{(2, -5), (1, 0), (2, -9), (3, 6)\}$
    $\{(2, 3), (2, 2)\}$; $\{(-5, 6), (8, 6), (-9, -5)\}$

Find $[g \circ h](x)$ and $[h \circ g](x)$. Justify each step. **See students' work.**

29. $g(x) = 4x$ **$8x - 4$;**
    $h(x) = 2x - 1$ **$8x - 1$**
30. $g(x) = -5x$ **$15x - 5$;**
    $h(x) = -3x + 1$ **$15x + 1$**
31. $g(x) = x + 2$ **$x^2 + 2$;**
    $h(x) = x^2$ **$x^2 + 4x + 4$**
32. $g(x) = x - 4$
    $h(x) = 3x^2$
33. $g(x) = 2x$
    $h(x) = x^3 + x^2 + x + 1$
34. $g(x) = x + 1$
    $h(x) = 2x^2 - 5x + 8$

32. $3x^2 - 4$; $3x^2 - 24x + 48$ 33. $2x^3 + 2x^2 + 2x + 2$; $8x^3 + 4x^2 + 2x + 1$

www.algebra2.com/self_check_quiz

About the Exercises...

Organization by Objective

- **Arithmetic Operations:** 17–22
- **Composition of Functions:** 23–46

Odd/Even Assignments

Exercises 17–46 are structured so that students practice the same concepts whether they are assigned odd or even problems.

Assignment Guide

Basic: 17–45 odd, 47, 48, 56–81

Average: 17–45 odd, 47, 48, 55–81

Advanced: 18–46 even, 49–54, 56–75 (optional: 76–81)

Answers

4. $4x + 9$; $2x - 1$; $3x^2 + 19x + 20$; $\frac{3x+4}{x+5}$, $x \neq -5$

5. $x^2 + x - 1$; $x^2 - x + 7$; $x^3 - 4x^2 + 3x - 12$; $\frac{x^2+3}{x-4}$, $x \neq 4$

16. Discount first, then coupon; sample answer: 25% of 49.99 is greater than 25% of 44.99.

17. $2x$; 18; $x^2 - 81$; $\frac{x+9}{x-9}$, $x \neq 9$

18. $6x + 6$; $-2x - 12$; $8x^2 + 6x - 27$; $\frac{2x-3}{4x+9}$, $x \neq -\frac{9}{4}$

19. $2x^2 - x + 8$; $2x^2 + x - 8$; $-2x^3 + 16x^2$; $\frac{2x^2}{8-x}$, $x \neq 8$

20. $x^2 + 8x + 15$; $x^2 + 4x + 3$; $2x^3 + 18x^2 + 54x + 54$; $\frac{x+3}{2}$, $x \neq -3$

21. $\frac{x^3+x^2-1}{x+1}$, $x \neq -1$; $\frac{x^3+x^2-2x-1}{x+1}$, $x \neq -1$; $x^2 - x$, $x \neq -1$; $\frac{x^3+x^2-x-1}{x}$, $x \neq 0$

22. $\frac{x^3+x^2-7x-15}{x+2}$, $x \neq -2$; $\frac{x^3+x^2-9x-9}{x+2}$, $x \neq -2$; $x^2 - 6x + 9$, $x \neq -2$; $x^2 + 4x + 4$, $x \neq -2, 3$

If $f(x) = 4x$, $g(x) = 2x - 1$, and $h(x) = x^2 + 1$, find each value.

35. $f[g(-1)]$ **−12**
36. $h[g(4)]$ **50**
37. $g[f(5)]$ **39**
38. $f[h(-4)]$ **68**
39. $g[g(7)]$ **25**
40. $f[f(-3)]$ **−48**
41. $h\left[f\left(\frac{1}{4}\right)\right]$ **2**
42. $g\left[h\left(-\frac{1}{2}\right)\right]$ **$1\frac{1}{2}$**
43. $[g \circ (f \circ h)](3)$ **79**
44. $[f \circ (h \circ g)](3)$ **104**
45. $[h \circ (g \circ f)](2)$ **226**
46. $[f \circ (g \circ h)](2)$ **36**

POPULATION GROWTH **For Exercises 47 and 48, use the following information.**
From 1990 to 1999, the number of births $b(x)$ in the U.S. can be modeled by the function $b(x) = -27x + 4103$, and the number of deaths $d(x)$ can be modeled by the function $d(x) = 23x + 2164$, where x is the number of years since 1990 and $b(x)$ and $d(x)$ are in thousands.

47. The net increase in population P is the number of births per year minus the number of deaths per year or $P = b - d$. Write an expression that can be used to model the population increase in the U.S. from 1990 to 1999 in function notation. **$P(x) = -50x + 1939$**

48. Assume that births and deaths continue at the same rates. Estimate the net increase in population in 2010. **939,000**

More About. . .

Shopping

Americans spent over $500 million on inline skates and equipment in 2000.

Source: National Sporting Goods Association

SHOPPING **For Exercises 49–51, use the following information.**
Liluye wants to buy a pair of inline skates that are on sale for 30% off the original price of $149. The sales tax is 5.75%.

49. Express the price of the inline skates after the discount and the price of the inline skates after the sales tax using function notation. Let x represent the price of the inline skates, $p(x)$ represent the price after the 30% discount, and $s(x)$ represent the price after the sales tax. **$p(x) = 0.70x$; $s(x) = 1.0575x$**

50. Which composition of functions represents the price of the inline skates, $p[s(x)]$ or $s[p(x)]$? Explain your reasoning.

50. $s[p(x)]$; The 30% would be taken off first, and then the sales tax would be calculated on this price.

51. How much will Liluye pay for the inline skates? **$110.30**

TEMPERATURE **For Exercises 52–54, use the following information.**
There are three temperature scales: Fahrenheit (°F), Celsius (°C), and Kelvin (K). The function $K(C) = C + 273$ can be used to convert Celsius temperatures to Kelvin. The function $C(F) = \frac{5}{9}(F - 32)$ can be used to convert Fahrenheit temperatures to Celsius.

52. Write a composition of functions that could be used to convert Fahrenheit temperatures to Kelvin. **$[K \circ C](F) = \frac{5}{9}(F - 32) + 273$**

53. Find the temperature in Kelvin for the boiling point of water and the freezing point of water if water boils at 212°F and freezes at 32°F. **373 K; 273 K**

54. While performing an experiment, Kimi found the temperature of a solution at different intervals. She needs to record the change in temperature in degrees Kelvin, but only has a thermometer with a Fahrenheit scale. What will she record when the temperature of the solution goes from 158°F to 256°F? **309.67 K**

55. **FINANCE** Kachina pays $50 each month on a credit card that charges 1.6% interest monthly. She has a balance of $700. The balance at the beginning of the nth month is given by $f(n) = f(n - 1) + 0.016\,f(n - 1) - 50$. Find the balance at the beginning of the first five months. No additional charges are made on the card. (*Hint:* $f(1) = 700$) **$700, $661.20, $621.78, $581.73, $541.04**

Study Guide and Intervention, p. 411 (shown) and p. 412

Arithmetic Operations

Operations with Functions	
	Sum $(f + g)(x) = f(x) + g(x)$
	Difference $(f - g)(x) = f(x) - g(x)$
	Product $(f \cdot g)(x) = f(x) \cdot g(x)$
	Quotient $\left(\frac{f}{g}\right)(x) = \frac{f(x)}{g(x)}, g(x) \neq 0$

Example **Find $(f + g)(x)$, $(f - g)(x)$, $(f \cdot g)(x)$, and $\left(\frac{f}{g}\right)(x)$ for $f(x) = x^2 + 3x - 4$ and $g(x) = 3x - 2$.**

$(f + g)(x) = f(x) + g(x)$ — Addition of functions
$= (x^2 + 3x - 4) + (3x - 2)$ — $f(x) = x^2 + 3x - 4$, $g(x) = 3x - 2$
$= x^2 + 6x - 6$ — Simplify.

$(f - g)(x) = f(x) - g(x)$ — Subtraction of functions
$= (x^2 + 3x - 4) - (3x - 2)$ — $f(x) = x^2 + 3x - 4$, $g(x) = 3x - 2$
$= x^2 - 2$ — Simplify.

$(f \cdot g)(x) = f(x) \cdot g(x)$ — Multiplication of functions
$= (x^2 + 3x - 4)(3x - 2)$ — $f(x) = x^2 + 3x - 4$, $g(x) = 3x - 2$
$= x^2(3x - 2) + 3x(3x - 2) - 4(3x - 2)$ — Distributive Property
$= 3x^3 - 2x^2 + 9x^2 - 6x - 12x + 8$ — Distributive Property
$= 3x^3 + 7x^2 - 18x + 8$ — Simplify.

$\left(\frac{f}{g}\right)(x) = \frac{f(x)}{g(x)}$ — Division of functions
$= \frac{x^2 + 3x - 4}{3x - 2}, x \neq \frac{2}{3}$ — $f(x) = x^2 + 3x - 4$ and $g(x) = 3x - 2$

Exercises

Find $(f + g)(x)$, $(f - g)(x)$, $(f \cdot g)(x)$, and $\left(\frac{f}{g}\right)(x)$ for each $f(x)$ and $g(x)$.

1. $f(x) = 8x - 3$; $g(x) = 4x + 5$
$12x + 2$; $4x - 8$; $32x^2 + 28x - 15$; $\frac{8x - 3}{4x + 5}, x \neq -\frac{5}{4}$

2. $f(x) = x^2 + x - 6$; $g(x) = x - 2$
$x^2 + 2x - 8$; $x^2 - 4$; $x^3 - x^2 - 8x + 12$; $x + 3, x \neq 2$

3. $f(x) = 3x^2 - x + 5$; $g(x) = 2x - 3$
$3x^2 + x + 2$; $3x^2 - 3x + 8$; $6x^3 - 11x^2 + 13x - 15$; $\frac{3x^2 - x + 5}{2x - 3}, x \neq \frac{3}{2}$

4. $f(x) = 2x - 1$; $g(x) = 3x^2 + 11x - 4$
$3x^2 + 13x - 5$; $-3x^2 - 9x + 3$; $6x^3 + 19x^2 - 19x + 4$; $\frac{2x - 1}{(3x - 1)(x + 4)}, x \neq \frac{1}{3}, -4$

5. $f(x) = x^2 - 1$; $g(x) = \frac{1}{x + 1}$
$x^2 - 1 + \frac{1}{x + 1}$; $x^2 - 1 - \frac{1}{x + 1}$; $x - 1$; $x^3 + x^2 - x - 1, x \neq -1$

Skills Practice, p. 413 and Practice, p. 414 (shown)

Find $(f + g)(x)$, $(f - g)(x)$, $(f \cdot g)(x)$, and $\left(\frac{f}{g}\right)(x)$ for each $f(x)$ and $g(x)$.

1. $f(x) = 2x + 1$
$g(x) = x - 3$
$3x - 2$; $x + 4$; $2x^2 - 5x - 3$; $\frac{2x + 1}{x - 3}, x \neq 3$

2. $f(x) = 8x^2$
$g(x) = \frac{1}{x^2}$
$\frac{8x^4 + 1}{x^2}, x \neq 0$; $\frac{8x^4 - 1}{x^2}, x \neq 0$; $8, x \neq 0$; $8x^4, x \neq 0$

3. $f(x) = x^2 + 7x + 12$
$g(x) = x^2 - 9$
$2x^2 + 7x + 3$; $7x + 21$; $x^4 + 7x^3 + 3x^2 - 63x - 108$; $\frac{x + 4}{x - 3}, x \neq \pm 3$

For each set of ordered pairs, find $f \circ g$ and $g \circ f$ if they exist.

4. $f = \{(-9, -1), (-1, 0), (3, 4)\}$
$g = \{(0, -9), (-1, 3), (4, -1)\}$
$\{(0, -1), (-1, 4), (4, 0)\}$; $\{(-9, 3), (-1, -9), (3, -1)\}$

5. $f = \{(-4, 3), (0, -2), (1, -2)\}$
$g = \{(-2, 0), (3, 1)\}$
$\{(-2, -2), (3, -2)\}$; $\{(-4, 1), (0, 0), (1, 0)\}$

6. $f = \{(-4, -5), (0, 3), (1, 6)\}$
$g = \{(6, 1), (-5, 0), (3, -4)\}$
$\{(6, 6), (-5, 3), (3, -5)\}$; $\{(-4, 0), (0, -4), (1, 1)\}$

7. $f = \{(0, -3), (1, -3), (6, 8)\}$
$g = \{(8, 2), (-3, 0), (-3, 1)\}$
does not exist; $\{(0, 0), (1, 0), (6, 2)\}$

Find $[g \circ h](x)$ and $[h \circ g](x)$.

8. $g(x) = 3x$
$h(x) = x - 4$
$3x - 12$; $3x - 4$

9. $g(x) = -8x$
$h(x) = 2x + 3$
$-16x - 24$; $-16x + 3$

10. $g(x) = x + 6$
$h(x) = 3x^2$
$3x^2 + 6$; $3x^2 + 36x + 108$

11. $g(x) = x + 3$
$h(x) = 2x^2$
$2x^2 + 3$; $2x^2 + 12x + 18$

12. $g(x) = -2x$
$h(x) = x^2 + 3x + 2$
$-2x^2 - 6x - 4$; $4x^2 - 6x + 2$

13. $g(x) = x - 2$
$h(x) = 3x^2 + 1$
$3x^2 - 1$; $3x^2 - 12x + 13$

If $f(x) = x^2$, $g(x) = 5x$, and $h(x) = x + 4$, find each value.

14. $f[g(1)]$ **25**
15. $g[h(-2)]$ **10**
16. $h[f(4)]$ **20**
17. $f[h(-9)]$ **25**
18. $h[g(-3)]$ **−11**
19. $g[f(8)]$ **320**
20. $h[f(20)]$ **404**
21. $[f \circ (h \circ g)](-1)$ **1**
22. $[f \circ (g \circ h)](4)$ **1600**

23. **BUSINESS** The function $f(x) = 1000 - 0.01x^2$ models the manufacturing cost per item when x items are produced, and $g(x) = 150 - 0.001x^2$ models the service cost per item. Write a function $C(x)$ for the total manufacturing and service cost per item.
$C(x) = 1150 - 0.011x^2$

24. **MEASUREMENT** The formula $f = \frac{n}{12}$ converts inches n to feet f, and $m = \frac{f}{5280}$ converts feet to miles m. Write a composition of functions that converts inches to miles.
$[m \circ f]n = \frac{n}{63{,}360}$

Reading to Learn Mathematics, p. 415 ELL

Pre-Activity Why is it important to combine functions in business?

Read the introduction to Lesson 7-7 at the top of page 383 in your textbook.

Describe two ways to calculate Ms. Coffmon's profit from the sale of 50 birdhouses. (Do not actually calculate her profit.) **Sample answer: 1. Find the revenue by substituting 50 for x in the expression $125x$. Next, find the cost by substituting 50 for x in the expression $65x + 5400$. Finally, subtract the cost from the revenue to find the profit. 2. Form the profit function $p(x) = r(x) - c(x) = 125x - (65x + 5400) = 60x - 5400$. Substitute 50 for x in the expression $60x - 5400$.**

Reading the Lesson

1. Determine whether each statement is *true* or *false*. (Remember that *true* means *always* true.)
   a. If f and g are polynomial functions, then $f + g$ is a polynomial function. **true**
   b. If f and g are polynomial functions, then $\frac{f}{g}$ is a polynomial function. **false**
   c. If f and g are polynomial functions, the domain of the function $f \cdot g$ is the set of all real numbers. **true**
   d. If $f(x) = 3x + 2$ and $g(x) = x - 4$, the domain of the function $\frac{f}{g}$ is the set of all real numbers. **false**
   e. If f and g are polynomial functions, then $(f \circ g)(x) = (g \circ f)(x)$. **false**
   f. If f and g are polynomial functions, then $(f \cdot g)(x) = (g \cdot f)(x)$ **true**

2. Let $f(x) = 2x - 5$ and $g(x) = x^2 + 1$.
   a. Explain in words how you would find $(f \circ g)(-3)$. (Do not actually do any calculations.) **Sample answer: Square −3 and add 1. Take the number you get, multiply it by 2, and subtract 5.**
   b. Explain in words how you would find $(g \circ f)(-3)$. (Do not actually do any calculations.) **Sample answer: Multiply −3 by 2 and subtract 5. Take the number you get, square it, and add 1.**

Helping You Remember

3. Some students have trouble remembering the correct order in which to apply the two original functions when evaluating a composite function. Write three sentences, each of which explains how to do this in a slightly different way. (Hint: Use the word *closest* in the first sentence, the words *inside* and *outside* in the second, and the words *left* and *right* in the third.) **Sample answer: 1. The function that is written closest to the variable is applied first. 2. Work from the inside to the outside. 3. Work from right to left.**

Enrichment, p. 416

Relative Maximum Values

The graph of $f(x) = x^3 - 6x - 9$ shows a relative maximum value somewhere between $f(-2)$ and $f(-1)$. You can obtain a closer approximation by comparing values such as those shown in the table.

To the nearest tenth a relative maximum value for $f(x)$ is -3.3.

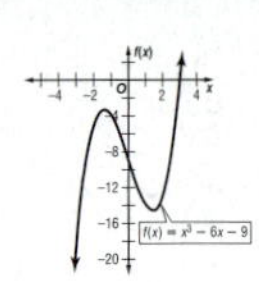

x	$f(x)$
−2	−5
−1.5	−3.375
−1.4	−3.344
−1.3	−3.397
−1	−4

Using a calculator to find points, graph each function. To the nearest tenth, find a relative maximum value of the function.

1. $f(x) = x(x^2 - 3)$ **rel. max. of 2.0**
2. $f(x) = x^3 - 3x - 3$ **rel. max. of −1.0**

56. **CRITICAL THINKING** If $f(0) = 4$ and $f(x + 1) = 3f(x) - 2$, find $f(4)$. **244**

57. **WRITING IN MATH** Answer the question that was posed at the beginning of the lesson. **See margin.**

Why is it important to combine functions in business?

Include the following in your answer:
- a description of how to write a new function that represents the profit, using the revenue and cost functions, and
- an explanation of the benefits of combining two functions into one function.

Standardized Test Practice

58. If $h(x) = 7x - 5$ and $g[h(x)] = 2x + 3$, then $g(x) =$ **A**

Ⓐ $\frac{2x + 31}{7}$. Ⓑ $-5x + 8$.

Ⓒ $5x - 8$. Ⓓ $\frac{2x + 26}{7}$.

59. If $f(x) = 4x^4 + 5x^3 - 3x^2 - 14x + 31$ and $g(x) = 7x^3 - 4x^2 + 5x - 42$, then $(f - g)(x) =$ **C**

Ⓐ $4x^4 + 12x^3 - 7x^2 - 9x - 11$. Ⓑ $4x^4 - 2x^3 - 7x^2 - 19x - 11$.

Ⓒ $4x^4 - 2x^3 + x^2 - 19x + 73$. Ⓓ $-3x^4 - 2x^3 - 7x^2 - 19x + 73$.

Maintain Your Skills

Mixed Review

List all of the possible rational zeros of each function. *(Lesson 7-6)*

60. $r(x) = x^2 - 6x + 8$ **$\pm 1, \pm 2, \pm 4, \pm 8$**

61. $f(x) = 4x^3 - 2x^2 + 6$

62. $g(x) = 9x^2 - 1$ **$\pm 1, \pm\frac{1}{3}, \pm\frac{1}{9}$**

Write a polynomial function of least degree with integral coefficients that has the given zeros. *(Lesson 7-5)*

63. 5, 3, −4

64. −3, −2, 8

65. $1, \frac{1}{2}, \frac{2}{3}$

66. $6, 2i$ **$x^3 - 6x^2 + 4x - 24$**

67. $3, 3 - 2i$ **$x^3 - 9x^2 + 31x - 39$**

68. $-5, 2, 1 - i$

69. **ELECTRONICS** There are three basic things to be considered in an electrical circuit: the flow of the electrical current I, the resistance to the flow Z called impedance, and electromotive force E called voltage. These quantities are related in the formula $E = I \cdot Z$. The current of a circuit is to be $35 - 40j$ amperes. Electrical engineers use the letter j to represent the imaginary unit. Find the impedance of the circuit if the voltage is to be $430 - 330j$ volts. *(Lesson 5-9)* **$10 + 2j$**

Find the inverse of each matrix, if it exists. *(Lesson 4-7)*

70. $\begin{bmatrix} 8 & 6 \\ 7 & 5 \end{bmatrix}$ **$-\frac{1}{2}\begin{bmatrix} 5 & -6 \\ -7 & 8 \end{bmatrix}$**

71. $\begin{bmatrix} 1 & 2 \\ 1 & 3 \end{bmatrix}$ **$\begin{bmatrix} 3 & -2 \\ -1 & 1 \end{bmatrix}$**

72. $\begin{bmatrix} 8 & 4 \\ 6 & 3 \end{bmatrix}$ **does not exist**

73. $\begin{bmatrix} -4 & 2 \\ 3 & -1 \end{bmatrix}$ **$-\frac{1}{2}\begin{bmatrix} -1 & -2 \\ -3 & -4 \end{bmatrix}$**

74. $\begin{bmatrix} 6 & -2 \\ 9 & -3 \end{bmatrix}$ **does not exist**

75. $\begin{bmatrix} 2 & 2 \\ 3 & -5 \end{bmatrix}$ **$-\frac{1}{16}\begin{bmatrix} -5 & -2 \\ -3 & 2 \end{bmatrix}$**

Getting Ready for the Next Lesson

PREREQUISITE SKILL **Solve each equation or formula for the specified variable.**
*(To review **solving equations for a variable**, see Lesson 1-3.)*

76. $2x - 3y = 6$, for x **$x = \frac{6 + 3y}{2}$**

77. $4x^2 - 5xy + 2 = 3$, for y **$y = \frac{1 - 4x^2}{-5x}$**

78. $3x + 7xy = -2$, for x **$x = \frac{-2}{3 + 7y}$**

79. $I = prt$, for t **$t = \frac{I}{pr}$**

80. $C = \frac{5}{9}(F - 32)$, for F **$F = \frac{9}{5}C + 32$**

81. $F = G\frac{Mm}{r^2}$, for m **$m = \frac{Fr^2}{GM}$**

Mixed Review answers:

61. $\pm 1, \pm\frac{1}{2}, \pm\frac{1}{4}, \pm 2, \pm 3, \pm\frac{3}{2}, \pm\frac{3}{4}, \pm 6$

63. $x^3 - 4x^2 - 17x + 60$

64. $x^3 - 3x^2 - 34x - 48$

65. $6x^3 - 13x^2 + 9x - 2$

68. $x^4 + x^3 - 14x^2 + 26x - 20$

4 Assess

Open-Ended Assessment

Modeling In some courses, a function $f(x)$ is modeled by a "machine" that accepts values for x as inputs and then outputs values for $f(x)$. Using this model, ask students to explain how the composition of two functions could be modeled by two such "machines" linked together. Also ask them to use the model to explain how the composition of two functions could be undefined for some initial input values.

Getting Ready for Lesson 7-8

PREREQUISITE SKILL In Lesson 7-8, students will find the inverse of a function. One method used will involve students solving an equation for a variable. Use Exercises 76–81 to determine your students' familiarity with solving equations for a variable.

Assessment Options

Quiz (Lessons 7-6 and 7-7) is available on p. 444 of the *Chapter 7 Resource Masters*.

Answer

57. Answers should include the following.
- Using the revenue and cost functions, a new function that represents the profit is $p(x) = r(c(x))$.
- The benefit of combining two functions into one function is that there are fewer steps to compute and it is less confusing to the general population of people reading the formulas.

7-8 Lesson Notes

1 Focus

5-Minute Check Transparency 7-8 Use as a quiz or review of Lesson 7-7.

Mathematical Background notes are available for this lesson on p. 344D.

How are inverse functions related to measurement conversions?

Ask students:

- When would you need to convert from SI units to customary units? **Sample answer: reading a recipe printed in a cookbook that was published in Canada**
- When would you need to convert between units within the customary system? **Sample answer: You might change pounds to ounces in order to estimate the cost per ounce of different grocery items.**
- When you are driving a car at a speed of 55 miles per hour, how many meters would you guess you are traveling each second? **Sample answer: 20 m/s**
- What is the calculated value for $f(55)$? **about 24.4 m/s**

7-8 Inverse Functions and Relations

What You'll Learn

- Find the inverse of a function or relation.
- Determine whether two functions or relations are inverses.

Vocabulary

- inverse relation
- inverse function
- identity function
- one-to-one

How are inverse functions related to measurement conversions?

Most scientific formulas involve measurements given in SI (International System) units. The SI units for speed are meters per second. However, the United States uses customary measurements such as miles per hour. To convert x miles per hour to an approximate equivalent in meters per second, you can evaluate

$$f(x) = \frac{x \text{ miles}}{1 \text{ hour}} \cdot \frac{1600 \text{ meters}}{1 \text{ mile}} \cdot \frac{1 \text{ hour}}{3600 \text{ seconds}} \text{ or } f(x) = \frac{4}{9}x.$$

To convert x meters per second to an approximate equivalent in miles per hour, you can evaluate

$$g(x) = \frac{x \text{ meters}}{1 \text{ second}} \cdot \frac{3600 \text{ seconds}}{1 \text{ hour}} \cdot \frac{1 \text{ mile}}{1600 \text{ meters}} \text{ or } g(x) = \frac{9}{4}x.$$

Notice that $f(x)$ multiplies a number by 4 and divides it by 9. The function $g(x)$ does the inverse operation of $f(x)$. It divides a number by 4 and multiplies it by 9. The functions $f(x) = \frac{4}{9}x$ and $g(x) = \frac{9}{4}x$ are inverses.

FIND INVERSES Recall that a relation is a set of ordered pairs. The **inverse relation** is the set of ordered pairs obtained by reversing the coordinates of each original ordered pair. The domain of a relation becomes the range of the inverse, and the range of a relation becomes the domain of the inverse.

Key Concept — *Inverse Relations*

- **Words** Two relations are inverse relations if and only if whenever one relation contains the element (a, b), the other relation contains the element (b, a).
- **Example** $Q = \{(1, 2), (3, 4), (5, 6)\}$ $S = \{(2, 1), (4, 3), (6, 5)\}$
  Q and S are inverse relations.

Example 1 *Find an Inverse Relation*

GEOMETRY The ordered pairs of the relation $\{(2, 1), (5, 1), (2, -4)\}$ are the coordinates of the vertices of a right triangle. Find the inverse of this relation and determine whether the resulting ordered pairs are also the vertices of a right triangle.

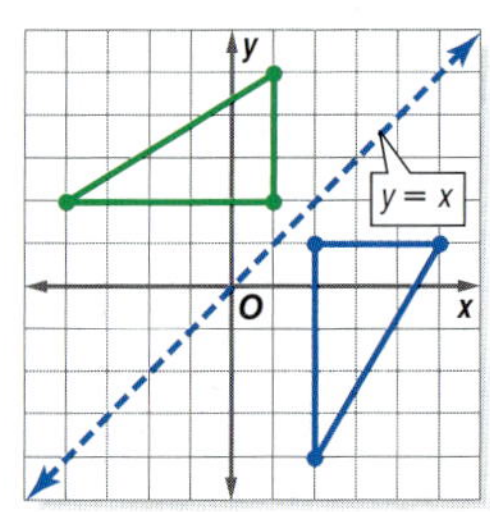

To find the inverse of this relation, reverse the coordinates of the ordered pairs.

The inverse of the relation is $\{(1, 2), (1, 5), (-4, 2)\}$.

Plotting the points shows that the ordered pairs also describe the vertices of a right triangle. Notice that the graphs of the relation and the inverse relation are reflections over the graph of $y = x$.

Resource Manager

Workbook and Reproducible Masters

Chapter 7 Resource Masters

- Study Guide and Intervention, pp. 417–418
- Skills Practice, p. 419
- Practice, p. 420
- Reading to Learn Mathematics, p. 421
- Enrichment, p. 422

School-to-Career Masters, p. 14

Teaching Algebra With Manipulatives Masters, pp. 256–257, 258

Transparencies

5-Minute Check Transparency 7-8

Answer Key Transparencies

Technology

Interactive Chalkboard

Study Tip

Reading Math

f^{-1} is read *f inverse* or *the inverse of f*. Note that -1 is not an exponent.

The ordered pairs of **inverse functions** are also related. We can write the inverse of function $f(x)$ as $f^{-1}(x)$.

Key Concept — *Property of Inverse Functions*

Suppose f and f^{-1} are inverse functions. Then, $f(a) = b$ if and only if $f^{-1}(b) = a$.

Let's look at the inverse functions $f(x) = x + 2$ and $f^{-1}(x) = x - 2$.

Evaluate $f(5)$.	Now, evaluate $f^{-1}(7)$.
$f(x) = x + 2$	$f^{-1}(x) = x - 2$
$f(5) = 5 + 2$ or 7	$f^{-1}(7) = 7 - 2$ or 5

Since $f(x)$ and $f^{-1}(x)$ are inverses, $f(5) = 7$ and $f^{-1}(7) = 5$. The inverse function can be found by exchanging the domain and range of the function.

Example 2 Find an Inverse Function

a. Find the inverse of $f(x) = \frac{x+6}{2}$.

Step 1 Replace $f(x)$ with y in the original equation.

$f(x) = \frac{x+6}{2} \rightarrow y = \frac{x+6}{2}$

Step 2 Interchange x and y.

$x = \frac{y+6}{2}$

Step 3 Solve for y.

$x = \frac{y+6}{2}$ Inverse

$2x = y + 6$ Multiply each side by 2.

$2x - 6 = y$ Subtract 6 from each side.

Step 4 Replace y with $f^{-1}(x)$.

$y = 2x - 6 \quad f^{-1}(x) = 2x - 6$

The inverse of $f(x) = \frac{x+6}{2}$ is $f^{-1}(x) = 2x - 6$.

b. Graph the function and its inverse.

Graph both functions on the coordinate plane. The graph of $f^{-1}(x) = 2x - 6$ is the reflection of the graph of $f(x) = \frac{x+6}{2}$ over the line $y = x$.

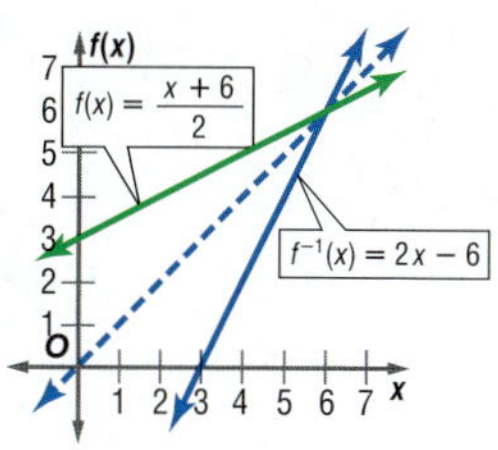

INVERSES OF RELATIONS AND FUNCTIONS You can determine whether two functions are inverses by finding both of their compositions. If both equal the $I(x) = x$, then the functions are inverse functions.

Key Concept — *Inverse Functions*

- **Words** Two functions f and g are inverse functions if and only if both of their compositions are the identity function.
- **Symbols** $[f \circ g](x) = x$ and $[g \circ f](x) = x$

www.algebra2.com/extra_examples

2 Teach

FIND INVERSES

In-Class Examples Power Point®

1 GEOMETRY The ordered pairs of the relation $\{(1, 3), (6, 3), (6, 0), (1, 0)\}$ are the coordinates of the vertices of a rectangle. Find the inverse of this relation and determine whether the resulting ordered pairs are also the coordinates of the vertices of a rectangle. **The inverse of the relation is $\{(3, 1), (3, 6), (0, 6), (0, 1)\}$. These ordered pairs also describe the vertices of a rectangle.**

2

a. Find the inverse of $f(x) = -\frac{1}{2}x + 1$. $f^{-1}(x) = -2x + 2$

b. Graph the function and its inverse.

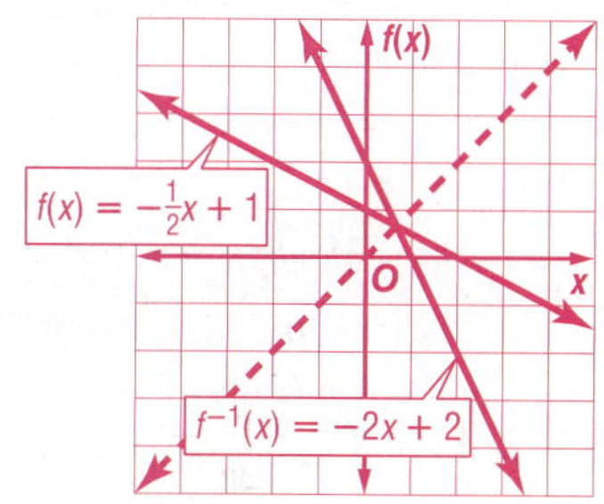

DAILY INTERVENTION — **Differentiated Instruction**

Logical After completing Example 3 on p. 392 and discussing how to determine if two functions are inverses, challenge students to find two functions, f and g, such that $f(g(x)) \neq g(f(x))$, with one of the two compositions having a value of x. You may wish to have students work in groups to brainstorm as they attempt this puzzler.

INVERSES OF RELATIONS AND FUNCTIONS

In-Class Example

Power Point®

3 Determine whether $f(x) = \frac{3}{4}x - 6$ and $g(x) = \frac{4}{3}x + 8$ are inverse functions. **The functions are inverses since both $[f \circ g](x)$ and $[g \circ f](x)$ equal x.**

Teaching Tip Have students verify the results of Example 3 by graphing the two functions on their graphing calculators and checking that the graphs are reflections over the line $y = x$.

3 Practice/Apply

Study Notebook

Have students—

- add the definitions/examples of the vocabulary terms to their Vocabulary Builder worksheets for Chapter 7.
- write the definitions of inverse relations and inverse functions.
- include any other item(s) that they find helpful in mastering the skills in this lesson.

About the Exercises...

Organization by Objective

- **Find Inverses:** 14–31
- **Inverses of Relations and Functions:** 32–37

Odd/Even Assignments

Exercises 14–37 are structured so that students practice the same concepts whether they are assigned odd or even problems.

Assignment Guide

Basic: 15–37 odd, 41, 44–61

Average: 15–37 odd, 41–61

Advanced: 14–36 even, 38–40, 42–55 (optional: 56–61)

Study Tip

Inverse Functions

Both compositions of $f(x)$ and $g(x)$ must be the identity function for $f(x)$ and $g(x)$ to be inverses. It is necessary to check them both.

Example 3 Verify Two Functions are Inverses

Determine whether $f(x) = 5x + 10$ and $g(x) = \frac{1}{5}x - 2$ are inverse functions.

Check to see if the compositions of $f(x)$ and $g(x)$ are identity functions.

$$\begin{aligned}[f \circ g](x) &= f[g(x)] \\ &= f\left(\tfrac{1}{5}x - 2\right) \\ &= 5\left(\tfrac{1}{5}x - 2\right) + 10 \\ &= x - 10 + 10 \\ &= x\end{aligned} \qquad \begin{aligned}[g \circ f](x) &= g[f(x)] \\ &= g(5x + 10) \\ &= \tfrac{1}{5}(5x + 10) - 2 \\ &= x + 2 - 2 \\ &= x\end{aligned}$$

The functions are inverses since both $[f \circ g](x)$ and $[g \circ f](x)$ equal x.

You can also determine whether two functions are inverse functions by graphing. The graphs of a function and its inverse are mirror images with respect to the graph of the identity function $I(x) = x$.

Algebra Activity

Inverses of Functions

- Use a full sheet of grid paper. Draw and label the x- and y-axes.
- Graph $y = 2x - 3$.
- On the same coordinate plane, graph $y = x$ as a dashed line.
- Place a geomirror so that the drawing edge is on the line $y = x$. Carefully plot the points that are part of the reflection of the original line. Draw a line through the points.

Analyze

1. What is the equation of the drawn line? **$y = \frac{x + 3}{2}$**
2. What is the relationship between the line $y = 2x - 3$ and the line that you drew? Justify your answer.
3. Try this activity with the function $y = |x|$. Is the inverse also a function? Explain. **No; the graph does not pass the vertical line test.**

2. They are inverses. $f(x) = 2x - 3$, $g(x) = \frac{x + 3}{2}$, $f[g(x)] = 2\left(\frac{x + 3}{2}\right) - 3 = x$, $g[f(x)] = \frac{(2x - 3) + 3}{2} = x$

When the inverse of a function is a function, then the original function is said to be **one-to-one**. To determine if the inverse of a function is a function, you can use the *horizontal line test*.

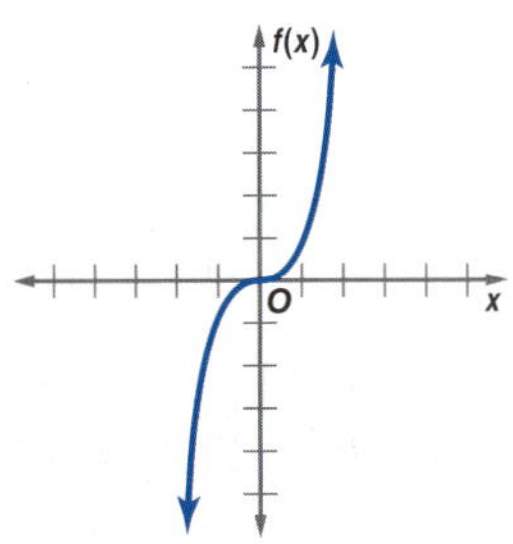

No horizontal line can be drawn so that it passes through more than one point. The inverse of this function is a function.

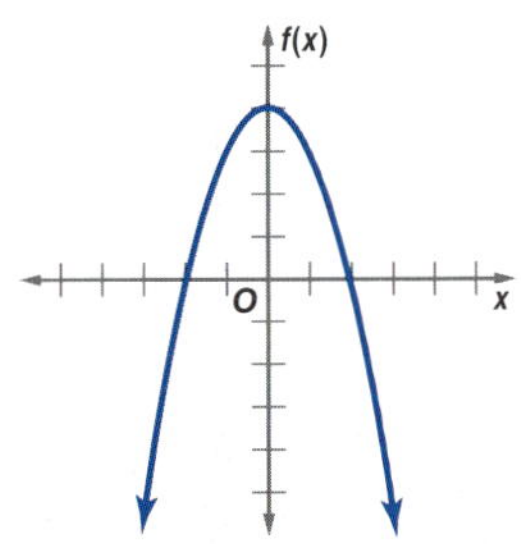

A horizontal line can be drawn that passes through more than one point. The inverse of this function is not a function.

Algebra Activity

Materials: grid paper, geomirror

- Watch for students who mistakenly place the geomirror on the graph of the function instead of the line $y = x$.
- Before students attempt Exercise 3, you may want to provide a brief review of the graphs of absolute value functions.

392 Chapter 7 Polynomial Functions

Check for Understanding

Concept Check

1. **Determine** whether $f(x) = 3x + 6$ and $g(x) = x - 2$ are inverses. **no**
2. **Explain** the steps you would take to find an inverse function.
3. **OPEN ENDED** Give an example of a function and its inverse. Verify that the two functions are inverses.
4. **Determine** the values of n for which $f(x) = x^n$ has an inverse that is a function. Assume that n is a whole number. **n is an odd whole number.**

2. Switch x and y in the equation and solve for y.
3. Sample answer: $f(x) = 2x$, $f^{-1}(x) = 0.5x$; $f[f^{-1}(x)] = f^{-1}[f(x)] = x$

Guided Practice

GUIDED PRACTICE KEY	
Exercises	Examples
5, 6	1
7–9, 12, 13	2
10, 11	3

Find the inverse of each relation.

5. $\{(2, 4), (-3, 1), (2, 8)\}$ **$\{(4, 2), (1, -3), (8, 2)\}$**
6. $\{(1, 3), (1, -1), (1, -3), (1, 1)\}$ **$\{(3, 1), (-1, 1), (-3, 1), (1, 1)\}$**

Find the inverse of each function. Then graph the function and its inverse.

7. $f(x) = -x$ **$f^{-1}(x) = -x$**
8. $g(x) = 3x + 1$ **$g^{-1}(x) = \frac{1}{3}x - \frac{1}{3}$**
9. $y = \frac{1}{2}x + 5$ **$y = 2x - 10$**

7–9. See pp. 407A–407H for graphs.

Determine whether each pair of functions are inverse functions.

10. $f(x) = x + 7$; $g(x) = x - 7$ **yes**
11. $g(x) = 3x - 2$; $f(x) = \frac{x - 2}{3}$ **no**

Application

PHYSICS **For Exercises 12 and 13, use the following information.**
The acceleration due to gravity is 9.8 meters per second squared (m/s^2). To convert to feet per second squared, you can use the following chain of operations:

$$\frac{9.8 \text{ m}}{s^2} \times \frac{100 \text{ cm}}{1 \text{ m}} \times \frac{1 \text{ in.}}{2.54 \text{ cm}} \times \frac{1 \text{ ft}}{12 \text{ in.}}$$

12. 32.2 ft/s²

12. Find the value of the acceleration due to gravity in feet per second squared.
13. An object is accelerating at 50 feet per second squared. How fast is it accelerating in meters per second squared? **15.24 m/s²**

Practice and Apply

Homework Help	
For Exercises	See Examples
14–19	1
20–31, 38–43	2
32–37	3

Extra Practice See page 844.

Find the inverse of each relation. **14–19. See margin.**

14. $\{(2, 6), (4, 5), (-3, -1)\}$
15. $\{(3, 8), (4, -2), (5, -3)\}$
16. $\{(7, -4), (3, 5), (-1, 4), (7, 5)\}$
17. $\{(-1, -2), (-3, -2), (-1, -4), (0, 6)\}$
18. $\{(6, 11), (-2, 7), (0, 3), (-5, 3)\}$
19. $\{(2, 8), (-6, 5), (8, 2), (5, -6)\}$

20–28. See pp. 407A–407H for graphs.

Find the inverse of each function. Then graph the function and its inverse.

20. $y = -3$ **$x = -3$**
21. $g(x) = -2x$
22. $f(x) = x - 5$
23. $g(x) = x + 4$
24. $f(x) = 3x + 3$
25. $y = -2x - 1$
26. $y = \frac{1}{3}x$ **$y = 3x$**
27. $f(x) = \frac{5}{8}x$
28. $f(x) = \frac{1}{3}x + 4$
29. $f(x) = \frac{4}{5}x - 7$
30. $g(x) = \frac{2x + 3}{6}$
31. $f(x) = \frac{7x - 4}{8}$

21. $g^{-1}(x) = -\frac{1}{2}x$
22. $f^{-1}(x) = x + 5$
23. $g^{-1}(x) = x - 4$
24. $f^{-1}(x) = \frac{1}{3}x - 1$
25. $y = -\frac{1}{2}x - \frac{1}{2}$
27. $f^{-1}(x) = \frac{8}{5}x$
28. $f^{-1}(x) = 3x - 12$
29–31. See pp. 407A–407H.

Determine whether each pair of functions are inverse functions.

32. $f(x) = x - 5$; $g(x) = x + 5$ **yes**
33. $f(x) = 3x + 4$; $g(x) = 3x - 4$ **no**
34. $f(x) = 6x + 2$; $g(x) = x - \frac{1}{3}$ **no**
35. $g(x) = 2x + 8$; $f(x) = \frac{1}{2}x - 4$ **yes**
36. $h(x) = 5x - 7$; $g(x) = \frac{1}{5}(x + 7)$ **yes**
37. $g(x) = 2x + 1$; $f(x) = \frac{x - 1}{2}$ **yes**

www.algebra2.com/self_check_quiz

Answers

14. $\{(6, 2), (5, 4), (-1, -3)\}$
15. $\{(8, 3), (-2, 4), (-3, 5)\}$
16. $\{(-4, 7), (5, 3), (4, -1), (5, 7)\}$
17. $\{(-2, -1), (-2, -3), (-4, -1), (6, 0)\}$
18. $\{(11, 6), (7, -2), (3, 0), (3, -5)\}$
19. $\{(8, 2), (5, -6), (2, 8), (-6, 5)\}$

Enrichment, p. 422

Miniature Golf

In miniature golf, the object of the game is to roll the golf ball into the hole in as few shots as possible. As in the diagram at the right, the hole is often placed so that a direct shot is impossible. Reflections can be used to help determine the direction that the ball should be rolled in order to score a hole-in-one.

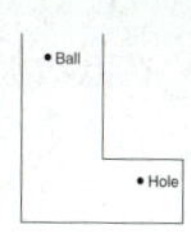

Example 1 **Using wall $\overline{EF}$, find the path to use to score a hole-in-one.**

Find the reflection image of the "hole" with respect to $\overline{EF}$ and label it H'. The intersection of $\overline{BH'}$ with wall $\overline{EF}$ is the point at which the shot should be directed.

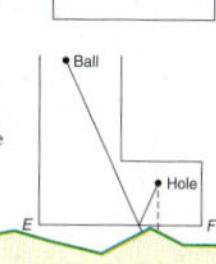

Study Guide and Intervention, p. 417 (shown) and p. 418

Find Inverses

Inverse Relations	Two relations are inverse relations if and only if whenever one relation contains the element (a, b), the other relation contains the element (b, a).
Property of Inverse Functions	Suppose f and f^{-1} are inverse functions. Then $f(a) = b$ if and only if $f^{-1}(b) = a$.

Example **Find the inverse of the function $f(x) = \frac{2}{5}x - \frac{1}{5}$. Then graph the function and its inverse.**

Step 1 Replace $f(x)$ with y in the original equation.
$f(x) = \frac{2}{5}x - \frac{1}{5} \rightarrow y = \frac{2}{5}x - \frac{1}{5}$

Step 2 Interchange x and y.
$x = \frac{2}{5}y - \frac{1}{5}$

Step 3 Solve for y.
$x = \frac{2}{5}y - \frac{1}{5}$ Inverse
$5x = 2y - 1$ Multiply each side by 5.
$5x + 1 = 2y$ Add 1 to each side.
$\frac{1}{2}(5x + 1) = y$ Divide each side by 2.

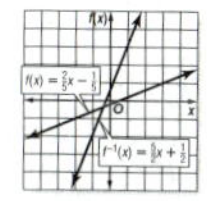

The inverse of $f(x) = \frac{2}{5}x - \frac{1}{5}$ is $f^{-1}(x) = \frac{1}{2}(5x + 1)$.

Exercises

Find the inverse of each function. Then graph the function and its inverse.

1. $f(x) = \frac{2}{3}x - 1$ **$f^{-1}(x) = \frac{3}{2}x + \frac{3}{2}$**

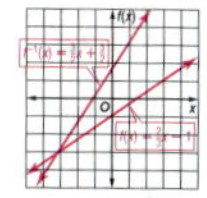

2. $f(x) = 2x - 3$ **$f^{-1}(x) = \frac{1}{2}x + \frac{3}{2}$**

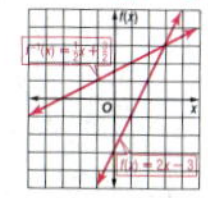

3. $f(x) = \frac{1}{4}x - 2$ **$f^{-1}(x) = 4x + 8$**

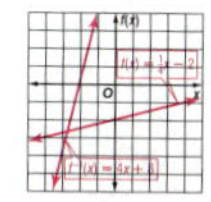

Skills Practice, p. 419 and Practice, p. 420 (shown)

Find the inverse of each relation.

1. $\{(0, 3), (4, 2), (5, -6)\}$ **$\{(3, 0), (2, 4), (-6, 5)\}$**
2. $\{(-5, 1), (-5, -1), (-5, 8)\}$ **$\{(1, -5), (-1, -5), (8, -5)\}$**
3. $\{(-3, -7), (0, -1), (5, 9), (7, 13)\}$ **$\{(-7, -3), (-1, 0), (9, 5), (13, 7)\}$**
4. $\{(8, -2), (10, 5), (12, 6), (14, 7)\}$ **$\{(-2, 8), (5, 10), (6, 12), (7, 14)\}$**
5. $\{(-5, -4), (1, 2), (3, 4), (7, 8)\}$ **$\{(-4, -5), (2, 1), (4, 3), (8, 7)\}$**
6. $\{(-3, 9), (-2, 4), (0, 0), (1, 1)\}$ **$\{(9, -3), (4, -2), (0, 0), (1, 1)\}$**

Find the inverse of each function. Then graph the function and its inverse.

7. $f(x) = \frac{3}{4}x$ **$f^{-1}(x) = \frac{4}{3}x$**

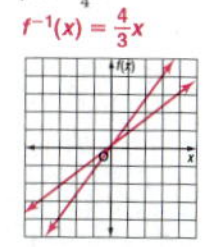

8. $g(x) = 3 + x$ **$g^{-1}(x) = x - 3$**

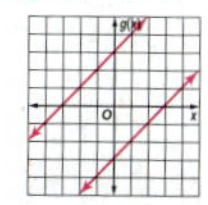

9. $y = 3x - 2$ **$y = \frac{x + 2}{3}$**

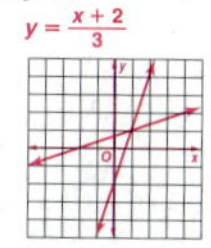

Determine whether each pair of functions are inverse functions.

10. $f(x) = x + 6$; $g(x) = x - 6$ **yes**
11. $f(x) = -4x + 1$; $g(x) = \frac{1}{4}(1 - x)$ **yes**
12. $g(x) = 13x - 13$; $h(x) = \frac{1}{13}x - 1$ **no**
13. $f(x) = 2x$; $g(x) = -2x$ **no**
14. $f(x) = \frac{6}{7}x$; $g(x) = \frac{7}{6}x$ **yes**
15. $g(x) = 2x - 8$; $h(x) = \frac{1}{2}x + 4$ **yes**

16. **MEASUREMENT** The points (63, 121), (71, 180), (67, 140), (65, 108), and (72, 165) give the weight in pounds as a function of height in inches for 5 students in a class. Give the points for these students that represent height as a function of weight. **(121, 63), (180, 71), (140, 67), (108, 65), (165, 72)**

REMODELING **For Exercises 17 and 18, use the following information.**
The Clearys are replacing the flooring in their 15 foot by 18 foot kitchen. The new flooring costs \$17.99 per square yard. The formula $f(x) = 9x$ converts square yards to square feet.

17. Find the inverse $f^{-1}(x)$. What is the significance of $f^{-1}(x)$ for the Clearys? **$f^{-1}(x) = \frac{x}{9}$; It will allow them to convert the square footage of their kitchen floor to square yards, so they can then calculate the cost of the new flooring.**
18. What will the new flooring cost the Cleary's? **\$539.70**

Reading to Learn Mathematics, p. 421 ELL

Pre-Activity **How are inverse functions related to measurement conversions?**

Read the introduction to Lesson 7-8 at the top of page 390 in your textbook.

A function multiplies a number by 3 and then adds 5 to the result. What does the inverse function do, and in what order? **Sample answer: It first subtracts 5 from the number and then divides the result by 3.**

Reading the Lesson

1. Complete each statement.
   a. If two relations are inverses, the domain of one relation is the **range** of the other.
   b. Suppose that $g(x)$ is a relation and that the point $(4, -2)$ is on its graph. Then a point on the graph of $g^{-1}(x)$ is **(−2, 4)**.
   c. The **horizontal line** test can be used on the graph of a function to determine whether the function has an inverse function.
   d. If you are given the graph of a function, you can find the graph of its inverse by reflecting the original graph over the line with equation **$y = x$**.
   e. If f and g are inverse functions, then $(f \circ g)(x) =$ **x** and $(g \circ f)(x) =$ **x**.
   f. A function has an inverse that is also a function only if the given function is **one-to-one**.
   g. Suppose that $h(x)$ is a function whose inverse is also a function. If $h(5) = 12$, then $h^{-1}(12) =$ **5**.
2. Assume that $f(x)$ is a one-to-one function defined by an algebraic equation. Write the four steps you would follow in order to find the equation for $f^{-1}(x)$.
   1. **Replace $f(x)$ with y in the original equation.**
   2. **Interchange x and y.**
   3. **Solve for y.**
   4. **Replace y with $f^{-1}(x)$.**

Helping You Remember

3. A good way to remember something new is to relate it to something you already know. How are the vertical and horizontal line tests related? **Sample answer: The vertical line test determines whether a relation is a function because the ordered pairs in a function can have no repeated x-values. The horizontal line test determines whether a function is one-to-one because a one-to-one function cannot have any repeated y-values.**

4 Assess

Open-Ended Assessment

Modeling On a large coordinate grid, have students model the graph of the identity function $f(x) = x$ using a length of string, a piece of raw spaghetti, or something similar. Then place a second length of string or spaghetti to model the graph of a function. Have students model the graph of the inverse of this function.

Getting Ready for Lesson 7-9

PREREQUISITE SKILL Students will graph square root functions and inequalities in Lesson 7-9. Students will need to solve radical equations. Use Exercises 56–61 to determine your students' familiarity with solving radical equations.

Answers

42. $C^{-1}(x) = \frac{9}{5}x + 32$; $C[C^{-1}(x)] = C^{-1}[C(x)] = x$

43. It can be used to convert Celsius to Fahrenheit.

44. Sample answer: $f(x) = x$ and $f^{-1}(x) = x$ or $f(x) = -x$ and $f^{-1}(x) = -x$

45. Inverses are used to convert between two units of measurement. Answers should include the following.
    - Even if it is not necessary, it is helpful to know the imperial units when given the metric units because most measurements in the U.S. are given in imperial units so it is easier to understand the quantities using our system.
    - To convert the speed of light from meters per second to miles per hour,

$$f(x) \approx \frac{3.0 \times 10^8 \text{ meters}}{1 \text{ second}} \cdot \frac{3600 \text{ seconds}}{1 \text{ hour}} \cdot \frac{1 \text{ mile}}{1600 \text{ meters}}$$

$$\approx 675{,}000{,}000 \text{ mi/hr}$$

NUMBER GAMES **For Exercises 38–40, use the following information.**
Damaso asked Sophia to choose a number between 1 and 20. He told her to add 7 to that number, multiply by 4, subtract 6, and divide by 2.

38. Write an equation that models this problem. $y = \frac{4(x + 7) - 6}{2}$
39. Find the inverse. $y = \frac{1}{2}x - \frac{11}{2}$
40. Sophia's final number was 35. What was her original number? **12**

41. **SALES** Sales associates at Electronics Unlimited earn \$8 an hour plus a 4% commission on the merchandise they sell. Write a function to describe their income, and find how much merchandise they must sell in order to earn \$500 in a 40-hour week. $I(m) = 320 + 0.04m$; **\$4500**

More About. . .

Temperature
The Fahrenheit temperature scale was established in 1724 by a physicist named Gabriel Daniel Fahrenheit. The Celsius temperature scale was established in the same year by an astronomer named Anders Celsius.
Source: www.infoplease.com

TEMPERATURE **For Exercises 42 and 43, use the following information.**
A formula for converting degrees Fahrenheit to Celsius is $C(x) = \frac{5}{9}(x - 32)$.

42. Find the inverse $C^{-1}(x)$. Show that $C(x)$ and $C^{-1}(x)$ are inverses.
43. Explain what purpose $C^{-1}(x)$ serves. **42–44. See margin.**

44. **CRITICAL THINKING** Give an example of a function that is its own inverse.

45. **WRITING IN MATH** Answer the question that was posed at the beginning of the lesson. **See margin.**

How are inverse functions related to measurement conversions?

Include the following items in your answer:
- an explanation of why you might want to know the customary units if you are given metric units even if it is not necessary for you to perform additional calculations, and
- a demonstration of how to convert the speed of light $c = 3.0 \times 10^8$ meters per second to miles per hour.

Standardized Test Practice

46. Which of the following is the inverse of the function $f(x) = \frac{3x - 5}{2}$? **A**

Ⓐ $g(x) = \frac{2x + 5}{3}$ Ⓑ $g(x) = \frac{3x + 5}{2}$ Ⓒ $g(x) = 2x + 5$ Ⓓ $g(x) = \frac{2x - 5}{3}$

47. For which of the following functions is the inverse also a function? **B**

I. $f(x) = x^3$ **II.** $f(x) = x^4$ **III.** $f(x) = -|x|$

Ⓐ I and II only Ⓑ I only Ⓒ I, II, and III Ⓓ III only

Maintain Your Skills

Mixed Review Find $[g \circ h](x)$ and $[h \circ g](x)$. *(Lesson 7-7)*

48. $g(x) = 4x$, $h(x) = x + 5$
49. $g(x) = 3x + 2$, $h(x) = 2x - 4$
50. $g(x) = x + 4$, $h(x) = x^2 - 3x - 28$

48. $g[h(x)] = 4x + 20$, $h[g(x)] = 4x + 5$
49. $g[h(x)] = 6x - 10$, $h[g(x)] = 6x$
50. $g[h(x)] = x^2 - 3x - 24$, $h[g(x)] = x^2 + 5x - 24$

Find all of the rational zeros of each function. *(Lesson 7-6)*

51. $f(x) = x^3 + 6x^2 - 13x - 42$ **−7, −2, 3**
52. $h(x) = 24x^3 - 86x^2 + 57x + 20$ $-\frac{1}{4}, \frac{4}{3}, \frac{5}{2}$

Evaluate each expression. *(Lesson 5-7)*

53. $16^{\frac{3}{2}}$ **64**
54. $64^{\frac{1}{3}} \cdot 64^{\frac{1}{2}}$ **32**
55. $\frac{3^{\frac{4}{3}}}{81^{\frac{1}{12}}}$ **3**

Getting Ready for the Next Lesson

PREREQUISITE SKILL **Solve each equation.**
*(To review **solving radical equations**, see Lesson 5-8.)*

56. $\sqrt{x} - 5 = -3$ **4**
57. $\sqrt{x + 4} = 11$ **117**
58. $12 - \sqrt{x} = -2$ **196**
59. $\sqrt{x - 5} = \sqrt{2x + 2}$ **−7**
60. $\sqrt{x - 3} = \sqrt{2} - \sqrt{x}$
61. $3 - \sqrt{x} = \sqrt{x - 6}$ $\frac{25}{4}$

60. ∅

7-9 Square Root Functions and Inequalities

What You'll Learn

- Graph and analyze square root functions.
- Graph square root inequalities.

Vocabulary

- square root function
- square root inequality

How are square root functions used in bridge design?

The Sunshine Skyway Bridge across Tampa Bay, Florida, is supported by 21 steel cables, each 9 inches in diameter. The amount of weight that a steel cable can support is given by $w = 8d^2$, where d is the diameter of the cable in inches and w is the weight in tons. If you need to know what diameter a steel cable should have to support a given weight, you can use the equation $d = \sqrt{\frac{w}{8}}$.

SQUARE ROOT FUNCTIONS If a function contains a square root of a variable, it is called a **square root function**. The inverse of a quadratic function is a square root function only if the range is restricted to nonnegative numbers.

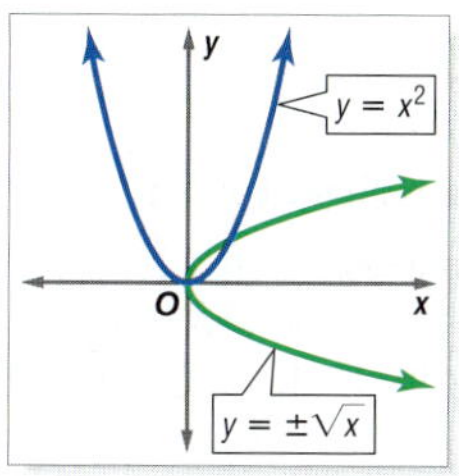

$y = \pm\sqrt{x}$ is not a function.

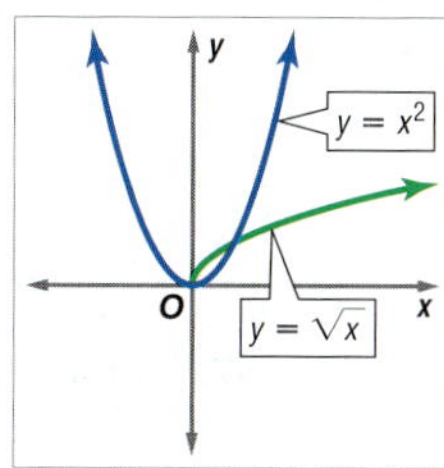

$y = \sqrt{x}$ is a function.

In order for a square root to be a real number, the radicand cannot be negative. When graphing a square root function, determine when the radicand would be negative and exclude those values from the domain.

Example 1 Graph a Square Root Function

Graph $y = \sqrt{3x + 4}$. State the domain, range, and x- and y-intercepts.

Since the radicand cannot be negative, identify the domain.

$3x + 4 \geq 0$ Write the expression inside the radicand as ≥ 0.

$x \geq -\frac{4}{3}$ Solve for x.

The x-intercept is $-\frac{4}{3}$.

Make a table of values and graph the function. From the graph, you can see that the domain is $x \geq -\frac{4}{3}$, and the range is $y \geq 0$. The y-intercept is 2.

x	y
$-\frac{4}{3}$	0
−1	1
0	2
2	3.2
4	4

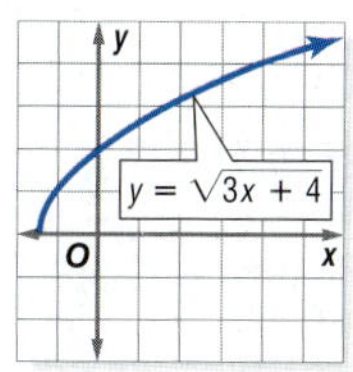

7-9 Lesson Notes

1 Focus

5-Minute Check Transparency 7-9 Use as a quiz or review of Lesson 7-8.

Mathematical Background notes are available for this lesson on p. 344D.

How are square root functions used in bridge design?

Ask students:

- Using the given data and formula, how much total weight can be supported by the bridge cables on the Sunshine Skyway Bridge? **13,608 tons**
- If the bridge had been designed to support 20,000 tons, how many 9-inch cables would have been used in the bridge design? **31 cables**

2 Teach

SQUARE ROOT FUNCTIONS

Teaching Tip In the graph on the right, point out the two graphs and lead students to see how the domain and range of the graph of $y = \sqrt{x}$ have been restricted.

Resource Manager

Workbook and Reproducible Masters

Chapter 7 Resource Masters

- Study Guide and Intervention, pp. 423–424
- Skills Practice, p. 425
- Practice, p. 426
- Reading to Learn Mathematics, p. 427
- Enrichment, p. 428
- Assessment, p. 444

Transparencies

5-Minute Check Transparency 7-9

Answer Key Transparencies

Technology

Interactive Chalkboard

In-Class Examples

Power Point®

1 Graph $y = \sqrt{\frac{3}{2}x - 1}$. State the domain, range, x- and y-intercepts.

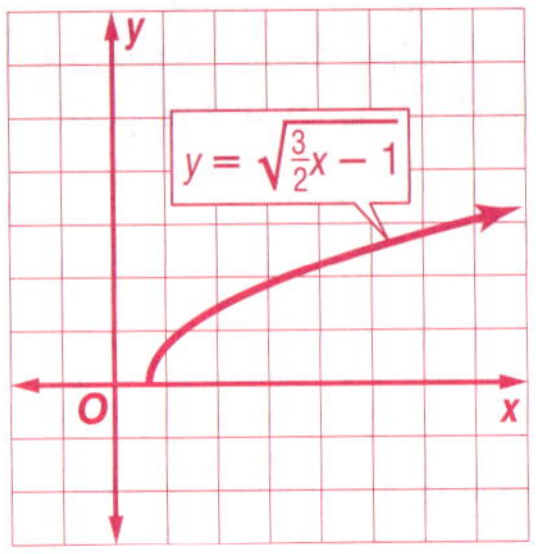

The domain is $x \geq \frac{2}{3}$, and the range is $y \geq 0$. The x-intercept is $\frac{2}{3}$. There is no y-intercept.

2 **PHYSICS** When an object is spinning in a circular path of radius 2 meters with velocity v, in meters per second, the centripetal acceleration a, in meters per second squared, is directed toward the center of the circle. The velocity v and acceleration a of the object are related by the function $v = \sqrt{2a}$.

a. Graph the function. State the domain and range. The domain is $a \geq 0$, and the range is $v \geq 0$.

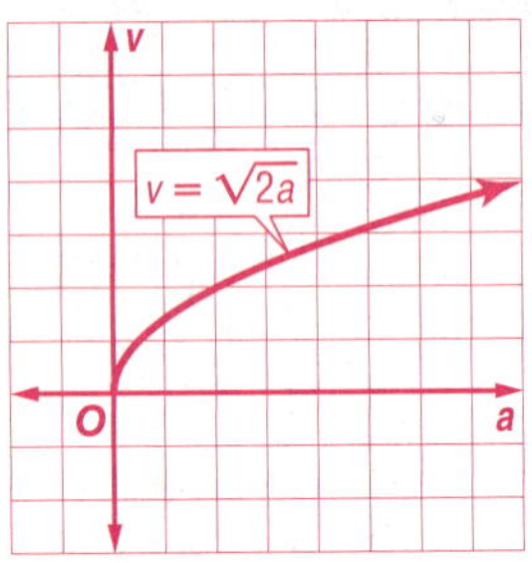

b. What would be the centripetal acceleration of an object spinning along the circular path with a velocity of 4 meters per second? 8 m/s^2

More About. . .

Submarines

Submarines were first used by The United States in 1776 during the Revolutionary War.

Source: www.infoplease.com

Example 2 Solve a Square Root Problem

SUBMARINES A lookout on a submarine is h feet above the surface of the water. The greatest distance d in miles that the lookout can see on a clear day is given by the square root of the quantity h multiplied by $\frac{3}{2}$.

a. Graph the function. State the domain and range.

The function is $d = \sqrt{\frac{3h}{2}}$. Make a table of values and graph the function.

h	d
0	0
2	$\sqrt{3}$ or 1.73
4	$\sqrt{6}$ or 2.45
6	$\sqrt{9}$ or 3.00
8	$\sqrt{12}$ or 3.46
10	$\sqrt{15}$ or 3.87

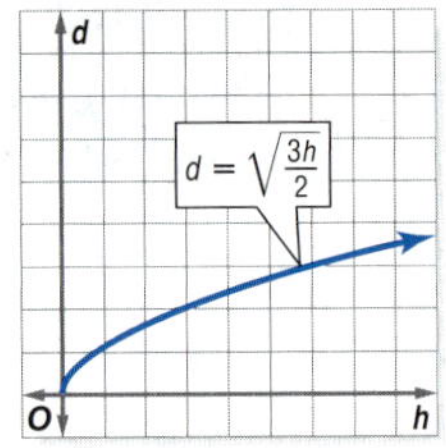

The domain is $h \geq 0$, and the range is $d \geq 0$.

b. A ship is 3 miles from a submarine. How high would the submarine have to raise its periscope in order to see the ship?

$d = \sqrt{\frac{3h}{2}}$ Original equation

$3 = \sqrt{\frac{3h}{2}}$ Replace d with 3.

$9 = \frac{3h}{2}$ Square each side.

$18 = 3h$ Multiply each side by 2.

$6 = h$ Divide each side by 3.

The periscope would have to be 6 feet above the water. Check this result on the graph.

Graphs of square root functions can be transformed just like quadratic functions.

Graphing Calculator Investigation

Square Root Functions

You can use a TI-83 Plus graphing calculator to graph square root functions. Use [2nd] [$\sqrt{\ }$] to enter the functions in the Y= list.

Think and Discuss 1–3. See pp. 407A–407H for graphs.

1. Graph $y = \sqrt{x}$, $y = \sqrt{x} + 1$, and $y = \sqrt{x} - 2$ in the viewing window $[-2, 8]$ by $[-4, 6]$. State the domain and range of each function and describe the similarities and differences among the graphs.
2. Graph $y = \sqrt{x}$, $y = \sqrt{2x}$, and $y = \sqrt{8x}$ in the viewing window [0, 10] by [0, 10]. State the domain and range of each function and describe the similarities and differences among the graphs.
3. Make a conjecture on how you could write an equation that translates the parent graph $y = \sqrt{x}$ to the left three units. Test your conjecture with the graphing calculator. $y = \sqrt{x + 3}$

1. D: $x \geq 0$, R: $y \geq 0$; D: $x \geq 0$, R: $y \geq 1$; D: $x \geq 0$, R: $y \geq -2$; Graphs are the same except they are translated vertically.
2. D: $x \geq 0$, R: $y \geq 0$; D: $x \geq 0$, R: $y \geq 0$; D: $x \geq 0$, R: $y \geq 0$; Graphs are the same except they get increasingly less steep.

Graphing Calculator Investigation

Square Roots Students who have worked with non-graphing calculators will likely be used to finding square roots by typing a value first and then pressing the square root key. On a graphing calculator, the square root key is pressed first, followed by the expression whose square root is to be found. Point out that the calculator treats the radical symbol like an open set of parentheses. If students want to graph $y = \sqrt{\frac{3}{2}x - 1}$, they need to enter a right parenthesis after the 1.

SQUARE ROOT INEQUALITIES

A **square root inequality** is an inequality involving square roots. You can use what you know about square root functions to graph square root inequalities.

Example 3 Graph a Square Root Inequality

a. Graph $y < \sqrt{2x - 6}$.

Graph the related equation $y = \sqrt{2x - 6}$. Since the boundary should not be included, the graph should be dashed.

The domain includes values for $x \geq 3$, so the graph is to the right of $x = 3$. Select a point and test its ordered pair.

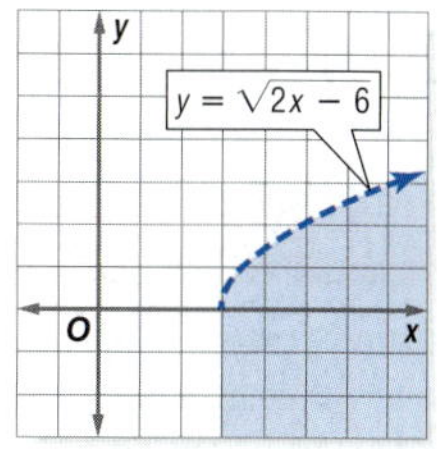

Test (4, 1).

$1 < \sqrt{2(4) - 6}$

$1 < \sqrt{2}$ true

Shade the region that includes the point (4, 1).

b. Graph $y \geq \sqrt{x + 1}$.

Graph the related equation $y = \sqrt{x + 1}$.

The domain includes values for $x \geq -1$, so the graph includes $x = -1$ and the values of x to the right of $x = -1$. Select a point and test its ordered pair.

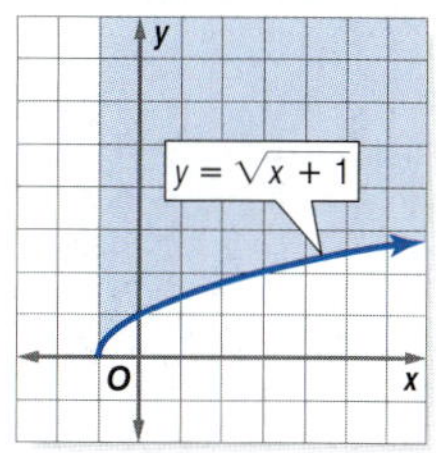

Test (2, 1).

$y \geq \sqrt{x + 1}$.

$1 \geq \sqrt{2 + 1}$

$1 \geq \sqrt{3}$ false

Shade the region that does not include (2, 1).

Check for Understanding

Concept Check

1–2. See margin.

1. **Explain** why the inverse of $y = 3x^2$ is not a square root function.
2. **Describe** the difference between the graphs of $y = \sqrt{x} - 4$ and $y = \sqrt{x - 4}$.
3. **OPEN ENDED** Write a square root function with a domain of $\{x \mid x \geq 2\}$. Sample answer: $y = \sqrt{2x - 4}$

Guided Practice

GUIDED PRACTICE KEY	
Exercises	Examples
4–7	1
8–11	3
12, 13	2

Graph each function. State the domain and range of the function.

4. $y = \sqrt{x} + 2$ D: $x \geq 0$, R: $y \geq 2$
5. $y = \sqrt{4x}$ D: $x \geq 0$; R: $y \geq 0$
6. $y = 3 - \sqrt{x}$ D: $x \geq 0$; R: $y \leq 3$
7. $y = \sqrt{x - 1} + 3$ D: $x \geq 1$; R: $y \geq 3$

4–7. See pp. 407A–407H for graphs.

Graph each inequality. 8–11. See pp. 407A–407H.

8. $y \leq \sqrt{x - 4} + 1$
9. $y > \sqrt{2x + 4}$
10. $y < 3 - \sqrt{5x + 1}$
11. $y \geq \sqrt{x + 2} - 1$

www.algebra2.com/extra_examples

SQUARE ROOT INEQUALITIES

In-Class Example

Power Point®

3 a. Graph $y > \sqrt{3x + 5}$.

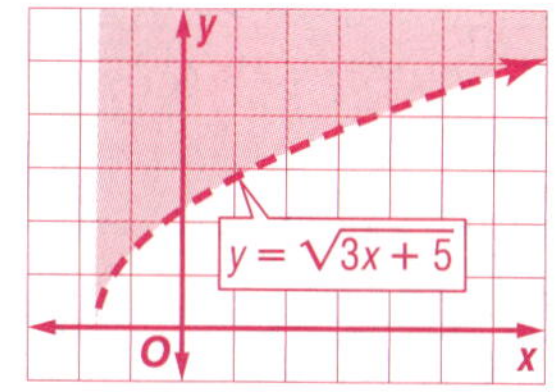

b. Graph $y \leq \sqrt{4 + \frac{3}{2}x}$.

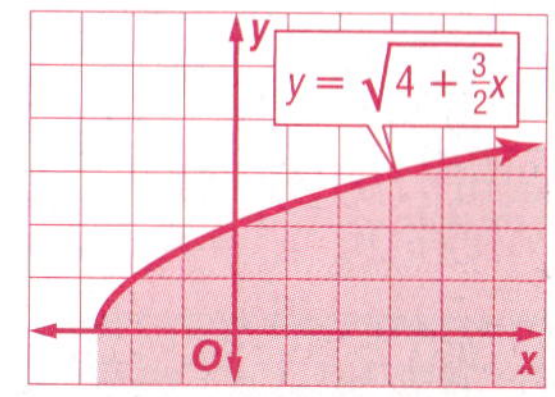

3 Practice/Apply

Study Notebook

Have students—

- complete the definitions/examples for the remaining terms on their Vocabulary Builder worksheets for Chapter 7.
- include any other item(s) that they find helpful in mastering the skills in this lesson.

Answers

1. In order for it to be a square root function, only the nonnegative range can be considered.
2. Both have the shape of the graph of $y = \sqrt{x}$, but $y = \sqrt{x} - 4$ is shifted down 4 units, and $y = \sqrt{x - 4}$ is shifted to the right 4 units.

DAILY INTERVENTION

Differentiated Instruction

Auditory/Musical Divide the class into groups of 4 to 6 students. Challenge each group to give themselves a rock or rap group name based on the vocabulary in this lesson, such as "The Intercepts." Have students write a musical verse about some of the key facts in the lesson, such as the domain, range, and intercepts of a graph.

Study Guide and Intervention, p. 423 (shown) and p. 424

Square Root Functions A function that contains the square root of a variable expression is a **square root function**.

Example Graph $y = \sqrt{3x - 2}$. State its domain and range.

Since the radicand cannot be negative, $3x - 2 \geq 0$ or $x \geq \frac{2}{3}$.

The x-intercept is $\frac{2}{3}$. The range is $y \geq 0$.

Make a table of values and graph the function.

x	y
$\frac{2}{3}$	0
1	1
2	2
3	$\sqrt{7}$

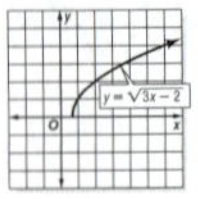

Exercises

Graph each function. State the domain and range of the function.

1. $y = \sqrt{2x}$ 2. $y = -3\sqrt{x}$ 3. $y = -\sqrt{\frac{x}{2}}$

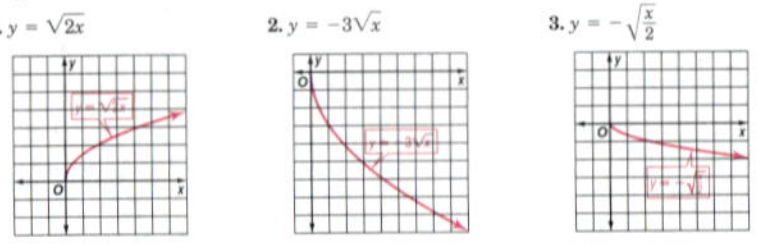

D: $x \geq 0$; R: $y \geq 0$ D: $x \geq 0$; R: $y \leq 0$ D: $x \geq 0$; R: $y \leq 0$

4. $y = 2\sqrt{x - 3}$ 5. $y = -\sqrt{2x - 3}$ 6. $y = \sqrt{2x + 5}$

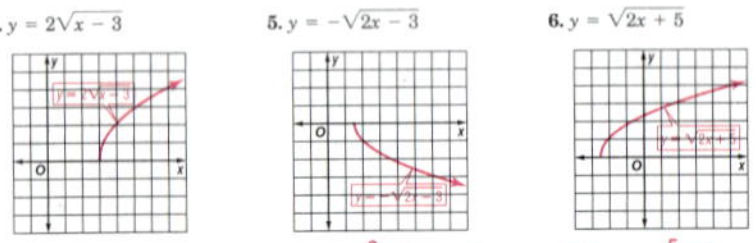

D: $x \geq 3$; R: $y \geq 0$ D: $x \geq \frac{3}{2}$; R: $y \leq 0$ D: $x \geq -\frac{5}{2}$; R: $y \geq 0$

Skills Practice, p. 425 and Practice, p. 426 (shown)

Graph each function. State the domain and range of each function.

1. $y = \sqrt{5x}$ 2. $y = -\sqrt{x - 1}$ 3. $y = 2\sqrt{x + 2}$

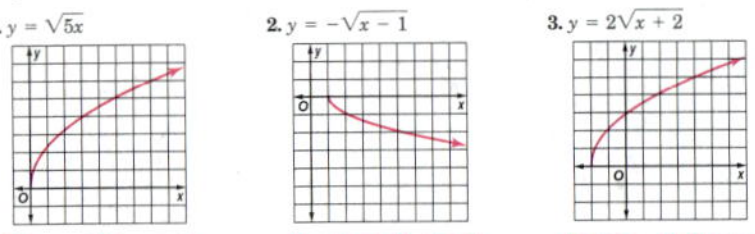

D: $x \geq 0$, R: $y \geq 0$ D: $x \geq 1$, R: $y \leq 0$ D: $x \geq -2$, R: $y \geq 0$

4. $y = \sqrt{3x - 4}$ 5. $y = \sqrt{x + 7} - 4$ 6. $y = 1 - \sqrt{2x + 3}$

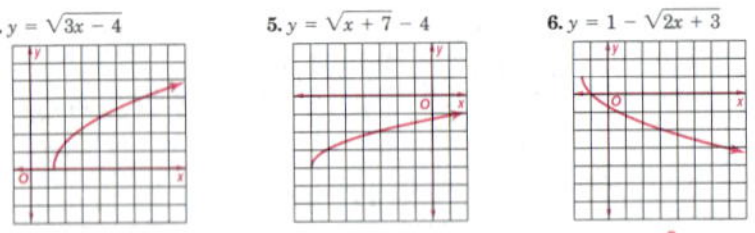

D: $x \geq \frac{4}{3}$, R: $y \geq 0$ D: $x \geq -7$, R: $y \geq -4$ D: $x \geq -\frac{3}{2}$, R: $y \leq 1$

Graph each inequality.

7. $y \geq -\sqrt{6x}$ 8. $y \leq \sqrt{x - 5} + 3$ 9. $y > -2\sqrt{3x + 2}$

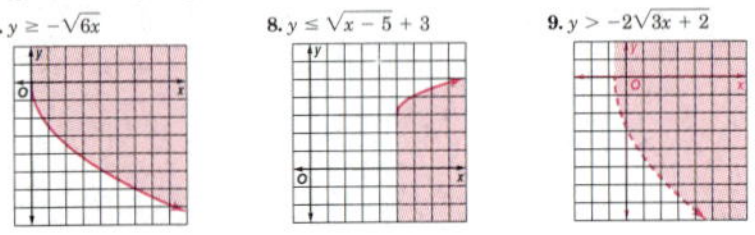

10. **ROLLER COASTERS** The velocity of a roller coaster as it moves down a hill is $v = \sqrt{v_0^2 + 64h}$, where v_0 is the initial velocity and h is the vertical drop in feet. If $v = 70$ feet per second and $v_0 = 8$ feet per second, find h. about 75.6 ft

11. **WEIGHT** Use the formula $d = \sqrt{\frac{3960^2 W_E}{W_s}} - 3960$, which relates distance from Earth d in miles to weight. If an astronaut's weight on Earth W_E is 148 pounds and in space W_s is 115 pounds, how far from Earth is the astronaut? about 532 mi

Reading to Learn Mathematics, p. 427 ELL

Pre-Activity How are square root functions used in bridge design?

Read the introduction to Lesson 7-9 at the top of page 395 in your textbook.

If the weight to be supported by a steel cable is doubled, should the diameter of the support cable also be doubled? If not, by what number should the diameter be multiplied?

no; $\sqrt{2}$

Reading the Lesson

1. Match each square root function from the list on the left with its domain and range from the list on the right.

a. $y = \sqrt{x}$ iv	i. domain: $x \geq 0$; range: $y \geq 3$
b. $y = \sqrt{x + 3}$ viii	ii. domain: $x \geq 0$; range: $y \leq 0$
c. $y = \sqrt{x} + 3$ i	iii. domain: $x \geq 0$; range: $y \leq -3$
d. $y = \sqrt{x - 3}$ v	iv. domain: $x \geq 0$; range: $y \geq 0$
e. $y = -\sqrt{x}$ ii	v. domain: $x \geq 3$; range: $y \geq 0$
f. $y = -\sqrt{x - 3}$ vii	vi. domain: $x \leq 3$; range: $y \geq 3$
g. $y = \sqrt{3 - x} + 3$ vi	vii. domain: $x \geq 3$; range: $y \leq 0$
h. $y = -\sqrt{x} - 3$ iii	viii. domain: $x \geq -3$; range: $y \geq 0$

2. The graph of the inequality $y \leq \sqrt{3x + 6}$ is a shaded region. Which of the following points lie inside this region?

(3, 0) (2, 4) (5, 2) (4, −2) (6, 6)

(3, 0), (5, 2), (4, −2)

Helping You Remember

3. A good way to remember something is to explain it to someone else. Suppose you are studying this lesson with a classmate who thinks that you cannot have square root functions because every positive real number has two square roots. How would you explain the idea of square root functions to your classmate?

Sample answer: To form a square root function, choose *either* the positive or negative square root. For example, $y = \sqrt{x}$ and $y = -\sqrt{x}$ are two separate functions.

Enrichment, p. 428

Reading Algebra

If two mathematical problems have basic structural similarities, they are said to be **analogous**. Using analogies is one way of discovering and proving new theorems.

The following numbered sentences discuss a three-dimensional analogy to the Pythagorean theorem.

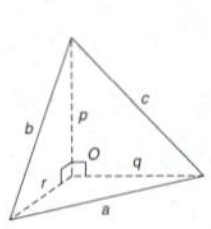

01 Consider a tetrahedron with three perpendicular faces that meet at vertex O.

02 Suppose you want to know how the areas A, B, and C of the three faces that meet at vertex O are related to the area D of the face opposite vertex O.

03 It is natural to expect a formula analogous to the Pythagorean theorem $z^2 = x^2 + y^2$, which is true for a similar situation in two dimensions.

04 To explore the three-dimensional case, you might guess a formula and then try to prove it.

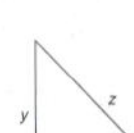

05 Two r... ble guesses are $D^3 = A^3 + B^3 + C^3$ and ...

Application

FIREFIGHTING For Exercises 12 and 13, use the following information.

When fighting a fire, the velocity v of water being pumped into the air is the square root of twice the product of the maximum height h and g, the acceleration due to gravity (32 ft/s²).

12. Determine an equation that will give the maximum height of the water as a function of its velocity. $h = \frac{v^2}{2g}$

13. The Coolville Fire Department must purchase a pump that is powerful enough to propel water 80 feet into the air. Will a pump that is advertised to project water with a velocity of 75 ft/s meet the fire department's need? Explain. **Yes; sample answer: the advertised pump will reach a maximum height of 87.9 ft.**

Practice and Apply

Homework Help

For Exercises	See Examples
14–25	1
26–31	3
32–34	2

Extra Practice See page 844.

14. D: $x \geq 0$, R: $y \geq 0$
15. D: $x \geq 0$, R: $y \leq 0$
16. D: $x \geq 0$, R: $y \leq 0$
17. D: $x \geq 0$, R: $y \geq 0$
18. D: $x \geq -2$, R: $y \geq 0$
19. D: $x \geq 7$, R: $y \geq 0$
20. D: $x \geq -0.5$, R: $y \leq 0$
21. D: $x \geq 0.6$, R: $y \geq 0$
22. D: $x \geq -6$, R: $y \geq -3$
23. D: $x \geq -4$, R: $y \leq 5$
24. D: $x \geq 2$, R: $y \geq 4$
25. D: $x \leq 0.75$, R: $y \geq 3$

Graph each function. State the domain and range of each function.

14. $y = \sqrt{3x}$ 15. $y = -\sqrt{5x}$ 16. $y = -4\sqrt{x}$

17. $y = \frac{1}{2}\sqrt{x}$ 18. $y = \sqrt{x + 2}$ 19. $y = \sqrt{x - 7}$

20. $y = -\sqrt{2x + 1}$ 21. $y = \sqrt{5x - 3}$ 22. $y = \sqrt{x + 6} - 3$

23. $y = 5 - \sqrt{x + 4}$ 24. $y = \sqrt{3x - 6} + 4$ 25. $y = 2\sqrt{3 - 4x} + 3$

14–25. See pp. 407A–407H for graphs.

Graph each inequality. **26–31. See pp. 407A–407H.**

26. $y \leq -6\sqrt{x}$ 27. $y < \sqrt{x + 5}$ 28. $y > \sqrt{2x + 8}$

29. $y \geq \sqrt{5x - 8}$ 30. $y \geq \sqrt{x - 3} + 4$ 31. $y < \sqrt{6x - 2} + 1$

32. **ROLLER COASTERS** The velocity of a roller coaster as it moves down a hill is $v = \sqrt{v_0^2 + 64h}$, where v_0 is the initial velocity and h is the vertical drop in feet. An engineer wants a new coaster to have a velocity of 90 feet per second when it reaches the bottom of the hill. If the initial velocity of the coaster at the top of the hill is 10 feet per second, how high should the engineer make the hill? **125 ft**

AEROSPACE For Exercises 33 and 34, use the following information.

The force due to gravity decreases with the square of the distance from the center of Earth. So, as an object moves farther from Earth, its weight decreases. The radius of Earth is approximately 3960 miles. The formula relating weight and distance is

$$r = \sqrt{\frac{3960^2 W_E}{W_S}} - 3960,$$ where W_E represents the weight of a body on Earth, W_S represents the weight of a body a certain distance from the center of Earth, and r represents the distance of an object above Earth's surface.

33. An astronaut weighs 140 pounds on Earth and 120 pounds in space. How far is he above Earth's surface? **317.29 mi**

34. An astronaut weighs 125 pounds on Earth. What is her weight in space if she is 99 miles above the surface of Earth? **119 lb**

More About. . .

Aerospace

The weight of a person is equal to the product of the person's mass and the acceleration due to Earth's gravity. Thus, as a person moves away from Earth, the person's weight decreases. However, mass remains constant.

35. **RESEARCH** Use the Internet or another resource to find the weights, on Earth, of several space shuttle astronauts and the average distance they were from Earth during their missions. Use this information to calculate their weights while in orbit. **See students' work.**

36. **CRITICAL THINKING** Recall how values of a, h, and k can affect the graph of a quadratic function of the form $y = a(x - h)^2 + k$. Describe how values of a, h, and k can affect the graph of a square root function of the form $y = a\sqrt{x - h} + k$. **See margin.**

37. **WRITING IN MATH** Answer the question that was posed at the beginning of the lesson. **See margin.**

How are square root functions used in bridge design?

Include the following in your answer:

- the weights for which a diameter less than 1 is reasonable, and
- the weight that the Sunshine Skyway Bridge can support.

Standardized Test Practice

38. What is the domain of $f(x) \geq \sqrt{5x - 3}$? **C**

(A) $\left\{x \mid x > \frac{3}{5}\right\}$ (B) $\left\{x \mid x > -\frac{3}{5}\right\}$ (C) $\left\{x \mid x \geq \frac{3}{5}\right\}$ (D) $\left\{x \mid x \geq -\frac{3}{5}\right\}$

39. Given the graph of the square root function at the right, which of the following must be true? **D**

I. The domain is all real numbers.

II. The function is $y = \sqrt{x} + 3.5$.

III. The range is $\{y \mid y \geq 3.5\}$.

(A) I only (B) I, II, and III
(C) II and III (D) III only

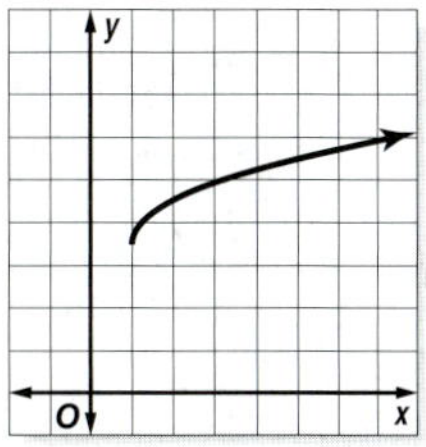

Maintain Your Skills

Mixed Review **Determine whether each pair of functions are inverse functions.** *(Lesson 7-8)*

40. $f(x) = 3x$ **yes**
$g(x) = \frac{1}{3}x$

41. $f(x) = 4x - 5$ **no**
$g(x) = \frac{1}{4}x - \frac{5}{16}$

42. $f(x) = \frac{3x + 2}{7}$ **yes**
$g(x) = \frac{7x - 2}{3}$

Find $(f + g)(x)$, $(f - g)(x)$, $(f \cdot g)(x)$, and $\left(\frac{f}{g}\right)(x)$ for each $f(x)$ and $g(x)$. *(Lesson 7-7)*

43. $f(x) = x + 5$
$g(x) = x - 3$

44. $f(x) = 10x - 20$
$g(x) = x - 2$

45. $f(x) = 4x^2 - 9$
$g(x) = \frac{1}{2x + 3}$

43–45. See margin.

46. **ENTERTAINMENT** A magician asked a member of his audience to choose any number. He said, "Multiply your number by 3. Add the sum of your number and 8 to that result. Now divide by the sum of your number and 2." The magician announced the final answer without asking the original number. What was the final answer? How did he know what it was? *(Lesson 5-4)*

46. 4; If x is your number, you can write the expression $\frac{3x + x + 8}{x + 2}$, which equals 4 after dividing the numerator and denominator by the GCF, $x + 2$.

Simplify. *(Lesson 5-2)*

47. $(x + 2)(2x - 8)$
$2x^2 - 4x - 16$

48. $(3p + 5)(2p - 4)$
$6p^2 - 2p - 20$

49. $(a^2 + a + 1)(a - 1)$
$a^3 - 1$

WebQuest Internet Project

Population Explosion

It is time to complete your project. Use the information and data you have gathered about the population to prepare a Web page. Be sure to include graphs, tables, and equations in the presentation.

www.algebra2.com/webquest

www.algebra2.com/self_check_quiz

About the Exercises...

Organization by Objective

- **Square Root Functions:** 14–25
- **Square Root Inequalities:** 26–31

Odd/Even Assignments

Exercises 14–31 are structured so that students practice the same concepts whether they are assigned odd or even problems.

Alert! Exercise 35 involves research on the Internet or other reference materials.

Assignment Guide

Basic: 15–31 odd, 33–49

Average: 15–31 odd, 33–49

Advanced: 14–32 even, 33–34, 36–49

4 Assess

Open-Ended Assessment

Writing Have students write a paragraph explaining why the domain and range of square root functions and square root inequalities must be restricted.

Assessment Options

Quiz (Lesson 7-9) is available on p. 444 of the *Chapter 7 Resource Masters.*

Answers

43. $2x + 2$; 8; $x^2 + 2x - 15$; $\frac{x + 5}{x - 3}$, $x \neq 3$

44. $11x - 22$; $9x - 18$; $10x^2 - 40x + 40$; 10, $x \neq 2$

45. $\frac{8x^3 + 12x^2 - 18x - 26}{2x + 3}$, $x \neq -\frac{3}{2}$; $\frac{8x^3 + 12x^2 - 18x - 28}{2x + 3}$, $x \neq -\frac{3}{2}$; $2x - 3$, $x \neq -\frac{3}{2}$; $8x^3 + 12x^2 - 18x - 27$, $x \neq -\frac{3}{2}$

Answers

36. If a is negative, the graph is reflected over the x-axis. The larger the value of a, the less steep the graph. If h is positive, the origin is translated to the right, and if h is negative, the origin is translated to the left. When k is positive, the origin is translated up, and when k is negative, the origin is translated down.

37. Square root functions are used in bridge design because the engineers must determine what diameter of steel cable needs to be used to support a bridge based on its weight. Answers should include the following.

- **Sample answer: When the weight to be supported is less than 8 tons.**
- **13,608 tons**

Chapter 7 Study Guide and Review

Vocabulary and Concept Check

- This alphabetical list of vocabulary terms in Chapter 7 includes a page reference where each term was introduced.
- **Assessment** A vocabulary test/review for Chapter 7 is available on p. 442 of the *Chapter 7 Resource Masters*.

Lesson-by-Lesson Review

For each lesson,

- the main ideas are summarized,
- additional examples review concepts, and
- practice exercises are provided.

Vocabulary PuzzleMaker

ELL The Vocabulary PuzzleMaker software improves students' mathematics vocabulary using four puzzle formats—crossword, scramble, word search using a word list, and word search using clues. Students can work on a computer screen or from a printed handout.

MindJogger Videoquizzes

ELL MindJogger Videoquizzes provide an alternative review of concepts presented in this chapter. Students work in teams in a game show format to gain points for correct answers. The questions are presented in three rounds.

Round 1 Concepts (5 questions)
Round 2 Skills (4 questions)
Round 3 Problem Solving (4 questions)

Chapter 7 Study Guide and Review

Vocabulary and Concept Check

Complex Conjugates Theorem (p. 374)
composition of functions (p. 384)
degree of a polynomial (p. 346)
depressed polynomial (p. 366)
Descartes' Rule of Signs (p. 372)
end behavior (p. 349)
Factor Theorem (p. 366)
Fundamental Theorem of Algebra (p. 371)
identity function (p. 391)
Integral Zero Theorem (p. 378)
inverse function (p. 391)
inverse relation (p. 390)
leading coefficients (p. 346)
Location Principle (p. 353)
one-to-one (p. 392)
polynomial function (p. 347)
polynomial in one variable (p. 346)
quadratic form (p. 360)
Rational Zero Theorem (p. 378)
relative maximum (p. 354)
relative minimum (p. 354)
Remainder Theorem (p. 365)
square root function (p. 395)
square root inequality (p. 397)
synthetic substitution (p. 365)

Choose the letter that best matches each statement or phrase.

1. A point on the graph of a polynomial function that has no other nearby points with lesser y-coordinates is a ______. **f**
2. The ______ is the coefficient of the term in a polynomial function with the highest degree. **d**
3. The ______ says that in any polynomial function, if an imaginary number is a zero of that function, then its conjugate is also a zero. **a**
4. When a polynomial is divided by one of its binomial factors, the quotient is called a(n) ______. **b**
5. $(x^2)^2 - 17(x^2) + 16 = 0$ is written in ______. **e**
6. $f(x) = 6x - 2$ and $g(x) = \frac{x+2}{6}$ are ______ since $[f \circ g](x)$ and $[g \circ f](x) = x$. **c**

a. Complex Conjugates Theorem
b. depressed polynomial
c. inverse functions
d. leading coefficient
e. quadratic form
f. relative minimum

Lesson-by-Lesson Review

7-1 Polynomial Functions

See pages 346–352.

Concept Summary

- The degree of a polynomial function in one variable is determined by the greatest exponent of its variable.

Example Find $p(a + 1)$ if $p(x) = 5x - x^2 + 3x^3$.

$p(a+1) = 5(a+1) - (a+1)^2 + 3(a+1)^3$ Replace x with $a + 1$.

$= 5a + 5 - (a^2 + 2a + 1) + 3(a^3 + 3a^2 + 3a + 1)$ Evaluate $5(a + 1)$, $(a + 1)^2$, and $3(a + 1)^3$.

$= 5a + 5 - a^2 - 2a - 1 + 3a^3 + 9a^2 + 9a + 3$

$= 3a^3 + 8a^2 + 12a + 7$ Simplify.

10. $21;\ x^2 + 2xh + h^2 + 5$ **11.** $20;\ x^2 + 2xh + h^2 - x - h$ **12.** $-129;\ 2x^3 + 6x^2h + 6xh^2 + 2h^3 - 1$

Exercises Find $p(-4)$ and $p(x + h)$ for each function.
See Examples 2 and 3 on pages 347 and 348.

7. $p(x) = x - 2$ **$-6;\ x + h - 2$**
8. $p(x) = -x + 4$ **$8;\ -x - h + 4$**
9. $p(x) = 6x + 3$ **$-21;\ 6x + 6h + 3$**
10. $p(x) = x^2 + 5$
11. $p(x) = x^2 - x$
12. $p(x) = 2x^3 - 1$

www.algebra2.com/vocabulary_review

FOLDABLES™ Study Organizer

For more information about Foldables, see *Teaching Mathematics with Foldables.*

Have students look through the chapter to make sure they have included notes and examples for each lesson in this chapter in their Foldable.

Encourage students to refer to their Foldables while completing the Study Guide and Review and to use them in preparing for the Chapter Test.

7-2 Graphing Polynomial Functions

See pages 353–358.

Concept Summary

- The Location Principle: Since zeros of a function are located at x-intercepts, there is also a zero between each pair of these zeros.
- Turning points of a function are called relative maxima and relative minima.

Example Graph $f(x) = x^4 - 2x^2 + 10x - 2$ by making a table of values.

Make a table of values for several values of x and plot the points. Connect the points with a smooth curve.

x	$f(x)$
−3	31
−2	−14
−1	−13
0	−2
1	7
2	26

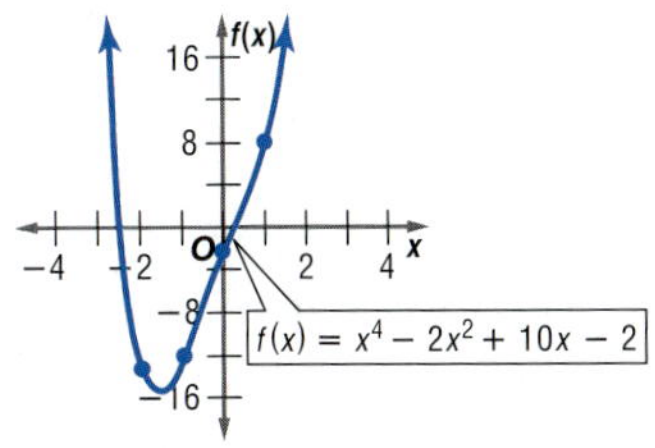

Exercises For Exercises 13–18, complete each of the following.

a. Graph each function by making a table of values.

b. Determine consecutive values of x between which each real zero is located.

c. Estimate the x-coordinates at which the relative maxima and relative minima occur. See Example 1 on page 353. **13–18. See margin.**

13. $h(x) = x^3 - 6x - 9$
14. $f(x) = x^4 + 7x + 1$
15. $p(x) = x^5 + x^4 - 2x^3 + 1$
16. $g(x) = x^3 - x^2 + 1$
17. $r(x) = 4x^3 + x^2 - 11x + 3$
18. $f(x) = x^3 + 4x^2 + x - 2$

7-3 Solving Equations Using Quadratic Techniques

See pages 360–364.

Concept Summary

- Solve polynomial equations by using quadratic techniques.

Example Solve $x^3 - 3x^2 - 54x = 0$.

$x^3 - 3x^2 - 54x = 0$ — Original equation

$x(x^2 - 3x - 54) = 0$ — Factor out the GCF.

$x(x - 9)(x + 6) = 0$ — Factor the trinomial.

$x = 0$ or $x - 9 = 0$ or $x + 6 = 0$ — Zero Product Property

$x = 0$ $\quad x = 9$ $\quad x = -6$

19. $\frac{5}{3}, -3, 0$
20. −8, 0, 5
21. $4, -2 \pm 2i\sqrt{3}$
23. 2, −2

Exercises Solve each equation. See Example 2 on page 361.

19. $3x^3 + 4x^2 - 15x = 0$
20. $m^4 + 3m^3 = 40m^2$
21. $a^3 - 64 = 0$
22. $r + 9\sqrt{r} = -8$ **∅**
23. $x^4 - 8x^2 + 16 = 0$
24. $x^{\frac{2}{3}} - 9x^{\frac{1}{3}} + 20 = 0$ **64, 125**

13a.

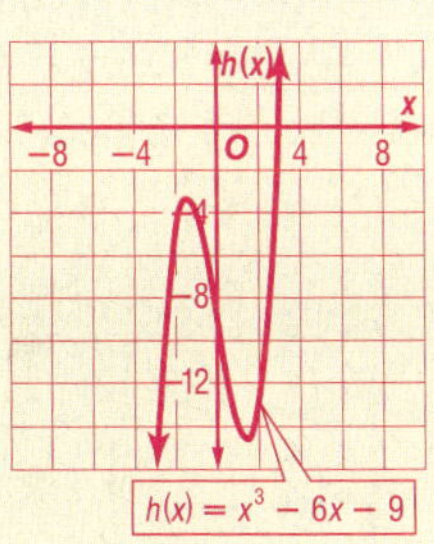

13b. at $x = 3$

13c. Sample answer: rel. max. at $x = -1.4$, rel. min. at $x = 1.4$

14a.

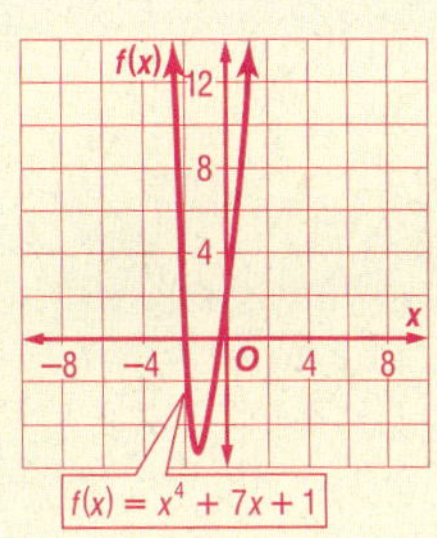

Study Guide and Review

Answers

14b. between −2 and −1, and between −1 and 0

14c. Sample answer: no rel. max., rel. min. at $x = -1.2$

15a.

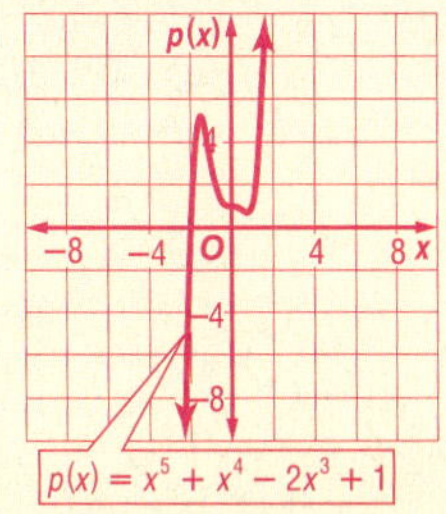

15b. between −2 and −3

15c. Sample answer: rel. max. at $x = 0$ $x = -1.6$, rel. min. at $x = 0.8$

16a.

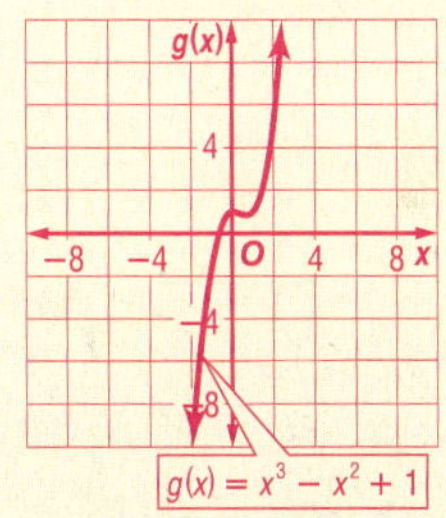

16b. between −1 and 0

16c. Sample answer: rel. max. at $x = 0$, rel. min. at $x = 0.7$

17a.

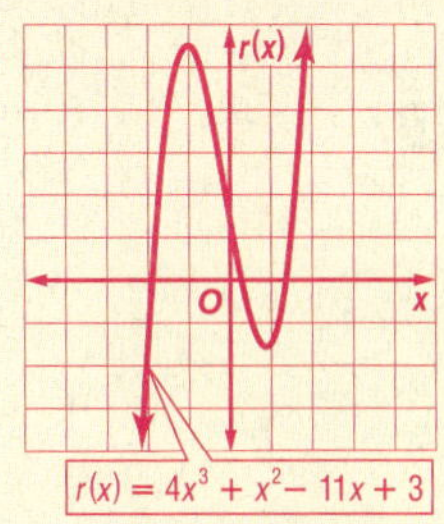

17b. between −2 and −1, between 0 and 1, and between 1 and 2

17c. Sample answer: rel. max. at $x = -1$, rel. min. at $x = 0.9$

18a.

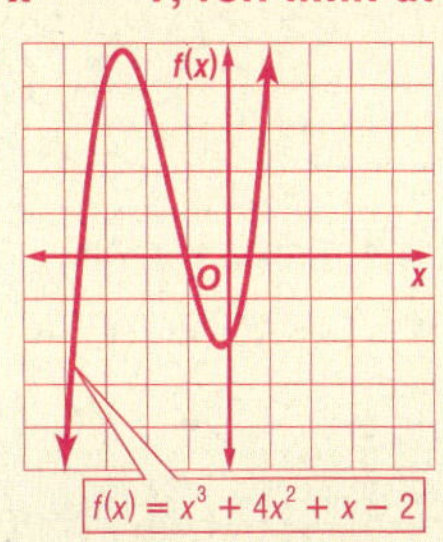

18b. between −4 and −3, at $x = -1$, and between 0 and 1

18c. Sample answer: rel. max. at $x = -2.5$, rel. min. at $x = -0.1$

7-4 The Remainder and Factor Theorems

See pages 365–370.

Concept Summary

- Remainder Theorem: If a polynomial $f(x)$ is divided by $x - a$, the remainder is the constant $f(a)$ and $f(x) = q(x) \cdot (x - a) + f(a)$ where $q(x)$ is a polynomial with degree one less than the degree of $f(x)$.
- Factor Theorem: $x - a$ is a factor of polynomial $f(x)$ if and only if $f(a) = 0$.

Example **Show that $x + 2$ is a factor of $x^3 - 2x^2 - 5x + 6$. Then find any remaining factors of the polynomial.**

$$\begin{array}{r|rrrr} -2 & 1 & -2 & -5 & 6 \\ & & -2 & 8 & -6 \\ \hline & 1 & -4 & 3 & 0 \end{array}$$

The remainder is 0, so $x + 2$ is a factor of $x^3 - 2x^2 - 5x + 6$. Since $x^3 - 2x^2 - 5x + 6 = (x + 2)(x^2 - 4x + 3)$, the factors of $x^3 - 2x^2 - 5x + 6$ are $(x + 2)(x - 3)(x - 1)$.

Exercises **Use synthetic substitution to find $f(3)$ and $f(-2)$ for each function.** *See Example 2 on page 367.* 26. 1, 16 27. 20, −20

25. $f(x) = x^2 - 5$ 4, −1
26. $f(x) = x^2 - 4x + 4$
27. $f(x) = x^3 - 3x^2 + 4x + 8$

Given a polynomial and one of its factors, find the remaining factors of the polynomial. Some factors may not be binomials. *See Example 3 on page 367.*

28. $x^3 + 5x^2 + 8x + 4;\ x + 1$
29. $x^3 + 4x^2 + 7x + 6;\ x + 2$ $x^2 + 2x + 3$

28. $x + 2,\ x + 2$

7-5 Roots and Zeros

See pages 371–377.

Concept Summary

- Fundamental Theorem of Algebra: Every polynomial equation with degree greater than zero has at least one root in the set of complex numbers.
- Use Descartes' Rule of Signs to determine types of zeros of polynomial functions.
- Complex Conjugates Theorem: If $a + bi$ is a zero of a polynomial function, then $a - bi$ is also a zero of the function.

Example **State the possible number of positive real zeros, negative real zeros, and imaginary zeros of $f(x) = 5x^4 + 6x^3 - 8x + 12$.**

Since $f(x)$ has two sign changes, there are 2 or 0 real positive zeros.

$f(-x) = 5x^4 - 6x^3 + 8x + 12$ Two sign changes → 0 or 2 negative real zeros

There are 0, 2, or 4 imaginary zeros. 30. 3 or 1; 1; 2 or 0
33. 3 or 1; 1; 0 or 2 34. 2 or 0; 2 or 0; 4, 2, or 0 35. 2 or 0; 2 or 0; 4, 2, or 0

Exercises **State the possible number of positive real zeros, negative real zeros, and imaginary zeros of each function.** *See Example 2 on page 373.*

30. $f(x) = 2x^4 - x^3 + 5x^2 + 3x - 9$
31. $f(x) = 7x^3 + 5x - 1$ 1; 0; 2
32. $f(x) = -4x^4 - x^2 - x + 1$ 1; 1; 2
33. $f(x) = 3x^4 - x^3 + 8x^2 + x - 7$
34. $f(x) = x^4 + x^3 - 7x + 1$
35. $f(x) = 2x^4 - 3x^3 - 2x^2 + 3$

402 Chapter 7 Polynomial Functions

7-6 Rational Zero Theorem

See pages 378–382.

Concept Summary

- Use the Rational Zero Theorem to find possible zeros of a polynomial function.
- Integral Zero Theorem: If the coefficients of a polynomial function are integers such that $a_0 = 1$ and $a_n \neq 0$, any rational zeros of the function must be factors of a_n.

Examples **Find all of the zeros of $f(x) = x^3 + 7x^2 - 36$.**

There are exactly three complex zeros.

There is exactly one positive real zero and two or zero negative real zeros.

The possible rational zeros are ±1, ±2, ±3, ±4, ±6, ±9, ±12, ±18, ±36.

$$\begin{array}{r|rrrr} 2 & 1 & 7 & 0 & -36 \\ & & 2 & 18 & 36 \\ \hline & 1 & 9 & 18 & 0 \end{array}$$

$$x^3 + 7x^2 - 36 = (x - 2)(x^2 + 9x + 18) = (x - 2)(x + 3)(x + 6)$$

Therefore, the zeros are 2, −3, and −6.

Exercises **Find all of the rational zeros of each function.** *See Example 3 on page 379.*

36. $f(x) = 2x^3 - 13x^2 + 17x + 12$ **$-\frac{1}{2}, 3, 4$**
37. $f(x) = x^4 + 5x^3 + 15x^2 + 19x + 8$
38. $f(x) = x^3 - 3x^2 - 10x + 24$ **−3, 2, 4**
39. $f(x) = x^4 - 4x^3 - 7x^2 + 34x - 24$
40. $f(x) = 2x^3 - 5x^2 - 28x + 15$
41. $f(x) = 2x^4 - 9x^3 + 2x^2 + 21x - 10$

37. −1, −1 39. 1, 2, 4, −3 40. −3, 5, $\frac{1}{2}$ 41. $\frac{1}{2}$, 2

7-7 Operations of Functions

See pages 383–389.

Concept Summary

Operation	Definition	Operation	Definition
Sum	$(f + g)(x) = f(x) + g(x)$	Quotient	$\left(\frac{f}{g}\right)(x) = \frac{f(x)}{g(x)}, g(x) \neq 0$
Difference	$(f - g)(x) = f(x) - g(x)$	Composition	$[f \circ g](x) = f[g(x)]$
Product	$(f \cdot g)(x) = f(x) \cdot g(x)$	—	—

Example **If $f(x) = x^2 - 2$ and $g(x) = 8x - 1$. Find $g[f(x)]$ and $f[g(x)]$.**

$g[f(x)] = 8(x^2 - 2) - 1$ — Replace $f(x)$ with $x^2 - 2$.

$= 8x^2 - 16 - 1$ — Multiply.

$= 8x^2 - 17$ — Simplify.

$f[g(x)] = (8x - 1)^2 - 2$ — Replace $g(x)$ with $8x - 1$.

$= 64x^2 - 16x + 1 - 2$ — Expand the binomial.

$= 64x^2 - 16x - 1$ — Simplify.

Exercises **Find $[g \circ h](x)$ and $[h \circ g](x)$.** *See Example 4 on page 385.*

42. $h(x) = 2x - 1$, $g(x) = 3x + 4$ **$6x + 1$; $6x + 7$**
43. $h(x) = x^2 + 2$, $g(x) = x - 3$
44. $h(x) = x^2 + 1$, $g(x) = -2x + 1$
45. $h(x) = -5x$, $g(x) = 3x - 5$
46. $h(x) = x^3$, $g(x) = x - 2$
47. $h(x) = x + 4$, $g(x) = |x|$ **$|x + 4|$; $|x| + 4$**

43. $x^2 - 1$; $x^2 - 6x + 11$ 44. $-2x^2 - 1$; $4x^2 - 4x + 2$ 45. $-15x - 5$; $-15x + 25$ 46. $x^3 - 2$; $x^3 - 6x^2 + 12x - 8$

Chapter 7 For More ...
• Extra Practice, see pages 842–844.
• Mixed Problem Solving, see page 868.

7-8 Inverse Functions and Relations

See pages 390–394.

Concept Summary

- Reverse the coordinates of ordered pairs to find the inverse of a relation.
- Two functions are inverses if and only if both of their compositions are the identity function. $[f \circ g](x) = x$ and $[g \circ f](x) = x$
- A function is one-to-one when the inverse of the function is a function.

Example **Find the inverse of $f(x) = -3x + 1$.**

Rewrite $f(x)$ as $y = -3x + 1$. Then interchange the variables and solve for y.

$x = -3y + 1$ Interchange the variables.

$3y = -x + 1$ Solve for y.

$y = \frac{-x + 1}{3}$ Divide each side by 3.

$f^{-1}(x) = \frac{-x + 1}{3}$ Rewrite in function notation.

Exercises **Find the inverse of each function. Then graph the function and its inverse.** *See Example 2 on page 391.* **48–53. See pp. 407A–407H.**

48. $f(x) = 3x - 4$ **49.** $f(x) = -2x - 3$ **50.** $g(x) = \frac{1}{3}x + 2$

51. $f(x) = \frac{-3x + 1}{2}$ **52.** $y = x^2$ **53.** $y = (2x + 3)^2$

7-9 Square Root Functions and Inequalities

See pages 395–399.

Concept Summary

- Graph square root inequalities in a similar manner as graphing square root equations.

Example **Graph $y = 2 + \sqrt{x - 1}$.**

x	y
1	2
2	3
3	$2 + \sqrt{2}$ or 3.4
4	$2 + \sqrt{3}$ or 3.7
5	4

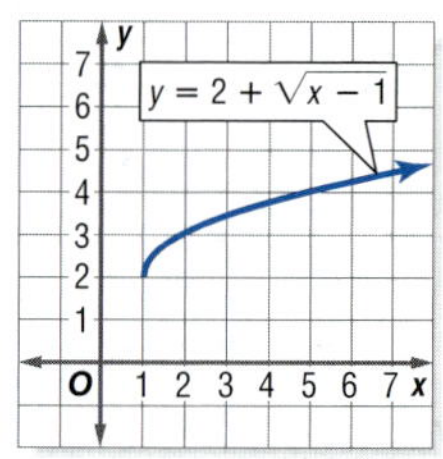

Exercises **Graph each function. State the domain and range of each function.** *See Examples 1 and 2 on pages 395 and 396.* **54–56. See margin.**

54. $y = \frac{1}{3}\sqrt{x + 2}$ **55.** $y = \sqrt{5x - 3}$ **56.** $y = 4 + 2\sqrt{x - 3}$

Graph each inequality. *See Example 3 on page 397.* **57–58. See margin.**

57. $y \geq \sqrt{x} - 2$ **58.** $y < \sqrt{4x - 5}$

Answers

54. D: $x \geq -2$, R: $y \geq 0$

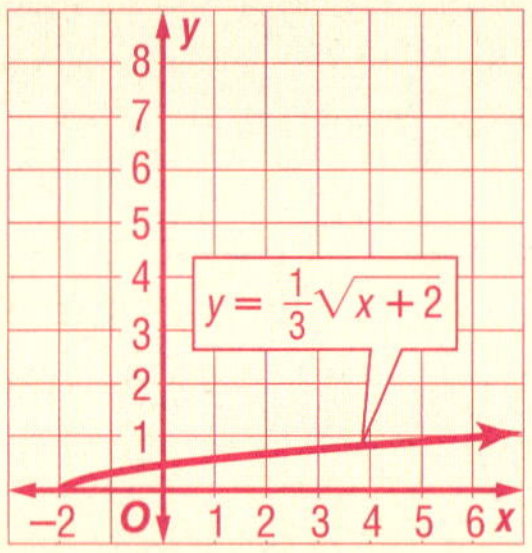

55. D: $x \geq \frac{3}{5}$, R: $y \geq 0$

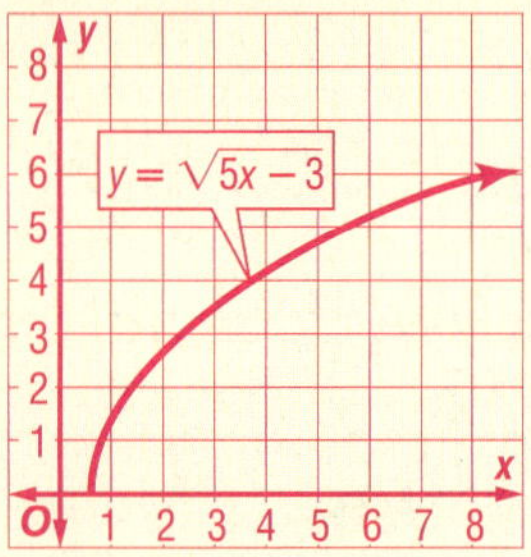

56. D: $x \geq 3$, R: $y \geq 4$

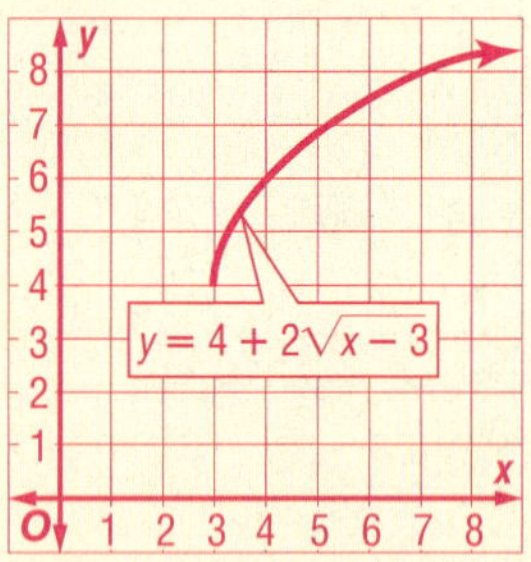

57.

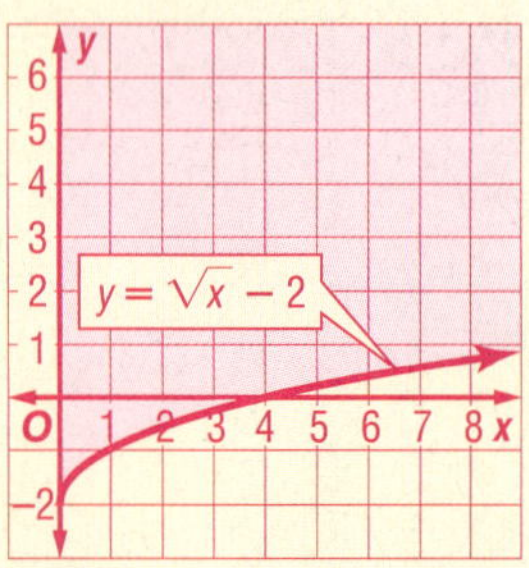

58.

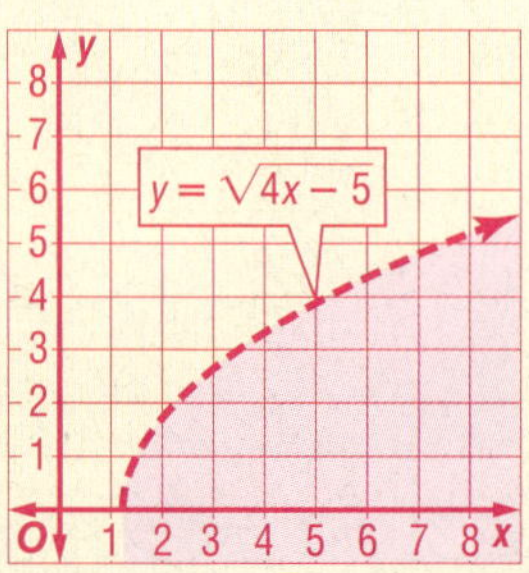

Answers (p. 405)

20. $x^2 + 2x - 1$

21. $-x^2 + 2x - 7$

22. $2x^3 - 4x^2 + 6x - 12$

23. $\frac{2x - 4}{x^2 + 3}$

24a. $A = 1000(1 + r)^6 + 1000(1 + r)^5 + 1000(1 + r)^4 + 1200(1 + r)^3 + 1200(1 + r)^2 + 2000(1 + r)$

404 Chapter 7 Polynomial Functions

Chapter 7 Practice Test

Vocabulary and Concepts

Match each statement with the term that it best describes.

1. $[f \circ g](x) = f[g(x)]$ **b**
2. $[f \circ g](x) = x$ and $[g \circ f](x) = x$ **c**
3. $(\sqrt{x})^2 - 2(\sqrt{x}) + 4 = 0$ **a**

a. quadratic form
b. composition of functions
c. inverse functions

Skills and Applications

For Exercises 4–7, complete each of the following. **4–7. See pp. 407A–407H.**

a. Graph each function by making a table of values.
b. Determine consecutive values of x between which each real zero is located.
c. Estimate the x-coordinates at which the relative maxima and relative minima occur.

4. $g(x) = x^3 + 6x^2 + 6x - 4$
5. $h(x) = x^4 + 6x^3 + 8x^2 - x$
6. $f(x) = x^3 + 3x^2 - 2x + 1$
7. $g(x) = x^4 - 2x^3 - 6x^2 + 8x + 5$

Solve each equation. **8. $0, -4 \pm \sqrt{34}$ 10. $\pm\sqrt{6}, \pm\sqrt{3}$**

8. $p^3 + 8p^2 = 18p$
9. $16x^4 - x^2 = 0$ **$0, \pm\frac{1}{4}$**
10. $r^4 - 9r^2 + 18 = 0$
11. $p^{\frac{3}{2}} - 8 = 0$ **4**

Given a polynomial and one of its factors, find the remaining factors of the polynomial. Some factors may not be binomials.

12. $x^3 - x^2 - 5x - 3; x + 1$ **$x - 3, x + 1$**
13. $x^3 + 8x + 24; x + 2$ **$x^2 - 2x + 12$**

State the possible number of positive real zeros, negative real zeros, and imaginary zeros for each function.

14. $f(x) = x^3 - x^2 - 14x + 24$ **2 or 0; 1; 0 or 2**
15. $f(x) = 2x^3 - x^2 + 16x - 5$ **3 or 1; 0; 0 or 2**

Find all of the rational zeros of each function.

16. $g(x) = x^3 - 3x^2 - 53x - 9$ **9**
17. $h(x) = x^4 + 2x^3 - 23x^2 + 2x - 24$ **−6, 4**

Determine whether each pair of functions are inverse functions.

18. $f(x) = 4x - 9, g(x) = \frac{x - 9}{4}$ **no**
19. $f(x) = \frac{1}{x + 2}, g(x) = \frac{1}{x} - 2$ **yes**

If $f(x) = 2x - 4$ and $g(x) = x^2 + 3$, find each value. **20–23. See margin.**

20. $(f + g)(x)$
21. $(f - g)(x)$
22. $(f \cdot g)(x)$
23. $\left(\frac{f}{g}\right)(x)$

24. **FINANCIAL PLANNING** Toshi will start college in six years. According to their plan, Toshi's parents will save \$1000 each year for the next three years. During the fourth and fifth years, they will save \$1200 each year. During the last year before he starts college, they will save \$2000.
 a. In the formula $A = P(1 + r)^t$, A = the balance, P = the amount invested, r = the interest rate, and t = the number of years the money has been invested. Use this formula to write a polynomial equation to describe the balance of the account when Toshi starts college. **See margin.**
 b. Find the balance of the account if the interest rate is 6%. **\$8916.76**

25. **STANDARDIZED TEST PRACTICE** Which value is included in the graph of $y < \sqrt{2x}$? **D**
 (A) (−2, −2) (B) (−1, −1) (C) (0, 0) (D) None of these

www.algebra2.com/chapter_test

Portfolio Suggestion

Introduction In mathematics, polynomial equations can be used to model many real-world problems. The solution to the polynomial equation provides a solution to the real-world problem.

Ask Students From your work in this chapter, select a real-world problem modeled by a polynomial equation and show how you solved it. Explain how the solution to the polynomial equation relates to the solution of the real-world problem. Place your work in your portfolio.

Chapter 7 Practice Test

Assessment Options

Vocabulary Test A vocabulary test/review for Chapter 7 can be found on p. 442 of the *Chapter 7 Resource Masters*.

Chapter Tests There are six Chapter 7 Tests and an Open-Ended Assessment task available in the *Chapter 7 Resource Masters*.

Chapter 7 Tests			
Form	**Type**	**Level**	**Pages**
1	MC	basic	429–430
2A	MC	average	431–432
2B	MC	average	433–434
2C	FR	average	435–436
2D	FR	average	437–438
3	FR	advanced	439–440

MC = multiple-choice questions
FR = free-response questions

Open-Ended Assessment Performance tasks for Chapter 7 can be found on p. 441 of the *Chapter 7 Resource Masters*. A sample scoring rubric for these tasks appears on p. A34.

Unit 2 Test A unit test/review can be found on pp. 449–450 of the *Chapter 7 Resource Masters*.

First Semester Test A test for Chapters 1–7 can be found on pp. 451–454 of the *Chapter 7 Resource Masters*.

ExamView® Pro

Use the networkable ExamView® Pro to:

- Create **multiple versions** of tests.
- Create **modified** tests for *Inclusion* students.
- **Edit** existing questions and **add** your own questions.
- Use built-in **state curriculum correlations** to create tests aligned with state standards.
- **Apply** art to your tests from a program bank of artwork.

Chapter 7 Standardized Test Practice

These two pages contain practice questions in the various formats that can be found on the most frequently given standardized tests.

A practice answer sheet for these two pages can be found on p. A1 of the *Chapter 7 Resource Masters*.

Standardized Test Practice Student Recording Sheet, p. A1

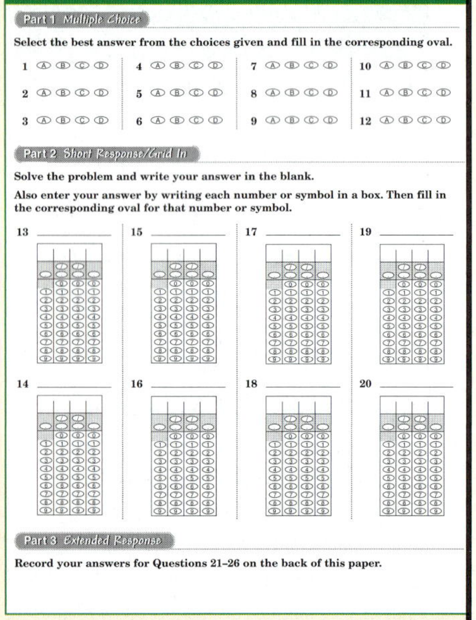

Part 1 Multiple Choice

Select the best answer from the choices given and fill in the corresponding oval.

1 Ⓐ Ⓑ Ⓒ Ⓓ 4 Ⓐ Ⓑ Ⓒ Ⓓ 7 Ⓐ Ⓑ Ⓒ Ⓓ 10 Ⓐ Ⓑ Ⓒ Ⓓ
2 Ⓐ Ⓑ Ⓒ Ⓓ 5 Ⓐ Ⓑ Ⓒ Ⓓ 8 Ⓐ Ⓑ Ⓒ Ⓓ 11 Ⓐ Ⓑ Ⓒ Ⓓ
3 Ⓐ Ⓑ Ⓒ Ⓓ 6 Ⓐ Ⓑ Ⓒ Ⓓ 9 Ⓐ Ⓑ Ⓒ Ⓓ 12 Ⓐ Ⓑ Ⓒ Ⓓ

Part 2 Short Response/Grid In

Solve the problem and write your answer in the blank.

Also enter your answer by writing each number or symbol in a box. Then fill in the corresponding oval for that number or symbol.

13 15 17 19

14 16 18 20

Part 3 Extended Response

Record your answers for Questions 21–26 on the back of this paper.

Additional Practice

See pp. 447–448 in the *Chapter 7 Resource Masters* for additional standardized test practice.

Chapter 7 Standardized Test Practice

Part 1 Multiple Choice

Record your answers on the answer sheet provided by your teacher or on a sheet of paper.

1. If $\frac{2}{p} - \frac{4}{p^2} = -\frac{2}{p^3}$, then what is the value of p? **B**

Ⓐ -1 Ⓑ 1 Ⓒ $-\frac{1}{2}$ Ⓓ $\frac{1}{2}$

2. There are n gallons of liquid available to fill a tank. After k gallons of the liquid have filled the tank, how do you represent in terms of n and k the percent of liquid that has filled the tank? **A**

Ⓐ $\frac{100k}{n}\%$ Ⓑ $\frac{n}{100k}\%$

Ⓒ $\frac{100n}{k}\%$ Ⓓ $\frac{n}{100(n-k)}\%$

3. How many different triangles have sides of lengths 4, 9 and s, where s is an integer and $4 < s < 9$? **D**

Ⓐ 0 Ⓑ 1 Ⓒ 2 Ⓓ 3

4. Triangles ABC and DEF are similar. The area of $\triangle ABC$ is 36 square units. What is the perimeter of $\triangle DEF$? **B**

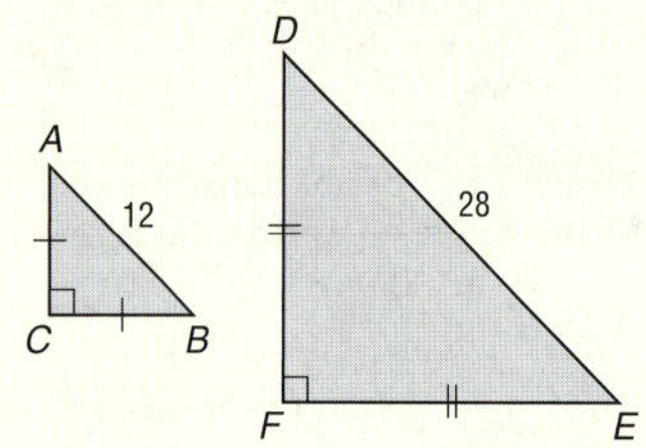

Ⓐ 56 units Ⓑ $28 + 28\sqrt{2}$ units

Ⓒ $56\sqrt{2}$ units Ⓓ $28 + 14\sqrt{2}$ units

5. If $2 - 3x > -1$ and $x + 5 > 0$, then x could equal each of the following *except* **A**

Ⓐ -5. Ⓑ -4. Ⓒ -2. Ⓓ 0.

6. What is the midpoint of the line segment whose endpoints are represented on the coordinate grid by the points $(-5, -3)$ and $(-1, 4)$? **B**

Ⓐ $\left(-3, -\frac{1}{2}\right)$ Ⓑ $\left(-3, \frac{1}{2}\right)$

Ⓒ $\left(-2, -\frac{7}{2}\right)$ Ⓓ $\left(-2, \frac{1}{2}\right)$

7. For all $n \neq 0$, what is the slope of the line passing through (n, k) and $(-n, -k)$? **D**

Ⓐ 0 Ⓑ 1 Ⓒ $\frac{n}{k}$ Ⓓ $\frac{k}{n}$

8. Which of the following is a quadratic equation in one variable? **B**

Ⓐ $3(x + 4) + 1 = 4x - 9$

Ⓑ $3x(x + 4) + 1 = 4x - 9$

Ⓒ $3x(x^2 + 4) + 1 = 4x - 9$

Ⓓ $y = 3x^2 + 8x + 10$

9. Simplify $\sqrt[4]{t^3} \cdot \sqrt[8]{t^2}$. **D**

Ⓐ $t^{\frac{3}{16}}$ Ⓑ $t^{\frac{1}{2}}$ Ⓒ $t^{\frac{3}{4}}$ Ⓓ t

10. Which of the following is a quadratic equation that has roots of $2\frac{1}{2}$ and $\frac{2}{3}$? **C**

Ⓐ $5x^2 + 11x - 7 = 0$

Ⓑ $5x^2 - 11x + 10 = 0$

Ⓒ $6x^2 - 19x + 10 = 0$

Ⓓ $6x^2 + 11x + 10 = 0$

11. If $f(x) = 3x - 5$ and $g(x) = 2 + x^2$, then what is equal to $f[g(2)]$? **D**

Ⓐ 3 Ⓑ 6 Ⓒ 12 Ⓓ 13

12. Which of the following is a zero of $f(x) = x^3 - 7x + 6$? **B**

Ⓐ -1 Ⓑ 2 Ⓒ 3 Ⓓ 6

ExamView® Pro

Special banks of standardized test questions similar to those on the SAT, ACT, TIMSS 12, NAEP 12, and state proficiency tests can be found on this CD-ROM.

406 Chapter 7 Polynomial Functions

Preparing for Standardized Tests
For test-taking strategies and more practice, see pages 877–892.

Part 2 Short Response/Grid In

Record your answers on the answer sheet provided by your teacher or on a sheet of paper.

13. A group of 34 people is to be divided into committees so that each person serves on exactly one committee. Each committee must have at least 3 members and not more than 5 members. If N represents the maximum number of committees that can be formed and n represents the minimum number of committees that can be formed, what is the value of $N - n$? **4**

14. Raisins selling for \$2.00 per pound are to be mixed with peanuts selling for \$3.00 per pound. How many pounds of peanuts are needed to produce a 20-pound mixture that sells for \$2.75 per pound? **15**

15. Jars X, Y, and Z each contain 10 marbles. What is the minimum number of marbles that must be transferred among the jars so that the ratio of the number of marbles in jar X to the number of marbles in jar Y to the number of marbles in jar Z is 1 : 2 : 3? **5**

16. If the area of $\triangle BCD$ is 40% of the area of $\triangle ABC$, what is the measure of $\overline{AD}$? **6**

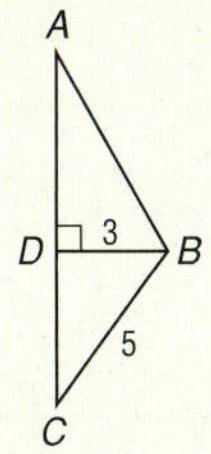

Test-Taking Tip

Questions 13, 15, and 18

Words such as *maximum*, *minimum*, *least*, and *greatest* indicate that a problem may involve an inequality. Take special care when simplifying inequalities that involve negative numbers.

www.algebra2.com/standardized_test

17. The mean of 15 scores is 82. If the mean of 7 of these scores is 78, what is the mean of the remaining 8 scores? **85.5**

18. If the measures of the sides of a triangle are 3, 8, and x and x is an integer, then what is the least possible perimeter of the triangle? **17**

19. If the operation ❖ is defined by the equation $x ❖ y = 3x - y$, what is the value of w in the equation $w ❖ 6 = 2 ❖ w$? **3**

20. In the figure below, $\ell \parallel m$. Find b. Assume that the figure is not drawn to scale. **100**

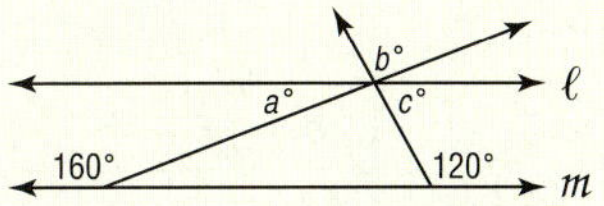

Part 3 Extended Response

Record your answers on a sheet of paper. Show your work.

For Exercises 21–25, use the polynomial function $f(x) = 3x^4 + 19x^3 + 7x^2 - 11x - 2$.

21. What is the degree of the function? **4**

22. Evaluate $f(1)$, $f(-2)$, and $f(2a)$. Show your work. **16; −56; $48a^4 + 152a^3 + 28a^2 - 22a - 2$**

23. State the number of possible positive real zeros, negative real zeros, and imaginary zeros of $f(x)$. **See margin.**

24. List all of the possible rational zeros of the function. **$\pm1, \pm2, \pm\frac{1}{3}, \pm\frac{2}{3}$**

25. Find all of the zeros of the function. **$-1, \frac{2}{3}, -3 \pm 2\sqrt{2}$**

26. Sketch the graphs of $f(x) = \frac{3x+1}{2}$ and $g(x) = \frac{2x-1}{3}$. Considering the graphs, describe the relationship between $f(x)$ and $g(x)$. Verify your conclusion. **See margin.**

Evaluating Extended Response Questions

Extended Response questions are graded by using a multilevel rubric that guides you in assessing a student's knowledge of a particular concept.

Goal: Determine the dimensions of a rectangular garden for a given perimeter.

Sample Scoring Rubric: The following rubric is a sample scoring device. You may wish to add more detail to this sample to meet your individual scoring needs.

Score	Criteria
4	A correct solution that is supported by well-developed, accurate explanations
3	A generally correct solution, but may contain minor flaws in reasoning or computation
2	A partially correct interpretation and/or solution to the problem
1	A correct solution with no supporting evidence or explanation
0	An incorrect solution indicating no mathematical understanding of the concept or task, or no solution is given

26. (continued from bottom wrap) Sample answer: The graphs are reflections over the line $y = x$, so it appears that the functions are inverses. To verify, find $f(g(x))$ and $g(f(x))$.

$$f(g(x)) = \frac{3\left(\frac{2x-1}{3}\right)+1}{2} = \frac{(2x-1)+1}{2} = \frac{2x}{2} = x;$$

$$g(f(x)) = \frac{2\left(\frac{3x+1}{2}\right)-1}{3} = \frac{(3x+1)-1}{3} = \frac{3x}{3} = x.$$

Since both $f(g(x))$ and $g(f(x)) = x$, the functions are inverses.

Answers

23. 1 positive real zero and 3 negative real zeros; or 1 positive real zero; 1 negative real zero and 2 imaginary zeros

26.

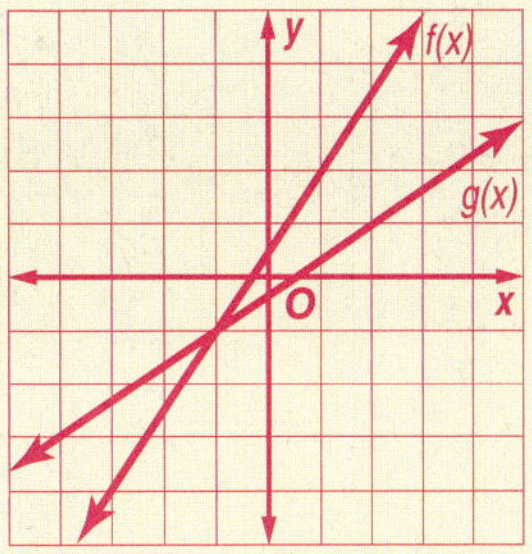

Pages 350–352, Lesson 7-1

68.

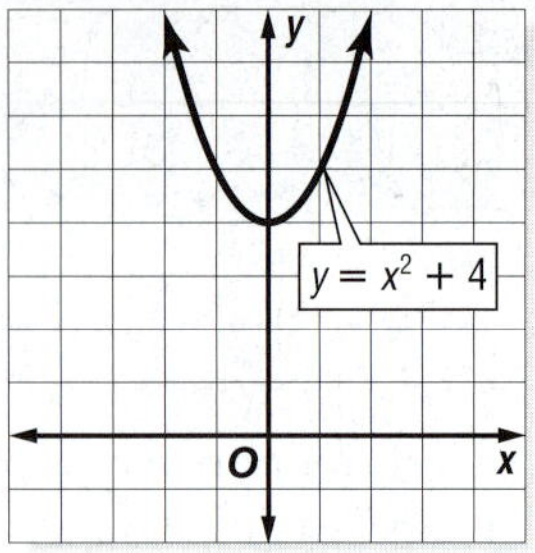

69.

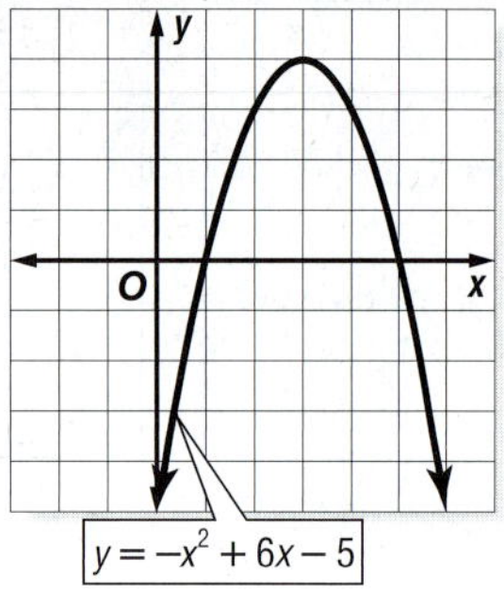

70.

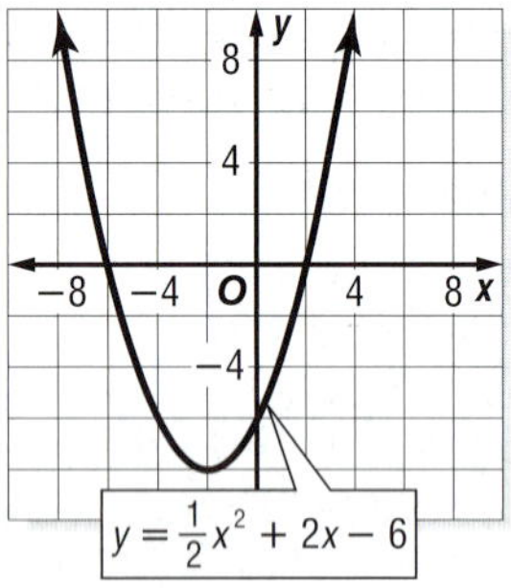

Page 356, Lesson 7-2
Graphing Calculator Investigation

1.

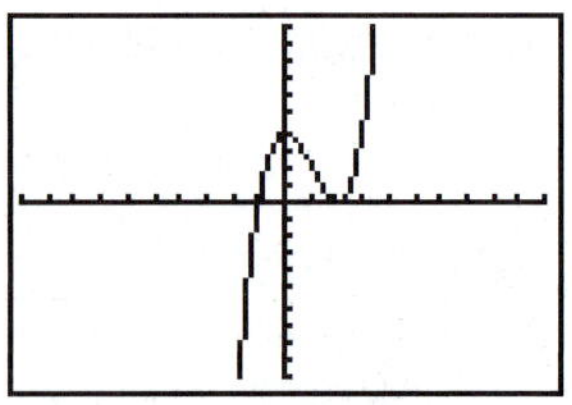

rel. max. at $x = 0$, rel. min. at $x = 2$

2. rel. max. at (0, 4) rel. min. at (2, 0)

3.

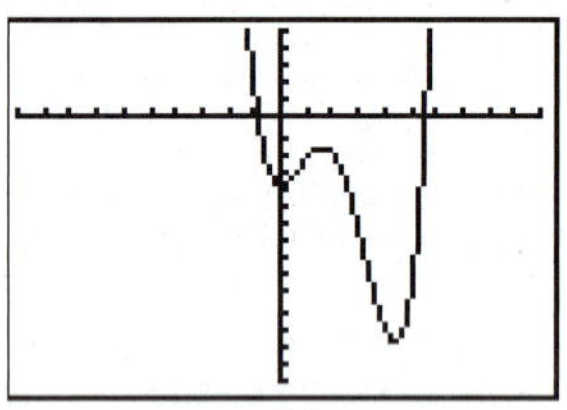

Sample answer: two rel. min. points at (0, −8) and (4.4, −25.8) and one rel. max. point at (1.6, −3.2)

Pages 356–358, Lesson 7-2

4.

x	$f(x)$
−3	−20
−2	0
−1	6
0	4
1	0
2	0
3	10

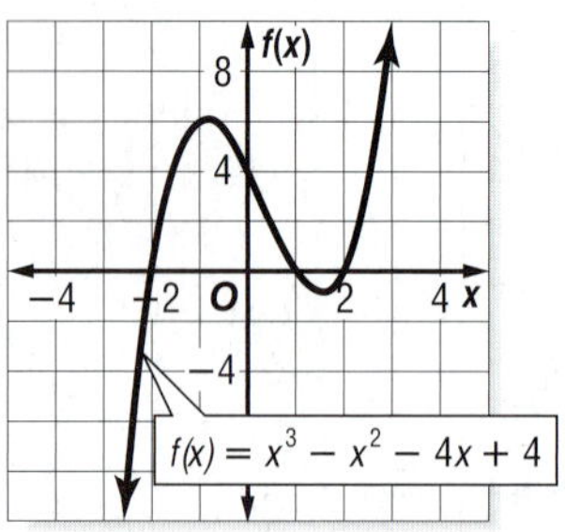

5.

x	$f(x)$
−3	20
−2	−9
−1	−2
0	5
1	0
2	−5
3	26

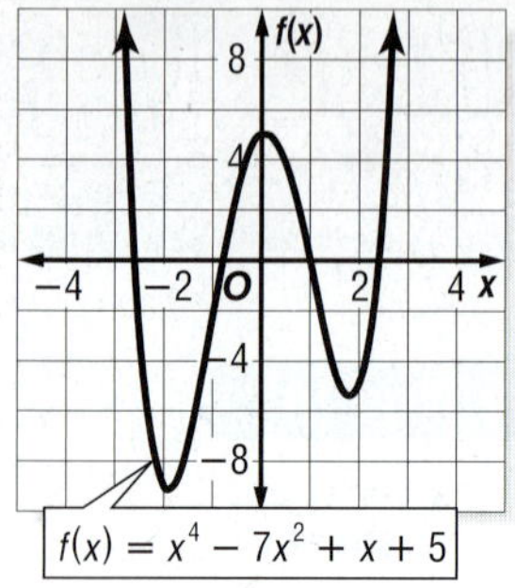

6.

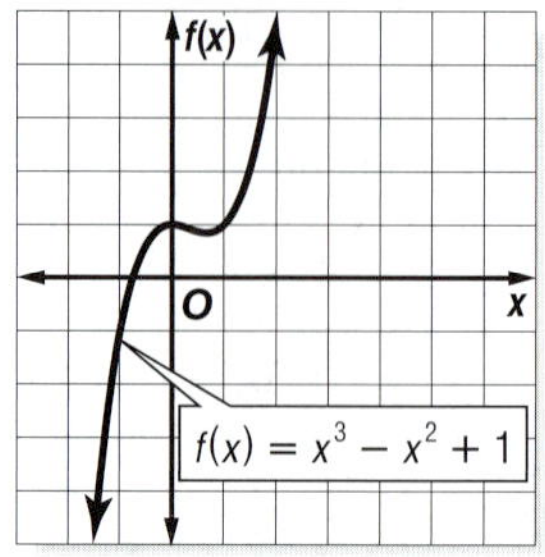

7.

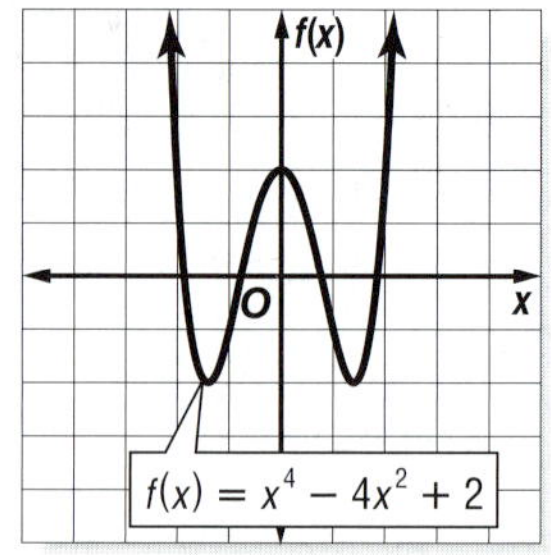

8.

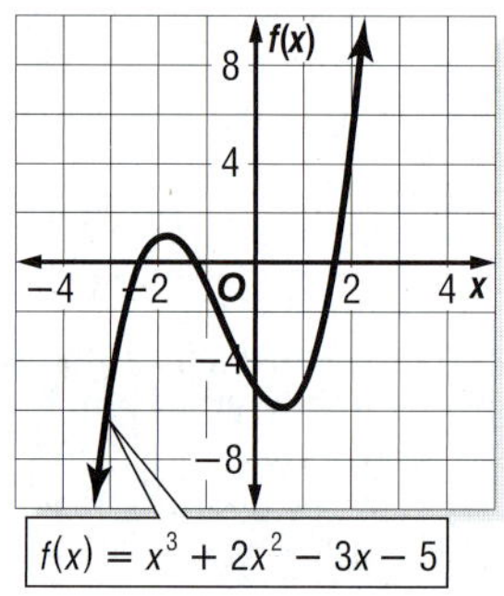

9.

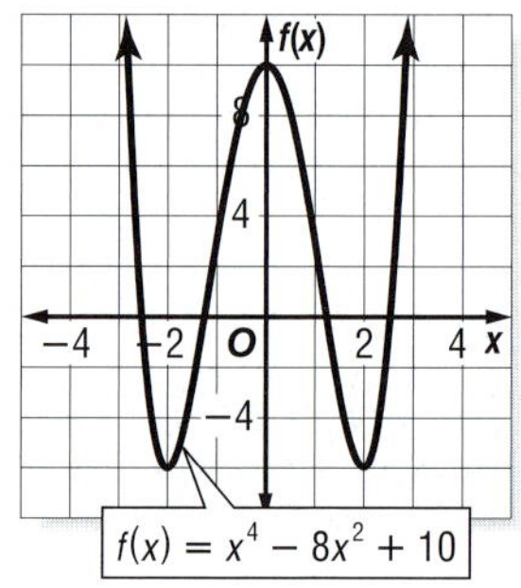

10.

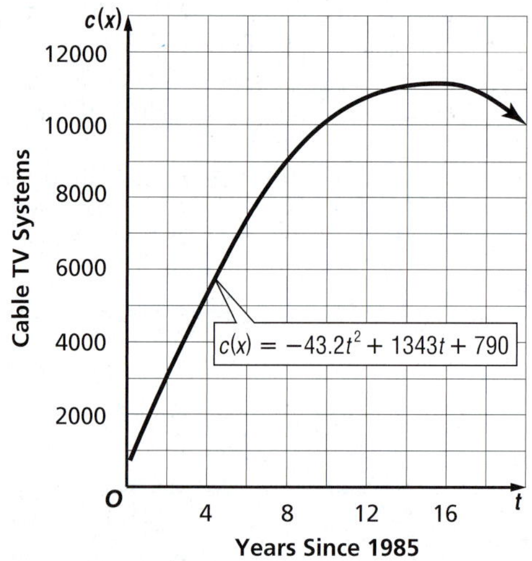

11. rel. max. between $x = 15$ and $x = 16$, and no rel. min.; $f(x) \to -\infty$ as $x \to -\infty$, $f(x) \to -\infty$ as $x \to +\infty$.

12. The number of cable TV systems rose steadily from 1985 to 2000. Then the number began to decline.

13a.

x	$f(x)$
−5	25
−4	0
−3	−9
−2	−8
−1	−3
0	0
1	−5
2	−24

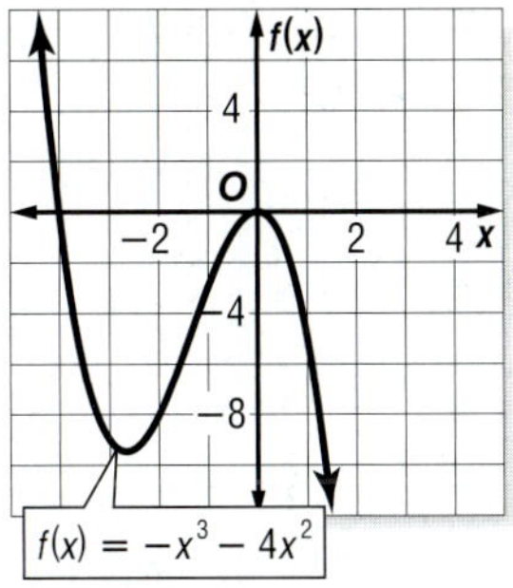

13b. at $x = -4$ and $x = 0$

13c. Sample answer: rel. max. at $x = 0$, rel. min. at $x = -3$

14a.

x	$f(x)$
−2	−10
−1	3
0	6
1	5
2	6
3	15
4	38

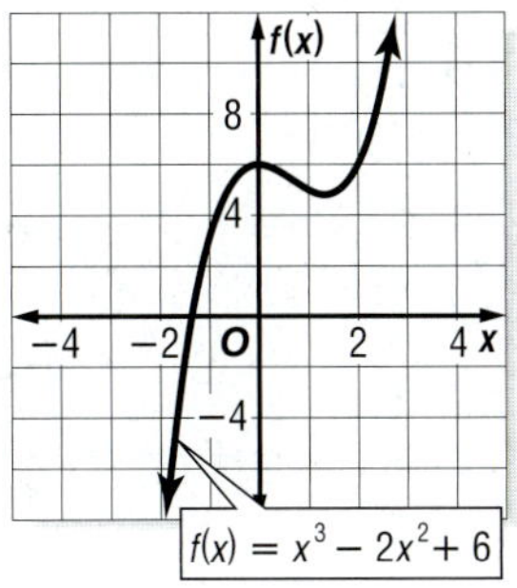

14b. between −2 and −1

14c. Sample answer: rel. max. at $x = 0$, rel. min. at $x = \frac{3}{2}$

15a.

x	$f(x)$
−2	−18
−1	−2
0	2
1	0
2	−2
3	2
4	18

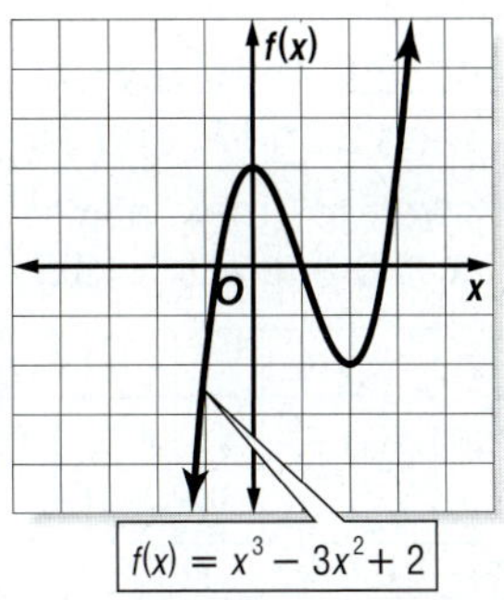

15b. at $x = 1$, between −1 and 0, and between 2 and 3

15c. Sample answer: rel. max. at $x = 0$, rel. min. at $x = 2$

16a.

x	$f(x)$
−5	−9
−4	7
−3	9
−2	3
−1	−5
0	−9
1	−3
2	19

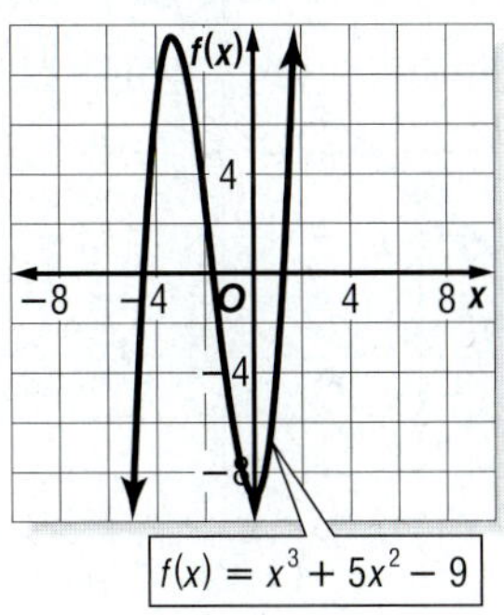

16b. between −5 and −4, between −2 and −1, and between 1 and 2

16c. Sample answer: rel. max. at $x = -3$, rel. min. at $x = 0$

17a.

x	$f(x)$
−1	75
0	16
1	−3
2	0
3	7
4	0
5	−39

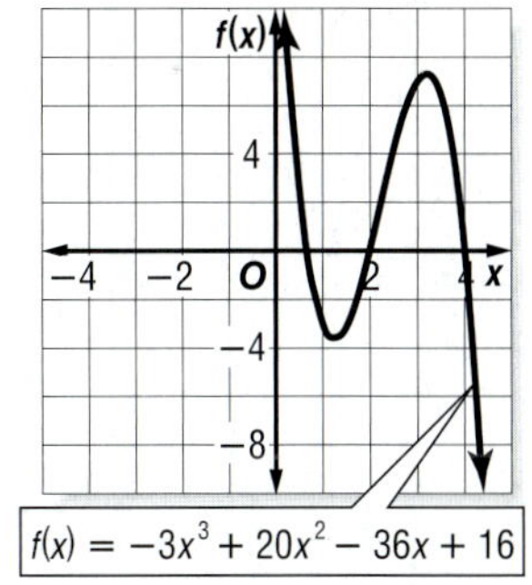

17b. between 0 and 1, at $x = 2$, and at $x = 4$

17c. Sample answer: rel. max. at $x = 3$, rel. min. at $x = 1$

18a.

x	$f(x)$
−2	−29
−1	−8
0	−1
1	−2
2	−5
3	−4
4	7
5	34

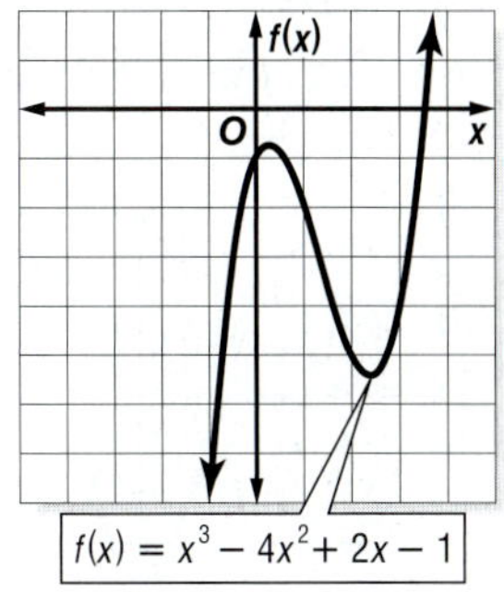

18b. between 3 and 4

18c. Sample answer: rel. max. at $x = 0.5$, rel. min. at $x = 2.5$

19a.

x	$f(x)$
−3	73
−2	8
−1	−7
0	−8
1	−7
2	8
3	73

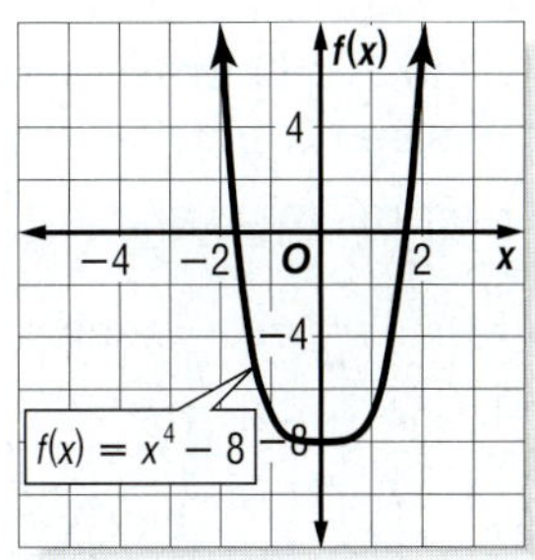

19b. between −2 and −1, and between 1 and 2

19c. Sample answer: no rel. max., rel. min. at $x = 0$

20a.

x	$f(x)$
−3	0
−2	−15
−1	0
0	9
1	0
2	−15
3	0
4	105

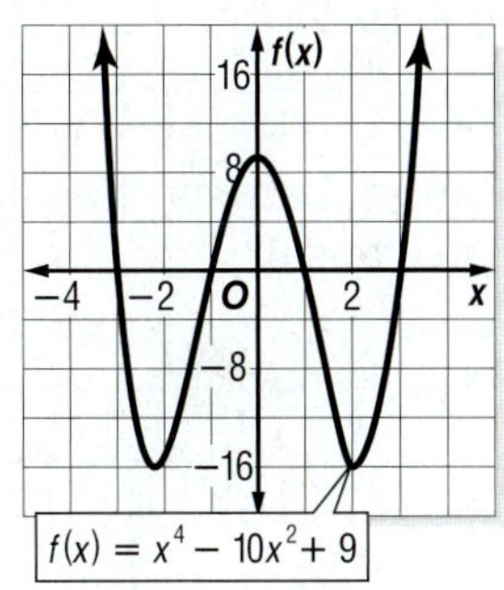

20b. at $x = -3$, $x = -1$, $x = 1$, and $x = 3$

20c. Sample answer: rel. max. at $x = 0$, rel. min. at $x = -2$ and $x = 2$

21a.

x	f(x)
−4	−169
−3	−31
−2	7
−1	5
0	−1
1	1
2	−1
3	−43

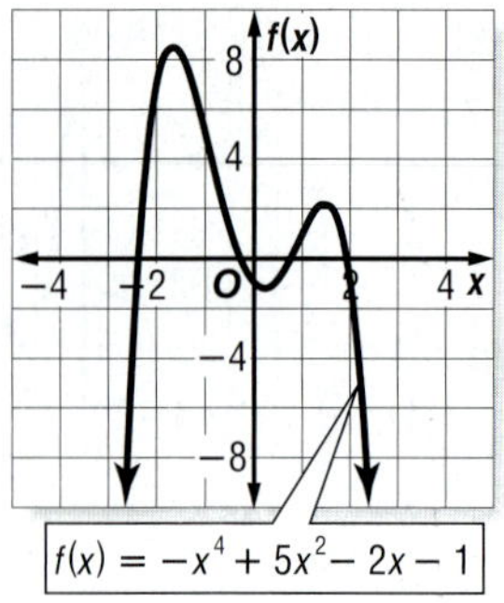

21b. between −3 and −2, between −1 and 0, between 0 and 1, and between 1 and 2

21c. Sample answer: rel. max. at $x = -2$ and at $x = 1.5$, rel. min. at $x = 0$

22a.

x	f(x)
−3	−39
−2	5
−1	3
0	−3
1	5
2	21
3	15
4	−67

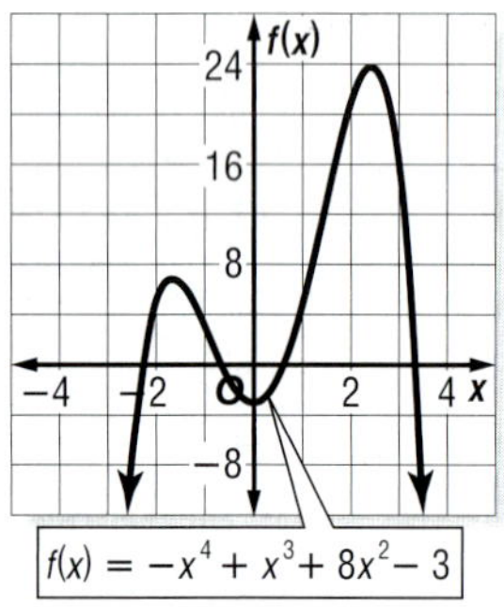

22b. between −3 and −2, between −1 and 0, between 0 and 1, and between 3 and 4

22c. Sample answer: rel. max. at $x = -1.5$ and at $x = 2.5$, rel. min. at $x = 0$

23a.

x	f(x)
−1	65
0	6
1	−1
2	2
3	−3
4	−10
5	11

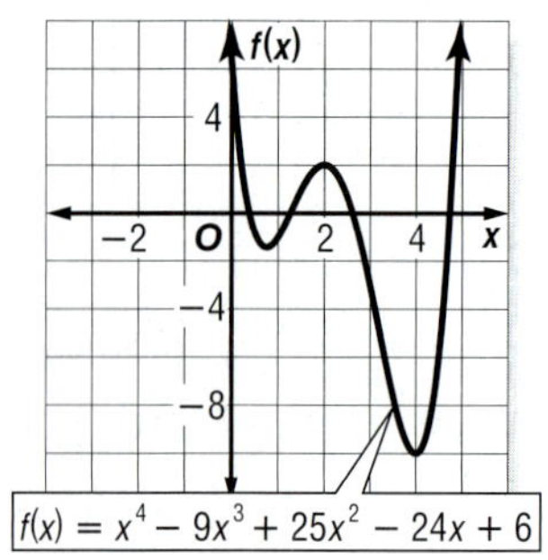

23b. between 0 and 1, between 1 and 2, between 2 and 3, and between 4 and 5

23c. Sample answer: rel. max. at $x = 2$, rel. min. at $x = 0.5$ and at $x = 4$

24a.

x	f(x)
−2	45
−1	−4
0	−5
1	−6
2	−7
3	40

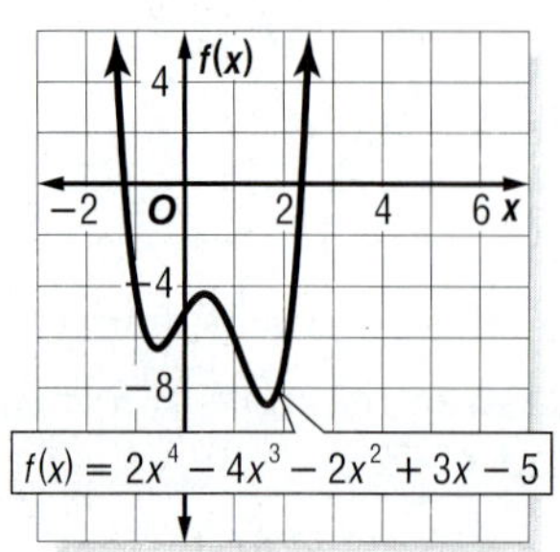

24b. between −2 and −1, and between 2 and 3

24c. Sample answer: rel. max. at $x = 0.5$, rel. min. at $x = -0.5$ and at $x = 1.5$

25a.

x	f(x)
−4	−77
−3	30
−2	7
−1	−2
0	3
1	−2
2	55

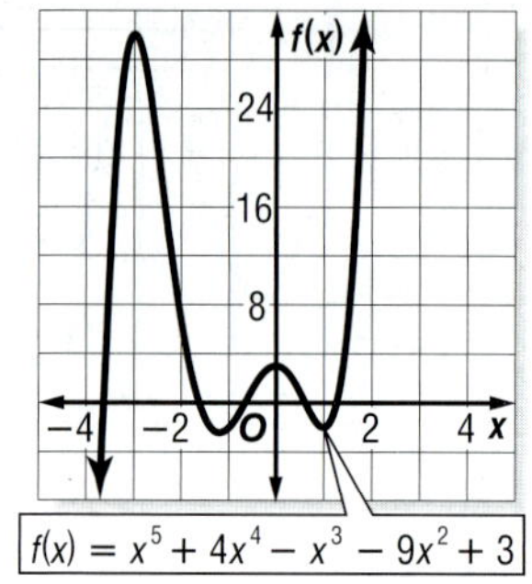

25b. between −4 and −3, between −2 and −1, between −1 and 0, between 0 and 1, and between 1 and 2

25c. Sample answer: rel. max. at $x = -3$ and at $x = 0$, rel. min. at $x = -1$ and at $x = 1$

26a.

x	f(x)
−2	−88
−1	5
0	−6
1	5
2	20
3	−3
4	−10
5	269

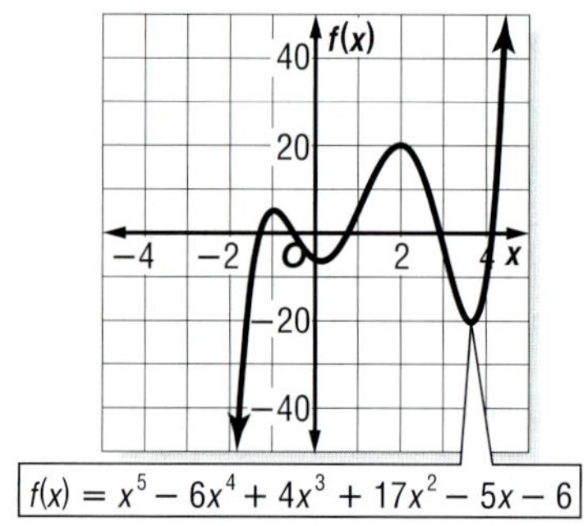

26b. between −2 and −1, between −1 and 0, between 0 and 1, between 2 and 3, and between 4 and 5

26c. Sample answer: rel. max. at $x = -1$ and at $x = 2$, rel. min. at $x = 0$ and at $x = 3.5$

31.

x	0	2	4	6	8	10	12	14	16	18	20
B(x)	25	34	40	45	50	54	59	64	68	71	71
G(x)	26	33	39	44	49	53	56	59	61	61	60

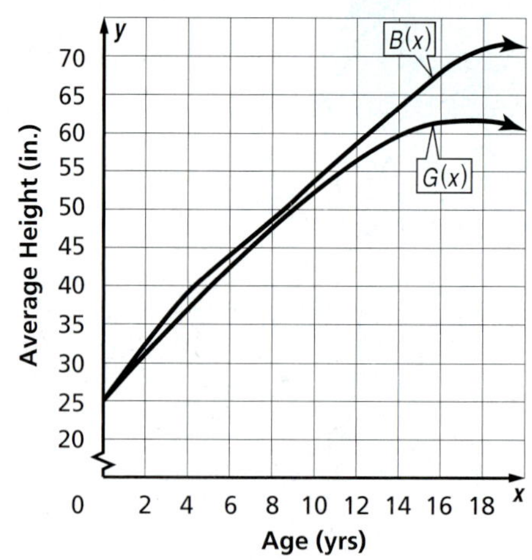

36.

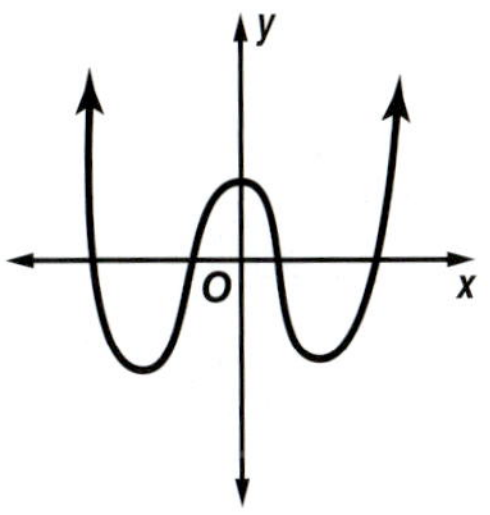

37.

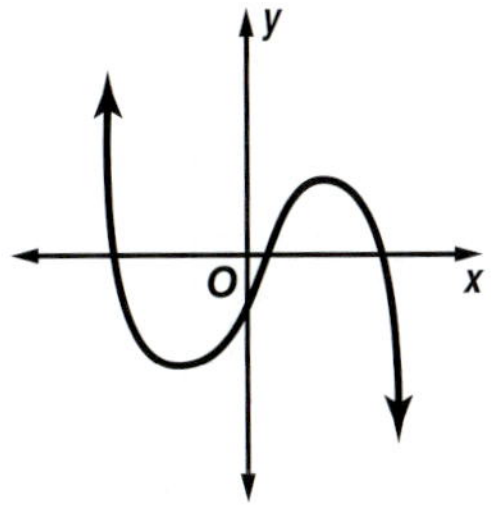

38.

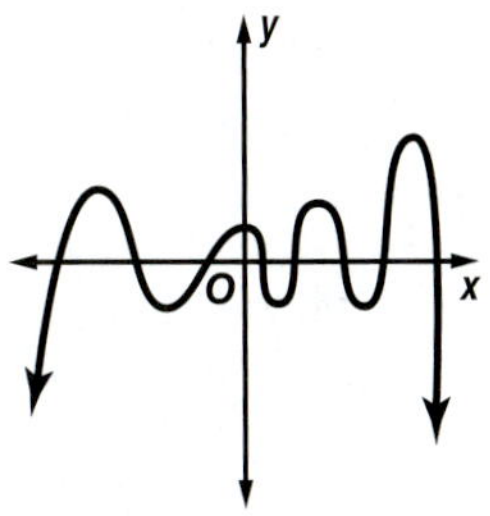

39.

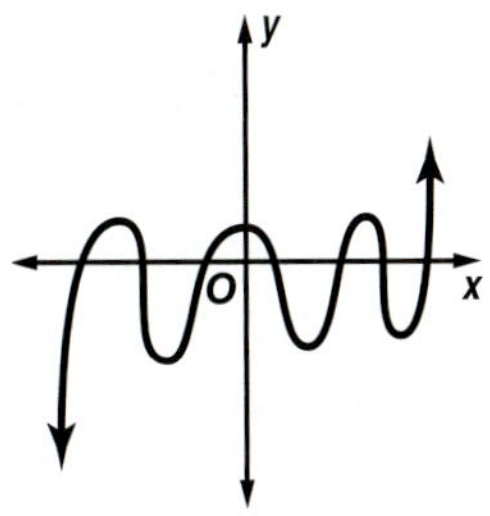

40. The turning points of a polynomial function that models a set of data can indicate fluctuations that may repeat. Answers should include the following.

- Polynomial equations best model data that contain turning points, rather than a constant increase or decrease like linear equations.
- To determine when the percentage of foreign-born citizens was at its highest, look for rel. max. of the graph, which is at $t = 5$. The lowest percentage is found at $t = 75$, the rel. min. of the graph.

Page 359, Follow-Up of Lesson 7-2 Graphing Calculator Investigation

1.

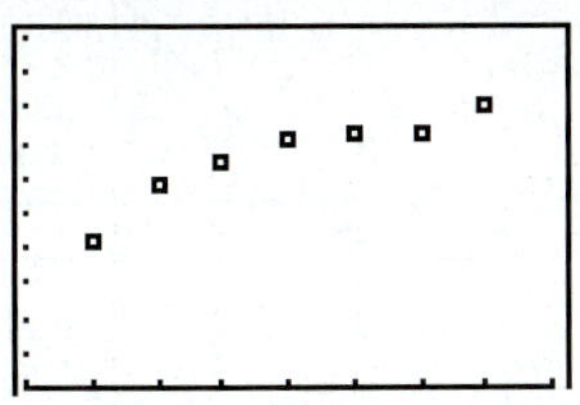

[1930, 2010] scl: 10 by [0, 200] scl: 20

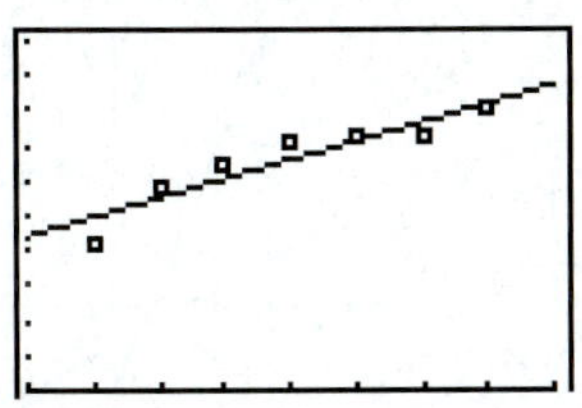

[1930, 2010] scl: 10 by [0, 200] scl: 20

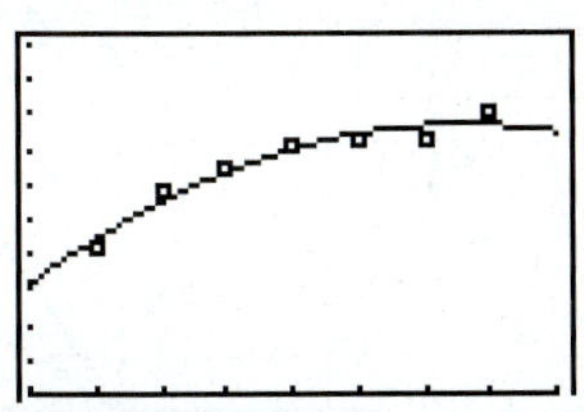

[1930, 2010] scl: 10 by [0, 200] scl: 20

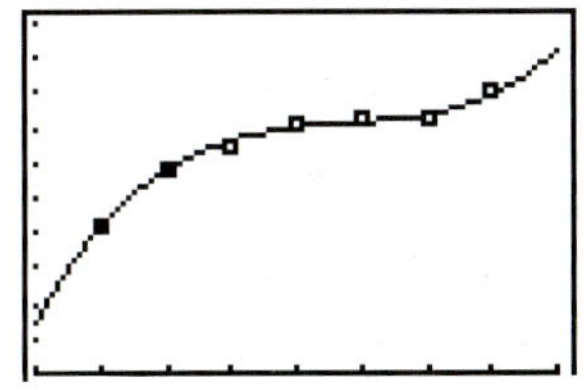

[1930, 2010] scl: 10 by [0, 200] scl: 20

Pages 362–364, Lesson 7-3

17. 0, −4, −3

18. 0, −1, −5

19. $-\sqrt{3}, \sqrt{3}, -i\sqrt{3}, i\sqrt{3}$

20. $0, -4, 4, -4i, 4i$

21. $2, -2, 2\sqrt{2}, -2\sqrt{2}$

22. $\sqrt{2}, -\sqrt{2}, 3, -3$

23. $-9, \frac{9 + 9i\sqrt{3}}{2}, \frac{9 - 9i\sqrt{3}}{2}$

24. $8, -4 + 4i\sqrt{3}, -4 - 4i\sqrt{3}$

25. 81, 625

26. −343, −64

27. 225, 16

28. 400

38. Answers should include the following.

- Solve the cubic equation $4x^3 + (-164x^2) + 1600x = 3600$ in order to determine the dimensions of the cut square if the desired volume is 3600 in^3. Solutions are 10 in. and $\frac{31 - \sqrt{601}}{2}$ in.
- There can be more than one square cut to produce the same volume because the height of the box is not specified and 3600 has a variety of different factors.

41.

x	f(x)
−2	−21
−1	−1
0	5
1	3
2	−1
3	−1
4	9
5	35

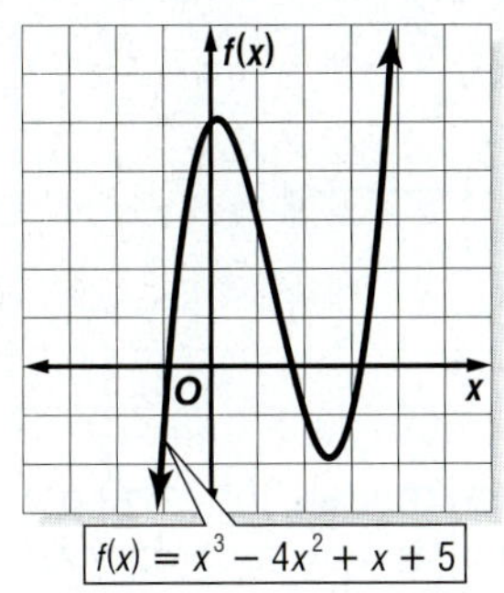

42.

x	f(x)
−1	15
0	−3
1	1
2	3
3	3
4	25

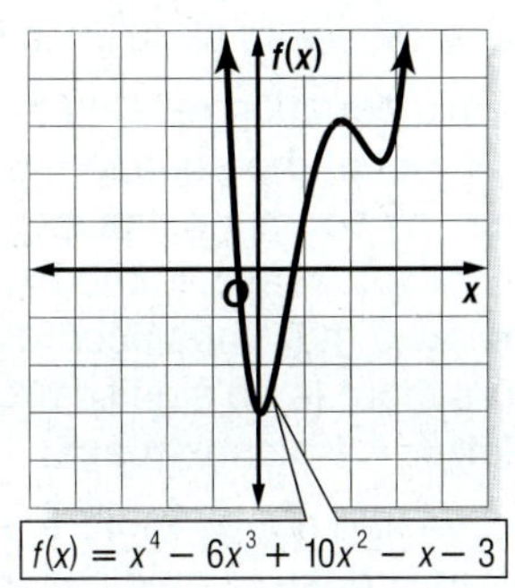

Page 364, Practice Quiz 1

3. Sample answer: maximum at $x = -2$, minimum at $x = 0.5$

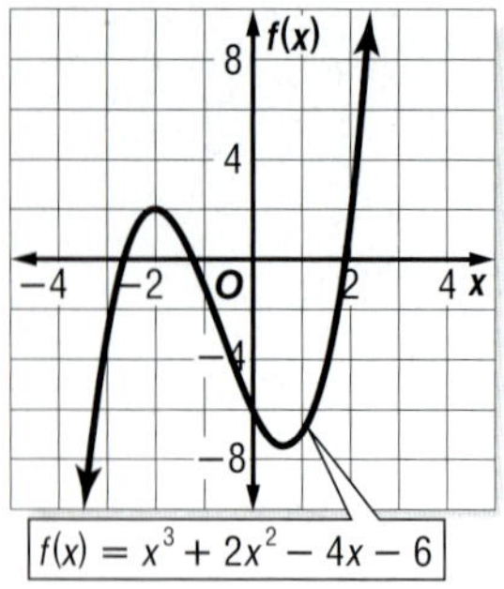

Pages 369–370, Lesson 7-4

32.

8	1	−4	−29	−24
		8	32	24
	1	4	3	0

37.

5	1	−14	69	−140	100
		5	−45	120	−100
	1	−9	24	−20	0

Pages 375–377, Lesson 7-5

41a.

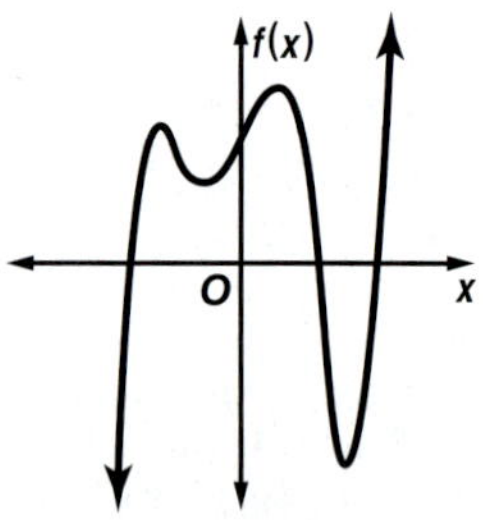

41b.

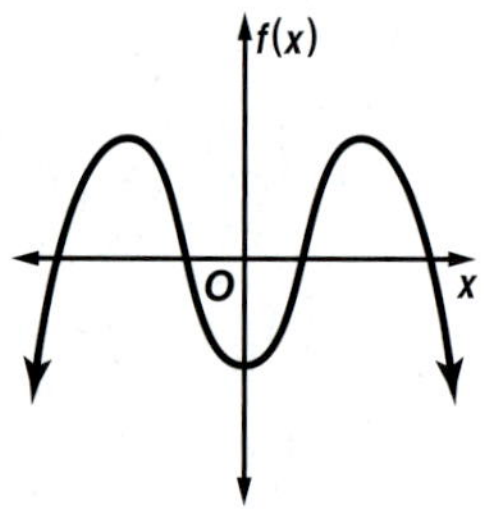

41c.

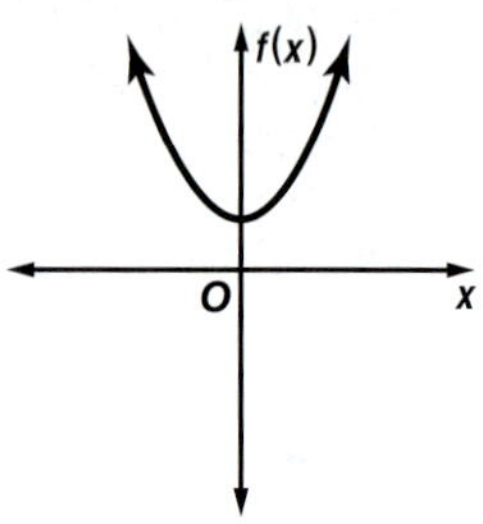

51. If the equation models the level of a medication in a patient's bloodstream, a doctor can use the roots of the equation to determine how often the patient should take the medication to maintain the necessary concentration in the body. Answers should include the following.

- A graph of this equation reveals that only the first positive real root of the equation, 5, has meaning for this situation, since the next positive real root occurs after the medication level in the bloodstream has dropped below 0 mg. Thus according to this model, after 5 hours there is no significant amount of medicine left in the bloodstream.
- The patient should not go more than 5 hours before taking their next dose of medication.

Pages 393–394, Lesson 7-8

7.

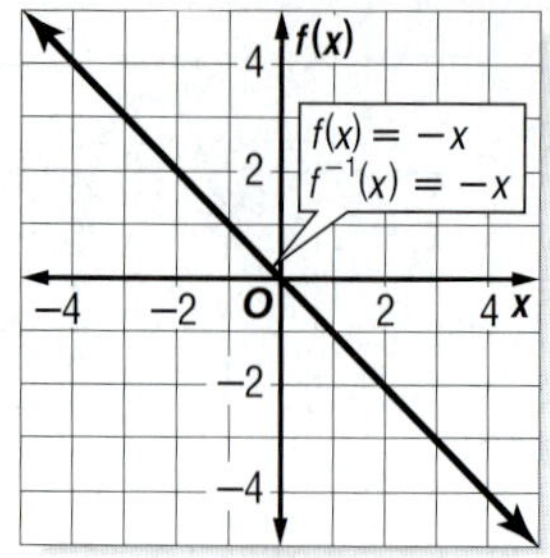

8.

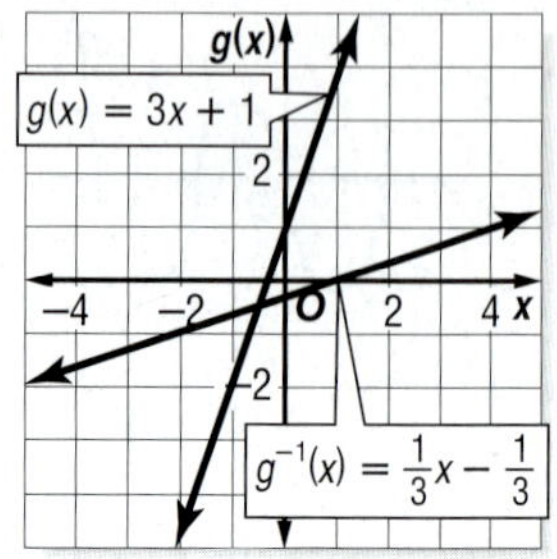

9.

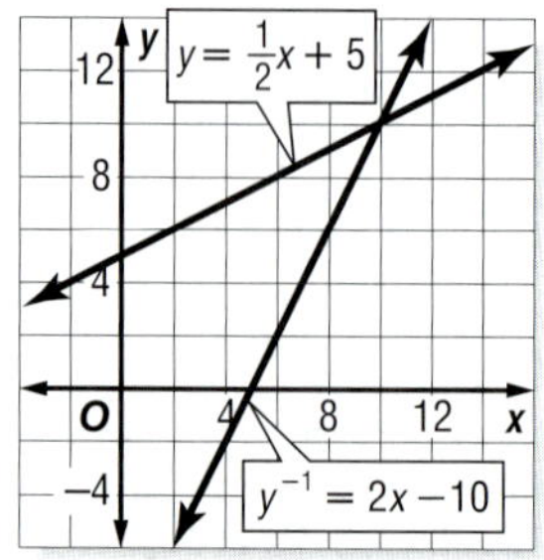

20.

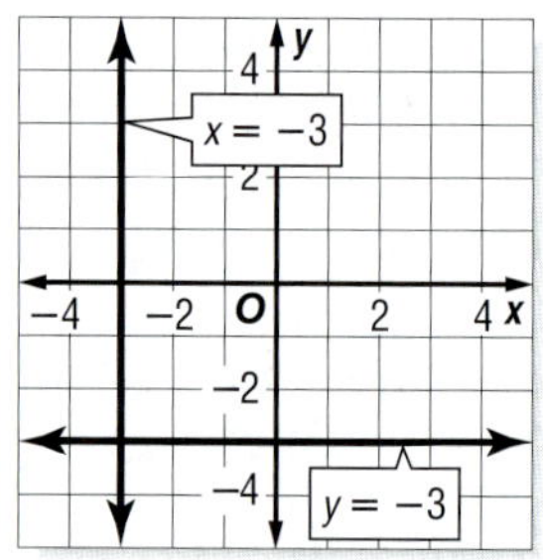

21.

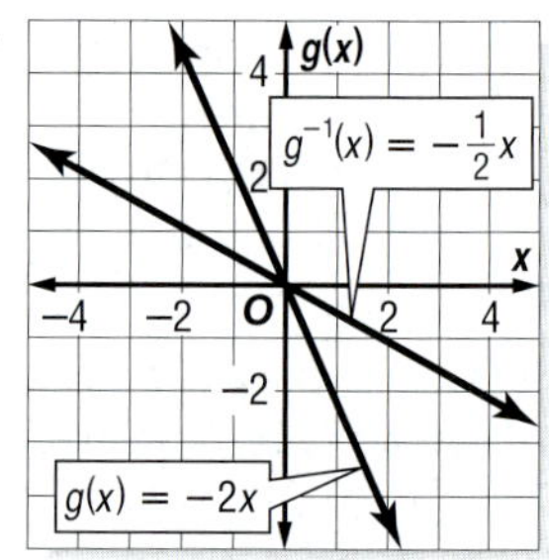

22.

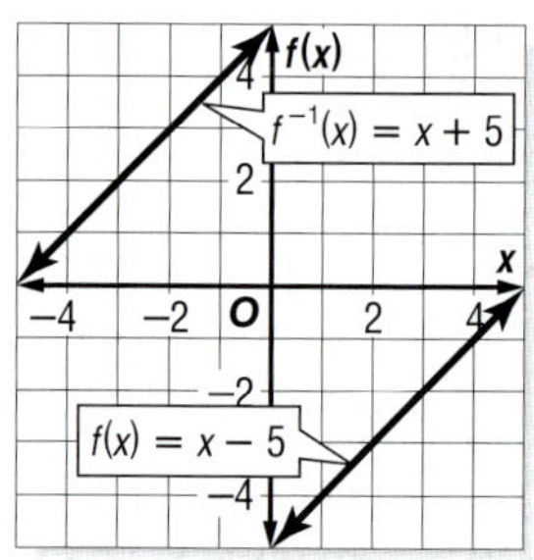

23.

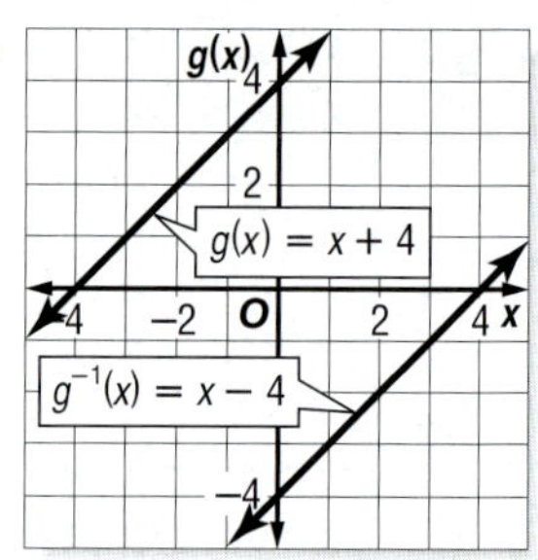

24.

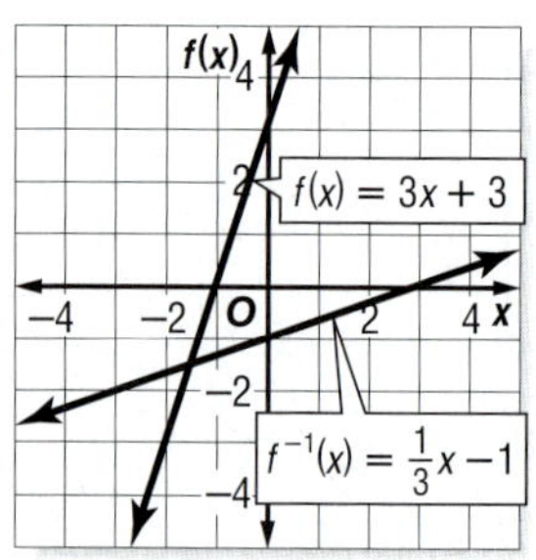

25.

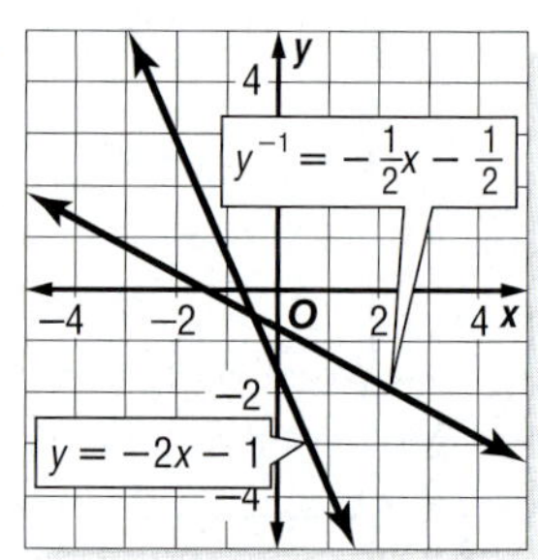

26.

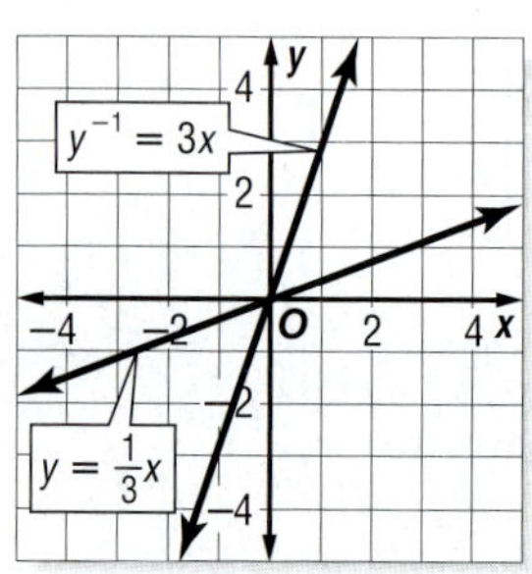

27.

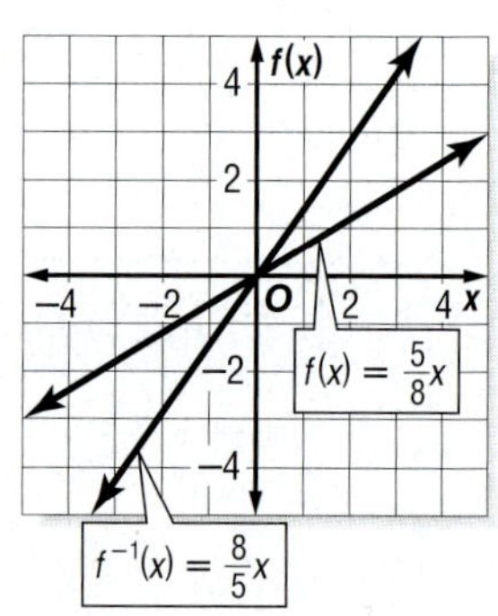

28.

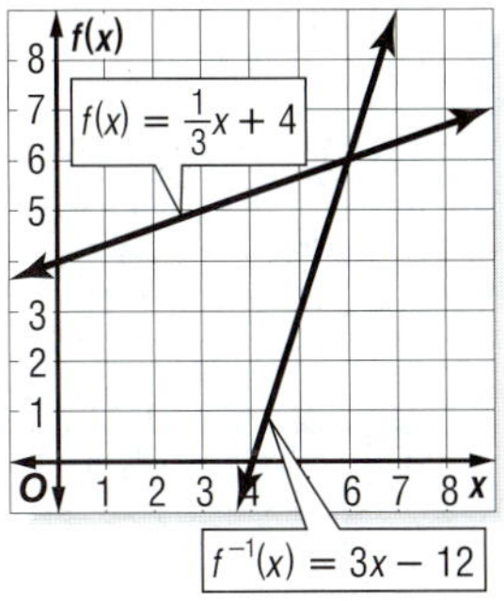

29. $f^{-1}(x) = \frac{5}{4}x + \frac{35}{4}$

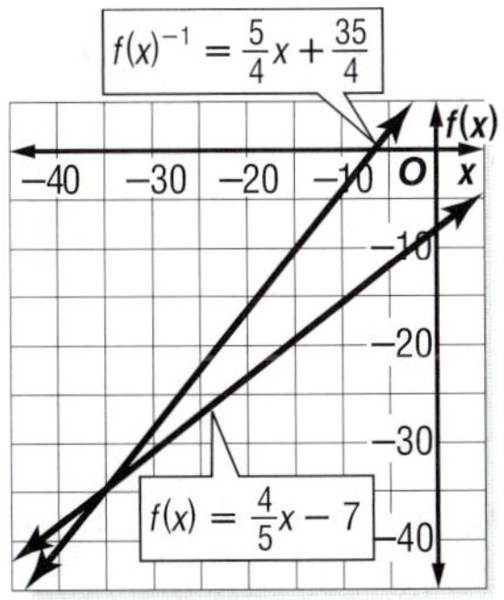

30. $g^{-1}(x) = 3x - \frac{3}{2}$

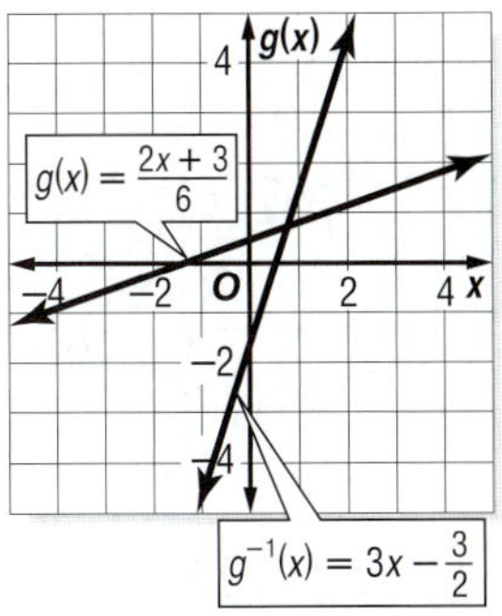

31. $f^{-1}(x) = \frac{8}{7}x + \frac{4}{7}$

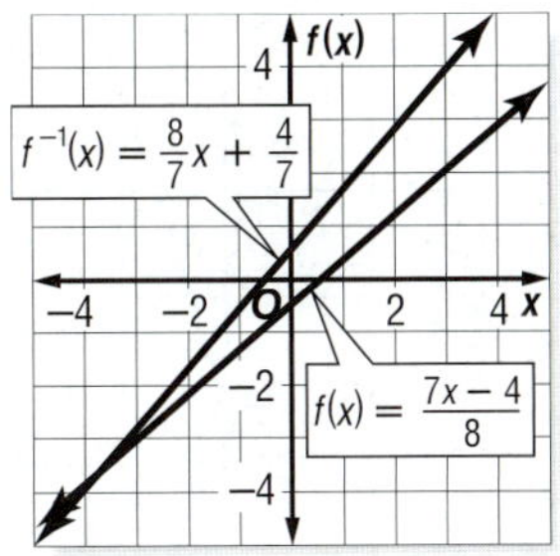

Page 396, Lesson 7-9
Graphing Calculator Investigation

1.

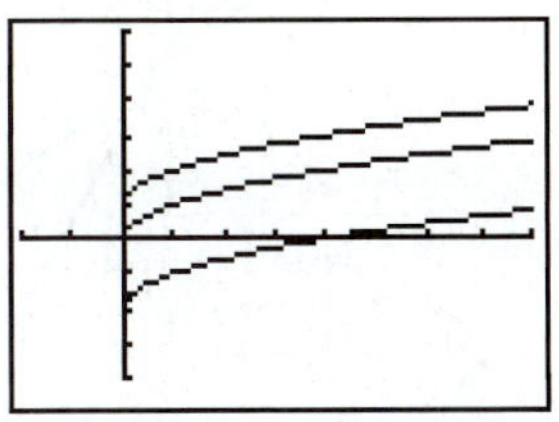

[−2, 8] scl: 1 by [−4, 6] scl: 1

2.

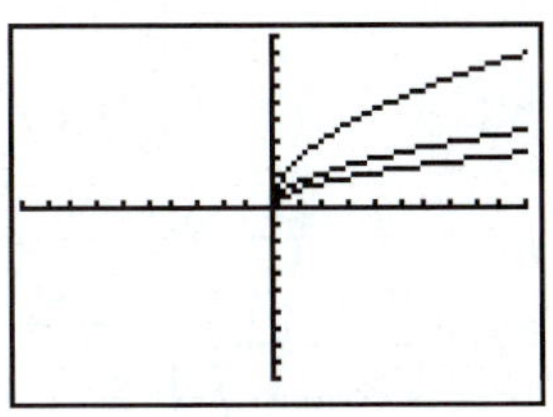

[−10, 10] scl: 1 by [−10, 10] scl: 1

3.

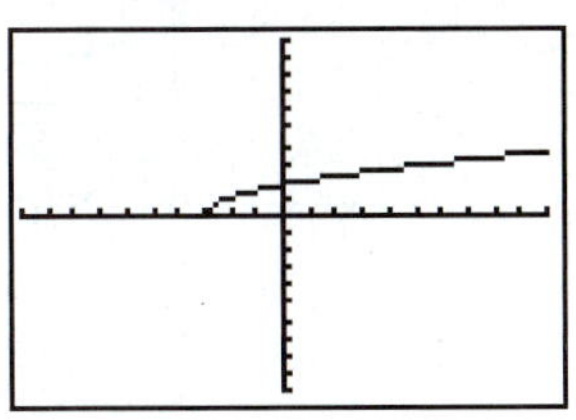

[−10, 10] scl: 1 by [−10, 10] scl: 1

Pages 397–399, Lesson 7-9

4.

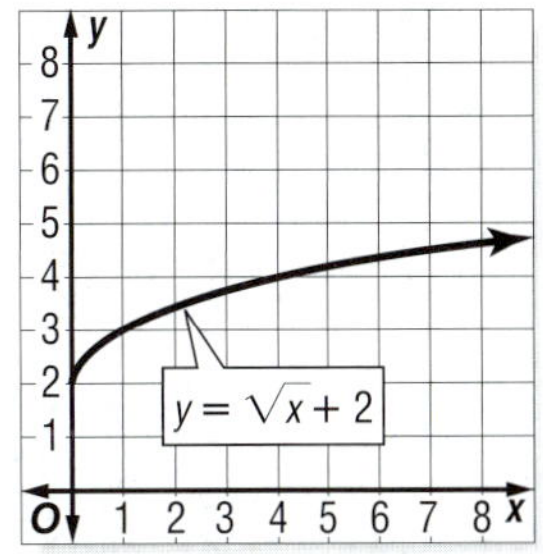

5.

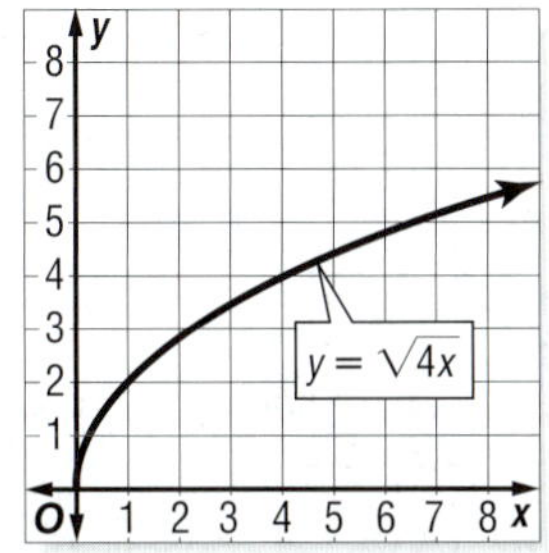

6.

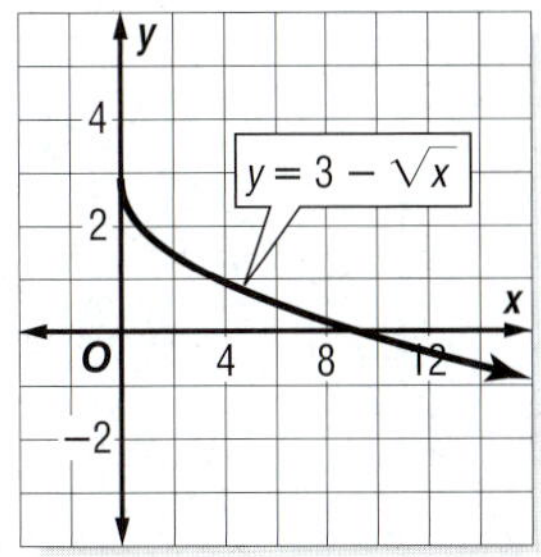

7.

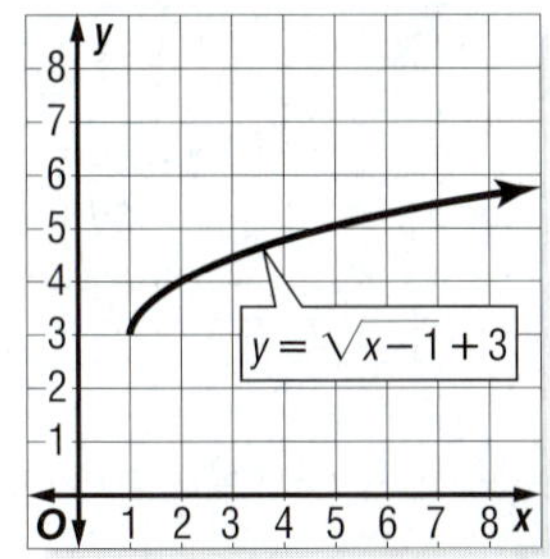

8.

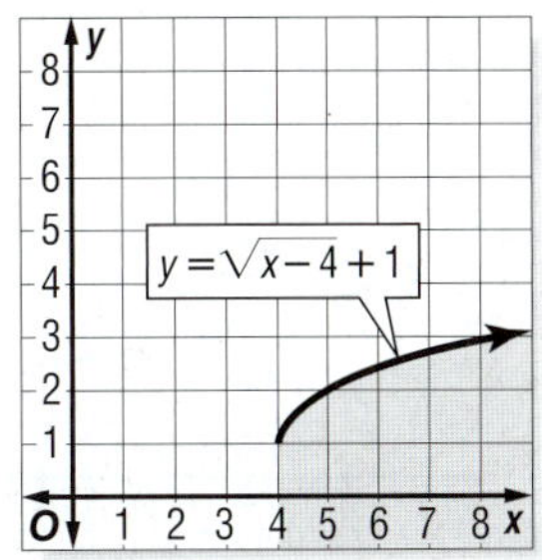

9.

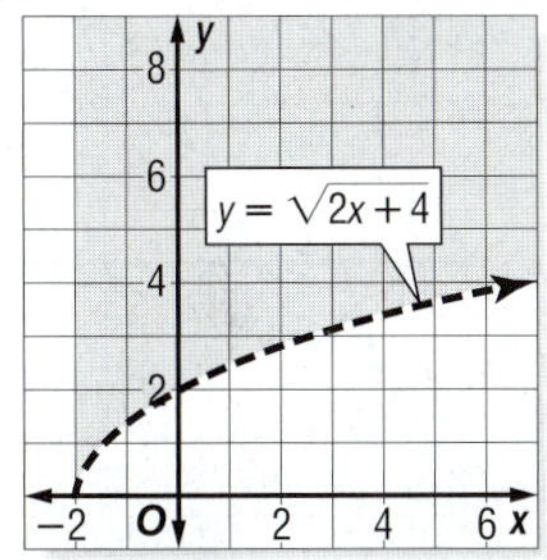

10.

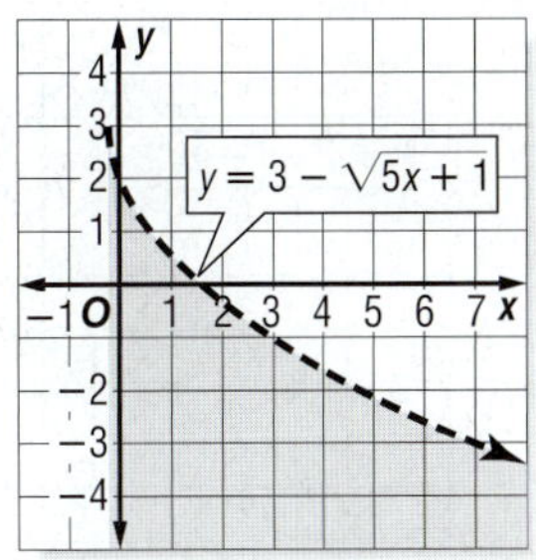

11.

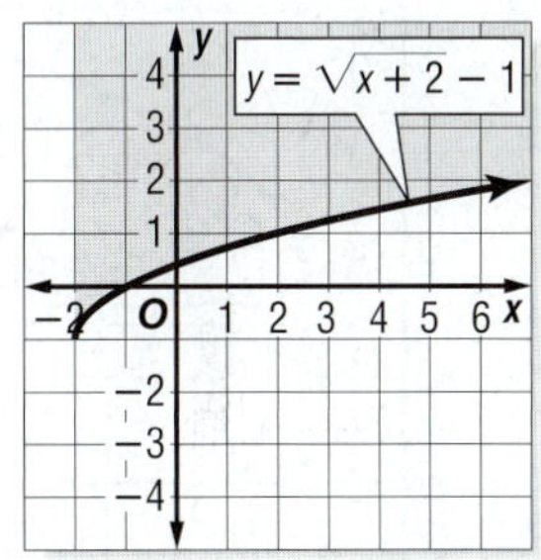

14.

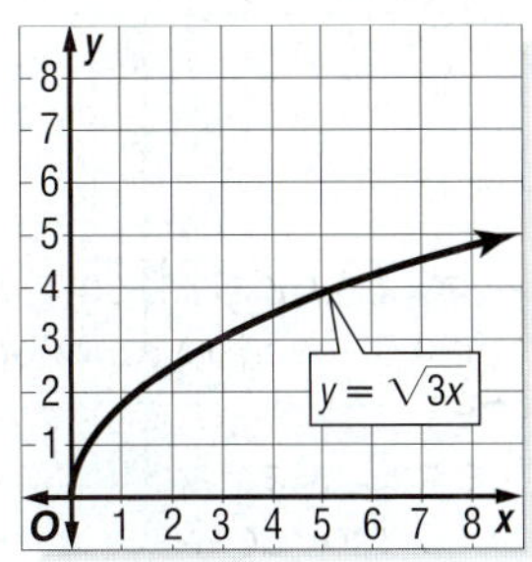

15.

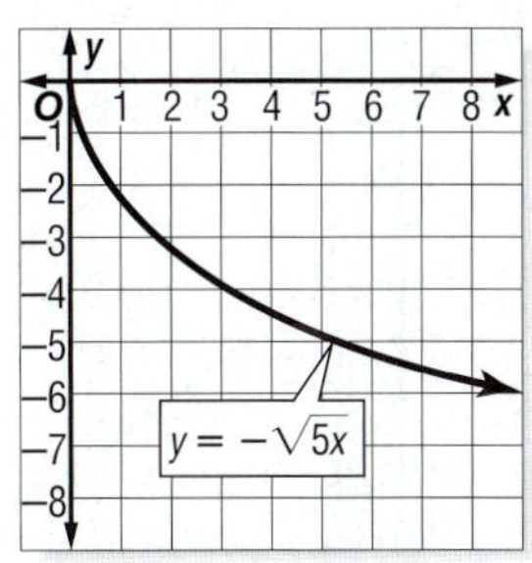

16.

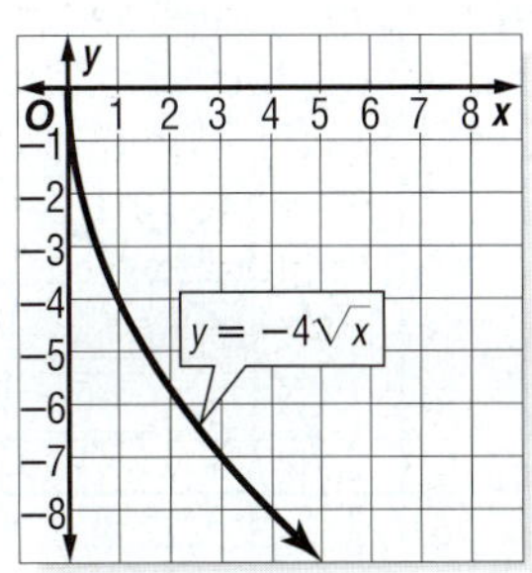

17.

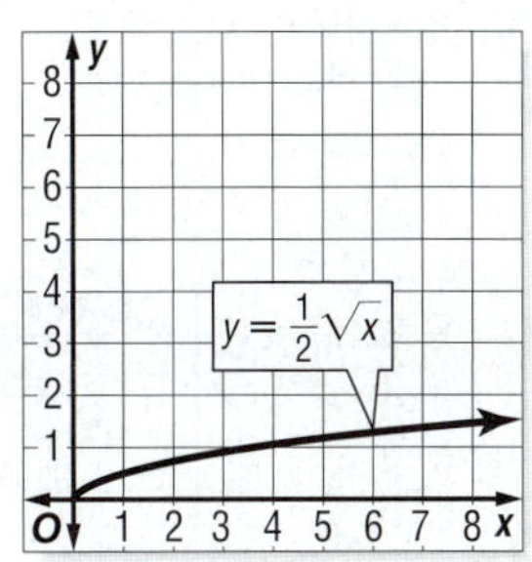

18.

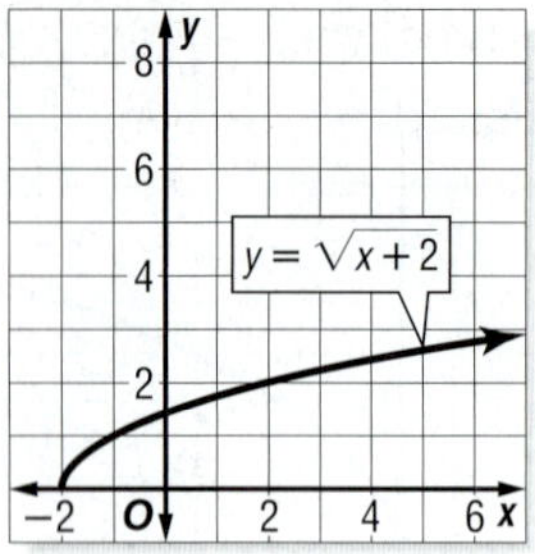

19.

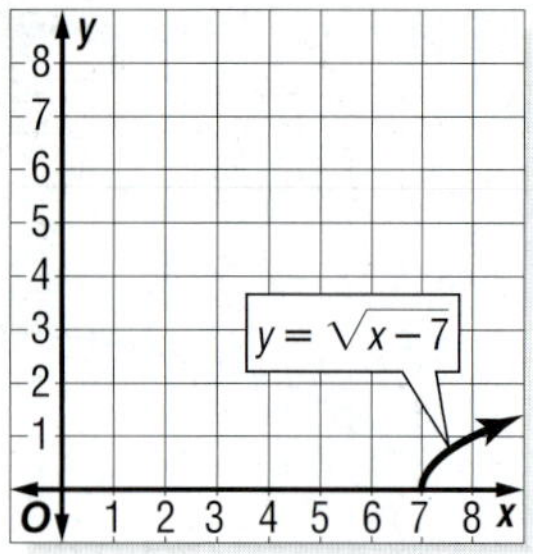

20.

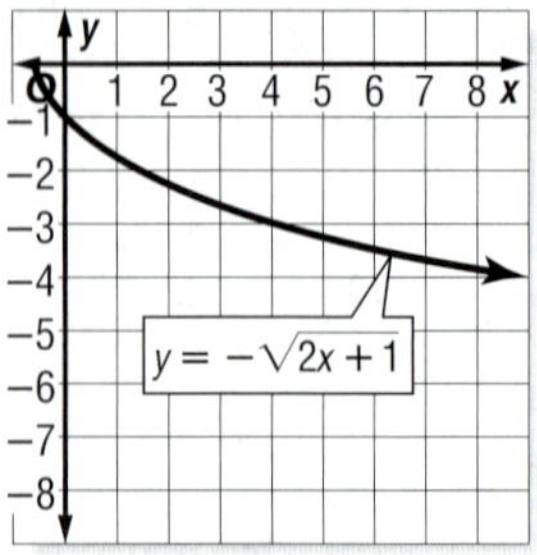

21.

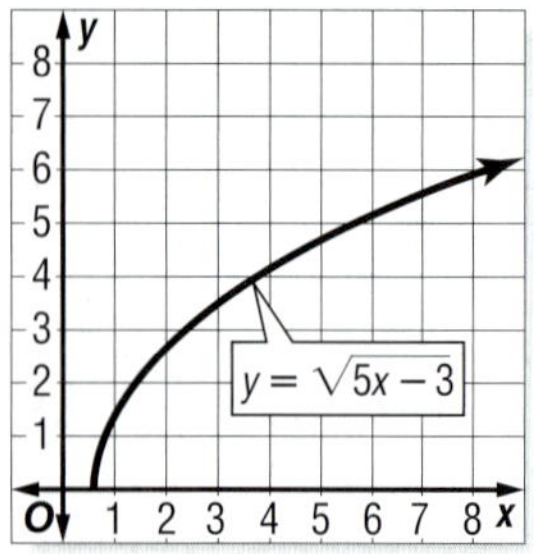

22.

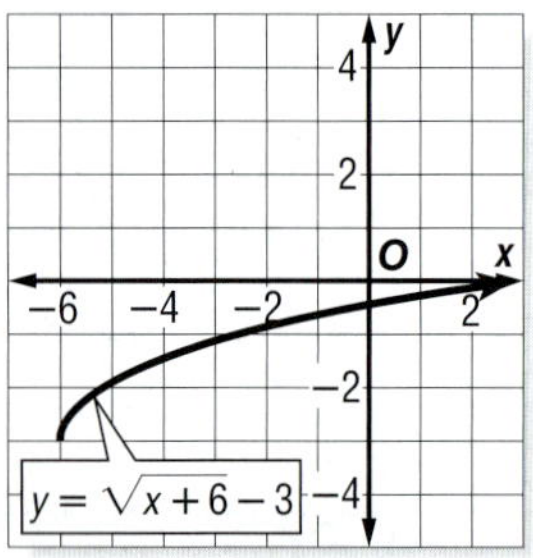

23.

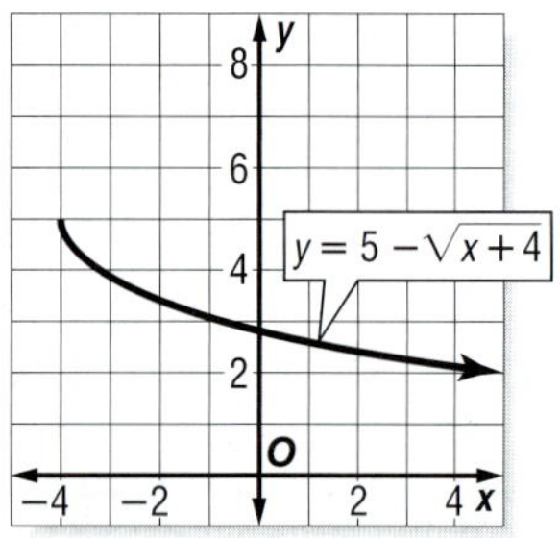

24.

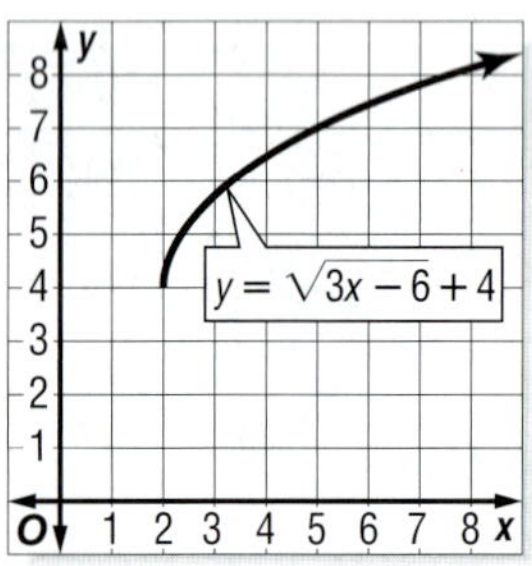

25.

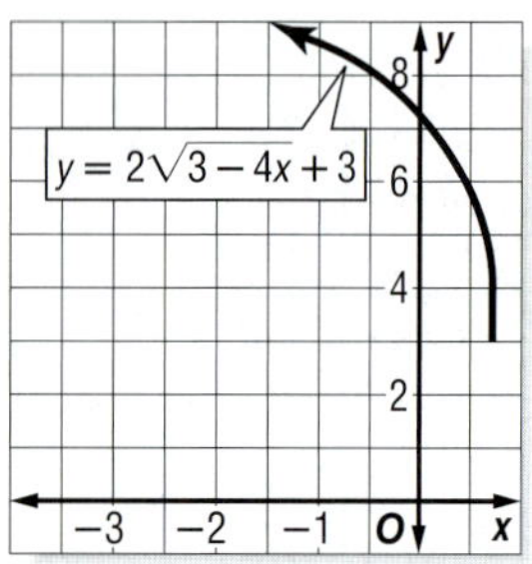

26.

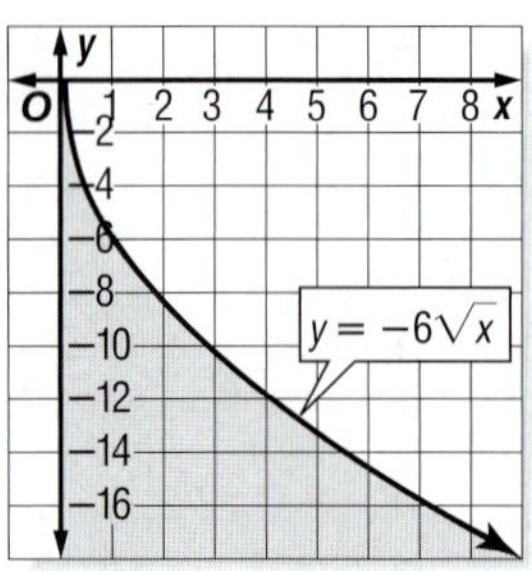

27.

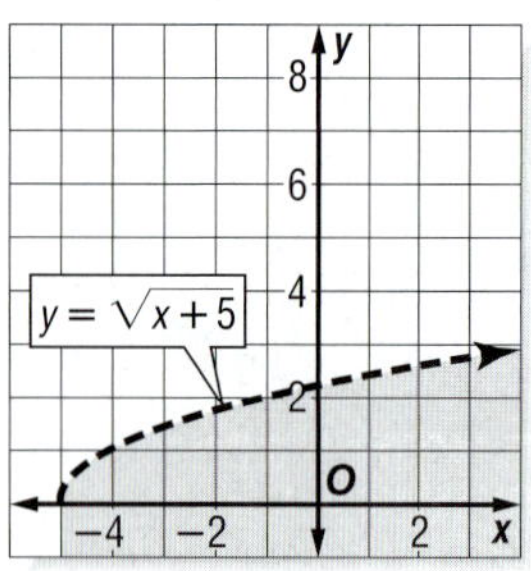

28.

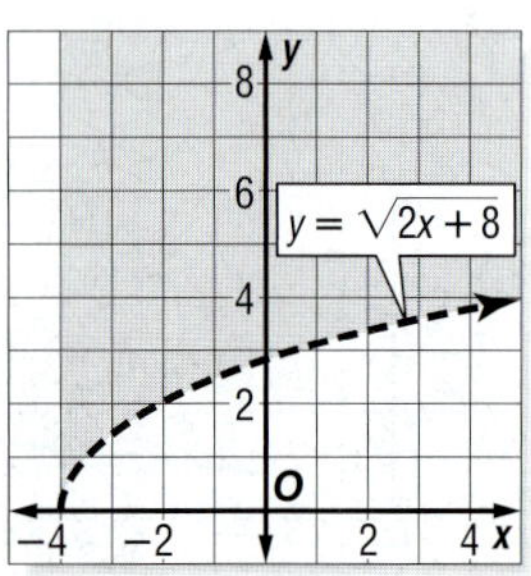

29.

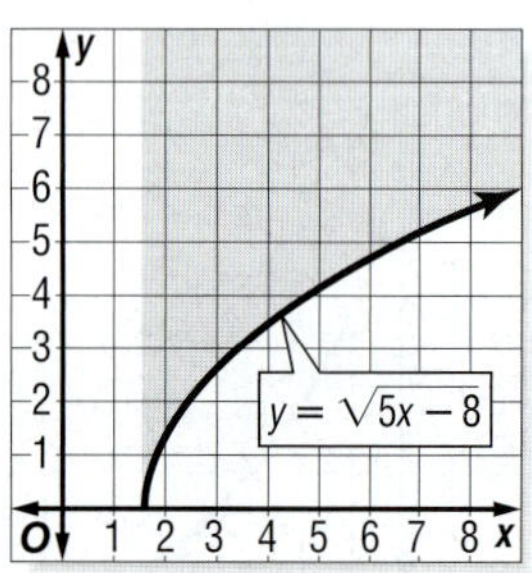

30.

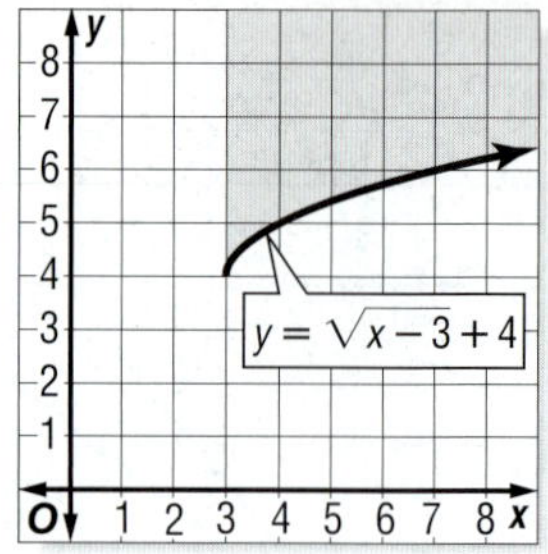

31.

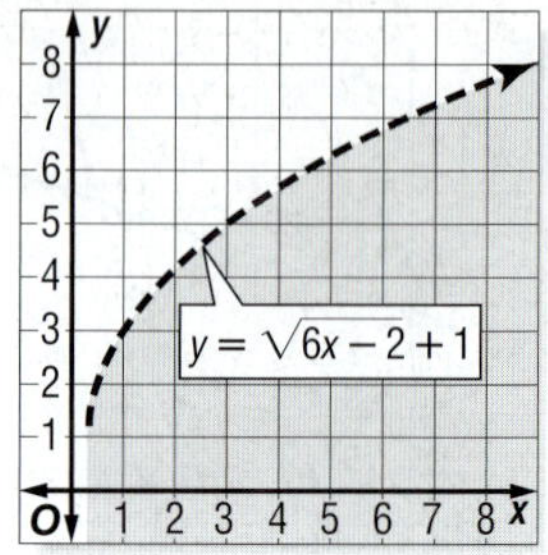

Pages 400–404, Chapter 7 Study Guide and Review

48. $f^{-1}(x) = \dfrac{x + 4}{3}$

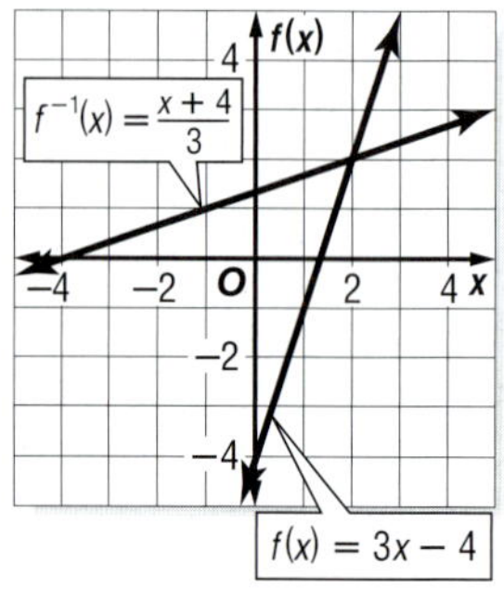

49. $f^{-1}(x) = \dfrac{-x - 3}{2}$

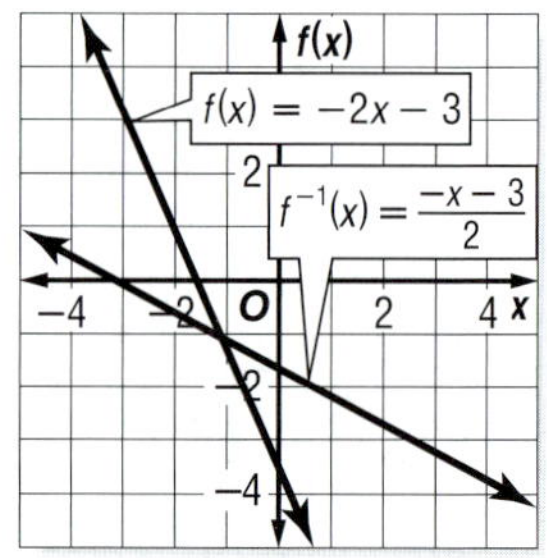

50. $g^{-1}(x) = 3x - 6$

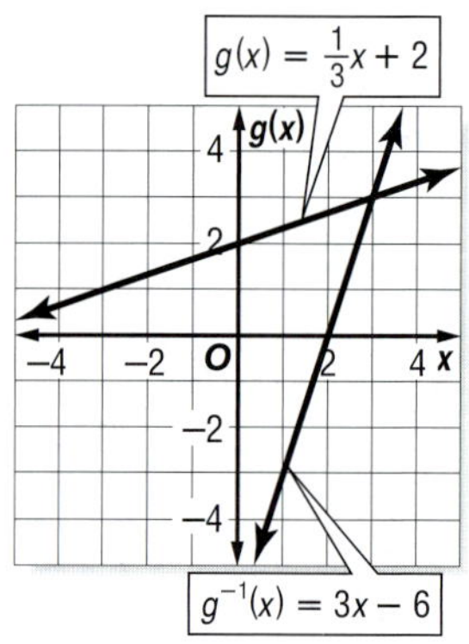

51. $f^{-1}(x) = \dfrac{2x - 1}{-3}$

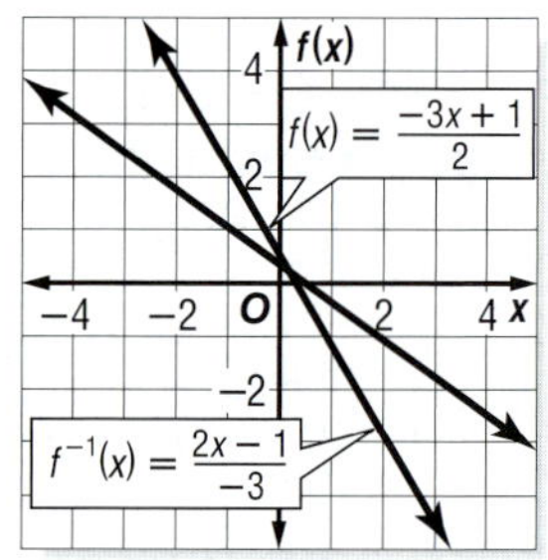

52. $y^{-1}(x) = \pm\sqrt{x}$

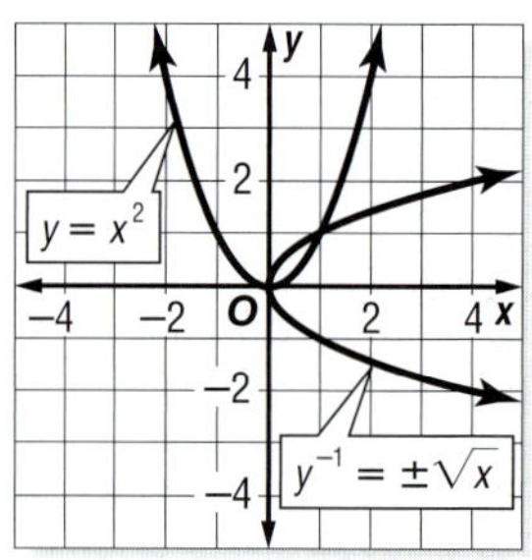

53. $y^{-1}(x) = \pm\frac{1}{2}\sqrt{x} - \frac{3}{2}$

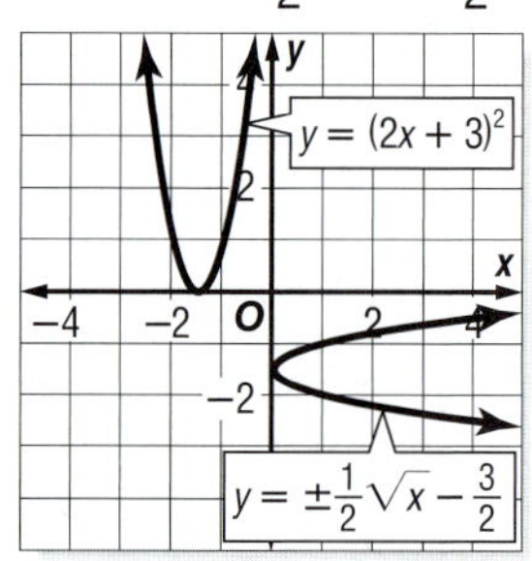

Page 405, Chapter 7 Practice Test

4a.

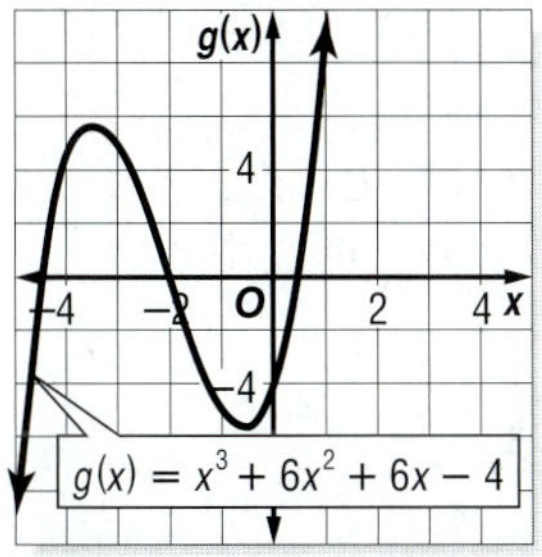

4b. between -5 and -4, zero at $x = -2$, between 0 and 1

4c. Sample answer: rel. max. at $x = -3.5$, rel. min. at $x = -0.5$

5a.

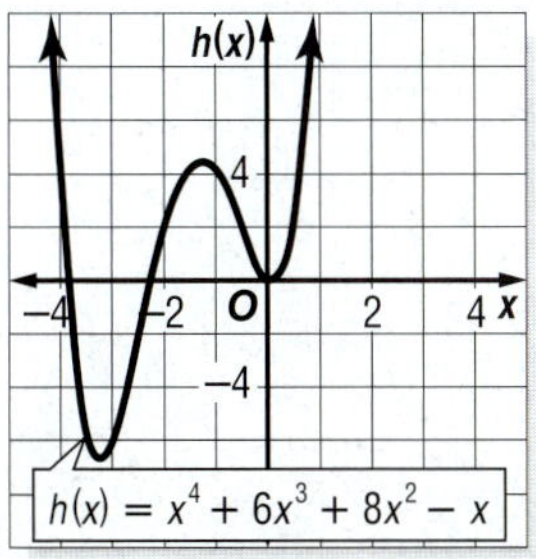

5b. between -4 and -3, between -3 and -2, zero at $x = 0$

5c. Sample answer: rel. max. at $x = -1$, rel. min. at $x = -3$ and at $x = 0$

6a.

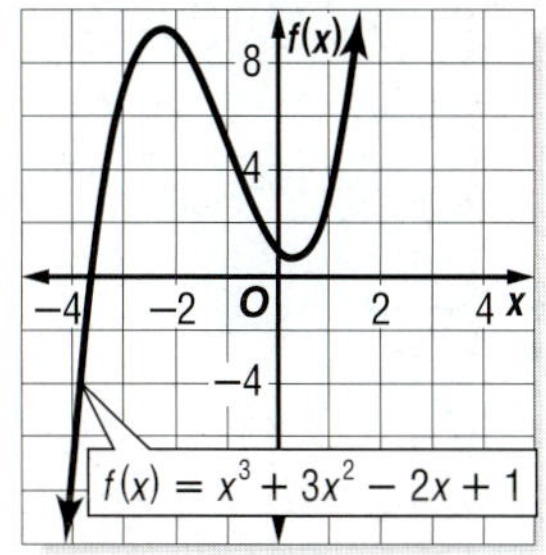

6b. between -4 and -3

6c. Sample answer: rel. max. at $x = -2.3$, rel. min. at $x = 0.3$

7a.

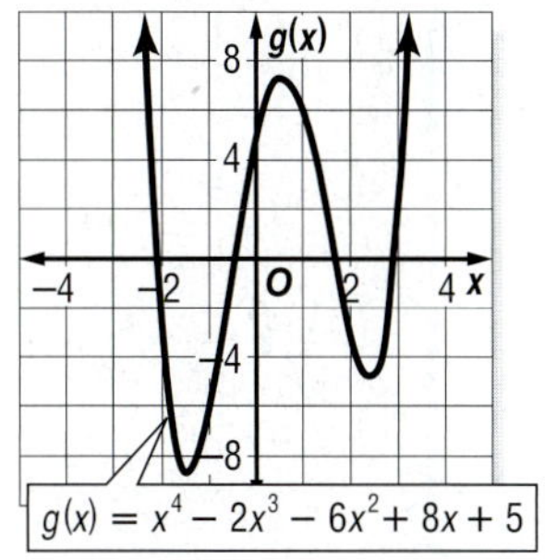

7b. between -3 and -2, between -1 and 0, between 1 and 2, between 2 and 3

7c. Sample answer: rel. max. at $x = 0.6$, rel. min. at $x = -1.5$ and at $x = 2.4$

UNIT 3 Notes

Introduction

At the beginning of this unit, students are reacquainted with the Midpoint and Distance Formulas before exploring conic sections. They then learn to combine rational expressions, which leads to graphing rational functions where they examine asymptotes and holes. This knowledge of functions is applied to direct, joint, and inverse variations.

The unit concludes with an investigation of exponential and logarithmic functions. Finally, logarithms with base *e* and natural logarithms are investigated and applied to real-world situations involving investigating growth and decay.

Assessment Options

Unit 3 Test Pages 629–630 of the *Chapter 10 Resource Masters* may be used as a test or review for Unit 3. This assessment contains both multiple-choice and short answer items.

ExamView® Pro

This CD-ROM can be used to create additional unit tests and review worksheets.

UNIT 3

Advanced Functions and Relations

You can use functions and relations to investigate events like earthquakes. In this unit, you will learn about conic sections, rational expressions and equations, and exponential and logarithmic functions.

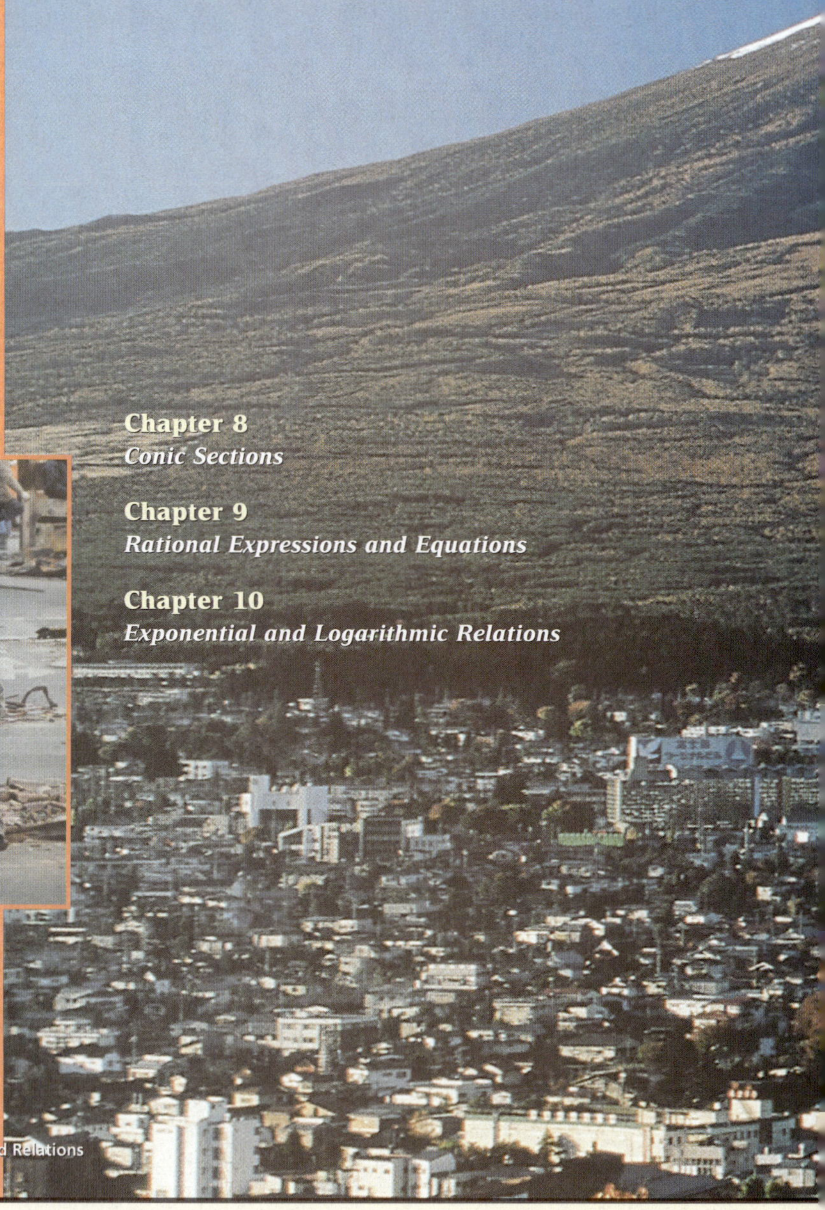

Chapter 8
Conic Sections

Chapter 9
Rational Expressions and Equations

Chapter 10
Exponential and Logarithmic Relations

WebQuest Internet Project

On Quake Anniversary, Japan Still Worries

Source: *USA TODAY*, January 16, 2001

"As Japan marks the sixth anniversary of the devastating Kobe earthquake this week, a different seismic threat is worrying the country: Mount Fuji. Researchers have measured a sudden increase of small earthquakes on the volcano, indicating there is movement of magma underneath its snowcapped, nearly symmetrical cone about 65 miles from Tokyo." In this project, you will explore how functions and relations are related to locating, measuring, and classifying earthquakes.

Log on to www.algebra2.com/webquest. Begin your WebQuest by reading the Task.

Then continue working on your WebQuest as you study Unit 3.

Lesson	8-3	9-5	10-1
Page	429	502	529

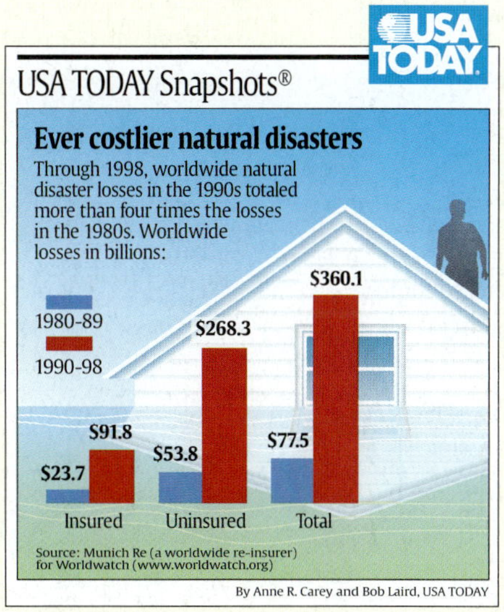

Teaching Suggestions

Have students study the USA TODAY Snapshot®.

- Ask students to revisit the population data from page 219 of Unit 2 and make a conjecture of how it relates to the increase in natural disaster losses provided here.
- Is the ratio of the losses for 1980–89 to the losses for 1990–98 greater for insured losses or for uninsured losses? **insured**
- Ask students to suggest what types of disasters, other than earthquakes, qualify as natural disasters and why the uninsured costs have grown so much.

Additional USA TODAY Snapshots® appearing in Unit 3:

Chapter 8	Staying in touch (p. 448)
Chapter 9	College high-tech spending (p. 492)
Chapter 10	July 4th can be loud. Be careful. (p. 535)
	Georgia led pecan production in 2000 (p. 565)

WebQuest Internet Project

Problem-Based Learning A WebQuest is an online project in which students do research on the Internet, gather data, and make presentations using word processing, graphing, page-making, or presentation software. In each chapter, students advance to the next step in their WebQuest. At the end of Chapter 10, the project culminates with a presentation of their findings.

Teaching suggestions and sample answers are available in the *WebQuest and Project Resources*.

Chapter 8

Conic Sections

Chapter Overview and Pacing

Year-long and two-year pacing: pages T20–T21.

LESSON OBJECTIVES

LESSON OBJECTIVES	PACING (days) Regular		PACING (days) Block	
	Basic/ Average	Advanced	Basic/ Average	Advanced
8-1 Midpoint and Distance Formulas *(pp. 412–418)* • Find the midpoint of a segment on the coordinate plane. • Find the distance between two points on the coordinate plane. ***Follow-Up:*** Midpoint and Distance Formulas in Three Dimensions	1	2 (with 8-1 Follow-Up)	0.5	1
8-2 Parabolas *(pp. 419–425)* • Write equations of parabolas in standard form. • Graph parabolas.	1	1	0.5	0.5
8-3 Circles *(pp. 426–431)* • Write equations of circles. • Graph circles.	1	1	0.5	0.5
8-4 Ellipses *(pp. 432–440)* ***Preview:*** Investigating Ellipses • Write equations of ellipses. • Graph ellipses.	2 (with 8-4 Preview)	2	1 (with 8-4 Preview)	1
8-5 Hyperpolas *(pp. 441–448)* • Write equations of hyperbolas. • Graph hyperbolas.	2	2	1	1
8-6 Conic Sections *(pp. 449–454)* • Write equations of conic sections in standard form. • Identify conic sections from their equations. ***Follow-Up:*** Conic Sections	1	2 (with 8-6 Follow-Up)	0.5	1
8-7 Solving Quadratic Systems *(pp. 455–460)* • Solve systems of quadratic equations algebraically and graphically. • Solve systems of quadratic inequalities graphically.	2	2	1	1
Study Guide and **Practice Test** *(pp. 461–467)* **Standardized Test Practice** *(pp. 468–469)*	1	1	0.5	0.5
Chapter Assessment	1	1	0.5	0.5
TOTAL	**12**	**14**	**6**	**7**

StudentWorks

*An electronic version of this chapter is available on **StudentWorks**™. This backpack solution CD-ROM allows students instant access to the Student Edition, lesson worksheet pages, and web resources.*

Timesaving Tools
TeacherWorks™
All-In-One Planner and Resource Center
See pages T5 and T21.

Chapter Resource Manager

Chapter 8 Resource Masters

Study Guide and Intervention	Practice (Skills and Average)	Reading to Learn Mathematics	Enrichment	Assessment	Applications*	5-Minute Check Transparencies	Interactive Chalkboard	Alge2PASS: Tutorial Plus (lessons)	Materials
455–456	457–458	459	460			8-1	8-1		(*Follow-Up:* shoe box or tissue box; wire, spaghetti, yarn, or thread)
461–462	463–464	465	466	511	GCS 42, SC 15, SM 119–122	8-2	8-2		wax paper, inch ruler, pen, posterboard, masking tape, yardsticks or meter sticks
467–468	469–470	471	472		GCS 41	8-3	8-3	15	
473–474	475–476	477	478	511, 513		8-4	8-4		(*Preview:* thumbtacks, cardboard, string, pencil, grid paper) grid paper, compass, index cards
479–480	481–482	483	484			8-5	8-5	16	posterboard
485–486	487–488	489	490	512		8-6	8-6		(*Follow-Up:* conic graph paper, concentric-circle graph paper)
491–492	493–494	495	496	512	SC 16	8-7	8-7		graphing calculator
				497–510, 514–516					

**Key to Abbreviations:* GCS = Graphing Calculator and Speadsheet Masters,
SC = School-to-Career Masters,
SM = Science and Mathematics Lab Manual

Chapter 8

Mathematical Connections and Background

Continuity of Instruction

Prior Knowledge

Students have found midpoints of segments and distances between points on the coordinate plane. In prior lessons they worked with equations for parabolas, and in previous courses they may have worked with equations for circles. They have also used graphs and algebraic techniques to explore systems of linear equations and inequalities.

This Chapter

This chapter begins with a review of the midpoint and distance formulas. Students then explore the characteristics and equations of the conic sections. They will study the effect that each number in the standard form of the equations has on the graph of the equations. Students also graph and solve systems of quadratic equations and inequalities.

Future Connections

This study of conic sections lays the foundation for future mathematics study in coordinate geometry. In future courses, students will apply their knowledge when they study parametric equations and the polar coordinate system.

8-1 Midpoint and Distance Formulas

In this lesson, students explore the formulas relating to line segments on a coordinate plane. The coordinates of the midpoint of a line segment are the means of the corresponding coordinates of the endpoints. The distance formula is an application of the Pythagorean Theorem. These two formulas are used often in the remaining lessons in the chapter as students investigate the general forms of the equations of conic sections.

8-2 Parabolas

A parabola is the set of all points in a plane that are the same distance from a given point called the *focus* and a given line called the *directrix*. Students will use this definition with the distance formula to derive the formula $y = a(x - h)^2 + k$, which is the standard form of the equation of a parabola. Students explore how the values a, h, and k are related to the parabola's vertex, axis of symmetry, focus, and directrix, and to whether it opens up or down. The coefficient a also is associated with a segment, called the *latus rectum*, which has endpoints on the parabola, contains the focus, and is perpendicular to the axis of symmetry.

8-3 Circles

A circle is the set of points in a plane that are a given distance (the radius) from a given point (the center). Using (h, k) as the given point and r as the given distance, the equation of a circle is $\sqrt{(x - h)^2 + (y - k)^2} = r$. Squaring both sides of that equation gives the standard form of the equation, $(x - h)^2 + (y - k)^2 = r^2$.

8-4 Ellipses

In this lesson students explore an ellipse, the set of points in a plane such that the sum of the distances from each point to two fixed points is constant. The lesson uses the definition of an ellipse and the distance formula to derive the standard form of an equation for an ellipse centered at the origin, $\frac{x^2}{a^2} + \frac{y^2}{b^2} = 1$. Students find the ellipse's foci, and they examine how the values of a and b determine the length of the major and minor axes and whether the direction of the major ellipse is horizontal or vertical. Also, students examine equations that have the form $\frac{(x - h)^2}{a^2} + \frac{(y - k)^2}{b^2} = 1$, exploring how such an equation represents an ellipse whose center is translated to the point (h, k).

8-5 Hyperbolas

This lesson begins by considering the distances between a general point and two fixed points, and defines a hyperbola as the set of all points in a plane for which the absolute value of the difference of those distances is constant. Then for a general point (x, y), two specific points, and a specific constant, the lesson finds the distances between (x, y) and each specific point, subtracts the distances, and equates that difference to the constant. The result is the standard form of the equation of a hyperbola centered at the origin, $\frac{x^2}{a^2} - \frac{y^2}{b^2} = 1$. Students identify the difference between the equations for ellipses and hyperbolas, and they examine the names of the parts of a hyperbola, its asymptotes $y = \pm\frac{b}{a}x$, its two branches, and whether the transverse axis is horizontal or vertical. They relate the variables a and b to the foci and vertices of the hyperbola, to the equations of the asymptotes, and to the lengths of the transverse and conjugate axes.

8-6 Conic Sections

This lesson presents the general quadratic equation for a conic section, $Ax^2 + Bxy + Cy^2 + Dx + Ey + F = 0$, and explores how that equation is related to the standard forms of the equations of the four conic sections. In one activity, students are given specific values for the coefficients A through F and manipulate the resulting equation until it is in the standard form for one of the four conic sections. In another activity, students analyze how the different relationships between the coefficients A and C determine whether a particular equation represents a parabola, circle, ellipse, or hyperbola. The lesson also reviews why the four curves are called conic sections; that is, how to slice a double cone with a plane to illustate a parabola, circle, ellipse, or hyperbola.

8-7 Solving Quadratic Systems

This lesson shows how to use graphing techniques to find the number of solutions to a quadratic system, and then how to use algebraic techniques to find those solutions. For a linear-quadratic system, a graph indicates whether the conic section and the line intersect in 0, 1, or 2 points; then substitution can be used as the first step in writing a one-variable equation from the two-variable system. For a quadratic-quadratic system, a graph indicates the number of solutions (0, 1, 2, 3, or 4); elimination can be used to generate a one-variable equation from the two-variable system. The lesson also explores systems of quadratic inequalities. By graphing the related equations, shading the appropriate regions, deciding when boundary lines are part of a solution region, and solving related systems of equations to find specific points, the solution to a system of quadratic inequalities can be illustrated graphically and described algebraically.

Additional mathematical information and teaching notes are available at www.algebra2.com/key_concepts.

Chapter 8

DAILY INTERVENTION and Assessment

	Type	Student Edition	Teacher Resources	Technology/Internet
INTERVENTION	Ongoing	Prerequisite Skills, pp. 411, 416, 425, 431, 440, 448, 452 Practice Quiz 1, p. 431 Practice Quiz 2, p. 448	5-Minute Check Transparencies Quizzes, *CRM* pp. 511–512 Mid-Chapter Test, *CRM* p. 513 Study Guide and Intervention, *CRM* pp. 455–456, 461–462, 467–468, 473–474, 479–480, 485–486, 491–492	Alge2PASS: Tutorial Plus www.algebra2.com/self_check_quiz www.algebra2.com/extra_examples
	Mixed Review	pp. 416, 425, 431, 440, 447, 452, 460	Cumulative Review, *CRM* p. 514	
	Error Analysis	Find the Error, pp. 423, 428	Find the Error, *TWE* pp. 423, 429 Unlocking Misconceptions, *TWE* pp. 420, 435, 442 Tips for New Teachers, *TWE* pp. 416, 440, 448	
	Standardized Test Practice	pp. 413, 414, 416, 425, 431, 439, 440, 446, 447, 452, 459, 468–469	*TWE* p. 413 Standardized Test Practice, *CRM* pp. 515–516	Standardized Test Practice CD-ROM www.algebra2.com/standardized_test
ASSESSMENT	Open-Ended Assessment	Writing in Math, pp. 416, 425, 430, 439, 447, 452, 459 Open Ended, pp. 414, 423, 437, 445, 450, 458	Modeling: *TWE* pp. 416, 452 Speaking: *TWE* pp. 431, 440, 460 Writing: *TWE* pp. 425, 448 Open-Ended Assessment, *CRM* p. 509	
	Chapter Assessment	Study Guide, pp. 461–466 Practice Test, p. 467	Multiple-Choice Tests (Forms 1, 2A, 2B), *CRM* pp. 497–502 Free-Response Tests (Forms 2C, 2D, 3), *CRM* pp. 503–508 Vocabulary Test/Review, *CRM* p. 510	ExamView® Pro (see below) MindJogger Videoquizzes www.algebra2.com/vocabulary_review www.algebra2.com/chapter_test

Key to Abbreviations: TWE = Teacher Wraparound Edition; CRM = Chapter Resource Masters

Intervention Technology

Alge2PASS: Tutorial Plus CD-ROM offers a complete, self-paced algebra curriculum.

Algebra 2 Lesson	Alge2PASS Lesson
8-3	**15** *Graphing Parabolas and Circles*
8-5	**16** *Graphing Ellipses and Hyperbolas*

ExamView® Pro

Use the networkable **ExamView® Pro** to:

- Create **multiple versions** of tests.
- Create **modified** tests for *Inclusion* students.
- **Edit** existing questions and **add** your own questions.
- Use built-in **state curriculum correlations** to create tests aligned with state standards.
- **Apply** art to your tests from a program bank of artwork.

For more information on Intervention and Assessment, see pp. T8–T11.

Reading and Writing in Mathematics

Glencoe Algebra 2 provides numerous opportunities to incorporate reading and writing into the mathematics classroom.

Student Edition

- Foldables Study Organizer, p. 411
- Concept Check questions require students to verbalize and write about what they have learned in the lesson. (pp. 414, 423, 428, 437, 445, 450, 458, 461)
- Writing in Math questions in every lesson, pp. 416, 425, 430, 439, 447, 452, 459
- Reading Study Tip, pp. 442, 449

Teacher Wraparound Edition

- Foldables Study Organizer, pp. 411, 461
- Study Notebook suggestions, pp. 414, 418, 423, 429, 432, 437, 445, 450, 454, 458
- Modeling activities, pp. 416, 452
- Speaking activities, pp. 431, 440, 460
- Writing activities, pp. 425, 448
- ELL Resources, pp. 410, 415, 424, 430, 439, 447, 451, 459, 461

Additional Resources

- Vocabulary Builder worksheets require students to define and give examples for key vocabulary terms as they progress through the chapter. (*Chapter 8 Resource Masters,* pp. vii-viii)
- Reading to Learn Mathematics master for each lesson (*Chapter 8 Resource Masters*, pp. 459, 465, 471, 477, 483, 489, 495)
- *Vocabulary PuzzleMaker* software creates crossword, jumble, and word search puzzles using vocabulary lists that you can customize.
- *Teaching Mathematics with Foldables* provides suggestions for promoting cognition and language.
- *Reading and Writing in the Mathematics Classroom*
- *WebQuest and Project Resources*

For more information on Reading and Writing in Mathematics, see pp. T6–T7.

ELL ENGLISH LANGUAGE LEARNERS

Lesson 8-1

Reading and Writing

Have students color-code the elements of one ordered pair with one color and the elements of the second ordered pair with another color. The color coding will allow them to easily identify the correct order in which to substitute the numbers in the Distance and Midpoint Formulas. Have students write an explanation in their own words, and if necessary in their own language, of how to find the midpoint of a segment and determine the distance between two points.

Lesson 8-2

Using Applications

Bring flashlights or car headlights to class to allow students to see how these objects use parabolic reflectors. Use mathematical terminology to discuss how the light bulbs are placed within the parabolic reflectors to achieve the most effective beam.

Lesson 8-5

Multisensory Activities

Have students view a hyperbola by placing a table lamp near the wall and observing the shape the light makes. You may wish to have groups of students place paper on the wall and sketch the shape of the light. Then determine a scale and axes to find an equation of the hyperbola formed.

Chapter

8 Notes

What You'll Learn

Have students read over the list of objectives and make a list of any words with which they are not familiar.

Why It's Important

Point out to students that this is only one of many reasons why each objective is important. Others are provided in the introduction to each lesson.

Lesson	NCTM Standards	Local Objectives
8-1	2, 3, 6, 8, 9, 10	
8-1 Follow-Up	3, 6, 8, 10	
8-2	2, 3, 6, 8, 9, 10	
8-3	2, 3, 6, 8, 9, 10	
8-4 Preview	3, 4, 7, 8	
8-4	2, 3, 6, 8, 9, 10	
8-5	2, 3, 6, 8, 9, 10	
8-6	2, 3, 6, 8, 9, 10	
8-6 Follow-Up	3, 7, 10	
8-7	2, 3, 6, 8, 9, 10	

Key to NCTM Standards:

1=Number & Operations, 2=Algebra, 3=Geometry, 4=Measurement, 5=Data Analysis & Probability, 6=Problem Solving, 7=Reasoning & Proof, 8=Communication, 9=Connections, 10=Representation

Chapter

8 Conic Sections

What You'll Learn

- **Lesson 8-1** Use the Midpoint and Distance Formulas.
- **Lessons 8-2 through 8-5** Write and graph equations of parabolas, circles, ellipses, and hyperbolas.
- **Lesson 8-6** Identify conic sections.
- **Lesson 8-7** Solve systems of quadratic equations and inequalities.

Key Vocabulary

- **parabola** (p. 419)
- **conic section** (p. 419)
- **circle** (p. 426)
- **ellipse** (p. 433)
- **hyperbola** (p. 441)

Why It's Important

Many planets, comets, and satellites have orbits in curves called *conic sections*. These curves include parabolas, circles, ellipses, and hyperbolas. The Moon's orbit is almost a perfect circle. *You will learn more about the orbits in Lessons 8-2 through 8-7.*

Vocabulary Builder

ELL

The Key Vocabulary list introduces students to some of the main vocabulary terms included in this chapter. For a more thorough vocabulary list with pronunciations of new words, give students the Vocabulary Builder worksheets found on pages vii and viii of the *Chapter 8 Resource Masters*. Encourage them to complete the definition of each term as they progress through the chapter. You may suggest that they add these sheets to their study notebooks for future reference when studying for the Chapter 8 test.

Getting Started

Prerequisite Skills To be successful in this chapter, you'll need to master these skills and be able to apply them in problem-solving situations. Review these skills before beginning Chapter 8.

For Lessons 8-2 through 8-6 — **Completing the Square**

Solve each equation by completing the square. *(For review, see Lesson 6-4.)*

1. $x^2 + 10x + 24 = 0$ $\{-4, -6\}$
2. $x^2 - 2x + 2 = 0$ $\{1 + i, 1 - i\}$
3. $2x^2 + 5x - 12 = 0$ $\left\{\frac{3}{2}, -4\right\}$

For Lessons 8-2 through 8-6 — **Translation Matrices**

A translation is given for each figure.

a. Write the vertex matrix for the given figure. 4a. $\begin{bmatrix} -2 & 3 & 1 & -4 \\ 1 & 1 & -2 & -2 \end{bmatrix}$

b. Write the translation matrix.

c. Find the coordinates in matrix form of the vertices of the translated figure.

(For review, see Lesson 4-4.)

4. 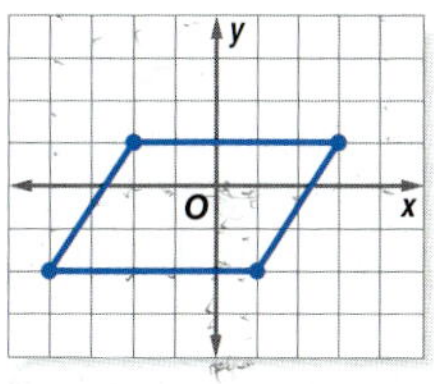

b. $\begin{bmatrix} -4 & -4 & -4 & -4 \\ 2 & 2 & 2 & 2 \end{bmatrix}$ c. $\begin{bmatrix} -6 & -1 & -3 & -8 \\ 3 & 3 & 0 & 0 \end{bmatrix}$

translated 4 units left and 2 units up

5. 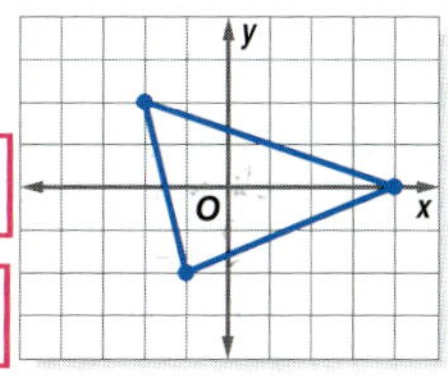

a. $\begin{bmatrix} -2 & 4 & -1 \\ 2 & 0 & -2 \end{bmatrix}$ b. $\begin{bmatrix} 5 & 5 & 5 \\ -3 & -3 & -3 \end{bmatrix}$ c. $\begin{bmatrix} 3 & 9 & 4 \\ -1 & -3 & -5 \end{bmatrix}$

translated 5 units right and 3 units down

For Lesson 8-7 — **Graph Linear Inequalities**

Graph each inequality. *(For review, see Lesson 2-7.)* **6–8. See margin.**

6. $y < x + 2$
7. $x + y \leq 3$
8. $2x - 3y > 6$

FOLDABLES™ Study Organizer

Conic Sections Make this Foldable to help you organize your notes. Begin with four sheets of grid paper and one piece of construction paper.

Step 1 Fold and Staple

Stack sheets of grid paper with edges $\frac{1}{2}$ inch apart. Fold top edges back. Staple to construction paper at top.

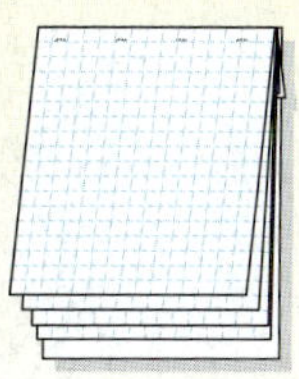

Step 2 Cut and Label

Cut grid paper in half lengthwise. Label tabs as shown.

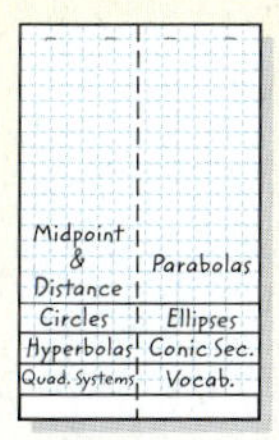

Reading and Writing As you read and study the chapter, use each tab to write notes, formulas, and examples for each conic section.

FOLDABLES™ Study Organizer

For more information about Foldables, see *Teaching Mathematics with Foldables.*

Expository Writing and Organizing Data After students make their Foldables, have them label each tab to correspond to a lesson in Chapter 8. Use the extra tab for vocabulary. Students use their Foldables to take notes, define terms, record concepts, and write examples. Ask students to use their notes to write expositions on conic sections so that someone who did not know or understand conic sections before will understand them after reading what students have written. Explain that textbooks are examples of expository writing.

Getting Started

This section provides a review of the basic concepts needed before beginning Chapter 8. Page references are included for additional student help.

Prerequisite Skills in the Getting Ready for the Next Lesson section at the end of each exercise set review a skill needed in the next lesson.

For Lesson	Prerequisite Skill
8-2	Completing the Square (p. 416)
8-3	Simplifying Radicals (p. 425)
8-4	Solving Quadratic Equations (p. 431)
8-5	Graphing Lines (p. 440)
8-6	Identifying Coefficients (p. 448)
8-7	Solving Systems of Linear Equations (p. 452)

Answers

6. 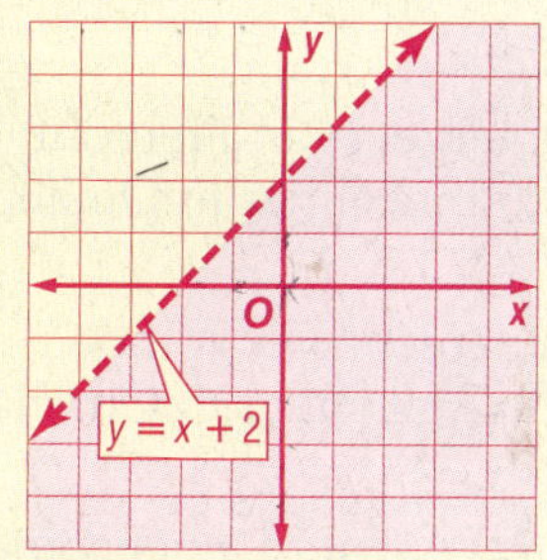

7. 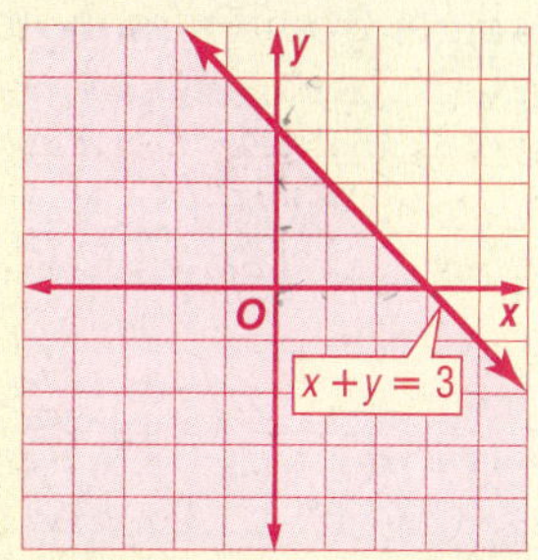

8. 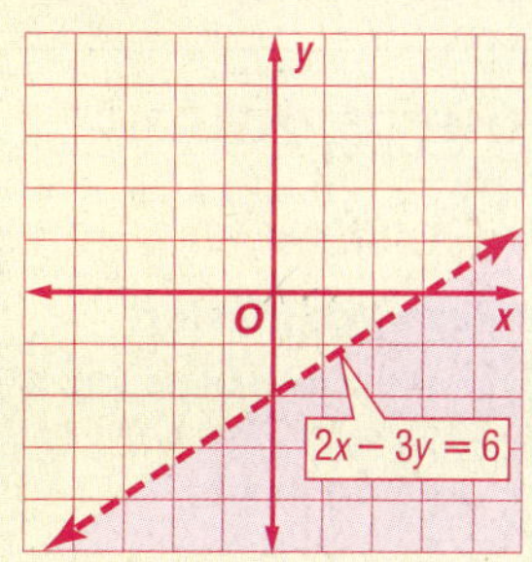

8-1 Lesson Notes

1 Focus

5-Minute Check Transparency 8-1 Use as a quiz or review of Chapter 7.

Mathematical Background notes are available for this lesson on p. 410C.

Building on Prior Knowledge

In Chapter 5, students simplified radical expressions. In this lesson, students will solve problems using the Pythagorean Theorem, which will require that they simplify radical expressions.

How are the Midpoint and Distance Formulas used in emergency medicine?

Ask students:

- Is an emergency in Fremont closer to Lincoln or to Omaha? **Omaha**
- Is an emergency in Wahoo closer to Lincoln or to Omaha? **They are about equally far.**
- What route would a helicopter follow to get from Fremont to Omaha? **Helicopters do not follow roads so they could fly directly from Fremont to Omaha.**

8-1 Midpoint and Distance Formulas

What You'll Learn

- Find the midpoint of a segment on the coordinate plane.
- Find the distance between two points on the coordinate plane.

How are the Midpoint and Distance Formulas used in emergency medicine?

A square grid is superimposed on a map of eastern Nebraska where emergency medical assistance by helicopter is available from both Lincoln and Omaha. Each side of a square represents 10 miles. You can use the formulas in this lesson to determine whether the site of an emergency is closer to Lincoln or to Omaha.

THE MIDPOINT FORMULA Recall that point M is the midpoint of segment PQ if M is between P and Q and $PM = MQ$. There is a formula for the coordinates of the midpoint of a segment in terms of the coordinates of the endpoints. *You will show that this formula is correct in Exercise 41.*

Study Tip

Midpoints
The coordinates of the midpoint are the means of the coordinates of the endpoints.

Key Concept — Midpoint Formula

- **Words** If a line segment has endpoints at (x_1, y_1) and (x_2, y_2), then the midpoint of the segment has coordinates $\left(\frac{x_1 + x_2}{2}, \frac{y_1 + y_2}{2}\right)$.
- **Symbols** midpoint $= \left(\frac{x_1 + x_2}{2}, \frac{y_1 + y_2}{2}\right)$
- **Model**

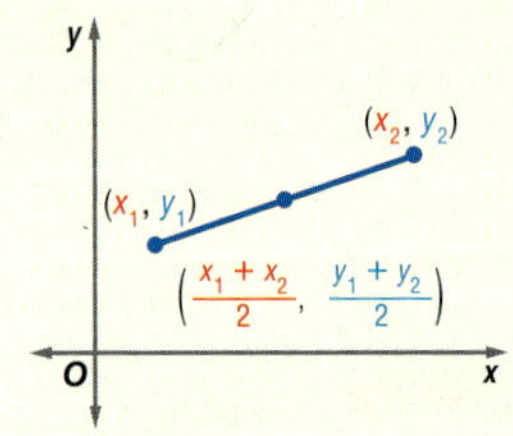

Example 1 Find a Midpoint

LANDSCAPING A landscape design includes two square flower beds and a sprinkler halfway between them. Find the coordinates of the sprinkler if the origin is at the lower left corner of the grid.

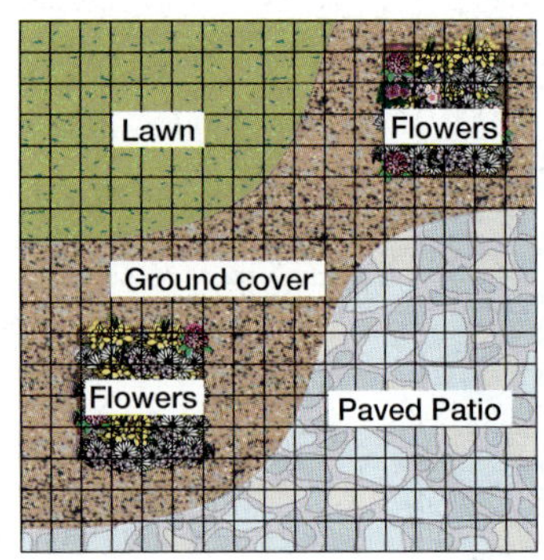

The centers of the flower beds are at (4, 5) and (14, 13). The sprinkler will be at the midpoint of the segment joining these points.

$$\left(\frac{x_1 + x_2}{2}, \frac{y_1 + y_2}{2}\right) = \left(\frac{4 + 14}{2}, \frac{5 + 13}{2}\right)$$

$$= \left(\frac{18}{2}, \frac{18}{2}\right) \text{ or } (9, 9)$$

The sprinkler will have coordinates (9, 9).

Resource Manager

Workbook and Reproducible Masters

Chapter 8 Resource Masters

- Study Guide and Intervention, pp. 455–456
- Skills Practice, p. 457
- Practice, p. 458
- Reading to Learn Mathematics, p. 459
- Enrichment, p. 460

Transparencies

5-Minute Check Transparency 8-1
Answer Key Transparencies

Technology

Interactive Chalkboard

THE DISTANCE FORMULA Recall that the distance between two points on a number line whose coordinates are a and b is $|a - b|$ or $|b - a|$. You can use this fact and the Pythagorean Theorem to derive a formula for the distance between two points on a coordinate plane.

Suppose (x_1, y_1) and (x_2, y_2) name two points. Draw a right triangle with vertices at these points and the point (x_1, y_2). The lengths of the legs of the right triangle are $|x_2 - x_1|$ and $|y_2 - y_1|$. Let d represent the distance between (x_1, y_1) and (x_2, y_2). Now use the Pythagorean Theorem.

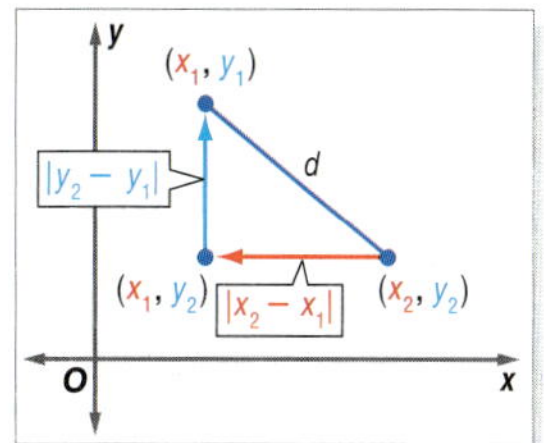

$c^2 = a^2 + b^2$	Pythagorean Theorem
$d^2 = \lvert x_2 - x_1\rvert^2 + \lvert y_2 - y_1\rvert^2$	Replace c with d, a with $\lvert x_2 - x_1\rvert$, and b with $\lvert y_2 - y_1\rvert$.
$d^2 = (x_2 - x_1)^2 + (y_2 - y_1)^2$	$\lvert x_2 - x_1\rvert^2 = (x_2 - x_1)^2$; $\lvert y_2 - y_1\rvert^2 = (y_2 - y_1)^2$
$d = \sqrt{(x_2 - x_1)^2 + (y_2 - y_1)^2}$	Find the nonnegative square root of each side.

Study Tip

Distance
In mathematics, distances are always positive.

Key Concept — *Distance Formula*

- **Words** The distance between two points with coordinates (x_1, y_1) and (x_2, y_2) is given by $d = \sqrt{(x_2 - x_1)^2 + (y_2 - y_1)^2}$.
- **Model**

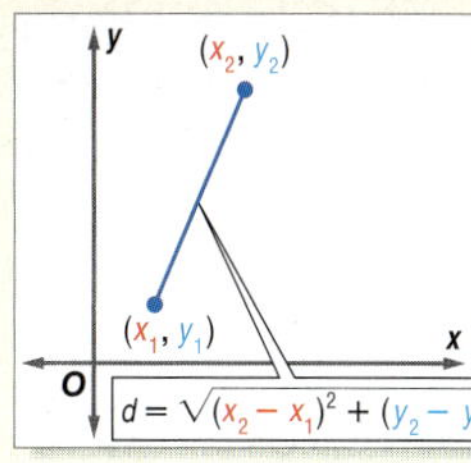

Example 2 Find the Distance Between Two Points

What is the distance between $A(-3, 6)$ and $B(4, -4)$?

$d = \sqrt{(x_2 - x_1)^2 + (y_2 - y_1)^2}$	Distance Formula
$= \sqrt{[4 - (-3)]^2 + (-4 - 6)^2}$	Let $(x_1, y_1) = (-3, 6)$ and $(x_2, y_2) = (4, -4)$.
$= \sqrt{7^2 + (-10)^2}$	Subtract.
$= \sqrt{49 + 100}$ or $\sqrt{149}$	Simplify.

The distance between the points is $\sqrt{149}$ units.

Standardized Test Practice
A B C D

Example 3 Find the Farthest Point

Multiple-Choice Test Item

> Which point is farthest from $(-1, 3)$?
>
> (A) (2, 4) (B) (−4, 1) (C) (0, 5) (D) (3, −2)

Read the Test Item

The word *farthest* refers to the greatest distance.

(continued on the next page)

www.algebra2.com/extra_examples

2 Teach

THE MIDPOINT FORMULA

In-Class Example

Power Point®

1 **COMPUTERS** A graphing program draws a line segment on a computer screen so that its ends are at (5, 2) and (7, 8). What are the coordinates of its midpoint? **(6, 5)**

Teaching Tip Have students draw a graph to check the coordinates.

THE DISTANCE FORMULA

In-Class Examples

Power Point®

2 What is the distance between $P(-1, 4)$ and $Q(2, -3)$? $\sqrt{58}$

Teaching Tip Ask students if the Distance Formula can be used when the two points are both on the same vertical line. **yes**

3 Which point is farthest from $(2, -3)$? **C**

A (0, 0) **B** (3, 2)
C (−3, 0) **D** (4, 1)

Teaching Tip Tell students that a distance in radical form, such as $\sqrt{41}$, is an exact mathematical representation of the value. However, if they are using this value to measure the length of some real object, such as a piece of wood, they can find an approximate value with a calculator. For example, $\sqrt{41} \approx 6.4$.

Standardized Test Practice
A B C D

Example 3 Point out that writing the steps in the calculations can be done very quickly, and helps to prevent careless errors in calculations with integers. This means that taking the time to write the steps is more efficient than trying to do all the calculations mentally.

Interactive Chalkboard
PowerPoint® Presentations

This CD-ROM is a customizable Microsoft® PowerPoint® presentation that includes:

- Step-by-step, dynamic solutions of each In-Class Example from the Teacher Wraparound Edition
- Additional, Your Turn exercises for each example
- The 5-Minute Check Transparencies
- Hot links to Glencoe Online Study Tools

3 Practice/Apply

Study Notebook

Have students—

- add the definitions/examples of the vocabulary terms to their Vocabulary Builder worksheets for Chapter 8.
- add the Test-Taking Tip for Example 3 to their list of tips to review before a test.
- include any other item(s) that they find helpful in mastering the skills in this lesson.

About the Exercises...

Organization by Objective

- **The Midpoint Formula:** 10–23
- **The Distance Formula:** 24–40

Odd/Even Assignments

Exercises 10–19 and 24–33 are structured so that students practice the same concepts whether they are assigned odd or even problems.

Alert! Exercises 22–23 require the Internet.

Assignment Guide

Basic: 11, 13, 21–23, 25–31 odd, 35, 38–39, 41–44, 47–59

Average: 11–19 odd, 21–23, 25–37 odd, 38–39, 41–44, 47–59 (optional: 45, 46)

Advanced: 10–20 even, 24–36 even, 38–53 (optional: 54–59)

Answer

1. Since the sum of the x-coordinates of the given points is negative, the x-coordinate of the midpoint is negative. Since the sum of the y-coordinates of the given points is positive, the y-coordinate of the midpoint is positive. Therefore, the midpoint is in Quadrant II.

Solve the Test Item

Use the Distance Formula to find the distance from $(-1, 3)$ to each point.

Distance to (2, 4)

$d = \sqrt{[2-(-1)]^2 + (4-3)^2}$
$= \sqrt{3^2 + 1^2}$ or $\sqrt{10}$

Distance to (−4, 1)

$d = \sqrt{[-4-(-1)]^2 + (1-3)^2}$
$= \sqrt{(-3)^2 + (-2)^2}$ or $\sqrt{13}$

Distance to (0, 5)

$d = \sqrt{[0-(-1)]^2 + (5-3)^2}$
$= \sqrt{1^2 + 2^2}$ or $\sqrt{5}$

Distance to (3, −2)

$d = \sqrt{[3-(-1)]^2 + (-2-3)^2}$
$= \sqrt{4^2 + (-5)^2}$ or $\sqrt{41}$

The greatest distance is $\sqrt{41}$ units. So, the farthest point from $(-1, 3)$ is $(3, -2)$. The answer is D.

Check for Understanding

Concept Check

1. **Explain** how you can determine in which quadrant the midpoint of the segment with endpoints at $(-6, 8)$ and $(4, 3)$ lies without actually calculating the coordinates. **See margin.**
2. **Identify** all of the points that are equidistant from the endpoints of a given segment. **all of the points on the perpendicular bisector of the segment**
3. **OPEN ENDED** Find two points that are $\sqrt{29}$ units apart. **Sample answer: (0, 0) and (5, 2)**

GUIDED PRACTICE KEY	
Exercises	Examples
4, 5	1
6–8	2, 3
9	3

Guided Practice

Find the midpoint of each line segment with endpoints at the given coordinates.

4. $(-5, 6), (1, 7)$ $\left(-2, \frac{13}{2}\right)$
5. $(8, 9), (-3, -4.5)$ **(2.5, 2.25)**

Find the distance between each pair of points with the given coordinates.

6. $(2, -4), (10, -10)$ **10 units**
7. $(7, 8), (-4, 9)$ $\sqrt{122}$ **units**
8. $(0.5, 1.4), (1.1, 2.9)$ $\sqrt{2.61}$ **units**

Standardized Test Practice

9. Which of the following points is closest to $(2, -4)$? **D**

Ⓐ (3, 1) Ⓑ (−2, 0) Ⓒ (1, 5) Ⓓ (4, −2)

★ indicates increased difficulty

Practice and Apply

Homework Help	
For Exercises	See Examples
10–23	1
24–40	2, 3

Extra Practice See page 845.

Find the midpoint of each line segment with endpoints at the given coordinates.

10. $(8, 3), (16, 7)$ **(12, 5)**
11. $(-5, 3), (-3, -7)$ **(−4, −2)**
12. $(6, -5), (-2, -7)$ **(2, −6)**
13. $(5, 9), (12, 18)$ $\left(\frac{17}{2}, \frac{27}{2}\right)$
14. ★ $(0.45, 7), (-0.3, -0.6)$ **(0.075, 3.2)**
15. ★ $(4.3, -2.1), (1.9, 7.5)$ **(3.1, 2.7)**
16. ★ $\left(\frac{1}{2}, -\frac{2}{3}\right), \left(\frac{1}{3}, \frac{1}{4}\right)$ $\left(\frac{5}{12}, -\frac{5}{24}\right)$
17. ★ $\left(\frac{1}{3}, \frac{3}{4}\right), \left(-\frac{1}{4}, \frac{1}{2}\right)$ $\left(\frac{1}{24}, \frac{5}{8}\right)$
18. **GEOMETRY** Triangle MNP has vertices $M(3, 5)$, $N(-2, 8)$, and $P(7, -4)$. Find the coordinates of the midpoint of each side. $\left(\frac{1}{2}, \frac{13}{2}\right), \left(\frac{5}{2}, 2\right), \left(5, \frac{1}{2}\right)$
19. ★ **GEOMETRY** Circle Q has a diameter $\overline{AB}$. If A is at $(-3, -5)$ and the center is at $(2, 3)$, find the coordinates of B. **(7, 11)**

DAILY INTERVENTION — Differentiated Instruction

Visual/Spatial Encourage students to relate coordinates to models and drawings of the lines and figures. Suggest that they construct and demonstrate models of some of the exercises to help other students who might have problems visualizing.

20. **REAL ESTATE** In John's town, the numbered streets and avenues form a grid. He belongs to a gym at the corner of 12th Street and 15th Avenue, and the deli where he works is at the corner of 4th Street and 5th Avenue. He wants to rent an apartment halfway between the two. In what area should he look? **around 8th Street and 10th Avenue**

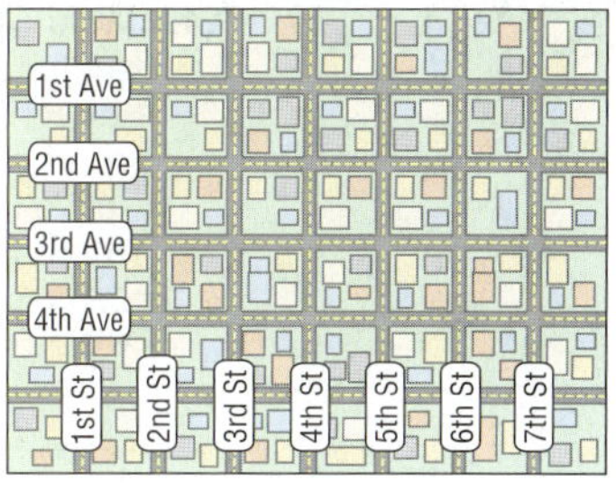

GEOGRAPHY **For Exercises 21–23, use the following information.**
The U.S. Geological Survey (USGS) has determined the official center of the continental United States.

21. Describe a method that might be used to approximate the geographical center of the continental United States. **See pp. 469A–469J.**

22. **RESEARCH** Use the Internet or other reference to look up the USGS geographical center of the continental United States. **near Lebanon, KS**

23. How does the location given by USGS compare to the result of your method? **See students' work.**

Find the distance between each pair of points with the given coordinates.

24. $(-4, 9), (1, -3)$ **13 units**

25. $(1, -14), (-6, 10)$ **25 units**

26. $(-4, -10), (-3, -11)$ **$\sqrt{2}$ units**

27. $(9, -2), (12, -14)$ **$3\sqrt{17}$ units**

28. $(0.23, 0.4), (0.68, -0.2)$ **0.75 unit**

29. $(2.3, -1.2), (-4.5, 3.7)$ **$\sqrt{70.25}$ units**

30. $\left(-3, -\frac{2}{11}\right), \left(5, \frac{9}{11}\right)$ **$\sqrt{65}$ units**

31. $\left(0, \frac{1}{5}\right), \left(\frac{3}{5}, -\frac{3}{5}\right)$ **1 unit**

★ 32. $(2\sqrt{3}, -5), (-3\sqrt{3}, 9)$ **$\sqrt{271}$ units**

★ 33. $\left(\frac{2\sqrt{3}}{3}, \frac{\sqrt{5}}{4}\right), \left(-\frac{2\sqrt{3}}{3}, \frac{\sqrt{5}}{2}\right)$

33. $\frac{\sqrt{813}}{12}$ units

34. **GEOMETRY** A circle has a radius with endpoints at (2, 5) and (−1, −4). Find the circumference and area of the circle. **$6\sqrt{10}\pi$ units, 90π units^2**

35. **GEOMETRY** Find the perimeter and area of the triangle shown at the right.

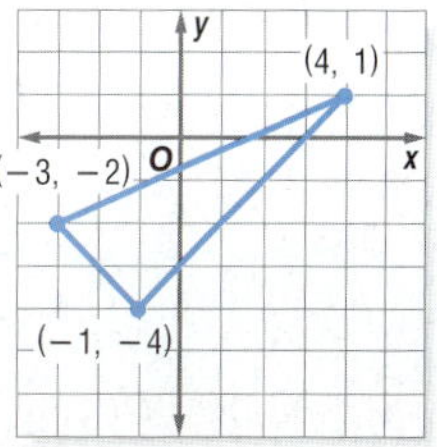

35. $7\sqrt{2} + \sqrt{58}$ units, 10 units^2

★ 36. **GEOMETRY** Quadrilateral *RSTV* has vertices $R(-4, 6)$, $S(4, 5)$, $T(6, 3)$, and $V(5, -8)$. Find the perimeter of the quadrilateral.

36. $\sqrt{65} + 2\sqrt{2} + \sqrt{122} + \sqrt{277}$ units

★ 37. **GEOMETRY** Triangle *CAT* has vertices $C(4, 9)$, $A(8, -9)$, and $T(-6, 5)$. *M* is the midpoint of $\overline{TA}$. Find the length of median $\overline{CM}$. (*Hint*: A median connects a vertex of a triangle to the midpoint of the opposite side.) **$\sqrt{130}$ units**

TRAVEL **For Exercises 38 and 39, use the figure at the right, where a grid is superimposed on a map of a portion of the state of Alabama.**

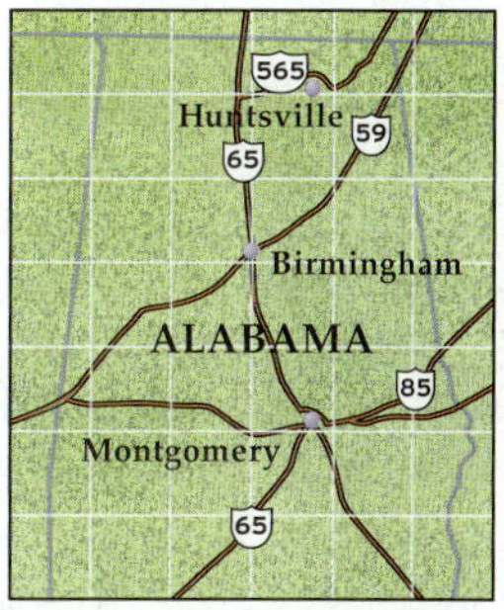

38. About how far is it from Birmingham to Montgomery if each unit on the grid represents 40 miles? **about 85 mi**

39. How long would it take a plane to fly from Huntsville to Montgomery if its average speed is 180 miles per hour? **about 0.9 h**

www.algebra2.com/self_check_quiz

Study Guide and Intervention, p. 455 (shown) and p. 456

The Midpoint Formula

Midpoint Formula	The midpoint *M* of a segment with endpoints (x_1, y_1) and (x_2, y_2) is $\left(\frac{x_1 + x_2}{2}, \frac{y_1 + y_2}{2}\right)$.

Example 1 **Find the midpoint of the line segment with endpoints at (4, −7) and (−2, 3).**

$$\left(\frac{x_1 + x_2}{2}, \frac{y_1 + y_2}{2}\right) = \left(\frac{4 + (-2)}{2}, \frac{-7 + 3}{2}\right) = \left(\frac{2}{2}, \frac{-4}{2}\right) \text{ or } (1, -2)$$

The midpoint of the segment is (1, −2).

Example 2 **A diameter $\overline{AB}$ of a circle has endpoints $A(5, -11)$ and $B(-7, 6)$. What are the coordinates of the center of the circle?**

The center of the circle is the midpoint of all of its diameters.

$$\left(\frac{x_1 + x_2}{2}, \frac{y_1 + y_2}{2}\right) = \left(\frac{5 + (-7)}{2}, \frac{-11 + 6}{2}\right) = \left(\frac{-2}{2}, \frac{-5}{2}\right) \text{ or } \left(-1, -2\frac{1}{2}\right)$$

The circle has center $\left(-1, -2\frac{1}{2}\right)$.

Exercises

Find the midpoint of each line segment with endpoints at the given coordinates.

1. (12, 7) and (−2, 11) **(5, 9)**
2. (−8, −3) and (10, 9) **(1, 3)**
3. (4, 15) and (10, 1) **(7, 8)**
4. (−3, −3) and (3, 3) **(0, 0)**
5. (15, 6) and (12, 14) **(13.5, 10)**
6. (22, −8) and (−10, 6) **(6, −1)**
7. (3, 5) and (−6, 11) **$\left(-\frac{3}{2}, 8\right)$**
8. (8, −15) and (−7, 13) **$\left(\frac{1}{2}, -1\right)$**
9. (2.5, −6.1) and (7.9, 13.7) **(5.2, 3.8)**
10. (−7, −6) and (−1, 24) **(−4, 9)**
11. (3, −10) and (30, −20) **$\left(\frac{33}{2}, -15\right)$**
12. (−9, 1.7) and (−11, 1.3) **(−10, 1.5)**
13. Segment $\overline{MN}$ has midpoint *P*. If *M* has coordinates (14, −3) and *P* has coordinates (−8, 6), what are the coordinates of *N*? **(−30, 15)**
14. Circle *R* has a diameter $\overline{ST}$. If *R* has coordinates (−4, −8) and *S* has coordinates (1, 4), what are the coordinates of *T*? **(−9, −20)**
15. Segment $\overline{AD}$ has midpoint *B*, and $\overline{BD}$ has midpoint *C*. If *A* has coordinates (−5, 4) and *C* has coordinates (10, 11), what are the coordinates of *B* and *D*? ***B* is $\left(5, 8\frac{2}{3}\right)$, *D* is $\left(15, 13\frac{1}{3}\right)$.**

Skills Practice, p. 457 and Practice, p. 458 (shown)

Find the midpoint of each line segment with endpoints at the given coordinates.

1. (8, −3), (−6, −11) **(1, −7)**
2. (−14, 5), (10, 6) **$\left(-2, \frac{11}{2}\right)$**
3. (−7, −6), (1, −2) **(−3, −4)**
4. (8, −2), (8, −8) **(8, −5)**
5. (9, −4), (1, −1) **$\left(5, -\frac{5}{2}\right)$**
6. (3, 3), (4, 9) **$\left(\frac{7}{2}, 6\right)$**
7. (4, −2), (3, −7) **$\left(\frac{7}{2}, -\frac{9}{2}\right)$**
8. (6, 7), (4, 4) **$\left(5, \frac{11}{2}\right)$**
9. (−4, −2), (−8, 2) **(−6, 0)**
10. (5, −2), (3, 7) **$\left(4, \frac{5}{2}\right)$**
11. (−6, 3), (−5, −7) **$\left(-\frac{11}{2}, -2\right)$**
12. (−9, −8), (8, 3) **$\left(-\frac{1}{2}, -\frac{5}{2}\right)$**
13. (2.6, −4.7), (8.4, 2.5) **(5.5, −1.1)**
14. $\left(-\frac{1}{3}, 6\right), \left(\frac{2}{3}, 4\right)$ **$\left(\frac{1}{6}, 5\right)$**
15. (−2.5, −4.2), (8.1, 4.2) **(2.8, 0)**
16. $\left(\frac{1}{8}, \frac{1}{2}\right), \left(-\frac{5}{8}, -\frac{1}{2}\right)$ **$\left(-\frac{1}{4}, 0\right)$**

Find the distance between each pair of points with the given coordinates.

17. (5, 2), (2, −2) **5 units**
18. (−2, −4), (4, 4) **10 units**
19. (−3, 8), (−1, −5) **$\sqrt{173}$ units**
20. (0, 1), (9, −6) **$\sqrt{130}$ units**
21. (−5, 6), (−6, 6) **1 unit**
22. (−3, 5), (12, −3) **17 units**
23. (−2, −3), (9, 3) **$\sqrt{157}$ units**
24. (−9, −8), (−7, 8) **$2\sqrt{65}$ units**
25. (9, 3), (9, −2) **5 units**
26. (−1, −7), (0, 6) **$\sqrt{170}$ units**
27. (10, −3), (−2, −8) **13 units**
28. (−0.5, −6), (1.5, 0) **$2\sqrt{10}$ units**
29. $\left(\frac{2}{5}, \frac{3}{5}\right), \left(1, \frac{7}{5}\right)$ **1 unit**
30. $(-4\sqrt{2}, -\sqrt{5}), (-5\sqrt{2}, 4\sqrt{5})$ **$\sqrt{127}$ units**
31. **GEOMETRY** Circle *O* has a diameter $\overline{AB}$. If *A* is at (−6, −2) and *B* is at (−3, 4), find the center of the circle and the length of its diameter. **$\left(-\frac{9}{2}, 1\right)$; $3\sqrt{5}$ units**
32. **GEOMETRY** Find the perimeter of a triangle with vertices at (1, −3), (−4, 9), and (−2, 1). **$18 + 2\sqrt{17}$ units**

Reading to Learn Mathematics, p. 459

ELL

Pre-Activity **How are the Midpoint and Distance Formulas used in emergency medicine?**

Read the introduction to Lesson 8-1 at the top of page 412 in your textbook.

How do you find distances on a road map?

Sample answer: Use the scale of miles on the map. You might also use a ruler.

Reading the Lesson

1. a. Write the coordinates of the midpoint of a segment with endpoints (x_1, y_1) and (x_2, y_2). **$\left(\frac{x_1 + x_2}{2}, \frac{y_1 + y_2}{2}\right)$**

   b. Explain how to find the midpoint of a segment if you know the coordinates of the endpoints. Do not use subscripts in your explanation.

   Sample answer: To find the *x*-coordinate of the midpoint, add the *x*-coordinates of the endpoints and divide by two. To find the *y*-coordinate of the midpoint, do the same with the *y*-coordinates of the endpoints.

2. a. Write an expression for the distance between two points with coordinates (x_1, y_1) and (x_2, y_2). **$\sqrt{(x_2 - x_1)^2 + (y_2 - y_1)^2}$**

   b. Explain how to find the distance between two points. Do not use subscripts in your explanation.

   Sample answer: Find the difference between the *x*-coordinates and square it. Find the difference between the *y*-coordinates and square it. Add the squares. Then find the square root of the sum.

3. Consider the segment connecting the points (−3, 5) and (9, 11).

   a. Find the midpoint of this segment. **(3, 8)**

   b. Find the length of the segment. Write your answer in simplified radical form. **$6\sqrt{5}$**

Helping You Remember

4. How can the "mid" in *midpoint* help you remember the midpoint formula?

   Sample answer: The *midpoint* is the point in the *middle* of a segment. It is halfway between the endpoints. The coordinates of the midpoint are found by finding the average of the two *x*-coordinates (add them and divide by 2) and the average of the two *y*-coordinates.

Enrichment, p. 460

Quadratic Form

Consider two methods for solving the following equation.

$$(y - 2)^2 - 5(y - 2) + 6 = 0$$

One way to solve the equation is to simplify first, then use factoring.

$$y^2 - 4y + 4 - 5y + 10 + 6 = 0$$
$$y^2 - 9y + 20 = 0$$
$$(y - 4)(y - 5) = 0$$

Thus, the solution set is {4, 5}.

Another way to solve the equation is first to replace $y - 2$ by a single variable. This will produce an equation that is easier to solve than the original equation. Let $t = y - 2$ and then solve the new equation.

$$(y - 2)^2 - 5(y - 2) + 6 = 0$$
$$t^2 - 5t + 6 = 0$$
$$(t - \ldots)(t - 3) = 0$$

4 Assess

Open-Ended Assessment

Modeling Have students plot two points on a coordinate grid and explain how to find the distance between them and the coordinates of the midpoint of the segment joining them.

Tips for New Teachers

Intervention Some students may have difficulty understanding and manipulating radicals. Take time to clear up any problems in this area before going on.

Getting Ready for Lesson 8-2

PREREQUISITE SKILL Students will analyze equations of parabolas in Lesson 8-2. When writing equations for parabolas into standard form, students frequently need to complete the square. Use Exercises 54–59 to determine your students' familiarity with completing squares.

Answers

46. -1; $\overleftrightarrow{AA'}$ is perpendicular to the line with equation $y = x$, which has slope 1.

47. $D = \{x \mid x \geq 2\}$, $R = \{y \mid y \geq 0\}$

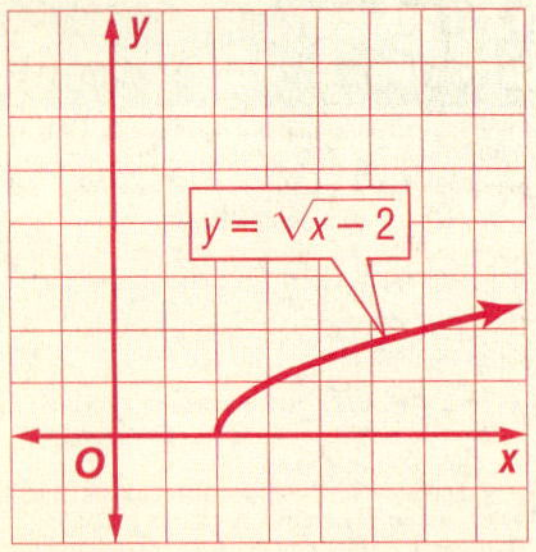

48. $D = \{x \mid x \geq 0\}$, $R = \{y \mid y \geq -1\}$

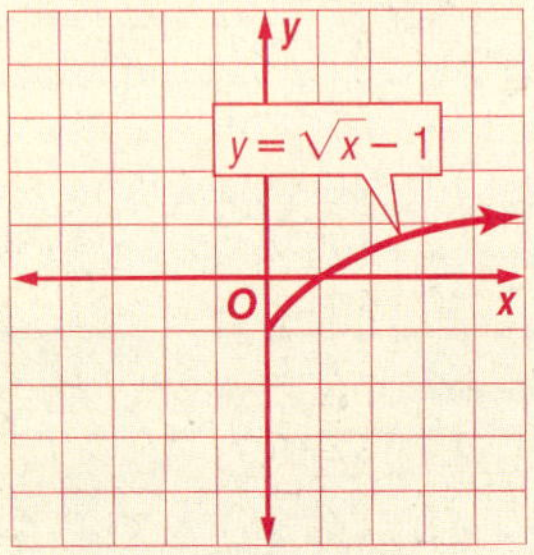

---

40. **WOODWORKING** A stage crew is making the set for a children's play. They want to make some gingerbread shapes out of some leftover squares of wood with sides measuring 1 foot. They can make taller shapes by cutting them out of the wood diagonally. To the nearest inch, how tall is the gingerbread shape in the drawing at the right? **14 in.**

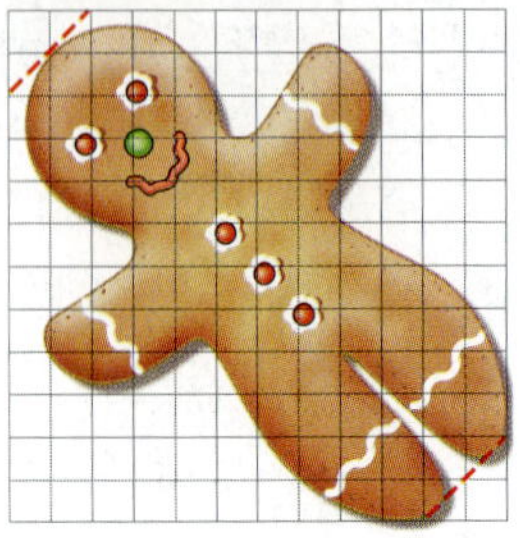

41. **CRITICAL THINKING** Verify the Midpoint Formula. (*Hint*: You must show that the formula gives the coordinates of a point on the line through the given endpoints and that the point is equidistant from the endpoints.) **See pp. 469A–469J.**

42. **WRITING IN MATH** Answer the question that was posed at the beginning of the lesson. **See pp. 469A–469J.**

How are the Midpoint and Distance Formulas used in emergency medicine?

Include the following in your answer:
- a few sentences explaining how to use the Distance Formula to approximate the distance between two cities on a map, and
- which city, Lincoln or Omaha, an emergency medical helicopter should be dispatched from to pick up a patient in Fremont.

Standardized Test Practice

43. What is the distance between the points $A(4, -2)$ and $B(-4, -8)$? **C**

Ⓐ 6 Ⓑ 8 Ⓒ 10 Ⓓ 14

44. Point $D(5, -1)$ is the midpoint of segment $\overline{CE}$. If point C has coordinates $(3, 2)$, then what are the coordinates of point E?

Ⓐ $(8, 1)$ Ⓑ $(7, -4)$ Ⓒ $(2, -3)$ Ⓓ $\left(4, \frac{1}{2}\right)$ **B**

Extending the Lesson

For Exercises 45 and 46, use the following information.
You can use midpoints and slope to describe some transformations. Suppose point A' is the image when point A is reflected over the line with equation $y = x$.

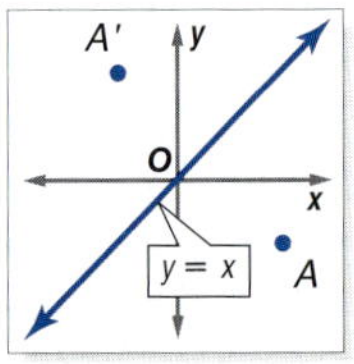

45. Where is the midpoint of $\overline{AA'}$? **45. on the line with equation $y = x$**

46. What is the slope of $\overleftrightarrow{AA'}$? Explain. **See margin.**

Maintain Your Skills

Mixed Review

Graph each function. State the domain and range. *(Lesson 7-9)* **47–49. See margin.**

47. $y = \sqrt{x - 2}$ 48. $y = \sqrt{x} - 1$ 49. $y = 2\sqrt{x} + 1$

50. Determine whether the functions $f(x) = x - 2$ and $g(x) = 2x$ are inverse functions. *(Lesson 7-8)* **no**

Simplify. *(Lesson 5-9)*

51. $(2 + 4i) + (-3 + 9i)$ **$-1 + 13i$**
52. $(4 - i) - (-2 + i)$ **$6 - 2i$**
53. $(1 - 2i)(2 + i)$ **$4 - 3i$**

Getting Ready for the Next Lesson

PREREQUISITE SKILL **Write each equation in the form $y = a(x - h)^2 + k$.**
(To review ***completing the square***, *see Lesson 6-4.)* **54–59. See margin.**

54. $y = x^2 + 6x + 9$
55. $y = x^2 - 4x + 1$
56. $y = 2x^2 + 20x + 50$
57. $y = 3x^2 - 6x + 5$
58. $y = -x^2 - 4x + 6$
59. $y = -3x^2 - 18x - 10$

---

49. $D = \{x \mid x \geq 0\}$, $R = \{y \mid y \geq 1\}$

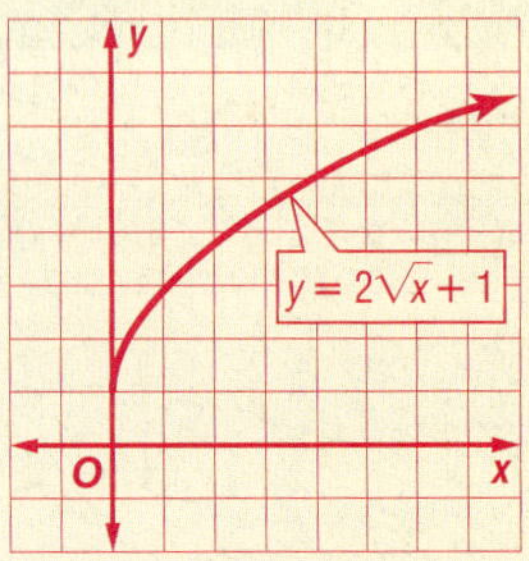

54. $y = (x + 3)^2$
55. $y = (x - 2)^2 - 3$
56. $y = 2(x + 5)^2$
57. $y = 3(x - 1)^2 + 2$
58. $y = -(x + 2)^2 + 10$
59. $y = -3(x + 3)^2 + 17$

Algebra Activity

A Follow-Up of Lesson 8-1

Midpoint and Distance Formulas in Three Dimensions

You can derive a formula for distance in three-dimensional space. It may seem that the formula would involve a cube root, but it actually involves a square root, similar to the formula in two dimensions.

Suppose (x_1, y_1, z_1) and (x_2, y_2, z_2) name two points in space. Draw the rectangular box that has opposite vertices at these points. The dimensions of the box are $|x_2 - x_1|$, $|y_2 - y_1|$, and $|z_2 - z_1|$. Let a be the length of a diagonal of the bottom of the box. By the Pythagorean Theorem, $a^2 = |x_2 - x_1|^2 + |y_2 - y_1|^2$.

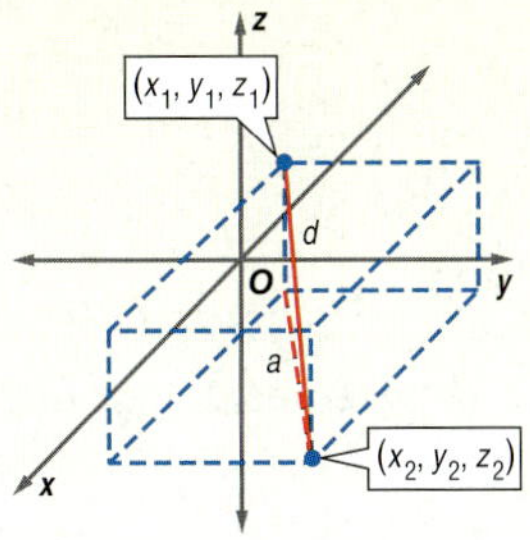

To find the distance d between (x_1, y_1, z_1) and (x_2, y_2, z_2), apply the Pythagorean Theorem to the right triangle whose legs are a diagonal of the bottom of the box and a vertical edge of the box.

$d^2 = a^2 + \|z_2 - z_1\|^2$	Pythagorean Theorem
$d^2 = \|x_2 - x_1\|^2 + \|y_2 - y_1\|^2 + \|z_2 - z_1\|^2$	$a^2 = \|x_2 - x_1\|^2 + \|y_2 - y_1\|^2$
$d^2 = (x_2 - x_1)^2 + (y_2 - y_1)^2 + (z_2 - z_1)^2$	$\|x_2 - x_1\|^2 = (x_2 - x_1)^2$, and so on
$d = \sqrt{(x_2 - x_1)^2 + (y_2 - y_1)^2 + (z_2 - z_1)^2}$	Take the square root of each side.

The distance d between the points with coordinates (x_1, y_1, z_1) and (x_2, y_2, z_2) is given by the formula $d = \sqrt{(x_2 - x_1)^2 + (y_2 - y_1)^2 + (z_2 - z_1)^2}$.

Example 1

Find the distance between (2, 0, −3) and (4, 2, 9).

$d = \sqrt{(x_2 - x_1)^2 + (y_2 - y_1)^2 + (z_2 - z_1)^2}$	Distance Formula
$= \sqrt{(4 - 2)^2 + (2 - 0)^2 + [9 - (-3)]^2}$	$(x_1, y_1, z_1) = (2, 0, -3)$ $(x_2, y_2, z_2) = (4, 2, 9)$
$= \sqrt{2^2 + 2^2 + 12^2}$	
$= \sqrt{152}$ or $2\sqrt{38}$	

The distance is $2\sqrt{38}$ or about 12.33 units.

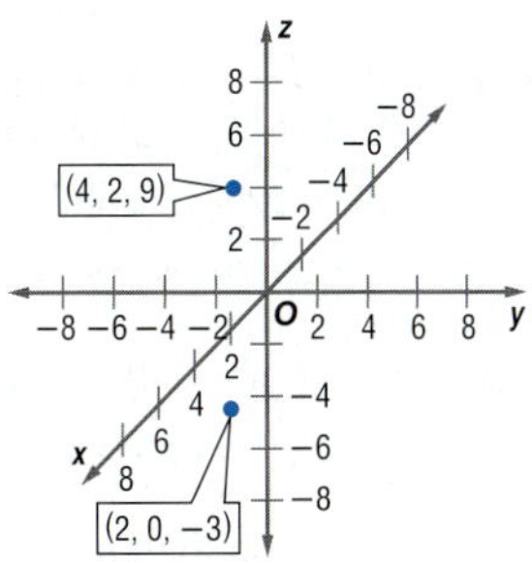

In three dimensions, the midpoint of the segment with coordinates (x_1, y_1, z_1) and (x_2, y_2, z_2) has coordinates $\left(\frac{x_1 + x_2}{2}, \frac{y_1 + y_2}{2}, \frac{z_1 + z_2}{2}\right)$. Notice how similar this is to the Midpoint Formula in two dimensions.

(continued on the next page)

Algebra Activity

A Follow-Up of Lesson 8-1

Getting Started

Objective To derive a formula for finding the distance between two points in three-dimensional space.

Materials
none

Teach

- To help students visualize this activity, use a box, such as a shoe box or tissue box, and mark points using the corners. Model the line between points with a piece of wire, dry spaghetti, yarn, or thread.
- Make sure that all students understand and pay attention to the difference between subscripts and superscripts.
- Ask students to explain how $\sqrt{152}$ simplifies to $2\sqrt{38}$.

Resource Manager

Teaching Algebra with Manipulatives

- p. 263 (student recording sheet)

Assess

In **Exercises 1–17**, students should

- find the distance between two points in a three-dimensional coordinate plane.
- find the midpoint of the segment between two points in a three-dimensional coordinate plane.
- use the midpoint and distance formulas to solve problems in a three-dimensional coordinate plane.

In **Exercises 12–15**, if students solve these problems correctly, they will have demonstrated their ability to use points in a three-dimensional coordinate plane.

Study Notebook

You may wish to have students summarize this activity and what they learned from it.

Answer

11. The distance between the points with coordinates (2, −4, 2) and (3, 1, 5) is $\sqrt{35}$ units. The distance between the points with coordinates (2, −4, 2) and (6, −3, −1) is $\sqrt{26}$ units. The distance between the points with coordinates (3, 1, 5) and (6, −3, −1) is $\sqrt{61}$ units. Since $(\sqrt{61})^2 = (\sqrt{35})^2 + (\sqrt{26})^2$, the triangle is a right triangle.

Example 2

Find the coordinates of the midpoint of the segment with endpoints (6, −5, 1) and (−2, 4, 0).

$$\left(\frac{x_1 + x_2}{2}, \frac{y_1 + y_2}{2}, \frac{z_1 + z_2}{2}\right) = \left(\frac{6 + (-2)}{2}, \frac{-5 + 4}{2}, \frac{1 + 0}{2}\right) \quad (x_1, y_1, z_1) = (6, -5, 1),\ (x_2, y_2, z_2) = (-2, 4, 0)$$

$$= \left(\frac{4}{2}, \frac{-1}{2}, \frac{1}{2}\right) \quad \text{Add.}$$

$$= \left(2, -\frac{1}{2}, \frac{1}{2}\right) \quad \text{Simplify.}$$

The midpoint has coordinates $\left(2, -\frac{1}{2}, \frac{1}{2}\right)$.

Exercises

Find the distance between each pair of points with the given coordinates.

1. (2, 4, 5), (1, 2, 3) **3 units**

2. (−1, 6, 2), (4, −3, 0) **$\sqrt{110}$ units**

3. (−2, 1, 7), (−2, 6, −3) **$5\sqrt{5}$ units**

4. (0, 7, −1), (−4, 1, 3) **$2\sqrt{17}$ units**

Find the midpoint of each line segment with endpoints at the given coordinates.

5. (2, 6, −1), (−4, 8, 5) **(−1, 7, 2)**

6. (4, −3, 2), (−2, 7, 6) **(1, 2, 4)**

7. (1, 3, 7), (−4, 2, −1) **$\left(-\frac{3}{2}, \frac{5}{2}, 3\right)$**

8. (2.3, −1.7, 0.6), (−2.7, 3.1, 1.8) **(−0.2, 0.7, 1.2)**

9. The coordinates of one endpoint of a segment are (4, −2, 3), and the coordinates of the midpoint are (3, 2, 5). Find the coordinates of the other endpoint. **(2, 6, 7)**

10. Two of the opposite vertices of a rectangular solid are at (4, 1, −1) and (2, 3, 5). Find the coordinates of the other six vertices. **(2, 3, −1), (4, 3, −1), (2, 1, −1), (4, 3, 5), (4, 1, 5), (2, 1, 5)**

11. Determine whether a triangle with vertices at (2, −4, 2), (3, 1, 5), and (6, −3, −1) is a right triangle. Explain. **Yes; see margin for explanation.**

The vertices of a rectangular solid are at (−2, 3, 2), (3, 3, 2), (3, 1, 2), (−2, 1, 2), (−2, 3, 6), (3, 3, 6), (3, 1, 6), and (−2, 1, 6).

12. Find the volume of the solid. **40 units^3**

13. Find the length of a diagonal of the solid. **$3\sqrt{5}$ units**

14. Show that the point with coordinates $\left(\frac{x_1 + x_2}{2}, \frac{y_1 + y_2}{2}, \frac{z_1 + z_2}{2}\right)$ is equidistant from the points with coordinates (x_1, y_1, z_1) and (x_2, y_2, z_2). **See pp. 469A–469J.**

15. Find the value of c so that the point with coordinates (2, 3, c) is $3\sqrt{6}$ units from the point with coordinates (−1, 0, 5). **−1 or 11**

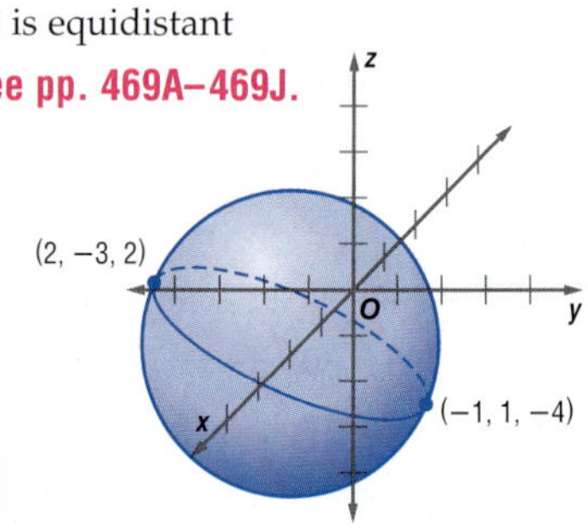

The endpoints of a diameter of a sphere are at (2, −3, 2) and (−1, 1, −4).

16. Find the length of a radius of the sphere. **$\frac{\sqrt{61}}{2}$**

17. Find the coordinates of the center of the sphere. **$\left(\frac{1}{2}, -1, -1\right)$**

418 Chapter 8 Conic Sections

8-2 Parabolas

What You'll Learn

- Write equations of parabolas in standard form.
- Graph parabolas.

Vocabulary

- parabola
- conic section
- focus
- directrix
- latus rectum

How are parabolas used in manufacturing?

A mirror or other reflective object in the shape of a parabola has the property that parallel incoming rays are all reflected to the same point. Or, if that point is the source of rays, the rays become parallel when they are reflected.

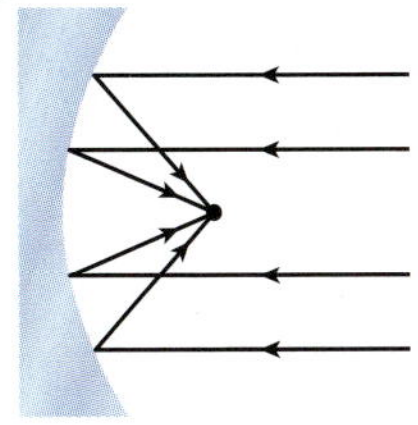

EQUATIONS OF PARABOLAS In Chapter 6, you learned that the graph of an equation of the form $y = ax^2 + bx + c$ is a **parabola**. A parabola can also be obtained by slicing a double cone on a slant as shown below on the left. Any figure that can be obtained by slicing a double cone is called a **conic section**. Other conic sections are also shown below.

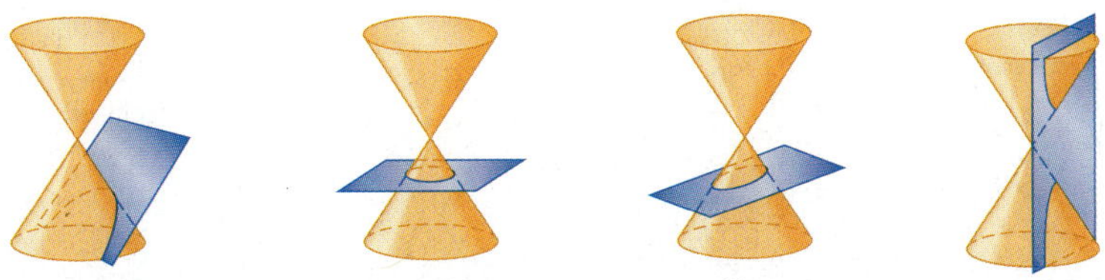

Study Tip

Focus of a Parabola
The focus is the special point referred to at the beginning of the lesson.

A parabola can also be defined as the set of all points in a plane that are the same distance from a given point called the **focus** and a given line called the **directrix**. The parabola at the right has its focus at (2, 3), and the equation of its directrix is $y = -1$. You can use the Distance Formula to find an equation of this parabola.

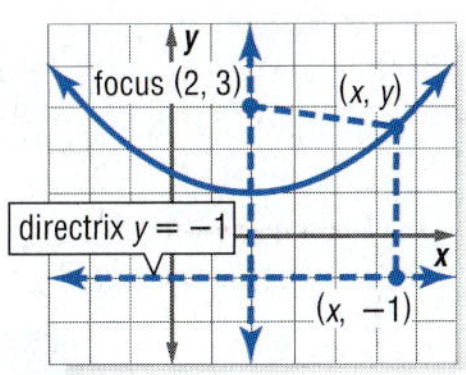

Let (x, y) be any point on this parabola. The distance from this point to the focus must be the same as the distance from this point to the directrix. The distance from a point to a line is measured along the perpendicular from the point to the line.

distance from (x, y) to $(2, 3)$ = distance from (x, y) to $(x, -1)$

$$\sqrt{(x-2)^2 + (y-3)^2} = \sqrt{(x-x)^2 + [y-(-1)]^2}$$

$$(x-2)^2 + (y-3)^2 = 0^2 + (y+1)^2 \quad \text{Square each side.}$$

$$(x-2)^2 + y^2 - 6y + 9 = y^2 + 2y + 1 \quad \text{Square } y - 3 \text{ and } y + 1.$$

$$(x-2)^2 + 8 = 8y \quad \text{Isolate the } y\text{-terms.}$$

$$\frac{1}{8}(x-2)^2 + 1 = y \quad \text{Divide each side by 8.}$$

8-2 Lesson Notes

1 Focus

5-Minute Check Transparency 8-2 Use as a quiz or review of Lesson 8-1.

Mathematical Background notes are available for this lesson on p. 410C.

Building on Prior Knowledge

In Chapter 6, students graphed quadratic functions. In this lesson, students will discover that parabolas are conic sections.

How are parabolas used in manufacturing?

Ask students:

- What are some applications in which reflecting light rays are important? **Sample answers: projecting slides and film; astronomy; cameras, and lenses**
- The rays from a lightbulb radiate in all directions, but the beam of a flashlight comes out in a straight line. Why? **Lead students to recognize that a parabolic mirror inside a flashlight reflects the rays so they are parallel.**

Resource Manager

Workbook and Reproducible Masters

Chapter 8 Resource Masters

- Study Guide and Intervention, pp. 461–462
- Skills Practice, p. 463
- Practice, p. 464
- Reading to Learn Mathematics, p. 465
- Enrichment, p. 466
- Assessment, p. 511

Graphing Calculator and Spreadsheet Masters, p. 42
School-to-Career Masters, p. 15
Science and Mathematics Lab Manual, pp. 119–122
Teaching Algebra With Manipulatives Masters, pp. 264, 265

Transparencies

5-Minute Check Transparency 8-2
Answer Key Transparencies

Technology

Interactive Chalkboard

2 Teach

EQUATIONS OF PARABOLAS

Teaching Tip Discuss with students what they already know and remember about parabolas, their shapes, and their equations.

Teaching Tip Remind students that the distance from a point to a line is measured by the perpendicular from the point to the line.

In-Class Example

Power Point®

1 Write $y = -x^2 - 2x + 3$ in standard form. Identify the vertex, axis of symmetry, and direction of opening of the parabola. **$y = -(x + 1)^2 + 4$; vertex: $(-1, 4)$; axis of symmetry: $x = -1$; opens downward**

Teaching Tip Point out that, while the standard form for a linear equation is $Ax + By = C$ (see Lesson 2-2), the standard form for the equation of a parabola is $y = a(x - h)^2 + k$.

An equation of the parabola with focus at (2, 3) and directrix with equation $y = -1$ is $y = \frac{1}{8}(x - 2)^2 + 1$. The equation of the *axis of symmetry* for this parabola is $x = 2$. The axis of symmetry intersects the parabola at a point called the *vertex*. The vertex is the point where the graph turns. The vertex of this parabola is at (2, 1). Since $\frac{1}{8}$ is positive, the parabola opens upward.

Any equation of the form $y = ax^2 + bx + c$ can be written in standard form.

Key Concept — *Equation of a Parabola*

The standard form of the equation of a parabola with vertex (h, k) and axis of symmetry $x = h$ is $y = a(x - h)^2 + k$.

- If $a > 0$, k is the minimum value of the related function and the parabola opens upward.
- If $a < 0$, k is the maximum value of the related function and the parabola opens downward.

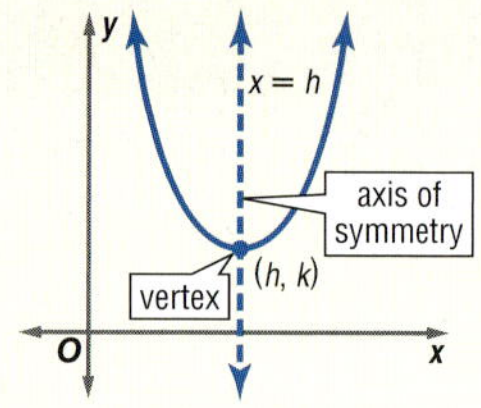

Example 1 *Analyze the Equation of a Parabola*

Write $y = 3x^2 + 24x + 50$ in standard form. Identify the vertex, axis of symmetry, and direction of opening of the parabola.

$y = 3x^2 + 24x + 50$	Original equation
$y = 3(x^2 + 8x) + 50$	Factor 3 from the x-terms.
$y = 3(x^2 + 8x + ■) + 50 - 3(■)$	Complete the square on the right side.
$y = 3(x^2 + 8x + 16) + 50 - 3(16)$	The 16 added when you complete the square is multiplied by 3.
$y = 3(x + 4)^2 + 2$	
$y = 3[x - (-4)]^2 + 2$	$(h, k) = (-4, 2)$

Study Tip

Look Back
To review **completing the square**, see Lesson 6-4.

The vertex of this parabola is located at $(-4, 2)$, and the equation of the axis of symmetry is $x = -4$. The parabola opens upward.

GRAPH PARABOLAS You can use symmetry and translations to graph parabolas. The equation $y = a(x - h)^2 + k$ can be obtained from $y = ax^2$ by replacing x with $x - h$ and y with $y - k$. Therefore, the graph of $y = a(x - h)^2 + k$ is the graph of $y = ax^2$ translated h units to the right and k units up.

Example 2 *Graph Parabolas*

Graph each equation.

a. $y = -2x^2$

For this equation, $h = 0$ and $k = 0$. The vertex is at the origin. Since the equation of the axis of symmetry is $x = 0$, substitute some small positive integers for x and find the corresponding y-values.

x	y
1	−2
2	−8
3	−18

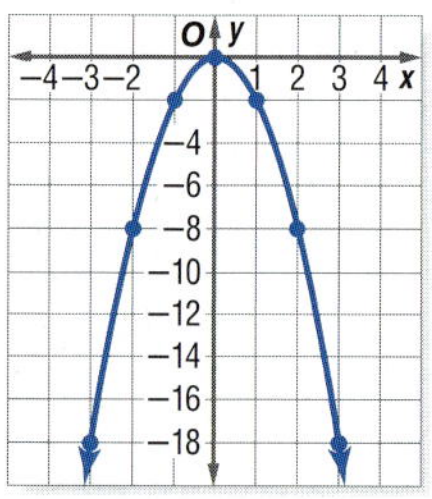

Since the graph is symmetric about the y-axis, the points at $(-1, -2)$, $(-2, -8)$, and $(-3, -18)$ are also on the parabola. Use all of these points to draw the graph.

Notice that each side of the graph is the reflection of the other side about the y-axis.

DAILY INTERVENTION

Unlocking Misconceptions

Some students may think that any curve can be called a parabola. Explain that only curves with a certain well-defined shape meet the definition of a parabola.

420 Chapter 8 Conic Sections

b. $y = -2(x - 2)^2 + 3$

The equation is of the form $y = a(x - h)^2 + k$, where $h = 2$ and $k = 3$. The graph of this equation is the graph of $y = -2x^2$ in part **a** translated 2 units to the right and up 3 units. The vertex is now at (2, 3).

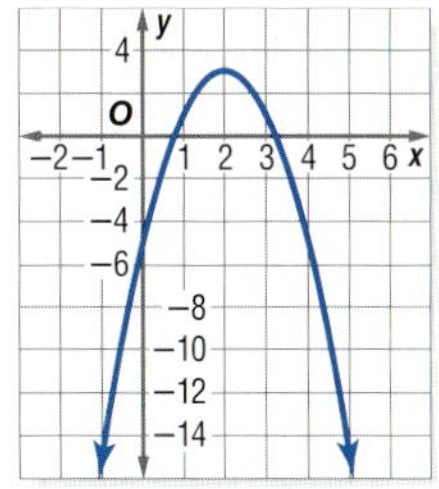

You can use paper folding to investigate the characteristics of a parabola.

Algebra Activity

Parabolas

Model

Step 1 Start with a sheet of wax paper that is about 15 inches long and 12 inches wide. Make a line that is perpendicular to the sides of the sheet by folding the sheet near one end. Open up the paper again. This line is the directrix. Mark a point about midway between the sides of the sheet so that the distance from the directrix is about 1 inch. This point is the focus.

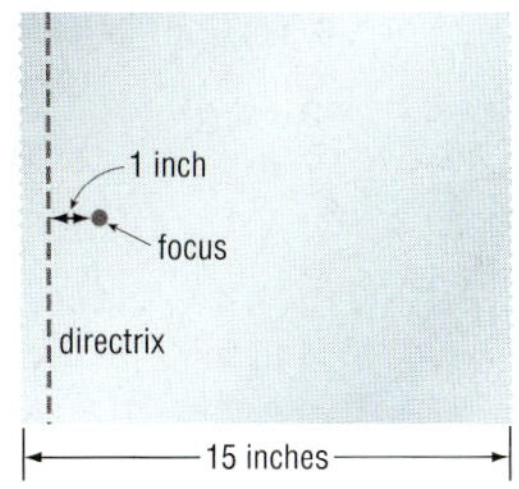

Put the focus on top of any point on the directrix and crease the paper. Make about 20 more creases by placing the focus on top of other points on the directrix. The lines form the outline of a parabola.

Step 2 Start with a new sheet of wax paper. Form another outline of a parabola with a focus that is about 3 inches from the directrix.

Step 3 On a new sheet of wax paper, form a third outline of a parabola with a focus that is about 5 inches from the directrix.

Analyze

Compare the shapes of the three parabolas. How does the distance between the focus and the directrix affect the shape of a parabola?

TEACHING TIP The line does not have to be perfectly perpendicular to the sides.

As the distance between the directrix and the focus increases, the parabola becomes wider.

The shape of a parabola and the distance between the focus and directrix depend on the value of a in the equation. The line segment through the focus of a parabola and perpendicular to the axis of symmetry is called the **latus rectum**. The endpoints of the latus rectum lie on the parabola.

In the figure at the right, the latus rectum is $\overline{AB}$. The length of the latus rectum of the parabola with equation $y = a(x - h)^2 + k$ is $\left|\frac{1}{a}\right|$ units. The endpoints of the latus rectum are $\left|\frac{1}{2a}\right|$ units from the focus.

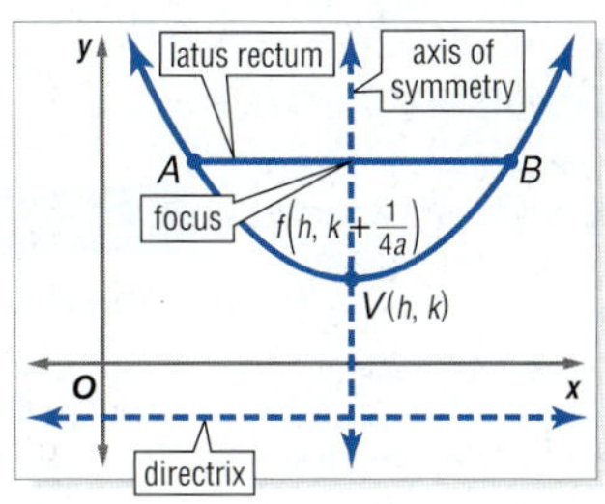

www.algebra2.com/extra_examples

GRAPH PARABOLAS

In-Class Example Power Point®

2 Graph each equation.

a. $y = 2x^2$

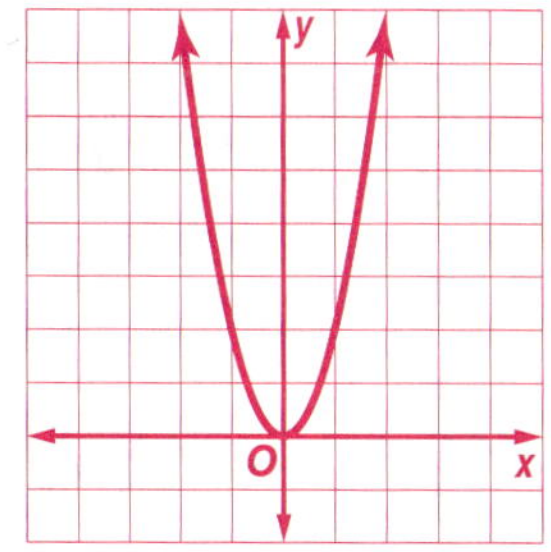

b. $y = 2(x - 1)^2 - 5$

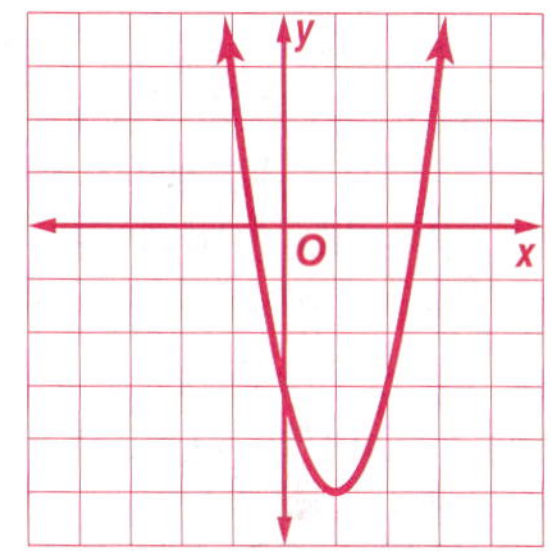

Teaching Tip Remind students that in $y = a(x - h)^2 + k$, the vertex is (h, k) and the axis of symmetry is $x = h$.

Algebra Activity

Materials: 3 sheets of wax paper about 15 inches by 12 inches, inch ruler

- You may wish to have students work in pairs. Caution students to avoid making unintentional creases as they fold the wax paper to find the focal point so that it is on the directrix.
- Suggest that students draw the directrix with a pen to make it easier to see.
- The creases that form the parabola may be easier to see against a dark background, such as a dark piece of posterboard.

In-Class Examples

Power Point®

3 Graph $x + y^2 = 4y - 1$.

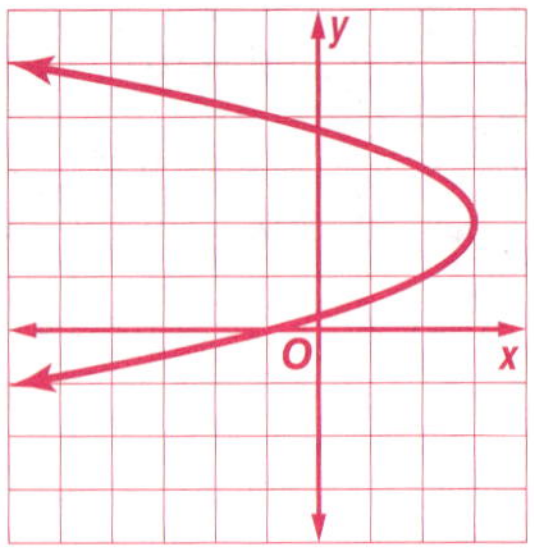

4 **BRIDGES** The 52 meter-long Hulme Arch Bridge in Manchester, England, is supported by cables suspended from a parabolic steel arch. The highest point of the arch is 25 meters above the bridge, and the focus of the arch is about 18 meters above the bridge.

a. Let the bridge be the x-axis, and let the y-axis pass through the vertex of the arch. Write an equation that models the arch. $y = -\frac{1}{28}x^2 + 25$

b. Graph the equation.

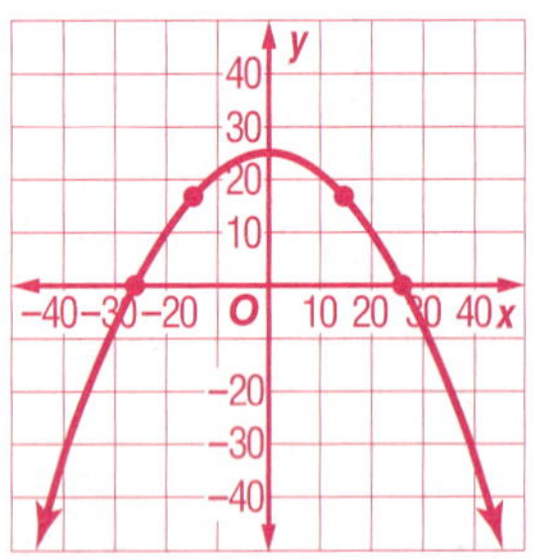

Teaching Tip Suggest that students make a rough sketch of the graph before they begin to find the details of the graph. They can do this by finding the vertex and then deciding which axis the graph wraps around and in which direction it opens.

Equations of parabolas with vertical axes of symmetry are of the form $y = a(x - h)^2 + k$ and are functions. Equations of parabolas with horizontal axes of symmetry are of the form $x = a(y - k)^2 + h$ and are not functions.

Concept Summary — Information About Parabolas

Form of Equation	$y = a(x - h)^2 + k$	$x = a(y - k)^2 + h$
Vertex	(h, k)	(h, k)
Axis of Symmetry	$x = h$	$y = k$
Focus	$\left(h, k + \frac{1}{4a}\right)$	$\left(h + \frac{1}{4a}, k\right)$
Directrix	$y = k - \frac{1}{4a}$	$x = h - \frac{1}{4a}$
Direction of Opening	upward if $a > 0$, downward if $a < 0$	right if $a > 0$, left if $a < 0$
Length of Latus Rectum	$\left\lvert\frac{1}{a}\right\rvert$ units	$\left\lvert\frac{1}{a}\right\rvert$ units

Example 3 *Graph an Equation Not in Standard Form*

Graph $4x - y^2 = 2y + 13$.

First, write the equation in the form $x = a(y - k)^2 + h$.

$4x - y^2 = 2y + 13$ — There is a y^2 term, so isolate the y and y^2 terms.

$4x = y^2 + 2y + 13$

$4x = (y^2 + 2y + \blacksquare) + 13 - \blacksquare$ — Complete the square.

$4x = (y^2 + 2y + 1) + 13 - 1$ — Add and subtract 1, since $\left(\frac{2}{2}\right)^2 = 1$.

$4x = (y + 1)^2 + 12$ — Write $y^2 + 2y + 1$ as a square.

$x = \frac{1}{4}(y + 1)^2 + 3$ — $(h, k) = (3, -1)$

Then use the following information to draw the graph.

vertex: $(3, -1)$

axis of symmetry: $y = -1$

focus: $\left(3 + \frac{1}{4\left(\frac{1}{4}\right)}, -1\right)$ or $(4, -1)$

directrix: $x = 3 - \frac{1}{4\left(\frac{1}{4}\right)}$ or 2

direction of opening: right, since $a > 0$

length of latus rectum: $\left\lvert\frac{1}{\frac{1}{4}}\right\rvert$ or 4 units

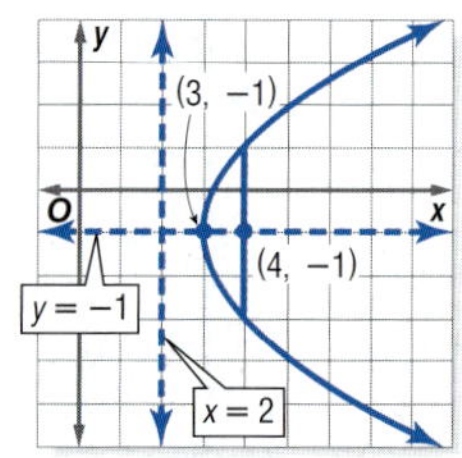

Remember that you can plot as many points as necessary to help you draw an accurate graph.

More About. . .

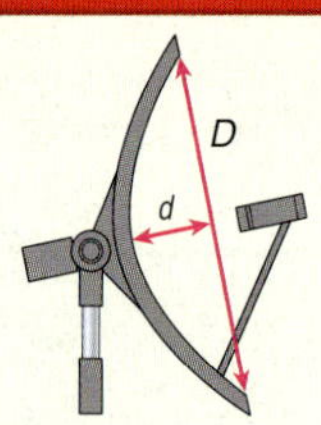

Satellite TV

The important characteristics of a satellite dish are the diameter D, depth d, and the ratio $\frac{f}{D}$, where f is the distance between the focus and the vertex. A typical dish has the values $D = 60$ cm, $d = 6.25$ cm, and $\frac{f}{D} = 0.6$.

Source: www.2000networks.com

Example 4 *Write and Graph an Equation for a Parabola*

SATELLITE TV **Satellite dishes have parabolic cross sections.**

a. Use the information at the left to write an equation that models a cross section of a satellite dish. Assume that the focus is at the origin and the parabola opens to the right.

First, solve for f. Since $\frac{f}{D} = 0.6$, and $D = 60$, $f = 0.6(60)$ or 36.

The focus is at $(0, 0)$, and the parabola opens to the right. So the vertex must be at $(-36, 0)$. Thus, $h = -36$ and $k = 0$. Use the x-coordinate of the focus to find a.

DAILY INTERVENTION — Differentiated Instruction

Kinesthetic Have students make a giant parabola. They can use masking tape to mark a focus point and a directrix line on the classroom floor. Then have them use yardsticks or meter sticks to find and mark, with colored masking tape, a set of points that are equidistant from the point and the line. These points trace a parabola.

$$-36 + \frac{1}{4a} = 0 \quad h = -36\text{; The } x\text{-coordinate of the focus is 0.}$$

$$\frac{1}{4a} = 36 \quad \text{Add 36 to each side.}$$

$$1 = 144a \quad \text{Multiply each side by } 4a.$$

$$\frac{1}{144} = a \quad \text{Divide each side by 144.}$$

An equation of the parabola is $x = \frac{1}{144}y^2 - 36$.

b. Graph the equation.

The length of the latus rectum is $\left|\frac{1}{\frac{1}{144}}\right|$ or 144 units, so the graph must pass through (0, 72) and (0, −72). According to the diameter and depth of the dish, the graph must pass through (−29.75, 30) and (−29.75, −30). Use these points and the information from part **a** to draw the graph.

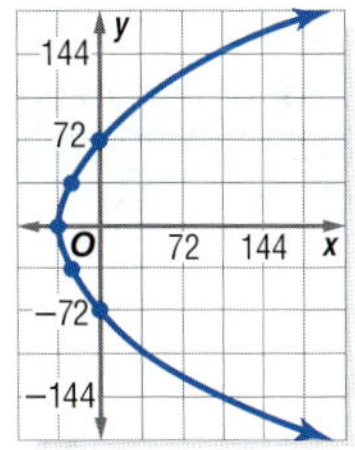

Check for Understanding

Concept Check

1. (3, −7), $\left(3, -6\frac{15}{16}\right)$, $x = 3$, $y = -7\frac{1}{16}$
2. Sample answer: $x = -y^2$

1. **Identify** the vertex, focus, axis of symmetry, and directrix of the graph of $y = 4(x - 3)^2 - 7$.
2. **OPEN ENDED** Write an equation for a parabola that opens to the left.
3. **FIND THE ERROR** Katie is finding the standard form of the equation $y = x^2 + 6x + 4$. What mistake did she make in her work? **When she added 9 to complete the square, she forgot to also subtract 9. The standard form is $y = (x + 3)^2 - 9 + 4$ or $y = (x + 3)^2 - 5$.**

$y = x^2 + 6x + 4$

$y = x^2 + 6x + 9 + 4$

$y = (x + 3)^2 + 4$

Guided Practice

4. Write $y = 2x^2 - 12x + 6$ in standard form. **$y = 2(x - 3)^2 - 12$**

GUIDED PRACTICE KEY	
Exercises	Examples
4	1
5–8	1–3
9, 10	2–4
11	4

Identify the coordinates of the vertex and focus, the equations of the axis of symmetry and directrix, and the direction of opening of the parabola with the given equation. Then find the length of the latus rectum and graph the parabola. **5–8. See pp. 469A–469J.**

5. $y = (x - 3)^2 - 4$
6. $y = 2(x + 7)^2 + 3$
7. $y = -3x^2 - 8x - 6$
8. $x = \frac{2}{3}y^2 - 6y + 12$

Write an equation for each parabola described below. Then draw the graph.

9–10. See margin for graphs.

9. focus (3, 8), directrix $y = 4$ **$y = \frac{1}{8}(x - 3)^2 + 6$**
10. vertex (5, −1), focus (3, −1) **$x = -\frac{1}{8}(y + 1)^2 + 5$**

Application

11. **COMMUNICATION** A microphone is placed at the focus of a parabolic reflector to collect sound for the television broadcast of a World Cup soccer game. Write an equation for the cross section, assuming that the focus is at the origin and the parabola opens to the right. **$x = \frac{1}{24}y^2 - 6$**

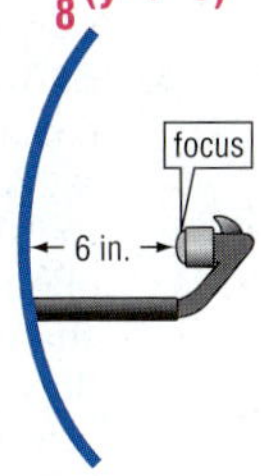

Answers

9.

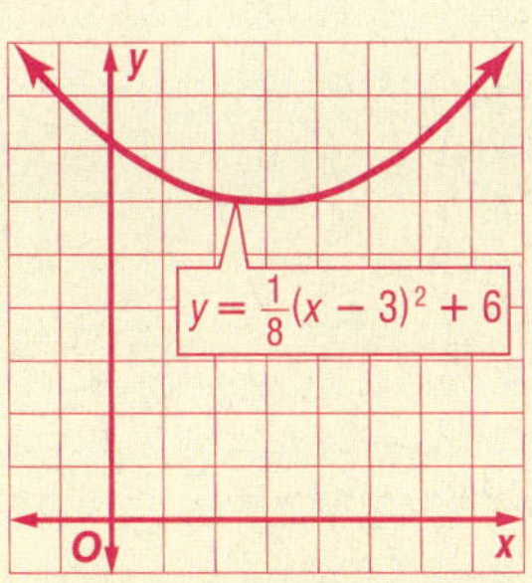

10.

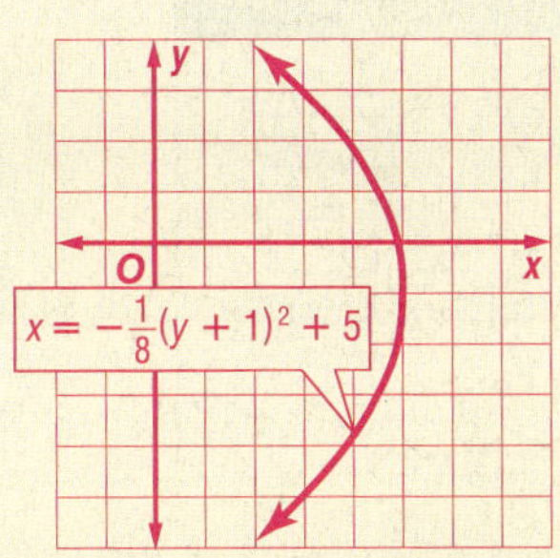

3 Practice/Apply

Study Notebook

Have students—

- add the definitions/examples of the vocabulary terms to their Vocabulary Builder worksheets for Chapter 8.
- copy the Concept Summary chart just before Example 3 into their notebooks and add some labeled sketches to illustrate.
- include any other item(s) that they find helpful in mastering the skills in this lesson.

DAILY INTERVENTION

FIND THE ERROR Remind students that any change they make to the form of an equation must not change the value, so when they add a quantity to complete the square they must also subtract that quantity or add it to the other side of the equation.

About the Exercises...

Organization by Objective

- **Equations of Parabolas:** 12–15, 35
- **Graph Parabolas:** 16–34, 36–45

Odd/Even Assignments

Exercises 12–29 and 36–41 are structured so that students practice the same concepts whether they are assigned odd or even problems.

Assignment Guide

Basic: 13, 17–27 odd, 31–45 odd, 46–62

Average: 13–45 odd, 46–62

Advanced: 12–46 even, 47–54 (optional: 55–62)

Study Guide and Intervention, p. 461 (shown) and p. 462

Equations of Parabolas A parabola is a curve consisting of all points in the coordinate plane that are the same distance from a given point (the **focus**) and a given line (the **directrix**). The following chart summarizes important information about parabolas.

Standard Form of Equation	$y = a(x - h)^2 + k$	$x = a(y - k)^2 + h$
Axis of Symmetry	$x = h$	$y = k$
Vertex	(h, k)	(h, k)
Focus	$\left(h, k + \frac{1}{4a}\right)$	$\left(h + \frac{1}{4a}, k\right)$
Directrix	$y = k - \frac{1}{4a}$	$x = h - \frac{1}{4a}$
Direction of Opening	upward if $a > 0$, downward if $a < 0$	right if $a > 0$, left if $a < 0$
Length of Latus Rectum	$\left\lvert\frac{1}{a}\right\rvert$ units	$\left\lvert\frac{1}{a}\right\rvert$ units

Example **Identify the coordinates of the vertex and focus, the equations of the axis of symmetry and directrix, and the direction of opening of the parabola with equation $y = 2x^2 - 12x - 25$.**

$y = 2x^2 - 12x - 25$ Original equation
$y = 2(x^2 - 6x) - 25$ Factor 2 from the x-terms.
$y = 2(x^2 - 6x + ■) - 25 - 2(■)$ Complete the square on the right side.
$y = 2(x^2 - 6x + 9) - 25 - 2(9)$ The 9 added to complete the square is multiplied by 2.
$y = 2(x - 3)^2 - 43$ Write in standard form.

The vertex of this parabola is located at $(3, -43)$, the focus is located at $\left(3, -42\frac{7}{8}\right)$, the equation of the axis of symmetry is $x = 3$, and the equation of the directrix is $y = -43\frac{1}{8}$. The parabola opens upward.

Exercises

Identify the coordinates of the vertex and focus, the equations of the axis of symmetry and directrix, and the direction of opening of the parabola with the given equation.

1. $y = x^2 + 6x - 4$ $(-3, -13)$, $\left(-3, -12\frac{3}{4}\right)$, $x = -3$, $y = -13\frac{1}{4}$, up
2. $y = 8x - 2x^2 + 10$ $(2, 18)$, $\left(2, 17\frac{1}{8}\right)$, $x = 2$, $y = 18\frac{1}{8}$, down
3. $x = y^2 - 8y + 6$ $(-10, 4)$, $\left(-9\frac{3}{4}, 4\right)$, $y = 4$, $x = -10\frac{1}{4}$; right

Write an equation of each parabola described below.

4. focus $(-2, 3)$, directrix $x = -2\frac{1}{12}$ $x = 6(y - 3)^2 - 2\frac{1}{24}$
5. vertex $(5, 1)$, focus $\left(4\frac{11}{12}, 1\right)$ $x = -3(y - 1)^2 + 5$

Skills Practice, p. 463 and Practice, p. 464 (shown)

Write each equation in standard form.

1. $y = 2x^2 - 12x + 19$ $y = 2(x - 3)^2 + 1$
2. $y = \frac{1}{2}x^2 + 3x + \frac{1}{2}$ $y = \frac{1}{2}[x - (-3)]^2 + (-4)$
3. $y = -3x^2 - 12x - 7$ $y = -3[x - (-2)]^2 + 5$

Identify the coordinates of the vertex and focus, the equations of the axis of symmetry and directrix, and the direction of opening of the parabola with the given equation. Then find the length of the latus rectum and graph the parabola.

4. $y = (x - 4)^2 + 3$

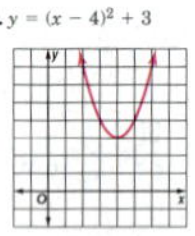

vertex: $(4, 3)$; focus: $\left(4, 3\frac{1}{4}\right)$; axis: $x = 4$; directrix: $y = 2\frac{3}{4}$; opens up; latus rectum: 1 unit

5. $x = -\frac{1}{3}y^2 + 1$

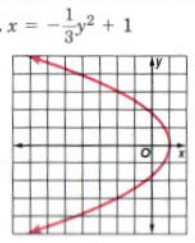

vertex: $(1, 0)$; focus: $\left(\frac{1}{4}, 0\right)$; axis: $y = 0$; directrix: $x = 1\frac{3}{4}$; opens left; latus rectum: 3 units

6. $x = 3(y + 1)^2 - 3$

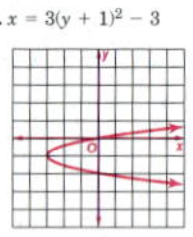

vertex: $(-3, -1)$; focus: $\left(-2\frac{11}{12}, -1\right)$; axis: $y = -1$; directrix: $x = -3\frac{1}{12}$; opens right; latus rectum: $\frac{1}{3}$ unit

Write an equation for each parabola described below. Then draw the graph.

7. vertex $(0, -4)$, focus $\left(0, -3\frac{7}{8}\right)$ $y = 2x^2 - 4$

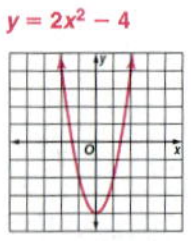

8. vertex $(-2, 1)$, directrix $x = -3$ $x = \frac{1}{4}(y - 1)^2 - 2$

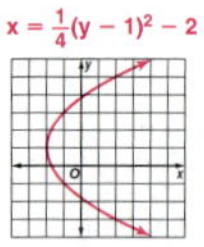

9. vertex $(1, 3)$, axis of symmetry $x = 1$, latus rectum: 2 units, $a < 0$ $y = -\frac{1}{2}(x - 1)^2 + 3$

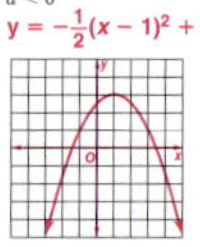

10. **TELEVISION** Write the equation in the form $y = ax^2$ for a satellite dish. Assume that the bottom of the upward-facing dish passes through $(0, 0)$ and that the distance from the bottom to the focus point is 8 inches. $y = \frac{1}{32}x^2$

Reading to Learn Mathematics, p. 465 ELL

Pre-Activity **How are parabolas used in manufacturing?**

Read the introduction to Lesson 8-2 at the top of page 419 in your textbook.

Name at least two reflective objects that might have the shape of a parabola.

Sample answer: telescope mirror, satellite dish

Reading the Lesson

1. In the parabola shown in the graph, the point $(2, -2)$ is called the __vertex__ and the point $(2, 0)$ is called the __focus__. The line $y = -4$ is called the __directrix__, and the line $x = 2$ is called the __axis of symmetry__.

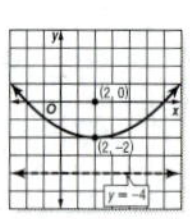

2. a. Write the standard form of the equation of a parabola that opens upward or downward. $y = a(x - h)^2 + k$
   b. The parabola opens downward if __$a < 0$__ and opens upward if __$a > 0$__. The equation of the axis of symmetry is __$x = h$__, and the coordinates of the vertex are __(h, k)__.
3. A parabola has equation $x = -\frac{1}{8}(y - 2)^2 + 4$. This parabola opens to the __left__. It has vertex __(4, 2)__ and focus __(2, 2)__. The directrix is __$x = 6$__. The length of the latus rectum is __8__ units.

Helping You Remember

4. How can the way in which you plot points in a rectangular coordinate system help you to remember what the sign of a tells you about the direction in which a parabola opens?
Sample answer: In plotting points, a positive x-coordinate tells you to move to the *right* and a negative x-coordinate tells you to move to the *left*. This is like a parabola whose equation is of the form "$x = ...$"; it opens to the *right* if $a > 0$ and to the *left* if $a < 0$. Likewise, a positive y-coordinate tells you to move *up* and a negative y-coordinate tells you to move *down*. This is like a parabola whose equation is of the form "$y = ...$"; it opens *upward* if $a > 0$ and *downward* if $a < 0$.

★ indicates increased difficulty

Practice and Apply

Homework Help

For Exercises	See Examples
12–15, 35	1
16–34	1–3
36–41	2–4
42–45	4a

Extra Practice See page 845.

Write each equation in standard form.

12. $y = x^2 - 6x + 11$ $y = (x - 3)^2 + 2$
13. $x = y^2 + 14y + 20$ $x = (y + 7)^2 - 29$
★14. $y = \frac{1}{2}x^2 + 12x - 8$ $y = \frac{1}{2}(x + 12)^2 - 80$
★15. $x = 3y^2 + 5y - 9$ $x = 3\left(y + \frac{5}{6}\right)^2 - 11\frac{1}{12}$

Identify the coordinates of the vertex and focus, the equations of the axis of symmetry and directrix, and the direction of opening of the parabola with the given equation. Then find the length of the latus rectum and graph the parabola. 16–29. See pp. 469A–469J.

16. $-6y = x^2$
17. $y^2 = 2x$
18. $3(y - 3) = (x + 6)^2$
19. $-2(y - 4) = (x - 1)^2$
20. $4(x - 2) = (y + 3)^2$
21. $(y - 8)^2 = -4(x - 4)$
22. $y = x^2 - 12x + 20$
23. $x = y^2 - 14y + 25$
24. $x = 5y^2 + 25y + 60$
25. $y = 3x^2 - 24x + 50$
26. $y = -2x^2 + 5x - 10$
27. $x = -4y^2 + 6y + 2$
★28. $y = \frac{1}{2}x^2 - 3x + \frac{19}{2}$
★29. $x = -\frac{1}{3}y^2 - 12y + 15$

For Exercises 30–34, use the equation $x = 3y^2 + 4y + 1$.

30. Draw the graph. See margin.
31. Find the x-intercept(s). 1
32. Find the y-intercept(s). -1 and $-\frac{1}{3}$
33. What is the equation of the axis of symmetry? $y = -\frac{2}{3}$
34. What are the coordinates of the vertex? $\left(-\frac{1}{3}, -\frac{2}{3}\right)$

35. **MANUFACTURING** The reflective surface in a flashlight has a parabolic cross section that can be modeled by $y = \frac{1}{3}x^2$, where x and y are in centimeters. How far from the vertex should the filament of the light bulb be located? 0.75 cm

36–41. See pp. 469A–469J for graphs.

Write an equation for each parabola described below. Then draw the graph.

36. vertex $(0, 1)$, focus $(0, 5)$ $y = \frac{1}{16}x^2 + 1$
37. vertex $(8, 6)$, focus $(2, 6)$
38. focus $(-4, -2)$, directrix $x = -8$
39. vertex $(1, 7)$, directrix $y = 3$
40. vertex $(-7, 4)$, axis of symmetry $x = -7$, measure of latus rectum 6, $a < 0$
41. vertex $(4, 3)$, axis of symmetry $y = 3$, measure of latus rectum 4, $a > 0$ $x = \frac{1}{4}(y - 3)^2 + 4$

37. $x = -\frac{1}{24}(y - 6)^2 + 8$
38. $x = \frac{1}{8}(y + 2)^2 - 6$
39. $y = \frac{1}{16}(x - 1)^2 + 7$
40. $y = -\frac{1}{6}(x + 7)^2 + 4$

42. Write an equation for the graph at the right. $y = \frac{2}{9}x^2 - 2$

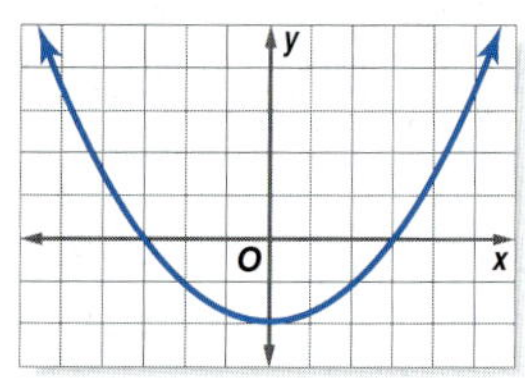

43. **BRIDGES** The Bayonne Bridge connects Staten Island, New York, to New Jersey. It has an arch in the shape of a parabola that opens downward. Write an equation of a parabola to model the arch, assuming that the origin is at the surface of the water, beneath the vertex of the arch. about $y = -0.00046x^2 + 325$

Enrichment, p. 466

Tangents to Parabolas

A line that intersects a parabola in exactly one point without crossing the curve is a **tangent** to the parabola. The point where a tangent line touches a parabola is the **point of tangency.** The line perpendicular to a tangent to a parabola at the point of tangency is called the **normal** to the parabola at that point. In the diagram, line ℓ is tangent to the parabola that is the graph of $y = x^2$ at $\left(\frac{3}{2}, \frac{9}{4}\right)$. The x-axis is tangent to the parabola at O, and the y-axis is the normal to the parabola at O.

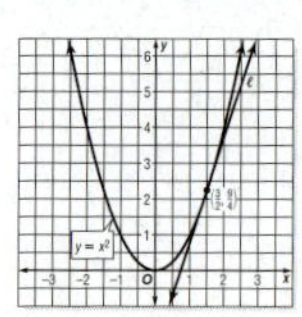

Solve each problem.

1. Find an equation for line ℓ in the diagram. *Hint:* A nonvertical line with an equation of the form $y = mx + b$ will be tangent to the graph of $y = x^2$ at

Answer

30.

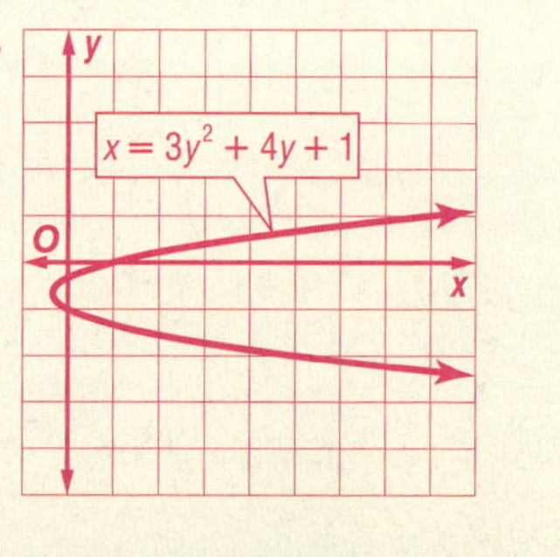

424 Chapter 8 Conic Sections

44. **SPORTS** When a ball is thrown or kicked, the path it travels is shaped like a parabola. Suppose a football is kicked from ground level, reaches a maximum height of 25 feet, and hits the ground 100 feet from where it was kicked. Assuming that the ball was kicked at the origin, write an equation of the parabola that models the flight of the ball. $y = -\frac{1}{100}(x - 50)^2 + 25$

45. **AEROSPACE** A spacecraft is in a circular orbit 150 kilometers above Earth. Once it attains the velocity needed to escape Earth's gravity, the spacecraft will follow a parabolic path with the center of Earth as the focus. Suppose the spacecraft reaches escape velocity above the North Pole. Write an equation to model the parabolic path of the spacecraft, assuming that the center of Earth is at the origin and the radius of Earth is 6400 kilometers. $y = -\frac{1}{26,200}x^2 + 6550$

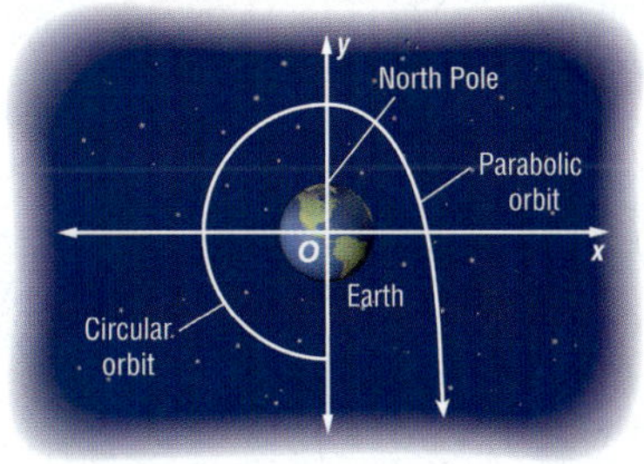

46. **CRITICAL THINKING** The parabola with equation $y = (x - 4)^2 + 3$ has its vertex at (4, 3) and passes through (5, 4). Find an equation of a different parabola with its vertex at (4, 3) that passes through (5, 4). $x = (y - 3)^2 + 4$

47. **WRITING IN MATH** Answer the question that was posed at the beginning of the lesson. **See margin.**

How are parabolas used in manufacturing?

Include the following in your answer:
- how you think the focus of a parabola got its name, and
- why a car headlight with a parabolic reflector is better than one with an unreflected light bulb.

Standardized Test Practice

48. Which equation has a graph that opens downward? **B**

Ⓐ $y = 3x^2 - 2$ Ⓑ $y = 2 - 3x^2$ Ⓒ $x = 3y^2 - 2$ Ⓓ $x = 2 - 3y^2$

49. Find the vertex of the parabola with equation $y = x^2 - 10x + 8$. **A**

Ⓐ (5, −17) Ⓑ (10, 8) Ⓒ (0, 8) Ⓓ (5, 8)

Maintain Your Skills

Mixed Review **Find the distance between each pair of points with the given coordinates.** *(Lesson 8-1)*

50. (7, 3), (−5, 8) **13 units** 51. (4, −1), (−2, 7) **10 units** 52. (−3, 1), (0, 6) $\sqrt{34}$ **units**

53. Graph $y \leq \sqrt{x + 1}$. *(Lesson 7-9)* **See margin.**

54. **HEALTH** Ty's heart rate is usually 120 beats per minute when he runs. If he runs for 2 hours every day, about how many times will his heart beat during the amount of time he exercises in two weeks? Express the answer in scientific notation. *(Lesson 5-1)* 2.016×10^5

Getting Ready for the Next Lesson **PREREQUISITE SKILL Simplify each radical expression.** *(To review* ***simplifying radicals****, see Lessons 5-5 and 5-6.)*

55. $\sqrt{16}$ **4** 56. $\sqrt{25}$ **5** 57. $\sqrt{81}$ **9** 58. $\sqrt{144}$ **12**
59. $\sqrt{12}$ $2\sqrt{3}$ 60. $\sqrt{18}$ $3\sqrt{2}$ 61. $\sqrt{48}$ $4\sqrt{3}$ 62. $\sqrt{72}$ $6\sqrt{2}$

www.algebra2.com/self_check_quiz

4 Assess

Open-Ended Assessment

Writing Have students draw and label four types of parabolas opening upward, downward, left, and right. Have them give the equation, vertex, axis of symmetry, focus, and directrix for each in terms of x, y, a, h, and k.

Getting Ready for Lesson 8-3

PREREQUISITE SKILL Students will write and analyze equations of circles in Lesson 8-3. Using the Distance Formula, students will simplify radicals to find the radii of circles. Exercises 55–62 should be used to determine your students' familiarity with simplifying radicals.

Assessment Options

Quiz (Lessons 8-1 and 8-2) is available on p. 511 of the *Chapter 8 Resource Masters*.

Answers

47. A parabolic reflector can be used to make a car headlight more effective. Answers should include the following.
- Reflected rays are *focused* at that point.
- The light from an unreflected bulb would shine in all directions. With a parabolic reflector, most of the light can be directed forward toward the road.

53.

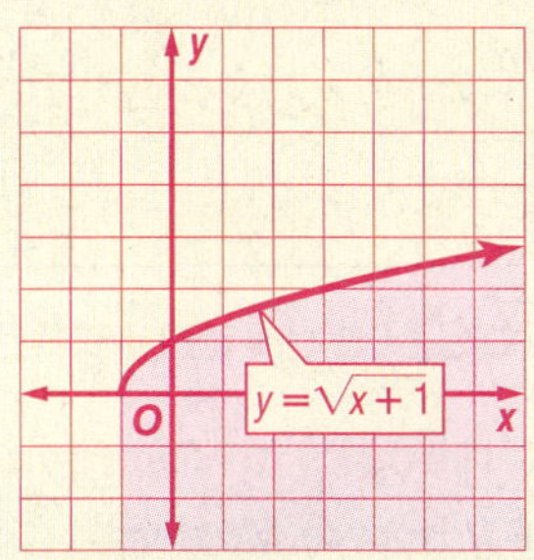

8-3 Lesson Notes

1 Focus

5-Minute Check Transparency 8-3 Use as a quiz or review of Lesson 8-2.

Mathematical Background notes are available for this lesson on p. 410C.

Why are circles important in air traffic control?

Ask students:

- Describe any radar antenna you may have seen on a boat or at an airport. How did the antenna move? **constantly turning in a circle**
- What part of a circle is the range of 45 to 70 miles describing? **radius**

8-3 Circles

What You'll Learn

- Write equations of circles.
- Graph circles.

Vocabulary

- circle
- center
- tangent

Why are circles important in air traffic control?

Radar equipment can be used to detect and locate objects that are too far away to be seen by the human eye. The radar systems at major airports can typically detect and track aircraft up to 45 to 70 miles in any direction from the airport. The boundary of the region that a radar system can monitor can be modeled by a circle.

EQUATIONS OF CIRCLES A **circle** is the set of all points in a plane that are equidistant from a given point in the plane, called the **center**. Any segment whose endpoints are the center and a point on the circle is a *radius* of the circle.

Assume that (x, y) are the coordinates of a point on the circle at the right. The center is at (h, k), and the radius is r. You can find an equation of the circle by using the Distance Formula.

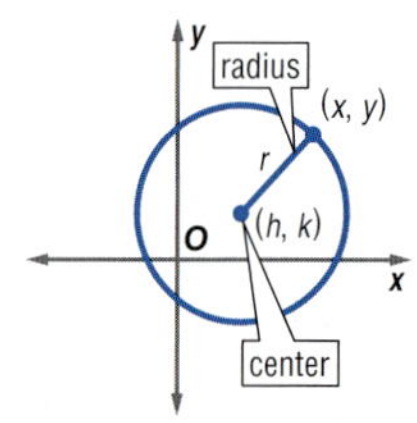

$$\sqrt{(x_2 - x_1)^2 + (y_2 - y_1)^2} = d \quad \text{Distance Formula}$$

$$\sqrt{(x - h)^2 + (y - k)^2} = r \quad (x_1, y_1) = (h, k),\ (x_2, y_2) = (x, y),\ d = r$$

$$(x - h)^2 + (y - k)^2 = r^2 \quad \text{Square each side.}$$

Key Concept — *Equation of a Circle*

The equation of a circle with center (h, k) and radius r units is $(x - h)^2 + (y - k)^2 = r^2$.

Example 1 Write an Equation Given the Center and Radius

NUCLEAR POWER In 1986, a nuclear reactor exploded at a power plant about 110 kilometers north and 15 kilometers west of Kiev. At first, officials evacuated people within 30 kilometers of the power plant. Write an equation to represent the boundary of the evacuated region if the origin of the coordinate system is at Kiev.

Since Kiev is at (0, 0), the power plant is at (−15, 110). The boundary of the evacuated region is the circle centered at (−15, 110) with radius 30 kilometers.

$$(x - h)^2 + (y - k)^2 = r^2 \quad \text{Equation of a circle}$$

$$[x - (-15)]^2 + (y - 110)^2 = 30^2 \quad (h, k) = (-15, 110),\ r = 30$$

$$(x + 15)^2 + (y - 110)^2 = 900 \quad \text{Simplify.}$$

The equation is $(x + 15)^2 + (y - 110)^2 = 900$.

Resource Manager

Workbook and Reproducible Masters

Chapter 8 Resource Masters
- Study Guide and Intervention, pp. 467–468
- Skills Practice, p. 469
- Practice, p. 470
- Reading to Learn Mathematics, p. 471
- Enrichment, p. 472

Graphing Calculator and Spreadsheet Masters, p. 41

Transparencies

5-Minute Check Transparency 8-3
Answer Key Transparencies

Technology

Alge2PASS: Tutorial Plus, Lesson 15
Interactive Chalkboard

Example 2 Write an Equation Given a Diameter

Write an equation for a circle if the endpoints of a diameter are at (5, 4) and (−2, −6).

Explore To write an equation of a circle, you must know the center and the radius.

Plan You can find the center of the circle by finding the midpoint of the diameter. Then you can find the radius of the circle by finding the distance from the center to one of the given points.

Solve First find the center of the circle.

$$(h, k) = \left(\frac{x_1 + x_2}{2}, \frac{y_1 + y_2}{2}\right) \quad \text{Midpoint Formula}$$

$$= \left(\frac{5 + (-2)}{2}, \frac{4 + (-6)}{2}\right) \quad (x_1, y_1) = (5, 4), (x_2, y_2) = (-2, -6)$$

$$= \left(\frac{3}{2}, \frac{-2}{2}\right) \quad \text{Add.}$$

$$= \left(\frac{3}{2}, -1\right) \quad \text{Simplify.}$$

Now find the radius.

$$r = \sqrt{(x_2 - x_1)^2 + (y_2 - y_1)^2} \quad \text{Distance Formula}$$

$$= \sqrt{\left(\frac{3}{2} - 5\right)^2 + (-1 - 4)^2} \quad (x_1, y_1) = (5, 4), (x_2, y_2) = \left(\frac{3}{2}, -1\right)$$

$$= \sqrt{\left(-\frac{7}{2}\right)^2 + (-5)^2} \quad \text{Subtract.}$$

$$= \sqrt{\frac{149}{4}} \quad \text{Simplify.}$$

The radius of the circle is $\sqrt{\frac{149}{4}}$ units, so $r^2 = \frac{149}{4}$.

Substitute h, k, and r^2 into the standard form of the equation of a circle.

An equation of the circle is $\left(x - \frac{3}{2}\right)^2 + [y - (-1)]^2 = \frac{149}{4}$ or $\left(x - \frac{3}{2}\right)^2 + (y + 1)^2 = \frac{149}{4}$.

Examine Each of the given points satisfies the equation, so the equation is reasonable.

A line in the plane of a circle can intersect the circle in zero, one, or two points. A line that intersects the circle in exactly one point is said to be **tangent** to the circle. The line and the circle are tangent to each other at this point.

Example 3 Write an Equation Given the Center and a Tangent

Write an equation for a circle with center at (−4, −3) that is tangent to the x-axis.

Sketch the circle. Since the circle is tangent to the x-axis, its radius is 3.

An equation of the circle is $(x + 4)^2 + (y + 3)^2 = 9$.

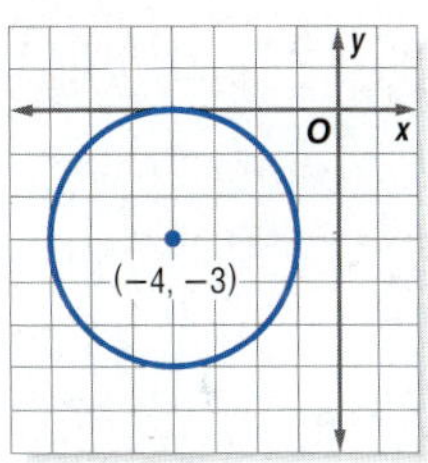

www.algebra2.com/extra_examples

2 Teach

EQUATIONS OF CIRCLES

In-Class Examples Power Point®

1 **LANDSCAPING** The plan for a park puts the center of a circular pond, of radius 0.6 miles, 2.5 miles east and 3.8 miles south of the park headquarters. Write an equation to represent the border of the pond, using the headquarters as the origin. $(x - 2.5)^2 + (y + 3.8)^2 = 0.36$

Teaching Tip Explain that h and k are the variables traditionally used, but other variables could be used just as well.

2 Write an equation for a circle if the endpoints of a diameter are at (2, 8) and (2, −2). $(x - 2)^2 + (y - 3)^2 = 25$

Teaching Tip Suggest that students draw a sketch showing the circle and the endpoints of the diameter to check their work.

3 Write an equation for a circle with center at (3, 5) that is tangent to the y-axis. $(x - 3)^2 + (y - 5)^2 = 9$

Teacher to Teacher

Nancy McKinney & Karen Rowe Camdenton H.S., Camdenton, MO

"While studying conic sections we have our students make a collage of the various shapes. They look in magazines, on the Internet, within computer print programs, etc."

GRAPH CIRCLES

In-Class Examples Power Point®

4 Find the center and radius of the circle with equation $x^2 + y^2 = 16$. Then graph the circle. **(0, 0); 4**

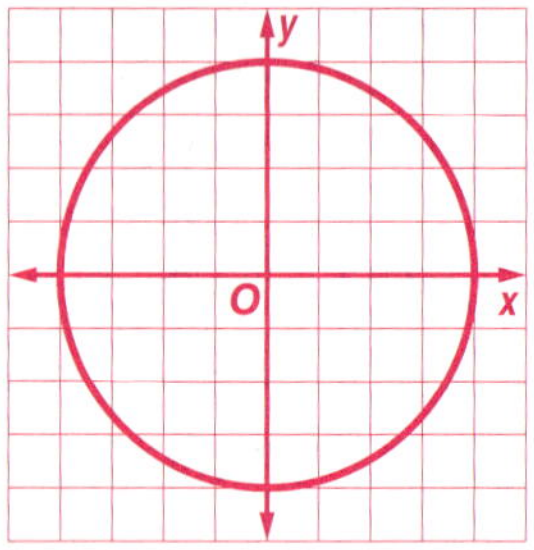

5 Find the center and radius of the circle with equation $x^2 + y^2 + 6x - 7 = 0$. Then graph the circle. **(−3, 0); 4**

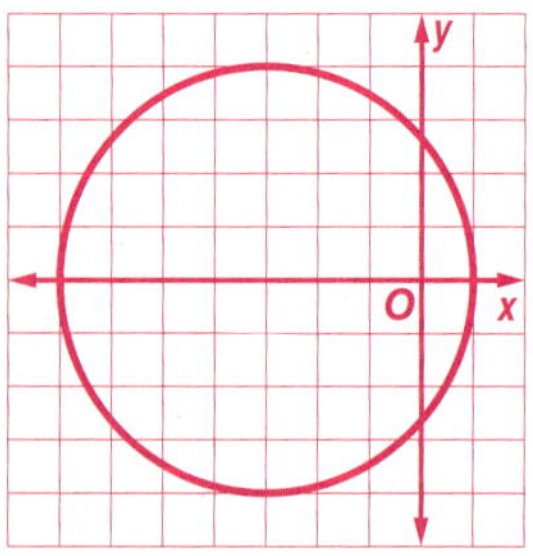

Answer

3. Lucy; 36 is the *square* of the radius, so the radius is 6 units.

GRAPH CIRCLES You can use completing the square, symmetry, and transformations to help you graph circles. The equation $(x - h)^2 + (y - k)^2 = r^2$ is obtained from the equation $x^2 + y^2 = r^2$ by replacing x with $x - h$ and y with $y - k$. So, the graph of $(x - h)^2 + (y - k)^2 = r^2$ is the graph of $x^2 + y^2 = r^2$ translated h units to the right and k units up.

Example 4 Graph an Equation in Standard Form

Find the center and radius of the circle with equation $x^2 + y^2 = 25$. Then graph the circle.

The center of the circle is at (0, 0), and the radius is 5.

x	y
0	5
3	4
4	3
5	0

The table lists some integer values for x and y that satisfy the equation.

Since the circle is centered at the origin, it is symmetric about the y-axis. Therefore, the points at (−3, 4), (−4, 3) and (−5, 0) lie on the graph.

The circle is also symmetric about the x-axis, so the points at (−4, −3), (−3, −4), (0, −5), (3, −4), and (4, −3) lie on the graph.

Graph all of these points and draw the circle that passes through them.

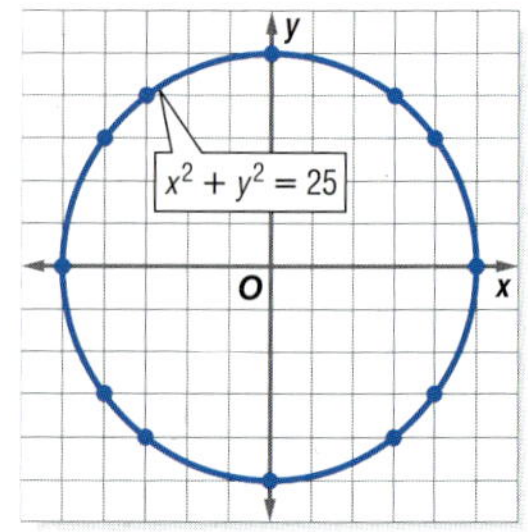

Example 5 Graph an Equation Not in Standard Form

Find the center and radius of the circle with equation $x^2 + y^2 - 4x + 8y - 5 = 0$. Then graph the circle.

Complete the squares.

$$x^2 + y^2 - 4x + 8y - 5 = 0$$

$$x^2 - 4x + \blacksquare + y^2 + 8y + \blacksquare = 5 + \blacksquare + \blacksquare$$

$$x^2 - 4x + 4 + y^2 + 8y + 16 = 5 + 4 + 16$$

$$(x - 2)^2 + (y + 4)^2 = 25$$

The center of the circle is at (2, −4), and the radius is 5. In the equation from Example 4, x has been replaced by $x - 2$, and y has been replaced by $y + 4$. The graph is the graph from Example 4 translated 2 units to the right and down 4 units.

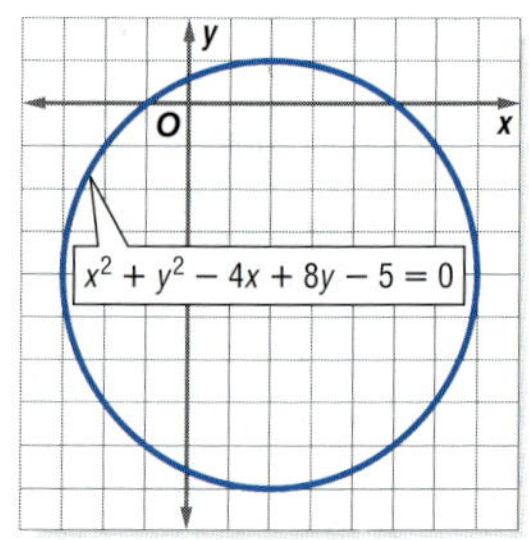

Check for Understanding

Concept Check

1. **OPEN ENDED** Write an equation for a circle with center at (6, −2).
2. Write $x^2 + y^2 + 6x - 2y - 54 = 0$ in standard form by completing the square. Describe the transformation that can be applied to the graph of $x^2 + y^2 = 64$ to obtain the graph of the given equation.
3. **FIND THE ERROR** Juwan says that the circle with equation $(x - 4)^2 + y^2 = 36$ has radius 36 units. Lucy says that the radius is 6 units. Who is correct? Explain. **See margin.**

1. Sample answer: $(x - 6)^2 + (y + 2)^2 = 16$

2. $(x + 3)^2 + (y - 1)^2 = 64$; left 3 units, up 1 unit

DAILY INTERVENTION

Differentiated Instruction

Naturalist Encourage students to find and describe objects in nature that are related to circles. Although circles in nature may not be mathematically perfect, they are often seen, as in the circle of leaves on a plant, or designs on an insect, or the shape of a flower, or ripples on a pond after a stone is tossed into the water.

Guided Practice

GUIDED PRACTICE KEY	
Exercises	Examples
4–7, 14, 15	1–3
8–13	4, 5

4. Write an equation for the graph at the right. $(x-3)^2+(y+1)^2=9$

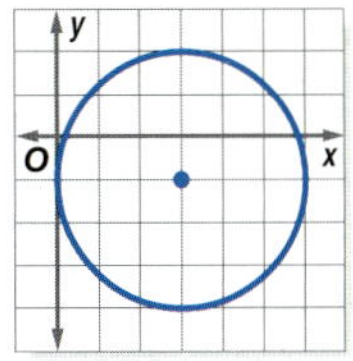

Write an equation for the circle that satisfies each set of conditions.

5. center $(-1, -5)$, radius 2 units $(x+1)^2+(y+5)^2=4$
6. endpoints of a diameter at $(-4, 1)$ and $(4, -5)$ $x^2+(y+2)^2=25$
7. center $(3, -7)$, tangent to the y-axis $(x-3)^2+(y+7)^2=9$

Find the center and radius of the circle with the given equation. Then graph the circle. 8–13. See pp. 469A–469J for graphs.

9. $(0, 14)$, $\sqrt{34}$ units

11. $\left(-\frac{2}{3}, \frac{1}{2}\right)$, $\frac{2\sqrt{2}}{3}$ unit

8. $(x-4)^2+(y-1)^2=9$ $(4, 1)$, 3 units
9. $x^2+(y-14)^2=34$
10. $(x-4)^2+y^2=\frac{16}{25}$ $(4, 0)$, $\frac{4}{5}$ unit
11. $\left(x+\frac{2}{3}\right)^2+\left(y-\frac{1}{2}\right)^2=\frac{8}{9}$
12. $x^2+y^2+8x-6y=0$ $(-4, 3)$, 5 units
13. $x^2+y^2+4x-8=0$ $(-2, 0)$, $2\sqrt{3}$ units

Application

AEROSPACE For Exercises 14 and 15, use the following information.
In order for a satellite to remain in a circular orbit above the same spot on Earth, the satellite must be 35,800 kilometers above the equator.

14. Write an equation for the orbit of the satellite. Use the center of Earth as the origin and 6400 kilometers for the radius of Earth. $x^2+y^2=42{,}200^2$
15. Draw a labeled sketch of Earth and the orbit to scale. See margin.

★ indicates increased difficulty

Practice and Apply

Homework Help

For Exercises	See Examples
16–29	1–3
30–48	4, 5

Extra Practice
See page 845.

Write an equation for each graph.

16.

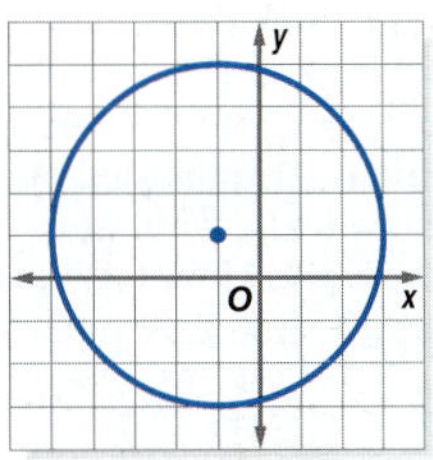

$(x+1)^2+(y-1)^2=16$

17.

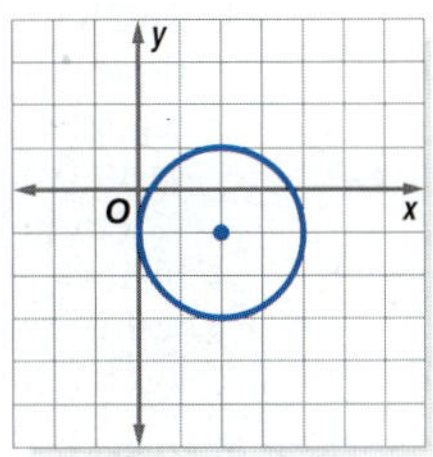

$(x-2)^2+(y+1)^2=4$

26–27. See margin.

Write an equation for the circle that satisfies each set of conditions.

18. center $(0, 3)$, radius 7 units $x^2+(y-3)^2=49$
19. center $(-8, 7)$, radius $\frac{1}{2}$ unit $(x+8)^2+(y-7)^2=\frac{1}{4}$
20. endpoints of a diameter at $(-5, 2)$ and $(3, 6)$ $(x+1)^2+(y-4)^2=20$
21. endpoints of a diameter at $(11, 18)$ and $(-13, -19)$ $(x+1)^2+\left(y+\frac{1}{2}\right)^2=\frac{1945}{4}$
22. center $(8, -9)$, passes through $(21, 22)$ $(x-8)^2+(y+9)^2=1130$
23. center $\left(-\sqrt{13}, 42\right)$, passes through the origin $\left(x+\sqrt{13}\right)^2+(y-42)^2=1777$
24. center at $(-8, -7)$, tangent to y-axis $(x+8)^2+(y+7)^2=64$
25. center at $(4, 2)$, tangent to x-axis $(x-4)^2+(y-2)^2=4$

★ 26. center in the first quadrant; tangent to $x=-3$, $x=5$, and the x-axis

★ 27. center in the second quadrant; tangent to $y=-1$, $y=9$, and the y-axis

www.algebra2.com/self_check_quiz

Answers

15.

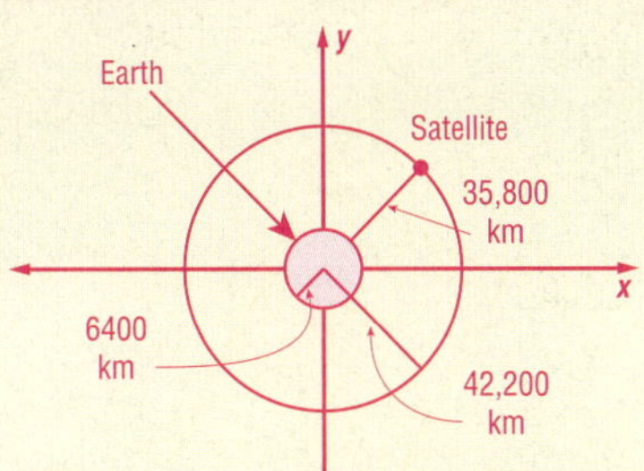

26. $(x-1)^2+(y-4)^2=16$
27. $(x+5)^2+(y-4)^2=25$

3 Practice/Apply

Study Notebook

Have students—

- add the definitions/examples of the vocabulary terms to their Vocabulary Builder worksheets for Chapter 8.
- include any other item(s) that they find helpful in mastering the skills in this lesson.

DAILY INTERVENTION

FIND THE ERROR Have students review the equation of a circle in standard form and notice that the right side of the equation is the square of the radius.

About the Exercises...

Organization by Objective

- **Equations of Circles:** 16–29
- **Graph Circles:** 30–48

Odd/Even Assignments

Exercises 16–27 and 30–47 are structured so that students practice the same concepts whether they are assigned odd or even problems.

Assignment Guide

Basic: 17–25 odd, 29–45 odd, 49–52, 57–71

Average: 17–47 odd, 49–52, 57–71 (optional: 53–56)

Advanced: 16–48 even, 49–65 (optional: 66–71)

All: Practice Quiz 1 (1–5)

Study Guide and Intervention, p. 467 (shown) and p. 468

Equations of Circles The equation of a circle with center (h, k) and radius r units is $(x - h)^2 + (y - k)^2 = r^2$.

Example **Write an equation for a circle if the endpoints of a diameter are at (−4, 5) and (6, −3).**

Use the midpoint formula to find the center of the circle.

$(h, k) = \left(\frac{x_1 + x_2}{2}, \frac{y_1 + y_2}{2}\right)$ Midpoint formula

$= \left(\frac{-4 + 6}{2}, \frac{5 + (-3)}{2}\right)$ $(x_1, y_1) = (-4, 5), (x_2, y_2) = (6, -3)$

$= \left(\frac{2}{2}, \frac{2}{2}\right)$ or $(1, 1)$ Simplify.

Use the coordinates of the center and one endpoint of the diameter to find the radius.

$r = \sqrt{(x_2 - x_1)^2 + (y_2 - y_1)^2}$ Distance formula

$r = \sqrt{(-4 - 1)^2 + (5 - 1)^2}$ $(x_1, y_1) = (1, 1), (x_2, y_2) = (-4, 5)$

$= \sqrt{(-5)^2 + 4^2} = \sqrt{41}$ Simplify.

The radius of the circle is $\sqrt{41}$, so $r^2 = 41$.

An equation of the circle is $(x - 1)^2 + (y - 1)^2 = 41$.

Exercises

Write an equation for the circle that satisfies each set of conditions.

1. center $(8, -3)$, radius 6 $(x - 8)^2 + (y + 3)^2 = 36$
2. center $(5, -6)$, radius 4 $(x - 5)^2 + (y + 6)^2 = 16$
3. center $(-5, 2)$, passes through $(-9, 6)$ $(x + 5)^2 + (y - 2)^2 = 32$
4. endpoints of a diameter at $(6, 6)$ and $(10, 12)$ $(x - 8)^2 + (y - 9)^2 = 13$
5. center $(3, 6)$, tangent to the x-axis $(x - 3)^2 + (y - 6)^2 = 36$
6. center $(-4, -7)$, tangent to $x = 2$ $(x + 4)^2 + (y + 7)^2 = 36$
7. center at $(-2, 8)$, tangent to $y = -4$ $(x + 2)^2 + (y - 8)^2 = 144$
8. center $(7, 7)$, passes through $(12, 9)$ $(x - 7)^2 + (y - 7)^2 = 29$
9. endpoints of a diameter are $(-4, -2)$ and $(8, 4)$ $(x - 2)^2 + (y - 1)^2 = 45$
10. endpoints of a diameter are $(-4, 3)$ and $(6, -8)$ $(x - 1)^2 + (y + 2.5)^2 = 55.25$

Skills Practice, p. 469 and Practice, p. 470 (shown)

Write an equation for the circle that satisfies each set of conditions.

1. center $(-4, 2)$, radius 8 units $(x + 4)^2 + (y - 2)^2 = 64$
2. center $(0, 0)$, radius 4 units $x^2 + y^2 = 16$
3. center $\left(-\frac{1}{4}, -\sqrt{3}\right)$, radius $5\sqrt{2}$ units $\left(x + \frac{1}{4}\right)^2 + (y + \sqrt{3})^2 = 50$
4. center $(2.5, 4.2)$, radius 0.9 unit $(x - 2.5)^2 + (y - 4.2)^2 = 0.81$
5. endpoints of a diameter at $(-2, -9)$ and $(0, -5)$ $(x + 1)^2 + (y + 7)^2 = 5$
6. center at $(-9, -12)$, passes through $(-4, -5)$ $(x + 9)^2 + (y + 12)^2 = 74$
7. center at $(-6, 5)$, tangent to x-axis $(x + 6)^2 + (y - 5)^2 = 25$

Find the center and radius of the circle with the given equation. Then graph the circle.

8. $(x + 3)^2 + y^2 = 16$ $(-3, 0)$, 4 units

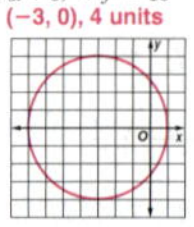

9. $3x^2 + 3y^2 = 12$ $(0, 0)$, 2 units

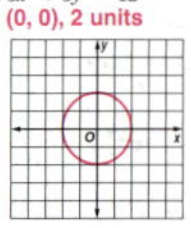

10. $x^2 + y^2 + 2x + 6y = 26$ $(-1, -3)$, 6 units

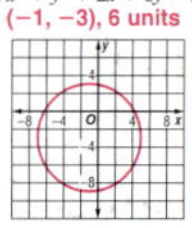

11. $(x - 1)^2 + y^2 + 4y = 12$ $(1, -2)$, 4 units

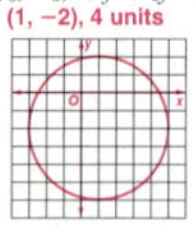

12. $x^2 - 6x + y^2 = 0$ $(3, 0)$, 3 units

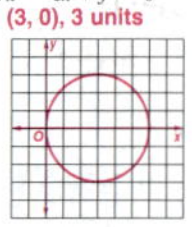

13. $x^2 + y^2 + 2x + 6y = -1$ $(-1, -3)$, 3 units

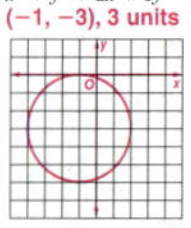

WEATHER For Exercises 14 and 15, use the following information.

On average, the circular eye of a hurricane is about 15 miles in diameter. Gale winds can affect an area up to 300 miles from the storm's center. In 1992, Hurricane Andrew devastated southern Florida. A satellite photo of Andrew's landfall showed the center of its eye on one coordinate system could be approximated by the point (80, 26).

14. Write an equation to represent a possible boundary of Andrew's eye. $(x - 80)^2 + (y - 26)^2 = 56.25$
15. Write an equation to represent a possible boundary of the area affected by gale winds. $(x - 80)^2 + (y - 26)^2 = 90{,}000$

Reading to Learn Mathematics, p. 471 ELL

Pre-Activity **Why are circles important in air traffic control?**

Read the introduction to Lesson 8-3 at the top of page 426 in your textbook.

A large home improvement chain is planning to enter a new metropolitan area and needs to select locations for its stores. Market research has shown that potential customers are willing to travel up to 12 miles to shop at one of their stores. How can circles help the managers decide where to place their store?

Sample answer: A store will draw customers who live inside a circle with center at the store and a radius of 12 miles. The management should select locations for which as many people as possible live within a circle of radius 12 miles around one of the stores.

Reading the Lesson

1. a. Write the equation of the circle with center (h, k) and radius r. $(x - h)^2 + (y - k)^2 = r^2$
   b. Write the equation of the circle with center $(4, -3)$ and radius 5. $(x - 4)^2 + (y + 3)^2 = 25$
   c. The circle with equation $(x + 8)^2 + y^2 = 121$ has center $\underline{(-8, 0)}$ and radius $\underline{11}$.
   d. The circle with equation $(x - 10)^2 + (y + 10)^2 = 1$ has center $\underline{(10, -10)}$ and radius $\underline{1}$.
2. a. In order to find center and radius of the circle with equation $x^2 + y^2 + 4x - 6y - 3 = 0$, it is necessary to $\underline{\text{complete the square}}$. Fill in the missing parts of this process.

$$x^2 + y^2 + 4x - 6y - 3 = 0$$
$$x^2 + y^2 + 4x - 6y = \underline{3}$$
$$x^2 + 4x + \underline{4} + y^2 - 6y + \underline{9} = \underline{3} + \underline{4} + \underline{9}$$
$$(x + \underline{2})^2 + (y - \underline{3})^2 = \underline{16}$$

   b. This circle has radius 4 and center at $\underline{(-2, 3)}$.

Helping You Remember

3. How can the distance formula help you to remember the equation of a circle? Sample answer: Write the distance formula. Replace (x_1, y_1) with (h, k) and (x_2, y_2) with (x, y). Replace d with r. Square both sides. Now you have the equation of a circle.

WebQuest

The epicenter of an earthquake can be located by using the equation of a circle. Visit www.algebra2.com/webquest to continue work on your WebQuest project.

28. **LANDSCAPING** The design of a garden is shown at the right. A pond is to be built in the center region. What is the equation of the largest circular pond centered at the origin that would fit within the walkways? $x^2 + y^2 = 18$

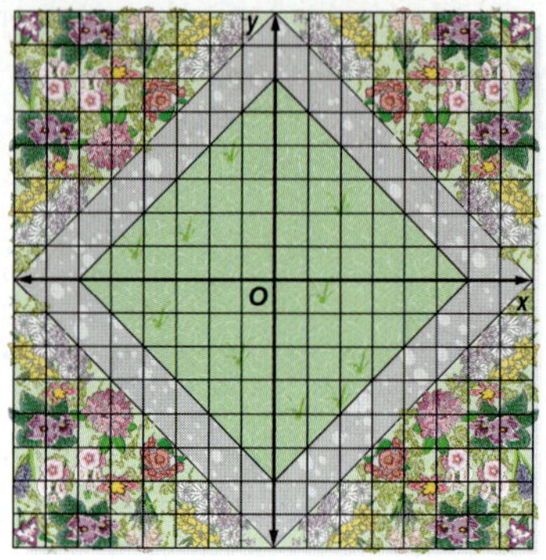

29. **EARTHQUAKES** The University of Southern California is located about 2.5 miles west and about 2.8 miles south of downtown Los Angeles. Suppose an earthquake occurs with its epicenter about 40 miles from the university. Assume that the origin of a coordinate plane is located at the center of downtown Los Angeles. Write an equation for the set of points that could be the epicenter of the earthquake. $(x + 2.5)^2 + (y + 2.8)^2 = 1600$

Find the center and radius of the circle with the given equation. Then graph the circle. 30–47. See pp. 469A–469J for graphs.

30. $x^2 + (y + 2)^2 = 4$ $(0, -2)$, 2 units
31. $x^2 + y^2 = 144$ $(0, 0)$, 12 units
32. $(x - 3)^2 + (y - 1)^2 = 25$ $(3, 1)$, 5 units
33. $(x + 3)^2 + (y + 7)^2 = 81$
34. $(x - 3)^2 + y^2 = 16$ $(3, 0)$, 4 units
35. $(x - 3)^2 + (y + 7)^2 = 50$
36. $(x + \sqrt{5})^2 + y^2 - 8y = 9$
37. $x^2 + (y - \sqrt{3})^2 + 4x = 25$
38. $x^2 + y^2 + 6y = -50 - 14x$
39. $x^2 + y^2 - 6y - 16 = 0$ $(0, 3)$, 5 units
40. $x^2 + y^2 + 2x - 10 = 0$
41. $x^2 + y^2 - 18x - 18y + 53 = 0$
42. $x^2 + y^2 + 9x - 8y + 4 = 0$
43. $x^2 + y^2 - 3x + 8y = 20$
44. $x^2 - 12x + 84 = -y^2 + 16y$
45. $x^2 + y^2 + 2x + 4y = 9$
46. ★ $3x^2 + 3y^2 + 12x - 6y + 9 = 0$ $(-2, 1)$, $\sqrt{2}$ unit
47. ★ $4x^2 + 4y^2 + 36y + 5 = 0$

33. $(-3, -7)$, 9 units
35. $(3, -7)$, $5\sqrt{2}$ units
36. $(-\sqrt{5}, 4)$, 5 units
37. $(-2, \sqrt{3})$, $\sqrt{29}$ units
38. $(-7, -3)$, $2\sqrt{2}$ units
40. $(-1, 0)$, $\sqrt{11}$ units
41. $(9, 9)$, $\sqrt{109}$ units
42. $\left(-\frac{9}{2}, 4\right)$, $\frac{\sqrt{129}}{2}$ units
43. $\left(\frac{3}{2}, -4\right)$, $\frac{3\sqrt{17}}{2}$ units
44. $(6, 8)$, 4 units
45. $(-1, -2)$, $\sqrt{14}$ units
47. $\left(0, -\frac{9}{2}\right)$, $\sqrt{19}$ units

48. **RADIO** The diagram at the right shows the relative locations of some cities in North Dakota. The x-axis represents Interstate 94. The scale is 1 unit = 30 miles. While driving west on the highway, Doralina is listening to a radio station in Minot. She estimates the range of the signal to be 120 miles. How far west of Bismarck will she be able to pick up the signal? about 109 mi

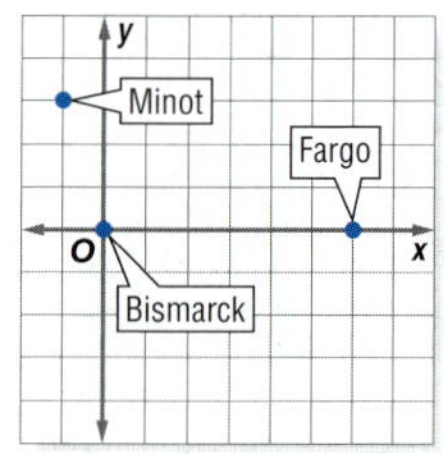

49. **CRITICAL THINKING** A circle has its center on the line with equation $y = 2x$. The circle passes through $(1, -3)$ and has a radius of $\sqrt{5}$ units. Write an equation of the circle. $(x + 1)^2 + (y + 2)^2 = 5$

50. **WRITING IN MATH** Answer the question that was posed at the beginning of the lesson. See margin.

Why are circles important in air traffic control?

Include the following in your answer:

- an equation of the circle that determines the boundary of the region where planes can be detected if the range of the radar is 50 miles and the radar is at the origin, and
- how an air traffic controller's job would be different for a region whose boundary is modeled by $x^2 + y^2 = 4900$ instead of $x^2 + y^2 = 1600$.

Enrichment, p. 472

Tangents to Circles

A line that intersects a circle in exactly one point is a **tangent** to the circle. In the diagram, line ℓ is tangent to the circle with equation $x^2 + y^2 = 25$ at the point whose coordinates are (3, 4).

A line is tangent to a circle at a point P on the circle if and only if the line is perpendicular to the radius from the center of the circle to point P. This fact enables you to find an equation of the tangent to a circle at a point P if you know an equation for the circle and the coordinates of P.

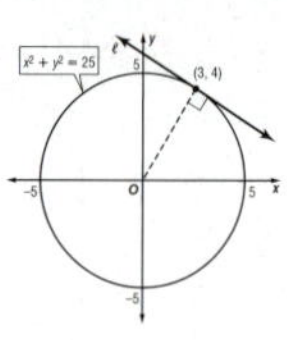

Use the diagram above to solve each problem.

1. What is [illegible]ope of the radius [illegible] point with coordinates (3, 4)? What is [illegible]

Standardized Test Practice

51. Find the radius of the circle with equation $x^2 + y^2 + 8x + 8y + 28 = 0$. **A**

Ⓐ 2 Ⓑ 4 Ⓒ 8 Ⓓ 28

52. Find the center of the circle with equation $x^2 + y^2 - 10x + 6y + 27 = 0$. **D**

Ⓐ (−10, 6) Ⓑ (1, 1) Ⓒ (10, −6) Ⓓ (5, −3)

Graphing Calculator

CIRCLES **For Exercises 53–56, use the following information.**
Since a circle is not the graph of a function, you cannot enter its equation directly into a graphing calculator. Instead, you must solve the equation for y. The result will contain a ± symbol, so you will have two functions.

53. Solve $(x + 3)^2 + y^2 = 16$ for y. $y = \pm\sqrt{16 - (x + 3)^2}$

54. What two functions should you enter to graph the given equation?

55. Graph $(x + 3)^2 + y^2 = 16$ on a graphing calculator. **See margin.**

56. Solve $(x + 3)^2 + y^2 = 16$ for x. What parts of the circle do the two expressions for x represent? **See margin.**

54. $y = \sqrt{16 - (x + 3)^2}$, $y = -\sqrt{16 - (x + 3)^2}$

Maintain Your Skills

Mixed Review

Identify the coordinates of the vertex and focus, the equations of the axis of symmetry and directrix, and the direction of opening of the parabola with the given equation. Then find the length of the latus rectum and graph the parabola. *(Lesson 8-2)* **57–59. See pp. 469A–469J.**

57. $x = -3y^2 + 1$ 58. $y + 2 = -(x - 3)^2$ 59. $y = x^2 + 4x$

Find the midpoint of the line segment with endpoints at the given coordinates. *(Lesson 8-1)*

60. (5, −7), (3, −1) **(4, −4)** 61. (2, −9), (−4, 5) **(−1, −2)** 62. (8, 0), (−5, 12) $\left(\frac{3}{2}, 6\right)$

Find all of the rational zeros for each function. *(Lesson 7-5)*

63. $f(x) = x^3 + 5x^2 + 2x - 8$ **−4, −2, 1** 64. $g(x) = 2x^3 - 9x^2 + 7x + 6$ $-\frac{1}{2}$, **2, 3**

65. **PHOTOGRAPHY** The perimeter of a rectangular picture is 86 inches. Twice the width exceeds the length by 2 inches. What are the dimensions of the picture? *(Lesson 3-2)* **28 in. by 15 in.**

Getting Ready for the Next Lesson

PREREQUISITE SKILL **Solve each equation. Assume that all variables are positive.** *(To review **solving quadratic equations**, see Lesson 6-4.)*

66. $c^2 = 13^2 - 5^2$ **12** 67. $c^2 = 10^2 - 8^2$ **6** 68. $(\sqrt{7})^2 = a^2 - 3^2$ **4**

69. $24^2 = a^2 - 7^2$ **25** 70. $4^2 = 6^2 - b^2$ $2\sqrt{5}$ 71. $(2\sqrt{14})^2 = 8^2 - b^2$ $2\sqrt{2}$

Practice Quiz 1

Lessons 8-1 through 8-3

Find the distance between each pair of points with the given coordinates. *(Lesson 8-1)*

1. (9, 5), (4, −7) **13 units** 2. (0, −5), (10, −3) $2\sqrt{26}$ **units**

Identify the coordinates of the vertex and focus, the equations of the axis of symmetry and directrix, and the direction of opening of the parabola with the given equation. Then find the length of the latus rectum and graph the parabola. *(Lesson 8-2)* **3–4. See pp. 469A–469J.**

3. $y^2 = 6x$ 4. $y = x^2 + 8x + 20$

5. Find the center and radius of the circle with equation $x^2 + (y - 4)^2 = 49$. Then graph the circle. *(Lesson 8-3)* **(0, 4), 7 units; see pp. 469A–469J for graph.**

4 Assess

Open-Ended Assessment

Speaking Have students explain how to tell from a given equation of a circle how the equation of a circle with its center at the origin and the same radius can be translated to give the graph of the given equation.

Getting Ready for Lesson 8-4

PREREQUISITE SKILL In the process of analyzing and simplifying equations of ellipses in Lesson 8-4, students will solve quadratic equations. Exercises 66–71 should be used to determine your students' familiarity with solving quadratic equations.

Assessment Options

Practice Quiz 1 The quiz provides students with a brief review of the concepts and skills in Lessons 8-1 through 8-3. Lesson numbers are given to the right of exercises or instruction lines so students can review concepts not yet mastered.

Answers

55.

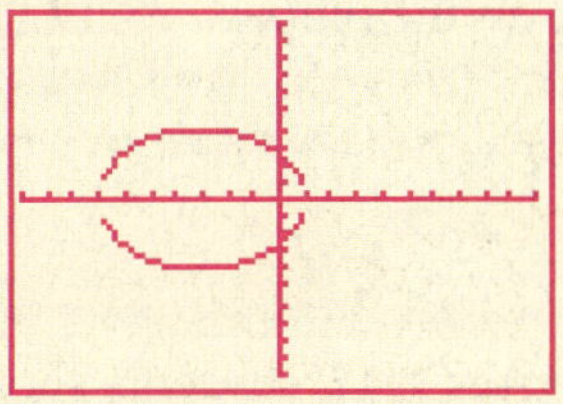

[−10, 10] scl:1 by [−10, 10] scl:1

56. $x = -3 \pm \sqrt{16 - y^2}$; The equations with the + symbol and − symbol represent the right and left halves of the circle, respectively.

Answer

50. A circle can be used to represent the limit at which planes can be detected by radar. Answers should include the following.

- $x^2 + y^2 = 2500$
- The region whose boundary is modeled by $x^2 + y^2 = 4900$ is larger, so there would be more planes to track.

Algebra Activity

A Preview of Lesson 8-4

Getting Started

Objective To derive an understanding of an ellipse as the set of points for which the sum of the distances from two fixed points is constant.

Materials

two thumbtacks	cardboard
string	pencil
grid paper	ruler

Teach

- It may be easier to manipulate the string and pencil if students work in pairs. Make sure that each member of the pair has the chance to draw an ellipse.
- Ask students to stop at one point in the curve they are drawing and ask what the total distance is from that point to one thumbtack plus the distance from the same point to the other thumbtack. **the length of the string minus the distance between the thumbtacks**
- Ask students what the total distance to the tacks is for any point on the ellipse. **the length of the string minus the distance between the thumbtacks**

Assess

In **Exercises 1–12**, students should

- be able to see what changes result from varying the positions of the thumbtacks and the length of the string.
- be able to predict what changes will result from varying the positions of the foci.

Study Notebook

You may wish to have students summarize this activity and what they learned from it.

Algebra Activity

A Preview of Lesson 8-4

Investigating Ellipses

Follow the steps below to construct another type of conic section.

Step 1 Place two thumbtacks in a piece of cardboard, about 1 foot apart.

Step 2 Tie a knot in a piece of string and loop it around the thumbtacks.

Step 3 Place your pencil in the string. Keep the string tight and draw a curve.

Step 4 Continue drawing until you return to your starting point.

The curve you have drawn is called an **ellipse**. The points where the thumbtacks are located are called the **foci** of the ellipse. *Foci* is the plural of *focus*.

Model and Analyze

Place a large piece of grid paper on a piece of cardboard. **1. See students' work.**

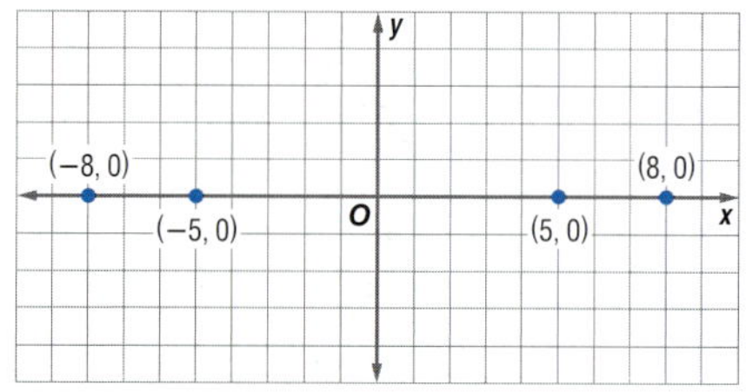

1. Place the thumbtacks at (8, 0) and (−8, 0). Choose a string long enough to loop around both thumbtacks. Draw an ellipse.

2. Repeat Exercise 1, but place the thumbtacks at (5, 0) and (−5, 0). Use the same loop of string and draw an ellipse. How does this ellipse compare to the one in Exercise 1? **See students' work; the ellipse is more circular.**

Place the thumbtacks at each set of points and draw an ellipse.
You may change the length of the loop of string if you like. **3–5. See students' work.**

3. (12, 0), (−12, 0) **4.** (2, 0), (−2, 0) **5.** (14, 4), (−10, 4)

Make a Conjecture

In Exercises 6–10, describe what happens to the shape of an ellipse when each change is made.

6. The thumbtacks are moved closer together. **The ellipse becomes more circular.**

7. The thumbtacks are moved farther apart. **The ellipse becomes more elongated.**

8. The length of the loop of string is increased. **The ellipse becomes larger.**

9. The thumbtacks are arranged vertically. **See pp. 469A–469J.**

10. One thumbtack is removed, and the string is looped around the remaining thumbtack. **The ellipse is a circle.**

11. Pick a point on one of the ellipses you have drawn. Use a ruler to measure the distances from that point to the points where the thumbtacks were located. Add the distances. Repeat for other points on the same ellipse. What relationship do you notice? **The sum of the distances is constant.**

12. Could this activity be done with a rubber band instead of a piece of string? Explain. **See pp. 469A–469J.**

Resource Manager

Teaching Algebra with Manipulatives

- p. 1 (master for grid paper)
- p. 24 (master for rulers)
- p. 266 (student recording sheet)

Glencoe Mathematics Classroom Manipulative Kit

- rulers

8-4 Ellipses

What You'll Learn

- Write equations of ellipses.
- Graph ellipses.

Vocabulary

- ellipse
- foci
- major axis
- minor axis
- center

Why are ellipses important in the study of the solar system?

Fascination with the sky has caused people to wonder, observe, and make conjectures about the planets since the beginning of history. Since the early 1600s, the orbits of the planets have been known to be ellipses with the Sun at a focus.

EQUATIONS OF ELLIPSES As you discovered in the Algebra Activity on page 432, an **ellipse** is the set of all points in a plane such that the sum of the distances from two fixed points is constant. The two fixed points are called the **foci** of the ellipse.

The ellipse at the right has foci at (5, 0) and (−5, 0). The distances from either of the x-intercepts to the foci are 2 units and 12 units, so the sum of the distances from any point with coordinates (x, y) on the ellipse to the foci is 14 units.

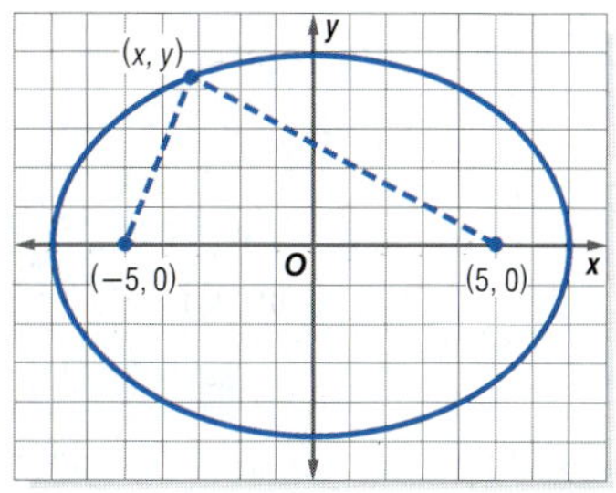

You can use the Distance Formula and the definition of an ellipse to find an equation of this ellipse.

$$\underbrace{\text{The distance between } (x, y) \text{ and } (-5, 0)} + \underbrace{\text{the distance between } (x, y) \text{ and } (5, 0)} = 14.$$

$$\sqrt{(x+5)^2 + y^2} + \sqrt{(x-5)^2 + y^2} = 14$$

$\sqrt{(x+5)^2 + y^2} = 14 - \sqrt{(x-5)^2 + y^2}$	Isolate the radicals.
$(x+5)^2 + y^2 = 196 - 28\sqrt{(x-5)^2 + y^2} + (x-5)^2 + y^2$	Square each side.
$x^2 + 10x + 25 + y^2 = 196 - 28\sqrt{(x-5)^2 + y^2} + x^2 - 10x + 25 + y^2$	
$20x - 196 = -28\sqrt{(x-5)^2 + y^2}$	Simplify.
$5x - 49 = -7\sqrt{(x-5)^2 + y^2}$	Divide each side by 4.
$25x^2 - 490x + 2401 = 49[(x-5)^2 + y^2]$	Square each side.
$25x^2 - 490x + 2401 = 49x^2 - 490x + 1225 + 49y^2$	Distributive Property
$-24x^2 - 49y^2 = -1176$	Simplify.
$\frac{x^2}{49} + \frac{y^2}{24} = 1$	Divide each side by −1176.

An equation for this ellipse is $\frac{x^2}{49} + \frac{y^2}{24} = 1$.

8-4 Lesson Notes

1 Focus

5-Minute Check Transparency 8-4 Use as a quiz or review of Lesson 8-3.

Mathematical Background notes are available for this lesson on p. 410C.

Why are ellipses important in the study of the solar system?

Ask students:

- What is the solar system? **The Sun and the group of objects orbiting around the Sun.**
- Does the Sun travel around Earth or does Earth travel around the Sun? **Earth travels around the Sun.**
- Describe some of the earlier conjectures about the Sun and Earth. **Sample answers: The Earth was the center of the solar system; the Earth was flat.**

Resource Manager

Workbook and Reproducible Masters

Chapter 8 Resource Masters

- Study Guide and Intervention, pp. 473–474
- Skills Practice, p. 475
- Practice, p. 476
- Reading to Learn Mathematics, p. 477
- Enrichment, p. 478
- Assessment, pp. 511, 513

Teaching Algebra With Manipulatives Masters, p. 267

Transparencies

5-Minute Check Transparency 8-4
Answer Key Transparencies

Technology

Interactive Chalkboard

2 Teach

EQUATIONS OF ELLIPSES

In-Class Example Power Point®

1 Write an equation for the ellipse shown.

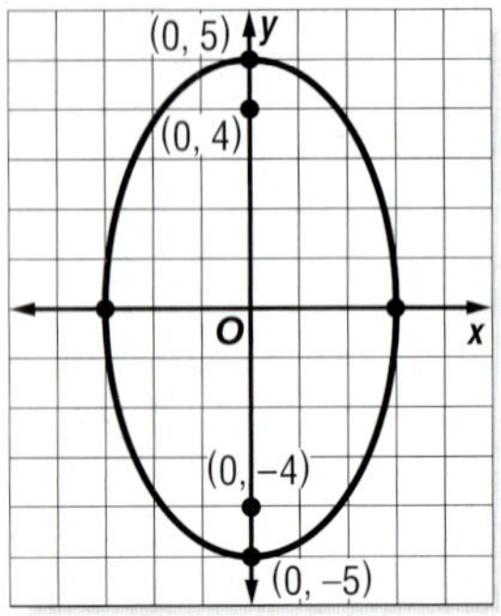

$\frac{y^2}{25} + \frac{x^2}{9} = 1$

Teaching Tip Lead a discussion to clarify how ellipses are different from circles and parabolas in their equations, foci, and other characteristics.

TEACHING TIP

Point out to students that *a*, not *c*, is the length of the hypotenuse.

Study Tip

Vertices of Ellipses

The endpoints of each axis are called the *vertices* of the ellipse.

Every ellipse has two axes of symmetry. The points at which the ellipse intersects its axes of symmetry determine two segments with endpoints on the ellipse called the **major axis** and the **minor axis**. The axes intersect at the **center** of the ellipse. The foci of an ellipse always lie on the major axis.

Study the ellipse at the right. The sum of the distances from the foci to any point on the ellipse is the same as the length of the major axis, or $2a$ units. The distance from the center to either focus is c units. By the Pythagorean Theorem, a, b, and c are related by the equation $c^2 = a^2 - b^2$. Notice that the x- and y-intercepts, $(\pm a, 0)$ and $(0, \pm b)$, satisfy the quadratic equation $\frac{x^2}{a^2} + \frac{y^2}{b^2} = 1$. This is the standard form of the equation of an ellipse with its center at the origin and a horizontal major axis.

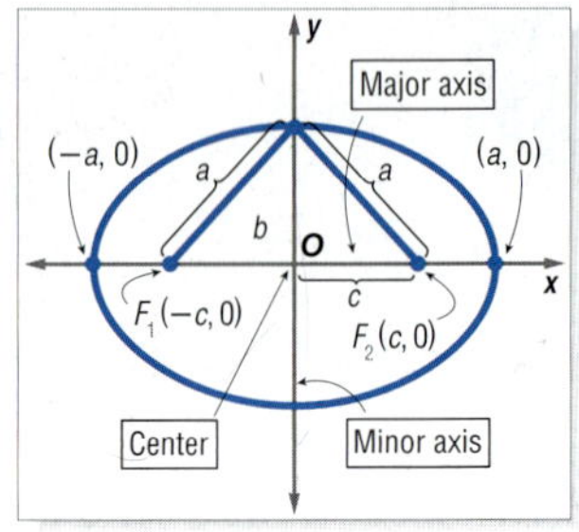

Key Concept	***Equations of Ellipses with Centers at the Origin***	
Standard Form of Equation	$\frac{x^2}{a^2} + \frac{y^2}{b^2} = 1$	$\frac{y^2}{a^2} + \frac{x^2}{b^2} = 1$
Direction of Major Axis	horizontal	vertical
Foci	$(c, 0), (-c, 0)$	$(0, c), (0, -c)$
Length of Major Axis	$2a$ units	$2a$ units
Length of Minor Axis	$2b$ units	$2b$ units

In either case, $a^2 \geq b^2$ and $c^2 = a^2 - b^2$. You can determine if the foci are on the x-axis or the y-axis by looking at the equation. If the x^2 term has the greater denominator, the foci are on the x-axis. If the y^2 term has the greater denominator, the foci are on the y-axis.

Example 1 *Write an Equation for a Graph*

Write an equation for the ellipse shown at the right.

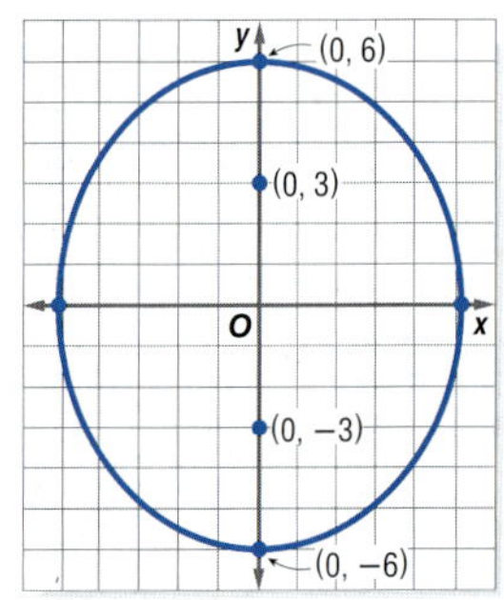

In order to write the equation for the ellipse, we need to find the values of a and b for the ellipse. We know that the length of the major axis of any ellipse is $2a$ units. In this ellipse, the length of the major axis is the distance between the points at (0, 6) and (0, −6). This distance is 12 units.

$2a = 12$ Length of major axis = 12

$a = 6$ Divide each side by 2.

The foci are located at (0, 3) and (0, −3), so $c = 3$. We can use the relationship between a, b, and c to determine the value of b.

$c^2 = a^2 - b^2$ Equation relating a, b, and c

$9 = 36 - b^2$ $c = 3$ and $a = 6$

$b^2 = 27$ Solve for b^2.

Since the major axis is vertical, substitute 36 for a^2 and 27 for b^2 in the form $\frac{y^2}{a^2} + \frac{x^2}{b^2} = 1$. An equation of the ellipse is $\frac{y^2}{36} + \frac{x^2}{27} = 1$.

DAILY INTERVENTION **Differentiated Instruction**

Auditory/Musical Have students describe some symbols in musical notation that use circles or ellipses.

434 Chapter 8 Conic Sections

Example 2 Write an Equation Given the Lengths of the Axes

MUSEUMS **In an ellipse, sound or light coming from one focus is reflected to the other focus. In a whispering gallery, a person can hear another person whisper from across the room if the two people are standing at the foci. The whispering gallery at the Museum of Science and Industry in Chicago has an elliptical cross section that is 13 feet 6 inches by 47 feet 4 inches.**

a. Write an equation to model this ellipse. Assume that the center is at the origin and the major axis is horizontal.

The length of the major axis is $47\frac{1}{3}$ or $\frac{142}{3}$ feet.

$2a = \frac{142}{3}$ Length of major axis $= \frac{142}{3}$

$a = \frac{71}{3}$ Divide each side by 2.

The length of the minor axis is $13\frac{1}{2}$ or $\frac{27}{2}$ feet.

$2b = \frac{27}{2}$ Length of minor axis $= \frac{27}{2}$

$b = \frac{27}{4}$ Divide each side by 2.

Substitute $a = \frac{71}{3}$ and $b = \frac{27}{4}$ into the form $\frac{x^2}{a^2} + \frac{y^2}{b^2} = 1$. An equation of the ellipse is $\frac{x^2}{\left(\frac{71}{3}\right)^2} + \frac{y^2}{\left(\frac{27}{4}\right)^2} = 1$.

b. How far apart are the points at which two people should stand to hear each other whisper?

People should stand at the two foci of the ellipse. The distance between the foci is $2c$ units.

$c^2 = a^2 - b^2$ Equation relating a, b, and c

$c = \sqrt{a^2 - b^2}$ Take the square root of each side.

$2c = 2\sqrt{a^2 - b^2}$ Multiply each side by 2.

$2c = 2\sqrt{\left(\frac{71}{3}\right)^2 - \left(\frac{27}{4}\right)^2}$ Substitute $a = \frac{71}{3}$ and $b = \frac{27}{4}$.

$2c \approx 45.37$ Use a calculator.

The points where two people should stand to hear each other whisper are about 45.37 feet or 45 feet 4 inches apart.

More About...

Museums

The whispering gallery at Chicago's Museum of Science and Industry has a parabolic dish at each focus to help collect sound.

Source: www.msichicago.org

GRAPH ELLIPSES As with circles, you can use completing the square, symmetry, and transformations to help graph ellipses. An ellipse with its center at the origin is represented by an equation of the form $\frac{x^2}{a^2} + \frac{y^2}{b^2} = 1$ or $\frac{y^2}{a^2} + \frac{x^2}{b^2} = 1$. The ellipse could be translated h units to the right and k units up. This would move the center to the point (h, k). Such a move would be equivalent to replacing x with $x - h$ and replacing y with $y - k$.

Key Concept: Equations of Ellipses with Centers at (h, k)

Standard Form of Equation	$\frac{(x-h)^2}{a^2} + \frac{(y-k)^2}{b^2} = 1$	$\frac{(y-k)^2}{a^2} + \frac{(x-h)^2}{b^2} = 1$
Direction of Major Axis	horizontal	vertical
Foci	$(h \pm c, k)$	$(h, k \pm c)$

www.algebra2.com/extra_examples

In-Class Example

Power Point®

2 SOUND A listener is standing in an elliptical room 150 feet wide and 320 feet long. When a speaker stands at one focus and whispers, the best place for the listener to stand is at the other focus.

a. Write an equation to model this ellipse, assuming the major axis is horizontal and the center is at the origin. $\frac{x^2}{160^2} + \frac{y^2}{75^2} = 1$

b. How far apart should the speaker and the listener be in this room? approximately 282.7 feet

Teaching Tip Suggest that students make a rough sketch of the situation in these problems.

DAILY INTERVENTION — Unlocking Misconceptions

- **Symmetry in Ellipses** Make sure that students understand that an ellipse has two axes of symmetry, the major axis and the minor axis.
- **Identifying Axes** Ask students how they can tell which is the major and which is the minor axis. **The major axis is longer.**

GRAPH ELLIPSES

In-Class Examples

Power Point®

3 Find the coordinates of the center and foci and the lengths of the major and minor axes of the ellipse with equation $\frac{x^2}{36} + \frac{y^2}{9} = 1$. Then graph the ellipse. **center: (0, 0); foci: $(3\sqrt{3}, 0)$, $(-3\sqrt{3}, 0)$; major axis: 12; minor axis: 6**

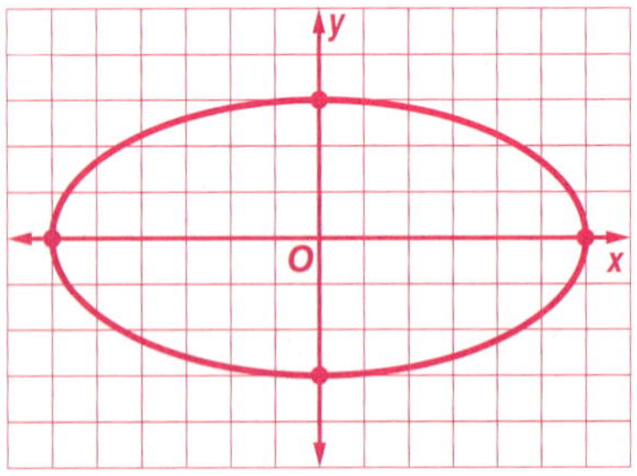

4 Find the coordinates of the center and foci and the lengths of the major and minor axes of the ellipse with equation $x^2 + 4y^2 - 6x - 16y - 11 = 0$. Then graph the ellipse. **center: (3, 2); foci: $(3\sqrt{3} + 3, 2)$, $(-3\sqrt{3} + 3, 2)$; major axis: 12; minor axis: 6**

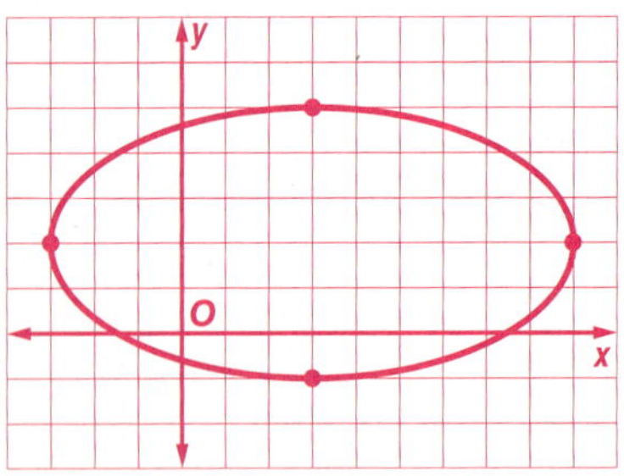

Example 3 Graph an Equation in Standard Form

Find the coordinates of the center and foci and the lengths of the major and minor axes of the ellipse with equation $\frac{x^2}{16} + \frac{y^2}{4} = 1$. Then graph the ellipse.

The center of this ellipse is at (0, 0).

Since $a^2 = 16$, $a = 4$. Since $b^2 = 4$, $b = 2$.

The length of the major axis is 2(4) or 8 units, and the length of the minor axis is 2(2) or 4 units. Since the x^2 term has the greater denominator, the major axis is horizontal.

$c^2 = a^2 - b^2$ Equation relating a, b, and c

$c^2 = 4^2 - 2^2$ or 12 $a = 4, b = 2$

$c = \sqrt{12}$ or $2\sqrt{3}$ Take the square root of each side.

The foci are at $(2\sqrt{3}, 0)$ and $(-2\sqrt{3}, 0)$.

You can use a calculator to find some approximate nonnegative values for x and y that satisfy the equation. Since the ellipse is centered at the origin, it is symmetric about the y-axis. Therefore, the points at (−4, 0), (−3, 1.3), (−2, 1.7), and (−1, 1.9) lie on the graph.

x	y
0	2.0
1	1.9
2	1.7
3	1.3
4	0.0

The ellipse is also symmetric about the x-axis, so the points at (−3, −1.3), (−2, −1.7), (−1, −1.9), (0, −2), (1, −1.9), (2, −1.7), and (3, −1.3) lie on the graph.

Graph the intercepts, (−4, 0), (4, 0), (0, 2), and (0, −2), and draw the ellipse that passes through them and the other points.

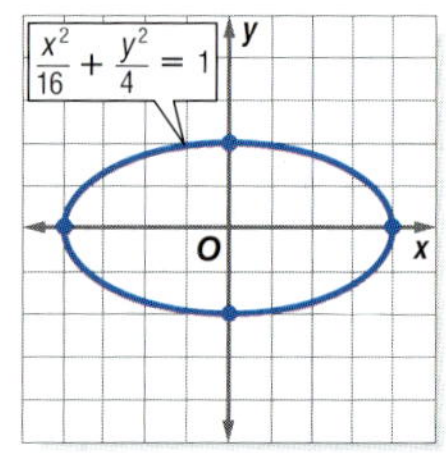

Study Tip

Graphing Calculator

You can graph an ellipse on a graphing calculator by first solving for y. Then graph the two equations that result on the same screen.

If you are given an equation of an ellipse that is not in standard form, write it in standard form first. This will make graphing the ellipse easier.

Example 4 Graph an Equation Not in Standard Form

Find the coordinates of the center and foci and the lengths of the major and minor axes of the ellipse with equation $x^2 + 4y^2 + 4x - 24y + 24 = 0$. Then graph the ellipse.

Complete the square for each variable to write this equation in standard form.

$x^2 + 4y^2 + 4x - 24y + 24 = 0$ Original equation

$(x^2 + 4x + \blacksquare) + 4(y^2 - 6y + \blacksquare) = -24 + \blacksquare + 4(\blacksquare)$ Complete the squares.

$(x^2 + 4x + 4) + 4(y^2 - 6y + 9) = -24 + 4 + 4(9)$ $\left(\frac{4}{2}\right)^2 = 4, \left(\frac{-6}{2}\right)^2 = 9$

$(x + 2)^2 + 4(y - 3)^2 = 16$ Write the trinomials as perfect squares.

$\frac{(x + 2)^2}{16} + \frac{(y - 3)^2}{4} = 1$ Divide each side by 16.

436 Chapter 8 Conic Sections

The graph of this ellipse is the graph from Example 3 translated 2 units to the left and up 3 units. The center is at $(-2, 3)$ and the foci are at $(-2 + 2\sqrt{3}, 0)$ and $(-2 - 2\sqrt{3}, 0)$. The length of the major axis is still 8 units, and the length of the minor axis is still 4 units.

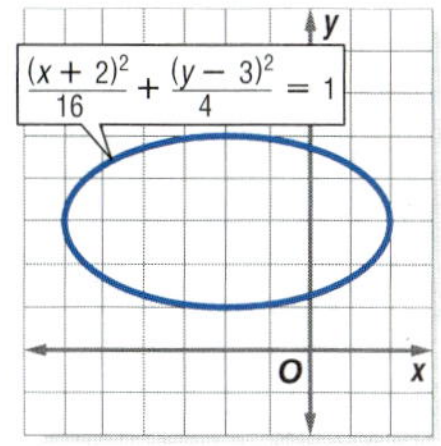

You can use a circle to locate the foci on the graph of a given ellipse.

Algebra Activity

Locating Foci

Step 1 Graph an ellipse so that its center is at the origin. Let the endpoints of the major axis be at $(-9, 0)$ and $(9, 0)$, and let the endpoints of the minor axis be at $(0, -5)$ and $(0, 5)$.

Step 2 Use a compass to draw a circle with center at $(0, 0)$ and radius 9 units.

Step 3 Draw the line with equation $y = 5$ and mark the points at which the line intersects the circle.

Step 4 Draw perpendicular lines from the points of intersection to the x-axis. The foci of the ellipse are located at the points where the perpendicular lines intersect the x-axis.

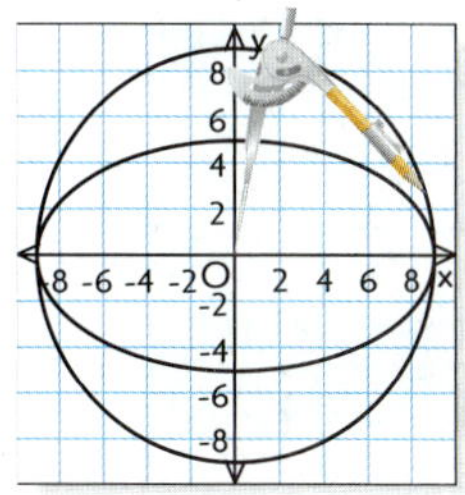

Make a Conjecture See students' work; see margin for explanation.

Draw another ellipse and locate its foci. Why does this method work?

Check for Understanding

Concept Check

2. See margin.

1. **Identify** the axes of symmetry of the ellipse at the right. $x = -1$, $y = 2$

2. **Explain** why a circle is a special case of an ellipse.

3. **OPEN ENDED** Write an equation for an ellipse with its center at $(2, -5)$ and a horizontal major axis.
Sample answer: $\frac{(x-2)^2}{4} + \frac{(y+5)^2}{1} = 1$

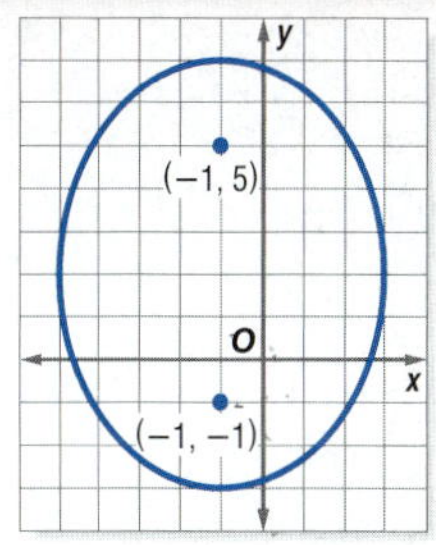

5. $\frac{(y+4)^2}{36} + \frac{(x-2)^2}{4} = 1$

Guided Practice

GUIDED PRACTICE KEY	
Exercises	Examples
4–6, 11	1, 2
7–10	3, 4

4. Write an equation for the ellipse shown at the right. $\frac{x^2}{36} + \frac{y^2}{20} = 1$

Write an equation for the ellipse that satisfies each set of conditions.

5. endpoints of major axis at $(2, 2)$ and $(2, -10)$, endpoints of minor axis at $(0, -4)$ and $(4, -4)$

6. endpoints of major axis at $(0, 10)$ and $(0, -10)$, foci at $(0, 8)$ and $(0, -8)$ $\frac{y^2}{100} + \frac{x^2}{36} = 1$

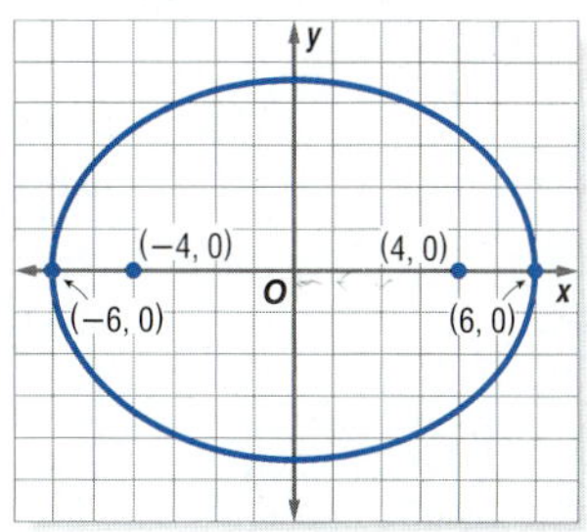

Algebra Activity

Materials: grid paper, compass, straightedge

- Suggest that students look at the point where the line intersects the circle (in Step 3) and think about how far that point is from the foci.
- Lead students to recall the relationship that they discovered in the Algebra Activity that was a preview of Lesson 8-4.

3 Practice/Apply

Study Notebook

Have students—

- add the definitions/examples of the vocabulary terms to their Vocabulary Builder worksheets for Chapter 8.
- copy both of the Key Concept summaries of ellipses into their notebooks, with labeled illustrations.
- include any other item(s) that they find helpful in mastering the skills in this lesson.

Answer

Algebra Activity

Let $(d, 0)$ be the coordinates of the point located on the positive x-axis. This point, the origin, and the point of intersection in the first quadrant of the circle and the ellipse form a right triangle. The length of the hypotenuse is the radius of the circle, which is half the length of the major axis of the ellipse, or a. One leg of the triangle has length d and the other has half the length of the minor axis of the ellipse, or b. By the Pythagorean Theorem, $a^2 = d^2 + b^2$ or $d^2 = a^2 - b^2$. Therefore, d satisfies the equation relating a, b, and c for an ellipse. Thus, one focus of the ellipse is at $(d, 0)$. By symmetry, the other focus is at $(-d, 0)$, which is the other point located by this method.

Answer

2. Let the equation of a circle be $(x - h)^2 + (y - k)^2 = r^2$. Divide each side by r^2 to get $\frac{(x-h)^2}{r^2} + \frac{(y-k)^2}{r^2} = 1$. This is the equation of an ellipse with a and b both equal to r. In other words, a circle is an ellipse whose major and minor axes are both diameters.

About the Exercises...

Organization by Objective

- Equations of Ellipses: 12–24
- Graph Ellipses: 25–38

Odd/Even Assignments

Exercises 12–21 and 27–38 are structured so that students practice the same concepts whether they are assigned odd or even problems.

Assignment Guide

Basic: 13–31 odd, 37, 39–42, 44–57

Average: 13–39 odd, 40–42, 44–57 (optional: 43)

Advanced: 12–38 even, 39–51 (optional: 52–57)

Find the coordinates of the center and foci and the lengths of the major and minor axes for the ellipse with the given equation. Then graph the ellipse.

7. $\frac{y^2}{18} + \frac{x^2}{9} = 1$

8. $\frac{(x-1)^2}{20} + \frac{(y+2)^2}{4} = 1$

9. $4x^2 + 8y^2 = 32$

10. $x^2 + 25y^2 - 8x + 100y + 91 = 0$

7–10. See margin.

Application

11. ASTRONOMY At its closest point, Mercury is 29.0 million miles from the center of the Sun. At its farthest point, Mercury is 43.8 million miles from the center of the Sun. Write an equation for the orbit of Mercury, assuming that the center of the orbit is the origin and the Sun lies on the x-axis.

about $\frac{x^2}{1.32 \times 10^{15}} + \frac{y^2}{1.27 \times 10^{15}} = 1$

★ indicates increased difficulty

Practice and Apply

Homework Help	
For Exercises	See Examples
12–24	1, 2
25–38	3, 4

Extra Practice
See page 846.

Write an equation for each ellipse. **12–15. See margin.**

12.

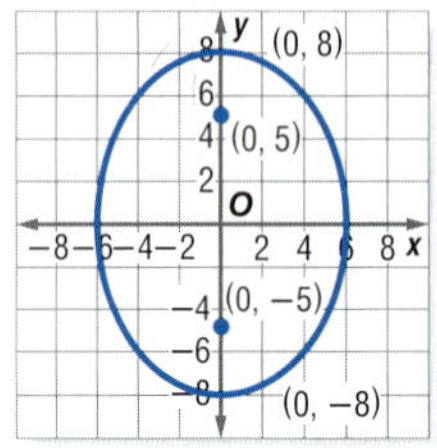

13.

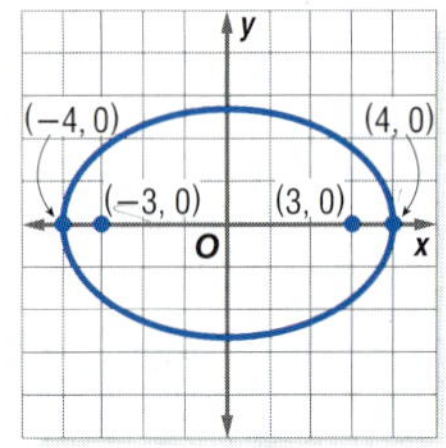

14.

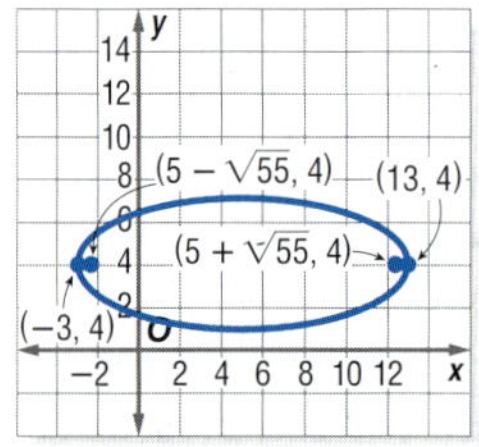

15.

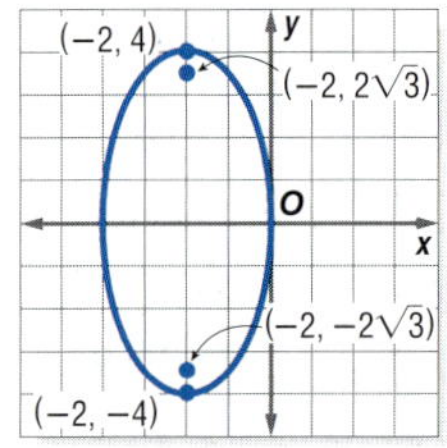

Write an equation for the ellipse that satisfies each set of conditions.

16. endpoints of major axis at (−11, 5) and (7, 5), endpoints of minor axis at (−2, 9) and (−2, 1)

17. endpoints of major axis at (2, 12) and (2, −4), endpoints of minor axis at (4, 4) and (0, 4)

18. major axis 20 units long and parallel to y-axis, minor axis 6 units long, center at (4, 2)

19. major axis 16 units long and parallel to x-axis, minor axis 9 units long, center at (5, 4)

20. endpoints of major axis at (10, 2) and (−8, 2), foci at (6, 2) and (−4, 2)

21. endpoints of minor axis at (0, 5) and (0, −5), foci at (12, 0) and (−12, 0)

16. $\frac{(x+2)^2}{81} + \frac{(y-5)^2}{16} = 1$

17. $\frac{(y-4)^2}{64} + \frac{(x-2)^2}{4} = 1$

18. $\frac{(y-2)^2}{100} + \frac{(x-4)^2}{9} = 1$

19. $\frac{(x-5)^2}{64} + \frac{(y-4)^2}{\frac{81}{4}} = 1$

20. $\frac{(x-1)^2}{81} + \frac{(y-2)^2}{56} = 1$

21. $\frac{x^2}{169} + \frac{y^2}{25} = 1$

22. INTERIOR DESIGN The rounded top of the window is the top half of an ellipse. Write an equation for the ellipse if the origin is at the midpoint of the bottom edge of the window. $\frac{x^2}{324} + \frac{y^2}{196} = 1$

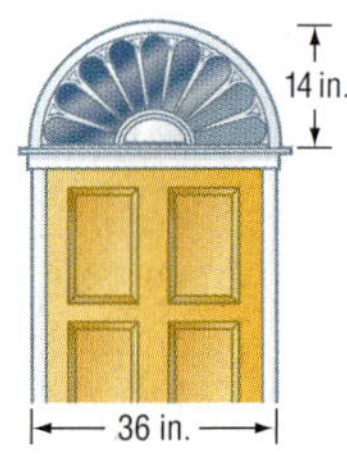

Answers

7. (0, 0); (0, ±3); $6\sqrt{2}$; 6

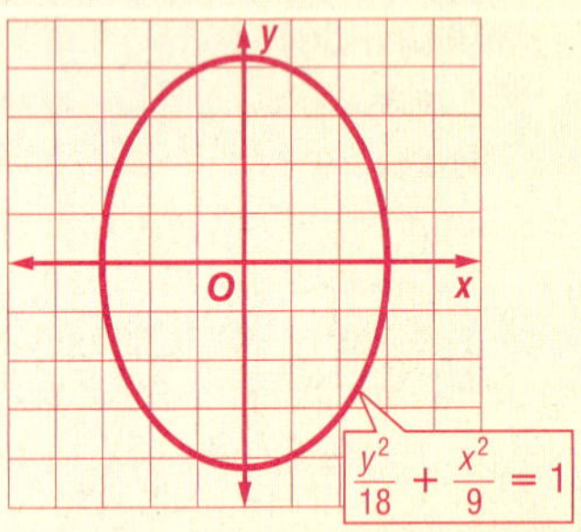

8. (1, −2); (5, −2), (−3, −2); $4\sqrt{5}$; 4

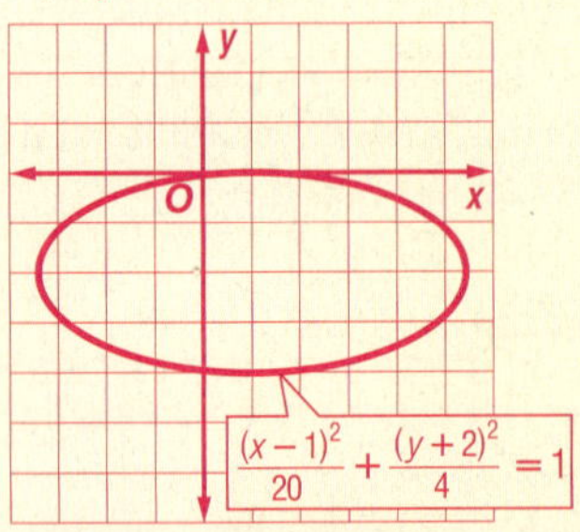

9. (0, 0); (±2, 0); $4\sqrt{2}$; 4

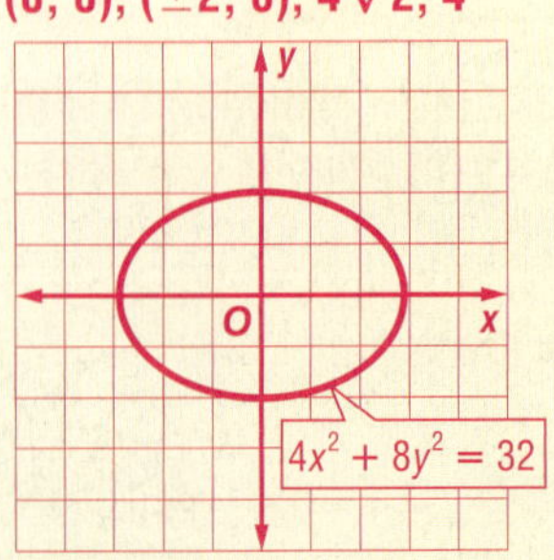

10. (4, −2); $(4 \pm 2\sqrt{6}, -2)$; 10; 2

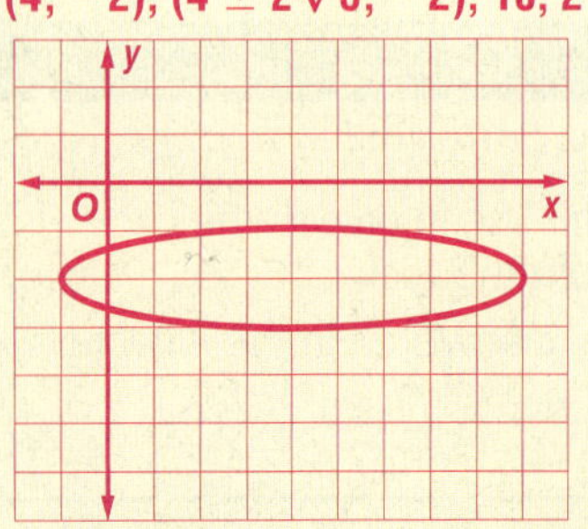

12. $\frac{y^2}{64} + \frac{x^2}{39} = 1$

13. $\frac{x^2}{16} + \frac{y^2}{7} = 1$

14. $\frac{(x-5)^2}{64} + \frac{(y-4)^2}{9} = 1$

15. $\frac{y^2}{16} + \frac{(x+2)^2}{4} = 1$

23. about $\frac{x^2}{2.02 \times 10^{16}} + \frac{y^2}{2.00 \times 10^{16}} = 1$

More About. . .

White House

The Ellipse, also known as President's Park South, has an area of about 16 acres.

Source: www.nps.gov

24. $\frac{x^2}{193{,}600} + \frac{y^2}{279{,}312.25} = 1$

29. $(-8, 2)$; $(-8 \pm 3\sqrt{7}, 2)$; 24; 18

30. $(5, -11)$; $(5, -11 \pm \sqrt{23})$; 24; 22

33. $(0, 0)$; $(0, \pm\sqrt{7})$; 8; 6

34. $(0, 0)$; $(\pm 3\sqrt{5}, 0)$; 18; 12

35. $(-3, 1)$; $(-3, 5)$, $(-3, -3)$; $4\sqrt{6}$; $4\sqrt{2}$

36. $(-2, 7)$; $(-2 \pm 4\sqrt{2}, 7)$; $4\sqrt{10}$; $4\sqrt{2}$

23. **ASTRONOMY** At its closest point, Mars is 128.5 million miles from the Sun. At its farthest point, Mars is 155.0 million miles from the Sun. Write an equation for the orbit of Mars. Assume that the center of the orbit is the origin, the Sun lies on the x-axis, and the radius of the Sun is 400,000 miles.

24. **WHITE HOUSE** There is an open area south of the White House known as the Ellipse. Write an equation to model the Ellipse. Assume that the origin is at the center of the Ellipse.

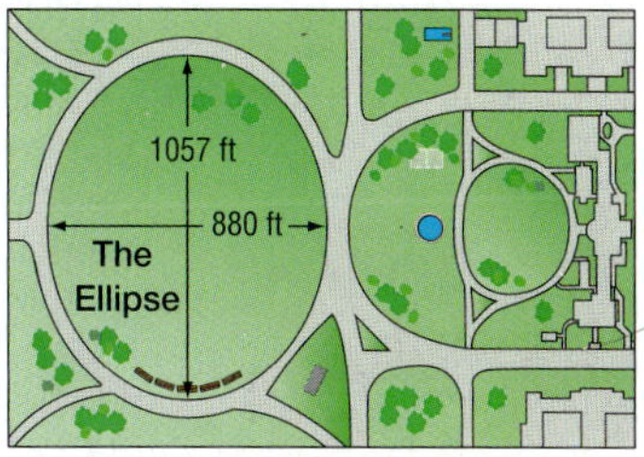

25. Write the equation $10x^2 + 2y^2 = 40$ in standard form. $\frac{y^2}{20} + \frac{x^2}{4} = 1$

26. What is the standard form of the equation $x^2 + 6y^2 - 2x + 12y - 23 = 0$? $\frac{(x-1)^2}{30} + \frac{(y+1)^2}{5} = 1$

27–38. See pp. 469A–469J for graphs.

Find the coordinates of the center and foci and the lengths of the major and minor axes for the ellipse with the given equation. Then graph the ellipse.

27. $\frac{y^2}{10} + \frac{x^2}{5} = 1$ $(0, 0)$; $(0, \pm\sqrt{5})$; $2\sqrt{10}$; $2\sqrt{5}$

28. $\frac{x^2}{25} + \frac{y^2}{9} = 1$ $(0, 0)$; $(\pm 4, 0)$; 10; 6

29. $\frac{(x+8)^2}{144} + \frac{(y-2)^2}{81} = 1$

30. $\frac{(y+11)^2}{144} + \frac{(x-5)^2}{121} = 1$

31. $3x^2 + 9y^2 = 27$ $(0, 0)$; $(\pm\sqrt{6}, 0)$; 6; $2\sqrt{3}$

32. $27x^2 + 9y^2 = 81$ $(0, 0)$; $(0, \pm\sqrt{6})$; 6; $2\sqrt{3}$

★ 33. $16x^2 + 9y^2 = 144$

34. $36x^2 + 81y^2 = 2916$

★ 35. $3x^2 + y^2 + 18x - 2y + 4 = 0$

★ 36. $x^2 + 5y^2 + 4x - 70y + 209 = 0$

37. $7x^2 + 3y^2 - 28x - 12y = -19$ $(2, 2)$; $(2, 4)$, $(2, 0)$; $2\sqrt{7}$; $2\sqrt{3}$

★ 38. $16x^2 + 25y^2 + 32x - 150y = 159$ $(-1, 3)$; $(2, 3)$, $(-4, 3)$; 10; 8

39. **CRITICAL THINKING** Find an equation for the ellipse with foci at $(\sqrt{3}, 0)$ and $(-\sqrt{3}, 0)$ that passes through $(0, 3)$. $\frac{x^2}{12} + \frac{y^2}{9} = 1$

40. **WRITING IN MATH** Answer the question that was posed at the beginning of the lesson. **See pp. 469A–469J.**

Why are ellipses important in the study of the solar system?

Include the following in your answer:
- why an equation that is an accurate model of the path of a planet might be useful, and
- the distance from the center of Earth's orbit to the center of the Sun given that the Sun is at a focus of the orbit of Earth. Use the information in the figure at the right.

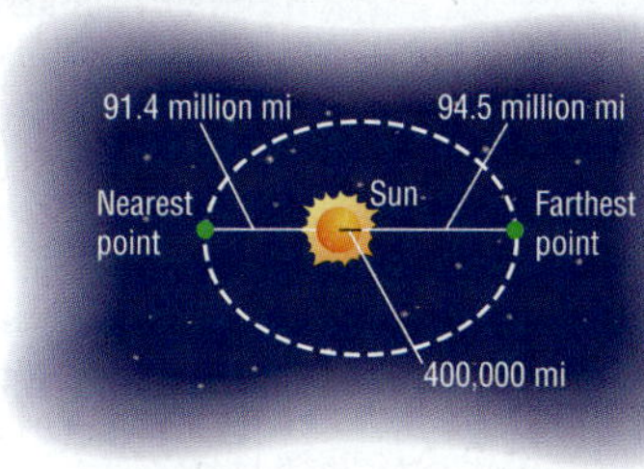

Standardized Test Practice

41. In the figure, A, B, and C are collinear. What is the measure of $\angle DBE$? **C**

Ⓐ 40° Ⓑ 65° Ⓒ 80° Ⓓ 100°

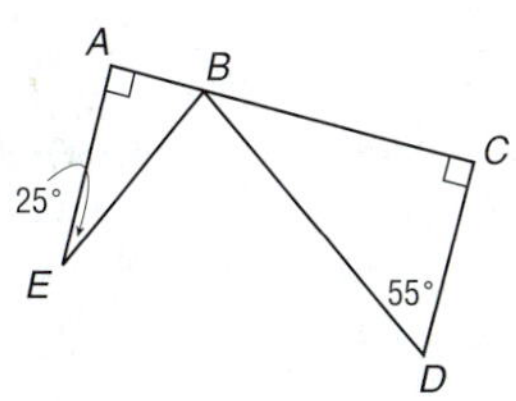

www.algebra2.com/self_check_quiz

Study Guide and Intervention, p. 473 (shown) and p. 474

Equations of Ellipses An **ellipse** is the set of all points in a plane such that the *sum* of the distances from two given points in the plane, called the foci, is constant. An ellipse has two axes of symmetry which contain the **major** and **minor axes**. In the table, the lengths a, b, and c are related by the formula $c^2 = a^2 - b^2$.

Standard Form of Equation	$\frac{(x-h)^2}{a^2} + \frac{(y-k)^2}{b^2} = 1$	$\frac{(y-k)^2}{a^2} + \frac{(x-h)^2}{b^2} = 1$
Center	(h, k)	(h, k)
Direction of Major Axis	Horizontal	Vertical
Foci	$(h + c, k)$, $(h - c, k)$	$(h, k - c)$, $(h, k + c)$
Length of Major Axis	$2a$ units	$2a$ units
Length of Minor Axis	$2b$ units	$2b$ units

Example **Write an equation for the ellipse shown.**

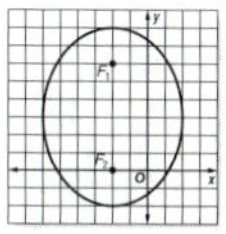

The length of the major axis is the distance between $(-2, -2)$ and $(-2, 8)$. This distance is 10 units.

$2a = 10$, so $a = 5$

The foci are located at $(-2, 6)$ and $(-2, 0)$, so $c = 3$.

$b^2 = a^2 - c^2$
$= 25 - 9$
$= 16$

The center of the ellipse is at $(-2, 3)$, so $h = -2$, $k = 3$, $a^2 = 25$, and $b^2 = 16$. The major axis is vertical.

An equation of the ellipse is $\frac{(y-3)^2}{25} + \frac{(x+2)^2}{16} = 1$.

Exercises

Write an equation for the ellipse that satisfies each set of conditions.

1. endpoints of major axis at $(-7, 2)$ and $(5, 2)$, endpoints of minor axis at $(-1, 0)$ and $(-1, 4)$ $\frac{(x+1)^2}{36} + \frac{(y-2)^2}{4} = 1$
2. major axis 8 units long and parallel to the x-axis, minor axis 2 units long, center at $(-2, -5)$ $\frac{(x+2)^2}{16} + (y+5)^2 = 1$
3. endpoints of major axis at $(-8, 4)$ and $(4, 4)$, foci at $(-3, 4)$ and $(-1, 4)$ $\frac{(x+2)^2}{36} + \frac{(y-4)^2}{35} = 1$
4. endpoints of major axis at $(3, 2)$ and $(3, -14)$, endpoints of minor axis at $(-1, -6)$ and $(7, -6)$ $\frac{(y+6)^2}{64} + \frac{(x-3)^2}{16} = 1$
5. minor axis 6 units long and parallel to the x-axis, major axis 12 units long, center at $(6, 1)$ $\frac{(y-1)^2}{36} + \frac{(x-6)^2}{9} = 1$

Skills Practice, p. 475 and Practice, p. 476 (shown)

Write an equation for each ellipse.

1. 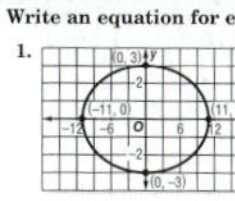

$\frac{x^2}{121} + \frac{y^2}{9} = 1$

2. 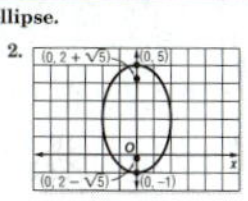

$\frac{(y-2)^2}{9} + \frac{x^2}{4} = 1$

3. 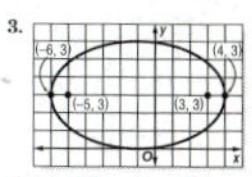

$\frac{(x+1)^2}{25} + \frac{(y-3)^2}{9} = 1$

Write an equation for the ellipse that satisfies each set of conditions.

4. endpoints of major axis at $(-9, 0)$ and $(9, 0)$, endpoints of minor axis at $(0, 3)$ and $(0, -3)$ $\frac{x^2}{81} + \frac{y^2}{9} = 1$
5. endpoints of major axis at $(4, 2)$ and $(4, -8)$, endpoints of minor axis at $(1, -3)$ and $(7, -3)$ $\frac{(y+3)^2}{25} + \frac{(x-4)^2}{9} = 1$
6. major axis 20 units long and parallel to x-axis, minor axis 10 units long, center at $(2, 1)$ $\frac{(x-2)^2}{100} + \frac{(y-1)^2}{25} = 1$
7. major axis 10 units long, minor axis 6 units long and parallel to x-axis, center at $(2, -4)$ $\frac{(y+4)^2}{25} + \frac{(x-2)^2}{9} = 1$
8. major axis 16 units long, center at $(0, 0)$, foci at $(0, 2\sqrt{15})$ and $(0, -2\sqrt{15})$ $\frac{y^2}{64} + \frac{x^2}{4} = 1$
9. endpoints of minor axis at $(0, 2)$ and $(0, -2)$, foci at $(-4, 0)$ and $(4, 0)$ $\frac{x^2}{20} + \frac{y^2}{4} = 1$

Find the coordinates of the center and foci and the lengths of the major and minor axes for the ellipse with the given equation. Then graph the ellipse.

10. $\frac{y^2}{16} + \frac{x^2}{9} = 1$ $(0, 0)$; $(0, \pm\sqrt{7})$; 8; 6

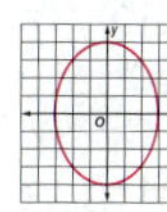

11. $\frac{(y-1)^2}{36} + \frac{(x-3)^2}{1} = 1$ $(3, 1)$; $(3, 1 \pm \sqrt{35})$; 12; 2

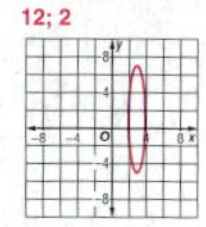

12. $\frac{(x+4)^2}{49} + \frac{(y+3)^2}{25} = 1$ $(-4, -3)$; $(-4 \pm 2\sqrt{6}, -3)$; 14; 10

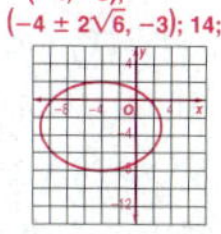

13. **SPORTS** An ice skater traces two congruent ellipses to form a figure eight. Assume that the center of the first loop is at the origin, with the second loop to its right. Write an equation to model the first loop if its major axis (along the x-axis) is 12 feet long and its minor axis is 6 feet long. Write another equation to model the second loop. $\frac{x^2}{36} + \frac{y^2}{9} = 1$; $\frac{(x-12)^2}{36} + \frac{y^2}{9} = 1$

Reading to Learn Mathematics, p. 477 ELL

Pre-Activity **Why are ellipses important in the study of the solar system?**

Read the introduction to Lesson 8-4 at the top of page 433 in your textbook.

Is the Earth always the same distance from the Sun? Explain your answer using the words *circle* and *ellipse*. No; if the Earth's orbit were a circle, it would always be the same distance from the Sun because every point on a circle is the same distance from the center. However, the Earth's orbit is an ellipse, and the points on an ellipse are not all the same distance from the center.

Reading the Lesson

1. An ellipse is the set of all points in a plane such that the ___sum___ of the distances from two fixed points is ___constant___. The two fixed points are called the ___foci___ of the ellipse.
2. Consider the ellipse with equation $\frac{x^2}{9} + \frac{y^2}{4} = 1$.
   a. For this equation, a = ___3___ and b = ___2___.
   b. Write an equation that relates the values of a, b, and c. $c^2 = a^2 - b^2$
   c. Find the value of c for this ellipse. $\sqrt{5}$
3. Consider the ellipses with equations $\frac{y^2}{25} + \frac{x^2}{16} = 1$ and $\frac{x^2}{9} + \frac{y^2}{4} = 1$. Complete the following table to describe characteristics of their graphs.

Standard Form of Equation	$\frac{y^2}{25} + \frac{x^2}{16} = 1$	$\frac{x^2}{9} + \frac{y^2}{4} = 1$
Direction of Major Axis	vertical	horizontal
Direction of Minor Axis	horizontal	vertical
Foci	$(0, 3)$, $(0, -3)$	$(\sqrt{5}, 0)$, $(-\sqrt{5}, 0)$
Length of Major Axis	10 units	6 units
Length of Minor Axis	8 units	4 units

Helping You Remember

4. Some students have trouble remembering the two standard forms for the equation of an ellipse. How can you remember which term comes first and where to place a and b in these equations? The x-axis is horizontal. If the major axis is horizontal, the first term is $\frac{x^2}{a^2}$. The y-axis is vertical. If the major axis is vertical, the first term is $\frac{y^2}{a^2}$. a is always the larger of the numbers a and b.

Enrichment, p. 478

Eccentricity

In an ellipse, the ratio $\frac{c}{d}$ is called the **eccentricity** and is denoted by the letter e. Eccentricity measures the elongation of an ellipse. The closer e is to 0, the more an ellipse looks like a circle. The closer e is to 1, the more elongated it is. Recall that the equation of an ellipse is $\frac{x^2}{a^2} + \frac{y^2}{b^2} = 1$ or $\frac{x^2}{b^2} + \frac{y^2}{a^2} = 1$ where a is the length of the major axis, and that $c = \sqrt{a^2 - b^2}$.

Find the eccentricity of each ellipse rounded to the nearest hundredth.

1. $\frac{x^2}{9} + \frac{y^2}{36} = 1$ 0.87
2. $\frac{x^2}{81} + \frac{y^2}{9} = 1$ 0.94
3. $\frac{x^2}{4} + \frac{y^2}{9} = 1$ 0.75

4 Assess

Open-Ended Assessment

Speaking Have students draw and label the various parts of an ellipse whose major axis is on the x-axis and one whose major axis is on the y-axis, and then explain the differences.

Tips for New Teachers

Intervention Suggest that students make a summary of the various quadratic graphs on a large index card so that they can easily refer to it as they solve problems. Clear up any questions they may have about the meaning of the various variables and vocabulary words.

Getting Ready for Lesson 8-5

PREREQUISITE SKILL In the process of graphing hyperbolas in Lesson 8-5, students will graph lines that represent asymptotes of the hyperbolas. Exercises 52–57 should be used to determine your students' familiarity with graphing lines.

Assessment Options

Quiz (Lessons 8-3 and 8-4) is available on p. 511 of the *Chapter 8 Resource Masters*.

Mid-Chapter Test (Lessons 8-1 through 8-4) is available on p. 513 of the *Chapter 8 Resource Masters*.

---

42. $\sqrt{25 + 144} =$ **B**

Ⓐ 7   Ⓑ 13   Ⓒ 17   Ⓓ 169

Extending the Lesson

43. **ASTRONOMY** In an ellipse, the ratio $\frac{c}{a}$ is called the **eccentricity** and is denoted by the letter e. Eccentricity measures the elongation of an ellipse. As shown in the graph at the right, the closer e is to 0, the more an ellipse looks like a circle. Pluto has the most eccentric orbit in our solar system with $e \approx 0.25$. Find an equation to model the orbit of Pluto, given that the length of the major axis is about 7.34 billion miles. Assume that the major axis is horizontal and that the center of the orbit is the origin.

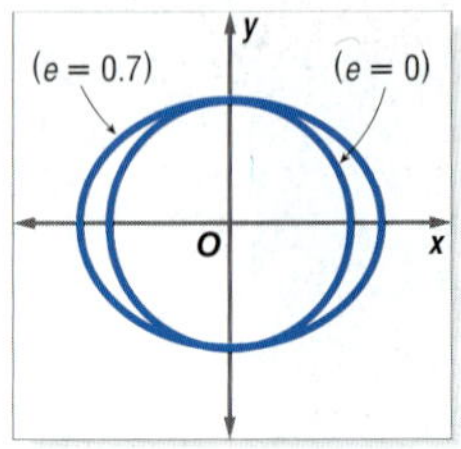

about $\frac{x^2}{1.35 \times 10^{19}} + \frac{y^2}{1.26 \times 10^{19}} = 1$

Maintain Your Skills

Mixed Review

Write an equation for the circle that satisfies each set of conditions. *(Lesson 8-3)*

44. center (3, −2), radius 5 units **$(x - 3)^2 + (y + 2)^2 = 25$**

45. endpoints of a diameter at (5, −9) and (3, 11) **$(x - 4)^2 + (y - 1)^2 = 101$**

46. center (−1, 0), passes through (2, −6) **$(x + 1)^2 + y^2 = 45$**

47. center (4, −1), tangent to y-axis **$(x - 4)^2 + (y + 1)^2 = 16$**

48. Write an equation of a parabola with vertex (3, 1) and focus $\left(3, 1\frac{1}{2}\right)$. Then draw the graph. *(Lesson 8-2)* **$y = \frac{1}{2}(x - 3)^2 + 1$; See margin for graph.**

MARRIAGE For Exercises 49–51, use the table at the right that shows the number of married Americans over the last few decades. *(Lesson 2-5)*

49. Draw a scatter plot in which x is the number of years since 1980. **49. See margin.**

50. Find a prediction equation. **50. Sample answer using (0, 104.6) and (10, 112.6): $y = 0.8x + 104.6$**

51. Predict the number of married Americans in 2010. **Sample answer: 128,600,000**

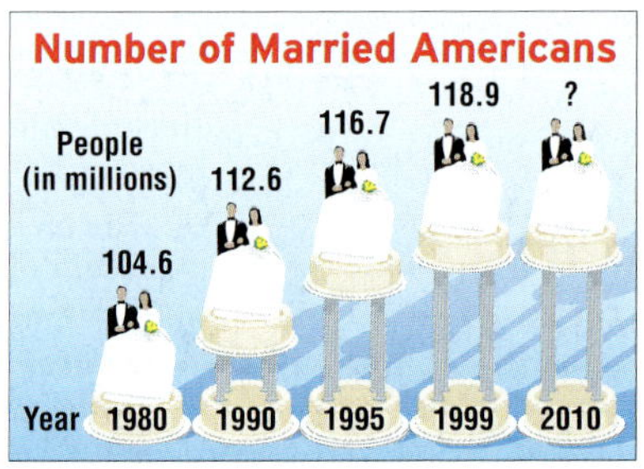

Source: U.S. Census Bureau

Online Research **Data Update** For the latest statistics on marriage and other characteristics of the population, visit www.algebra2.com/data_update to learn more.

Getting Ready for the Next Lesson

PREREQUISITE SKILL Graph the line with the given equation.
*(To review **graphing lines**, see Lessons 2-1, 2-2, and 2-3.)* **52–57. See pp. 469A–469J.**

52. $y = 2x$ 53. $y = -2x$ 54. $y = -\frac{1}{2}x$

55. $y = \frac{1}{2}x$ 56. $y + 2 = 2(x - 1)$ 57. $y + 2 = -2(x - 1)$

Answers

48.

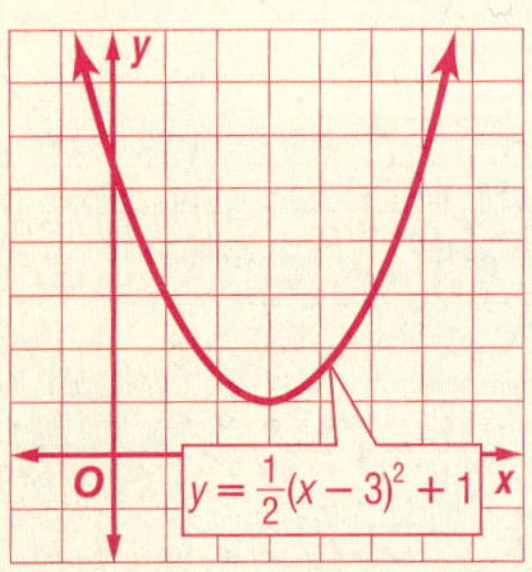

49.

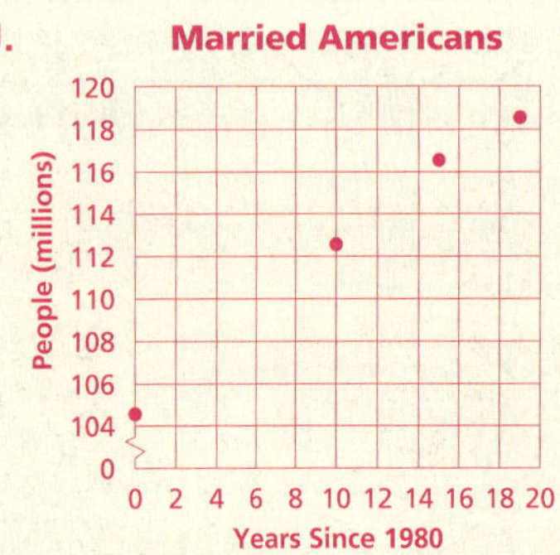

8-5 Hyperbolas

What You'll Learn

- Write equations of hyperbolas.
- Graph hyperbolas.

How are hyperbolas different from parabolas?

A hyperbola is a conic section with the property that rays directed toward one focus are reflected toward the other focus. Notice that, unlike the other conic sections, a hyperbola has two branches.

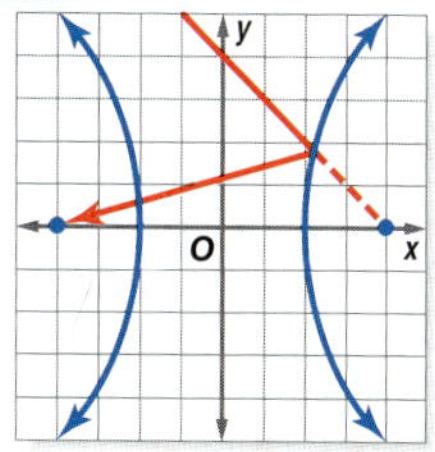

Vocabulary

- hyperbola
- foci
- center
- vertex
- asymptote
- transverse axis
- conjugate axis

EQUATIONS OF HYPERBOLAS

A **hyperbola** is the set of all points in a plane such that the absolute value of the difference of the distances from two fixed points, called the **foci**, is constant.

The hyperbola at the right has foci at (0, 3) and (0, −3). The distances from either of the y-intercepts to the foci are 1 unit and 5 units, so the difference of the distances from any point with coordinates (x, y) on the hyperbola to the foci is 4 or −4 units, depending on the order in which you subtract.

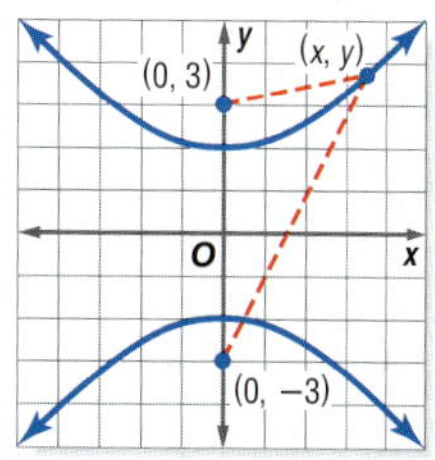

You can use the Distance Formula and the definition of a hyperbola to find an equation of this hyperbola.

The distance between (x, y) and $(0, 3)$ − the distance between (x, y) and $(0, -3)$ = ±4.

$$\sqrt{x^2 + (y - 3)^2} - \sqrt{x^2 + (y + 3)^2} = \pm 4$$

$$\sqrt{x^2 + (y - 3)^2} = \pm 4 + \sqrt{x^2 + (y + 3)^2} \quad \text{Isolate the radicals.}$$

$$x^2 + (y - 3)^2 = 16 \pm 8\sqrt{x^2 + (y + 3)^2} + x^2 + (y + 3)^2 \quad \text{Square each side.}$$

$$x^2 + y^2 - 6y + 9 = 16 \pm 8\sqrt{x^2 + (y + 3)^2} + x^2 + y^2 + 6y + 9$$

$$-12y - 16 = \pm 8\sqrt{x^2 + (y + 3)^2} \quad \text{Simplify.}$$

$$3y + 4 = \pm 2\sqrt{x^2 + (y + 3)^2} \quad \text{Divide each side by } -4.$$

$$9y^2 + 24y + 16 = 4[x^2 + (y + 3)^2] \quad \text{Square each side.}$$

$$9y^2 + 24y + 16 = 4x^2 + 4y^2 + 24y + 36 \quad \text{Distributive Property}$$

$$5y^2 - 4x^2 = 20 \quad \text{Simplify.}$$

$$\frac{y^2}{4} - \frac{x^2}{5} = 1 \quad \text{Divide each side by 20.}$$

An equation of this hyperbola is $\frac{y^2}{4} - \frac{x^2}{5} = 1$.

8-5 Lesson Notes

1 Focus

5-Minute Check Transparency 8-5 Use as a quiz or review of Lesson 8-4.

Mathematical Background notes are available for this lesson on p. 410D.

How are hyperbolas different from parabolas?

Ask students:

- Why are parabolas, circles, ellipses, and hyperbolas called conic sections? **They are cross sections that result when a plane intersects a double right circular cone.**
- Where is a ray directed toward one focus and reflected toward the other? **inside or between the two branches of the hyperbola**

Resource Manager

Workbook and Reproducible Masters

Chapter 8 Resource Masters

- Study Guide and Intervention, pp. 479–480
- Skills Practice, p. 481
- Practice, p. 482
- Reading to Learn Mathematics, p. 483
- Enrichment, p. 484

Transparencies

5-Minute Check Transparency 8-5
Real-World Transparency 8
Answer Key Transparencies

Technology

Alge2PASS: Tutorial Plus, Lesson 16
Interactive Chalkboard

2 Teach

EQUATIONS OF HYPERBOLAS

In-Class Examples

Power Point®

1 Write an equation for the hyperbola shown.

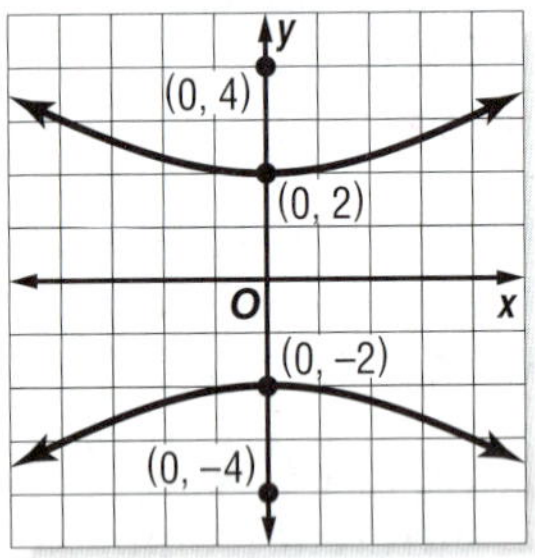

$\frac{y^2}{4} - \frac{x^2}{12} = 1$

2 **NAVIGATION** A ship notes that the difference of its distance from two LORAN stations that are located at (−70, 0) and (70, 0) is 70 nautical miles. Write an equation for the hyperbola on which the ship lies.

$\frac{x^2}{1225} - \frac{y^2}{3675} = 1$

Teaching Tip Tell students to make a rough sketch of the situation in problems such as these.

The diagram below shows the parts of a hyperbola.

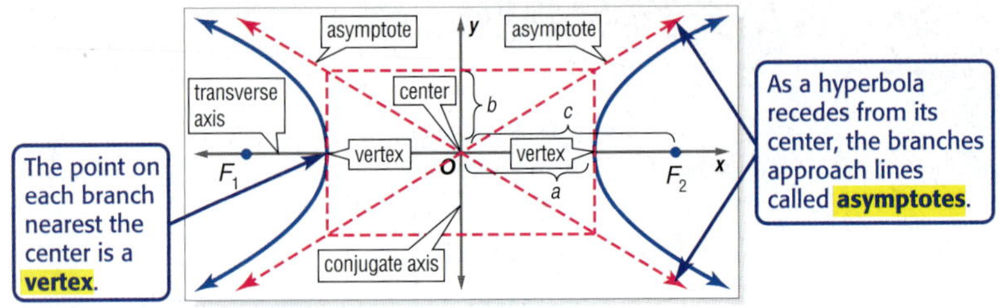

A hyperbola has some similarities to an ellipse. The distance from the **center** to a vertex is a units. The distance from the center to a focus is c units. There are two axes of symmetry. The **transverse axis** is a segment of length $2a$ whose endpoints are the vertices of the hyperbola. The **conjugate axis** is a segment of length $2b$ units that is perpendicular to the transverse axis at the center. The values of a, b, and c are related differently for a hyperbola than for an ellipse. For a hyperbola, $c^2 = a^2 + b^2$. The table below summarizes many of the properties of hyperbolas with centers at the origin.

Study Tip

Reading Math In the standard form of a hyperbola, the squared terms are subtracted (−). For an ellipse, they are added (+).

Key Concept *Equations of Hyperbolas with Centers at the Origin*

Standard Form of Equation	$\frac{x^2}{a^2} - \frac{y^2}{b^2} = 1$	$\frac{y^2}{a^2} - \frac{x^2}{b^2} = 1$
Direction of Transverse Axis	horizontal	vertical
Foci	$(c, 0), (-c, 0)$	$(0, c), (0, -c)$
Vertices	$(a, 0), (-a, 0)$	$(0, a), (0, -a)$
Length of Transverse Axis	$2a$ units	$2a$ units
Length of Conjugate Axis	$2b$ units	$2b$ units
Equations of Asymptotes	$y = \pm\frac{b}{a}x$	$y = \pm\frac{a}{b}x$

Example 1 *Write an Equation for a Graph*

Write an equation for the hyperbola shown at the right.

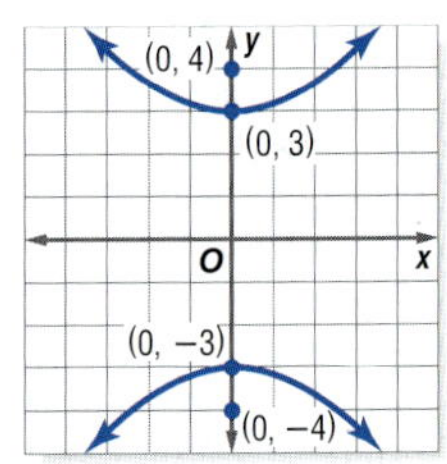

The center is the midpoint of the segment connecting the vertices, or (0, 0).

The value of a is the distance from the center to a vertex, or 3 units. The value of c is the distance from the center to a focus, or 4 units.

$c^2 = a^2 + b^2$	Equation relating a, b, and c for a hyperbola
$4^2 = 3^2 + b^2$	$c = 4, a = 3$
$16 = 9 + b^2$	Evaluate the squares.
$7 = b^2$	Solve for b^2.

Since the transverse axis is vertical, the equation is of the form $\frac{y^2}{a^2} - \frac{x^2}{b^2} = 1$.

Substitute the values for a^2 and b^2. An equation of the hyperbola is $\frac{y^2}{9} - \frac{x^2}{7} = 1$.

DAILY INTERVENTION **Unlocking Misconceptions**

Some students may think that a hyperbola has the shape of two parabolas. Explain that this is not true, and encourage students to draw a parabola on thin paper and place it over a hyperbola to see that the shapes of these curves are different.

442 Chapter 8 Conic Sections

Example 2 Write an Equation Given the Foci and Transverse Axis

NAVIGATION **The LORAN navigational system is based on hyperbolas. Two stations send out signals at the same time. A ship notes the difference in the times at which it receives the signals. The ship is on a hyperbola with the stations at the foci. Suppose a ship determines that the difference of its distances from two stations is 50 nautical miles. The stations are 100 nautical miles apart. Write an equation for a hyperbola on which the ship lies if the stations are at (−50, 0) and (50, 0).**

First, draw a figure. By studying either of the x-intercepts, you can see that the difference of the distances from any point on the hyperbola to the stations at the foci is the same as the length of the transverse axis, or $2a$. Therefore, $2a = 50$, or $a = 25$. According to the coordinates of the foci, $c = 50$.

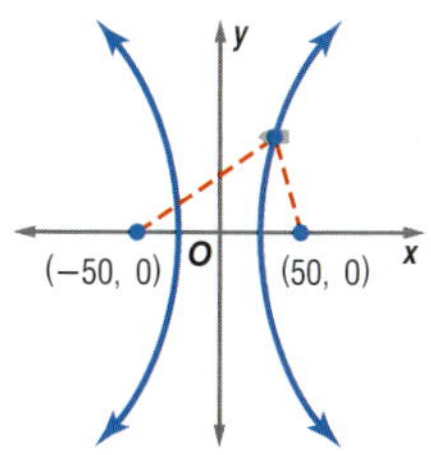

Use the values of a and c to determine the value of b for this hyperbola.

$c^2 = a^2 + b^2$	Equation relating a, b, and c for a hyperbola
$50^2 = 25^2 + b^2$	$c = 50$, $a = 25$
$2500 = 625 + b^2$	Evaluate the squares.
$1875 = b^2$	Solve for b^2.

Since the transverse axis is horizontal, the equation is of the form $\frac{x^2}{a^2} - \frac{y^2}{b^2} = 1$.

Substitute the values for a^2 and b^2. An equation of the hyperbola is $\frac{x^2}{625} - \frac{y^2}{1875} = 1$.

GRAPH HYPERBOLAS So far, you have studied hyperbolas that are centered at the origin. A hyperbola may be translated so that its center is at (h, k). This corresponds to replacing x by $x - h$ and y by $y - k$ in both the equation of the hyperbola and the equations of the asymptotes.

Key Concept: Equations of Hyperbolas with Centers at (h, k)

Standard Form of Equation	$\frac{(x-h)^2}{a^2} - \frac{(y-k)^2}{b^2} = 1$	$\frac{(y-k)^2}{a^2} - \frac{(x-h)^2}{b^2} = 1$
Direction of Transverse Axis	horizontal	vertical
Equations of Asymptotes	$y - k = \pm\frac{b}{a}(x - h)$	$y - k = \pm\frac{a}{b}(x - h)$

It is easier to graph a hyperbola if the asymptotes are drawn first. To graph the asymptotes, use the values of a and b to draw a rectangle with dimensions $2a$ and $2b$. The diagonals of the rectangle should intersect at the center of the hyperbola. The asymptotes will contain the diagonals of the rectangle.

Example 3 Graph an Equation in Standard Form

Find the coordinates of the vertices and foci and the equations of the asymptotes for the hyperbola with equation $\frac{x^2}{9} - \frac{y^2}{4} = 1$. Then graph the hyperbola.

The center of this hyperbola is at the origin. According to the equation, $a^2 = 9$ and $b^2 = 4$, so $a = 3$ and $b = 2$. The coordinates of the vertices are (3, 0) and (−3, 0).

(continued on the next page)

www.algebra2.com/extra_examples

More About. . .

Navigation

LORAN stands for *Long Range Navigation*. The LORAN system is generally accurate to within 0.25 nautical mile.

Source: U.S. Coast Guard

GRAPH HYPERBOLAS

In-Class Example

Power Point®

3 Find the coordinates of the vertices and foci and the equations of the asymptotes for the hyperbola with equation $x^2 - y^2 = 1$. Then graph the hyperbola.
vertices: (−1, 0), (1, 0); foci: $(\sqrt{2}, 0)$, $(-\sqrt{2}, 0)$; asymptotes: $y = x$, $y = -x$

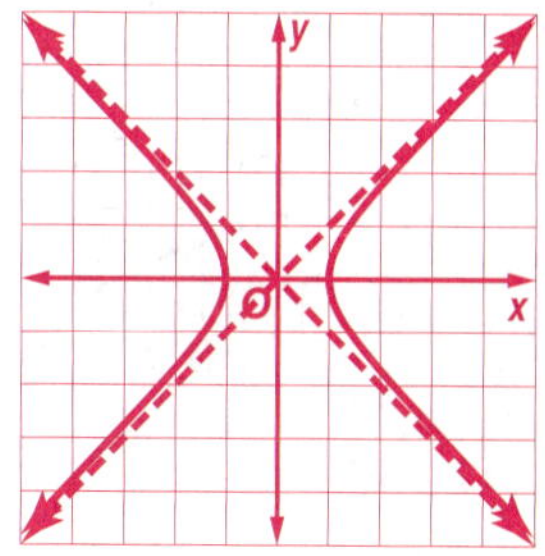

DAILY INTERVENTION — Differentiated Instruction

Logical Have students create a chart for a classroom poster that summarizes all the facts and equations for each type of conic section, with illustrations.

In-Class Example

PowerPoint®

4 Find the coordinates of the vertices and foci and the equations of the asymptotes for the hyperbola with equation $x^2 - y^2 + 6x + 10y - 17 = 0$. Then graph the hyperbola. **vertices: $(-4, 5)$, $(-2, 5)$; foci: $(\sqrt{2} - 3, 5)$, $(-\sqrt{2} - 3, 5)$; asymptotes: $y = x + 8$, $y = -x + 2$**

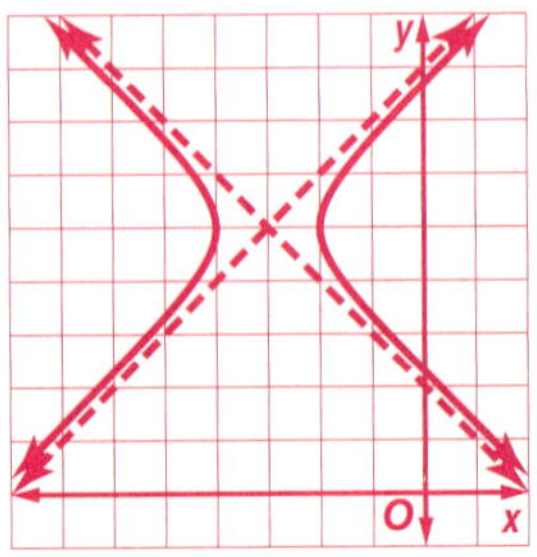

Teaching Tip Suggest that students make a list of the values of a, b, c, h, and k. This will help avoid confusions about the signs of h and k.

$c^2 = a^2 + b^2$ Equation relating a, b, and c for a hyperbola

$c^2 = 3^2 + 2^2$ $a = 3$, $b = 2$

$c^2 = 13$ Simplify.

$c = \sqrt{13}$ Take the square root of each side.

The foci are at $(\sqrt{13}, 0)$ and $(-\sqrt{13}, 0)$.

The equations of the asymptotes are $y = \pm\frac{b}{a}x$ or $y = \pm\frac{2}{3}x$.

You can use a calculator to find some approximate nonnegative values for x and y that satisfy the equation. Since the hyperbola is centered at the origin, it is symmetric about the y-axis. Therefore, the points at $(-8, 4.9)$, $(-7, 4.2)$, $(-6, 3.5)$, $(-5, 2.7)$, $(-4, 1.8)$, and $(-3, 0)$ lie on the graph.

x	y
3	0
4	1.8
5	2.7
6	3.5
7	4.2
8	4.9

The hyperbola is also symmetric about the x-axis, so the points at $(-8, -4.9)$, $(-7, -4.2)$, $(-6, -3.5)$, $(-5, -2.7)$, $(-4, -1.8)$, $(4, -1.8)$, $(5, -2.7)$, $(6, -3.5)$, $(7, -4.2)$, and $(8, -4.9)$ also lie on the graph.

Draw a 6-unit by 4-unit rectangle. The asymptotes contain the diagonals of the rectangle. Graph the vertices, which, in this case, are the x-intercepts. Use the asymptotes as a guide to draw the hyperbola that passes through the vertices and the other points. The graph does not intersect the asymptotes.

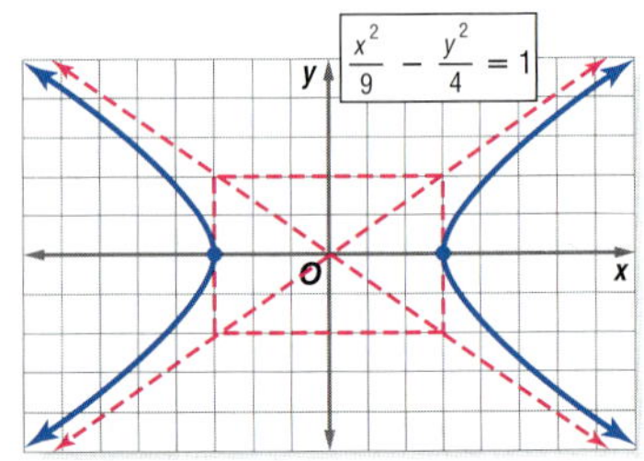

Study Tip

Graphing Calculator

You can graph a hyperbola on a graphing calculator. Similar to an ellipse, first solve the equation for y. Then graph the two equations that result on the same screen.

When graphing a hyperbola given an equation that is not in standard form, begin by rewriting the equation in standard form.

Example 4 Graph an Equation Not in Standard Form

Find the coordinates of the vertices and foci and the equations of the asymptotes for the hyperbola with equation $4x^2 - 9y^2 - 32x - 18y + 19 = 0$. Then graph the hyperbola.

Complete the square for each variable to write this equation in standard form.

$4x^2 - 9y^2 - 32x - 18y + 19 = 0$ Original equation

$4(x^2 - 8x + \blacksquare) - 9(y^2 + 2y + \blacksquare) = -19 + 4(\blacksquare) - 9(\blacksquare)$ Complete the squares.

$4(x^2 - 8x + 16) - 9(y^2 + 2y + 1) = -19 + 4(16) - 9(1)$

$4(x - 4)^2 - 9(y + 1)^2 = 36$ Write the trinomials as perfect squares.

$\frac{(x-4)^2}{9} - \frac{(y+1)^2}{4} = 1$ Divide each side by 36.

The graph of this hyperbola is the graph from Example 3 translated 4 units to the right and down 1 unit. The vertices are at $(7, -1)$ and $(1, -1)$, and the foci are at $(4 + \sqrt{13}, -1)$ and $(4 - \sqrt{13}, -1)$. The equations of the asymptotes are $y + 1 = \pm\frac{2}{3}(x - 4)$.

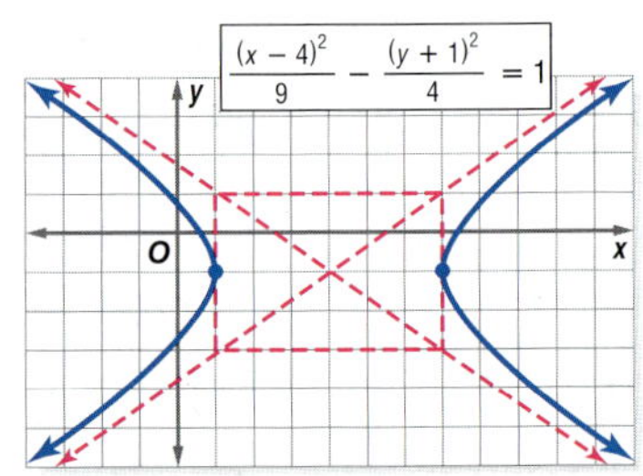

444 Chapter 8 Conic Sections

Check for Understanding

Concept Check

1. **Determine** whether the statement is *sometimes*, *always*, or *never* true. *The graph of a hyperbola is symmetric about the x-axis.* **sometimes**
2. **Describe** how the graph of $y^2 - \frac{x^2}{k^2} = 1$ changes as k increases.
3. **OPEN ENDED** Find a counterexample to the following statement. *If the equation of a hyperbola is* $\frac{x^2}{a^2} - \frac{y^2}{b^2} = 1$, *then* $a^2 \geq b^2$. **Sample answer:** $\frac{x^2}{4} - \frac{y^2}{9} = 1$

2. As k increases, the branches of the hyperbola become wider.

Guided Practice

GUIDED PRACTICE KEY	
Exercises	Examples
4, 5	1, 2
6–10	3, 4

4. Write an equation for the hyperbola shown at the right.
5. A hyperbola has foci at (4, 0) and (−4, 0). The value of a is 1. Write an equation for the hyperbola. $\frac{x^2}{1} - \frac{y^2}{15} = 1$

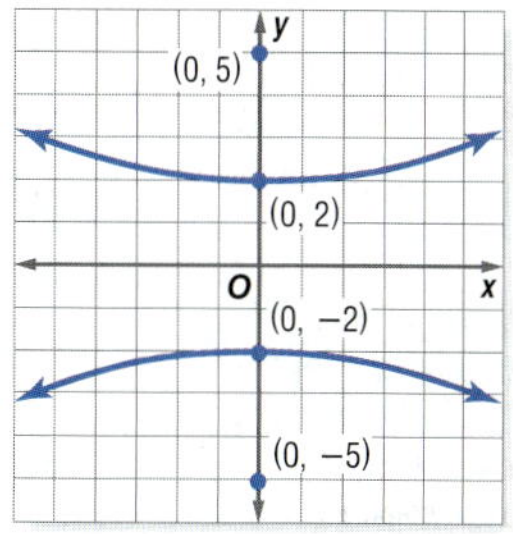

4. $\frac{y^2}{4} - \frac{x^2}{21} = 1$

6–9. See margin.

Find the coordinates of the vertices and foci and the equations of the asymptotes for the hyperbola with the given equation. Then graph the hyperbola.

6. $\frac{y^2}{18} - \frac{x^2}{20} = 1$
7. $\frac{(y + 6)^2}{20} - \frac{(x - 1)^2}{25} = 1$
8. $x^2 - 36y^2 = 36$
9. $5x^2 - 4y^2 - 40x - 16y - 36 = 0$

Application

10. **ASTRONOMY** Comets that pass by Earth only once may follow hyperbolic paths. Suppose a comet's path is modeled by a branch of the hyperbola with equation $\frac{y^2}{225} - \frac{x^2}{400} = 1$. Find the coordinates of the vertices and foci and the equations of the asymptotes for the hyperbola. Then graph the hyperbola.

10. (0, ±15); (0, ±25); $y = \pm\frac{3}{4}x$; See margin for graph.

★ indicates increased difficulty

Practice and Apply

Homework Help	
For Exercises	See Examples
11–20, 35	1, 2
21–34	3, 4
36–38	

Extra Practice
See page 846.

Write an equation for each hyperbola.

11.

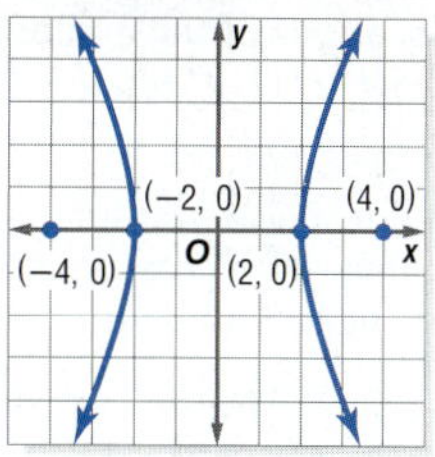

$\frac{x^2}{4} - \frac{y^2}{12} = 1$

12.

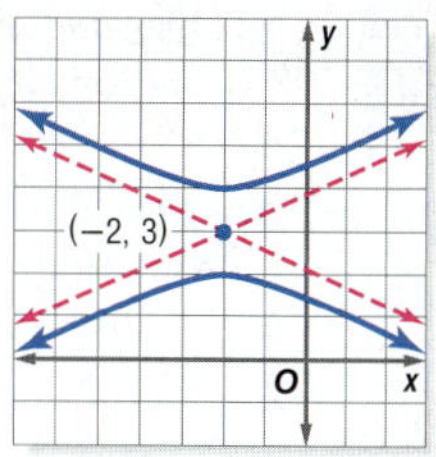

$\frac{(y - 3)^2}{1} - \frac{(x + 2)^2}{4} = 1$

13.

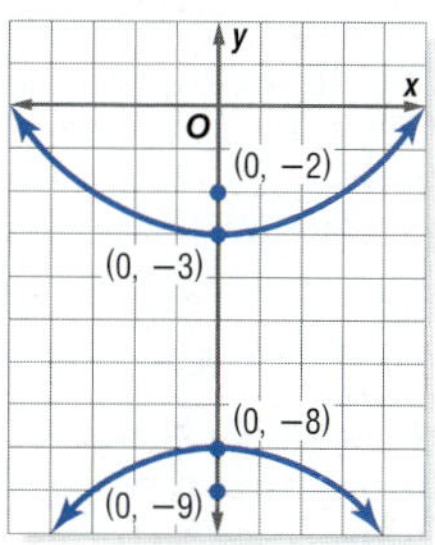

$\frac{\left(y + \frac{11}{2}\right)^2}{\frac{25}{4}} - \frac{x^2}{6} = 1$

14.

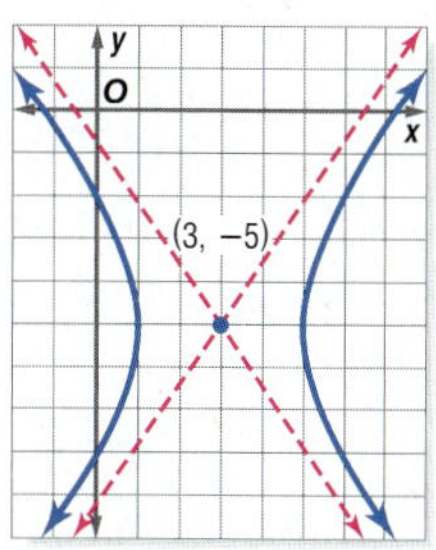

$\frac{(x - 3)^2}{4} - \frac{(y + 5)^2}{9} = 1$

www.algebra2.com/self_check_quiz

3 Practice/Apply

Study Notebook

Have students—
- add the definitions/examples of the vocabulary terms to their Vocabulary Builder worksheets for Chapter 8.
- copy both of the Key Concept summaries of hyperbolas into their notebooks, with labeled illustrations.
- include any other item(s) that they find helpful in mastering the skills in this lesson.

Answers

6. $(0, \pm 3\sqrt{2})$; $(0, \pm\sqrt{38})$; $y = \pm\frac{3\sqrt{10}}{10}x$

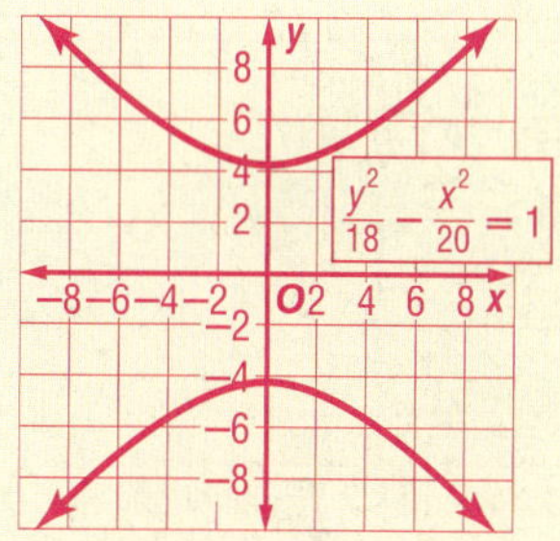

7. $(1, -6 \pm 2\sqrt{5})$; $(1, -6 \pm 3\sqrt{5})$; $y + 6 = \pm\frac{2\sqrt{5}}{5}(x - 1)$

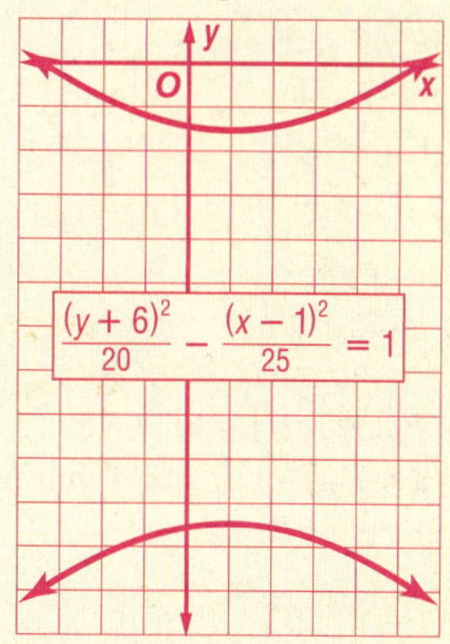

8. $(\pm 6, 0)$; $(\pm\sqrt{37}, 0)$; $y = \pm\frac{1}{6}x$

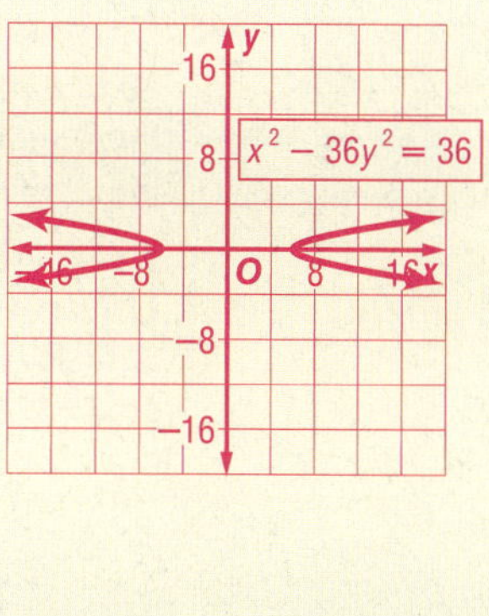

9. $(4 \pm 2\sqrt{5}, -2)$; $(4 \pm 3\sqrt{5}, -2)$; $y + 2 = \pm\frac{\sqrt{5}}{2}(x - 4)$

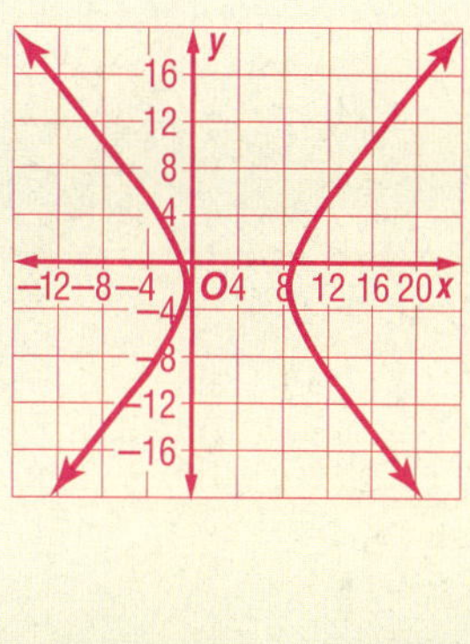

10.

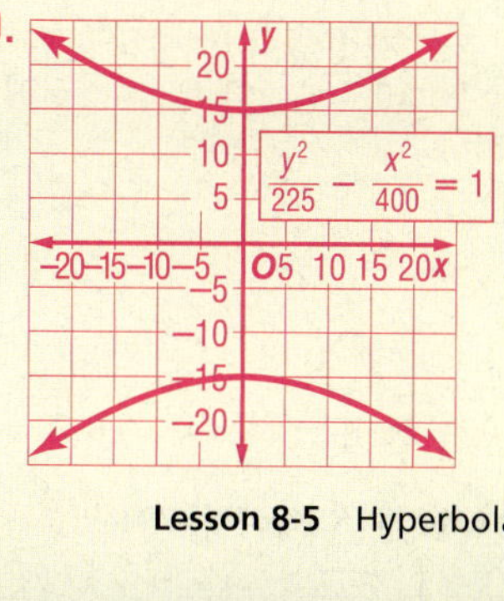

About the Exercises...

Organization by Objective

- **Equations of Hyperbolas:** 11–20, 35
- **Graph Hyperbolas:** 21–34, 36–38

Odd/Even Assignments

Exercises 11–34 are structured so that students practice the same concepts whether they are assigned odd or even problems.

Assignment Guide

Basic: 11–31 odd, 35–38, 40–42, 47–49

Average: 11–35 odd, 36–42, 47–49 (optional: 43–46)

Advanced: 12–34 even, 35, 36, 38–46 (optional: 47–49)

All: Practice Quiz 2 (1–5)

16. $\frac{y^2}{16} - \frac{x^2}{49} = 1$

19. $\frac{x^2}{16} - \frac{y^2}{9} = 1$

23. $(0, \pm 4)$; $(0, \pm\sqrt{41})$; $y = \pm\frac{4}{5}x$

24. $(\pm 3, 0)$; $(\pm\sqrt{34}, 0)$; $y = \pm\frac{5}{3}x$

25. $(\pm\sqrt{2}, 0)$; $(\pm\sqrt{3}, 0)$; $y = \pm\frac{\sqrt{2}}{2}x$

26. $(\pm 2, 0)$; $(\pm 2\sqrt{2}, 0)$; $y = \pm x$

27. $(0, \pm 6)$; $(0, \pm 3\sqrt{5})$; $y = \pm 2x$

28. $(0, \pm\sqrt{2})$; $(0, \pm 2\sqrt{2})$; $y = \pm\frac{\sqrt{3}}{3}x$

Write an equation for the hyperbola that satisfies each set of conditions.

15. vertices $(-5, 0)$ and $(5, 0)$, conjugate axis of length 12 units $\frac{x^2}{25} - \frac{y^2}{36} = 1$
16. vertices $(0, -4)$ and $(0, 4)$, conjugate axis of length 14 units
17. vertices $(9, -3)$ and $(-5, -3)$, foci $(2 \pm \sqrt{53}, -3)$ $\frac{(x-2)^2}{49} - \frac{(y+3)^2}{4} = 1$
18. vertices $(-4, 1)$ and $(-4, 9)$, foci $(-4, 5 \pm \sqrt{97})$ $\frac{(y-5)^2}{16} - \frac{(x+4)^2}{81} = 1$
19. Find an equation for a hyperbola centered at the origin with a horizontal transverse axis of length 8 units and a conjugate axis of length 6 units.
20. What is an equation for the hyperbola centered at the origin with a vertical transverse axis of length 12 units and a conjugate axis of length 4 units? $\frac{y^2}{36} - \frac{x^2}{4} = 1$

21–28. See pp. 469A–469J for graphs.

Find the coordinates of the vertices and foci and the equations of the asymptotes for the hyperbola with the given equation. Then graph the hyperbola.

21. $\frac{x^2}{81} - \frac{y^2}{49} = 1$ $(\pm 9, 0)$; $(\pm\sqrt{130}, 0)$; $y = \pm\frac{7}{9}x$
22. $\frac{y^2}{36} - \frac{x^2}{4} = 1$ $(0, \pm 6)$; $(0, \pm 2\sqrt{10})$; $y = \pm 3x$
23. $\frac{y^2}{16} - \frac{x^2}{25} = 1$
24. $\frac{x^2}{9} - \frac{y^2}{25} = 1$
25. $x^2 - 2y^2 = 2$
26. $x^2 - y^2 = 4$
27. $y^2 = 36 + 4x^2$
28. $6y^2 = 2x^2 + 12$
29. $\frac{(y-4)^2}{16} - \frac{(x+2)^2}{9} = 1$
30. $\frac{(y-3)^2}{25} - \frac{(x-2)^2}{16} = 1$
31. $\frac{(x+1)^2}{4} - \frac{(y+3)^2}{9} = 1$
32. $\frac{(x+6)^2}{36} - \frac{(y+3)^2}{9} = 1$
33. ★ $y^2 - 3x^2 + 6y + 6x - 18 = 0$
34. ★ $4x^2 - 25y^2 - 8x - 96 = 0$

29–34. pp. 469A–469J.

Career Choices

Forester

Foresters work for private companies or governments to protect and manage forest land. They also supervise the planting of trees and use controlled burning to clear weeds, brush, and logging debris.

Online Research For information about a career as a forester, visit: www.algebra2.com/careers

FORESTRY For Exercises 35 and 36, use the following information.
A forester at an outpost and another forester at the primary station both heard an explosion. The outpost and the primary station are 6 kilometers apart.

35. If one forester heard the explosion 6 seconds before the other, write an equation that describes all the possible locations of the explosion. Place the two forester stations on the x-axis with the midpoint between the stations at the origin. The transverse axis is horizontal. (*Hint*: The speed of sound is about 0.35 kilometer per second.)
36. Draw a sketch of the possible locations of the explosion. Include the ranger stations in the drawing.

35–36. See margin.

37. **STRUCTURAL DESIGN** An architect's design for a building includes some large pillars with cross sections in the shape of hyperbolas. The curves can be modeled by the equation $\frac{x^2}{0.25} - \frac{y^2}{9} = 1$, where the units are in meters. If the pillars are 4 meters tall, find the width of the top of each pillar and the width of each pillar at the narrowest point in the middle. Round to the nearest centimeter. 120 cm, 100 cm

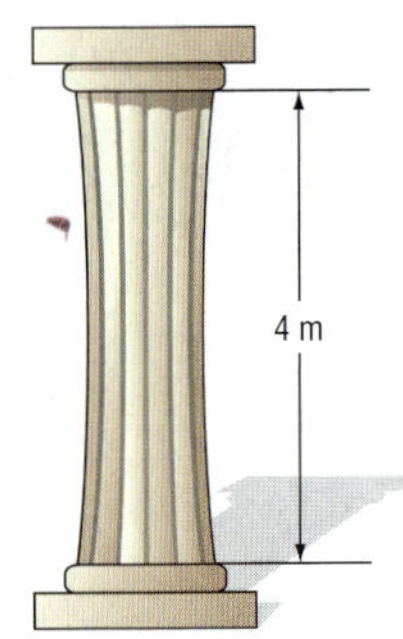

38. **CRITICAL THINKING** A hyperbola with a horizontal transverse axis contains the point at (4, 3). The equations of the asymptotes are $y - x = 1$ and $y + x = 5$. Write the equation for the hyperbola. $\frac{(x-2)^2}{4} - \frac{(y-3)^2}{4} = 1$

Answers

35. $\frac{x^2}{1.1025} - \frac{y^2}{7.8975} = 1$

36.

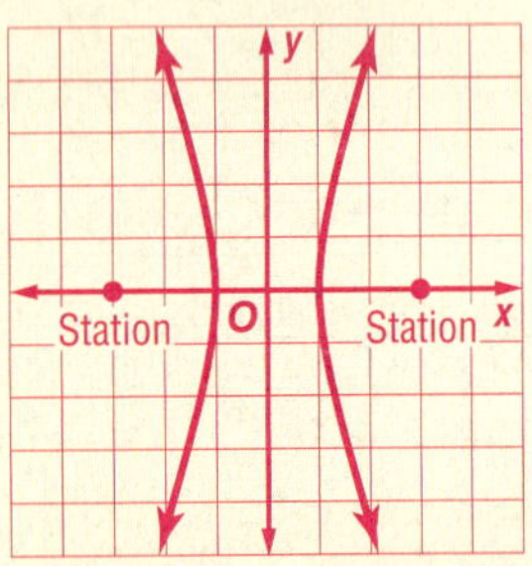

40. Hyperbolas and parabolas have different graphs and different reflective properties. Answers should include the following.
    - Hyperbolas have two branches, two foci, and two vertices. Parabolas have only one branch, one focus, and one vertex. Hyperbolas have asymptotes, but parabolas do not.
    - Hyperbolas reflect rays directed at one focus toward the other focus. Parabolas reflect parallel incoming rays toward the only focus.

43.

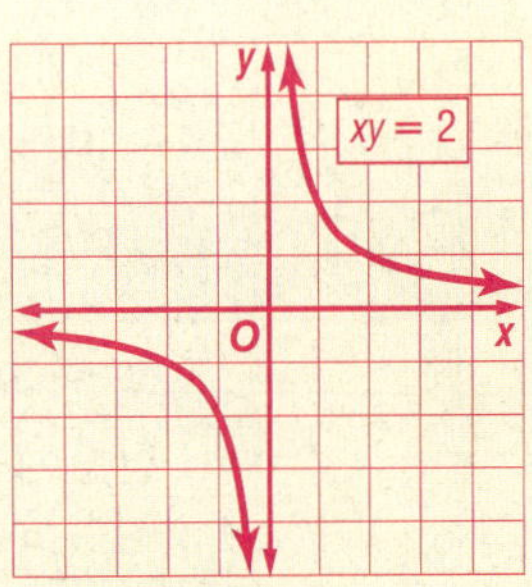

45.

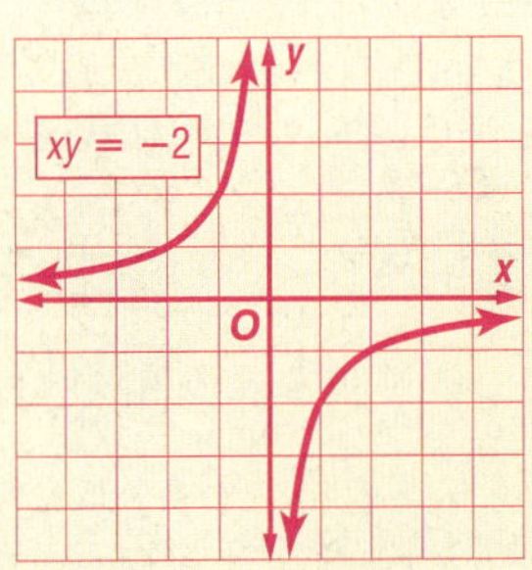

46. The graph of $xy = -2$ can be obtained by reflecting the graph of $xy = 2$ over the x-axis or over the y-axis. The graph of $xy = -2$ can also be obtained by rotating the graph of $xy = 2$ by 90°.

39. about 47.32 ft

★ 39. **PHOTOGRAPHY** A curved mirror is placed in a store for a wide-angle view of the room. The right-hand branch of $\frac{x^2}{1} - \frac{y^2}{3} = 1$ models the curvature of the mirror. A small security camera is placed so that all of the 2-foot diameter of the mirror is visible. If the back of the room lies on $x = -18$, what width of the back of the room is visible to the camera? (*Hint*: Find the equations of the lines through the focus and each edge of the mirror.)

40. WRITING IN MATH Answer the question that was posed at the beginning of the lesson. **See margin.**

How are hyperbolas different from parabolas?

Include the following in your answer:

- differences in the graphs of hyperbolas and parabolas, and
- differences in the reflective properties of hyperbolas and parabolas.

Standardized Test Practice

41. A leg of an isosceles right triangle has a length of 5 units. What is the length of the hypotenuse? **C**

(A) $\frac{5\sqrt{2}}{2}$ units (B) 5 units (C) $5\sqrt{2}$ units (D) 10 units

42. In the figure, what is the sum of the slopes of $\overline{AB}$ and $\overline{AC}$? **B**

(A) −1 (B) 0 (C) 1 (D) 8

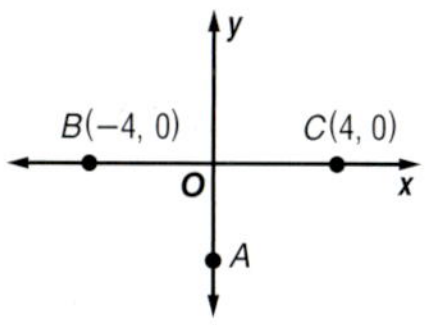

Extending the Lesson

A hyperbola with asymptotes that are not perpendicular is called a **nonrectangular hyperbola**. Most of the hyperbolas you have studied so far are nonrectangular. A **rectangular hyperbola** has perpendicular asymptotes. For example, the graph of $x^2 - y^2 = 1$ is a rectangular hyperbola. The graphs of equations of the form $xy = c$, where c is a constant, are rectangular hyperbolas with the coordinate axes as their asymptotes.

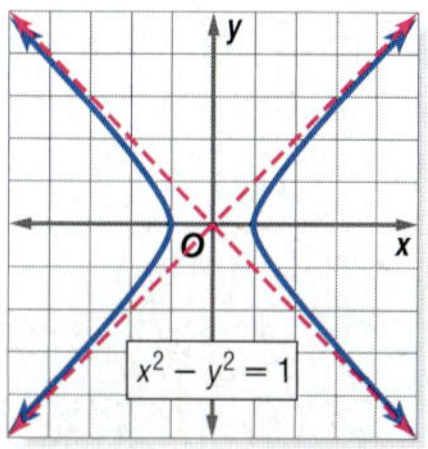

For Exercises 43 and 44, consider the equation $xy = 2$.

43. Plot some points and use them to graph the equation. Be sure to consider negative values for the variables. **See margin.**

44. $(\sqrt{2}, \sqrt{2})$, $(-\sqrt{2}, -\sqrt{2})$

44. Find the coordinates of the vertices of the graph of the equation.

45. Graph $xy = -2$. **See margin.**

46. Describe the transformations that can be applied to the graph of $xy = 2$ to obtain the graph of $xy = -2$. **See margin.**

Maintain Your Skills

Mixed Review

47–49. See margin.

Write an equation for the ellipse that satisfies each set of conditions. *(Lesson 8-4)*

47. endpoints of major axis at (1, 2) and (9, 2), endpoints of minor axis at (5, 1) and (5, 3)

48. major axis 8 units long and parallel to y-axis, minor axis 6 units long, center at (−3, 1)

49. foci at (5, 4) and (−3, 4), major axis 10 units long

Answers

47. $\frac{(x-5)^2}{16} + \frac{(y-2)^2}{1} = 1$

48. $\frac{(y-1)^2}{16} + \frac{(x+3)^2}{9} = 1$

49. $\frac{(x-1)^2}{25} + \frac{(y-4)^2}{9} = 1$

Enrichment, p. 484

Rectangular Hyperbolas

A **rectangular hyperbola** is a hyperbola with perpendicular asymptotes. For example, the graph of $x^2 - y^2 = 1$ is a rectangular hyperbola. A hyperbola with asymptotes that are not perpendicular is called a **nonrectangular hyperbola**. The graphs of equations of the form $xy = c$, where c is a constant, are rectangular hyperbolas.

Make a table of values and plot points to graph each rectangular hyperbola below. Be sure to consider negative values for the variables. See students' tables.

1. $xy = -4$

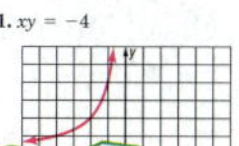

2. $xy = 3$

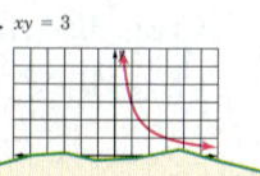

Study Guide and Intervention, p. 479 (shown) and p. 480

Equations of Hyperbolas A **hyperbola** is the set of all points in a plane such that the absolute value of the *difference* of the distances from any point on the hyperbola to any two given points in the plane, called the **foci**, is constant.

In the table, the lengths a, b, and c are related by the formula $c^2 = a^2 + b^2$.

Standard Form of Equation	$\frac{(x-h)^2}{a^2} - \frac{(y-k)^2}{b^2} = 1$	$\frac{(y-k)^2}{a^2} - \frac{(x-h)^2}{b^2} = 1$
Equations of the Asymptotes	$y - k = \pm\frac{b}{a}(x - h)$	$y - k = \pm\frac{a}{b}(x - h)$
Transverse Axis	Horizontal	Vertical
Foci	$(h - c, k), (h + c, k)$	$(h, k - c), (h, k + c)$
Vertices	$(h - a, k), (h + a, k)$	$(h, k - a), (h, k + a)$

Example **Write an equation for the hyperbola with vertices (−2, 1) and (6, 1) and foci (−4, 1) and (8, 1).**

Use a sketch to orient the hyperbola correctly. The center of the hyperbola is the midpoint of the segment joining the two vertices. The center is $(\frac{-2+6}{2}, 1)$, or (2, 1). The value of a is the distance from the center to a vertex, so $a = 4$. The value of c is the distance from the center to a focus, so $c = 6$.

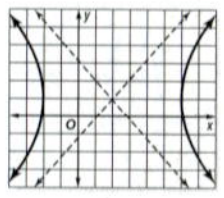

$c^2 = a^2 + b^2$
$6^2 = 4^2 + b^2$
$b^2 = 36 - 16 = 20$

Use h, k, a^2, and b^2 to write an equation of the hyperbola.

$\frac{(x-2)^2}{16} - \frac{(y-1)^2}{20} = 1$

Exercises

Write an equation for the hyperbola that satisfies each set of conditions.

1. vertices (−7, 0) and (7, 0), conjugate axis of length 10 $\frac{x^2}{49} - \frac{y^2}{25} = 1$
2. vertices (−2, −3) and (4, −3), foci (−5, −3) and (7, −3) $\frac{(x-1)^2}{9} - \frac{(y+3)^2}{27} = 1$
3. vertices (4, 3) and (4, −5), conjugate axis of length 4 $\frac{(y+1)^2}{16} - \frac{(x-4)^2}{4} = 1$
4. vertices (−8, 0) and (8, 0), equation of asymptotes $y = \pm\frac{1}{6}x$ $\frac{x^2}{64} - \frac{9y^2}{16} = 1$
5. vertices (−4, 6) and (−4, −2), foci (−4, 10) and (−4, −6) $\frac{(y-2)^2}{16} - \frac{(x+4)^2}{48} = 1$

Skills Practice, p. 481 and Practice, p. 482 (shown)

Write an equation for each hyperbola.

1.

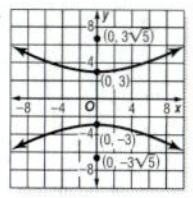

$\frac{y^2}{9} - \frac{x^2}{36} = 1$

2.

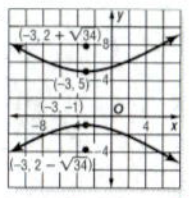

$\frac{(y-2)^2}{9} - \frac{(x+3)^2}{25} = 1$

3.

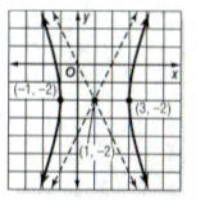

$\frac{(x-1)^2}{4} - \frac{(y+2)^2}{16} = 1$

Write an equation for the hyperbola that satisfies each set of conditions.

4. vertices (0, 7) and (0, −7), conjugate axis of length 18 units $\frac{y^2}{49} - \frac{x^2}{81} = 1$
5. vertices (1, −1) and (1, −9), conjugate axis of length 6 units $\frac{(y+5)^2}{16} - \frac{(x-1)^2}{9} = 1$
6. vertices (−5, 0) and (5, 0), foci $(\pm\sqrt{26}, 0)$ $\frac{x^2}{25} - \frac{y^2}{1} = 1$
7. vertices (1, 1) and (1, −3), foci $(1, -1 \pm \sqrt{5})$ $\frac{(y+1)^2}{4} - \frac{(x-1)^2}{1} = 1$

Find the coordinates of the vertices and foci and the equations of the asymptotes for the hyperbola with the given equation. Then graph the hyperbola.

8. $\frac{y^2}{16} - \frac{x^2}{4} = 1$
$(0, \pm 4)$; $(0, \pm 2\sqrt{5})$; $y = \pm 2x$

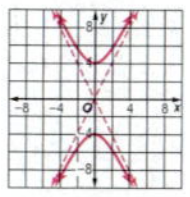

9. $\frac{(y-2)^2}{1} - \frac{(x-1)^2}{4} = 1$
(1, 3), (1, 1); $(1, 2 \pm \sqrt{5})$; $y - 2 = \pm\frac{1}{2}(x - 1)$

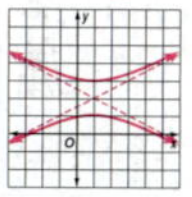

10. $\frac{(y+2)^2}{4} - \frac{(x-3)^2}{4} = 1$
(3, 0), (3, −4); $(3, -2 \pm 2\sqrt{2})$; $y + 2 = \pm(x - 3)$

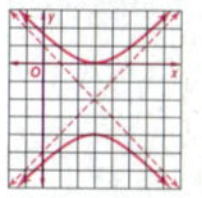

11. **ASTRONOMY** Astronomers use special X-ray telescopes to observe the sources of celestial X rays. Some X-ray telescopes are fitted with a metal mirror in the shape of a hyperbola, which reflects the X rays to a focus. Suppose the vertices of such a mirror are located at (−3, 0) and (3, 0), and one focus is located at (5, 0). Write an equation that models the hyperbola formed by the mirror. $\frac{x^2}{9} - \frac{y^2}{16} = 1$

Reading to Learn Mathematics, p. 483

ELL

Pre-Activity **How are hyperbolas different from parabolas?**

Read the introduction to Lesson 8-5 at the top of page 441 in your textbook.

Look at the sketch of a hyperbola in the introduction to this lesson. List three ways in which hyperbolas are different from parabolas.
Sample answer: A hyperbola has two branches, while a parabola is one continuous curve. A hyperbola has two foci, while a parabola has one focus. A hyperbola has two vertices, while a parabola has one vertex.

Reading the Lesson

1. The graph at the right shows the hyperbola whose equation in standard form is $\frac{x^2}{16} - \frac{y^2}{9} = 1$.

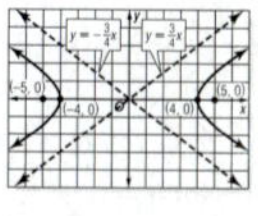

The point (0, 0) is the ___center___ of the hyperbola.

The points (4, 0) and (−4, 0) are the ___vertices___ of the hyperbola.

The points (5, 0) and (−5, 0) are the ___foci___ of the hyperbola.

The segment connecting (4, 0) and (−4, 0) is called the ___transverse___ axis.

The segment connecting (0, 3) and (0, −3) is called the ___conjugate___ axis.

The lines $y = \frac{3}{4}x$ and $y = -\frac{3}{4}x$ are called the ___asymptotes___.

2. Study the hyperbola graphed at the right.

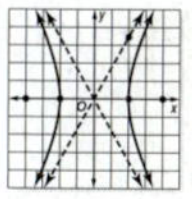

The center is ___(0, 0)___.

The value of a is ___2___.

The value of c is ___4___.

To find b^2, solve the equation ___c^2___ = ___a^2___ + ___b^2___.

The equation in standard form for this hyperbola is ___$\frac{x^2}{4} - \frac{y^2}{12} = 1$___.

Helping You Remember

3. What is an easy way to remember the equation relating the values of a, b, and c for a hyperbola? This equation looks just like the Pythagorean Theorem, although the variables represent different lengths in a hyperbola than in a right triangle.

4 Assess

Open-Ended Assessment

Writing Have students write a paragraph that explains how hyperbolas are related to other conic sections, and how they are alike and different from the others.

Tips for New Teachers

Intervention To diagnose confusion, ask students to draw a quick sketch, without labels, of each of the conic sections and to name the shape of each of the graphs they drew.

Getting Ready for Lesson 8-6

PREREQUISITE SKILL In Lesson 8-6, students will learn how to identify conic sections from their equations, including comparing specific coefficients within equations. Exercises 58–63 should be used to determine your students' familiarity with identifying coefficients.

Assessment Options

Practice Quiz 2 The quiz provides students with a brief review of the concepts and skills in Lessons 8-4 and 8-5. Lesson numbers are given to the right of exercises or instruction lines so students can review concepts not yet mastered.

Answer

50.

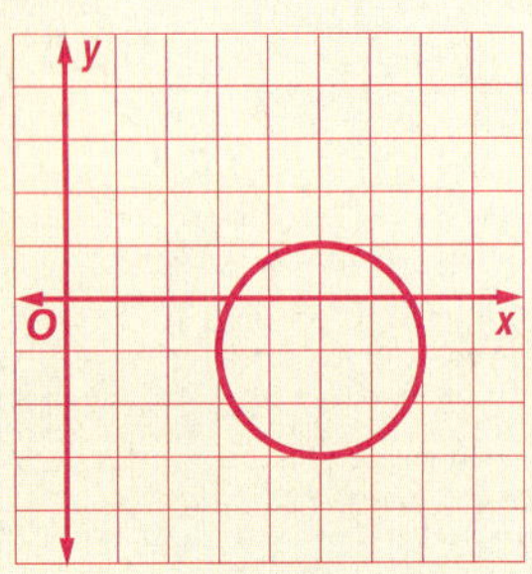

---

50. Find the center and radius of the circle with equation $x^2 + y^2 - 10x + 2y + 22 = 0$. Then graph the circle. *(Lesson 8-3)* **(5, −1), 2 units; See margin for graph.**

Solve each equation by factoring. *(Lesson 6-2)*

51. $x^2 + 6x + 8 = 0$ **−4, −2**

52. $2q^2 + 11q = 21$ **$-7, \frac{3}{2}$**

Perform the indicated operations, if possible. *(Lesson 4-5)*

53. $\begin{bmatrix} 2 & -1 \\ 0 & 5 \end{bmatrix} \cdot \begin{bmatrix} -3 & 2 \\ 1 & 4 \end{bmatrix}$ **$\begin{bmatrix} -7 & 0 \\ 5 & 20 \end{bmatrix}$**

54. $[1 \quad -3] \cdot \begin{bmatrix} 4 & -2 & 1 \\ -3 & 2 & 0 \end{bmatrix}$ **[13 −8 1]**

55. about 5,330,000 subscribers per year

55. **PAGERS** Refer to the graph at the right. What was the average rate of change of the number of pager subscribers from 1996 to 1999? *(Lesson 2-3)*

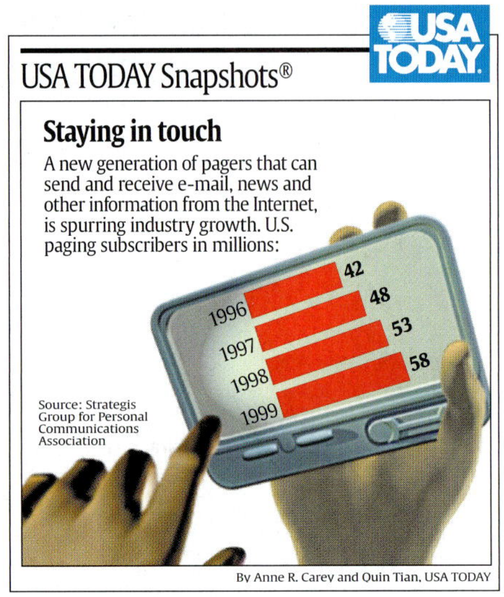

56. Solve $|2x + 1| = 9$. *(Lesson 1-4)* **−5, 4**

57. Simplify $7x + 8y + 9y - 5x$. *(Lesson 1-2)* **$2x + 17y$**

Getting Ready for the Next Lesson

PREREQUISITE SKILL **Each equation is of the form $Ax^2 + Bxy + Cy^2 + Dx + Ey + F = 0$. Identify the values of A, B, and C.**
*(To review **coefficients**, see Lesson 5-1.)*

58. $2x^2 + 3xy - 5y^2 = 0$ **2, 3, −5**

59. $x^2 - 2xy + 9y^2 = 0$ **1, −2, 9**

60. $-3x^2 + xy + 2y^2 + 4x - 7y = 0$

61. $5x^2 - 2y^2 + 5x - y = 0$ **5, 0, −2**

62. $x^2 - 4x + 5y + 2 = 0$ **1, 0, 0**

63. $xy - 2x - 3y + 6 = 0$ **0, 1, 0**

60. −3, 1, 2

Practice Quiz 2 — *Lessons 8-4 and 8-5*

1. Write an equation of the ellipse with foci at (3, 8) and (3, −6) and endpoints of the major axis at (3, −8) and (3, 10). *(Lesson 8-4)* **$\frac{(y-1)^2}{81} + \frac{(x-3)^2}{32} = 1$**

2–3. See pp. 469A–469J for graphs.

Find the coordinates of the center and foci and the lengths of the major and minor axes of the ellipse with the given equation. Then graph the ellipse. *(Lesson 8-4)*

2. $\frac{(x-4)^2}{9} + \frac{(y+2)^2}{1} = 1$ **$(4, -2)$; $(4 \pm 2\sqrt{2}, -2)$; 6; 2**

3. $16x^2 + 5y^2 + 32x - 10y - 59 = 0$ **$(-1, 1)$; $(-1, 1 \pm \sqrt{11})$; 8; $2\sqrt{5}$**

Write an equation for the hyperbola that satisfies each set of conditions. *(Lesson 8-5)*

4. vertices (−3, 0) and (3, 0), conjugate axis of length 8 units **$\frac{x^2}{9} - \frac{y^2}{16} = 1$**

5. vertices (−2, 2) and (6, 2), foci $(2 \pm \sqrt{21}, 2)$ **5. $\frac{(x-2)^2}{16} - \frac{(y-2)^2}{5} = 1$**

Online Lesson Plans

USA TODAY Education's Online site offers resources and interactive features connected to each day's newspaper. *Experience TODAY*, USA TODAY's daily lesson plan, is available on the site and delivered daily to subscribers. This plan provides instruction for integrating USA TODAY graphics and key editorial features into your mathematics classroom. Log on to www.education.usatoday.com.

448 Chapter 8 Conic Sections

8-6 Conic Sections

What You'll Learn

- Write equations of conic sections in standard form.
- Identify conic sections from their equations.

How can you use a flashlight to make conic sections?

Recall that parabolas, circles, ellipses, and hyperbolas are called conic sections because they are the cross sections formed when a double cone is sliced by a plane. You can use a flashlight and a flat surface to make patterns in the shapes of conic sections.

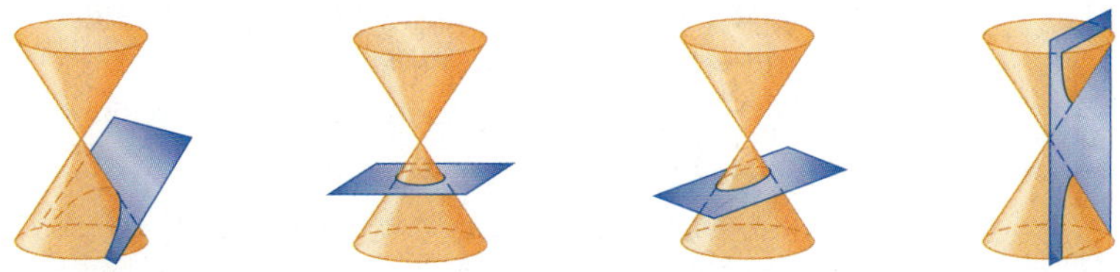

parabola circle ellipse hyperbola

STANDARD FORM The equation of any conic section can be written in the form of the general quadratic equation $Ax^2 + Bxy + Cy^2 + Dx + Ey + F = 0$, where A, B, and C are not all zero. If you are given an equation in this general form, you can complete the square to write the equation in one of the standard forms you have learned.

Study Tip

Reading Math
In this lesson, the word *ellipse* means an ellipse that is not a circle.

Concept Summary — Standard Form of Conic Sections

Conic Section	Standard Form of Equation
Parabola	$y = a(x - h)^2 + k$ or $x = a(y - k)^2 + h$
Circle	$(x - h)^2 + (y - k)^2 = r^2$
Ellipse	$\frac{(x-h)^2}{a^2} + \frac{(y-k)^2}{b^2} = 1$ or $\frac{(y-k)^2}{a^2} + \frac{(x-h)^2}{b^2} = 1$, $a \neq b$
Hyperbola	$\frac{(x-h)^2}{a^2} - \frac{(y-k)^2}{b^2} = 1$ or $\frac{(y-k)^2}{a^2} - \frac{(x-h)^2}{b^2} = 1$

Example 1 Rewrite an Equation of a Conic Section

Write the equation $x^2 + 4y^2 - 6x - 7 = 0$ in standard form. State whether the graph of the equation is a *parabola, circle, ellipse,* or *hyperbola.* Then graph the equation.

Write the equation in standard form.

$x^2 + 4y^2 - 6x - 7 = 0$	Original equation
$x^2 - 6x + \blacksquare + 4y^2 = 7 + \blacksquare$	Isolate terms.
$x^2 - 6x + 9 + 4y^2 = 7 + 9$	Complete the square.
$(x - 3)^2 + 4y^2 = 16$	$x^2 - 6x + 9 = (x - 3)^2$
$\frac{(x-3)^2}{16} + \frac{y^2}{4} = 1$	Divide each side by 16.

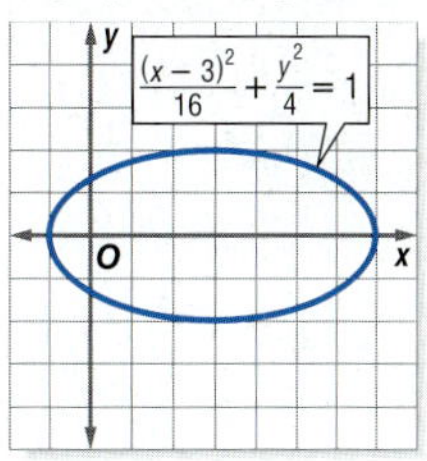

The graph of the equation is an ellipse with its center at (3, 0).

www.algebra2.com/extra_examples

8-6 Lesson Notes

1 Focus

5-Minute Check Transparency 8-6 Use as a quiz or review of Lesson 8-5.

Mathematical Background notes are available for this lesson on p. 410D.

Building on Prior Knowledge

In Chapter 6, students learned how to analyze graphs of quadratic equations and rewrite the equations in different forms. In this lesson, students will use similar techniques to analyze graphs of conic sections.

How can you use a flashlight to make conic sections?

Ask students:

- Describe the plane that forms a hyperbola. **perpendicular to the base of the cone**
- Describe the plane that forms a parabola. **parallel to the slant height of the cone**

Resource Manager

Workbook and Reproducible Masters

Chapter 8 Resource Masters
- Study Guide and Intervention, pp. 485–486
- Skills Practice, p. 487
- Practice, p. 488
- Reading to Learn Mathematics, p. 489
- Enrichment, p. 490
- Assessment, p. 512

Teaching Algebra With Manipulatives Masters, pp. 268–269

Transparencies

5-Minute Check Transparency 8-6
Answer Key Transparencies

Technology

Interactive Chalkboard

2 Teach

STANDARD FORM

In-Class Example

Power Point®

1 Write the equation $y^2 = 18 - 2x^2$ in standard form. State whether the graph of the equation is a *parabola*, *circle*, *ellipse*, or *hyperbola*. Then graph the equation. $\frac{x^2}{9} + \frac{y^2}{18} = 1$; **ellipse**

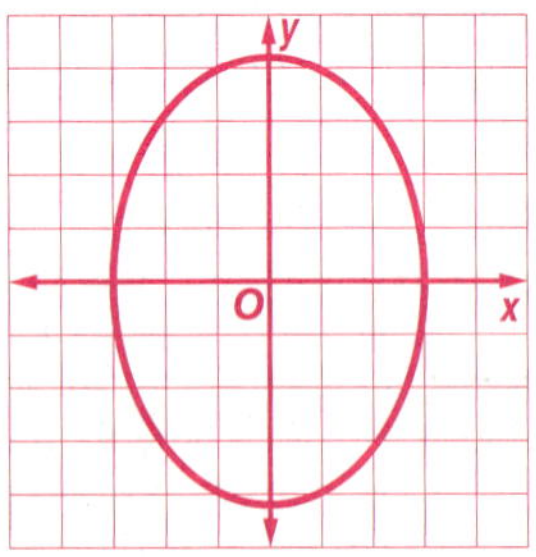

IDENTIFY CONIC SECTIONS

In-Class Example

Power Point®

2 Without writing the equation in standard form, state whether the graph of the equation is a *parabola*, *circle*, *ellipse*, or *hyperbola*.

a. $3y^2 - x^2 - 9 = 0$ **hyperbola**

b. $2x^2 + 2y^2 + 16x - 20y = -32$ **circle**

c. $y^2 - 2x - 4y + 10 = 0$ **parabola**

3 Practice/Apply

Study Notebook

Have students—
- add the definitions/examples of the vocabulary terms to their Vocabulary Builder worksheets for Chapter 8.
- include any other item(s) that they find helpful in mastering the skills in this lesson.

IDENTIFY CONIC SECTIONS Instead of writing the equation in standard form, you can determine what type of conic section an equation of the form $Ax^2 + Bxy + Cy^2 + Dx + Ey + F = 0$, where $B = 0$, represents by looking at A and C.

Concept Summary — *Identifying Conic Sections*

Conic Section	Relationship of *A* and *C*
Parabola	$A = 0$ or $C = 0$, but not both.
Circle	$A = C$
Ellipse	A and C have the same sign and $A \neq C$.
Hyperbola	A and C have opposite signs.

Example 2 Analyze an Equation of a Conic Section

Without writing the equation in standard form, state whether the graph of each equation is a *parabola, circle, ellipse,* or *hyperbola*.

a. $y^2 - 2x^2 - 4x - 4y - 4 = 0$

$A = -2$ and $C = 1$. Since A and C have opposite signs, the graph is a hyperbola.

b. $4x^2 + 4y^2 + 20x - 12y + 30 = 0$

$A = 4$ and $C = 4$. Since $A = C$, the graph is a circle.

c. $y^2 - 3x + 6y + 12 = 0$

$C = 1$. Since there is no x^2 term, $A = 0$. The graph is a parabola.

Check for Understanding

Concept Check

1. **OPEN ENDED** Write an equation of the form $Ax^2 + Bxy + Cy^2 + Dx + Ey + F = 0$, where $A = 2$, that represents a circle. **Sample answer: $2x^2 + 2y^2 - 1 = 0$**
2. **Write** the general quadratic equation for which $A = 2$, $B = 0$, $C = 0$, $D = -4$, $E = 7$, and $F = 1$. **$2x^2 - 4x + 7y + 1 = 0$**
3. **Explain** why the graph of $x^2 + y^2 - 4x + 2y + 5 = 0$ is a single point. **See margin.**

Guided Practice

GUIDED PRACTICE KEY	
Exercises	Examples
4–7, 10, 11	1
8, 9	2

Write each equation in standard form. State whether the graph of the equation is a *parabola, circle, ellipse,* or *hyperbola*. Then graph the equation.

4. $y = x^2 + 3x + 1$ **parabola**
5. $y^2 - 2x^2 - 16 = 0$ **hyperbola**
6. $x^2 + y^2 = x + 2$ **circle**
7. $x^2 + 4y^2 + 2x - 24y + 33 = 0$ **ellipse**

4–7. See pp. 469A–469J for equations and graphs.

Without writing the equation in standard form, state whether the graph of each equation is a *parabola, circle, ellipse,* or *hyperbola*.

8. $y^2 - x - 10y + 34 = 0$ **parabola**
9. $3x^2 + 2y^2 + 12x - 28y + 104 = 0$ **ellipse**

Application

AVIATION **For Exercises 10 and 11, use the following information.**
When an airplane flies faster than the speed of sound, it produces a shock wave in the shape of a cone. Suppose the shock wave intersects the ground in a curve that can be modeled by $x^2 - 14x + 4 = 9y^2 - 36y$.

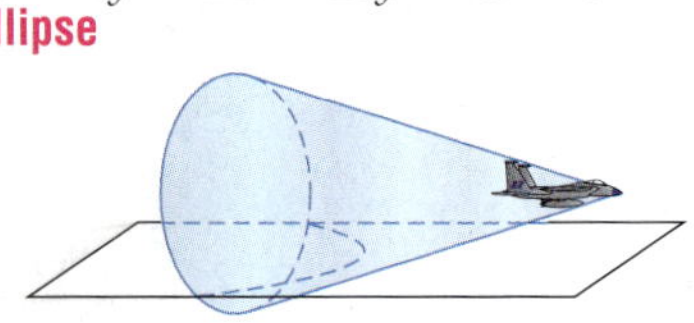

10. Identify the shape of the curve. **hyperbola**
11. Graph the equation. **See pp. 469A–469J.**

DAILY INTERVENTION

Differentiated Instruction

Intrapersonal Encourage students to make a list of the techniques and hints that they use as they answer questions like the ones in this lesson. Invite students to share their techniques with the class.

450 Chapter 8 Conic Sections

★ indicates increased difficulty

Practice and Apply

Homework Help

For Exercises	See Examples
12–32	1
33–43	2

Extra Practice
See page 846.

12–29. See pp. 469A–469J for equations and graphs.

Write each equation in standard form. State whether the graph of the equation is a *parabola, circle, ellipse,* or *hyperbola.* Then graph the equation.

12. $6x^2 + 6y^2 = 162$ circle
13. $4x^2 + 2y^2 = 8$ ellipse
14. $x^2 = 8y$ parabola
15. $4y^2 - x^2 + 4 = 0$ hyperbola
16. $(x - 1)^2 - 9(y - 4)^2 = 36$ hyperbola
17. $y + 4 = (x - 2)^2$ parabola
18. $(y - 4)^2 = 9(x - 4)$ parabola
19. $x^2 + y^2 + 4x - 6y = -4$ circle
20. $x^2 + y^2 + 6y + 13 = 40$ circle
21. $x^2 - y^2 + 8x = 16$ hyperbola
22. $x^2 + 2y^2 = 2x + 8$ ellipse
23. $x^2 - 8y + y^2 + 11 = 0$ circle
24. $9y^2 + 18y = 25x^2 + 216$ hyperbola
25. $3x^2 + 4y^2 + 8y = 8$ ellipse
26. $x^2 + 4y^2 - 11 = 2(4y - x)$ ellipse
27. $y + x^2 = -(8x + 23)$ parabola
★ 28. $6x^2 - 24x - 5y^2 - 10y - 11 = 0$ hyperbola
★ 29. $25y^2 + 9x^2 - 50y - 54x = 119$ ellipse

30. **ASTRONOMY** The orbits of comets follow paths in the shapes of conic sections. For example, Halley's Comet follows an elliptical orbit with the Sun located at one focus. What type(s) of orbit(s) pass by the Sun only once? parabolas and hyperbolas

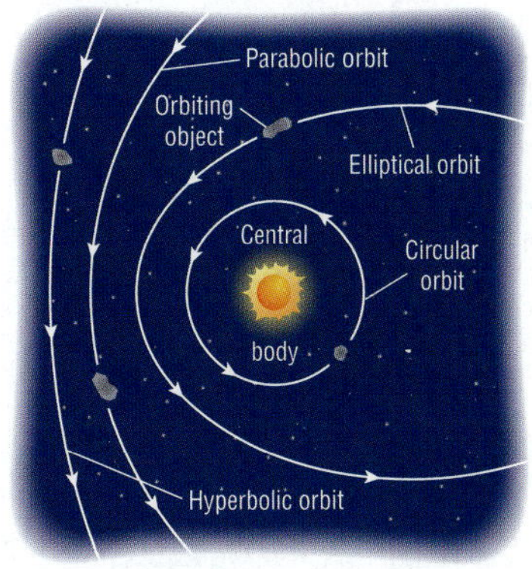

WATER **For Exercises 31 and 32, use the following information.**
If two stones are thrown into a lake at different points, the points of intersection of the resulting ripples will follow a conic section. Suppose the conic section has the equation $x^2 - 2y^2 - 2x - 5 = 0$.

31. Identify the shape of the curve. hyperbola
32. Graph the equation. See pp. 469A–469J.

Without writing the equation in standard form, state whether the graph of each equation is a *parabola, circle, ellipse,* or *hyperbola.*

33. $x^2 + y^2 - 8x - 6y + 5 = 0$ circle
34. $3x^2 - 2y^2 + 32y - 134 = 0$ hyperbola
35. $y^2 + 18y - 2x = -84$ parabola
36. $7x^2 - 28x + 4y^2 + 8y = -4$ ellipse
★ 37. $5x^2 + 6x - 4y = x^2 - y^2 - 2x$ ellipse
★ 38. $2x^2 + 12x + 18 - y^2 = 3(2 - y^2) + 4y$ circle
39. Identify the shape of the graph of the equation $2x^2 + 3x - 4y + 2 = 0$. parabola
40. What type of conic section is represented by the equation $y^2 - 6y = x^2 - 8$? hyperbola

For Exercises 41–43, match each equation below with the situation that it could represent.

a. $9x^2 + 4y^2 - 36 = 0$
b. $0.004x^2 - x + y - 3 = 0$
c. $x^2 + y^2 - 20x + 30y - 75 = 0$

41. **SPORTS** the flight of a baseball b
42. **PHOTOGRAPHY** the oval opening in a picture frame a
43. **GEOGRAPHY** the set of all points that are 20 miles from a landmark c

www.algebra2.com/self_check_quiz

Answer

3. The standard form of the equation is $(x - 2)^2 + (y + 1)^2 = 0$. This is an equation of a circle centered at $(2, -1)$ with radius 0. In other words, $(2, -1)$ is the only point that satisfies the equation.

Enrichment, p. 490

Loci

A *locus* (plural, *loci*) is the set of all points, and only those points, that satisfy a given set of conditions. In geometry, figures often are defined as loci. For example, a circle is the locus of points of a plane that are a given distance from a given point. The definition leads naturally to an equation whose graph is the curve described.

Example **Write an equation of the locus of points that are the same distance from (3, 4) and $y = -4$.**

Recognizing that the locus is a parabola with focus (3, 4) and directrix $y = -4$, you can find that $h = 3$, $k = 0$, and $a = 4$ where (h, k) is the vertex and 4 units is the distance from the vertex to both the focus and directrix.

Thus, an equation for the parabola is $y = \frac{1}{16}(x - 3)^2$.

The problem also may be approached analytically as follows:

Let (x, y) be a point of the locus.

The distance from (3, 4) to (x, y) = the distance from $y = -4$ to (x, y)

Study Guide and Intervention, p. 485 (shown) and p. 486

Standard Form Any conic section in the coordinate plane can be described by an equation of the form

$Ax^2 + Bxy + Cy^2 + Dx + Ey + F = 0$, where A, B, and C are not all zero.

One way to tell what kind of conic section an equation represents is to rearrange terms and complete the square, if necessary, to get one of the standard forms from an earlier lesson. This method is especially useful if you are going to graph the equation.

Example **Write the equation $3x^2 - 4y^2 - 30x - 8y + 59 = 0$ in standard form. State whether the graph of the equation is a *parabola, circle, ellipse,* or *hyperbola.***

$3x^2 - 4y^2 - 30x - 8y + 59 = 0$	Original equation
$3x^2 - 30x - 4y^2 - 8y = -59$	Isolate terms.
$3(x^2 - 10x + ■) - 4(y^2 + 2y + ■) = -59 + ■ + ■$	Factor out common multiples.
$3(x^2 - 10x + 25) - 4(y^2 + 2y + 1) = -59 + 3(25) + (-4)(1)$	Complete the squares.
$3(x - 5)^2 - 4(y + 1)^2 = 12$	Simplify.
$\frac{(x-5)^2}{4} - \frac{(y+1)^2}{3} = 1$	Divide each side by 12.

The graph of the equation is a hyperbola with its center at $(5, -1)$. The length of the transverse axis is 4 units and the length of the conjugate axis is $2\sqrt{3}$ units.

Exercises

Write each equation in standard form. State whether the graph of the equation is a *parabola, circle, ellipse,* or *hyperbola.*

1. $x^2 + y^2 - 6x + 4y + 3 = 0$ $(x - 3)^2 + (y + 2)^2 = 10$; circle
2. $x^2 + 2y^2 + 6x - 20y + 53 = 0$ $\frac{(x+3)^2}{6} + \frac{(y-5)^2}{3} = 1$; ellipse
3. $6x^2 - 60x - y + 161 = 0$ $y = 6(x - 5)^2 + 11$; parabola
4. $x^2 + y^2 - 4x - 14y + 29 = 0$ $(x - 2)^2 + (y - 7)^2 = 24$; circle
5. $6x^2 - 5y^2 + 24x + 20y - 56 = 0$ $\frac{(x+2)^2}{10} - \frac{(y-2)^2}{12} = 1$; hyperbola
6. $3y^2 + x - 24y + 46 = 0$ $x = -3(y - 4)^2 + 2$; parabola
7. $x^2 - 4y^2 - 16x + 24y - 36 = 0$ $\frac{(x-8)^2}{64} - \frac{(y-3)^2}{16} = 1$; hyperbola
8. $x^2 + 2y^2 + 8x + 4y + 2 = 0$ $\frac{(x+4)^2}{16} + \frac{(y+1)^2}{8} = 1$; ellipse
9. $4x^2 + 48x + y + 158 = 0$ $y = -4(x + 6)^2 - 14$; parabola
10. $3x^2 + y^2 - 48x - 4y + 184 = 0$ $\frac{(x-8)^2}{4} + \frac{(y-2)^2}{12} = 1$; ellipse
11. $-3x^2 + 2y^2 - 18x + 20y + 5 = 0$ $\frac{(y+5)^2}{9} - \frac{(x+3)^2}{6} = 1$; hyperbola
12. $x^2 + y^2 + 8x + 2y + 8 = 0$ $(x + 4)^2 + (y + 1)^2 = 9$; circle

Skills Practice, p. 487 and Practice, p. 488 (shown)

Write each equation in standard form. State whether the graph of the equation is a *parabola, circle, ellipse,* or *hyperbola.* Then graph the equation.

1. $y^2 = -3x$ parabola $x = -\frac{1}{3}y^2$

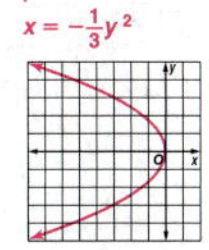

2. $x^2 + y^2 + 6x = 7$ circle $(x + 3)^2 + y^2 = 16$

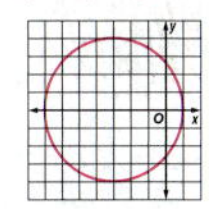

3. $5x^2 - 6y^2 - 30x - 12y = -9$ hyperbola $\frac{(x-3)^2}{6} - \frac{(y+1)^2}{5} = 1$

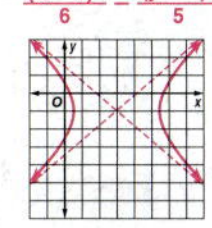

4. $196y^2 = 1225 - 100x^2$ ellipse $\frac{x^2}{12.25} + \frac{y^2}{6.25} = 1$

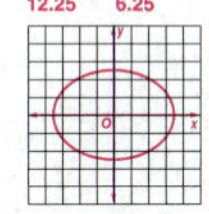

5. $3x^2 = 9 - 3y^2 - 6y$ circle $x^2 + (y + 1)^2 = 4$

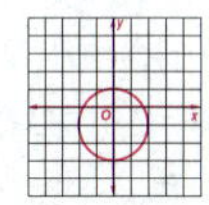

6. $9x^2 + y^2 + 54x - 6y = -81$ ellipse $\frac{(x+3)^2}{1} + \frac{(y-3)^2}{9} = 1$

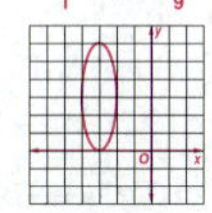

Without writing the equation in standard form, state whether the graph of each equation is a *parabola, circle, ellipse,* or *hyperbola.*

7. $6x^2 + 6y^2 = 36$ circle
8. $4x^2 - y^2 = 16$ hyperbola
9. $9x^2 + 16y^2 - 64y - 80 = 0$ ellipse
10. $5x^2 + 5y^2 - 45 = 0$ circle
11. $x^2 + 2x = y$ parabola
12. $4y^2 - 36x^2 + 4x - 144 = 0$ hyperbola
13. **ASTRONOMY** A satellite travels in an hyperbolic orbit. It reaches the vertex of its orbit at (5, 0) and then travels along a path that gets closer and closer to the line $y = \frac{2}{5}x$. Write an equation that describes the path of the satellite if the center of its hyperbolic orbit is at (0, 0). $\frac{x^2}{25} - \frac{y^2}{4} = 1$

Reading to Learn Mathematics, p. 489 ELL

Pre-Activity **How can you use a flashlight to make conic sections?**

Read the introduction to Lesson 8-6 at the top of page 449 in your textbook.

The figures in the introduction show how a plane can slice a double cone to form the conic sections. Name the conic section that is formed if the plane slices the double cone in each of the following ways:

- The plane is parallel to the base of the double cone and slices through one of the cones that form the double cone. circle
- The plane is perpendicular to the base of the double cone and slices through both of the cones that form the double cone. hyperbola

Reading the Lesson

1. Name the conic section that is the graph of each of the following equations. Give the coordinates of the vertex if the conic section is a parabola and of the center if it is a circle, an ellipse, or a hyperbola.
   a. $\frac{(x-3)^2}{36} + \frac{(y+5)^2}{15} = 1$ ellipse; (3, −5)
   b. $x = -2(y + 1)^2 + 7$ parabola; (7, −1)
   c. $(x - 5)^2 - (y + 5)^2 = 1$ hyperbola; (5, −5)
   d. $(x + 6)^2 + (y - 2)^2 = 1$ circle; (−6, 2)
2. Each of the following is the equation of a conic section. For each equation, identify the values of A and C. Then, without writing the equation in standard form, state whether the graph of each equation is a *parabola, circle, ellipse,* or *hyperbola.*
   a. $2x^2 + y^2 - 6x + 8y + 12 = 0$ A = 2; C = 1; type of graph: ellipse
   b. $2x^2 + 3x - 2y - 5 = 0$ A = 2; C = 0; type of graph: parabola
   c. $5x^2 + 10x + 5y^2 - 20y + 1 = 0$ A = 5; C = 5; type of graph: circle
   d. $x^2 - y^2 + 4x + 2y - 5 = 0$ A = 1; C = −1; type of graph: hyperbola

Helping You Remember

3. What is an easy way to recognize that an equation represents a parabola rather than one of the other conic sections?
   If the equation has an x^2 term and y term but no y^2 term, then the graph is a parabola. Likewise, if the equation has a y^2 term and x term but no x^2 term, then the graph is a parabola.

About the Exercises...

Organization by Objective

- **Standard Form:** 12–32
- **Identify Conic Sections:** 33–43

Odd/Even Assignments

Exercises 12–29 and 33–38 are structured so that students practice the same concepts whether they are assigned odd or even problems.

Assignment Guide

Basic: 13–27 odd, 31–35 odd, 39–43 odd, 44–48, 50–59

Average: 13–43 odd, 44–48, 50–59 (optional: 49)

Advanced: 12–30 even, 31, 32–44 even, 45–56 (optional: 57–59)

4 Assess

Open-Ended Assessment

Modeling Have students use paper folding to construct double right circular cones. Use them to demonstrate the various conic sections.

Getting Ready for Lesson 8-7

PREREQUISITE SKILL In Lesson 8-7, students will solve systems of quadratic equations. Students should be sure they can solve systems of simpler linear equations before continuing. Exercises 57–59 should be used to determine your students' familiarity with solving systems of linear equations.

Assessment Options

Quiz (Lessons 8-5 and 8-6) is available on p. 512 of the *Chapter 8 Resource Masters*.

CRITICAL THINKING **For Exercises 44 and 45, use the following information.**

The graph of an equation of the form $\frac{x^2}{a^2} - \frac{y^2}{b^2} = 0$ is a special case of a hyperbola.

44. Identify the graph of such an equation. **2 intersecting lines**

45. Explain how to obtain such a set of points by slicing a double cone with a plane.

45. The plane should be vertical and contain the axis of the double cone.

46. WRITING IN MATH Answer the question that was posed at the beginning of the lesson. **See pp. 469A - 469J.**

How can you use a flashlight to make conic sections?

Include the following in your answer:

- an explanation of how you could point the flashlight at a ceiling or wall to make a circle, and
- an explanation of how you could point the flashlight to make a branch of a hyperbola.

Standardized Test Practice

47. Which conic section is not symmetric about the y-axis? **D**

(A) $x^2 - y + 3 = 0$ (B) $y^2 - x^2 - 1 = 0$

(C) $6x^2 + y^2 - 6 = 0$ (D) $x^2 + y^2 - 2x - 3 = 0$

48. What is the equation of the graph at the right? **C**

(A) $y = x^2 + 1$ (B) $y - x = 1$

(C) $y^2 - x^2 = 1$ (D) $x^2 + y^2 = 1$

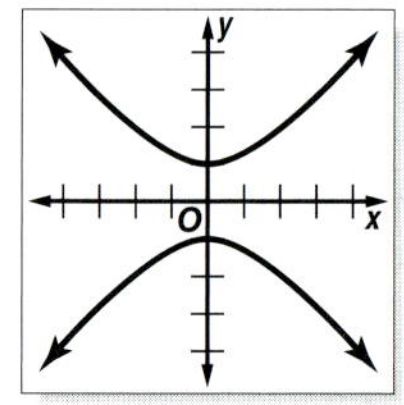

Extending the Lesson

49. Refer to Exercise 43 on page 440. Eccentricity can be studied for conic sections other than ellipses. The expression for the eccentricity of a hyperbola is $\frac{c}{a}$, just as for an ellipse. The eccentricity of a parabola is 1. Find inequalities for the eccentricities of noncircular ellipses and hyperbolas, respectively. $0 < e < 1$, $e > 1$

Maintain Your Skills

Mixed Review

Write an equation of the hyperbola that satisfies each set of conditions. *(Lesson 8-5)*

50. vertices (5, 10) and (5, −2), conjugate axis of length 8 units

50. $\frac{(y-4)^2}{36} - \frac{(x-5)^2}{16} = 1$

51. vertices (6, −6) and (0, −6), foci $(3 \pm \sqrt{13}, -6)$ $\frac{(x-3)^2}{9} - \frac{(y+6)^2}{4} = 1$

52. Find the coordinates of the center and foci and the lengths of the major and minor axes of the ellipse with equation $4x^2 + 9y^2 - 24x + 72y + 144 = 0$. Then graph the ellipse. *(Lesson 8-4)* $(3, -4)$; $(3 \pm \sqrt{5}, -4)$; 6; 4; **See pp. 469A-469J for graph.**

Simplify. Assume that no variable equals 0. *(Lesson 5-1)*

53. $(x^3)^4$ x^{12}

54. $(m^5n^{-3})^2m^2n^7$ $m^{12}n$

55. $\frac{x^2y^{-3}}{x^{-5}y}$ $\frac{x^7}{y^4}$

56. **HEALTH** The prediction equation $y = 205 - 0.5x$ relates a person's maximum heart rate for exercise y and age x. Use the equation to find the maximum heart rate for an 18-year-old. *(Lesson 2-5)* **196 beats per min**

Getting Ready for the Next Lesson

PREREQUISITE SKILL **Solve each system of equations.**

(To review ***solving systems of linear equations****, see Lesson 3-2.)*

57. $y = x + 4$
$2x + y = 10$ **(2, 6)**

58. $4x + y = 14$
$4x - y = 10$ **(3, 2)**

59. $x + 5y = 10$
$3x - 2y = -4$ **(0, 2)**

Teacher to Teacher

Judie Campbell Derry Area H.S., Derry, PA

"As a final project, I have students make a poster with 6 different types of graphs, explaining how each was derived, and then have them label each part of the graph. An explanation of each graph is also required."

Algebra Activity

A Follow-Up of Lesson 8-6

Conic Sections

Recall that a parabola is the set of all points that are equidistant from the focus and the directrix.

You can draw a parabola based on this definition by using special conic graph paper. This graph paper contains a series of concentric circles equally spaced from each other and a series of parallel lines tangent to the circles.

Number the circles consecutively beginning with the smallest circle. Number the lines with consecutive integers as shown in the sample at the right. Be sure that line 1 is tangent to circle 1.

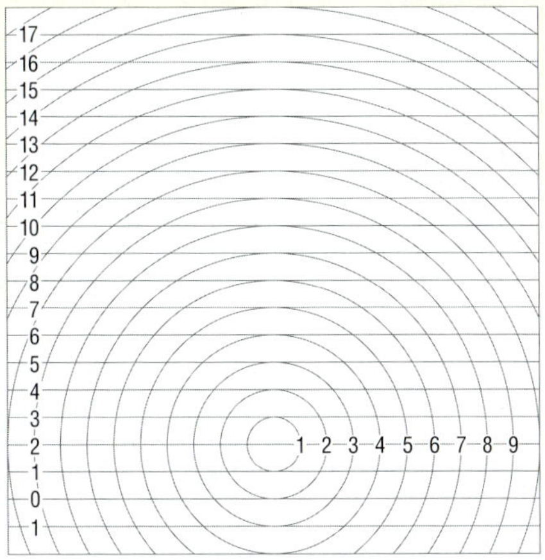

Activity 1

Mark the point at the intersection of circle 1 and line 1. Mark both points that are on line 2 and circle 2. Continue this process, marking both points on line 3 and circle 3, and so on. Then connect the points with a smooth curve.

Look at the diagram at the right. What shape is the graph? Note that every point on the graph is equidistant from the center of the small circle and the line labeled 0. The center of the small circle is the focus of the parabola, and line 0 is the directrix.

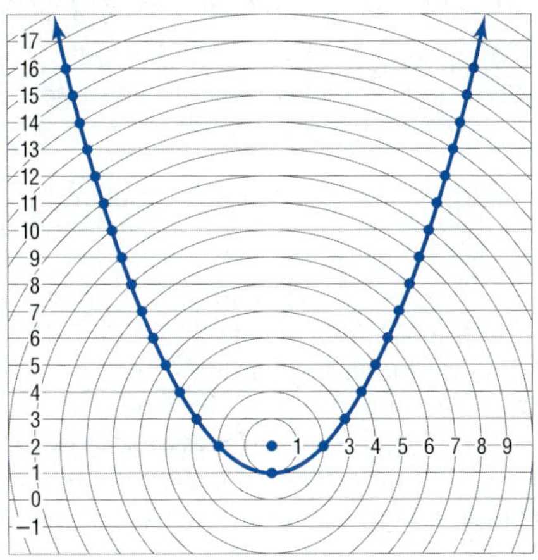

Resource Manager

Teaching Algebra with Manipulatives

- pp. 8–9 (masters for conic paper)
- p. 270 (student recording sheet)

Algebra Activity

A Follow-Up of Lesson 8-6

Getting Started

Objective To illustrate definitions of the conic sections by graphing them on conic graph paper.

Materials
conic graph paper

Teach

- Explain to students that there are many different kinds of graph paper (logarithmic, polar, and isometric, for example) in addition to the familiar rectangular grid. Each type helps visualize various mathematical relationships.
- Some students may have some visual difficulties with this kind of graph. Suggest that students move a pointer (finger or pencil) to keep track of where they are.
- To correct numbering errors, have students work in pairs to check the numbering before graphing.

Assess

In **Exercises 1–3,** students should

- use conic graphing paper.
- see how the distance definitions of the conic sections define their shapes.

Study Notebook

You may wish to have students summarize this activity and what they learned from it.

Answers

1. The points on each graph are equidistant from the focus and the directrix.

2a. There are no intersecting circles whose sum is less than 10.

2b. The ellipses become more circular; the ellipses become more oblong.

3. Each branch of the hyperbola becomes more narrow and the vertices become farther apart; each branch of the hyperbola becomes wider and the vertices become closer.

Algebra Activity

Activity 2

An ellipse is the set of points such that the sum of the distances from two fixed points is constant. The two fixed points are called the foci.

- Use graph paper like that shown. It contains two small circles and a series of concentric circles from each. The concentric circles are tangent to each other as shown.
- Choose the constant 13. Mark the points at the intersections of circle 9 and circle 4, because $9 + 4 = 13$. Continue this process until you have marked the intersection of all circles whose sum is 13.
- Connect the points to form a smooth curve. The curve is an ellipse whose foci are the centers of the two small circles on the graph paper.

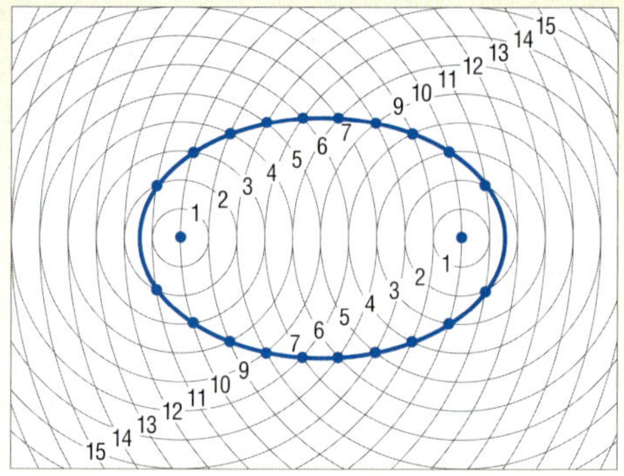

Activity 3

A hyperbola is the set of points such that the difference of the distances from two fixed points is constant. The two fixed points are called the foci.

- Use the same type of graph paper that you used for the ellipse in Activity 2. Choose the constant 7. Mark the points at the intersections of circle 9 and circle 2, because $9 - 2 = 7$. Continue this process until you have marked the intersections of all circles whose difference in radius is 7.
- Connect the points to form a hyperbola.

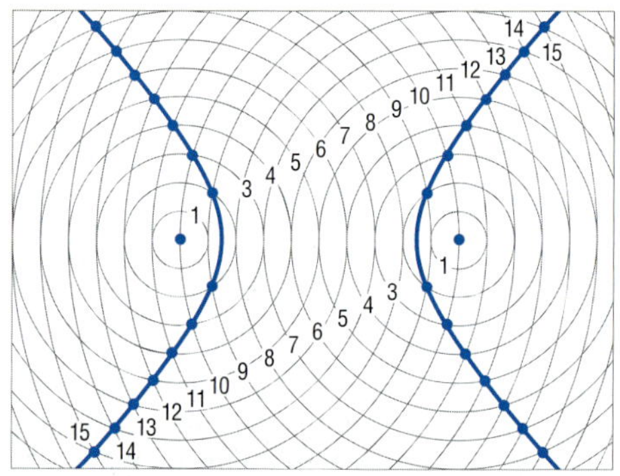

Model and Analyze 1–3. See margin.

1. Use the type of graph paper you used in Activity 1. Mark the intersection of line 0 and circle 2. Then mark the two points on line 1 and circle 3, the two points on line 2 and circle 4, and so on. Draw the new parabola. Continue this process and make as many parabolas as you can on one sheet of the graph paper. The focus is always the center of the small circle. Why are the resulting graphs parabolas?
2. In Activity 2, you drew an ellipse such that the sum of the distances from two fixed points was 13. Choose 10, 11, 12, 14, and so on for that sum, and draw as many ellipses as you can on one piece of the graph paper.
   a. Why can you not start with 9 as the sum?
   b. What happens as the sum increases? decreases?
3. In Activity 3, you drew a hyperbola such that the difference of the distances from two fixed points was 7. Choose other numbers and draw as many hyperbolas as you can on one piece of graph paper. What happens as the difference increases? decreases?

454 Chapter 8 Conic Sections

8-7 Solving Quadratic Systems

What You'll Learn

- Solve systems of quadratic equations algebraically and graphically.
- Solve systems of quadratic inequalities graphically.

How do systems of equations apply to video games?

Computer software often uses a coordinate system to keep track of the locations of objects on the screen. Suppose an enemy space station is located at the center of the screen, which is the origin in a coordinate system. The space station is surrounded by a circular force field of radius 50 units. If the spaceship you control is flying toward the center along the line with equation $y = 3x$, the point where the ship hits the force field is a solution of a system of equations.

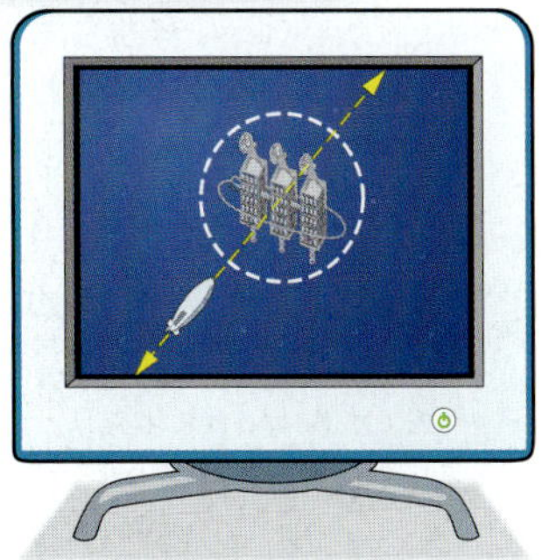

SYSTEMS OF QUADRATIC EQUATIONS If the graphs of a system of equations are a conic section and a line, the system may have zero, one, or two solutions. Some of the possible situations are shown below.

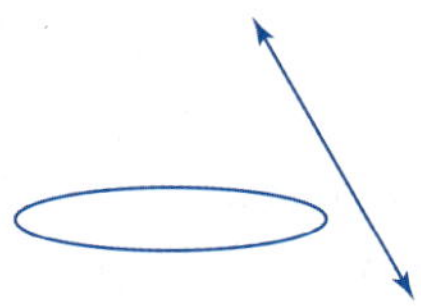

no solutions

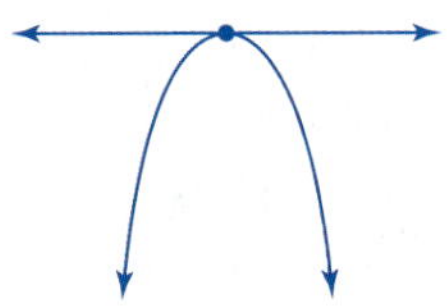

one solution

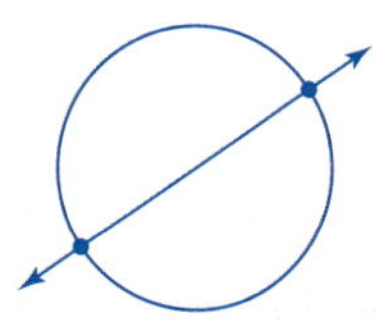

two solutions

You have solved systems of linear equations graphically and algebraically. You can use similar methods to solve systems involving quadratic equations.

Example 1 Linear-Quadratic System

Solve the system of equations.

$$x^2 - 4y^2 = 9$$
$$4y - x = 3$$

You can use a graphing calculator to help visualize the relationships of the graphs of the equations and predict the number of solutions.

Solve each equation for y to obtain $y = \pm\frac{\sqrt{x^2 - 9}}{2}$ and $y = \frac{1}{4}x + \frac{3}{4}$. Enter the functions $y = \frac{\sqrt{x^2 - 9}}{2}$, $y = -\frac{\sqrt{x^2 - 9}}{2}$, and $y = \frac{1}{4}x + \frac{3}{4}$ on the Y= screen. The graph indicates that the hyperbola and line intersect in two points. So the system has two solutions.

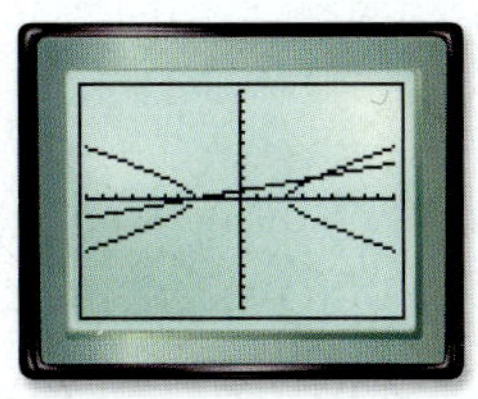

[−10, 10] scl: 1 by [−10, 10] scl: 1

(continued on the next page)

8-7 Lesson Notes

1 Focus

5-Minute Check Transparency 8-7 Use as a quiz or review of Lesson 8-6.

Mathematical Background notes are available for this lesson on p. 410D.

How do systems of equations apply to video games?

Ask students:

- Describe the graph of $y = 3x$. **a line through the origin and steeper than the 45 degree line in quadrants I and III**
- Describe the graph of the force field. **a circle with a radius of 50 and a center at the center of the screen**

Resource Manager

Workbook and Reproducible Masters

Chapter 8 Resource Masters

- Study Guide and Intervention, pp. 491–492
- Skills Practice, p. 493
- Practice, p. 494
- Reading to Learn Mathematics, p. 495
- Enrichment, p. 496
- Assessment, p. 512

School-to-Career Masters, p. 16

Transparencies

5-Minute Check Transparency 8-7
Answer Key Transparencies

Technology

Interactive Chalkboard
Multimedia Applications

2 Teach

SYSTEMS OF QUADRATIC EQUATIONS

In-Class Examples

Power Point®

1 Solve the system of equations.
$4x^2 - 16y^2 = 25$
$2y + x = 2$
$\left(\frac{41}{16}, -\frac{9}{32}\right)$

Teaching Tip Point out that there are combinations of graphs other than those shown just before Example 2 which are possible for each number of solutions. For example, the ellipse in the third figure could be shifted to the right to intersect the circle in four points.

2 Solve the system of equations.
$x^2 + y^2 = 16$
$4x^2 + y^2 = 23$
$\left(\frac{\sqrt{21}}{3}, \frac{\sqrt{123}}{3}\right), \left(-\frac{\sqrt{21}}{3}, \frac{\sqrt{123}}{3}\right),$
$\left(\frac{\sqrt{21}}{3}, -\frac{\sqrt{123}}{3}\right),$
$\left(-\frac{\sqrt{21}}{3}, -\frac{\sqrt{123}}{3}\right)$

Use substitution to solve the system. First rewrite $4y - x = 3$ as $x = 4y - 3$.

$x^2 - 4y^2 = 9$	First equation in the system
$(4y - 3)^2 - 4y^2 = 9$	Substitute $4y - 3$ for x.
$12y^2 - 24y = 0$	Simplify.
$y^2 - 2y = 0$	Divide each side by 12.
$y(y - 2) = 0$	Factor.
$y = 0$ or $y - 2 = 0$	Zero Product Property
$y = 2$	Solve for y.

Now solve for x.

$x = 4y - 3$	$x = 4y - 3$	Equation for x in terms of y
$= 4(0) - 3$	$= 4(2) - 3$	Substitute the y values.
$= -3$	$= 5$	Simplify.

The solutions of the system are $(-3, 0)$ and $(5, 2)$. Based on the graph, these solutions are reasonable.

If the graphs of a system of equations are two conic sections, the system may have zero, one, two, three, or four solutions. Some of the possible situations are shown below.

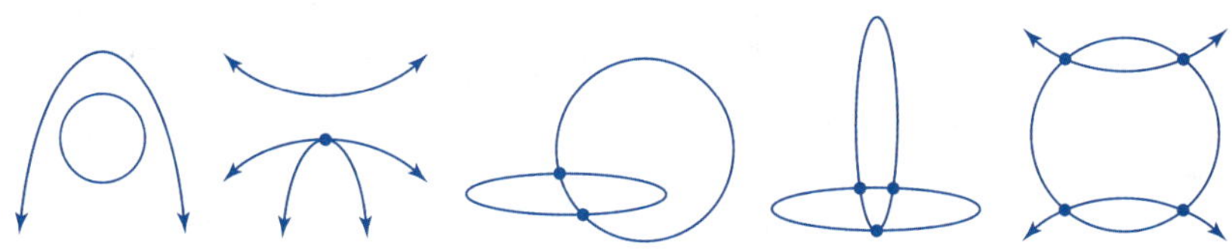

no solutions   one solution   two solutions   three solutions   four solutions

Example 2 Quadratic-Quadratic System

Solve the system of equations.
$y^2 = 13 - x^2$
$x^2 + 4y^2 = 25$

A graphing calculator indicates that the circle and ellipse intersect in four points. So, this system has four solutions.

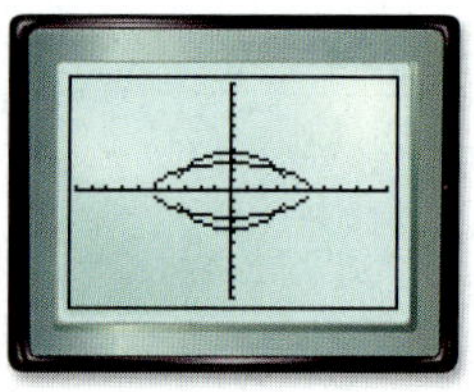

[−10, 10] scl: 1 by [−10, 10] scl: 1

Use the elimination method to solve the system.

$y^2 = 13 - x^2$
$x^2 + 4y^2 = 25$

$-x^2 - y^2 = -13$	Rewrite the first original equation.
$(+)\ x^2 + 4y^2 = 25$	Second original equation
$3y^2 = 12$	Add.
$y^2 = 4$	Divide each side by 3.
$y = \pm 2$	Take the square root of each side.

Study Tip

Graphing Calculators
If you use ZSquare on the ZOOM menu, the graph of the first equation will look like a circle.

DAILY INTERVENTION **Differentiated Instruction**

Logical Challenge students to sketch as many different possibilities as they can think of to add to the figures just before Example 2.

Substitute 2 and -2 for y in either of the original equations and solve for x.

$x^2 + 4y^2 = 25$	$x^2 + 4y^2 = 25$	Second original equation
$x^2 + 4(2)^2 = 25$	$x^2 + 4(-2)^2 = 25$	Substitute for y.
$x^2 = 9$	$x^2 = 9$	Subtract 16 from each side.
$x = \pm 3$	$x = \pm 3$	Take the square root of each side.

The solutions are $(3, 2)$, $(-3, 2)$, $(-3, -2)$, and $(3, -2)$.

A graphing calculator can be used to approximate the solutions of a system of equations.

Graphing Calculator Investigation

Quadratic Systems

The calculator screen shows the graphs of two circles.

Think and Discuss

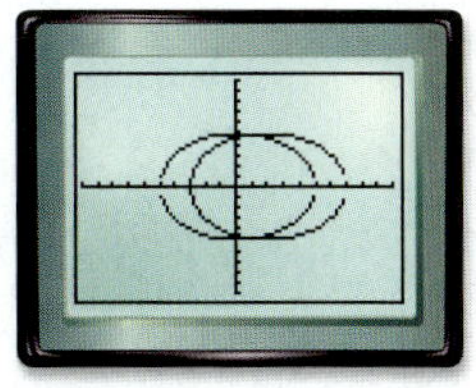

$[-10, 10]$ scl: 1 by $[-10, 10]$ scl: 1

1. Write the system of equations represented by the graph. $x^2 + y^2 = 25;\ (x - 2)^2 + y^2 = 25$
2. Enter the equations into a TI-83 Plus and use the intersect feature on the CALC menu to solve the system. Round to the nearest hundredth.
3. Solve the system algebraically. $(1, \pm 2\sqrt{6})$
4. Can you always find the exact solution of a system using a graphing calculator? Explain.

Use a graphing calculator to solve each system of equations. Round to the nearest hundredth.

5. $y = x + 2$, $x^2 + y^2 = 9$ **(0.87, 2.87), (−2.87, −0.87)**
6. $3x^2 + y^2 = 11$, $y = x^2 + x + 1$ **(−1.57, 1.90), (0.96, 2.87)**

2. (1, ±4.90)

4. No; a calculator only gives decimal approximations. If the solution involves irrational numbers or unfamiliar fractions, you may not be able to recognize them.

SYSTEMS OF QUADRATIC INEQUALITIES

You have learned how to solve systems of linear inequalities by graphing. Systems of quadratic inequalities are also solved by graphing.

The graph of an inequality involving a parabola, circle, or ellipse is either the interior or the exterior of the conic section. The graph of an inequality involving a hyperbola is either the region between the branches or the two regions inside the branches. As with linear inequalities, examine the inequality symbol to determine whether to include the boundary.

Example 3 System of Quadratic Inequalities

Solve the system of inequalities by graphing.

$y \le x^2 - 2$
$x^2 + y^2 < 16$

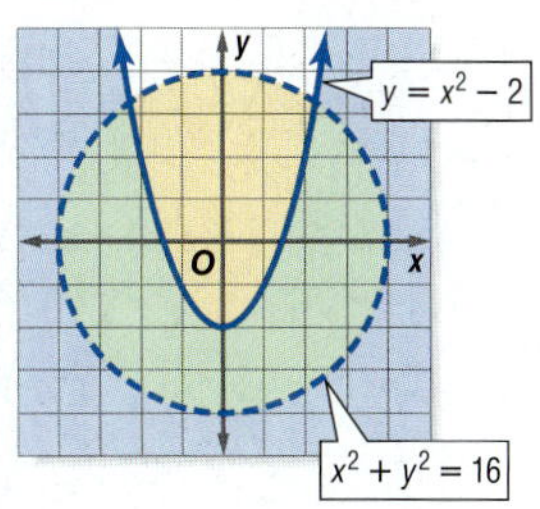

The graph of $y \le x^2 - 2$ is the parabola $y = x^2 - 2$ and the region outside or below it. This region is shaded blue.

The graph of $x^2 + y^2 < 16$ is the interior of the circle $x^2 + y^2 = 16$. This region is shaded yellow.

The intersection of these regions, shaded green, represents the solution of the system of inequalities.

Study Tip

Graphing Quadratic Inequalities

If you are unsure about which region to shade, you can test one or more points, as you did with linear inequalities.

www.algebra2.com/extra_examples

SYSTEMS OF QUADRATIC INEQUALITIES

In-Class Example Power Point®

3 Solve the system of inequalities by graphing.

$y \ge x^2 + 1$
$x^2 + y^2 \le 9$

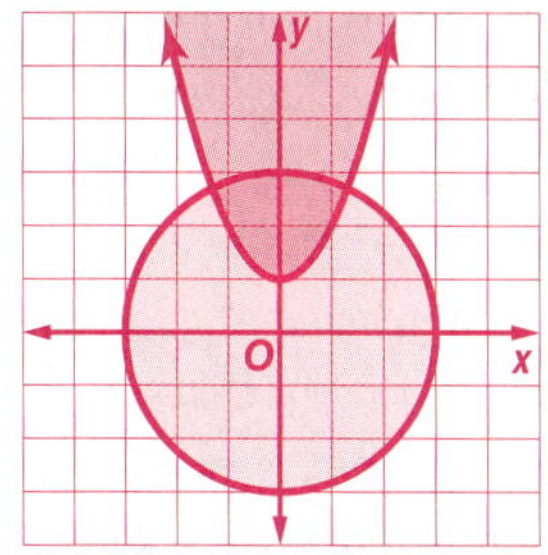

Teaching Tip Remind students that it is always a good idea to check a point inside the solution area as well as one outside that area.

Graphing Calculator Investigation

- **Rounding** Discuss the fact that a coordinate such as $\sqrt{3}$ is an exact symbol for this irrational number, while the decimal value used for a graph, 1.73, is an approximation rounded to the nearest hundredth.
- **Viewing Window** Remind students that the circles will not appear circular on the calculator screen unless the viewing window ranges have been set for a square grid.

3 Practice/Apply

Study Notebook

Have students—

- complete the definitions/examples of the remaining terms on their Vocabulary Builder worksheets for Chapter 8.
- summarize what they learned about graphing systems of quadratic equations and quadratic inequalities.
- include any other item(s) that they find helpful in mastering the skills in this lesson.

About the Exercises...

Organization by Objective

- **Systems of Quadratic Equations:** 11–31
- **Systems of Quadratic Inequalities:** 32–37

Odd/Even Assignments

Exercises 11–26 and 32–37 are structured so that students practice the same concepts whether they are assigned odd or even problems.

Assignment Guide

Basic: 11–21 odd, 25–27, 29–31, 33–37 odd, 38–45, 52–72

Average: 11–27 odd, 29–31, 33, 35, 37–45, 52–72 (optional: 46–51)

Advanced: 12–30 even, 31, 32–38 even, 39–72

Check for Understanding

Concept Check

1. See margin for graphs.

1. **Graph** each system of equations. Use the graph to solve the system.
   a. $4x - 3y = 0$, $x^2 + y^2 = 25$ **(−3, −4), (3, 4)**
   b. $y = 5 - x^2$, $y = 2x^2 + 2$ **(±1, 4)**
2. **Sketch** a parabola and an ellipse that intersect at exactly three points. **See margin.**
3. **OPEN ENDED** Write a system of quadratic equations for which (2, 6) is a solution. **Sample answer: $x^2 + y^2 = 40$, $y = x^2 + x$**

Guided Practice

GUIDED PRACTICE KEY	
Exercises	Examples
4, 5	1
6, 7, 10	2
8, 9	3

Find the exact solution(s) of each system of equations.

4. $y = 5$, $y^2 = x^2 + 9$ **(±4, 5)**
5. $y - x = 1$, $x^2 + y^2 = 25$ **(−4, −3), (3, 4)**
6. $3x = 8y^2$, $8y^2 - 2x^2 = 16$ **no solution**
7. $5x^2 + y^2 = 30$, $9x^2 - y^2 = -16$ **(1, ±5), (−1, ±5)**

Solve each system of inequalities by graphing. **8–9. See pp. 469A–469J.**

8. $x + y < 4$, $9x^2 - 4y^2 \geq 36$
9. $x^2 + y^2 < 25$, $4x^2 - 9y^2 < 36$

Application

10. **EARTHQUAKES** In a coordinate system where a unit represents one mile, the epicenter of an earthquake was determined to be 50 miles from a station at the origin. It was also 40 miles from a station at (0, 30) and 13 miles from a station at (35, 18). Where was the epicenter located? **(40, 30)**

★ indicates increased difficulty

Practice and Apply

Homework Help	
For Exercises	See Examples
11–16, 25 29	1
17–24, 26–28, 30, 31	2
32–37	3

Extra Practice See page 847.

Find the exact solution(s) of each system of equations.

11. $y = x + 2$, $y = x^2$ **(2, 4), (−1, 1)**
12. $y = x + 3$, $y = 2x^2$ **$\left(\frac{3}{2}, \frac{9}{2}\right)$, (−1, 2)**
13. $x^2 + y^2 = 36$, $y = x + 2$ **$(-1 + \sqrt{17}, 1 + \sqrt{17})$, $(-1 - \sqrt{17}, 1 - \sqrt{17})$**
14. $y^2 + x^2 = 9$, $y = 7 - x$ **no solution**
15. $\frac{x^2}{30} + \frac{y^2}{6} = 1$, $x = y$ **$(\sqrt{5}, \sqrt{5})$, $(-\sqrt{5}, -\sqrt{5})$**
16. $\frac{x^2}{36} - \frac{y^2}{4} = 1$, $x = y$ **no solution**
17. $4x + y^2 = 20$, $4x^2 + y^2 = 100$ **(5, 0), (−4, ±6)**
18. $y + x^2 = 3$, $x^2 + 4y^2 = 36$ **(0, 3), $\left(\pm\frac{\sqrt{23}}{2}, -\frac{11}{4}\right)$**
19. $x^2 + y^2 = 64$, $x^2 + 64y^2 = 64$ **(±8, 0)**
20. $y^2 + x^2 = 25$, $y^2 + 9x^2 = 25$ **(0, ±5)**
21. $y^2 = x^2 - 25$, $x^2 - y^2 = 7$ **no solution**
22. $y^2 = x^2 - 7$, $x^2 + y^2 = 25$ **(4, ±3), (−4, ±3)**
23. ★ $2x^2 + 8y^2 + 8x - 48y + 30 = 0$, $2x^2 - 8y^2 = -48y + 90$ **(−5, 5), (−5, 1), (3, 3)**
24. ★ $3x^2 - 20y^2 - 12x + 80y - 96 = 0$, $3x^2 + 20y^2 = 80y + 48$ **(6, 3), (6, 1), (−4, 4), (−4, 0)**
25. Where do the graphs of the equations $y = 2x + 1$ and $2x^2 + y^2 = 11$ intersect?

25. **$\left(-\frac{5}{3}, -\frac{7}{3}\right)$, (1, 3)**

26. What are the coordinates of the points that lie on the graphs of both $x^2 + y^2 = 25$ and $2x^2 + 3y^2 = 66$? **(3, ±4), (−3, ±4)**
27. **ROCKETS** Two rockets are launched at the same time, but from different heights. The height y in feet of one rocket after t seconds is given by $y = -16t^2 + 150t + 5$. The height of the other rocket is given by $y = -16t^2 + 160t$. After how many seconds are the rockets at the same height? **0.5 s**

Answers

1a.

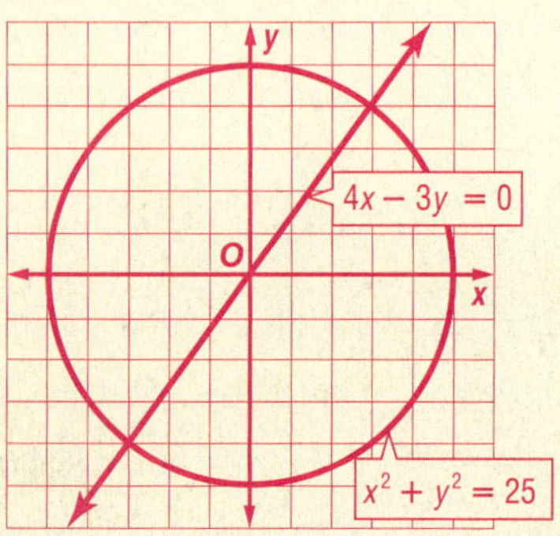

1b.

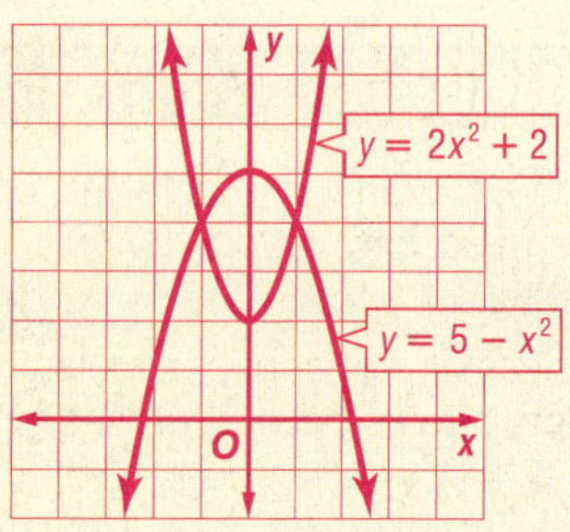

2. The vertex of the parabola is on the ellipse. The parabola opens toward the interior of the ellipse and is narrow enough to intersect the ellipse in two other points. Thus, there are exactly three points of intersection.

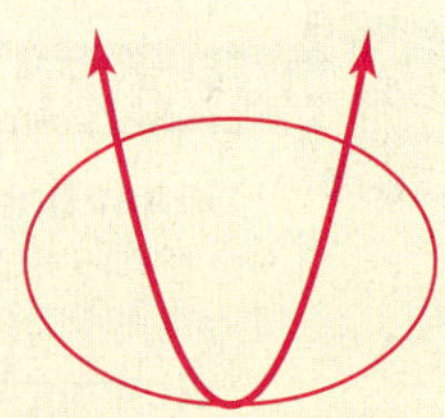

★ 28. **ADVERTISING** The corporate logo for an automobile manufacturer is shown at the right. Write a system of three equations to model this logo.
Sample answer:
$$\frac{x^2}{36}+\frac{y^2}{16}=1,\ \frac{x^2}{16}+\frac{(y-2)^2}{4}=1,\ \frac{x^2}{2}+\frac{y^2}{16}=1$$

29. $\left(\frac{40-24\sqrt{5}}{5},\ \frac{45-12\sqrt{5}}{5}\right)$

29. **MIRRORS** A hyperbolic mirror is a mirror in the shape of one branch of a hyperbola. Such a mirror reflects light rays directed at one focus toward the other focus. Suppose a hyperbolic mirror is modeled by the upper branch of the hyperbola with equation $\frac{y^2}{9}-\frac{x^2}{16}=1$. A light source is located at $(-10, 0)$. Where should the light from the source hit the mirror so that the light will be reflected to $(0, -5)$?

More About. . .

Astronomy

The astronomical unit (AU) is the mean distance between Earth and the Sun. One AU is about 93 million miles or 150 million kilometers.

Source: www.infoplease.com

ASTRONOMY For Exercises 30 and 31, use the following information.

The orbit of Pluto can be modeled by the equation $\frac{x^2}{39.5^2}+\frac{y^2}{38.3^2}=1$, where the units are astronomical units. Suppose a comet is following a path modeled by the equation $x = y^2 + 20$.

30. Find the point(s) of intersection of the orbits of Pluto and the comet. Round to the nearest tenth. **(39.2, ±4.4)**

31. Will the comet necessarily hit Pluto? Explain. **No; the comet and Pluto may not be at either point of intersection at the same time.**

Solve each system of inequalities by graphing. **32–37. See pp. 469A–469J.**

32. $x + 2y > 1$ $x^2 + y^2 \le 25$	33. $x + y \le 2$ $4x^2 - y^2 \ge 4$	34. $x^2 + y^2 \ge 4$ $4y^2 + 9x^2 \le 36$
35. $x^2 + y^2 < 36$ $4x^2 + 9y^2 > 36$	36. $y^2 < x$ $x^2 - 4y^2 < 16$	37. $x^2 \le y$ $y^2 - x^2 \ge 4$

CRITICAL THINKING For Exercises 38–42, find all values of k for which the system of equations has the given number of solutions. If no values of k meet the condition, write *none*. **38. $k < -3$, $-2 < k < 2$, or $k > 3$**

$$x^2 + y^2 = k^2 \qquad \frac{x^2}{9}+\frac{y^2}{4}=1$$

40. $k = \pm 2$ or $k = \pm 3$

38. no solutions
39. one solution **none**
40. two solutions
41. three solutions **none**
42. four solutions **$-3 < k < -2$ or $2 < k < 3$**

43. **WRITING IN MATH** Answer the question that was posed at the beginning of the lesson. **See pp. 469A–469J.**

How do systems of equations apply to video games?

Include the following in your answer:
- a linear-quadratic system of equations that applies to this situation,
- an explanation of how you know that the spaceship is headed directly toward the center of the screen, and
- the coordinates of the point at which the spaceship will hit the force field, assuming that the spaceship moves from the bottom of the screen toward the center.

Standardized Test Practice

44. If $\boxed{x}$ is defined to be $x^2 - 4x$ for all numbers x, which of the following is the greatest? **A**

(A) $\boxed{0}$ (B) $\boxed{1}$ (C) $\boxed{2}$ (D) $\boxed{3}$

45. How many three-digit numbers are divisible by 3? **B**

(A) 299 (B) 300 (C) 301 (D) 302

www.algebra2.com/self_check_quiz

Study Guide and Intervention, p. 491 (shown) and p. 492

Systems of Quadratic Equations Like systems of linear equations, systems of quadratic equations can be solved by substitution and elimination. If the graphs are a conic section and a line, the system will have 0, 1, or 2 solutions. If the graphs are two conic sections, the system will have 0, 1, 2, 3, or 4 solutions.

Example **Solve the system of equations.** $y = x^2 - 2x - 15$, $x + y = -3$

Rewrite the second equation as $y = -x - 3$ and substitute into the first equation.

$-x - 3 = x^2 - 2x - 15$
$0 = x^2 - x - 12$ Add $x + 3$ to each side.
$0 = (x - 4)(x + 3)$ Factor.

Use the Zero Product property to get $x = 4$ or $x = -3$.

Substitute these values for x in $x + y = -3$:

$4 + y = -3$ or $-3 + y = -3$
$y = -7$ $\quad y = 0$

The solutions are $(4, -7)$ and $(-3, 0)$.

Exercises

Find the exact solution(s) of each system of equations.

1. $y = x^2 - 5$, $y = x - 3$ **(2, −1), (−1, −4)**
2. $x^2 + (y - 5)^2 = 25$, $y = -x^2$ **(0, 0)**
3. $x^2 + (y - 5)^2 = 25$, $y = x^2$ **(0, 0), (3, 9), (−3, 9)**
4. $x^2 + y^2 = 9$, $x^2 + y = 3$ **$(0, 3), (\sqrt{5}, -2), (-\sqrt{5}, -2)$**
5. $x^2 - y^2 = 1$, $x^2 + y^2 = 16$ **$\left(\frac{\sqrt{34}}{2}, \frac{\sqrt{30}}{2}\right), \left(\frac{\sqrt{34}}{2}, -\frac{\sqrt{30}}{2}\right), \left(-\frac{\sqrt{34}}{2}, \frac{\sqrt{30}}{2}\right), \left(-\frac{\sqrt{34}}{2}, -\frac{\sqrt{30}}{2}\right)$**
6. $y = x - 3$, $x = y^2 - 4$ **$\left(\frac{7+\sqrt{29}}{2}, \frac{1+\sqrt{29}}{2}\right), \left(\frac{7-\sqrt{29}}{2}, \frac{1-\sqrt{29}}{2}\right)$**

Skills Practice, p. 493 and Practice, p. 494 (shown)

Find the exact solution(s) of each system of equations.

1. $(x - 2)^2 + y^2 = 5$, $x - y = 1$ **(0, −1), (3, 2)**
2. $x = 2(y + 1)^2 - 6$, $x + y = 3$ **(2, 1), (6.5, −3.5)**
3. $y^2 - 3x^2 = 6$, $y = 2x - 1$ **(−1, −3), (5, 9)**
4. $x^2 + 2y^2 = 1$, $y = -x + 1$ **$(1, 0), \left(\frac{1}{3}, \frac{2}{3}\right)$**
5. $4y^2 - 9x^2 = 36$, $4x^2 - 9y^2 = 36$ **no solution**
6. $y = x^2 - 3$, $x^2 + y^2 = 9$ **$(0, -3), (\pm\sqrt{5}, 2)$**
7. $x^2 + y^2 = 25$, $4y = 3x$ **(4, 3), (−4, −3)**
8. $y^2 = 10 - 6x^2$, $4y^2 = 40 - 2x^2$ **$(0, \pm\sqrt{10})$**
9. $x^2 + y^2 = 25$, $x = 3y - 5$ **(−5, 0), (4, 3)**
10. $4x^2 + 9y^2 = 36$, $2x^2 - 9y^2 = 18$ **(±3, 0)**
11. $x = -(y - 3)^2 + 2$, $x = (y - 3)^2 + 3$ **no solution**
12. $\frac{x^2}{9} - \frac{y^2}{16} = 1$, $x^2 + y^2 = 9$ **(±3, 0)**
13. $25x^2 + 4y^2 = 100$, $x = -\frac{5}{2}$ **no solution**
14. $x^2 + y^2 = 4$, $\frac{x^2}{4} + \frac{y^2}{8} = 1$ **(±2, 0)**
15. $x^2 - y^2 = 3$, $y^2 - x^2 = 3$ **no solution**
16. $\frac{x^2}{7} + \frac{y^2}{7} = 1$, $3x^2 - y^2 = 9$ **$(\pm 2, \pm\sqrt{3})$**
17. $x + 2y = 3$, $x^2 + y^2 = 9$ **$(3, 0), \left(-\frac{9}{5}, \frac{12}{5}\right)$**
18. $x^2 + y^2 = 64$, $x^2 - y^2 = 8$ **$(\pm 6, \pm 2\sqrt{7})$**

Solve each system of inequalities by graphing.

19. $y \ge x^2$, $y > -x + 2$

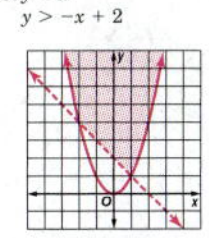

20. $x^2 + y^2 < 36$, $x^2 + y^2 \ge 16$

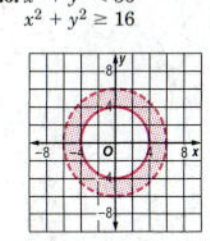

21. $\frac{(y-3)^2}{16} + \frac{(x+2)^2}{4} \le 1$, $(x + 1)^2 + (y - 2)^2 \le 4$

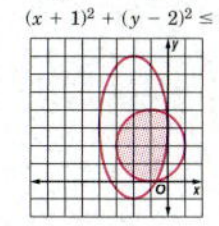

22. **GEOMETRY** The top of an iron gate is shaped like half an ellipse with two congruent segments from the center of the ellipse to the ellipse as shown. Assume that the center of the ellipse is at (0, 0). If the ellipse can be modeled by the equation $x^2 + 4y^2 = 4$ for $y \ge 0$ and the two congruent segments can be modeled by $y = \frac{\sqrt{3}}{2}x$ and $y = -\frac{\sqrt{3}}{2}x$, what are the coordinates of points A and B?

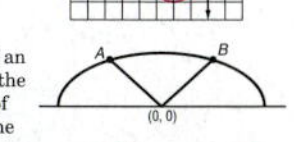

$\left(-1, \frac{\sqrt{3}}{2}\right)$ and $\left(1, \frac{\sqrt{3}}{2}\right)$

Reading to Learn Mathematics, p. 495 ELL

Pre-Activity **How do systems of equations apply to video games?**

Read the introduction to Lesson 8-7 at the top of page 455 in your textbook.

The figure in your textbook shows that the spaceship hits the circular force field in two points. Is it possible for the spaceship to hit the force field in either fewer or more than two points? State all possibilities and explain how these could happen. **Sample answer: The spaceship could hit the force field in zero points if the spaceship missed the force field all together. The spaceship could also hit the force field in one point if the spaceship just touched the edge of the force field.**

Reading the Lesson

1. Draw a sketch to illustrate each of the following possibilities.

a. a parabola and a line that intersect in 2 points

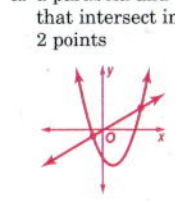

b. an ellipse and a circle that intersect in 4 points

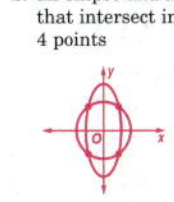

c. a hyperbola and a line that intersect in 1 point

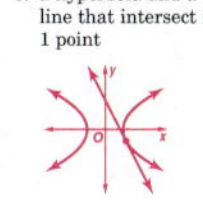

2. Consider the following system of equations.
$x^2 = 3 + y^2$
$2x^2 + 3y^2 = 11$

a. What kind of conic section is the graph of the first equation? **hyperbola**

b. What kind of conic section is the graph of the second equation? **ellipse**

c. Based on your answers to parts a and b, what are the possible numbers of solutions that this system could have? **0, 1, 2, 3, or 4**

Helping You Remember

3. Suppose that the graph of a quadratic inequality is a region whose boundary is a circle. How can you remember whether to shade the interior or the exterior of the circle? **Sample answer: The solutions of an inequality of the form $x^2 + y^2 < r^2$ are all points that are *less than* r units from the origin, so the graph is the *interior* of the circle. The solutions of an inequality of the form $x^2 + y^2 > r^2$ are the points that are *more than* r units from the origin, so the graph is the *exterior* of the circle.**

Enrichment, p. 496

Graphing Quadratic Equations with xy-Terms

You can use a graphing calculator to examine graphs of quadratic equations that contain xy-terms.

Example **Use a graphing calculator to display the graph of $x^2 + xy + y^2 = 4$.**

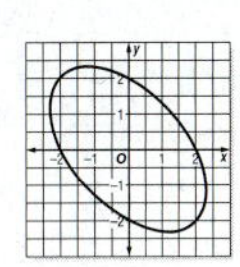

Solve the equation for y in terms of x by using the quadratic formula.

$y^2 + xy + (x^2 - 4) = 0$

To use the formula, let $a = 1$, $b = x$, and $c = (x^2 - 4)$.

$y = \frac{-x \pm \sqrt{x^2 - 4(1)(x^2 - 4)}}{2}$

$y = \frac{-x \pm \sqrt{16 - 3x^2}}{2}$

...on the...

4 Assess

Open-Ended Assessment

Speaking Have students use a sketch to explain how they know which regions to shade for systems of quadratic inequalities.

Assessment Options

Quiz (Lesson 8-7) is available on p. 512 of the *Chapter 8 Resource Masters*.

Answers

46. Sample answer: $y = x^2$, $x = (y - 2)^2$

47. Sample answer: $x^2 + y^2 = 36$, $\frac{(x+2)^2}{16} - \frac{y^2}{4} = 1$

48. Sample answer: $x^2 + y^2 = 100$, $\frac{x^2}{16} + \frac{y^2}{4} = 1$

49. Sample answer: $x^2 + y^2 = 81$, $\frac{x^2}{4} + \frac{y^2}{100} = 1$

50. Sample answer: $\frac{x^2}{64} + \frac{y^2}{36} = 1$, $\frac{x^2}{64} - \frac{y^2}{36} = 1$

52. $(x + 2)^2 + (y + 1)^2 = 11$, circle

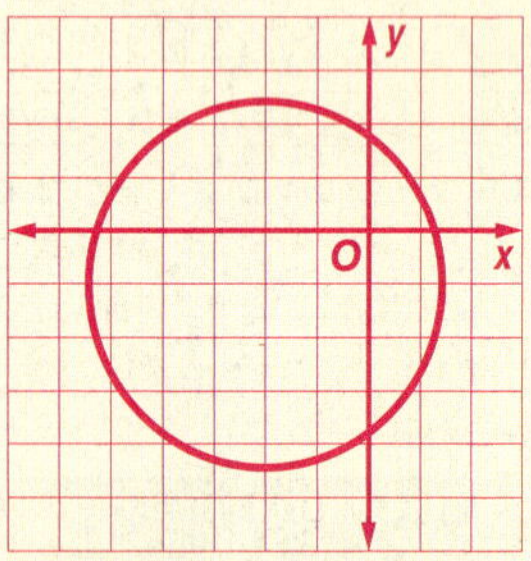

53. $\frac{(y-3)^2}{9} + \frac{x^2}{4} = 1$, ellipse

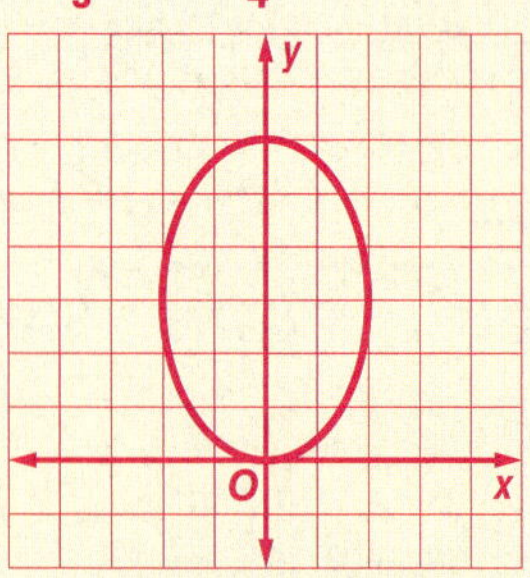

Graphing Calculator

SYSTEMS OF EQUATIONS **Write a system of equations that satisfies each condition. Use a graphing calculator to verify that you are correct.**

46–50. See margin for sample answers.

46. two parabolas that intersect in two points
47. a hyperbola and a circle that intersect in three points
48. a circle and an ellipse that do not intersect
49. a circle and an ellipse that intersect in four points
50. a hyperbola and an ellipse that intersect in two points
51. two circles that intersect in three points **impossible**

Maintain Your Skills

Mixed Review **Write each equation in standard form. State whether the graph of the equation is a *parabola, circle, ellipse,* or *hyperbola*. Then graph the equation.** *(Lesson 8-6)*

52. $x^2 + y^2 + 4x + 2y - 6 = 0$ 53. $9x^2 + 4y^2 - 24y = 0$

52–53. See margin.

54. Find the coordinates of the vertices and foci and the equations of the asymptotes of the hyperbola with the equation $6y^2 - 2x^2 = 24$. Then graph the hyperbola. *(Lesson 8-5)* **See margin.**

Solve each equation by factoring. *(Lesson 6-5)*

55. $x^2 + 7x = 0$ **−7, 0** 56. $x^2 - 3x = 0$ **0, 3** 57. $21 = x^2 + 4x$ **−7, 3**

58. $35 = -2x + x^2$ **7, −5** 59. $9x^2 + 24x = -16$ $-\frac{4}{3}$ 60. $8x^2 + 2x = 3$ $-\frac{3}{4}, \frac{1}{2}$

For Exercises 61 and 62, complete parts a–c for each quadratic equation. *(Lesson 6-5)*

a. Find the value of the discriminant.

b. Describe the number and type of roots.

c. Find the exact solutions by using the Quadratic Formula.

61. $5x^2 = 2$ 62. $-3x^2 + 6x - 7 = 0$

61a. 40
61b. two real, irrational
61c. $\pm\frac{\sqrt{10}}{5}$
62a. −48
62b. two imaginary
62c. $1 \pm \frac{2i\sqrt{3}}{3}$

Simplify. *(Lesson 5-9)*

63. $(3 + 2i) - (1 - 7i)$ **$2 + 9i$** 64. $(8 - i)(4 - 3i)$ **$29 - 28i$** 65. $\frac{2 + 3i}{1 + 2i}$ $\frac{8}{5} - \frac{1}{5}i$

66. **CHEMISTRY** The mass of a proton is about 1.67×10^{-27} kilogram. The mass of an electron is about 9.11×10^{-31} kilogram. About how many times as massive as an electron is a proton? *(Lesson 5-1)* **about 1830 times**

Evaluate each determinant. *(Lesson 4-3)*

67. $\begin{vmatrix} 2 & -3 \\ 2 & 0 \end{vmatrix}$ **6** 68. $\begin{vmatrix} -4 & -2 \\ 5 & 3 \end{vmatrix}$ **−2** 69. $\begin{vmatrix} 2 & 1 & -2 \\ 4 & 0 & 3 \\ -3 & 1 & 7 \end{vmatrix}$ **−51**

70. Solve the system of equations. *(Lesson 3-5)* **(5, 3, 7)**

$r + s + t = 15$
$r + t = 12$
$s + t = 10$

Write an equation in slope-intercept form for each graph. *(Lesson 2-4)*

71. 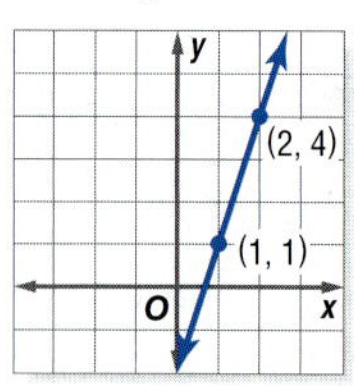

$y = 3x - 2$

72. 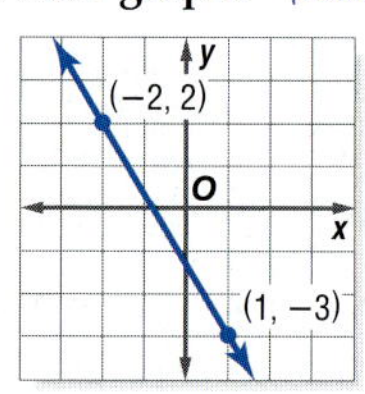

$y = -\frac{5}{3}x - \frac{4}{3}$

---

54. $(0, \pm 2)$; $(0, \pm 4)$; $y = \pm\frac{\sqrt{3}}{3}x$

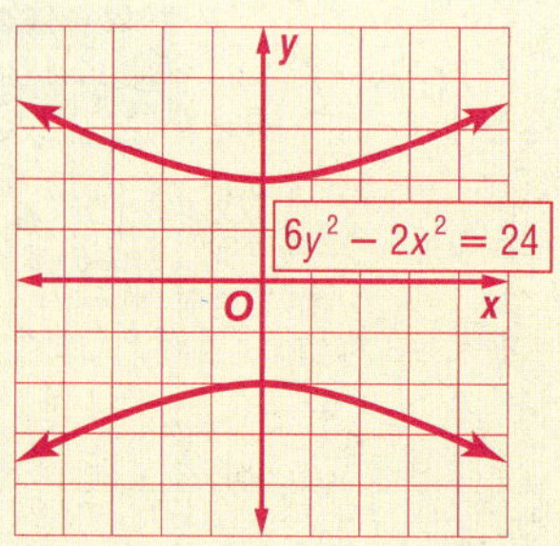

Chapter 8 Study Guide and Review

Vocabulary and Concept Check

asymptote (p. 442)	Distance Formula (p. 413)	major axis (p. 434)
center of a circle (p. 426)	ellipse (p. 433)	Midpoint Formula (p. 412)
center of a hyperbola (p. 442)	foci of a hyperbola (p. 441)	minor axis (p. 434)
center of an ellipse (p. 434)	foci of an ellipse (p. 433)	parabola (p. 419)
circle (p. 426)	focus of a parabola (p. 419)	tangent (p. 427)
conic section (p. 419)	hyperbola (p. 441)	transverse axis (p. 442)
conjugate axis (p. 442)	latus rectum (p. 421)	vertex of a hyperbola (p. 442)
directrix (p. 419)		

Tell whether each statement is *true* or *false*. If the statement is false, correct it to make it true. 4, 6, 8, 9. See pp. 469A–469J for correct statements.

1. An ellipse is the set of all points in a plane such that the sum of the distances from two given points in the plane, called the foci, is constant. true
2. The major axis is the longer of the two axes of symmetry of an ellipse. true
3. The formula used to find the distance between two points in a coordinate plane is $d = \sqrt{(x_2 - x_1)^2 + (y_2 - y_1)^2}$. true
4. A parabola is the set of all points that are the same distance from a given point called the directrix and a given line called the focus. false
5. The radius is the distance from the center of a circle to any point on the circle. true
6. The conjugate axis of a hyperbola is a line segment parallel to the transverse axis. false
7. A conic section is formed by slicing a double cone by a plane. true
8. A hyperbola is the set of all points in a plane such that the absolute value of the sum of the distances from any point on the hyperbola to two given points is constant. false
9. The midpoint formula is given by $\left(\frac{x_1 - x_2}{2}, \frac{y_1 - y_2}{2}\right)$. false
10. The set of all points in a plane that are equidistant from a given point in a plane, called the center, forms a circle. true

Lesson-by-Lesson Review

8-1 Midpoint and Distance Formulas

See pages 412–416.

Concept Summary

- Midpoint Formula: $M = \left(\frac{x_1 + x_2}{2}, \frac{y_1 + y_2}{2}\right)$
- Distance Formula: $d = \sqrt{(x_2 - x_1)^2 + (y_2 - y_1)^2}$

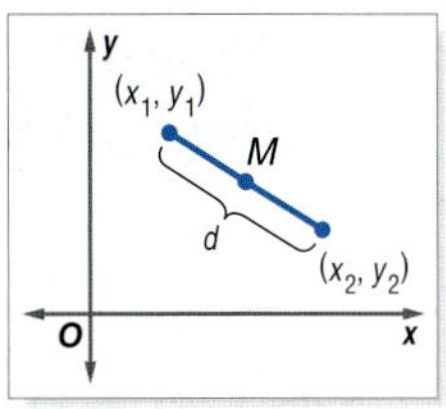

Examples

1 **Find the midpoint of a segment whose endpoints are at (−5, 9) and (11, −1).**

$\left(\frac{x_1 + x_2}{2}, \frac{y_1 + y_2}{2}\right) = \left(\frac{-5 + 11}{2}, \frac{9 + (-1)}{2}\right)$ Let $(x_1, y_1) = (-5, 9)$ and $(x_2, y_2) = (11, -1)$.

$= \left(\frac{6}{2}, \frac{8}{2}\right)$ or (3, 4) Simplify.

www.algebra2.com/vocabulary_review

Chapter 8 Study Guide and Review

Vocabulary and Concept Check

- This alphabetical list of vocabulary terms in Chapter 8 includes a page reference where each term was introduced.
- **Assessment** A vocabulary test/review for Chapter 8 is available on p. 510 of the *Chapter 8 Resource Masters*.

Lesson-by-Lesson Review

For each lesson,

- the main ideas are summarized,
- additional examples review concepts, and
- practice exercises are provided.

Vocabulary PuzzleMaker

ELL The Vocabulary PuzzleMaker software improves students' mathematics vocabulary using four puzzle formats—crossword, scramble, word search using a word list, and word search using clues. Students can work on a computer screen or from a printed handout.

MindJogger Videoquizzes

ELL MindJogger Videoquizzes provide an alternative review of concepts presented in this chapter. Students work in teams in a game show format to gain points for correct answers. The questions are presented in three rounds.

Round 1 Concepts (5 questions)
Round 2 Skills (4 questions)
Round 3 Problem Solving (4 questions)

FOLDABLES™ Study Organizer

For more information about Foldables, see *Teaching Mathematics with Foldables.*

As students review their Foldable for this chapter, ask them if they have used titles and headings to make it clear what the topic is for each section they write. Ask volunteers for some possible headings within a lesson, and for titles that might tie the whole chapter together. Encourage students to use both their own informal phrasing and the correct mathematical terminology of the textbook as they write.

Encourage students to refer to their Foldables while completing the Study Guide and Review and to use them in preparing for the Chapter Test.

Answers

17. (1, 1); (1, 4); $x = 1$; $y = -2$; upward; 12 units

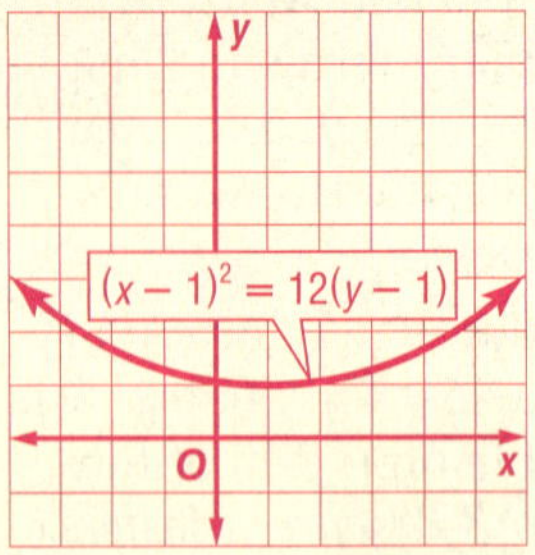

18. $(3, -6)$; $\left(3, -5\frac{63}{64}\right)$; $x = 3$; $y = -6\frac{1}{64}$; upward; $\frac{1}{16}$ unit

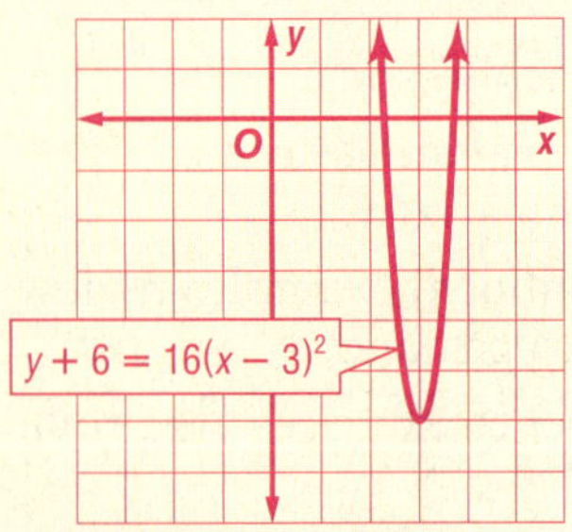

19. $(4, -2)$; $(4, -4)$; $x = 4$; $y = 0$; downward; 8 units

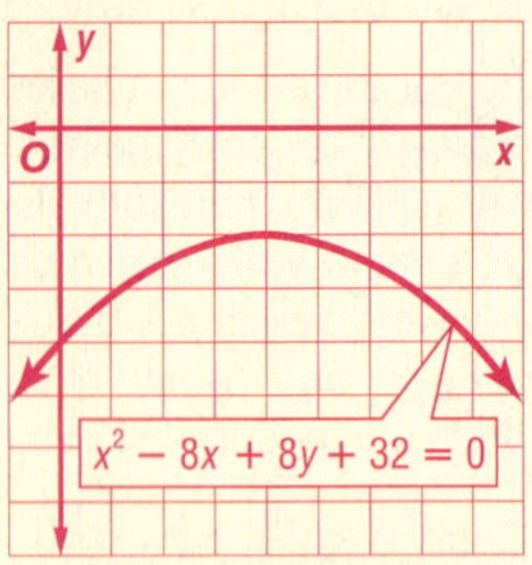

20. $(0, 0)$; $\left(\frac{1}{64}, 0\right)$; $y = 0$; $x = -\frac{1}{64}$; right; $\frac{1}{16}$ unit

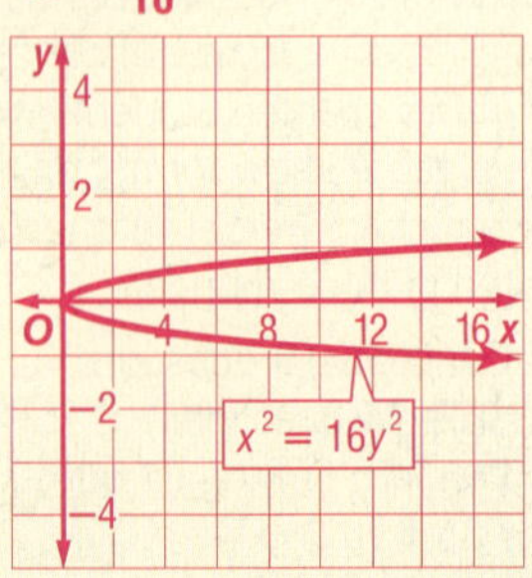

2 Find the distance between $P(6, -4)$ and $Q(-3, 8)$.

$d = \sqrt{(x_2 - x_1)^2 + (y_2 - y_1)^2}$ — Distance Formula

$= \sqrt{(-3 - 6)^2 + [8 - (-4)]^2}$ — Let $(x_1, y_1) = (6, -4)$ and $(x_2, y_2) = (-3, 8)$.

$= \sqrt{81 + 144}$ — Subtract.

$= \sqrt{225}$ or 15 units — Simplify.

Exercises **Find the midpoint of the line segment with endpoints at the given coordinates.** *See Example 1 on page 412.*

11. (1, 2), (4, 6) $\left(\frac{5}{2}, 4\right)$ **12.** (−8, 0), (−2, 3) $\left(-5, \frac{3}{2}\right)$ **13.** $\left(\frac{3}{5}, -\frac{7}{4}\right), \left(\frac{1}{4}, -\frac{2}{5}\right)$ $\left(\frac{17}{40}, -\frac{43}{40}\right)$

Find the distance between each pair of points with the given coordinates.
See Examples 2 and 3 on pages 413 and 414.

14. (−2, 10), (−2, 13) 3 units **15.** (8, 5), (−9, 4) $\sqrt{290}$ units **16.** (7, −3), (1, 2) $\sqrt{61}$ units

8-2 Parabolas

See pages 419–425.

Concept Summary

Parabolas		
Standard Form	$y = a(x - h)^2 + k$	$x = a(y - k)^2 + h$
Vertex	(h, k)	(h, k)
Axis of Symmetry	$x = h$	$y = k$
Focus	$\left(h, k + \frac{1}{4a}\right)$	$\left(h + \frac{1}{4a}, k\right)$
Directrix	$y = k - \frac{1}{4a}$	$x = h - \frac{1}{4a}$

Example Graph $4y - x^2 = 14x - 27$.

First write the equation in the form $y = a(x - h)^2 + k$.

$4y - x^2 = 14x - 27$ — Original equation

$4y = x^2 + 14x - 27$ — Isolate the terms with x.

$4y = (x^2 + 14x + ■) - 27 - ■$ — Complete the square.

$4y = (x^2 + 14x + 49) - 27 - 49$ — Add and subtract 49, since $\left(\frac{14}{2}\right)^2 = 49$.

$4y = (x + 7)^2 - 76$ — $x^2 + 14x + 49 = (x + 7)^2$

$y = \frac{1}{4}(x + 7)^2 - 19$ — Divide each side by 4.

vertex: $(-7, -19)$ axis of symmetry: $x = -7$

focus: $\left(-7, -19 + \frac{1}{4\left(\frac{1}{4}\right)}\right)$ or $(-7, -18)$

directrix: $y = -19 - \frac{1}{4\left(\frac{1}{4}\right)}$ or $y = -20$

direction of opening: upward since $a > 0$

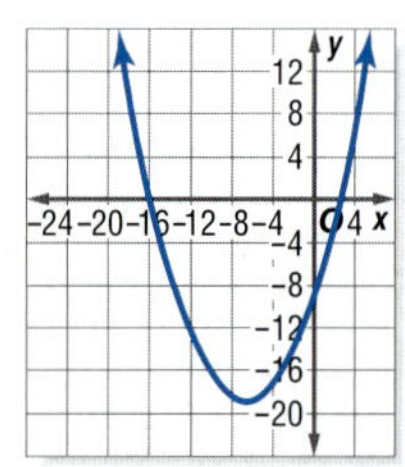

462 Chapter 8 Conic Sections

Exercises Identify the coordinates of the vertex and focus, the equations of the axis of symmetry and directrix, and the direction of opening of the parabola with the given equation. Then find the length of the latus rectum and graph the parabola. *See Examples 2–4 on pages 420–423.* **17–21. See margin.**

17. $(x-1)^2 = 12(y-1)$
18. $y + 6 = 16(x-3)^2$
19. $x^2 - 8x + 8y + 32 = 0$
20. $x = 16y^2$
21. Write an equation for a parabola with vertex (0, 1) and focus (0, −1). Then graph the parabola. *See Example 4 on pages 422 and 423.*

8-3 Circles

See pages 426–431.

Concept Summary

- The equation of a circle with center (h, k) and radius r can be written in the form $(x-h)^2 + (y-k)^2 = r^2$.

Example Graph $x^2 + y^2 + 8x - 24y + 16 = 0$.

First write the equation in the form $(x-h)^2 + (y-k)^2 = r^2$.

$x^2 + y^2 + 8x - 24y + 16 = 0$	Original equation
$x^2 + 8x + ■ + y^2 - 24y + ■ = -16 + ■ + ■$	Complete the squares.
$x^2 + 8x + 16 + y^2 - 24y + 144 = -16 + 16 + 144$	$\left(\frac{8}{2}\right)^2 = 16, \left(\frac{-24}{2}\right)^2 = 144$
$(x+4)^2 + (y-12)^2 = 144$	Write the trinomials as squares.

The center of the circle is at (−4, 12) and the radius is 12.

Now draw the graph.

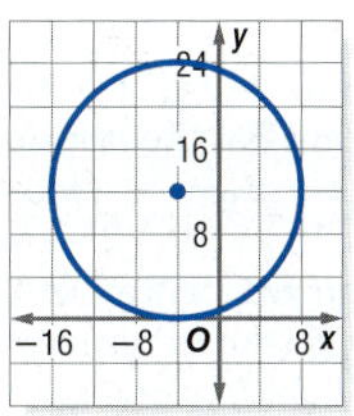

Exercises Write an equation for the circle that satisfies each set of conditions. *See Example 1 on page 426.*

22. center (2, −3), radius 5 units $(x-2)^2 + (y+3)^2 = 25$
23. center (−4, 0), radius $\frac{3}{4}$ unit $(x+4)^2 + y^2 = \frac{9}{16}$
24. endpoints of a diameter at (9, 4) and (−3, −2) $(x-3)^2 + (y-1)^2 = 45$
25. center at (−1, 2), tangent to x-axis $(x+1)^2 + (y-2)^2 = 4$

Find the center and radius of the circle with the given equation. Then graph the circle. *See Examples 4 and 5 on page 428.* **26–29. See margin for graphs.**

26. $x^2 + y^2 = 169$ **(0, 0), 13 units**
27. $(x+5)^2 + (y-11)^2 = 49$ **(−5, 11), 7 units**
28. $x^2 + y^2 - 6x + 16y - 152 = 0$ **(3, −8), 15 units**
29. $x^2 + y^2 + 6x - 2y - 15 = 0$ **(−3, 1), 5 units**

Answers

21. $y = -\frac{1}{8}x^2 + 1$

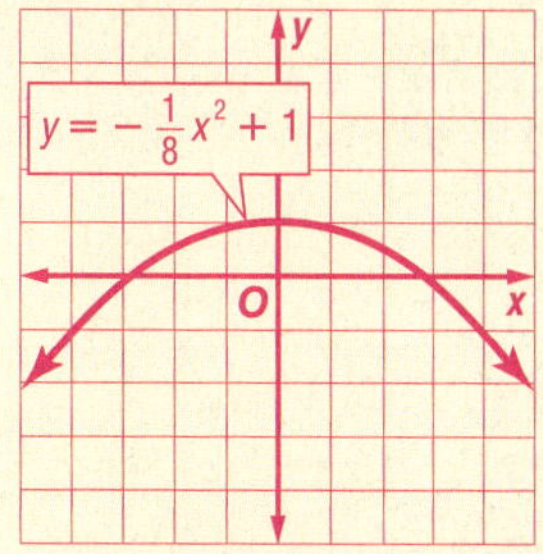

26.

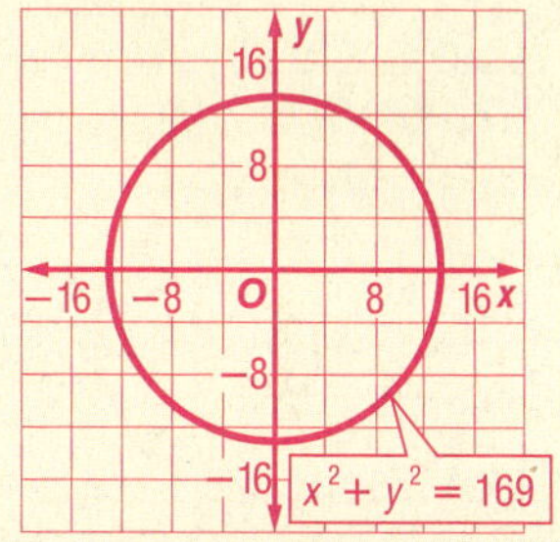

27.

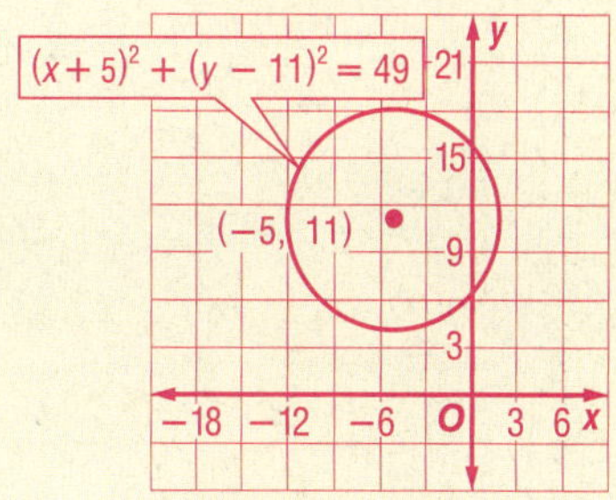

28.

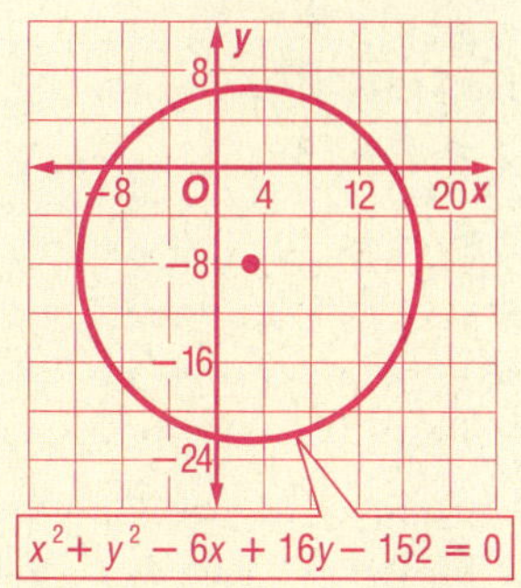

29.

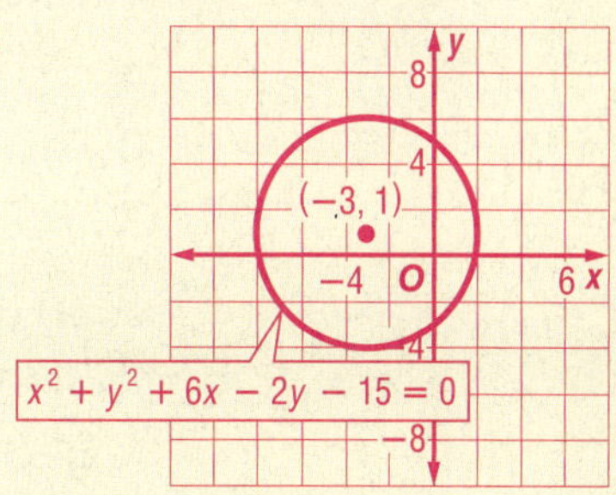

Answers

31. (0, 0); (0, ±3); 10; 8

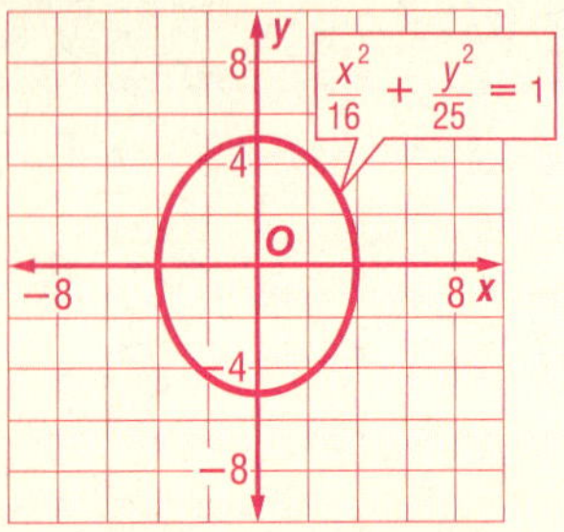

32. (−2, 3); $(-2 \pm \sqrt{7}, 3)$; 8; 6

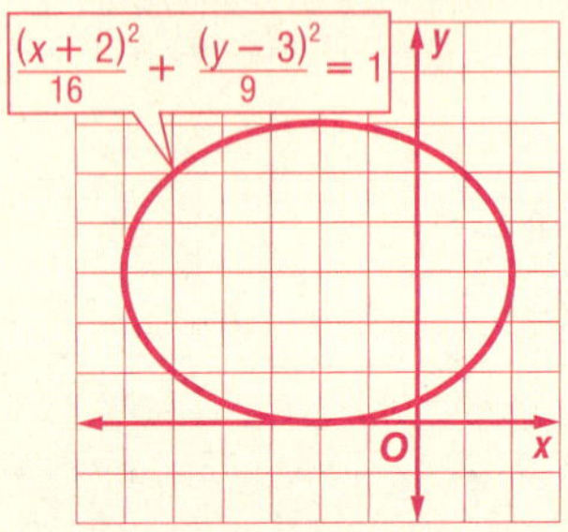

33. (1, −2); $(1 \pm \sqrt{3}, -2)$; 4; 2

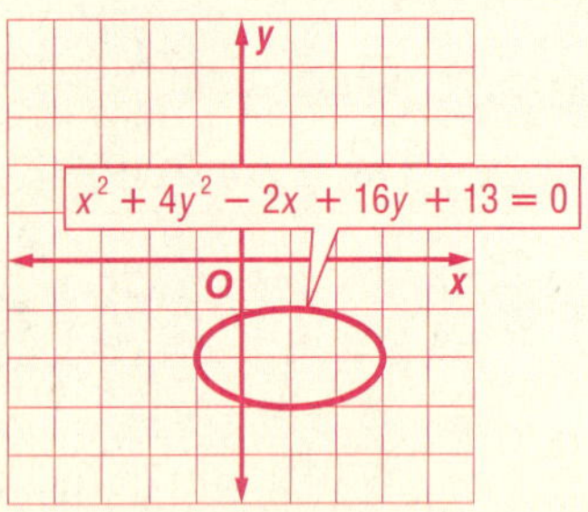

8-4 Ellipses

See pages 433–440.

Concept Summary

Ellipses		
Standard Form of Equation	$\frac{(x-h)^2}{a^2} + \frac{(y-k)^2}{b^2} = 1$	$\frac{(y-k)^2}{a^2} + \frac{(x-h)^2}{b^2} = 1$
Direction of Major Axis	horizontal	vertical

Example **Graph $x^2 + 3y^2 - 16x + 24y + 31 = 0$.**

First write the equation in standard form by completing the squares.

$x^2 + 3y^2 - 16x + 24y + 31 = 0$	Original equation
$x^2 - 16x + ■ + 3(y^2 + 8y + ■) = -31 + ■ + 3(■)$	Complete the squares.
$x^2 - 16x + 64 + 3(y^2 + 8y + 16) = -31 + 64 + 3(16)$	$\left(\frac{-16}{2}\right)^2 = 64, \left(\frac{8}{2}\right)^2 = 16$
$(x - 8)^2 + 3(y + 4)^2 = 81$	Write the trinomials as squares.
$\frac{(x-8)^2}{81} + \frac{(y+4)^2}{27} = 1$	Divide each side by 81.

The center of the ellipse is at (8, −4). The length of the major axis is 18, and the length of the minor axis is $6\sqrt{3}$.

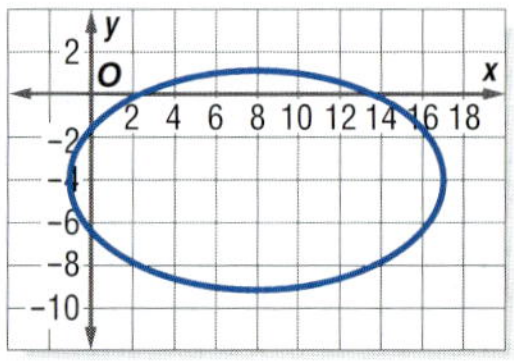

Exercises

30. Write an equation for the ellipse with endpoints of the major axis at (4, 1) and (−6, 1) and endpoints of the minor axis at (−1, 3) and (−1, −1).
See Examples 1 and 2 on pages 434 and 435. $\frac{(x+1)^2}{25} + \frac{(y-1)^2}{4} = 1$

Find the coordinates of the center and foci and the lengths of the major and minor axes for the ellipse with the given equation. Then graph the ellipse.
See Examples 3 and 4 on pages 436 and 437. **31–33. See margin.**

31. $\frac{x^2}{16} + \frac{y^2}{25} = 1$ 32. $\frac{(x+2)^2}{16} + \frac{(y-3)^2}{9} = 1$ 33. $x^2 + 4y^2 - 2x + 16y + 13 = 0$

8-5 Hyperbolas

See pages 441–448.

Concept Summary

Hyperbolas		
Standard Form	$\frac{(x-h)^2}{a^2} - \frac{(y-k)^2}{b^2} = 1$	$\frac{(y-k)^2}{a^2} - \frac{(x-h)^2}{b^2} = 1$
Transverse Axis	horizontal	vertical
Asymptotes	$y - k = \pm\frac{b}{a}(x - h)$	$y - k = \pm\frac{a}{b}(x - h)$

Example **Graph $9x^2 - 4y^2 + 18x + 32y - 91 = 0$.**

Complete the square for each variable to write this equation in standard form.

$9x^2 - 4y^2 + 18x + 32y - 91 = 0$	Original equation
$9(x^2 + 2x + \blacksquare) - 4(y^2 - 8y + \blacksquare) = 91 + 9(\blacksquare) - 4(\blacksquare)$	Complete the squares.
$9(x^2 + 2x + 1) - 4(y^2 - 8y + 16) = 91 + 9(1) - 4(16)$	$\left(\frac{2}{2}\right)^2 = 1, \left(\frac{-8}{2}\right)^2 = 16$
$9(x + 1)^2 - 4(y - 4)^2 = 36$	Write the trinomials as squares.
$\frac{(x+1)^2}{4} - \frac{(y-4)^2}{9} = 1$	Divide each side by 36.

The center is at $(-1, 4)$. The vertices are at $(-3, 4)$ and $(1, 4)$ and the foci are at $\left(-1 \pm \sqrt{13}, 4\right)$. The equations of the asymptotes are $y - 4 = \pm\frac{3}{2}(x + 1)$.

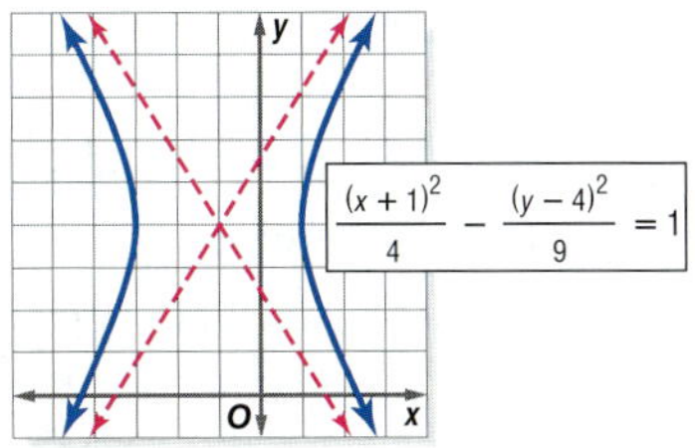

Exercises

34. Write an equation for a hyperbola that has vertices at (2, 5) and (2, 1) and a conjugate axis of length 6 units. *See Example 1 on page 442.* $\frac{(y-3)^2}{4} - \frac{(x-2)^2}{9} = 1$

Find the coordinates of the vertices and foci and the equations of the asymptotes for the hyperbola with the given equation. Then graph the hyperbola.
See Examples 3 and 4 on pages 443 and 444. **35–38. See margin.**

35. $\frac{y^2}{4} - \frac{x^2}{9} = 1$

36. $\frac{(x-2)^2}{1} - \frac{(y+1)^2}{9} = 1$

37. $9y^2 - 16x^2 = 144$

38. $16x^2 - 25y^2 - 64x - 336 = 0$

8-6 Conic Sections

See pages 449–452.

Concept Summary

- Conic sections can be identified directly from their equations of the form $Ax^2 + Bxy + Cy^2 + Dx + Ey + F = 0$, assuming $B = 0$.

Conic Section	Relationship of *A* and *C*
Parabola	$A = 0$ or $C = 0$, but not both.
Circle	$A = C$
Ellipse	*A* and *C* have the same sign and $A \neq C$.
Hyperbola	*A* and *C* have opposite signs.

Example **Without writing the equation in standard form, state whether the graph of $4x^2 + 9y^2 + 16x - 18y - 11 = 0$ is a *parabola, circle, ellipse,* or *hyperbola*.**

In this equation, $A = 4$ and $C = 9$. Since *A* and *C* are both positive and $A \neq C$, the graph is an ellipse.

Answers

35. $(0, \pm 2)$; $\left(0, \pm\sqrt{13}\right)$; $y = \pm\frac{2}{3}x$

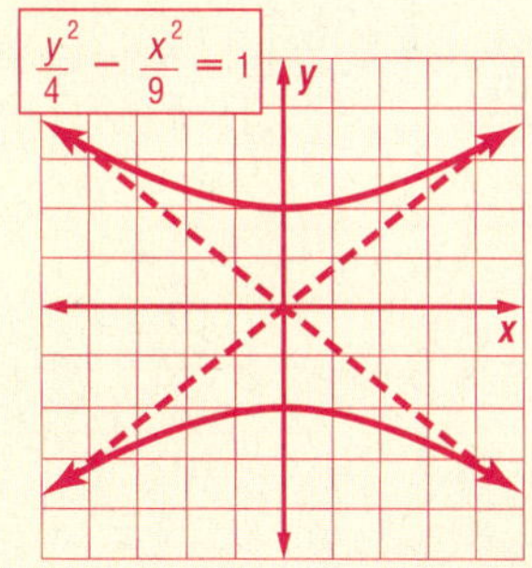

36. $(1, -1), (3, -1)$; $\left(2 \pm \sqrt{10}, -1\right)$; $y + 1 = \pm 3(x - 2)$

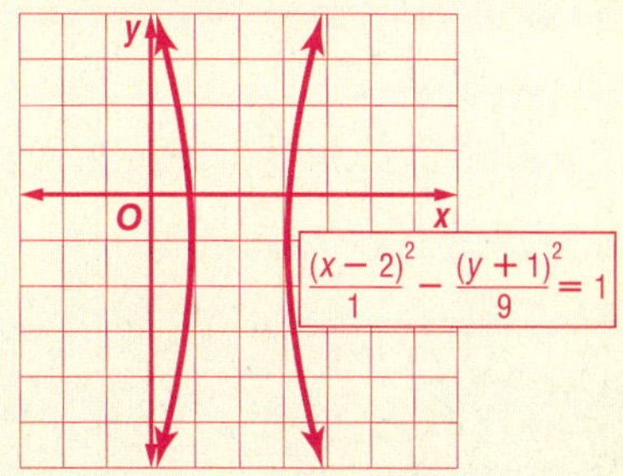

37. $(0, \pm 4)$; $(0, \pm 5)$; $y \pm\frac{4}{3}x$

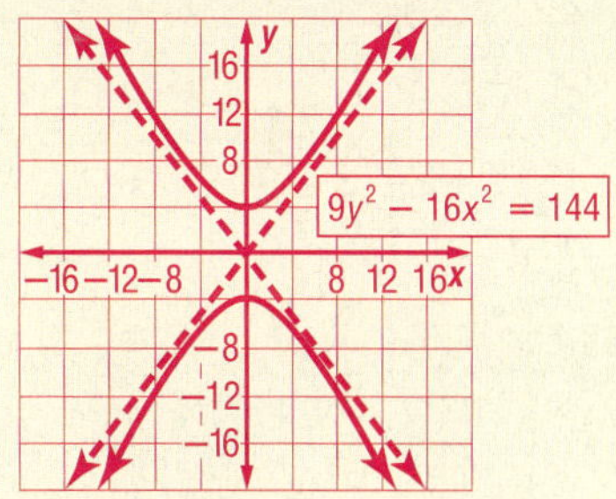

38. $(-3, 0), (7, 0)$; $\left(2 \pm \sqrt{41}, 0\right)$; $y = \pm\frac{4}{5}(x - 2)$

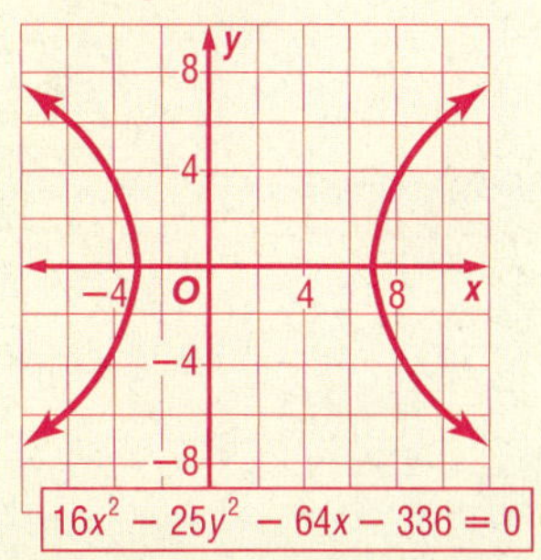

Chapter 8 For More ...
• Extra Practice, see pages 845–847.
• Mixed Problem Solving, see page 869.

Exercises Write each equation in standard form. State whether the graph of the equation is a *parabola*, *circle*, *ellipse*, or *hyperbola*. Then graph the equation. *See Example 1 on page 449.* **39–42. See margin for equations and graphs.**

39. $x^2 + 4x - y = 0$ **parabola**
40. $9x^2 + 4y^2 = 36$ **ellipse**
41. $-4x^2 + y^2 + 8x - 8 = 0$ **hyperbola**
42. $x^2 + y^2 - 4x - 6y + 4 = 0$ **circle**

Without writing the equation in standard form, state whether the graph of each equation is a *parabola*, *circle*, *ellipse*, or *hyperbola*. *See Example 2 on page 450.*

43. $7x^2 + 9y^2 = 63$ **ellipse**
44. $x^2 - 8x + 16 = 6y$ **parabola**
45. $x^2 + 4x + y^2 - 285 = 0$ **circle**
46. $5y^2 + 2y + 4x - 13x^2 = 81$ **hyperbola**

8-7 Solving Quadratic Systems

See pages 455–460.

Concept Summary

- Systems of quadratic equations can be solved using substitution and elimination.
- A system of quadratic equations can have zero, one, two, three, or four solutions.

Example **Solve the system of equations.**
$x^2 + y^2 + 2x - 12y + 12 = 0$
$y + x = 0$

Use substitution to solve the system.

First, rewrite $y + x = 0$ as $y = -x$.

$x^2 + y^2 + 2x - 12y + 12 = 0$ First original equation

$x^2 + (-x)^2 + 2x - 12(-x) + 12 = 0$ Substitute $-x$ for y.

$2x^2 + 14x + 12 = 0$ Simplify.

$x^2 + 7x + 6 = 0$ Divide each side by 2.

$(x + 6)(x + 1) = 0$ Factor.

$x + 6 = 0$ or $x + 1 = 0$ Zero Product Property.

$x = -6$ $\quad x = -1$ Solve for x.

Now solve for y.

$y = -x$ $\quad y = -x$ Equation for y in terms of x

$= -(-6)$ or 6 $\quad = -(-1)$ or 1 Substitute the x values.

The solutions of the system are $(-6, 6)$ and $(-1, 1)$.

Exercises Find the exact solution(s) of each system of equations. *See Examples 1 and 2 on pages 455–457.*

47. $x^2 + y^2 - 18x + 24y + 200 = 0$
$4x + 3y = 0$ **(6, −8), (12, −16)**
48. $4x^2 + y^2 = 16$
$x^2 + 2y^2 = 4$ **(±2, 0)**

49–50. See margin.
Solve each system of inequalities by graphing. *See Example 3 on page 457.*

49. $y < x$
$y > x^2 - 4$
50. $x^2 + y^2 \le 9$
$x^2 + 4y^2 \le 16$

Answers

39. $y = (x + 2)^2 - 4$

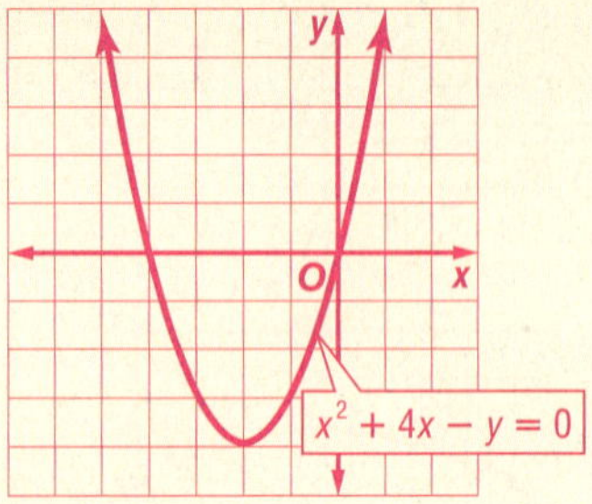

40. $\frac{y^2}{9} + \frac{x^2}{4} = 1$

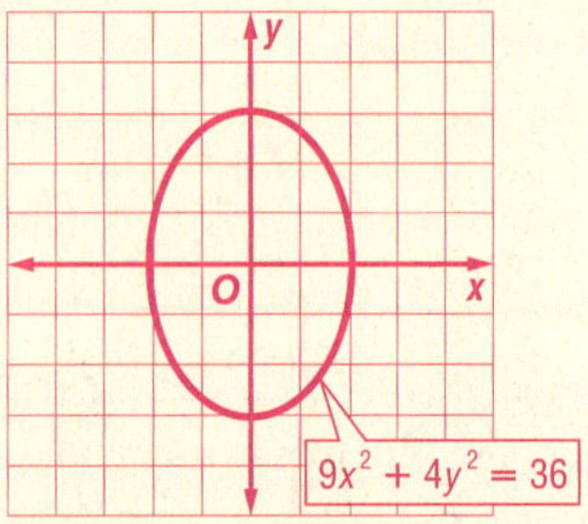

41. $\frac{y^2}{4} - \frac{(x-1)^2}{1} = 1$

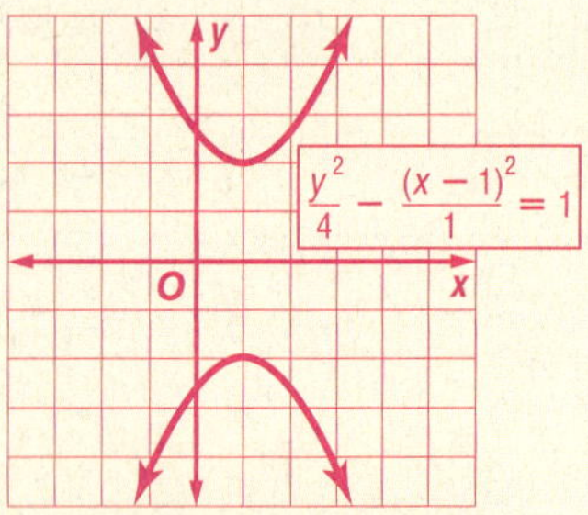

42. $(x - 2)^2 + (y - 3)^2 = 9$

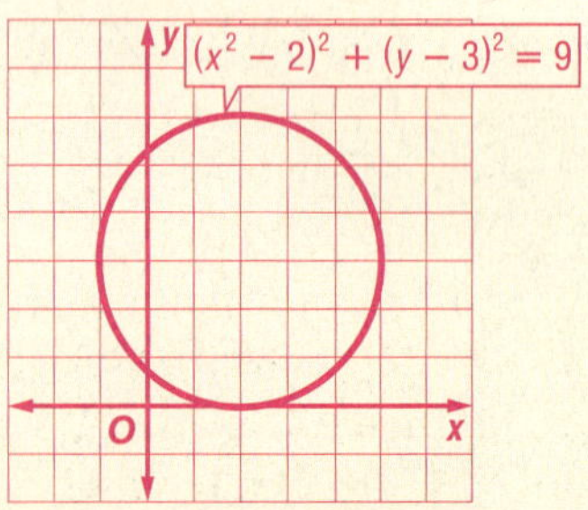

49.

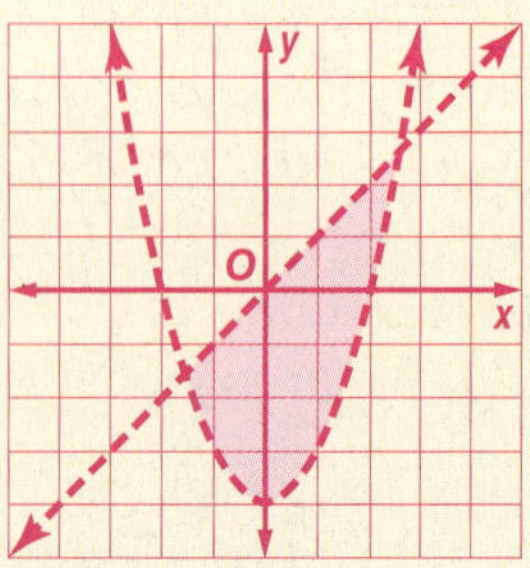

50.

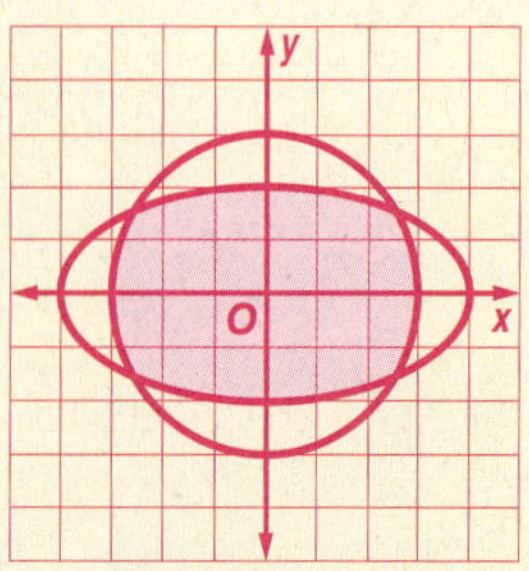

466 Chapter 8 Conic Sections

Chapter 8 Practice Test

Vocabulary and Concepts

Choose the letter that best matches each description.

1. the set of all points in a plane that are the same distance from a given point, the focus, and a given line, the directrix **b**
2. the set of all points in a plane such that the absolute value of the difference of the distances from two fixed points, the foci, is constant **c**
3. the set of all points in a plane such that the sum of the distances from two fixed points, the foci, is constant **a**

a. ellipse
b. parabola
c. hyperbola

Skills and Applications

Find the midpoint of the line segment with endpoints at the given coordinates.

4. $(7, 1), (-5, 9)$ **$(1, 5)$**
5. $\left(\frac{3}{8}, -1\right), \left(-\frac{8}{5}, 2\right)$ **$\left(-\frac{49}{80}, \frac{1}{2}\right)$**
6. $(-13, 0), (-1, -8)$ **$(-7, -4)$**

Find the distance between each pair of points with the given coordinates.

7. $(-6, 7), (3, 2)$ **$\sqrt{106}$ units**
8. $\left(\frac{1}{2}, \frac{5}{2}\right), \left(-\frac{3}{4}, -\frac{11}{4}\right)$ **$\frac{\sqrt{466}}{4}$ units**
9. $(8, -1), (8, -9)$ **8 units**

State whether the graph of each equation is a *parabola, circle, ellipse,* or *hyperbola*. Then graph the equation. **10–19. See pp. 469A–469J for graphs.**

10. $x^2 + 4y^2 = 25$ **ellipse**
11. $y = 4x^2 + 1$ **parabola**
12. $x^2 = 36 - y^2$ **circle**
13. $(x + 4)^2 = 7(y + 5)$ **parabola**
14. $4x^2 - 26y^2 + 10 = 0$ **hyperbola**
15. $25x^2 + 49y^2 = 1225$ **ellipse**
16. $-(y^2 - 24) = x^2 + 10x$ **circle**
17. $5x^2 - y^2 = 49$ **hyperbola**
18. $\frac{1}{3}x^2 - 4 = y$ **parabola**
19. $\frac{y^2}{9} - \frac{x^2}{25} = 1$ **hyperbola**

20. **TUNNELS** The opening of a tunnel is in the shape of a semielliptical arch. The arch is 60 feet wide and 40 feet high. Find the height of the arch 12 feet from the edge of the tunnel. **32 ft**

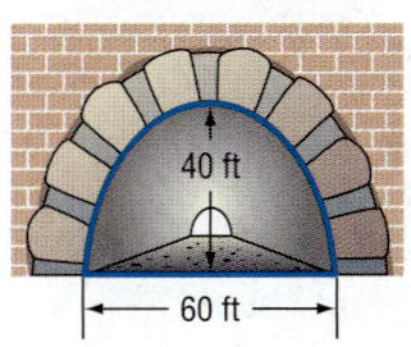

Find the exact solution(s) of each system of equations.

21. $x^2 + y^2 = 100$
$y = 2 - x$ **$(-6, 8), (8, -6)$**
22. $x^2 + 2y^2 = 6$
$x + y = 1$ **$\left(-\frac{2}{3}, \frac{5}{3}\right), (2, -1)$**
23. $x^2 - y^2 - 12x + 12y = 36$
$x^2 + y^2 - 12x - 12y + 36 = 0$ **$(0, 6), (12, 6)$**
24. Solve the system of inequalities by graphing. **See pp. 469A–469J.**
$x^2 - y^2 \geq 1$
$x^2 + y^2 \leq 16$

25. **STANDARDIZED TEST PRACTICE** Which is *not* the equation of a parabola? **C**
Ⓐ $y = 2x^2 + 4x - 9$
Ⓑ $3x + 2y^2 + y + 1 = 0$
Ⓒ $x^2 + 2y^2 + 8y = 8$
Ⓓ $x = \frac{1}{2}(y - 1)^2 + 5$

www.algebra2.com/chapter_test

Chapter 8 Practice Test

Assessment Options

Vocabulary Test A vocabulary test/review for Chapter 8 can be found on p. 510 of the *Chapter 8 Resource Masters.*

Chapter Tests There are six Chapter 8 Tests and an Open-Ended Assessment task available in the *Chapter 8 Resource Masters.*

Chapter 8 Tests			
Form	**Type**	**Level**	**Pages**
1	MC	basic	497–498
2A	MC	average	499–500
2B	MC	average	501–502
2C	FR	average	503–504
2D	FR	average	505–506
3	FR	advanced	507–508

MC = multiple-choice questions
FR = free-response questions

Open-Ended Assessment
Performance tasks for Chapter 8 can be found on p. 509 of the *Chapter 8 Resource Masters*. A sample scoring rubric for these tasks appears on p. A28.

ExamView® Pro

Use the networkable **ExamView® Pro** to:

- Create **multiple versions** of tests.
- Create **modified** tests for *Inclusion* students.
- **Edit** existing questions and **add** your own questions.
- Use built-in **state curriculum correlations** to create tests aligned with state standards.
- **Apply** art to your tests from a program bank of artwork.

Portfolio Suggestion

Introduction In this chapter you have worked with four different conic sections.

Ask Students Describe what you have learned about conic sections and how they are related. List and compare their standard equations. Then draw each conic section on a separate coordinate grid, labeling important characteristics, such as center, foci, vertex, axes of symmetry, and so on. Place these drawings in your portfolio.

These two pages contain practice questions in the various formats that can be found on the most frequently given standardized tests.

A practice answer sheet for these two pages can be found on p. A1 of the *Chapter 8 Resource Masters*.

Standardized Test Practice
Student Recording Sheet, p. A1

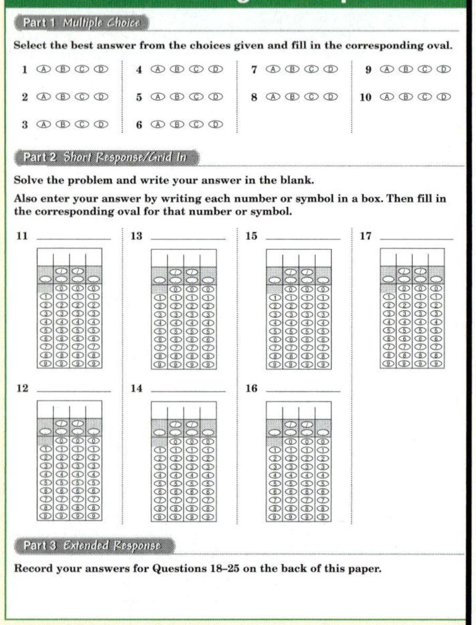
Part 1 Multiple Choice
Select the best answer from the choices given and fill in the corresponding oval.
1 Ⓐ Ⓑ Ⓒ Ⓓ 4 Ⓐ Ⓑ Ⓒ Ⓓ 7 Ⓐ Ⓑ Ⓒ Ⓓ 9 Ⓐ Ⓑ Ⓒ Ⓓ
2 Ⓐ Ⓑ Ⓒ Ⓓ 5 Ⓐ Ⓑ Ⓒ Ⓓ 8 Ⓐ Ⓑ Ⓒ Ⓓ 10 Ⓐ Ⓑ Ⓒ Ⓓ
3 Ⓐ Ⓑ Ⓒ Ⓓ 6 Ⓐ Ⓑ Ⓒ Ⓓ
Part 2 Short Response/Grid In
Solve the problem and write your answer in the blank.
Also enter your answer by writing each number or symbol in a box. Then fill in the corresponding oval for that number or symbol.
11 13 15 17
12 14 16
Part 3 Extended Response
Record your answers for Questions 18–25 on the back of this paper.

Additional Practice

See pp. 515–516 in the *Chapter 8 Resource Masters* for additional standardized test practice.

Chapter 8 Standardized Test Practice

Part 1 Multiple Choice

Record your answers on the answer sheet provided by your teacher or on a sheet of paper.

1. The product of a prime number and a composite number must be **B**

Ⓐ prime. Ⓑ composite.
Ⓒ even. Ⓓ negative.

2. In 1990, the population of Clayton was 54,200, and the population of Montrose was 47,500. By 2000, the population of each city had decreased by exactly 5%. How many more people lived in Clayton than in Montrose in 2000? **C**

Ⓐ 335 Ⓑ 5085
Ⓒ 6365 Ⓓ 6700

3. If 4% of n is equal to 40% of p, then n is what percent of $10p$? **C**

Ⓐ $\frac{1}{1000}\%$ Ⓑ 10%
Ⓒ 100% Ⓓ 1000%

4. Leroy bought m magazines at d dollars per magazine and p paperback books at $2d + 1$ dollars per book. Which of the following represents the total amount Leroy spent? **A**

Ⓐ $d(m + 2p) + p$ Ⓑ $(m + p)(3d + 1)$
Ⓒ $md + 2pd + 1$ Ⓓ $pd(m + 2)$

Test-Taking Tip

Questions 3, 4
In problems with variables, you can substitute values to try to eliminate some of the answer choices. For example, in Question 3, choose a value for n and compute the corresponding value of p. Then find $\frac{n}{10p}$ to answer the question.

5. What is the midpoint of the line segment whose endpoints are at $(-5, -3)$ and $(-1, 4)$? **B**

Ⓐ $\left(-3, -\frac{1}{2}\right)$ Ⓑ $\left(-3, \frac{1}{2}\right)$
Ⓒ $\left(-2, \frac{7}{2}\right)$ Ⓓ $\left(-2, \frac{1}{2}\right)$

6. Point $M(-2, 3)$ is the midpoint of line segment NP. If point N has coordinates $(-7, 1)$, then what are the coordinates of point P? **D**

Ⓐ $(-5, 2)$ Ⓑ $(-4, 6)$
Ⓒ $\left(-\frac{9}{2}, 2\right)$ Ⓓ $(3, 5)$

7. Which equation's graph is a parabola? **D**

Ⓐ $3x^2 - 2y^2 = 10$
Ⓑ $4x^2 + 3y^2 = 20$
Ⓒ $2x^2 + 2y^2 = 15$
Ⓓ $3x^2 + 4y = 8$

8. What is the center of the circle with equation $x^2 + y^2 - 4x + 6y - 9 = 0$? **C**

Ⓐ $(-4, 6)$ Ⓑ $(-2, 3)$
Ⓒ $(2, -3)$ Ⓓ $(3, 3)$

9. What is the distance between the points shown in the graph? **B**

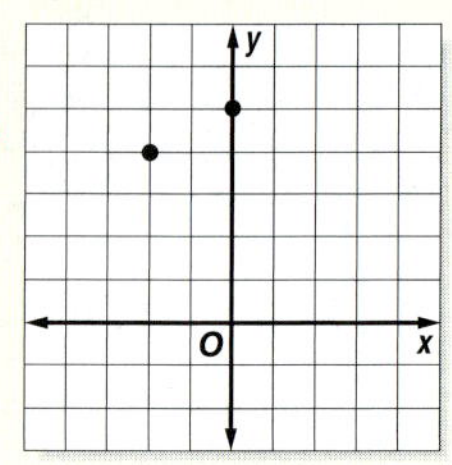

Ⓐ $\sqrt{3}$ units
Ⓑ $\sqrt{5}$ units
Ⓒ 3 units
Ⓓ $\sqrt{17}$ units

10. The median of seven test scores is 52, the mode is 64, the lowest score is 40, and the average is 53. If the scores are integers, what is the greatest possible test score? **A**

Ⓐ 68 Ⓑ 72 Ⓒ 76 Ⓓ 84

ExamView® Pro

Special banks of standardized test questions similar to those on the SAT, ACT, TIMSS 12, NAEP 12, and state proficiency tests can be found on this CD-ROM.

Preparing for Standardized Tests
For test-taking strategies and more practice, see pages 877–892.

Part 2 Short Response/Grid In

Record your answers on the answer sheet provided by your teacher or on a sheet of paper.

11. What is the least positive integer p for which $2^{2p} + 3$ is not a prime number? **4**

12. The ratio of cars to SUVs in a parking lot is 4 to 5. After 6 cars leave the parking lot, the ratio of cars to SUVs becomes 1 to 2. How many SUVs are in the parking lot? **20**

13. Each dimension of a rectangular box is an integer greater than 1. If the area of one side of the box is 27 square units and the area of another side is 12 square units, what is the volume of the box in cubic units? **108**

14. Let the operation * be defined as $a * b = 2ab - (a + b)$. If $4 * x = 10$, then what is the value of x? **2**

15. If the slope of line PQ in the figure is $\frac{1}{4}$, what is the area of quadrilateral $OPQR$? **14**

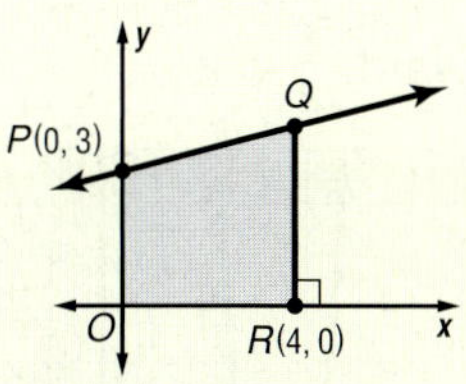

16. In the figure, the slope of line ℓ is $\frac{5}{4}$, and the slope of line k is $\frac{3}{8}$. What is the distance from point A to point B? **7/2 or 3.5**

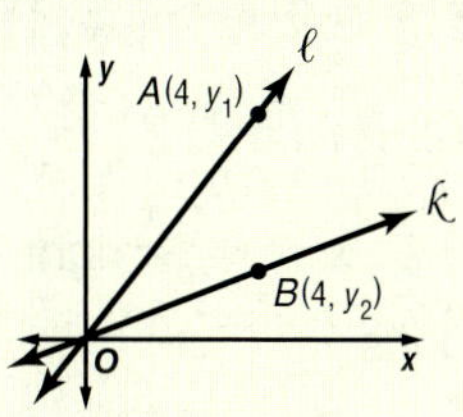

17. If $(2x - 3)(4x + n) = ax^2 + bx - 15$ for all values of x, what is the value of $a + b$? **6**

www.algebra2.com/standardized_test

Part 3 Extended Response

Record your answers on a sheet of paper. Show your work.

For Exercises 18–20, use the following information.

A certain map of New York City's Central Park shows the Loeb Boathouse at (4, 7), Belvedere Castle at (3, 9), and the center of the Great Lawn at (3, 12). Each grid on the map represents 0.1 mile in the park.

18. How far apart are the Loeb Boathouse and Belvedere Castle? Round to the nearest hundredth of a mile. **0.22 mi**

19. Loeb Boathouse is halfway between the center of the Great Lawn and the statue of Balto. Where is the statue of Balto on this map? **(5, 2)**

20. The Great Lawn is approximately shaped like an ellipse. The major axis is 0.2 mile and the minor axis is 0.14 mile. If the major axis is vertical, write an equation to model the Great Lawn. $\frac{(y-12)^2}{1} + \frac{(x-3)^2}{0.49} = 1$

21. A fountain sprays water from a jet at ground level. The water reaches a maximum height of 84 inches and hits the ground 18 inches away from the jet. Write an equation of the parabola that models the path of the water. Assume that the jet is at the origin. **See margin.**

For Exercises 22–25, use the following information.

The endpoints of a diameter of a circle are at $(-1, 0)$, and $(5, -8)$.

22. What are the coordinates of the center of the circle? Explain your method. **(2, −4); The center is the midpoint of a diameter.**

23. Find the radius of the circle. Explain your method. **5; The radius is the distance between the center and any point on the circle.**

24. Write an equation of the circle. $(x - 2)^2 + (y + 4)^2 = 25$

25. Find the point(s) at which the circle intersects the line with equation $y = x - 7$. **(6, −1), (−1, −8)**

Evaluating Extended Response Questions

Extended Response questions are graded by using a multilevel rubric that guides you in assessing a student's knowledge of a particular concept.

Goal: Determine the dimensions of a rectangular garden for a given perimeter.

Sample Scoring Rubric: The following rubric is a sample scoring device. You may wish to add more detail to this sample to meet your individual scoring needs.

Score	Criteria
4	A correct solution that is supported by well-developed, accurate explanations
3	A generally correct solution, but may contain minor flaws in reasoning or computation
2	A partially correct interpretation and/or solution to the problem
1	A correct solution with no supporting evidence or explanation
0	An incorrect solution indicating no mathematical understanding of the concept or task, or no solution is given

Pages 415–416, Lesson 8-1

21. Sample answer: Draw several line segments across the U.S. One should go from the northeast corner to the southwest corner; another should go from the southeast corner to the northwest corner; another should go across the middle of the U.S. from east to west; and so on. Find the midpoints of these segments. Locate a point to represent all of these midpoints.

41. The slope of the line through (x_1, y_1) and (x_2, y_2) is $\frac{y_2 - y_1}{x_2 - x_1}$ and the point-slope form of the equation of the line is $y - y_1 = \frac{y_2 - y_1}{x_2 - x_1}(x - x_1)$. Substitute $\left(\frac{x_1 + x_2}{2}, \frac{y_1 + y_2}{2}\right)$ into this equation. The left side is $\frac{y_1 + y_2}{2} - y_1$ or $\frac{y_2 - y_1}{2}$. The right side is $\frac{y_2 - y_1}{x_2 - x_1}\left(\frac{x_1 + x_2}{2} - x_1\right) = \frac{y_2 - y_1}{x_2 - x_1}\left(\frac{x_2 - x_1}{2}\right)$ or $\frac{y_2 - y_1}{2}$. Therefore, the point with coordinates $\left(\frac{x_1 + x_2}{2}, \frac{y_1 + y_2}{2}\right)$ lies on the line through (x_1, y_1) and (x_2, y_2). The distance from $\left(\frac{x_1 + x_2}{2}, \frac{y_1 + y_2}{2}\right)$ to (x_1, y_1) is $\sqrt{\left(x_1 - \frac{x_1 + x_2}{2}\right)^2 + \left(y_1 - \frac{y_1 + y_2}{2}\right)^2}$ or $\sqrt{\left(\frac{x_1 - x_2}{2}\right)^2 + \left(\frac{y_1 - y_2}{2}\right)^2}$. The distance from $\left(\frac{x_1 + x_2}{2}, \frac{y_1 + y_2}{2}\right)$ to (x_2, y_2) is $\sqrt{\left(x_2 - \frac{x_1 + x_2}{2}\right)^2 + \left(y_2 - \frac{y_1 + y_2}{2}\right)^2} = \sqrt{\left(\frac{x_2 - x_1}{2}\right)^2 + \left(\frac{y_2 - y_1}{2}\right)^2}$ or $\sqrt{\left(\frac{x_1 - x_2}{2}\right)^2 + \left(\frac{y_1 - y_2}{2}\right)^2}$. Therefore the point with coordinates $\left(\frac{x_1 + x_2}{2}, \frac{y_1 + y_2}{2}\right)$ is equidistant from (x_1, y_1) and (x_2, y_2).

42. The formulas can be used to decide from which location an emergency squad should be dispatched. Answers should include the following.

- Most maps have a superimposed grid. Think of the grid as a coordinate system and assign approximate coordinates to the two cities. Then use the Distance Formula to find the distance between the points with those coordinates.
- Suppose the bottom left of the grid is the origin. Then the coordinates of Lincoln are about (0.7, 0.2); the coordinates of Omaha are about (4.4, 3.9); and the coordinates of Fremont are about (1.7, 4.6). The distance from Omaha to Fremont is about $10\sqrt{(1.7 - 4.4)^2 + (4.6 - 3.9)^2}$ or about 28 miles. The distance from Lincoln to Fremont is about $10\sqrt{(1.7 - 0.7)^2 + (4.6 - 0.2)^2}$ or about 45 miles. Since Omaha is closer than Lincoln, the helicopter should be dispatched from Omaha.

Page 418, Follow-Up of Lesson 8-1
Algebra Activity

14. The distance between the points with coordinates $\left(\frac{x_1 + x_2}{2}, \frac{y_1 + y_2}{2}, \frac{z_1 + z_2}{2}\right)$ and (x_1, y_1, z_1) is $\sqrt{\left(x_1 - \frac{x_1 + x_2}{2}\right)^2 + \left(y_1 - \frac{y_1 + y_2}{2}\right)^2 + \left(z_1 - \frac{z_1 + z_2}{2}\right)^2}$ or $\sqrt{\left(\frac{x_1 - x_2}{2}\right)^2 + \left(\frac{y_1 - y_2}{2}\right)^2 + \left(\frac{z_1 - z_2}{2}\right)^2}$ units. The distance between the points with coordinates $\left(\frac{x_1 + x_2}{2}, \frac{y_1 + y_2}{2}, \frac{z_1 + z_2}{2}\right)$ and (x_2, y_2, z_2) is $\sqrt{\left(x_2 - \frac{x_1 + x_2}{2}\right)^2 + \left(y_2 - \frac{y_1 + y_2}{2}\right)^2 + \left(z_2 - \frac{z_1 + z_2}{2}\right)^2}$ or $\sqrt{\left(\frac{x_2 - x_1}{2}\right)^2 + \left(\frac{y_2 - y_1}{2}\right)^2 + \left(\frac{z_2 - z_1}{2}\right)^2}$ units. Since $\sqrt{\left(\frac{x_2 - x_1}{2}\right)^2 + \left(\frac{y_2 - y_1}{2}\right)^2 + \left(\frac{z_2 - z_1}{2}\right)^2} = \sqrt{\left(\frac{x_1 - x_2}{2}\right)^2 + \left(\frac{y_1 - y_2}{2}\right)^2 + \left(\frac{z_1 - z_2}{2}\right)^2}$, the distances are equal.

Pages 423–424, Lesson 8-2

5. $(3, -4)$, $\left(3, -3\frac{3}{4}\right)$, $x = 3$, $y = -4\frac{1}{4}$, upward, 1 unit

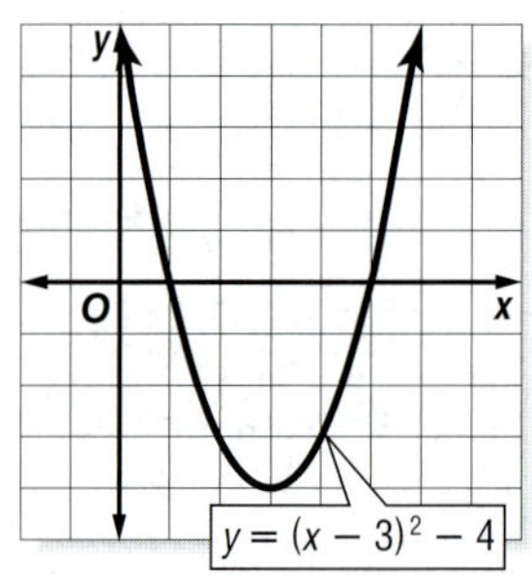

6. $(-7, 3)$, $\left(-7, 3\frac{1}{8}\right)$, $x = -7$, $y = 2\frac{7}{8}$, upward, $\frac{1}{2}$ unit

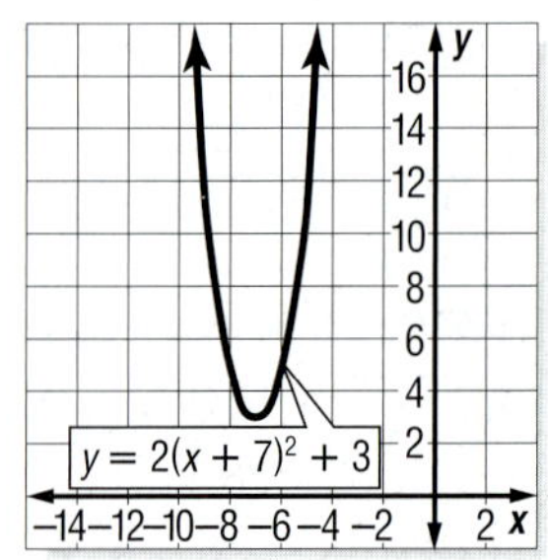

7. $\left(-\frac{4}{3}, -\frac{2}{3}\right)$, $\left(-\frac{4}{3}, -\frac{3}{4}\right)$, $x = -\frac{4}{3}$, $y = -\frac{7}{12}$, downward, $\frac{1}{3}$ unit

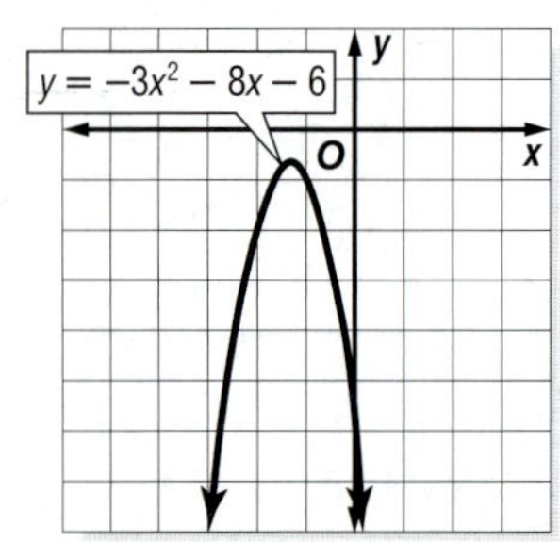

8. $\left(-\frac{3}{2}, \frac{9}{2}\right)$, $\left(-\frac{9}{8}, \frac{9}{2}\right)$, $y = \frac{9}{2}$, $x = -\frac{15}{8}$, right, $\frac{3}{2}$ units

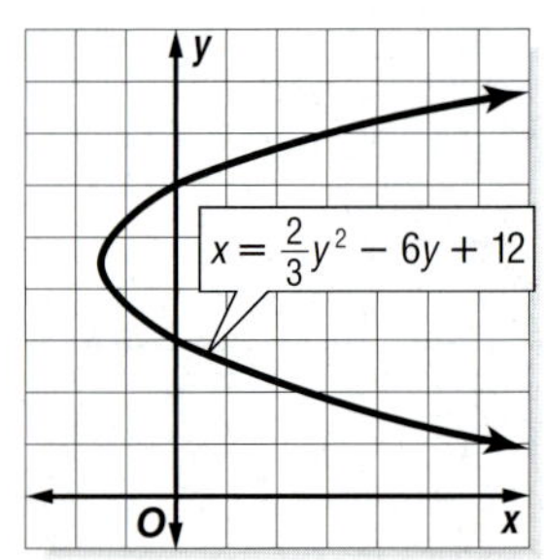

16. $(0, 0), \left(0, -\frac{3}{2}\right), x = 0,$ $y = \frac{3}{2}$, downward, 6 units

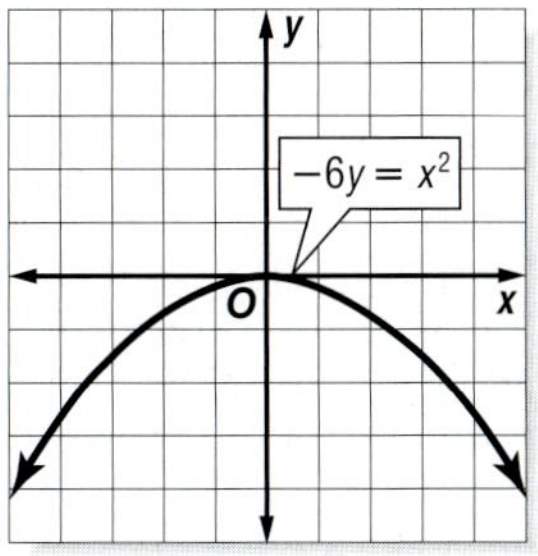

17. $(0, 0), \left(\frac{1}{2}, 0\right), y = 0,$ $x = -\frac{1}{2}$, right, 2 units

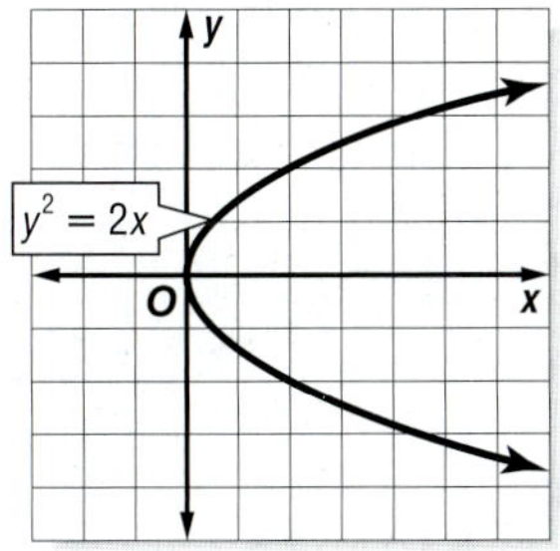

18. $(-6, 3), \left(-6, 3\frac{3}{4}\right),$ $x = -6, y = 2\frac{1}{4}$, upward, 3 units

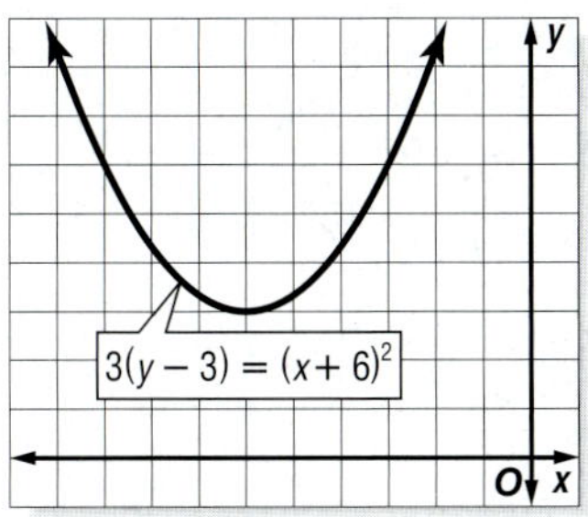

19. $(1, 4), \left(1, 3\frac{1}{2}\right), x = 1,$ $y = 4\frac{1}{2}$, downward, 2 units

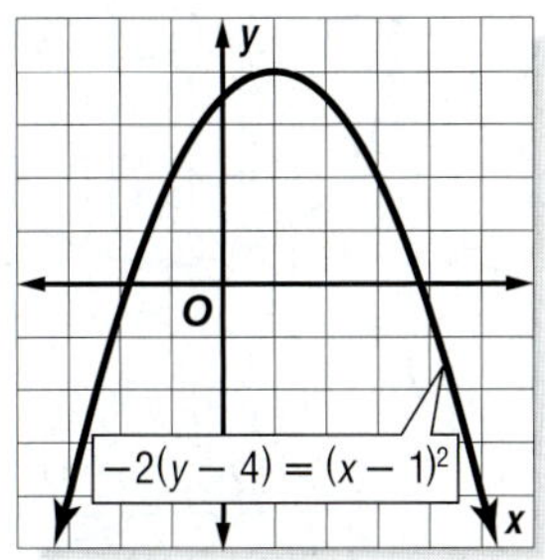

20. $(2, -3), (3, -3), y = -3,$ $x = 1$, right, 4 units

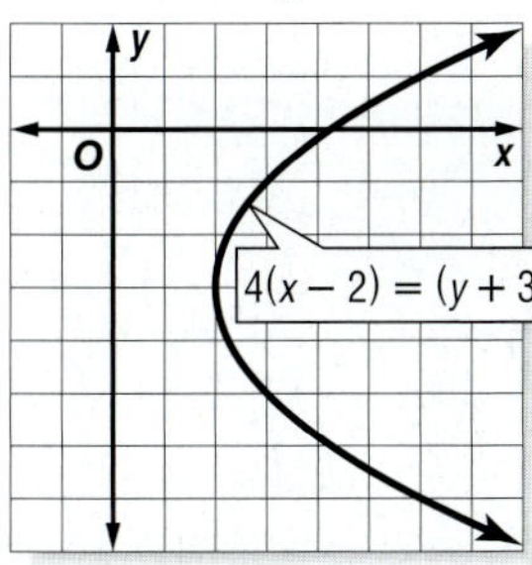

21. $(4, 8), (3, 8), y = 8,$ $x = 5$, left, 4 units

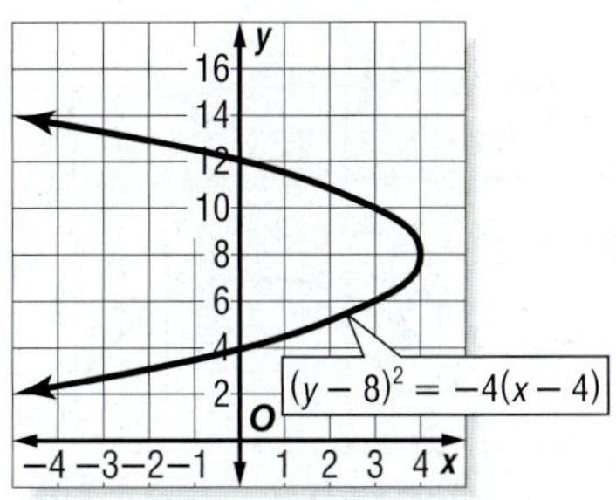

22. $(6, -16), \left(6, -15\frac{3}{4}\right),$ $x = 6, y = -16\frac{1}{4}$, upward, 1 unit

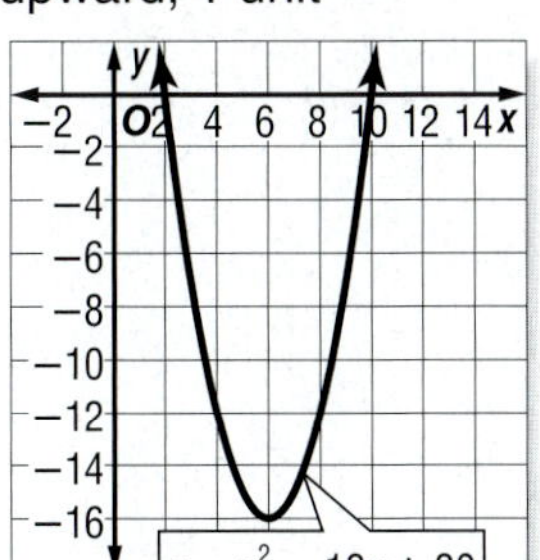

23. $(-24, 7), \left(-23\frac{3}{4}, 7\right),$ $y = 7, x = -24\frac{1}{4}$, right, 1 unit

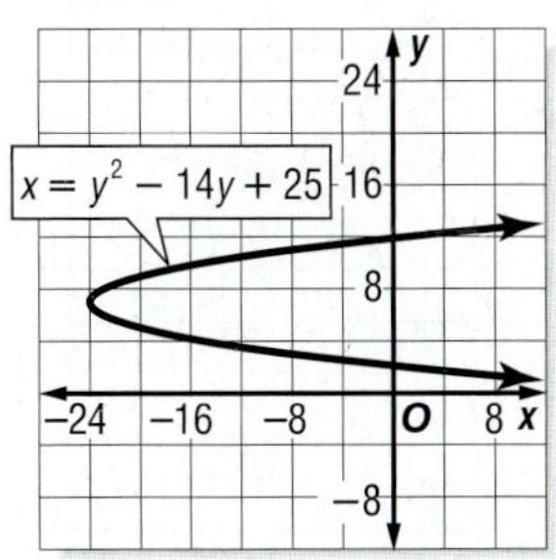

24. $\left(\frac{115}{4}, -\frac{5}{2}\right), \left(\frac{144}{5}, -\frac{5}{2}\right),$ $y = -\frac{5}{2}, x = \frac{287}{10}$, right, $\frac{1}{5}$ unit

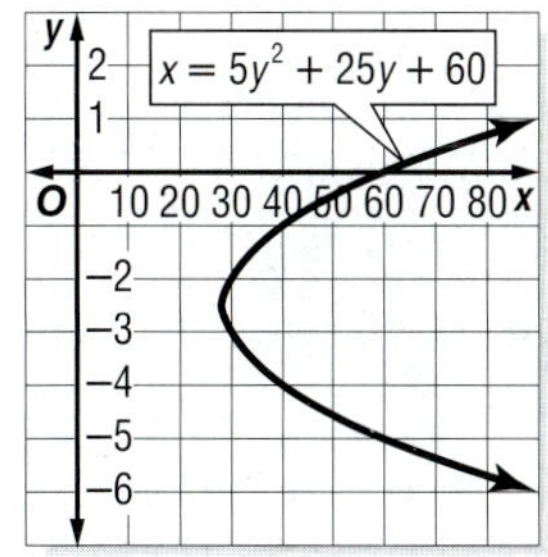

25. $(4, 2), \left(4, 2\frac{1}{12}\right), x = 4,$ $y = 1\frac{11}{12}$, upward, $\frac{1}{3}$ unit

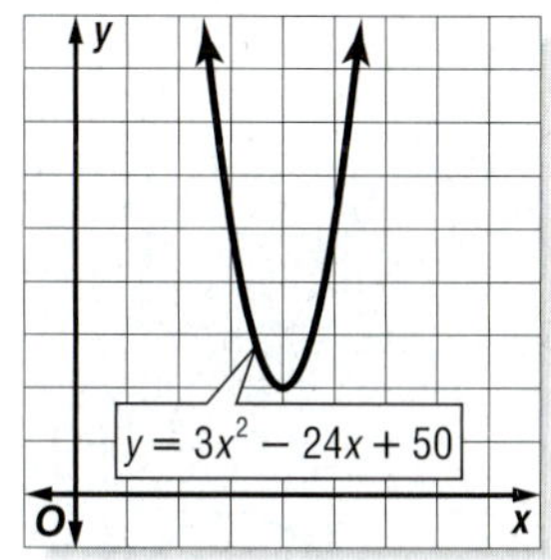

26. $\left(\frac{5}{4}, -\frac{55}{8}\right), \left(\frac{5}{4}, -7\right),$ $x = \frac{5}{4}, y = -\frac{27}{4},$ downward, $\frac{1}{2}$ unit

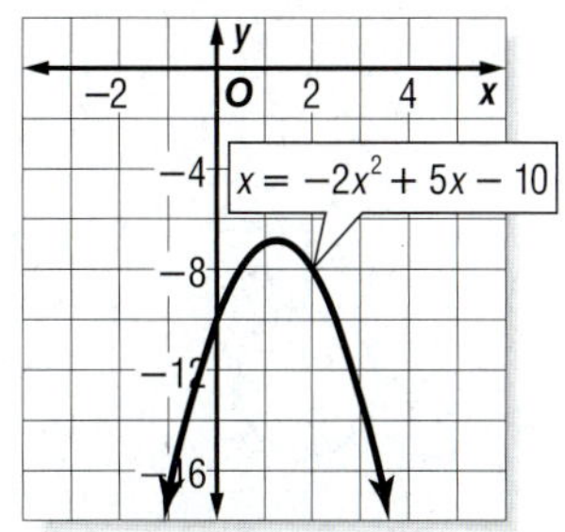

27. $\left(\frac{17}{4}, \frac{3}{4}\right), \left(\frac{67}{16}, \frac{3}{4}\right), y = \frac{3}{4},$ $x = \frac{69}{16}$, left, $\frac{1}{4}$ unit

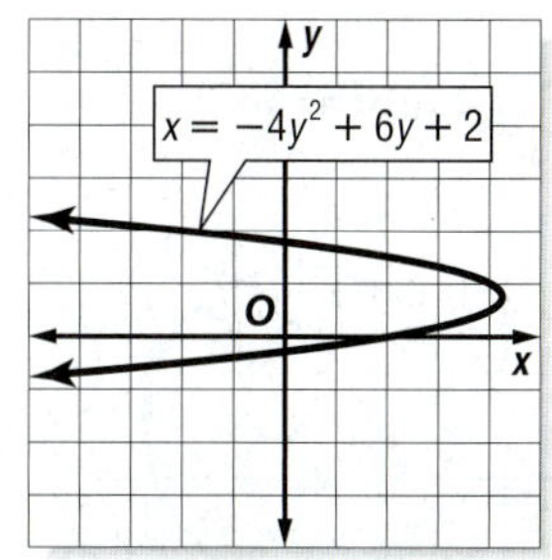

28. $(3, 5), \left(3, 5\frac{1}{2}\right), x = 3,$ $y = 4\frac{1}{2}$, upward, 2 units

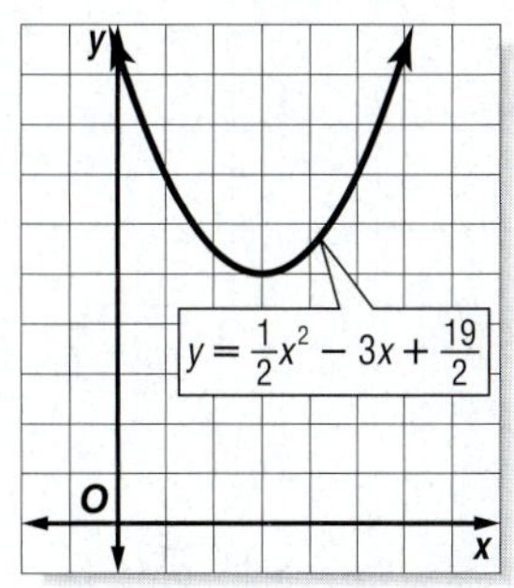

29. $(123, -18),$ $\left(122\frac{1}{4}, -18\right), y = -18,$ $x = 123\frac{3}{4}$, left, 3 units

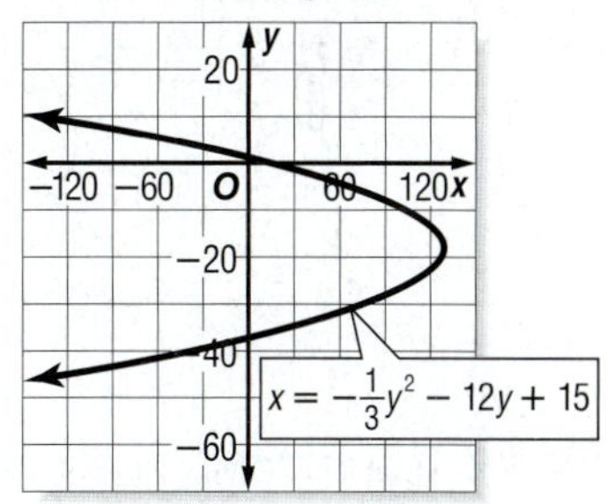

36.

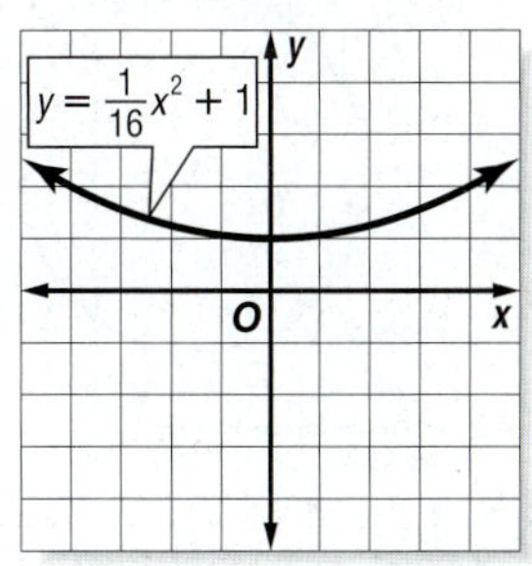

37.

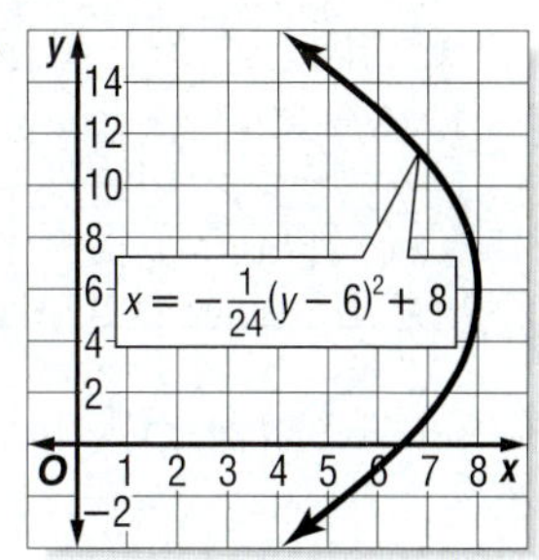

38.

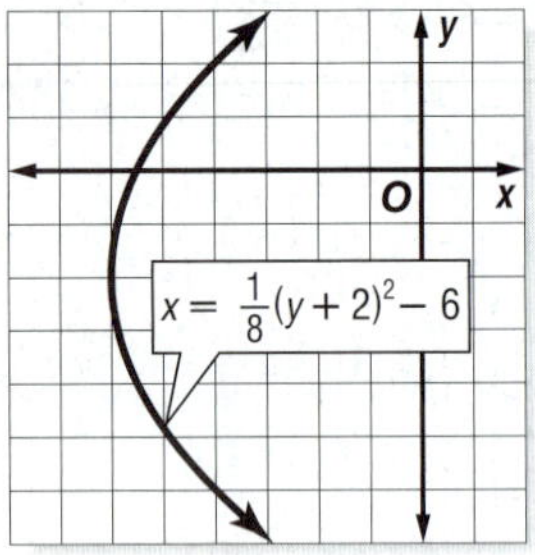

39.

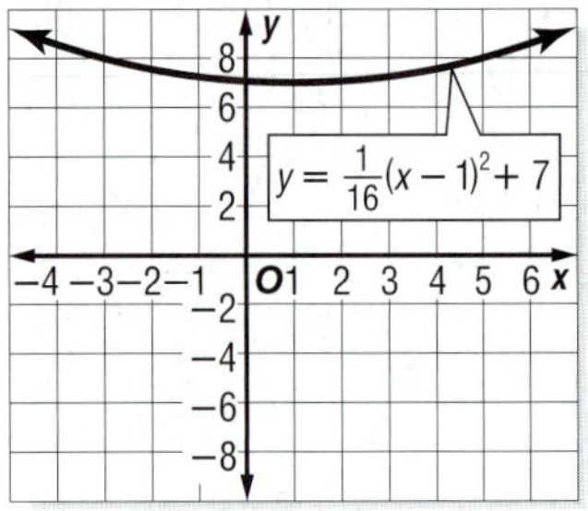

40.

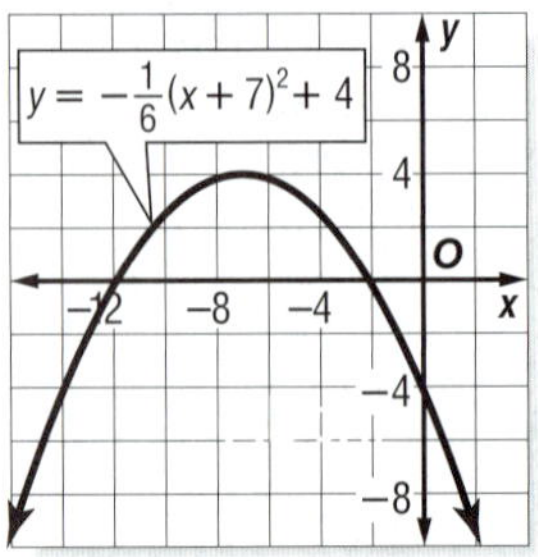

41.

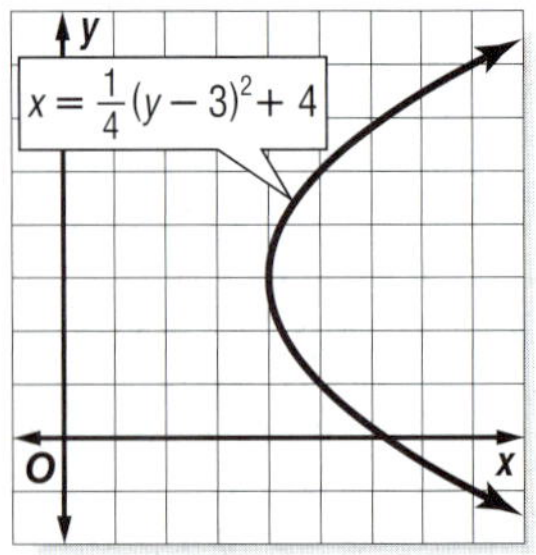

Pages 429–431, Lesson 8-3

8.

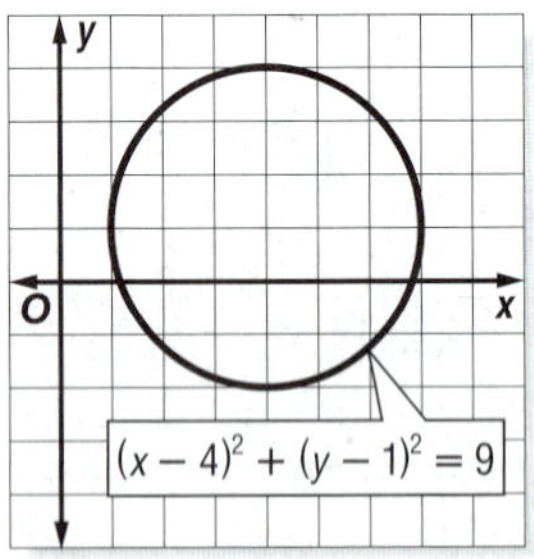

9.

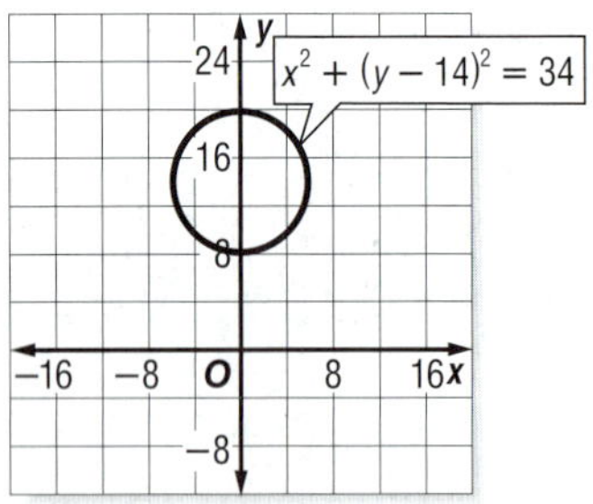

10.

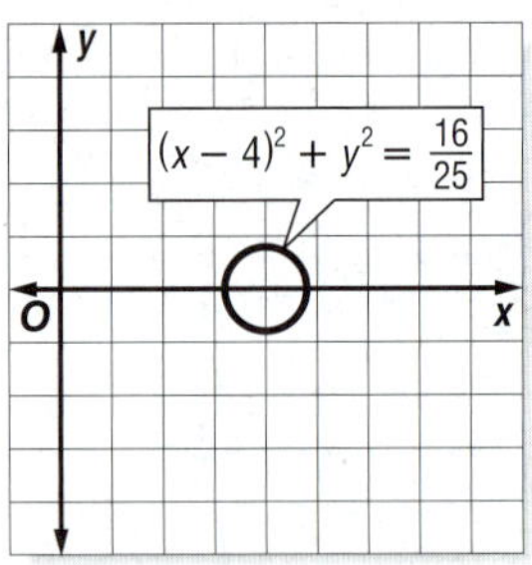

11.

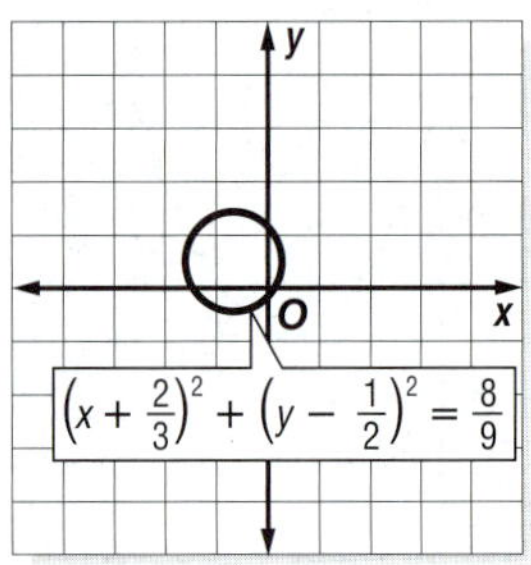

12.

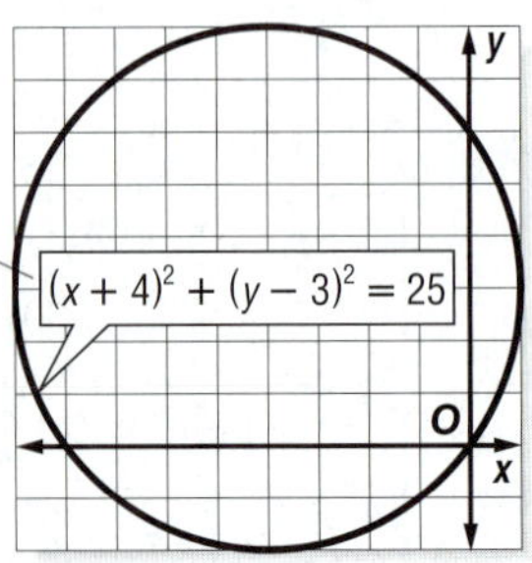

13.

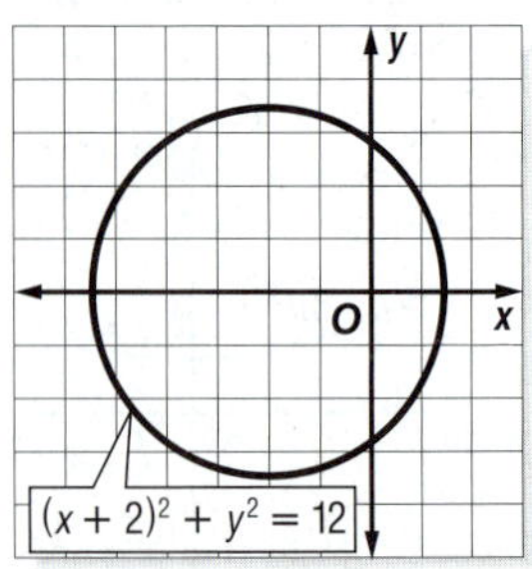

30.

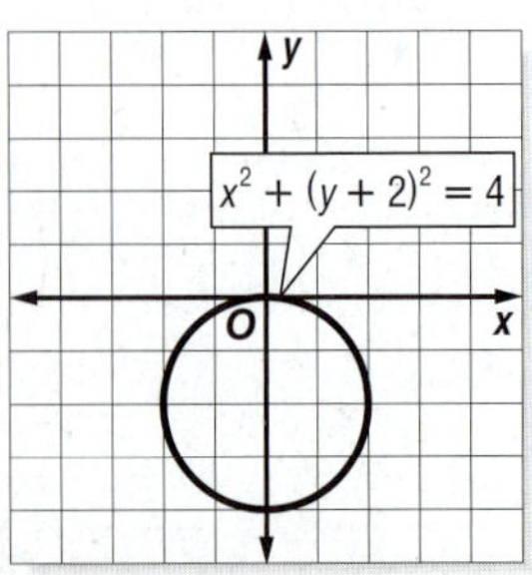

31.

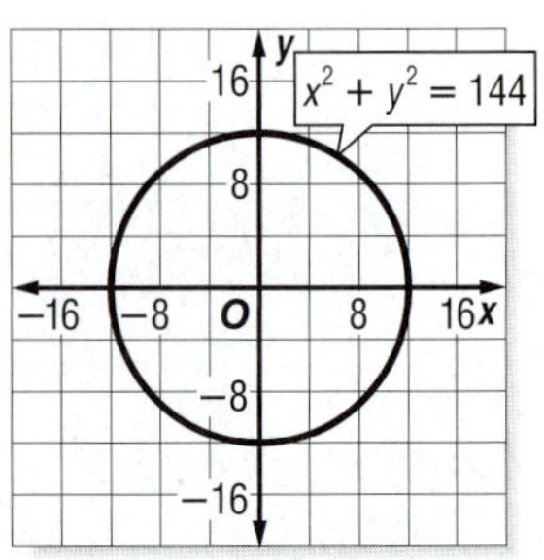

32.

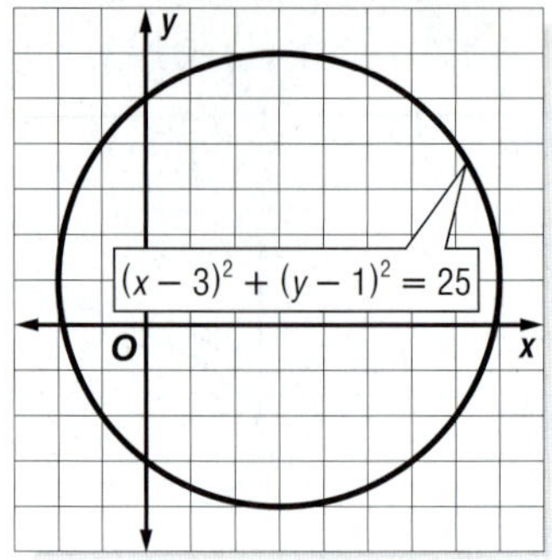

33.

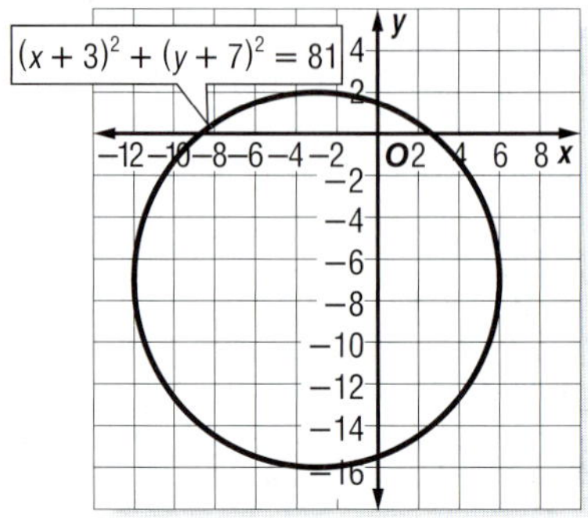

34.

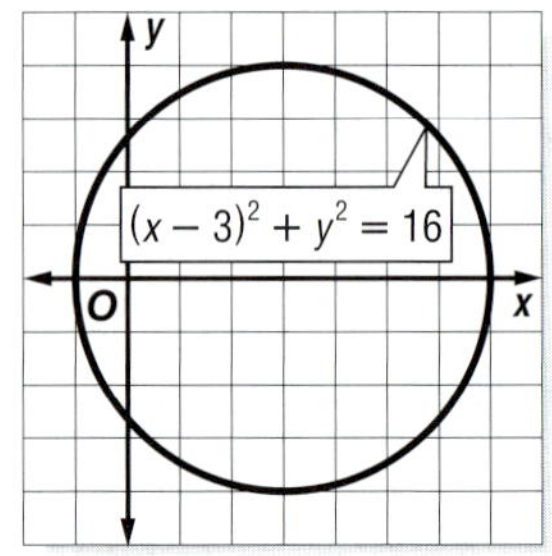

35.

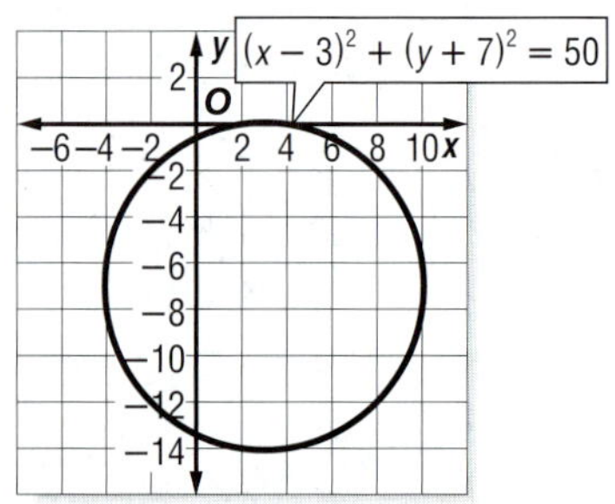

36.

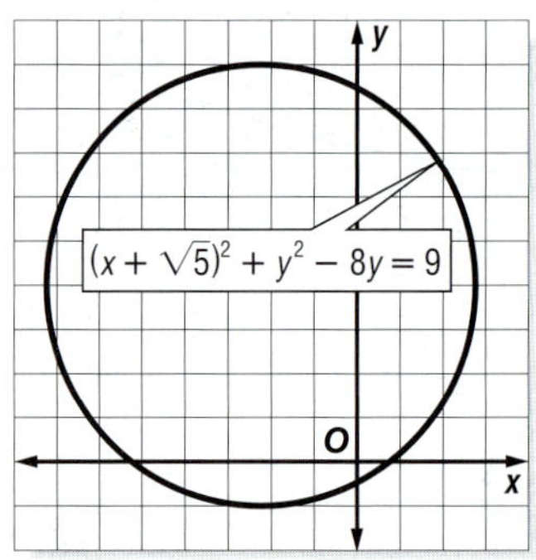

37.

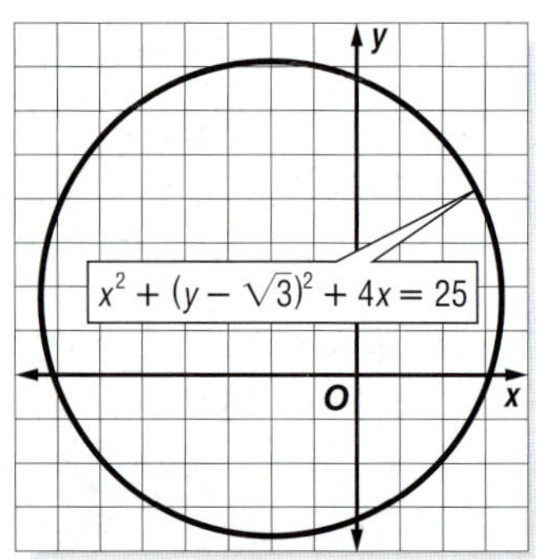

38.

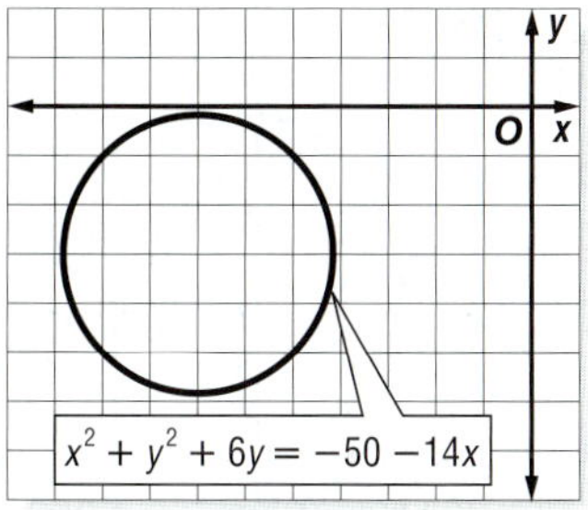

39.

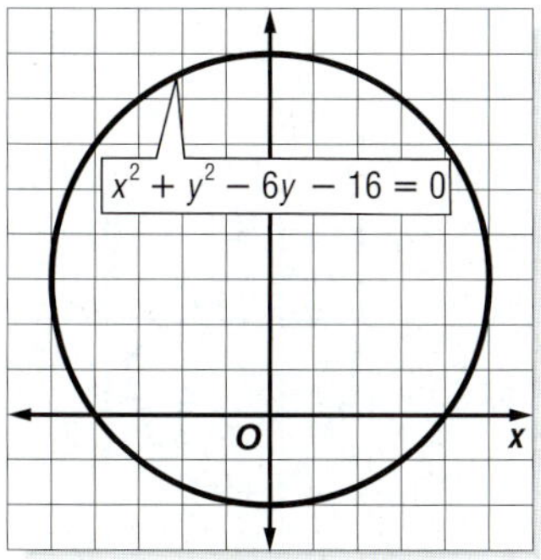

40.

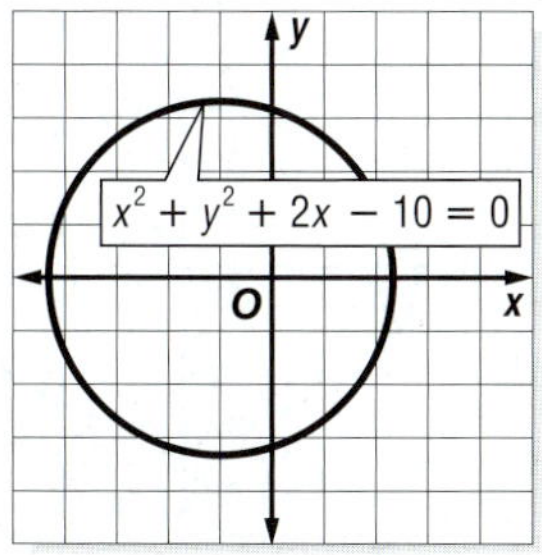

41.

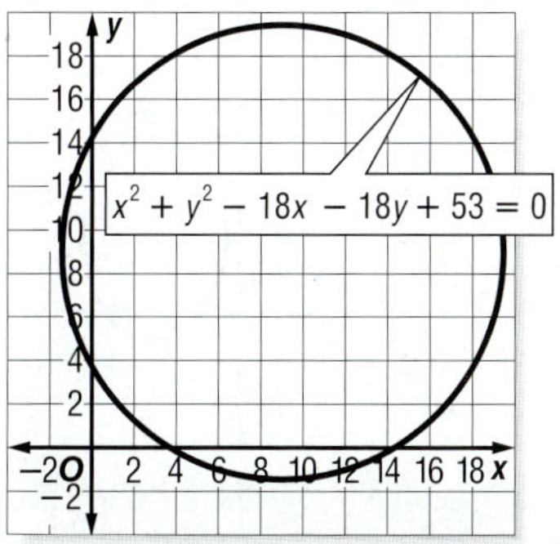

42.

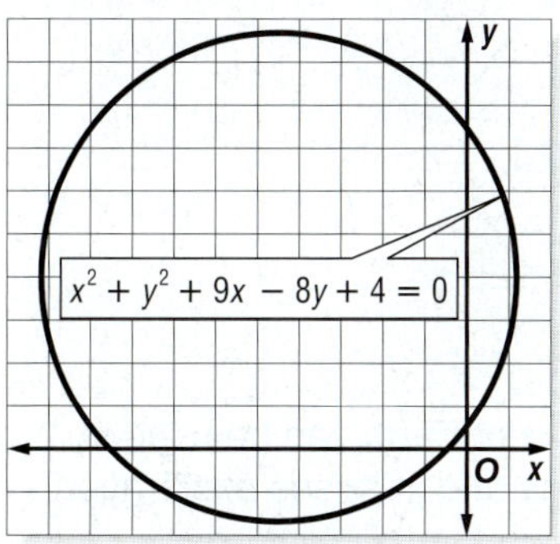

43.

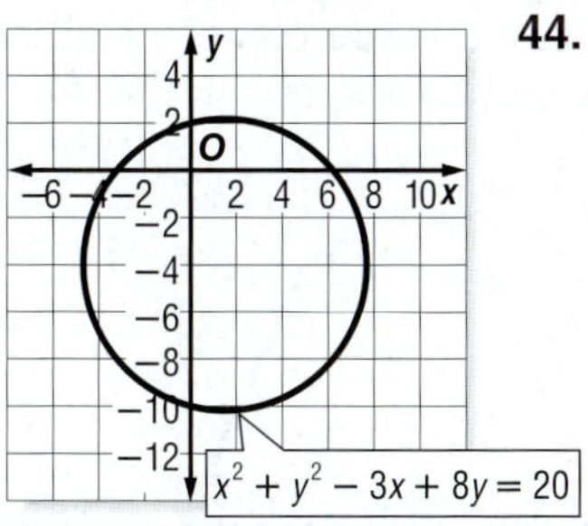

44.

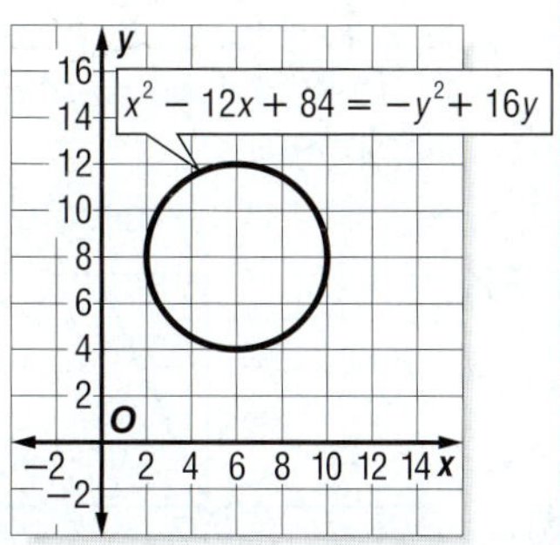

45.

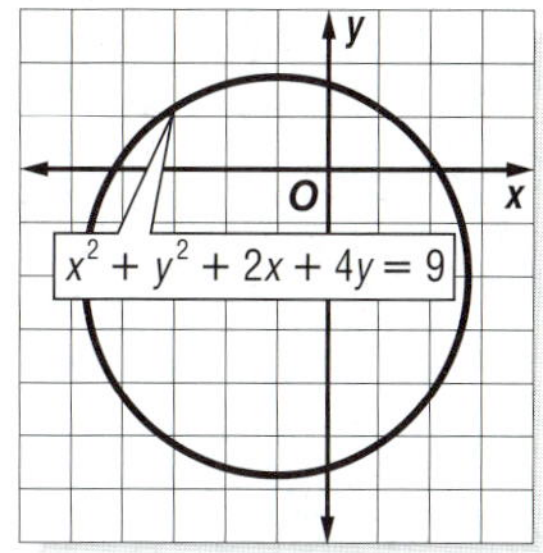

46.

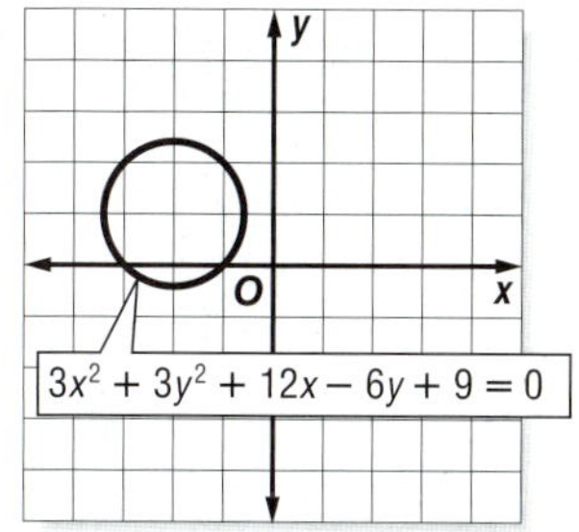

47.

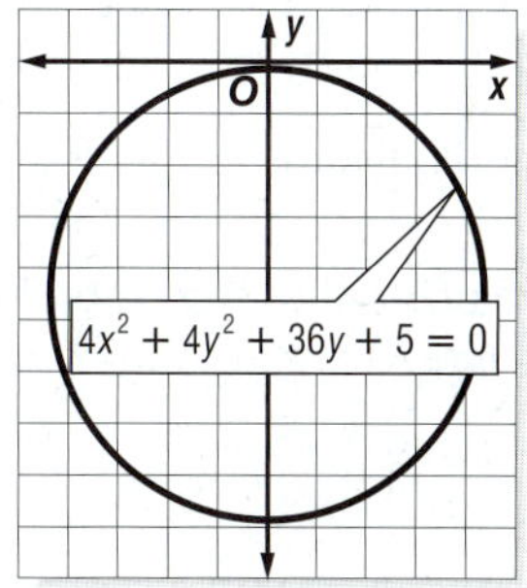

57. $(1, 0)$, $\left(\frac{11}{12}, 0\right)$, $y = 0$, $x = 1\frac{1}{12}$, left, $\frac{1}{3}$ unit

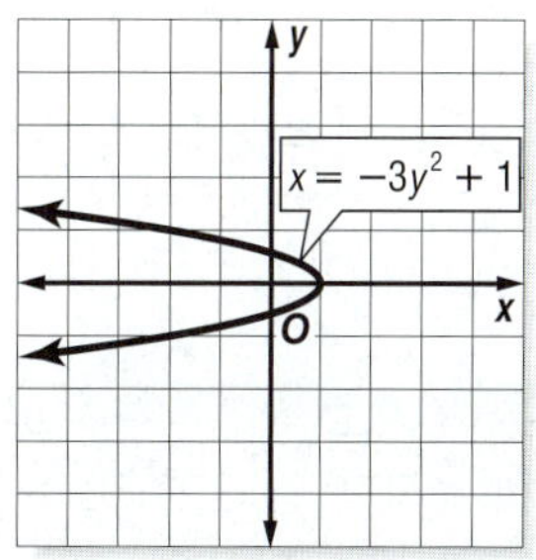

58. $(3, -2)$, $\left(3, -2\frac{1}{4}\right)$, $x = 3$, $y = -1\frac{3}{4}$, downward, 1 unit

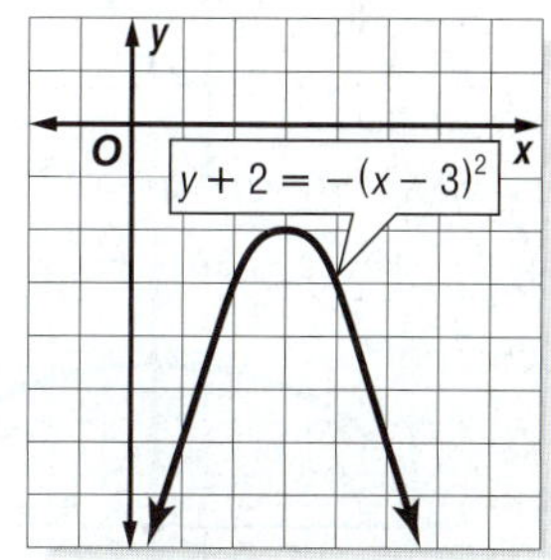

59. $(-2, -4)$, $\left(-2, -3\frac{3}{4}\right)$, $x = -2$, $y = -4\frac{1}{4}$, upward, 1 unit

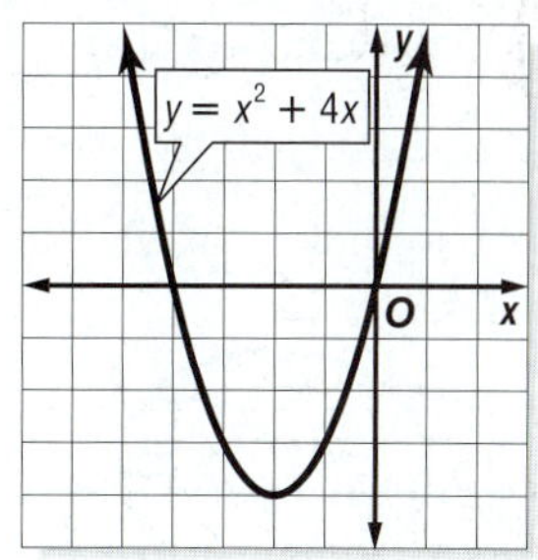

Page 431, Practice Quiz 1

3. $(0, 0)$, $(1\frac{1}{2}, 0)$, $y = 0$, $x = -1\frac{1}{2}$, right, 6 units

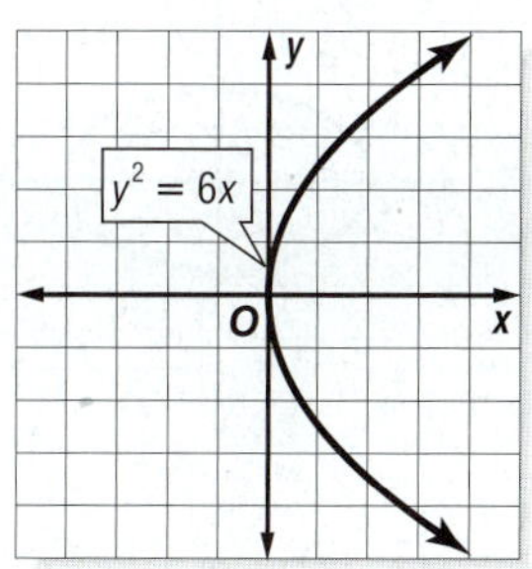

4. $(-4, 4)$, $\left(-4, 4\frac{1}{4}\right)$, $x = -4$, $y = 3\frac{3}{4}$, upward, 1 unit

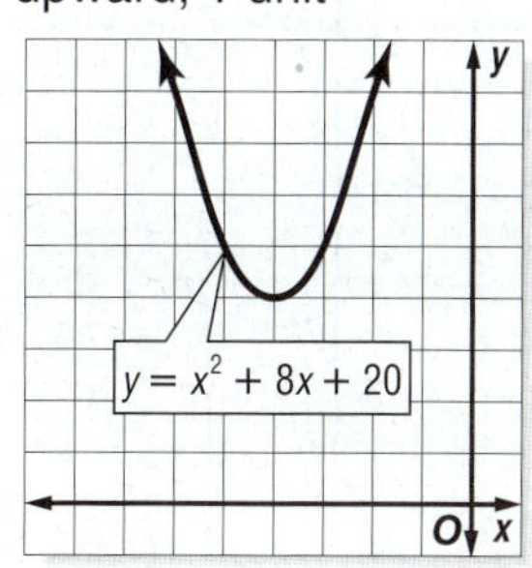

5.

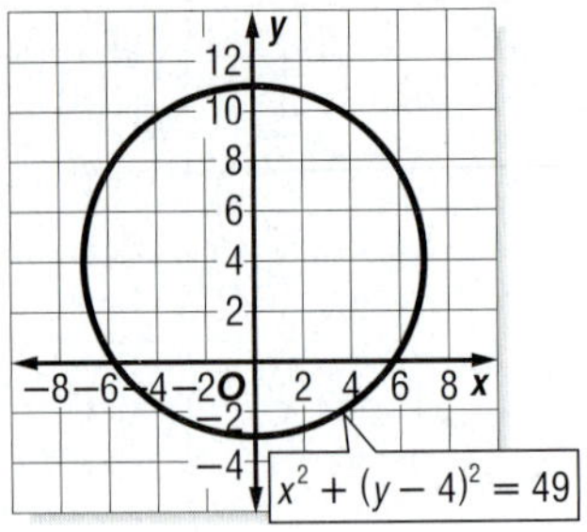

Page 432, Preview of Lesson 8-4
Algebra Activity

9. The ellipse is longer in the vertical direction than in the horizontal direction.

12. No; a rubber band might stretch so that the sum of the distances to the thumbtacks would not be constant.

Pages 438–440, Lesson 8-4

27.

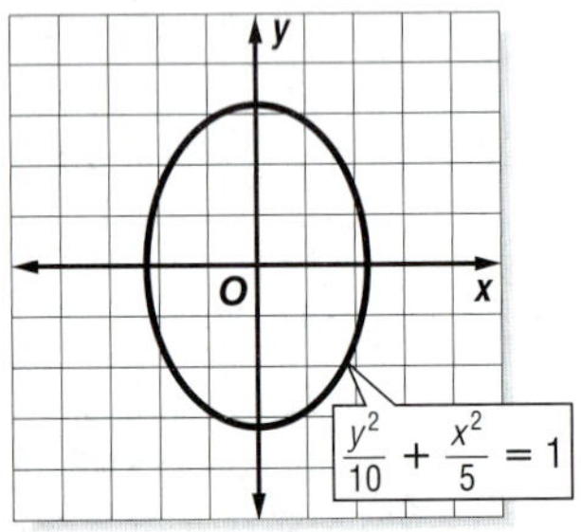

28.

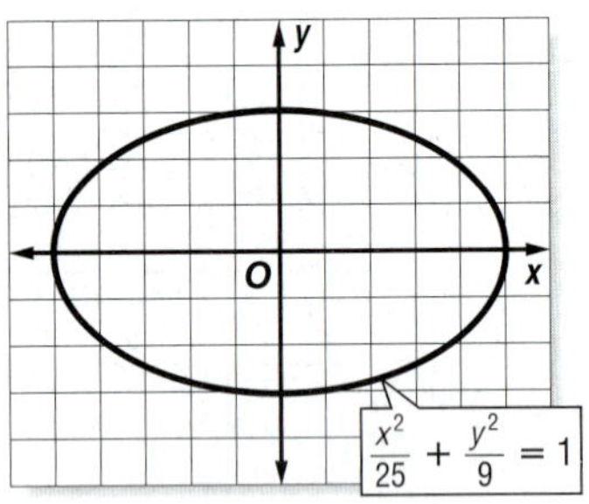

29.

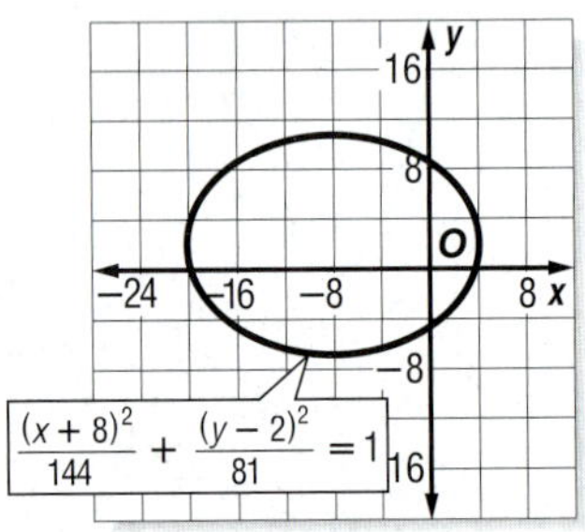

30.

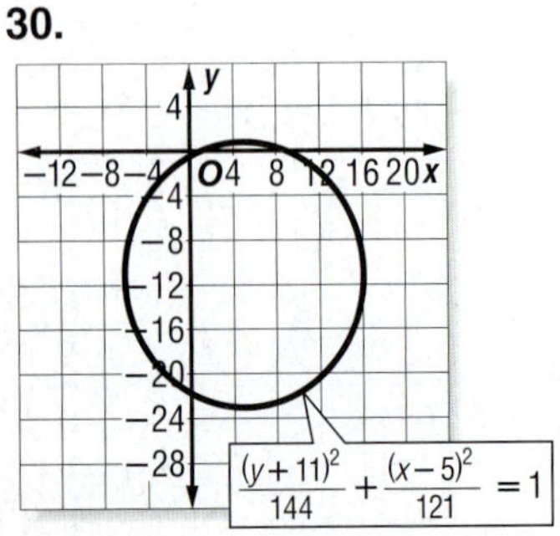

31.

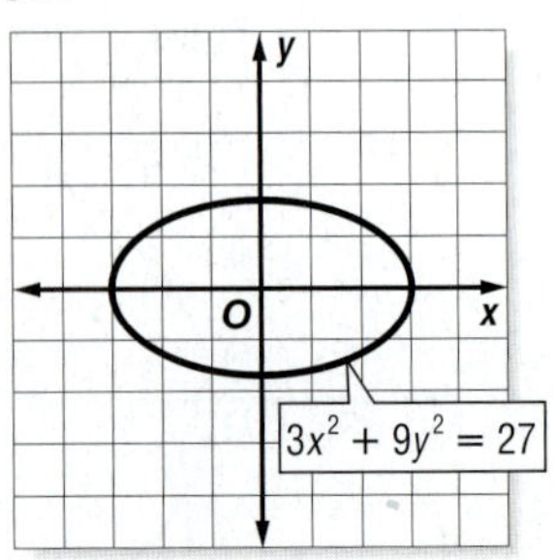

32.

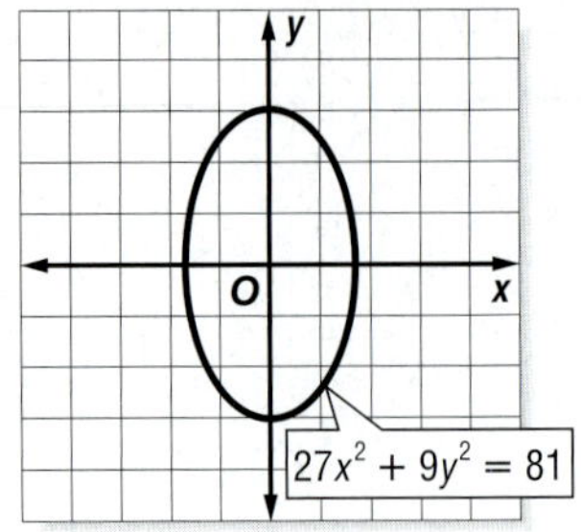

33.

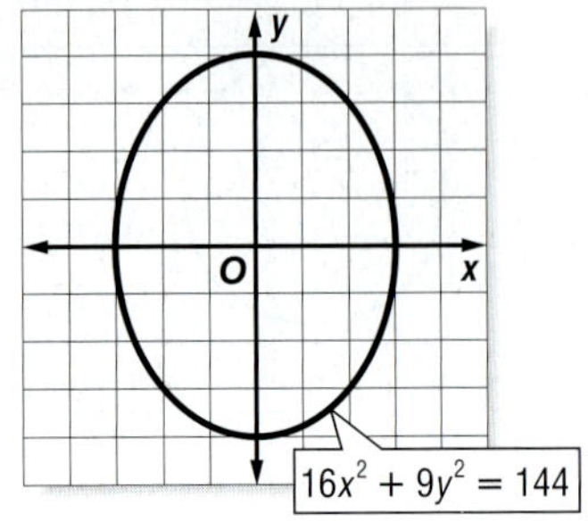

34.

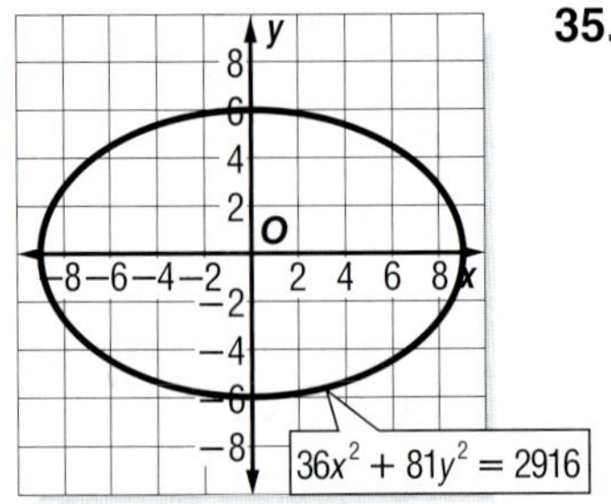

35.

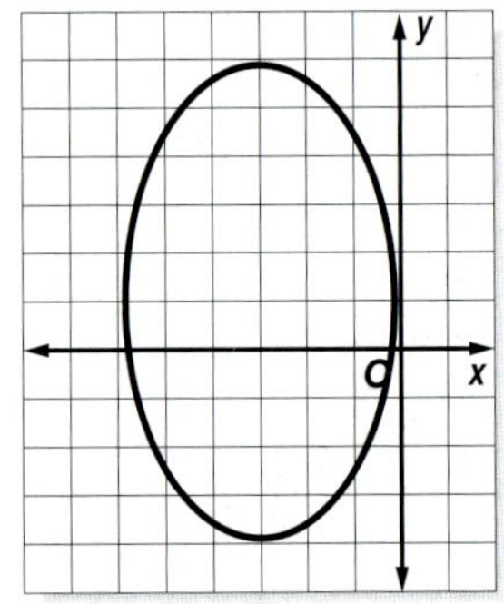

36.

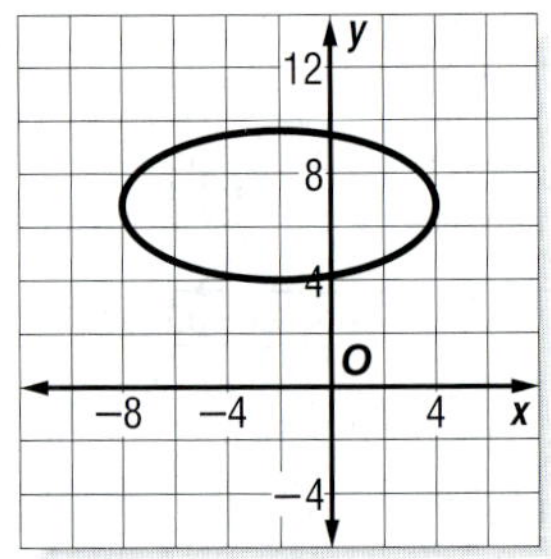

37.

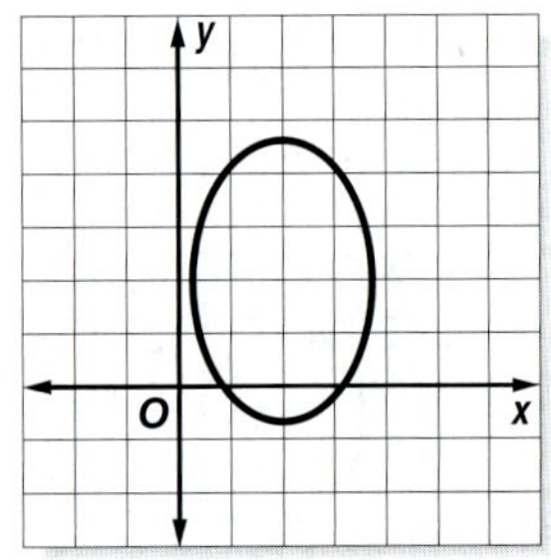

38.

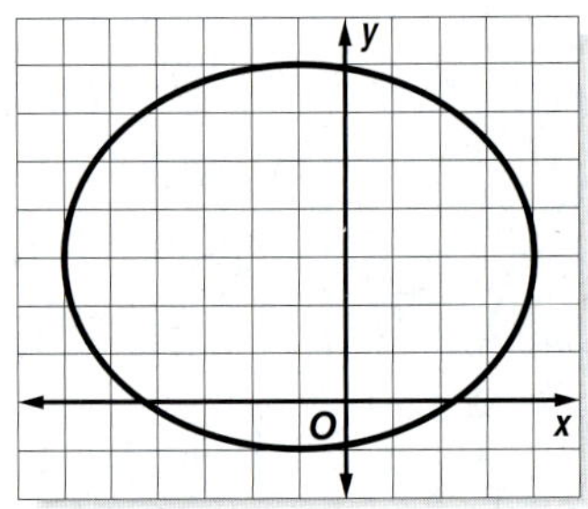

40. Knowledge of the orbit of Earth can be used in predicting the seasons and in space exploration. Answers should include the following.

- Knowledge of the path of another planet would be needed if we wanted to send a spacecraft to that planet.
- 1.55 million miles

52.

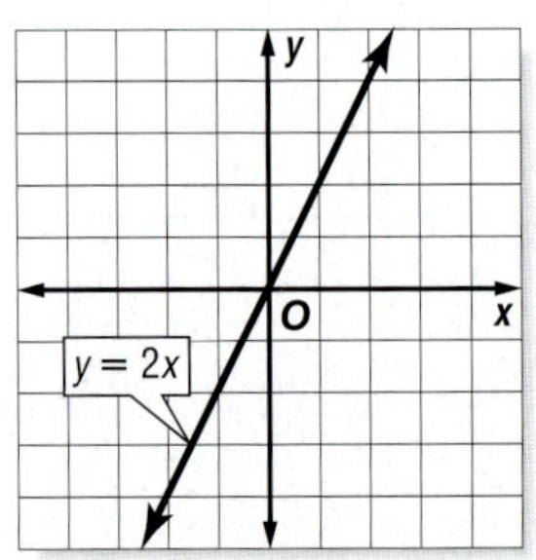

53.

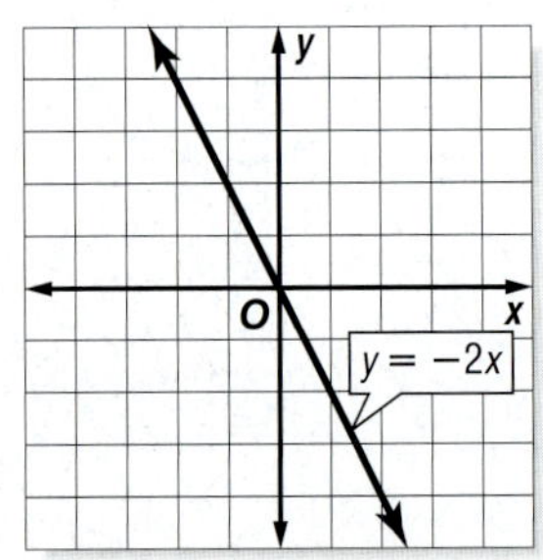

54.

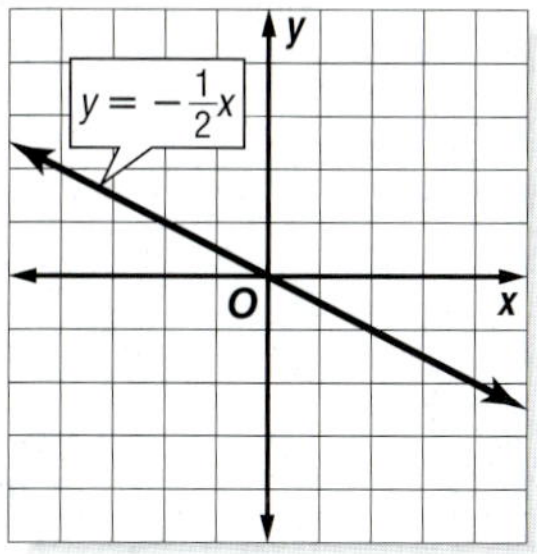

55.

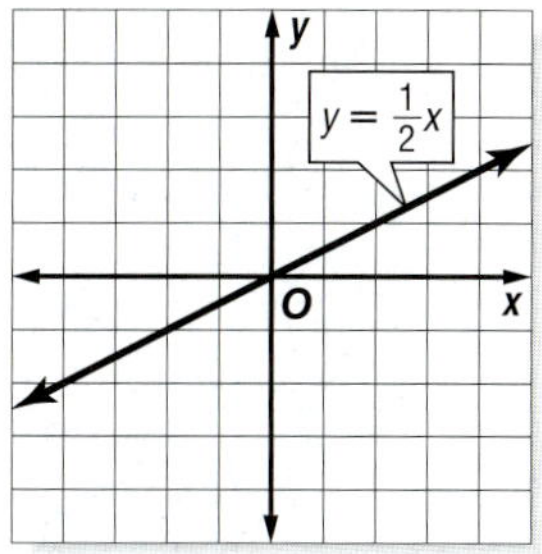

56.

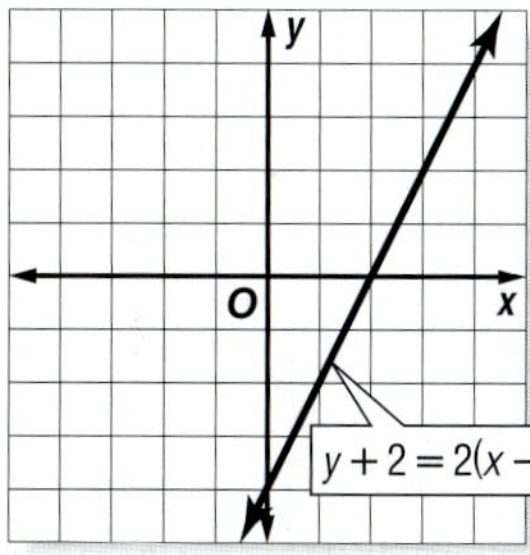

57.

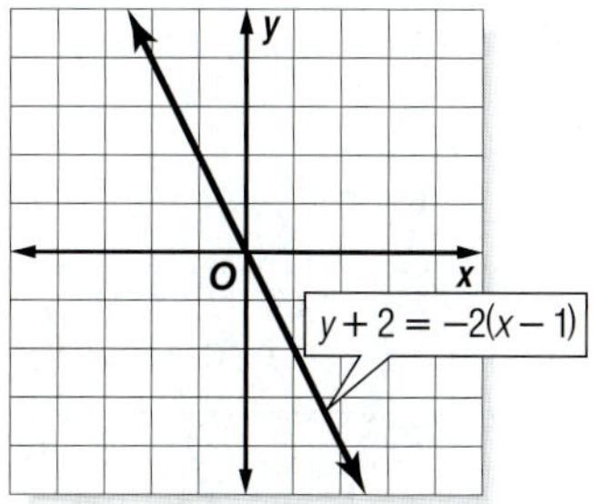

Page 446, Lesson 8-5

21.

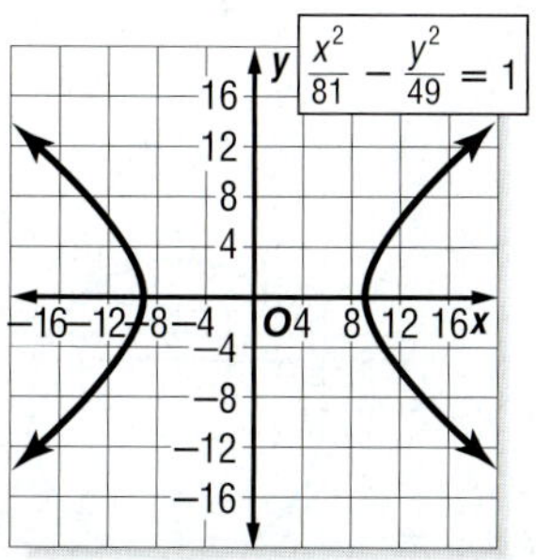

22.

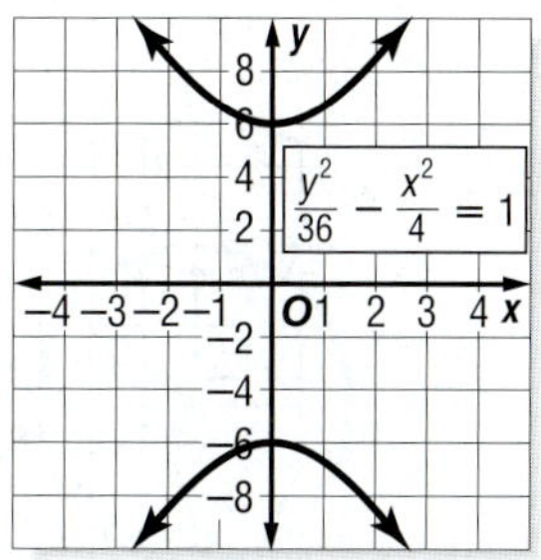

23.

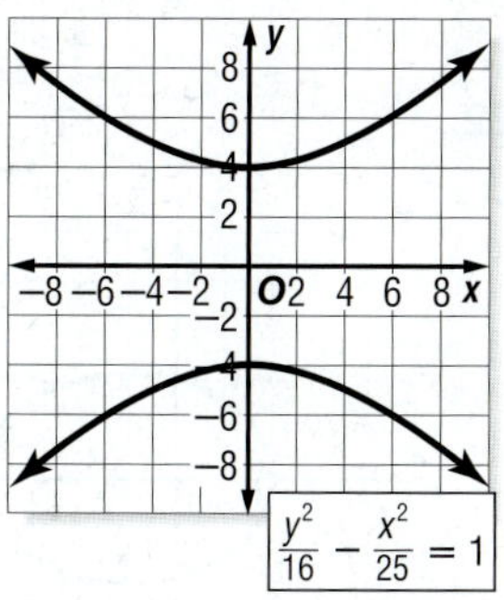

24.

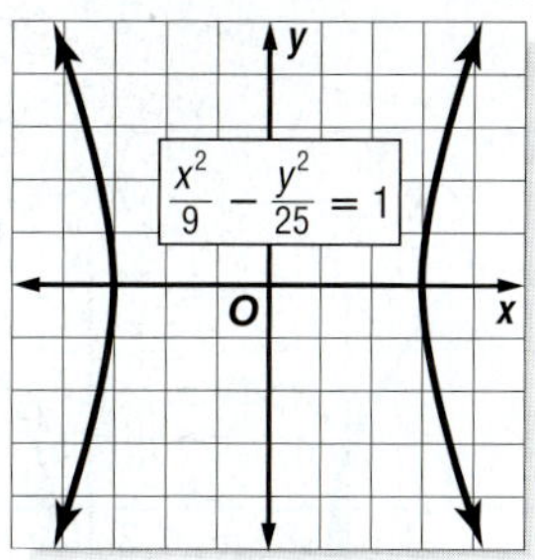

25.

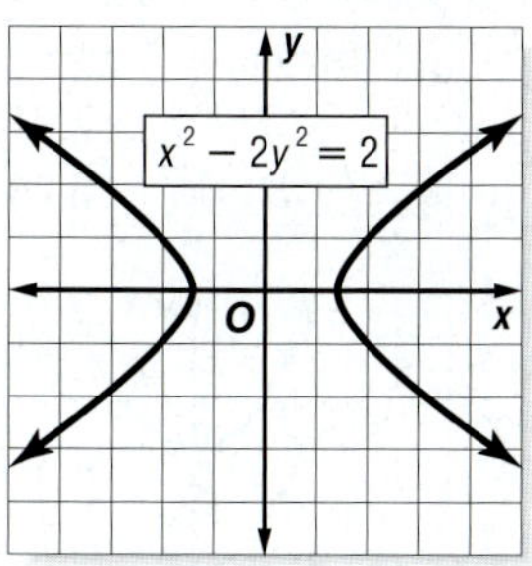

26.

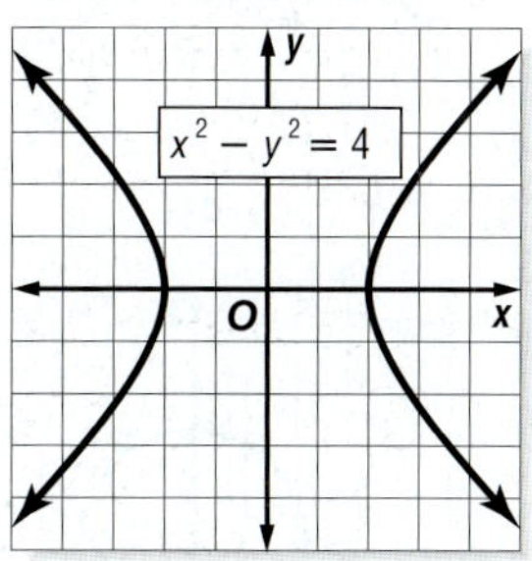

27.

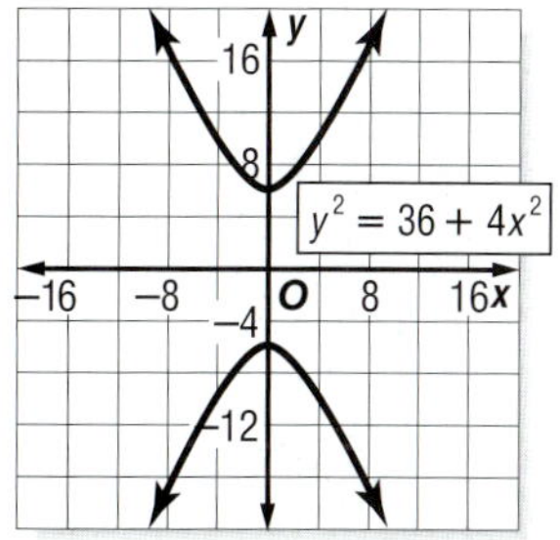

28.

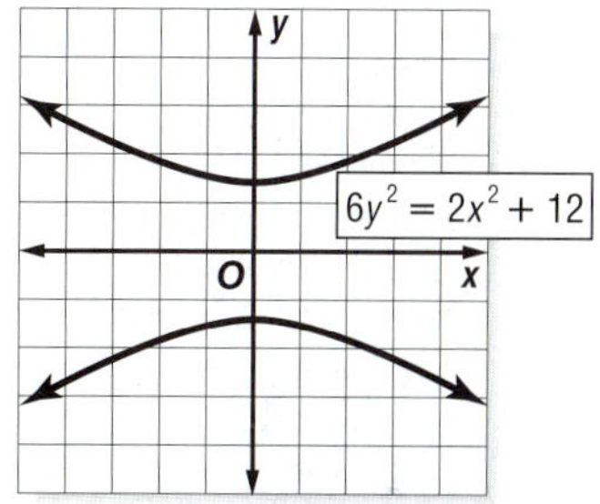

29. $(-2, 0), (-2, 8)$;
$(-2, -1), (-2, 9)$;
$y - 4 = \pm\frac{4}{3}(x + 2)$

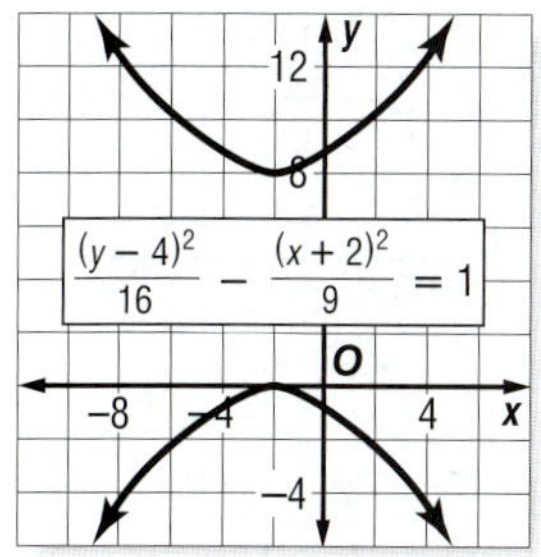

30. $(2, -2), (2, 8)$;
$(2, 3 \pm \sqrt{41})$;
$y - 3 = \pm\frac{5}{4}(x - 2)$

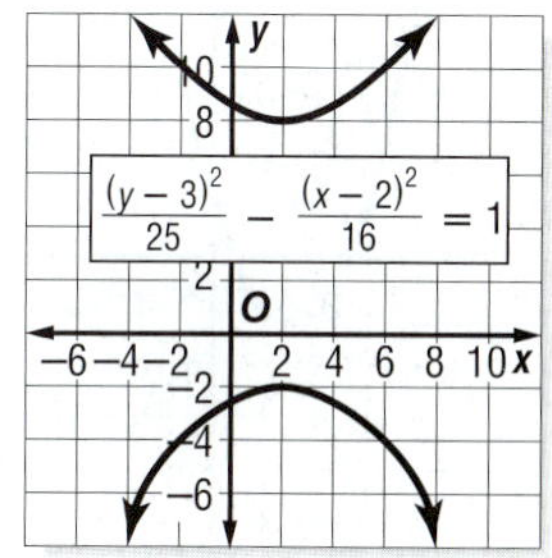

31. $(-3, -3), (1, -3)$;
$(-1 \pm \sqrt{13}, -3)$;
$y + 3 = \pm\frac{3}{2}(x + 1)$

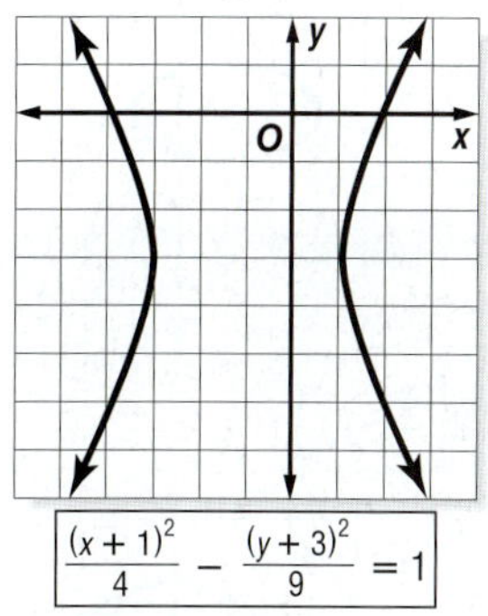

32. $(-12, -3), (0, -3)$;
$(-6 \pm 3\sqrt{5}, -3)$;
$y + 3 = \pm\frac{1}{2}(x + 6)$

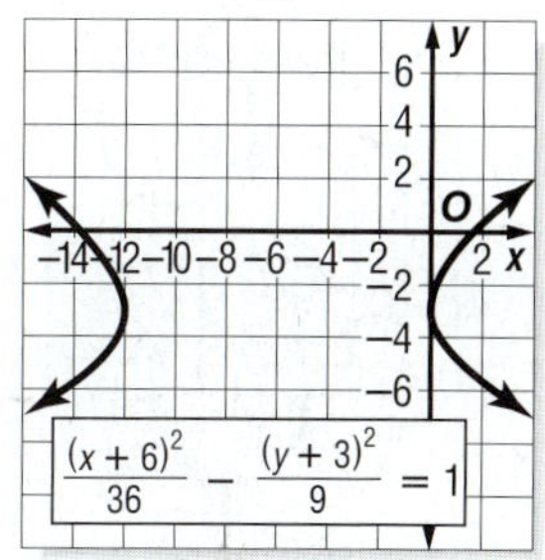

33. $(1, -3 \pm 2\sqrt{6})$;
$(1, -3 \pm 4\sqrt{2})$;
$y + 3 = \pm\sqrt{3}(x - 1)$

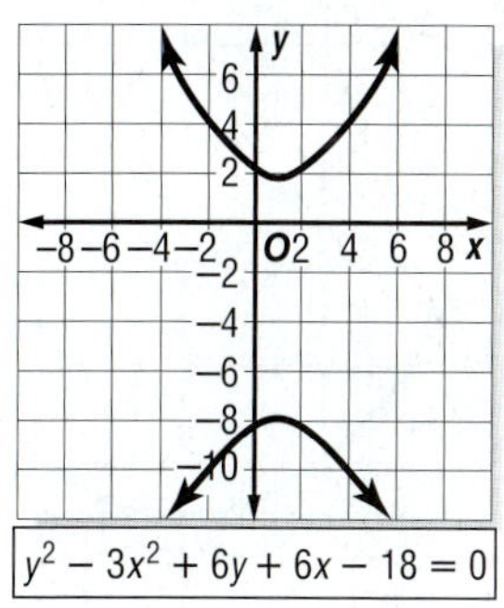

34. $(-4, 0), (6, 0)$;
$(1 \pm \sqrt{29}, 0)$;
$y = \pm\frac{2}{5}(x - 1)$

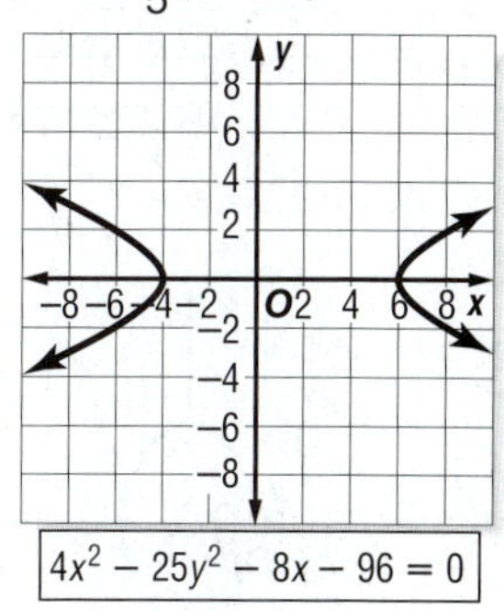

Page 448, Practice Quiz 2

2.

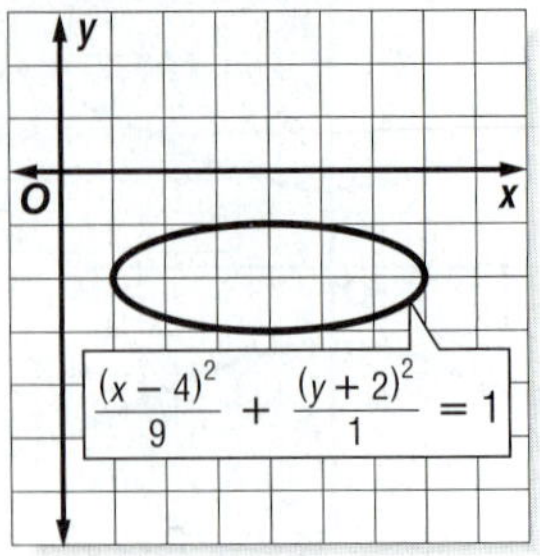

3.

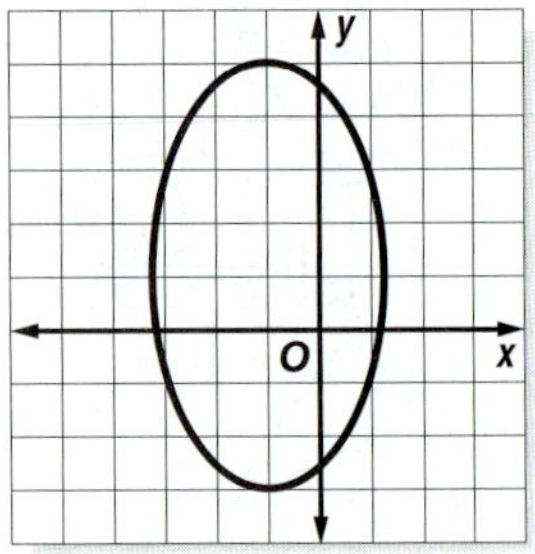

Pages 450–452, Lesson 8-6

4. $y = \left(x + \frac{3}{2}\right)^2 - \frac{5}{4}$

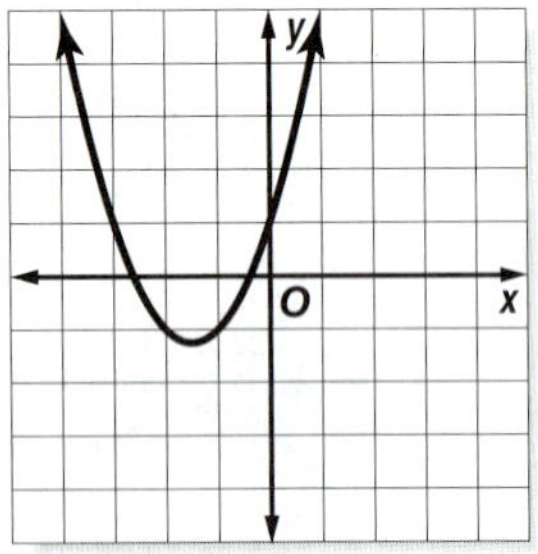

5. $\frac{y^2}{16} - \frac{x^2}{8} = 1$

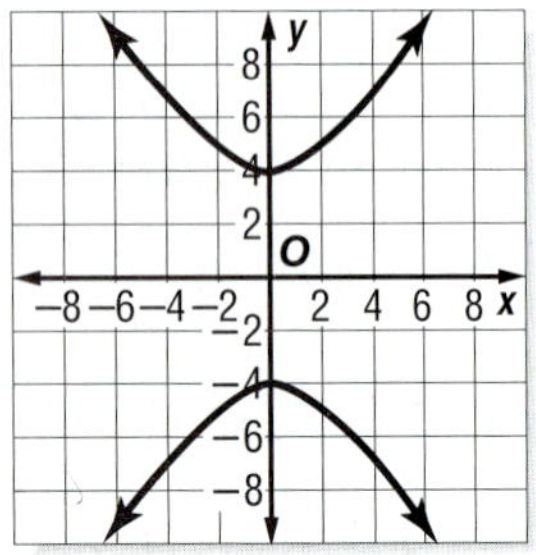

6. $\left(x - \frac{1}{2}\right)^2 + y^2 = \frac{9}{4}$

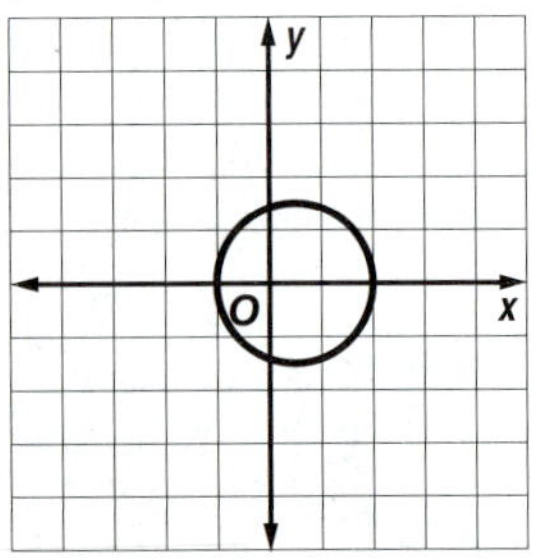

7. $\frac{(x+1)^2}{4} + \frac{(y-3)^2}{1} = 1$

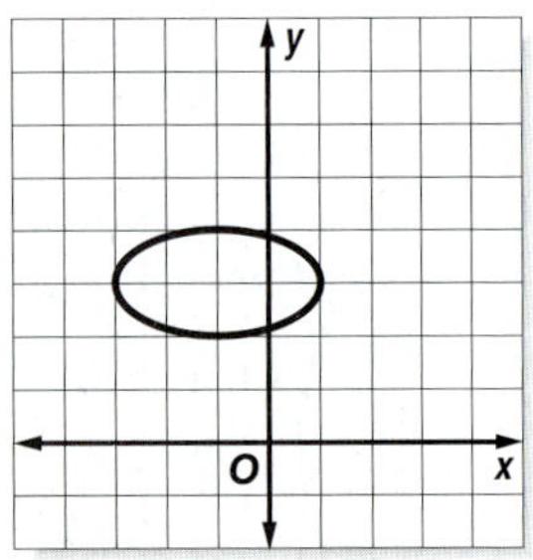

11.

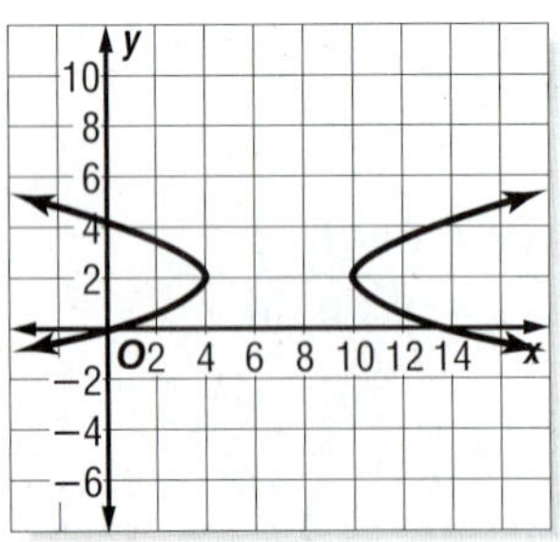

12. $x^2 + y^2 = 27$

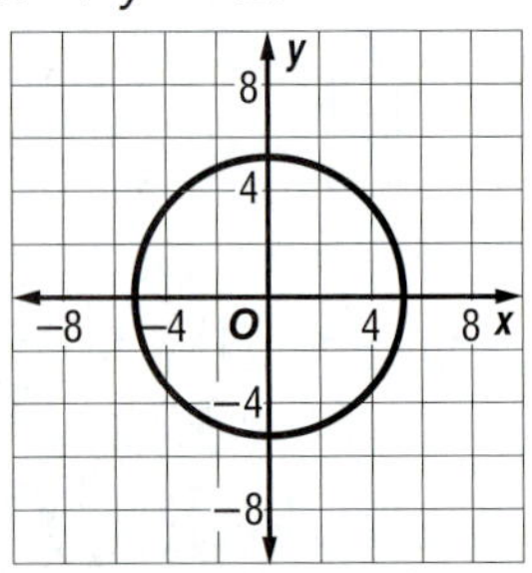

13. $\frac{y^2}{4} + \frac{x^2}{2} = 1$

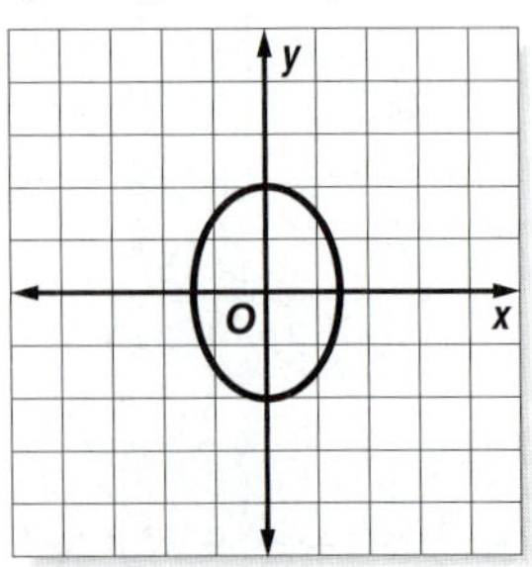

14. $y = \frac{1}{8}x^2$

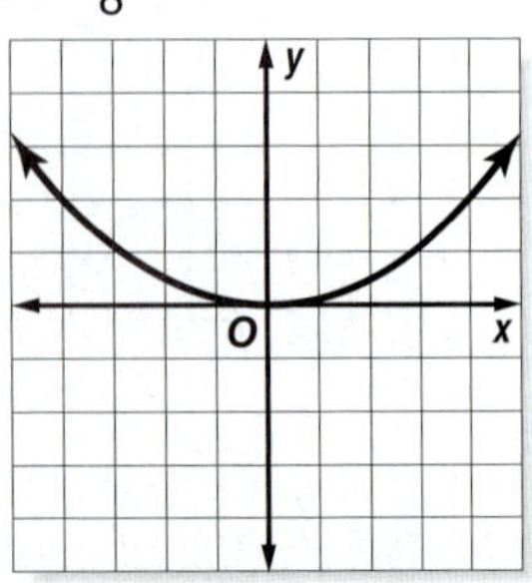

15. $\frac{x^2}{4} - \frac{y^2}{1} = 1$

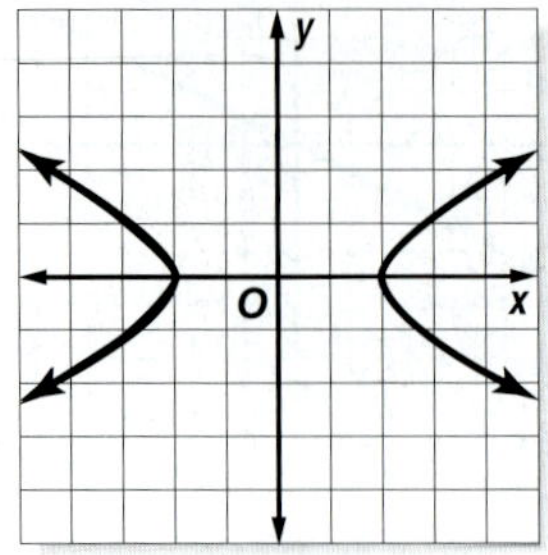

16. $\frac{(x-1)^2}{36} - \frac{(y-4)^2}{4} = 1$

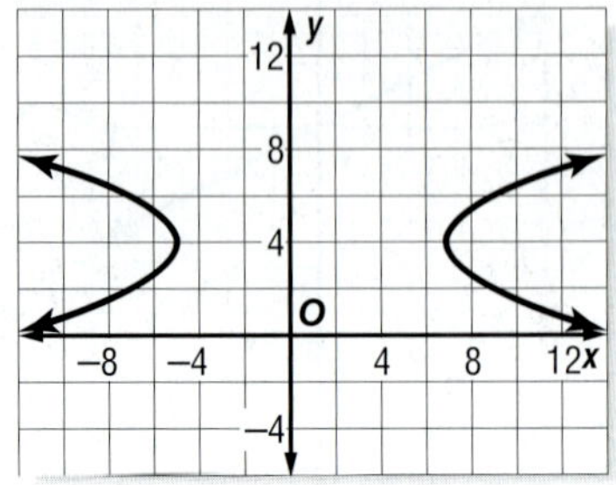

17. $y = (x-2)^2 - 4$

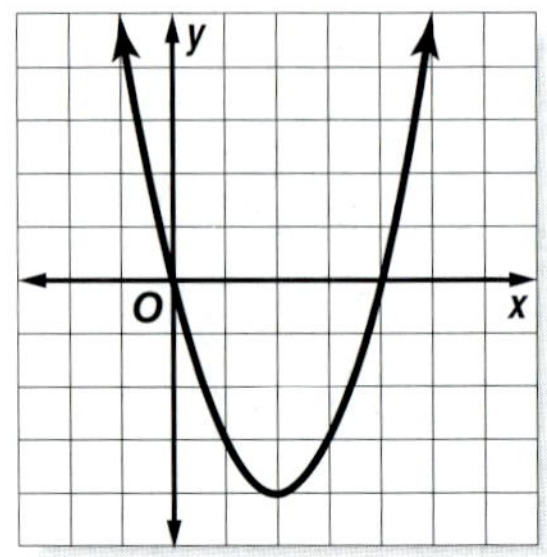

18. $x = \frac{1}{9}(y-4)^2 + 4$

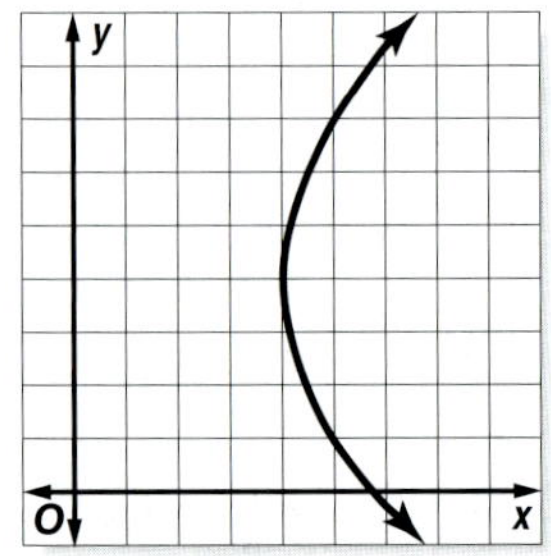

19. $(x+2)^2 + (y-3)^2 = 9$

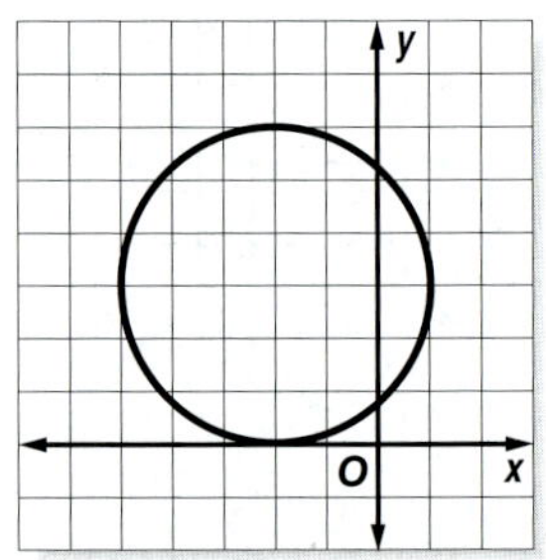

20. $x^2 + (y+3)^2 = 36$

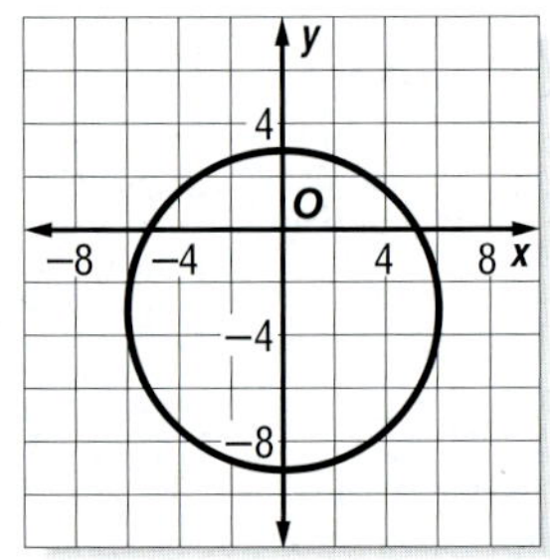

21. $\frac{(x+4)^2}{32} - \frac{y^2}{32} = 1$

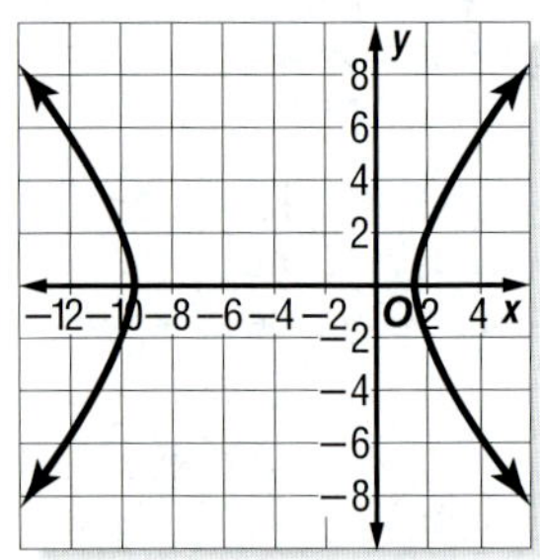

22. $\frac{(x-1)^2}{9} + \frac{y^2}{\frac{9}{2}} = 1$

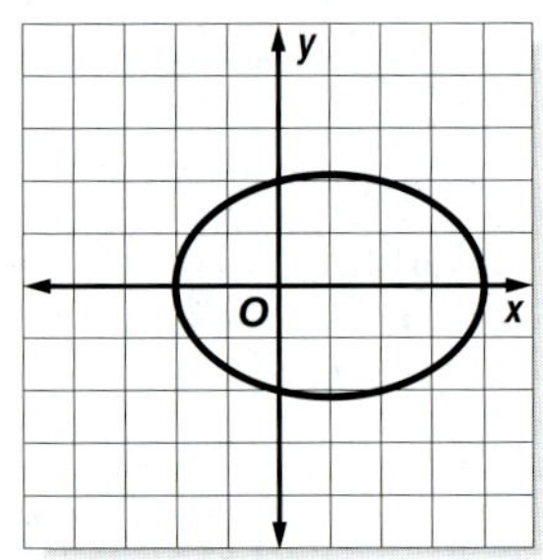

23. $x^2 + (y-4)^2 = 5$

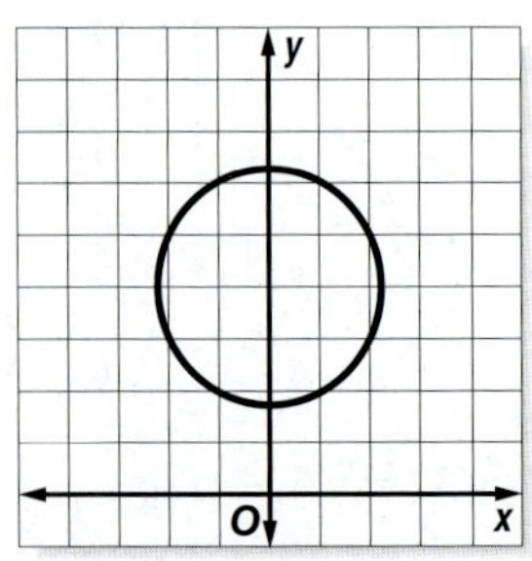

24. $\frac{(y+1)^2}{25} - \frac{x^2}{9} = 1$

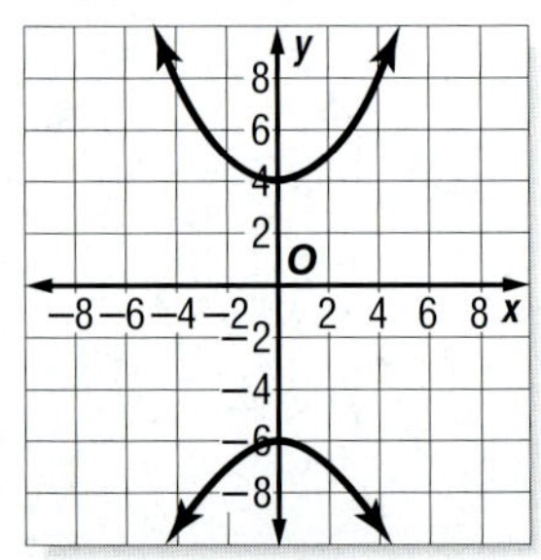

25. $\frac{x^2}{4} + \frac{(y+1)^2}{3} = 1$

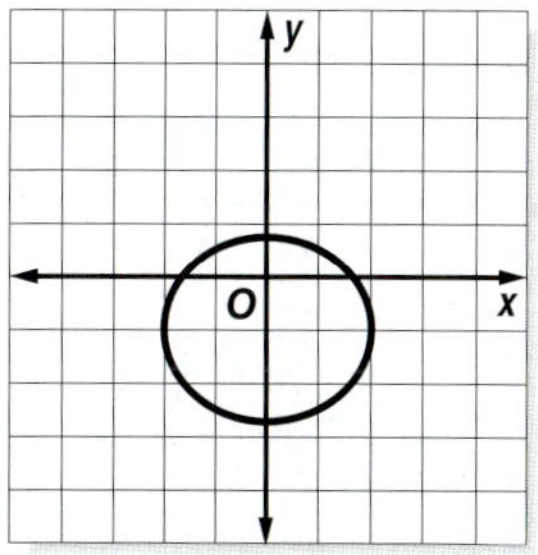

26. $\frac{(x+1)^2}{16} + \frac{(y-1)^2}{4} = 1$

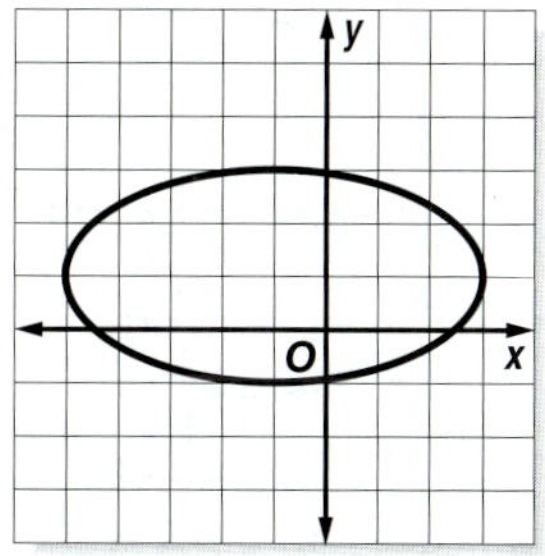

27. $y = -(x+4)^2 - 7$

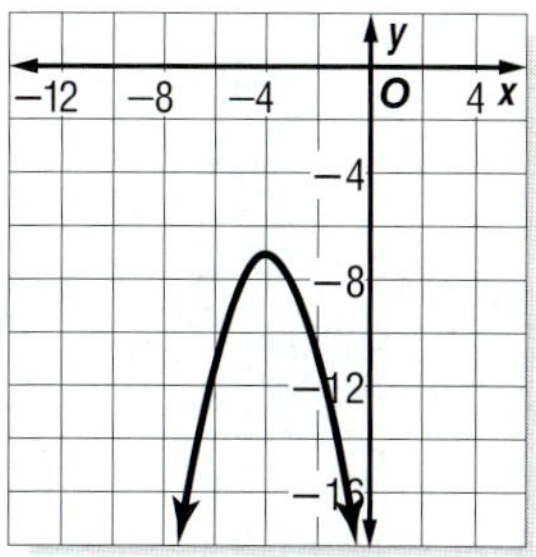

28. $\frac{(x-2)^2}{5} - \frac{(y+1)^2}{6} = 1$

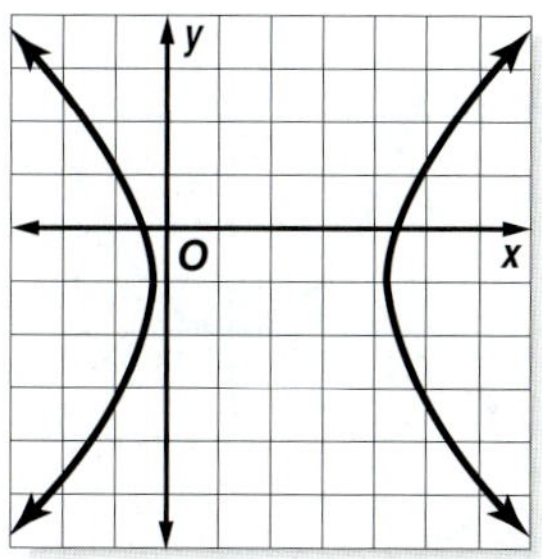

29. $\frac{(x-3)^2}{25} + \frac{(y-1)^2}{9} = 1$

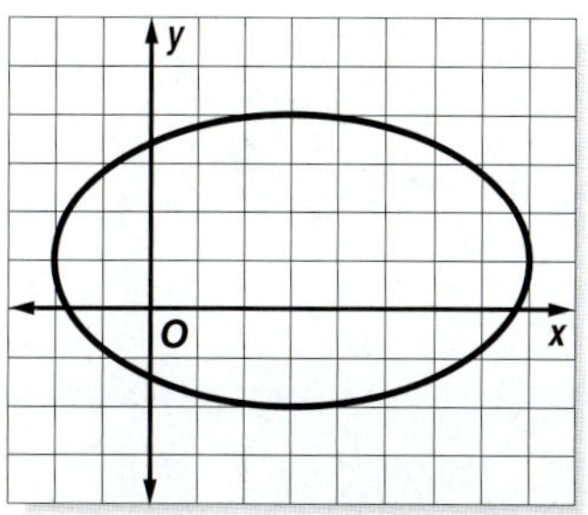

32.

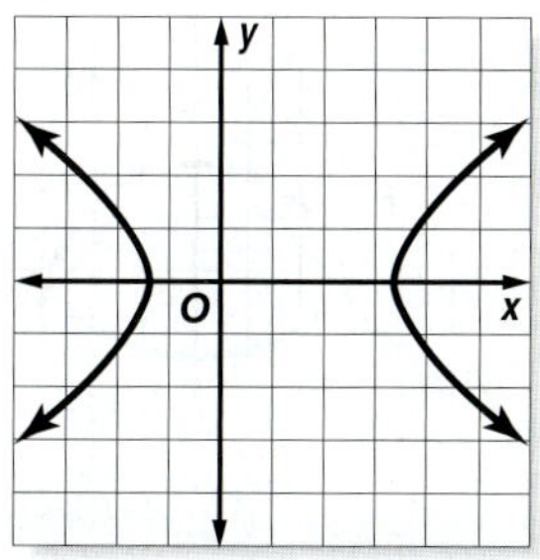

46. If you point a flashlight at a flat surface, you can make different conic sections by varying the angle at which you point the flashlight. Answers should include the following.

- Point the flashlight directly at a ceiling or wall. The light from the flashlight is in the shape of a cone and the ceiling or wall acts as a plane perpendicular to the axis of the cone.
- Hold the flashlight close to a wall and point it directly vertically toward the ceiling. A branch of a hyperbola will appear on the wall. In this case, the wall acts as a plane parallel to the axis of the cone.

52.

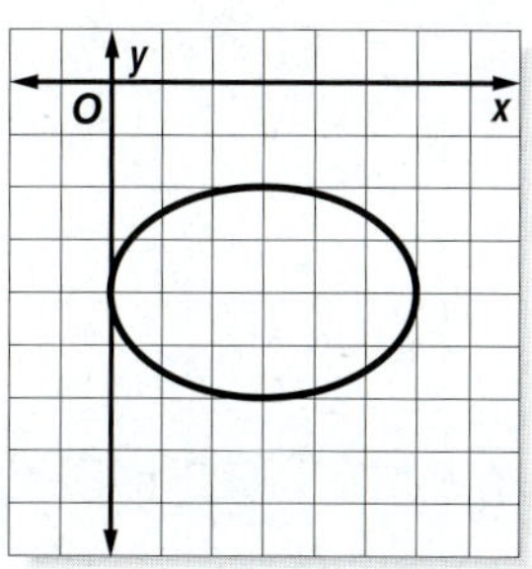

Pages 458–459, Lesson 8-7

8.

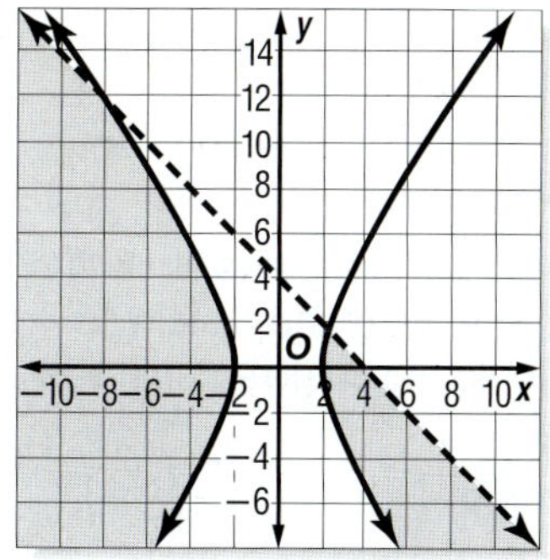

9.

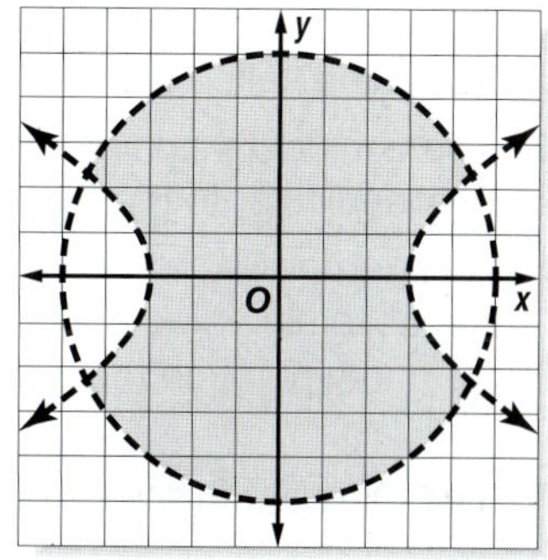

32.

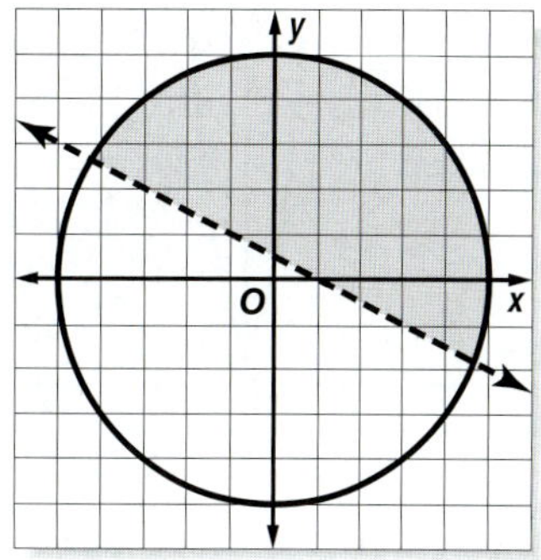

33.

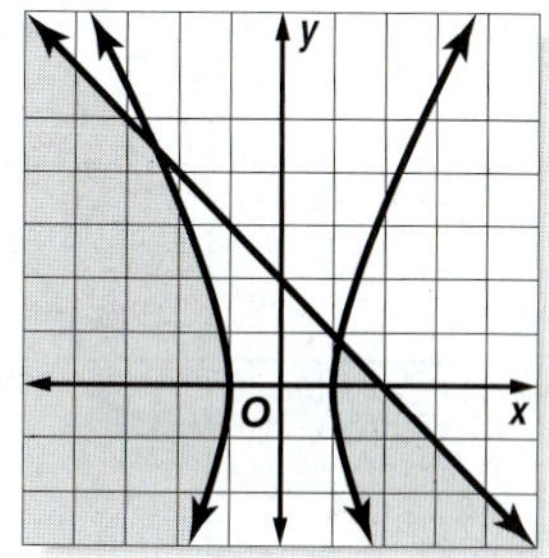

34.

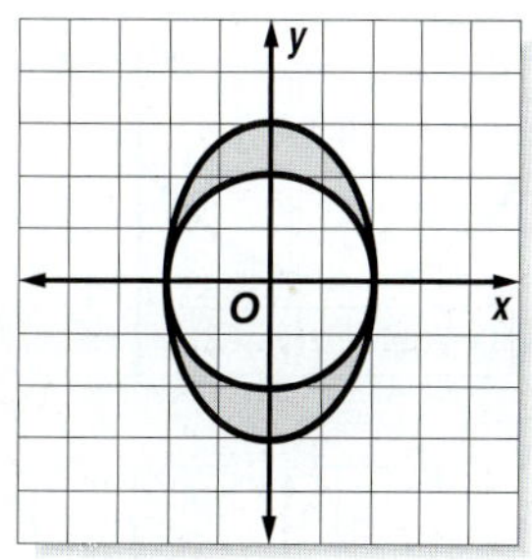

35.

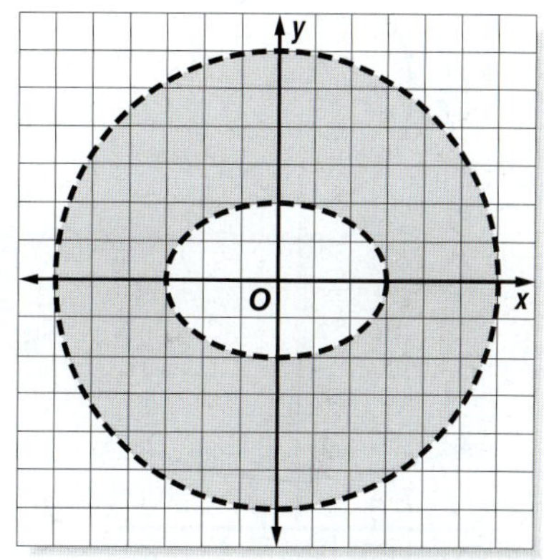

36.

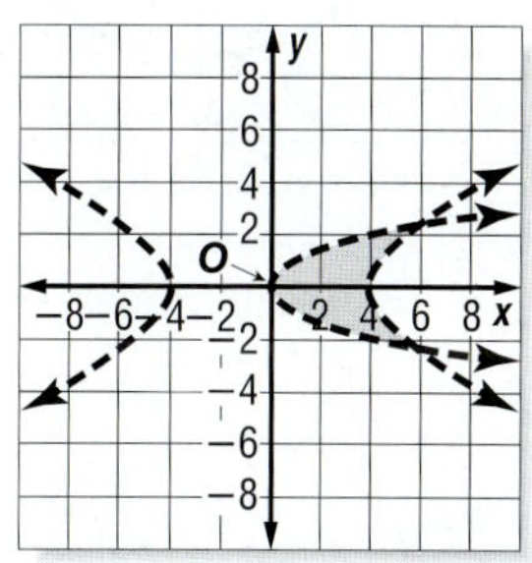

37.

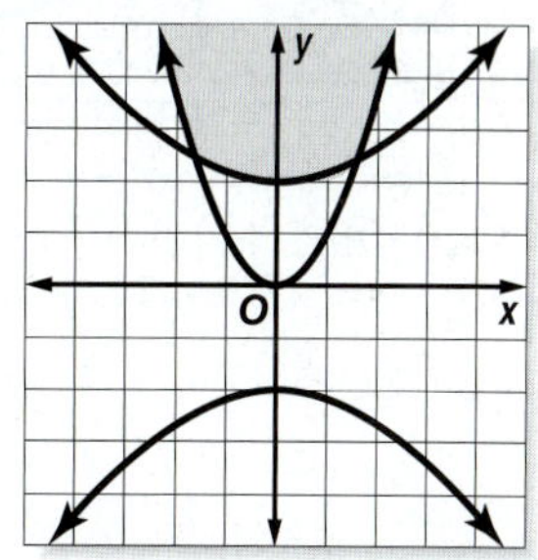

43. Systems of equations can be used to represent the locations and/or paths of objects on the screen. Answers should include the following.

- $y = 3x$, $x^2 + y^2 = 2500$
- The y-intercept of the graph of the equation $y = 3x$ is 0, so the path of the spaceship contains the origin.
- $(-5\sqrt{10}, -15\sqrt{10})$ or about $(-15.81, -47.43)$

Page 461, Chapter 8 Study Guide and Review

4. A parabola is the set of all points that are the same distance from a given point called the focus and a given line called the directrix.
6. The conjugate axis of a hyperbola is a line segment perpendicular to the transverse axis.
8. A hyperbola is the set of all points in a plane such that the absolute value of the difference of the distances from any point on the hyperbola to two given points is constant.
9. The midpoint formula is given by $\left(\frac{x_1 + x_2}{2}, \frac{y_1 + y_2}{2}\right)$.

Page 467, Chapter 8 Practice Test

10.
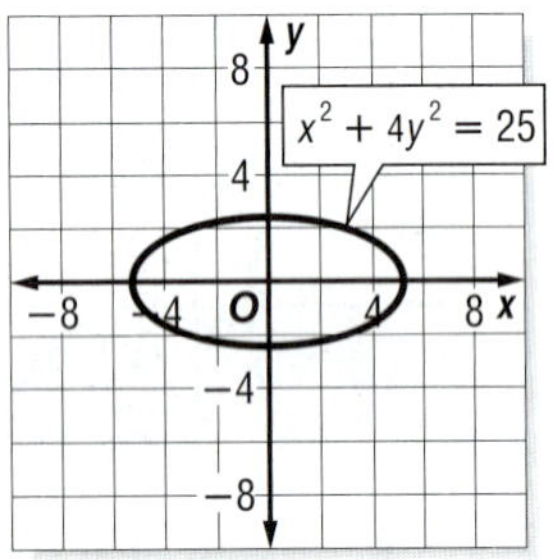

11.
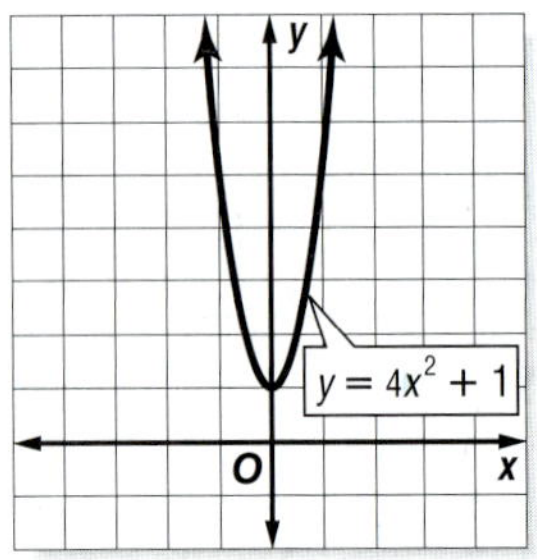

12.
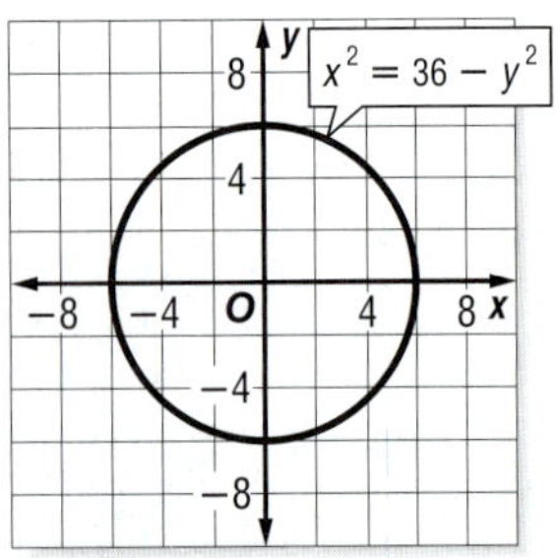

13.
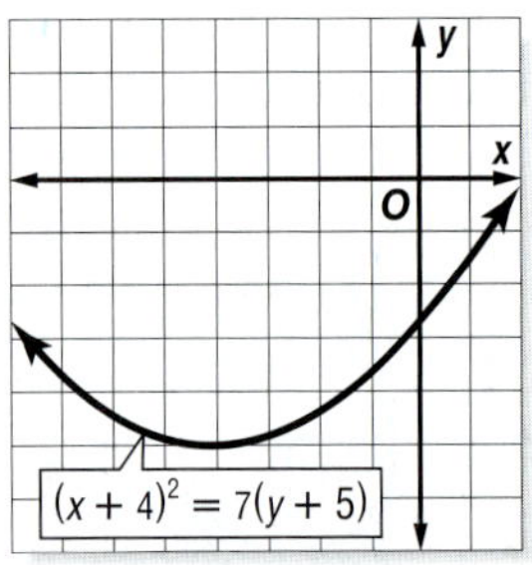

14.
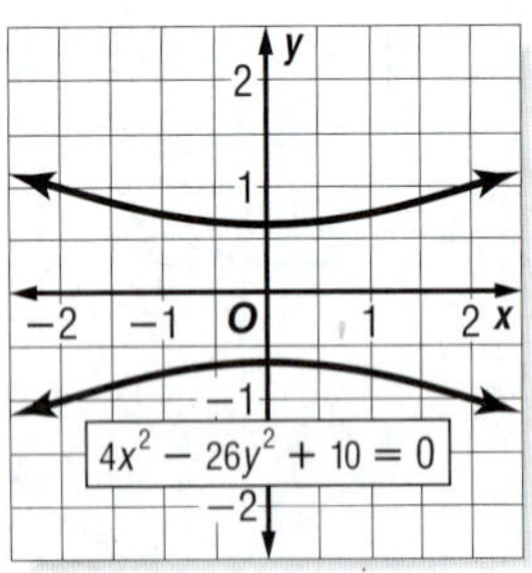

15.
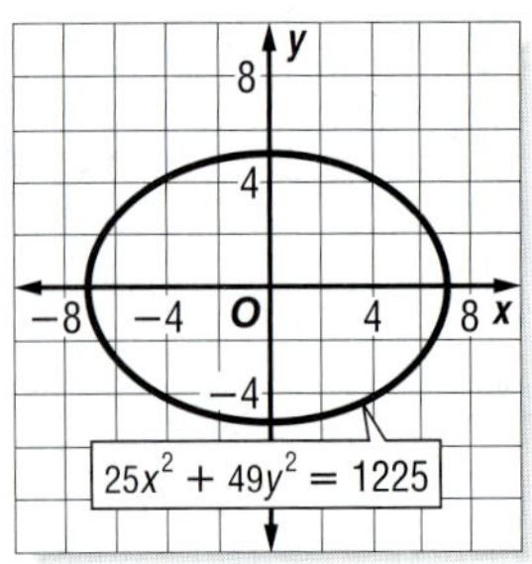

16.
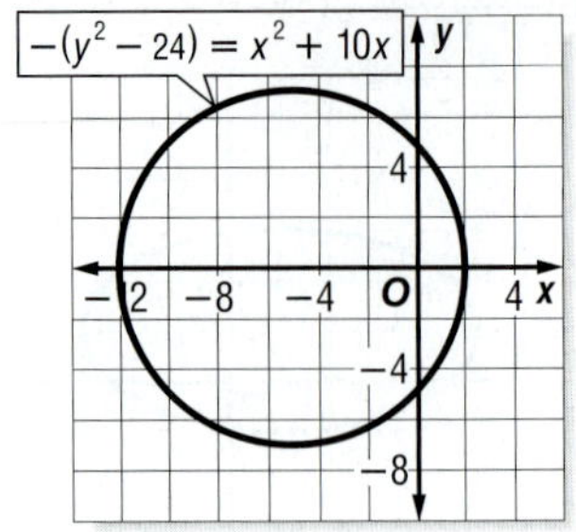

17.
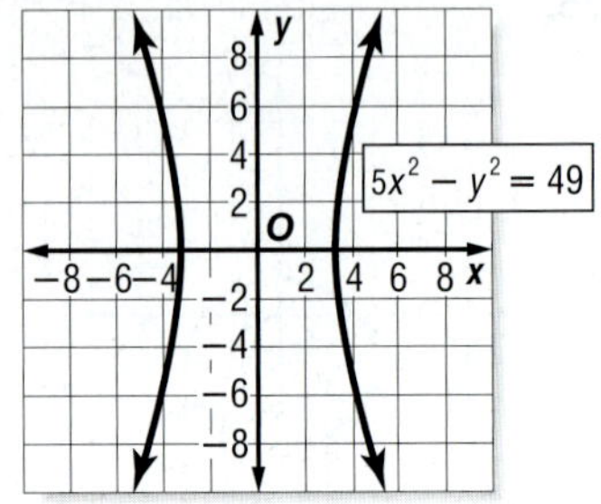

18.
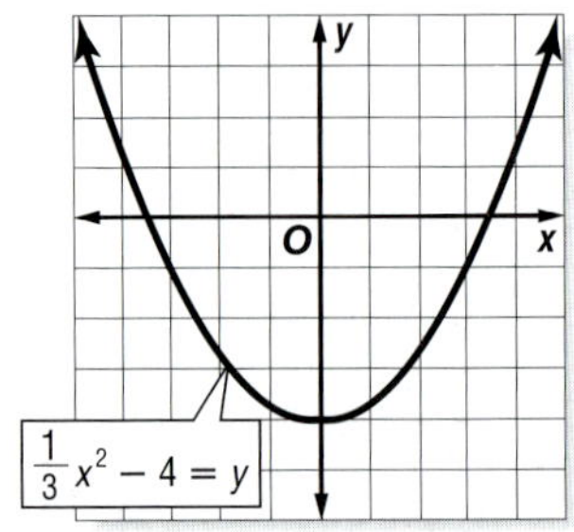

19.
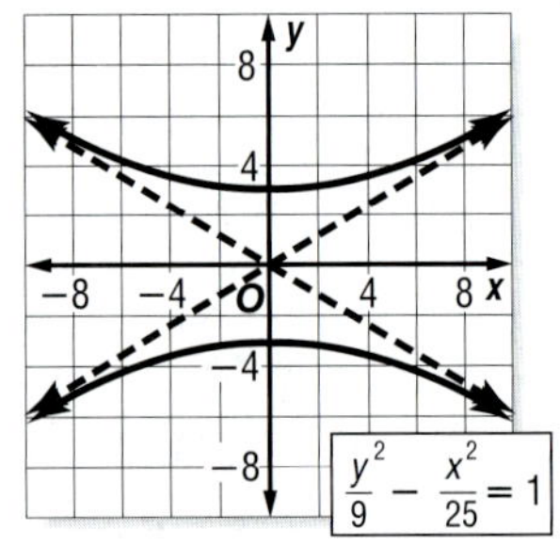

24.
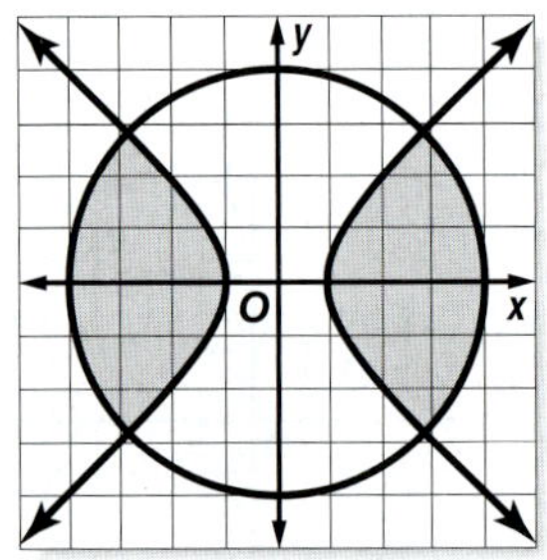

Notes

Chapter 9

Rational Expressions and Equations

Chapter Overview and Pacing

Year-long and two-year pacing: pages T20–T21.

LESSON OBJECTIVES

LESSON OBJECTIVES	PACING (days) Regular: Basic/Average	Regular: Advanced	Block: Basic/Average	Block: Advanced
9-1 Multiplying and Dividing Rational Expressions *(pp. 472–478)* • Simplify rational expressions. • Simplify complex fractions.	1	1	0.5	0.5
9-2 Adding and Subtracting Rational Expressions *(pp. 479–484)* • Determine the LCM of polynomials. • Add and subtract rational expressions.	2	2	1	1
9-3 Graphing Rational Functions *(pp. 485–491)* • Determine the vertical asymptotes and the point discontinuity for the graphs of rational functions. • Graph rational functions. ***Follow-Up:*** Graphing Rational Functions	1	2 (with 9-3 Follow-Up)	0.5	1
9-4 Direct, Joint, and Inverse Variation *(pp. 492–498)* • Recognize and solve direct and joint variation problems. • Recognize and solve inverse variation problems.	2	2	1	1
9-5 Classes of Functions *(pp. 499–504)* • Identify graphs as different types of functions. • Identify equations as different types of functions.	1	1	0.5	0.5
9-6 Solving Rational Equations and Inequalities *(pp. 505–512)* • Solve rational equations. • Solve rational inequalities. ***Follow-Up:*** Solving Rational Equations by Graphing	2	3 (with 9-6 Follow-Up)	1.5	1
Study Guide and Practice Test *(pp. 513–517)* **Standardized Test Practice** *(pp. 518–519)*	1	1	0.5	0.5
Chapter Assessment	1	1	0.5	0.5
TOTAL	**11**	**13**	**6**	**6**

StudentWorks

*An electronic version of this chapter is available on **StudentWorks**™. This backpack solution CD-ROM allows students instant access to the Student Edition, lesson worksheet pages, and web resources.*

Timesaving Tools
TeacherWorks™
All-In-One Planner and Resource Center
See pages T5 and T21.

Chapter Resource Manager

Chapter 9 Resource Masters: Study Guide and Intervention	Practice (Skills and Average)	Reading to Learn Mathematics	Enrichment	Assessment	Applications*	5-Minute Check Transparencies	Interactive Chalkboard	Alge2PASS: Tutorial Plus (lessons)	Materials
517–518	519–520	521	522		SC 17	9-1	9-1	17	
523–524	525–526	527	528	567		9-2	9-2	18	
529–530	531–532	533	534	567, 569	GCS 43	9-3	9-3		balance, metric measuring cup, graph paper (*Follow-Up:* graphing calculator)
535–536	537–538	539	540		GCS 44, SC 18, SM 123–126	9-4	9-4		
541–542	543–544	545	546	568		9-5	9-5		string, grid paper
547–548	549–550	551	552	568		9-6	9-6		(*Follow-Up:* graphing calculator)
				553–566, 570–572					

Key to Abbreviations: GCS = Graphing Calculator and Speadsheet Masters,
SC = School-to-Career Masters,
SM = Science and Mathematics Lab Manual

Chapter 9

Mathematical Connections and Background

Continuity of Instruction

Prior Knowledge

Students have simplified rational numbers, written equivalent rational numbers, and found least common denominators. They have graphed functions from tables of values and they have explored functions whose graphs are lines or other shapes. Also, they have solved linear and polynomial equations.

This Chapter

Students extend basic arithmetic operations to rational expressions and extend solving equations to rational equations and inequalities. They graph rational functions and identify discontinuities in the graphs. They investigate equations that represent direct, inverse, and joint variation. They look at graphs of various shapes, including continuous curves, discontinuous curves, and lines, and associate each graph with a specific kind of function.

Future Connections

Students will extend operations on rational expressions to finding powers and roots of rational expressions. They will continue to study situations involving direct or inverse variation, and they will extend the study of joint variation to variables with exponents. Also, they will continue to see how the shape of a graph is used to classify the function represented by the graph.

9-1 Multiplying and Dividing Rational Expressions

In this lesson students look at some familiar ideas from fractions and apply them to rational expressions. One idea is that a fraction is undefined if the denominator is zero. To extend that idea, students examine polynomials that are the denominators of rational expressions and use their skill in factoring to identify values of variables for which the denominator would be zero. Students extend another idea to rational expressions, simplifying fractions, by finding common factors in the numerator and denominator, and replacing the quotient of those factors with 1. As another extension of simplifying, students identify factors of the form $a - b$ and $b - a$ in the numerator and denominator of a rational expression, and replace the quotient of those factors with -1. The ideas of multiplying or dividing fractions and simplifying complex fractions have direct extensions. To multiply two rational expressions, students divide the product of the numerators by the product of the denominators; to divide by a rational expression, they multiply by the reciprocal of that expression; and to simplify a complex fraction involving rational expressions, they rewrite it and treat it as a division expression.

9-2 Adding and Subtracting Rational Expressions

Students continue to look at familiar ideas from fractions and extend those ideas to rational expressions. For the idea of a least common multiple, students factor two or more polynomials. They write each factor of either of the given polynomials, with each factor having an exponent that indicates the maximum number of times that factor appears in any one of the given polynomials. Another familiar idea is writing two fractions as equivalent fractions with a common denominator. To extend this idea to rational expressions, students find the LCM of the given denominators and rewrite each rational expression as an equivalent expression whose denominator is that LCM. The ideas of adding and subtracting fractions are extended to rational expressions by writing the rational expressions with a common denominator (again, the common denominator is the LCM of the given denominators) and then adding or subtracting the numerators.

9-3 Graphing Rational Functions

In this lesson (and the next two) students use graphs to examine properties of functions. This lesson introduces graphs of rational functions, which are functions whose numerator and denominator are both polynomials. To look at values of x for which the denominator and thus the function is undefined, students study two kinds of rational functions. In one kind, the denominator is a factor of the numerator; for example, $f(x) = \frac{(x + 1)^5}{x + 1}$ or $g(x) = \frac{(2x + 1)(3x - 2)(5x + 3)}{(2x + 1)(5x + 3)}$. These functions can be simplified, and the graph of the simplified function is a continuous curve. However, there are one or more values of the variable for which the denominator of the original polynomial is zero. These values, called point discontinuities, represent places when the function is undefined. Students explore these types of functions by reducing the function, graphing the reduced function, and identifying the "holes" in the graph. Another kind of rational expression is one in which the entire denominator is not a factor of the numerator. For these functions, each value of the variable for which the denominator is zero is associated with a vertical asymptote on the graph. Students explore these functions by identifying all the vertical asymptotes, and then using tables of values and the asymptotes to graph the function.

9-4 Direct, Joint, and Inverse Variation

In this lesson students examine graphs for two-variable equations that represent two types of relationships. For variables x and y and constant k, the relationship $y = kx$ (also written as $\frac{y}{x} = k$) is called direct variation. The graph of a direct variation is a line; the line goes through the origin (0, 0) and has slope k. The relationship $y = \frac{k}{x}$ (or $xy = k$) is called inverse variation; its graph is a hyperbola. Students explore direct and inverse variation by finding the constant or a missing value of a variable for a given type of variation or by stating the type of variation for a given graph or set of values. Students also explore the relationship $y = kxz$ among the variables x, y, and z and constant k. Students explore this type of variation, called joint variation, by using given values to find the value of the constant or a missing value of a variable.

9-5 Classes of Functions

In this lesson students organize information learned previously about graphs and functions. Given the graph of a line, they describe that line as the graph of a constant function, a direct variation function, or the identity function; if a line has a hole, they identify it as the graph of one type of rational function. Given a graph that is a continuous curve, they describe that curve as the graph of a quadratic function or a square root function. Also, they relate V-shaped graphs to absolute value functions and relate discontinuous graphs to greatest integer functions, inverse variation functions, and rational functions.

9-6 Solving Rational Equations and Inequalities

In this lesson students return to the topics of the chapter's first two lessons and solve equations involving rational expressions. In general, the first step is to multiply both sides of the equation by the least common denominator of all the denominators. The result is to rewrite the original equation as an equation with no denominators; the new equation can be solved using familiar methods for solving linear or polynomial equations. Students apply these methods to several kinds of word problems. One kind is "work problems," in which one complete job is the sum of partial jobs, each partial job being the quotient of some number of time units divided by a per-unit rate. Another kind is "rate problems," in which a total duration is the sum of two smaller durations, each one being the quotient of distance divided by rate. Also in this lesson students explore rational inequalities by finding values that make the denominator equal to 0, solving a related equation, and identifying intervals on the number line.

Additional mathematical information and teaching notes are available at www.algebra2.com/key_concepts.

Chapter 9

DAILY INTERVENTION and Assessment

	Type	Student Edition	Teacher Resources	Technology/Internet
INTERVENTION	Ongoing	Prerequisite Skills, pp. 471, 478, 484, 490, 498, 504 Practice Quiz 1, p. 484 Practice Quiz 2, p. 498	5-Minute Check Transparencies Quizzes, *CRM* pp. 567–568 Mid-Chapter Test, *CRM* p. 569 Study Guide and Intervention, *CRM* pp. 517–518, 523–524, 529–530, 535–536, 541–542, 547–548	Alge2PASS: Tutorial Plus www.algebra2.com/self_check_quiz www.algebra2.com/extra_examples
	Mixed Review	pp. 478, 484, 490, 498, 504, 511	Cumulative Review, *CRM* p. 570	
	Error Analysis	Find the Error, pp. 481, 509	Find the Error, *TWE* pp. 481, 509 Unlocking Misconceptions, *TWE* pp. 474, 486, 494 Tips for New Teachers, *TWE* pp. 478, 484, 487, 498, 504, 511	
ASSESSMENT	Standardized Test Practice	pp. 473, 476, 478, 484, 490, 498, 503, 504, 511, 517, 518–519	*TWE* p. 473 Standardized Test Practice, *CRM* pp. 571–572	Standardized Test Practice CD-ROM www.algebra2.com/standardized_test
	Open-Ended Assessment	Writing in Math, pp. 477, 484, 490, 498, 503, 511 Open Ended, pp. 476, 478, 482, 488, 495, 501, 509	Modeling: *TWE* pp. 498, 504 Speaking: *TWE* p. 478 Writing: *TWE* pp. 484, 490, 511 Open-Ended Assessment, *CRM* p. 565	
	Chapter Assessment	Study Guide, pp. 513–516 Practice Test, p. 517	Multiple-Choice Tests (Forms 1, 2A, 2B), *CRM* pp. 553–558 Free-Response Tests (Forms 2C, 2D, 3), *CRM* pp. 559–564 Vocabulary Test/Review, *CRM* p. 566	ExamView® Pro (see below) MindJogger Videoquizzes www.algebra2.com/vocabulary_review www.algebra2.com/chapter_test

Key to Abbreviations: TWE = Teacher Wraparound Edition; CRM = Chapter Resource Masters

Intervention Technology

Alge2PASS: Tutorial Plus CD-ROM offers a complete, self-paced algebra curriculum.

Algebra 2 Lesson	Alge2PASS Lesson
9-1	**17** *Simplifying Rational Expressions*
9-2	**18** *Operations with Rational Functions*

ExamView® Pro

Use the networkable **ExamView® Pro** to:

- Create **multiple versions** of tests.
- Create **modified** tests for *Inclusion* students.
- **Edit** existing questions and **add** your own questions.
- Use built-in **state curriculum correlations** to create tests aligned with state standards.
- **Apply** art to your tests from a program bank of artwork.

For more information on Intervention and Assessment, see pp. T8–T11.

Reading and Writing in Mathematics

Glencoe Algebra 2 provides numerous opportunities to incorporate reading and writing into the mathematics classroom.

Student Edition

- Foldables Study Organizer, p. 471
- Concept Check questions require students to verbalize and write about what they have learned in the lesson. (pp. 476, 481, 488, 495, 501, 509, 513)
- Writing in Math questions in every lesson, pp. 477, 484, 490, 498, 503, 511
- WebQuest, p. 502

Teacher Wraparound Edition

- Foldables Study Organizer, pp. 471, 513
- Study Notebook suggestions, pp. 476, 481, 488, 495, 501, 509
- Modeling activities, pp. 498, 504
- Speaking activities, p. 478
- Writing activities, pp. 484, 490, 511
- ELL Resources, pp. 470, 477, 483, 489, 496, 503, 510, 513

Additional Resources

- Vocabulary Builder worksheets require students to define and give examples for key vocabulary terms as they progress through the chapter. (*Chapter 9 Resource Masters*, pp. vii-viii)
- Reading to Learn Mathematics master for each lesson (*Chapter 9 Resource Masters*, pp. 521, 527, 533, 539, 545, 551)
- *Vocabulary PuzzleMaker* software creates crossword, jumble, and word search puzzles using vocabulary lists that you can customize.
- *Teaching Mathematics with Foldables* provides suggestions for promoting cognition and language.
- *Reading and Writing in the Mathematics Classroom*
- *WebQuest and Project Resources*

For more information on Reading and Writing in Mathematics, see pp. T6–T7.

PROJECT CRISS℠ Study Skill

Students can create their own representation of the concepts they are learning by using a free-form map. Free-form mapping is done as a cooperative activity after students have read about and discussed a lesson or chapter. Give groups of students large sheets of paper and colored markers. Allow them to decide the important ideas of the lesson to represent on their map. Then present their ideas on paper with words pictures, and diagrams.

The free-form map at the right represents the concepts in Lesson 9-4.

Direct, Joint, and Inverse Variation

Direct	Inverse	Joint
• $y = kx$ • graph passes thru (0, 0) • Sample graphs: 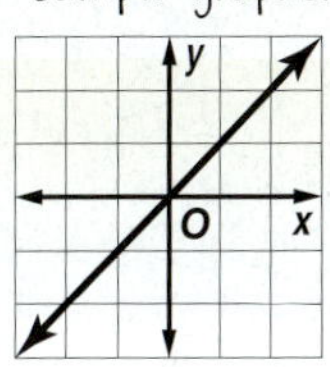 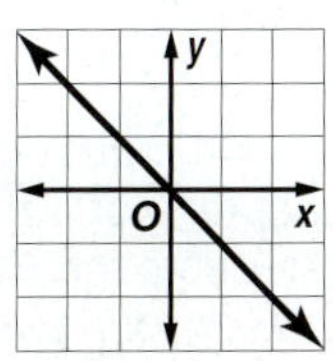	• $xy = k$ • graph is a hyperbola, often restricted to Quadrant I for real data • Sample graph: 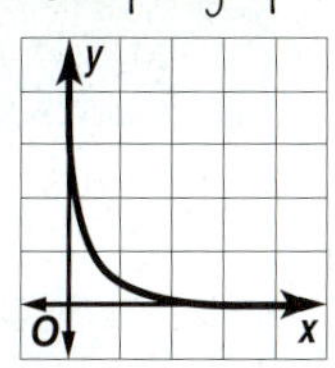	• $y = kxz$ • Sample application: the area of a triangle varies jointly as its base and height, $A = \frac{1}{2}bh$

CReating Independence Through Student-Owned Strategies

Chapter 9 Notes

What You'll Learn

Have students read over the list of objectives and make a list of any words with which they are not familiar.

Why It's Important

Point out to students that this is only one of many reasons why each objective is important. Others are provided in the introduction to each lesson.

Lesson	NCTM Standards	Local Objectives
9-1	2, 6, 8, 9, 10	
9-2	2, 6, 8, 9, 10	
9-3	2, 6, 8, 9, 10	
9-3 Follow-Up	2, 6, 8	
9-4	1, 2, 6, 8, 9, 10	
9-5	2, 6, 8, 9, 10	
9-6	2, 6, 8, 9, 10	
9-6 Follow-Up	2, 6	

Key to NCTM Standards:

1=Number & Operations, 2=Algebra, 3=Geometry, 4=Measurement, 5=Data Analysis & Probability, 6=Problem Solving, 7=Reasoning & Proof, 8=Communication, 9=Connections, 10=Representation

Chapter 9 Rational Expressions and Equations

What You'll Learn

- **Lessons 9-1 and 9-2** Simplify rational expressions.
- **Lesson 9-3** Graph rational functions.
- **Lesson 9-4** Solve direct, joint, and inverse variation problems.
- **Lesson 9-5** Identify graphs and equations as different types of functions.
- **Lesson 9-6** Solve rational equations and inequalities.

Key Vocabulary

- rational expression (p. 472)
- asymptote (p. 485)
- point discontinuity (p. 485)
- direct variation (p. 492)
- inverse variation (p. 493)

Why It's Important

Rational expressions, functions, and equations can be used to solve problems involving mixtures, photography, electricity, medicine, and travel, to name a few. Direct, joint, and inverse variation are important applications of rational expressions. For example, scuba divers can use direct variation to determine the amount of pressure at various depths. *You will learn how to determine the amount of pressure exerted on the ears of a diver in Lesson 9-4.*

Vocabulary Builder ELL

The Key Vocabulary list introduces students to some of the main vocabulary terms included in this chapter. For a more thorough vocabulary list with pronunciations of new words, give students the Vocabulary Builder worksheets found on pages vii and viii of the *Chapter 9 Resource Masters*. Encourage them to complete the definition of each term as they progress through the chapter. You may suggest that they add these sheets to their study notebooks for future reference when studying for the Chapter 9 test.

Getting Started

Prerequisite Skills To be successful in this chapter, you'll need to master these skills and be able to apply them in problem-solving situations. Review these skills before beginning Chapter 9.

For Lesson 9-1 — Solve Equations with Rational Numbers

Solve each equation. Write your answer in simplest form. *(For review, see Lesson 1-3.)*

1. $\frac{8}{5}x = \frac{4}{15}$ **$\frac{1}{6}$**
2. $\frac{27}{14}t = \frac{6}{7}$ **$\frac{4}{9}$**
3. $\frac{3}{10} = \frac{12}{25}a$ **$\frac{5}{8}$**
4. $\frac{6}{7} = 9m$ **$\frac{2}{21}$**
5. $\frac{9}{8}b = 18$ **16**
6. $\frac{6}{7}s = \frac{3}{4}$ **$\frac{7}{8}$**
7. $\frac{1}{3}r = \frac{5}{6}$ **$2\frac{1}{2}$**
8. $\frac{2}{3}n = 7$ **$10\frac{1}{2}$**
9. $\frac{4}{5}r = \frac{5}{6}$ **$1\frac{1}{24}$**

For Lesson 9-3 — Determine Asymptotes and Graph Equations

Draw the asymptotes and graph each hyperbola. *(For review, see Lesson 8-5.)*

10. $\frac{(x-3)^2}{4} - \frac{(y+5)^2}{9} = 1$
11. $\frac{y^2}{4} - \frac{(x+4)^2}{1} = 1$
12. $\frac{(x+2)^2}{4} - \frac{(y-3)^2}{25} = 1$

10–12. See margin.

For Lesson 9-4 — Solve Proportions

Solve each proportion.

13. $\frac{3}{4} = \frac{r}{16}$ **12**
14. $\frac{8}{16} = \frac{5}{y}$ **10**
15. $\frac{6}{8} = \frac{m}{20}$ **15**
16. $\frac{t}{3} = \frac{5}{24}$ **$\frac{5}{8}$**
17. $\frac{5}{a} = \frac{6}{18}$ **15**
18. $\frac{3}{4} = \frac{b}{6}$ **$4\frac{1}{2}$**
19. $\frac{v}{9} = \frac{12}{18}$ **6**
20. $\frac{7}{p} = \frac{1}{4}$ **28**
21. $\frac{2}{5} = \frac{3}{z}$ **$7\frac{1}{2}$**
22. $\frac{16}{18} = \frac{6}{x}$ **6.75**
23. $\frac{12}{14} = \frac{3}{2}$ **21**
24. $\frac{5}{7} = \frac{1.5}{t}$ **2.1**

FOLDABLES™ Study Organizer

Rational Expressions and Equations Make this Foldable to help you organize your notes. Begin with a sheet of plain $8\frac{1}{2}$" by 11" paper.

Step 1 Fold

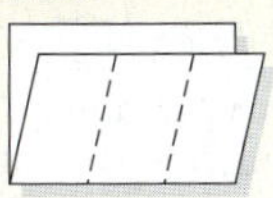

Fold in half lengthwise leaving a $1\frac{1}{2}$" margin at the top. Fold again in thirds.

Step 2 Cut and Label

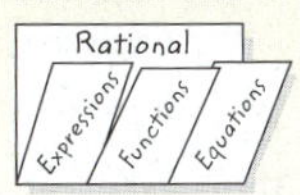

Open. Cut along the second folds to make three tabs. Label as shown.

Reading and Writing As you read and study the chapter, write notes and examples for each concept under the tabs.

FOLDABLES™ Study Organizer

For more information about Foldables, see *Teaching Mathematics with Foldables.*

Organization of Data with a Concept Map Concept maps are visual study guides that allow students to view main ideas or key words and use them to recall and organize what they know and what they have learned. Begin by writing *Rational* on the base of the Foldable and the words *Expressions*, *Functions*, and *Equations* on the tabs of the concept map. Under the tabs of their Foldable, have students take notes, define terms, record concepts, and write examples. Students can check their responses and memory by reviewing their notes under the tabs.

Getting Started

This section provides a review of the basic concepts needed before beginning Chapter 9. Page references are included for additional student help.

Prerequisite Skills in the Getting Ready for the Next Lesson section at the end of each exercise set review a skill needed in the next lesson.

For Lesson	Prerequisite Skill
9-2	Solving Equations (p. 478)
9-3	Graphing Hyperbolas (p. 484)
9-4	Solving Proportions (p. 490)
9-5	Special Functions (p. 498)
9-6	Least Common Multiples of Polynomials (p. 504)

Answers

10.

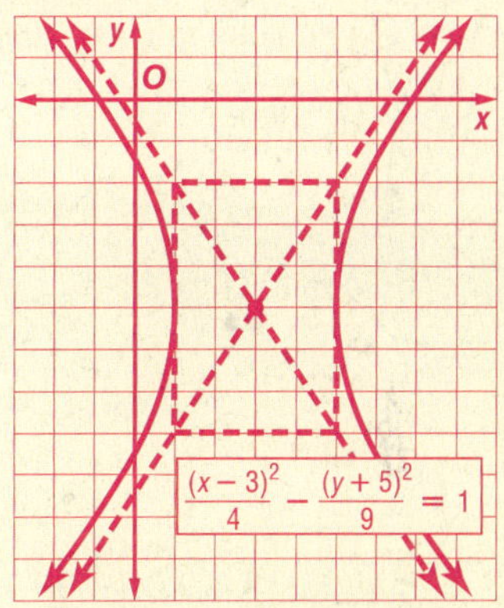

11.

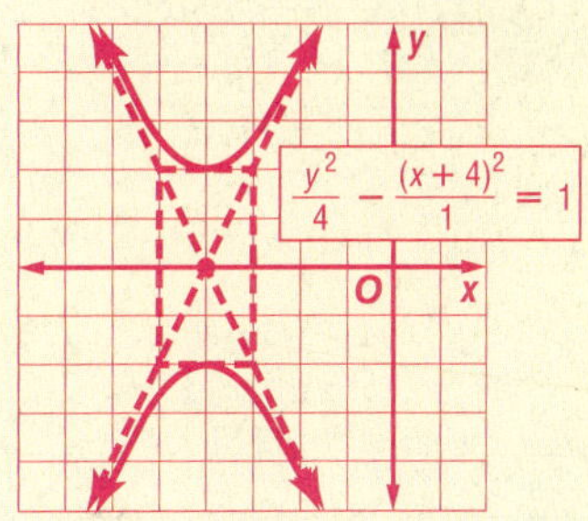

12.

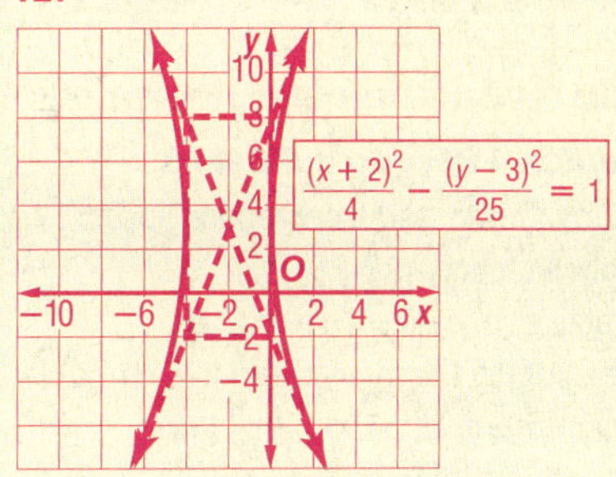

9-1 Lesson Notes

1 Focus

5-Minute Check Transparency 9-1 Use as a quiz or review of Chapter 8.

Mathematical Background notes are available for this lesson on p. 470C.

How are rational expressions used in mixtures?

Ask students:

- How can the term *rational expression* help you recall what it means? **The word "rational" contains the word "ratio."**
- What does it mean to say that 6 is the GCF of 12 and 30? **It is the greatest integer that divides into both 12 and 30 without a remainder.**

9-1 Multiplying and Dividing Rational Expressions

What You'll Learn

- Simplify rational expressions.
- Simplify complex fractions.

Vocabulary

- rational expression
- complex fraction

How are rational expressions used in mixtures?

The Goodie Shoppe sells candy and nuts by the pound. One of their items is a mixture of peanuts and cashews. This mixture is made with 8 pounds of peanuts and 5 pounds of cashews. Therefore, $\frac{8}{8+5}$ or $\frac{8}{13}$ of the mixture is peanuts. If the store manager adds an additional x pounds of peanuts to the mixture, then $\frac{8+x}{13+x}$ of the mixture will be peanuts.

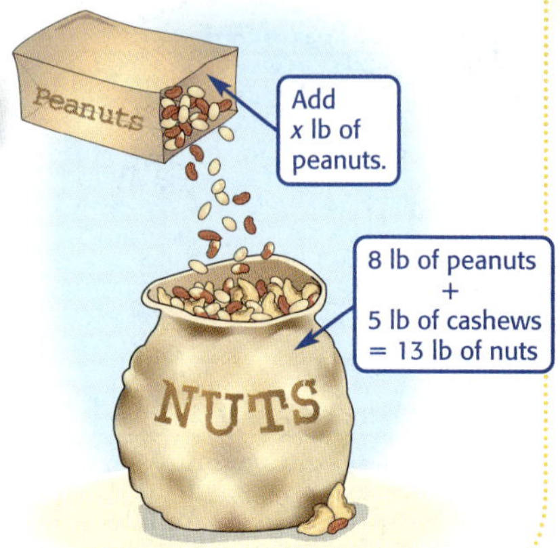

SIMPLIFY RATIONAL EXPRESSIONS A ratio of two polynomial expressions such as $\frac{8+x}{13+x}$ is called a **rational expression**. Because variables in algebra represent real numbers, operations with rational numbers and rational expressions are similar.

To write a fraction in simplest form, you divide both the numerator and denominator by their greatest common factor (GCF). To simplify a rational expression, you use similar properties.

Example 1 Simplify a Rational Expression

a. Simplify $\frac{2x(x-5)}{(x-5)(x^2-1)}$.

Look for common factors.

$$\frac{2x(x-5)}{(x-5)(x^2-1)} = \frac{2x}{x^2-1} \cdot \frac{\overset{1}{\cancel{x-5}}}{\underset{1}{\cancel{x-5}}} \quad \text{How is this similar to simplifying } \tfrac{10}{15}?$$

$$= \frac{2x}{x^2-1} \quad \text{Simplify.}$$

b. Under what conditions is this expression undefined?

Just as with a fraction, a rational expression is undefined if the denominator is equal to 0. To find when this expression is undefined, completely factor the original denominator.

$$\frac{2x(x-5)}{(x-5)(x^2-1)} = \frac{2x(x-5)}{(x-5)(x-1)(x+1)} \quad x^2-1=(x-1)(x+1)$$

The values that would make the denominator equal to 0 are 5, 1, or -1. So the expression is undefined when $x = 5$, $x = 1$, or $x = -1$. These numbers are called *excluded values*.

Resource Manager

Workbook and Reproducible Masters

Chapter 9 Resource Masters
- Study Guide and Intervention, pp. 517–518
- Skills Practice, p. 519
- Practice, p. 520
- Reading to Learn Mathematics, p. 521
- Enrichment, p. 522

School-to-Career Masters, p. 17
Teaching Algebra With Manipulatives Masters, p. 272

Transparencies

5-Minute Check Transparency 9-1
Answer Key Transparencies

Technology

Alge2PASS: Tutorial Plus, Lesson 17
Interactive Chalkboard

Standardized Test Practice

Example 2 Use the Process of Elimination

Multiple-Choice Test Item

For what value(s) of x is $\frac{x^2 + x - 12}{x^2 + 7x + 12}$ undefined?

Ⓐ $-4, -3$ Ⓑ -4 Ⓒ 0 Ⓓ $-4, 3$

Read the Test Item

You want to determine which values of x make the denominator equal to 0.

Solve the Test Item

Look at the possible answers. Notice that if x equals 0 or a positive number, $x^2 + 7x + 12$ must be greater than 0. Therefore, you can eliminate choices C and D. Since both choices A and B contain -4, determine whether the denominator equals 0 when $x = -3$.

$$x^2 + 7x + 12 = (-3)^2 + 7(-3) + 12 \quad x = -3$$
$$= 9 - 21 + 12 \quad \text{Multiply.}$$
$$= 0 \quad \text{Simplify.}$$

Since the denominator equals 0 when $x = -3$, the answer is A.

Test-Taking Tip

Sometimes you can save time by looking at the possible answers and eliminating choices, rather than actually evaluating an expression or solving an equation.

Sometimes you can factor out -1 in the numerator or denominator to help simplify rational expressions.

Example 3 Simplify by Factoring Out –1

Simplify $\frac{z^2w - z^2}{z^3 - z^3w}$.

$$\frac{z^2w - z^2}{z^3 - z^3w} = \frac{z^2(w-1)}{z^3(1-w)} \quad \text{Factor the numerator and the denominator.}$$
$$= \frac{\overset{1}{\cancel{z^2}}(-1)\overset{1}{\cancel{(1-w)}}}{\underset{z}{\cancel{z^3}}\underset{1}{\cancel{(1-w)}}} \quad w - 1 = -(-w + 1) \text{ or } -1(1 - w)$$
$$= \frac{-1}{z} \text{ or } -\frac{1}{z} \quad \text{Simplify.}$$

Remember that to multiply two fractions, you first multiply the numerators and then multiply the denominators. To divide two fractions, you multiply by the multiplicative inverse, or reciprocal, of the divisor.

Multiplication

$$\frac{5}{6} \cdot \frac{4}{15} = \frac{\overset{1}{\cancel{5}} \cdot \overset{1}{\cancel{2}} \cdot 2}{\underset{1}{\cancel{2}} \cdot 3 \cdot 3 \cdot \underset{1}{\cancel{5}}}$$
$$= \frac{2}{3 \cdot 3} \text{ or } \frac{2}{9}$$

Division

$$\frac{3}{7} \div \frac{9}{14} = \frac{3}{7} \cdot \frac{14}{9}$$
$$= \frac{\overset{1}{\cancel{3}} \cdot 2 \cdot \overset{1}{\cancel{7}}}{\underset{1}{\cancel{7}} \cdot \underset{1}{\cancel{3}} \cdot 3}$$
$$= \frac{2}{3}$$

The same procedures are used for multiplying and dividing rational expressions.

www.algebra2.com/extra_examples

2 Teach

SIMPLIFY RATIONAL EXPRESSIONS

In-Class Examples

Power Point®

1 a. Simplify $\frac{3y(y + 7)}{(y + 7)(y^2 - 9)}$. $\frac{3y}{y^2 - 9}$

b. Under what conditions is this expression undefined? when $y = -7$, $y = -3$, or $y = 3$

Teaching Tip Ask students to explain why division by zero is undefined.

2 For what value(s) of p is $\frac{p^2 + 2p - 3}{p^2 - 2p - 15}$ undefined? B

A 5 **B** $-3, 5$

C $3, -5$ **D** $5, 1, -3$

3 Simplify $\frac{a^4b - 2a^4}{2a^3 - a^3b}$. $-a$

Teaching Tip Point out that rational expressions are usually used without specifically excluding those values that make the expression undefined. It is understood that only those values for which the expression has meaning are included.

Interactive Chalkboard

PowerPoint® Presentations

This CD-ROM is a customizable Microsoft® PowerPoint® presentation that includes:

- Step-by-step, dynamic solutions of each In-Class Example from the Teacher Wraparound Edition
- Additional, Your Turn exercises for each example
- The 5-Minute Check Transparencies
- Hot links to Glencoe Online Study Tools

Standardized Test Practice

Example 2 Make sure students know to study only the denominator to determine the values that make the expression undefined. In this example, the numerator is irrelevant.

In-Class Examples

Power Point®

4 Simplify each expression.

a. $\frac{8x}{21y^3} \cdot \frac{7y^2}{16x^3}$ $\frac{1}{6x^2y}$

b. $\frac{5a^4c}{12b} \cdot \frac{24bc^2}{15a^3b^2}$ $\frac{2ac^3}{3b^2}$

5 Simplify $\frac{10ps^2}{3c^2d} \div \frac{5ps}{6c^2d^2}$. $4ds$

Teaching Tip To help students understand why division is equivalent to multiplying by the reciprocal, discuss simple examples such as this: dividing 18 marbles between 2 people means that each person gets one-half, or 9, of the marbles.

Key Concept — *Rational Expressions*

Multiplying Rational Expressions

- **Words** To multiply two rational expressions, multiply the numerators and the denominators.
- **Symbols** For all rational expressions $\frac{a}{b}$ and $\frac{c}{d}$, $\frac{a}{b} \cdot \frac{c}{d} = \frac{ac}{bd}$, if $b \neq 0$ and $d \neq 0$.

Dividing Rational Expressions

- **Words** To divide two rational expressions, multiply by the reciprocal of the divisor.
- **Symbols** For all rational expressions $\frac{a}{b}$ and $\frac{c}{d}$, $\frac{a}{b} \div \frac{c}{d} = \frac{a}{b} \cdot \frac{d}{c} = \frac{ad}{bc}$, if $b \neq 0$, $c \neq 0$, and $d \neq 0$.

The following examples show how these rules are used with rational expressions.

Study Tip

Alternative Method

When multiplying rational expressions, you can multiply first and then divide by the common factors. For instance, in Example 4,

$\frac{4a}{5b} \cdot \frac{15b^2}{16a^3} = \frac{60ab^2}{80a^3b}$.

Now divide the numerator and the denominator by the common factors.

$\frac{\overset{3}{\cancel{60}}\,\overset{1}{\cancel{a}}\,\overset{b}{\cancel{b^2}}}{\underset{4}{\cancel{80}}\,\underset{a^2}{\cancel{a^3}}\,\underset{1}{\cancel{b}}} = \frac{3b}{4a^2}$

Example 4 *Multiply Rational Expressions*

Simplify each expression.

a. $\frac{4a}{5b} \cdot \frac{15b^2}{16a^3}$

$$\frac{4a}{5b} \cdot \frac{15b^2}{16a^3} = \frac{\overset{1}{\cancel{2}} \cdot \overset{1}{\cancel{2}} \cdot \overset{1}{\cancel{a}} \cdot 3 \cdot \overset{1}{\cancel{5}} \cdot \overset{1}{\cancel{b}} \cdot b}{\underset{1}{\cancel{5}} \cdot \underset{1}{\cancel{b}} \cdot \underset{1}{\cancel{2}} \cdot \underset{1}{\cancel{2}} \cdot 2 \cdot 2 \cdot \underset{1}{\cancel{a}} \cdot a \cdot a} \quad \text{Factor.}$$

$$= \frac{3 \cdot b}{2 \cdot 2 \cdot a \cdot a} \quad \text{Simplify.}$$

$$= \frac{3b}{4a^2} \quad \text{Simplify.}$$

b. $\frac{8t^2s}{5r^2} \cdot \frac{15sr}{12t^3s^2}$

$$\frac{8t^2s}{5r^2} \cdot \frac{15sr}{12t^3s^2} = \frac{\overset{1}{\cancel{2}} \cdot \overset{1}{\cancel{2}} \cdot 2 \cdot \overset{1}{\cancel{t}} \cdot \overset{1}{\cancel{t}} \cdot \overset{1}{\cancel{s}} \cdot \overset{1}{\cancel{3}} \cdot \overset{1}{\cancel{5}} \cdot \overset{1}{\cancel{s}} \cdot \overset{1}{\cancel{r}}}{\underset{1}{\cancel{5}} \cdot \underset{1}{\cancel{r}} \cdot r \cdot \underset{1}{\cancel{2}} \cdot \underset{1}{\cancel{2}} \cdot \underset{1}{\cancel{3}} \cdot \underset{1}{\cancel{t}} \cdot \underset{1}{\cancel{t}} \cdot t \cdot \underset{1}{\cancel{s}} \cdot \underset{1}{\cancel{s}}} \quad \text{Factor.}$$

$$= \frac{2}{rt} \quad \text{Simplify.}$$

Example 5 *Divide Rational Expressions*

Simplify $\frac{4x^2y}{15a^3b^3} \div \frac{2xy^2}{5ab^3}$.

$$\frac{4x^2y}{15a^3b^3} \div \frac{2xy^2}{5ab^3} = \frac{4x^2y}{15a^3b^3} \cdot \frac{5ab^3}{2xy^2} \quad \text{Multiply by the reciprocal of the divisor.}$$

$$= \frac{\overset{1}{\cancel{2}} \cdot 2 \cdot \overset{1}{\cancel{x}} \cdot x \cdot \overset{1}{\cancel{y}} \cdot \overset{1}{\cancel{5}} \cdot \overset{1}{\cancel{a}} \cdot \overset{1}{\cancel{b}} \cdot \overset{1}{\cancel{b}} \cdot \overset{1}{\cancel{b}}}{3 \cdot \underset{1}{\cancel{5}} \cdot \underset{1}{\cancel{a}} \cdot a \cdot a \cdot \underset{1}{\cancel{b}} \cdot \underset{1}{\cancel{b}} \cdot \underset{1}{\cancel{b}} \cdot \underset{1}{\cancel{2}} \cdot \underset{1}{\cancel{x}} \cdot \underset{1}{\cancel{y}} \cdot y} \quad \text{Factor.}$$

$$= \frac{2 \cdot x}{3 \cdot a \cdot a \cdot y} \quad \text{Simplify.}$$

$$= \frac{2x}{3a^2y} \quad \text{Simplify.}$$

DAILY INTERVENTION — Unlocking Misconceptions

- **Simplifying the Quotient of Opposites** Help students understand why the quotient of $(x - y)$ and $(y - x)$ is -1 by pointing out that these two expressions are opposites (or additive inverses) just as are 2 and -2.
- **Division by Zero** By definition, $\frac{a}{b} = c$ if $a = bc$. If students think $\frac{6}{0} = 0$, use the definition to show $\frac{6}{0} = 0$ if $6 = 0 \cdot 0$, which is false.

474 Chapter 9 Rational Expressions and Equations

These same steps are followed when the rational expressions contain numerators and denominators that are polynomials.

Example 6 Polynomials in the Numerator and Denominator

Simplify each expression.

a. $\frac{x^2+2x-8}{x^2+4x+3} \cdot \frac{3x+3}{x-2}$

$\frac{x^2+2x-8}{x^2+4x+3} \cdot \frac{3x+3}{x-2} = \frac{(x+4)\cancel{(x-2)}}{(x+3)\cancel{(x+1)}} \cdot \frac{3\cancel{(x+1)}}{\cancel{(x-2)}}$ Factor.

$= \frac{3(x+4)}{(x+3)}$ Simplify.

$= \frac{3x+12}{x+3}$ Simplify.

b. $\frac{a+2}{a+3} \div \frac{a^2+a-12}{a^2-9}$

$\frac{a+2}{a+3} \div \frac{a^2+a-12}{a^2-9} = \frac{a+2}{a+3} \cdot \frac{a^2-9}{a^2+a-12}$ Multiply by the reciprocal of the divisor.

$= \frac{(a+2)\cancel{(a+3)}\cancel{(a-3)}}{\cancel{(a+3)}(a+4)\cancel{(a-3)}}$ Factor.

$= \frac{a+2}{a+4}$ Simplify.

Study Tip

Factor First
As in Example 6, sometimes you must factor the numerator and/or the denominator first before you can simplify a quotient of rational expressions.

TEACHING TIP

Have students replace a with a value such as 2. Then simplify both $\frac{a+2}{a+3} \div \frac{a^2+a-12}{a^2-9}$ and $\frac{a+2}{a+4}$. The values are the same.

SIMPLIFY COMPLEX FRACTIONS

A **complex fraction** is a rational expression whose numerator and/or denominator contains a rational expression. The expressions below are complex fractions.

$$\frac{\frac{a}{5}}{3b} \qquad \frac{\frac{3}{t}}{t+5} \qquad \frac{\frac{m^2-9}{8}}{\frac{3-m}{12}} \qquad \frac{\frac{1}{p}+2}{\frac{3}{p}-4}$$

Remember that a fraction is nothing more than a way to express a division problem. For example, $\frac{2}{5}$ can be expressed as $2 \div 5$. So to simplify any complex fraction, rewrite it as a division expression and use the rules for division.

Example 7 Simplify a Complex Fraction

Simplify $\frac{\frac{r^2}{r^2-25s^2}}{\frac{r}{5s-r}}$.

$\frac{\frac{r^2}{r^2-25s^2}}{\frac{r}{5s-r}} = \frac{r^2}{r^2-25s^2} \div \frac{r}{5s-r}$ Express as a division expression.

$= \frac{r^2}{r^2-25s^2} \cdot \frac{5s-r}{r}$ Multiply by the reciprocal of the divisor.

$= \frac{\cancel{r} \cdot r(-1)\cancel{(r-5s)}}{(r+5s)\cancel{(r-5s)}\cancel{r}}$ Factor.

$= \frac{-r}{r+5s}$ or $-\frac{r}{r+5s}$ Simplify.

In-Class Example Power Point®

6 Simplify each expression.

a. $\frac{k-3}{k+1} \div \frac{k^2-4k+3}{1-k^2}$. -1

b. $\frac{2d+6}{d^2+d-2} \div \frac{d+3}{d^2+3d+2}$ $\frac{2(d+1)}{d-1}$

SIMPLIFY COMPLEX FRACTIONS

In-Class Example Power Point®

7 Simplify $\frac{\frac{x^2}{9x^2-4y^2}}{\frac{x^3}{2y-3x}}$ $-\frac{1}{x(3x+2y)}$ or $\frac{-1}{x(3x+2y)}$

3 Practice/Apply

Study Notebook

Have students—

- add the definitions/examples of the vocabulary terms to their Vocabulary Builder worksheets for Chapter 9.
- add the Key Concepts in this lesson to their notebook, adding their own examples for each one.
- add the Test-Taking Tip to their list of test-taking tips for review as they prepare for standardized tests.
- include any other item(s) that they find helpful in mastering the skills in this lesson.

About the Exercises...

Organization by Objective

- **Simplify Rational Expressions:** 14–35
- **Simplify Complex Fractions:** 36–41

Odd/Even Assignments

Exercises 14–43 and 46–47 are structured so that students practice the same concepts whether they are assigned odd or even problems.

Assignment Guide

Basic: 15–37 odd, 43, 47–70

Average: 15–43 odd, 44, 45, 47–70

Advanced: 14–42 even, 44–46, 48–64 (optional: 65–70)

Answer

2. To multiply rational numbers or rational expressions, you multiply the numerators and multiply the denominators. To divide rational numbers or rational expressions, you multiply by the reciprocal of the divisor. In either case, you can reduce your answer by dividing the numerator and the denominator of the results by any common factors.

Check for Understanding

Concept Check

1. **OPEN ENDED** Write two rational expressions that are equivalent. **1. Sample answer:** $\frac{4}{6}, \frac{4(x+2)}{6(x+2)}$
2. **Explain** how multiplication and division of rational expressions are similar to multiplication and division of rational numbers. **See margin.**
3. **Determine** whether $\frac{2d+5}{3d+5} = \frac{2}{3}$ is *sometimes*, *always*, or *never* true. Explain. **Never; solving the equation using cross products leads to 15 = 10, which is never true.**

Guided Practice Simplify each expression.

GUIDED PRACTICE KEY	
Exercises	**Examples**
4–6	1, 3
7–10	4–6
11, 12	7
13	2

4. $\frac{45mn^3}{20n^7}$ $\frac{9m}{4n^4}$
5. $\frac{a+b}{a^2-b^2}$ $\frac{1}{a-b}$
6. $\frac{6y^3-9y^2}{2y^2+5y-12}$ $\frac{3y^2}{y+4}$
7. $\frac{2a^2}{5b^2c} \cdot \frac{3bc^2}{8a^2}$ $\frac{3c}{20b}$
8. $\frac{35}{16x^2} \div \frac{21}{4x}$ $\frac{5}{12x}$
9. $\frac{3t+6}{7t-7} \cdot \frac{14t-14}{5t+10}$ $\frac{6}{5}$
10. $\frac{12p^2+6p-6}{4(p+1)^2} \div \frac{6p-3}{2p+10}$ $\frac{p+5}{p+1}$
11. $\frac{\frac{c^3d^3}{a}}{\frac{xc^2d}{ax^2}}$ cd^2x
12. $\frac{\frac{2y}{y^2-4}}{\frac{3}{y^2-4y+4}}$ $\frac{2y(y-2)}{3(y+2)}$

Standardized Test Practice Ⓐ Ⓑ Ⓒ Ⓓ

13. Identify all of the values of y for which the expression $\frac{y-4}{y^2-4y-12}$ is undefined. **D**

Ⓐ −2, 4, 6 Ⓑ −6, −4, 2 Ⓒ −2, 0, 6 Ⓓ −2, 6

★ indicates increased difficulty

Practice and Apply

Homework Help	
For Exercises	**See Examples**
14–21	1, 3
22–35	4–6
36–41	7
42, 43, 50	2

Extra Practice See page 847.

Simplify each expression.

14. $\frac{30bc}{12b^2}$ $\frac{5c}{2b}$
15. $\frac{-3mn^4}{21m^2n^2}$ $-\frac{n^2}{7m}$
16. $\frac{(-3x^2y)^3}{9x^2y^2}$ $-3x^4y$
17. $\frac{(-2rs^2)^2}{12r^2s^3}$ $\frac{s}{3}$
18. $\frac{5t-5}{t^2-1}$ $\frac{5}{t+1}$
19. $\frac{c+5}{2c+10}$ $\frac{1}{2}$
20. $\frac{y^2+4y+4}{3y^2+5y-2}$ $\frac{y+2}{3y-1}$
21. $\frac{a^2+2a+1}{2a^2+3a+1}$ $\frac{a+1}{2a+1}$
22. $\frac{3xyz}{4xz} \cdot \frac{6x^2}{3y^2}$ $\frac{3x^2}{2y}$
23. $\frac{-4ab}{21c} \cdot \frac{14c^2}{18a^2}$ $-\frac{4bc}{27a}$
24. $\frac{3}{5d} \div \left(\frac{-9}{15df}\right)$ $-f$
25. $\frac{p^3}{2q} \div \frac{-p}{4q}$ $-2p^2$
26. $\frac{2x^3y}{z^5} \div \left(\frac{4xy}{z^3}\right)^2$ $\frac{xz}{8y}$
27. $\frac{xy}{a^3} \div \left(\frac{xy}{ab}\right)^3$ $\frac{b^3}{x^2y^2}$
28. $\frac{3t^2}{t+2} \cdot \frac{t+2}{t^2}$ 3
29. $\frac{4w+4}{3} \cdot \frac{1}{w+1}$ $\frac{4}{3}$
30. $\frac{4t^2-4}{9(t+1)^2} \cdot \frac{3t+3}{2t-2}$ $\frac{2}{3}$
31. $\frac{3p-21}{p^2-49} \cdot \frac{p^2+7p}{3p}$ 1
32. $\frac{5x^2+10x-75}{4x^2-24x-28} \cdot \frac{2x^2-10x-28}{x^2+7x+10}$ **32.** $\frac{5(x-3)}{2(x+1)}$
33. $\frac{w^2-11w+24}{w^2-18w+80} \cdot \frac{w^2-15w+50}{w^2-9w+20}$ $\frac{w-3}{w-4}$
34. $\frac{r^2+2r-8}{r^2+4r+3} \div \frac{r-2}{3r+3}$ $\frac{3(r+4)}{r+3}$
35. $\frac{a^2+2a-15}{a-3} \div \frac{a^2-4}{2}$ $\frac{2(a+5)}{(a-2)(a+2)}$

DAILY INTERVENTION — Differentiated Instruction

Intrapersonal Have students think about what aspects of multiplying and dividing rational expressions they find most challenging. Have them write a paragraph explaining why, and what steps they can take to help their challenges or confusions.

★ 36. $\dfrac{\frac{m^3}{3n}}{-\frac{m^4}{9n^2}}$ $-\dfrac{3n}{m}$

37. $\dfrac{\frac{p^3}{2q}}{\frac{-p^2}{4q}}$ $-2p$

38. $\dfrac{\frac{m+n}{5}}{\frac{m^2+n^2}{5}}$ $\dfrac{m+n}{m^2+n^2}$

★ 39. $\dfrac{\frac{x+y}{2x-y}}{\frac{x+y}{2x+y}}$ $\dfrac{2x+y}{2x-y}$

40. $\dfrac{\frac{6y^2-6}{8y^2+8y}}{\frac{3y-3}{4y^2+4y}}$ $y+1$

41. $\dfrac{\frac{5x^2-5x-30}{45-15x}}{\frac{6+x-x^2}{4x-12}}$ $\dfrac{4}{3}$

42. Under what conditions is $\dfrac{2d(d+1)}{(d+1)(d^2-4)}$ undefined? **$d = -2, -1,$ or 2**

43. Under what conditions is $\dfrac{a^2+ab+b^2}{a^2-b^2}$ undefined? **$a = -b$ or b**

More About. . .

Basketball

After graduating from the U.S. Naval Academy, David Robinson became the NBA Rookie of the Year in 1990. He has played basketball in 3 different Olympic Games.

Source: NBA

BASKETBALL **For Exercises 44 and 45, use the following information.**
At the end of the 2000–2001 season, David Robinson had made 6827 field goals out of 13,129 attempts during his NBA career.

44. Write a fraction to represent the ratio of the number of career field goals made to career field goals attempted by David Robinson at the end of the 2000–2001 season. $\dfrac{6827}{13{,}129}$

45. Suppose David Robinson attempted a field goals and made m field goals during the 2001–2002 season. Write a rational expression to represent the number of career field goals made to the number of career field goals attempted at the end of the 2001–2002 season. $\dfrac{6827+m}{13{,}129+a}$

Online Research **Data Update** What are the current scoring statistics of your favorite NBA player? Visit www.algebra2.com/data_update to learn more.

46. **GEOMETRY** A parallelogram with an area of $6x^2 - 7x - 5$ square units has a base of $3x - 5$ units. Determine the height of the parallelogram. **$2x + 1$ units**

47. **GEOMETRY** Parallelogram L has an area of $3x^2 + 10x + 3$ square meters and a height of $3x + 1$ meters. Parallelogram M has an area of $2x^2 - 13x + 20$ square meters and a height of $x - 4$ meters. Find the area of rectangle N.
$(2x^2 + x - 15)$ m^2

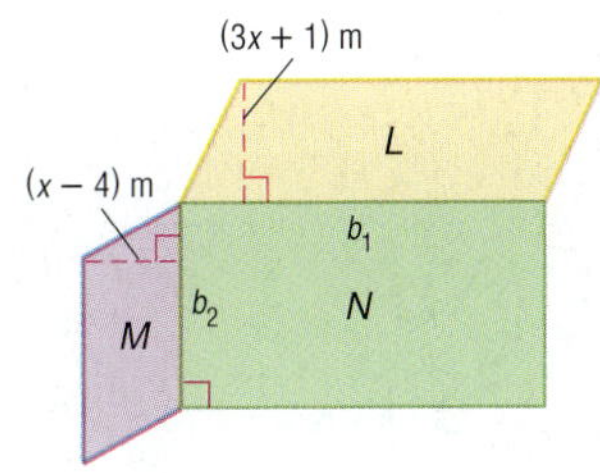

48. **CRITICAL THINKING** Simplify $\dfrac{(a^2-5a+6)^{-1}}{(a-2)^{-2}} \div \dfrac{(a-3)^{-1}}{(a-2)^{-2}}$. $\dfrac{1}{a-2}$

49. **WRITING IN MATH** Answer the question that was posed at the beginning of the lesson. **See pp. 519A–519D.**

How are rational expressions used in mixtures?

Include the following in your answer:

- an explanation of how to determine whether the rational expression representing the nut mixture is in simplest form, and
- an example of a mixture problem that could be represented by $\dfrac{8+x}{13+x+y}$.

www.algebra2.com/self_check_quiz

Study Guide and Intervention, p. 517 (shown) and p. 518

Simplify Rational Expressions A ratio of two polynomial expressions is a **rational expression**. To simplify a rational expression, divide both the numerator and the denominator by their greatest common factor (GCF).

Multiplying Rational Expressions	For all rational expressions $\frac{a}{b}$ and $\frac{c}{d}$, $\frac{a}{b} \cdot \frac{c}{d} = \frac{ac}{bd}$, if $b \neq 0$ and $d \neq 0$.
Dividing Rational Expressions	For all rational expressions $\frac{a}{b}$ and $\frac{c}{d}$, $\frac{a}{b} \div \frac{c}{d} = \frac{ad}{bc}$, if $b \neq 0$, $c \neq 0$, and $d \neq 0$.

Example **Simplify each expression.**

a. $\dfrac{24a^5b^2}{(2ab)^4}$

$\dfrac{24a^5b^2}{(2ab)^4} = \dfrac{2 \cdot 2 \cdot 2 \cdot 3 \cdot a \cdot a \cdot a \cdot a \cdot a \cdot b \cdot b}{2 \cdot 2 \cdot 2 \cdot 2 \cdot a \cdot a \cdot a \cdot a \cdot b \cdot b \cdot b \cdot b} = \dfrac{3a}{2b^2}$

b. $\dfrac{3r^2s^3}{5t^4} \cdot \dfrac{20t^2}{9r^3s}$

$\dfrac{3r^2s^3}{5t^4} \cdot \dfrac{20t^2}{9r^3s} = \dfrac{3 \cdot r \cdot r \cdot s \cdot s \cdot s \cdot 2 \cdot 2 \cdot 5 \cdot t \cdot t}{5 \cdot t \cdot t \cdot t \cdot t \cdot 3 \cdot 3 \cdot r \cdot r \cdot r \cdot s} = \dfrac{2 \cdot 2 \cdot s \cdot s}{3 \cdot r \cdot t \cdot t} = \dfrac{4s^2}{3rt^2}$

c. $\dfrac{x^2+8x+16}{2x-2} \div \dfrac{x^2+2x-8}{x-1}$

$\dfrac{x^2+8x+16}{2x-2} \div \dfrac{x^2+2x-8}{x-1} = \dfrac{x^2+8x+16}{2x-2} \cdot \dfrac{x-1}{x^2+2x-8}$

$= \dfrac{(x+4)(x+4)(x-1)}{2(x-1)(x-2)(x+4)} = \dfrac{x+4}{2(x-2)}$

Exercises

Simplify each expression.

1. $\dfrac{(-2ab^2)^3}{20ab^4}$ $-\dfrac{2a^2b^2}{5}$
2. $\dfrac{4x^2-12x+9}{9-6x}$ $\dfrac{3-2x}{3}$
3. $\dfrac{x^2+x-6}{x^2-6x-27}$ $\dfrac{x-2}{x-9}$
4. $\dfrac{3m^3-3m}{6m^4} \cdot \dfrac{4m^5}{m+1}$ $2m^2(m-1)$
5. $\dfrac{c^2-3c}{c^2-25} \cdot \dfrac{c^2+4c-5}{c^2-4c+3}$ $\dfrac{c}{c-5}$
6. $\dfrac{(m-3)^2}{m^2-6m+9} \cdot \dfrac{m^3-9m}{m^2-9}$ m
7. $\dfrac{6xy^4}{25z^3} \div \dfrac{18xz^2}{5y}$ $\dfrac{y^5}{15z^5}$
8. $\dfrac{16p^2-8p+1}{14p^4} \div \dfrac{4p^2+7p-2}{7p^5}$ $\dfrac{p(4p-1)}{2(p+2)}$
9. $\dfrac{2m-1}{m^2-3m-10} \div \dfrac{4m^2-1}{4m+8}$ $\dfrac{4}{(2m+1)(m-5)}$

Skills Practice, p. 519 and Practice, p. 520 (shown)

Simplify each expression.

1. $\dfrac{9a^2b^3}{27a^4b^4c}$ $\dfrac{1}{3a^2bc}$
2. $\dfrac{(2m^3n^2)^3}{-18m^5n^4}$ $-\dfrac{4m^4n^2}{9}$
3. $\dfrac{10y^2+15y}{35y^2-5y}$ $\dfrac{2y+3}{7y-1}$
4. $\dfrac{2k^2-k-15}{k^2-9}$ $\dfrac{2k+5}{k+3}$
5. $\dfrac{25-v^2}{3v^2-13v-10}$ $-\dfrac{v+5}{3v+2}$
6. $\dfrac{x^4+x^3-2x^2}{x^4-x^3}$ $\dfrac{x+2}{x}$
7. $\dfrac{-2u^3y}{15xz^5} \cdot \dfrac{25x^3}{14u^2y^2}$ $-\dfrac{5ux^2}{21yz^5}$
8. $\dfrac{a+y}{6} \cdot \dfrac{4}{y+a}$ $\dfrac{2}{3}$
9. $\dfrac{n^5}{n-6} \cdot \dfrac{n^2-6n}{n^8}$ $\dfrac{1}{n^2}$
10. $\dfrac{a-y}{w+n} \cdot \dfrac{w^2-n^2}{y-a}$ $n-w$
11. $\dfrac{x^2-5x-24}{6x+2x^2} \cdot \dfrac{5x^2}{8-x}$ $-\dfrac{5x}{2}$
12. $\dfrac{x-5}{10x-2} \cdot \dfrac{25x^2-1}{x^2-10x+25}$ $\dfrac{5x+1}{2(x-5)}$
13. $\dfrac{a^5y^3}{wy^7} \div \dfrac{a^3w^2}{w^5y^2}$ $\dfrac{a^2w^2}{y^2}$
14. $\left(\dfrac{2xy}{w^2}\right)^3 \div \dfrac{24x^2}{w^5}$ $\dfrac{xy^3}{3w}$
15. $\dfrac{x+y}{6} \div \dfrac{x^2-y^2}{3}$ $\dfrac{1}{2(x-y)}$
16. $\dfrac{3x+6}{x^2-9} \div \dfrac{6x^2+12x}{4x+12}$ $\dfrac{2}{x(x-3)}$
17. $\dfrac{2s^2-7s-15}{(s+4)^2} \div \dfrac{s^2-10s+25}{s+4}$ $\dfrac{2s+3}{(s+4)(s-5)}$
18. $\dfrac{9-a^2}{a^2+5a+6} \div \dfrac{2a-6}{5a+10}$ $-\dfrac{5}{2}$
19. $\dfrac{\frac{2x+1}{x}}{\frac{4-x}{x}}$ $\dfrac{2x+1}{4-x}$
20. $\dfrac{\frac{x^2-9}{4}}{\frac{3-x}{8}}$ $-2(x+3)$
21. $\dfrac{\frac{x^3+2^3}{x^2-2x}}{\frac{(x+2)^3}{x^2+4x+4}}$ $\dfrac{x^2-2x+4}{x(x-2)}$

22. **GEOMETRY** A right triangle with an area of $x^2 - 4$ square units has a leg that measures $2x + 4$ units. Determine the length of the other leg of the triangle. **$x - 2$ units**

23. **GEOMETRY** A rectangular pyramid has a base area of $\dfrac{x^2+3x-10}{2x}$ square centimeters and a height of $\dfrac{x^2-3x}{x^2-5x+6}$ centimeters. Write a rational expression to describe the volume of the rectangular pyramid. $\dfrac{x+5}{6}$ cm^3

Reading to Learn Mathematics, p. 521 ELL

Pre-Activity **How are rational expressions used in mixtures?**

Read the introduction to Lesson 9-1 at the top of page 472 in your textbook.

- Suppose that the Goodie Shoppe also sells a candy mixture of chocolate mints and caramels. If this mixture is made with 4 pounds of chocolate mints and 3 pounds of caramels, then $\frac{4}{7}$ of the mixture is mints and $\frac{3}{7}$ of the mixture is caramels.
- If the store manager adds another y pounds of mints to the mixture, what fraction of the mixture will be mints? $\dfrac{4+y}{7+y}$

Reading the Lesson

1. a. In order to simplify a rational number or rational expression, **factor** the numerator and **denominator** and divide both of them by their **greatest common factor**.
   b. A rational expression is undefined when its **denominator** is equal to **0**. To find the values that make the expression undefined, completely **factor** the original **denominator** and set each factor equal to **0**.
2. a. To multiply two rational expressions, **multiply** the **numerators** and multiply the denominators.
   b. To divide two rational expressions, **multiply** by the **reciprocal** of the **divisor**.
3. a. Which of the following expressions are complex fractions? **ii, iv, v**

   i. $\dfrac{7}{12}$ ii. $\dfrac{\frac{3}{8}}{\frac{5}{16}}$ iii. $\dfrac{r+5}{r-5}$ iv. $\dfrac{\frac{z+1}{z}}{z}$ v. $\dfrac{\frac{r^2-25}{9}}{\frac{r+5}{3}}$

   b. Does a complex fraction express a multiplication or division problem? **division** How is multiplication used in simplifying a complex fraction? **Sample answer: To divide the numerator of the complex fraction by the denominator, multiply the numerator by the reciprocal of the denominator.**

Helping You Remember

4. One way to remember something new is to see how it is similar to something you already know. How can your knowledge of division of fractions in arithmetic help you to understand how to divide rational expressions? **Sample answer: To divide rational expressions, multiply the first expression by the reciprocal of the second. This is the same "invert and multiply" process that is used when dividing arithmetic fractions.**

Enrichment, p. 522

Reading Algebra

In mathematics, the term *group* has a special meaning. The following numbered sentences discuss the idea of group and one interesting example of a group.

01 To be a group, a set of elements and a binary operation must satisfy four conditions: the set must be closed under the operation, the operation must be associative, there must be an identity element, and every element must have an inverse.

02 The following six functions form a group under the operation of composition of functions: $f_1(x) = x$, $f_2(x) = \frac{1}{x}$, $f_3(x) = 1 - x$, $f_4(x) = \frac{(x-1)}{x}$, $f_5(x) = \frac{x}{(x-1)}$, and $f_6(x) = \frac{1}{(1-x)}$.

03 This group is an example of a noncommutative group. For example, $f_3 \circ f_2 = f_4$, but $f_2 \circ f_3 = f_6$.

04 Some experimentation with this group will show that the identity element is f_1.

4 Assess

Open-Ended Assessment

Speaking Have students explain the procedures and cautions for multiplying and dividing rational expressions, demonstrating with examples.

Tips for New Teachers

Intervention Encourage students who are having difficulty with these problems to use several steps, writing each one below the previous one, and keeping each line equivalent to the one above. Caution them to make only one change per step.

Getting Ready for Lesson 9-2

PREREQUISITE SKILL Students will add and subtract rational expressions in Lesson 9-2. As with equations containing fractions, students will find common denominators, combine like terms, and simplify equations. Use Exercises 65–70 to determine your students' familiarity with solving equations containing fractions.

Answers

54.

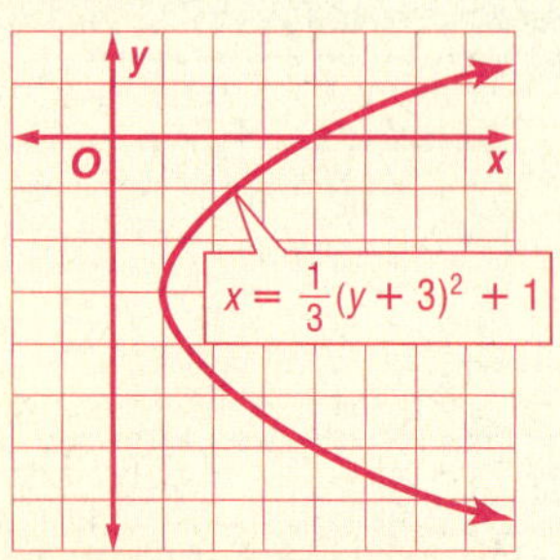

55.

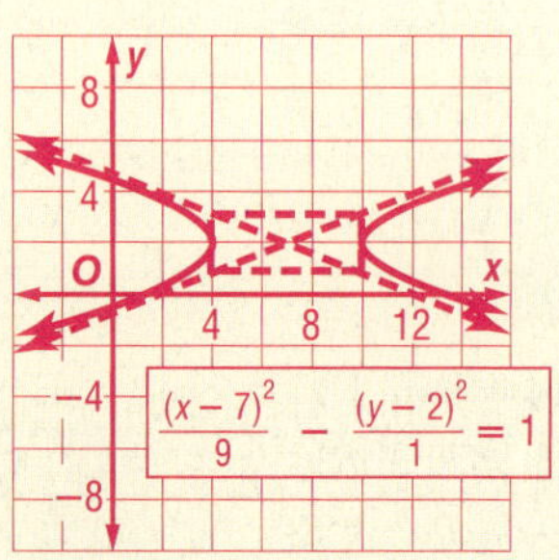

Standardized Test Practice

50. For what value(s) of x is the expression $\frac{4x}{x^2 - x}$ undefined? **C**

Ⓐ $-1, 1$ Ⓑ $-1, 0, 1$ Ⓒ $0, 1$ Ⓓ 0 Ⓔ $1, 2$

51. Compare the quantity in Column A and the quantity in Column B. Then determine whether: **A**

Ⓐ the quantity in Column A is greater,
Ⓑ the quantity in Column B is greater,
Ⓒ the two quantities are equal, or
Ⓓ the relationship cannot be determined from the information given.

Column A	Column B
$\frac{a^2 + 3a - 10}{a - 2}$	$\frac{a^2 + a - 6}{a + 3}$

Maintain Your Skills

Mixed Review

Find the exact solution(s) of each system of equations. *(Lesson 8–7)*

52. $x^2 + 2y^2 = 33$
$x^2 + y^2 - 19 = 2x$
$(-1, \pm 4), (5, \pm 2)$

53. $x^2 + 2y^2 = 33$
$x^2 - y^2 = 9$
$(\pm\sqrt{17}, \pm 2\sqrt{2})$

Write each equation in standard form. State whether the graph of the equation is a *parabola, circle, ellipse,* or *hyperbola*. Then graph the equation. *(Lesson 8–6)*

54. $y^2 - 3x + 6y + 12 = 0$

55. $x^2 - 14x + 4 = 9y^2 - 36y$

54–55. See margin for graphs.

54. $x = \frac{1}{3}(y + 3)^2 + 1$; parabola

55. $\frac{(x-7)^2}{9} - \frac{(y-2)^2}{1} = 1$; hyperbola

Determine whether each graph represents an odd-degree function or an even-degree function. Then state how many real zeros each function has. *(Lesson 7–1)*

56.

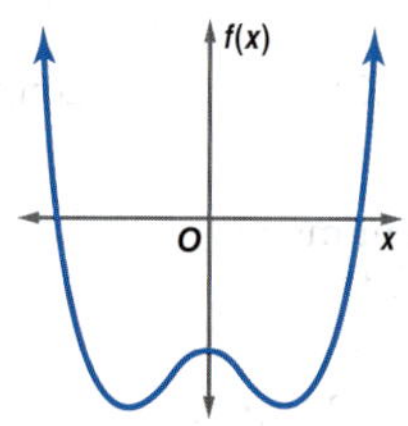

even; 2

57.

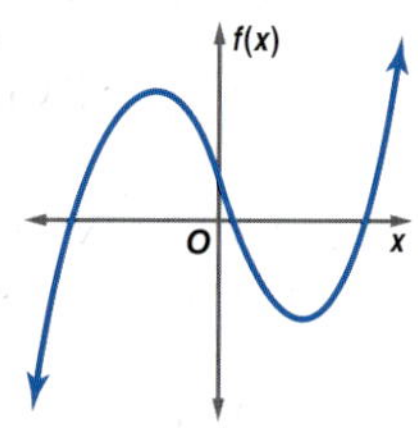

odd; 3

58.

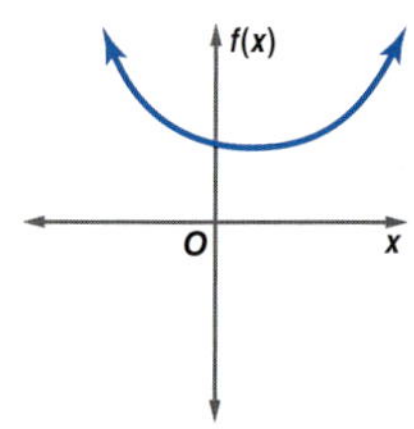

even; 0

Solve each equation by factoring. *(Lesson 6–3)*

59. $r^2 - 3r = 4$ **$\{-1, 4\}$** **60.** $18u^2 - 3u = 1$ **$\left\{-\frac{1}{6}, \frac{1}{3}\right\}$** **61.** $d^2 - 5d = 0$ **$\{0, 5\}$**

62. ASTRONOMY Earth is an average 1.496×10^8 kilometers from the Sun. If light travels 3×10^5 kilometers per second, how long does it take sunlight to reach Earth? *(Lesson 5–1)* **4.99×10^2 s or about 8 min 19 s**

Solve each equation. *(Lesson 1–4)*

63. $|2x + 7| + 5 = 0$ **$\varnothing$**

64. $5|3x - 4| = x + 1$ **$\left\{\frac{3}{2}, \frac{19}{16}\right\}$**

Getting Ready for the Next Lesson

PREREQUISITE SKILL Solve each equation. *(To review **solving equations,** see Lesson 1–3.)*

65. $\frac{2}{3} + x = -\frac{4}{9}$ **$-1\frac{1}{9}$** **66.** $x + \frac{5}{8} = -\frac{5}{6}$ **$-1\frac{11}{24}$** **67.** $x - \frac{3}{5} = \frac{2}{3}$ **$1\frac{4}{15}$**

68. $x + \frac{3}{16} = -\frac{1}{2}$ **$-\frac{11}{16}$** **69.** $x - \frac{1}{6} = -\frac{7}{9}$ **$-\frac{11}{18}$** **70.** $x - \frac{3}{8} = -\frac{5}{24}$ **$\frac{1}{6}$**

478 Chapter 9 Rational Expressions and Equations

9-2 Adding and Subtracting Rational Expressions

What You'll Learn

- Determine the LCM of polynomials.
- Add and subtract rational expressions.

How is subtraction of rational expressions used in photography?

To take sharp, clear pictures, a photographer must focus the camera precisely. The distance from the object to the lens p and the distance from the lens to the film q must be accurately calculated to ensure a sharp image. The focal length of the lens is f.

The formula $\frac{1}{q} = \frac{1}{f} - \frac{1}{p}$ can be used to determine how far the film should be placed from the lens to create a perfect photograph.

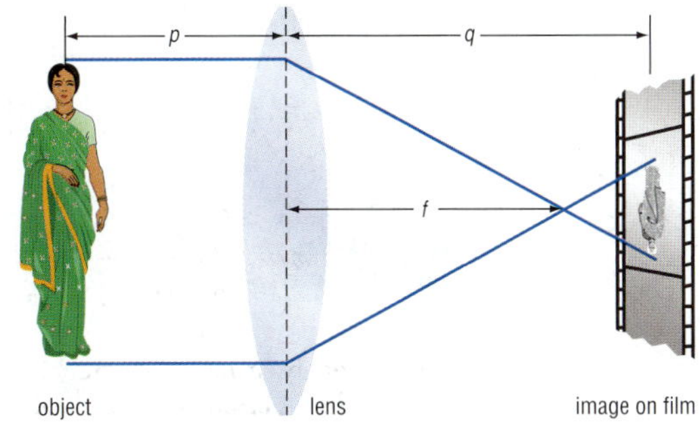

LCM OF POLYNOMIALS To find $\frac{5}{6} - \frac{1}{4}$ or $\frac{1}{f} - \frac{1}{p}$, you must first find the least common denominator (LCD). The LCD is the least common multiple (LCM) of the denominators.

To find the LCM of two or more numbers or polynomials, factor each number or polynomial. The LCM contains *each* factor the greatest number of times it appears as a factor.

LCM of 6 and 4	**LCM of $a^2 - 6a + 9$ and $a^2 + a - 12$**
$6 = 2 \cdot 3$	$a^2 - 6a + 9 = (a - 3)^2$
$4 = 2^2$	$a^2 + a - 12 = (a - 3)(a + 4)$
LCM $= 2^2 \cdot 3$ or 12	LCM $= (a - 3)^2(a + 4)$

Example 1 LCM of Monomials

Find the LCM of $18r^2s^5$, $24r^3st^2$, and $15s^3t$.

$18r^2s^5 = 2 \cdot 3^2 \cdot r^2 \cdot s^5$ — Factor the first monomial.

$24r^3st^2 = 2^3 \cdot 3 \cdot r^3 \cdot s \cdot t^2$ — Factor the second monomial.

$15s^3t = 3 \cdot 5 \cdot s^3 \cdot t$ — Factor the third monomial.

LCM $= 2^3 \cdot 3^2 \cdot 5 \cdot r^3 \cdot s^5 \cdot t^2$
$= 360r^3s^5t^2$ — Use each factor the greatest number of times it appears as a factor and simplify.

9-2 Lesson Notes

1 Focus

5-Minute Check Transparency 9-2 Use as a quiz or review of Lesson 9-1.

Mathematical Background notes are available for this lesson on p. 470C.

How is subtraction of rational expressions used in photography?

Ask students:

- If you drew a box to represent the camera, which of the variables shown would describe dimensions within the camera? ***f* and *q***
- In a camera, what is a typical value for q? **Sample answer: 0.5 in.**

Resource Manager

Workbook and Reproducible Masters

Chapter 9 Resource Masters

- Study Guide and Intervention, pp. 523–524
- Skills Practice, p. 525
- Practice, p. 526
- Reading to Learn Mathematics, p. 527
- Enrichment, p. 528
- Assessment, p. 567

Transparencies

5-Minute Check Transparency 9-2
Answer Key Transparencies

Technology

Alge2PASS: Tutorial Plus, Lesson 18
Interactive Chalkboard

2 Teach

LCM OF POLYNOMIALS

In-Class Examples Power Point®

1 Find the LCM of $15a^2bc^3$, $16b^5c^2$, and $20a^3c^6$. $240a^3b^5c^6$

2 Find the LCM of $x^3 - x^2 - 2x$ and $x^2 - 4x + 4$. $x(x + 1)(x - 2)^2$

ADD AND SUBTRACT RATIONAL EXPRESSIONS

In-Class Examples Power Point®

3 Simplify $\frac{5a^2}{6b} + \frac{9}{14a^2b^2}$. $\frac{35a^4b + 27}{42a^2b^2}$

Teaching Tip Have students discuss the differences between procedures for adding and multiplying fractions.

4 Simplify $\frac{x + 10}{3x - 15} - \frac{3x + 15}{6x - 30}$. $-\frac{1}{6}$

Example 2 *LCM of Polynomials*

Find the LCM of $p^3 + 5p^2 + 6p$ and $p^2 + 6p + 9$.

$p^3 + 5p^2 + 6p = p(p + 2)(p + 3)$ Factor the first polynomial.

$p^2 + 6p + 9 = (p + 3)^2$ Factor the second polynomial.

$\text{LCM} = p(p + 2)(p + 3)^2$ Use each factor the greatest number of times it appears as a factor.

ADD AND SUBTRACT RATIONAL EXPRESSIONS As with fractions, to add or subtract rational expressions, you must have common denominators.

Specific Case		General Case
$\frac{2}{3} + \frac{3}{5} = \frac{2 \cdot 5}{3 \cdot 5} + \frac{3 \cdot 3}{5 \cdot 3}$	Find equivalent fractions that have a common denominator.	$\frac{a}{c} + \frac{b}{d} = \frac{a \cdot d}{c \cdot d} + \frac{b \cdot c}{d \cdot c}$
$= \frac{10}{15} + \frac{9}{15}$	Simplify each numerator and denominator.	$= \frac{ad}{cd} + \frac{bc}{cd}$
$= \frac{19}{15}$	Add the numerators.	$= \frac{ad + bc}{cd}$

Example 3 *Monomial Denominators*

Simplify $\frac{7x}{15y^2} + \frac{y}{18xy}$.

$\frac{7x}{15y^2} + \frac{y}{18xy} = \frac{7x \cdot 6x}{15y^2 \cdot 6x} + \frac{y \cdot 5y}{18xy \cdot 5y}$ The LCD is $90xy^2$. Find equivalent fractions that have this denominator.

$= \frac{42x^2}{90xy^2} + \frac{5y^2}{90xy^2}$ Simplify each numerator and denominator.

$= \frac{42x^2 + 5y^2}{90xy^2}$ Add the numerators.

Example 4 *Polynomial Denominators*

Simplify $\frac{w + 12}{4w - 16} - \frac{w + 4}{2w - 8}$.

$\frac{w + 12}{4w - 16} - \frac{w + 4}{2w - 8} = \frac{w + 12}{4(w - 4)} - \frac{w + 4}{2(w - 4)}$ Factor the denominators.

$= \frac{w + 12}{4(w - 4)} - \frac{(w + 4)(2)}{2(w - 4)(2)}$ The LCD is $4(w - 4)$.

$= \frac{(w + 12) - (2)(w + 4)}{4(w - 4)}$ Subtract the numerators.

$= \frac{w + 12 - 2w - 8}{4(w - 4)}$ Distributive Property

$= \frac{-w + 4}{4(w - 4)}$ Combine like terms.

$= \frac{-1\cancel{(w - 4)}^{1}}{4\cancel{(w - 4)}_{1}}$ or $-\frac{1}{4}$ Simplify.

Study Tip

Common Factors

Sometimes when you simplify the numerator, the polynomial contains a factor common to the denominator. Thus, the rational expression can be further simplified.

Sometimes simplifying complex fractions involves adding or subtracting rational expressions. One way to simplify a complex fraction is to simplify the numerator and the denominator separately, and then simplify the resulting expressions.

Example 5 Simplify Complex Fractions

Simplify $\dfrac{\frac{1}{x} - \frac{1}{y}}{1 + \frac{1}{x}}$.

TEACHING TIP Point out that $\dfrac{\frac{1}{x} + \frac{1}{y}}{1 + \frac{1}{x}}$ can also be simplified by multiplying the numerator and denominator by *xy*.

$$\frac{\frac{1}{x} - \frac{1}{y}}{1 + \frac{1}{x}} = \frac{\frac{y}{xy} - \frac{x}{xy}}{\frac{x}{x} + \frac{1}{x}}$$ The LCD of the numerator is *xy*. The LCD of the denominator is *x*.

$$= \frac{\frac{y - x}{xy}}{\frac{x + 1}{x}}$$ Simplify the numerator and denominator.

$$= \frac{y - x}{xy} \div \frac{x + 1}{x}$$ Write as a division expression.

$$= \frac{y - x}{\cancel{x}y} \cdot \frac{\cancel{x}}{x + 1}$$ Multiply by the reciprocal of the divisor.

$$= \frac{y - x}{y(x + 1)} \text{ or } \frac{y - x}{xy + y}$$ Simplify.

Example 6 Use a Complex Fraction to Solve a Problem

COORDINATE GEOMETRY Find the slope of the line that passes through $A\left(\frac{2}{p}, \frac{1}{2}\right)$ and $B\left(\frac{1}{3}, \frac{3}{p}\right)$.

$$m = \frac{y_2 - y_1}{x_2 - x_1}$$ Definition of slope

$$= \frac{\frac{3}{p} - \frac{1}{2}}{\frac{1}{3} - \frac{2}{p}}$$ $y_2 = \frac{3}{p}, y_1 = \frac{1}{2}, x_2 = \frac{1}{3}$, and $x_1 = \frac{2}{p}$

$$= \frac{\frac{6 - p}{2p}}{\frac{p - 6}{3p}}$$ The LCD of the numerator is 2*p*. The LCD of the denominator is 3*p*.

$$= \frac{6 - p}{2p} \div \frac{p - 6}{3p}$$ Write as a division expression.

$$= \frac{\overset{-1}{\cancel{6 - p}}}{\underset{1}{2\cancel{p}}} \cdot \frac{\overset{1}{3\cancel{p}}}{\underset{1}{\cancel{p - 6}}} \text{ or } -\frac{3}{2}$$ The slope is $-\frac{3}{2}$.

Study Tip

Check Your Solution

You can check your answer be letting *p* equal any nonzero number, say 1. Use the definition of slope to find the slope of the line through the points.

Check for Understanding

Concept Check

1. FIND THE ERROR Catalina and Yong-Chan are simplifying $\frac{x}{a} - \frac{x}{b}$.

Catalina

$\frac{x}{a} - \frac{x}{b} = \frac{bx}{ab} - \frac{ax}{ab}$

$= \frac{bx - ax}{ab}$

Yong-Chan

$\frac{x}{a} - \frac{x}{b} = \frac{x}{a - b}$

Who is correct? Explain your reasoning.

1. Catalina; you need a common denominator, not a common numerator, to subtract two rational expressions.

www.algebra2.com/extra_examples

In-Class Examples

Power Point®

5 Simplify $\dfrac{\frac{1}{a} + \frac{1}{b}}{\frac{1}{b} - 1}$. $\frac{a + b}{a(1 - b)}$ or $\frac{a + b}{a - ab}$

6 COORDINATE GEOMETRY Find the slope of the line that passes through $P\left(\frac{3}{k}, \frac{1}{3}\right)$ and $Q\left(\frac{1}{2}, \frac{2}{k}\right)$. $-\frac{2}{3}$

Teaching Tip Remind students that the slope of a line is the change in *y* divided by the change in *x*, or the rise over the run.

3 Practice/Apply

Study Notebook

Have students—

- add the definitions/examples of the vocabulary terms to their Vocabulary Builder worksheets for Chapter 9.
- include any item(s) that they find helpful in mastering the skills in this lesson.

DAILY INTERVENTION

FIND THE ERROR One way to find the error is to substitute values for the variables. With $x = 4$, $a = 5$, and $b = 3$, $\frac{x}{a} - \frac{x}{b}$ becomes $\frac{4}{5} - \frac{4}{3}$. Since the first fraction is less than 1 and the second is greater than 1, the result must be negative, which means that the answer $\frac{x}{a - b} = \frac{4}{2}$ or 2 cannot be correct.

DAILY INTERVENTION

Differentiated Instruction

Interpersonal Have students work with a partner, one in the role of coach and the other in the role of athlete. The athlete works the problem, using steps and explaining the thinking, while the coach listens and watches for errors, correcting as necessary. Then the partners exchange roles.

About the Exercises...

Organization by Objective

- **LCM of Polynomials:** 14–21
- **Add and Subtract Rational Expressions:** 22–49

Odd/Even Assignments

Exercises 14–43 are structured so that students practice the same concepts whether they are assigned odd or even problems.

Assignment Guide

Basic: 15–37 odd, 50–61

Average: 15–43 odd, 49–61

Advanced: 14–42 even, 44–48, 50–58 (optional: 59–61)

All: Practice Quiz 1 (1–10)

Answers

3a. Since a, b, and c are factors of abc, abc is always a common denominator of $\frac{1}{a} + \frac{1}{b} + \frac{1}{c}$.

3b. If a, b, and c have no common factors, then abc is the LCD of $\frac{1}{a} + \frac{1}{b} + \frac{1}{c}$.

3c. If a and b have no common factors and c is a factor of ab, then ab is the LCD of $\frac{1}{a} + \frac{1}{b} + \frac{1}{c}$.

3d. If a and c are factors of b, then b is the LCD of $\frac{1}{a} + \frac{1}{b} + \frac{1}{c}$.

3e. Since $\frac{1}{a} + \frac{1}{b} + \frac{1}{c} = \frac{bc}{abc} + \frac{ac}{abc} + \frac{ab}{abc}$, the sum is always $\frac{bc + ac + ab}{abc}$.

---

2. Sample answer: $d^2 - d$, $d + 1$

2. **OPEN ENDED** Write two polynomials that have a LCM of $d^3 - d$.

★ 3. Consider $\frac{1}{a} + \frac{1}{b} + \frac{1}{c}$ if a, b, and c are real numbers. Determine whether each statement is *sometimes*, *always*, or *never* true. Explain your answer.
   a. abc is a common denominator. **always** a–e. See margin for explanations.
   b. abc is the LCD. **sometimes**
   c. ab is the LCD. **sometimes**
   d. b is the LCD. **sometimes**
   e. The sum is $\frac{bc + ac + ab}{abc}$. **always**

Guided Practice

GUIDED PRACTICE KEY	
Exercises	**Examples**
4–6	1, 2
7–11	3, 4
12	5
13	6

Find the LCM of each set of polynomials.

4. $12y^2$, $6x^2$ $\mathbf{12x^2y^2}$
5. $16ab^3$, $5b^2a^2$, $20ac$ $\mathbf{80a^2b^3c}$
6. $x^2 - 2x$, $x^2 - 4$ $\mathbf{x(x - 2)(x + 2)}$

Simplify each expression.

7. $\frac{2}{x^2y} - \frac{x}{y}$ $\frac{2 - x^3}{x^2y}$
8. $\frac{7a}{15b^2} + \frac{b}{18ab}$ $\frac{42a^2 + 5b^2}{90ab^2}$
9. $\frac{5}{3m} - \frac{2}{7m} - \frac{1}{2m}$ $\frac{37}{42m}$
10. $\frac{6}{d^2 + 4d + 4} + \frac{5}{d + 2}$ $\frac{5d + 16}{(d + 2)^2}$
11. $\frac{a}{a^2 - a - 20} + \frac{2}{a + 4}$ $\frac{3a - 10}{(a - 5)(a + 4)}$
12. $\frac{x + \frac{x}{3}}{x - \frac{x}{6}}$ $\frac{8}{5}$

Application

13. **GEOMETRY** Find the perimeter of the quadrilateral. Express in simplest form. $\frac{13x^2 + 4x - 9}{2x(x - 1)(x + 1)}$ units

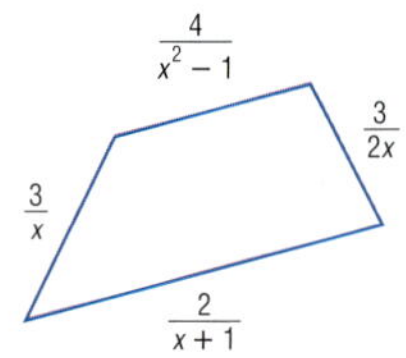

★ indicates increased difficulty

Practice and Apply

Homework Help	
For Exercises	**See Examples**
14–21	1, 2
22–39	3, 4
40–43	5
44–49	6

Extra Practice See page 847.

Find the LCM of each set of polynomials.

14. $10s^2$, $35s^2t^2$ $\mathbf{70s^2t^2}$
15. $36x^2y$, $20xyz$ $\mathbf{180x^2yz}$
16. $14a^3$, $15bc^3$, $12b^3$ $\mathbf{420a^3b^3c^3}$
17. $9p^2q^3$, $6pq^4$, $4p^3$ $\mathbf{36p^3q^4}$
18. $4w - 12$, $2w - 6$ $\mathbf{4(w - 3)}$
19. $x^2 - y^2$, $x^3 + x^2y$ $\mathbf{x^2(x - y)(x + y)}$
20. $2t^2 + t - 3$, $2t^2 + 5t + 3$ $\mathbf{(2t + 3)(t - 1)(t + 1)}$
21. $n^2 - 7n + 12$, $n^2 - 2n - 8$ $\mathbf{(n - 4)(n - 3)(n + 2)}$

Simplify each expression.

22. $\frac{6}{ab} + \frac{8}{a}$ $\frac{6 + 8b}{ab}$
23. $\frac{5}{6v} + \frac{7}{4v}$ $\frac{31}{12v}$
24. $\frac{5}{r} + 7$ $\frac{5 + 7r}{r}$
25. $\frac{2x}{3y} + 5$ $\frac{2x + 15y}{3y}$
26. $\frac{3x}{4y^2} - \frac{y}{6x}$ $\frac{9x^2 - 2y^3}{12xy^2}$
27. $\frac{5}{a^2b} - \frac{7a}{5b^2}$ $\frac{25b - 7a^3}{5a^2b^2}$
28. $\frac{3}{4q} - \frac{2}{5q} - \frac{1}{2q}$ $-\frac{3}{20q}$
29. $\frac{11}{9} - \frac{7}{2w} - \frac{6}{5w}$ $\frac{110w - 423}{90w}$
30. $\frac{7}{y - 8} - \frac{6}{8 - y}$ $\frac{13}{y - 8}$
31. $\frac{a}{a - 4} - \frac{3}{4 - a}$ $\frac{a + 3}{a - 4}$
32. $\frac{m}{m^2 - 4} + \frac{2}{3m + 6}$ $\frac{5m - 4}{3(m + 2)(m - 2)}$
33. $\frac{y}{y + 3} - \frac{6y}{y^2 - 9}$ $\frac{y(y - 9)}{(y + 3)(y - 3)}$

482 Chapter 9 Rational Expressions and Equations

35. $\frac{-8d + 20}{(d - 4)(d + 4)(d - 2)}$

36. $\frac{-4h + 15}{(h - 4)(h - 5)^2}$

37. $\frac{x^2 - 6}{(x + 2)^2(x + 3)}$

39. $\frac{2y^2 + y - 4}{(y - 1)(y - 2)}$

34. $\frac{5}{x^2 - 3x - 28} + \frac{7}{2x - 14}$ $\frac{7x + 38}{2(x - 7)(x + 4)}$

35. $\frac{d - 4}{d^2 + 2d - 8} - \frac{d + 2}{d^2 - 16}$

36. $\frac{1}{h^2 - 9h + 20} - \frac{5}{h^2 - 10h + 25}$

37. $\frac{x}{x^2 + 5x + 6} - \frac{2}{x^2 + 4x + 4}$

★ 38. $\frac{m^2 + n^2}{m^2 - n^2} + \frac{m}{n - m} + \frac{n}{m + n}$ 0

39. $\frac{y + 1}{y - 1} + \frac{y + 2}{y - 2} + \frac{y}{y^2 - 3y + 2}$

★ 40. $\frac{\frac{1}{b + 2} + \frac{1}{b - 5}}{\frac{2b^2 - b - 3}{b^2 - 3b - 10}}$ $\frac{1}{b + 1}$

41. $\frac{(x + y)\left(\frac{1}{x} - \frac{1}{y}\right)}{(x - y)\left(\frac{1}{x} + \frac{1}{y}\right)}$ -1

★ 42. Write $\left(\frac{2s}{2s + 1} - 1\right) \div \left(1 + \frac{2s}{1 - 2s}\right)$ in simplest form. $\frac{2s - 1}{2s + 1}$

★ 43. What is the simplest form of $\left(3 + \frac{5}{a + 2}\right) \div \left(3 - \frac{10}{a + 7}\right)$? $\frac{a + 7}{a + 2}$

ELECTRICITY **For Exercises 44 and 45, use the following information.**
In an electrical circuit, if two resistors with resistance R_1 and R_2 are connected in parallel as shown, the relationship between these resistances and the resulting combination resistance R is $\frac{1}{R} = \frac{1}{R_1} + \frac{1}{R_2}$.

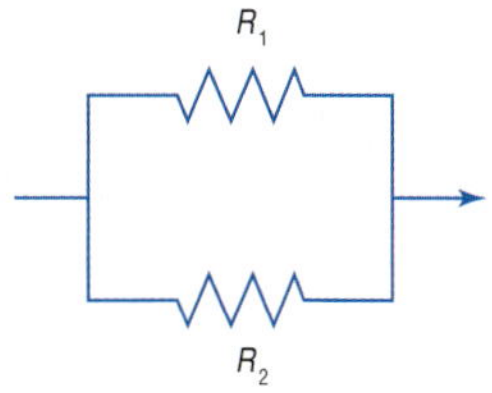

44. If R_1 is x ohms and R_2 is 4 ohms less than twice x ohms, write an expression for $\frac{1}{R}$. $\frac{3x - 4}{2x(x - 2)}$

45. Find the effective resistance of a 30-ohm resistor and a 20-ohm resistor that are connected in parallel. 12 ohms

More About. . .

Bicycling

The Tour de France is the most popular bicycle road race. It lasts 24 days and covers 2500 miles.

Source: *World Book Encyclopedia*

BICYCLING **For Exercises 46–48, use the following information.**
Jalisa is competing in a 48-mile bicycle race. She travels half the distance at one rate. The rest of the distance, she travels 4 miles per hour slower.

46. If x represents the faster pace in miles per hour, write an expression that represents the time spent at that pace. $\frac{24}{x}$ h

47. Write an expression for the amount of time spent at the slower pace. $\frac{24}{x - 4}$ h

48. Write an expression for the amount of time Jalisa needed to complete the race. $\frac{48(x - 2)}{x(x - 4)}$ h

49. **MAGNETS** For a bar magnet, the magnetic field strength H at a point P along the axis of the magnet is $H = \frac{m}{2L(d - L)^2} - \frac{m}{2L(d + L)^2}$. Write a simpler expression for H. $\frac{2md}{(d - L)^2(d + L)^2}$ or $\frac{2md}{(d^2 - L^2)^2}$

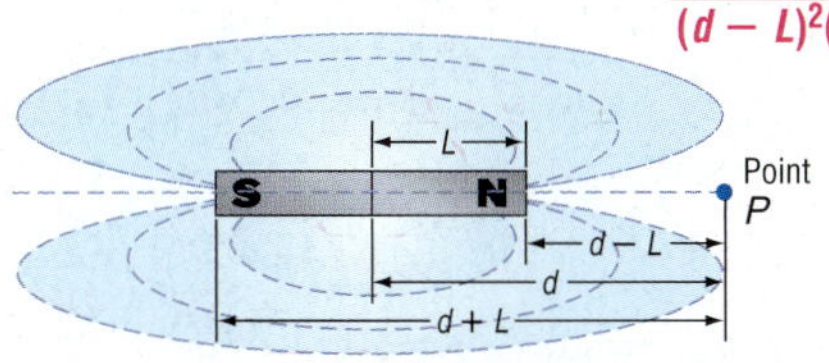

50. **CRITICAL THINKING** Find two rational expressions whose sum is $\frac{2x - 1}{(x + 1)(x - 2)}$. Sample answer: $\frac{1}{x + 1}$, $\frac{1}{x - 2}$

www.algebra2.com/self_check_quiz

Study Guide and Intervention, p. 523 (shown) and p. 524

LCM of Polynomials To find the least common multiple of two or more polynomials, factor each expression. The LCM contains each factor the greatest number of times it appears as a factor.

Example 1 **Find the LCM of $16p^2q^3r$, $40pq^4r^2$, and $15p^3r^4$.**

$16p^2q^3r = 2^4 \cdot p^2 \cdot q^3 \cdot r$
$40pq^4r^2 = 2^3 \cdot 5 \cdot p \cdot q^4 \cdot r^2$
$15p^3r^4 = 3 \cdot 5 \cdot p^3 \cdot r^4$
$LCM = 2^4 \cdot 3 \cdot 5 \cdot p^3 \cdot q^4 \cdot r^4$
$= 240p^3q^4r^4$

Example 2 **Find the LCM of $3m^2 - 3m - 6$ and $4m^2 + 12m - 40$.**

$3m^2 - 3m - 6 = 3(m + 1)(m - 2)$
$4m^2 + 12m - 40 = 4(m - 2)(m + 5)$
$LCM = 12(m + 1)(m - 2)(m + 5)$

Exercises

Find the LCM of each set of polynomials.

1. $14ab^2, 42bc^3, 18a^2c$ $126a^2b^2c^3$
2. $8cdf^3, 28c^2f, 35d^4f^2$ $280c^2d^4f^3$
3. $65x^4y, 10x^2y^2, 26y^4$ $130x^4y^4$
4. $11mn^5, 18m^2n^3, 20mn^4$ $1980m^2n^5$
5. $15a^4b, 50a^2b^2, 40b^8$ $600a^4b^8$
6. $24p^7q, 30p^2q^2, 45pq^3$ $360p^7q^3$
7. $39b^2c^2, 52b^4c, 12c^3$ $156b^4c^3$
8. $12xy^4, 42x^2y, 30x^2y^3$ $420x^2y^4$
9. $56stv^2, 24s^2v^2, 70t^3v^3$ $840s^2t^3v^3$
10. $x^2 + 3x, 10x^2 + 25x - 15$ $5x(x + 3)(2x - 1)$
11. $9x^2 - 12x + 4, 3x^2 + 10x - 8$ $(3x - 2)^2(x + 4)$
12. $22x^2 + 66x - 220, 4x^2 - 16$ $44(x - 2)(x + 2)(x + 5)$
13. $8x^2 - 36x - 20, 2x^2 + 2x - 60$ $4(x - 5)(x + 6)(2x + 1)$
14. $5x^2 - 125, 5x^2 + 24x - 5$ $5(x - 5)(x + 5)(5x - 1)$
15. $3x^2 - 18x + 27, 2x^3 - 4x^2 - 6x$ $6x(x - 3)^2(x + 1)$
16. $45x^2 - 6x - 3, 45x^2 - 5$ $15(5x + 1)(3x - 1)(3x + 1)$
17. $x^3 + 4x^2 - x - 4, x^2 + 2x - 3$ $(x - 1)(x + 1)(x + 3)(x + 4)$
18. $54x^3 - 24x, 12x^2 - 26x + 12$ $6x(3x + 2)(3x - 2)(2x - 3)$

Skills Practice, p. 525 and Practice, p. 526 (shown)

Find the LCM of each set of polynomials.

1. x^2y, xy^3 x^2y^3
2. a^2b^3c, abc^4 $a^2b^3c^4$
3. $x + 1, x + 3$ $(x + 1)(x + 3)$
4. $g - 1, g^2 + 3g - 4$ $(g - 1)(g + 4)$
5. $2r + 2, r^2 + r, r + 1$ $2r(r + 1)$
6. $3, 4w + 2, 4w^2 - 1$ $6(2w + 1)(2w - 1)$
7. $x^2 + 2x - 8, x + 4$ $(x + 4)(x - 2)$
8. $x^2 - x - 6, x^2 + 6x + 8$ $(x + 2)(x + 4)(x - 3)$
9. $d^2 + 6d + 9, 2(d^2 - 9)$ $2(d - 3)(d + 3)^2$

Simplify each expression.

10. $\frac{5}{6ab} - \frac{7}{8a}$ $\frac{20 - 21b}{24ab}$
11. $\frac{5}{12x^4y} - \frac{1}{5x^2y^3}$ $\frac{25y^2 - 12x^2}{60x^4y^3}$
12. $\frac{1}{6c^2d} + \frac{3}{4cd^3}$ $\frac{2d^2 + 9c}{12c^2d^3}$
13. $\frac{4m}{3mn} + 2$ $\frac{2(2 + 3n)}{3n}$
14. $2x - 5 - \frac{x - 8}{x + 4}$ $\frac{2(x + 3)(x - 2)}{x + 4}$
15. $\frac{4}{a - 3} + \frac{9}{a - 5}$ $\frac{13a - 47}{(a - 3)(a - 5)}$
16. $\frac{16}{x^2 - 16} + \frac{2}{x + 4}$ $\frac{2}{x - 4}$
17. $\frac{2 - 5m}{m - 9} + \frac{4m - 5}{9 - m}$ $\frac{7 - 9m}{m - 9}$
18. $\frac{y - 5}{y^2 - 3y - 10} + \frac{y}{y^2 + y - 2}$ $\frac{2y - 1}{(y + 2)(y - 1)}$
19. $\frac{5}{2x - 12} - \frac{20}{x^2 - 4x - 12}$ $\frac{5}{2(x + 2)}$
20. $\frac{2p - 3}{p^2 - 5p + 6} - \frac{5}{p^2 - 9}$ $\frac{2p^2 - 2p + 1}{(p - 2)(p + 3)(p - 3)}$
21. $\frac{1}{5n} - \frac{3}{4} + \frac{7}{10n}$ $\frac{3(6 - 5n)}{20n}$
22. $\frac{2a}{a - 3} - \frac{2a}{a + 3} + \frac{36}{a^2 - 9}$ $\frac{12}{a - 3}$
23. $\frac{\frac{2}{x - y} + \frac{1}{x + y}}{\frac{1}{x - y}}$ $\frac{3x + y}{x + y}$
24. $\frac{\frac{r + 6}{r} - \frac{1}{r + 2}}{\frac{r^2 + 4r + 3}{r^2 + 2r}}$ $\frac{r + 4}{r + 1}$

25. **GEOMETRY** The expressions $\frac{5x}{2}$, $\frac{20}{x + 4}$, and $\frac{10}{x - 4}$ represent the lengths of the sides of a triangle. Write a simplified expression for the perimeter of the triangle. $\frac{5(x^3 - 4x - 16)}{2(x - 4)(x + 4)}$

26. **KAYAKING** Mai is kayaking on a river that has a current of 2 miles per hour. If r represents her rate in calm water, then $r + 2$ represents her rate with the current, and $r - 2$ represents her rate against the current. Mai kayaks 2 miles downstream and then back to her starting point. Use the formula for time, $t = \frac{d}{r}$, where d is the distance, to write a simplified expression for the total time it takes Mai to complete the trip. $\frac{4r}{(r + 2)(r - 2)}$ h

Reading to Learn Mathematics, p. 527 ELL

Pre-Activity **How is subtraction of rational expressions used in photography?**

Read the introduction to Lesson 9-2 at the top of page 479 in your textbook.

A person is standing 5 feet from a camera that has a lens with a focal length of 3 feet. Write an equation that you could solve to find how far the film should be from the lens to get a perfectly focused photograph.
$\frac{1}{q} = \frac{1}{3} - \frac{1}{5}$

Reading the Lesson

1. a. In work with rational expressions, LCD stands for least common denominator and LCM stands for least common multiple. The LCD is the LCM of the denominators.
b. To find the LCM of two or more numbers or polynomials, factor each number or polynomial. The LCM contains each factor the greatest number of times it appears as a factor.
2. To add $\frac{x^2 - 3}{x^2 - 5x + 6}$ and $\frac{x - 4}{x^3 - 4x^2 + 4x}$, you should first factor the denominator of each fraction. Then use the factorizations to find the LCM of $x^2 - 5x + 6$ and $x^3 - 4x^2 + 4x$. This is the LCD for the two fractions.
3. When you add or subtract fractions, you often need to rewrite the fractions as equivalent fractions. You do this so that the resulting equivalent fractions will each have a denominator equal to the LCD of the original fractions.
4. To add or subtract two fractions that have the same denominator, you add or subtract their numerators and keep the same denominator.
5. The sum or difference of two rational expressions should be written as a polynomial or as a fraction in simplest form.

Helping You Remember

6. Some students have trouble remembering whether a common denominator is needed to add and subtract rational expressions or to multiply and divide them. How can your knowledge of working with fractions in arithmetic help you remember this?
Sample answer: In arithmetic, a common denominator is needed to add and subtract fractions, but not to multiply and divide them. The situation is the same for rational expressions.

Enrichment, p. 528

Superellipses

The circle and the ellipse are members of an interesting family of curves that were first studied by the French physicist and mathematician Gabriel Lamé (1795–1870). The general equation for the family is

$\left|\frac{x}{a}\right|^n + \left|\frac{y}{b}\right|^n = 1$, with $a \neq 0$, $b \neq 0$, and $n > 0$.

For even values of n greater than 2, the curves are called **superellipses.**

1. Consider two curves that are *not* superellipses. Graph each equation on the grid at the right. State the type of curve produced each time.
a. $\left|\frac{x}{2}\right|^2 + \left|\frac{y}{2}\right|^2 = 1$ circle
b. $\left|\frac{x}{3}\right|^2 + \left|\frac{y}{2}\right|^2 = 1$ ellipse

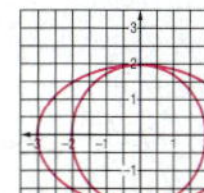

4 Assess

Open-Ended Assessment

Writing Have students write their own problems of the types in this lesson by beginning with an answer and working backward to create a problem.

Tips for New Teachers

Intervention The skills for combining and simplifying done in this lesson are used extensively in algebra. Take time to clear up student errors and misconceptions before proceeding.

Getting Ready for Lesson 9-3

PREREQUISITE SKILL Students will graph rational functions using asymptotes in Lesson 9-3. In previous course material, students graphed hyperbolas by using asymptotes and will apply these skills to graphing rational functions. Use Exercises 59–61 to determine your students' familiarity with graphing hyperbolas.

Assessment Options

Practice Quiz 1 The quiz provides students with a brief review of the concepts and skills in Lessons 9-1 and 9-2. Lesson numbers are given to the right of exercises or instruction lines so students can review concepts not yet mastered.

Quiz (Lessons 9-1 and 9-2) is available on p. 567 of the *Chapter 9 Resource Masters*.

51. WRITING IN MATH Answer the question that was posed at the beginning of the lesson. See pp. 519A–519D.

How is subtraction of rational expressions used in photography?
Include the following in your answer:
- an explanation of how to subtract rational expressions, and
- an equation that could be used to find the distance between the lens and the film if the focal length of the lens is 10 centimeters and the distance between the lens and the object is 60 centimeters.

Standardized Test Practice

52. For all $t \neq 5$, $\frac{t^2 - 25}{3t - 15} =$ **B**

(A) $\frac{t-5}{3}$. (B) $\frac{t+5}{3}$. (C) $t - 5$. (D) $t + 5$. (E) $\frac{t-5}{t-3}$.

53. What is the sum of $\frac{x-y}{5}$ and $\frac{x+y}{4}$? **C**

(A) $\frac{9x+9y}{20}$ (B) $\frac{x+9y}{20}$ (C) $\frac{9x+y}{20}$ (D) $\frac{9x-y}{20}$ (E) $\frac{x-9y}{20}$

Maintain Your Skills

Mixed Review

Simplify each expression. *(Lesson 9-1)*

54. $\frac{9x^2y^3}{(5xyz)^2} \div \frac{(3xy)^3}{20x^2y}$ $\frac{4}{15xyz^2}$

55. $\frac{5a^2 - 20}{2a + 2} \cdot \frac{4a}{10a - 20}$ $\frac{a(a+2)}{a+1}$

Solve each system of inequalities by graphing. *(Lesson 8-7)* 56–57. See pp. 519A–519D.

56. $9x^2 + y^2 < 81$
$x^2 + y^2 \geq 16$

57. $(y - 3)^2 \geq x + 2$
$x^2 \leq y + 4$

58. **GARDENS** Helene Jonson has a rectangular garden 25 feet by 50 feet. She wants to increase the garden on all sides by an equal amount. If the area of the garden is to be increased by 400 square feet, by how much should each dimension be increased? *(Lesson 6-4)* **2.5 ft**

Getting Ready for the Next Lesson

PREREQUISITE SKILL **Draw the asymptotes and graph each hyperbola.**
*(To review **graphing hyperbolas**, see Lesson 8-5.)* 59–61. See pp. 519A–519D.

59. $\frac{x^2}{16} - \frac{y^2}{20} = 1$

60. $\frac{y^2}{49} - \frac{x^2}{25} = 1$

61. $\frac{(x+2)^2}{16} - \frac{(y-5)^2}{25} = 1$

Practice Quiz 1 — Lessons 9-1 and 9-2

Simplify each expression. *(Lesson 9-1)*

1. $\frac{t^2 - t - 6}{t^2 - 6t + 9}$ $\frac{t+2}{t-3}$

2. $\frac{3ab^3}{8a^2b} \cdot \frac{4ac}{9b^4}$ $\frac{c}{6b^2}$

3. $-\frac{4}{8x} \div \frac{16}{xy^2}$ $-\frac{y^2}{32}$

4. $\frac{48}{6a + 42} \cdot \frac{7a + 49}{16}$ $\frac{7}{2}$

5. $\frac{w^2 + 5w + 4}{6} \div \frac{w + 1}{18w + 24}$ $(w + 4)(3w + 4)$

6. $\frac{\frac{x^2 + x}{x + 1}}{\frac{x}{x - 1}}$ $x - 1$

Simplify each expression. *(Lesson 9-2)*

7. $\frac{4a + 2}{a + b} + \frac{1}{-b - a}$ $\frac{4a+1}{a+b}$

8. $\frac{2x}{5ab^3} + \frac{4y}{3a^2b^2}$ $\frac{6ax + 20by}{15a^2b^3}$

9. $\frac{5}{n + 6} - \frac{4}{n - 1}$ $\frac{n - 29}{(n+6)(n-1)}$

10. $\frac{x - 5}{2x - 6} - \frac{x - 7}{4x - 12}$ $\frac{1}{4}$

484 Chapter 9 Rational Expressions and Equations

9-3 Graphing Rational Functions

What You'll Learn

- Determine the vertical asymptotes and the point discontinuity for the graphs of rational functions.
- Graph rational functions.

Vocabulary

- rational function
- continuity
- asymptote
- point discontinuity

How can rational functions be used when buying a group gift?

A group of students want to get their favorite teacher, Mr. Salgado, a retirement gift. They plan to get him a gift certificate for a weekend package at a lodge in a state park. The certificate costs \$150. If c represents the cost for each student and s represents the number of students, then $c = \frac{150}{s}$.

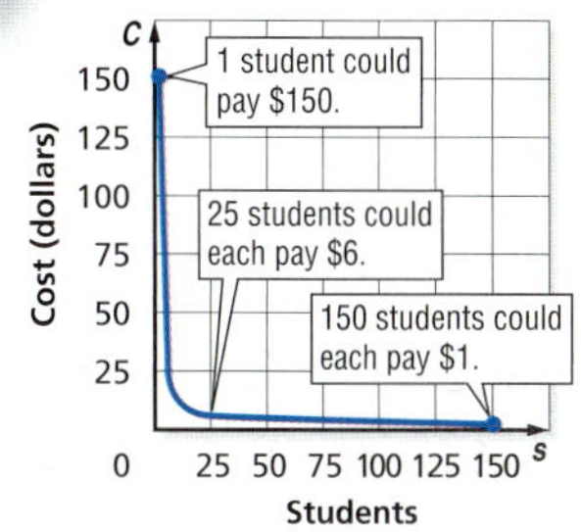

VERTICAL ASYMPTOTES AND POINT DISCONTINUITY The function $c = \frac{150}{s}$ is an example of a rational function. A **rational function** is an equation of the form $f(x) = \frac{p(x)}{q(x)}$, where $p(x)$ and $q(x)$ are polynomial functions and $q(x) \neq 0$. Here are other examples of rational functions.

$$f(x) = \frac{x}{x+3} \qquad g(x) = \frac{5}{x-6} \qquad h(x) = \frac{x+4}{(x-1)(x+4)}$$

No denominator in a rational function can be zero because division by zero is not defined. In the examples above, the functions are not defined at $x = -3$, $x = 6$, and $x = 1$ and $x = -4$, respectively.

The graphs of rational functions may have breaks in **continuity**. This means that, unlike polynomial functions, which can be traced with a pencil never leaving the paper, not all rational functions are traceable. Breaks in continuity can appear as a vertical **asymptote** or as a **point discontinuity**. Recall that an asymptote is a line that the graph of the function approaches, but never crosses. Point discontinuity is like a hole in a graph.

Study Tip

Look Back
To review **asymptotes**, see Lesson 8-5.

Key Concept — Vertical Asymptotes

Property	Words	Example	Model
Vertical Asymptote	If the rational expression of a function is written in simplest form and the function is undefined for $x = a$, then $x = a$ is a vertical asymptote.	For $f(x) = \frac{x}{x-3}$, $x = 3$ is a vertical asymptote.	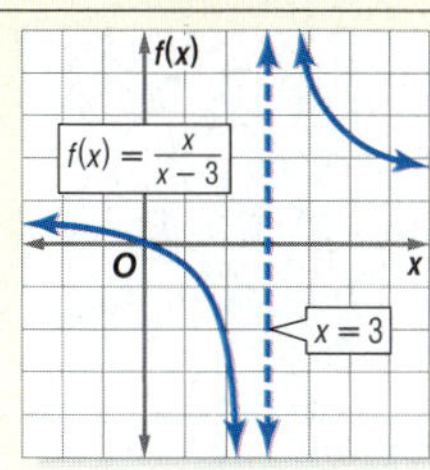

9-3 Lesson Notes

1 Focus

5-Minute Check Transparency 9-3 Use as a quiz or review of Lesson 9-2.

Mathematical Background notes are available for this lesson on p. 470D.

Building on Prior Knowledge

In Chapter 7, students learned to graph polynomial equations. In this lesson, they will apply the same skills to graphing rational functions.

How can rational functions be used when buying a group gift?

Ask students:

- What does the cost for one student depend on? **the number of students who participate**
- What happens to the value of c as the value of s increases? **It decreases.**

Resource Manager

Workbook and Reproducible Masters

Chapter 9 Resource Masters

- Study Guide and Intervention, pp. 529–530
- Skills Practice, p. 531
- Practice, p. 532
- Reading to Learn Mathematics, p. 533
- Enrichment, p. 534
- Assessment, pp. 567, 569

Graphing Calculator and Spreadsheet Masters, p. 43

Teaching Algebra With Manipulatives Masters, p. 273

Transparencies

5-Minute Check Transparency 9-3
Answer Key Transparencies

Technology

Interactive Chalkboard

2 Teach

VERTICAL ASYMPTOTES AND POINT DISCONTINUITY

In-Class Example Power Point®

1 Determine the equations of any vertical asymptotes and the values of x for any holes in the graph of

$$f(x) = \frac{x^2 - 4}{x^2 + 5x + 6}.$$

$x = -3$ is a vertical asymptote and $x = -2$ represents a hole in the graph.

GRAPH RATIONAL FUNCTIONS

In-Class Example Power Point®

2 Graph $f(x) = \frac{x}{x + 1}$.

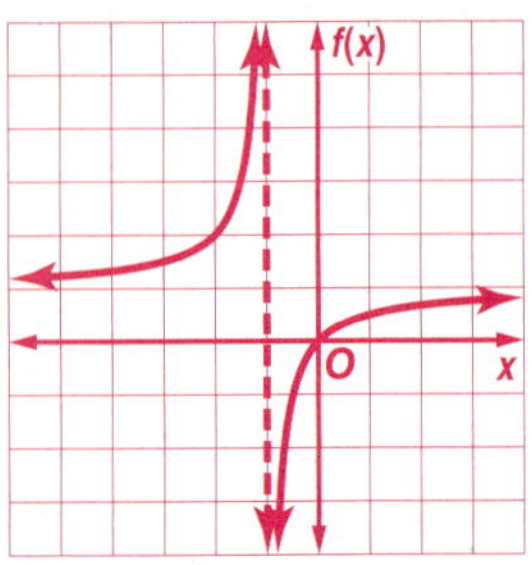

Teaching Tip Suggest that students choose a large unit on their grid paper and estimate points to the nearest tenth. Point out that they will probably not be able to see the shape of the graph as a whole unless they use a graphing calculator or computer program.

Key Concept — *Point Discontinuity*

Property	Words	Example	Model
Point Discontinuity	If the original function is undefined for $x = a$ but the rational expression of the function in simplest form is defined for $x = a$, then there is a hole in the graph at $x = a$.	$f(x) = \frac{(x + 2)(x - 1)}{x + 2}$ can be simplified to $f(x) = x - 1$. So, $x = -2$ represents a hole in the graph.	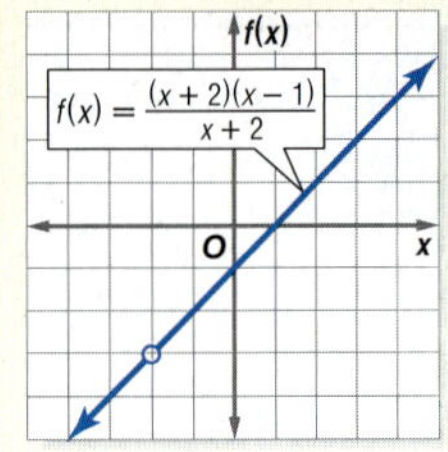

Example 1 Vertical Asymptotes and Point Discontinuity

Determine the equations of any vertical asymptotes and the values of x for any holes in the graph of $f(x) = \frac{x^2 - 1}{x^2 - 6x + 5}$.

First factor the numerator and denominator of the rational expression.

$$\frac{x^2 - 1}{x^2 - 6x + 5} = \frac{(x - 1)(x + 1)}{(x - 1)(x - 5)}$$

The function is undefined for $x = 1$ and $x = 5$. Since $\frac{\overset{1}{\cancel{(x - 1)}}(x + 1)}{\underset{1}{\cancel{(x - 1)}}(x - 5)} = \frac{x + 1}{x - 5}$,

$x = 5$ is a vertical asymptote, and $x = 1$ represents a hole in the graph.

GRAPH RATIONAL FUNCTIONS You can use what you know about vertical asymptotes and point discontinuity to graph rational functions.

Example 2 Graph with a Vertical Asymptote

Graph $f(x) = \frac{x}{x - 2}$.

The function is undefined for $x = 2$. Since $\frac{x}{x - 2}$ is in simplest form, $x = 2$ is a vertical asymptote. Draw the vertical asymptote. Make a table of values. Plot the points and draw the graph.

x	$f(x)$
−50	0.96154
−20	0.90909
−10	0.83333
−2	0.5
−1	0.33333
0	0
1	−1
3	3
4	2
5	1.6667
10	1.25
20	1.1111
50	1.0417

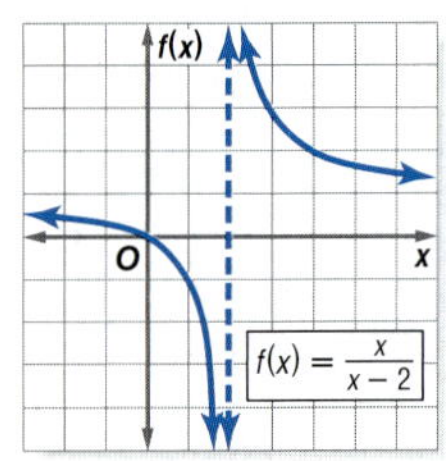

As $|x|$ increases, it appears that the y values of the function get closer and closer to 1. The line with the equation $f(x) = 1$ is a horizontal asymptote of the function.

Study Tip

Graphing Rational Functions
Finding the x- and y-intercepts is often useful when graphing rational functions.

DAILY INTERVENTION — **Unlocking Misconceptions**

Asymptotes Students should understand that a graph continues to approach an asymptote and gets closer and closer to that value, but never reaches it. This is an abstract mathematical idea that cannot be represented accurately with any form of visual illustration.

As you have learned, graphs of rational functions may have point discontinuity rather than vertical asymptotes. The graphs of these functions appear to have holes. These holes are usually shown as circles on graphs.

Example 3 Graph with Point Discontinuity

Graph $f(x) = \frac{x^2 - 9}{x + 3}$.

Notice that $\frac{x^2 - 9}{x + 3} = \frac{(x + 3)(x - 3)}{x + 3}$ or $x - 3$.

Therefore, the graph of $f(x) = \frac{x^2 - 9}{x + 3}$ is the graph of $f(x) = x - 3$ with a hole at $x = -3$.

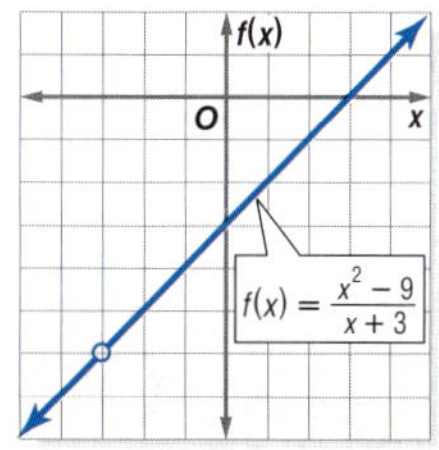

Many real-life situations can be described by using rational functions.

Algebra Activity

Rational Functions

The density of a material can be expressed as $D = \frac{m}{V}$, where m is the mass of the material in grams and V is the volume in cubic centimeters. By finding the volume and density of 200 grams of each liquid, you can draw a graph of the function $D = \frac{200}{V}$.

Collect the Data

- Use a balance and metric measuring cups to find the volumes of 200 grams of different liquids such as water, cooking oil, isopropyl alcohol, sugar water, and salt water.
- Use $D = \frac{m}{V}$ to find the density of each liquid.

Analyze the Data

1. Graph the data by plotting the points (volume, density) on a graph. Connect the points. See pp. 519A–519D.
2. From the graph, find the asymptotes. $x = 0$, $y = 0$

In the real world, sometimes values on the graph of a rational function are not meaningful.

Example 4 Use Graphs of Rational Functions

TRANSPORTATION A train travels at one velocity V_1 for a given amount of time t_1 and then another velocity V_2 for a different amount of time t_2. The average velocity is given by $V = \frac{V_1t_1 + V_2t_2}{t_1 + t_2}$.

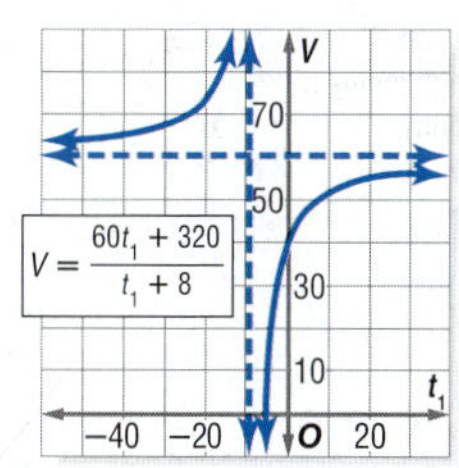

a. Let t_1 be the independent variable and let V be the dependent variable. Draw the graph if $V_1 = 60$ miles per hour, $V_2 = 40$ miles per hour, and $t_2 = 8$ hours.

The function is $V = \frac{60t_1 + 40(8)}{t_1 + 8}$ or $V = \frac{60t_1 + 320}{t_1 + 8}$. The vertical asymptote is $t_1 = -8$. Graph the vertical asymptote and the function. Notice that the horizontal asymptote is $V = 60$.

www.algebra2.com/extra_examples

In-Class Examples

Power Point®

3 Graph $f(x) = \frac{x^2 - 4}{x - 2}$.

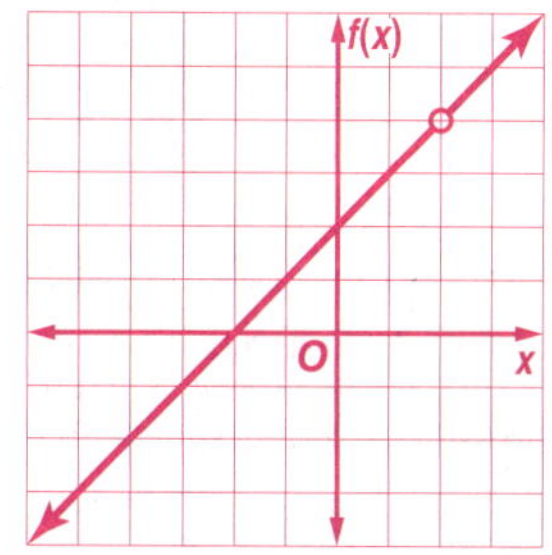

Teaching Tip Since the discontinuity is only one point (and a mathematical point has no dimensions), suggest that students draw a circle on their graphs to indicate the discontinuity.

4 **TRANSPORTATION** Use the situation and formula given in Example 4.

a. Draw the graph if $V_1 = 50$ miles per hour, $V_2 = 30$ miles per hour, and $t_2 = 1$ hour.

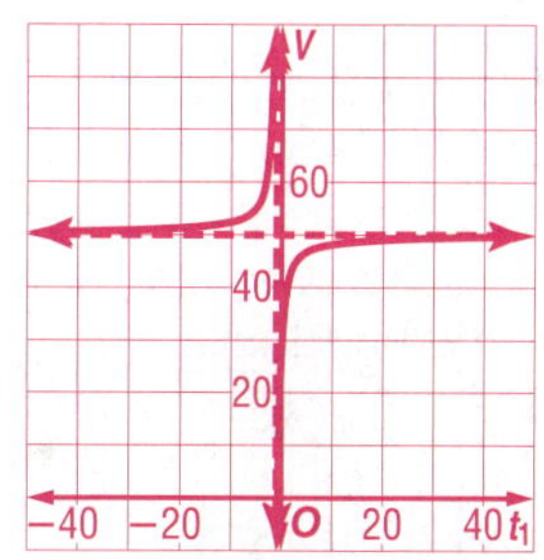

b. What is the V-intercept of the graph? **30**

c. What values of t_1 and V are meaningful in the context of the problem? **Positive values of t_1 and values of V between 30 and 50 are meaningful.**

Algebra Activity

Materials: balance, metric measuring cups, different liquids, graph paper

- Choose liquids that are quite different in density. If you make sugar or salt water, dissolve as much of the substance as you can in it.
- Differences in volume for each 200 grams will be easier to read if you use the smallest measuring cup that holds the amount.

Tips for New Teachers

Intervention To make sure the situation in Example 4 is meaningful, ask a student to explain the situation as a story without using letter names for variables. For example, the story might begin "A train travels 40 miles per hour as it goes through towns. Eight hours of its total trip are spent going through towns."

3 Practice/Apply

Study Notebook

Have students—

- add the definitions/examples of the vocabulary terms to their Vocabulary Builder worksheets for Chapter 9.
- include any other item(s) that they find helpful in mastering the skills in this lesson.

About the Exercises...

Organization by Objective

- **Vertical Asymptotes and Point Discontinuity:** 16–21
- **Graph Rational Functions:** 22–45, 47–50

Odd/Even Assignments

Exercises 16–39 are structured so that students practice the same concepts whether they are assigned odd or even problems.

Assignment Guide

Basic: 17–39 odd, 46, 51–66

Average: 17–39 odd, 40–42, 46–66

Advanced: 16–38 even, 40–62 (optional: 63–66)

Answer

2. Each of the graphs is a straight line passing through $(-5, 0)$ and $(0, 5)$. However, the graph of $f(x) = \frac{(x-1)(x+5)}{x-1}$ has a hole at $(1, 6)$, and the graph of $g(x) = x + 5$ does not have a hole.

---

b. What is the V-intercept of the graph?

The V-intercept is 40.

c. What values of t_1 and V are meaningful in the context of the problem?

In the problem context, time and velocity are positive values. Therefore, only values of t_1 greater than 0 and values of V between 40 and 60 are meaningful.

Check for Understanding

Concept Check

1. Sample answer: $f(x) = \frac{1}{(x+5)(x-2)}$

1. **OPEN ENDED** Write a function whose graph has two vertical asymptotes located at $x = -5$ and $x = 2$.
2. **Compare and contrast** the graphs of $f(x) = \frac{(x-1)(x+5)}{x-1}$ and $g(x) = x + 5$. See margin.
3. **Describe** the graph at the right. Include the equations of any asymptotes, the x values of any holes, and the x- and y-intercepts.
   $x = 2$ and $y = 0$ are asymptotes of the graph. The y-intercept is 0.5 and there is no x-intercept because $y = 0$ is an asymptote.

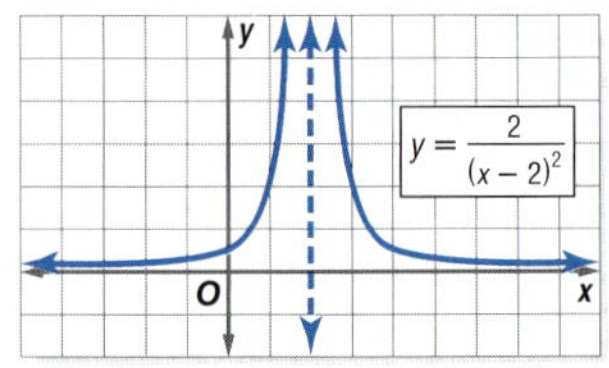

Guided Practice

GUIDED PRACTICE KEY	
Exercises	**Examples**
4, 5	1
6–11	2, 3
12–15	4

Determine the equations of any vertical asymptotes and the values of x for any holes in the graph of each rational function.

4. $f(x) = \frac{3}{x^2 - 4x + 4}$ asymptote: $x = 2$
5. $f(x) = \frac{x-1}{x^2 + 4x - 5}$ asymptote: $x = -5$; hole: $x = 1$

Graph each rational function. 6–11. See pp. 519A–519D.

6. $f(x) = \frac{x}{x+1}$
7. $f(x) = \frac{6}{(x-2)(x+3)}$
8. $f(x) = \frac{x^2 - 25}{x - 5}$
9. $f(x) = \frac{x-5}{x+1}$
10. $f(x) = \frac{4}{(x-1)^2}$
11. $f(x) = \frac{x+2}{x^2 - x - 6}$

Application

MEDICINE **For Exercises 12–15, use the following information.**
For certain medicines, health care professionals may use Young's Rule, $C = \frac{y}{y+12} \cdot D$, to estimate the proper dosage for a child when the adult dosage is known. In this equation, C represents the child's dose, D represents the adult dose, and y represents the child's age in years.

12. Use Young's Rule to estimate the dosage of amoxicillin for an eight-year-old child if the adult dosage is 250 milligrams. 100 mg
13. Graph $C = \frac{y}{y+12}$. See pp. 519A–519D.
14. Give the equations of any asymptotes and y- and C-intercepts of the graph. $y = -12$, $C = 1$; 0; 0
15. What values of y and C are meaningful in the context of the problem? $y > 0$ and $0 < C < 1$

DAILY INTERVENTION — **Differentiated Instruction**

Visual/Spatial Have students graph one of the examples from the lesson with colors on a large sheet of posterboard, to clearly show how a graph approaches but never reaches an asymptote or how a graph may have a hole in it for a certain value of the variable. Display the results in the classroom.

488 Chapter 9 Rational Expressions and Equations

★ indicates increased difficulty

Practice and Apply

Homework Help

For Exercises	See Examples
16–21	1
22–39	2, 3
40–50	4

Extra Practice
See page 849.

19. asymptote: $x = -1$; hole: $x = 5$

Determine the equations of any vertical asymptotes and the values of x for any holes in the graph of each rational function.

16. $f(x) = \frac{2}{x^2 - 5x + 6}$ asymptotes: $x = 2$, $x = 3$

17. $f(x) = \frac{4}{x^2 + 2x - 8}$ asymptotes: $x = -4$, $x = 2$

18. $f(x) = \frac{x + 3}{x^2 + 7x + 12}$ asymptote: $x = -4$; hole: $x = -3$

19. $f(x) = \frac{x - 5}{x^2 - 4x - 5}$

20. $f(x) = \frac{x^2 - 8x + 16}{x - 4}$ hole: $x = 4$

21. $f(x) = \frac{x^2 - 3x + 2}{x - 1}$ hole: $x = 1$

Graph each rational function. 22–39. See pp. 519A–519D.

22. $f(x) = \frac{1}{x}$
23. $f(x) = \frac{3}{x}$
24. $f(x) = \frac{1}{x + 2}$
25. $f(x) = \frac{-5}{x + 1}$
26. $f(x) = \frac{x}{x - 3}$
27. $f(x) = \frac{5x}{x + 1}$
28. $f(x) = \frac{-3}{(x - 2)^2}$
29. $f(x) = \frac{1}{(x + 3)^2}$
30. $f(x) = \frac{x + 4}{x - 1}$
31. $f(x) = \frac{x - 1}{x - 3}$
32. $f(x) = \frac{x^2 - 36}{x + 6}$
33. $f(x) = \frac{x^2 - 1}{x - 1}$
34. $f(x) = \frac{3}{(x - 1)(x + 5)}$
35. $f(x) = \frac{-1}{(x + 2)(x - 3)}$
36. $f(x) = \frac{x}{x^2 - 1}$
37. $f(x) = \frac{x - 1}{x^2 - 4}$
38. $f(x) = \frac{6}{(x - 6)^2}$
39. $f(x) = \frac{1}{(x + 2)^2}$

More About...

History

Mathematician Maria Gaetana Agnesi was one of the greatest scholars of all time. Born in Milan, Italy, in 1718, she mastered Greek, Hebrew, and several modern languages by the age of 11.

Source: *A History of Mathematics*

HISTORY For Exercises 40–42, use the following information.

In Maria Gaetana Agnesi's book *Analytical Institutions*, Agnesi discussed the characteristics of the equation $x^2y = a^2(a - y)$, whose graph is called the "curve of Agnesi." This equation can be expressed as $y = \frac{a^3}{x^2 + a^2}$.

40. Graph $f(x) = \frac{a^3}{x^2 + a^2}$ if $a = 4$. See pp. 519A–519D.

41. Describe the graph.

42. **Make a conjecture** about the shape of the graph of $f(x) = \frac{a^3}{x^2 + a^2}$ if $a = -4$. Explain your reasoning. See pp. 519A–519D.

41. The graph is bell-shaped with a horizontal asymptote at $f(x) = 0$.

AUTO SAFETY For Exercises 43–45, use the following information.

When a car has a front-end collision, the objects in the car (including passengers) keep moving forward until the impact occurs. After impact, objects are repelled. Seat belts and airbags limit how far you are jolted forward. The formula for the velocity at which you are thrown backward is $V_f = \frac{(m_1 - m_2)v_i}{m_1 + m_2}$, where m_1 and m_2 are masses of the two objects meeting and v_i is the initial velocity. 43. See pp. 519A–519D.

43. Let m_1 be the independent variable, and let V_f be the dependent variable. Graph the function if $m_2 = 7$ kilograms and $v_i = 5$ meters per second.

44. Give the equation of the vertical asymptote and the m_1- and V_f-intercepts of the graph. $m_1 = -7$; 7; -5

45. Find the value of V_f when the value of m_1 is 5 kilograms. about -0.83 m/s

46. Sample answers: $f(x) = \frac{x + 2}{(x + 2)(x - 3)}$, $f(x) = \frac{2(x + 2)}{(x + 2)(x - 3)}$, $f(x) = \frac{5(x + 2)}{(x + 2)(x - 3)}$

46. **CRITICAL THINKING** Write three rational functions that have a vertical asymptote at $x = 3$ and a hole at $x = -2$.

www.algebra2.com/self_check_quiz

Study Guide and Intervention, p. 529 (shown) and p. 530

Vertical Asymptotes and Point Discontinuity

Rational Function	an equation of the form $f(x) = \frac{p(x)}{q(x)}$, where $p(x)$ and $q(x)$ are polynomial expressions and $q(x) \neq 0$
Vertical Asymptote of the Graph of a Rational Function	An asymptote is a line that the graph of a function approaches, but never crosses. If the simplified form of the related rational expression is undefined for $x = a$, then $x = a$ is a vertical asymptote.
Point Discontinuity of the Graph of a Rational Function	Point discontinuity is like a hole in a graph. If the original related expression is undefined for $x = a$ but the simplified expression is defined for $x = a$, then there is a hole in the graph at $x = a$.

Example Determine the equations of any vertical asymptotes and the values of x for any holes in the graph of $f(x) = \frac{4x^2 + x - 3}{x^2 - 1}$.

First factor the numerator and the denominator of the rational expression.

$f(x) = \frac{4x^2 + x - 3}{x^2 - 1} = \frac{(4x - 3)(x + 1)}{(x + 1)(x - 1)}$

The function is undefined for $x = 1$ and $x = -1$.

Since $\frac{(4x - 3)(x + 1)}{(x + 1)(x - 1)} = \frac{4x - 3}{x - 1}$, $x = 1$ is a vertical asymptote. The simplified expression is defined for $x = -1$, so this value represents a hole in the graph.

Exercises

Determine the equations of any vertical asymptotes and the values of x for any holes in the graph of each rational function.

1. $f(x) = \frac{4}{x^2 + 3x - 10}$ asymptotes: $x = 2$, $x = -5$
2. $f(x) = \frac{2x^2 - x - 10}{2x - 5}$ hole: $x = \frac{5}{2}$
3. $f(x) = \frac{x^2 - x - 12}{x^2 - 4x}$ asymptote: $x = 0$; hole $x = 4$
4. $f(x) = \frac{3x - 1}{3x^2 + 5x - 2}$ asymptote: $x = -2$; hole: $x = \frac{1}{3}$
5. $f(x) = \frac{x^2 - 6x - 7}{x^2 + 6x - 7}$ asymptotes: $x = 1$, $x = -7$
6. $f(x) = \frac{3x^2 - 5x - 2}{x + 3}$ asymptote: $x = -3$
7. $f(x) = \frac{x + 1}{x^2 - 6x + 5}$ asymptotes: $x = 1$, $x = 5$
8. $f(x) = \frac{2x^2 - x - 3}{2x^2 + 3x - 9}$ asymptote: $x = -3$; hole: $x = \frac{3}{2}$
9. $f(x) = \frac{x^3 - 2x^2 - 5x + 6}{x^2 - 4x + 3}$ holes: $x = 1$, $x = 3$

Skills Practice, p. 531 and Practice, p. 532 (shown)

Determine the equations of any vertical asymptotes and the values of x for any holes in the graph of each rational function.

1. $f(x) = \frac{6}{x^2 + 3x - 10}$ asymptotes: $x = 2$, $x = -5$
2. $f(x) = \frac{x - 7}{x^2 - 10x + 21}$ asymptote: $x = 3$; hole: $x = 7$
3. $f(x) = \frac{x - 2}{x^2 + 4x + 4}$ asymptote: $x = -2$
4. $f(x) = \frac{x^2 - 100}{x + 10}$ hole: $x = -10$
5. $f(x) = \frac{x^2 - 2x - 24}{x - 6}$ hole: $x = 6$
6. $f(x) = \frac{x^2 + 9x + 20}{x + 5}$ hole: $x = -5$

Graph each rational function.

7. $f(x) = \frac{-4}{x - 2}$
8. $f(x) = \frac{x - 3}{x - 2}$
9. $f(x) = \frac{3x}{(x + 3)^2}$

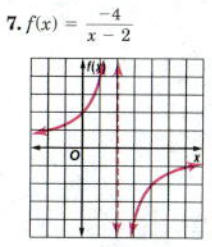
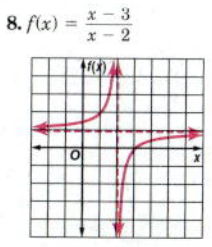
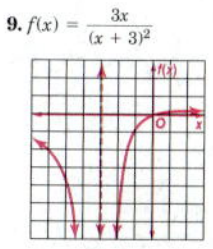

10. **PAINTING** Working alone, Tawa can give the shed a coat of paint in 6 hours. It takes her father x hours working alone to give the shed a coat of paint. The equation $f(x) = \frac{6 + x}{6x}$ describes the portion of the job Tawa and her father working together can complete in 1 hour. Graph $f(x) = \frac{6 + x}{6x}$ for $x \geq 0$, $y \geq 0$. If Tawa's father can complete the job in 4 hours alone, what portion of the job can they complete together in 1 hour? $\frac{5}{12}$

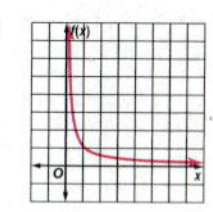

11. **LIGHT** The relationship between the illumination an object receives from a light source of I foot-candles and the square of the distance d in feet of the object from the source can be modeled by $I(d) = \frac{4500}{d^2}$. Graph the function $I(d) = \frac{4500}{d^2}$ for $0 \leq I \leq 80$ and $0 \leq d \leq 80$. What is the illumination in foot-candles that the object receives at a distance of 20 feet from the light source? 11.25 foot-candles

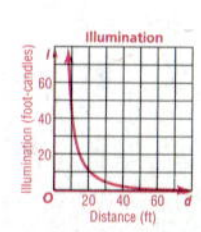

Reading to Learn Mathematics, p. 533

ELL

Pre-Activity How can rational functions be used when buying a group gift?

Read the introduction to Lesson 9-3 at the top of page 485 in your textbook.

- If 15 students contribute to the gift, how much would each of them pay? $10
- If each student pays $5, how many students contributed? 30 students

Reading the Lesson

1. Which of the following are rational functions? A and C
   A. $f(x) = \frac{1}{x - 5}$ B. $g(x) = \sqrt{x}$ C. $h(x) = \frac{x^2 - 25}{x^2 + 6x + 9}$

2. a. Graphs of rational functions may have breaks in continuity. These may occur as vertical asymptotes or as point discontinuities.
   b. The graphs of two rational functions are shown below.

I. 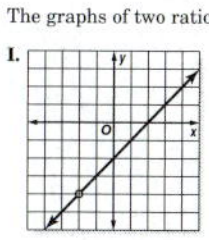 II. 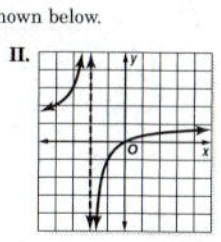

Graph I has a point discontinuity at $x = -2$.
Graph II has a vertical asymptote at $x = -2$.
Match each function with its graph above.
$f(x) = \frac{x}{x + 2}$ II $g(x) = \frac{x^2 - 4}{x + 2}$ I

Helping You Remember

3. One way to remember something new is to see how it is related to something you already know. How can knowing that division by zero is undefined help you to remember how to find the places where a rational function has a point discontinuity or an asymptote?

Sample answer: A point discontinuity or vertical asymptote occurs where the function is undefined, that is, where the denominator of the related rational expression is equal to 0. Therefore, set the denominator equal to zero and solve for the variable.

Enrichment, p. 534

Graphing with Addition of y-Coordinates

Equations of parabolas, ellipses, and hyperbolas that are "tipped" with respect to the x- and y-axes are more difficult to graph than the equations you have been studying.

Often, however, you can use the graphs of two simpler equations to graph a more complicated equation. For example, the graph of the ellipse in the diagram at the right is obtained by adding the y-coordinate of each point on the circle and the y-coordinate of the corresponding point of the line.

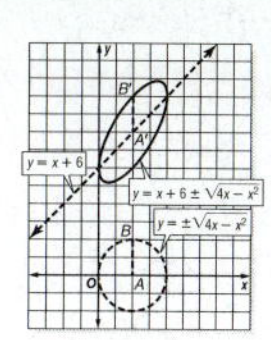

Graph each equation. State the type of curve for each graph.

4 Assess

Open-Ended Assessment

Writing Have students write their own examples of rational functions and graph them, showing discontinuities.

Getting Ready for Lesson 9-4

BASIC SKILL Students will write and solve direct, joint, and inverse variation problems in Lesson 9-4. This will include students writing and solving proportions that relate the values in the variation. Use Exercises 63–66 to determine your students' familiarity with solving proportions.

Assessment Options

Quiz (Lesson 9-3) is available on p. 567 of the *Chapter 9 Resource Masters*.

Mid-Chapter Test (Lessons 9-1 through 9-3) is available on p. 569 of the *Chapter 9 Resource Masters*.

BASKETBALL **For Exercises 47–50, use the following information.**
Zonta plays basketball for Centerville High School. So far this season, she has made 6 out of 10 free throws. She is determined to improve her free-throw percentage. If she can make x consecutive free throws, her free-throw percentage can be determined using $P(x) = \frac{6 + x}{10 + x}$.

47. Graph the function. **See margin.**

48. What part of the graph is meaningful in the context of the problem?

★ 49. Describe the meaning of the y-intercept.

★ 50. What is the equation of the horizontal asymptote? Explain its meaning with respect to Zonta's shooting percentage.

48. the part in the first quadrant

49. It represents her original free-throw percentage of 60%.

50. $y = 1$; this represents 100% which she cannot achieve because she has already missed 4 free throws.

51. **WRITING IN MATH** Answer the question that was posed at the beginning of the lesson. **See margin.**

How can rational functions be used when buying a group gift?

Include the following in your answer:
- a complete graph of the function $c = \frac{150}{s}$ with asymptotes, and
- an explanation of why only part of the graph is meaningful in the context of the problem.

Standardized Test Practice

52. Which set is the domain of the function graphed at the right? **A**

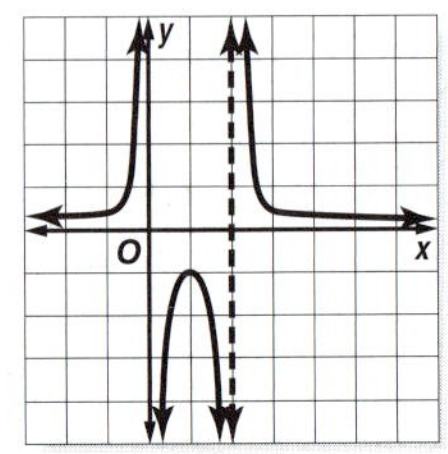

Ⓐ $\{x \mid x \neq 0, 2\}$
Ⓑ $\{x \mid x \neq -2, 0\}$
Ⓒ $\{x \mid x < 4\}$
Ⓓ $\{x \mid x > -4\}$

53. Which set is the range of the function $y = \frac{x^2 + 8}{2}$? **B**

Ⓐ $\{y \mid y \neq \pm 2\sqrt{2}\}$ Ⓑ $\{y \mid y \geq 4\}$
Ⓒ $\{y \mid y \geq 0\}$ Ⓓ $\{y \mid y \leq 0\}$

Maintain Your Skills

Mixed Review

Simplify each expression. *(Lessons 9-2 and 9-1)*

54. $\frac{3m + 2}{m + n} + \frac{4}{2m + 2n}$ 55. $\frac{5}{x + 3} - \frac{2}{x - 2}$ 56. $\frac{2w - 4}{w + 3} \div \frac{2w + 6}{5}$

54. $\frac{3m + 4}{m + n}$

55. $\frac{3x - 16}{(x + 3)(x - 2)}$

56. $\frac{5(w - 2)}{(w + 3)^2}$

Find the coordinates of the center and the radius of the circle with the given equation. Then graph the circle. *(Lesson 8-3)* **57–58. See margin for graphs.**

57. $(x - 6)^2 + (y - 2)^2 = 25$ **(6, 2); 5** 58. $x^2 + y^2 + 4x = 9$ **(−2, 0); $\sqrt{13}$**

59. **ART** Joyce Jackson purchases works of art for an art gallery. Two years ago, she bought a painting for \$20,000, and last year, she bought one for \$35,000. If paintings appreciate 14% per year, how much are the two paintings worth now? *(Lesson 7-1)* **\$65,892**

Solve each equation by completing the square. *(Lesson 6-4)*

60. $x^2 + 8x + 20 = 0$ **$\{-4 \pm 2i\}$** 61. $x^2 + 2x - 120 = 0$ **$\{-12, 10\}$** 62. $x^2 + 7x - 17 = 0$ **$\left\{\frac{-7 \pm 3\sqrt{13}}{2}\right\}$**

Getting Ready for the Next Lesson

BASIC SKILL **Solve each proportion.**

63. $\frac{16}{v} = \frac{32}{9}$ **4.5** 64. $\frac{7}{25} = \frac{a}{5}$ **1.4** 65. $\frac{6}{15} = \frac{8}{s}$ **20** 66. $\frac{b}{9} = \frac{40}{30}$ **12**

Answers

47.

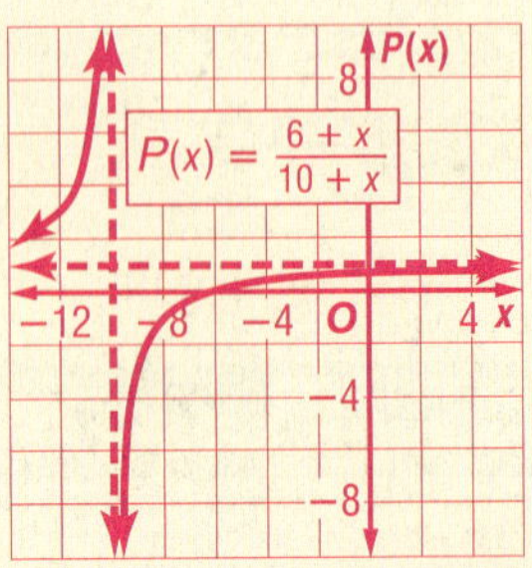

51. A rational function can be used to determine how much each person owes if the cost of the gift is known and the number of people sharing the cost is s. Answers should include the following.

•

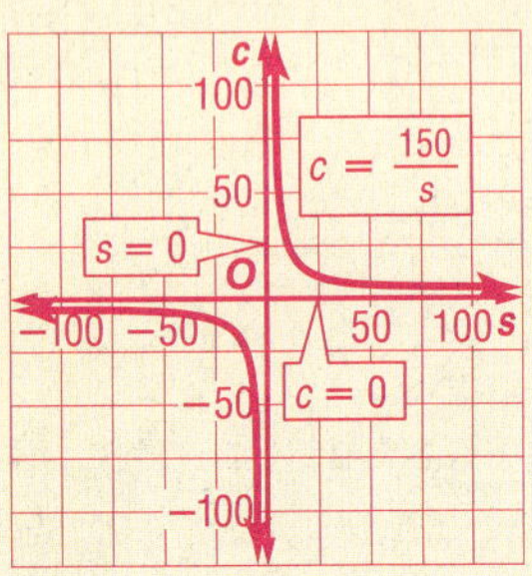

• Only the portion in the first quadrant is significant in the real world because there cannot be a negative number of people nor a negative amount of money owed for the gift.

57.

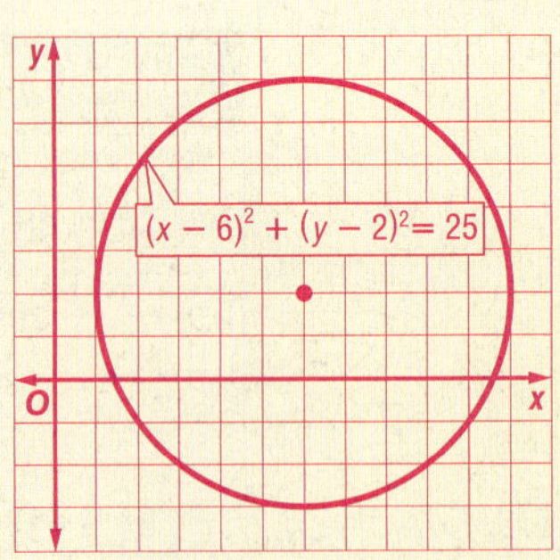

58.

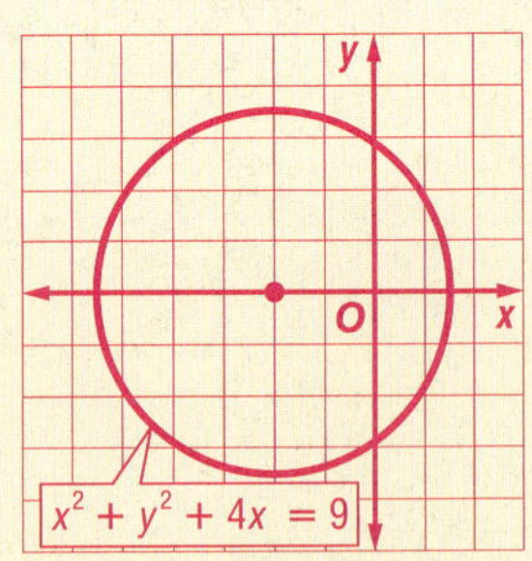

Graphing Rational Functions

A TI-83 Plus graphing calculator can be used to explore the graphs of rational functions. These graphs have some features that never appear in the graphs of polynomial functions.

Example 1 **Graph $y = \frac{8x - 5}{2x}$ in the standard viewing window. Find the equations of any asymptotes.**

- Enter the equation in the Y= list.

KEYSTROKES: [Y=] [(] 8 [X,T,θ,n] [−] 5 [)] [÷] [(] 2 [X,T,θ,n] [)] [ZOOM] 6

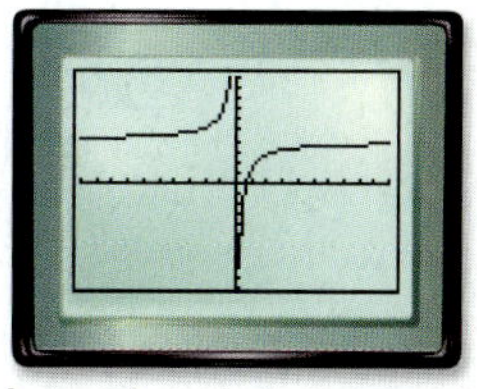

[−10, 10] scl: 1 by [−10, 10] scl: 1

By looking at the equation, we can determine that if $x = 0$, the function is undefined. The equation of the vertical asymptote is $x = 0$. Notice what happens to the y values as x grows larger and as x gets smaller. The y values approach 4. So, the equation for the horizontal asymptote is $y = 4$.

Example 2 **Graph $y = \frac{x^2 - 16}{x + 4}$ in the window [−5, 4.4] by [−10, 2] with scale factors of 1.**

- Because the function is not continuous, put the calculator in dot mode.

KEYSTROKES: [MODE] [▼] [▼] [▼] [▼] [▶] [ENTER]

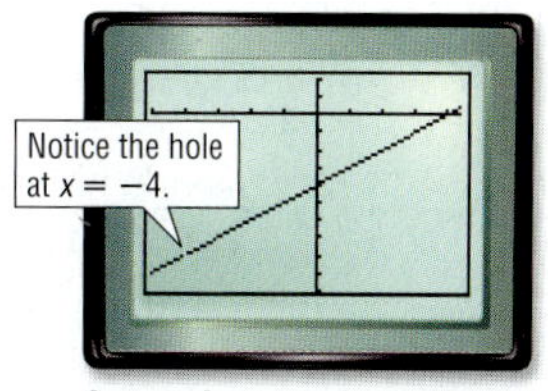

[−5, 4.4] scl: 1 by [−10, 2] scl: 1

This graph looks like a line with a break in continuity at $x = -4$. This happens because the denominator is 0 when $x = -4$. Therefore, the function is undefined when $x = -4$.

If you TRACE along the graph, when you come to $x = -4$, you will see that there is no corresponding y value.

Exercises 1–6. See pp. 519A–519D for graphs.

Use a graphing calculator to graph each function. Be sure to show a complete graph. Draw the graph on a sheet of paper. Write the x-coordinates of any points of discontinuity and/or the equations of any asymptotes. 2. $x = -2$, $y = 1$ 4. $x = 2$, $y = \frac{2}{3}$ 5. $x = 1$, $y = 4$

1. $f(x) = \frac{1}{x}$ $x = 0$, $y = 0$
2. $f(x) = \frac{x}{x + 2}$
3. $f(x) = \frac{2}{x - 4}$ $x = 4$, $y = 0$
4. $f(x) = \frac{2x}{3x - 6}$
5. $f(x) = \frac{4x + 2}{x - 1}$
6. $f(x) = \frac{x^2 - 9}{x + 3}$ point discontinuity at $x = -3$
7. Which graph(s) has point discontinuity? 6
8. Describe functions that have point discontinuity. See margin.

www.algebra2.com/other_calculator_keystrokes

Graphing Calculator Investigation

A Follow-Up of Lesson 9-3

Getting Started

Graphing Window For the examples, students should use the settings shown below the diagrams. For all the exercises, a good window is [−10, 10] scl: 1 by [−10, 10] scl: 1.

Graph Style Students may find it instructive to experiment with the graph style. They can begin by using the usual line style. This is the style when the icon to the left of the equation on the Y= list is a backslash. The best alternate style to use is path style. The icon for this style is a small numeral 0 with a short minus sign attached to the left side of the 0.

Teach

Suggest that students try graphing Example 2 in Connected mode as well as Dot mode. Ask them which way makes it easier to see the discontinuity.

Assess

Ask: How can you use the graphing calculator to check the exact value where a discontinuity occurs? Use [TRACE] to find where there is no y value.

Answer

8. rational functions where a value of the function is not defined, but the rational expression in simplest form is defined for that value

9-4 Lesson Notes

1 Focus

5-Minute Check Transparency 9-4 Use as a quiz or review of Lesson 9-3.

Mathematical Background notes are available for this lesson on p. 470D.

How is variation used to find the total cost given the unit cost?

Ask students:

- If the number of students increases, what happens to the value of the total spending? **It increases.**
- If the number of students decreases, what happens to the value of the total spending? **It decreases.**

9-4 Direct, Joint, and Inverse Variation

What You'll Learn

- Recognize and solve direct and joint variation problems.
- Recognize and solve inverse variation problems.

Vocabulary

- direct variation
- constant of variation
- joint variation
- inverse variation

How is variation used to find the total cost given the unit cost?

The total high-tech spending t of an average public college can be found by using the equation $t = 149s$, where s is the number of students.

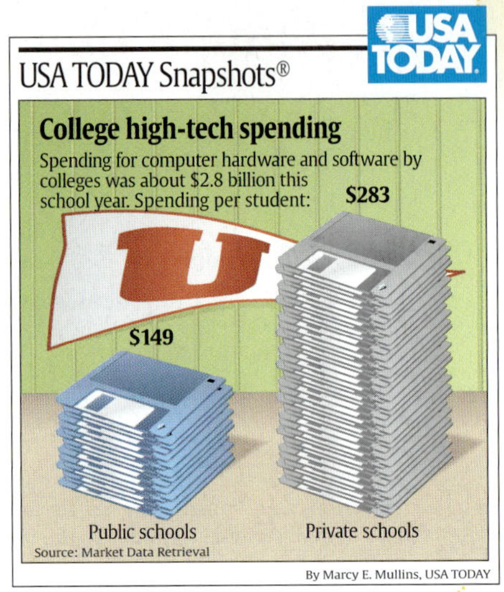

DIRECT VARIATION AND JOINT VARIATION The relationship given by $t = 149s$ is an example of direct variation. A **direct variation** can be expressed in the form $y = kx$. The k in this equation is a constant and is called the **constant of variation**.

Notice that the graph of $t = 149s$ is a straight line through the origin. An equation of a direct variation is a special case of an equation written in slope-intercept form, $y = mx + b$. When $m = k$ and $b = 0$, $y = mx + b$ becomes $y = kx$. So the slope of a direct variation equation is its constant of variation.

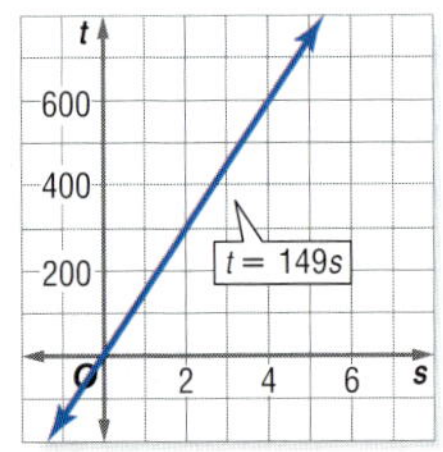

To express a direct variation, we say that y varies directly as x. In other words, as x increases, y increases or decreases at a constant rate.

> **Key Concept** — *Direct Variation*
>
> y varies directly as x if there is some nonzero constant k such that $y = kx$.
>
> k is called the constant of variation.

If you know that y varies directly as x and one set of values, you can use a proportion to find the other set of corresponding values.

$$y_1 = kx_1 \quad \text{and} \quad y_2 = kx_2$$

$$\frac{y_1}{x_1} = k \qquad \frac{y_2}{x_2} = k$$

Therefore, $\frac{y_1}{x_1} = \frac{y_2}{x_2}$.

Resource Manager

Workbook and Reproducible Masters

Chapter 9 Resource Masters

- Study Guide and Intervention, pp. 535–536
- Skills Practice, p. 537
- Practice, p. 538
- Reading to Learn Mathematics, p. 539
- Enrichment, p. 540

Graphing Calculator and Spreadsheet Masters, p. 44

School-to-Career Masters, p. 18

Science and Mathematics Lab Manual, pp. 123–126

Transparencies

5-Minute Check Transparency 9-4

Real-World Transparency 9

Answer Key Transparencies

Technology

Interactive Chalkboard

Using the properties of equality, you can find many other proportions that relate these same x and y values.

Example 1 Direct Variation

If y varies directly as x and $y = 12$ when $x = -3$, find y when $x = 16$.

Use a proportion that relates the values.

$\frac{y_1}{x_1} = \frac{y_2}{x_2}$ Direct proportion

$\frac{12}{-3} = \frac{y_2}{16}$ $y_1 = 12, x_1 = -3$, and $x_2 = 16$

$16(12) = -3(y_2)$ Cross multiply.

$192 = -3y_2$ Simplify.

$-64 = y_2$ Divide each side by -3.

When $x = 16$, the value of y is -64.

Another type of variation is joint variation. **Joint variation** occurs when one quantity varies directly as the product of two or more other quantities.

Key Concept *Joint Variation*

y varies jointly as x and z if there is some number k such that $y = kxz$, where $k \neq 0$.

If you know y varies jointly as x and z and one set of values, you can use a proportion to find the other set of corresponding values.

$$y_1 = kx_1z_1 \quad \text{and} \quad y_2 = kx_2z_2$$

$$\frac{y_1}{x_1z_1} = k \qquad \frac{y_2}{x_2z_2} = k$$

Therefore, $\frac{y_1}{x_1z_1} = \frac{y_2}{x_2z_2}$.

Example 2 Joint Variation

Suppose y varies jointly as x and z. Find y when $x = 8$ and $z = 3$, if $y = 16$ when $z = 2$ and $x = 5$.

Use a proportion that relates the values.

$\frac{y_1}{x_1z_1} = \frac{y_2}{x_2z_2}$ Joint variation

$\frac{16}{5(2)} = \frac{y_2}{8(3)}$ $y_1 = 16, x_1 = 5, z_1 = 2, x_2 = 8$, and $z_2 = 3$

$8(3)(16) = 5(2)(y_2)$ Cross multiply.

$384 = 10y_2$ Simplify.

$38.4 = y_2$ Divide each side by 10.

When $x = 8$ and $z = 3$, the value of y is 38.4.

INVERSE VARIATION Another type of variation is inverse variation. For two quantities with **inverse variation**, as one quantity increases, the other quantity decreases. For example, speed and time for a fixed distance vary inversely with each other. When you travel to a particular location, as your speed increases, the time it takes to arrive at that location decreases.

www.algebra2.com/extra_examples

2 Teach

DIRECT VARIATION AND JOINT VARIATION

In-Class Examples Power Point®

1. If y varies directly as x and $y = -15$ when $x = 5$, find y when $x = 3$. **−9**

2. Suppose y varies jointly as x and z. Find y when $x = 10$ and $z = 5$, if $y = 12$ when $z = 8$ and $x = 3$. **25**

Teaching Tip Discuss with students what happens to the constant of variation in the proportions used to solve these examples.

INVERSE VARIATION

In-Class Examples

Power Point®

3 If a varies inversely as b and $a = -6$ when $b = 2$, find a when $b = -7$. $\frac{12}{7}$

4 **SPACE** The next closest planet to the Sun after Mercury is Venus, which is about 67 million miles away. How much larger would the diameter of the Sun appear on Venus than on Earth? **about 1.39 times as large as it appears from Earth**

Teaching Tip To understand the situation in the problem, some students may find it useful to make a sketch showing the relative distances from the Sun to Earth, Mercury, and Venus.

TEACHING TIP

In Example 3, students may wish to solve the problem by using the equation $r_1t_1 = r_2t_2$.

More About. . .

Space

Mercury is about 36 million miles from the Sun, making it the closest planet to the Sun. Its proximity to the Sun causes its temperature to be as high as 800°F.

Source: *World Book Encyclopedia*

Key Concept — *Inverse Variation*

y varies inversely as x if there is some nonzero constant k such that

$$xy = k \text{ or } y = \frac{k}{x} \text{ where } x \neq 0 \text{ and } y \neq 0.$$

Suppose y varies inversely as x such that $xy = 6$ or $y = \frac{6}{x}$. The graph of this equation is shown at the right. Note that in this case, k is a positive value 6, so as the values of x increase, the values of y decrease.

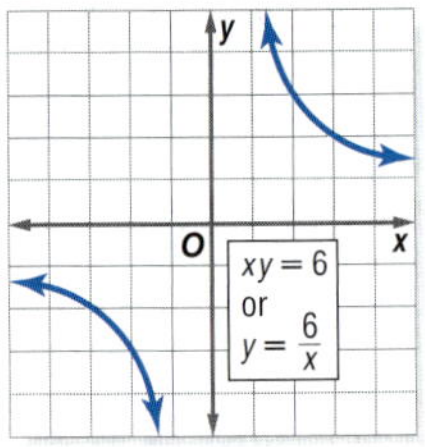

Just as with direct variation and joint variation, a proportion can be used with inverse variation to solve problems where some quantities are known. The following proportion is only one of several that can be formed.

$x_1y_1 = k$ and $x_2y_2 = k$

$x_1y_1 = x_2y_2$ — Substitution Property of Equality

$\frac{x_1}{y_2} = \frac{x_2}{y_1}$ — Divide each side by y_1y_2.

Example 3 *Inverse Variation*

If r varies inversely as t and $r = 18$ when $t = -3$, find r when $t = -11$.

Use a proportion that relates the values.

$\frac{r_1}{t_2} = \frac{r_2}{t_1}$ — Inverse variation

$\frac{18}{-11} = \frac{r_2}{-3}$ — $r_1 = 18$, $t_1 = -3$, and $t_2 = -11$

$18(-3) = -11(r_2)$ — Cross multiply.

$-54 = -11r_2$ — Simplify.

$4\frac{10}{11} = r_2$ — Divide each side by -11.

When $t = -11$, the value of r is $4\frac{10}{11}$.

Example 4 *Use Inverse Variation*

SPACE The apparent length of an object is inversely proportional to one's distance from the object. Earth is about 93 million miles from the Sun. Use the information at the left to find how much larger the diameter of the Sun would appear on Mercury than on Earth.

Explore You know that the apparent diameter of the Sun varies inversely with the distance from the Sun. You also know Mercury's distance from the Sun and Earth's distance from the Sun. You want to determine how much larger the diameter of the Sun appears on Mercury than on Earth.

Plan Let the apparent diameter of the Sun from Earth equal 1 unit and the apparent diameter of the Sun from Mercury equal m. Then use a proportion that relates the values.

DAILY INTERVENTION — **Unlocking Misconceptions**

Direct and Inverse Variation Help students understand the difference between the two types of variation by using the example of gas in the tank of a car, distance, and driving time. The amount of distance increases as the driving time increases (direct). The amount of gas decreases as the driving time increases (inverse).

Solve

$$\frac{\text{distance from Mercury}}{\text{apparent diameter from Earth}} = \frac{\text{distance from Earth}}{\text{apparent diameter from Mercury}}$$ Inverse variation

$$\frac{36 \text{ million miles}}{1 \text{ unit}} = \frac{93 \text{ million miles}}{m \text{ units}}$$ Substitution

$$(36 \text{ million miles})(m \text{ units}) = (93 \text{ million miles})(1 \text{ unit})$$ Cross multiply.

$$m = \frac{(93 \text{ million miles})(1 \text{ unit})}{36 \text{ million miles}}$$ Divide each side by 36 million miles.

$$m \approx 2.58 \text{ units}$$ Simplify.

Examine Since the distance between the Sun and Earth is between 2 and 3 times the distance between the Sun and Mercury, the answer seems reasonable. From Mercury, the diameter of the Sun will appear about 2.58 times as large as it appears from Earth.

Check for Understanding

Concept Check

1. **Determine** whether each graph represents a *direct* or an *inverse* variation.

a. 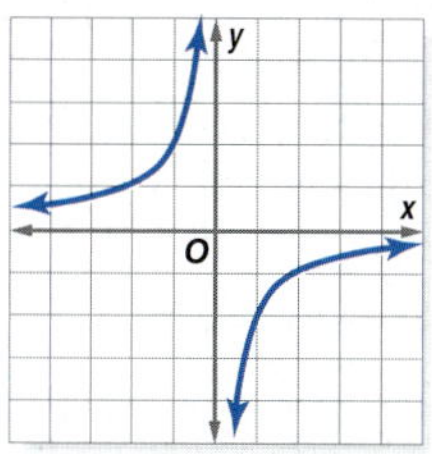

inverse

b. 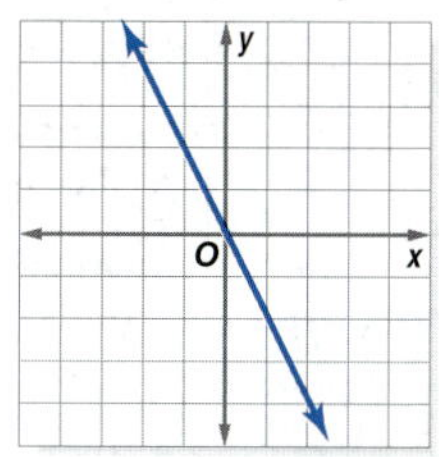

direct

2. **Compare and contrast** $y = 5x$ and $y = -5x$.

2. Both are examples of direct variation. For $y = 5x$, y increases as x increases. For $y = -5x$, y decreases as x increases.

3. **OPEN ENDED** Describe two quantities in real life that vary directly with each other and two quantities that vary inversely with each other. See margin.

Guided Practice

GUIDED PRACTICE KEY	
Exercises	Examples
4–9	1–3
10–13	4

State whether each equation represents a *direct, joint,* or *inverse* variation. Then name the constant of variation.

4. $ab = 20$ inverse; 20
5. $\frac{y}{x} = -0.5$ direct; −0.5
6. $A = \frac{1}{2}bh$ joint; $\frac{1}{2}$

Find each value.

7. If y varies directly as x and $y = 18$ when $x = 15$, find y when $x = 20$. 24
8. Suppose y varies jointly as x and z. Find y when $x = 9$ and $z = -5$, if $y = -90$ when $z = 15$ and $x = -6$. −45
9. If y varies inversely as x and $y = -14$ when $x = 12$, find x when $y = 21$. −8

Application

SWIMMING **For Exercises 10–13, use the following information.**

When a person swims underwater, the pressure in his or her ears varies directly with the depth at which he or she is swimming.

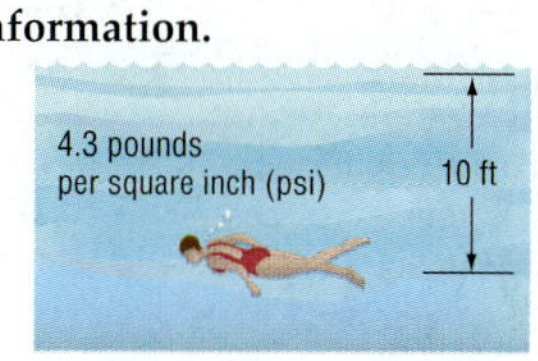

10. Write an equation of direct variation that represents this situation. $P = 0.43d$
11. Find the pressure at 60 feet. 25.8 psi
12. It is unsafe for amateur divers to swim where the water pressure is more than 65 pounds per square inch. How deep can an amateur diver safely swim?
13. Make a table showing the number of pounds of pressure at various depths of water. Use the data to draw a graph of pressure versus depth. See margin.

12. about 150 ft

3 Practice/Apply

Study Notebook

Have students—

- add the definitions/examples of the vocabulary terms to their Vocabulary Builder worksheets for Chapter 9.
- write the names and some examples from their lives for direct, joint, and inverse variation.
- include any other item(s) that they find helpful in mastering the skills in this lesson.

Answers

3. Sample answers: wages and hours worked, total cost and number of pounds of apples purchased; distances traveled and amount of gas remaining in the tank, distance of an object and the size it appears

13.

Depth (ft)	Pressure (psi)
0	0
1	0.43
2	0.86
3	1.29
4	1.72

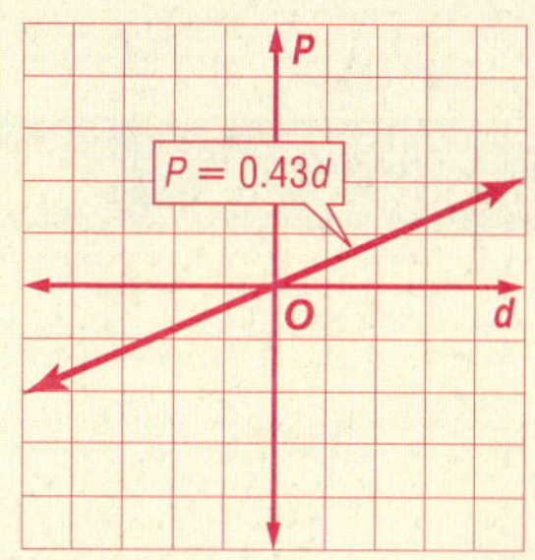

DAILY INTERVENTION

Differentiated Instruction

Auditory/Musical Have students find various kinds of variation in the sounds made by musical instruments. Suggest that they investigate the length and size of guitar strings relative to their vibrations, and the length and diameter of the columns of air used in wind and brass instruments for various notes.

★ indicates increased difficulty

Practice and Apply

Homework Help

For Exercises	See Examples
14–37	1–3
38–53	4

Extra Practice
See page 848.

State whether each equation represents a *direct, joint,* or *inverse* variation. Then name the constant of variation. **14. direct; 1.5 16. inverse; −18**

14. $\frac{n}{m} = 1.5$

15. $a = 5bc$ **joint; 5**

16. $vw = -18$

17. $3 = \frac{a}{b}$ **direct; 3**

18. $p = \frac{12}{q}$ **inverse; 12**

19. $y = -7x$ **direct; −7**

20. $V = \frac{1}{3}Bh$ **joint; $\frac{1}{3}$**

21. $\frac{2.5}{t} = s$ **inverse; 2.5**

22. **CHEMISTRY** Boyle's Law states that when a sample of gas is kept at a constant temperature, the volume varies inversely with the pressure exerted on it. Write an equation for Boyle's Law that expresses the variation in volume V as a function of pressure P. $V = \frac{k}{P}$

23. **CHEMISTRY** Charles' Law states that when a sample of gas is kept at a constant pressure, its volume V will increase as the temperature t increases. Write an equation for Charles' Law that expresses volume as a function. $V = kt$

24. **GEOMETRY** How does the circumference of a circle vary with respect to its radius? What is the constant of variation? **directly; 2π**

25. **TRAVEL** A map is scaled so that 3 centimeters represents 45 kilometers. How far apart are two towns if they are 7.9 centimeters apart on the map? **118.5 km**

Find each value.

26. If y varies directly as x and $y = 15$ when $x = 3$, find y when $x = 12$. **60**

27. If y varies directly as x and $y = 8$ when $x = 6$, find y when $x = 15$. **20**

28. Suppose y varies jointly as x and z. Find y when $x = 2$ and $z = 27$, if $y = 192$ when $x = 8$ and $z = 6$. **216**

29. If y varies jointly as x and z and $y = 80$ when $x = 5$ and $z = 8$, find y when $x = 16$ and $z = 2$. **64**

30. If y varies inversely as x and $y = 5$ when $x = 10$, find y when $x = 2$. **25**

31. If y varies inversely as x and $y = 16$ when $x = 5$, find y when $x = 20$. **4**

32. If y varies inversely as x and $y = 2$ when $x = 25$, find x when $y = 40$. **1.25**

33. If y varies inversely as x and $y = 4$ when $x = 12$, find y when $x = 5$. **9.6**

34. If y varies directly as x and $y = 9$ when $x = -15$, find y when $x = 21$. **−12.6**

35. If y varies directly as x and $x = 6$ when $y = 0.5$, find y when $x = 10$. **0.83**

★ 36. Suppose y varies jointly as x and z. Find y when $x = \frac{1}{2}$ and $z = 6$, if $y = 45$ when $x = 6$ and $z = 10$. **$2\frac{1}{4}$**

★ 37. If y varies jointly as x and z and $y = \frac{1}{8}$ when $x = \frac{1}{2}$ and $z = 3$, find y when $x = 6$ and $z = \frac{1}{3}$. **$\frac{1}{6}$**

38. **WORK** Paul drove from his house to work at an average speed of 40 miles per hour. The drive took him 15 minutes. If the drive home took him 20 minutes and he used the same route in reverse, what was his average speed going home? **30 mph**

39. **WATER SUPPLY** Many areas of Northern California depend on the snowpack of the Sierra Nevada Mountains for their water supply. If 250 cubic centimeters of snow will melt to 28 cubic centimeters of water, how much water does 900 cubic centimeters of snow produce? **100.8 cm³**

Career Choices

Travel Agent

Travel agents give advice and make arrangements for transportation, accommodations, and recreation. For international travel, they also provide information on customs and currency exchange.

Online Research For information about a career as a travel agent, visit: www.algebra2.com/careers

Study Guide and Intervention, p. 535 (shown) and p. 536

Direct Variation and Joint Variation

Direct Variation	y varies directly as x if there is some nonzero constant k such that $y = kx$. k is called the constant of variation.
Joint Variation	y varies jointly as x and z if there is some number k such that $y = kxz$, where $x \neq 0$ and $z \neq 0$.

Example **Find each value.**

a. If y varies directly as x and $y = 16$ when $x = 4$, find x when $y = 20$.

$\frac{y_1}{x_1} = \frac{y_2}{x_2}$ Direct proportion

$\frac{16}{4} = \frac{20}{x_2}$ $y_1 = 16$, $x_1 = 4$, and $y_2 = 20$

$16x_2 = (20)(4)$ Cross multiply.

$x_2 = 5$ Simplify.

The value of x is 5 when y is 20.

b. If y varies jointly as x and z and $y = 10$ when $x = 2$ and $z = 4$, find y when $x = 4$ and $z = 3$.

$\frac{y_1}{x_1z_1} = \frac{y_2}{x_2z_2}$ Joint variation

$\frac{10}{2 \cdot 4} = \frac{y_2}{4 \cdot 3}$ $y_1 = 10$, $x_1 = 2$, $z_1 = 4$, $x_2 = 4$, and $z_2 = 3$

$120 = 8y_2$ Simplify.

$y_2 = 15$ Divide each side by 8.

The value of y is 15 when $x = 4$ and $z = 3$.

Exercises

Find each value.

1. If y varies directly as x and $y = 9$ when $x = 6$, find y when $x = 8$. **12**
2. If y varies directly as x and $y = 16$ when $x = 36$, find y when $x = 54$. **24**
3. If y varies directly as x and $x = 15$ when $y = 5$, find x when $y = 9$. **27**
4. If y varies directly as x and $x = 33$ when $y = 22$, find x when $y = 32$. **48**
5. Suppose y varies jointly as x and z. Find y when $x = 5$ and $z = 3$, if $y = 18$ when $x = 3$ and $z = 2$. **45**
6. Suppose y varies jointly as x and z. Find y when $x = 6$ and $z = 8$, if $y = 6$ when $x = 4$ and $z = 2$. **36**
7. Suppose y varies jointly as x and z. Find y when $x = 4$ and $z = 11$, if $y = 60$ when $x = 3$ and $z = 5$. **176**
8. Suppose y varies jointly as x and z. Find y when $x = 5$ and $z = 2$, if $y = 84$ when $x = 4$ and $z = 7$. **30**
9. If y varies directly as x and $y = 14$ when $x = 35$, find y when $x = 12$. **4.8**
10. If y varies directly as x and $x = 200$ when $y = 50$, find x when $y = 1000$. **4000**
11. If y varies directly as x and $y = 39$ when $x = 52$, find y when $x = 22$. **16.5**
12. If y varies directly as x and $x = 60$ when $y = 75$, find x when $y = 42$. **33.6**
13. Suppose y varies jointly as x and z. Find y when $x = 6$ and $z = 11$, if $y = 120$ when $x = 5$ and $z = 12$. **132**
14. Suppose y varies jointly as x and z. Find y when $x = 5$ and $z = 10$, if $y = 12$ when $x = 8$ and $z = 6$. **12.5**
15. Suppose y varies jointly as x and z. Find y when $x = 7$ and $z = 18$, if $y = 351$ when $x = 6$ and $z = 13$. **567**
16. Suppose y varies jointly as x and z. Find y when $x = 5$ and $z = 27$, if $y = 480$ when $x = 9$ and $z = 20$. **360**

Skills Practice, p. 537 and Practice, p. 538 (shown)

State whether each equation represents a *direct, joint,* or *inverse* variation. Then name the constant of variation.

1. $u = 8wz$ **joint; 8**
2. $p = 4s$ **direct; 4**
3. $L = \frac{5}{k}$ **inverse; 5**
4. $xy = 4.5$ **inverse; 4.5**
5. $\frac{C}{d} = \pi$ **direct; π**
6. $2d = mn$ **joint; $\frac{1}{2}$**
7. $\frac{1.25}{g} = h$ **inverse; 1.25**
8. $y = \frac{3}{4x}$ **inverse; $\frac{3}{4}$**

Find each value.

9. If y varies directly as x and $y = 8$ when $x = 2$, find y when $x = 6$. **24**
10. If y varies directly as x and $y = -16$ when $x = 6$, find x when $y = -4$. **1.5**
11. If y varies directly as x and $y = 132$ when $x = 11$, find y when $x = 33$. **396**
12. If y varies directly as x and $y = 7$ when $x = 1.5$, find y when $x = 4$. **$\frac{56}{3}$**
13. If y varies jointly as x and z and $y = 24$ when $x = 2$ and $z = 1$, find y when $x = 12$ and $z = 2$. **288**
14. If y varies jointly as x and z and $y = 60$ when $x = 3$ and $z = 4$, find y when $x = 6$ and $z = 8$. **240**
15. If y varies jointly as x and z and $y = 12$ when $x = -2$ and $z = 3$, find y when $x = 4$ and $z = -1$. **8**
16. If y varies inversely as x and $y = 16$ when $x = 4$, find y when $x = 3$. **$\frac{64}{3}$**
17. If y varies inversely as x and $y = 3$ when $x = 5$, find x when $y = 2.5$. **6**
18. If y varies inversely as x and $y = -18$ when $x = 6$, find y when $x = 5$. **−21.6**
19. If y varies directly as x and $y = 5$ when $x = 0.4$, find x when $y = 37.5$. **3**
20. **GASES** The volume V of a gas varies inversely as its pressure P. If $V = 80$ cubic centimeters when $P = 2000$ millimeters of mercury, find V when $P = 320$ millimeters of mercury. **500 cm³**
21. **SPRINGS** The length S that a spring will stretch varies directly with the weight F that is attached to the spring. If a spring stretches 20 inches with 25 pounds attached, how far will it stretch with 15 pounds attached? **12 in.**
22. **GEOMETRY** The area A of a trapezoid varies jointly as its height and the sum of its bases. If the area is 480 square meters when the height is 20 meters and the bases are 28 meters and 20 meters, what is the area of a trapezoid when its height is 8 meters and its bases are 10 meters and 15 meters? **100 m²**

Reading to Learn Mathematics, p. 539 ELL

Pre-Activity **How is variation used to find the total cost given the unit cost?**

Read the introduction to Lesson 9-4 at the top of page 492 in your textbook.

- For each additional student who enrolls in a public college, the total high-tech spending will **increase** (increase/decrease) by **$149**.
- For each decrease in enrollment of 100 students in a public college, the total high-tech spending will **decrease** (increase/decrease) by **$14,900**.

Reading the Lesson

1. Write an equation to represent each of the following variation statements. Use k as the constant of variation.
   a. m varies inversely as n. $m = \frac{k}{n}$
   b. s varies directly as r. $s = kr$
   c. t varies jointly as p and q. $t = kpq$
2. Which type of variation, direct or inverse, is represented by each graph?

a. 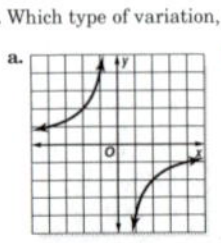 **inverse**

b. 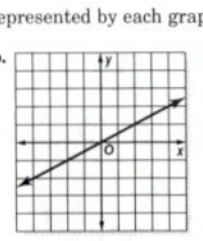 **direct**

Helping You Remember

3. How can your knowledge of the equation of the slope-intercept form of the equation of a line help you remember the equation for direct variation?

Sample answer: The graph of an equation expressing direct variation is a line. The slope-intercept form of the equation of a line is $y = mx + b$. In direct variation, if one of the quantities is 0, the other quantity is also 0, so $b = 0$ and the line goes through the origin. The equation of a line through the origin is $y = mx$, where m is the slope. This is the same as the equation for direct variation with $k = m$.

Enrichment, p. 540

Expansions of Rational Expressions

Many rational expressions can be transformed into **power series.** A power series is an infinite series of the form $A + Bx + Cx^2 + Dx^3 + \ldots$. The rational expression and the power series normally can be said to have the same values only for certain values of x. For example, the following equation holds only for values of x such that $-1 < x < 1$.

$$\frac{1}{1-x} = 1 + x + x^2 + x^3 + \ldots \text{ for } -1 < x < 1$$

Example **Expand $\frac{2 + 3x}{1 + x + x^2}$ in ascending powers of x.**

Assume that the expression equals a series of the form $A + Bx + Cx^2 + Dx^3 + \ldots$. Then multiply both sides of the equation by the denominator $1 + x + x^2$.

$$\frac{2 + 3x}{1 + x + x^2} = A + Bx + Cx^2 + Dx^3 + \ldots$$

$$2 + 3x = (1 + x + x^2)(A + Bx + Cx^2 + Dx^3 + \ldots)$$

$2 + 3x = A + Bx + Cx^2 + Dx^3 +$

496 Chapter 9 Rational Expressions and Equations

40. **RESEARCH** According to Johannes Kepler's third law of planetary motion, the ratio of the square of a planet's period of revolution around the Sun to the cube of its mean distance from the Sun is constant for all planets. Verify that this is true for at least three planets. **See students' work.**

More About. . .

Biology

In order to sustain itself in its cold habitat, a Siberian tiger requires 20 pounds of meat per day.

Source: *Wildlife Fact File*

BIOLOGY **For Exercises 41–43, use the information at the left.**

41. Write an equation to represent the amount of meat needed to sustain s Siberian tigers for d days. $m = 20sd$

42. Is your equation in Exercise 41 a *direct, joint,* or *inverse* variation? **joint**

43. How much meat do three Siberian tigers need for the month of January? **1860 lb**

LAUGHTER **For Exercises 44–46, use the following information.**
According to *The Columbus Dispatch*, the average American laughs 15 times per day.

44. Write an equation to represent the average number of laughs produced by m household members during a period of d days. $\ell = 15md$

45. Is your equation in Exercise 44 a *direct, joint,* or *inverse* variation? **joint**

46. Assume that members of your household laugh the same number of times each day as the average American. How many times would the members of your household laugh in a week? **See students' work.**

ARCHITECTURE **For Exercises 47–49, use the following information.**
When designing buildings such as theaters, auditoriums, or museums architects have to consider how sound travels. Sound intensity I is inversely proportional to the square of the distance from the sound source d.

★47. Write an equation that represents this situation. $I = \frac{k}{d^2}$

★48. If d is the independent variable and I is the dependent variable, graph the equation from Exercise 47 when $k = 16$. **See margin.**

★49. If a person in a theater moves to a seat twice as far from the speakers, compare the new sound intensity to that of the original.

49. The sound will be heard $\frac{1}{4}$ as intensely.

Study Tip

Combined Variation
Many applied problems involve a combination of direct, inverse, and joint variation. This is called *combined variation.*

TELECOMMUNICATIONS **For Exercises 50–53, use the following information.**
It has been found that the average number of daily phone calls C between two cities is directly proportional to the product of the populations P_1 and P_2 of two cities and inversely proportional to the square of the distance d between the cities. That is, $C = \frac{kP_1P_2}{d^2}$.

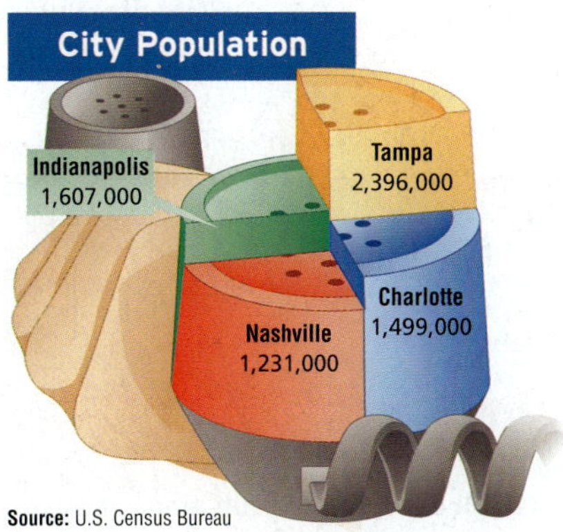

Source: U.S. Census Bureau

★50. The distance between Nashville and Charlotte is about 425 miles. If the average number of daily phone calls between the cities is 204,000, find the value of k and write the equation of variation. Round to the nearest hundredth.

50. 0.02; $C = \frac{0.02P_1P_2}{d^2}$

★51. Nashville is about 680 miles from Tampa. Find the average number of daily phone calls between them. **about 127,572 calls**

★52. The average daily phone calls between Indianapolis and Charlotte is 133,380. Find the distance between Indianapolis and Charlotte. **about 601 mi**

★53. Could you use this formula to find the populations or the average number of phone calls between two adjoining cities? Explain. **no;** $d \neq 0$

54. **CRITICAL THINKING** Write a real-world problem that involves a joint variation. Solve the problem.

54. Sample answer: If the average student spends \$2.50 for lunch in the school cafeteria, write an equation to represent the amount s students will spend for lunch in d days. How much will 30 students spend in a week? $a = 2.50sd$**; \$375**

www.algebra2.com/self_check_quiz

About the Exercises...

Organization by Objective
- **Direct Variation and Joint Variation:** 23–29, 34–37, 39–46
- **Inverse Variation:** 22, 30–33, 38, 47–49

Odd/Even Assignments
Exercises 14–39 are structured so that students practice the same concepts whether they are assigned odd or even problems.

Alert! Exercise 40 requires reference materials for planetary data.

Assignment Guide

Basic: 15–35 odd, 39, 41–43, 54–73

Average: 15–39 odd, 44–49, 54–73

Advanced: 14–40 even, 44–67 (optional: 68–73)

All: Practice Quiz 2 (1–5)

Answer

48.

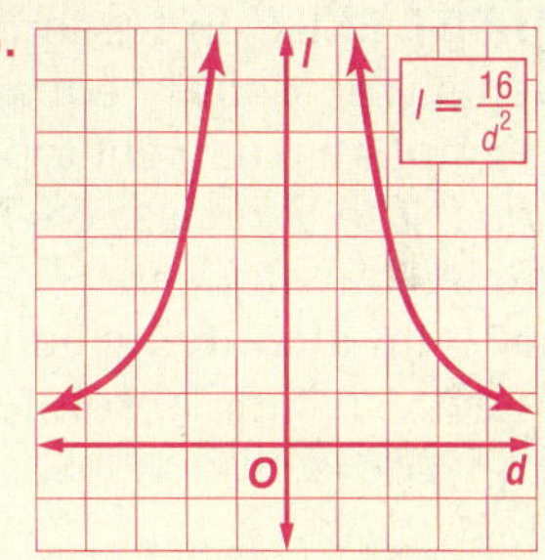

Teacher to Teacher

Susan Nelson — Spring H.S., Spring, TX

"I have my students do a data gathering activity called Rotations where we have the student do a regression for the diameter of a lid versus the number of rotations it takes to move across a fixed length of masking tape."

4 Assess

Open-Ended Assessment

Modeling Have students write equations for various variations in their life (time spent studying, hours of sleep, and so on). Ask them to write examples and explain them.

Tips for New Teachers

Intervention Make sure that students understand the essential differences between direct and inverse variation. For example, ask them if the number of candles on a birthday cake varies directly or indirectly with the age of the birthday person. Then ask how the length of the candle remaining varies with the time the candle has burned.

Getting Ready for Lesson 9-5

PREREQUISITE SKILL In Lesson 9-5, students will identify equations and graphs as different types of functions. Use Exercises 68–73 to determine your students' familiarity with identifying equations as step, constant, absolute value, or piecewise functions.

Assessment Options

Practice Quiz 2 The quiz provides students with a brief review of the concepts and skills in Lessons 9-3 and 9-4. Lesson numbers are given to the right of exercises or instruction lines so students can review concepts not yet mastered.

---

55. **WRITING IN MATH** Answer the question that was posed at the beginning of the lesson. **See margin.**

How is variation used to find the total cost given the unit cost?

Include the following in your answer:
- an explanation of why the equation for the total cost is a direct variation, and
- a problem involving unit cost and total cost of an item and its solution.

Standardized Test Practice

56. If the ratio of $2a$ to $3b$ is 4 to 5, what is the ratio of $5a$ to $4b$? **D**

Ⓐ $\frac{4}{3}$ Ⓑ $\frac{3}{4}$ Ⓒ $\frac{9}{8}$ Ⓓ $\frac{3}{2}$

57. Suppose b varies inversely as the square of a. If a is multiplied by 9, which of the following is true for the value of b? **C**

Ⓐ It is multiplied by $\frac{1}{3}$. Ⓑ It is multiplied by $\frac{1}{9}$.

Ⓒ It is multiplied by $\frac{1}{81}$. Ⓓ It is multiplied by 3.

Maintain Your Skills

Mixed Review Determine the equations of any vertical asymptotes and the values of x for any holes in the graph of each rational function. *(Lesson 9-3)*

58. $f(x) = \frac{x+1}{x^2-1}$ **asymp.: $x = 1$; hole: $x = -1$**
59. $f(x) = \frac{x+3}{x^2+x-12}$ **asymp.: $x = -4$, $x = 3$**
60. $f(x) = \frac{x^2+4x+3}{x+3}$ **hole: $x = -3$**

Simplify each expression. *(Lesson 9-2)*

61. $\frac{3x}{x-y} + \frac{4x}{y-x}$ **$\frac{x}{y-x}$**
62. $\frac{t}{t+2} - \frac{2}{t^2-4}$
63. $\frac{m - \frac{1}{m}}{1 + \frac{4}{m} - \frac{5}{m^2}}$ **$\frac{m(m+1)}{m+5}$**

62. $\frac{t^2-2t-2}{(t+2)(t-2)}$

64. **ASTRONOMY** The distance from Earth to the Sun is approximately 93,000,000 miles. Write this number in scientific notation. *(Lesson 5-1)* **9.3×10^7**

State the slope and the y-intercept of the graph of each equation. *(Lesson 2-4)*

65. $y = 0.4x + 1.2$ **0.4; 1.2**
66. $2y = 6x + 14$ **3; 7**
67. $3x + 5y = 15$ **$-\frac{3}{5}$; 3**

Getting Ready for the Next Lesson **PREREQUISITE SKILL** Identify each function as S for step, C for constant, A for absolute value, or P for piecewise. *(To review* ***special functions****, see Lesson 2-6.)*

68. $h(x) = \frac{2}{3}$ **C**
69. $g(x) = 3|x|$ **A**
70. $f(x) = [\![2x]\!]$ **S**
71. $f(x) = \begin{cases} 1 \text{ if } x > 0 \\ -1 \text{ if } x \le 0 \end{cases}$ **P**
72. $h(x) = |x - 2|$ **A**
73. $g(x) = -3$ **C**

Practice Quiz 2 — Lessons 9-3 and 9-4

Graph each rational function. *(Lesson 9-3)* **1–2. See pp. 519A–519D.**

1. $f(x) = \frac{x-1}{x-4}$
2. $f(x) = \frac{-2}{x^2-6x+9}$

Find each value. *(Lesson 9-4)*

3. If y varies inversely as x and $x = 14$ when $y = 7$, find x when $y = 2$. **49**
4. If y varies directly as x and $y = 1$ when $x = 5$, find y when $x = 22$. **4.4**
5. If y varies jointly as x and z and $y = 80$ when $x = 25$ and $z = 4$, find y when $x = 20$ and $z = 7$. **112**

Answer

55. A direct variation can be used to determine the total cost when the cost per unit is known. Answers should include the following.
- Since the total cost T is the cost per unit u times the number of units n or $T = un$, the relationship is a direct variation. In this equation u is the constant of variation.
- Sample answer: The school store sells pencils for 20¢ each. John wants to buy 5 pencils. What is the total cost of the pencils? ($1.00)

9-5 Classes of Functions

What You'll Learn

- Identify graphs as different types of functions.
- Identify equations as different types of functions.

How can graphs of functions be used to determine a person's weight on a different planet?

The purpose of the 2001 Mars Odyssey Mission is to study conditions on Mars. The findings will help NASA prepare for a possible mission with human explorers. The graph at the right compares a person's weight on Earth with his or her weight on Mars. This graph represents a direct variation, which you studied in the previous lesson.

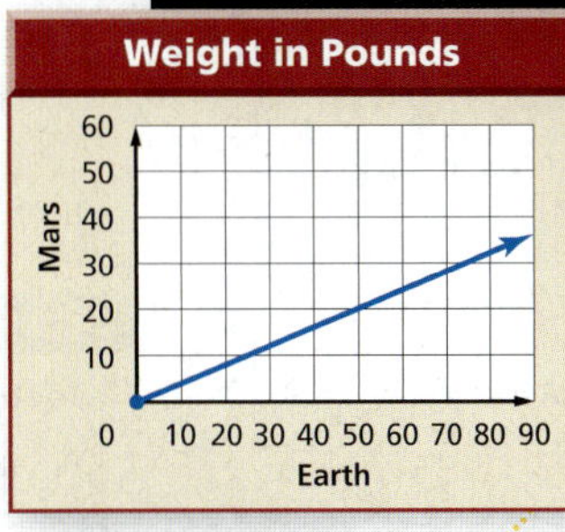

IDENTIFY GRAPHS In this book, you have studied several types of graphs representing special functions. The following is a summary of these graphs.

Concept Summary — *Special Functions*

Constant Function

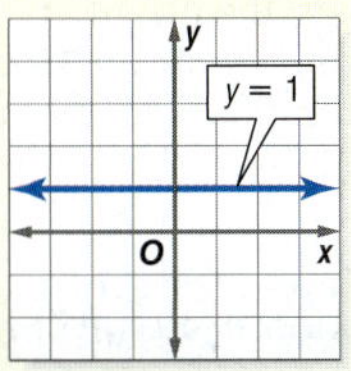

The general equation of a constant function is $y = a$, where a is any number. Its graph is a horizontal line that crosses the y-axis at a.

Direct Variation Function

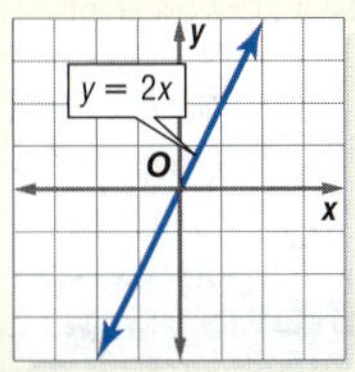

The general equation of a direct variation function is $y = ax$, where a is a nonzero constant. Its graph is a line that passes through the origin and is neither horizontal nor vertical.

Identity Function

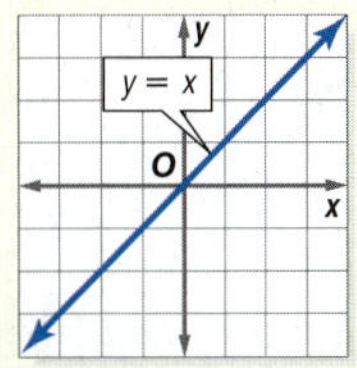

The identity function $y = x$ is a special case of the direct variation function in which the constant is 1. Its graph passes through all points with coordinates (a, a).

Greatest Integer Function

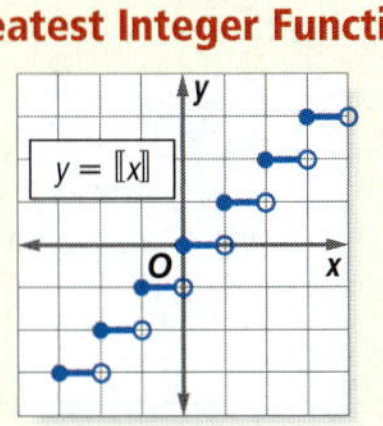

If an equation includes an expression inside the greatest integer symbol, the function is a greatest integer function. Its graph looks like steps.

Absolute Value Function

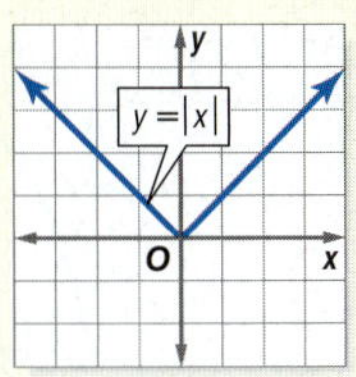

An equation with a direct variation expression inside absolute value symbols is an absolute value function. Its graph is in the shape of a V.

Quadratic Function

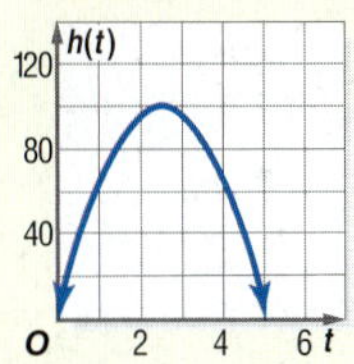

The general equation of a quadratic function is $y = ax^2 + bx + c$, where $a \neq 0$. Its graph is a parabola.

(continued on the next page)

9-5 Lesson Notes

1 Focus

5-Minute Check Transparency 9-5 Use as a quiz or review of Lesson 9-4.

Mathematical Background notes are available for this lesson on p. 470D.

Building on Prior Knowledge

In previous course material, students have learned about different kinds of functions. In this lesson, students will revisit different functions and group them into logical categories based on their characteristics.

How can graphs of functions be used to determine a person's weight on a different planet?

Ask students:

- According to the graph, what is the approximate weight on Mars of a person who weighs 50 pounds on Earth? **about 20 lb**
- According to the graph, what is the approximate weight on Earth of a person who would weigh 30 pounds on Mars? **about 75 lb**

Resource Manager

Workbook and Reproducible Masters

Chapter 9 Resource Masters

- Study Guide and Intervention, pp. 541–542
- Skills Practice, p. 543
- Practice, p. 544
- Reading to Learn Mathematics, p. 545
- Enrichment, p. 546
- Assessment, p. 568

Transparencies

5-Minute Check Transparency 9-5
Answer Key Transparencies

Technology

Interactive Chalkboard

2 Teach

IDENTIFY GRAPHS

In-Class Example

1 Identify the type of function represented by each graph.

a.

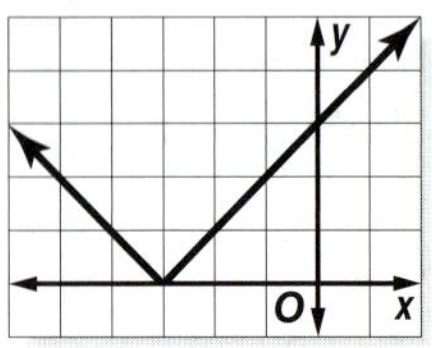

absolute value function

b.

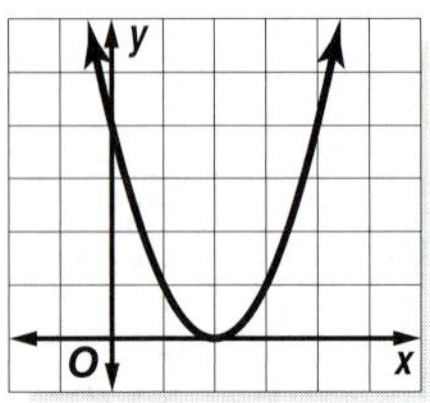

quadratic function

IDENTIFY EQUATIONS

In-Class Example

2 **SHIPPING CHARGES** A chart gives the shipping rates for an Internet company. They charge \$3.50 to ship less than 1 pound, \$3.95 for 1 pound and over up to 2 pounds, and \$5.20 for 2 pounds and over up to 3 pounds. Which graph depicts these rates? **c, the step or greatest integer function**

a.

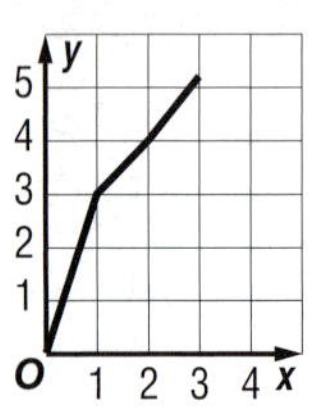

b.

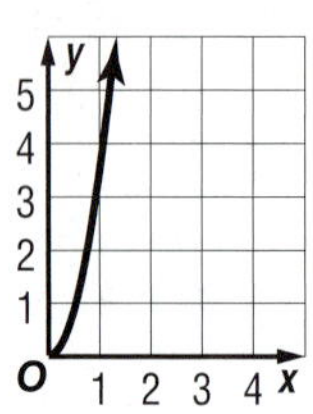

c.

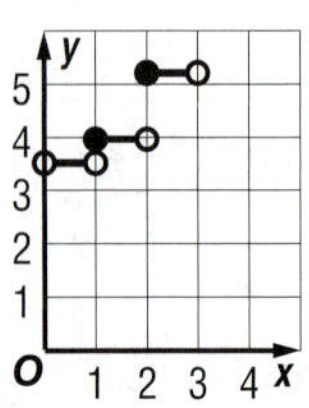

Concept Summary — Special Functions

Square Root Function	Rational Function	Inverse Variation Function
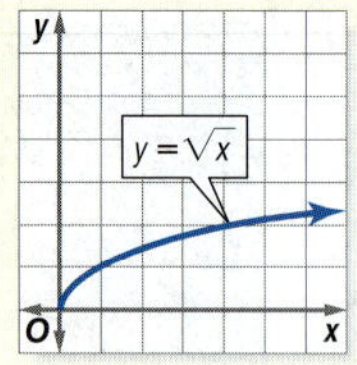	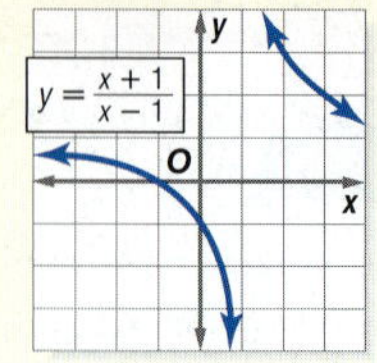	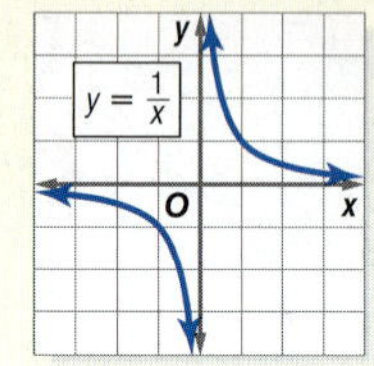
If an equation includes an expression inside the radical sign, the function is a square root function. Its graph is a curve that starts at a point and continues in only one direction.	The general equation for a rational function is $y = \frac{p(x)}{q(x)}$, where $p(x)$ and $q(x)$ are polynomial functions. Its graph has one or more asymptotes and/or holes.	The inverse variation function $y = \frac{a}{x}$ is a special case of the rational function where $p(x)$ is a constant and $q(x) = x$. Its graph has two asymptotes, $x = 0$ and $y = 0$.

Example 1 Identify a Function Given the Graph

Identify the type of function represented by each graph.

a.

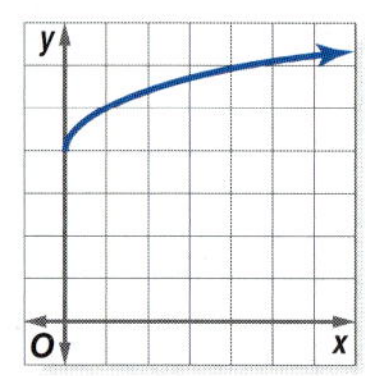

The graph has a starting point and curves in one direction. The graph represents a square root function.

b.

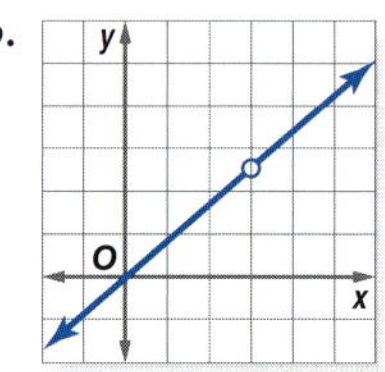

The graph appears to be a direct variation since it is a straight line passing through the origin. However, the hole indicates that it represents a rational function.

IDENTIFY EQUATIONS If you can identify an equation as a type of function, you can determine the shape of the graph.

Example 2 Match Equation with Graph

ROCKETRY **Emily launched a toy rocket from ground level. The height above the ground level h, in feet, after t seconds is given by the formula $h(t) = -16t^2 + 80t$. Which graph depicts the height of the rocket during its flight?**

a.

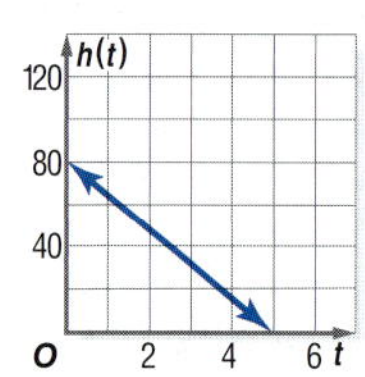

b.

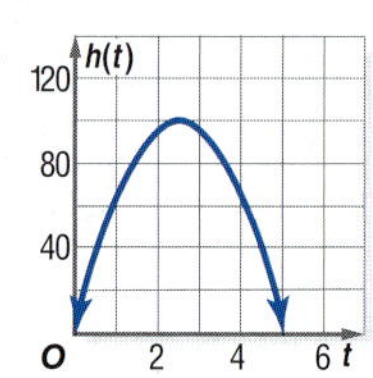

c.

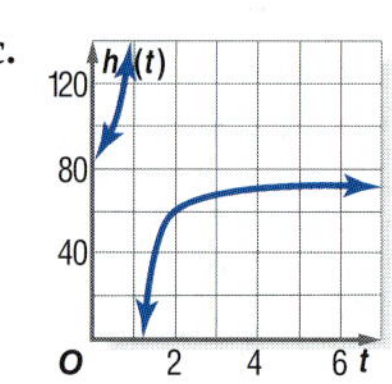

The function includes a second-degree polynomial. Therefore, it is a quadratic function, and its graph is a parabola. Graph b is the only parabola. Therefore, the answer is graph **b**.

Teacher to Teacher

Deedee S. Adams — Oxford H.S., Oxford, AL

"I have my students play Simon Says by having them all stand and graph different types of functions with their arms. Students sit down if they don't illustrate the correct graph."

500 Chapter 9 Rational Expressions and Equations

Sometimes recognizing an equation as a specific type of function can help you graph the function.

Example 3 Identify a Function Given its Equation

Identify the type of function represented by each equation. Then graph the equation.

a. $y = |x| - 1$

Since the equation includes an expression inside absolute value symbols, it is an absolute value function. Therefore, the graph will be in the shape of a V. Determine some points on the graph and use what you know about graphs of absolute value functions to graph the function.

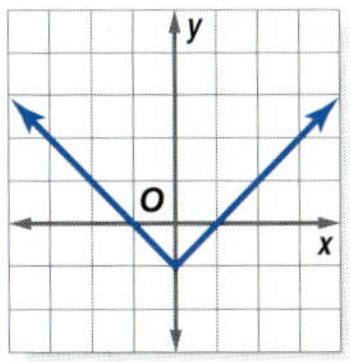

b. $y = -\frac{2}{3}x$

The function is in the form $y = ax$, where $a = -\frac{2}{3}$. Therefore, it is a direct variation function. The graph passes through the origin and has a slope of $-\frac{2}{3}$.

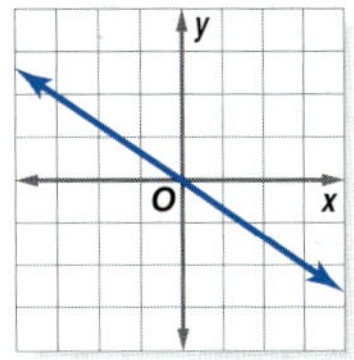

Check for Understanding

Concept Check

1. **OPEN ENDED** Find a counterexample to the statement *All functions are continuous.* Describe your function. **See margin.**
2. **Name** three special functions whose graphs are straight lines. Give an example of each function. **constant ($y = 1$), direct variation ($y = 2x$), identity ($y = x$)**
3. **Describe** the graph of $y = [\![x + 2]\!]$.

Guided Practice

GUIDED PRACTICE KEY	
Exercises	Examples
4–6	1
7, 8, 12	2
9–11	3

3. The equation is a greatest integer function. The graph looks like a series of steps.

Identify the type of function represented by each graph.

4. 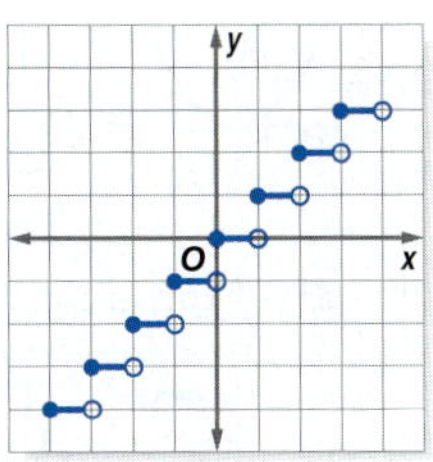

greatest integer

5. 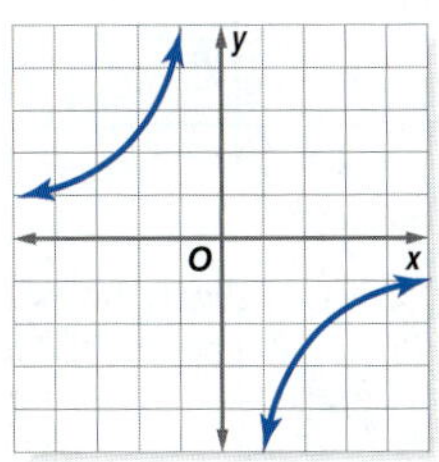

inverse variation or rational

6. 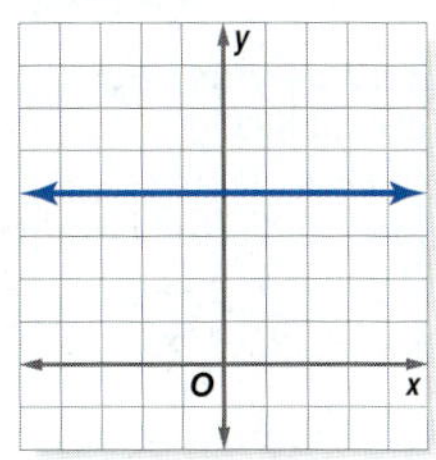

constant

Match each graph with an equation at the right.

7. 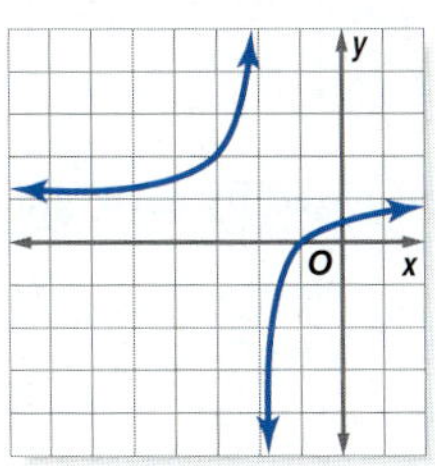

c

8. 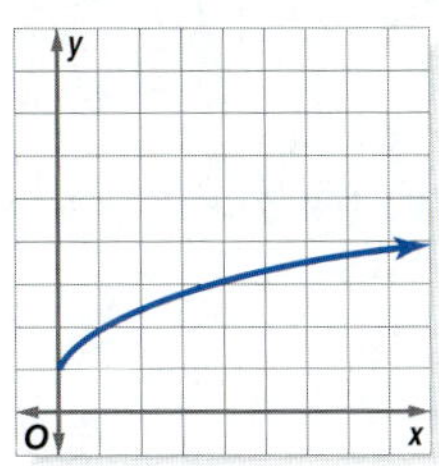

b

a. $y = x^2 + 2x + 3$
b. $y = \sqrt{x + 1}$
c. $y = \frac{x + 1}{x + 2}$
d. $y = [\![2x]\!]$

www.algebra2.com/extra_examples

In-Class Example (PowerPoint®)

3 Identify the type of function represented by each equation. Then graph the equation.

a. $y = -3$ **constant function**

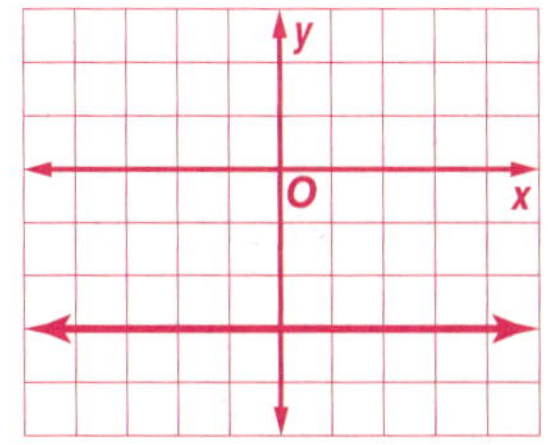

b. $y = \sqrt{9x}$ **square root function**

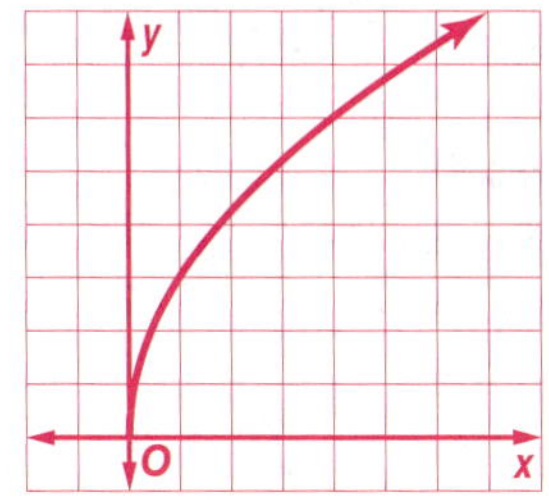

3 Practice/Apply

Study Notebook

Have students—

- add the definitions/examples of the vocabulary terms to their Vocabulary Builder worksheets for Chapter 9.
- add sketches to illustrate each special function graph.
- include any other item(s) that they find helpful in mastering the skills in this lesson.

Answer

1. Sample answer:

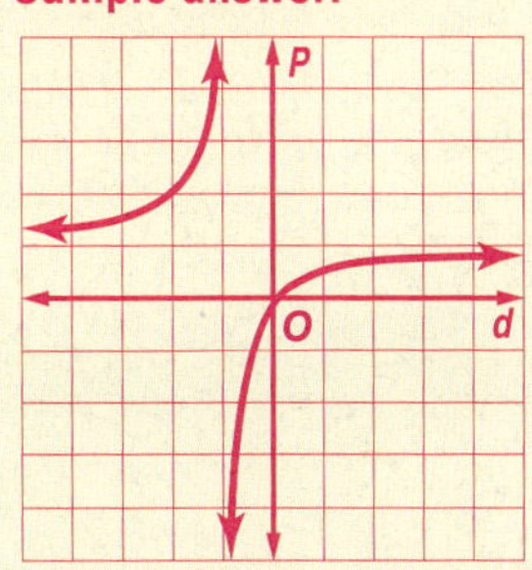

This graph is a rational function. It has an asymptote at $x = -1$.

DAILY INTERVENTION — Differentiated Instruction

Interpersonal Have students work with a partner or in small groups to do quick sketches of graphs and identify the type of function the graph could represent. Have each group make a list of the identifying characteristics of the graph; then ask groups to exchange and compare their lists.

About the Exercises...

Organization by Objective

- **Identify Graphs:** 13–22
- **Identify Equations:** 23–34

Odd/Even Assignments

Exercises 13–30 are structured so that students practice the same concepts whether they are assigned odd or even problems.

Assignment Guide

Basic: 13–29 odd, 31–33, 37–61

Average: 13–29 odd, 31–33, 37–61

Advanced: 14–30 even, 31–55 (optional: 56–61)

Answers

9.

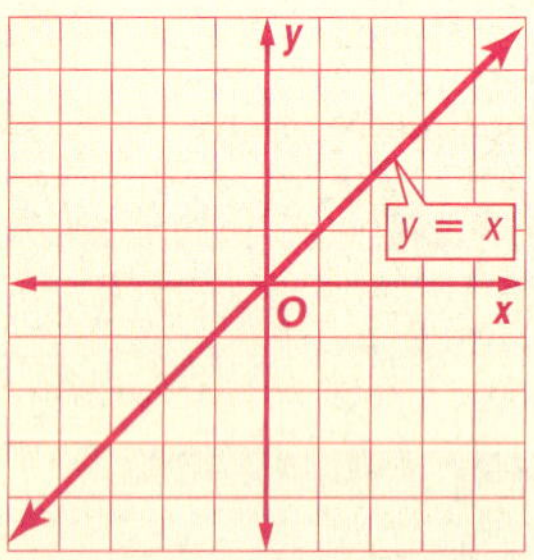

10.

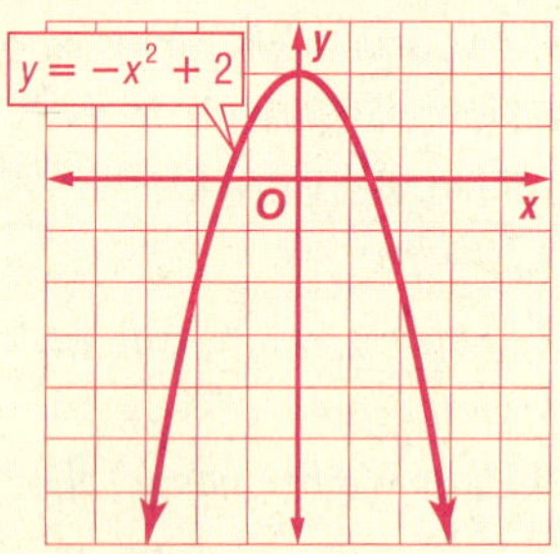

11.

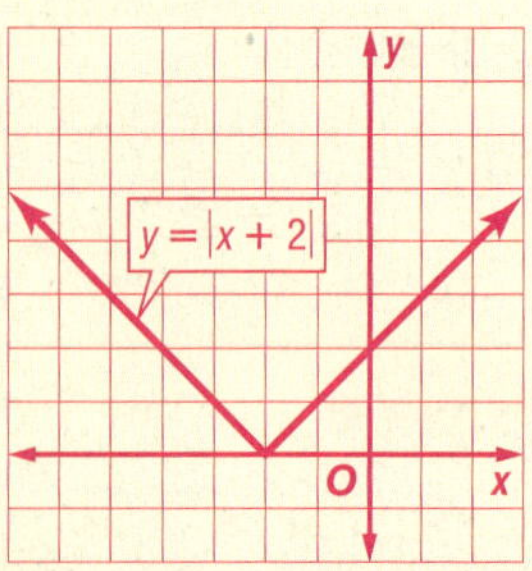

9–11. See margin for graphs.

Identify the type of function represented by each equation. Then graph the equation.

9. $y = x$ identity or direct variation
10. $y = -x^2 + 2$ quadratic
11. $y = |x + 2|$ absolute value

Application

12. **GEOMETRY** Write the equation for the area of a circle. Identify the equation as a type of function. Describe the graph of the function. $A = \pi r^2$; quadratic; the graph is a parabola.

★ indicates increased difficulty

Practice and Apply

Homework Help

For Exercises	See Examples
13–18	1
19–22, 31–36	2
23–30	3

Extra Practice See page 848.

Identify the type of function represented by each graph.

13.

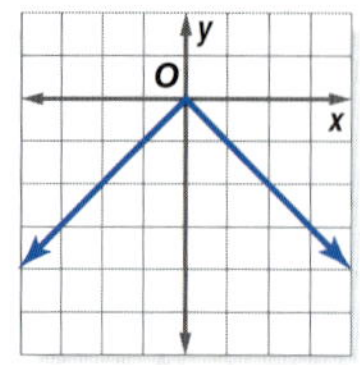

absolute value

14.

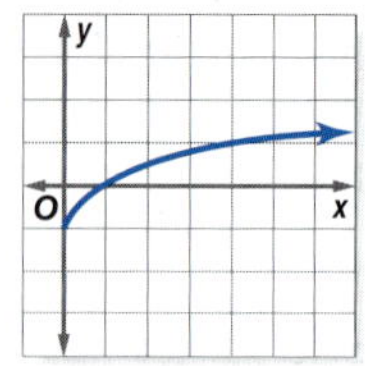

square root

15.

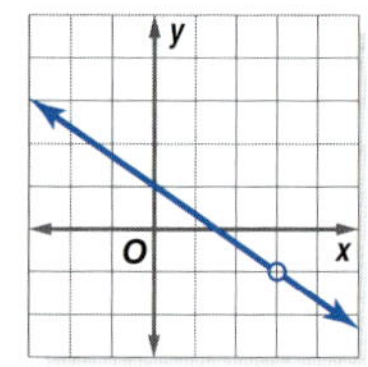

rational

16.

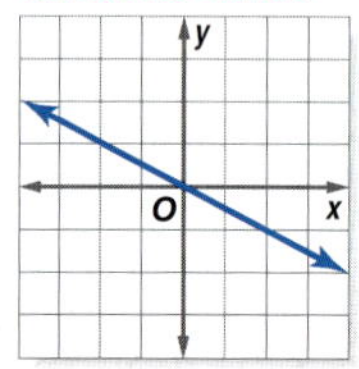

direct variation

17.

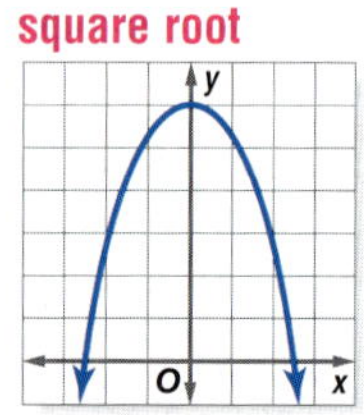

quadratic

18.

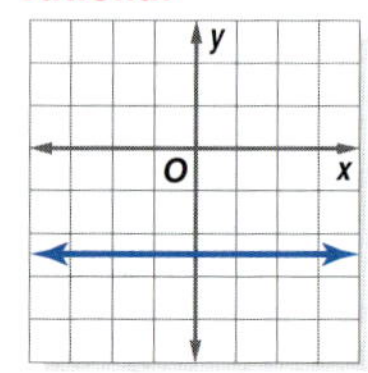

constant

Match each graph with an equation at the right.

19.

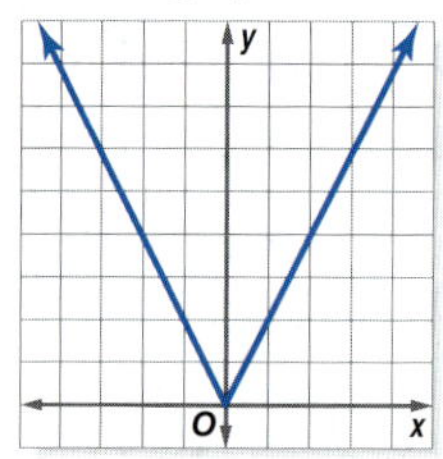

b

20.

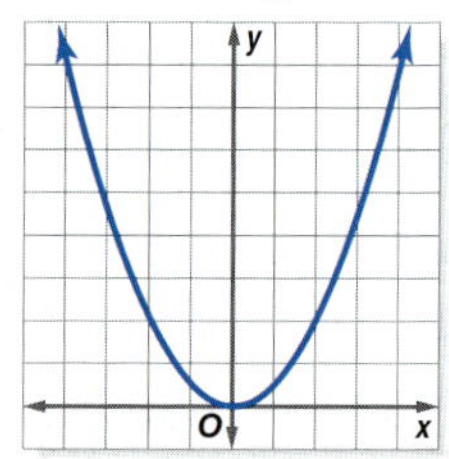

e

21.

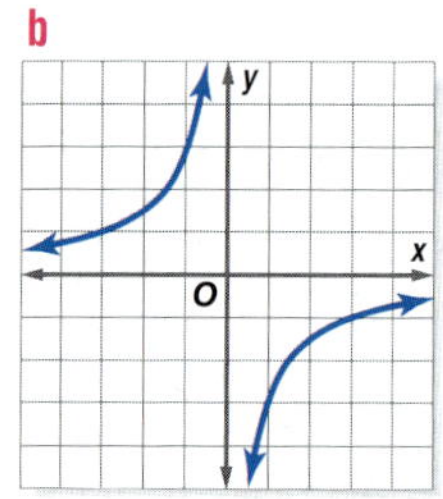

g

22.

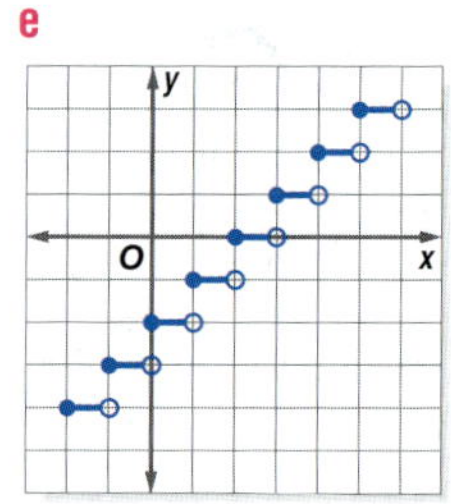

a

a. $y = [\![x]\!] - 2$
b. $y = 2|x|$
c. $y = 2\sqrt{x}$
d. $y = -3x$
e. $y = 0.5x^2$
f. $y = -\frac{3}{x+1}$
g. $y = -\frac{3}{x}$

WebQuest

You can use functions to determine the relationship between primary and secondary earthquake waves. Visit www.algebra2.com/webquest to continue work on your WebQuest project.

23–30. See pp. 519A–519D for graphs.
23. constant
24. direct variation
25. square root
26. inverse variation or rational
27. rational
28. greatest integer
29. absolute value
30. quadratic

Identify the type of function represented by each equation. Then graph the equation.

23. $y = -1.5$
24. $y = 2.5x$
25. $y = \sqrt{9x}$
26. $y = \frac{4}{x}$
27. $y = \frac{x^2 - 1}{x - 1}$
28. $y = 3[\![x]\!]$
29. $y = |2x|$
30. $y = 2x^2$

35.

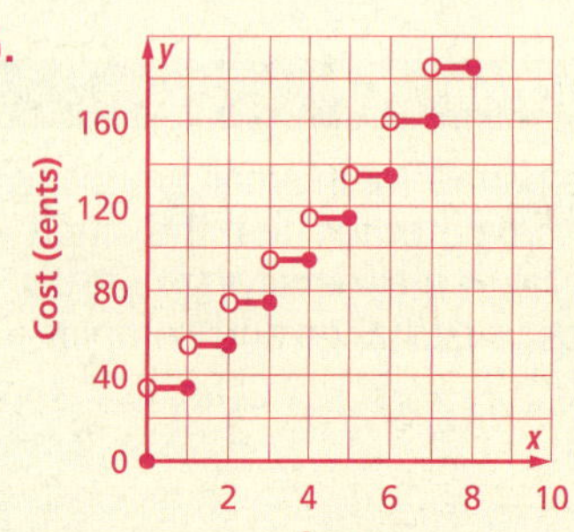

36. The graph is similar to the graph of the greatest integer function because both graphs look like a series of steps. In the graph of the postage rates, the solid dots are on the right and the circles are on the left. However, in the greatest integer function, the circles are on the right and the solid dots are on the left.

502 Chapter 9 Rational Expressions and Equations

HEALTH **For Exercises 31–33, use the following information.**
A woman painting a room will burn an average of 4.5 Calories per minute.

31. Write an equation for the number of Calories burned in m minutes. **$C = 4.5m$**
32. Identify the equation in Exercise 31 as a type of function. **direct variation**
33. Describe the graph of the function. **a line slanting to the right and passing through the origin**

34. **ARCHITECTURE** The shape of the Gateway Arch of the Jefferson National Expansion Memorial in St. Louis, Missouri, resembles the graph of the function $f(x) = -0.00635x^2 + 4.0005x - 0.07875$, where x is in feet. Describe the shape of the Gateway Arch. **similar to a parabola**

More About. . .

Architecture

The Gateway Arch is 630 feet high and is the tallest monument in the United States.

Source: *World Book Encyclopedia*

MAIL **For Exercises 35 and 36, use the following information.**
In 2001, the cost to mail a first-class letter was 34¢ for any weight up to and including 1 ounce. Each additional ounce or part of an ounce added 21¢ to the cost.

★35. Make a graph showing the postal rates to mail any letter from 0 to 8 ounces. **See margin.**

★36. Compare your graph in Exercise 35 to the graph of the greatest integer function. **See margin.**

37. **CRITICAL THINKING** Identify each table of values as a type of function.

a.

x	$f(x)$
−5	7
−3	5
−1	3
0	2
1	3
3	5
5	7
7	9

absolute value

b.

x	$f(x)$
−5	24
−3	8
−1	0
0	−1
1	0
3	8
5	24
7	48

quadratic

c.

x	$f(x)$
−1.3	−1
−1.7	−1
0	1
0.8	1
0.9	1
1	2
1.5	2
2.3	3

greatest integer

d.

x	$f(x)$
−5	undefined
−3	undefined
−1	undefined
0	0
1	1
4	2
9	3
16	4

square root

38. **WRITING IN MATH** Answer the question that was posed at the beginning of the lesson. **See pp. 519A–519D.**

How can graphs of functions be used to determine a person's weight on a different planet?

Include the following in your answer:
- an explanation of why the graph comparing weight on Earth and Mars represents a direct variation function, and
- an equation and a graph comparing a person's weight on Earth and Venus if a person's weight on Venus is 0.9 of his or her weight on Earth.

Standardized Test Practice Ⓐ Ⓑ Ⓒ Ⓓ

39. The curve at the right could be part of the graph of which function? **C**

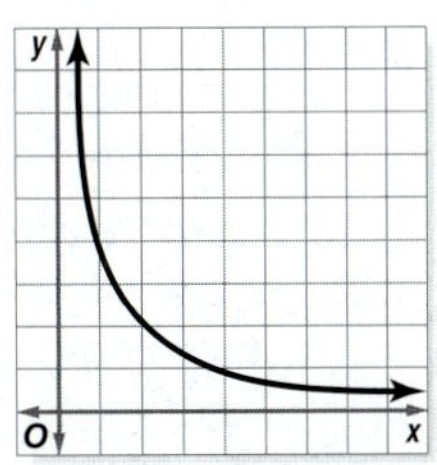

Ⓐ $y = \sqrt{x}$
Ⓑ $y = x^2 - 5x + 4$
Ⓒ $xy = 4$
Ⓓ $y = -x + 20$

www.algebra2.com/self_check_quiz

Study Guide and Intervention, p. 541 (shown) and p. 542

Identify Graphs You should be familiar with the graphs of the following functions.

Function	Description of Graph
Constant	a horizontal line that crosses the y-axis at a
Direct Variation	a line that passes through the origin and is neither horizontal nor vertical
Identity	a line that passes through the point (a, a), where a is any real number
Greatest Integer	a step function
Absolute Value	V-shaped graph
Quadratic	a parabola
Square Root	a curve that starts at a point and curves in only one direction
Rational	a graph with one or more asymptotes and/or holes
Inverse Variation	a graph with 2 curved branches and 2 asymptotes, $x = 0$ and $y = 0$ (special case of rational function)

Exercises

Identify the function represented by each graph.

1. 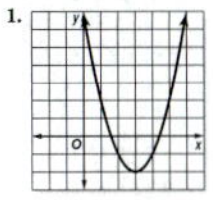 **quadratic**
2. 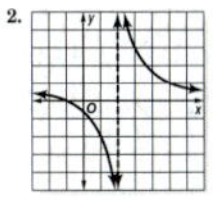 **rational**
3. 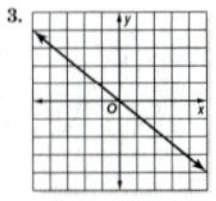 **direct variation**
4. 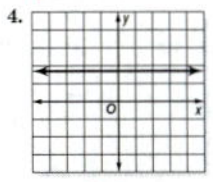 **constant**
5. 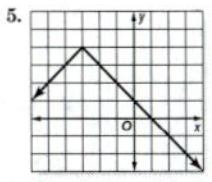 **absolute value**
6. 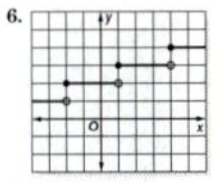 **greatest integer**
7. 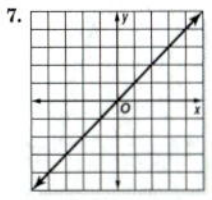 **identity**
8. 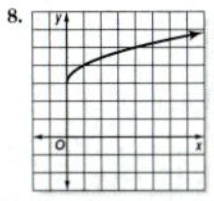 **square root**
9. 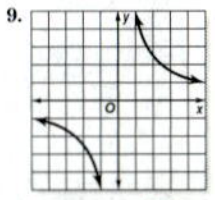 **inverse variation**

Skills Practice, p. 543 and Practice, p. 544 (shown)

Identify the type of function represented by each graph.

1. 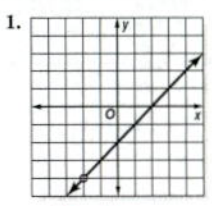 **rational**
2. 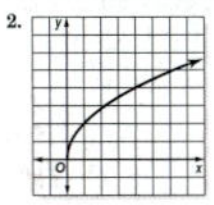 **square root**
3. 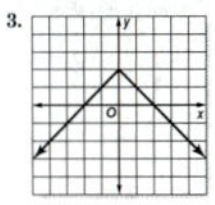 **absolute value**

Match each graph with an equation below.

A. $y = |2x + 1|$ B. $y = [\![2x + 1]\!]$ C. $y = \frac{x - 3}{2}$ D. $y = \sqrt{-x}$

4. 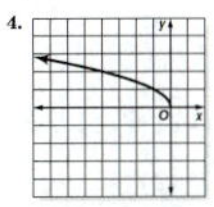 **D**
5. 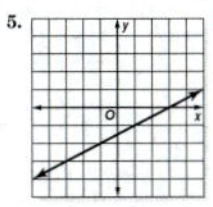 **C**
6. 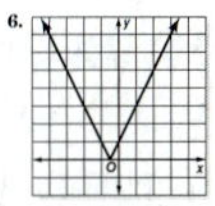 **A**

Identify the type of function represented by each equation. Then graph the equation.

7. $y = -3$ 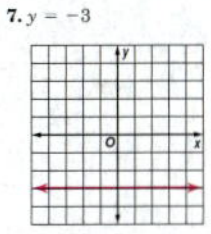 **constant**
8. $y = 2x^2 + 1$ 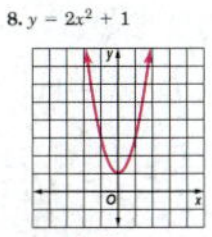 **quadratic**
9. $y = \frac{x^2 + 5x + 6}{x + 2}$ 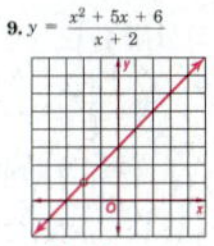 **rational**

10. BUSINESS A startup company uses the function $P = 1.3x^2 + 3x - 7$ to predict its profit or loss during its first 7 years of operation. Describe the shape of the graph of the function. **The graph is U-shaped; it is a parabola.**

11. PARKING A parking lot charges \$10 to park for the first day or part of a day. After that, it charges an additional \$8 per day or part of a day. Describe the graph and find the cost of parking for $6\frac{1}{2}$ days. **The graph looks like a series of steps, similar to a greatest integer function, but with open circles on the left and closed circles on the right; \$58.**

Reading to Learn Mathematics, p. 545 ELL

Pre-Activity **How can graphs of functions be used to determine a person's weight on a different planet?**

Read the introduction to Lesson 9-5 at the top of page 499 in your textbook.

- Based on the graph, estimate the weight on Mars of a child who weighs 40 pounds on Earth. **about 15 pounds**
- Although the graph does not extend far enough to the right to read it directly from the graph, use the weight you found above and your knowledge that this graph represents direct variation to estimate the weight on Mars of a woman who weighs 120 pounds on Earth. **about 45 pounds**

Reading the Lesson

1. Match each graph below with the type of function it represents. Some types may be used more than once and others not at all.

I. square root II. quadratic III. absolute value IV. rational
V. greatest integer VI. constant VII. identity

a. 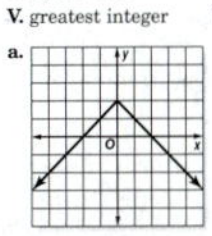 **III**
b. 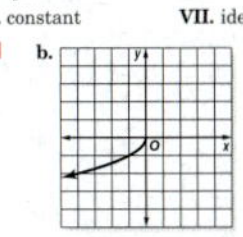 **I**
c. 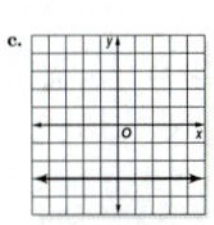 **VI**
d. 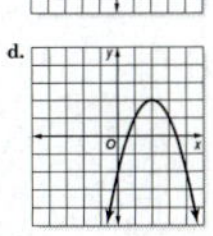 **II**
e. 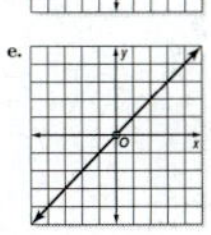 **IV**
f. 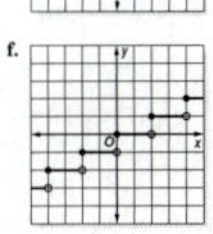 **V**

Helping You Remember

2. How can the symbolic definition of absolute value that you learned in Lesson 1-4 help you to remember the graph of the function $f(x) = |x|$? **Sample answer: Using the definition of absolute value, $f(x) = x$ if $x \geq 0$ and $f(x) = -x$ if $x < 0$. Therefore, the graph is made up of pieces of two lines, one with slope 1 and one with slope −1, meeting at the origin. This forms a V-shaped graph with "vertex" at the origin.**

Enrichment, p. 546

Partial Fractions

It is sometimes an advantage to rewrite a rational expression as the sum of two or more fractions. For example, you might do this in a calculus course while carrying out a procedure called integration.

You can resolve a rational expression into partial fractions if two conditions are met:

(1) The degree of the numerator must be less than the degree of the denominator; and
(2) The factors of the denominator must be known.

Example **Resolve $\frac{3}{x^3 + 1}$ into partial fractions.**

The denominator has two factors, a linear factor, $x + 1$, and a quadratic factor, $x^2 - x + 1$. Start by writing the following equation. Notice that the degree of the numerators of each partial fraction is less than its denominator.

4 Assess

Open-Ended Assessment

Modeling Have students use string on a coordinate grid to model some of the nine different types of functions in this lesson. Ask them to give an example of an equation that might have that sort of graph.

Tips for New Teachers

Intervention Help students associate the graphs and their functions by grouping the 9 types into 2 groups, those which involve straight lines and those which involve curves.

Getting Ready for Lesson 9-6

PREREQUISITE SKILL Students will solve rational equations in Lesson 9-6. These equations often contain fractions that are simplified by finding the LCD. Use Exercises 56–61 to determine your students' familiarity with finding LCMs of polynomials.

Assessment Options

Quiz (Lessons 9-4 and 9-5) is available on p. 568 of the *Chapter 9 Resource Masters.*

Answers

45. $(8, -1)$; $\left(8, -\frac{7}{8}\right)$; $x = 8$; $y = -1\frac{1}{8}$; up; $\frac{1}{2}$ unit

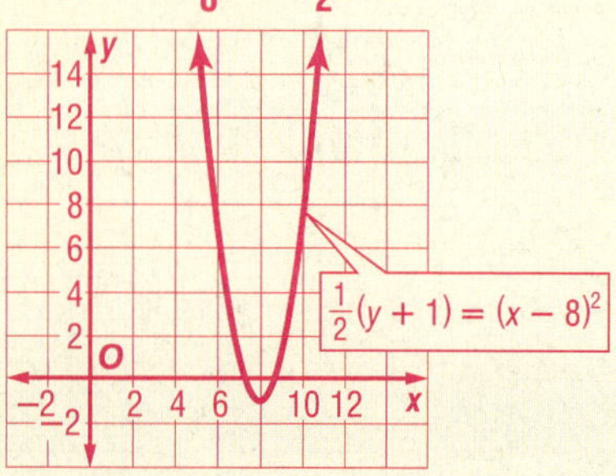

46. $\left(-3\frac{1}{4}, 1\right)$, $\left(-2\frac{1}{4}, 1\right)$; $y = 1$; $x = -4\frac{1}{4}$; right; 4 units

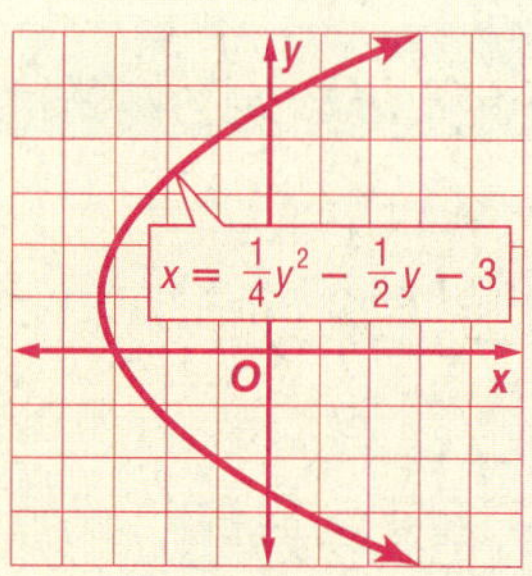

47. $(5, -4)$; $\left(5\frac{3}{4}, -4\right)$; $y = -4$; $x = 4\frac{1}{4}$; right; 3 units

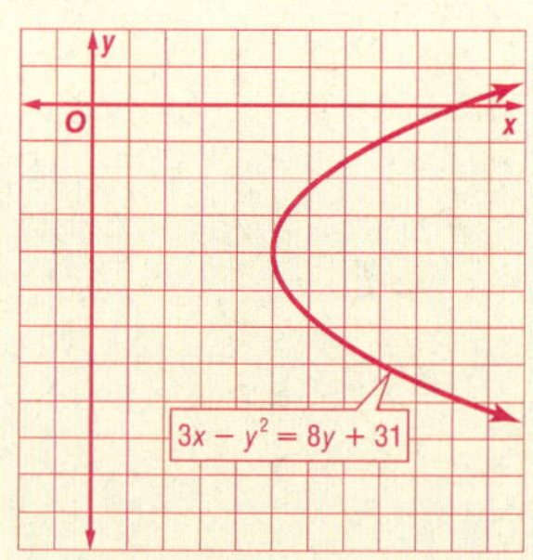

40. If $g(x) = [\![x]\!]$, which of the following is the graph of $g\left(\frac{x}{2}\right) + 2$? **D**

Ⓐ 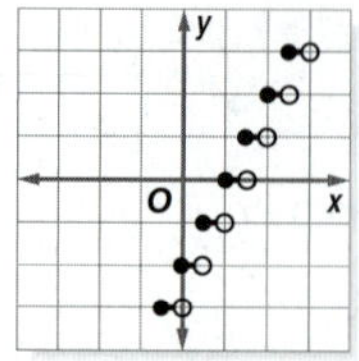

Ⓑ 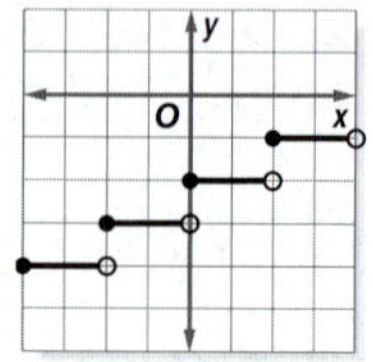

Ⓒ 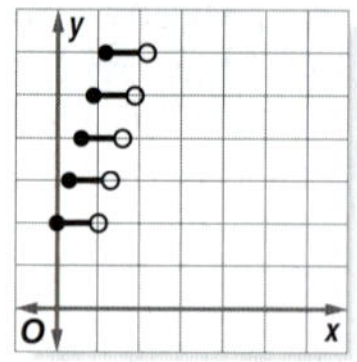

Ⓓ 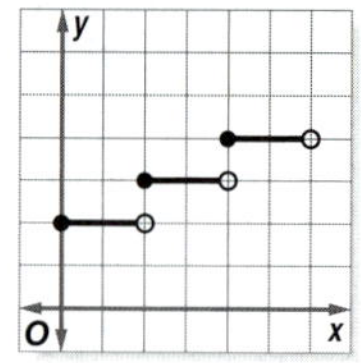

Maintain Your Skills

Mixed Review

41. If x varies directly as y and $y = \frac{1}{5}$ when $x = 11$, find x when $y = \frac{2}{5}$. *(Lesson 9–4)* **22**

Graph each rational function. *(Lesson 9-3)* **42–44. See pp. 519A–519D.**

42. $f(x) = \frac{3}{x + 2}$
43. $f(x) = \frac{8}{(x - 1)(x + 3)}$
44. $f(x) = \frac{x^2 - 5x + 4}{x - 4}$

Identify the coordinates of the vertex and focus, the equations of the axis of symmetry and directrix, and the direction of opening of the parabola with the given equation. Then find the length of the latus rectum and graph the parabola. *(Lesson 8-2)* **45–47. See margin.**

45. $\frac{1}{2}(y + 1) = (x - 8)^2$
46. $x = \frac{1}{4}y^2 - \frac{1}{2}y - 3$
47. $3x - y^2 = 8y + 31$

Find each product, if possible. *(Lesson 4-3)*

48. $\begin{bmatrix} 3 & -5 \\ 2 & 7 \end{bmatrix} \cdot \begin{bmatrix} 5 & 1 & -3 \\ 8 & -4 & 9 \end{bmatrix}$ **48.** $\begin{bmatrix} -25 & 23 & -54 \\ 66 & -26 & 57 \end{bmatrix}$

49. $\begin{bmatrix} 4 & -1 & 6 \\ 1 & 5 & -8 \end{bmatrix} \cdot \begin{bmatrix} 1 & 3 \\ 9 & -6 \end{bmatrix}$ **impossible**

Solve each system of equations by using either substitution or elimination. *(Lesson 3-2)*

50. $3x + 5y = -4$; $2x - 3y = 29$ **(7, −5)**
51. $3a - 2b = -3$; $3a + b = 3$ $\left(\frac{1}{3}, 2\right)$
52. $3s - 2t = 10$; $4s + t = 6$ **(2, −2)**

Determine the value of r so that a line through the points with the given coordinates has the given slope. *(Lesson 2-3)*

53. $(r, 2)$, $(4, -6)$; slope $= \frac{-8}{3}$ **1**
54. $(r, 6)$, $(8, 4)$; slope $= \frac{1}{2}$ **12**

55. Evaluate $[(-7 + 4) \times 5 - 2] \div 6$. *(Lesson 1-1)* $-\frac{17}{6}$

Getting Ready for the Next Lesson

PREREQUISITE SKILL Find the LCM of each set of polynomials.
(To review ***least common multiples of polynomials,*** *see Lesson 9-2.)*

56. $15ab^2c$, $6a^3$, $4bc^2$ **56.** $60a^3b^2c^2$
57. $9x^3$, $5xy^2$, $15x^2y^3$ **57.** $45x^3y^3$
58. $5d - 10$, $3d - 6$ **58.** $15(d - 2)$
59. $x^2 - y^2$, $3x + 3y$ $3(x - y)(x + y)$
60. $a^2 - 2a - 3$, $a^2 - a - 6$ $(a - 3)(a + 1)(a + 2)$
61. $2t^2 - 9t - 5$, $t^2 + t - 30$ $(t - 5)(t + 6)(2t + 1)$

504 Chapter 9 Rational Expressions and Equations

9-6 Solving Rational Equations and Inequalities

What You'll Learn

- Solve rational equations.
- Solve rational inequalities.

Vocabulary

- rational equation
- rational inequality

How are rational equations used to solve problems involving unit price?

The Coast to Coast Phone Company advertises 5¢ a minute for long-distance calls. However, it also charges a monthly fee of \$5. If the customer has x minutes in long distance calls last month, the bill in cents will be $500 + 5x$. The actual cost per minute is $\frac{500 + 5x}{x}$. To find how many long-distance minutes a person would need to make the actual cost per minute 6¢, you would need to solve the equation $\frac{500 + 5x}{x} = 6$.

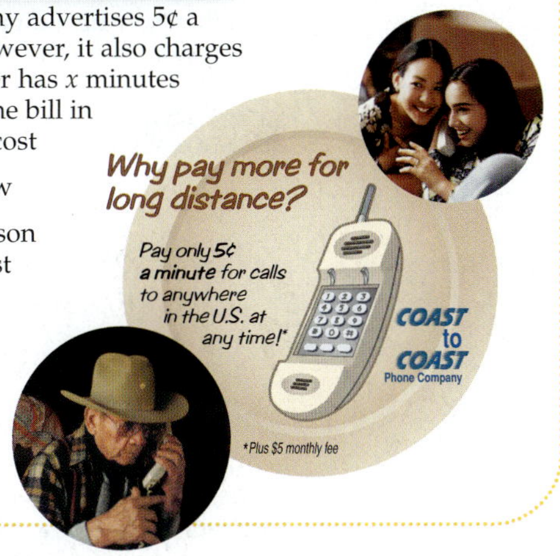

SOLVE RATIONAL EQUATIONS The equation $\frac{500 + 5x}{x} = 6$ is an example of a rational equation. In general, any equation that contains one or more rational expressions is called a **rational equation**.

Rational equations are easier to solve if the fractions are eliminated. You can eliminate the fractions by multiplying each side of the equation by the least common denominator (LCD). Remember that when you multiply each side by the LCD, each term on each side must be multiplied by the LCD.

Example 1 Solve a Rational Equation

Solve $\frac{9}{28} + \frac{3}{z + 2} = \frac{3}{4}$. Check your solution.

The LCD for the three denominators is $28(z + 2)$.

$$\frac{9}{28} + \frac{3}{z+2} = \frac{3}{4}$$ Original equation

$$28(z+2)\left(\frac{9}{28} + \frac{3}{z+2}\right) = 28(z+2)\left(\frac{3}{4}\right)$$ Multiply each side by $28(z + 2)$.

$$\overset{1}{\cancel{28}}(z+2)\left(\frac{9}{\underset{1}{\cancel{28}}}\right) + 28\overset{1}{\cancel{(z+2)}}\left(\frac{3}{\underset{1}{\cancel{z+2}}}\right) = \overset{7}{\cancel{28}}(z+2)\left(\frac{3}{\underset{1}{\cancel{4}}}\right)$$ Distributive Property

$$(9z + 18) + 84 = 21z + 42$$ Simplify.

$$9z + 102 = 21z + 42$$ Simplify.

$$60 = 12z$$ Subtract $9z$ and 42 from each side.

$$5 = z$$ Divide each side by 12.

9-6 Lesson Notes

1 Focus

5-Minute Check Transparency 9-6 Use as a quiz or review of Lesson 9-5.

Mathematical Background notes are available for this lesson on p. 470D.

Building on Prior Knowledge

In Chapter 1, students reviewed techniques for solving linear equations and inequalities. In this lesson, students will apply those same techniques to solving rational equations and inequalities.

How are rational equations used to solve problems involving unit price?

Ask students:

- Why does the equation use 500 instead of 5 for the monthly fee? **The fee and the per minute cost are both expressed in cents.**
- If a person makes 100 minutes of calls for a given month, how much did the monthly fee add to the per minute cost for these calls? **The fee adds 5 cents per minute.**

Resource Manager

Workbook and Reproducible Masters

Chapter 9 Resource Masters

- Study Guide and Intervention, pp. 547–548
- Skills Practice, p. 549
- Practice, p. 550
- Reading to Learn Mathematics, p. 551
- Enrichment, p. 552
- Assessment, p. 568

Transparencies

5-Minute Check Transparency 9-6
Answer Key Transparencies

Technology

Interactive Chalkboard
Multimedia Applications

2 Teach

SOLVE RATIONAL EQUATIONS

In-Class Examples

1 Solve $\frac{5}{24} + \frac{2}{3 - x} = \frac{1}{4}$. Check your solution. $x = -45$

Teaching Tip Discuss with students that, when checking a solution, it is not possible to simply substitute the solution for the variable and then multiply both sides of the equation by the LCD because the truth of the equation is being tested; it cannot be assumed.

2 Solve
$\frac{p^2 - p + 1}{p + 1} = \frac{p^2 - 7}{p^2 - 1} + p$.
Check your solution. $p = 2$

Teaching Tip Remind students that solutions must always be checked in the original equation, rather than in any of the steps of the solution.

CHECK

$\frac{9}{28} + \frac{3}{z + 2} = \frac{3}{4}$ Original equation

$\frac{9}{28} + \frac{3}{5 + 2} \stackrel{?}{=} \frac{3}{4}$ $z = 5$

$\frac{9}{28} + \frac{3}{7} \stackrel{?}{=} \frac{3}{4}$ Simplify.

$\frac{9}{28} + \frac{12}{28} \stackrel{?}{=} \frac{3}{4}$ Simplify.

$\frac{3}{4} = \frac{3}{4}$ ✓ The solution is correct.

The solution is 5.

When solving a rational equation, any possible solution that results in a zero in the denominator must be excluded from your list of solutions.

Example 2 Elimination of a Possible Solution

Solve $r + \frac{r^2 - 5}{r^2 - 1} = \frac{r^2 + r + 2}{r + 1}$. Check your solution.

The LCD is $(r^2 - 1)$.

$r + \frac{r^2 - 5}{r^2 - 1} = \frac{r^2 + r + 2}{r + 1}$ Original equation

$(r^2 - 1)\left(r + \frac{r^2 - 5}{r^2 - 1}\right) = (r^2 - 1)\left(\frac{r^2 + r + 2}{r + 1}\right)$ Multiply each side by the LCD, $(r^2 - 1)$.

$(r^2 - 1)r + \overset{1}{\cancel{(r^2 - 1)}}\left(\frac{r^2 - 5}{\underset{1}{\cancel{r^2 - 1}}}\right) = \overset{(r - 1)}{\cancel{(r^2 - 1)}}\left(\frac{r^2 + r + 2}{\underset{1}{\cancel{r + 1}}}\right)$ Distributive Property

$(r^3 - r) + (r^2 - 5) = (r - 1)(r^2 + r + 2)$ Simplify.

$r^3 + r^2 - r - 5 = r^3 + r - 2$ Simplify.

$r^2 - 2r - 3 = 0$ Subtract $(r^3 + r - 2)$ from each side.

$(r - 3)(r + 1) = 0$ Factor.

$r - 3 = 0$ or $r + 1 = 0$ Zero Product Property

$r = 3$ $\quad$ $r = -1$

CHECK

$r + \frac{r^2 - 5}{r^2 - 1} = \frac{r^2 + r + 2}{r + 1}$ Original equation

$3 + \frac{3^2 - 5}{3^2 - 1} \stackrel{?}{=} \frac{3^2 + 3 + 2}{3 + 1}$ $r = 3$

$3 + \frac{4}{8} \stackrel{?}{=} \frac{14}{4}$ Simplify.

$\frac{7}{2} = \frac{7}{2}$ ✓

$r + \frac{r^2 - 5}{r^2 - 1} = \frac{r^2 + r + 2}{r + 1}$ Original equation

$-1 + \frac{(-1)^2 - 5}{(-1)^2 - 1} \stackrel{?}{=} \frac{(-1)^2 + (-1) + 2}{-1 + 1}$ $r = -1$

$-1 + \frac{-4}{0} \stackrel{?}{=} \frac{2}{0}$ Simplify.

Since $r = -1$ results in a zero in the denominator, eliminate -1 from the list of solutions.

The solution is 3.

Study Tip

Extraneous Solutions
Multiplying each side of an equation by the LCD of rational expressions can yield results that are not solutions of the original equation. These solutions are called *extraneous solutions*.

506 Chapter 9 Rational Expressions and Equations

Some real-world problems can be solved with rational equations.

Example 3 Work Problem

TUNNELS **When building the Chunnel, the English and French each started drilling on opposite sides of the English Channel. The two sections became one in 1990. The French used more advanced drilling machinery than the English. Suppose the English could drill the Chunnel in 6.2 years and the French could drill it in 5.8 years. How long would it have taken the two countries to drill the tunnel?**

In 1 year, the English could complete $\frac{1}{6.2}$ of the tunnel.

In 2 years, the English could complete $\frac{1}{6.2} \cdot 2$ or $\frac{2}{6.2}$ of the tunnel.

In t years, the English could complete $\frac{1}{6.2} \cdot t$ or $\frac{t}{6.2}$ of the tunnel.

Likewise, in t years, the French could complete $\frac{1}{5.8} \cdot t$ or $\frac{t}{5.8}$ of the tunnel.

Together, they completed the whole tunnel.

Part completed by the English	plus	part completed by the French	equals	entire tunnel.
$\frac{t}{6.2}$	$+$	$\frac{t}{5.8}$	$=$	1

Solve the equation.

$$\frac{t}{6.2} + \frac{t}{5.8} = 1 \quad \text{Original equation}$$

$$17.98\left(\frac{t}{6.2} + \frac{t}{5.8}\right) = 17.98(1) \quad \text{Multiply each side by 17.98.}$$

$$17.98\left(\frac{t}{6.2}\right) + 17.98\left(\frac{t}{5.8}\right) = 17.98 \quad \text{Distributive Property}$$

$$2.9t + 3.1t = 17.98 \quad \text{Simplify.}$$

$$6t = 17.98 \quad \text{Simplify.}$$

$$t \approx 3.00 \quad \text{Divide each side by 6.}$$

It would have taken about 3 years to build the Chunnel.

More About. . .

Tunnels

The Chunnel is a tunnel under the English Channel that connects England with France. It is 32 miles long with 23 miles of the tunnel under water.

Source: www.pbs.org

Rate problems frequently involve rational equations.

Example 4 Rate Problem

NAVIGATION **The speed of the current in the Puget sound is 5 miles per hour. A barge travels 26 miles with the current and returns in $10\frac{2}{3}$ hours. What is the speed of the barge in still water?**

WORDS The formula that relates distance, time, and rate is $d = rt$ or $\frac{d}{r} = t$.

VARIABLES Let r be the speed of the barge in still water. Then the speed of the barge with the current is $r + 5$, and the speed of the barge against the current is $r - 5$.

	Time going with the current	plus	time going against the current	equals	total time.
EQUATION	$\frac{26}{r+5}$	$+$	$\frac{26}{r-5}$	$=$	$10\frac{2}{3}$

(continued on the next page)

www.algebra2.com/extra_examples

In-Class Examples

Power Point®

3 **MOWING LAWNS** Tim and Ashley mow lawns together. Tim working alone could complete the job in 4.5 hours, and Ashley could complete it alone in 3.7 hours. How long does it take to complete the job when they work together? **about 2 h**

Teaching Tip Suggest to students that when the problem involves completing part of a job and working together, students think about what part of the work gets done in one day, or one year—whatever is one unit of the time.

4 **SWIMMING** Janine swims for 5 hours in a stream that has a current of 1 mile per hour. She leaves her dock and swims upstream for 2 miles and then back to her dock. What is her swimming speed in still water? **about 1.5 mi/h**

Teaching Tip Make sure that students understand the difference between the solutions to the quadratic equation (of which there are two) and the solution to the problem (of which there is only one).

Lesson 9-6 Solving Rational Equations and Inequalities 507

SOLVE RATIONAL INEQUALITIES

In-Class Example Power Point®

5 Solve $\frac{1}{3s} + \frac{2}{9s} < \frac{2}{3}$. $s < 0$ or $s > \frac{5}{6}$

Teaching Tip Suggest that students also verify whether the boundary indicated by the solution of the equation is or is not in the solution set of the inequality.

Study Tip

Look Back
To review the **Quadratic Formula,** see Lesson 6-5.

Solve the equation.

$$\frac{26}{r+5} + \frac{26}{r-5} = 10\frac{2}{3} \quad \text{Original equation}$$

$$3(r^2-25)\left(\frac{26}{r+5} + \frac{26}{r-5}\right) = 3(r^2-25)\left(10\frac{2}{3}\right) \quad \text{Multiply each side by } 3(r^2-25).$$

$$3(r^2-25)\left(\frac{26}{r+5}\right) + 3(r^2-25)\left(\frac{26}{r-5}\right) = 3(r^2-25)\left(\frac{32}{3}\right) \quad \text{Distributive Property}$$

$$(78r - 390) + (78r + 390) = 32r^2 - 800 \quad \text{Simplify.}$$

$$156r = 32r^2 - 800 \quad \text{Simplify.}$$

$$0 = 32r^2 - 156r - 800 \quad \text{Subtract } 156r \text{ from each side.}$$

$$0 = 8r^2 - 39r - 200 \quad \text{Divide each side by 4.}$$

Use the Quadratic Formula to solve for r.

$$x = \frac{-b \pm \sqrt{b^2 - 4ac}}{2a} \quad \text{Quadratic Formula}$$

$$r = \frac{-(-39) \pm \sqrt{(-39)^2 - 4(8)(-200)}}{2(8)} \quad x = r, a = 8, b = -39, \text{ and } c = -200$$

$$r = \frac{39 \pm \sqrt{7921}}{16} \quad \text{Simplify.}$$

$$r = \frac{39 \pm 89}{16} \quad \text{Simplify.}$$

$$r = 8 \text{ or } -3.125 \quad \text{Simplify.}$$

Since the speed must be positive, the answer is 8 miles per hour.

SOLVE RATIONAL INEQUALITIES Inequalities that contain one or more rational expressions are called **rational inequalities**. To solve rational inequalities, complete the following steps.

Step 1 State the excluded values.

Step 2 Solve the related equation.

Step 3 Use the values determined in Steps 1 and 2 to divide a number line into regions. Test a value in each region to determine which regions satisfy the original inequality.

Example 5 Solve a Rational Inequality

Solve $\frac{1}{4a} + \frac{5}{8a} > \frac{1}{2}$.

Step 1 Values that make a denominator equal to 0 are excluded from the domain. For this inequality, the excluded value is 0.

Step 2 Solve the related equation.

$$\frac{1}{4a} + \frac{5}{8a} = \frac{1}{2} \quad \text{Related equation}$$

$$8a\left(\frac{1}{4a} + \frac{5}{8a}\right) = 8a\left(\frac{1}{2}\right) \quad \text{Multiply each side by } 8a.$$

$$2 + 5 = 4a \quad \text{Simplify.}$$

$$7 = 4a \quad \text{Add.}$$

$$1\frac{3}{4} = a \quad \text{Divide each side by 4.}$$

508 Chapter 9 Rational Expressions and Equations

Step 3 Draw vertical lines at the excluded value and at the solution to separate the number line into regions.

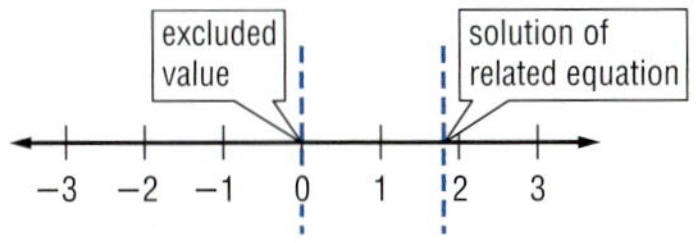

Now test a sample value in each region to determine if the values in the region satisfy the inequality.

Test $a = -1$.	Test $a = 1$.	Test $a = 2$.
$\frac{1}{4(-1)} + \frac{5}{8(-1)} \overset{?}{>} \frac{1}{2}$	$\frac{1}{4(1)} + \frac{5}{8(1)} \overset{?}{>} \frac{1}{2}$	$\frac{1}{4(2)} + \frac{5}{8(2)} \overset{?}{>} \frac{1}{2}$
$-\frac{1}{4} - \frac{5}{8} \overset{?}{>} \frac{1}{2}$	$\frac{1}{4} + \frac{5}{8} \overset{?}{>} \frac{1}{2}$	$\frac{1}{8} + \frac{5}{16} \overset{?}{>} \frac{1}{2}$
$-\frac{7}{8} \not> \frac{1}{2}$	$\frac{7}{8} > \frac{1}{2}$ ✓	$\frac{7}{16} \not> \frac{1}{2}$
$a < 0$ is *not* a solution.	$0 < a < 1\frac{3}{4}$ is a solution.	$a > 1\frac{3}{4}$ is *not* a solution.

The solution is $0 < a < 1\frac{3}{4}$.

Check for Understanding

Concept Check

1. **OPEN ENDED** Write a rational equation that can be solved by first multiplying each side by $5(a + 2)$.
2. State the expression by which you would multiply each side of $\frac{x}{x+4} + \frac{1}{2} = 1$ in order to solve the equation. What value(s) of x cannot be a solution?
3. **FIND THE ERROR** Jeff and Dustin are solving $2 - \frac{3}{a} = \frac{2}{3}$.

Jeff	Dustin
$2 - \frac{3}{a} = \frac{2}{3}$	$2 - \frac{3}{a} = \frac{2}{3}$
$6a - 9 = 2a$	$2 - 9 = 2a$
$4a = 9$	$-7 = 2a$
$a = 2.25$	$-3.5 = a$

Who is correct? Explain your reasoning.

1. Sample answer: $\frac{1}{5} + \frac{2}{a+2} = 1$

2. $2(x + 4)$; -4

3. Jeff; when Dustin multiplied by $3a$, he forgot to multiply the 2 by $3a$.

Guided Practice

Solve each equation or inequality. Check your solutions.

4. $\frac{2}{d} + \frac{1}{4} = \frac{11}{12}$ **3**
5. $t + \frac{12}{t} - 8 = 0$ **2, 6**
6. $\frac{1}{x-1} + \frac{2}{x} = 0$ $\frac{2}{3}$
7. $\frac{12}{v^2 - 16} - \frac{24}{v - 4} = 3$ **−6, −2**
8. $\frac{4}{c+2} > 1$ $-2 < c < 2$
9. $\frac{1}{3v} + \frac{1}{4v} < \frac{1}{2}$ $v < 0$ or $v > 1\frac{1}{6}$

GUIDED PRACTICE KEY	
Exercises	Examples
4–9	1, 2, 5
10	3, 4

Application

10. **WORK** A bricklayer can build a wall of a certain size in 5 hours. Another bricklayer can do the same job in 4 hours. If the bricklayers work together, how long would it take to do the job? $2\frac{2}{9}$ **h**

DAILY INTERVENTION — Differentiated Instruction

Logical Have students think about the difference between "pure" mathematics, such as solving an equation, and "applied" mathematics, such as solving a real-world problem. Ask them to list some ways in which these two are alike and some ways in which they are different.

3 Practice/Apply

Study Notebook

Have students—

- complete the definitions/examples for the remaining terms on their Vocabulary Builder worksheets for Chapter 9.
- add the steps for solving rational inequalities given in this lesson to their notebooks, along with an example from their work.
- write a list of cautions, or checks, that must be done as part of working with problems such as those in the lesson.
- include any other item(s) that they find helpful in mastering the skills in this lesson.

DAILY INTERVENTION

FIND THE ERROR Ask students what value of a can be excluded at the beginning of the problem. **The value of a cannot be 0.**

About the Exercises...

Organization by Objective

- **Solve Rational Equations:** 11–14, 17, 18, 23–39
- **Solve Rational Inequalities:** 15, 16, 19–22

Odd/Even Assignments

Exercises 11–32 are structured so that students practice the same concepts whether they are assigned odd or even problems.

Assignment Guide

Basic: 11–27 odd, 31, 33, 40–54

Average: 11–33 odd, 37–54

Advanced: 12–32 even, 34–36, 38–54

Lesson 9-6 Solving Rational Equations and Inequalities 509

★ indicates increased difficulty

Practice and Apply

Homework Help

For Exercises	See Examples
11–30	1, 2, 5
31–39	3, 4

Extra Practice
See page 849.

19. $t < 0$ or $t > 3$
22. $p < 0$ or $p > 2\frac{1}{2}$

Solve each equation or inequality. Check your solutions.

11. $\frac{y}{y+1} = \frac{2}{3}$ **2**
12. $\frac{p}{p-2} = \frac{2}{5}$ $-\frac{4}{3}$
13. $s + 5 = \frac{6}{s}$ **−6, 1**
14. $a + 1 = \frac{6}{a}$ **−3, 2**
15. $\frac{7}{a+1} > 7$ $-1 < a < 0$
16. $\frac{10}{m+1} > 5$ $-1 < m < 1$
17. $\frac{9}{t-3} = \frac{t-4}{t-3} + \frac{1}{4}$ **11**
18. $\frac{w}{w-1} + w = \frac{4w-3}{w-1}$ **3**
19. $5 + \frac{1}{t} > \frac{16}{t}$
20. $7 - \frac{2}{b} < \frac{5}{b}$ $0 < b < 1$
21. $\frac{2}{3y} + \frac{5}{6y} > \frac{3}{4}$ $0 < y < 2$
22. $\frac{1}{2p} + \frac{3}{4p} < \frac{1}{2}$
23. $\frac{b-4}{b-2} = \frac{b-2}{b+2} + \frac{1}{b-2}$ **14**
24. $\frac{4n^2}{n^2-9} - \frac{2n}{n+3} = \frac{3}{n-3}$ $\frac{3}{2}$
25. $\frac{1}{d+4} = \frac{2}{d^2+3d-4} - \frac{1}{1-d}$ $\varnothing$
26. $\frac{2}{y+2} - \frac{y}{2-y} = \frac{y^2+4}{y^2-4}$ $\varnothing$
27. $\frac{3}{b^2+5b+6} + \frac{b-1}{b+2} = \frac{7}{b+3}$ **7**
28. $\frac{1}{n-2} = \frac{2n+1}{n^2+2n-8} + \frac{2}{n+4}$ $\frac{7}{3}$
★ 29. $\frac{2q}{2q+3} - \frac{2q}{2q-3} = 1$ $\frac{-3 \pm 3\sqrt{2}}{2}$
30. $\frac{4}{z-2} - \frac{z+6}{z+1} = 1$ $\frac{1 \pm \sqrt{145}}{4}$

31. **NUMBER THEORY** The ratio of 8 less than a number to 28 more than that number is 2 to 5. What is the number? **32**

32. **NUMBER THEORY** The sum of a number and 8 times its reciprocal is 6. Find the number(s). **2 or 4**

33. **ACTIVITIES** The band has 30 more members than the school chorale. If each group had 10 more members, the ratio of their membership would be 3:2. How many members are in each group? **band, 80 members; chorale, 50 members**

PHYSICS **For Exercises 34 and 35, use the following information.**
The distance a spring stretches is related to the mass attached to the spring. This is represented by $d = km$, where d is the distance, m is the mass, and k is the spring constant. When two springs with spring constants k_1 and k_2 are attached in a series, the resulting spring constant k is found by the equation $\frac{1}{k} = \frac{1}{k_1} + \frac{1}{k_2}$.

34. If one spring with constant of 12 centimeters per gram is attached in a series with another spring with constant of 8 centimeters per gram, find the resultant spring constant. **4.8 cm/g**

35. If a 5-gram object is hung from the series of springs, how far will the springs stretch? **24 cm**

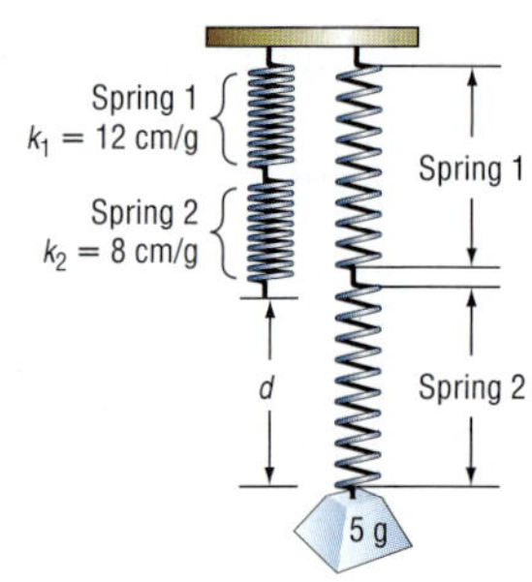

36. **CYCLING** On a particular day, the wind added 3 kilometers per hour to Alfonso's rate when he was cycling with the wind and subtracted 3 kilometers per hour from his rate on his return trip. Alfonso found that in the same amount of time he could cycle 36 kilometers with the wind, he could go only 24 kilometers against the wind. What is his normal bicycling speed with no wind? **15 km/h**

Study Guide and Intervention, p. 547 (shown) and p. 548

Solve Rational Equations A **rational equation** contains one or more rational expressions. To solve a rational equation, first multiply each side by the least common denominator of all of the denominators. Be sure to exclude any solution that would produce a denominator of zero.

Example **Solve $\frac{9}{10} + \frac{2}{x+1} = \frac{2}{5}$.**

$\frac{9}{10} + \frac{2}{x+1} = \frac{2}{5}$ Original equation
$10(x+1)\left(\frac{9}{10} + \frac{2}{x+1}\right) = 10(x+1)\left(\frac{2}{5}\right)$ Multiply each side by $10(x + 1)$.
$9(x+1) + 2(10) = 4(x+1)$ Multiply.
$9x + 9 + 20 = 4x + 4$ Distributive Property
$5x = -25$ Subtract 4x and 29 from each side.
$x = -5$ Divide each side by 5.

Check
$\frac{9}{10} + \frac{2}{x+1} = \frac{2}{5}$ Original equation
$\frac{9}{10} + \frac{2}{-5+1} \stackrel{?}{=} \frac{2}{5}$ $x = -5$
$\frac{9}{10} + \frac{2}{-4} \stackrel{?}{=} \frac{2}{5}$ Simplify.
$\frac{18}{20} - \frac{10}{20} \stackrel{?}{=} \frac{2}{5}$ Simplify.
$\frac{8}{20} \stackrel{?}{=} \frac{2}{5}$ Simplify.
$\frac{2}{5} = \frac{2}{5}$

Exercises

Solve each equation.

1. $\frac{2y}{3} - \frac{y+3}{6} = 2$ **5**
2. $\frac{4t-3}{5} - \frac{4-2t}{3} = 1$ **2**
3. $\frac{2x+1}{3} - \frac{x-5}{4} = \frac{1}{2}$ $-\frac{13}{5}$
4. $\frac{3m+2}{5m} + \frac{2m-1}{2m} = 4$ $-\frac{1}{24}$
5. $\frac{4}{x-1} = \frac{x+1}{12}$ ± 7
6. $\frac{x}{x-2} + \frac{4}{x-2} = 10$ $\frac{8}{3}$

7. **NAVIGATION** The current in a river is 6 miles per hour. In her motorboat Marissa can travel 12 miles upstream or 16 miles downstream in the same amount of time. What is the speed of her motorboat in still water? **42 mph**

8. **WORK** Adam, Bethany, and Carlos own a painting company. To paint a particular house alone, Adam estimates that it would take him 4 days, Bethany estimates $5\frac{1}{2}$ days, and Carlos 6 days. If these estimates are accurate, how long should it take the three of them to paint the house if they work together? **about $1\frac{2}{3}$ days**

Skills Practice, p. 549 and Practice, p. 550 (shown)

Solve each equation or inequality. Check your solutions.

1. $\frac{12}{x} + \frac{3}{4} = \frac{3}{2}$ **16**
2. $\frac{x}{x-1} - 1 = \frac{x}{2}$ **−1, 2**
3. $\frac{p+10}{p^2-2} = \frac{4}{p}$ $-\frac{2}{3}, 4$
4. $\frac{s}{s+2} + s = \frac{5s+8}{s+2}$ **4**
5. $\frac{5}{y-5} = \frac{y}{y-5} - 1$ **all reals except 5**
6. $\frac{1}{3x-2} + \frac{5}{x} = 0$ $\frac{5}{8}$
7. $\frac{5}{t} < \frac{9}{2t+1}$ $t < -5$ or $-\frac{1}{2} < t < 0$
8. $\frac{1}{2h} + \frac{5}{h} = \frac{3}{h-1}$ $\frac{11}{5}$
9. $\frac{4}{w-2} = \frac{-1}{w+3}$ **−2**
10. $5 - \frac{3}{a} < \frac{7}{a}$ $0 < a < 2$
11. $\frac{4}{5x} + \frac{1}{10} < \frac{3}{2x}$ $0 < x < 7$
12. $8 + \frac{3}{y} > \frac{19}{y}$ $y < 0$ or $y > 2$
13. $\frac{4}{p} + \frac{1}{3p} < \frac{1}{5}$ $p < 0$ or $p > \frac{65}{3}$
14. $\frac{6}{x-1} = \frac{4}{x-2} + \frac{2}{x+1}$ $\varnothing$
15. $g + \frac{g}{g-2} = \frac{2}{g-2}$ **−1**
16. $b + \frac{2b}{b-1} = 1 - \frac{b-3}{b-1}$ **−2**
17. $2 = \frac{x+2}{x-3} + \frac{x-2}{x-6}$ $\frac{14}{3}$
18. $5 - \frac{3d+2}{d-1} = \frac{2d-4}{d+2}$ **6**
19. $\frac{1}{n+2} + \frac{1}{n-2} = \frac{3}{n^2-4}$ $\frac{3}{2}$
20. $\frac{c+1}{c-3} = 4 - \frac{12}{c^2-2c-3}$ $-\frac{5}{3}, 5$
21. $\frac{3}{k-3} + \frac{4}{k-4} = \frac{25}{k^2-7k+12}$ **7**
22. $\frac{4v}{v-1} - \frac{5v}{v-2} = \frac{2}{v^2-3v+2}$ **−1, −2**
23. $\frac{y}{y+2} + \frac{7}{y-5} = \frac{14}{y^2-3y-10}$ **0**
24. $\frac{x^2+4}{x^2-4} + \frac{x}{2-x} = \frac{2}{x+2}$ $\varnothing$
25. $\frac{r}{r+4} + \frac{4}{r-4} = \frac{r^2+16}{r^2-16}$ **all reals except −4 and 4**
26. $3 = \frac{6a-1}{2a+7} + \frac{22}{a+5}$ **−2**

27. **BASKETBALL** Kiana has made 9 of 19 free throws so far this season. Her goal is to make 60% of her free throws. If Kiana makes her next x free throws in a row, the function $f(x) = \frac{9+x}{19+x}$ represents Kiana's new ratio of free throws made. How many successful free throws in a row will raise Kiana's percent made to 60%? **6**

28. **OPTICS** The lens equation $\frac{1}{p} + \frac{1}{q} = \frac{1}{f}$ relates the distance p of an object from a lens, the distance q of the image of the object from the lens, and the focal length f of the lens. What is the distance of an object from a lens if the image of the object is 5 centimeters from the lens and the focal length of the lens is 4 centimeters? **20 cm**

Reading to Learn Mathematics, p. 551 ELL

Pre-Activity **How are rational equations used to solve problems involving unit price?**

Read the introduction to Lesson 9-6 at the top of page 505 in your textbook.

- If you increase total number of minutes of long-distance calls from March to April, will your long-distance phone bill increase or decrease? **increase**
- Will your actual cost per minute increase or decrease? **decrease**

Reading the Lesson

1. When solving a rational equation, any possible solution that results in 0 in the denominator must be excluded from the list of solutions.

2. Suppose that on a quiz you are asked to solve the rational inequality $\frac{3}{z+2} - \frac{6}{z} > 0$. Complete the steps of the solution.

Step 1 The excluded values are **−2** and **0**.

Step 2 The related equation is $\frac{3}{z+2} - \frac{6}{z} = 0$.

To solve this equation, multiply both sides by the LCD, which is $z(z + 2)$. Solving this equation will show that the only solution is −4.

Step 3 Divide a number line into **4** regions using the excluded values and the solution of the related equation. Draw dashed vertical lines on the number line below to show these regions.

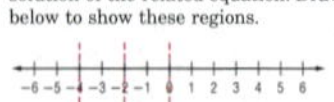

Consider the following values of $\frac{3}{z+2} - \frac{6}{z}$ for various test values of z.

If $z = -5$, $\frac{3}{z+2} - \frac{6}{z} = 0.2$.
If $z = -3$, $\frac{3}{z+2} - \frac{6}{z} = -1$.
If $z = -1$, $\frac{3}{z+2} - \frac{6}{z} = 9$.
If $z = 1$, $\frac{3}{z+2} - \frac{6}{z} = -5$.

Using this information and your number line, write the solution of the inequality.
$z < -4$ or $-2 < z < 0$

Helping You Remember

3. How are the processes of adding rational expressions with different denominators and of solving rational expressions alike, and how are they different? **Sample answer: They are alike because both use the LCD of all the rational expressions in the problem. They are different because in an addition problem, the LCD remains after the fractions are added, while in solving a rational equation, the LCD is eliminated.**

Enrichment, p. 552

Limits

Sequences of numbers with a rational expression for the general term often approach some number as a finite limit. For example, the reciprocals of the positive integers approach 0 as n gets larger and larger. This is written using the notation shown below. The symbol ∞ stands for infinity and $n \to \infty$ means that n is getting larger and larger, or "n goes to infinity."

$1, \frac{1}{2}, \frac{1}{3}, \frac{1}{4}, \ldots, \frac{1}{n}, \ldots$ $\lim_{n \to \infty} \frac{1}{n} = 0$

Example **Find $\lim_{n \to \infty} \frac{n^2}{(n+1)^2}$**

It is not immediately apparent whether the sequence approaches a limit or not. But notice what happens if we divide the numerator and denominator of the general term by n^2.

510 Chapter 9 Rational Expressions and Equations

Career Choices

Chemist Many chemists work for manufacturers developing products or doing quality control to ensure the products meet industry and government standards.

Online Research For information about a career as a chemist, visit: www.algebra2.com/careers

37. **CHEMISTRY** Kiara adds an 80% acid solution to 5 milliliters of solution that is 20% acid. The function that represents the percent of acid in the resulting solution is $f(x) = \frac{5(0.20) + x(0.80)}{5 + x}$, where x is the amount of 80% solution added. How much 80% solution should be added to create a solution that is 50% acid? **5 mL**

STATISTICS **For Exercises 38 and 39, use the following information.**
A number x is the *harmonic mean* of y and z if $\frac{1}{x}$ is the average of $\frac{1}{y}$ and $\frac{1}{z}$.

★38. Find y if $x = 8$ and $z = 20$. **5**

★39. Find x if $y = 5$ and $z = 8$. **6.15**

40. **CRITICAL THINKING** Solve for a if $\frac{1}{a} - \frac{1}{b} = c$. $\frac{b}{bc + 1}$

41. **WRITING IN MATH** Answer the question that was posed at the beginning of the lesson. **See margin.**

How are rational equations used to solve problems involving unit price?

Include the following in your answer:
- an explanation of how to solve $\frac{500 + 5x}{x} = 6$, and
- the reason why the actual price per minute could never be 5¢.

Standardized Test Practice

42. If $T = \frac{4st}{s - t}$, what is the value of s when $t = 5$ and $T = 40$? **B**

(A) 20 (B) 10 (C) 5 (D) 2

43. Amanda wanted to determine the average of her 6 test scores. She added the scores correctly to get T, but divided by 7 instead of 6. Her average was 12 less than the actual average. Which equation could be used to determine the value of T? **C**

(A) $6T + 12 = 7T$
(B) $\frac{T}{7} = \frac{T - 12}{6}$
(C) $\frac{T}{7} + 12 = \frac{T}{6}$
(D) $\frac{T}{6} = \frac{T - 12}{7}$

Maintain Your Skills

Mixed Review **Identify the type of function represented by each equation. Then graph the equation.** *(Lesson 9-5)* **44–46. See margin for graphs.**

44. $y = 2x^2 + 1$ **quad.**
45. $y = 2\sqrt{x}$ **sq. root**
46. $y = 0.8x$ **direct var.**

47. If y varies inversely as x and $y = 24$ when $x = 9$, find y when $x = 6$. *(Lesson 9-4)* **36**

48. If y varies directly as x and $y = 9$ when $x = 4$, find y when $x = 15$. *(Lesson 9-4)* **33.75**

Find the distance between each pair of points with the given coordinates. *(Lesson 8-1)*

49. $(-5, 7), (9, -11)$ $2\sqrt{130}$
50. $(3, 5), (7, 3)$ $2\sqrt{5}$
51. $(-1, 3), (-5, -8)$ $\sqrt{137}$

Solve each inequality. *(Lesson 6-7)*

52. $(x + 11)(x - 3) > 0$
53. $x^2 - 4x \leq 0$
54. $2b^2 - b < 6$

52. $\{x \mid x < -11 \text{ or } x > 3\}$
53. $\{x \mid 0 \leq x \leq 4\}$
54. $\left\{b \mid -1\frac{1}{2} < b < 2\right\}$

www.algebra2.com/self_check_quiz

41. If something has a general fee and cost per unit, rational equations can be used to determine how many units a person must buy in order for the actual unit price to be a given number. Answers should include the following.
- To solve $\frac{500 + 5x}{x} = 6$, multiply each side of the equation by x to eliminate the rational expression. Then subtract $5x$ from each side. Therefore, $500 = x$. A person would need to make 500 minutes of long distance minutes to make the actual unit price 6¢.
- Since the cost is 5¢ per minute plus \$5.00 per month, the actual cost per minute could never be 5¢ or less.

4 Assess

Open-Ended Assessment

Writing Have students write their own real-world problems similar to some they have seen in this lesson, but using their own data. Then solve them.

Tips for New Teachers

Intervention Make sure every student is clear about why some values must be excluded, even though they appear as solutions in the course of working a problem. Point out that multiplying each side of an equation by the variable may introduce extraneous roots.

Assessment Options

Quiz (Lesson 9-6) is available on p. 568 of the *Chapter 9 Resource Masters.*

Answers

44.

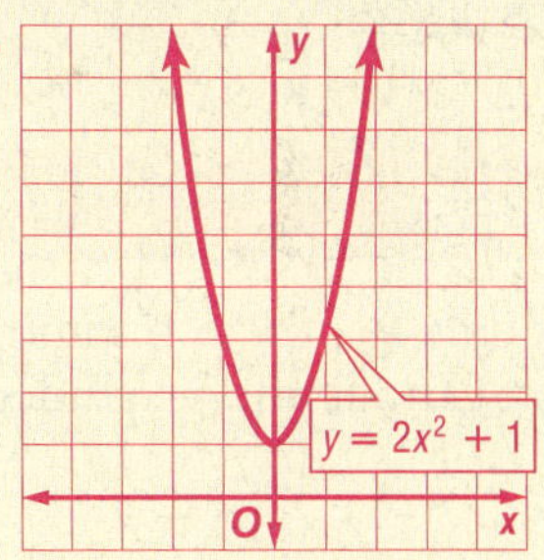

45.

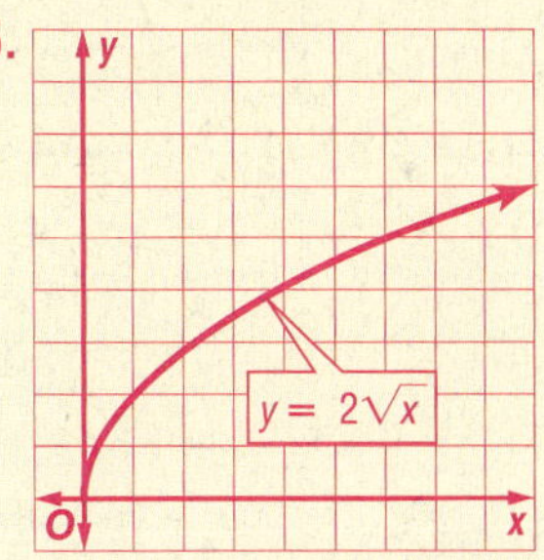

46.

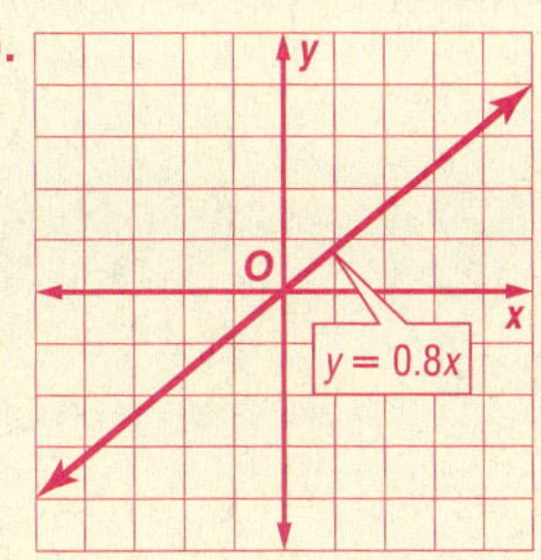

Graphing Calculator Investigation

A Follow-Up of Lesson 9-6

Getting Started

Know Your Calculator To see the vertical asymptote for the graph of y_1, students should check to see that the calculator is in Connected mode.

Using Parentheses When students enter functions on the Y= list, they should use parentheses around any numerator or denominator that is not a single number or variable.

Teach

After they enter the two equations, have students enter the CALC menu and select 5:Intersect. Then have them move the cursor and press [ENTER] to identify each of the graphs. In response to Guess?, move the cursor to an estimated point of intersection and press [ENTER].

Assess

Ask students to describe some ways you can identify excluded values.

Graphing Calculator Investigation

A Follow-Up of Lesson 9-6

Solving Rational Equations by Graphing

You can use a graphing calculator to solve rational equations. You need to graph both sides of the equation and locate the point(s) of intersection. You can also use a graphing calculator to confirm solutions that you have found algebraically.

Example

Use a graphing calculator to solve $\frac{4}{x+1} = \frac{3}{2}$.

- First, rewrite as two functions, $y_1 = \frac{4}{x+1}$ and $y_2 = \frac{3}{2}$.
- Next, graph the two functions on your calculator.

KEYSTROKES: [Y=] 4 [÷] [(] [X,T,θ,n] [+] 1 [)] [▼] 3 [÷] 2 [ZOOM] 6

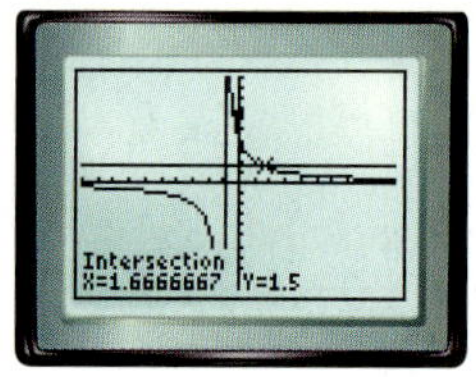

[−10, 10] scl: 1 by [−10, 10] scl: 1

Notice that because the calculator is in connected mode, a vertical line is shown connecting the two branches of the hyperbola. This line is not part of the graph.

- Next, locate the point(s) of intersection.

KEYSTROKES: [2nd] [CALC] 5

Select one graph and press [ENTER]. Select the other graph, press [ENTER], and press [ENTER] again. The solution is $1\frac{2}{3}$. Check this solution by substitution.

Exercises

Use a graphing calculator to solve each equation.

1. $\frac{1}{x} + \frac{1}{2} = \frac{2}{x}$ **2**
2. $\frac{1}{x-4} = \frac{2}{x-2}$ **6**
3. $\frac{4}{x} = \frac{6}{x^2}$ **1.5**
4. $\frac{1}{1-x} = 1 - \frac{x}{x-1}$ **all real numbers except 1**
5. $\frac{1}{x+4} = \frac{2}{x^2+3x-4} - \frac{1}{1-x}$ **no real solution**
6. $\frac{1}{x-1} + \frac{1}{x+2} = \frac{1}{2}$ **−1, 4**

Solve each equation algebraically. Then, confirm your solution(s) using a graphing calculator.

7. $\frac{3}{x} + \frac{7}{x} = 9$ **$1\frac{1}{9}$**
8. $\frac{1}{x-1} + \frac{2}{x} = 0$ **$\frac{2}{3}$**
9. $1 + \frac{5}{x-1} = \frac{7}{6}$ **31**
10. $\frac{1}{x^2-1} = \frac{2}{x^2+x-2}$ **0**
11. $\frac{6}{x^2+2x} - \frac{x+1}{x+2} = \frac{2}{x}$
12. $\frac{3}{x^2+5x+6} + \frac{x-1}{x+2} = \frac{7}{x+3}$ **7**

11. $\frac{-3 \pm \sqrt{17}}{2}$ or about −3.56 and 0.56

512 Chapter 9 Rational Expressions and Equations

www.algebra2.com/other_calculator_keystrokes

Chapter 9 Study Guide and Review

Vocabulary and Concept Check

asymptote (p. 485)	direct variation (p. 492)	rational equation (p. 505)
complex fraction (p. 475)	inverse variation (p. 493)	rational expression (p. 472)
constant of variation (p. 492)	joint variation (p. 493)	rational function (p. 485)
continuity (p. 485)	point discontinuity (p. 485)	rational inequality (p. 508)

State whether each sentence is *true* or *false*. If false, replace the underlined word or number to make a true sentence.

1. The equation $y = \frac{x^2 - 1}{x + 1}$ has a(n) <u>*asymptote*</u> at $x = -1$. **false; point discontinuity**
2. The equation $y = 3x$ is an example of a <u>*direct*</u> variation equation. **true**
3. The equation $y = \frac{x^2}{x + 1}$ is a(n) <u>*polynomial*</u> equation. **false; rational**
4. The graph of $y = \frac{4}{x - 4}$ has a(n) <u>*variation*</u> at $x = 4$. **false; asymptote**
5. The equation $b = \frac{2}{a}$ is a(n) <u>*inverse*</u> variation equation. **true**
6. On the graph of $y = \frac{x - 5}{x + 2}$, there is a break in continuity at $x = \underline{2}$. **false; −2**

Lesson-by-Lesson Review

9-1 Multiplying and Dividing Rational Expressions

See pages 472–478.

Concept Summary

- Multiplying and dividing rational expressions is similar to multiplying and dividing fractions.

Examples

1 Simplify $\frac{3x}{2y} \cdot \frac{8y^3}{6x^2}$.

$$\frac{3x}{2y} \cdot \frac{8y^3}{6x^2} = \frac{\overset{1}{\cancel{3}} \cdot \overset{1}{\cancel{x}} \cdot \overset{1}{\cancel{2}} \cdot \overset{1}{\cancel{2}} \cdot 2 \cdot \overset{1}{\cancel{y}} \cdot y \cdot y}{\underset{1}{\cancel{2}} \cdot \underset{1}{\cancel{y}} \cdot \underset{1}{\cancel{2}} \cdot \underset{1}{\cancel{3}} \cdot \underset{1}{\cancel{x}} \cdot x}$$

$$= \frac{2y^2}{x}$$

2 Simplify $\frac{p^2 + 7p}{3p} \div \frac{49 - p^2}{3p - 21}$.

$$\frac{p^2 + 7p}{3p} \div \frac{49 - p^2}{3p - 21} = \frac{p^2 + 7p}{3p} \cdot \frac{3p - 21}{49 - p^2}$$

$$= \frac{\overset{1}{\cancel{p}}\overset{1}{\cancel{(7 + p)}}}{\underset{1}{\cancel{3}}\underset{1}{\cancel{p}}} \cdot \frac{\overset{-1}{\cancel{-3}}\overset{1}{\cancel{(7 - p)}}}{\underset{1}{\cancel{(7 + p)}}\underset{1}{\cancel{(7 - p)}}}$$

$$= -1$$

Exercises Simplify each expression. *See Examples 4–7 on pages 474 and 475.*

7. $\frac{-4ab}{21c} \cdot \frac{14c^2}{22a^2}$ $\mathbf{-\frac{4bc}{33a}}$

8. $\frac{a^2 - b^2}{6b} \div \frac{a + b}{36b^2}$ $\mathbf{6b(a - b)}$

9. $\frac{y^2 - y - 12}{y + 2} \div \frac{y - 4}{y^2 - 4y - 12}$ $\mathbf{(y + 3)(y - 6)}$

10. $\frac{\frac{x^2 + 7x + 10}{x + 2}}{\frac{x^2 + 2x - 15}{x + 2}}$ $\mathbf{\frac{x + 2}{x - 3}}$

11. $\frac{\frac{1}{n^2 - 6n + 9}}{\frac{n + 3}{2n^2 - 18}}$ $\mathbf{\frac{2}{n - 3}}$

12. $\frac{x^2 + 3x - 10}{x^2 + 8x + 15} \cdot \frac{x^2 + 5x + 6}{x^2 + 4x + 4}$

12. $\mathbf{\frac{x - 2}{x + 2}}$

www.algebra2.com/vocabulary_review

FOLDABLES™ Study Organizer

For more information about Foldables, see *Teaching Mathematics with Foldables.*

number or variable.

After they enter the two equations, have students enter the CALC menu and select **5:Intersect**. Then have them move the

Chapter 9 Study Guide and Review

Vocabulary and Concept Check

- This alphabetical list of vocabulary terms in Chapter 9 includes a page reference where each term was introduced.
- **Assessment** A vocabulary test/review for Chapter 9 is available on p. 566 of the *Chapter 9 Resource Masters*.

Lesson-by-Lesson Review

For each lesson,

- the main ideas are summarized,
- additional examples review concepts, and
- practice exercises are provided.

Vocabulary PuzzleMaker

ELL The Vocabulary PuzzleMaker software improves students' mathematics vocabulary using four puzzle formats—crossword, scramble, word search using a word list, and word search using clues. Students can work on a computer screen or from a printed handout.

MindJogger Videoquizzes

ELL MindJogger Videoquizzes provide an alternative review of concepts presented in this chapter. Students work in teams in a game show format to gain points for correct answers. The questions are presented in three rounds.

Round 1 Concepts (5 questions)
Round 2 Skills (4 questions)
Round 3 Problem Solving (4 questions)

Answers

19.

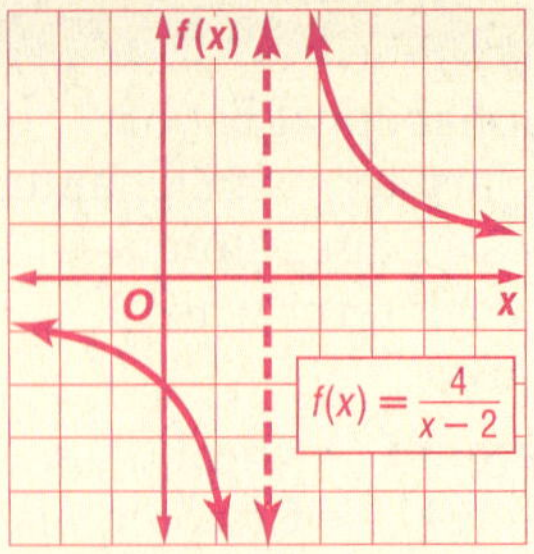

20.

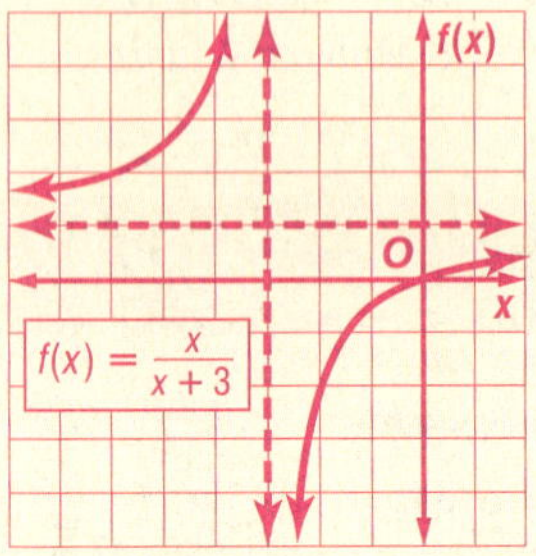

21.

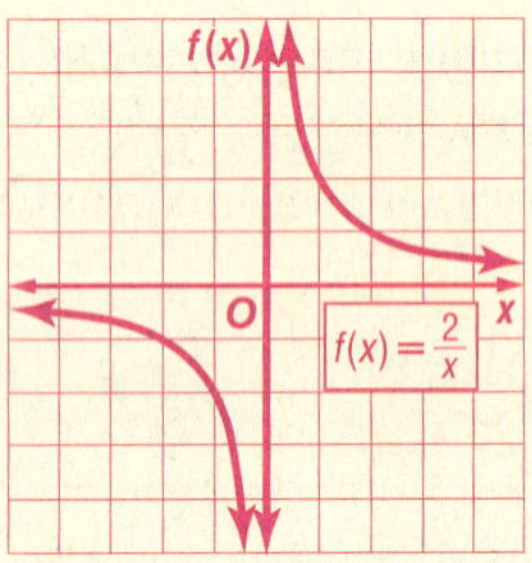

22.

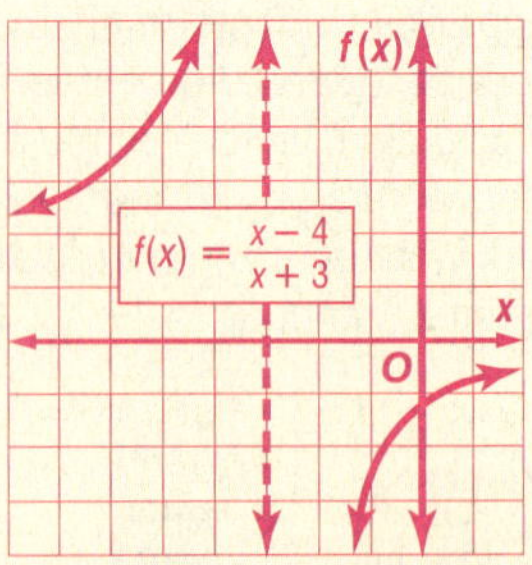

23.

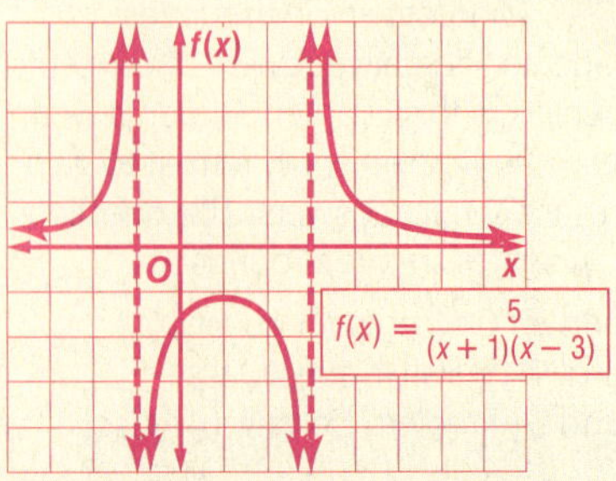

24.

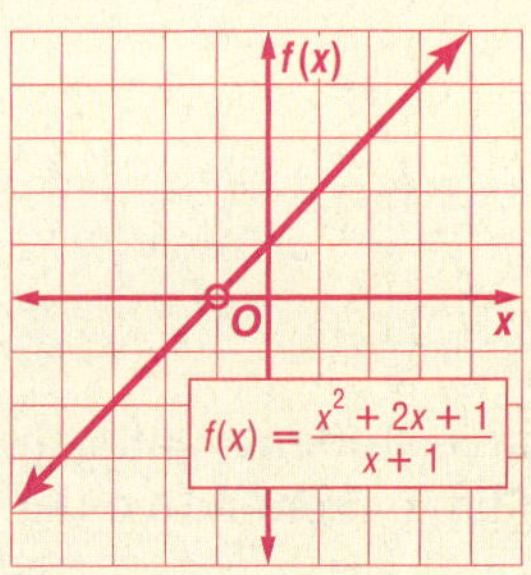

9-2 Adding and Subtracting Rational Expressions

See pages 479–484.

Concept Summary

- To add or subtract rational expressions, find a common denominator.
- To simplify complex fractions, simplify the numerator and the denominator separately, and then simplify the resulting expression.

Example Simplify $\frac{14}{x+y} - \frac{9x}{x^2 - y^2}$.

$$\frac{14}{x+y} - \frac{9x}{x^2-y^2} = \frac{14}{x+y} - \frac{9x}{(x+y)(x-y)} \quad \text{Factor the denominators.}$$

$$= \frac{14(x-y)}{(x+y)(x-y)} - \frac{9x}{(x+y)(x-y)} \quad \text{The LCD is } (x+y)(x-y).$$

$$= \frac{14(x-y) - 9x}{(x+y)(x-y)} \quad \text{Subtract the numerators.}$$

$$= \frac{14x - 14y - 9x}{(x+y)(x-y)} \quad \text{Distributive Property}$$

$$= \frac{5x - 14y}{(x+y)(x-y)} \quad \text{Simplify.}$$

Exercises Simplify each expression. *See Examples 3 and 4 on page 480.*

13. $\frac{x+2}{x-5} + 6$ **$\frac{7(x-4)}{x-5}$**

14. $\frac{x-1}{x^2-1} + \frac{2}{5x+5}$ **$\frac{7}{5(x+1)}$**

15. $\frac{7}{y} - \frac{2}{3y}$ **$\frac{19}{3y}$**

16. $\frac{7}{y-2} - \frac{11}{2-y}$ **$\frac{18}{y-2}$**

17. $\frac{3}{4b} - \frac{2}{5b} - \frac{1}{2b}$ **$-\frac{3}{20b}$**

18. $\frac{m+3}{m^2-6m+9} - \frac{8m-24}{9-m^2}$ **$\frac{3(3m^2-14m+27)}{(m+3)(m-3)^2}$**

9-3 Graphing Rational Functions

See pages 485–490.

Concept Summary

- Functions are undefined at any x value where the denominator is zero.
- An asymptote is a line that the graph of the function approaches, but never crosses.

Example Graph $f(x) = \frac{5}{x(x+4)}$.

The function is undefined for $x = 0$ and $x = -4$. Since $\frac{5}{x(x+4)}$ is in simplest form, $x = 0$ and $x = -4$ are vertical asymptotes. Draw the two asymptotes and sketch the graph.

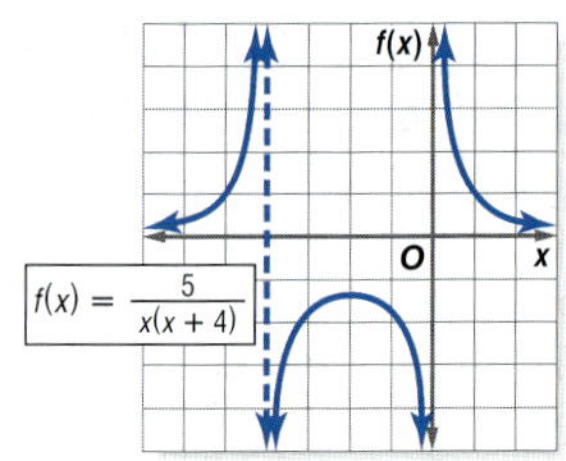

19–24. See margin.

Exercises Graph each rational function. *See Examples 2–4 on pages 486–488.*

19. $f(x) = \frac{4}{x-2}$

20. $f(x) = \frac{x}{x+3}$

21. $f(x) = \frac{2}{x}$

22. $f(x) = \frac{x-4}{x+3}$

23. $f(x) = \frac{5}{(x+1)(x-3)}$

24. $f(x) = \frac{x^2+2x+1}{x+1}$

514 Chapter 9 Rational Expressions and Equations

9-4 Direct, Joint, and Inverse Variation

See pages 492–498.

Concept Summary

- Direct Variation: There is a nonzero constant k such that $y = kx$.
- Joint Variation: There is a number k such that $y = kxz$, where $x \neq 0$ and $z \neq 0$.
- Inverse Variation: There is a nonzero constant k such that $xy = k$ or $y = \frac{k}{x}$.

Example **If y varies inversely as x and $x = 14$ when $y = -6$, find x when $y = -11$.**

$\frac{x_1}{y_2} = \frac{x_2}{y_1}$ Inverse variation

$\frac{14}{-11} = \frac{x_2}{-6}$ $x_1 = 14, y_1 = -6, y_2 = -11$

$14(-6) = -11(x_2)$ Cross multiply.

$-84 = -11x_2$ Simplify.

$7\frac{7}{11} = x_2$ When $y = -11$, the value of x is $7\frac{7}{11}$.

Exercises **Find each value.** *See Examples 1–3 on pages 493 and 494.*

25. If y varies directly as x and $y = 21$ when $x = 7$, find x when $y = -5$. $-1\frac{2}{3}$

26. If y varies inversely as x and $y = 9$ when $x = 2.5$, find y when $x = -0.6$. -37.5

27. If y varies inversely as x and $x = 28$ when $y = 18$, find x when $y = 63$. 8

28. If y varies directly as x and $x = 28$ when $y = 18$, find x when $y = 63$. 98

29. If y varies jointly as x and z and $x = 2$ and $z = 4$ when $y = 16$, find y when $x = 5$ and $z = 8$. 80

9-5 Classes of Functions

See pages 499–504.

Concept Summary

The following is a list of special functions.

- constant function
- direct variation function
- identity function
- greatest integer function
- absolute value function
- quadratic function
- square root function
- rational function
- inverse variation function

Examples **Identify the type of function represented by each graph.**

1

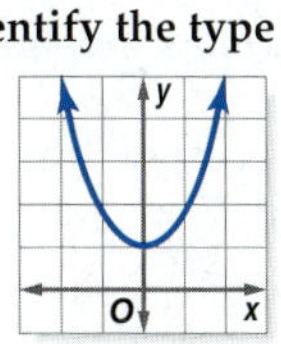

The graph has a parabolic shape, therefore it is a quadratic function.

2

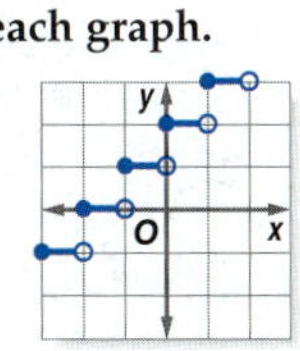

The graph has a stair-step pattern, therefore it is a greatest integer function.

Chapter 9 Study Guide and Review 515

Chapter 9 For More ...
- Extra Practice, see pages 847–849.
- Mixed Problem Solving, see page 870.

Exercises **Identify the type of function represented by each graph.**
See Example 1 on page 500.

30. 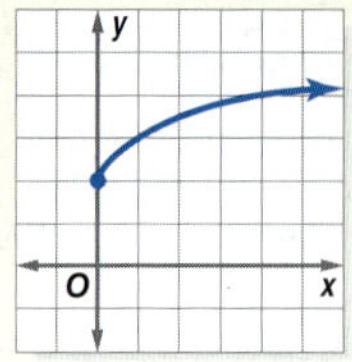

31. 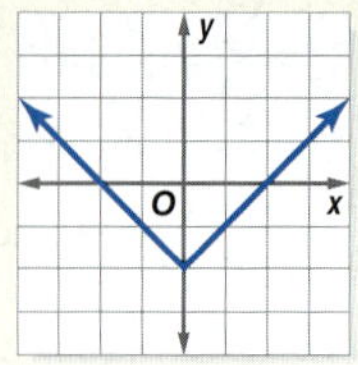

32. 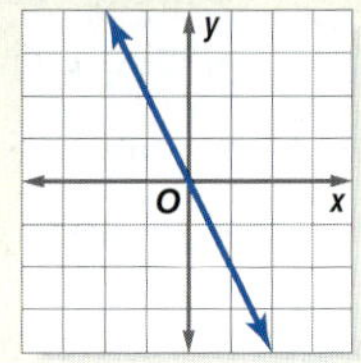

square root   absolute value   direct variation

9-6 Solving Rational Equations and Inequalities

See pages 505–511.

Concept Summary

- Eliminate fractions in rational equations by multiplying each side of the equation by the LCD.
- Possible solutions to a rational equation must exclude values that result in zero in the denominator.
- To solve rational inequalities, find the excluded values, solve the related equation, and use these values to divide a number line into regions. Then test a value in each region to determine which regions satisfy the original inequality.

Example **Solve $\frac{1}{x-1} + \frac{2}{x} = 0$.**

The LCD is $x(x - 1)$.

$\frac{1}{x-1} + \frac{2}{x} = 0$	Original equation
$x(x-1)\left(\frac{1}{x-1} + \frac{2}{x}\right) = x(x-1)(0)$	Multiply each side by $x(x - 1)$.
$x(x-1)\left(\frac{1}{x-1}\right) + x(x-1)\left(\frac{2}{x}\right) = x(x-1)(0)$	Distributive Property
$1(x) + 2(x-1) = 0$	Simplify.
$x + 2x - 2 = 0$	Distributive Property
$3x - 2 = 0$	Simplify.
$3x = 2$	Add 2 to each side.
$x = \frac{2}{3}$	Divide each side by 3.

The solution is $\frac{2}{3}$.

Exercises **Solve each equation or inequality. Check your solutions.**
See Examples 1, 2, and 5 on pages 505, 506, 508, and 509.

33. $\frac{3}{y} + \frac{7}{y} = 9$ $1\frac{1}{9}$

34. $1 + \frac{5}{y-1} = \frac{7}{6}$ 31

35. $\frac{3x+2}{4} = \frac{9}{4} - \frac{3-2x}{6}$ 3

36. $\frac{1}{r^2-1} = \frac{2}{r^2+r-2}$ 0

37. $\frac{x}{x^2-1} + \frac{2}{x+1} = 1 + \frac{1}{2x-2}$ $1\frac{1}{2}$

38. $\frac{1}{3b} - \frac{3}{4b} > \frac{1}{6}$ $-2\frac{1}{2} < b < 0$

Answers

Practice Test

12. 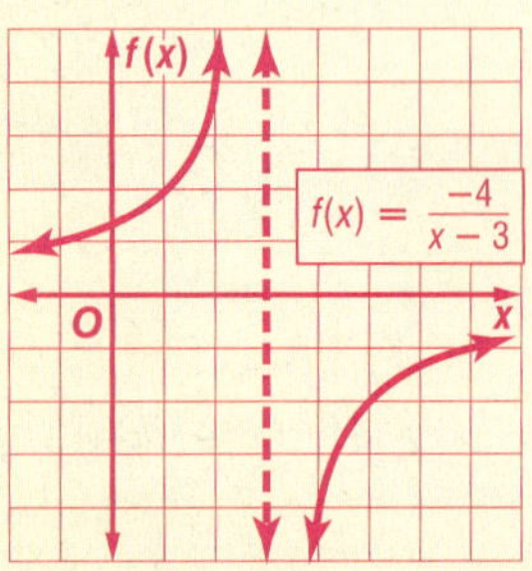

13. 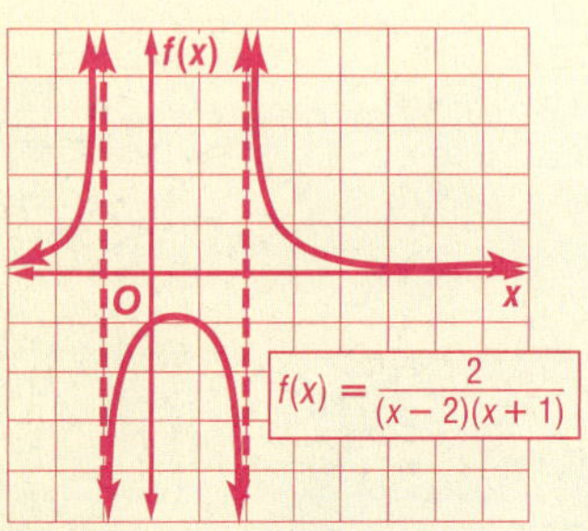

516 Chapter 9 Rational Expressions and Equations

Chapter 9 Practice Test

Vocabulary and Concepts

Match each example with the correct term.

1. $y = 4xz$ **c**
2. $y = 5x$ **b**
3. $y = \frac{7}{x}$ **a**

a. inverse variation equation
b. direct variation equation
c. joint variation equation

Skills and Applications

Simplify each expression.

4. $\frac{a^2 - ab}{3a} \div \frac{a - b}{15b^2}$ **$5b^2$**

5. $\frac{x^2 - y^2}{y^2} \cdot \frac{y^3}{y - x}$ **$-y(x + y)$**

6. $\frac{x^2 - 2x + 1}{y - 5} \div \frac{x - 1}{y^2 - 25}$ **$(x - 1)(y + 5)$**

7. $\dfrac{\frac{x^2 - 1}{x^2 - 3x - 10}}{\frac{x^2 + 3x + 2}{x^2 - 12x + 35}}$ **$\frac{(x - 1)(x - 7)}{(x + 2)^2}$**

8. $\frac{x - 2}{x - 1} + \frac{6}{7x - 7}$ **$\frac{7x - 8}{7(x - 1)}$**

9. $\frac{x}{x^2 - 9} + \frac{1}{2x + 6}$ **$\frac{3(x - 1)}{2(x - 3)(x + 3)}$**

Identify the type of function represented by each graph.

10. 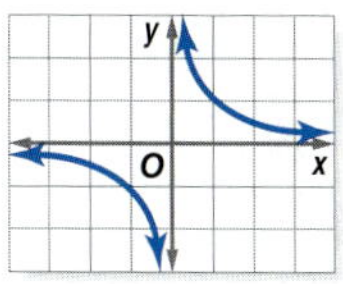 **inverse variation**

11. 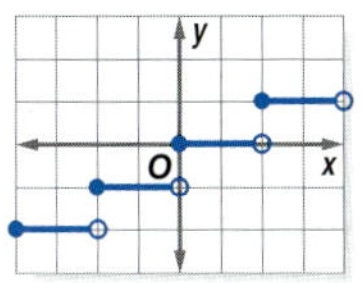 **greatest integer**

Graph each rational function. **12–13. See margin.**

12. $f(x) = \frac{-4}{x - 3}$

13. $f(x) = \frac{2}{(x - 2)(x + 1)}$

Solve each equation or inequality.

14. $\frac{2}{x - 1} = 4 - \frac{x}{x - 1}$ **2**

15. $\frac{9}{28} + \frac{3}{z + 2} = \frac{3}{4}$ **5**

16. $5 + \frac{3}{t} > -\frac{2}{t}$ **$t < -1$ or $t > 0$**

17. $x + \frac{12}{x} - 8 = 0$ **2, 6**

18. $\frac{5}{6} - \frac{2m}{2m + 3} = \frac{19}{6}$ **$-1\frac{1}{20}$**

19. $\frac{x - 3}{2x} = \frac{x - 2}{2x + 1} - \frac{1}{2}$ **$\pm\frac{\sqrt{6}}{2}$**

20. If y varies inversely as x and $y = 9$ when $x = -\frac{2}{3}$, find x when $y = -7$. **$\frac{6}{7}$**

21. If g varies directly as w and $g = 10$ when $w = -3$, find w when $g = 4$. **$-1\frac{1}{5}$**

22. Suppose y varies jointly as x and z. If $x = 10$ when $y = 250$ and $z = 5$, find x when $y = 2.5$ and $z = 4.5$. **$\frac{1}{9}$**

23. **AUTO MAINTENANCE** When air is pumped into a tire, the pressure required varies inversely as the volume of the air. If the pressure is 30 pounds per square inch when the volume is 140 cubic inches, find the pressure when the volume is 100 cubic inches. **42 lb/in^2**

24. **ELECTRICITY** The current I in a circuit varies inversely with the resistance R.
 a. Use the table at the right to write an equation relating the current and the resistance. **$I = \frac{6}{R}$**
 b. What is the constant of variation? **6**

I	0.5	1.0	1.5	2.0	2.5	3.0	5.0
R	12.0	6.0	4.0	3.0	2.4	2.0	1.2

25. **STANDARDIZED TEST PRACTICE** If $m = \frac{1}{x}$, $n = 7m$, $p = \frac{1}{n}$, $q = 14p$, and $r = \frac{1}{\frac{1}{2}q}$, find x. **D**

 (A) r (B) q (C) p (D) $\frac{1}{r}$ (E) $\frac{1}{q}$

www.algebra2.com/chapter_test

Portfolio Suggestion

Introduction In this lesson, you have been working with several kinds of variation.

Ask Students Write an application problem for each of the three types of variations (direct, inverse, and joint), and show your steps for each problem. Then describe how your problems are modeled by each of the variations, and how your answers relate to the solutions for the problems.

Chapter 9 Practice Test

Assessment Options

Vocabulary Test A vocabulary test/review for Chapter 9 can be found on p. 566 of the *Chapter 9 Resource Masters*.

Chapter Tests There are six Chapter 9 Tests and an Open-Ended Assessment task available in the *Chapter 9 Resource Masters*.

Chapter 9 Tests			
Form	**Type**	**Level**	**Pages**
1	MC	basic	553–554
2A	MC	average	555–556
2B	MC	average	557–558
2C	FR	average	559–560
2D	FR	average	561–562
3	FR	advanced	563–564

MC = multiple-choice questions
FR = free-response questions

Open-Ended Assessment
Performance tasks for Chapter 9 can be found on p. 565 of the *Chapter 9 Resource Masters*. A sample scoring rubric for these tasks appears on p. A25.

ExamView® Pro

Use the networkable **ExamView® Pro** to:

- Create **multiple versions** of tests.
- Create **modified** tests for *Inclusion* students.
- **Edit** existing questions and **add** your own questions.
- Use built-in **state curriculum correlations** to create tests aligned with state standards.
- **Apply** art to your tests from a program bank of artwork.

Chapter 9 Standardized Test Practice

These two pages contain practice questions in the various formats that can be found on the most frequently given standardized tests.

A practice answer sheet for these two pages can be found on p. A1 of the *Chapter 9 Resource Masters*.

Standardized Test Practice Student Recording Sheet, p. A1

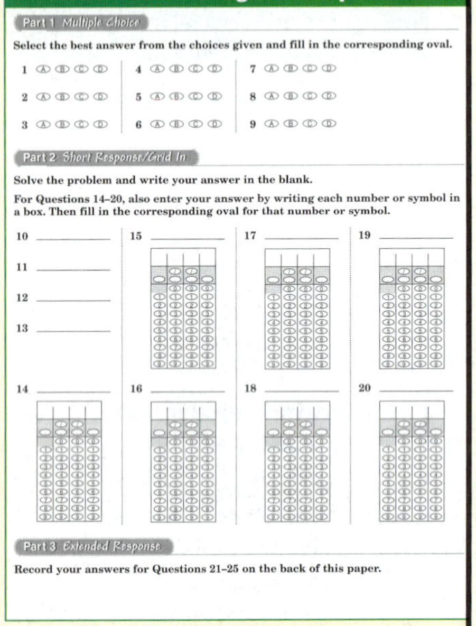
Part 1 Multiple Choice
Select the best answer from the choices given and fill in the corresponding oval.
Part 2 Short Response/Grid In
Solve the problem and write your answer in the blank.
For Questions 14–20, also enter your answer by writing each number or symbol in a box. Then fill in the corresponding oval for that number or symbol.
Part 3 Extended Response
Record your answers for Questions 21–25 on the back of this paper.

Additional Practice

See pp. 571–572 in the *Chapter 9 Resource Masters* for additional standardized test practice.

Chapter 9 Standardized Test Practice

Part 1 Multiple Choice

Record your answers on the answer sheet provided by your teacher or on a sheet of paper.

1. Best Bikes has 5000 bikes in stock on May 1. By the end of May, 40 percent of the bikes have been sold. By the end of June, 40 percent of the remaining bikes have been sold. How many bikes remain unsold? **C**

Ⓐ 1000 Ⓑ 1200
Ⓒ 1800 Ⓓ 2000

2. In $\triangle ABC$, if AB is equal to 8, then BC is equal to **C**

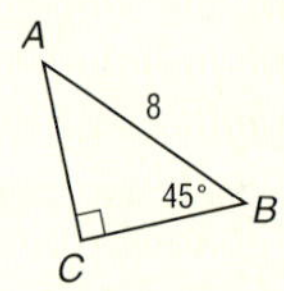

Ⓐ $\frac{\sqrt{2}}{8}$. Ⓑ 4.
Ⓒ $4\sqrt{2}$. Ⓓ 8.

3. In the figure, the slope of $\overline{AC}$ is $-\frac{1}{3}$ and $m\angle C = 30°$. What is the length of $\overline{BC}$? **D**

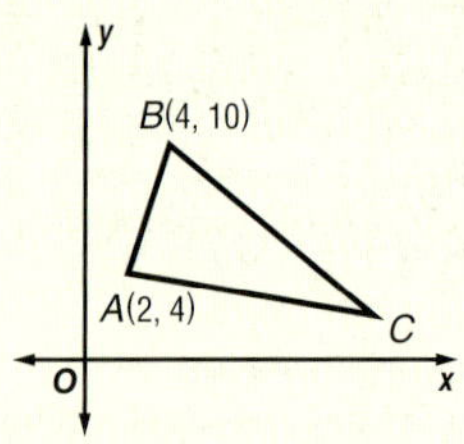

Ⓐ $\sqrt{10}$ Ⓑ $2\sqrt{10}$
Ⓒ $3\sqrt{10}$ Ⓓ $4\sqrt{10}$

4. Given that $-|2 - 4k| = -14$, which of the following could be k? **B**

Ⓐ 5 Ⓑ 4
Ⓒ 3 Ⓓ 2

5. In a hardware store, n nails cost c cents. Which of the following expresses the cost of k nails? **B**

Ⓐ nck Ⓑ $\frac{kc}{n}$
Ⓒ $n + \frac{k}{c}$ Ⓓ $n + \frac{c}{n}$

6. If $5w + 3 \leq w - 9$, then **D**

Ⓐ $w \leq 3$. Ⓑ $w \geq 3$.
Ⓒ $w \leq 12$. Ⓓ $w \leq -3$.

7. The graphs show a driver's distance d from a designated point as a function of time t. The driver passed the designated point at 60 mph and continued at that speed for 2 hours. Then she slowed to 50 mph for 1 hour. She stopped for gas and lunch for 1 hour and then drove at 60 mph for 1 hour. Which graph best represents this trip? **B**

Ⓐ
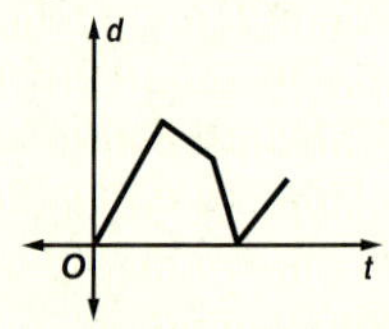

Ⓑ
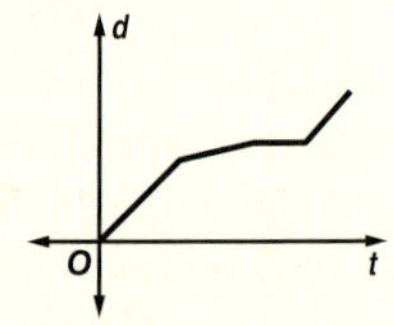

Ⓒ
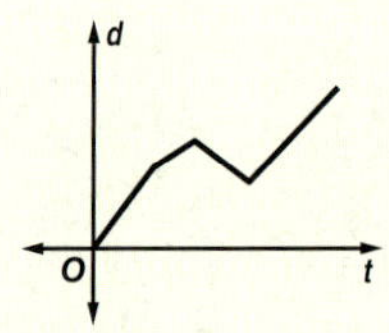

Ⓓ
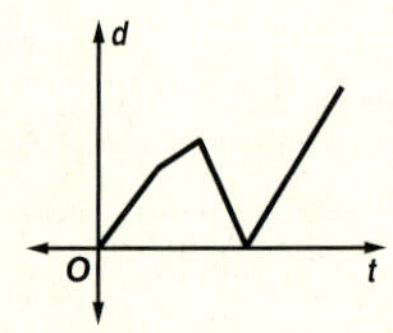

8. Which equation has roots of $-2n$, $2n$, and 2? **D**

Ⓐ $2x^2 - 8n^2 = 0$
Ⓑ $8n^2 - 2x^2 = 0$
Ⓒ $x^3 - 2x^2 - 4n^2x - 8n^2 = 0$
Ⓓ $x^3 - 2x^2 - 4n^2x + 8n^2 = 0$

9. What point is on the graph of $y - x^2 = 2$ and has a y-coordinate of 5? **A**

Ⓐ $(-\sqrt{3}, 5)$ Ⓑ $(\sqrt{7}, 5)$
Ⓒ $(5, \sqrt{3})$ Ⓓ $(3, 5)$

ExamView® Pro

Special banks of standardized test questions similar to those on the SAT, ACT, TIMSS 12, NAEP 12, and state proficiency tests can be found on this CD-ROM.

Preparing for Standardized Tests
For test-taking strategies and more practice, see pages 877–892.

Part 2 Short Response/Grid In

Record your answers on the answer sheet provided by your teacher or on a sheet of paper.

10. In the figure, what is the equation of the circle Q that is circumscribed around the square $ABCD$? $(x + 5)^2 + y^2 = 18$

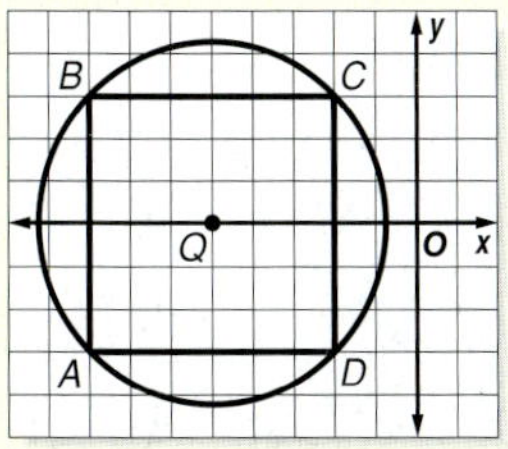

11. Find one possible value for k such that k is an integer between 20 and 40 that has a remainder of 2 when it is divided by 3 and that has a remainder of 2 when divided by 4. **26 or 38**

12. The coordinates of the vertices of a triangle are $(2, -4)$, $(10, -4)$, and (a, b). If the area of the triangle is 36 square units, what is a possible value for b? **−13 or 5**

13. If $(x + 2)(x - 3) = 6$, what is a possible value of x? **−3 or 4**

14. If the average of five consecutive even integers is 76, what is the greatest of these integers? **80**

15. In May, Hank's Camping Supply Store sold 45 tents. In June, it sold 90 tents. What is the percent increase in the number of tents sold? **100**

16. If $2^{n-4} = 64$, what is the value of n? **10**

17. If $xy = 5$ and $x^2 + y^2 = 20$, what is the value of $(x + y)^2$? **30**

18. If $\frac{2}{a} - \frac{8}{a^2} = \frac{-8}{a^3}$, then what is the value of a? **2**

19. If $\sqrt[x]{80} = 2\sqrt[x]{5}$, what is the value of x? **4**

20. What is the y-intercept of the graph of $3x + 2 = 4y - 6$? **2**

Part 3 Extended Response

Record your answers on a sheet of paper. Show your work. **21–22. See margin.**

21. Write an expression that is undefined when $x = 1$ or $x = -1$. Justify your answer.

22. Identify the type of function represented by the graph below. Then write an equation for the function. Explain your answer.

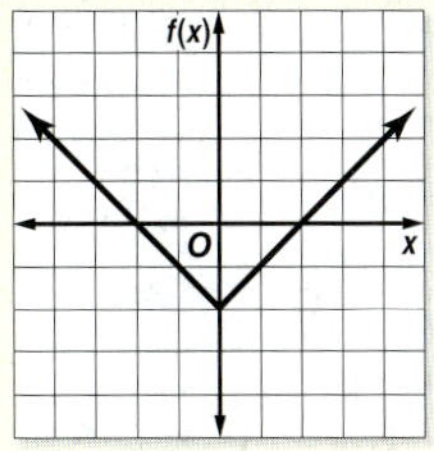

23. If 100 feet of a certain type of cable weighs 12 pounds, how much do 3 yards of the same cable weigh? **1.08 lb**

For Exercises 24 and 25, use the following information. **24. See margin.**

A gear that is 8 inches in diameter turns a smaller gear that is 3 inches in diameter.

24. Does this situation represent a direct or inverse variation? Explain your reasoning.

25. If the larger gear makes 36 revolutions, how many revolutions does the smaller gear make in that time? **96**

Test-Taking Tip

Questions 23–25
When solving problems involving variation, pay careful attention to the units of measure in your calculations. Check that the units in the fractions of a proportion for a direct variation problem are the same in the numerators and the denominators.

www.algebra2.com/standardized_test

Evaluating Extended Response Questions

Extended Response questions are graded by using a multilevel rubric that guides you in assessing a student's knowledge of a particular concept.

Goal: Determine the dimensions of a rectangular garden for a given perimeter.

Sample Scoring Rubric: The following rubric is a sample scoring device. You may wish to add more detail to this sample to meet your individual scoring needs.

Score	Criteria
4	A correct solution that is supported by well-developed, accurate explanations
3	A generally correct solution, but may contain minor flaws in reasoning or computation
2	A partially correct interpretation and/or solution to the problem
1	A correct solution with no supporting evidence or explanation
0	An incorrect solution indicating no mathematical understanding of the concept or task, or no solution is given

Answers

21. Sample answer: $\frac{x+2}{x^2-1}$; 1 and −1 make the denominator 0 and division by 0 is undefined .

22. Absolute value; $f(x) = |x| - 2$; absolute value functions have v-shaped graphs. The graph of $f(x) = |x|$ is translated down 2 units for $f(x) = |x| - 2$.

24. Inverse; the smaller gear has fewer teeth and makes more revolutions.

Chapter 9 Standardized Test Practice 519

Page 477, Lesson 9-1

49. A rational expression can be used to express the fraction of a nut mixture that is peanuts. Answers should include the following.

- The rational expression $\frac{8 + x}{13 + x}$ is in simplest form because the numerator and the denominator have no common factors.
- Sample answer: $\frac{8 + x}{13 + x + y}$ could be used to represent the fraction that is peanuts if x pounds of peanuts and y pounds of cashews were added to the original mixture.

Page 484, Lesson 9-2

51. Subtraction of rational expressions can be used to determine the distance between the lens and the film if the focal length of the lens and the distance between the lens and the object are known. Answers should include the following.

- To subtract rational expressions, first find a common denominator. Then, write each fraction as an equivalent fraction with the common denominator. Subtract the numerators and place the difference over the common denominator. If possible, reduce the answer.
- $\frac{1}{q} = \frac{1}{10} - \frac{1}{60}$ could be used to determine the distance between the lens and the film if the focal length of the lens is 10 cm and the distance between the lens and the object is 60 cm.

56.

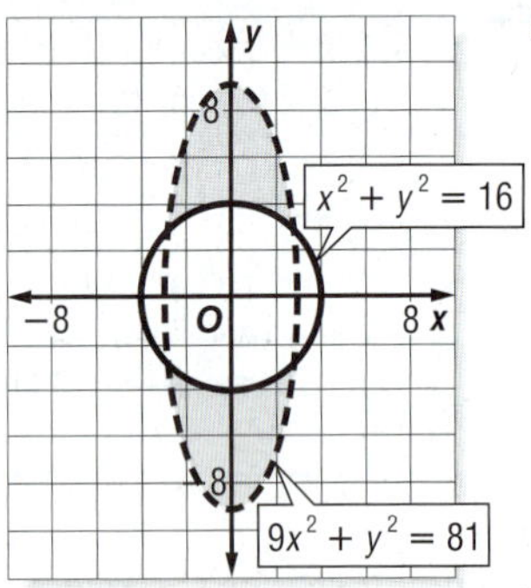

57.

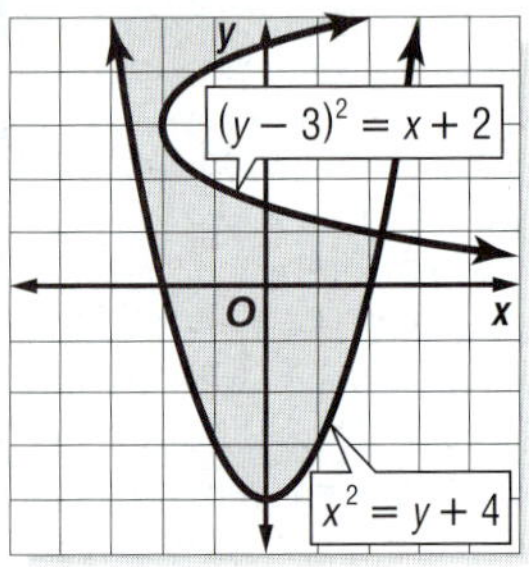

59.

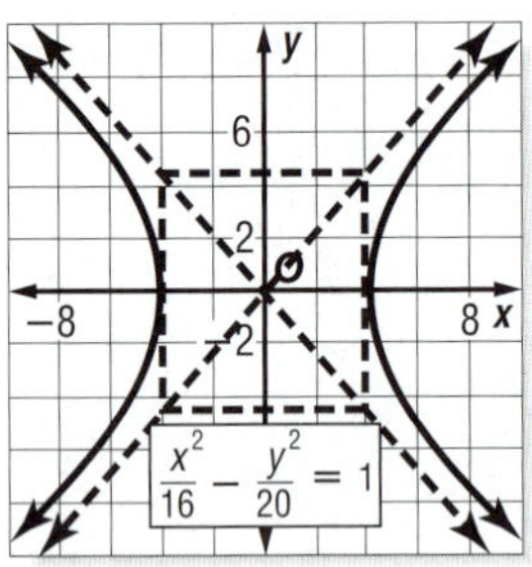

60.

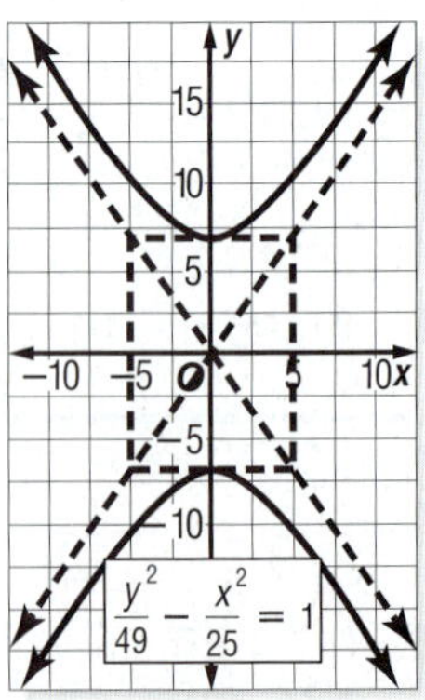

61.

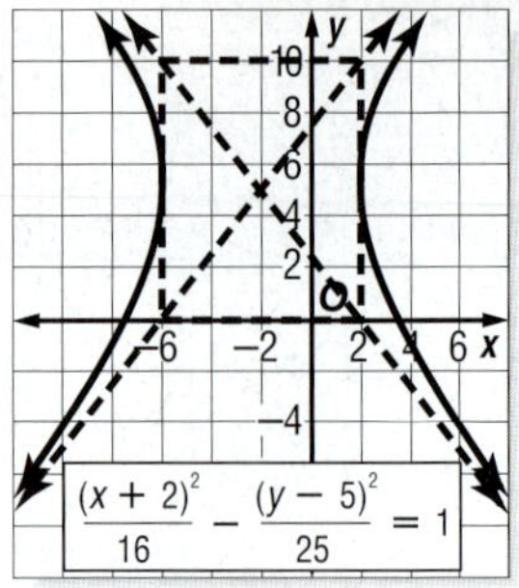

Page 487, Algebra Activity

1.

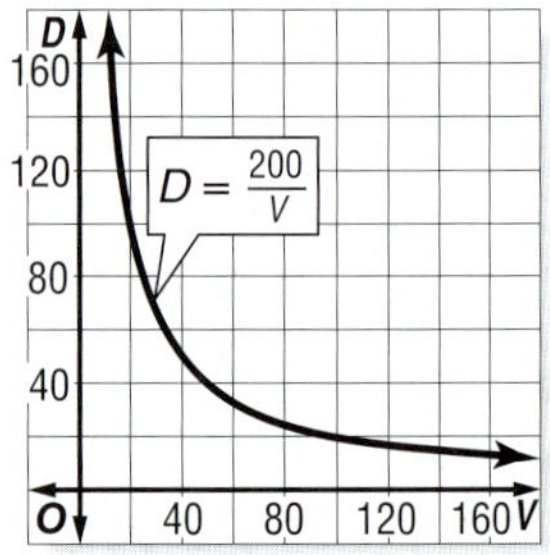

Pages 488–489, Lesson 9-3

6.

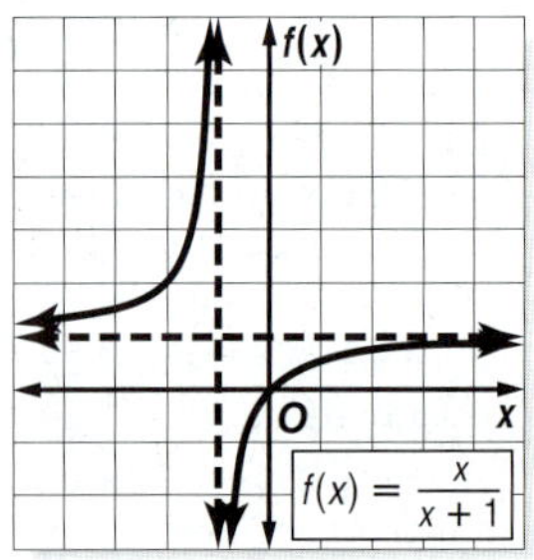

7.

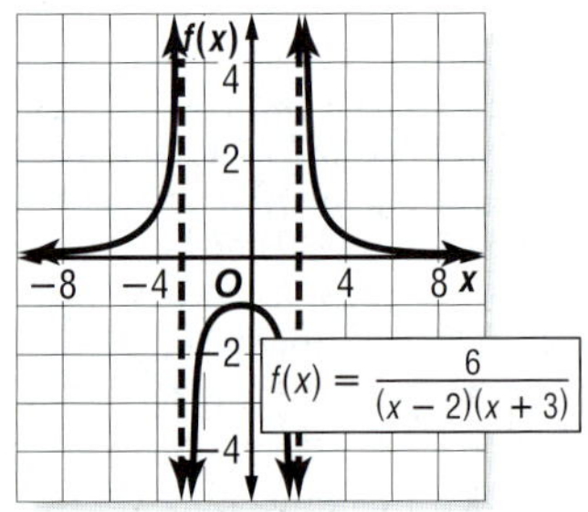

8.

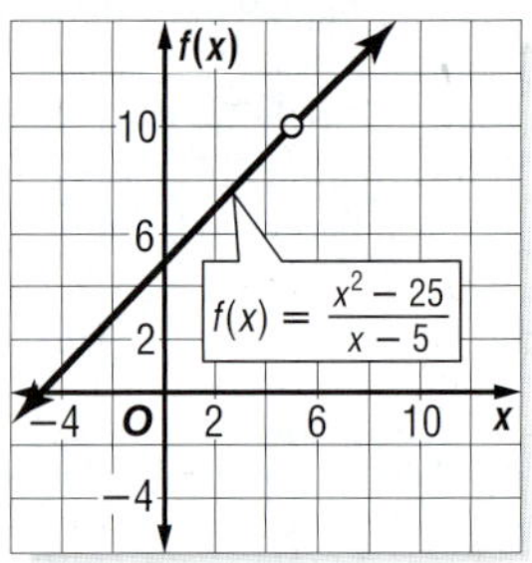

9.

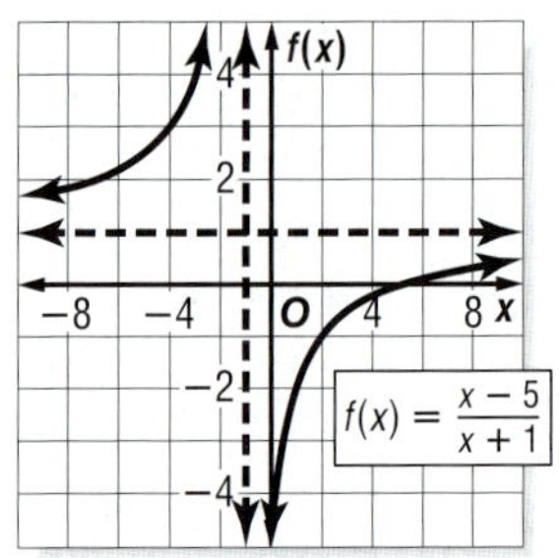

10.

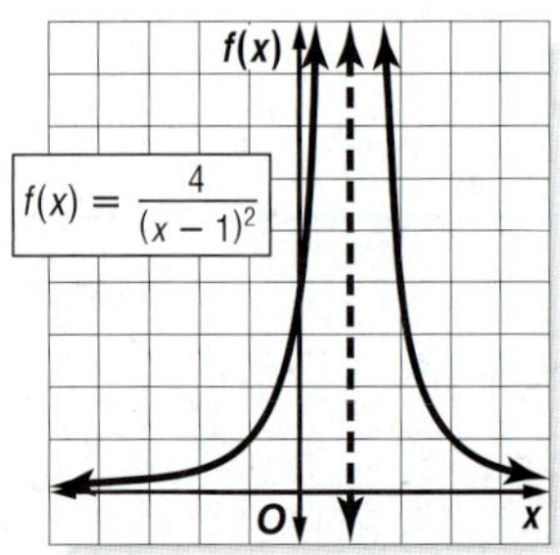

11.

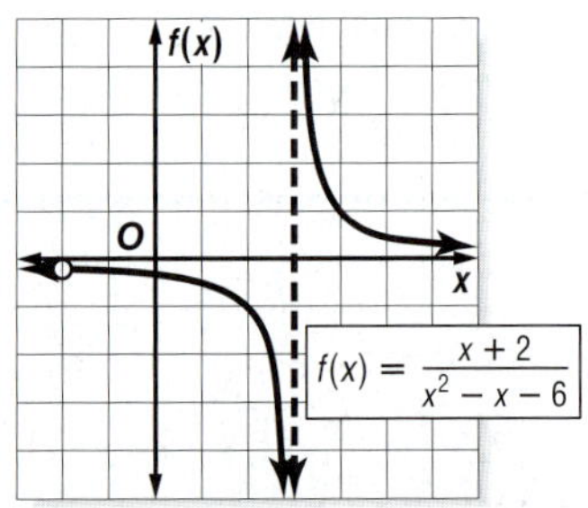

13.

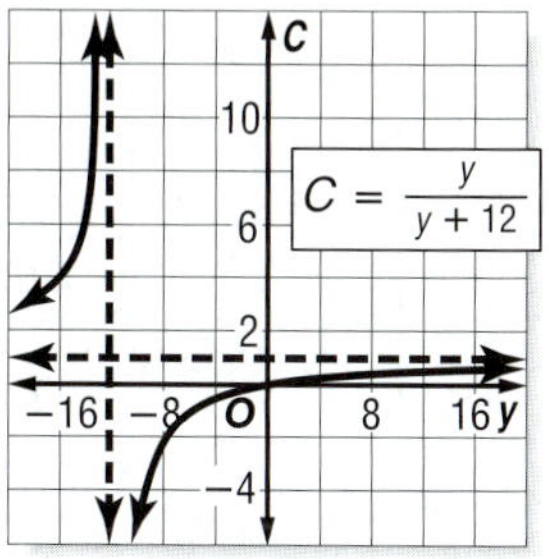

22.

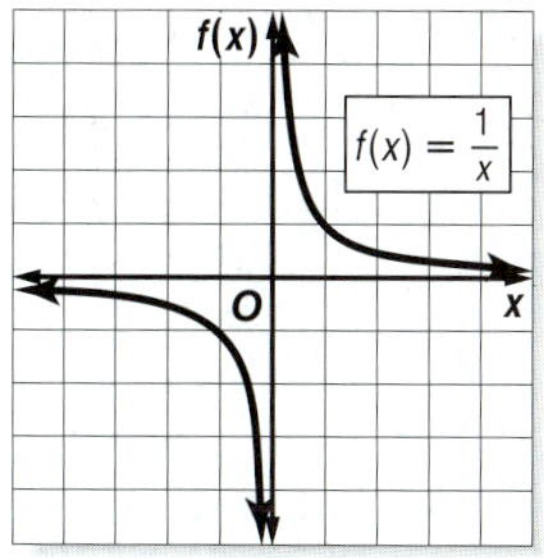

23.

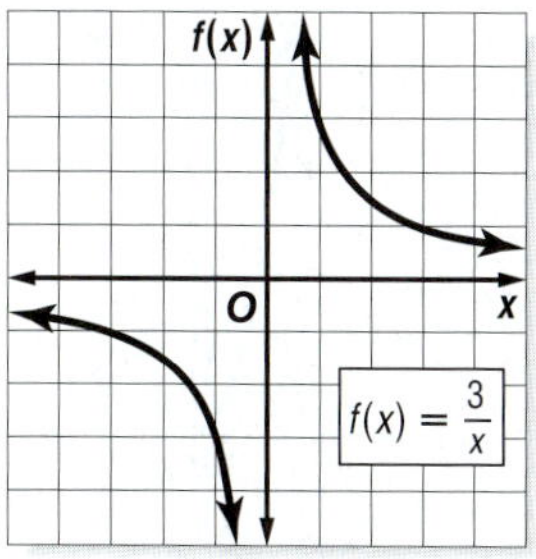

24.

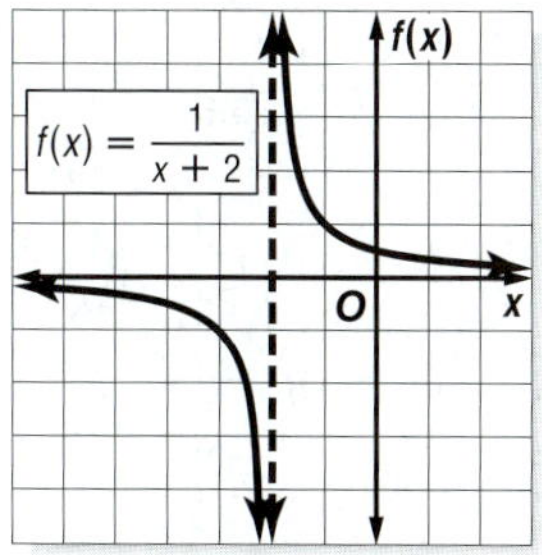

25.

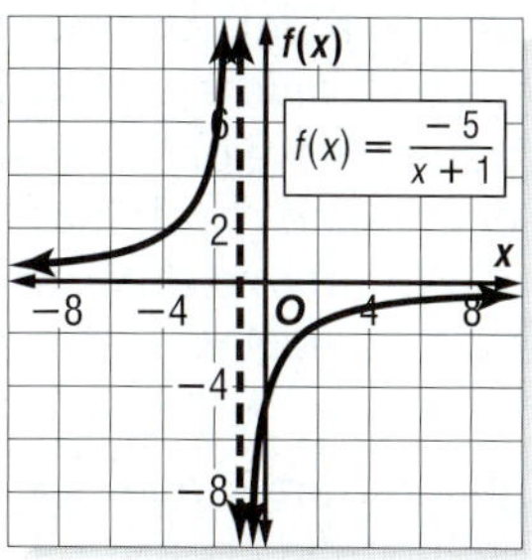

26.

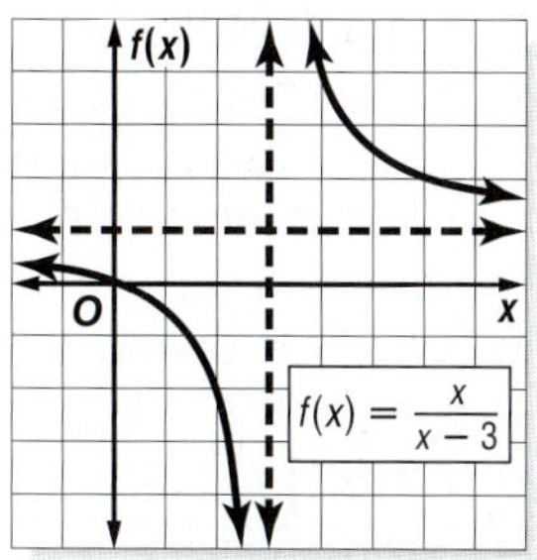

27.

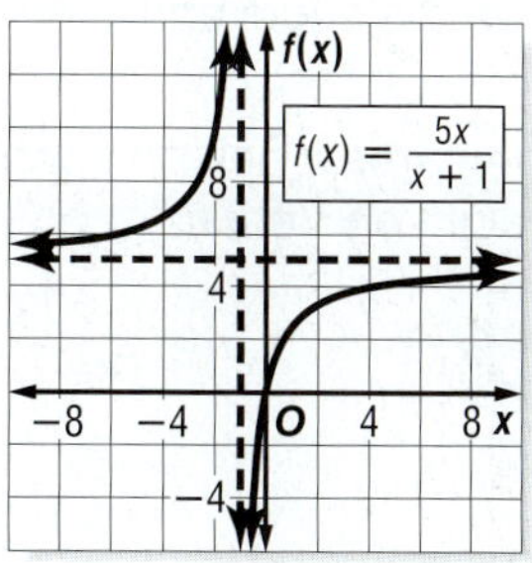

28.

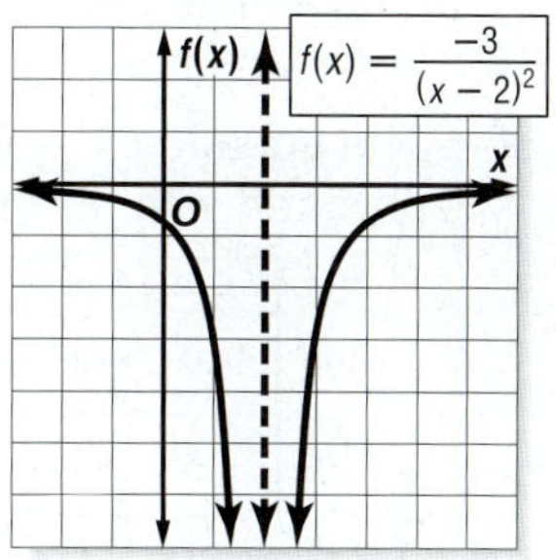

29.

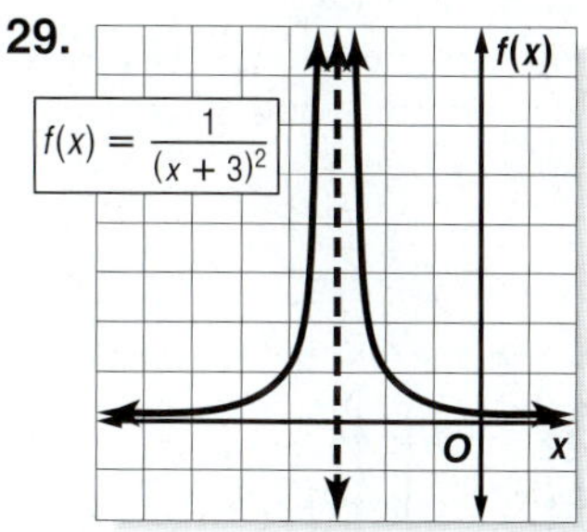

30.

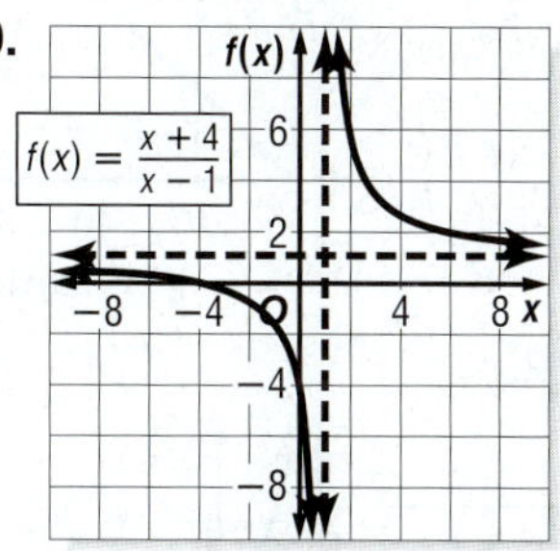

31.

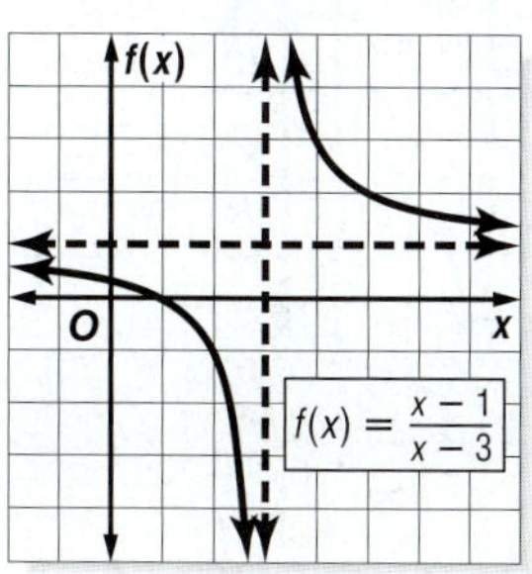

32.

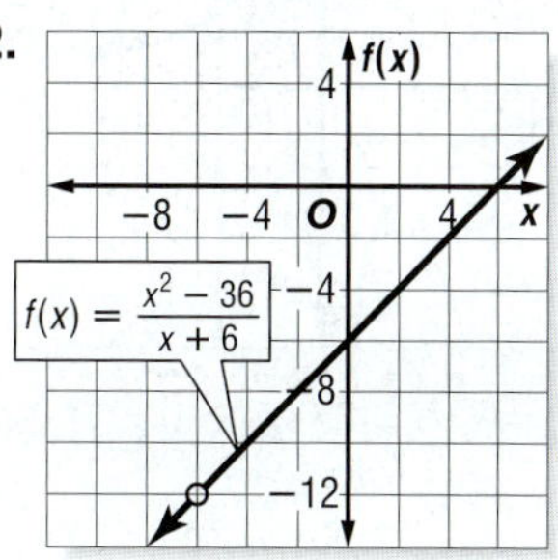

33.

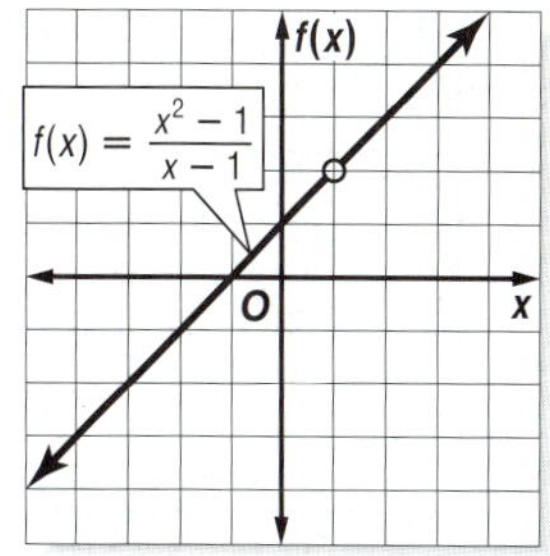

34.

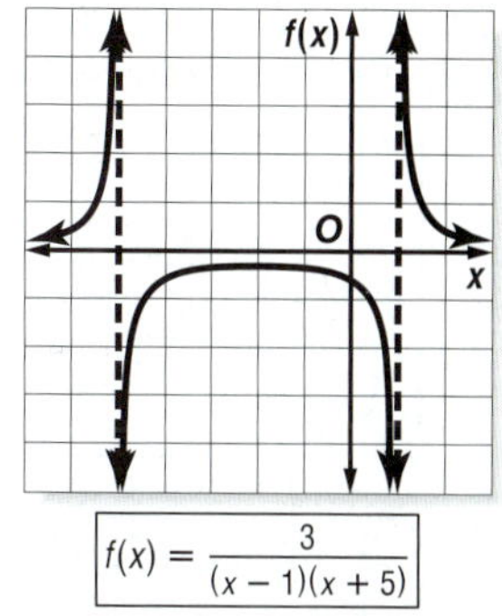

35.

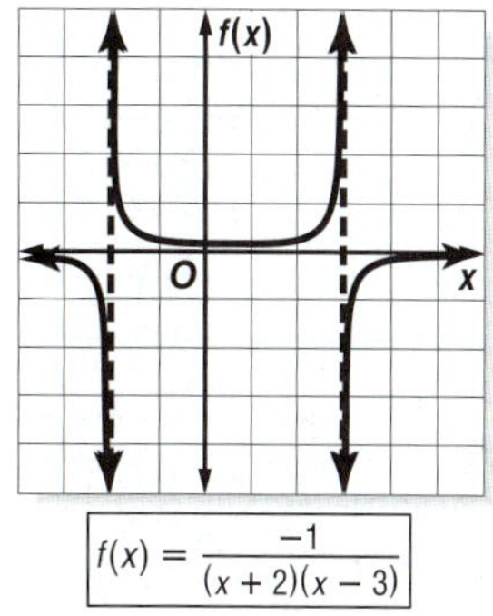

36.

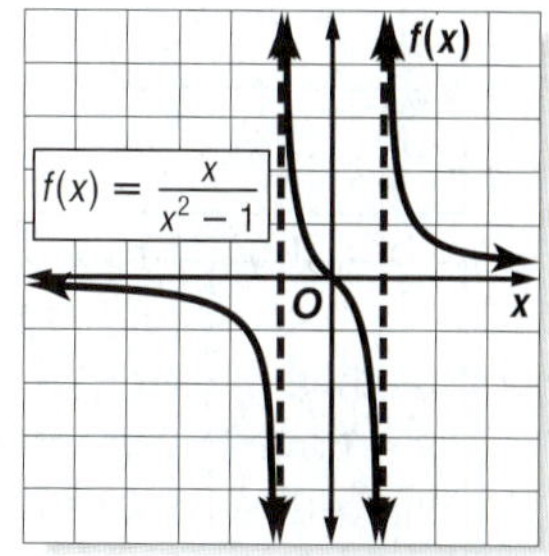

37.

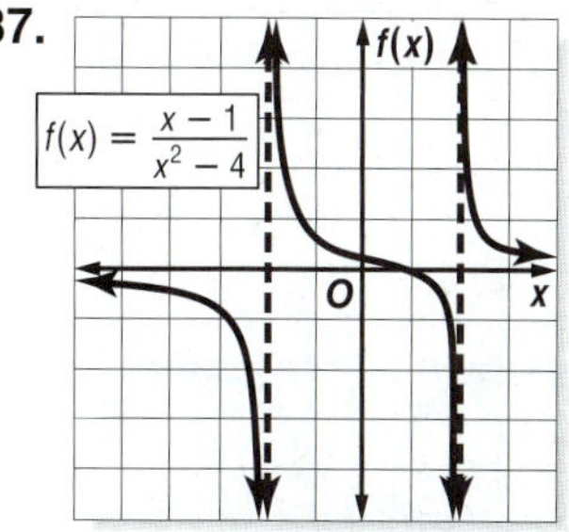

38.

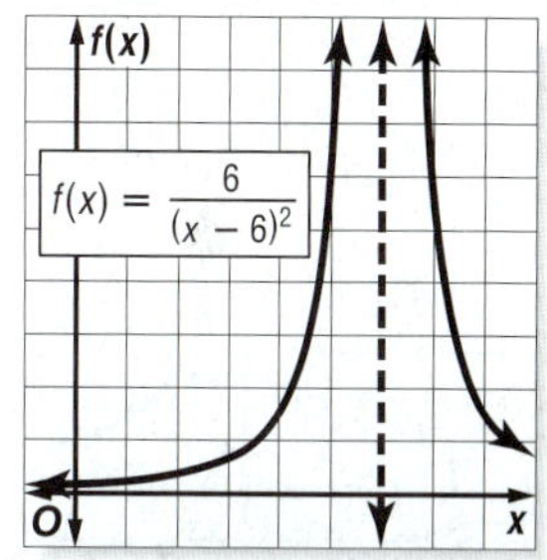

39.

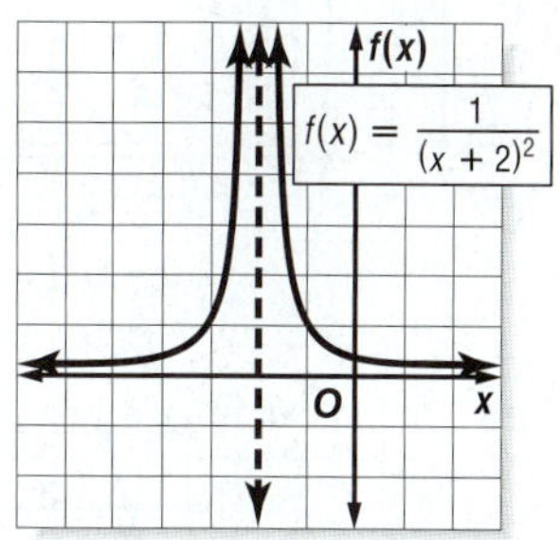

40.

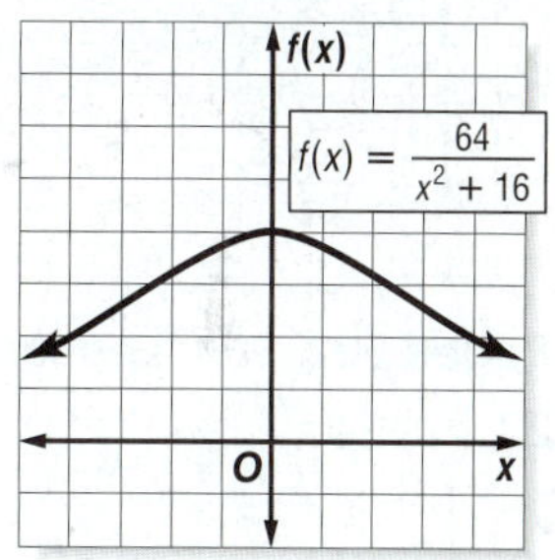

42. Since $\frac{-64}{x^2 + 16} = -\left(\frac{64}{x^2 + 16}\right)$, the graph of $f(x) = \frac{-64}{x^2 + 16}$ would be a reflection of the graph of $f(x) = \frac{64}{x^2 + 16}$ over the x-axis.

43.

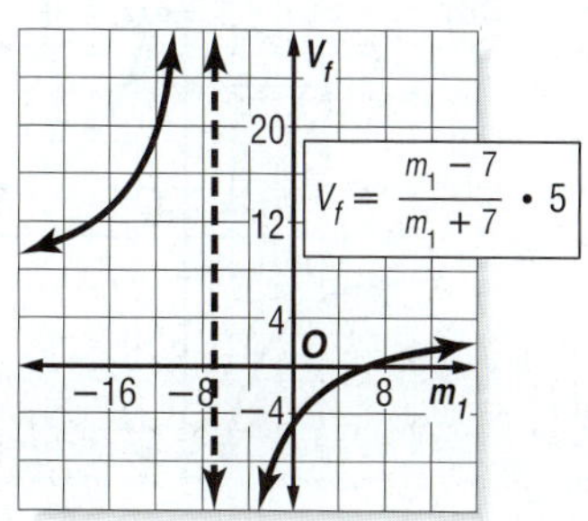

Page 491, Follow-Up of Lesson 9-3
Graphing Calculator Investigation

1. 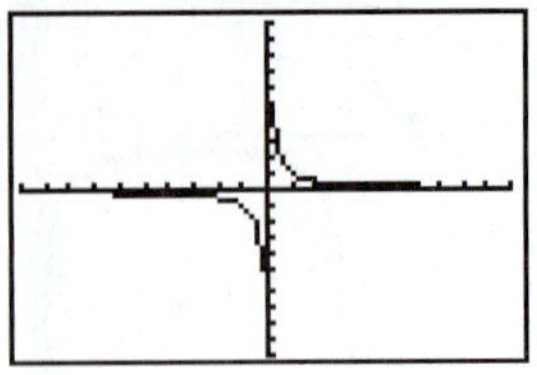

2. 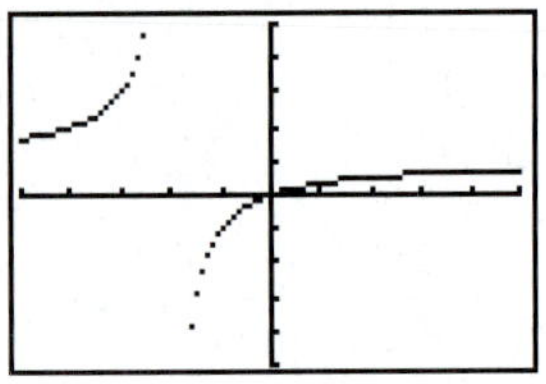

3. 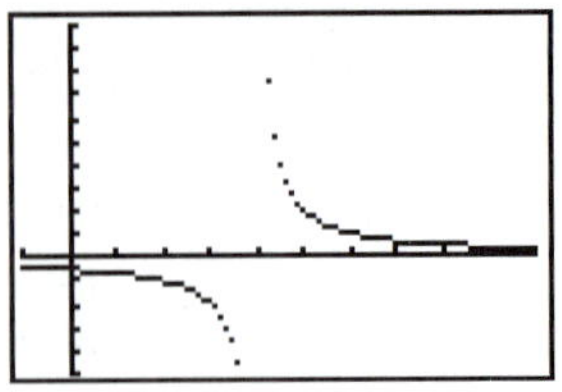

4. 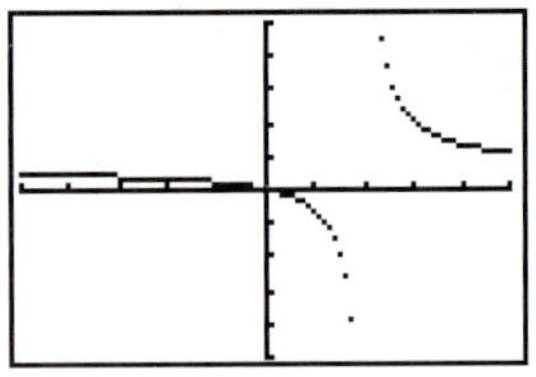

5. 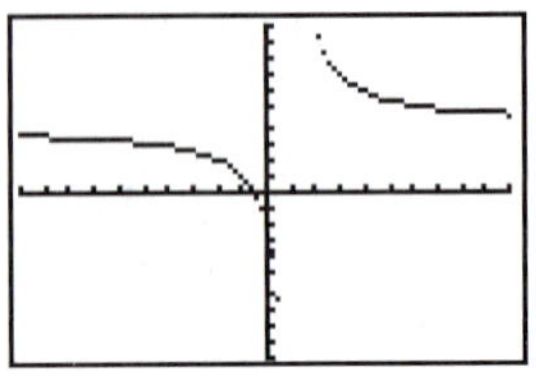

6. 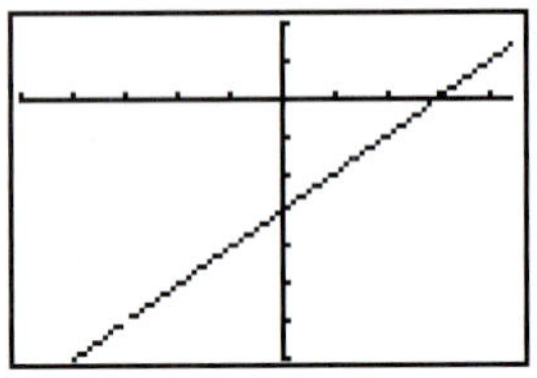

Page 498, Practice Quiz 2

1. 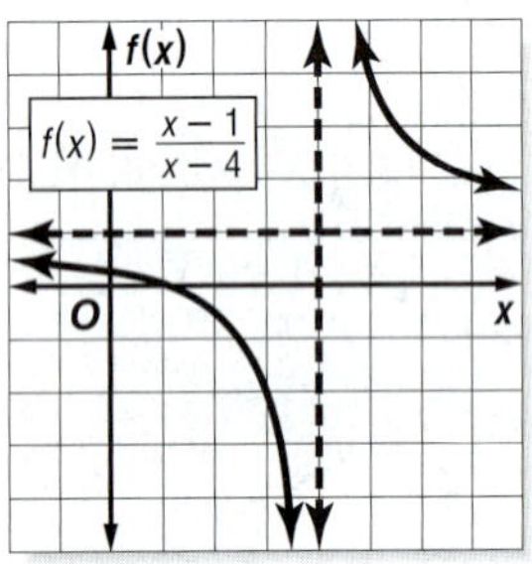

2. 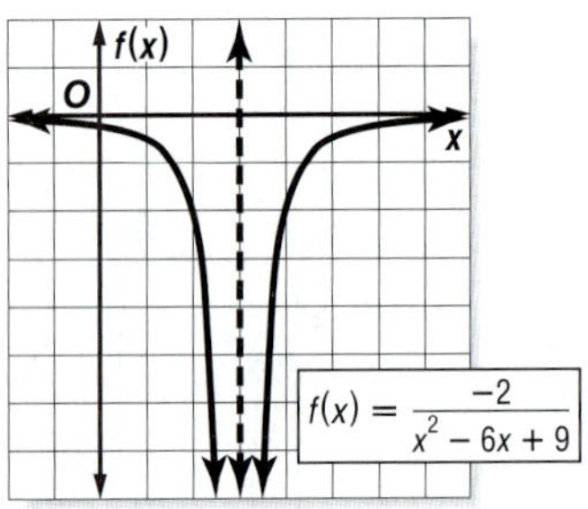

Pages 502–504, Lesson 9-5

23. 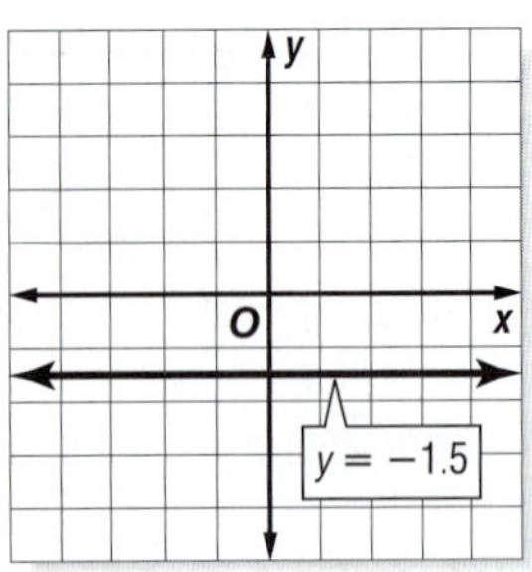

24. 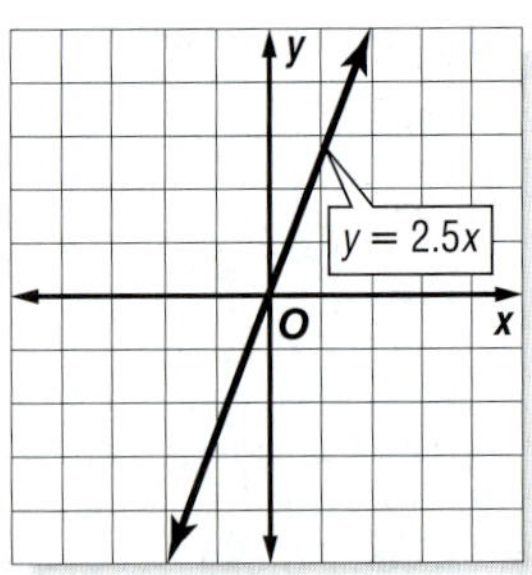

25. 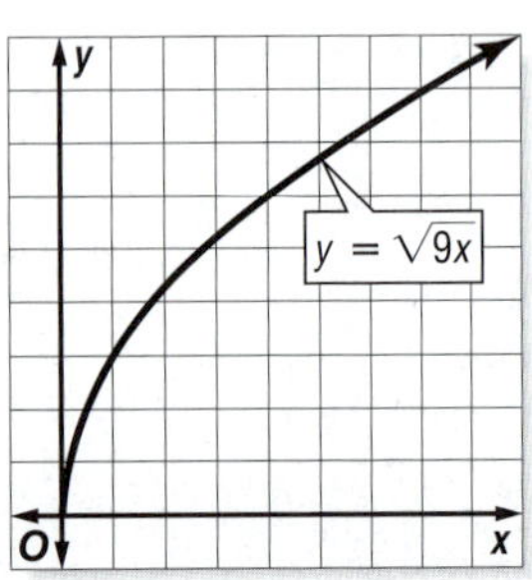

26. 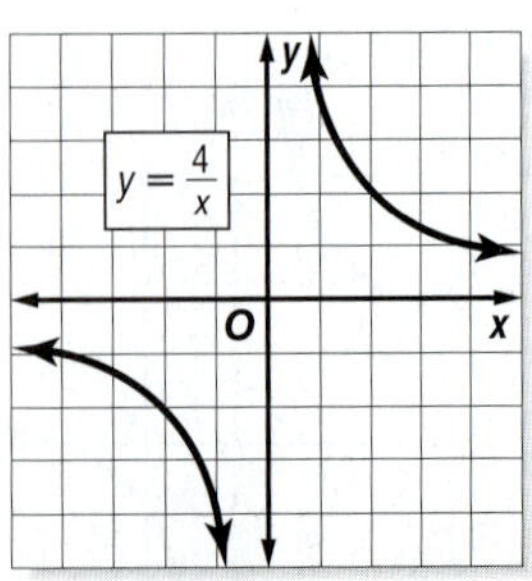

27. 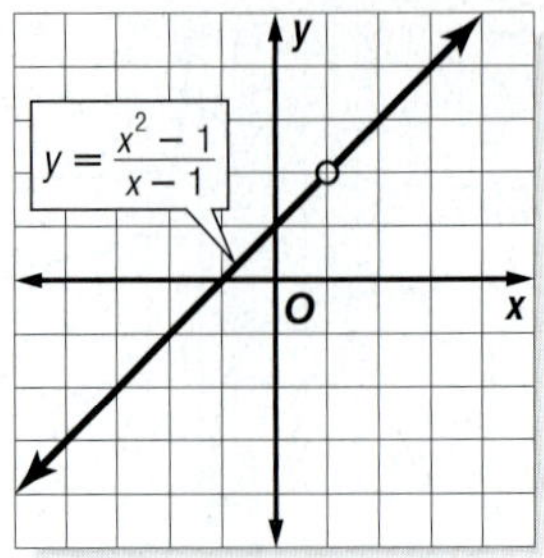

28. 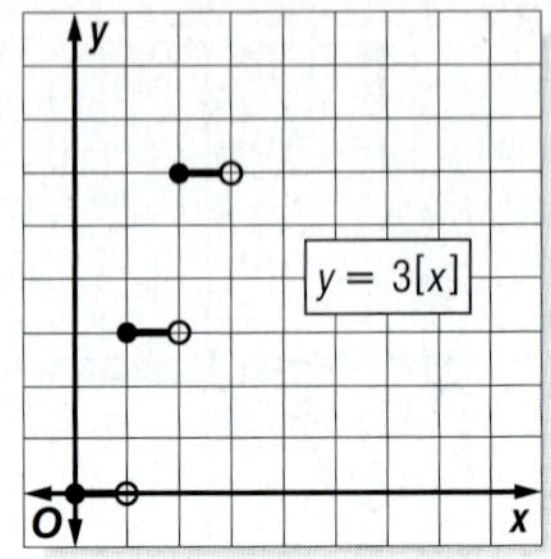

29. 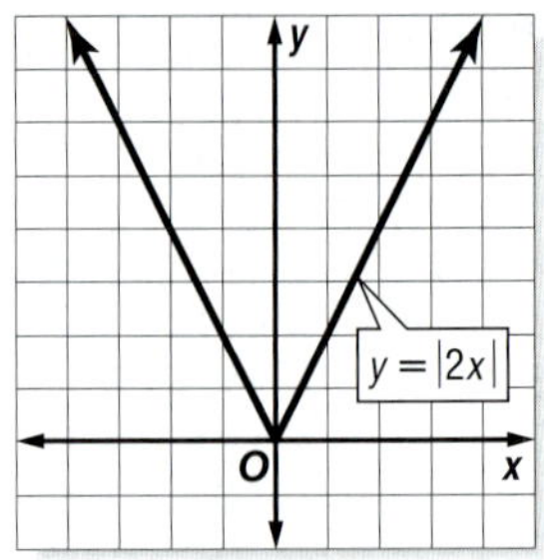

30. 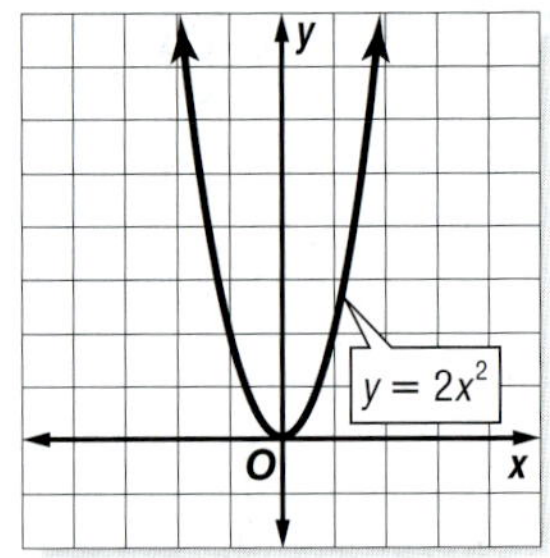

38. A graph of the function that relates a person's weight on Earth with his or her weight on a different planet can be used to determine a person's weight on the other planet by finding the point on the graph that corresponds with the weight on Earth and determining the value on the other planet's axis. Answers should include the following.

- The graph comparing weight on Earth and Mars represents a direct variation function because it is a straight line passing through the origin and is neither horizontal nor vertical.
- The equation $V = 0.9E$ compares a person's weight on Earth with his or her weight on Venus.

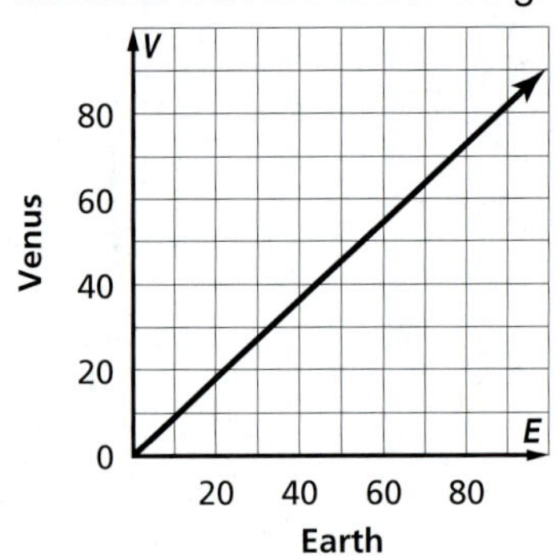

42. 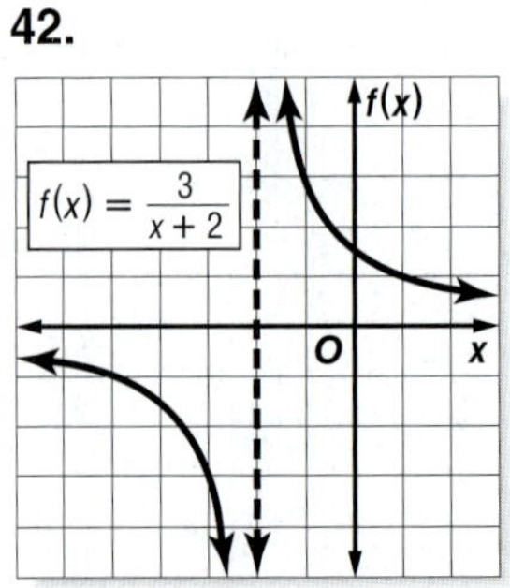

43. 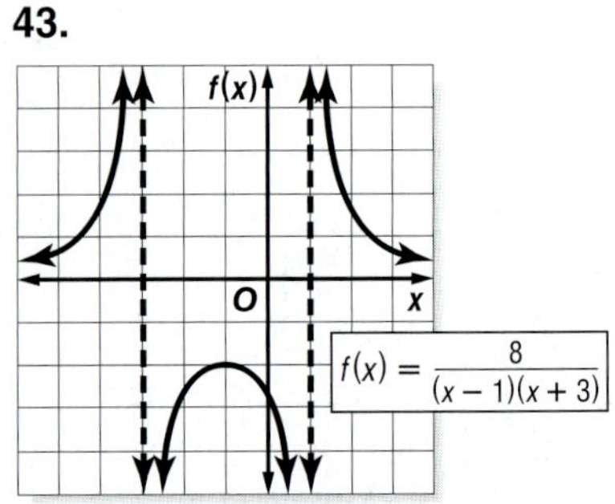

44. 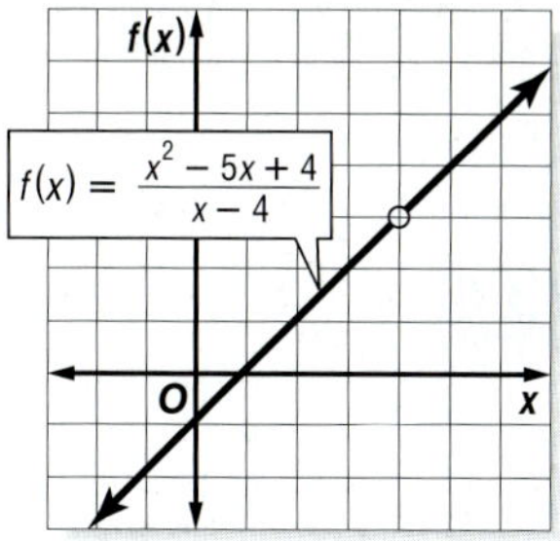

Additional Answers for Chapter 9

Chapter 10 Exponential and Logarithmic Relations

Chapter Overview and Pacing

Year-long and two-year pacing: pages T20–T21.

LESSON OBJECTIVES	PACING (days)			
	Regular		Block	
	Basic/ Average	Advanced	Basic/ Average	Advanced
10-1 Exponential Functions *(pp. 522–530)* ***Preview:*** Investigating Exponential Functions • Graph exponential functions. • Solve exponential equations and inequalities.	1	1	0.5	0.5
10-2 Logarithms and Logarithmic Functions *(pp. 531–540)* • Evaluate logarithmic expressions. • Solve logarithmic equations and inequalities. ***Follow-Up:*** Modeling Real-World Data: Curve Fitting	2	2 (with 10-2 Follow-Up)	1	1
10-3 Properties of Logarithms *(pp. 541–546)* • Simplify and evaluate expressions using the properties of logarithms. • Solve logarithmic equations using the properties of logarithms.	1	1	0.5	0.5
10-4 Common Logarithms *(pp. 547–553)* • Solve exponential equations and inequalities using common logarithms. • Evaluate logarithmic expressions using the Change of Base Formula. ***Follow-Up:*** Solving Exponential and Logarithmic Equations and Inequalities	1	1	0.5	0.5
10-5 Base *e* and Natural Logarithms *(pp. 554–559)* • Evaluate expressions involving the natural base and natural logarithms. • Solve exponential equations and inequalities using natural logarithms.	2 (with 10-4 Follow-Up)	2 (with 10-4 Follow-Up)	1 (with 10-4 Follow-Up)	1 (with 10-4 Follow-Up)
10-6 Exponential Growth and Decay *(pp. 560–565)* • Use logarithms to solve problems involving exponential decay. • Use logarithms to solve problems involving exponential growth.	1	1	0.5	0.5
Study Guide and **Practice Test** *(pp. 566–571)* **Standardized Test Practice** *(pp. 572–573)*	1	1	0.5	0.5
Chapter Assessment	1	1	0.5	0.5
TOTAL	**10**	**10**	**5**	**5**

*An electronic version of this chapter is available on **StudentWorks™**. This backpack solution CD-ROM allows students instant access to the Student Edition, lesson worksheet pages, and web resources.*

Timesaving Tools
TeacherWorks™
All-In-One Planner and Resource Center
See pages T5 and T21.

Chapter Resource Manager

Study Guide and Intervention	Practice (Skills and Average)	Reading to Learn Mathematics	Enrichment	Assessment	Applications*	5-Minute Check Transparencies	Interactive Chalkboard	Alge2PASS: Tutorial Plus (lessons)	Materials
Chapter 10 Resource Masters									
573–574	575–576	577	578		GCS 45	10-1	10-1		(*Preview:* paper, scissors, grid paper, calculator) graphing calculator, grid paper, string
579–580	581–582	583	584	623	SC 19	10-2	10-2		posterboard (*Follow-Up:* graphing calculator, grid paper)
585–586	587–588	589	590	623, 625		10-3	10-3		
591–592	593–594	595	596			10-4	10-4		(*Follow-Up:* graphing calculator)
597–598	599–600	601	602	624	SM 127–132	10-5	10-5	19	plastic coins, paper currency
603–604	605–606	607	608	624	GCS 46, SC 20	10-6	10-6		
				609–622, 626–628					

**Key to Abbreviations:* GCS = Graphing Calculator and Speadsheet Masters,
SC = School-to-Career Masters,
SM = Science and Mathematics Lab Manual

Chapter 10

Mathematical Connections and Background

Continuity of Instruction

Prior Knowledge

Students have worked with exponents in many situations, including performing calculations, manipulating expressions, and applying properties. They have explored properties of inverses for operations and for functions, and they have solved many kinds of equations and inequalities.

This Chapter

Students are introduced to the term logarithm to solve for a variable that appears as an exponent. They explore the relationship between exponents and logarithms, and they use logarithms with two special bases, base 10 or common logarithms, and base *e* or natural logarithms. They apply the Change of Base Formula to rewrite a logarithm using a different base, and they apply appropriate formulas to solve problems involving exponential growth and exponential decay.

Future Connections

Students will continue to look at properties of and relationships between exponents and logarithms. They will apply formulas for exponential growth and exponential decay in science courses and in consumer situations. The natural-base exponential function will have an important role in precalculus and calculus topics.

10-1 Exponential Functions

Examine the list of characteristics for an exponential function on page 524. The first characteristic states that an exponential function is continuous and one-to-one. The term *continuous* means that the function can be traced without lifting your pencil. The term *one-to-one* means that a horizontal line passing through the graph will intersect no more than one point on the graph. This characteristic is important for the development of the logarithmic function in Lesson 10-2, since only one-to-one functions can have inverses. The second characteristic listed is that the domain of the function is the set of all real numbers. This property is important because it means that $3^{\sqrt{5}}$ has meaning, since $\sqrt{5}$ is a real number and part of the domain of $y = 3^x$. The third and fourth characteristics of an exponential function are related. The x-axis is a *horizontal asymptote* of the graph of an exponential function. This means that the graph of this function approaches the horizontal line $x = 0$, getting closer and closer to this line but never crossing it. This restricts the graph of an exponential function to either Quadrants I and II, when a is positive, or to Quadrants III and IV, when a is negative. In terms of the range of the function, this means that when a is positive, all y values of the function will be positive, and when a is negative, all y values of the function will be negative. These two properties will also be important when considering the inverse of the exponential function. The last two properties are useful for graphing and writing exponential functions.

10-2 Logarithms and Logarithmic Functions

In the equation $y = \log_b x$, y is referred to as the *logarithm*, b is the *base*, and x is sometimes referred to as the *argument*. The definition of a logarithm given on page 532 indicates that a logarithm is an exponent.

When solving logarithmic equations and inequalities, it is important to remember that a defining characteristic of a logarithmic function is that its domain is the set of all *positive* numbers. This means that the logarithm of 0 or of a negative number for any base is undefined. It is very important to check possible solutions to logarithmic equations in the original equation, to be sure that they would not result in taking the logarithm of 0 or a negative number. For logarithmic inequalities, this fact will exclude not just one value from the solution set, but a range of values. In Example 8 on page 534, since the original inequality asks for the values $\log_{10}(3x - 4)$ and $\log_{10}(x + 6)$, we must solve two inequalities, $3x - 4 \leq 0$ and $x + 6 \leq 0$, to find what values must be excluded from the solution set we found using the Property of Inequality for Logarithmic Functions. Excluding the values such that $x \leq \frac{4}{3}$ and $x \leq -6$, the solution set is all x such that the following three inequalities are all satisfied: $x > \frac{4}{3}$, $x > -6$, and $x < 5$. To simplify this compound inequality, sketch all three inequalities, as shown below, and find where all three intersect.

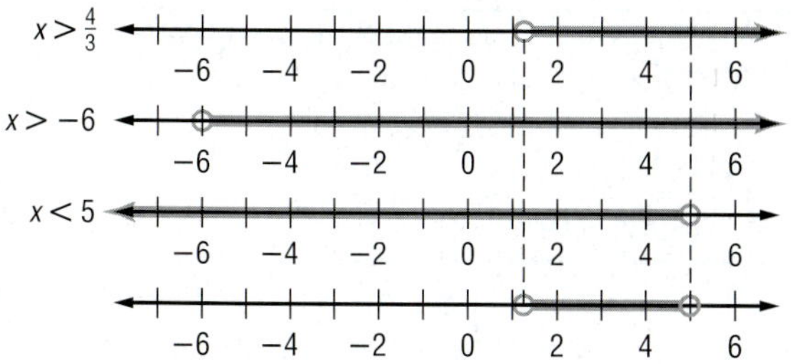

The final number line shows that the solution set is the compound inequality $\frac{4}{3} < x < 5$.

10-3 Properties of Logarithms

The word logarithm is actually a contraction of "logical arithmetic." Logarithms were invented to make computation easier. Using logarithms, multiplication changes to addition, according to the Product Property of Logarithms, and division changes to subtraction, according to the Quotient Property of Logarithms. This is illustrated in Examples 1, 2, and 4 of Lesson 10-3. In these examples, students are given the approximate value of specific logarithms. Before the invention of the scientific calculator, these values took a good deal of time to compute. Rather than use the same arduous process to compute each and every logarithm one encountered, the properties of logarithms allowed the use of a relative few logarithmic values to compute others.

10-4 Common Logarithms

Before the invention of the scientific calculator, the appendices of algebra texts contained extensive tables of common logarithms of numbers. In order to read these tables, you had to understand the parts of a logarithm. Every logarithm has two parts, the *characteristic* and the *mantissa*. A mantissa is the logarithm of a number between 1 and 10. When the original number is expressed in scientific notation, the characteristic is the power of 10.

$645{,}000 = 6.45 \cdot 10^5$	Scientific notation
$\log 645{,}000 = \log (6.45 \cdot 10^5)$	Property of Equality for Log Functions
$= \log 6.45 + \log 10^5$	Product Property
$= \log 6.45 + 5$	Inverse Property of Exponents and Logs
$\approx 0.8096 + 5$	$\log 6.45 \approx 0.8096$
≈ 5.8096	Simplify.

$\log 645{,}000 \approx 5.8096$

↑ characteristic ↑ mantissa

10-5 Base *e* and Natural Logarithms

Exponentiation, which is the inverse operation of taking a logarithm, is sometimes referred to as finding the *antilogarithm*. That is, if $\log x = a$ then $x = \text{antilog } a$. Since antilogarithms mean the same operation as exponentiation, it follows that to find the antilogarithm of a common logarithm, you would use 2nd [10^x] on a graphing calculator. To find the antilogarithm of a natural logarithm, antiln a, you would use 2nd [e^x].

10-6 Exponential Growth and Decay

It is important to note that the variable r in the exponential decay formula $y = a(1 - r)^t$ and the variable k in the alternate exponential decay formula $y = ae^{-kt}$ are not equivalent. In a problem where a decay factor is given or asked for, the formula $y = a(1 - r)^t$ should be used and not the formula $y = ae^{-kt}$. The same is true of the exponential growth formulas $y = a(1 + r)^t$ and $y = ae^{kt}$.

Additional mathematical information and teaching notes are available at www.algebra2.com/key_concepts.

Chapter 10

DAILY INTERVENTION and Assessment

	Type	Student Edition	Teacher Resources	Technology/Internet
INTERVENTION	Ongoing	Prerequisite Skills, pp. 521, 530, 538, 546, 551, 559 Practice Quiz 1, p. 538 Practice Quiz 2, p. 559	5-Minute Check Transparencies Quizzes, *CRM* pp. 623–624 Mid-Chapter Test, *CRM* p. 625 Study Guide and Intervention, *CRM* pp. 573–574, 579–580, 585–586, 591–592, 597–598, 603–604	Alge2PASS: Tutorial Plus www.algebra2.com/self_check_quiz www.algebra2.com/extra_examples
	Mixed Review	pp. 531, 538, 546, 551, 559, 565	Cumulative Review, *CRM* p. 626	
	Error Analysis	Find the Error, pp. 535, 544, 557 Common Misconceptions, p. 523	Find the Error, *TWE* pp. 535, 544, 557 Unlocking Misconceptions, *TWE* pp. 542, 548 Tips for New Teachers, *TWE* p. 534	
	Standardized Test Practice	pp. 530, 537, 538, 546, 551, 559, 562, 563, 564, 572–573	*TWE* p. 562 Standardized Test Practice, *CRM* pp. 627–628	Standardized Test Practice CD-ROM www.algebra2.com/standardized_test
ASSESSMENT	Open-Ended Assessment	Writing in Math, pp. 530, 537, 546, 551, 559, 564 Open Ended, pp. 527, 535, 544, 549, 557, 563	Modeling: *TWE* pp. 530, 565 Speaking: *TWE* pp. 546, 559 Writing: *TWE* pp. 538, 551 Open-Ended Assessment, *CRM* p. 621	
	Chapter Assessment	Study Guide, pp. 566–570 Practice Test, p. 571	Multiple-Choice Tests (Forms 1, 2A, 2B), *CRM* pp. 609–614 Free-Response Tests (Forms 2C, 2D, 3), *CRM* pp. 615–620 Vocabulary Test/Review, *CRM* p. 622	ExamView® Pro (see below) MindJogger Videoquizzes www.algebra2.com/vocabulary_review www.algebra2.com/chapter_test

Key to Abbreviations: TWE = Teacher Wraparound Edition; CRM = Chapter Resource Masters

Intervention Technology

Alge2PASS: Tutorial Plus CD-ROM offers a complete, self-paced algebra curriculum.

Algebra 2 Lesson	Alge2PASS Lesson
10-5	19 *Exponential and Logarithmic Functions*

ExamView® Pro

Use the networkable **ExamView® Pro** to:

- Create **multiple versions** of tests.
- Create **modified** tests for *Inclusion* students.
- **Edit** existing questions and **add** your own questions.
- Use built-in **state curriculum correlations** to create tests aligned with state standards.
- **Apply** art to your tests from a program bank of artwork.

For more information on Intervention and Assessment, see pp. T8–T11.

Reading and Writing in Mathematics

Glencoe Algebra 2 provides numerous opportunities to incorporate reading and writing into the mathematics classroom.

Student Edition

- Foldables Study Organizer, p. 521
- Concept Check questions require students to verbalize and write about what they have learned in the lesson. (pp. 527, 535, 544, 549, 557, 563, 566)
- Writing in Math questions in every lesson, pp. 530, 537, 546, 551, 559, 564
- WebQuest, pp. 529, 565

Teacher Wraparound Edition

- Foldables Study Organizer, pp. 521, 566
- Study Notebook suggestions, pp. 522, 527, 535, 544, 549, 557, 563
- Modeling activities, pp. 530, 565
- Speaking activities, pp. 546, 559
- Writing activities, pp. 538, 551
- ELL Resources, pp. 520, 529, 537, 545, 550, 558, 564, 566

Additional Resources

- Vocabulary Builder worksheets require students to define and give examples for key vocabulary terms as they progress through the chapter. (*Chapter 10 Resource Masters*, pp. vii-viii)
- Reading to Learn Mathematics master for each lesson (*Chapter 10 Resource Masters*, pp. 577, 583, 589, 595, 601, 607)
- *Vocabulary PuzzleMaker* software creates crossword, jumble, and word search puzzles using vocabulary lists that you can customize.
- *Teaching Mathematics with Foldables* provides suggestions for promoting cognition and language.
- *Reading and Writing in the Mathematics Classroom*
- *WebQuest and Project Resources*

For more information on Reading and Writing in Mathematics, see pp. T6–T7.

ELL ENGLISH LANGUAGE LEARNERS

Lesson 10-1

Using Applications

Show students a copy of a family tree. Have students discuss the number of parents, grandparents, and great-grandparents a person has. Lead students in a discussion of using a pattern to find an equation that can represent the number of direct ancestors a person has *x* generations ago.

Lesson 10-6

Multisensory Activities

Give each group of students a "superball" and a meter stick. Have them experiment to find the percent of the previous height that is reached on the first bounce when the ball is dropped. Suggest that they drop the ball from a height of 1 meter and measure the height of the first bounce for at least 3 trials and average the results. Then have each group write an equation for the height of the ball on the *n*th bounce. For example, $h = 8(0.9^n)$ represents the height of the *n*th bounce if the ball rebounds to 90% of its previous height. Then discuss how to find the height of the 12th bounce using their equations.

Chapter 10 Notes

What You'll Learn

Have students read over the list of objectives and make a list of any words with which they are not familiar.

Why It's Important

Point out to students that this is only one of many reasons why each objective is important. Others are provided in the introduction to each lesson.

Lesson	NCTM Standards	Local Objectives
10-1 Preview	1, 2, 3, 6, 7, 8, 10	
10-1	1, 2, 3, 4, 6, 8, 9, 10	
10-2	1, 2, 3, 4, 6, 7, 8, 9	
10-2 Follow-Up	1, 2, 3, 5, 6, 8, 10	
10-3	1, 2, 4, 6, 7, 8, 9	
10-4	1, 2, 4, 6, 8, 9	
10-4 Follow-Up	1, 2, 3	
10-5	1, 2, 3, 4, 6, 7, 8, 9	
10-6	1, 2, 4, 6, 8, 9	

Key to NCTM Standards:

1=Number & Operations, 2=Algebra, 3=Geometry, 4=Measurement, 5=Data Analysis & Probability, 6=Problem Solving, 7=Reasoning & Proof, 8=Communication, 9=Connections, 10=Representation

Chapter 10 Exponential and Logarithmic Relations

What You'll Learn

- **Lessons 10-1 through 10-3** Simplify exponential and logarithmic expressions.
- **Lessons 10-1, 10-4, and 10-5** Solve exponential equations and inequalities.
- **Lessons 10-2 and 10-3** Solve logarithmic equations and inequalities.
- **Lesson 10-6** Solve problems involving exponential growth and decay.

Key Vocabulary

- exponential growth (p. 524)
- exponential decay (p. 524)
- logarithm (p. 531)
- common logarithm (p. 547)
- natural logarithm (p. 554)

Why It's Important

Exponential functions are often used to model problems involving growth and decay. Logarithms can also be used to solve such problems. *You will learn how a declining farm population can be modeled by an exponential function in Lesson 10-1.*

Vocabulary Builder

ELL

The Key Vocabulary list introduces students to some of the main vocabulary terms included in this chapter. For a more thorough vocabulary list with pronunciations of new words, give students the Vocabulary Builder worksheets found on pages vii and viii of the *Chapter 10 Resource Masters*. Encourage them to complete the definition of each term as they progress through the chapter. You may suggest that they add these sheets to their study notebooks for future reference when studying for the Chapter 10 test.

Getting Started

Prerequisite Skills To be successful in this chapter, you'll need to master these skills and be able to apply them in problem-solving situations. Review these skills before beginning Chapter 10.

Lessons 10-1 through 10-3 — **Multiply and Divide Monomials**

Simplify. Assume that no variable equals 0. *(For review, see Lesson 5-1.)*

1. $x^5 \cdot x \cdot x^6$ x^{12}
2. $(3ab^4c^2)^3$ $27a^3b^{12}c^6$
3. $\frac{-36x^7y^4z^3}{21x^4y^9z^4}$ $-\frac{12x^3}{7y^5z}$
4. $\left(\frac{4ab^2}{64b^3c}\right)^2$ $\frac{a^2}{256b^2c^2}$

Lessons 10-2 and 10-3 — **Solve Inequalities**

Solve each inequality. *(For review, see Lesson 1-5)*

5. $a + 4 < -10$ $a < -14$
6. $-5n \leq 15$ $n \geq -3$
7. $3y + 2 \geq -4$ $y \geq -2$
8. $15 - x > 9$ $x < 6$

Lessons 10-2 and 10-3 — **Inverse Functions**

Find the inverse of each function. Then graph the function and its inverse. 9. $f^{-1}(x) = -\frac{1}{2}x$
(For review, see Lesson 7-8.) 9–12. See pp. 573A–573D for graphs.

9. $f(x) = -2x$
10. $f(x) = 3x - 2$ $f^{-1}(x) = \frac{x+2}{3}$
11. $f(x) = -x + 1$ $f^{-1}(x) = -x + 1$
12. $f(x) = \frac{x-4}{3}$ $f^{-1}(x) = 3x + 4$

Lessons 10-2 and 10-3 — **Composition of Functions**

Find $g[h(x)]$ and $h[g(x)]$. *(For review, see Lesson 7-7.)*

13. $h(x) = 3x + 4$ $g[h(x)] = 3x + 2$
    $g(x) = x - 2$ $h[g(x)] = 3x - 2$
14. $h(x) = 2x - 7$ $g[h(x)] = 10x - 35$
    $g(x) = 5x$ $h[g(x)] = 10x - 7$
15. $h(x) = x - 4$ $g[h(x)] = x^2 - 8x + 16$
    $g(x) = x^2$ $h[g(x)] = x^2 - 4$
16. $h(x) = 4x + 1$ $g[h(x)] = -8x - 5$
    $g(x) = -2x - 3$ $h[g(x)] = -8x - 11$

FOLDABLES™ Study Organizer

Exponential and Logarithmic Relations Make this Foldable to help you organize your notes. Begin with four sheets of grid paper.

Step 1 **Fold and Cut**

Fold in half along the width. On the first two sheets, cut along the fold at the ends. On the second two sheets, cut in the center of the fold as shown.

First Sheets

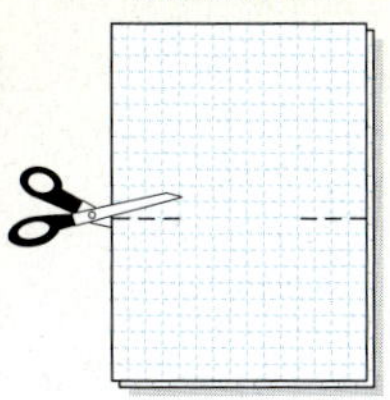

Second Sheets

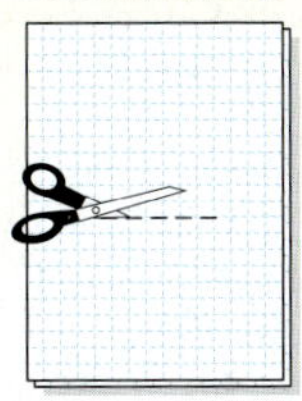

Step 2 **Fold and Label**

Insert first sheets through second sheets and align the folds. Label the pages with lesson numbers.

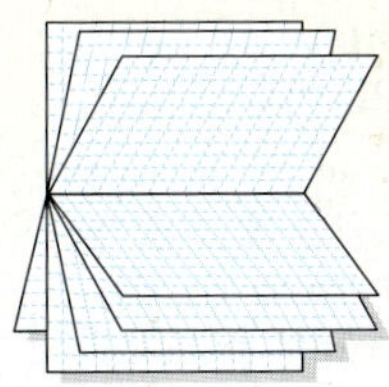

Reading and Writing As you read and study the chapter, fill the journal with notes, diagrams, and examples for each lesson.

Getting Started

This section provides a review of the basic concepts needed before beginning Chapter 10. Page references are included for additional student help.

Prerequisite Skills in the Getting Ready for the Next Lesson section at the end of each exercise set review a skill needed in the next lesson.

For Lesson	Prerequisite Skill
10-2	Composition of Functions (p. 530)
10-3	Multiplying and Dividing Monomials (p. 538)
10-4	Solving Logarithmic Equations and Inequalities (p. 546)
10-5	Logarithmic Equations (p. 551)
10-6	Exponential Equations and Inequalities (p. 559)

FOLDABLES™ Study Organizer

For more information about Foldables, see *Teaching Mathematics with Foldables.*

Organization of Data and Journal Writing After students make their Foldable journals, have them label two pages for each lesson in Chapter 10. Writers' journals can be used by students to record the direction and progress of learning, to describe positive and negative experiences during learning, to write about personal associations and experiences called to mind during learning, and to list examples of ways in which new knowledge has or will be used in their daily life, as well as take notes, record key concepts, and write examples.

Algebra Activity

A Preview of Lesson 10-1

Getting Started

Objective Use paper stacking to explore an exponential function.

Materials
notebook paper
scissors
grid paper

Teach

- You may wish to do the example as a demonstration while students complete the table on the chalkboard.
- Students may recognize that the y value is doubled for each successive cut, but they may have to be led to realizing that this can be written in the form 2^x.
- Show students how to connect the points with a smooth curve, rather than connecting each pair of points with a straight line.

Assess

Have students work in small groups for **Exercises 1–9**. Observe students' work to determine if they are able to write the function in **Exercise 5**. Students should conclude after **Exercise 9** that exponential functions can increase faster than seems reasonable.

Algebra Activity

A Preview of Lesson 10-1

Investigating Exponential Functions

Collect the Data

Step 1 Cut a sheet of notebook paper in half.

Step 2 Stack the two halves, one on top of the other.

Step 3 Make a table like the one below and record the number of sheets of paper you have in the stack after one cut.

Number of Cuts	Number of Sheets
0	1
1	2
2	4

Step 4 Cut the two stacked sheets in half, placing the resulting pieces in a single stack. Record the number of sheets of paper in the new stack after 2 cuts.

Step 5 Continue cutting the stack in half, each time putting the resulting piles in a single stack and recording the number of sheets in the stack. Stop when the resulting stack is too thick to cut.

Analyze the Data

1. Write a list of ordered pairs (x, y), where x is the number of cuts and y is the number of sheets in the stack. Notice that the list starts with the ordered pair $(0, 1)$, which represents the single sheet of paper before any cuts were made. **1. (0, 1), (1, 2), (2, 4), (3, 8), (4, 16), ...**
2. Continue the list, beyond the point where you stopped cutting, until you reach the ordered pair for 7 cuts. Explain how you calculated the last y values for your list, after you had stopped cutting. **2. (5, 32), (6, 64), (7, 128); The y value is found by raising 2 to the number of cuts.**
3. Plot the ordered pairs in your list on a coordinate grid. Be sure to choose a scale for the y-axis so that you can plot all of the points. **See pp. 573A–573D.**
4. Describe the pattern of the points you have plotted. Do they lie on a straight line? **The points do not lie in a straight line. The slope increases as the x values increase.**

Make a Conjecture

5. Write a function that expresses y as a function of x. $y = 2^x$
6. Use a calculator to evaluate the function you wrote in Exercise 5 for $x = 8$ and $x = 9$. Does it give the correct number of sheets in the stack after 8 and 9 cuts? **256, 512; yes**
7. Notebook paper usually stacks about 500 sheets to the inch. How thick would your stack of paper be if you had been able to make 9 cuts? **about 1 in.**
8. Suppose each cut takes about 5 seconds. If you had been able to keep cutting, you would have made 36 cuts in three minutes. At 500 sheets to the inch, make a conjecture as to how thick you think the stack would be after 36 cuts. **Sample answer: 1 million ft**
9. Use your function from Exercise 5 to calculate the thickness of your stack after 36 cuts. Write your answer in miles. **2169 mi**

Resource Manager

Teaching Algebra with Manipulatives
- p. 1 (grid paper)
- p. 275 (student recording sheet)

Glencoe Mathematics Classroom Manipulative Kit
- scissors
- coordinate grid stamp

Study Notebook

You may wish to have students summarize this activity and what they learned from it.

522 Chapter 10 Exponential and Logarithmic Relations

10-1 Exponential Functions

What You'll Learn

- Graph exponential functions.
- Solve exponential equations and inequalities.

Vocabulary

- exponential function
- exponential growth
- exponential decay
- exponential equation
- exponential inequality

How does an exponential function describe tournament play?

The NCAA women's basketball tournament begins with 64 teams and consists of 6 rounds of play. The winners of the first round play against each other in the second round. The winners then move from the Sweet Sixteen to the Elite Eight to the Final Four and finally to the Championship Game.

The number of teams y that compete in a tournament of x rounds is $y = 2^x$.

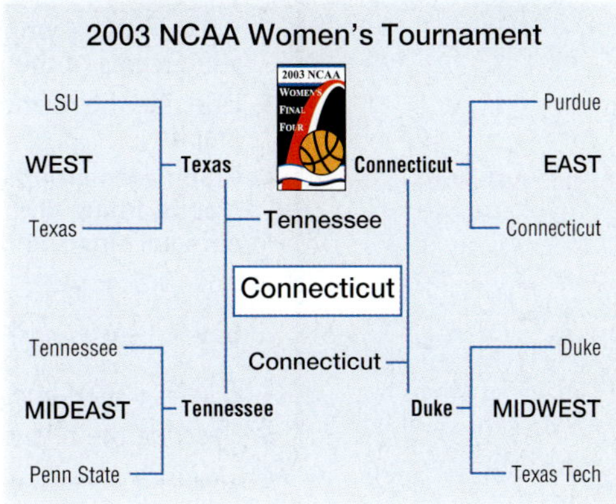

Study Tip

Common Misconception

Be sure not to confuse polynomial functions and exponential functions. While $y = x^2$ and $y = 2^x$ each have an exponent, $y = x^2$ is a polynomial function and $y = 2^x$ is an exponential function.

EXPONENTIAL FUNCTIONS In an exponential function like $y = 2^x$, the base is a constant, and the exponent is a variable. Let's examine the graph of $y = 2^x$.

Example 1 Graph an Exponential Function

Sketch the graph of $y = 2^x$. Then state the function's domain and range.

Make a table of values. Connect the points to sketch a smooth curve.

x	$y = 2^x$
-3	$2^{-3} = \frac{1}{8}$
-2	$2^{-2} = \frac{1}{4}$
-1	$2^{-1} = \frac{1}{2}$
0	$2^0 = 1$
$\frac{1}{2}$	$2^{\frac{1}{2}} = \sqrt{2}$
1	$2^1 = 2$
2	$2^2 = 4$
3	$2^3 = 8$

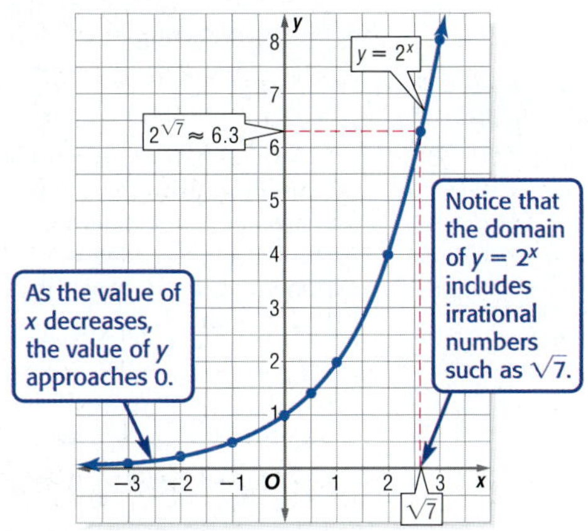

The domain is all real numbers, while the range is all positive numbers.

10-1 Lesson Notes

1 Focus

5-Minute Check Transparency 10-1 Use as a quiz or review of Chapter 9.

Mathematical Background notes are available for this lesson on p. 520C.

Building on Prior Knowledge

Ask students where they have heard the term *exponential* before and what they think it might mean. Students may have heard terms like *exponential growth* on a television news program and they might think that exponential means "enormous." Use students' answers to introduce the concept of exponential functions.

How does an exponential function describe tournament play?

Ask students:

- How many winners are there in the first round of the tournament? **32**
- After each round, how has the number of teams changed? **The number of teams remaining after each round is half the number of teams that played in that round.**
- If the tournament field was reduced to 32 teams, how many basketball games would have to be played by the tournament's winning team? **5 games**

Resource Manager

Workbook and Reproducible Masters

Chapter 10 Resource Masters

- Study Guide and Intervention, pp. 573–574
- Skills Practice, p. 575
- Practice, p. 576
- Reading to Learn Mathematics, p. 577
- Enrichment, p. 578

Graphing Calculator and Spreadsheet Masters, p. 45

Teaching Algebra With Manipulatives Masters, pp. 276–277

Transparencies

5-Minute Check Transparency 10-1
Answer Key Transparencies

Technology

Interactive Chalkboard

2 Teach

EXPONENTIAL FUNCTIONS

In-Class Example

Power Point®

Teaching Tip Watch for students who do not understand why the graph in Example 1 cannot simply be modeled by a quadratic, cubic, or quartic function. Point out that the graphs of $y = x^2$, $y = x^3$, and $y = x^4$ all pass through the point (0, 0) and not through the point (0, 1).

1 Sketch the graph of $y = 4^x$. Then state the function's domain and range.

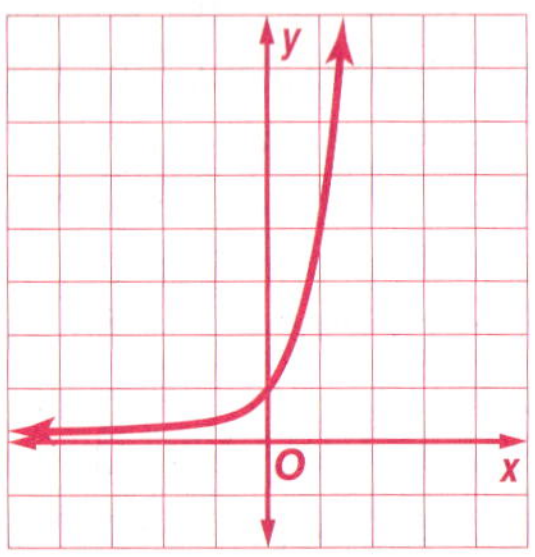

The domain is all real numbers, while the range is all positive numbers.

Answers

1. The shapes of the graphs are the same.
2. The asymptote for each graph is the x-axis and the y-intercept for each graph is 1.
3. The graphs are reflections of each other over the y-axis.
5. The graphs are reflections of each other over the x-axis.

You can use a TI-83 Plus graphing calculator to look at the graph of two other exponential functions, $y = 3^x$ and $y = \left(\frac{1}{3}\right)^x$.

Graphing Calculator Investigation

Families of Exponential Functions

The calculator screen shows the graphs of $y = 3^x$ and $y = \left(\frac{1}{3}\right)^x$.

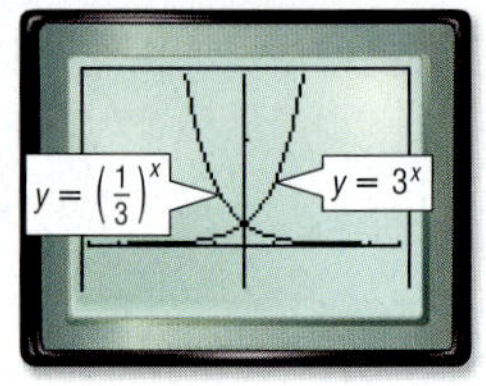

[−5, 5] scl: 1 by [−2, 8] scl: 1

Think and Discuss **1–3,5. See margin.**

1. How do the shapes of the graphs compare?
2. How do the asymptotes and y-intercepts of the graphs compare?
3. Describe the relationship between the graphs.
4. Graph each group of functions on the same screen. Then compare the graphs, listing both similarities and differences in shape, asymptotes, domain, range, and y-intercepts. **4. See pp. 573A–573D.**
   a. $y = 2^x$, $y = 3^x$, and $y = 4^x$
   b. $y = \left(\frac{1}{2}\right)^x$, $y = \left(\frac{1}{3}\right)^x$, and $y = \left(\frac{1}{4}\right)^x$
   c. $y = -3(2)^x$ and $y = 3(2)^x$; $y = -1(2)^x$ and $y = 2^x$.
5. Describe the relationship between the graphs of $y = -1(2)^x$ and $y = 2^x$.

Study Tip

Look Back
To review **continuous functions**, see page 62, Exercises 60 and 61. To review **one-to-one functions**, see Lesson 2-1.

In general, an equation of the form $y = ab^x$, where $a \neq 0$, $b > 0$, and $b \neq 1$, is called an **exponential function** with base b. Exponential functions have the following characteristics.

1. The function is continuous and one-to-one.
2. The domain is the set of all real numbers.
3. The x-axis is an asymptote of the graph.
4. The range is the set of all positive numbers if $a > 0$ and all negative numbers if $a < 0$.
5. The graph contains the point $(0, a)$. That is, the y-intercept is a.
6. The graphs of $y = ab^x$ and $y = a\left(\frac{1}{b}\right)^x$ are reflections across the y-axis.

Study Tip

Exponential Growth and Decay
Notice that the graph of an exponential growth function *rises* from left to right. The graph of an exponential decay function *falls* from left to right.

There are two types of exponential functions: **exponential growth** and **exponential decay**. The base of an exponential growth function is a number greater than one. The base of an exponential decay function is a number between 0 and 1.

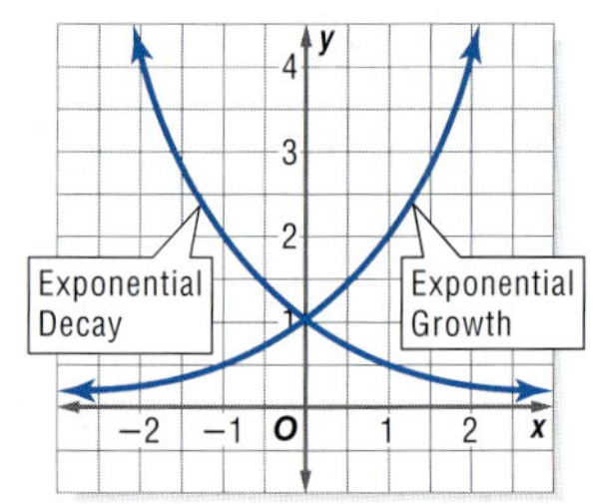

Key Concept — *Exponential Growth and Decay*

- If $a > 0$ and $b > 1$, the function $y = ab^x$ represents exponential growth.
- If $a > 0$ and $0 < b < 1$, the function $y = ab^x$ represents exponential decay.

Graphing Calculator Investigation

Families of Exponential Functions Have students begin by graphing the two functions separately, so they recognize that the two curves shown in the book are two distinct graphs. Students are used to seeing the U-shaped graphs of polynomial functions and might have difficulty separating the graphs visually. Also, a reminder about the meaning of the term *asymptotes* may be helpful for many students.

Example 2 Identify Exponential Growth and Decay

Determine whether each function represents exponential *growth* or *decay*.

Function	Exponential Growth or Decay?
a. $y = \left(\frac{1}{5}\right)^x$	The function represents exponential decay, since the base, $\frac{1}{5}$, is between 0 and 1.
b. $y = 3(4)^x$	The function represents exponential growth, since the base, 4, is greater than 1.
c. $y = 7(1.2)^x$	The function represents exponential growth, since the base, 1.2, is greater than 1.

Exponential functions are frequently used to model the growth or decay of a population. You can use the y-intercept and one other point on the graph to write the equation of an exponential function.

Example 3 Write an Exponential Function

FARMING **In 1983, there were 102,000 farms in Minnesota, but by 1998, this number had dropped to 80,000.**

a. Write an exponential function of the form $y = ab^x$ that could be used to model the farm population y of Minnesota. Write the function in terms of x, the number of years since 1983.

For 1983, the time x equals 0, and the initial population y is 102,000. Thus, the y-intercept, and value of a, is 102,000.

For 1998, the time x equals 1998 − 1983 or 15, and the population y is 80,000. Substitute these values and the value of a into an exponential function to approximate the value of b.

$y = ab^x$	Exponential function
$80{,}000 = 102{,}000b^{15}$	Replace x with 15, y with 80,000, and a with 102,000.
$0.78 \approx b^{15}$	Divide each side by 102,000.
$\sqrt[15]{0.78} \approx b$	Take the 15th root of each side.

To find the 15th root of 0.78, use selection 5: $\sqrt[x]{\ }$ under the MATH menu on the TI-83 Plus.

KEYSTROKES: 15 [MATH] 5 0.78 [ENTER] .9835723396

An equation that models the farm population of Minnesota from 1983 to 1998 is $y = 102{,}000(0.98)^x$.

b. Suppose the number of farms in Minnesota continues to decline at the same rate. Estimate the number of farms in 2010.

For 2010, the time x equals 2010 − 1983 or 27.

$y = 102{,}000(0.98)^x$	Modeling equation
$y = 102{,}000(0.98)^{27}$	Replace x with 27.
$y \approx 59{,}115$	Use a calculator.

The farm population in Minnesota will be about 59,115 in 2010.

More About. . .

Farming

In 1999, 47% of the net farm income in the United States was from direct government payments. The USDA has set a goal of reducing this percent to 14% by 2005.

Source: USDA

TEACHING TIP

In Example 3, one of the given points is the y-intercept. You may wish to give your students a challenge problem in which any two points are given and students use a system of equations to find the equation of the exponential function.

www.algebra2.com/extra_examples

In-Class Examples

PowerPoint®

2 Determine whether each function represents exponential *growth* or *decay*.

a. $y = (0.7)^x$ **The function represents exponential decay, since the base, 0.7, is between 0 and 1.**

b. $y = \frac{1}{2}(3)^x$ **The function represents exponential growth, since the base, 3, is greater than 1.**

c. $y = 10\left(\frac{4}{3}\right)^x$ **The function represents exponential growth, since the base, $\frac{4}{3}$, is greater than 1.**

3 **CELLULAR PHONES** In December of 1990, there were 5,283,000 cellular telephone subscribers in the United States. By December of 2000, this number had risen to 109,478, 000. **Source:** Cellular Telecommunications Industry Association

a. Write an exponential function of the form $y = ab^x$ that could be used to model the number of cellular telephone subscribers y in the U.S. Write the function in terms of x, the number of years since 1990. $\boldsymbol{y = 5{,}283{,}000(1.35)^x}$

b. Suppose the number of cellular telephone subscribers continues to increase at the same rate. Estimate the number of U.S. subscribers in 2010. **about 2,136,000,000 subscribers**

Teacher to Teacher

David S. Daniels — Longmeadow H.S., Longmeadow, MA

"As a lead-in activity for exponential functions, have students flip 50 pennies and count the number of heads. Then have students remove those pennies that landed on heads and repeat the activity. Students should record their results and make a plot of the trial number versus the number of heads counted in that trial. The graph will model that of $y = \left(\frac{1}{2}\right)^x$."

Interactive Chalkboard

PowerPoint® Presentations

This CD-ROM is a customizable Microsoft® PowerPoint® presentation that includes:

- Step-by-step, dynamic solutions of each In-Class Example from the Teacher Wraparound Edition
- Additional, Your Turn exercises for each example
- The 5-Minute Check Transparencies
- Hot links to Glencoe Online Study Tools

EXPONENTIAL EQUATIONS AND INEQUALITIES

In-Class Examples

Power Point®

4 Simplify each expression.

a. $5^{\sqrt{3}} \div 5^{\sqrt{2}}$ $5^{\sqrt{3}-\sqrt{2}}$

b. $\left(6^{\sqrt{5}}\right)^{\sqrt{6}}$ $6^{\sqrt{30}}$

5 Solve each equation.

a. $4^{9n-2} = 256$ $n = \frac{2}{3}$

b. $3^{5x} = 9^{2x-1}$ $x = -2$

Study Tip

Look Back

To review **Properties of Power**, see Lesson 5-1.

EXPONENTIAL EQUATIONS AND INEQUALITIES Since the domain of an exponential function includes irrational numbers such as $\sqrt{2}$, all the properties of rational exponents apply to irrational exponents.

Example 4 Simplify Expressions with Irrational Exponents

Simplify each expression.

a. $2^{\sqrt{5}} \cdot 2^{\sqrt{3}}$

$2^{\sqrt{5}} \cdot 2^{\sqrt{3}} = 2^{\sqrt{5}+\sqrt{3}}$ Product of Powers

b. $\left(7^{\sqrt{2}}\right)^{\sqrt{3}}$

$\left(7^{\sqrt{2}}\right)^{\sqrt{3}} = 7^{\sqrt{2}\cdot\sqrt{3}}$ Power of a Power

$= 7^{\sqrt{6}}$ Product of Radicals

The following property is useful for solving exponential equations. **Exponential equations** are equations in which variables occur as exponents.

Key Concept — Property of Equality for Exponential Functions

- **Symbols** If b is a positive number other than 1, then $b^x = b^y$ if and only if $x = y$.
- **Example** If $2^x = 2^8$, then $x = 8$.

Example 5 Solve Exponential Equations

Solve each equation.

a. $3^{2n+1} = 81$

$3^{2n+1} = 81$	Original equation
$3^{2n+1} = 3^4$	Rewrite 81 as 3^4 so each side has the same base.
$2n + 1 = 4$	Property of Equality for Exponential Functions
$2n = 3$	Subtract 1 from each side.
$n = \frac{3}{2}$	Divide each side by 2.

The solution is $\frac{3}{2}$.

CHECK

$3^{2n+1} = 81$	Original equation
$3^{2\left(\frac{3}{2}\right)+1} \stackrel{?}{=} 81$	Substitute $\frac{3}{2}$ for n.
$3^4 \stackrel{?}{=} 81$	Simplify.
$81 = 81$ ✓	Simplify.

b. $4^{2x} = 8^{x-1}$

$4^{2x} = 8^{x-1}$	Original equation
$(2^2)^{2x} = (2^3)^{x-1}$	Rewrite each side with a base of 2.
$2^{4x} = 2^{3(x-1)}$	Power of a Power
$4x = 3(x - 1)$	Property of Equality for Exponential Functions
$4x = 3x - 3$	Distributive Property
$x = -3$	Subtract $3x$ from each side.

The solution is -3.

526 Chapter 10 Exponential and Logarithmic Relations

The following property is useful for solving inequalities involving exponential functions or **exponential inequalities**.

Key Concept *Property of Inequality for Exponential Functions*

- **Symbols** If $b > 1$, then $b^x > b^y$ if and only if $x > y$, and $b^x < b^y$ if and only if $x < y$.
- **Example** If $5^x < 5^4$, then $x < 4$.

This property also holds for ≤ and ≥.

Example 6 Solve Exponential Inequalities

Solve $4^{3p-1} > \frac{1}{256}$.

$4^{3p-1} > \frac{1}{256}$ Original inequality

$4^{3p-1} > 4^{-4}$ Rewrite $\frac{1}{256}$ as $\frac{1}{4^4}$ or 4^{-4} so each side has the same base.

$3p - 1 > -4$ Property of Inequality for Exponential Functions

$3p > -3$ Add 1 to each side.

$p > -1$ Divide each side by 3.

The solution set is $p > -1$.

CHECK Test a value of p greater than -1; for example, $p = 0$.

$4^{3p-1} > \frac{1}{256}$ Original inequality

$4^{3(0)-1} \overset{?}{>} \frac{1}{256}$ Replace p with 0.

$4^{-1} \overset{?}{>} \frac{1}{256}$ Simplify.

$\frac{1}{4} > \frac{1}{256}$ ✓ $a^{-1} = \frac{1}{a}$

Check for Understanding

Concept Check 1. **OPEN ENDED** Give an example of a value of b for which $y = b^x$ represents exponential decay. **Sample answer: 0.8**

2a. quadratic
2b. exponential
2c. linear
2d. exponential

2. Identify each function as *linear, quadratic,* or *exponential*.

a. $y = 3x^2$ b. $y = 4(3)^x$ c. $y = 2x + 4$ d. $y = 4(0.2)^x + 1$

Match each function with its graph.

3. $y = 5^x$ **c** 4. $y = 2(5)^x$ **a** 5. $y = \left(\frac{1}{5}\right)^x$ **b**

a.
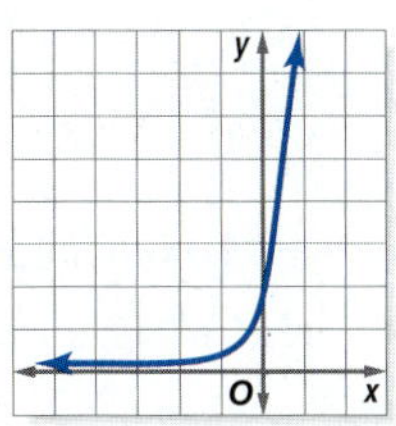

b.
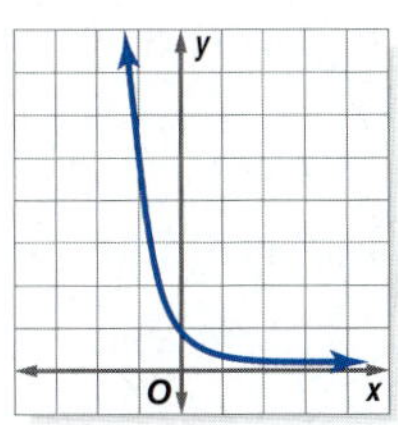

c.
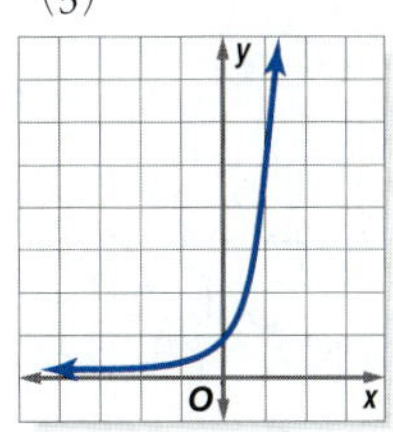

Guided Practice **Sketch the graph of each function. Then state the function's domain and range.**

6–7. See margin.

6. $y = 3(4)^x$ 7. $y = 2\left(\frac{1}{3}\right)^x$

In-Class Example Power Point®

6 Solve $5^{3-2k} > \frac{1}{625}$. $k < \frac{7}{2}$

3 Practice/Apply

Study Notebook

Have students—

- add the definitions/examples of the vocabulary terms to their Vocabulary Builder worksheets for Chapter 10.
- include examples of exponential growth and decay graphs and equations.
- include any other item(s) that they find helpful in mastering the skills in this lesson.

Answers

6. D = {x | x is all real numbers.}, R = {y | y > 0}

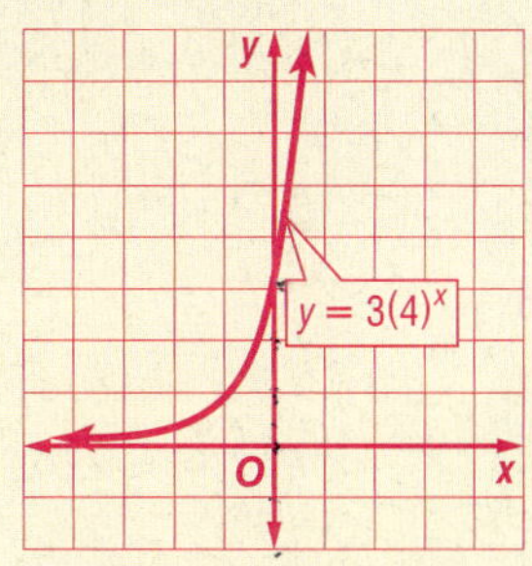

7. D = {x | x is all real numbers.}, R = {y | y > 0}

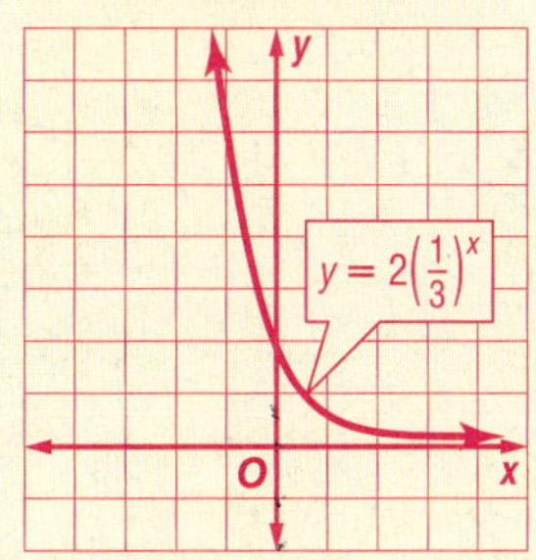

DAILY INTERVENTION

Differentiated Instruction

Auditory/Musical Going around the room, have students count by ones beginning at 2, with each student calling out one number. Instruct them to record the number they called as n. Then have students find n^2 and 2^n. Now go around the room again and ask students to state their value of n^2 (for a class of 30 students, the recited numbers are all the squares from 4 to 961). Now have students state their values of 2^n (for a class of 30, the recited numbers are all the powers of 2 from 4 to 2^{31} or about 2×10^9).

About the Exercises...

Organization by Objective

- **Exponential Functions:** 21–38, 57–61
- **Exponential Equations and Inequalities:** 37–56, 62–66

Odd/Even Assignments

Exercises 21–56 are structured so that students practice the same concepts whether they are assigned odd or even problems.

Alert! Exercise 61 involves research on the Internet or other reference materials. Exercises 71–75 require the use of graphing calculators.

Assignment Guide

Basic: 21, 23, 27–53 odd, 57–61, 68–70, 76–89

Average: 21–55 odd, 59–64, 67–70, 76–89 (optional: 71–75)

Advanced: 22–56 even, 61–86 (optional: 87–89)

GUIDED PRACTICE KEY	
Exercises	Examples
6, 7	1
8–10	2
11, 12, 19, 20	3
13–15	4
16–18	5, 6

Determine whether each function represents exponential *growth* or *decay*.

8. $y = 2(7)^x$ **growth** **9.** $y = (0.5)^x$ **decay** **10.** $y = 0.3(5)^x$ **growth**

Write an exponential function whose graph passes through the given points.

11. (0, 3) and (−1, 6) $y = 3\left(\frac{1}{2}\right)^x$ **12.** (0, −18) and (−2, −2) $y = -18(3)^x$

Simplify each expression.

13. $2^{\sqrt{7}} \cdot 2^{\sqrt{7}}$ $2^{2\sqrt{7}}$ or $4^{\sqrt{7}}$ **14.** $(a^{\pi})^4$ $a^{4\pi}$ **15.** $81^{\sqrt{2}} \div 3^{\sqrt{2}}$ $3^{3\sqrt{2}}$ or $27^{\sqrt{2}}$

Solve each equation or inequality. Check your solution.

16. $2^{n+4} = \frac{1}{32}$ **−9** **17.** $5^{2x+3} \leq 125$ $x \leq 0$ **18.** $9^{2y-1} = 27^y$ **2**

Application

ANIMAL CONTROL **For Exercises 19 and 20, use the following information.** During the 19th century, rabbits were brought to Australia. Since the rabbits had no natural enemies on that continent, their population increased rapidly. Suppose there were 65,000 rabbits in Australia in 1865 and 2,500,000 in 1867.

19. Write an exponential function that could be used to model the rabbit population y in Australia. Write the function in terms of x, the number of years since 1865. $y = 65{,}000(6.20)^x$

20. Assume that the rabbit population continued to grow at that rate. Estimate the Australian rabbit population in 1872. **22,890,495,000**

★ indicates increased difficulty

Practice and Apply

Homework Help	
For Exercises	See Examples
21–26	1
27–32	2
33–38, 57–66	3
39–44	4
45–56	5, 6

Extra Practice
See page 849.

Sketch the graph of each function. Then state the function's domain and range. **21–26. See pp. 573A–573D.**

21. $y = 2(3)^x$ **22.** $y = 5(2)^x$ **23.** $y = 0.5(4)^x$

24. $y = 4\left(\frac{1}{3}\right)^x$ ★ **25.** $y = -\left(\frac{1}{5}\right)^x$ ★ **26.** $y = -2.5(5)^x$

Determine whether each function represents exponential *growth* or *decay*.

27. $y = 10(3.5)^x$ **growth** **28.** $y = 2(4)^x$ **growth** **29.** $y = 0.4\left(\frac{1}{3}\right)^x$ **decay**

30. $y = 3\left(\frac{5}{2}\right)^x$ **growth** **31.** $y = 30^{-x}$ **decay** **32.** $y = 0.2(5)^{-x}$ **decay**

Write an exponential function whose graph passes through the given points.

33. (0, −2) and (−2, −32) $y = -2\left(\frac{1}{4}\right)^x$ **34.** (0, 3) and (1, 15) $y = 3(5)^x$

35. (0, 7) and (2, 63) $y = 7(3)^x$ **36.** (0, −5) and (−3, −135) $y = -5\left(\frac{1}{3}\right)^x$

37. (0, 0.2) and (4, 51.2) $y = 0.2(4)^x$ **38.** (0, −0.3) and (5, −9.6) $y = -0.3(2)^x$

Simplify each expression.

39. $\left(5^{\sqrt{2}}\right)^{\sqrt{8}}$ 5^4 or 625 **40.** $\left(x^{\sqrt{5}}\right)^{\sqrt{3}}$ $x^{\sqrt{15}}$ **41.** $7^{\sqrt{2}} \cdot 7^{3\sqrt{2}}$ $7^{4\sqrt{2}}$

42. $y^{3\sqrt{3}} \div y^{\sqrt{3}}$ $y^{2\sqrt{3}}$ **43.** $n^2 \cdot n^{\pi}$ $n^{2+\pi}$ **44.** $64^{\pi} \div 2^{\pi}$ $2^{5\pi}$

Solve each equation or inequality. Check your solution. **54.** $p \geq -2$

45. $3^{n-2} = 27$ **5** **46.** $2^{3x+5} = 128$ $\frac{2}{3}$ **47.** $5^{n-3} = \frac{1}{25}$ **1**

48. $2^{2n} \leq \frac{1}{16}$ $n \leq -2$ **49.** $\left(\frac{1}{9}\right)^m = 81^{m+4}$ $-\frac{8}{3}$ **50.** $\left(\frac{1}{7}\right)^{y-3} = 343$ **0**

51. $16^n < 8^{n+1}$ $n < 3$ **52.** $10^{x-1} = 100^{2x-3}$ $\frac{5}{3}$ **53.** $36^{2p} = 216^{p-1}$ **−3**

54. $32^{5p+2} \geq 16^{5p}$ ★ **55.** $3^{5x} \cdot 81^{1-x} = 9^{x-3}$ **10** **56.** $49^x = 7^{x^2-15}$ **−3, 5**

Answers (p. 529)

60. 9.67 million; 17.62 million; 32.12 million; These answers are in close agreement with the actual populations in those years.

61. 2144.97 million; 281.42 million; No, the growth rate has slowed considerably. The population in 2000 was much smaller than the equation predicts it would be.

528 Chapter 10 Exponential and Logarithmic Relations

WebQuest

The magnitude of an earthquake can be represented by an exponential equation. Visit www.algebra2.com/webquest to continue work on your WebQuest project.

BIOLOGY **For Exercises 57 and 58, use the following information.**
The number of bacteria in a colony is growing exponentially.

57. Write an exponential function to model the population y of bacteria x hours after 2 P.M. $y = 100(6.32)^x$

58. How many bacteria were there at 7 P.M. that day? about 1,008,290

POPULATION **For Exercises 59–61, use the following information.**
Every ten years, the Bureau of the Census counts the number of people living in the United States. In 1790, the population of the U.S. was 3.93 million. By 1800, this number had grown to 5.31 million.

59. Write an exponential function that could be used to model the U.S. population y in millions for 1790 to 1800. Write the equation in terms of x, the number of decades x since 1790. $y = 3.93(1.35)^x$

60. Assume that the U.S. population continued to grow at that rate. Estimate the population for the years 1820, 1840, and 1860. Then compare your estimates with the actual population for those years, which were 9.64, 17.06, and 31.44 million, respectively. See margin.

61. **RESEARCH** Estimate the population of the U.S. in 2000. Then use the Internet or other reference to find the actual population of the U.S. in 2000. Has the population of the U.S. continued to grow at the same rate at which it was growing in the early 1800s? Explain. See margin.

62. Exponential; the base, $1 + \frac{r}{n}$, is fixed, but the exponent, nt, is variable since the time t can vary.

More About. . .

Computers

Since computers were invented, computational speed has multiplied by a factor of 4 about every three years.

Source: www.wired.com

MONEY **For Exercises 62–64, use the following information.**
Suppose you deposit a principal amount of P dollars in a bank account that pays compound interest. If the annual interest rate is r (expressed as a decimal) and the bank makes interest payments n times every year, the amount of money A you would have after t years is given by $A(t) = P\left(1 + \frac{r}{n}\right)^{nt}$.

62. If the principal, interest rate, and number of interest payments are known, what type of function is $A(t) = P\left(1 + \frac{r}{n}\right)^{nt}$? Explain your reasoning.

63. Write an equation giving the amount of money you would have after t years if you deposit $1000 into an account paying 4% annual interest compounded quarterly (four times per year). $A(t) = 1000(1.01)^{4t}$

64. Find the account balance after 20 years. $2216.72

COMPUTERS **For Exercises 65 and 66, use the information at the left.**

65. If a typical computer operates with a computational speed s today, write an expression for the speed at which you can expect an equivalent computer to operate after x three-year periods. $s \cdot 4^x$

★ 66. Suppose your computer operates with a processor speed of 600 megahertz and you want a computer that can operate at 4800 megahertz. If a computer with that speed is currently unavailable for home use, how long can you expect to wait until you can buy such a computer? 1.5 three-year periods or 4.5 yr

★ 67. **CRITICAL THINKING** Decide whether the following statement is *sometimes*, *always*, or *never* true. Explain your reasoning.

For a positive base b other than 1, $b^x > b^y$ if and only if $x > y$.

67. Sometimes; true when $b > 1$, but false when $b < 1$.

Enrichment, p. 578

Finding Solutions of $x^y = y^x$

Perhaps you have noticed that if x and y are interchanged in equations such as $x = y$ and $xy = 1$, the resulting equation is equivalent to the original equation. The same is true of the equation $x^y = y^x$. However, finding solutions of $x^y = y^x$ and drawing its graph is not a simple process.

Solve each problem. Assume that x and y are positive real numbers.

1. If $a > 0$, will (a, a) be a solution of $x^y = y^x$? Justify your answer.
Yes, since $a^a = a^a$ must be true (Reflexive Prop. of Equality).

2. If $c > 0$, $d > 0$, and (c, d) is a solution of $x^y = y^x$, will (d, c) also be a solution? Justify your answer.
Yes; replacing x with d, y with c gives $d^c = c^d$; but if (c, d) is a solution, $c^d = d^c$. So, by the Symmetric Property of Equality, $d^c = c^d$ is true.

Study Guide and Intervention, p. 573 (shown) and p. 574

Exponential Functions An **exponential function** has the form $y = ab^x$, where $a \neq 0$, $b > 0$, and $b \neq 1$.

Properties of an Exponential Function	1. The function is continuous and one-to-one. 2. The domain is the set of all real numbers. 3. The x-axis is the asymptote of the graph. 4. The range is the set of all positive numbers if $a > 0$ and all negative numbers if $a < 0$. 5. The graph contains the point $(0, a)$.
Exponential Growth and Decay	If $a > 0$ and $b > 1$, the function $y = ab^x$ represents exponential growth. If $a > 0$ and $0 < b < 1$, the function $y = ab^x$ represents exponential decay.

Example 1 **Sketch the graph of $y = 0.1(4)^x$. Then state the function's domain and range.**

Make a table of values. Connect the points to form a smooth curve.

x	−1	0	1	2	3
y	0.025	0.1	0.4	1.6	6.4

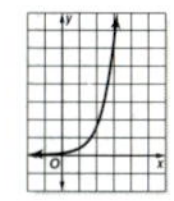

The domain of the function is all real numbers, while the range is the set of all positive real numbers.

Example 2 **Determine whether each function represents exponential *growth* or *decay*.**

a. $y = 0.5(2)^x$
exponential growth, since the base, 2, is greater than 1

b. $y = -2.8(2)^x$
neither, since −2.8, the value of a is less than 0.

c. $y = 1.1(0.5)^x$
exponential decay, since the base, 0.5, is between 0 and 1

Exercises

Sketch the graph of each function. Then state the function's domain and range.

1. $y = 3(2)^x$

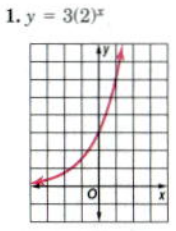

Domain: all real numbers; Range: all positive real numbers

2. $y = -2\left(\frac{1}{4}\right)^x$

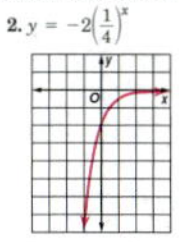

Domain: all real numbers; Range: all negative real numbers

3. $y = 0.25(5)^x$

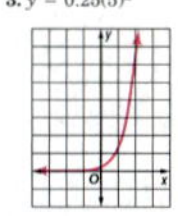

Domain: all real numbers; Range: all positive real numbers

Determine whether each function represents exponential growth or decay.

4. $y = 0.3(1.2)^x$ growth
5. $y = -5\left(\frac{4}{5}\right)^x$ neither
6. $y = 3(10)^{-x}$ decay

Skills Practice, p. 575 and Practice, p. 576 (shown)

Sketch the graph of each function. Then state the function's domain and range.

1. $y = 1.5(2)^x$

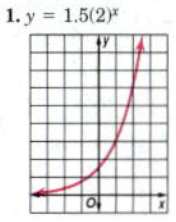

domain: all real numbers; range: all positive numbers

2. $y = 4(3)^x$

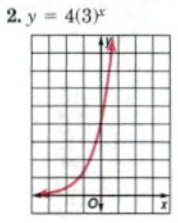

domain: all real numbers; range: all positive numbers

3. $y = 3(0.5)^x$

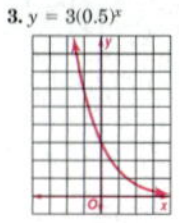

domain: all real numbers; range: all positive numbers

Determine whether each function represents exponential *growth* or *decay*.

4. $y = 5(0.6)^x$ decay
5. $y = 0.1(2)^x$ growth
6. $y = 5 \cdot 4^{-x}$ decay

Write an exponential function whose graph passes through the given points.

7. (0, 1) and (−1, 4) $y = \left(\frac{1}{4}\right)^x$
8. (0, 2) and (1, 10) $y = 2(5)^x$
9. (0, −3) and (1, −1.5) $y = -3(0.5)^x$
10. (0, 0.8) and (1, 1.6) $y = 0.8(2)^x$
11. (0, −0.4) and (2, −10) $y = -0.4(5)^x$
12. (0, π) and (3, 8π) $y = \pi(2)^x$

Simplify each expression.

13. $(2^{\sqrt{2}})^{\sqrt{8}}$ 16
14. $(n^{\sqrt{3}})^{\sqrt{75}}$ n^{15}
15. $y^{\sqrt{6}} \cdot y^{5\sqrt{6}}$ $y^{6\sqrt{6}}$
16. $13^{\sqrt{6}} \cdot 13^{\sqrt{24}}$ $13^{3\sqrt{6}}$
17. $n^3 \div n^{\pi}$ $n^{3-\pi}$
18. $125^{\sqrt{11}} \div 5^{\sqrt{11}}$ $5^{2\sqrt{11}}$

Solve each equation or inequality. Check your solution.

19. $3^{3x-5} > 81$ $x > 3$
20. $7^{6x} = 7^{2x-20}$ −5
21. $3^{6n-5} < 9^{4n-3}$ $n > \frac{1}{2}$
22. $9^{2x-1} = 27^{x+4}$ 14
23. $2^{3n-1} \geq \left(\frac{1}{8}\right)^n$ $n \geq \frac{1}{6}$
24. $16^{4n-1} = 128^{2n+1}$ $\frac{11}{2}$

BIOLOGY For Exercises 25 and 26, use the following information.
The initial number of bacteria in a culture is 12,000. The number after 3 days is 96,000.

25. Write an exponential function to model the population y of bacteria after x days.
$y = 12{,}000(2)^x$

26. How many bacteria are there after 6 days? 768,000

27. **EDUCATION** A college with a graduating class of 4000 students in the year 2002 predicts that it will have a graduating class of 4862 in 4 years. Write an exponential function to model the number of students y in the graduating class t years after 2002.
$y = 4000(1.05)^t$

Reading to Learn Mathematics, p. 577 ELL

Pre-Activity **How does an exponential function describe tournament play?**

Read the introduction to Lesson 10-1 at the top of page 523 in your textbook.

How many rounds of play would be needed for a tournament with 100 players? 7

Reading the Lesson

1. Indicate whether each of the following statements about the exponential function $y = 10^x$ is *true* or *false*.
   a. The domain is the set of all positive real numbers. false
   b. The y-intercept is 1. true
   c. The function is one-to-one. true
   d. The y-axis is an asymptote of the graph. false
   e. The range is the set of all real numbers. false

2. Determine whether each function represents exponential *growth* or *decay*.
   a. $y = 0.2(3)^x$. growth
   b. $y = 3\left(\frac{2}{5}\right)^x$. decay
   c. $y = 0.4(1.01)^x$. growth

3. Supply the reason for each step in the following solution of an exponential equation.

$9^{2x-1} = 27x$	Original equation
$(3^2)^{2x-1} = (3^3)^x$	Rewrite each side with a base of 3.
$3^{2(2x-1)} = 3^{3x}$	Power of a Power
$2(2x - 1) = 3x$	Property of Equality for Exponential Functions
$4x - 2 = 3x$	Distributive Property
$x - 2 = 0$	Subtract 3x from each side.
$x = 2$	Add 2 to each side.

Helping You Remember

4. One way to remember that polynomial functions and exponential functions are different is to contrast the polynomial function $y = x^2$ and the exponential function $y = 2^x$. Tell at least three ways they are different.
Sample answer: In $y = x^2$, the variable x is a base, but in $y = 2^x$, the variable x is an exponent. The graph of $y = x^2$ is symmetric with respect to the y-axis, but the graph of $y = 2^x$ is not. The graph of $y = x^2$ touches the x-axis at (0, 0), but the graph of $y = 2^x$ has the x-axis as an asymptote. You can compute the value of $y = x^2$ mentally for $x = 100$, but you cannot compute the value of $y = 2^x$ mentally for $x = 100$.

4 Assess

Open-Ended Assessment

Modeling Give students a sheet of grid paper and a length of string. Have students model the graph of the equation $y = \left(\frac{1}{2}\right)^x$. Have them check their model by graphing the equation on a graphing calculator.

Getting Ready for Lesson 10-2

PREREQUISITE SKILL In Lesson 10-2, students will evaluate logarithmic expressions. Because logarithmic and exponential functions are inverses of each other, their composites are the identity function. Students must be familiar with compositions of functions in order to evaluate these inverse functions. Use Exercises 87–89 to determine your students' familiarity with composition of functions.

Answers

75. For $h > 0$, the graph of $y = 2^x$ is translated $|h|$ units to the right. For $h < 0$, the graph of $y = 2^x$ is translated $|h|$ units to the left. For $k > 0$, the graph of $y = 2^x$ is translated $|k|$ units up. For $k < 0$, the graph of $y = 2^x$ is translated $|k|$ units down.

80.

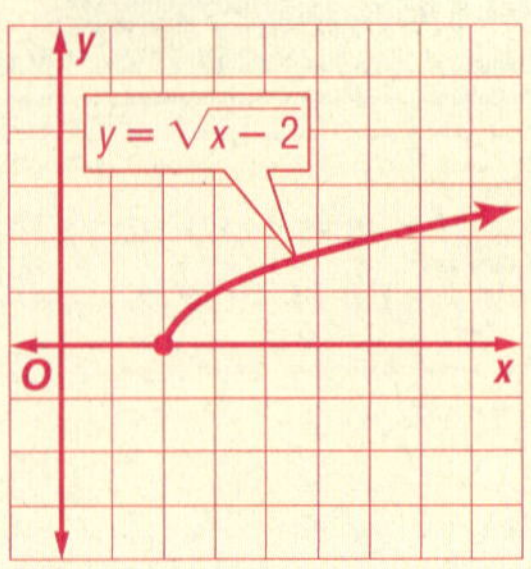

81.

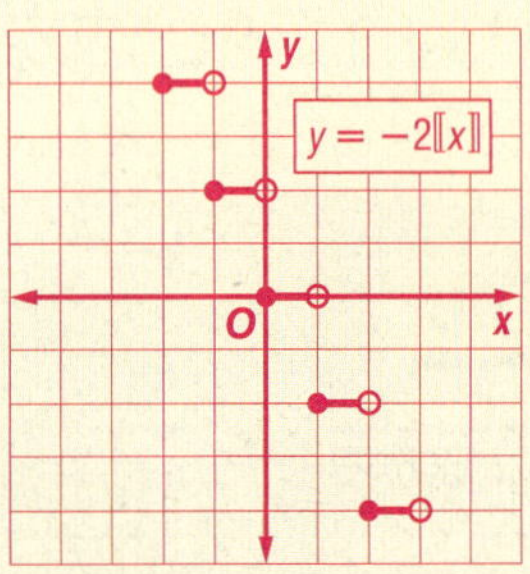

82.

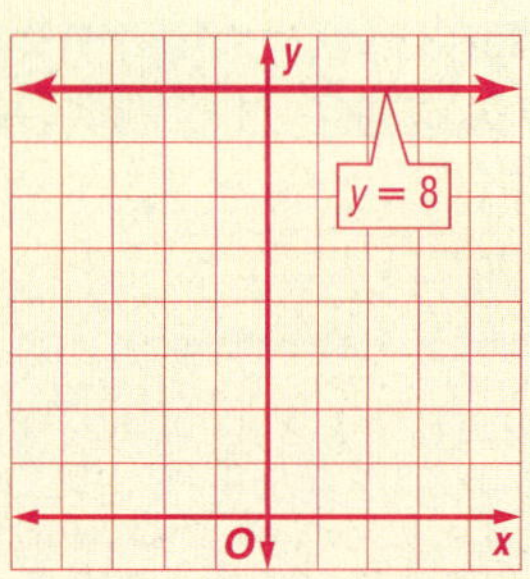

87. $g[h(x)] = 2x - 6$; $h[g(x)] = 2x - 11$

88. $g[h(x)] = x^2 + 6x + 9$; $h[g(x)] = x^2 + 3$

89. $g[h(x)] = -2x - 2$; $h[g(x)] = -2x + 11$

---

68. **WRITING IN MATH** Answer the question that was posed at the beginning of the lesson. See pp. 573A–573D.

How does an exponential function describe tournament play?

Include the following in your answer:
- an explanation of how you could use the equation $y = 2^x$ to determine the number of rounds of tournament play for 128 teams, and
- an example of an inappropriate number of teams for tournament play with an explanation as to why this number would be inappropriate.

Standardized Test Practice

69. If $4^{x+2} = 48$, then $4^x =$ **A**

(A) 3.0. (B) 6.4. (C) 6.9. (D) 12.0. (E) 24.0.

70. **GRID IN** Suppose you deposit \$500 in an account paying 4.5% interest compounded semiannually. Find the dollar value of the account rounded to the nearest penny after 10 years. **780.25**

Graphing Calculator

FAMILIES OF GRAPHS Graph each pair of functions on the same screen. Then compare the graphs, listing both similarities and differences in shape, asymptotes, domain, range, and y-intercepts. **71–74. See pp. 573A–573D.**

71. $y = 2^x$ and $y = 2^x + 3$

72. $y = 3^x$ and $y = 3^{x+1}$

73. $y = \left(\frac{1}{5}\right)^x$ and $y = \left(\frac{1}{5}\right)^{x-2}$

74. $y = \left(\frac{1}{4}\right)^x$ and $y = \left(\frac{1}{4}\right)^x - 1$

75. Describe the effect of changing the values of h and k in the equation $y = 2^{x-h} + k$. **See margin.**

Maintain Your Skills

Mixed Review

Solve each equation or inequality. Check your solutions. *(Lesson 9-6)*

76. $\frac{15}{p} + p = 16$ **1, 15**

77. $\frac{s-3}{s+4} = \frac{6}{s^2-16}$ **1, 6**

78. $\frac{2a-5}{a-9} + \frac{a}{a+9} = \frac{-6}{a^2-81}$ $-\frac{13}{3}$, **3**

79. $\frac{x-2}{x} < \frac{x-4}{x-6}$ $0 < x < 3$ **or** $x > 6$

Identify each equation as a type of function. Then graph the equation. *(Lesson 9-5)*

80–82. See margin for graphs.

80. $y = \sqrt{x-2}$ **square root**

81. $y = -2⟦x⟧$ **greatest integer**

82. $y = 8$ **constant**

Find the inverse of each matrix, if it exists. *(Lesson 4-7)*

83. $\begin{bmatrix} 1 & 0 \\ 0 & 1 \end{bmatrix}$ $\begin{bmatrix} 1 & 0 \\ 0 & 1 \end{bmatrix}$

84. $\begin{bmatrix} 2 & 4 \\ 5 & 10 \end{bmatrix}$ **does not exist**

85. $\begin{bmatrix} -5 & 6 \\ -11 & 3 \end{bmatrix}$ $\frac{1}{51}\begin{bmatrix} 3 & -6 \\ 11 & -5 \end{bmatrix}$

86. **ENERGY** A circular cell must deliver 18 watts of energy. If each square centimeter of the cell that is in sunlight produces 0.01 watt of energy, how long must the radius of the cell be? *(Lesson 5-8)* **about 23.94 cm**

Getting Ready for the Next Lesson

PREREQUISITE SKILL Find $g[h(x)]$ and $h[g(x)]$.
*(To review **composition of functions**, see Lesson 7-7.)* **87–89. See margin.**

87. $h(x) = 2x - 1$
$g(x) = x - 5$

88. $h(x) = x + 3$
$g(x) = x^2$

89. $h(x) = 2x + 5$
$g(x) = -x + 3$

530 Chapter 10 Exponential and Logarithmic Relations

10-2 Logarithms and Logarithmic Functions

What You'll Learn

- Evaluate logarithmic expressions.
- Solve logarithmic equations and inequalities.

Vocabulary

- logarithm
- logarithmic function
- logarithmic equation
- logarithmic inequality

Why is a logarithmic scale used to measure sound?

Many scientific measurements have such an enormous range of possible values that it makes sense to write them as powers of 10 and simply keep track of their exponents. For example, the loudness of sound is measured in units called *decibels*. The graph shows the relative intensities and decibel measures of common sounds.

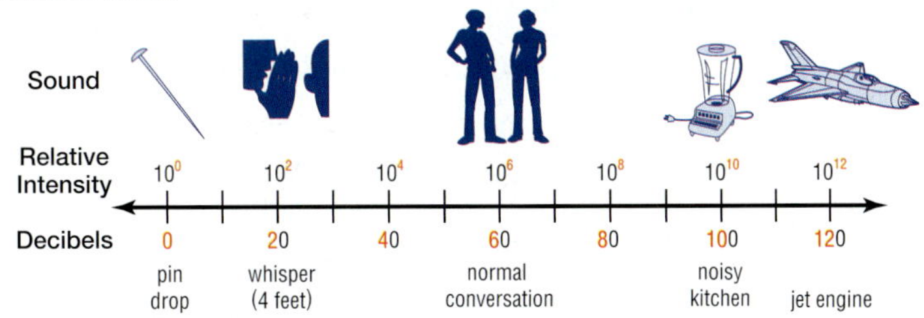

The decibel measure of the loudness of a sound is the exponent or logarithm of its relative intensity multiplied by 10.

LOGARITHMIC FUNCTIONS AND EXPRESSIONS To better understand what is meant by a logarithm, let's look at the graph of $y = 2^x$ and its inverse. Since exponential functions are one-to-one, the inverse of $y = 2^x$ exists and is also a function. Recall that you can graph the inverse of a function by interchanging the x and y values in the ordered pairs of the function.

Study Tip

Look Back
To review **inverse functions**, see Lesson 7-8.

$y = 2^x$	
x	y
-3	$\frac{1}{8}$
-2	$\frac{1}{4}$
-1	$\frac{1}{2}$
0	1
1	2
2	4
3	8

$x = 2^y$	
x	y
$\frac{1}{8}$	-3
$\frac{1}{4}$	-2
$\frac{1}{2}$	-1
1	0
2	1
4	2
8	3

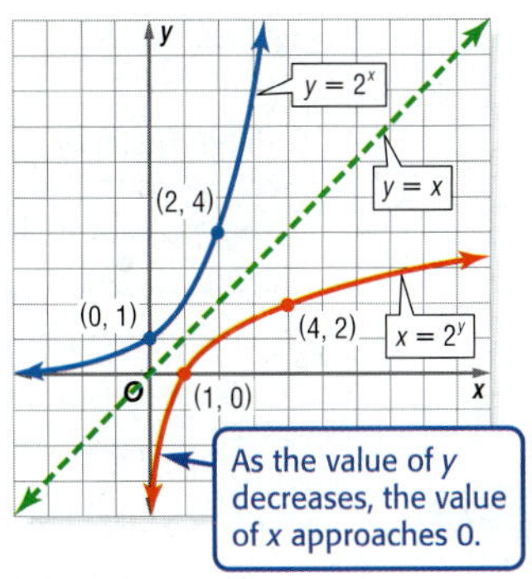

The inverse of $y = 2^x$ can be defined as $x = 2^y$. Notice that the graphs of these two functions are reflections of each other over the line $y = x$.

In general, the inverse of $y = b^x$ is $x = b^y$. In $x = b^y$, y is called the **logarithm** of x. It is usually written as $y = \log_b x$ and is read *y equals log base b of x*.

10-2 Lesson Notes

1 Focus

5-Minute Check Transparency 10-2 Use as a quiz or review of Lesson 10-1.

Mathematical Background notes are available for this lesson on p. 520C.

Why is a logarithmic scale used to measure sound?

Ask students:

- On the number line shown, the scale along the bottom is 10 decibels per tick mark. What do you notice about the scale along the top for relative intensity? **The scale is not uniform; the relative intensity at the first tick mark is 10, at the second it is 100, at the third it is 1000, and so on.**
- If you draw a number line with a uniform scale whose tick marks are labeled from 0 to 10^{12}, what number is at the midpoint between 0 to 10^{12}? **5×10^{11}**
- Where does the point 1 million appear on your number line? **very close to the point for 0**
- Where does the point 100 appear on your number line? **very, very close to the point for 0**
- What problem arises with trying to represent the relative intensities on a standard number line? **Sample answer: The lesser intensities are so close together near 0 on the number line that they are difficult to represent accurately.**

Resource Manager

Workbook and Reproducible Masters

Chapter 10 Resource Masters

- Study Guide and Intervention, pp. 579–580
- Skills Practice, p. 581
- Practice, p. 582
- Reading to Learn Mathematics, p. 583
- Enrichment, p. 584
- Assessment, p. 623

School-to-Career Masters, p. 19

Transparencies

5-Minute Check Transparency 10-2
Answer Key Transparencies

Technology

Interactive Chalkboard

2 Teach

LOGARITHMIC FUNCTIONS AND EXPRESSIONS

Teaching Tip After discussing the definition of logarithm at the bottom of p. 531, write the equation $y = 2x$ on the chalkboard and ask students to rewrite the equation with x in terms of y. $\left(x = \frac{1}{2}y\right)$ Repeat this for the equation $y = x^2$. $(x = \pm\sqrt{y})$ Now write the equation $y = 2^x$ on the chalkboard and ask students to rewrite this equation with x in terms of y. This will likely have students stymied. Explain that the rewritten equation is $x = \log_2 y$.

In-Class Examples

Power Point®

1 Write each equation in exponential form.

a. $\log_3 9 = 2$ **$3^2 = 9$**

b. $\log_{10} \frac{1}{100} = -2$ **$10^{-2} = \frac{1}{100}$**

2 Write each equation in logarithmic form.

a. $5^3 = 125$ **$\log_5 125 = 3$**

b. $27^{\frac{1}{3}} = 3$ **$\log_{27} 3 = \frac{1}{3}$**

3 Evaluate $\log_3 243$. **5**

Key Concept — *Logarithm with Base b*

- **Words** Let b and x be positive numbers, $b \neq 1$. The *logarithm of x with base b* is denoted $\log_b x$ and is defined as the exponent y that makes the equation $b^y = x$ true.
- **Symbols** Suppose $b > 0$ and $b \neq 1$. For $x > 0$, there is a number y such that $\log_b x = y$ if and only if $b^y = x$.

Example 1 Logarithmic to Exponential Form

Write each equation in exponential form.

a. $\log_8 1 = 0$

$\log_8 1 = 0 \rightarrow 1 = 8^0$

b. $\log_2 \frac{1}{16} = -4$

$\log_2 \frac{1}{16} = -4 \rightarrow \frac{1}{16} = 2^{-4}$

Example 2 Exponential to Logarithmic Form

Write each equation in logarithmic form.

a. $10^3 = 1000$

$10^3 = 1000 \rightarrow \log_{10} 1000 = 3$

b. $9^{\frac{1}{2}} = 3$

$9^{\frac{1}{2}} = 3 \rightarrow \log_9 3 = \frac{1}{2}$

You can use the definition of logarithm to find the value of a logarithmic expression.

Example 3 Evaluate Logarithmic Expressions

Evaluate $\log_2 64$.

$\log_2 64 = y$ Let the logarithm equal y.

$64 = 2^y$ Definition of logarithm

$2^6 = 2^y$ $64 = 2^6$

$6 = y$ Property of Equality for Exponential Functions

So, $\log_2 64 = 6$.

The function $y = \log_b x$, where $b > 0$ and $b \neq 1$, is called a **logarithmic function**. As shown in the graph on the previous page, this function is the inverse of the exponential function $y = b^x$ and has the following characteristics.

1. The function is continuous and one-to-one.
2. The domain is the set of all positive real numbers.
3. The y-axis is an asymptote of the graph.
4. The range is the set of all real numbers.
5. The graph contains the point (1, 0). That is, the x-intercept is 1.

Study Tip

Look Back
To review **composition of functions**, see Lesson 7-7.

Since the exponential function $f(x) = b^x$ and the logarithmic function $g(x) = \log_b x$ are inverses of each other, their composites are the identity function. That is, $f[g(x)] = x$ and $g[f(x)] = x$.

$$f[g(x)] = x \qquad g[f(x)] = x$$
$$f(\log_b x) = x \qquad g(b^x) = x$$
$$b^{\log_b x} = x \qquad \log_b b^x = x$$

Thus, if their bases are the same, exponential and logarithmic functions "undo" each other. You can use this inverse property of exponents and logarithms to simplify expressions.

Example 4 Inverse Property of Exponents and Logarithms

Evaluate each expression.

a. $\log_6 6^8$

$\log_6 6^8 = 8$ $\log_b b^x = x$

b. $3^{\log_3 (4x - 1)}$

$3^{\log_3 (4x-1)} = 4x - 1$ $b^{\log_b x} = x$

SOLVE LOGARITHMIC EQUATIONS AND INEQUALITIES

A **logarithmic equation** is an equation that contains one or more logarithms. You can use the definition of a logarithm to help you solve logarithmic equations.

Example 5 Solve a Logarithmic Equation

Solve $\log_4 n = \frac{5}{2}$.

$\log_4 n = \frac{5}{2}$ Original equation

$n = 4^{\frac{5}{2}}$ Definition of logarithm

$n = (2^2)^{\frac{5}{2}}$ $4 = 2^2$

$n = 2^5$ Power of a Power

$n = 32$ Simplify.

A **logarithmic inequality** is an inequality that involves logarithms. In the case of inequalities, the following property is helpful.

Key Concept: Logarithmic to Exponential Inequality

- **Symbols** If $b > 1$, $x > 0$, and $\log_b x > y$, then $x > b^y$.
  If $b > 1$, $x > 0$, and $\log_b x < y$, then $0 < x < b^y$.
- **Examples** $\log_2 x > 3$, $x > 2^3$ $\log_3 x < 5$, $0 < x < 3^5$

Example 6 Solve a Logarithmic Inequality

Solve $\log_5 x < 2$. Check your solution.

$\log_5 x < 2$ Original inequality

$0 < x < 5^2$ Logarithmic to exponential inequality

$0 < x < 25$ Simplify.

The solution set is $\{x \mid 0 < x < 25\}$.

CHECK Try 5 to see if it satisfies the inequality.

$\log_5 x < 2$ Original inequality

$\log_5 5 \overset{?}{<} 2$ Substitute 5 for x.

$1 < 2$ ✓ $\log_5 5 = 1$ because $5^1 = 5$.

Study Tip

Special Values

If $b > 0$ and $b \neq 1$, then the following statements are true.

- $\log_b b = 1$ because $b^1 = b$.
- $\log_b 1 = 0$ because $b^0 = 1$.

www.algebra2.com/extra_examples

In-Class Example (PowerPoint®)

4 Evaluate each expression.

a. $\log_9 9^2$ **2**

b. $7^{\log_7 (x^2 - 1)}$ $x^2 - 1$

SOLVE LOGARITHMIC EQUATIONS AND INEQUALITIES

In-Class Examples (PowerPoint®)

5 Solve $\log_8 n = \frac{4}{3}$. **16**

6 Solve $\log_6 x > 3$. Check your solution. $\{x \mid x > 216\}$

In-Class Examples

7 Solve $\log_4 x^2 = \log_4 (4x - 3)$. Check your solution. **1, 3**

8 Solve $\log_7 (2x + 8) > \log_7 (x + 5)$. Check your solution. $x > -3$

Tips for New Teachers

Intervention Students have not covered logarithmic functions before and are likely to find them confusing. Expect students to need extra time to absorb the material in this lesson before continuing with the rest of the chapter.

Use the following property to solve logarithmic equations that have logarithms with the same base on each side.

Key Concept — Property of Equality for Logarithmic Functions

- **Symbols** If b is a positive number other than 1, then $\log_b x = \log_b y$ if and only if $x = y$.
- **Example** If $\log_7 x = \log_7 3$, then $x = 3$.

Example 7 Solve Equations with Logarithms on Each Side

Solve $\log_5 (p^2 - 2) = \log_5 p$. Check your solution.

$\log_5 (p^2 - 2) = \log_5 p$	Original equation
$p^2 - 2 = p$	Property of Equality for Logarithmic Functions
$p^2 - p - 2 = 0$	Subtract p from each side.
$(p - 2)(p + 1) = 0$	Factor.
$p - 2 = 0$ or $p + 1 = 0$	Zero Product Property
$p = 2$ $\quad$ $p = -1$	Solve each equation.

CHECK Substitute each value into the original equation.

$\log_5 (2^2 - 2) \stackrel{?}{=} \log_5 2$	Substitute 2 for p.
$\log_5 2 = \log_5 2$ ✓	Simplify.
$\log_5 [(-1)^2 - 2] \stackrel{?}{=} \log_5 (-1)$	Substitute -1 for p.

Since $\log_5 (-1)$ is undefined, -1 is an *extraneous* solution and must be eliminated. Thus, the solution is 2.

Study Tip

Extraneous Solutions
The domain of a logarithmic function does not include negative values. For this reason, be sure to check for extraneous solutions of logarithmic equations.

Use the following property to solve logarithmic inequalities that have the same base on each side. Exclude values from your solution set that would result in taking the logarithm of a number less than or equal to zero in the original inequality.

Key Concept — Property of Inequality for Logarithmic Functions

- **Symbols** If $b > 1$, then $\log_b x > \log_b y$ if and only if $x > y$, and $\log_b x < \log_b y$ if and only if $x < y$.
- **Example** If $\log_2 x > \log_2 9$, then $x > 9$.

This property also holds for $\leq$ and $\geq$.

Example 8 Solve Inequalities with Logarithms on Each Side

Solve $\log_{10} (3x - 4) < \log_{10} (x + 6)$. Check your solution.

$\log_{10} (3x - 4) < \log_{10} (x + 6)$	Original inequality
$3x - 4 < x + 6$	Property of Inequality for Logarithmic Functions
$2x < 10$	Addition and Subtraction Properties of Inequalities
$x < 5$	Divide each side by 2.

We must exclude from this solution all values of x such that $3x - 4 \leq 0$ or $x + 6 \leq 0$. Thus, the solution set is $x > \frac{4}{3}$, $x > -6$, and $x < 5$. This compound inequality simplifies to $\frac{4}{3} < x < 5$.

Study Tip

Look back
To review **compound inequalities**, see Lesson 1-6.

DAILY INTERVENTION — Differentiated Instruction

Visual/Spatial Have students create colorful posters showing several equivalent exponential and logarithmic equations, such as $2^3 = 8$ and $3 = \log_2 8$. Suggest that students use a different color for each of the digits 2, 3, and 8 to help them visualize the relative locations of the digits in the pairs of equations.

Check for Understanding

Concept Check

1. **OPEN ENDED** Give an example of an exponential equation and its related logarithmic equation. **Sample answer: $x = 5^y$ and $y = \log_5 x$**

2. **Describe** the relationship between $y = 3^x$ and $y = \log_3 x$. **They are inverses.**

3. **FIND THE ERROR** Paul and Scott are solving $\log_3 x = 9$.

Paul	Scott
$\log_3 x = 9$	$\log_3 x = 9$
$3^x = 9$	$x = 3^9$
$3^x = 3^2$	$x = 19{,}683$
$x = 2$	

Who is correct? Explain your reasoning. **Scott; see margin for explanation.**

Guided Practice

GUIDED PRACTICE KEY	
Exercises	Examples
4, 5	1
6, 7	2
8–11	3
12–17	4–7
18–20	4

Write each equation in logarithmic form.

4. $5^4 = 625$ **$\log_5 625 = 4$**
5. $7^{-2} = \frac{1}{49}$ **$\log_7 \frac{1}{49} = -2$**

Write each equation in exponential form.

6. $\log_3 81 = 4$ **$3^4 = 81$**
7. $\log_{36} 6 = \frac{1}{2}$ **$36^{\frac{1}{2}} = 6$**

Evaluate each expression.

8. $\log_4 256$ **4**
9. $\log_2 \frac{1}{8}$ **−3**
10. $3^{\log_3 21}$ **21**
11. $\log_5 5^{-1}$ **−1**

Solve each equation or inequality. Check your solutions.

12. $\log_9 x = \frac{3}{2}$ **27**
13. $\log_{\frac{1}{10}} x = -3$ **1000**
14. $\log_3 (2x - 1) \leq 2$ **$\frac{1}{2} < x \leq 5$**
15. $\log_5 (3x - 1) = \log_5 2x^2$ **$\frac{1}{2}$, 1**
16. $\log_2 (3x - 5) > \log_2 (x + 7)$ **$x > 6$**
17. $\log_b 9 = 2$ **3**

Application

SOUND **For Exercises 18–20, use the following information.**
An equation for loudness L, in decibels, is $L = 10 \log_{10} R$, where R is the relative intensity of the sound.

18. Solve $130 = 10 \log_{10} R$ to find the relative intensity of a fireworks display with a loudness of 130 decibels. **10^{13}**
19. Solve $75 = 10 \log_{10} R$ to find the relative intensity of a concert with a loudness of 75 decibels. **$10^{7.5}$**
20. How many times more intense is the fireworks display than the concert? In other words, find the ratio of their intensities. **$10^{5.5}$ or about 316,228 times**

3 Practice/Apply

Study Notebook

Have students—
- add the definitions/examples of the vocabulary terms to their Vocabulary Builder worksheets for Chapter 10.
- include examples of how to write logarithms in exponential form.
- include any other item(s) that they find helpful in mastering the skills in this lesson.

DAILY INTERVENTION **FIND THE ERROR**

Review converting logarithms to exponential form. Also note that, according to Paul, $\log_3 x = 3^x$, which cannot be true.

Answer

3. The value of a logarithmic equation, 9, is the exponent of the equivalent exponential equation, and the base of the logarithmic expression, 3, is the base of the exponential equation. Thus $x = 3^9$ or 19,683.

Online Lesson Plans

USA TODAY Education's Online site offers resources and interactive features connected to each day's newspaper. *Experience TODAY*, USA TODAY's daily lesson plan, is available on the site and delivered daily to subscribers. This plan provides instruction for integrating USA TODAY graphics and key editorial features into your mathematics classroom. Log on to www.education.usatoday.com.

★ indicates increased difficulty

Practice and Apply

Homework Help	
For Exercises	See Examples
21–26	1
27–32	2
33–46	3
47–62	4–7
63–65	4
68–70	5

Extra Practice
See page 849.

Write each equation in logarithmic form. **23.** $\log_5 \frac{1}{125} = -3$

21. $8^3 = 512$ $\log_8 512 = 3$
22. $3^3 = 27$ $\log_3 27 = 3$
23. $5^{-3} = \frac{1}{125}$
24. $\left(\frac{1}{3}\right)^{-2} = 9$ $\log_{\frac{1}{3}} 9 = -2$
25. $100^{\frac{1}{2}} = 10$ $\log_{100} 10 = \frac{1}{2}$
26. $2401^{\frac{1}{4}} = 7$ $\log_{2401} 7 = \frac{1}{4}$

Write each equation in exponential form. **29.** $4^{-1} = \frac{1}{4}$

27. $\log_5 125 = 3$ $5^3 = 125$
28. $\log_{13} 169 = 2$ $13^2 = 169$
29. $\log_4 \frac{1}{4} = -1$
30. $\log_{100} \frac{1}{10} = -\frac{1}{2}$ $100^{-\frac{1}{2}} = \frac{1}{10}$
31. $\log_8 4 = \frac{2}{3}$ $8^{\frac{2}{3}} = 4$
32. $\log_{\frac{1}{5}} 25 = -2$ $\left(\frac{1}{5}\right)^{-2} = 25$

Evaluate each expression.

33. $\log_2 16$ **4**
34. $\log_{12} 144$ **2**
35. $\log_{16} 4$ $\frac{1}{2}$
36. $\log_9 243$ $\frac{5}{2}$
37. $\log_2 \frac{1}{32}$ -5
38. $\log_3 \frac{1}{81}$ -4
39. $\log_5 5^7$ **7**
40. $2^{\log_2 45}$ **45**
41. $\log_{11} 11^{(n-5)}$ $n - 5$
42. $6^{\log_6 (3x+2)}$ $3x + 2$
★ 43. $\log_{10} 0.001$ -3
★ 44. $\log_4 16^x$ $2x$

WORLD RECORDS For Exercises 45 and 46, use the information given for Exercises 18–20 to find the relative intensity of each sound. **Source:** *The Guinness Book of Records*

45. The loudest animal sounds are the low-frequency pulses made by blue whales when they communicate. These pulses have been measured up to 188 decibels. $10^{18.8}$

46. The loudest insect is the African cicada. It produces a calling song that measures 106.7 decibels at a distance of 50 centimeters. $10^{10.67}$

Solve each equation or inequality. Check your solutions.

47. $\log_9 x = 2$ **81**
48. $\log_2 c > 8$ $c > 256$
49. $\log_{64} y \le \frac{1}{2}$ $0 < y \le 8$
50. $\log_{25} n = \frac{3}{2}$ **125**
51. $\log_{\frac{1}{7}} x = -1$ **7**
52. $\log_{\frac{1}{3}} p < 0$ $p > 1$
53. $\log_2 (3x - 8) \ge 6$ $x \ge 24$
54. $\log_{10} (x^2 + 1) = 1$ ± 3
55. $\log_b 64 = 3$ **4**
56. $\log_b 121 = 2$ **11**
57. $\log_5 5^{6n+1} = 13$ **2**
58. $\log_5 x = \frac{1}{2}$ $\sqrt{5}$
59. $\log_6 (2x - 3) = \log_6 (x + 2)$ **5**
60. $\log_2 (4y - 10) \ge \log_2 (y - 1)$ $y \ge 3$

★ 61. $\log_{10} (a^2 - 6) > \log_{10} a$ $a > 3$
★ 62. $\log_7 (x^2 + 36) = \log_7 100$ ± 8

Show that each statement is true. **63–65. See margin.**

★ 63. $\log_5 25 = 2 \log_5 5$
★ 64. $\log_{16} 2 \cdot \log_2 16 = 1$
★ 65. $\log_7 [\log_3 (\log_2 8)] = 0$

About the Exercises...

Organization by Objective

- **Logarithmic Functions and Expressions:** 21–46, 66–71
- **Solve Logarithmic Equations and Inequalities:** 47–65

Odd/Even Assignments
Exercises 21–62 are structured so that students practice the same concepts whether they are assigned odd or even problems.

Assignment Guide

Basic: 21–41 odd, 45–59 odd, 71–90

Average: 21–67 odd, 68, 69, 71–90

Advanced: 22–66 even, 68–84 (optional: 85–90)

All: Practice Quiz 1 (1–10)

Answers

63.
$\log_5 25 \stackrel{?}{=} 2 \log_5 5$	Original equation
$\log_5 5^2 \stackrel{?}{=} 2 \log_5 5^1$	$25 = 5^2$ and $5 = 5^1$
$2 \stackrel{?}{=} 2(1)$	Inverse Property of Exponents and Logarithms
$2 = 2$ ✓	Simplify.

64.
$\log_{16} 2 \cdot \log_2 16 \stackrel{?}{=} 1$	Original equation
$\log_{16} 16^{\frac{1}{4}} \cdot \log_2 2^4 \stackrel{?}{=} 1$	$2 = 16^{\frac{1}{4}}$ and $16 = 2^4$
$\frac{1}{4}(4) \stackrel{?}{=} 1$	Inverse Property of Exponents and Logarithms
$1 = 1$ ✓	

65.
$\log_7 [\log_3 (\log_2 8)] \stackrel{?}{=} 0$	Original equation
$\log_7 [\log_3 (\log_2 2^3)] \stackrel{?}{=} 0$	$8 = 2^3$
$\log_7 (\log_3 3) \stackrel{?}{=} 0$	Inverse Property of Exponents and Logarithms
$\log_7 (\log_3 3^1) \stackrel{?}{=} 0$	$3 = 3^1$
$\log_7 1 \stackrel{?}{=} 0$	Inverse Property of Exponents and Logarithms
$\log_7 7^0 \stackrel{?}{=} 0$	$1 = 7^0$
$0 = 0$ ✓	Inverse Property of Exponents and Logarithms

66–67. See pp. 573A–573D.

66. a. Sketch the graphs of $y = \log_{\frac{1}{2}} x$ and $y = \left(\frac{1}{2}\right)^x$ on the same axes.

b. Describe the relationship between the graphs.

★ 67. a. Sketch the graphs of $y = \log_2 x + 3$, $y = \log_2 x - 4$, $y = \log_2 (x - 1)$, and $y = \log_2 (x + 2)$.

b. Describe this family of graphs in terms of its parent graph $y = \log_2 x$.

EARTHQUAKE **For Exercises 68 and 69, use the following information.**
The magnitude of an earthquake is measured on a logarithmic scale called the Richter scale. The magnitude M is given by $M = \log_{10} x$, where x represents the amplitude of the seismic wave causing ground motion.

68. How many times as great is the amplitude caused by an earthquake with a Richter scale rating of 7 as an aftershock with a Richter scale rating of 4? 10^3 or 1000 as times great

69. How many times as great was the motion caused by the 1906 San Francisco earthquake that measured 8.3 on the Richter scale as that caused by the 2001 Bhuj, India, earthquake that measured 6.9? $10^{1.4}$ or about 25 times as great

More About. . .

Earthquake
The Loma Prieta earthquake measured 7.1 on the Richter scale and interrupted the 1989 World Series in San Francisco.
Source: U.S. Geological Survey

70. **NOISE ORDINANCE** A proposed city ordinance will make it illegal to create sound in a residential area that exceeds 72 decibels during the day and 55 decibels during the night. How many times more intense is the noise level allowed during the day than at night? $10^{1.7}$ or about 50 times

71. **CRITICAL THINKING** The value of $\log_2 5$ is between two consecutive integers. Name these integers and explain how you determined them. 2 and 3; Sample answer: 5 is between 2^2 and 2^3.

72. **CRITICAL THINKING** Using the definition of a logarithmic function where $y = \log_b x$, explain why the base b cannot equal 1. All powers of 1 are 1, so the inverse of $y = 1^x$ is not a function.

73. *WRITING IN MATH* Answer the question that was posed at the beginning of the lesson. See pp. 573A–573D.

Why is a logarithmic scale used to measure sound?

Include the following in your answer:

- the relative intensities of a pin drop, a whisper, normal conversation, kitchen noise, and a jet engine written in scientific notation,
- a plot of each of these relative intensities on the scale shown below, and

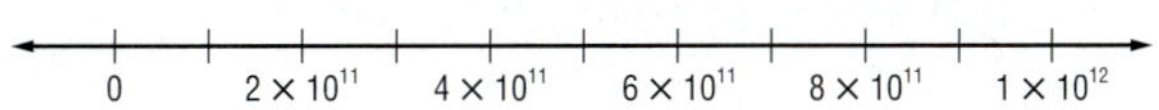

- an explanation as to why the logarithmic scale might be preferred over the scale shown above.

Standardized Test Practice

74. What is the equation of the function graphed at the right? B

(A) $y = 2(3)^x$

(B) $y = 2\left(\frac{1}{3}\right)^x$

(C) $y = 3\left(\frac{1}{2}\right)^x$

(D) $y = 3(2)^x$

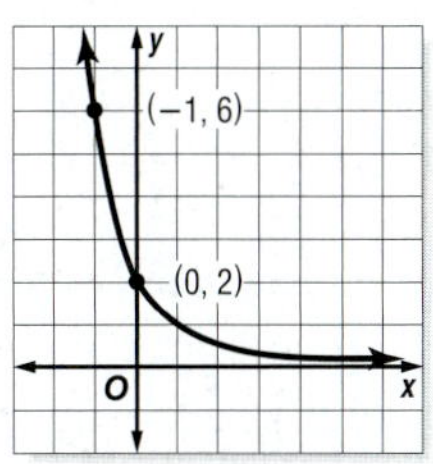

www.algebra2.com/self_check_quiz

Study Guide and Intervention, p. 579 (shown) and p. 580

Logarithmic Functions and Expressions

Definition of Logarithm with Base b	Let b and x be positive numbers, $b \neq 1$. The logarithm of x with base b is denoted $\log_b x$ and is defined as the exponent y that makes the equation $b^y = x$ true.

The inverse of the exponential function $y = b^x$ is the **logarithmic function** $x = b^y$. This function is usually written as $y = \log_b x$.

Properties of Logarithmic Functions	1. The function is continuous and one-to-one. 2. The domain is the set of all positive real numbers. 3. The y-axis is an asymptote of the graph. 4. The range is the set of all real numbers. 5. The graph contains the point (0, 1).

Example 1 **Write an exponential equation equivalent to $\log_3 243 = 5$.**
$3^5 = 243$

Example 2 **Write a logarithmic equation equivalent to $6^{-3} = \frac{1}{216}$.**
$\log_6 \frac{1}{216} = -3$

Example 3 **Evaluate $\log_8 16$.**
$8^{\frac{4}{3}} = 16$, so $\log_8 16 = \frac{4}{3}$.

Exercises

Write each equation in logarithmic form.

1. $2^7 = 128$ $\log_2 128 = 7$
2. $3^{-4} = \frac{1}{81}$ $\log_3 \frac{1}{81} = -4$
3. $\left(\frac{1}{7}\right)^3 = \frac{1}{343}$ $\log_{\frac{1}{7}} \frac{1}{343} = 3$

Write each equation in exponential form.

4. $\log_{15} 225 = 2$ $15^2 = 225$
5. $\log_3 \frac{1}{27} = -3$ $3^{-3} = \frac{1}{27}$
6. $\log_4 32 = \frac{5}{2}$ $4^{\frac{5}{2}} = 32$

Evaluate each expression.

7. $\log_4 64$ 3
8. $\log_2 64$ 6
9. $\log_{100} 100{,}000$ 2.5
10. $\log_5 625$ 4
11. $\log_{27} 81$ $\frac{4}{3}$
12. $\log_{25} 5$ $\frac{1}{2}$
13. $\log_2 \frac{1}{128}$ −7
14. $\log_{10} 0.00001$ −5
15. $\log_4 \frac{1}{32}$ −2.5

Skills Practice, p. 581 and Practice, p. 582 (shown)

Write each equation in logarithmic form.

1. $5^3 = 125$ $\log_5 125 = 3$
2. $7^0 = 1$ $\log_7 1 = 0$
3. $3^4 = 81$ $\log_3 81 = 4$
4. $3^{-4} = \frac{1}{81}$ $\log_3 \frac{1}{81} = -4$
5. $\left(\frac{1}{4}\right)^3 = \frac{1}{64}$ $\log_{\frac{1}{4}} \frac{1}{64} = 3$
6. $7776^{\frac{1}{5}} = 6$ $\log_{7776} 6 = \frac{1}{5}$

Write each equation in exponential form.

7. $\log_6 216 = 3$ $6^3 = 216$
8. $\log_2 64 = 6$ $2^6 = 64$
9. $\log_3 \frac{1}{81} = -4$ $3^{-4} = \frac{1}{81}$
10. $\log_{10} 0.00001 = -5$ $10^{-5} = 0.00001$
11. $\log_{25} 5 = \frac{1}{2}$ $25^{\frac{1}{2}} = 5$
12. $\log_{32} 8 = \frac{3}{5}$ $32^{\frac{3}{5}} = 8$

Evaluate each expression.

13. $\log_3 81$ 4
14. $\log_{10} 0.0001$ −4
15. $\log_2 \frac{1}{16}$ −4
16. $\log_{\frac{1}{3}} 27$ −3
17. $\log_9 1$ 0
18. $\log_8 4$ $\frac{2}{3}$
19. $\log_7 \frac{1}{49}$ −2
20. $\log_6 6^4$ 4
21. $\log_3 \frac{1}{3}$ −1
22. $\log_4 \frac{1}{256}$ −4
23. $\log_9 9^{(n+1)}$ $n + 1$
24. $2^{\log_2 32}$ 32

Solve each equation or inequality. Check your solutions.

25. $\log_{10} n = -3$ $\frac{1}{1000}$
26. $\log_4 x > 3$ $x > 64$
27. $\log_4 x = \frac{3}{2}$ 8
28. $\log_{\frac{1}{5}} x = -3$ 125
29. $\log_7 q < 0$ $0 < q < 1$
30. $\log_6 (2y + 8) \geq 2$ $y \geq 14$
31. $\log_y 16 = -4$ $\frac{1}{2}$
32. $\log_n \frac{1}{8} = -3$ 2
33. $\log_b 1024 = 5$ 4
34. $\log_8 (3x + 7) < \log_8 (7x + 4)$ $x > \frac{3}{4}$
35. $\log_7 (8x + 20) = \log_7 (x + 6)$ −2
36. $\log_3 (x^2 - 2) = \log_3 x$ 2

37. **SOUND** Sounds that reach levels of 130 decibels or more are painful to humans. What is the relative intensity of 130 decibels? 10^{13}

38. **INVESTING** Maria invests \$1000 in a savings account that pays 8% interest compounded annually. The value of the account A at the end of five years can be determined from the equation $\log A = \log[1000(1 + 0.08)^5]$. Find the value of A to the nearest dollar. \$1469

Reading to Learn Mathematics, p. 583 ELL

Pre-Activity **Why is a logarithmic scale used to measure sound?**

Read the introduction to Lesson 10-2 at the top of page 531 in your textbook.

How many times louder than a whisper is normal conversation? 10^4 or 10,000 times

Reading the Lesson

1. a. Write an exponential equation that is equivalent to $\log_3 81 = 4$. $3^4 = 81$
   b. Write a logarithmic equation that is equivalent to $25^{-\frac{1}{2}} = \frac{1}{5}$. $\log_{25} \frac{1}{5} = -\frac{1}{2}$
   c. Write an exponential equation that is equivalent to $\log_4 1 = 0$. $4^0 = 1$
   d. Write a logarithmic equation that is equivalent to $10^{-3} = 0.001$. $\log_{10} 0.001 = -3$
   e. What is the inverse of the function $y = 5^x$? $y = \log_5 x$
   f. What is the inverse of the function $y = \log_{10} x$? $y = 10^x$

2. Match each function with its graph.
   a. $y = 3^x$ IV
   b. $y = \log_3 x$ I
   c. $y = \left(\frac{1}{3}\right)^x$ II

I.
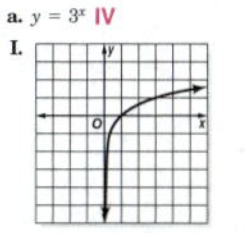

II.
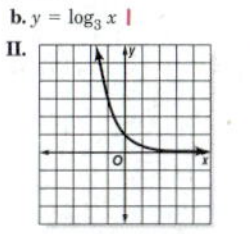

III.
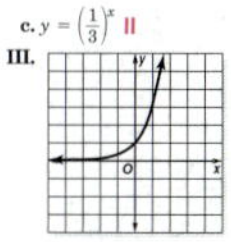

3. Indicate whether each of the following statements about the exponential function $y = \log_5 x$ is *true* or *false*.
   a. The y-axis is an asymptote of the graph. true
   b. The domain is the set of all real numbers. false
   c. The graph contains the point (5, 0). false
   d. The range is the set of all real numbers. true
   e. The y-intercept is 1. false

Helping You Remember

4. An important skill needed for working with logarithms is changing an equation between logarithmic and exponential forms. Using the words *base*, *exponent*, and *logarithm*, describe an easy way to remember and apply the part of the definition of logarithm that says, "$\log_b x = y$ if and only if $b^y = x$." Sample answer: In these equations, b stands for *base*. In log form, b is the subscript, and in exponential form, b is the number that is raised to a power. A *logarithm* is an *exponent*, so y, which is the log in the first equation, becomes the exponent in the second equation.

Enrichment, p. 584

Musical Relationships

The frequencies of notes in a musical scale that are one octave apart are related by an exponential equation. For the eight C notes on a piano, the equation is $C_n = C_1 2^{n-1}$, where C_n represents the frequency of note C_n.

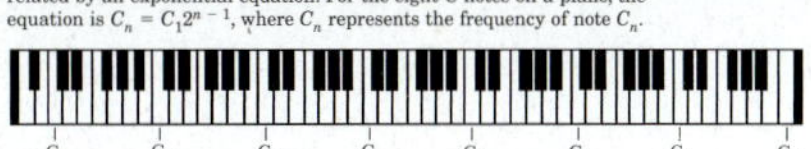

1. Find the relationship between C_1 and C_2. $C_2 = 2C_1$
2. Find the relationship between C_1 and C_4. $C_4 = 8C_1$

The frequencies of consecutive notes are related by a common ratio r. The general equation is $f_n = f_1 r^{n-1}$.

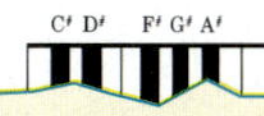

3. If the frequency of middle C is 261.6 cycles per second

4 Assess

Open-Ended Assessment

Writing Have students write a step-by-step explanation of the procedure for solving a logarithmic equation such as $\log_8 n = \frac{7}{3}$.

Getting Ready for Lesson 10-3

PREREQUISITE SKILL In Lesson 10-3, students will evaluate expressions using the properties of logarithms. Because these properties are related to exponential properties, students must be familiar with exponential properties when multiplying or dividing terms with like bases. Use Exercises 85–90 to determine your students' familiarity with multiplying and dividing monomials.

Assessment Options

Practice Quiz 1 The quiz provides students with a brief review of the concepts and skills in Lessons 10-1 and 10-2. Lesson numbers are given to the right of the exercises or instruction lines so students can review concepts not yet mastered.

Quiz (Lessons 10-1 and 10-2) is available on p. 623 of the *Chapter 10 Resource Masters.*

---

75. In the figure at the right, if $y = \frac{2}{7}x$ and $z = 3w$, then $x =$ **D**

Ⓐ 14. Ⓑ 20.

Ⓒ 28. Ⓓ 35.

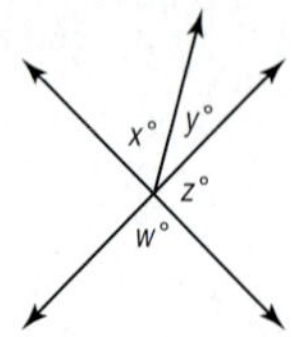

Maintain Your Skills

Mixed Review

Simplify each expression. *(Lesson 10-1)*

76. $x^{\sqrt{6}} \cdot x^{\sqrt{6}}$ $x^{2\sqrt{6}}$

77. $(b^{\sqrt{6}})^{\sqrt{24}}$ b^{12}

Solve each equation. Check your solutions. *(Lesson 9-6)* **79.** $-3, \frac{14}{5}$

78. $\frac{2x+1}{x} - \frac{x+1}{x-4} = \frac{-20}{x^2-4x}$ $\varnothing$

79. $\frac{2a-5}{a-9} - \frac{a-3}{3a+2} = \frac{5}{3a^2-25a-18}$

Solve each equation by using the method of your choice. Find exact solutions. *(Lesson 6-5)*

80. $9y^2 = 49$ $\pm\frac{7}{3}$

81. $2p^2 = 5p + 6$ $\frac{5 \pm \sqrt{73}}{4}$

Simplify each expression. *(Lesson 9-2)* **83.** $\frac{6x-58}{(x-3)(x+3)(x+7)}$

82. $\frac{3}{2y} + \frac{4}{3y} - \frac{7}{5y}$ $\frac{43}{30y}$

83. $\frac{x-7}{x^2-9} - \frac{x-3}{x^2+10x+21}$

84. BANKING Donna Bowers has \$4000 she wants to save in the bank. A certificate of deposit (CD) earns 8% annual interest, while a regular savings account earns 3% annual interest. Ms. Bowers doesn't want to tie up all her money in a CD, but she has decided she wants to earn \$240 in interest for the year. How much money should she put in to each type of account? (*Hint:* Use Cramer's Rule.) *(Lesson 4-4)* **\$2400, CD; \$1600, savings**

Getting Ready for the Next Lesson

PREREQUISITE SKILL **Simplify. Assume that no variable equals zero.**
*(To review **multiplying and dividing monomials**, see Lesson 5-1.)*

85. $x^4 \cdot x^6$ x^{10}

86. $(y^3)^8$ y^{24}

87. $(2a^2b)^3$ $8a^6b^3$

88. $\frac{a^4n^7}{a^3n}$ an^6

89. $\frac{x^5yz^2}{x^2y^3z^5}$ $\frac{x^3}{y^2z^3}$

90. $\left(\frac{b^7}{a^4}\right)^0$ 1

Practice Quiz 1 — Lessons 10-1 and 10-2

1. Determine whether $5(1.2)^x$ represents exponential *growth* or *decay*. *(Lesson 10-1)* **growth**

2. Write an exponential function whose graph passes through (0, 2) and (2, 32). $y = 2(4)^x$

3. Write an equivalent logarithmic equation for $4^6 = 4096$. *(Lesson 10-2)* $\log_4 4096 = 6$

4. Write an equivalent exponential equation for $\log_9 27 = \frac{3}{2}$. *(Lesson 10-2)* $9^{\frac{3}{2}} = 27$

Evaluate each expression. *(Lesson 10-2)*

5. $\log_8 16$ $\frac{4}{3}$

6. $\log_4 4^{15}$ 15

Solve each equation or inequality. Check your solutions. *(Lessons 10-1 and 10-2)*

7. $3^{4x} = 3^{3-x}$ $\frac{3}{5}$

8. $3^{2n} \leq \frac{1}{9}$ $n \leq -1$

9. $\log_2 (x + 6) > 5$ $x > 26$

10. $\log_5 (4x - 1) = \log_5 (3x + 2)$ 3

538 Chapter 10 Exponential and Logarithmic Relations

Graphing Calculator Investigation

A Follow-Up of Lesson 10-2

Modeling Real-World Data: Curve Fitting

We are often confronted with data for which we need to find an equation that best fits the information. We can find exponential and logarithmic functions of best fit using a TI-83 Plus graphing calculator.

Example

The population per square mile in the United States has changed dramatically over a period of years. The table shows the number of people per square mile for several years.

U.S. Population Density			
Year	People per square mile	Year	People per square mile
1790	4.5	1900	21.5
1800	6.1	1910	26.0
1810	4.3	1920	29.9
1820	5.5	1930	34.7
1830	7.4	1940	37.2
1840	9.8	1950	42.6
1850	7.9	1960	50.6
1860	10.6	1970	57.5
1870	10.9	1980	64.0
1880	14.2	1990	70.3
1890	17.8	2000	80.0

Source: Northeast-Midwest Institute

a. Use a graphing calculator to enter the data and draw a scatter plot that shows how the number of people per square mile is related to the year.

Step 1 Enter the year into L1 and the people per square mile into L2.

KEYSTROKES: *See pages 87 and 88 to review how to enter lists.*

Be sure to clear the Y= list. Use the ▶ key to move the cursor from L1 to L2.

Step 2 Draw the scatter plot.

KEYSTROKES: *See pages 87 and 88 to review how to graph a scatter plot.*

Make sure that Plot 1 is on, the scatter plot is chosen, Xlist is L1, and Ylist is L2. Use the viewing window [1780, 2020] with a scale factor of 10 by [0, 115] with a scale factor of 5.

We see from the graph that the equation that best fits the data is a curve. Based on the shape of the curve, try an exponential model.

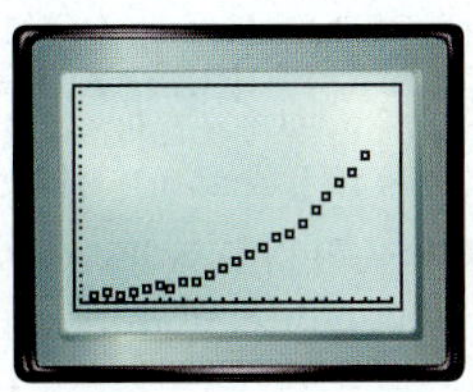

[1780, 2020] scl: 10 by [0, 115] scl: 5

Step 3 To determine the exponential equation that best fits the data, use the exponential regression feature of the calculator.

KEYSTROKES: STAT ▶ 0 2nd [L1] , 2nd [L2] ENTER

The equation is $y = 1.835122 \times 10^{-11}(1.014700091)^x$.

(continued on the next page)

www.algebra2.com/other_calculator_keystrokes

Graphing Calculator Investigation

A Follow-Up of Lesson 10-2

Getting Started

Turning Off Stat Plots Before Step 1, students should use the keystrokes 2nd [STAT PLOT] and check that both plot 2 and plot 3 are turned off.

Diagnostics Display Students should have the calculator set to DiagnosticOn. To set the calculator for diagnostics, use 2nd [CATALOG], move the cursor down to DiagnosticOn, and press ENTER twice.

Teach

- When students begin the exercises, they should clear lists L1 and L2. They should also enter appropriate settings for the graphing window.
- Point out that the table of data is arranged in two "double" columns.
- Suggest that students compare their graphs to the one shown.
- Have students estimate the population density in 2010 and 2050. How soon will the population density be twice what it was in 2000? **about 2040**
- If you have time, consider extending this activity into a discussion of how life in the future will be different as the result of the increasing population density. Ask students to think about the effect on transportation, housing, crime rates, and so on. You may wish to team-teach with a social studies teacher.

The calculator also reports an r value of 0.991887235. Recall that this number is a correlation coefficient that indicates how well the equation fits the data. A perfect fit would be $r = 1$. Therefore, we can conclude that this equation is a pretty good fit for the data.

To check this equation visually, overlap the graph of the equation with the scatter plot.

KEYSTROKES: [Y=] [VARS] 5 [▶] [▶] 1 [GRAPH]

The *residual* is the difference between actual and predicted data. The predicted population per square mile in 2000 using this model was 86.9 (To calculate, press [2nd] [CALC] 1 2000 [ENTER].) So the residual for 2000 was 80.0 − 86.9 or −6.9.

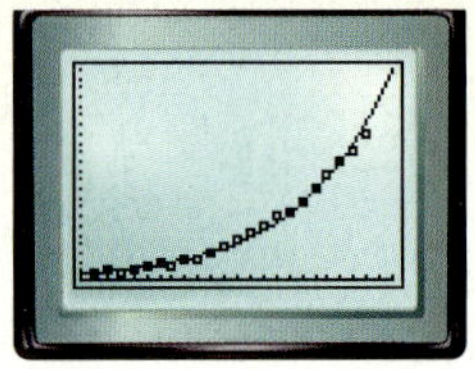

[1780, 2020] scl: 10 by [0, 115] scl: 5

b. If this trend continues, what will be the population per square mile in 2010?

To determine the population per square mile in 2010, from the graphics screen, find the value of y when $x = 2010$.

KEYSTROKES: [2nd] [CALC] 1 2010 [ENTER]

[1780, 2020] scl: 10 by [0, 115] scl: 5

The calculator returns a value of approximately 100.6. If this trend continues, in 2010, there will be approximately 100.6 people per square mile.

Exercises

In 1985, Erika received $30 from her aunt and uncle for her seventh birthday. Her father deposited it into a bank account for her. Both Erika and her father forgot about the money and made no further deposits or withdrawals. The table shows the account balance for several years.

Elapsed Time (years)	Balance
0	$30.00
5	$41.10
10	$56.31
15	$77.16
20	$105.71
25	$144.83
30	$198.43

1. Use a graphing calculator to draw a scatter plot for the data. **See margin.**
2. Calculate and graph the curve of best fit that shows how the elapsed time is related to the balance. Use ExpReg for this exercise. **See margin.**
3. Write the equation of best fit. $y = 29.99908551(1.06500135)^x$
4. Write a sentence that describes the fit of the graph to the data. **This equation is a good fit because $r \approx 1$.**
5. Based on the graph, estimate the balance in 41 years. Check this using the CALC value. **After 41 years she will have approximately $397.**
6. Do you think there are any other types of equations that would be good models for these data? Why or why not? **A quadratic equation might be a good model for this example because the shape is close to a portion of a parabola.**

Assess

In **Exercise 2**, make sure students can explain why their equation of best fit is a good choice. In **Exercise 3**, students' answers may vary slightly. When you discuss **Exercise 6**, you may want to ask for any ideas students have about how to use the calculator to judge the relative merits of various models (quadratic, cubic, quartic, and exponential).

Answers

1.

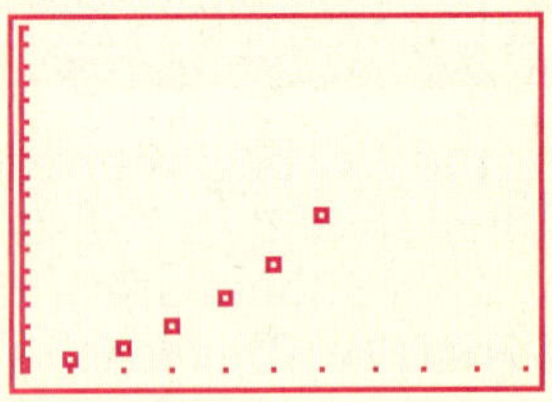

[0, 50] scl: 5 by [30, 400] scl: 20

2.

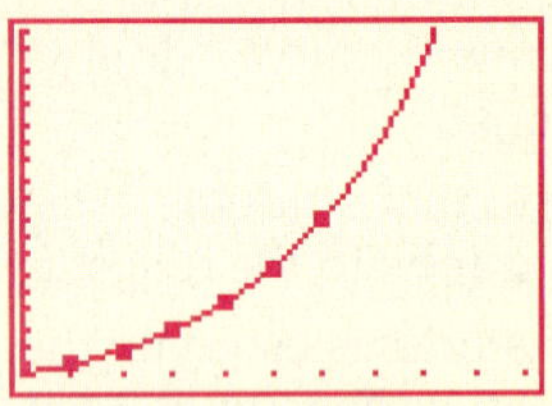

[0, 50] scl: 5 by [30, 400] scl: 20

540 Chapter 10 Exponential and Logarithmic Relations

10-3 Properties of Logarithms

What You'll Learn

- Simplify and evaluate expressions using the properties of logarithms.
- Solve logarithmic equations using the properties of logarithms.

How are the properties of exponents and logarithms related?

In Lesson 5-1, you learned that the product of powers is the sum of their exponents.

$$9 \cdot 81 = 3^2 \cdot 3^4 \text{ or } 3^{2+4}$$

In Lesson 10-2, you learned that logarithms *are* exponents, so you might expect that a similar property applies to logarithms. Let's consider a specific case. Does $\log_3 (9 \cdot 81) = \log_3 9 + \log_3 81$?

$\log_3 (9 \cdot 81) = \log_3 (3^2 \cdot 3^4)$ Replace 9 with 3^2 and 81 with 3^4.

$= \log_3 3^{(2+4)}$ Product of Powers

$= 2 + 4$ or 6 Inverse property of exponents and logarithms

$\log_3 9 + \log_3 81 = \log_3 3^2 + \log_3 3^4$ Replace 9 with 3^2 and 81 with 3^4.

$= 2 + 4$ or 6 Inverse property of exponents and logarithms

So, $\log_3 (9 \cdot 81) = \log_3 9 + \log_3 81$.

PROPERTIES OF LOGARITHMS Since logarithms are exponents, the properties of logarithms can be derived from the properties of exponents. The example above and other similar examples suggest the following property of logarithms.

Key Concept — Product Property of Logarithms

- **Words** The logarithm of a product is the sum of the logarithms of its factors.
- **Symbols** For all positive numbers m, n, and b, where $b \neq 1$, $\log_b mn = \log_b m + \log_b n$.
- **Example** $\log_3 (4)(7) = \log_3 4 + \log_3 7$

To show that this property is true, let $b^x = m$ and $b^y = n$. Then, using the definition of logarithm, $x = \log_b m$ and $y = \log_b n$.

$b^x b^y = mn$

$b^{x+y} = mn$ Product of Powers

$\log_b b^{x+y} = \log_b mn$ Property of Equality for Logarithmic Functions

$x + y = \log_b mn$ Inverse Property of Exponents and Logarithms

$\log_b m + \log_b n = \log_b mn$ Replace x with $\log_b m$ and y with $\log_b n$.

You can use the Product Property of Logarithms to approximate logarithmic expressions.

10-3 Lesson Notes

1 Focus

5-Minute Check Transparency 10-3 Use as a quiz or review of Lesson 10-2.

Mathematical Background notes are available for this lesson on p. 520D.

How are the properties of exponents and logarithms related?

Ask students:

- How do you know that logarithms are exponents? **Sample answer: The logarithm of a number is equal to the power (or exponent) when the number is rewritten in exponential form.**
- Since $\log_3 (9 \cdot 81) = \log_3 729$, how could $\log_3 729$ have been used in the justification that $\log_3 (9 \cdot 81) = \log_3 9 + \log_3 81$? **After stating that $\log_3 (9 \cdot 81) = \log_3 729$, then the statements $\log_3 729 = \log_3 3^6$ and $\log_3 3^6 = 6$ could be used to justify that $\log_3 (9 \cdot 81) = 6$.**

Resource Manager

Workbook and Reproducible Masters

Chapter 10 Resource Masters

- Study Guide and Intervention, pp. 585–586
- Skills Practice, p. 587
- Practice, p. 588
- Reading to Learn Mathematics, p. 589
- Enrichment, p. 590
- Assessment, pp. 623, 625

Transparencies

5-Minute Check Transparency 10-3
Answer Key Transparencies

Technology

Interactive Chalkboard

2 Teach

PROPERTIES OF LOGARITHMS

In-Class Examples

Power Point®

Teaching Tip When discussing the Product Property of Logarithms, point out that the logarithms used in the example $(\log_3 (4)(7) = \log_3 4 + \log_3 7)$ show the property applies to all logarithms and not just those that can be simplified. Be sure students did not get this impression from the earlier example where it was shown that $\log_3 (9 \cdot 81) = \log_3 9 + \log_3 81$.

1 Use $\log_5 2 \approx 0.4307$ to approximate the value of $\log_5 250$. **3.4307**

Teaching Tip Some students may wonder how the approximation for $\log_2 3$ was determined since on most calculators the log button calculates only logarithms of base 10. State that $\log_2 3 = \frac{\log_{10} 3}{\log_{10} 2}$, which can be evaluated using a calculator. Stress that this procedure will be formally discussed in Lesson 10-4.

2 Use $\log_6 8 \approx 1.1606$ and $\log_6 32 \approx 1.9343$ to approximate the value of $\log_6 4$. **0.7737**

3 **SOUND** The sound made by a lawnmower has a relative intensity of 10^9 or 90 decibels. Would the sound of ten lawnmowers running at that same intensity be ten times as loud or 900 decibels? Explain your reasoning. **No; the sound of ten lawnmowers is perceived to be only 10 decibels louder than the sound of one lawnmower, or 100 decibels.**

TEACHING TIP

The value of R in Example 3 is determined by finding the ratio of the intensity I of the sound in watts per square meter to the intensity I_0 of 10^{-12} watts per square meter. The intensity I_0 corresponds to the threshold of hearing. Thus a formula that relates the intensity of a sound in watts per square meter to its loudness in decibels is $L = 10 \log_{10} \frac{I}{I_0}$.

Career Choices

Sound Technician

Sound technicians produce movie sound tracks in motion picture production studios, control the sound of live events such as concerts, or record music in a recording studio.

Online Research For information about a career as a sound technician, visit: www.algebra2.com/careers

Example 1 Use the Product Property

Use $\log_2 3 \approx 1.5850$ to approximate the value of $\log_2 48$.

$\log_2 48 = \log_2 (2^4 \cdot 3)$ — Replace 48 with $16 \cdot 3$ or $2^4 \cdot 3$.

$= \log_2 2^4 + \log_2 3$ — Product Property

$= 4 + \log_2 3$ — Inverse Property of Exponents and Logarithms

$\approx 4 + 1.5850$ or 5.5850 — Replace $\log_2 3$ with 1.5850.

Thus, $\log_2 48$ is approximately 5.5850.

Recall that the quotient of powers is found by subtracting exponents. The property for the logarithm of a quotient is similar.

Key Concept — Quotient Property of Logarithms

- **Words** The logarithm of a quotient is the difference of the logarithms of the numerator and the denominator.
- **Symbols** For all positive numbers m, n, and b, where $b \neq 1$, $\log_b \frac{m}{n} = \log_b m - \log_b n$.

You will show that this property is true in Exercise 47.

Example 2 Use the Quotient Property

Use $\log_3 5 \approx 1.4650$ and $\log_3 20 \approx 2.7268$ to approximate $\log_3 4$.

$\log_3 4 = \log_3 \frac{20}{5}$ — Replace 4 with the quotient $\frac{20}{5}$.

$= \log_3 20 - \log_3 5$ — Quotient Property

$\approx 2.7268 - 1.4650$ or 1.2618 — $\log_3 20 = 2.7268$ and $\log_3 5 = 1.4650$

Thus, $\log_3 4$ is approximately 1.2618.

CHECK Using the definition of logarithm and a calculator, $3^{1.2618} \approx 4$. ✓

Example 3 Use Properties of Logarithms

SOUND **The loudness L of a sound in decibels is given by $L = 10 \log_{10} R$, where R is the sound's relative intensity. Suppose one person talks with a relative intensity of 10^6 or 60 decibels. Would the sound of ten people each talking at that same intensity be ten times as loud or 600 decibels? Explain your reasoning.**

Let L_1 be the loudness of one person talking. → $L_1 = 10 \log_{10} 10^6$

Let L_2 be the loudness of ten people talking. → $L_2 = 10 \log_{10} (10 \cdot 10^6)$

Then the increase in loudness is $L_2 - L_1$.

$L_2 - L_1 = 10 \log_{10} (10 \cdot 10^6) - 10 \log_{10} 10^6$ — Substitute for L_1 and L_2.

$= 10(\log_{10} 10 + \log_{10} 10^6) - 10 \log_{10} 10^6$ — Product Property

$= 10 \log_{10} 10 + 10 \log_{10} 10^6 - 10 \log_{10} 10^6$ — Distributive Property

$= 10 \log_{10} 10$ — Subtract.

$= 10(1)$ or 10 — Inverse Property of Exponents and Logarithms

The sound of ten people talking is perceived by the human ear to be only about 10 decibels louder than the sound of one person talking, or 70 decibels.

DAILY INTERVENTION — Unlocking Misconceptions

Power Property After you have discussed the Power Property of Logarithms on p. 543, clarify that the property works for logarithms because they are equivalent to exponents. Stress that students should not read a statement such as $\log_2 5^3 = 3 \log_2 5$ and conclude that $5^3 = 3 \times 5$.

Recall that the power of a power is found by multiplying exponents. The property for the logarithm of a power is similar.

Key Concept *Power Property of Logarithms*

- **Words** The logarithm of a power is the product of the logarithm and the exponent.
- **Symbols** For any real number p and positive numbers m and b, where $b \neq 1$, $\log_b m^p = p \log_b m$.

You will show that this property is true in Exercise 50.

Example 4 Power Property of Logarithms

Given $\log_4 6 \approx 1.2925$, approximate the value of $\log_4 36$.

$\log_4 36 = \log_4 6^2$ Replace 36 with 6^2.

$= 2 \log_4 6$ Power Property

$\approx 2(1.2925)$ or 2.585 Replace $\log_4 6$ with 1.2925.

SOLVE LOGARITHMIC EQUATIONS You can use the properties of logarithms to solve equations involving logarithms.

Example 5 Solve Equations Using Properties of Logarithms

Solve each equation.

a. $3 \log_5 x - \log_5 4 = \log_5 16$

$3 \log_5 x - \log_5 4 = \log_5 16$ Original equation

$\log_5 x^3 - \log_5 4 = \log_5 16$ Power Property

$\log_5 \frac{x^3}{4} = \log_5 16$ Quotient Property

$\frac{x^3}{4} = 16$ Property of Equality for Logarithmic Functions

$x^3 = 64$ Multiply each side by 4.

$x = 4$ Take the cube root of each side.

The solution is 4.

b. $\log_4 x + \log_4 (x - 6) = 2$

$\log_4 x + \log_4 (x - 6) = 2$ Original equation

$\log_4 x(x - 6) = 2$ Product Property

$x(x - 6) = 4^2$ Definition of logarithm

$x^2 - 6x - 16 = 0$ Subtract 16 from each side.

$(x - 8)(x + 2) = 0$ Factor.

$x - 8 = 0$ or $x + 2 = 0$ Zero Product Property

$x = 8$ $\quad x = -2$ Solve each equation.

CHECK Substitute each value into the original equation.

$\log_4 8 + \log_4 (8 - 6) \stackrel{?}{=} 2$

$\log_4 8 + \log_4 2 \stackrel{?}{=} 2$

$\log_4 (8 \cdot 2) \stackrel{?}{=} 2$

$\log_4 16 \stackrel{?}{=} 2$

$2 = 2$ ✓

$\log_4 (-2) + \log_4 (-2 - 6) \stackrel{?}{=} 2$

$\log_4 (-2) + \log_4 (-8) \stackrel{?}{=} 2$

Since $\log_4 (-2)$ and $\log_4 (-8)$ are undefined, -2 is an extraneous solution and must be eliminated.

The only solution is 8.

Study Tip

Checking Solutions

It is wise to check all solutions to see if they are valid since the domain of a logarithmic function is not the complete set of real numbers.

www.algebra2.com/extra_examples

In-Class Example Power Point®

4 Given $\log_5 6 \approx 1.1133$, approximate the value of $\log_5 216$. **3.3399**

SOLVE LOGARITHMIC EQUATIONS

In-Class Example Power Point®

5 Solve each equation.

a. $4 \log_2 x - \log_2 5 = \log_2 125$ **5**

b. $\log_8 x + \log_8 (x - 12) = 2$ **16**

DAILY INTERVENTION **Differentiated Instruction**

Interpersonal Right after discussing Example 5, have pairs of students rework both parts of the example together without looking at the solution in the text. Have the partners take turns explaining the solution steps to each other.

3 Practice/Apply

Study Notebook

Have students—
- add the definitions/examples of the vocabulary terms to their Vocabulary Builder worksheets for Chapter 10.
- summarize the properties of logarithms they learned in this lesson.
- include any other item(s) that they find helpful in mastering the skills in this lesson.

DAILY INTERVENTION **FIND THE ERROR** When discussing the error made by Clemente, remind students that logarithms are exponents. Adding $\log_7 6 + \log_7 3$ as $\log_7 (6 + 3)$ is similar to saying that $x^2 + x^3 = x^{2+3}$ or x^5, which students should recognize as being untrue because x^2 and x^3 are unlike terms.

About the Exercises...

Organization by Objective
- **Properties of Logarithms:** 13–20, 37–46
- **Solve Logarithmic Equations:** 21–34

Odd/Even Assignments
Exercises 13–34 are structured so that students practice the same concepts whether they are assigned odd or even problems.

Alert! Exercise 46 involves research on the Internet or other reference materials.

Assignment Guide

Basic: 13–17 odd, 21–31 odd, 35–40, 47–66

Average: 13–33 odd, 35–43, 47–66

Advanced: 14–34 even, 35, 36, 41–62 (optional: 63–66)

Check for Understanding

Concept Check

1. properties of exponents
2. Sample answer: $2\log_3 x + \log_3 5$; $\log_3 5x^2$

1. **Name** the properties that are used to derive the properties of logarithms.
2. **OPEN ENDED** Write an expression that can be simplified by using two or more properties of logarithms. Then simplify it.
3. **FIND THE ERROR** Umeko and Clemente are simplifying $\log_7 6 + \log_7 3 - \log_7 2$.

Umeko	Clemente
$\log_7 6 + \log_7 3 - \log_7 2$	$\log_7 6 + \log_7 3 - \log_7 2$
$= \log_7 18 - \log_7 2$	$= \log_7 9 - \log_7 2$
$= \log_7 9$	$= \log_7 7$ or 1

Who is correct? Explain your reasoning. **Umeko; see margin for explanation.**

Guided Practice

GUIDED PRACTICE KEY

Exercises	Examples
4–6	1, 2, 4
7–10	5
11, 12	3

Use $\log_3 2 \approx 0.6310$ and $\log_3 7 \approx 1.7712$ to approximate the value of each expression.

4. $\log_3 \frac{7}{2}$ **1.1402**
5. $\log_3 18$ **2.6310**
6. $\log_3 \frac{2}{3}$ **−0.3690**

Solve each equation. Check your solutions.

7. $\log_3 42 - \log_3 n = \log_3 7$ **6**
8. $\log_2 3x + \log_2 5 = \log_2 30$ **2**
9. $2\log_5 x = \log_5 9$ **3**
10. $\log_{10} a + \log_{10} (a + 21) = 2$ **4**

Application

MEDICINE **For Exercises 11 and 12, use the following information.**
The pH of a person's blood is given by $\text{pH} = 6.1 + \log_{10} B - \log_{10} C$, where B is the concentration of bicarbonate, which is a base, in the blood and C is the concentration of carbonic acid in the blood. **11. $\text{pH} = 6.1 + \log_{10} \frac{B}{C}$**

11. Use the Quotient Property of Logarithms to simplify the formula for blood pH.
12. Most people have a blood pH of 7.4. What is the approximate ratio of bicarbonate to carbonic acid for blood with this pH? **20:1**

★ indicates increased difficulty

Practice and Apply

Homework Help

For Exercises	See Examples
13–20	1, 2, 4
21–34	5
37–45	3

Extra Practice See page 850.

Use $\log_5 2 \approx 0.4307$ and $\log_5 3 \approx 0.6826$ to approximate the value of each expression.

13. $\log_5 9$ **1.3652**
14. $\log_5 8$ **1.2921**
15. $\log_5 \frac{2}{3}$ **−0.2519**
16. $\log_5 \frac{3}{2}$ **0.2519**
17. $\log_5 50$ **2.4307**
18. $\log_5 30$ **2.1133**
★19. $\log_5 0.5$ **−0.4307**
★20. $\log_5 \frac{10}{9}$ **0.0655**

Solve each equation. Check your solutions.

21. $\log_3 5 + \log_3 x = \log_3 10$ **2**
22. $\log_4 a + \log_4 9 = \log_4 27$ **3**
23. $\log_{10} 16 - \log_{10} 2t = \log_{10} 2$ **4**
24. $\log_7 24 - \log_7 (y + 5) = \log_7 8$ **−2**
25. $\log_2 n = \frac{1}{4}\log_2 16 + \frac{1}{2}\log_2 49$ **14**
26. $2\log_{10} 6 - \frac{1}{3}\log_{10} 27 = \log_{10} x$ **12**
27. $\log_{10} z + \log_{10} (z + 3) = 1$ **2**
28. $\log_6 (a^2 + 2) + \log_6 2 = 2$ **±4**
29. $\log_2 (12b - 21) - \log_2 (b^2 - 3) = 2$ **∅**
30. $\log_2 (y + 2) - \log_2 (y - 2) = 1$ **6**
31. $\log_3 0.1 + 2\log_3 x = \log_3 2 + \log_3 5$ **10**
32. $\log_5 64 - \log_5 \frac{8}{3} + \log_5 2 = \log_5 4p$ **12**

Answer

3. Clemente incorrectly applied the product and quotient properties of logarithms.
$\log_7 6 + \log_7 3 = \log_7 (6 \cdot 3)$ or $\log_7 18$ Product Property of Logarithms
$\log_7 18 - \log_7 2 = \log_7 (18 \div 2)$ or $\log_7 9$ Quotient Property of Logarithms

544 Chapter 10 Exponential and Logarithmic Relations

35. False; $\log_2 (2^2 + 2^3) = \log_2 12$, $\log_2 2^2 + \log_2 2^3 = 2 + 3$ or 5, and $\log_2 12 \neq 5$ since $2^5 \neq 12$.

Solve for n. $34.\ \frac{1}{2}(x - 1)$

★ 33. $\log_a 4n - 2 \log_a x = \log_a x$ $\frac{x^3}{4}$

★ 34. $\log_b 8 + 3 \log_b n = 3 \log_b (x - 1)$

CRITICAL THINKING **Tell whether each statement is *true* or *false*. If true, show that it is true. If false, give a counterexample.**

35. For all positive numbers m, n, and b, where $b \neq 1$, $\log_b (m + n) = \log_b m + \log_b n$.

36. For all positive numbers m, n, x, and b, where $b \neq 1$, $n \log_b x + m \log_b x = (n + m) \log_b x$. See pp. 573A–573D.

37. **EARTHQUAKES** The great Alaskan earthquake in 1964 was about 100 times more intense than the Loma Prieta earthquake in San Francisco in 1989. Find the difference in the Richter scale magnitudes of the earthquakes. 2

39. about 0.4214 kilocalorie per gram

40. about 0.8429 kilocalorie per gram

BIOLOGY **For Exercises 38–40, use the following information.**
The energy E (in kilocalories per gram molecule) needed to transport a substance from the outside to the inside of a living cell is given by $E = 1.4(\log_{10} C_2 - \log_{10} C_1)$, where C_1 is the concentration of the substance outside the cell and C_2 is the concentration inside the cell.

38. Express the value of E as one logarithm. $E = 1.4 \log \frac{C_2}{C_1}$

39. Suppose the concentration of a substance inside the cell is twice the concentration outside the cell. How much energy is needed to transport the substance on the outside of the cell to the inside? (Use $\log_{10} 2 \approx 0.3010$.)

40. Suppose the concentration of a substance inside the cell is four times the concentration outside the cell. How much energy is needed to transport the substance from the outside of the cell to the inside?

More About. . .

Star Light

The Greek astronomer Hipparchus made the first known catalog of stars. He listed the brightness of each star on a scale of 1 to 6, the brightest being 1. With no telescope, he could only see stars as dim as the 6th magnitude.

Source: NASA

SOUND **For Exercises 41–43, use the formula for the loudness of sound in Example 3 on page 542. Use $\log_{10} 2 \approx 0.3010$ and $\log_{10} 3 \approx 0.47712$.**

41. A certain sound has a relative intensity of R. By how many decibels does the sound increase when the intensity is doubled? 3

42. A certain sound has a relative intensity of R. By how many decibels does the sound decrease when the intensity is halved? 3

★ 43. A stadium containing 10,000 cheering people can produce a crowd noise of about 90 decibels. If every one cheers with the same relative intensity, how much noise, in decibels, is a crowd of 30,000 people capable of producing? Explain your reasoning. About 95 decibels; see pp. 573A–573D for explanation.

STAR LIGHT **For Exercises 44–46, use the following information.**
The brightness, or apparent magnitude, m of a star or planet is given by the formula $m = 6 - 2.5 \log_{10} \frac{L}{L_0}$, where L is the amount of light coming to Earth from the star or planet and L_0 is the amount of light from a sixth magnitude star.

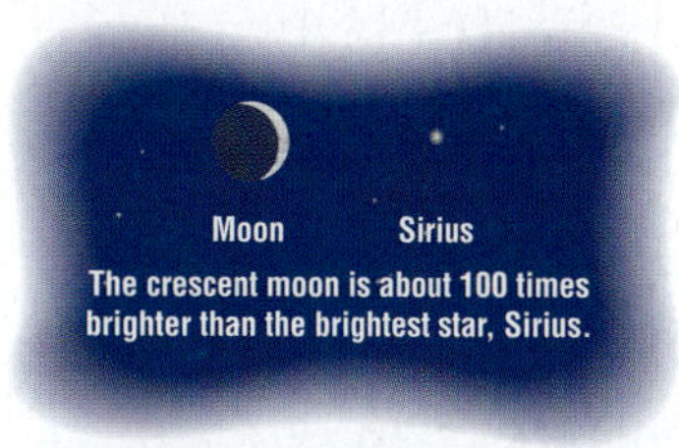

The crescent moon is about 100 times brighter than the brightest star, Sirius.

★ 44. Find the difference in the magnitudes of Sirius and the crescent moon. 5

★ 45. Find the difference in the magnitudes of Saturn and Neptune. 7.5

46. **RESEARCH** Use the Internet or other reference to find the magnitude of the dimmest stars that we can now see with ground-based telescopes. about 22

Saturn, as seen from Earth, is 1000 times brighter than Neptune.

www.algebra2.com/self_check_quiz

Study Guide and Intervention, p. 585 (shown) and p. 586

Properties of Logarithms Properties of exponents can be used to develop the following properties of logarithms.

Product Property of Logarithms	For all positive numbers m, n, and b, where $b \neq 1$, $\log_b mn = \log_b m + \log_b n$.
Quotient Property of Logarithms	For all positive numbers m, n, and b, where $b \neq 1$, $\log_b \frac{m}{n} = \log_b m - \log_b n$.
Power Property of Logarithms	For any real number p and positive numbers m and b, where $b \neq 1$, $\log_b m^p = p \log_b m$.

Example **Use $\log_3 28 \approx 3.0331$ and $\log_3 4 \approx 1.2619$ to approximate the value of each expression.**

a. $\log_3 36$

$\log_3 36 = \log_3 (3^2 \cdot 4) = \log_3 3^2 + \log_3 4 = 2 + \log_3 4 \approx 2 + 1.2619 \approx 3.2619$

b. $\log_3 7$

$\log_3 7 = \log_3 \left(\frac{28}{4}\right) = \log_3 28 - \log_3 4 \approx 3.0331 - 1.2619 \approx 1.7712$

c. $\log_3 256$

$\log_3 256 = \log_3 (4^4) = 4 \cdot \log_3 4 \approx 4(1.2619) \approx 5.0476$

Exercises

Use $\log_{12} 3 \approx 0.4421$ and $\log_{12} 7 \approx 0.7831$ to evaluate each expression.

1. $\log_{12} 21$ 1.2252
2. $\log_{12} \frac{7}{3}$ 0.3410
3. $\log_{12} 49$ 1.5662
4. $\log_{12} 36$ 1.4421
5. $\log_{12} 63$ 1.6673
6. $\log_{12} \frac{27}{49}$ −0.2399
7. $\log_{12} \frac{81}{49}$ 0.2022
8. $\log_{12} 16{,}807$ 3.9155
9. $\log_{12} 441$ 2.4504

Use $\log_5 3 \approx 0.6826$ and $\log_5 4 \approx 0.8614$ to evaluate each expression.

10. $\log_5 12$ 1.5440
11. $\log_5 100$ 2.8614
12. $\log_5 0.75$ −0.1788
13. $\log_5 144$ 3.0880
14. $\log_5 \frac{27}{16}$ 0.3250
15. $\log_5 375$ 3.6826
16. $\log_5 1.\overline{3}$ 0.1788
17. $\log_5 \frac{9}{16}$ −0.3576
18. $\log_5 \frac{81}{5}$ 1.7304

Skills Practice, p. 587 and Practice, p. 588 (shown)

Use $\log_{10} 5 \approx 0.6990$ and $\log_{10} 7 \approx 0.8451$ to approximate the value of each expression.

1. $\log_{10} 35$ 1.5441
2. $\log_{10} 25$ 1.3980
3. $\log_{10} \frac{7}{5}$ 0.1461
4. $\log_{10} \frac{5}{7}$ −0.1461
5. $\log_{10} 245$ 2.3892
6. $\log_{10} 175$ 2.2431
7. $\log_{10} 0.2$ −0.6990
8. $\log_{10} \frac{25}{7}$ 0.5529

Solve each equation. Check your solutions.

9. $\log_7 n = \frac{2}{3} \log_7 8$ 4
10. $\log_{10} u = \frac{3}{2} \log_{10} 4$ 8
11. $\log_6 x + \log_6 9 = \log_6 54$ 6
12. $\log_8 48 - \log_8 w = \log_8 4$ 12
13. $\log_9 (3u + 14) - \log_9 5 = \log_9 2u$ 2
14. $4 \log_2 x + \log_2 5 = \log_2 405$ 3
15. $\log_3 y = -\log_3 16 + \frac{1}{3} \log_3 64$ $\frac{1}{4}$
16. $\log_2 d = 5 \log_2 2 - \log_2 8$ 4
17. $\log_{10} (3m - 5) + \log_{10} m = \log_{10} 2$ 2
18. $\log_{10} (b + 3) + \log_{10} b = \log_{10} 4$ 1
19. $\log_8 (t + 10) - \log_8 (t - 1) = \log_8 12$ 2
20. $\log_3 (a + 3) + \log_3 (a + 2) = \log_3 6$ 0
21. $\log_{10} (r + 4) - \log_{10} r = \log_{10} (r + 1)$ 2
22. $\log_4 (x^2 - 4) - \log_4 (x + 2) = \log_4 1$ 3
23. $\log_{10} 4 + \log_{10} w = 2$ 25
24. $\log_8 (n - 3) + \log_8 (n + 4) = 1$ 4
25. $3 \log_5 (x^2 + 9) - 6 = 0$ ±4
26. $\log_{16} (9x + 5) - \log_{16} (x^2 - 1) = \frac{1}{2}$ 3
27. $\log_6 (2x - 5) + 1 = \log_6 (7x + 10)$ 8
28. $\log_2 (5y + 2) - 1 = \log_2 (1 - 2y)$ 0
29. $\log_{10} (c^2 - 1) - 2 = \log_{10} (c + 1)$ 101
30. $\log_7 x + 2 \log_7 x - \log_7 3 = \log_7 72$ 6

31. **SOUND** The loudness L of a sound in decibels is given by $L = 10 \log_{10} R$, where R is the sound's relative intensity. If the intensity of a certain sound is tripled, by how many decibels does the sound increase? about 4.8 db

32. **EARTHQUAKES** An earthquake rated at 3.5 on the Richter scale is felt by many people, and an earthquake rated at 4.5 may cause local damage. The Richter scale magnitude reading m is given by $m = \log_{10} x$, where x represents the amplitude of the seismic wave causing ground motion. How many times greater is the amplitude of an earthquake that measures 4.5 on the Richter scale than one that measures 3.5? 10 times

Reading to Learn Mathematics, p. 589 ELL

Pre-Activity **How are the properties of exponents and logarithms related?**

Read the introduction to Lesson 10-3 at the top of page 541 in your textbook.

Find the value of $\log_5 125$. 3 Find the value of $\log_5 5$. 1

Find the value of $\log_5 (125 \div 5)$. 2

Which of the following statements is true? B

A. $\log_5 (125 \div 5) = (\log_5 125) \div (\log_5 5)$

B. $\log_5 (125 \div 5) = \log_5 125 - \log_5 5$

Reading the Lesson

1. Each of the properties of logarithms can be stated in words or in symbols. Complete the statements of these properties in words.
   a. The logarithm of a quotient is the difference of the logarithms of the numerator and the denominator.
   b. The logarithm of a power is the product of the logarithm of the base and the exponent.
   c. The logarithm of a product is the sum of the logarithms of its factors.

2. State whether each of the following equations is *true* or *false*. If the statement is true, name the property of logarithms that is illustrated.
   a. $\log_3 10 = \log_3 30 - \log_3 3$ true; Quotient Property
   b. $\log_4 12 = \log_4 4 + \log_4 8$ false
   c. $\log_2 81 = 2 \log_2 9$ true; Power Property
   d. $\log_8 30 = \log_8 5 \cdot \log_8 6$ false

3. The algebraic process of solving the equation $\log_2 x + \log_2 (x + 2) = 3$ leads to "$x = -4$ or $x = 2$." Does this mean that both −4 and 2 are solutions of the logarithmic equation? Explain your reasoning. Sample answer: No; 2 is a solution because it checks: $\log_2 2 + \log_2 (2 + 2) = \log_2 2 + \log_2 4 = 1 + 2 = 3$. However, because $\log_2 (-4)$ and $\log_2 (-2)$ are undefined, −4 is an extraneous solution and must be eliminated. The only solution is 2.

Helping You Remember

4. A good way to remember something is to relate it something you already know. Use words to explain how the Product Property for exponents can help you remember the product property for logarithms. Sample answer: When you *multiply* two numbers or expressions with the same base, you *add* the exponents and keep the *same base*. Logarithms are exponents, so to find the logarithm of a *product*, you *add* the logarithms of the factors, keeping the *same base*.

Answer

43. $L = 10 \log_{10} R$, where L is the loudness of the sound in decibels and R is the relative intensity of the sound. Since the crowd increased by a factor of 3, we assume that the intensity also increases by a factor of 3. Thus, we need to find the loudness of $3R$.
$L = 10 \log_{10} 3R$; $L = 10(\log_{10} 3 + \log_{10} R)$
$L = 10 \log_{10} 3 + 10 \log_{10} R$;
$L \approx 10(0.4771) + 90$; $L \approx 4.771 + 90$ or about 95

Enrichment, p. 590

Spirals

Consider an angle in standard position with its vertex at a point O called the pole. Its initial side is on a coordinatized axis called the *polar axis*. A point P on the terminal side of the angle is named by the *polar coordinates* (r, θ), where r is the directed distance of the point from O and θ is the measure of the angle. Graphs in this system may be drawn on polar coordinate paper such as the kind shown below.

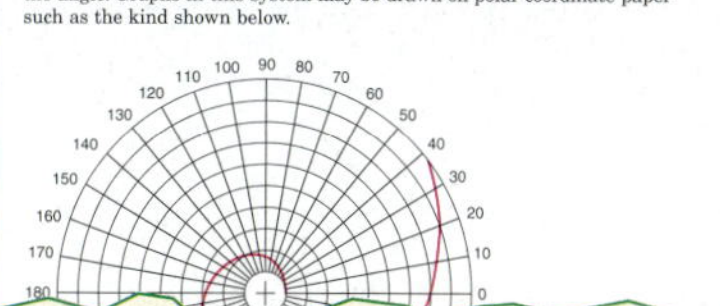

4 Assess

Open-Ended Assessment

Speaking Ask students to explain the Product Property, Quotient Property, and Power Property of Logarithms in their own words. Encourage them to use specific examples for clarification.

Getting Ready for Lesson 10-4

PREREQUISITE SKILL Students will use common logarithms to solve exponential equations and inequalities in Lesson 10-4. The solution techniques involve using the skills they learned when solving logarithmic equations and inequalities. Use Exercises 63–66 to determine your students' familiarity with solving logarithmic equations and inequalities.

Assessment Options

Quiz (Lesson 10-3) is available on p. 623 of the *Chapter 10 Resource Masters*.

Mid-Chapter Test (Lessons 10-1 through 10-3) is available on p. 625 of the *Chapter 10 Resource Masters*.

47. **CRITICAL THINKING** Use the properties of exponents to prove the Quotient Property of Logarithms. **See margin.**

48. **WRITING IN MATH** Answer the question that was posed at the beginning of the lesson. **See pp. 573A–573D.**

How are the properties of exponents and logarithms related?

Include the following in your answer:

- examples like the one shown at the beginning of the lesson illustrating the Quotient Property and Power Property of Logarithms, and
- an explanation of the similarity between one property of exponents and its related property of logarithms.

Standardized Test Practice

49. Simplify $2 \log_5 12 - \log_5 8 - 2 \log_5 3$. **A**

Ⓐ $\log_5 2$ Ⓑ $\log_5 3$ Ⓒ $\log_5 0.5$ Ⓓ 1

50. **SHORT RESPONSE** Show that $\log_b m^p = p \log_b m$ for any real number p and positive number m and b, where $b \neq 1$. **See margin.**

Maintain Your Skills

Mixed Review

Evaluate each expression. *(Lesson 10-2)*

51. $\log_3 81$ **4**
52. $\log_9 \frac{1}{729}$ **−3**
53. $\log_7 7^{2x}$ **2x**

Solve each equation or inequality. Check your solutions. *(Lesson 10-1)*

54. $3^{5n+3} = 3^{33}$ **6**
55. $7^a = 49^{-4}$ **−8**
56. $3^{d+4} > 9^d$ **$d < 4$**

Determine whether each graph represents an odd-degree polynomial function or an even-degree polynomial function. Then state how many real zeros each function has. *(Lesson 7-1)*

57. **odd; 3**

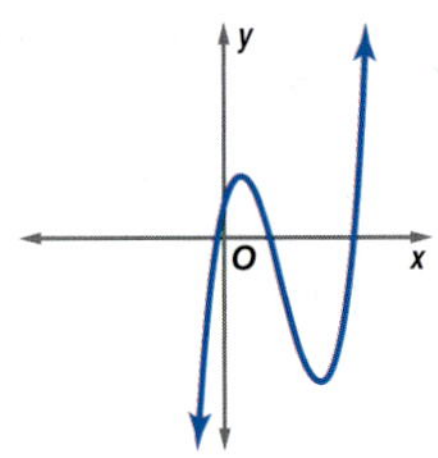

58. **even; 4**

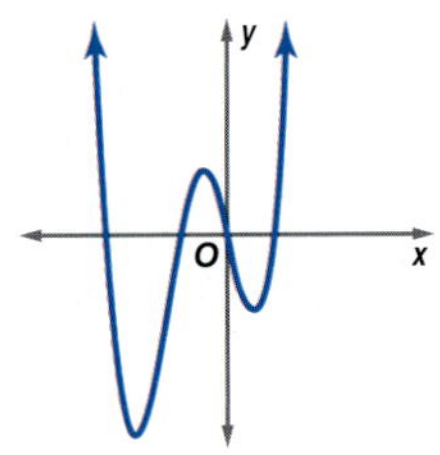

Simplify each expression. *(Lesson 9-1)*

59. $\frac{39a^3b^4}{13a^4b^3}$ $\mathbf{\frac{3b}{a}}$
60. $\frac{k+3}{5kl} \cdot \frac{10kl}{k+3}$ **2**
61. $\frac{5y - 15z}{42x^2} \div \frac{y - 3z}{14x}$ $\mathbf{\frac{5}{3x}}$

62. **PHYSICS** If a stone is dropped from a cliff, the equation $t = \frac{1}{4}\sqrt{d}$ represents the time t in seconds that it takes for the stone to reach the ground. If d represents the distance in feet that the stone falls, find how long it would take for a stone to fall from a 150-foot cliff. *(Lesson 5-6)* **3.06 s**

Getting Ready for the Next Lesson

PREREQUISITE SKILL Solve each equation or inequality. Check your solutions.
*(To review **solving logarithmic equations and inequalities**, see Lesson 10-2.)*

63. $\log_3 x = \log_3 (2x - 1)$ **1**
64. $\log_{10} 2^x = \log_{10} 32$ **5**
65. $\log_2 3x > \log_2 5$ $\mathbf{x > \frac{5}{3}}$
66. $\log_5 (4x + 3) < \log_5 11$ $\mathbf{-\frac{3}{4} < x < 2}$

Answers

47. Let $b^x = m$ and $b^y = n$. Then $\log_b m = x$ and $\log_b n = y$.

$$\frac{b^x}{b^y} = \frac{m}{n}$$

$$b^{x-y} = \frac{m}{n} \quad \text{Quotient Property}$$

$$\log_b b^{x-y} = \log_b \frac{m}{n} \quad \text{Property of Equality for Logarithmic Equations}$$

$$x - y = \log_b \frac{m}{n} \quad \text{Inverse Property of Exponents and Logarithms}$$

$$\log_b m - \log_b n = \log_b \frac{m}{n} \quad \text{Replace } x \text{ with } \log_b m \text{ and } y \text{ with } \log_b n.$$

50. Let $b^x = m$, then $\log_b m = x$.

$$(b^x)^p = m^p$$

$$b^{xp} = m^p \quad \text{Product of Powers}$$

$$\log_b b^{xp} = \log_b m^p \quad \text{Property of Equality for Logarithmic Equations}$$

$$xp = \log_b m^p \quad \text{Inverse Property of Exponents and Logarithms}$$

$$p \log_b m = \log_b m^p \quad \text{Replace } x \text{ with } \log_b m.$$

546 Chapter 10 Exponential and Logarithmic Relations

10-4 Common Logarithms

What You'll Learn

- Solve exponential equations and inequalities using common logarithms.
- Evaluate logarithmic expressions using the Change of Base Formula.

Vocabulary

- common logarithm
- Change of Base Formula

Why is a logarithmic scale used to measure acidity?

The pH level of a substance measures its acidity. A low pH indicates an acid solution while a high pH indicates a basic solution. The pH levels of some common substances are shown.

Acidity of Common Substances

Substance	pH Level
Battery acid	1.0
Sauerkraut	3.5
Tomatoes	4.2
Black Coffee	5.0
Milk	6.4
Distilled Water	7.0
Eggs	7.8
Milk of magnesia	10.0

The pH level of a substance is given by pH $= -\log_{10}[H+]$, where $H+$ is the substance's hydrogen ion concentration in moles per liter. Another way of writing this formula is pH $= -\log[H+]$.

COMMON LOGARITHMS You have seen that the base 10 logarithm function, $y = \log_{10} x$, is used in many applications. Base 10 logarithms are called **common logarithms**. Common logarithms are usually written without the subscript 10.

$$\log_{10} x = \log x, x > 0$$

Most calculators have a LOG key for evaluating common logarithms.

Study Tip

Technology
Nongraphing scientific calculators often require entering the number followed by the function, for example, 3 LOG.

Example 1 Find Common Logarithms

Use a calculator to evaluate each expression to four decimal places.

a. **log 3** KEYSTROKES: LOG 3 ENTER .4771212547 about 0.4771

b. **log 0.2** KEYSTROKES: LOG 0.2 ENTER −.6989700043 about −0.6990

Sometimes an application of logarithms requires that you use the inverse of logarithms, or exponentiation.

$$10^{\log x} = x$$

Example 2 Solve Logarithmic Equations Using Exponentiation

EARTHQUAKES **The amount of energy E, in ergs, that an earthquake releases is related to its Richter scale magnitude M by the equation $\log E = 11.8 + 1.5M$. The Chilean earthquake of 1960 measured 8.5 on the Richter scale. How much energy was released?**

$\log E = 11.8 + 1.5M$	Write the formula.
$\log E = 11.8 + 1.5(8.5)$	Replace M with 8.5.
$\log E = 24.55$	Simplify.
$10^{\log E} = 10^{24.55}$	Write each side using exponents and base 10.
$E = 10^{24.55}$	Inverse Property of Exponents and Logarithms
$E \approx 3.55 \times 10^{24}$	Use a calculator.

The amount of energy released by this earthquake was about 3.55×10^{24} ergs.

10-4 Lesson Notes

1 Focus

5-Minute Check Transparency 10-4 Use as a quiz or review of Lesson 10-3.

Mathematical Background notes are available for this lesson on p. 520D.

Why is a logarithmic scale used to measure acidity?

Ask students:

- What does pH level measure? the acidity of a substance
- Where have you heard of pH levels? Sample answer: in soap and shampoo commercials
- Distilled water has a neutral pH. That is, it is neither acidic nor basic. Use the chart to determine the pH level for a neutral substance. 7.0

2 Teach

COMMON LOGARITHMS

Teaching Tip Stress that when the base of a logarithm is not shown, the base is assumed to be 10.

Resource Manager

Workbook and Reproducible Masters

Chapter 10 Resource Masters

- Study Guide and Intervention, pp. 591–592
- Skills Practice, p. 593
- Practice, p. 594
- Reading to Learn Mathematics, p. 595
- Enrichment, p. 596

Transparencies

5-Minute Check Transparency 10-4
Real-World Transparency 10
Answer Key Transparencies

Technology

Interactive Chalkboard

In-Class Examples

1 Use a calculator to evaluate each expression to four decimal places.

a. log 6 **about 0.7782**

b. log 0.35 **about −0.4559**

2 **EARTHQUAKE** Refer to Example 2. The San Fernando Valley earthquake of 1994 measured 6.6 on the Richter scale. How much energy did this earthquake release? **about 5.01×10^{21} ergs**

Teaching Tip After discussing In-Class Example 2, have students compare the Richter scale magnitudes of the Chilean and San Fernando Valley earthquakes. **$8.5 \div 6.6 \approx 1.29$; the Chilean magnitude was about 29% greater.** Then have them compare the energy released by the Chilean earthquake to the energy released by the San Fernando Valley earthquake. **$3.55 \times 10^{24} \div 5.01 \times 10^{21} \approx 708.58$; the Chilean earthquake released more than 6 times as much energy.** Point out that these results demonstrate the nonlinear nature of the equation that models the amount of energy released.

3 Solve $5^x = 62$. **about 2.5643**

4 Solve $2^{7x} > 3^{5x-3}$. **$\{x \mid x < 5.1415\}$**

Study Tip

Using Logarithms
When you use the Property for Logarithmic Functions as in the second step of Example 3, this is sometimes referred to as *taking the logarithm of each side.*

Example 3 Solve Exponential Equations Using Logarithms

Solve $3^x = 11$.

$3^x = 11$ — Original equation

$\log 3^x = \log 11$ — Property of Equality for Logarithmic Functions

$x \log 3 = \log 11$ — Power Property of Logarithms

$x = \dfrac{\log 11}{\log 3}$ — Divide each side by log 3.

$x \approx \dfrac{1.0414}{0.4771}$ — Use a calculator.

$x \approx 2.1828$ — The solution is approximately 2.1828.

CHECK You can check this answer using a calculator or by using estimation. Since $3^2 = 9$ and $3^3 = 27$, the value of x is between 2 and 3. In addition, the value of x should be closer to 2 than 3, since 11 is closer to 9 than 27. Thus, 2.1828 is a reasonable solution. ✓

Example 4 Solve Exponential Inequalities Using Logarithms

Solve $5^{3y} < 8^{y-1}$.

$5^{3y} < 8^{y-1}$ — Original inequality

$\log 5^{3y} < \log 8^{y-1}$ — Property of Inequality for Logarithmic Functions

$3y \log 5 < (y-1) \log 8$ — Power Property of Logarithms

$3y \log 5 < y \log 8 - \log 8$ — Distributive Property

$3y \log 5 - y \log 8 < -\log 8$ — Subtract $y \log 8$ from each side.

$y(3 \log 5 - \log 8) < -\log 8$ — Distributive Property

$y < \dfrac{-\log 8}{3 \log 5 - \log 8}$ — Divide each side by $3 \log 5 - \log 8$.

$y < \dfrac{-(0.9031)}{3(0.6990) - 0.9031}$ — Use a calculator.

$y < -0.7564$ — The solution set is $\{y \mid y < -0.7564\}$.

CHECK Test $y = -1$.

$5^{3y} < 8^{y-1}$ — Original inequality

$5^{3(-1)} < 8^{(-1)-1}$ — Replace y with 1.

$5^{-3} < 8^{-2}$ — Simplify.

$\dfrac{1}{125} < \dfrac{1}{64}$ ✓ — Negative Exponent Property

CHANGE OF BASE FORMULA The **Change of Base Formula** allows you to write equivalent logarithmic expressions that have different bases.

Key Concept — Change of Base Formula

- **Symbols** For all positive numbers, a, b and n, where $a \neq 1$ and $b \neq 1$,

  $\log_a n = \dfrac{\log_b n}{\log_b a}$ ← log base b of original number / ← log base b of old base

- **Example** $\log_5 12 = \dfrac{\log_{10} 12}{\log_{10} 5}$

DAILY INTERVENTION — Unlocking Misconceptions

Change of Base As you discuss the Change of Base Formula, point out that the base b that students are changing to does not have to be 10. Any base could be used; however, b is most commonly 10 because this allows for the logarithms to be evaluated with a calculator.

To prove this formula, let $\log_a n = x$.

$a^x = n$	Definition of logarithm
$\log_b a^x = \log_b n$	Property of Equality for Logarithms
$x \log_b a = \log_b n$	Power Property of Logarithms
$x = \dfrac{\log_b n}{\log_b a}$	Divide each side by $\log_b a$.
$\log_a n = \dfrac{\log_b n}{\log_b a}$	Replace x with $\log_a n$.

This formula makes it possible to evaluate a logarithmic expression of any base by translating the expression into one that involves common logarithms.

Example 5 Change of Base Formula

Express $\log_4 25$ in terms of common logarithms. Then approximate its value to four decimal places.

$\log_4 25 = \dfrac{\log_{10} 25}{\log_{10} 4}$ Change of Base Formula

≈ 2.3219 Use a calculator.

The value of $\log_4 25$ is approximately 2.3219.

Check for Understanding

Concept Check

1. **Name** the base used by the calculator LOG key. What are these logarithms called? **10; common logarithms**
2. **OPEN ENDED** Give an example of an exponential equation requiring the use of logarithms to solve. Then solve your equation.
3. **Explain** why you must use the Change of Base Formula to find the value of $\log_2 7$ on a calculator. **A calculator is not programmed to find base 2 logarithms.**

2. Sample answer: $5^x = 2$; $x \approx 0.4307$

Guided Practice

GUIDED PRACTICE KEY

Exercises	Examples
4–6	1
7–12	3, 4
13–15	5
16	2

Use a calculator to evaluate each expression to four decimal places.

4. $\log 4$ **0.6021**
5. $\log 23$ **1.3617**
6. $\log 0.5$ **−0.3010**

Solve each equation or inequality. Round to four decimal places. **12. $\{p \mid p \leq 4.8188\}$**

7. $9^x = 45$ **1.7325**
8. $4^{5n} > 30$ **$\{n \mid n > 0.4907\}$**
9. $3.1^{a-3} = 9.42$ **4.9824**
10. $11^{x^2} = 25.4$ **±1.1615**
11. $7^{t-2} = 5^t$ **11.5665**
12. $4^{p-1} \leq 3^p$

Express each logarithm in terms of common logarithms. Then approximate its value to four decimal places.

13. $\log_7 5$ **$\dfrac{\log 5}{\log 7}$; 0.8271**
14. $\log_3 42$ **$\dfrac{\log 42}{\log 3}$; 3.4022**
15. $\log_2 9$ **$\dfrac{\log 9}{\log 2}$; 3.1699**

Application

16. **DIET** Sandra's doctor has told her to avoid foods with a pH that is less than 4.5. What is the hydrogen ion concentration of foods Sandra is allowed to eat? Use the information at the beginning of the lesson. **at least 0.00003 mole per liter**

★ **indicates increased difficulty**

Practice and Apply

Use a calculator to evaluate each expression to four decimal places.

17. $\log 5$ **0.6990**
18. $\log 12$ **1.0792**
19. $\log 7.2$ **0.8573**
20. $\log 2.3$ **0.3617**
21. $\log 0.8$ **−0.0969**
22. $\log 0.03$ **−1.5229**

www.algebra2.com/extra_examples

CHANGE OF BASE FORMULA

In-Class Example

Power Point®

5 Express $\log_3 18$ in terms of common logarithms. Then approximate its value to four decimal places. **$\log_3 18 = \dfrac{\log_{10} 18}{\log_{10} 3}$; $\log_3 18 \approx 2.6309$**

3 Practice/Apply

Study Notebook

Have students—

- add the definitions/examples of the vocabulary terms to their Vocabulary Builder worksheets for Chapter 10.
- include an example of an exponential inequality that they solved, and an example showing how to use the Change of Base Formula.
- include any other item(s) that they find helpful in mastering the skills in this lesson.

About the Exercises...

Organization by Objective

- **Common Logarithms:** 17–44, 51–55
- **Change of Base Formula:** 45–50

Odd/Even Assignments

Exercises 17–52 are structured so that students practice the same concepts whether they are assigned odd or even problems.

Assignment Guide

Basic: 17–41 odd, 45, 47, 51, 56, 59–77

Average: 17–51 odd, 56–77

Advanced: 18–52 even, 53–74 (optional: 72–77)

DAILY INTERVENTION

Differentiated Instruction

Naturalist Have interested students research earthquakes and their Richter scale measurements. Have them calculate the amount of energy released by three earthquakes they found to be of interest. Have students compare the energy released to the amount of destruction caused and share their findings with the class.

Homework Help

For Exercises	See Examples
17–22	1
23–44, 53–57	3, 4
45–50	5
51–55	2

Extra Practice
See page 850.

ACIDITY For Exercises 23–26, use the information at the beginning of the lesson to find the pH of each substance given its concentration of hydrogen ions.

23. ammonia: $[H+] = 1 \times 10^{-11}$ mole per liter **11**
24. vinegar: $[H+] = 6.3 \times 10^{-3}$ mole per liter **2.2**
25. lemon juice: $[H+] = 7.9 \times 10^{-3}$ mole per liter **2.1**
26. orange juice: $[H+] = 3.16 \times 10^{-4}$ mole per liter **3.5**

Solve each equation or inequality. Round to four decimal places.

27. $6^x \geq 42$ **$\{x \mid x \geq 2.0860\}$**
28. $5^x = 52$ **2.4550**
29. $8^{2a} < 124$ **$\{a \mid a < 1.1590\}$**
30. $4^{3p} = 10$ **0.5537**
31. $3^{n+2} = 14.5$ **0.4341**
32. $9^{z-4} = 6.28$ **4.8362**
33. $8.2^{n-3} = 42.5$ **4.7820**
34. $2.1^{t-5} = 9.32$ **8.0086**
35. $20^{x^2} = 70$ **±1.1909**
36. $2^{x^2-3} = 15$ **±2.6281**
37. $8^{2n} > 52^{4n+3}$ **$\{n \mid n < -1.0178\}$**
38. $2^{2x+3} = 3^{3x}$ **1.0890**
39. $16^{d-4} = 3^{3-d}$ **3.7162**
40. $7^{p+2} \leq 13^{5-p}$ **$\{p \mid p \leq 1.9803\}$**
41. $5^{5y-2} = 2^{2y+1}$ **0.5873**
42. $8^{2x-5} = 5^{x+1}$ **4.7095**
★ 43. $2^n = \sqrt{3^{n-2}}$ **−7.6377**
★ 44. $4^x = \sqrt{5^{x+2}}$ **2.7674**

Express each logarithm in terms of common logarithms. Then approximate its value to four decimal places.

45. $\log_2 13$
46. $\log_5 20$
47. $\log_7 3$
48. $\log_3 8$
★ 49. $\log_4 (1.6)^2$
★ 50. $\log_6 \sqrt{5}$

45. $\frac{\log 13}{\log 2} \approx 3.7004$
46. $\frac{\log 20}{\log 5} \approx 1.8614$
47. $\frac{\log 3}{\log 7} \approx 0.5646$
48. $\frac{\log 8}{\log 3} \approx 1.8928$
49. $\frac{2 \log 1.6}{\log 4} \approx 0.6781$
50. $\frac{0.5 \log 5}{\log 6} \approx 0.4491$

For Exercises 51 and 52, use the information presented at the beginning of the lesson.

51. **POLLUTION** The acidity of water determines the toxic effects of runoff into streams from industrial or agricultural areas. A pH range of 6.0 to 9.0 appears to provide protection for freshwater fish. What is this range in terms of the water's hydrogen ion concentration? **between 0.000000001 and 0.000001 mole per liter**

52. **BUILDING DESIGN** The 1971 Sylmar earthquake in Los Angeles had a Richter scale magnitude of 6.3. Suppose an architect has designed a building strong enough to withstand an earthquake 50 times as intense as the Sylmar quake. Find the magnitude of the strongest quake this building is designed to withstand. **8**

ASTRONOMY For Exercises 53–55, use the following information.
Some stars appear bright only because they are very close to us. Absolute magnitude M is a measure of how bright a star would appear if it were 10 parsecs, about 32 light years, away from Earth. A lower magnitude indicates a brighter star. Absolute magnitude is given by $M = m + 5 - 5 \log d$, where d is the star's distance from Earth measured in parsecs and m is its apparent magnitude.

53. Sirius and Vega are two of the brightest stars in Earth's sky. The apparent magnitude of Sirius is −1.44 and of Vega is 0.03. Which star appears brighter?
54. Sirius is 2.64 parsecs from Earth while Vega is 7.76 parsecs from Earth. Find the absolute magnitude of each star. **Sirius: 1.45, Vega: 0.58**
55. Which star is actually brighter? That is, which has a lower absolute magnitude? **Vega**
56. **CRITICAL THINKING**
    a. Without using a calculator, find the value of $\log_2 8$ and $\log_8 2$. **$3; \frac{1}{3}$**
    b. Without using a calculator, find the value of $\log_9 27$ and $\log_{27} 9$. **$\frac{3}{2}; \frac{2}{3}$**
    c. Make and prove a conjecture as to the relationship between $\log_a b$ and $\log_b a$. **See margin.**

53. Sirius

Study Guide and Intervention, p. 591 (shown) and p. 592

Common Logarithms Base 10 logarithms are called **common logarithms**. The expression $\log_{10} x$ is usually written without the subscript as $\log x$. Use the LOG key on your calculator to evaluate common logarithms.

The relation between exponents and logarithms gives the following identity.

Inverse Property of Logarithms and Exponents	$10^{\log x} = x$

Example 1 **Evaluate log 50 to four decimal places.**
Use the LOG key on your calculator. To four decimal places, log 50 = 1.6990.

Example 2 **Solve $3^{2x+1} = 12$.**

$3^{2x+1} = 12$ Original equation
$\log 3^{2x+1} = \log 12$ Property of Equality for Logarithms
$(2x+1)\log 3 = \log 12$ Power Property of Logarithms
$2x + 1 = \frac{\log 12}{\log 3}$ Divide each side by log 3.
$2x = \frac{\log 12}{\log 3} - 1$ Subtract 1 from each side.
$x = \frac{1}{2}\left(\frac{\log 12}{\log 3} - 1\right)$ Multiply each side by $\frac{1}{2}$.
$x \approx 0.6309$

Exercises

Use a calculator to evaluate each expression to four decimal places.

1. log 18 **1.2553**
2. log 39 **1.5911**
3. log 120 **2.0792**
4. log 5.8 **0.7634**
5. log 42.3 **1.6263**
6. log 0.003 **−2.5229**

Solve each equation or inequality. Round to four decimal places.

7. $4^{3x} = 12$ **0.5975**
8. $6^{x+2} = 18$ **−0.3869**
9. $5^{4x-2} = 120$ **1.2437**
10. $7^{3x-1} \geq 21$ **$\{x \mid x \geq 0.8549\}$**
11. $2.4^{x+4} = 30$ **−0.1150**
12. $6.5^{2x} \geq 200$ **$\{x \mid x \geq 1.4153\}$**
13. $3.6^{4x-1} = 85.4$ **1.1180**
14. $2^{x+5} = 3^{x-2}$ **13.9666**
15. $9^{3x} = 4^{5x+2}$ **−8.1595**
16. $6^{x-5} = 2^{7x+3}$ **−3.6069**

Skills Practice, p. 593 and Practice, p. 594 (shown)

Use a calculator to evaluate each expression to four decimal places.

1. log 101 **2.0043**
2. log 2.2 **0.3424**
3. log 0.05 **−1.3010**

Use the formula pH = −log[H+] to find the pH of each substance given its concentration of hydrogen ions.

4. milk: $[H+] = 2.51 \times 10^{-7}$ mole per liter **6.6**
5. acid rain: $[H+] = 2.51 \times 10^{-6}$ mole per liter **5.6**
6. black coffee: $[H+] = 1.0 \times 10^{-5}$ mole per liter **5.0**
7. milk of magnesia: $[H+] = 3.16 \times 10^{-11}$ mole per liter **10.5**

Solve each equation or inequality. Round to four decimal places.

8. $2^x < 25$ **$\{x \mid x < 4.6439\}$**
9. $5^a = 120$ **2.9746**
10. $6^z = 45.6$ **2.1319**
11. $9^m \geq 100$ **$\{m \mid m \geq 2.0959\}$**
12. $3.5^x = 47.9$ **3.0885**
13. $8.2^y = 64.5$ **1.9802**
14. $2^{b+1} \leq 7.31$ **$\{b \mid b \leq 1.8699\}$**
15. $4^{2x} = 27$ **1.1887**
16. $2^{a-4} = 82.1$ **10.3593**
17. $9^{z-2} > 38$ **$\{z \mid z > 3.6555\}$**
18. $5^{w+3} = 17$ **−1.2396**
19. $30^{x^2} = 50$ **±1.0725**
20. $5^{x^2-3} = 72$ **±2.3785**
21. $4^{2x} = 9^{x+1}$ **3.8188**
22. $2^{n+1} = 5^{2n-1}$ **0.9117**

Express each logarithm in terms of common logarithms. Then approximate its value to four decimal places.

23. $\log_5 12$ **$\frac{\log_{10} 12}{\log_{10} 5}$; 1.5440**
24. $\log_8 32$ **$\frac{\log_{10} 32}{\log_{10} 8}$; 1.6667**
25. $\log_{11} 9$ **$\frac{\log_{10} 9}{\log_{10} 11}$; 0.9163**
26. $\log_2 18$ **$\frac{\log_{10} 18}{\log_{10} 2}$; 4.1699**
27. $\log_9 6$ **$\frac{\log_{10} 6}{\log_{10} 9}$; 0.8155**
28. $\log_7 \sqrt{8}$ **$\frac{\log_{10} 8}{2 \log_{10} 7}$; 0.5343**

29. **HORTICULTURE** Siberian irises flourish when the concentration of hydrogen ions [H+] in the soil is not less than 1.58×10^{-8} mole per liter. What is the pH of the soil in which these irises will flourish? **7.8 or less**

30. **ACIDITY** The pH of vinegar is 2.9 and the pH of milk is 6.6. How many times greater is the hydrogen ion concentration of vinegar than of milk? **about 5000**

31. **BIOLOGY** There are initially 1000 bacteria in a culture. The number of bacteria doubles each hour. The number of bacteria N present after t hours is $N = 1000(2)^t$. How long will it take the culture to increase to 50,000 bacteria? **about 5.6 h**

32. **SOUND** An equation for loudness L in decibels is given by $L = 10 \log R$, where R is the sound's relative intensity. An air-raid siren can reach 150 decibels and jet engine noise can reach 120 decibels. How many times greater is the relative intensity of the air-raid siren than that of the jet engine noise? **1000**

Reading to Learn Mathematics, p. 595 ELL

Pre-Activity **Why is a logarithmic scale used to measure acidity?**

Read the introduction to Lesson 10-4 at the top of page 547 in your textbook.

Which substance is more acidic, milk or tomatoes?
tomatoes

Reading the Lesson

1. Rhonda used the following keystrokes to enter an expression on her graphing calculator:
   LOG 17) ENTER
   The calculator returned the result 1.230448921.
   Which of the following conclusions are correct? **a, c, and d**
   a. The base 10 logarithm of 17 is about 1.2304.
   b. The base 17 logarithm of 10 is about 1.2304.
   c. The common logarithm of 17 is about 1.230449.
   d. $10^{1.230448921}$ is very close to 17.
   e. The common logarithm of 17 is exactly 1.230448921.

2. Match each expression from the first column with an expression from the second column that has the same value.
   a. $\log_2 2$ **iv** — i. $\log_4 1$
   b. $\log 12$ **iii** — ii. $\log_2 8$
   c. $\log_3 1$ **i** — iii. $\log_{10} 12$
   d. $\log_5 \frac{1}{5}$ **v** — iv. $\log_5 5$
   e. $\log 1000$ **ii** — v. $\log 0.1$

3. Calculators do not have keys for finding base 8 logarithms directly. However, you can use a calculator to find $\log_8 20$ if you apply the **change of base** formula.
   Which of the following expressions are equal to $\log_8 20$? **B and C**
   A. $\log_{20} 8$ B. $\frac{\log_{10} 20}{\log_{10} 8}$ C. $\frac{\log 20}{\log 8}$ D. $\frac{\log 8}{\log 20}$

Helping You Remember

4. Sometimes it is easier to remember a formula if you can state it in words. State the change of base formula in words. **Sample answer: To change the logarithm of a number from one base to another, divide the log of the original number in the old base by the log of the new base in the old base.**

Enrichment, p. 596

The Slide Rule

Before the invention of electronic calculators, computations were often performed on a slide rule. A slide rule is based on the idea of logarithms. It has two movable rods labeled with C and D scales. Each of the scales is logarithmic.

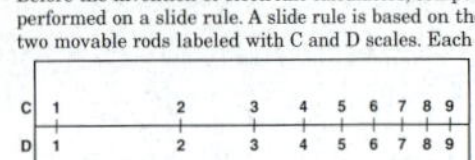

To multiply 2×3 on a slide rule, move the C rod to the right as shown below. You can find 2×3 by adding log 2 to log 3, and the slide rule adds the lengths for you. The distance you get is 0.778, or the logarithm of 6.

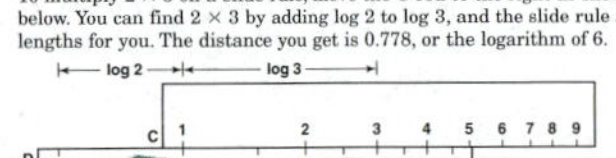

550 Chapter 10 Exponential and Logarithmic Relations

MONEY **For Exercises 57 and 58, use the following information.**
If you deposit P dollars into a bank account paying an annual interest rate r (expressed as a decimal), with n interest payments each year, the amount A you would have after t years is $A = P\left(1 + \frac{r}{n}\right)^{nt}$. Marta places $100 in a savings account earning 6% annual interest, compounded quarterly.

57. If Marta adds no more money to the account, how long will it take the money in the account to reach $125? **about 3.75 yr or 3 yr 9 mo**

58. How long will it take for Marta's money to double? **about 11.64 yr or 11 yr 8 mo**

59. WRITING IN MATH Answer the question that was posed at the beginning of the lesson. **See margin.**

Why is a logarithmic scale used to measure acidity?

Include the following in your answer:
- the hydrogen ion concentration of three substances listed in the table, and
- an explanation as to why it is important to be able to distinguish between a hydrogen ion concentration of 0.00001 mole per liter and 0.0001 mole per liter.

Standardized Test Practice

60. The expression $\log (x - 5)$ is undefined for all values of x such that **A**
   (A) $x \leq 5$
   (B) $x > 5$
   (C) $x \leq 0$
   (D) $x > 1$

61. If $2^4 = 3^x$, then what is the value of x? **C**
   (A) 0.63 (B) 2.34
   (C) 2.52 (D) 4

Maintain Your Skills

Mixed Review

Use $\log_7 2 \approx 0.3562$ and $\log_7 3 \approx 0.5646$ to approximate the value of each expression. *(Lesson 10-3)*

62. $\log_7 16$ **1.4248** 63. $\log_7 27$ **1.6938** 64. $\log_7 36$ **1.8416**

Solve each equation or inequality. Check your solutions. *(Lesson 10-2)*

65. $\log_4 r = 3$ **64** 66. $\log_8 z \leq -2$ 67. $\log_3 (4x - 5) = 5$ **62**

66. $\left\{z \mid 0 < z \leq \frac{1}{64}\right\}$

68. Use synthetic substitution to find $f(-2)$ for $f(x) = x^3 + 6x - 2$. *(Lesson 7-4)* **−22**

Factor completely. If the polynomial is not factorable, write prime. *(Lesson 5-4)*

69. $3d^2 + 2d - 8$ **$(d + 2)(3d - 4)$** 70. $42pq - 35p + 18q - 15$ **$(7p + 3)(6q - 5)$** 71. $13xyz + 3x^2z + 4k$ **prime**

Getting Ready for the Next Lesson

PREREQUISITE SKILLS **Write an equivalent exponential equation.**
*(For review of **logarithmic equations**, see Lesson 10-2.)*

72. $\log_2 3 = x$ **$2^x = 3$** 73. $\log_3 x = 2$ **$3^2 = x$** 74. $\log_5 125 = 3$ **$5^3 = 125$**

Write an equivalent logarithmic equation.
*(For review of **logarithmic equations**, see Lesson 10-2.)*

75. $5^x = 45$ **$\log_5 45 = x$** 76. $7^3 = x$ **$\log_7 x = 3$** 77. $b^y = x$ **$\log_b x = y$**

www.algebra2.com/self_check_quiz

4 Assess

Open-Ended Assessment

Writing Ask students to explain in writing what it means to use the Change of Base Formula. They should include comments about why this formula is useful.

Getting Ready for Lesson 10-5

PREREQUISITE SKILL In Lesson 10-5, students will solve exponential equations and inequalities using natural logarithms and the skills they learned solving common logarithmic equations and inequalities. Students should be confident when converting between exponential and logarithmic equations before proceeding. Use Exercises 72–77 to determine your students' familiarity with converting between exponential and logarithmic equations.

Answers

56c. conjecture: $\log_a b = \frac{1}{\log_b a}$;
proof:

$\log_a b \stackrel{?}{=} \frac{1}{\log_b a}$	Original statement
$\frac{\log_b b}{\log_b a} \stackrel{?}{=} \frac{1}{\log_b a}$	Change of Base Formula
$\frac{1}{\log_b a} = \frac{1}{\log_b a}$ ✓	Inverse Property of Exponents and Logarithms

59. Comparisons between substances of different acidities are more easily distinguished on a logarithmic scale. Answers should include the following.
- Sample answer:
  Tomatoes: 6.3×10^{-5} mole per liter
  Milk: 3.98×10^{-7} mole per liter
  Eggs: 1.58×10^{-8} mole per liter
- Those measurements correspond to pH measurements of 5 and 4, indicating a weak acid and a stronger acid. On the logarithmic scale we can see the difference in these acids, whereas on a normal scale, these hydrogen ion concentrations would appear nearly the same. For someone who has to watch the acidity of the foods they eat, this could be the difference between an enjoyable meal and heartburn.

Graphing Calculator Investigation

A Follow-Up of Lesson 10-4

Getting Started

Using Parentheses In Step 1 of Example 1, remind students that they must also use parentheses around the fraction $\frac{1}{2}$.

Teach

- Before discussing Example 1, use a simple equation such as $2x = 6$ to show students how the equation can be solved by graphing. Graph the equations $y = 2x$ and $y = 6$ and then identify the point of intersection of the graphs.
- Ask students why it is necessary in Step 1 to enter the equations using parentheses around the exponents.
- Have students substitute the solution to Example 1 into the original equation to verify that it is correct.
- In Example 2, make sure students understand why the equations must be rewritten using the Change of Base Formula.
- Students can find the solution set for Example 2 without using the shading options. Simply have them use the **intersect** feature, noting that the graph of Y_1 intersects or is above the graph of Y_2 at and to the right of $x = 0.5$.

Graphing Calculator

A Follow-Up of Lesson 10-4

Solving Exponential and Logarithmic Equations and Inequalities

You can use a TI-83 Plus graphing calculator to solve exponential and logarithmic equations and inequalities. This can be done by graphing each side of the equation separately and using the **intersect** feature on the calculator.

Example 1

Solve $2^{3x-9} = \left(\frac{1}{2}\right)^{x-3}$ by graphing.

Step 1 *Graph each side of the equation.*

- Graph each side of the equation as a separate function. Enter 2^{3x-9} as Y1. Enter $\left(\frac{1}{2}\right)^{x-3}$ as Y2. Be sure to include the added parentheses around each exponent. Then graph the two equations.

KEYSTROKES: *See pages 87 and 88 to review graphing equations.*

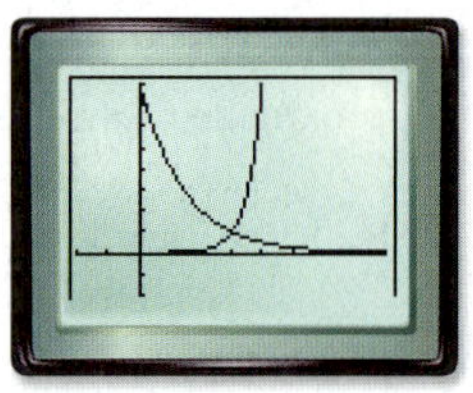

[−2, 8] scl: 1 by [−2, 8] scl: 1

Step 2 *Use the intersect feature.*

- You can use the **intersect** feature on the CALC menu to approximate the ordered pair of the point at which the curves cross.

KEYSTROKES: *See page 115 to review how to use the intersect feature.*

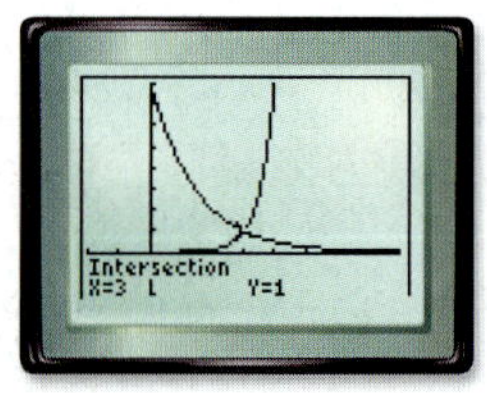

[−2, 8] scl: 1 by [−2, 8] scl: 1

The calculator screen shows that the x-coordinate of the point at which the curves cross is 3. Therefore, the solution of the equation is 3.

The TI-83 Plus has $y = \log_{10} x$ as a built-in function. Enter [Y=] [LOG] [X,T,θ,n] [GRAPH] to view this graph. To graph logarithmic functions with bases other than 10, you must use the Change of Base Formula,

$$\log_a n = \frac{\log_b n}{\log_b a}.$$

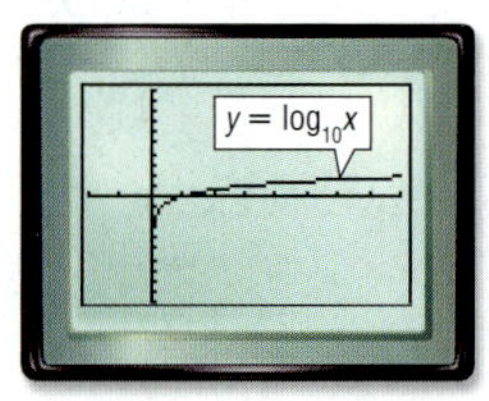

[−2, 8] scl: 1 by [−5, 5] scl: 1

For example, $\log_3 x = \frac{\log_{10} x}{\log_{10} 3}$, so to graph $y = \log_3 x$ you must enter [LOG] [X,T,θ,n] [)] [÷] [LOG] 3 [)] as Y1.

 www.algebra2.com/other_calculator_keystrokes

552 Chapter 10 Exponential and Logarithmic Relations

Investigation

Example 2

Solve $\log_2 2x \geq \log_{\frac{1}{2}} 2x$ by graphing.

Step 1 *Rewrite the problem as a system of common logarithmic inequalities.*

- The first inequality is $\log_2 2x \geq y$ or $y \leq \log_2 2x$. The second inequality is $y \geq \log_{\frac{1}{2}} 2x$.
- Use the Change of Base Formula to create equations that can be entered into the calculator.

$$\log_2 2x = \frac{\log 2x}{\log 2} \qquad \log_{\frac{1}{2}} 2x = \frac{\log 2x}{\log \frac{1}{2}}$$

Thus, the two inequalities are $y \leq \frac{\log 2x}{\log 2}$ and $y \geq \frac{\log 2x}{\log \frac{1}{2}}$.

Step 2 *Enter the first inequality.*

- Enter $y \leq \frac{\log 2x}{\log 2}$ as Y1. Since the inequality includes *less than*, shade below the curve.

KEYSTROKES: [Y=] [LOG] 2 [X,T,θ,n] [)] [÷] [LOG] 2 [)]

Use the arrow and [ENTER] keys to choose the shade below icon, ◣.

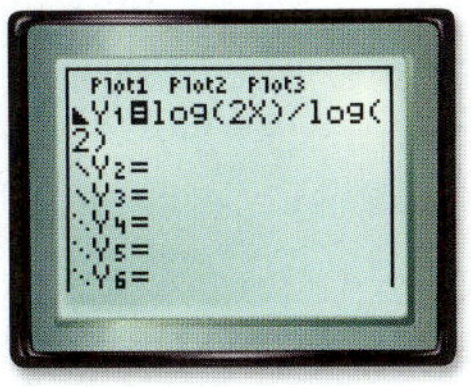

Step 3 *Enter the second inequality.*

- Enter $y \geq \frac{\log 2x}{\log \frac{1}{2}}$ as Y2. Since the inequality includes *greater than*, shade above the curve.

KEYSTROKES: [LOG] 2 [X,T,θ,n] [)] [÷] [LOG] 1 [÷] 2 [)] [GRAPH]

Use the arrow and [ENTER] keys to choose the shade above icon, ◥.

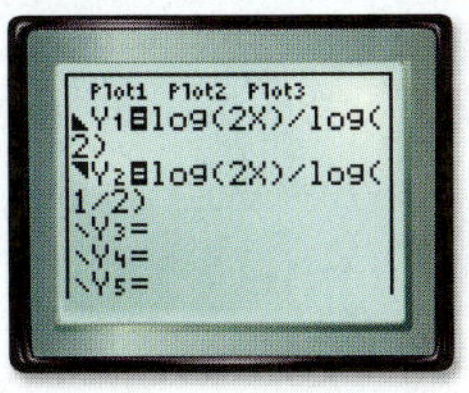

Step 4 *Graph the inequalities.*

KEYSTROKES: [GRAPH]

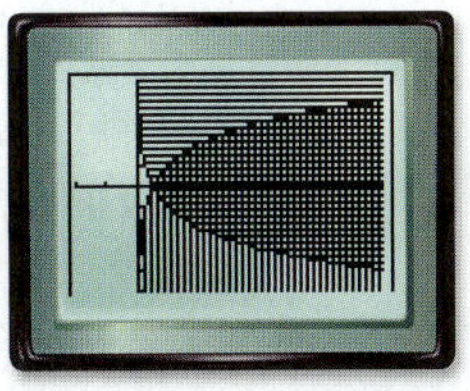

[−2, 8] scl: 1 by [−5, 5] scl: 1

The x values of the points in the region where the shadings overlap is the solution set of the original inequality. Using the calculator's intersect feature, you can conclude that the solution set is $\{x \mid x \geq 0.5\}$.

Exercises Solve each equation or inequality by graphing.

1. $3.5^{x+2} = 1.75^{x+3}$ **−1.2**
2. $-3^{x+4} = -0.5^{2x+3}$ **−2.6**
3. $6^{2-x} - 4 = -0.25^{x-2.5}$ **1.8**
4. $3^x - 4 = 5^{\frac{x}{2}}$ **2**
5. $\log_2 3x = \log_3 (2x + 2)$ **0.7**
6. $2^{x-2} \geq 0.5^{x-3}$ **$x \geq 2.5$**
7. $\log_3 (3x - 5) \geq \log_3 (x + 7)$ **$x \geq 6$**
8. $5^{x+3} \leq 2^{x+4}$ **$x \leq -2.24$**
9. $\log_2 2x \leq \log_4 (x + 3)$ **$0 < x \leq 1$**

Assess

In **Exercise 9**, check that students record the inequalities in the solution set correctly. In particular, students must include the fact that x must be greater than 0.

Graphing Calculator Investigation Solving Exponential and Logarithmic Equations and Inequalities 553

10-5 Lesson Notes

1 Focus

5-Minute Check Transparency 10-5 Use as a quiz or review of Lesson 10-4.

Mathematical Background notes are available for this lesson on p. 520D.

How is the natural base e used in banking?

Ask students:

- What is the formula calculating? **the amount of money in the account**
- Why does the interest increase as the time between compounding periods decreases? **Sample answer: Interest is earned not just on the initial \$1 but also on the total interest that has accrued. As the compounding occurs more often, the amount of money earning interest grows faster.**

10-5 Base e and Natural Logarithms

What You'll Learn

- Evaluate expressions involving the natural base and natural logarithms.
- Solve exponential equations and inequalities using natural logarithms.

Vocabulary

- natural base, e
- natural base exponential function
- natural logarithm
- natural logarithmic function

How is the natural base e used in banking?

Suppose a bank compounds interest on accounts *continuously*, that is, with no waiting time between interest payments.

To develop an equation to determine continuously compounded interest, examine what happens to the value A of an account for increasingly larger numbers of compounding periods n. Use a principal P of \$1, an interest rate r of 100% or 1, and time t of 1 year.

Continuously Compounded Interest

n	$A = P\left(1 + \frac{r}{n}\right)^{nt}$	A
1 (yearly)	$1\left(1 + \frac{1}{1}\right)^{1(1)}$	2
4 (quarterly)	$1\left(1 + \frac{1}{4}\right)^{4(1)}$	2.4414...
12 (monthly)	$1\left(1 + \frac{1}{12}\right)^{12(1)}$	2.6130...
365 (daily)	$1\left(1 + \frac{1}{365}\right)^{365(1)}$	2.7145...
8760 (hourly)	$1\left(1 + \frac{1}{8760}\right)^{8760(1)}$	2.7181...

BASE e AND NATURAL LOGARITHMS In the table above, as n increases, the expression $1\left(1 + \frac{1}{n}\right)^{n(1)}$ or $\left(1 + \frac{1}{n}\right)^{n}$ approaches the irrational number 2.71828... . This number is referred to as the **natural base, e**.

An exponential function with base e is called a **natural base exponential function**. The graph of $y = e^x$ is shown at the right. Natural base exponential functions are used extensively in science to model quantities that grow and decay continuously.

Most calculators have an e^x function for evaluating natural base expressions.

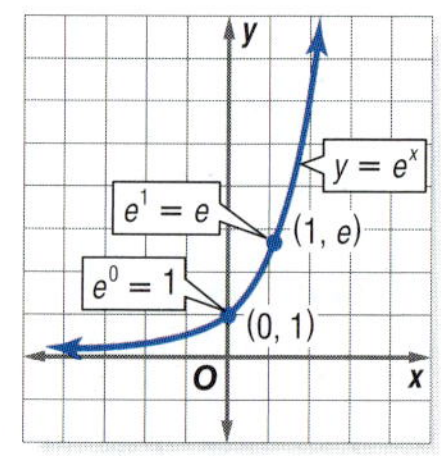

Study Tip

Simplifying Expressions with e
You can simplify expressions involving e in the same manner in which you simplify expressions involving π.
Examples:
- $\pi^2 \cdot \pi^3 = \pi^5$
- $e^2 \cdot e^3 = e^5$

Example 1 Evaluate Natural Base Expressions

Use a calculator to evaluate each expression to four decimal places.

a. e^2 KEYSTROKES: 2nd [e^x] 2 ENTER 7.389056099 about 7.3891

b. $e^{-1.3}$ KEYSTROKES: 2nd [e^x] -1.3 ENTER .272531793 about 0.2725

The logarithm with base e is called the **natural logarithm**, sometimes denoted by $\log_e x$, but more often abbreviated $\ln x$. The **natural logarithmic function**, $y = \ln x$, is the inverse of the natural base exponential function, $y = e^x$. The graph of these two functions shows that $\ln 1 = 0$ and $\ln e = 1$.

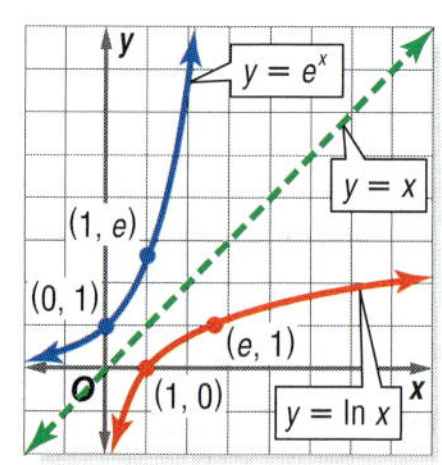

Resource Manager

Workbook and Reproducible Masters

Chapter 10 Resource Masters
- Study Guide and Intervention, pp. 597–598
- Skills Practice, p. 599
- Practice, p. 600
- Reading to Learn Mathematics, p. 601
- Enrichment, p. 602
- Assessment, p. 624

Science and Mathematics Lab Manual, pp. 127–132

Transparencies
5-Minute Check Transparency 10-5
Answer Key Transparencies

Technology
Alge2PASS: Tutorial Plus, Lesson 19
Interactive Chalkboard

Most calculators have an LN key for evaluating natural logarithms.

Example 2 Evaluate Natural Logarithmic Expressions

Use a calculator to evaluate each expression to four decimal places.

a. $\ln 4$ KEYSTROKES: LN 4 ENTER 1.386294361 about 1.3863

b. $\ln 0.05$ KEYSTROKES: LN 0.05 ENTER −2.995732274 about −2.9957

You can write an equivalent base e exponential equation for a natural logarithmic equation and vice versa by using the fact that $\ln x = \log_e x$.

Example 3 Write Equivalent Expressions

Write an equivalent exponential or logarithmic equation.

a. $e^x = 5$

$e^x = 5 \rightarrow \log_e 5 = x$

$\ln 5 = x$

b. $\ln x \approx 0.6931$

$\ln x \approx 0.6931 \rightarrow \log_e x \approx 0.6931$

$x \approx e^{0.6931}$

Since the natural base function and the natural logarithmic function are inverses, these two functions can be used to "undo" each other.

$$e^{\ln x} = x \qquad \ln e^x = x$$

Example 4 Inverse Property of Base e and Natural Logarithms

Evaluate each expression.

a. $e^{\ln 7}$

$e^{\ln 7} = 7$

b. $\ln e^{4x+3}$

$\ln e^{4x+3} = 4x + 3$

EQUATIONS AND INEQUALITIES WITH e AND ln Equations and inequalities involving base e are easier to solve using natural logarithms than using common logarithms. All of the properties of logarithms that you have learned apply to natural logarithms as well.

Example 5 Solve Base e Equations

Solve $5e^{-x} - 7 = 2$.

$5e^{-x} - 7 = 2$	Original equation
$5e^{-x} = 9$	Add 7 to each side.
$e^{-x} = \frac{9}{5}$	Divide each side by 5.
$\ln e^{-x} = \ln \frac{9}{5}$	Property of Equality for Logarithms
$-x = \ln \frac{9}{5}$	Inverse Property of Exponents and Logarithms
$x = -\ln \frac{9}{5}$	Divide each side by −1.
$x \approx -0.5878$	Use a calculator.

The solution is about −0.5878.

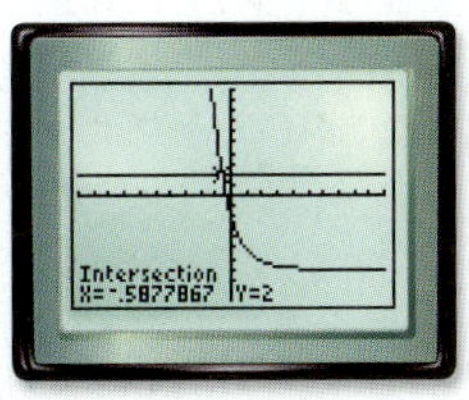

CHECK You can check this value by substituting −0.5878 into the original equation or by finding the intersection of the graphs of $y = 5e^{-x} - 7$ and $y = 2$.

www.algebra2.com/extra_examples

2 Teach

BASE e AND NATURAL LOGARITHMS

Teaching Tip Stress that e is a constant like π, and not a variable like x or y.

In-Class Examples Power Point®

1 Use a calculator to evaluate each expression to four decimal places.

a. $e^{0.5}$ **about 1.6487**

b. e^{-8} **about 0.0003**

2 Use a calculator to evaluate each expression to four decimal places.

a. $\ln 3$ **about 1.0986**

b. $\ln \frac{1}{4}$ **about −1.3863**

3 Write an equivalent exponential or logarithmic equation.

a. $e^x = 23$ **$\ln 23 = x$**

b. $\ln x \approx 1.2528$ **$x \approx e^{1.2528}$**

4 Evaluate each expression.

a. $e^{\ln 21}$ **21**

b. $\ln e^{x^2 - 1}$ **$x^2 - 1$**

EQUATIONS AND INEQUALITIES WITH e AND ln

In-Class Example Power Point®

5 Solve $3e^{-2x} + 4 = 10$. **$x \approx -0.3466$**

Lesson 10-5 Base e and Natural Logarithms 555

In-Class Examples

Power Point®

6 **SAVINGS** Suppose you deposit \$700 into an account paying 6% annual interest, compounded continuously.

a. What is the balance after 8 years? **\$1131.25**

b. How long will it take for the balance in your account to reach at least \$2000? **at least 17.5 years**

7 Solve each equation or inequality.

a. $\ln 3x = 0.5$ **about 0.5496**

b. $\ln (2x - 3) < 2.5$ **$1.5 < x < 7.5912$**

Study Tip

Continuously Compounded Interest
Although no banks actually pay interest compounded continuously, the equation $A = Pe^{rt}$ is so accurate in computing the amount of money for quarterly compounding, or daily compounding, that it is often used for this purpose.

When interest is compounded continuously, the amount A in an account after t years is found using the formula $A = Pe^{rt}$, where P is the amount of principal and r is the annual interest rate.

Example 6 Solve Base e Inequalities

SAVINGS **Suppose you deposit \$1000 in an account paying 5% annual interest, compounded continuously.**

a. What is the balance after 10 years?

$A = Pe^{rt}$	Continuous compounding formula
$= 1000e^{(0.05)(10)}$	Replace P with 1000, r with 0.05, and t with 10.
$= 1000e^{0.5}$	Simplify.
≈ 1648.72	Use a calculator.

The balance after 10 years would be \$1648.72.

b. How long will it take for the balance in your account to reach at least \$1500?

$$\underbrace{\text{The balance}}\ \underbrace{\text{is at least}}\ \underbrace{\$1500.}$$

$A \geq 1500$	Write an inequality.
$1000e^{(0.05)t} \geq 1500$	Replace A with $1000e^{(0.05)t}$.
$e^{(0.05)t} \geq 1.5$	Divide each side by 1000.
$\ln e^{(0.05)t} \geq \ln 1.5$	Property of Equality for Logarithms
$0.05t \geq \ln 1.5$	Inverse Property of Exponents and Logarithms
$t \geq \frac{\ln 1.5}{0.05}$	Divide each side by 0.05.
$t \geq 8.11$	Use a calculator.

It will take at least 8.11 years for the balance to reach \$1500.

Study Tip

Equations with ln
As with other logarithmic equations, remember to check for extraneous solutions.

Example 7 Solve Natural Log Equations and Inequalities

Solve each equation or inequality.

a. $\ln 5x = 4$

$\ln 5x = 4$	Original equation
$e^{\ln 5x} = e^4$	Write each side using exponents and base e.
$5x = e^4$	Inverse Property of Exponents and Logarithms
$x = \frac{e^4}{5}$	Divide each side by 5.
$x \approx 10.9196$	Use a calculator.

The solution is 10.9196. Check this solution using substitution or graphing.

b. $\ln (x - 1) > -2$

$\ln (x - 1) > -2$	Original inequality
$e^{\ln (x - 1)} > e^{-2}$	Write each side using exponents and base e.
$x - 1 > e^{-2}$	Inverse Property of Exponents and Logarithms
$x > e^{-2} + 1$	Add 1 to each side.
$x > 1.1353$	Use a calculator.

The solution is all numbers greater than about 1.1353. Check this solution using substitution.

DAILY INTERVENTION

Differentiated Instruction

Kinesthetic Using plastic coins and paper currency, have pairs of students begin with \$10, choose an interest rate, and calculate how much they will have after 5, 10, 15, and 20 years. After each calculation, have students model the amount with their money to help them visualize the growth over time.

556 Chapter 10 Exponential and Logarithmic Relations

Check for Understanding

Concept Check

1. **Name** the base of natural logarithms. **the number e**
2. **OPEN ENDED** Give an example of an exponential equation that requires using natural logarithms instead of common logarithms to solve. **Sample answer: $e^x = 8$**
3. **FIND THE ERROR** Colby and Elsu are solving $\ln 4x = 5$.

3. Elsu; Colby tried to write each side as a power of 10. Since the base of the natural logarithmic function is e, he should have written each side as a power of e; $10^{\ln 4x} \neq 4x$.

Colby

$\ln 4x = 5$

$10^{\ln 4x} = 10^5$

$4x = 100{,}000$

$x = 25{,}000$

Elsu

$\ln 4x = 5$

$e^{\ln 4x} = e^5$

$4x = e^5$

$x = \frac{e^5}{4}$

$x \approx 37.1033$

Who is correct? Explain your reasoning.

Guided Practice

GUIDED PRACTICE KEY	
Exercises	Examples
4–7	1, 2
8, 9	3
10, 11	4
12–17	5–7
18, 19	5

Use a calculator to evaluate each expression to four decimal places.

4. e^6 **403.4288**
5. $e^{-3.4}$ **0.0334**
6. $\ln 1.2$ **0.1823**
7. $\ln 0.1$ **−2.3026**

Write an equivalent exponential or logarithmic equation.

8. $e^x = 4$ **$x = \ln 4$**
9. $\ln 1 = 0$ **$e^0 = 1$**

Evaluate each expression.

10. $e^{\ln 3}$ **3**
11. $\ln e^{5x}$ **$5x$**

Solve each equation or inequality. **15. $0 < x < 403.4288$**

12. $e^x > 30$ **$x > 3.4012$**
13. $2e^x - 5 = 1$ **1.0986**
14. $3 + e^{-2x} = 8$ **−0.8047**
15. $\ln x < 6$
16. $2 \ln 3x + 1 = 5$ **2.4630**
17. $\ln x^2 = 9$ **±90.0171**

Application

ALTITUDE **For Exercises 18 and 19, use the following information.**
The altimeter in an airplane gives the altitude or height h (in feet) of a plane above sea level by measuring the outside air pressure P (in kilopascals). The height and air pressure are related by the model $P = 101.3\, e^{-\frac{h}{26{,}200}}$.

18. $h = -26{,}200 \ln \frac{P}{101.3}$

18. Find a formula for the height in terms of the outside air pressure.
19. Use the formula you found in Exercise 18 to approximate the height of a plane above sea level when the outside air pressure is 57 kilopascals. **about 15,066 ft**

★ **indicates increased difficulty**

Practice and Apply

Homework Help	
For Exercises	See Examples
20–29	1, 2
30–33	3
34–37	4
38–53	5–7
54–57	6
58–61	3, 5

Extra Practice
See page 850.

Use a calculator to evaluate each expression to four decimal places.

20. e^4 **54.5982**
21. e^5 **148.4132**
22. $e^{-1.2}$ **0.3012**
23. $e^{0.5}$ **1.6487**
24. $\ln 3$ **1.0986**
25. $\ln 10$ **2.3026**
26. $\ln 5.42$ **1.6901**
27. $\ln 0.03$ **−3.5066**

28. **SAVINGS** If you deposit \$150 in a savings account paying 4% interest compounded continuously, how much money will you have after 5 years? Use the formula presented in Example 6. **\$183.21**

29. **PHYSICS** The equation $\ln \frac{I_0}{I} = 0.014d$ relates the intensity of light at a depth of d centimeters of water I with the intensity in the atmosphere I_0. Find the depth of the water where the intensity of light is half the intensity of the light in the atmosphere. **about 49.5 cm**

www.algebra2.com/self_check_quiz

3 Practice/Apply

Study Notebook

Have students—

- add the definitions/examples of the vocabulary terms to their Vocabulary Builder worksheets for Chapter 10.
- include examples of how to evaluate expressions containing natural logarithms.
- include any other item(s) that they find helpful in mastering the skills in this lesson.

DAILY INTERVENTION **FIND THE ERROR**

Make sure students can identify what Colby did incorrectly. Point out that raising both terms to base 10 is not incorrect but the step that follows incorrectly states that $10^{\ln 4x} = 4x$.

About the Exercises...

Organization by Objective

- **Base e and Natural Logarithms:** 20–37
- **Equations and Inequalities with e and ln:** 38–61

Odd/Even Assignments

Exercises 20–53 are structured so that students practice the same concepts whether they are assigned odd or even problems.

Assignment Guide

Basic: 21–51 odd, 54–59, 62–80

Average: 21–53 odd, 54–59, 62–80

Advanced: 20–52 even, 54–74 (optional: 75–80)

All: Practice Quiz 2 (1–5)

Write an equivalent exponential or logarithmic equation.

30. $e^{-x} = 5$ $-x = \ln 5$
31. $e^2 = 6x$ $2 = \ln 6x$
32. $\ln e = 1$ $e^1 = e$
33. $\ln 5.2 = x$ $e^x = 5.2$

Evaluate each expression.

34. $e^{\ln 0.2}$ **0.2**
35. $e^{\ln y}$ y
36. $\ln e^{-4x}$ $-4x$
37. $\ln e^{45}$ **45**

Solve each equation or inequality.

38. $3e^x + 1 = 5$ **0.2877**
39. $2e^x - 1 = 0$ **−0.6931**
40. $e^x < 4.5$ $x < 1.5041$
41. $e^x > 1.6$ $x > 0.4700$
42. $-3e^{4x} + 11 = 2$ **0.2747**
43. $8 + 3e^{3x} = 26$ **0.5973**
44. $e^{5x} \geq 25$ $x \geq 0.6438$
45. $e^{-2x} \leq 7$ $x \geq -0.9730$
46. $\ln 2x = 4$ **27.2991**
47. $\ln 3x = 5$ **49.4711**
48. $\ln (x + 1) = 1$ **1.7183**
49. $\ln (x - 7) = 2$ **14.3891**
50. $\ln x + \ln 3x = 12$ **232.9197**
51. $\ln 4x + \ln x = 9$ **45.0086**

★ 52. $\ln (x^2 + 12) = \ln x + \ln 8$ **2, 6**

★ 53. $\ln x + \ln (x + 4) = \ln 5$ **1**

More About. . .

Money

To determine the doubling time on an account paying an interest rate r that is compounded *annually*, investors use the "Rule of 72." Thus, the amount of time needed for the money in an account paying 6% interest compounded annually to double is $\frac{72}{6}$ or 12 years.

Source: www.datachimp.com

MONEY For Exercises 54–57, use the formula for continuously compounded interest found in Example 6. **55.** $t = \frac{100 \ln 2}{r}$

54. If you deposit \$100 in an account paying 3.5% interest compounded continuously, how long will it take for your money to double? **about 19.8 yr**

55. Suppose you deposit A dollars in an account paying an interest rate r as a percent, compounded continuously. Write an equation giving the time t needed for your money to double, or the *doubling time*.

56. Explain why the equation you found in Exercise 55 might be referred to as the "Rule of 70." $100 \ln 2 \approx 70$

57. **MAKE A CONJECTURE** State a rule that could be used to approximate the amount of time t needed to triple the amount of money in a savings account paying r percent interest compounded continuously. $t = \frac{110}{r}$

POPULATION For Exercises 58 and 59, use the following information.
In 2000, the world's population was about 6 billion. If the world's population continues to grow at a constant rate, the future population P, in billions, can be predicted by $P = 6e^{0.02t}$, where t is the time in years since 2000. **58. about 7.33 billion**

58. According to this model, what will the world's population be in 2010?

59. Some experts have estimated that the world's food supply can support a population of, at most, 18 billion. According to this model, for how many more years will the world's population remain at 18 billion or less? **about 55 yr**

Online Research **Data Update** What is the current world population? Visit www.algebra2.com/data_update to learn more.

RUMORS For Exercises 60 and 61, use the following information.
The number of people H who have heard a rumor can be approximated by $H = \frac{P}{1 + (P - S)e^{-0.35t}}$, where P is the total population, S is the number of people who start the rumor, and t is the time in minutes. Suppose two students start a rumor that the principal will let everyone out of school one hour early that day.

60. If there are 1600 students in the school, how many students will have heard the rumor after 10 minutes? **about 32 students**

61. How much time will pass before half of the students have heard the rumor? **about 21 min**

62. **CRITICAL THINKING** Determine whether the following statement is *sometimes*, *always*, or *never* true. Explain your reasoning. **Always; see pp. 573A-573D.**

For all positive numbers x and y, $\frac{\log x}{\log y} = \frac{\ln x}{\ln y}$.

Study Guide and Intervention, p. 597 (shown) and p. 598

Base e and Natural Logarithms The irrational number $e \approx 2.71828...$ often occurs as the base for exponential and logarithmic functions that describe real-world phenomena.

Natural Base e	As n increases, $\left(1 + \frac{1}{n}\right)^n$ approaches $e \approx 2.71828....$ $\ln x = \log_e x$

The functions $y = e^x$ and $y = \ln x$ are inverse functions.

Inverse Property of Base e and Natural Logarithms	$e^{\ln x} = x$ $\ln e^x = x$

Natural base expressions can be evaluated using the e^x and ln keys on your calculator.

Example 1 **Evaluate ln 1685.**
Use a calculator.
$\ln 1685 \approx 7.4295$

Example 2 **Write a logarithmic equation equivalent to $e^{2x} = 7$.**
$e^{2x} = 7 \rightarrow \log_e 7 = 2x$ or $2x = \ln 7$

Example 3 **Evaluate ln e^{18}.**
Use the Inverse Property of Base e and Natural Logarithms.
$\ln e^{18} = 18$

Exercises

Use a calculator to evaluate each expression to four decimal places.

1. ln 732 **6.5958**
2. ln 84,350 **11.3427**
3. ln 0.735 **−0.3079**
4. ln 100 **4.6052**
5. ln 0.0824 **−2.4962**
6. ln 2.388 **0.8705**
7. ln 128,245 **11.7617**
8. ln 0.00614 **−5.0929**

Write an equivalent exponential or logarithmic equation.

9. $e^{15} = x$ $\ln x = 15$
10. $e^{3x} = 45$ $3x = \ln 45$
11. $\ln 20 = x$ $e^x = 20$
12. $\ln x = 8$ $x = e^8$
13. $e^{-5x} = 0.2$ $-5x = \ln 0.2$
14. $\ln (4x) = 9.6$ $4x = e^{9.6}$
15. $e^{8.2} = 10x$ $\ln 10x = 8.2$
16. $\ln 0.0002 = x$ $e^x = 0.0002$

Evaluate each expression.

17. $\ln e^3$ **3**
18. $e^{\ln 42}$ **42**
19. $e^{\ln 0.5}$ **0.5**
20. $\ln e^{16.2}$ **16.2**

Skills Practice, p. 599 and Practice, p. 600 (shown)

Use a calculator to evaluate each expression to four decimal places.

1. $e^{1.5}$ **4.4817**
2. ln 8 **2.0794**
3. ln 3.2 **1.1632**
4. $e^{-0.6}$ **0.5488**
5. $e^{4.2}$ **66.6863**
6. ln 1 **0**
7. $e^{-2.5}$ **0.0821**
8. ln 0.037 **−3.2968**

Write an equivalent exponential or logarithmic equation.

9. $\ln 50 = x$ $e^x = 50$
10. $\ln 36 = 2x$ $e^{2x} = 36$
11. $\ln 6 \approx 1.7918$ $e^{1.7918} \approx 6$
12. $\ln 9.3 \approx 2.2300$ $e^{2.2300} \approx 9.3$
13. $e^x = 8$ $x = \ln 8$
14. $e^5 = 10x$ $5 = \ln 10x$
15. $e^{-x} = 4$ $x = -\ln 4$
16. $e^2 = x + 1$ $2 = \ln (x + 1)$

Evaluate each expression.

17. $e^{\ln 12}$ **12**
18. $e^{\ln 3x}$ $3x$
19. $\ln e^{-1}$ **−1**
20. $\ln e^{-2y}$ $-2y$

Solve each equation or inequality.

21. $e^x < 9$ $\{x \mid x < 2.1972\}$
22. $e^{-x} = 31$ **−3.4340**
23. $e^x = 1.1$ **0.0953**
24. $e^x = 5.8$ **1.7579**
25. $2e^x - 3 = 1$ **0.6931**
26. $5e^x + 1 \geq 7$ $\{x \mid x \geq 0.1823\}$
27. $4 + e^x = 19$ **2.7081**
28. $-3e^x + 10 < 8$ $\{x \mid x > -0.4055\}$
29. $e^{3x} = 8$ **0.6931**
30. $e^{-4x} = 5$ **−0.4024**
31. $e^{0.5x} = 6$ **3.5835**
32. $2e^{5x} = 24$ **0.4970**
33. $e^{2x} + 1 = 55$ **1.9945**
34. $e^{3x} - 5 = 32$ **1.2036**
35. $9 + e^{2x} = 10$ **0**
36. $e^{-3x} + 7 \geq 15$ $\{x \mid x \leq -0.6931\}$
37. $\ln 4x = 3$ **5.0214**
38. $\ln (-2x) = 7$ **−548.3166**
39. $\ln 2.5x = 10$ **8810.5863**
40. $\ln (x - 6) = 1$ **8.7183**
41. $\ln (x + 2) = 3$ **18.0855**
42. $\ln (x + 3) = 5$ **145.4132**
43. $\ln 3x + \ln 2x = 9$ **36.7493**
44. $\ln 5x + \ln x = 7$ **14.8097**

INVESTING **For Exercises 45 and 46, use the formula for continuously compounded interest, $A = Pe^{rt}$, where P is the principal, r is the annual interest rate, and t is the time in years.**

45. If Sarita deposits \$1000 in an account paying 3.4% annual interest compounded continuously, what is the balance in the account after 5 years? **\$1185.30**

46. How long will it take the balance in Sarita's account to reach \$2000? **about 20.4 yr**

47. **RADIOACTIVE DECAY** The amount of a radioactive substance y that remains after t years is given by the equation $y = ae^{kt}$, where a is the initial amount present and k is the decay constant for the radioactive substance. If $a = 100$, $y = 50$, and $k = -0.035$, find t. **about 19.8 yr**

Reading to Learn Mathematics, p. 601 ELL

Pre-Activity **How is the natural base e used in banking?**

Read the introduction to Lesson 10-5 at the top of page 554 in your textbook.

Suppose that you deposit \$675 in a savings account that pays an annual interest rate of 5%. In each case listed below, indicate which method of compounding would result in more money in your account at the end of one year.

a. annual compounding or monthly compounding **monthly**
b. quarterly compounding or daily compounding **daily**
c. daily compounding or continuous compounding **continuous**

Reading the Lesson

1. Jagdish entered the following keystrokes in his calculator:

LN 5) ENTER

The calculator returned the result 1.609437912. Which of the following conclusions are correct? **d and f**

a. The common logarithm of 5 is about 1.6094.
b. The natural logarithm of 5 is exactly 1.609437912.
c. The base 5 logarithm of e is about 1.6094.
d. The natural logarithm of 5 is about 1.609438.
e. $10^{1.609437912}$ is very close to 5.
f. $e^{1.609437912}$ is very close to 5.

2. Match each expression from the first column with its value in the second column. Some choices may be used more than once or not at all.

a. $e^{\ln 5}$ **IV**	I. 1
b. $\ln 1$ **V**	II. 10
c. $e^{\ln e}$ **VI**	III. −1
d. $\ln e^5$ **IV**	IV. 5
e. $\ln e$ **I**	V. 0
f. $\ln \left(\frac{1}{e}\right)$ **III**	VI. e

Helping You Remember

3. A good way to remember something is to explain it to someone else. Suppose that you are studying with a classmate who is puzzled when asked to evaluate $\ln e^3$. How would you explain to him an easy way to figure this out? **Sample answer: *ln* means *natural log*. The natural log of e^3 is the power to which you raise e to get e^3. This is obviously 3.**

Enrichment, p. 602

Approximations for π and e

The following expression can be used to approximate e. If greater and greater values of n are used, the value of the expression approximates e more and more closely.

$$\left(1 + \frac{1}{n}\right)^n$$

Another way to approximate e is to use this infinite sum. The greater the value of n, the closer the approximation.

$$e = 1 + 1 + \frac{1}{2} + \frac{1}{2 \cdot 3} + \frac{1}{2 \cdot 3 \cdot 4} + \ldots + \frac{1}{2 \cdot 3 \cdot 4 \cdot \ldots \cdot n} + \ldots$$

In a similar manner, π can be approximated using an infinite product discovered by the English mathematician John Wallis (1616–1703).

$$\frac{\pi}{2} = \frac{2}{1} \cdot \frac{2}{3} \cdot \frac{4}{3} \cdot \frac{4}{5} \cdot \frac{6}{5} \cdot \frac{6}{7} \cdot \ldots \cdot \frac{2n}{2n - 1} \cdot \frac{2n}{2n + 1} \ldots$$

Solve each problem.

[illegible] with an e^x key [illegible] to 7 decimal places. **2.7182818**

63. **WRITING IN MATH** Answer the question that was posed at the beginning of the lesson. **See margin.**

How is the natural base *e* used in banking?

Include the following in your answer:
- an explanation of how to calculate the value of an account whose interest is compounded continuously, and
- an explanation of how to use natural logarithms to find when the account will have a specified value.

Standardized Test Practice

64. If $e^x \neq 1$ and $e^{x^2} = \frac{1}{(\sqrt{2})^x}$, what is the value of x? **B**

(A) −1.41 (B) −0.35 (C) 1.00 (D) 1.10

65. **SHORT RESPONSE** The population of a certain country can be modeled by the equation $P(t) = 40\,e^{0.02t}$, where P is the population in millions and t is the number of years since 1900. When will the population be 100 million, 200 million, and 400 million? What do you notice about these time periods? **1946, 1981, 2015; It takes between 34 and 35 years for the population to double.**

Maintain Your Skills

Mixed Review

Express each logarithm in terms of common logarithms. Then approximate its value to four decimal places. *(Lesson 10-4)*

66. $\log_4 68$ 67. $\log_6 0.047$ 68. $\log_{50} 23$

66. $\frac{\log 68}{\log 4} = 3.0437$
67. $\frac{\log 0.047}{\log 6} = -1.7065$
68. $\frac{\log 23}{\log 50} = 0.8015$

Solve each equation. Check your solutions. *(Lesson 10-3)*

69. $\log_3(a + 3) + \log_3(a - 3) = \log_3 16$ **5** 70. $\log_{11} 2 + 2\log_{11} x = \log_{11} 32$ **4**

State whether each equation represents a *direct*, *joint*, or *inverse* variation. Then name the constant of variation. *(Lesson 9-4)*

71. $mn = 4$ **inverse, 4** 72. $\frac{a}{b} = c$ **joint, 1** 73. $y = -7x$ **direct, −7**

74. **COMMUNICATION** A microphone is placed at the focus of a parabolic reflector to collect sounds for the television broadcast of a football game. The focus of the parabola that is the cross section of the reflector is 5 inches from the vertex. The latus rectum is 20 inches long. Assuming that the focus is at the origin and the parabola opens to the right, write the equation of the cross section. *(Lesson 8-2)* $x = \frac{1}{20}y^2 - 5$

Getting Ready for the Next Lesson

PREREQUISITE SKILL **Solve each equation or inequality.**
*(To review **exponential equations and inequalities**, see Lesson 10-1.)*

75. $2^x = 10$ **3.32** 76. $5^x = 12$ **1.54** 77. $6^x = 13$ **1.43**

78. $2(1 + 0.1)^x = 50$ **33.77** 79. $10(1 + 0.25)^x = 200$ **13.43** 80. $400(1 - 0.2)^x = 50$ **9.32**

Practice Quiz 2

Lessons 10-3 through 10-5

1. Express $\log_4 5$ in terms of common logarithms. Then approximate its value to four decimal places. *(Lesson 10-4)* $\frac{\log 5}{\log 4}$; **1.1610**
2. Write an equivalent exponential equation for $\ln 3x = 2$. *(Lesson 10-5)* $e^2 = 3x$

Solve each equation or inequality. *(Lesson 10-3 through 10-5)*

3. $\log_2(9x + 5) = 2 + \log_2(x^2 - 1)$ **3** 4. $2^{x-3} > 5$ $x > 5.3219$ 5. $2e^x - 1 = 7$ **1.3863**

4 Assess

Open-Ended Assessment

Speaking Ask students to explain how evaluating expressions involving base *e* and natural logarithms is similar to evaluating expressions involving common logarithms and base 10, and also how they differ.

Getting Ready for Lesson 10-6

PREREQUISITE SKILL Students will encounter exponential growth and decay problems in Lesson 10-6. They will be required to solve exponential equations and inequalities. Use Exercises 75–80 to determine your students' familiarity with solving exponential equations and inequalities.

Assessment Options

Practice Quiz 2 The quiz provides students with a brief review of the concepts and skills in Lessons 10-3 through 10-5. Lesson numbers are given to the right of the exercises or instruction lines so students can review concepts not yet mastered.

Quiz (Lessons 10-4 and 10-5) is available on p. 624 of the *Chapter 10 Resource Masters*.

Answer

63. The number *e* is used in the formula for continuously compounded interest, $A = Pe^{rt}$. Although no banks actually pay interest compounded continually, the equation is so accurate in computing the amount of money for quarterly compounding, or daily compounding, that it is often used for this purpose. Answers should include the following.
- If you know the annual interest rate *r* and the principal *P*, the value of the account after *t* years is calculated by multiplying *P* times *e* raised to the *r* times *t* power. Use a calculator to find the value of e^{rt}.
- If you know the value *A* you wish the account to achieve, the principal *P*, and the annual interest rate *r*, the time *t* needed to achieve this value is found by first taking the natural logarithm of *A* minus the natural logarithm of *P*. Then, divide this quantity by *r*.

10-6 Lesson Notes

1 Focus

5-Minute Check Transparency 10-6 Use as a quiz or review of Lesson 10-5.

Mathematical Background notes are available for this lesson on p. 520D.

How can you determine the current value of your car?

Ask students:

- What kinds of items increase in value? **Sample answer: artwork, some trading cards, some collectibles**
- During which year does the car depreciate the most? **first year**
- How can the amount of depreciation be different each year when the percent of decrease is always the same? **In the first year, the car depreciates 16% of its value at the beginning of that year (when it was new). This is when its value is greatest, so the amount of depreciation is also the greatest during this year. The amount of depreciation decreases each year because the value of the car at the beginning of each year is less than it was at the beginning of the previous year.**

10-6 Exponential Growth and Decay

What You'll Learn

- Use logarithms to solve problems involving exponential decay.
- Use logarithms to solve problems involving exponential growth.

Vocabulary

- rate of decay
- rate of growth

How can you determine the current value of your car?

Certain assets, like homes, can *appreciate* or increase in value over time. Others, like cars, *depreciate* or decrease in value with time. Suppose you buy a car for $22,000 and the value of the car decreases by 16% each year. The table shows the value of the car each year for up to 5 years after it was purchased.

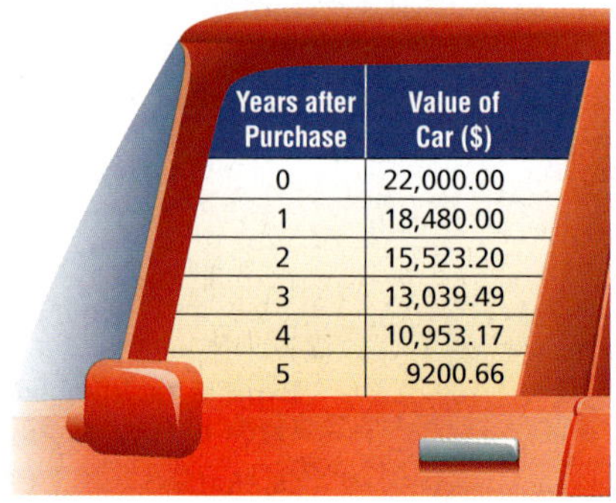

Years after Purchase	Value of Car ($)
0	22,000.00
1	18,480.00
2	15,523.20
3	13,039.49
4	10,953.17
5	9200.66

EXPONENTIAL DECAY The depreciation of the value of a car is an example of exponential decay. When a quantity *decreases* by a fixed percent each year, or other period of time, the amount y of that quantity after t years is given by $y = a(1 - r)^t$, where a is the initial amount and r is the percent of decrease expressed as a decimal. The percent of decrease r is also referred to as the **rate of decay**.

Study Tip

Rate of Change Remember to rewrite the rate of change as a decimal before using it in the formula.

Example 1 Exponential Decay of the Form $y = a(1 - r)^t$

CAFFEINE A cup of coffee contains 130 milligrams of caffeine. If caffeine is eliminated from the body at a rate of 11% per hour, how long will it take for half of this caffeine to be eliminated from a person's body?

Explore The problem gives the amount of caffeine consumed and the rate at which the caffeine is eliminated. It asks you to find the time it will take for half of the caffeine to be eliminated from a person's body.

Plan Use the formula $y = a(1 - r)^t$. Let t be the number of hours since drinking the coffee. The amount remaining y is half of 130 or 65.

Solve

$y = a(1 - r)^t$ Exponential decay formula

$65 = 130(1 - 0.11)^t$ Replace y with 65, a with 130, and r with 11% or 0.11.

$0.5 = (0.89)^t$ Divide each side by 130.

$\log 0.5 = \log (0.89)^t$ Property of Equality for Logarithms

$\log 0.5 = t \log (0.89)$ Power Property for Logarithms

$\frac{\log 0.5}{\log 0.89} = t$ Divide each side by log 0.89.

$5.9480 \approx t$ Use a calculator.

Resource Manager

Workbook and Reproducible Masters

Chapter 10 Resource Masters

- Study Guide and Intervention, pp. 603–604
- Skills Practice, p. 605
- Practice, p. 606
- Reading to Learn Mathematics, p. 607
- Enrichment, p. 608
- Assessment, p. 624

Graphing Calculator and Spreadsheet Masters, p. 46

School-to-Career Masters, p. 20

Teaching Algebra With Manipulatives Masters, p. 278

Transparencies

5-Minute Check Transparency 10-6
Answer Key Transparencies

Technology

Interactive Chalkboard
Multimedia Applications

It will take approximately 6 hours for half of the caffeine to be eliminated from a person's body.

Examine Use the formula to find how much of the original 130 milligrams of caffeine would remain after 6 hours.

$y = a(1 - r)^t$ Exponential decay formula

$y = 130(1 - 0.11)^6$ Replace a with 130, r with 0.11, and t with 6.

$y \approx 64.6$ Use a calculator.

Half of 130 is 65, so the answer seems reasonable.

Another model for exponential decay is given by $y = ae^{-kt}$, where k is a constant. This is the model preferred by scientists. Use this model to solve problems involving radioactive decay.

Example 2 Exponential Decay of the Form $y = ae^{-kt}$

PALEONTOLOGY The *half-life* of a radioactive substance is the time it takes for half of the atoms of the substance to become disintegrated. All life on Earth contains the radioactive element Carbon-14, which decays continuously at a fixed rate. The half-life of Carbon-14 is 5760 years. That is, every 5760 years half of a mass of Carbon-14 decays away.

a. What is the value of k for Carbon-14?

To determine the constant k for Carbon-14, let a be the initial amount of the substance. The amount y that remains after 5760 years is then represented by $\frac{1}{2}a$ or $0.5a$.

$y = ae^{-kt}$ Exponential decay formula

$0.5a = ae^{-k(5760)}$ Replace y with $0.5a$ and t with 5760.

$0.5 = e^{-5760k}$ Divide each side by a.

$\ln 0.5 = \ln e^{-5760k}$ Property of Equality for Logarithmic Functions

$\ln 0.5 = -5760k$ Inverse Property of Exponents and Logarithms

$\frac{\ln 0.5}{-5760} = k$ Divide each side by -5760.

$0.00012 \approx k$ Use a calculator.

The constant for Carbon-14 is 0.00012. Thus, the equation for the decay of Carbon-14 is $y = ae^{-0.00012t}$, where t is given in years.

b. A paleontologist examining the bones of a woolly mammoth estimates that they contain only 3% as much Carbon-14 as they would have contained when the animal was alive. How long ago did the mammoth die?

Let a be the initial amount of Carbon-14 in the animal's body. Then the amount y that remains after t years is 3% of a or $0.03a$.

$y = ae^{-0.00012t}$ Formula for the decay of Carbon-14

$0.03a = ae^{-0.00012t}$ Replace y with $0.03a$.

$0.03 = e^{-0.00012t}$ Divide each side by a.

$\ln 0.03 = \ln e^{-0.00012t}$ Property of Equality for Logarithms

$\ln 0.03 = -0.00012t$ Inverse Property of Exponents and Logarithms

$\frac{\ln 0.03}{-0.00012} = t$ Divide each side by -0.00012.

$29{,}221 \approx t$ Use a calculator.

The mammoth lived about 29,000 years ago.

Career Choices

Paleontologist

Paleontologists study fossils found in geological formations. They use these fossils to trace the evolution of plant and animal life and the geologic history of Earth.

Online Research For information about a career as a paleontologist, visit: www.algebra2.com/careers

Source: U.S. Department of Labor

www.algebra2.com/extra_examples

2 Teach

EXPONENTIAL DECAY

Teaching Tip In Example 1, point out that you are calculating how long until half the caffeine has been eliminated, which also means half the caffeine remains. If the value to be found is something other than half, students must be careful that they use the formula correctly.

In-Class Examples

Power Point®

1 **CAFFEINE** Refer to Example 1. How long will it take for 90% of this caffeine to be eliminated from a person's body? **about 20 h**

2 **GEOLOGY** The half-life of Sodium-22 is 2.6 years.

a. What is the value of k for Sodium-22? **about 0.2666**

b. A geologist examining a meteorite estimates that it contains only about 10% as much Sodium-22 as it would have contained when it reached Earth's surface. How long ago did the meteorite reach the surface of Earth? **about 9 years ago**

DAILY INTERVENTION

Differentiated Instruction

Logical Have students work in pairs or small groups. Ask them to examine the growth and decay formulas used in Examples 1–4 and to discuss how the equations are related. In particular, ask them to discuss how they can identify which equations are used for exponential decay situations (minus/negative sign) and which are used for exponential growth.

EXPONENTIAL GROWTH

In-Class Examples Power Point®

3 The population of a city of one million is increasing at a rate of 3% per year. If the population continues to grow at this rate, in how many years will the population have doubled? **D**

A 4 years **B** 5 years

C 20 years **D** 23 years

4 **POPULATION** As of 2000, Nigeria had an estimated population of 127 million people and the United States had an estimated population of 278 million people. The populations of Nigeria and the United States can be modeled by $N(t) = 127e^{0.026t}$ and $U(t) = 278e^{0.009t}$, respectively. According to these models, when will Nigeria's population be more than the population of the United States? **after 46 years or in 2046**

EXPONENTIAL GROWTH When a quantity *increases* by a fixed percent each time period, the amount y of that quantity after t time periods is given by $y = a(1 + r)^t$, where a is the initial amount and r is the percent of increase expressed as a decimal. The percent of increase r is also referred to as the **rate of growth**.

Standardized Test Practice

Example 3 Exponential Growth of the Form $y = a(1 + r)^t$

Multiple-Choice Test Item

> In 1910, the population of a city was 120,000. Since then, the population has increased by exactly 1.5% per year. If the population continues to grow at this rate, what will the population be in 2010?
>
> (A) 138,000 (B) 531,845
>
> (C) 1,063,690 (D) 1.4×10^{11}

Test-Taking Tip

To change a percent to a decimal, drop the percent symbol and move the decimal point two places to the left.

1.5% = 0.015

Read the Test Item

You need to find the population of the city 2010 − 1910 or 100 years later. Since the population is growing at a fixed percent each year, use the formula $y = a(1 + r)^t$.

Solve the Test Item

$y = a(1 + r)^t$	Exponential growth formula
$y = 120{,}000(1 + 0.015)^{100}$	Replace a = 120,000, r with 0.015, and t with 2010 − 1910 or 100.
$y = 120{,}000(1.015)^{100}$	Simplify.
$y \approx 531{,}845.48$	Use a calculator.

The answer is B.

Another model for exponential growth, preferred by scientists, is $y = ae^{kt}$, where k is a constant. Use this model to find the constant k.

Example 4 Exponential Growth of the Form $y = ae^{kt}$

POPULATION **As of 2000, China was the world's most populous country, with an estimated population of 1.26 billion people. The second most populous country was India, with 1.01 billion. The populations of India and China can be modeled by $I(t) = 1.01e^{0.015t}$ and $C(t) = 1.26e^{0.009t}$, respectively. According to these models, when will India's population be more than China's?**

You want to find t such that $I(t) > C(t)$.

$I(t) > C(t)$	
$1.01e^{0.015t} > 1.26e^{0.009t}$	Replace $I(t)$ with $1.01e^{0.015t}$ and $C(t)$ with $1.26e^{0.009t}$
$\ln 1.01e^{0.015t} > \ln 1.26e^{0.009t}$	Property of Inequality for Logarithms
$\ln 1.01 + \ln e^{0.015t} > \ln 1.26 + \ln e^{0.009t}$	Product Property of Logarithms
$\ln 1.01 + 0.015t > \ln 1.26 + 0.009t$	Inverse Property of Exponents and Logarithms
$0.006t > \ln 1.26 - \ln 1.01$	Subtract $0.009t$ from each side.
$t > \dfrac{\ln 1.26 - \ln 1.01}{0.006}$	Divide each side by 0.006.
$t > 36.86$	Use a calculator.

After 37 years or in 2037, India will be the most populous country in the world.

Standardized Test Practice

Example 3 On all standardized tests, students should look to identify any answer choices that can be logically eliminated. In Example 3, students can quickly determine that since 1% of 120,000 is 1200 and therefore 1.5% is 1800, over the 100 years from 1920 to 2010 the city's population will have increased by more than 1800(100) or 180,000 people. So Choice A is much too low. Choice D can also be eliminated, because 1.4×10^{11} written in standard notation is 140,000,000,000 (which is 140 billion). That's more than the population of the entire planet! So, the answer must be either Choice B or Choice C.

562 Chapter 10 Exponential and Logarithmic Relations

Check for Understanding

Concept Check

1. **Write** a general formula for exponential growth and decay where r is the percent of change. $y = a(1 + r)^t$, where $r > 0$ represents exponential growth and $r < 0$ represents exponential decay
2. **Explain** how to solve $y = (1 + r)^t$ for t. See margin.
3. **OPEN ENDED** Give an example of a quantity that grows or decays at a fixed rate. Sample answer: money in a bank

Guided Practice

SPACE **For Exercises 4–6, use the following information.**
A radioisotope is used as a power source for a satellite. The power output P (in watts) is given by $P = 50e^{-\frac{t}{250}}$, where t is the time in days.

GUIDED PRACTICE KEY	
Exercises	Examples
4–6	2
7, 8	4
9	1, 3

4. Is the formula for power output an example of exponential growth or decay? Explain your reasoning. Decay; the exponent is negative.
5. Find the power available after 100 days. about 33.5 watts
6. Ten watts of power are required to operate the equipment in the satellite. How long can the satellite continue to operate? about 402 days

POPULATION GROWTH **For Exercises 7 and 8, use the following information.**
The city of Raleigh, North Carolina, grew from a population of 212,000 in 1990 to a population of 259,000 in 1998.

7. Write an exponential growth equation of the form $y = ae^{kt}$ for Raleigh, where t is the number of years after 1990. $y = 212{,}000e^{0.025t}$
8. Use your equation to predict the population of Raleigh in 2010. about 349,529 people

Standardized Test Practice

9. Suppose the weight of a bar of soap decreases by 2.5% each time it is used. If the bar weighs 95 grams when it is new, what is its weight to the nearest gram after 15 uses? C

   (A) 57.5 g (B) 59.4 g (C) 65 g (D) 93 g

★ indicates increased difficulty

Practice and Apply

Homework Help	
For Exercises	See Examples
10	1
12–14,	2
11, 17–20	3
15, 16	4

Extra Practice
See page 851.

10. **COMPUTERS** Zeus Industries bought a computer for \$2500. It is expected to depreciate at a rate of 20% per year. What will the value of the computer be in 2 years? \$1600
11. **REAL ESTATE** The Martins bought a condominium for \$85,000. Assuming that the value of the condo will appreciate at most 5% a year, how much will the condo be worth in 5 years? at most \$108,483.93
12. **MEDICINE** Radioactive iodine is used to determine the health of the thyroid gland. It decays according to the equation $y = ae^{-0.0856t}$, where t is in days. Find the half-life of this substance. about 8.1 days
13. **PALEONTOLOGY** A paleontologist finds a bone that might be a dinosaur bone. In the laboratory, she finds that the Carbon-14 found in the bone is $\frac{1}{12}$ of that found in living bone tissue. Could this bone have belonged to a dinosaur? Explain your reasoning. (*Hint:* The dinosaurs lived from 220 million years ago to 63 million years ago.) No; the bone is only about 21,000 years old, and dinosaurs died out 63,000,000 years ago.
14. **ANTHROPOLOGY** An anthropologist finds there is so little remaining Carbon-14 in a prehistoric bone that instruments cannot measure it. This means that there is less than 0.5% of the amount of Carbon-14 the bones would have contained when the person was alive. How long ago did the person die?

14. more than 44,000 years ago

www.algebra2.com/self_check_quiz

3 Practice/Apply

Study Notebook

Have students—
- complete the definitions/examples for the remaining terms on their Vocabulary Builder worksheets for Chapter 10.
- record the formulas for exponential growth and decay.
- include any other item(s) that they find helpful in mastering the skills in this lesson.

About the Exercises...

Organization by Objective
- **Exponential Decay:** 10–20
- **Exponential Growth:** 10–20

Assignment Guide

Basic: 11, 13, 17, 18, 21–40

Average: 11, 13, 15–18, 21–40

Advanced: 10–14 even, 15, 16, 19–40

Answer

2. Take the common logarithm of each side, use the Power Property to write $\log (1 + r)^t$ as $t \log (1 + r)$, and then divide each side by the quantity $\log (1 + r)$.

Study Guide and Intervention, p. 603 (shown) and p. 604

Exponential Decay Depreciation of value and radioactive decay are examples of **exponential decay**. When a quantity decreases by a fixed percent each time period, the amount of the quantity after t time periods is given by $y = a(1 - r)^t$, where a is the initial amount and r is the percent decrease expressed as a decimal.

Another exponential decay model often used by scientists is $y = ae^{-kt}$, where k is a constant.

Example **CONSUMER PRICES As technology advances, the price of many technological devices such as scientific calculators and camcorders goes down. One brand of hand-held organizer sells for $89.**

a. If its price decreases by 6% per year, how much will it cost after 5 years?

Use the exponential decay model with initial amount $89, percent decrease 0.06, and time 5 years.

$y = a(1 - r)^t$ Exponential decay formula

$y = 89(1 - 0.06)^5$ $a = 89, r = 0.06, t = 5$

$y = \$65.32$

After 5 years the price will be $65.32.

b. After how many years will its price be $50?

To find when the price will be $50, again use the exponential decay formula and solve for t.

$y = a(1 - r)^t$ Exponential decay formula

$50 = 89(1 - 0.06)^t$ $y = 50, a = 89, r = 0.06$

$\frac{50}{89} = (0.94)^t$ Divide each side by 89.

$\log\left(\frac{50}{89}\right) = \log(0.94)^t$ Property of Equality for Logarithms

$\log\left(\frac{50}{89}\right) = t \log 0.94$ Power Property

$t = \frac{\log\left(\frac{50}{89}\right)}{\log 0.94}$ Divide each side by log 0.94.

$t \approx 9.3$

The price will be $50 after about 9.3 years.

Exercises

1. **BUSINESS** A furniture store is closing out its business. Each week the owner lowers prices by 25%. After how many weeks will the sale price of a $500 item drop below $100? **6 weeks**

CARBON DATING Use the formula $y = ae^{-0.00012t}$, where a is the initial amount of Carbon-14, t is the number of years ago the animal lived, and y is the remaining amount after t years.

2. How old is a fossil remain that has lost 95% of its Carbon-14? **about 25,000 years old**
3. How old is a skeleton that has 95% of its Carbon-14 remaining? **about 427.5 years old**

Skills Practice, p. 605 and Practice, p. 606 (shown)

Solve each problem.

1. **INVESTING** The formula $A = P\left(1 + \frac{r}{2}\right)^{2t}$ gives the value of an investment after t years in an account that earns an annual interest rate r compounded twice a year. Suppose $500 is invested at 6% annual interest compounded twice a year. In how many years will the investment be worth $1000? **about 11.7 yr**
2. **BACTERIA** How many hours will it take a culture of bacteria to increase from 20 to 2000 if the growth rate per hour is 85%? **about 7.5 h**
3. **RADIOACTIVE DECAY** A radioactive substance has a half-life of 32 years. Find the constant k in the decay formula for the substance. **about 0.02166**
4. **DEPRECIATION** A piece of machinery valued at $250,000 depreciates at a fixed rate of 12% per year. After how many years will the value have depreciated to $100,000? **about 7.2 yr**
5. **INFLATION** For Dave to buy a new car comparably equipped to the one he bought 8 years ago would cost $12,500. Since Dave bought the car, the inflation rate for cars like his has been at an average annual rate of 5.1%. If Dave originally paid $8400 for the car, how long ago did he buy it? **about 8 yr**
6. **RADIOACTIVE DECAY** Cobalt, an element used to make alloys, has several isotopes. One of these, cobalt-60, is radioactive and has a half-life of 5.7 years. Cobalt-60 is used to trace the path of nonradioactive substances in a system. What is the value of k for Cobalt-60? **about 0.1216**
7. **WHALES** Modern whales appeared 5–10 million years ago. The vertebrae of a whale discovered by paleontologists contain roughly 0.25% as much carbon-14 as they would have contained when the whale was alive. How long ago did the whale die? Use $k = 0.00012$. **about 50,000 yr**
8. **POPULATION** The population of rabbits in an area is modeled by the growth equation $P(t) = 8e^{0.26t}$, where P is in thousands and t is in years. How long will it take for the population to reach 25,000? **about 4.4 yr**
9. **DEPRECIATION** A computer system depreciates at an average rate of 4% per month. If the value of the computer system was originally $12,000, in how many months is it worth $7350? **about 12 mo**
10. **BIOLOGY** In a laboratory, a culture increases from 30 to 195 organisms in 5 hours. What is the hourly growth rate in the growth formula $y = a(1 + r)^t$? **about 45.4%**

Reading to Learn Mathematics, p. 607 ELL

Pre-Activity **How can you determine the current value of your car?**

Read the introduction to Lesson 10-6 at the top of page 560 in your textbook.

- Between which two years shown in the table did the car depreciate by the greatest amount?
  between years 0 and 1
- Describe two ways to calculate the value of the car 6 years after it was purchased. (Do not actually calculate the value.)
  Sample answer: 1. Multiply $9200.66 by 0.16 and subtract the result from $9200.66. 2. Multiply $9200.66 by 0.84.

Reading the Lesson

1. State whether each situation is an example of exponential *growth* or *decay*.
   a. A city had 42,000 residents in 1980 and 128,000 residents in 2000. **growth**
   b. Raul compared the value of his car when he bought it new to the value when he traded ';lpit in six years later. **decay**
   c. A paleontologist compared the amount of carbon-14 in the skeleton of an animal when it died to the amount 300 years later. **decay**
   d. Maria deposited $750 in a savings account paying 4.5% annual interest compounded quarterly. She did not make any withdrawals or further deposits. She compared the balance in her passbook immediately after she opened the account to the balance 3 years later. **growth**
2. State whether each equation represents exponential growth or decay.
   a. $y = 5e^{0.15t}$ **growth** b. $y = 1000(1 - 0.05)^t$ **decay**
   c. $y = 0.3e^{-1200t}$ **decay** d. $y = 2(1 + 0.0001)^t$ **growth**

Helping You Remember

3. Visualizing their graphs is often a good way to remember the difference between mathematical equations. How can your knowledge of the graphs of exponential equations from Lesson 10-1 help you to remember that equations of the form $y = a(1 + r)^t$ represent exponential growth, while equations of the form $y = a(1 - r)^t$ represent exponential decay?
   Sample answer: If $a > 0$, the graph of $y = ab^x$ is always increasing if $b > 1$ and is always decreasing if $0 < b < 1$. Since r is always a positive number, if $b = 1 + r$, the base will be greater than 1 and the function will be increasing (growth), while if $b = 1 - r$, the base will be less than 1 and the function will be decreasing (decay).

More About. . .

Olympics

The women's high jump competition first took place in the USA in 1895, but it did not become an Olympic event until 1928.

Source: www.princeton.edu

BIOLOGY **For Exercises 15 and 16, use the following information.**
Bacteria usually reproduce by a process known as *binary fission*. In this type of reproduction, one bacterium divides, forming two bacteria. Under ideal conditions, some bacteria reproduce every 20 minutes. **15. about 0.0347**

15. Find the constant k for this type of bacteria under ideal conditions.
16. Write the equation for modeling the exponential growth of this bacterium. $y = ae^{0.0347t}$

ECONOMICS **For Exercises 17 and 18, use the following information.**
The annual Gross Domestic Product (GDP) of a country is the value of all of the goods and services produced in the country during a year. During the period 1985–1999, the Gross Domestic Product of the United States grew about 3.2% per year, measured in 1996 dollars. In 1985, the GDP was $5717 billion.

17. Assuming this rate of growth continues, what will the GDP of the United States be in the year 2010? **$12,565 billion**
18. In what year will the GDP reach $20 trillion? **about 2025**
19. ★ **OLYMPICS** In 1928, when the high jump was first introduced as a women's sport at the Olympic Games, the winning women's jump was 62.5 inches, while the winning men's jump was 76.5 inches. Since then, the winning jump for women has increased by about 0.38% per year, while the winning jump for men has increased at a slower rate, 0.3%. If these rates continue, when will the women's winning high jump be higher than the men's? **after the year 2182**
20. ★ **HOME OWNERSHIP** The Mendes family bought a new house 10 years ago for $120,000. The house is now worth $191,000. Assuming a steady rate of growth, what was the yearly rate of appreciation? **4.8%**
21. **CRITICAL THINKING** The half-life of Radium is 1620 years. When will a 20-gram sample of Radium be completely gone? Explain your reasoning. **Never; theoretically, the amount left will always be half of the previous amount.**
22. *WRITING IN MATH* Answer the question that was posed at the beginning of the lesson. **See margin.**

How can you determine the current value of your car?

Include the following in your answer:

- a description of how to find the percent decrease in the value of the car each year, and
- a description of how to find the value of a car for any given year when the rate of depreciation is known.

Standardized Test Practice

23. **SHORT RESPONSE** An artist creates a sculpture out of salt that weighs 2000 pounds. If the sculpture loses 3.5% of its mass each year to erosion, after how many years will the statue weigh less than 1000 pounds? **about 19.5 yr**
24. The curve shown at the right represents a portion of the graph of which function? **D**

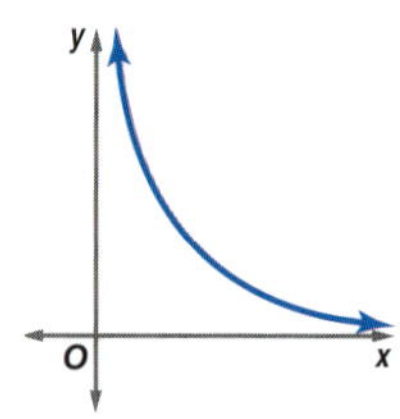

Ⓐ $y = 50 - x$ Ⓑ $y = \log x$

Ⓒ $y = e^{-x}$ Ⓓ $xy = 5$

Enrichment, p. 608

Effective Annual Yield

When interest is compounded more than once per year, the effective annual yield is higher than the annual interest rate. The effective annual yield, E, is the interest rate that would give the same amount of interest if the interest were compounded once per year. If P dollars are invested for one year, the value of the investment at the end of the year is $A = P(1 + E)$. If P dollars are invested for one year at a nominal rate r compounded n times per year, the value of the investment at the end of the year is $A = P\left(1 + \frac{r}{n}\right)^n$. Setting the amounts equal and solving for E will produce a formula for the effective annual yield.

$$P(1 + E) = P\left(1 + \frac{r}{n}\right)^n$$

$$1 + E = \left(1 + \frac{r}{n}\right)^n$$

$$E = \left(1 + \frac{r}{n}\right)^n - 1$$

Maintain Your Skills

Mixed Review **Write an equivalent exponential or logarithmic equation.** *(Lesson 10-5)*

25. $e^3 = y$ **$\ln y = 3$**
26. $e^{4n-2} = 29$ **$\ln 29 = 4n - 2$**
27. $\ln 4 + 2 \ln x = 8$ **$4x^2 = e^8$**

Solve each equation or inequality. Round to four decimal places. *(Lesson 10-4)*

28. $16^x = 70$ **1.5323**
29. $2^{3p} > 1000$ **$p > 3.3219$**
30. $\log_b 81 = 2$ **9**

BUSINESS **For Exercises 31–33, use the following information.**
The board of a small corporation decided that 8% of the annual profits would be divided among the six managers of the corporation. There are two sales managers and four nonsales managers. Fifty percent of the amount would be split equally among all six managers. The other 50% would be split among the four nonsales managers. Let p represent the annual profits of the corporation. *(Lesson 9-2)*

31. Write an expression to represent the share of the profits each nonsales manager will receive. **$\frac{0.5(0.08p)}{6} + \frac{0.5(0.08p)}{4}$**
32. Simplify this expression. **$\frac{p}{60}$**
33. Write an expression in simplest form to represent the share of the profits each sales manager will receive. **$\frac{p}{150}$**

Without writing the equation in standard form, state whether the graph of each equation is a *parabola, circle, ellipse,* or *hyperbola.* *(Lesson 8-6)*

34. $4y^2 - 3x^2 + 8y - 24x = 50$ **hyperbola**
35. $7x^2 - 42x + 6y^2 - 24y = -45$ **ellipse**
36. $y^2 + 3x - 8y = 4$ **parabola**
37. $x^2 + y^2 - 6x + 2y + 5 = 0$ **circle**

AGRICULTURE **For Exercises 38–40, use the graph at the right.** *(Lesson 5-1)*

38. Write the number of pounds of pecans produced by U.S. growers in 2000 in scientific notation. **2.06×10^8**
39. Write the number of pounds of pecans produced by the state of Georgia in 2000 in scientific notation. **8×10^7**
40. What percent of the overall pecan production for 2000 can be attributed to Georgia? **about 38.8%**

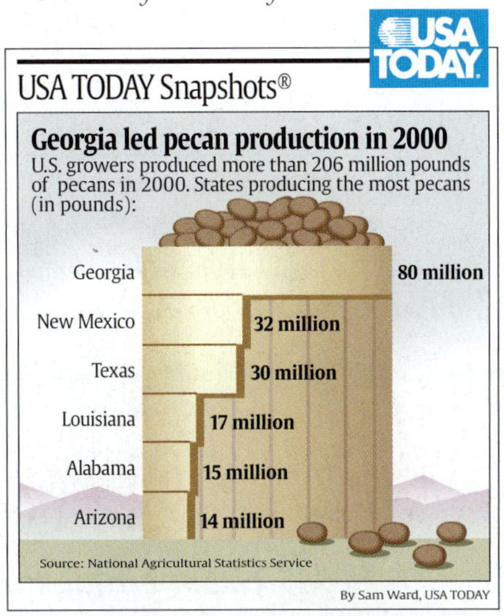

WebQuest Internet Project

On Quake Anniversary, Japan Still Worries

It is time to complete your project. Use the information and data you have gathered about earthquakes to prepare a research report or Web page. Be sure to include graphs, tables, diagrams, and any calculations you need for the earthquake you chose.

www.algebra2.com/webquest

4 Assess

Open-Ended Assessment

Modeling Using manipulatives, ask students to demonstrate why the amount of compound interest earned annually increases each year. Have students relate this to their understanding of exponential growth.

Assessment Options

Quiz (Lesson 10-6) is available on p. 624 of the *Chapter 10 Resource Masters*.

Answer

22. Answers should include the following.
    - Find the absolute value of the difference between the price of the car for two consecutive years. Then divide this difference by the price of the car for the earlier year.
    - Find 1 minus the rate of decrease in the value of the car as a decimal. Raise this value to the number of years it has been since the car was purchased, and then multiply by the original value of the car.

Online Lesson Plans

USA TODAY Education's Online site offers resources and interactive features connected to each day's newspaper. *Experience TODAY*, USA TODAY's daily lesson plan, is available on the site and delivered daily to subscribers. This plan provides instruction for integrating USA TODAY graphics and key editorial features into your mathematics classroom. Log on to www.education.usatoday.com.

Chapter 10 Study Guide and Review

Vocabulary and Concept Check

- This alphabetical list of vocabulary terms in Chapter 10 includes a page reference where each term was introduced.
- **Assessment** A vocabulary test/review for Chapter 10 is available on p. 622 of the *Chapter 10 Resource Masters.*

Lesson-by-Lesson Review

For each lesson,
- the main ideas are summarized,
- additional examples review concepts, and
- practice exercises are provided.

Vocabulary PuzzleMaker

ELL The Vocabulary PuzzleMaker software improves students' mathematics vocabulary using four puzzle formats—crossword, scramble, word search using a word list, and word search using clues. Students can work on a computer screen or from a printed handout.

MindJogger Videoquizzes

ELL MindJogger Videoquizzes provide an alternative review of concepts presented in this chapter. Students work in teams in a game show format to gain points for correct answers. The questions are presented in three rounds.

Round 1 Concepts (5 questions)
Round 2 Skills (4 questions)
Round 3 Problem Solving (4 questions)

Chapter 10 Study Guide and Review

Vocabulary and Concept Check

Change of Base Formula (p. 548)
common logarithm (p. 547)
exponential decay (p. 524)
exponential equation (p. 526)
exponential function (p. 524)
exponential growth (p. 524)
exponential inequality (p. 527)
logarithm (p. 531)
logarithmic equation (p. 533)
logarithmic function (p. 532)
logarithmic inequality (p. 533)
natural base, *e* (p. 554)
natural base exponential function (p. 554)
natural logarithm (p. 554)
natural logarithmic function (p. 554)
Power Property of Logarithms (p. 543)
Product Property of Logarithms (p. 541)
Property of Equality for Exponential Functions (p. 526)
Property of Equality for Logarithmic Functions (p. 534)
Property of Inequality for Exponential Functions (p. 527)
Property of Inequality for Logarithmic Functions (p. 534)
Quotient Property of Logarithms (p. 542)
rate of decay (p. 560)
rate of growth (p. 562)

State whether each sentence is *true* or *false*. If false, replace the underlined word(s) to make a true statement.

1. If $24^{2y+3} = 24^{y-4}$, then $2y + 3 = y - 4$ by the *Property of Equality for Exponential Functions*. **true**
2. The number of bacteria in a petri dish over time is an example of *exponential decay*. **false; exponential growth**
3. The *natural logarithm* is the inverse of the exponential function with base 10.
4. The *Power Property of Logarithms* shows that $\ln 9 < \ln 81$.
5. If a savings account yields 2% interest per year, then 2% is the *rate of growth*.
6. Radioactive half-life is an example of *exponential decay*. **true**
7. The inverse of an exponential function is a *composite function*.
8. The *Quotient Property of Logarithms* is shown by $\log_4 2x = \log_4 2 + \log_4 x$.
9. The function $f(x) = 2(5)^x$ is an example of a *quadratic function*. **false; exponential function**

3. false; common logarithm
4. false; Property of Inequality for Logarithms
5. true
7. false; logarithmic function
8. false; Product Property of Logarithms

Lesson-by-Lesson Review

10-1 Exponential Functions

See pages 523–530.

Concept Summary

- An exponential function is in the form $y = ab^x$, where $a \neq 0$, $b > 0$, and $b \neq 1$.
- The function $y = ab^x$ represents exponential growth for $a > 0$ and $b > 1$, and exponential decay for $a > 0$ and $0 < b < 1$.
- Property of Equality for Exponential Functions: If b is a positive number other than 1, then $b^x = b^y$ if and only if $x = y$.
- Property of Inequality for Exponential Functions: If $b > 1$, then $b^x > b^y$ if and only if $x > y$, and $b^x < b^y$ if and only if $x < y$.

www.algebra2.com/vocabulary_review

FOLDABLES™ Study Organizer

For more information about Foldables, see *Teaching Mathematics with Foldables.*

Have students look through the chapter to make sure they have included notes and examples for each lesson in this chapter in their Foldable.

Encourage students to refer to their Foldables while completing the Study Guide and Review and to use them in preparing for the Chapter Test.

Study Guide and Review

Example Solve $64 = 2^{3n+1}$ for n.

$64 = 2^{3n+1}$ Original equation

$2^6 = 2^{3n+1}$ Rewrite 64 as 2^6 so each side has the same base.

$6 = 3n + 1$ Property of Equality for Exponential Functions

$\frac{5}{3} = n$ The solution is $\frac{5}{3}$.

Exercises **Determine whether each function represents exponential** ***growth*** **or** ***decay.*** *See Example 2 on page 525.*

10. $y = 5(0.7)^x$ **decay**

11. $y = \frac{1}{3}(4)^x$ **growth**

Write an exponential function whose graph passes through the given points. *See Example 3 on page 525.*

12. (0, −2) and (3, −54) $y = -2(3)^x$

13. (0, 7) and (1, 1.4) $y = 7\left(\frac{1}{5}\right)^x$

Solve each equation or inequality. *See Examples 5 and 6 on pages 526 and 527.*

14. $9^x = \frac{1}{81}$ -2

15. $2^{6x} = 4^{5x+2}$ -1

16. $49^{3p+1} = 7^{2p-5}$ $-\frac{7}{4}$

17. $9^{x^2} \leq 27^{x^2-2}$ $x \leq -\sqrt{6}$ or $x \geq \sqrt{6}$

10-2 Logarithms and Logarithmic Functions

See pages 531–538.

Concept Summary

- Suppose $b > 0$ and $b \neq 1$. For $x > 0$, there is a number y such that $\log_b x = y$ if and only if $b^y = x$.
- Logarithmic to exponential inequality: If $b > 1$, $x > 0$, and $\log_b x > y$, then $x > b^y$.

  If $b > 1$, $x > 0$, and $\log_b x < y$, then $0 < x < b^y$.
- Property of Equality for Logarithmic Functions: If b is a positive number other than 1, then $\log_b x = \log_b y$ if and only if $x = y$.
- Property of Inequality for Logarithmic Functions: If $b > 1$, then $\log_b x > \log_b y$ if and only if $x > y$, and $\log_b x < \log_b y$ if and only if $x < y$.

Examples

$\log_7 x = 2 \rightarrow 7^2 = x$

$\log_2 x > 5 \rightarrow x > 2^5$

$\log_3 x < 4 \rightarrow 0 < x < 3^4$

If $\log_5 x = \log_5 6$, then $x = 6$.

If $\log_4 x > \log_4 10$, then $x > 10$.

Examples **1** Solve $\log_9 n > \frac{3}{2}$.

$\log_9 n > \frac{3}{2}$ Original inequality

$n > 9^{\frac{3}{2}}$ Logarithmic to exponential inequality

$n > (3^2)^{\frac{3}{2}}$ $9 = 3^2$

$n > 3^3$ Power of a Power

$n > 27$ Simplify.

Chapter 10 Study Guide and Review 567

2 Solve $\log_3 12 = \log_3 2x$.

$\log_3 12 = \log_3 2x$	Original equation
$12 = 2x$	Property of Equality for Logarithmic Functions
$6 = x$	Divide each side by 2.

Exercises Write each equation in logarithmic form. *See Example 1 on page 532.*

18. $7^3 = 343$ $\log_7 343 = 3$ 19. $5^{-2} = \frac{1}{25}$ $\log_5 \frac{1}{25} = -2$ 20. $4^{\frac{3}{2}} = 8$ $\log_4 8 = \frac{3}{2}$

Write each equation in exponential form. *See Example 2 on page 532.* **23.** $6^{-2} = \frac{1}{36}$

21. $\log_4 64 = 3$ $4^3 = 64$ 22. $\log_8 2 = \frac{1}{3}$ $8^{\frac{1}{3}} = 2$ 23. $\log_6 \frac{1}{36} = -2$

Evaluate each expression. *See Examples 3 and 4 on pages 532 and 533.*

24. $4^{\log_4 9}$ 9 25. $\log_7 7^{-5}$ -5 26. $\log_{81} 3$ $\frac{1}{4}$ 27. $\log_{13} 169$ 2

Solve each equation or inequality. *See Examples 5–8 on pages 533 and 534.*

28. $\log_4 x = \frac{1}{2}$ 2
29. $\log_{81} 729 = x$ $\frac{3}{2}$
30. $\log_b 9 = 2$ 3
31. $\log_8 (3y - 1) < \log_8 (y + 5)$ $\frac{1}{3} < y < 3$
32. $\log_5 12 < \log_5 (5x - 3)$ $x > 3$
33. $\log_8 (x^2 + x) = \log_8 12$ $-4, 3$

10-3 Properties of Logarithms

See pages 541–546.

Concept Summary

- The logarithm of a product is the sum of the logarithms of its factors.
- The logarithm of a quotient is the difference of the logarithms of the numerator and the denominator.
- The logarithm of a power is the product of the logarithm and the exponent.

Example Use $\log_{12} 9 \approx 0.884$ and $\log_{12} 18 \approx 1.163$ to approximate the value of $\log_{12} 2$.

$\log_{12} 2 = \log_{12} \frac{18}{9}$	Replace 2 with $\frac{18}{9}$.
$= \log_{12} 18 - \log_{12} 9$	Quotient Property
$\approx 1.163 - 0.884$ or 0.279	Replace $\log_{12} 9$ with 0.884 and $\log_{12} 18$ with 1.163.

Exercises Use $\log_9 7 \approx 0.8856$ and $\log_9 4 \approx 0.6309$ to approximate the value of each expression. *See Examples 1 and 2 on page 542.*

34. $\log_9 28$ 1.5165 35. $\log_9 49$ 1.7712 36. $\log_9 144$ 2.2618

Solve each equation. *See Example 5 on page 543.*

37. $\log_2 y = \frac{1}{3} \log_2 27$ 3
38. $\log_5 7 + \frac{1}{2} \log_5 4 = \log_5 x$ 14
39. $2 \log_2 x - \log_2 (x + 3) = 2$ 6
40. $\log_3 x - \log_3 4 = \log_3 12$ 48
41. $\log_6 48 - \log_6 \frac{16}{5} + \log_6 5 = \log_6 5x$ 15
42. $\log_7 m = \frac{1}{3} \log_7 64 + \frac{1}{2} \log_7 121$ 44

568 Chapter 10 Exponential and Logarithmic Relations

Study Guide and Review

10-4 Common Logarithms

See pages 547–551.

Concept Summary

- Base 10 logarithms are called common logarithms and are usually written without the subscript 10: $\log_{10} x = \log x$.
- You use the inverse of logarithms, or exponentiation, to solve equations or inequalities involving common logarithms: $10^{\log x} = x$.
- The Change of Base Formula: $\log_a n = \dfrac{\log_b n}{\log_b a}$ ← log base b original number; ← log base b old base

Example **Solve $5^x = 7$.**

$5^x = 7$	Original equation
$\log 5^x = \log 7$	Property of Equality for Logarithmic Functions
$x \log 5 = \log 7$	Power Property of Logarithms
$x = \dfrac{\log 7}{\log 5}$	Divide each side by log 5.
$x \approx \dfrac{0.8451}{0.6990}$ or 1.2090	Use a calculator.

Exercises **Solve each equation or inequality. Round to four decimal places.**
See Examples 3 and 4 on page 548.

43. $2^x = 53$ **5.7279**
44. $2.3^{x^2} = 66.6$ **±2.2452**
45. $3^{4x-7} < 4^{2x+3}$ **$x < 7.3059$**
46. $6^{3y} = 8^{y-1}$ **−0.6309**
47. $12^{x-5} \geq 9.32$ **$x \geq 5.8983$**
48. $2.1^{x-5} = 9.32$ **8.0086**

Express each logarithm in terms of common logarithms. Then approximate its value to four decimal places. *See Example 5 on page 549.*

49. $\log_4 11$ $\dfrac{\log 11}{\log 4}$; **1.7297**
50. $\log_2 15$ $\dfrac{\log 15}{\log 2}$; **3.9069**
51. $\log_{20} 1000$ $\dfrac{\log 1000}{\log 20}$; **2.3059**

10-5 Base e and Natural Logarithms

See pages 554–559.

Concept Summary

- You can write an equivalent base e exponential equation for a natural logarithmic equation and vice versa by using the fact that $\ln x = \log_e x$.
- Since the natural base function and the natural logarithmic function are inverses, these two functions can be used to "undo" each other. $e^{\ln x} = x$ and $\ln e^x = x$

Example **Solve $\ln (x + 4) > 5$.**

$\ln (x + 4) > 5$	Original inequality
$e^{\ln (x + 4)} > e^5$	Write each side using exponents and base e.
$x + 4 > e^5$	Inverse Property of Exponents and Logarithms
$x > e^5 - 4$	Subtract 4 from each side.
$x > 144.4132$	Use a calculator.

Chapter 10 Study Guide and Review 569

Chapter 10 For More ...
• Extra Practice, see pages 849–851.
• Mixed Problem Solving, see page 871.

Exercises **Write an equivalent exponential or logarithmic equation.**
See Example 3 on page 555.

52. $e^x = 6$ **$\ln 6 = x$**

53. $\ln 7.4 = x$ **$e^x = 7.4$**

Evaluate each expression. *See Example 4 on page 555.*

54. $e^{\ln 12}$ **12**

55. $\ln e^{7x}$ **$7x$**

Solve each equation or inequality.
See Examples 5 and 7 on pages 555 and 556.

56. $2e^x - 4 = 1$ **0.9163**

57. $e^x > 3.2$ **$x > 1.1632$**

58. $-4e^{2x} + 15 = 7$ **0.3466**

59. $\ln 3x \leq 5$ **$0 < x \leq 49.4711$**

60. $\ln (x - 10) = 0.5$ **11.6487**

61. $\ln x + \ln 4x = 10$ **74.2066**

10-6 Exponential Growth and Decay

See pages 560–565.

Concept Summary

- Exponential decay: $y = a(1 - r)^t$ or $y = ae^{-kt}$
- Exponential growth: $y = a(1 + r)^t$ or $y = ae^{kt}$

Example BIOLOGY **A certain culture of bacteria will grow from 500 to 4000 bacteria in 1.5 hours. Find the constant k for the growth formula. Use $y = ae^{kt}$.**

$y = ae^{kt}$	Exponential growth formula
$4000 = 500e^{k(1.5)}$	Replace y with 4000, a with 500, and t with 1.5.
$8 = e^{1.5k}$	Divide each side by 500.
$\ln 8 = \ln e^{1.5k}$	Property of Equality for Logarithmic Functions
$\ln 8 = 1.5k$	Inverse Property of Exponents and Logarithms
$\frac{\ln 8}{1.5} = k$	Divide each side by 1.5.
$1.3863 \approx k$	Use a calculator.

Exercises *See Examples 1–4 on pages 560–562.*

62. **BUSINESS** Able Industries bought a fax machine for $250. It is expected to depreciate at a rate of 25% per year. What will be the value of the fax machine in 3 years? **$105.47**

63. **BIOLOGY** For a certain strain of bacteria, k is 0.872 when t is measured in days. How long will it take 9 bacteria to increase to 738 bacteria? **5.05 days**

64. **CHEMISTRY** Radium-226 decomposes radioactively. Its half-life, the time it takes for half of the sample to decompose, is 1800 years. Find the constant k in the decay formula for this compound. **about 0.000385**

65. **POPULATION** The population of a city 10 years ago was 45,600. Since then, the population has increased at a steady rate each year. If the population is currently 64,800, find the annual rate of growth for this city. **about 3.6%**

570 Chapter 10 Exponential and Logarithmic Relations

Practice Test

Vocabulary and Concepts

Choose the term that best completes each sentence.

1. The equation $y = 0.3(4)^x$ is an exponential (*growth*, *decay*) function.
2. The logarithm of a quotient is the (*sum*, *difference*) of the logarithms of the numerator and the denominator.
3. The base of a natural logarithm is (*10*, *e*).

Skills and Applications

4. Write $3^7 = 2187$ in logarithmic form. $\log_3 2187 = 7$
5. Write $\log_8 16 = \frac{4}{3}$ in exponential form. $8^{\frac{4}{3}} = 16$
6. Write an exponential function whose graph passes through (0, 0.4) and (2, 6.4). $y = 0.4(4)^x$
7. Express $\log_3 5$ in terms of common logarithms. $\frac{\log 5}{\log 3}$
8. Evaluate $\log_2 \frac{1}{32}$. -5

Use $\log_4 7 \approx 1.4037$ and $\log_4 3 \approx 0.7925$ to approximate the value of each expression.

9. $\log_4 21$ 2.1962
10. $\log_4 \frac{7}{12}$ -0.3888

Simplify each expression.

11. $\left(3^{\sqrt{8}}\right)^{\sqrt{2}}$ 81
12. $81^{\sqrt{5}} \div 3^{\sqrt{5}}$ $3^{3\sqrt{5}}$

Solve each equation or inequality. Round to four decimal places if necessary. 17. 108 19. 2, 6 22. 15

13. $2^{x-3} = \frac{1}{16}$ -1
14. $27^{2p+1} = 3^{4p-1}$ -2
15. $\log_2 x < 7$ $0 < x < 128$
16. $\log_m 144 = -2$ $\frac{1}{12}$
17. $\log_3 x - 2\log_3 2 = 3\log_3 3$
18. $\log_9 (x + 4) + \log_9 (x - 4) = 1$ 5
19. $\log_5 (8y - 7) = \log_5 (y^2 + 5)$
20. $\log_3 3^{(4x-1)} = 15$ 4
21. $7.6^{x-1} = 431$ 3.9910
22. $\log_2 5 + \frac{1}{3}\log_2 27 = \log_2 x$
23. $3^x = 5^{x-1}$ 3.1507
24. $4^{2x-3} = 9^{x+3}$ 18.6848
25. $e^{3y} > 6$ $y > 0.5973$
26. $2e^{3x} + 5 = 11$ 0.3662
27. $\ln 3x - \ln 15 = 2$ 36.9453

COINS **For Exercises 28 and 29, use the following information.**
You buy a commemorative coin for \$25. The value of the coin increases 3.25% per year.

28. How much will the coin be worth in 15 years? \$40.39
29. After how many years will the coin have doubled in value? 22
30. **STANDARDIZED TEST PRACTICE** Which equation represents the graph at the right? B

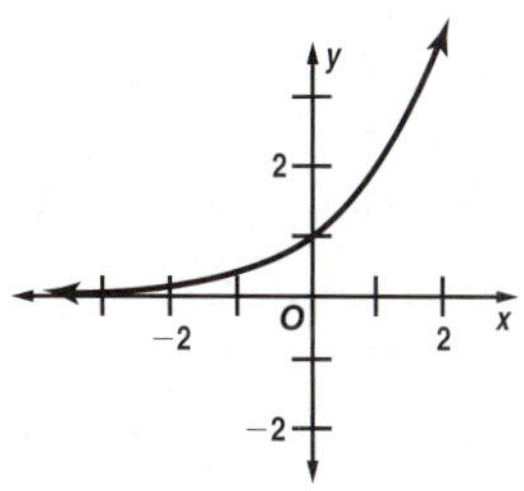

Ⓐ $y = x^2 + 2$
Ⓑ $y = 2^x$
Ⓒ $y = \log_2 x$
Ⓓ $y = 2^x + 1$

www.algebra2.com/chapter_test

Portfolio Suggestion

Introduction In mathematics, exponential functions can be used to model real-world problems. The solution to the exponential function provides a solution to the real-world problem.

Ask Students Find a real-world problem modeled by an exponential function from your work in this chapter and show how you solved it. Explain how the function models the real-world situation and what could be gained by understanding the real-world problem better. Place your work in your portfolio.

Chapter 10 Practice Test

Assessment Options

Vocabulary Test A vocabulary test/review for Chapter 10 can be found on p. 622 of the *Chapter 10 Resource Masters*.

Chapter Tests There are six Chapter 10 Tests and an Open-Ended Assessment task available in the *Chapter 10 Resource Masters*.

Chapter 10 Tests			
Form	Type	Level	Pages
1	MC	basic	609–610
2A	MC	average	611–612
2B	MC	average	613–614
2C	FR	average	615–616
2D	FR	average	617–618
3	FR	advanced	619–620

MC = multiple-choice questions
FR = free-response questions

Open-Ended Assessment
Performance tasks for Chapter 10 can be found on p. 621 of the *Chapter 10 Resource Masters*. A sample scoring rubric for these tasks appears on p. A25.

Unit 3 Test A unit test/review can be found on pp. 629–630 of the *Chapter 10 Resource Masters*.

ExamView® Pro

Use the networkable **ExamView® Pro** to:

- Create **multiple versions** of tests.
- Create **modified** tests for *Inclusion* students.
- **Edit** existing questions and **add** your own questions.
- Use built-in **state curriculum correlations** to create tests aligned with state standards.
- **Apply** art to your tests from a program bank of artwork.

Chapter 10 Standardized Test Practice

These two pages contain practice questions in the various formats that can be found on the most frequently given standardized tests.

A practice answer sheet for these two pages can be found on p. A1 of the *Chapter 10 Resource Masters*.

Standardized Test Practice Student Recording Sheet, p. A1

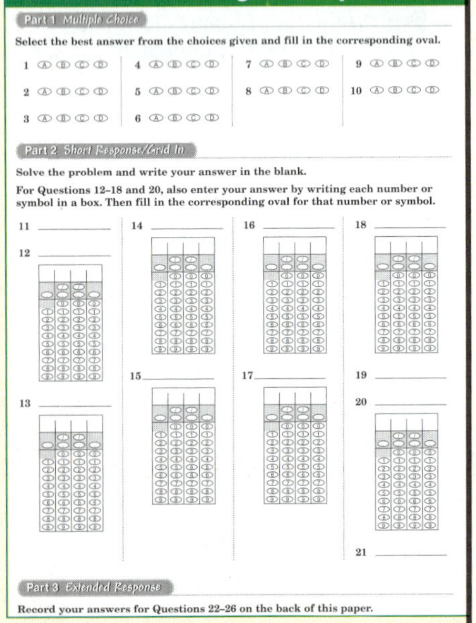
Part 1 Multiple Choice

Select the best answer from the choices given and fill in the corresponding oval.

Part 2 Short Response/Grid In

Solve the problem and write your answer in the blank.

For Questions 12–18 and 20, also enter your answer by writing each number or symbol in a box. Then fill in the corresponding oval for that number or symbol.

Part 3 Extended Response

Record your answers for Questions 22–26 on the back of this paper.

Additional Practice

See pp. 627–628 in the *Chapter 10 Resource Masters* for additional standardized test practice.

Chapter 10 Standardized Test Practice

Part 1 Multiple Choice

Record your answers on the answer sheet provided by your teacher or on a sheet of paper.

1. The arc shown is part of a circle. Find the area of the shaded region. **B**

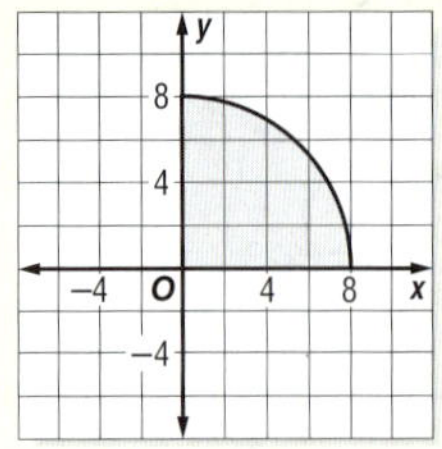

Ⓐ 8π units^2

Ⓑ 16π units^2

Ⓒ 32π units^2

Ⓓ 64π units^2

2. If line ℓ is parallel to line m in the figure below, what is the value of x? **D**

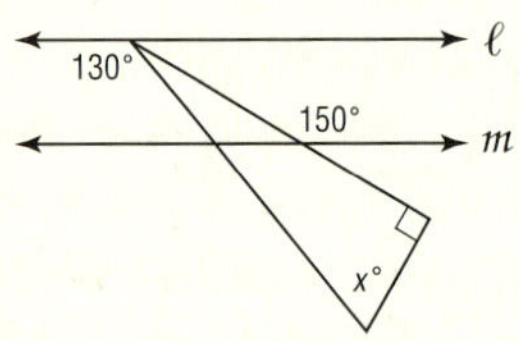

Ⓐ 40 Ⓑ 50

Ⓒ 60 Ⓓ 70

3. According to the graph, what was the percent of increase in sales from 1998 to 2000? **D**

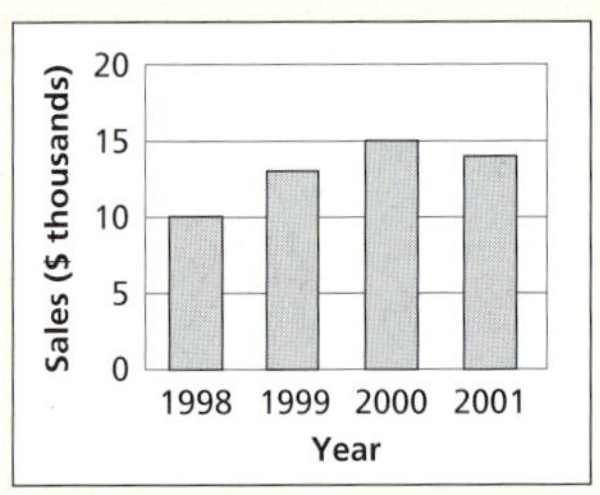

Ⓐ 5% Ⓑ 15% Ⓒ 25% Ⓓ 50%

4. What is the x-intercept of the line described by the equation $y = 2x + 5$? **B**

Ⓐ -5 Ⓑ $-\frac{5}{2}$

Ⓒ 0 Ⓓ $\frac{5}{2}$

5. $\frac{(xy)^2z^0}{y^2x^3} =$ **D**

Ⓐ $\frac{1}{x^2y}$ Ⓑ $\frac{z}{x^2}$

Ⓒ $\frac{z}{x}$ Ⓓ $\frac{1}{x}$

6. If $\frac{v^2 - 36}{6 - v} = 10$, then $v =$ **A**

Ⓐ -16. Ⓑ -4.

Ⓒ 4. Ⓓ 8.

7. The expression $\frac{1}{3}\sqrt{45}$ is equivalent to **A**

Ⓐ $\sqrt{5}$. Ⓑ $3\sqrt{5}$.

Ⓒ 5. Ⓓ 15.

8. What are all the values for x such that $x^2 < 3x + 18$? **B**

Ⓐ $x < -3$ Ⓑ $-3 < x < 6$

Ⓒ $x > -3$ Ⓓ $x < 6$

9. If $f(x) = 2x^3 - 18x$, what are all the values of x at which $f(x) = 0$? **B**

Ⓐ 0, 3 Ⓑ -3, 0, 3

Ⓒ -6, 0, 6 Ⓓ -3, 2, 3

10. Which of the following is equal to $\frac{17.5(10^{-2})}{500(10^{-4})}$? **D**

Ⓐ $0.035(10^{-2})$ Ⓑ $0.35(10^{-2})$

Ⓒ $0.0035(10^2)$ Ⓓ $0.035(10^2)$

Test-Taking Tip

Question 7

You can use estimates to help you eliminate answer choices. For example, in Question 7, you can estimate that $\frac{1}{3}\sqrt{45}$ is less than $\frac{1}{3}\sqrt{49}$, which is $\frac{7}{3}$ or $2\frac{1}{3}$. Eliminate choices C and D.

ExamView® Pro

Special banks of standardized test questions similar to those on the SAT, ACT, TIMSS 12, NAEP 12, and state proficiency tests can be found on this CD-ROM.

Preparing for Standardized Tests
For test-taking strategies and more practice, see pages 877–892.

Part 2 Short Response/Grid In

Record your answers on the answer sheet provided by your teacher or on a sheet of paper.

11. If the outer diameter of a cylindrical tank is 62.46 centimeters and the inner diameter is 53.32 centimeters, what is the thickness of the tank? **4.57 cm**

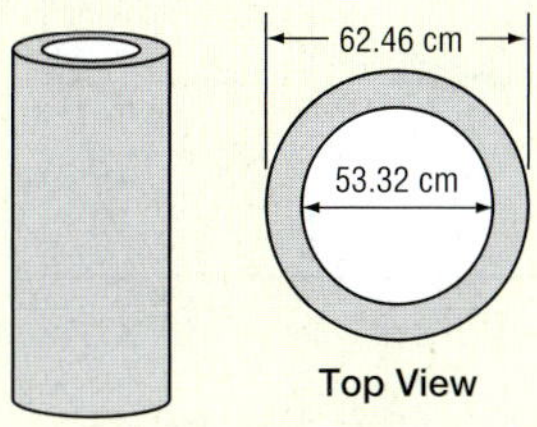

12. What number added to 80% of itself is equal to 45? **25**

13. Of 200 families surveyed, 95% have at least one TV and 60% of those with TVs have more than 2 TVs. If 50 families have exactly 2 TVs, how many families have exactly 1 TV? **26**

14. In the figure, if $ED = 8$, what is the measure of line segment AE? **2**

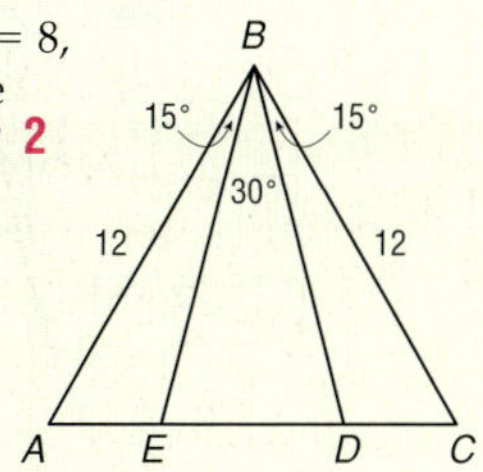

15. If $a \leftrightarrow b$ is defined as $a - b + ab$, find the value of $4 \leftrightarrow 2$. **10**

16. If $6(m + k) = 26 + 4(m + k)$, what is the value of $m + k$? **13**

17. If $y = 1 - x^2$ and $-3 \leq x \leq 1$, what number is found by subtracting the *least* possible value of y from the *greatest* possible value of y? **9**

www.algebra2.com/standardized_test

18. If $f(x) = (x - \pi)(x - 3)(x - e)$, what is the difference between the greatest and least roots of $f(x)$? Round to the nearest hundredth. **.42**

19. Suppose you deposit $700 in an account that pays 2.5% interest compounded quarterly. What will the dollar value of the account be in 5 years? Round to the nearest penny. **$792.90**

20. Solve $128^{x+1} = 64^{3x}$ for x. $\mathbf{\frac{7}{11}}$

21. State the domain and range of the function $y = \log_5 (x + 1)$. **domain: $x > -1$, range: all real numbers**

Part 3 Extended Response

Record your answers on a sheet of paper. Show your work.

For Exercises 22–25, use the following information.

In $y = 8.430\,(1.058)^x$, x represents the number of years since 1990, and y represents the approximate number of millions of Americans 7 years of age and older who participated in backpacking two or more times that year.

22. Describe how the number of millions of Americans who go backpacking is changing over time. **growing exponentially**

23. What do the numbers 8.430 and 1.058 represent? **See margin.**

24. About how many Americans went backpacking in 1996? **11.8 million**

25. In what year would you expect the number of Americans who participated in backpacking to reach 20 million for the first time? **2005**

26. Sketch the graphs of $f(x) = 2^x$ and $g(x) = \log_2 x$. **See margin.**

a. Study the graphs and describe the relationship between $f(x)$ and $g(x)$.

b. Specify the domain and range of $g(x)$.

26a. The graphs are reflections across the line $y = x$, so these are inverse functions.

26b. domain: $x > 0$, range: all real numbers

Evaluating Extended Response Questions

Extended Response questions are graded by using a multilevel rubric that guides you in assessing a student's knowledge of a particular concept.

Goal: Determine the dimensions of a rectangular garden for a given perimeter.

Sample Scoring Rubric: The following rubric is a sample scoring device. You may wish to add more detail to this sample to meet your individual scoring needs.

Score	Criteria
4	A correct solution that is supported by well-developed, accurate explanations
3	A generally correct solution, but may contain minor flaws in reasoning or computation
2	A partially correct interpretation and/or solution to the problem
1	A correct solution with no supporting evidence or explanation
0	An incorrect solution indicating no mathematical understanding of the concept or task, or no solution is given

Answers

23. 8.430 is the number of millions of backpackers in 1990 and 1.058 is 1 plus the rate at which the number of backpackers is growing each year.

26.

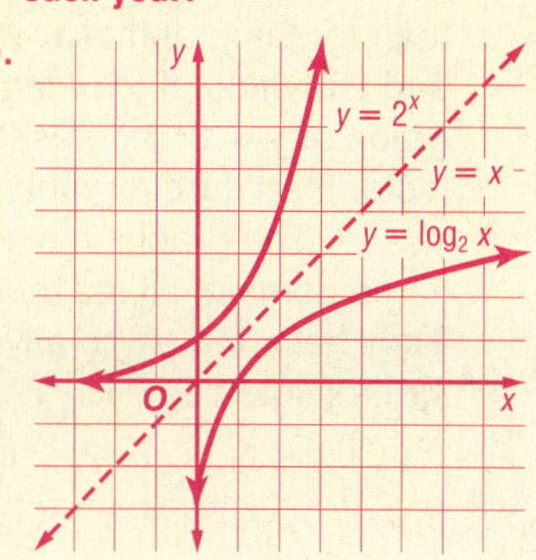

Page 521, Chapter 10 Getting Started

9.
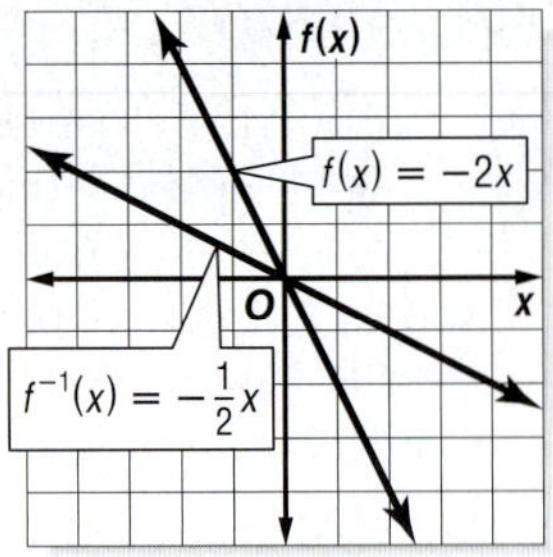

10.
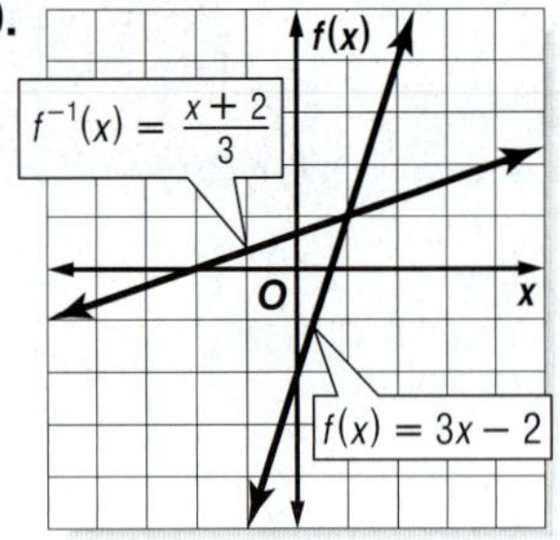

11.
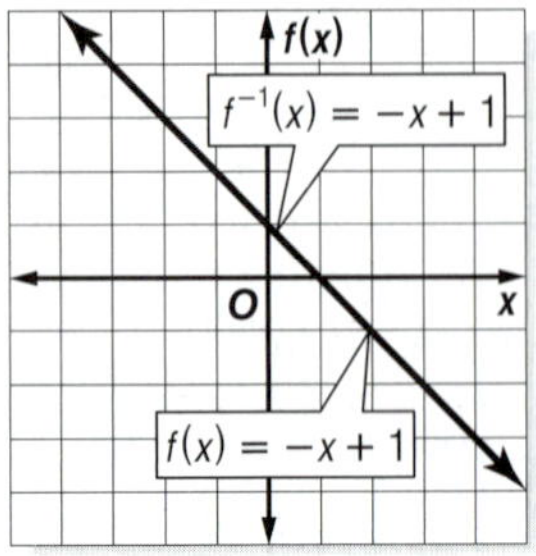

12.
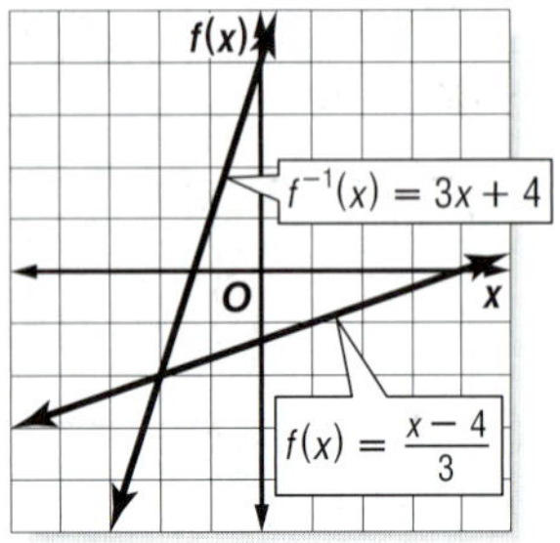

Page 522, Preview of Lesson 10-1 Algebra Activity

3. Sample graph:
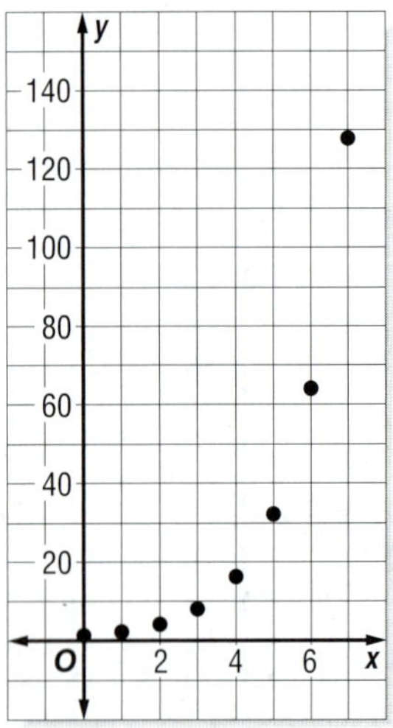

Page 524, Lesson 10-1 Graphing Calculator Investigation

4a. As the value of x increases, the value of y for the graph of $y = 4^x$ increases faster than for the graph of $y = 3^x$, and the value of y for the graph of $y = 3^x$ increases faster than for the graph of $y = 2^x$. The graphs have the same domain, all real numbers, and range, $y > 0$. They have the same asymptote, the x-axis, and the same y-intercept, 1.

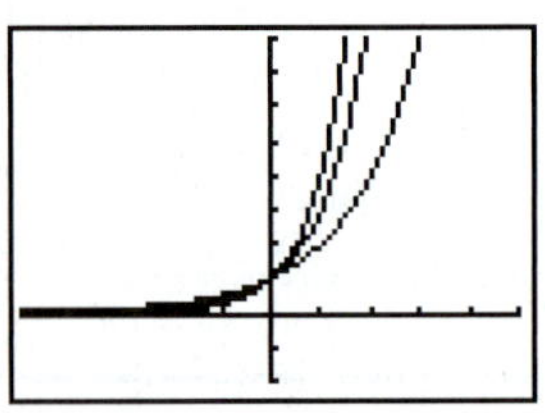
[−5, 5] scl: 1 by [−2, 8] scl: 1

4b. As the value of x increases, the value of y for the graph of $y = \left(\frac{1}{3}\right)^x$ decreases faster than for the graph of $y = \left(\frac{1}{2}\right)^x$, and the value of y for the graph of $y = \left(\frac{1}{4}\right)^x$ decreases faster than for the graph of $y = \left(\frac{1}{3}\right)^x$. The graphs have the same domain, all real numbers, and range, $y > 0$. They have the same asymptote, the x-axis, and the same y-intercept, 1.

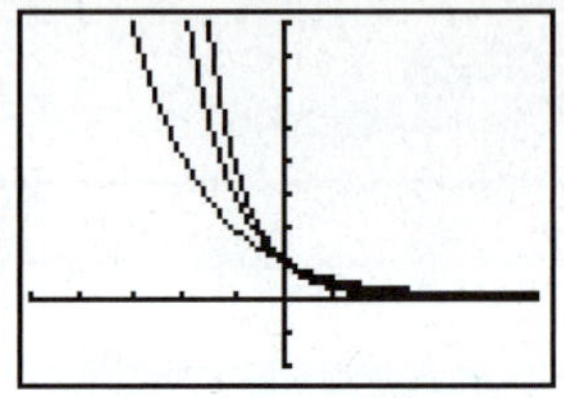
[−5, 5] scl: 1 by [−2, 8] scl: 1

4c. The graph of $y = 3(2)^x$ moves down and to the right more quickly than the graph of $y = -1(2)^x$. The graph of $y = 3(2)^x$ moves up and to the right more quickly than the graph of $y = 2^x$. All of the graphs have the same domain, all real numbers, and asymptote, the x-axis, but the range of $y = -3(2)^x$ and $y = -1(2)^x$ is $y < 0$, while the range of $y = 2^x$ and $y = 3(2)^x$ is $y > 0$. The y-intercept of $y = -3(2)^x$ is −3, of $y = -1(2)^x$ is −1, of $y = 2^x$ is 1, and of $y = 3(2)^x$ is 3.

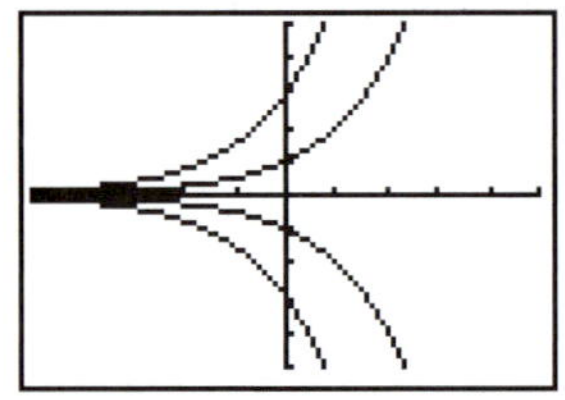
[−5, 5] scl: 1 by [−5, 5] scl: 1

Pages 528–530, Lesson 10-1

21. D = {x|x is all real numbers.}, R = {y|$y > 0$}

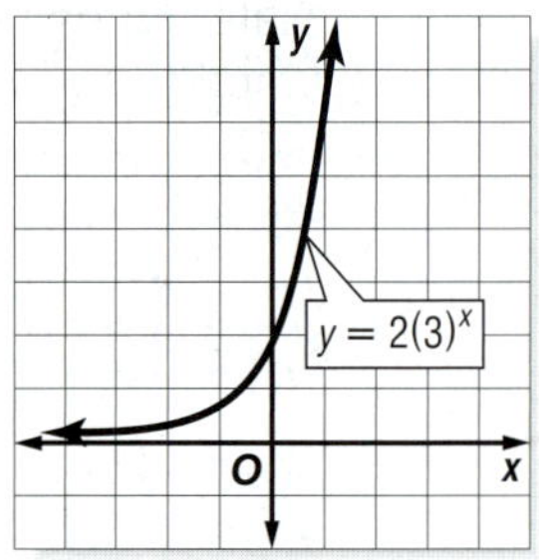

22. D = {x|x is all real numbers.}, R = {y|$y > 0$}

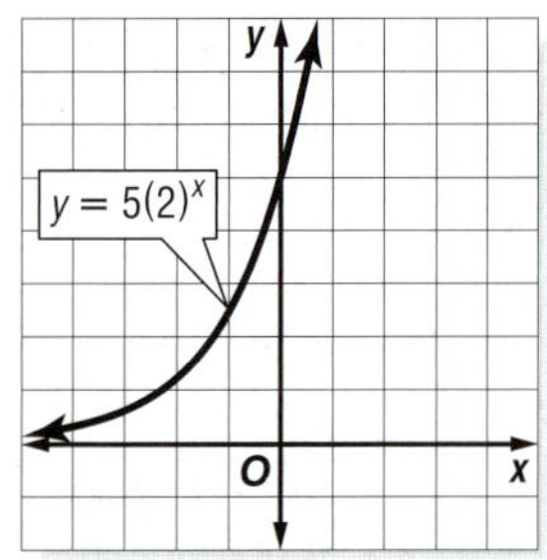

23. D = {x|x is all real numbers.}, R = {y|$y > 0$}

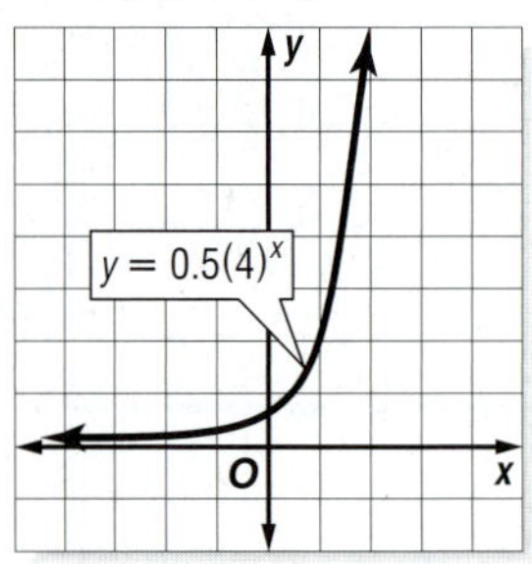

24. D = {x|x is all real numbers.}, R = {y|$y > 0$}

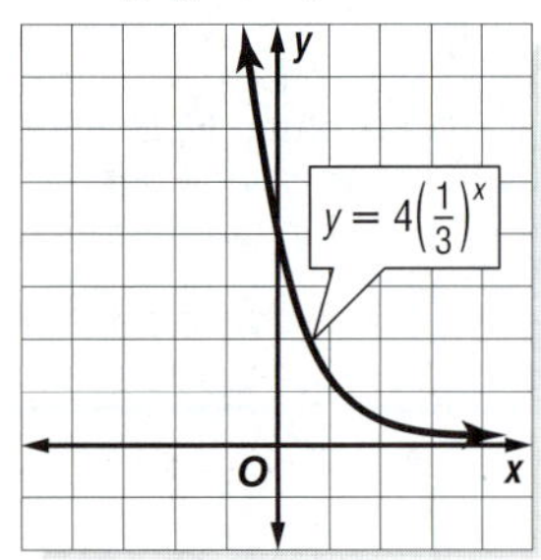

25. D = {$x|x$ is all real numbers.}, R = {$y|y < 0$}

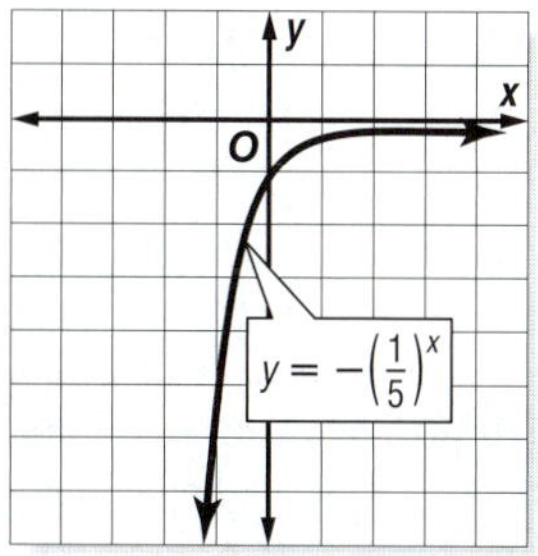

26. D = {$x|x$ is all real numbers.}, R = {$y|y < 0$}

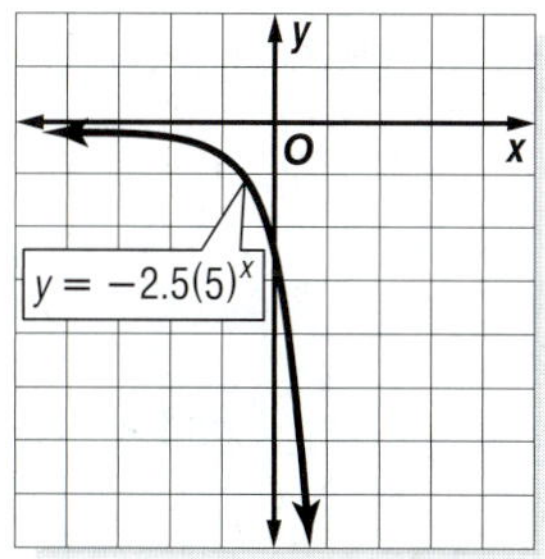

68. The number of teams y that could compete in a tournament in which x rounds are played can be expressed as $y = 2^x$. The 2 teams that make it to the final round got there as a result of winning games played with 2 other teams, for a total of $2 \cdot 2 = 2^2$ or 4 games played in the previous round or semifinal round. Answers should include the following.

- Rewrite 128 as a power of 2, 2^7. Substitute 2^7 for y in the equation $y = 2^x$. Then, using the Property of Equality for Exponents, x must be 7. Therefore, 128 teams would need to play 7 rounds of tournament play.
- Sample answer: 52 would be an inappropriate number of teams to play in this type of tournament because 52 is not a power of 2.

71.

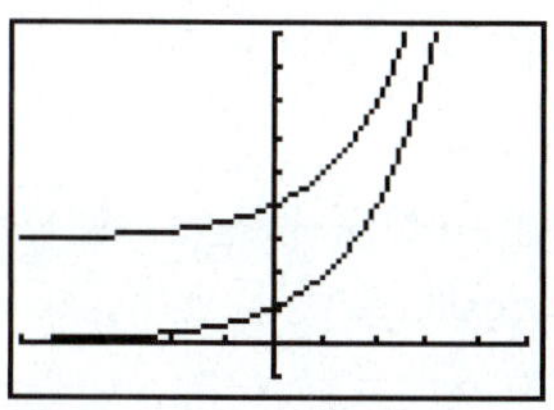

[−5, 5] scl: 1 by [−1, 9] scl: 1

The graphs have the same shape. The graph of $y = 2^x + 3$ is the graph of $y = 2^x$ translated three units up. The asymptote for the graph of $y = 2^x$ is the line $y = 0$ and for $y = 2^x + 3$ is the line $y = 3$. The graphs have the same domain, all real numbers, but the range of $y = 2^x$ is $y > 0$ and the range of $y = 2^x + 3$ is $y > 3$. The y-intercept of the graph of $y = 2^x$ is 1 and for the graph of $y = 2^x + 3$ is 4.

72.

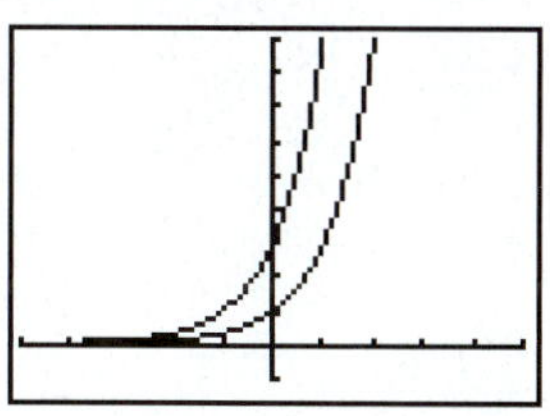

[−5, 5] scl: 1 by [−1, 9] scl: 1

The graphs have the same shape. The graph of $y = 3^{x+1}$ is the graph of $y = 3^x$ translated one unit to the left. The asymptote for the graph of $y = 3^x$ and for $y = 3^{x+1}$ is the line $y = 0$. The graphs have the same domain, all real numbers, and range, $y > 0$. The y-intercept of the graph of $y = 3^x$ is 1 and for the graph of $y = 3^{x+1}$ is 3.

73.

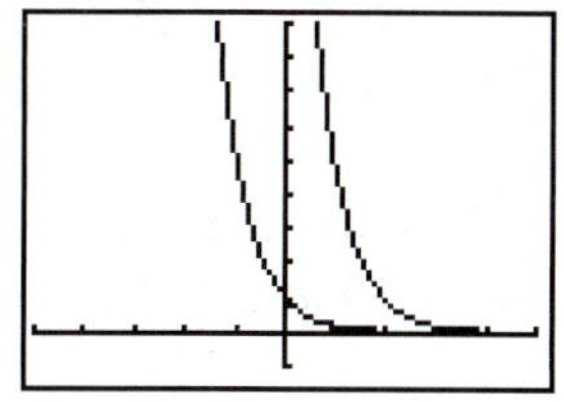

[−5, 5] scl: 1 by [−1, 9] scl: 1

The graphs have the same shape. The graph of $y = \left(\frac{1}{5}\right)^{x-2}$ is the graph of $y = \left(\frac{1}{5}\right)^x$ translated two units to the right. The asymptote for the graph of $y = \left(\frac{1}{5}\right)^x$ and for $y = \left(\frac{1}{5}\right)^{x-2}$ is the line $y = 0$. The graphs have the same domain, all real numbers, and range, $y > 0$. The y-intercept of the graph of $y = \left(\frac{1}{5}\right)^x$ is 1 and for the graph of $y = \left(\frac{1}{5}\right)^{x-2}$ is 25.

74.

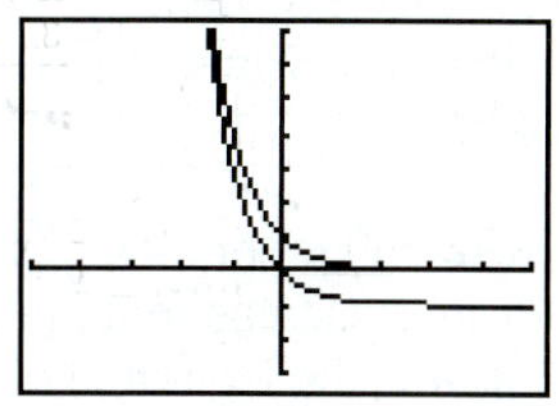

[−5, 5] scl: 1 by [−3, 7] scl: 1

The graphs have the same shape. The graph of $y = \left(\frac{1}{4}\right)^x - 1$ is the graph of $y = \left(\frac{1}{4}\right)^x$ translated one unit down. The asymptote for the graph of $y = \left(\frac{1}{4}\right)^x$ is the line $y = 0$ and for the graph of $y = \left(\frac{1}{4}\right)^x - 1$ is the line $y = -1$. The graphs have the same domain, all real numbers, but the range of $y = \left(\frac{1}{4}\right)^x$ is $y > 0$ and of $y = \left(\frac{1}{4}\right)^x - 1$ is $y > -1$. The y-intercept of the graph of $y = \left(\frac{1}{4}\right)^x$ is 1 and for the graph of $y = \left(\frac{1}{4}\right)^x - 1$ is 0.

Page 537, Lesson 10-2

66a.

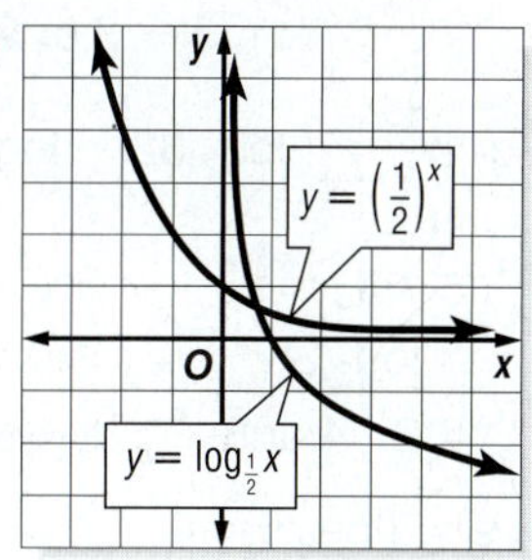

66b. The graphs are reflections of each other over the line $y = x$.

67a.

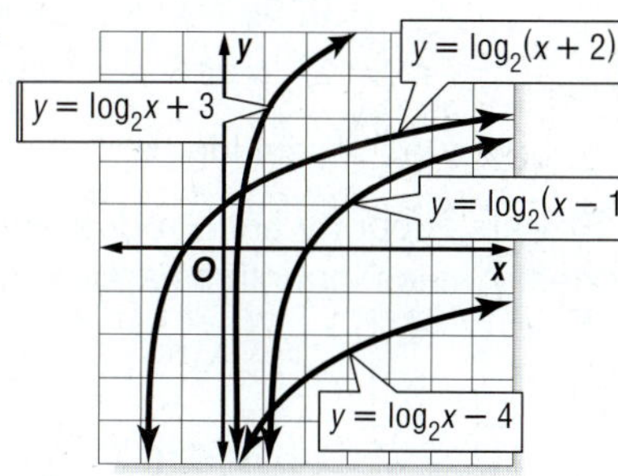

67b. The graph of $y = \log_2 x + 3$ is the graph of $y = \log_2 x$ translated 3 units up. The graph of $y = \log_2 x - 4$ is the graph of $y = \log_2 x$ translated 4 units down. The graph of $\log_2 (x - 1)$ is the graph of $y = \log_2 x$ translated 1 unit to the right. The graph of $\log_2 (x + 2)$ is the graph of $y = \log_2 x$ translated 2 units to the left.

73. A logarithmic scale illustrates that values next to each other vary by a factor of 10. Answers should include the following.

- Pin drop: 1×10^0; Whisper: 1×10^2; Normal conversation: 1×10^6; Kitchen noise: 1×10^{10}; Jet engine: 1×10^{12}
- 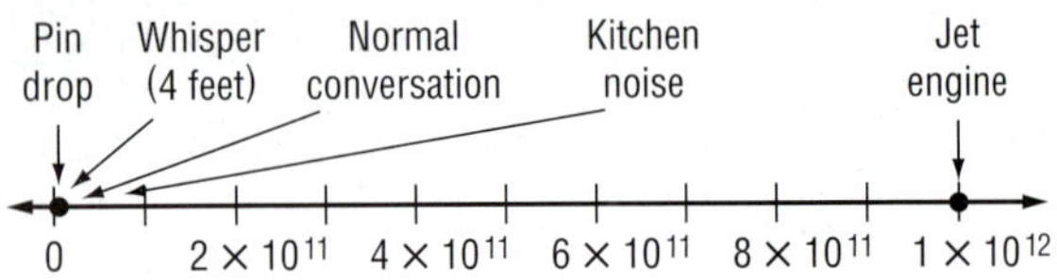

- On the scale shown above, the sound of a pin drop and the sound of normal conversation appear not to differ by much at all, when in fact they do differ in terms of the loudness we perceive. The first scale shows this difference more clearly.

Pages 545–546, Lesson 10-3

36. true;

$n \log_b x + m \log_b x \stackrel{?}{=} (n + m) \log_b x$

$\log_b x^n + \log_b x^m \stackrel{?}{=} (n + m) \log_b x$ Power Property of Logarithms

$\log_b (x^n \cdot x^m) \stackrel{?}{=} (n + m) \log_b x$ Product Property of Logarithms

$\log_b (x^{n+m}) \stackrel{?}{=} (n + m) \log_b x$ Product of Powers Property

$(n + m) \log_b x = (n + m) \log_b x$ ✓ Power Property of Logarithms

48. Since logarithms are exponents, the properties of logarithms are similar to the properties of exponents. The Product Property states that to multiply two powers that have the same base, add the exponents. Similarly, the logarithm of a product is the sum of the logarithms of its factors. The Quotient Property states that to divide two powers that have the same base, subtract their exponents. Similarly, the logarithm of a quotient is the difference of the logarithms of the numerator and the denominator. The Power Property states that to find the power of a power, multiply the exponents. Similarly, the logarithm of a power is the product of the logarithm and the exponent. Answers should include the following.

- Quotient Property: $\log_2 \left(\frac{32}{8}\right) = \log_2 \left(\frac{2^5}{2^3}\right)$ Replace 32 with 2^5 and 8 with 2^3.

  $= \log_2 2^{(5-3)}$ Quotient of Powers

  $= 5 - 3$ or 2 Inverse Property of Exponents and Logarithms

  $\log_2 32 - \log_2 8 = \log_2 2^5 - \log_2 2^3$ Replace 32 with 2^5 and 8 with 2^3.

  $= 5 - 3$ or 2 Inverse Property of Exponents and Logarithms

  So, $\log_2 \left(\frac{32}{8}\right) = \log_2 32 - \log_2 8$.

  Power Property: $\log_3 9^4 = \log_3 (3^2)^4$ Replace 9 with 3^2.

  $= \log_3 3^{(2 \cdot 4)}$ Power of a Power

  $= 2 \cdot 4$ or 8 Inverse Property of Exponents and Logarithms

  $4 \log_3 9 = (\log_3 9) \cdot 4$ Commutative Property (×)

  $= (\log_3 3^2) \cdot 4$ Replace 9 with 3^2.

  $= 2 \cdot 4$ or 8 Inverse Property of Exponents and Logarithms

  So, $\log_3 9^4 = 4 \log_3 9$.

- The Product of Powers Property and Product Property of Logarithms both involve the addition of exponents, since logarithms are exponents.

Page 558, Lesson 10-5

62. $\frac{\log x}{\log y} \stackrel{?}{=} \frac{\ln x}{\ln y}$ Original statement

$\frac{\log x}{\log y} \stackrel{?}{=} \frac{\frac{\log x}{\log e}}{\frac{\log y}{\log e}}$ Change of Base Formula

$\frac{\log x}{\log y} \stackrel{?}{=} \frac{\log x}{\log e} \cdot \frac{\log e}{\log y}$ Multiply $\frac{\log x}{\log e}$ by the reciprocal of $\frac{\log y}{\log e}$.

$\frac{\log x}{\log y} = \frac{\log x}{\log y}$ Simplify.

Notes

UNIT 4

Notes

Introduction

In this unit, students explore various topics of discrete mathematics, including arithmetic and geometric sequences and series, as well as recursion and fractals. They also apply the Binomial Theorem, and prove statements using mathematical induction.

The unit concludes with an investigation of probability and statistics, including permutations, combinations, and the normal distribution. Finally, students apply their mathematical skills in a simulation, as well as to sampling situations and to testing hypotheses.

Assessment Options

Unit 4 Test Pages 773–774 of the *Chapter 12 Resource Masters* may be used as a test or review for Unit 4. This assessment contains both multiple-choice and short answer items.

ExamView® Pro

This CD-ROM can be used to create additional unit tests and review worksheets.

UNIT 4

Discrete Mathematics

Discrete mathematics is the branch of mathematics that involves finite or discontinuous quantities. In this unit, you will learn about sequences, series, probability, and statistics.

Richard Kaye
Professor of Mathematics
University of Birmingham

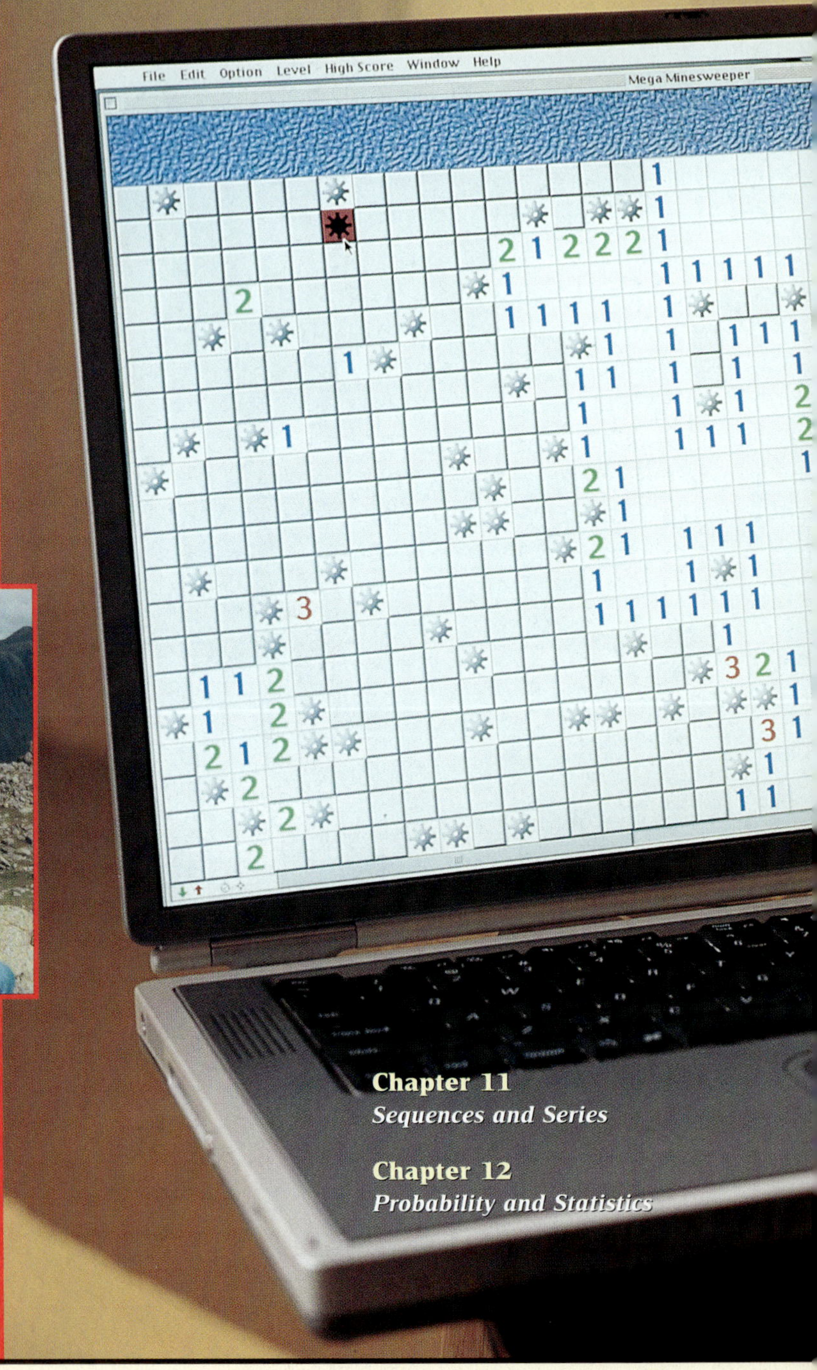

Chapter 11
Sequences and Series

Chapter 12
Probability and Statistics

WebQuest Internet Project

'Minesweeper': Secret to Age-Old Puzzle?

Source: *USA TODAY*, November 3, 2000

"*Minesweeper*, a seemingly simple game included on most personal computers, could help mathematicians crack one of the field's most intriguing problems. The buzz began after Richard Kaye, a mathematics professor at the University of Birmingham in England, started playing *Minesweeper*. After playing the game steadily for a few weeks, Kaye realized that *Minesweeper*, if played on a much larger grid, has the same mathematical characteristics as other problems deemed insolvable." In this project, you will research a mathematician of the past and his or her role in the development of discrete mathematics.

Log on to www.algebra2.com/webquest. Begin your WebQuest by reading the Task.

Then continue working on your WebQuest as you study Unit 4.

Lesson	11-7	12-1
Page	616	635

USA TODAY Education Teaching Suggestions

Have students study the USA TODAY Snapshot®.

- Ask students to name some historical mathematicians (such as Pythagoras, Fibonacci, and Descartes).
- Have students compare their own use of a PC at home to the percents shown in the graph.
- Point out to students that use of the Internet is quickly becoming the main tool for doing research on a given topic.

Additional USA TODAY Snapshots® appearing in Unit 4:

Chapter 11 Yosemite visitors peak in '96 (p. 604)

Chapter 12 Getting ready for bed (p. 658)

WebQuest Internet Project

Problem-Based Learning A WebQuest is an online project in which students do research on the Internet, gather data, and make presentations using word processing, graphing, page-making, or presentation software. In each chapter, students advance to the next step in their WebQuest. At the end of Chapter 12, the project culminates with a presentation of their findings.

Teaching suggestions and sample answers are available in the *WebQuest and Project Resources*.

Unit 4 Discrete Mathematics 575

Chapter 11

Sequences and Series

Chapter Overview and Pacing

Year-long and two-year pacing: pages T20–T21.

LESSON OBJECTIVES	PACING (days) Regular		Block	
	Basic/ Average	Advanced	Basic/ Average	Advanced
11-1 Arithmetic Sequences *(pp. 578–582)* • Use arithmetic sequences. • Find arithmetic means.	1	1	0.5	0.5
11-2 Arithmetic Series *(pp. 583–587)* • Find sums of arithmetic series. • Use sigma notation.	1	1	0.5	0.5
11-3 Geometric Sequences *(pp. 588–592)* • Use geometric sequences. • Find geometric means.	1	1	0.5	0.5
11-4 Geometric Series *(pp. 593–598)* ***Preview:*** Limits • Find sums of geometric series. • Find specific terms of geometric series.	2 (with 11-4 Preview)	1	0.5	0.5
11-5 Infinite Geometric Series *(pp. 599–604)* • Find the sum of an infinite geometric series. • Write repeating decimals as fractions.	1	2 (with 11-6 Preview)	0.5	1
11-6 Recursion and Special Sequences *(pp. 605–611)* ***Preview:*** Amortizing Loans • Recognize and use special sequences. • Iterate functions. ***Follow-Up:*** Fractals	2	3 (with 11-6 Follow-Up	1	1
11-7 The Binomial Theorem *(pp. 612–617)* • Use Pascal's triangle to expand powers of binomials. • Use the Binomial Theorem to expand powers of binomials.	1	1	0.5	0.5
11-8 Proof and Mathematical Induction *(pp. 618–621)* • Prove statements by using mathematical induction. • Disprove statements by finding a counterexample.	2	1	1	0.5
Study Guide and **Practice Test** *(pp. 622–627)* **Standardized Test Practice** *(pp. 628–629)*	1	1	0.5	0.5
Chapter Assessment	1	1	0.5	0.5
TOTAL	13	13	6	6

StudentWorks

An electronic version of this chapter is available on ***StudentWorks™****. This backpack solution CD-ROM allows students instant access to the Student Edition, lesson worksheet pages, and web resources.*

Timesaving Tools

TeacherWorks™

All-In-One Planner and Resource Center

See pages T5 and T21.

Chapter Resource Manager

Study Guide and Intervention	Practice (Skills and Average)	Reading to Learn Mathematics	Enrichment	Assessment	Applications*	5-Minute Check Transparencies	Interactive Chalkboard	Alge2PASS: Tutorial Plus (lessons)	Materials
631–632	633–634	635	636			11-1	11-1	20	isometric dot paper, wooden or plastic cubes
637–638	639–640	641	642	693	GCS 48, SM 133–138	11-2	11-2		graphing calculator
643–644	645–646	647	648		SC 21	11-3	11-3		
649–650	651–652	653	654	693, 695	SC 22	11-4	11-4	21	(*Preview:* graphing calculator)
655–656	657–658	659	660			11-5	11-5		
661–662	663–664	665	666	694	GCS 47	11-6	11-6		(*Preview:* spreadsheet software) penny, nickel, dime, cardboard (*Follow-Up:* isometric dot paper)
667–668	669–670	671	672			11-7	11-7		colored pens or pencils
673–674	675–676	677	678	694		11-8	11-8		
				679–692, 696–698					

Key to Abbreviations: GCS = Graphing Calculator and Speadsheet Masters,
SC = School-to-Career Masters,
SM = Science and Mathematics Lab Manual

Chapter 11

Mathematical Connections and Background

Continuity of Instruction

Prior Knowledge

Students have used formulas for area, volume, and other attributes, and they have used notation such as subscripts, superscripts, and factorials. They have used the output of one function as input for another function, and they have explored proof by deriving particular properties from other properties.

This Chapter

Students study and apply formulas for arithmetic sequences and series, for finite geometric sequences and series, and for infinite geometric series. They use sigma notation and factorial notation to write concise forms of formulas, especially the formula for the Binomial Theorem. Students continue to use formulas and notation as they explore two-part recursive formulas and the three-step process of mathematical induction.

Future Connections

Students will continue to use subscripts, factorials and sigma notation in later math topics, and they will continue to see sequences and series. They will translate between recursive formulas and non-recursive formulas ("explicit formulas"), and they frequently will revisit the powerful idea of mathematical induction.

11-1 Arithmetic Sequences

Throughout this chapter students study notation and formulas. They see how notation allows a formula to be written in a concise form, and they investigate how one formula can be related to or contain another formula. The first lesson uses arithmetic sequences to explore notation and formulas. In an *arithmetic sequence,* each term after the first is found by adding a constant, called the *common difference,* to the previous term. (Students may need to be told that when "arithmetic" is used as an adjective, the accent is on the next-to-last syllable.) Subscripts indicate a particular order for terms, and the formula $a_n = a_1 + (n - 1)d$ is seen as a concise way to represent the nth term.

11-2 Arithmetic Series

This lesson begins with two ideas, that the average of the first n terms in an arithmetic sequence is the mean of the first and nth terms, and that the sum of the first n terms is the number n times the average of these terms. This leads to two formulas for the sum S_n of the terms of an arithmetic series. For one formula, a_1 and $a_1 + (n - 1)d$ are used as the first and nth terms; a different formula uses a_1 and a_n as the first and nth terms. As in the previous lesson, students investigate the formulas by solving problems that ask them to find the value S_n, n, a_1, a_n, or d from given information. The lesson also introduces sigma notation. For example, $\sum_{n=0}^{6}(x^2 + 2)$ represents a sum of terms where x is replaced by each of the seven values 0, 1, 2, 3, 4, 5, and 6. So $\sum_{n=0}^{6}(x^2 + 2) = (0^2 + 2) + (1^2 + 2) + (2^2 + 2) + (3^2 + 2) + (4^2 + 2) + (5^2 + 2) + (6^2 + 2)$. The value of this series is 105.

11-3 Geometric Sequences

This lesson introduces sequences whose terms have a constant ratio r; that is, each successive term is the product of r and the previous term. Using r to represent that common ratio, the formula $a_n = a_1 \cdot r^{n-1}$, which includes both subscripts and superscripts, is a concise way to represent the nth term. After geometric means are described as numbers that form a geometric sequence, students use the formula for the nth term of a geometric sequence to find a given number of geometric means between two given numbers. A *geometric mean* is a number or numbers that are missing terms between two nonsuccessive terms of a geometric sequence.

11-4 Geometric Series

The formula for the sum of the first n terms of a geometric series is derived by using several ideas, each expressed concisely with subscripts and exponents.

(1) The $(n + 1)$st term of a geometric sequence is $a_1 r^n$.

(2) If you multiply the 1st through nth terms of a geometric series by the common ratio, the result is the 2nd through $(n + 1)$st terms.

(3) The difference $S_n - rS_n$ can be written as the equation $S_n - rS_n = a_1 - a_1 r^n$.

Dividing each side of that equation by $1 - r$ results in a formula for S_n, the sum of an geometric series. Another formula for S_n can be derived by substituting $a_n r$ for $a_1 r^n$.

11-5 Infinite Geometric Series

A formula for an infinite series can be simpler than a formula for a finite series. If the common ratio r is such that $|r| < 1$, then as n get larger the value of r^n approaches 0. As a result, the sum of an infinite geometric series can be written as a formula that has no exponents, and the sum is completely determined by the first term and the common ratio. Repeating decimals can be expressed by an infinite geometric series as well as by a fraction.

11-6 Recursion and Special Sequences

This lesson introduces a different kind of formula for the terms of a sequence. The formulas have two parts. One part gives specific values for the first one or more terms of the sequence. The second part describes the "next" term as a function of previous terms. (The formulas from the earlier lessons are often called explicit formulas.) The lesson also introduces a situation in which a function rule and the first term of a sequence are given. Each term of the sequence, as an input value, yields the next term in the sequence. This situation is called iterating a function or generating a sequence using iteration.

11-7 The Binomial Theorem

Previously learned formulas are used to develop the Binomial Theorem. The expansion of the expression $(a + b)^n$ for nonnegative values of n involves finding the coefficient and the exponents for a and b for each term. The expansions show many patterns: the sum of the exponents for a and b is n; the coefficients are the entries in Pascal's triangle; and the coefficients are functions of the exponents.

One way to write the Binomial Theorem is to describe each coefficient as a fraction. Another way is to use factorial notation for the coefficients. And another way, using sigma notation as well as factorial notation, illustrates how notation can provide a concise way to write a complex formula. Students should be familiar with factorials, finding powers of monomials, and using sigma notation.

11-8 Proof and Mathematical Induction

Mathematical induction is a powerful idea in mathematics. Much of higher mathematics uses this principle for verification of conjectures. This lesson examines series, divisibility, and finding a counterexample to show that a formula is not true. Students can relate mathematical induction to the two-part recursive formulas of a previous lesson. One part of mathematical induction is to show that a particular property is true for a particular number (often for the number 1). The second part is to prove that if the property holds for some positive integer, then the property holds for the "next" integer. Completing both parts is a proof that the property holds for all positive integers.

Another way to look at mathematical induction is to consider the set S of positive integers for which some property is true. If you can show that 1 is in S and, for any integer k in S, that the integer following k is in S, then S contains all positive integers. Since S is the set of integers for which the property is true, then the property is true for all positive integers.

Additional mathematical information and teaching notes are available at www.algebra2.com/key_concepts.

Chapter 11

DAILY INTERVENTION and Assessment

	Type	Student Edition	Teacher Resources	Technology/Internet
INTERVENTION	Ongoing	Prerequisite Skills, pp. 577, 582, 587, 592, 598, 604, 610, 617 Practice Quiz 1, p. 592 Practice Quiz 2, p. 617	5-Minute Check Transparencies Quizzes, *CRM* pp. 693–694 Mid-Chapter Test, *CRM* p. 695 Study Guide and Intervention, *CRM* pp. 631–632, 637–638, 643–644, 649–650, 655–656, 661–662, 667–668, 673–674	Alge2PASS: Tutorial Plus www.algebra2.com/self_check_quiz www.algebra2.com/extra_examples
INTERVENTION	Mixed Review	pp. 582, 587, 592, 598, 604, 610, 617, 621	Cumulative Review, *CRM* p. 696	
INTERVENTION	Error Analysis	Find the Error, pp. 590, 602	Find the Error, *TWE* pp. 590, 602 Unlocking Misconceptions, *TWE* pp. 579, 600 Tips for New Teachers, *TWE* pp. 582, 587, 592, 598, 604, 610, 617, 620	
ASSESSMENT	Standardized Test Practice	pp. 582, 587, 588, 591, 592, 598, 603, 610, 616, 621, 627, 628–629	*TWE* p. 589 Standardized Test Practice, *CRM* pp. 697–698	Standardized Test Practice CD-ROM www.algebra2.com/standardized_test
ASSESSMENT	Open-Ended Assessment	Writing in Math, pp. 582, 587, 592, 598, 603, 610, 616, 621 Open Ended, pp. 580, 586, 590, 596, 602, 608, 615, 619	Modeling: *TWE* p. 592 Speaking: *TWE* pp. 582, 587, 610 Writing: *TWE* pp. 598, 604, 617 Open-Ended Assessment, *CRM* p. 691	
ASSESSMENT	Chapter Assessment	Study Guide, pp. 622–626 Practice Test, p. 627	Multiple-Choice Tests (Forms 1, 2A, 2B), *CRM* pp. 679–684 Free-Response Tests (Forms 2C, 2D, 3), *CRM* pp. 685–690 Vocabulary Test/Review, *CRM* p. 692	ExamView® Pro (see below) MindJogger Videoquizzes www.algebra2.com/vocabulary_review www.algebra2.com/chapter_test

Key to Abbreviations: TWE = Teacher Wraparound Edition; CRM = Chapter Resource Masters

Intervention Technology

Alge2PASS: Tutorial Plus CD-ROM offers a complete, self-paced algebra curriculum.

Algebra 2 Lesson	Alge2PASS Lesson
11-1	**20** *Finding the Missing Number in a Sequence*
11-4	**21** *Sequences and Series*

ExamView® Pro

Use the networkable **ExamView® Pro** to:

- Create **multiple versions** of tests.
- Create **modified** tests for *Inclusion* students.
- **Edit** existing questions and **add** your own questions.
- Use built-in **state curriculum correlations** to create tests aligned with state standards.
- **Apply** art to your tests from a program bank of artwork.

For more information on Intervention and Assessment, see pp. T8–T11.

Reading and Writing in Mathematics

Glencoe Algebra 2 provides numerous opportunities to incorporate reading and writing into the mathematics classroom.

Student Edition

- Foldables Study Organizer, p. 577
- Concept Check questions require students to verbalize and write about what they have learned in the lesson. (pp. 580, 586, 590, 596, 602, 608, 615, 619, 622)
- Writing in Math questions in every lesson, pp. 582, 587, 592, 598, 603, 610, 616, 621
- Reading Study Tip, pp. 606, 619
- WebQuest, p. 616

Teacher Wraparound Edition

- Foldables Study Organizer, pp. 577, 622
- Study Notebook suggestions, pp. 580, 585, 590, 596, 602, 605, 608, 611, 615, 619
- Modeling activities, p. 592
- Speaking activities, pp. 582, 587, 610
- Writing activities, pp. 598, 604, 617
- Differentiated Instruction, (Verbal/Linguistic), p. 615
- ELL Resources, pp. 576, 581, 586, 591, 597, 603, 609, 615, 616, 621, 622

Additional Resources

- Vocabulary Builder worksheets require students to define and give examples for key vocabulary terms as they progress through the chapter. (*Chapter 11 Resource Masters,* pp. vii-viii)
- Reading to Learn Mathematics master for each lesson (*Chapter 11 Resource Masters*, pp. 635, 641, 647, 653, 659, 665, 671, 677)
- *Vocabulary PuzzleMaker* software creates crossword, jumble, and word search puzzles using vocabulary lists that you can customize.
- *Teaching Mathematics with Foldables* provides suggestions for promoting cognition and language.
- *Reading and Writing in the Mathematics Classroom*
- *WebQuest and Project Resources*

For more information on Reading and Writing in Mathematics, see pp. T6–T7.

PROJECT CRISS[SM] Study Skill

Students have used Venn diagrams in their mathematics studies in previous courses. Using Venn diagrams to compare and contrast concepts may be helpful as they assimilate new knowledge. Encourage students to look for the relationships between math concepts that are taught together. These relationships can be represented using Venn diagrams. Students can brainstorm ideas for Venn diagrams individually or as a cooperative learning activity.

The Venn diagram at the right represents the concepts of arithmetic and geometric sequences (Lessons 11-1 and 11-3).

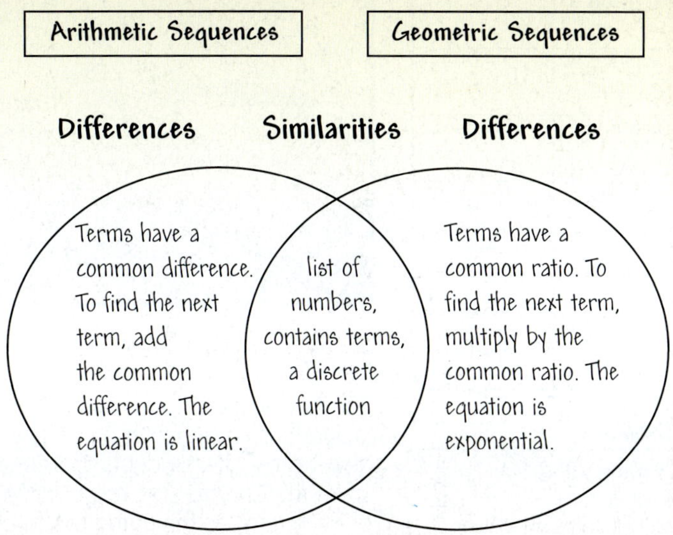

CReating **I**ndependence Through **S**tudent-Owned **S**trategies

Chapter 11 Notes

What You'll Learn

Have students read over the list of objectives and make a list of any words with which they are not familiar.

Why It's Important

Point out to students that this is only one of many reasons why each objective is important. Others are provided in the introduction to each lesson.

Lesson	NCTM Standards	Local Objectives
11-1	1, 6, 7, 8, 9	
11-2	1, 6, 7, 8, 9, 10	
11-3	1, 2, 6, 7, 8, 9, 10	
11-4 Preview	1, 6	
11-4	1, 2, 6, 7, 8, 9, 10	
11-5	1, 2, 6, 7, 8, 9, 10	
11-6 Preview	1, 6, 8, 9, 10	
11-6	1, 2, 6, 7, 8, 9, 10	
11-6 Follow-Up	1, 3, 4, 6, 7, 8	
11-7	1, 2, 6, 8, 9, 10	
11-8	2, 6, 7, 8	

Key to NCTM Standards:

1=Number & Operations, 2=Algebra, 3=Geometry, 4=Measurement, 5=Data Analysis & Probability, 6=Problem Solving, 7=Reasoning & Proof, 8=Communication, 9=Connections, 10=Representation

Chapter 11 Sequences and Series

What You'll Learn

- **Lessons 11-1 through 11-5** Use arithmetic and geometric sequences and series.
- **Lesson 11-6** Use special sequences and iterate functions.
- **Lesson 11-7** Expand powers by using the Binomial Theorem.
- **Lesson 11-8** Prove statements by using mathematical induction.

Key Vocabulary

- arithmetic sequence (p. 578)
- arithmetic series (p. 583)
- sigma notation (p. 585)
- geometric sequence (p. 588)
- geometric series (p. 594)

Why It's Important

Many number patterns found in nature and used in business can be modeled by sequences, which are lists of numbers. Some sequences are classified by the method used to predict the next term from the previous term(s). When the terms of a sequence are added, a series is formed. *In Lesson 11-2, you will learn how the number of seats in the rows of an amphitheater can be modeled using a series.*

Vocabulary Builder ELL

The Key Vocabulary list introduces students to some of the main vocabulary terms included in this chapter. For a more thorough vocabulary list with pronunciations of new words, give students the Vocabulary Builder worksheets found on pages vii and viii of the *Chapter 11 Resource Masters*. Encourage them to complete the definition of each term as they progress through the chapter. You may suggest that they add these sheets to their study notebooks for future reference when studying for the Chapter 11 test.

Getting Started

Prerequisite Skills To be successful in this chapter, you'll need to master these skills and be able to apply them in problem-solving situations. Review these skills before beginning Chapter 11.

For Lessons 11-1 and 11-3 — Solve Equations

Solve each equation. *(For review, see Lessons 1-3 and 5-5.)*

1. $36 = 12 + 4x$ **6**
2. $-40 = 10 + 5x$ **−10**
3. $12 - 3x = 27$ **−5**
4. $162 = 2x^4$ **±3**
5. $\frac{1}{8} = 4x^5$ $\mathbf{\frac{1}{2}}$
6. $3x^3 + 4 = -20$ **−2**

For Lessons 11-1 and 11-5 — Graph Functions

Graph each function. *(For review, see Lesson 2-1.)* **7–10. See pp. 629A–629F.**

7. $\{(1, 1), (2, 3), (3, 5), (4, 7), (5, 9)\}$
8. $\{(1, -20), (2, -16), (3, -12), (4, -8), (5, -4)\}$
9. $\left\{(1, 64), (2, 16), (3, 4), (4, 1), \left(5, \frac{1}{4}\right)\right\}$
10. $\left\{(1, 2), (2, 3), \left(3, \frac{7}{2}\right), \left(4, \frac{15}{4}\right), \left(5, \frac{31}{8}\right)\right\}$

For Lessons 11-1 through 11-5, 11-8 — Evaluate Expressions

Evaluate each expression for the given value(s) of the variable(s). *(For review, see Lesson 1-1.)*

11. $x + (y - 1)z$ if $x = 3$, $y = 8$, and $z = 2$ **17**
12. $\frac{x}{2}(y + z)$ if $x = 10$, $y = 3$, and $z = 25$ **140**
13. $a \cdot b^{c-1}$ if $a = 2$, $b = \frac{1}{2}$, and $c = 7$ $\mathbf{\frac{1}{32}}$
14. $\frac{a(1 - bc)^2}{1 - b}$ if $a = -2$, $b = 3$, and $c = 5$ **196**
15. $\frac{a}{1 - b}$ if $a = \frac{1}{2}$, and $b = \frac{1}{6}$ $\mathbf{\frac{3}{5}}$
16. $\frac{n(n + 1)}{2}$ if $n = 10$ **55**

FOLDABLES™ Study Organizer

Sequences and Series Make this Foldable to help you organize your notes. Begin with one sheet of 11" by 17" paper and four sheets of notebook paper.

Step 1 Fold and Cut

Fold the short sides of the 11" by 17" paper to meet in the middle.

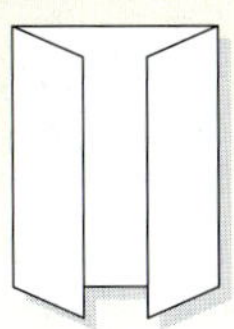

Step 2 Staple and Label

Fold the notebook paper in half lengthwise. Insert two sheets of notebook paper in each tab and staple the edges. Label with lesson numbers.

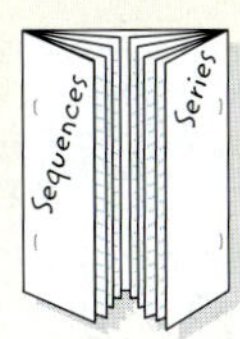

Reading and Writing As you read and study the chapter, fill the journal with examples for each lesson.

Getting Started

This section provides a review of the basic concepts needed before beginning Chapter 11. Page references are included for additional student help.

Prerequisite Skills in the Getting Ready for the Next Lesson section at the end of each exercise set review a skill needed in the next lesson.

For Lesson	Prerequisite Skill
11-2	Evaluating Expressions (p. 582)
11-3	Evaluating Expressions (p. 587)
11-4	Evaluating Expressions (p. 592)
11-5	Evaluating Expressions (p. 598)
11-6	Evaluating Functions (p. 604)
11-8	Evaluating Expressions (p. 617)

FOLDABLES™ Study Organizer

For more information about Foldables, see *Teaching Mathematics with Foldables.*

Questioning and Organizing Data Before beginning each lesson, ask students to preview each lesson and write several questions about what they see on each of the lesson tabs of their Foldable. Encourage students to write different types of questions including factual, open-ended, analytical, and test-like questions. As students read and work through the lesson, ask them to record the answers to their questions in their journal. Students can add questions to their Foldable that arise during reading, taking notes, or doing homework.

11-1 Lesson Notes

1 Focus

5-Minute Check Transparency 11-1 Use as a quiz or review of Chapter 10.

Mathematical Background notes are available for this lesson on p. 576C.

How are arithmetic sequences related to roofing?

Ask students:

- What other sequences have you seen before? **Answers will vary, but some may recall the Fibonacci sequence: 1, 1, 2, 3, 5, 8, 13, 21, ….**
- How can you find the next 5 numbers in the shingles sequence? **Add 1 to each successive row.**

11-1 Arithmetic Sequences

What You'll Learn

- Use arithmetic sequences.
- Find arithmetic means.

Vocabulary

- sequence
- term
- arithmetic sequence
- common difference
- arithmetic means

How are arithmetic sequences related to roofing?

A roofer is nailing shingles to the roof of a house in overlapping rows. There are three shingles in the top row. Since the roof widens from top to bottom, one additional shingle is needed in each successive row.

Row	1	2	3	4	5	6	7
Shingles	3	4	5	6	7	8	9

Study Tip

Sequences
The numbers in a sequence may not be ordered. For example, the numbers 33, 25, 36, 40, 36, 66, 63, 50, ... are a sequence that represents the number of home runs Sammy Sosa hit in each year beginning with 1993.

ARITHMETIC SEQUENCES The numbers 3, 4, 5, 6, …, representing the number of shingles in each row, are an example of a sequence of numbers. A **sequence** is a list of numbers in a particular order. Each number in a sequence is called a **term**. The first term is symbolized by a_1, the second term is symbolized by a_2, and so on.

The graph represents the information from the table above. A sequence is a discrete function whose domain is the set of positive integers.

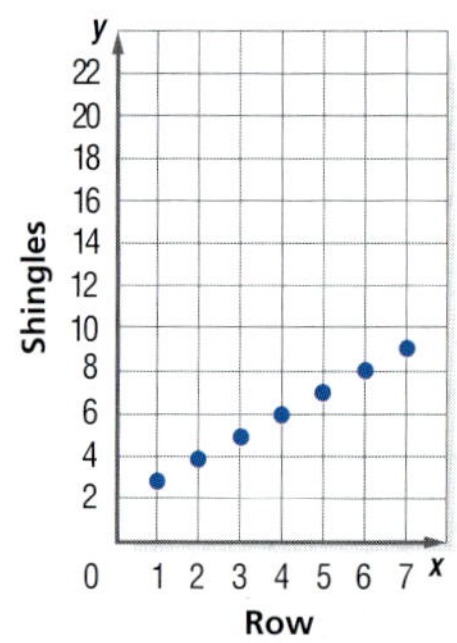

Many sequences have patterns. For example, in the sequence above for the number of shingles, each term can be found by adding 1 to the previous term. A sequence of this type is called an arithmetic sequence. An **arithmetic sequence** is a sequence in which each term after the first is found by adding a constant, called the **common difference** d, to the previous term.

Example 1 *Find the Next Terms*

Find the next four terms of the arithmetic sequence 55, 49, 43, … .

Find the common difference d by subtracting two consecutive terms.

$49 - 55 = -6$ and $43 - 49 = -6$ So, $d = -6$.

Now add -6 to the third term of the sequence, and then continue adding -6 until the next four terms are found.

$$43 \xrightarrow{+(-6)} 37 \xrightarrow{+(-6)} 31 \xrightarrow{+(-6)} 25 \xrightarrow{+(-6)} 19$$

The next four terms of the sequence are 37, 31, 25, and 19.

It is possible to develop a formula for each term of an arithmetic sequence in terms of the first term a_1 and the common difference d. Consider Example 1.

Resource Manager

Workbook and Reproducible Masters

Chapter 11 Resource Masters
- Study Guide and Intervention, pp. 631–632
- Skills Practice, p. 633
- Practice, p. 634
- Reading to Learn Mathematics, p. 635
- Enrichment, p. 636

Teaching Algebra With Manipulatives Masters, pp. 282, 283

Transparencies

5-Minute Check Transparency 11-1
Answer Key Transparencies

Technology

Alge2PASS: Tutorial Plus, Lesson 20
Interactive Chalkboard

Sequence	numbers	55	49	43	37	...	
	symbols	a_1	a_2	a_3	a_4	...	a_n
Expressed in Terms of d and the First Term	numbers	$55 + 0(-6)$	$55 + 1(-6)$	$55 + 2(-6)$	$55 + 3(-6)$	...	$55 + (n - 1)(-6)$
	symbols	$a_1 + 0 \cdot d$	$a_1 + 1 \cdot d$	$a_1 + 2 \cdot d$	$a_1 + 3 \cdot d$	...	$a_1 + (n - 1)d$

The following formula generalizes this pattern for any arithmetic sequence.

Key Concept ***nth Term of an Arithmetic Sequence***

The nth term a_n of an arithmetic sequence with first term a_1 and common difference d is given by

$$a_n = a_1 + (n - 1)d,$$

where n is any positive integer.

More About. . .

Construction

The table below shows typical costs for a construction company to rent a crane for one, two, three, or four months.

Months	Cost ($)
1	75,000
2	90,000
3	105,000
4	120,000

Source: www.howstuffworks.com

Example 2 Find a Particular Term

CONSTRUCTION Refer to the information at the left. Assuming that the arithmetic sequence continues, how much would it cost to rent the crane for twelve months?

Explore Since the difference between any two successive costs is $15,000, the costs form an arithmetic sequence with common difference 15,000.

Plan You can use the formula for the nth term of an arithmetic sequence with $a_1 = 75{,}000$ and $d = 15{,}000$ to find a_{12}, the cost for twelve months.

Solve

$a_n = a_1 + (n - 1)d$ Formula for nth term

$a_{12} = 75{,}000 + (12 - 1)15{,}000$ $n = 12, a_1 = 75{,}000, d = 15{,}000$

$a_{12} = 240{,}000$ Simplify.

It would cost $240,000 to rent the crane for twelve months.

Examine You can find terms of the sequence by adding 15,000. a_5 through a_{12} are 135,000, 150,000, 165,000, 180,000, 195,000, 210,000, 225,000, and 240,000. Therefore, $240,000 is correct.

Example 3 Write an Equation for the nth Term

Write an equation for the nth term of the arithmetic sequence 8, 17, 26, 35, ...

In this sequence, $a_1 = 8$ and $d = 9$. Use the nth term formula to write an equation.

$a_n = a_1 + (n - 1)d$ Formula for nth term

$a_n = 8 + (n - 1)9$ $a_1 = 8, d = 9$

$a_n = 8 + 9n - 9$ Distributive Property

$a_n = 9n - 1$ Simplify.

An equation is $a_n = 9n - 1$.

www.algebra2.com/extra_examples

DAILY INTERVENTION

Unlocking Misconceptions

Subscripts Make sure that all students understand that the subscript in a_n names a term and that it is not an exponent.

2 Teach

ARITHMETIC SEQUENCES

In-Class Examples Power Point®

1 Find the next four terms of the arithmetic sequence $-8, -6, -4, \ldots$. **$-2, 0, 2, 4$**

2 **CONSTRUCTION** Use the information in Example 2 to find the cost to rent the crane for 24 months. **$420,000**

Teaching Tip Ask students to use the formula for the nth Term of an Arithmetic Sequence to show why doubling n does not result in doubling a_n.

3 Write an equation for the nth term of the arithmetic sequence $-8, -6, -4, \ldots$. **$a_n = 2n - 10$**

Teaching Tip Ask students to explain why the equation of an arithmetic sequence is always linear. **There is no power greater than 1, and two variables are not multiplied together.**

Interactive Chalkboard
PowerPoint® Presentations

This CD-ROM is a customizable Microsoft® PowerPoint® presentation that includes:

- Step-by-step, dynamic solutions of each In-Class Example from the Teacher Wraparound Edition
- Additional, Your Turn exercises for each example
- The 5-Minute Check Transparencies
- Hot links to Glencoe Online Study Tools

ARITHMETIC MEANS

In-Class Example

Power Point®

4 Find the three arithmetic means between 21 and 45. **27, 33, 39**

Teaching Tip Ask students to create their own questions similar to Example 4. Lead them to see that, for the value of d to be an integer, the number of arithmetic means must evenly divide the difference between the first and last terms given.

3 Practice/Apply

Study Notebook

Have students—
- add the definitions/examples of the vocabulary terms to their Vocabulary Builder worksheets for Chapter 11.
- include any other item(s) that they find helpful in mastering the skills in this lesson.

About the Exercises...

Organization by Objective
- **Arithmetic Sequences:** 15–51
- **Arithmetic Means:** 52–55

Odd/Even Assignments
Exercises 15–40, 43–48, and 52–55 are structured so that students practice the same concepts whether they are assigned odd or even problems.

Assignment Guide

Basic: 15, 17, 23, 25, 29, 31, 33, 37–43 odd, 47, 49–51, 53, 55–67

Average: 15–47 odd, 49–51, 53, 55–67

Advanced: 16–48 even, 49–51, 52, 54, 56–64 (optional: 65–67)

Algebra Activity

Arithmetic Sequences

Study the figures below. The length of an edge of each cube is 1 centimeter.

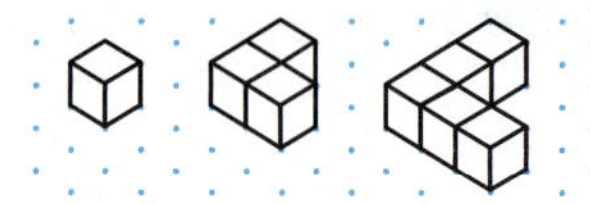

Model and Analyze 1. See margin.

1. Based on the pattern, draw the fourth figure on a piece of isometric dot paper.
2. Find the volumes of the four figures. 1 cm^3, 3 cm^3, 5 cm^3, 7 cm^3
3. Suppose the number of cubes in the pattern continues. Write an equation that gives the volume of Figure n. $V_n = 2n - 1$
4. What would the volume of the twelfth figure be? 23 cm^3

ARITHMETIC MEANS Sometimes you are given two terms of a sequence, but they are not successive terms of that sequence. The terms between any two nonsuccessive terms of an arithmetic sequence are called **arithmetic means**. In the sequence below, 41, 52, and 63 are the three arithmetic means between 30 and 74.

19, 30, 41, 52, 63, 74, 85, 96, …

3 arithmetic means between 30 and 74

Example 4 Find Arithmetic Means

Find the four arithmetic means between 16 and 91.

You can use the nth term formula to find the common difference. In the sequence 16, __?__, __?__, __?__, __?__, 91, …, a_1 is 16 and a_6 is 91.

$a_n = a_1 + (n - 1)d$	Formula for the nth term
$a_6 = 16 + (6 - 1)d$	$n = 6, a_1 = 16$
$91 = 16 + 5d$	$a_6 = 91$
$75 = 5d$	Subtract 16 from each side.
$15 = d$	Divide each side by 5.

Now use the value of d to find the four arithmetic means.

16 → 31 → 46 → 61 → 76 (each + 15)

The arithmetic means are 31, 46, 61, and 76. **CHECK** $76 + 15 = 91$ ✓

Study Tip

Alternate Method
You may prefer this method. The four means will be $16 + d$, $16 + 2d$, $16 + 3d$, and $16 + 4d$. The common difference is $d = 91 - (16 + 4d)$ or $d = 15$.

Check for Understanding

Concept Check

1. **Explain** why the sequence 4, 5, 7, 10, 14, … is not arithmetic. **See margin.**
2. **Find** the 15th term in the arithmetic sequence −3, 4, 11, 18, … . **95**
3. **OPEN ENDED** Write an arithmetic sequence with common difference −5.

3. Sample answer: 1, −4, −9, −14, …

Guided Practice

Find the next four terms of each arithmetic sequence.

4. 12, 16, 20, … **24, 28, 32, 36**
5. 3, 1, −1, … **−3, −5, −7, −9**

Find the first five terms of each arithmetic sequence described.

6. $a_1 = 5, d = 3$ **5, 8, 11, 14, 17**
7. $a_1 = 14, d = -2$ **14, 12, 10, 8, 6**

Algebra Activity

Materials: isometric dot paper
- Point out that this activity does not stack cubes but keeps them on a plane.
- Suggest that students explore by repeating this activity using a different initial arrangement of three cubes.

Answer

Algebra Activity

1.

GUIDED PRACTICE KEY	
Exercises	Examples
4–7	1
8–11, 14	2
12	3
13	4

12. $a_n = 11n - 37$

8. Find a_{13} for the arithmetic sequence $-17, -12, -7, \ldots$. **43**

Find the indicated term of each arithmetic sequence.

9. $a_1 = 3, d = -5, n = 24$ **−112**
10. $a_1 = -5, d = 7, n = 13$ **79**
11. Complete: 68 is the __?__th term of the arithmetic sequence $-2, 3, 8, \ldots$. **15**
12. Write an equation for the nth term of the arithmetic sequence $-26, -15, -4, 7, \ldots$.
13. Find the three arithmetic means between 44 and 92. **56, 68, 80**

Application

14. **ENTERTAINMENT** A basketball team has a halftime promotion where a fan gets to shoot a 3-pointer to try to win a jackpot. The jackpot starts at $5000 for the first game and increases $500 each time there is no winner. Ken has tickets to the fifteenth game of the season. How much will the jackpot be for that game if no one wins by then? **$12,000**

★ indicates increased difficulty

Practice and Apply

Homework Help	
For Exercises	See Examples
15–28, 49	1
29–45, 51	2
46–48, 50	3
52–55	4

Extra Practice
See page 851.

Find the next four terms of each arithmetic sequence.

15. $9, 16, 23, \ldots$ **30, 37, 44, 51**
16. $31, 24, 17, \ldots$ **10, 3, −4, −11**
17. $-6, -2, 2, \ldots$ **6, 10, 14, 18**
18. $-8, -5, -2, \ldots$ **1, 4, 7, 10**

★19. $\frac{1}{3}, 1, \frac{5}{3}, \ldots$ $\frac{7}{3}, 3, \frac{11}{3}, \frac{13}{3}$

★20. $\frac{18}{5}, \frac{16}{5}, \frac{14}{5}, \ldots$ $\frac{12}{5}, 2, \frac{8}{5}, \frac{6}{5}$

★21. $6.7, 6.3, 5.9, \ldots$ **5.5, 5.1, 4.7, 4.3**

★22. $1.3, 3.8, 6.3, \ldots$ **8.8, 11.3, 13.8, 16.3**

Find the first five terms of each arithmetic sequence described.

23. $a_1 = 2, d = 13$ **2, 15, 28, 41, 54**
24. $a_1 = 41, d = 5$ **41, 46, 51, 56, 61**
25. $a_1 = 6, d = -4$ **6, 2, −2, −6, −10**
26. $a_1 = 12, d = -3$ **12, 9, 6, 3, 0**

★27. $a_1 = \frac{4}{3}, d = -\frac{1}{3}$ $\frac{4}{3}, 1, \frac{2}{3}, \frac{1}{3}, 0$

★28. $a_1 = \frac{5}{8}, d = \frac{3}{8}$ $\frac{5}{8}, 1, \frac{11}{8}, \frac{7}{4}, \frac{17}{8}$

29. Find a_8 if $a_n = 4 + 3n$. **28**
30. If $a_n = 1 - 5n$, what is a_{10}? **−49**

Find the indicated term of each arithmetic sequence.

31. $a_1 = 3, d = 7, n = 14$ **94**
32. $a_1 = -4, d = -9, n = 20$ **−175**
33. $a_1 = 35, d = 3, n = 101$ **335**
34. $a_1 = 20, d = 4, n = 81$ **340**

★35. $a_1 = 5, d = \frac{1}{3}, n = 12$ $\frac{26}{3}$

★36. $a_1 = \frac{5}{2}, d = -\frac{3}{2}, n = 11$ $-\frac{25}{2}$

37. a_{12} for $-17, -13, -9, \ldots$ **27**
38. a_{12} for $8, 3, -2, \ldots$ **−47**
39. a_{21} for $121, 118, 115, \ldots$ **61**
40. a_{43} for $5, 9, 13, 17, \ldots$ **173**
41. **GEOLOGY** Geologists estimate that the continents of Europe and North America are drifting apart at a rate of an average of 12 miles every 1 million years, or about 0.75 inch per year. If the continents continue to drift apart at that rate, how many inches will they drift in 50 years? (*Hint*: $a_1 = 0.75$) **37.5 in.**
42. **TOWER OF PISA** To prove that objects of different weights fall at the same rate, Galileo dropped two objects with different weights from the Leaning Tower of Pisa in Italy. The objects hit the ground at the same time. When an object is dropped from a tall building, it falls about 16 feet in the first second, 48 feet in the second second, and 80 feet in the third second, regardless of its weight. How many feet would an object fall in the tenth second? **304 ft**

www.algebra2.com/self_check_quiz

Answer

1. The differences between the terms are not constant.

Enrichment, p. 636

Fibonacci Sequence

Leonardo Fibonacci first discovered the sequence of numbers named for him while studying rabbits. He wanted to know how many pairs of rabbits would be produced in n months, starting with a single pair of newborn rabbits. He made the following assumptions.

1. Newborn rabbits become adults in one month.
2. Each pair of rabbits produces one pair each month.
3. No rabbits die.

Let F_n represent the number of pairs of rabbits at the end of n months. If you begin with one pair of newborn rabbits, $F_0 = F_1 = 1$. This pair of rabbits would produce one pair at the end of the second month, so $F_2 = 1 + 1$, or 2. At the end of the third month, the first pair of rabbits would produce another pair. Thus, $F_3 = 2 + 1$, or 3.

The chart below shows the number of rabbits each month for several months.

Study Guide and Intervention, p. 631 (shown) and p. 632

Arithmetic Sequences An **arithmetic sequence** is a sequence of numbers in which each **term** after the first term is found by adding the **common difference** to the preceding term.

nth Term of an Arithmetic Sequence	$a_n = a_1 + (n - 1)d$, where a_1 is the first term, d is the common difference, and n is any positive integer

Example 1 **Find the next four terms of the arithmetic sequence 7, 11, 15, ...**

Find the common difference by subtracting two consecutive terms.

$11 - 7 = 4$ and $15 - 11 = 4$, so $d = 4$.

Now add 4 to the third term of the sequence, and then continue adding 4 until the four terms are found. The next four terms of the sequence are 19, 23, 27, and 31.

Example 2 **Find the thirteenth term of the arithmetic sequence with $a_1 = 21$ and $d = -6$.**

Use the formula for the nth term of an arithmetic sequence with $a_1 = 21, n = 13$, and $d = -6$.

$a_n = a_1 + (n - 1)d$ Formula for nth term
$a_{13} = 21 + (13 - 1)(-6)$ $n = 13, a_1 = 21, d = -6$
$a_{13} = -51$ Simplify.

The thirteenth term is −51.

Example 3 **Write an equation for the nth term of the arithmetic sequence −14, −5, 4, 13, ...**

In this sequence $a_1 = -14$ and $d = 9$. Use the formula for a_n to write an equation.

$a_n = a_1 + (n - 1)d$ Formula for the nth term
$= -14 + (n - 1)9$ $a_1 = -14, d = 9$
$= -14 + 9n - 9$ Distributive Property
$= 9n - 23$ Simplify.

Exercises

Find the next four terms of each arithmetic sequence.

1. 106, 111, 116, ... **121, 126, 131, 136**
2. −28, −31, −34, ... **−37, −40, −43, −46**
3. 207, 194, 181, ... **168, 155, 142, 129**

Find the first five terms of each arithmetic sequence described.

4. $a_1 = 101, d = 9$ **101, 110, 119, 128, 137**
5. $a_1 = -60, d = 4$ **−60, −56, −52, −48, −44**
6. $a_1 = 210, d = -40$ **210, 170, 130, 90, 50**

Find the indicated term of each arithmetic sequence.

7. $a_1 = 4, d = 6, n = 14$ **82**
8. $a_1 = -4, d = -2, n = 12$ **−26**
9. $a_1 = 80, d = -8, n = 21$ **−80**
10. a_{10} for 0, −3, −6, −9, ... **−27**

Write an equation for the nth term of each arithmetic sequence.

11. 18, 25, 32, 39, ... **$7n + 11$**
12. −110, −85, −60, −35, ... **$25n - 135$**
13. 6.2, 8.1, 10.0, 11.9, ... **$1.9n + 4.3$**

Skills Practice, p. 633 and Practice, p. 634 (shown)

Find the next four terms of each arithmetic sequence.

1. 5, 8, 11, ... **14, 17, 20, 23**
2. −4, −6, −8, ... **−10, −12, −14, −16**
3. 100, 93, 86, ... **79, 72, 65, 58**
4. −24, −19, −14, ... **−9, −4, 1, 6**
5. $\frac{7}{2}, 6, \frac{17}{2}, 11, \ldots$ $\frac{27}{2}, 16, \frac{37}{2}, 21$
6. 4.8, 4.1, 3.4, ... **2.7, 2, 1.3, 0.6**

Find the first five terms of each arithmetic sequence described.

7. $a_1 = 7, d = 7$ **7, 14, 21, 28, 35**
8. $a_1 = -8, d = 2$ **−8, −6, −4, −2, 0**
9. $a_1 = -12, d = -4$ **−12, −16, −20, −24, −28**
10. $a_1 = \frac{1}{2}, d = \frac{1}{2}$ $\frac{1}{2}, 1, \frac{3}{2}, 2, \frac{5}{2}$
11. $a_1 = -\frac{5}{6}, d = -\frac{1}{3}$ $-\frac{5}{6}, -\frac{7}{6}, -\frac{3}{2}, -\frac{11}{6}, -\frac{13}{6}$
12. $a_1 = 10.2, d = -5.8$ **10.2, 4.4, −1.4, −7.2, −13**

Find the indicated term of each arithmetic sequence.

13. $a_1 = 5, d = 3, n = 10$ **32**
14. $a_1 = 9, d = 3, n = 29$ **93**
15. a_{18} for −6, −7, −8, ... **−23**
16. a_{37} for 124, 119, 114, ... **−56**
17. $a_1 = \frac{9}{5}, d = -\frac{3}{5}, n = 10$ $-\frac{18}{5}$
18. $a_1 = 14.25, d = 0.15, n = 31$ **18.75**

Complete the statement for each arithmetic sequence.

19. 166 is the __?__th term of 30, 34, 38, ... **35**
20. 2 is the __?__th term of $\frac{3}{5}, \frac{4}{5}, 1, \ldots$ **8**

Write an equation for the nth term of each arithmetic sequence.

21. −5, −3, −1, 1, ... **$a_n = 2n - 7$**
22. −8, −11, −14, −17, ... **$a_n = -3n - 5$**
23. 1, −1, −3, −5, ... **$a_n = -2n + 3$**
24. −5, 3, 11, 19, ... **$a_n = 8n - 13$**

Find the arithmetic means in each sequence.

25. −5, __?__, __?__, __?__, 11 **−1, 3, 7**
26. 82, __?__, __?__, __?__, 18 **66, 50, 34**
27. **EDUCATION** Trevor Koba has opened an English Language School in Isehara, Japan. He began with 26 students. If he enrolls 3 new students each week, in how many weeks will he have 101 students? **26 wk**
28. **SALARIES** Yolanda interviewed for a job that promised her a starting salary of $32,000 with a $1250 raise at the end of each year. What will her salary be during her sixth year if she accepts the job? **$38,250**

Reading to Learn Mathematics, p. 635 ELL

Pre-Activity **How are arithmetic sequences related to roofing?**

Read the introduction to Lesson 11-1 at the top of page 578 in your textbook.

Describe how you would find the number of shingles needed for the fifteenth row. (Do not actually calculate this number.) Explain why your method will give the correct answer. **Sample answer: Add 3 times 14 to 2. This works because the first row has 2 shingles and 3 more are added 14 times to go from the first row to the fifteenth row.**

Reading the Lesson

1. Consider the formula $a_n = a_1 + (n - 1)d$.
   a. What is this formula used to find? **a particular term of an arithmetic sequence**
   b. What do each of the following represent?
      a_n: **the nth term**
      a_1: **the first term**
      n: **a positive integer that indicates which term you are finding**
      d: **the common difference**
2. Consider the equation $a_n = -3n + 5$.
   a. What does this equation represent? **Sample answer: It gives the nth term of an arithmetic sequence with first term 2 and common difference −3.**
   b. Is the graph of this equation a straight line? Explain your answer. **Sample answer: No; the graph is a set of points that fall on a line, but the points do not fill the line.**
   c. The functions represented by the equations $a_n = -3n + 5$ and $f(x) = -3x + 5$ are alike in that they have the same formula. How are they different? **Sample answer: They have different domains. The domain of the first function is the set of positive integers. The domain of the second function is the set of all real numbers.**

Helping You Remember

3. A good way to remember something is to explain it to someone else. Suppose that your classmate Shala has trouble remembering the formula $a_n = a_1 + (n - 1)d$ correctly. She thinks that the formula should be $a_n = a_1 + nd$. How would you explain to her that she should use $(n - 1)d$ rather than nd in the formula? **Sample answer: Each term after the first in an arithmetic sequence is found by adding d to the previous term. You would add d once to get to the second term, twice to get to the third term, and so on. So d is added $n - 1$ times, not n times, to get the nth term.**

4 Assess

Open-Ended Assessment

Speaking Have students explain what an arithmetic sequence is and how to find a specified term without repeatedly adding the common difference.

Tips for New Teachers

Intervention Make sure students understand that an arithmetic sequence is a list of numbers that share a certain characteristic, but not all lists of numbers are arithmetic sequences. This will prepare them for Lesson 11-3 on geometric sequences.

Getting Ready for Lesson 11-2

PREREQUISITE SKILL Students will use sigma notation in Lesson 11-2. This will involve their evaluating variable expressions for different values as they find values in a series. Use Exercises 65–67 to determine your students' familiarity with evaluating variable expressions for given values.

Answer

57. Arithmetic sequences can be used to model the numbers of shingles in the rows on a section of roof. Answers should include the following.
   - One additional shingle is needed in each successive row.
   - One method is to successively add 1 to the terms of the sequence: $a_8 = 9 + 1$ or 10, $a_9 = 10 + 1$ or 11, $a_{10} = 11 + 1$ or 12, $a_{11} = 12 + 1$ or 13, $a_{12} = 13 + 1$ or 14, $a_{13} = 14 + 1$ or 15, $a_{14} = 15 + 1$ or 16, $a_{15} = 16 + 1$ or 17. Another method is to use the formula for the nth term: $a_{15} = 3 + (15 - 1)1$ or 17.

Complete the statement for each arithmetic sequence.

43. 170 is the __?__ term of $-4, 2, 8, \ldots$. **30th**

44. 124 is the __?__ term of $-2, 5, 12, \ldots$. **19th**

★45. -14 is the __?__ term of $2\frac{1}{5}, 2, 1\frac{4}{5}, \ldots$. **82nd**

Write an equation for the nth term of each arithmetic sequence.

46. $7, 16, 25, 34, \ldots$ $a_n = 9n - 2$

47. $18, 11, 4, -3, \ldots$ $a_n = -7n + 25$

48. $-3, -5, -7, -9, \ldots$ $a_n = -2n - 1$

GEOMETRY For Exercises 49–51, refer to the first three arrays of numbers below.

49. Make drawings to find the next three numbers in this pattern. **49. 13, 17, 21; See pp. 629A–629F for drawings.**

50. Write an equation representing the nth number in this pattern. **50. $p_n = 4n - 3$**

51. Is 397 a number in this pattern? Explain. **51. Yes; it corresponds to $n = 100$.**

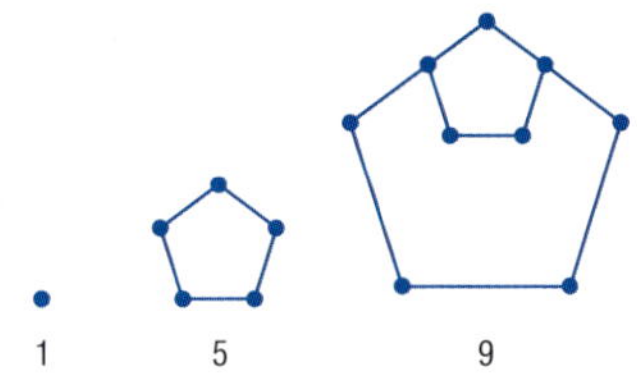

Find the arithmetic means in each sequence.

52. $55, \_\_?\_\_, \_\_?\_\_, \_\_?\_\_, 115$ **70, 85, 100**

53. $10, \_\_?\_\_, \_\_?\_\_, -8$ **4, −2**

54. $-8, \_\_?\_\_, \_\_?\_\_, \_\_?\_\_, \_\_?\_\_, 7$ **54. −5, −2, 1, 4**

55. $3, \_\_?\_\_, \_\_?\_\_, \_\_?\_\_, \_\_?\_\_, \_\_?\_\_, 27$ **55. 7, 11, 15, 19, 23**

56. **CRITICAL THINKING** The numbers x, y, and z are the first three terms of an arithmetic sequence. Express z in terms of x and y. $z = 2y - x$

57. **WRITING IN MATH** Answer the question that was posed at the beginning of the lesson. **See margin.**

How are arithmetic sequences related to roofing?

Include the following in your answer:
- the words that indicate that the numbers of shingles in the rows form an arithmetic sequence, and
- explanations of at least two ways to find the number of shingles in the fifteenth row.

Standardized Test Practice

58. What number follows 20 in this arithmetic sequence? **B**

$8, 11, 14, 17, 20, \ldots$

Ⓐ 5 Ⓑ 23 Ⓒ 26 Ⓓ 29

59. Find the first term in the arithmetic sequence. **B**

$\_\_\_\_, 8\frac{1}{3}, 7, 5\frac{2}{3}, 4\frac{1}{3}, \ldots$

Ⓐ 3 Ⓑ $9\frac{2}{3}$ Ⓒ $10\frac{1}{3}$ Ⓓ 11

Maintain Your Skills

Mixed Review

60. **COMPUTERS** Suppose a computer that costs \$3000 new is only worth \$600 after 3 years. What is the average annual rate of depreciation? *(Lesson 10-6)* **about 41.5%**

Solve each equation. *(Lesson 10-5)*

61. $3e^x - 2 = 0$ **−0.4055**

62. $e^{3x} = 4$ **0.4621**

63. $\ln(x + 2) = 5$ **146.4132**

64. If y varies directly as x and $y = 5$ when $x = 2$, find y when $x = 6$. *(Lesson 9-4)* **15**

Getting Ready for the Next Lesson

PREREQUISITE SKILL **Evaluate each expression for the given values of the variable.** *(To review **evaluating expressions**, see Lesson 1-1.)*

65. $3n - 1$; $n = 1, 2, 3, 4$ **2, 5, 8, 11**

66. $6 - j$; $j = 1, 2, 3, 4$ **5, 4, 3, 2**

67. $4m + 7$; $m = 1, 2, 3, 4, 5$ **11, 15, 19, 23, 27**

582 Chapter 11 Sequences and Series

DAILY INTERVENTION **Differentiated Instruction**

Kinesthetic Have students use wooden or plastic cubes (or ones they make themselves out of paper with a net for a cube drawn on it) to model the Algebra Activity in this lesson.

11-2 Arithmetic Series

What You'll Learn

- Find sums of arithmetic series.
- Use sigma notation.

Vocabulary

- series
- arithmetic series
- sigma notation
- index of summation

How do arithmetic series apply to amphitheaters?

The first amphitheaters were built for contests between gladiators. Modern amphitheaters are usually used for the performing arts. Amphitheaters generally get wider as the distance from the stage increases.

Suppose a small amphitheater can seat 18 people in the first row and each row can seat 4 more people than the previous row.

Study Tip

Indicated Sum
The sum of a series is the result when the terms of the series are added. An *indicated sum* is the expression that illustrates the series, which includes the terms + or −.

ARITHMETIC SERIES The numbers of seats in the rows of the amphitheater form an arithmetic sequence. To find the number of people who could sit in the first four rows, add the first four terms of the sequence. That sum is 18 + 22 + 26 + 30 or 96. A **series** is an indicated sum of the terms of a sequence. Since 18, 22, 26, 30 is an arithmetic sequence, 18 + 22 + 26 + 30 is an **arithmetic series**. Below are some more arithmetic sequences and the corresponding arithmetic series.

Arithmetic Sequence	Arithmetic Series
$-9, -3, 3$	$-9 + (-3) + 3$
$\frac{3}{8}, \frac{8}{8}, \frac{13}{8}, \frac{18}{8}$	$\frac{3}{8} + \frac{8}{8} + \frac{13}{8} + \frac{18}{8}$

S_n represents the sum of the first n terms of a series. For example, S_4 is the sum of the first four terms.

To develop a formula for the sum of any arithmetic series, consider the series below.

$$S_9 = 4 + 11 + 18 + 25 + 32 + 39 + 46 + 53 + 60$$

Write S_9 in two different orders and add the two equations.

$$\begin{aligned} S_9 &= 4 + 11 + 18 + 25 + 32 + 39 + 46 + 53 + 60 \\ (+)\ S_9 &= 60 + 53 + 46 + 39 + 32 + 25 + 18 + 11 + 4 \\ \hline 2S_9 &= 64 + 64 + 64 + 64 + 64 + 64 + 64 + 64 + 64 \\ 2S_9 &= 9(64) \\ S_9 &= \frac{9}{2}(64) \end{aligned}$$

Note that the sum had 9 terms.

The first and last terms of the sum are 64.

An arithmetic series S_n has n terms, and the sum of the first and last terms is $a_1 + a_n$. Thus, the formula $S_n = \frac{n}{2}(a_1 + a_n)$ represents the sum of any arithmetic series.

Key Concept — *Sum of an Arithmetic Series*

The sum S_n of the first n terms of an arithmetic series is given by

$$S_n = \frac{n}{2}[2a_1 + (n - 1)d] \text{ or } S_n = \frac{n}{2}(a_1 + a_n).$$

11-2 Lesson Notes

1 Focus

5-Minute Check Transparency 11-2 Use as a quiz or review of Lesson 11-1.

Mathematical Background notes are available for this lesson on p. 576C.

How do arithmetic series apply to amphitheaters?

Ask students:

- What is the value of a_1 in the sequence of seats? **18**
- What is the value of d? **4**
- How would you determine the number of people who could be seated in 15 rows? **Answers will vary.**

Resource Manager

Workbook and Reproducible Masters

Chapter 11 Resource Masters
- Study Guide and Intervention, pp. 637–638
- Skills Practice, p. 639
- Practice, p. 640
- Reading to Learn Mathematics, p. 641
- Enrichment, p. 642
- Assessment, p. 693

Graphing Calculator and Spreadsheet Masters, p. 48
Science and Mathematics Lab Manual, pp. 133–138

Transparencies
5-Minute Check Transparency 11-2
Answer Key Transparencies

Technology
Interactive Chalkboard

2 Teach

ARITHMETIC SERIES

In-Class Examples

Power Point®

1 Find the sum of the first 20 even numbers, beginning with 2. **420**

Teaching Tip Discuss the difference between a sequence and a series, and ask students to suggest ways to remember which is which.

2 **RADIO** Refer to Example 2 in the Student Edition. Suppose the radio station decided to give away another \$124,000 during the next month, using the same plan. How much should they give away on the first day of September, rounded to the nearest cent? **\$2683.33**

Teaching Tip Remind students that September is one day shorter than August.

3 Find the first four terms of an arithmetic series in which $a_1 = 14$, $a_n = 29$, and $S_n = 129$. **14, 17, 20, 23**

Answer

Graphing Calculator Investigation

1. The index of summation is always replaced by specific values, so the letter that is used does not affect the value of the sum.

Example 1 Find the Sum of an Arithmetic Series

Find the sum of the first 100 positive integers.

The series is $1 + 2 + 3 + \ldots + 100$. Since you can see that $a_1 = 1$, $a_{100} = 100$, and $d = 1$, you can use either sum formula for this series.

Method 1		Method 2
$S_n = \frac{n}{2}(a_1 + a_n)$	Sum formula	$S_n = \frac{n}{2}[2a_1 + (n-1)d]$
$S_{100} = \frac{100}{2}(1 + 100)$	$n = 100, a_1 = 1, a_{100} = 100, d = 1$	$S_{100} = \frac{100}{2}[2(1) + (100-1)1]$
$S_{100} = 50(101)$	Simplify.	$S_{100} = 50(101)$
$S_{100} = 5050$	Multiply.	$S_{100} = 5050$

The sum of the first 100 positive integers is 5050.

Example 2 Find the First Term

RADIO **A radio station considered giving away \$4000 every day in the month of August for a total of \$124,000. Instead, they decided to increase the amount given away every day while still giving away the same total amount. If they want to increase the amount by \$100 each day, how much should they give away the first day?**

You know the values of n, S_n, and d. Use the sum formula that contains d.

$S_n = \frac{n}{2}[2a_1 + (n-1)d]$	Sum formula
$S_{31} = \frac{31}{2}[2a_1 + (31-1)100]$	$n = 31, d = 100$
$124{,}000 = \frac{31}{2}(2a_1 + 3000)$	$S_{31} = 124{,}000$
$8000 = 2a_1 + 3000$	Multiply each side by $\frac{2}{31}$.
$5000 = 2a_1$	Subtract 3000 from each side.
$2500 = a_1$	Divide each side by 2.

The radio station should give away \$2500 the first day.

More About. . .

Radio

99.0% of teens ages 12–17 listen to the radio at least once a week. 79.1% listen at least once a day.

Source: Radio Advertising Bureau

Sometimes it is necessary to use both a sum formula and the formula for the nth term to solve a problem.

Example 3 Find the First Three Terms

Find the first three terms of an arithmetic series in which $a_1 = 9$, $a_n = 105$, and $S_n = 741$.

Step 1 Since you know a_1, a_n, and S_n, use $S_n = \frac{n}{2}(a_1 + a_n)$ to find n.

$$S_n = \frac{n}{2}(a_1 + a_n)$$
$$741 = \frac{n}{2}(9 + 105)$$
$$741 = 57n$$
$$13 = n$$

Step 2 Find d.

$$a_n = a_1 + (n-1)d$$
$$105 = 9 + (13-1)d$$
$$96 = 12d$$
$$8 = d$$

Step 3 Use d to determine a_2 and a_3.

$a_2 = 9 + 8$ or 17 $\quad$ $a_3 = 17 + 8$ or 25

The first three terms are 9, 17, and 25.

584 Chapter 11 Sequences and Series

SIGMA NOTATION

Writing out a series can be time-consuming and lengthy. For convenience, there is a more concise notation called **sigma notation**. The series $3 + 6 + 9 + 12 + \ldots + 30$ can be expressed as $\sum_{n=1}^{10} 3n$. This expression is read *the sum of 3n as n goes from 1 to 10.*

$$\sum_{n=1}^{10} 3n$$

last value of n → 10

formula for the terms of the series → $3n$

first value of n → $n = 1$

The variable, in this case n, is called the **index of summation**.

To generate the terms of a series given in sigma notation, successively replace the index of summation with consecutive integers between the first and last values of the index, inclusive. For the series above, the values of n are 1, 2, 3, and so on, through 10.

Study Tip

Sigma Notation

There are many ways to represent a given series.

$$\sum_{r=4}^{9}(r-3) = \sum_{s=2}^{7}(s-1) = \sum_{j=0}^{5}(j+1)$$

Example 4 Evaluate a Sum in Sigma Notation

Evaluate $\sum_{j=5}^{8}(3j - 4)$.

Method 1

Find the terms by replacing j with 5, 6, 7, and 8. Then add.

$$\sum_{j=5}^{8}(3j-4) = [3(5) - 4] + [3(6) - 4] + [3(7) - 4] + [3(8) - 4]$$
$$= 11 + 14 + 17 + 20$$
$$= 62$$

Method 2

Since the sum is an arithmetic series, use the formula $S_n = \frac{n}{2}(a_1 + a_n)$.

There are 4 terms, $a_1 = 3(5) - 4$ or 11, and $a_4 = 3(8) - 4$ or 20.

$$S_4 = \frac{4}{2}(11 + 20)$$
$$S_4 = 62$$

The sum of the series is 62.

You can use the sum and sequence features on a graphing calculator to find the sum of a series.

Study Tip

Graphing Calculators

On the TI-83 Plus, sum(is located on the LIST MATH menu. The function seq(is located on the LIST OPS menu.

Graphing Calculator Investigation

Sums of Series

The calculator screen shows the evaluation of $\sum_{N=2}^{10}(5N - 2)$. The first four entries for seq(are

- the formula for the general term of the series,
- the index of summation,
- the first value of the index, and
- the last value of the index, respectively.

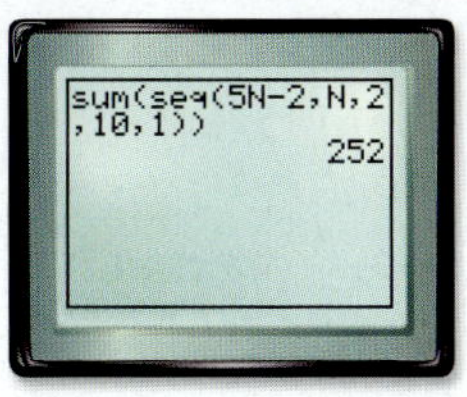

The last entry is always 1 for the types of series that we are considering.

Think and Discuss

1. Explain why you can use any letter for the index of summation.
2. Evaluate $\sum_{n=1}^{8}(2n - 1)$ and $\sum_{j=5}^{12}(2j - 9)$. **Make a conjecture** as to their relationship and explain why you think it is true.

1. See margin.

2. 64; They represent the same series. Any series can be written in many ways using sigma notation.

www.algebra2.com/extra_examples

Graphing Calculator Investigation

When the calculator is in **Seq** mode, the variable will automatically be n rather than x. To select **Seq** mode, press MODE, move the cursor down to **FUNC** and over to **Seq** and press ENTER.

SIGMA NOTATION

In-Class Example Power Point®

4 Evaluate $\sum_{k=3}^{10}(2k + 1)$. **112**

Teaching Tip Help students become comfortable with sigma notation by having them read aloud several expressions written in this notation. Explain that sigma is simply the upper case letter S in the Greek alphabet. Ask them what other mathematical notation uses Greek letters. **Sample answer: π**

3 Practice/Apply

Study Notebook

Have students—

- add the definitions/examples of the vocabulary terms to their Vocabulary Builder worksheets for Chapter 11.
- keep a list of study tips for the graphing calculator, including the one in this lesson.
- include any other item(s) that they find helpful in mastering the skills in this lesson.

About the Exercises...

Organization by Objective

- **Arithmetic Series:** 15–32, 39–45
- **Sigma Notation:** 33–38

Odd/Even Assignments

Exercises 15–26, 29–38, and 41–44 are structured so that students practice the same concepts whether they are assigned odd or even problems.

Assignment Guide

Basic: 15–23 odd, 27–35 odd, 39–45 odd, 46–50, 54–65

Average: 15–45 odd, 46–50, 54–65 (optional: 51–53)

Advanced: 16–46 even, 47–62 (optional: 63–65)

Study Guide and Intervention, p. 637 (shown) and p. 638

Arithmetic Series An **arithmetic series** is the sum of consecutive terms of an arithmetic sequence.

Sum of an Arithmetic Series	The sum S_n of the first n terms of an arithmetic series is given by the formula $S_n = \frac{n}{2}[2a_1 + (n-1)d]$ or $S_n = \frac{n}{2}(a_1 + a_n)$

Example 1 **Find S_n for the arithmetic series with $a_1 = 14$, $a_n = 101$, and $n = 30$.**

Use the sum formula for an arithmetic series.

$S_n = \frac{n}{2}(a_1 + a_n)$ Sum formula

$S_{30} = \frac{30}{2}(14 + 101)$ $n = 30, a_1 = 14, a_n = 101$

$= 15(115)$ Simplify.

$= 1725$ Multiply.

The sum of the series is 1725.

Example 2 **Find the sum of all positive odd integers less than 180.**

The series is $1 + 3 + 5 + \ldots + 179$.

Find n using the formula for the nth term of an arithmetic sequence.

$a_n = a_1 + (n-1)d$ Formula for nth term

$179 = 1 + (n-1)2$ $a_n = 179, a_1 = 1, d = 2$

$179 = 2n - 1$ Simplify.

$180 = 2n$ Add 1 to each side.

$n = 90$ Divide each side by 2.

Then use the sum formula for an arithmetic series.

$S_n = \frac{n}{2}(a_1 + a_n)$ Sum formula

$S_{90} = \frac{90}{2}(1 + 179)$ $n = 90, a_1 = 1, a_n = 179$

$= 45(180)$ Simplify.

$= 8100$ Multiply.

The sum of all positive odd integers less than 180 is 8100.

Exercises

Find S_n for each arithmetic series described.

1. $a_1 = 12, a_n = 100, n = 12$ **672**
2. $a_1 = 50, a_n = -50, n = 15$ **0**
3. $a_1 = 60, a_n = -136, n = 50$ **−1900**
4. $a_1 = 20, d = 4, a_n = 112$ **1584**
5. $a_1 = 180, d = -8, a_n = 68$ **1860**
6. $a_1 = -8, d = -7, a_n = -71$ **−395**
7. $a_1 = 42, n = 8, d = 6$ **504**
8. $a_1 = 4, n = 20, d = 2\frac{1}{2}$ **555**
9. $a_1 = 32, n = 27, d = 3$ **1917**

Find the sum of each arithmetic series.

10. $8 + 6 + 4 + \ldots + -10$ **−10**
11. $16 + 22 + 28 + \ldots + 112$ **1088**
12. $-45 + (-41) + (-37) + \ldots + 35$ **−105**

Find the first three terms of each arithmetic series described.

13. $a_1 = 12, a_n = 174, S_n = 1767$ **12, 21, 30**
14. $a_1 = 80, a_n = -115, S_n = -245$ **80, 65, 50**
15. $a_1 = 6.2, a_n = 12.6, S_n = 84.6$ **6.2, 7.0, 7.8**

Skills Practice, p. 639 and Practice, p. 640 (shown)

Find S_n for each arithmetic series described.

1. $a_1 = 16, a_n = 98, n = 13$ **741**
2. $a_1 = 3, a_n = 36, n = 12$ **234**
3. $a_1 = -5, a_n = -26, n = 8$ **−124**
4. $a_1 = 5, n = 10, a_n = -13$ **−40**
5. $a_1 = 6, n = 15, a_n = -22$ **−120**
6. $a_1 = -20, n = 25, a_n = 148$ **1600**
7. $a_1 = 13, d = -6, n = 21$ **−987**
8. $a_1 = 5, d = 4, n = 11$ **275**
9. $a_1 = 5, d = 2, a_n = 33$ **285**
10. $a_1 = -121, d = 3, a_n = 5$ **−2494**
11. $d = 0.4, n = 10, a_n = 3.8$ **20**
12. $d = -\frac{2}{3}, n = 16, a_n = 44$ **784**

Find the sum of each arithmetic series.

13. $5 + 7 + 9 + 11 + \ldots + 27$ **192**
14. $-4 + 1 + 6 + 11 + \ldots + 91$ **870**
15. $13 + 20 + 27 + \ldots + 272$ **5415**
16. $89 + 86 + 83 + 80 + \ldots + 20$ **1308**
17. $\sum_{n=1}^{4}(1 - 2n)$ **−16**
18. $\sum_{j=1}^{6}(5 + 3n)$ **93**
19. $\sum_{n=1}^{5}(9 - 4n)$ **−15**
20. $\sum_{k=4}^{10}(2k + 1)$ **105**
21. $\sum_{n=3}^{8}(5n - 10)$ **105**
22. $\sum_{n=1}^{101}(4 - 4n)$ **−20,200**

Find the first three terms of each arithmetic series described.

23. $a_1 = 14, a_n = -85, S_n = -1207$ **14, 11, 8**
24. $a_1 = 1, a_n = 19, S_n = 100$ **1, 3, 5**
25. $n = 16, a_n = 15, S_n = -120$ **−30, −27, −24**
26. $n = 15, a_n = 5\frac{4}{5}, S_n = 45$ **$\frac{1}{5}, \frac{3}{5}, 1$**
27. **STACKING** A health club rolls its towels and stacks them in layers on a shelf. Each layer of towels has one less towel than the layer below it. If there are 20 towels on the bottom layer and one towel on the top layer, how many towels are stacked on the shelf? **210 towels**
28. **BUSINESS** A merchant places $1 in a jackpot on August 1, then draws the name of a regular customer. If the customer is present, he or she wins the $1 in the jackpot. If the customer is not present, the merchant adds $2 to the jackpot on August 2 and draws another name. Each day the merchant adds an amount equal to the day of the month. If the first person to win the jackpot wins $496, on what day of the month was her or his name drawn? **August 31**

Reading to Learn Mathematics, p. 641 ELL

Pre-Activity **How do arithmetic series apply to amphitheaters?**

Read the introduction to Lesson 11-2 at the top of page 583 in your textbook.

Suppose that an amphitheater can seat 50 people in the first row and that each row thereafter can seat 9 more people than the previous row. Using the vocabulary of arithmetic sequences, describe how you would find the number of people who could be seated in the first 10 rows. (Do not actually calculate the sum.) **Sample answer: Find the first 10 terms of an arithmetic sequence with first term 50 and common difference 9. Then add these 10 terms.**

Reading the Lesson

1. What is the relationship between an arithmetic sequence and the corresponding arithmetic series? **Sample answer: An arithmetic sequence is a list of terms with a common difference between successive terms. The corresponding arithmetic series is the sum of the terms of the sequence.**
2. Consider the formula $S_n = \frac{n}{2}(a_1 + a_n)$. Explain the meaning of this formula in words. **Sample answer: To find the sum of the first n terms of an arithmetic sequence, find half the number of terms you are adding. Multiply this number by the sum of the first term and the nth term.**
3. a. What is the purpose of sigma notation? **Sample answer: to write a series in a concise form**

   b. Consider the expression $\sum_{i=2}^{12}(4i - 2)$.

   This form of writing a sum is called **sigma notation**.

   The variable i is called the **index of summation**.

   The first value of i is **2**.

   The last value of i is **12**.

   How would you read this expression? **The sum of $4i - 2$ as i goes from 2 to 12.**

Helping You Remember

4. A good way to remember something is to relate it to something you already know. How can your knowledge of how to find the average of two numbers help you remember the formula $S_n = \frac{n}{2}(a_1 + a_n)$? **Sample answer: Rewrite the formula as $S_n = n \cdot \frac{a_1 + a_n}{2}$. The average of the first and last terms is given by the expression $\frac{a_1 + a_n}{2}$. The sum of the first n terms is the average of the first and last terms multiplied by the number of terms.**

Check for Understanding

Concept Check **1–3. See margin.**

1. **Explain** the difference between a sequence and a series.
2. **OPEN ENDED** Write an arithmetic series for which $S_5 = 10$.
3. **OPEN ENDED** Write the series $7 + 10 + 13 + 16$ using sigma notation.

Guided Practice

GUIDED PRACTICE KEY	
Exercises	**Examples**
4–9	1, 2
10, 11	4
12, 13	3
14	1

Find S_n for each arithmetic series described.

4. $a_1 = 4, a_n = 100, n = 25$ **1300**
5. $a_1 = 40, n = 20, d = -3$ **230**
6. $a_1 = 132, d = -4, a_n = 52$ **1932**
7. $d = 5, n = 16, a_n = 72$ **552**

Find the sum of each arithmetic series.

8. $5 + 11 + 17 + \ldots + 95$ **800**
9. $38 + 35 + 32 + \ldots + 2$ **260**
10. $\sum_{n=1}^{7}(2n + 1)$ **63**
11. $\sum_{k=3}^{7}(3k + 4)$ **95**

Find the first three terms of each arithmetic series described.

12. $a_1 = 11, a_n = 110, S_n = 726$ **11, 20, 29**
13. $n = 8, a_n = 36, S_n = 120$ **−6, 0, 6**

Application

14. **WORLD CULTURES** The African-American festival of *Kwanzaa* includes a ritual involving candles. The first night, a candle is lit and then blown out. The second night, a new candle and the candle from the previous night are lit and blown out. This pattern of lighting a new candle and relighting all the candles from the previous nights is continued for seven nights. Use a formula from this lesson to find the total number of candle lightings during the festival. **28**

★ indicates increased difficulty

Practice and Apply

Homework Help	
For Exercises	**See Examples**
15–32, 39, 40, 45	1, 2
33–38	4
31–43	3

Extra Practice See page 851.

Find S_n for each arithmetic series described.

15. $a_1 = 7, a_n = 79, n = 8$ **344**
16. $a_1 = 58, a_n = -7, n = 26$ **663**
17. $a_1 = 43, n = 19, a_n = 115$ **1501**
18. $a_1 = 76, n = 21, a_n = 176$ **2646**
19. $a_1 = 7, d = -2, n = 9$ **−9**
20. $a_1 = 3, d = -4, n = 8$ **−88**
21. $a_1 = 5, d = \frac{1}{2}, n = 13$ **104**
22. $a_1 = 12, d = \frac{1}{3}, n = 13$ **182**
23. $d = -3, n = 21, a_n = -64$ **−714**
24. $d = 7, n = 18, a_n = 72$ **225**
25. ★ $d = \frac{1}{5}, n = 10, a_n = \frac{23}{10}$ **14**
26. ★ $d = -\frac{1}{4}, n = 20, a_n = -\frac{53}{12}$ **$-\frac{245}{6}$**

27. **TOYS** Jamila is making a triangular wall with building blocks. The top row has one block, the second row has three, the third has five, and so on. How many rows can she make with a set of 100 blocks? **10 rows**

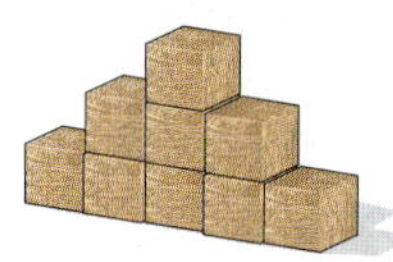

28. **CONSTRUCTION** A construction company will be fined for each day it is late completing its current project. The daily fine will be $4000 for the first day and will increase by $1000 each day. Based on its budget, the company can only afford $60,000 in total fines. What is the maximum number of days it can be late? **8 days**

Find the sum of each arithmetic series.

29. $6 + 13 + 20 + 27 + \ldots + 97$ **721**
30. $7 + 14 + 21 + 28 + \ldots + 98$ **735**
31. $34 + 30 + 26 + \ldots + 2$ **162**
32. $16 + 10 + 4 + \ldots + (-50)$ **−204**
33. $\sum_{n=1}^{6}(2n + 11)$ **108**
34. $\sum_{n=1}^{5}(2 - 3n)$ **−35**
35. $\sum_{k=7}^{11}(42 - 9k)$ **−195**
36. $\sum_{t=19}^{23}(5t - 3)$ **510**
37. ★ $\sum_{i=1}^{300}(7i - 3)$ **315,150**
38. ★ $\sum_{k=1}^{150}(11 + 2k)$ **24,300**

Enrichment, p. 642

Geometric Puzzlers

For the problems on this page, you will need to use the Pythagorean Theorem and the formulas for the area of a triangle and a trapezoid.

1. A rectangle measures 5 by 12 units. The upper left corner is cut off as shown in the diagram.

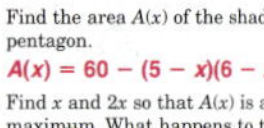

   a. Find the area $A(x)$ of the shaded pentagon. **$A(x) = 60 - (5 - x)(6 - x)$**

   b. Find x and $2x$ so that $A(x)$ is a maximum. What happens to the ...le?

2. A triangle with sides of lengths a, a, and b is isosceles. Two triangles are cut off so that the remaining pentagon has five equal sides of length x. The value of x can be found using this equation.

   $(2b - a)x^2 + (4a^2 - b^2)(2x - a) = 0$

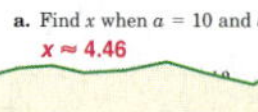

   a. Find x when $a = 10$ and $b = 12$. **$x \approx 4.46$**

Answers

1. In a series, the terms are added. In a sequence, they are not.
2. Sample answer: $0 + 1 + 2 + 3 + 4$
3. Sample answer: $\sum_{n=1}^{4}(3n + 4)$

39. Find the sum of the first 1000 positive even integers. **1,001,000**

★ 40. What is the sum of the multiples of 3 between 3 and 999, inclusive? **166,833**

Find the first three terms of each arithmetic series described.

41. $a_1 = 17, a_n = 197, S_n = 2247$ | 42. $a_1 = -13, a_n = 427, S_n = 18{,}423$

43. $n = 31, a_n = 78, S_n = 1023$ | 44. $n = 19, a_n = 103, S_n = 1102$

41. 17, 26, 35
42. −13, −8, −3
43. −12, −9, −6
44. 13, 18, 23

45. **AEROSPACE** On the Moon, a falling object falls just 2.65 feet in the first second after being dropped. Each second it falls 5.3 feet farther than in the previous second. How far would an object fall in the first ten seconds after being dropped? **265 ft**

CRITICAL THINKING **State whether each statement is *true* or *false*. Explain.**

46. Doubling each term in an arithmetic series will double the sum.

47. Doubling the number of terms in an arithmetic series, but keeping the first term and common difference the same, will double the sum.

46. True; for any series, $2a_1 + 2a_2 + 2a_3 + \ldots + 2a_n = 2(a_1 + a_2 + a_3 + \ldots + a_n)$.
47. False; for example, $7 + 10 + 13 + 16 = 46$, but $7 + 10 + 13 + 16 + 19 + 22 + 25 + 28 = 140$.

48. **WRITING IN MATH** Answer the question that was posed at the beginning of the lesson. **See pp. 629A–629F.**

How do arithmetic series apply to amphitheaters?

Include the following in your answer:
- explanations of what the sequence and the series that can be formed from the given numbers represent, and
- two ways to find the amphitheater capacity if it has ten rows of seats.

Standardized Test Practice

49. $18 + 22 + 26 + 30 + \ldots + 50 = ?$ **C**

Ⓐ 146 Ⓑ 272 Ⓒ 306 Ⓓ 340

50. The angles of a triangle form an arithmetic sequence. If the smallest angle measures 36°, what is the measure of the largest angle? **C**

Ⓐ 60° Ⓑ 72°
Ⓒ 84° Ⓓ 144°

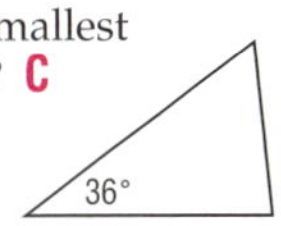

Graphing Calculator

Use a graphing calculator to find the sum of each arithmetic series.

51. $\sum_{n=21}^{75} (2n + 5)$ **5555** 52. $\sum_{n=10}^{50} (3n - 1)$ **3649** 53. $\sum_{n=20}^{60} (4n + 3)$ **6683**

Maintain Your Skills

Mixed Review **Find the indicated term of each arithmetic sequence.** *(Lesson 11-1)*

54. $a_1 = 46, d = 5, n = 14$ **111** 55. $a_1 = 12, d = -7, n = 22$ **−135**

56. **RADIOACTIVITY** The decay of Radon-222 can be modeled by the equation $y = ae^{-0.1813t}$, where t is measured in days. What is the half-life of Radon-222? *(Lesson 10-6)* **about 3.82 days**

Solve each equation by completing the square. *(Lesson 6-4)*

57. $x^2 + 9x + 20.25 = 0$ 58. $9x^2 + 96x + 256 = 0$ 59. $x^2 - 3x - 20 = 0$

57. $-\frac{9}{2}$
58. $-\frac{16}{3}$
59. $\frac{3 \pm \sqrt{89}}{2}$

Simplify. *(Lesson 5-6)*

60. $5\sqrt{3} - 4\sqrt{3}$ $\sqrt{3}$ 61. $\sqrt{26} \cdot \sqrt{39} \cdot \sqrt{14}$ $26\sqrt{21}$ 62. $(\sqrt{10} - \sqrt{6})(\sqrt{5} + \sqrt{3})$ $2\sqrt{2}$

Getting Ready for the Next Lesson **PREREQUISITE SKILL** **Evaluate the expression $a \cdot b^{n-1}$ for the given values of a, b, and n.** *(To review **evaluating expressions**, see Lesson 1-1.)*

63. $a = 1, b = 2, n = 5$ **16** 64. $a = 2, b = -3, n = 4$ **−54** 65. $a = 18, b = \frac{1}{3}, n = 6$ $\frac{2}{27}$

www.algebra2.com/self_check_quiz

4 Assess

Open-Ended Assessment

Speaking Have students explain what sigma notation means, and why it is a useful way to write a series.

Tips for New Teachers

Intervention Make sure that all students can demonstrate understanding of sigma notation by asking them to write out the terms of a series described in sigma notation.

Getting Ready for Lesson 11-3

PREREQUISITE SKILL Students will find terms in geometric sequences in Lesson 11-3. This will involve their evaluating variable expressions for different values as they find values in a sequence. Use Exercises 63–65 to determine your students' familiarity with evaluating variable expressions for given values.

Assessment Options

Quiz (Lessons 11-1 and 11-2) is available on p. 693 of the *Chapter 11 Resource Masters*.

DAILY INTERVENTION **Differentiated Instruction**

Auditory/Musical Have musical students explore and explain how the keys from octave to octave on a piano might relate to a sequence such as A_1, A_2, A_3.

11-3 Lesson Notes

1 Focus

5-Minute Check Transparency 11-3 Use as a quiz or review of Lesson 11-2.

Mathematical Background notes are available for this lesson on p. 576C.

How do geometric sequences apply to a bouncing ball?

Ask students:

- Why is this sequence not an arithmetic sequence? **There is no common difference between terms.**
- Compare a common difference and a common ratio. **The first involves addition; the second, multiplication.**

11-3 Geometric Sequences

What You'll Learn

- Use geometric sequences.
- Find geometric means.

Vocabulary

- geometric sequence
- common ratio
- geometric means

How do geometric sequences apply to a bouncing ball?

If you have ever bounced a ball, you know that when you drop it, it never rebounds to the height from which you dropped it. Suppose a ball is dropped from a height of three feet, and each time it falls, it rebounds to 60% of the height from which it fell. The heights of the ball's rebounds form a sequence.

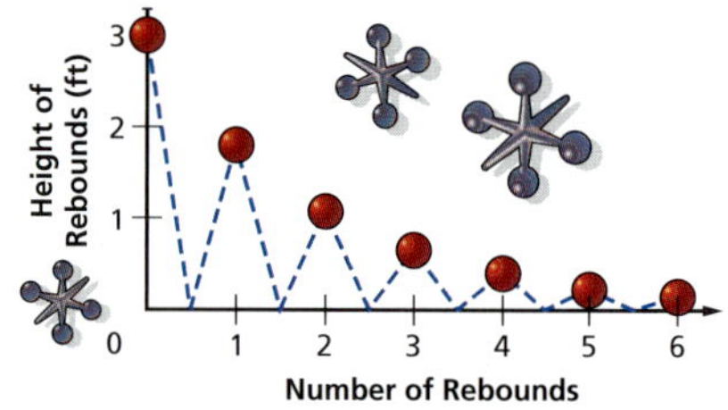

GEOMETRIC SEQUENCES The height of the first rebound of the ball is 3(0.6) or 1.8 feet. The height of the second rebound is 1.8(0.6) or 1.08 feet. The height of the third rebound is 1.08(0.6) or 0.648 feet. The sequence of heights, 1.8, 1.08, 0.648, …, is an example of a **geometric sequence**. A geometric sequence is a sequence in which each term after the first is found by multiplying the previous term by a constant r called the **common ratio**.

As with an arithmetic sequence, you can label the terms of a geometric sequence as a_1, a_2, a_3, and so on. The nth term is a_n and the previous term is a_{n-1}. So, $a_n = r(a_{n-1})$. Thus, $r = \frac{a_n}{a_{n-1}}$. That is, the common ratio can be found by dividing any term by its previous term.

Standardized Test Practice

Example 1 *Find the Next Term*

Multiple-Choice Test Item

Find the missing term in the geometric sequence: 8, 20, 50, 125, ___.			
(A) 75	(B) 200	(C) 250	(D) 312.5

Test-Taking Tip
Since the terms of this sequence are increasing, the missing term must be greater than 125. You can immediately eliminate 75 as a possible answer.

Read the Test Item

Since $\frac{20}{8} = 2.5$, $\frac{50}{20} = 2.5$, and $\frac{125}{50} = 2.5$, the sequence has a common ratio of 2.5.

Solve the Test Item

To find the missing term, multiply the last given term by 2.5: 125(2.5) = 312.5.

The answer is D.

You have seen that each term of a geometric sequence can be expressed in terms of r and its previous term. It is also possible to develop a formula that expresses each term of a geometric sequence in terms of r and the first term a_1. Study the patterns shown in the table on the next page for the sequence 2, 6, 18, 54, … .

Resource Manager

Workbook and Reproducible Masters

Chapter 11 Resource Masters

- Study Guide and Intervention, pp. 643–644
- Skills Practice, p. 645
- Practice, p. 646
- Reading to Learn Mathematics, p. 647
- Enrichment, p. 648

School-to-Career Masters, p. 21

Transparencies

5-Minute Check Transparency 11-3
Answer Key Transparencies

Technology

Interactive Chalkboard

Sequence	numbers	2	6	18	54	...	
	symbols	a_1	a_2	a_3	a_4	...	a_n
Expressed in Terms of r and the Previous Term	numbers	2	2(3)	6(3)	18(3)	...	
	symbols	a_1	$a_1 \cdot r$	$a_2 \cdot r$	$a_3 \cdot r$	...	$a_{n-1} \cdot r$
Expressed in Terms of r and the First Term	numbers	2	2(3)	2(9)	2(27)	...	
		$2(3^0)$	$2(3^1)$	$2(3^2)$	$2(3^3)$	...	
	symbols	$a_1 \cdot r^0$	$a_1 \cdot r^1$	$a_1 \cdot r^2$	$a_1 \cdot r^3$	...	$a_1 \cdot r^{n-1}$

The three entries in the last column of the table all describe the nth term of a geometric sequence. This leads us to the following formula for finding the nth term of a geometric sequence.

Key Concept *nth Term of a Geometric Sequence*

The nth term a_n of a geometric sequence with first term a_1 and common ratio r is given by

$$a_n = a_1 \cdot r^{n-1},$$

where n is any positive integer.

Example 2 Find a Particular Term

Find the eighth term of a geometric sequence for which $a_1 = -3$ and $r = -2$.

$a_n = a_1 \cdot r^{n-1}$ Formula for nth term

$a_8 = (-3) \cdot (-2)^{8-1}$ $n = 8, a_1 = -3, r = -2$

$a_8 = (-3) \cdot (-128)$ $(-2)^7 = -128$

$a_8 = 384$ Multiply.

The eighth term is 384.

Example 3 Write an Equation for the nth Term

Write an equation for the nth term of the geometric sequence 3, 12, 48, 192, … .

In this sequence, $a_1 = 3$ and $r = 4$. Use the nth term formula to write an equation.

$a_n = a_1 \cdot r^{n-1}$ Formula for nth term

$a_n = 3 \cdot 4^{n-1}$ $a_1 = 3, r = 4$

An equation is $a_n = 3 \cdot 4^{n-1}$.

You can also use the formula for the nth term if you know the common ratio and one term of a geometric sequence, but not the first term.

Example 4 Find a Term Given the Fourth Term and the Ratio

Find the tenth term of a geometric sequence for which $a_4 = 108$ and $r = 3$.

First, find the value of a_1.

$a_n = a_1 \cdot r^{n-1}$ Formula for nth term

$a_4 = a_1 \cdot 3^{4-1}$ $n = 4, r = 3$

$108 = 27a_1$ $a_4 = 108$

$4 = a_1$ Divide each side by 27.

Now find a_{10}.

$a_n = a_1 \cdot r^{n-1}$ Formula for nth term

$a_{10} = 4 \cdot 3^{10-1}$ $n = 10, a_1 = 4, r = 3$

$a_{10} = 78{,}732$ Use a calculator.

The tenth term is 78,732.

www.algebra2.com/extra_examples

2 Teach

GEOMETRIC SEQUENCES

In-Class Examples Power Point®

1 Find the missing term in the geometric sequence: 324, 108, 36, 12, ____. **B**

A 972 **B** 4

C 0 **D** −12

Teaching Tip Discuss the fact that when a sequence has three consecutive terms that are decreasing (or increasing), it will continue to do so.

2 Find the sixth term of a geometric sequence for which $a_1 = -3$ and $r = -2$. $a_6 = 96$

3 Write an equation for the nth term of the geometric sequence 5, 10, 20, 40, … . $a_n = 5 \cdot 2^{n-1}$

Teaching Tip Encourage students to begin a geometric sequence problem by writing the known values for each of the variables n, a, and r.

4 Find the seventh term of a geometric sequence for which $a_3 = 96$ and $r = 2$. **1536**

Teaching Tip Emphasize the importance of writing every step of the calculations as an equation, so that each numeric value found during the process is clearly identified.

Teacher to Teacher

Holly K. Plunkett University H.S., Morgantown, WV

"I have my students investigate the problem presented at the beginning of this lesson using a CBL."

Standardized Test Practice

A B C D

Example 1 In discussing the Test-Taking Tip for Example 1, point out that a geometric sequence with a negative common ratio is neither increasing nor decreasing.

GEOMETRIC MEANS

In-Class Example Power Point®

5 Find three geometric means between 3.12 and 49.92. **6.24, 12.48, 24.96 or −6.24, 12.48, −24.96**

3 Practice/Apply

Study Notebook

Have students—
- add the definitions/examples of the vocabulary terms to their Vocabulary Builder worksheets for Chapter 11.
- include any other item(s) that they find helpful in mastering the skills in this lesson.

DAILY INTERVENTION

FIND THE ERROR Help students see that if the first term is greater than 1, then a decreasing sequence must have a common ratio less than 1.

About the Exercises...

Organization by Objective
- **Geometric Sequences:** 13–42
- **Geometric Means:** 43–46

Odd/Even Assignments
Exercises 13–24, 27–36, and 39–46 are structured so that students practice the same concepts whether they are assigned odd or even problems.

Alert! Exercise 47 requires the Internet or other reference materials.

Assignment Guide

Basic: 13, 15, 21–29 odd, 33–47 odd, 48–61

Average: 13–47 odd, 48–61

Advanced: 14–46 even, 47–58 (optional: 59–61)

All: Practice Quiz 1 (1–5)

GEOMETRIC MEANS In Lesson 11-1, you learned that missing terms between two nonsuccessive terms in an arithmetic sequence are called *arithmetic means*. Similarly, the missing terms(s) between two nonsuccessive terms of a geometric sequence are called **geometric means**. For example, 6, 18, and 54 are three geometric means between 2 and 162 in the sequence 2, 6, 18, 54, 162, … . You can use the common ratio to find the geometric means in a given sequence.

Study Tip

Alternate Method
You may prefer this method. The three means will be $2.25r$, $2.25r^2$, and $2.25r^3$. Then the common ratio is $r = \frac{576}{2.25r^3}$ or $r^4 = \frac{576}{2.25}$. Thus, $r = 4$.

Example 5 Find Geometric Means

Find three geometric means between 2.25 and 576.

Use the *n*th term formula to find the value of r. In the sequence 2.25, __?__, __?__, __?__, 576, a_1 is 2.25 and a_5 is 576.

$a_n = a_1 \cdot r^{n-1}$	Formula for *n*th term
$a_5 = 2.25 \cdot r^{5-1}$	$n = 5$, $a_1 = 2.25$
$576 = 2.25r^4$	$a_5 = 576$
$256 = r^4$	Divide each side by 2.25.
$\pm 4 = r$	Take the fourth root of each side.

There are two possible common ratios, so there are two possible sets of geometric means. Use each value of r to find three geometric means.

$r = 4$	$r = -4$
$a_2 = 2.25(4)$ or 9	$a_2 = 2.25(-4)$ or -9
$a_3 = 9(4)$ or 36	$a_3 = -9(-4)$ or 36
$a_4 = 36(4)$ or 144	$a_4 = 36(-4)$ or -144

The geometric means are 9, 36, and 144, or −9, 36, and −144.

Check for Understanding

Concept Check

1. **Decide** whether each sequence is *arithmetic* or *geometric*. Explain.
   a. 1, −2, 4, −8, …
   b. 1, −2, −5, −8, …

   1a. Geometric; the terms have a common ratio of −2.
   1b. Arithmetic; the terms have a common difference of −3.

2. **OPEN ENDED** Write a geometric sequence with a common ratio of $\frac{2}{3}$. **2. Sample answer: 1, $\frac{2}{3}$, $\frac{4}{9}$, $\frac{8}{27}$, …**

3. **FIND THE ERROR** Marika and Lori are finding the seventh term of the geometric sequence 9, 3, 1, … .

   Marika
   $r = \frac{3}{9}$ or $\frac{1}{3}$
   $a_7 = 9\left(\frac{1}{3}\right)^{7-1}$
   $= \frac{1}{81}$

   Lori
   $r = \frac{9}{3}$ or 3
   $a_7 = 9 \cdot 3^{7-1}$
   $= 6561$

   Who is correct? Explain your reasoning.
   Marika; Lori divided in the wrong order when finding r.

Guided Practice **Find the next two terms of each geometric sequence.**

4. 20, 30, 45, … **67.5, 101.25**
5. $-\frac{1}{4}, \frac{1}{2}, -1, \ldots$ **2, −4**
6. Find the first five terms of the geometric sequence for which $a_1 = -2$ and $r = 3$. **−2, −6, −18, −54, −162**

DAILY INTERVENTION

Differentiated Instruction

Interpersonal Have students in small groups discuss any confusions they may have about the language, formulas, and definitions for arithmetic and geometric sequences and series. Suggest that they help each other organize their notes and thinking to make these topics clear.

GUIDED PRACTICE KEY	
Exercises	Examples
4–6, 12	1
8, 9	2
10	3
7	4
11	5

7. Find a_9 for the geometric sequence 60, 30, 15, … . $\frac{15}{64}$

Find the indicated term of each geometric sequence.

8. $a_1 = 7, r = 2, n = 4$ **56**

9. $a_3 = 32, r = -0.5, n = 6$ **−4**

10. Write an equation for the nth term of the geometric sequence 4, 8, 16, … .

11. Find two geometric means between 1 and 27. **3, 9**

10. $a_n = 4 \cdot 2^{n-1}$

Standardized Test Practice

12. Find the missing term in the geometric sequence: $\frac{9}{4}, \frac{3}{4}, \frac{1}{4}, \frac{1}{12}$, ____. **A**

Ⓐ $\frac{1}{36}$ Ⓑ $\frac{1}{20}$ Ⓒ $\frac{1}{6}$ Ⓓ $\frac{1}{3}$

★ indicates increased difficulty

Practice and Apply

Homework Help	
For Exercises	See Examples
13–24	1
25–30, 33–38, 47, 48	2
31, 32	4
39–42	3
43–46	5

Extra Practice
See page 852.

Find the next two terms of each geometric sequence.

13. 405, 135, 45, … **15, 5**

14. 81, 108, 144, … **192, 256**

15. 16, 24, 36, … **54, 81**

16. 162, 108, 72, … **48, 32**

★ 17. $\frac{5}{2}, \frac{5}{3}, \frac{10}{9}, \ldots$ $\frac{20}{27}, \frac{40}{81}$

★ 18. $\frac{1}{3}, \frac{5}{6}, \frac{25}{12}, \ldots$ $\frac{125}{24}, \frac{625}{48}$

★ 19. 1.25, −1.5, 1.8, … **−2.16, 2.592**

★ 20. 1.4, −3.5, 8.75, … **−21.875, 54.6875**

Find the first five terms of each geometric sequence described.

21. $a_1 = 2, r = -3$ **2, −6, 18, −54, 162**

22. $a_1 = 1, r = 4$ **1, 4, 16, 64, 256**

23. $a_1 = 243, r = \frac{1}{3}$ **243, 81, 27, 9, 3**

24. $a_1 = 576, r = -\frac{1}{2}$ **576, −288, 144, −72, 36**

25. Find a_7 if $a_n = 12\left(\frac{1}{2}\right)^{n-1}$. $\frac{3}{16}$

26. If $a_n = \frac{1}{3} \cdot 6^{n-1}$, what is a_6? **2592**

Find the indicated term of each geometric sequence.

27. $a_1 = \frac{1}{3}, r = 3, n = 8$ **729**

28. $a_1 = \frac{1}{64}, r = 4, n = 9$ **1024**

29. $a_1 = 16{,}807, r = \frac{3}{7}, n = 6$ **243**

30. $a_1 = 4096, r = \frac{1}{4}, n = 8$ $\frac{1}{4}$

★ 31. $a_4 = 16, r = 0.5, n = 8$ **1**

★ 32. $a_6 = 3, r = 2, n = 12$ **192**

33. a_9 for $\frac{1}{5}, 1, 5, \ldots$ **78,125**

34. a_7 for $\frac{1}{32}, \frac{1}{16}, \frac{1}{8}, \ldots$ **2**

35. a_8 for 4, −12, 36, … **−8748**

36. a_6 for 540, 90, 15, … $\frac{5}{72}$

37. **ART** A one-ton ice sculpture is melting so that it loses one-fifth of its weight per hour. How much of the sculpture will be left after five hours? Write the answer in pounds. **655.36 lb**

38. **SALARIES** Geraldo's current salary is $40,000 per year. His annual pay raise is always a percent of his salary at the time. What would his salary be if he got four consecutive 4% increases? **$46,794.34**

Write an equation for the nth term of each geometric sequence.

39. 36, 12, 4, … $a_n = 36\left(\frac{1}{3}\right)^{n-1}$

40. 64, 16, 4, … $a_n = 64\left(\frac{1}{4}\right)^{n-1}$

41. −2, 10, −50, … $a_n = -2(-5)^{n-1}$

42. 4, −12, 36, … $a_n = 4(-3)^{n-1}$

Find the geometric means in each sequence. **46. 6, 12, 24, 48**

43. 9, __?__, __?__, __?__, 144

44. 4, __?__, __?__, __?__, 324

45. 32, __?__, __?__, __?__, __?__, 1

46. 3, __?__, __?__, __?__, __?__, 96

43. ±18, 36, ±72
44. ±12, 36, ±108
45. 16, 8, 4, 2

www.algebra2.com/self_check_quiz

Study Guide and Intervention, p. 643 (shown) and p. 644

Geometric Sequences A **geometric sequence** is a sequence in which each term after the first is the product of the previous term and a constant called the **constant ratio**.

nth Term of a Geometric Sequence	$a_n = a_1 \cdot r^{n-1}$, where a_1 is the first term, r is the common ratio, and n is any positive integer

Example 1 **Find the next two terms of the geometric sequence 1200, 480, 192, … .**

Since $\frac{480}{1200} = 0.4$ and $\frac{192}{480} = 0.4$, the sequence has a common ratio of 0.4. The next two terms in the sequence are 192(0.4) = 76.8 and 76.8(0.4) = 30.72.

Example 2 **Write an equation for the nth term of the geometric sequence 3.6, 10.8, 32.4, … .**

In this sequence $a_1 = 3.6$ and $r = 3$. Use the nth term formula to write an equation.

$a_n = a_1 \cdot r^{n-1}$ Formula for nth term
$= 3.6 \cdot 3^{n-1}$ $a_1 = 3.6, r = 3$

An equation for the nth term is $a_n = 3.6 \cdot 3^{n-1}$.

Exercises

Find the next two terms of each geometric sequence.

1. 6, 12, 24, … **48, 96**
2. 180, 60, 20, … $\frac{20}{3}, \frac{20}{9}$
3. 2000, −1000, 500, … **−250, 125**
4. 0.8, −2.4, 7.2, … **−21.6, 64.8**
5. 80, 60, 45, … **33.75, 25.3125**
6. 3, 16.5, 90.75, … **499.125, 2745.1875**

Find the first five terms of each geometric sequence described.

7. $a_1 = \frac{1}{9}, r = 3$ $\frac{1}{9}, \frac{1}{3}$, 1, 3, 9
8. $a_1 = 240, r = -\frac{3}{4}$ 240, −180, 135, $-101\frac{1}{4}, 75\frac{15}{16}$
9. $a_1 = 10, r = \frac{5}{2}$ 10, 25, $62\frac{1}{2}, 156\frac{1}{4}, 390\frac{5}{8}$

Find the indicated term of each geometric sequence.

10. $a_1 = -10, r = 4, n = 2$ **−40**
11. $a_1 = -6, r = -\frac{1}{2}, n = 8$ $\frac{3}{64}$
12. $a_3 = 9, r = -3, n = 7$ **729**
13. $a_4 = 16, r = 2, n = 10$ **1024**
14. $a_4 = -54, r = -3, n = 6$ **−486**
15. $a_1 = 8, r = \frac{2}{3}, n = 5$ $\frac{128}{81}$

Write an equation for the nth term of each geometric sequence.

16. 500, 350, 245, … $500 \cdot 0.7^{n-1}$
17. 8, 32, 128, … $8 \cdot 4^{n-1}$
18. 11, −24.2, 53.24, … $11 \cdot (-2.2)^{n-1}$

Skills Practice, p. 645 and Practice, p. 646 (shown)

Find the next two terms of each geometric sequence.

1. −15, −30, −60, … **−120, −240**
2. 80, 40, 20, … **10, 5**
3. 90, 30, 10, … $\frac{10}{3}, \frac{10}{9}$
4. −1458, 486, −162, … **54, −18**
5. $\frac{1}{4}, \frac{3}{8}, \frac{9}{16}, \ldots$ $\frac{27}{32}, \frac{81}{64}$
6. 216, 144, 96, … 64, $\frac{128}{3}$

Find the first five terms of each geometric sequence described.

7. $a_1 = -1, r = -3$ **−1, 3, −9, 27, −81**
8. $a_1 = 7, r = -4$ **7, −28, 112, −448, 1792**
9. $a_1 = -\frac{1}{3}, r = 2$ $-\frac{1}{3}, -\frac{2}{3}, -\frac{4}{3}, -\frac{8}{3}, -\frac{16}{3}$
10. $a_1 = 12, r = \frac{2}{3}$ 12, 8, $\frac{16}{3}, \frac{32}{9}, \frac{64}{27}$

Find the indicated term of each geometric sequence.

11. $a_1 = 5, r = 3, n = 6$ **1215**
12. $a_1 = 20, r = -3, n = 6$ **−4860**
13. $a_1 = -4, r = -2, n = 10$ **2048**
14. a_8 for $-\frac{1}{250}, -\frac{1}{50}, -\frac{1}{10}, \ldots$ $-\frac{625}{2}$
15. a_{12} for 96, 48, 24, … $\frac{3}{64}$
16. $a_1 = 8, r = \frac{1}{2}, n = 9$ $\frac{1}{32}$
17. $a_1 = -3125, r = -\frac{1}{5}, n = 9$ $-\frac{1}{125}$
18. $a_1 = 3, r = \frac{1}{10}, n = 8$ $\frac{3}{10{,}000{,}000}$

Write an equation for the nth term of each geometric sequence.

19. 1, 4, 16, … $a_n = (4)^{n-1}$
20. −1, −5, −25, … $a_n = -1(5)^{n-1}$
21. $1, \frac{1}{2}, \frac{1}{4}, \ldots$ $a_n = \left(\frac{1}{2}\right)^{n-1}$
22. −3, −6, −12, … $a_n = -3(2)^{n-1}$
23. 7, −14, 28, … $a_n = 7(-2)^{n-1}$
24. −5, −30, −180, … $a_n = -5(6)^{n-1}$

Find the geometric means in each sequence.

25. 3, __?__, __?__, __?__, 768 **12, 48, 192**
26. 5, __?__, __?__, __?__, 1280 **±20, 80, ±320**
27. 144, __?__, __?__, __?__, 9 **±72, 36, ±18**
28. 37,500, __?__, __?__, __?__, __?__, −12 **−7500, 1500, −300, 60**

29. **BIOLOGY** A culture initially contains 200 bacteria. If the number of bacteria doubles every 2 hours, how many bacteria will be in the culture at the end of 12 hours? **12,800**

30. **LIGHT** If each foot of water in a lake screens out 60% of the light above, what percent of the light passes through 5 feet of water? **1.024%**

31. **INVESTING** Raul invests $1000 in a savings account that earns 5% interest compounded annually. How much money will he have in the account at the end of 5 years? **$1276.28**

Reading to Learn Mathematics, p. 647 ELL

Pre-Activity **How do geometric sequences apply to a bouncing ball?**

Read the introduction to Lesson 11-3 at the top of page 588 in your textbook.

Suppose that you drop a ball from a height of 4 feet, and that each time it falls, it bounces back to 74% of the height from which it fell. Describe how would you find the height of the third bounce. (Do not actually calculate the height of the bounce.)

Sample answer: Multiply 4 by 0.74 three times.

Reading the Lesson

1. Explain the difference between an arithmetic sequence and a geometric sequence.
Sample answer: In an arithmetic sequence, each term after the first is found by adding the common difference to the previous term. In a geometric sequence, each term after the first is found by multiplying the previous term by the common ratio.

2. Consider the formula $a_n = a_1 \cdot r^{n-1}$.
   a. What is this formula used to find? **a particular term of a geometric sequence**
   b. What do each of the following represent?
   a_n: **the nth term**
   a_1: **the first term**
   r: **the common ratio**
   n: **a positive integer that indicates which term you are finding**

3. a. In the sequence 5, 8, 11, 14, 17, 20, the numbers 8, 11, 14, and 17 are **arithmetic means** between 5 and 20.
   b. In the sequence 12, 4, $\frac{4}{3}, \frac{4}{9}, \frac{4}{27}$, the numbers 4, $\frac{4}{3}$, and $\frac{4}{9}$ are **geometric means** between 12 and $\frac{4}{27}$.

Helping You Remember

4. Suppose that your classmate Ricardo has trouble remembering the formula $a_n = a_1 \cdot r^{n-1}$ correctly. He thinks that the formula should be $a_n = a_1 \cdot r^n$. How would you explain to him that he should use r^{n-1} rather than r^n in the formula?
Sample answer: Each term after the first in a geometric sequence is found by multiplying the previous term by r. There are $n - 1$ terms before the nth term, so you would need to multiply by r a total of $n - 1$ times, not n times, to get the nth term.

Enrichment, p. 648

Half the Distance

Suppose you are 200 feet from a fixed point, P. Suppose that you are able to move to the halfway point in one minute, to the next halfway point one minute after that, and so on.

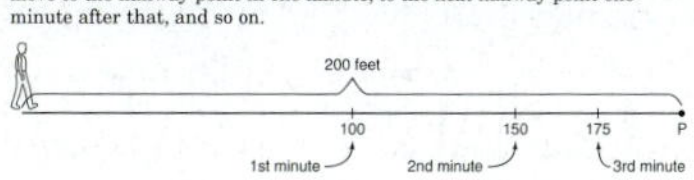

An interesting sequence results because according to the problem, you never actually reach the point P, although you do get arbitrarily close to it.

You can compute how long it will take to get within some specified small distance of the point. On a calculator, you enter the distance to be covered and then count the number of successive divisions by 2 necessary to get within the desired distance.

4 Assess

Open-Ended Assessment

Modeling With manipulatives or sketches, have students use various geometric elements (for example, numbers of sides and diagonals) to model problems involving arithmetic and geometric sequences.

Tips for New Teachers

Intervention Make sure that students understand the difference between arithmetic and geometric sequences by asking them to create a simple example of each one.

Getting Ready for Lesson 11-4

PREREQUISITE SKILL Students will find the sum of the first n terms of geometric series in Lesson 11-4. This will involve evaluating rational expressions for different values. Use Exercises 59–61 to determine your students' familiarity with evaluating rational expressions.

Assessment Options

Practice Quiz 1 The quiz provides students with a brief review of the concepts and skills in Lessons 11-1 through 11-3. Lesson numbers are given to the right of exercises or instruction lines so students can review concepts not yet mastered.

MEDICINE For Exercises 47 and 48, use the following information.
Iodine-131 is a radioactive element used to study the thyroid gland.

47. **RESEARCH** Use the Internet or other resource to find the *half-life* of Iodine-131, rounded to the nearest day. This is the amount of time it takes for half of a sample of Iodine-131 to decay into another element. **8 days**

48. How much of an 80-milligram sample of Iodine-131 would be left after 32 days? **5 mg**

CRITICAL THINKING **Determine whether each statement is *true* or *false*. If true, explain. If false, provide a counterexample.**

49. Every sequence is either arithmetic or geometric.

50. There is no sequence that is both arithmetic and geometric.

49. False; the sequence 1, 4, 9, 16, ..., for example, is neither arithmetic nor geometric.

50. False, the sequence 1, 1, 1, 1, ..., for example, is arithmetic ($d = 0$) and geometric ($r = 1$).

51. **WRITING IN MATH** Answer the question that was posed at the beginning of the lesson. **See margin.**

How do geometric sequences apply to a bouncing ball?

Include the following in your answer:
- the first five terms of the sequence of heights from which the ball falls, and
- any similarities or differences in the sequences for the heights the ball rebounds and the heights from which the ball falls.

Standardized Test Practice

52. Find the missing term in the geometric sequence: −5, 10, −20, 40, ___. **A**

Ⓐ −80 Ⓑ −35 Ⓒ 80 Ⓓ 100

53. What is the tenth term in the geometric sequence: 144, 72, 36, 18, …? **C**

Ⓐ 0 Ⓑ $\frac{9}{64}$ Ⓒ $\frac{9}{32}$ Ⓓ $\frac{9}{16}$

Maintain Your Skills

Mixed Review Find S_n for each arithmetic series described. *(Lesson 11-2)*

54. $a_1 = 11, a_n = 44, n = 23$ **632.5**

55. $a_1 = -5, d = 3, n = 14$ **203**

Find the arithmetic means in each sequence. *(Lesson 11-1)*

56. 15, _?_, _?_, 27 **19, 23**

57. −8, _?_, _?_, _?_, −24

57. −12, −16, −20

58. **GEOMETRY** Find the perimeter of a triangle with vertices at (2, 4), (−1, 3) and (1, −3). *(Lesson 8-1)* **$5\sqrt{2} + 3\sqrt{10}$ units**

Getting Ready for the Next Lesson **PREREQUISITE SKILL** Evaluate each expression. *(To review **expressions**, see Lesson 1-1.)*

59. $\frac{1-2^7}{1-2}$ **127**

60. $\frac{1-\left(\frac{1}{2}\right)^6}{1-\left(\frac{1}{2}\right)}$ **$\frac{63}{32}$**

61. $\frac{1-\left(-\frac{1}{3}\right)^5}{1-\left(-\frac{1}{3}\right)}$ **$\frac{61}{81}$**

Practice Quiz 1 — Lessons 11-1 through 11-3

Find the indicated term of each arithmetic sequence. *(Lesson 11-1)*

1. $a_1 = 7, d = 3, n = 14$ **46**

2. $a_1 = 2, d = \frac{1}{2}, n = 8$ **$\frac{11}{2}$**

Find the sum of each arithmetic series described. *(Lesson 11-2)*

3. $a_1 = 5, a_n = 29, n = 11$ **187**

4. 6 + 12 + 18 + … + 96 **816**

5. Find a_7 for the geometric sequence 729, −243, 81, … . *(Lesson 11-3)* **1**

Answer

51. The heights of the bounces of a ball and the heights from which a bouncing ball falls each form geometric sequences. Answers should include the following.
- 3, 1.8, 1.08, 0.648, 0.3888
- The common ratios are the same, but the first terms are different. The sequence of heights from which the ball falls is the sequence of heights of the bounces with the term 3 inserted at the beginning.

Graphing Calculator Investigation

A Preview of Lesson 11-4

Limits

You may have noticed that in some geometric sequences, the later the term in the sequence, the closer the value is to 0. Another way to describe this is that as n increases, a_n approaches 0. The value that the terms of a sequence approach, in this case 0, is called the **limit** of the sequence. Other types of infinite sequences may also have limits. If the terms of a sequence do not approach a unique value, we say that the limit of the sequence does not exist.

Find the limit of the geometric sequence $1, \frac{1}{3}, \frac{1}{9}, \ldots$.

Step 1 *Enter the sequence.*

- The formula for this sequence is $a_n = \left(\frac{1}{3}\right)^{n-1}$.
- Position the cursor on L1 in the STAT EDIT Edit … screen and enter the formula seq(N,N,1,10,1). This generates the values 1, 2, …, 10 of the index N.
- Position the cursor on L2 and enter the formula seq((1/3)^(N-1),N,1,10,1). This generates the first ten terms of the sequence.

  KEYSTROKES: *Review sequences in the Graphing Calculator Investigation on page 585.*

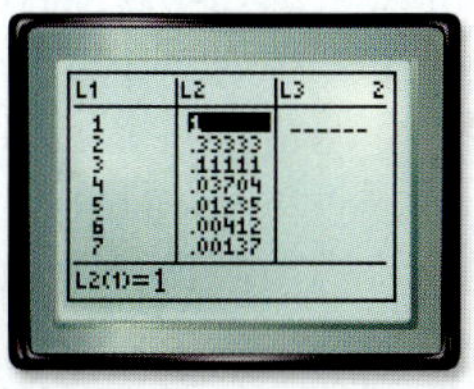

Notice that as n increases, the terms of the given sequence get closer and closer to 0. If you scroll down, you can see that for $n \geq 8$ the terms are so close to 0 that the calculator expresses them in scientific notation. This suggests that the limit of the sequence is 0.

Step 2 *Graph the sequence.*

- Use a STAT PLOT to graph the sequence. Use L1 as the Xlist and L2 as the Ylist.

  KEYSTROKES: *Review STAT PLOTs on page 87.*

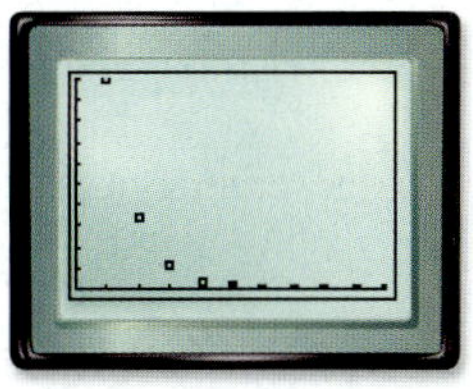

[0, 10] scl: 1 by [0, 1] scl: 0.1

The graph also shows that, as n increases, the terms approach 0. In fact, for $n \geq 6$, the marks appear to lie on the horizontal axis. This strongly suggests that the limit of the sequence is 0.

Exercises

Use a graphing calculator to find the limit, if it exists, of each sequence.

1. $a_n = \left(\frac{1}{2}\right)^n$ **0**

2. $a_n = \left(-\frac{1}{2}\right)^n$ **0**

3. $a_n = 4^n$ **does not exist**

4. $a_n = \frac{1}{n^2}$ **0**

5. $a_n = \frac{2^n}{2^n + 1}$ **1**

6. $a_n = \frac{n^2}{n + 1}$ **does not exist**

www.algebra2.com/other_calculator_keystrokes

Graphing Calculator Investigation

A Preview of Lesson 11-4

Getting Started

Entering Sequences To enter the formula seq (N,N,1,10,1) in Step 1, use the keystrokes 2nd [LIST] ▶ 5 ALPHA [N] , ALPHA [N] , 1 , 10 , 1) . Follow a similar procedure to enter the formula for L2.

Graphing Sequences Stat plots for sequences are graphed in the same way as any other stat plot. It is essential that lists L1 and L2 contain the same number of elements.

Graphing Window The x-axis settings are determined by the values in L1. The y-axis settings are determined by the values in L2.

Teach

Ask: Does the sequence 3, 9, 27, 81, … have a limit? **no** Does 0.1, 0.01, 0.001, 0.0001, …? **yes**

Assess

Ask the students:

- Describe the graph in the example in terms of asymptotes. **The graph has the x-axis as an asymptote.**
- Does every decreasing geometric sequence have a limit? Explain. **No; A sequence such as −2, −4, −8, −16, … is decreasing and has no limit.**

11-4 Lesson Notes

1 Focus

5-Minute Check Transparency 11-4 Use as a quiz or review of Lesson 11-3.

Mathematical Background notes are available for this lesson on p. 576D.

How is e-mailing a joke like a geometric series?

Ask students:

- How many people have read your joke at the end of Monday? **3** at the end of Tuesday? **12** at the end of Wednesday? **39** at the end of Thursday? **120**

11-4 Geometric Series

What You'll Learn

- Find sums of geometric series.
- Find specific terms of geometric series.

Vocabulary

- geometric series

How is e-mailing a joke like a geometric series?

Suppose you e-mail a joke to three friends on Monday. Each of those friends sends the joke on to three of their friends on Tuesday. Each person who receives the joke on Tuesday sends it to three more people on Wednesday, and so on.

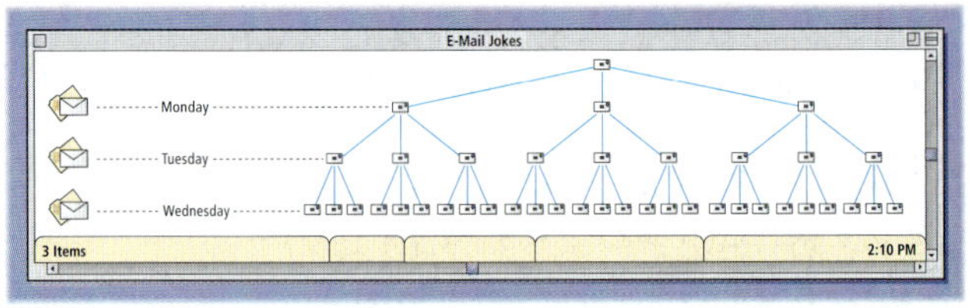

GEOMETRIC SERIES Notice that every day, the number of people who read your joke is three times the number that read it the day before. By Sunday, the number of people, including yourself, who have read the joke is 1 + 3 + 9 + 27 + 81 + 243 + 729 + 2187 or 3280!

The numbers 1, 3, 9, 27, 81, 243, 729, and 2187 form a geometric sequence in which $a_1 = 1$ and $r = 3$. Since 1, 3, 9, 27, 81, 243, 729, 2187 is a geometric sequence, 1 + 3 + 9 + 27 + 81 + 243 + 729 + 2187 is called a **geometric series**. Below are some more examples of geometric sequences and their corresponding geometric series.

Geometric Sequences	Geometric Series
1, 2, 4, 8, 16	1 + 2 + 4 + 8 + 16
4, −12, 36	4 + (−12) + 36
$5, 1, \frac{1}{5}, \frac{1}{25}$	$5 + 1 + \frac{1}{5} + \frac{1}{25}$

To develop a formula for the sum of a geometric series, consider the series given in the e-mail situation above.

$$\begin{aligned} S_8 &= 1 + 3 + 9 + 27 + 81 + 243 + 729 + 2187 \\ (-)\ 3S_8 &= \quad\ \ 3 + 9 + 27 + 81 + 243 + 729 + 2187 + 6561 \\ \hline (1-3)S_8 &= 1 + 0 + 0 + 0 + 0 + 0 + 0 + 0 - 6561 \end{aligned}$$

$$S_8 = \frac{1 - 6561}{1 - 3} \text{ or } 3280$$

first term in series

last term in series multiplied by common ratio; in this case, a_9

common ratio

Study Tip

Terms of Geometric Sequences
Remember that a_9 can also be written as a_1r^8.

The expression for S_8 can be written as $S_8 = \frac{a_1 - a_1r^8}{1 - r}$. A rational expression like this can be used to find the sum of any geometric series.

Resource Manager

Workbook and Reproducible Masters

Chapter 11 Resource Masters
- Study Guide and Intervention, pp. 649–650
- Skills Practice, p. 651
- Practice, p. 652
- Reading to Learn Mathematics, p. 653
- Enrichment, p. 654
- Assessment, pp. 693, 695

School-to-Career Masters, p. 22

Transparencies

5-Minute Check Transparency 11-4
Answer Key Transparencies

Technology

Alge2PASS: Tutorial Plus, Lesson 21
Interactive Chalkboard

Key Concept — Sum of a Geometric Series

The sum S_n of the first n terms of a geometric series is given by

$$S_n = \frac{a_1 - a_1 r^n}{1 - r} \text{ or } S_n = \frac{a_1(1 - r^n)}{1 - r}, \text{ where } r \neq 1.$$

You cannot use the formula for the sum with a geometric series for which $r = 1$ because division by 0 would result. In a geometric series with $r = 1$, the terms are constant. For example, $4 + 4 + 4 + \ldots + 4$ is such a series. In general, the sum of n terms of a geometric series with $r = 1$ is $n \cdot a^1$.

Example 1 Find the Sum of the First n Terms

GENEALOGY In the book *Roots*, author Alex Haley traced his family history back many generations to the time one of his ancestors was brought to America from Africa. If you could trace your family back for 15 generations, starting with your parents, how many ancestors would there be?

Counting your two parents, four grandparents, eight great-grandparents, and so on gives you a geometric series with $a_1 = 2$, $r = 2$, and $n = 15$.

$S_n = \frac{a_1(1 - r^n)}{1 - r}$ Sum formula

$S_{15} = \frac{2(1 - 2^{15})}{1 - 2}$ $n = 15, a_1 = 2, r = 2$

$S_{15} = 65{,}534$ Use a calculator.

Going back 15 generations, you have 65,534 ancestors.

As with arithmetic series, you can use sigma notation to represent geometric series.

Example 2 Evaluate a Sum Written in Sigma Notation

Evaluate $\sum_{n=1}^{6} 5 \cdot 2^{n-1}$.

Method 1

Find the terms by replacing n with 1, 2, 3, 4, 5, and 6. Then add.

$$\begin{aligned}\sum_{n=1}^{6} 5 \cdot 2^{n-1} &= 5(2^{1-1}) + 5(2^{2-1}) + 5(2^{3-1}) + 5(2^{4-1}) + 5(2^{5-1}) + 5(2^{6-1}) \\ &= 5(1) + 5(2) + 5(4) + 5(8) + 5(16) + 5(32) \\ &= 5 + 10 + 20 + 40 + 80 + 160 \\ &= 315\end{aligned}$$

Method 2

Since the sum is a geometric series, you can use the formula

$S_n = \frac{a_1(1 - r^n)}{1 - r}$.

$S_6 = \frac{5(1 - 2^6)}{1 - 2}$ $n = 6, a_1 = 5, r = 2$

$S_6 = \frac{5(-63)}{-1}$ $2^6 = 64$

$S_6 = 315$ Simplify.

The sum of the series is 315.

How can you find the sum of a geometric series if you know the first and last terms and the common ratio, but not the number of terms? Remember the formula for the nth term of a geometric sequence or series, $a_n = a_1 \cdot r^{n-1}$. You can use this formula to find an expression involving r^n.

$a_n = a_1 \cdot r^{n-1}$ Formula for nth term

$a_n \cdot r = a_1 \cdot r^{n-1} \cdot r$ Multiply each side by r.

$a_n \cdot r = a_1 \cdot r^n$ $r^{n-1} \cdot r^1 = r^{n-1+1}$ or r^n

More About. . .

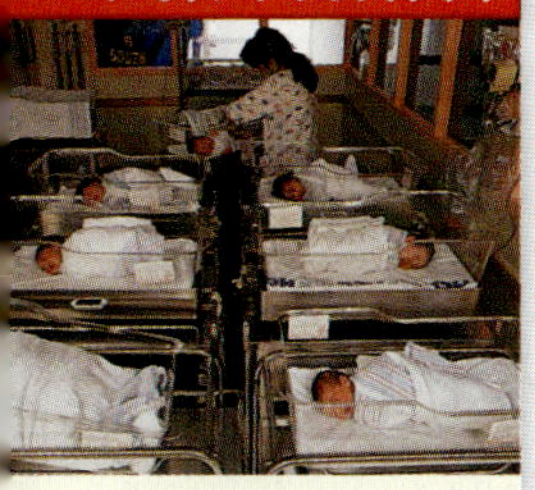

Genealogy

When he died in 1992, Samuel Must of Fryburg, Pennsylvania, had a record 824 living descendants.

Source: *The Guinness Book of Records*

www.algebra2.com/extra_examples

2 Teach

GEOMETRIC SERIES

In-Class Examples

Power Point®

Teaching Tip Ask students to explain the difference between counting direct ancestors, as in Example 1, and counting living descendants. Point out that this example counts only direct biological parents, not taking into consideration step-parents, adoptive parents, aunts, uncles, and so on. Discuss how the counting process might change if this assumption were not made.

1 **GENEALOGY** Use the information in Example 2. How many direct ancestors would a person have after 8 generations? **510**

Teaching Tip Review the basic ideas by asking students to explain the difference between a sequence and a series. Ask them to read Example 2 aloud to be sure they can interpret the sigma notation correctly.

2 Evaluate $\sum_{n=1}^{12} 3 \cdot 2^{n-1}$. **12,285**

In-Class Example

Power Point®

3 Find the sum of a geometric series for which $a_1 = 7776$, $a_n = 6$, and $r = -\frac{1}{6}$. **6666**

SPECIFIC TERMS

In-Class Example

Power Point®

4 Find a_1 in a geometric series for which $S_8 = 765$ and $r = 2$. **3**

3 Practice/Apply

Study Notebook

Have students—

- *add the definitions/examples of the vocabulary terms to their Vocabulary Builder worksheets for Chapter 11.*
- *include any other item(s) that they find helpful in mastering the skills in this lesson.*

About the Exercises...

Organization by Objective

- **Geometric Series:** 15–40, 47
- **Specific Terms:** 41–46

Odd/Even Assignments

Exercises 15–46 are structured so that students practice the same concepts whether they are assigned odd or even problems.

Alert! A graphing calculator is needed for Exercises 52–54.

Assignment Guide

Basic: 15–25 odd, 29–37 odd, 41, 43, 47–51, 55–67

Average: 15–47 odd, 48–51, 55–67 (optional: 52–54)

Advanced: 16–48 even, 49–61 (optional: 62–67)

Now substitute $a_n \cdot r$ for $a_1 \cdot r^n$ in the formula for the sum of a geometric series. The result is $S_n = \frac{a_1 - a_n r}{1 - r}$.

Example 3 Use the Alternate Formula for a Sum

Find the sum of a geometric series for which $a_1 = 15{,}625$, $a_n = -5$, and $r = -\frac{1}{5}$.

Since you do not know the value of n, use the formula derived above.

$S_n = \frac{a_1 - a_n r}{1 - r}$ Alternate sum formula

$= \frac{15{,}625 - (-5)\left(-\frac{1}{5}\right)}{1 - \left(-\frac{1}{5}\right)}$ $a_1 = 15{,}625$, $a_n = -5$, $r = -\frac{1}{5}$

$= \frac{15{,}624}{\frac{6}{5}}$ or 13,020 Simplify.

SPECIFIC TERMS You can use the formula for the sum of a geometric series to help find a particular term of the series.

Example 4 Find the First Term of a Series

Find a_1 in a geometric series for which $S_8 = 39{,}360$ and $r = 3$.

$S_n = \frac{a_1(1 - r^n)}{1 - r}$ Sum formula

$39{,}360 = \frac{a_1(1 - 3^8)}{1 - 3}$ $S_8 = 39{,}360$; $r = 3$; $n = 8$

$39{,}360 = \frac{-6560a_1}{-2}$ Subtract.

$39{,}360 = 3280a_1$ Divide.

$12 = a_1$ Divide each side by 3280.

The first term of the series is 12.

Check for Understanding

Concept Check

1. **OPEN ENDED** Write a geometric series for which $r = \frac{1}{2}$ and $n = 4$.
   1. Sample answer: $4 + 2 + 1 + \frac{1}{2}$
2. **Explain**, using geometric series, why the polynomial $1 + x + x^2 + x^3$ can be written as $\frac{x^4 - 1}{x - 1}$, assuming $x \neq 1$. **See margin.**
3. **Explain** how to write the series 2 + 12 + 72 + 432 + 2592 using sigma notation. **See pp. 629A–629F.**

Guided Practice

GUIDED PRACTICE KEY	
Exercises	**Examples**
4, 5	3
6–9, 14	1
10, 11	2
12, 13	4

Find S_n for each geometric series described.

4. $a_1 = 12$, $a_5 = 972$, $r = -3$ **732**
5. $a_1 = 3$, $a_n = 46{,}875$, $r = -5$ **39,063**
6. $a_1 = 5$, $r = 2$, $n = 14$ **81,915**
7. $a_1 = 243$, $r = -\frac{2}{3}$, $n = 5$ **165**

Find the sum of each geometric series.

8. 54 + 36 + 24 + 16 + ... to 6 terms **8.** $\frac{1330}{9}$
9. 3 − 6 + 12 − ... to 7 terms **129**
10. $\sum_{n=1}^{5} \frac{1}{4} \cdot 2^{n-1}$ $\frac{31}{4}$
11. $\sum_{n=1}^{7} 81\left(\frac{1}{3}\right)^{n-1}$ $\frac{1093}{9}$

DAILY INTERVENTION

Differentiated Instruction

Naturalist Have students research how biologists and ecologists use geometric series in their work to count and predict the population changes for various organisms.

Find the indicated term for each geometric series described.

12. $S_n = \frac{381}{64}, r = \frac{1}{2}, n = 7; a_1$ **3**

13. $S_n = 33, a_n = 48, r = -2; a_1$ **3**

Application 14. **WEATHER** Heavy rain caused a river to rise. The river rose three inches the first day, and each additional day it rose twice as much as the previous day. How much did the river rise in five days? **93 in. or 7 ft 9 in.**

★ indicates increased difficulty

Practice and Apply

Homework Help	
For Exercises	See Examples
15–34, 47	1, 3
35–40	2
41–46	4

Extra Practice
See page 852.

Find S_n for each geometric series described.

15. $a_1 = 2, a_6 = 486, r = 3$ **728**

16. $a_1 = 3, a_8 = 384, r = 2$ **765**

17. $a_1 = 1296, a_n = 1, r = -\frac{1}{6}$ **1111**

18. $a_1 = 343, a_n = -1, r = -\frac{1}{7}$ **300**

19. $a_1 = 4, r = -3, n = 5$ **244**

20. $a_1 = 5, r = 3, n = 12$ **1,328,600**

21. $a_1 = 2401, r = -\frac{1}{7}, n = 5$ **2101**

22. $a_1 = 625, r = \frac{3}{5}, n = 5$ **1441**

23. $a_1 = 162, r = \frac{1}{3}, n = 6$ $\frac{728}{3}$

24. $a_1 = 80, r = -\frac{1}{2}, n = 7$ $\frac{215}{4}$

25. $a_1 = 625, r = 0.4, n = 8$ **1040.984**

26. $a_1 = 4, r = 0.5, n = 8$ **7.96875**

★27. $a_2 = -36, a_5 = 972, n = 7$ **6564**

★28. $a_3 = -36, a_6 = -972, n = 10$ **−118,096**

29. **HEALTH** Contagious diseases can spread very quickly. Suppose five people are ill during the first week of an epidemic and that each person who is ill spreads the disease to four people by the end of the next week. By the end of the tenth week of the epidemic, how many people have been affected by the illness? **1,747,625**

30. **LEGENDS** There is a legend of a king who wanted to reward a boy for a good deed. The king gave the boy a choice. He could have $1,000,000 at once, or he could be rewarded daily for a 30-day month, with one penny on the first day, two pennies on the second day, and so on, receiving twice as many pennies each day as the previous day. How much would the second option be worth? **$10,737,418.23**

Find the sum of each geometric series.

31. $4096 - 512 + 64 - \ldots$ to 5 terms **3641**

32. $7 + 21 + 63 + \ldots$ to 10 terms **206,668**

33. $\frac{1}{16} + \frac{1}{4} + 1 + \ldots$ to 7 terms $\frac{5461}{16}$

34. $\frac{1}{9} - \frac{1}{3} + 1 - \ldots$ to 6 terms $-\frac{182}{9}$

35. $\sum_{n=1}^{9} 5 \cdot 2^{n-1}$ **2555**

36. $\sum_{n=1}^{6} 2(-3)^{n-1}$ **−364**

37. $\sum_{n=1}^{7} 144\left(-\frac{1}{2}\right)^{n-1}$ $\frac{387}{4}$

38. $\sum_{n=1}^{8} 64\left(\frac{3}{4}\right)^{n-1}$ $\frac{58{,}975}{256}$

★39. $\sum_{n=1}^{20} 3 \cdot 2^{n-1}$ **3,145,725**

★40. $\sum_{n=1}^{16} 4 \cdot 3^{n-1}$ **86,093,440**

Find the indicated term for each geometric series described.

41. $S_n = 165, a_n = 48, r = -\frac{2}{3}; a_1$ **243**

42. $S_n = 688, a_n = 16, r = -\frac{1}{2}; a_1$ **1024**

43. $S_n = -364, r = -3, n = 6; a_1$ **2**

44. $S_n = 1530, r = 2, n = 8; a_1$ **6**

★45. $S_n = 315, r = 0.5, n = 6; a_2$ **80**

★46. $S_n = 249.92, r = 0.2, n = 5, a_3$ **8**

47. **LANDSCAPING** Rob is helping his dad install a fence. He is using a sledgehammer to drive the pointed fence posts into the ground. On his first swing, he drives a post five inches into the ground. Since the soil is denser the deeper he drives, on each swing after the first, he can only drive the post 30% as far into the ground as he did on the previous swing. How far has he driven the post into the ground after five swings? **about 7.13 in.**

www.algebra2.com/self_check_quiz

Answer

2. The polynomial is a geometric series with first term 1, common ratio x, and 4 terms. The sum is $\frac{1(1 - x^4)}{1 - x} = \frac{x^4 - 1}{x - 1}$.

Enrichment, p. 654

Annuities

An annuity is a fixed amount of money payable at given intervals. For example, suppose you wanted to set up a trust fund so that $30,000 could be withdrawn each year for 14 years before the money ran out. Assume the money can be invested at 9%.

You must find the amount of money that needs to be invested. Call this amount A. After the third payment, the amount left is

$1.09[1.09A - 30{,}000(1 + 1.09)] - 30{,}000 = 1.09^2A - 30{,}000(1 + 1.09 + 1.09^2)$.

The results are summarized in the table below.

Payment Number	Number of Dollars Left After Payment
1	$A - 30{,}000$
2	$1.09A - 30{,}000(1 + 1.09)$
3	$1.09^2A - 30{,}000(1 + 1.09 + 1.09^2)$

1. Use the pattern shown in the table to find the number of dollars left after the fourth payment. $1.09^3A - 30{,}000(1 + 1.09 + 1.09^2 + 1.09^3)$

Study Guide and Intervention, p. 649 (shown) and p. 650

Geometric Series A **geometric series** is the indicated sum of consecutive terms of a geometric sequence.

Sum of a Geometric Series	The sum S_n of the first n terms of a geometric series is given by $S_n = \frac{a_1(1 - r^n)}{1 - r}$ or $S_n = \frac{a_1 - a_1r^n}{1 - r}$, where $r \neq 1$.

Example 1 **Find the sum of the first four terms of the geometric sequence for which $a_1 = 120$ and $r = \frac{1}{3}$.**

$S_n = \frac{a_1(1 - r^n)}{1 - r}$ Sum formula

$S_4 = \frac{120\left(1 - \left(\frac{1}{3}\right)^4\right)}{1 - \frac{1}{3}}$ $n = 4, a_1 = 120, r = \frac{1}{3}$

≈ 177.78 Use a calculator.

The sum of the series is 177.78.

Example 2 **Find the sum of the geometric series $\sum_{j=1}^{7} 4 \cdot 3^{j-2}$.**

Since the sum is a geometric series, you can use the sum formula.

$S_n = \frac{a_1(1 - r^n)}{1 - r}$ Sum formula

$S_7 = \frac{\frac{4}{3}(1 - 3^7)}{1 - 3}$ $n = 7, a_1 = \frac{4}{3}, r = 3$

≈ 1457.33 Use a calculator.

The sum of the series is 1457.33.

Exercises

Find S_n for each geometric series described.

1. $a_1 = 2, a_n = 486, r = 3$ **728**
2. $a_1 = 1200, a_n = 75, r = \frac{1}{2}$ **2325**
3. $a_1 = \frac{1}{25}, a_n = 125, r = 5$ **156.24**
4. $a_1 = 3, r = \frac{1}{3}, n = 4$ **4.44**
5. $a_1 = 2, r = 6, n = 4$ **518**
6. $a_1 = 2, r = 4, n = 6$ **2730**
7. $a_1 = 100, r = -\frac{1}{2}, n = 5$ **68.75**
8. $a_3 = 20, a_6 = 160, n = 8$ **1275**
9. $a_4 = 16, a_7 = 1024, n = 10$ **87,381.25**

Find the sum of each geometric series.

10. $6 + 18 + 54 + \ldots$ to 6 terms **2184**
11. $\frac{1}{4} + \frac{1}{2} + 1 + \ldots$ to 10 terms **255.75**
12. $\sum_{j=4}^{8} 2^j$ **496**
13. $\sum_{k=1}^{7} 3 \cdot 2^{k-1}$ **381**

Skills Practice, p. 651 and Practice, p. 652 (shown)

Find S_n for each geometric series described.

1. $a_1 = 2, a_6 = 64, r = 2$ **126**
2. $a_1 = 160, a_6 = 5, r = \frac{1}{2}$ **315**
3. $a_1 = -3, a_n = -192, r = -2$ **−129**
4. $a_1 = -81, a_n = -16, r = -\frac{2}{3}$ **−55**
5. $a_1 = -3, a_n = 3072, r = -4$ **2457**
6. $a_1 = 54, a_6 = \frac{2}{9}, r = \frac{1}{3}$ $\frac{728}{9}$
7. $a_1 = 5, r = 3, n = 9$ **49,205**
8. $a_1 = -6, r = -1, n = 21$ **−6**
9. $a_1 = -6, r = -3, n = 7$ **−3282**
10. $a_1 = -9, r = \frac{2}{3}, n = 4$ $-\frac{65}{3}$
11. $a_1 = \frac{1}{3}, r = 3, n = 10$ $\frac{29{,}524}{3}$
12. $a_1 = 16, r = -1.5, n = 6$ **−66.5**

Find the sum of each geometric series.

13. $162 + 54 + 18 + \ldots$ to 6 terms $\frac{728}{3}$
14. $2 + 4 + 8 + \ldots$ to 8 terms **510**
15. $64 - 96 + 144 - \ldots$ to 7 terms **463**
16. $\frac{1}{9} - \frac{1}{3} + 1 - \ldots$ to 6 terms $-\frac{182}{9}$
17. $\sum_{n=1}^{8} (-3)^{n-1}$ **−1640**
18. $\sum_{n=1}^{9} 5(-2)^{n-1}$ **855**
19. $\sum_{n=1}^{5} -1(4)^{n-1}$ **−341**
20. $\sum_{n=1}^{6} \left(\frac{1}{2}\right)^{n-1}$ $\frac{63}{32}$
21. $\sum_{n=1}^{10} 2560\left(\frac{1}{2}\right)^{n-1}$ **5115**
22. $\sum_{n=1}^{4} 9\left(\frac{2}{3}\right)^{n-1}$ $\frac{65}{3}$

Find the indicated term for each geometric series described.

23. $S_n = 1023, a_n = 768, r = 4; a_1$ **3**
24. $S_n = 10{,}160, a_n = 5120, r = 2; a_1$ **80**
25. $S_n = -1365, n = 12, r = -2; a_1$ **1**
26. $S_n = 665, n = 6, r = 1.5; a_1$ **32**

27. **CONSTRUCTION** A pile driver drives a post 27 inches into the ground on its first hit. Each additional hit drives the post $\frac{2}{3}$ the distance of the prior hit. Find the total distance the post has been driven after 5 hits. $70\frac{1}{3}$ **in.**

28. **COMMUNICATIONS** Hugh Moore e-mails a joke to 5 friends on Sunday morning. Each of these friends e-mails the joke to 5 of her or his friends on Monday morning, and so on. Assuming no duplication, how many people will have heard the joke by the end of Saturday, not including Hugh? **97,655 people**

Reading to Learn Mathematics, p. 653 ELL

Pre-Activity **How is e-mailing a joke like a geometric series?**

Read the introduction to Lesson 11-4 at the top of page 594 in your textbook.

- Suppose that you e-mail the joke on Monday to five friends, rather than three, and that each of those friends e-mails it to five friends on Tuesday, and so on. Write a sum that shows that total number of people, including yourself, who will have read the joke by Thursday. (Write out the sum using plus signs rather than sigma notation. Do not actually find the sum.) **1 + 5 + 25 + 125**
- Use exponents to rewrite the sum you found above. (Use an exponent in each term, and use the same base for all terms.) $5^0 + 5^1 + 5^2 + 5^3$

Reading the Lesson

1. Consider the formula $S_n = \frac{a_1(1 - r^n)}{1 - r}$.
   a. What is this formula used to find? **the sum of the first n terms of a geometric series**
   b. What do each of the following represent?
      S_n: **the sum of the first n terms**
      a_1: **the first term**
      r: **the common ratio**
   c. Suppose that you want to use the formula to evaluate $3 - 1 + \frac{1}{3} - \frac{1}{9} + \frac{1}{27}$. Indicate the values you would substitute into the formula in order to find S_n. (Do not actually calculate the sum.)
      $n =$ **5** $a_1 =$ **3** $r =$ $-\frac{1}{3}$ $r^n =$ $\left(-\frac{1}{3}\right)^5$ or $-\frac{1}{243}$
   d. Suppose that you want to use the formula to evaluate the sum $\sum_{n=1}^{6} 8(-2)^{n-1}$. Indicate the values you would substitute into the formula in order to find S_n. (Do not actually calculate the sum.)
      $n =$ **6** $a_1 =$ **8** $r =$ **−2** $r^n =$ $(-2)^6$ or 64

Helping You Remember

2. This lesson includes three formulas for the sum of the first n terms of a geometric series. All of these formulas have the same denominator and have the restriction $r \neq 1$. How can this restriction help you to remember the denominator in the formulas? **Sample answer: If $r = 1$, then $r - 1 = 0$. Because division by 0 is undefined, a formula with $r - 1$ in the denominator will not apply when $r = 1$.**

Lesson 11-4 Geometric Series 597

4 Assess

Open-Ended Assessment

Writing Have students make a chart that compares and contrasts arithmetic and geometric sequences and series, explaining what the variables represent in each formula.

Tips for New Teachers

Intervention Make sure that students can read the notation used in the various formulas and that they understand what each variable and subscript means.

Getting Ready for Lesson 11-5

PREREQUISITE SKILL Students will find the sum of infinite geometric series in Lesson 11-5. This will involve their evaluating rational expressions for different values. Use Exercises 62–67 to determine your students' familiarity with evaluating rational expressions for given values.

Assessment Options

Quiz (Lessons 11-3 and 11-4) is available on p. 693 of the *Chapter 11 Resource Masters*.

Mid-Chapter Test (Lessons 11-1 through 11-4) is available on p. 695 of the *Chapter 11 Resource Masters*.

48. If the first term and common ratio of a geometric series are integers, then all the terms of the series are integers. Therefore, the sum of the series is an integer.

48. **CRITICAL THINKING** If a_1 and r are integers, explain why the value of $\frac{a_1 - a_1r^n}{1 - r}$ must also be an integer.

49. **WRITING IN MATH** Answer the question that was posed at the beginning of the lesson. See margin.

How is e-mailing a joke like a geometric series?

Include the following in your answer:
- how the related geometric series would change if each person e-mailed the joke on to four people instead of three, and
- how the situation could be changed to make it better to use a formula than to add terms.

Standardized Test Practice

50. The first term of a geometric series is −1, and the common ratio is −3. How many terms are in the series if its sum is 182? **A**

(A) 6 (B) 7 (C) 8 (D) 9

51. What is the first term in a geometric series with ten terms, a common ratio of 0.5, and a sum of 511.5? **C**

(A) 64 (B) 128 (C) 256 (D) 512

Graphing Calculator

Use a graphing calculator to find the sum of each geometric series.

52. $\sum_{n=1}^{20} 3(-2)^{n-1}$ **−1,048,575**

53. $\sum_{n=1}^{15} 2\left(\frac{1}{2}\right)^{n-1}$ **3.99987793**

54. $\sum_{n=1}^{10} 5(0.2)^{n-1}$ **6.24999936**

Maintain Your Skills

Mixed Review

55. $\pm\frac{1}{4}, \frac{3}{2}, \pm 9$

56. $-3, -\frac{9}{2}, -\frac{27}{4}, -\frac{81}{8}$

Find the geometric means in each sequence. *(Lesson 11-3)*

55. $\frac{1}{24}$, ___?___, ___?___, ___?___, 54

56. −2, ___?___, ___?___, ___?___, ___?___, $-\frac{243}{16}$

Find the sum of each arithmetic series. *(Lesson 11-2)*

57. 50 + 44 + 38 + … + 8 **232**

58. $\sum_{n=1}^{12} (2n + 3)$ **192**

ENTERTAINMENT For Exercises 59–61, use the table that shows the number of drive-in movie screens in the United States for 1995–2000. *(Lesson 2-5)*

Year	1995	1996	1997	1998	1999	2000
Screens	848	826	815	750	737	637

Source: National Association of Theatre Owners

59. See margin.

60. Sample answer using (1, 826) and (3, 750): $y = -38x + 864$

59. Draw a scatter plot, in which x is the number of years since 1995.

60. Find a prediction equation.

61. Predict the number of screens in 2010. **Sample answer: 294**

Online Research **Data Update** For the latest statistics on the movie industry, visit: www.algebra2.com/data_update

Getting Ready for the Next Lesson

PREREQUISITE SKILL Evaluate $\frac{a}{1 - b}$ for the given values of a and b.

*(To review **evaluating expressions**, see Lesson 1-1.)*

62. $a = 1, b = \frac{1}{2}$ **2**

63. $a = 3, b = -\frac{1}{2}$ **2**

64. $a = \frac{1}{3}, b = -\frac{1}{3}$ $\frac{1}{4}$

65. $a = \frac{1}{2}, b = \frac{1}{4}$ $\frac{2}{3}$

66. $a = -1, b = 0.5$ **−2**

67. $a = 0.9, b = -0.5$ **0.6**

Answers

49. If the number of people that each person sends the joke to is constant, then the total number of people who have seen the joke is the sum of a geometric series. Answers should include the following.
- The common ratio would change from 3 to 4.
- Increase the number of days that the joke circulates so that it is inconvenient to find and add all the terms of the series.

59. **Drive-In Movie Screens**

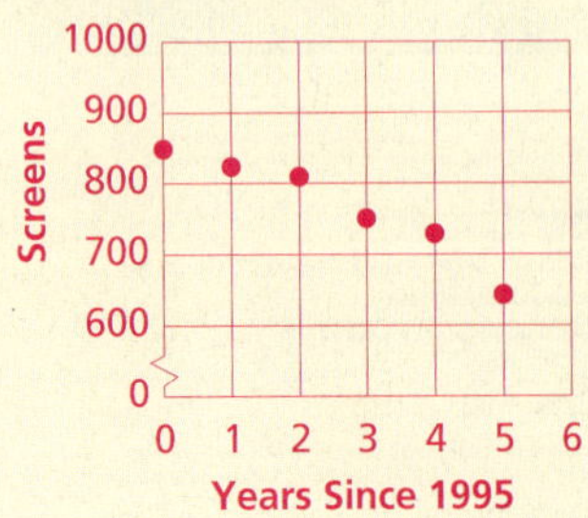

11-5 Infinite Geometric Series

What You'll Learn

- Find the sum of an infinite geometric series.
- Write repeating decimals as fractions.

Vocabulary

- infinite geometric series
- partial sum
- convergent series

How does an infinite geometric series apply to a bouncing ball?

Refer to the beginning of Lesson 11-3. Suppose you wrote a geometric series to find the sum of the heights of the rebounds of the ball. The series would have no last term because theoretically there is no last bounce of the ball. For every rebound of the ball, there is another rebound, 60% as high. Such a geometric series is called an **infinite geometric series**.

In the Bleachers By Steve Moore

"And that, ladies and gentlemen, is the way the ball bounces."

INFINITE GEOMETRIC SERIES Consider the infinite geometric series $\frac{1}{2} + \frac{1}{4} + \frac{1}{8} + \frac{1}{16} + \ldots$. You have already learned how to find the sum S_n of the first n terms of a geometric series. For an infinite series, S_n is called a **partial sum** of the series. The table and graph show some values of S_n.

n	S_n
1	$\frac{1}{2}$ or 0.5
2	$\frac{3}{4}$ or 0.75
3	$\frac{7}{8}$ or 0.875
4	$\frac{15}{16}$ or 0.9375
5	$\frac{31}{32}$ or 0.96875
6	$\frac{63}{64}$ or 0.984375
7	$\frac{127}{128}$ or 0.9921875

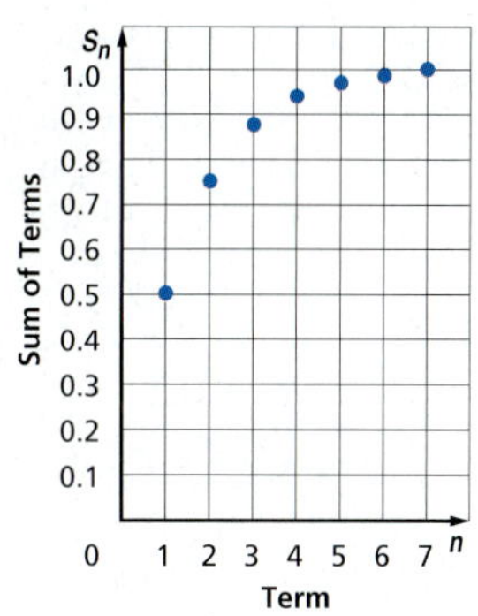

Notice that as n increases, the partial sums level off and approach a limit of 1. This leveling-off behavior is characteristic of infinite geometric series for which $|r| < 1$.

Study Tip

Absolute Value
Recall that $|r| < 1$ means $-1 < r < 1$.

11-5 Lesson Notes

1 Focus

5-Minute Check Transparency 11-5 Use as a quiz or review of Lesson 11-4.

Mathematical Background notes are available for this lesson on p. 576D.

Building on Prior Knowledge

In Lesson 11-4, students worked with geometric series that had a specific number of terms. In this lesson, students extend these skills to finding the sum of an infinite geometric series.

How does an infinite geometric series apply to a bouncing ball?

Ask students:

- Why might someone find this cartoon amusing? **Answers will vary.**
- What is the difference between what is happening theoretically and what really happens with the ball? **Answers will vary.**

Resource Manager

Workbook and Reproducible Masters

Chapter 11 Resource Masters

- Study Guide and Intervention, pp. 655–656
- Skills Practice, p. 657
- Practice, p. 658
- Reading to Learn Mathematics, p. 659
- Enrichment, p. 660

Transparencies

5-Minute Check Transparency 11-5
Answer Key Transparencies

Technology

Interactive Chalkboard
Multimedia Applications

2 Teach

INFINITE GEOMETRIC SERIES

In-Class Example

Power Point®

1 Find the sum of each infinite geometric series, if it exists.

a. $-\frac{4}{3} + 4 - 12 + 36 - 108 + \ldots$ **no sum**

b. $3 - \frac{3}{2} + \frac{3}{4} - \frac{3}{8} + \ldots$ **2**

Teaching Tip To help students understand when an infinite geometric series has a sum, lead students to make a generalization about the size of a product of a number and a fraction between −1 and 1. **The absolute value of such a product will always be less than the absolute value of the original number.**

Let's look at the formula for the sum of a finite geometric series and use it to find a formula for the sum of an infinite geometric series.

$$S_n = \frac{a_1 - a_1r^n}{1 - r} \quad \text{Sum of first } n \text{ terms}$$

$$= \frac{a_1}{1 - r} - \frac{a_1r^n}{1 - r} \quad \text{Write the fraction as a difference of fractions.}$$

If $-1 < r < 1$, the value of r^n will approach 0 as n increases. Therefore, the partial sums of an infinite geometric series will approach $\frac{a_1}{1-r} - \frac{a_1(0)}{1-r}$ or $\frac{a_1}{1-r}$. This expression gives the sum of an infinite geometric series. An infinite series that has a sum is called a **convergent series**.

Key Concept — *Sum of an Infinite Geometric Series*

The sum S of an infinite geometric series with $-1 < r < 1$ is given by

$$S = \frac{a_1}{1 - r}.$$

Study Tip

Formula for Sum if $-1 < r < 1$

To convince yourself of this formula, make a table of the first ten partial sums of the geometric series with $r = \frac{1}{2}$ and $a_1 = 100$.

Term Number	Term	Partial Sum
1	100	100
2	50	150
3	25	175
⋮	⋮	⋮
10		

Complete the table and compare the sum that the series is approaching to that obtained by using the formula.

An infinite geometric series for which $|r| \geq 1$ does not have a sum. Consider the series $1 + 3 + 9 + 27 + 81 + \ldots$. In this series, $a_1 = 1$ and $r = 3$. The table shows some of the partial sums of this series. As n increases, S_n rapidly increases and has no limit. That is, the partial sums do not approach a particular value.

n	S_n
5	121
10	29,524
15	7,174,453
20	1,743,392,200

Example 1 *Sum of an Infinite Geometric Series*

Find the sum of each infinite geometric series, if it exists.

a. $\frac{1}{2} + \frac{3}{8} + \frac{9}{32} + \ldots$

First, find the value of r to determine if the sum exists.

$a_1 = \frac{1}{2}$ and $a_2 = \frac{3}{8}$, so $r = \frac{\frac{3}{8}}{\frac{1}{2}}$ or $\frac{3}{4}$. Since $\left|\frac{3}{4}\right| < 1$, the sum exists.

Now use the formula for the sum of an infinite geometric series.

$$S = \frac{a_1}{1 - r} \quad \text{Sum formula}$$

$$= \frac{\frac{1}{2}}{1 - \frac{3}{4}} \quad a_1 = \frac{1}{2}, r = \frac{3}{4}$$

$$= \frac{\frac{1}{2}}{\frac{1}{4}} \text{ or } 2 \quad \text{Simplify.}$$

The sum of the series is 2.

b. $1 - 2 + 4 - 8 + \ldots$

$a_1 = 1$ and $a_2 = -2$, so $r = \frac{-2}{1}$ or -2. Since $|-2| \geq 1$, the sum does not exist.

DAILY INTERVENTION — Unlocking Misconceptions

Absolute Value Make sure students can explain why $|r| < 1$ can also be written as $-1 < r < 1$. Graphing this inequality on a number line may help students understand what is meant by these two different mathematical notations.

In Lessons 11-2 and 11-4, we used sigma notation to represent finite series. You can also use sigma notation to represent infinite series. An *infinity symbol* ∞ is placed above the Σ to indicate that a series is infinite.

Example 2 Infinite Series in Sigma Notation

Evaluate $\sum_{n=1}^{\infty} 24\left(-\frac{1}{5}\right)^{n-1}$.

In this infinite geometric series, $a_1 = 24$ and $r = -\frac{1}{5}$.

$S = \frac{a_1}{1 - r}$ Sum formula

$= \frac{24}{1 - \left(-\frac{1}{5}\right)}$ $a_1 = 24, r = -\frac{1}{5}$

$= \frac{24}{\frac{6}{5}}$ or 20 Simplify.

Thus, $\sum_{n=1}^{\infty} 24\left(-\frac{1}{5}\right)^{n-1} = 20$.

REPEATING DECIMALS The formula for the sum of an infinite geometric series can be used to write a repeating decimal as a fraction. Remember that decimals with bar notation such as $0.\overline{2}$ and $0.\overline{47}$ represent 0.222222… and 0.474747…, respectively. Each of these expressions can be written as an infinite geometric series.

Example 3 Write a Repeating Decimal as a Fraction

Write $0.\overline{39}$ as a fraction.

Method 1

Write the repeating decimal as a sum.

$0.\overline{39} = 0.393939\ldots$

$= 0.39 + 0.0039 + 0.000039 + \ldots$

$= \frac{39}{100} + \frac{39}{10{,}000} + \frac{39}{1{,}000{,}000} + \ldots$

In this series, $a_1 = \frac{39}{100}$ and $r = \frac{1}{100}$.

$S = \frac{a_1}{1 - r}$ Sum formula

$= \frac{\frac{39}{100}}{1 - \frac{1}{100}}$ $a_1 = \frac{39}{100}, r = \frac{1}{100}$

$= \frac{\frac{39}{100}}{\frac{99}{100}}$ Subtract.

$= \frac{39}{99}$ or $\frac{13}{33}$ Simplify.

Thus, $0.\overline{39} = \frac{13}{33}$.

Method 2

$S = 0.\overline{39}$ Label the given decimal.

$S = 0.393939\ldots$ Repeating decimal

$100S = 39.393939\ldots$ Multiply each side by 100.

$99S = 39$ Subtract the second equation from the third.

$S = \frac{39}{99}$ or $\frac{13}{33}$ Divide each side by 99.

www.algebra2.com/extra_examples

In-Class Example Power Point®

2 Evaluate $\sum_{n=1}^{\infty} 5\left(\frac{1}{2}\right)^{n-1}$. **10**

Teaching Tip Ask students to write a few terms of the series in Example 2 to make sure they know how to read the notation.

REPEATING DECIMALS

In-Class Example Power Point®

3 Write $0.\overline{25}$ as a fraction. $\frac{25}{99}$

DAILY INTERVENTION **Differentiated Instruction**

Logical Have students research and read about the famous mathematical puzzle called Zeno's paradox. Have them discuss this story of the tortoise's race in terms of the content of this lesson.

3 Practice/Apply

Study Notebook

Have students—

- add the definitions/examples of the vocabulary terms to their Vocabulary Builder worksheets for Chapter 11.
- include any other item(s) that they find helpful in mastering the skills in this lesson.

DAILY INTERVENTION **FIND THE ERROR** Help students realize that, although $\frac{a_1}{1 - r}$ may have a value, that value represents the sum of an infinite geometric series *only* when $|r| < 1$.

About the Exercises...

Organization by Objective

- **Infinite Geometric Series:** 14–39
- **Repeating Decimals:** 40–47

Odd/Even Assignments

Exercises 14–31 and 36–47 are structured so that students practice the same concepts whether they are assigned odd or even problems.

Assignment Guide

Basic: 15–33 odd, 34, 35, 41–45 odd, 48–75

Average: 15–33 odd, 34, 35–47 odd, 48–75

Advanced: 14–32 even, 33, 34–48 even, 49–69 (optional: 70–75)

Check for Understanding

Concept Check

1. **OPEN ENDED** Write the series $\frac{1}{2} + \frac{1}{4} + \frac{1}{8} + \frac{1}{16} + \ldots$ using sigma notation.

   1. Sample answer: $\sum_{n=1}^{\infty}\left(\frac{1}{2}\right)^n$

2. Explain why $0.999999\ldots = 1$. **See margin.**
3. **FIND THE ERROR** Miguel and Beth are discussing the series $-\frac{1}{3} + \frac{4}{9} - \frac{16}{27} + \ldots$. Miguel says that the sum of the series is $-\frac{1}{7}$. Beth says that the series does not have a sum. Who is correct? Explain your reasoning. **Beth; see margin for explanation.**

   Miguel

   $$S = \frac{-\frac{1}{3}}{1-\left(-\frac{4}{3}\right)} = -\frac{1}{7}$$

Guided Practice

GUIDED PRACTICE KEY	
Exercises	**Examples**
4–8, 13	1
9	2
10–12	3

Find the sum of each infinite geometric series, if it exists.

4. $a_1 = 36, r = \frac{2}{3}$ **108**
5. $a_1 = 18, r = -1.5$ **does not exist**
6. $16 + 24 + 36 + \ldots$ **does not exist**
7. $\frac{1}{4} + \frac{1}{6} + \frac{2}{18} + \ldots$ $\mathbf{\frac{3}{4}}$
8. $6 - 2.4 + 0.96 - \ldots$ $\mathbf{\frac{30}{7}}$
9. $\sum_{n=1}^{\infty} 40\left(\frac{3}{5}\right)^{n-1}$ **100**

Write each repeating decimal as a fraction.

10. $0.\overline{5}$ $\mathbf{\frac{5}{9}}$
11. $0.\overline{73}$ $\mathbf{\frac{73}{99}}$
12. $0.\overline{175}$ $\mathbf{\frac{175}{999}}$

Application

13. **CLOCKS** Jasmine's old grandfather clock is broken. When she tries to set the pendulum in motion by holding it against the side of the clock and letting it go, it first swings 24 centimeters to the other side, then 18 centimeters back, then 13.5 centimeters, and so on. What is the total distance that the pendulum swings? **96 cm**

★ indicates increased difficulty

Practice and Apply

Homework Help	
For Exercises	**See Examples**
14–27, 32–39	1
28–31	2
40–47	3

Extra Practice See page 852.

Find the sum of each infinite geometric series, if it exists.

14. $a_1 = 4, r = \frac{5}{7}$ **14**
15. $a_1 = 14, r = \frac{7}{3}$
16. $a_1 = 12, r = -0.6$ **7.5**
17. $a_1 = 18, r = 0.6$ **45**
18. $16 + 12 + 9 + \ldots$ **64**
19. $-8 - 4 - 2 - \ldots$ **−16**
20. $12 - 18 + 24 - \ldots$
21. $18 - 12 + 8 - \ldots$ $\mathbf{\frac{54}{5}}$
22. $1 + \frac{2}{3} + \frac{4}{9} + \ldots$ **3**
23. $\frac{5}{3} + \frac{25}{3} + \frac{125}{3} + \ldots$
24. $\frac{5}{3} - \frac{10}{9} + \frac{20}{27} - \ldots$ **1**
25. $\frac{3}{2} - \frac{3}{4} + \frac{3}{8} - \ldots$ **1**
26. $3 + 1.8 + 1.08 + \ldots$ **7.5**
27. $1 - 0.5 + 0.25 - \ldots$ $\mathbf{\frac{2}{3}}$
28. $\sum_{n=1}^{\infty} 48\left(\frac{2}{3}\right)^{n-1}$ **144**
29. $\sum_{n=1}^{\infty} \left(\frac{3}{8}\right)\left(\frac{3}{4}\right)^{n-1}$ $\mathbf{\frac{3}{2}}$
30. $\sum_{n=1}^{\infty} 3(0.5)^{n-1}$ **6**
31. $\sum_{n=1}^{\infty} (1.5)(0.25)^{n-1}$ **2**

15. does not exist
20. does not exist
23. does not exist

32. **CHILD'S PLAY** Kimimela's little sister likes to swing at the playground. Yesterday, Kimimela pulled the swing back and let it go. The swing traveled a distance of 9 feet before heading back the other way. Each swing afterward was only 70% as long as the previous one. Find the total distance the swing traveled. **30 ft**

Answers

2. $0.999999\ldots$ can be written as the infinite geometric series $\frac{9}{10} + \frac{9}{100} + \frac{9}{1000} + \ldots$. The first term of this series is $\frac{9}{10}$ and the common ratio is $\frac{1}{10}$, so the sum is $\frac{\frac{9}{10}}{1 - \frac{1}{10}}$ or 1.
3. The common ratio for the infinite geometric series is $-\frac{4}{3}$. Since $\left|-\frac{4}{3}\right| \geq 1$, the series does not have a sum and the formula $S = \frac{a_1}{1 - r}$ does not apply.

GEOMETRY For Exercises 33 and 34, refer to square $ABCD$, which has a perimeter of 40 centimeters.

If the midpoints of the sides are connected, a smaller square results. Suppose the process of connecting midpoints of sides and drawing new squares is continued indefinitely.

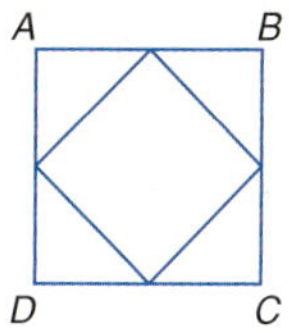

33. Write an infinite geometric series to represent the sum of the perimeters of all of the squares. $40 + 20\sqrt{2} + 20 + \ldots$

34. Find the sum of the perimeters of all of the squares. $80 + 40\sqrt{2}$ or about 136.6 cm

35. **AVIATION** A hot-air balloon rises 90 feet in its first minute of flight. In each succeeding minute, it rises only 90% as far as it did during the preceding minute. What is the final height of the balloon? **900 ft**

★ 36. The sum of an infinite geometric series is 81, and its common ratio is $\frac{2}{3}$. Find the first three terms of the series. **27, 18, 12**

★ 37. The sum of an infinite geometric series is 125, and the value of r is 0.4. Find the first three terms of the series. **75, 30, 12**

★ 38. The common ratio of an infinite geometric series is $\frac{11}{16}$, and its sum is $76\frac{4}{5}$. Find the first four terms of the series. $24, 16\frac{1}{2}, 11\frac{11}{32}, 7\frac{409}{512}$

★ 39. The first term of an infinite geometric series is -8, and its sum is $-13\frac{1}{3}$. Find the first four terms of the series. $-8, -3\frac{1}{5}, -1\frac{7}{25}, -\frac{64}{125}$

More About. . .

Aviation

The largest hot-air balloon ever flown had a capacity of 2.6 million cubic feet.

Source: *The Guinness Book of Records*

Write each repeating decimal as a fraction.

40. $0.\overline{7}$ $\frac{7}{9}$ 41. $0.\overline{1}$ $\frac{1}{9}$ 42. $0.\overline{36}$ $\frac{4}{11}$ 43. $0.\overline{82}$ $\frac{82}{99}$

44. $0.\overline{246}$ $\frac{82}{333}$ 45. $0.\overline{427}$ $\frac{427}{999}$ ★ 46. $0.4\overline{5}$ $\frac{5}{11}$ ★ 47. $0.2\overline{31}$ $\frac{229}{990}$

48. **CRITICAL THINKING** Derive the formula for the sum of an infinite geometric series by using the technique in Lessons 11-2 and 11-4. That is, write an equation for the sum S of a general infinite geometric series, multiply each side of the equation by r, and subtract equations. See pp. 629A–629F.

49. **WRITING IN MATH** Answer the question that was posed at the beginning of the lesson. See pp. 629A–629F.

How does an infinite geometric series apply to a bouncing ball?

Include the following in your answer:

- some formulas you might expect to see on the chalkboard if the character in the comic strip really was discussing a bouncing ball, and
- an explanation of how to find the total distance traveled, both up and down, by the bouncing ball described at the beginning of Lesson 11-3.

Standardized Test Practice

50. What is the sum of an infinite geometric series with a first term of 6 and a common ratio of $\frac{1}{2}$? **D**

Ⓐ 3 Ⓑ 4 Ⓒ 9 Ⓓ 12

51. $2 + \frac{2}{3} + \frac{2}{9} + \frac{2}{27} + \ldots =$ **C**

Ⓐ $\frac{3}{2}$ Ⓑ $\frac{80}{27}$ Ⓒ 3 Ⓓ does not exist

www.algebra2.com/self_check_quiz

Study Guide and Intervention, p. 655 (shown) and p. 656

Infinite Geometric Series A geometric series that does not end is called an **infinite geometric series**. Some infinite geometric series have sums, but others do not because the **partial sums** increase without approaching a limiting value.

Sum of an Infinite Geometric Series	$S = \frac{a_1}{1 - r}$ for $-1 < r < 1$. If $\lvert r \rvert \geq 1$, the infinite geometric series does not have a sum.

Example **Find the sum of each infinite geometric series, if it exists.**

a. $75 + 15 + 3 + \ldots$

First, find the value of r to determine if the sum exists. $a_1 = 75$ and $a_2 = 15$, so $r = \frac{15}{75}$ or $\frac{1}{5}$. Since $\left|\frac{1}{5}\right| < 1$, the sum exists. Now use the formula for the sum of an infinite geometric series.

$S = \frac{a_1}{1 - r}$ Sum formula

$= \frac{75}{1 - \frac{1}{5}}$ $a_1 = 75, r = \frac{1}{5}$

$= \frac{75}{\frac{4}{5}}$ or 93.75 Simplify.

The sum of the series is 93.75.

b. $\sum_{n=1}^{\infty} 48\left(-\frac{1}{3}\right)^{n-1}$

In this infinite geometric series, $a_1 = 48$ and $r = -\frac{1}{3}$.

$S = \frac{a_1}{1 - r}$ Sum formula

$= \frac{48}{1 - \left(-\frac{1}{3}\right)}$ $a_1 = 48, r = -\frac{1}{3}$

$= \frac{48}{\frac{4}{3}}$ or 36 Simplify.

Thus $\sum_{n=1}^{\infty} 48\left(-\frac{1}{3}\right)^{n-1} = 36$.

Exercises

Find the sum of each infinite geometric series, if it exists.

1. $a_1 = -7, r = \frac{5}{8}$ $-18\frac{2}{3}$
2. $1 + \frac{5}{4} + \frac{25}{16} + \ldots$ does not exist
3. $a_1 = 4, r = \frac{1}{2}$ 8
4. $\frac{2}{9} + \frac{5}{27} + \frac{25}{162} + \ldots$ $1\frac{1}{3}$
5. $15 + 10 + 6\frac{2}{3} + \ldots$ 45
6. $18 - 9 + 4\frac{1}{2} - 2\frac{1}{4} + \ldots$ 12
7. $\frac{1}{10} + \frac{1}{20} + \frac{1}{40} + \ldots$ $\frac{1}{5}$
8. $1000 + 800 + 640 + \ldots$ 5000
9. $6 - 12 + 24 - 48 + \ldots$ does not exist
10. $\sum_{n=1}^{\infty} 50\left(\frac{4}{5}\right)^{n-1}$ 250
11. $\sum_{k=1}^{\infty} 22\left(-\frac{1}{2}\right)^{k-1}$ $14\frac{2}{3}$
12. $\sum_{s=1}^{\infty} 24\left(\frac{7}{12}\right)^{s-1}$ $57\frac{3}{5}$

Skills Practice, p. 657 and Practice, p. 658 (shown)

Find the sum of each infinite geometric series, if it exists.

1. $a_1 = 35, r = \frac{2}{7}$ 49
2. $a_1 = 26, r = \frac{1}{2}$ 52
3. $a_1 = 98, r = -\frac{3}{4}$ 56
4. $a_1 = 42, r = \frac{6}{5}$ does not exist
5. $a_1 = 112, r = -\frac{3}{5}$ 70
6. $a_1 = 500, r = \frac{1}{5}$ 625
7. $a_1 = 135, r = -\frac{1}{2}$ 90
8. $18 - 6 + 2 - \ldots$ $\frac{27}{2}$
9. $2 + 6 + 18 + \ldots$ does not exist
10. $6 + 4 + \frac{8}{3} + \ldots$ 18
11. $\frac{4}{25} + \frac{2}{5} + 1 + \ldots$ does not exist
12. $10 + 1 + 0.1 + \ldots$ $\frac{100}{9}$
13. $100 + 20 + 4 + \ldots$ 125
14. $-270 + 135 - 67.5 + \ldots$ -180
15. $0.5 + 0.25 + 0.125 + \ldots$ 1
16. $\frac{7}{10} + \frac{7}{100} + \frac{7}{1000} + \ldots$ $\frac{7}{9}$
17. $0.8 + 0.08 + 0.008 + \ldots$ $\frac{8}{9}$
18. $\frac{1}{12} - \frac{1}{6} + \frac{1}{3} - \ldots$ does not exist
19. $3 + \frac{9}{7} + \frac{27}{49} + \ldots$ $\frac{21}{4}$
20. $0.3 - 0.003 + 0.00003 - \ldots$ $\frac{30}{101}$
21. $0.06 + 0.006 + 0.0006 + \ldots$ $\frac{1}{15}$
22. $\frac{2}{3} - 2 + 6 - \ldots$ does not exist
23. $\sum_{n=1}^{\infty} 3\left(\frac{1}{4}\right)^{n-1}$ 4
24. $\sum_{n=1}^{\infty} \frac{2}{3}\left(-\frac{3}{4}\right)^{n-1}$ $\frac{8}{21}$
25. $\sum_{n=1}^{\infty} 18\left(\frac{2}{3}\right)^{n-1}$ 54
26. $\sum_{n=1}^{\infty} 5(-0.1)^{n-1}$ $\frac{50}{11}$

Write each repeating decimal as a fraction.

27. $0.\overline{6}$ $\frac{2}{3}$
28. $0.\overline{09}$ $\frac{1}{11}$
29. $0.\overline{43}$ $\frac{43}{99}$
30. $0.\overline{27}$ $\frac{3}{11}$
31. $0.\overline{243}$ $\frac{9}{37}$
32. $0.\overline{84}$ $\frac{28}{33}$
33. $0.\overline{990}$ $\frac{110}{111}$
34. $0.\overline{150}$ $\frac{50}{333}$

35. **PENDULUMS** On its first swing, a pendulum travels 8 feet. On each successive swing, the pendulum travels $\frac{4}{5}$ the distance of its previous swing. What is the total distance traveled by the pendulum when it stops swinging? 40 ft

36. **ELASTICITY** A ball dropped from a height of 10 feet bounces back $\frac{9}{10}$ of that distance. With each successive bounce, the ball continues to reach $\frac{9}{10}$ of its previous height. What is the total vertical distance (both up and down) traveled by the ball when it stops bouncing? (*Hint*: Add the total distance the ball falls to the total distance it rises.) 190 ft

Reading to Learn Mathematics, p. 659 ELL

Pre-Activity **How does an infinite geometric series apply to a bouncing ball?**

Read the introduction to Lesson 11-5 at the top of page 599 in your textbook.

Note the following powers of 0.6: $0.6^1 = 0.6$; $0.6^2 = 0.36$; $0.6^3 = 0.216$; $0.6^4 = 0.1296$; $0.6^5 = 0.07776$; $0.6^6 = 0.046656$; $0.6^7 = 0.0279936$. If a ball is dropped from a height of 10 feet and bounces back to 60% of its previous height on each bounce, after how many bounces will it bounce back to a height of less than 1 foot? 5 bounces

Reading the Lesson

1. Consider the formula $S = \frac{a_1}{1 - r}$.
   a. What is the formula used to find? the sum of an infinite geometric series
   b. What do each of the following represent?
      S: the sum
      a_1: the first term
      r: the common ratio
   c. For what values of r does an infinite geometric sequence have a sum? $-1 < r < 1$
   d. Rewrite your answer for part d as an absolute value inequality. $|r| < 1$

2. For each of the following geometric series, give the values of a_1 and r. Then state whether the sum of the series exists. (Do not actually find the sum.)
   a. $\frac{2}{3} + \frac{2}{9} + \frac{2}{27} + \ldots$ $a_1 =$ $\frac{2}{3}$ $r =$ $\frac{1}{3}$ Does the sum exist? yes
   b. $2 - 1 + \frac{1}{2} - \frac{1}{4} + \ldots$ $a_1 =$ 2 $r =$ $-\frac{1}{2}$ Does the sum exist? yes
   c. $\sum_{i=1}^{\infty} 3^i$ $a_1 =$ 3 $r =$ 3 Does the sum exist? no

Helping You Remember

3. One good way to remember something is to relate it to something you already know. How can you use the formula $S_n = \frac{a_1(1 - r^n)}{1 - r}$ that you learned in Lesson 11-4 for finding the sum of a geometric series to help you remember the formula for finding the sum of an infinite geometric series? Sample answer: If $-1 < r < 1$, then as n gets large, r^n approaches 0, so $1 - r^n$ approaches 1. Therefore, S_n approaches $\frac{a_1 \cdot 1}{1 - r}$, or $\frac{a_1}{1 - r}$.

Enrichment, p. 660

Convergence and Divergence

Convergence and divergence are terms that relate to the existence of a sum of an infinite series. If a sum exists, the series is convergent. If not, the series is divergent. Consider the series $12 + 3 + \frac{3}{4} + \frac{3}{16} + \ldots$. This is a geometric series with $r = \frac{1}{4}$. The sum is given by the formula $S = \frac{a_1}{1 - r}$. Thus, the sum is $12 \div \frac{3}{4}$ or 16. This series is convergent since a sum exists. Notice that the first two terms have a sum of 15. As more terms are added, the sum comes closer (or converges) to 16.

Recall that a geometric series has a sum if and only if $-1 < r < 1$. Thus, a geometric series is convergent if r is between -1 and 1, and divergent if r has another value. An infinite arithmetic series cannot have a sum unless all of the terms are equal to zero.

Example **Determine whether each series is convergent or divergent.**

a. $2 + 5 + 8 + 11 + \ldots$ divergent

4 Assess

Open-Ended Assessment

Writing Have students write their own examples of an infinite geometric series—one that has a sum and one that does not. Have them also write an example of a repeating decimal and then express it as a fraction.

Tips for New Teachers

Intervention Make sure that students can read the notation used in the various formulas and that they understand what each variable and subscript means.

Getting Ready for Lesson 11-6

PREREQUISITE SKILL Students will use recursive formulas in Lesson 11-6. This will involve their evaluating functions for given values. Use Exercises 70–75 to determine your students' familiarity with evaluating functions for given values.

Maintain Your Skills

Mixed Review

Find S_n for each geometric series described. *(Lesson 11-4)*

52. $a_1 = 1, a_6 = -243, r = -3$ **−182**

53. $a_1 = 72, r = \frac{1}{3}, n = 7$ $\mathbf{\frac{8744}{81}}$

54. **PHYSICS** A vacuum pump removes 20% of the air from a container with each stroke of its piston. What percent of the original air remains after five strokes of the piston? *(Lesson 11-3)* **32.768%**

Solve each equation or inequality. Check your solution. *(Lesson 10-1)*

55. $6^x = 216$ **3**

56. $2^{2x} = \frac{1}{8}$ $\mathbf{-\frac{3}{2}}$

57. $3^{x-2} \geq 27$ $\mathbf{x \geq 5}$

Simplify each expression. *(Lesson 9-2)*

58. $\frac{-2}{ab} + \frac{5}{a^2}$ $\mathbf{\frac{-2a + 5b}{a^2b}}$

59. $\frac{1}{x-3} - \frac{2}{x+1}$

60. $\frac{1}{x^2 + 6x + 8} + \frac{3}{x+4}$

59. $\mathbf{\frac{-x + 7}{(x-3)(x+1)}}$

60. $\mathbf{\frac{3x + 7}{(x+4)(x+2)}}$

Write an equation for the circle that satisfies each set of conditions. *(Lesson 8-3)*

61. center (2, 4), radius 6 $\mathbf{(x-2)^2 + (y-4)^2 = 36}$

62. endpoints of a diameter at (7, 3) and (−1, −5) $\mathbf{(x-3)^2 + (y+1)^2 = 32}$

Find all the zeros of each function. *(Lesson 7-5)*

63. $f(x) = 8x^3 - 36x^2 + 22x + 21$

64. $g(x) = 12x^4 + 4x^3 - 3x^2 - x$

63. $\mathbf{-\frac{1}{2}, \frac{3}{2}, \frac{7}{2}}$

64. $\mathbf{-\frac{1}{2}, -\frac{1}{3}, 0, \frac{1}{2}}$

Write a quadratic equation with the given roots. Write the equation in the form $ax^2 + bx + c = 0$, where a, b, and c are integers. *(Lesson 6-3)*

65. 6, −6 $\mathbf{x^2 - 36 = 0}$

66. −2, −7 $\mathbf{x^2 + 9x + 14 = 0}$

67. 6, 4 $\mathbf{x^2 - 10x + 24 = 0}$

RECREATION **For Exercises 68 and 69, refer to the graph at the right.** *(Lesson 2-3)*

68. Find the average rate of change of the number of visitors to Yosemite National Park from 1996 to 1999. **about −180,724 visitors per year**

69. Was the number of visitors increasing or decreasing from 1996 to 1999? **The number of visitors was decreasing.**

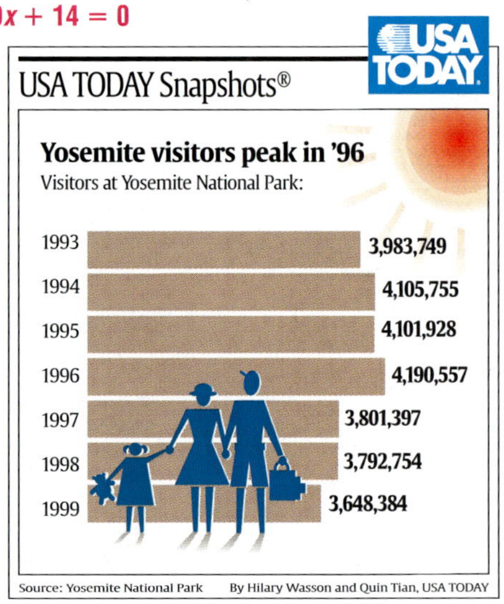

Getting Ready for the Next Lesson

PREREQUISITE SKILL **Find each function value.**
*(To review **evaluating functions**, see Lesson 2-1.)*

70. $f(x) = 2x, f(1)$ **2**

71. $g(x) = 3x - 3, g(2)$ **3**

72. $h(x) = -2x + 2, h(0)$ **2**

73. $f(x) = 3x - 1, f\left(\frac{1}{2}\right)$ $\mathbf{\frac{1}{2}}$

74. $g(x) = x^2, g(2)$ **4**

75. $h(x) = 2x^2 - 4, h(0)$ **−4**

Online Lesson Plans

USA TODAY Education

USA TODAY Education's Online site offers resources and interactive features connected to each day's newspaper. *Experience TODAY*, USA TODAY's daily lesson plan, is available on the site and delivered daily to subscribers. This plan provides instruction for integrating USA TODAY graphics and key editorial features into your mathematics classroom. Log on to www.education.usatoday.com.

604 Chapter 11 Sequences and Series

Spreadsheet Investigation

A Preview of Lesson 11-6

Amortizing Loans

When a payment is made on a loan, part of the payment is used to cover the interest that has accumulated since the last payment. The rest is used to reduce the *principal*, or original amount of the loan. This process is called *amortization*. You can use a spreadsheet to analyze the payments, interest, and balance on a loan. A table that shows this kind of information is called an *amortization schedule*.

Example

Marisela just bought a new sofa for $495. The store is letting her make monthly payments of $43.29 at an interest rate of 9% for one year. How much will she still owe after six months?

Every month, the interest on the remaining balance will be $\frac{9\%}{12}$ or 0.75%. You can find the balance after a payment by multiplying the balance after the previous payment by 1 + 0.0075 or 1.0075 and then subtracting 43.29.

In a spreadsheet, use the column of numbers for the number of payments and use column B for the balance. Enter the interest rate and monthly payment in cells in column A so that they can be easily updated if the information changes.

The spreadsheet at the right shows the formulas for the balances after each of the first six payments. After six months, Marisela still owes $253.04.

Loans

	A	B
1	Interest rate	=495*(1+A2)-A5
2	0.0075	=B1*(1+A2)-A5
3		=B2*(1+A2)-A5
4	Monthly payment	=B3*(1+A2)-A5
5	43.29	=B4*(1+A2)-A5
6		=B5*(1+A2)-A5
7		

Sheet1 / Sh

Exercises

1. Let b_n be the balance left on Marisela's loan after n months. Write an equation relating b_n and b_{n+1}. **$b_{n+1} = 1.0075b_n - 43.29$**
2. Payments at the beginning of a loan go more toward interest than payments at the end. What percent of Marisela's loan remains to be paid after half a year? **about 51%**
3. Extend the spreadsheet to the whole year. What is the balance after 12 payments? Why is it not 0? **About −$0.01; the balance is not exactly 0 due to rounding.**
4. Suppose Marisela decides to pay $50 every month. How long would it take her to pay off the loan? **11 months**
5. Suppose that, based on how much she can afford, Marisela will pay a variable amount each month in addition to the $43.29. Explain how the flexibility of a spreadsheet can be used to adapt to this situation. **See margin.**
6. Jamie has a three-year, $12,000 car loan. The annual interest rate is 6%, and his monthly payment is $365.06. After twelve months, he receives an inheritance which he wants to use to pay off the loan. How much does he owe at that point? **$8236.91**

Spreadsheet Investigation

A Preview of Lesson 11-6

Getting Started

Objective To discover how a spreadsheet can express a relationship in which the calculation of the value of the next term involves using the value of the previous term.

Teach

Ask students why the balance is multiplied by 1.0075 rather than 0.75%. **to find the balance plus the interest**

Assess

In **Exercises 1–5**, students should

- be able to relate the list of payments to the series and sequences they have been studying.
- be able to use a spreadsheet to make a flexible table of payments.

Study Notebook

You may wish to have students summarize this activity and what they learned from it.

Answer

5. **Changing the monthly payment only requires editing the amount subtracted in the formula in each cell.**

Spreadsheet Investigation Amortizing Loans 605

11-6 Lesson Notes

1 Focus

5-Minute Check Transparency 11-6 Use as a quiz or review of Lesson 11-5.

Mathematical Background notes are available for this lesson on p. 576D.

How is the Fibonacci sequence illustrated in nature?

Ask students:

- What number follows 5 in the Fibonacci sequence? **8**
- Is the Fibonacci sequence an arithmetic sequence? **no** Is it a geometric sequence? **no** Explain. **There is no common difference and no common ratio.**

11-6 Recursion and Special Sequences

What You'll Learn

- Recognize and use special sequences.
- Iterate functions.

Vocabulary

- Fibonacci sequence
- recursive formula
- iteration

How is the Fibonacci sequence illustrated in nature?

A shoot on a sneezewort plant must grow for two months before it is strong enough to put out another shoot. After that, it puts out at least one shoot every month.

Month	1	2	3	4	5
Shoots	1	1	2	3	5

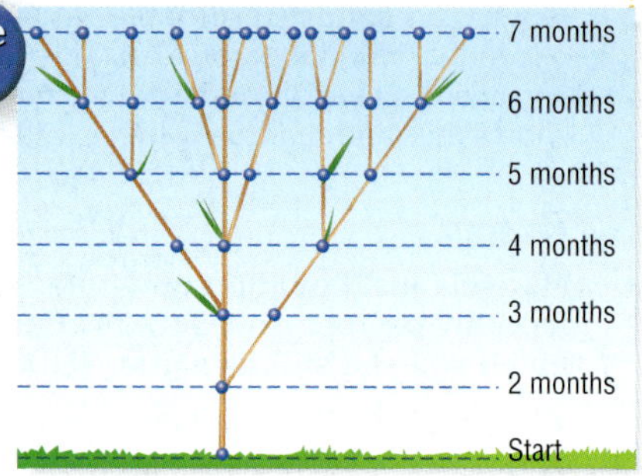

SPECIAL SEQUENCES Notice that the sequence 1, 1, 2, 3, 5, 8, 13, … has a pattern. Each term in the sequence is the sum of the two previous terms. For example, $8 = 3 + 5$ and $13 = 5 + 8$. This sequence is called the **Fibonacci sequence**, and it is found in many places in nature.

first term	a_1		1
second term	a_2		1
third term	a_3	$a_1 + a_2$	$1 + 1 = 2$
fourth term	a_4	$a_2 + a_3$	$1 + 2 = 3$
fifth term	a_5	$a_3 + a_4$	$2 + 3 = 5$
⋮	⋮	⋮	
nth term	a_n	$a_{n-2} + a_{n-1}$	

Study Tip

Reading Math

A recursive formula is often called a *recursive relation* or a *recurrence relation*.

The formula $a_n = a_{n-2} + a_{n-1}$ is an example of a **recursive formula**. This means that each term is formulated from one or more previous terms. To be able to use a recursive formula, you must be given the value(s) of the first term(s) so that you can start the sequence and then use the formula to generate the rest of the terms.

TEACHING TIP

$a_n = a_1 + (n - 1)d$ and $a_n = a_1r^{n-1}$ are not recursive. These sequences are determined by the number of the term n rather than by the preceding term.

Example 1 Use a Recursive Formula

Find the first five terms of the sequence in which $a_1 = 4$ and $a_{n+1} = 3a_n - 2$, $n \geq 1$.

$a_{n+1} = 3a_n - 2$ Recursive formula

$a_{1+1} = 3a_1 - 2$ $n = 1$
$a_2 = 3(4) - 2$ or 10 $a_1 = 4$

$a_{2+1} = 3a_2 - 2$ $n = 2$
$a_3 = 3(10) - 2$ or 28 $a_2 = 10$

$a_{3+1} = 3a_3 - 2$ $n = 3$
$a_4 = 3(28) - 2$ or 82 $a_3 = 28$

$a_{4+1} = 3a_4 - 2$ $n = 4$
$a_5 = 3(82) - 2$ or 244 $a_4 = 82$

The first five terms of the sequence are 4, 10, 28, 82, and 244.

Resource Manager

Workbook and Reproducible Masters

Chapter 11 Resource Masters

- Study Guide and Intervention, pp. 661–662
- Skills Practice, p. 663
- Practice, p. 664
- Reading to Learn Mathematics, p. 665
- Enrichment, p. 666
- Assessment, p. 694

Graphing Calculator and Spreadsheet Masters, p. 47

Teaching Algebra With Manipulatives Masters, pp. 285, 286–287

Transparencies

5-Minute Check Transparency 11-6
Real-World Transparency 11
Answer Key Transparencies

Technology

Interactive Chalkboard

Example 2 Find and Use a Recursive Formula

GARDENING **Mr. Yazaki discovered that there were 225 dandelions in his garden on the first Saturday of spring. He had time to pull out 100, but by the next Saturday, there were twice as many as he had left. Each Saturday in spring, he removed 100 dandelions, only to find that the number of remaining dandelions had doubled by the following Saturday.**

a. Write a recursive formula for the number of dandelions Mr. Yazaki finds in his garden each Saturday.

Let d_n represent the number of dandelions at the beginning of the nth Saturday. Mr. Yazaki will pull 100 of these out of his garden, leaving $d_n - 100$. The number d_{n+1} of dandelions the next Saturday will be twice this number. So, $d_{n+1} = 2(d_n - 100)$ or $2d_n - 200$.

b. Find the number of dandelions Mr. Yazaki would find on the fifth Saturday.

On the first Saturday, there were 225 dandelions, so $d_1 = 225$.

$d_{n+1} = 2d_n - 200$ Recursive formula

$d_{1+1} = 2d_1 - 200$ $n = 1$
$d_2 = 2(225) - 200$ or 250

$d_{2+1} = 2d_2 - 200$ $n = 2$
$d_3 = 2(250) - 200$ or 300

$d_{3+1} = 2d_3 - 200$ $n = 3$
$d_4 = 2(300) - 200$ or 400

$d_{4+1} = 2d_4 - 200$ $n = 4$
$d_5 = 2(400) - 200$ or 600

On the fifth Saturday, there would be 600 dandelions in Mr. Yazaki's garden.

You can use sequences to analyze some games.

Algebra Activity

Special Sequences

The object of the *Towers of Hanoi* game is to move a stack of n coins from one position to another in the fewest number a_n of moves with these rules.

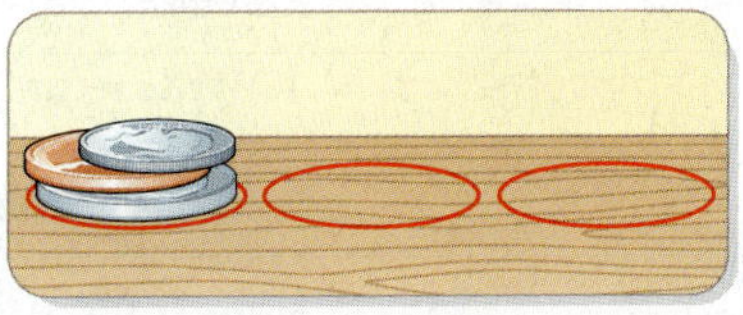

- You may only move one coin at a time.
- A coin must be placed on top of another coin, not underneath.
- A smaller coin may be placed on top of a larger coin, but not vice versa. For example, a penny may not be placed on top of a dime.

Model and Analyze

1. Draw three circles on a sheet of paper, as shown above. Place a penny on the first circle. What is the least number of moves required to get the penny to the second circle? 1
2. Place a nickel and a penny on the first circle, with the penny on top. What is the least number of moves that you can make to get the stack to another circle? (Remember, a nickel cannot be placed on top of a penny.) 3
3. Place a nickel, penny, and dime on the first circle. What is the least number of moves that you can take to get the stack to another circle? 7

Make a Conjecture

4. Place a quarter, nickel, penny, and dime on the first circle. Experiment to find the least number of moves needed to get the stack to another circle. Make a conjecture about a formula for the minimum number a_n of moves required to move a stack of n coins. 15; $a_n = 2^n - 1$

www.algebra2.com/extra_examples

2 Teach

SPECIAL SEQUENCES

In-Class Examples Power Point®

1 Find the first five terms of the sequence in which $a_1 = 5$ and $a_{n+1} = 2a_n + 7, n \geq 1$. **5, 17, 41, 89, 185**

Teaching Tip Make sure students understand that you use the value of one term to find the value of the next term.

2 **BIOLOGY** Dr. Elliot is growing cells in lab dishes. She starts with 108 cells Monday morning and then removes 20 of these for her experiment. By Tuesday the remaining cells have multiplied by 1.5. She again removes 20. This pattern repeats each day in the week.

a. Write a recursive formula for the number of cells Dr. Elliot finds each day before she removes any. $c_{n+1} = 1.5(c_n - 20)$ or $c_{n+1} = 1.5c_n - 30$

b. Find the number of cells she will find on Friday morning. **303**

Algebra Activity

Materials: compass, penny, nickel, dime, quarter

- Tell students that according to Martin Gardner, in *The Scientific American Book of Mathematical Puzzles & Diversions*, the "Tower of Hanoi was invented by the French mathematician Edouard Lucas and sold as a toy in 1883."
- The toy usually has 3 pegs, with a tower of 8 disks on one peg. The task is to transfer all 8 disks to one of the vacant pegs, using the rules in the activity, in the fewest possible moves.

ITERATION

In-Class Example

Power Point®

3 Find the first three iterates x_1, x_2, x_3 of the function $f(x) = 3x - 1$ for an initial value of $x_0 = 5$. **14, 41, 122**

3 Practice/Apply

Study Notebook

Have students—
- add the definitions/examples of the vocabulary terms to their Vocabulary Builder worksheets for Chapter 11.
- include any other item(s) that they find helpful in mastering the skills in this lesson.

About the Exercises...

Organization by Objective
- **Special Sequences:** 13–30
- **Iteration:** 31–39

Odd/Even Assignments

Exercises 13–24 and 31–38 are structured so that students practice the same concepts whether they are assigned odd or even problems.

Assignment Guide

Basic: 13–19 odd, 23, 25–30, 31–35 odd, 39–55

Average: 13–25 odd, 26–30, 31–37 odd, 39–55

Advanced: 14–24 even, 25–30, 32–40 even, 41–49 (optional: 50–55)

Study Tip

Look Back
To review **composition of functions**, see Lesson 7-7.

ITERATION

Iteration is the process of composing a function with itself repeatedly. For example, if you compose a function with itself once, the result is $f \circ f(x)$ or $f(f(x))$. If you compose a function with itself two times, the result is $f \circ f \circ f(x)$ or $f(f(f(x)))$, and so on.

You can use iteration to recursively generate a sequence. Start with an initial value x_0. Let $x_1 = f(x_0)$, $x_2 = f(x_1)$ or $f(f(x_0))$, $x_3 = f(x_2)$ or $f(f(f(x_0)))$, and so on.

Example 3 Iterate a Function

Find the first three iterates x_1, x_2, x_3 of the function $f(x) = 2x + 3$ for an initial value of $x_0 = 1$.

To find the first iterate x_1, find the value of the function for $x_0 = 1$.

$x_1 = f(x_0)$ — Iterate the function.
$= f(1)$ — $x_0 = 1$
$= 2(1) + 3$ or 5 — Simplify.

To find the second iterate x_2, substitute x_1 for x.

$x_2 = f(x_1)$ — Iterate the function.
$= f(5)$ — $x_1 = 5$
$= 2(5) + 3$ or 13 — Simplify.

Substitute x_2 for x to find the third iterate.

$x_3 = f(x_2)$ — Iterate the function.
$= f(13)$ — $x_2 = 13$
$= 2(13) + 3$ or 29 — Simplify.

Therefore, 1, 5, 13, 29 is an example of a sequence generated using iteration.

Check for Understanding

Concept Check

1. Write recursive formulas for the nth terms of arithmetic and geometric sequences. $a_n = a_{n-1} + d$; $a_n = r \cdot a_{n-1}$
2. **OPEN ENDED** Write a recursive formula for a sequence whose first three terms are 1, 1, and 3. **Sample answer:** $a_n = 2a_{n-1} + a_{n-2}$
3. State whether the statement $x_n \neq x_{n-1}$ is *sometimes*, *always*, or *never* true if $x_n = f(x_{n-1})$. Explain. **Sometimes; see margin for explanation.**

Guided Practice

GUIDED PRACTICE KEY	
Exercises	**Examples**
4–7	1, 2
8–10	3
11, 12	2

Find the first five terms of each sequence. **5. −3, −2, 0, 3, 7**

4. $a_1 = 12, a_{n+1} = a_n - 3$ **12, 9, 6, 3, 0**
5. $a_1 = -3, a_{n+1} = a_n + n$
6. $a_1 = 0, a_{n+1} = -2a_n - 4$ **0, −4, 4, −12, 20**
7. $a_1 = 1, a_2 = 2, a_{n+2} = 4a_{n+1} - 3a_n$ **1, 2, 5, 14, 41**

Find the first three iterates of each function for the given initial value.

8. $f(x) = 3x - 4, x_0 = 3$ **5, 11, 29**
9. $f(x) = -2x + 5, x_0 = 2$ **1, 3, −1**
10. $f(x) = x^2 + 2, x_0 = -1$ **3, 11, 123**

Application

BANKING For Exercises 11 and 12, use the following information.

Rita has deposited $1000 in a bank account. At the end of each year, the bank posts interest to her account in the amount of 5% of the balance, but then takes out a $10 annual fee.

11. Let b_0 be the amount Rita deposited. Write a recursive equation for the balance b_n in her account at the end of n years. $b_n = 1.05b_{n-1} - 10$
12. Find the balance in the account after four years. **$1172.41**

DAILY INTERVENTION

Differentiated Instruction

Kinesthetic Have students make and play a Tower of Hanoi game as described in the notes for the Algebra Activity. Students can cut 8 cardboard squares of graduated sizes and move them between three circles to represent the pegs. Students can also do additional research about this classic puzzle.

★ indicates increased difficulty

Practice and Apply

Homework Help

For Exercises	See Examples
13–30	1–2
31–39	3

Extra Practice
See page 853.

Find the first five terms of each sequence.

13. $a_1 = -6, a_{n+1} = a_n + 3$
14. $a_1 = 13, a_{n+1} = a_n + 5$
15. $a_1 = 2, a_{n+1} = a_n - n$
16. $a_1 = 6, a_{n+1} = a_n + n + 3$
17. $a_1 = 9, a_{n+1} = 2a_n - 4$
18. $a_1 = 4, a_{n+1} = 3a_n - 6$
19. $a_1 = -1, a_2 = 5, a_{n+1} = a_n + a_{n-1}$
20. $a_1 = 4, a_2 = -3, a_{n+2} = a_{n+1} + 2a_n$
★ 21. $a_1 = \frac{7}{2}, a_{n+1} = \frac{n}{n+1} \cdot a_n$
★ 22. $a_1 = \frac{3}{4}, a_{n+1} = \frac{n^2+1}{n} \cdot a_n$

13. −6, −3, 0, 3, 6
14. 13, 18, 23, 28, 33
15. 2, 1, −1, −4, −8
16. 6, 10, 15, 21, 28
17. 9, 14, 24, 44, 84
18. 4, 6, 12, 30, 84
19. −1, 5, 4, 9, 13
20. 4, −3, 5, −1, 9
21. $\frac{7}{2}, \frac{7}{4}, \frac{7}{6}, \frac{7}{8}, \frac{7}{10}$
22. $\frac{3}{4}, \frac{3}{2}, \frac{15}{4}, \frac{25}{2}, \frac{425}{8}$

23. If $a_0 = 7$ and $a_{n+1} = a_n + 12$ for $n \geq 0$, find the value of a_5. **67**

24. If $a_0 = 1$ and $a_{n+1} = -2.1$ for $n \geq 0$, then what is the value of a_4? **−2.1**

GEOMETRY For Exercises 25 and 26, use the following information.
Join two 1-unit by 1-unit squares to form a rectangle. Next, draw a larger square along a long side of the rectangle. Continue this process of drawing a square along a long side of the rectangle formed at the previous step.

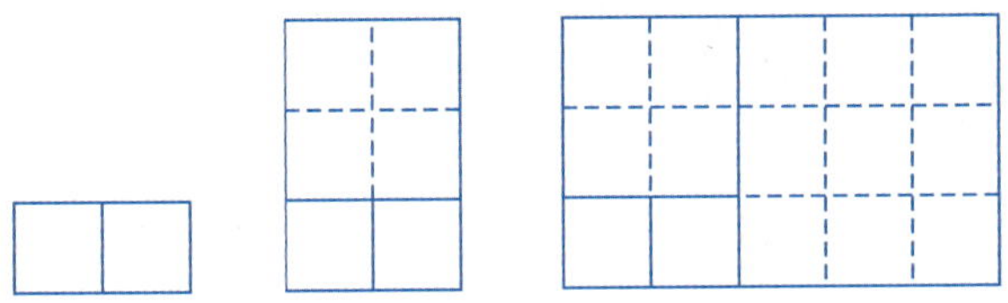

Step 1 Step 2 Step 3

25. Write the sequence of the lengths of the sides of the squares you added at each step. Begin the sequence with the lengths of the sides of the two original squares. **1, 1, 2, 3, 5, ...**

26. Identify the sequence in Exercise 25. **the Fibonacci sequence**

27. **LOANS** The Cruz family is taking out a mortgage loan for \$100,000 to buy a house. Their monthly payment is \$678.79. The recursive formula $b_n = 1.006\, b_{n-1} - 678.79$ describes the balance left on the loan after n payments. Find the balances of the loan after each of the first eight payments.

GEOMETRY For Exercises 28–30, study the triangular numbers shown below.

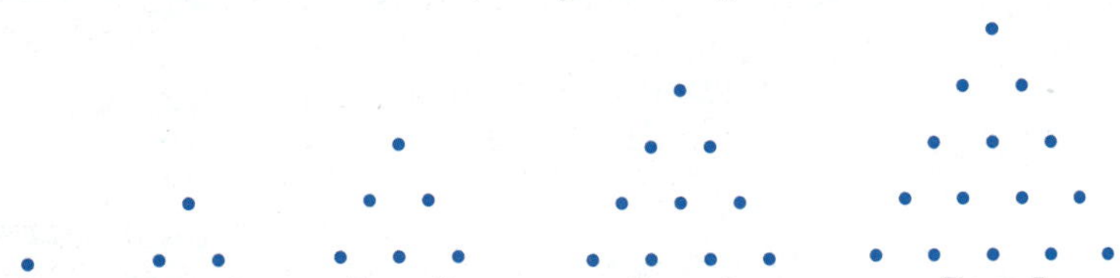

Figure 1 Figure 2 Figure 3 Figure 4 Figure 5

28. Write a sequence of the first five triangular numbers. **1, 3, 6, 10, 15**

29. Write a recursive formula for the nth triangular number t_n. $t_n = t_{n-1} + n$

30. What is the 200th triangular number? **20,100**

27. \$99,921.21, \$99,841.95, \$99,762.21, \$99,681.99, \$99,601.29, \$99,520.11, \$99,438.44, \$99,356.28

Career Choices

Real Estate Agent
Most real estate agents are independent business-people who earn their income from commission.

Online Research
To learn more about a career in real estate, visit: www.algebra2.com/careers

www.algebra2.com/self_check_quiz

Study Guide and Intervention, p. 661 (shown) and p. 662

Special Sequences In a **recursive formula**, each succeeding term is formulated from one or more previous terms. A recursive formula for a sequence has two parts:

1. the value(s) of the first term(s), and
2. an equation that shows how to find each term from the term(s) before it.

Example **Find the first five terms of the sequence in which $a_1 = 6, a_2 = 10$, and $a_n = 2a_{n-2}$ for $n \geq 3$.**

$a_1 = 6$
$a_2 = 10$
$a_3 = 2a_1 = 2(6) = 12$
$a_4 = 2a_2 = 2(10) = 20$
$a_5 = 2a_3 = 2(12) = 24$

The first five terms of the sequence are 6, 10, 12, 20, 24.

Exercises

Find the first five terms of each sequence.

1. $a_1 = 1, a_2 = 1, a_n = 2(a_{n-1} + a_{n-2}), n \geq 3$ **1, 1, 4, 10, 28**
2. $a_1 = 1, a_n = \frac{1}{1 + a_{n-1}}, n \geq 2$ **1, $\frac{1}{2}, \frac{2}{3}, \frac{3}{5}, \frac{5}{8}$**
3. $a_1 = 3, a_n = a_{n-1} + 2(n-2), n \geq 2$ **3, 3, 5, 9, 15**
4. $a_1 = 5, a_n = a_{n-1} + 2, n \geq 2$ **5, 7, 9, 11, 13**
5. $a_1 = 1, a_n = (n-1)a_{n-1}, n \geq 2$ **1, 1, 2, 6, 24**
6. $a_1 = 7, a_n = 4a_{n-1} - 1, n \geq 2$ **7, 27, 107, 427, 1707**
7. $a_1 = 3, a_2 = 4, a_n = 2a_{n-2} + 3a_{n-1}, n \geq 3$ **3, 4, 18, 62, 222**
8. $a_1 = 0.5, a_n = a_{n-1} + 2n, n \geq 2$ **0.5, 4.5, 10.5, 18.5, 28.5**
9. $a_1 = 8, a_2 = 10, a_n = \frac{a_{n-2}}{a_{n-1}}, n \geq 3$ **8, 10, 0.8, 12.5, 0.064**
10. $a_1 = 100, a_n = \frac{a_{n-1}}{n}, n \geq 2$ **100, 50, $\frac{50}{3}, \frac{50}{12}, \frac{50}{60}$**

Skills Practice, p. 663 and Practice, p. 664 (shown)

Find the first five terms of each sequence.

1. $a_1 = 3, a_{n+1} = a_n + 5$ **3, 8, 13, 18, 23**
2. $a_1 = -7, a_{n+1} = a_n + 8$ **−7, 1, 9, 17, 25**
3. $a_1 = -3, a_{n+1} = 3a_n + 2$ **−3, −7, −19, −55, −163**
4. $a_1 = -8, a_{n+1} = 10 - a_n$ **−8, 18, −8, 18, −8**
5. $a_1 = 4, a_{n+1} = n - a_n$ **4, −3, 5, −2, 6**
6. $a_1 = -3, a_{n+1} = 3a_n$ **−3, −9, −27, −81, −243**
7. $a_1 = 4, a_{n+1} = -3a_n + 4$ **4, −8, 28, −80, 244**
8. $a_1 = 2, a_{n+1} = -4a_n - 5$ **2, −13, 47, −193, 767**
9. $a_1 = 3, a_2 = 1, a_{n+1} = a_n - a_{n-1}$ **3, 1, −2, −3, −1**
10. $a_1 = -1, a_2 = 5, a_{n+1} = 4a_{n-1} - a_n$ **−1, 5, −9, 29, −65**
11. $a_1 = 2, a_2 = -3, a_{n+1} = 5a_n - 8a_{n-1}$ **2, −3, −31, −131, −407**
12. $a_1 = -2, a_2 = 1, a_{n+1} = -2a_n + 6a_{n-1}$ **−2, 1, −14, 34, −152**

Find the first three iterates of each function for the given initial value.

13. $f(x) = 3x + 4, x_0 = -1$ **1, 7, 25**
14. $f(x) = 10x + 2, x_0 = -1$ **−8, −78, −778**
15. $f(x) = 8 + 3x, x_0 = 1$ **11, 41, 131**
16. $f(x) = 8 - x, x_0 = -3$ **11, −3, 11**
17. $f(x) = 4x + 5, x_0 = -1$ **1, 9, 41**
18. $f(x) = 5(x + 3), x_0 = -2$ **5, 40, 215**
19. $f(x) = -8x + 9, x_0 = 1$ **1, 1, 1**
20. $f(x) = -4x^2, x_0 = -1$ **−4; −64; −16,384**
21. $f(x) = x^2 - 1, x_0 = 3$ **8, 63, 3968**
22. $f(x) = 2x^2; x_0 = 5$ **50; 5000; 50,000,000**

23. **INFLATION** Iterating the function $c(x) = 1.05x$ gives the future cost of an item at a constant 5% inflation rate. Find the cost of a \$2000 ring in five years at 5% inflation. **\$2552.56**

FRACTALS For Exercises 24–27, use the following information.
Replacing each side of the square shown with the combination of segments below it gives the figure to its right.

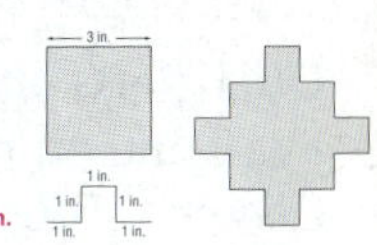

24. What is the perimeter of the original square? **12 in.**
25. What is the perimeter of the new shape? **20 in.**
26. If you repeat the process by replacing each side of the new shape by a proportional combination of 5 segments, what will the perimeter of the third shape be? **$33\frac{1}{3}$ in.**
27. What function $f(x)$ can you iterate to find the perimeter of each successive shape if you continue this process? $f(x) = \frac{5}{3}x$

Reading to Learn Mathematics, p. 665 ELL

Pre-Activity **How is the Fibonacci sequence illustrated in nature?**

Read the introduction to Lesson 11-6 at the top of page 606 in your textbook.

What are the next three numbers in the sequence that gives the number of shoots corresponding to each month? **8, 13, 21**

Reading the Lesson

1. Consider the sequence in which $a_1 = 4$ and $a_n = 2a_{n-1} + 5$.
   a. Explain why this is a *recursive* formula. **Sample answer: Each term is found from the value of the previous term.**
   b. Explain in your own words how to find the first four terms of this sequence. (Do not actually find any terms after the first.) **Sample answer: The first term is 4. To find the second term, double the first term and add 5. To find the third term, double the second term and add 5. To find the fourth term, double the third term and add 5.**
   c. What happens to the terms of this sequence as n increases? **Sample answer: They keep getting larger and larger.**
2. Consider the function $f(x) = 3x - 1$ with an initial value of $x_0 = 2$.
   a. What does it mean to *iterate* this function? **to compose the function with itself repeatedly**
   b. Fill in the blanks to find the first three iterates. The blanks that follow the letter x are for subscripts.
   $x_1 = f(x_{\underline{0}}) = f(\underline{2}) = 3(\underline{2}) - 1 = \underline{6} - 1 = \underline{5}$
   $x_2 = f(x_{\underline{1}}) = f(\underline{5}) = 3(\underline{5}) - 1 = \underline{14}$
   $x_3 = f(x_{\underline{2}}) = f(\underline{14}) = 3(\underline{14}) - 1 = \underline{41}$
   c. As this process continues, what happens to the values of the iterates? **Sample answer: They keep getting larger and larger.**

Helping You Remember

3. Use a dictionary to find the meanings of the words *recurrent* and *iterate*. How can the meanings of these words help you to remember the meaning of the mathematical terms *recursive* and *iteration*? How are these ideas related? **Sample answer: *Recurrent* means happening repeatedly, while *iterate* means to repeat a process or operation. A *recursive* formula is used repeatedly to find the value of one term of a sequence based on the previous term. *Iteration* means to compose a function with it self repeatedly. Both ideas have to do with repetition—doing the same thing over and over again.**

Answers

3. If $f(x) = x^2$ and $x_1 = 2$, then $x_2 = 2^2$ or 4, so $x_2 \neq x_1$. But, if $x_1 = 1$, then $x_2 = 1$, so $x_2 = x_1$.

Enrichment, p. 666

Continued Fractions

The fraction below is an example of a continued fraction. Note that each fraction in the continued fraction has a numerator of 1.

$$2 + \cfrac{1}{3 + \cfrac{1}{4 + \frac{1}{5}}}$$

Example 1 **Evaluate the continued fraction above. Start at the bottom and work your way up.**

Step 1: $4 + \frac{1}{5} = \frac{20}{5} + \frac{1}{5} = \frac{21}{5}$
Step 2: $\frac{1}{\frac{21}{5}} = \frac{5}{21}$
Step 3: $3 + \frac{5}{21} = \frac{63}{21} + \frac{5}{21} = \frac{68}{21}$
Step 4: $\frac{1}{68} = \frac{21}{}$

Example 2 **Change $\frac{25}{11}$ into a continued fraction.**

Follow the steps.

Step 1: $\frac{25}{11} = \frac{22}{11} + \frac{3}{11} = 2 + \frac{3}{11}$
Step 2: $\frac{3}{11} = \frac{1}{\frac{11}{3}}$
Step 3: $\frac{11}{3} = \frac{9}{3} + \frac{2}{3} = 3 + \frac{2}{3}$
Step 4: $\frac{2}{3} = \frac{1}{3}$

4 Assess

Open-Ended Assessment

Speaking Have students explain, with examples, what it means to say that a formula or a function is recursive.

Tips for New Teachers

Intervention Make sure that students understand the language used in this lesson, particularly *iteration*, *iterative*, and *iterate*.

Getting Ready for Lesson 11-7

BASIC SKILL Students will use the Binomial Theorem in Lesson 11-7. This will involve their simplifying factorial expressions. Use Exercises 50–55 to determine your students' familiarity with evaluating the kinds of expressions they will encounter when simplifying factorials.

Assessment Options

Quiz (Lessons 11-5 and 11-6) is available on p. 694 of the *Chapter 11 Resource Masters.*

Answer

41. Under certain conditions, the Fibonacci sequence can be used to model the number of shoots on a plant. Answers should include the following.
    - The 13th term of the sequence is 233, so there are 233 shoots on the plant during the 13th month.
    - The Fibonacci sequence is not arithmetic because the differences (0, 1, 1, 2, ...) of the terms are not constant. The Fibonacci sequence is not geometric because the ratios $\left(1, 2, \frac{3}{2}, \ldots\right)$ of the terms are not constant.

Find the first three iterates of each function for the given initial value.

31. $f(x) = 9x - 2, x_0 = 2$ **16, 142, 1276**
32. $f(x) = 4x - 3, x_0 = 2$ **5, 17, 65**
33. $f(x) = 3x + 5, x_0 = -4$ **−7, −16, −43**
34. $f(x) = 5x + 1, x_0 = -1$ **−4, −19, −94**
35. $f(x) = 2x^2 - 5, x_0 = -1$ **−3, 13, 333**
36. $f(x) = 3x^2 - 4, x_0 = 1$ **−1, −1, −1**

★37. $f(x) = 2x^2 + 2x + 1, x_0 = \frac{1}{2}$

★38. $f(x) = 3x^2 - 3x + 2, x_0 = \frac{1}{3}$

37. $\frac{5}{2}, \frac{37}{2}, \frac{1445}{2}$

38. $\frac{4}{3}, \frac{10}{3}, \frac{76}{3}$

39. **ECONOMICS** If the rate of inflation is 2%, the cost of an item in future years can be found by iterating the function $c(x) = 1.02x$. Find the cost of a \$70 portable stereo in four years if the rate of inflation remains constant. **\$75.78**

40. **CRITICAL THINKING** Are there a function $f(x)$ and an initial value x_0 such that the first three iterates, in order, are 4, 4, and 7? If so, state such a function and initial value. If not, explain.

40. No; according to the first two iterates, $f(4) = 4$. According to the second and third iterates, $f(4) = 7$. Since $f(x)$ is a function, it cannot have two values when $x = 4$.

41. **WRITING IN MATH** Answer the question that was posed at the beginning of the lesson. **See margin.**

    How is the Fibonacci sequence illustrated in nature?

    Include the following in your answer:
    - the 13th term in the Fibonacci sequence, with an explanation of what it tells you about the plant described, and
    - an explanation of why the Fibonacci sequence is neither arithmetic nor geometric.

Standardized Test Practice

42. If a is positive, what percent of $4a$ is 8? **D**

    (A) $\frac{a}{100}\%$ (B) $\frac{a}{2}\%$ (C) $\frac{8}{a}\%$ (D) $\frac{200}{a}\%$

43. The figure at the right is made of three concentric semicircles. What is the total area of the shaded regions? **C**

    (A) 4π units^2 (B) 10π units^2

    (C) 12π units^2 (D) 20π units^2

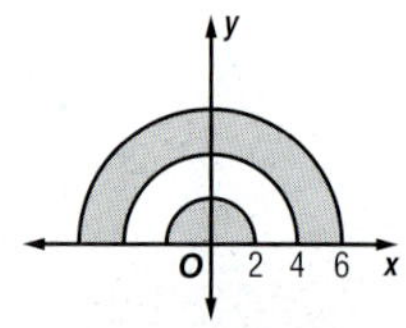

Maintain Your Skills

Mixed Review

Find the sum of each infinite geometric series, if it exists. *(Lesson 11-5)*

44. $9 + 6 + 4 + \ldots$ **27**
45. $\frac{1}{8} + \frac{1}{32} + \frac{1}{128} + \ldots$ $\frac{1}{6}$
46. $4 - \frac{8}{3} + \frac{16}{9} + \ldots$ $\frac{12}{5}$

Find the sum of each geometric series. *(Lesson 11-4)*

47. $2 - 10 + 50 - \ldots$ to 6 terms **−5208**
48. $3 + 1 + \frac{1}{3} + \ldots$ to 7 terms $\frac{1093}{243}$

49. **GEOMETRY** The area of rectangle $ABCD$ is $6x^2 + 38x + 56$ square units. Its width is $2x + 8$ units. What is the length of the rectangle? *(Lesson 5-3)* **$3x + 7$ units**

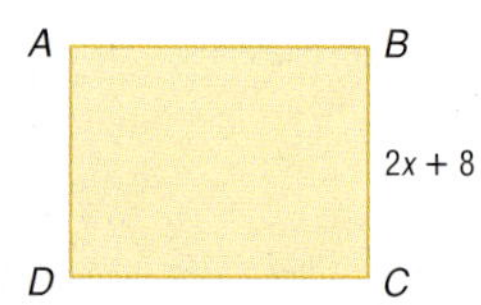

Getting Ready for the Next Lesson

BASIC SKILL **Evaluate each expression.** **51. 5040**

50. $5 \cdot 4 \cdot 3 \cdot 2 \cdot 1$ **120**
51. $7 \cdot 6 \cdot 5 \cdot 4 \cdot 3 \cdot 2 \cdot 1$
52. $\frac{4 \cdot 3}{2 \cdot 1}$ **6**
53. $\frac{6 \cdot 5 \cdot 4}{3 \cdot 2 \cdot 1}$ **20**
54. $\frac{9 \cdot 8 \cdot 7 \cdot 6}{4 \cdot 3 \cdot 2 \cdot 1}$ **126**
55. $\frac{10 \cdot 9 \cdot 8 \cdot 7 \cdot 6 \cdot 5}{6 \cdot 5 \cdot 4 \cdot 3 \cdot 2 \cdot 1}$ **210**

610 Chapter 11 Sequences and Series

Algebra Activity

A Follow-Up of Lesson 11-6

Fractals

Fractals are sets of points that often involve intricate geometric shapes. Many fractals have the property that when small parts are magnified, the detail of the fractal is not lost. In other words, the magnified part is made up of smaller copies of itself. Such fractals can be constructed recursively.

You can use isometric dot paper to draw stages of the construction of a fractal called the *von Koch snowflake.*

Stage 1 Draw an equilateral triangle with sides of length 9 units on the dot paper.

Stage 2 Now remove the middle third of each side of the triangle from Stage 1 and draw the other two sides of an equilateral triangle pointing outward.

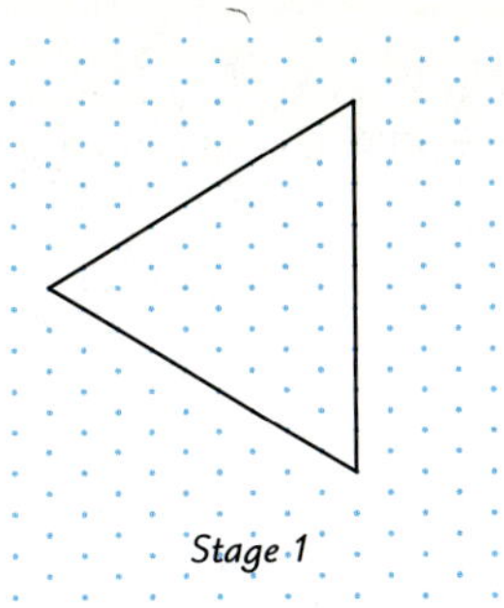

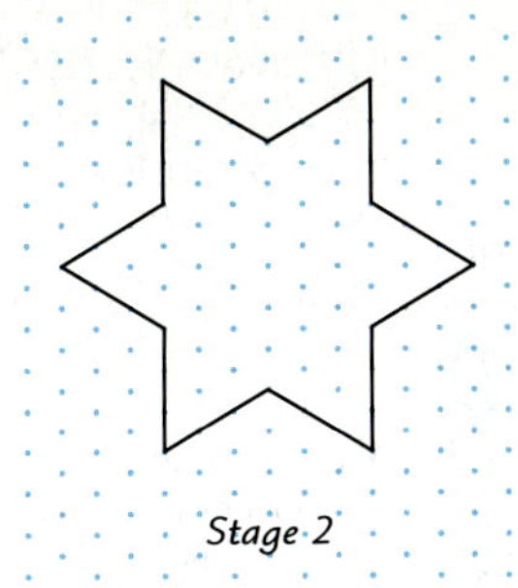

Imagine continuing this process indefinitely. The von Koch snowflake is the shape that these stages approach.

Model and Analyze

3. $s_n = 3 \cdot 4^{n-1}$, $\ell_n = 9\left(\frac{1}{3}\right)^{n-1}$, $P_n = 27\left(\frac{4}{3}\right)^{n-1}$

1. Copy and complete the table. Draw Stage 3, if necessary.

Stage	1	2	3	4
Number of Segments	3	12	48	192
Length of each Segment	9	3	1	$\frac{1}{3}$
Perimeter	27	36	48	64

2. $s_n = 4s_{n-1}$, $\ell_n = \frac{1}{3}\ell_{n-1}$, $P_n = \frac{4}{3}P_{n-1}$

2. Write recursive formulas for the number s_n of segments in Stage n, the length ℓ_n of each segment in Stage n, and the perimeter P_n of Stage n.
3. Write nonrecursive formulas for s_n, ℓ_n, and P_n.
4. What is the perimeter of the von Koch snowflake? Explain. **See pp. 629A–629F.**
5. Explain why the area of the von Koch snowflake can be represented by the infinite series $\frac{81\sqrt{3}}{4} + \frac{27\sqrt{3}}{4} + 3\sqrt{3} + \frac{4\sqrt{3}}{3} + \ldots$. **See pp. 629A–629F.**
6. Find the sum of the series in Exercise 5. Explain your steps. **See pp. 629A–629F.**
7. Do you think the results of Exercises 4 and 6 are contradictory? Explain. **See pp. 629A–629F.**

Resource Manager

Teaching Algebra with Manipulatives
- p. 19 (isometric dot paper)
- p. 284 (student recording sheet)

Glencoe Mathematics Classroom Manipulative Kit
- isometric dot grid stamp

Algebra Activity

A Follow-Up of Lesson 11-6

Getting Started

Objective To apply iterations to various aspects of the Koch snowflake fractal.

Materials
isometric dot paper

Teach

- Have students explore the difference between using dot paper and graph paper in terms of counting the units in the perimeter.
- Tell students that one of the characteristics of the Koch snowflake is that the area of the interior is finite but its perimeter is infinite. Invite them to explore fractals using the many related sites on the Internet.

Assess

In **Exercises 1–3,** students should
- be able to see the iterative nature of these data.
- make the generalizations that will form the parts of the formulas.

In **Exercises 4–7,** students should
- apply the formulas of the previous lessons.
- understand that fractals have mathematical characteristics that distinguish them from ordinary polygons.

Study Notebook

You may wish to have students summarize this activity and what they learned from it.

11-7 Lesson Notes

1 Focus

5-Minute Check Transparency 11-7 Use as a quiz or review of Lesson 11-6.

Mathematical Background notes are available for this lesson on p. 576D.

How does a power of a binomial describe the numbers of boys and girls in a family?

Ask students:

- In this problem, does the order make a difference, or is any family with 2 girls and 2 boys the same as any other? **Order does make a difference.**
- What does b^2g^2 represent in the triangle shown? **Any sequence with 2 boys and 2 girls.**

11-7 The Binomial Theorem

What You'll Learn

- Use Pascal's triangle to expand powers of binomials.
- Use the Binomial Theorem to expand powers of binomials.

Vocabulary

- Pascal's triangle
- Binomial Theorem
- factorial

How does a power of a binomial describe the numbers of boys and girls in a family?

According to the U.S. Census Bureau, ten percent of families have three or more children. If a family has four children, there are six sequences of births of boys and girls that result in two boys and two girls. These sequences are listed below.

BBGG BGBG BGGB GBBG GBGB GGBB

PASCAL'S TRIANGLE You can use the coefficients in powers of binomials to count the number of possible sequences in situations such as the one above. Remember that a binomial is a polynomial with two terms. Expand a few powers of the binomial $b + g$.

$$(b+g)^0 = 1b^0g^0$$
$$(b+g)^1 = 1b^1g^0 + 1b^0g^1$$
$$(b+g)^2 = 1b^2g^0 + 2b^1g^1 + 1b^0g^1$$
$$(b+g)^3 = 1b^3g^0 + 3b^2g^1 + 3b^1g^2 + 1b^0g^3$$
$$(b+g)^4 = 1b^4g^0 + 4b^3g^1 + 6b^2g^2 + 4b^1g^3 + 1b^0g^4$$

The coefficient 6 of the b^2g^2 term in the expansion of $(b + g)^4$ gives the number of sequences of births that result in two boys and two girls. As another example, the coefficient 4 of the b^1g^3 term gives the number of sequences with one boy and 3 girls.

More About. . .

Pascal's Triangle

Although he did not discover it, Pascal's triangle is named for the French mathematician Blaise Pascal (1623–1662).

Here are some patterns that can be seen in any binomial expansion of the form $(a + b)^n$.

1. There are $n + 1$ terms.
2. The exponent n of $(a + b)^n$ is the exponent of a in the first term and the exponent of b in the last term.
3. In successive terms, the exponent of a decreases by one, and the exponent of b increases by one.
4. The sum of the exponents in each term is n.
5. The coefficients are symmetric. They increase at the beginning of the expansion and decrease at the end.

The coefficients form a pattern that is often displayed in a triangular formation. This is known as **Pascal's triangle**. Notice that each row begins and ends with 1. Each coefficient is the sum of the two coefficients above it in the previous row.

$(a+b)^0$	1
$(a+b)^1$	1 1
$(a+b)^2$	1 2 1
$(a+b)^3$	1 3 3 1
$(a+b)^4$	1 4 6 4 1
$(a+b)^5$	1 5 10 10 5 1

Resource Manager

Workbook and Reproducible Masters

Chapter 11 Resource Masters

- Study Guide and Intervention, pp. 667–668
- Skills Practice, p. 669
- Practice, p. 670
- Reading to Learn Mathematics, p. 671
- Enrichment, p. 672

Transparencies

5-Minute Check Transparency 11-7
Answer Key Transparencies

Technology

Interactive Chalkboard

Example 1 Use Pascal's Triangle

Expand $(x + y)^7$.

Write two more rows of Pascal's triangle.

$$\begin{array}{ccccccccccccccc} & 1 & & 6 & & 15 & & 20 & & 15 & & 6 & & 1 & \\ 1 & & 7 & & 21 & & 35 & & 35 & & 21 & & 7 & & 1 \end{array}$$

Use the patterns of a binomial expansion and the coefficients to write the expansion of $(x + y)^7$.

$$(x + y)^7 = 1x^7y^0 + 7x^6y^1 + 21x^5y^2 + 35x^4y^3 + 35x^3y^4 + 21x^2y^5 + 7x^1y^6 + 1x^0y^7$$
$$= x^7 + 7x^6y + 21x^5y^2 + 35x^4y^3 + 35x^3y^4 + 21x^2y^5 + 7xy^6 + y^7$$

THE BINOMIAL THEOREM Another way to show the coefficients in a binomial expansion is to write them in terms of the previous coefficients.

$(a + b)^0$ $\quad 1$

$(a + b)^1$ $\quad 1 \quad \frac{1}{1}$

$(a + b)^2$ $\quad 1 \quad \frac{2}{1} \quad \frac{2 \cdot 1}{1 \cdot 2}$

$(a + b)^3$ $\quad 1 \quad \frac{3}{1} \quad \frac{3 \cdot 2}{1 \cdot 2} \quad \frac{3 \cdot 2 \cdot 1}{1 \cdot 2 \cdot 3}$

$(a + b)^4$ $\quad 1 \quad \frac{4}{1} \quad \frac{4 \cdot 3}{1 \cdot 2} \quad \frac{4 \cdot 3 \cdot 2}{1 \cdot 2 \cdot 3} \quad \frac{4 \cdot 3 \cdot 2 \cdot 1}{1 \cdot 2 \cdot 3 \cdot 4}$

Eliminate common factors that are shown in color.

This pattern provides the coefficients of $(a + b)^n$ for any nonnegative integer n. The pattern is summarized in the **Binomial Theorem**.

Key Concept — Binomial Theorem

If n is a nonnegative integer, then

$$(a + b)^n = 1a^nb^0 + \frac{n}{1}a^{n-1}b^1 + \frac{n(n-1)}{1 \cdot 2}a^{n-2}b^2 + \frac{n(n-1)(n-2)}{1 \cdot 2 \cdot 3}a^{n-3}b^3 + \ldots + 1a^0b^n.$$

Example 2 Use the Binomial Theorem

Expand $(a - b)^6$.

The expansion will have seven terms. Use the sequence $1, \frac{6}{1}, \frac{6 \cdot 5}{1 \cdot 2}, \frac{6 \cdot 5 \cdot 4}{1 \cdot 2 \cdot 3}$ to find the coefficients for the first four terms. Then use symmetry to find the remaining coefficients.

$$(a - b)^6 = 1a^6(-b)^0 + \frac{6}{1}a^5(-b)1 + \frac{6 \cdot 5}{1 \cdot 2}a^4(-b)^2 + \frac{6 \cdot 5 \cdot 4}{1 \cdot 2 \cdot 3}a^3(-b)^3 + \ldots + 1a^0(-b)^6$$
$$= a^6 - 6a^5b + 15a^4b^2 - 20a^3b^3 + 15a^2b^4 - 6ab^5 + b^6$$

Notice that in terms having the same coefficients, the exponents are reversed, as in $15a^4b^2$ and $15a^2b^4$.

Study Tip

Graphing Calculators

On a TI-83 Plus, the factorial symbol, !, is located on the MATH PRB menu.

The factors in the coefficients of binomial expansions involve special products called **factorials**. For example, the product $4 \cdot 3 \cdot 2 \cdot 1$ is written 4! and is read *4 factorial*. In general, if n is a positive integer, then $n! = n(n - 1)(n - 2)(n - 3) \ldots 2 \cdot 1$. *By definition, 0! = 1.*

www.algebra2.com/extra_examples

2 Teach

PASCAL'S TRIANGLE

In-Class Example Power Point®

1 Expand $(p + q)^5$. $p^5 + 5p^4q + 10p^3q^2 + 10p^2q^3 + 5pq^4 + q^5$

THE BINOMIAL THEOREM

In-Class Example Power Point®

2 Expand $(t - s)^8$. $t^8 - 8t^7s + 28t^6s^2 - 56t^5s^3 + 70t^4s^4 - 56t^3s^5 + 28t^2s^6 - 8ts^7 + s^8$

Teaching Tip Have students discuss each of the various patterns in these examples to make sure they see what happens with coefficients, exponents, and signs.

In-Class Examples

Power Point®

3 Evaluate $\frac{6!}{2!4!}$. **15**

Teaching Tip Encourage students to write the factors and simplify before they calculate.

4 Expand $(3x - y)^4$. $81x^4 - 108x^3y + 54x^2y^2 - 12xy^3 + y^4$

Teaching Tip Make sure students understand that $0! = 1$ by definition, and also that $1! = 1$.

Example 3 Factorials

Evaluate $\frac{8!}{3!5!}$.

$$\frac{8!}{3!5!} = \frac{8 \cdot 7 \cdot 6 \cdot 5 \cdot 4 \cdot 3 \cdot 2 \cdot 1}{3 \cdot 2 \cdot 1 \cdot 5 \cdot 4 \cdot 3 \cdot 2 \cdot 1}$$

Note that $8! = 8 \cdot 7 \cdot 6 \cdot 5!$, so $\frac{8!}{3!5!} = \frac{8 \cdot 7 \cdot 6 \cdot 5!}{3!5!}$ or $\frac{8 \cdot 7 \cdot 6}{3 \cdot 2 \cdot 1}$

$$= \frac{8 \cdot 7 \cdot 6}{3 \cdot 2 \cdot 1} \text{ or } 56$$

Study Tip

Missing Steps

If you don't understand a step like $\frac{6 \cdot 5 \cdot 4}{1 \cdot 2 \cdot 3} = \frac{6!}{3!3!}$, work it out on a piece of scrap paper.

$$\frac{6 \cdot 5 \cdot 4}{1 \cdot 2 \cdot 3} = \frac{6 \cdot 5 \cdot 4 \cdot 3!}{1 \cdot 2 \cdot 3 \cdot 3!} = \frac{6!}{3!3!}$$

An expression such as $\frac{6 \cdot 5 \cdot 4}{1 \cdot 2 \cdot 3}$ in Example 2 can be written as a quotient of factorials. In this case, $\frac{6 \cdot 5 \cdot 4}{1 \cdot 2 \cdot 3} = \frac{6!}{3!3!}$. Using this idea, you can rewrite the expansion of $(a + b)^6$ using factorials.

$$(a + b)^6 = \frac{6!}{6!0!}a^6b^0 + \frac{6!}{5!1!}a^5b^1 + \frac{6!}{4!2!}a^4b^2 + \frac{6!}{3!3!}a^3b^3 + \frac{6!}{2!4!}a^2b^4 + \frac{6!}{1!5!}a^1b^5 + \frac{6!}{0!6!}a^0b^6$$

You can also write this series using sigma notation.

$$(a + b)^6 = \sum_{k=0}^{6} \frac{6!}{(6 - k)!k!}a^{6-k}b^k$$

In general, the Binomial Theorem can be written both in factorial notation and in sigma notation.

Key Concept — *Binomial Theorem, Factorial Form*

$$(a + b)^n = \frac{n!}{n!0!}a^nb^0 + \frac{n!}{(n-1)!1!}a^{n-1}b^1 + \frac{n!}{(n-2)!2!}a^{n-2}b^2 + \ldots + \frac{n!}{0!n!}a^0b^n$$

$$= \sum_{k=0}^{n} \frac{n!}{(n-k)!k!}a^{n-k}b^k$$

Example 4 Use a Factorial Form of the Binomial Theorem

Expand $(2x + y)^5$.

$$(2x + y)^5 = \sum_{k=0}^{5} \frac{5!}{(5-k)!k!}(2x)^{5-k}y^k \quad \text{Binomial Theorem, factorial form}$$

$$= \frac{5!}{5!0!}(2x)^5y^0 + \frac{5!}{4!1!}(2x)^4y^1 + \frac{5!}{3!2!}(2x)^3y^2 + \frac{5!}{2!3!}(2x)^2y^3 + \frac{5!}{1!4!}(2x)^1y^4 + \frac{5!}{0!5!}(2x)^0y^5 \quad \text{Let } k = 0, 1, 2, 3, 4, \text{ and } 5.$$

$$= \frac{5 \cdot 4 \cdot 3 \cdot 2 \cdot 1}{5 \cdot 4 \cdot 3 \cdot 2 \cdot 1 \cdot 1}(2x)^5 + \frac{5 \cdot 4 \cdot 3 \cdot 2 \cdot 1}{4 \cdot 3 \cdot 2 \cdot 1 \cdot 1}(2x)^4 y + \frac{5 \cdot 4 \cdot 3 \cdot 2 \cdot 1}{3 \cdot 2 \cdot 1 \cdot 2 \cdot 1}(2x)^3 y^2 + \frac{5 \cdot 4 \cdot 3 \cdot 2 \cdot 1}{2 \cdot 1 \cdot 3 \cdot 2 \cdot 1}(2x)^2 y^3 + \frac{5 \cdot 4 \cdot 3 \cdot 2 \cdot 1}{1 \cdot 4 \cdot 3 \cdot 2 \cdot 1}(2x)y^4 + \frac{5 \cdot 4 \cdot 3 \cdot 2 \cdot 1}{1 \cdot 5 \cdot 4 \cdot 3 \cdot 2 \cdot 1}y^5$$

$$= 32x^5 + 80x^4y + 80x^3y^2 + 40x^2y^3 + 10xy^4 + y^5 \quad \text{Simplify.}$$

TEACHING TIP

A common mistake students make is forgetting to evaluate $(2x)^5$ as 2^5x^5. Inserting the line $1(32x^5) + 5(16x^4)y + 10(8x^3)y^2 + 10(4x^2)y^3 + 5(2x)y^4 + y^5$ might help.

Sometimes you need to know only a particular term of a binomial expansion. Note that when the Binomial Theorem is written in sigma notation, $k = 0$ for the first term, $k = 1$ for the second term, and so on. In general, the value of k is always one less than the number of the term you are finding.

Answers

7. $p^5 + 5p^4q + 10p^3q^2 + 10p^2q^3 + 5pq^4 + q^5$
8. $t^6 + 12t^5 + 60t^4 + 160t^3 + 240t^2 + 192t + 64$
9. $x^4 - 12x^3y + 54x^2y^2 - 108xy^3 + 81y^4$
19. $a^3 - 3a^2b + 3ab^2 - b^3$
20. $m^4 + 4m^3n + 6m^2n^2 + 4mn^3 + n^4$
21. $r^8 + 8r^7s + 28r^6s^2 + 56r^5s^3 + 70r^4s^4 + 56r^3s^5 + 28r^2s^6 + 8rs^7 + s^8$
22. $m^5 - 5m^4a + 10m^3a^2 - 10m^2a^3 + 5ma^4 - a^5$
23. $x^5 + 15x^4 + 90x^3 + 270x^2 + 405x + 243$
24. $a^4 - 8a^3 + 24a^2 - 32a + 16$
25. $16b^4 - 32b^3x + 24b^2x^2 - 8bx^3 + x^4$
26. $64a^6 + 192a^5b + 240a^4b^2 + 160a^3b^3 + 60a^2b^4 + 12ab^5 + b^6$
27. $243x^5 - 810x^4y + 1080x^3y^2 - 720x^2y^3 + 240xy^4 - 32y^5$
28. $81x^4 + 216x^3y + 216x^2y^2 + 96xy^3 + 16y^4$
29. $\frac{a^5}{32} + \frac{5a^4}{8} + 5a^3 + 20a^2 + 40a + 32$
30. $243 + 135m + 30m^2 + \frac{10m^3}{3} + \frac{5m^4}{27} + \frac{m^5}{243}$

Example 5 Find a Particular Term

Find the fifth term in the expansion of $(p + q)^{10}$.

First, use the Binomial Theorem to write the expansion in sigma notation.

$$(p + q)^{10} = \sum_{k=0}^{10} \frac{10!}{(10 - k)!k!} p^{10 - k} q^k$$

In the fifth term, $k = 4$.

$$\frac{10!}{(10-k)!k!} p^{10 - k} q^k = \frac{10!}{(10-4)!4!} p^{10 - 4} q^4 \quad k = 4$$

$$= \frac{10 \cdot 9 \cdot 8 \cdot 7}{4 \cdot 3 \cdot 2 \cdot 1} p^6 q^4 \quad \frac{10!}{6!4!} = \frac{10 \cdot 9 \cdot 8 \cdot 7 \cdot 6!}{6!4!} \text{ or } \frac{10 \cdot 9 \cdot 8 \cdot 7}{4 \cdot 3 \cdot 2 \cdot 1}$$

$$= 210p^6q^4 \quad \text{Simplify.}$$

In-Class Example

Power Point®

5 Find the fourth term in the expansion of $(a + 3b)^4$. $108ab^3$

3 Practice/Apply

Check for Understanding

Concept Check

1. **List** the coefficients in the row of Pascal's triangle corresponding to $n = 8$.
2. **Identify** the coefficient of $a^{n-1}b$ in the expansion of $(a + b)^n$. n
3. **OPEN ENDED** Write a power of a binomial for which the first term of the expansion is $625x^4$. **Sample answer: $(5x + y)^4$**

1. 1, 8, 28, 56, 70, 56, 28, 8, 1

Guided Practice

GUIDED PRACTICE KEY	
Exercises	Examples
4–6	3
7–9, 12	1, 2, 4
10, 11	5

Evaluate each expression.

4. $8!$ **40,320**
5. $\frac{13!}{9!}$ **17,160**
6. $\frac{12!}{2!10!}$ **66**

Expand each power. **7–9. See margin.**

7. $(p + q)^5$
8. $(t + 2)^6$
9. $(x - 3y)^4$

Find the indicated term of each expansion.

10. fourth term of $(a + b)^8$ $56a^5b^3$
11. fifth term of $(2a + 3b)^{10}$ $1{,}088{,}640a^6b^4$

Application

12. **SCHOOL** Mr. Hopkins is giving a five-question true-false quiz. How many ways could a student answer the questions with three trues and two falses? **10**

★ indicates increased difficulty

Practice and Apply

Homework Help	
For Exercises	See Examples
13–18	3
19–33	1, 2, 4
34–41	5

Extra Practice See page 853.

Evaluate each expression.

13. $9!$ **362,880**
14. $13!$ **6,227,020,800**
15. $\frac{9!}{7!}$ **72**
16. $\frac{7!}{4!}$ **210**
17. $\frac{12!}{8!4!}$ **495**
18. $\frac{14!}{5!9!}$ **2002**

Expand each power. **19-30. See margin.**

19. $(a - b)^3$
20. $(m + n)^4$
21. $(r + s)^8$
22. $(m - a)^5$
23. $(x + 3)^5$
24. $(a - 2)^4$
25. $(2b - x)^4$
26. $(2a + b)^6$
27. $(3x - 2y)^5$
28. $(3x + 2y)^4$
★ 29. $\left(\frac{a}{2} + 2\right)^5$
★ 30. $\left(3 + \frac{m}{3}\right)^5$

31. **GEOMETRY** Write an expanded expression for the volume of the cube at the right. $27x^3 + 54x^2 + 36x + 8 \text{ cm}^3$

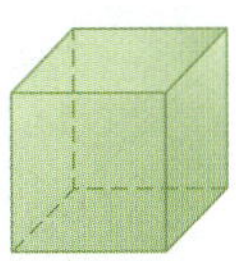

3x + 2 cm

www.algebra2.com/self_check_quiz

Study Notebook

Have students—
- add the definitions/examples of the vocabulary terms to their Vocabulary Builder worksheets for Chapter 11.
- add the Study Tip on p. 613 to their list of tips about the graphing calculator.
- include any other item(s) that they find helpful in mastering the skills in this lesson.

About the Exercises...

Organization by Objective
- **Pascal's Triangle:** 19–22, 34–41
- **The Binomial Theorem:** 13–18, 23–33

Odd/Even Assignments
Exercises 13–30 and 34–41 are structured so that students practice the same concepts whether they are assigned odd or even problems.

Assignment Guide

Basic: 13–27 odd, 31–39 odd, 42–62

Average: 13–39 odd, 41–62

Advanced: 14–40 even, 42–58 (optional: 59–62)

All: Practice Quiz 2 (1–10)

DAILY INTERVENTION — Differentiated Instruction — ELL

Verbal/Linguistic Have pairs of students work together to make up a jingle or a poem that describes the patterns in the Binomial Theorem. The poem should describe at least three of the five patterns in the Binomial Theorem described on p. 612 of the Student Edition.

WebQuest

Pascal's triangle displays many patterns. Visit www.algebra2.com/webquest to continue work on your WebQuest project.

32. **GAMES** The diagram shows the board for a game in which ball bearings are dropped down a chute. A pattern of nails and dividers causes the bearings to take various paths to the sections at the bottom. For each section, how many paths through the board lead to that section? **1, 4, 6, 4, 1**

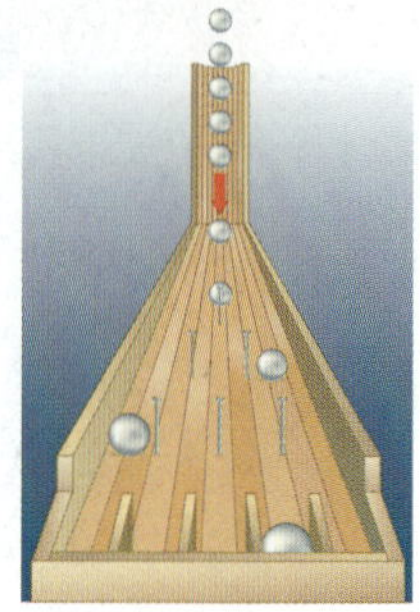

33. **INTRAMURALS** Ofelia is taking ten shots in the intramural free-throw shooting competition. How many sequences of makes and misses are there that result in her making eight shots and missing two? **45**

Find the indicated term of each expansion.

34. sixth term of $(x - y)^9$ $-126x^4y^5$
35. seventh term of $(x + y)^{12}$ $924x^6y^6$
36. fourth term of $(x + 2)^7$ $280x^4$
37. fifth term of $(a - 3)^8$ $5670a^4$
38. fifth term of $(2a + 3b)^{10}$ $1{,}088{,}640a^6b^4$
39. fourth term of $(2x + 3y)^9$ $145{,}152x^6y^3$
★ 40. fourth term of $\left(x + \frac{1}{3}\right)^7$ $\frac{35}{27}x^4$
★ 41. sixth term of $\left(x - \frac{1}{2}\right)^{10}$ $-\frac{63}{8}x^5$

42. **CRITICAL THINKING** Explain why $\frac{12!}{7!5!} + \frac{12!}{6!6!} = \frac{13!}{7!6!}$ without finding the value of any of the expressions. **See pp. 629A–629F.**

43. **WRITING IN MATH** Answer the question that was posed at the beginning of the lesson. **See pp. 629A–629F.**

How does a power of a binomial describe the numbers of boys and girls in a family?

Include the following in your answer:

- the expansion of $(b + g)^5$ and what it tells you about sequences of births of boys and girls in families with five children, and
- an explanation of how to find a formula for the number of sequences of births that have exactly k girls in a family of n children.

Standardized Test Practice

44. Which of the following represents the values of x that are solutions of the inequality $x^2 < x + 20$? **D**

(A) $x > -4$
(B) $x < 5$
(C) $-5 < x < 4$
(D) $-4 < x < 5$

45. If four lines intersect as shown in the figure at the right, $x + y =$ **C**

(A) 70.
(B) 115.
(C) 140.
(D) It cannot be determined from the information given.

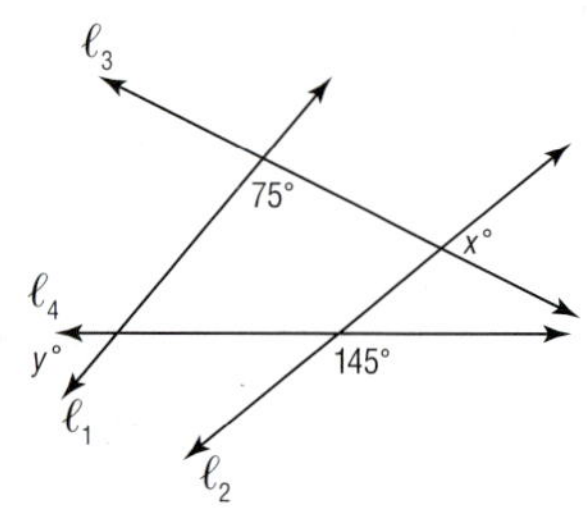

Study Guide and Intervention, p. 667 (shown) and p. 668

Pascal's Triangle **Pascal's triangle** is the pattern of coefficients of powers of binomials displayed in triangular form. Each row begins and ends with 1 and each coefficient is the sum of the two coefficients above it in the previous row.

Pascal's Triangle		
	$(a + b)^0$	1
	$(a + b)^1$	1 1
	$(a + b)^2$	1 2 1
	$(a + b)^3$	1 3 3 1
	$(a + b)^4$	1 4 6 4 1
	$(a + b)^5$	1 5 10 10 5 1

Example **Use Pascal's triangle to find the number of possible sequences consisting of 3 *a*s and 2 *b*s.**

The coefficient 10 of the a^3b^2-term in the expansion of $(a + b)^5$ gives the number of sequences that result in three *a*s and two *b*s.

Exercises

Expand each power using Pascal's triangle.

1. $(a + 5)^4$ $a^4 + 20a^3 + 150a^2 + 500a + 625$
2. $(x - 2y)^6$ $x^6 - 12x^5y + 60x^4y^2 - 160x^3y^3 + 240x^2y^4 - 192xy^5 + 64y^6$
3. $(j - 3k)^5$ $j^5 - 15j^4k + 90j^3k^2 - 270j^2k^3 + 405jk^4 - 243k^5$
4. $(2s + t)^7$ $128s^7 + 448s^6t + 672s^5t^2 + 560s^4t^3 + 280s^3t^4 + 84s^2t^5 + 14st^6 + t^7$
5. $(2p + 3q)^6$ $64p^6 + 576p^5q + 2160p^4q^2 + 4320p^3q^3 + 4860p^2q^4 + 2916pq^5 + 729q^6$
6. $\left(a - \frac{b}{2}\right)^4$ $a^4 - 2a^3b + \frac{3}{2}a^2b^2 - \frac{1}{2}ab^3 + \frac{1}{16}b^4$
7. Ray tosses a coin 15 times. How many different sequences of tosses could result in 4 heads and 11 tails? **1365**
8. There are 9 true/false questions on a quiz. If twice as many of the statements are true as false, how many different sequences of true/false answers are possible? **84**

Skills Practice, p. 669 and Practice, p. 670 (shown)

Evaluate each expression.

1. $7!$ 5040
2. $11!$ 39,916,800
3. $\frac{9!}{5!}$ 3024
4. $\frac{20!}{18!}$ 380
5. $\frac{8!}{6!2!}$ 28
6. $\frac{8!}{5!3!}$ 56
7. $\frac{12!}{6!6!}$ 924
8. $\frac{41!}{3!38!}$ 10,660

Expand each power.

9. $(n + v)^5$ $n^5 + 5n^4v + 10n^3v^2 + 10n^2v^3 + 5nv^4 + v^5$
10. $(x - y)^4$ $x^4 - 4x^3y + 6x^2y^2 - 4xy^3 + y^4$
11. $(x + y)^6$ $x^6 + 6x^5y + 15x^4y^2 + 20x^3y^3 + 15x^2y^4 + 6xy^5 + y^6$
12. $(r + 3)^5$ $r^5 + 15r^4 + 90r^3 + 270r^2 + 405r + 243$
13. $(m - 5)^5$ $m^5 - 25m^4 + 250m^3 - 1250m^2 + 3125m - 3125$
14. $(x + 4)^4$ $x^4 + 16x^3 + 96x^2 + 256x + 256$
15. $(3x + y)^4$ $81x^4 + 108x^3y + 54x^2y^2 + 12xy^3 + y^4$
16. $(2m - y)^4$ $16m^4 - 32m^3y + 24m^2y^2 - 8my^3 + y^4$
17. $(w - 3z)^3$ $w^3 - 9w^2z + 27wz^2 - 27z^3$
18. $(2d + 3)^6$ $64d^6 + 576d^5 + 2160d^4 + 4320d^3 + 4860d^2 + 2916d + 729$
19. $(x + 2y)^5$ $x^5 + 10x^4y + 40x^3y^2 + 80x^2y^3 + 80xy^4 + 32y^5$
20. $(2x - y)^5$ $32x^5 - 80x^4y + 80x^3y^2 - 40x^2y^3 + 10xy^4 - y^5$
21. $(a - 3b)^4$ $a^4 - 12a^3b + 54a^2b^2 - 108ab^3 + 81b^4$
22. $(3 - 2z)^4$ $16z^4 - 96z^3 + 216z^2 - 216z + 81$
23. $(3m - 4n)^3$ $27m^3 - 108m^2n + 144mn^2 - 64n^3$
24. $(5x - 2y)^4$ $625x^4 - 1000x^3y + 600x^2y^2 - 160xy^3 + 16y^4$

Find the indicated term of each expansion.

25. seventh term of $(a + b)^{10}$ $210a^4b^6$
26. sixth term of $(m - n)^{10}$ $-252m^5n^5$
27. ninth term of $(r - s)^{14}$ $3003r^6s^8$
28. tenth term of $(2x + y)^{12}$ $1760x^3y^9$
29. fourth term of $(x - 3y)^6$ $-540x^3y^3$
30. fifth term of $(2x - 1)^9$ $4032x^5$
31. **GEOMETRY** How many line segments can be drawn between ten points, no three of which are collinear, if you use exactly two of the ten points to draw each segment? **45**
32. **PROBABILITY** If you toss a coin 4 times, how many different sequences of tosses will give exactly 3 heads and 1 tail or exactly 1 head and 3 tails? **8**

Reading to Learn Mathematics, p. 671 ELL

Pre-Activity **How does a power of a binomial describe the numbers of boys and girls in a family?**

Read the introduction to Lesson 11-7 at the top of page 612 in your textbook.

- If a family has four children, list the sequences of births of girls and boys that result in three girls and one boy. **BGGG GBGG GGBG GGGB**
- Describe a way to figure out how many such sequences there are without listing them. **Sample answer: The boy could be the first, second, third, or fourth child, so there are four sequences with three girls and one boy.**

Reading the Lesson

1. Consider the expansion of $(w + z)^5$.
   a. How many terms does this expansion have? **6**
   b. In the second term of the expansion, what is the exponent of w? **4**
   What is the exponent of z? **1**
   What is the coefficient of the second term? **5**
   c. In the fourth term of the expansion, what is the exponent of w? **2**
   What is the exponent of z? **3**
   What is the coefficient of the fourth term? **10**
   d. What is the last term of this expansion? z^5
2. a. State the definition of a *factorial* in your own words. (Do not use mathematical symbols in your definition.) **Sample answer: The factorial of any positive integer is the product of that integer and all the smaller integers down to one. The factorial of zero is one.**
   b. Write out the product that you would use to calculate 10!. (Do not actually calculate the product.) $10 \cdot 9 \cdot 8 \cdot 7 \cdot 6 \cdot 5 \cdot 4 \cdot 3 \cdot 2 \cdot 1$
   c. Write an expression involving factorials that could be used to find the coefficient of the third term of the expansion of $(m - n)^6$. (Do not actually calculate the coefficient.) $\frac{6!}{4!2!}$

Helping You Remember

3. Without using Pascal's triangle or factorials, what is an easy way to remember the first two and last two coefficients for the terms of the binomial expansion of $(a + b)^n$? **Sample answer: The first and last coefficients are always 1. The second and next-to-last coefficients are always *n*, the power to which the binomial is being raised.**

Enrichment, p. 672

Patterns in Pascal's Triangle

You have learned that the coefficients in the expansion of $(x + y)^n$ yield a number pyramid called **Pascal's triangle.**

Row	
Row 1 →	1
Row 2 →	1 1
Row 3 →	1 2 1
Row 4 →	1 3 4 1
Row 5 →	1 4 6 4 1
Row 6 →	1 5 10 10 5 1
Row 7 →	1 6 15 20 15 6 1

As many rows can be added to the bottom of the pyramid as you please.

This activity explores some of the interesting properties of this famous number pyramid.

1. Pick a row of Pascal's triangle.
   a. What is the sum of all the numbers in all the rows *above* the row
   See studen

Maintain Your Skills

Mixed Review Find the first five terms of each sequence. *(Lesson 11-6)*

46. $a_1 = 7, a_{n+1} = a_n - 2$ **7, 5, 3, 1, −1** 47. $a_1 = 3, a_{n+1} = 2a_n - 1$ **3, 5, 9, 17, 33**

48. **CLOCKS** The spring in Juanita's old grandfather clock is broken. When you try to set the pendulum in motion by holding it against the wall of the clock and letting go, it follows a swing pattern of 25 centimeters, 20 centimeters, 16 centimeters, and so on until it comes to rest. What is the total distance the pendulum swings before coming to rest? *(Lesson 11-5)* **125 cm**

Express each logarithm in terms of common logarithms. Then approximate its value to four decimal places. *(Lesson 10-4)*

49. $\log_2 5$ $\frac{\log 5}{\log 2}$; **2.3219** 50. $\log_3 10$ $\frac{1}{\log 3}$; **2.0959** 51. $\log_5 8$ $\frac{\log 8}{\log 5}$; **1.2920**

Determine any vertical asymptotes and holes in the graph of each rational function. *(Lesson 9-3)* **54. hole: $x = -3$**

52. $f(x) = \frac{1}{x^2 + 5x + 6}$ 53. $f(x) = \frac{x + 2}{x^2 + 3x - 4}$ 54. $f(x) = \frac{x^2 + 4x + 3}{x + 3}$

52. asymptotes: $x = -2, x = -3$

53. asymptotes: $x = -4, x = 1$

Without writing the equation in standard form, state whether the graph of each equation is a *parabola*, *circle*, *ellipse*, or *hyperbola*. *(Lesson 8-6)*

55. $x^2 - 6x - y^2 - 3 = 0$ **hyperbola** 56. $4y - x + y^2 = 1$ **parabola**

Determine whether each pair of functions are inverse functions. *(Lesson 7-8)*

57. $f(x) = x + 3$, $g(x) = x - 3$ **yes** 58. $f(x) = 2x + 1$, $g(x) = \frac{x + 1}{2}$ **no**

Getting Ready for the Next Lesson **PREREQUISITE SKILL** State whether each statement is *true* or *false* when $n = 1$. Explain. *(To review **evaluating expressions**, see Lesson 1-1.)*

59–62. See margin for explanations.

59. $1 = \frac{n(n + 1)}{2}$ **true** 60. $1 = \frac{(n + 1)(2n + 1)}{2}$ **false**

61. $1 = \frac{n^2(n + 1)^2}{4}$ **true** 62. $3^n - 1$ is even. **true**

Practice Quiz 2 — *Lessons 11-4 through 11-7*

Find the sum of each geometric series. *(Lessons 11-4 and 11-5)*

1. $a_1 = 5, r = 3, n = 12$ **1,328,600** 2. $\sum_{n=1}^{6} 2(-3)^{n-1}$ **−364**

3. $\sum_{n=1}^{\infty} 8\left(\frac{2}{3}\right)^{n-1}$ **24** 4. $5 + 1 + \frac{1}{5} + \ldots$ $\frac{25}{4}$

Find the first five terms of each sequence. *(Lesson 11-6)*

5. $a_1 = 1, a_{n+1} = 2a_n + 3$ **1, 5, 13, 29, 61** 6. $a_1 = 2, a_{n+1} = a_n + 2n$ **2, 4, 8, 14, 22**

7. Find the first three iterates of the function $f(x) = -3x + 2$ for an initial value of $x_0 = -1$. *(Lesson 11-6)* **5, −13, 41**

Expand each power. *(Lesson 11-7)* **8. $243x^5 + 405x^4y + 270x^3y^2 + 90x^2y^3 + 15xy^4 + y^5$**

8. $(3x + y)^5$ 9. $(a + 2)^6$ **$a^6 + 12a^5 + 60a^4 + 160a^3 + 240a^2 + 192a + 64$**

10. Find the fifth term of the expansion of $(2a + b)^9$. *(Lesson 11-7)* **$4032a^5b^4$**

4 Assess

Open-Ended Assessment

Writing Have students write their own example of a binomial expansion, using colored pens or pencils to emphasize the patterns.

Tips for New Teachers

Intervention Invite students to share their confusions about this material in small groups or with a partner in order to clear up any problems.

Getting Ready for Lesson 11-8

PREREQUISITE SKILL Students will prove statements using mathematical induction in Lesson 11-8. This will include their showing that a statement is true for $n = 1$ by evaluating an equation for that value. Use Exercises 59–62 to determine your students' familiarity with evaluating equations for a given value.

Assessment Options

Practice Quiz 2 The quiz provides students with a brief review of the concepts and skills in Lessons 11-4 through 11-7. Lesson numbers are given to the right of the exercises or instruction lines so students can review concepts not yet mastered.

Answers

59. $\frac{1(1 + 1)}{2} = \frac{1(2)}{2}$ or 1

60. $\frac{(1 + 1)(2 \cdot 1 + 1)}{2} = \frac{2(3)}{2}$ or 3

61. $\frac{1^2(1 + 1)^2}{4} = \frac{1(4)}{4}$ or 1

62. $3^1 - 1 = 2$, which is even

11-8 Lesson Notes

1 Focus

5-Minute Check Transparency 11-8 Use as a quiz or review of Lesson 11-7.

Mathematical Background notes are available for this lesson on p. 576D.

How does the concept of a ladder help you prove statements about numbers?

Ask students:

Why is it not enough to prove only Step 2 and Step 3? **Steps 2 and 3 prove the statement for the next integer, given that it is true for some integer. You must prove the statement for some specific value of n in order to prove that the statement is true for any value of k that is greater than or equal to n.**

11-8 Proof and Mathematical Induction

What You'll Learn

- Prove statements by using mathematical induction.
- Disprove statements by finding a counterexample.

How does the concept of a ladder help you prove statements about numbers?

Imagine the positive integers as a ladder that goes upward forever. You know that you cannot leap to the top of the ladder, but you can stand on the first step, and no matter which step you are on, you can always climb one step higher. Is there any step you cannot reach?

Vocabulary

- mathematical induction
- inductive hypothesis

MATHEMATICAL INDUCTION **Mathematical induction** is used to prove statements about positive integers. An induction proof consists of three steps.

Study Tip

Step 1
In many cases, it will be helpful to let $n = 1$.

Key Concept — Mathematical Induction

Step 1 Show that the statement is true for some integer n.

Step 2 Assume that the statement is true for some positive integer k, where $k \geq n$. This assumption is called the **inductive hypothesis**.

Step 3 Show that the statement is true for the next integer $k + 1$.

Example 1 Summation Formula

Prove that the sum of the squares of the first n positive integers is $\frac{n(n+1)(2n+1)}{6}$. That is, prove that $1^2 + 2^2 + 3^2 + \ldots + n^2 = \frac{n(n+1)(2n+1)}{6}$.

Step 1 When $n = 1$, the left side of the given equation is 1^2 or 1. The right side is $\frac{1(1+1)[2(1)+1]}{6}$ or 1. Thus, the equation is true for $n = 1$.

Step 2 Assume $1^2 + 2^2 + 3^2 + \ldots + k^2 = \frac{k(k+1)(2k+1)}{6}$ for a positive integer k.

Step 3 Show that the given equation is true for $n = k + 1$.

$$1^2 + 2^2 + 3^2 + \ldots + k^2 + (k+1)^2 = \frac{k(k+1)(2k+1)}{6} + (k+1)^2 \quad \text{Add } (k+1)^2 \text{ to each side.}$$

$$= \frac{k(k+1)(2k+1) + 6(k+1)^2}{6} \quad \text{Add.}$$

$$= \frac{(k+1)[k(2k+1) + 6(k+1)]}{6} \quad \text{Factor.}$$

$$= \frac{(k+1)[2k^2 + 7k + 6]}{6} \quad \text{Simplify.}$$

$$= \frac{(k+1)(k+2)(2k+3)}{6} \quad \text{Factor.}$$

$$= \frac{(k+1)[(k+1)+1][2(k+1)+1]}{6}$$

Resource Manager

Workbook and Reproducible Masters

Chapter 11 Resource Masters

- Study Guide and Intervention, pp. 673–674
- Skills Practice, p. 675
- Practice, p. 676
- Reading to Learn Mathematics, p. 677
- Enrichment, p. 678
- Assessment, p. 694

Transparencies

5-Minute Check Transparency 11-8
Answer Key Transparencies

Technology

Interactive Chalkboard

The last expression on page 618 is the right side of the equation to be proved, where n has been replaced by $k + 1$. Thus, the equation is true for $n = k + 1$.

This proves that $1^2 + 2^2 + 3^2 + \ldots + n^2 = \frac{n(n+1)(2n+1)}{6}$ for all positive integers n.

Example 2 Divisibility

Prove that $7^n - 1$ is divisible by 6 for all positive integers n.

Step 1 When $n = 1$, $7^n - 1 = 7^1 - 1$ or 6. Since 6 is divisible by 6, the statement is true for $n = 1$.

Step 2 Assume that $7^k - 1$ is divisible by 6 for some positive integer k. This means that there is a whole number r such that $7^k - 1 = 6r$.

Step 3 Show that the statement is true for $n = k + 1$.

$7^k - 1 = 6r$ Inductive hypothesis

$7^k = 6r + 1$ Add 1 to each side.

$7(7^k) = 7(6r + 1)$ Multiply each side by 7.

$7^{k+1} = 42r + 7$ Simplify.

$7^{k+1} - 1 = 42r + 6$ Subtract 1 from each side.

$7^{k+1} - 1 = 6(7r + 1)$ Factor.

Since r is a whole number, $7r + 1$ is a whole number. Therefore, $7^{k+1} - 1$ is divisible by 6. Thus, the statement is true for $n = k + 1$.

This proves that $7^n - 1$ is divisible by 6 for all positive integers n.

TEACHING TIP

Point out that, since k is a positive integer, $7^k - 1$ is a positive integer. Therefore, r must also be a positive integer. This type of reasoning will help students analyze the factorizations that they obtain in Exercises 7, 19, and 20.

COUNTEREXAMPLES

Of course, not every formula that you can write is true. A formula that works for a few positive integers may not work for *every* positive integer. You can show that a formula is not true by finding a *counterexample*. This often involves trial and error.

Study Tip

Reading Math

One of the meanings of *counter* is *to oppose*, so a counterexample is an example that opposes a hypothesis.

Example 3 Counterexample

Find a counterexample for the formula $1^4 + 2^4 + 3^4 + \ldots + n^4 = 1 + (4n - 4)^2$.

Check the first few positive integers.

n	Left Side of Formula	Right Side of Formula	
1	1^4 or 1	$1 + [4(1) - 4]^2 = 1 + 0^2$ or 1	true
2	$1^4 + 2^4 = 1 + 16$ or 17	$1 + [4(2) - 4]^2 = 1 + 4^2$ or 17	true
3	$1^4 + 2^4 + 3^4 = 1 + 16 + 81$ or 98	$1 + [4(3) - 4]^2 = 1 + 64$ or 65	false

The value $n = 3$ is a counterexample for the formula.

Check for Understanding

Concept Check

1–2. See pp. 629A–629F.

1. **Describe** some of the types of statements that can be proved by using mathematical induction.

2. **Explain** the difference between mathematical induction and a counterexample.

3. **OPEN ENDED** Write an expression of the form $b^n - 1$ that is divisible by 2 for all positive integers n. **Sample answer: $3^n - 1$**

www.algebra2.com/extra_examples

2 Teach

MATHEMATICAL INDUCTION

In-Class Examples Power Point®

1 Prove that $1 + 3 + 5 + \ldots + (2n - 1) = n^2$. **Step 1: When $n = 1$, the left side of the given equation is 1. The right side is 1^2. Since $1 = 1^2$, the equation is true for $n = 1$. Step 2: Assume $1 + 3 + 5 + \ldots + (2k - 1) = k^2$ for a positive integer k. Step 3: Does $1 + 3 + 5 + \ldots + [2(k + 1) - 1] = (k + 1)^2$? Yes, the left side simplifies to $k^2 + 2k + 1$ or $(k + 1)^2$ which is equal to the right side.**

2 Prove that $6^n - 1$ is divisible by 5 for all positive integers n. **Proof uses steps similar to those in Example 2 in the Student Edition.**

COUNTEREXAMPLES

In-Class Example Power Point®

3 Find a counterexample for the formula that $n^2 + n + 5$ is always a prime number for any positive integer n. **$n = 4$**

3 Practice/Apply

Study Notebook

Have students—
- complete the definitions/examples for the remaining terms on their Vocabulary Builder worksheets for Chapter 11.
- include any other item(s) that they find helpful in mastering the skills in this lesson.

DAILY INTERVENTION

Differentiated Instruction

Visual/Spatial Have students demonstrate proof by induction by laying out a "train" of dominoes. Have them relate the steps in an inductive proof to the requirements that (1) the first domino must fall and (2) if any one domino falls, the next one must fall.

About the Exercises...

Organization by Objective
- **Mathematical Induction:** 11–24
- **Counterexamples:** 25–30

Odd/Even Assignments
Exercises 11–20 and 25–30 are structured so that students practice the same concepts whether they are assigned odd or even problems.

Assignment Guide

Basic: 11–27 odd, 31–42
Average: 11–29 odd, 31–42
Advanced: 12–30 even, 31–42

4 Assess

Open-Ended Assessment

Speaking Have students explain how you can prove or disprove statements by using induction and counterexamples.

Tips for New Teachers

Intervention Use simple examples to help students understand the principles of an inductive proof before they get involved in elaborate calculations.

Assessment Options

Quiz (Lessons 11-7 and 11-8) is available on p. 694 of the *Chapter 11 Resource Masters*.

Guided Practice

GUIDED PRACTICE KEY	
Exercises	Examples
4, 5, 10	1
6, 7	2
8, 9	3

Prove that each statement is true for all positive integers. 4–7. See pp. 629A–629F.

4. $1 + 2 + 3 + \ldots + n = \frac{n(n+1)}{2}$
5. $\frac{1}{2} + \frac{1}{2^2} + \frac{1}{2^3} + \ldots + \frac{1}{2^n} = 1 - \frac{1}{2^n}$
6. $4^n - 1$ is divisible by 3.
7. $5^n + 3$ is divisible by 4.

Find a counterexample for each statement.

8. $1 + 2 + 3 + \ldots + n = n^2$ Sample answer: $n = 2$
9. $2^n + 2n$ is divisible by 4. Sample answer: $n = 3$

Application

10. **PARTIES** Suppose that each time a new guest arrives at a party, he or she shakes hands with each person already at the party. Prove that after n guests have arrived, a total of $\frac{n(n-1)}{2}$ handshakes have taken place. See pp. 629A–629F.

★ indicates increased difficulty

Practice and Apply

Homework Help	
For Exercises	See Examples
11–23, 31	1
24	1, 2
25–30	3

Extra Practice See page 853.

Prove that each statement is true for all positive integers.

11. $1 + 5 + 9 + \ldots + (4n - 3) = n(2n - 1)$ 11–20. See pp. 629A–629F.
12. $2 + 5 + 8 + \ldots + (3n - 1) = \frac{n(3n+1)}{2}$
13. $1^3 + 2^3 + 3^3 + \ldots + n^3 = \frac{n^2(n+1)^2}{4}$
14. $1^2 + 3^2 + 5^2 + \ldots + (2n-1)^2 = \frac{n(2n-1)(2n+1)}{3}$
15. $\frac{1}{3} + \frac{1}{3^2} + \frac{1}{3^3} + \ldots + \frac{1}{3^n} = \frac{1}{2}\left(1 - \frac{1}{3^n}\right)$
16. $\frac{1}{4} + \frac{1}{4^2} + \frac{1}{4^3} + \ldots + \frac{1}{4^n} = \frac{1}{3}\left(1 - \frac{1}{4^n}\right)$
17. $8^n - 1$ is divisible by 7.
18. $9^n - 1$ is divisible by 8.
19. $12^n + 10$ is divisible by 11.
20. $13^n + 11$ is divisible by 12.

21. **ARCHITECTURE** A memorial being constructed in a city park will be a brick wall, with a top row of six gold-plated bricks engraved with the names of six local war veterans. Each row has two more bricks than the row above it. Prove that the number of bricks in the top n rows is $n^2 + 5n$. See pp. 629A–629F.

22. **GEOMETRIC SERIES** Use mathematical induction to prove the formula $a_1 + a_1r + a_1r^2 + \ldots + a_1r^{n-1} = \frac{a_1(1 - r^n)}{1 - r}$ for the sum of a finite geometric series.

23. **ARITHMETIC SERIES** Use mathematical induction to prove the formula $a_1 + (a_1 + d) + (a_1 + 2d) + \ldots + [a_1 + (n-1)d] = \frac{n}{2}[2a_1 + (n-1)d]$ for the sum of an arithmetic series.

22–23. See pp. 629A–629F.

24. **PUZZLES** Show that a 2^n by 2^n checkerboard with the top right square missing can always be covered by nonoverlapping L-shaped tiles like the one at the right. See pp. 629A–629F.

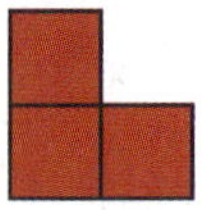

Answer

31. Write 7^n as $(6 + 1)^n$. Then use the Binomial Theorem.

$$7^n - 1 = (6+1)^n - 1$$
$$= 6^n + n \cdot 6^{n-1} + \frac{n(n-1)}{2}6^{n-2} + \ldots + n \cdot 6 + 1 - 1$$
$$= 6^n + n \cdot 6^{n-1} + \frac{n(n-1)}{2}6^{n-2} + \ldots + n \cdot 6$$

Since each term in the last expression is divisible by 6, the whole expression is divisible by 6. Thus, $7^n - 1$ is divisible by 6.

Find a counterexample for each statement.

25. $1^2 + 2^2 + 3^2 + \ldots + n^2 = \frac{n(3n-1)}{2}$ **Sample answer: $n = 3$**

26. $1^3 + 3^3 + 5^3 + \ldots + (2n-1)^3 = 12n^3 - 23n^2 + 12n$ **Sample answer: $n = 4$**

27. $3^n + 1$ is divisible by 4. **Sample answer: $n = 2$**

28. $2^n + 2n^2$ is divisible by 4. **Sample answer: $n = 3$**

★ 29. $n^2 - n + 11$ is prime. **Sample answer: $n = 11$**

★ 30. $n^2 + n + 41$ is prime. **Sample answer: $n = 41$**

31. See margin.

31. **CRITICAL THINKING** Refer to Example 2. Explain how to use the Binomial Theorem to show that $7^n - 1$ is divisible by 6 for all positive integers n.

32. **WRITING IN MATH** Answer the question that was posed at the beginning of the lesson. **See pp. 629A–629F.**

How does the concept of a ladder help you prove statements about numbers?

Include the following in your answer:

- an explanation of which part of an inductive proof corresponds to stepping onto the bottom step of the ladder, and
- an explanation of which part of an inductive proof corresponds to climbing from one step on the ladder to the next.

Standardized Test Practice (A) (B) (C) (D)

33. $\dfrac{x - \frac{4}{x}}{1 - \frac{4}{x} + \frac{4}{x^2}} =$ **C**

(A) $\frac{x}{x-2}$ (B) $\frac{x^2+2}{x-2}$ (C) $\frac{x^2+2x}{x-2}$ (D) $\frac{x^2+2x}{(x-2)^2}$

34. $PQRS$ is a square. Find the ratio of the length of diagonal $\overline{QS}$ to the length of side $\overline{RS}$. **A**

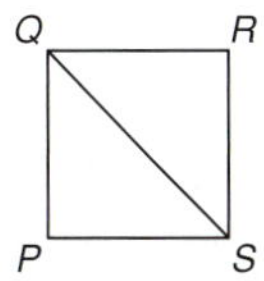

(A) $\sqrt{2}$

(B) 1

(C) $\frac{\sqrt{2}}{2}$

(D) The relationship cannot be determined from the given information.

Maintain Your Skills

Mixed Review Expand each power. *(Lesson 11-7)* **35–37. See margin.**

35. $(x + y)^6$ 36. $(a - b)^7$ 37. $(2x + y)^8$

Find the first three iterates of each function for the given initial value. *(Lesson 11-6)*

38. $f(x) = 3x - 2,\ x_0 = 2$ **4, 10, 28** 39. $f(x) = 4x^2 - 2,\ x_0 = 1$ **2, 14, 782**

40. **BIOLOGY** Suppose an amoeba divides into two amoebas once every hour. How long would it take for a single amoeba to become a colony of 4096 amoebas? *(Lesson 10-2)* **12 h**

Solve each equation. Check your solutions. *(Lesson 9-6)*

41. $\frac{1}{y+1} - \frac{3}{y-3} = 2$ **0, 1** 42. $\frac{6}{a-7} = \frac{a-49}{a^2-7a} + \frac{1}{a}$ **−14**

www.algebra2.com/self_check_quiz

Answers

35. $x^6 + 6x^5y + 15x^4y^2 + 20x^3y^3 + 15x^2y^4 + 6xy^5 + y^6$

36. $a^7 - 7a^6b + 21a^5b^2 - 35a^4b^3 + 35a^3b^4 - 21a^2b^5 + 7ab^6 - b^7$

37. $256x^8 + 1024x^7y + 1792x^6y^2 + 1792x^5y^3 + 1120x^4y^4 + 448x^3y^5 + 112x^2y^6 + 16xy^7 + y^8$

Enrichment, p. 678

Proof by Induction

Mathematical induction is a useful tool when you want to prove that a statement is true for all natural numbers.

The three steps in using induction are:

1. Prove that the statement is true for $n = 1$.
2. Prove that if the statement is true for the natural number n, it must also be true for $n + 1$.
3. Conclude that the statement is true for all natural numbers.

Follow the steps to complete each proof.

Theorem A: The sum of the first n odd natural numbers is equal to n^2.

1. Show that the theorem is true for $n = 1$.
$1 = (1)^2$

2. Suppose $1 + 3 + 5 + \ldots + (2n - 1) = n^2$. Show that

Study Guide and Intervention, p. 673 (shown) and p. 674

Mathematical Induction Mathematical induction is a method of proof used to prove statements about positive integers.

Mathematical Induction Proof	Step 1 Show that the statement is true for some integer n. Step 2 Assume that the statement is true for some positive integer k where $k \ge n$. This assumption is called the **inductive hypothesis**. Step 3 Show that the statement is true for the next integer $k + 1$.

Example **Prove that $5 + 11 + 17 + \ldots + (6n - 1) = 3n^2 + 2n$.**

Step 1 When $n = 1$, the left side of the given equation is $6(1) - 1 = 5$. The right side is $3(1)^2 + 2(1) = 5$. Thus the equation is true for $n = 1$.

Step 2 Assume that $5 + 11 + 17 + \ldots + (6k - 1) = 3k^2 + 2k$ for some positive integer k.

Step 3 Show that the equation is true for $n = k + 1$. First, add $[6(k + 1) - 1]$ to each side.

$$\begin{aligned} 5 + 11 + 17 + \ldots + (6k-1) + [6(k+1) - 1] &= 3k^2 + 2k + [6(k+1) - 1] \\ &= 3k^2 + 2k + 6k + 5 && \text{Add.} \\ &= 3k^2 + 6k + 3 + 2k + 2 && \text{Rewrite.} \\ &= 3(k^2 + 2k + 1) + 2(k + 1) && \text{Factor.} \\ &= 3(k+1)^2 + 2(k+1) && \text{Factor.} \end{aligned}$$

The last expression above is the right side of the equation to be proved, where n has been replaced by $k + 1$. Thus the equation is true for $n = k + 1$.

This proves that $5 + 11 + 17 + \ldots + (6n - 1) = 3n^2 + 2n$ for all positive integers n.

Exercises

Prove that each statement is true for all positive integers.

1. $3 + 7 + 11 + \ldots + (4n - 1) = 2n^2 + n$.

Step 1 The statement is true for $n = 1$ since $4(1) - 1 = 3$ and $2(1)^2 + 1 = 3$.
Step 2 Assume that $3 + 7 + 11 + \ldots + (4k - 1) = 2k^2 + k$ for some positive integer k.
Step 3 Adding the $(k + 1)$st term to each side from step 2, we get $3 + 7 + 11 + \ldots + (4k - 1) + [4(k + 1) - 1] = 2k^2 + k + [4(k + 1) - 1]$.
Simplifying the right side of the equation gives $2(k + 1)^2 + (k + 1)$, which is the statement to be proved.

2. $500 + 100 + 20 + \ldots + 4 \cdot 5^{4-n} = 625\left(1 - \frac{1}{5^n}\right)$.

Step 1 The statement is true for $n = 1$, since $4 \cdot 5^{4-1} = 4 \cdot 5^3 = 500$ and $625\left(1 - \frac{1}{5^1}\right) = \frac{4}{5}(625) = 500$.
Step 2 Assume that $500 + 100 + 20 + \ldots + 4 \cdot 5^{4-k} = 625\left(1 - \frac{1}{5^k}\right)$ for some positive integer k.
Step 3 Adding the $(k + 1)$st term to each side from step 2 and simplifying gives $500 + 100 + 20 + \ldots + 4 \cdot 5^{4-k} + 4 \cdot 5^{3-k} = 625\left(1 - \frac{1}{5^k}\right) + 4 \cdot 5^{3-k} = 625\left(1 - \frac{1}{5^{k+1}}\right)$, which is the statement to be proved.

Skills Practice, p. 675 and Practice, p. 676 (shown)

Prove that each statement is true for all positive integers.

1. $1 + 2 + 4 + 8 + \ldots + 2^{n-1} = 2^n - 1$

Step 1: When $n = 1$, then $2^{n-1} = 2^{1-1} = 2^0 = 1 = 2^1 - 1$. So, the equation is true for $n = 1$.
Step 2: Assume that $1 + 2 + 4 + 8 + \ldots + 2^{k-1} = 2^k - 1$ for some positive integer k.
Step 3: Show that the given equation is true for $n = k + 1$.
$1 + 2 + 4 + 8 + \ldots + 2^{k-1} + 2^{(k+1)-1} = (2^k - 1) + 2^{(k+1)-1}$
$= 2^k - 1 + 2^k = 2 \cdot 2^k - 1 = 2^{k+1} - 1$
So, $1 + 2 + 4 + 8 + \ldots + 2^{n-1} = 2^n - 1$ for all positive integers n.

2. $1 + 4 + 9 + \ldots + n^2 = \frac{n(n+1)(2n+1)}{6}$

Step 1: When $n = 1$, $n^2 = 1^2 = 1 = \frac{1(1+1)(2 \cdot 1 + 1)}{6}$; true for $n = 1$.
Step 2: Assume that $1 + 4 + 9 + \ldots + k^2 = \frac{k(k+1)(2k+1)}{6}$ for some positive integer k.
Step 3: Show that the given equation is true for $n = k + 1$.

$$\begin{aligned} 1 + 4 + 9 + \ldots + k^2 + (k+1)^2 &= \frac{k(k+1)(2k+1)}{6} + (k+1)^2 \\ &= \frac{k(k+1)(2k+1)}{6} + \frac{6(k+1)^2}{6} = \frac{(k+1)[k(2k+1) + 6(k+1)]}{6} \\ &= \frac{(k+1)(2k^2 + 7k + 6)}{6} = \frac{(k+1)[(k+2)(2k+3)]}{6} \\ &= \frac{(k+1)[(k+1)+1][2(k+1)+1]}{6} \end{aligned}$$

So, $1 + 4 + 9 + \ldots + n^2 = \frac{n(n+1)(2n+1)}{6}$ for all positive integers n.

3. $18^n - 1$ is a multiple of 17.

Step 1: When $n = 1$, $18^n - 1 = 18 - 1$ or 17; true for $n = 1$.
Step 2: Assume that $18^k - 1$ is divisible by 17 for some positive integer k. This means that there is a whole number r such that $18^k - 1 = 17r$.
Step 3: Show that the statement is true for $n = k + 1$.
$18^k - 1 = 17r$, so $18^k = 17r + 1$, and $18(18^k) = 18(17r + 1)$. This is equivalent to $18^{k+1} = 306r + 18$, so $18^{k+1} - 1 = 306r + 17$, and $18^{k+1} - 1 = 17(18r + 1)$.
Since r is a whole number, $18r + 1$ is a whole number, and $18^{k+1} - 1$ is divisible by 17. The statement is true for $n = k + 1$. So, $18^n - 1$ is divisible by 17 for all positive integers n.

Find a counterexample for each statement.

4. $1 + 4 + 7 + \ldots + (3n - 2) = n^3 - n^2 + 1$
Sample answer: $n = 3$

5. $5^n - 2n - 3$ is divisible by 3.
Sample answer: $n = 3$

6. $1 + 3 + 5 + \ldots + (2n - 1) = \frac{n^2 + 3n - 2}{2}$
Sample answer: $n = 3$

7. $1^3 + 2^3 + 3^3 + \ldots + n^3 = n^4 - n^3 + 1$
Sample answer: $n = 3$

Reading to Learn Mathematics, p. 677 ELL

Pre-Activity **How does the concept of a ladder help you prove statements about numbers?**

Read the introduction to Lesson 11-8 at the top of page 618 in your textbook.

What are two ways in which a ladder could be constructed so that you could not reach every step of the ladder?

Sample answer: 1. The first step could be too far off the ground for you to climb on it. 2. The steps could be too far apart for you to go up from one step to the next.

Reading the Lesson

1. Fill in the blanks to describe the three steps in a proof by mathematical induction.

Step 1 Show that the statement is __true__ for the number __1__.

Step 2 Assume that the statement is __true__ for some positive __integer__ k.
This assumption is called the __inductive hypothesis__.

Step 3 Show that the statement is __true__ for the next integer __$k + 1$__.

2. Suppose that you wanted to prove that the following statement is true for all positive integers.

$$3 + 6 + 9 + \ldots + 3n = \frac{3n(n+1)}{2}$$

a. Which of the following statements shows that the statement is true for $n = 1$? **ii**

i. $3 = \frac{3 \cdot 2 + 1}{2}$ ii. $3 = \frac{3 \cdot 1 \cdot 2}{2}$ iii. $3 = \frac{3 + 1 + 2}{2}$

b. Which of the following is the statement for $n = k + 1$? **iv**

i. $3 + 6 + 9 + \ldots + 3^k = \frac{3k(k+1)}{2}$

ii. $3 + 6 + 9 + \ldots + 3^{k+1} = \frac{3k(k+1)}{2}$

iii. $3 + 6 + 9 + \ldots + 3^{k+1} = 3(k+1)(k+2)$

iv. $3 + 6 + 9 + \ldots + 3(k+1) = \frac{3(k+1)(k+2)}{2}$

Helping You Remember

3. Many students confuse the roles of n and k in a proof by mathematical induction. What is a good way to remember the difference in the ways these variables are used in such a proof?
Sample answer: The letter n stands for "number" and is used as a variable to represent *any* natural number. The letter k is used to represent a particular value of n.

Chapter 11 Study Guide and Review

Vocabulary and Concept Check

- This alphabetical list of vocabulary terms in Chapter 11 includes a page reference where each term was introduced.
- **Assessment** A vocabulary test/review for Chapter 11 is available on p. 692 of the *Chapter 11 Resource Masters*.

Lesson-by-Lesson Review

For each lesson,

- the main ideas are summarized,
- additional examples review concepts, and
- practice exercises are provided.

Vocabulary PuzzleMaker

ELL The Vocabulary PuzzleMaker software improves students' mathematics vocabulary using four puzzle formats—crossword, scramble, word search using a word list, and word search using clues. Students can work on a computer screen or from a printed handout.

MindJogger Videoquizzes

ELL MindJogger Videoquizzes provide an alternative review of concepts presented in this chapter. Students work in teams in a game show format to gain points for correct answers. The questions are presented in three rounds.

Round 1 Concepts (5 questions)
Round 2 Skills (4 questions)
Round 3 Problem Solving (4 questions)

Chapter 11 Study Guide and Review

Vocabulary and Concept Check

arithmetic means (p. 580)	Fibonacci sequence (p. 606)	mathematical induction (p. 618)
arithmetic sequence (p. 578)	geometric means (p. 590)	partial sum (p. 599)
arithmetic series (p. 583)	geometric sequence (p. 588)	Pascal's triangle (p. 612)
Binomial Theorem (p. 613)	geometric series (p. 594)	recursive formula (p. 606)
common difference (p. 578)	index of summation (p. 585)	sequence (p. 578)
common ratio (p. 588)	inductive hypothesis (p. 618)	series (p. 583)
convergent series (p. 600)	infinite geometric series (p. 599)	sigma notation (p. 585)
factorial (p. 613)	iteration (p. 608)	term (p. 578)

Choose the term from the list above that best completes each statement.

1. A(n) ____________ of an infinite series is the sum of a certain number of terms. **partial sum**
2. If a sequence has a common ratio, then it is a(n) ____________. **geometric sequence**
3. Using _________, the series $2 + 5 + 8 + 11 + 14$ can be written as $\sum_{n=1}^{5}(3n - 1)$. **sigma notation**
4. Eleven and 17 are the two ____ between 5 and 23 in the sequence 5, 11, 17, 23. **arithmetic means**
5. Using the _________, $(a - 2)^4$ can be expanded to $a^4 - 8a^3 + 24a^2 - 32a + 16$. **Binomial Theorem**
6. The _________ of the sequence $3, 2, \frac{4}{3}, \frac{8}{9}, \frac{16}{27}$ is $\frac{2}{3}$. **common ratio**
7. The _______ $11 + 16.5 + 22 + 27.5 + 33$ has a sum of 110. **arithmetic series**
8. A(n) ____ is expressed as $n! = n(n - 1)(n - 2) \ldots 2 \cdot 1$. **factorial**

Lesson-by-Lesson Review

11-1 Arithmetic Sequences

See pages 578–582.

Concept Summary

- An arithmetic sequence is formed by adding a constant to each term to get the next term.
- The *n*th term a_n of an arithmetic sequence with first term a_1 and common difference d is given by $a_n = a_1 + (n - 1)d$, where n is any positive integer.

Examples

1 **Find the 12th term of an arithmetic sequence if $a_1 = -17$ and $d = 4$.**

$a_n = a_1 + (n - 1)d$ Formula for the *n*th term

$a_{12} = -17 + (12 - 1)4$ $n = 12, a_1 = -17, d = 4$

$a_{12} = 27$ Simplify.

2 **Find the two arithmetic means between 4 and 25.**

$a_n = a_1 + (n - 1)d$ Formula for the *n*th term

$a_4 = 4 + (4 - 1)d$ $n = 4, a_1 = 4$

$25 = 4 + 3d$ $a_4 = 25$

$7 = d$ The arithmetic means are 4 + 7 or 11 and 11 + 7 or 18.

www.algebra2.com/vocabulary_review

FOLDABLES™ Study Organizer

For more information about Foldables, see *Teaching Mathematics with Foldables.*

Students are usually interested in doing well on various kinds of tests. One way to achieve this goal is by writing their own questions about the material. Have student volunteers read some of their questions. Have other student volunteers answer, and have the writer of the question comment on the answer. Ask students to use what they have learned in this discussion to revise their own Foldables.

Encourage students to refer to their Foldables while completing the Study Guide and Review and to use them in preparing for the Chapter Test.

Exercises **Find the indicated term of each arithmetic sequence.** *See Example 2 on p. 579.*

9. $a_1 = 6, d = 8, n = 5$ **38**
10. $a_1 = -5, d = 7, n = 22$ **142**
11. $a_1 = 5, d = -2, n = 9$ **−11**
12. $a_1 = -2, d = -3, n = 15$ **−44**

Find the arithmetic means in each sequence. *See Example 4 on page 580.* **15. 6, 3, 0, −3**

13. −7, __?__, __?__, __?__, 9 **−3, 1, 5**
14. 12, __?__, __?__, 4 $\frac{28}{3}, \frac{20}{3}$
15. 9, __?__, __?__, __?__, __?__, −6
16. 56, __?__, __?__, __?__, 28 **49, 42, 35**

11-2 Arithmetic Series

See pages 583–587.

Concept Summary

- The sum S_n of the first n terms of an arithmetic series is given by $S_n = \frac{n}{2}[2a_1 + (n-1)d]$ or $S_n = \frac{n}{2}(a_1 + a_n)$.

Example **Find S_n for the arithmetic series with $a_1 = 34$, $a_n = 2$, and $n = 9$.**

$S_n = \frac{n}{2}(a_1 + a_n)$ Sum formula

$S_9 = \frac{9}{2}(34 + 2)$ $n = 9, a_1 = 34, a_n = 2$

$S_9 = 162$ Simplify.

Exercises **Find S_n for each arithmetic series.** *See Examples on pages 584 and 585.*

17. $a_1 = 12, a_n = 117, n = 36$ **2322**
18. $4 + 10 + 16 + \ldots + 106$ **990**
19. $10 + 4 + (-2) + \ldots + (-50)$ **−220**
20. $\sum_{n=2}^{13}(3n + 1)$ **282**

11-3 Geometric Sequences

See pages 588–592.

Concept Summary

- A geometric sequence is one in which each term after the first is found by multiplying the previous term by a common ratio.
- The nth term a_n of a geometric sequence with first term a_1 and common ratio r is given by $a_n = a_1 \cdot r^{n-1}$, where n is any positive integer.

Examples

1 Find the fifth term of a geometric sequence for which $a_1 = 7$ and $r = 3$.

$a_n = a_1 \cdot r^{n-1}$ Formula for nth term

$a_5 = 7 \cdot 3^{5-1}$ $n = 5, a_1 = 7, r = 3$

$a_5 = 567$ The fifth term is 567.

2 Find two geometric means between 1 and 8.

$a_n = a_1 \cdot r^{n-1}$ Formula for nth term

$a_4 = 1 \cdot r^{4-1}$ $n = 4$ and $a_1 = 1$

$8 = r^3$ $a_4 = 8$

$2 = r$ The geometric means are 1(2) or 2 and 2(2) or 4.

Exercises Find the indicated term of each geometric sequence.
See Example 2 on page 589.

21. $a_1 = 2, r = 2, n = 5$ **32**
22. $a_1 = 7, r = 2, n = 4$ **56**
23. $a_1 = 243, r = -\frac{1}{3}, n = 5$ **3**
24. a_6 for $\frac{2}{3}, \frac{4}{3}, \frac{8}{3}, \ldots$ $\frac{64}{3}$

Find the geometric means in each sequence. *See Example 5 on page 590.*

25. 3, _?_, _?_, 24 **6, 12**
26. 7.5, _?_, _?_, _?_, 120 **±15, 30, ±60**
27. 8, _?_, _?_, _?_, _?_, $\frac{1}{4}$ **4, 2, 1,** $\frac{1}{2}$
28. 5, _?_, _?_, _?_, 80 **±10, 20, ±40**

11-4 Geometric Series

See pages 594–598.

Concept Summary

- The sum S_n of the first n terms of a geometric series is given by

$$S_n = \frac{a_1(1 - r^n)}{1 - r} \text{ or } S_n = \frac{a_1 - a_1 r^n}{1 - r}, \text{ where } r \neq 1.$$

Example **Find the sum of a geometric series for which $a_1 = 7$, $r = 3$, and $n = 14$.**

$S_n = \frac{a_1 - a_1 r^n}{1 - r}$ Sum formula

$S_{14} = \frac{7 - 7 \cdot 3^{14}}{1 - 3}$ $n = 14, a_1 = 7, r = 3$

$S_{14} = 16{,}740{,}388$ Use a calculator.

Exercises Find S_n for each geometric series. *See Examples 1 and 3 on pages 595 and 596.*

29. $a_1 = 12, r = 3, n = 5$ **1452**
30. $4 - 2 + 1 - \ldots$ to 6 terms $\frac{21}{8}$
31. $256 + 192 + 144 + \ldots$ to 7 terms $\frac{14{,}197}{16}$
32. $\sum_{n=1}^{5}\left(-\frac{1}{2}\right)^{n-1}$ $\frac{11}{16}$

11-5 Infinite Geometric Series

See pages 599–604.

Concept Summary

- The sum S of an infinite geometric series with $-1 < r < 1$ is given by $S = \frac{a_1}{1 - r}$.

Example **Find the sum of the infinite geometric series for which $a_1 = 18$ and $r = -\frac{2}{7}$.**

$S = \frac{a_1}{1 - r}$ Sum formula

$= \frac{18}{1 - \left(-\frac{2}{7}\right)}$ $a_1 = 18, r = -\frac{2}{7}$

$= \frac{18}{\frac{9}{7}}$ or 14 Simplify.

624 Chapter 11 Sequences and Series

Exercises Find the sum of each infinite geometric series, if it exists.
See Example 1 on page 600. **34. does not exist**

33. $a_1 = 6, r = \frac{11}{12}$ **72**
34. $\frac{1}{8} - \frac{3}{16} + \frac{9}{32} - \frac{27}{64} + \ldots$
35. $\sum_{n=1}^{\infty} -2\left(-\frac{5}{8}\right)^{n-1}$ **$-\frac{16}{13}$**

11-6 Recursion and Special Sequences

See pages 606–610.

Concept Summary

- In a recursive formula, each term is formulated from one or more previous terms.
- Iteration is the process of composing a function with itself repeatedly.

Examples

1 **Find the first five terms of the sequence in which $a_1 = 2$ and $a_{n+1} = 2a_n - 1$.**

$a_{n+1} = 2a_n - 1$ Recursive formula

$a_{1+1} = 2a_1 - 1$ $n = 1$
$a_2 = 2(2) - 1$ or 3 $a_1 = 2$
$a_{2+1} = 2a_2 - 1$ $n = 2$
$a_3 = 2(3) - 1$ or 5 $a_2 = 3$

$a_{3+1} = 2a_3 - 1$ $n = 3$
$a_4 = 2(5) - 1$ or 9 $a_3 = 5$
$a_{4+1} = 2a_4 - 1$ $n = 4$
$a_5 = 2(9) - 1$ or 17 $a_4 = 9$

The first five terms of the sequence are 2, 3, 5, 9, and 17.

2 **Find the first three iterates of $f(x) = -5x - 1$ for an initial value of $x_0 = -1$.**

$x_1 = f(x_0)$
$= f(-1)$
$= -5(-1) - 1$ or 4

$x_2 = f(x_1)$
$= f(4)$
$= -5(4) - 1$ or -21

$x_3 = f(x_2)$
$= f(-21)$
$= -5(-21) - 1$ or 104

The first three iterates are 4, −21, and 104.

Exercises Find the first five terms of each sequence. *See Example 1 on page 606.*

36. $a_1 = -2, a_{n+1} = a_n + 5$
37. $a_1 = 3, a_{n+1} = 4a_n - 10$
38. $a_1 = 2, a_{n+1} = a_n + 3n$ **2, 5, 11, 20, 32**
39. $a_1 = 1, a_2 = 3, a_{n+2} = a_{n+1} + a_n$

36. −2, 3, 8, 13, 18 37. 3, 2, −2, −18, −82 39. 1, 3, 4, 7, 11

Find the first three iterates of each function for the given initial value.
See Example 3 on page 608. **43. −1, 4, −31**

40. $f(x) = -2x + 3, x_0 = 1$ **1, 1, 1**
41. $f(x) = 7x - 4, x_0 = 2$ **10, 66, 458**
42. $f(x) = x^2 - 6, x_0 = -1$ **−5, 19, 355**
43. $f(x) = -2x^2 - x + 5, x_0 = -2$

11-7 The Binomial Theorem

See pages 612–617.

Concept Summary

- Pascal's triangle can be used to find the coefficients in a binomial expansion.
- The Binomial Theorem: $(a + b)^n = \sum_{k=0}^{n} \frac{n!}{(n-k)!k!} a^{n-k} b^k$

Chapter 11 Study Guide and Review 625

Chapter 11 For More ...

• Extra Practice, see pages 851–855.
• Mixed Problem Solving, see page 872.

Answers

44. $x^3 + 3x^2y + 3xy^2 + y^3$

45. $x^4 - 8x^3 + 24x^2 - 32x + 16$

46. $243r^5 + 405r^4s + 270r^3s^2 + 90r^2s^3 + 15rs^4 + s^5$

49. Step 1: When $n = 1$, the left side of the given equation is 1. The right side is $2^1 - 1$ or 1, so the equation is true for $n = 1$.

Step 2: Assume $1 + 2 + 4 + \ldots + 2^{k-1} = 2^k - 1$ for some positive integer k.

Step 3: $1 + 2 + 4 + \ldots + 2^{k-1} + 2^{(k+1)-1}$
$= 2^k - 1 + 2^{(k+1)-1}$
$= 2^k - 1 + 2^k$
$= 2 \cdot 2^k - 1$
$= 2^{k+1} - 1$

The last expression is the right side of the equation to be proved, where $n = k + 1$. Thus, the equation is true for $n = k + 1$.

Therefore, $1 + 2 + 4 + \ldots + 2^{n-1} = 2^n - 1$ for all positive integers n.

50. Step 1: $6^1 - 1 = 5$, which is divisible by 5. The statement is true for $n = 1$.

Step 2: Assume that $6^k - 1$ is divisible by 5 for some positive integer k. This means that $6^k - 1 = 5r$ for some whole number r.

Step 3:
$$\begin{aligned} 6^k - 1 &= 5r \\ 6^k &= 5r + 1 \\ 6(6^k) &= 6(5r + 1) \\ 6^{k+1} &= 30r + 6 \\ 6^{k+1} - 1 &= 30r + 5 \\ 6^{k+1} - 1 &= 5(6r + 1) \end{aligned}$$

Since r is a whole number, $6r + 1$ is a whole number. Thus, $6^{k+1} - 1$ is divisible by 5, so the statement is true for $n = k + 1$.

Therefore, $6^n - 1$ is divisible by 5 for all positive integers n.

Example Expand $(a - 2b)^4$.

$$(a - 2b)^4 = \sum_{k=0}^{4} \frac{4!}{(4-k)!k!} a^{4-k}(-2b)^k \quad \text{Binomial Theorem}$$

$$= \frac{4!}{4!0!}a^4(-2b)^0 + \frac{4!}{3!1!}a^3(-2b)^1 + \frac{4!}{2!2!}a^2(-2b)^2 + \frac{4!}{1!3!}a^1(-2b)^3 + \frac{4!}{0!4!}a^0(-2b)^4$$

$$= a^4 - 8a^3b + 24a^2b^2 - 32ab^3 + 16b^4 \quad \text{Simplify.}$$

Exercises Expand each power. *See Examples 1, 2, and 4 on pages 613 and 614.*

44. $(x + y)^3$ 45. $(x - 2)^4$ 46. $(3r + s)^5$

44–46. See margin. 48. $-13{,}107{,}200x^9$

Find the indicated term of each expansion. *See Example 5 on page 615.*

47. fourth term of $(x + 2y)^6$ $160x^3y^3$ 48. second term of $(4x - 5)^{10}$

11-8 Proof and Mathematical Induction

See pages 618–621.

Concept Summary

- Mathematical induction is a method of proof used to prove statements about the positive integers.

Example Prove $1 + 5 + 25 + \ldots + 5^{n-1} = \frac{1}{4}(5^n - 1)$ for all positive integers n.

Step 1 When $n = 1$, the left side of the given equation is 1. The right side is $\frac{1}{4}(5^1 - 1)$ or 1. Thus, the equation is true for $n = 1$.

Step 2 Assume that $1 + 5 + 25 + \ldots + 5^{k-1} = \frac{1}{4}(5^k - 1)$ for some positive integer k.

Step 3 Show that the given equation is true for $n = k + 1$.

$$\begin{aligned} 1 + 5 + 25 + \ldots + 5^{k-1} + 5^{(k+1)-1} &= \tfrac{1}{4}(5^k - 1) + 5^{(k+1)-1} && \text{Add } 5^{(k+1)-1} \text{ to each side.} \\ &= \tfrac{1}{4}(5^k - 1) + 5^k && \text{Simplify the exponent.} \\ &= \frac{5^k - 1 + 4 \cdot 5^k}{4} && \text{Common denominator} \\ &= \frac{5 \cdot 5^k - 1}{4} && \text{Distributive Property} \\ &= \tfrac{1}{4}(5^{k+1} - 1) && 5 \cdot 5^k = 5^{k+1} \end{aligned}$$

The last expression above is the right side of the equation to be proved, where n has been replaced by $k + 1$. Thus, the equation is true for $n = k + 1$.

This proves that $1 + 5 + 25 + \ldots + 5^{n-1} = \frac{1}{4}(5^n - 1)$ for all positive integers n.

Exercises Prove that each statement is true for all positive integers. *See Examples 1 and 2 on pages 618 and 619.* 49–50. See margin.

49. $1 + 2 + 4 + \ldots + 2^{n-1} = 2^n - 1$ 50. $6^n - 1$ is divisible by 5.

626 Chapter 11 Sequences and Series

Chapter 11 Practice Test

Vocabulary and Concepts

Choose the correct term to complete each sentence.

1. A sequence in which each term after the first is found by adding a constant to the previous term is called a(n) (*arithmetic*, *geometric*) sequence.
2. A (*Fibonacci sequence*, *series*) is a sum of terms of a sequence.
3. (*Pascal's triangle*, *Recursive formulas*) and the Binomial Theorem can be used to expand powers of binomials.

Skills and Applications

4. Find the next four terms of the arithmetic sequence 42, 37, 32, … . **27, 22, 17, 12**
5. Find the 27th term of an arithmetic sequence for which $a_1 = 2$ and $d = 6$. **158**
6. Find the three arithmetic means between −4 and 16. **1, 6, 11**
7. Find the sum of the arithmetic series for which $a_1 = 7$, $n = 31$, and $a_n = 127$. **2077**
8. Find the next two terms of the geometric sequence $\frac{1}{81}, \frac{1}{27}, \frac{1}{9}, \ldots$. **$\frac{1}{3}$, 1**
9. Find the sixth term of the geometric sequence for which $a_1 = 5$ and $r = -2$. **−160**
10. Find the two geometric means between 7 and 189. **21, 63**
11. Find the sum of the geometric series for which $a_1 = 125$, $r = \frac{2}{5}$, and $n = 4$. **203**

Find the sum of each series, if it exists. **13. does not exist**

12. $\sum_{k=3}^{15}(14 - 2k)$ **−52**
13. $\sum_{n=1}^{\infty}\frac{1}{3}(-2)^{n-1}$
14. $91 + 85 + 79 + \ldots + (-29)$ **651**
15. $12 + (-6) + 3 - \frac{3}{2} + \ldots$ **8**

Find the first five terms of each sequence.

16. $a_1 = 1, a_{n+1} = a_n + 3$ **1, 4, 7, 10, 13**
17. $a_1 = -3, a_{n+1} = a_n + n^2$ **−3, −2, 2, 11, 27**

18. Find the first three iterates of $f(x) = x^2 - 3x$ for an initial value of $x_0 = 1$. **−2, 10, 70**
19. Expand $(2s - 3t)^5$. **$32s^5 - 240s^4t + 720s^3t^2 - 1080s^2t^3 + 810st^4 - 243t^5$**
20. Find the third term of the expansion of $(x + y)^{10}$. **$45x^8y^2$**

Prove that each statement is true for all positive integers. **21–22. See pp. 629A–629F.**

21. $1 + 3 + 5 + \ldots + (2n - 1) = n^2$
22. $14^n - 1$ is divisible by 13.

23. **DESIGN** A landscaper is designing a wall of white brick and red brick. The pattern starts with 20 red bricks on the bottom row. Each row above it contains 3 fewer red bricks than the preceding row. If the top row contains no red bricks, how many rows are there and how many red bricks were used? **8 rows, 77 red bricks**
24. **RECREATION** One minute after it is released, a gas-filled balloon has risen 100 feet. In each succeeding minute, the balloon rises only 50% as far as it rose in the previous minute. How far will the balloon rise in 5 minutes? **193.75 ft**
25. **STANDARDIZED TEST PRACTICE** Find the next term in the geometric sequence $8, 6, \frac{9}{2}, \frac{27}{8}, \ldots$. **D**

(A) $\frac{11}{8}$ (B) $\frac{27}{16}$ (C) $\frac{9}{4}$ (D) $\frac{81}{32}$

www.algebra2.com/chapter_test

Portfolio Suggestion

Introduction Throughout this course, you have been working in groups to solve problems.

Ask Students What roles do you play in the group?

- Do you help to keep your group on task? ask questions? just listen and copy down answers?
- List some ways you are a good group member and some ways you could do better.

Place your responses in your portfolio.

Chapter 11 Practice Test

Assessment Options

Vocabulary Test A vocabulary test/review for Chapter 11 can be found on p. 692 of the *Chapter 11 Resource Masters*.

Chapter Tests There are six Chapter 11 Tests and an Open-Ended Assessment task available in the *Chapter 11 Resource Masters*.

Chapter 11 Tests			
Form	**Type**	**Level**	**Pages**
1	MC	basic	679–680
2A	MC	average	681–682
2B	MC	average	683–684
2C	FR	average	685–686
2D	FR	average	687–688
3	FR	advanced	689–690

MC = multiple-choice questions
FR = free-response questions

Open-Ended Assessment Performance tasks for Chapter 11 can be found on p. 691 of the *Chapter 11 Resource Masters*. A sample scoring rubric for these tasks appears on p. A31.

ExamView® Pro

Use the networkable **ExamView® Pro** to:

- Create **multiple versions** of tests.
- Create **modified** tests for *Inclusion* students.
- **Edit** existing questions and **add** your own questions.
- Use built-in **state curriculum correlations** to create tests aligned with state standards.
- **Apply** art to your tests from a program bank of artwork.

Chapter 11 Standardized Test Practice

These two pages contain practice questions in the various formats that can be found on the most frequently given standardized tests.

A practice answer sheet for these two pages can be found on p. A1 of the *Chapter 11 Resource Masters*.

Standardized Test Practice Student Recording Sheet, p. A1

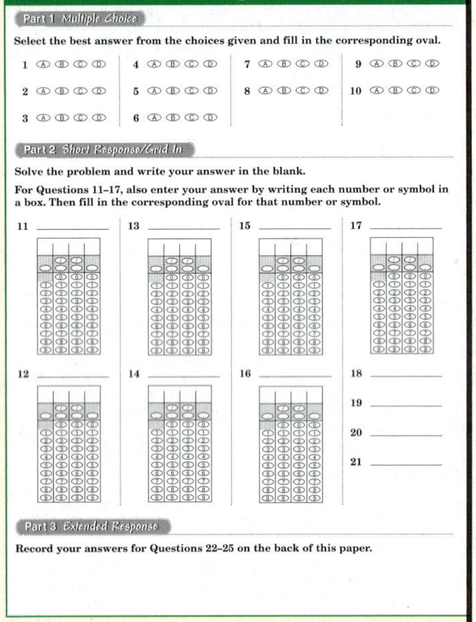
Part 1 Multiple Choice

Select the best answer from the choices given and fill in the corresponding oval.

Part 2 Short Response/Grid In

Solve the problem and write your answer in the blank.

For Questions 11–17, also enter your answer by writing each number or symbol in a box. Then fill in the corresponding oval for that number or symbol.

Part 3 Extended Response

Record your answers for Questions 22–25 on the back of this paper.

Additional Practice

See pp. 697–698 in the *Chapter 11 Resource Masters* for additional standardized test practice.

Chapter 11 Standardized Test Practice

Part 1 Multiple Choice

Record your answers on the answer sheet provided by your teacher or on a sheet of paper.

1. For all positive integers, let $\boxed{n} = n + g$, where g is the greatest factor of n, and $g < n$. If $\boxed{18} = x$, then $\boxed{x} =$ **D**

Ⓐ 9. Ⓑ 8.

Ⓒ 27. Ⓓ 36.

2. If p is positive, what percent of $6p$ is 12? **D**

Ⓐ $\frac{p}{100}\%$ Ⓑ $\frac{p}{2}\%$

Ⓒ $\frac{12}{p}\%$ Ⓓ $\frac{200}{p}\%$

3. A box is 12 units tall, 6 units long, and 8 units wide. A designer is creating a new box that must have the same volume as the first box. If the length and width of the new box are each 50% greater than the length and width of the first box, about how many units tall will the new box be? **A**

Ⓐ 5.3 Ⓑ 6.8

Ⓒ 7.1 Ⓓ 8.5

4. Which of the following statements must be true when $0 < m < 1$? **A**

I $\frac{\sqrt{m}}{m} > 1$ **II** $4m < 1$ **III** $m^2 - m^3 < 0$

Ⓐ I only

Ⓑ III only

Ⓒ I and II only

Ⓓ I, II, and III

5. If $3kx - \frac{4s}{t} = 3ky$, then $x - y = ?$ **D**

Ⓐ $-\frac{4s}{3kt}$ Ⓑ $\frac{-4s}{t} + \frac{1}{3k}$

Ⓒ $\frac{4s}{3t} - k$ Ⓓ $\frac{4s}{3kt}$

6. For all $n \neq 0$, what is the slope of the line passing through $(3n, -k)$ and $(-n, -k)$? **A**

Ⓐ 0 Ⓑ $\frac{k}{2n}$

Ⓒ $\frac{2n}{k}$ Ⓓ undefined

7. Which is the graph of the equation $x^2 + (y - 4)^2 = 20$? **C**

Ⓐ line Ⓑ parabola

Ⓒ circle Ⓓ ellipse

8. $\dfrac{x - \frac{9}{x}}{1 - \frac{6}{x} + \frac{9}{x^2}} =$ **C**

Ⓐ $\frac{x}{x - 3}$ Ⓑ $\frac{x^2 + 3}{x - 3}$

Ⓒ $\frac{x^2 + 3x}{x - 3}$ Ⓓ $\frac{x^2 + 3x}{(x - 3)^2}$

9. What is the sum of the positive even factors of 30? **C**

Ⓐ 18 Ⓑ 30

Ⓒ 48 Ⓓ 72

10. If ℓ_1 is parallel to ℓ_2 in the figure, what is the value of x? **D**

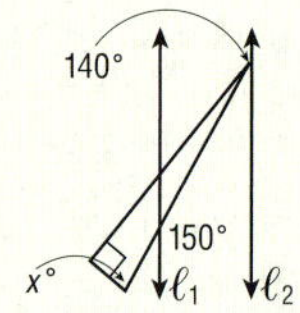

Ⓐ 30

Ⓑ 40

Ⓒ 70

Ⓓ 80

Test-Taking Tip

Question 5

Some questions ask you to find the value of an expression. It is often not necessary to find the value of each variable in the expression. For example, to answer Question 5, it is not necessary to find the values of x and y. Isolate the expression $x - y$ on one side of the equation.

628 **Chapter 11** Standardized Test Practice

ExamView® Pro

Special banks of standardized test questions similar to those on the SAT, ACT, TIMSS 12, NAEP 12, and state proficiency tests can be found on this CD-ROM.

Preparing for Standardized Tests
For test-taking strategies and more practice, see pages 877–892.

Part 2 Short Response/Grid In

Record your answers on the answer sheet provided by your teacher or on a sheet of paper.

11. $\begin{array}{r} AA \\ +\ BB \\ \hline CC \end{array}$

If A, B, and C are each digits and $A = 3B$, then what is one possible value of C? **0, 4, or 8**

12. In the figure, each arc is a semicircle. If B is the midpoint of $\overline{AD}$ and C is the midpoint of $\overline{BD}$, what is the ratio of the area of the semicircle $\overset{\frown}{CD}$ to the area of the semicircle $\overset{\frown}{AD}$? **1/16**

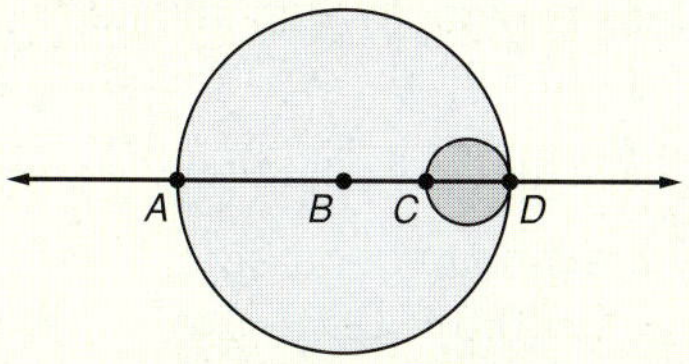

13. Two people are 17.5 miles apart. They begin to walk toward each other along a straight line at the same time. One walks at the rate of 4 miles per hour, and the other walks at the rate of 3 miles per hour. In how many hours will they meet? **2.5 or 5/2**

14. If $\frac{x+y}{x} = \frac{5}{4}$, then $\frac{y}{x} =$ **1/4 or .25**

15. A car's gasoline tank is $\frac{1}{2}$ full. After adding 7 gallons of gas, the gauge shows that the tank is $\frac{3}{4}$ full. How many gallons does the tank hold? **28**

16. If $a = 15 - b$, what is the value of $3a + 3b$? **45**

17. If $x^9 = \frac{45}{y}$ and $x^7 = \frac{1}{5y}$, and $x > 0$, what is the value of x? **15**

18. The rates at a parking garage are shown. How much will it cost to park in the garage for $4\frac{1}{2}$ hours? **$6.50**

PARKING RATES
First hour..............$2.50
Each additional hour or part of an hour.....$1.00

19. Evaluate. $\sum_{n=1}^{6}(1 - 4n)$. **−78**

20. Does an infinite geometric series with a first term of 36 and a common ratio of $\frac{1}{4}$ have a sum? If the series has a sum, find the sum. **yes, 48**

21. What is the last term in the expansion of $(x + 2y)^6$? $\mathbf{64y^6}$

Part 3 Extended Response

Record your answers on a sheet of paper. Show your work.

For Exercises 22–25, use the following information.

The amount of a certain medication remaining in the body is reduced by 20% each hour after the medication is given. The prescription that Dr. Barr gave Kim called for a 100-milligram dose of medication.

22. Is the amount of medication in Kim's body each hour after she takes it modeled by an arithmetic sequence or a geometric sequence? Explain. **See margin.**

23. Write an equation to find the amount of medication A in Kim's body each hour h after she takes it. $\mathbf{A = 100(0.8)^{h-1}}$

24. How many milligrams of medication remain in Kim's body 4 hours after she takes a 100-milligram dose? If necessary, round to the nearest tenth of a milligram. **51.2 mg**

25. If Dr. Barr wants Kim to have at least 30 milligrams of medication in her body at all times, after how many hours should Kim take a second dose of medication? Explain your method of solution. **See margin.**

www.algebra2.com/standardized_test

Answers

22. Geometric; each term is found by multipling the previous term by 0.8.

25. 6 hours; There are about 32.8 mg remaining in the body after 6 hours and about 26.2 mg after 7 hours, so taking another dose after 6 hours keeps the level above 30 mg.

Evaluating Extended Response Questions

Extended Response questions are graded by using a multilevel rubric that guides you in assessing a student's knowledge of a particular concept.

Goal: Determine the dimensions of a rectangular garden for a given perimeter.

Sample Scoring Rubric: The following rubric is a sample scoring device. You may wish to add more detail to this sample to meet your individual scoring needs.

Score	Criteria
4	A correct solution that is supported by well-developed, accurate explanations
3	A generally correct solution, but may contain minor flaws in reasoning or computation
2	A partially correct interpretation and/or solution to the problem
1	A correct solution with no supporting evidence or explanation
0	An incorrect solution indicating no mathematical understanding of the concept or task, or no solution is given

Page 577, Chapter 11 Getting Started

7.

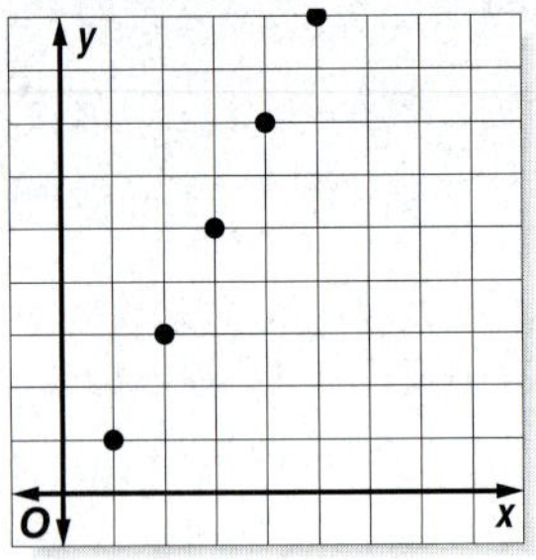

8.

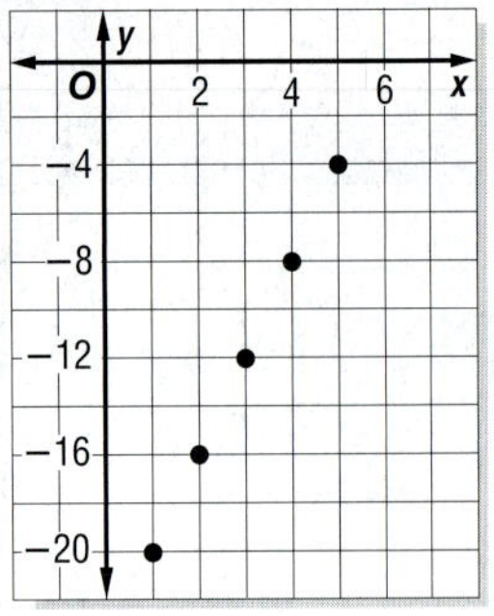

9.

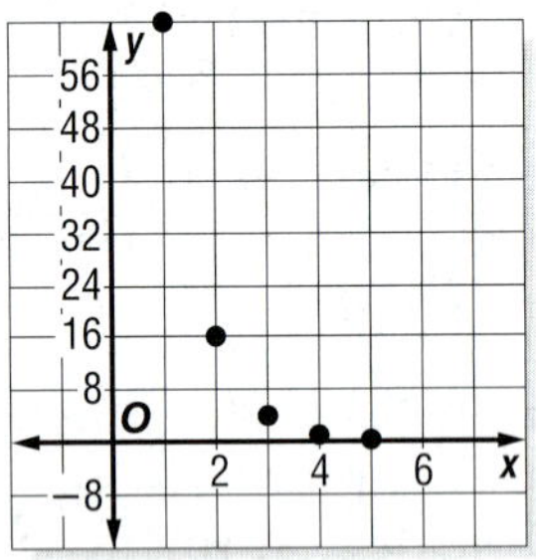

10.

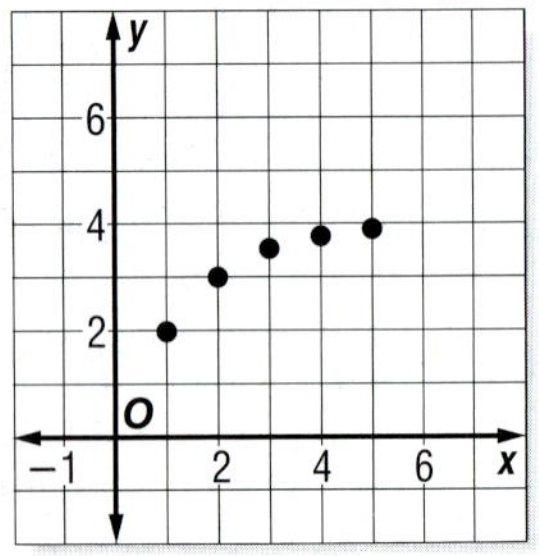

Page 582, Lesson 11-1

49.

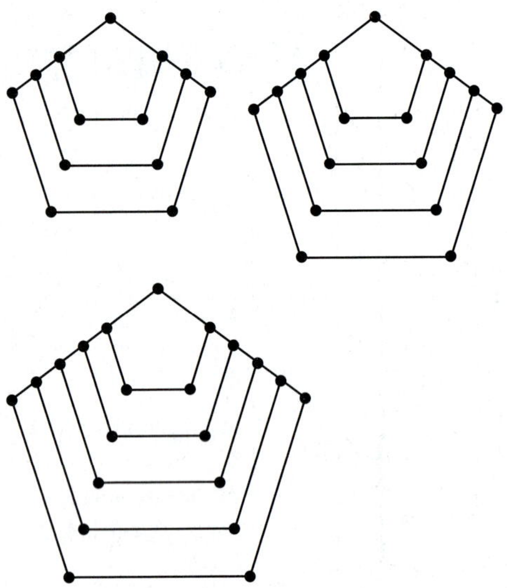

Page 587, Lesson 11-2

48. Arithmetic series can be used to find the seating capacity of an amphitheater. Answers should include the following.

- The sequence represents the numbers of seats in the rows. The sum of the first n terms of the series is the seating capacity of the first n rows.
- One method is to write out the terms and add them: $18 + 22 + 26 + 30 + 34 + 38 + 42 + 46 + 50 + 54 = 360$. Another method is to use the formula $S_n = \frac{n}{2}[2a_1 + (n - 1)d]$:

$$S_{10} = \frac{10}{2}[2(18) + (10 - 1)4] \text{ or } 360.$$

Page 596, Lesson 11-4

3. Sample answer: The first term is $a_1 = 2$. Divide the second term by the first to find that the common ratio is $r = 6$. Therefore, the nth term of the series is given by $2 \cdot 6^{n-1}$. There are five terms, so the series can be written as $\sum_{n=1}^{5} 2 \cdot 6^{n-1}$.

Page 603, Lesson 11-5

48.

$$\begin{aligned} S &= a_1 + a_1r + a_1r^2 + a_1r^3 + \ldots \\ (-)\ rS &= \phantom{a_1 + {}} a_1r + a_1r^2 + a_1r^3 + a_1r^4 + \ldots \\ \hline S - rS &= a_1 + 0 + 0 + 0 + 0 + \ldots \\ S(1 - r) &= a_1 \\ S &= \frac{a_1}{1 - r} \end{aligned}$$

49. The total distance that a ball bounces, both up and down, can be found by adding the sums of two infinite geometric series. Answers should include the following.

- $a_n = a_1 \cdot r^{n-1}$, $S_n = \frac{a_1(1 - r^n)}{1 - r}$, or $S = \frac{a_1}{1 - r}$
- The total distance the ball falls is given by the infinite geometric series $3 + 3(0.6) + 3(0.6)^2 + \ldots$. The sum of this series is $\frac{3}{1 - 0.6}$ or 7.5. The total distance the ball bounces up is given by the infinite geometric series $1.8(0.6) + 1.8(0.6)^2 + 1.8(0.6)^3 + \ldots$. The sum of this series is $\frac{1.8(0.6)}{1 - 0.6}$ or 2.7. Thus, the total distance the ball travels is $7.5 + 2.7$ or 10.2 feet.

Page 611, Follow-Up of Lesson 11-6
Algebra Activity

4. The von Koch snowflake has infinite perimeter. As n increases, the perimeter P_n of Stage n increases without bound. That is, the limit of $27\left(\frac{4}{3}\right)^{n-1}$ is ∞.

5. Stage 1 is an equilateral triangle with sides of length 9 units, so its area is $\frac{81\sqrt{3}}{4}$ units^2. Each subsequent stage encloses $3 \cdot 4^{n-2}$ additional equilateral triangular regions of area $\frac{81\sqrt{3}}{4 \cdot 3^{2n-2}}$ units^2. Thus, the additional area at each stage is $3 \cdot 4^{n-2} \cdot \frac{81\sqrt{3}}{4 \cdot 3^{2n-2}}$ or $\frac{4^{n-3}\sqrt{3}}{3^{2n-7}}$ units^2. This is the general term of the series for $n \geq 2$.

6. Beginning with the second term, the terms of the series in Exercise 5 form an infinite geometric series with common ratio $\frac{4}{9}$. Therefore, the sum of the whole series in Exercise 5 is $\frac{81\sqrt{3}}{4} + \frac{\frac{27\sqrt{3}}{4}}{1 - \frac{4}{9}}$ or $\frac{162\sqrt{3}}{5}$. The area of the von Koch snowflake is $\frac{162\sqrt{3}}{5}$ units^2.

7. Sample answer: No, they show that it is possible for a figure with infinite perimeter to enclose only a finite amount of area.

Page 616, Lesson 11-7

42. $\frac{12!}{7!5!}$ and $\frac{12!}{6!6!}$ represent the sixth and seventh entries in the row for $n = 12$ in Pascal's triangle. $\frac{13!}{7!6!}$ represents the seventh entry in the row for $n = 13$.

Since $\frac{13!}{7!6!}$ is below $\frac{12!}{7!5!}$ and $\frac{12!}{6!6!}$ in Pascal's triangle, $\frac{12!}{7!5!} + \frac{12!}{6!6!} = \frac{13!}{7!6!}$.

43. The coefficients in a binomial expansion give the numbers of sequences of births resulting in given numbers of boys and girls. Answers should include the following.

- $(b + g)^5 = b^5 + 5b^4g + 10b^3g^2 + 10b^2g^3 + 5bg^4 + g^5$; There is one sequence of births with all five boys, five sequences with four boys and one girl, ten sequences with three boys and two girls, ten sequences with two boys and three girls, five sequences with one boy and four girls, and one sequence with all five girls.
- The number of sequences of births that have exactly k girls in a family of n children is the coefficient of $b^{n-k}g^k$ in the expansion of $(b + g)^n$. According to the Binomial Theorem, this coefficient is $\frac{n!}{(n-k)!k!}$.

Pages 619–621, Lesson 11-8

1. Sample answers: formulas for the sums of powers of the first n positive integers and statements that expressions involving exponents of n are divisible by certain numbers

2. Mathematical induction is used to show that a statement is true. A counterexample is used to show that a statement is false.

4. **Step 1:** When $n = 1$, the left side of the given equation is 1. The right side is $\frac{1(1+1)}{2}$ or 1, so the equation is true for $n = 1$.

Step 2: Assume $1 + 2 + 3 + \ldots + k = \frac{k(k+1)}{2}$ for some positive integer k.

Step 3: $1 + 2 + 3 + \ldots + k + (k + 1)$

$$= \frac{k(k+1)}{2} + (k+1)$$
$$= \frac{k(k+1) + 2(k+1)}{2}$$
$$= \frac{(k+1)(k+2)}{2}$$

The last expression is the right side of the equation to be proved, where $n = k + 1$. Thus, the equation is true for $n = k + 1$.

Therefore, $1 + 2 + 3 + \ldots + n = \frac{n(n+1)}{2}$ for all positive integers n.

5. **Step 1:** When $n = 1$, the left side of the given equation is $\frac{1}{2}$. The right side is $1 - \frac{1}{2}$ or $\frac{1}{2}$, so the equation is true for $n = 1$.

Step 2: Assume $\frac{1}{2} + \frac{1}{2^2} + \frac{1}{2^3} + \ldots + \frac{1}{2^k} = 1 - \frac{1}{2^k}$ for some positive integer k.

Step 3:

$$\frac{1}{2} + \frac{1}{2^2} + \frac{1}{2^3} + \ldots + \frac{1}{2^k} + \frac{1}{2^{k+1}} = 1 - \frac{1}{2^k} + \frac{1}{2^{k+1}}$$
$$= 1 - \frac{2}{2^{k+1}} + \frac{1}{2^{k+1}}$$
$$= 1 - \frac{1}{2^{k+1}}$$

The last expression is the right side of the equation to be proved, where $n = k + 1$. Thus, the equation is true for $n = k + 1$.

Therefore, $\frac{1}{2} + \frac{1}{2^2} + \frac{1}{2^3} + \ldots + \frac{1}{2^n} = 1 - \frac{1}{2^n}$ for all positive integers n.

6. **Step 1:** $4^1 - 1 = 3$, which is divisible by 3. The statement is true for $n = 1$.

Step 2: Assume that $4^k - 1$ is divisible by 3 for some positive integer k. This means that $4^k - 1 = 3r$ for some whole number r.

Step 3:

$$4^k - 1 = 3r$$
$$4^k = 3r + 1$$
$$4^{k+1} = 12r + 4$$
$$4^{k+1} - 1 = 12r + 3$$
$$4^{k+1} - 1 = 3(4r + 1)$$

Since r is a whole number, $4r + 1$ is a whole number. Thus, $4^{k+1} - 1$ is divisible by 3, so the statement is true for $n = k + 1$. Therefore, $4^n - 1$ is divisible by 3 for all positive integers n.

7. **Step 1:** $5^1 + 3 = 8$, which is divisible by 4. The statement is true for $n = 1$.

Step 2: Assume that $5^k + 3$ is divisible by 4 for some positive integer k. This means that $5k + 3 = 4r$ for some positive integer r.

Step 3:

$$5^k + 3 = 4r$$
$$5^k = 4r - 3$$
$$5^{k+1} = 20r - 15$$
$$5^{k+1} + 3 = 20r - 12$$
$$5^{k+1} + 3 = 4(5r - 3)$$

Since r is a positive integer, $5r - 3$ is a positive integer. Thus, $5^{k+1} + 3$ is divisible by 4, so the statement is true for $n = k + 1$.

Therefore, $5^n + 3$ is divisible by 4 for all positive integers n.

10. **Step 1:** After the first guest has arrived, no handshakes have taken place. $\frac{1(1-1)}{2} = 0$, so the formula is correct for $n = 1$.

Step 2: Assume that after k guests have arrived, a total of $\frac{k(k-1)}{2}$ handshakes have take place, for some positive integer k.

Step 3: When the $(k + 1)$st guest arrives, he or she shakes hands with the k guests already there, so the total number of handshakes that have then taken place is $\frac{k(k-1)}{2} + k$.

$$\frac{k(k-1)}{2} + k = \frac{k(k-1) + 2k}{2}$$
$$= \frac{k[(k-1) + 2]}{2}$$
$$= \frac{k(k+1)}{2} \text{ or } \frac{(k+1)k}{2}$$

Additional Answers for Chapter 11

The last expression is the formula to be proved, where $n = k + 1$. Thus, the formula is true for $n = k + 1$.

Therefore, the total number of handshakes is $\frac{n(n-1)}{2}$ for all positive integers n.

11. **Step 1:** When $n = 1$, the left side of the given equation is 1. The right side is $1[2(1) - 1]$ or 1, so the equation is true for $n = 1$.

Step 2: Assume $1 + 5 + 9 + \ldots + (4k - 3) = k(2k - 1)$ for some positive integer k.

Step 3: $1 + 5 + 9 + \ldots + (4k - 3) + [4(k + 1) - 3]$

$$= k(2k - 1) + [4(k + 1) - 3]$$
$$= 2k^2 - k + 4k + 4 - 3$$
$$= 2k^2 + 3k + 1$$
$$= (k + 1)(2k + 1)$$
$$= (k + 1)[2(k + 1) - 1]$$

The last expression is the right side of the equation to be proved, where $n = k + 1$. Thus, the equation is true for $n = k + 1$.

Therefore, $1 + 5 + 9 + \ldots + (4n - 3) = n(2n - 1)$ for all positive integers n.

12. **Step 1:** When $n = 1$, the left side of the given equation is 2. The right side is $\frac{1[3(1) + 1]}{2}$ or 2, so the equation is true for $n = 1$.

Step 2: Assume $2 + 5 + 8 + \ldots + (3k - 1) = \frac{k(3k + 1)}{2}$ for some positive integer k.

Step 3: $2 + 5 + 8 + \ldots + (3k - 1) + [3(k + 1) - 1]$

$$= \frac{k(3k + 1)}{2} + [3(k + 1) - 1]$$
$$= \frac{k(3k + 1) + 2[3(k + 1) - 1]}{2}$$
$$= \frac{3k^2 + k + 6k + 6 - 2}{2}$$
$$= \frac{3k^2 + 7k + 4}{2}$$
$$= \frac{(k + 1)(3k + 4)}{2}$$
$$= \frac{(k + 1)[(3(k + 1) + 1]}{2}$$

The last expression is the right side of the equation to be proved, where $n = k + 1$. Thus, the equation is true for $n = k + 1$.

Therefore, $2 + 5 + 8 + \ldots + (3n - 1) = \frac{n(3n + 1)}{2}$ for all positive integers n.

13. **Step 1:** When $n = 1$, the left side of the given equation is 1^3 or 1. The right side is $\frac{1^2(1 + 1)^2}{4}$ or 1, so the equation is true for $n = 1$.

Step 2: Assume $1^3 + 2^3 + 3^3 + \ldots + k^3 = \frac{k^2(k + 1)^2}{4}$ for some positive integer k.

Step 3: $1^3 + 2^3 + 3^3 + \ldots + k^3 + (k + 1)^3$

$$= \frac{k^2(k + 1)^2}{4} + (k + 1)^3$$
$$= \frac{k^2(k + 1)^2 + 4(k + 1)^3}{4}$$
$$= \frac{(k + 1)^2[k^2 + 4(k + 1)]}{4}$$
$$= \frac{(k + 1)^2(k^2 + 4k + 4)}{4}$$
$$= \frac{(k + 1)^2(k + 2)^2}{4}$$
$$= \frac{(k + 1)^2[(k + 1) + 1]^2}{4}$$

The last expression is the right side of the equation to be proved, where $n = k + 1$. Thus, the equation is true for $n = k + 1$.

Therefore, $1^3 + 2^3 + 3^3 + \ldots + n^3 = \frac{n^2(n + 1)^2}{4}$ for all positive integers n.

14. **Step 1:** When $n = 1$, the left side of the given equation is 1^2 or 1. The right side is $\frac{1[2(1) - 1][2(1) + 1]}{3}$ or 1, so the equation is true for $n = 1$.

Step 2: Assume $1^2 + 3^2 + 5^2 + \ldots + (2k - 1)^2 = \frac{k(2k - 1)(2k + 1)}{3}$ for some positive integer k.

Step 3: $1^2 + 3^2 + 5^2 + \ldots + (2k - 1)^2 + [2(k + 1) - 1]^2$

$$= \frac{k(2k - 1)(2k + 1)}{3} + [2(k + 1) - 1]^2$$
$$= \frac{k(2k - 1)(2k + 1) + 3(2k + 1)^2}{3}$$
$$= \frac{(2k + 1)[k(2k - 1) + 3(2k + 1)]}{3}$$
$$= \frac{(2k + 1)(2k^2 - k + 6k + 3)}{3}$$
$$= \frac{(2k + 1)(2k^2 + 5k + 3)}{3}$$
$$= \frac{(2k + 1)(k + 1)(2k + 3)}{3}$$
$$= \frac{(k + 1)[2(k + 1) - 1][2(k + 1) + 1]}{3}$$

The last expression is the right side of the equation to be proved, where $n = k + 1$. Thus, the equation is true for $n = k + 1$.

Therefore, $1^2 + 3^2 + 5^2 + \ldots + (2n - 1)^2 = \frac{n(2n - 1)(2n + 1)}{3}$ for all positive integers n.

15. **Step 1:** When $n = 1$, the left side of the given equation is $\frac{1}{3}$. The right side is $\frac{1}{2}\left(1 - \frac{1}{3}\right)$ or $\frac{1}{3}$, so the equation is true for $n = 1$.

Step 2: Assume $\frac{1}{3} + \frac{1}{3^2} + \frac{1}{3^3} + \ldots + \frac{1}{3^k} = \frac{1}{2}\left(1 - \frac{1}{3^k}\right)$ for some positive integer k.

Step 3: $\frac{1}{3} + \frac{1}{3^2} + \frac{1}{3^3} + \ldots + \frac{1}{3^k} + \frac{1}{3^{k+1}}$

$$= \frac{1}{2}\left(1 - \frac{1}{3^k}\right) + \frac{1}{3^{k+1}}$$
$$= \frac{1}{2} - \frac{1}{2 \cdot 3^k} + \frac{1}{3^{k+1}}$$
$$= \frac{3^{k+1} - 3 + 2}{2 \cdot 3^{k+1}}$$
$$= \frac{3^{k+1} - 1}{2 \cdot 3^{k+1}}$$
$$= \frac{1}{2}\left(\frac{3^{k+1} - 1}{3^{k+1}}\right)$$
$$= \frac{1}{2}\left(1 - \frac{1}{3^{k+1}}\right)$$

The last expression is the right side of the equation to be proved, where $n = k + 1$. Thus, the equation is true for $n = k + 1$.

Therefore, $\frac{1}{3} + \frac{1}{3^2} + \frac{1}{3^3} + \ldots + \frac{1}{3^n} = \frac{1}{2}\left(1 - \frac{1}{3^n}\right)$ for all positive integers n.

16. **Step 1:** When $n = 1$, the left side of the given equation is $\frac{1}{4}$. The right side is $\frac{1}{3}\left(1 - \frac{1}{4}\right)$ or $\frac{1}{4}$, so the equation is true for $n = 1$.

Step 2: Assume $\frac{1}{4} + \frac{1}{4^2} + \frac{1}{4^3} + \ldots + \frac{1}{4^k} = \frac{1}{3}\left(1 - \frac{1}{4^k}\right)$ for some positive integer k.

Step 3: $\frac{1}{4} + \frac{1}{4^2} + \frac{1}{4^3} + \ldots + \frac{1}{4^k} + \frac{1}{4^{k+1}}$

$$= \frac{1}{3}\left(1 - \frac{1}{4^k}\right) + \frac{1}{4^{k+1}}$$
$$= \frac{1}{3} - \frac{1}{3 \cdot 4^k} + \frac{1}{4^{k+1}}$$
$$= \frac{4^{k+1} - 4 + 3}{3 \cdot 4^{k+1}}$$
$$= \frac{4^{k+1} - 1}{3 \cdot 4^{k+1}}$$
$$= \frac{1}{3}\left(\frac{4^{k+1} - 1}{4^{k+1}}\right)$$
$$= \frac{1}{3}\left(1 - \frac{1}{4^{k+1}}\right)$$

The last expression is the right side of the equation to be proved, where $n = k + 1$. Thus, the equation is true for $n = k + 1$.

Therefore, $\frac{1}{4} + \frac{1}{4^2} + \frac{1}{4^3} + \ldots + \frac{1}{4^n} = \frac{1}{3}\left(1 - \frac{1}{4^n}\right)$ for all positive integers n.

17. **Step 1:** $8^1 - 1 = 7$, which is divisible by 7. The statement is true for $n = 1$.

Step 2: Assume that $8^k - 1$ is divisible by 7 for some positive integer k. This means that $8^k - 1 = 7r$ for some whole number r.

Step 3: $8^k - 1 = 7r$

$$8^k = 7r + 1$$
$$8^{k+1} = 56r + 8$$
$$8^{k+1} - 1 = 56r + 7$$
$$8^{k+1} - 1 = 7(8r + 1)$$

Since r is a whole number, $8r + 1$ is a whole number. Thus, $8^{k+1} - 1$ is divisible by 7, so the statement is true for $n = k + 1$.

Therefore, $8^n - 1$ is divisible by 7 for all positive integers n.

18. **Step 1:** $9^1 - 1 = 8$, which is divisible by 8. The statement is true for $n = 1$.

Step 2: Assume that $9^k - 1$ is divisible by 8 for some positive integer k. This means that $9^k - 1 = 8r$ for some whole number r.

Step 3: $9^k - 1 = 8r$

$$9^k = 8r + 1$$
$$9^{k+1} = 72r + 9$$
$$9^{k+1} - 1 = 72r + 8$$
$$9^{k+1} - 1 = 8(9r + 1)$$

Since r is a whole number, $9r + 1$ is a whole number. Thus, $9^{k+1} - 1$ is divisible by 8, so the statement is true for $n = k + 1$.

Therefore, $9^n - 1$ is divisible by 8 for all positive integers n.

19. **Step 1:** $12^1 + 10 = 22$, which is divisible by 11. The statement is true for $n = 1$.

Step 2: Assume that $12^k + 10$ is divisible by 11 for some positive integer k. This means that $12^k + 10 = 11r$ for some positive integer r.

Step 3: $12^k + 10 = 11r$

$$12^k = 11r - 10$$
$$12^{k+1} = 132r - 120$$
$$12^{k+1} + 10 = 132r - 110$$
$$12^{k+1} + 10 = 11(12r - 10)$$

Since r is a positive integer, $12r - 10$ is a positive integer. Thus, $12^{k+1} + 10$ is divisible by 11, so the statement is true for $n = k + 1$.

Therefore, $12^n + 10$ is divisible by 11 for all positive integers n.

20. **Step 1:** $13^1 + 11 = 24$, which is divisible by 12. The statement is true for $n = 1$.

Step 2: Assume that $13^k + 11$ is divisible by 12 for some positive integer k. This means that $13^k + 11 = 12r$ for some positive integer r.

Step 3: $13^k + 11 = 12r$

$$13^k = 12r - 11$$
$$13^{k+1} = 156r - 143$$
$$13^{k+1} + 11 = 156r - 132$$
$$13^{k+1} + 11 = 12(13r - 11)$$

Since r is a positive integer, $13r - 11$ is a positive integer. Thus, $13^{k+1} + 11$ is divisible by 12, so the statement is true for $n = k + 1$.

Therefore, $13^n + 11$ is divisible by 12 for all positive integers n.

21. **Step 1:** There are 6 bricks in the top row, and $1^2 + 5(1) = 6$, so the formula is true for $n = 1$.

Step 2: Assume that there are $k^2 + 5k$ bricks in the top k rows for some positive integer k.

Step 3: Since each row has 2 more bricks than the one above, the numbers of bricks in the rows form an arithmetic sequence. The number of bricks in the $(k + 1)$st row is $6 + [(k + 1) - 1](2)$ or $2k + 6$. Then the number of bricks in the top $k + 1$ rows is $k^2 + 5k + (2k + 6)$ or $k^2 + 7k + 6$.

$k^2 + 7k + 6 = (k + 1)^2 + 5(k + 1)$, which is the formula to be proved, where $n = k + 1$. Thus, the formula is true for $n = k + 1$.

Therefore, the number of bricks in the top n rows is $n^2 + 5n$ for all positive integers n.

22. Step 1: When $n = 1$, the left side of the given equation is a_1. The right side is $\frac{a_1(1 - r^1)}{1 - r}$ or a_1, so the equation is true for $n = 1$.

Step 2: Assume $a_1 + a_1r + a_1r^2 + \ldots + a_1r^{k-1} = \frac{a_1(1 - r^k)}{1 - r}$ for some positive integer k.

Step 3: $a_1 + a_1r + a_1r^2 + \ldots + a_1r^{k-1} + a_1r^k$

$$= \frac{a_1(1 - r^k)}{1 - r} + a_1r^k$$
$$= \frac{a_1(1 - r^k) + (1 - r)a_1r^k}{1 - r}$$
$$= \frac{a_1 - a_1r^k + a_1r^k - a_1r^{k+1}}{1 - r}$$
$$= \frac{a_1(1 - r^{k+1})}{1 - r}$$

The last expression is the right side of the equation to be proved, where $n = k + 1$. Thus, the equation is true for $n = k + 1$.

Therefore, $a_1 + a_1r + a_1r^2 + \ldots + a_1r^{n-1} = \frac{a_1(1 - r^n)}{1 - r}$ for all positive integers n.

23. Step 1: When $n = 1$, the left side of the given equation is a_1. The right side is $\frac{1}{2}[2a_1 + (1 - 1)d]$ or a_1, so the equation is true for $n = 1$.

Step 2: Assume $a_1 + (a_1 + d) + (a_1 + 2d) + \ldots + [a_1 + (k - 1)d] = \frac{k}{2}[2a_1 + (k - 1)d]$ for some positive integer k.

Step 3: $a_1 + (a_1 + d) + (a_1 + 2d) + \ldots + [a_1 + (k - 1)d] + [a_1 + (k + 1 - 1)d]$

$$= \frac{k}{2}[2a_1 + (k - 1)d] + [a_1 + (k + 1 - 1)d]$$
$$= \frac{k}{2}[2a_1 + (k - 1)d] + a_1 + kd$$
$$= \frac{k[2a_1 + (k - 1)d] + 2(a_1 + kd)}{2}$$
$$= \frac{k \cdot 2a_1 + (k^2 - k)d + 2a_1 + 2kd}{2}$$
$$= \frac{(k + 1)2a_1 + (k^2 - k + 2k)d}{2}$$
$$= \frac{(k + 1)2a_1 + k(k + 1)d}{2}$$
$$= \frac{k + 1}{2}(2a_1 + kd)$$
$$= \frac{k + 1}{2}[2a_1 + (k + 1 - 1)d]$$

The last expression is the right side of the formula to be proved, where $n = k + 1$. Thus, the formula is true for $n = k + 1$.

Therefore, $a_1 + (a_1 + d) + (a_1 + 2d) + \ldots + [a_1 + (n - 1)d] = \frac{n}{2}[2a_1 + (n - 1)d]$ for all positive integers n.

24. Step 1: The figure below shows how to cover a 2^1 by 2^1 board, so the statement is true for $n = 1$.

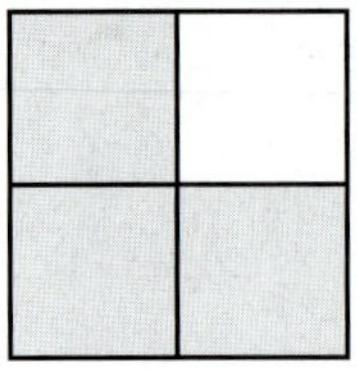

Step 2: Assume that a 2^k by 2^k board can be covered for some positive integer k.

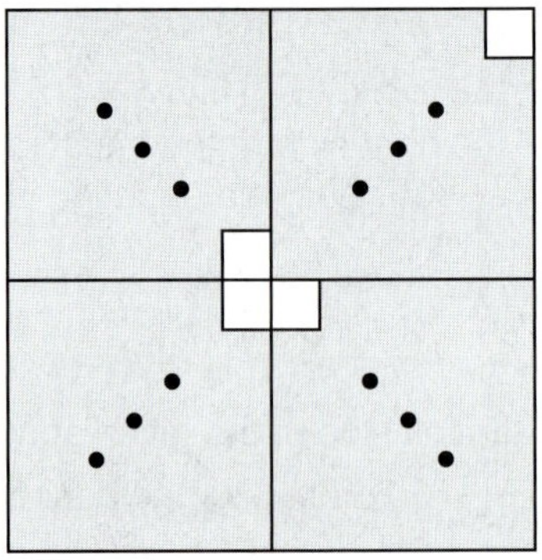

Step 3: Divide a 2^{k+1} by 2^{k+1} board into four quadrants. By the inductive hypothesis, the first quadrant can be covered. Rotate the design that covers Quadrant I 90° clockwise and use it to cover Quadrant II. Use the design that covers Quadrant I to cover Quadrant III. Rotate the design that covers Quadrant I 90° counterclockwise and use it to cover Quadrant IV. This leaves three empty squares near the center of the board, as shown. Use one more L-shaped tile to cover these 3 squares. Thus, a 2^{k+1} by 2^{k+1} board can be covered. The statement is true for $n = k + 1$.

Therefore, a 2^n by 2^n checkerboard with the top right square missing can be covered for all positive integers n.

32. An analogy can be made between mathematical induction and a ladder with the positive integers on the steps. Answers should include the following.

- Showing that the statement is true for $n = 1$ (Step 1).
- Assuming that the statement is true for some positive integer k and showing that it is true for $k + 1$ (Steps 2 and 3).

Page 627, Chapter 11 Practice Test

21. Step 1: When $n = 1$, the left side of the given equation is 1. The right side is 1^2 or 1, so the equation is true for $n = 1$.

Step 2: Assume $1 + 3 + 5 + \ldots + (2k - 1) = k^2$ for some positive integer k.

Step 3: $1 + 3 + 5 + \ldots + (2k - 1) + [2(k + 1) - 1]$

$$= k^2 + [2(k + 1) - 1]$$
$$= k^2 + 2k + 2 - 1$$
$$= k^2 + 2k + 1$$
$$= (k + 1)^2$$

The last expression is the right side of the equation to be proved, where $n = k + 1$. Thus, the equation is true for $n = k + 1$.

Therefore, $1 + 3 + 5 + \ldots + (2n - 1) = n^2$ for all positive integers n.

22. **Step 1:** $14^1 - 1 = 13$, which is divisible by 13. The statement is true for $n = 1$.

Step 2: Assume that $14^k - 1$ is divisible by 13 for some positive integer k. This means that $14^k - 1 = 13r$ for some whole number r.

Step 3: $14^k - 1 = 13r$

$$14^k = 13r + 1$$
$$14^{k+1} = 182r + 14$$
$$14^{k+1} - 1 = 182r + 13$$
$$14^{k+1} - 1 = 13(14r + 1)$$

Since r is a whole number, $14r + 1$ is a whole number. Thus, $14^{k+1} - 1$ is divisible by 13, so the statement is true for $n = k + 1$.

Therefore, $14^n - 1$ is divisible by 13 for all positive integers n.

Chapter 12 Probability and Statistics

Chapter Overview and Pacing

Year-long and two-year pacing: pages T20–T21.

LESSON OBJECTIVES

LESSON OBJECTIVES	PACING (days) Regular Basic/Average	Regular Advanced	Block Basic/Average	Block Advanced
12-1 The Counting Principle *(pp. 632–637)* • Solve problems involving independent events. • Solve problems involving dependent events.	1	1	0.5	0.5
12-2 Permutations and Combinations *(pp. 638–643)* • Solve problems involving linear permutations. • Solve problems involving combinations.	1	1	0.5	0.5
12-3 Probability *(pp. 644–650)* • Find the probability and odds of events. • Create and use graphs of probability distributions.	1	1	0.5	0.5
12-4 Multiplying Probabilities *(pp. 651–657)* • Find the probability of two independent events. • Find the probability of two dependent events.	1	1	0.5	0.5
12-5 Adding Probabilities *(pp. 658–663)* • Find the probability of mutually exclusive events. • Find the probability of inclusive events.	1	1	0.5	0.5
12-6 Statistical Measures *(pp. 664–670)* • Use measures of central tendency to represent a set of data. • Find measures of variation for a set of data.	1	1	0.5	0.5
12-7 The Normal Distribution *(pp. 671–675)* • Determine whether a set of data appears to be normally distributed or skewed. • Solve problems involving normally distributed data.	2	1	1.5	0.5
12-8 Binomial Experiments *(pp. 676–681)* • Use binomial expansions to find probabilities. • Find probabilities for binomial experiments. ***Follow-Up:*** Simulations	2	2 (with 12-8 (Follow-Up)	1	1.5 (with 12-8 (Follow-Up)
12-9 Sampling and Error *(pp. 682–686)* • Determine whether a sample is unbiased. • Find margins of sampling error. ***Follow-Up:*** Testing Hypotheses	1	2 (with 12-9 (Follow-Up)	0.5	1
Study Guide and **Practice Test** *(pp. 687–693)* **Standardized Test Practice** *(pp. 694–695)*	1	1	0.5	0.5
Chapter Assessment	1	1	0.5	0.5
TOTAL	13	13	7	7

StudentWorks

*An electronic version of this chapter is available on **StudentWorks**™. This backpack solution CD-ROM allows students instant access to the Student Edition, lesson worksheet pages, and web resources.*

Timesaving Tools

TeacherWorks™

All-In-One Planner and Resource Center

See pages T5 and T21.

Chapter Resource Manager

Study Guide and Intervention	Practice (Skills and Average)	Reading to Learn Mathematics	Enrichment	Assessment	Applications*	5-Minute Check Transparencies	Interactive Chalkboard	Alge2PASS: Tutorial Plus (lessons)	Materials
CHAPTER 12 RESOURCE MASTERS									
699–700	701–702	703	704			12-1	12-1		restaurant menu
705–706	707–708	709	710		GCS 50	12-2	12-2	22	index cards
711–712	713–714	715	716	767		12-3	12-3		
717–718	719–720	721	722		SC 23	12-4	12-4	23	compass, protractor
723–724	725–726	727	728	767, 769	GCS 49	12-5	12-5		colored chips, index cards
729–730	731–732	733	734		SM 115–118	12-6	12-6		graphing calculator
735–736	737–738	739	740	768		12-7	12-7		measuring tape
741–742	743–744	745	746		SC 24	12-8	12-8		ball (*Follow-Up:* die, tally sheet, grid paper)
747–748	749–750	751	752	768	SM 57–62	12-9	12-9		(*Follow-Up:* ruler)
				753–766, 770–772					

**Key to Abbreviations:* GCS = Graphing Calculator and Speadsheet Masters,
SC = School-to-Career Masters,
SM = Science and Mathematics Lab Manual

Chapter 12

Mathematical Connections and Background

Continuity of Instruction

Prior Knowledge

Some of the notation used in this chapter will be familiar, including factorials and binomial expansion. Also, some of the content will be familiar, including simple probability, relative frequency, and finding means, medians, and modes.

This Chapter

Students learn to represent counting situations using permutations and combinations. They describe the likelihood of single events using odds and probability, and they calculate probabilities for pairs of dependent or independent events, mutually exclusive or inclusive events, and binomial experiments. They calculate the central tendency and variation of data sets by calculating means, medians, variance, and standard deviations, and they explore normal distributions, skewed distributions, and sampling error.

Future Connections

Students will continue to use permutations, combinations, and probabilities in their math classes. They will study the mathematical underpinnings of statistical ideas in later math courses, and they will apply those statistical ideas in courses on behavioral science, psychology, economics, and many other fields.

12-1 The Counting Principle

In this lesson students investigate the Fundamental Counting Principle. The Fundamental Counting Principle states that the total number of options for a succession of choices is the product of the number of options for the individual choices. Students use exponents and factorials to express answers to counting problems.

12-2 Permutations and Combinations

The real-world situations in this lesson involve selecting some number of objects from a larger group of objects. If the order of selection is one of the attributes that differentiates among the selected objects, then the selection is called a permutation. If the order does not differentiate among the selected objects, then the selection is called a combination. As students analyze and apply the formulas for permutations, they consider situations in which some of the items in the large group are duplicates. Students also explore the relationship between permutations and combinations, which can be represented by the formula $C(n, r) = \frac{P(n, r)}{r!}$.

12-3 Probability

In this lesson, students analyze the likelihood that a particular event will happen. The likelihood of an event can be described in terms of odds and probability. Some of the mathematical properties of these expressions are that the odds of success and the odds of failure for any given event are reciprocals, that each probability is a number between 0 and 1, inclusive, and that if you add the probability of success and the probability of failure for any given event, the sum is 1. As students explore these descriptions of likelihood, they compare the probabilities for all the events in a sample space. They investigate the probabilities by looking at tables of probability distributions and by graphing those distributions as relative-frequency histograms.

12-4 Multiplying Probabilities

In this lesson, students consider the likelihood that two events will both happen and determine how that likelihood is related to the probabilities of the separate events. If two events *A* and *B* are independent, then the probability that both *A* and *B* occur is the product of the individual probabilities. If the two events are dependent, then the probability of both occurring is the product of the probability of *A* occurring times the probability of *B* occurring given that *A* occurred. Students explore problems in which they calculate values of $P(A)$, $P(B)$, and $P(B \text{ following } A)$, and use those values to calculate the value $P(A \text{ and } B)$.

12-5 Adding Probabilities

This lesson considers the likelihood that at least one of two events will happen, and relates that likelihood to the probabilities of the separate events. If it is not possible that two events *A* and *B* both occur, then *A* and *B* are mutually exclusive and $P(A \text{ or } B)$ is $P(A) + P(B)$. If two events are not mutually exclusive, then $P(A \text{ or } B)$ is the probability that *A* will happen, plus the probability that *B* will happen, minus the probability that both will happen. Formulas can clarify the relationship between mutually exclusive and inclusive events. Starting with $P(A \text{ or } B) = P(A) + P(B) - P(A \text{ and } B)$, if $P(A \text{ and } B) = 0$ then the events cannot both happen, so they are mutually exclusive. In that case, $P(A \text{ or } B) = P(A) + P(B)$.

12-6 Statistical Measures

Students investigate how the values of a data set are distributed. They will choose the most appropriate measure of central tendency for a given set of data. For the dispersion of the data, they find the *variance* by using a formula whose key step is to look at how the individual data values differ from the mean of the set. They also calculate the *standard deviation*, which is the square root of the variance.

12-7 The Normal Distribution

For a large data set, the heights of the bars of a relative-frequency histogram can be replaced with a curve. A curve is a normal distribution curve if the probability distribution curve is symmetric and the mean, median, and mode are indicated by the peak of the curve. Another condition for a distribution to be normal involves the percent of data values that are within one, two, or three standard deviations of the mean. A data set with a long tail above the mean is positively skewed, while a data set with a long tail below the mean is negatively skewed.

12-8 Binomial Experiments

One or more terms of the binomial expansion $(p + q)^n$ can be used to calculate the probability for a binomial experiment. In a binomial experiment there are exactly two outcomes for each trial, there is a fixed number of trials, each trial is independent, and the probability of success or failure is the same for each trial. Tossing a coin five times is an example of a binomial experiment because each of these conditions is met.

12-9 Sampling and Error

In this lesson, students investigate sampling. They discuss how the response from a sample reflects what the responses might be from the entire population. If everyone in the population has an equal chance to be in the sample, then the sample is called an unbiased or random sample. For unbiased samples, students will describe the difference between sample and population responses by calculating the margin of sampling error (ME). If some percent p of people in a sample answer a question in a particular way, then for that question the percent of the population expected to answer the same way will be in the interval $p \pm ME$. A formula lets students calculate the ME based on the sample size and the value of p.

Additional mathematical information and teaching notes are available at www.algebra2.com/key_concepts.

Chapter 12

DAILY INTERVENTION and Assessment

	Type	Student Edition	Teacher Resources	Technology/Internet
INTERVENTION	Ongoing	Prerequisite Skills, pp. 631, 637, 643, 650, 657, 663, 670, 675, 680 Practice Quiz 1, p. 650 Practice Quiz 2, p. 670	5-Minute Check Transparencies Quizzes, *CRM* pp. 767–768 Mid-Chapter Test, *CRM* p. 769 Study Guide and Intervention, *CRM* pp. 699–700, 705–706, 711–712, 717–718, 723–724, 729–730, 735–736, 741–742, 747–748	Alge2PASS: Tutorial Plus www.algebra2.com/self_check_quiz www.algebra2.com/extra_examples
	Mixed Review	pp. 637, 643, 650, 657, 663, 670, 675, 681, 685	Cumulative Review, *CRM* p. 770	
	Error Analysis	Find the Error, pp. 654, 660 Common Misconceptions, p. 659	Find the Error, *TWE* pp. 654, 660 Unlocking Misconceptions, *TWE* p. 639 Tips for New Teachers, *TWE* pp. 648, 668	
ASSESSMENT	Standardized Test Practice	pp. 633, 634, 636, 642, 649, 657, 662, 669, 675, 680, 685, 693, 694–695	*TWE* p. 633 Standardized Test Practice, *CRM* pp. 771–772	Standardized Test Practice CD-ROM www.algebra2.com/standardized_test
	Open-Ended Assessment	Writing in Math, pp. 636, 642, 649, 657, 662, 669, 675, 679, 685 Open Ended, pp. 634, 641, 647, 654, 660, 666, 673, 678, 683	Modeling: *TWE* pp. 650, 663 Speaking: *TWE* pp. 643, 657, 680, 684 Writing: *TWE* pp. 637, 670, 675 Open-Ended Assessment, *CRM* p. 765	
	Chapter Assessment	Study Guide, pp. 687–692 Practice Test, p. 693	Multiple-Choice Tests (Forms 1, 2A, 2B), *CRM* pp. 753–758 Free-Response Tests (Forms 2C, 2D, 3), *CRM* pp. 759–764 Vocabulary Test/Review, *CRM* p. 766	ExamView® Pro (see below) MindJogger Videoquizzes www.algebra2.com/vocabulary_review www.algebra2.com/chapter_test

Key to Abbreviations: TWE = Teacher Wraparound Edition; CRM = Chapter Resource Masters

Intervention Technology

Alge2PASS: Tutorial Plus CD-ROM offers a complete, self-paced algebra curriculum.

Algebra 2 Lesson	Alge2PASS Lesson
12-2	**22** *Combinations and Permutations*
12-4	**23** *Integration: Introduction to Probability*

ExamView® Pro

Use the networkable **ExamView® Pro** to:

- Create **multiple versions** of tests.
- Create **modified** tests for *Inclusion* students.
- **Edit** existing questions and **add** your own questions.
- Use built-in **state curriculum correlations** to create tests aligned with state standards.
- **Apply** art to your tests from a program bank of artwork.

For more information on Intervention and Assessment, see pp. T8–T11.

Reading and Writing in Mathematics

Glencoe Algebra 2 provides numerous opportunities to incorporate reading and writing into the mathematics classroom.

Student Edition

- Foldables Study Organizer, p. 631
- Concept Check questions require students to verbalize and write about what they have learned in the lesson. (pp. 634, 641, 647, 654, 660, 666, 673, 678, 683, 687)
- Writing in Math questions in every lesson, pp. 636, 642, 649, 657, 662, 669, 675, 679, 685
- Reading Study Tip, pp. 633, 638, 644, 646, 665, 669
- WebQuest, pp. 635, 685

Teacher Wraparound Edition

- Foldables Study Organizer, pp. 631, 687
- Study Notebook suggestions, pp. 635, 641, 647, 654, 660, 667, 673, 678, 681, 684, 686
- Modeling activities, pp. 650, 663
- Speaking activities, pp. 643, 657, 680, 684
- Writing activities, pp. 637, 670, 675
- Differentiated Instruction, (Verbal/Linguistic), p. 683
- ELL Resources, pp. 630, 636, 642, 649, 656, 662, 669, 674, 679, 683, 685, 687

Additional Resources

- Vocabulary Builder worksheets require students to define and give examples for key vocabulary terms as they progress through the chapter. (*Chapter 12 Resource Masters*, pp. vii-viii)
- Reading to Learn Mathematics master for each lesson (*Chapter 12 Resource Masters*, pp. 703, 709, 715, 721, 727, 733, 739, 745, 751)
- *Vocabulary PuzzleMaker* software creates crossword, jumble, and word search puzzles using vocabulary lists that you can customize.
- *Teaching Mathematics with Foldables* provides suggestions for promoting cognition and language.
- *Reading and Writing in the Mathematics Classroom*
- *WebQuest and Project Resources*

For more information on Reading and Writing in Mathematics, see pp. T6–T7.

ELL ENGLISH LANGUAGE LEARNERS

Lesson 12-1

Reading and Writing

Give groups of students a list of courses offered in your school. Have students discuss choosing courses for their schedule. Are the events independent or dependent? Have them write an explanation defending their answer.

Lesson 12-2

Using Manipulatives

Give groups of students actual books of different types or flashcards showing book titles and subjects. For example, you might give a group 3 novels, 2 nonfiction books, and 1 biography. Have the students use the books to model permutations and combinations. Ask questions like, "In how many ways can the books be arranged on a shelf?" and "In how many ways can I choose to read 1 novel and 1 nonfiction book?"

Lesson 12-6

Using Applications

Give groups of students newspapers or magazines. Have students find uses of data in the media. Have them discuss how they think the data were gathered and what statistical measures were used to analyze the data. For example, if an article states an "average," what statistical measure was used and why?

Chapter 12 Notes

What You'll Learn

Have students read over the list of objectives and make a list of any words with which they are not familiar.

Why It's Important

Point out to students that this is only one of many reasons why each objective is important. Others are provided in the introduction to each lesson.

Lesson	NCTM Standards	Local Objectives
12-1	1, 5, 6, 8, 9, 10	
12-2	1, 5, 6, 8, 9, 10	
12-3	1, 5, 6, 8, 9, 10	
12-4	1, 5, 6, 8, 9, 10	
12-5	1, 5, 6, 8, 9, 10	
12-6	1, 5, 6, 8, 9, 10	
12-7	1, 5, 6, 8, 9, 10	
12-8	1, 5, 6, 8, 9, 10	
12-8 Follow-Up	1, 5, 6, 9, 10	
12-9	1, 5, 6, 8, 9, 10	
12-9 Follow-Up	5, 7, 8, 9, 10	

Key to NCTM Standards:

1=Number & Operations, 2=Algebra, 3=Geometry, 4=Measurement, 5=Data Analysis & Probability, 6=Problem Solving, 7=Reasoning & Proof, 8=Communication, 9=Connections, 10=Representation

Chapter 12 Probability and Statistics

What You'll Learn

- **Lessons 12-1 and 12-2** Solve problems involving independent events, dependent events, permutations, and combinations.
- **Lessons 12-3, 12-4, 12-5, and 12-8** Find probability and odds.
- **Lesson 12-6** Find statistical measures.
- **Lesson 12-7** Use the normal distribution.
- **Lesson 12-9** Determine whether a sample is unbiased.

Key Vocabulary

- permutation (p. 638)
- combination (p. 640)
- probability (p. 644)
- measures of central tendency (p. 664)
- measures of variation (p. 665)

Why It's Important

Being able to analyze data is an important skill for every citizen. Business decision-makers rely on statistical measures to ensure quality products, medical researchers test and design new treatments by performing experiments with sample populations, and sports coaches use probabilities to design a winning team.

Each day during a presidential election campaign, journalists report the results of public opinion polls. Pollsters must make sure that the sample they choose accurately represents all of the voters. *You will investigate how opinion polls are used in political campaigns in Lesson 12-9.*

Vocabulary Builder

ELL

The Key Vocabulary list introduces students to some of the main vocabulary terms included in this chapter. For a more thorough vocabulary list with pronunciations of new words, give students the Vocabulary Builder worksheets found on pages vii and viii of the *Chapter 12 Resource Masters*. Encourage them to complete the definition of each term as they progress through the chapter. You may suggest that they add these sheets to their study notebooks for future reference when studying for the Chapter 12 test.

Getting Started

Prerequisite Skills To be successful in this chapter, you'll need to master these skills and be able to apply them in problem-solving situations. Review these skills before beginning Chapter 12.

For Lesson 12-3 — Find Simple Probability

Find each probability if a die is rolled once. **6.** $\frac{5}{6}$

1. $P(2)$ $\frac{1}{6}$
2. $P(5)$ $\frac{1}{6}$
3. P(even number) $\frac{1}{2}$
4. P(odd number) $\frac{1}{2}$
5. P(numbers less than 5) $\frac{2}{3}$
6. P(numbers greater than 1)

For Lesson 12-6 — Box-and-Whisker Plots

Make a box-and-whisker plot for each set of data. *(For review, see pages 826 and 827.)*

7. {24, 32, 38, 38, 26, 33, 37, 39, 23, 31, 40, 21} **7–10. See margin.**
8. {25, 46, 31, 53, 39, 59, 48, 43, 68, 64, 29}
9. {51, 69, 46, 27, 60, 53, 55, 39, 81, 54, 46, 23}
10. {13.6, 15.1, 14.9, 15.7, 16.0, 14.1, 16.3, 14.3, 13.8}

For Lesson 12-6 — Evaluate Expressions

Evaluate $\sqrt{\frac{(a-b)^2+(c-b)^2}{d}}$ for each set of values. *(For review, see Lesson 5-6.)*

11. $a = 4, b = 7, c = 1, d = 5$ **3**
12. $a = 2, b = 6, c = 9, d = 5$ $\sqrt{5}$
13. $a = 5, b = 1, c = 7, d = 4$ $\sqrt{13}$
14. $a = 3, b = 4, c = 11, d = 10$ $\sqrt{5}$

For Lesson 12-8 — Expand Binomials

15. $a^3 + 3a^2b + 3ab^2 + b^3$

Expand each binomial. *(For review, see Lesson 5-2.)* **16.** $c^4 + 4c^3d + 6c^2d^2 + 4cd^3 + d^4$

15. $(a + b)^3$
16. $(c + d)^4$
17. $(m - n)^5$
18. $(x + y)^6$

17. $m^5 - 5m^4n + 10m^3n^2 - 10m^2n^3 + 5mn^4 - n^5$

18. $x^6 + 6x^5y + 15x^4y^2 + 20x^3y^3 + 15x^2y^4 + 6xy^5 + y^6$

FOLDABLES™ Study Organizer

Probability and Statistics Make this Foldable to help you organize your notes. Begin with one sheet of 11" by 17" paper.

Step 1 Fold

Fold 2" tabs on each of the short sides.

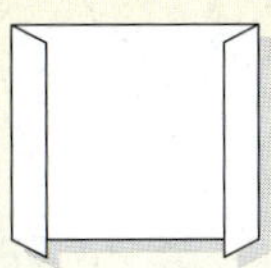

Step 2 Fold and Cut

Then fold in half in both directions. Open and cut as shown.

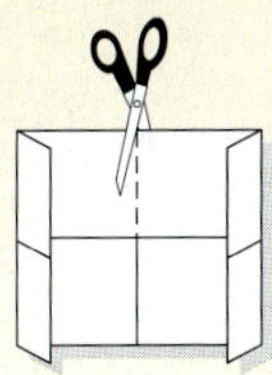

Step 3 Staple and Label

Refold along the width. Staple each pocket. Label pockets as *The Counting Principle, Permutations and Combinations, Probability,* and *Statistics.*

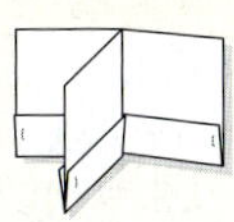

Reading and Writing As you read and study the chapter, you can write notes and examples on index cards and store the cards in the Foldable pockets.

Getting Started

This section provides a review of the basic concepts needed before beginning Chapter 12. Page references are included for additional student help.

Prerequisite Skills in the Getting Ready for the Next Lesson section at the end of each exercise set review a skill needed in the next lesson.

For Lesson	Prerequisite Skill
12-2	Factorials (p. 637)
12-3	Evaluating Expressions (p. 643)
12-6	Mean, Median, Mode, and Range (p. 663)
12-8	Binomial Expansions (p. 675)
12-9	Radical Expressions (p. 680)

Answers

7.

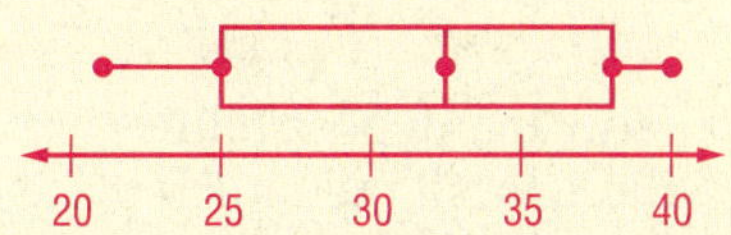

8.

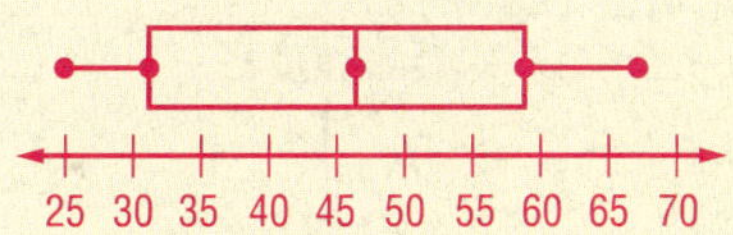

9.

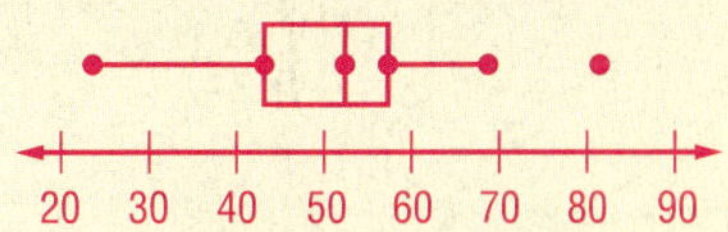

10.

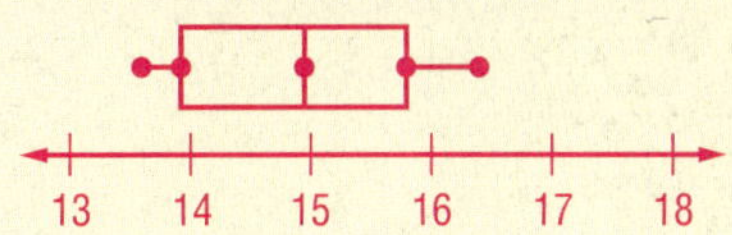

FOLDABLES™ Study Organizer

For more information about Foldables, see *Teaching Mathematics with Foldables.*

Organization of Data and Statistics in Writing After students make their Foldable, have them label the four pockets with the key topics of this chapter—The Counting Principle, Permutations and Combinations, Probability, and Statistics. Throughout the chapter, students might record examples of probability and statistics they see in everyday print (newspapers, magazines, and advertisements). They should note how writers use statistics to prove or disprove points of view and discuss the ethical responsibilities writers have when using statistics.

12-1 Lesson Notes

1 Focus

5-Minute Check Transparency 12-1 Use as a quiz or review of Chapter 11.

Mathematical Background notes are available for this lesson on p. 630C.

How can you count the maximum number of license plates a state can issue?

Ask students:

- How many letters are there on the license plate? how many digits? **3; 3**
- How many possibilities are there to fill the first place on this plate? **26 (assuming all letters are possibilities)**
- How many possibilities are there to fill the fourth place on this plate? **10 (assuming all digits are possibilities)**

12-1 The Counting Principle

What You'll Learn

- Solve problems involving independent events.
- Solve problems involving dependent events.

Vocabulary

- outcomes
- sample space
- event
- independent events
- Fundamental Counting Principle
- dependent events

How can you count the maximum number of license plates a state can issue?

Most states have letters and digits on their license plates. The number of possible plates is too great to count by listing all of the possibilities. It is much more efficient to count the number of possibilities by using the Fundamental Counting Principle.

INDEPENDENT EVENTS An **outcome** is the result of a single trial. For example, the trial of flipping a coin once has two outcomes: head or tail. The set of all possible outcomes is called the **sample space**. An **event** consists of one or more outcomes of a trial. The choices of letters and digits to be put on a license plate are called **independent events** because each letter or digit chosen does *not* affect the choices for the others.

For situations in which the number of choices leads to a small number of total possibilities, you can use a tree diagram or a table to count them.

Example 1 Independent Events

FOOD A sandwich cart offers customers a choice of hamburger, chicken, or fish on either a plain or a sesame seed bun. How many different combinations of meat and a bun are possible?

First, note that the choice of the type of meat does not affect the choice of the type of bun, so these events are independent.

Method 1 Tree Diagram

Let H represent hamburger, C, chicken, F, fish, P, plain, and S, sesame seed. Make a tree diagram in which the first row shows the choice of meat and the second row shows the choice of bun.

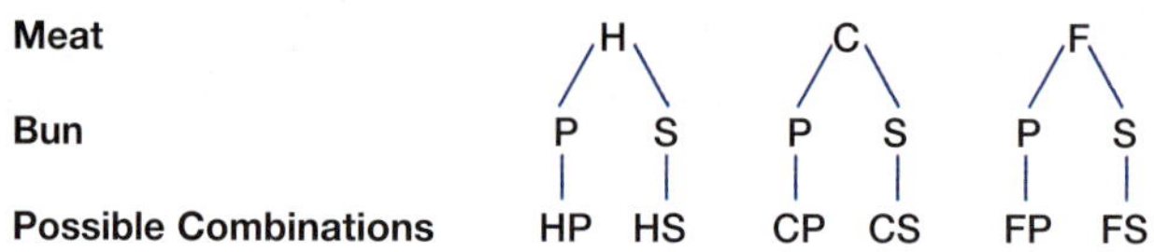

There are six possible outcomes.

Method 2 Make a Table

Make a table in which each row represents a type of meat and each column represents a type of bun.

This method also shows that there are six outcomes.

		Bun	
		Plain	Sesame
Meat	Hamburger	HP	HS
	Chicken	CP	CS
	Fish	FP	FS

Resource Manager

Workbook and Reproducible Masters

Chapter 12 Resource Masters

- Study Guide and Intervention, pp. 699–700
- Skills Practice, p. 701
- Practice, p. 702
- Reading to Learn Mathematics, p. 703
- Enrichment, p. 704

Transparencies

5-Minute Check Transparency 12-1
Answer Key Transparencies

Technology

Interactive Chalkboard

Notice that in Example 1, there are 3 ways to choose the type of meat, 2 ways to choose the type of bun, and $3 \cdot 2$ or 6 total ways to choose a combination of the two. This illustrates the **Fundamental Counting Principle**.

Key Concept — Fundamental Counting Principle

- **Words** If event M can occur in m ways and event N can occur in n ways, then event M followed by event N can occur in $m \cdot n$ ways.
- **Example** If event M can occur in 2 ways and event N can occur in 3 ways, then M followed by N can occur in $2 \cdot 3$ or 6 ways.

This rule can be extended to any number of events.

Standardized Test Practice

Example 2 Fundamental Counting Principle

Multiple-Choice Test Item

Kim won a contest on a radio station. The prize was a restaurant gift certificate and tickets to a sporting event. She can select one of three different restaurants and tickets to a football, baseball, basketball, or hockey game. How many different ways can she select a restaurant followed by a sporting event?

(A) 7 (B) 12 (C) 15 (D) 16

Read the Test Item

Her choice of a restaurant does not affect her choice of a sporting event, so these events are independent.

Solve the Test Item

There are 3 ways she can choose a restaurant and there are 4 ways she can choose the sporting event. By the Fundamental Counting Principle, there are $3 \cdot 4$ or 12 total ways she can choose her two prizes. The answer is B.

Test-Taking Tip
Remember that you can check your answer by making a tree diagram or a table showing the outcomes.

The Fundamental Counting Principle can be used to count the number of outcomes possible for any number of successive events.

Example 3 More than Two Independent Events

COMMUNICATION Many answering machines allow owners to call home and get their messages by entering a 3-digit code. How many codes are possible?

The choice of any digit does not affect the other two digits, so the choices of the digits are independent events.

There are 10 possible first digits in the code, 10 possible second digits, and 10 possible third digits. So, there are $10 \cdot 10 \cdot 10$ or 1000 possible different code numbers.

DEPENDENT EVENTS Some situations involve dependent events. With **dependent events**, the outcome of one event *does* affect the outcome of another event. The Fundamental Counting Principle applies to dependent events as well as independent events.

www.algebra2.com/extra_examples

2 Teach

INDEPENDENT EVENTS

In-Class Examples Power Point®

1. A sandwich menu offers customers a choice of white, wheat, or rye bread with one spread chosen from butter, mustard, or mayonnaise. How many different combinations of bread and spread are possible? **9**

Teaching Tip Make sure students know how to read a tree diagram so that they can identify the possibilities.

2. For their vacation, the Murray family is choosing a trip to the beach or to the mountains. They can select their transportation from a car, plane, or train. How many different ways can they select a destination followed by a means of transportation? **C**

A 2 **B** 5
C 6 **D** 9

3. How many codes are possible if the code is just two digits? **100**

Standardized Test Practice **Example 2** Have students draw tree diagrams to show the possible prize outcomes. Make sure students recognize that which restaurant is chosen has no affect on the choice of sporting event Kim attends.

Interactive Chalkboard PowerPoint® Presentations

This CD-ROM is a customizable Microsoft® PowerPoint® presentation that includes:

- Step-by-step, dynamic solutions of each In-Class Example from the Teacher Wraparound Edition
- Additional, Your Turn exercises for each example
- The 5-Minute Check Transparencies
- Hot links to Glencoe Online Study Tools

DEPENDENT EVENTS

In-Class Example PowerPoint®

4 Refer to the table in Example 4 in the Student Edition. How many different schedules could a student have who is planning to take only 4 different classes? **24**

Study Tip

Look Back
To review **factorials**, see Lesson 11-7.

Example 4 Dependent Events

SCHOOL Charlita wants to take 6 different classes next year. Assuming that each class is offered each period, how many different schedules could she have?

When Charlita schedules a given class for a given period, she cannot schedule that class for any other period. Therefore, the choices of which class to schedule each period are dependent events.

There are 6 classes Charlita can take during first period. That leaves 5 classes she can take second period. After she chooses which classes to take the first two periods, there are 4 remaining choices for third period, and so on.

Period	1st	2nd	3rd	4th	5th	6th
Number of Choices	6	5	4	3	2	1

There are $6 \cdot 5 \cdot 4 \cdot 3 \cdot 2 \cdot 1$ or 720 schedules that Charlita could have.
Note that $6 \cdot 5 \cdot 4 \cdot 3 \cdot 2 \cdot 1 = 6!$.

Concept Summary — Independent and Dependent Events

- **Words** If the outcome of an event does *not* affect the outcome of another event, the two events are *independent*.
- **Example** Tossing a coin and rolling a die are independent events.

- **Words** If the outcome of an event *does* affect the outcome of another event, the two events are *dependent*.
- **Example** Taking a piece of candy from a jar and then taking a second piece without replacing the first are dependent events because taking the first piece affects what is available to be taken next.

Check for Understanding

Concept Check

1. **List** the possible outcomes when a coin is tossed three times. Use H for heads and T for tails. **HHH, HHT, HTH, HTT, THH, THT, TTH, TTT**
2. **OPEN ENDED** Describe a situation in which you can use the Fundamental Counting Principle to show that there are 18 total possibilities.
3. **Explain** how choosing to buy a car or a pickup truck and then selecting the color of the vehicle could be dependent events. **The available colors for the car could be different from those for the truck.**

2. Sample answer: buying a shirt that comes in 3 sizes and 6 colors

Guided Practice

GUIDED PRACTICE KEY	
Exercises	Examples
4–9	1–4

State whether the events are *independent* or *dependent*.

4. choosing the color and size of a pair of shoes **independent**
5. choosing the winner and runner-up at a dog show **dependent**

Solve each problem.

6. An ice cream shop offers a choice of two types of cones and 15 flavors of ice cream. How many different 1-scoop ice cream cones can a customer order? **30**
7. Lance's math quiz has eight true-false questions. How many different choices for giving answers to the eight questions are possible? **256**
8. For a college application, Macawi must select one of five topics on which to write a short essay. She must also select a different topic from the list for a longer essay. How many ways can she choose the topics for the two essays? **20**

Standardized Test Practice

9. A bookshelf holds 4 different biographies and 5 different mystery novels. How many ways can one book of each type be selected? **D**
   (A) 1 (B) 9 (C) 10 (D) 20

DAILY INTERVENTION

Differentiated Instruction

Interpersonal Have students work in pairs or small groups. Give each group a menu from a neighborhood restaurant, or have them design a brief menu. Then ask each group to use their menu to write, and answer, four problems similar to Examples 1 through 4. Have groups exchange problems and solve.

★ indicates increased difficulty

Practice and Apply

Homework Help

For Exercises	See Examples
10–23, 25–27	1–4

Extra Practice See page 854.

12. independent

State whether the events are *independent* or *dependent*.

10. choosing a president, vice president, secretary, and treasurer for Student Council, assuming that a person can hold only one office **dependent**
11. selecting a fiction book and a nonfiction book at the library **independent**
12. Each of six people guess the total number of points scored in a basketball game. Each person writes down his or her guess without telling what it is.
13. The letters A through Z are written on pieces of paper and placed in a jar. Four of them are selected one after the other without replacing any of them. **dependent**

Solve each problem.

14. Tim wants to buy one of three different albums he sees in a music store. Each is available on tape and on CD. From how many combinations of album and format does he have to choose? **6**
15. A video store has 8 new releases this week. Each is available on videotape and on DVD. How many ways can a customer choose a new release and a format to rent? **16**
16. Carlos has homework to do in math, chemistry, and English. How many ways can he choose the order in which to do his homework? **6**
17. The menu for a banquet has a choice of 2 types of salad, 5 main courses, and 3 desserts. How many ways can a salad, main course, and dessert be selected to form a meal? **30**
18. A golf club manufacturer makes drivers with 4 different shaft lengths, 3 different lofts, 2 different grips, and 2 different club head materials. How many different combinations are possible? **48**
19. Each question on a five-question multiple-choice quiz has answer choices labeled A, B, C, and D. How many different ways can a student answer the five questions? **1024**
20. ★ How many ways can six different books be arranged on a shelf if one of the books is a dictionary and it must be on an end? **240**
21. ★ In how many orders can eight actors be listed in the opening credits of a movie if the leading actor must be listed first or last? **10,080**

WebQuest

You can use the Fundamental Counting Principle to list possible outcomes in games. Visit www.algebra2.com/webquest to continue work on your WebQuest project.

22. **PASSWORDS** Abby is registering at a Web site. She must select a password containing 6 numerals to be able to use the site. How many passwords are allowed if no digit may be used more than once? **151,200**
23. **ENTERTAINMENT** Solve the problem in the comic strip below. Assume that the books are all different. **362,880**

Peanuts®

24. **CRITICAL THINKING** The members of the Math Club need to elect a president and a vice-president. They determine that there are a total of 272 ways that they can fill the positions with two different members. How many people are in the Math Club? **17**

www.algebra2.com/self_check_quiz

3 Practice/Apply

Study Notebook

Have students—
- add the definitions/examples of the vocabulary terms to their Vocabulary Builder worksheets for Chapter 12.
- include their own examples of both independent and dependent events.
- include any other item(s) that they find helpful in mastering the skills in this lesson.

About the Exercises...

Organization by Objective
- **Independent Events:** 11, 12
- **Dependent Events:** 10, 13

Odd/Even Assignments
Exercises 10–21 are structured so that students practice the same concepts whether they are assigned odd or even problems.

Alert! Exercise 28 involves research on the Internet or other reference materials.

Assignment Guide

Basic: 11–19 odd, 23–25, 29–31, 34–63

Average: 11–25 odd, 29–31, 34–63 (optional: 32, 33)

Advanced: 10–24 even, 26–55 (optional: 56–63)

Lesson 12-1 The Counting Principle 635

Study Guide and Intervention, p. 699 (shown) and p. 700

Independent Events If the outcome of one event does not affect the outcome of another event and vice versa, the events are called **independent events**.

Fundamental Counting Principle	If event M can occur in m ways and is followed by event N that can occur in n ways, then the event M followed by the event N can occur in $m \cdot n$ ways.

Example **FOOD For the Breakfast Special at the Country Pantry, customers can choose their eggs scrambled, fried, or poached, whole wheat or white toast, and either orange, apple, tomato, or grapefruit juice. How many different Breakfast Specials can a customer order?**

A customer's choice of eggs does not affect his or her choice of toast or juice, so the events are independent. There are 3 ways to choose eggs, 2 ways to choose toast, and 4 ways to choose juice. By the Fundamental Counting Principle, there are $3 \cdot 2 \cdot 4$ or 24 ways to choose the Breakfast Special.

Exercises

Solve each problem.

1. The Palace of Pizza offers small, medium, or large pizzas with 14 different toppings available. How many different one-topping pizzas do they serve? **42**
2. The letters A, B, C, and D are used to form four-letter passwords for entering a computer file. How many passwords are possible if letters can be repeated? **256**
3. A restaurant serves 5 main dishes, 3 salads, and 4 desserts. How many different meals could be ordered if each has a main dish, a salad, and a dessert? **60**
4. Marissa brought 8 T-shirts and 6 pairs of shorts to summer camp. How many different outfits consisting of a T-shirt and a pair of shorts does she have? **48**
5. There are 6 different packages available for school pictures. The studio offers 5 different backgrounds and 2 different finishes. How many different options are available? **60**
6. How many 5-digit even numbers can be formed using the digits 4, 6, 7, 2, 8 if digits can be repeated? **2500**
7. How many license plate numbers consisting of three letters followed by three numbers are possible when repetition is allowed? **17,576,000**
8. How many 4-digit positive even integers are there? **4500**

Skills Practice, p. 701 and Practice, p. 702 (shown)

State whether the events are *independent* or *dependent*.

1. choosing an ice cream flavor and choosing a topping for the ice cream **independent**
2. choosing an offensive player of the game and a defensive player of the game in a professional football game **independent**
3. From 15 entries in an art contest, a camp counselor chooses first, second, and third place winners. **dependent**
4. Jillian is selecting two more courses for her block schedule next semester. She must select one of three morning history classes and one of two afternoon math classes. **independent**

Solve each problem.

5. A briefcase lock has 3 rotating cylinders, each containing 10 digits. How many numerical codes are possible? **1000**
6. A golf club manufacturer makes irons with 7 different shaft lengths, 3 different grips, 5 different lies, and 2 different club head materials. How many different combinations are offered? **210**
7. There are five different routes that a commuter can take from her home to the office. In how many ways can she make a round trip if she uses a different route coming than going? **20**
8. In how many ways can the four call letters of a radio station be arranged if the first letter must be W or K and no letters repeat? **27,600**
9. How many 7-digit phone numbers can be formed if the first digit cannot be 0 or 1, and any digit can be repeated? **8,000,000**
10. How many 7-digit phone numbers can be formed if the first digit cannot be 0, and any digit can be repeated? **9,000,000**
11. How many 7-digit phone numbers can be formed if the first digit cannot be 0 or 1, and if no digit can be repeated? **483,840**
12. How many 7-digit phone numbers can be formed if the first digit cannot be 0, and if no digit can be repeated? **544,320**
13. How many 6-character passwords can be formed if the first character is a digit and the remaining 5 characters are letters that can be repeated? **118,813,760**
14. How many 6-character passwords can be formed if the first and last characters are digits and the remaining characters are letters? Assume that any character can be repeated. **45,697,600**

Reading to Learn Mathematics, p. 703 ELL

Pre-Activity **How can you count the maximum number of license plates a state can issue?**

Read the introduction to Lesson 12-1 at the top of page 632 in your textbook.

Assume that all Florida license plates have three letters followed by three digits, and that there are no rules against using the same letter or number more than once. How many choices are there for each letter? for each digit? **26; 10**

Reading the Lesson

1. Shamim is signing up for her classes. Most of her classes are required, but she has two electives. For her arts class, she can chose between Art, Band, Chorus, or Drama. For her language class, she can choose between French, German, and Spanish.

   a. To organize her choices, Shamim decides to make a tree diagram. Let A, B, C, and D represent Art, Band, Chorus, and Drama, and F, G, and S represent French, German, and Spanish. Complete the following diagram.

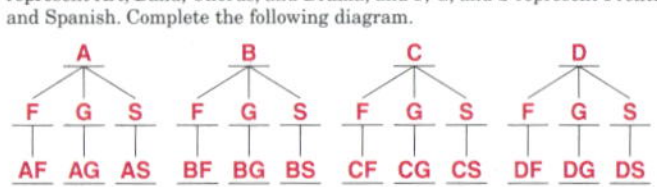

   b. How could Shamim have found the number of possible combinations without making a tree diagram? **Sample answer: Multiply the number of choices for her arts class by the number of choices for her language class: 3 × 4 = 12.**

2. A jar contains 6 red marbles, 4 blue marbles, and 3 yellow marbles. Indicate whether the events described are *dependent* or *independent*.

   a. A marble is drawn out of the jar and is not replaced. A second marble is drawn. **dependent**

   b. A marble is drawn out of the jar and is put back in. The jar is shaken. A second marble is drawn. **independent**

Helping You Remember

3. One definition of *independent* is "not determined or influenced by someone or something else." How can this definition help you remember the difference between *independent* and *dependent* events? **Sample answer: If the outcome of one event does *not* affect or influence the outcome of another, the events are independent. If the outcome of one event *does* affect or influence the outcome of another, the events are dependent.**

25. **HOME SECURITY** How many different 5-digit codes are possible using the keypad shown at the right if the first digit cannot be 0 and no digit may be used more than once? **27,216**

More About. . .

Area Codes

Before 1995, area codes had the following format.

(XYZ)

X = 2, 3, …, or 9

Y = 0 or 1

Z = 0, 1, 2, …, or 9

Source: www.nanpa.com

AREA CODES **For Exercises 26 and 27, refer to the information about telephone area codes at the left.**

26. How many area codes were possible before 1995? **160**
27. In 1995, the restriction on the middle digit was removed, allowing any digit in that position. How many total codes were possible after this change was made? **800**
28. **RESEARCH** Use the Internet or other resource to find the configuration of letters and numbers on license plates in your state. Then find the number of possible plates. **See students' work.**
29. **WRITING IN MATH** Answer the question that was posed at the beginning of the lesson. **See margin.**

    How can you count the maximum number of license plates a state can issue?

    Include the following in your answer:
    - an explanation of how to use the Fundamental Counting Principle to find the number of different license plates in a state such as Florida, which has 3 letters followed by 3 numbers, and
    - a way that a state can increase the number of possible plates without increasing the length of the plate number.

Standardized Test Practice

30. How many numbers between 100 and 999, inclusive, have 7 in the tens place? **A**

    (A) 90 (B) 100 (C) 110 (D) 120

31. A coin is tossed four times. How many possible sequences of heads or tails are possible? **C**

    (A) 4 (B) 8 (C) 16 (D) 32

Extending the Lesson

For Exercises 32 and 33, use the following information.
A **finite graph** is a collection of points, called **vertices**, and segments, called **edges**, connecting the vertices. For example, the graph shown at the right has 4 vertices and 2 edges.

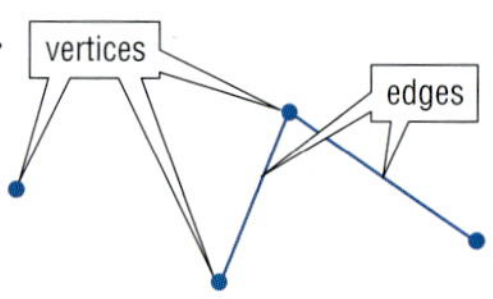

32. Suppose a graph has 10 vertices and each pair of vertices is connected by exactly one edge. Find the number of edges in the graph. (*Hint:* If you use the Fundamental Counting Principle, be sure to count each edge only once.) **45**
33. **TRANSPORTATION** The table shows the distances in miles of the roads between some towns. Draw a graph in which the vertices represent the towns and the edges are labeled with the lengths of the roads. Use your graph to find the length of the shortest route from Greenville to Red Rock. **20 mi**

Route	Miles
Greenville to Roseburg	14
Greenville to Bluemont	12
Greenville to Whiteston	9
Roseburg to Bluemont	8
Bluemont to Whiteston	5
Roseburg to Red Rock	7
Bluemont to Red Rock	9
Whiteston to Red Rock	11

636 Chapter 12 Probability and Statistics

Enrichment, p. 704

Tree Diagrams and the Power Rule

If you flip a coin once, there are two possible outcomes: heads showing (H) or tails showing (T). The tree diagram to the right shows the four (2^2) possible outcomes if you flip a coin twice.

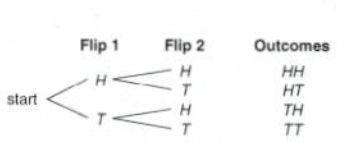

Example 1 **Draw a tree diagram to show all the possible outcomes for flipping a coin three times. List the outcomes.**

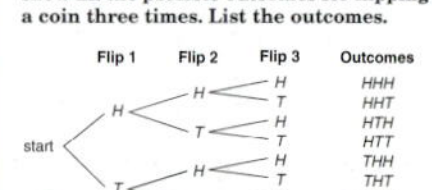

Example 2 **In a cup there are a red, a blue, and a yellow marble. How many possible outcomes are there if you draw one marble at random, replace it, and then draw another?**

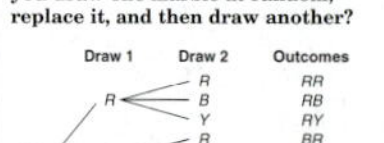

Maintain Your Skills

Mixed Review

34. Prove that $4 + 7 + 10 \cdots + (3n + 1) = \frac{n(3n+5)}{2}$ for all positive integers n. *(Lesson 11-8)* **See pp. 695A–695B.**

Find the indicated term of each expansion. *(Lesson 11-7)*

35. third term of $(x + y)^8$ **$28x^6y^2$**
36. fifth term of $(2a - b)^7$ **$280a^3b^4$**

Evaluate each expression. *(Lesson 10-2)*

37. $\log_2 128$ **7**
38. $\log_3 243$ **5**
39. $\log_9 3$ **$\frac{1}{2}$**

Simplify each expression. *(Lesson 9-1)*

40. $-\frac{x^2 - y^2}{x + y} \cdot \frac{1}{x - y}$ **-1**
41. $\dfrac{\frac{x^2}{x^2 - 25y^2}}{\frac{x}{5y - x}}$ **$-\frac{x}{x + 5y}$**

42. **CARTOGRAPHY** Edison is located at (9, 3) in the coordinate system on a road map. Kettering is located at (12, 5) on the same map. Each side of a square on the map represents 10 miles. To the nearest mile, what is the distance between Edison and Kettering? *(Lesson 8-1)* **36 mi**

Solve each equation. *(Lesson 7-3)*

43. $x^4 - 5x^2 + 4 = 0$ **$\pm 1, \pm 2$**
44. $y^4 + 4y^3 + 4y^2 = 0$ **$0, -2$**

Write an equation of the form $y = a(x - h)^2 + k$ for the parabola with the given vertex that passes through the given point. *(Lesson 6-6)*

45. vertex (3, 2)
    point (5, 6)
    $y = (x - 3)^2 + 2$
46. vertex (−1, 4)
    point (−2, 2)
    $y = -2(x + 1)^2 + 4$
47. vertex (0, 8)
    point (4, 0)
    $y = -\frac{1}{2}x^2 + 8$

Solve each equation. *(Lesson 5-8)*

48. $\sqrt{2x + 1} = 3$ **4**
49. $3 + \sqrt{x + 1} = 5$ **3**
50. $\sqrt{x} + \sqrt{x + 5} = 5$ **4**

Find the inverse of each matrix, if it exists. *(Lesson 4-7)*

51. $\begin{bmatrix} 3 & 1 \\ -4 & 1 \end{bmatrix}$ **$\frac{1}{7}\begin{bmatrix} 1 & -1 \\ 4 & 3 \end{bmatrix}$**
52. $\begin{bmatrix} 4 & -5 \\ 2 & -1 \end{bmatrix}$ **$\frac{1}{6}\begin{bmatrix} -1 & 5 \\ -2 & 4 \end{bmatrix}$**
53. $\begin{bmatrix} -3 & 2 \\ -6 & 4 \end{bmatrix}$ **no inverse exists**

Write an equation in slope-intercept form for each graph. *(Lesson 2-4)*

54.
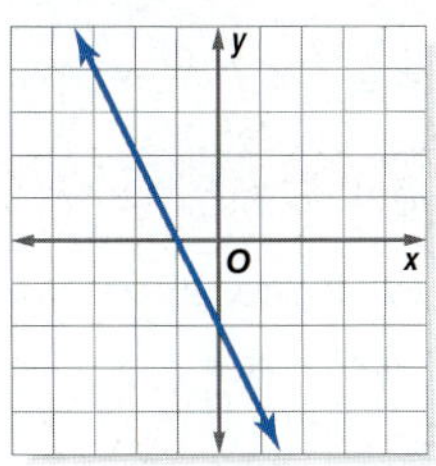

$y = -2x - 2$

55.
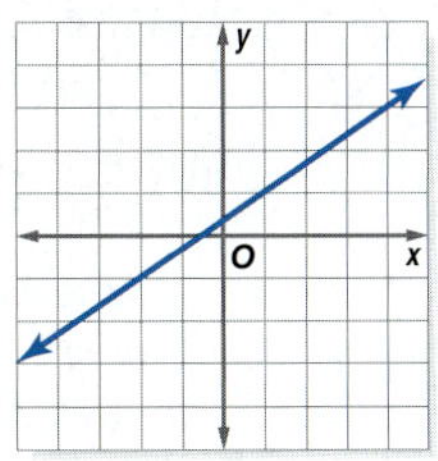

$y = \frac{2}{3}x + \frac{1}{3}$

Getting Ready for the Next Lesson

PREREQUISITE SKILL Evaluate each expression.
*(To review **factorials**, see Lesson 11-7.)*

56. $\frac{5!}{2!}$ **60**
57. $\frac{6!}{4!}$ **30**
58. $\frac{7!}{3!}$ **840**
59. $\frac{6!}{1!}$ **720**
60. $\frac{4!}{2!2!}$ **6**
61. $\frac{6!}{2!4!}$ **15**
62. $\frac{8!}{3!5!}$ **56**
63. $\frac{5!}{5!0!}$ **1**

4 Assess

Open-Ended Assessment

Writing Ask students to write a brief explanation of the difference between independent and dependent events, and to give several examples for each.

Getting Ready for Lesson 12-2

PREREQUISITE SKILL Lesson 12-2 presents solving problems involving permutations and combinations. Students will use their familiarity with evaluating expressions involving factorials as they apply formulas for permutations and combinations. Exercises 56–63 should be used to determine your students' familiarity with evaluating factorials.

Answer

29. The maximum number of license plates is a product with factors of 26s and 10s, depending on how many letters are used and how many digits are used. Answers should include the following.
    - There are 26 choices for the first letter, 26 for the second, and 26 for the third. There are 10 choices for the first number, 10 for the second, and 10 for the third. By the Fundamental Counting Principle, there are $26^3 \cdot 10^3$ or 17,576,000 possible license plates.
    - Replace positions containing numbers with letters.

12-2 Lesson Notes

1 Focus

5-Minute Check Transparency 12-2 Use as a quiz or review of Lesson 12-1.

Mathematical Background notes are available for this lesson on p. 630C.

How do permutations and combinations apply to softball?

Ask students:

- Is a lineup or batting order for the first batters of A, B, C, and D different from a lineup of D, C, B, A? **Yes, the order matters.**
- Is the number of ways, 840, equal to either 7! or 4!? **no**
- How could you write $7 \cdot 6 \cdot 5 \cdot 4$ as an expression in terms of 7! and 4!? $\frac{7!}{(7-4)!}$

12-2 Permutations and Combinations

What You'll Learn

- Solve problems involving linear permutations.
- Solve problems involving combinations.

How do permutations and combinations apply to softball?

When the manager of a softball team fills out her team's lineup card before the game, the order in which she fills in the names is important because it determines the order in which the players will bat.

Suppose she has 7 possible players in mind for the top 4 spots in the lineup. You know from the Fundamental Counting Principle that there are $7 \cdot 6 \cdot 5 \cdot 4$ or 840 ways that she could assign players to the top 4 spots.

Vocabulary

- permutation
- linear permutation
- combination

PERMUTATIONS When a group of objects or people are arranged in a certain order, the arrangement is called a **permutation**. In a permutation, the *order* of the objects is very important. The arrangement of objects or people in a line is called a **linear permutation**.

Notice that $7 \cdot 6 \cdot 5 \cdot 4$ is the product of the first 4 factors of 7!. You can rewrite this product in terms of 7!.

$$7 \cdot 6 \cdot 5 \cdot 4 = 7 \cdot 6 \cdot 5 \cdot 4 \cdot \frac{3 \cdot 2 \cdot 1}{3 \cdot 2 \cdot 1} \quad \text{Multiply by } \frac{3 \cdot 2 \cdot 1}{3 \cdot 2 \cdot 1} \text{ or } 1.$$

$$= \frac{7 \cdot 6 \cdot 5 \cdot 4 \cdot 3 \cdot 2 \cdot 1}{3 \cdot 2 \cdot 1} \text{ or } \frac{7!}{3!} \quad 7! = 7 \cdot 6 \cdot 5 \cdot 4 \cdot 3 \cdot 2 \cdot 1 \text{ and } 3! = 3 \cdot 2 \cdot 1$$

Notice that 3! is the same as $(7-4)!$.

Study Tip

Reading Math
The expression $P(n, r)$ is read *the number of permutations of n objects taken r at a time*. It is sometimes written as $_nP_r$.

The number of ways to arrange 7 people or objects taken 4 at a time is written $P(7, 4)$. The expression for the softball lineup above is a case of the following formula.

Key Concept *Permutations*

The number of permutations of n distinct objects taken r at a time is given by

$$P(n, r) = \frac{n!}{(n-r)!}.$$

Example 1 *Permutation*

FIGURE SKATING There are 10 finalists in a figure skating competition. How many ways can gold, silver, and bronze medals be awarded?

Since each winner will receive a different medal, order is important. You must find the number of permutations of 10 things taken 3 at a time.

Resource Manager

Workbook and Reproducible Masters

Chapter 12 Resource Masters

- Study Guide and Intervention, pp. 705–706
- Skills Practice, p. 707
- Practice, p. 708
- Reading to Learn Mathematics, p. 709
- Enrichment, p. 710

Graphing Calculator and Spreadsheet Masters, p. 50

Transparencies

5-Minute Check Transparency 12-2
Answer Key Transparencies

Technology

Alge2PASS: Tutorial Plus, Lesson 22
Interactive Chalkboard

$$P(n, r) = \frac{n!}{(n-r)!}$$ Permutation formula

$$P(10, 3) = \frac{10!}{(10 - 3!)}$$ $n = 10, r = 3$

$$= \frac{10!}{7!}$$ Simplify.

$$= \frac{10 \cdot 9 \cdot 8 \cdot \cancel{7}^{1} \cdot \cancel{6}^{1} \cdot \cancel{5}^{1} \cdot \cancel{4}^{1} \cdot \cancel{3}^{1} \cdot \cancel{2}^{1} \cdot \cancel{1}^{1}}{\cancel{7}_{1} \cdot \cancel{6}_{1} \cdot \cancel{5}_{1} \cdot \cancel{4}_{1} \cdot \cancel{3}_{1} \cdot \cancel{2}_{1} \cdot \cancel{1}_{1}} \text{ or } 720$$ Divide by common factors.

The gold, silver, and bronze medals can be awarded in 720 ways.

Notice that in Example 1, all of the factors of $(n - r)!$ are also factors of $n!$. Instead of writing all of the factors, you can also evaluate the expression in the following way.

$$\frac{10!}{(10-3)!} = \frac{10!}{7!}$$ Simplify.

$$= \frac{10 \cdot 9 \cdot 8 \cdot 7!}{7!}$$ $\frac{7!}{7!} = 1$

$$= 10 \cdot 9 \cdot 8 \text{ or } 720$$ Multiply.

Suppose you want to rearrange the letters of the word *geometry* to see if you can make a different word. If the two *e*'s were not identical, the eight letters in the word could be arranged in $P(8, 8)$ or $8!$ ways. To account for the identical *e*'s, divide $P(8, 8)$ or 40,320 by the number of arrangements of *e*. The two *e*'s can be arranged in $P(2, 2)$ or $2!$ ways.

$$\frac{P(8, 8)}{P(2, 2)} = \frac{8!}{2!}$$ Divide.

$$= \frac{8 \cdot 7 \cdot 6 \cdot 5 \cdot 4 \cdot 3 \cdot 2!}{2!} \text{ or } 20{,}160$$ Simplify.

Thus, there are 20,160 ways to arrange the letters in *geometry*.

When some letters or objects are alike, use the rule below to find the number of permutations.

Key Concept — ***Permutations with Repetitions***

The number of permutations of n objects of which p are alike and q are alike is $\frac{n!}{p!q!}$.

This rule can be extended to any number of objects that are repeated.

Example 2 *Permutation with Repetition*

How many different ways can the letters of the word *MISSISSIPPI* be arranged?

The second, fifth, eighth, and eleventh letters are each I.

The third, fourth, sixth, and seventh letters are each S.

The ninth and tenth letters are each P.

You need to find the number of permutations of 11 letters of which 4 of one letter, 4 of another letter, and 2 of another letter are the same.

$$\frac{11!}{4!4!2!} = \frac{11 \cdot 10 \cdot 9 \cdot 8 \cdot 7 \cdot 6 \cdot 5 \cdot 4!}{4!4!2!} \text{ or } 34{,}650$$

There are 34,650 ways to arrange the letters.

www.algebra2.com/extra_examples

2 Teach

PERMUTATIONS

In-Class Examples Power Point®

1 Eight people enter the Best Pie contest. How many ways can blue, red, and green ribbons be awarded? **336**

2 How many different ways can the letters of the word *BANANA* be arranged? **60**

DAILY INTERVENTION — **Unlocking Misconceptions**

Discuss the notation for $P(n, r)$. Students might reasonably think that this expression could be an ordered pair, or function notation. However, when the P is used, the expression is probably for permutation notation, which can also be written as ${}_nP_r$.

Lesson 12-2 Permutations and Combinations 639

COMBINATIONS

In-Class Examples

Power Point®

3 Five cousins at a family reunion decide that three of them will go to pick up a pizza. How many ways can they choose the three people to go? **10**

4 Six cards are drawn from a standard deck of cards. How many hands consist of two hearts and four spades? **55,770**

Study Tip

Permutations and Combinations

- If order in an arrangement *is* important, the arrangement is a *permutation*.
- If order is *not* important, the arrangement is a *combination*.

COMBINATIONS

An arrangement or selection of objects in which order is *not* important is called a **combination**. The number of combinations of n objects taken r at a time is written $C(n, r)$. *It is sometimes written* ${}_nC_r$.

You know that there are $P(n, r)$ ways to select r objects from a group of n if the order is important. There are $r!$ ways to order the r objects that are selected, so there are $r!$ permutations that are all the same combination. Therefore,

$$C(n, r) = \frac{P(n, r)}{r!} \text{ or } \frac{n!}{(n-r)!r!}.$$

Key Concept — Combinations

The number of combinations of n distinct objects taken r at a time is given by

$$C(n, r) = \frac{n!}{(n-r)!r!}.$$

Example 3 Combination

A group of seven students working on a project needs to choose two from their group to present the group's report to the class. How many ways can they choose the two students?

Since the order they choose the students is not important, you must find the number of combinations of 7 students taken 2 at a time.

$C(n, r) = \dfrac{n!}{(n-r)!r!}$ Combination formula

$C(7, 2) = \dfrac{7!}{(7-2)!2!}$ $n = 7$ and $r = 2$

$= \dfrac{7!}{5!2!}$ or 21 Simplify.

There are 21 possible ways to choose the two students.

In more complicated situations, you may need to multiply combinations and/or permutations.

Study Tip

Deck of Cards

In this text, a *standard deck of cards* always means a deck of 52 playing cards. There are 4 suits—clubs (black), diamonds (red), hearts (red), and spades (black)—with 13 cards in each suit.

Example 4 Multiple Events

Five cards are drawn from a standard deck of cards. How many hands consist of three clubs and two diamonds?

By the Fundamental Counting Principle, you can multiply the number of ways to select three clubs and the number of ways to select two diamonds.

Only the cards in the hand matter, not the order in which they were drawn, so use combinations.

$C(13, 3)$ Three of 13 clubs are to be drawn.

$C(13, 2)$ Two of 13 diamonds are to be drawn.

$C(13, 3) \cdot C(13, 2) = \dfrac{13!}{(13-3)!3!} \cdot \dfrac{13!}{(13-2)!2!}$ Combination formula

$= \dfrac{13!}{10!3!} \cdot \dfrac{13!}{11!2!}$ Subtract.

$= 286 \cdot 78$ or 22,308 Simplify.

There are 22,308 hands consisting of 3 clubs and 2 diamonds.

DAILY INTERVENTION — Differentiated Instruction

Visual/Spatial Have students model the various problems by writing letters, names, or other labels on index cards. After students have tried to model and tally possible combinations, they will soon realize that the formulas save lots of time.

640 Chapter 12 Probability and Statistics

Check for Understanding

Concept Check

1. **OPEN ENDED** Describe a situation in which the number of outcomes is given by $P(6, 3)$. **See margin.**
2. **Show** that $C(n, n - r) = C(n, r)$. **See margin.**
3. **Determine** whether the statement $C(n, r) = P(n, r)$ is *sometimes*, *always*, or *never* true. Explain your reasoning. **Sometimes; the statement is true when $r = 1$.**

Guided Practice

GUIDED PRACTICE KEY	
Exercises	Examples
4, 5	1
6, 7, 11	3
8, 9	1, 3
10	2

Evaluate each expression.

4. $P(5, 3)$ **60**
5. $P(6, 3)$ **120**
6. $C(4, 2)$ **6**
7. $C(6, 1)$ **6**

Determine whether each situation involves a *permutation* or a *combination*. Then find the number of possibilities.

8. choosing 2 different pizza toppings from a list of 6 **combination; 15**
9. seven shoppers in line at a checkout counter **permutation; 5040**
10. an arrangement of the letters in the word *intercept* **permutation; 90,720**

Application

11. **SCHOOL** The principal at Cobb County High School wants to start a mentoring group. He needs to narrow his choice of students to be mentored to six from a group of nine. How many ways can a group of six be selected? **84**

★ indicates increased difficulty

Practice and Apply

Homework Help	
For Exercises	See Examples
12–15	1
16–19	3
20, 21, 32–35	4
22–31	1–3

Extra Practice
See page 854.

Evaluate each expression.

12. $P(8, 2)$ **56**
13. $P(9, 1)$ **9**
14. $P(7, 5)$ **2520**
15. $P(12, 6)$ **665,280**
16. $C(5, 2)$ **10**
17. $C(8, 4)$ **70**
18. $C(12, 7)$ **792**
19. $C(10, 4)$ **210**
20. ★ $C(12, 4) \cdot C(8, 3)$ **27,720**
21. ★ $C(9, 3) \cdot C(6, 2)$ **1260**

Determine whether each situation involves a *permutation* or a *combination*. Then find the number of possibilities. **22. permutation; 5040 26. combination; 220**

22. the winner and first, second, and third runners-up in a contest with 10 finalists
23. selecting two of eight employees to attend a business seminar **combination; 28**
24. an arrangement of the letters in the word *algebra* **permutation; 2520**
25. placing an algebra book, a geometry book, a chemistry book, an English book, and a health book on a shelf **permutation; 120**
26. selecting nine books to check out of the library from a reading list of twelve
27. an arrangement of the letters in the word *parallel* **permutation; 3360**
28. choosing two CDs to buy from ten that are on sale **combination; 45**
29. selecting three of fifteen flavors of ice cream at the grocery store **combination; 455**
30. **MOVIES** The manager of a four-screen movie theater is deciding which of 12 available movies to show. The screens are in rooms with different seating capacities. How many ways can he show four different movies on the screens? **11,880**
31. **LANGUAGES** How many different arrangements of the letters of the Hawaiian word *aloha* are possible? **60**
32. **GOVERNMENT** How many ways can five members of the 100-member United States Senate be chosen to be put on a committee? **75,287,520**

www.algebra2.com/self_check_quiz

3 Practice/Apply

Study Notebook

Have students—
- add the definitions/examples of the vocabulary terms to their Vocabulary Builder worksheets for Chapter 12.
- include their own examples for different kinds of permutations and combinations.
- include any other item(s) that they find helpful in mastering the skills in this lesson.

About the Exercises...

Organization by Objective
- **Permutations:** 12–15, 22, 24, 25, 27
- **Combinations:** 16–21, 23, 26, 28, 29

Odd/Even Assignments
Exercises 12–29 are structured so that students practice the same concepts whether they are assigned odd or even problems.

Assignment Guide

Basic: 13–19 odd, 23–31 odd, 37–40, 44–72

Average: 13–35 odd, 37–40, 44–72 (optional: 41–43)

Advanced: 12–36 even, 37–68 (optional: 69–72)

Answers

1. Sample answer: There are six people in a contest. How many ways can the first, second, and third prizes be awarded?
2. $C(n, n - r)$
$= \frac{n!}{[n - (n - r)]!(n - r)!}$
$= \frac{n!}{r!(n - r)!}$
$= \frac{n!}{(n - r)!r!}$
$= C(n, r)$

Teacher to Teacher

Harry Rattien Townsend Harris H.S. at Queens College, Flushing, NY

I use the following mnemonic device to help my students remember the difference between permutations and combinations.

Permutation → place   Combination → choose

Study Guide and Intervention, p. 705 (shown) and p. 706

Permutations When a group of objects or people are arranged in a certain order, the arrangement is called a **permutation**.

Permutations	The number of permutations of n distinct objects taken r at a time is given by $P(n, r) = \frac{n!}{(n-r)!}$.
Permutations with Repetitions	The number of permutations of n objects of which p are alike and q are alike is $\frac{n!}{p!q!}$.

The rule for permutations with repetitions can be extended to any number of objects that are repeated.

Example **From a list of 20 books, each student must choose 4 books for book reports. The first report is a traditional book report, the second a poster, the third a newspaper interview with one of the characters, and the fourth a timeline of the plot. How many different orderings of books can be chosen?**

Since each book report has a different format, order is important. You must find the number of permutations of 20 objects taken 4 at a time.

$P(n, r) = \frac{n!}{(n-r)!}$ — Permutation formula

$P(20, 4) = \frac{20!}{(20-4)!}$ — $n = 20, r = 4$

$= \frac{20!}{16!}$ — Simplify.

$= \frac{20 \cdot 19 \cdot 18 \cdot 17 \cdot \overset{1}{\cancel{16}} \cdot \overset{1}{\cancel{15}} \cdot \ldots \cdot \overset{1}{\cancel{1}}}{\underset{1}{\cancel{16}} \cdot \underset{1}{\cancel{15}} \cdot \ldots \cdot \underset{1}{\cancel{1}}}$ — Divide by common factors.

$= 116{,}280$

Books for the book reports can be chosen 116,280 ways.

Exercises

Evaluate each expression.

1. $P(6, 3)$ 120
2. $P(8, 5)$ 6720
3. $P(9, 4)$ 3024
4. $P(11, 6)$ 332,640

How many different ways can the letters of each word be arranged?

5. MOM 3
6. MONDAY 720
7. STEREO 360

8. **SCHOOL** The high school chorus has been practicing 12 songs, but there is time for only 5 of them at the spring concert. How may different orderings of 5 songs are possible? 95,040

Skills Practice, p. 707 and Practice, p. 708 (shown)

Evaluate each expression.

1. $P(8, 6)$ 20,160
2. $P(9, 7)$ 181,440
3. $P(3, 3)$ 6
4. $P(4, 3)$ 24
5. $P(4, 1)$ 4
6. $P(7, 2)$ 42
7. $C(8, 2)$ 28
8. $C(11, 3)$ 165
9. $C(20, 18)$ 190
10. $C(9, 9)$ 1
11. $C(3, 1)$ 3
12. $C(9, 3) \cdot C(6, 2)$ 1260

Determine whether each situation involves a *permutation* or a *combination*. Then find the number of possibilities.

13. selecting a 4-person bobsled team from a group of 9 athletes
combination; 126
14. an arrangement of the letters in the word *Canada*
permutation; 120
15. arranging 4 charms on a bracelet that has a clasp, a front, and a back
permutation; 24
16. selecting 3 desserts from 10 desserts that are displayed on a dessert cart in a restaurant
combination; 120
17. an arrangement of the letters in the word *annually*
permutation; 5040
18. forming a 2-person sales team from a group of 12 salespeople
combination; 66
19. making 5-sided polygons by choosing any 5 of 11 points located on a circle to be the vertices
combination; 462
20. seating 5 men and 5 women alternately in a row, beginning with a woman
permutation; 14,400
21. **STUDENT GROUPS** Farmington High is planning its academic festival. All math classes will send 2 representatives to compete in the math bowl. How many different groups of students can be chosen from a class of 16 students? 120
22. **PHOTOGRAPHY** A photographer is taking pictures of a bride and groom and their 6 attendants. If she takes photographs of 3 people in a group, how many different groups can she photograph? 56
23. **AIRLINES** An airline is hiring 5 flight attendants. If 8 people apply for the job, how many different groups of 5 attendants can the airline hire? 56
24. **SUBSCRIPTIONS** A school librarian would like to buy subscriptions to 7 new magazines. Her budget, however, will allow her to buy only 4 new subscriptions. How many different groups of 4 magazines can she choose from the 7 magazines? 35

Reading to Learn Mathematics, p. 709 ELL

Pre-Activity **How do permutations and combinations apply to softball?**

Read the introduction to Lesson 12-2 at the top of page 638 in your textbook.

Suppose that 20 students enter a math contest. In how many ways can first, second, and third places be awarded? (Write your answer as a product. Do not calculate the product.) $20 \cdot 19 \cdot 18$

Reading the Lesson

1. Indicate whether each situation involves a *permutation* or a *combination*.
   a. choosing five students from a class to work on a special project **combination**
   b. arranging five pictures in a row on a wall **permutation**
   c. drawing a hand of 13 cards from a 52-card deck **combination**
   d. arranging the letters of the word *algebra* **permutation**
2. Write an expression that can be used to calculate each of the following.
   a. number of combinations of n distinct objects taken r at a time $\frac{n!}{(n-r)!r!}$
   b. number of permutations of n objects of which p are alike and q are alike $\frac{n!}{p!q!}$
   c. number of permutations of n distinct objects taken r at a time $\frac{n!}{(n-r)!}$
3. Five cards are drawn from a standard deck of cards. Suppose you are asked to determine how many possible hands consist of one heart, two diamonds, and two spades.
   a. Which of the following would you use to solve this problem: *Fundamental Counting Principle*, *permutations*, or *combinations*? (More than one of these may apply.)

      Fundamental Counting Principle, combinations
   b. Write an expression that involves the notation $P(n, r)$ and/or $C(n, r)$ that you would use to solve this problem. (Do not do any calculations.)

      $C(13, 1) \cdot C(13, 2) \cdot C(13, 2)$

Helping You Remember

4. Many students have trouble knowing when to use permutations and when to use combinations to solve counting problems. How can the idea of *order* help you to remember the difference between permutations and combinations?

Sample answer: A permutation is an arrangement of objects in which order is important. A combination is a selection of objects in which order is *not* important.

More About. . .

Card Games

Hanafuda cards are often called "flower cards" because each suit is depicted by a different flower. Each flower is representative of the calendar month in which the flower blooms.

Source: www.gamesdomain.com

★ 33. How many ways can a hand of five cards consisting of four cards from one suit and one card from another suit be drawn from a standard deck of cards? **111,540**

★ 34. How many ways can a hand of five cards consisting of three cards from one suit and two cards from another suit be drawn from a standard deck of cards? **267,696**

35. **LOTTERIES** In a multi-state lottery, the player must guess which five of forty nine white balls numbered from 1 to 49 will be drawn. The order in which the balls are drawn does not matter. The player must also guess which one of forty-two red balls numbered from 1 to 42 will be drawn. How many ways can the player fill out a lottery ticket? **80,089,128**

36. **CARD GAMES** *Hanafuda* is a Japanese game that uses a deck of cards made up of 12 suits, with each suit having four cards. How many 7-card hands can be formed so that 3 are from one suit and 4 are from another? **528**

37. **CRITICAL THINKING** Show that $C(n-1, r) + C(n-1, r-1) = C(n, r)$. **See pp. 695A–695B.**

38. WRITING IN MATH Answer the question that was posed at the beginning of the lesson. **See pp. 695A–695B.**

How do permutations and combinations apply to softball?

Include the following in your answer:
- an explanation of how to find the number of 9-person lineups that are possible, and
- an explanation of how many ways there are to choose 9 players if 16 players show up for a game.

Standardized Test Practice

39. How many ways can eight runners in an Olympic race finish in first, second, and third places? **D**

Ⓐ 8 Ⓑ 24 Ⓒ 56 Ⓓ 336

40. How many diagonals can be drawn in the pentagon? **A**

Ⓐ 5 Ⓑ 10
Ⓒ 15 Ⓓ 20

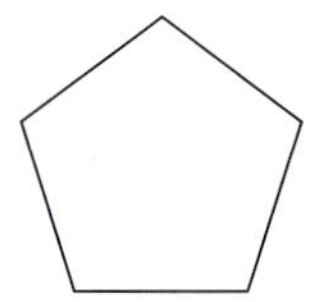

Extending the Lesson

When n distinct objects are arranged in a circle, there are n ways that the arrangement can be rotated to obtain an arrangement that is really the same as the original. For example, the two arrangements of three objects shown below are the same. Therefore, the number of **circular permutations** of n distinct objects is $\frac{n!}{n}$ or $(n-1)!$ *Note that the keys are not turned over.*

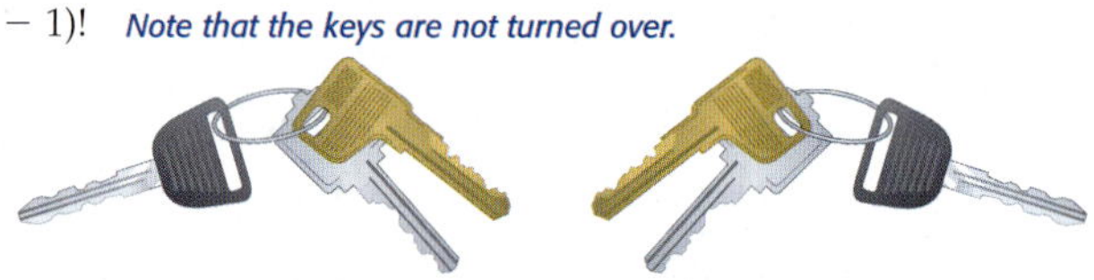

Find the number of possibilities for each situation.

41. a basketball huddle of 5 players **24**
42. four different dishes on a revolving tray in the middle of a table at a Chinese restaurant **6**
43. six quarters with designs from six different states arranged in a circle on top of your desk **120**

Enrichment, p. 710

Combinations and Pascal's Triangle

Pascal's triangle is a special array of numbers invented by Blaise Pascal (1623–1662). The values in Pascal's triangle can be found using the combinations shown below.

1. Evaluate the expression in each cell of the triangle.

C(1,0) 1 | C(1,1) 1

C(2,0) 1 | C(2,1) 2 | C(2,2) 1

C(3,0) 1 | C(3,1) 3 | C(3,2) 3 | C(3,3) 1

C(4,1) | C(4,2) | C(4,3) | C(4,4)

Maintain Your Skills

Mixed Review

44. Darius can do his homework in pencil or pen, using lined or unlined paper, and on one or both sides of each page. How many ways can he prepare his homework? *(Lesson 12-1)* **8**

45. A customer in an ice cream shop can order a sundae with a choice of 10 flavors of ice cream, a choice of 4 flavors of sauce, and with or without a cherry on top. How many different sundaes are possible? *(Lesson 12-1)* **80**

Find a counterexample to each statement. *(Lesson 11-8)*

46. $1 + 2 + 3 + \ldots + n = 2n - 1$

47. $5^n + 1$ is divisible by 6.

46. Sample answer: $n = 3$
47. Sample answer: $n = 2$

Solve each equation or inequality. *(Lesson 10-5)*

48. $3e^x + 1 = 2$ **−1.0986**

49. $e^{2x} > 5$ **$x > 0.8047$**

50. $\ln (x - 1) = 3$ **21.0855**

51. **CONSTRUCTION** A painter works on a job for 10 days and is then joined by an associate. Together they finish the job in 6 more days. The associate could have done the job in 30 days. How long would it have taken the painter to do the job alone? *(Lesson 9-6)* **20 days**

Write an equation for each ellipse. *(Lesson 8-4)*

52.

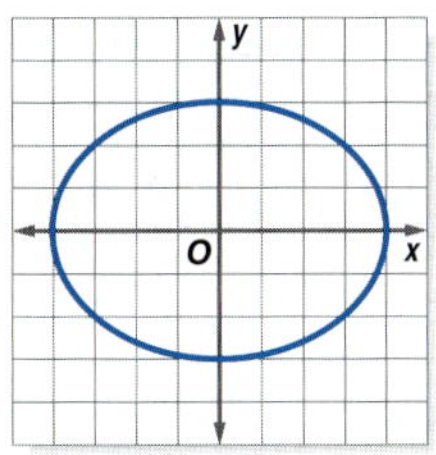

53.

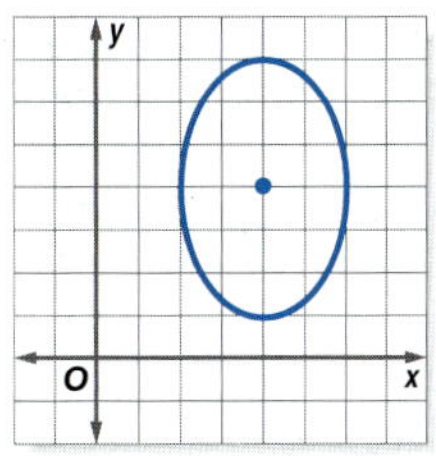

52. $\frac{x^2}{16} + \frac{y^2}{9} = 1$
53. $\frac{(y-4)^2}{9} + \frac{(x-4)^2}{4} = 1$

Find $p(-1)$ and $p(5)$ for each function. *(Lesson 7-1)*

54. $p(x) = \frac{1}{2}x^2 + 3x - 1$ **$-\frac{7}{2}; \frac{53}{2}$**

55. $p(x) = x^4 - 4x^3 + 2x - 7$ **−4; 128**

Solve each equation by factoring. *(Lesson 6-3)*

56. $x^2 - 16 = 0$ **{−4, 4}**

57. $x^2 - 3x - 10 = 0$ **{−2, 5}**

58. $3x^2 + 8x - 3 = 0$ **$\left\{-3, \frac{1}{3}\right\}$**

Simplify. *(Lesson 5-6)*

59. $\sqrt{128}$ **$8\sqrt{2}$**

60. $\sqrt{3x^6y^4}$ **$|x^3|y^2\sqrt{3}$**

61. $\sqrt{20} + 2\sqrt{45} - \sqrt{80}$ **$4\sqrt{5}$**

Solve each system of equations by using inverse matrices. *(Lesson 4-8)*

62. $x + 2y = 5$
$3x - 3y = -12$ **(−1, 3)**

63. $5a + 2b = 4$
$-3a + b = 2$ **(0, 2)**

Find the slope of the line that passes through each pair of points. *(Lesson 2-3)*

64. (2, 1), (5, −3) **$-\frac{4}{3}$**

65. (0, 4), (7, −2) **$-\frac{6}{7}$**

66. (5, 3), (2, 3) **0**

Solve each equation. Check your solutions. *(Lesson 1-4)*

67. $|x - 4| = 11$ **{−7, 15}**

68. $|2x + 2| = -3$ **∅**

Getting Ready for the Next Lesson

PREREQUISITE SKILL **Evaluate the expression $\frac{x}{x + y}$ for the given values of x and y.** *(To review* ***evaluating expressions****, see Lesson 1-1.)*

69. $x = 3, y = 2$ **$\frac{3}{5}$**

70. $x = 4, y = 4$ **$\frac{1}{2}$**

71. $x = 2, y = 8$ **$\frac{1}{5}$**

72. $x = 5, y = 10$ **$\frac{1}{3}$**

4 Assess

Open-Ended Assessment

Speaking Ask students to work with a partner. One writes an expression, such as C(3, 2), and hands it to the other, who reads the notation aloud (for example, "the number of combinations of 3 things taken 2 at a time") and calculates the value. **3** The partners discuss and correct this value as necessary. Then they exchange roles.

Getting Ready for Lesson 12-3

PREREQUISITE SKILL Lesson 12-3 presents finding the probability and odds of events. Students will use their familiarity with evaluating rational expressions as they apply probability formulas. Exercises 69–72 should be used to determine your students' familiarity with evaluating rational expressions.

12-3 Lesson Notes

1 Focus

5-Minute Check Transparency 12-3 Use as a quiz or review of Lesson 12-2.

Mathematical Background notes are available for this lesson on p. 630C.

What do probability and odds tell you about life's risks?

Ask students:

- On average, out of 750,000 people, how many will be struck by lightning each year? **1**
- If there are 260 million people in the United States, how many people will be struck by lightning each year? **about 347**
- Does probability say anything about where or why an event occurs? **no**

12-3 Probability

What You'll Learn

- Find the probability and odds of events.
- Create and use graphs of probability distributions.

Vocabulary

- probability
- success
- failure
- random
- odds
- random variable
- probability distribution
- uniform distribution
- relative-frequency histogram

What do probability and odds tell you about life's risks?

The risk of getting struck by lightning in any given year is 1 in 750,000. The chances of surviving a lightning strike are 3 in 4. These risks and chances are a way of describing the probability of an event. The **probability** of an event is a ratio that measures the chances of the event occurring.

PROBABILITY AND ODDS

Mathematicians often use tossing of coins and rolling of dice to illustrate probability. When you toss a coin, there are only two possible outcomes—heads or tails. A desired outcome is called a **success**. Any other outcome is called a **failure**.

Key Concept *Probability of Success and Failure*

If an event can succeed in s ways and fail in f ways, then the probabilities of success, $P(S)$, and of failure, $P(F)$, are as follows.

$$P(S) = \frac{s}{s+f} \qquad P(F) = \frac{f}{s+f}$$

The probability of an event occurring is always between 0 and 1, inclusive. The closer the probability of an event is to 1, the more likely the event is to occur. The closer the probability of an event is to 0, the less likely the event is to occur.

Study Tip

Reading Math
When P is followed by an event in parentheses, P stands for *probability*. When there are two numbers in parentheses, P stands for *permutations*.

Example 1 Probability

When two coins are tossed, what is the probability that both are tails?

You can use a tree diagram to find the sample space.

First coin	H		T	
Second coin	H	T	H	T
Possible outcomes	HH	HT	TH	TT

There are 4 possible outcomes. You can confirm this using the Fundamental Counting Principle. There are 2 possible results for the first coin and 2 for the second coin, so there are $2 \cdot 2$ or 4 possible outcomes. Only one of these outcomes, TT, is a success, so $s = 1$. The other three outcomes are failures, so $f = 3$.

$P(\text{two tails}) = \frac{s}{s+f}$ Probability formula

$= \frac{1}{1+3}$ or $\frac{1}{4}$ $s = 1, f = 3$

The probability of tossing two heads is $\frac{1}{4}$. *This probability can also be written as a decimal, 0.25, or as a percent, 25%.*

Resource Manager

Workbook and Reproducible Masters

Chapter 12 Resource Masters
- Study Guide and Intervention, pp. 711–712
- Skills Practice, p. 713
- Practice, p. 714
- Reading to Learn Mathematics, p. 715
- Enrichment, p. 716
- Assessment, p. 767

Transparencies

5-Minute Check Transparency 12-3
Answer Key Transparencies

Technology

Interactive Chalkboard

In more complicated situations, you may need to use permutations and/or combinations to count the outcomes. When all outcomes have an equally likely chance of occurring, we say that the outcomes occur at **random**.

Example 2 Probability with Combinations

Monifa has a collection of 32 CDs—18 R&B and 14 rap. As she is leaving for a trip, she randomly chooses 6 CDs to take with her. What is the probability that she selects 3 R&B and 3 rap?

Step 1 Determine how many 6-CD selections meet the conditions.

$C(18, 3)$ Select 3 R&B CDs. Their order does not matter.
$C(14, 3)$ Select 3 rap CDs.

Step 2 Use the Fundamental Counting Principle to find the number of successes.

$$C(18, 3) \cdot C(14, 3) = \frac{18!}{15!3!} \cdot \frac{14!}{11!3!} \text{ or } 297{,}024$$

Step 3 Find the total number, $s + f$, of possible 6-CD selections.

$$C(32, 6) = \frac{32!}{26!6!} \text{ or } 906{,}192 \qquad s + f = 906{,}192$$

Step 4 Determine the probability.

$$P(\text{3 R\&B CDs and 3 rap CDs}) = \frac{s}{s+f} \qquad \text{Probability formula}$$
$$= \frac{297{,}024}{906{,}192} \qquad \text{Substitute.}$$
$$\approx 0.32777 \qquad \text{Use a calculator.}$$

The probability of selecting 3 R&B CDs and 3 rap CDs is about 0.32777 or 33%.

Another way to measure the chance of an event occurring is with odds. The **odds** that an event will occur can be expressed as the ratio of the number of successes to the number of failures.

Key Concept — Odds

The odds that an event will occur can be expressed as the ratio of the number of ways it can succeed to the number of ways it can fail. If an event can succeed in s ways and fail in f ways, then the odds of success and of failure are as follows.

Odds of success = $s:f$ Odds of failure = $f:s$

Example 3 Odds

LIFE EXPECTANCY According to the U.S. National Center for Health Statistics, the chances of a male born in 1990 living to be at least 65 years of age are about 3 in 4. For females, the chances are about 17 in 20.

a. What are the odds of a male living to be at least 65?

Three out of four males will live to be at least 65, so the number of successes (living to 65) is 3. The number of failures is $4 - 3$ or 1.

$$\text{odds of a male living to 65} = s:f \qquad \text{Odds formula}$$
$$= 3:1 \qquad s = 3, f = 1$$

The odds of a male living to at least 65 are 3:1.

www.algebra2.com/extra_examples

2 Teach

PROBABILITY AND ODDS

In-Class Examples Power Point®

1. When three coins are tossed, what is the probability that all three are heads? **$\frac{1}{8}$ or 12.5%**

2. Roman has a collection of 26 books—16 are fiction and 10 are nonfiction. He randomly chooses 8 books to take with him on vacation. What is the probability that he chooses 4 fiction and 4 nonfiction? **0.24464 or 24.5%**

3. Using the statistics in Example 3 in the Student Edition, what are the odds that a male born in 1990 will die before age 65? **1:3** a female born in 1990? **3:17**

PROBABILITY DISTRIBUTIONS

In-Class Example Power Point®

4 Use the table and graph in Example 4 in the Student Edition.

a. Use the graph to determine which outcomes are least likely. What is their probability? **The least likely outcomes are 2 and 12, with a probability of $\frac{1}{36}$ for each.**

b. Use the table to find $P(S = 11)$. What other sum has the same probability? **The probability of a sum of 11 is $\frac{1}{18}$, which is the same as that for a sum of 3.**

c. What are the odds of rolling a sum of 5? **1:8**

b. What are the odds of a female living to be at least 65?

Seventeen out of twenty females will live to be at least 65, so the number of successes in this case is 17. The number of failures is 20 − 17 or 3.

odds of a female living to be 65 $= s{:}f$ — Odds formula

$= 17{:}3$ — $s = 17, f = 3$

The odds of a female living to at least 65 are 17:3.

PROBABILITY DISTRIBUTIONS Many experiments, such as rolling a die, have numerical outcomes. A **random variable** is a variable whose value is the numerical outcome of a random event. For example, when rolling a die we can let the random variable D represent the number showing on the die. Then D can equal 1, 2, 3, 4, 5, or 6. A **probability distribution** for a particular random variable is a function that maps the sample space to the probabilities of the outcomes in the sample space. The table below illustrates the probability distribution for rolling a die. *A distribution like this one where all of the probabilities are the same is called a* ***uniform distribution***.

Study Tip

Reading Math
The notation $P(X = n)$ is used with random variables. $P(D = 4) = \frac{1}{6}$ is read *the probability that D equals 4 is one sixth.*

D = Roll	1	2	3	4	5	6
Probability	$\frac{1}{6}$	$\frac{1}{6}$	$\frac{1}{6}$	$\frac{1}{6}$	$\frac{1}{6}$	$\frac{1}{6}$

$P(D = 4) = \frac{1}{6}$

To help visualize a probability distribution, you can use a table of probabilities or a graph, called a **relative-frequency histogram**.

Example 4 Probability Distribution

Suppose two dice are rolled. The table and the relative-frequency histogram show the distribution of the sum of the numbers rolled.

S = Sum	2	3	4	5	6	7	8	9	10	11	12
Probability	$\frac{1}{36}$	$\frac{1}{18}$	$\frac{1}{12}$	$\frac{1}{9}$	$\frac{5}{36}$	$\frac{1}{6}$	$\frac{5}{36}$	$\frac{1}{9}$	$\frac{1}{12}$	$\frac{1}{18}$	$\frac{1}{36}$

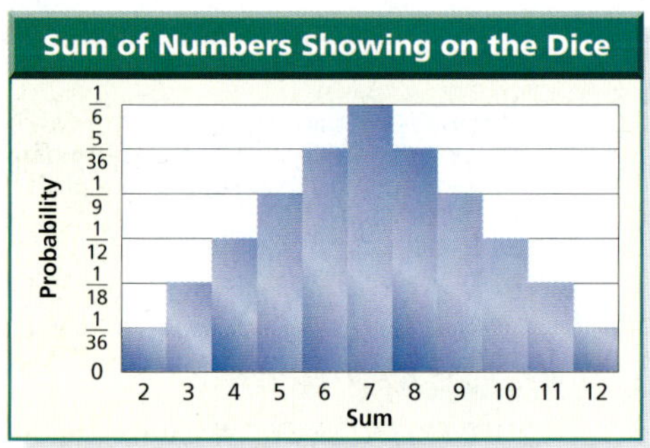

a. Use the graph to determine which outcome is most likely. What is its probability?

The most likely outcome is a sum of 7, and its probability is $\frac{1}{6}$.

b. Use the table to find $P(S = 9)$. What other sum has the same probability?

According to the table, the probability of a sum of 9 is $\frac{1}{9}$. The other outcome with a probability of $\frac{1}{9}$ is 5.

DAILY INTERVENTION **Differentiated Instruction**

Naturalist Ask students to find examples outside the classroom of odds and probabilities, perhaps from statistics on natural disasters or weather reports. Have them share these examples with the class.

646 Chapter 12 Probability and Statistics

c. **What are the odds of rolling a sum of 7?**

Step 1 Identify s and f.

$P(\text{rolling a 7}) = \frac{1}{6}$

$= \frac{s}{s+f}$ $s = 1, f = 5$

Step 2 Find the odds.

Odds $= s:f$

$= 1:5$

So, the odds of rolling a sum of 7 are 1:5.

Check for Understanding

Concept Check

1. **OPEN ENDED** Describe an event that has a probability of 0 and an event that has a probability of 1. **See margin.**
2. **Write** the probability of an event whose odds are 3:2. $\frac{3}{5}$
3. **Verify** the probabilities given for sums of 2 and 3 in Example 4. **See margin.**

Guided Practice

GUIDED PRACTICE KEY	
Exercises	**Examples**
4–6	2
7–12	3
13, 14	4
15–18	1

Suppose you select 2 letters at random from the word *compute*. Find each probability.

4. $P(\text{2 vowels})$ $\frac{1}{7}$
5. $P(\text{2 consonants})$ $\frac{2}{7}$
6. $P(\text{1 vowel, 1 consonant})$ **6.** $\frac{4}{7}$

Find the odds of an event occurring, given the probability of the event.

7. $\frac{8}{9}$ **8:1**
8. $\frac{1}{6}$ **1:5**
9. $\frac{2}{9}$ **2:7**

Find the probability of an event occurring, given the odds of the event.

10. 6:5 $\frac{6}{11}$
11. 10:1 $\frac{10}{11}$
12. 2:5 $\frac{2}{7}$

The table and the relative-frequency histogram show the distribution of the number of heads when 3 coins are tossed. Find each probability.

H = Heads	0	1	2	3
Probability	$\frac{1}{8}$	$\frac{3}{8}$	$\frac{3}{8}$	$\frac{1}{8}$

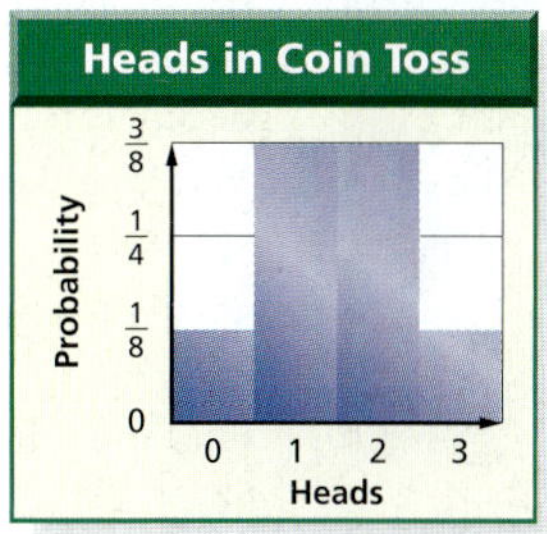

13. $P(H = 0)$ $\frac{1}{8}$
14. $P(H = 2)$ $\frac{3}{8}$

Application

GEOGRAPHY **For Exercises 15–18, find each probability if a state is chosen at random from the 50 states.**

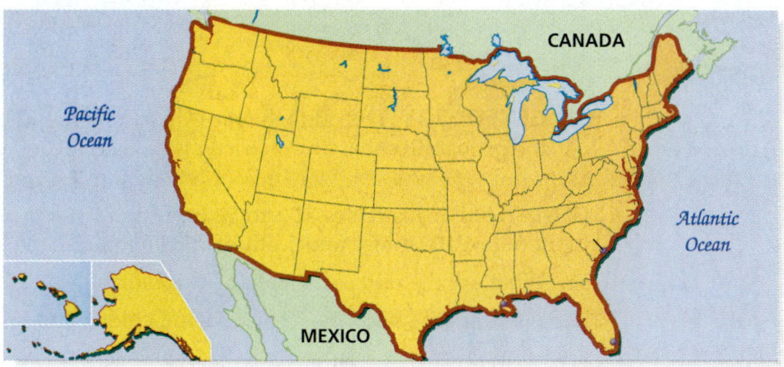

15. $P(\text{next to the Pacific Ocean})$ $\frac{1}{10}$
16. $P(\text{has at least five neighboring states})$ $\frac{21}{50}$
17. $P(\text{borders Mexico})$ $\frac{2}{25}$
18. $P(\text{is surrounded by water})$ $\frac{1}{50}$

3 Practice/Apply

Study Notebook

Have students—

- add the definitions/examples of the vocabulary terms to their Vocabulary Builder worksheets for Chapter 12.
- include any other item(s) that they find helpful in mastering the skills in this lesson.

About the Exercises…

Organization by Objective

- **Probability and Odds:** 19–53
- **Probability Distributions:** 55–60

Odd/Even Assignments

Exercises 19–29, 34–49, and 55–60 are structured so that students practice the same concepts whether they are assigned odd or even problems.

Assignment Guide

Basic: 19–59 odd, 62–65, 70–82

Average: 19–61 odd, 62–65, 70–82 (optional: 66–69)

Advanced: 20–60 even, 61–78 (optional: 79–83)

All: Practice Quiz 1 (1–10)

Answers

1. Sample answer: The event *July comes before June* has a probability of 0. The event *June comes before July* has a probability of 1.
3. There are 6 · 6 or 36 possible outcomes for the two dice. Only 1 outcome, 1 and 1, results in a sum of 2, so $P(2) = \frac{1}{36}$. There are 2 outcomes, 1 and 2 as well as 2 and 1, that result in a sum of 3, so $P(3) = \frac{2}{36}$ or $\frac{1}{18}$.

Tips for New Teachers

Intervention Students may be confused about the nature of odds and probability. Lead students in a discussion about the difference between theoretical and experimental probability. If you have a state lottery, this may be an opportunity to examine mistaken beliefs about chance. Be sensitive to the fact that some students may have cultural or familial prohibitions against cards, dice, or gambling of any kind. Explain that historically the laws of probability were actually developed in the context of gambling, but they are now used in many other ways, including medicine and meteorology.

★ indicates increased difficulty

Practice and Apply

Homework Help

For Exercises	See Examples
19–33, 54	1, 2
34–53	3
55–60	4

Extra Practice
See page 854.

Ebony has 4 male kittens and 7 female kittens. She picks up 2 kittens to give to a friend. Find the probability of each selection.

19. P(2 male) $\frac{6}{55}$
20. P(2 female) $\frac{21}{55}$
21. P(1 of each) $\frac{28}{55}$

Bob is moving and all of his CDs are mixed up in a box. Twelve CDs are rock, eight are jazz, and five are classical. If he reaches in the box and selects them at random, find each probability.

22. P(3 jazz) $\frac{14}{575}$
23. P(3 rock) $\frac{11}{115}$
24. P(1 classical, 2 jazz) $\frac{7}{115}$
25. P(2 classical, 1 rock) $\frac{6}{115}$
26. P(1 jazz, 2 rock) $\frac{132}{575}$
27. P(1 classical, 1 jazz, 1 rock) $\frac{24}{115}$
28. P(2 rock, 2 classical) $\frac{6}{115}$
29. P(2 jazz, 1 reggae) 0

30. **LOTTERIES** The state of Florida has a lottery in which 6 numbers out of 53 are drawn at random. What is the probability of a given ticket matching all 6 numbers in any order? $\frac{1}{22,957,480}$

More About. . .

Entrance Tests

In addition to the MCAT, most medical schools require applicants to have had one year each of biology, physics, and English, and two years of chemistry in college.

ENTRANCE TESTS **For Exercises 31–33, use the table that shows the college majors of the students who took the Medical College Admission Test (MCAT) in April 2000.**
If a student taking the test were randomly selected, find each probability. Express as decimals rounded to the nearest thousandth.

Major	Students
biological sciences	15,819
humanities	963
math or statistics	179
physical sciences	2770
social sciences	2482
specialized health sciences	1431
other	1761

31. P(math or statistics) 0.007
32. P(biological sciences) 0.623
33. P(physical sciences) 0.109

Find the odds of an event occurring, given the probability of the event.

34. $\frac{1}{2}$ 1:1
35. $\frac{3}{8}$ 3:5
36. $\frac{11}{12}$ 11:1
37. $\frac{5}{8}$ 5:3
38. $\frac{4}{7}$ 4:3
39. $\frac{1}{5}$ 1:4
40. $\frac{4}{11}$ 4:7
41. $\frac{3}{4}$ 3:1

Find the probability of an event occurring, given the odds of the event.

42. 6:1 $\frac{6}{7}$
43. 3:7 $\frac{3}{10}$
44. 5:6 $\frac{5}{11}$
45. 4:5 $\frac{4}{9}$
46. 9:8 $\frac{9}{17}$
47. 1:8 $\frac{1}{9}$
48. 7:9 $\frac{7}{16}$
49. 3:2 $\frac{3}{5}$

50. **GENEOLOGY** The odds that an American is of English ancestry are 1:9. What is the probability that an American is of English ancestry? $\frac{1}{10}$

GENETICS **For Exercises 51 and 52, use the following information.**
Eight out of 100 males and 1 out of 1000 females have some form of color blindness.

51. What are the odds of a male being color-blind? 2:23
52. What are the odds of a female being color-blind? 1:999

53. **EDUCATION** Josefina's guidance counselor estimates that the probability she will get a college scholarship is $\frac{4}{5}$. What are the odds that she will *not* earn a scholarship? 1:4

Answer

63. Probability and odds are good tools for assessing risk. Answers should include the following.

- P(struck by lightning) $= \frac{s}{s+f} = \frac{1}{750,000}$, so Odds = 1:(750,000 − 1) or 1:749,999.
  P(surviving a lightning strike) $= \frac{s}{s+f} = \frac{3}{4}$, so Odds = 3:(4 − 3) or 3:1.
- In this case, success is being struck by lightning or surviving the lightning strike. Failure is not being struck by lightning or not surviving the lightning strike.

★54. **CARD GAMES** The game of euchre is played using only the 9s, 10s, jacks, queens, kings, and aces from a standard deck of cards. Find the probability of being dealt a 5-card euchre hand containing all four suits. $\frac{540}{1771}$

Three students are selected at random from a group of 3 sophomores and 3 juniors. The table and relative-frequency histogram show the distribution of the number of sophomores chosen. Find each probability.

Sophomores	0	1	2	3
Probability	$\frac{1}{20}$	$\frac{9}{20}$	$\frac{9}{20}$	$\frac{1}{20}$

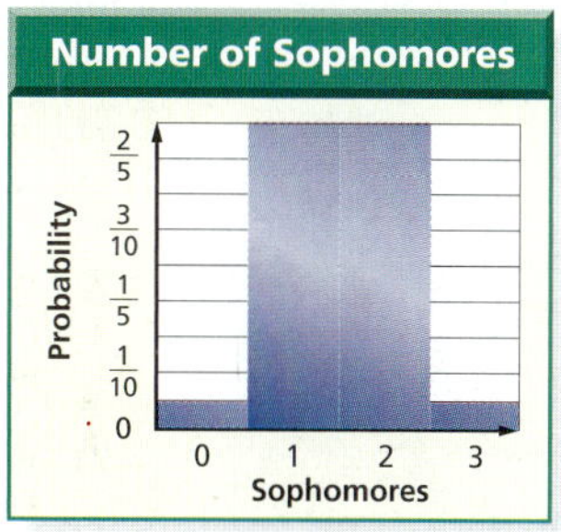

55. P(0 sophomores) $\frac{1}{20}$
56. P(1 sophomore) $\frac{9}{20}$
57. P(2 sophomores) $\frac{9}{20}$
58. P(3 sophomores) $\frac{1}{20}$
59. P(2 juniors) $\frac{9}{20}$
60. P(1 junior) $\frac{9}{20}$

★61. **WRITING** Josh types the 5 entries in the bibliography of his term paper in random order, forgetting that they should be in alphabetical order by author. What is the probability that he actually typed them in alphabetical order? $\frac{1}{120}$

62. **CRITICAL THINKING** Find the probability that a point chosen at random in the figure is in the shaded region. Write your answer in terms of π. $\frac{\pi - 1}{\pi}$

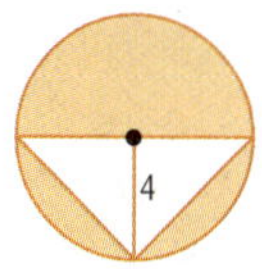

63. **WRITING IN MATH** Answer the question that was posed at the beginning of the lesson. **See margin.**

What do probability and odds tell you about life's risks?

Include the following in your answer:
- the odds of being struck by lightning and surviving the lightning strike, and
- a description of the meaning of *success* and *failure* in this case.

Standardized Test Practice

64. $\frac{6!}{2!} = ?$ **C**

Ⓐ 3 Ⓑ 60 Ⓒ 360 Ⓓ 720

65. A jar contains 4 red marbles, 3 green marbles, and 2 blue marbles. If a marble is drawn at random, what is the probability that it is not green? **D**

Ⓐ $\frac{2}{9}$ Ⓑ $\frac{1}{3}$ Ⓒ $\frac{4}{9}$ Ⓓ $\frac{2}{3}$

Extending the Lesson

Theoretical probability is determined using mathematical methods and assumptions about the fairness of coins, dice, and so on. **Experimental probability** is determined by performing experiments and observing the outcomes.

Determine whether each probability is *theoretical* or *experimental*. Then find the probability.

66. Two dice are rolled. What is the probability that the sum will be 12? **66. theoretical; $\frac{1}{36}$**

67. A baseball player has 126 hits in 410 at-bats this season. What is the probability that he gets a hit in his next at-bat? **experimental; about 0.307**

68. A bird watcher observes that 5 out of 25 birds in a garden are red. What is the probability that the next bird to fly into the garden will be red? **experimental; $\frac{1}{5}$**

69. A hand of 2 cards is dealt from a standard deck of cards. What is the probability that both cards are clubs? **theoretical; $\frac{1}{17}$**

www.algebra2.com/self_check_quiz

Study Guide and Intervention, p. 711 (shown) and p. 712

Probability and Odds In probability, a desired outcome is called a **success**; any other outcome is called a **failure**.

Probability of Success and Failure	If an event can succeed in s ways and fail in f ways, then the probabilities of success, $P(S)$, and of failure, $P(F)$, are as follows. $P(S) = \frac{s}{s+f}$ and $P(F) = \frac{f}{s+f}$.
Definition of Odds	If an event can succeed in s ways and fail in f ways, then the odds of success and of failure are as follows. Odds of success = $s:f$ Odds of failure = $f:s$

Example 1 **When 3 coins are tossed, what is the probability that at least 2 are heads?**

You can use a tree diagram to find the sample space.

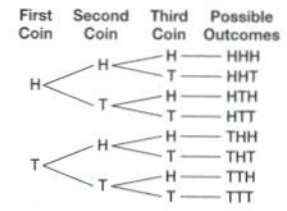

Of the 8 possible outcomes, 4 have at least 2 heads. So the probability of tossing at least 2 heads is $\frac{4}{8}$ or $\frac{1}{2}$.

Example 2 **What is the probability of picking 4 fiction books and 2 biographies from a best-seller list that consists of 12 fiction books and 6 biographies?**

By the Fundamental Counting Principle, the number of successes is $C(12, 4) \cdot C(6, 2)$. The total number of selections, $s + f$, of 6 books is $C(18, 6)$.

$P(\text{4 fiction, 2 biography}) = \frac{C(12, 4) \cdot C(6, 2)}{C(18, 6)}$ or about 0.40

The probability of selecting 4 fiction books and 2 biographies is about 40%.

Exercises

Find the odds of an event occurring, given the probability of the event.

1. $\frac{3}{7}$ **3:4** 2. $\frac{4}{5}$ **4:1** 3. $\frac{2}{13}$ **2:11** 4. $\frac{1}{15}$ **1:14**

Find the probability of an event occurring, given the odds of the event.

5. 10:1 $\frac{10}{11}$ 6. 2:5 $\frac{2}{7}$ 7. 4:9 $\frac{4}{13}$ 8. 8:3 $\frac{8}{11}$

One bag of candy contains 15 red candies, 10 yellow candies, and 6 green candies. Find the probability of each selection.

9. picking a red candy $\frac{15}{31}$
10. not picking a yellow candy $\frac{21}{31}$
11. picking a green candy $\frac{6}{31}$
12. not picking a red candy $\frac{16}{31}$

Skills Practice, p. 713 and Practice, p. 714 (shown)

A bag contains 1 green, 4 red, and 5 yellow balls. Two balls are selected at random. Find the probability of each selection.

1. P(2 red) $\frac{2}{15}$
2. P(1 red and 1 yellow) $\frac{4}{9}$
3. P(1 green and 1 yellow) $\frac{1}{9}$
4. P(2 green) **0**
5. P(2 red and 1 yellow) **0**
6. P(1 red and 1 green) $\frac{4}{45}$

A bank contains 3 pennies, 8 nickels, 4 dimes, and 10 quarters. Two coins are selected at random. Find the probability of each selection.

7. P(2 pennies) $\frac{1}{100}$
8. P(2 dimes) $\frac{1}{50}$
9. P(1 nickel and 1 dime) $\frac{8}{75}$
10. P(1 quarter and 1 penny) $\frac{1}{10}$
11. P(1 quarter and 1 nickel) $\frac{4}{15}$
12. P(2 dimes and 1 quarter) **0**

Henrico visits a home decorating store to choose wallpapers for his new house. The store has 28 books of wallpaper samples, including 10 books of WallPride samples and 18 books of Deluxe Wall Coverings samples. The store will allow Henrico to bring 4 books home for a few days so he can decide which wallpapers he wants to buy. If Henrico randomly chooses 4 books to bring home, find the probability of each selection.

13. P(4 WallPride) $\frac{2}{195}$
14. P(2 WallPride and 2 Deluxe) $\frac{153}{455}$
15. P(1 WallPride and 3 Deluxe) $\frac{544}{1365}$
16. P(3 WallPride and 1 Deluxe) $\frac{48}{455}$

For Exercises 17–20, use the table that shows the range of verbal SAT scores for freshmen at a small liberal arts college. If a freshman student is chosen at random, find each probability. Express as decimals rounded to the nearest thousandth.

Range	400–449	450–499	500–549	550–559	600–649	650+
Number of Students	129	275	438	602	620	412

17. P(400–449) **0.052**
18. P(550–559) **0.243**
19. P(at least 650) **0.166**

Find the odds of an event occurring, given the probability of the event.

20. $\frac{4}{11}$ **4:7** 21. $\frac{12}{13}$ **12:1** 22. $\frac{5}{99}$ **5:94** 23. $\frac{1}{1000}$ **1:999**
24. $\frac{5}{16}$ **5:11** 25. $\frac{3}{95}$ **3:92** 26. $\frac{9}{70}$ **9:61** 27. $\frac{8}{15}$ **8:7**

Find the probability of an event occurring, given the odds of the event.

28. 2:23 $\frac{2}{25}$ 29. 2:5 $\frac{2}{7}$ 30. 15:1 $\frac{15}{16}$ 31. 9:7 $\frac{9}{16}$
32. 11:14 $\frac{11}{25}$ 33. 1000:1 $\frac{1000}{1001}$ 34. 12:17 $\frac{12}{29}$ 35. 8:13 $\frac{8}{21}$

Reading to Learn Mathematics, p. 715 ELL

Pre-Activity **What do probability and odds tell you about life's risks?**

Read the introduction to Lesson 12-3 at the top of page 644 in your textbook.

What is the probability that a person will *not* be struck by lightning in a given year? $\frac{749,999}{750,000}$

Reading the Lesson

1. Indicate whether each of the following statements is *true* or *false*.
   a. If an event can never occur, its probability is a negative number. **false**
   b. If an event is certain to happen, its probability is 1. **true**
   c. If an event can succeed in s ways and fail in f ways, then the probability of success is $\frac{s}{f}$. **false**
   d. If an event can succeed in s ways and fail in f ways, then the odds against the event are $s:f$. **false**
   e. A probability distribution is a function in which the domain is the sample space of an experiment. **true**
2. A weather forecast says that the chance of rain tomorrow is 40%.
   a. Write the probability that it will rain tomorrow as a fraction in lowest terms. $\frac{2}{5}$
   b. Write the probability that it will not rain tomorrow as a fraction in lowest terms. $\frac{3}{5}$
   c. What are the odds in favor of rain? **2:3**
   d. What are the odds against rain? **3:2**
3. Refer to the table in Example 4 on page 646 in your textbook.
   a. What other sum has the same probability as a sum of 11? **3**
   b. What are the odds of rolling a sum of 8? **5:31**
   c. What are the odds against rolling a sum of 9? **8:1**

Helping You Remember

4. A good way to remember something is to explain it to someone else. Suppose that your friend Roberto is having trouble remembering the difference between probability and odds. What would you tell him to help him remember this easily?

Sample answer: Probability gives the ratio of successes to the total number of outcomes, while odds gives the ratio of successes to failures.

Enrichment, p. 716

Geometric Probability

If a dart, thrown at random, hits the triangular board shown at the right, what is the chance that it will hit the shaded region? This chance, also called a probability, can be determined by comparing the area of the shaded region to the area of the board. This ratio indicates what fraction of the tosses should hit in the shaded region.

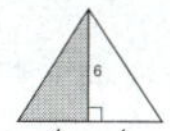

$$\frac{\text{area of shaded region}}{\text{area of triangular board}} = \frac{\frac{1}{2}(4)(6)}{\frac{1}{2}(8)(6)} = \frac{12}{24} \text{ or } \frac{1}{2}$$

In general, if S is a subregion of some region R, then the probability, $P(S)$, that a point, chosen at random, belongs to subregion S is given by the following.

$$P(S) = \frac{\text{area of subregion } S}{\text{are of region } R}$$

4 Assess

Open-Ended Assessment

Modeling Have students create a simple probability experiment using manipulatives and classroom objects. Have students first calculate the probability and then perform the experiment to verify their calculations.

Assessment Options

Practice Quiz 1 The quiz provides students with a brief review of the concepts and skills in Lessons 12-1 through 12-3. Lesson numbers are given to the right of exercises or instruction lines so students can review concepts not yet mastered.

Quiz (Lessons 12-1 through 12-3) is available on p. 767 of the *Chapter 12 Resource Masters*.

Getting Ready for Lesson 12-4

BASIC SKILL Lesson 12-4 presents finding the probability of two events. Students will use their familiarity with multiplying fractions as they calculate probabilities. Exercises 79–83 should be used to determine your students' familiarity with multiplying rational expressions.

Maintain Your Skills

Mixed Review

Determine whether each situation involves a *permutation* or a *combination*. Then find the number of possibilities. *(Lesson 12-2)*

70. arranging 5 different books on a shelf **permutation; 120**

71. arranging the letters of the word *arrange* **permutation; 1260**

72. picking 3 apples from the last 7 remaining at the grocery store **combination; 35**

73. A mail-order computer company offers a choice of 4 amounts of memory, 2 sizes of hard drives, and 2 sizes of monitors. How many different systems are available to a customer? *(Lesson 12-1)* **16**

74. How many ways can 4 different gifts be placed into 4 different gift bags if each bag gets exactly 1 gift? *(Lesson 12-1)* **24**

Identify the type of function represented by each graph. *(Lesson 9-5)*

75. 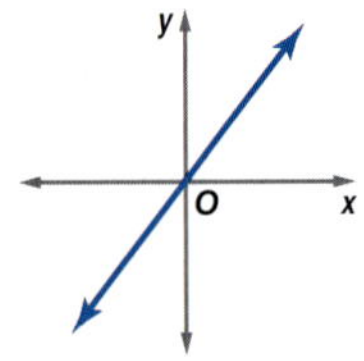

direct variation

76. 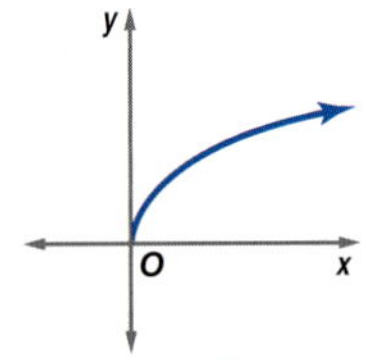

square root

Solve each matrix equation. *(Lesson 4-1)*

77. $[x \quad y] = [y \quad 4]$ **(4, 4)**

78. $\begin{bmatrix} 3y \\ 2x \end{bmatrix} = \begin{bmatrix} x+8 \\ y-x \end{bmatrix}$ **(1, 3)**

Getting Ready for the Next Lesson

BASIC SKILL **Find each product if $a = \frac{3}{5}$, $b = \frac{2}{7}$, $c = \frac{3}{4}$, and $d = \frac{1}{3}$.**

79. ab $\frac{6}{35}$

80. bc $\frac{3}{14}$

81. cd $\frac{1}{4}$

82. bd $\frac{2}{21}$

83. ac $\frac{9}{20}$

Practice Quiz 1 — Lessons 12-1 through 12-3

1. At the Burger Bungalow, you can order your hamburger with or without cheese, with or without onions or pickles, and either rare, medium, or well-done. How many different ways can you order your hamburger? *(Lesson 12-1)* **24**

2. For a particular model of car, a dealer offers 3 sizes of engines, 2 types of stereos, 18 body colors, and 7 upholstery colors. How many different possibilities are available for that model? *(Lesson 12-1)* **756**

3. How many codes consisting of a letter followed by 3 digits can be made if no digit can be used more than once? *(Lesson 12-1)* **18,720**

Evaluate each expression. *(Lesson 12-2)*

4. $P(12, 3)$ **1320**

5. $C(8, 3)$ **56**

Determine whether each situation involves a *permutation* or a *combination*. Then find the number of possibilities. *(Lesson 12-2)*

6. 8 cars in a row parked next to a curb **permutation; 40,320**

7. a hand of 6 cards from a standard deck of cards **combination; 20,358,520**

Two cards are drawn from a standard deck of cards. Find each probability. *(Lesson 12-3)*

8. P(2 aces) $\frac{1}{221}$

9. P(1 heart, 1 club) $\frac{13}{102}$

10. P(1 queen, 1 king) $\frac{8}{663}$

12-4 Multiplying Probabilities

What You'll Learn

- Find the probability of two independent events.
- Find the probability of two dependent events.

Vocabulary

- area diagram

How does probability apply to basketball?

Reggie Miller of the Indiana Pacers is one of the best free-throw shooters in the National Basketball Association. The table shows the five highest season free-throw statistics of his career. For any year, you can determine the probability that Miller will make two free throws in a row based on the probability of his making one free throw.

Season	FT%
1990–91	91.8
1993–94	90.8
1998–99	91.5
1999–00	92.9
2000–01	92.8

Source: *Sporting News*

PROBABILITY OF INDEPENDENT EVENTS In a situation with two events like shooting a free throw and then shooting another one, you can find the probability of *both* events occurring if you know the probability of each event occurring. You can use an **area diagram** to model the probability of the two events occurring at the same time.

Algebra Activity

Area Diagrams

Suppose there are 1 red and 3 blue paper clips in one drawer and 1 gold and 2 silver paper clips in another drawer. The area diagram represents the probabilities of choosing one colored paper clip and one metallic paper clip if one of each is chosen at random. For example, rectangle A represents drawing 1 silver clip and 1 blue clip.

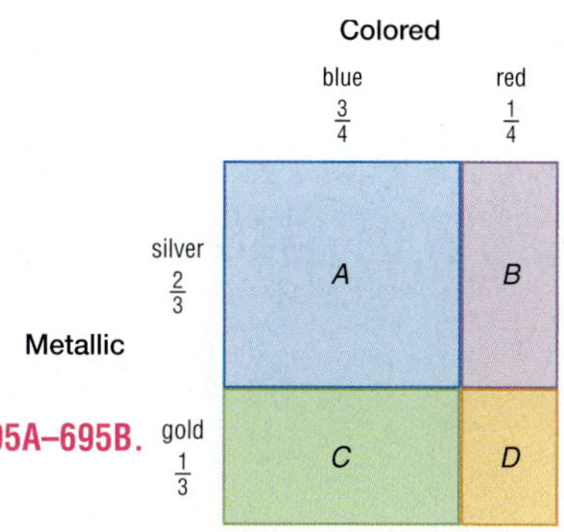

Model and Analyze 1, 4. See pp. 695A–695B.

1. Find the areas of rectangles A, B, C, and D, and explain what each area represents. 2. $\frac{1}{6}$ 3. 1; 1; 1; The sum of the probabilities must be 1.
2. What is the probability of choosing a red paper clip and a silver paper clip?
3. What are the length and width of the whole square? What is the area? Why does the area need to have this value?
4. Make an area diagram that represents the probability of each outcome if you spin each spinner once. Label the diagram and describe what the area of each rectangle represents.

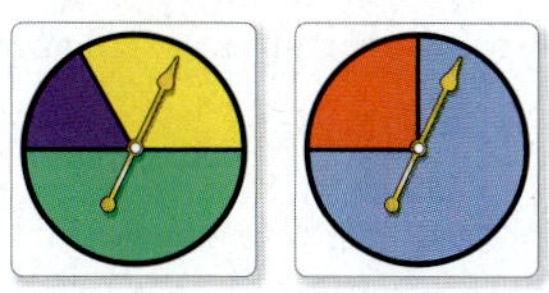

12-4 Lesson Notes

1 Focus

5-Minute Check Transparency 12-4 Use as a quiz or review of Lesson 12-3.

Mathematical Background notes are available for this lesson on p. 630D.

How does probability apply to basketball?

Ask students:

- Ask a volunteer to explain what a free throw is, for the benefit of any students who might not be familiar with the game of basketball.
- Based on the information in this table, out of 10 free throws, how many would you expect Miller to make in the 2001-2002 season? 9

Resource Manager

Workbook and Reproducible Masters

Chapter 12 Resource Masters
- Study Guide and Intervention, pp. 717–718
- Skills Practice, p. 719
- Practice, p. 720
- Reading to Learn Mathematics, p. 721
- Enrichment, p. 722

School-to-Career Masters, p. 23

Teaching Algebra With Manipulatives Masters, pp. 291, 292–293

Transparencies

5-Minute Check Transparency 12-4
Answer Key Transparencies

Technology

Alge2PASS: Tutorial Plus, Lesson 23
Interactive Chalkboard

2 Teach

PROBABILITY OF INDEPENDENT EVENTS

In-Class Examples

Power Point®

1 Gerardo has 9 dimes and 7 pennies in his pocket. He randomly selects one coin, looks at it, and replaces it. He then randomly selects another coin. What is the probability that both of the coins he selects are dimes? $\frac{81}{256}$

2 When three dice are rolled, what is the probability that two dice show a 5 and the third die shows an even number? $\frac{1}{72}$

Teaching Tip To verify that students understand the notation in the Key Concept box, have them read aloud the expression $P(A \text{ and } B) = P(A) \cdot P(B)$ and ask them to explain it.

In Exercise 4 of the activity, spinning one spinner has no effect on the second spinner. These events are independent.

Key Concept — ***Probability of Two Independent Events***

If two events, A and B, are independent, then the probability of both events occurring is $P(A \text{ and } B) = P(A) \cdot P(B)$.

This formula can be applied to any number of independent events.

Example 1 Two Independent Events

At a picnic, Julio reaches into an ice-filled cooler containing 8 regular soft drinks and 5 diet soft drinks. He removes a can, then decides he is not really thirsty, and puts it back. What is the probability that Julio and the next person to reach into the cooler both randomly select a regular soft drink?

Explore These events are independent since Julio replaced the can that he removed. The outcome of the second person's selection is not affected by Julio's selection.

Plan Since there are 13 cans, the probability of each person's getting a regular soft drink is $\frac{8}{13}$.

Solve $P(\text{both regular}) = P(\text{regular}) \cdot P(\text{regular})$ Probability of independent events

$= \frac{8}{13} \cdot \frac{8}{13}$ or $\frac{64}{169}$ Substitute and multiply.

The probability that both people select a regular soft drink is $\frac{64}{169}$ or about 0.38.

Examine You can verify this result by making a tree diagram that includes probabilities. Let R stand for regular and D stand for diet.

$P(R, R) = \frac{8}{13} \cdot \frac{8}{13}$

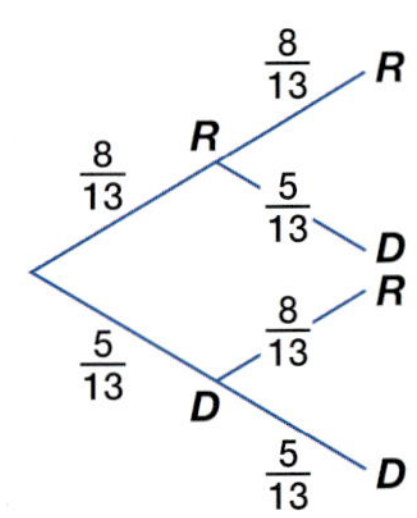

Study Tip

Alternative Method

You could use the Fundamental Counting Principle to find the number of successes and the number of total outcomes.

both regular $= 8 \cdot 8$ or 64

total outcomes $= 13 \cdot 13$ or 169

So, $P(\text{both reg.}) = \frac{64}{169}$.

The formula for the probability of independent events can be extended to any number of independent events.

Example 2 Three Independent Events

In a board game, three dice are rolled to determine the number of moves for the players. What is the probability that the first die shows a 6, the second die shows a 6, and the third die does not?

Let A be the event that the first die shows a 6. → $P(A) = \frac{1}{6}$

Let B be the event that the second die shows a 6. → $P(B) = \frac{1}{6}$

Let C be the event that the third die does *not* show a 6. → $P(C) = \frac{5}{6}$

Algebra Activity

Materials (optional): paper clips in red, blue, gold, and silver; spinner with circle whose segments can be changed

Suggest that students use the least common denominator of the probabilities to choose the length of the side of the square for their area diagram. For example, when representing probabilities of $\frac{1}{6}$, $\frac{1}{2}$, and $\frac{1}{3}$, a square with sides of 6 centimeters works well.

$$P(A, B, \text{ and } C) = P(A) \cdot P(B) \cdot P(C)$$ Probability of independent events

$$= \frac{1}{6} \cdot \frac{1}{6} \cdot \frac{5}{6} \text{ or } \frac{5}{216}$$ Substitute and multiply.

The probability that the first and second dice show a 6 and the third die does not is $\frac{5}{36}$.

PROBABILITY OF DEPENDENT EVENTS In Example 1, what is the probability that both people select a regular soft drink if Julio does not put his back in the cooler? In this case, the two events are dependent because the outcome of the first event affects the outcome of the second event.

Study Tip

Conditional Probability
The event of getting a regular soft drink the second time *given* that Julio got a regular soft drink the first time is called a *conditional probability*.

First selection	**Second selection**
$P(\text{regular}) = \frac{8}{13}$	$P(\text{regular}) = \frac{7}{12}$

Notice that when Julio removes his can, there is not only one fewer regular soft drink but also one fewer drink in the cooler.

$$P(\text{both regular}) = P(\text{regular}) \cdot P(\text{regular following regular})$$

$$= \frac{8}{13} \cdot \frac{7}{12} \text{ or } \frac{14}{39}$$ Substitute and multiply.

The probability that both people select a regular soft drink is $\frac{14}{39}$ or about 0.36.

Key Concept — *Probability of Two Dependent Events*

If two events, A and B, are dependent, then the probability of both events occurring is $P(A \text{ and } B) = P(A) \cdot P(B \text{ following } A)$.

This formula can be extended to any number of dependent events.

Example 3 Two Dependent Events

The host of a game show is drawing chips from a bag to determine the prizes for which contestants will play. Of the 10 chips in the bag, 6 show *television*, 3 show *vacation*, and 1 shows *car*. If the host draws the chips at random and does not replace them, find each probability.

Because the first chip is not replaced, the events are dependent. Let T represent a television, V a vacation, and C a car.

a. a vacation, then a car

$$P(V, \text{ then } C) = P(V) \cdot P(C \text{ following } V)$$ Dependent events

$$= \frac{3}{10} \cdot \frac{1}{9} \text{ or } \frac{1}{30}$$ After the first chip is drawn, there are 9 left.

The probability of a vacation and then a car is $\frac{1}{30}$ or about 0.03.

b. two televisions

$$P(T, \text{ then } T) = P(T) \cdot P(T \text{ following } T)$$ Dependent events

$$= \frac{6}{10} \cdot \frac{5}{9} \text{ or } \frac{1}{3}$$ If the first chip shows television, then 5 of the remaining 9 show television.

The probability of the host drawing two televisions is $\frac{1}{3}$.

www.algebra2.com/extra_examples

PROBABILITY OF DEPENDENT EVENTS

In-Class Example Power Point®

3 Refer to Example 3 in the Student Edition. The next week, the host of the game show draws from a bag of 20 chips, of which 11 say *computer*, 8 say *trip*, and 1 says *truck*. Drawing at random and without replacement, find each of the following probabilities.

a. a computer, then a truck $\frac{11}{380}$ or about 0.03

b. two trips $\frac{14}{95}$ or about 0.15

Lesson 12-4 Multiplying Probabilities 653

In-Class Example

Power Point®

4 Three cards are drawn from a standard deck of cards without replacement. Find the probability of drawing a heart, another heart, and a spade in that order. $\frac{13}{850}$ or about 0.015

3 Practice/Apply

Study Notebook

Have students—

- add the definitions/examples of the vocabulary terms to their Vocabulary Builder worksheets for Chapter 12.
- include any other item(s) that they find helpful in mastering the skills in this lesson.

DAILY INTERVENTION **FIND THE ERROR** Ask students to describe a situation in which Tabitha would be correct. Sample answer: Once a number is rolled with the die, that number roll is considered invalid and the die must be rolled again until a valid number is rolled.

Example 4 Three Dependent Events

Three cards are drawn from a standard deck of cards without replacement. Find the probability of drawing a diamond, a club, and another diamond in that order.

Since the cards are not replaced, the events are dependent. Let D represent a diamond and C a club.

$$P(D, C, D) = P(D) \cdot P(C \text{ following } D) \cdot P(D \text{ following } D \text{ and } C)$$

$$= \frac{13}{52} \cdot \frac{13}{51} \cdot \frac{12}{50} \text{ or } \frac{13}{850}$$ If the first two cards are a diamond and a club, then 12 of the remaining cards are diamonds.

The probability is $\frac{13}{850}$ or about 0.015.

Check for Understanding

Concept Check

1. **OPEN ENDED** Describe two real-life events that are dependent.
2. **Write** a formula for $P(A, B, C, \text{and } D)$ if A, B, C, and D are independent.
3. **FIND THE ERROR** Mario and Tabitha are calculating the probability of getting a 4 and then a 2 if they roll a die twice.

Mario	Tabitha
$P(4, \text{then } 2) = \frac{1}{6} \cdot \frac{1}{6} = \frac{1}{36}$	$P(4, \text{then } 2) = \frac{1}{6} \cdot \frac{1}{5} = \frac{1}{30}$

Who is correct? Explain your reasoning. Mario; the probabilities of rolling a 4 and rolling a 2 are both $\frac{1}{6}$.

1. Sample answer: putting on your socks, and then your shoes
2. $P(A, B, C, \text{and } D) = P(A) \cdot P(B) \cdot P(C) \cdot P(D)$

Guided Practice

GUIDED PRACTICE KEY	
Exercises	Examples
4, 5, 9, 12	1, 2
6–8, 10, 13	3
11	4

A die is rolled twice. Find each probability.

4. $P(5, \text{then } 1)$ $\frac{1}{36}$
5. $P(\text{two even numbers})$ $\frac{1}{4}$

Two cards are drawn from a standard deck of cards. Find each probability if no replacement occurs.

6. $P(\text{two hearts})$ $\frac{1}{17}$
7. $P(\text{ace, then king})$ $\frac{4}{663}$

There are 8 action, 3 romantic comedy, and 5 children's DVDs on a shelf. Suppose two DVDs are selected at random from the shelf. Find each probability.

8. $P(2 \text{ action DVDs})$, if no replacement occurs $\frac{7}{30}$
9. $P(2 \text{ action DVDs})$, if replacement occurs $\frac{1}{4}$
10. $P(\text{a romantic comedy DVD, then a children's DVD})$, if no replacement occurs $\frac{1}{16}$

Determine whether the events are *independent* or *dependent*. Then find the probability. 11. dependent; $\frac{21}{220}$

11. Yana has 7 blue pens, 3 black pens, and 2 red pens in his desk drawer. If he selects three pens at random with no replacement, what is the probability that he will first select a blue pen, then a black pen, and then another blue pen?
12. A black die and a white die are rolled. What is the probability that a 3 shows on the black die and a 5 shows on the white die? independent; $\frac{1}{36}$

DAILY INTERVENTION

Differentiated Instruction

Naturalist Have students investigate how probability can be used to report the results of Mendel's famous experiments with seeds, and how it is used today by botanists who are developing desired characteristics in flowers and vegetables.

Application

13. **ELECTIONS** Tami, Sonia, Malik, and Roger are the four candidates for student council president. If their names are placed in random order on the ballot, what is the probability that Malik's name will be first on the ballot followed by Sonia's name second? $\frac{1}{12}$

★ indicates increased difficulty

Practice and Apply

Homework Help	
For Exercises	See Examples
14–19, 36–39, 44–46	1, 2
20–29	1, 3
30–35	1–4
40–43	3

Extra Practice
See page 855.

A die is rolled twice. Find each probability.

14. $P(2, \text{then } 3)$ $\frac{1}{36}$
15. $P(\text{no 6s})$ $\frac{25}{36}$
16. $P(\text{two 4s})$ $\frac{1}{36}$
17. $P(1, \text{then any number})$ $\frac{1}{6}$
18. $P(\text{two of the same number})$ $\frac{1}{6}$
19. $P(\text{two different numbers})$ $\frac{5}{6}$

The tiles *A*, *B*, *G*, *I*, *M*, *R*, and *S* of a word game are placed face down in the lid of the game. If two tiles are chosen at random, find each probability.

20. $P(R, \text{then } S)$, if no replacement occurs $\frac{1}{42}$
21. $P(A, \text{then } M)$, if replacement occurs $\frac{1}{49}$
22. $P(\text{2 consonants})$, if replacement occurs $\frac{25}{49}$
23. $P(\text{2 consonants})$, if no replacement occurs $\frac{10}{21}$
24. ★ $P(B, \text{then } D)$, if replacement occurs 0
25. ★ $P(\text{selecting the same letter twice})$, if no replacement occurs 0

Ashley takes her 3-year-old brother Alex into an antique shop. There are 4 statues, 3 picture frames, and 3 vases on a shelf. Alex accidentally knocks 2 items off the shelf and breaks them. Find each probability.

26. $P(\text{breaking 2 vases})$ $\frac{1}{15}$
27. $P(\text{breaking 2 statues})$ $\frac{2}{15}$
28. $P(\text{breaking a picture frame, then a vase})$ $\frac{1}{10}$
29. $P(\text{breaking a statue, then a picture frame})$ $\frac{2}{15}$

Determine whether the events are *independent* or *dependent*. Then find the probability.

30. There are 3 miniature chocolate bars and 5 peanut butter cups in a candy dish. Judie chooses 2 of them at random. What is the probability that she chooses 2 miniature chocolate bars? dependent; $\frac{3}{28}$
31. A bowl contains 4 peaches and 5 apricots. Maxine randomly selects one, puts it back, and then randomly selects another. What is the probability that both selections were apricots? independent; $\frac{25}{81}$
32. A bag contains 7 red, 4 blue, and 6 yellow marbles. If 3 marbles are selected in succession, what is the probability of selecting blue, then yellow, then red, if replacement occurs each time? independent; $\frac{168}{4913}$
33. Joe's wallet contains three \$1 bills, four \$5 bills, and two \$10 bills. If he selects three bills in succession, find the probability of selecting a \$10 bill, then a \$5 bill, and then a \$1 bill if the bills are not replaced. dependent; $\frac{1}{21}$
34. What is the probability of getting heads each time if a coin is tossed 5 times?
35. ★ When Diego plays his favorite video game, the odds are 3 to 4 that he will reach the highest level of the game. What is the probability that he will reach the highest level each of the next four times he plays? dependent; $\frac{81}{2401}$

34. independent; $\frac{1}{32}$

www.algebra2.com/self_check_quiz

About the Exercises...

Organization by Objective

- **Probability of Independent Events:** 14–19, 21, 22, 24, 28, 29, 31, 32, 34
- **Probability of Dependent Events:** 20, 23, 25–27, 30, 33, 35, 40–43

Odd/Even Assignments

Exercises 14–35 are structured so that students practice the same concepts whether they are assigned odd or even problems.

Assignment Guide

Basic: 15–23 odd, 27–33 odd, 45, 50–77

Average: 15–35 odd, 41–45 odd, 50–77

Advanced: 14–34 even, 36–39, 40–46 even, 47–71 (optional: 72–77)

For Exercises 36–39, suppose you spin the spinner twice.

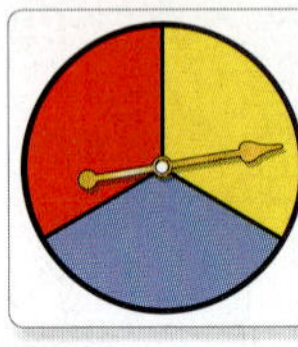

36. Sketch a tree diagram showing all of the possibilities. Use it to find the probability of spinning red and then blue. **$\frac{1}{9}$; See pp. 695A–695B for diagram.**

37. Sketch an area diagram of the outcomes. Shade the region on your area diagram corresponding to getting the same color twice. **See pp. 695A–695B.**

38. What is the probability that you get the same color on both spins? **$\frac{1}{9}$**

39. If you spin the same color twice, what is the probability that the color is red? **$\frac{1}{3}$**

Find each probability if 13 cards are drawn from a standard deck of cards and no replacement occurs.

★ 40. P(all clubs) **$\frac{1}{635{,}013{,}559{,}600}$**

★ 41. P(all black cards) **$\frac{19}{1{,}160{,}054}$**

★ 42. P(all one suit) **$\frac{1}{158{,}753{,}389{,}900}$**

★ 43. P(no aces) **$\frac{6327}{20{,}825}$**

44. **UTILITIES** A city water system includes a sequence of 4 pumps as shown below. Water enters the system at point A, is pumped through the system by pumps at locations 1, 2, 3, and 4, and exits the system at point B.

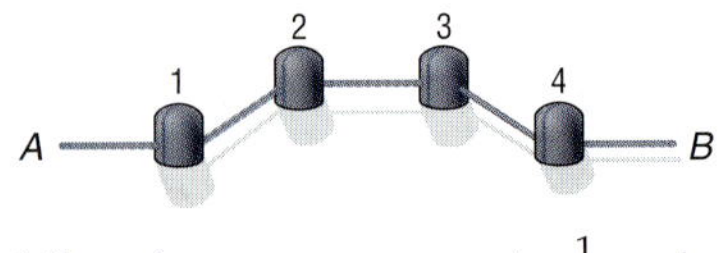

If the probability of failure for any one pump is $\frac{1}{100}$, what is the probability that water will flow all the way through the system from A to B? **$\left(\frac{99}{100}\right)^4$ or about 0.96**

45. **SPELLING** Suppose a contestant in a spelling bee has a 93% chance of spelling any given word correctly. What is the probability that he or she spells the first five words in a bee correctly and then misspells the sixth word? **about 4.87%**

★ 46. **LITERATURE** The following quote is from *The Mirror Crack'd,* which was written by Agatha Christie in 1962.

> "I think you're begging the question," said Haydock, "and I can see looming ahead one of those terrible exercises in probability where six men have white hats and six men have black hats and you have to work it out by mathematics how likely it is that the hats will get mixed up and in what proportion. If you start thinking about things like that, you would go round the bend. Let me assure you of that!"

If the twelve hats are all mixed up and each man randomly chooses a hat, what is the probability that the first three men get their own hats? Assume that no replacement occurs. **$\frac{1}{1320}$**

For Exercises 47–49, use the following information.

You have a bag containing 10 marbles. In this problem, a *cycle* means that you draw a marble, record its color, and put it back.

47. You go through the cycle 10 times. If you do not record any black marbles, can you conclude that there are no black marbles in the bag? **no**

48. Can you conclude that there are none if you repeat the cycle 50 times? **no**

49. How many times do you have to repeat the cycle to be certain that there are no black marbles in the bag? Explain your reasoning. **See margin.**

50. **CRITICAL THINKING** If one bulb in a string of holiday lights fails to work, the whole string will not light. If each bulb in a set has a 99.5% chance of working, what is the maximum number of lights that can be strung together with at least a 90% chance of the whole string lighting? **21**

More About. . .

Spelling

The National Spelling Bee has been held every year since 1925, except for 1943-1945. Of the first 76 champions, 42 were girls and 34 were boys.

Source: www.spellingbee.com

Study Guide and Intervention, p. 717 (shown) and p. 718

Probability of Independent Events

Probability of Two Independent Events	If two events, A and B, are independent, then the probability of both occurring is $P(A \text{ and } B) = P(A) \cdot P(B)$.

Example **In a board game each player has 3 different-colored markers. To move around the board the player first spins a spinner to determine which piece can be moved. He or she then rolls a die to determine how many spaces that colored piece should move. On a given turn what is the probability that a player will be able to move the yellow piece more than 2 spaces?**

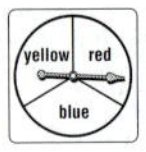

Let A be the event that the spinner lands on yellow, and let B be the event that the die shows a number greater than 2. The probability of A is $\frac{1}{3}$, and the probability of B is $\frac{2}{3}$.

$P(A \text{ and } B) = P(A) \cdot P(B)$ Probability of independent events

$= \frac{1}{3} \cdot \frac{2}{3}$ or $\frac{2}{9}$ Substitute and multiply.

The probability that the player can move the yellow piece more than 2 spaces is $\frac{2}{9}$.

Exercises

A die is rolled 3 times. Find the probability of each event.

1. a 1 is rolled, then a 2, then a 3 **$\frac{1}{216}$**
2. a 1 or a 2 is rolled, then a 3, then a 5 or a 6 **$\frac{1}{54}$**
3. 2 odd numbers are rolled, then a 6 **$\frac{1}{24}$**
4. a number less than 3 is rolled, then a 3, then a number greater than 3 **$\frac{1}{36}$**
5. A box contains 5 triangles, 6 circles, and 4 squares. If a figure is removed, replaced, and a second figure is picked, what is the probability that a triangle and then a circle will be picked? **$\frac{2}{15}$ or about 0.13**
6. A bag contains 5 red marbles and 4 white marbles. A marble is selected from the bag, then replaced, and a second selection is made. What is the probability of selecting 2 red marbles? **$\frac{25}{81}$ or about 0.31**
7. A jar contains 7 lemon jawbreakers, 3 cherry jawbreakers, and 8 rainbow jawbreakers. What is the probability of selecting 2 lemon jawbreakers in succession providing the jawbreaker drawn first is then replaced before the second is drawn? **$\frac{49}{324}$ or about 0.15**

Skills Practice, p. 719 and Practice, p. 720 (shown)

A die is rolled three times. Find each probability.

1. P(three 4s) **$\frac{1}{216}$**
2. P(no 4s) **$\frac{125}{216}$**
3. P(2, then 3, then 1) **$\frac{1}{216}$**
4. P(three different even numbers) **$\frac{1}{36}$**
5. P(any number, then 5, then 5) **$\frac{1}{36}$**
6. P(even number, then odd number, then 1) **$\frac{1}{24}$**

There are 3 nickels, 2 dimes, and 5 quarters in a purse. Three coins are selected in succession at random. Find the probability.

7. P(nickel, then dime, then quarter), if no replacement occurs **$\frac{1}{24}$**
8. P(nickel, then dime, then quarter), if replacement occurs **$\frac{3}{100}$**
9. P(2 nickels, then 1 quarter), if no replacement occurs **$\frac{1}{24}$**
10. P(3 dimes), if replacement occurs **$\frac{1}{125}$**
11. P(3 dimes), if no replacement occurs **0**

For Exercises 12 and 13, determine whether the events are *independent* or *dependent*. Then find each probability.

12. Serena is creating a painting. She wants to use 2 more colors. She chooses randomly from 6 shades of red, 10 shades of green, 4 shades of yellow, 4 shades of purple, and 6 shades of blue. What is the probability that she chooses 2 shades of green? **dependent; $\frac{3}{29}$**
13. Kershel's mother is shopping at a bakery. The owner offers Kershel a cookie from a jar containing 22 chocolate chip cookies, 18 sugar cookies, and 15 oatmeal cookies. Without looking, Kershel selects one, drops it back in, and then randomly selects another. What is the probability that neither selection was a chocolate chip cookie? **independent; $\frac{9}{25}$**
14. **METEOROLOGY** The Fadeeva's are planning a 3-day vacation to the mountains. A long-range forecast reports that the probability of rain each day is 10%. Assuming that the daily probabilities of rain are independent, what is the probability that there is no rain on the first two days, but that it rains on the third day? **$\frac{81}{1000}$**

RANDOM NUMBERS For Exercises 15 and 16, use the following information.

Anita has a list of 20 jobs around the house to do, and plans to do 3 of them today. She assigns each job a number from 1 to 20, and sets her calculator to generate random numbers from 1 to 20, which can reoccur. Of the jobs, 3 are outside, and the rest are inside.

15. Sketch a tree diagram showing all of the possibilities that the first three numbers generated correspond to inside jobs or outside jobs. Use it to find the probability that the first two numbers correspond to inside jobs, and the third to an outside job. **0.108375**
16. What is the probability that the number generated corresponds to an outside job three times in a row? **0.003375**

Reading to Learn Mathematics, p. 721 ELL

Pre-Activity **How does probability apply to basketball?**

Read the introduction to Lesson 12-4 at the top of page 651 in your textbook.

Write the probability that Reggie Miller made a free-throw shot during the 1998–99 season as a fraction in lowest terms. (Your answer should not include a decimal.) **$\frac{183}{200}$**

Reading the Lesson

1. A bag contains 4 yellow balls, 5 red balls, 1 white ball, and 2 black balls. A ball is drawn from the bag and is not replaced. A second ball is drawn.
   a. Let Y be the event "first ball is yellow" and B be the event "second ball is black." Are these events *independent* or *dependent*? **dependent**
   b. Tell which formula you would use to find the probability that the first ball is yellow and the second ball is black. **C**
      A. $P(Y \text{ and } B) = \frac{P(Y)}{P(Y) + P(B)}$
      B. $P(Y \text{ and } B) = P(Y) \cdot P(B)$
      C. $P(Y \text{ and } B) = P(Y) \cdot P(B \text{ following } Y)$
   c. Which equation shows the correct calculation of this probability? **B**
      A. $\frac{1}{3} + \frac{2}{11} = \frac{17}{33}$
      B. $\frac{1}{3} \cdot \frac{2}{11} = \frac{2}{33}$
      C. $\frac{1}{3} + \frac{1}{6} = \frac{1}{2}$
      D. $\frac{1}{3} \cdot \frac{1}{6} = \frac{1}{18}$
   d. Which equation shows the correct calculation of the probability that if three balls are drawn in succession without replacement, all three will be red? **B**
      A. $\frac{5}{12} \cdot \frac{5}{12} \cdot \frac{5}{12} = \frac{125}{1728}$
      B. $\frac{5}{12} \cdot \frac{4}{11} \cdot \frac{3}{10} = \frac{1}{22}$
      C. $\frac{5}{12} + \frac{4}{11} + \frac{3}{10} = \frac{713}{660}$

Helping You Remember

2. Some students have trouble remembering a lot of formulas, so they try to keep the number of formulas they have to know to a minimum. Can you learn just one formula that will allow you to find probabilities for both independent and dependent events? Explain your reasoning. **Sample answer: Just remember the formula for dependent events: $P(A \text{ and } B) = P(A) \cdot P(B \text{ following } A)$. When the events are independent, $P(B \text{ following } A) = P(B)$, so the formula for dependent events simplifies to $P(A \text{ and } B) = P(A) \cdot P(B)$, which is the correct formula for independent events.**

656 Chapter 12 Probability and Statistics

Enrichment, p. 722

Conditional Probability

Suppose a pair of dice is thrown. It is known that the sum is greater than seven. Find the probability that the dice match.

The probability of an event given the occurrence of another event is called *conditional probability*. The conditional probability of event A, the dice match, given event B, their sum is greater than seven, is denoted $P(A/B)$.

There are 15 sums greater than seven and there are 36 possible pairs altogether.

$P(B) = \frac{15}{36}$

There are three matching pairs greater than seven.

$P(A \text{ and } B) = \frac{3}{36}$

$P(A/B) = \frac{P(A \text{ and } B)}{P(B)}$

$P(A/B) = \frac{\frac{3}{36}}{\frac{15}{36}}$ or $\frac{1}{5}$

...bability is $\frac{1}{}$...

Answer

49. **Sample answer: As the number of trials increases, the results become more reliable. However, you cannot be absolutely certain that there are no black marbles in the bag without looking at all of the marbles.**

51. **WRITING IN MATH** Answer the question that was posed at the beginning of the lesson. See pp. 695A–695B.

How does probability apply to basketball?

Include the following in your answer:

- an explanation of how a value such as one of those in the table at the beginning of the lesson could be used to find the chances of Reggie Miller making 0, 1, or 2 of 2 successive free throws, assuming the 2 free throws are independent, and
- a possible psychological reason why 2 free throws on the same trip to the foul line might not be independent.

Standardized Test Practice

52. The spinner is spun four times. What is the probability that the spinner lands on 2 each time? **D**

(A) $\frac{1}{2}$ (B) $\frac{1}{4}$

(C) $\frac{1}{16}$ (D) $\frac{1}{256}$

53. A coin is tossed and a die is rolled. What is the probability of a head and a 3? **C**

(A) $\frac{1}{4}$ (B) $\frac{1}{8}$ (C) $\frac{1}{12}$ (D) $\frac{1}{24}$

Maintain Your Skills

Mixed Review

A gumball machine contains 7 red, 8 orange, 9 purple, 7 white, and 5 yellow gumballs. Tyson buys 3 gumballs. Find each probability, assuming that the machine dispenses the gumballs at random. *(Lesson 12-3)*

54. P(3 red) $\frac{1}{204}$

55. P(2 white, 1 purple) $\frac{3}{340}$

56. P(1 purple, 1 orange, 1 yellow) $\frac{6}{119}$

57. **PHOTOGRAPHY** A photographer is taking a picture of a bride and groom together with 6 attendants. How many ways can he arrange the 8 people in a row if the bride and groom stand in the middle? *(Lesson 12-2)* **1440 ways**

Solve each equation. Check your solutions. *(Lesson 10-3)*

58. $\log_5 5 + \log_5 x = \log_5 30$ **6**

59. $\log_{16} c - 2\log_{16} 3 = \log_{16} 4$ **36**

Given a polynomial and one of its factors, find the remaining factors of the polynomial. Some factors may not be binomials. *(Lesson 7-4)*

60. $x^3 - x^2 - 10x + 6; x + 3$ — **60.** $x^2 - 4x + 2$

61. $x^3 - 7x^2 + 12x; x - 3$ — **61.** $x, x - 4$

Graph each inequality. *(Lesson 6-7)* **62–64. See margin.**

62. $y \le x^2 + x - 2$

63. $y < x^2 - 4$

64. $y > x^2 - 3x$

Simplify. *(Lesson 5-5)*

65. $\sqrt{(153)^2}$ **153**

66. $\sqrt[3]{-729}$ **−9**

67. $\sqrt[16]{b^{16}}$ $|b|$

68. $\sqrt{25a^8b^6}$ $5a^4|b^3|$

Solve each system of equations. *(Lesson 3-2)*

69. $z = 4y - 2$, $z = -y + 3$ **(1, 2)**

70. $j - k = 4$, $2j + k = 35$ **(13, 9)**

71. $3x + 1 = -y - 1$, $2y = -4x$ **(−2, 4)**

Getting Ready for the Next Lesson

BASIC SKILL Find each sum if $a = \frac{1}{2}$, $b = \frac{1}{6}$, $c = \frac{2}{3}$, and $d = \frac{3}{4}$.

72. $a + b$ $\frac{2}{3}$

73. $b + c$ $\frac{5}{6}$

74. $a + d$ $\frac{5}{4}$

75. $b + d$ $\frac{11}{12}$

76. $c + a$ $1\frac{1}{6}$

77. $c + d$ $1\frac{5}{12}$

4 Assess

Open-Ended Assessment

Speaking Ask students working in small groups to write an original problem (using objects or situations at school) that involves two selections. Have them create two versions, one with and one without replacement. Ask each group to present their problems to the class and to lead a discussion to compare the two solutions.

Getting Ready for Lesson 12-5

BASIC SKILL Lesson 12-5 presents finding the probability of mutually exclusive events. Students will use their familiarity with adding fractions as they calculate these probabilities. Exercises 72–77 should be used to determine your students' familiarity with adding fractions.

Answers

62.

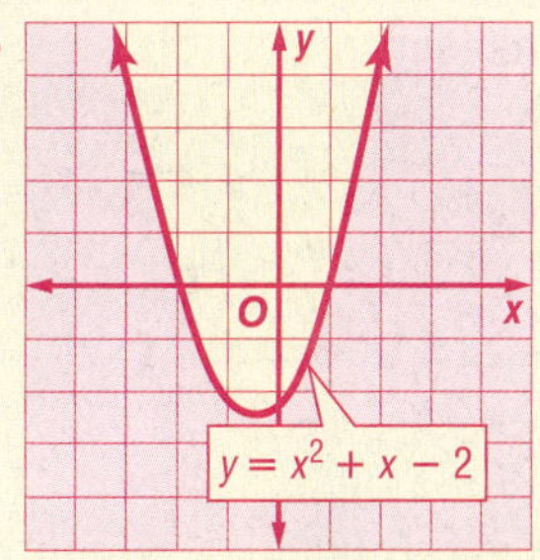

63.

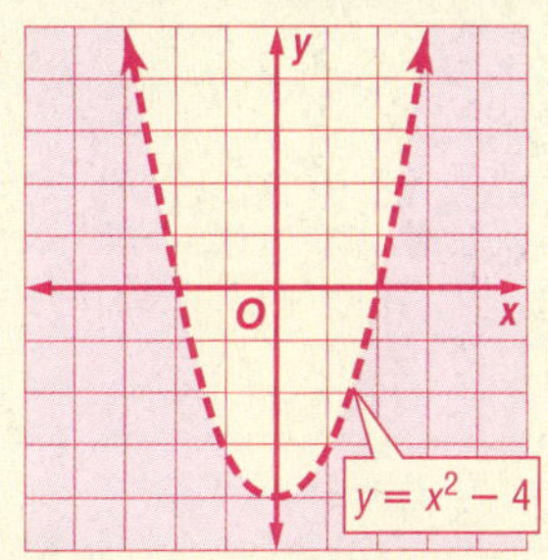

64.

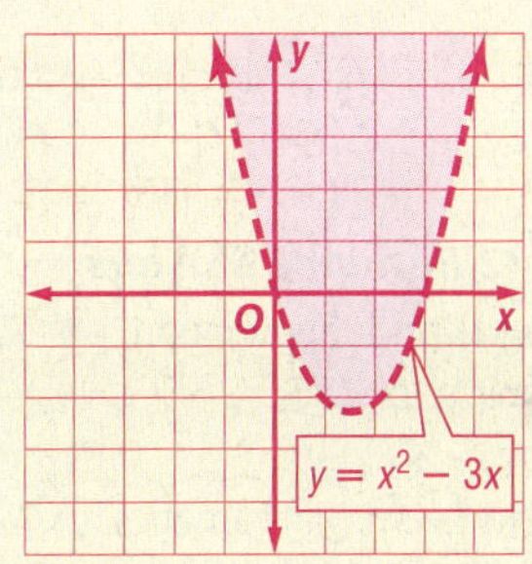

12-5 Lesson Notes

1 Focus

5-Minute Check Transparency 12-5 Use as a quiz or review of Lesson 12-4.

Mathematical Background notes are available for this lesson on p. 630D.

How does probability apply to your personal habits?

Ask students:

- Which of these activities would have the greatest probability of being reported by a randomly selected person? **brushing teeth**
- Which of these activities would have the least probability of being reported by a randomly selected person? **Preparing clothes and taking medication have the same least probability.**

12-5 Adding Probabilities

What You'll Learn

- Find the probability of mutually exclusive events.
- Find the probability of inclusive events.

Vocabulary

- simple event
- compound event
- mutually exclusive events
- inclusive events

How does probability apply to your personal habits?

The graph shows the results of a survey about bedtime rituals. Determining the probability that a randomly selected person reads a book or brushes his or her teeth before going to bed requires adding probabilities.

MUTUALLY EXCLUSIVE EVENTS When you roll a die, an event such as rolling a 1 is called a **simple event** because it consists of only one event. An event that consists of two or more simple events is called a **compound event**. For example, the event of rolling an odd number or a number greater than 5 is a compound event because it consists of the simple events rolling a 1, rolling a 3, rolling a 5, or rolling a 6.

When there are two events, it is important to understand how they are related before finding the probability of one or the other event occurring. Suppose you draw a card from a standard deck of cards. What is the probability of drawing a 2 or an ace? Since a card cannot be both a 2 *and* an ace, these are called **mutually exclusive events**. That is, the two events cannot occur at the same time. The probability of drawing a 2 or an ace is found by adding their individual probabilities.

$$P(2 \text{ or ace}) = P(2) + P(\text{ace}) \quad \text{Add probabilities.}$$
$$= \frac{4}{52} + \frac{4}{52} \quad \text{There are 4 twos and 4 aces in a deck.}$$
$$= \frac{8}{52} \text{ or } \frac{2}{13} \quad \text{Simplify.}$$

The probability of drawing a 2 or an ace is $\frac{2}{13}$.

Key Concept — Probability of Mutually Exclusive Events

- **Words** If two events, A and B, are mutually exclusive, then the probability that A or B occurs is the sum of their probabilities.
- **Symbols** $P(A \text{ or } B) = P(A) + P(B)$

This formula can be extended to any number of mutually exclusive events.

Resource Manager

Workbook and Reproducible Masters

Chapter 12 Resource Masters
- Study Guide and Intervention, pp. 723–724
- Skills Practice, p. 725
- Practice, p. 726
- Reading to Learn Mathematics, p. 727
- Enrichment, p. 728
- Assessment, pp. 767, 769

Graphing Calculator and Spreadsheet Masters, p. 49

Transparencies

5-Minute Check Transparency 12-5
Answer Key Transparencies

Technology

Interactive Chalkboard
Multimedia Applications

Example 1 Two Mutually Exclusive Events

Keisha has a stack of 8 baseball cards, 5 basketball cards, and 6 soccer cards. If she selects a card at random from the stack, what is the probability that it is a baseball or a soccer card?

These are mutually exclusive events, since the card cannot be both a baseball card *and* a soccer card. Note that there is a total of 19 cards.

$P(\text{baseball or soccer}) = P(\text{baseball}) + P(\text{soccer})$ Mutually exclusive events

$= \frac{8}{19} + \frac{6}{19}$ or $\frac{14}{19}$ Substitute and add.

The probability that Keisha selects a baseball or a soccer card is $\frac{14}{19}$.

Example 2 Three Mutually Exclusive Events

There are 7 girls and 6 boys on the junior class homecoming committee. A subcommittee of 4 people is being chosen at random to decide the theme for the class float. What is the probability that the subcommittee will have at least 2 girls?

At least 2 girls means that the subcommittee may have 2, 3, or 4 girls. It is not possible to select a group of 2 girls, a group of 3 girls, and a group of 4 girls all in the same 4-member subcommittee, so the events are mutually exclusive. Add the probabilities of each type of committee.

$P(\text{at least 2 girls}) = P(\text{2 girls}) + P(\text{3 girls}) + P(\text{4 girls})$

2 girls, 2 boys; 3 girls, 1 boy; 4 girls, 0 boys

$= \frac{C(7, 2) \cdot C(6, 2)}{C(13, 4)} + \frac{C(7, 3) \cdot C(6, 1)}{C(13, 4)} + \frac{C(7, 4) \cdot C(6, 0)}{C(13, 4)}$

$= \frac{315}{715} + \frac{210}{715} + \frac{35}{715}$ or $\frac{112}{143}$ Simplify.

The probability of at least 2 girls on the subcommittee is $\frac{112}{143}$ or about 0.78.

Study Tip

Choosing a Committee
$C(13, 4)$ refers to choosing 4 subcommittee members from 13 committee members. Since order does not matter, the number of combinations is found.

INCLUSIVE EVENTS What is the probability of drawing a queen or a diamond from a standard deck of cards? Since it is possible to draw a card that is both a queen and a diamond, these events are *not* mutually exclusive. These are called **inclusive events**.

P(queen)	P(diamond)	P(diamond, queen)
$\frac{4}{52}$	$\frac{13}{52}$	$\frac{1}{52}$
1 queen in each suit	diamonds	queen of diamonds

In the first two fractions above, the probability of drawing the queen of diamonds is counted twice, once for a queen and once for a diamond. To find the correct probability, you must subtract P(queen of diamonds) from the sum of the first two probabilities.

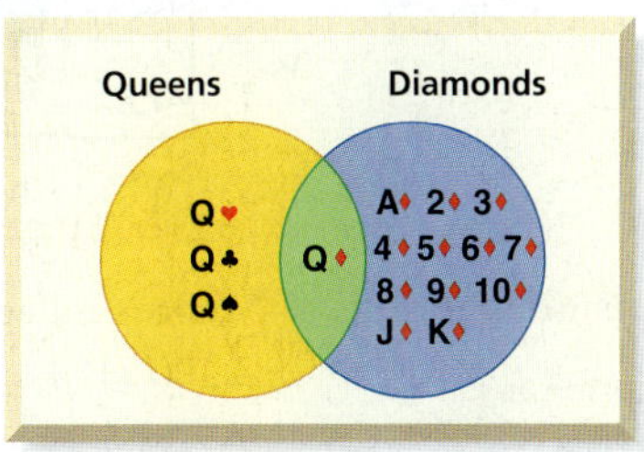

Study Tip

Common Misconception
In mathematics, unlike everyday language, the expression *A* or *B* allows the possibility of both *A* and *B* occurring.

www.algebra2.com/extra_examples

2 Teach

MUTUALLY EXCLUSIVE EVENTS

In-Class Examples PowerPoint®

1. Sylvia has a stack of playing cards consisting of 10 hearts, 8 spades, and 7 clubs. If she selects a card at random from this stack, what is the probability that it is a heart or a club? $\frac{17}{25}$

2. The Film Club makes a list of 9 comedies and 5 adventure movies they want to see. They plan to select 4 titles at random to show this semester. What is the probability that at least two of the films they select are comedies? $\frac{906}{1001}$ or about 0.91

Online Lesson Plans

USA TODAY Education

USA TODAY Education's Online site offers resources and interactive features connected to each day's newspaper. *Experience TODAY*, USA TODAY's daily lesson plan, is available on the site and delivered daily to subscribers. This plan provides instruction for integrating USA TODAY graphics and key editorial features into your mathematics classroom. Log on to www.education.usatoday.com.

INCLUSIVE EVENTS

In-Class Example Power Point®

3 There are 2400 subscribers to an Internet service provider. Of these, 1200 own Brand A computers, 500 own Brand B, and 100 own both A and B. What is the probability that a subscriber selected at random owns either Brand A or Brand B? $\frac{2}{3}$

3 Practice/Apply

Study Notebook

Have students—
- add the definitions/examples of the vocabulary terms to their Vocabulary Builder worksheets for Chapter 12.
- add a representative problem for each of the probability situations in this lesson.
- include any other item(s) that they find helpful in mastering the skills in this lesson.

DAILY INTERVENTION **FIND THE ERROR** Discuss whether the two events are inclusive or exclusive. Some students may feel that rain Saturday reduces the chance of rain on Sunday. Some students may think the two events are independent. Encourage interested students to research the science of weather forecasting.

$$P(\text{queen or diamond}) = P(\text{queen}) + P(\text{diamond}) - P(\text{queen of diamonds})$$
$$= \frac{4}{52} + \frac{13}{52} - \frac{1}{52} \text{ or } \frac{4}{13}$$

The probability of drawing a queen or a diamond is $\frac{4}{13}$.

Key Concept *Probability of Inclusive Events*

- **Words** If two events, A and B, are inclusive, then the probability that A or B occurs is the sum of their probabilities decreased by the probability of both occurring.
- **Symbols** $P(A \text{ or } B) = P(A) + P(B) - P(A \text{ and } B)$

Example 3 *Inclusive Events*

EDUCATION **The enrollment at Southburg High School is 1400. Suppose 550 students take French, 700 take algebra, and 400 take both French and algebra. What is the probability that a student selected at random takes French or algebra?**

Since some students take both French and algebra, the events are inclusive.

$P(\text{French}) = \frac{550}{1400}$ $\quad P(\text{algebra}) = \frac{700}{1400}$ $\quad P(\text{French and algebra}) = \frac{400}{1400}$

$$P(\text{French or algebra}) = P(\text{French}) + P(\text{algebra}) - P(\text{French and algebra})$$
$$= \frac{550}{1400} + \frac{700}{1400} - \frac{400}{1400} \text{ or } \frac{17}{28} \quad \text{Substitute and simplify.}$$

The probability that a student selected at random takes French or algebra is $\frac{17}{28}$.

Check for Understanding

Concept Check

1. **OPEN ENDED** Describe two mutually exclusive events and two inclusive events. **1. Sample answer: mutually exclusive events: tossing a coin and rolling a die; inclusive events: drawing a 7 and a diamond from a standard deck of cards**
2. **Draw** a Venn diagram to illustrate Example 3. **See margin.**
3. **FIND THE ERROR** Refer to the comic below.

The Born Loser®

Why is the weather forecaster's prediction incorrect? **The events are not mutually exclusive, so the chance of rain is less than 100%.**

Guided Practice **A die is rolled. Find each probability.**

4. $P(1 \text{ or } 6)$ $\frac{1}{3}$
5. $P(\text{at least } 5)$ $\frac{1}{3}$
6. $P(\text{less than } 3)$ $\frac{1}{3}$
7. $P(\text{prime})$ $\frac{1}{2}$
8. $P(\text{even or prime})$ $\frac{5}{6}$
9. $P(\text{multiple of 2 or 3})$ $\frac{2}{3}$

DAILY INTERVENTION **Differentiated Instruction**

Intrapersonal Have students reflect on the definitions, skills, and formulas they have learned in these first five lessons on probability. Ask them to write an entry in their notes that describes their reaction to this topic in general, and to indicate which kinds of problems they find the most interesting, and which they find the most challenging.

GUIDED PRACTICE KEY	
Exercises	Examples
4–7, 12	1, 2
8–11	3

A card is drawn from a standard deck of cards. Determine whether the events are *mutually exclusive* or *inclusive*. Then find the probability.

10. P(6 or king) mutually exclusive; $\frac{2}{13}$

11. P(queen or spade) inclusive; $\frac{4}{13}$

Application

12. **SCHOOL** There are 8 girls and 8 boys on the student senate. Three of the students are seniors. What is the probability that a person selected from the student senate is not a senior? $\frac{13}{16}$

★ indicates increased difficulty

Practice and Apply

Homework Help	
For Exercises	See Examples
13–22, 33–42	1, 2
23–26	1–3
27–32, 43–46	3

Extra Practice See page 855.

See page 855.

18. $\frac{105}{143}$ 21. $\frac{38}{143}$

Lisa has 9 rings in her jewelry box. Five are gold and 4 are silver. If she randomly selects 3 rings to wear to a party, find each probability.

13. P(2 silver or 2 gold) $\frac{5}{6}$

14. P(all gold or all silver) $\frac{1}{6}$

15. P(at least 2 gold) $\frac{25}{42}$

16. P(at least 1 silver) $\frac{37}{42}$

Seven girls and six boys walk into a video store at the same time. There are five salespeople available to help them. Find the probability that the salespeople will first help the given numbers of girls and boys.

17. P(4 girls or 4 boys) $\frac{35}{143}$

18. P(3 girls or 3 boys)

★ 19. P(all girls or all boys) $\frac{3}{143}$

20. P(at least 3 girls) $\frac{84}{143}$

21. P(at least 4 girls or at least 4 boys)

★ 22. P(at least 2 boys) $\frac{32}{39}$

For Exercises 23–26, determine whether the events are *mutually exclusive* or *inclusive*. Then find the probability. 24. inclusive; $\frac{1}{2}$

23. There are 3 literature books, 4 algebra books, and 2 biology books on a shelf. If a book is randomly selected, what is the probability of selecting a literature book or an algebra book? mutually exclusive; $\frac{7}{9}$

24. A die is rolled. What is the probability of rolling a 5 or a number greater than 3?

25. In the Math Club, 7 of the 20 girls are seniors, and 4 of the 14 boys are seniors. What is the probability of randomly selecting a boy or a senior to represent the Math Club at a statewide math contest? inclusive; $\frac{21}{34}$

26. A card is drawn from a standard deck of cards. What is the probability of drawing an ace or a face card? (*Hint:* A face card is a jack, queen, or king.) mutually exclusive; $\frac{4}{13}$

27. One tile with each letter of the alphabet is placed in a bag, and one is drawn at random. What is the probability of selecting a vowel or a letter from the word *equation*? $\frac{4}{13}$

28. Each of the numbers from 1 to 30 is written on a card and placed in a bag. If one card is drawn at random, what is the probability that the number is a multiple of 2 or a multiple of 3? $\frac{2}{3}$

Two cards are drawn from a standard deck of cards. Find each probability.

29. P(both kings or both black) $\frac{55}{221}$

30. P(both kings or both face cards) $\frac{11}{221}$

31. P(both face cards or both red) $\frac{188}{663}$

32. P(both either red or a king) $\frac{63}{221}$

More About. . .

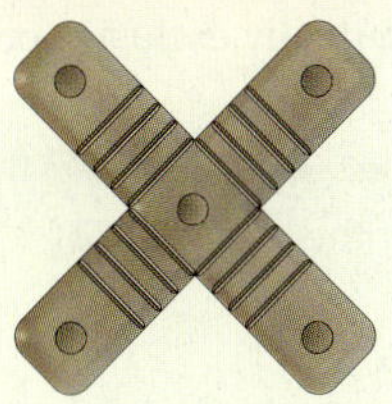

World Cultures

Totolospi is a Hopi game of chance. The players use cane dice, which have both a flat side and a round side, and a counting board inscribed in stone.

WORLD CULTURES **For Exercises 33–36, refer to the information at the left.** When tossing 3 cane dice, if three round sides land up, the player advances 2 lines. If three flat sides land up, the player advances 1 line. If a combination is thrown, the player loses a turn. Find each probability.

33. P(advancing 2 lines) $\frac{1}{8}$

34. P(advancing 1 line) $\frac{1}{8}$

35. P(advancing at least 1 line) $\frac{1}{4}$

36. P(losing a turn) $\frac{3}{4}$

www.algebra2.com/self_check_quiz

About the Exercises...

Organization by Objective

- **Mutually Exclusive Events:** 13–23, 26, 33–42
- **Inclusive Events:** 24, 25, 27–32, 43–46

Odd/Even Assignments

Exercises 13–42 are structured so that students practice the same concepts whether they are assigned odd or even problems.

Assignment Guide

Basic: 13–17 odd, 21–39 odd, 47–75

Average: 13–43 odd, 47–75

Advanced: 14–42 even, 44–69 (optional: 70–75)

Answer

2.

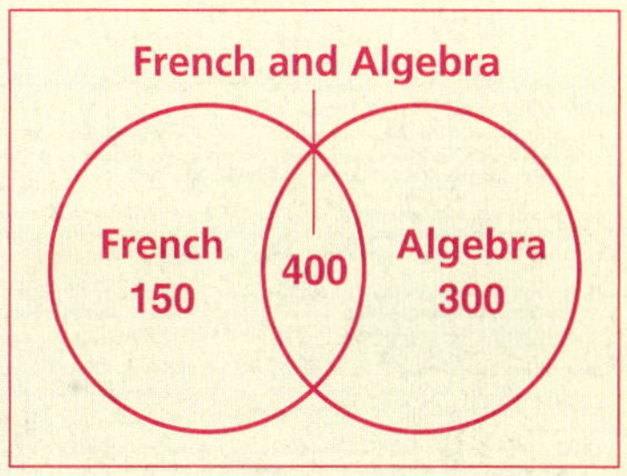

Lesson 12-5 Adding Probabilities 661

Study Guide and Intervention, p. 723 (shown) and p. 724

Mutually Exclusive Events Events that cannot occur at the same time are called mutually exclusive events.

Probability of Mutually Exclusive Events	If two events, A and B, are mutually exclusive, then $P(A \text{ or } B) = P(A) + P(B)$.

This formula can be extended to any number of mutually exclusive events.

Example 1 **To choose an afternoon activity, summer campers pull slips of paper out of a hat. Today there are 25 slips for a nature walk, 35 slips for swimming, and 30 slips for arts and crafts. What is the probability that a camper will pull a slip for a nature walk or for swimming?**

These are mutually exclusive events. Note that there is a total of 90 slips.

$$P(\text{nature walk or swimming}) = P(\text{nature walk}) + P(\text{swimming}) = \frac{25}{90} + \frac{35}{90} \text{ or } \frac{2}{3}$$

The probability of a camper's pulling out a slip for a nature walk or for swimming is $\frac{2}{3}$.

Example 2 **By the time one tent of 6 campers gets to the front of the line, there are only 10 nature walk slips and 15 swimming slips left. What is the probability that more than 4 of the 6 campers will choose a swimming slip?**

$$P(\text{more than 4 swimmers}) = P(\text{5 swimmers}) + P(\text{6 swimmers}) = \frac{C(10, 1) \cdot C(15, 5)}{C(25, 6)} + \frac{C(10, 0) \cdot C(15, 6)}{C(25, 6)} \approx 0.2$$

The probability of more than 4 of the campers swimming is about 0.2.

Exercises

Find each probability.

1. A bag contains 45 dyed eggs: 15 yellow, 12 green, and 18 red. What is the probability of selecting a green or a red egg? $\frac{2}{3}$
2. The letters from the words LOVE and LIVE are placed on cards and put in a box. What is the probability of selecting an L or an O from the box? $\frac{3}{8}$
3. A pair of dice is rolled, and the two numbers are added. What is the probability that the sum is either a 5 or a 7? $\frac{5}{18}$ or about 0.28
4. A bowl has 10 whole wheat crackers, 16 sesame crackers, and 14 rye crisps. If a person picks a cracker at random, what is the probability of picking either a sesame cracker or a rye crisp? $\frac{3}{4}$
5. An art box contains 12 colored pencils and 20 pastels. If 5 drawing implements are chosen at random, what is the probability that at least 4 of them are pastels? about 0.37

Skills Practice, p. 725 and Practice, p. 726 (shown)

An urn contains 7 white marbles and 5 blue marbles. Four marbles are selected without replacement. Find each probability.

1. P(4 white or 4 blue) $\frac{8}{99}$
2. P(exactly 3 white) $\frac{35}{99}$
3. P(at least 3 white) $\frac{14}{33}$
4. P(fewer than 3 white) $\frac{19}{33}$
5. P(3 white or 3 blue) $\frac{49}{99}$
6. P(no white or no blue) $\frac{8}{99}$

Jason and Maria are playing a board game in which three dice are tossed to determine a player's move. Find each probability.

7. P(two 5s) $\frac{5}{72}$
8. P(three 5s) $\frac{1}{216}$
9. P(at least two 5s) $\frac{2}{27}$
10. P(no 5s) $\frac{125}{216}$
11. P(one 5) $\frac{25}{72}$
12. P(one 5 or two 5s) $\frac{5}{12}$

Determine whether the events are *mutually exclusive* or *inclusive*. Then find the probability.

13. A clerk chooses 4 CD players at random for floor displays from a shipment of 24 CD players. If 15 of the players have a blue case and the rest have a red case, what is the probability of choosing 4 players with a blue case or 4 players with a red case? mutual. exclus.; $\frac{71}{506}$
14. A department store employs 28 high school students, all juniors and seniors. Six of the 12 seniors are females and 12 of the juniors are males. One student employee is chosen at random. What is the probability of selecting a senior or a female? inclusive; $\frac{4}{7}$
15. A restaurant has 5 pieces of apple pie, 4 pieces of chocolate cream pie, and 3 pieces of blueberry pie. If Janine selects a piece of pie at random for dessert, what is the probability that she selects either apple or chocolate cream? mutually exclusive; $\frac{3}{4}$
16. At a statewide meeting, there are 20 school superintendents, 13 principals, and 6 assistant principals. If one of these people is chosen at random, what is the probability that he or she is either a principal or an assistant principal? mutually exclusive; $\frac{19}{39}$
17. An airline has one bank of 13 telephones at a reservations office. Of the 13 operators who work there, 8 take reservations for domestic flights and 5 take reservations for international flights. Seven of the operators taking domestic reservations and 3 of the operators taking international reservations are female. If an operator is chosen at random, what is the probability that the person chosen takes domestic reservations or is a male? inclusive; $\frac{10}{13}$
18. **MUSIC** Forty senior citizens were surveyed about their music preferences. The results are displayed in the Venn diagram. If a senior citizen from the survey group is selected at random, what is the probability that he or she likes only country and western music? What is the probability that he or she likes classical and/or country, but not 1940's pop? $\frac{3}{20}$; $\frac{2}{5}$

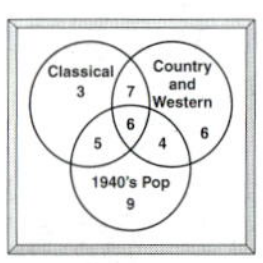

Reading to Learn Mathematics, p. 727 ELL

Pre-Activity **How does probability apply to your personal habits?**

Read the introduction to Lesson 12-5 at the top of page 658 in your textbook.

Why do the percentages shown on the bar graph add up to more than 100%? Sample answer: Many people do more than one of the listed bedtime rituals.

Reading the Lesson

1. Indicate whether the events in each pair are *inclusive* or *mutually exclusive*.
   a. Q: drawing a queen from a standard deck of cards
      D: drawing a diamond from a standard deck of cards inclusive
   b. J: drawing a jack from a standard deck of cards
      K: drawing a king from a standard deck of cards mutually exclusive
2. Marla took a quiz on this lesson that contained the following problem.
   Each of the integers from 1 through 25 is written on a slip of paper and placed in an envelope. If one slip is drawn at random, what is the probability that it is odd or a multiple of 5?
   Here is Marla's work.
   $P(\text{odd}) = \frac{13}{25}$ $\quad P(\text{multiple of } 5) = \frac{5}{25} \text{ or } \frac{1}{5}$
   $$P(\text{odd or multiple of } 5) = P(\text{odd}) + P(\text{multiple of } 5) = \frac{13}{25} + \frac{5}{25} = \frac{18}{25}$$
   a. Why is Marla's work incorrect? Sample answer: Marla used the formula for mutually exclusive events, but the events are inclusive. She should use the formula for inclusive events so that the odd multiples of 5 will not be counted twice.
   b. Show the corrected work.
   $$P(\text{odd or multiple of } 5) = P(\text{odd}) + P(\text{multiple of } 5) - P(\text{odd multiple of } 5) = \frac{13}{25} + \frac{5}{25} - \frac{3}{25} = \frac{15}{25} = \frac{3}{5}$$

Helping You Remember

3. Some students have trouble remembering a lot of formulas, so they try to keep the number of formulas they have to know to a minimum. Can you learn just one formula that will allow you to find probabilities for both mutually exclusive and inclusive events? Explain your reasoning. Sample answer: Just remember the formula for inclusive events: $P(A \text{ or } B) = P(A) + P(B) - P(A \text{ and } B)$. When the events are mutually exclusive, $P(A \text{ and } B) = 0$, so the formula for inclusive events simplifies to $P(A \text{ and } B) = P(A) + P(B)$, which is the correct formula for mutually exclusive events.

For Exercises 37–42, use the following information.

Each of the numbers 1 through 30 is written on a table tennis ball and placed in a wire cage. Each of the numbers 20 through 45 is written on a table tennis ball and placed in a different wire cage. One ball is chosen at random from each spinning cage. Find each probability.

37. P(each is a 25) $\frac{1}{780}$
38. P(neither is a 20) $\frac{145}{156}$
39. P(exactly one is a 30) $\frac{9}{130}$
40. P(exactly one is a 40) $\frac{1}{26}$
★ 41. P(the numbers are equal) $\frac{11}{780}$
★ 42. P(the sum is 30) $\frac{1}{78}$

More About. . .

Recycling

The United States recycles 28% of its waste.

Source: The U.S. Environmental Protection Agency

43. **RECYCLING** In one community, 300 people were surveyed to see if they would participate in a curbside recycling program. Of those surveyed, 134 said they would recycle aluminum cans, and 108 said they would recycle glass. Of those, 62 said they would recycle both. What is the probability that a randomly selected member of the community would recycle aluminum or glass? $\frac{3}{5}$

SCHOOL **For Exercises 44–46, use the Venn diagram that shows the number of participants in extracurricular activities for a junior class of 324 students.** Determine each probability if a student is selected at random from the class.

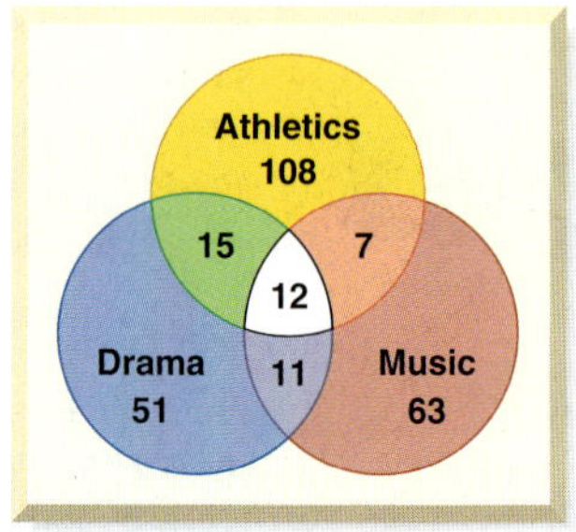

44. P(drama or music) $\frac{53}{108}$
45. P(drama or athletics) $\frac{17}{27}$
★ 46. P(athletics and drama, or music and athletics) $\frac{17}{162}$

47. **CRITICAL THINKING** Consider the following probability equation.

$$P(A \text{ and } B) = P(A) + P(B) - P(A \text{ or } B)$$

A textbook gives this equation for events A and B that are mutually exclusive or inclusive. Is this correct? Explain. **See margin.**

48. **WRITING IN MATH** Answer the question that was posed at the beginning of the lesson. **See pp. 695A–695B.**

How does probability apply to your personal habits?

Include the following in your answer:

- an explanation of whether the events listed in the graphic are mutually exclusive or inclusive, and
- an explanation of how to determine the probability that a randomly selected person reads a book or brushes his or her teeth before going to bed if in a survey of 2000 people, 600 said that they do both.

Standardized Test Practice

Ⓐ Ⓑ Ⓒ Ⓓ

49. In a jar of red and white gumballs, the ratio of white gumballs to red gumballs is 5:4. If the jar contains a total of 180 gumballs, how many of them are red? **C**

Ⓐ 45 Ⓑ 64 Ⓒ 80 Ⓓ 100

50. $\langle x \rangle = \frac{1}{2}x$ if x is composite. $\langle x \rangle = 2x$ if x is prime. What is the value of $\langle 7 \rangle + \langle 18 \rangle$? **A**

Ⓐ 23 Ⓑ 46 Ⓒ 50 Ⓓ 64

Enrichment, p. 728

Probability and Tic-Tac-Toe

What would be the chances of winning at tic-tac-toe if it were turned into a game of pure chance? To find out, the nine cells of the tic-tac-toe board are numbered from 1 to 9 and nine chips (also numbered from 1 to 9) are put into a bag. Player A draws a chip at random and enters an X in the corresponding cell. Player B does the same and enters an O.

To solve the problem, assume that both players draw all their chips without looking and all X and O entries are made at the same time. There are four possible outcomes: a draw, A wins, B wins, and either A or B can win.

There are 16 arrangements that result in a draw. Reflections and rotations must be counted as shown below.

```
o x o        x o x        o o x
x o x  4     o o x  4     x x o  8
x o x        x x o        o x x
```

There are 36 arrangements in which either player may win because both players have winning triples.

Maintain Your Skills

Mixed Review A die is rolled three times. Find each probability. *(Lesson 12-4)*

51. $P(1, \text{then } 2, \text{then } 3)$ $\frac{1}{216}$
52. $P(\text{no } 4\text{s})$ $\frac{125}{216}$
53. $P(\text{three } 1\text{s})$ $\frac{1}{216}$
54. $P(\text{three even numbers})$ $\frac{1}{8}$

Find the odds of an event occurring, given the probability of the event. *(Lesson 12-3)*

55. $\frac{4}{5}$ 4:1
56. $\frac{1}{9}$ 1:8
57. $\frac{2}{7}$ 2:5
58. $\frac{5}{8}$ 5:3

Find the sum of each series. *(Lessons 11-2 and 11-4)*

59. $2 + 4 + 8 + \cdots + 128$ 254
60. $\sum_{n=1}^{3}(5n - 2)$ 24

Find the exact solution(s) of each system of equations. *(Lesson 8-7)*

61. $y = -10$
    $y^2 = x^2 + 36$ $(\pm 8, -10)$
62. $x^2 = 144$
    $x^2 + y^2 = 169$ $(\pm 12, \pm 5)$

63. Use the graph of the polynomial function at the right to determine at least one binomial factor of the polynomial. Then find all factors of the polynomial. *(Lesson 7-4)*
    $(x + 1)^2(x - 1)(x^2 + 1)$

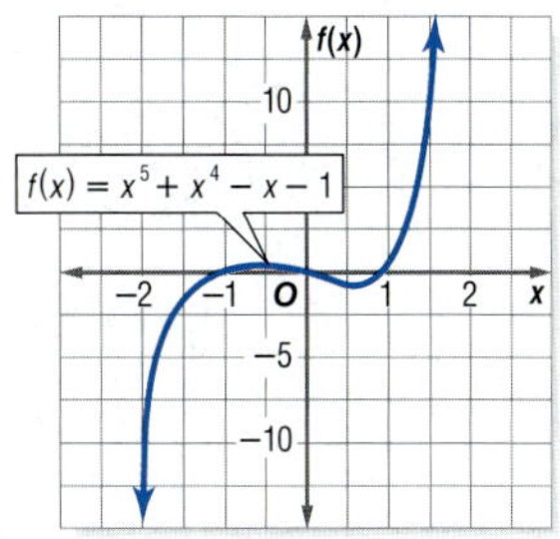

Find the maxima and minima of each function. Round to the nearest hundredth. *(Lesson 6-2)*

64. $f(x) = x^3 + 2x^2 - 5$
65. $f(x) = x^3 + 3x^2 + 2x + 1$

64. min: $(0, -5)$; max: $(-1.33, -3.81)$
65. min: $(-0.42, 0.62)$; max: $(-1.58, 1.38)$

Graph each system of inequalities. Name the coordinates of the vertices of the feasible region. Find the maximum and minimum values of the given function for this region. *(Lesson 3-4)* 66–67. See margin for graphs.

66. $y \geq x - 2$
    $x \geq 0$
    $y \leq 2 - x$
    $f(x, y) = 3x + y$
    $(0, 2), (2, 0), (0, -2)$; max: $f(2, 0) = 6$; min: $f(0, -2) = -2$
67. $y \geq 2x - 3$
    $1 \leq x \leq 3$
    $y \leq x + 2$
    $f(x, y) = x + 4y$
    $(1, 3), (1, -1), (3, 3), (3, 5)$; max: $f(3, 5) = 23$; min: $f(1, -1) = -3$

SPEED SKATING For Exercises 68 and 69, use the following information.
In the 1988 Winter Olympics, Bonnie Blair set a world record for women's speed skating by skating approximately 12.79 meters per second in the 500-meter race. *(Lesson 2-6)*

68. Suppose she could maintain that speed. Write an equation that represents how far she could travel in t seconds. $d = 12.79t$
69. What type of equation is the one in Exercise 68? direct variation

Getting Ready for the Next Lesson **PREREQUISITE SKILL** Find the mean, median, mode, and range for each set of data. Round to the nearest hundredth, if necessary. 70–75. See margin.
*(To review **mean**, **median**, **mode**, and **range**, see pages 822 and 823.)*

70. 298, 256, 399, 388, 276
71. 3, 75, 58, 7, 34
72. 4.8, 5.7, 2.1, 2.1, 4.8, 2.1
73. 80, 50, 65, 55, 70, 65, 75, 50
74. 61, 89, 93, 102, 45, 89
75. 13.3, 15.4, 12.5, 10.7

4 Assess

Open-Ended Assessment

Modeling Ask students to use colored chips, index cards, and other objects to design and model two probability problems—one that involves mutually exclusive events and the other inclusive events. Have students explain their problems to a partner.

Assessment Options

Quiz (Lessons 12-4 and 12-5) is available on p. 767 of the *Chapter 12 Resource Masters.*

Mid-Chapter Test (Lessons 12-1 through 12-5) is available on p. 769 of the *Chapter 12 Resource Masters.*

Getting Ready for Lesson 12-6

PREREQUISITE SKILL Lesson 12-6 presents using measures of central tendency and variation for a set of data. Students will use their familiarity with mean, median, mode, and range as they calculate standard deviation. Exercises 70–75 should be used to determine your students' familiarity with finding mean, median, mode, and range for a set of values.

Answers

67.

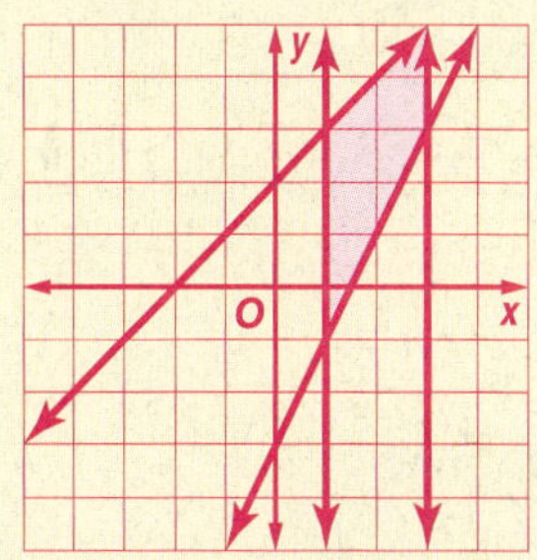

70. 323.4, 298, no mode, 143
71. 35.4, 34, no mode, 72
72. 3.6, 3.45, 2.1, 3.6
73. 63.75, 65, 50 and 65, 30
74. 79.83, 89, 89, 57
75. 12.98, 12.9, no mode, 4.7

47. Subtracting $P(A \text{ and } B)$ from each side and adding $P(A \text{ or } B)$ to each side results in the equation $P(A \text{ or } B) = P(A) + P(B) - P(A \text{ and } B)$. This is the equation for the probability of inclusive events. If A and B are mutually exclusive, then $P(A \text{ and } B) = 0$, so the equation simplifies to $P(A \text{ or } B) = P(A) + P(B)$, which is the equation for the probability of mutually exclusive events. Therefore, the equation is correct in either case.

66.

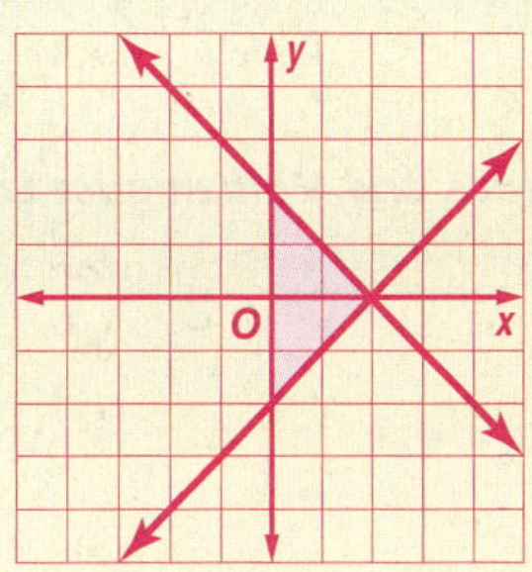

12-6 Lesson Notes

1 Focus

5-Minute Check Transparency 12-6 Use as a quiz or review of Lesson 12-5.

Mathematical Background notes are available for this lesson on p. 630D.

What statistics should a teacher tell the class after a test?

Ask students:

- Why is it helpful to put a list in order when studying data? **Sample answer: If the data are in order, it is much easier to find the lowest value, median, mode, and highest value.**
- What observations can you make about this data without doing any calculations, or using only mental math? **Sample answers may include: greatest and least values (94 to 19) and the range (75), as well as the fact that 19 and 34 seem to be outliers**

12-6 Statistical Measures

What You'll Learn

- Use measures of central tendency to represent a set of data.
- Find measures of variation for a set of data.

What statistics should a teacher tell the class after a test?

Vocabulary

- univariate data
- measure of central tendency
- measure of variation
- dispersion
- variance
- standard deviation

On Mr. Dent's most recent Algebra 2 test, his students earned these scores.

72	70	77	76	90	68	81	86	34	94
71	84	89	67	19	85	75	66	80	94

When his students ask how they did on the test, which measure of central tendency should Mr. Dent use to describe the scores?

MEASURES OF CENTRAL TENDENCY Data with one variable, such as the test scores, is called **univariate data**. Sometimes it is convenient to have one number that describes a set of univariate data. This number is called a **measure of central tendency**, because it represents the center or middle of the data. The most commonly used measures of central tendency are the *mean*, *median*, and *mode*.

When deciding which measure of central tendency to use to represent a set of data, look closely at the data itself.

Study Tip

Look Back
To review **outliers**, see Lesson 2-5.

Concept Summary *Measures of Tendency*

Use	When . . .
mean	the data are spread out, and you want an average of the values.
median	the data contain outliers.
mode	the data are tightly clustered around one or two values.

Example 1 *Choose a Measure of Central Tendency*

SWEEPSTAKES **A sweepstakes offers a first prize of \$10,000, two second prizes of \$100, and one hundred third prizes of \$10.**

a. Which measure of central tendency best represents the available prizes?

Since 100 of the 103 prizes are \$10, the mode (\$10) best represents the available prizes. Notice that in this case the median is the same as the mode.

b. Which measure of central tendency would the organizers of the sweepstakes be most likely to use in their advertising?

The organizers would be most likely to use the mean (about \$109) to make people think they had a better chance of winning more money.

Resource Manager

Workbook and Reproducible Masters

Chapter 12 Resource Masters
- Study Guide and Intervention, pp. 729–730
- Skills Practice, p. 731
- Practice, p. 732
- Reading to Learn Mathematics, p. 733
- Enrichment, p. 734

Science and Mathematics Lab Manual, pp. 115–118

Transparencies

5-Minute Check Transparency 12-6
Answer Key Transparencies

Technology

Interactive Chalkboard

MEASURES OF VARIATION

Measures of variation or **dispersion** measure how spread out or scattered a set of data is. The simplest measure of variation to calculate is the *range*, the difference between the greatest and the least values in a set of data. Variance and standard deviation are measures of variation that indicate how much the data values differ from the mean.

Study Tip

Reading Math
The symbol σ is the lower case Greek letter *sigma*. $\bar{x}$ is read *x bar*.

To find the **variance** σ^2 of a set of data, follow these steps.

1. Find the mean, $\bar{x}$.
2. Find the difference between each value in the set of data and the mean.
3. Square each difference.
4. Find the mean of the squares.

The **standard deviation** σ is the square root of the variance.

TEACHING TIP
In this text, assume that students are being asked to find the standard deviation of a population, for which the formula has been given.

Key Concept — *Standard Deviation*

If a set of data consists of the n values $x_1, x_2, \ldots, x_n$ and has mean $\bar{x}$, then the standard deviation σ is given by the following formula.

$$\sigma = \sqrt{\frac{(x_1 - \bar{x})^2 + (x_2 - \bar{x})^2 + \cdots + (x_n - \bar{x})^2}{n}}$$

Example 2 Standard Deviation

STATES **The table shows the populations in millions of 11 eastern states as of the 2000 Census. Find the variance and standard deviation of the data to the nearest tenth.**

State	Population	State	Population	State	Population
NY	19.0	MD	5.3	RI	1.0
PA	12.3	CT	3.4	DE	0.8
NJ	8.4	ME	1.3	VT	0.6
MA	6.3	NH	1.2	—	—

Source: U.S. Census Bureau

Step 1 Find the mean. Add the data and divide by the number of items.

$$\bar{x} = \frac{19.0 + 12.3 + 8.4 + 6.3 + 5.3 + 3.4 + 1.3 + 1.2 + 1.0 + 0.8 + 0.6}{11}$$

$\approx 5.4\overline{18}$ The mean is about 5.4 people.

Step 2 Find the variance.

$$\sigma^2 = \frac{(x_1 - \bar{x})^2 + (x_2 - \bar{x})^2 + \cdots + (x_n - \bar{x})^2}{n}$$ Variance formula

$$\approx \frac{(19.0 - 5.4)^2 + (12.3 - 5.4)^2 + \cdots + (0.8 - 5.4)^2 + (0.6 - 5.4)^2}{11}$$

$$\approx \frac{344.4}{11}$$ Simplify.

≈ 31.309 The variance is about 31.3 people.

Step 3 Find the standard deviation.

$\sigma^2 \approx 31.3$ Take the square root of each side.

$\sigma \approx 5.594640292$ The standard deviation is about 5.6 people.

www.algebra2.com/extra_examples

2 Teach

MEASURES OF CENTRAL TENDENCY

In-Class Example Power Point®

1 A new Internet company has 3 employees who are paid \$300,000, 10 who are paid \$100,000, and 60 who are paid \$50,000.

a. Which measure of central tendency best represents the pay at this company? **mode or median**

b. Which measure of central tendency would recruiters for this company be most likely to use to attract job applicants? **mean**

MEASURES OF VARIATION

In-Class Example Power Point®

2 **RIVERS** This table shows the length in thousands of miles of some of the longest rivers in the world. Find the standard deviation for these data. **1.05**

River	Length (thousands of miles)
Nile	4.16
Amazon	4.08
Missouri	2.35
Rio Grande	1.90
Danube	1.78

Teaching Tip Explain to students that the standard deviation is a number representing the typical or representative variation for the data items in that set. It tells how far a data value will typically be from the mean of the entire data set.

Teaching Tip Tell students that, for a normal distribution, 68.3% of the data is always within one standard deviation of the mean; 95.4% is always within two standard deviations, and 99.7% is always within three standard deviations, because of the way standard deviation is defined.

Answers

2. Sample answer: The variance of the set {0, 1} is 0.25 and the standard deviation is 0.5.

3. $\sigma = \sqrt{\frac{1}{n}\sum_{i=1}^{n}(x_i - \bar{x})^2}$

Most of the members of a set of data are within 1 standard deviation of the mean. The populations of the states in Example 2 can be broken down as shown below.

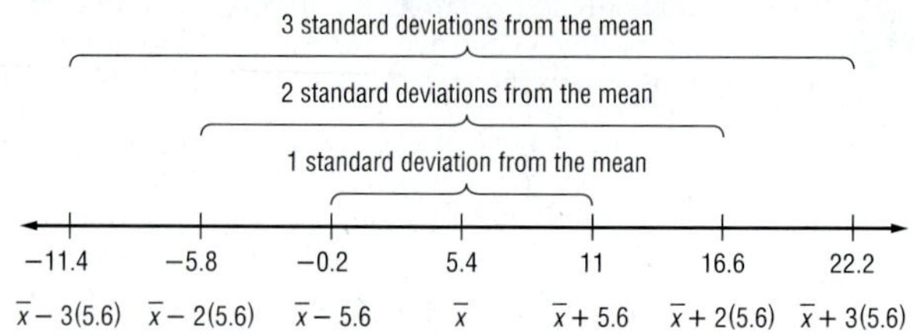

Looking at the original data, you can see that most of the states' populations were between 2.4 million and 20.2 million. That is, the majority of members of the data set were within 1 standard deviation of the mean.

You can use a TI-83 Plus graphing calculator to find statistics for the data in Example 2.

Graphing Calculator Investigation

One-Variable Statistics

The TI-83 Plus can compute a set of one-variable statistics from a list of data. These statistics include the mean, variance, and standard deviation. Enter the data into L1.

KEYSTROKES: STAT ENTER 19.0 ENTER 12.3 ENTER ...

Then use STAT ▶ 1 ENTER to show the statistics. The mean $\bar{x}$ is about 5.4, the sum of the values $\sum x$ is 59.6, the standard deviation σx is about 5.6, and there are $n = 11$ data items. If you scroll down, you will see the least value (minX = .6), the three quartiles (1, 3.4, and 8.4), and the greatest value (maxX = 19).

Think and Discuss

1. Find the variance of the data set. **about 31.31**
2. Enter the data set in list L1 but without the outlier 19.0. What are the new mean, median, and standard deviation? **4.06, 2.35, about 3.8**
3. Did the mean or median change less when the outlier was deleted? **median**

Check for Understanding

Concept Check

1. **OPEN ENDED** Give a sample set of data with a variance and standard deviation of 0. **Sample answer: {10, 10, 10, 10, 10, 10}**
2. **Find a counterexample** for the following statement. **See margin.**
   The standard deviation of a set of data is always less than the variance.
3. **Write** the formula for standard deviation using sigma notation. (*Hint:* To review sigma notation, see Lesson 11-5.) **See margin.**

GUIDED PRACTICE KEY	
Exercises	Examples
4–6	2
7, 8	1

Guided Practice

Find the variance and standard deviation of each set of data to the nearest tenth.

4. {48, 36, 40, 29, 45, 51, 38, 47, 39, 37} **40, 6.3**
5. {321, 322, 323, 324, 325, 326, 327, 328, 329, 330} **8.3, 2.9**
6. {43, 56, 78, 81, 47, 42, 34, 22, 78, 98, 38, 46, 54, 67, 58, 92, 55} **424.3, 20.6**

Graphing Calculator Investigation

Statistics On the TI-83 Plus calculator, press STAT and 1 (to select Edit) followed by ENTER to enter the values in L1. Then press STAT ▶ 1 and ENTER to display the statistics for L1. For Exercise 2, students can edit the list in L1 to delete 19.0. Then recalculate the statistics.

Application

EDUCATION **For Exercises 7 and 8, use the following information.**
The table below shows the amounts of money spent on education per student in 1998 in two regions of the United States.

Pacific States		Southwest Central States	
State	Expenditures per Student ($)	State	Expenditures per Student ($)
Alaska	10,650	Texas	6291
California	5345	Arkansas	5222
Washington	6488	Louisiana	5194
Oregon	6719	Oklahoma	4634

Source: National Education Association

7. Find the mean for each region. **$7300.50, $5335.25**
8. For which region is the mean more representative of the data? Explain.

8. The mean is more representative for the southwest central states because the data for the Pacific states contains the most extreme value, $10,650.

★ indicates increased difficulty

Practice and Apply

Homework Help

For Exercises	See Examples
17–26	1
9–16, 27–33	2

Extra Practice
See page 855.

Find the variance and standard deviation of each set of data to the nearest tenth.

9. {400, 300, 325, 275, 425, 375, 350} **2500, 50**
10. {5, 4, 5, 5, 5, 5, 6, 6, 6, 6, 7, 7, 7, 7, 8, 9} **1.6, 1.3**
11. {2.4, 5.6, 1.9, 7.1, 4.3, 2.7, 4.6, 1.8, 2.4} **3.1, 1.7**
12. {4.3, 6.4, 2.9, 3.1, 8.7, 2.8, 3.6, 1.9, 7.2} **4.8, 2.2**
13. {234, 345, 123, 368, 279, 876, 456, 235, 333, 444} **37,691.2, 194.1**
14. {13, 14, 15, 16, 17, 18, 19, 20, 21, 23, 67, 56, 34, 99, 44, 55} **569.4, 23.9**

★ 15.

Stem	Leaf
4	4 5 6 7 7
5	3 5 6 7 8 9
6	7 7 8 9 9 9 *4|5 = 45*

82.9, 9.1

★ 16.

Stem	Leaf
5	7 7 7 8 9
6	3 4 5 5 6 7
7	2 3 4 5 6 *6|3 = 63*

43.6, 6.6

18. The median seems to represent the center of the data.

BASKETBALL **For Exercises 17 and 18, use the following information.**
The table below shows the rebounding totals for the 2003 Los Angeles Sparks.

141	231	220	126	175	55	29	68	19	32	19	21	6	23	1

Source: WNBA

17. Find the mean, median, and mode of the data to the nearest tenth. **77.7; 32; 19**
18. Which measure of central tendency best represents the data? Explain.

More About. . .

Basketball

Chamique Holdsclaw of the Washington Mystics led the Women's National Basketball Association in rebounding in 2003 with 284 rebounds in 26 games, an average of about 10.9 rebounds per game.
Source: WNBA

Online Research **Data Update** For the latest rebounding statistics for both women's and men's professional basketball, visit: www.algebra2.com/data_update

EDUCATION **For Exercises 19 and 20, use the following information. The Millersburg school board is negotiating a pay raise with the teacher's union. Three of the administrators have salaries of $80,000 each. However, a majority of the teachers have salaries of about $35,000 per year.**

19. You are a member of the school board and would like to show that the current salaries are reasonable. Would you quote the mean, median, or mode as the "average" salary to justify your claim? Explain. **Mean; it is highest.**
20. You are the head of the teacher's union and maintain that a pay raise is in order. Which of the mean, median, or mode would you quote to justify your claim? Explain your reasoning. **See margin.**

www.algebra2.com/self_check_quiz

3 Practice/Apply

Study Notebook

Have students—
- add the definitions/examples of the vocabulary terms to their Vocabulary Builder worksheets for Chapter 12.
- include any other item(s) that they find helpful in mastering the skills in this lesson.

About the Exercises...

Organization by Objective
- **Measures of Central Tendency:** 17–26
- **Measures of Variation:** 9–16, 27–33

Odd/Even Assignments
Exercises 9–16 are structured so that students practice the same concepts whether they are assigned odd or even problems.

Assignment Guide

Basic: 9–13 odd, 17, 18, 21–23, 27–30, 34 –41, 45–64

Average: 9–15 odd, 19–23, 31–41, 45–64 (optional: 42–44)

Advanced: 10–16 even, 19, 20, 24–26, 31–58 (optional: 59–64)

All: Practice Quiz 2 (1–10)

Answer

20. Mode; it is lower and is what most employees make. It reflects the most representative worker.

DAILY INTERVENTION **Differentiated Instruction**

Interpersonal Assign students to partners, one who is fairly new to the graphing calculator and the other who is confident in calculator skills. Have them work together to address difficulties using the calculator.

Tips for New Teachers

Intervention Scientific calculators, as well as graphing calculators, have special keys and functions that can be used to find mean, median, and standard deviation. Since the scientific calculator costs only a fraction of the graphing calculator, more students may have their own calculator of this type.

Answers

34. Different scales are used on the vertical axes.
35. Sample answer: The first graph might be used by a sales manager to show a salesperson that he or she does not deserve a big raise. It appears that sales are steady but not increasing fast enough to warrant a big raise.
36. Sample answer: The second graph might be shown by the company owner to a prospective buyer of the company. It looks like there is a dramatic rise in sales.
39. The statistic(s) that best represent a set of test scores depends on the distribution of the particular set of scores. Answers should include the following.
    - mean, 73.9; median, 76.5; mode, 94
    - The mode is not representative at all because it is the highest score. The median is more representative than the mean because it is influenced less than the mean by the two very low scores of 34 and 19.
    - Each measure is increased by 5.
44. The mean deviations would be greater for the greater standard deviation and lower for the groups of data that have the smaller standard deviation.

22. Mode; it is the least expensive price.
23. Mean or median; they are nearly equal and are more representative of the prices than the mode.

ADVERTISING **For Exercises 21–23, use the following information.**
A camera store placed an ad in the newspaper showing five digital cameras for sale. The ad says, "Our digital cameras average $695." The prices of the digital cameras are $1200, $999, $1499, $895, $695, $1100, $1300, and $695.

21. Find the mean, median, and mode of the prices. **$1047.88, $1049.50, $695**
22. Which measure is the store using in its ad? Why did they choose this measure?
23. As a consumer, which measure would you want to see advertised? Explain.

More About. . .

Shopping
While the Mall of America does not have the most gross leasable area, it is the largest fully enclosed retail and entertainment complex in the United States.
Source: Mall of America

SHOPPING MALLS **For Exercises 24–26, use the following information.**
The table lists the areas of some large shopping malls in the United States.

	Mall	Gross Leasable Area (ft^2)
1	Del Amo Fashion Center, Torrance, CA	3,000,000
2	South Coast Plaza/Crystal Court, Costa Mesa, CA	2,918,236
3	Mall of America, Bloomington, MN	2,472,500
4	Lakewood Center Mall, Lakewood, CA	2,390,000
5	Roosevelt Field Mall, Garden City, NY	2,300,000
6	Gurnee Mills, Gurnee, IL	2,200,000
7	The Galleria, Houston, TX	2,100,000
8	Randall Park Mall, North Randall, OH	2,097,416
9	Oakbrook Shopping Center, Oak Brook, IL	2,006,688
10	Sawgrass Mills, Sunrise, FL	2,000,000
10	The Woodlands Mall, The Woodlands, TX	2,000,000
10	Woodfield, Schaumburg, IL	2,000,000

Source: Blackburn Marketing Service

24. Find the mean, median, and mode of the gross leasable areas.
25. You are a realtor who is trying to lease mall space in different areas of the country to a large retailer. Which measure would you talk about if the customer felt that the malls were too large for his store? Explain. **Mode; it is lowest.**
26. Which measure would you talk about if the customer had a large inventory? Explain. **Mean; it is highest.**

24. 2,290,403; 2,150,000; 2,000,000

FOOTBALL **For Exercises 27–30, use the weights in pounds of the starting offensive linemen of the football teams from three high schools.**

Jackson	Washington	King
170, 165, 140, 188, 195	144, 177, 215, 225, 197	166, 175, 196, 206, 219

27. Find the standard deviation of the weights for Jackson High. **19.3**
28. Find the standard deviation of the weights for Washington High. **28.9**
29. Find the standard deviation of the weights for King High **19.5**
30. Which team had the most variation in weights? How do you think this variation will impact their play? **Washington; see students' work.**

SCHOOL **For Exercises 31–33, use the frequency table at the right that shows the scores on a multiple-choice test.**

Score	Frequency
90	3
85	2
80	3
75	7
70	6
65	4

31. Find the variance and standard deviation of the scores.
★ 32. What percent of the scores are within one standard deviation of the mean? **64%**
★ 33. What percent of the scores are within two standard deviations of the mean? **100%**

31. 59.8, 7.7

For Exercises 34–36, consider the two graphs below. **34–36. See margin.**

34. Explain why the graphs made from the same data look different.

35. Describe a situation where the first graph might be used.

36. Describe a situation where the second graph might be used.

CRITICAL THINKING For Exercises 37 and 38, consider the two sets of data.

$A = \{1, 2, 2, 2, 2, 3, 3, 3, 3, 4\}$, $B = \{1, 1, 2, 2, 2, 3, 3, 3, 4, 4\}$

37. Find the mean, median, variance, and standard deviation of each set of data.

37. A: 2.5, 2.5, 0.7, 0.8; B: 2.5, 2.5, 1.1, 1.0

38. Explain how you can tell which histogram below goes with each data set without counting the frequencies in the sets.

38. The first histogram is lower in the middle and higher on the ends, so it represents data that are more spread out. Since set B has the greater standard deviation, set B corresponds to the first histogram and set A corresponds to the second.

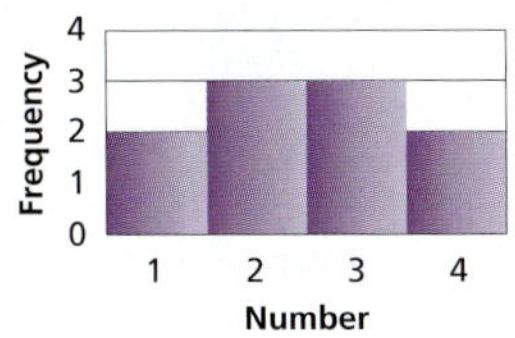

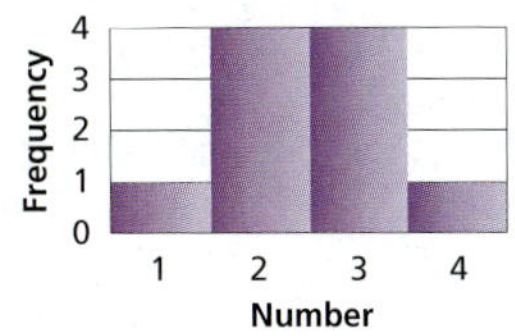

39. WRITING IN MATH Answer the question that was posed at the beginning of the lesson. **See margin.**

What statistics should a teacher tell the class after a test?

Include the following in your answer:

- the mean, median, and mode of the given data set,
- which measure of central tendency best represents the test scores and why, and
- how the measures of central tendency are affected if Mr. Dent adds 5 points to each score.

Standardized Test Practice

40. What is the mean of the numbers represented by $x + 1$, $3x - 2$, and $2x - 5$? **A**

Ⓐ $2x - 2$ Ⓑ $\frac{6x - 7}{3}$ Ⓒ $\frac{x + 1}{3}$ Ⓓ $x + 4$

41. Manuel got scores of 92, 85, and 84 on three successive tests. What score must he get on a fourth test in order to have an average of 90? **D**

Ⓐ 96 Ⓑ 97 Ⓒ 98 Ⓓ 99

Extending the Lesson

Mean deviation is another method of dispersion. It is the mean of the deviations of the data from the mean of the data. If a set of data consists of n values $x_1, x_2, \ldots, x_n$ and has mean $\bar{x}$, then the mean deviation is given by the following formula.

$$MD = \frac{|x_1 - \bar{x}| + |x_2 - \bar{x}| + \cdots + |x_n - \bar{x}|}{n} \text{ or } \frac{1}{n}\sum_{i=1}^{n} |x_i - \bar{x}|$$

Study Tip

Reading Math

Mean deviation is also sometimes called *mean absolute deviation.*

Find the mean deviation of each set of data to the nearest tenth.

42. $\{95, 91, 88, 86\}$ **3**

43. $\{10.4, 11.4, 16.2, 14.9, 13.5\}$ **1.9**

44. Suppose two sets of data have the same mean and different standard deviations. Describe their mean deviations. **See margin.**

Enrichment, p. 734

Probabilities in Genetics

Genes are the units which transmit hereditary traits. The possible forms which a gene may take, **dominant** and **recessive,** are called **alleles.** A particular trait is determined by two alleles, one from the female parent and one from the male parent. If an organism has the trait which is dominant, it may have either two dominant alleles or one dominant and one recessive allele. If the organism has the trait which is recessive, it must have two recessive alleles.

Example **Consider a plant in which tall stems, *T*, are dominant to short stems, *t*. What is the probability of obtaining a long-stemmed plant if two long-stemmed plants both with the genetic formula *Tt* are crossed?**

	T	*t*
T	*TT*	*Tt*

A *Punnett square* is a chart used to determine the possible combinations of characteristics among offspring.

Study Guide and Intervention, p. 729 (shown) and p. 730

Measures of Central Tendency

	Use	When
Measures of Central Tendency	mean	the data are spread out and you want an average of values
	median	the data contain outliers
	mode	the data are tightly clustered around one or two values

Example **Find the mean, median, and mode of the following set of data: {42, 39, 35, 40, 38, 35, 45}.**

To find the mean, add the values and divide by the number of values.

$\text{mean} = \frac{42 + 39 + 35 + 40 + 38 + 35 + 45}{7} \approx 39.14.$

To find the median, arrange the values in ascending or descending order and choose the middle value. (If there is an even number of values, find the mean of the two middle values.) In this case, the median is 39.

To find the mode, take the most common value. In this case, the mode is 35.

Exercises

Find the mean, median, and mode of each set of data. Round to the nearest hundredth, if necessary.

1. {238, 261, 245, 249, 255, 262, 241, 245} **249.5; 247; 245**

2. {9, 13, 8, 10, 11, 9, 12, 16, 10, 9} **10.7; 10; 9**

3. {120, 108, 145, 129, 102, 132, 134, 118, 108, 142} **123.8; 124.5; 108**

4. {68, 54, 73, 58, 63, 72, 65, 70, 61} **64.89; 65; no mode**

5. {34, 49, 42, 38, 40, 45, 34, 28, 43, 30} **38.3; 39; 34**

6. The table at the right shows the populations of the six New England capitals. Which would be the most appropriate measure of central tendency to represent the data? Explain why and find that value. **There is no mode. The population of Boston is an outlier and would raise the mean too high. The median, 79,500, would be the best choice.**

City	Population (rounded to the nearest 1000)
Augusta, ME	19,000
Boston, MA	589,000
Concord, NH	37,000
Hartford, CT	122,000
Montpelier, VT	8,000
Providence, RI	174,000

Source: www.factfinder.census.gov

Skills Practice, p. 731 and Practice, p. 732 (shown)

Find the variance and standard deviation of each set of data to the nearest tenth.

1. {47, 61, 93, 22, 82, 22, 37} **673.1, 25.9**

2. {10, 10, 54, 39, 96, 91, 91, 18} **1228.6, 35.1**

3. {1, 2, 2, 3, 3, 3, 4, 4, 4, 4, 5, 5, 5, 5, 5} **1.6, 1.2**

4. {1100, 725, 850, 335, 700, 800, 950} **49,150.0; 221.7**

5. {3.4, 7.1, 8.5, 5.1, 4.7, 6.3, 9.9, 8.4, 3.6} **4.7, 2.2**

6. {2.8, 0.5, 1.9, 0.8, 1.9, 1.5, 3.3, 2.6, 0.7, 2.5} **0.8, 0.9**

7. HEALTH CARE Eight physicians with 15 patients on a hospital floor see these patients an average of 18 minutes a day. The 22 nurses on the same floor see the patients an average of 3 hours a day. As a hospital administrator, would you quote the mean, median, or mode as an indicator of the amount of daily medical attention the patients on this floor receive? Explain. **Either the median or the mode; they are equal and higher than the mean, which is lowered by the smaller amount of time the physicians spend with the patients.**

For Exercises 8–10, use the frequency table that shows the percent of public school teachers in the U. S. in 1999 who used computers or the Internet at school for various administrative and teaching activities.

Activity	Percent Using Computer or Internet
Create instructional materials	39
Administrative record keeping	34
Communicate with colleagues	23
Gather information for planning lessons	16
Multimedia classroom presentations	8
Access research and best practices for teaching	8
Communicate with parents or students	8
Access model lesson plans	6

Source: National Assessment of Educational Progress

8. Find the mean, median, and mode of the data. **17.75%, 12%, 8%**

9. Suppose you believe teachers use computers or the Internet too infrequently. Which measure would you quote as the "average?" Explain. **Mode; it is lowest.**

10. Suppose you believe teachers use computers or the Internet too often. Which measure would you quote as the "average?" Explain. **Mean; it is highest.**

For Exercises 11 and 12, use the frequency table that shows the number of games played by 24 American League baseball players between opening day, 2001 and September 8, 2001.

No. of Games	Frequency
141	4
140	3
139	4
138	5
137	2
136	3
135	3

Source: Major League Baseball

11. Find the mean, median, mode, and standard deviation of the number of games played to the nearest tenth. **138.2, 138; 138, 2.0**

12. For how many players is the number of games within one standard deviation of the mean? **14**

Reading to Learn Mathematics, p. 733 ELL

Pre-Activity **What statistics should a teacher tell the class after a test?**

Read the introduction to Lesson 12-6 at the top of page 664 in your textbook.

There is more than one way to give an "average" score for this test. Three measures of central tendency for these scores are 94, 76.5 and 73.9. Can you tell which of these is the mean, the median, and the mode without doing any calculations? Explain your answer.

Sample answer: Yes. The mode must be one of the scores, so it must be an integer. The median must be either one of the scores or halfway between two of the scores, so it must be an integer or a decimal ending with .5. Therefore, 94 is the mode, 76.5 is the median, and 73.9 is the mean.

Reading the Lesson

1. Match each measure with one of the six descriptions of how to find measures of central tendency and variation.

a. median **vi** b. mode **i** c. range **iv**

d. variance **iii** e. mean **ii** f. standard deviation **v**

i. Find the most commonly occurring values or values in a set of data.

ii. Add the data and divide by the number of items.

iii. Find the mean of the squares of the differences between each value in the set of data and the mean.

iv. Find the difference between the largest and smallest values in the set of data.

v. Take the positive square root of the variance.

vi. If there is an odd number of items in a set of data, take the middle one. If there is an even number of items, add the two middle items and divide by 2.

Helping You Remember

2. It is usually easier to remember a complicated procedure if you break it down into steps. Write the procedure for finding the standard deviation for a set of data in a series of brief, numbered steps.

Sample answer:
1. Find the mean.
2. Find the difference between each value and the mean.
3. Square each difference.
4. Find the mean of the squares.
5. Take the positive square root.

Lesson 12-6 Statistical Measures 669

4 Assess

Open-Ended Assessment

Writing Ask students to write a brief explanation of what standard deviation is and how it can be used. Have them include at least one example.

Assessment Options

Practice Quiz 2 The quiz provides students with a brief review of the concepts and skills in Lessons 12-4 through 12-6. Lesson numbers are given to the right of exercises or instruction lines so students can review concepts not yet mastered.

Getting Ready for Lesson 12-7

BASIC SKILL Lesson 12-7 presents data that are normally distributed. Students will use percents to calculate ranges of data within a normal distribution. Exercises 59–64 should be used to determine your students' familiarity with finding percents.

Maintain Your Skills

Mixed Review **Determine whether the events are mutually exclusive or inclusive. Then find the probability.** *(Lesson 12-5)*

45. A card is drawn from a standard deck of cards. What is the probability that it is a 5 or a spade? **inclusive; $\frac{4}{13}$**

46. A jar of change contains 5 quarters, 8 dimes, 10 nickels, and 19 pennies. If a coin is pulled from the jar at random, what is the probability that it is a nickel or a dime? **mutually exclusive; $\frac{3}{7}$**

Two cards are drawn from a standard deck of cards. Find each probability. *(Lesson 12-4)*

47. P(ace, then king) if replacement occurs $\frac{1}{169}$

48. P(ace, then king) if no replacement occurs $\frac{4}{663}$

49. P(heart, then club) if no replacement occurs $\frac{13}{204}$

50. P(heart, then club) if replacement occurs $\frac{1}{16}$

51. Find the coordinates of the vertices and foci and the slopes of the asymptotes for the hyperbola given by $\frac{y^2}{81} - \frac{x^2}{25} = 1$. *(Lesson 8-5)* **$(0, \pm 9)$; $(0, \pm\sqrt{106})$; $\pm\frac{9}{5}$**

If $f(x) = x - 7$, $g(x) = 4x^2$, and $h(x) = 2x + 1$, find each value. *(Lesson 7-7)*

52. $f[g(-1)]$ **−3**

53. $h[f(15)]$ **17**

54. $f \circ h(2)$ **−2**

55. **BUSINESS** The Energy Booster Company keeps their stock of Health Aid liquid in a rectangular tank whose sides measure $x - 1$ centimeters, $x + 3$ centimeters, and $x - 2$ centimeters. Suppose they would like to bottle their Health Aid in $x - 3$ containers of the same size. How much liquid in cubic centimeters will remain unbottled? *(Lesson 7-2)* **12 cm^3**

Use Cramer's Rule to solve each system of equations. *(Lesson 4-6)*

56. $2x + 6y = 28$; $-x - 4y = -20$ **(−4, 6)**

57. $7c - 3d = -8$; $4c + d = 9$ **(1, 5)**

58. $m - 2n = -7$; $-3m + n = -4$ **(3, 5)**

Getting Ready for the Next Lesson **BASIC SKILL** **Find each percent.**

59. 68% of 200 **136**

60. 68% of 500 **340**

61. 95% of 400 **380**

62. 95% of 500 **475**

63. 99% of 400 **396**

64. 99% of 500 **495**

Practice Quiz 2 — Lessons 12-4 through 12-6

A bag contains 5 red marbles, 3 green marbles, and 2 blue marbles. Two marbles are drawn at random from the bag. Find each probability. *(Lesson 12-4)*

1. P(red, then green) if replacement occurs $\frac{3}{20}$

2. P(red, then green) if no replacement occurs $\frac{1}{6}$

3. P(2 red) if no replacement occurs $\frac{2}{9}$

4. P(2 red) if replacement occurs $\frac{1}{4}$

A twelve-sided die has sides numbered 1 through 12. The die is rolled once. Find each probability. *(Lesson 12-5)*

5. P(4 or 5) $\frac{1}{6}$

6. P(even or a multiple of 3) $\frac{2}{3}$

7. P(odd or a multiple of 4) $\frac{3}{4}$

Find the variance and standard deviation of each set of data to the nearest tenth. *(Lesson 12-6)*

8. {5, 8, 2, 9, 4} **6.6, 2.6**

9. {16, 22, 18, 31, 25, 22} **23.6, 4.9**

10. {425, 400, 395, 415, 420} **134.0, 11.6**

670 Chapter 12 Probability and Statistics

12-7 The Normal Distribution

What You'll Learn

- Determine whether a set of data appears to be normally distributed or skewed.
- Solve problems involving normally distributed data.

Vocabulary

- discrete probability distribution
- continuous probability distribution
- normal distribution
- skewed distribution

How are the heights of professional athletes distributed?

The frequency table below lists the heights of the 2001 Baltimore Ravens. The table shows the heights of the players, but it does not show how these heights compare to the height of an average player. To make that comparison, you can determine how the heights are distributed.

Height (in.)	67	69	70	71	72	73	74	75	76	77	80
Frequency	1	1	4	4	10	6	6	8	7	5	1

Source: www.ravenszone.net

NORMAL AND SKEWED DISTRIBUTIONS The probability distributions you have studied thus far are **discrete probability distributions** because they have only a finite number of possible values. A discrete probability distribution can be represented by a histogram. For a **continuous probability distribution**, the outcome can be any value in an interval of real numbers. Continuous probability distributions are represented by curves instead of histograms.

The curve at the right represents a continuous probability distribution. Notice that the curve is symmetric. Such a curve is often called a *bell curve*. Many distributions with symmetric curves or histograms are **normal distributions**.

Normal Distribution

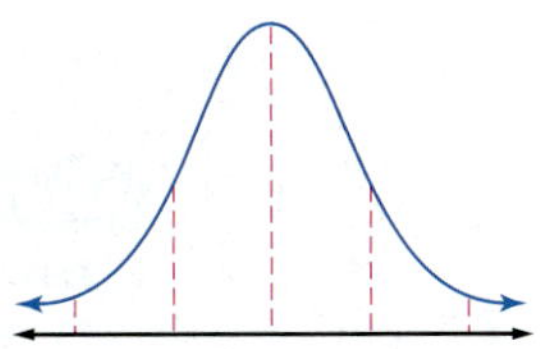

Study Tip

Skewed Distributions

In a positively skewed distribution, the long tail is in the positive direction. These are sometimes said to be *skewed to the right*. In a negatively skewed distribution, the long tail is in the negative direction. These are sometimes said to be *skewed to the left*.

A curve or histogram that is not symmetric represents a **skewed distribution**. For example, the distribution for a curve that is high at the left and has a tail to the right is said to be *positively skewed*. Similarly, the distribution for a curve that is high at the right and has a tail to the left is said to be *negatively skewed*.

Positively Skewed

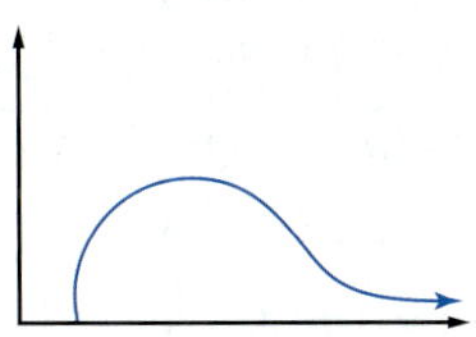

Negatively Skewed

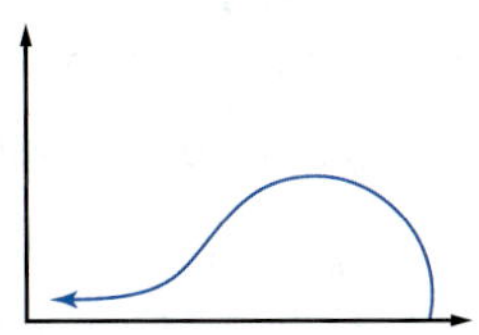

12-7 Lesson Notes

1 Focus

5-Minute Check Transparency 12-7 Use as a quiz or review of Lesson 12-6.

Mathematical Background notes are available for this lesson on p. 630D.

How are the heights of professional athletes distributed?

Ask students:

- What is the greatest height listed in feet and inches? **6 ft 8 in.**
- How many players were exactly 6 feet tall? **10**

Resource Manager

Workbook and Reproducible Masters

Chapter 12 Resource Masters

- Study Guide and Intervention, pp. 735–736
- Skills Practice, p. 737
- Practice, p. 738
- Reading to Learn Mathematics, p. 739
- Enrichment, p. 740
- Assessment, p. 768

Transparencies

5-Minute Check Transparency 12-7
Real-World Transparency 12
Answer Key Transparencies

Technology

Interactive Chalkboard

2 Teach

NORMAL AND SKEWED DISTRIBUTIONS

In-Class Example Power Point®

1 Determine whether the data {31, 37, 35, 36, 34, 36, 32, 36, 33, 32, 34, 34, 35, 34} appear to be *positively skewed*, *negatively skewed*, or *normally distributed*. **normally distributed**

USE NORMAL DISTRIBUTIONS

In-Class Example Power Point®

2 Students counted the number of candies in 100 small packages. They found that the number of candies per package was normally distributed, with a mean of 23 candies per package and a standard deviation of 1 piece of candy.

a. About how many packages had between 24 and 22 candies? **about 68 packages**

b. What is the probability that a package selected at random had more than 25 candies? **about 2.5%**

Study Tip

Normal Distribution

If you randomly select an item from data that are normally distributed, the probability that the one you pick will be within one standard deviation of the mean is 0.68. If you do this 1000 times, about 680 of those picked will be within one standard deviation of the mean.

Example 1 Classify a Data Distribution

Determine whether the data {14, 15, 11, 13, 13, 14, 15, 14, 12, 13, 14, 15} appear to be *positively skewed, negatively skewed,* or *normally distributed.*

Make a frequency table for the data. Then use the table to make a histogram.

Value	11	12	13	14	15
Frequency	1	1	3	4	3

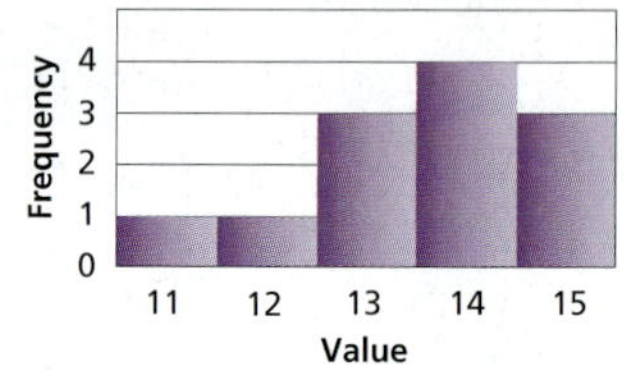

Since the histogram is high at the right and has a tail to the left, the data are negatively skewed.

USE NORMAL DISTRIBUTIONS Normal distributions occur quite frequently in real life. Standardized test scores, the lengths of newborn babies, the useful life and size of manufactured items, and production levels can all be represented by normal distributions. In all of these cases, the number of data values must be large for the distribution to be approximately normal.

Key Concept — *Normal Distribution*

Normal distributions have these properties.

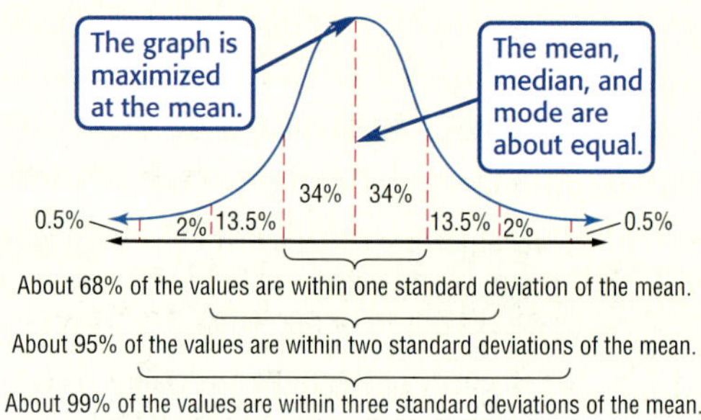

Example 2 Normal Distribution

PHYSIOLOGY The reaction times for a hand-eye coordination test administered to 1800 teenagers are normally distributed with a mean of 0.35 second and a standard deviation of 0.05 second.

a. About how many teens had reaction times between 0.25 and 0.45 second?

Draw a normal curve. Label the mean and the mean plus or minus multiples of the standard deviation.

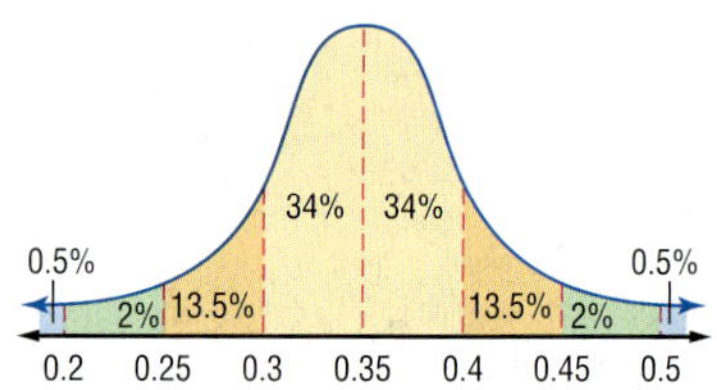

The values 0.25 and 0.45 are 2 standard deviations *below and above* the mean, respectively. Therefore, about 95% of the data are between 0.25 and 0.45.

$1800 \times 95\% = 1710$ Multiply 1800 by 0.95.

About 1710 of the teenagers had reaction times between 0.25 and 0.45 second.

DAILY INTERVENTION — **Differentiated Instruction**

Kinesthetic Ask students to measure carefully the distance around the wrists of 15 classmates to the nearest tenth of a centimeter and find the mean and standard deviation for their data. Then have them determine if this data is normally distributed, or positively or negatively skewed.

b. What is the probability that a teenager selected at random had a reaction time greater than 0.4 second?

The value 0.4 is one standard deviation above the mean. You know that about 100% − 68% or 32% of the data are more than one standard deviation away from the mean. By the symmetry of the normal curve, half of 32%, or 16%, of the data are more than one standard deviation above the mean.

The probability that a teenager selected at random had a reaction time greater than 0.4 second is about 16% or 0.16.

Check for Understanding

Concept Check

1. **OPEN ENDED** Sketch a positively skewed graph. Describe a situation in which you would expect data to be distributed this way. **See margin.**
2. **Compare and contrast** the means and standard deviations of the graphs.

See margin.

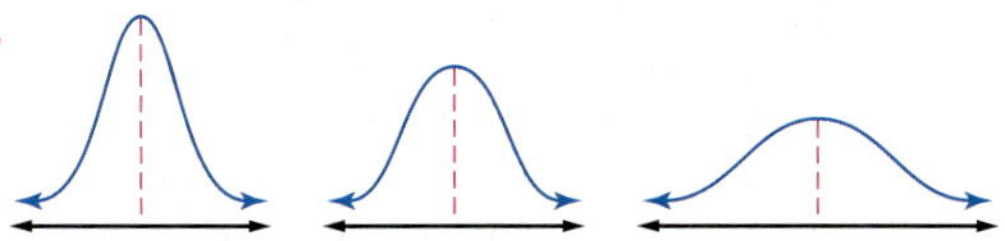

3. **Explain** how to find what percent of a set of normally distributed data is more than 3 standard deviations above the mean. **See margin.**

Guided Practice

GUIDED PRACTICE KEY	
Exercises	Examples
4	1
5–11	2

4. The table at the right shows female mathematics SAT scores in 2000. Determine whether the data appear to be *positively skewed*, *negatively skewed*, or *normally distributed*. **normally distributed**

Score	Percent of Females
200–299	3
300–399	14
400–499	33
500–599	31
600–699	15
700–800	4

Source: www.collegeboard.org

For Exercises 5–7, use the following information.
Mrs. Sung gave a test in her trigonometry class. The scores were normally distributed with a mean of 85 and a standard deviation of 3.

5. What percent would you expect to score between 82 and 88? **68%**
6. What percent would you expect to score between 88 and 91? **13.5%**
7. What is the probability that a student chosen at random scored between 79 and 91? **95%**

Application

QUALITY CONTROL For Exercises 8–11, use the following information.
The useful life of a radial tire is normally distributed with a mean of 30,000 miles and a standard deviation of 5000 miles. The company makes 10,000 tires a month.

8. About how many tires will last between 25,000 and 35,000 miles? **6800**
9. About how many tires will last more than 40,000 miles? **250**
10. About how many tires will last less than 25,000 miles? **1600**
11. What is the probability that if you buy a radial tire at random, it will last between 20,000 and 35,000 miles? **81.5%**

www.algebra2.com/extra_examples

3 Practice/Apply

Study Notebook

Have students—
- add the definitions/examples of the vocabulary terms to their Vocabulary Builder worksheets for Chapter 12.
- draw graphs of normally distributed, positively skewed, and negatively skewed sets of data.
- include any other item(s) that they find helpful in mastering the skills in this lesson.

About the Exercises...

Organization by Objective
- **Normal and Skewed Distributions:** 12–14
- **Use Normal Distributions:** 15–26

Odd/Even Assignments
Exercises 12, 13, and 15–26 are structured so that students practice the same concepts whether they are assigned odd or even problems.

Assignment Guide

Basic: 13, 15–21, 27–44
Average: 13, 22–44
Advanced: 12, 14, 22–41 (optional: 42–44)

Answers

1. Sample answer:

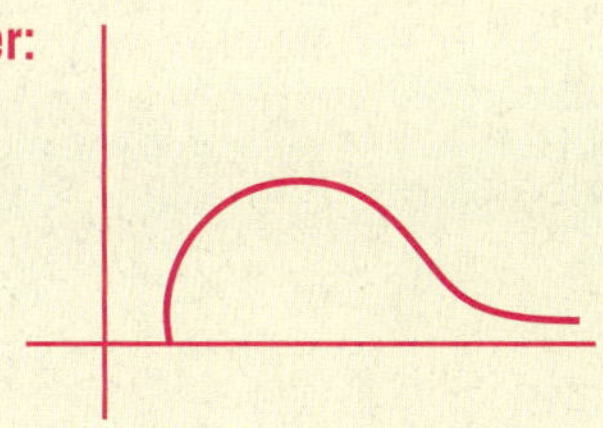

the use of cassettes since CDs were introduced

2. The mean of the three graphs is the same, but the standard deviations are different. The first graph has the least standard deviation, the standard deviation of the middle graph is slightly greater, and the standard deviation of the last graph is greatest.

3. Since 99% of the data is within 3 standard deviations of the mean, 1% of the data is more than 3 standard deviations from the mean. By symmetry, half of this, or 0.5%, is more than 3 standard deviations above the mean.

Study Guide and Intervention, p. 735 (shown) and p. 736

Normal and Skewed Distributions A **continuous probability** distribution is represented by a curve.

Types of Continuous Distributions

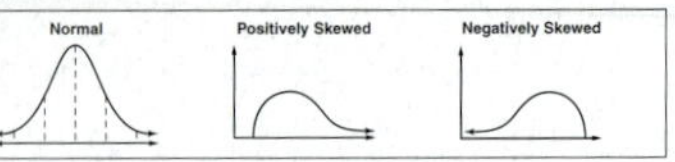

Example **Determine whether the data below appear to be *positively skewed*, *negatively skewed*, or *normally distributed*.**
{100, 120, 110, 100, 110, 80, 100, 90, 100, 120, 100, 90, 110, 100, 90, 80, 100, 90}

Make a frequency table for the data.

Value	80	90	100	110	120
Frequency	2	4	7	3	2

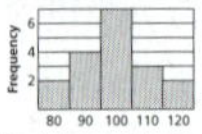

Then use the data to make a histogram.
Since the histogram is roughly symmetric, the data appear to be normally distributed.

Exercises

Determine whether the data in each table appear to be *positively skewed*, *negatively skewed*, or *normally distributed*. Make a histogram of the data.

1. {27, 24, 29, 25, 27, 22, 24, 25, 29, 24, 25, 22, 27, 24, 22, 25, 24, 22} positively skewed

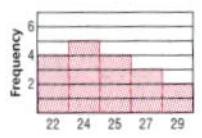

2.

Shoe Size	4	5	6	7	8	9	10
No. of Students	1	2	4	8	5	1	2

normally distributed

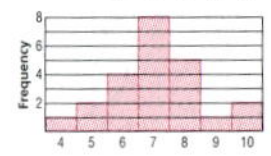

3.

Housing Price	No. of Houses Sold
less than \$100,000	0
\$100,00–\$120,000	1
\$121,00–\$140,000	3
\$141,00–\$160,000	7
\$161,00–\$180,000	8
\$181,00–\$200,000	6
over \$200,000	12

negatively skewed

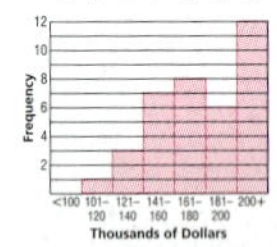

Skills Practice, p. 737 and Practice, p. 738 (shown)

Determine whether the data in each table appear to be *positively skewed*, *negatively skewed*, or *normally distributed*.

1.

Time Spent at a Museum Exhibit	
Minutes	Frequency
0–25	27
26–50	46
51–75	89
75–100	57
100+	24

normally distributed

2.

Average Age of High School Principals	
Age in Years	Number
31–35	3
36–40	8
41–45	15
46–50	32
51–55	40
56–60	38
60+	4

negatively skewed

For Exercises 3 and 4, use the frequency table that shows the number of hours worked per week by 100 high school seniors.

Hours	Number of Students
0–8	30
9–17	45
18–25	20
26+	5

3. Make a histogram of the data.

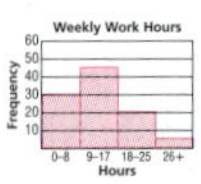

4. Do the data appear to be *positively skewed*, *negatively skewed*, or *normally distributed*? Explain. Positively skewed; the histogram is high at the left and has a tail to the right.

TESTING **For Exercises 5–10, use the following information.**
The scores on a test administered to prospective employees are normally distributed with a mean of 100 and a standard deviation of 15.

5. About what percent of the scores are between 70 and 130? 95%
6. About what percent of the scores are between 85 and 130? 81.5%
7. About what percent of the scores are over 115? 16%
8. About what percent of the scores are lower than 85 or higher than 115? 32%
9. If 80 people take the test, how many would you expect to score higher than 130? 2
10. If 75 people take the test, how many would you expect to score lower than 85? 12
11. TEMPERATURE The daily July surface temperature of a lake at a resort has a mean of 82° and a standard deviation of 4.2°. If you prefer to swim when the temperature is at least 77.8°, about what percent of the days does the temperature meet your preference? 84%

Reading to Learn Mathematics, p. 739 ELL

Pre-Activity **How are the heights of professional athletes distributed?**

Read the introduction to Lesson 12-7 at the top of page 671 in your textbook.

There were 53 players on the team and the mean height was approximately 73.6. About what fraction of the players' heights are between 72 and 75, inclusive? Sample answer: about $\frac{2}{3}$

Reading the Lesson

1. Indicate whether each of the following statements is *true* or *false*.
   a. In a continuous probability distribution, there is a finite number of possible outcomes. false
   b. Every normal distribution can be represented by a bell curve. true
   c. A distribution that is represented by a curve that is high at the left and has a tail to the right is negatively skewed. false
   d. A normal distribution is an example of a skewed distribution. false

2. Ms. Rose gave the same quiz to her two geometry classes. She recorded the following scores.

First-period class:

Score	0	1	2	3	4	5	6	7	8	9	10
Frequency	1	0	1	0	3	4	5	7	4	3	2

Fifth-period class:

Score	0	1	2	3	4	5	6	7	8	9	10
Frequency	0	0	0	0	3	4	9	7	6	1	0

In each class, 30 students took the quiz. The mean score for each class was 6.4. Which set of scores has the greater standard deviation? (Answer this question without doing any calculations.) Explain your answer.

First period class; sample answer: The scores are more spread out from the mean than for the fifth period class.

Helping You Remember

3. Many students have trouble remembering how to determine if a curve represents a distribution that is *positively skewed* or *negatively skewed*. What is an easy way to remember this?

Sample answer: Follow the tail! If the tail is on the right (positive direction), the distribution is positively skewed. If the tail is on the left (negative direction), the distribution is negatively skewed.

Practice and Apply

Homework Help

For Exercises	See Examples
12–14	1
15–26	2

Extra Practice
See page 856.

Determine whether the data in each table appear to be *positively skewed*, *negatively skewed*, or *normally distributed*.

12.

U.S. Population	
Age	Percent
0–19	28.7
20–39	29.3
40–59	25.5
60–79	13.3
80–99	3.2
100+	0.0

Source: U.S. Census Bureau

positively skewed

13.

Record Low Temperatures in the 50 States	
Temperature (°F)	Number of States
−80 to −65	4
−64 to −49	12
−48 to −33	19
−32 to −17	12
−16 to −1	2
0 to 15	1

Source: *The World Almanac*

13. normally distributed

14. **SCHOOL** The frequency table at the right shows the grade-point averages (GPAs) of the juniors at Stanhope High School. Do the data appear to be *positively skewed*, *negatively skewed*, or *normally distributed*? Explain. **Negatively skewed; the histogram is high at the right and has a tail to the left.**

GPA	Frequency
0.0–0.4	4
0.5–0.9	4
1.0–1.4	2
1.5–1.9	32
2.0–2.4	96
2.5–2.9	91
3.0–3.4	110
3.5–4.0	75

FOOD **For Exercises 15–18, use the following information.**
The shelf life of a particular dairy product is normally distributed with a mean of 12 days and a standard deviation of 3.0 days.

15. About what percent of the products last between 9 and 15 days? **68%**
16. About what percent of the products last between 12 and 15 days? **34%**
17. About what percent of the products last less than 3 days? **0.5%**
18. About what percent of the products last more than 15 days? **16%**

VENDING **For Exercises 19–21, use the following information.**
The vending machine in the school cafeteria usually dispenses about 6 ounces of soft drink. Lately, it is not working properly, and the variability of how much of the soft drink it dispenses has been getting greater. The amounts are normally distributed with a standard deviation of 0.2 ounce.

19. What percent of the time will you get more than 6 ounces of soft drink? **50%**
20. What percent of the time will you get less than 6 ounces of soft drink? **50%**
21. What percent of the time will you get between 5.6 and 6.4 ounces of soft drink? **95%**

MANUFACTURING **For Exercises 22–24, use the following information.**
A company manufactures 1000 CDs per hour that are supposed to be 120 millimeters in diameter. These CDs are made for drives 122 millimeters wide. The sizes of CDs made by this company are normally distributed with a standard deviation of 1 millimeter. **22. 50%**

22. What percent of the CDs would you expect to be greater than 120 millimeters?
23. In one hour, how many CDs would you expect to be between 119 and 122 millimeters? **815**
24. About how many CDs per hour will be too large to fit in the drives? **25**

Enrichment, p. 740

Street Networks: Finding All Possible Routes

A section of a city is laid out in square blocks. Going north from the intersection of First Avenue and First Street, the avenues are 1st, 2nd, 3rd, and so on. Going east, the streets are numbered in the same way.

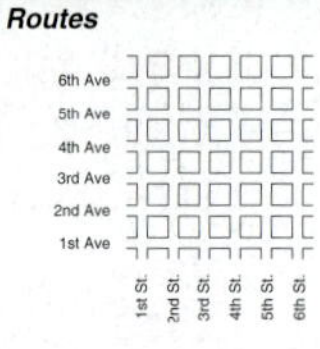

Factorials can be used to find the number, $r(e, n)$, of different routes between two intersections. The formula is shown below.

$$r(e, n) = \frac{[(e - 1) + (n - 1)]!}{(e - 1)!(n - 1)!}$$

The number of streets going east is e; the number of avenues going north is n.

The following problems examine the possible routes from one location to another. Assume that you never use a route that is unnecessarily long. Assume that $e \geq 1$ and $n \geq 1$.

Answer

27. The mean would increase by 25; the standard deviation would not change; and the graph would be translated 25 units to the right.

More About. . .

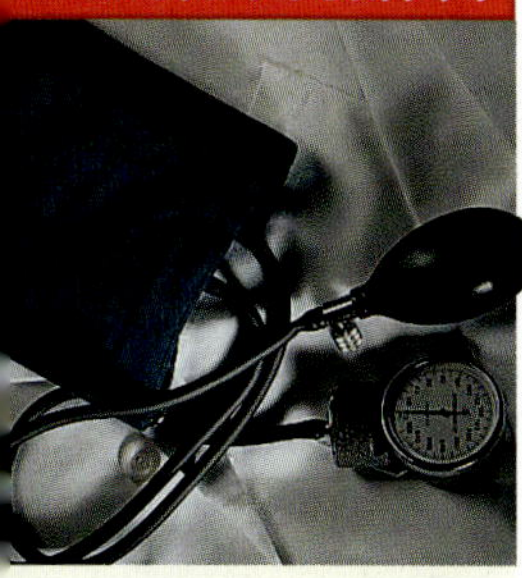

Health

A systolic blood pressure below 130 is normal and between 130 and 139 is "high normal."

Source: National Institutes of Health

HEALTH **For Exercises 25 and 26, use the following information.**
A recent study showed that the systolic blood pressure of high school students ages 14–17 is normally distributed with a mean of 120 and a standard deviation of 12. Suppose a high school has 800 students.

25. About what percent of the students have blood pressures below 108? **16%**

26. About how many students have blood pressures between 108 and 144? **652**

27. **CRITICAL THINKING** The graphing calculator screen shows the graph of a normal distribution for a large set of test scores whose mean is 500 and whose standard deviation is 100. If every test score in the data set were increased by 25 points, describe how the mean, standard deviation, and graph of the data would change. **See margin.**

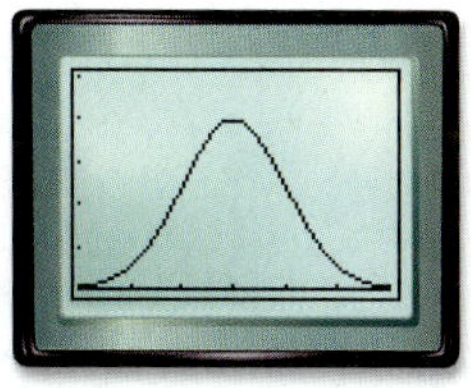

[200, 800] scl: 100 by [0, 0.005] scl: 0.001

28. **WRITING IN MATH** Answer the question that was posed at the beginning of the lesson. **See margin.**

How are the heights of professional athletes distributed?

Include the following items in your answer:

- a histogram of the given data, and
- an explanation of whether you think the data are normally distributed.

Standardized Test Practice

29. If $x + y = 5$ and $xy = 6$, what is the value of $x^2 + y^2$? **A**

(A) 13 (B) 17 (C) 25 (D) 37

30. Which of the following is not the square of a rational number? **D**

(A) 0.04 (B) 0.16 (C) $\frac{4}{9}$ (D) $\frac{2}{3}$

Maintain Your Skills

Mixed Review

Find the variance and standard deviation of each set of data to the nearest tenth. *(Lesson 12-6)*

31. {7, 16, 9, 4, 12, 3, 9, 4} **17.5, 4.2**

32. {12, 14, 28, 19, 11, 7, 10} **42.5, 6.5**

A card is drawn from a standard deck of cards. Find each probability. *(Lesson 12-5)*

33. P(jack or queen) $\frac{2}{13}$

34. P(ace or heart) $\frac{4}{13}$

35. P(2 or face card) $\frac{4}{13}$

Find all of the rational zeros for each function. *(Lesson 7-6)* **39.** $\frac{1}{4}$, **1**

36. $f(x) = x^3 + 4x^2 - 5x$ **−5, 0, 1**

37. $p(x) = x^3 - 3x^2 - 10x + 24$ **−3, 2, 4**

38. $h(x) = x^4 - 2x^2 + 1$ **1, −1**

39. $f(x) = 4x^4 - 13x^3 - 13x^2 + 28x - 6$

METEOROLOGY **For excercises 40 and 41, use the following information.**
Weather forecasters can determine the approximate time that a thunderstorm will last if they know the diameter d of the storm in miles. The time t in hours can be found by using the formula $216t^2 = d^3$. *(Lesson 6-2)*

40. Graph $y = 216t^2 - 5^3$ and use it to estimate how long a thunderstorm will last if its diameter is 5 miles. **See margin for graph; about 45 min.**

41. Find how long a thunderstorm will last if its diameter is 5 miles and compare this time with your estimate in Exercise 40. **0.76 h**

Getting Ready for the Next Lesson

PREREQUISITE SKILL **Find the indicated term of each expression.** **42.** $21a^5b^2$
(For review of ***binomial expansions****, see Lesson 5-2.)* **43.** $56c^5d^3$ **44.** $126x^5y^4$

42. third term of $(a + b)^7$

43. fourth term of $(c + d)^8$

44. fifth term of $(x + y)^9$

www.algebra2.com/self_check_quiz

4 Assess

Open-Ended Assessment

Writing Have students in small groups discuss the meaning of standard deviation and normally distributed and skewed data. Ask them to write their own definitions for these terms in informal, but accurate, language.

Assessment Options

Quiz (Lessons 12-6 and 12-7) is available on p. 768 of the *Chapter 12 Resource Masters.*

Getting Ready for Lesson 12-8

PREREQUISITE SKILL Lesson 12-8 presents finding probabilities by using binomial expansions. Students will use their familiarity with finding terms of a binomial expansion as they determine probabilities. Exercises 42–44 should be used to determine your students' familiarity with finding a specified term of a binomial expansion.

Answer

40.

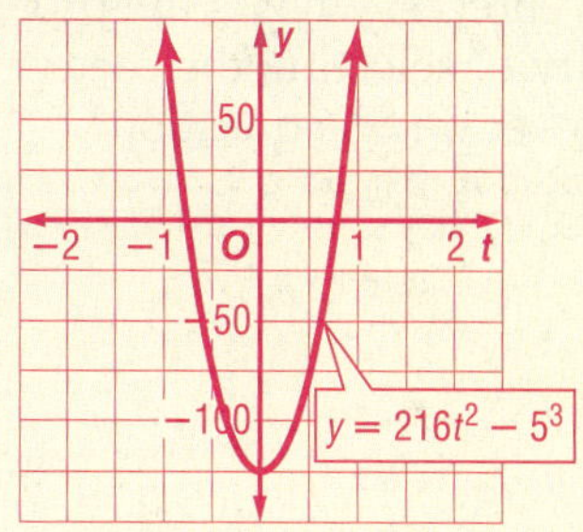

Answer

28. If a large enough group of athletes is studied, some of the characteristics may be normally distributed; others may have skewed distributions. Answers should include the following.

- See graph at the right.
- Since the histogram has two peaks, the data may not be normally distributed. This may be due to players who play certain positions tending to be of similar large sizes while players who play the other positions tend to be of similar smaller sizes.

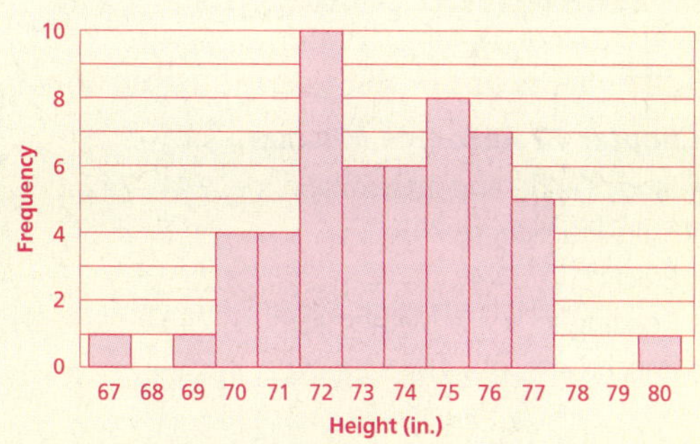

12-8 Lesson Notes

1 Focus

5-Minute Check Transparency 12-8 Use as a quiz or review of Lesson 12-7.

Mathematical Background notes are available for this lesson on p. 630D.

Building on Prior Knowledge

In Chapter 11, students learned to use the Binomial Theorem. In this lesson, students will use the Binomial Theorem to find probabilities.

How can you determine whether guessing is worth it?

Ask students:

- How many choices are there for each question? **4**
- If you guess at random, without being able to eliminate any of the choices, what is the probability of selecting the correct answer on one question? **1 out of 4 or 25%**

12-8 Binomial Experiments

What You'll Learn

- Use binomial expansions to find probabilities.
- Find probabilities for binomial experiments.

Vocabulary

- binomial experiment

How can you determine whether guessing is worth it?

What is the probability of getting exactly 4 questions correct on a 5-question multiple-choice quiz if you guess at every question?

Study Tip

Look Back
To review the **Binomial Theorem**, see Lesson 11-7.

BINOMIAL EXPANSIONS You can use the Binomial Theorem to find probabilities in certain situations where there are two possible outcomes. The 5 possible ways of getting 4 questions right r and 1 question wrong w are shown at the right. This chart shows the combination of 5 things (answer choices) taken 4 at a time (right answers) or $C(5, 4)$.

w	r	r	r	r
r	w	r	r	r
r	r	w	r	r
r	r	r	w	r
r	r	r	r	w

The terms of the binomial expansion of $(r + w)^5$ can be used to find the probabilities of each combination of right and wrong.

$$(r + w)^5 = r^5 + 5r^4w + 10r^3w^2 + 10r^2w^3 + 5rw^4 + w^5$$

Coefficient	Term	Meaning
$C(5, 5) = 1$	r^5	1 way to get all 5 questions right
$C(5, 4) = 5$	$5r^4w$	5 ways to get 4 questions right and 1 question wrong
$C(5, 3) = 10$	$10r^3w^2$	10 ways to get 3 questions right and 2 questions wrong
$C(5, 2) = 10$	$10r^2w^3$	10 ways to get 2 questions right and 3 questions wrong
$C(5, 1) = 5$	$5rw^4$	5 ways to get 1 question right and 4 questions wrong
$C(5, 0) = 1$	w^5	1 way to get all 5 questions wrong

The probability of getting a question right that you guessed on is $\frac{1}{4}$. So, the probability of getting the question wrong is $\frac{3}{4}$. To find the probability of getting 4 questions right and 1 question wrong, substitute $\frac{1}{4}$ for r and $\frac{3}{4}$ for w in the term $5r^4w$.

$$P(\text{4 right, 1 wrong}) = 5r^4w$$

$$= 5\left(\frac{1}{4}\right)^4\left(\frac{3}{4}\right) \quad r = \frac{1}{4}, w = \frac{3}{4}$$

$$= \frac{15}{1024} \quad \text{Multiply.}$$

The probability of getting exactly 4 questions correct is $\frac{15}{1024}$ or about 1.5%.

Resource Manager

Workbook and Reproducible Masters

Chapter 12 Resource Masters
- Study Guide and Intervention, pp. 741–742
- Skills Practice, p. 743
- Practice, p. 744
- Reading to Learn Mathematics, p. 745
- Enrichment, p. 746

School-to-Career Masters, p. 24
Teaching Algebra With Manipulatives Masters, p. 294

Transparencies

5-Minute Check Transparency 12-8
Answer Key Transparencies

Technology

Interactive Chalkboard

Example 1 Binomial Theorem

If a family has 4 children, what is the probability that they have 3 boys and 1 girl?

There are two possible outcomes for the gender of each of their children: boy or girl. The probability of a boy b is $\frac{1}{2}$, and the probability of a girl g is $\frac{1}{2}$.

$(b + g)^4 = b^4 + 4b^3g + 6b^2g^2 + 4bg^3 + g^4$

The term $4b^3g$ represents 3 boys and 1 girl.

$P(\text{3 boys, 1 girl}) = 4b^3g$

$= 4\left(\frac{1}{2}\right)^3\left(\frac{1}{2}\right)$ $\quad b = \frac{1}{2}, g = \frac{1}{2}$

$= \frac{1}{4}$ $\quad$ Multiply.

The probability of 3 boys and 1 girl is $\frac{1}{4}$ or 25%.

BINOMIAL EXPERIMENTS Problems like Example 1 that can be solved using binomial expansion are called **binomial experiments**.

Key Concept — Binomial Experiments

A binomial experiment exists if and only if all of these conditions occur.

- There are exactly two possible outcomes for each trial.
- There is a fixed number of trials.
- The trials are independent.
- The probabilities for each trial are the same.

A binomial experiment is sometimes called a *Bernoulli experiment*.

Suppose that in the application at the beginning of the lesson, the first 3 questions are answered correctly. Then the last 2 are answered incorrectly. The probability of this occurring is $\frac{1}{4} \cdot \frac{1}{4} \cdot \frac{1}{4} \cdot \frac{3}{4} \cdot \frac{3}{4}$ or $\left(\frac{1}{4}\right)^3\left(\frac{3}{4}\right)^2$. In general, there are $C(5, 3)$ ways to arrange 3 correct answers among the 5 questions, so the probability of exactly 3 correct answers is given by $C(5, 3)\left(\frac{1}{4}\right)^3\left(\frac{3}{4}\right)^2$.

More About. . .

Sports

The National Hockey League record for most goals in a game by one player is seven. A player has scored five or more goals in a game 53 times in league history.

Source: NHL

Example 2 Binomial Experiment

SPORTS Suppose that when hockey star Jaromir Jagr takes a shot, he has a $\frac{1}{7}$ probability of scoring a goal. He takes 6 shots in a game one night.

a. What is the probability that he will score exactly 2 goals?

The probability that he scores a goal on a given shot is $\frac{1}{7}$. The probability that he does not score on a given shot is $\frac{6}{7}$. There are $C(6, 2)$ ways to choose the 2 shots that score.

$P(\text{2 goals}) = C(6, 2)\left(\frac{1}{7}\right)^2\left(\frac{6}{7}\right)^4$ $\quad$ If he scores on 2 shots, he fails to score on 4 shots.

$= \frac{6 \cdot 5}{2}\left(\frac{1}{7}\right)^2\left(\frac{6}{7}\right)^4$ $\quad C(6, 2) = \frac{6!}{4!2!}$

$= \frac{19{,}440}{117{,}649}$ $\quad$ Simplify.

The probability that Jagr will score exactly 2 goals is $\frac{19{,}440}{117{,}649}$ or about 0.17.

www.algebra2.com/extra_examples

2 Teach

BINOMIAL EXPANSIONS

In-Class Example Power Point®

Teaching Tip This example assumes that the chance for having a boy is 1 out of 2. Actually, from a biological standpoint, this is not quite accurate. In the U.S., about 1050 males are born for each 1000 females.

1 If a family has 4 children, what is the probability that they have 2 girls and 2 boys? **37.5%**

BINOMIAL EXPERIMENTS

In-Class Example Power Point®

2 A report said that approximately 1 out of 6 cars sold in a certain year was green. Suppose a salesperson sells 7 cars per week.

a. What is the probability that this salesperson will sell exactly 3 green cars in a week? **about 0.078**

b. What is the probability that this salesperson will sell at least 3 green cars in a week? **about 0.096**

DAILY INTERVENTION — Differentiated Instruction

Kinesthetic Have students in small groups do a binomial experiment by tossing a ball into the wastebasket about 20 times to establish the probability of scoring a goal. Then have them find the probability that they will score exactly 4 goals in 8 tries.

3 Practice/Apply

Study Notebook

Have students—

- add the definitions/examples of the vocabulary terms to their Vocabulary Builder worksheets for Chapter 12.
- include any other item(s) that they find helpful in mastering the skills in this lesson.

About the Exercises...

Organization by Objective

- **Binomial Expansions:** 12–37
- **Binomial Experiments:** 12–37

Odd/Even Assignments

Exercises 12–33 are structured so that students practice the same concepts whether they are assigned odd or even problems.

Alert! Exercises 42–43 require a graphing calculator.

Assignment Guide

Basic: 13–31 odd, 35, 38–41, 44–56

Average: 13–35 odd, 36–41, 44–56 (optional: 42, 43)

Advanced: 12–34 even, 36–50 (optional: 51–56)

b. What is the probability that he will score at least 2 goals?

Instead of adding the probabilities of getting exactly 2, 3, 4, 5, and 6 goals, it is easier to subtract the probabilities of getting exactly 0 or 1 goal from 1.

$$\begin{aligned} P(\text{at least 2 goals}) &= 1 - P(0 \text{ goals}) - P(1 \text{ goal}) \\ &= 1 - C(6, 0)\left(\frac{1}{7}\right)^0\left(\frac{6}{7}\right)^6 - C(6, 1)\left(\frac{1}{7}\right)^1\left(\frac{6}{7}\right)^5 \\ &= 1 - \frac{46{,}656}{117{,}649} - \frac{46{,}656}{117{,}649} \quad \text{Simplify.} \\ &= \frac{24{,}337}{117{,}649} \quad \text{Subtract.} \end{aligned}$$

The probability that Jagr will score at least 2 goals is $\frac{24{,}337}{117{,}649}$ or about 0.21.

Check for Understanding

Concept Check

1. Sample answer: In a 5-card hand, what is the probability that at least 2 cards are hearts?
2. RRRWW, RRWRW, RRWWR, RWRRW, RWRWR, RWWRR, WRRRW, WRRWR, WRWRR, WWRRR

1. **OPEN ENDED** Describe a situation for which the P(2 or more) can be found by using a binomial expansion.
2. Refer to the application at the beginning of the lesson. List the possible sequences of 3 right answers and 2 wrong answers.
3. **Explain** why each experiment is not binomial.
   a. rolling a die and recording whether a 1, 2, 3, 4, 5, or 6 comes up **3a. Each trial has more than two possible outcomes.**
   b. tossing a coin repeatedly until it comes up heads **The number of trials is not fixed.**
   c. removing marbles from a bag and recording whether each one is black or white, if no replacement occurs **The trials are not independent.**

Guided Practice

GUIDED PRACTICE KEY	
Exercises	Examples
4–11	1, 2

Find each probability if a coin is tossed 3 times.

4. $P(\text{exactly 2 heads})$ $\frac{3}{8}$
5. $P(0 \text{ heads})$ $\frac{1}{8}$
6. $P(\text{at least 1 head})$ $\frac{7}{8}$

Four cards are drawn from a standard deck of cards. Each card is replaced before the next one is drawn. Find each probability.

7. $P(4 \text{ jacks})$ $\frac{1}{28{,}561}$
8. $P(\text{exactly 3 jacks})$ $\frac{48}{28{,}561}$
9. $P(\text{at most 1 jack})$ $\frac{27{,}648}{28{,}561}$

Application

SPORTS **For Exercises 10 and 11, use the following information.**
Jessica Mendoza of Stanford University was the 2000 NCAA women's softball batting leader with an average of .475. This means that the probability of her getting a hit in a given at-bat was 0.475.

10. Find the probability of her getting 4 hits in 4 at-bats. **about 0.05**
11. Find the probability of her getting exactly 2 hits in 4 at-bats. **about 0.37**

★ indicates increased difficulty

Practice and Apply

Homework Help

For Exercises	See Examples
12–37	1, 2

Extra Practice
See page 856.

Find each probability if a coin is tossed 4 times.

12. $P(4 \text{ tails})$ $\frac{1}{16}$
13. $P(0 \text{ tails})$ $\frac{1}{16}$
14. $P(\text{exactly 2 tails})$ $\frac{3}{8}$
15. $P(\text{exactly 1 tail})$ $\frac{1}{4}$
16. $P(\text{at least 3 tails})$ $\frac{5}{16}$
17. $P(\text{at most 2 tails})$ $\frac{11}{16}$

Find each probability if a die is rolled 5 times.

18. $P(\text{exactly one 5})$ $\frac{3125}{7776}$
19. $P(\text{exactly three 5s})$ $\frac{125}{3888}$
20. $P(\text{at most two 5s})$ $\frac{625}{648}$
21. $P(\text{at least three 5s})$ $\frac{23}{648}$

678 Chapter 12 Probability and Statistics

As an apartment manager, Jackie Thomas is responsible for showing prospective renters different models of apartments. When showing a model, the probability that she selects the correct key from her set is $\frac{1}{4}$. If she shows 5 models in a day, find each probability.

22. P(never the correct key) $\frac{243}{1024}$
23. P(always the correct key) $\frac{1}{1024}$
24. P(correct exactly 4 times) $\frac{15}{1024}$
25. P(correct exactly 2 times) $\frac{135}{512}$
26. P(no more than 2 times correct) $\frac{459}{512}$
27. P(at least 3 times correct) $\frac{53}{512}$

Prisana guesses at all 10 true/false questions on her history test. Find each probability.

28. P(exactly 6 correct) $\frac{105}{512}$
29. P(exactly 4 correct) $\frac{105}{512}$
30. P(at most half correct) $\frac{319}{512}$
31. P(at least half correct) $\frac{319}{512}$

If a thumbtack is dropped, the probability of it landing point-up is 0.4. If 12 tacks are dropped, find each probability.

★ 32. P(at least 9 points up) **about 0.02**
★ 33. P(at most 4 points up) **about 0.44**

34. **CARS** According to a recent survey, about 1 in 3 new cars is leased rather than bought. What is the probability that 3 of 7 randomly-selected new cars are leased? $\frac{560}{2187}$

35. **INTERNET** In 2001, it was estimated that 32.5% of U.S. adults use the Internet. What is the probability that exactly 2 out of 5 randomly-selected U.S. adults use the Internet? **about 0.32**

More About. . .

Internet

The word *Internet* was virtually unknown until the mid-1980s. By 1997, 19 million Americans were using the Internet. That number tripled in 1998 and passed 100 million in 1999.

Source: UCLA

WORLD CULTURES **For Exercises 36 and 37, use the following information.**
The Cayuga Indians played a game of chance called *Dish*, in which they used 6 flattened peach stones blackened on one side. They placed the peach stones in a wooden bowl and tossed them. The winner was the first person to get a prearranged number of points. The table below shows the points that were given for each toss. Assume that each face (black or neutral) of each stone has an equal chance of showing up.

36. Copy and complete the table by finding the probability of each outcome.

37. Find the probability that a player gets at least 1 point for a toss. $\frac{7}{32}$

Outcome	Points	Probability
6 black	5	1/64
5 black, 1 neutral	1	3/32
4 black, 2 neutral	0	15/64
3 black, 3 neutral	0	5/16
2 black, 4 neutral	0	15/64
1 black, 5 neutral	1	3/32
6 neutral	5	1/64

38. **CRITICAL THINKING** Write an expression for the probability of exactly m successes in n trials of a binomial experiment where the probability of success in a given trial is p. $C(n, m)p^m(1 - p)^{n - m}$

39. **WRITING IN MATH** Answer the question that was posed at the beginning of the lesson. **See pp. 695A–695B.**

How can you determine whether guessing is worth it?

Include the following in your answer:
- an explanation of how to find the probability of getting any number of questions right on a 5-question multiple-choice quiz, and
- the probability of each score.

www.algebra2.com/self_check_quiz

Study Guide and Intervention, p. 741 (shown) and p. 742

Binomial Expansions For situations with only 2 possible outcomes, you can use the Binomial Theorem to find probabilities. The coefficients of terms in a binomial expansion can be found by using combinations.

Example **What is the probability that 3 coins show heads and 3 show tails when 6 coins are tossed?**

There are 2 possible outcomes that are equally likely: heads (H) and tails (T). The tosses of 6 coins are independent events. When $(H + T)^6$ is expanded, the term containing H^3T^3, which represents 3 heads and 3 tails, is used to get the desired probability. By the Binomial Theorem the coefficient of H^3T^3 is $C(6, 3)$.

$$P(\text{3 heads, 3 tails}) = \frac{6!}{3!3!}\left(\frac{1}{2}\right)^3\left(\frac{1}{2}\right)^3 \quad P(H) = \frac{1}{2} \text{ and } P(T) = \frac{1}{2}$$
$$= \frac{20}{64}$$
$$= \frac{5}{16}$$

The probability of getting 3 heads and 3 tails is $\frac{5}{16}$ or 0.3125.

Exercises

Find each probability if a coin is tossed 8 times.

1. P(exactly 5 heads) **about 22%**
2. P(exactly 2 heads) **about 11%**
3. P(even number of heads) **50%**
4. P(at least 6 heads) **about 14%**

Mike guesses on all 10 questions of a true-false test. If the answers true and false are evenly distributed, find each probability.

5. Mike gets exactly 8 correct answers. $\frac{45}{1024}$ **or 0.044**
6. Mike gets at most 3 correct answers. $\frac{11}{64}$ **or 0.172**
7. A die is tossed 4 times. What is the probability of tossing exactly two sixes? $\frac{25}{216}$ **or 0.116**

Skills Practice, p. 743 and Practice, p. 744 (shown)

Find each probability if a coin is tossed 6 times.

1. P(exactly 3 tails) $\frac{5}{16}$
2. P(exactly 5 tails) $\frac{3}{32}$
3. P(0 tails) $\frac{1}{64}$
4. P(at least 4 heads) $\frac{11}{32}$
5. P(at least 4 tails) $\frac{11}{32}$
6. P(at most 2 tails) $\frac{11}{32}$

The probability of Chris making a free throw is $\frac{2}{3}$. If she shoots 5 times, find each probability.

7. P(all missed) $\frac{1}{243}$
8. P(all made) $\frac{32}{243}$
9. P(exactly 2 made) $\frac{40}{243}$
10. P(exactly 1 missed) $\frac{80}{243}$
11. P(at least 3 made) $\frac{64}{81}$
12. P(at most 2 made) $\frac{17}{81}$

When Tarin and Sam play a certain board game, the probability that Tarin will win a game is $\frac{3}{4}$. If they play 5 games, find each probability.

13. P(Sam wins only once) $\frac{405}{1024}$
14. P(Tarin wins exactly twice) $\frac{45}{512}$
15. P(Sam wins exactly 3 games) $\frac{45}{512}$
16. P(Sam wins at least 1 game) $\frac{781}{1024}$
17. P(Tarin wins at least 3 games) $\frac{459}{512}$
18. P(Tarin wins at most 2 games) $\frac{53}{512}$

19. **SAFETY** In August 2001, the American Automobile Association reported that 73% of Americans use seat belts. In a random selection of 10 Americans in 2001, what is the probability that exactly half of them use seat belts? Source: AAA **about 7.5%**

HEALTH **For Exercises 20 and 21, use the following information.**
In 2001, the American Heart Association reported that 50 percent of the Americans who receive heart transplants are ages 50–64 and 20 percent are ages 35–49. Source: American Heart Association

20. In a randomly selected group of 10 heart transplant recipients, what is the probability that at least 8 of them are ages 50–64? $\frac{7}{128}$

21. In a randomly selected group of 5 heart transplant recipients, what is the probability that 2 of them are ages 35–49? $\frac{128}{625}$

Reading to Learn Mathematics, p. 745 ELL

Pre-Activity **How can you determine whether guessing is worth it?**

Read the introduction to Lesson 12-8 at the top of page 676 in your textbook.

Suppose you are taking a 50-question multiple-choice test in which there are 5 answer choices for each question. You are told that no points will be deducted for wrong answers. Should you guess the answers to the questions you do not know? Explain your reasoning. **Sample answer: Yes; the probability of guessing the right answer to a question is $\frac{1}{5}$, so you have a chance to get some points by guessing, and you have nothing to lose.**

Reading the Lesson

1. Indicate whether each of the following is a *binomial experiment* or *not a binomial experiment*. If the experiment is not a binomial experiment, explain why.
   a. A fair coin is tossed 10 times and "heads" or "tails" is recorded each time. **binomial experiment**
   b. A pair of dice is thrown 5 times and the sum of the numbers that come up is recorded each time. **Not a binomial experiment; there are more than two possible outcomes for each trial.**
   c. There are 5 red marbles and 6 blue marbles in a bag. One marble is drawn from the bag and its color recorded. The marble is not put back in the bag. A second marble is drawn and its color recorded. **Not a binomial experiment; the trials are not independent (or, the probabilities for the two trials are not the same).**
   d. There are 5 red marbles and 6 blue marbles in a bag. One marble is drawn from the bag and its color recorded. The marble is put back in the bag. A second marble is drawn and its color recorded. **binomial experiment**

2. Len randomly guesses the answers to all 6 multiple-choice questions on his chemistry test. Each question has 5 choices. Which of the following expressions gives the probability that he will get at least 4 of the answers correct? **B**
   A. $P(6, 4)\left(\frac{1}{5}\right)^4\left(\frac{4}{5}\right)^2 + P(6, 5)\left(\frac{1}{5}\right)^5\left(\frac{4}{5}\right)^1 + P(6, 6)\left(\frac{1}{5}\right)^6\left(\frac{4}{5}\right)^0$
   B. $C(6, 4)\left(\frac{1}{5}\right)^4\left(\frac{4}{5}\right)^2 + C(6, 5)\left(\frac{1}{5}\right)^5\left(\frac{4}{5}\right)^1 + C(6, 6)\left(\frac{1}{5}\right)^6\left(\frac{4}{5}\right)^0$
   C. $C(6, 4)\left(\frac{1}{5}\right)^2\left(\frac{4}{5}\right)^4 + C(6, 5)\left(\frac{1}{5}\right)^1\left(\frac{4}{5}\right)^5 + C(6, 6)\left(\frac{1}{5}\right)^0\left(\frac{4}{5}\right)^6$

Helping You Remember

3. Some students have trouble remembering how to calculate binomial probabilities. What is an easy way to remember which numbers to put into an expression like $C(6, 4)\left(\frac{1}{5}\right)^2\left(\frac{4}{5}\right)^4$? **Sample answer: The binomial coefficient is $C(n, r)$, where n is the number of trials and r is the number of successes. The probability of success is raised to the rth power and the probability of failure is raised to the $(n - r)$th power.**

Enrichment, p. 746

Misuses of Statistics

Statistics can be misleading. Graphs for a set of data can look very different from one another. Compare the following graphs.

Source: U.S. Department of Labor

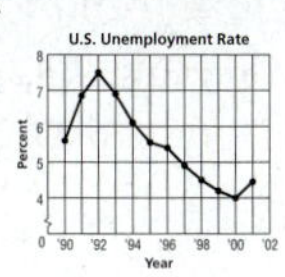

Source: U.S. Department of Labor

Notice that the two graphs show the same data, but the spacing in the vertical and horizontal scales differs. Scales can be cramped or spread out to make a graph that gives a certain impression. Which graph would you use to

4 Assess

Open-Ended Assessment

Speaking Ask students to use their own families, for example, 2 boys and a girl, and find the probabilities for that particular group of siblings. Then have students explain the steps they used.

Getting Ready for Lesson 12-9

PREREQUISITE SKILL Lesson 12-9 presents finding sources of bias and sampling error. Students will use their familiarity with evaluating radical expressions as they find the margin of error. Exercises 51–54 should be used to determine your students' familiarity with finding the value of a radical expression.

Answers

48.

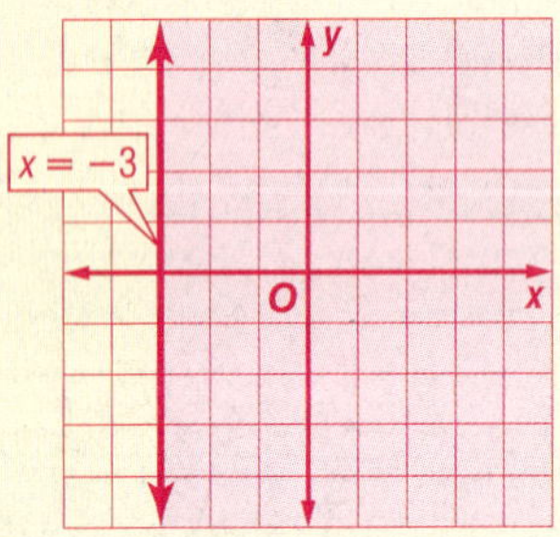

49.

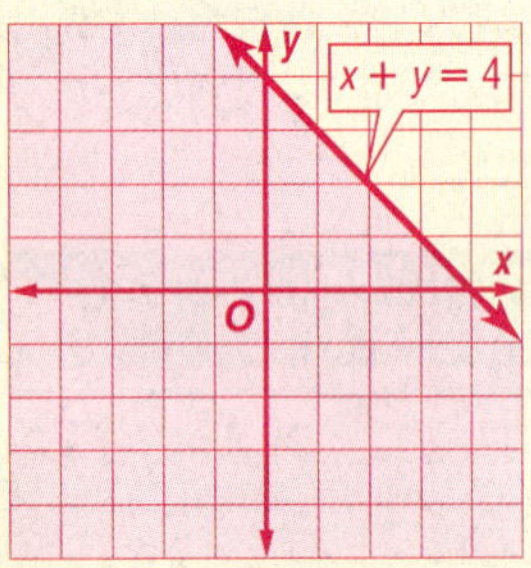

50.

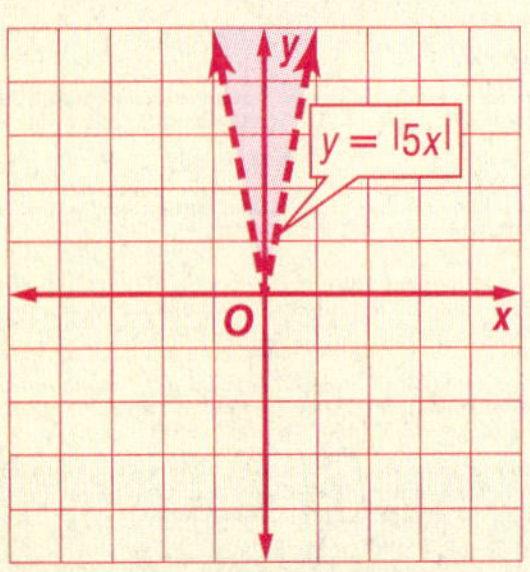

Standardized Test Practice

40. GRID IN In the figure, if $DE = 2$, what is the sum of the area of $\triangle ABE$ and the area of $\triangle BCD$? **2**

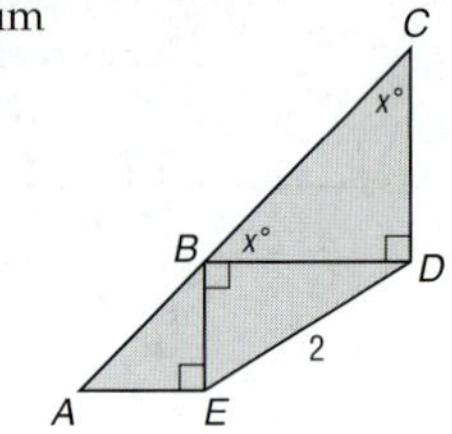

41. What is the net result if a discount of 5% is applied to a bill of \$340.60? **B**

Ⓐ \$306.54 Ⓑ \$323.57

Ⓒ \$335.60 Ⓓ \$357.63

Graphing Calculator

BINOMIAL DISTRIBUTION You can use a TI-83 Plus to investigate the graph of a binomial distribution.

Step 1 Enter the number of trials in L1. Start with 10 trials.

KEYSTROKES: [STAT] 1 [▲] [2nd] [LIST] [▶] 5 [X,T,θ,n] [,] [X,T,θ,n] [,] 0 [,] 10 [)] [ENTER]

Step 2 Calculate the probability of success for each trial in L2.

KEYSTROKES: [▶] [▲] [2nd] [DISTR] 0 10 [,] .5 [,] [2nd] [L1] [)] [ENTER]

Step 3 Graph the histogram.

KEYSTROKES: [2nd] [STATPLOT]

Use the arrow and [ENTER] keys to choose ON, the histogram, L1 as the Xlist, and L2 as the frequency. Use the window [0, 10] scl:1 by [0, 0.5] scl: 0.1.

42. Replace the 10 in the keystrokes for steps 1 and 2 to graph the binomial distribution for several values of n less than or equal to 47. You may have to adjust your viewing window to see all of the histogram. Make sure Xscl is 1. **42. See students' work.**

43. What type of distribution does the binomial distribution start to resemble as n increases? **normal distribution**

Maintain Your Skills

Mixed Review

For Exercises 44–46, use the following information.
A set of 400 test scores is normally distributed with a mean of 75 and a standard deviation of 8. *(Lesson 12-7)*

44. What percent of the test scores lie between 67 and 83? **68%**

45. How many of the test scores are greater than 91? **10**

46. What is the probability that a randomly-selected score is less than 67? **16%**

47. A salesperson had sales of \$11,000, \$15,000, \$11,000, \$16,000, \$12,000, and \$12,000 in the last six months. Which measure of central tendency would he be likely to use to represent these data when he talks with his supervisor? Explain. *(Lesson 12-6)* **Mean; it is highest.**

Graph each inequality. *(Lesson 2-7)* **48–50. See margin.**

48. $x \geq -3$ **49.** $x + y \leq 4$ **50.** $y > |5x|$

Getting Ready for the Next Lesson

PREREQUISITE SKILL Evaluate $2\sqrt{\frac{p(1-p)}{n}}$ for the given values of p and n. Round to the nearest thousandth, if necessary. *(For review of* ***radical expressions****, see Lesson 5-6.)*

51. $p = 0.5, n = 100$ **0.1** **52.** $p = 0.5, n = 400$ **0.05**

53. $p = 0.25, n = 500$ **0.039** **54.** $p = 0.75, n = 1000$ **0.027**

55. $p = 0.3, n = 500$ **0.041** **56.** $p = 0.6, n = 1000$ **0.031**

Answers (p. 681)

5. The class results should be better since it is a much larger set of data.

7. Sample answer: Put 20 marbles—5 red, 3, yellow, 3 blue, 3 green, 3 orange, and 3 black—into a bag. The red will represent Amazing Amy, and the other colors will represent each of the other prizes.

Algebra Activity

A Follow-Up of Lesson 12-8

Simulations

A **simulation** uses a probability experiment to mimic a real-life situation. You can use a simulation to solve the following problem about **expected value**.

A brand of cereal is offering one of six different prizes in every box. If the prizes are equally and randomly distributed within the cereal boxes, how many boxes, on average, would you have to buy in order to get a complete set of the six prizes?

Collect the Data

Work in pairs or small groups to complete steps 1 through 4.

Step 1 Use the six numbers on a die to represent the six different prizes.

Step 2 Roll the die and record which prize was in the first box of cereal. Use a tally sheet like the one shown.

Step 3 Continue to roll the die and record the prize number until you have a complete set of prizes. Stop as soon as you have a complete set. This is the end of one trial in your simulation. Record the number of boxes required for this trial.

Step 4 Repeat steps 1, 2, and 3 until your group has carried out 25 trials. Use a new tally sheet for each trial.

Simulation Tally Sheet	
Prize Number	**Boxes Purchased**
1	
2	
3	
4	
5	
6	
Total Needed	

Analyze the Data 1–2. See pp. 695A–695B.

1. Create two different statistical graphs of the data collected for 25 trials.
2. Determine the mean, median, maximum, minimum, and standard deviation of the total number of boxes needed in the 25 trials.
3. Combine the small-group results and determine the mean, median, maximum, minimum, and standard deviation of the number of boxes required for all the trials conducted by the class. **See students' work.**

Make a Conjecture

4. If you carry out 25 additional trials, will your results be the same as in the first 25 trials? Explain. **Probably not; the outcomes of the trials are random since you are rolling a die.**
5. Should the small-group results or the class results give a better idea of the average number of boxes required to get a complete set of superheroes? Explain. **See margin.**
6. If there were 8 superheroes instead of 6, would you need to buy more boxes of cereal or fewer boxes of cereal on average? **more**
7. What if one of the 6 prizes was more common than the other 5? For instance, suppose that one prize, Amazing Amy, appears in 25% of all the boxes and the other 5 prizes are equally and randomly distributed among the remaining 75% of the boxes? Design and carry out a new simulation to predict the average number of boxes you would need to buy to get a complete set. Include some measures of central tendency and dispersion with your data. **See margin.**

Algebra Activity

A Follow-Up of Lesson 12-8

Getting Started

Objective Simulate a real-life situation, collect data, and do a statistical analysis.

Materials
one die for each group

Teach

- Ask students why rolling a die can simulate this problem. **because it has 6 random outcomes**
- Ask students before they collect their data if they would expect every group in the class to have the same results. **Probably not, since you are finding experimental and not theoretical probabilities.**
- Have students complete the simulation to collect data and then complete Exercises 1–7.

Assess

- In **Exercises 1–3**, students should be able to collect and organize data in a usable form and find various statistical measures.
- In **Exercises 4–7**, students should conclude that the greater the number of trials, the closer the experimental probabilities will be to the theoretical probabilities. They should also recognize that changes in the parameters of the experiment affect the outcomes.

Resource Manager

Teaching Algebra with Manipulatives
- p. 22 (master for die patterns)
- p. 295 (student recording sheet)

Glencoe Mathematics Classroom Manipulative Kit
- dice

Study Notebook

You may wish to have students summarize this activity and what they learned from it.

12-9 Lesson Notes

1 Focus

5-Minute Check Transparency 12-9 Use as a quiz or review of Lesson 12-8.

Mathematical Background notes are available for this lesson on p. 630D.

How are opinion polls used in political campaigns?

Ask students:

- Do the results of this poll indicate beyond doubt that Bush will be the victor? **No, the 7% undecided could be the deciding margin in the actual election.**
- What is the difference between the Other category and the Undecided category? **Those in the Other category are probably going to vote for a candidate other than Bush or Gore, while those in the Undecided category might be added to the total for one of those two.**

12-9 Sampling and Error

What You'll Learn

- Determine whether a sample is unbiased.
- Find margins of sampling error.

How are opinion polls used in political campaigns?

About a month before the 2000 presidential election, Mason-Dixon Polling & Research surveyed the preferences of Florida voters. The results shown were published in the *Orlando Sentinel*.

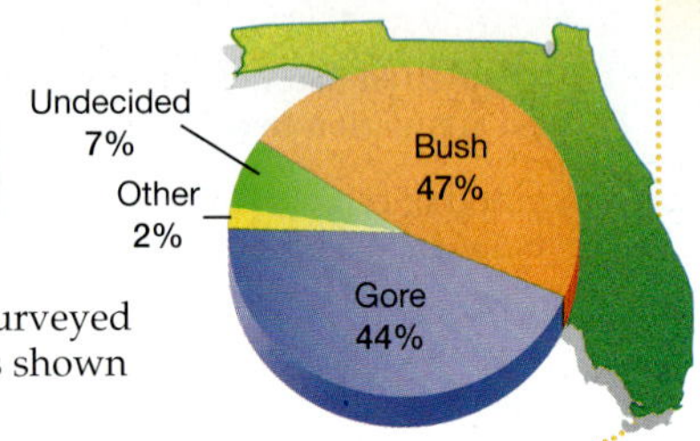

Vocabulary

- unbiased sample
- margin of sampling error

BIAS When polling organizations want to find how the public feels about an issue, they do not have the time or money to ask everyone. Instead, they obtain their results by polling a small portion of the population. To be sure that the results are representative of the population, they need to make sure that this portion is a random or **unbiased sample** of the population. A sample of size n is random when every possible sample of size n has an equal chance of being selected.

Example 1 Biased and Unbiased Samples

State whether each method would produce a random sample. Explain.

a. asking every tenth person coming out of a health club how many times a week they exercise to determine how often people in the city exercise

This would not result in a random sample because the people surveyed would probably exercise more often than the average person.

b. surveying people going into an Italian restaurant to find out people's favorite type of food

This would probably not result in a random sample because the people surveyed would probably be more likely than others to prefer Italian food.

Study Tip

Law of Large Numbers
The principle that as sample size increases, the closer the experimental probability is to the theoretical probability is called the *Law of Large Numbers.*

MARGIN OF ERROR As the size of a sample increases, it more accurately reflects the population. If you sampled only three people and two prefer Brand A, you could say, "Two out of three people chose Brand A over any other brand," but you may not be giving a true picture of how the total population would respond. The **margin of sampling error (ME)** gives a limit on the difference between how a sample responds and how the total population would respond.

Key Concept *Margin of Sampling Error*

If the percent of people in a sample responding in a certain way is p and the size of the sample is n, then 95% of the time, the percent of the population responding in that same way will be between $p - ME$ and $p + ME$, where

$$ME = 2\sqrt{\frac{p(1-p)}{n}}.$$

That is, the probability is 0.95 that $p \pm ME$ will contain the true population results.

Resource Manager

Workbook and Reproducible Masters

Chapter 12 Resource Masters

- Study Guide and Intervention, pp. 747–748
- Skills Practice, p. 749
- Practice, p. 750
- Reading to Learn Mathematics, p. 751
- Enrichment, p. 752
- Assessment, p. 768

Science and Mathematics Lab Manual, pp. 57–62

Transparencies

5-Minute Check Transparency 12-9
Answer Key Transparencies

Technology

Interactive Chalkboard

Example 2 Find a Margin of Error

In a survey of 1000 randomly selected adults, 37% answered "yes" to a particular question. What is the margin of error?

$ME = 2\sqrt{\frac{p(1-p)}{n}}$ Formula for margin of sampling error

$= 2\sqrt{\frac{0.37(1-0.37)}{1000}}$ p = 37% or 0.37, n = 1000

≈ 0.030535 Use a calculator.

The margin of error is about 3%. This means that there is a 95% chance that the percent of people in the whole population who would answer "yes" is between 37 − 3 or 34% and 37 + 3 or 40%.

Published survey results often include the margin of error for the data. You can use this information to determinine the sample size.

Example 3 Analyze a Margin of Error

HEALTH **In a recent Gallup Poll, 25% of the people surveyed said they had smoked cigarettes in the past week. The margin of error was 3%.**

a. What does the 3% indicate about the results?

The 3% means that the probability is 95% that the percent of people in the population who had smoked cigarettes in the past week was between 25 − 3 or 22% and 25 + 3 or 28%.

b. How many people were surveyed?

$ME = 2\sqrt{\frac{p(1-p)}{n}}$ Formula for margin of sampling error

$0.03 = 2\sqrt{\frac{0.25(1-0.25)}{n}}$ ME = 0.03, p = 0.25

$0.015 = \sqrt{\frac{0.25(0.75)}{n}}$ Divide each side by 2.

$0.000225 = \frac{0.25(0.75)}{n}$ Square each side.

$n = \frac{0.25(0.75)}{0.000225}$ Multiply by n and divide by 0.000225.

$n \approx 833.33$ Use a calculator.

About 833 people were surveyed.

More About. . .

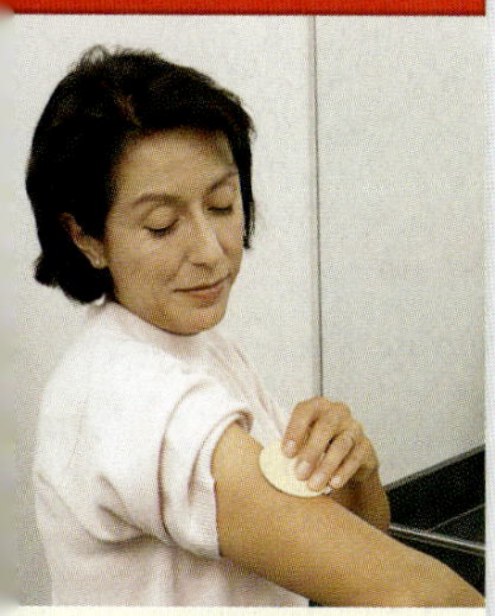

Health

The percent of smokers in the United States population declined from 38.7% in 1985 to 25.8% in 1999. New therapies, like the nicotine patch, are helping more people to quit.

Source: U.S. Department of Health and Human Services

Check for Understanding

Concept Check

1–3. See pp. 695A–695B.

1. **Describe** how sampling techniques can influence the results of a survey.
2. **OPEN ENDED** Give an example of a good sample and a bad sample. Explain your reasoning.
3. **Explain** what happens to the margin of sampling error when the size of the sample n increases. Why does this happen?

www.algebra2.com/extra_examples

2 Teach

BIAS

In-Class Example Power Point®

1 State whether each method would produce a random sample. Explain.

a. surveying people going into an action movie to find out the most popular kind of movie **No; they will most likely think action movies are the most popular kind of movie.**

b. calling every 10th person on the list of subscribers to a newspaper to ask about the quality of the delivery service **Yes; no obvious bias exists in calling every 10th subscriber.**

MARGIN OF ERROR

In-Class Examples Power Point®

2 In a survey of 100 randomly selected adults, 37% answered "yes" to a particular question. What is the margin of error? **0.09656 or about 10%**

3 **HEALTH** In an earlier survey, 30% of the people surveyed said they had smoked cigarettes in the past week. The margin of error was 2%.

a. What does the 2% indicate about the results? **There is a 95% chance that the percent of people in the population who had smoked cigarettes in the past week was between 28% and 32%.**

b. How many people were surveyed? **2100**

DAILY INTERVENTION Differentiated Instruction ELL

Verbal/Linguistic Have students in small groups design a survey question and practice asking it in such a way that there is bias built into the tone of voice and facial expression of the questioner. Then have them try out the question on other groups to see if they get a high percentage of the answer that the bias is designed to elicit.

3 Practice/Apply

Study Notebook

Have students—
- complete the definitions/examples for the remaining terms on their Vocabulary Builder worksheets for Chapter 12.
- include any other item(s) that they find helpful in mastering the skills in this lesson.

About the Exercises...

Organization by Objective
- **Bias:** 11–14
- **Margin of Error:** 15–28

Odd/Even Assignments
Exercises 11–24 are structured so that students practice the same concepts whether they are assigned odd or even problems.

Assignment Guide

Basic: 11–25 odd, 28–38
Average: 11–27 odd, 28–38
Advanced: 12–26 even, 28–38

4 Assess

Open-Ended Assessment

Speaking Ask students to explain why a larger sample will result in a lower margin of error, if the percent stays the same.

Assessment Options

Quiz (Lessons 12-8 and 12-9) is available on p. 768 of the *Chapter 12 Resource Masters.*

Guided Practice

GUIDED PRACTICE KEY	
Exercises	Examples
4, 5	1
6–8	2
9, 10	3

Determine whether each situation would produce a random sample. Write *yes* or *no* and explain your answer.

4. the government sending a tax survey to everyone whose social security number ends in a particular digit **Yes; last digits of social security numbers are random.**
5. surveying students in the honors chemistry classes to determine the average time students in your school study each week **No; these students probably study more than average.**

For Exercises 6–8, find the margin of sampling error to the nearest percent.

6. $p = 72\%$, $n = 100$ **about 9%**
7. $p = 31\%$, $n = 500$ **about 4%**
8. In a survey of 520 randomly-selected high school students, 68% of those surveyed stated that they were involved in extracurricular activities at their school. **about 4%**

Application

MEDIA **For Exercises 9 and 10, use the following information.**
According to a survey in *American Demographics,* 77% of Americans age 12 or older said they listen to the radio every day. Suppose the survey had a margin of error of 5%.

9. What does the 5% indicate about the results? **See margin.**
10. How many people were surveyed? **about 283**

Practice and Apply

Homework Help	
For Exercises	See Examples
11–14	1
15–26	2
27–28	3

Extra Practice
See page 856.

Determine whether each situation would produce a random sample. Write *yes* or *no* and explain your answer. **11–14. See margin for explanations.**

11. pointing with your pencil at a class list with your eyes closed as a way to find a sample of students in your class **no**
12. putting the names of all seniors in a hat, then drawing names from the hat to select a sample of seniors **yes**
13. calling every twentieth person listed in the telephone book to determine which political candidate is favored **yes**
14. finding the heights of all the boys in a freshman physical education class to determine the average height of all the boys in your school **no**

For Exercises 15–24, find the margin of sampling error to the nearest percent.

15. $p = 81\%$, $n = 100$
16. $p = 16\%$, $n = 400$
17. $p = 54\%$, $n = 500$
18. $p = 48\%$, $n = 1000$ **about 3%**
19. $p = 33\%$, $n = 1000$ **about 3%**
20. $p = 67\%$, $n = 1500$ **about 2%**
21. A poll asked people to name the most serious problem facing the country. Forty-six percent of the 800 randomly selected people said crime. **about 4%**
22. Although skim milk has as much calcium as whole milk, only 33% of 2406 adults surveyed in *Shape* magazine said skim milk is a good calcium source.
23. Three hundred sixty-seven of 425 high school students said pizza was their favorite food in the school cafeteria. **about 3%**
24. Nine hundred thirty-four of 2150 subscribers to a particular newspaper said their favorite sport was football. **about 2%**
25. **ECONOMICS** In a poll conducted by ABC News, 83% of the 1020 people surveyed said they supported raising the minimum wage. What was the margin of error? **about 2%**

15. about 8%
16. about 4%
17. about 4%
22. about 2%

Answers

9. The probability is 0.95 that the percent of Americans ages 12 and older who listen to the radio every day is between 72% and 82%.
11. You would tend to point toward the middle of the page.
12. All seniors would have the same chance of being selected.
13. A wide variety of people would be called since almost everyone has a phone.
14. Freshmen are more likely than older students to be still growing, so a sample of freshmen would not give representative heights for the whole school.

Career Choices

Physician

Physicians diagnose illnesses and prescribe and administer treatment.

Online Research For information about a career as a physician, visit: www.algebra2.com/careers

26. **PHYSICIANS** In a recent Harris Poll, 61% of the 1010 people surveyed said they considered being a physician to be a very prestigious occupation. What was the margin of error? **about 3%**

27. **SHOPPING** According to a Gallup Poll, 33% of shoppers planned to spend \$1000 or more during a recent holiday season. The margin of error was 3%. How many people were surveyed? **about 983**

28. **CRITICAL THINKING** One hundred people were asked a yes-or-no question in an opinion poll. How many said "yes" if the margin of error was 9.6%? **36 or 64**

29. **WRITING IN MATH** Answer the question that was posed at the beginning of the lesson. **See pp. 695A–695B.**

How are opinion polls used in political campaigns?

Include the following in your answer:

- a description of how a candidate could use statistics from opinion polls to determine where to make campaign stops,
- the margin of error for Bush if 807 people were surveyed, and
- an explanation of how to use the margin of error to determine the range of percent of Florida voters who favored Bush.

Standardized Test Practice

30. In rectangle $ABCD$, what is $x + y$ in terms of z? **A**

Ⓐ $90 + z$ Ⓑ $190 - z$

Ⓒ $180 + z$ Ⓓ $270 - z$

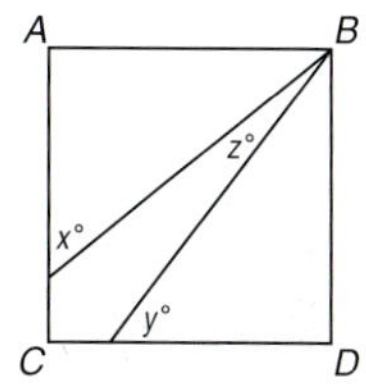

31. If $xy^{-2} + y^{-1} = y^{-2}$, then the value of x *cannot* equal which of the following? **C**

Ⓐ -1 Ⓑ 0 Ⓒ 1 Ⓓ 2

Maintain Your Skills

Mixed Review **A student guesses at all 5 questions on a true-false quiz. Find each probability.** *(Lesson 12-8)*

32. P(all 5 correct) $\frac{1}{32}$ 33. P(exactly 4 correct) $\frac{5}{32}$ 34. P(at least 3 correct) $\frac{1}{2}$

A set of 250 data values is normally distributed with a mean of 50 and a standard deviation of 5.5. *(Lesson 12-7)*

35. What percent of the data lies between 39 and 61? **95%**

36. How many data values are less than 55.5? **210**

37. What is the probability that a data value selected at random is greater than 39?

37. 97.5%

38. Given $x^3 - 3x^2 - 4x + 12$ and one of its factors $x + 2$, find the remaining factors of the polynomial. *(Lesson 7-4)* $x - 2, x - 3$

WebQuest Internet Project

'Minesweeper': Secret to Age-Old Puzzle?

It is time to complete your project. Use the information and data you have gathered about the history of mathematics to prepare a presentation or web page. Be sure to include transparencies and a sample mathematics problem or idea in the presentation.

www.algebra2.com/webquest

www.algebra2.com/extra_examples

Study Guide and Intervention, p. 747 (shown) and p. 748

Bias A sample of size n is random (or **unbiased**) when every possible sample of size n has an equal chance of being selected. If a sample is biased, then information obtained from it may not be reliable.

Example **To find out how people in the U.S. feel about mass transit, people at a commuter train station are asked their opinion. Does this situation represent a random sample?**

No; the sample includes only people who actually use a mass-transit facility. The sample does not include people who ride bikes, drive cars, or walk.

Exercises

Determine whether each situation would produce a random sample. Write *yes* or *no* and explain your answer.

1. asking people in Phoenix, Arizona, about rainfall to determine the average rainfall for the United States **No; it rains less in Phoenix than most places in the U.S.**
2. obtaining the names of tree types in North America by surveying all of the U.S. National Forests **Yes; there are National Forests in about every state in the U.S.**
3. surveying every tenth person who enters the mall to find out about music preferences in that part of the country **Yes; mall customers should be fairly representative in terms of music tastes.**
4. interviewing country club members to determine the average number of televisions per household in the community **No; country club members would tend to be more affluent and thus not a representative sample of the community.**
5. surveying all students whose ID numbers end in 4 about their grades and career counseling needs **Yes; ID numbers are probably assigned alphabetically or by some other method not connected to students' grades or counseling needs.**
6. surveying parents at a day care facility about their preferences for brands of baby food for a marketing campaign **Yes; choice of a daycare facility would probably not influence baby food preferences.**
7. asking people in a library about the number of magazines to which they subscribe in order to describe the reading habits of a town **No; library visitors tend to read more than most citizens.**

Skills Practice, p. 749 and Practice, p. 750 (shown)

Determine whether each situation would produce a random sample. Write *yes* or *no* and explain your answer.

1. calling every twentieth registered voter to determine whether people own or rent their homes in your community **No; registered voters may be more likely to be homeowners, causing the survey to underrepresent renters.**
2. predicting local election results by polling people in every twentieth residence in all the different neighborhoods of your community **Yes; since all neighborhoods are represented proportionally, the views of the community should as a whole should be well represented.**
3. to find out why not many students are using the library, a school's librarian gives a questionnaire to every tenth student entering the library **No; she is polling only the students who are coming to the library, and will obtain no input from those who aren't using the library.**
4. testing overall performance of tires on interstate highways only **No; for overall performance, tires should be tested on many kinds of surfaces, and under many types of conditions.**
5. selecting every 50th hamburger from a fast-food restaurant chain and determining its fat content to assess the fat content of hamburgers served in fast-food restaurant chains throughout the country **No; the selected hamburgers are a random sample of the hamburgers served in one chain, and may represent the fat content for that chain, but will not necessarily represent the fat content of hamburgers served in other fast-food restaurant chains.**
6. assigning all shift workers in a manufacturing plant a unique identification number, and then placing the numbers in a hat and drawing 30 at random to determine the annual average salary of the workers **Yes; because the numbers are randomly chosen from among all shift workers, all workers have the same chance of being selected.**

Find the margin of sampling error to the nearest percent.

7. $p = 26\%, n = 100$ **about 9%** 8. $p = 55\%, n = 100$ **about 10%** 9. $p = 75\%, n = 500$ **about 4%**

10. $p = 14\%, n = 500$ **about 3%** 11. $p = 96\%, n = 1000$ **about 1%** 12. $p = 21\%, n = 1000$ **about 3%**

13. $p = 34\%, n = 1000$ **about 3%** 14. $p = 49\%, n = 1500$ **about 3%** 15. $p = 65\%, n = 1500$ **about 2%**

16. **COMPUTING** According to a poll of 500 teenagers, 43% said that they use a personal computer at home. What is the margin of sampling error? **about 4%**
17. **TRUST** A survey of 605 people, ages 13–33, shows that 68% trust their parents more than their best friends to tell them the truth. What is the margin of sampling error? **about 4%**
18. **PRODUCTIVITY** A study by the University of Illinois in 1995 showed an increase in productivity by 10% of the employees who wore headsets and listened to music of their choice while they were working. The margin of sampling error for the study was about 7%. How many employees participated in the study? **about 76**

Reading to Learn Mathematics, p. 751 ELL

Pre-Activity **How are opinion polls used in political campaigns?**

Read the introduction to Lesson 12-9 at the top of page 682 in your textbook.

Do you think the results of the survey about the presidential preferences demonstrates that Bush was actually ahead in Florida a month before the election? If there is not enough information given to determine this, list at least two questions you would ask about the survey that would help you determine the significance of the survey. **Sample answer: There is not enough information to tell. 1. How many people were surveyed? 2. How was the sample for the survey selected? 3. What is the margin of error for this survey?**

Reading the Lesson

1. Determine whether each situation would produce a random sample. Write *yes* or *no* and explain your answer.
   a. asking all the customers at five restaurants on the same evening how many times a month they eat dinner in restaurants to determine how often the average American eats dinner in a restaurants **No; people surveyed at a restaurant might be likely to eat dinner in restaurants more often than other people.**
   b. putting the names of all seniors at your high school in a hat and then drawing 20 names for a survey to find out where seniors would like to hold their prom **Yes; every senior would have an equal chance of being chosen for the survey.**
2. A survey determined that 58% of registered voters in the United States support increased federal spending for education. The margin of error for this survey is 4%. Explain in your own words what this tells you about the actual percentage of registered voters who support increased spending for education. **Sample answer: There is a 95% chance that the actual percentage of voters supporting increased federal spending for education is between 54% and 62%.**

Helping You Remember

3. The formula for margin of sampling error may be tricky to remember. A good way to start is to think about the variables that must be included in the formula. What are these variables, and what do they represent? What is an easy way to remember which variable goes in the denominator in the formula? **Sample answer: *p* is the probability of a certain response and *n* is the sample size. The larger the sample size, the smaller the margin of error, so *n* must go in the denominator since dividing by a larger number gives a smaller number. The square root of a smaller number is a smaller number, and twice the square root of a smaller number is a smaller number.**

Enrichment, p. 752

Shapes of Distribution Curves

Graphs of frequency distributions can be described as either symmetric or skewed.

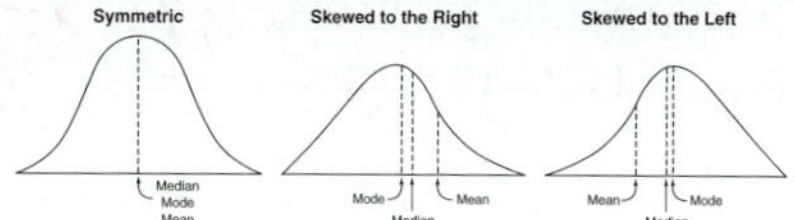

In a distribution skewed to the right, there are a larger number of high values. The long "tail" extends to the right.

In a distribution skewed to the left, there are a larger number of low values. The "tail" extends to the left.

Algebra Activity

A Follow-Up of Lesson 12-9

Getting Started

Objective State hypotheses for conjectures and design an experiment to test a hypothesis.

Materials
ruler
stopwatch

Teach

- Ask students why the tested hypothesis is called the null hypothesis. **Because it is often stated in the form "there is no (or null) difference."**
- Make sure students know how to use the stopwatches before beginning the experiment.
- Have students complete the simulation to collect data and then complete Exercises 1–4.

Assess

In **Exercises 1–3**, students should
- state the null hypothesis saying that there is no difference.
- state the alternative hypothesis saying that there is a difference.

In **Exercise 4**, students should
- design an experiment that they could carry out.
- restate the hypothesis so that it is in the form of a null hypothesis.

Algebra Activity

A Follow-Up of Lesson 12-9

Testing Hypotheses

A **hypothesis** is a statement to be tested. Testing a hypothesis to determine whether it is supported by the data involves five steps.

Step 1 State the hypothesis. The statement should include a *null hypothesis*, which is the hypothesis to be tested, and an *alternative hypothesis*.

Step 2 Design the experiment.

Step 3 Conduct the experiment and collect the data.

Step 4 Evaluate the data. Decide whether to reject the null hypothesis.

Step 5 Summarize the results.

Test the following hypothesis.

People react to sound and touch at the same rate.

You can measure reaction time by having someone drop a ruler and then having someone else catch it between their fingers. The distance the ruler falls will depend on their reaction time. Half of the class will investigate the time it takes to react when someone is told the ruler has dropped. The other half will measure the time it takes to react when the catcher is alerted by touch.

Step 1 The null hypothesis H_0 and alternative hypothesis H_1 are as follows.
These statements often use $=, \neq, <, >, \geq$, and $\leq$.
- H_0: reaction time to sound = reaction time to touch
- H_1: reaction time to sound $\neq$ reaction time to touch

Step 2 You will need to decide the height from which the ruler is dropped, the position of the person catching the ruler, the number of practice runs, and whether to use one try or the average of several tries.

Step 3 Conduct the experiment in each group and record the results.

Step 4 Organize the results so that they can be compared.

Step 5 Based on the results, do you think the hypothesis is true? Explain.

Analyze

State the null and alternative hypotheses for each conjecture. **1–3. See pp. 695A–695B.**

1. A teacher feels that playing classical music during a math test will cause the test scores to change (either up or down). In the past, the average test score was 73.
2. An engineer thinks that the mean number of defects can be decreased by using robots on an assembly line. Currently, there are 18 defects for every 1000 items.
3. A researcher is concerned that a new medicine will cause pulse rates to rise dangerously. The mean pulse rate for the population is 82 beats per minute.
4. **MAKE A CONJECTURE** Design and conduct an experiment to test the following hypothesis. Interpret the data and present your results. *Pulse rates increase 20% after moderate exercise.* **See students' work.**

Resource Manager

Teaching Algebra with Manipulatives
- p. 24 (master for rulers)
- p. 296 (student recording sheet)

Glencoe Mathematics Classroom Manipulative Kit
- rulers
- stopwatches

Study Notebook

You may wish to have students summarize this activity and what they learned from it.

Chapter 12 Study Guide and Review

Vocabulary and Concept Check

- area diagram (p. 651)
- binomial experiment (p. 677)
- combination (p. 640)
- compound event (p. 658)
- continuous probability distribution (p. 671)
- dependent events (p. 633)
- discrete probability distributions (p. 671)
- event (p. 632)
- failure (p. 644)
- Fundamental Counting Principle (p. 633)
- inclusive events (p. 659)
- independent events (p. 632)
- linear permutation (p. 638)
- margin of sampling error (p. 682)
- measure of central tendency (p. 664)
- measure of variation (p. 665)
- mutually exclusive events (p. 658)
- normal distribution (p. 671)
- odds (p. 645)
- outcome (p. 632)
- permutation (p. 638)
- probability (p. 644)
- probability distribution (p. 646)
- random (p. 645)
- random variable (p. 646)
- relative-frequency histogram (p. 646)
- sample space (p. 632)
- simple event (p. 658)
- skewed distribution (p. 671)
- standard deviation (p. 665)
- success (p. 644)
- unbiased sample (p. 682)
- uniform distribution (p. 646)
- univariate data (p. 664)
- variance (p. 665)

Choose the letter of the term that best matches each statement or phrase.

1. the ratio of the number of ways an event can succeed to the number of possible outcomes c
2. an arrangement of objects in which order does not matter b
3. two or more events in which the outcome of one event affects the outcome of another event a
4. a sample in which every member of the population has an equal chance to be selected g
5. an arrangement of objects in which order matters d
6. two events in which the outcome can never be the same e
7. the ratio of the number of ways an event can succeed to the number of ways it can fail f

a. dependent events
b. combination
c. probability
d. permutation
e. mutually exclusive events
f. odds
g. unbiased sample

Lesson-by-Lesson Review

12-1 *The Counting Principle*

See pages 632–637.

Concept Summary

- Fundamental Counting Principle: If event M can occur in m ways and event N can occur in n ways, then event M followed by event N can occur in $m \cdot n$ ways.
- Independent Events: The outcome of one event does *not* affect the outcome of another.
- Dependent Events: The outcome of one event *does* affect the outcome of another.

Example

How many different license plates are possible with two letters followed by three digits?

There are 26 possibilities for each letter. There are 10 possibilities, the digits 0–9, for each number. Thus, the number of possible license plates is as follows.

$26 \cdot 26 \cdot 10 \cdot 10 \cdot 10 = 26^2 \cdot 10^3$ or 676,000

www.algebra2.com/vocabulary_review

FOLDABLES™ Study Organizer

For more information about Foldables, see *Teaching Mathematics with Foldables.*

Remind students to review the Foldable and make sure that the lists of terms, concepts, and examples are complete. Have student volunteers share some of the printed examples of statistics that they found. Ask them to check over their notes and examples about probability and statistics to see if they wish to add any further information about the uses and misuses of statistics in the world around them.

Encourage students to refer to their Foldables while completing the Study Guide and Review and to use them in preparing for the Chapter Test.

Chapter 12 Study Guide and Review

Vocabulary and Concept Check

- This alphabetical list of vocabulary terms in Chapter 12 includes a page reference where each term was introduced.
- **Assessment** A vocabulary test/review for Chapter 12 is available on p. 766 of the *Chapter 12 Resource Masters.*

Lesson-by-Lesson Review

For each lesson,

- the main ideas are summarized,
- additional examples review concepts, and
- practice exercises are provided.

Vocabulary PuzzleMaker

ELL The Vocabulary PuzzleMaker software improves students' mathematics vocabulary using four puzzle formats—crossword, scramble, word search using a word list, and word search using clues. Students can work on a computer screen or from a printed handout.

MindJogger Videoquizzes

ELL MindJogger Videoquizzes provide an alternative review of concepts presented in this chapter. Students work in teams in a game show format to gain points for correct answers. The questions are presented in three rounds.

Round 1 Concepts (5 questions)
Round 2 Skills (4 questions)
Round 3 Problem Solving (4 questions)

Exercises **Solve each problem.** *See Examples 2 and 3 on page 633.*

8. The letters a, c, e, g, i, and k are used to form 6-letter passwords for a movie theater security system. How many passwords can be formed if the letters can be used more than once in any given password? **46,656 passwords**
9. How many 4-digit personal identification codes can be formed if each numeral can only be used once? **5040 codes**

12-2 Permutations and Combinations

See pages 638–643.

Concept Summary

- In a permutation, the order of objects is important.
- In a combination, the order of objects is not important.

Example

A basket contains 3 apples, 6 oranges, 7 pears, and 9 peaches. How many ways can 1 apple, 2 oranges, 6 pears, and 2 peaches be selected?

This involves the product of four combinations, one for each type of fruit.

$$C(3, 1) \cdot C(6, 2) \cdot C(7, 6) \cdot C(9, 2) = \frac{3!}{(3-1)!1!} \cdot \frac{6!}{(6-2)!2!} \cdot \frac{7!}{(7-6)!6!} \cdot \frac{9!}{(9-2)!2!}$$

$$= 3 \cdot 15 \cdot 7 \cdot 36 \text{ or } 11{,}340$$

There are 11,340 different ways to choose the fruit from the basket.

Exercises **Solve each problem.** *See Example 4 on page 640.*

10. A committee of 3 is selected from Jillian, Miles, Mark, and Nikia. How many committees contain 2 boys and 1 girl? **2**
11. Five cards are drawn from a standard deck of cards. How many different hands consist of four queens and one king? **4**
12. A box of pencils contains 4 red, 2 white, and 3 blue pencils. How many different ways can 2 red, 1 white, and 1 blue pencil be selected? **36**

12-3 Probability

See pages 644–650.

Concept Summary

- $P(\text{success}) = \frac{s}{s+f}$; $P(\text{failure}) = \frac{f}{s+f}$
- odds of success $= s:f$; odds of failure $= f:s$

Example

A bag of golf tees contains 23 red, 19 blue, 16 yellow, 21 green, 11 orange, 19 white, and 17 black tees. What is the probability that if you choose a tee from the bag at random, you will choose a green tee?

There are 21 ways to choose a green tee and 23 + 19 + 16 + 11 + 19 + 17 or 105 ways not to choose a green tee. So, s is 21 and f is 105.

$$P(\text{green tee}) = \frac{s}{s+f}$$

$$= \frac{21}{21+105} \text{ or } \frac{1}{6}$$ The probability is 1 out of 6 or about 16.7%.

688 Chapter 12 Probability and Statistics

Exercises Find the odds of an event occurring, given the probability of the event.
See Example 3 on pages 645 and 646.

13. $\frac{1}{4}$ 1:3 **14.** $\frac{5}{8}$ 5:3 **15.** $\frac{7}{12}$ 7:5 **16.** $\frac{3}{7}$ 3:4 **17.** $\frac{2}{5}$ 2:3

18. The table shows the distribution of the number of heads occurring when four coins are tossed. Find $P(H = 3)$. $\frac{1}{4}$
See Example 4 on page 646.

H = Heads	0	1	2	3	4
Probability	$\frac{1}{16}$	$\frac{1}{4}$	$\frac{3}{8}$	$\frac{1}{4}$	$\frac{1}{16}$

12-4 Multiplying Probabilities

See pages 651–657.

Concept Summary

- Probability of two independent events: $P(A \text{ and } B) = P(A) \cdot P(B)$
- Probability of two dependent events: $P(A \text{ and } B) = P(A) \cdot P(B \text{ following } A)$

Example

There are 3 dimes, 2 quarters, and 5 nickels in Langston's pocket. If he reaches in and selects three coins at random without replacing any of them, what is the probability that he will choose a dime d, then a quarter q, then a nickel n?

Because the outcomes of the first and second choices affect the later choices, these are dependent events.

$P(d, \text{ then } q, \text{ then } n) = \frac{3}{10} \cdot \frac{2}{9} \cdot \frac{5}{8}$ or $\frac{1}{24}$ The probability is $\frac{1}{24}$ or about 4.2%.

Exercises Determine whether the events are *independent* or *dependent*. Then find the probability. *See Examples 1–4 on pages 652 and 654.* 19. independent; $\frac{1}{36}$

19. Two dice are rolled. What is the probability that each die shows a 4?

20. dependent; $\frac{13}{204}$ **20.** Two cards are drawn from a standard deck of cards without replacement. Find the probability of drawing a heart and a club, in that order.

21. Luz has 2 red, 2 white, and 3 blue marbles in a cup. If she draws two marbles at random and does not replace the first one, find the probability of a white marble and then a blue marble. dependent; $\frac{1}{7}$

12-5 Adding Probabilities

See pages 658–663.

Concept Summary

- Probability of mutually exclusive events: $P(A \text{ or } B) = P(A) + P(B)$
- Probability of inclusive events: $P(A \text{ or } B) = P(A) + P(B) - P(A \text{ and } B)$

Example

Trish has four \$1 bills and six \$5 bills. She takes three bills from her wallet at random. What is the probability that Trish will select at least two \$1 bills?

$$P(\text{at least two \$1 bills}) = P(\text{two \$1, one \$5}) + P(\text{three \$1, no \$5})$$

$$= \frac{C(4, 2) \cdot C(6, 1)}{C(10, 3)} + \frac{C(4, 3) \cdot C(6, 0)}{C(10, 3)}$$

$$= \frac{\frac{4! \cdot 6!}{(4-2)!2!(6-1)!1!}}{C(10, 3)} + \frac{\frac{4! \cdot 6!}{(4-3)!3!(6-0)!0!}}{C(10, 3)}$$

$$= \frac{36}{120} + \frac{4}{120} \text{ or } \frac{1}{3}$$ The probability is $\frac{1}{3}$ or about 0.333.

Chapter 12 Study Guide and Review 689

Exercises **Determine whether the events are *mutually exclusive* or *inclusive*. Then find the probability.** *See Examples 1–3 on pages 659 and 660.*

22. There are 5 English, 2 math, and 3 chemistry books on a shelf. If a book is randomly selected, what is the probability of selecting a math book or a chemistry book? **mutually exclusive; $\frac{1}{2}$** **23. mutually exclusive; $\frac{2}{3}$**
23. A die is rolled. What is the probability of rolling a 6 or a number less than 4?
24. A die is rolled. What is the probability of rolling a 6 or a number greater than 4?
25. A card is drawn from a standard deck of cards. What is the probability of drawing a king or a red card? **inclusive; $\frac{7}{13}$** **24. inclusive; $\frac{1}{3}$**

12-6 Statistical Measures

See pages 664–670.

Concept Summary

- To represent a set of data, use the mean if the data are spread out and you want an average of the values, the median when the data contain outliers, or the mode when the data are tightly clustered around one or two values.
- Standard deviation for n values:

$$\sigma = \sqrt{\frac{(x_1 - \overline{x})^2 + (x_2 - \overline{x})^2 + \cdots + (x_n - \overline{x})^2}{n}}, \ \overline{x} \text{ is the mean}$$

Example **Find the variance and standard deviation for {100, 156, 158, 159, 162, 165, 170, 190}.**

Step 1 Find the mean.

$$\overline{x} = \frac{100 + 156 + 158 + 159 + 162 + 165 + 170 + 190}{8}$$ Add the data and divide by the number of items.

$$= \frac{1260}{8}$$

$$= 157.5$$

Step 2 Find the variance.

$$\sigma^2 = \frac{(x_1 - \overline{x})^2 + (x_2 - \overline{x})^2 + \cdots + (x_n - \overline{x})^2}{n}$$

$$= \frac{(100 - 157.5)^2 + (156 - 157.5)^2 + \cdots + (170 - 157.5)^2 + (190 - 157.5)^2}{8}$$

$$= \frac{4600}{8}$$ Simplify.

$$= 575$$ Use a calculator.

Step 3 Find the standard deviation.

$\sigma^2 = 575$ Take the square root of each side.

$\sigma \approx 23.98$ Use a calculator.

Exercises **Find the variance and standard deviation of each set of data to the nearest tenth.** *See Examples 1 and 2 on pages 664 and 665.*

26. {56, 56, 57, 58, 58, 58, 59, 61} **2.4, 1.5**
27. {302, 310, 331, 298, 348, 305, 314, 284, 321, 337} **341.0, 18.5**
28. {3.4, 4.2, 8.6, 5.1, 3.6, 2.8, 7.1, 4.4, 5.2, 5.6} **2.8, 1.7**

690 Chapter 12 Probability and Statistics

Study Guide and Review

12-7 The Normal Distribution

See pages 671–675.

Concept Summary

Normal distributions have these properties.

- The graph is maximized and the data are symmetric at the mean.
- The mean, median, and mode are about equal.
- About 68% of the values are within one standard deviation of the mean.
- About 95% of the values are within two standard deviations of the mean.
- About 99% of the values are within three standard deviations of the mean.

Example **Mr. Byrum gave an exam to his 30 Algebra 2 students at the end of the first semester. The scores were normally distributed with a mean score of 78 and a standard deviation of 6.**

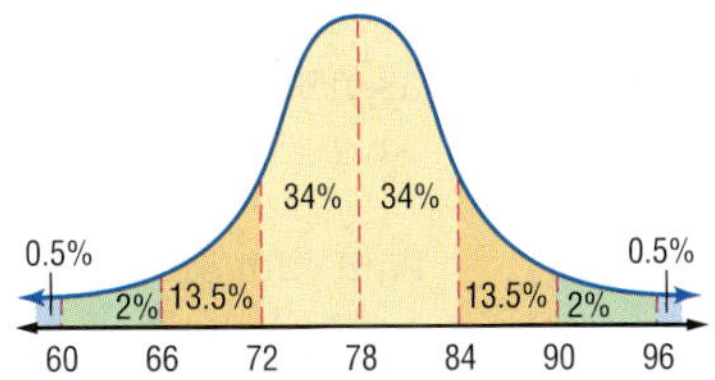

a. What percent of the class would you expect to have scored between 72 and 84?

Since 72 and 84 are 1 standard deviation to the left and right of the mean, respectively, 34% + 34% or 68% of the students scored within this range.

b. What percent of the class would you expect to have scored between 90 and 96?

90 to 96 on the test includes 2% of the students.

c. Approximately how many students scored between 84 and 90?

84 to 90 on the test includes 13.5% of the students. $0.135 \times 30 = 4$ students

d. Approximately how many students scored between 72 and 84?

34% + 34% or 68% of the students scored between 72 and 84.

$0.68 \times 30 = 20$ students

Exercises **For Exercises 29–32, use the following information.**

The utility bills in a city of 5000 households are normally distributed with a mean of \$180 and a standard deviation of \$16. *See Example 2 on pages 672 and 673.*

29. About how many utility bills were between \$164 and \$196? **3400**

30. About how many bills were more than \$212? **125**

31. About how many bills were less than \$164? **800**

32. What is the probability that a household selected at random will have a utility bill between \$164 and \$180? **34%**

12-8 Binomial Experiments

See pages 676–680.

Concept Summary

A binomial experiment exists if and only if all of these conditions occur.

- There are exactly two possible outcomes for each trial.
- There is a fixed number of trials.
- The trials are independent.
- The possibilities for each trial are the same.

Chapter 12 Study Guide and Review 691

Chapter 12 For More ...
• Extra Practice, see pages 854–856.
• Mixed Problem Solving, see page 837.

Example **To practice for a jigsaw puzzle competition, Laura and Julian completed four jigsaw puzzles. The probability that Laura places the last piece is $\frac{3}{5}$, and the probability that Julian places the last piece is $\frac{2}{5}$. What is the probability that Laura will place the last piece of at least two puzzles?**

$P = L^4 + 4L^3J + 6L^2J^2$ — P(last piece in 4) + P(last piece in 3) + P(last piece in 2)

$= \left(\frac{3}{5}\right)^4 + 4\left(\frac{3}{5}\right)^3\left(\frac{2}{5}\right) + 6\left(\frac{3}{5}\right)^2\left(\frac{2}{5}\right)^2$ — $L = \frac{3}{5}, J = \frac{2}{5}$

$= \frac{81}{625} + \frac{216}{625} + \frac{216}{625}$ or 0.8208 — The probability is 82.08%.

Exercises *See Example 2 on pages 677 and 678.*

33. Find the probability of getting 7 heads in 8 tosses of a coin. $\frac{1}{32}$
34. Find the probability that a family with seven children has exactly five boys. $\frac{21}{128}$

Find each probability if a die is rolled twelve times.

35. P(twelve 3s) $\frac{1}{2,176,782,336}$
36. P(exactly one 3) $\frac{48,828,125}{181,398,528}$
37. P(six 3s) $\frac{14,437,500}{2,176,782,336}$

12-9 Sampling and Error

See pages 682–685.

Concept Summary

- Margin of sampling error: $ME = 2\sqrt{\frac{p(1-p)}{n}}$ if the percent of people in a sample responding in a certain way is p and the size of the sample is n

Example **In a survey taken at a local high school, 75% of the student body stated that they thought school lunches should be free. This survey had a margin of error of 2%. How many people were surveyed?**

$ME = 2\sqrt{\frac{p(1-p)}{n}}$ — Formula for margin of sampling error

$0.02 = 2\sqrt{\frac{0.75(1-0.75)}{n}}$ — $ME = 0.02$, $p = 0.75$

$0.01 = \sqrt{\frac{0.75(1-0.75)}{n}}$ — Divide each side by 2.

$0.0001 = \frac{0.75(0.25)}{n}$ — Square each side of the equation.

$n = \frac{0.75(0.25)}{0.0001}$ — Multiply each side by n and divide each side by 0.0001.

$n = 1875$ — There were about 1875 people in the survey.

Exercises

38. In a poll asking people to name their most valued freedom, 51% of the randomly selected people said it was the freedom of speech. Find the margin of sampling error if 625 people were randomly selected. *See Example 2 on page 683.* **about 4%**
39. According to a recent survey of mothers with children who play sports, 63% of them would prefer that their children not play football. Suppose the margin of error is 4.5%. How many mothers were surveyed? *See Example 3 on page 683.* **460 mothers**

692 Chapter 12 Probability and Statistics

Chapter 12

Practice Test

Vocabulary and Concepts

Match the following terms and descriptions.

1. data are symmetric about the mean **c**
2. variance and standard deviation **b**
3. mode, median, mean **a**

a.	measures of central tendency
b.	measures of variation
c.	normal distribution

Skills and Applications

Evaluate each expression.

4. $P(7, 3)$ **210**
5. $C(7, 3)$ **35**
6. $P(13, 5)$ **154,440**

Solve each problem. **10. 31,824**

7. How many ways can 9 bowling balls be arranged on the upper rack of a bowling ball rack? **362,880 ways**
8. How many different outfits can be made if you choose 1 each from 11 skirts, 9 blouses, 3 belts, and 7 pairs of shoes? **2079 outfits**
9. How many ways can the letters of the word *probability* be arranged? **9,979,200 ways**
10. How many different soccer teams consisting of 11 players can be formed from 18 players?
11. In a row of 10 parking spaces in a parking lot, how many ways can 4 cars park? **5040 ways**
12. Eleven points are equally spaced on a circle. How many ways can 5 of these points be chosen as the vertices of a pentagon? **462 pentagons**
13. A number is drawn at random from a hat that contains all the numbers from 1 to 100. What is the probability that the number is less than sixteen? $\frac{3}{20}$
14. Two cards are drawn in succession from a standard deck of cards without replacement. What is the probability that both cards are greater than 2 and less than 9? $\frac{46}{221}$
15. A shipment of ten television sets contains 3 defective sets. How many ways can a hospital purchase 4 of these sets and receive at least 2 of the defective sets? **70 ways**
16. While shooting arrows, William Tell can hit an apple 9 out of 10 times. What is the probability that he will hit it exactly 4 out of 7 times? $\frac{45{,}927}{2{,}000{,}000}$
17. Ten people are going on a camping trip in 3 cars that hold 5, 2, and 4 passengers, respectively. How many ways is it possible to transport the 10 people to their campsite? **6930 ways**
18. From a box containing 5 white golf balls and 3 red golf balls, 3 golf balls are drawn in succession, each being replaced in the box before the next draw is made. What is the probability that all 3 golf balls are the same color? $\frac{19}{64}$

For Exercises 19–21, use the following information.
In a ten-question multiple-choice test with four choices for each question, a student who was not prepared guesses on each item. Find each probability.

19. six questions correct $\frac{8505}{524{,}288}$
20. at least eight questions correct $\frac{109}{262{,}144}$
21. fewer than eight questions correct $\frac{262{,}035}{262{,}144}$

22. **STANDARDIZED TEST PRACTICE** Lila throws a die and writes down the number showing. If she throws the number cube again, what is the probability that the second throw will have the same number showing as the first throw? **D**

Ⓐ $\frac{1}{2}$ Ⓑ $\frac{1}{3}$ Ⓒ $\frac{1}{4}$ Ⓓ $\frac{1}{6}$

www.algebra2.com/chapter_test

Portfolio Suggestion

Introduction The Fundamental Counting Principle, permutations and combinations, probability, and statistics may have been new topics for many students. These topics are used in many ways in almost all fields of employment.

Ask Students Which was your favorite word problem from this chapter? Put the problem in your portfolio and write a note that explains why it is your favorite. Add a brief conjecture about how you might be able to use the topics of this chapter in a future career that you might like to have.

Chapter 12 Practice Test

Assessment Options

Vocabulary Test A vocabulary test/review for Chapter 12 can be found on p. 766 of the *Chapter 12 Resource Masters*.

Chapter Tests There are six Chapter 12 Tests and an Open-Ended Assessment task available in the *Chapter 12 Resource Masters*.

Chapter 12 Tests			
Form	**Type**	**Level**	**Pages**
1	MC	basic	753–754
2A	MC	average	755–756
2B	MC	average	757–758
2C	FR	average	759–760
2D	FR	average	761–762
3	FR	advanced	763–764

MC = multiple-choice questions
FR = free-response questions

Open-Ended Assessment
Performance tasks for Chapter 12 can be found on p. 765 of the *Chapter 12 Resource Masters*. A sample scoring rubric for these tasks appears on p. A34.

Unit 4 Test A unit test/review can be found on pp. 773–774 of the *Chapter 12 Resource Masters*.

ExamView® Pro

Use the networkable **ExamView® Pro** to:

- Create **multiple versions** of tests.
- Create **modified** tests for *Inclusion* students.
- **Edit** existing questions and **add** your own questions.
- Use built-in **state curriculum correlations** to create tests aligned with state standards.
- **Apply** art to your tests from a program bank of artwork.

Chapter 12 Standardized Test Practice

These two pages contain practice questions in the various formats that can be found on the most frequently given standardized tests.

A practice answer sheet for these two pages can be found on p. A1 of the *Chapter 12 Resource Masters*.

Standardized Test Practice Student Recording Sheet, p. A1

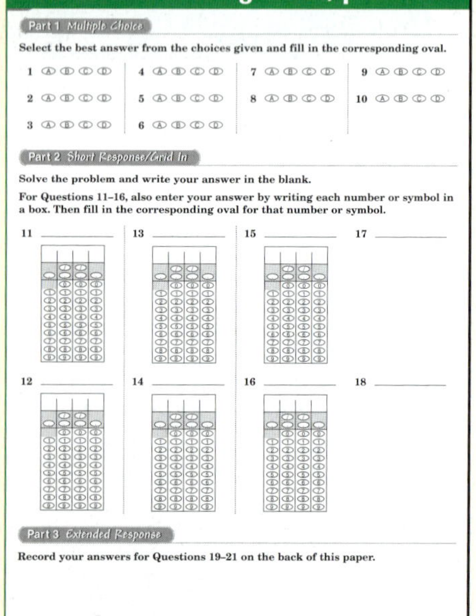
Part 1 Multiple Choice

Select the best answer from the choices given and fill in the corresponding oval.

1 Ⓐ Ⓑ Ⓒ Ⓓ 2 Ⓐ Ⓑ Ⓒ Ⓓ 3 Ⓐ Ⓑ Ⓒ Ⓓ 4 Ⓐ Ⓑ Ⓒ Ⓓ 5 Ⓐ Ⓑ Ⓒ Ⓓ 6 Ⓐ Ⓑ Ⓒ Ⓓ 7 Ⓐ Ⓑ Ⓒ Ⓓ 8 Ⓐ Ⓑ Ⓒ Ⓓ 9 Ⓐ Ⓑ Ⓒ Ⓓ 10 Ⓐ Ⓑ Ⓒ Ⓓ

Part 2 Short Response/Grid In

Solve the problem and write your answer in the blank.

For Questions 11–16, also enter your answer by writing each number or symbol in a box. Then fill in the corresponding oval for that number or symbol.

11 ____ 13 ____ 15 ____ 17 ____

12 ____ 14 ____ 16 ____ 18 ____

Part 3 Extended Response

Record your answers for Questions 19–21 on the back of this paper.

Teaching Tip In Question 8, students may want to write a list of the odd numbers in the set that are divisible by 3.

Additional Practice

See pp. 771–772 in the *Chapter 12 Resource Masters* for additional standardized test practice.

Chapter 12 Standardized Test Practice

Part 1 Multiple Choice

Record your answers on the answer sheet provided by your teacher or on a sheet of paper.

1. In a jar of red and green gumdrops, the ratio of red gumdrops to green gumdrops is 7 to 3. If the jar contains a total of 150 gumdrops, how many gumdrops are green? **C**

Ⓐ 21 Ⓑ 30

Ⓒ 45 Ⓓ 105

2. $\langle x \rangle = \frac{1}{2}x$ if x is composite and $\langle x \rangle = 2x$ if x is prime. What is the value of $\langle 16 \rangle + \langle 11 \rangle$? **B**

Ⓐ 10 Ⓑ 30

Ⓒ 54 Ⓓ 60

3. In rhombus *ABCD*, which of the following are true? **D**

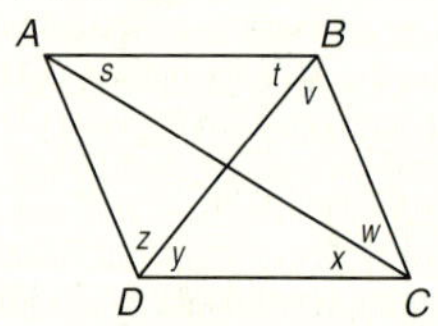

I. $\angle s$ and $\angle x$ are congruent.

II. $\angle t$ and $\angle v$ are congruent.

III. $\angle z$ and $\angle t$ are congruent.

Ⓐ I only

Ⓑ II only

Ⓒ I and II only

Ⓓ I, II, and III

4. What is the area of an isosceles right triangle with hypotenuse $3\sqrt{2}$ units? **B**

Ⓐ $1.5\sqrt{2}$ units^2

Ⓑ 4.5 units^2

Ⓒ 9 units^2

Ⓓ $6 + 3\sqrt{2}$ units^2

5. What is the solution set for $t(t + 7) = 18$? **D**

Ⓐ $\{-2, 9\}$

Ⓑ $\{-3, 6\}$

Ⓒ $\{0, 18\}$

Ⓓ $\{-9, 2\}$

6. The equation $3x - 8 = 5x^2 - y$ represents which of the following conic sections? **B**

Ⓐ hyperbola

Ⓑ parabola

Ⓒ circle

Ⓓ ellipse

7. If the equations $x^2 + y^2 = 16$ and $y = x^2 + 4$ are graphed on the same coordinate plane, how many points of intersection exist? **B**

Ⓐ none

Ⓑ one

Ⓒ two

Ⓓ three

8. A number is chosen at random from the set $\{1, 2, 3, \ldots 20\}$. What is the probability that the number is odd and divisible by 3? **A**

Ⓐ $\frac{3}{20}$ Ⓑ $\frac{3}{10}$

Ⓒ $\frac{7}{20}$ Ⓓ $\frac{13}{20}$

9. What is the least positive integer that is divisible by 3, 4, 5, and 6? **A**

Ⓐ 60 Ⓑ 180

Ⓒ 240 Ⓓ 360

10. If $4y - 5x + 6xy - 50 = 0$ and $x + 7 = 13$, then what is $y + 5$? **C**

Ⓐ 2 Ⓑ 6

Ⓒ 7 Ⓓ 11

ExamView® Pro

Special banks of standardized test questions similar to those on the SAT, ACT, TIMSS 12, NAEP 12, and state proficiency tests can be found on this CD-ROM.

Preparing for Standardized Tests
For test-taking strategies and more practice, see pages 877–892.

Part 2 Short Response/Grid In

Record your answers on the answer sheet provided by your teacher or on a sheet of paper.

11. In a high school, 250 students take math and 50 students take art. If there are 280 students enrolled in the school and they all take at least one of these courses, how many students take both math and art? **20**

12. If $20 < y < 30$ and x and y are both integers, what is the greatest possible value for x? **79**

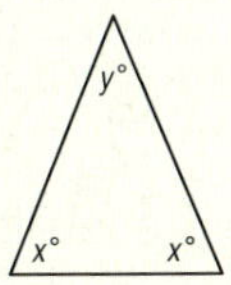

13. Four numbers are selected at random. Their average (arithmetic mean) is 45. The fourth number selected is 34. What is the sum of the other three numbers? **146**

14. If one half of an even positive integer and three fourths of the next greater even integer have a sum of 24, what is the mean of the two integers? **19**

15. Shane has six tiles, each of which has one of the letters A, B, C, D, E, or F on it. If one of the letters must be A and the last letter must be F, how many different arrangements of three letters (such as ADF) can Shane create with these titles? **8**

> **Test-Taking Tip** Ⓐ Ⓑ Ⓒ Ⓓ
>
> **Question 10**
> When answering questions, read carefully and make sure that you know exactly what the question is asking you to find. For example, if you only find the value of y in Question 10, you have not solved the problem. You need to find the value of $y + 5$.

www.algebra2.com/standardized_test

16. The lunch special at Wally's Hot Dog Stand offers a sandwich, snack, and a drink for $3.99. How many different lunch combinations can be ordered? **36**

> **Wally's lunch specials**
>
> **Sandwiches:** hotdog, a bratwurst, or Italian sausage
>
> **Snacks:** potato chips, pretzels, or corn chips
>
> **Drinks:** diet soda, regular cola, lemon-lime bottled soda, or bottled water

17. A gumball machine contains 84 gumballs. Of the gumballs, 19 are yellow, 32 are red, and 33 are green. Each gumball sold is selected at random. If a yellow gumball is sold, what is the probability that the next gumball is also yellow? $\frac{18}{83}$

18. Esteban is the place kicker on his high school football team. Based on past experience, the probability that Esteban will succeed in kicking an extra point is $\frac{9}{10}$. What is the probability that he will succeed on exactly three of the four extra point attempts in a game? $\frac{729}{2500}$ **or 0.2916**

Part 3 Extended Response

Record your answers on a sheet of paper. Show your work.

For Exercises 19-21, use the following information.

Twenty-five students took an English examination and received the following scores.

76, 62, 78, 69, 88, 82, 79, 70, 81, 87, 90, 71, 75, 78, 88, 83, 91, 93, 95, 78, 80, 82, 86, 89, 79

19. Find the mean, median, mode, and standard deviation of the data. If necessary, round to the nearest hundredth. **81.2, 81, 78, 7.93**

20. How many students scored within one standard deviation of the mean? **17**

21. Do the results of the examination approximate a normal distribution? Justify your answer. **See margin.**

Evaluating Extended Response Questions

Extended Response questions are graded by using a multilevel rubric that guides you in assessing a student's knowledge of a particular concept.

Goal: Determine the dimensions of a rectangular garden for a given perimeter.

Sample Scoring Rubric: The following rubric is a sample scoring device. You may wish to add more detail to this sample to meet your individual scoring needs.

Score	Criteria
4	A correct solution that is supported by well-developed, accurate explanations
3	A generally correct solution, but may contain minor flaws in reasoning or computation
2	A partially correct interpretation and/or solution to the problem
1	A correct solution with no supporting evidence or explanation
0	An incorrect solution indicating no mathematical understanding of the concept or task, or no solution is given

Answer

21. Yes; 68% of the students are within one standard deviation of the mean. That is consistent with a normal distribution.

Chapter 12 Standardized Test Practice 695

Page 637, Lesson 12-1

34. Step 1: When $n = 1$, the left side of the given equation is 4. The right side is $\frac{1[3(1) + 5]}{2}$ or 4, so the equation is true for $n = 1$.

Step 2: Assume $4 + 7 + 10 + \ldots + (3k + 1) = \frac{k(3k + 5)}{2}$ for some positive integer k.

Step 3: $4 + 7 + 10 + \ldots + (3k + 1) + [3(k + 1) + 1]$

$$= \frac{k(3k + 5)}{2} + [3(k + 1) + 1]$$
$$= \frac{k(3k + 5) + 2[3(k + 1) + 1]}{2}$$
$$= \frac{3k^2 + 5k + 6k + 6 + 2}{2}$$
$$= \frac{3k^2 + 11k + 8}{2}$$
$$= \frac{(k + 1)(3k + 8)}{2}$$
$$= \frac{(k + 1)[3(k + 1) + 5]}{2}$$

The last expression is the right side of the equation to be proved, where $n = k + 1$. Thus, the equation is true for $n = k + 1$.

Therefore, $4 + 7 + 10 + \ldots + (3n + 1) = \frac{n(3n + 5)}{2}$ for all positive integers n.

Page 642, Lesson 12-2

37. $C(n - 1, r) + C(n - 1, r - 1)$

$$= \frac{(n-1)!}{(n-1-r)!r!} + \frac{(n-1)!}{[n-1-(r-1)]!(r-1)!}$$
$$= \frac{(n-1)!}{(n-r-1)!r!} + \frac{(n-1)!}{(n-r)!(r-1)!}$$
$$= \frac{(n-1)!}{(n-r-1)!r!} \cdot \frac{n-r}{n-r} + \frac{(n-1)!}{(n-r)!(r-1)!} \cdot \frac{r}{r}$$
$$= \frac{(n-1)!(n-r)}{(n-r)!r!} + \frac{(n-1)!r}{(n-r)!r!}$$
$$= \frac{(n-1)!(n-r+r)}{(n-r)!r!}$$
$$= \frac{(n-1)!n}{(n-r)!r!}$$
$$= \frac{n!}{(n-r)!r!}$$
$$= C(n, r)$$

38. Permutations and combinations can be used to find the number of different lineups. Answers should include the following.

- There are 9! different 9-person lineups available: 9 choices for the first player, 8 choices for the second player, 7 for the third player, and so on. So, there are 362,880 different lineups.
- There are $C(16, 9)$ ways to choose 9 players from 16: $C(16, 9) = \frac{16!}{7!9!}$ or 11,440.

Page 651, Algebra Activity

1. The area of rectangle A is $\frac{1}{2}$; it represents the probability of drawing a silver clip and a blue clip. The area of rectangle B is $\frac{1}{6}$; it represents the probability of drawing a silver clip and a red clip. The area of rectangle C is $\frac{1}{4}$; it represents the probability of drawing a gold clip and a blue clip. The area of rectangle D is $\frac{1}{12}$; it represents the probability of drawing a gold clip and a red clip.

4.

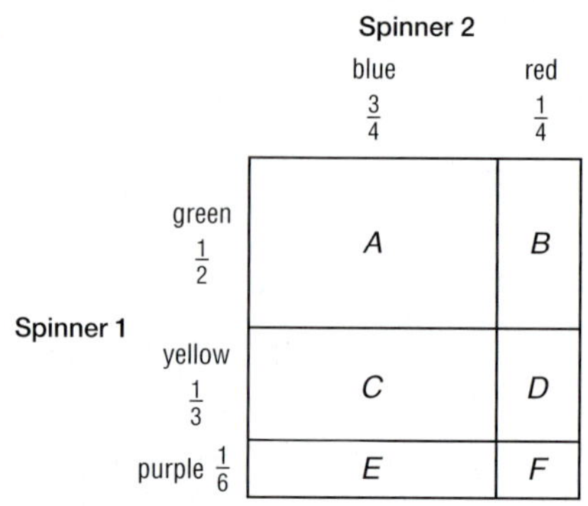

The area of rectangle *A* represents the probability of spinning green and blue. The area of rectangle *B* represents the probability of spinning green and red. The area of rectangle *C* represents the probability of spinning yellow and blue. The area of rectangle *D* represents the probability of spinning yellow and red. The area of rectangle *E* represents the probability of spinning purple and blue. The area of rectangle *F* represents the probability of spinning purple and red.

Page 656, Lesson 12-4

36.

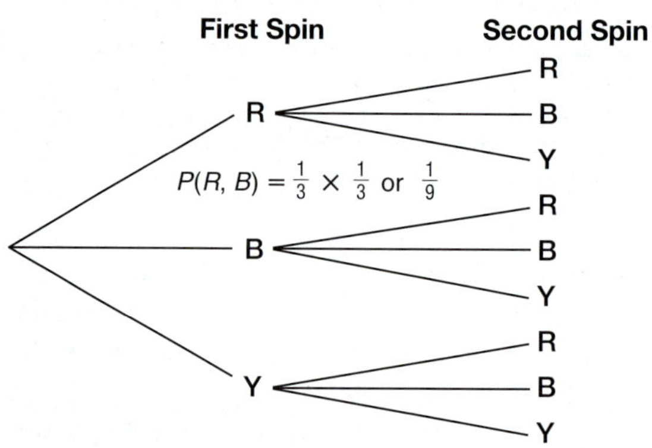

37.

		First Spin blue $\frac{1}{3}$	First Spin yellow $\frac{1}{3}$	First Spin red $\frac{1}{3}$
Second Spin	blue $\frac{1}{3}$	BB $\frac{1}{9}$	BY $\frac{1}{9}$	BR $\frac{1}{9}$
	yellow $\frac{1}{3}$	YB $\frac{1}{9}$	YY $\frac{1}{9}$	YR $\frac{1}{9}$
	red $\frac{1}{3}$	RB $\frac{1}{9}$	RY $\frac{1}{9}$	RR $\frac{1}{9}$

51. Probability can be used to analyze the chances of a player making 0, 1, or 2 free throws when he or she goes to the foul line to shoot 2 free throws. Answers should include the following.

- One of the decimals in the table could be used as the value of p, the probability that a player makes a given free throw. The probability that a player misses both free throws is $(1 - p)(1 - p)$ or $(1 - p)^2$. The probability that a player makes both free throws is $p \cdot p$ or p^2. Since the sum of the probabilities of all the possible outcomes is 1, the probability that a player makes exactly 1 of the 2 free throws is $1 - (1 - p)^2 - p^2$ or $2p(1 - p)$.
- The result of the first free throw could affect the player's confidence on the second free throw. For example, if the player makes the first free throw, the probability of he or she making the second free throw might increase. Or, if the player misses the first free throw, the probability that he or she makes the second free throw might decrease.

Page 662, Lesson 12-5

48. Probability can be used to estimate the percents of people who do the same things before going to bed. Answers should include the following.

- The events are inclusive because some people brush their teeth and set their alarm. Also, you know that the events are inclusive because the sum of the percents is not 100%.
- According to the information in the text and the table, $P(\text{read book}) = \frac{38}{100}$ and $P(\text{brush teeth}) = \frac{81}{100}$. Since the events are inclusive,
  $P(\text{read book or brush teeth}) =$
  $P(\text{read book}) + P(\text{brush teeth}) -$
  $P(\text{read book and brush teeth}) =$
  $\frac{38}{100} + \frac{81}{100} - \frac{600}{2000} = \frac{89}{100}$.

Page 679, Lesson 12-8

39. Getting a right answer and a wrong answer are the outcomes of a binomial experiment. The probability is far greater that guessing will result in a low grade than in a high grade. Answers should include the following.

- Use $(r + w)^5 = r^5 + 5r^4w + 10r^3w^2 + 10r^2w^3 + 5rw^4 + w^5$ and the chart on page 676 to determine the probabilities of each combination of right and wrong.
- $P(\text{5 right})$: $r^5 = \left(\frac{1}{4}\right)^5 = \frac{1}{1024}$ or about 0.098%;
  $P(\text{4 right, 1 wrong})$: $\frac{15}{1024}$ or about 1.5%;
  $P(\text{3 right, 2 wrong})$: $10r^3w^2 = 10\left(\frac{1}{4}\right)^3\left(\frac{3}{4}\right)^2 = \frac{45}{512}$ or about 8.8%;
  $P(\text{3 wrong, 2 right})$: $10r^2w^3 = 10\left(\frac{1}{4}\right)^2\left(\frac{3}{4}\right)^3 = \frac{135}{512}$ or about 26.4%;
  $P(\text{4 wrong, 1 right})$: $5rw^4 = 5\left(\frac{1}{4}\right)\left(\frac{3}{4}\right)^4 = \frac{405}{1024}$ or about 39.6%;
  $P(\text{5 wrong})$: $w^5 = \left(\frac{3}{4}\right)^5 = \frac{243}{1024}$ or about 23.7%.

Page 681, Follow-Up of Lesson 12-8
Algebra Activity

1. Sample answer:

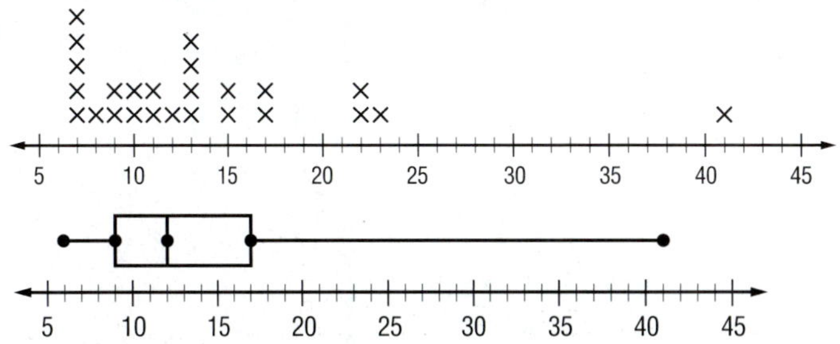

2. Sample answer: mean = 13.56; median = 12; maximum = 41; minimum = 7; standard deviation ≈ 7.3.

Pages 683–685, Lesson 12-9

1. Sample answer: If a sample is not random, the results of a survey may not be valid.
2. Sample answer for good sample: doing a random telephone poll to rate the mayor's performance; sample answer for bad sample: conducting a survey on how much the average person reads at a bookstore
3. The margin of sampling error decreases when the size of the sample n increases. As n increases, $\frac{p(1 - p)}{n}$ decreases.

29. A political candidate can use the statistics from an opinion poll to analyze his or her standing and to help plan the rest of the campaign. Answers should include the following.

- The candidate could decide to skip areas where he or she is way ahead or way behind, and concentrate on areas where the polls indicate the race is close.
- about 3.5%
- The margin of error indicates that with a probability of 0.95 the percent of the Florida population that favored Bush was between 43.5% and 50.5%. The margin of error for Gore was also about 3.5%, so with probability 0.95 the percent that favored Gore was between 40.5% and 47.5%. Therefore, it was possible that the percent of the Florida population that favored Bush was less than the percent that favored Gore.

Page 686, Follow-Up of Lesson 12-9
Algebra Activity

1. H_0: playing classical music during a math test, average test score ≠ 73
   H_1: playing classical music during a math test, average test score = 73
2. H_0: using robots on an assembly line, mean number of defects per 1000 items < 18
   H_1: using robots on an assembly line, mean number of defects per 1000 items ≥ 18
3. H_0: taking medication, mean pulse rate for the population > 82 beats per minute
   H_1: taking medication, mean pulse rate for the population ≤ 82 beats per minute

UNIT 5 Notes

Introduction

In this unit, students investigate the six trigonometric functions, both as ratios in right triangles and as circular functions, which they graph. The Law of Sines and the Law of Cosines are used to solve problems, as are the inverse trigonometric functions.

The unit concludes with lessons in which students verify and use trigonometric identities, and solve trigonometric equations.

Assessment Options

Unit 5 Test Pages 899–900 of the *Chapter 14 Resource Masters* may be used as a test or review for Unit 5. This assessment contains both multiple-choice and short answer items.

ExamView® Pro

This CD-ROM can be used to create additional unit tests and review worksheets.

UNIT 5 Trigonometry

Trigonometry is used in navigation, physics, and construction, among other fields. In this unit, you will learn about trigonometric functions, graphs, and identities.

Chapter 13
Trigonometric Functions

Chapter 14
Trigonometric Graphs and Identities

WebQuest Internet Project

Trig Class Angles for Lessons in Lit

Source: *USA TODAY*, November 21, 2000

"The groans from the trigonometry students immediately told teacher Michael Buchanan what the class thought of his idea to read Homer Hickam's *October Sky*. In the story, in order to accomplish what they would like, the kids had to teach themselves trig, calculus, and physics." In this project, you will research applications of trigonometry as it applies to a possible career for you.

Log on to www.algebra2.com/webquest. Begin your WebQuest by reading the Task.

Then continue working on your WebQuest as you study Unit 5.

Lesson	13-1	14-2
Page	708	775

Teaching Suggestions

Have students study the USA TODAY Snapshot®.

- Ask students to discuss their own reading habits, and whether for recreational reading they prefer fictional or non-fictional material.
- Ask students to share any experience they have with teaching themselves a subject or skill.
- Point out to students that good reading skills have a high positive correlation with academic and vocational success.

Additional USA TODAY Snapshots® appearing in Unit 5:

Chapter 13 Sports radio extends coverage area (p. 715)

Chapter 14 Strongest quakes in 20th century (p. 797)

WebQuest Internet Project

Problem-Based Learning A WebQuest is an online project in which students do research on the Internet, gather data, and make presentations using word processing, graphing, page-making, or presentation software. In each chapter, students advance to the next step in their WebQuest. At the end of Chapter 14, the project culminates with a presentation of their findings.

Teaching suggestions and sample answers are available in the *WebQuest and Project Resources.*

Chapter 13

Trigonometric Functions
Chapter Overview and Pacing

Year-long and two-year pacing: pages T20–T21.

LESSON OBJECTIVES

Lesson Objectives	PACING (days) Regular Basic/Average	Regular Advanced	Block Basic/Average	Block Advanced
13-1 Right Triangle Trigonometry *(pp. 700–708)* ***Preview:*** Special Right Triangles • Find values of trigonometric functions for acute angles. • Solve problems involving right triangles.	2 (with 13-1 Preview)	2 (with 13-1 Preview)	1	1
13-2 Angles and Angle Measure *(pp. 709–716)* • Change radian measure to degree measure and vice versa. • Identify coterminal angles. ***Follow-Up:*** Investigating Regular Polygons Using Trigonometry	1	1	0.5	0.5
13-3 Trigonometric Functions of General Angles *(pp. 717–724)* • Find values of trigonometric functions for general angles. • Use reference angles to find values of trigonometric functions.	2 (with 13-2 Follow-Up)	2 (with 13-2 Follow-Up)	1	1
13-4 Law of Sines *(pp. 725–732)* • Solve problems by using the Law of Sines. • Determine whether a triangle has one, two, or no solutions.	2	2	1	1
13-5 Law of Cosines *(pp. 733–738)* • Solve problems by using the Law of Cosines. • Determine whether a triangle can be solved by first using the Law of Sines or the Law of Cosines.	2	2	1	1
13-6 Circular Functions *(pp. 739–745)* • Define and use the trigonometric functions based on the unit circle. • Find the exact values of trigonometric functions of angles.	2	2	1	1
13-7 Inverse Trigonometric Functions *(pp. 746–751)* • Solve equations by using inverse trigonometric functions. • Find values of expressions involving trigonometric functions.	1	1	0.5	0.5
Study Guide and **Practice Test** *(pp. 752–757)* **Standardized Test Practice** *(pp. 758–759)*	1	1	0.5	0.5
Chapter Assessment	1	1	1	0.5
TOTAL	**14**	**14**	**7.5**	**7**

An electronic version of this chapter is available on ***StudentWorks***™. *This backpack solution CD-ROM allows students instant access to the Student Edition, lesson worksheet pages, and web resources.*

Timesaving Tools
TeacherWorks™
All-In-One Planner and Resource Center
See pages T5 and T21.

Chapter Resource Manager

Chapter 13 Resource Masters: Study Guide and Intervention	Practice (Skills and Average)	Reading to Learn Mathematics	Enrichment	Assessment	Applications*	5-Minute Check Transparencies	Interactive Chalkboard	Alge2PASS: Tutorial Plus (lessons)	Materials
775–776	777–778	779	780		GCS 52, SM 37–40	13-1	13-1		(*Preview:* spreadsheet program)
781–782	783–784	785	786	831		13-2	13-2		yardsticks (*Follow-Up:* compass, ruler, protractor, calculator)
787–788	789–790	791	792			13-3	13-3	24, 25	
793–794	795–796	797	798	831, 833	GCS 51, SC 25	13-4	13-4		coffee stirrer sticks
799–800	801–802	803	804		SC 26	13-5	13-5	26	
805–806	807–808	809	810	832		13-6	13-6		graphing calculator, geoboard and rubber bands
811–812	813–814	815	816	832		13-7	13-7		
				817–830, 834–836					

**Key to Abbreviations:* GCS = Graphing Calculator and Speadsheet Masters,
SC = School-to-Career Masters,
SM = Science and Mathematics Lab Manual

Chapter 13

Mathematical Connections and Background

Continuity of Instruction

Prior Knowledge

Students have used the Pythagorean Theorem. They are familiar with applying formulas to solve problems. Also, they introduced new notation for inverse functions when they explored logarithmic functions, and they restricted domains when they found inverses for functions such as $y = x^2$.

This Chapter

Students explore trigonometric functions, first for acute angles in right triangles, then for angles in standard form, and also for points on the unit circle. They derive and use the Law of Sines and the Law of Cosines as applications of trigonometric functions, and they develop inverses for the sine, cosine, and tangent functions.

Future Connections

Students' exploration of trigonometric functions and periodic functions continues in the following chapter. There they will explore amplitude and frequency for periodic functions and will look at translations of their graphs. They will develop and use trigonometric functions for sums or differences of angles and they will solve equations involving trigonometric terms.

13-1 Right Triangle Trigonometry

To solve many of the problems in this lesson, students will need to remember the Pythagorean Theorem, which states that the sum of the squares of the legs of a right triangle equals the square of the hypotenuse. The *hypotenuse* is the side directly across from the right angle. It is also the longest side. The *legs* of a right triangle are the two shorter sides. If the legs of a right triangle have measures a and b and the hypotenuse has a measure of c, then $a^2 + b^2 = c^2$.

When solving real-world problems involving right triangles and trigonometric ratios, first determine what information is given about the triangle's sides and angles. Then determine how the given side or sides of the triangle relate to the given angle or the angle to be considered. In other words, determine if a given side is the hypotenuse, side opposite, or side adjacent to the angle. For example, if the problem gives an angle measure and the measure of the side adjacent to this angle and asks you to find the side opposite this angle, use the tangent ratio.

13-2 Angles and Angle Measure

This lesson will introduce students to concept of negative angle measure. It is important to note that an angle measuring $-210°$ is not less than an angle measuring $210°$. The negative angle measure indicates that the direction of the rotation is clockwise instead of counterclockwise.

Remember that in trigonometry an angle is a measure of rotation. One complete rotation, or turn, measures $360°$, the number of degrees in a circle. In the real world, objects can rotate about a fixed point many times. Angles measuring more than $360°$ are used to describe these rotations. To draw such an angle, first subtract $360°$ from the angle measure and continue doing so until you arrive at a measure that is less than or equal to $360°$. Draw this angle and then use an arrow spiraling from the angle's initial side, through as many $360°$-increments as you subtracted, until finally reaching the angle's terminal side.

13-3 Trigonometric Functions of General Angles

In Lesson 13-1, students find the exact values of the six trigonometric functions for angles measuring less than $90°$, since the angles other than the right angle in a right triangle are both acute angles. Right triangle trigonometry is also used to define the values of the trigonometric functions for angles other than acute angles. From a point $P(x, y)$ on the terminal side of any angle, draw a segment perpendicular, meeting at a right angle, to the x-axis. A right triangle is formed with one side measuring x units, another side measuring y units, and the hypotenuse measuring r units. The value of r can be found using the Pythagorean Theorem, $r = \sqrt{x^2 + y^2}$. To determine the values of the six trigonometric functions for the original angle θ, find the value of each function for the angle θ' that is formed by the terminal side and the x-axis. It is important to note that the trigonometric values of angles other than acute angles can be negative.

13-4 Law of Sines

Examine the chart on page 727. In each drawing where A is acute, the positions of the horizontal segment and segment b are fixed, thus allowing the measure of A to remain constant. The measure of segment b in each drawing is also constant. The position of segment a, however, can change like a door on a hinge. Notice also that the length of a in each diagram is different and often compared to the value $b \sin A$.

In the first diagram in the first row, the length of a is less than the value $b \sin A$. In other words, side a is too short to form a right triangle. When a is too short to form a right triangle, no triangle can be formed. When a equals $b \sin A$, a right triangle is formed, as in the second diagram. In the first diagram in the second row, a is greater than $b \sin A$ but still less than b. In other words, side a is too long to form a right triangle, but can still rotate on its hinge and meet the opposite side to form either an obtuse or an acute triangle. If a is greater than or equal to b, side a is again too long to form a right triangle, but is now also too long to rotate back towards $\angle A$. Instead, it can meet the opposite side in only one place, as shown in the second diagram in the second row.

13-5 Law of Cosines

The Law of Cosines involves three sides and one angle of a triangle. The key step in deriving the Law of Cosines is to consider an altitude of length h intersecting a side and dividing it into two parts. Using x to represent one of the two parts, the Pythagorean Theorem gives an equation in terms of x and the three side lengths. Then x can be replaced with an expression involving one side and the cosine of one angle. The result is an expression of the Law of Cosines as three equations:

$$a^2 = b^2 + c^2 - 2bc \cos A$$
$$b^2 = a^2 + c^2 - 2ac \cos B$$
$$c^2 = a^2 + b^2 - 2ab \cos C$$

The Law of Cosines lets you calculate measures for all the angles and all the sides of a triangle if you are given (1) the length of two sides and the measure of the included angle, or (2) the lengths of three sides. The Law of Sines can be used if the measures of two angles and one side are known (ASA, AAS) or if the measures of two sides and the non-included angle are known (SSA, but there may be 0, 1, or 2 triangles).

13-6 Circular Functions

To memorize the unit circle presented at the bottom of page 740, use the following technique. First memorize the coordinates for the angles in the first quadrant. Notice that all of these coordinates contain values having a denominator of 2, that the x- and y-coordinates for 45° are identical, $\frac{\sqrt{2}}{2}$, and that the x- and y-coordinates for 30° and 60° are reversed. With this quadrant memorized, the coordinates for the other quadrants can be obtained using the signs of x and y in each quadrant and the symmetry of a circle.

The coordinates of the quadrantal angles, 0°, 90°, 180°, and 270°, are easily remembered by recalling that a unit circle has a radius of 1 unit. For example, since 0° is located on the x-axis, its coordinates are (1, 0).

Once the unit circle is completed, it can be used as a reference to obtain the sine or cosine of any of the angles listed. One need only remember that the x-coordinate given for an angle measure is the cosine of the angle and the y-coordinate is the sine of the angle. In other words $(x, y) = (\cos \theta, \sin \theta)$.

13-7 Inverse Trigonometric Functions

In this lesson, students limit the domain of the sine, cosine, and tangent functions, and define an inverse for each function. The graphs of those functions do not pass the horizontal line test, so they would not have inverses. However, the section of the sine and tangent graphs between −180° and 180° and the section of the cosine graph between 0° and 180° do pass the horizontal line test, so inverse functions can be identified for those domains. The function $y = \text{Sin } x$ is the restricted-domain sine function. Its inverse is $x = \text{Sin}^{-1}\, y$ or $x = \text{Arcsin } y$. The function $y = \text{Cos } x$ is the restricted-domain cosine function. Its inverse is $x = \text{Cos}^{-1}\, y$ or $x = \text{Arccos } y$. The function $y = \text{Tan } x$ represents the restricted-domain tangent function. Its inverse function is $x = \text{Tan}^{-1}\, y$ or $x = \text{Arctan } y$.

Additional mathematical information and teaching notes are available at www.algebra2.com/key_concepts.

Chapter 13 DAILY INTERVENTION and Assessment

	Type	Student Edition	Teacher Resources	Technology/Internet
INTERVENTION	Ongoing	Prerequisite Skills, pp. 699, 708, 715, 724, 732, 738, 745 Practice Quiz 1, p. 715 Practice Quiz 2, p. 738	5-Minute Check Transparencies Quizzes, *CRM* pp. 831–832 Mid-Chapter Test, *CRM* p. 833 Study Guide and Intervention, *CRM* pp. 775–776, 781–782, 787–788, 793–794, 799–800, 805–806, 811–812	Alge2PASS: Tutorial Plus www.algebra2.com/self_check_quiz www.algebra2.com/extra_examples
	Mixed Review	pp. 708, 714, 724, 732, 738, 745, 751	Cumulative Review, *CRM* p. 834	
	Error Analysis	Find the Error, pp. 730, 735 Common Misconceptions, p. 703	Find the Error, *TWE* pp. 730, 735 Unlocking Misconceptions, *TWE* pp. 718, 726 Tips for New Teachers, *TWE* pp. 703, 711	
	Standardized Test Practice	pp. 702, 706, 708, 714, 724, 732, 737, 738, 745, 751, 757, 758–759	*TWE* p. 702 Standardized Test Practice, *CRM* pp. 835–836	Standardized Test Practice CD-ROM www.algebra2.com/standardized_test
ASSESSMENT	Open-Ended Assessment	Writing in Math, pp. 708, 714, 724, 732, 737, 744, 751 Open Ended, pp. 706, 712, 722, 729, 736, 742, 749	Modeling: *TWE* pp. 708, 732, 745 Speaking: *TWE* pp. 715, 751 Writing: *TWE* pp. 724, 738 Open-Ended Assessment, *CRM* p. 829	
	Chapter Assessment	Study Guide, pp. 752–756 Practice Test, p. 757	Multiple-Choice Tests (Forms 1, 2A, 2B), *CRM* pp. 817–822 Free-Response Tests (Forms 2C, 2D, 3), *CRM* pp. 823–828 Vocabulary Test/Review, *CRM* p. 830	ExamView® Pro (see below) MindJogger Videoquizzes www.algebra2.com/vocabulary_review www.algebra2.com/chapter_test

Key to Abbreviations: TWE = Teacher Wraparound Edition; CRM = Chapter Resource Masters

Intervention Technology

Alge2PASS: Tutorial Plus CD-ROM offers a complete, self-paced algebra curriculum.

Algebra 2 Lesson	Alge2PASS Lesson
13-3	**24** *Trigonometric Functions of Coterminal Angles*
13-3	**25** *Trigonometric Functions of Acute and Quadrantal Angles*
13-5	**26** *Trigonometry II*

ExamView® Pro

Use the networkable **ExamView® Pro** to:

- Create **multiple versions** of tests.
- Create **modified** tests for *Inclusion* students.
- **Edit** existing questions and **add** your own questions.
- Use built-in **state curriculum correlations** to create tests aligned with state standards.
- **Apply** art to your tests from a program bank of artwork.

For more information on Intervention and Assessment, see pp. T8–T11.

Reading and Writing in Mathematics

Glencoe Algebra 2 provides numerous opportunities to incorporate reading and writing into the mathematics classroom.

Student Edition

- Foldables Study Organizer, p. 699
- Concept Check questions require students to verbalize and write about what they have learned in the lesson. (pp. 706, 712, 722, 729, 735, 742, 749, 752)
- Writing in Math questions in every lesson, pp. 708, 714, 724, 732, 737, 744, 751
- Reading Study Tip, pp. 701, 709, 711, 718, 740
- WebQuest, p. 708

Teacher Wraparound Edition

- Foldables Study Organizer, pp. 699, 752
- Study Notebook suggestions, pp. 706, 713, 716, 722, 730, 735, 743, 749
- Modeling activities, pp. 708, 732, 745
- Speaking activities, pp. 715, 751
- Writing activities, pp. 724, 738
- Differentiated Instruction, (Verbal/Linguistic), p. 735
- ELL Resources, pp. 698, 707, 714, 723, 731, 735, 737, 744, 750, 752

Additional Resources

- Vocabulary Builder worksheets require students to define and give examples for key vocabulary terms as they progress through the chapter. (*Chapter 13 Resource Masters,* pp. vii-viii)
- Reading to Learn Mathematics master for each lesson (*Chapter 13 Resource Masters*, pp. 779, 785, 791, 797, 803, 809, 815)
- *Vocabulary PuzzleMaker* software creates crossword, jumble, and word search puzzles using vocabulary lists that you can customize.
- *Teaching Mathematics with Foldables* provides suggestions for promoting cognition and language.
- *Reading and Writing in the Mathematics Classroom*
- *WebQuest and Project Resources*

For more information on Reading and Writing in Mathematics, see pp. T6–T7.

PROJECT CRISS℠ Study Skill

Students may find that making study cards is helpful when they are learning many new terms and concepts. Making the cards aids in learning because it requires summarizing the information. Then the cards can be used as students review when preparing for tests. The cards shown at the right describe some of the concepts of right triangle trigonometry that students study in Lesson 13-1. A name and sketch are on the front of the card and the relationship is on the back of the card.

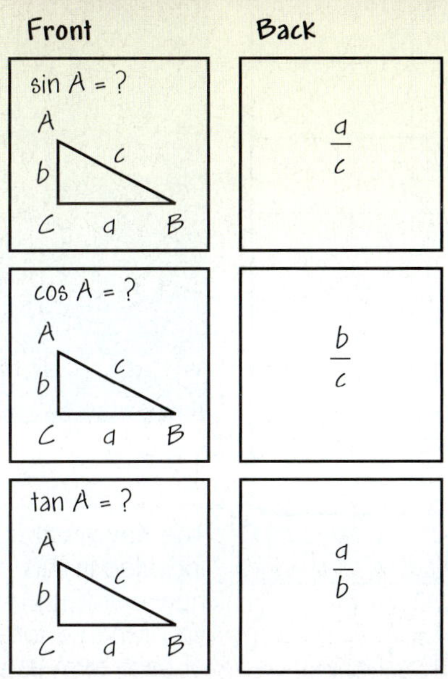

CReating **I**ndependence Through **S**tudent-Owned **S**trategies

Chapter 13 Notes

What You'll Learn

Have students read over the list of objectives and make a list of any words with which they are not familiar.

Why It's Important

Point out to students that this is only one of many reasons why each objective is important. Others are provided in the introduction to each lesson.

Lesson	NCTM Standards	Local Objectives
13-1 Preview	2, 3, 6	
13-1	2, 3, 6, 8, 9, 10	
13-2	1, 3, 4, 6, 8, 9, 10	
13-2 Follow-Up	1, 3, 4, 6, 8	
13-3	1, 3, 4, 6, 8, 9, 10	
13-4	1, 3, 4, 6, 8, 9, 10	
13-5	1, 3, 4, 6, 8, 9, 10	
13-6	1, 3, 4, 6, 8, 9, 10	
13-7	1, 3, 4, 6, 8, 9, 10	

Key to NCTM Standards:

1=Number & Operations, 2=Algebra, 3=Geometry, 4=Measurement, 5=Data Analysis & Probability, 6=Problem Solving, 7=Reasoning & Proof, 8=Communication, 9=Connections, 10=Representation

Chapter 13 Trigonometric Functions

What You'll Learn

- **Lessons 13-1, 13-2, 13-3, 13-6, and 13-7** Find values of trigonometric functions.
- **Lessons 13-1, 13-4, and 13-5** Solve problems by using right triangle trigonometry.
- **Lessons 13-4 and 13-5** Solve triangles by using the Law of Sines and Law of Cosines.

Key Vocabulary

- solve a right triangle (p. 704)
- radian (p. 710)
- Law of Sines (p. 726)
- Law of Cosines (p. 733)
- circular function (p. 740)

Why It's Important

Trigonometry is the study of the relationships among the angles and sides of right triangles. One of the many real-world applications of trigonometric functions involves solving problems using indirect measurement. For example, surveyors use a trigonometric function to find the heights of buildings. *You will learn how architects who design fountains use a trigonometric function to aim the water jets in Lesson 13-7.*

698 Chapter 13 Trigonometric Functions

Vocabulary Builder

ELL

The Key Vocabulary list introduces students to some of the main vocabulary terms included in this chapter. For a more thorough vocabulary list with pronunciations of new words, give students the Vocabulary Builder worksheets found on pages vii and viii of the *Chapter 13 Resource Masters*. Encourage them to complete the definition of each term as they progress through the chapter. You may suggest that they add these sheets to their study notebooks for future reference when studying for the Chapter 13 test.

Getting Started

Prerequisite Skills To be successful in this chapter, you'll need to master these skills and be able to apply them in problem-solving situations. Review these skills before beginning Chapter 13.

For Lessons 13-1 and 13-3 — Pythagorean Theorem

Find the value of x to the nearest tenth. *(For review, see pages 820 and 821.)*

1. **10**
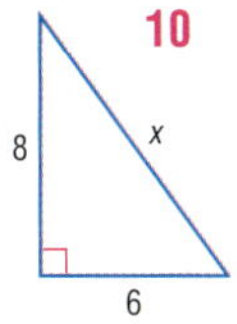

2. **10.3**
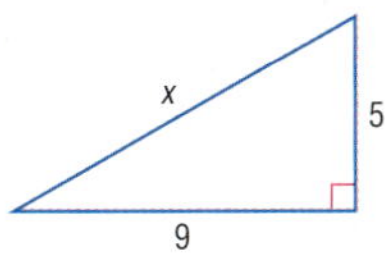

3. **16.7**
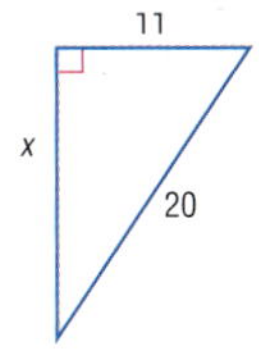

4. **21.8**
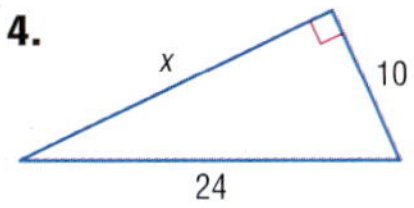

For Lesson 13-1 — 45°-45°-90° and 30°-60°-90° Triangles

Find each missing measure. Write all radicals in simplest form.

5. **$x = 7,\ y = 7\sqrt{2}$**
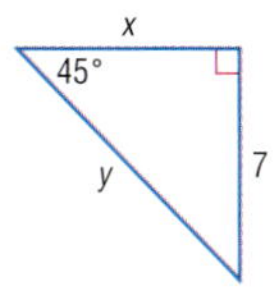

6. **$x = \frac{21\sqrt{2}}{2}$**
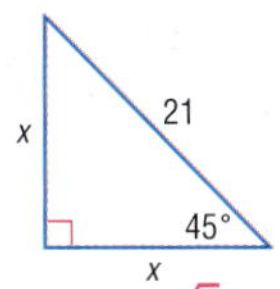

7. **$x = 4\sqrt{3},\ y = 8$**
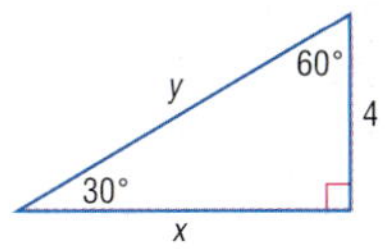

8. **$x = 3\sqrt{3},\ y = 6\sqrt{3}$**
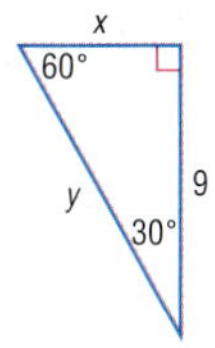

For Lesson 13-7 — Inverse Functions

Find the inverse of each function. Then graph the function and its inverse. *(For review, see Lesson 7-8.)* **9–12. See pp. 759A–759D for graphs.**

9. $f(x) = x + 3$ **$f^{-1}(x) = x - 3$**

10. $f(x) = \frac{x-2}{5}$ **$f^{-1}(x) = 5x + 2$**

11. $f(x) = x^2 - 4$ **$f^{-1}(x) = \pm\sqrt{x+4}$**

12. $f(x) = -7x - 9$ **$f^{-1}(x) = \frac{-x-9}{7}$**

FOLDABLES™ Study Organizer

Trigonometric Functions Make this Foldable to help you organize your notes. Begin with one sheet of construction paper and two pieces of grid paper.

Step 1 **Fold and Cut**

Stack and fold on the diagonal. Cut to form a triangular stack.

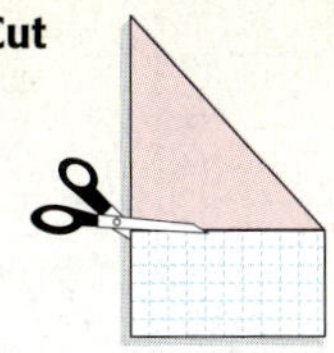

Step 2 **Staple and Label**

Staple the edge to form a booklet.

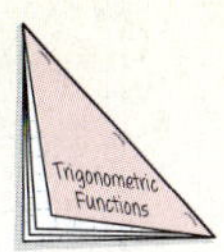

Reading and Writing As you read and study the chapter, you can write notes, draw diagrams, and record formulas on the grid paper.

Getting Started

This section provides a review of the basic concepts needed before beginning Chapter 13. Page references are included for additional student help.

Prerequisite Skills in the Getting Ready for the Next Lesson section at the end of each exercise set review a skill needed in the next lesson.

For Lesson	Prerequisite Skill
13-2	Dimensional Analysis (p. 708)
13-3	Rationalizing Denominators (p. 715)
13-4	Solving Equations with Trigonometric Functions (p. 724)
13-5	Solving Equations with Trigonometric Functions (p. 732)
13-6	Coterminal Angles (p. 738)
13-7	Finding Angle Measures (p. 745)

FOLDABLES™ Study Organizer

For more information about Foldables, see *Teaching Mathematics with Foldables.*

Organization of Data: Vocabulary and Visuals Have students use their right triangle journals to practice writing concise definitions in their own words and to design visuals that present the information introduced in the lesson in a concrete, easy-to-study format. Encourage students to clearly label their visuals and write captions when needed. Students can use this study guide to review what they know and apply it to what they are currently learning.

Spreadsheet Investigation

A Preview of Lesson 13-1

Getting Started

Cell Format Students can format the cells before they start to enter the data by using the Format Cells command. Format columns C, E, and F for numbers with 8 decimal places. Format columns A, B, and D for integers.

Teach

- Make sure students understand how to enter the formulas shown for columns C through F.
- Have students practice their skills in using a spreadsheet by entering the data in the example.
- Have students complete Exercises 1–3.
- To extend this investigation, ask students to explore the effect on the ratios if $a = b$ when a and b are rational numbers (such as 3.6) instead of integers.

Assess

Ask students:

- What formula could you have used in column B for the 45°-45°-90° triangle instead of entering the data? in column C for the 30°-60°-90° triangle? **$b = a$; $c = 2a$**
- Compare the 30°-60°-90° triangles. What is the same? What is different? **The angle measures are all the same, but the side measures are different. The triangles are similar but not congruent.**

Answer

3. **All of the ratios of side b to side a are approximately 1.73. All of the ratios of side b to side c are approximately 0.87. All of the ratios of side a to side c are 0.5.**

Special Right Triangles

You can use a computer spreadsheet program to investigate the relationships among the ratios of the side measures of special right triangles.

Example

The legs of a 45°-45°-90° triangle, a and b, are equal in measure. Use a spreadsheet to investigate the dimensions of 45°-45°-90° triangles. What patterns do you observe in the ratios of the side measures of these triangles?

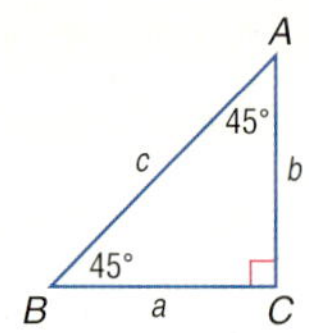

=SQRT(A2^2 + B2^2) =B2/A2 =B2/C2 =A2/C2

45-45-90 Triangles

	A	B	C	D	E	F
1	a	b	c	b/a	b/c	a/c
2	1	1	1.41421356	1	0.70710678	0.70710678
3	2	2	2.82842712	1	0.70710678	0.70710678
4	3	3	4.24264069	1	0.70710678	0.70710678
5	4	4	5.65685425	1	0.70710678	0.70710678
6	5	5	7.07106781	1	0.70710678	0.70710678
7						

Sheet1 / Sheet2 / S

The spreadsheet shows the formula that will calculate the length of side c. The formula uses the Pythagorean Theorem in the form $c = \sqrt{a^2 + b^2}$. Since 45°-45°-90° triangles share the same angle measures, the triangles listed in the spreadsheet are all similar triangles. Notice that all of the ratios of side b to side a are 1. All of the ratios of side b to side c and of side a to side c are approximately 0.71.

Exercises

For Exercises 1–3, use the spreadsheet below for 30°-60°-90° triangles.

If the measure of one leg of a right triangle and the measure of the hypotenuse are in a ratio of 1 to 2, then the acute angles of the triangle measure 30° and 60°.

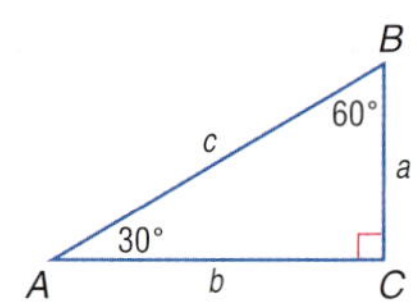

30-60-90 Triangles

	A	B	C	D	E	F
1	a	b	c	b/a	b/c	a/c
2	1	=SQRT(C2^2-A2^2)	2	=B2/A2	=B2/C2	=A2/C2
3	2	=SQRT(C3^2-A3^2)	4	=B3/A3	=B3/C3	=A3/C3
4	3	=SQRT(C4^2-A4^2)	6	=B4/A4	=B4/C4	=A4/C4
5	4	=SQRT(C5^2-A5^2)	8	=B5/A5	=B5/C5	=A5/C5
6	5	=SQRT(C6^2-A6^2)	10	=B6/A6	=B6/C6	=A6/C6
7						

Sheet1 / Sheet2 / Sheet3

2. The triangles are all similar.

1. Copy and complete the spreadsheet above.
2. Describe the relationship among the 30°-60°-90° triangles whose dimensions are given.
3. What patterns do you observe in the ratios of the side measures of these triangles? **See margin.**

700 Chapter 13 Trigonometric Functions

13-1 Right Triangle Trigonometry

What You'll Learn

- Find values of trigonometric functions for acute angles.
- Solve problems involving right triangles.

Vocabulary

- trigonometry
- trigonometric functions
- sine
- cosine
- tangent
- cosecant
- secant
- cotangent
- solve a right triangle
- angle of elevation
- angle of depression

How is trigonometry used in building construction?

The Americans with Disabilities Act (ADA) provides regulations designed to make public buildings accessible to all. Under this act, the slope of an entrance ramp designed for those with mobility disabilities must not exceed a ratio of 1 to 12. This means that for every 12 units of horizontal run, the ramp can rise or fall no more than 1 unit.

When viewed from the side, a ramp forms a right triangle. The slope of the ramp can be described by the *tangent* of the angle the ramp makes with the ground. In this example, the tangent of angle A is $\frac{1}{12}$.

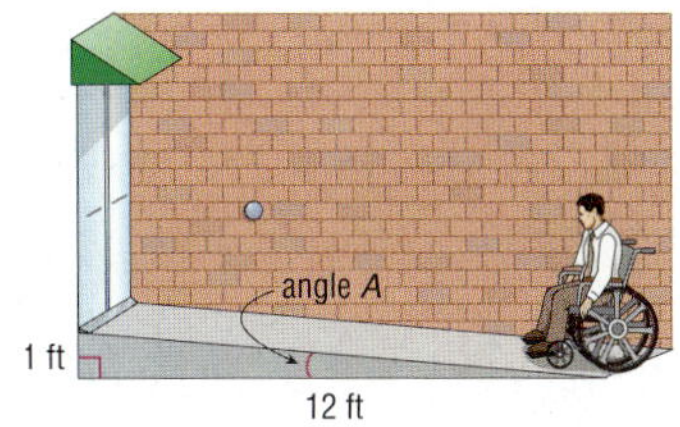

Study Tip

Reading Math
The word *trigonometry* is derived from two Greek words—*trigon* meaning triangle and *metra* meaning measurement.

TRIGONOMETRIC VALUES The tangent of an angle is one of the ratios used in trigonometry. **Trigonometry** is the study of the relationships among the angles and sides of a right triangle.

Consider right triangle ABC in which the measure of acute angle A is identified by the Greek letter *theta*, θ. The sides of the triangle are the *hypotenuse*, the *leg opposite* θ, and the *leg adjacent to* θ.

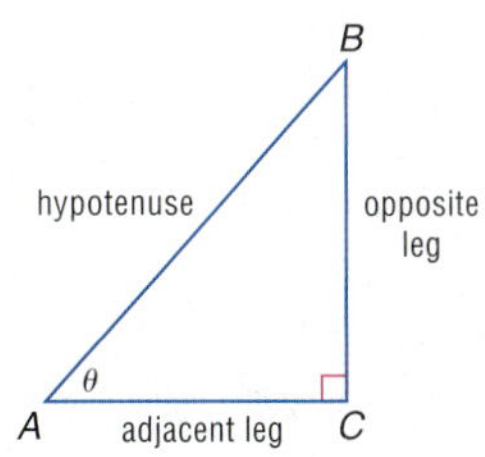

Using these sides, you can define six **trigonometric functions**: **sine**, **cosine**, **tangent**, **cosecant**, **secant**, and **cotangent**. These functions are abbreviated sin, cos, tan, sec, csc, and cot, respectively.

Key Concept — Trigonometric Functions

If θ is the measure of an acute angle of a right triangle, *opp* is the measure of the leg opposite θ, *adj* is the measure of the leg adjacent to θ, and *hyp* is the measure of the hypotenuse, then the following are true.

$$\sin\theta = \frac{\text{opp}}{\text{hyp}} \qquad \cos\theta = \frac{\text{adj}}{\text{hyp}} \qquad \tan\theta = \frac{\text{opp}}{\text{adj}}$$

$$\csc\theta = \frac{\text{hyp}}{\text{opp}} \qquad \sec\theta = \frac{\text{hyp}}{\text{adj}} \qquad \cot\theta = \frac{\text{adj}}{\text{opp}}$$

Notice that the sine, cosine, and tangent functions are reciprocals of the cosecant, secant, and cotangent functions, respectively. Thus, the following are also true.

$$\csc\theta = \frac{1}{\sin\theta} \qquad \sec\theta = \frac{1}{\cos\theta} \qquad \cot\theta = \frac{1}{\tan\theta}$$

13-1 Lesson Notes

1 Focus

5-Minute Check Transparency 13-1 Use as a quiz or review of Chapter 12.

Mathematical Background notes are available for this lesson on p. 698C.

Building on Prior Knowledge

In previous courses, students learned about the Pythagorean Theorem. In this lesson, students will apply their knowledge to solve triangles.

How is trigonometry used in building construction?

Ask students:

- To meet the required ratio, how long would a ramp have to be to rise from the ground to a door that is 2 feet above ground level? **24 ft**
- Does a ramp with a ratio of 0.08 meet the requirement? **yes**

Resource Manager

Workbook and Reproducible Masters

Chapter 13 Resource Masters

- Study Guide and Intervention, pp. 775–776
- Skills Practice, p. 777
- Practice, p. 778
- Reading to Learn Mathematics, p. 779
- Enrichment, p. 780

Graphing Calculator and Spreadsheet Masters, p. 52

Science and Mathematics Lab Manual, pp. 37–40

Teaching Algebra With Manipulatives Masters, pp. 299–300

Transparencies

5-Minute Check Transparency 13-1
Answer Key Transparencies

Technology

Interactive Chalkboard

2 Teach

TRIGONOMETRIC VALUES

In-Class Examples

Power Point®

Teaching Tip Discuss with students the relationship between sin A and cos C in the figure. Remind students that A and C are complementary angles.

1 Find the values of the six trigonometric functions for angle G.

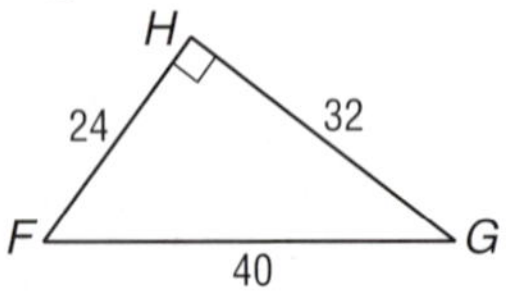

$\sin G = \frac{3}{5}$, $\cos G = \frac{4}{5}$, $\tan G = \frac{3}{4}$, $\cot G = \frac{4}{3}$, $\sec G = \frac{5}{4}$, $\csc G = \frac{5}{3}$

2 If $\tan A = \frac{5}{3}$, find the value of $\csc A$. **D**

A $\frac{3}{5}$ **B** $\frac{4}{3}$

C $\sqrt{34}$ **D** $\frac{\sqrt{34}}{5}$

This CD-ROM is a customizable Microsoft® PowerPoint® presentation that includes:

- Step-by-step, dynamic solutions of each In-Class Example from the Teacher Wraparound Edition
- Additional, Your Turn exercises for each example
- The 5-Minute Check Transparencies
- Hot links to Glencoe Online Study Tools

Study Tip

Memorize Trigonometric Ratios

SOH-CAH-TOA is a mnemonic device for remembering the first letter of each word in the ratios for sine, cosine, and tangent.

$\sin \theta = \frac{\text{opp}}{\text{hyp}}$

$\cos \theta = \frac{\text{adj}}{\text{hyp}}$

$\tan \theta = \frac{\text{opp}}{\text{adj}}$

The domain of each of these trigonometric functions is the set of all acute angles θ of a right triangle. The values of the functions depend only on the measure of θ and not on the size of the right triangle. For example, consider $\sin \theta$ in the figure at the right.

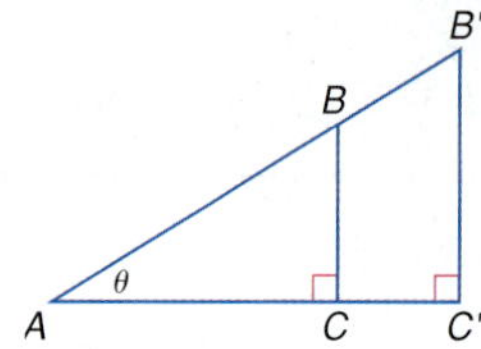

Using $\triangle ABC$: $\sin \theta = \frac{BC}{AB}$

Using $\triangle AB'C'$: $\sin \theta = \frac{B'C'}{AB'}$

The right triangles are similar because they share angle θ. Since they are similar, the ratios of corresponding sides are equal. That is, $\frac{BC}{AB} = \frac{B'C'}{AB'}$. Therefore, you will find the same value for $\sin \theta$ regardless of which triangle you use.

Example 1 Find Trigonometric Values

Find the values of the six trigonometric functions for angle θ.

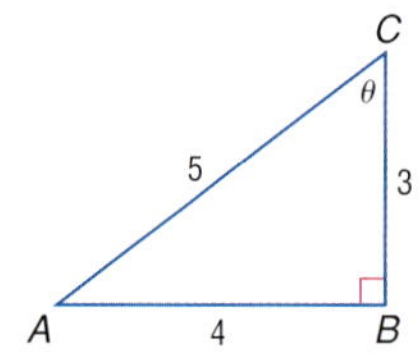

For this triangle, the leg opposite θ is $\overline{AB}$, and the leg adjacent to θ is $\overline{CB}$. Recall that the hypotenuse is always the longest side of a right triangle, in this case $\overline{AC}$.

Use opp = 4, adj = 3, and hyp = 5 to write each trigonometric ratio.

$\sin \theta = \frac{\text{opp}}{\text{hyp}} = \frac{4}{5}$ $\cos \theta = \frac{\text{adj}}{\text{hyp}} = \frac{3}{5}$ $\tan \theta = \frac{\text{opp}}{\text{adj}} = \frac{4}{3}$

$\csc \theta = \frac{\text{hyp}}{\text{opp}} = \frac{5}{4}$ $\sec \theta = \frac{\text{hyp}}{\text{adj}} = \frac{5}{3}$ $\cot \theta = \frac{\text{adj}}{\text{opp}} = \frac{3}{4}$

Throughout Unit 5, a capital letter will be used to denote both a vertex of a triangle and the measure of the angle at that vertex. The same letter in lowercase will be used to denote the side opposite that angle and its measure.

Standardized Test Practice

Example 2 Use One Trigonometric Ratio to Find Another

Multiple-Choice Test Item

If $\cos A = \frac{2}{5}$, find the value of $\tan A$.
(A) $\frac{5}{2}$ (B) $\frac{2\sqrt{21}}{21}$ (C) $\frac{\sqrt{21}}{2}$ (D) $\sqrt{21}$

Test-Taking Tip

Whenever necessary or helpful, draw a diagram of the situation.

Read the Test Item

Begin by drawing a right triangle and labeling one acute angle A. Since $\cos \theta = \frac{\text{adj}}{\text{hyp}}$ and $\cos A = \frac{2}{5}$ in this case, label the adjacent leg 2 and the hypotenuse 5.

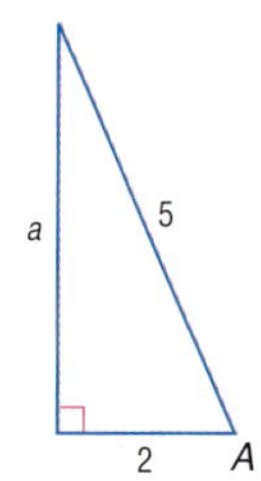

Solve the Test Item

Use the Pythagorean Theorem to find a.

$a^2 + b^2 = c^2$ Pythagorean Theorem

$a^2 + 2^2 = 5^2$ Replace b with 2 and c with 5.

$a^2 + 4 = 25$ Simplify.

$a^2 = 21$ Subtract 4 from each side.

$a = \sqrt{21}$ Take the square root of each side.

Standardized Test Practice

Example 2 Another approach would be to write $\cos A = \frac{2}{5} = \frac{\text{adj}}{\text{hyp}}$. Then write $\tan A = \frac{\text{opp}}{\text{adj}}$. This suggests looking for an answer choice that has a denominator of 2.

Now find tan A.

$\tan A = \frac{\text{opp}}{\text{adj}}$ Tangent ratio

$= \frac{\sqrt{21}}{2}$ Replace *opp* with $\sqrt{21}$ and *adj* with 2.

The answer is C.

Angles that measure 30°, 45°, and 60° occur frequently in trigonometry. The table below gives the values of the six trigonometric functions for these angles. To remember these values, use the properties of 30°-60°-90° and 45°-45°-90° triangles.

Key Concept — Trigonometric Values for Special Angles

30°-60°-90° Triangle

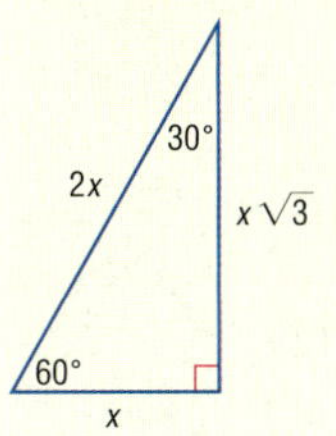

45°-45°-90° Triangle

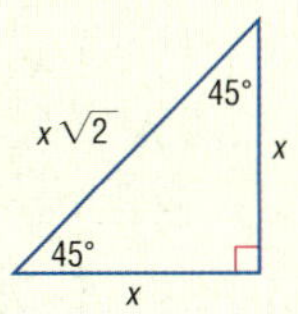

θ	$\sin\theta$	$\cos\theta$	$\tan\theta$	$\csc\theta$	$\sec\theta$	$\cot\theta$
30°	$\frac{1}{2}$	$\frac{\sqrt{3}}{2}$	$\frac{\sqrt{3}}{3}$	2	$\frac{2\sqrt{3}}{3}$	$\sqrt{3}$
45°	$\frac{\sqrt{2}}{2}$	$\frac{\sqrt{2}}{2}$	1	$\sqrt{2}$	$\sqrt{2}$	1
60°	$\frac{\sqrt{3}}{2}$	$\frac{1}{2}$	$\sqrt{3}$	$\frac{2\sqrt{3}}{3}$	2	$\frac{\sqrt{3}}{3}$

You will verify some of these values in Exercises 27 and 28.

RIGHT TRIANGLE PROBLEMS You can use trigonometric functions to solve problems involving right triangles.

Example 3 Find a Missing Side Length of a Right Triangle

Write an equation involving sin, cos, or tan that can be used to find the value of x. Then solve the equation. Round to the nearest tenth.

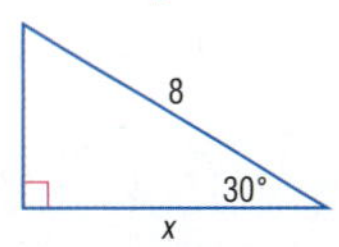

The measure of the hypotenuse is 8. The side with the missing length is *adjacent* to the angle measuring 30°. The trigonometric function relating the adjacent side of a right triangle and the hypotenuse is the cosine function.

$\cos\theta = \frac{\text{adj}}{\text{hyp}}$ cosine ratio

$\cos 30° = \frac{x}{8}$ Replace θ with 30°, *adj* with x, and *hyp* with 8.

$\frac{\sqrt{3}}{2} = \frac{x}{8}$ $\cos 30° = \frac{\sqrt{3}}{2}$.

$4\sqrt{3} = x$ Multiply each side by 8.

The value of x is $4\sqrt{3}$ or about 6.9.

Study Tip

Common Misconception

The $\cos^{-1} x$ on a graphing calculator does *not* find $\frac{1}{\cos x}$. To find sec x or $\frac{1}{\cos x}$, find cos x and then use the [x^{-1}] key.

A calculator can be used to find the value of trigonometric functions for *any* angle, not just the special angles mentioned. Use [SIN], [COS], and [TAN] for sine, cosine, and tangent. Use these keys and the reciprocal key, [x^{-1}], for cosecant, secant, and cotangent. Be sure your calculator is in degree mode.

www.algebra2.com/extra_examples

RIGHT TRIANGLE PROBLEMS

In-Class Example Power Point®

3 Write an equation involving sin, cos, or tan that can be used to find the value of x. Then solve the equation. Round to the nearest tenth.

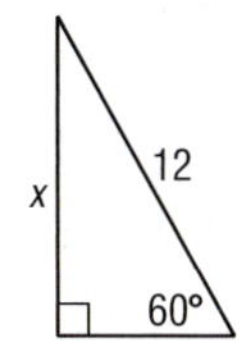

$\sin 60° = \frac{x}{12}$; $6\sqrt{3}$ or about 10.4

Teaching Tip On a scientific calculator (in contrast to a graphing calculator), the sequence may be to enter the angle measure, such as 20, first, and then press the [TAN] key.

Tips for New Teachers

Intervention The lengths of the two special triangles in the Key Concept box on this page are critical. Have students learn how to recreate these triangles themselves from memory. From the triangles, they can generate the entire table shown in the Key Concept box.

In-Class Examples

Power Point®

4 Solve $\triangle XYZ$. Round measures of sides to the nearest tenth and measures of angles to the nearest degree.

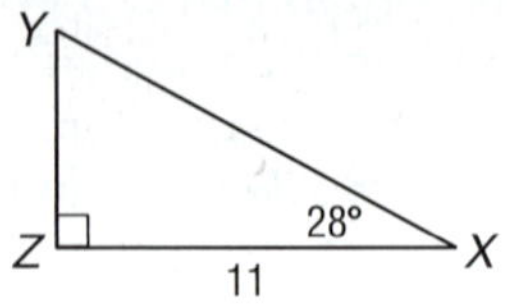

$Y = 62°$, $x \approx 5.8$, $z \approx 12.5$

5 Solve $\triangle ABC$. Round measures of sides to the nearest tenth and measures of angles to the nearest degree.

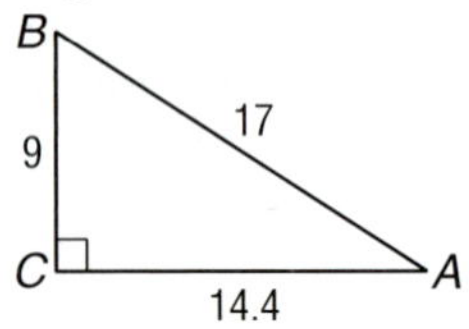

$A \approx 32°$, $B \approx 58°$

Concept Check

Solving Right Triangles What is the minimum information you have to have about a right triangle to solve it? **two of the following: measure of an acute angle, length of a leg, length of the hypotenuse**

Here are some calculator examples.

cos 46° **KEYSTROKES:** [COS] 46 [ENTER] .6946583705

cot 20° **KEYSTROKES:** [TAN] 20 [ENTER] [x^{-1}] [ENTER] 2.747477419

If you know the measures of any two sides of a right triangle or the measures of one side and one acute angle, you can determine the measures of all the sides and angles of the triangle. This process of finding the missing measures is known as **solving a right triangle**.

Study Tip

Error in Measurement

The value of z in Example 4 is found using the secant instead of using the Pythagorean Theorem. This is because the secant uses values given in the problem rather than calculated values.

Example 4 Solve a Right Triangle

Solve $\triangle XYZ$. Round measures of sides to the nearest tenth and measures of angles to the nearest degree.

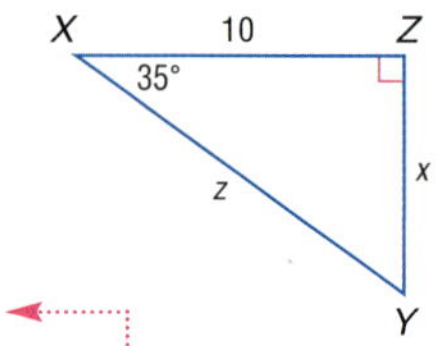

You know the measures of one side, one acute angle, and the right angle. You need to find x, z, and Y.

Find x and z.

$$\tan 35° = \frac{x}{10} \qquad \sec 35° = \frac{z}{10}$$

$$10 \tan 35° = x \qquad \frac{1}{\cos 35°} = \frac{z}{10}$$

$$7.0 \approx x \qquad \frac{10}{\cos 35°} = z$$

$$12.2 \approx z$$

Find Y.

$35° + Y = 90°$ Angles X and Y are complementary.

$Y = 55°$ Solve for Y.

Therefore, $Y = 55°$, $x \approx 7.0$, and $z \approx 12.2$.

TEACHING TIP

An alternative method that is equally accurate is to solve the equation $\cos 35° = \frac{10}{z}$.

Use the inverse capabilities of your calculator to find the measure of an angle when one of its trigonometric ratios is known. For example, use the SIN^{-1} function to find the measure of an angle when the sine of the angle is known. *You will learn more about inverses of trigonometric functions in Lesson 13-7.*

Example 5 Find Missing Angle Measures of Right Triangles

Solve $\triangle ABC$. Round measures of sides to the nearest tenth and measures of angles to the nearest degree.

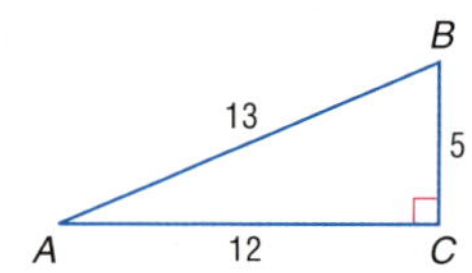

You know the measures of the sides. You need to find A and B.

Find A. $\sin A = \frac{5}{13}$ $\quad \sin A = \frac{\text{opp}}{\text{hyp}}$

Use a calculator and the SIN^{-1} function to find the angle whose sine is $\frac{5}{13}$.

KEYSTROKES: [2nd] [SIN^{-1}] 5 [÷] 13 [)] [ENTER] 22.61986495

To the nearest degree, $A \approx 23°$.

Find B. $23° + B \approx 90°$ Angles A and B are complementary.

$B \approx 67°$ Solve for B.

Therefore, $A \approx 23°$ and $B \approx 67°$.

DAILY INTERVENTION

Differentiated Instruction

Visual/Spatial Have students use a stack of books and a notebook to model a ramp and investigate how steep the ramp needs to be for a toy car to roll down it without being pushed. Have them report their results in terms of the trigonometric functions of a right triangle.

704 Chapter 13 Trigonometric Functions

Trigonometry has many practical applications. Among the most important is the ability to find distances or lengths that either cannot be measured directly or are not easily measured directly.

Example 6 Indirect Measurement

BRIDGE CONSTRUCTION **In order to construct a bridge across a river, the width of the river at that location must be determined. Suppose a stake is planted on one side of the river directly across from a second stake on the opposite side. At a distance 50 meters to the left of the stake, an angle of 82° is measured between the two stakes. Find the width of the river.**

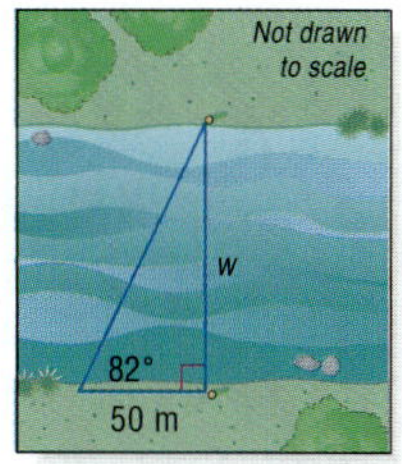

Let w represent the width of the river at that location. Write an equation using a trigonometric function that involves the ratio of the distance w and 50.

$$\tan 82° = \frac{w}{50} \qquad \tan \theta = \frac{\text{opp}}{\text{adj}}$$

$$50 \tan 82° = w \qquad \text{Multiply each side by 50.}$$

$$355.8 \approx w \qquad \text{Use a calculator.}$$

The width of the river is about 355.8 meters.

Some applications of trigonometry use an angle of elevation or depression. In the figure at the right, the angle formed by the line of sight from the observer and a line parallel to the ground is called the **angle of elevation**. The angle formed by the line of sight from the plane and a line parallel to the ground is called the **angle of depression**.

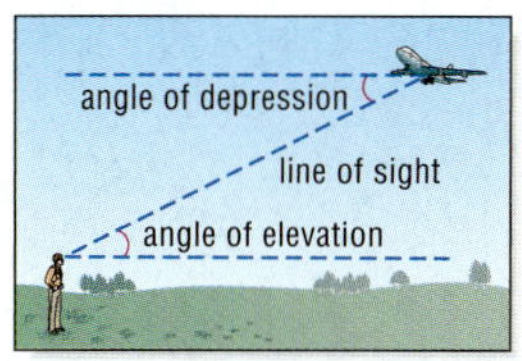

The angle of elevation and the angle of depression are congruent since they are alternate interior angles of parallel lines.

More About. . .

Skiing

The average annual snowfall in Snowbird, Utah, is 500 inches. The longest designated run there is Chip's Run, at 2.5 miles.

Source: www.utahskiing.com

Example 7 Use an Angle of Elevation

SKIING **The Aerial run in Snowbird, Utah, has an angle of elevation of 20.2°. Its vertical drop is 2900 feet. Estimate the length of this run.**

Let ℓ represent the length of the run. Write an equation using a trigonometric function that involves the ratio of ℓ and 2900.

$$\sin 20.2° = \frac{2900}{\ell} \qquad \sin \theta = \frac{\text{opp}}{\text{hyp}}$$

$$\ell = \frac{2900}{\sin 20.2°} \qquad \text{Solve for } \ell.$$

$$\ell \approx 8398.5 \qquad \text{Use a calculator.}$$

The length of the run is about 8399 feet.

In-Class Examples

Power Point®

6 **BRIDGE CONSTRUCTION** Suppose, in a situation similar to that of Example 6 in the Student Edition, the angle was measured at a distance 30 meters away from the stake, and found to be 55°. Find the width of the river. **about 42.8 m**

7 **SKIING** A run has an angle of elevation of 15.7° and a vertical drop of 1800 feet. Estimate the length of this run. **about 6652 ft**

Lesson 13-1 Right Triangle Trigonometry 705

3 Practice/Apply

Study Notebook

Have students—

- add the definitions/examples of the vocabulary terms to their Vocabulary Builder worksheets for Chapter 13.
- include any other item(s) that they find helpful in mastering the skills in this lesson.

About the Exercises...

Organization by Objective

- **Trigonometric Values:** 15–20, 27, 28
- **Right Triangle Problems:** 21–26, 29–46

Odd/Even Assignments

Exercises 15–42 are structured so that students practice the same concepts whether they are assigned odd or even problems.

Assignment Guide

Basic: 15, 17, 21–37 odd, 41, 47–62

Average: 15–41 odd, 43–45, 47–62

Advanced: 16–42 even, 43–58 (optional: 59–62)

Check for Understanding

Concept Check

1. **Define** the word *trigonometry*. **1–3. See margin.**
2. **OPEN ENDED** Draw a right triangle. Label one of its acute angles θ. Then, label the hypotenuse, the leg adjacent to θ, and the leg opposite θ.
3. **Find a counterexample** to the following statement.
   It is always possible to solve a right triangle.

Guided Practice

GUIDED PRACTICE KEY	
Exercises	Examples
4–6	1
7	3
8, 11, 12	5
9, 10	4
13	6, 7
14	2

Find the values of the six trigonometric functions for angle θ. **4–6. See margin.**

4. 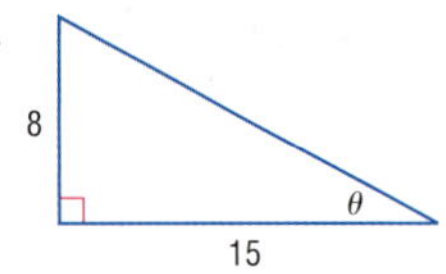

5. 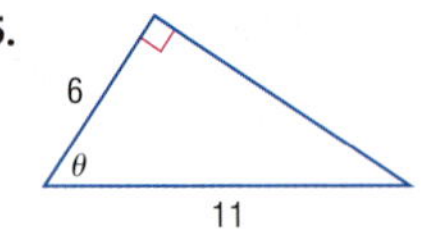

6. 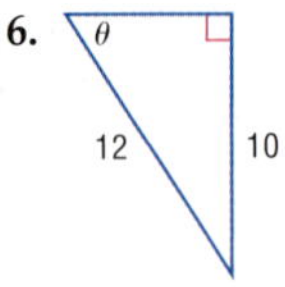

Write an equation involving sin, cos, or tan that can be used to find x. Then solve the equation. Round measures of sides to the nearest tenth and angles to the nearest degree.

7. 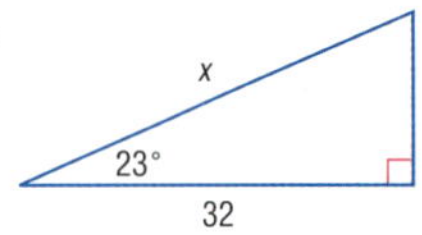

$\cos 23° = \frac{32}{x}$; $x \approx 34.8$

8. 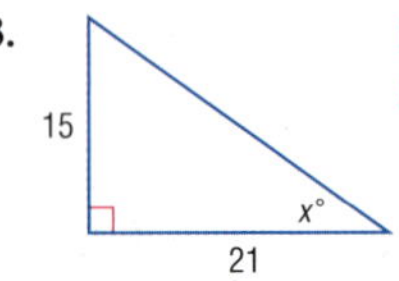

$\tan x° = \frac{15}{21}$; $x \approx 36$

Solve $\triangle ABC$ by using the given measurements. Round measures of sides to the nearest tenth and measures of angles to the nearest degree.

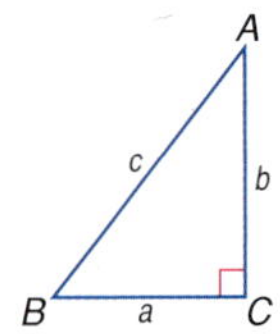

9. $A = 45°, b = 6$
10. $B = 56°, c = 16$
11. $b = 7, c = 18$ $a \approx 16.6$, $A \approx 67°$, $B \approx 23°$
12. $a = 14, b = 13$ $c \approx 19.1$, $A \approx 47°$, $B \approx 43°$

9. $B = 45°$, $a = 6$, $c \approx 8.5$
10. $A = 34°$, $a \approx 8.9$, $b \approx 13.3$

13. **AVIATION** When landing, a jet will average a 3° angle of descent. What is the altitude x, to the nearest foot, of a jet on final descent as it passes over an airport beacon 6 miles from the start of the runway? **1660 ft**

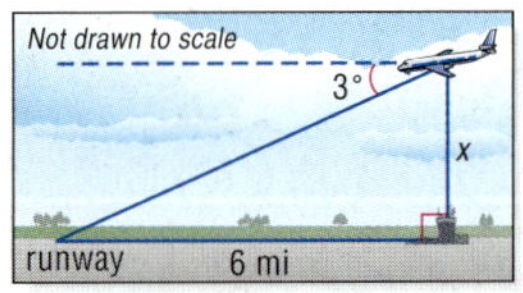

Standardized Test Practice

14. If $\tan \theta = 3$, find the value of $\sin \theta$. **B**

(A) $\frac{3}{10}$ (B) $\frac{3\sqrt{10}}{10}$ (C) $\frac{10}{3}$ (D) $\frac{1}{3}$

★ indicates increased difficulty

Practice and Apply

Homework Help	
For Exercises	See Examples
15–20	1
21–24, 27, 28	3
25, 26, 37–40	5
29–36	4
41–46	6, 7
49	2

Extra Practice See page 857.

Find the values of the six trigonometric functions for angle θ. **15–20. See pp. 759A–759D.**

15. 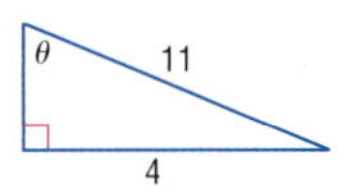

16. 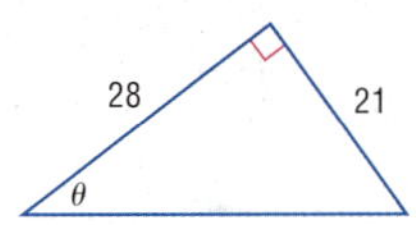

17. 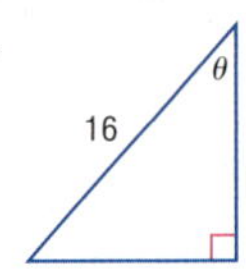

18. 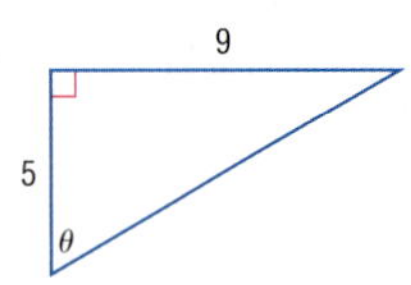

★ 19. 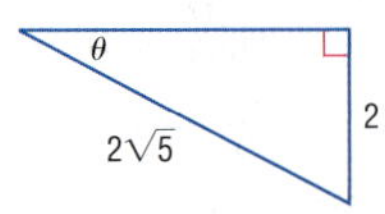

★ 20. 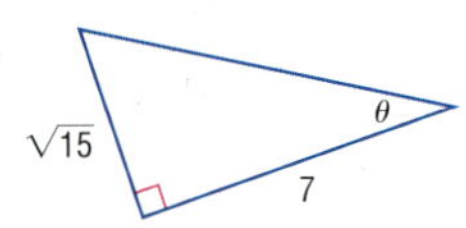

Answers

1. Trigonometry is the study of the relationships between the angles and sides of a right triangle.
2. 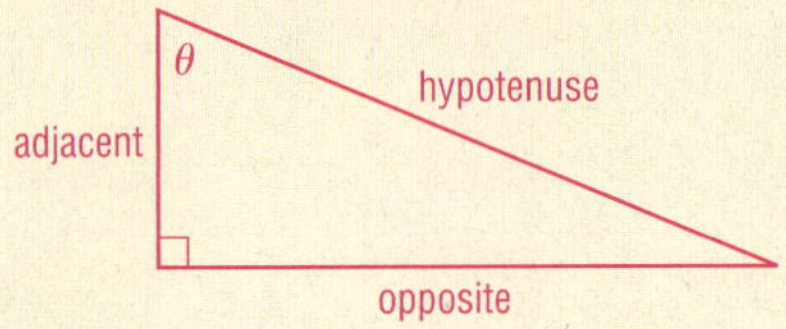

3. Given only the measures of the angles of a right triangle, you cannot find the measures of its sides.
4. $\sin \theta = \frac{8}{17}$; $\cos \theta = \frac{15}{17}$; $\tan \theta = \frac{8}{15}$; $\csc \theta = \frac{17}{8}$; $\sec \theta = \frac{17}{15}$; $\cot \theta = \frac{15}{8}$
5. $\sin \theta = \frac{\sqrt{85}}{11}$; $\cos \theta = \frac{6}{11}$; $\tan \theta = \frac{\sqrt{85}}{6}$; $\csc \theta = \frac{11\sqrt{85}}{85}$; $\sec \theta = \frac{11}{6}$; $\cot \theta = \frac{6\sqrt{85}}{85}$
6. $\sin \theta = \frac{5}{6}$; $\cos \theta = \frac{\sqrt{11}}{6}$; $\tan \theta = \frac{5\sqrt{11}}{11}$; $\csc \theta = \frac{6}{5}$; $\sec \theta = \frac{6\sqrt{11}}{11}$; $\cot \theta = \frac{\sqrt{11}}{5}$

Write an equation involving sin, cos, or tan that can be used to find x. Then solve the equation. Round measures of sides to the nearest tenth and measures of angles to the nearest degree.

21. $\tan 30° = \frac{x}{10}$; $x \approx 5.8$
22. $\cos 60° = \frac{3}{x}$; $x = 6$
23. $\sin 54° = \frac{17.8}{x}$; $x \approx 22.0$
24. $\tan 17.5° = \frac{x}{23.7}$; $x \approx 7.5$
25. $\cos x° = \frac{15}{36}$; $x \approx 65$
26. $\sin x° = \frac{16}{22}$; $x \approx 47$
27–28. See pp. 759A–759D.

21.

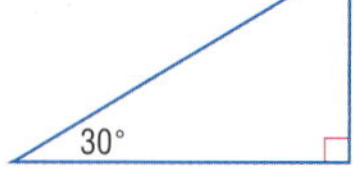

22.

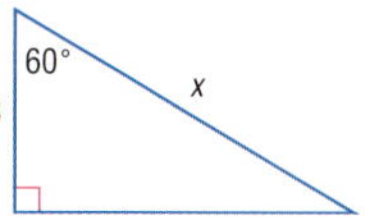

23.

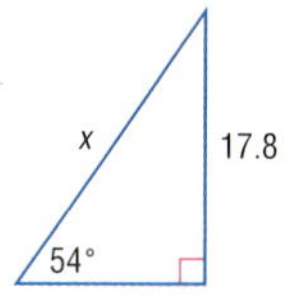

24.

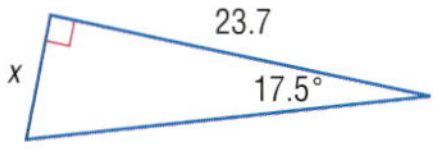

25.

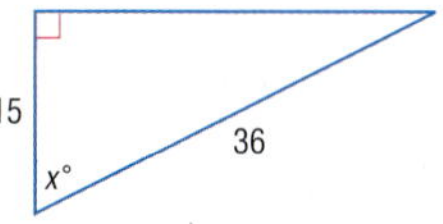

26.

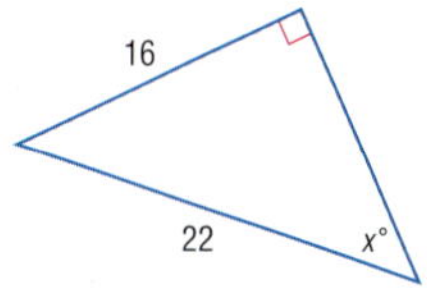

27. Using the 30°-60°-90° triangle shown on page 703, verify each value.

a. $\sin 30° = \frac{1}{2}$ b. $\cos 30° = \frac{\sqrt{3}}{2}$ c. $\sin 60° = \frac{\sqrt{3}}{2}$

28. Using the 45°-45°-90° triangle shown on page 703, verify each value.

a. $\sin 45° = \frac{\sqrt{2}}{2}$ b. $\cos 45° = \frac{\sqrt{2}}{2}$ c. $\tan 45° = 1$

Solve $\triangle ABC$ by using the given measurements. Round measures of sides to the nearest tenth and measures of angles to the nearest degree. 29–40. See pp. 759A–759D.

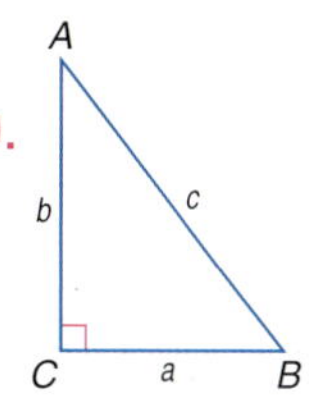

29. $A = 16°, c = 14$ **30.** $B = 27°, b = 7$

31. $A = 34°, a = 10$ **32.** $B = 15°, c = 25$

33. $B = 30°, b = 11$ **34.** $A = 45°, c = 7\sqrt{2}$

35. $B = 18°, a = \sqrt{15}$ **36.** $A = 10°, b = 15$

37. $b = 6, c = 13$ **38.** $a = 4, c = 9$

★ **39.** $\tan B = \frac{7}{8}, b = 7$ ★ **40.** $\sin A = \frac{1}{3}, a = 5$

Career Choices

Surveyor

Land surveyors manage survey parties that measure distances, directions, and angles between points, lines, and contours on Earth's surface.

Online Research
For information about a career as a surveyor, visit: www.algebra2.com/careers

41. TRAVEL In a sightseeing boat near the base of the Horseshoe Falls at Niagara Falls, a passenger estimates the angle of elevation to the top of the falls to be 30°. If the Horseshoe Falls are 173 feet high, what is the distance from the boat to the base of the falls? about 300 ft

42. SURVEYING A surveyor stands 100 feet from a building and sights the top of the building at a 55° angle of elevation. Find the height of the building. about 142.8 ft

EXERCISE For Exercises 43 and 44, use the following information.

A preprogrammed workout on a treadmill consists of intervals walking at various rates and angles of incline. A 1% incline means 1 unit of vertical rise for every 100 units of horizontal run.

43. At what angle, with respect to the horizontal, is the treadmill bed when set at a 10% incline? Round to the nearest degree. about 6°

44. If the treadmill bed is 40 inches long, what is the vertical rise when set at an 8% incline? about 3.2 in.

★ **45. GEOMETRY** Find the area of the regular hexagon with point O as its center. (*Hint*: First find the value of x.) 93.53 units^2

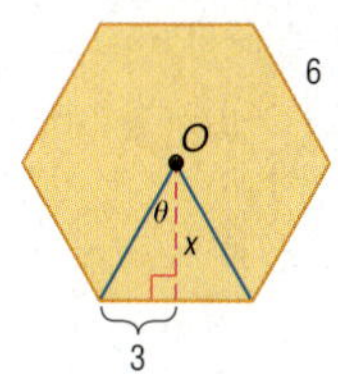

www.algebra2.com/self_check_quiz

Study Guide and Intervention, p. 775 (shown) and p. 776

Trigonometric Values

Trigonometric Functions: If θ is the measure of an acute angle of a right triangle, *opp* is the measure of the leg opposite θ, *adj* is the measure of the leg adjacent to θ, and *hyp* is the measure of the hypotenuse, then the following are true.

$\sin\theta = \frac{opp}{hyp}$ $\cos\theta = \frac{adj}{hyp}$ $\tan\theta = \frac{opp}{adj}$
$\csc\theta = \frac{hyp}{opp}$ $\sec\theta = \frac{hyp}{adj}$ $\cot\theta = \frac{adj}{opp}$

Example Find the values of the six trigonometric functions for angle θ.

Use the Pythagorean Theorem to find x, the measure of the leg opposite θ.

$x^2 + 7^2 = 9^2$ Pythagorean Theorem
$x^2 + 49 = 81$ Simplify.
$x^2 = 32$ Subtract 49 from each side.
$x = \sqrt{32}$ or $4\sqrt{2}$ Take the square root of each side.

Use opp = $4\sqrt{2}$, adj = 7, and hyp = 9 to write each trigonometric ratio.

$\sin\theta = \frac{4\sqrt{2}}{9}$ $\cos\theta = \frac{7}{9}$ $\tan\theta = \frac{4\sqrt{2}}{7}$ $\csc\theta = \frac{9\sqrt{2}}{8}$ $\sec\theta = \frac{9}{7}$ $\cot\theta = \frac{7\sqrt{2}}{8}$

Exercises

Find the values of the six trigonometric functions for angle θ.

1. $\sin\theta = \frac{5}{13}$; $\cos\theta = \frac{12}{13}$; $\tan\theta = \frac{5}{12}$; $\csc\theta = \frac{13}{5}$; $\sec\theta = \frac{13}{12}$; $\cot\theta = \frac{12}{5}$
2. $\sin\theta = \frac{4}{5}$; $\cos\theta = \frac{3}{5}$; $\tan\theta = \frac{4}{3}$; $\csc\theta = \frac{5}{4}$; $\sec\theta = \frac{5}{3}$; $\cot\theta = \frac{3}{4}$
3. $\sin\theta = \frac{8}{17}$; $\cos\theta = \frac{15}{17}$; $\tan\theta = \frac{8}{15}$; $\csc\theta = \frac{17}{8}$; $\sec\theta = \frac{17}{15}$; $\cot\theta = \frac{15}{8}$
4. $\sin\theta = \frac{\sqrt{2}}{2}$; $\cos\theta = \frac{\sqrt{2}}{2}$; $\tan\theta = 1$; $\csc\theta = \sqrt{2}$; $\sec\theta = \sqrt{2}$; $\cot\theta = 1$
5. $\sin\theta = \frac{\sqrt{3}}{2}$; $\cos\theta = \frac{1}{2}$; $\tan\theta = \sqrt{3}$; $\csc\theta = \frac{2\sqrt{3}}{3}$; $\sec\theta = 2$; $\cot\theta = \frac{\sqrt{3}}{3}$
6. $\sin\theta = \frac{5\sqrt{61}}{61}$; $\cos\theta = \frac{6\sqrt{61}}{61}$; $\tan\theta = \frac{5}{6}$; $\csc\theta = \frac{\sqrt{61}}{5}$; $\sec\theta = \frac{\sqrt{61}}{6}$; $\cot\theta = \frac{6}{5}$

Skills Practice, p. 777 and Practice, p. 778 (shown)

Find the values of the six trigonometric functions for angle θ.

1. $\sin\theta = \frac{15}{17}$, $\cos\theta = \frac{8}{17}$, $\tan\theta = \frac{15}{8}$, $\csc\theta = \frac{17}{15}$, $\sec\theta = \frac{17}{8}$, $\cot\theta = \frac{8}{15}$
2. $\sin\theta = \frac{5}{11}$, $\cos\theta = \frac{4\sqrt{6}}{11}$, $\tan\theta = \frac{5\sqrt{6}}{24}$, $\csc\theta = \frac{11}{5}$, $\sec\theta = \frac{11\sqrt{6}}{24}$, $\cot\theta = \frac{4\sqrt{6}}{5}$
3. $\sin\theta = \frac{1}{2}$, $\cos\theta = \frac{\sqrt{3}}{2}$, $\tan\theta = \frac{\sqrt{3}}{3}$, $\csc\theta = 2$, $\sec\theta = \frac{2\sqrt{3}}{3}$, $\cot\theta = \sqrt{3}$

Write an equation involving sin, cos, or tan that can be used to find x. Then solve the equation. Round measures of sides to the nearest tenth and measures of angles to the nearest degree.

4. $\tan 30° = \frac{x}{7}$, $x \approx 4.0$
5. $\sin 20° = \frac{x}{32}$, $x \approx 10.9$
6. $\tan 49° = \frac{17}{x}$, $x \approx 14.8$
7. $\cos 41° = \frac{28}{x}$, $x \approx 37.1$
8. $\tan x° = \frac{19.2}{17}$, $x \approx 48$
9. $\sin x° = \frac{7}{15.3}$, $x \approx 27$

Solve $\triangle ABC$ by using the given measurements. Round measures of sides to the nearest tenth and measures of angles to the nearest degree.

10. $A = 35°, a = 12$ — $b \approx 17.1$, $c \approx 20.9$, $B = 55°$
11. $B = 71°, b = 25$ — $a \approx 8.6$, $c \approx 26.4$, $A = 19°$
12. $B = 36°, c = 8$ — $a \approx 6.5$, $b \approx 4.7$, $A = 54°$
13. $a = 4, b = 7$ — $c \approx 8.1$, $A \approx 30°$, $B \approx 60°$
14. $A = 17°, c = 3.2$ — $a \approx 0.9$, $b \approx 3.1$, $B = 73°$
15. $b = 52, c = 95$ — $a \approx 79.5$, $A \approx 33°$, $B \approx 57°$

16. **SURVEYING** John stands 150 meters from a water tower and sights the top at an angle of elevation of 36°. How tall is the tower? Round to the nearest meter. 109 m

Reading to Learn Mathematics, p. 779 ELL

Pre-Activity How is trigonometry used in building construction?

Read the introduction to Lesson 13-1 at the top of page 701 in your textbook.

If a different ramp is built so that the angle shown in the figure has a tangent of $\frac{1}{14}$, will this ramp meet, exceed, or fail to meet ADA regulations? exceed

Reading the Lesson

1. Refer to the triangle at the right. Match each trigonometric function with the correct ratio.

i. $\frac{r}{t}$ ii. $\frac{r}{s}$ iii. $\frac{t}{s}$ iv. $\frac{s}{t}$ v. $\frac{s}{r}$ vi. $\frac{t}{r}$

a. $\sin\theta$ iv b. $\tan\theta$ v c. $\sec\theta$ iii
d. $\cot\theta$ ii e. $\cos\theta$ i f. $\csc\theta$ vi

2. Refer to the Key Concept box on page 703 in your textbook. Use the drawings of the 30°-60°-90° triangle and 45°-45°-90° triangle and/or the table to complete the following.

a. The tangent of 45° and the cotangent of 45° are equal.
b. The sine of 30° is equal to the cosine of 60°.
c. The sine and cosine of 45° are equal.
d. The reciprocal of the cosecant of 60° is the sine of 60°.
e. The reciprocal of the cosine of 30° is the cosecant of 60°.
f. The reciprocal of the tangent of 60° is the tangent of 30°.

Helping You Remember

3. In studying trigonometry, it is important for you to know the relationships between the lengths of the sides of a 30°-60°-90° triangle. If you remember just one fact about this triangle, you will always be able to figure out the lengths of all the sides. What fact can you use, and why is it enough?

Sample answer: The shorter leg is half as long as the hypotenuse. You can use the Pythagorean Theorem to find the length of the longer leg.

Enrichment, p. 780

The Angle of Repose

Suppose you place a block of wood on an inclined plane, as shown at the right. If the angle, θ, at which the plane is inclined from the horizontal is very small, the block will not move. If you increase the angle, the block will eventually overcome the force of friction and start to slide down the plane.

At the instant the block begins to slide, the angle formed by the plane is called the angle of friction, or the angle of repose.

For situations in which the block and plane are smooth but unlubricated, the angle of repose depends *only* on the types of materials in the block and the plane. The angle is independent of the area of contact between the two surfaces and of the weight of the block.

The drawing at the right shows how to use vectors to find a coefficient of friction. This coefficient varies with ...

$F = W\sin\theta$ $N = W\cos\theta$

4 Assess

Open-Ended Assessment

Modeling Have students design a ramp, specifying the angle and how far the end of the ramp is from the place on the ground where it begins. Have them draw a picture and show steps to find the length of the ramp. Does the ramp meet ADA requirements?

Getting Ready for Lesson 13-2

PREREQUISITE SKILL Lesson 13-2 presents changing between measuring angles in radians and in degrees. Students will use their familiarity with converting units as they write the measures of angles in both radians and degrees. Exercises 59–62 should be used to determine your students' familiarity with dimensional analysis.

WebQuest

You can use the tangent ratio to determine the maximum height of a rocket. Visit www.algebra2.com/webquest to continue work on your WebQuest project.

46. **GEOLOGY** A geologist measured a 40° angle of elevation to the top of a mountain. After moving 0.5 kilometer farther away, the angle of elevation was 34°. How high is the top of the mountain? (*Hint*: Write a system of equations in two variables.) **about 1.7 km high**

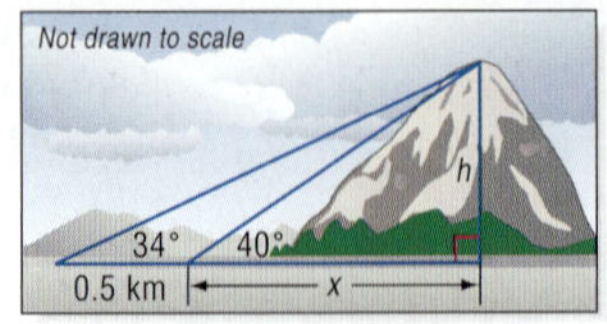

47. **CRITICAL THINKING** Explain why the sine and cosine of an acute angle are never greater than 1, but the tangent of an acute angle may be greater than 1. **See margin.**

48. **WRITING IN MATH** Answer the question that was posed at the beginning of the lesson. **See margin.**

How is trigonometry used in building construction?

Include the following in your answer:
- an explanation as to why the ratio of vertical rise to horizontal run on an entrance ramp is the tangent of the angle the ramp makes with the horizontal, and
- an explanation of how an architect can use the tangent ratio to ensure that all the ramps he or she designs meet the ADA requirement.

Standardized Test Practice

49. If the secant of an angle θ is $\frac{25}{7}$, what is the sine of angle θ? **C**

Ⓐ $\frac{5}{25}$ Ⓑ $\frac{7}{25}$ Ⓒ $\frac{24}{25}$ Ⓓ $\frac{25}{7}$

50. **GRID IN** The tailgate of a moving truck is 2 feet above the ground. The incline of the ramp used for loading the truck is 15° as shown. Find the length of the ramp to the nearest tenth of a foot. **7.7**

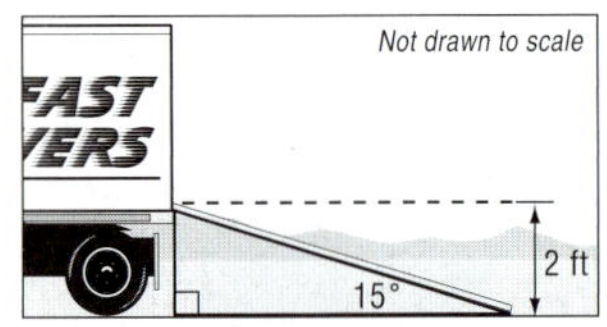

Maintain Your Skills

Mixed Review

Determine whether each situation would produce a random sample. Write *yes* or *no* and explain your answer. *(Lesson 12-9)* **51–52. See margin for explanation.**

51. surveying band members to find the most popular type of music at your school **no**

52. surveying people coming into a post office to find out what color cars are most popular **yes**

Find each probability if a coin is tossed 4 times. *(Lesson 12-8)*

53. P(exactly 2 heads) $\frac{3}{8}$

54. P(4 heads) $\frac{1}{16}$

55. P(at least 1 head) $\frac{15}{16}$

Solve each equation. *(Lesson 7-3)*

56. $y^4 - 64 = 0$ $\{\pm 2\sqrt{2}, \pm 2i\sqrt{2}\}$

57. $x^5 - 5x^3 + 4x = 0$ $\{-2, -1, 0, 1, 2\}$

58. $d + \sqrt{d} - 132 = 0$ $\{121\}$

Getting Ready for the Next Lesson

PREREQUISITE SKILL Find each product. Include the appropriate units with your answer. *(To review **dimensional analysis**, see Lesson 5-1.)*

59. 5 gallons $\left(\frac{4 \text{ quarts}}{1 \text{ gallon}}\right)$ **20 qt**

60. 6.8 miles $\left(\frac{5280 \text{ feet}}{1 \text{ mile}}\right)$ **35,904 ft**

61. $\left(\frac{2 \text{ square meters}}{5 \text{ dollars}}\right)$ 30 dollars **12 m^2**

62. $\left(\frac{4 \text{ liters}}{5 \text{ minutes}}\right)$ 60 minutes **48 L**

Answers

47. The sine and cosine ratios of acute angles of right triangles each have the longest measure of the triangle, the hypotenuse, as their denominator. A fraction whose denominator is greater than its numerator is less than 1. The tangent ratio of an acute angle of a right triangle does not involve the measure of the hypotenuse, $\frac{\text{opp}}{\text{adj}}$. If the measure of the opposite side is greater than the measure of the adjacent side, the tangent ratio is greater than 1. If the measure of the opposite side is less than the measure of the adjacent side, the tangent ratio is less than 1.

48. When construction involves right triangles, including building ramps, designing buildings, or surveying land before building, trigonometry is likely to be used. Answers should include the following.
- If you view the ramp from the side then the vertical rise is opposite the angle that the ramp makes with the horizontal. Similarly, the horizontal run is the adjacent side. So the tangent of the angle is the ratio of the rise to the run or the slope of the ramp.
- Given the ratio of the slope of ramp, you can find the angle of inclination by calculating $\tan^{-1}$ of this ratio.

51. Band members may be more likely to like the same kinds of music.

52. This sample is random since different kinds of people go to the post office.

13-2 Angles and Angle Measure

What You'll Learn

- Change radian measure to degree measure and vice versa.
- Identify coterminal angles.

How can angles be used to describe circular motion?

The Ferris wheel at Navy Pier in Chicago has a 140-foot diameter and 40 gondolas equally spaced around its circumference. The average angular velocity ω of one of the gondolas is given by $\omega = \frac{\theta}{t}$, where θ is the angle through which the gondola has revolved after a specified amount of time t. For example, if a gondola revolves through an angle of 225° in 40 seconds, then its average angular velocity is 225° ÷ 40 or about 5.6° per second.

Vocabulary

- initial side
- terminal side
- standard position
- unit circle
- radian
- coterminal angles

TEACHING TIP

ω is the lowercase Greek letter omega.

ANGLE MEASUREMENT What does an angle measuring 225° look like? In Lesson 13-1, you worked only with acute angles, those measuring between 0° and 90°, but angles can have *any* real number measurement.

On a coordinate plane, an angle may be generated by the rotation of two rays that share a fixed endpoint at the origin. One ray, called the **initial side** of the angle, is fixed along the positive x-axis. The other ray, called the **terminal side** of the angle, can rotate about the center. An angle positioned so that its vertex is at the origin and its initial side is along the positive x-axis is said to be in **standard position**.

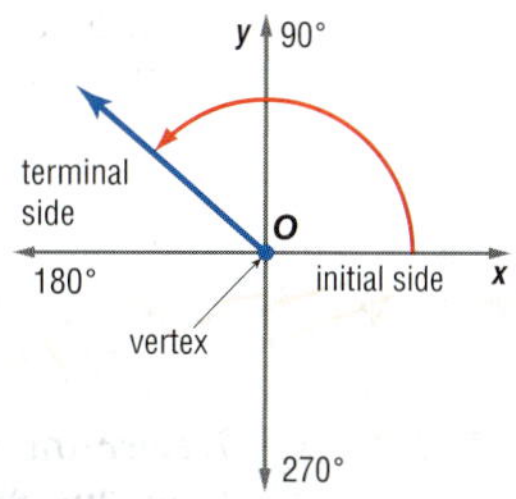

Study Tip

Reading Math

In trigonometry, an angle is sometimes referred to as an *angle of rotation*.

The measure of an angle is determined by the amount and direction of rotation from the initial side to the terminal side.

Positive Angle Measure
counterclockwise

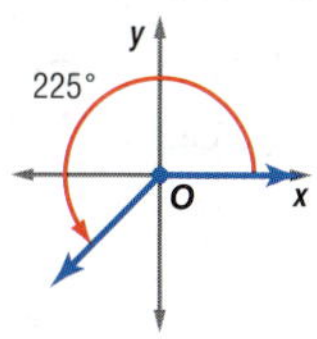

Negative Angle Measure
clockwise

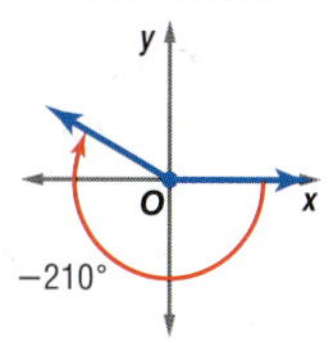

When terminal sides rotate, they may sometimes make one or more revolutions. An angle whose terminal side has made exactly one revolution has a measure of 360°.

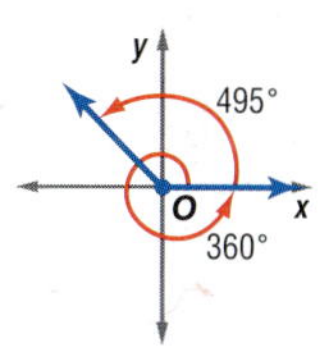

TEACHING TIP

Why are there 360° in one revolution instead of 100° or 1000°? The answer may lie in ancient Babylon where mathematicians developed a number system based on 60. For example, there were 60 bushels in a *mana* and 60 *mana* in a *talent*. Today, we further subdivide 1 degree into 60 minutes (60′) and 1 minute into 60 seconds (60″).

13-2 Lesson Notes

1 Focus

5-Minute Check Transparency 13-2 Use as a quiz or review of Lesson 13-1.

Mathematical Background notes are available for this lesson on p. 698C.

How can angles be used to describe circular motion?

Ask students:

- What is the space between one gondola and the next around the circumference? **11 ft**
- How many degrees of the circle are between one gondola and the next? **9°**
- What is the radius of the Ferris wheel? **70 ft**

Resource Manager

Workbook and Reproducible Masters

Chapter 13 Resource Masters

- Study Guide and Intervention, pp. 781–782
- Skills Practice, p. 783
- Practice, p. 784
- Reading to Learn Mathematics, p. 785
- Enrichment, p. 786
- Assessment, p. 831

Transparencies

5-Minute Check Transparency 13-2
Answer Key Transparencies

Technology

Interactive Chalkboard

2 Teach

ANGLE MEASUREMENT

In-Class Example

Power Point®

Teaching Tip Discuss with students the importance of marking the drawings of angles with the arrows as shown, and labeling the degrees.

1 Draw an angle with the given measure in standard position.

a. 210°

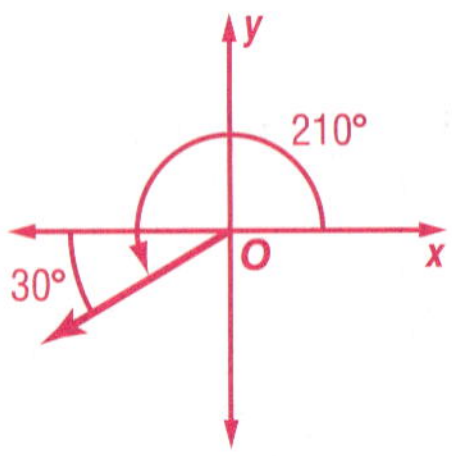

b. −45°

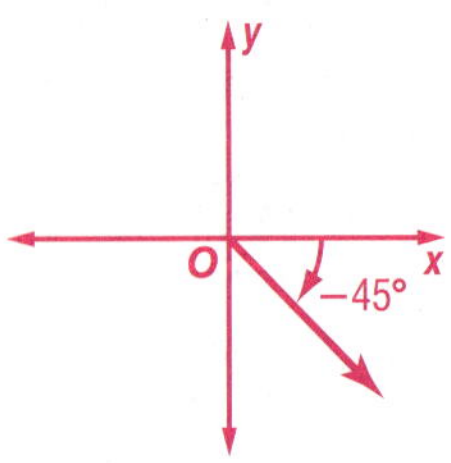

c. 540°

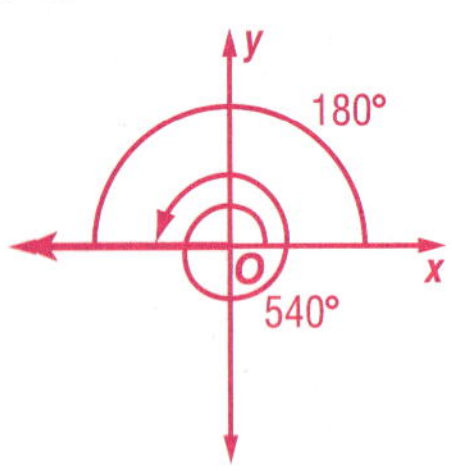

Example 1 Draw an Angle in Standard Position

Draw an angle with the given measure in standard position.

a. 240° $240° = 180° + 60°$
Draw the terminal side of the angle 60° counterclockwise past the negative x-axis.

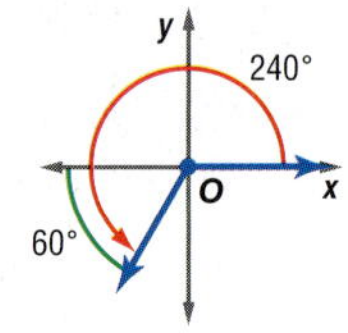

b. −30° The angle is negative.
Draw the terminal side of the angle 30° clockwise from the positive x-axis.

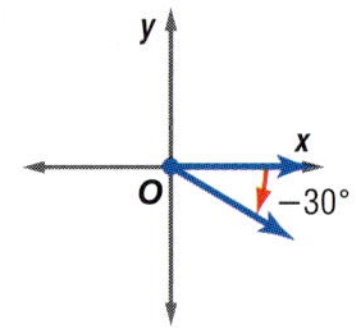

c. 450° $450° = 360° + 90°$
Draw the terminal side of the angle 90° counterclockwise past the positive x-axis.

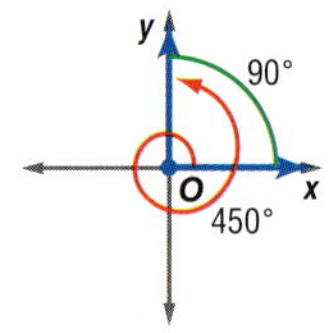

Another unit used to measure angles is a radian. The definition of a radian is based on the concept of a **unit circle**, which is a circle of radius 1 unit whose center is at the origin of a coordinate system. One **radian** is the measure of an angle θ in standard position whose rays intercept an arc of length 1 unit on the unit circle.

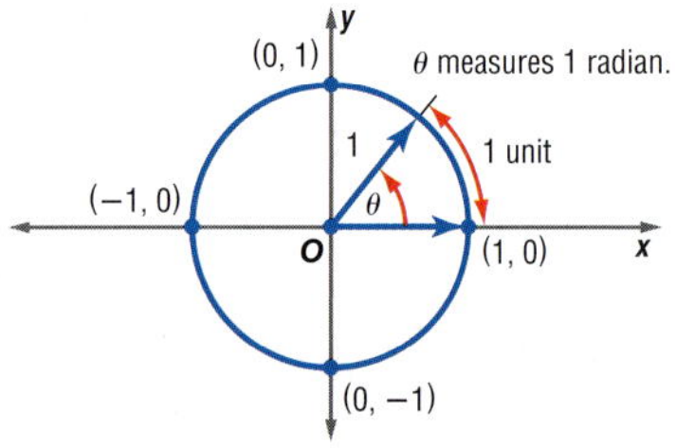

Study Tip

Radian Measure
As with degrees, the measure of an angle in radians is positive if its rotation is counterclockwise. The measure is negative if the rotation is clockwise.

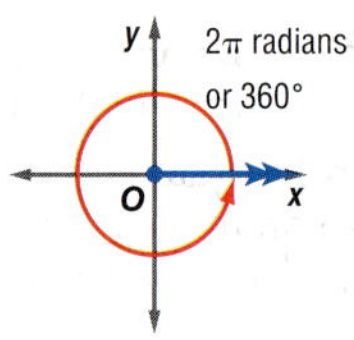

The circumference of any circle is $2\pi r$, where r is the radius measure. So the circumference of a unit circle is $2\pi(1)$ or 2π units. Therefore, an angle representing one complete revolution of the circle measures 2π radians. This same angle measures 360°. Therefore, the following equation is true.

$$2\pi \text{ radians} = 360°$$

To change angle measures from radians to degrees or vice versa, solve the equation above in terms of both units.

$2\pi \text{ radians} = 360°$	$2\pi \text{ radians} = 360°$
$\dfrac{2\pi \text{ radians}}{2\pi} = \dfrac{360°}{2\pi}$	$\dfrac{2\pi \text{ radians}}{360} = \dfrac{360°}{360}$
$1 \text{ radian} = \dfrac{180°}{\pi}$	$\dfrac{\pi \text{ radians}}{180} = 1°$
1 radian is about 57 degrees.	1 degree is about 0.0175 radian.

These equations suggest a method for converting between radian and degree measure.

710 Chapter 13 Trigonometric Functions

Key Concept — Radian and Degree Measure

- To rewrite the radian measure of an angle in degrees, multiply the number of radians by $\frac{180°}{\pi \text{ radians}}$.
- To rewrite the degree measure of an angle in radians, multiply the number of degrees by $\frac{\pi \text{ radians}}{180°}$.

Example 2 Convert Between Degree and Radian Measure

Rewrite the degree measure in radians and the radian measure in degrees.

a. 60°

$$60° = 60°\left(\frac{\pi \text{ radians}}{180°}\right)$$

$$= \frac{60\pi}{180} \text{ radians or } \frac{\pi}{3}$$

b. $-\frac{7\pi}{4}$

$$-\frac{7\pi}{4} = \left(-\frac{7\pi}{4} \text{ radians}\right)\left(\frac{180°}{\pi \text{ radians}}\right)$$

$$= -\frac{1260°}{4} \text{ or } -315°$$

Study Tip

Reading Math

The word *radian* is usually omitted when angles are expressed in radian measure. Thus, when no units are given for an angle measure, radian measure is implied.

You will find it useful to learn equivalent degree and radian measures for the special angles shown in the diagram at the right. This diagram is more easily learned by memorizing the equivalent degree and radian measures for the first quadrant and for 90°. All of the other special angles are multiples of these angles.

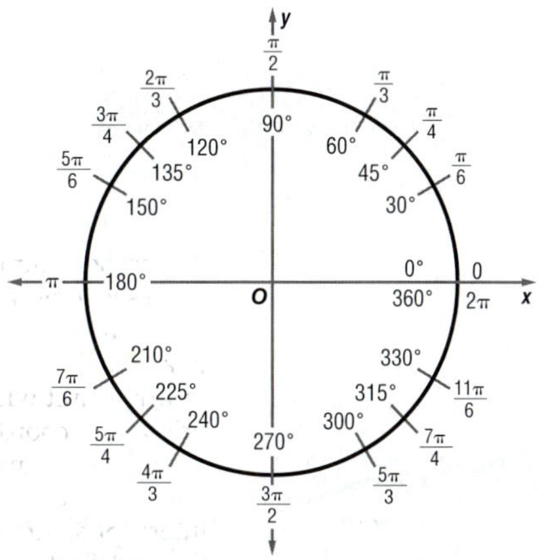

Example 3 Measure an Angle in Degrees and Radians

TIME **Find both the degree and radian measures of the angle through which the hour hand on a clock rotates from 1:00 P.M. to 3:00 P.M.**

The numbers on a clock divide it into 12 equal parts with 12 equal angles. The angle from 1 to 3 on the clock represents $\frac{2}{12}$ or $\frac{1}{6}$ of a complete rotation of 360°. $\frac{1}{6}$ of 360° is 60°.

Since the rotation is clockwise, the angle through which the hour hand rotates is negative. Therefore, the angle measures −60°.

60° has an equivalent radian measure of $\frac{\pi}{3}$. So the equivalent radian measure of −60° is $-\frac{\pi}{3}$.

COTERMINAL ANGLES If you graph a 405° angle and a 45° angle in standard position on the same coordinate plane, you will notice that the terminal side of the 405° angle is the same as the terminal side of the 45° angle. When two angles in standard position have the same terminal sides, they are called **coterminal angles**.

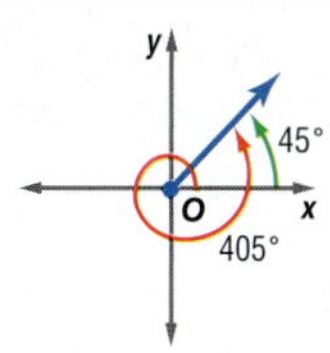

In-Class Examples

2 Rewrite the degree measure in radians and the radian measure in degrees.

a. 30° $\frac{\pi}{6}$

b. $-\frac{5\pi}{3}$ **−300°**

3 Find both the degree and radian measures of the angle through which the hour hand on a clock rotates from 6 P.M. to 7 P.M. **−30° or** $-\frac{\pi}{6}$

Tips for New Teachers

Intervention Students must be completely certain they know the definitions of the sine, cosine, and tangent ratios before they proceed with the rest of the chapter material. Have students create mnemonics, make color-coded index cards, or use other methods to help them review the ratios until they are confident of them.

Lesson 13-2 Angles and Angle Measure 711

COTERMINAL ANGLES

In-Class Example

Power Point®

4 Find one angle with positive measure and one angle with negative measure coterminal with each angle.

a. 210° **Sample answers: 570°, −150°**

b. $\frac{7\pi}{3}$ **Sample answers:** $\frac{13\pi}{3}$, $-\frac{5\pi}{3}$

Answers

2. In a circle of radius *r* units, one radian is the measure of an angle whose rays intercept an arc length of *r* units.

3.

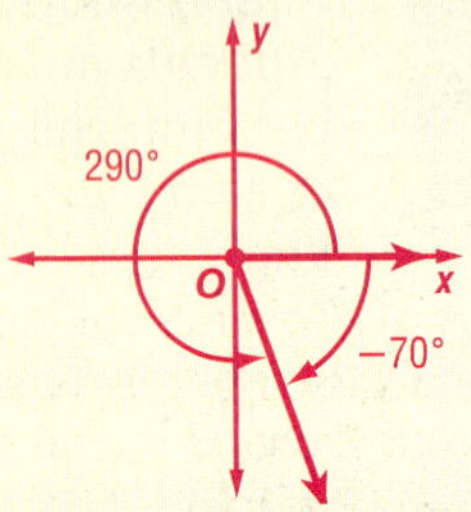

4.

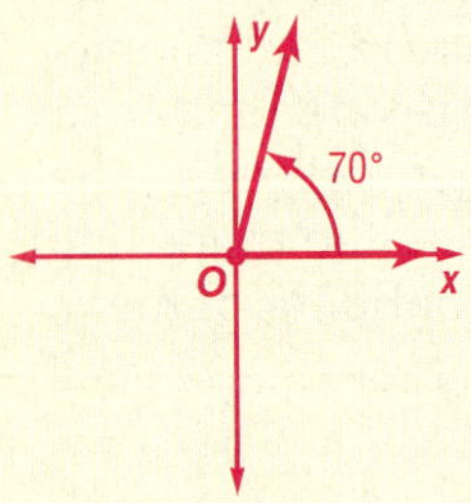

5.

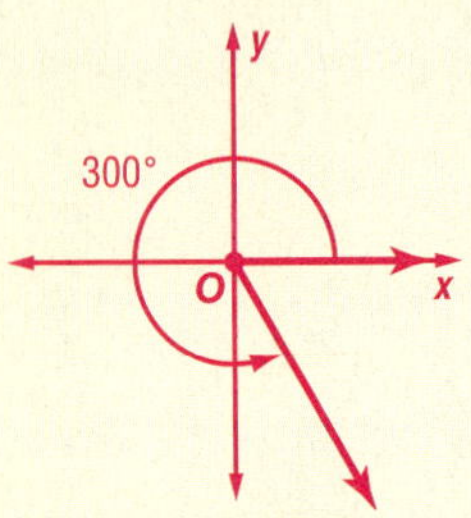

6.

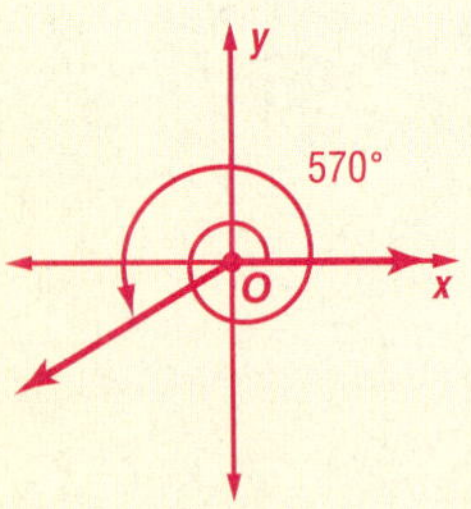

7.

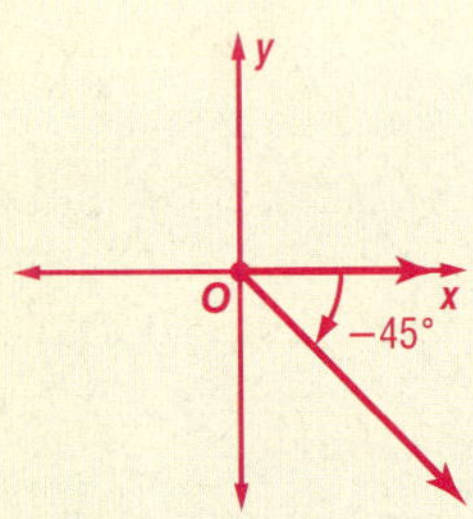

Notice that 405° − 45° = 360°. In degree measure, coterminal angles differ by an integral multiple of 360°. You can find an angle that is coterminal to a given angle by adding or subtracting a multiple of 360°. In radian measure, a coterminal angle is found by adding or subtracting a multiple of 2π.

Example 4 Find Coterminal Angles

Find one angle with positive measure and one angle with negative measure coterminal with each angle.

a. 240°

A positive angle is 240° + 360° or 600°.

A negative angle is 240° − 360° or −120°.

b. $\frac{9\pi}{4}$

A positive angle is $\frac{9\pi}{4} + 2\pi$ or $\frac{17\pi}{4}$. $\quad \frac{9\pi}{4} + \frac{8\pi}{4} = \frac{17\pi}{4}$

A negative angle is $\frac{9\pi}{4} - 2(2\pi)$ or $-\frac{7\pi}{4}$. $\quad \frac{9\pi}{4} + \left(-\frac{16\pi}{4}\right) = -\frac{7\pi}{4}$

Study Tip

Coterminal Angles
Notice in Example 4b that it is necessary to subtract a multiple of 2π to find a coterminal angle with negative measure.

Check for Understanding

Concept Check

1. **Name** the set of numbers to which angle measures belong. **reals**
2. **Define** the term radian. **See margin.**
3. **OPEN ENDED** Draw and label an example of an angle with negative measure in standard position. Then find an angle with positive measure that is coterminal with this angle. **See margin.**

Guided Practice

GUIDED PRACTICE KEY	
Exercises	**Examples**
4–7	1
8–13	2
14–16	4
17, 18	3

Draw an angle with the given measure in standard position. 4–7. See margin.

4. 70° 5. 300° 6. 570° 7. −45°

Rewrite each degree measure in radians and each radian measure in degrees.

8. 130° $\frac{13\pi}{18}$ 9. −10° $-\frac{\pi}{18}$ 10. 485° $\frac{97\pi}{36}$

11. $\frac{3\pi}{4}$ **135°** 12. $-\frac{\pi}{6}$ **−30°** 13. $\frac{19\pi}{3}$ **1140°**

Find one angle with positive measure and one angle with negative measure coterminal with each angle. 14–16. Sample answers are given.

14. 60° **420°, −300°** 15. 425° **785°, −295°** 16. $\frac{\pi}{3}$ $\frac{7\pi}{3}$, $-\frac{5\pi}{3}$

Application

ASTRONOMY For Exercises 17 and 18, use the following information.
Earth rotates on its axis once every 24 hours.

17. How long does it take Earth to rotate through an angle of 315°? **21 h**
18. How long does it take Earth to rotate through an angle of $\frac{\pi}{6}$? **2 h**

★ indicates increased difficulty

Practice and Apply

Draw an angle with the given measure in standard position. 19–26. See pp. 759A–759D.

19. 235° 20. 270° 21. 790° 22. 380°
23. −150° 24. −50° ★25. π ★26. $-\frac{2\pi}{3}$

DAILY INTERVENTION — Differentiated Instruction

Kinesthetic Have students work with a partner so that one person models an angle with outstretched arms or with two pencils or yardsticks. The other partner then names a positive and negative angle, less than or more than a full circle, that is coterminal with the modeled angle.

Homework Help

For Exercises	See Examples
19–26	1
27–42	2
43–54	4
55–59	3

Extra Practice
See page 857.

41. $\frac{1620}{\pi} \approx 515.7°$

42. $\frac{540}{\pi} \approx 171.9°$

Rewrite each degree measure in radians and each radian measure in degrees.

27. 120° $\frac{2\pi}{3}$ 28. 60° $\frac{\pi}{3}$ 29. −15° $-\frac{\pi}{12}$ 30. −225° $-\frac{5\pi}{4}$

31. 660° $\frac{11\pi}{3}$ 32. 570° $\frac{19\pi}{6}$ 33. 158° $\frac{79\pi}{90}$ 34. 260° $\frac{13\pi}{9}$

35. $\frac{5\pi}{6}$ **150°** 36. $\frac{11\pi}{4}$ **495°** 37. $-\frac{\pi}{4}$ **−45°** 38. $-\frac{\pi}{3}$ **−60°**

39. $\frac{29\pi}{4}$ **1305°** 40. $\frac{17\pi}{6}$ **510°** ★41. 9 ★42. 3

Find one angle with positive measure and one angle with negative measure coterminal with each angle. **43–54. Sample answers are given.**

43. 225° **585°, −135°** 44. 30° **390°, −330°** 45. −15° **345°, −375°**

46. −140° **220°, −500°** 47. 368° **8°, −352°** 48. 760° **400°, −320°**

49. $\frac{3\pi}{4}$ $\frac{11\pi}{4}, -\frac{5\pi}{4}$ 50. $\frac{7\pi}{6}$ $\frac{19\pi}{6}, -\frac{5\pi}{6}$ 51. $-\frac{5\pi}{4}$ $\frac{3\pi}{4}, -\frac{13\pi}{4}$

52. $-\frac{2\pi}{3}$ $\frac{4\pi}{3}, -\frac{8\pi}{3}$ 53. $\frac{9\pi}{2}$ $\frac{13\pi}{2}, -\frac{3\pi}{2}$ 54. $\frac{17\pi}{4}$ $\frac{25\pi}{4}, -\frac{7\pi}{4}$

55. **DRIVING** Some sport-utility vehicles (SUVs) use 15-inch radius wheels. When driven 40 miles per hour, determine the measure of the angle through which a point on the wheel travels every second. Round to both the nearest degree and nearest radian. **2689° per second; 47 radians per second**

More About. . .

Driving

A leading U.S. automaker plans to build a hybrid sport-utility vehicle in the near future that will use an electric motor to boost fuel efficiency and reduce polluting emissions.

Source: *The Dallas Morning News*

GEOMETRY **For Exercises 56 and 57, use the following information.**

A *sector* is a region of a circle that is bounded by a central angle θ and its intercepted arc. The area A of a sector with radius r and central angle θ is given by $A = \frac{1}{2}r^2\theta$, where θ is measured in radians.

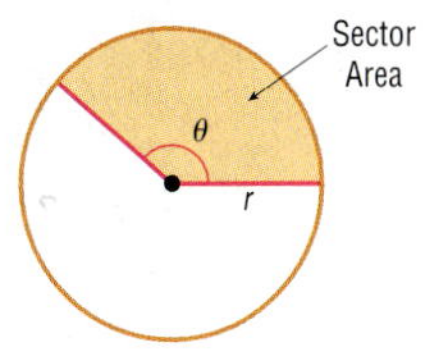

56. Find the area of a sector with a central angle of $\frac{4\pi}{3}$ radians in a circle whose radius measures 10 inches. **209.4 in^2**

57. Find the area of a sector with a central angle of 150° in a circle whose radius measures 12 meters. **about 188.5 m^2**

58. **ENTERTAINMENT** Suppose the gondolas on the Navy Pier Ferris wheel were numbered from 1 through 40 consecutively in a counterclockwise fashion. If you were sitting in gondola number 3 and the wheel were to rotate counterclockwise through $\frac{47\pi}{10}$ radians, which gondola used to be in the position that you are in now? **number 17**

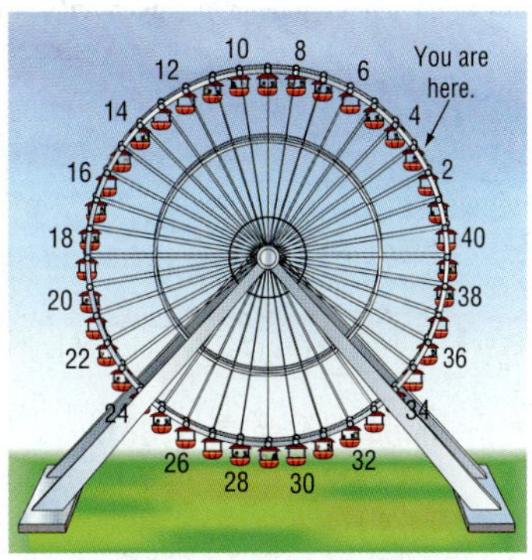

★59. **CARS** Use the Area of a Sector Formula in Exercises 56 and 57 to find the area swept by the rear windshield wiper of the car shown at the right. **about 640.88 in^2**

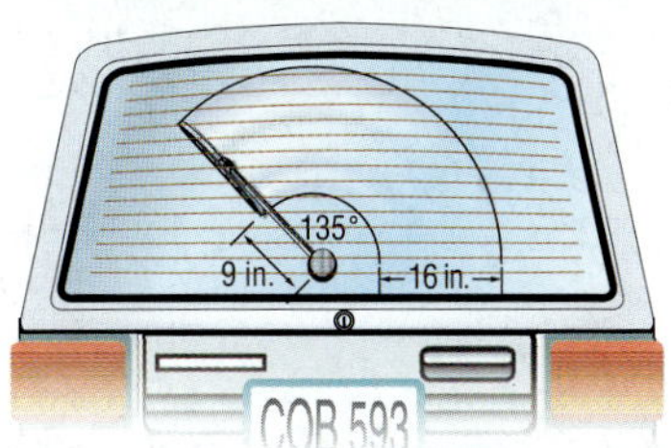

www.algebra2.com/self_check_quiz

3 Practice/Apply

Study Notebook

Have students—

- add the definitions/examples of the vocabulary terms to their Vocabulary Builder worksheets for Chapter 13.
- add the diagram of radian and degree measures on p. 711 to their notebook.
- include any other item(s) that they find helpful in mastering the skills in this lesson.

About the Exercises...

Organization by Objective

- **Angle Measurement:** 19–42, 55–59
- **Coterminal Angles:** 43–54

Odd/Even Assignments

Exercises 19–54 are structured so that students practice the same concepts whether they are assigned odd or even problems.

Assignment Guide

Basic: 19–23 odd, 27–39 odd, 43–57 odd, 60–81

Average: 19–59 odd, 60–81

Advanced: 20–58 even, 60–75 (optional: 76–81)

All: Practice Quiz 1 (1–10)

Study Guide and Intervention, p. 781 (shown) and p. 782

Angle Measurement An angle is determined by two rays. The degree measure of an angle is described by the amount and direction of rotation from the **initial side** along the positive x-axis to the **terminal side**. A counterclockwise rotation is associated with positive angle measure and a clockwise rotation is associated with negative angle measure. An angle can also be measured in **radians**.

Radian and Degree Measure	To rewrite the radian measure of an angle in degrees, multiply the number of radians by $\frac{180°}{\pi \text{ radians}}$. To rewrite the degree measure of an angle in radians, multiply the number of degrees by $\frac{\pi \text{ radians}}{180°}$.

Example 1 **Draw an angle with measure 290° in standard notation.**

The negative y-axis represents a positive rotation of 270°. To generate an angle of 290°, rotate the terminal side 20° more in the counterclockwise direction.

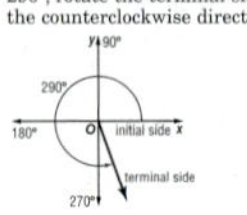

Example 2 **Rewrite the degree measure in radians and the radian measure in degrees.**

a. 45°

$45° = 45°\left(\frac{\pi \text{ radians}}{180°}\right) = \frac{\pi}{4}$ radians

b. $\frac{5\pi}{3}$ radians

$\frac{5\pi}{3}$ radians $= \frac{5\pi}{3}\left(\frac{180°}{\pi}\right) = 300°$

Exercises

Draw an angle with the given measure in standard position.

1. 160°
2. $-\frac{5\pi}{4}$
3. 400°

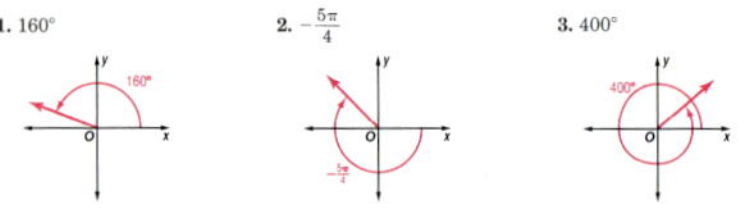

Rewrite each degree measure in radians and each radian measure in degrees.

4. 140° $\frac{7\pi}{9}$
5. −860° $-\frac{43\pi}{9}$
6. $-\frac{3\pi}{5}$ −108°
7. $\frac{11\pi}{3}$ 660°

Skills Practice, p. 783 and Practice, p. 784 (shown)

Draw an angle with the given measure in standard position.

1. 210°
2. 305°
3. 580°
4. 135°
5. −450°
6. −560°

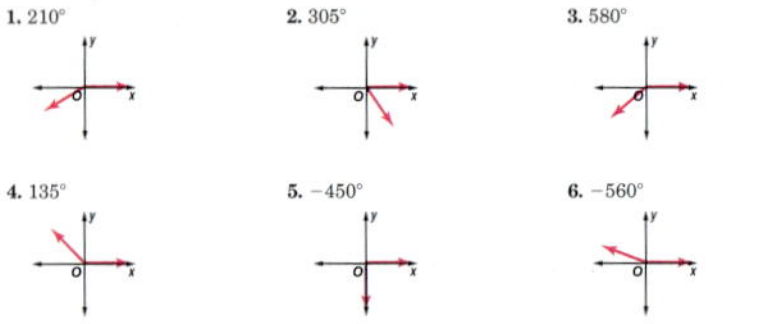

Rewrite each degree measure in radians and each radian measure in degrees.

7. 18° $\frac{\pi}{10}$
8. 6° $\frac{\pi}{30}$
9. 870° $\frac{29\pi}{6}$
10. 347° $\frac{347\pi}{180}$
11. −72° $-\frac{2\pi}{5}$
12. −820° $-\frac{41\pi}{9}$
13. −250° $-\frac{25\pi}{18}$
14. −165° $-\frac{11\pi}{12}$
15. 4π 720°
16. $\frac{5\pi}{2}$ 450°
17. $\frac{13\pi}{5}$ 468°
18. $\frac{13\pi}{30}$ 78°
19. $-\frac{9\pi}{2}$ −810°
20. $-\frac{7\pi}{12}$ −105°
21. $-\frac{3\pi}{8}$ −67.5°
22. $-\frac{3\pi}{16}$ −33.75°

Find one angle with positive measure and one angle with negative measure coterminal with each angle. 23–34. Sample answers are given.

23. 65° 425°, −295°
24. 80° 440°, −280°
25. 285° 645°, −75°
26. 110° 470°, −250°
27. −37° 323°, −397°
28. −93° 267°, −453°
29. $\frac{2\pi}{5}$ $\frac{12\pi}{5}, -\frac{8\pi}{5}$
30. $\frac{5\pi}{6}$ $\frac{17\pi}{6}, -\frac{7\pi}{6}$
31. $\frac{17\pi}{6}$ $\frac{29\pi}{6}, -\frac{7\pi}{6}$
32. $-\frac{3\pi}{2}$ $\frac{\pi}{2}, -\frac{7\pi}{2}$
33. $-\frac{\pi}{4}$ $\frac{7\pi}{4}, -\frac{9\pi}{4}$
34. $-\frac{5\pi}{12}$ $\frac{19\pi}{12}, -\frac{29\pi}{12}$

35. **TIME** Find both the degree and radian measures of the angle through which the hour hand on a clock rotates from 5 A.M. to 10 A.M. −150°; $-\frac{5\pi}{6}$

36. **ROTATION** A truck with 16-inch radius wheels is driven at 77 feet per second (52.5 miles per hour). Find the measure of the angle through which a point on the outside of the wheel travels each second. Round to the nearest degree and nearest radian. 3309°/s; 58 radians/s

Reading to Learn Mathematics, p. 785

ELL

Pre-Activity **How can angles be used to describe circular motion?**

Read the introduction to Lesson 13-2 at the top of page 709 in your textbook.

If a gondola revolves through a complete revolution in one minute, what is its angular velocity in degrees per second? 6° per second

Reading the Lesson

1. Match each degree measure with the corresponding radian measure on the right.

a. 30° v	i. $\frac{2\pi}{3}$
b. 90° ii	ii. $\frac{\pi}{2}$
c. 120° i	iii. $\frac{7\pi}{6}$
d. 135° vi	iv. π
e. 180° iv	v. $\frac{\pi}{6}$
f. 210° iii	vi. $\frac{3\pi}{4}$

2. The sine of 30° is $\frac{1}{2}$ and the sine of 150° is also $\frac{1}{2}$. Does this mean that 30° and 150° are coterminal angles? Explain your reasoning. Sample answer: No; the terminal side of a 30° angle is in Quadrant I, while the terminal side of a 150° angle is in Quadrant II.

3. Describe how to find two angles that are coterminal with an angle of 155°, one with positive measure and one with negative measure. (Do not actually calculate these angles.) Sample answer: Positive angle: Add 360° to 155°. Negative angle: Subtract 360° from 155°.

4. Describe how to find two angles that are coterminal with an angle of $\frac{5\pi}{3}$, one positive and one negative. (Do not actually calculate these angles.) Sample answer: Positive angle: Add 2π to $\frac{5\pi}{3}$. Negative angle: Subtract 2π from $\frac{5\pi}{3}$.

Helping You Remember

5. How can you use what you know about the circumference of a circle to remember how to convert between radian and degree measure? Sample answer: The circumference of a circle is given by the formula $C = 2\pi r$, so the circumference of a circle with radius 1 is 2π. In degree measure, one complete circle is 360°. So 2π radians = 360° and π radians = 180°.

60. **CRITICAL THINKING** If (a, b) is on a circle that has radius r and center at the origin, prove that each of the following points is also on this circle.

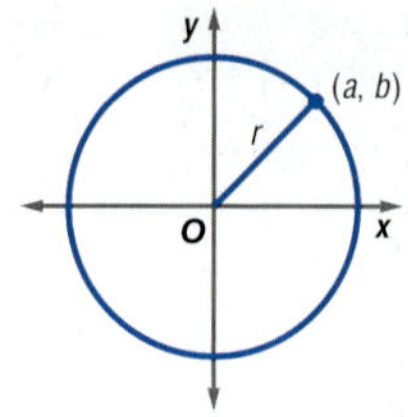

a. $(a, -b)$ $a^2 + (-b)^2 = a^2 + b^2 = 1$

b. (b, a) $b^2 + a^2 = a^2 + b^2 = 1$

c. $(b, -a)$ $b^2 + (-a)^2 = a^2 + b^2 = 1$

61. **WRITING IN MATH** Answer the question that was posed at the beginning of the lesson. See pp. 759A–759D.

How can angles be used to describe circular motion?

Include the following in your answer:

- an explanation of the significance of angles of more than 180° in terms of circular motion,
- an explanation of the significance of angles with negative measure in terms of circular motion, and
- an interpretation of a rate of more than 360° per minute.

Standardized Test Practice Ⓐ Ⓑ Ⓒ Ⓓ

62. Choose the radian measure that is equal to 56°. C

Ⓐ $\frac{\pi}{15}$ Ⓑ $\frac{7\pi}{45}$

Ⓒ $\frac{14\pi}{45}$ Ⓓ $\frac{\pi}{3}$

63. Angular velocity is defined by the equation $\omega = \frac{\theta}{t}$, where θ is usually expressed in radians and t represents time. Find the angular velocity in radians per second of a point on a bicycle tire if it completes 2 revolutions in 3 seconds. D

Ⓐ $\frac{\pi}{3}$ Ⓑ $\frac{\pi}{2}$

Ⓒ $\frac{2\pi}{3}$ Ⓓ $\frac{4\pi}{3}$

Maintain Your Skills

Mixed Review **Solve $\triangle ABC$ by using the given measurements. Round measures of sides to the nearest tenth and measures of angles to the nearest degree.** *(Lesson 13-1)*

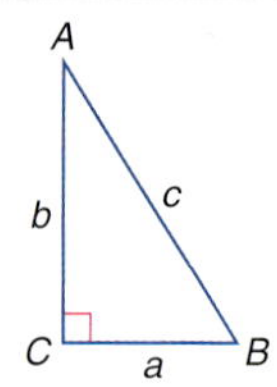

64. $A = 34°$, $b = 5$
65. $B = 68°$, $b = 14.7$
66. $B = 55°$, $c = 16$
67. $a = 0.4$, $b = 0.4\sqrt{3}$

64. $a \approx 3.4$, $c \approx 6.0$, $B = 56°$
65. $A = 22°$, $a \approx 5.9$, $c \approx 15.9$
66. $A = 35°$, $a \approx 9.2$, $b \approx 13.1$
67. $c = 0.8$, $A = 30°$, $B = 60°$

Find the margin of sampling error. *(Lesson 12-9)*

68. $p = 72\%$, $n = 100$ about 8.98%
69. $p = 50\%$, $n = 200$ about 7.07%

714 Chapter 13 Trigonometric Functions

Enrichment, p. 786

Making and Using a Hypsometer

A **hypsometer** is a device that can be used to measure the height of an object. To construct your own hypsometer, you will need a rectangular piece of heavy cardboard that is at least 7 cm by 10 cm, a straw, transparent tape, a string about 20 cm long, and a small weight that can be attached to the string.

Mark off 1-cm increments along one short side and one long side of the cardboard. Tape the straw to the other short side. Then attach the weight to one end of the string, and attach the other end of the string to one corner of the cardboard, as shown in the figure below. The diagram below shows how your hypsometer should look.

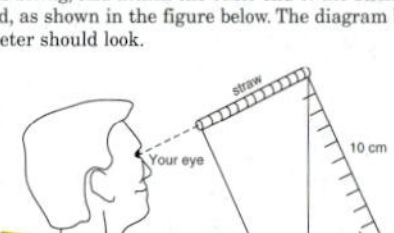

Determine whether each situation involves a *permutation* or a *combination*. Then find the number of possibilities. *(Lesson 12-2)*

70. permutation, 17,100,720

70. choosing an arrangement of 5 CDs from your 30 favorite CDs

71. choosing 3 different types of snack foods out of 7 at the store to take on a trip **combination, 35**

Find $[g \circ h](x)$ and $[h \circ g](x)$. *(Lesson 7-7)*

72. $[g \circ h](x) = 6x - 8$, $[h \circ g](x) = 6x - 4$

73. $[g \circ h](x) = 4x^2 - 6x + 23$, $[h \circ g](x) = 8x^2 + 34x + 44$

72. $g(x) = 2x$
$h(x) = 3x - 4$

73. $g(x) = 2x + 5$
$h(x) = 2x^2 - 3x + 9$

For Exercises 74 and 75, use the graph at the right.
The number of sports radio stations can be modeled by $R(x) = 7.8x^2 + 16.6x + 95.8$, where x is the number of years since 1996. *(Lesson 7-5)*

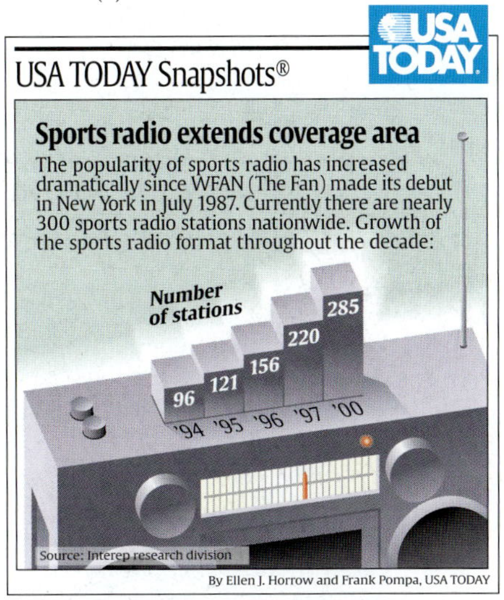

74. Use synthetic substitution to estimate the number of sports radio stations for 2006. **1041.8**

75. Evaluate $R(12)$. What does this value represent? **1418.2 or about 1418; the number of sports radio stations in 2008**

Getting Ready for the Next Lesson

PREREQUISITE SKILL **Simplify each expression.**
*(To review **rationalizing denominators**, see Lesson 5-6.)*

76. $\frac{2}{\sqrt{3}}$ $\frac{2\sqrt{3}}{3}$

77. $\frac{3}{\sqrt{5}}$ $\frac{3\sqrt{5}}{5}$

78. $\frac{4}{\sqrt{6}}$ $\frac{2\sqrt{6}}{3}$

79. $\frac{5}{\sqrt{10}}$ $\frac{\sqrt{10}}{2}$

80. $\frac{\sqrt{7}}{\sqrt{2}}$ $\frac{\sqrt{14}}{2}$

81. $\frac{\sqrt{5}}{\sqrt{8}}$ $\frac{\sqrt{10}}{4}$

Practice Quiz 1 — *Lessons 13-1 and 13-2*

Solve $\triangle ABC$ by using the given measurements. Round measures of sides to the nearest tenth and measures of angles to the nearest degree. *(Lesson 13-1)*

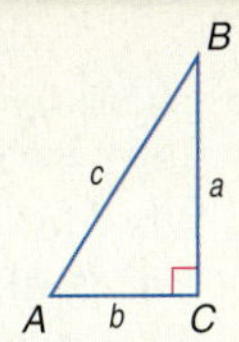

1. $A = 48°$, $b = 12$ **$B = 42°$, $a \approx 13.3$, $c \approx 17.9$**

2. $a = 18$, $c = 21$ **$A \approx 59°$, $B \approx 31°$, $b \approx 10.8$**

3. Draw an angle measuring $-60°$ in standard position. *(Lesson 13-1)* **See margin.**

4. Find the values of the six trigonometric functions for angle θ in the triangle at the right. *(Lesson 13-1)* **See margin.**

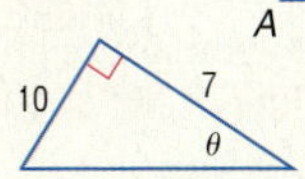

Rewrite each degree measure in radians and each radian measure in degrees. *(Lesson 13-2)*

5. $190°$ $\frac{19\pi}{18}$

6. $450°$ $\frac{5\pi}{2}$

7. $\frac{7\pi}{6}$ $210°$

8. $-\frac{11\pi}{5}$ $-396°$

Find one angle with positive measure and one angle with negative measure coterminal with each angle. *(Lesson 13-2)*

9. $-55°$ **305°; −415°**

10. $\frac{11\pi}{3}$ $\frac{5\pi}{3}, -\frac{\pi}{3}$

Online Lesson Plans

USA TODAY Education's Online site offers resources and interactive features connected to each day's newspaper. *Experience TODAY*, USA TODAY's daily lesson plan, is available on the site and delivered daily to subscribers. This plan provides instruction for integrating USA TODAY graphics and key editorial features into your mathematics classroom. Log on to www.education.usatoday.com.

4 Assess

Open-Ended Assessment

Speaking Have students work in small groups to decide on an informal explanation of what a radian is and what coterminal angles are. Then have a reporter from each group share that explanation with the whole class.

Getting Ready for Lesson 13-3

PREREQUISITE SKILL Lesson 13-3 presents finding the trigonometric functions for general angles. Students will use their familiarity with rationalizing denominators as they find values of trigonometric functions. Exercises 76–81 should be used to determine your students' familiarity with rationalizing denominators.

Assessment Options

Practice Quiz 1 The quiz provides students with a brief review of the concepts and skills in Lessons 13-1 and 13-2. Lesson numbers are given to the right of exercises or instruction lines so students can review concepts not yet mastered.

Quiz (Lessons 13-1 and 13-2) is available on p. 831 of the *Chapter 13 Resource Masters*.

Answers

3.

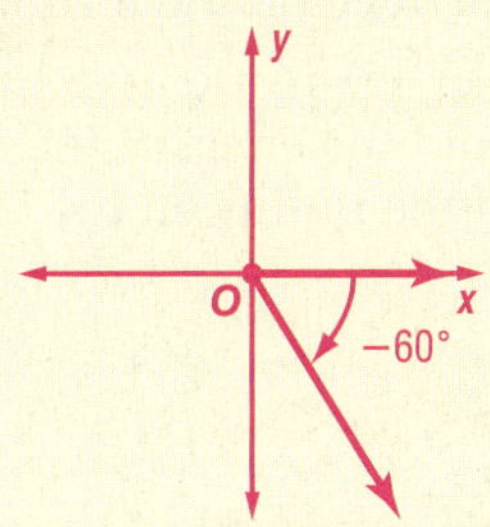

4. $\sin \theta = \frac{10\sqrt{149}}{149}$; $\cos \theta = \frac{7\sqrt{149}}{149}$; $\tan \theta = \frac{10}{7}$; $\csc \theta = \frac{\sqrt{149}}{10}$; $\sec \theta = \frac{\sqrt{149}}{7}$; $\cot \theta = \frac{7}{10}$

Algebra Activity

A Follow-Up of Lesson 13-2

Getting Started

Objective Investigate measures in regular polygons using trigonometry.

Materials
compass
straightedge
protractor

Teach

- Have students copy the figure and use a colored pencil to show which line is the apothem.
- Make sure the students draw circles with a radius (not a diameter) of 1 inch.
- If it is available, you may want to have students use computer software to draw the inscribed regular polygons.

Assess

In **Exercises 1–3**, students should
- be able to see the pattern in the table.
- understand how the apothem changes as the number of sides in the polygon increases.

In **Exercises 4–7**, students should
- be able to develop the formula for the apothem.
- understand the effect of the length of the radius on the formula.

Study Notebook

You may wish to have students summarize this activity and what they learned from it.

Algebra Activity

A Follow-Up of Lesson 13-2

Investigating Regular Polygons Using Trigonometry

Collect the Data

- Use a compass to draw a circle with a radius of one inch. Inscribe an equilateral triangle inside of the circle. To do this, use a protractor to measure three angles of 120° at the center of the circle, since $\frac{360°}{3} = 120°$. Then connect the points where the sides of the angles intersect the circle using a straightedge.
- The **apothem** of a regular polygon is a segment that is drawn from the center of the polygon perpendicular to a side of the polygon. Use the cosine of angle θ to find the length of an apothem, labeled a in the diagram below.

Analyze the Data

1. Make a table like the one shown below and record the length of the apothem of the equilateral triangle.

Number of Sides, n	θ	a
3	60	**0.50**
4	45	**0.71**
5	**36**	**0.81**
6	**30**	**0.87**
7	**≈26**	**0.90**
8	**22.5**	**0.92**
9	**20**	**0.94**
10	**18**	**0.95**

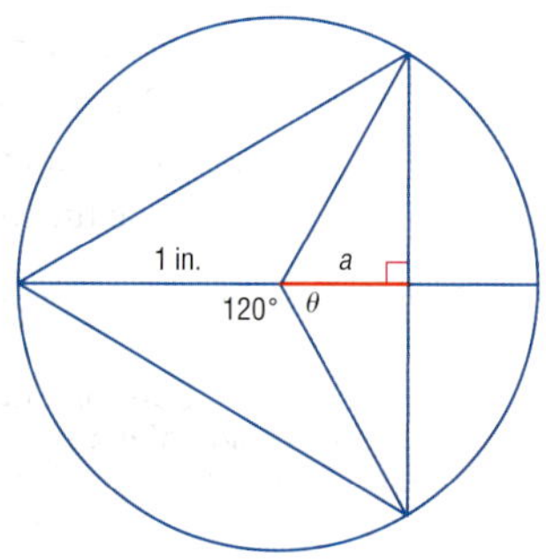

Inscribe each regular polygon named in the table in a circle of radius one inch. Copy and complete the table.

2. What do you notice about the measure of θ as the number of sides of the inscribed polygon increases? **The measure of θ decreases.**

3. What do you notice about the values of a?
The length of the apothem increases as the number of sides increases.

Make a Conjecture

4. Suppose you inscribe a 20-sided regular polygon inside a circle. Find the measure of angle θ. **9°**

5. Write a formula that gives the measure of angle θ for a polygon with n sides. **$\theta = 360° \div 2n$ or $\theta = 180° \div n$**

6. Write a formula that gives the length of the apothem of a regular polygon inscribed in a circle of radius one inch. **$a = \cos\theta$**

7. How would the formula you wrote in Exercise 6 change if the radius of the circle was not one inch? **See pp. 759A–759D.**

716 Chapter 13 Trigonometric Functions

Resource Manager

Teaching Algebra with Manipulatives
- p. 23 (master for protractors)
- p. 301 (student recording sheet)

Glencoe Mathematics Classroom Manipulative Kit
- protractors
- rulers
- compasses

13-3 Trigonometric Functions of General Angles

What You'll Learn

- Find values of trigonometric functions for general angles.
- Use reference angles to find values of trigonometric functions.

Vocabulary

- quadrantal angle
- reference angle

How can you model the position of riders on a skycoaster?

A skycoaster consists of a large arch from which two steel cables hang and are attached to riders suited together in a harness. A third cable, coming from a larger tower behind the arch, is attached with a ripcord. Riders are hoisted to the top of the larger tower, pull the ripcord, and then plunge toward Earth. They swing through the arch, reaching speeds of more than 60 miles per hour. After the first several swings of a certain skycoaster, the angle θ of the riders from the center of the arch is given by $\theta = 0.2 \cos (1.6t)$, where t is the time in seconds after leaving the bottom of their swing.

TRIGONOMETRIC FUNCTIONS AND GENERAL ANGLES In Lesson 13-1, you found values of trigonometric functions whose domains were the set of all acute angles, angles between 0 and $\frac{\pi}{2}$, of a right triangle. For $t > 0$ in the equation above, you must find the cosine of an angle greater than $\frac{\pi}{2}$. In this lesson, we will extend the domain of trigonometric functions to include angles of *any* measure.

Key Concept — *Trigonometric Functions, θ in Standard Position*

Let θ be an angle in standard position and let $P(x, y)$ be a point on the terminal side of θ. Using the Pythagorean Theorem, the distance r from the origin to P is given by $r = \sqrt{x^2 + y^2}$. The trigonometric functions of an angle in standard position may be defined as follows.

$\sin \theta = \frac{y}{r}$ $\quad\cos \theta = \frac{x}{r}$ $\quad\tan \theta = \frac{y}{x}, x \neq 0$

$\csc \theta = \frac{r}{y}, y \neq 0$ $\quad\sec \theta = \frac{r}{x}, x \neq 0$ $\quad\cot \theta = \frac{x}{y}, y \neq 0$

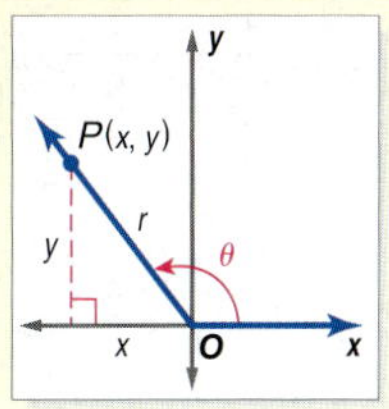

Example 1 *Evaluate Trigonometric Functions for a Given Point*

Find the exact values of the six trigonometric functions of θ if the terminal side of θ contains the point (5, −12).

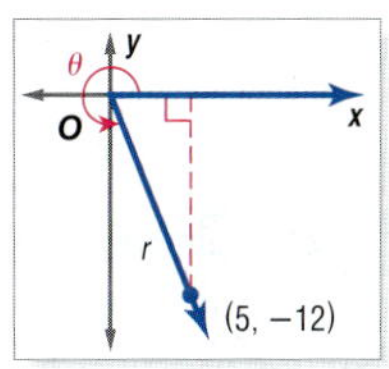

From the coordinates given, you know that $x = 5$ and $y = -12$. Use the Pythagorean Theorem to find r.

(continued on the next page)

13-3 Lesson Notes

1 Focus

5-Minute Check Transparency 13-3 Use as a quiz or review of Lesson 13-2.

Mathematical Background notes are available for this lesson on p. 698C.

How can you model the position of riders on a skycoaster?

Ask students:

- What happens to the size of the swing (as measured by θ) as the time after the plunge increases? **The arc of the swing decreases, eventually to zero.**
- Is the t in the formula the same as the time that has elapsed since the rider plunged? **no**

Resource Manager

Workbook and Reproducible Masters

Chapter 13 Resource Masters

- Study Guide and Intervention, pp. 787–788
- Skills Practice, p. 789
- Practice, p. 790
- Reading to Learn Mathematics, p. 791
- Enrichment, p. 792

Transparencies

5-Minute Check Transparency 13-3
Real-World Transparency 13
Answer Key Transparencies

Technology

Alge2PASS: Tutorial Plus, Lessons 24 and 25
Interactive Chalkboard

2 Teach

TRIGONOMETRIC FUNCTIONS AND GENERAL ANGLES

In-Class Examples Power Point®

1 Find the exact values of the six trigonometric functions of θ if the terminal side of θ contains the point $(8, -15)$.
$\sin\theta = -\frac{15}{17}$; $\cos\theta = \frac{8}{17}$; $\tan\theta = -\frac{15}{8}$; $\csc\theta = -\frac{17}{15}$; $\sec\theta = \frac{17}{8}$; $\cot\theta = -\frac{8}{15}$

Teaching Tip When finding functions for an angle, students may find it helpful to sketch an angle in standard position and drop a perpendicular to the *x*-axis to form a right triangle. Then they can see that, as the angle increases from 90° to 180°, the *y* value approaches 0, and the values of the other two sides approach each other.

2 Find the values of the six trigonometric functions for an angle in standard position that measures 180°.
$\sin\theta = 0$; $\cos\theta = -1$; $\tan\theta = 0$; $\csc\theta$ is undefined; $\sec\theta = -1$; $\cot\theta$ is undefined.

$r = \sqrt{x^2 + y^2}$ Pythagorean Theorem

$= \sqrt{5^2 + (-12)^2}$ Replace x with 5 and y with -12.

$= \sqrt{169}$ or 13 Simplify.

Now, use $x = 5$, $y = -12$, and $r = 13$ to write the ratios.

$\sin\theta = \frac{y}{r} = \frac{-12}{13}$ or $-\frac{12}{13}$ $\quad$ $\cos\theta = \frac{x}{r} = \frac{5}{13}$ $\quad$ $\tan\theta = \frac{y}{x} = \frac{-12}{5}$ or $-\frac{12}{5}$

$\csc\theta = \frac{r}{y} = \frac{13}{-12}$ or $-\frac{13}{12}$ $\quad$ $\sec\theta = \frac{r}{x} = \frac{13}{5}$ $\quad$ $\cot\theta = \frac{x}{y} = \frac{5}{-12}$ or $-\frac{5}{12}$

If the terminal side of angle θ lies on one of the axes, θ is called a **quadrantal angle**. The quadrantal angles are 0°, 90°, 180°, and 270°. Notice that for these angles either x or y is equal to 0. Since division by zero is undefined, two of the trigonometric values are undefined for each quadrantal angle.

Key Concept *Quadrantal Angles*

$\theta = 0°$ or 0 radians

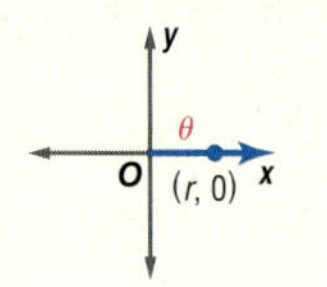

$\theta = 90°$ or $\frac{\pi}{2}$

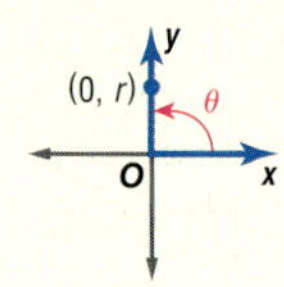

$\theta = 180°$ or π radians

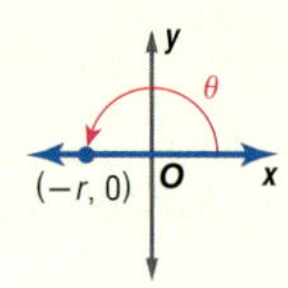

$\theta = 270°$ or $\frac{3\pi}{2}$ radians

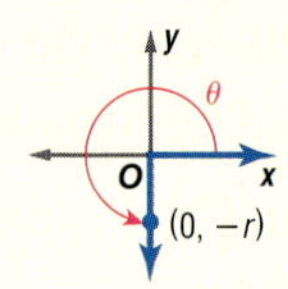

Example 2 Quadrantal Angles

Find the values of the six trigonometric functions for an angle in standard position that measures 270°.

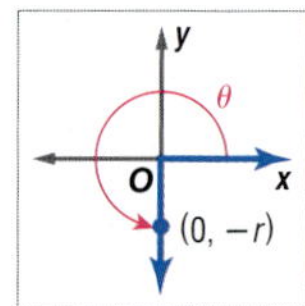

When $\theta = 270°$, $x = 0$ and $y = -r$.

$\sin\theta = \frac{y}{r} = \frac{-r}{r}$ or -1 $\quad$ $\cos\theta = \frac{x}{r} = \frac{0}{r}$ or 0 $\quad$ $\tan\theta = \frac{y}{x} = \frac{-r}{0}$ or undefined

$\csc\theta = \frac{r}{y} = \frac{r}{-r}$ or -1 $\quad$ $\sec\theta = \frac{r}{x} = \frac{r}{0}$ or undefined $\quad$ $\cot\theta = \frac{x}{y} = \frac{0}{-r}$ or 0

REFERENCE ANGLES To find the values of trigonometric functions of angles greater than 90° (or less than 0°), you need to know how to find the measures of reference angles. If θ is a nonquadrantal angle in standard position, its **reference angle**, θ', is defined as the acute angle formed by the terminal side of θ and the x-axis.

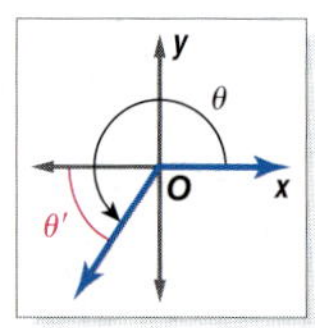

Study Tip

Reading Math
θ' is read *theta prime*.

DAILY INTERVENTION **Unlocking Misconceptions**

In Example 1, watch for students who think that *r* must be a negative number any time either *x* or *y* has a negative value.

You can use the rule below to find the reference angle for any nonquadrantal angle θ where $0° < \theta < 360°$ (or $0 < \theta < 2\pi$).

Key Concept — *Reference Angle Rule*

For any nonquadrantal angle θ, $0° < \theta < 360°$ (or $0 < \theta < 2\pi$), its reference angle θ' is defined as follows.

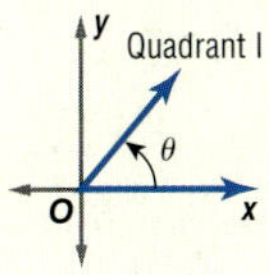

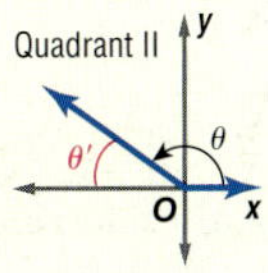

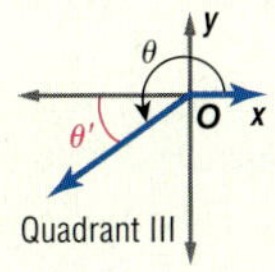

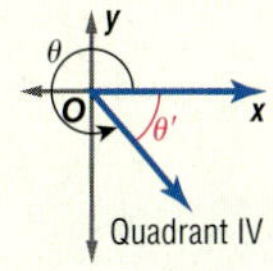

Quadrant I	Quadrant II	Quadrant III	Quadrant IV
$\theta' = \theta$	$\theta' = 180° - \theta$ ($\theta' = \pi - \theta$)	$\theta' = \theta - 180°$ ($\theta' = \theta - \pi$)	$\theta' = 360° - \theta$ ($\theta' = 2\pi - \theta$)

If the measure of θ is greater than 360° or less than 0°, its reference angle can be found by associating it with a coterminal angle of positive measure between 0° and 360°.

Example 3 Find the Reference Angle for a Given Angle

Sketch each angle. Then find its reference angle.

a. **300°**

Because the terminal side of 300° lies in Quadrant IV, the reference angle is $360° - 300°$ or $60°$.

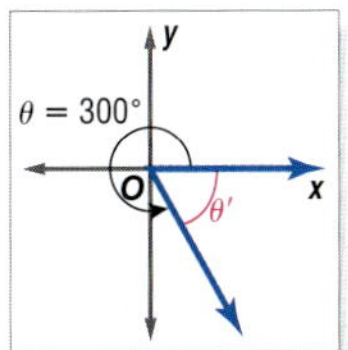

b. $-\frac{2\pi}{3}$

A coterminal angle of $-\frac{2\pi}{3}$ is $2\pi - \frac{2\pi}{3}$ or $\frac{4\pi}{3}$.

Because the terminal side of this angle lies in Quadrant III, the reference angle is $\frac{4\pi}{3} - \pi$ or $\frac{\pi}{3}$.

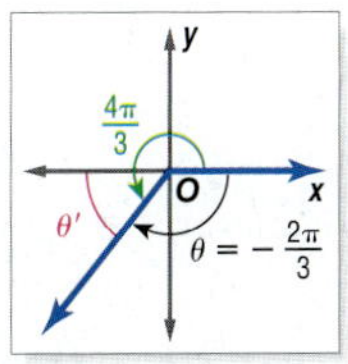

To use the reference angle θ' to find a trigonometric value of θ, you need to know the sign of that function for an angle θ. From the function definitions, these signs are determined by x and y, since r is always positive. Thus, the sign of each trigonometric function is determined by the quadrant in which the terminal side of θ lies.

The chart below summarizes the signs of the trigonometric functions for each quadrant.

Function	Quadrant I	II	III	IV
$\sin\theta$ or $\csc\theta$	+	+	−	−
$\cos\theta$ or $\sec\theta$	+	−	−	+
$\tan\theta$ or $\cot\theta$	+	−	+	−

TEACHING TIP

For example, because $\sin\theta = \frac{y}{r}$ and r is always positive, $\sin\theta$ is positive wherever $y > 0$, which is in Quadrants I and II.

www.algebra2.com/extra_examples

REFERENCE ANGLES

In-Class Example Power Point®

3 Sketch each angle. Then find its reference angle.

a. 330° **30°**

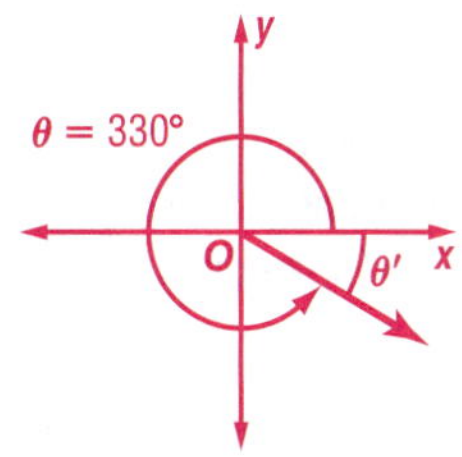

b. $-\frac{5\pi}{6}$ $\frac{\pi}{6}$

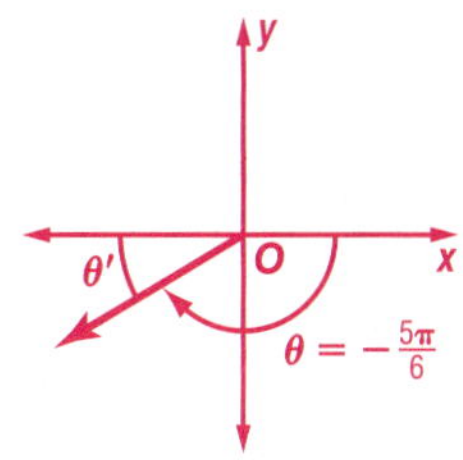

Teacher to Teacher

David S. Daniels — Longmeadow H.S., Longmeadow, MA

"Have students measure the distance from the tip of the minute hand on the classroom clock to the ceiling at 5 minute or 30° intervals, starting at the top of an hour. Then have students plot the data and describe the graph. Students will discover that the distance from the ceiling is a function of the number of degrees through which the minute hand has turned."

In-Class Examples

Power Point®

4 Find the exact value of each trigonometric function.

a. sin 135° $\frac{\sqrt{2}}{2}$

b. cot $\frac{7\pi}{3}$ $\frac{\sqrt{3}}{3}$

5 Suppose θ is an angle in standard position whose terminal side is in Quadrant III and $\csc \theta = -\frac{5}{3}$. Find the exact values of the remaining five trigonometric functions of θ.
$\sin \theta = -\frac{3}{5}$; $\cos \theta = -\frac{4}{5}$; $\tan \theta = \frac{3}{4}$; $\sec \theta = -\frac{5}{4}$; $\cot \theta = \frac{4}{3}$

Study Tip

Look Back
To review **trigonometric values of angles measuring 30°, 45°, and 60°**, see Lesson 13-1.

Use the following steps to find the value of a trigonometric function of any angle θ.

Step 1 Find the reference angle θ'.

Step 2 Find the value of the trigonometric function for θ'.

Step 3 Using the quadrant in which the terminal side of θ lies, determine the sign of the trigonometric function value of θ.

Example 4 Use a Reference Angle to Find a Trigonometric Value

Find the exact value of each trigonometric function.

a. sin 120°

Because the terminal side of 120° lies in Quadrant II, the reference angle θ' is 180° − 120° or 60°. The sine function is positive in Quadrant II, so sin 120° = sin 60° or $\frac{\sqrt{3}}{2}$.

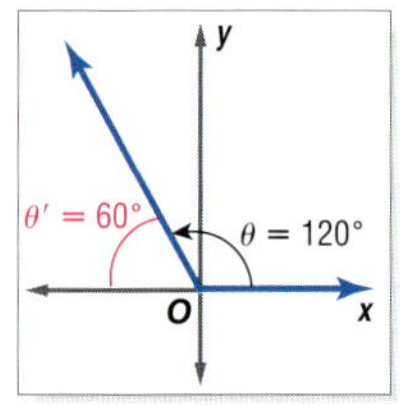

b. $\cot \frac{7\pi}{4}$

Because the terminal side of $\frac{7\pi}{4}$ lies in Quadrant IV, the reference angle θ' is $2\pi - \frac{7\pi}{4}$ or $\frac{\pi}{4}$. The cotangent function is negative in Quadrant IV.

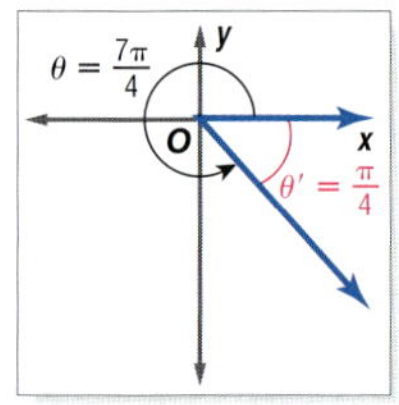

$\cot \frac{7\pi}{4} = -\cot \frac{\pi}{4}$

$= -\cot 45°$ $\quad \frac{\pi}{4}$ radians = 45°

$= -1$ $\quad$ cot 45° = 1

If you know the quadrant that contains the terminal side of θ in standard position and the exact value of one trigonometric function of θ, you can find the values of the other trigonometric functions of θ using the function definitions.

Example 5 Quadrant and One Trigonometric Value of θ

Suppose θ is an angle in standard position whose terminal side is in Quadrant III and $\sec \theta = -\frac{4}{3}$. Find the exact values of the remaining five trigonometric functions of θ.

Draw a diagram of this angle, labeling a point $P(x, y)$ on the terminal side of θ. Use the definition of secant to find the values of x and r.

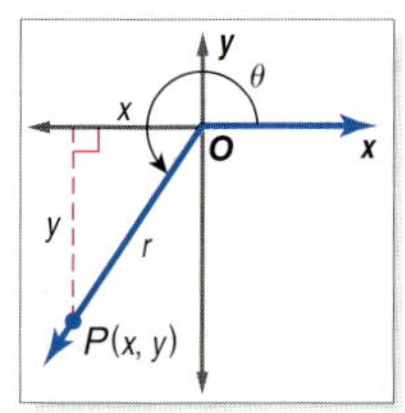

$\sec \theta = -\frac{4}{3}$ Given

$\frac{r}{x} = -\frac{4}{3}$ Definition of secant

Since x is negative in Quadrant III and r is always positive, $x = -3$ and $r = 4$. Use these values and the Pythagorean Theorem to find y.

DAILY INTERVENTION **Differentiated Instruction**

Auditory/Musical Have students work in small groups to create a jingle, song, rap, or short poem to help them remember the basic equivalences between angle measures in degrees and radians.

$x^2 + y^2 = r^2$ — Pythagorean Theorem

$(-3)^2 + y^2 = 4^2$ — Replace x with -3 and r with 4.

$y^2 = 16 - 9$ — Simplify. Then subtract 9 from each side.

$y = \pm\sqrt{7}$ — Simplify. Then take the square root of each side.

$y = -\sqrt{7}$ — y is negative in Quadrant III.

Use $x = -3$, $y = -\sqrt{7}$, and $r = 4$ to write the remaining trigonometric ratios.

$\sin\theta = \frac{y}{r} = \frac{-\sqrt{7}}{4}$ $\qquad$ $\cos\theta = \frac{x}{r} = \frac{-3}{4}$ $\qquad$ $\tan\theta = \frac{y}{x} = \frac{-\sqrt{7}}{-3}$ or $\frac{\sqrt{7}}{3}$

$\csc\theta = \frac{r}{y} = -\frac{4}{\sqrt{7}}$ or $-\frac{4\sqrt{7}}{7}$ $\qquad$ $\cot\theta = \frac{x}{y} = \frac{-3}{-\sqrt{7}}$ or $\frac{3\sqrt{7}}{7}$

Just as an exact point on the terminal side of an angle can be used to find trigonometric function values, trigonometric function values can be used to find the exact coordinates of a point on the terminal side of an angle.

Example 6 Find Coordinates Given a Radius and an Angle

ROBOTICS In a robotics competition, a robotic arm 4 meters long is to pick up an object at point A and release it into a container at point B. The robot's owner programs the arm to rotate through an angle of precisely 135° to accomplish this task. What is the new position of the object relative to the pivot point O?

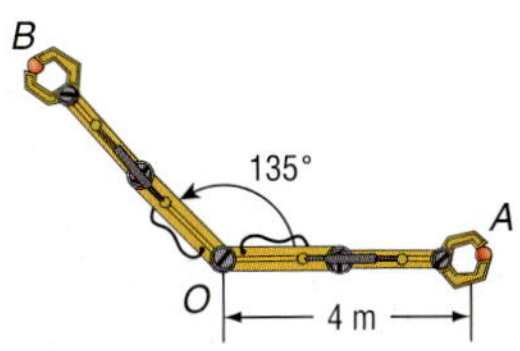

With the pivot point at the origin and the angle through which the arm rotates in standard position, point A has coordinates (4, 0). The reference angle θ' for 135° is 180° − 135° or 45°.

Let the position of point B have coordinates (x, y). Then, use the definitions of sine and cosine to find the value of x and y. The value of r is the length of the robotic arm, 4 meters. Because B is in Quadrant II, the cosine of 135° is negative.

$\cos 135° = \frac{x}{r}$	cosine ratio	$\sin 135° = \frac{y}{r}$	sine ratio
$-\cos 45° = \frac{x}{4}$	180° − 135° = 45°	$\sin 45° = \frac{y}{4}$	180° − 35° = 45°
$-\frac{\sqrt{2}}{2} = \frac{x}{4}$	$\cos 45° = \frac{\sqrt{2}}{2}$	$\frac{\sqrt{2}}{2} = \frac{y}{4}$	$\sin 45° = \frac{\sqrt{2}}{2}$
$-2\sqrt{2} = x$	Solve for x.	$2\sqrt{2} = y$	Solve for y.

The exact coordinates of B are $(-2\sqrt{2}, 2\sqrt{2})$. Since $2\sqrt{2}$ is about 2.83, the object is about 2.83 meters to the left of the pivot point and about 2.83 meters in front of the pivot point.

More About. . .

Robotics

RoboCup is an annual event in which teams from all over the world compete in a series of soccer matches in various classes according to the size and intellectual capacity of their robot. The robots are programmed to react to the ball and communicate with each other.

Source: www.robocup.org

In-Class Example

Power Point®

6 ROBOTICS Use the figure for Example 6 in the Student Edition to find the new position of the object relative to the pivot point for a robotic arm that is 3 meters long and that rotates through an angle of 150°.

$\left(-\frac{3\sqrt{3}}{2}, \frac{3}{2}\right)$ or about 2.60 meters to the left of O and 1.5 meters in front of O

3 Practice/Apply

Study Notebook

Have students—

- add the definitions/examples of the vocabulary terms to their Vocabulary Builder worksheets for Chapter 13.
- add diagrams explaining quadrantal and reference angles.
- include any other item(s) that they find helpful in mastering the skills in this lesson.

About the Exercises...

Organization by Objective

- **Trigonometric Functions and General Angles:** 17–24
- **Reference Angles:** 25–52

Odd/Even Assignments

Exercises 17–52 are structured so that students practice the same concepts whether they are assigned odd or even problems.

Assignment Guide

Basic: 17–21 odd, 25–51 odd, 56–77

Average: 17–51 odd, 55, 56–77

Advanced: 18–52 even, 53, 54, 56–71 (optional: 72–77)

Answers

3. To find the value of a trigonometric function of θ, where θ is greater than 90°, find the value of the trigonometric function for θ', then use the quadrant in which the terminal side of θ lies to determine the sign of the trigonometric function value of θ.

4. $\sin\theta = \frac{8}{17}$, $\cos\theta = -\frac{15}{17}$, $\tan\theta = -\frac{8}{15}$, $\csc\theta = \frac{17}{8}$, $\sec\theta = -\frac{17}{15}$, $\cot\theta = -\frac{15}{8}$

Check for Understanding

Concept Check

1. False; $\sec 0° = \frac{r}{r}$ or 1 and $\tan 0° = \frac{0}{r}$ or 0.

2. Sample answer: 190°

1. **Determine** whether the following statement is *true* or *false*. If true, explain your reasoning. If false, give a counterexample.

   The values of the secant and tangent functions for any quadrantal angle are undefined.

2. **OPEN ENDED** Give an example of an angle whose sine is negative.

3. **Explain** how the reference angle θ' is used to find the value of a trigonometric function of θ, where θ is greater than 90°. **See margin.**

Guided Practice

GUIDED PRACTICE KEY	
Exercises	**Examples**
4–6	1
7–9	3
10–13	2, 4
14–16	5

Find the exact values of the six trigonometric functions of θ if the terminal side of θ in standard position contains the given point. **4–6. See margin.**

4. $(-15, 8)$
5. $(-3, 0)$
6. $(4, 4)$

Sketch each angle. Then find its reference angle.

7. 235° **55°**
8. $\frac{7\pi}{4}$ $\frac{\pi}{4}$
9. −240° **60°**

7–9. See pp. 759A–759D for sketches.

Find the exact value of each trigonometric function.

10. $\sin 300°$ $-\frac{\sqrt{3}}{2}$
11. $\cos 180°$ **−1**
12. $\tan \frac{5\pi}{3}$ $-\sqrt{3}$
13. $\sec \frac{7\pi}{6}$ $-\frac{2\sqrt{3}}{3}$

Suppose θ is an angle in standard position whose terminal side is in the given quadrant. For each function, find the exact values of the remaining five trigonometric functions of θ. **14–15. See pp. 759A–759D.**

14. $\cos\theta = -\frac{1}{2}$, Quadrant II
15. $\cot\theta = -\frac{\sqrt{2}}{2}$, Quadrant IV

Application

16. **BASKETBALL** The maximum height H in feet that a basketball reaches after being shot is given by the formula $H = \frac{V_0^2 (\sin\theta)^2}{64}$, where V_0 represents the initial velocity in feet per second, θ represents the degree measure of the angle that the path of the basketball makes with the ground. Find the maximum height reached by a ball shot with an initial velocity of 30 feet per second at an angle of 70°. **about 12.4 ft**

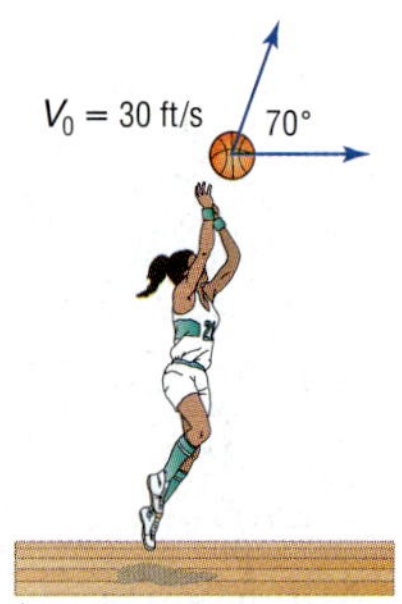

★ indicates increased difficulty

Practice and Apply

Homework Help

For Exercises	See Examples
17–24	1
25–32	3
33–46	2, 4
47–52	5
53–55	6

Extra Practice See page 857.

Find the exact values of the six trigonometric functions of θ if the terminal side of θ in standard position contains the given point. **17–24. See pp. 759A–759D.**

17. $(7, 24)$
18. $(2, 1)$
19. $(5, -8)$
20. $(4, -3)$
21. $(0, -6)$
22. $(-1, 0)$
★ 23. $(\sqrt{2}, -\sqrt{2})$
★ 24. $(-\sqrt{3}, -\sqrt{6})$

25–32. See pp. 759A–759D for sketches.

Sketch each angle. Then find its reference angle.

25. 315° **45°**
26. 240° **60°**
27. −210° **30°**
28. −125° **55°**
29. $\frac{5\pi}{4}$ $\frac{\pi}{4}$
30. $\frac{5\pi}{6}$ $\frac{\pi}{6}$
31. $\frac{13\pi}{7}$ $\frac{\pi}{7}$
32. $-\frac{2\pi}{3}$ $\frac{\pi}{3}$

5. $\sin\theta = 0$, $\cos\theta = -1$, $\tan\theta = 0$, $\csc\theta =$ undefined, $\sec\theta = -1$, $\cot\theta =$ undefined

6. $\sin\theta = \frac{\sqrt{2}}{2}$, $\cos\theta = \frac{\sqrt{2}}{2}$, $\tan\theta = 1$, $\csc\theta = \sqrt{2}$, $\sec\theta = \sqrt{2}$, $\cot\theta = 1$

14. $\sin\theta = \frac{\sqrt{3}}{2}$, $\tan\theta = -\sqrt{3}$, $\csc\theta = \frac{2\sqrt{3}}{3}$, $\sec\theta = -2$, $\cot\theta = -\frac{\sqrt{3}}{3}$

15. $\sin\theta = -\frac{\sqrt{6}}{3}$, $\cos\theta = \frac{\sqrt{3}}{3}$, $\tan\theta = -\sqrt{2}$, $\csc\theta = -\frac{\sqrt{6}}{2}$, $\sec\theta = \sqrt{3}$

37. undefined

47. $\sin \theta = -\frac{4}{5}$, $\tan \theta = -\frac{4}{3}$, $\csc \theta = -\frac{5}{4}$, $\sec \theta = \frac{5}{3}$, $\cot \theta = -\frac{3}{4}$

More About. . .

Baseball

If a major league pitcher throws a pitch at 95 miles per hour, it takes only about 4 tenths of a second for the ball to travel the 60 feet, 6 inches from the pitcher's mound to home plate. In that time, the hitter must decide whether to swing at the ball and if so, when to swing.

Source: www.exploratorium.edu

48. $\sin \theta = \frac{\sqrt{26}}{26}$, $\cos \theta = -\frac{5\sqrt{26}}{26}$, $\csc \theta = \sqrt{26}$, $\sec \theta = -\frac{\sqrt{26}}{5}$, $\cot \theta = -5$

49. $\cos \theta = -\frac{2\sqrt{2}}{3}$, $\tan \theta = -\frac{\sqrt{2}}{4}$, $\csc \theta = 3$, $\sec \theta = -\frac{3\sqrt{2}}{4}$, $\cot \theta = -2\sqrt{2}$

Find the exact value of each trigonometric function.

33. $\sin 240°$ $-\frac{\sqrt{3}}{2}$
34. $\sec 120°$ -2
35. $\tan 300°$ $-\sqrt{3}$
36. $\cot 510°$ $-\sqrt{3}$
37. $\csc 5400°$
38. $\cos \frac{11\pi}{3}$ $\frac{1}{2}$
39. $\cot\left(-\frac{5\pi}{6}\right)$ $\sqrt{3}$
40. $\sin \frac{3\pi}{4}$ $\frac{\sqrt{2}}{2}$
41. $\sec \frac{3\pi}{2}$ undefined
42. $\csc \frac{17\pi}{6}$ 2
43. $\cos(-30°)$ $\frac{\sqrt{3}}{2}$
44. $\tan\left(-\frac{5\pi}{4}\right)$ -1

45. **SKYCOASTING** Mikhail and Anya visit a local amusement park to ride a skycoaster. After the first several swings, the angle the skycoaster makes with the vertical is modeled by $\theta = 0.2 \cos \pi t$, with θ measured in radians and t measured in seconds. Determine the measure of the angle for $t = 0, 0.5, 1, 1.5, 2, 2.5$, and 3 in both radians and degrees. **0.2, 0, −0.2, 0, 0.2, 0, and −0.2; or about 11.5°, 0°, −11.5°, 0°, 11.5°, 0°, and −11.5°**

46. **NAVIGATION** Ships and airplanes measure distance in nautical miles. The formula 1 nautical mile $= 6077 - 31 \cos 2\theta$ feet, where θ is the latitude in degrees, can be used to find the approximate length of a nautical mile at a certain latitude. Find the length of a nautical mile where the latitude is 60°. **6092.5 ft**

Suppose θ is an angle in standard position whose terminal side is in the given quadrant. For each function, find the exact values of the remaining five trigonometric functions of θ. **50–52. See pp. 759A–759D.**

47. $\cos \theta = \frac{3}{5}$, Quadrant IV
48. $\tan \theta = -\frac{1}{5}$, Quadrant II
49. $\sin \theta = \frac{1}{3}$, Quadrant II
50. $\cot \theta = \frac{1}{2}$, Quadrant III
51. $\sec \theta = -\sqrt{10}$, Quadrant III
52. $\csc \theta = -5$, Quadrant IV

BASEBALL **For Exercises 53 and 54, use the following information.**

The formula $R = \frac{V_0^2 \sin 2\theta}{32}$ gives the distance of a baseball that is hit at an initial velocity of V_0 feet per second at an angle of θ with the ground.

53. If the ball was hit with an initial velocity of 80 feet per second at an angle of 30°, how far was it hit? **about 173.2 ft**

54. Which angle will result in the greatest distance? Explain your reasoning. **45°; 2 × 45° or 90° yields the greatest value for sin 2θ.**

55. **CAROUSELS** Anthony's little brother gets on a carousel that is 8 meters in diameter. At the start of the ride, his brother is 3 meters from the fence to the ride. How far will his brother be from the fence after the carousel rotates 240°? **9 meters**

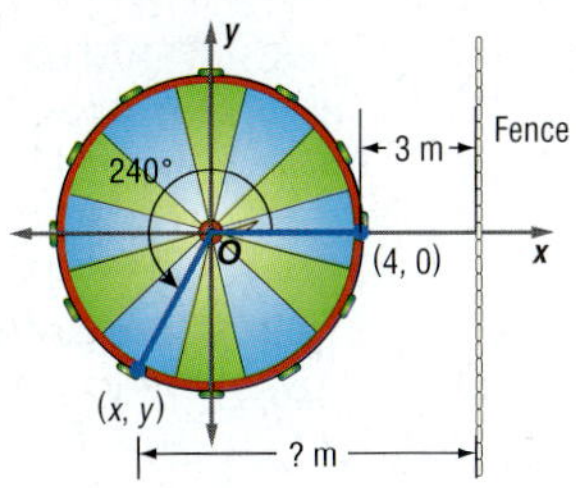

Online Research **Data Update** What is the diameter of the world's largest carousel? Visit www.algebra2.com/data_update to learn more.

CRITICAL THINKING **Suppose θ is an angle in standard position with the given conditions. State the quadrant(s) in which the terminal side of θ lies.**

56. $\sin \theta > 0$ **I, II**
57. $\sin \theta > 0$, $\cos \theta < 0$ **II**
58. $\tan \theta > 0$, $\cos \theta < 0$ **III**

www.algebra2.com/self_check_quiz

Study Guide and Intervention, p. 787 (shown) and p. 788

Trigonometric Functions and General Angles

Trigonometric Functions, θ in Standard Position	Let θ be an angle in standard position and let $P(x, y)$ be a point on the terminal side of θ. By the Pythagorean Theorem, the distance r from the origin is given by $r = \sqrt{x^2 + y^2}$. The trigonometric functions of an angle in standard position may be defined as follows. $\sin \theta = \frac{y}{r}$ $\cos \theta = \frac{x}{r}$ $\tan \theta = \frac{y}{x}$ $\csc \theta = \frac{r}{y}$ $\sec \theta = \frac{r}{x}$ $\cot \theta = \frac{x}{y}$

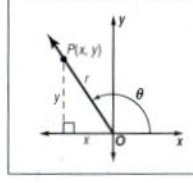

Example **Find the exact values of the six trigonometric functions of θ if the terminal side of θ contains the point $(-5, 5\sqrt{2})$.**

You know that $x = -5$ and $y = 5\sqrt{2}$. You need to find r.

$r = \sqrt{x^2 + y^2}$ Pythagorean Theorem

$= \sqrt{(-5)^2 + (5\sqrt{2})^2}$ Replace x with −5 and y with $5\sqrt{2}$.

$= \sqrt{75}$ or $5\sqrt{3}$

Now use $x = -5$, $y = 5\sqrt{2}$, and $r = 5\sqrt{3}$ to write the ratios.

$\sin \theta = \frac{y}{r} = \frac{5\sqrt{2}}{5\sqrt{3}} = \frac{\sqrt{6}}{3}$ $\cos \theta = \frac{x}{r} = \frac{-5}{5\sqrt{3}} = -\frac{\sqrt{3}}{3}$ $\tan \theta = \frac{y}{x} = \frac{5\sqrt{2}}{-5} = -\sqrt{2}$

$\csc \theta = \frac{r}{y} = \frac{5\sqrt{3}}{5\sqrt{2}} = \frac{\sqrt{6}}{2}$ $\sec \theta = \frac{r}{x} = \frac{5\sqrt{3}}{-5} = -\sqrt{3}$ $\cot \theta = \frac{x}{y} = \frac{-5}{5\sqrt{2}} = -\frac{\sqrt{2}}{2}$

Exercises

Find the exact values of the six trigonometric functions of θ if the terminal side of θ in standard position contains the given point.

1. (8, 4) $\sin \theta = \frac{\sqrt{5}}{5}$, $\cos \theta = \frac{2\sqrt{5}}{5}$, $\tan \theta = \frac{1}{2}$, $\csc \theta = \sqrt{5}$, $\sec \theta = \frac{\sqrt{5}}{2}$, $\cot \theta = 2$

2. $(4, 4\sqrt{3})$ $\sin \theta = \frac{\sqrt{3}}{2}$, $\cos \theta = \frac{1}{2}$, $\tan \theta = \sqrt{3}$, $\csc \theta = \frac{2\sqrt{3}}{3}$, $\sec \theta = 2$, $\cot \theta = \frac{\sqrt{3}}{3}$

3. (0, −4) $\sin \theta = -1$, $\cos \theta = 0$, $\tan \theta$ undefined, $\csc \theta = -1$, $\sec \theta$ undefined, $\cot \theta = 0$

4. (6, 2) $\sin \theta = \frac{\sqrt{10}}{10}$, $\cos \theta = \frac{3\sqrt{10}}{10}$, $\tan \theta = \frac{1}{3}$, $\csc \theta = \sqrt{10}$, $\sec \theta = \frac{\sqrt{10}}{3}$, $\cot \theta = 3$

Skills Practice, p. 789 and Practice, p. 790 (shown)

Find the exact values of the six trigonometric functions of θ if the terminal side of θ in standard position contains the given point.

1. (6, 8) $\sin \theta = \frac{4}{5}$, $\cos \theta = \frac{3}{5}$, $\tan \theta = \frac{4}{3}$, $\csc \theta = \frac{5}{4}$, $\sec \theta = \frac{5}{3}$, $\cot \theta = \frac{3}{4}$

2. (−20, 21) $\sin \theta = \frac{21}{29}$, $\cos \theta = -\frac{20}{29}$, $\tan \theta = -\frac{21}{20}$, $\csc \theta = \frac{29}{21}$, $\sec \theta = -\frac{29}{20}$, $\cot \theta = -\frac{20}{21}$

3. (−2, −5) $\sin \theta = -\frac{5\sqrt{29}}{29}$, $\cos \theta = -\frac{2\sqrt{29}}{29}$, $\tan \theta = \frac{5}{2}$, $\csc \theta = -\frac{\sqrt{29}}{5}$, $\sec \theta = -\frac{\sqrt{29}}{2}$, $\cot \theta = \frac{2}{5}$

Find the reference angle for the angle with the given measure.

4. 236° 56°
5. $\frac{13\pi}{8}$ $\frac{3\pi}{8}$
6. −210° 30°
7. $-\frac{7\pi}{4}$ $\frac{\pi}{4}$

Find the exact value of each trigonometric function.

8. $\tan 135°$ -1
9. $\cot 210°$ $\sqrt{3}$
10. $\cot(-90°)$ 0
11. $\cos 405°$ $\frac{\sqrt{2}}{2}$
12. $\tan \frac{5\pi}{3}$ $-\sqrt{3}$
13. $\csc\left(-\frac{3\pi}{4}\right)$ $-\sqrt{2}$
14. $\cot 2\pi$ undefined
15. $\tan \frac{13\pi}{6}$ $\frac{\sqrt{3}}{3}$

Suppose θ is an angle in standard position whose terminal side is in the given quadrant. For each function, find the exact values of the remaining five trigonometric functions of θ.

16. $\tan \theta = -\frac{12}{5}$, Quadrant IV $\sin \theta = -\frac{12}{13}$, $\cos \theta = \frac{5}{13}$, $\csc \theta = -\frac{13}{12}$, $\sec \theta = \frac{13}{5}$, $\cot \theta = -\frac{5}{12}$

17. $\sin \theta = \frac{2}{3}$, Quadrant III $\cos \theta = -\frac{\sqrt{5}}{3}$, $\tan \theta = -\frac{2\sqrt{5}}{5}$, $\csc \theta = \frac{3}{2}$, $\sec \theta = -\frac{3\sqrt{5}}{5}$, $\cot \theta = -\frac{\sqrt{5}}{2}$

18. **LIGHT** Light rays that "bounce off" a surface are *reflected* by the surface. If the surface is partially transparent, some of the light rays are bent or *refracted* as they pass from the air through the material. The angles of reflection θ_1 and of refraction θ_2 in the diagram at the right are related by the equation $\sin \theta_1 = n \sin \theta_2$. If $\theta_1 = 60°$ and $n = \sqrt{3}$, find the measure of θ_2. 30°

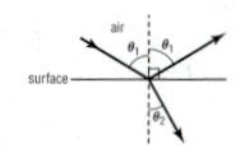

19. **FORCE** A cable running from the top of a utility pole to the ground exerts a horizontal pull of 800 Newtons and a vertical pull of $800\sqrt{3}$ Newtons. What is the sine of the angle θ between the cable and the ground? What is the measure of this angle? $\frac{\sqrt{3}}{2}$; 60°

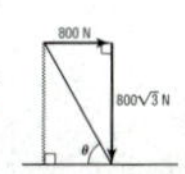

Reading to Learn Mathematics, p. 791 ELL

Pre-Activity **How can you model the position of riders on a skycoaster?**

Read the introduction to Lesson 13-3 at the top of page 717 in your textbook.

- What does $t = 0$ represent in this application? Sample answer: the time when the riders leave the bottom of their swing
- Do negative values of t make sense in this application? Explain your answer. Sample answer: No; $t = 0$ represents the starting time, so the value of t cannot be less than 0.

Reading the Lesson

1. Suppose θ is an angle in standard position, $P(x, y)$ is a point on the terminal side of θ, and the distance from the origin to P is r. Determine whether each of the following statements is *true* or *false*.
   a. The value of r can be found by using either the Pythagorean Theorem or the distance formula. true
   b. $\cos \theta = \frac{x}{r}$ true
   c. $\csc \theta$ is defined if $y \neq 0$. true
   d. $\tan \theta$ is undefined if $y = 0$. false
   e. $\sin \theta$ is defined for every value of θ. true

2. Let θ be an angle measured in degrees. Match the quadrant of θ from the first column with the description of how to find the reference angle for θ from the second column.
   a. Quadrant III ii
   b. Quadrant IV i
   c. Quadrant II iv
   d. Quadrant I iii

   i. Subtract θ from 360°.
   ii. Subtract 180° from θ.
   iii. θ is its own reference angle.
   iv. Subtract θ from 180°.

Helping You Remember

3. The chart on page 719 in your textbook summarizes the signs of the six trigonometric functions in the four quadrants. Since reciprocals always have the same sign, you only need to remember where the sine, cosine, and tangent are positive. How can you remember this with a simple diagram?

Sample answer:

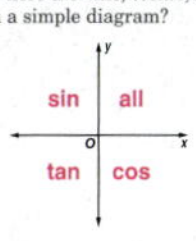

Enrichment, p. 792

Areas of Polygons and Circles

A regular polygon has sides of equal length and angles of equal measure. A regular polygon can be inscribed in or circumscribed about a circle. For n-sided regular polygons, the following area formulas can be used.

Area of circle $A_C = \pi r^2$

Area of inscribed polygon $A_I = \frac{nr^2}{2} \times \sin \frac{360°}{n}$

Area of circumscribed polygon $A_C = nr^2 \times \tan \frac{180°}{n}$

Use a calculator to complete the chart below for a unit circle (a circle of radius 1).

	Area of ...ribed	Area of Circle	Area of	Area of Polygon

59. **WRITING IN MATH** Answer the question that was posed at the beginning of the lesson. **See margin.**

How can you model the position of riders on a skycoaster?

Include the following in your answer:

- an explanation of how you could use the cosine of the angle θ and the length of the cable from which they swing to find the horizontal position of a person on a skycoaster relative to the center of the arch, and
- an explanation of how you would use the angle θ, the height of the tower, and the length of the cable to find the height of riders from the ground.

Standardized Test Practice

60. If the cotangent of angle θ is 1, then the tangent of angle θ is **C**

Ⓐ -1. Ⓑ 0. Ⓒ 1. Ⓓ 3.

61. **SHORT RESPONSE** Find the exact coordinates of point P, which is located at the intersection of a circle of radius 5 and the terminal side of angle θ measuring $\frac{5\pi}{3}$. $\left(\frac{5}{2}, -\frac{5\sqrt{3}}{2}\right)$

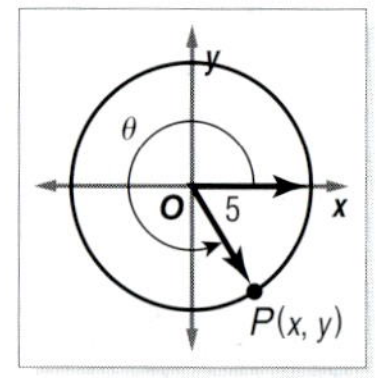

Maintain Your Skills

Mixed Review

Rewrite each degree measure in radians and each radian measure in degrees. *(Lesson 13-2)*

62. $90°$ $\frac{\pi}{2}$ 63. $\frac{5\pi}{3}$ **300°** 64. 5 $\frac{900}{\pi} \approx 286.5°$

Write an equation involving sin, cos, or tan that can be used to find x. Then solve the equation. Round measures of sides to the nearest tenth and measures of angles to the nearest degree. *(Lesson 13-1)*

65.

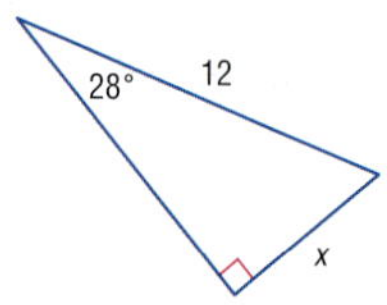

$\sin 28° = \frac{x}{12}$, 5.6

66.

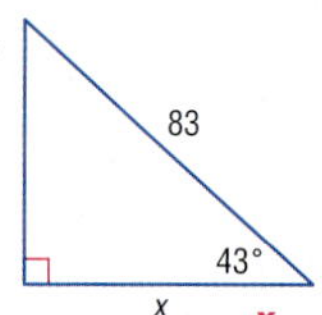

$\cos 43° = \frac{x}{83}$, 60.7

67.

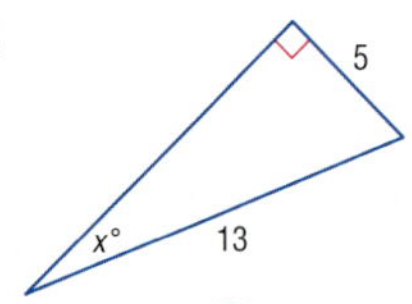

$\sin x° = \frac{5}{13}$, 23

68. **LITERATURE** In one of *Grimm's Fairy Tales*, Rumpelstiltskin has the ability to spin straw into gold. Suppose on the first day, he spun 5 pieces of straw into gold, and each day thereafter he spun twice as much. How many pieces of straw would he have spun into gold by the end of the week? *(Lesson 11-4)* **635**

Use Cramer's Rule to solve each system of equations. *(Lesson 4-6)*

69. $3x - 4y = 13$, $-2x + 5y = -4$ **(7, 2)**

70. $5x + 7y = 1$, $3x + 5y = 3$ **(−4, 3)**

71. $2x + 3y = -2$, $-6x + y = -34$ **(5, −4)**

Getting Ready for the Next Lesson

PREREQUISITE SKILL **Solve each equation. Round to the nearest tenth.**
(To review ***solving equations with trigonometric functions****, see Lesson 13-1.)*

72. $\frac{a}{\sin 32°} = \frac{8}{\sin 65°}$ **4.7** 73. $\frac{b}{\sin 45°} = \frac{21}{\sin 100°}$ **15.1** 74. $\frac{c}{\sin 60°} = \frac{3}{\sin 75°}$ **2.7**

75. $\frac{\sin A}{14} = \frac{\sin 104°}{25}$ **32.9°** 76. $\frac{\sin B}{3} = \frac{\sin 55°}{7}$ **20.6°** 77. $\frac{\sin C}{10} = \frac{\sin 35°}{9}$ **39.6°**

4 Assess

Open-Ended Assessment

Writing Have students choose an angle measure, draw an angle with that measure in standard position, find the reference angle, and give the values of all 6 of the trigonometric functions, in both degrees and radians. Then display some of these in the classroom.

Getting Ready for Lesson 13-4

PREREQUISITE SKILL Lesson 13-4 presents the Law of Sines which will require students to solve equations involving trigonometric functions as they apply the Law of Sines. Exercises 72–77 should be used to determine your students' familiarity with solving equations with trigonometric functions.

Answer

59. Answers should include the following.

- The cosine of any angle is defined as $\frac{x}{r}$, where x is the x-coordinate of any point on the terminal ray of the angle and r is the distance from the origin to that point. This means that for angles with terminal sides to the left of the y-axis, the cosine is negative, and those with terminal sides to the right of the y-axis, the cosine is positive. Therefore the cosine function can be used to model real-world data that oscillate between being positive and negative.
- If we knew the length of the cable we could find the vertical distance from the top of the tower to the rider. Then if we knew the height of the tower we could subtract from it the vertical distance calculated previously. This will leave the height of the rider from the ground.

724 Chapter 13 Trigonometric Functions

13-4 Law of Sines

What You'll Learn

- Solve problems by using the Law of Sines.
- Determine whether a triangle has one, two, or no solutions.

Vocabulary

- Law of Sines

How can trigonometry be used to find the area of a triangle?

You know how to find the area of a triangle when the base and the height are known. Using this formula, the area of $\triangle ABC$ below is $\frac{1}{2}ch$. If the height h of this triangle were not known, you could still find the area given the measures of angle A and the length of side b.

$$\sin A = \frac{h}{b} \rightarrow h = b \sin A$$

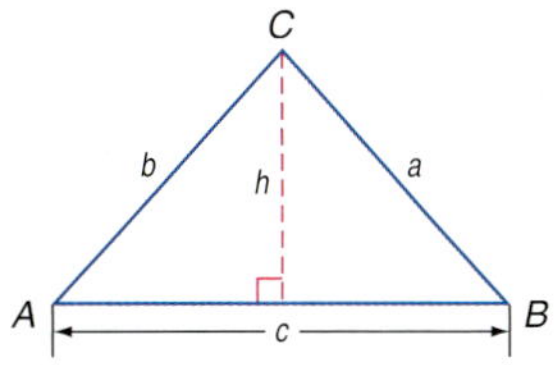

By combining this equation with the area formula, you can find a new formula for the area of the triangle.

$$\text{Area} = \frac{1}{2}ch \rightarrow \text{Area} = \frac{1}{2}c(b \sin A)$$

LAW OF SINES You can find two other formulas for the area of the triangle above in a similar way. These formulas, summarized below, allow you to find the area of any triangle when you know the measures of two sides and the included angle.

Key Concept — *Area of a Triangle*

- **Words** The area of a triangle is one half the product of the lengths of two sides and the sine of their included angle.
- **Symbols** area $= \frac{1}{2}bc \sin A$

  area $= \frac{1}{2}ac \sin B$

  area $= \frac{1}{2}ab \sin C$

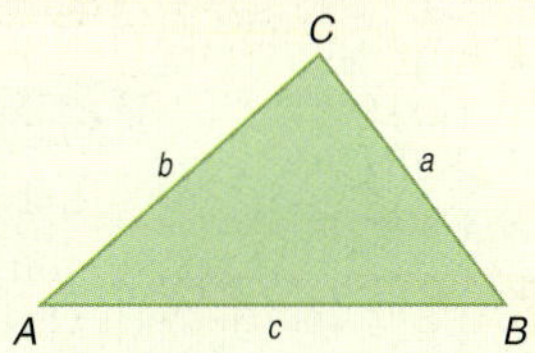

Example 1 Find the Area of a Triangle

Find the area of $\triangle ABC$ to the nearest tenth.

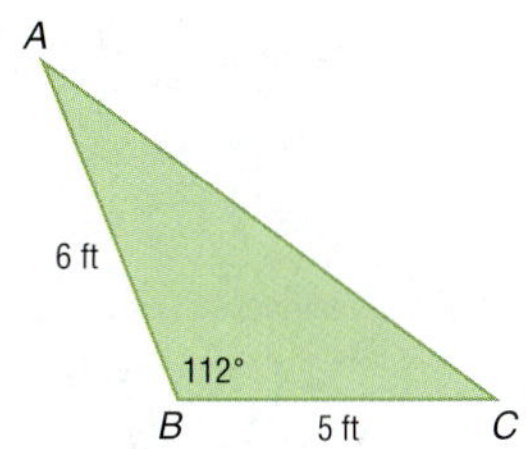

In this triangle, $a = 5$, $c = 6$, and $B = 112°$. Choose the second formula because you know the values of its variables.

$\text{Area} = \frac{1}{2}ac \sin B$	Area formula
$= \frac{1}{2}(5)(6) \sin 112°$	Replace a with 5, c with 6, and B with 112°.
≈ 13.9	Use a calculator.

To the nearest tenth, the area is 13.9 square feet.

13-4 Lesson Notes

1 Focus

5-Minute Check Transparency 13-4 Use as a quiz or review of Lesson 13-3.

Mathematical Background notes are available for this lesson on p. 698D.

How can trigonometry be used to find the area of a triangle?

Ask students:

- What kind of triangles does the height separate the triangle into? **right triangles**
- Does the size of angle C make any difference in this derivation of the area formula? **no**

Resource Manager

Workbook and Reproducible Masters

Chapter 13 Resource Masters

- Study Guide and Intervention, pp. 793–794
- Skills Practice, p. 795
- Practice, p. 796
- Reading to Learn Mathematics, p. 797
- Enrichment, p. 798
- Assessment, pp. 831, 833

Graphing Calculator and Spreadsheet Masters, p. 51

School-to-Career Masters, p. 25

Transparencies

5-Minute Check Transparency 13-4

Answer Key Transparencies

Technology

Interactive Chalkboard

2 Teach

LAW OF SINES

In-Class Examples Power Point®

Teaching Tip In Example 1, ask students if they knew only the values for sides c and a, and for angle A, could they use the formula to find the area of the triangle? **No, to use the formula, the angle must be included in the known sides.**

1 Find the area of $\triangle ABC$ to the nearest tenth. **3.8 cm²**

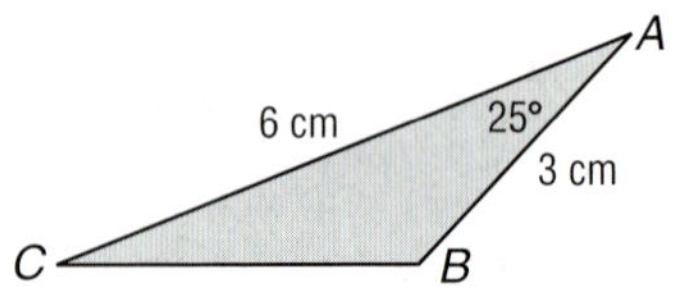

2 Solve $\triangle ABC$.

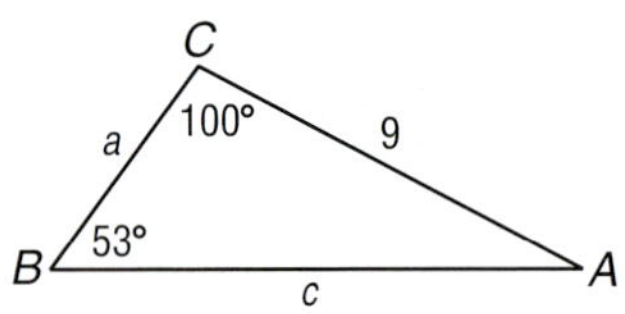

$A = 27°$, $a \approx 5.1$, $c \approx 11.1$

All of the area formulas for $\triangle ABC$ represent the area of the same triangle. So, $\frac{1}{2}bc \sin A$, $\frac{1}{2}ac \sin B$, and $\frac{1}{2}ab \sin C$ are all equal. You can use this fact to derive the **Law of Sines**.

$$\frac{1}{2}bc \sin A = \frac{1}{2}ac \sin B = \frac{1}{2}ab \sin C$$ Set area formulas equal to each other.

$$\frac{\frac{1}{2}bc \sin A}{\frac{1}{2}abc} = \frac{\frac{1}{2}ac \sin B}{\frac{1}{2}abc} = \frac{\frac{1}{2}ab \sin C}{\frac{1}{2}abc}$$ Divide each expression by $\frac{1}{2}abc$.

$$\frac{\sin A}{a} = \frac{\sin B}{b} = \frac{\sin C}{c}$$ Simplify.

Study Tip

Alternate Representations
The Law of Sines may also be written as $\frac{a}{\sin A} = \frac{b}{\sin B} = \frac{c}{\sin C}$.

Key Concept — *Law of Sines*

Let $\triangle ABC$ be any triangle with a, b, and c representing the measures of sides opposite angles with measurements A, B, and C respectively. Then,

$$\frac{\sin A}{a} = \frac{\sin B}{b} = \frac{\sin C}{c}.$$

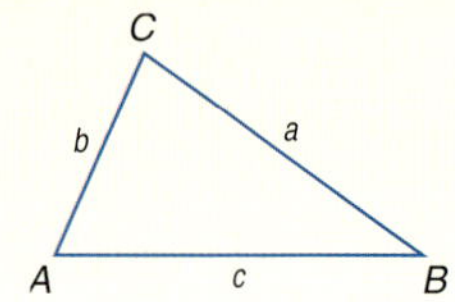

The Law of Sines can be used to write three different equations.

$$\frac{\sin A}{a} = \frac{\sin B}{b} \quad \text{or} \quad \frac{\sin B}{b} = \frac{\sin C}{c} \quad \text{or} \quad \frac{\sin A}{a} = \frac{\sin C}{c}$$

In Lesson 13-1, you learned how to solve right triangles. To solve *any* triangle, you can apply the Law of Sines if you know

- the measures of two angles and any side or
- the measures of two sides and the angle opposite one of them.

TEACHING TIP
You may wish to use the abbreviation AAS when referring to a triangle where the measure of two angles and a nonincluded side are known and ASA when the measure of two angles and their included side are known. Similarly, you can use SSA to refer to a triangle where the measures of two sides and the angle opposite one of them is known.

Example 2 *Solve a Triangle Given Two Angles and a Side*

Solve $\triangle ABC$.

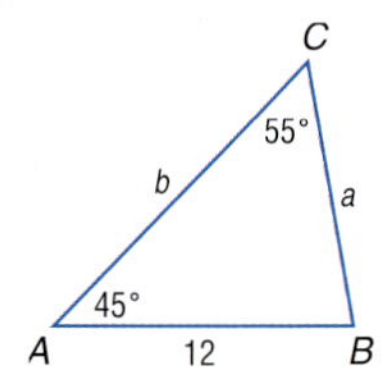

You are given the measures of two angles and a side. First, find the measure of the third angle.

$45° + 55° + B = 180°$ The sum of the angle measures of a triangle is 180°.

$B = 80°$ $180 - (45 + 55) = 80$

Now use the Law of Sines to find a and b. Write two equations, each with one variable.

$\frac{\sin A}{a} = \frac{\sin C}{c}$	Law of Sines	$\frac{\sin B}{b} = \frac{\sin C}{c}$
$\frac{\sin 45°}{a} = \frac{\sin 55°}{12}$	Replace A with 45°, B with 80°, C with 55°, and c with 12.	$\frac{\sin 80°}{b} = \frac{\sin 55°}{12}$
$a = \frac{12 \sin 45°}{\sin 55°}$	Solve for the variable.	$b = \frac{12 \sin 80°}{\sin 55°}$
$a \approx 10.4$	Use a calculator.	$b \approx 14.4$

Therefore, $B = 80°$, $a \approx 10.4$, and $b \approx 14.4$.

DAILY INTERVENTION — **Unlocking Misconceptions**

Students may think the Law of Sines only works for right triangles. Clarify that this formula works for any triangle, as does the Law of Cosines explored in Lesson 13-5.

ONE, TWO, OR NO SOLUTIONS When solving a triangle, you must analyze the data you are given to determine whether there is a solution. For example, if you are given the measures of two angles and a side, as in Example 2, the triangle has a unique solution. However, if you are given the measures of two sides and the angle opposite one of them, a single solution may not exist. One of the following will be true.

- No triangle exists, and there is no solution.
- Exactly one triangle exists, and there is one solution.
- Two triangles exist, and there are two solutions.

Key Concept — Possible Triangles Given Two Sides and One Opposite Angle

Suppose you are given a, b, and A for a triangle.

A Is Acute ($A < 90°$).

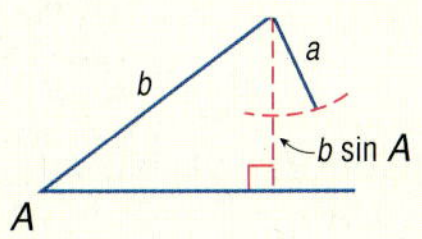

$a < b \sin A$
no solution

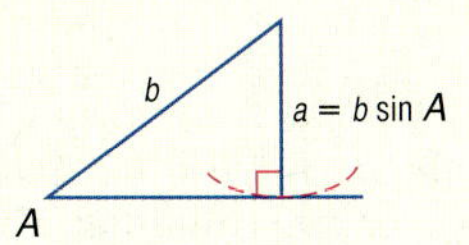

$a = b \sin A$
one solution

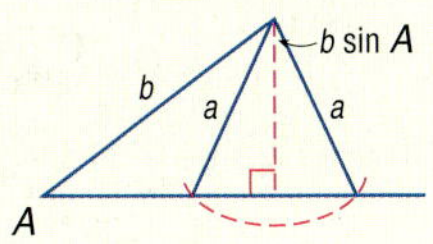

$b > a > b \sin A$
two solutions

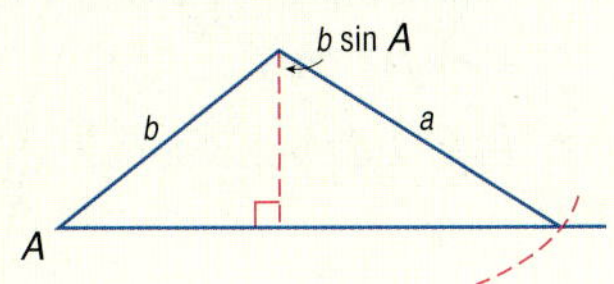

$a \geq b$
one solution

A Is Right or Obtuse ($A \geq 90°$).

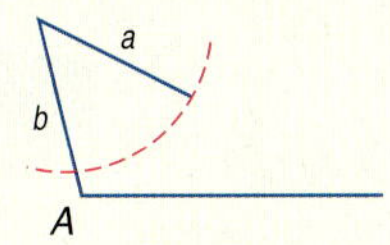

$a \leq b$
no solution

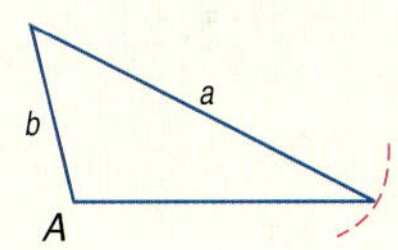

$a > b$
one solution

Example 3 One Solution

In $\triangle ABC$, $A = 118°$, $a = 20$, and $b = 17$. Determine whether $\triangle ABC$ has *no* solution, *one* solution, or *two* solutions. Then solve $\triangle ABC$.

Because angle A is obtuse and $a > b$, you know that one solution exists.

Make a sketch and then use the Law of Sines to find B.

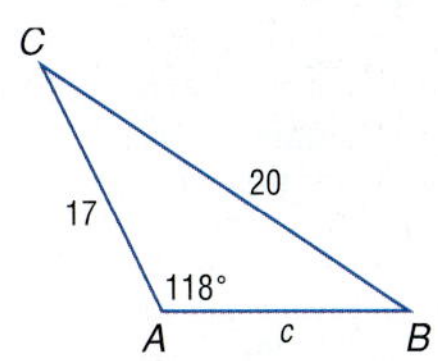

$\frac{\sin B}{17} = \frac{\sin 118°}{20}$ Law of Sines

$\sin B = \frac{17 \sin 118°}{20}$ Multiply each side by 17.

$\sin B \approx 0.7505$ Use a calculator.

$B \approx 49°$ Use the $\sin^{-1}$ function.

The measure of angle C is approximately $180 - (118 + 49)$ or $13°$.

Use the Law of Sines again to find c.

$\frac{\sin 13}{c} = \frac{\sin 118°}{20}$ Law of Sines

$c = \frac{20 \sin 13°}{\sin 118°}$ or about 5.1 Use a calculator.

Therefore, $B \approx 49°$, $C \approx 13°$, and $c \approx 5.1$.

www.algebra2.com/extra_examples

ONE, TWO, OR NO SOLUTIONS

In-Class Example

Power Point®

3 In $\triangle ABC$, $A = 25°$, $a = 13$, and $b = 12$. Determine whether $\triangle ABC$ has *no* solution, *one* solution, or *two* solutions. Then solve $\triangle ABC$.
one; $B \approx 23°$, $C \approx 132°$; $c \approx 22.9$

Lesson 13-4 Law of Sines 727

In-Class Examples

Power Point®

4 In $\triangle ABC$, $A = 125°$, $a = 35$, and $b = 32$. Determine whether $\triangle ABC$ has *no* solution, *one* solution, or *two* solutions. Then solve $\triangle ABC$.
no

5 In $\triangle ABC$, $A = 25°$, $a = 5$, and $b = 10$. Determine whether $\triangle ABC$ has *no* solution, *one* solution, or *two* solutions. Then solve $\triangle ABC$.
two; $B \approx 58°$, $C \approx 97°$, $c \approx 11.7$; $B \approx 122°$, $C \approx 33°$, $c \approx 6.4$

Study Tip

A Is Acute

We compare $b \sin A$ to a because $b \sin A$ is the minimum distance from C to $\overline{AB}$ when A is acute.

Example 4 No Solution

In $\triangle ABC$, $A = 50°$, $a = 5$, and $b = 9$. Determine whether $\triangle ABC$ has *no* solution, *one* solution, or *two* solutions. Then solve $\triangle ABC$.

Since angle A is acute, find $b \sin A$ and compare it with a.

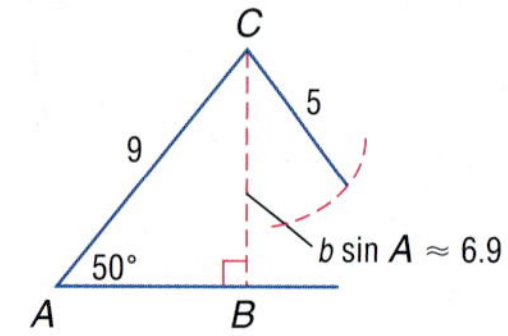

$b \sin A = 9 \sin 50°$ Replace b with 9 and A with 50°.

≈ 6.9 Use a calculator.

Since $5 < 6.9$, there is no solution.

TEACHING TIP

You can have students verify this solution using the Law of Sines.

$$\frac{\sin 50°}{5} = \frac{\sin B}{9}$$

$$\sin B \approx 1.379$$

Since a sine ratio cannot exceed 1, there is no solution.

When two solutions for a triangle exist, it is called the *ambiguous case.*

Example 5 Two Solutions

In $\triangle ABC$, $A = 39°$, $a = 10$, and $b = 14$. Determine whether $\triangle ABC$ has *no* solution, *one* solution, or *two* solutions. Then solve $\triangle ABC$.

Since angle A is acute, find $b \sin A$ and compare it with a.

$b \sin A = 14 \sin 39°$ Replace b with 14 and A with 39°.

≈ 8.81 Use a calculator.

Since $14 > 10 > 8.81$, there are two solutions. Thus, there are two possible triangles to be solved.

Case 1 Acute Angle B

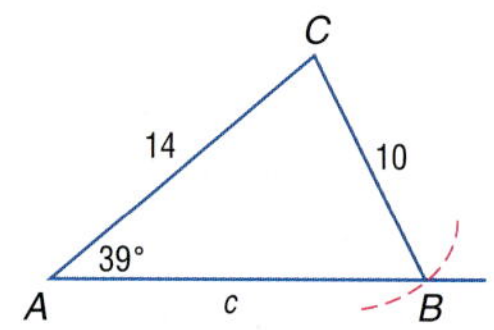

First, use the Law of Sines to find B.

$$\frac{\sin B}{14} = \frac{\sin 39°}{10}$$

$$\sin B = \frac{14 \sin 39°}{10}$$

$$\sin B = 0.8810$$

$$B \approx 62°$$

The measure of angle C is approximately $180 - (39 + 62)$ or $79°$.

Use the Law of Sines again to find c.

$$\frac{\sin 79°}{c} = \frac{\sin 39°}{10}$$

$$c = \frac{10 \sin 79°}{\sin 39°}$$

$$c \approx 15.6$$

Therefore, $B \approx 62°$, $C \approx 79°$, and $c \approx 15.6$.

Case 2 Obtuse Angle B

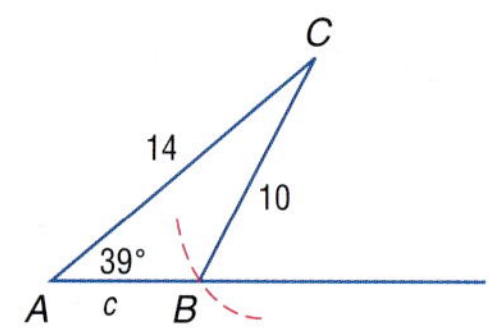

To find B, you need to find an obtuse angle whose sine is also 0.8810. To do this, subtract the angle given by your calculator, 62°, from 180°. So B is approximately $180 - 62$ or $118°$.

The measure of angle C is approximately $180 - (39 + 118)$ or $23°$.

Use the Law of Sines to find c.

$$\frac{\sin 23°}{c} = \frac{\sin 39°}{10}$$

$$c = \frac{10 \sin 23°}{\sin 39°}$$

$$c \approx 6.2$$

Therefore, $B \approx 118°$, $C \approx 23°$, and $c \approx 6.2$.

Study Tip

Alternate Method

Another way to find the obtuse angle in Case 2 of Example 5 is to notice in the figure below that $\triangle CBB'$ is isosceles. Since the base angles of an isosceles triangle are always congruent and $m\angle B' = 62°$, $m\angle CBB' = 62°$. Also, $\angle ABC$ and $m\angle CBB'$ are supplementary. Therefore, $m\angle ABC = 180° - 62°$ or $118°$.

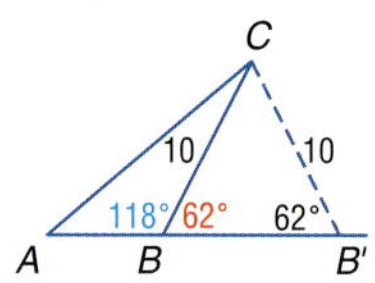

DAILY INTERVENTION

Differentiated Instruction

Intrapersonal Have students write a journal entry about which example they found the most challenging and why. Ask them to include any questions they still have about the lesson.

Example 6 Use the Law of Sines to Solve a Problem

LIGHTHOUSES **A lighthouse is located on a rock at a certain distance from a straight shore. The light revolves counterclockwise at a steady rate of one revolution per minute. As the beam revolves, it strikes a point on the shore that is 2000 feet from the lighthouse. Three seconds later, the light strikes a point 750 feet further down the shore. To the nearest foot, how far is the lighthouse from the shore?**

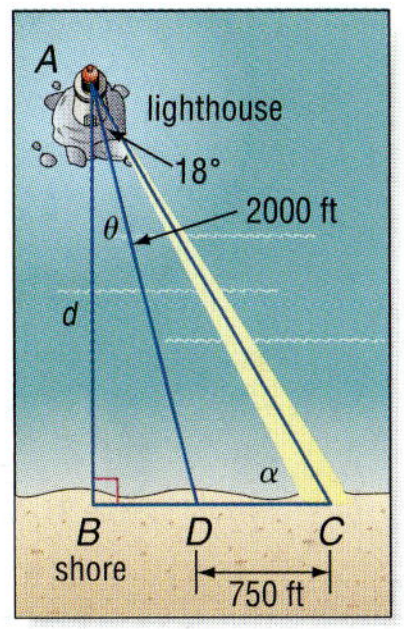

Because the lighthouse makes one revolution every 60 seconds, the angle through which the light revolves in 3 seconds is $\frac{3}{60}(360°)$ or 18°.

Use the Law of Sines to find the measure of angle α.

$\frac{\sin \alpha}{2000} = \frac{\sin 18°}{750}$ Law of Sines

$\sin \alpha = \frac{2000 \sin 18°}{750}$ Multiply each side by 2000.

$\sin \alpha \approx 0.8240$ Use a calculator.

$\alpha \approx 55°$ Use the $\sin^{-1}$ function.

Use this angle measure to find the measure of angle θ. Since $\triangle ABC$ is a right triangle, the measures of angle α and $\angle BAC$ are complementary.

$\alpha + m\angle BAC = 90°$ Angles α and $\angle BAC$ are complementary.

$55° + (\theta + 18°) \approx 90°$ $\alpha \approx 55°$ and $m\angle BAC = \theta + 18°$

$\theta + 73° \approx 90°$ Simplify.

$\theta \approx 17°$ Solve for θ.

To find the distance from the lighthouse to the shore, solve $\triangle ABD$ for d.

$\cos \theta = \frac{AB}{AD}$ Cosine ratio

$\cos 17° \approx \frac{d}{2000}$ $\theta = 17°$ and $AD = 2000$

$d \approx 2000 \cos 17°$ Solve for d.

$d \approx 1913$ Use a calculator.

The distance from the lighthouse to the shore, to the nearest foot, is 1913 feet. This answer is reasonable since 1913 is less than 2000.

More About. . .

Lighthouses

Standing 208 feet tall, the Cape Hatteras Lighthouse in North Carolina is the tallest lighthouse in the United States.

Source: www.oldcapehatteraslighthouse.com

TEACHING TIP

Point out to students that the Greek letter *alpha*, α, can be used to denote the measure of an angle.

Check for Understanding

Concept Check

1. **Determine** whether the following statement is *sometimes*, *always* or *never* true. Explain your reasoning.
   If given the measure of two sides of a triangle and the angle opposite one of them, you will be able to find a unique solution.

   1. Sometimes; only if when A is acute, $a = b \sin A$ or $a > b$ and if when A is obtuse, $a > b$.

2. **OPEN ENDED** Give an example of a triangle that has two solutions by listing measures for A, a, and b, where a and b are in centimeters. Then draw both cases using a ruler and protractor. **Sample answer: $A = 42°$, $a = 2.6$ cm, $b = 3.2$ cm; See margin for drawings.**

In-Class Example

Power Point®

6 **LIGHTHOUSES** Refer to Example 6 in the Student Edition. Suppose a different lighthouse has a beam that revolves at the same rate (one revolution per minute) but the beam strikes a point on the shore that is 1840 feet from the lighthouse. Two seconds later, the light strikes a point 500 feet farther down the shore. To the nearest foot, how far is the lighthouse from the shore? **1625 ft**

3 Practice/Apply

Study Notebook

Have students—

- add the definitions/examples of the vocabulary terms to their Vocabulary Builder worksheets for Chapter 13.
- include any other item(s) that they find helpful in mastering the skills in this lesson.

Answer

2.

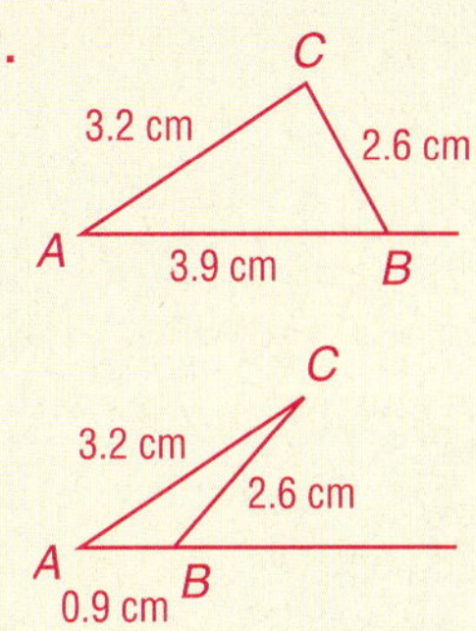

DAILY INTERVENTION **FIND THE ERROR** Have students sketch a triangle and label sides a, b, and angle A on the sketch. Make sure students label angle A opposite side a.

About the Exercises...

Organization by Objective

- **Law of Sines:** 14–27, 38–41
- **One, Two, or No Solutions:** 28–37

Odd/Even Assignments

Exercises 14–37 are structured so that students practice the same concepts whether they are assigned odd or even problems.

Assignment Guide

Basic: 15–37 odd, 42–58

Average: 15–37 odd, 42–58

Advanced: 14–40 even, 41–54 (optional: 55–58)

3. **FIND THE ERROR** Dulce and Gabe are finding the area of $\triangle ABC$ for $A = 64°$, $a = 15$ meters, and $b = 8$ meters using the sine function.

Dulce	Gabe
$\text{Area} = \frac{1}{2}(15)(8)\sin 64°$	$\text{Area} = \frac{1}{2}(15)(8)\sin 87.4°$
$\approx 53.9 \text{ m}^2$	$\approx 59.9 \text{ m}^2$

Who is correct? Explain your reasoning. **Gabe; see margin for explanation.**

Guided Practice

GUIDED PRACTICE KEY	
Exercises	**Examples**
4, 5	1
6, 7	2
8–12	3–5
13	6

Find the area of $\triangle ABC$ to the nearest tenth.

4. **57.5 in²**

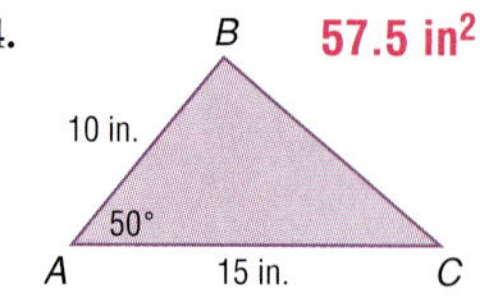

5. **6.4 cm²**

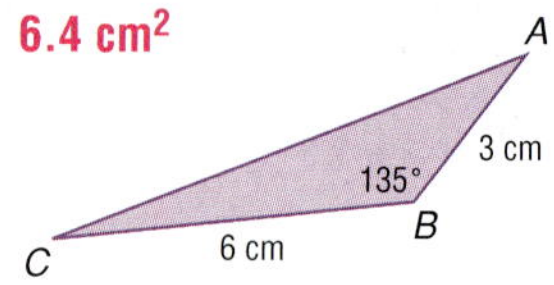

Solve each triangle. Round measures of sides to the nearest tenth and measures of angles to the nearest degree. **6–8. See margin.**

6.

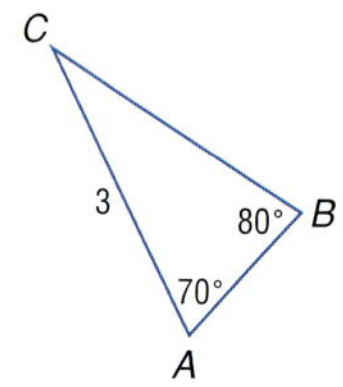

7.

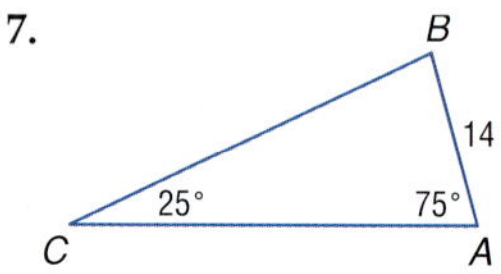

8.

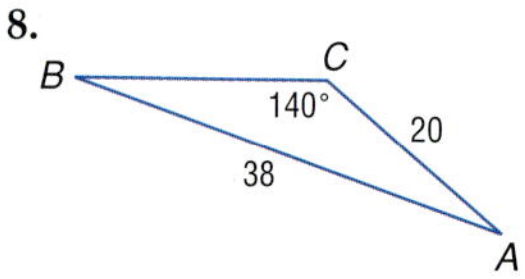

Determine whether each triangle has *no* solution, *one* solution, or *two* solutions. Then solve each triangle. Round measures of sides to the nearest tenth and measures of angles to the nearest degree.

9. $A = 123°, a = 12, b = 23$ **no**
10. $A = 30°, a = 3, b = 4$
11. $A = 55°, a = 10, b = 5$ **one; $B \approx 24°$, $C \approx 101°$, $c \approx 12.0$**
12. $A = 145°, a = 18, b = 10$ **one; $B \approx 19°$, $C \approx 16°$, $c \approx 8.9$**

10. two; $B \approx 42°$, $C \approx 108°$, $c \approx 5.7$; $B \approx 138°$, $C \approx 12°$, $c \approx 1.2$

Application

13. **WOODWORKING** Latisha is constructing a triangular brace from three beams of wood. She is to join the 6-meter beam to the 7-meter beam so the angle opposite the 7-meter beam measures 75°. To what length should Latisha cut the third beam in order to form the triangular brace? Round to the nearest tenth. **5.5 m**

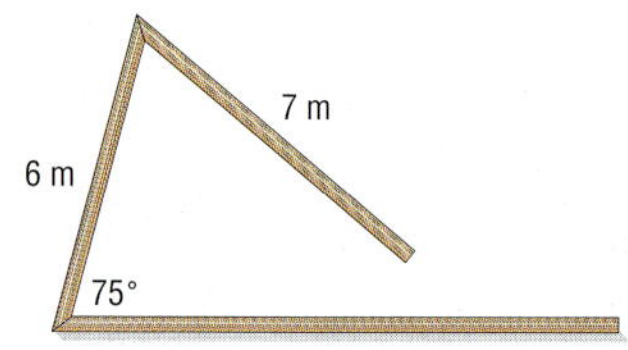

★ indicates increased difficulty

Practice and Apply

Homework Help	
For Exercises	**See Examples**
14–19	1
20–37	2–5
38–41	6

Extra Practice See page 858.

Find the area of $\triangle ABC$ to the nearest tenth.

14. **43.1 m²**

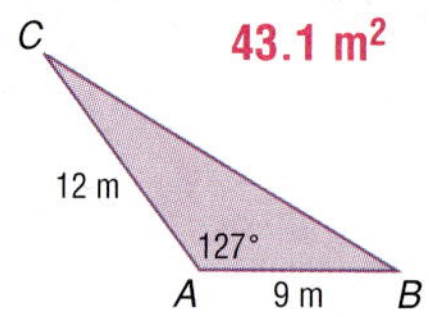

15. **19.5 yd²**

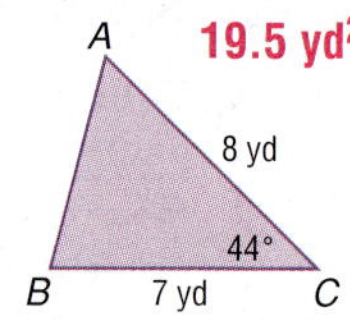

16. $B = 85°, c = 23 \text{ ft}, a = 50 \text{ ft}$ **572.8 ft²**
17. $A = 60°, b = 12 \text{ cm}, c = 12 \text{ cm}$ **62.4 cm²**
18. $C = 136°, a = 3 \text{ m}, b = 4 \text{ m}$ **4.2 m²**
19. $B = 32°, a = 11 \text{ mi}, c = 5 \text{ mi}$ **14.6 mi²**

Answers

3. Gabe; $\frac{\sin 64°}{15} = \frac{\sin B}{8}$

$\sin B = \frac{\sin 64°}{15}$

$B \approx 28.6°$

$m\angle C \approx 180° - (64° + 28.6°)$

$\approx 87.4°$

$\text{Area} = \frac{1}{2}ab \sin C$

$\approx \frac{1}{2}(15)(8) \sin 87.4°$

$\approx 59.9 \text{ m}^2$

6. $C = 30°$, $a \approx 2.9$, $c \approx 1.5$
7. $B = 80°$, $a \approx 32.0$, $b \approx 32.6$
8. $B \approx 20°$, $A \approx 20°$, $a \approx 20.2$
28. no
29. one; $B \approx 36°$, $C \approx 45°$, $c \approx 1.8$
30. two; $B \approx 72°$, $C \approx 75°$, $c \approx 3.5$; $B \approx 108°$, $C \approx 39°$, $c \approx 2.3$
31. no
32. one; $B = 90°$, $C = 60°$, $c \approx 24.2$
33. one; $B \approx 18°$, $C \approx 101°$, $c \approx 25.8$
34. two; $B \approx 56°$, $C \approx 72°$, $c \approx 229.3$; $B \approx 124°$, $C \approx 4°$, $c \approx 16.8$
35. two; $B \approx 85°$, $C \approx 15°$, $c \approx 2.4$; $B \approx 95°$, $C \approx 5°$, $c \approx 0.8$
36. one; $B \approx 23°$, $C \approx 129°$, $c \approx 14.1$
37. two; $B \approx 65°$, $C \approx 68°$, $c \approx 84.9$; $B \approx 115°$, $C \approx 18°$, $c \approx 28.3$

Solve each triangle. Round measures of sides to the nearest tenth and measures of angles to the nearest degree.

20. 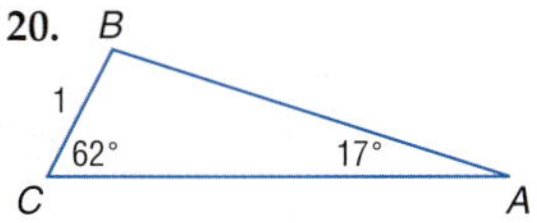

$B = 101°$, $c \approx 3.0$, $b \approx 3.4$

21. 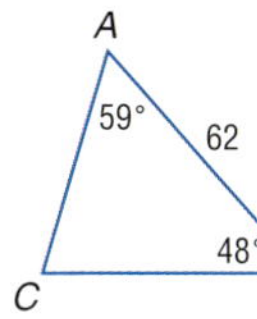

$C = 73°$, $a \approx 55.6$, $b \approx 48.2$

22. 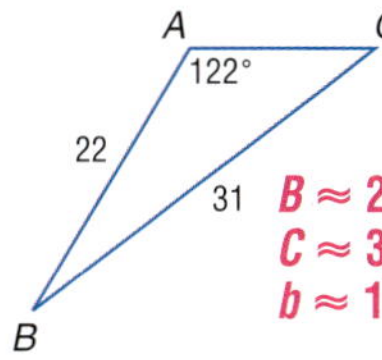

$B \approx 21°$, $C \approx 37°$, $b \approx 13.1$

23. 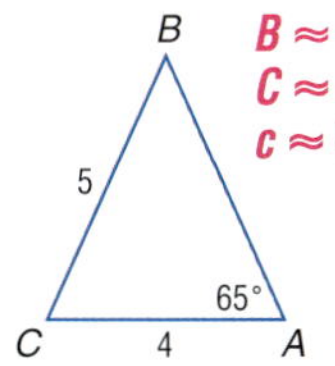

$B \approx 46°$, $C \approx 69°$, $c \approx 5.1$

24. 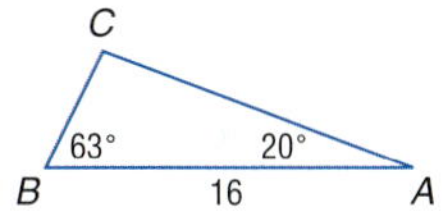

$C = 97°$, $a \approx 5.5$, $b \approx 14.4$

25. 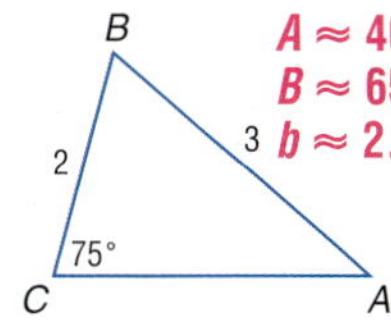

$A \approx 40°$, $B \approx 65°$, $b \approx 2.8$

26. $A = 50°, a = 2.5, c = 3$
$C \approx 67°$, $B \approx 63°$, $b \approx 2.9$

27. $B = 18°, C = 142°, b = 20$
$A = 20°$, $a \approx 22.1$, $c \approx 39.8$

Determine whether each triangle has *no* solution, *one* solution, or *two* solutions. Then solve each triangle. Round measures of sides to the nearest tenth and measures of angles to the nearest degree. **28–37. See margin.**

28. $A = 124°, a = 1, b = 2$

29. $A = 99°, a = 2.5, b = 1.5$

30. $A = 33°, a = 2, b = 3.5$

31. $A = 68°, a = 3, b = 5$

32. $A = 30°, a = 14, b = 28$

33. $A = 61°, a = 23, b = 8$

34. $A = 52°, a = 190, b = 200$

35. $A = 80°, a = 9, b = 9.1$

36. $A = 28°, a = 8.5, b = 7.2$

37. $A = 47°, a = 67, b = 83$

38. RADIO A radio station providing local tourist information has its transmitter on Beacon Road, 8 miles from where it intersects with the interstate highway. If the radio station has a range of 5 miles, between what two distances from the intersection can cars on the interstate tune in to hear this information? **4.6 and 8.5 mi**

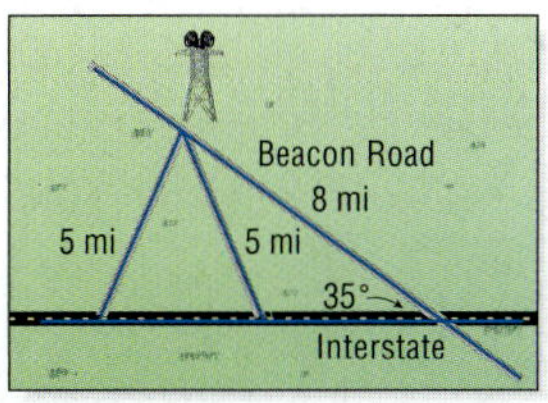

More About...

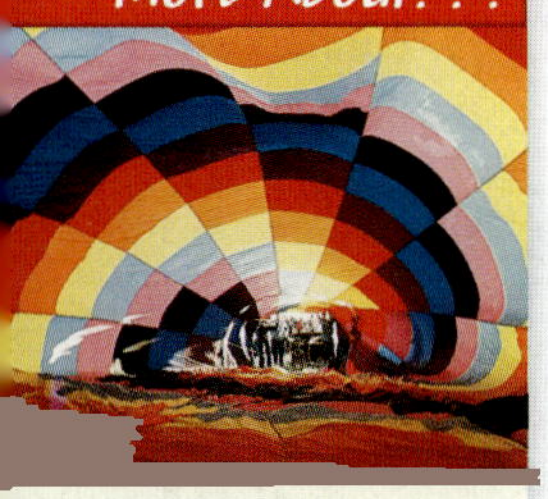

Ballooning

Hot-air balloons range in size from approximately 54,000 cubic feet to over 250,000 cubic feet.

Source: www.unicorn-ballon.com

39. FORESTRY Two forest rangers, 12 miles from each other on a straight service road, both sight an illegal bonfire away from the road. Using their radios to communicate with each other, they determine that the fire is between them. The first ranger's line of sight to the fire makes an angle of 38° with the road, and the second ranger's line of sight to the fire makes a 63° angle with the road. How far is the fire from each ranger? **7.5 mi from Ranger B, 10.9 mi from Ranger A**

40. BALLOONING As a hot-air balloon crosses over a straight portion of interstate highway, its pilot eyes two consecutive mileposts on the same side of the balloon. When viewing the mileposts the angles of depression are 64° and 7°. How high is the balloon to the nearest foot? **690 ft**

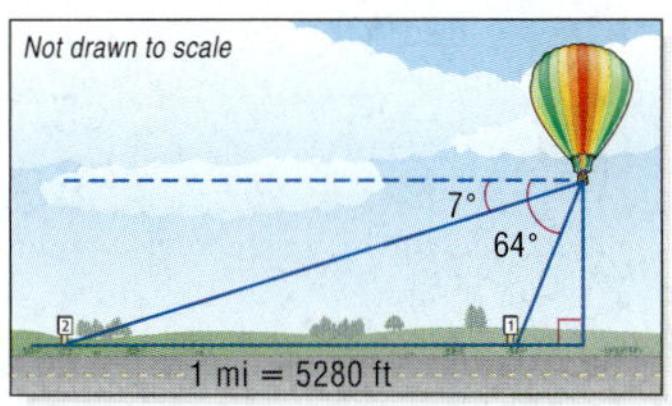

www.algebra2.com/self_check_quiz

Study Guide and Intervention, p. 793 (shown) and p. 794

Law of Sines The area of any triangle is one half the product of the lengths of two sides and the sine of the included angle.

Area of a Triangle	area $= \frac{1}{2}bc \sin A$ area $= \frac{1}{2}ac \sin B$ area $= \frac{1}{2}ab \sin C$

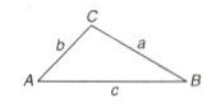

You can use the Law of Sines to solve any triangle if you know the measures of two angles and any side, or the measures of two sides and the angle opposite one of them.

Law of Sines	$\frac{\sin A}{a} = \frac{\sin B}{b} = \frac{\sin C}{c}$

Example 1 **Find the area of $\triangle ABC$ if $a = 10$, $b = 14$, and $C = 40°$.**

Area $= \frac{1}{2}ab \sin C$ — Area formula

$= \frac{1}{2}(10)(14)\sin 40°$ — Replace *a*, *b*, and *C*.

≈ 44.9951 — Use a calculator.

The area of the triangle is approximately 45 square units.

Example 2 **If $a = 12$, $b = 9$, and $A = 28°$, find B.**

$\frac{\sin A}{a} = \frac{\sin B}{b}$ — Law of Sines

$\frac{\sin 28°}{12} = \frac{\sin B}{9}$ — Replace *A*, *a*, and *b*.

$\sin B = \frac{9 \sin 28°}{12}$ — Solve for sin *B*.

$\sin B \approx 0.3521$ — Use a calculator.

$B \approx 20.62°$ — Use the $\sin^{-1}$ function.

Exercises

Find the area of $\triangle ABC$ to the nearest tenth.

1. 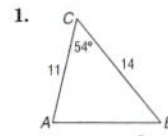

62.3 units²

2. 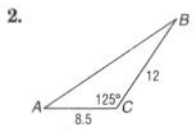

41.8 units²

3. 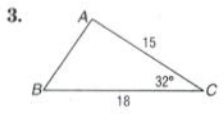

71.5 units²

Solve each triangle. Round measures of sides to the nearest tenth and measures of angles to the nearest degree.

4. $B = 42°, C = 68°, a = 10$ — $A = 70°$, $b \approx 7.1$, $c \approx 9.9$

5. $A = 40°, B = 14°, a = 52$ — $b \approx 19.6$, $c \approx 65.4$, $C = 126°$

6. $A = 15°, B = 50°, b = 36$ — $C = 115°$, $a \approx 12.2$, $c \approx 42.6$

Skills Practice, p. 795 and Practice, p. 796 (shown)

Find the area of $\triangle ABC$ to the nearest tenth.

1. 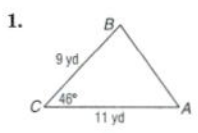

35.6 yd²

2. 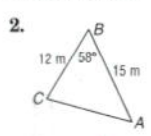

76.3 m²

3. 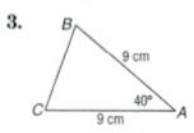

26.0 cm²

4. $C = 32°, a = 12.6$ m, $b = 8.9$ m — 29.7 m²

5. $B = 27°, a = 14.9$ cm, $c = 18.6$ cm — 62.9 cm²

6. $A = 17.4°, b = 12$ km, $c = 14$ km — 25.1 km²

7. $A = 34°, b = 19.4$ ft, $c = 8.6$ ft — 46.6 ft²

Solve each triangle. Round measures of sides to the nearest tenth and measures of angles to the nearest degree.

8. $A = 50°, B = 30°, c = 9$ — $C = 100°$, $a \approx 7.0$, $b \approx 4.6$

9. $A = 56°, B = 38°, a = 12$ — $C = 86°$, $b \approx 8.9$, $c \approx 14.4$

10. $A = 80°, C = 14°, a = 40$ — $B = 86°$, $b \approx 40.5$, $c \approx 9.8$

11. $B = 47°, C = 112°, b = 13$ — $A = 21°$, $a \approx 6.4$, $c \approx 16.5$

12. $A = 72°, a = 8, c = 6$ — $B \approx 62°$, $C \approx 46°$, $b \approx 7.5$

13. $A = 25°, C = 107°, b = 12$ — $B = 48°$, $a \approx 6.8$, $c \approx 15.4$

Determine whether each triangle has *no* solution, *one* solution, or *two* solutions. Then solve each triangle. Round measures of sides to the nearest tenth and measures of angles to the nearest degree.

14. $A = 29°, a = 6, b = 13$ no solution

15. $A = 70°, a = 25, b = 20$ one solution; $B \approx 49°$, $C \approx 61°$, $c \approx 23.3$

16. $A = 113°, a = 21, b = 25$ no solution

17. $A = 110°, a = 20, b = 8$ one solution; $B \approx 22°$, $C \approx 48°$, $c \approx 15.8$

18. $A = 66°, a = 12, b = 7$ one solution; $B \approx 32°$, $C \approx 82°$, $c \approx 13.0$

19. $A = 54°, a = 5, b = 8$ no solution

20. $A = 45°, a = 15, b = 18$ two solutions; $B \approx 58°$, $C \approx 77°$, $c \approx 20.7$; $B \approx 122°$, $C \approx 13°$, $c \approx 4.8$

21. $A = 60°, a = 4\sqrt{3}, b = 8$ one solution; $B = 90°$, $C = 30°$, $c \approx 23.3$

22. **WILDLIFE** Sarah Phillips, an officer for the Department of Fisheries and Wildlife, checks boaters on a lake to make sure they do not disturb two osprey nesting sites. She leaves a dock and heads due north in her boat to the first nesting site. From here, she turns 5° north of due west and travels an additional 2.14 miles to the second nesting site. She then travels 6.7 miles directly back to the dock. How far from the dock is the first osprey nesting site? Round to the nearest tenth. 6.2 mi

Reading to Learn Mathematics, p. 797 ELL

Pre-Activity **How can trigonometry be used to find the area of a triangle?**

Read the introduction to Lesson 13-4 at the top of page 725 in your textbook.

What happens when the formula Area $= \frac{1}{2}ab \sin C$ is applied to a right triangle in which C is the right angle? **Sample answer: The formula gives Area $= \frac{1}{2}ab \sin 90° = \frac{1}{2}ab \cdot 1 = \frac{1}{2}ab$, which is the same as the result from using the formula Area $= \frac{1}{2}$(base)(height).**

Reading the Lesson

1. In each case below, the measures of three parts of a triangle are given. For each case, write the formula you would use to find the area of the triangle. Show the formulas with specific values substituted, but do not actually calculate the area. If there is not enough information provided to find the area of the triangle by using the area formulas on page 725 in your textbook and without finding other parts of the triangle first, explain why.

   a. $A = 48°, b = 9, c = 5$ — $\frac{1}{2}(9)(5) \sin 48°$

   b. $a = 15, b = 15, C = 120°$ — $\frac{1}{2}(15)(15) \sin 120°$

   c. $b = 16, c = 10, B = 120°$ — Not enough information; B is not the included angle between the two given sides.

2. Tell whether the equation must be true based on the Law of Sines. Write *yes* or *no*.

   a. $\frac{\sin A}{b} = \frac{\sin B}{a}$ no

   b. $\frac{b}{\sin B} = \frac{c}{\sin C}$ yes

   c. $a \sin C = c \sin A$ yes

   d. $b = \frac{a \sin A}{\sin B}$ no

3. Determine whether $\triangle ABC$ has *no solution*, *one solution*, or *two solutions*. Do not try to solve the triangle.

   a. $a = 20, A = 30°, B = 70°$ — one solution

   b. $A = 55°, b = 5, a = 3$ ($b \sin A \approx 4.1$) — no solution

   c. $c = 12, A = 100°, a = 30$ — one solution

   d. $C = 27°, b = 23.5, c = 17.5$ ($b \sin C \approx 10.7$) — two solutions

Helping You Remember

4. Suppose that you are taking a quiz and cannot remember whether the formula for the area of a triangle is Area $= \frac{1}{2}ab \cos C$ or Area $= \frac{1}{2}ab \sin C$. How can you quickly remember which of these is correct? **Sample answer: The formula has to work when C is a right angle. The formula cannot contain $\cos C$ because $\cos 90° = 0$ and this would make the area of a right triangle be 0.**

Enrichment, p. 798

Navigation

The bearing of a boat is an angle showing the direction the boat is heading. Often, the angle is measured from north, but it can be measured from any of the four compass directions. At the right, the bearing of the boat is 155°. Or, it can be described as 25° east of south (S25°E).

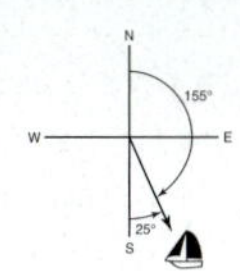

Example **A boat A sights the lighthouse B in the direction N65°E and the spire of a church C in the direction S75°E. According to the map, B is 7 miles from C in the direction N30°W. In order for A to avoid running aground, find the bearing it should keep to pass B at 4 miles distance.**

In $\triangle ABC$, $\angle\alpha = 180° - 65° - 75°$ or $40°$

$\angle C = 180° - 30° - (180° - 75°)$

$= 45°$

$a = 7$ miles

4 Assess

Open-Ended Assessment

Modeling Have students work in small groups to build models with coffee stirrer sticks (or similar objects) to illustrate and explain the various number of solutions that are possible for triangles.

Assessment Options

Quiz (Lessons 13-3 and 13-4) is available on p. 831 of the *Chapter 13 Resource Masters.*

Mid-Chapter Test (Lessons 13-1 through 13-4) is available on p. 833 of the *Chapter 13 Resource Masters.*

Getting Ready for Lesson 13-5

PREREQUISITE SKILL Lesson 13-5 presents the Law of Cosines. Students will use their familiarity with solving equations involving trigonometric functions as they apply the Law of Cosines. Exercises 55–58 should be used to determine your students' familiarity with solving equations with trigonometric functions.

★ **41. NAVIGATION** Two fishing boats, A and B, are anchored 4500 feet apart in open water. A plane flies at a constant speed in a straight path directly over the two boats, maintaining a constant altitude. At one point during the flight, the angle of depression to A is 85°, and the angle of depression to B is 25°. Ten seconds later the plane has passed over A and spots B at a 35° angle of depression. How fast is the plane flying? **107 mph**

42. CRITICAL THINKING Given $\triangle ABC$, if $a = 20$ and $B = 47°$, then determine all possible values of b so that the triangle has

a. two solutions. **$14.63 < b < 20$**

b. one solution. **$b = 14.63$ or $b \geq 20$**

c. no solutions. **$b < 14.63$**

43. WRITING IN MATH Answer the question that was posed at the beginning of the lesson. **See margin.**

How can trigonometry be used to find the area of a triangle?

Include the following in your answer:

- the conditions that would indicate that trigonometry is needed to find the area of a triangle,
- an example of a real-world situation in which you would need trigonometry to find the area of a triangle, and
- a derivation of one of the other two area formulas.

Standardized Test Practice

44. Which of the following is the perimeter of the triangle shown? **D**

Ⓐ 49.0 cm Ⓑ 66.0 cm

Ⓒ 91.4 cm Ⓓ 93.2 cm

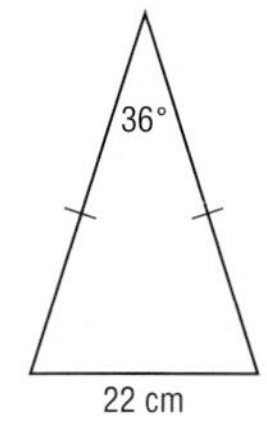

45. SHORT RESPONSE The longest side of a triangle is 67 inches. Two angles have measures of 47° and 55°. Solve the triangle. **$B = 78°$, $a \approx 50.1$, $c \approx 56.1$**

Maintain Your Skills

Mixed Review

Find the exact value of each trigonometric function. *(Lesson 13-3)*

46. $\cos 30°$ **$\frac{\sqrt{3}}{2}$**

47. $\cot\left(\frac{\pi}{3}\right)$ **$\frac{\sqrt{3}}{3}$**

48. $\csc\left(\frac{\pi}{4}\right)$ **$\sqrt{2}$**

Find one angle with positive measure and one angle with negative measure coterminal with each angle. *(Lesson 13-2)*

49. 300° **660°, −60°**

50. 47° **407°, −313°**

51. $\frac{5\pi}{6}$ **$\frac{17\pi}{6}$, $-\frac{7\pi}{6}$**

Two cards are drawn from a deck of cards. Find each probability. *(Lesson 12-5)*

52. P(both 5s or both spades) **$\frac{1}{17}$**

53. P(both 7s or both red) **$\frac{55}{221}$**

54. AERONAUTICS A rocket rises 20 feet in the first second, 60 feet in the second second, and 100 feet in the third second. If it continues at this rate, how many feet will it rise in the 20th second? *(Lesson 11-1)* **780 ft**

Getting Ready for the Next Lesson

PREREQUISITE SKILL Solve each equation. Round to the nearest tenth.

*(To review **solving equations with trigonometric functions**, see Lesson 13-1.)*

55. $a^2 = 3^2 + 5^2 - 2(3)(5) \cos 85°$ **5.6**

56. $c^2 = 12^2 + 10^2 - 2(12)(10) \cos 40°$ **7.8**

57. $7^2 = 11^2 + 9^2 - 2(11)(9) \cos B°$ **39.4°**

58. $13^2 = 8^2 + 6^2 - 2(8)(6) \cos A°$ **136.0°**

Answer

43. Answers should include the following.

- If the height of the triangle is not given, but the measure of two sides and their included angle are given, then the formula for the area of a triangle using the sine function should be used.
- You might use this formula to find the area of a triangular piece of land, since it might be easier to measure two sides and use surveying equipment to measure the included angle than to measure the perpendicular distance from one vertex to its opposite side.
- The area of $\triangle ABC$ is $\frac{1}{2}ah$.

  $\sin B = \frac{h}{c}$ or $c = c \sin B$

  Area $= \frac{1}{2}ah$ or Area $= \frac{1}{2}a(c \sin B)$

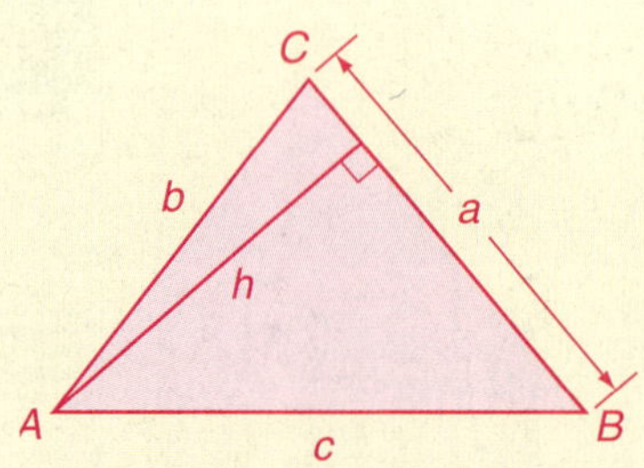

13-5 Law of Cosines

What You'll Learn

- Solve problems by using the Law of Cosines.
- Determine whether a triangle can be solved by first using the Law of Sines or the Law of Cosines.

Vocabulary

- Law of Cosines

How can you determine the angle at which to install a satellite dish?

The GE-3 satellite is in a *geosynchronous orbit* about Earth, meaning that it circles Earth once each day. As a result, the satellite appears to remain stationary over one point on the equator. A receiving dish for the satellite can be directed at one spot in the sky. The satellite orbits 35,786 kilometers above the equator at 87°W longitude. The city of Valparaiso, Indiana, is located at approximately 87°W longitude and 41.5°N latitude.

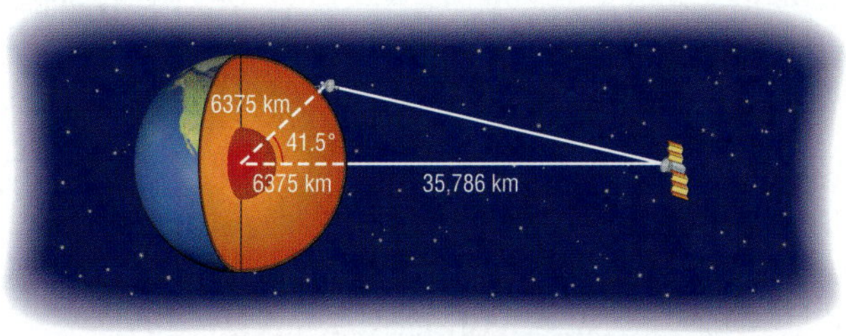

Knowing the radius of Earth to be about 6375 kilometers, a satellite dish installer can use trigonometry to determine the angle at which to direct the receiver.

LAW OF COSINES Problems such as this, in which you know the measures of two sides and the included angle of a triangle, cannot be solved using the Law of Sines. You can solve problems such as this by using the **Law of Cosines**.

To derive the Law of Cosines, consider $\triangle ABC$. What relationship exists between a, b, c, and A?

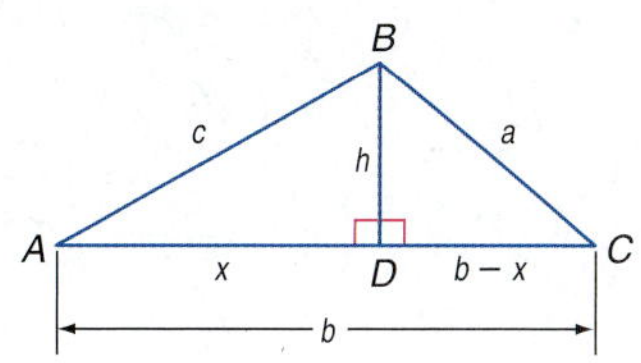

$a^2 = (b - x)^2 + h^2$	Use the Pythagorean Theorem for $\triangle DBC$.
$= b^2 - 2bx + x^2 + h^2$	Expand $(b - x)^2$.
$= b^2 - 2bx + c^2$	In $\triangle ADB$, $c^2 = x^2 + h^2$.
$= b^2 - 2b(c \cos A) + c^2$	$\cos A = \frac{x}{c}$, so $x = c \cos A$.
$= b^2 + c^2 - 2bc \cos A$	Commutative Property

Key Concept — Law of Cosines

Let $\triangle ABC$ be any triangle with a, b, and c representing the measures of sides, and opposite angles with measures A, B, and C, respectively. Then the following equations are true.

$$a^2 = b^2 + c^2 - 2bc \cos A$$
$$b^2 = a^2 + c^2 - 2ac \cos B$$
$$c^2 = a^2 + b^2 - 2ab \cos C$$

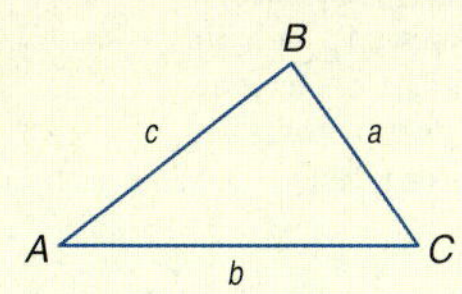

13-5 Lesson Notes

1 Focus

5-Minute Check Transparency 13-5 Use as a quiz or review of Lesson 13-4.

Mathematical Background notes are available for this lesson on p. 698D.

How can you determine the angle at which to install a satellite dish?

Ask students:

- Why does the satellite appear to remain stationary over one point on the equator? **because the satellite and Earth are turning at the same rate**
- What is the zero point for measuring longitude? **The zero meridian of longitude passes through Greenwich, England.**

Resource Manager

Workbook and Reproducible Masters

Chapter 13 Resource Masters

- Study Guide and Intervention, pp. 799–800
- Skills Practice, p. 801
- Practice, p. 802
- Reading to Learn Mathematics, p. 803
- Enrichment, p. 804

School-to-Career Masters, p. 26

Transparencies

5-Minute Check Transparency 13-5
Answer Key Transparencies

Technology

Alge2PASS: Tutorial Plus, Lesson 26
Interactive Chalkboard

2 Teach

LAW OF COSINES

In-Class Examples Power Point®

1 Solve $\triangle ABC$.

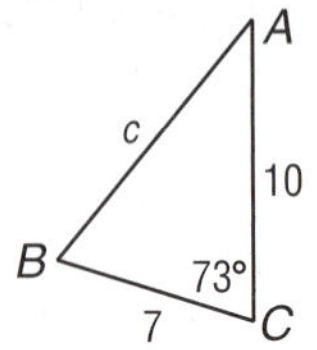

$A \approx 40°$; $B \approx 67°$; $c \approx 10.4$

2 Solve $\triangle ABC$.

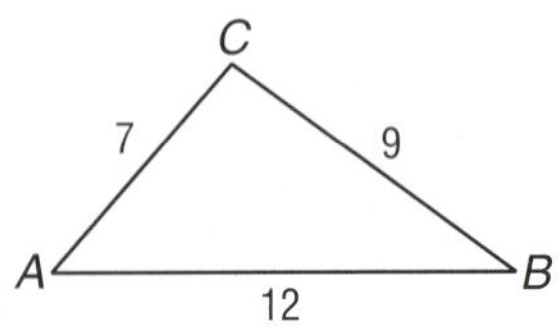

$A \approx 48°$, $B \approx 35°$, $C \approx 97°$

TEACHING TIP

You may wish to use the abbreviation SAS when referring to a triangle where the measures of the sides and the included angle are known and SSS when the measure of three sides are known.

Study Tip

Alternate Method

After finding the measure of c in Example 1, the Law of Cosines could be used again to find a second angle.

Study Tip

Sides and Angles

When solving triangles, remember that the angle with the greatest measure is always opposite the longest side. The angle with the least measure is always opposite the shortest side.

You can apply the Law of Cosines to a triangle if you know

- the measures of two sides and the included angle, or
- the measures of three sides.

Example 1 Solve a Triangle Given Two Sides and Included Angle

Solve $\triangle ABC$.

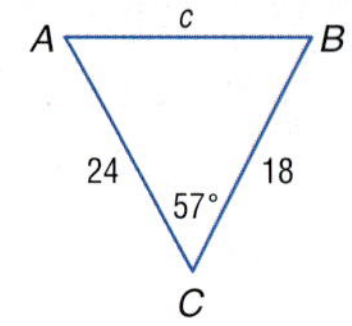

You are given the measures of two sides and the included angle. Begin by using the Law of Cosines to determine c.

$c^2 = a^2 + b^2 - 2ab \cos C$ — Law of Cosines

$c^2 = 18^2 + 24^2 - 2(18)(24) \cos 57°$ — $a = 18$, $b = 24$, and $C = 57°$

$c^2 \approx 429.4$ — Simplify using a calculator.

$c \approx 20.7$ — Take the square root of each side.

Next, you can use the Law of Sines to find the measure of angle A.

$\frac{\sin A}{a} = \frac{\sin C}{c}$ — Law of Sines

$\frac{\sin A}{18} \approx \frac{\sin 57°}{20.7}$ — $a = 18$, $C = 57°$, and $c \approx 20.7$

$\sin A \approx \frac{18 \sin 57°}{20.7}$ — Multiply each side by 18.

$\sin A \approx 0.7293$ — Use a calculator.

$A \approx 47°$ — Use the $\sin^{-1}$ function.

The measure of the angle B is approximately $180° - (57° + 47°)$ or $76°$. Therefore, $c \approx 20.7$, $A \approx 47°$, and $B \approx 76°$.

Example 2 Solve a Triangle Given Three Sides

Solve $\triangle ABC$.

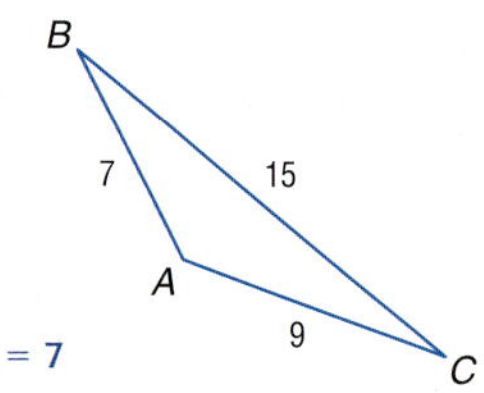

You are given the measures of three sides. Use the Law of Cosines to find the measure of the largest angle first, angle A.

$a^2 = b^2 + c^2 - 2bc \cos A$ — Law of Cosines

$15^2 = 9^2 + 7^2 - 2(9)(7) \cos A$ — $a = 15$, $b = 9$, and $c = 7$

$15^2 - 9^2 - 7^2 = -2(9)(7) \cos A$ — Subtract 9^2 and 7^2 from each side.

$\frac{15^2 - 9^2 - 7^2}{-2(9)(7)} = \cos A$ — Divide each side by $-2(9)(7)$.

$-0.7540 \approx \cos A$ — Use a calculator.

$139° \approx A$ — Use the $\cos^{-1}$ function.

You can use the Law of Sines to find the measure of angle B.

$\frac{\sin B}{b} = \frac{\sin A}{a}$ — Law of Sines

$\frac{\sin B}{9} \approx \frac{\sin 139°}{15}$ — $b = 9$, $A \approx 139°$, and $a = 15$

$\sin B \approx \frac{9 \sin 139°}{15}$ — Multiply each side by 9.

$\sin B \approx 0.3936$ — Use a calculator.

$B \approx 23°$ — Use the $\sin^{-1}$ function.

The measure of the angle C is approximately $180° - (139° + 23°)$ or $18°$. Therefore, $A \approx 139°$, $B \approx 23°$, and $C \approx 18°$.

734 Chapter 13 Trigonometric Functions

CHOOSE THE METHOD To solve a triangle that is *oblique*, or having no right angle, you need to know the measure of at least one side and any two other parts. If the triangle has a solution, then you must decide whether to begin solving by using the Law of Sines or by using the Law of Cosines. Use the chart below to help you choose.

TEACHING TIP
Remind students that when given two sides and an angle opposite one of them, they must first decide whether the triangle has one solution, two solutions, or no solution.

Concept Summary — Solving an Oblique Triangle

Given	Begin by Using
two angles and any side	Law of Sines
two sides and an angle opposite one of them	Law of Sines
two sides and their included angle	Law of Cosines
three sides	Law of Cosines

More About. . .

Emergency Medicine

Medical evacuation (Medevac) helicopters provide quick transportation from areas that are difficult to reach by any other means. These helicopters can cover long distances and are primary emergency vehicles in locations where there are few hospitals.

Source: The Helicopter Education Center

Example 3 Apply the Law of Cosines

EMERGENCY MEDICINE A medical rescue helicopter has flown from its home base at point C to pick up an accident victim at point A and then from there to the hospital at point B. The pilot needs to know how far he is now from his home base so he can decide whether to refuel before returning. How far is the hospital from the helicopter's base?

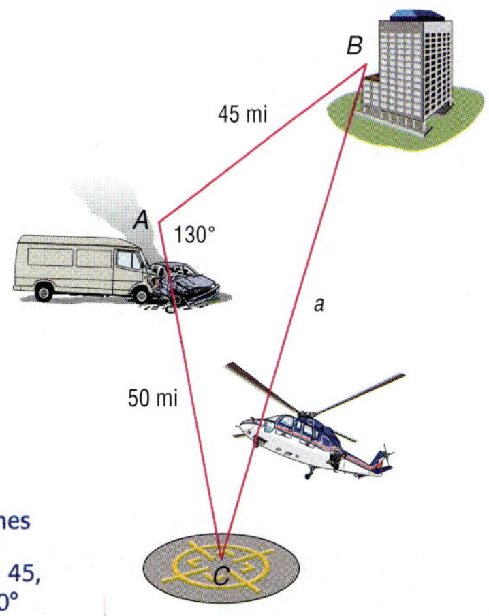

You are given the measures of two sides and their included angle, so use the Law of Cosines to find a.

$a^2 = b^2 + c^2 - 2bc \cos A$	Law of Cosines
$a^2 = 50^2 + 45^2 - 2(50)(45) \cos 130°$	$b = 50$, $c = 45$, and $A = 130°$
$a^2 \approx 7417.5$	Use a calculator to simplify.
$a \approx 86.1$	Take the square root of each side.

The distance between the hospital and the helicopter base is approximately 86.1 miles.

Check for Understanding

Concept Check

1. **FIND THE ERROR** Mateo and Amy are deciding which method, the Law of Sines or the Law of Cosines, should be used first to solve $\triangle ABC$.

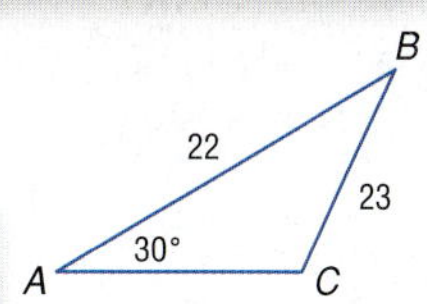

Mateo	Amy
Begin by using the Law of Sines, since you are given two sides and an angle opposite one of them.	Begin by using the Law of Cosines, since you are given two sides and their included angle.

Who is correct? Explain your reasoning.

1. Mateo; the angle given is not between the two sides; therefore the Law of Sines should be used.

www.algebra2.com/extra_examples

CHOOSE THE METHOD

In-Class Example Power Point®

3 EMERGENCY MEDICINE Refer to Example 3 in the Student Edition. A helicopter flies 55 miles from its base at point C to an accident at point B and then 35 miles to the hospital at point A. Angle B equals 42°. How far will the helicopter have to fly to return to its base from the hospital?

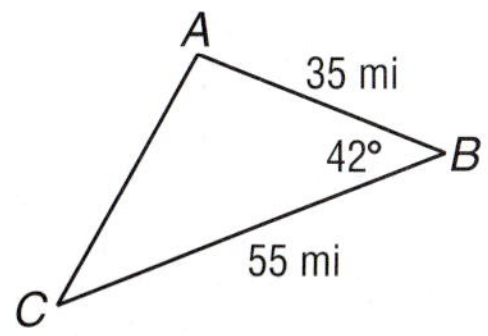

about 37.27 mi

3 Practice/Apply

Study Notebook

Have students—

- add the definitions/examples of the vocabulary terms to their Vocabulary Builder worksheets for Chapter 13.
- include any other item(s) that they find helpful in mastering the skills in this lesson.

DAILY INTERVENTION **FIND THE ERROR** Watch for students who think the angle is included. Review the definition of included angles with them.

Verbal/Linguistic Have students

DAILY INTERVENTION **Differentiated Instruction** ELL

discuss in small groups how to choose which method to use when solving a triangle. Have them compare their approaches and develop a brief explanation to help others decide. Then have each group share their conclusions with the class.

About the Exercises...

Organization by Objective

- **Law of Cosines:** 10–33
- **Choose the Method:** 10–27

Odd/Even Assignments

Exercises 10–27 are structured so that students practice the same concepts whether they are assigned odd or even problems.

Assignment Guide

Basic: 11–25 odd, 31, 34–37, 41–54

Average: 11–27 odd, 31, 33–37, 41–54 (optional: 38–40)

Advanced: 10–26 even, 28–30, 32–48 (optional: 49–54)

All: Practice Quiz 2 (1–5)

Answers

2a. Use the Law of Cosines to find the measure of one angle. Then use the Law of Sines or the Law of Cosines to find the measure of a second angle. Finally, subtract the sum of these two angles from 180° to find the measure of the third angle.

2b. Use the Law of Cosines to find the measure of the third side. Then use the Law of Sines or the Law of Cosines to find the measure of a second angle. Finally, subtract the sum of these two angles from 180° to find the measure of the third angle.

3. Sample answer:

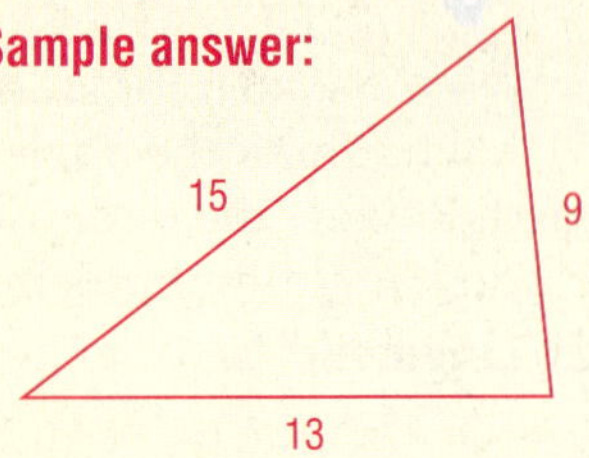

2. **Explain** how to solve a triangle by using the Law of Cosines if the lengths of
   a. three sides are known. **See margin.**
   b. two sides and the measure of the angle between them are known.
3. **OPEN ENDED** Give an example of a triangle that can be solved by first using the Law of Cosines. **See margin.**

Guided Practice

Determine whether each triangle should be solved by beginning with the Law of Sines or Law of Cosines. Then solve each triangle. Round measures of sides to the nearest tenth and measures of angles to the nearest degree.

GUIDED PRACTICE KEY	
Exercises	Examples
4–7	1, 2
8, 9	3

4. 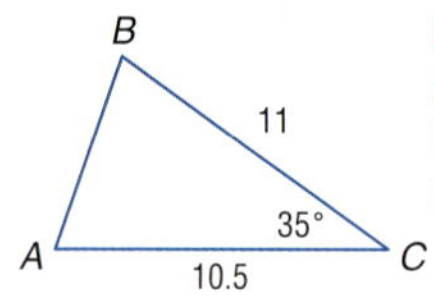

cosines; $A \approx 77°$, $B \approx 68°$, $c \approx 6.5$

5. 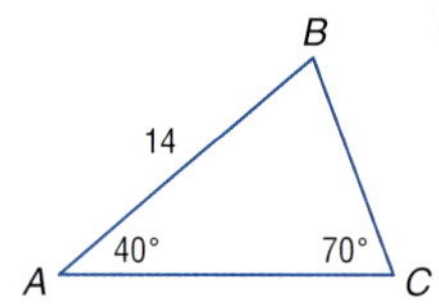

sines; $B = 70°$, $a \approx 9.6$, $b = 14$

6. $A = 42°, b = 57, a = 63$ **sines; $C \approx 101°$, $B \approx 37°$, $c \approx 92.5$**
7. $a = 5, b = 12, c = 13$ **cosines; $A \approx 23°$, $B \approx 67°$, $C = 90°$**

Application

BASEBALL **For Exercises 8 and 9, use the following information.**

In Australian baseball, the bases lie at the vertices of a square 27.5 meters on a side and the pitcher's mound is 18 meters from home plate.

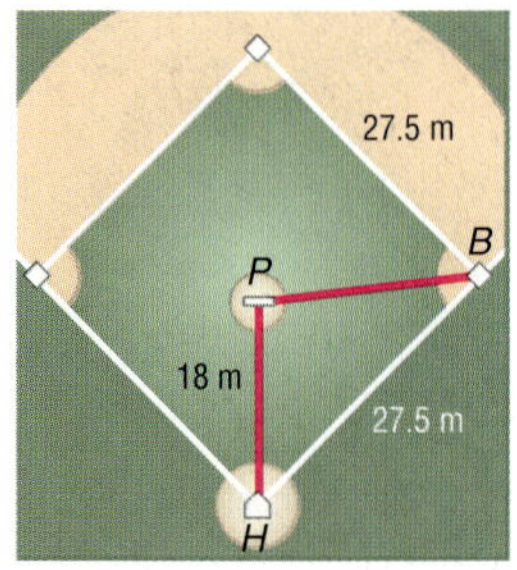

8. Find the distance from the pitcher's mound to first base. **19.5 m**
9. Find the angle between home plate, the pitcher's mound, and first base. **94.3°**

★ indicates increased difficulty

Practice and Apply

Homework Help	
For Exercises	See Examples
10–27	1, 2
28–33	3

Extra Practice
See page 858.

Determine whether each triangle should be solved by beginning with the Law of Sines or Law of Cosines. Then solve each triangle. Round measures of sides to the nearest tenth and measures of angles to the nearest degree. **10–27. See margin.**

10. 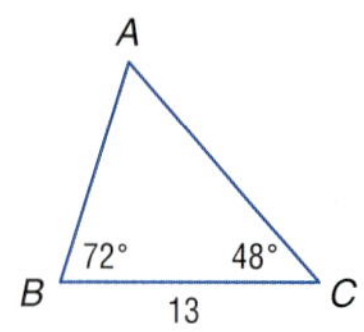

11. 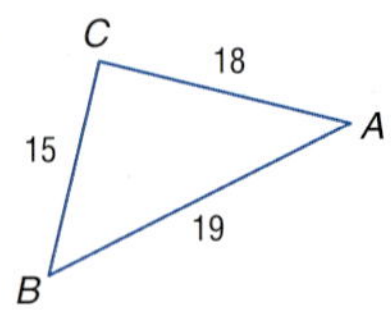

12. 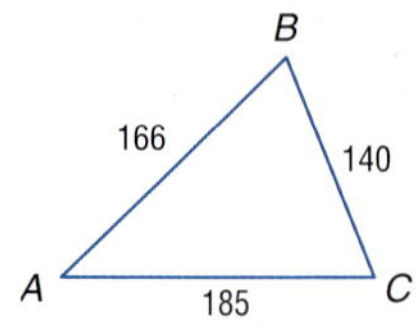

13. 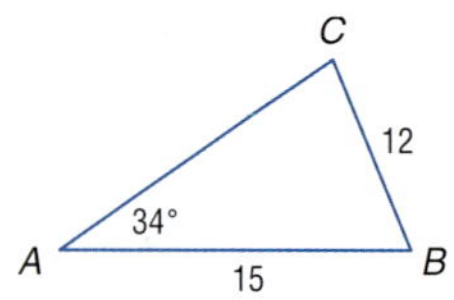

14. 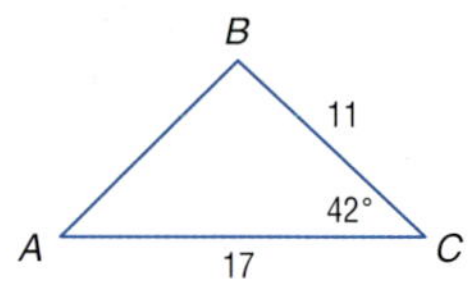

15. 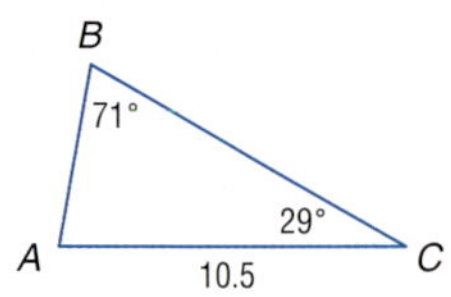

16. $a = 20, c = 24, B = 47°$
17. $a = 345, b = 648, c = 442$
18. $A = 36°, a = 10, b = 19$
19. $A = 25°, B = 78°, a = 13.7$
20. $a = 21.5, b = 16.7, c = 10.3$
21. $a = 16, b = 24, c = 41$
22. $a = 8, b = 24, c = 18$
23. $B = 19°, a = 51, c = 61$
24. $A = 56°, B = 22°, a = 12.2$
25. $a = 4, b = 8, c = 5$
26. $a = 21.5, b = 13, C = 38°$

★27. $A = 40°, b = 7, a = 6$

10. sines; $A = 60°$, $b \approx 14.3$, $c \approx 11.2$
11. cosines; $A \approx 48°$, $B \approx 63°$, $C \approx 70°$
12. cosines; $A \approx 47°$, $B \approx 74°$, $C \approx 60°$
13. sines; $B \approx 102°$, $C \approx 44°$, $b \approx 21.0$
14. cosines; $A \approx 57°$, $B \approx 82°$, $c \approx 11.5$
15. sines; $A = 80°$, $a \approx 10.9$, $c \approx 5.4$
16. cosines; $A \approx 55°$, $C \approx 78°$, $b \approx 17.9$

see answers 11, 12, 14, 22, 27

17. cosines; $A \approx 30°$, $B \approx 110°$, $C \approx 40°$
18. no
19. sines; $C \approx 77°$, $b \approx 31.7$, $c \approx 31.6$
20. cosines; $A \approx 103°$, $B \approx 49°$, $C \approx 28°$
21. no
22. cosines; $A \approx 15°$, $B \approx 130°$, $C \approx 35°$
23. cosines; $A \approx 52°$, $C \approx 109°$, $b \approx 21.0$
24. sines; $C = 102°$, $b \approx 5.5$, $c \approx 14.4$
25. cosines; $A \approx 24°$, $B \approx 125°$, $C \approx 31°$
26. cosines; $A \approx 107°$, $B \approx 35°$, $c \approx 13.8$
27. sines; $B \approx 1°$, $C \approx 139°$, $c \approx 6.3$

736 Chapter 13 Trigonometric Functions

More About. . .

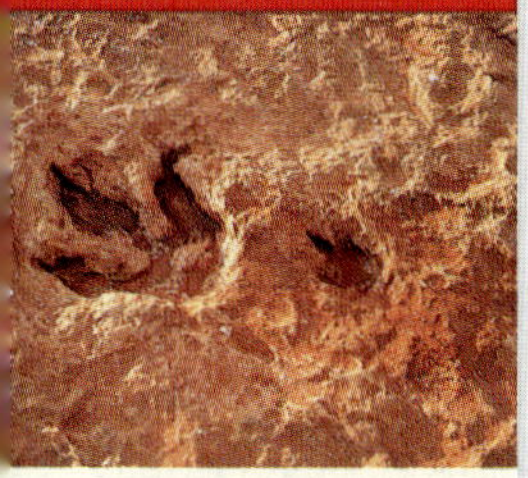

Dinosaurs

At digs such as the one at the Glen Rose formation in Texas, anthropologists study the footprints made by dinosaurs millions of years ago. *Locomoter* parameters, such as pace and stride, taken from these prints can be used to describe how a dinosaur once moved.

Source: Mid-America Paleontology Society

DINOSAURS For Exercises 28–30, use the diagram at the right.

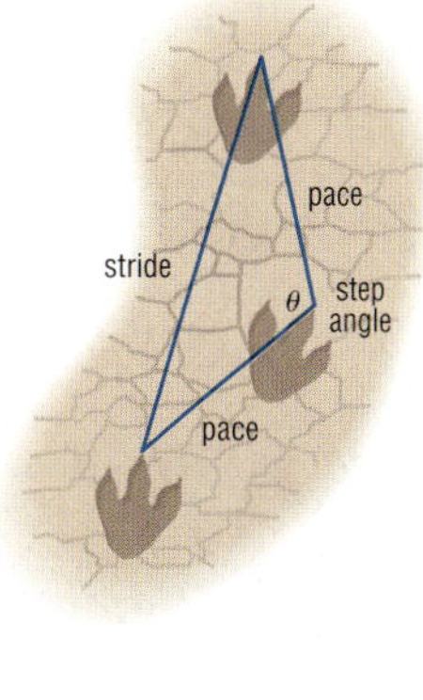

28. An anthropologist examining the footprints made by a bipedal (two-footed) dinosaur finds that the dinosaur's average pace was about 1.60 meters and average stride was about 3.15 meters. Find the step angle θ for this dinosaur. **about 159.7°**

29. Find the step angle θ made by the hindfeet of a herbivorous dinosaur whose pace averages 1.78 meters and stride averages 2.73 meters. **about 100.1°**

30. An efficient walker has a step angle that approaches 180°, meaning that the animal minimizes "zig-zag" motion while maximizing forward motion. What can you tell about the motion of each dinosaur from its step angle?

30. Since the step angle for the carnivore is closer to 180°, it appears as though the carnivore made more forward progress with each step than the herbivore did.

31. **GEOMETRY** In rhombus $ABCD$, the measure of $\angle ADC$ is 52°. Find the measures of diagonals $\overline{AC}$ and $\overline{DB}$ to the nearest tenth. **4.4 cm, 9.0 cm**

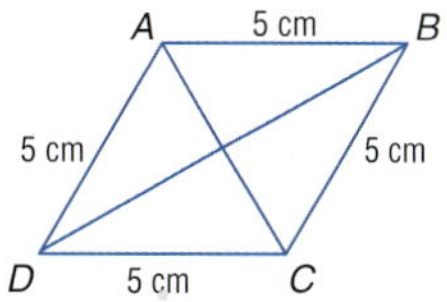

32. **SURVEYING** Two sides of a triangular plot of land have lengths of 425 feet and 550 feet. The measure of the angle between those sides is 44.5°. Find the perimeter and area of the plot. **about 1362 ft; about 81,919 ft²**

★ 33. **AVIATION** A pilot typically flies a route from Bloomington to Rockford, covering a distance of 117 miles. In order to avoid a storm, the pilot first flies from Bloomington to Peoria, a distance of 42 miles, then turns the plane and flies 108 miles on to Rockford. Through what angle did the pilot turn the plane over Peoria? **91.6°**

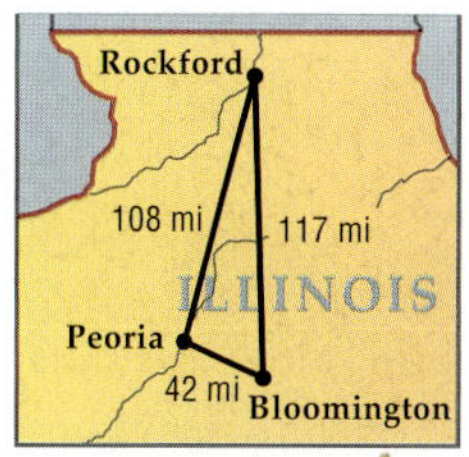

34. **CRITICAL THINKING** Explain how the Pythagorean Theorem is a special case of the Law of Cosines.

34. Since cos 90° = 0, $a^2 = b^2 + c^2 - 2bc \cos A$ becomes $a^2 = b^2 + c^2$.

35. **WRITING IN MATH** Answer the question that was posed at the beginning of the lesson. **See pp. 759A–759D.**

How can you determine the angle at which to install a satellite dish?

Include the following in your answer:

- a description of the conditions under which you can use the Law of Cosines to solve a triangle, and
- given the latitude of a point on Earth's surface, an explanation of how you can determine the angle at which to install a satellite dish at the same longitude.

Standardized Test Practice

36. In $\triangle DEF$, what is the value of θ to the nearest degree? **B**

(A) 26° (B) 74°
(C) 80° (D) 141°

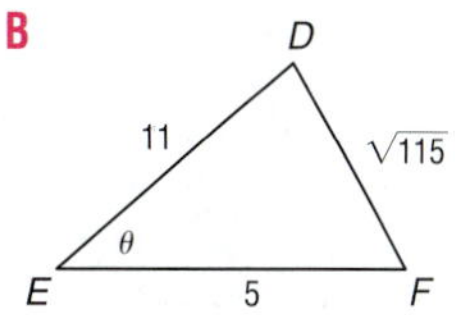

www.algebra2.com/self_check_quiz

Enrichment, p. 804

The Law of Cosines and the Pythagorean Theorem

The law of cosines bears strong similarities to the Pythagorean theorem. According to the law of cosines, if two sides of a triangle have lengths a and b and if the angle between them has a measure of $x°$, then the length, y, of the third side of the triangle can be found by using the equation

$y^2 = a^2 + b^2 - 2ab \cos x°$.

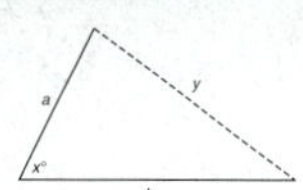

Answer the following questions to clarify the relationship between the law of cosines and the Pythagorean theorem.

1. If the value of $x°$ becomes less and less, what number is $\cos x°$ close to? **1**

2. If the value of $x°$ is very close to zero but then increases, what happens to $\cos x°$ as $x°$ approaches 90°? **decreases, approaches 0**

Study Guide and Intervention, p. 799 (shown) and p. 800

Law of Cosines

Law of Cosines	Let $\triangle ABC$ be any triangle with a, b, and c representing the measures of the sides, and opposite angles with measures A, B, and C, respectively. Then the following equations are true. $a^2 = b^2 + c^2 - 2bc \cos A$; $b^2 = a^2 + c^2 - 2ac \cos B$; $c^2 = a^2 + b^2 - 2ab \cos C$

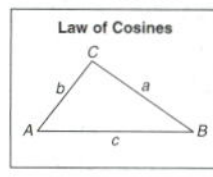

You can use the Law of Cosines to solve any triangle if you know the measures of two sides and the included angle, or the measures of three sides.

Example Solve $\triangle ABC$.

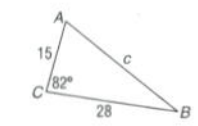

You are given the measures of two sides and the included angle. Begin by using the Law of Cosines to determine c.

$c^2 = a^2 + b^2 - 2ab \cos C$
$c^2 = 28^2 + 15^2 - 2(28)(15)\cos 82°$
$c^2 \approx 892.09$
$c \approx 29.9$

Next you can use the Law of Sines to find the measure of angle A.

$\frac{\sin A}{a} = \frac{\sin C}{c}$
$\frac{\sin A}{28} \approx \frac{\sin 82°}{29.9}$
$\sin A \approx 0.9273$
$A \approx 68°$

The measure of B is about $180° - (82° + 68°)$ or about 30°.

Exercises

Solve each triangle described below. Round measures of sides to the nearest tenth and angles to the nearest degree.

1. $a = 14, c = 20, B = 38°$ **$b \approx 12.4, A \approx 44°, C \approx 98°$**
2. $A = 60°, c = 17, b = 12$ **$a \approx 15.1, B \approx 43°, C \approx 77°$**
3. $a = 4, b = 6, c = 3$ **$A \approx 36°, B \approx 118°, C \approx 26°$**
4. $A = 103°, b = 31, c = 52$ **$a \approx 66, B \approx 27°, C \approx 50°$**
5. $a = 15, b = 26, C = 132°$ **$c \approx 38, A \approx 17°, B \approx 31°$**
6. $a = 31, b = 52, c = 43$ **$A \approx 36°, B \approx 88°, C \approx 56°$**

Skills Practice, p. 801 and Practice, p. 802 (shown)

Determine whether each triangle should be solved by beginning with the Law of Sines or Law of Cosines. Then solve each triangle. Round measures of sides to the nearest tenth and measures of angles to the nearest degree.

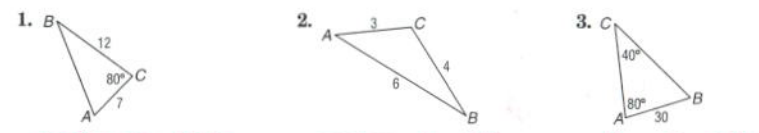

1. **cosines; $c \approx 12.8$, $A \approx 67°, B \approx 33°$**
2. **cosines; $A \approx 36°$, $B \approx 26°, C \approx 117°$**
3. **sines; $B = 60°$, $a \approx 46.0, b \approx 40.4$**
4. $a = 16, b = 20, C = 54°$ **cosines; $A \approx 51°, B \approx 75°, c \approx 16.7$**
5. $B = 71°, c = 6, a = 11$ **cosines; $A \approx 77°, C \approx 32°, b \approx 10.7$**
6. $A = 37°, a = 20, b = 18$ **sines; $B \approx 33°, C \approx 110°, c \approx 31.2$**
7. $C = 35°, a = 18, b = 24$ **cosines; $A \approx 48°, B \approx 97°, c \approx 13.9$**
8. $a = 8, b = 6, c = 9$ **cosines; $A \approx 61°, B \approx 41°, C \approx 79°$**
9. $A = 23°, b = 10, c = 12$ **cosines; $B \approx 54°, C \approx 103°, a \approx 4.8$**
10. $a = 4, b = 5, c = 8$ **cosines; $A \approx 24°, B \approx 31°, C \approx 125°$**
11. $B = 46.6°, C = 112°, b = 13$ **sines; $A = 21°, a \approx 6.5, c \approx 16.6$**
12. $A = 46.3°, a = 35, b = 30$ **sines; $B \approx 38°, C \approx 95°, c \approx 48.2$**
13. $a = 16.4, b = 21.1, c = 18.5$ **cosines; $A \approx 48°, B \approx 74°, C \approx 57°$**
14. $C = 43.5°, b = 8, c = 6$ **sines; $A \approx 70°, B \approx 67°, a \approx 8.2$**
15. $A = 78.3°, b = 7, c = 11$ **cosines; $B \approx 36°, C \approx 66°, a \approx 11.8$**

16. **SATELLITES** Two radar stations 2.4 miles apart are tracking an airplane. The straight-line distance between Station A and the plane is 7.4 miles. The straight-line distance between Station B and the plane is 6.9 miles. What is the angle of elevation from Station A to the plane? Round to the nearest degree. **69°**

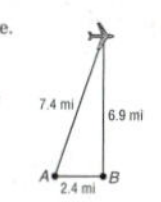

17. **DRAFTING** Marion is using a computer-aided drafting program to produce a drawing for a client. She begins a triangle by drawing a segment 4.2 inches long from point A to point B. From B, she moves 42° degrees counterclockwise from the segment connecting A and B and draws a second segment that is 6.4 inches long, ending at point C. To the nearest tenth, how long is the segment from C to A? **9.9 in.**

Reading to Learn Mathematics, p. 803 ELL

Pre-Activity **How can you determine the angle at which to install a satellite dish?**

Read the introduction to Lesson 13-5 at the top of page 733 in your textbook.

One side of the triangle in the figure is not labeled with a length. What does the length of this side represent? Is this length greater than or less than the distance from the satellite to the equator?

the distance from the satellite to Valparaiso; greater than

Reading the Lesson

1. Each of the following equations can be changed into a correct statement of the Law of Cosines by making one change. In each case, indicate what change should be made to make the statement correct.

   a. $b^2 = a^2 + c^2 + 2ac \cos B$ **Change the second + to −.**
   b. $a^2 = b^2 + c^2 - 2bc \sin A$ **Change sin A to cos A.**
   c. $c = a^2 + b^2 - 2ab \cos C$ **Change c to c^2.**
   d. $a^2 = b^2 - c^2 - 2bc \cos A$ **Change the first − to +.**

2. Suppose that you are asked to solve $\triangle ABC$ given the following information about the sides and angles of the triangle. In each case, indicate whether you would begin by using the *Law of Sines* or the *Law of Cosines*.

   a. $a = 8, b = 7, c = 6$ **Law of Cosines**
   b. $b = 9.5, A = 72°, B = 39°$ **Law of Sines**
   c. $C = 123°, b = 22.95, a = 34.35$ **Law of Cosines**

Helping You Remember

3. It is often easier to remember a complicated procedure if you can break it down into small steps. Describe in your own words how to use the Law of Cosines to find the length of one side of a triangle if you know the lengths of the other two sides and the measure of the included angle. Use numbered steps. (You may use mathematical terms, but do not use any mathematical symbols.)

   Sample answer: 1. Square each of the lengths of the two known sides. 2. Add these squares. 3. Find the cosine of the included angle. 4. Multiply this cosine by two times the product of the lengths of the two known sides. 5. Subtract the product from the sum. 6. Take the positive square root of the result.

4 Assess

Open-Ended Assessment

Writing Have students sketch and label some of the parts of three triangles—one that they would choose to solve using the Law of Cosines, one that they would choose to solve using the Law of Sines, and one that they would use both laws to solve.

Assessment Options

Practice Quiz 2 The quiz provides students with a brief review of the concepts and skills in Lessons 13-3 through 13-5. Lesson numbers are given to the right of exercises or instruction lines so students can review concepts not yet mastered.

Getting Ready for Lesson 13-6

PREREQUISITE SKILL Lesson 13-6 presents the unit circle approach to trigonometric functions. Students will use their familiarity with coterminal angles as they work with the unit circle. Exercises 49–54 should be used to determine your students' familiarity with finding coterminal angles.

Answers

40. By finding the measure of angle C in one step using the Law of Cosines, only the given information was used. By finding this angle measure using the Law of Cosines and then the Law of Sines, a calculated value that was not exact was introduced.

43. $\sin\theta = \frac{12}{13}$, $\cos\theta = \frac{5}{13}$, $\tan\theta = \frac{12}{5}$, $\csc\theta = \frac{13}{12}$, $\sec\theta = \frac{13}{5}$, $\cot\theta = \frac{5}{12}$

44. $\sin\theta = \frac{7\sqrt{65}}{65}$, $\cos\theta = \frac{4\sqrt{65}}{65}$, $\tan\theta = \frac{7}{4}$, $\csc\theta = \frac{\sqrt{65}}{7}$, $\sec\theta = \frac{\sqrt{65}}{4}$, $\cot\theta = \frac{4}{7}$

37. Two trucks, A and B, start from the intersection C of two straight roads at the same time. Truck A is traveling twice as fast as truck B and after 4 hours, the two trucks are 350 miles apart. Find the approximate speed of truck B in miles per hour. **A**

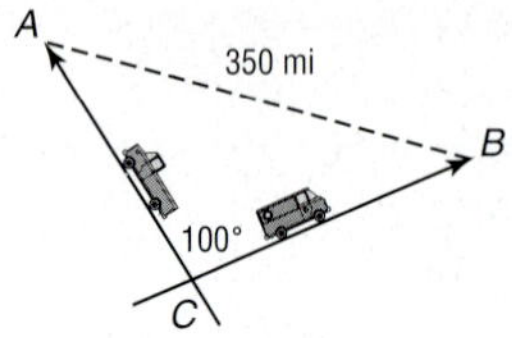

(A) 35 (B) 37 (C) 57 (D) 73

Extending the Lesson

ERROR IN MEASUREMENT For Exercises 38–40, use the following information. Consider $\triangle ABC$, in which $a = 17$, $b = 8$, and $c = 20$.

38. Find the measure of angle C in one step using the Law of Cosines. Round to the nearest tenth. **100.0°**

39. Find the measure of angle C in two steps using the Law of Cosines and then the Law of Sines. Round to the nearest tenth. **Sample answer: 100.2°**

40. Explain why your answers for Exercises 38 and 39 are different. Which answer gives you the better approximation for the measure of angle C? **See margin for explanation; 100.0°.**

Maintain Your Skills

Mixed Review

Determine whether each triangle has *no* solution, *one* solution, or *two* solutions. Then solve each triangle. Round measures of sides to the nearest tenth and measures of angles to the nearest degree. *(Lesson 13-4)*

41. $A = 55°, a = 8, b = 7$ **one; $B \approx 46°$, $C \approx 79°$, $c \approx 9.6$**

42. $A = 70°, a = 7, b = 10$ **no solution**

Find the exact values of the six trigonometric functions of θ if the terminal side of θ in standard position contains the given point. *(Lesson 13-3)* **43–45. See margin.**

43. $(5, 12)$ **44.** $(4, 7)$ **45.** $(\sqrt{10}, \sqrt{6})$

Solve each equation or inequality. *(Lesson 10-5)*

46. $e^x + 5 = 9$ **1.3863**

47. $4e^x - 3 > -1$ **$\{x \mid x > -0.6931\}$**

48. $\ln(x + 3) = 2$ **4.3891**

Getting Ready for the Next Lesson

PREREQUISITE SKILL **Find one angle with positive measure and one angle with negative measure coterminal with each angle.**
*(To review **coterminal angles**, see Lesson 13-2.)*

49. 45° **405°, −315°** **50.** 30° **390°, −330°** **51.** 180° **540°, −180°**

52. $\frac{\pi}{2}$ **$\frac{5\pi}{2}, -\frac{3\pi}{2}$** **53.** $\frac{7\pi}{6}$ **$\frac{19\pi}{6}, -\frac{5\pi}{6}$** **54.** $\frac{4\pi}{3}$ **$\frac{10\pi}{3}, -\frac{2\pi}{3}$**

Practice Quiz 2 — Lessons 13-3 through 13-5

1. Find the exact value of the six trigonometric functions of θ if the terminal side of θ in standard position contains the point $(-2, 3)$. *(Lesson 13-3)* **See margin.**

2. Find the exact value of $\csc \frac{5\pi}{3}$. *(Lesson 13-3)* **$-\frac{2\sqrt{3}}{3}$**

3. Find the area of $\triangle DEF$ to the nearest tenth. *(Lesson 13-4)* **27.7 m²**

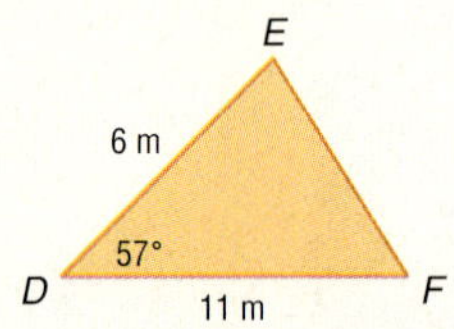

4. Determine whether $\triangle ABC$, with $A = 22°$, $a = 15$, and $b = 18$, has *no* solution, *one* solution, or *two* solutions. Then solve the triangle, if possible. Round measures of sides to the nearest tenth and measures of angles to the nearest degree. *(Lesson 13-4)* **See margin.**

5. Determine whether $\triangle ABC$, with $b = 11$, $c = 14$, and $A = 78°$, should be solved by beginning with the Law of Sines or Law of Cosines. Then solve the triangle. Round measures of sides to the nearest tenth and measures of angles to the nearest degree. *(Lesson 13-5)* **cosines; $c \approx 15.9$, $C \approx 59°$, $B \approx 43°$**

738 Chapter 13 Trigonometric Functions

45. $\sin\theta = \frac{\sqrt{6}}{4}$, $\cos\theta = \frac{\sqrt{10}}{4}$, $\tan\theta = \frac{\sqrt{15}}{5}$, $\csc\theta = \frac{2\sqrt{6}}{3}$, $\sec\theta = \frac{2\sqrt{10}}{5}$, $\cot\theta = \frac{\sqrt{15}}{3}$

Answer (Practice Quiz 2)

1. $\sin\theta = \frac{3\sqrt{13}}{13}$; $\cos\theta = -\frac{2\sqrt{13}}{13}$; $\tan\theta = -\frac{3}{2}$; $\csc\theta = \frac{\sqrt{13}}{3}$; $\sec\theta = -\frac{\sqrt{13}}{2}$; $\cot\theta = -\frac{2}{3}$

4. two; $B \approx 27°$; $C \approx 131°$; $c \approx 30.1$; $B \approx 153°$; $C \approx 5°$; $c \approx 3.3$

13-6 Circular Functions

What You'll Learn

- Define and use the trigonometric functions based on the unit circle.
- Find the exact values of trigonometric functions of angles.

Vocabulary

- circular function
- periodic
- period

How can you model annual temperature fluctuations?

The average high temperatures, in degrees Fahrenheit, for Barrow, Alaska, are given in the table at the right. With January assigned a value of 1, February a value of 2, March a value of 3, and so on, these data can be graphed as shown below. This pattern of temperature fluctuations repeats after a period of 12 months.

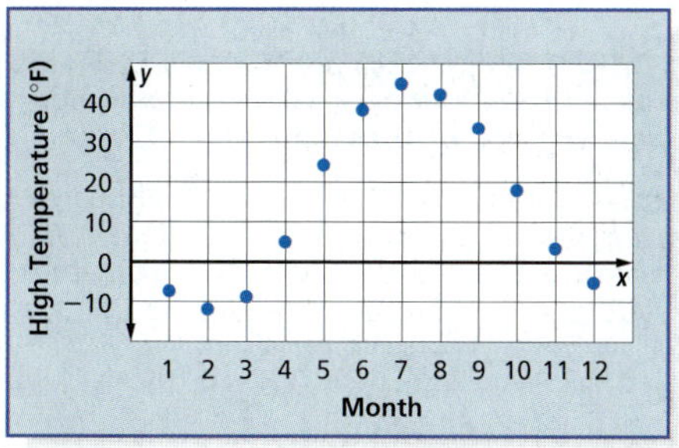

BARROW, ALASKA

MONTH	HIGH TEMP. (°F)
Jan	-7.4
Feb	-11.8
March	-9.0
April	4.7
May	24.2
June	38.3
July	45.0
Aug	42.3
Sept	33.8
Oct	18.1
Nov	3.5
Dec	-5.2

Source: www.met.utah.edu

UNIT CIRCLE DEFINITIONS From your work with reference angles, you know that the values of trigonometric functions also repeat. For example, sin 30° and sin 150° have the same value, $\frac{1}{2}$. In this lesson, we will further generalize the trigonometric functions by defining them in terms of the unit circle.

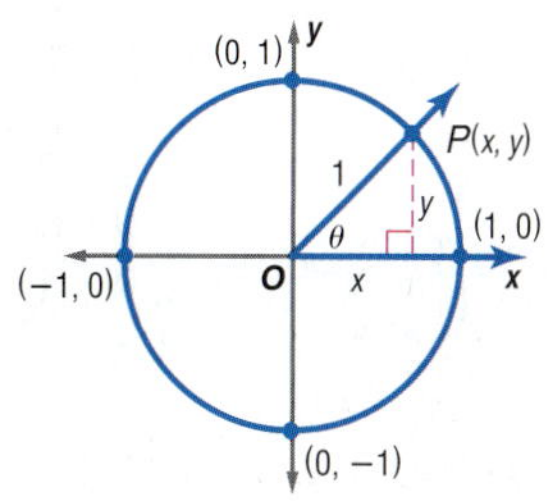

Consider an angle θ in standard position. The terminal side of the angle intersects the unit circle at a unique point, $P(x, y)$. Recall that $\sin \theta = \frac{y}{r}$ and $\cos \theta = \frac{x}{r}$. Since $P(x, y)$ is on the unit circle, $r = 1$. Therefore, $\sin \theta = y$ and $\cos \theta = x$.

Key Concept — *Definition of Sine and Cosine*

- **Words** If the terminal side of an angle θ in standard position intersects the unit circle at $P(x, y)$, then $\cos \theta = x$ and $\sin \theta = y$. Therefore, the coordinates of P can be written as $P(\cos \theta, \sin \theta)$.
- **Model**

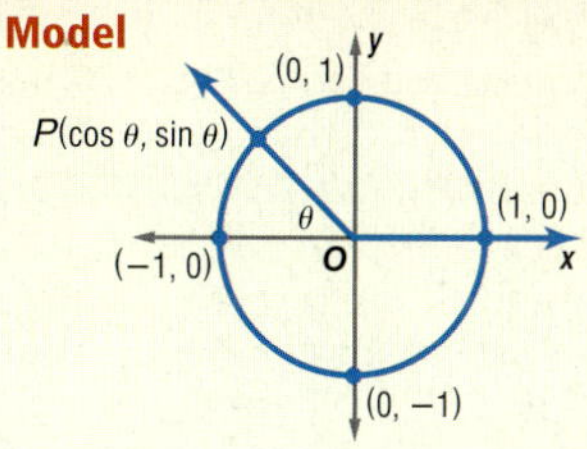

13-6 Lesson Notes

1 Focus

5-Minute Check Transparency 13-6 Use as a quiz or review of Lesson 13-5.

Mathematical Background notes are available for this lesson on p. 698D.

How can you model annual temperature fluctuations?

Ask students:

- Describe the change in the average high temperature from February to July. **It is increasing.**
- Describe the change in the average high temperature from August to December. **It is decreasing.**
- Does the pattern of change in temperature seem to fall along a straight line? **no**
- What curve does the pattern of change in temperature seem to suggest? **a sine or cosine curve**

Resource Manager

Workbook and Reproducible Masters

Chapter 13 Resource Masters

- Study Guide and Intervention, pp. 805–806
- Skills Practice, p. 807
- Practice, p. 808
- Reading to Learn Mathematics, p. 809
- Enrichment, p. 810
- Assessment, p. 832

Teaching Algebra With Manipulatives Masters, p. 302

Transparencies

5-Minute Check Transparency 13-6
Answer Key Transparencies

Technology

Interactive Chalkboard

2 Teach

UNIT CIRCLE DEFINITIONS

In-Class Example

Power Point®

1 Given an angle θ in standard position, if $P\left(\frac{\sqrt{7}}{4}, \frac{3}{4}\right)$ lies on the terminal side of θ and on the unit circle, find $\sin \theta$ and $\cos \theta$.

$\sin \theta = \frac{3}{4}$, $\cos \theta = \frac{\sqrt{7}}{4}$

Study Tip

Reading Math

To help you remember that $x = \cos \theta$ and $y = \sin \theta$, notice that alphabetically x comes before y and cosine comes before sine.

Since there is exactly one point $P(x, y)$ for any angle θ, the relations $\cos \theta = x$ and $\sin \theta = y$ are functions of θ. Because they are both defined using a unit circle, they are often called **circular functions**.

Example 1 Find Sine and Cosine Given Point on Unit Circle

Given an angle θ in standard position, if $P\left(\frac{2\sqrt{2}}{3}, -\frac{1}{3}\right)$ lies on the terminal side and on the unit circle, find $\sin \theta$ and $\cos \theta$.

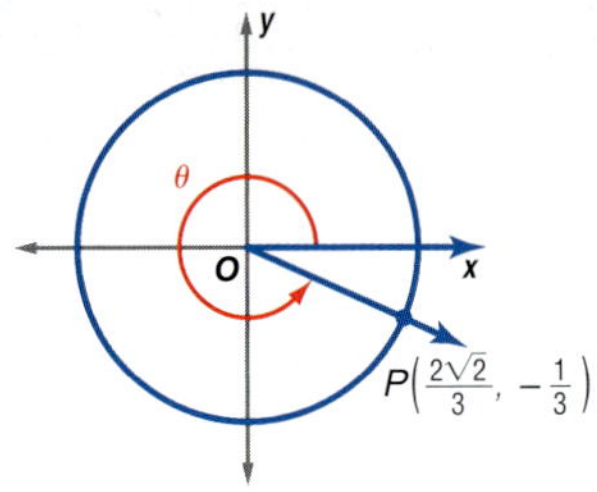

$P\left(\frac{2\sqrt{2}}{3}, -\frac{1}{3}\right) = P(\cos \theta, \sin \theta)$,

so $\sin \theta = -\frac{1}{3}$ and $\cos \theta = \frac{2\sqrt{2}}{3}$.

In the Investigation below, you will explore the behavior of the sine and cosine functions on the unit circle.

Graphing Calculator Investigation

Sine and Cosine on the Unit Circle

Press [MODE] on a TI-83 Plus and highlight Degree and Par. Then use the following range values to set up a viewing window: TMIN = 0, TMAX = 360, TSTEP = 15, XMIN = −2.4, XMAX = 2.35, XSCL = 0.5, YMIN = −1.5, YMAX = 1.55, YSCL = 0.5. Press [Y=] to define the unit circle with $X_{1T} = \cos T$ and $Y_{1T} = \sin T$. Press [GRAPH]. Use the [TRACE] function to move around the circle.

Think and Discuss

1. What does T represent? What does the x value represent? What does the y value represent? **the angle θ; $\cos \theta$; $\sin \theta$**
2. Determine the sine and cosine of the angles whose terminal sides lie at 0°, 90°, 180°, and 270°.
3. How do the values of sine change as you move around the unit circle? How do the values of cosine change?

2. $\sin 0° = 0$, $\cos 0° = 1$, $\sin 90° = 1$, $\cos 90° = 0$, $\sin 180° = 0$, $\cos 180° = -1$, $\sin 270° = -1$, $\cos 270° = 0$

3. For $0° \le \theta \le 90°$, $\sin \theta$ increases from 0 to 1. For $90° \le \theta \le 270°$, $\sin \theta$ decreases from 1 to −1. For $270° \le \theta \le 360°$, $\sin \theta$ increases from −1 to 0. For $0° \le \theta \le 180°$, $\cos \theta$ decreases from 1 to −1. For $180° \le \theta \le 270°$, $\cos \theta$ increases from −1 to 1.

The exact values of the sine and cosine functions for specific angles are summarized using the definition of sine and cosine on the unit circle below.

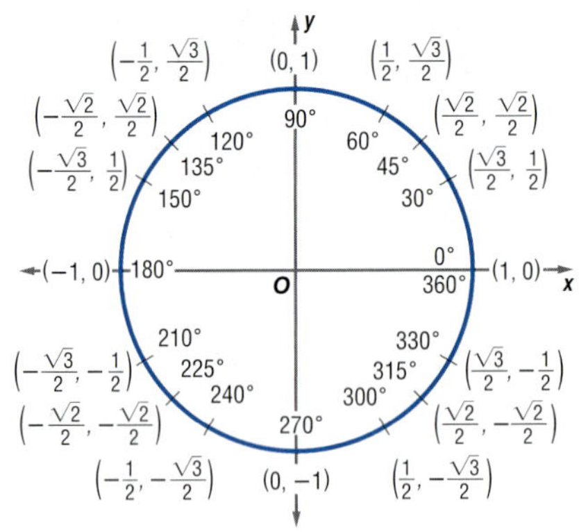

Graphing Calculator Investigation

Sine and Cosine on the Unit Circle Follow the given steps for the TI-83 calculator also. Point out that, after the mode and window have been defined, from the **Y=** screen, pressing the [X,T,θ,*n*] key will automatically enter T. After pressing [TRACE], use [▶] to move around the circle counterclockwise. Call attention to the information that appears on the screen as the cursor moves around the circle.

This same information is presented on the graphs of the sine and cosine functions below, where the horizontal axis shows the values of θ and the vertical axis shows the values of $\sin \theta$ or $\cos \theta$.

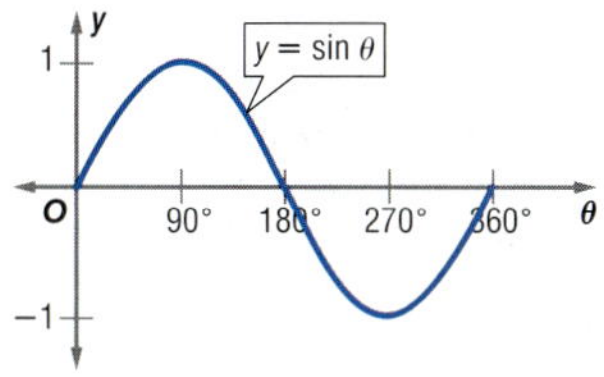

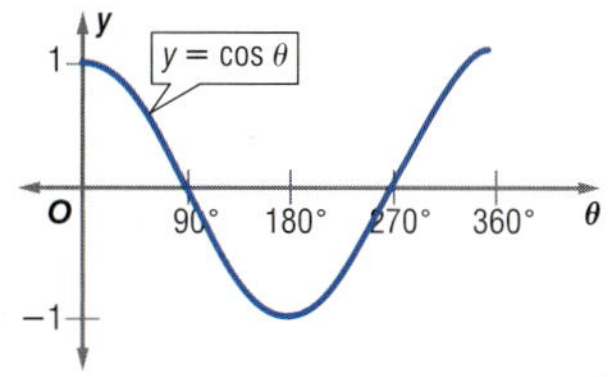

PERIODIC FUNCTIONS Notice in the graph above that the values of sine for the coterminal angles 0° and 360° are both 0. The values of cosine for these angles are both 1. Every 360° or 2π radians, the sine and cosine functions repeat their values. So, we can say that the sine and cosine functions are **periodic**, each having a **period** of 360° or 2π radians.

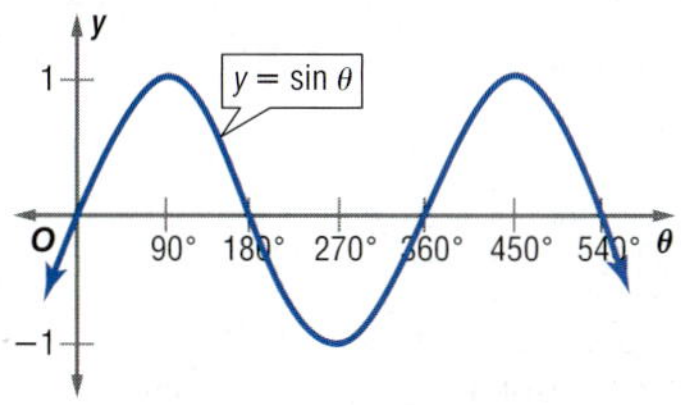

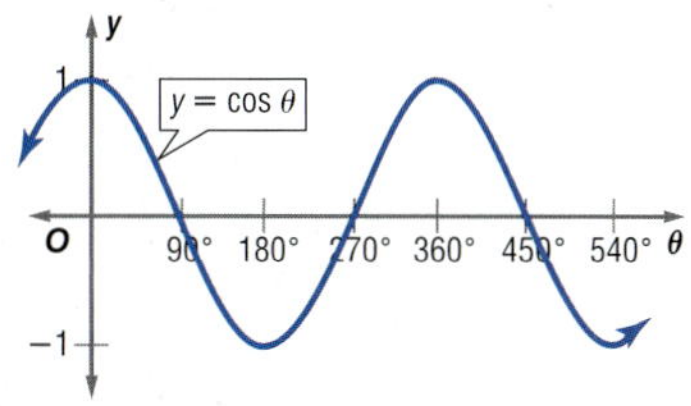

Key Concept — *Periodic Function*

A function is called periodic if there is a number a such that $f(x) = f(x + a)$ for all x in the domain of the function. The least positive value of a for which $f(x) = f(x + a)$ is called the period of the function.

For the sine and cosine functions, $\cos (x + 360°) = \cos x$, and $\sin (x + 360°) = \sin x$. In radian measure, $\cos (x + 2\pi) = \cos x$, and $\sin (x + 2\pi) = \sin x$. Therefore, the period of the sine and cosine functions is 360° or 2π.

Example 2 Find the Value of a Trigonometric Function

Find the exact value of each function.

a. $\cos 675°$

$$\cos 675° = \cos (315° + 360°)$$
$$= \cos 315°$$
$$= \frac{\sqrt{2}}{2}$$

b. $\sin \left(-\frac{5\pi}{6}\right)$

$$\sin \left(-\frac{5\pi}{6}\right) = \sin \left(-\frac{5\pi}{6} + 2\pi\right)$$
$$= \sin \frac{7\pi}{6}$$
$$= -\frac{1}{2}$$

When you look at the graph of a periodic function, you will see a repeating pattern: a shape that repeats over and over as you move to the right on the x-axis. The period is the distance along the x-axis from the beginning of the pattern to the point at which it begins again.

www.algebra2.com/extra_examples

PERIODIC FUNCTIONS

In-Class Example

Power Point®

2 Find the exact value of each function.

a. $\cos 690°$ $\frac{\sqrt{3}}{2}$

b. $\sin \left(-\frac{3\pi}{4}\right)$ $-\frac{\sqrt{2}}{2}$

Lesson 13-6 Circular Functions 741

In-Class Example Power Point®

3 **FERRIS WHEEL** On another Ferris wheel, the diameter is 42 feet, and it travels at a rate of 3 revolutions per minute.

a. Identify the period of this function. **20 seconds**

b. Make a graph in which the horizontal axis represents the time t in seconds and the vertical axis represents the height h in feet in relation to the starting point.

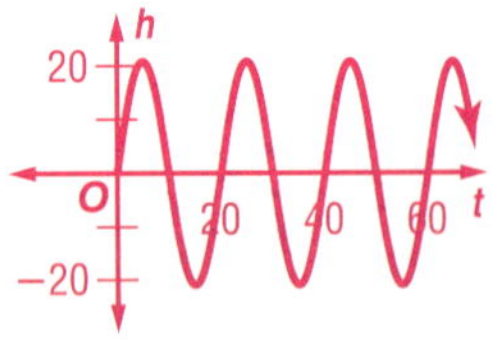

Answer

1. The terminal side of the angle θ in standard position must intersect the unit circle at $P(x, y)$.

Many real-world situations have characteristics that can be described with periodic functions.

Example 3 Find the Value of a Trigonometric Function

FERRIS WHEEL **As you ride a Ferris wheel, the height that you are above the ground varies periodically as a function of time. Consider the height of the center of the wheel to be the starting point. A particular wheel has a diameter of 38 feet and travels at a rate of 4 revolutions per minute.**

a. Identify the period of this function.

Since the wheel makes 4 complete counterclockwise rotations every minute, the period is the time it takes to complete one rotation, which is $\frac{1}{4}$ of a minute or 15 seconds.

b. Make a graph in which the horizontal axis represents the time t in seconds and the vertical axis represents the height h in feet in relation to the starting point.

Since the diameter of the wheel is 38 feet, the wheel reaches a maximum height of $\frac{38}{2}$ or 19 feet above the starting point and a minimum of 19 feet below the starting point.

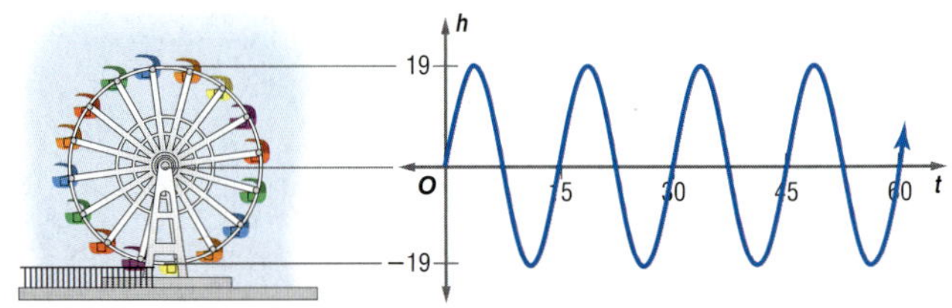

Check for Understanding

Concept Check

1. State the conditions under which $\cos \theta = x$ and $\sin \theta = y$. **See margin.**
2. **OPEN ENDED** Give an example of a situation that could be described by a periodic function. Then state the period of the function.
3. **Compare and contrast** the graphs of the sine and cosine functions on page 741. **Sample answer: The graphs have the same shape, but cross the x-axis at different points.**

2. Sample answer: the motion of the minute hand on a clock; 60 s

Guided Practice

GUIDED PRACTICE KEY	
Exercises	Examples
4, 5	1
6, 7	2
8–10	3

If the given point P is located on the unit circle, find $\sin \theta$ and $\cos \theta$.

4. $P\left(\frac{5}{13}, -\frac{12}{13}\right)$ **$\sin \theta = -\frac{12}{13}$; $\cos \theta = \frac{5}{13}$**
5. $P\left(\frac{\sqrt{2}}{2}, \frac{\sqrt{2}}{2}\right)$ **$\sin \theta = \frac{\sqrt{2}}{2}$; $\cos \theta = \frac{\sqrt{2}}{2}$**

Find the exact value of each function.

6. $\sin -240°$ **$\frac{\sqrt{3}}{2}$**
7. $\cos \frac{10\pi}{3}$ **$-\frac{1}{2}$**

8. Determine the period of the function that is graphed below. **720°**

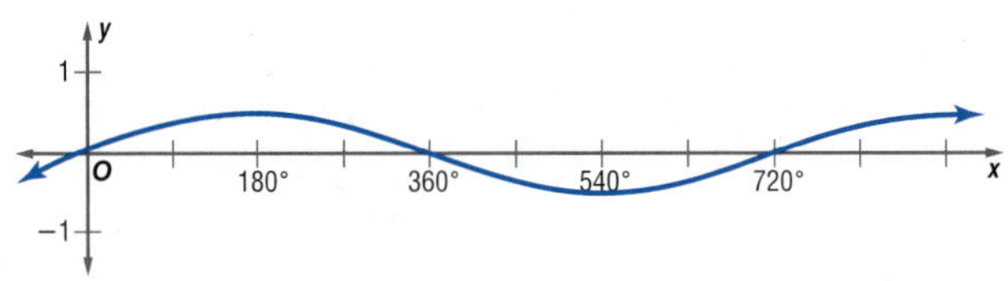

DAILY INTERVENTION

Differentiated Instruction

Naturalist Have students research various kinds of circular calendars, such as those used by the Mayans, to predict the weather and determine the best time for planting crops.

742 Chapter 13 Trigonometric Functions

Application **PHYSICS** **For Exercises 9 and 10, use the following information.**
The motion of a weight on a spring varies periodically as a function of time. Suppose you pull the weight down 3 inches from its equilibrium point and then release it. It bounces above the equilibrium point and then returns below the equilibrium point in 2 seconds.

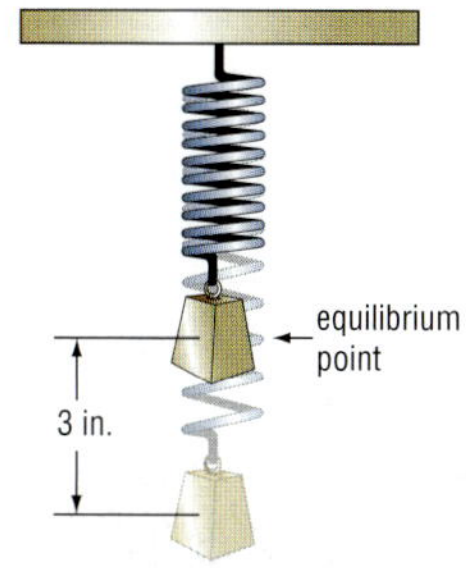

9. Find the period of this function. **2 s**
10. Graph the height of the spring as a function of time. **See margin.**

★ indicates increased difficulty

Practice and Apply

Homework Help

For Exercises	See Examples
11–16	1
17–28	2
29–42	3

Extra Practice
See page 858.

The given point P is located on the unit circle. Find $\sin \theta$ and $\cos \theta$.

11. $P\left(-\frac{3}{5}, \frac{4}{5}\right)$
12. $P\left(-\frac{12}{13}, -\frac{5}{13}\right)$
13. $P\left(\frac{8}{17}, \frac{15}{17}\right)$
14. $P\left(\frac{\sqrt{3}}{2}, -\frac{1}{2}\right)$
15. $P\left(-\frac{1}{2}, \frac{\sqrt{3}}{2}\right)$
16. $P(0.6, 0.8)$

11. $\sin \theta = \frac{4}{5}$; $\cos \theta = -\frac{3}{5}$
12. $\sin \theta = -\frac{5}{13}$; $\cos \theta = -\frac{12}{13}$
13. $\sin \theta = \frac{15}{17}$; $\cos \theta = \frac{8}{17}$
14. $\sin \theta = -\frac{1}{2}$; $\cos \theta = \frac{\sqrt{3}}{2}$
15. $\sin \theta = \frac{\sqrt{3}}{2}$; $\cos \theta = -\frac{1}{2}$
16. $\sin \theta = 0.8$; $\cos \theta = 0.6$

Find the exact value of each function.

17. $\sin 690°$ $-\frac{1}{2}$
18. $\cos 750°$ $\frac{\sqrt{3}}{2}$
19. $\cos 5\pi$ -1
20. $\sin\left(\frac{14\pi}{6}\right)$ $\frac{\sqrt{3}}{2}$
21. $\sin\left(-\frac{3\pi}{2}\right)$ 1
22. $\cos(-225°)$ $-\frac{\sqrt{2}}{2}$
★ 23. $\frac{\cos 60° + \sin 30°}{4}$ $\frac{1}{4}$
★ 24. $3(\sin 60°)(\cos 30°)$ $\frac{9}{4}$
★ 25. $\sin 30° - \sin 60°$ $\frac{1-\sqrt{3}}{2}$
★ 26. $\frac{4 \cos 330° + 2 \sin 60°}{3}$ $\sqrt{3}$
★ 27. $12(\sin 150°)(\cos 150°)$ $-3\sqrt{3}$
★ 28. $(\sin 30°)^2 + (\cos 30°)^2$ 1

Determine the period of each function.

29. **6**

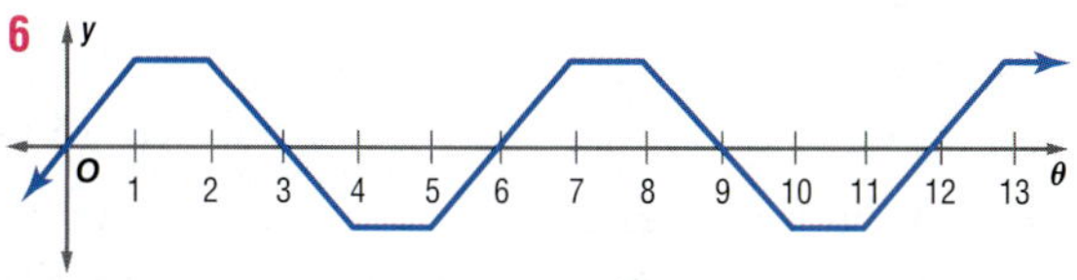

30. **9**

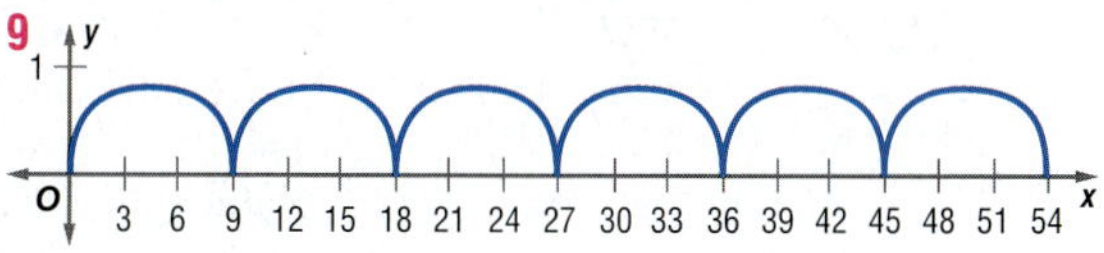

31. **2π**

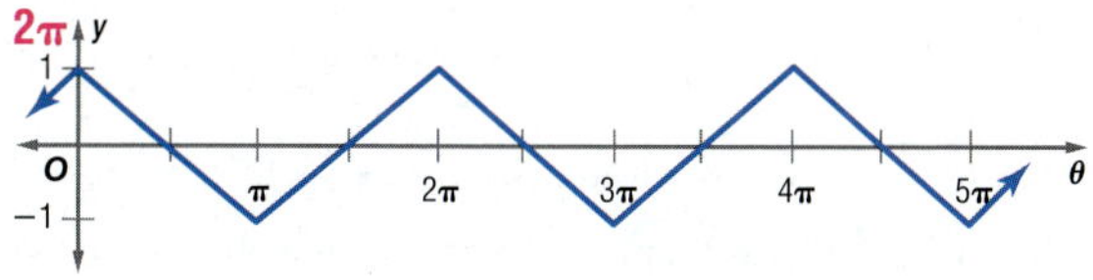

32. **8**

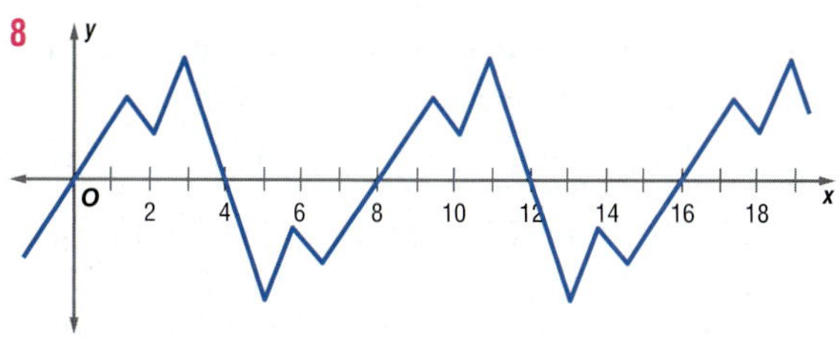

www.algebra2.com/self_check_quiz

3 Practice/Apply

Study Notebook

Have students—
- add the definitions/examples of the vocabulary terms to their Vocabulary Builder worksheets for Chapter 13.
- include any other item(s) that they find helpful in mastering the skills in this lesson.

About the Exercises...

Organization by Objective
- **Unit Circle Definitions:** 11–16
- **Periodic Functions:** 17–42

Odd/Even Assignments
Exercises 11–32 are structured so that students practice the same concepts whether they are assigned odd or even problems.

Assignment Guide

Basic: 11–21 odd, 29, 31, 33, 34, 43–69
Average: 11–31 odd, 33–35, 43–69
Advanced: 12–32 even, 33, 34, 36–63 (optional: 64–69)

Answer

10.

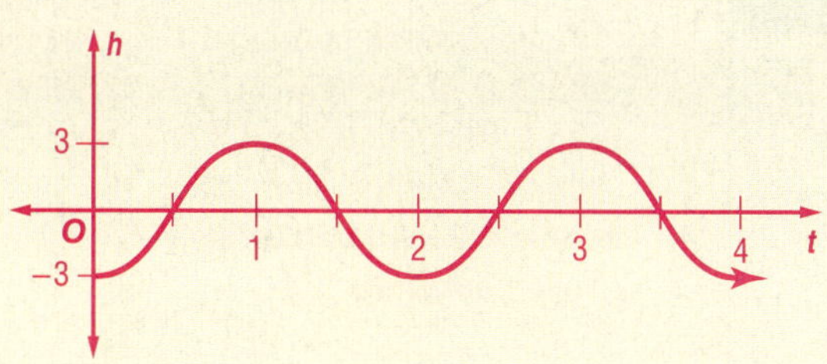

GUITAR For Exercises 33 and 34, use the following information.
When a guitar string is plucked, it is displaced from a fixed point in the middle of the string and vibrates back and forth, producing a musical tone. The exact tone depends on the frequency, or number of cycles per second, that the string vibrates. To produce an A, the frequency is 440 cycles per second, or 440 hertz (Hz).

33. Find the period of this function. $\frac{1}{440}$ s

34. Graph the height of the fixed point on the string from its resting position as a function of time. Let the maximum distance above the resting position have a value of 1 unit and the minimum distance below this position have a value of 1 unit. See pp. 759A–759D.

More About. . .

Guitar

Most guitars have six strings. The frequency at which one of these strings vibrates is controlled by the length of the string, the amount of tension on the string, the weight of the string, and springiness of the strings' material.

Source: www.howstuffworks.com

35. **GEOMETRY** A regular hexagon is inscribed in a unit circle centered at the origin. If one vertex of the hexagon is at (1, 0), find the exact coordinates of the remaining vertices.

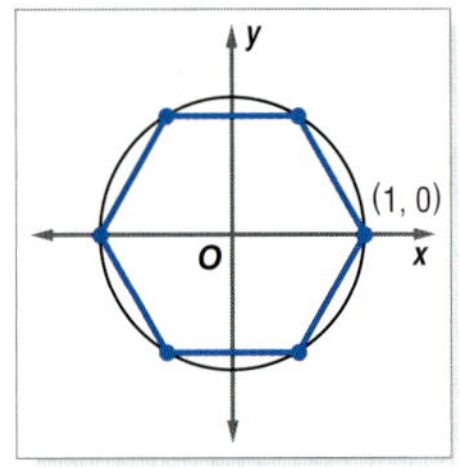

$\left(\frac{1}{2}, \frac{\sqrt{3}}{2}\right), \left(-\frac{1}{2}, \frac{\sqrt{3}}{2}\right), (-1, 0), \left(-\frac{1}{2}, -\frac{\sqrt{3}}{2}\right), \left(\frac{1}{2}, -\frac{\sqrt{3}}{2}\right)$

36. **BIOLOGY** In a certain area of forested land, the population of rabbits R increases and decreases periodically throughout the year. If the population can be modeled by $R = 425 + 200 \sin\left[\frac{\pi}{365}(d - 60)\right]$, where d represents the dth day of the year, describe what happens to the population throughout the year. See margin.

SLOPE For Exercises 37–42, use the following information.
Suppose the terminal side of an angle θ in standard position intersects the unit circle at $P(x, y)$.

37. What is the slope of $\overline{OP}$? $\frac{y}{x}$

38. Which of the six trigonometric functions is equal to the slope of $\overline{OP}$? $\tan \theta$

39. What is the slope of any line perpendicular to $\overline{OP}$? $-\frac{x}{y}$

40. Which of the six trigonometric functions is equal to the slope of any line perpendicular to $\overline{OP}$? $-\cot \theta$

41. Find the slope of $\overline{OP}$ when $\theta = 60°$. $\sqrt{3}$

42. If $\theta = 60°$, find the slope of the line tangent to circle O at point P. $-\frac{\sqrt{3}}{3}$

43. **CRITICAL THINKING** Determine the domain and range of the functions $y = \sin \theta$ and $y = \cos \theta$. sine: D = {all reals}, R = $\{-1 \le y \le 1\}$; cosine: D = {all reals}, R = $\{-1 \le y \le 1\}$

44. **WRITING IN MATH** Answer the question that was posed at the beginning of the lesson. See margin.

How can you model annual temperature fluctuations?

Include the following in your answer:
- a description of how the sine and cosine functions are similar to annual temperature fluctuations, and
- if the formula for the temperature T in degrees Fahrenheit of a city t months into the year is given by $T = 50 + 25 \sin\left(\frac{\pi}{6}t\right)$, explain how to find the average temperature and the maximum and minimum predicted over the year.

Study Guide and Intervention, p. 805 (shown) and p. 806

Unit Circle Definitions

Definition of Sine and Cosine	If the terminal side of an angle θ in standard position intersects the unit circle at $P(x, y)$, then $\cos \theta = x$ and $\sin \theta = y$. Therefore, the coordinates of P can be written as $P(\cos \theta, \sin \theta)$.

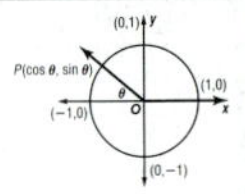

Example Given an angle θ in standard position, if $P\left(-\frac{5}{6}, \frac{\sqrt{11}}{6}\right)$ lies on the terminal side and on the unit circle, find $\sin \theta$ and $\cos \theta$.
$P\left(-\frac{5}{6}, \frac{\sqrt{11}}{6}\right) = P(\cos \theta, \sin \theta)$, so $\sin \theta = \frac{\sqrt{11}}{6}$ and $\cos \theta = -\frac{5}{6}$.

Exercises

If θ is an angle in standard position and if the given point P is located on the terminal side of θ and on the unit circle, find $\sin \theta$ and $\cos \theta$.

1. $P\left(-\frac{\sqrt{3}}{2}, \frac{1}{2}\right)$ $\sin \theta = \frac{1}{2}, \cos \theta = -\frac{\sqrt{3}}{2}$
2. $P(0, -1)$ $\sin \theta = -1, \cos \theta = 0$
3. $P\left(-\frac{2}{3}, \frac{\sqrt{5}}{3}\right)$ $\sin \theta = \frac{\sqrt{5}}{3}, \cos \theta = -\frac{2}{3}$
4. $P\left(-\frac{4}{5}, -\frac{3}{5}\right)$ $\sin \theta = -\frac{3}{5}, \cos \theta = -\frac{4}{5}$
5. $P\left(\frac{1}{6}, -\frac{\sqrt{35}}{6}\right)$ $\sin \theta = -\frac{\sqrt{35}}{6}, \cos \theta = \frac{1}{6}$
6. $P\left(\frac{\sqrt{7}}{4}, \frac{3}{4}\right)$ $\sin \theta = \frac{3}{4}, \cos \theta = \frac{\sqrt{7}}{4}$
7. P is on the terminal side of $\theta = 45°$. $\sin \theta = \frac{\sqrt{2}}{2}, \cos \theta = \frac{\sqrt{2}}{2}$
8. P is on the terminal side of $\theta = 120°$. $\sin \theta = \frac{\sqrt{3}}{2}, \cos \theta = -\frac{1}{2}$
9. P is on the terminal side of $\theta = 240°$. $\sin \theta = -\frac{\sqrt{3}}{2}, \cos \theta = -\frac{1}{2}$
10. P is on the terminal side of $\theta = 330°$. $\sin \theta = -\frac{1}{2}, \cos \theta = \frac{\sqrt{3}}{2}$

Skills Practice, p. 807 and Practice, p. 808 (shown)

The given point P is located on the unit circle. Find $\sin \theta$ and $\cos \theta$.

1. $P\left(-\frac{1}{2}, \frac{\sqrt{3}}{2}\right)$ $\sin \theta = \frac{\sqrt{3}}{2}, \cos \theta = -\frac{1}{2}$
2. $P\left(\frac{20}{29}, -\frac{21}{29}\right)$ $\sin \theta = -\frac{21}{29}, \cos \theta = \frac{20}{29}$
3. $P(0.8, 0.6)$ $\sin \theta = 0.6, \cos \theta = 0.8$
4. $P(0, -1)$ $\sin \theta = -1, \cos \theta = 0$
5. $P\left(-\frac{\sqrt{2}}{2}, -\frac{\sqrt{2}}{2}\right)$ $\sin \theta = -\frac{\sqrt{2}}{2}, \cos \theta = -\frac{\sqrt{2}}{2}$
6. $P\left(\frac{\sqrt{3}}{2}, \frac{1}{2}\right)$ $\sin \theta = \frac{1}{2}, \cos \theta = \frac{\sqrt{3}}{2}$

Find the exact value of each function.

7. $\cos \frac{7\pi}{4}$ $\frac{\sqrt{2}}{2}$
8. $\sin(-30°)$ $-\frac{1}{2}$
9. $\sin\left(-\frac{2\pi}{3}\right)$ $-\frac{\sqrt{3}}{2}$
10. $\cos(-330°)$ $\frac{\sqrt{3}}{2}$
11. $\cos 600°$ $-\frac{1}{2}$
12. $\sin \frac{9\pi}{2}$ 1
13. $\cos 7\pi$ -1
14. $\cos\left(-\frac{11\pi}{4}\right)$ $-\frac{\sqrt{2}}{2}$
15. $\sin(-225°)$ $\frac{\sqrt{2}}{2}$
16. $\sin 585°$ $-\frac{\sqrt{2}}{2}$
17. $\cos\left(-\frac{10\pi}{3}\right)$ $-\frac{1}{2}$
18. $\sin 840°$ $\frac{\sqrt{3}}{2}$

Determine the period of each function.

19.
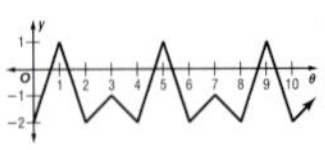

4

20.
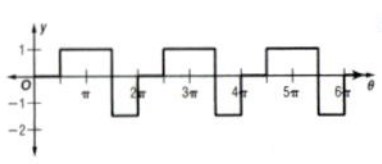

2π

21. **FERRIS WHEELS** A Ferris wheel with a diameter of 100 feet completes 2.5 revolutions per minute. What is the period of the function that describes the height of a seat on the outside edge of the Ferris Wheel as a function of time? 24 s

Reading to Learn Mathematics, p. 809 ELL

Pre-Activity **How can you model annual temperature fluctuations?**

Read the introduction to Lesson 13-6 at the top of page 739 in your textbook.
- If the graph in your textbook is continued, what month will $x = 17$ represent? May of the following year
- About what do you expect the average high temperature to be for that month? 24.2°F
- Will this be exactly the average high temperature for that month? Explain your answer. Sample answer: No; temperatures vary from year to year.

Reading the Lesson

1. Use the unit circle on page 740 in your textbook to find the exact values of each expression.
   a. $\cos 45°$ $\frac{\sqrt{2}}{2}$ b. $\sin 150°$ $\frac{1}{2}$ c. $\sin 240°$ $-\frac{\sqrt{3}}{2}$
   d. $\sin 315°$ $-\frac{\sqrt{2}}{2}$ e. $\cos 270°$ 0 f. $\sin 210°$ $-\frac{1}{2}$
   g. $\cos 0°$ 1 h. $\sin 180°$ 0 i. $\cos 330°$ $\frac{\sqrt{3}}{2}$
2. Tell whether each function is periodic. Write *yes* or *no*.
   a. $y = 2x$ no b. $y = x^2$ no c. $y = \cos x$ yes d. $y = |x|$ no
3. Find the period of each function by examining its graph.
   a.
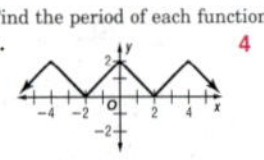

   4
   b.
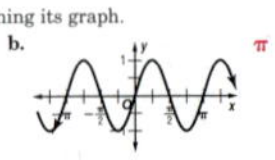

   π
   c.
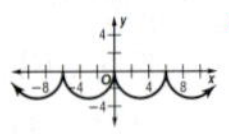

   6

Helping You Remember

4. What is an easy way to remember the periods of the sine and cosine functions in radian measure? Sample answer: The period of both functions is 2π, which is the circumference of the unit circle.

Enrichment, p. 810

Polar Coordinates

Consider an angle in standard position with its vertex at a point O called the *pole*. Its initial side is on a coordinated axis called the *polar axis*. A point P on the terminal side of the angle is named by the *polar coordinates* (r, θ) where r is the directed distance of the point from O and θ is the measure of the angle.

Graphs in this system may be drawn on polar coordinate paper such as the kind shown at the right.

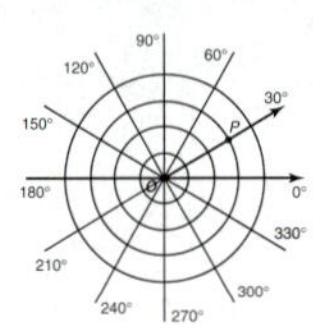

The polar coordinates of a point are not unique. For example, (3, 30°) names point P as well as (3, 390°). Another name for P is (−3, 210°). Can you see why? The coordinates of the pole are $(0, \theta)$ where θ may be any angle.

Example Draw the graph of the function $r = \cos \theta$. Make a table of convenient values for θ ... points

Answer

36. The population is around 425 near the 60th day of the year. It rises to around 625 in May/June. It falls to around 425 again by August/September. It continues to drop to around 225 in November/December.

744 Chapter 13 Trigonometric Functions

Standardized Test Practice

45. If $\triangle ABC$ is an equilateral triangle, what is the length of $\overline{AD}$, in units? **A**

Ⓐ $5\sqrt{2}$
Ⓑ 5
Ⓒ $10\sqrt{2}$
Ⓓ 10

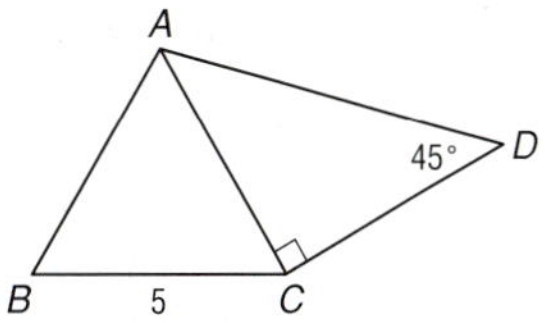

46. **SHORT RESPONSE** What is the exact value of tan 1830°? $\frac{\sqrt{3}}{3}$

Maintain Your Skills

Mixed Review

Determine whether each triangle should be solved by beginning with the Law of Sines or Law of Cosines. Then solve each triangle. Round measures of sides to the nearest tenth and measures of angles to the nearest degree. *(Lesson 13-5)*

47.

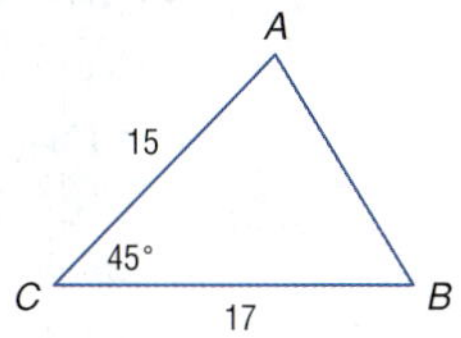

cosines; $c \approx 12.4$, $B \approx 59°$, $A \approx 76°$

48.

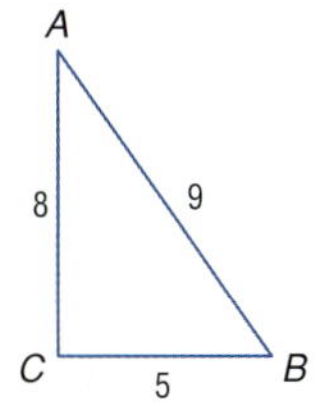

cosines; $A \approx 34°$, $B \approx 62°$, $C \approx 84°$

Find the area of $\triangle ABC$. Round to the nearest tenth. *(Lesson 13-4)*

49. $a = 11$ in., $c = 5$ in., $B = 79°$ **27.0 in²**
50. $b = 4$ m, $c = 7$ m, $A = 63°$ **12.5 m²**

BULBS **For Exercises 51–56, use the following information.**
The lifetimes of 10,000 light bulbs are normally distributed. The mean lifetime is 300 days, and the standard deviation is 40 days. *(Lesson 12-7)*

51. How many light bulbs will last between 260 and 340 days? **6800**
52. How many light bulbs will last between 220 and 380 days? **9500**
53. How many light bulbs will last fewer than 300 days? **5000**
54. How many light bulbs will last more than 300 days? **5000**
55. How many light bulbs will last more than 380 days? **250**
56. How many light bulbs will last fewer than 180 days? **50**

Find the sum of each infinite geometric series, if it exists. *(Lesson 11-5)*

57. $a_1 = 3, r = 1.2$ **does not exist**
58. $16, 4, 1, \frac{1}{4}, \ldots$ $\frac{64}{3}$
59. $\sum_{n=1}^{\infty} 13(-0.625)^{n-1}$ **8**

Use synthetic division to find each quotient. *(Lesson 5-3)*

60. $(4x^2 - 13x + 10) \div (x - 2)$ $4x - 5$
61. $(2x^2 + 21x + 54) \div (x + 6)$ $2x + 9$
62. $(5y^3 + y^2 - 7) \div (y + 1)$ $5y^2 - 4y + 4 - \frac{11}{y+1}$
63. $(2y^2 + y - 16) \div (y - 3)$ $2y + 7 + \frac{5}{y-3}$

Getting Ready for the Next Lesson

PREREQUISITE SKILL **Find each value of θ. Round to the nearest degree.**
*(To review **finding angle measures**, see Lesson 13-1.)*

64. $\sin \theta = 0.3420$ **20°**
65. $\cos \theta = -0.3420$ **110°**
66. $\tan \theta = 3.2709$ **73°**
67. $\tan \theta = 5.6713$ **80°**
68. $\sin \theta = 0.8290$ **56°**
69. $\cos \theta = 0.0175$ **89°**

4 Assess

Open-Ended Assessment

Modeling Have students use a geoboard to make a model of a regular octagon inscribed in a unit circle, similar to that shown in Exercise 35, and find the exact coordinates of the vertices.

Assessment Options

Quiz (Lessons 13-5 and 13-6) is available on p. 832 of the *Chapter 13 Resource Masters*.

Getting Ready for Lesson 13-7

PREREQUISITE SKILL Lesson 13-7 presents inverse trigonometric functions. Students will use their familiarity with finding angle measures as they solve equations with inverse trigonometric functions. Exercises 64–69 should be used to determine your students' familiarity with finding angle measures.

Answer

44. Answers should include the following.

- Over the course of one period both the sine and cosine function attain their maximum value once and their minimum value once. From the maximum to the minimum the functions decrease slowly at first, then decrease more quickly and return to a slow rate of change as they come into the minimum. Similarly, the functions rise slowly from their minimum. They begin to increase more rapidly as they pass the halfway point, and then begin to rise more slowly as they increase into the maximum. Annual temperature fluctuations behave in exactly the same manner.
- The maximum value of the sine function is 1 so the maximum temperature would be 50 + 25(1) or 75° F. Similarly, the minimum value would be 50 + 25(−1) or 25° F. The average temperature over this time period occurs when the sine function takes on a value of 0. In this case that would be 50° F.

13-7 Lesson Notes

1 Focus

5-Minute Check Transparency 13-7 Use as a quiz or review of Lesson 13-6.

Mathematical Background notes are available for this lesson on p. 698D.

How are inverse trigonometric functions used in road design?

Ask students:

- Where else have you seen banking on curves used? **Sample answers: skating, skateboarding, and other sports**
- The force of gravity is 32 feet per second per second. Where does gravity appear in the formula? **in the denominator**

13-7 Inverse Trigonometric Functions

What You'll Learn

- Solve equations by using inverse trigonometric functions.
- Find values of expressions involving trigonometric functions.

Vocabulary

- principal values
- Arcsine function
- Arccosine function
- Arctangent function

How are inverse trigonometric functions used in road design?

When a car travels a curve on a horizontal road, the friction between the tires and the road keeps the car on the road. Above a certain speed, however, the force of friction will not be great enough to hold the car in the curve. For this reason, civil engineers design banked curves.

The proper banking angle θ for a car making a turn of radius r feet at a velocity v in feet per second is given by the equation $\tan \theta = \frac{v^2}{32r}$. In order to determine the appropriate value of θ for a specific curve, you need to know the radius of the curve, the maximum allowable velocity of cars making the curve, and how to determine the angle θ given the value of its tangent.

SOLVE EQUATIONS USING INVERSES Sometimes the value of a trigonometric function for an angle is known and it is necessary to find the measure of the angle. The concept of inverse functions can be applied to find the inverse of trigonometric functions.

In Lesson 8-8, you learned that the inverse of a function is the relation in which all the values of x and y are reversed. The graphs of $y = \sin x$ and its inverse, $x = \sin y$, are shown below.

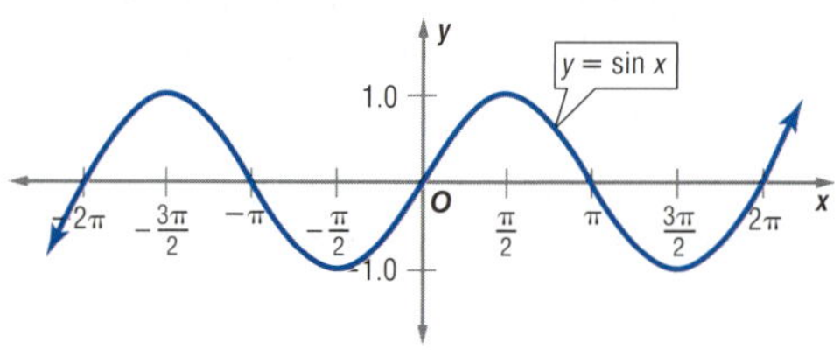

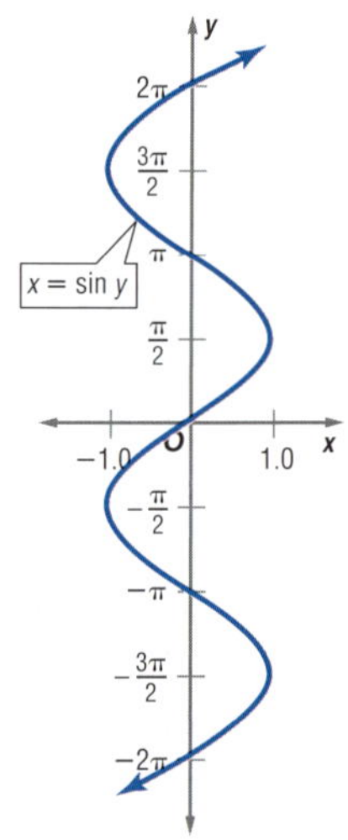

Notice that the inverse is not a function, since it fails the vertical line test. None of the inverses of the trigonometric functions are functions.

We must restrict the domain of trigonometric functions so that their inverses are functions. The values in these restricted domains are called **principal values**. Capital letters are used to distinguish trigonometric functions with restricted domains from the usual trigonometric functions.

Resource Manager

Workbook and Reproducible Masters

Chapter 13 Resource Masters

- Study Guide and Intervention, pp. 811–812
- Skills Practice, p. 813
- Practice, p. 814
- Reading to Learn Mathematics, p. 815
- Enrichment, p. 816
- Assessment, p. 832

Transparencies

5-Minute Check Transparency 13-7
Answer Key Transparencies

Technology

Interactive Chalkboard
Multimedia Applications

Key Concept — *Principal Values of Sine, Cosine, and Tangent*

$y = \text{Sin } x$ if and only if $y = \sin x$ and $-\frac{\pi}{2} \leq x \leq \frac{\pi}{2}$.

$y = \text{Cos } x$ if and only if $y = \cos x$ and $0 \leq x \leq \pi$.

$y = \text{Tan } x$ if and only if $y = \tan x$ and $-\frac{\pi}{2} < x < \frac{\pi}{2}$.

The inverse of the Sine function is called the **Arcsine function** and is symbolized by $\mathbf{Sin^{-1}}$ or **Arcsin**. The Arcsine function has the following characteristics.

- Its domain is the set of real numbers from −1 to 1.
- Its range is the set of angle measures from $-\frac{\pi}{2} \leq x \leq \frac{\pi}{2}$.
- $\text{Sin } x = y$ if and only if $\text{Sin}^{-1} y = x$.
- $[\text{Sin}^{-1} \circ \text{Sin}](x) = [\text{Sin} \circ \text{Sin}^{-1}](x) = x$.

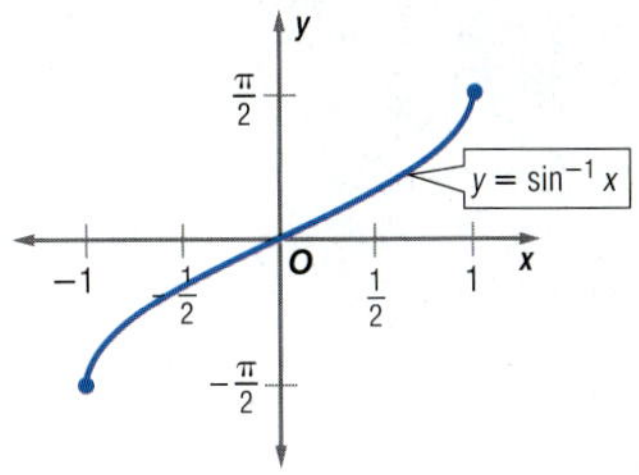

Study Tip

Look Back

To review **composition of functions**, see Lesson 8-7.

The definitions of the Arccosine and Arctangent functions are similar to the definition of the Arcsine function.

Concept Summary — *Inverse Sine, Cosine, and Tangent*

- Given $y = \text{Sin } x$, the inverse Sine function is defined by $y = \text{Sin}^{-1} x$ or $y = \text{Arcsin } x$.
- Given $y = \text{Cos } x$, the inverse Cosine function is defined by $y = \text{Cos}^{-1} x$ or $y = \text{Arccos } x$.
- Given $y = \text{Tan } x$, the inverse Tangent function is defined by $y = \text{Tan}^{-1} x$ or $y = \text{Arctan } x$.

The expressions in each row of the table below are equivalent. You can use these expressions to rewrite and solve trigonometric equations.

$y = \text{Sin } x$	$x = \text{Sin}^{-1} y$	$x = \text{Arcsin } y$
$y = \text{Cos } x$	$x = \text{Cos}^{-1} y$	$x = \text{Arccos } y$
$y = \text{Tan } x$	$x = \text{Tan}^{-1} y$	$x = \text{Arctan } y$

Example 1 — *Solve an Equation*

Solve $\text{Sin } x = \frac{\sqrt{3}}{2}$ by finding the value of x to the nearest degree.

If $\text{Sin } x = \frac{\sqrt{3}}{2}$, then x is the least value whose sine is $\frac{\sqrt{3}}{2}$. So, $x = \text{Arcsin } \frac{\sqrt{3}}{2}$.

Use a calculator to find x.

KEYSTROKES: [2nd] [SIN⁻¹] [2nd] [√] 3 [ )] [÷] 2 [ )] [ENTER] 60

Therefore, $x = 60°$.

www.algebra2.com/extra_examples

2 Teach

SOLVE EQUATIONS USING INVERSES

In-Class Example — Power Point®

Teaching Tip It may help students to understand the nature of inverse functions if they read arcsin x and $\sin^{-1} x$ as "the angle whose sine is x."

1 Solve $\text{Sin } x = \frac{\sqrt{2}}{2}$ by finding the value of x to the nearest degree. $x = 45°$

Lesson 13-7 Inverse Trigonometric Functions 747

In-Class Example

2 **DRAWBRIDGE** For the drawbridge shown in Example 2 in the Student Edition, what is the minimum angle θ, to the nearest degree, to which each leaf should open so that a ship that is 100 feet wide will fit? **52°**

TRIGONOMETRIC VALUES

In-Class Example

Teaching Tip Remind students to first put their calculators in Radian mode.

3 Find each value. Write angle measures in radians. Round to the nearest hundredth.

a. Arcsin $\frac{\sqrt{2}}{2}$ **about 0.79 radians**

b. $\tan\left(\text{Cos}^{-1}\frac{4}{5}\right)$ **0.75 radians**

Many application problems involve finding the inverse of a trigonometric function.

Example 2 Apply an Inverse to Solve a Problem

DRAWBRIDGE **Each leaf of a certain double-leaf drawbridge is 130 feet long. If an 80-foot wide ship needs to pass through the bridge, what is the minimum angle θ, to the nearest degree, which each leaf of the bridge should open so that the ship will fit?**

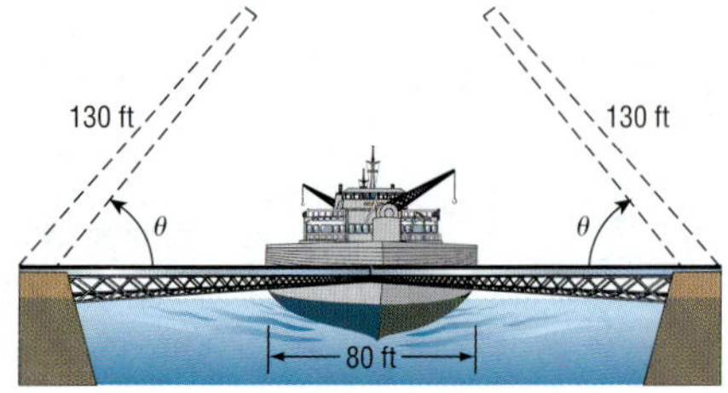

When the two parts of the bridge are in their lowered position, the bridge spans 130 + 130 or 260 feet. In order for the ship to fit, the distance between the leaves must be at least 80 feet.

This leaves a horizontal distance of $\frac{260 - 80}{2}$ or 90 feet from the pivot point of each leaf to the ship as shown in the diagram at the right.

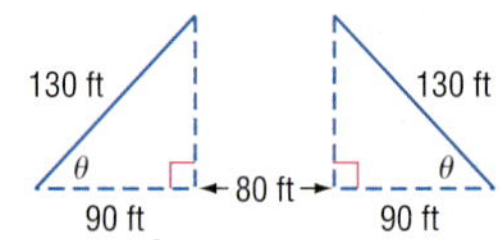

To find the measure of angle θ, use the cosine ratio for right triangles.

$\cos\theta = \frac{adj}{hyp}$ — Cosine ratio

$\cos\theta = \frac{90}{130}$ — Replace *adj* with 90 and *hyp* with 130.

$\theta = \cos^{-1}\left(\frac{90}{130}\right)$ — Inverse cosine function

$\theta \approx 46.2°$ — Use a calculator.

Thus, the minimum angle through which each leaf of the bridge should open is 47°.

More About. . .

Drawbridges

Bascule bridges have spans (leaves) that pivot upward utilizing gears, motors, and counterweights.

Source: www.multnomah.lib.or.us

TRIGONOMETRIC VALUES You can use a calculator to find the values of trigonometric expressions.

Example 3 Find a Trigonometric Value

Find each value. Write angle measures in radians. Round to the nearest hundredth.

a. ArcSin $\frac{\sqrt{3}}{2}$

KEYSTROKES: [2nd] [SIN⁻¹] [2nd] [√] 3 [)] [÷] 2 [)] [ENTER] 1.047197551

Therefore, ArcSin $\frac{\sqrt{3}}{2} \approx 1.05$ radians.

b. $\tan\left(\text{Cos}^{-1}\frac{6}{7}\right)$

KEYSTROKES: [TAN] [2nd] [COS⁻¹] 6 [÷] 7 [)] [ENTER] .6009252126

Therefore, $\tan\left(\text{Cos}^{-1}\frac{6}{7}\right) \approx 0.60$.

Study Tip

Angle Measure

Remember that when evaluating an inverse trigonometric function the result is an angle measure.

DAILY INTERVENTION

Differentiated Instruction

Visual/Spatial Ask students to find Arcsin 2. If they use a calculator, suggest that they study the graph of $y = \sin x$ to explain why an error message was the result. **The graph of $y = \sin x$ has no y values greater than 1 or less than −1.**

Check for Understanding

Concept Check

2. Sample answer: $\text{Cos } 45° = \frac{\sqrt{2}}{2}$; $\text{Cos}^{-1} \frac{\sqrt{2}}{2} = 45°$

1. **Explain** how you know when the domain of a trigonometric function is restricted. **Restricted domains are denoted with a capital letter.**
2. **OPEN ENDED** Write an equation giving the value of the Cosine function for an angle measure in its domain. Then, write your equation in the form of an inverse function.
3. Describe how $y = \text{Cos } x$ and $y = \text{Arccos } x$ are related. **They are inverses of each other.**

Guided Practice

GUIDED PRACTICE KEY	
Exercises	Examples
4–7	1
8–13	3
14	2

Write each equation in the form of an inverse function.

4. $\text{Tan } \theta = x$ **$\theta = \text{Arctan } x$**
5. $\text{Cos } \alpha = 0.5$ **$\alpha = \text{Arccos } 0.5$**

Solve each equation by finding the value of x to the nearest degree.

6. $x = \text{Cos}^{-1} \frac{\sqrt{2}}{2}$ **45°**
7. $\text{Arctan } 0 = x$ **0°**

Find each value. Write angle measures in radians. Round to the nearest hundredth.

8. $\text{Tan}^{-1}\left(-\frac{\sqrt{3}}{3}\right)$ **$8.\ -\frac{\pi}{6} \approx -0.52$**
9. $\text{Cos}^{-1}(-1)$ **$\pi \approx 3.14$**
10. $\cos\left(\text{Cos}^{-1} \frac{2}{9}\right)$ **0.22**
11. $\sin\left(\text{Sin}^{-1} \frac{3}{4}\right)$ **0.75**
12. $\sin\left(\text{Cos}^{-1} \frac{3}{4}\right)$ **0.66**
13. $\tan\left(\text{Sin}^{-1} \frac{1}{2}\right)$ **0.58**

Application

14. **ARCHITECTURE** The support for a roof is shaped like two right triangles as shown at the right. Find θ. **30°**

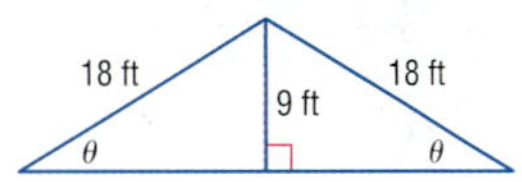

Practice and Apply

Homework Help	
For Exercises	See Examples
15–26	1
27–42	3
43–48	2

Extra Practice See page 859.

Write each equation in the form of an inverse function. **15–20. See margin.**

15. $\alpha = \text{Sin } \beta$
16. $\text{Tan } a = b$
17. $\text{Cos } y = x$
18. $\text{Sin } 30° = \frac{1}{2}$
19. $\text{Cos } 45° = y$
20. $-\frac{4}{3} = \text{Tan } x$

Solve each equation by finding the value of x to the nearest degree.

21. $x = \text{Cos}^{-1} \frac{1}{2}$ **60°**
22. $\text{Sin}^{-1} \frac{1}{2} = x$ **30°**
23. $\text{Arctan } 1 = x$ **45°**
24. $x = \text{Arctan } \frac{\sqrt{3}}{3}$ **30°**
25. $x = \text{Sin}^{-1}$ **45°**
26. $x = \text{Cos}^{-1} 0$ **90°**

Find each value. Write angle measures in radians. Round to the nearest hundredth.

27. $\text{Cos}^{-1}\left(-\frac{1}{2}\right)$ **2.09**
28. $\text{Sin}^{-1} \frac{\pi}{2}$ **28. does not exist**
29. $\text{Arctan } \frac{\sqrt{3}}{3}$ **0.52**
30. $\text{Arccos } \frac{\sqrt{3}}{2}$ **0.52**
31. $\sin\left(\text{Sin}^{-1} \frac{1}{2}\right)$ **0.5**
32. $\cot\left(\text{Sin}^{-1} \frac{5}{6}\right)$ **0.66**
33. $\tan\left(\text{Cos}^{-1} \frac{6}{7}\right)$ **0.60**
34. $\sin\left(\text{Arctan } \frac{\sqrt{3}}{3}\right)$ **0.5**
35. $\cos\left(\text{Arcsin } \frac{3}{5}\right)$ **0.8**
36. $\cot\left(\text{Sin}^{-1} \frac{7}{9}\right)$ **0.81**
37. $\cos\left(\text{Tan}^{-1} \sqrt{3}\right)$ **0.5**
38. $\tan(\text{Arctan } 3)$ **3**
39. $\cos\left[\text{Arccos}\left(-\frac{1}{2}\right)\right]$ **−0.5**
40. $\text{Sin}^{-1}\left(\tan \frac{\pi}{4}\right)$ **1.57**
41. $\cos\left(\text{Cos}^{-1} \frac{\sqrt{2}}{2} - \frac{\pi}{2}\right)$ **41. 0.71**
42. $\text{Cos}^{-1}(\text{Sin}^{-1} 90)$ **does not exist**
43. $\sin\left(2\ \text{Cos}^{-1} \frac{3}{5}\right)$ **0.96**
44. $\sin\left(2\ \text{Sin}^{-1} \frac{1}{2}\right)$ **0.87**

www.algebra2.com/self_check_quiz

3 Practice/Apply

Study Notebook

Have students—
- complete the definitions/examples for the remaining terms on their Vocabulary Builder worksheets for Chapter 13.
- include any other item(s) that they find helpful in mastering the skills in this lesson.

About the Exercises…

Organization by Objective
- **Solve Equations Using Inverses:** 15–26, 45–48
- **Trigonometric Values:** 27–44

Odd/Even Assignments
Exercises 15–44 are structured so that students practice the same concepts whether they are assigned odd or even problems.

Assignment Guide

Basic: 15–45 odd, 49–54, 58–66

Average: 15–47 odd, 49–54, 58–66 (optional: 55–57)

Advanced: 16–48 even, 49–66

Answers

15. $\beta = \text{Arcsin } \alpha$
16. $a = \text{Arctan } b$
17. $y = \text{Arccos } x$
18. $30° = \text{Arcsin } \frac{1}{2}$
19. $\text{Arccos } y = 45°$
20. $\text{Arctan}\left(-\frac{4}{3}\right) = x$

Study Guide and Intervention, p. 811 (shown) and p. 812

Solve Equations Using Inverses If the domains of trigonometric functions are restricted to their **principal values**, then their inverses are also functions.

Principal Values of Sine, Cosine, and Tangent	$y = \text{Sin } x$ if and only if $y = \sin x$ and $-\frac{\pi}{2} \le x \le \frac{\pi}{2}$. $y = \text{Cos } x$ if and only if $y = \cos x$ and $0 \le x \le \pi$. $y = \text{Tan } x$ if and only if $y = \tan x$ and $-\frac{\pi}{2} \le x \le \frac{\pi}{2}$.
Inverse Sine, Cosine, and Tangent	Given $y = \text{Sin } x$, the inverse Sine function is defined by $y = \text{Sin}^{-1} x$ or $y = \text{Arcsin } x$. Given $y = \text{Cos } x$, the inverse Cosine function is defined by $y = \text{Cos}^{-1} x$ or $y = \text{Arccos } x$. Given $y = \text{Tan } x$, the inverse Tangent function is given by $y = \text{Tan}^{-1} x$ or $y = \text{Arctan } x$.

Example 1 **Solve $x = \text{Sin}^{-1}\left(\frac{\sqrt{3}}{2}\right)$.**

If $x = \text{Sin}^{-1}\left(\frac{\sqrt{3}}{2}\right)$, then $\text{Sin } x = \frac{\sqrt{3}}{2}$ and $-\frac{\pi}{2} \le x \le \frac{\pi}{2}$.

The only x that satisfies both criteria is $x = \frac{\pi}{3}$ or 60°.

Example 2 **Solve $\text{Arctan}\left(-\frac{\sqrt{3}}{3}\right) = x$.**

If $x = \text{Arctan}\left(-\frac{\sqrt{3}}{3}\right)$, then $\text{Tan } x = -\frac{\sqrt{3}}{3}$ and $-\frac{\pi}{2} \le x \le \frac{\pi}{2}$.

The only x that satisfies both criteria is $-\frac{\pi}{6}$ or −30°.

Exercises

Solve each equation by finding the value of x to the nearest degree.

1. $\text{Cos}^{-1}\left(-\frac{\sqrt{3}}{2}\right) = x$ **150°**
2. $x = \text{Sin}^{-1} \frac{\sqrt{3}}{2}$ **60°**
3. $x = \text{Arccos}(-0.8)$ **143°**
4. $x = \text{Arctan } \sqrt{3}$ **60°**
5. $x = \text{Arccos}\left(-\frac{\sqrt{2}}{2}\right)$ **135°**
6. $x = \text{Tan}^{-1}(-1)$ **−45°**
7. $\text{Sin}^{-1} 0.45 = x$ **27°**
8. $x = \text{Arcsin}\left(-\frac{\sqrt{3}}{2}\right)$ **−60°**
9. $x = \text{Arccos}\left(-\frac{1}{2}\right)$ **120°**
10. $\text{Cos}^{-1}(-0.2) = x$ **102°**
11. $x = \text{Tan}^{-1}(-\sqrt{3})$ **−60°**
12. $x = \text{Arcsin } 0.3$ **17°**
13. $x = \text{Tan}^{-1}(15)$ **86°**
14. $x = \text{Cos}^{-1} 1$ **0**
15. $\text{Arctan}^{-1}(-3) = x$ **−72°**
16. $x = \text{Sin}^{-1}(-0.9)$ **−64°**
17. $\text{Arccos}^{-1} 0.15$ **81°**
18. $x = \text{Tan}^{-1} 0.2$ **11°**

Skills Practice, p. 813 and Practice, p. 814 (shown)

Write each equation in the form of an inverse function.

1. $\beta = \cos \alpha$ — **$\alpha = \cos^{-1} \beta$**
2. $\tan \beta = \alpha$ — **$\beta = \tan^{-1} \alpha$**
3. $y = \tan 120°$ — **$120° = \tan^{-1} y$**
4. $-\frac{1}{2} = \cos x$ — **$x = \cos^{-1}\left(-\frac{1}{2}\right)$**
5. $\sin \frac{2\pi}{3} = \frac{\sqrt{3}}{2}$ — **$\sin^{-1} \frac{\sqrt{3}}{2} = \frac{2\pi}{3}$**
6. $\cos \frac{\pi}{3} = \frac{1}{2}$ — **$\cos^{-1} \frac{1}{2} = \frac{\pi}{3}$**

Solve each equation by finding the value of x to the nearest degree.

7. $\text{Arcsin } 1 = x$ **90°**
8. $\text{Cos}^{-1} \frac{\sqrt{3}}{2} = x$ **30°**
9. $x = \tan^{-1}\left(-\frac{\sqrt{3}}{3}\right)$ **−30°**
10. $x = \text{Arccos} \frac{\sqrt{2}}{2}$ **45°**
11. $x = \text{Arctan}(-\sqrt{3})$ **−60°**
12. $\text{Sin}^{-1}\left(-\frac{1}{2}\right) = x$ **−30°**

Find each value. Write angle measures in radians. Round to the nearest hundredth.

13. $\text{Cos}^{-1}\left(-\frac{\sqrt{3}}{2}\right)$ **2.62 radians**
14. $\text{Sin}^{-1}\left(-\frac{\sqrt{2}}{2}\right)$ **−0.79 radians**
15. $\text{Arctan}\left(-\frac{\sqrt{3}}{3}\right)$ **−0.52 radians**
16. $\tan\left(\text{Cos}^{-1} \frac{1}{2}\right)$ **1.73**
17. $\cos\left[\text{Sin}^{-1}\left(-\frac{3}{5}\right)\right]$ **0.8**
18. $\cos[\text{Arctan}(-1)]$ **0.71**
19. $\tan\left(\sin^{-1} \frac{12}{13}\right)$ **2.4**
20. $\sin\left(\text{Arctan} \frac{\sqrt{3}}{3}\right)$ **0.5**
21. $\text{Cos}^{-1}\left(\tan \frac{3\pi}{4}\right)$ **3.14 radians**
22. $\text{Sin}^{-1}\left(\cos \frac{\pi}{3}\right)$ **0.52 radians**
23. $\sin\left(2 \text{ Cos}^{-1} \frac{15}{17}\right)$ **0.83**
24. $\cos\left(2 \text{ Sin}^{-1} \frac{\sqrt{3}}{2}\right)$ **−0.5**

25. **PULLEYS** The equation $x = \cos^{-1} 0.95$ describes the angle through which pulley A moves, and $y = \cos^{-1} 0.17$ describes the angle through which pulley B moves. Both angles are greater than 270° and less than 360°. Which pulley moves through a greater angle? **pulley *A***

26. **FLYWHEELS** The equation $y = \text{Arctan } 1$ describes the counterclockwise angle through which a flywheel rotates in 1 millisecond. Through how many degrees has the flywheel rotated after 25 milliseconds? **1125°**

Reading to Learn Mathematics, p. 815 ELL

Pre-Activity **How are inverse trigonometric functions used in road design?**

Read the introduction to Lesson 13-7 at the top of page 746 in your textbook.

Suppose you are given specific values for v and r. What feature of your graphing calculator could you use to find the approximate measure of the banking angle θ? **Sample answer: the TABLE feature**

Reading the Lesson

1. Indicate whether each statement is *true* or *false*.
   a. The domain of the function $y = \sin x$ is the set of all real numbers. **true**
   b. The domain of the function $y = \text{Cos } x$ is $0 \le x \le \pi$. **true**
   c. The range of the function $y = \text{Tan } x$ is $-1 \le y \le 1$. **false**
   d. The domain of the function $y = \text{Cos}^{-1} x$ is $-\frac{\pi}{2} \le x \le \frac{\pi}{2}$. **false**
   e. The domain of the function $y = \text{Tan}^{-1} x$ is the set of all real numbers. **true**
   f. The range of the function $y = \text{Arcsin } x$ is $0 \le x \le \pi$. **false**
2. Answer each question in your own words.
   a. What is the difference between the functions $y = \sin x$ and the function $y = \text{Sin } x$? **Sample answer: The domain of $y = \sin x$ is the set of all real numbers, while the domain of $y = \text{Sin } x$ is restricted to $-\frac{\pi}{2} \le x \le \frac{\pi}{2}$.**
   b. Why is it necessary to restrict the domains of the trigonometric functions in order to define their inverses? **Sample answer: Only one-to-one functions have inverses. None of the six basic trigonometric functions is one-to-one, but related one-to-one functions can be formed if the domains are restricted in certain ways.**

Helping You Remember

3. What is a good way to remember the domains of the functions $y = \text{Sin } x$, $y = \text{Cos } x$, and $y = \text{Tan } x$, which are also the range of the functions $y = \text{Arcsin } x$, $y = \text{Arccos } x$, and $y = \text{Arctan } x$? (You may want to draw a diagram.) **Sample answer: Each restricted domain must include an interval of numbers for which the function values are positive and one for which they are negative.**

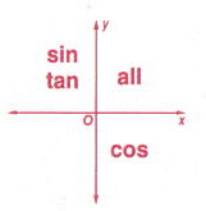

47. No; with this point on the terminal side of the throwing angle θ, the measure of θ is found by solving the equation $\tan \theta = \frac{17}{18}$. Thus $\theta = \tan^{-1} \frac{17}{18}$ or about 43.4°, which is greater than the 40° requirement.

More About. . .

Track and Field

The shot is a metal sphere that can be made out of solid iron. Shot putters stand inside a seven-foot circle and must "put" the shot from the shoulder with one hand.

Source: www.coolrunning.com.au

45. **TRAVEL** The cruise ship *Reno* sailed due west 24 miles before turning south. When the *Reno* became disabled and radioed for help, the rescue boat found that the fastest route to her covered a distance of 48 miles. The cosine of the angle at which the rescue boat should sail is 0.5. Find the angle θ, to the nearest tenth of a degree, at which the rescue boat should travel to aid the *Reno*. **60° south of west**

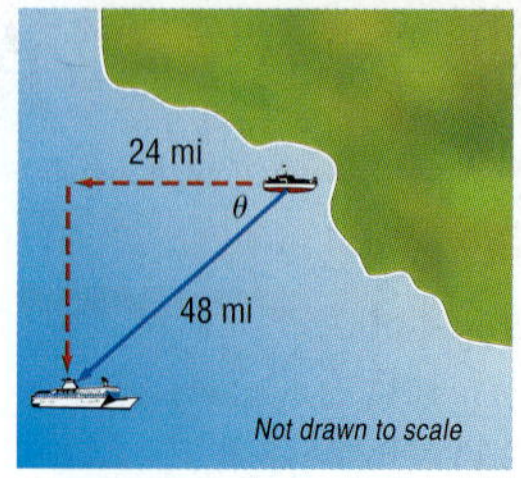

46. **FOUNTAINS** Architects who design fountains know that both the height and distance that a water jet will project is dependent on the angle θ at which the water is aimed. For a given angle θ, the ratio of the maximum height H of the parabolic arc to the horizontal distance D it travels is given by $\frac{H}{D} = \frac{1}{4} \tan \theta$. Find the value of θ, to the nearest degree, that will cause the arc to go twice as high as it travels horizontally. **83°**

47. **TRACK AND FIELD** When a shot put is thrown, it must land in a 40° sector. Consider a coordinate system in which the vertex of the sector is at the origin and one side lies along the x-axis. If an athlete puts the shot so that it lands at a point with coordinates (18, 17), did the shot land in the required region? Explain your reasoning.

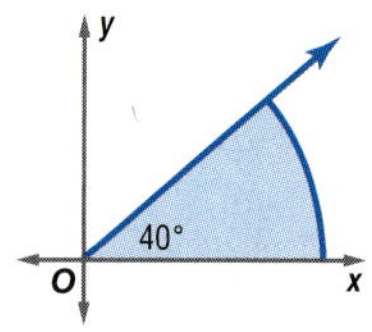

48. **OPTICS** You may have polarized sunglasses that eliminate glare by polarizing the light. When light is polarized, all of the waves are traveling in parallel planes. Suppose horizontally-polarized light with intensity I_0 strikes a polarizing filter with its axis at an angle of θ with the horizontal. The intensity of the transmitted light I_t and θ are related by the equation $\cos \theta = \sqrt{\frac{I_t}{I_0}}$. If one fourth of the polarized light is transmitted through the lens, what angle does the transmission axis of the filter make with the horizontal? **60°**

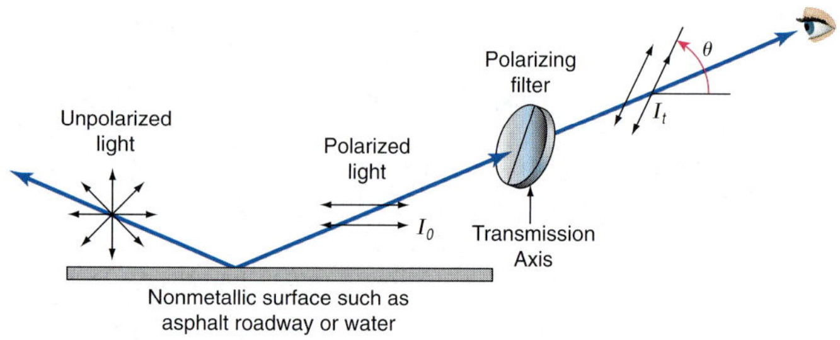

CRITICAL THINKING **For Exercises 49–51, use the following information.**

If the graph of the line $y = mx + b$ intersects the x-axis such that an angle of θ is formed with the positive x-axis, then $\tan \theta = m$.

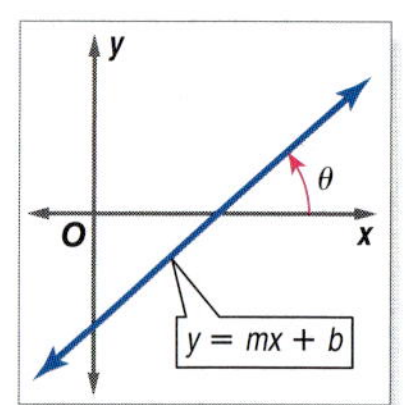

49. Find the acute angle that the graph of $3x + 5y = 7$ makes with the positive x-axis to the nearest degree. **31°**
50. Determine the obtuse angle formed at the intersection of the graphs of $2x + 5y = 8$ and $6x - y = -8$. State the measure of the angle to the nearest degree. **102°**
51. Explain why this relationship, $\tan \theta = m$, holds true. **See margin.**

Enrichment, p. 816

Snell's Law

Snell's Law describes what happens to a ray of light that passes from air into water or some other substance. In the figure, the ray starts at the left and makes an angle of incidence θ with the surface.

Part of the ray is reflected, creating an angle of reflection θ. The rest of the ray is bent, or refracted, as it passes through the other medium. This creates angle θ'.

The angle of incidence equals the angle of reflection.

The angles of incidence and refraction are related by Snell's Law:

$\sin \theta = k \sin \theta'$

The constant k is called the index of refraction.

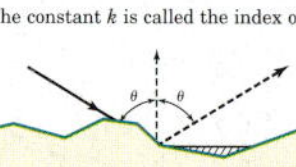

k	Substance
1.33	Water
1.36	Ethyl alcohol

52. WRITING IN MATH Answer the question that was posed at the beginning of the lesson. **See margin.**

How are inverse trigonometric functions used in road design?

Include the following in your answer:

- a few sentences describing how to determine the banking angle for a road, and
- a description of what would have to be done to a road if the speed limit were increased and the banking angle was not changed.

Standardized Test Practice

53. **GRID IN** Find the angle of depression θ between the shallow end and the deep end of the swimming pool to the nearest degree. **37°**

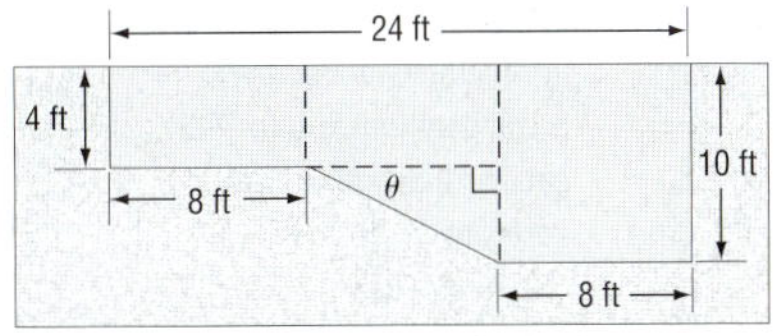

Side View of Swimming Pool

54. If $\sin \theta = \frac{2}{3}$ and $-90° \leq \theta \leq 90°$, then $\cos 2\theta =$ **D**

(A) $-\frac{1}{9}$. (B) $-\frac{1}{3}$. (C) $\frac{1}{3}$. (D) $\frac{1}{9}$. (E) 1.

Graphing Calculator

ADDITION OF TRIGONOMETRIC INVERSES Consider the function $y = \text{Sin}^{-1} x + \text{Cos}^{-1} x$.

55. Copy and complete the table below by evaluating y for each value of x.

x	0	$\frac{1}{2}$	$\frac{\sqrt{2}}{2}$	$\frac{\sqrt{3}}{2}$	1	$-\frac{1}{2}$	$-\frac{\sqrt{2}}{2}$	$-\frac{\sqrt{3}}{2}$	-1
y	$\frac{\pi}{2}$	$\frac{\pi}{2}$	$\frac{\pi}{2}$	$\frac{\pi}{2}$	$\frac{\pi}{2}$	$\frac{\pi}{2}$	$\frac{\pi}{2}$	$\frac{\pi}{2}$	$\frac{\pi}{2}$

56. $\text{Sin}^{-1} x + \text{Cos}^{-1} x = \frac{\pi}{2}$ **for all values of** x

56. Make a conjecture about the function $y = \text{Sin}^{-1} x + \text{Cos}^{-1} x$.

57. Considering only positive values of x, provide an explanation of why your conjecture might be true. **See margin.**

Maintain Your Skills

Mixed Review

Find the exact value of each function. *(Lesson 13-6)*

58. $\sin -660°$ $\frac{\sqrt{3}}{2}$

59. $\cos 25\pi$ **−1**

60. $(\sin 135°)^2 + (\cos -675°)^2$ **1**

Determine whether each triangle should be solved by beginning with the Law of Sines or Law of Cosines. Then solve each triangle. Round measures of sides to the nearest tenth and measures of angles to the nearest degree. *(Lesson 13-5)*

61. $a = 3.1, b = 5.8, A = 30°$

61. sines; $B \approx 69°$, $C \approx 81°$, $c \approx 6.1$ **or** $B \approx 111°$, $C \approx 39°$, $c \approx 3.9$

62. $a = 9, b = 40, c = 41$ **cosines;** $A \approx 13°$, $B \approx 77°$, $C \approx 90°$

Use synthetic substitution to find $f(3)$ and $f(-4)$ for each function. *(Lesson 7-4)*

63. $f(x) = 5x^2 + 6x - 17$ **46, 39**

64. $f(x) = -3x^2 + 2x - 1$ **−22, −57**

65. $f(x) = 4x^2 - 10x + 5$ **11, 109**

66. **PHYSICS** A toy rocket is fired upward from the top of a 200-foot tower at a velocity of 80 feet per second. The height of the rocket t seconds after firing is given by the formula $h(t) = -16t^2 + 80t + 200$. Find the time at which the rocket reaches its maximum height of 300 feet. *(Lesson 6-5)* **2.5 s**

51. Suppose $P(x_1, y_1)$ and $Q(x_2, y_2)$ lie on the line $y = mx + b$. Then $m = \frac{y_2 - y_1}{x_2 - x_1}$. The tangent of the angle θ the line makes with the positive x-axis is equal to the ratio $\frac{\text{opp}}{\text{adj}}$ or $\frac{y_2 - y_1}{x_2 - x_1}$. Thus $\tan \theta = m$.

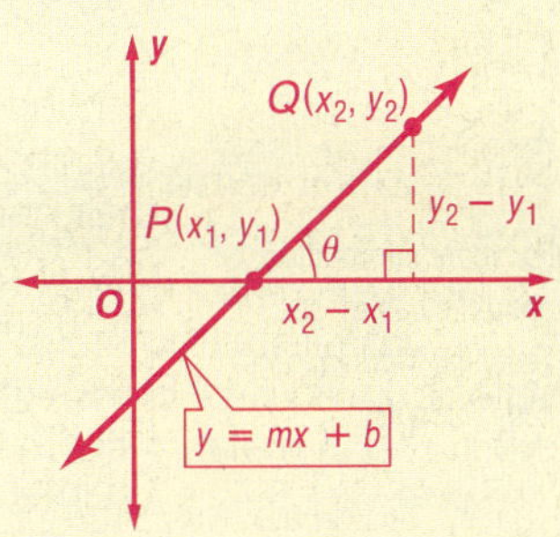

4 Assess

Open-Ended Assessment

Speaking Have students write several problems of their own using both types of notation for inverse trigonometric functions. Then ask them to read their problems aloud and explain how to solve each one.

Assessment Options

Quiz (Lesson 13-7) is available on p. 832 of the *Chapter 13 Resource Masters*.

Answers

52. Trigonometry is used to determine proper banking angles. Answers should include the following.

- **Knowing the velocity of the cars to be traveling on a road and the radius of the curve to be built, then the banking angle can be determined. First find the ratio of the square of the velocity to the product of the acceleration due to gravity and the radius of the curve. Then determine the angle that had this ratio as its tangent. This will be the banking angle for the turn.**
- **If the speed limit were increased and the banking angle remained the same, then in order to maintain a safe road the curvature would have to be decreased. That is, the radius of the curve would also have to increase, which would make the road less curved.**

57. From a right triangle perspective, if an acute angle θ has a given sine, say x, then the complementary angle $\frac{\pi}{2} - \theta$ has that same value as its cosine. This can be verified by looking at a right triangle. Therefore, the sum of the angle whose sine is x and the angle whose cosine is x should be $\frac{\pi}{2}$.

Chapter 13 Study Guide and Review

Vocabulary and Concept Check

- This alphabetical list of vocabulary terms in Chapter 13 includes a page reference where each term was introduced.
- **Assessment** A vocabulary test/review for Chapter 13 is available on p. 830 of the *Chapter 13 Resource Masters.*

Lesson-by-Lesson Review

For each lesson,

- the main ideas are summarized,
- additional examples review concepts, and
- practice exercises are provided.

Vocabulary PuzzleMaker

ELL The Vocabulary PuzzleMaker software improves students' mathematics vocabulary using four puzzle formats—crossword, scramble, word search using a word list, and word search using clues. Students can work on a computer screen or from a printed handout.

MindJogger Videoquizzes

ELL MindJogger Videoquizzes provide an alternative review of concepts presented in this chapter. Students work in teams in a game show format to gain points for correct answers. The questions are presented in three rounds.

Round 1 Concepts (5 questions)
Round 2 Skills (4 questions)
Round 3 Problem Solving (4 questions)

Chapter 13 Study Guide and Review

Vocabulary and Concept Check

angle of depression (p. 705)
angle of elevation (p. 705)
arccosine function (p. 747)
arcsine function (p. 747)
arctangent function (p. 747)
circular function (p. 740)
cosecant (p. 701)
cosine (p. 701)
cotangent (p. 701)
coterminal angles (p. 711)
initial side (p. 709)
law of cosines (p. 733)
law of sines (p. 726)
period (p. 741)
periodic (p. 741)
principal values (p. 746)
quadrantal angles (p. 718)
radian (p. 710)
reference angle (p. 718)
secant (p. 701)
sine (p. 701)
solve a right triangle (p. 704)
standard position (p. 709)
tangent (p. 701)
terminal side (p. 709)
trigonometric functions (p. 701)
trigonometry (p. 701)
unit circle (p. 710)

State whether each sentence is *true* or *false*. If false, replace the underlined word(s) or number to make a true sentence.

1. When two angles in standard position have the same terminal side, they are called quadrantal angles. **false, coterminal**
2. The Law of Sines is used to solve a triangle when the measure of two angles and the measure of any side are known. **true**
3. Trigonometric functions can be defined by using a unit circle. **true**
4. For all values of θ, $\csc \theta = \frac{1}{\cos \theta}$. **false; sec θ**
5. A radian is the measure of an angle on the unit circle where the rays of the angle intercept an arc with length 1 unit. **true**
6. If the measures of three sides of a triangle are known, then the Law of Sines can be used to solve the triangle. **false; Law of Cosines**
7. An angle measuring $60°$ is a quadrantal angle. **False; see margin.**
8. For all values of x, $\cos(x + 180°) = \cos x$. **false; 360°**
9. In a coordinate plane, the initial side of an angle is the ray that rotates about the center. **false; terminal**

Lesson-by-Lesson Review

13-1 Right Triangle Trigonometry

See pages 701–708.

Concept Summary

θ: measure of an acute angle of a right triangle
opp: measure of the leg opposite θ
adj: measure of the leg adjacent to θ
hyp: measure of the hypotenuse

$\sin \theta = \frac{opp}{hyp}$ $\quad \cos \theta = \frac{adj}{hyp}$ $\quad \tan \theta = \frac{opp}{adj}$

$\csc \theta = \frac{hyp}{opp}$ $\quad \sec \theta = \frac{hyp}{adj}$ $\quad \cot \theta = \frac{adj}{opp}$

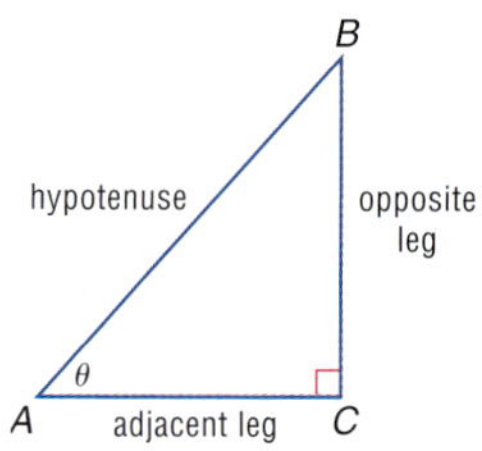

www.algebra2.com/vocabulary_review

FOLDABLES™ Study Organizer

For more information about Foldables, see *Teaching Mathematics with Foldables.*

Remind students to review the Foldable and make sure that their definitions are accurate and complete. Ask them to check over their notes to make sure they have a diagram for each concept and application that they have worked with in this chapter. Have them make any needed additions.

Encourage students to refer to their Foldables while completing the Study Guide and Review and to use them in preparing for the Chapter Test.

Example **Solve $\triangle ABC$. Round measures of sides to the nearest tenth and measures of angles to the nearest degree.**

Find a.

$a^2 + b^2 = c^2$ Pythagorean Theorem

$a^2 + 11^2 = 14^2$ $b = 11$ and $c = 14$

$a = \sqrt{14^2 - 11^2}$ Solve for a.

$a \approx 8.7$ Use a calculator.

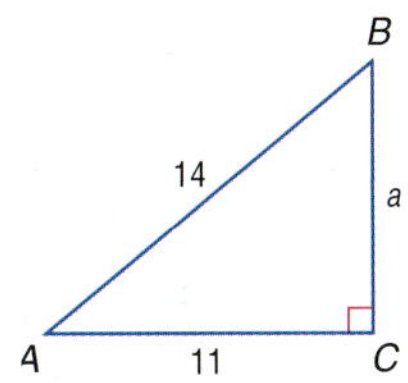

Find A. $\cos A = \frac{11}{14}$ $\cos A = \frac{\text{adj}}{\text{hyp}}$

Use a calculator to find the angle whose cosine is $\frac{11}{14}$.

KEYSTROKES: 2nd [COS^{-1}] 11 ÷ 14) ENTER 38.2132107

To the nearest degree, $A \approx 38°$.

Find B. $38° + B \approx 90°$ Angles A and B are complementary.

$B \approx 52°$ Solve for B.

Therefore, $a \approx 8.7$, $A \approx 38°$, and $B \approx 52°$.

10–15. See margin.

Exercises **Solve $\triangle ABC$ by using the given measurements. Round measures of sides to the nearest tenth and measures of angles to the nearest degree.** *See Examples 4 and 5 on page 704.*

10. $c = 16, a = 7$

11. $A = 25°, c = 6$

12. $B = 45°, c = 12$

13. $B = 83°, b = \sqrt{31}$

14. $a = 9, B = 49°$

15. $\cos A = \frac{1}{4}, a = 4$

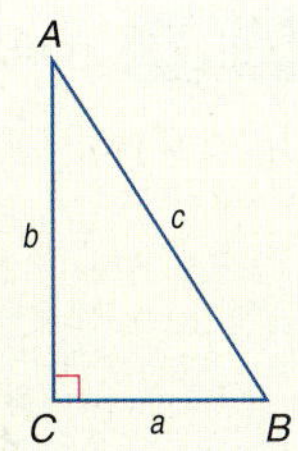

13-2 Angles and Angle Measure

See pages 709–715.

Concept Summary

- An angle in standard position has its vertex at the origin and its initial side along the positive x-axis.
- The measure of an angle is determined by the amount of rotation from the initial side to the terminal side. If the rotation is in a counterclockwise direction, the measure of the angle is positive. If the rotation is in a clockwise direction, the measure of the angle is negative.

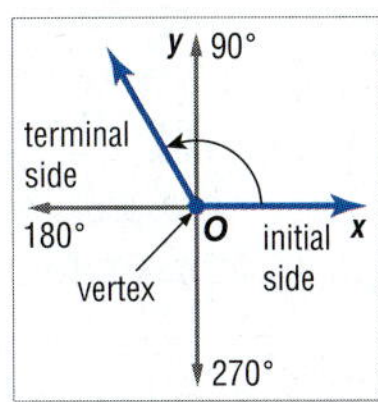

Examples **Rewrite the degree measure in radians and the radian measure in degrees.**

1 240°

$240° = 240°\left(\frac{\pi \text{ radians}}{180°}\right)$

$= \frac{240\pi}{180}$ radians or $\frac{4\pi}{3}$

2 $\frac{\pi}{12}$

$\frac{\pi}{12} = \left(\frac{\pi}{12} \text{ radians}\right)\left(\frac{180°}{\pi \text{ radians}}\right)$

$= \frac{180°}{12}$ or 15°

Answers

7. 0°, 90°, 180°, 270°, 360°, or any multiple of these
10. $A \approx 26°$, $B \approx 64°$, $b \approx 14.4$
11. $B = 65°$, $a \approx 2.5$, $b \approx 5.4$
12. $A = 45°$, $a \approx 8.5$, $b \approx 8.5$
13. $A = 7°$, $a \approx 0.7$, $c \approx 5.6$
14. $A = 41°$, $b \approx 10.4$, $c \approx 13.7$
15. $A \approx 76°$, $B \approx 14°$, $b \approx 1.0$, $c \approx 4.1$

Answers

24. $\sin\theta = \frac{5\sqrt{29}}{29}$, $\cos\theta = \frac{2\sqrt{29}}{29}$, $\tan\theta = \frac{5}{2}$, $\csc\theta = \frac{\sqrt{29}}{5}$, $\sec\theta = \frac{\sqrt{29}}{2}$, $\cot\theta = \frac{2}{5}$

25. $\sin\theta = -\frac{8}{17}$, $\cos\theta = \frac{15}{17}$, $\tan\theta = -\frac{8}{15}$, $\csc\theta = -\frac{17}{8}$, $\sec\theta = \frac{17}{15}$, $\cot\theta = -\frac{15}{8}$

Exercises **Rewrite each degree measure in radians and each radian measure in degrees.** *See Example 2 on page 711.*

16. 255° $\frac{17\pi}{12}$ 17. −210° $-\frac{7\pi}{6}$ 18. $\frac{7\pi}{4}$ 315° 19. -4π −720°

Find one angle with positive measure and one angle with negative measure coterminal with each angle. *See Example 4 on page 712.*

20. 205° 565°, −155° 21. −40° 320°, −400° 22. $\frac{4\pi}{3}$ $\frac{10\pi}{3}$, $-\frac{2\pi}{3}$ 23. $-\frac{7\pi}{4}$ $\frac{\pi}{4}$, $-\frac{15\pi}{4}$

13-3 Trigonometric Functions of General Angles

See pages 717–724.

Concept Summary

- You can find the exact values of the six trigonometric functions of θ given the coordinates of a point $P(x, y)$ on the terminal side of the angle.

$\sin\theta = \frac{y}{r}$ $\cos\theta = \frac{x}{r}$ $\tan\theta = \frac{y}{x}, x \neq 0$

$\csc\theta = \frac{r}{y}, y \neq 0$ $\sec\theta = \frac{r}{x}, x \neq 0$ $\cot\theta = \frac{x}{y}, y \neq 0$

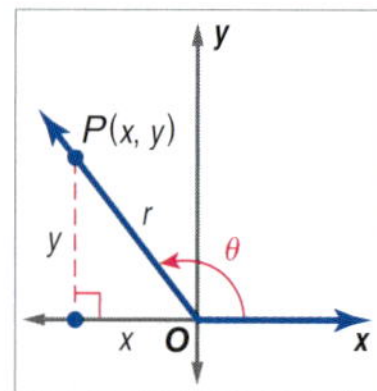

Example **Find the exact value of cos 150°.**

Because the terminal side of 150° lies in Quadrant II, the reference angle θ' is 180° − 150° or 30°. The cosine function is negative in Quadrant II, so cos 150° = −cos 30° or $-\frac{\sqrt{3}}{2}$.

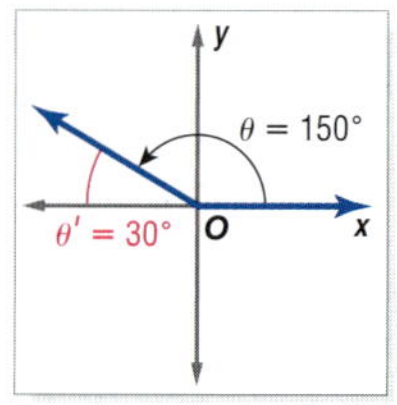

Exercises **Find the exact value of the six trigonometric functions of θ if the terminal side of θ in standard position contains the given point.** *See Example 1 on pages 717 and 718.* **24–25. See margin.**

24. $P(2, 5)$ 25. $P(15, -8)$

Find the exact value of each trigonometric function. *See Example 4 on page 720.*

26. cos 3π −1 27. tan 120° $-\sqrt{3}$ 28. $\sin\frac{5\pi}{4}$ $-\frac{\sqrt{2}}{2}$ 29. sec (−30°) $\frac{2\sqrt{3}}{3}$

13-4 Law of Sines

See pages 725–732.

Concept Summary

- You can find the area of $\triangle ABC$ if the measures of two sides and their included angle are known.

area $= \frac{1}{2}bc \sin A$ area $= \frac{1}{2}ac \sin B$ area $= \frac{1}{2}ab \sin C$

- Law of Sines: $\frac{\sin A}{a} = \frac{\sin B}{b} = \frac{\sin C}{c}$

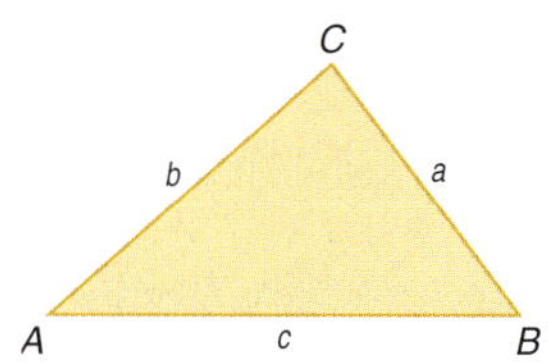

754 Chapter 13 Trigonometric Functions

Example Solve $\triangle ABC$.

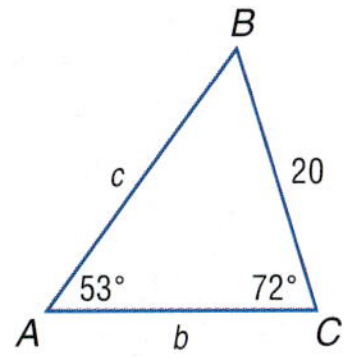

First, find the measure of the third angle.

$53° + 72° + B = 180°$ The sum of the angle measures is 180°.

$B = 55°$ $180 - (53 + 72) = 55$

Now use the Law of Sines to find b and c. Write two equations, each with one variable.

$\frac{\sin A}{a} = \frac{\sin C}{c}$	Law of Sines	$\frac{\sin B}{b} = \frac{\sin A}{a}$
$\frac{\sin 53°}{20} = \frac{\sin 72°}{c}$	Replace A with 53°, B with 55°, C with 72°, and a with 20.	$\frac{\sin 55°}{b} = \frac{\sin 53°}{20}$
$c = \frac{20 \sin 72°}{\sin 53°}$	Solve for the variable.	$b = \frac{20 \sin 55°}{\sin 53°}$
$c \approx 23.8$	Use a calculator.	$b \approx 20.5$

Therefore, $B = 55°$, $b \approx 20.5$, and $c \approx 23.8$.

31, 32, 34, 35. See margin.

Exercises **Determine whether each triangle has *no* solution, *one* solution, or *two* solutions. Then solve each triangle. Round measures of sides to the nearest tenth and measures of angles to the nearest degree.** *See Examples 3–5 on pages 727 and 728.*

30. $a = 24, b = 36, A = 64°$ **no**

31. $A = 40°, b = 10, a = 8$

32. $b = 10, c = 15, C = 66°$

33. $A = 82°, a = 9, b = 12$ **no**

34. $A = 105°, a = 18, b = 14$

35. $B = 46°, C = 83°, b = 65$

13-5 Law of Cosines

See pages 733–738.

Concept Summary

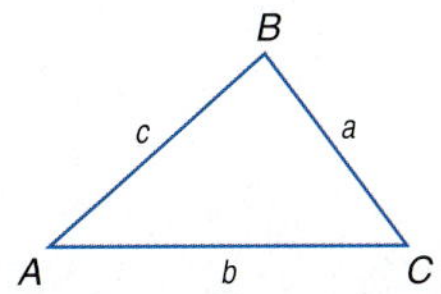

- Law of Cosines: $a^2 = b^2 + c^2 - 2bc \cos A$
  $b^2 = a^2 + c^2 - 2ac \cos B$
  $c^2 = a^2 + b^2 - 2ab \cos C$

Example **Solve $\triangle ABC$ for $A = 62°$, $b = 15$, and $c = 12$.**

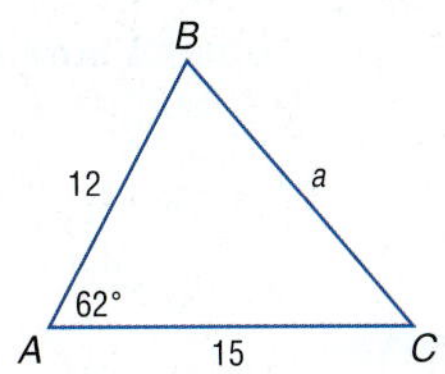

You are given the measure of two sides and the included angle. Begin by drawing a diagram and using the Law of Cosines to determine a.

$a^2 = b^2 + c^2 - 2bc \cos A$ Law of Cosines

$a^2 = 15^2 + 12^2 - 2(15)(12) \cos 62°$ $b = 15, c = 12$, and $A = 62°$

$a^2 \approx 200$ Simplify.

$a \approx 14.1$ Take the square root of each side.

Next, you can use the Law of Sines to find the measure of angle C.

$\frac{\sin 62°}{14.1} \approx \frac{\sin C}{12}$ Law of Sines

$\sin C \approx \frac{12 \sin 62°}{14.1}$ or about 48.7° Use a calculator.

The measure of the angle B is approximately $180 - (62 + 48.7)$ or 69.3°.

Therefore, $a \approx 14.1$, $C \approx 48.7°$, $B \approx 69.3°$.

Answers

31. two; $B \approx 53°$, $C \approx 87°$, $c \approx 12.4$; $B \approx 127°$, $C \approx 13°$, $c \approx 3.0$

32. one; $B = 38°$, $A = 76°$, $a = 16.0$

34. one; $B = 49°$, $C = 26°$, $c = 8.2$

35. one; $A = 51°$, $a = 70.2$, $c = 89.7$

Chapter 13 Study Guide and Review 755

Chapter 13 **For More ...**
- Extra Practice, see pages 857–859.
- Mixed Problem Solving, see page 874.

Answers

36. cosines; 4.6, $A \approx 84°$, $B \approx 61°$
37. sines; $C = 105°$, $a \approx 28.3$, $c \approx 38.6$
38. cosines; $A \approx 46°$, $B \approx 58°$, $C \approx 77°$
39. cosines; $A \approx 34°$, $B \approx 81°$, $c \approx 6.4$
40. sines; $B \approx 52°$, $C \approx 92°$, $c \approx 10.2$; $B \approx 128°$, $C \approx 16°$, $c \approx 2.7$
41. cosines; $B \approx 26°$, $C \approx 125°$, $a \approx 8.3$

Exercises Determine whether each triangle should be solved by beginning with the Law of Sines or Law of Cosines. Then solve each triangle. Round measures of sides to the nearest tenth and measures of angles to the nearest degree. *See Examples 1 and 2 on pages 734 and 735.* **36–41. See margin.**

36. 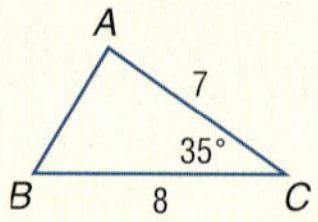

37. 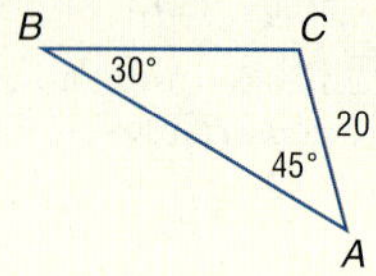

38. 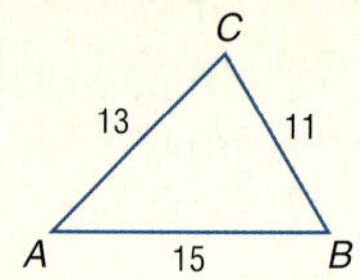

39. $C = 65°, a = 4, b = 7$
40. $A = 36°, a = 6, b = 8$
41. $b = 7.6, c = 14.1, A = 29°$

13-6 Circular Functions

See pages 739–745.

Concept Summary

- If the terminal side of an angle θ in standard position intersects the unit circle at $P(x, y)$, then $\cos \theta = x$ and $\sin \theta = y$. Therefore, the coordinates of P can be written as $P(\cos \theta, \sin \theta)$.

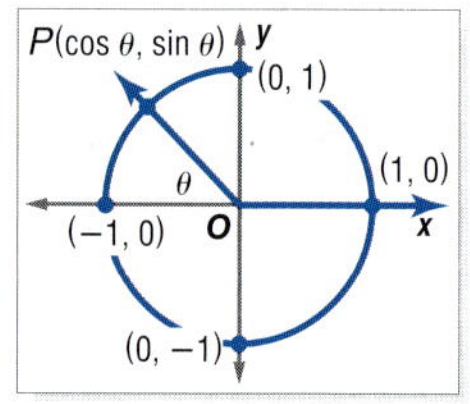

Example Find the exact value of $\cos\left(-\frac{7\pi}{4}\right)$.

$$\cos\left(-\frac{7\pi}{4}\right) = \cos\left(-\frac{7\pi}{4} + 2\pi\right) = \cos\frac{\pi}{4} \text{ or } \frac{\sqrt{2}}{2}$$

Exercises Find the exact value of each function. *See Example 2 on page 741.*

42. $\sin(-150°)$ $-\frac{1}{2}$
43. $\cos 300°$ $\frac{1}{2}$
44. $(\sin 45°)(\sin 225°)$ $-\frac{1}{2}$
45. $\sin\frac{5\pi}{4}$ $-\frac{\sqrt{2}}{2}$
46. $(\sin 30°)^2 + (\cos 30°)^2$ 1
47. $\frac{4\cos 150° + 2\sin 300°}{3}$ $-\sqrt{3}$

13-7 Inverse Trigonometric Functions

See pages 746–751.

Concept Summary

- $y = \text{Sin}\, x$ if and only if $y = \sin x$ and $-\frac{\pi}{2} \le x \le \frac{\pi}{2}$.
- $y = \text{Cos}\, x$ if and only if $y = \cos x$ and $0 \le x \le \pi$.
- $y = \text{Tan}\, x$ if and only if $y = \tan x$ and $-\frac{\pi}{2} < x < \frac{\pi}{2}$.

Example Find the value of $\text{Cos}^{-1}\left[\tan\left(-\frac{\pi}{6}\right)\right]$ in radians. Round to the nearest hundredth.

KEYSTROKES: [2nd] [COS^{-1}] [TAN] [(−)] [2nd] [π] [÷] 6 [)] [)] [ENTER] 2.186276035

Therefore, $\text{Cos}^{-1}\left[\tan\left(-\frac{\pi}{6}\right)\right] \approx 2.19$ radians.

Exercises Find each value. Write angle measures in radians. Round to the nearest hundredth. *See Example 3 on page 748.* **48. −1.57 50. 0.75**

48. $\text{Sin}^{-1}(-1)$
49. $\text{Tan}^{-1}\sqrt{3}$ 1.05
50. $\tan\left(\text{Arcsin}\frac{3}{5}\right)$
51. $\cos(\text{Sin}^{-1} 1)$ 0

Answer (Practice Test)

1. 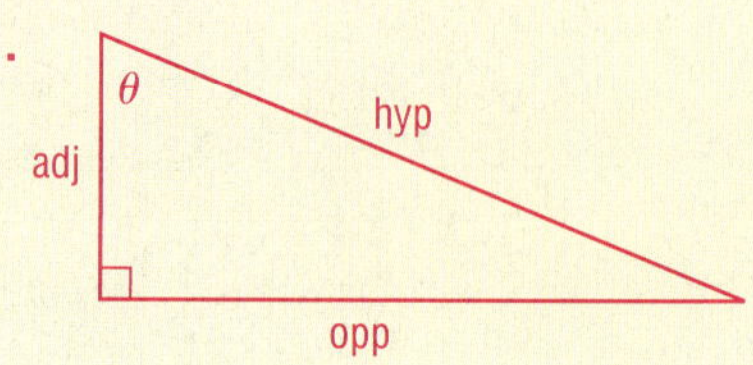

Chapter 13 Practice Test

Vocabulary and Concepts

1. **Draw** a right triangle and label one of the acute angles θ. Then label the hypotenuse *hyp*, the side opposite θ *opp*, and the side adjacent θ *adj*. **See margin.**
2. **State** the Law of Sines for $\triangle ABC$. $\frac{\sin A}{a} = \frac{\sin B}{b} = \frac{\sin C}{c}$
3. **Describe** a situation in which you would solve a triangle by first applying the Law of Cosines. **Sample answer: when the measures of two sides and the included angle are given**

Skills and Applications

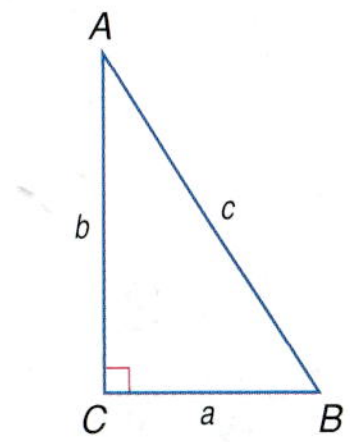

Solve $\triangle ABC$ by using the given measurements. Round measures of sides to the nearest tenth and measures of angles to the nearest degree.

4. $a = 7, A = 49°$ **$b \approx 6.1, c \approx 9.3, B = 41°$**
5. $B = 75°, b = 6$
6. $A = 22°, c = 8$ **$a \approx 3.0, b \approx 7.4, B = 68°$**
7. $a = 7, c = 16$

5. $a \approx 1.6, c \approx 6.2, A = 15°$ 7. $b \approx 14.4, A \approx 26°, B \approx 64°$

Rewrite each degree measure in radians and each radian measure in degrees.

8. $275°$ **$\frac{55\pi}{36}$**
9. $-\frac{\pi}{6}$ **$-30°$**
10. $\frac{11\pi}{2}$ **$990°$**
11. $330°$ **$\frac{11\pi}{6}$**
12. $-600°$ **$-\frac{10\pi}{3}$**
13. $-\frac{7\pi}{4}$ **$-315°$**

Find the exact value of each expression. Write angle measures in degrees.

14. $\cos(-120°)$ **$-\frac{1}{2}$**
15. $\sin \frac{7\pi}{4}$ **$-\frac{\sqrt{2}}{2}$**
16. $\cot 300°$ **$-\frac{\sqrt{3}}{3}$**
17. $\sec\left(-\frac{7\pi}{6}\right)$ **$-\frac{2\sqrt{3}}{3}$**
18. $\text{Sin}^{-1}\left(-\frac{\sqrt{3}}{2}\right)$ **$-60°$**
19. Arctan 1 **$45°$**
20. $\tan 135°$ **-1**
21. $\csc \frac{5\pi}{6}$ **2**

22. Determine the number of possible solutions for a triangle in which $A = 40°, b = 10$, and $a = 14$. If a solution exists, solve the triangle. Round measures of sides to the nearest tenth and measures of angles to the nearest degree. **one; $B \approx 27°, C \approx 113°, c \approx 20.1$**

23. Suppose θ is an angle in standard position whose terminal side lies in Quadrant II. Find the exact values of the remaining five trigonometric functions for θ for $\cos \theta = -\frac{\sqrt{3}}{2}$. **$\sin \theta = \frac{1}{2}, \tan \theta = -\frac{\sqrt{3}}{3}, \sec \theta = -\frac{2\sqrt{3}}{3}, \csc \theta = 2, \cot \theta = -\sqrt{3}$**

24. **GEOLOGY** From the top of the cliff, a geologist spots a dry riverbed. The measurement of the angle of depression to the riverbed is 70°. The cliff is 50 meters high. How far is the riverbed from the base of the cliff? **18.2 m**

25. **STANDARDIZED TEST PRACTICE** Triangle *ABC* has a right angle at *C*, angle $B = 30°$, and $BC = 6$. Find the area of triangle *ABC*. **C**

Ⓐ 6 units^2 Ⓑ $\sqrt{3}$ units^2 Ⓒ $6\sqrt{3}$ units^2 Ⓓ 12 units^2

www.algebra2.com/chapter_test

Portfolio Suggestion

Introduction In this chapter, you studied a number of different approaches to naming and measuring angles and their trigonometric functions.

Ask Students Of the seven lessons in this chapter, pick the one that you are still having some trouble understanding. Describe what questions you still have about this lesson. Explain how this lesson might have been explained in a way that would be clearer to you. Place this in your portfolio.

Chapter 13 Practice Test

Assessment Options

Vocabulary Test A vocabulary test/review for Chapter 13 can be found on p. 830 of the *Chapter 13 Resource Masters.*

Chapter Tests There are six Chapter 13 Tests and an Open-Ended Assessment task available in the *Chapter 13 Resource Masters.*

Chapter 13 Tests

Form	Type	Level	Pages
1	MC	basic	817–818
2A	MC	average	819–820
2B	MC	average	821–822
2C	FR	average	823–824
2D	FR	average	825–826
3	FR	advanced	827–828

MC = multiple-choice questions
FR = free-response questions

Open-Ended Assessment Performance tasks for Chapter 13 can be found on p. 829 of the *Chapter 13 Resource Masters.* A sample scoring rubric for these tasks appears on p. A28.

ExamView® Pro

Use the networkable **ExamView® Pro** to:

- Create **multiple versions** of tests.
- Create **modified** tests for *Inclusion* students.
- **Edit** existing questions and **add** your own questions.
- Use built-in **state curriculum correlations** to create tests aligned with state standards.
- **Apply** art to your tests from a program bank of artwork.

Chapter 13 Standardized Test Practice

These two pages contain practice questions in the various formats that can be found on the most frequently given standardized tests.

A practice answer sheet for these two pages can be found on p. A1 of the *Chapter 13 Resource Masters*.

Standardized Test Practice Student Recording Sheet, p. A1

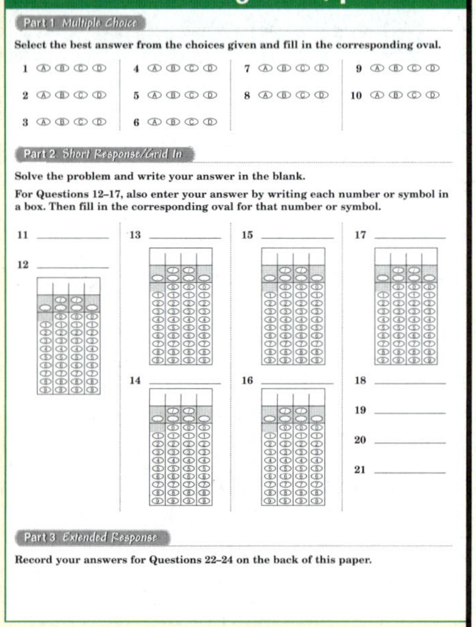

Part 1 Multiple Choice

Select the best answer from the choices given and fill in the corresponding oval.

1 Ⓐ Ⓑ Ⓒ Ⓓ 4 Ⓐ Ⓑ Ⓒ Ⓓ 7 Ⓐ Ⓑ Ⓒ Ⓓ 9 Ⓐ Ⓑ Ⓒ Ⓓ
2 Ⓐ Ⓑ Ⓒ Ⓓ 5 Ⓐ Ⓑ Ⓒ Ⓓ 8 Ⓐ Ⓑ Ⓒ Ⓓ 10 Ⓐ Ⓑ Ⓒ Ⓓ
3 Ⓐ Ⓑ Ⓒ Ⓓ 6 Ⓐ Ⓑ Ⓒ Ⓓ

Part 2 Short Response/Grid In

Solve the problem and write your answer in the blank.

For Questions 12–17, also enter your answer by writing each number or symbol in a box. Then fill in the corresponding oval for that number or symbol.

11 ____ 12 ____ 13 14 15 16 17 18 ____ 19 ____ 20 ____ 21 ____

Part 3 Extended Response

Record your answers for Questions 22–24 on the back of this paper.

Teaching Tip In Questions 7 and 9, students may find it helpful to make sketches to increase their understanding of the problem.

Additional Practice

See pp. 835–836 in the *Chapter 13 Resource Masters* for additional standardized test practice.

Chapter 13 Standardized Test Practice

Part 1 Multiple Choice

Record your answers on the answer sheet provided by your teacher or on a sheet of paper.

1. If $3n + k = 30$ and n is a positive even integer, then which of the following statements must be true? **C**

I. k is divisible by 3.
II. k is an even integer.
III. k is less than 20.

Ⓐ I only Ⓑ II only
Ⓒ I and II only Ⓓ I, II, and III

2. If $4x^2 + 5x = 80$ and $4x^2 - 5y = 30$, then what is the value of $6x + 6y$? **C**

Ⓐ 10 Ⓑ 50 Ⓒ 60 Ⓓ 110

3. If $a = b + cb$, then what does $\frac{b}{a}$ equal in terms of c? **B**

Ⓐ $\frac{1}{c}$ Ⓑ $\frac{1}{1 + c}$
Ⓒ $1 - c$ Ⓓ $1 + c$

4. What is the value of $\sum_{n=1}^{5} 3n^2$? **D**

Ⓐ 55 Ⓑ 58
Ⓒ 75 Ⓓ 165

5. There are 16 green marbles, 2 red marbles, and 6 yellow marbles in a jar. How many yellow marbles need to be added to the jar in order to double the probability of selecting a yellow marble? **D**

Ⓐ 4 Ⓑ 6 Ⓒ 8 Ⓓ 12

Test-Taking Tip

Questions 1–10

The answer choices to multiple-choice questions can provide clues to help you solve a problem. In Question 5, you can add the values in the answer choices to the number of yellow marbles and the total number of marbles to find which is the correct answer.

6. From a lookout point on a cliff above a lake, the angle of depression to a boat on the water is 12°. The boat is 3 kilometers from the shore just below the cliff. What is the height of the cliff from the surface of the water to the lookout point? **D**

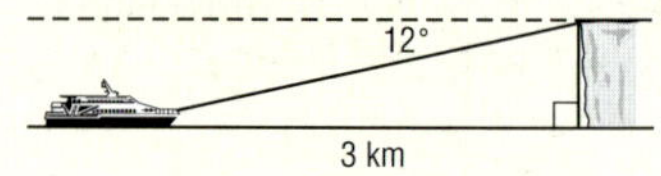

Ⓐ $\frac{3}{\sin 12°}$ Ⓑ $\frac{3}{\tan 12°}$
Ⓒ $\frac{3}{\cos 12°}$ Ⓓ $3 \tan 12°$

7. If $x + y = 90°$ and x and y are positive, then $\frac{\cos x}{\sin y} =$ **C**

Ⓐ 0. Ⓑ $\frac{1}{2}$.
Ⓒ 1. Ⓓ cannot be determined

8. A child flying a kite holds the string 4 feet above the ground. The taut string is 40 feet long and makes an angle of 35° with the horizontal. How high is the kite off the ground? **A**

Ⓐ $4 + 40 \sin 35°$ Ⓑ $4 + 40 \cos 35°$
Ⓒ $4 + 40 \tan 35°$ Ⓓ $4 + \frac{40}{\sin 35°}$

9. If $\sin \theta = -\frac{1}{2}$ and $180° < \theta < 270°$, then $\theta =$ **B**

Ⓐ 200°. Ⓑ 210°.
Ⓒ 225°. Ⓓ 240°.

10. If $\cos \theta = \frac{8}{17}$ and the terminal side of the angle is in quadrant IV, then $\sin \theta =$ **C**

Ⓐ $-\frac{15}{8}$. Ⓑ $-\frac{17}{15}$.
Ⓒ $-\frac{15}{17}$. Ⓓ $\frac{15}{17}$.

ExamView® Pro

Special banks of standardized test questions similar to those on the SAT, ACT, TIMSS 12, NAEP 12, and state proficiency tests can be found on this CD-ROM.

Preparing for Standardized Tests
For test-taking strategies and more practice, see pages 877–892.

Part 2 Short Response/Grid In

Record your answers on the answer sheet provided by your teacher or on a sheet of paper.

11. The length, width, and height of the rectangular box illustrated below are each integers greater than 1. If the area of *ABCD* is 18 square units and the area of *CDEF* is 21 square units, what is the volume of the box? **126 units^3**

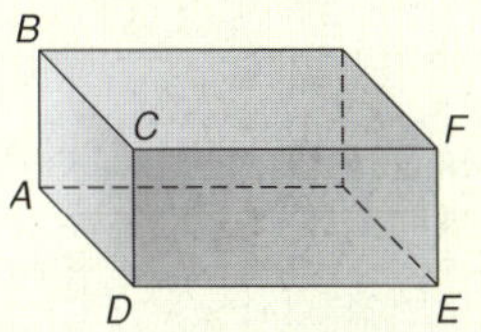

12. When six consecutive integers are multiplied, their product is 0. What is their greatest possible sum? **15**

13. The average (arithmetic mean) score for the 25 players on a team is n. Their scores range from 60 to 100, inclusive. The average score of 20 of the players is 70. What is the difference between the greatest and least possible values of n? **8**

14. The variables a, b, c, d, and e are integers in a sequence, where $a = 2$ and $b = 12$. To find the next term, double the last term and add that result to one less than the next-to-last term. For example, $c = 25$, because $2(12) = 24$, $2 - 1 = 1$, and $24 + 1 = 25$. What is the value of e? **146**

15. In the figure, if $t = 2v$, what is the value of x? **150**

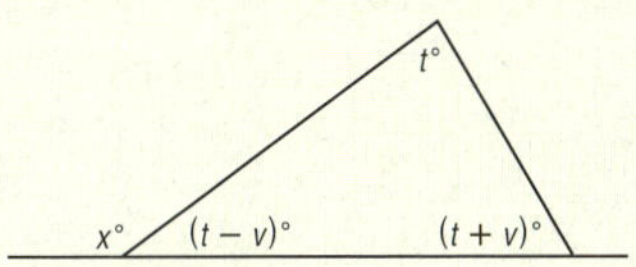

16. If $b = 4$, then what is the value of a in the equations below? **5**

$3a + 4b + 2c = 33$
$2b + 4c = 12$

17. At the head table at a banquet, 3 men and 3 women sit in a row. In how many ways can the row be arranged so that the men and women alternate? **72**

www.algebra2.com/standardized_test

18. If $\csc \theta > 0$ and $\cos \theta < 0$, in which quadrant does the terminal side of angle θ lie? **II**

19. In $\triangle ABC$, if $AC = 15$, $BC = 10$, and $m\angle A = 35°$, what type of angle could angle C be? **acute or obtuse**

20. Through how many radians does the minute hand of a clock turn in 18 minutes? $\frac{3\pi}{5}$

21. The terminal side of an angle θ passes through the point at $\left(-\frac{1}{2}, \frac{\sqrt{3}}{2}\right)$ on a unit circle. If θ is in standard position, name a possible value of θ. **120° or $\frac{2\pi}{3}$**

Part 3 Extended Response

Record your answers on a sheet of paper. Show your work.

For Exercises 22–24, use the following information.

A boater at sea uses a sextant to sight an angle of elevation of 42° to the top of a lighthouse. After the ship travels 330 feet directly toward the lighthouse, she makes another sighting and the new angle of elevation is 49°. The ship's charts show that water becomes too shallow for the boat 125 feet from the base of the lighthouse.

22. Tell whether you can find the distance from the boat to the lighthouse by using the Law of Sines or the Law of Cosines. Justify your answer. **See margin.**

23. Find the distance from the boat to the base of the lighthouse. Show each step and round to the nearest foot. **1189 ft**

24. If the boat is traveling at 44 feet per second, how long does the boater have to change course before the boat runs aground in the shallow water? Round to the nearest tenth of a second. **24.2 s**

Evaluating Extended Response Questions

Extended Response questions are graded by using a multilevel rubric that guides you in assessing a student's knowledge of a particular concept.

Goal: Determine the dimensions of a rectangular garden for a given perimeter.

Sample Scoring Rubric: The following rubric is a sample scoring device. You may wish to add more detail to this sample to meet your individual scoring needs.

Score	Criteria
4	A correct solution that is supported by well-developed, accurate explanations
3	A generally correct solution, but may contain minor flaws in reasoning or computation
2	A partially correct interpretation and/or solution to the problem
1	A correct solution with no supporting evidence or explanation
0	An incorrect solution indicating no mathematical understanding of the concept or task, or no solution is given

Answer

22. Law of sines; the measures of two angles and the included side are known.

Chapter 13 Standardized Test Practice 759

Page 699, Chapter 13 Getting Started

9.

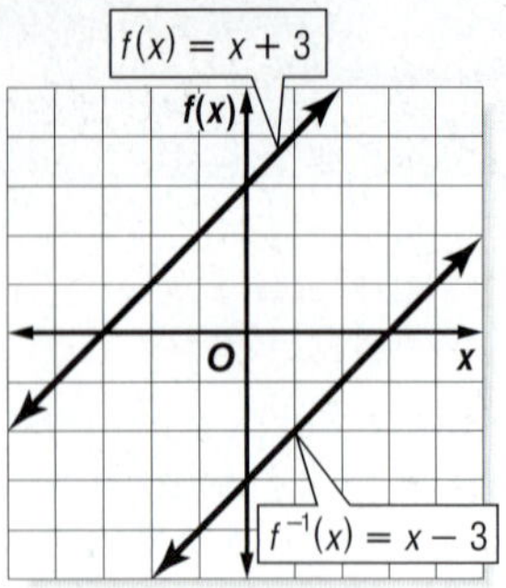

10.

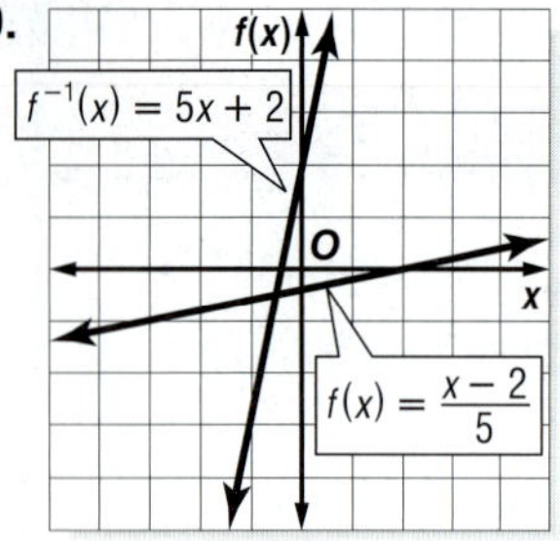

11.

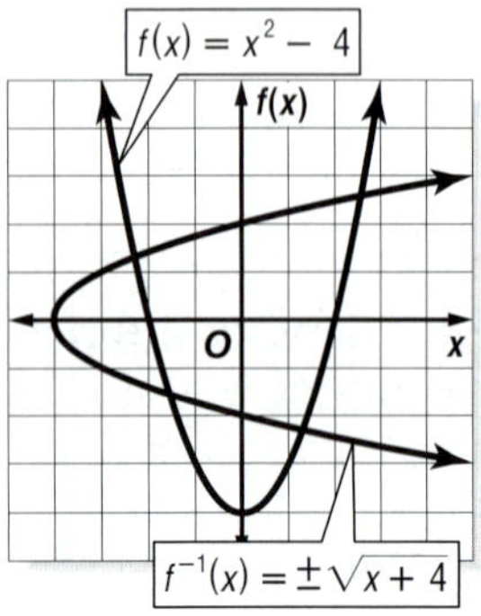

12.

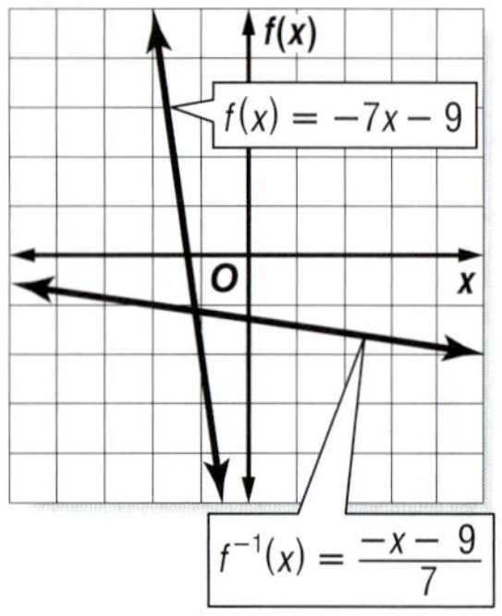

Pages 706–708, Lesson 13-1

15. $\sin\theta = \frac{4}{11}$; $\cos\theta = \frac{\sqrt{105}}{11}$; $\tan\theta = \frac{4\sqrt{105}}{105}$; $\csc\theta = \frac{11}{4}$; $\sec\theta = \frac{11\sqrt{105}}{105}$; $\cot\theta = \frac{\sqrt{105}}{4}$

16. $\sin\theta = \frac{3}{5}$; $\cos\theta = \frac{4}{5}$; $\tan\theta = \frac{3}{4}$; $\csc\theta = \frac{5}{3}$; $\sec\theta = \frac{5}{4}$; $\cot\theta = \frac{4}{3}$

17. $\sin\theta = \frac{\sqrt{7}}{4}$; $\cos\theta = \frac{3}{4}$; $\tan\theta = \frac{\sqrt{7}}{3}$; $\csc\theta = \frac{4\sqrt{7}}{7}$; $\sec\theta = \frac{4}{3}$; $\cot\theta = \frac{3\sqrt{7}}{7}$

18. $\sin\theta = \frac{9\sqrt{106}}{106}$; $\cos\theta = \frac{5\sqrt{106}}{106}$; $\tan\theta = \frac{9}{5}$; $\csc\theta = \frac{\sqrt{106}}{9}$; $\sec\theta = \frac{\sqrt{106}}{5}$; $\cot\theta = \frac{5}{9}$

19. $\sin\theta = \frac{\sqrt{5}}{5}$; $\cos\theta = \frac{2\sqrt{5}}{5}$; $\tan\theta = \frac{1}{2}$; $\csc\theta = \sqrt{5}$; $\sec\theta = \frac{\sqrt{5}}{2}$; $\cot\theta = 2$

20. $\sin\theta = \frac{\sqrt{15}}{8}$; $\cos\theta = \frac{7}{8}$; $\tan\theta = \frac{\sqrt{15}}{7}$; $\csc\theta = \frac{8\sqrt{15}}{15}$; $\sec\theta = \frac{8}{7}$; $\cot\theta = \frac{7\sqrt{15}}{15}$

27a. $\sin 30° = \frac{\text{opp}}{\text{hyp}}$ sine ratio

$\sin 30° = \frac{x}{2x}$ Replace opp with x and hyp with $2x$.

$\sin 30° = \frac{1}{2}$ Simplify.

27b. $\cos 30° = \frac{\text{adj}}{\text{hyp}}$ cosine ratio

$\cos 30° = \frac{\sqrt{3}x}{2x}$ Replace adj with $\sqrt{3}x$ and hyp with $2x$.

$\cos 30° = \frac{\sqrt{3}}{2}$ Simplify.

27c. $\sin 60° = \frac{\text{opp}}{\text{hyp}}$ sine ratio

$\sin 60° = \frac{\sqrt{3}x}{2x}$ Replace opp with $\sqrt{3}x$ and hyp with $2x$.

$\sin 60° = \frac{\sqrt{3}}{2}$ Simplify.

28a. $\sin 45° = \frac{\text{opp}}{\text{hyp}}$ sine ratio

$\sin 45° = \frac{x}{\sqrt{2}x}$ Replace opp with x and hyp with $\sqrt{2}x$.

$\sin 45° = \frac{1}{\sqrt{2}}$ Simplify.

$\sin 45° = \frac{\sqrt{2}}{2}$ Rationalize the denominator.

28b. $\cos 45° = \frac{\text{adj}}{\text{hyp}}$ cosine ratio

$\cos 45° = \frac{x}{\sqrt{2}x}$ Replace adj with x and hyp with $\sqrt{2}x$.

$\cos 45° = \frac{1}{\sqrt{2}}$ Simplify.

$\cos 45° = \frac{\sqrt{2}}{2}$ Rationalize the denominator.

28c. $\tan 45° = \frac{\text{opp}}{\text{adj}}$ tangent ratio

$\tan 45° = \frac{x}{x}$ Replace opp with x and adj with x.

$\tan 45° = 1$ Simplify.

29. $B = 74°$, $a \approx 3.9$, $b \approx 13.5$

30. $A = 63°$, $a \approx 13.7$, $c \approx 15.4$

31. $B = 56°$, $b \approx 14.8$, $c \approx 17.9$

32. $A = 75°$, $a \approx 24.1$, $b \approx 6.5$

33. $A = 60°$, $a \approx 19.1$, $c = 22$

34. $B = 45°$, $a = 7$, $b = 7$

35. $A = 72°$, $b \approx 1.3$, $c \approx 4.1$

36. $B = 80°$, $a \approx 2.6$, $c \approx 15.2$

37. $A \approx 63°$, $B \approx 27°$, $a \approx 11.5$

38. $A \approx 26°$, $B \approx 64°$, $b \approx 8.1$

39. $A \approx 49°$, $B \approx 41°$, $a = 8$, $c \approx 10.6$

40. $A \approx 19°$, $B \approx 71°$, $b \approx 14.1$, $c = 15$

Pages 712–714, Lesson 13-2

19.

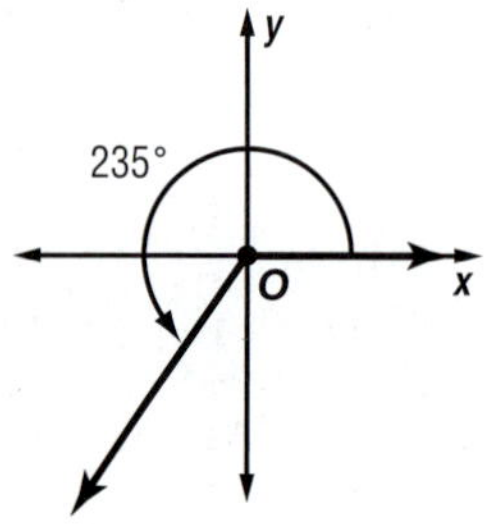

20.

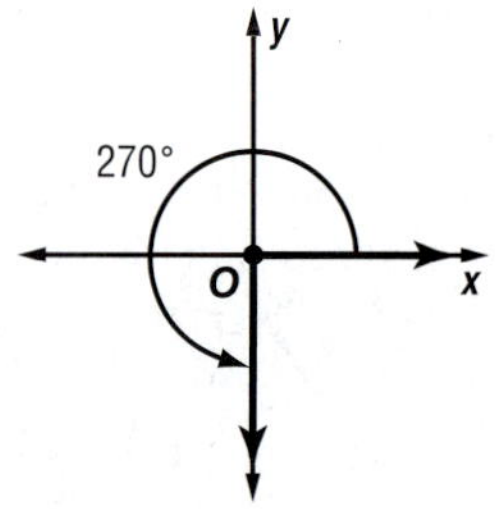

21.

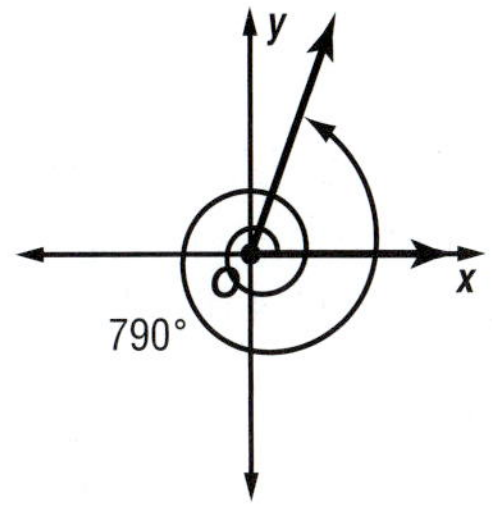

22.

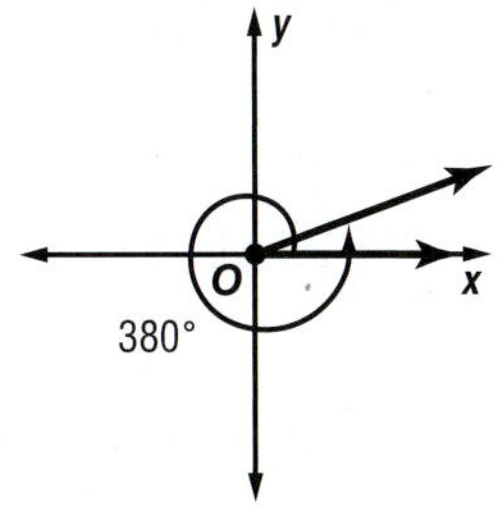

23.

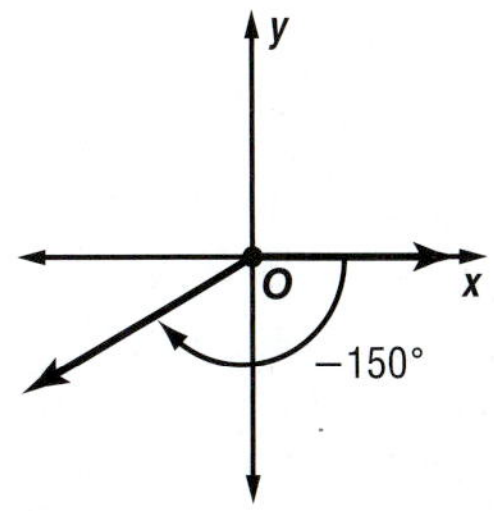

24.

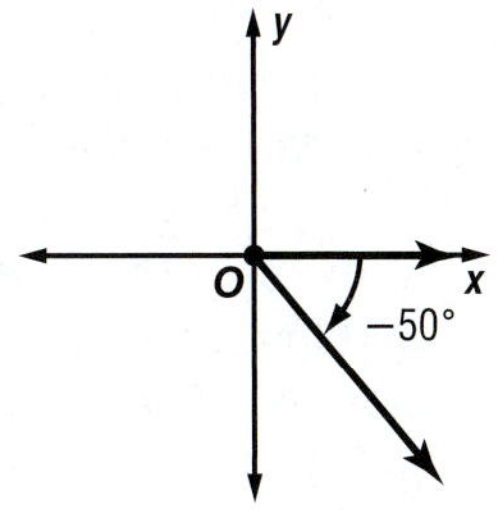

25.

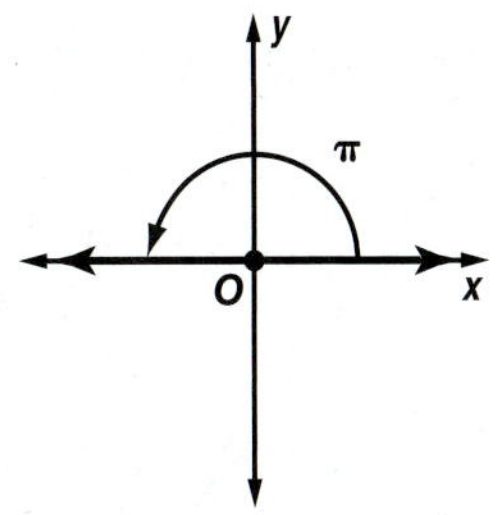

26.

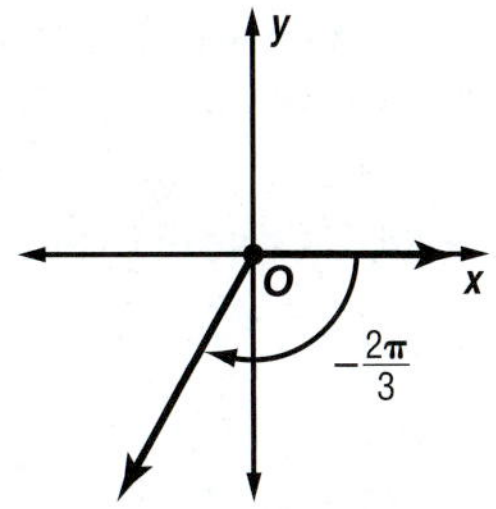

61. Student answers should include the following.

- An angle with a measure of more than 180° gives an indication of motion in a circular path that ended at a point more than halfway around the circle from where it started.
- Negative angles convey the same meaning as positive angles, but in an opposite direction. The standard convention is that negative angles represent rotations in a clockwise direction.
- Rates over 360° per minute indicate that an object is rotating or revolving more than one revolution per minute.

Page 716, Follow-Up of Lesson 13-2
Algebra Activity

7. To find the length of the apothem, you need to write this equation: $\cos \theta = \frac{a}{\text{length of radius}}$. If the radius is 1, then $\cos \theta = a$. If the radius is not 1, then $a =$ length of radius $\cdot \cos \theta$.

Pages 722–723, Lesson 13-3

7.

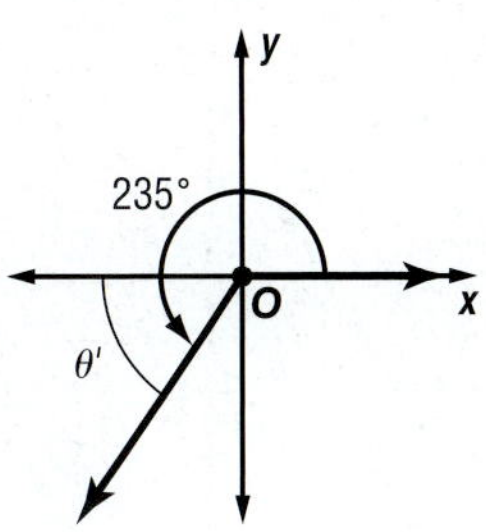

8.

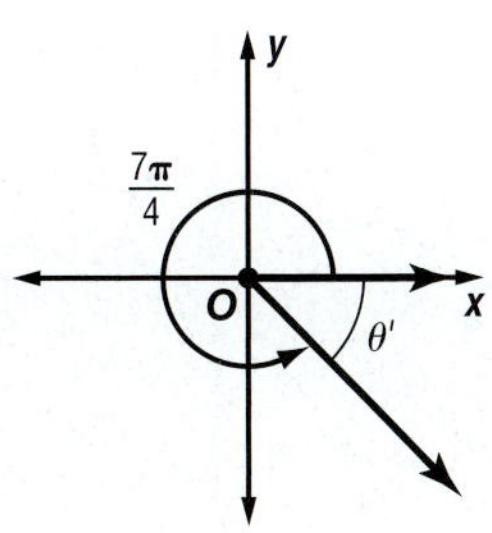

9.

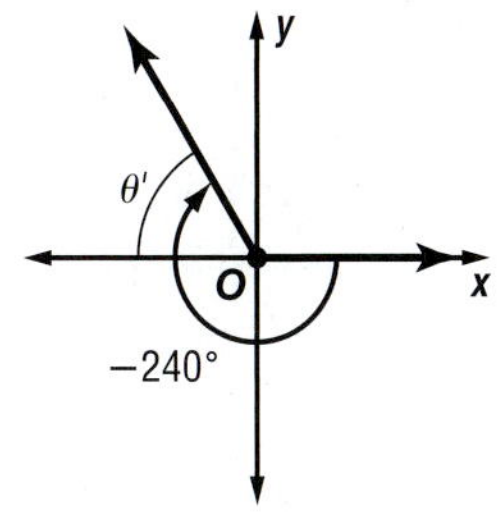

17. $\sin \theta = \frac{24}{25}$, $\cos \theta = \frac{7}{25}$, $\tan \theta = \frac{24}{7}$, $\csc \theta = \frac{25}{24}$, $\sec \theta = \frac{25}{7}$, $\cot \theta = \frac{7}{24}$

18. $\sin \theta = \frac{\sqrt{5}}{5}$, $\cos \theta = \frac{2\sqrt{5}}{5}$, $\tan \theta = \frac{1}{2}$, $\csc \theta = \sqrt{5}$, $\sec \theta = \frac{\sqrt{5}}{2}$, $\cot \theta = 2$

19. $\sin \theta = -\frac{8\sqrt{89}}{89}$, $\cos \theta = \frac{5\sqrt{89}}{89}$, $\tan \theta = -\frac{8}{5}$, $\csc \theta = -\frac{\sqrt{89}}{8}$, $\sec \theta = \frac{\sqrt{89}}{5}$, $\cot \theta = -\frac{5}{8}$

20. $\sin \theta = -\frac{3}{5}$, $\cos \theta = \frac{4}{5}$, $\tan \theta = -\frac{3}{4}$, $\csc \theta = -\frac{5}{3}$, $\sec \theta = \frac{5}{4}$, $\cot \theta = -\frac{4}{3}$

21. $\sin \theta = -1$, $\cos \theta = 0$, $\tan \theta =$ undefined, $\csc \theta = -1$, $\sec \theta =$ undefined, $\cot \theta = 0$

22. $\sin \theta = 0$, $\cos \theta = -1$, $\tan \theta = 0$, $\csc \theta =$ undefined, $\sec \theta = -1$, $\cot \theta =$ undefined

23. $\sin \theta = -\frac{\sqrt{2}}{2}$, $\cos \theta = \frac{\sqrt{2}}{2}$, $\tan \theta = -1$, $\csc \theta = -\sqrt{2}$, $\sec \theta = \sqrt{2}$, $\cot \theta = -1$

24. $\sin \theta = -\frac{\sqrt{6}}{3}$, $\cos \theta = -\frac{\sqrt{3}}{3}$, $\tan \theta = \sqrt{2}$, $\csc \theta = -\frac{\sqrt{6}}{2}$, $\sec \theta = -\sqrt{3}$, $\cot \theta = \frac{\sqrt{2}}{2}$

25.

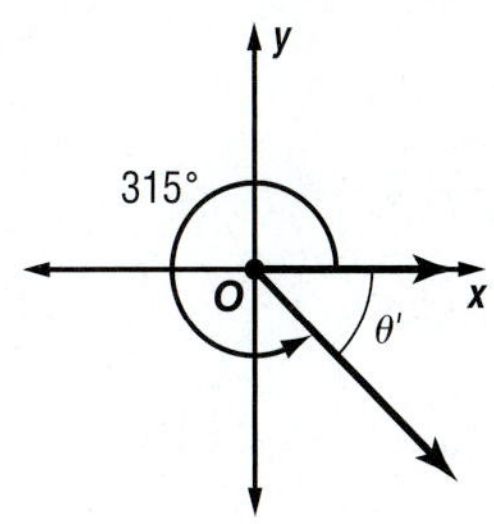

26.

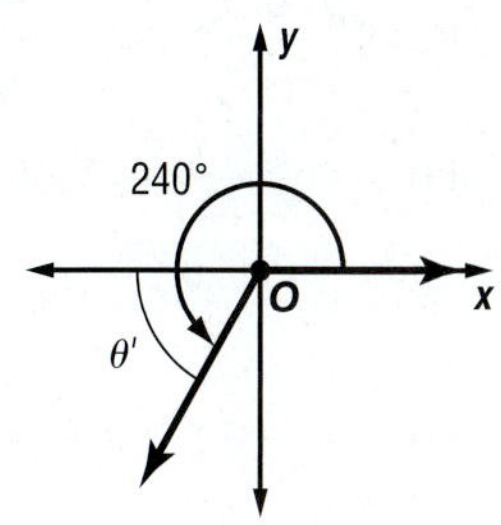

27.

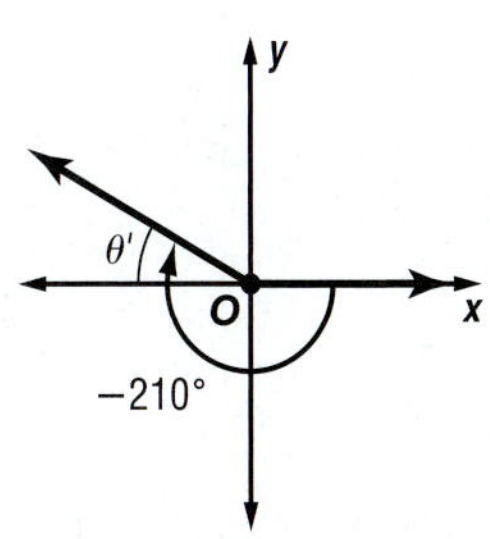

28.

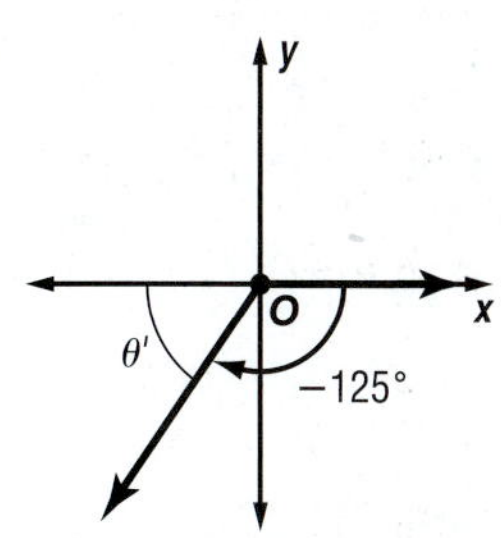

29.

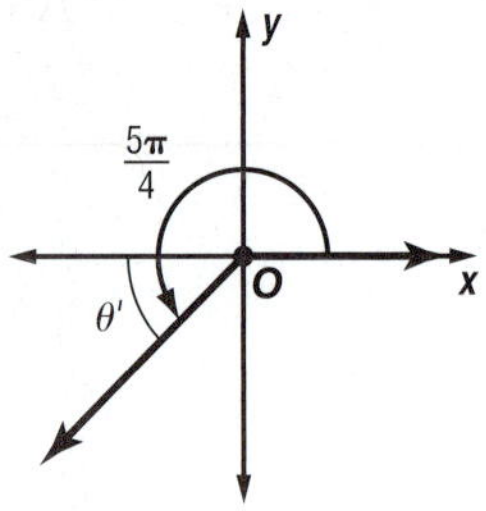

30.

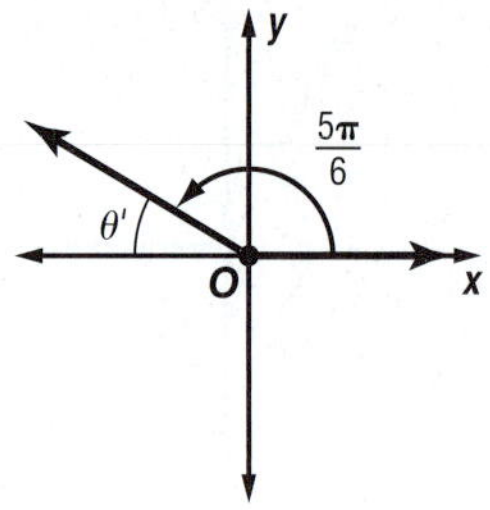

31.

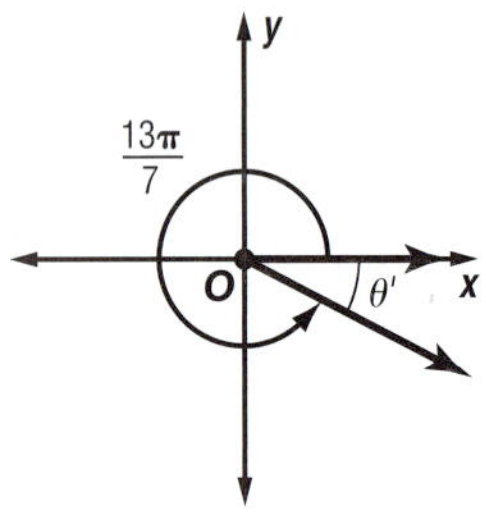

32.

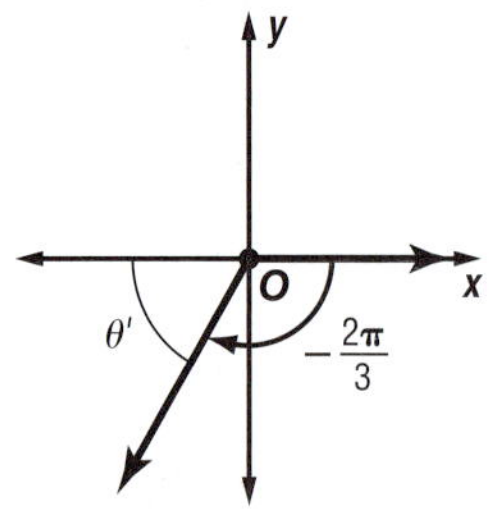

50. $\sin\theta = -\frac{2\sqrt{5}}{5}$, $\cos\theta = -\frac{\sqrt{5}}{5}$, $\tan\theta = 2$,
$\csc\theta = -\frac{\sqrt{5}}{2}$, $\sec\theta = -\sqrt{5}$

51. $\sin\theta = -\frac{3\sqrt{10}}{10}$, $\cos\theta = -\frac{\sqrt{10}}{10}$, $\tan\theta = 3$,
$\csc\theta = -\frac{\sqrt{10}}{3}$, $\cot\theta = \frac{1}{3}$

52. $\sin\theta = -\frac{1}{5}$, $\cos\theta = \frac{2\sqrt{6}}{5}$, $\tan\theta = -\frac{\sqrt{6}}{12}$,
$\sec\theta = \frac{5\sqrt{6}}{12}$, $\cot\theta = -2\sqrt{6}$

Pages 736–737, Lesson 13-5

35. Answers should include the following.

- The Law of Cosines can be used when you know all three sides of a triangle or when you know two sides and the included angle. It can even be used with two sides and the nonincluded angle. This set of conditions leaves a quadratic equation to be solved. It may have one, two, or no solution just like the SSA case with the Law of Sines.
- Given the latitude of a point on the surface of Earth, you can use the radius of the Earth and the orbiting height of a satellite in geosynchronous orbit to create a triangle. This triangle will have two known sides and the measure of the included angle. Find the third side using the Law of Cosines and then use the Law of Sines to determine the angles of the triangle. Subtract 90 degrees from the angle with its vertex on Earth's surface to find the angle at which to aim the receiver dish.

Page 744, Lesson 13-6

34.

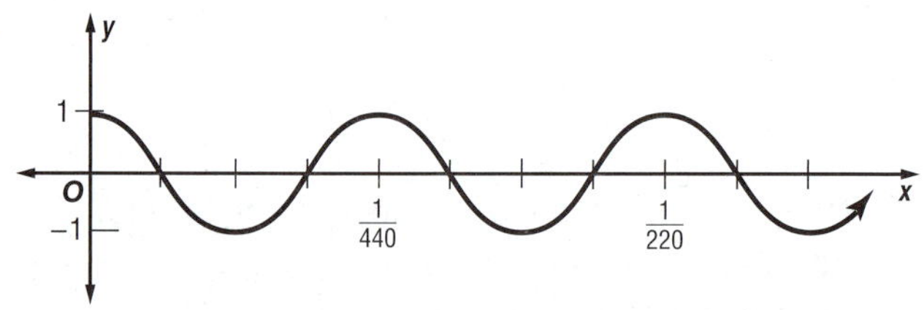

Notes

Chapter 14

Trigonometric Graphs and Identities

Chapter Overview and Pacing

Year-long and two-year pacing: pages T20–T21.

LESSON OBJECTIVES	PACING (days)			
	Regular		Block	
	Basic/ Average	Advanced	Basic/ Average	Advanced
14-1 Graphing Trigonometric Functions *(pp. 762–768)* • Graph trigonometric functions. • Find the amplitude and period of variation of the sine, cosine, and tangent functions.	optional	2	optional	1
14-2 Translations of Trigonometric Graphs *(pp. 769–776)* • Graph horizontal translations of trigonometric graphs and find phase shifts. • Graph vertical translations of trigonometric graphs.	optional	2	optional	1
14-3 Trigonometric Identities *(pp. 777–781)* • Use identities to find trigonometric values. • Use trigonometric identities to simplify expressions.	optional	2	optional	1
14-4 Verifying Trigonometric Identities *(pp. 782–785)* • Verify trigonometric identities by transforming one side of an equation into the form of the other side. • Verify trigonometric identities by transforming each side of the equation into the same form.	optional	2	optional	1
14-5 Sum and Difference of Angles Formulas *(pp. 786–790)* • Find values of sine and cosine involving sum and difference formulas. • Verify identities by using sum and difference formulas.	optional	1	optional	0.5
14-6 Double-Angle and Half-Angle Formulas *(pp. 791–797)* • Find values of sine and cosine involving double-angle formulas. • Find values of sine and cosine involving half-angle formulas.	optional	1	optional	0.5
14-7 Solving Trigonometric Equations *(pp. 798–804)* ***Preview:*** Solving Trigonometric Equations • Solve trigonometric equations. • Use trigonometric equations to solve real-world problems.	optional	2 (with 14-7 Preview)	optional	1 (with 14-7 Preview)
Study Guide and **Practice Test** *(pp. 805–809)* **Standardized Test Practice** *(pp. 810–811)*	optional	1	optional	0.5
Chapter Assessment	optional	1	optional	1
TOTAL	0	14	0	7.5

StudentWorks

An electronic version of this chapter is available on ***StudentWorks™****. This backpack solution CD-ROM allows students instant access to the Student Edition, lesson worksheet pages, and web resources.*

Timesaving Tools

TeacherWorks™

All-In-One Planner and Resource Center

See pages T5 and T21.

Chapter Resource Manager

Chapter 14 Resource Masters

Study Guide and Intervention	Practice (Skills and Average)	Reading to Learn Mathematics	Enrichment	Assessment	Applications*	5-Minute Check Transparencies	Interactive Chalkboard	Alge2PASS: Tutorial Plus (lessons)	Materials
837–838	839–840	841	842			14-1	14-1		graphing calculator, posterboard
843–844	845–846	847	848	893	GCS 53, SC 27	14-2	14-2	27	graphing calculator, grid paper, string, masking tape, rope
849–850	851–852	853	854		GCS 54	14-3	14-3		
855–856	857–858	859	860	893, 895		14-4	14-4	28	
861–862	863–864	865	866			14-5	14-5		
867–868	869–870	871	872	894		14-6	14-6		
873–874	875–876	877	878	894	SC 28, SM 145–148	14-7	14-7		(*Preview:* graphing calculator)
				879–892, 896–898					

**Key to Abbreviations:* GCS = Graphing Calculator and Speadsheet Masters,
SC = School-to-Career Masters,
SM = Science and Mathematics Lab Manual

Chapter 14

Mathematical Connections and Background

Continuity of Instruction

Prior Knowledge

In the previous chapter students investigated the six trigonometric functions and worked with angles measured in degrees or radians, as rays in standard position, and as points on the unit circle. Also, they explored periodicity and inverse trigonometric functions.

This Chapter

Students use the trigonometric functions to explore amplitude and period, and they investigate phase shifts and vertical shifts in the graphs of trigonometric functions. They learn how to verify and use trigonometric identities, including sum and difference formulas, double-angle formulas, and half-angle formulas. Finally they solve trigonometric equations using the ideas of factoring, the zero product property, trigonometric inverses, and periodic behavior.

Future Connections

Students will continue to study amplitude, period, and frequency for trigonometric and other periodic functions. In later math courses they will use trigonometric formulas and identities, and frequently when they study graphs they will analyze vertical and horizontal translations of graphs and how those changes are related to changes in the algebraic description of the graph.

14-1 Graphing Trigonometric Functions

This chapter continues the extensive investigation of trigonometric functions from the previous chapter. This lesson focuses on graphs of the trigonometric functions. Each of the sine, cosine, and tangent functions repeats a pattern of values, or is periodic. For the sine and cosine functions, the period is 2π radians or 360°, while for the tangent function the period is π radians or 180°. For periodic functions the distance between a horizontal center line and the maximum or minimum value is called the amplitude of the graph. For $y = \sin x$ and $y = \cos x$, the horizontal center line is the x-axis and the maximum and minimum values are ± 1, so the amplitude is 1.

The lesson also describes these properties algebraically. For the sine function $y = a \sin b\theta$ and the cosine function $y = a \cos b\theta$, the period is $2\pi \div |b|$ and the amplitude is $|a|$. For $y = a \tan b\theta$, the tangent function, the period is $\pi \div |b|$. The tangent function has no finite maximum or minimum, so amplitude is not defined for the tangent function.

14-2 Translations of Trigonometric Graphs

In this lesson students explore the graphs of $y = \sin \theta$ and $y = \cos \theta$. More specifically, they use the functions $y = a \sin b(\theta - h) + k$ and $y = a \cos b(\theta - h) + k$ and see how changing each of the values a, b, h, and k affects the graph. Also, they explore how to sketch a graph for a given set of values of the four variables.

One change in the graph of a periodic function is to move the horizontal center line of the graph. When $k = 0$ the horizontal center line is the x-axis and the vertical shift is zero. A positive value of k represents a vertical shift upward while a negative value of k represents a downward vertical shift. The amplitude is determined by the value of $|a|$, so the maximum values of the function are $|a|$ units above k and the minimum values of the function are $|a|$ units below k. A third change is the period of the function. The expression for the length of one period has the variable b in the denominator, so as the value of $|b|$ increases the period of the function decreases. The fourth variable, h, is associated with the phase shift of the function. If h is positive, the entire graph is shifted to the right; if h is negative, the entire graph is shifted to the left. Students also use the equation $y = a \tan b(\theta - h) + k$ and explore phase shifts, periods, and vertical shifts for the tangent function.

14-3 Trigonometric Identities

This lesson and the next three deal with trigonometric identities. Students learn the definition of an identity, and they work with arguments that are half of a given angle, twice a given angle, or the sum or difference of two given angles. In this lesson students work with the definitions of the six trigonometric functions in terms of x, y, and r. By dividing each side of $x^2 + y^2 = r^2$ by r^2, y^2, or x^2, the results are three identities called the Pythagorean Identities. For other identities, called the Reciprocal Identities, students note that the definitions for sine and cosecant, for cosine and secant, and for tangent and cotangent are reciprocals. Also, they see that the ratios $\sin\theta \div \cos\theta$ and $\cos\theta \div \sin\theta$ can be simplified to $\tan\theta$ and $\cot\theta$, respectively, resulting in two identities called the Quotient Identities. Students explore how to use identities to simplify trigonometric expressions, and they use identities to evaluate a complicated trigonometric expression for a given argument.

14-4 Verifying Trigonometric Identities

In this lesson students continue exploring how to identify and use trigonometric identities. For each equation, the goal is to transform each side, replacing expressions with equivalent expressions, until the two sides are identical. There are several approaches for writing equivalent expressions. First, students can make substitutions using the Pythagorean Identities. Second, they can use the Distributive Property to factor an expression or to collect like terms. Third, they can transform a term by multiplying the term by an expression equivalent to 1. And fourth, they can rewrite all the trigonometric functions in terms of $\sin\theta$ or $\cos\theta$ by using the Quotient and Reciprocal Identities. Students also relate trigonometric identities to graphs, using a graphing calculator to show that the expressions on each side of a trigonometric identity have the same graph.

14-5 Sum and Difference of Angles Formulas

Students derive and then use formulas for rewriting the two-variable functions $\sin(\alpha \pm \beta)$ and $\cos(\alpha \pm \beta)$ in terms of the one-variable functions $\sin\alpha$, $\sin\beta$, $\cos\alpha$, and $\cos\beta$. The derivation of the difference formula for the cosine function begins with the two ordered pairs on the unit circle that correspond to two angles α and β. The distance d between the two points can be found using the distance formula, or it can be found as the distance between the point (1, 0) and the coordinates of the point on the unit circle associated with angle $(\alpha - \beta)$. After equating the two expressions for d, algebraic manipulation gives an expression for the two-variable function $\cos(\alpha - \beta)$ in terms of one-variable functions. Students use the formulas to find exact values for particular trigonometric expressions. They also use the formulas in problems such as verifying that the equation $\sin(180° + \theta) = -\sin\theta$ is an identity.

14-6 Double-Angle and Half-Angle Formulas

Students begin with the formulas for $\sin(\alpha + \beta)$ and $\cos(\alpha + \beta)$ and replace both α and β with θ. The results, called the Double-Angle Formulas, are equations in which each of $\sin 2\theta$ and $\cos 2\theta$ is expressed in terms of $\sin\theta$ and $\cos\theta$. Then students use an algebraic technique and let α represent 2θ (so $\frac{\alpha}{2}$ represents θ), and derive formulas for $\sin\frac{\alpha}{2}$ and $\cos\frac{\alpha}{2}$ in terms of $\sin\alpha$ and $\cos\alpha$. The two formulas are called the Half-Angle Formulas. Students use the Half-Angle and Double-Angle Formulas, along with other formulas, to find exact values for particular trigonometric expressions. They also substitute the formulas in equations to verify trigonometric identities.

14-7 Solving Trigonometric Equations

In this last lesson of the two-chapter investigation of trigonometric functions, students solve trigonometric equations and review some of the important general ideas of algebra. The first step in solving a trigonometric equation is to use factoring, the zero product property, and identities to rewrite a complicated equation as a string of simpler trigonometric equations. The second step is to use trigonometric inverses to isolate the variable; that is, to solve an equation such as $\cos\theta = 0.5$ for θ. The third step is to use ideas of periodicity to include all the occurrences of that value. Students solve trigonometric equations for arguments measured in degrees or in radians, and they use trigonometric equations and their solutions to solve statements of real-world problems.

Additional mathematical information and teaching notes are available at www.algebra2.com/key_concepts.

Chapter 14

DAILY INTERVENTION and Assessment

	Type	Student Edition	Teacher Resources	Technology/Internet
INTERVENTION	Ongoing	Prerequisite Skills, pp. 761, 768, 776, 781, 785, 790, 797 Practice Quiz 1, p. 781 Practice Quiz 2, p. 797	5-Minute Check Transparencies Quizzes, *CRM* pp. 893–894 Mid-Chapter Test, *CRM* p. 895 Study Guide and Intervention, *CRM* pp. 837–838, 843–844, 849–850, 855–856, 861–862, 867–868, 873–874	Alge2PASS: Tutorial Plus www.algebra2.com/self_check_quiz www.algebra2.com/extra_examples
INTERVENTION	Mixed Review	pp. 768, 776, 781, 785, 790, 797, 804	Cumulative Review, *CRM* p. 896	
INTERVENTION	Error Analysis	Find the Error, p. 766 Common Misconceptions, p. 782	Find the Error, *TWE* p. 766 Tips for New Teachers, *TWE* p. 793	
	Standardized Test Practice	pp. 768, 776, 781, 783, 784, 785, 790, 796, 804, 809, 810–811	*TWE* p. 783 Standardized Test Practice, *CRM* pp. 897–898	Standardized Test Practice CD-ROM www.algebra2.com/standardized_test
ASSESSMENT	Open-Ended Assessment	Writing in Math, pp. 768, 776, 781, 785, 790, 796, 804 Open Ended, pp. 766, 774, 779, 784, 788, 794, 802	Modeling: *TWE* pp. 768, 790 Speaking: *TWE* pp. 781, 784, 803 Writing: *TWE* pp. 776, 797 Open-Ended Assessment, *CRM* p. 891	
ASSESSMENT	Chapter Assessment	Study Guide, pp. 805–808 Practice Test, p. 809	Multiple-Choice Tests (Forms 1, 2A, 2B), *CRM* pp. 879–884 Free-Response Tests (Forms 2C, 2D, 3), *CRM* pp. 885–890 Vocabulary Test/Review, *CRM* p. 892	ExamView® Pro (see below) MindJogger Videoquizzes www.algebra2.com/vocabulary_review www.algebra2.com/chapter_test

Key to Abbreviations: TWE = Teacher Wraparound Edition; CRM = Chapter Resource Masters

Intervention Technology

Alge2PASS: Tutorial Plus CD-ROM offers a complete, self-paced algebra curriculum.

Algebra 2 Lesson	Alge2PASS Lesson
14-2	**27** *Graphing Trigonometric Functions*
14-4	**28** *Trigonometric Identities*

ExamView® Pro

Use the networkable **ExamView® Pro** to:

- Create **multiple versions** of tests.
- Create **modified** tests for *Inclusion* students.
- **Edit** existing questions and **add** your own questions.
- Use built-in **state curriculum correlations** to create tests aligned with state standards.
- **Apply** art to your tests from a program bank of artwork.

For more information on Intervention and Assessment, see pp. T8–T11.

Reading and Writing in Mathematics

Glencoe Algebra 2 provides numerous opportunities to incorporate reading and writing into the mathematics classroom.

Student Edition

- Foldables Study Organizer, p. 761
- Concept Check questions require students to verbalize and write about what they have learned in the lesson. (pp. 766, 774, 779, 784, 788, 794, 802, 805)
- Writing in Math questions in every lesson, pp. 768, 776, 781, 785, 790, 796, 804
- Reading Study Tip, pp. 786, 788
- WebQuest, pp. 775, 804

Teacher Wraparound Edition

- Foldables Study Organizer, pp. 761, 805
- Study Notebook suggestions, pp. 766, 774, 779, 783, 788, 794, 802
- Modeling activities, pp. 768, 790
- Speaking activities, pp. 781, 784, 803
- Writing activities, pp. 776, 797
- ELL Resources, pp. 760, 767, 775, 780, 785, 789, 796, 803, 805

Additional Resources

- Vocabulary Builder worksheets require students to define and give examples for key vocabulary terms as they progress through the chapter. (*Chapter 14 Resource Masters,* pp. vii-viii)
- Reading to Learn Mathematics master for each lesson (*Chapter 14 Resource Masters*, pp. 841, 847, 853, 859, 865, 871, 877)
- *Vocabulary PuzzleMaker* software creates crossword, jumble, and word search puzzles using vocabulary lists that you can customize.
- *Teaching Mathematics with Foldables* provides suggestions for promoting cognition and language.
- *Reading and Writing in the Mathematics Classroom*
- *WebQuest and Project Resources*

For more information on Reading and Writing in Mathematics, see pp. T6–T7.

ELL ENGLISH LANGUAGE LEARNERS

Lesson 14-1

Flexible Groups

Have groups of students brainstorm a list of characteristics of the graphs of $y = \sin x$ and $y = \cos x$. Suggest that their list include characteristics such as domain, range, period, x- and y-intercepts, and maximum and minimum values. Then compare and contrast the characteristics of the graphs of the two functions.

Lesson 14-3

Language Experience Approach to Illustrations

Discuss the term *identity* with students. What are the mathematical and real-life meanings of the word? How are these meanings related? How can knowing the real-life meanings help you in learning about mathematical identities?

Chapter 14 Notes

What You'll Learn

Have students read over the list of objectives and make a list of any words with which they are not familiar.

Why It's Important

Point out to students that this is only one of many reasons why each objective is important. Others are provided in the introduction to each lesson.

Chapter 14 Trigonometric Graphs and Identities

What You'll Learn

- **Lessons 14-1 and 14-2** Graph trigonometric functions and determine period, amplitude, phase shifts, and vertical shifts.
- **Lessons 14-3 and 14-4** Use and verify trigonometric identities.
- **Lessons 14-5 and 14-6** Use sum and difference formulas and double- and half-angle formulas.
- **Lesson 14-7** Solve trigonometric equations.

Key Vocabulary

- amplitude (p. 763)
- phase shift (p. 769)
- vertical shift (p. 771)
- trigonometric identity (p. 777)
- trigonometric equation (p. 799)

Why It's Important

Some equations contain one or more trigonometric functions. It is important to know how to simplify trigonometric expressions to solve these equations. Trigonometric functions can be used to model many real-world applications, such as music. *You will learn how a trigonometric function can be used to describe music in Lesson 14-6.*

Lesson	NCTM Standards	Local Objectives
14-1	1, 2, 6, 7, 8, 9, 10	
14-2	1, 2, 3, 4, 6, 8, 9, 10	
14-3	1, 2, 6, 8, 9, 10	
14-4	2, 8	
14-5	1, 2, 6, 7, 8, 9, 10	
14-6	1, 2, 7, 8, 9, 10	
14-7 Preview	2, 10	
14-7	1, 2, 6, 7, 8, 9, 10	

Key to NCTM Standards:

1=Number & Operations, 2=Algebra, 3=Geometry, 4=Measurement, 5=Data Analysis & Probability, 6=Problem Solving, 7=Reasoning & Proof, 8=Communication, 9=Connections, 10=Representation

Vocabulary Builder ELL

The Key Vocabulary list introduces students to some of the main vocabulary terms included in this chapter. For a more thorough vocabulary list with pronunciations of new words, give students the Vocabulary Builder worksheets found on pages vii and viii of the *Chapter 14 Resource Masters*. Encourage them to complete the definition of each term as they progress through the chapter. You may suggest that they add these sheets to their study notebooks for future reference when studying for the Chapter 14 test.

760 Chapter 14 Trigonometric Graphs and Identities

Getting Started

Prerequisite Skills To be successful in this chapter, you'll need to master these skills and be able to apply them in problem-solving situations. Review these skills before beginning Chapter 14.

For Lessons 14-1 and 14-2 — Trigonometric Values

Find the exact value of each trigonometric function. *(For review, see Lesson 13-3.)*

1. $\sin 135°$ $\frac{\sqrt{2}}{2}$ **2.** $\tan 315°$ -1 **3.** $\cos 90°$ 0 **4.** $\tan 45°$ 1

5. $\sin \frac{5\pi}{4}$ $-\frac{\sqrt{2}}{2}$ **6.** $\cos \frac{7\pi}{6}$ $-\frac{\sqrt{3}}{2}$ **7.** $\sin \frac{11\pi}{6}$ $-\frac{1}{2}$ **8.** $\tan \frac{3\pi}{2}$ not defined

For Lessons 14-3, 14-5, and 14-6 — Circular Functions

Find the exact value of each trigonometric function. *(For review, see Lesson 13-6.)*

9. $\cos(-150°)$ $-\frac{\sqrt{3}}{2}$ **10.** $\sin 510°$ $\frac{1}{2}$ **11.** $\cot \frac{9\pi}{4}$ 1 **12.** $\sec \frac{13\pi}{6}$ $\frac{2\sqrt{3}}{3}$

13. $\tan\left(-\frac{3\pi}{2}\right)$ not defined **14.** $\csc(-720°)$ not defined **15.** $\cos \frac{7\pi}{3}$ $\frac{1}{2}$ **16.** $\tan \frac{8\pi}{3}$ $-\sqrt{3}$

For Lesson 14-4 — Factor Polynomials

Factor completely. If the polynomial is not factorable, write *prime*. *(For review, see Lesson 5-4.)*

17. $-15x^2 - 5x$ $-5x(3x + 1)$ **18.** $2x^4 - 4x^2$ $2x^2(x^2 - 2)$ **19.** $x^3 + 4$ prime

20. $x^2 - 6x + 8$ $(x - 4)(x - 2)$ **21.** $2x^2 - 3x - 2$ $(2x + 1)(x - 2)$ **22.** $3x^3 - 2x^2 - x$ $x(3x + 1)(x - 1)$

For Lesson 14-7 — Solve Quadratic Equations

Solve each equation by factoring. *(For review, see Lesson 6-3.)*

23. $x^2 - 5x - 24 = 0$ $8, -3$ **24.** $x^2 - 2x - 48 = 0$ $8, -6$ **25.** $x^2 + 3x - 40 = 0$ $-8, 5$

26. $x^2 - 12x = 0$ $0, 12$ **27.** $-2x^2 - 11x - 12 = 0$ $-4, -\frac{3}{2}$ **28.** $x^2 - 16 = 0$ $-4, 4$

FOLDABLES™ Study Organizer

Trigonometric Graphs and Identities Make this Foldable to help you organize your notes. Begin with eight sheets of grid paper.

Step 1 Staple

Staple the stack of grid paper along the top to form a booklet.

Step 2 Cut and Label

Cut seven lines from the bottom of the top sheet, six lines from the second sheet, and so on. Label with lesson numbers as shown.

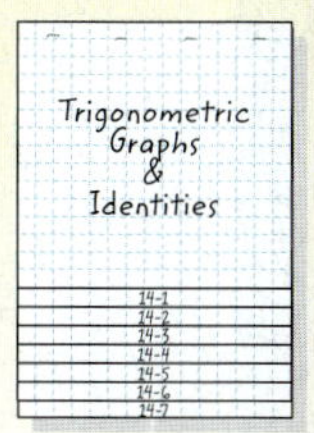

Reading and Writing As you read and study the chapter, use each page to write notes and to graph examples for each lesson.

Getting Started

This section provides a review of the basic concepts needed before beginning Chapter 14. Page references are included for additional student help.

Prerequisite Skills in the Getting Ready for the Next Lesson section at the end of each exercise set review a skill needed in the next lesson.

For Lesson	Prerequisite Skill
14-2	Graphs of Quadratic Functions (p. 768)
14-3	Reference Angles (p. 776)
14-4	Properties of Equality (p. 781)
14-5	Simplifying Radical Expressions (p. 785)
14-6	Solving Equations Using the Square Root Property (p. 790)
14-7	Solving Equations Using the Zero Product Property (p. 797)

FOLDABLES™ Study Organizer

For more information about Foldables, see *Teaching Mathematics with Foldables.*

Writing Instructions and Sequencing Data After students make their Foldable, have them label each tab to correspond to a lesson in this chapter. Students use their Foldable to take notes, define terms, record concepts, and write examples. After each lesson, ask students to write a set of instructions on how to do something presented in the lesson. For example, a student might write instructions for graphing trigonometric functions. Have students follow their own instructions to check them for accuracy. Use their notes and textbook to make needed revisions.

14-1 Lesson Notes

1 Focus

5-Minute Check Transparency 14-1 Use as a quiz or review of Chapter 13.

Mathematical Background notes are available for this lesson on p. 760C.

Building on Prior Knowledge

In Chapter 13, students learned the sine, cosine, and tangent of 30°, 45°, and 60° angles. In this lesson, students will learn the sine, cosine, and tangent of angles with other measures and use all of these values to create graphs of the six trigonometric functions.

Why can you predict the behavior of tides?

Ask students:

- Why would you need to know the times for the high and low tides? **Sample answer: roads might be flooded at high tide**
- How is the period of a tide defined? **Sample answer: the length of time between two high tides (or two low tides)**
- What is a tidal range? **Sample answer: the difference in water level between low tide and high tide**

14-1 Graphing Trigonometric Functions

What You'll Learn

- Graph trigonometric functions.
- Find the amplitude and period of variation of the sine, cosine, and tangent functions.

Vocabulary

- amplitude

Why can you predict the behavior of tides?

The rise and fall of tides can have great impact on the communities and ecosystems that depend upon them. One type of tide is a semidiurnal tide. This means that bodies of water, like the Atlantic Ocean, have two high tides and two low tides a day. Because tides are periodic, they behave the same way each day.

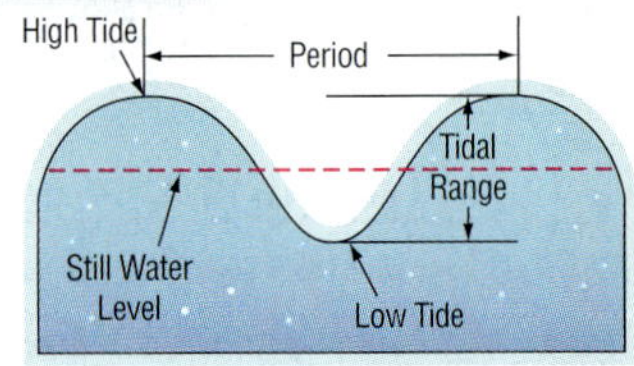

GRAPH TRIGONOMETRIC FUNCTIONS The diagram below illustrates the water level as a function of time for a body of water with semidiurnal tides.

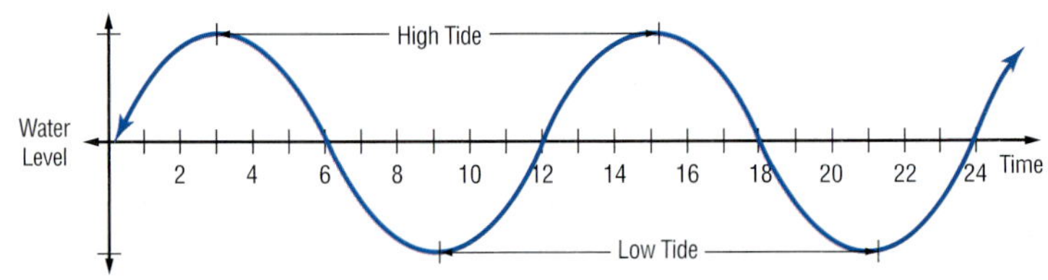

Study Tip

Look Back
To review **period** and **periodic functions**, see Lesson 13-6.

In each cycle of high and low tides, the pattern repeats itself. Recall that a function whose graph repeats a basic pattern is called a *periodic function*.

To find the period, start from any point on the graph and proceed to the right until the pattern begins to repeat. The simplest approach is to begin at the origin. Notice that after about 12 hours the graph begins to repeat. Thus, the period of the function is about 12 hours.

To graph the periodic functions $y = \sin\theta$, $y = \cos\theta$, or $y = \tan\theta$, use values of θ expressed either in degrees or radians. Ordered pairs for points on these graphs are of the form $(\theta, \sin\theta)$, $(\theta, \cos\theta)$, and $(\theta, \tan\theta)$, respectively.

θ	0°	30°	45°	60°	90°	120°	135°	150°	180°	210°	225°	240°	270°	300°	315°	330°	360°
$\sin\theta$	0	$\frac{1}{2}$	$\frac{\sqrt{2}}{2}$	$\frac{\sqrt{3}}{2}$	1	$\frac{\sqrt{3}}{2}$	$\frac{\sqrt{2}}{2}$	$\frac{1}{2}$	0	$-\frac{1}{2}$	$-\frac{\sqrt{2}}{2}$	$-\frac{\sqrt{3}}{2}$	−1	$-\frac{\sqrt{3}}{2}$	$-\frac{\sqrt{2}}{2}$	$-\frac{1}{2}$	0
nearest tenth	0	0.5	0.7	0.9	1	0.9	0.7	0.5	0	−0.5	−0.7	−0.9	−1	−0.9	−0.7	−0.5	0
$\cos\theta$	1	$\frac{\sqrt{3}}{2}$	$\frac{\sqrt{2}}{2}$	$\frac{1}{2}$	0	$-\frac{1}{2}$	$-\frac{\sqrt{2}}{2}$	$-\frac{\sqrt{3}}{2}$	−1	$-\frac{\sqrt{3}}{2}$	$-\frac{\sqrt{2}}{2}$	$-\frac{1}{2}$	0	$\frac{1}{2}$	$\frac{\sqrt{2}}{2}$	$\frac{\sqrt{3}}{2}$	1
nearest tenth	1	0.9	0.7	0.5	0	−0.5	−0.7	−0.9	−1	−0.9	−0.7	−0.5	0	0.5	0.7	0.9	1
$\tan\theta$	0	$\frac{\sqrt{3}}{3}$	1	$\sqrt{3}$	nd	$-\sqrt{3}$	−1	$-\frac{\sqrt{3}}{3}$	0	$\frac{\sqrt{3}}{3}$	1	$\sqrt{3}$	nd	$-\sqrt{3}$	−1	$-\frac{\sqrt{3}}{3}$	0
nearest tenth	0	0.6	1	1.7	nd	−1.7	−1	−0.6	0	0.6	1	1.7	nd	−1.7	−1	−0.6	0
θ	0	$\frac{\pi}{6}$	$\frac{\pi}{4}$	$\frac{\pi}{3}$	$\frac{\pi}{2}$	$\frac{2\pi}{3}$	$\frac{3\pi}{4}$	$\frac{5\pi}{6}$	π	$\frac{7\pi}{6}$	$\frac{5\pi}{4}$	$\frac{4\pi}{3}$	$\frac{3\pi}{2}$	$\frac{5\pi}{3}$	$\frac{7\pi}{4}$	$\frac{11\pi}{6}$	2π

nd = not defined

Resource Manager

Workbook and Reproducible Masters

Chapter 14 Resource Masters

- Study Guide and Intervention, pp. 837–838
- Skills Practice, p. 839
- Practice, p. 840
- Reading to Learn Mathematics, p. 841
- Enrichment, p. 842

Transparencies

5-Minute Check Transparency 14-1
Real-World Transparency 14
Answer Key Transparencies

Technology

Interactive Chalkboard
Multimedia Applications

After plotting several points, complete the graphs of $y = \sin\theta$ and $y = \cos\theta$ by connecting the points with a smooth, continuous curve. Recall from Chapter 13 that each of these functions has a period of 360° or 2π radians. That is, the graph of each function repeats itself every 360° or 2π radians.

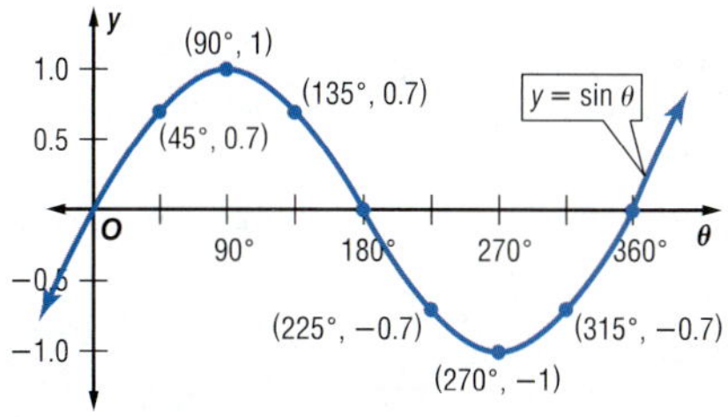

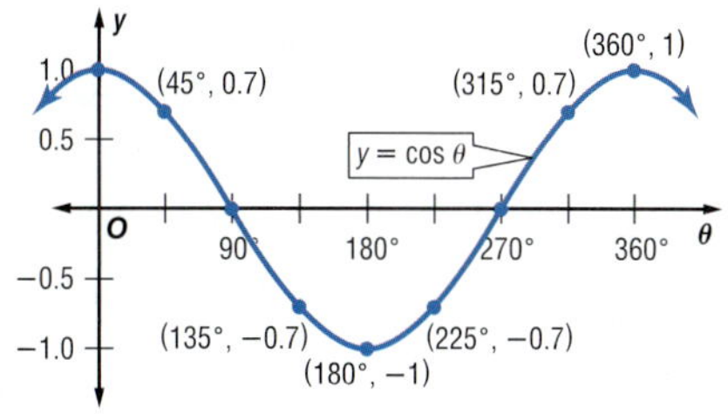

Notice that both the sine and cosine have a maximum value of 1 and a minimum value of −1. The **amplitude** of the graph of a periodic function is the absolute value of half the difference between its maximum value and its minimum value. So, for both the sine and cosine functions, the amplitude of their graphs is $\left|\frac{1-(-1)}{2}\right|$ or 1.

The graph of the tangent function can also be drawn by plotting points. By examining the values for $\tan\theta$ in the table, you can see that the tangent function is not defined for 90°, 270°, …, 90° + $k \cdot 180°$, where k is an integer. The graph is separated by vertical asymptotes whose x-intercepts are the values for which $y = \tan\theta$ is not defined.

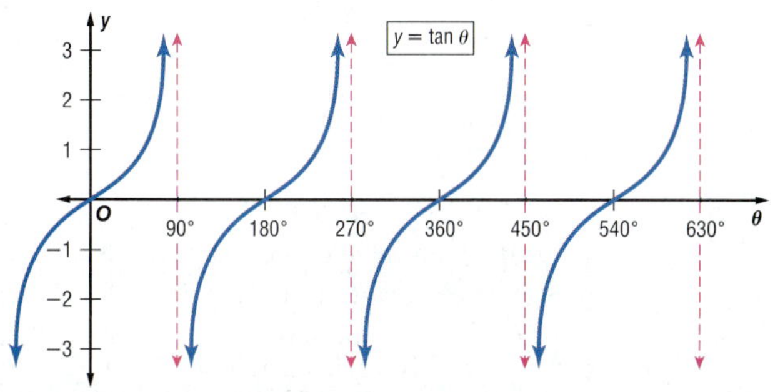

The period of the tangent function is 180° or π radians. Since the tangent function has no maximum or minimum value, it has no amplitude.

The graphs of the secant, cosecant, and cotangent functions are shown below. Compare them to the graphs of the cosine, sine, and tangent functions, which are shown in red.

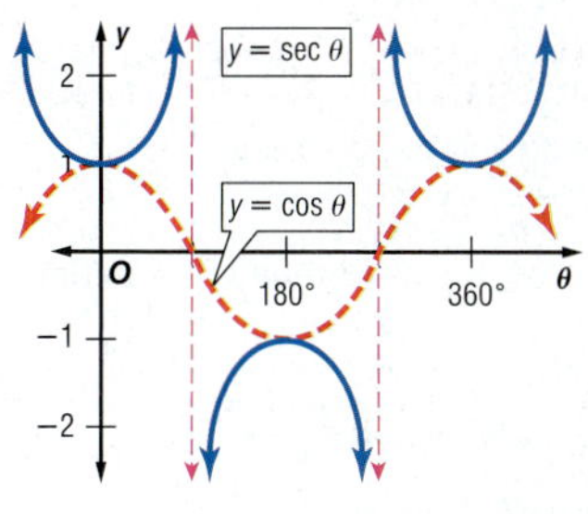

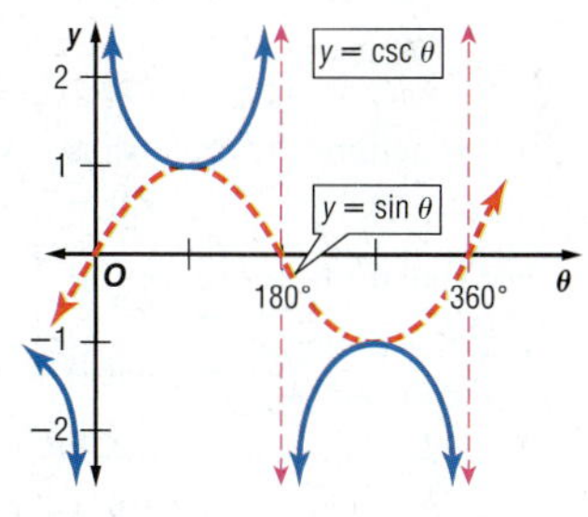

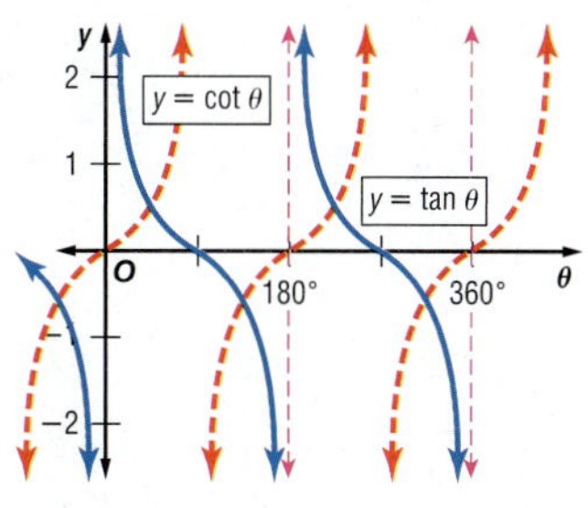

Notice that the period of the secant and cosecant functions is 360° or 2π radians. The period of the cotangent is 180° or π radians. Since none of these functions have a maximum or minimum value, they have no amplitude.

2 Teach

GRAPH TRIGONOMETRIC FUNCTIONS

Teaching Tip Have students brainstorm to make a list of real-world phenomena that fluctuate in a regular periodic pattern. (Examples: temperatures, the number of people in a mall over the course of a week)

Teaching Tip You might wish to have students draw their own graphs of the sine, cosine, and tangent functions using the data from the table on p. 762. Their graphs can be compared to those shown on p. 763. This immediate feedback can be beneficial in helping students acquire the skills for graphing trigonometric functions, which are more complicated than most of the graphing students have done to this point.

Teaching Tip Review how the sine and cosine functions are related in a right triangle.

DAILY INTERVENTION — **Differentiated Instruction**

Visual/Spatial Have groups of students make posters showing sketches of the graphs of the six trigonometric functions. Encourage students to color-code the key features of all the graphs, such as period, amplitude, asymptotes, and so on.

Lesson 14-1 Graphing Trigonometric Functions 763

VARIATIONS OF TRIGONOMETRIC FUNCTIONS

Concept Check

After discussing the Key Concept box about amplitudes and periods, ask: What is an equation involving the sine function with a period of 90° and an amplitude of $\frac{1}{2}$? **One of the following: $y = \frac{1}{2}\sin 4\theta$, $y = \frac{1}{2}\sin(-4\theta)$, $y = -\frac{1}{2}\sin 4\theta$, or $y = -\frac{1}{2}\sin(-4\theta)$**

VARIATIONS OF TRIGONOMETRIC FUNCTIONS Just as with other functions, a trigonometric function can be used to form a family of graphs by changing the period and amplitude.

Graphing Calculator Investigation

Period and Amplitude

On a TI-83 Plus graphing calculator, set the MODE to degrees.

Think and Discuss

1. Graph $y = \sin x$ and $y = \sin 2x$. What is the maximum value of each function? **1**
2. How many times does each function reach a maximum value? **2, 4**
3. Graph $y = \sin\left(\frac{x}{2}\right)$. What is the maximum value of this function? How many times does this function reach its maximum value? **1; 1**

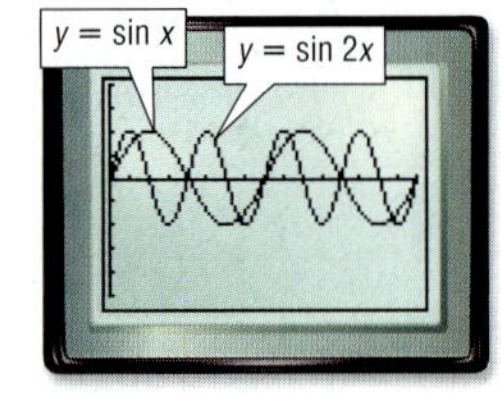

[0, 720] scl: 45 by [−2.5, 2.5] scl: 0.5

4. Use the equations $y = \sin bx$ and $y = \cos bx$. Repeat Exercises 1–3 for maximum values and the other values of b. What conjecture can you make about the effect of b on the maximum values and the periods of these functions? **The greater the value of b, the smaller the period. b has no effect on the maximum value.**
5. Graph $y = \sin x$ and $y = 2 \sin x$. What is the maximum value of each function? What is the period of each function? **1, 2; 360°**
6. Graph $y = \frac{1}{2}\sin x$. What is the maximum value of this function? What is the period of this function? **$\frac{1}{2}$; 360°**

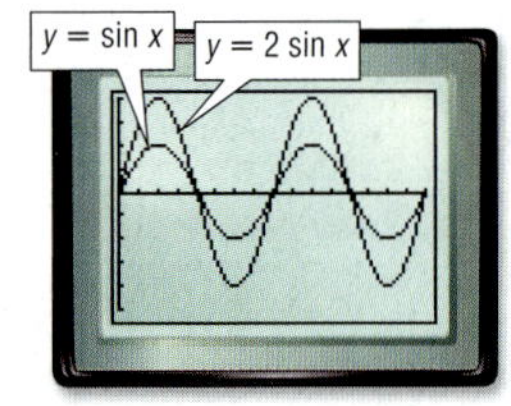

[0, 720] scl: 45 by [−2.5, 2.5] scl: 0.5

7. Use the equations $y = a \sin x$ and $y = a \cos x$. Repeat Exercises 5 and 6 for other values of a. What conjecture can you make about the effect of a on the amplitudes and periods of $y = a \sin x$ and $y = a \cos x$?

Study Tip

Amplitude and Period

Note that the amplitude affects the graph along the vertical axis and the period affects it along the horizontal axis.

7. When a is positive, the amplitude is a. When a is negative, then amplitude is $|a|$. a has no effect on the period.

The results of the investigation suggest the following generalization.

Key Concept — *Amplitudes and Periods*

- **Words** For functions of the form $y = a \sin b\theta$ and $y = a \cos b\theta$, the amplitude is $|a|$, and the period is $\frac{360°}{|b|}$ or $\frac{2\pi}{|b|}$.
  For functions of the form $y = a \tan b\theta$, the amplitude is not defined, and the period is $\frac{180°}{|b|}$ or $\frac{\pi}{|b|}$.
- **Examples**

$y = 3 \sin 4\theta$	amplitude 3 and period $\frac{360°}{4}$ or 90°
$y = -6 \cos 5\theta$	amplitude $|-6|$ or 6 and period $\frac{2\pi}{5}$
$y = 2 \tan \frac{1}{3}\theta$	no amplitude and period 3π

Graphing Calculator Investigation

Graphing Trigonometric Functions To set the calculator for degrees, press MODE and move the cursor to highlight **DEGREE** and press ENTER. Also, be sure to have students clear the **Y=** lists before beginning Exercise 1.

You can use the amplitude and period of a trigonometric function to help you graph the function.

Example 1 Graph Trigonometric Functions

Find the amplitude and period of each function. Then graph the function.

a. $y = \cos 3\theta$

First, find the amplitude.

$|a| = |1|$ The coefficient of $\cos 3\theta$ is 1.

Next, find the period.

$\frac{360°}{|b|} = \frac{360°}{|3|}$ $b = 3$

$= 120°$

Use the amplitude and period to graph the function.

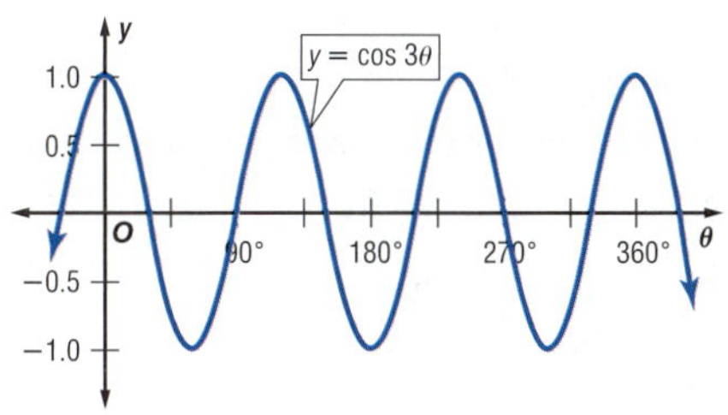

b. $y = \frac{1}{4}\sin\theta$

Amplitude: $|a| = \left|\frac{1}{4}\right|$

$= \frac{1}{4}$

Period: $\frac{360°}{|b|} = \frac{360°}{|1|}$

$= 360°$

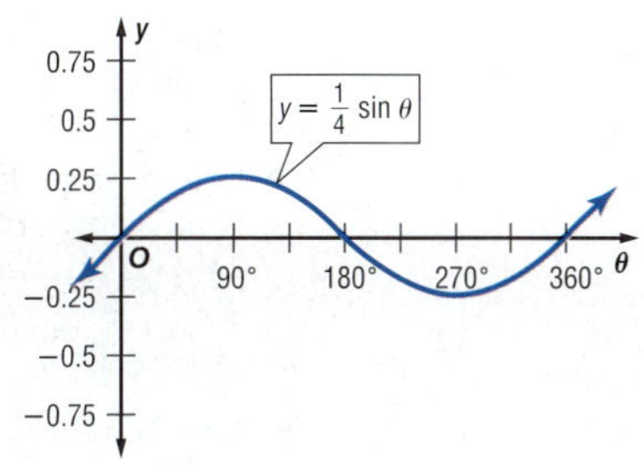

c. $y = \frac{1}{2}\sin\left(-\frac{1}{3}\theta\right)$

Amplitude: $|a| = \left|\frac{1}{2}\right|$

$= \frac{1}{2}$

Period: $\frac{2\pi}{|b|} = \frac{2\pi}{\left|-\frac{1}{3}\right|}$

$= 6\pi$

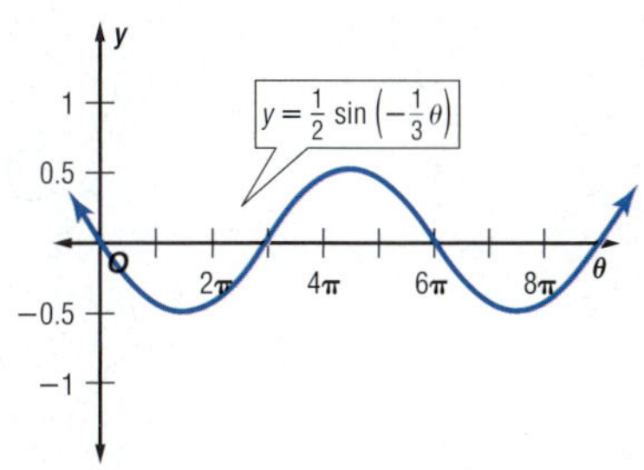

www.algebra2.com/extra_examples

In-Class Example

Power Point®

1 Find the amplitude and period of each function. Then graph the function.

a. $y = \sin\frac{1}{3}\theta$

ampl: 1; period: 1080°

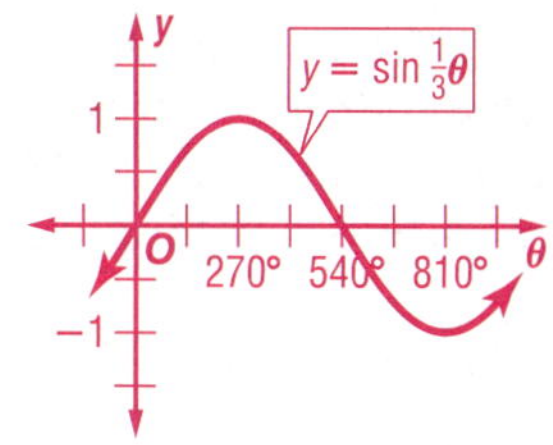

b. $y = \frac{1}{3}\cos\theta$

ampl: $\frac{1}{3}$; period: 360°

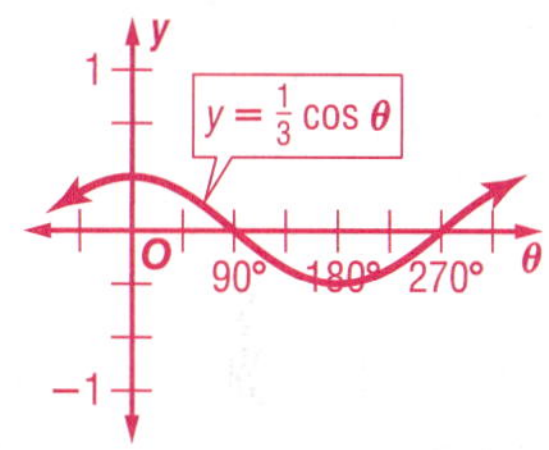

c. $y = 2\sin\frac{1}{4}\theta$

ampl: 2; period: 1440°

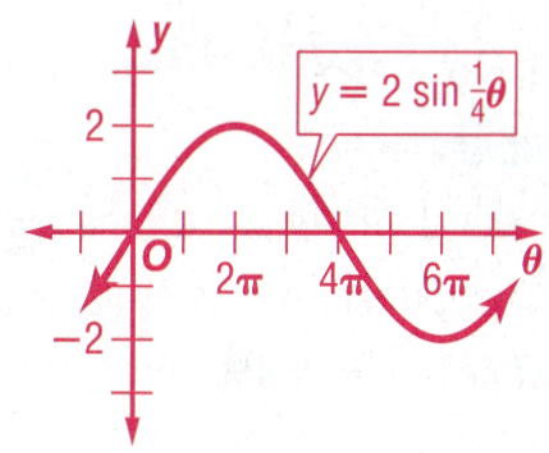

Teacher to Teacher

Berchie Holliday — Author, Cincinnati, OH

"I have my students construct a unit circle on a coordinate plane with a toothpick length radius. They mark every 15°. Students form right triangles inside the circle and break toothpicks to match the lengths of each vertical leg. (See Figure 1 at the right.) They transfer each leg to its appropriate degree mark on a second x-axis and place a dot at the top of each toothpick. (See Figure 2.) Finally, students connect the dots with a smooth curve."

Figure 1

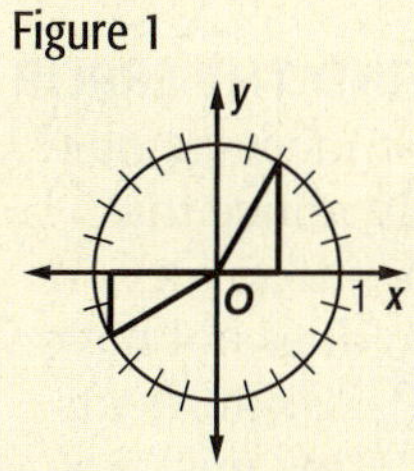

Figure 2

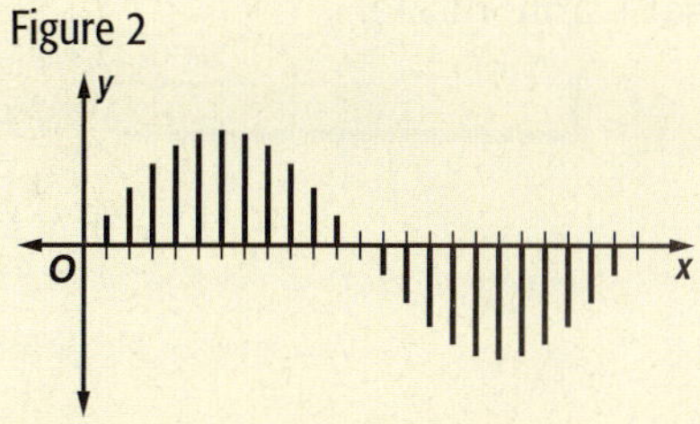

In-Class Example

Power Point®

2 OCEANOGRAPHY Refer to the application at the beginning of the lesson. The tidal range in the Bay of Fundy in Canada measures 50 feet.

a. Write a function to represent the height h of the tide. Assume that the tide is at equilibrium at $t = 0$ and that the high tide is beginning. $y = 25 \sin \frac{\pi}{6}t$

b. Graph the tide function.

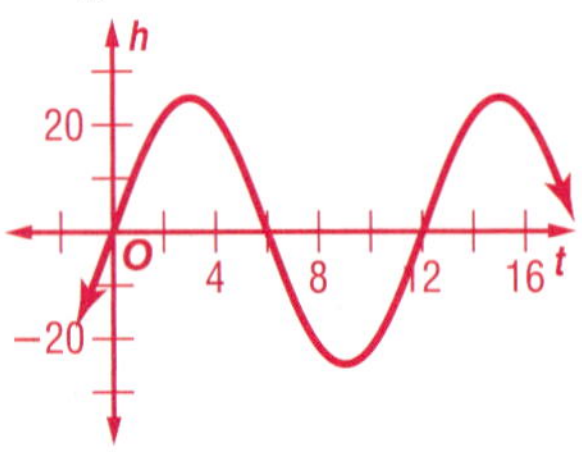

You can use trigonometric functions to describe real-world situations.

Example 2 Use Trigonometric Functions

OCEANOGRAPHY **Refer to the application at the beginning of the lesson. Suppose the tidal range of a city on the Atlantic coast is 18 feet. A tide is at *equilibrium* when it is at its normal level, halfway between its highest and lowest points.**

a. Write a function to represent the height h of the tide. Assume that the tide is at equilibrium at $t = 0$ and that the high tide is beginning.

Since the height of the tide is 0 at $t = 0$, use the sine function $h = a \sin bt$, where a is the amplitude of the tide and t is the time in hours.

Find the amplitude. The difference between high tide and low tide is the tidal range or 18 feet.

$a = \frac{18}{2}$ or 9

Find the value of b. Each tide cycle lasts about 12 hours.

$\frac{2\pi}{|b|} = 12$ period $= \frac{2\pi}{|b|}$

$b = \frac{2\pi}{12}$ or $\frac{\pi}{6}$ Solve for b.

Thus, an equation to represent the height of the tide is $h = 9 \sin \frac{\pi}{6}t$.

b. Graph the tide function.

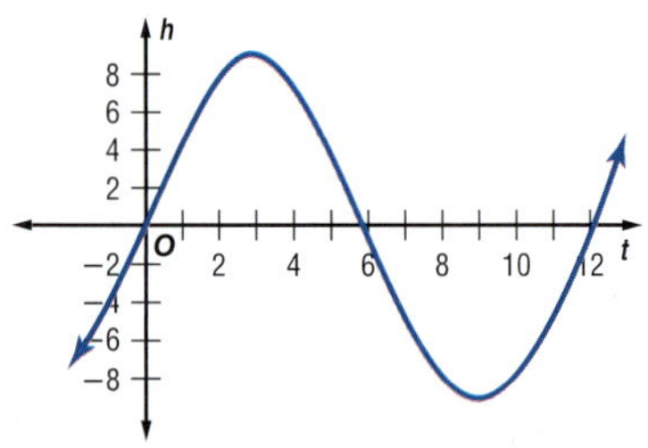

More About. . .

Oceanography

Lake Superior has one of the smallest tidal ranges. It can be measured in inches, while the tidal range in the Bay of Funday in Canada measures up to 50 feet.

Source: Office of Naval Research

3 Practice/Apply

Study Notebook

Have students—

- add the definitions/examples of the vocabulary terms to their Vocabulary Builder worksheets for Chapter 14.
- include any other item(s) that they find helpful in mastering the skills in this lesson.

DAILY INTERVENTION **FIND THE ERROR** Students should quickly notice that Dante must be incorrect because his graph does not have an amplitude of 3. Ask students if they can identify the mistake that Dante made.

Check for Understanding

Concept Check

1. **OPEN ENDED** Explain why $y = \tan \theta$ has no amplitude.
2. **Explain** what it means to say that the period of a function is 180°.
3. **FIND THE ERROR** Dante and Jamile graphed $y = 3 \cos \frac{2}{3}\theta$.

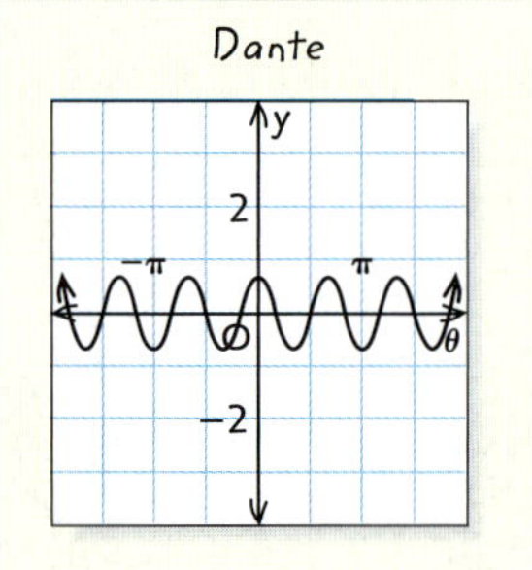

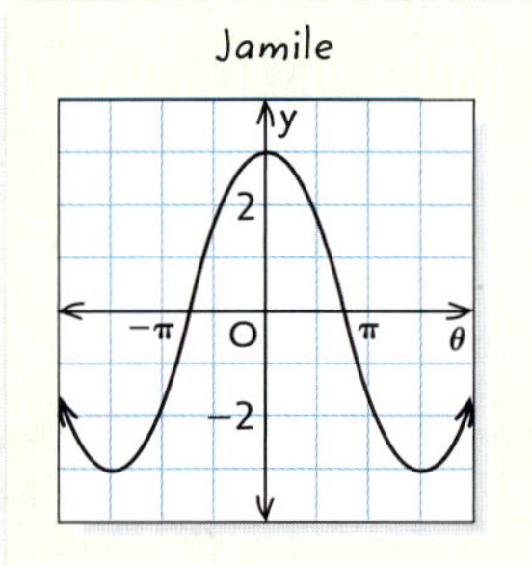

Who is correct? Explain your reasoning. **Jamile; the amplitude is 3, and the period is 3π.**

1. Sample answer: Amplitude is half the difference beween the maximum and minimum values of a graph; $y = \tan \theta$ has no maximum or minimum value.

2. Sample answer: The graph repeats itself every 180°.

Answers

33.

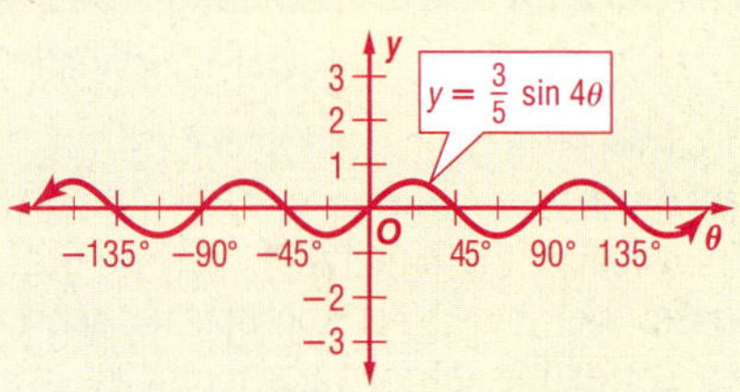

34.

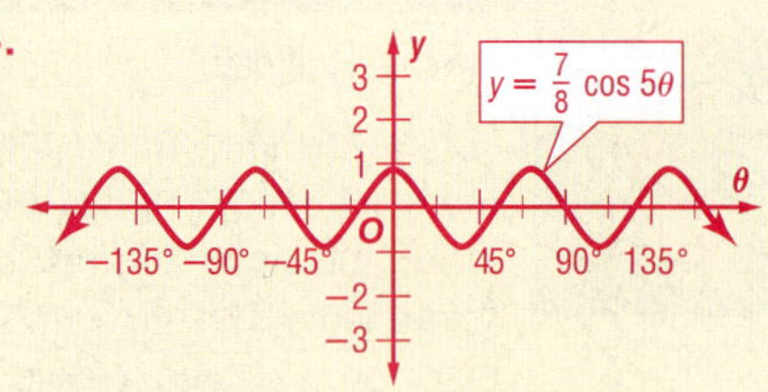

766 Chapter 14 Trigonometric Graphs and Identities

Guided Practice Find the amplitude, if it exists, and period of each function. Then graph each function. **4–12. See pp. 811A–811N.**

GUIDED PRACTICE KEY	
Exercises	Examples
4–12	1
13, 14	2

4. $y = \frac{1}{2}\sin\theta$
5. $y = 2\sin\theta$
6. $y = \frac{2}{3}\cos\theta$
7. $y = \frac{1}{4}\tan\theta$
8. $y = \csc 2\theta$
9. $y = 4\sin 2\theta$
10. $y = 4\cos\frac{3}{4}\theta$
11. $y = \frac{1}{2}\sec 3\theta$
12. $y = \frac{3}{4}\cos\frac{1}{2}\theta$

Application

BIOLOGY For Exercises 13 and 14, use the following information.
In a certain wildlife refuge, the population of field mice can be modeled by $y = 3000 + 1250\sin\frac{\pi}{6}t$, where y represents the number of mice and t represents the number of months past March 1 of a given year.

13. Determine the period of the function. What does this period represent? **13. 12 months; Sample answer: The pattern in the population will repeat itself every 12 months.**
14. What is the maximum number of mice and when does this occur? **4250; June 1**

★ indicates increased difficulty

Practice and Apply

Homework Help	
For Exercises	See Examples
15–35	1
36–41	2

Extra Practice
See page 859.

Find the amplitude, if it exists, and period of each function. Then graph each function. **15–32. See pp. 811A–811N.**

15. $y = 3\sin\theta$
16. $y = 5\cos\theta$
17. $y = 2\csc\theta$
18. $y = 2\tan\theta$
19. $y = \frac{1}{5}\sin\theta$
20. $y = \frac{1}{3}\sec\theta$
21. $y = \sin 4\theta$
22. $y = \sin 2\theta$
23. $y = \sec 3\theta$
24. $y = \cot 5\theta$
25. $y = 4\tan\frac{1}{3}\theta$
26. $y = 2\cot\frac{1}{2}\theta$
27. $y = 6\sin\frac{2}{3}\theta$
28. $y = 3\cos\frac{1}{2}\theta$
29. $y = 3\csc\frac{1}{2}\theta$
30. $y = \frac{1}{2}\cot 2\theta$
31. $2y = \tan\theta$
32. $\frac{3}{4}y = \frac{2}{3}\sin\frac{3}{5}\theta$

★ 33. Draw a graph of a sine function with an amplitude of $\frac{3}{5}$ and a period of 90°. Then write an equation for the function. **See margin for graph; $y = \frac{3}{5}\sin 4\theta$.**

★ 34. Draw a graph of a cosine function with an amplitude of $\frac{7}{8}$ and a period of $\frac{2\pi}{5}$. Then write an equation for the function. **See margin for graph; $y = \frac{7}{8}\cos 5\theta$.**

35. **COMMUNICATIONS** The carrier wave for a certain FM radio station can be modeled by the equation $y = A\sin(10^7 \cdot 2\pi t)$, where A is the amplitude of the wave and t is the time in seconds. Determine the period of the carrier wave. $\frac{1}{10^7}$

PHYSICS For Exercises 36 and 37, use the following information.
The sound wave produced by a tuning fork can be modeled by a sine function.

36. If the amplitude of the sine function is 0.25, write the equations for tuning forks that resonate with a frequency of 64, 256, and 512 Hertz. **36. $y = 0.25\sin 128\pi t$, $y = 0.25\sin 512\pi t$, $y = 0.25\sin 1024\pi t$**
37. How do the periods of the tuning forks compare? **See margin.**
38. **CRITICAL THINKING** A function is called *even* if the graphs of $y = f(x)$ and $y = f(-x)$ are the same. A function is *odd* if $f(-x) = -f(x)$. The graphs of even functions have reflection symmetry with respect to the y-axis and the graphs of odd functions have rotation symmetry with respect to the origin. Which of the six trigonometric functions are even and which are odd? Justify your answer with a graph of each function. **See margin.**

www.algebra2.com/self_check_quiz

Answers

37. **Sample answer: The amplitudes are the same. As the frequency increases, the period decreases.**
38. **even: $f(x) = \cos x$, $f(x) = \sec x$; odd: $f(x) = \sin x$, $f(x) = \tan x$, $f(x) = \csc x$, $f(x) = \cot x$; see pp. 811A–811N for graphs.**

Study Guide and Intervention, p. 837 (shown) and p. 838

Graph Trigonometric Functions To graph a trigonometric function, make a table of values for known degree measures (0°, 30°, 45°, 60°, 90°, and so on). Round function values to the nearest tenth, and plot the points. Then connect the points with a smooth, continuous curve. The *period* of the sine, cosine, secant, and cosecant functions is 360° or 2π radians.

Amplitude of a Function	The **amplitude** of the graph of a periodic function is the absolute value of half the difference between its maximum and minimum values.

Example **Graph $y = \sin\theta$ for $-360° \le \theta \le 0°$.**
First make a table of values.

θ	−360°	−330°	−315°	−300°	−270°	−240°	−225°	−210°	−180°
$\sin\theta$	0	$\frac{1}{2}$	$\frac{\sqrt{2}}{2}$	$\frac{\sqrt{3}}{2}$	1	$\frac{\sqrt{3}}{2}$	$\frac{\sqrt{2}}{2}$	$\frac{1}{2}$	0
θ	−150°	−135°	−120°	−90°	−60°	−45°	−30°	0°	
$\sin\theta$	$-\frac{1}{2}$	$-\frac{\sqrt{2}}{2}$	$-\frac{\sqrt{3}}{2}$	−1	$-\frac{\sqrt{3}}{2}$	$-\frac{\sqrt{2}}{2}$	$-\frac{1}{2}$	0	

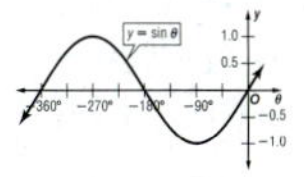

Exercises

Graph the following functions for the given domain.

1. $\cos\theta$, $-360° \le \theta \le 0°$
2. $\tan\theta$, $-2\pi \le \theta \le 0$

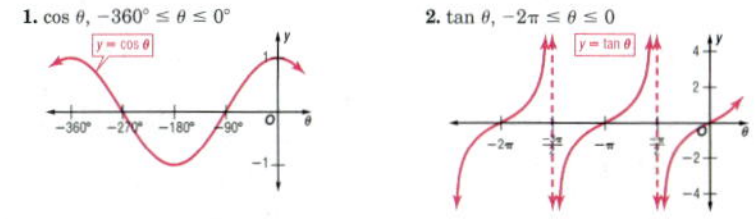

What is the amplitude of each function?

3. **4**
4. **8**

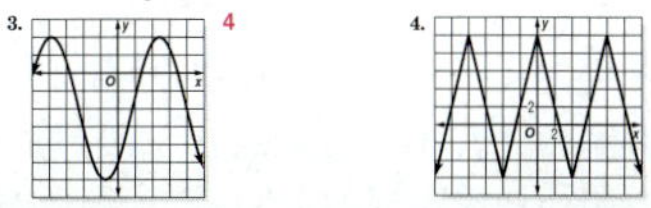

Skills Practice, p. 839 and Practice, p. 840 (shown)

Find the amplitude, if it exists, and period of each function. Then graph each function.

1. $y = -4\sin\theta$ **4; 360°**
2. $y = \cot\frac{1}{2}\theta$ **no amplitude; 360°**
3. $y = \cos 5\theta$ **1; 72°**
4. $y = \csc\frac{3}{4}\theta$ **no amplitude; 480°**
5. $y = 2\tan\frac{1}{2}\theta$ **no amplitude; 360°**
6. $2y = \sin\theta$ **$\frac{1}{2}$; 360°**

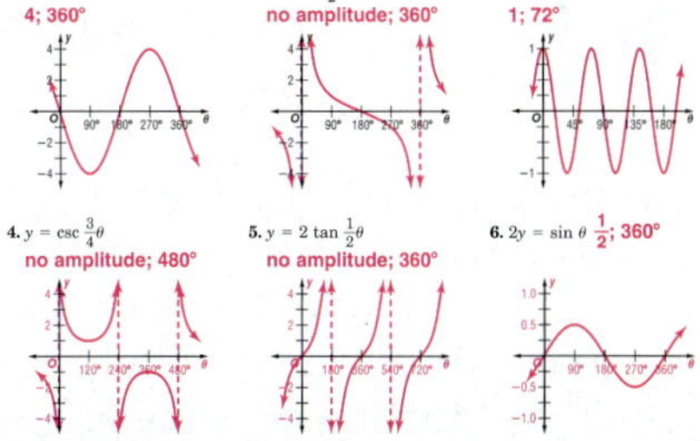

FORCE For Exercises 7 and 8, use the following information.
An anchoring cable exerts a force of 500 Newtons on a pole. The force has the horizontal and vertical components F_x and F_y. (A force of one Newton (N), is the force that gives an acceleration of 1 m/sec² to a mass of 1 kg.)

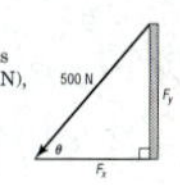

7. The function $F_x = 500\cos\theta$ describes the relationship between the angle θ and the horizontal force. What are the amplitude and period of this function? **500; 360°**
8. The function $F_y = 500\sin\theta$ describes the relationship between the angle θ and the vertical force. What are the amplitude and period of this function? **500; 360°**

WEATHER For Exercises 9 and 10, use the following information.
The function $y = 60 + 25\sin\frac{\pi}{6}t$, where t is in months and $t = 0$ corresponds to April 15, models the average high temperature in degrees Fahrenheit in Centerville.

9. Determine the period of this function. What does this period represent? **12; a calendar year**
10. What is the maximum high temperature and when does this occur? **85°F; July 15**

Reading to Learn Mathematics, p. 841 ELL

Pre-Activity **Why can you predict the behavior of tides?**
Read the introduction to Lesson 14-1 at the top of page 762 in your textbook.
Consider the tides of the Atlantic Ocean as a function of time. Approximately what is the period of this function? **12 hours**

Reading the Lesson

1. Determine whether each statement is *true* or *false*.
   a. The period of a function is the distance between the maximum and minimum points. **false**
   b. The amplitude of a function is the difference between its maximum and minimum values. **false**
   c. The amplitude of the function $y = \sin\theta$ is 2π. **false**
   d. The function $y = \cot\theta$ has no amplitude. **true**
   e. The period of the function $y = \sec\theta$ is π. **false**
   f. The amplitude of the function $y = 2\cos\theta$ is 4. **false**
   g. The function $y = \sin 2\theta$ has a period of π. **true**
   h. The period of the function $y = \cot 3\theta$ is $\frac{\pi}{3}$. **true**
   i. The amplitude of the function $y = -5\sin\theta$ is −5. **false**
   j. The period of the function $y = \csc\frac{1}{4}\theta$ is 4π. **false**
   k. The graph of the function $y = \sin\theta$ has no asymptotes. **true**
   l. The graph of the function $y = \tan\theta$ has an asymptote at $\theta = 180°$. **false**
   m. When $\theta = 360°$, the values of $\cos\theta$ and $\sec\theta$ are equal. **true**
   n. When $\theta = 270°$, $\cot\theta$ is undefined. **false**
   o. When $\theta = 180°$, $\csc\theta$ is undefined. **true**

Helping You Remember

2. What is an easy way to remember the periods of $y = a\sin b\theta$ and $y = a\cos b\theta$? **Sample answer: The period of the functions $y = \sin\theta$ and $y = \cos\theta$ is 360° or 2π. Divide 360° or 2π by the absolute value of the coefficient of θ, depending on whether you want to find the period in degrees or in radians.**

Enrichment, p. 842

Blueprints

Interpreting blueprints requires the ability to select and use trigonometric functions and geometric properties. The figure below represents a plan for an improvement to a roof. The metal fitting shown makes a 30° angle with the horizontal. The vertices of the geometric shapes are *not* labeled in these plans. Relevant information must be selected and the appropriate function used to find the unknown measures.

Example **Find the unknown measures in the figure at the right.**

The measures x and y are the legs of a right triangle.

The measure of the hypotenuse is $\frac{15}{16}$ in. + $\frac{5}{16}$ in. or $\frac{20}{16}$ in.

Roofing Improvement

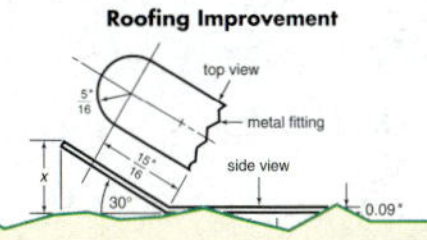

About the Exercises...

Organization by Objective
- **Graph Trigonometric Functions:** 15–34
- **Variations of Trigonometric Functions:** 15–34

Odd/Even Assignments
Exercises 15–34 are structured so that students practice the same concepts whether they are assigned odd or even problems.

Assignment Guide

Basic: 15–31 odd, 35–56

Average: 15–35 odd, 36–56

Advanced: 16–34 even, 36–52 (optional: 53–56)

4 Assess

Open-Ended Assessment

Modeling Provide students with a coordinate grid and a length of string. Give students one of the six basic trigonometric functions and have them model the graph using the string. Have students check their model with a graphing calculator.

Getting Ready for Lesson 14-2

PREREQUISITE SKILL Students will graph horizontal and vertical translations of trigonometric functions in Lesson 14-2. Students will apply what they learned about families of quadratic functions. Use Exercises 53–56 to determine your students' familiarity with the graphs of families of functions.

BOATING **For Exercises 39–41, use the following information.**
A marker buoy off the coast of Gulfport, Mississippi, bobs up and down with the waves. The distance between the highest and lowest point is 4 feet. The buoy moves from its highest point to its lowest point and back to its highest point every 10 seconds.

39. Write an equation for the motion of the buoy. Assume that it is at equilibrium at $t = 0$ and that it is on the way up from the normal water level. $y = 2 \sin \frac{\pi}{5}t$

40. Draw a graph showing the height of the buoy as a function of time. **See margin.**

41. What is the height of the buoy after 12 seconds? **about 1.9 ft**

42. *WRITING IN MATH* Answer the question that was posed at the beginning of the lesson. **See margin.**

Why can you predict the behavior of tides?

Include the following in your answer:
- an explanation of why certain tidal characteristics follow the patterns seen in the graph of the sine function, and
- a description of how to determine the amplitude of a function using the maximum and minimum values.

Standardized Test Practice

43. What is the period of $f(x) = \frac{1}{2} \cos 3x$? **A**

Ⓐ 120° Ⓑ 180° Ⓒ 360° Ⓓ 720°

44. Identify the equation of the graphed function. **C**

Ⓐ $y = \frac{1}{2} \sin 4\theta$

Ⓑ $y = 2 \sin \frac{1}{4}\theta$

Ⓒ $y = \frac{1}{4} \sin 2\theta$

Ⓓ $y = 4 \sin \frac{1}{2}\theta$

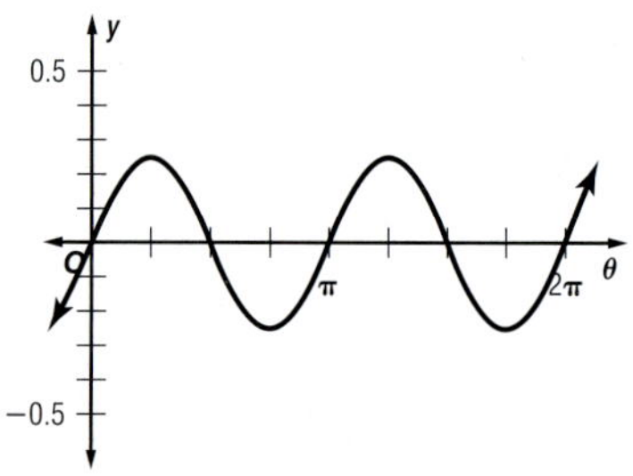

Maintain Your Skills

Mixed Review

Solve each equation. *(Lesson 13-7)*

45. $x = \text{Sin}^{-1} 1$ **90°**

46. $\text{Arcsin}(-1) = y$ **−90°**

47. $\text{Arccos} \frac{\sqrt{2}}{2} = x$ **45°**

Find the exact value of each function. *(Lesson 13-6)*

48. $\sin 390°$ $\frac{1}{2}$

49. $\sin(-315°)$ $\frac{\sqrt{2}}{2}$

50. $\cos 405°$ $\frac{\sqrt{2}}{2}$

51. **PROBABILITY** There are 8 girls and 8 boys on the Faculty Advisory Board. Three are juniors. Find the probability of selecting a boy or a girl from the committee who is not a junior. *(Lesson 12-5)* $\frac{13}{16}$

52. Find the first five terms of the sequence in which $a_1 = 3, a_{n+1} = 2a_n + 5$. *(Lesson 11-5)* **3, 11, 27, 59, 123**

Getting Ready for the Next Lesson

53–56. See pp. 811A–811N.

PREREQUISITE SKILL **Graph each pair of functions on the same set of axes.**
*(To review **graphs of quadratic functions**, see Lesson 6-6.)*

53. $y = x^2, y = 3x^2$

54. $y = 3x^2, y = 3x^2 - 4$

55. $y = 2x^2, y = 2(x + 1)^2$

56. $y = x^2 + 2, y = (x - 3)^2 + 2$

Answers

40.

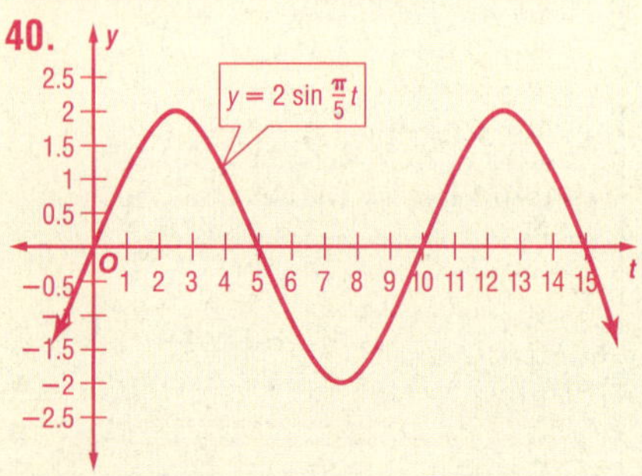

42. Sample answer: Tides display periodic behavior. This means that their pattern repeats at regular intervals. Answers should include the following information.
- Tides rise and fall in a periodic manner, similar to the sine function.
- In $f(x) = a \sin bx$, the amplitude is the absolute value of a.

14-2 Translations of Trigonometric Graphs

What You'll Learn

- Graph horizontal translations of trigonometric graphs and find phase shifts.
- Graph vertical translations of trigonometric graphs.

Vocabulary

- phase shift
- vertical shift
- midline

How can translations of trigonometric graphs be used to show animal populations?

In predator-prey ecosystems, the number of predators and the number of prey tend to vary in a periodic manner. In a certain region with coyotes as predators and rabbits as prey, the rabbit population R can be modeled by the equation $R = 1200 + 250 \sin \frac{1}{2}\pi t$, where t is the time in years since January 1, 2001.

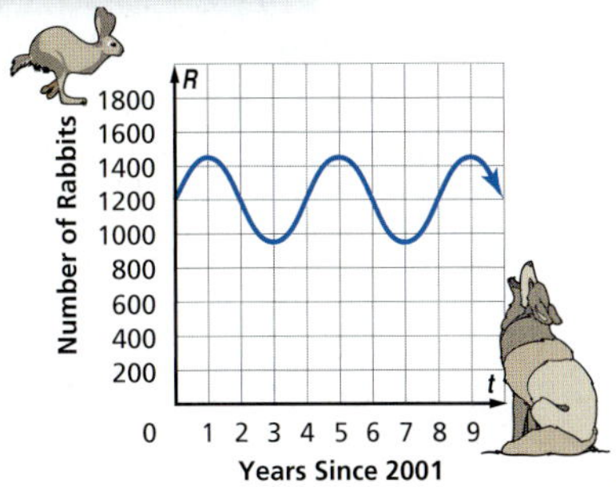

HORIZONTAL TRANSLATIONS Recall that a translation is a type of transformation in which the image is identical to the preimage in all aspects except its location on the coordinate plane. A horizontal translation shifts to the left or right, and not upward or downward.

Graphing Calculator Investigation

Horizontal Translations

On a TI-83 Plus, set the MODE to degrees.

Think and Discuss 1–3. See margin.

1. Graph $y = \sin x$ and $y = \sin (x - 30)$. How do the two graphs compare?
2. Graph $y = \sin (x + 60)$. How does this graph compare to the other two?
3. What conjecture can you make about the effect of h in the function $y = \sin (x - h)$?
4. Test your conjecture on the following pairs of graphs.
   - $y = \cos x$ and $y = \cos (x + 30)$ See pp. 811A–811N for graphs;
   - $y = \tan x$ and $y = \tan (x - 45)$ the conjecture holds.
   - $y = \sec x$ and $y = \sec (x + 75)$

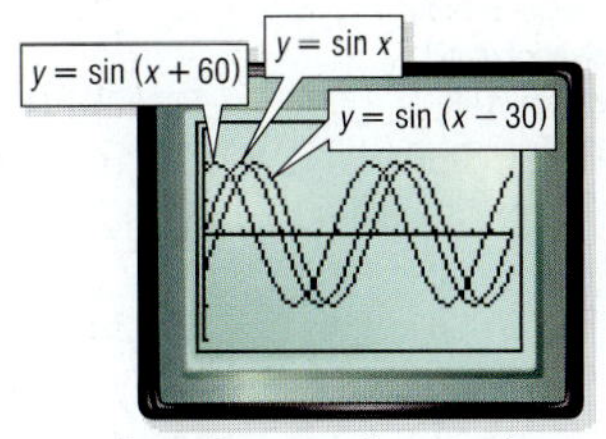

[0, 720] scl: 45 by [−1.5, 1.5] scl: 0.5

TEACHING TIP

For Exercise 4, point out that since the calculator has no preprogrammed button for the secant function, they will need to graph $y = \frac{1}{\cos x}$.

Notice that when a constant is added to an angle measure in a trigonometric function, the graph is shifted to the left or to the right. If (x, y) are coordinates of $y = \sin x$, then $(x \pm h, y)$ are coordinates of $y = \sin (x \mp h)$. A horizontal translation of a trigonometric function is called a **phase shift**.

14-2 Lesson Notes

1 Focus

5-Minute Check Transparency 14-2 Use as a quiz or review of Lesson 14-1.

Mathematical Background notes are available for this lesson on page 760C.

How can translations of trigonometric graphs be used to show animal populations?

Ask students:

- Why might the two animal populations vary? **Sample answer: As the number of predators increases, more prey are eaten and there are fewer prey left. Lower numbers of prey means increased competition for food by the predator species, so the number of predators decreases as their food supply diminishes.**
- What are the minimum and maximum rabbit populations shown by the graph? **950 rabbits, 1450 rabbits**
- How could you find the range of the population without calculating the maximum and minimum values or graphing the function? **The range, 500, is twice the amplitude, 250, of the graph.**

Answers

See p. 770 for Graphing Calculator answers.

Resource Manager

Workbook and Reproducible Masters

Chapter 14 Resource Masters

- Study Guide and Intervention, pp. 843–844
- Skills Practice, p. 845
- Practice, p. 846
- Reading to Learn Mathematics, p. 847
- Enrichment, p. 848
- Assessment, p. 893

Graphing Calculator and Spreadsheet Masters, p. 53

School-to-Career Masters, p. 27

Transparencies

5-Minute Check Transparency 14-2

Answer Key Transparencies

Technology

Alge2PASS: Tutorial Plus, Lesson 27

Interactive Chalkboard

2 Teach

Building on Prior Knowledge

In Lesson 6-6, students learned about the translations of graphs of quadratic functions. In this lesson, students will use similar techniques to study the translations of graphs of the trigonometric functions.

HORIZONTAL TRANSLATIONS

In-Class Example Power Point®

1 State the amplitude, period, and phase shift for each function. Then graph the function.

a. $y = 2\sin(\theta + 20°)$ **amplitude: 2; period: 360°; phase shift: 20° left**

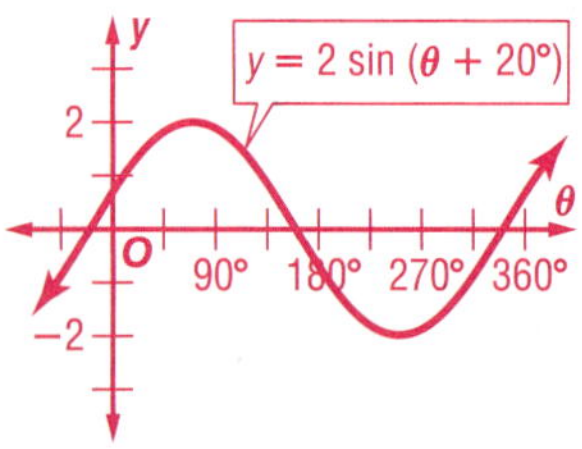

b. $y = \frac{1}{2}\cos\left(\theta - \frac{\pi}{6}\right)$

amplitude: $\frac{1}{2}$; period: 2π; phase shift: $\frac{\pi}{6}$ right

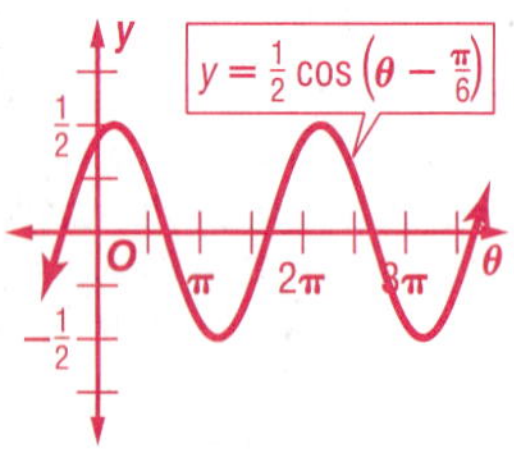

Answers (p. 769)

Graphing Calculator Investigation

1. The graph of $y = \sin(x - 30)$ is shifted 30° to the right of the graph of $y = \sin x$.
2. The graph of $y = \sin(x + 60)$ is shifted 60° to the left of the graph of $y = \sin x$.
3. Sample answer: When h is positive the graph shifts right h units. When h is negative the graph shifts left h units.

Key Concept *Phase Shift*

- **Words:** The phase shift of the functions $y = a \sin b(\theta - h)$, $y = a \cos b(\theta - h)$, and $y = a \tan b(\theta - h)$ is h, where $b > 0$.

  If $h > 0$, the shift is to the right. If $h < 0$, the shift is to the left.

- **Models:**

Sine

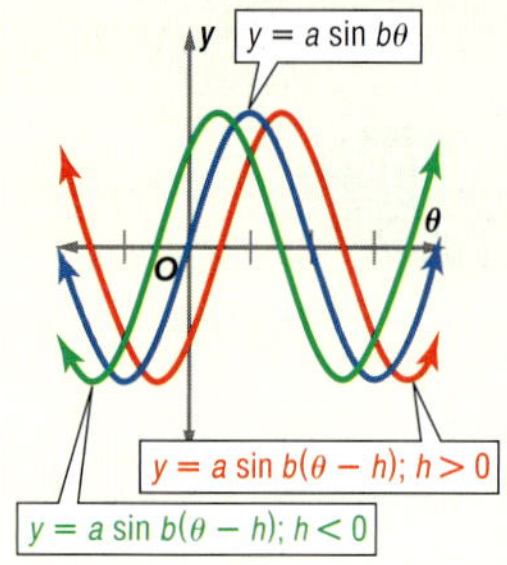

Cosine

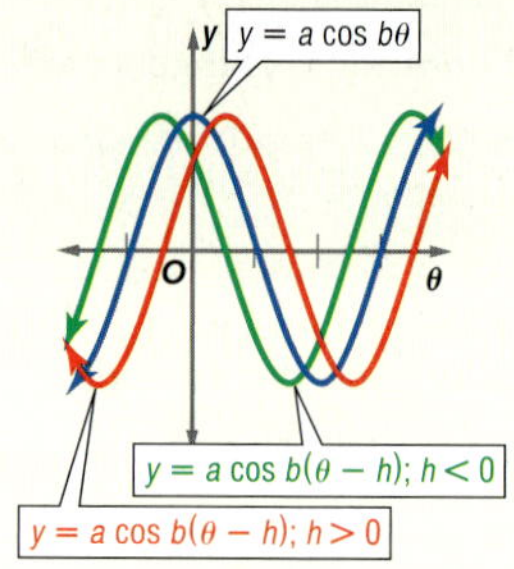

Tangent

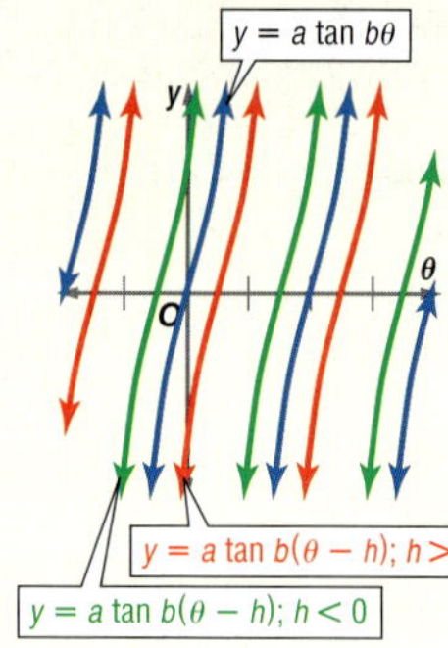

The secant, cosecant, and cotangent can be graphed using the same rules.

Study Tip

Verifying a Graph

After drawing the graph of a trigonometric function, select values of θ and evaluate them in the equation to verify your graph.

Example 1 Graph Horizontal Translations

State the amplitude, period, and phase shift for each function. Then graph the function.

a. $y = \cos(\theta - 60°)$

Since $a = 1$ and $b = 1$, the amplitude and period of the function are the same as $y = \cos\theta$. However, $h = 60°$, so the phase shift is 60°. Because $h > 0$, the parent graph is shifted to the right.

To graph $y = \cos(\theta - 60°)$, consider the graph of $y = \cos\theta$. Graph this function and then shift the graph 60° to the right. The graph $y = \cos(\theta - 60°)$ is the graph of $y = \cos\theta$ shifted to the right.

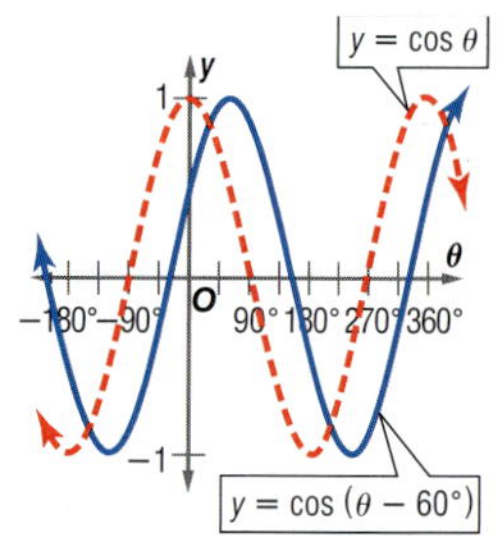

b. $y = 2\sin\left(\theta + \frac{\pi}{4}\right)$

Amplitude: $a = |2|$ or 2

Period: $\frac{2\pi}{|b|} = \frac{2\pi}{|1|}$ or 2π

Phase Shift: $h = -\frac{\pi}{4}$ $\left(\theta + \frac{\pi}{4}\right) = \theta - \left(-\frac{\pi}{4}\right)$

The phase shift is to the left since $-\frac{\pi}{4} < 0$.

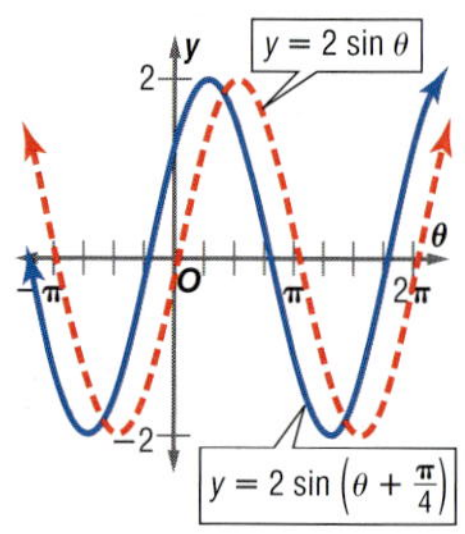

Graphing Calculator Investigation

Graphing the Secant Function The graph of the secant function should look like a pattern of U shapes, alternately opening upward and downward. Students might want to extend the activity by investigating the graphs of cosecant (csc) and cotangent (cot). The graphing calculator does not have a button for graphing either of these functions so students should enter Y= 1/sin x to graph the cosecant function, and they should enter Y= 1/tan x to graph the cotangent function.

770 Chapter 14 Trigonometric Graphs and Identities

Study Tip

Look Back
Pay close attention to trigonometric functions for the placement of parentheses. Note that $\sin(\theta + x) \neq \sin\theta + x$. The expression on the left represents a phase shift while the expression on the right represents a vertical shift.

VERTICAL TRANSLATIONS In Chapter 6, you learned that the graph of $y = x^2 + 4$ is a vertical translation of the parent graph of $y = x^2$. Similarly, graphs of trigonometric functions can be translated vertically through a **vertical shift**.

When a constant is added to a trigonometric function, the graph is shifted upward or downward. If (x, y) are coordinates of $y = \sin x$, then $(x, y \pm k)$ are coordinates of $y = \sin x \pm k$.

A new horizontal axis called the **midline** becomes the reference line about which the graph oscillates. For the graph of $y = \sin\theta + k$, the midline is the graph of $y = k$.

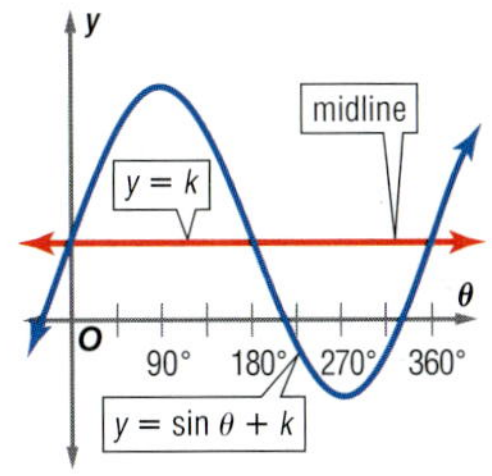

Key Concept — Vertical Shift

- **Words** The vertical shift of the functions $y = a \sin b(\theta - h) + k$, $y = a \cos b(\theta - h) + k$, and $y = a \tan b(\theta - h) + k$ is k.

  If $k > 0$, the shift is up. If $k < 0$, the shift is down. The midline is $y = k$.

- **Models:**

Sine

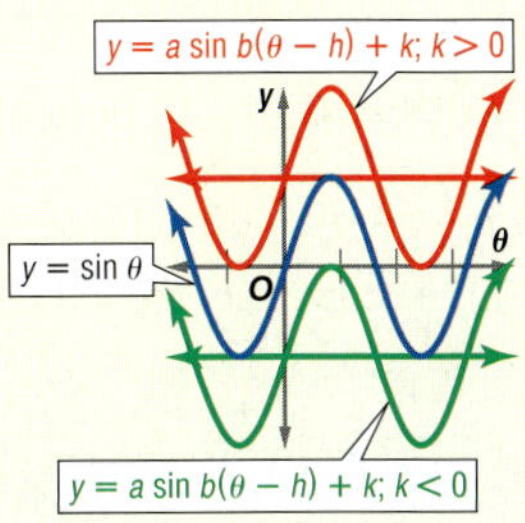

Cosine

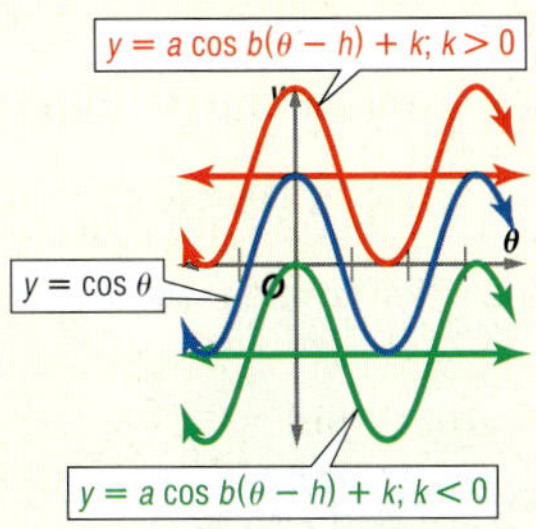

Tangent

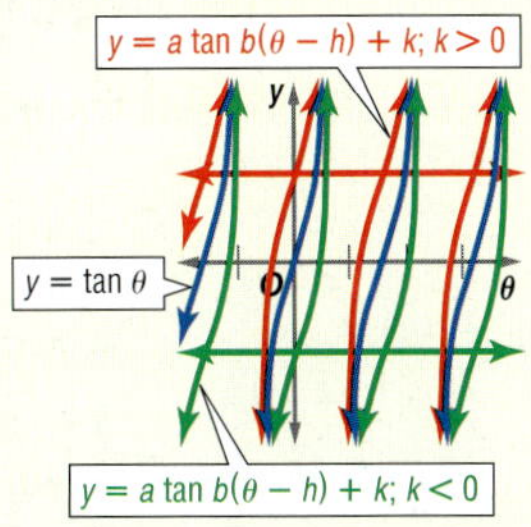

The secant, cosecant, and cotangent can be graphed using the same rules.

Example 2 Graph Vertical Translations

State the vertical shift, equation of the midline, amplitude, and period for each function. Then graph the function.

a. $y = \tan\theta - 2$

Since $\tan\theta - 2 = \tan\theta + (-2)$, $k = -2$, and the vertical shift is -2. Draw the midline, $y = -2$. The tangent function has no amplitude and the period is the same as that of $\tan\theta$.

Draw the graph of the function relative to the midline.

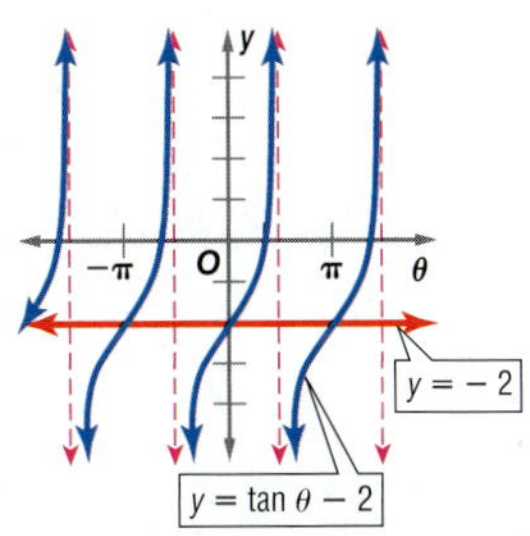

VERTICAL TRANSLATIONS

Teaching Tip Make sure students are clear about the difference between the terms *vertical* and *horizontal*. Students can use the words *vertigo* (a sense of dizziness some people experience when looking down from a great height) and *horizon* to help link vertical and horizontal to their relative direction.

In-Class Example

Power Point®

2 State the vertical shift, equation of the midline, amplitude, and period for each function. Then graph the function.

a. $y = 2\sin\theta - 1$
vertical shift: -1; midline: $y = -1$; amplitude: 2; period: 2π

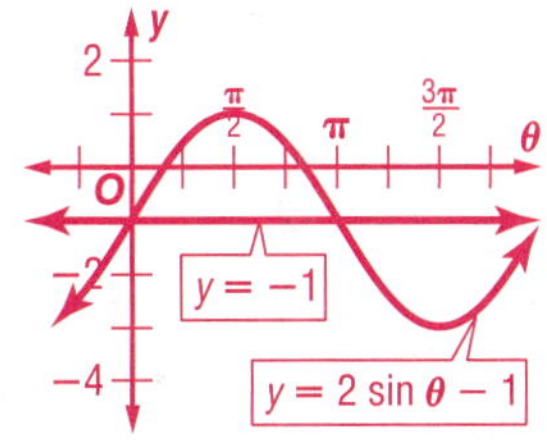

b. $y = \frac{1}{2}\cos\theta + 3$
vertical shift: $+3$; midline: $y = 3$; amplitude: $\frac{1}{2}$; period: 2π

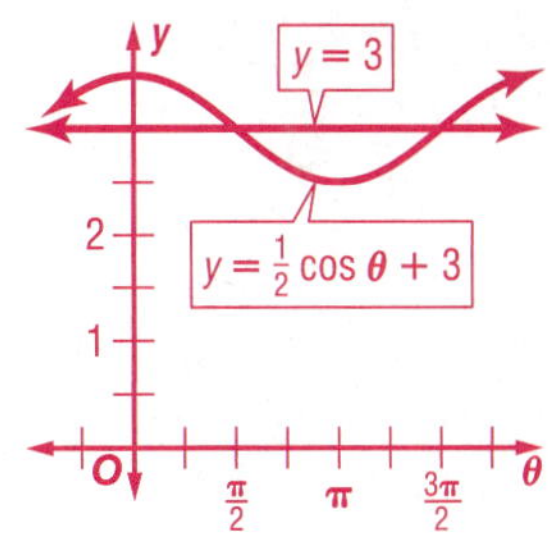

Interactive Chalkboard PowerPoint® Presentations

This CD-ROM is a customizable Microsoft® PowerPoint® presentation that includes:

- Step-by-step, dynamic solutions of each In-Class Example from the Teacher Wraparound Edition
- Additional, Your Turn exercises for each example
- The 5-Minute Check Transparencies
- Hot links to Glencoe Online Study Tools

Teaching Tip Suggest that students write the general form of the sine, cosine, and tangent functions given in the Key Concept box on p. 771 at the top of separate index cards. Using the steps listed in the Concept Summary on p. 772, have students identify the parts of the equation used in each step to find the information listed in each step. They can then use their index cards to work Example 3.

In-Class Example

Power Point®

3 State the vertical shift, amplitude, period, and phase shift of $y = 3\sin\left[2\left(\theta - \frac{\pi}{2}\right)\right] + 4$. Then graph the function.

The vertical shift is +4. The amplitude is 3. The period is π. The phase shift is $\frac{\pi}{2}$ to the right.

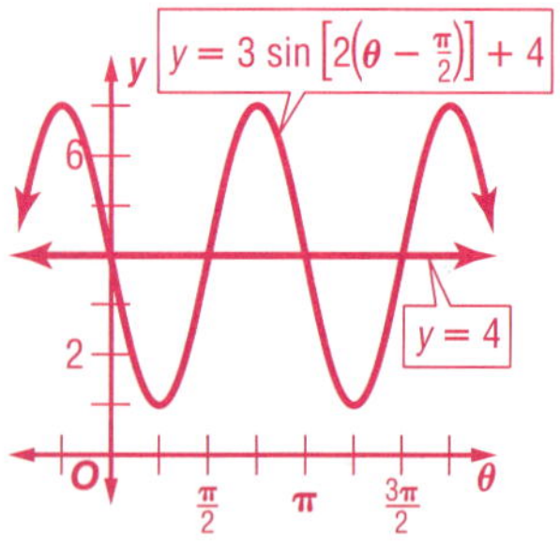

Study Tip

Look Back

It may be helpful to first graph the parent graph $y = \sin\theta$ in one color. Then apply the vertical shift and graph the function in another color. Then apply the change in amplitude and graph the function in the final color.

b. $y = \frac{1}{2}\sin\theta + 1$

Vertical shift: $k = 1$, so the midline is the graph of $y = 1$.

Amplitude: $|a| = \left|\frac{1}{2}\right|$ or $\frac{1}{2}$

Period: $\frac{2\pi}{|b|} = 2\pi$

Since the amplitude of the function is $\frac{1}{2}$, draw dashed lines parallel to the midline that are $\frac{1}{2}$ unit above and below the midline. Then draw the sine curve.

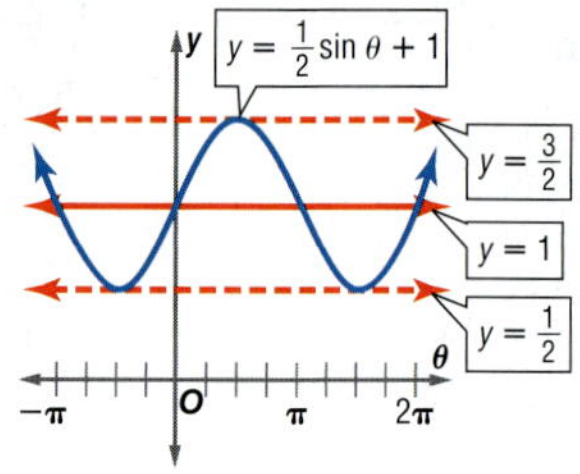

In general, use the following steps to graph any trigonometric function.

Concept Summary — Graphing Trigonometric Functions

Step 1 Determine the vertical shift, and graph the midline.

Step 2 Determine the amplitude, if it exists. Use dashed lines to indicate the maximum and minimum values of the function.

Step 3 Determine the period of the function and graph the appropriate function.

Step 4 Determine the phase shift and translate the graph accordingly.

Example 3 Graph Transformations

State the vertical shift, amplitude, period, and phase shift of $y = 4\cos\left[\frac{1}{2}\left(\theta - \frac{\pi}{3}\right)\right] - 6$. Then graph the function.

The function is written in the form $y = a\cos[b(\theta - h)] + k$. Identify the values of k, a, b, and h.

$k = -6$, so the vertical shift is -6.

$a = 4$, so the amplitude is $|4|$ or 4.

$b = \frac{1}{2}$, so the period is $\frac{2\pi}{\left|\frac{1}{2}\right|}$ or 4π.

$h = \frac{\pi}{3}$, so the phase shift is $\frac{\pi}{3}$ to the right.

Then graph the function.

Step 1 The vertical shift is -6. Graph the midline $y = -6$.

Step 2 The amplitude is 4. Draw dashed lines 4 units above and below the midline at $y = -2$ and $y = -10$.

Step 3 The period is 4π, so the graph will be stretched. Graph $y = 4\cos\frac{1}{2}\theta - 6$ using the midline as a reference.

Step 4 Shift the graph $\frac{\pi}{3}$ to the right.

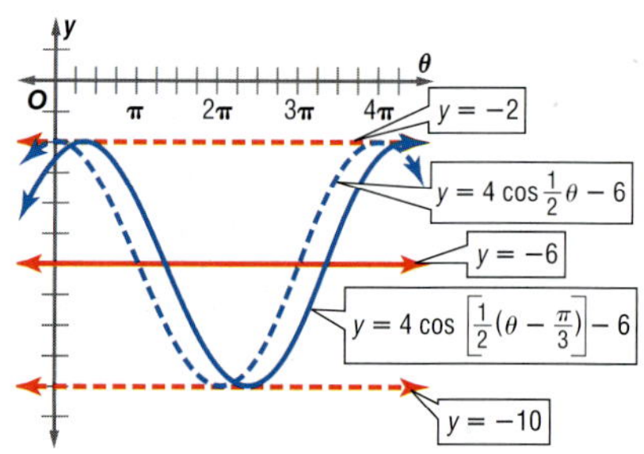

772 Chapter 14 Trigonometric Graphs and Identities

You can use information about amplitude, period, and translations of trigonometric functions to model real-world applications.

Example 4 Use Translations to Solve a Problem

HEALTH Suppose a person's resting blood pressure is 120 over 80. This means that the blood pressure oscillates between a maximum of 120 and a minimum of 80. If this person's resting heart rate is 60 beats per minute, write a sine function that represents the blood pressure for t seconds. Then graph the function.

Explore You know that the function is periodic and can be modeled using sine.

Plan Let P represent blood pressure and let t represent time in seconds. Use the equation $P = a \sin [b(t - h)] + k$.

Solve

- Write the equation for the midline. Since the maximum is 120 and the minimum is 80, the midline lies halfway between these values.

$$P = \frac{120 + 80}{2} \text{ or } 100$$

- Determine the amplitude by finding the difference between the midline value and the maximum and minimum values.

$$a = |120 - 100| \qquad a = |80 - 100|$$
$$= |20| \text{ or } 20 \qquad = |-20| \text{ or } 20$$

Thus, $a = 20$.

- Determine the period of the function and solve for b. Recall that the period of a function can be found using the expression $\frac{2\pi}{|b|}$. Since the heart rate is 60 beats per minute, there is one heartbeat, or cycle, per second. So, the period is 1 second.

$1 = \frac{2\pi}{|b|}$ Write an equation.

$|b| = 2\pi$ Multiply each side by $|b|$.

$b = \pm 2\pi$ Solve.

For this example, let $b = 2\pi$. The use of the positive or negative value depends upon whether you begin a cycle with a maximum value (positive) or a minimum value (negative).

- There is no phase shift, so $h = 0$. So, the equation is $P = 20 \sin 2\pi t + 100$.
- Graph the function.

Step 1 Draw the midline $P = 100$.

Step 2 Draw maximum and minimum reference lines.

Step 3 Use the period to draw the graph of the function.

Step 4 There is no phase shift.

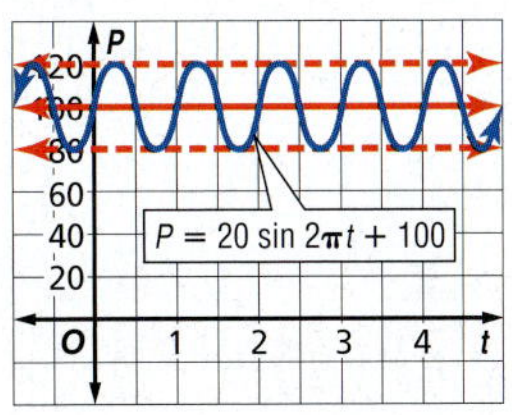

Examine Notice that each cycle begins at the midline, rises to 120, drops to 80, and then returns to the midline. This represents the blood pressure of 120 over 80 for one heartbeat. Since each cycle lasts 1 second, there will be 60 cycles, or heartbeats, in 1 minute. Therefore, the graph accurately represents the information.

More About. . .

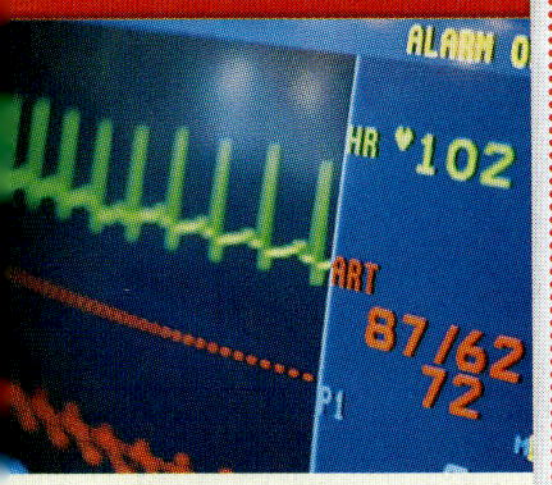

Health

Blood pressure can change from minute to minute and can be affected by the slightest of movements, such as tapping your fingers or crossing your arms.

Source: American Heart Association

In-Class Example

Power Point®

4 HEALTH Refer to Example 4 in the Student Edition. Write a sine function that represents the blood pressure for t seconds of a person with a resting blood pressure of 120 over 90 and a heart rate of 75 beats per minute. Then graph the function.

$P = 15 \sin \frac{5\pi}{2} t + 105$

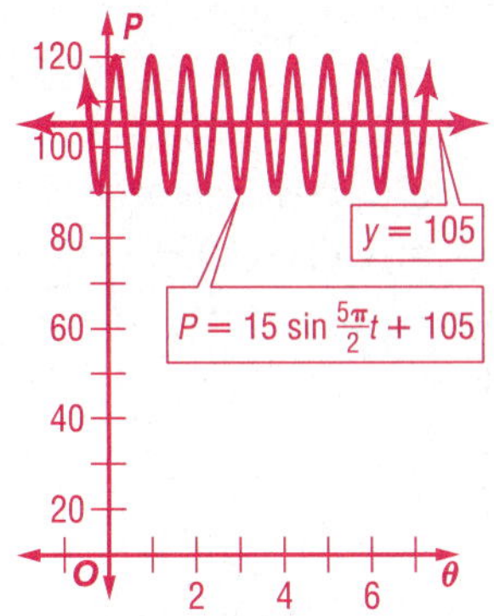

DAILY INTERVENTION — Differentiated Instruction

Kinesthetic Make coordinate axes with masking tape on the classroom floor. Give students at least 15 feet of rope and have them stand along the x-axis, positioning the rope to model the graph of $y = \sin x$. As you call out equations of functions whose graphs are horizontal phase shifts of the graph of $y = \sin x$, students can step left or right to model the translated graph. Similarly, call out functions whose graphs are vertical shifts of the graph of $y = \sin x$.

3 Practice/Apply

Study Notebook

Have students—
- add the definitions/examples of the vocabulary terms to their Vocabulary Builder worksheets for Chapter 14.
- record the Key Concepts box on p. 771 and the Concept Summary on p. 772.
- include any other item(s) that they find helpful in mastering the skills in this lesson.

About the Exercises…

Organization by Objective
- **Horizontal Translations:** 19–24, 33–42
- **Vertical Translations:** 25–42

Odd/Even Assignments
Exercises 19–42 are structured so that students practice the same concepts whether they are assigned odd or even problems.

Assignment Guide

Basic: 19–29 odd, 33–43 odd, 44–73

Average: 19–43 odd, 44–73

Advanced: 20–42 even, 44–46, 48–65 (optional: 66–73)

Check for Understanding

Concept Check

1–3. See margin.

1. **Identify** the vertical shift, amplitude, period, and phase shift of the graph of $y = 3\cos(2x - 90°) + 15$.
2. **Define** the midline of a trigonometric graph.
3. **OPEN ENDED** Write the equation of a trigonometric function with a phase shift of $-45°$.

Guided Practice

GUIDED PRACTICE KEY	
Exercises	Examples
4–7	1
8–11	2
12–15	3
16–18	4

State the amplitude, period, and phase shift for each function. Then graph the function. 4–15. See pp. 811A–811N for graphs.

4. $y = \sin\left(\theta - \frac{\pi}{2}\right)$ 1; 2π; $\frac{\pi}{2}$
5. $y = \tan(\theta + 60°)$ no amplitude; 180°; −60°
6. $y = \cos(\theta - 45°)$ 1; 360°; 45°
7. $y = \sec\left(\theta + \frac{\pi}{3}\right)$ no amplitude; 2π; $-\frac{\pi}{3}$

State the vertical shift, equation of the midline, amplitude, and period for each function. Then graph the function.

8. $y = \cos\theta + \frac{1}{4}$ $\frac{1}{4}$; $y = \frac{1}{4}$; 1; 360°
9. $y = \sec\theta - 5$ −5; $y = -5$; no amplitude; 360°
10. $y = \tan\theta + 4$ 4; $y = 4$; no amplitude; 180°
11. $y = \sin\theta + 0.25$ 0.25; $y = 0.25$; 1; 360°

State the vertical shift, amplitude, period, and phase shift of each function. Then graph the function. 12. 10; 3; 180°; 30° 13. −6; no amplitude; 60°; −45°

12. $y = 3\sin[2(\theta - 30°)] + 10$
13. $y = 2\cot(3\theta + 135°) - 6$
14. $y = \frac{1}{2}\sec\left[4\left(\theta - \frac{\pi}{4}\right)\right] + 1$ 1; no amplitude; $\frac{\pi}{2}$; $\frac{\pi}{4}$
15. $y = \frac{2}{3}\cos\left[\frac{1}{2}\left(\theta + \frac{\pi}{6}\right)\right] - 2$ −2; $\frac{2}{3}$; 4π; $-\frac{\pi}{6}$

Application

PHYSICS For Exercises 16–18, use the following information.
A weight is attached to a spring and suspended from the ceiling. At equilibrium, the weight is located 4 feet above the floor. The weight is pulled down 1 foot and released.

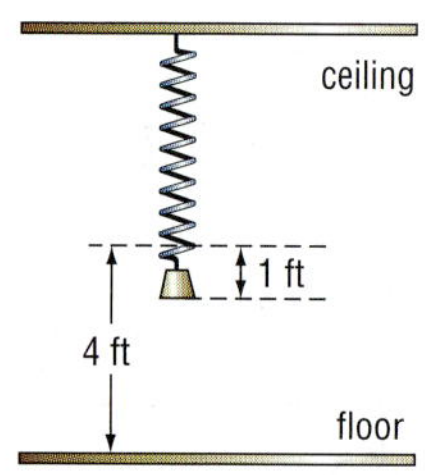

16. Determine the vertical shift, amplitude, and period of a function that represents the height of the weight above the floor if the weight returns to its lowest position every 4 seconds. 4; 1; 4 s
17. Write the equation for the height h of the weight above the floor as a function of time t seconds. $h = 4 - \cos\frac{\pi}{2}t$ or $h = 4 - \cos 90°t$
18. Draw a graph of the function you wrote in Exercise 17. See pp. 811A–811N.

★ indicates increased difficulty

Practice and Apply

State the amplitude, period, and phase shift for each function. Then graph the function. 20. no amplitude; 180°; 30°

19–24. See pp. 811A–811N for graphs.

19. $y = \cos(\theta + 90°)$ 1; 360°; −90°
20. $y = \cot(\theta - 30°)$
21. $y = \sin\left(\theta - \frac{\pi}{4}\right)$ 1; 2π; $\frac{\pi}{4}$
22. $y = \cos\left(\theta + \frac{\pi}{3}\right)$ 1; 2π; $-\frac{\pi}{3}$
23. $y = \frac{1}{4}\tan(\theta + 22.5°)$ no amplitude; 180°; −22.5°
24. $y = 3\sin(\theta - 75°)$ 3; 360°; 75°

Answers

1. vertical shift: 15; amplitude: 3; period: 180°; phase shift: 45°
2. The midline of a trigonometric function is the line about which the graph of the function oscillates after a vertical shift.
3. Sample answer: $y = \sin(\theta + 45°)$

31.

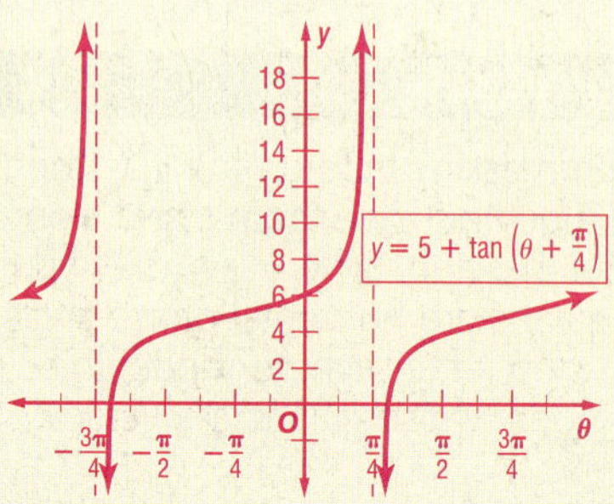

32.

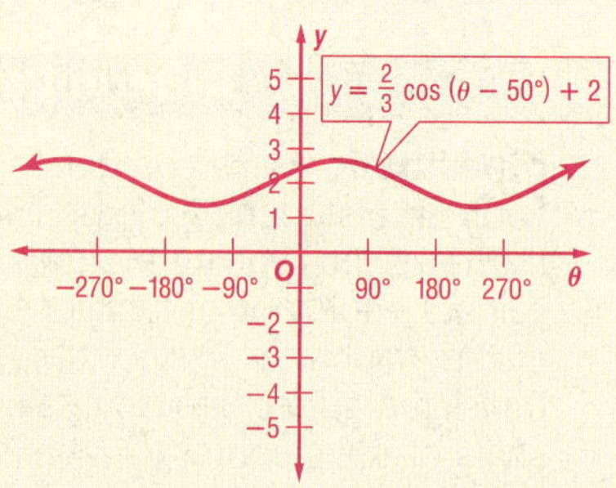

Homework Help	
For Exercises	See Examples
19–24	1
25–30	2
31, 32	1, 2
33–42	3
37–40	4

Extra Practice
See page 859.

State the vertical shift, equation of the midline, amplitude, and period for each function. Then graph the function. **25–30. See pp. 811A–811N for graphs.**

25. $y = \sin\theta - 1$ **−1; y = −1; 1; 360°**
26. $y = \sec\theta + 2$ **2; y = 2; no amplitude; 360°**
27. $y = \cos\theta - 5$ **−5; y = −5; 1; 360°**
28. $y = \csc\theta - \frac{3}{4}$
29. $y = \frac{1}{2}\sin\theta + \frac{1}{2}$ **$\frac{1}{2}$; $y = \frac{1}{2}$; $\frac{1}{2}$; 360°**
30. $y = 6\cos\theta + 1.5$ **1.5; y = 1.5; 6; 360°**

★ 31. Graph $y = 5 + \tan\left(\theta + \frac{\pi}{4}\right)$. Describe the transformation to the parent graph $y = \tan\theta$. **See margin for graph; translation $\frac{\pi}{4}$ units left and 5 units up.**

★ 32. Draw a graph of the function $y = \frac{2}{3}\cos(\theta - 50°) + 2$. How does this graph compare to the graph of $y = \cos\theta$? **See margin for graph; translation 50° right and 2 units up with an amplitude of $\frac{2}{3}$ unit.**

28. $-\frac{3}{4}$; $y = -\frac{3}{4}$; **no amplitude; 360°**

State the vertical shift, amplitude, period, and phase shift of each function. Then graph the function. **33–42. See pp. 811A–811N.**

33. $y = 2\sin[3(\theta - 45°)] + 1$
34. $y = 4\cos[2(\theta + 30°)] - 5$
35. $y = 3\csc\left[\frac{1}{2}(\theta + 60°)\right] - 3.5$
36. $y = 6\cot\left[\frac{2}{3}(\theta - 90°)\right] + 0.75$
37. $y = \frac{1}{4}\cos(2\theta - 150°) + 1$
38. $y = \frac{2}{5}\tan(6\theta + 135°) - 4$
39. $y = 3 + 2\sin\left[2\left(\theta + \frac{\pi}{4}\right)\right]$
40. $y = 4 + 5\sec\left[\frac{1}{3}\left(\theta + \frac{2\pi}{3}\right)\right]$
41. Graph $y = 3 - \frac{1}{2}\cos\theta$ and $y = 3 - \frac{1}{2}\cos(\theta + \pi)$. How do the graphs compare?
42. Compare the graphs of $y = -\sin\left[\frac{1}{4}\left(\theta - \frac{\pi}{2}\right)\right]$ and $y = \cos\left[\frac{1}{4}\left(\theta + \frac{3\pi}{2}\right)\right]$.

43. **MUSIC** When represented on an oscilloscope, the note A above middle C has period of $\frac{1}{440}$. Which of the following can be an equation for an oscilloscope graph of this note? The amplitude of the graph is K. **c**
    a. $y = K\sin 220\pi t$ b. $y = K\sin 440\pi t$ c. $y = K\sin 880\pi t$

WebQuest
Translations of trigonometric graphs can be used to describe temperature trends. Visit www.algebra2.com/webquest to continue work on your WebQuest project.

ZOOLOGY For Exercises 44–46, use the following information.
The population of predators and prey in a closed ecological system tends to vary periodically over time. In a certain system, the population of owls O can be represented by $O = 150 + 30\sin\left(\frac{\pi}{10}t\right)$ where t is the time in years since January 1, 2001. In that same system, the population of mice M can be represented by $M = 600 + 300\sin\left(\frac{\pi}{10}t + \frac{\pi}{20}\right)$. **44. 180; 5 yr**

44. Find the maximum number of owls. After how many years does this occur?
45. What is the minimum number of mice? How long does it take for the population of mice to reach this level? **300; 14.5 yr**
46. Why would the maximum owl population follow behind the population of mice? **See margin.**

47. $h = 9 + 6\sin\left[\frac{\pi}{9}(t - 1.5)\right]$

47. **TIDES** The height of the water in a harbor rose to a maximum height of 15 feet at 6:00 P.M. and then dropped to a minimum level of 3 feet by 3:00 A.M. Assume that the water level can be modeled by the sine function. Write an equation that represents the height h of the water t hours after noon on the first day.

Online Research **Data Update** Use the Internet or another resource to find tide data for a location of your choice. Write a sine function to represent your data. Then graph the function. Visit www.algebra2.com/data_update to learn more.

www.algebra2.com/self_check_quiz

Answer

46. Sample answer: When the prey (mouse) population is at its greatest the predator will consume more and the predator population will grow while the prey population falls.

Study Guide and Intervention, p. 843 (shown) and p. 844

Horizontal Translations When a constant is subtracted from the angle measure in a trigonometric function, a **phase shift** of the graph results.

Phase Shift	The horizontal phase shift of the graphs of the functions $y = a\sin b(\theta - h)$, $y = a\cos b(\theta - h)$, and $y = a\tan b(\theta - h)$ is h, where $b > 0$. If $h > 0$, the shift is to the right. If $h < 0$, the shift is to the left.

Example **State the amplitude, period, and phase shift for $y = \frac{1}{2}\cos 3\left(\theta - \frac{\pi}{2}\right)$. Then graph the function.**

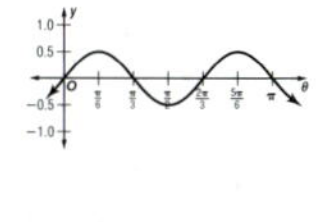

Amplitude: $a = \left|\frac{1}{2}\right|$ or $\frac{1}{2}$
Period: $\frac{2\pi}{|b|} = \frac{2\pi}{|3|}$ or $\frac{2\pi}{3}$
Phase Shift: $h = \frac{\pi}{2}$
The phase shift is to the right since $\frac{\pi}{2} > 0$.

Exercises

State the amplitude, period, and phase shift for each function. Then graph the function.

1. $y = 2\sin(\theta + 60°)$ **2; 360°; 60° to the left**

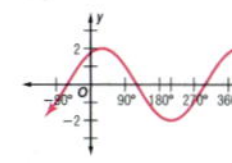

2. $y = \tan\left(\theta - \frac{\pi}{2}\right)$ **no amplitude; π; $\frac{\pi}{2}$ to the right**

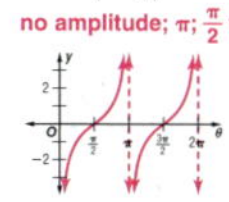

3. $y = 3\cos(\theta - 45°)$ **3; 360°; 45° to the right**

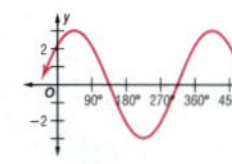

4. $y = \frac{1}{2}\sin 3\left(\theta - \frac{\pi}{3}\right)$ **$\frac{1}{2}$; $\frac{2\pi}{3}$; $\frac{\pi}{3}$ to the right**

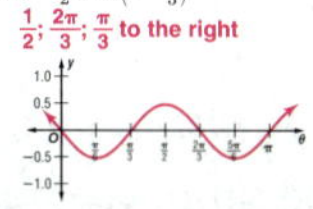

Skills Practice, p. 845 and Practice, p. 846 (shown)

State the vertical shift, amplitude, period, and phase shift for each function. Then graph the function.

1. $y = \frac{1}{2}\tan\left(\theta - \frac{\pi}{2}\right)$ **no vertical shift; no amplitude; π; $\frac{\pi}{2}$**

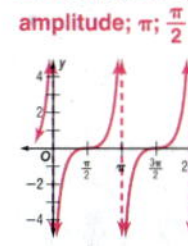

2. $y = 2\cos(\theta + 30°) + 3$ **3; 2; 360; −30°**

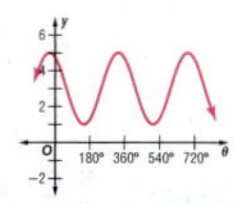

3. $y = 3\csc(2\theta + 60°) - 2.5$ **−2.5; no amplitude; 180°; −60°**

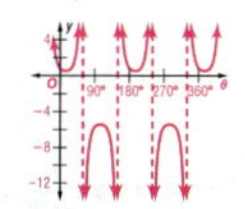

ECOLOGY For Exercises 4–6, use the following information.
The population of an insect species in a stand of trees follows the growth cycle of a particular tree species. The insect population can be modeled by the function $y = 40 + 30\sin 6t$, where t is the number of years since the stand was first cut in November, 1920.

4. How often does the insect population reach its maximum level? **every 60 yr**
5. When did the population last reach its maximum? **1995**
6. What condition in the stand do you think corresponds with a minimum insect population? **Sample answer: The species on which the insect feeds has been cut.**

BLOOD PRESSURE For Exercises 7–9, use the following information.
Jason's blood pressure is 110 over 70, meaning that the pressure oscillates between a maximum of 110 and a minimum of 70. Jason's heart rate is 45 beats per minute. The function that represents Jason's blood pressure P can be modeled using a sine function with no phase shift.

7. Find the amplitude, midline, and period in seconds of the function. **20; P = 90; $1\frac{1}{3}$ s**
8. Write a function that represents Jason's blood pressure P after t seconds. **P = 20 sin 270t + 90**
9. Graph the function.

Jason's Blood Pressure

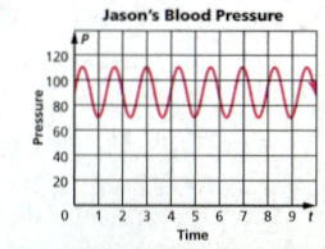

Reading to Learn Mathematics, p. 847

ELL

Pre-Activity **How can translations of trigonometric graphs be used to show animal populations?**

Read the introduction to Lesson 14-2 at the top of page 769 in your textbook.

According to the model given in your textbook, what would be the estimated rabbit population for January 1, 2005? **1200**

Reading the Lesson

1. Determine whether the graph of each function represents a shift of the parent function *to the left*, *to the right*, *upward*, or *downward*. (Do not actually graph the functions.)
   a. $y = \sin(\theta + 90°)$ **to the left**
   b. $y = \sin\theta + 3$ **upward**
   c. $y = \cos\left(\theta - \frac{\pi}{3}\right)$ **to the right**
   d. $y = \tan\theta - 4$ **downward**
2. Determine whether the graph of each function has an *amplitude change*, *period change*, *phase shift*, or *vertical shift* compared to the graph of the parent function. (More than one of these may apply to each function. Do not actually graph the functions.)
   a. $y = 3\sin\left(\theta + \frac{5\pi}{6}\right)$ **amplitude change and phase shift**
   b. $y = \cos(2\theta + 70°)$ **period change and phase shift**
   c. $y = -4\cos 3\theta$ **amplitude change and period change**
   d. $y = \sec\frac{1}{2}\theta + 3$ **period change and vertical shift**
   e. $y = \tan\left(\theta - \frac{\pi}{4}\right) - 1$ **phase shift and vertical shift**
   f. $y = 2\sin\left(\frac{1}{3}\theta + \frac{\pi}{6}\right) - 4$ **amplitude change, period change, phase shift, and vertical shift**

Helping You Remember

3. Many students have trouble remembering which of the functions $y = \sin(\theta + \alpha)$ and $y = \sin(\theta - \alpha)$ represents a shift to the left and which represents a shift to the right. Using $\alpha = 45°$, explain a good way to remember which is which.

Sample answer: Although sine curves are infinitely repeating periodic graphs, think of y = sin x starting a period or cycle at (0, 0). Then y = sin (θ + 45°) "starts early" at (−45°), a shift of 45° to the left, while y = sin (θ − 45°) "starts late" at 45°, a shift of 45° to the right.

Enrichment, p. 848

Translating Graphs of Trigonometric Functions

Three graphs are shown at the right:
$y = 3\sin 2\theta$
$y = 3\sin 2(\theta - 30°)$
$y + 4 = 3\sin 2\theta$

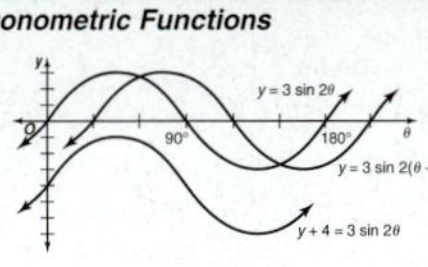

Replacing θ with $(\theta - 30°)$ translates the graph to the right. Replacing y with $y + 4$ translates the graph 4 units down.

Example **Graph one cycle of $y = 6\cos(5\theta + 80°) + 2$.**

Step 1 Transform the equation into the form $y - k = a\cos b(\theta - h)$.
$y - 2 = 6\cos 5(\theta + 16°)$

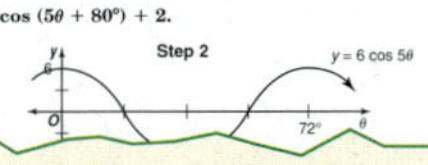

4 Assess

Open-Ended Assessment

Writing Have students write a summary explaining how to use the equation for a trigonometric function to identify how its graph is shifted vertically and/or horizontally from its parent graph.

Getting Ready for Lesson 14-3

PREREQUISITE SKILL Students will find the value of a trigonometric function in Lesson 14-3. Students should be familiar with values of the sine, cosine, and tangent functions for various reference angles in order to determine the sign of the value. Use Exercises 66–73 to determine your students' familiarity with reference angles.

Assessment Options

Quiz (Lessons 14-1 and 14-2) is available on p. 893 of the *Chapter 14 Resource Masters*.

Answers

52. amplitude: does not exist; period: 360° or 2π

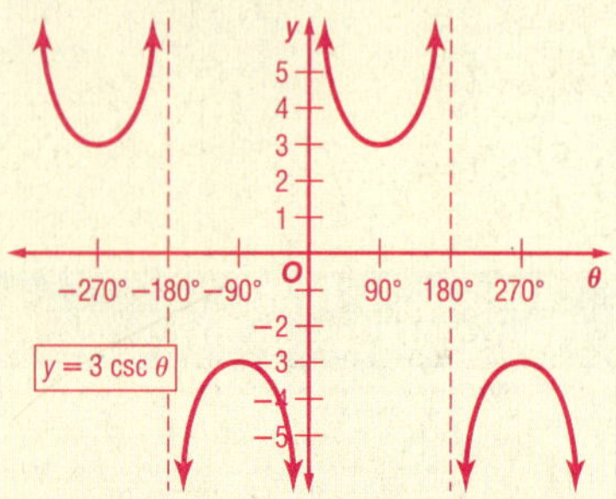

53. amplitude: 1; period: 720° or 4π

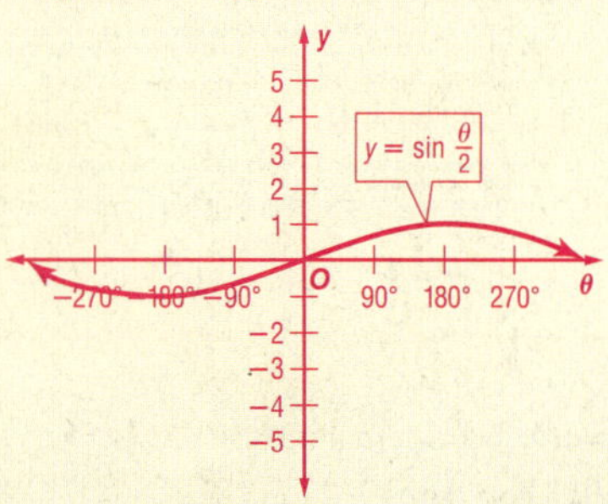

54. amplitude: does not exist; period: 270° or $\frac{3\pi}{2}$

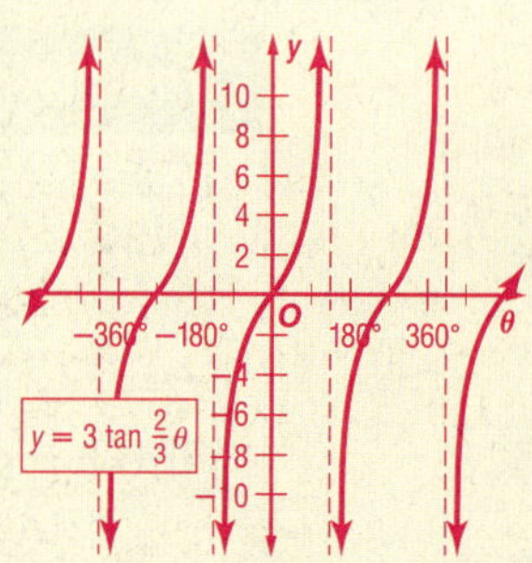

63. $\frac{5a - 13}{(a-2)(a-3)}$

65. $\frac{3y^2 + 10y + 5}{2(y-5)(y+3)}$

48. **CRITICAL THINKING** The graph of $y = \cot \theta$ is a transformation of the graph of $y = \tan \theta$. Determine a, b, and h so that $\cot \theta = a \tan [b(\theta - h)]$ for all values of θ for which each function is defined. **$a = -1$, $b = 1$, $h = \frac{\pi}{2}$**

49. **WRITING IN MATH** Answer the question that was posed at the beginning of the lesson. **See pp. 811A–811N.**

How can translations of trigonometric graphs be used to show animal populations?

Include the following in your answer:
- a description of what each number in the equation $R = 1200 + 250 \sin \frac{1}{2}\pi t$ represents, and
- a comparison of the graphs of $y = a \cos bx$, $y = a \cos bx + k$, and $y = a \cos [b(x - h)]$.

Standardized Test Practice

50. Which equation is represented by the graph? **B**

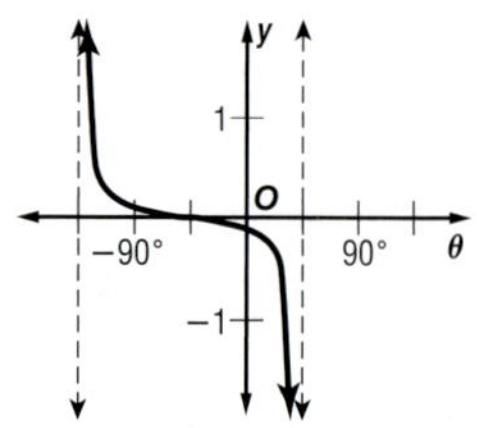

(A) $y = \cot (\theta + 45°)$
(B) $y = \cot (\theta - 45°)$
(C) $y = \tan (\theta + 45°)$
(D) $y = \tan (\theta - 45°)$

51. Identify the equation for a sine function of period 90°, after a phase shift 20° to the left. **D**

(A) $y = \sin [0.25(\theta - 20°)]$
(B) $y = \sin [4(\theta - 20°)]$
(C) $y = \sin [0.25(\theta + 20°)]$
(D) $y = \sin [4(\theta + 20°)]$

Maintain Your Skills

Mixed Review **Find the amplitude, if it exists, and period of each function. Then graph each function.** *(Lesson 14-1)* **52–54. See margin.**

52. $y = 3 \csc \theta$
53. $y = \sin \frac{\theta}{2}$
54. $y = 3 \tan \frac{2}{3}\theta$

Find each value. *(Lesson 13-7)*

55. $\sin \left(\text{Cos}^{-1} \frac{2}{3}\right)$ **0.75**
56. $\cos \left(\text{Cos}^{-1} \frac{4}{7}\right)$ **0.57**
57. $\text{Sin}^{-1} \left(\sin \frac{5}{6}\right)$ **0.83**
58. $\cos \left(\text{Tan}^{-1} \frac{3}{4}\right)$ **0.8**

59. **GEOMETRY** Find the total number of diagonals that can be drawn in a decagon. *(Lesson 12-2)* **35**

Solve each equation. Round to the nearest hundredth. *(Lesson 10-4)*

60. $4^x = 24$ **2.29**
61. $4.3^{3x+1} = 78.5$ **0.66**
62. $7^{x-2} = 53^{-x}$ **0.66**

Simplify each expression. *(Lesson 9-4)*

63. $\frac{3}{a-2} + \frac{2}{a-3}$
64. $\frac{w + 12}{4w - 16} - \frac{w + 4}{2w - 8}$ **$-\frac{1}{4}$**
65. $\frac{3y + 1}{2y - 10} + \frac{1}{y^2 - 2y - 15}$

Getting Ready for the Next Lesson **PREREQUISITE SKILL** **Find the value of each function.**
*(To review **reference angles**, see Lesson 13-3.)*

66. $\cos 150°$ **$-\frac{\sqrt{3}}{2}$**
67. $\tan 135°$ **−1**
68. $\sin \frac{3\pi}{2}$ **−1**
69. $\cos \left(-\frac{\pi}{3}\right)$ **$\frac{1}{2}$**
70. $\sin (-\pi)$ **0**
71. $\tan \left(-\frac{5\pi}{6}\right)$ **$\frac{\sqrt{3}}{3}$**
72. $\cos 225°$ **$-\frac{\sqrt{2}}{2}$**
73. $\tan 405°$ **1**

14-3 Trigonometric Identities

What You'll Learn

- Use identities to find trigonometric values.
- Use trigonometric identities to simplify expressions.

Vocabulary

- trigonometric identity

How can trigonometry be used to model the path of a baseball?

A model for the height of a baseball after it is hit as a function of time can be determined using trigonometry. If the ball is hit with an initial velocity of v feet per second at an angle of θ from the horizontal, then the height h of the ball after t seconds can be represented by $h = \left(\frac{-16}{v^2 \cos^2 \theta}\right)t^2 + \left(\frac{\sin \theta}{\cos \theta}\right)t + h_0$, where h_0 is the height of the ball in feet the moment it is hit.

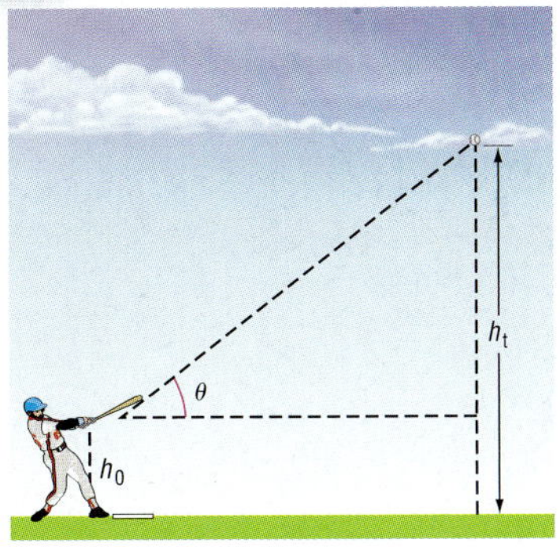

FIND TRIGONOMETRIC VALUES In the equation above, the second term $\left(\frac{\sin \theta}{\cos \theta}\right)t$ can also be written as $(\tan \theta)t$. $\left(\frac{\sin \theta}{\cos \theta}\right)t = (\tan \theta)t$ is an example of a trigonometric identity. A **trigonometric identity** is an equation involving trigonometric functions that is true for all values for which every expression in the equation is defined.

The identity $\tan \theta = \frac{\sin \theta}{\cos \theta}$ is true except for angle measures such as 90°, 270°, 450°, …, 90° + 180° · k. The cosine of each of these angle measures is 0, so none of the expressions tan 90°, tan 270°, tan 450°, and so on, are defined. An identity similar to this is $\cot \theta = \frac{\cos \theta}{\sin \theta}$.

These identities are sometimes called *quotient identities*. These and other basic trigonometric identities are listed below.

Key Concept — *Basic Trigonometric Identities*

Quotient Identities	$\tan \theta = \frac{\sin \theta}{\cos \theta}$	$\cot \theta = \frac{\cos \theta}{\sin \theta}$	
Reciprocal Identities	$\csc \theta = \frac{1}{\sin \theta}$	$\sec \theta = \frac{1}{\cos \theta}$	$\cot \theta = \frac{1}{\tan \theta}$
Pythagorean Identities	$\cos^2 \theta + \sin^2 \theta = 1$ $\tan^2 \theta + 1 = \sec^2 \theta$ $\cot^2 \theta + 1 = \csc^2 \theta$		

You can use trigonometric identities to find values of trigonometric functions.

14-3 Lesson Notes

1 Focus

5-Minute Check Transparency 14-3 Use as a quiz or review of Lesson 14-2.

Mathematical Background notes are available for this lesson on page 760D.

How can trigonometry be used to model the path of a baseball?

Ask students:

- What assumptions are made when using this function as a model for the height of the ball above the ground? **Sample answers: There is no wind; there is no friction as the ball passes through the air; the ball does not hit an obstruction.**
- To hit the ball the farthest horizontal distance, is it better for the angle θ to be a large angle or a small angle? **Answers will vary. Ignoring friction, an angle measure of 45° will result in the farthest horizontal distance.**
- What might be a reasonable value for h_0? **Sample answer: 3 ft**

Resource Manager

Workbook and Reproducible Masters

Chapter 14 Resource Masters

- Study Guide and Intervention, pp. 849–850
- Skills Practice, p. 851
- Practice, p. 852
- Reading to Learn Mathematics, p. 853
- Enrichment, p. 854

Graphing Calculator and Spreadsheet Masters, p. 54

Transparencies

5-Minute Check Transparency 14-3
Answer Key Transparencies

Technology

Interactive Chalkboard

2 Teach

FIND TRIGONOMETRIC VALUES

Teaching Tip You can use the familiar definitions of sine, cosine, and tangent as the ratios of the opposite side, adjacent side, and hypotenuse of a right triangle to show why $\frac{\sin\theta}{\cos\theta} = \tan\theta$.

In-Class Example — Power Point®

1

a. Find $\tan\theta$ if $\sec\theta = -2$ and $180° < \theta < 270°$. $\tan\theta = \sqrt{3}$

b. Find $\sin\theta$ if $\cos\theta = -\frac{1}{2}$ and $90° < \theta < 180°$. $\sin\theta = \frac{\sqrt{3}}{2}$

SIMPLIFY EXPRESSIONS

In-Class Example — Power Point®

2 Simplify $\sin\theta(\csc\theta - \sin\theta)$. $\cos^2\theta$

Example 1 Find a Value of a Trigonometric Function

a. Find $\cos\theta$ if $\sin\theta = -\frac{3}{5}$ and $90° < \theta < 180°$.

$\cos^2\theta + \sin^2\theta = 1$	Trigonometric identity
$\cos^2\theta = 1 - \sin^2\theta$	Subtract $\sin^2\theta$ from each side.
$\cos^2\theta = 1 - \left(\frac{3}{5}\right)^2$	Substitute $\frac{3}{5}$ for $\sin\theta$.
$\cos^2\theta = 1 - \frac{9}{25}$	Square $\frac{3}{5}$.
$\cos^2\theta = \frac{16}{25}$	Subtract.
$\cos\theta = \pm\frac{4}{5}$	Take the square root of each side.

Since θ is in the second quadrant, $\cos\theta$ is negative. Thus, $\cos\theta = -\frac{4}{5}$.

b. Find $\csc\theta$ if $\cot\theta = -\frac{1}{4}$ and $270° < \theta < 360°$.

$\cot^2\theta + 1 = \csc^2\theta$	Trigonometric identity
$\left(-\frac{1}{4}\right)^2 + 1 = \csc^2\theta$	Substitute $-\frac{1}{4}$ for $\cot\theta$.
$\frac{1}{16} + 1 = \csc^2\theta$	Square $-\frac{1}{4}$.
$\frac{17}{16} = \csc^2\theta$	Add.
$\pm\frac{\sqrt{17}}{4} = \csc\theta$	Take the square root of each side.

Since θ is in the fourth quadrant, $\csc\theta$ is negative. Thus, $\csc\theta = -\frac{\sqrt{17}}{4}$.

SIMPLIFY EXPRESSIONS Trigonometric identities can also be used to simplify expressions containing trigonometric functions. Simplifying an expression that contains trigonometric functions means that the expression is written as a numerical value or in terms of a single trigonometric function, if possible.

Example 2 Simplify an Expression

Simplify $\frac{\csc^2\theta - \cot^2\theta}{\cos\theta}$.

$\frac{\csc^2\theta - \cot^2\theta}{\cos\theta} = \frac{\frac{1}{\sin^2\theta} - \frac{\cos^2\theta}{\sin^2\theta}}{\cos\theta}$	$\csc^2\theta = \frac{1}{\sin^2\theta}$, $\cot^2\theta = \frac{\cos^2\theta}{\sin^2\theta}$
$= \frac{\frac{1-\cos^2\theta}{\sin^2\theta}}{\cos\theta}$	Add.
$= \frac{\frac{\sin^2\theta}{\sin^2\theta}}{\cos\theta}$	$1 - \cos^2\theta = \sin^2\theta$
$= \frac{1}{\cos\theta}$	$\frac{\sin^2\theta}{\sin^2\theta} = 1$
$= \sec\theta$	$\frac{1}{\cos\theta} = \sec\theta$

DAILY INTERVENTION — Differentiated Instruction

Logical Have students work in groups of three. Ask each group to choose one of the identities in the Key Concept box on p. 777 and work together to demonstrate that it is true. Students should verify their results using the definitions of sine, cosine, and tangent in terms of the sides of a right triangle.

Example 3 Simplify and Use an Expression

BASEBALL **Refer to the application at the beginning of the lesson. Rewrite the equation in terms of tan θ.**

$$h = \left(\frac{-16}{v^2 \cos^2 \theta}\right)t^2 + \left(\frac{\sin \theta}{\cos \theta}\right)t + h_0 \quad \text{Original equation}$$

$$= \frac{-16}{v^2}\left(\frac{1}{\cos^2 \theta}\right)t^2 + \left(\frac{\sin \theta}{\cos \theta}\right)t + h_0 \quad \text{Factor.}$$

$$= \frac{-16}{v^2}\left(\frac{1}{\cos^2 \theta}\right)t^2 + (\tan \theta)t + h_0 \quad \frac{\sin \theta}{\cos \theta} = \tan \theta$$

$$= \frac{-16}{v^2}(\sec^2 \theta)t^2 + (\tan \theta)t + h_0 \quad \text{Since } \frac{1}{\cos \theta} = \sec \theta, \frac{1}{\cos^2 \theta} = \sec^2 \theta.$$

$$= \frac{-16}{v^2}(1 + \tan^2 \theta)t^2 + (\tan \theta)t + h_0 \quad \sec^2 \theta = 1 + \tan^2 \theta$$

Thus, $\left(\frac{-16}{v^2 \cos^2 \theta}\right)t^2 + \left(\frac{\sin \theta}{\cos \theta}\right)t + h_0 = \frac{-16}{v^2}(1 + \tan^2 \theta)t^2 + (\tan \theta)t + h_0$.

In-Class Example

Power Point®

3 **BASEBALL** Refer to the application at the beginning of the lesson. Rewrite the equation in terms of sec θ.

$h = \frac{-16}{v^2}(\sec^2 \theta)t^2 + \left(\sqrt{\sec^2 \theta - 1}\right)t + h_0$

Check for Understanding

Concept Check

1–3. See margin.

1. **Describe** how you can determine the quadrant in which the terminal side of angle α lies if $\sin \alpha = -\frac{1}{4}$.
2. **Explain** why the Pythagorean identities are so named.
3. **OPEN ENDED** Explain what it means to simplify a trigonometric expression.

Guided Practice

GUIDED PRACTICE KEY	
Exercises	Examples
4–7	1
9–11	2
12	3

Find the value of each expression.

4. $\tan \theta$, if $\sin \theta = \frac{1}{2}$; $90° \leq \theta < 180°$ $-\frac{\sqrt{3}}{3}$
5. $\csc \theta$, if $\cos \theta = -\frac{3}{5}$; $180° \leq \theta < 270°$ $-\frac{5}{4}$
6. $\cos \theta$, if $\sin \theta = \frac{4}{5}$; $0° \leq \theta < 90°$ $\frac{3}{5}$
7. $\sec \theta$, if $\tan \theta = -1$; $270° < \theta < 360°$ $\sqrt{2}$

Simplify each expression.

8. $\csc \theta \cos \theta \tan \theta$ **1**
9. $\sec^2 \theta - 1$ $\tan^2 \theta$
10. $\frac{\tan \theta}{\sin \theta}$ $\sec \theta$
11. $\sin \theta (1 + \cot^2 \theta)$ $\csc \theta$

Application

12. **PHYSICAL SCIENCE** When a person moves along a circular path, the body leans away from a vertical position. The nonnegative acute angle that the body makes with the vertical is called the *angle of inclination* and is represented by the equation $\tan \theta = \frac{v^2}{gR}$, where R is the radius of the circular path, v is the speed of the person in meters per second, and g is the acceleration due to gravity, 9.8 meters per second squared. Write an equivalent expression using $\sin \theta$ and $\cos \theta$. $\sin \theta = \cos \theta \frac{v^2}{gR}$

3 Practice/Apply

Study Notebook

Have students—

- add the definitions/examples of the vocabulary terms to their Vocabulary Builder worksheets for Chapter 14.
- record the basic trigonometric identities from the Key Concept box on p. 777.
- include any other item(s) that they find helpful in mastering the skills in this lesson.

Practice and Apply

Practice and Apply

15. $-\sqrt{5}$
16. $2\sqrt{2}$

Find the value of each expression.

13. $\tan \theta$, if $\cot \theta = 2$; $0° \leq \theta < 90°$ $\frac{1}{2}$
14. $\sin \theta$, if $\cos \theta = \frac{2}{3}$; $0° \leq \theta < 90°$ $\frac{\sqrt{5}}{3}$
15. $\sec \theta$, if $\tan \theta = -2$; $90° < \theta < 180°$
16. $\tan \theta$, if $\sec \theta = -3$; $180° < \theta < 270°$

www.algebra2.com/extra_examples

About the Exercises...

Organization by Objective

- **Find Trigonometric Values:** 13–24
- **Simplify Expressions:** 25–36

Odd/Even Assignments

Exercises 13–36 are structured so that students practice the same concepts whether they are assigned odd or even problems.

Assignment Guide

Basic: 13–35 odd, 37–41, 43–58

Average: 13–35 odd, 37–41, 43–58

Advanced: 14–36 even, 37–54 (optional: 55–58)

All: Practice Quiz 1 (1–5)

Answers

1. Sample answer: The sine function is negative in the third and fourth quadrants. Therefore, the terminal side of the angle must lie in one of those two quadrants.
2. Sample answer: Pythagorean identities are derived by applying the Pythagorean Theorem to trigonometric concepts.
3. Sample answer: Simplifying a trigonometric expression means writing the expression as a numerical value or in terms of a single trigonometric function, if possible.

Study Guide and Intervention, p. 849 (shown) and p. 850

Find Trigonometric Values A **trigonometric identity** is an equation involving trigonometric functions that is true for all values for which every expression in the equation is defined.

Basic Trigonometric Identities				
	Quotient Identities	$\tan\theta = \frac{\sin\theta}{\cos\theta}$	$\cot\theta = \frac{\cos\theta}{\sin\theta}$	
	Reciprocal Identities	$\csc\theta = \frac{1}{\sin\theta}$	$\sec\theta = \frac{1}{\cos\theta}$	$\cot\theta = \frac{1}{\tan\theta}$
	Pythagorean Identities	$\cos^2\theta + \sin^2\theta = 1$	$\tan^2\theta + 1 = \sec^2\theta$	$\cot^2\theta + 1 = \csc^2\theta$

Example **Find the value of cot θ if $\csc\theta = -\frac{11}{5}$; $180° < \theta < 270°$.**

$\cot^2\theta + 1 = \csc^2\theta$ Trigonometric identity

$\cot^2\theta + 1 = \left(-\frac{11}{5}\right)^2$ Substitute $-\frac{11}{5}$ for csc θ.

$\cot^2\theta + 1 = \frac{121}{25}$ Square $-\frac{11}{5}$.

$\cot^2\theta = \frac{96}{25}$ Subtract 1 from each side.

$\cot\theta = \pm\frac{4\sqrt{6}}{5}$ Take the square root of each side.

Since θ is in the third quadrant, cot θ is positive, Thus $\cot\theta = \frac{4\sqrt{6}}{5}$.

Exercises

Find the value of each expression.

1. tan θ, if cot θ = 4; $180° < \theta < 270°$ $\frac{1}{4}$
2. csc θ, if $\cos\theta = \frac{\sqrt{3}}{2}$; $0° \le \theta < 90°$ 2
3. cos θ, if $\sin\theta = \frac{3}{5}$; $0° \le \theta < 90°$ $\frac{4}{5}$
4. sec θ, if $\sin\theta = \frac{1}{3}$; $0° \le \theta < 90°$ $\frac{3\sqrt{2}}{4}$
5. cos θ, if $\tan\theta = -\frac{4}{3}$; $90° < \theta < 180°$ $-\frac{3}{5}$
6. tan θ, if $\sin\theta = \frac{3}{7}$; $0° \le \theta < 90°$ $\frac{3\sqrt{10}}{20}$
7. sec θ, if $\cos\theta = -\frac{7}{8}$; $90° < \theta < 180°$ $-\frac{8}{7}$
8. sin θ, if $\cos\theta = \frac{6}{7}$; $270° \le \theta < 360°$ $-\frac{\sqrt{13}}{7}$
9. cot θ, if $\csc\theta = \frac{12}{5}$; $90° < \theta < 180°$ $-\frac{\sqrt{119}}{5}$
10. sin θ, if $\csc\theta = -\frac{9}{4}$; $270° < \theta < 360°$ $-\frac{4}{9}$

Skills Practice, p. 851 and Practice, p. 852 (shown)

Find the value of each expression.

1. sin θ, if $\cos\theta = \frac{5}{13}$ and $0° \le \theta < 90°$ $\frac{12}{13}$
2. sec θ, if $\sin\theta = -\frac{15}{17}$ and $180° < \theta < 270°$ $-\frac{17}{8}$
3. cot θ, if $\cos\theta = \frac{3}{10}$ and $270° < \theta < 360°$ $-\frac{3\sqrt{91}}{91}$
4. sin θ, if $\cot\theta = \frac{1}{2}$ and $0° \le \theta < 90°$ $\frac{2\sqrt{5}}{5}$
5. cot θ, if $\csc\theta = -\frac{3}{2}$ and $180° < \theta < 270°$ $\frac{\sqrt{5}}{2}$
6. sec θ, if csc θ = −8 and $270° < \theta < 360°$ $\frac{8\sqrt{7}}{21}$
7. sec θ, if tan θ = 4 and $180° < \theta < 270°$ $-\sqrt{17}$
8. sin θ, if $\tan\theta = -\frac{1}{2}$ and $270° < \theta < 360°$ $-\frac{\sqrt{5}}{5}$
9. cot θ, if $\tan\theta = \frac{2}{5}$ and $0° \le \theta < 90°$ $\frac{5}{2}$
10. cot θ, if $\cos\theta = \frac{1}{3}$ and $270° < \theta < 360°$ $-\frac{\sqrt{2}}{4}$

Simplify each expression.

11. $\csc\theta\tan\theta$ sec x
12. $\frac{\sin^2\theta}{\tan^2\theta}$ $\cos^2\theta$
13. $\sin^2\theta\cot^2\theta$ $\cos^2\theta$
14. $\cot^2\theta + 1$ $\csc^2\theta$
15. $\frac{\csc^2\theta - \cot^2\theta}{1 - \cos^2\theta}$ $\csc^2\theta$
16. $\frac{\csc\theta - \sin\theta}{\cos\theta}$ cot θ
17. $\sin\theta + \cos\theta\cot\theta$ csc θ
18. $\frac{\cos\theta}{1 - \sin\theta} - \frac{\cos\theta}{1 + \sin\theta}$ 2 tan θ
19. $\sec^2\theta\cos^2\theta - \tan^2\theta$ $\sec^2\theta$
20. **AERIAL PHOTOGRAPHY** The illustration shows a plane taking an aerial photograph of point *A*. Because the point is directly below the plane, there is no distortion in the image. For any point *B* not directly below the plane, however, the increase in distance creates distortion in the photograph. This is because as the distance from the camera to the point being photographed increases, the exposure of the film reduces by $(\sin\theta)(\csc\theta - \sin\theta)$. Express $(\sin\theta)(\csc\theta - \sin\theta)$ in terms of cos θ only. $\cos^2\theta$

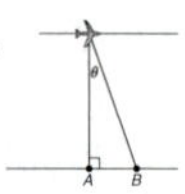

21. **TSUNAMIS** The equation $y = a\sin\theta t$ represents the height of the waves passing a buoy at a time *t* in seconds. Express *a* in terms of csc θt. $a = y\csc\theta t$

Reading to Learn Mathematics, p. 853 ELL

Pre-Activity **How can trigonometry be used to model the path of a baseball?**

Read the introduction to Lesson 14-3 at the top of page 777 in your textbook.

Suppose that a baseball is hit from home plate with an initial velocity of 58 feet per second at an angle of 36° with the horizontal from an initial height of 5 feet. Show the equation that you would use to find the height of the ball 10 seconds after the ball is hit. (Show the formula with the appropriate numbers substituted, but do not do any calculations.)

$h = \left(\frac{-16}{58^2\cos^2 36°}\right)10^2 + \left(\frac{\sin 36°}{\cos 36°}\right)10 + 5$

Reading the Lesson

1. Match each expression from the list on the left with an expression from the list on the right that is equal to it for all values for which each expression is defined. (Some of the expressions from the list on the right may be used more than once or not at all.)

a. $\sec^2\theta - \tan^2\theta$ iii	i. $\frac{1}{\sin\theta}$
b. $\cot^2\theta + 1$ v	ii. tan θ
c. $\frac{\sin\theta}{\cos\theta}$ ii	iii. 1
d. $\sin^2\theta + \cos^2\theta$ iii	iv. sec θ
e. csc θ i	v. $\csc^2\theta$
f. $\frac{1}{\cos\theta}$ iv	vi. cot θ
g. $\frac{\cos\theta}{\sin\theta}$ vi	

2. Write an identity that you could use to find each of the indicated trigonometric values and tell whether that value is positive or negative. (Do not actually find the values.)
   a. tan θ, if $\sin\theta = -\frac{4}{5}$ and $180° < \theta < 270°$ $\tan\theta = \frac{\sin\theta}{\cos\theta}$; positive
   b. sec θ, if tan θ = −3 and $90° < \theta < 180°$ $\tan^2\theta + 1 = \sec^2\theta$; negative

Helping You Remember

3. A good way to remember something new is to relate it to something you already know. How can you use the unit circle definitions of the sine and cosine that you learned in Chapter 13 to help you remember the Pythagorean identity $\cos^2\theta + \sin^2\theta = 1$? **Sample answer: On a unit circle, $x = \cos\theta$ and $y = \sin\theta$. The equation of the unit circle is $x^2 + y^2 = 1$, so this is equivalent to the equation $\cos^2\theta + \sin^2\theta = 1$.**

Homework Help

For Exercises	See Examples
13–24	1
25–36	2
37–43	3

Extra Practice
See page 860.

20. $-\frac{3\sqrt{5}}{5}$

22. $\frac{4}{5}$

More About. . .

Amusement Parks

The oldest operational carousel in the United States is the Flying Horse Carousel at Martha's Vineyard, Massachusetts.

Source: Martha's Vineyard Preservation Trust

Find the value of each expression.

17. csc θ, if $\cos\theta = -\frac{3}{5}$; $90° < \theta < 180°$ $\frac{5}{4}$
18. cos θ, if $\sec\theta = \frac{5}{3}$; $270° < \theta < 360°$ $\frac{3}{5}$
19. cos θ, if $\sin\theta = \frac{1}{2}$; $0° \le \theta < 90°$ $\frac{\sqrt{3}}{2}$
20. csc θ, if $\cos\theta = -\frac{2}{3}$; $180° < \theta < 270°$
21. tan θ, if $\cos\theta = \frac{4}{5}$; $0° \le \theta < 90°$ $\frac{3}{4}$
22. cos θ, if $\csc\theta = -\frac{5}{3}$; $270° < \theta < 360°$
23. sec θ, if $\sin\theta = \frac{3}{4}$; $90° < \theta < 180°$ $-\frac{4\sqrt{7}}{7}$
24. sin θ, if tan θ = 4; $180° < \theta < 270°$ $-\frac{4\sqrt{17}}{17}$

Simplify each expression.

25. $\cos\theta\csc\theta$ **cot θ**
26. $\tan\theta\cot\theta$ **1**
27. $\sin\theta\cot\theta$ **cos θ**
28. $\cos\theta\tan\theta$ **sin θ**
29. $2(\csc^2\theta - \cot^2\theta)$ **2**
30. $3(\tan^2\theta - \sec^2\theta)$ **−3**
31. $\frac{\cos\theta\csc\theta}{\tan\theta}$ $\cot^2\theta$
32. $\frac{\sin\theta\csc\theta}{\cot\theta}$ **tan θ**
33. $\frac{1 - \cos^2\theta}{\sin^2\theta}$ **1**
34. $\frac{1 - \sin^2\theta}{\sin^2\theta}$ $\cot^2\theta$
35. $\frac{\sin^2\theta + \cos^2\theta}{\sin^2\theta}$ $\csc^2\theta$
36. $\frac{\tan^2\theta - \sin^2\theta}{\tan^2\theta\sin^2\theta}$ **1**

AMUSEMENT PARKS For Exercises 37–39, use the following information.
Suppose a child is riding on a merry-go-round and is seated on an outside horse. The diameter of the merry-go-round is 16 meters.

37. If the sine of the angle of inclination of the child is $\frac{1}{5}$, what is the angle of inclination made by the child? Refer to Exercise 12 for information on the angle of inclination. **about 11.5°**
38. What is the velocity of the merry-go-round? **about 4 m/s**
39. If the speed of the merry-go-round is 3.6 meters per second, what is the value of the angle of inclination of a rider? **about 9.4°**

LIGHTING For Exercises 40 and 41, use the following information.
The amount of light that a source provides to a surface is called the illuminance. The illuminance *E* in foot candles on a surface is related to the distance *R* in feet from the light source. The formula $\sec\theta = \frac{I}{ER^2}$, where *I* is the intensity of the light source measured in candles and θ is the angle between the light beam and a line perpendicular to the surface, can be used in situations in which lighting is important.

40. Solve the formula in terms of *E*. $E = \frac{I\cos\theta}{R^2}$
41. Is the equation in Exercise 40 equivalent to $R^2 = \frac{I\tan\theta\cos\theta}{E}$? Explain. **No; $R^2 = \frac{I\tan\theta\cos\theta}{E}$ simplifies to $E = \frac{I\sin\theta}{R^2}$.**

ELECTRONICS For Exercises 42 and 43, use the following information.
When an alternating current of frequency *f* and a peak current *I* pass through a resistance *R*, then the power delivered to the resistance at time *t* seconds is $P = I^2R - I^2R\cos^2(2ft\pi)$.

42. Write an expression for the power in terms of $\sin^2 2ft\pi$. $P = I^2R\sin^2 2\pi ft$
43. Write an expression for the power in terms of $\tan^2 2ft\pi$. $P = I^2R - \frac{I^2R}{1 + \tan^2 2\pi ft}$
44. **CRITICAL THINKING** If $\tan\beta = \frac{3}{4}$, find $\frac{\sin\beta\sec\beta}{\cot\beta}$. $\frac{9}{16}$

Enrichment, p. 854

Planetary Orbits

The orbit of a planet around the sun is an ellipse with the sun at one focus. Let the pole of a polar coordinate system be that focus and the polar axis be toward the other focus. The polar equation of an ellipse is

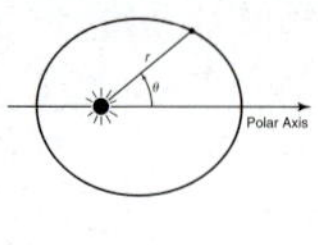

$r = \frac{2ep}{1 - e\cos\theta}$. Since $2p = \frac{b^2}{c}$ and $b^2 = a^2 - c^2$,

$2p = \frac{a^2 - c^2}{c} = \frac{a^2}{c}\left(1 - \frac{c^2}{a^2}\right)$. Because $e = \frac{c}{a}$,

$2p = a\left(\frac{a}{c}\right)\left(1 - \left(\frac{c}{a}\right)^2\right) = a\left(\frac{1}{e}\right)(1 - e^2)$.

Therefore $2ep = a(1 - e^2)$. Substituting into the polar equation of an ellipse yields an equation that is useful for finding distances from the planet to the sun.

$r = \frac{a(1 - e^2)}{1 - e\cos\theta}$

780 Chapter 14 Trigonometric Graphs and Identities

45. **WRITING IN MATH** Answer the question that was posed at the beginning of the lesson. **See margin.**

How can trigonometry be used to model the path of a baseball?

Include the following in your answer:

- an explanation of why the equation at the beginning of the lesson is the same as $y = \frac{-16 \sec^2 \theta}{v^2}x^2 + (\tan \theta)x + h_0$, and
- examples of how you might use this equation for other situations.

Standardized Test Practice

46. If $\sin x = m$ and $0 < x < 90°$, then $\tan x =$ **B**

(A) $\frac{1}{m^2}$. (B) $\frac{m}{\sqrt{1 - m^2}}$. (C) $\frac{1 - m^2}{m}$. (D) $\frac{m}{1 - m^2}$.

47. $\frac{1}{1 + \sin x} + \frac{1}{1 - \sin x} =$ **A**

(A) $2 \sec^2 x$ (B) $-\sec^2 x$ (C) $2 \csc^2 x$ (D) $-\csc^2 x$

Maintain Your Skills

Mixed Review

48–49. See margin for graphs.

State the vertical shift, equation of the midline, amplitude, and period for each function. Then graph the function. *(Lesson 14-2)*

48. $y = \sin \theta - 1$ **−1; y = −1; 1; 360°**
49. $y = \tan \theta + 12$ **12; y = 12; no amplitude; 180°**

Find the amplitude, if it exists, and period of each function. Then graph each function. *(Lesson 14-1)* **50–52. See pp. 811A–811N.**

50. $y = \csc 2\theta$
51. $y = \cos 3\theta$
52. $y = \frac{1}{3} \cot 5\theta$

53. Find the sum of a geometric series for which $a_1 = 48$, $a_n = 3$, and $r = \frac{1}{2}$. *(Lesson 11-4)* **93**

54. Write an equation of a parabola with focus at (11, −1) and whose directrix is $y = 2$. *(Lesson 8-2)* $y = -\frac{1}{6}(x - 11)^2 + \frac{1}{2}$

Getting Ready for the Next Lesson

PREREQUISITE SKILL **Name the property illustrated by each statement.**
*(To review **properties of equality**, see Lesson 1-3.)* **55. Symmetric (=)**

55. If $4 + 8 = 12$, then $12 = 4 + 8$.
56. If $7 + s = 21$, then $s = 14$. **Subt. (=)**
57. If $4x = 16$, then $12x = 48$. **Multiplication (=)**
58. If $q + (8 + 5) = 32$, then $q + 13 = 32$. **Substitution (=)**

Practice Quiz 1 — *Lessons 14-1 through 14-3*

1. Find the amplitude and period of $y = \frac{3}{4} \sin \frac{1}{2}\theta$. Then graph the function. *(Lesson 14-1)* $\frac{3}{4}$, **720° or 4π** **1–2. See pp. 811A–811N for graphs.**
2. State the vertical shift, amplitude, period, and phase shift for $y = 2 \cos\left[\frac{1}{4}\left(\theta - \frac{\pi}{4}\right)\right] - 5$. Then graph the function. *(Lesson 14-2)* **−5, 2, 8π,** $\frac{\pi}{4}$

Find the value of each expression. *(Lesson 14-3)*

3. $\cos \theta$, if $\sin \theta = \frac{4}{5}$; $90° < \theta < 180°$ $-\frac{3}{5}$
4. $\csc \theta$, if $\cot \theta = -\frac{2}{3}$; $270° < \theta < 360°$ $-\frac{\sqrt{13}}{3}$
5. $\sec \theta$, if $\tan \theta = \frac{1}{2}$; $0° < \theta < 90°$ $\frac{\sqrt{5}}{2}$

www.algebra2.com/self_check_quiz

4 Assess

Open-Ended Assessment

Speaking Have students explain how to rewrite and simplify expressions using trigonometric identities, using an example they worked during the lesson.

Getting Ready for Lesson 14-4

PREREQUISITE SKILL Students will verify trigonometric identities in Lesson 14-4 by rewriting them using the properties of equality. Use Exercises 55–58 to determine your students' familiarity with the properties of equality.

Assessment Options

Practice Quiz 1 The quiz provides students with a brief review of the concepts and skills in Lessons 14-1 through 14-3. Lesson numbers are given to the right of the exercises or instruction lines so students can review concepts not yet mastered.

Answer

45. Sample answer: You can use equations to find the height and the horizontal distance of a baseball after it has been hit. The equations involve using the initial angle the ball makes with the ground with the sine function. Answers should include the following information.

- **Both equations are quadratic in nature with a leading negative coefficient. Thus, both are inverted parabolas which model the path of a baseball.**
- **model rockets, hitting a golf ball, kicking a rock**

Answers

48.

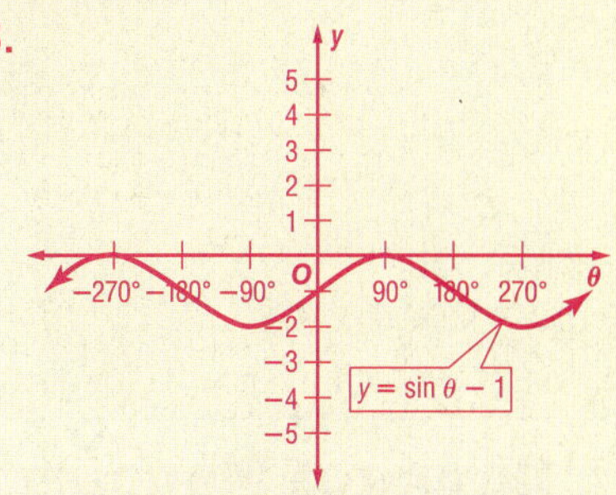

49.

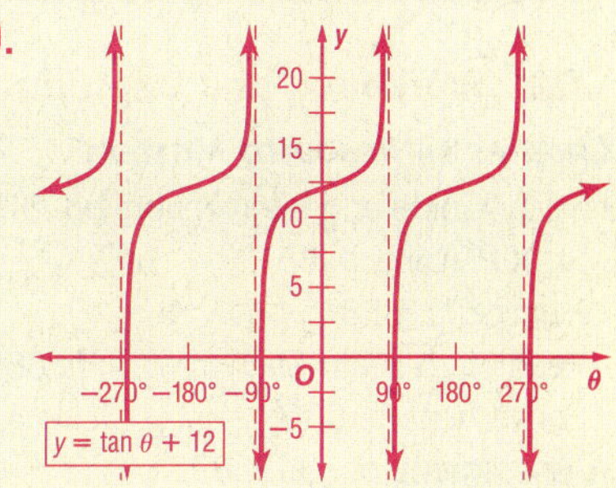

14-4 Lesson Notes

1 Focus

5-Minute Check Transparency 14-4 Use as a quiz or review of Lesson 14-3.

Mathematical Background notes are available for this lesson on p. 760D.

How can you verify trigonometric identities?

Ask students:

- Can you give an example of two functions in x whose values are equal for some values of x but not all? **Sample answer: $y = x$ and $y = |x|$ are the same for $x \geq 0$ but not for $x < 0$.**
- Why is it not sufficient to show that two functions have equal values for specific values of x when trying to justify that two functions are equivalent? **The functions might have equal values for some values of x but not others.**

14-4 Verifying Trigonometric Identities

What You'll Learn

- Verify trigonometric identities by transforming one side of an equation into the form of the other side.
- Verify trigonometric identities by transforming each side of the equation into the same form.

How can you verify trigonometric identities?

Examine the graphs of $y = \tan^2 \theta - \sin^2 \theta$ and $y = \tan^2 \theta \sin^2 \theta$. Recall that when the graphs of two functions coincide, the functions are equivalent. However, the graphs only show a limited range of solutions. It is not sufficient to show some values of θ and conclude that the statement is true for all values of θ. In order to show that the equation $\tan^2 \theta - \sin^2 \theta = \tan^2 \theta \sin^2 \theta$ for all values of θ, you must consider the general case.

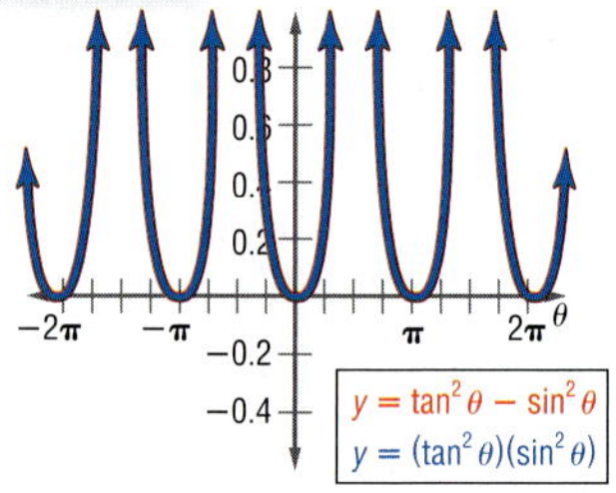

Study Tip

Common Misconception
You cannot perform operations to the quantities from each side of an unverified identity as you do with equations. Until an identity is verified it is not considered an equation, so the properties of equality do not apply.

TRANSFORM ONE SIDE OF AN EQUATION You can use the basic trigonometric identities along with the definitions of the trigonometric functions to verify identities. For example, if you wish to show that $\tan^2 \theta - \sin^2 \theta = \tan^2 \theta \sin^2 \theta$ is an identity, you need to show that it is true for all values of θ.

Verifying an identity is like checking the solution of an equation. You must simplify one or both sides of an equation *separately* until they are the same. In many cases, it is easier to work with only one side of an equation. You may choose either side, but it is often easier to begin with the more complicated side of the equation. Transform that expression into the form of the simpler side.

Example 1 Transform One Side of an Equation

Verify that $\tan^2 \theta - \sin^2 \theta = \tan^2 \theta \sin^2 \theta$ is an identity.

Transform the left side.

$\tan^2 \theta - \sin^2 \theta \stackrel{?}{=} \tan^2 \theta \sin^2 \theta$	Original equation
$\frac{\sin^2 \theta}{\cos^2 \theta} - \sin^2 \theta \stackrel{?}{=} \tan^2 \theta \sin^2 \theta$	$\tan^2 \theta = \frac{\sin^2 \theta}{\cos^2 \theta}$
$\frac{\sin^2 \theta}{\cos^2 \theta} - \frac{\sin^2 \theta \cos^2 \theta}{\cos^2 \theta} \stackrel{?}{=} \tan^2 \theta \sin^2 \theta$	Rewrite using the LCD, $\cos^2 \theta$.
$\frac{\sin^2 \theta - \sin^2 \theta \cos^2 \theta}{\cos^2 \theta} \stackrel{?}{=} \tan^2 \theta \sin^2 \theta$	Subtract.
$\frac{\sin^2 \theta (1 - \cos^2 \theta)}{\cos^2 \theta} \stackrel{?}{=} \tan^2 \theta \sin^2 \theta$	Factor.
$\frac{\sin^2 \theta \sin^2 \theta}{\cos^2 \theta} \stackrel{?}{=} \tan^2 \theta \sin^2 \theta$	$1 - \cos^2 \theta = \sin^2 \theta$
$\frac{\sin^2 \theta}{\cos^2 \theta} \cdot \frac{\sin^2 \theta}{1} \stackrel{?}{=} \tan^2 \theta \sin^2 \theta$	$\frac{ab}{c} = \frac{a}{c} \cdot \frac{b}{1}$
$\tan^2 \theta \sin^2 \theta = \tan^2 \theta \sin^2 \theta$	$\frac{\sin^2 \theta}{\cos^2 \theta} = \tan \theta$

Resource Manager

Workbook and Reproducible Masters

Chapter 14 Resource Masters
- Study Guide and Intervention, pp. 855–856
- Skills Practice, p. 857
- Practice, p. 858
- Reading to Learn Mathematics, p. 859
- Enrichment, p. 860
- Assessment, pp. 893, 895

Transparencies

5-Minute Check Transparency 14-4
Answer Key Transparencies

Technology

Alge2PASS: Tutorial Plus, Lesson 28
Interactive Chalkboard

Standardized Test Practice
A B C D

Example 2 Find an Equivalent Expression

Multiple-Choice Test Item

$\sin \theta \left(\frac{1}{\sin \theta} - \frac{\cos \theta}{\cot \theta}\right) =$

(A) $\cos \theta$ (B) $\sin \theta$ (C) $\cos^2 \theta$ (D) $\sin^2 \theta$

Read the Test Item

Find an expression that is equal to the given expression.

Solve the Test Item

Write a trigonometric identity by using the basic trigonometric identities and the definitions of trigonometric functions to transform the given expression to match one of the choices.

$$\sin \theta \left(\frac{1}{\sin \theta} - \frac{\cos \theta}{\cot \theta}\right) = \sin \theta \left(\frac{1}{\sin \theta} - \frac{\cos \theta}{\frac{\cos \theta}{\sin \theta}}\right) \quad \cot \theta = \frac{\cos \theta}{\sin \theta}$$

$$= \sin \theta \left(\frac{1}{\sin \theta} - \frac{\cos \theta \sin \theta}{\cos \theta}\right) \quad \text{Simplify.}$$

$$= \sin \theta \left(\frac{1}{\sin \theta} - \sin \theta\right) \quad \text{Simplify.}$$

$$= 1 - \sin^2 \theta \quad \text{Distributive property.}$$

$$= \cos^2 \theta \quad 1 - \sin^2 \theta = \cos^2 \theta$$

Since $\sin \theta \left(\frac{1}{\sin \theta} - \frac{\cos \theta}{\cot \theta}\right) = \cos^2 \theta$, the answer is C.

Test-Taking Tip

Verify your answer by choosing values for θ. Then evaluate the original expression and compare to your answer choice.

TRANSFORM BOTH SIDES OF AN EQUATION Sometimes it is easier to transform both sides of an equation separately into a common form. The following suggestions may be helpful as you verify trigonometric identities.

- Substitute one or more basic trigonometric identities to simplify an expression.
- Factor or multiply to simplify an expression.
- Multiply both the numerator and denominator by the same trigonometric expression.
- Write both sides of the identity in terms of sine and cosine only. Then simplify each side as much as possible.

Example 3 Verify by Transforming Both Sides

Verify that $\sec^2 \theta - \tan^2 \theta = \tan \theta \cot \theta$ is an identity.

$$\sec^2 \theta - \tan^2 \theta \stackrel{?}{=} \tan \theta \cot \theta \quad \text{Original equation}$$

$$\frac{1}{\cos^2 \theta} - \frac{\sin^2 \theta}{\cos^2 \theta} \stackrel{?}{=} \frac{\sin \theta}{\cos \theta} \cdot \frac{\cos \theta}{\sin \theta} \quad \text{Express all terms using sine and cosine.}$$

$$\frac{1 - \sin^2 \theta}{\cos^2 \theta} \stackrel{?}{=} 1 \quad \text{Subtract on the left. Multiply on the right.}$$

$$\frac{\cos^2 \theta}{\cos^2 \theta} \stackrel{?}{=} 1 \quad 1 - \sin^2 \theta = \cos^2 \theta$$

$$1 = 1 \quad \text{Simplify the left side.}$$

www.algebra2.com/extra_examples

2 Teach

TRANSFORM ONE SIDE OF AN EQUATION

In-Class Examples Power Point®

1 Verify that $\csc \theta \cos \theta \tan \theta = 1$ is an identity.

$$\csc \theta \cos \theta \tan \theta \stackrel{?}{=} 1$$

$$\frac{1}{\sin \theta} \cdot \cos \theta \cdot \frac{\sin \theta}{\cos \theta} \stackrel{?}{=} 1$$

$$1 = 1$$

2 $\frac{\csc \theta}{\cos \theta} - \tan \theta =$ **A**

A $\cot \theta$ **B** $\frac{1 - \sin \theta}{\cos^2 \theta}$

C 0 **D** $\cos^2 \theta$

TRANSFORM BOTH SIDES OF AN EQUATION

In-Class Example Power Point®

3 Verify that $\csc \theta + \sec \theta = \frac{1 + \cot \theta}{\cos \theta}$ is an identity.

$$\csc \theta + \sec \theta \stackrel{?}{=} \frac{1 + \cot \theta}{\cos \theta}$$

$$\frac{1}{\sin \theta} + \frac{1}{\cos \theta} \stackrel{?}{=} \frac{1 + \frac{\cos \theta}{\sin \theta}}{\cos \theta}$$

$$\frac{\cos \theta + \sin \theta}{\sin \theta \cos \theta} \stackrel{?}{=} \frac{\sin \theta\left(1 + \frac{\cos \theta}{\sin \theta}\right)}{\sin \theta \cos \theta}$$

$$\frac{\cos \theta + \sin \theta}{\sin \theta \cos \theta} = \frac{\sin \theta + \cos \theta}{\sin \theta \cos \theta}$$

3 Practice/Apply

Study Notebook

Have students—

- add the definitions/examples of the vocabulary terms to their Vocabulary Builder worksheets for Chapter 14.
- include any other item(s) that they find helpful in mastering the skills in this lesson.

Standardized Test Practice
A B C D

Example 2 Urge students to read slowly and carefully so they do not mistake cot for cos or csc for cos. In addition, students may find the correct answer, $\cos^2 \theta$, but choose choice A because they did not notice the difference between $\cos^2 \theta$ and $\cos \theta$.

Check for Understanding

Concept Check

1–3. See pp. 811A–811N.

1. **Explain** the steps used to verify the identity $\sin\theta \tan\theta = \sec\theta - \cos\theta$.
2. **Describe** the various methods you can use to show that two trigonometric expressions form an identity.
3. **OPEN ENDED** Write a trigonometric equation that is not an identity. Explain how you know it is not an identity.

Guided Practice

GUIDED PRACTICE KEY	
Exercises	Examples
4, 6–10	1, 2
5	3

Verify that each of the following is an identity. **4–9. See pp. 811A–811N.**

4. $\tan\theta(\cot\theta + \tan\theta) = \sec^2\theta$
5. $\tan^2\theta\cos^2\theta = 1 - \cos^2\theta$
6. $\dfrac{\cos^2\theta}{1 - \sin\theta} = 1 + \sin\theta$
7. $\dfrac{1 + \tan^2\theta}{\csc^2\theta} = \tan^2\theta$
8. $\dfrac{\sin\theta}{\sec\theta} = \dfrac{1}{\tan\theta + \cot\theta}$
9. $\dfrac{\sec\theta + 1}{\tan\theta} = \dfrac{\tan\theta}{\sec\theta - 1}$

Standardized Test Practice

10. Which expression is equivalent to $\dfrac{\sec\theta + \csc\theta}{1 + \tan\theta}$? **D**

Ⓐ $\sin\theta$ Ⓑ $\cos\theta$ Ⓒ $\tan\theta$ Ⓓ $\csc\theta$

Practice and Apply

Homework Help	
For Exercises	See Examples
11–24, 26–32	1, 2
25	3

Extra Practice
See page 860.

Verify that each of the following is an identity. **11–28. See pp. 811A–811N.**

11. $\cos^2\theta + \tan^2\theta\cos^2\theta = 1$
12. $\cot\theta(\cot\theta + \tan\theta) = \csc^2\theta$
13. $1 + \sec^2\theta\sin^2\theta = \sec^2\theta$
14. $\sin\theta\sec\theta\cot\theta = 1$
15. $\dfrac{1 - \cos\theta}{1 + \cos\theta} = (\csc\theta - \cot\theta)^2$
16. $\dfrac{1 - 2\cos^2\theta}{\sin\theta\cos\theta} = \tan\theta - \cot\theta$
17. $\cot\theta\csc\theta = \dfrac{\cot\theta + \csc\theta}{\sin\theta + \tan\theta}$
18. $\sin\theta + \cos\theta = \dfrac{1 + \tan\theta}{\sec\theta}$
19. $\dfrac{\sec\theta}{\sin\theta} - \dfrac{\sin\theta}{\cos\theta} = \cot\theta$
20. $\dfrac{\sin\theta}{1 - \cos\theta} + \dfrac{1 - \cos\theta}{\sin\theta} = 2\csc\theta$
21. $\dfrac{1 + \sin\theta}{\sin\theta} = \dfrac{\cot^2\theta}{\csc\theta - 1}$
22. $\dfrac{1 + \tan\theta}{1 + \cot\theta} = \dfrac{\sin\theta}{\cos\theta}$
23. $\dfrac{1}{\sec^2\theta} + \dfrac{1}{\csc^2\theta} = 1$
24. $1 + \dfrac{1}{\cos\theta} = \dfrac{\tan^2\theta}{\sec\theta - 1}$
25. $1 - \tan^4\theta = 2\sec^2\theta - \sec^4\theta$
26. $\cos^4\theta - \sin^4\theta = \cos^2\theta - \sin^2\theta$
27. $\dfrac{1 - \cos\theta}{\sin\theta} = \dfrac{\sin\theta}{1 + \cos\theta}$
28. $\dfrac{\cos\theta}{1 + \sin\theta} + \dfrac{\cos\theta}{1 - \sin\theta} = 2\sec\theta$
29. Verify that $\tan\theta\sin\theta\cos\theta\csc^2\theta = 1$ is an identity. **See pp. 811A–811N.**
30. Show that $1 + \cos\theta$ and $\dfrac{\sin^2\theta}{1 - \cos\theta}$ form an identity. **See pp. 811A–811N.**

PHYSICS **For Exercises 31 and 32, use the following information.**
If an object is propelled from ground level, the maximum height that it reaches is given by $h = \dfrac{v^2\sin^2\theta}{2g}$, where θ is the angle between the ground and the initial path of the object, v is the object's initial velocity, and g is the acceleration due to gravity, 9.8 meters per second squared.

31. Verify the identity $\dfrac{v^2\sin^2\theta}{2g} = \dfrac{v^2\tan^2\theta}{2g\sec^2\theta}$. **See pp. 811A–811N.**
32. A model rocket is launched with an initial velocity of 110 meters per second at an angle of 80° with the ground. Find the maximum height of the rocket. **598.7 m**

About the Exercises...

Organization by Objective

- **Transform One Side of an Equation:** 11–24, 26–30
- **Transform Both Sides of an Equation:** 25

Odd/Even Assignments
Exercises 11–30 are structured so that students practice the same concepts whether they are assigned odd or even problems.

Alert! Exercises 37–42 require a graphing calculator.

Assignment Guide

Basic: 11–31 odd, 33–36, 43–54

Average: 11–31 odd, 33–36, 43–54 (optional: 37–42)

Advanced: 12–30 even, 31–50 (optional: 51–54)

4 *Assess*

Open-Ended Assessment

Speaking Have students explain some of the techniques they have seen or used to verify trigonometric identities in this lesson.

Getting Ready for Lesson 14-5

PREREQUISITE SKILL Students will simplify radical expressions in the process of using the Sum and Difference of Angles Formulas in Lesson 14-5. Use Exercises 51–54 to determine your students' familiarity with simplifying radical expressions.

Assessment Options

Quiz (Lessons 14-3 and 14-4) is available on p. 893 of the *Chapter 14 Resource Masters*.

Mid-Chapter Test (Lessons 14-1 through 14-4) is available on p. 895 of the *Chapter 14 Resource Masters*.

DAILY INTERVENTION

Differentiated Instruction

Interpersonal Have groups or pairs of students work together to verify some of the identities in Exercises 4–9. Have students record the techniques they found helpful. Ask students to compare their list of techniques to the list of suggestions given above Example 3 on p. 783.

784 Chapter 14 Trigonometric Graphs and Identities

33. **CRITICAL THINKING** Present a logical argument for why the identity $\sin^{-1} x + \cos^{-1} x = \frac{\pi}{2}$ is true when $0 \le x \le 1$. **See margin.**

34. **WRITING IN MATH** Answer the question that was posed at the beginning of the lesson. **See pp. 811A–811N.**

How can you verify trigonometric identities?

Include the following in your answer:

- an explanation of why you cannot perform operations to each side of an unverified identity,
- an explanation of how you can tell if two expressions are equivalent, and
- an explanation of why you cannot use the graphs of two equations to verify an identity.

Standardized Test Practice

35. Which of the following is not equivalent to $\cos\theta$? **D**

(A) $\dfrac{\cos\theta}{\cos^2\theta + \sin^2\theta}$ (B) $\dfrac{1 - \sin^2\theta}{\cos\theta}$ (C) $\cot\theta \sin\theta$ (D) $\tan\theta \csc\theta$

36. Which of the following is equivalent to $\sin\theta + \cot\theta\cos\theta$? **B**

(A) $2\sin\theta$ (B) $\dfrac{1}{\sin\theta}$ (C) $\cos^2\theta$ (D) $\dfrac{\sin\theta + \cos\theta}{\sin^2\theta}$

Graphing Calculator

37–42. See pp. 811A–811N for graphs.

VERIFYING TRIGONOMETRIC IDENTITIES You can determine whether or not an equation may be a trigonometric identity by graphing the expressions on either side of the equals sign as two separate functions. If the graphs do not match, then the equation is not an identity. If the two graphs do coincide, the equation *might* be an identity. The equation has to be verified algebraically to ensure that it is an identity.

Determine whether each of the following *may be* or *is not* an identity.

37. $\cot x + \tan x = \csc x \cot x$ **is not**

38. $\sec^2 x - 1 = \sin^2 x \sec^2 x$ **may be**

39. $(1 + \sin x)(1 - \sin x) = \cos^2 x$ **may be**

40. $\dfrac{1}{\sec x \tan x} = \csc x - \sin x$ **may be**

41. $\dfrac{\sec^2 x}{\tan x} = \sec x \csc x$ **may be**

42. $\dfrac{1}{\sec x} + \dfrac{1}{\csc x} = 1$ **is not**

Maintain Your Skills

Mixed Review

Find the value of each expression. *(Lesson 14-3)*

43. $\sec\theta$, if $\tan\theta = \frac{1}{2}$; $0° < \theta < 90°$ $\mathbf{\frac{\sqrt{5}}{2}}$

44. $\cos\theta$, if $\sin\theta = -\frac{2}{3}$; $180° < \theta < 270°$

45. $\csc\theta$, if $\cot\theta = -\frac{7}{12}$; $90° < \theta < 180°$

46. $\sin\theta$, if $\cos\theta = \frac{3}{4}$; $270° < \theta < 360°$

44. $-\frac{\sqrt{5}}{3}$

45. $\frac{\sqrt{193}}{12}$

46. $-\frac{\sqrt{7}}{4}$

49. $3;\ 2\pi;\ -\frac{\pi}{2}$

53. $\frac{\sqrt{6} + 2\sqrt{2}}{4}$

54. $\frac{2 - \sqrt{3}}{4}$

State the amplitude, period, and phase shift of each function. Then graph each function. *(Lesson 14-2)* **47–49. See pp. 811A–811N for graphs.**

47. $y = \cos(\theta - 30°)$ **1; 360°; 30°**

48. $y = \sin(\theta - 45°)$ **1; 360°; 45°**

49. $y = 3\cos\left(\theta + \frac{\pi}{2}\right)$

50. What is the probability that an event occurs if the odds of the event occurring are 5:1? *(Lesson 12-4)* $\mathbf{\frac{5}{6}}$

Getting Ready for the Next Lesson

PREREQUISITE SKILL **Simplify each expression.**

*(To review **simplifying radical expressions**, see Lesson 5-6.)*

51. $\frac{\sqrt{3}}{2} \cdot \frac{\sqrt{2}}{2}$ $\mathbf{\frac{\sqrt{6}}{4}}$

52. $\frac{1}{2} \cdot \frac{\sqrt{2}}{2}$ $\mathbf{\frac{\sqrt{2}}{4}}$

53. $\frac{\sqrt{6}}{4} + \frac{\sqrt{2}}{2}$

54. $\frac{1}{2} - \frac{\sqrt{3}}{4}$

www.algebra2.com/self_check_quiz

Answer

33. Sample answer: Consider a right triangle *ABC* with right angle at *C*. If an angle, say *A*, has a sine of *x*, then angle *B* must have a cosine of *x*. Since *A* and *B* are both in a right triangle and neither is the right angle, their sum must be $\frac{\pi}{2}$.

Enrichment, p. 860

Heron's Formula

Heron's formula can be used to find the area of a triangle if you know the lengths of the three sides. Consider any triangle ABC. Let K represent the area of $\triangle ABC$. Then

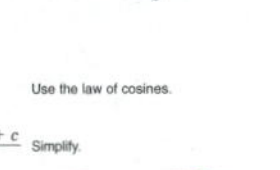

$K = \frac{1}{2}bc \sin A$

$K^2 = \frac{b^2c^2 \sin^2 A}{4}$ Square both sides.

$= \frac{b^2c^2(1 - \cos^2 A)}{4}$

$= \frac{b^2c^2(1 + \cos A)(1 - \cos A)}{4}$

$= \frac{b^2c^2}{4}\left(1 + \frac{b^2 + c^2 - a^2}{2bc}\right)\left(1 - \frac{b^2 + c^2 - a^2}{2bc}\right)$ Use the law of cosines.

$= \frac{b + c + a}{2} \cdot \frac{b + c - a}{2} \cdot \frac{a + b - c}{2} \cdot \frac{a - b + c}{2}$ Simplify.

Study Guide and Intervention, p. 855 (shown) and p. 856

Transform One Side of an Equation Use the basic trigonometric identities along with the definitions of the trigonometric functions to verify trigonometric identities. Often it is easier to begin with the more complicated side of the equation and transform that expression into the form of the simpler side.

Example **Verify that each of the following is an identity.**

a. $\dfrac{\sin\theta}{\cot\theta} - \sec\theta = -\cos\theta$

Transform the left side.

$\frac{\sin\theta}{\cot\theta} - \sec\theta \stackrel{?}{=} -\cos\theta$

$\frac{\sin\theta}{\frac{\cos\theta}{\sin\theta}} - \frac{1}{\cos\theta} \stackrel{?}{=} -\cos\theta$

$\frac{\sin^2\theta}{\cos\theta} - \frac{1}{\cos\theta} \stackrel{?}{=} -\cos\theta$

$\frac{\sin^2 - 1}{\cos\theta} \stackrel{?}{=} -\cos\theta$

$\frac{-\cos^2\theta}{\cos\theta} \stackrel{?}{=} -\cos\theta$

$-\cos\theta = -\cos\theta$

b. $\dfrac{\tan\theta}{\csc\theta} + \cos\theta = \sec\theta$

Transform the left side.

$\frac{\tan\theta}{\csc\theta} + \cos\theta \stackrel{?}{=} \sec\theta$

$\frac{\frac{\sin\theta}{\cos\theta}}{\frac{1}{\sin\theta}} + \cos\theta \stackrel{?}{=} \sec\theta$

$\frac{\sin^2\theta}{\cos\theta} + \cos\theta \stackrel{?}{=} \sec\theta$

$\frac{\sin^2\theta + \cos^2\theta}{\cos\theta} \stackrel{?}{=} \sec\theta$

$\frac{1}{\cos\theta} \stackrel{?}{=} \sec\theta$

$\sec\theta = \sec\theta$

Exercises

Verify that each of the following is an identity.

1. $1 + \csc^2\theta \cdot \cos^2\theta = \csc^2\theta$

$1 + \frac{1}{\sin^2\theta} \cdot \cos^2\theta \stackrel{?}{=} \csc^2\theta$

$\frac{\sin^2\theta + \cos^2\theta}{\sin^2\theta} \stackrel{?}{=} \csc^2\theta$

$\frac{1}{\sin^2\theta} \stackrel{?}{=} \csc^2\theta$

$\csc^2\theta = \csc^2\theta$

2. $\dfrac{\sin\theta}{1 - \cos\theta} - \dfrac{\cot\theta}{1 + \cos\theta} = \dfrac{1 - \cos^3\theta}{\sin^3\theta}$

$\frac{\sin\theta(1 + \cos\theta) - \frac{\cos\theta}{\sin\theta}(1 - \cos\theta)}{(1 - \cos\theta)(1 + \cos\theta)} \stackrel{?}{=} \frac{1 - \cos^3\theta}{\sin^3\theta}$

$\frac{\sin\theta + \sin\theta \cdot \cos\theta - \frac{\cos\theta}{\sin\theta} + \frac{\cos^2\theta}{\sin\theta}}{1 - \cos^2\theta} \stackrel{?}{=} \frac{1 - \cos^3\theta}{\sin^3\theta}$

$\frac{\frac{\sin^2\theta + \sin^2\theta\cos\theta - \cos\theta + \cos^2\theta}{\sin\theta}}{\sin^2\theta} \stackrel{?}{=} \frac{1 - \cos^3\theta}{\sin^3\theta}$

$\frac{\sin^2\theta + \cos^2\theta + \cos\theta(\sin^2\theta - 1)}{\sin^3\theta} \stackrel{?}{=} \frac{1 - \cos^3\theta}{\sin^3\theta}$

$\frac{1 + \cos\theta(-\cos^2\theta)}{\sin^3\theta} \stackrel{?}{=} \frac{1 - \cos^3\theta}{\sin^3\theta}$

$\frac{1 - \cos^3\theta}{\sin^3\theta} = \frac{1 - \cos^3\theta}{\sin^3\theta}$

Skills Practice, p. 857 and Practice, p. 858 (shown)

Verify that each of the following is an identity.

1. $\dfrac{\sin^2\theta + \cos^2\theta}{\cos^2\theta} = \sec^2\theta$

$\frac{\sin^2\theta + \cos^2\theta}{\cos^2\theta} \stackrel{?}{=} \sec^2\theta$

$\frac{1}{\cos^2\theta} \stackrel{?}{=} \sec^2\theta$

$\sec^2\theta = \sec^2\theta$

2. $\dfrac{\cos^2\theta}{1 - \sin^2\theta} = 1$

$\frac{\cos^2\theta}{1 - \sin^2\theta} \stackrel{?}{=} 1$

$\frac{\cos^2\theta}{\cos^2\theta} \stackrel{?}{=} 1$

$1 = 1$

3. $(1 + \sin\theta)(1 - \sin\theta) = \cos^2\theta$

$(1 + \sin\theta)(1 - \sin\theta) \stackrel{?}{=} \cos^2\theta$

$1 - \sin^2\theta \stackrel{?}{=} \cos^2\theta$

$\cos^2\theta = \cos^2\theta$

4. $\tan^4\theta + 2\tan^2\theta + 1 = \sec^4\theta$

$\tan^4\theta + 2\tan^2\theta + 1 \stackrel{?}{=} \sec^4\theta$

$(\tan^2\theta + 1)^2 \stackrel{?}{=} \sec^4\theta$

$(\sec^2\theta)^2 \stackrel{?}{=} \sec^4\theta$

$\sec^4\theta = \sec^4\theta$

5. $\cos^2\theta\cot^2\theta = \cot^2\theta - \cos^2\theta$

$\cos^2\theta\cot^2\theta \stackrel{?}{=} \cot^2\theta - \cos^2\theta$

$\cos^2\theta\cot^2\theta \stackrel{?}{=} \frac{\cos^2\theta}{\sin^2\theta} - \cos^2\theta$

$\cos^2\theta\cot^2\theta \stackrel{?}{=} \frac{\cos^2\theta - \cos^2\theta\sin^2\theta}{\sin^2\theta}$

$\cos^2\theta\cot^2\theta \stackrel{?}{=} \frac{(\cos^2\theta)(1 - \sin^2\theta)}{\sin^2\theta}$

$\cos^2\theta\cot^2\theta \stackrel{?}{=} \frac{\cos^2\theta}{1} \cdot \frac{\cos^2\theta}{\sin^2\theta}$

$\cos^2\theta\cot^2\theta = \cos^2\theta\cot^2\theta$

6. $(\sin^2\theta)(\csc^2\theta + \sec^2\theta) = \sec^2\theta$

$(\sin^2\theta)(\csc^2\theta + \sec^2\theta) \stackrel{?}{=} \sec^2\theta$

$(\sin^2\theta)\left(\frac{1}{\sin^2\theta} + \frac{1}{\cos^2\theta}\right) \stackrel{?}{=} \sec^2\theta$

$1 + \frac{\sin^2\theta}{\cos^2\theta} \stackrel{?}{=} \sec^2\theta$

$1 + \tan^2\theta \stackrel{?}{=} \sec^2\theta$

$\sec^2\theta = \sec^2\theta$

7. **PROJECTILES** The square of the initial velocity of an object launched from the ground is $v^2 = \frac{2gh}{\sin^2\theta}$, where θ is the angle between the ground and the initial path, h is the maximum height reached, and g is the acceleration due to gravity. Verify the identity $\frac{2gh}{\sin^2\theta} = \frac{2gh\sec^2\theta}{\sec^2\theta - 1}$.

$\frac{2gh}{\sin^2\theta} = \frac{2gh}{1 - \cos^2\theta} = \frac{2gh}{1 - \frac{1}{\sec^2\theta}} = \frac{2gh}{\frac{\sec^2\theta - 1}{\sec^2\theta}} = \frac{2gh\sec^2\theta}{\sec^2\theta - 1}$

8. **LIGHT** The intensity of a light source measured in candles is given by $I = ER^2 \sec\theta$, where E is the illuminance in foot candles on a surface, R is the distance in feet from the light source, and θ is the angle between the light beam and a line perpendicular to the surface. Verify the identity $ER^2(1 + \tan^2\theta)\cos\theta = ER^2\sec\theta$.

$ER^2(1 + \tan^2\theta)\cos\theta = ER^2\sec^2\theta\cos\theta = ER^2\sec^2\theta \cdot \frac{1}{\sec\theta} = ER^2\sec\theta$

Reading to Learn Mathematics, p. 859 ELL

Pre-Activity **How can you verify trigonometric identities?**

Read the introduction to Lesson 14-4 at the top of page 782 in your textbook.

For $\theta = -\pi$, 0, or π, $\sin\theta = \sin 2\theta$. Does this mean that $\sin\theta = \sin 2\theta$ is an identity? Explain your reasoning. **Sample answer: No; an identity is an equation that is true for *all* values of a variable for which the functions involved are defined, not just *some* values. If $\theta = \frac{\pi}{4}$, $\sin\theta = \frac{\sqrt{2}}{2}$, and $\sin 2\theta = 1$.**

Reading the Lesson

1. Determine whether each equation is an *identity* or *not an identity*.

a. $\frac{1}{\sin^2\theta} - \frac{1}{\tan^2\theta} = 1$ **identity**

b. $\frac{\cos\theta}{\sin\theta\tan\theta}$ **not an identity**

c. $\frac{\sin\theta}{\cos\theta} + \frac{\cos\theta}{\sin\theta} = \cos\theta\sin\theta$ **not an identity**

d. $\cos^2\theta(\tan^2\theta + 1) = 1$ **identity**

e. $\frac{\sin^2\theta}{\cos^2\theta} + \sin\theta\csc\theta = \sec^2\theta$ **identity**

f. $\frac{1}{1 - \sin\theta} + \frac{1}{1 + \sin\theta} = 2\cos^2\theta$ **not an identity**

g. $\tan^2\theta\cos^2\theta = \frac{1}{\csc^2\theta}$ **identity**

h. $\frac{\sin\theta}{\sec\theta} = \frac{1}{\tan\theta} + \frac{1}{\cot\theta}$ **not an identity**

2. Which of the following is *not* permitted when verifying an identity? **B**

A. simplifying one side of the identity to match the other side

B. cross multiplying if the identity is a proportion

C. simplifying each side of the identity separately to get the same expression on both sides

Helping You Remember

3. Many students have trouble knowing where to start in verifying a trigonometric identity. What is a simple rule that you can remember that you can always use if you don't see a quicker approach? **Sample answer: Write both sides in terms of sines and cosines. Then simplify each side as much as possible.**

14-5 Lesson Notes

1 Focus

5-Minute Check Transparency 14-5 Use as a quiz or review of Lesson 14-4.

Mathematical Background notes are available for this lesson on p. 760D.

How are the sum and difference formulas used to describe communication interference?

Ask students:

- Where else have you heard the term *interference*? **Sample answer: television and radio**
- At its peak, how does the amplitude of the combined wave compare to the amplitude of the initial two waves? **The amplitude of the combined wave is the sum of the amplitudes of the two initial waves.**
- Why does the combined wave cross the x-axis at a point where neither of the two initial waves are crossing the axis? **The combined wave is the sum of the other two waves. It crosses the x-axis at points where one of the initial waves is above the x-axis and the other wave is an equal distance below the x-axis.**

14-5 Sum and Difference of Angles Formulas

What You'll Learn

- Find values of sine and cosine involving sum and difference formulas.
- Verify identities by using sum and difference formulas.

How are the sum and difference formulas used to describe communication interference?

Have you ever been talking on a cell phone and temporarily lost the signal? Radio waves that pass through the same place at the same time cause interference. *Constructive interference* occurs when two waves combine to have a greater amplitude than either of the component waves. *Destructive interference* occurs when the component waves combine to have a smaller amplitude.

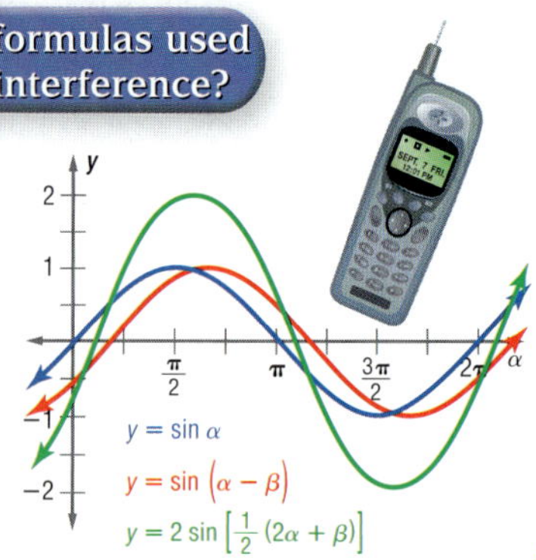

Study Tips

Reading Math

The Greek letter *beta*, β, can be used to denote the measure of an angle.

It is important to realize that $\sin(\alpha \pm \beta)$ is not the same as $\sin \alpha \pm \sin \beta$.

SUM AND DIFFERENCE FORMULAS

Notice that the third equation shown above involves the sum of α and β. It is often helpful to use formulas for the trigonometric values of the difference or sum of two angles. For example, you could find sin 15° by evaluating sin (60° − 45°). Formulas can be developed that can be used to evaluate expressions like $\sin(\alpha - \beta)$ or $\cos(\alpha + \beta)$.

The figure at the right shows two angles α and β in standard position on the unit circle. Use the Distance Formula to find d, where $(x_1, y_1) = (\cos \beta, \sin \beta)$ and $(x_2, y_2) = (\cos \alpha, \sin \alpha)$.

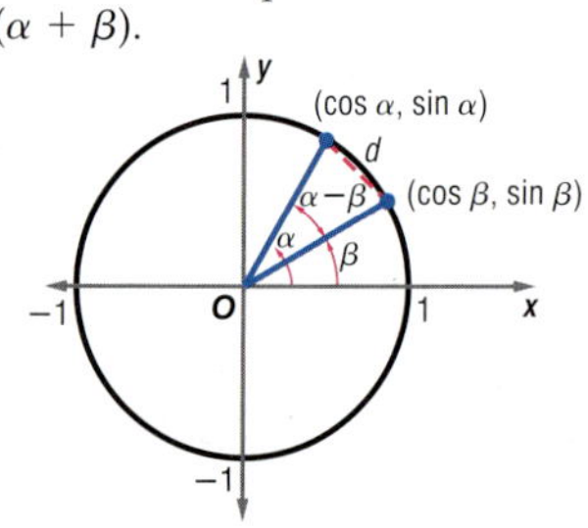

$$d = \sqrt{(\cos \alpha - \cos \beta)^2 + (\sin \alpha - \sin \beta)^2}$$
$$d^2 = (\cos \alpha - \cos \beta)^2 + (\sin \alpha - \sin \beta)^2$$
$$d^2 = (\cos^2 \alpha - 2\cos \alpha \cos \beta + \cos^2 \beta) + (\sin^2 \alpha - 2\sin \alpha \sin \beta + \sin^2 \beta)$$
$$d^2 = \cos^2 \alpha + \sin^2 \alpha + \cos^2 \beta + \sin^2 \beta - 2\cos \alpha \cos \beta - 2\sin \alpha \sin \beta$$
$$d^2 = 1 + 1 - 2\cos \alpha \cos \beta - 2\sin \alpha \sin \beta \quad \sin^2 \alpha + \cos^2 \alpha = 1 \text{ and}$$
$$d^2 = 2 - 2\cos \alpha \cos \beta - 2\sin \alpha \sin \beta \quad \sin^2 \beta + \cos^2 \beta = 1$$

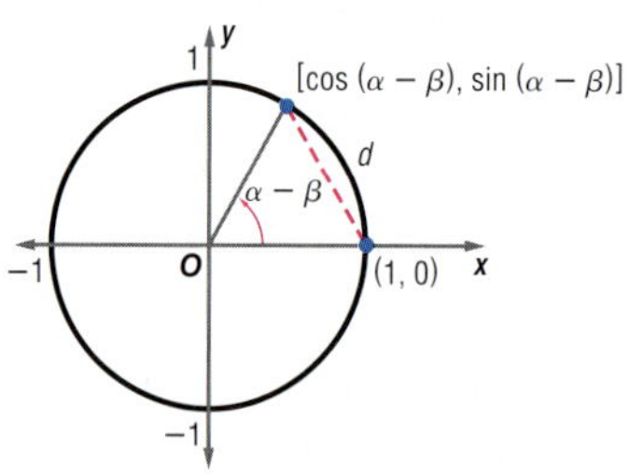

Now find the value of d^2 when the angle having measure $\alpha - \beta$ is in standard position on the unit circle, as shown in the figure at the left.

$$d = \sqrt{[\cos(\alpha - \beta) - 1]^2 + [\sin(\alpha - \beta) - 0]^2}$$
$$d^2 = [\cos(\alpha - \beta) - 1]^2 + [\sin(\alpha - \beta) - 0]^2$$
$$= [\cos^2(\alpha - \beta) - 2\cos(\alpha - \beta) + 1] + \sin^2(\alpha - \beta)$$
$$= \cos^2(\alpha - \beta) + \sin^2(\alpha - \beta) - 2\cos(\alpha - \beta) + 1$$
$$= 1 - 2\cos(\alpha - \beta) + 1$$
$$= 2 - 2\cos(\alpha - \beta)$$

Resource Manager

Workbook and Reproducible Masters

Chapter 14 Resource Masters

- Study Guide and Intervention, pp. 861–862
- Skills Practice, p. 863
- Practice, p. 864
- Reading to Learn Mathematics, p. 865
- Enrichment, p. 866

Transparencies

5-Minute Check Transparency 14-5
Answer Key Transparencies

Technology

Interactive Chalkboard

By equating the two expressions for d^2, you can find a formula for $\cos(\alpha - \beta)$.

$$d^2 = d^2$$

$$2 - 2\cos(\alpha - \beta) = 2 - 2\cos\alpha\cos\beta - 2\sin\alpha\sin\beta$$

$$-1 + \cos(\alpha - \beta) = -1 + \cos\alpha\cos\beta + \sin\alpha\sin\beta \quad \text{Divide each side by } -2.$$

$$\cos(\alpha - \beta) = \cos\alpha\cos\beta + \sin\alpha\sin\beta \quad \text{Add 1 to each side.}$$

Use the formula for $\cos(\alpha - \beta)$ to find a formula for $\cos(\alpha + \beta)$.

$$\cos(\alpha - \beta) = \cos[\alpha - (-\beta)]$$
$$= \cos\alpha\cos(-\beta) + \sin\alpha\sin(-\beta)$$
$$= \cos\alpha\cos\beta - \sin\alpha\sin\beta \quad \cos(-\beta) = \cos\beta;\ \sin(-\beta) = -\sin\beta$$

You can use a similar method to find formulas for $\sin(\alpha + \beta)$ and $\sin(\alpha - \beta)$.

Key Concept *Sum and Difference of Angles Formulas*

The following identities hold true for all values of α and β.

$$\cos(\alpha \pm \beta) = \cos\alpha\cos\beta \mp \sin\alpha\sin\beta$$

$$\sin(\alpha \pm \beta) = \sin\alpha\cos\beta \pm \cos\alpha\sin\beta$$

Notice the symbol $\mp$ in the formula for $\cos(\alpha \pm \beta)$. It means "minus or plus." In the cosine formula, when the sign on the left side of the equation is plus, the sign on the right side is minus; when the sign on the left side is minus, the sign on the right side is plus. The signs match each other in the sine formula.

Example 1 Use Sum and Difference of Angles Formulas

Find the exact value of each expression.

a. cos 75°

Use the formula $\cos(\alpha + \beta) = \cos\alpha\cos\beta - \sin\alpha\sin\beta$.

$$\cos 75° = \cos(30° + 45°) \quad \alpha = 30°, \beta = 45°$$
$$= \cos 30°\cos 45° - \sin 30°\sin 45°$$
$$= \left(\frac{\sqrt{3}}{2}\cdot\frac{\sqrt{2}}{2}\right) - \left(\frac{1}{2}\cdot\frac{\sqrt{2}}{2}\right) \quad \text{Evaluate each expression.}$$
$$= \frac{\sqrt{6}}{4} - \frac{\sqrt{2}}{4} \quad \text{Multiply.}$$
$$= \frac{\sqrt{6} - \sqrt{2}}{4} \quad \text{Simplify.}$$

b. sin (−210°)

Use the formula $\sin(\alpha - \beta) = \sin\alpha\cos\beta - \cos\alpha\sin\beta$.

$$\sin(-210°) = \sin(60° - 270°) \quad \alpha = 60°, \beta = 270°$$
$$= \sin 60°\cos 270° - \cos 60°\sin 270°$$
$$= \left(\frac{\sqrt{3}}{2}\right)(0) - \left(\frac{1}{2}\right)(-1) \quad \text{Evaluate each expression.}$$
$$= 0 - \left(-\frac{1}{2}\right) \quad \text{Multiply.}$$
$$= \frac{1}{2} \quad \text{Simplify.}$$

www.algebra2.com/extra_examples

2 Teach

Building on Prior Knowledge

The Distance Formula was first discussed and used back in Lesson 8-1. In this lesson, students will learn how to apply the Distance Formula to derive the Sum and Difference of Angles Formulas.

SUM AND DIFFERENCE FORMULAS

Teaching Tip Explain that sin 15° can be found by evaluating sin (60° − 45°) because the exact values of sin 60° and sin 45° are known. Stress that using a difference such as sin (90° − 75°) is ineffective because sin 75° is not easily computed or remembered.

Reading Tip Have the entire class read aloud the paragraph directly below the Key Concept box, about the minus and plus signs. Stress that its use in the formula for the cosine of a sum or difference indicates that the sign on the right side of the identity is the opposite of the sign on the left side.

In-Class Examples

Power Point®

1 Find the exact value of each expression.

a. sin 75° $\frac{\sqrt{2} + \sqrt{6}}{4}$

b. cos (−75°) $\frac{\sqrt{6} - \sqrt{2}}{4}$

2 Refer to Example 2 in the Student Edition. Use the difference of angles formula to determine the amount of light energy in Raleigh, North Carolina, located at a latitude of 35.8° N. **The maximum light energy per square foot is 0.9770*E*.**

DAILY INTERVENTION **Differentiated Instruction**

Naturalist Have students apply what they know about geography, or have them conduct research, to find out whether the light energy per square foot would be increasing or decreasing as you travel toward the equator. Have students research the latitude of your city, or a city that interests them, and repeat Example 2 for that city.

VERIFY IDENTITIES

In-Class Example

Power Point®

3 Verify that each of the following is an identity.

a. $\cos(90° - \theta) = \sin\theta$

$\cos(90° - \theta) \stackrel{?}{=} \sin\theta$
$\cos 90° \cos\theta + \sin 90° \sin\theta \stackrel{?}{=} \sin\theta$
$0 \cos\theta + 1 \sin\theta \stackrel{?}{=} \sin\theta$
$\sin\theta = \sin\theta$

b. $\cos(180° - \theta) = -\cos\theta$

$\cos(180° - \theta) \stackrel{?}{=} -\cos\theta$
$\cos 180° \cos\theta + \sin 180° \sin\theta \stackrel{?}{=} -\cos\theta$
$-1 \cos\theta + 0 \sin\theta \stackrel{?}{=} -\cos\theta$
$-\cos\theta = -\cos\theta$

3 Practice/Apply

Study Notebook

Have students—
- add the definitions/examples of the vocabulary terms to their Vocabulary Builder worksheets for Chapter 14.
- record the sum and difference formulas.
- include any other item(s) that they find helpful in mastering the skills in this lesson.

About the Exercises...

Organization by Objective
- **Sum and Difference Formulas:** 14–27
- **Verify Identities:** 28–39

Odd/Even Assignments
Exercises 14–39 are structured so that students practice the same concepts whether they are assigned odd or even problems.

Assignment Guide

Basic: 15–25 odd, 29–37 odd, 40, 41–45 odd, 46–74

Average: 15–39 odd, 40, 41–45 odd, 46–74

Advanced: 14–40 even, 41–66 (optional: 67–74)

Example 2 Use Sum and Difference Formulas to Solve a Problem

PHYSICS **On June 22, the maximum amount of light energy falling on a square foot of ground at a location in the northern hemisphere is given by $E \sin(113.5° - \phi)$, where ϕ is the latitude of the location and E is the amount of light energy when the Sun is directly overhead. Use the difference of angles formula to determine the amount of light energy in Rochester, New York, located at a latitude of 43.1° N.**

Use the difference formula for sine.

$$\begin{aligned}\sin(113.5° - \phi) &= \sin 113.5° \cos\phi - \cos 113.5° \sin\phi \\ &= \sin 113.5° \cos 43.1° - \cos 113.5° \sin 43.1° \\ &= 0.9171 \cdot 0.7302 - (-0.3987) \cdot 0.6833 \\ &= 0.9420\end{aligned}$$

In Rochester, New York, the maximum light energy per square foot is $0.9420E$.

Study Tip

Reading Math
The symbol ϕ is the lowercase Greek letter *phi*.

VERIFY IDENTITIES You can also use the sum and difference formulas to verify identities.

Example 3 Verify Identities

Verify that each of the following is an identity.

a. $\sin(180° + \theta) = -\sin\theta$

$\sin(180° + \theta) \stackrel{?}{=} -\sin\theta$	Original equation
$\sin 180° \cos\theta + \cos 180° \sin\theta \stackrel{?}{=} -\sin\theta$	Sum of angles formula
$0 \cos\theta + (-1) \sin\theta \stackrel{?}{=} -\sin\theta$	Evaluate each expression.
$-\sin\theta = -\sin\theta$	Simplify.

b. $\cos(180° + \theta) = -\cos\theta$

$\cos(180° + \theta) \stackrel{?}{=} -\cos\theta$	Original equation
$\cos 180° \cos\theta - \sin 180° \sin\theta \stackrel{?}{=} -\cos\theta$	Sum of angles formula
$(-1) \cos\theta - 0 \sin\theta \stackrel{?}{=} -\cos\theta$	Evaluate each expression.
$-\cos\theta = -\cos\theta$	Simplify.

Check for Understanding

Concept Check

1–2. See margin.
3. Sometimes; sample answer: The cosine function can equal 1.

1. **Determine** whether $\sin(\alpha + \beta) = \sin\alpha + \sin\beta$ is an identity.
2. **Describe** a method for finding the exact value of sin 105°. Then find the value.
3. **OPEN ENDED** Determine whether $\cos(\alpha - \beta) < 1$ is *sometimes*, *always*, or *never* true. Explain your reasoning.

Guided Practice

GUIDED PRACTICE KEY	
Exercises	Examples
4–9, 13	1
10–12	3

Find the exact value of each expression.

4. sin 75° $\frac{\sqrt{6}+\sqrt{2}}{4}$
5. sin 165° $\frac{\sqrt{6}-\sqrt{2}}{4}$
6. cos 255° $\frac{\sqrt{2}-\sqrt{6}}{4}$
7. cos (−30°) $\frac{\sqrt{3}}{2}$
8. sin (−240°) $\frac{\sqrt{3}}{2}$
9. cos (−120°) $-\frac{1}{2}$

Verify that each of the following is an identity. 10–12. See pp 811A–811N.

10. $\cos(270° - \theta) = -\sin\theta$
11. $\sin\left(\theta + \frac{\pi}{2}\right) = \cos\theta$
12. $\sin(\theta + 30°) + \cos(\theta + 60°) = \cos\theta$

Answers

1. $\sin(\alpha + \beta) \stackrel{?}{=} \sin\alpha + \sin\beta$

$\sin\alpha \cos\beta + \cos\alpha \sin\beta \neq \sin\alpha + \sin\beta$

2. Use the formula $\sin(\alpha + \beta) = \sin\alpha \cos\beta + \cos\alpha \sin\beta$. Since $\sin 105° = \sin(60° + 45°)$, replace α with 60° and β with 45° to get $\sin 60° \cos 45° + \cos 60° \sin 45°$. By finding the sum of the products of the values, the result is $\frac{\sqrt{6}+\sqrt{2}}{4}$ or about 0.9659.

Application 13. **GEOMETRY** Determine the exact value of $\tan \alpha$. $\frac{5-\sqrt{3}}{1+5\sqrt{3}}$

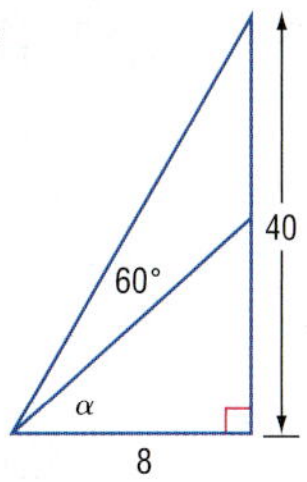

★ indicates increased difficulty

Practice and Apply

Homework Help

For Exercises	See Examples
14–27	1
28–39	3
40, 41, 43–46	2

Extra Practice See page 860.

Find the exact value of each expression.

14. $\sin 135°$ $\frac{\sqrt{2}}{2}$
15. $\cos 105°$ $\frac{\sqrt{2}-\sqrt{6}}{4}$
16. $\sin 285°$ $\frac{-\sqrt{6}-\sqrt{2}}{4}$
17. $\cos 165°$
18. $\cos 195°$
19. $\sin 255°$
20. $\cos 225°$ $-\frac{\sqrt{2}}{2}$
21. $\sin 315°$ $-\frac{\sqrt{2}}{2}$
22. $\sin(-15°)$ $\frac{\sqrt{2}-\sqrt{6}}{4}$
23. $\cos(-45°)$
24. $\cos(-150°)$
25. $\sin(-165°)$

★ 26. What is the exact value of $\sin 75° - \sin 15°$? $\frac{\sqrt{2}}{2}$

★ 27. Find the exact value of $\cos 105° + \cos 225°$. $\frac{-\sqrt{6}-\sqrt{2}}{4}$

17. $\frac{-\sqrt{6}-\sqrt{2}}{4}$

18. $\frac{-\sqrt{6}-\sqrt{2}}{4}$

19. $\frac{-\sqrt{6}-\sqrt{2}}{4}$

23. $\frac{\sqrt{2}}{2}$ 24. $-\frac{\sqrt{3}}{2}$

25. $\frac{\sqrt{2}-\sqrt{6}}{4}$

Verify that each of the following is an identity. 28–39. See pp. 811A–811N.

28. $\sin(270° - \theta) = -\cos\theta$
29. $\cos(90° + \theta) = -\sin\theta$
30. $\cos(90° - \theta) = \sin\theta$
31. $\sin(90° - \theta) = \cos\theta$
32. $\sin\left(\theta + \frac{3\pi}{2}\right) = -\cos\theta$
33. $\cos(\pi - \theta) = -\cos\theta$
34. $\cos(2\pi + \theta) = \cos\theta$
35. $\sin(\pi - \theta) = \sin\theta$
36. $\sin(60° + \theta) + \sin(60° - \theta) = \sqrt{3}\cos\theta$
37. $\sin\left(\theta + \frac{\pi}{3}\right) - \cos\left(\theta + \frac{\pi}{6}\right) = \sin\theta$

★ 38. $\sin(\alpha + \beta)\sin(\alpha - \beta) = \sin^2\alpha - \sin^2\beta$

★ 39. $\cos(\alpha + \beta) = \frac{1 - \tan\alpha\tan\beta}{\sec\alpha\sec\beta}$

COMMUNICATION **For Exercises 40 and 41, use the following information.**
A radio transmitter sends out two signals, one for voice communication and another for data. Suppose the equation of the voice wave is $v = 10\sin(2t - 30°)$ and the equation of the data wave is $d = 10\cos(2t + 60°)$.

40. Draw a graph of the waves when they are combined. See margin.

41. Refer to the application at the beginning of the lesson. What type of interference results? Explain. **Destructive; the resulting graph has a smaller amplitude than the two initial graphs.**

More About. . .

Physics

In the northern hemisphere, the day with the least number of hours of daylight is December 21 or 22, the first day of winter.

Source: www.infoplease.com

PHYSICS **For Exercises 42–45, use the following information.**
On December 22, the maximum amount of light energy that falls on a square foot of ground at a certain location is given by $E \sin(113.5° + \phi)$, where ϕ is the latitude of the location. Use the sum of angles formula to find the amount of light energy, in terms of E, for each location. **42. 0.3681 *E*** **43. 0.4179 *E***

42. Salem, OR (Latitude: 44.9° N)
43. Chicago, IL (Latitude: 41.8° N)
44. Charleston, SC (Latitude: 28.5°N) **0.6157 *E***
45. San Diego, CA (Latitude 32.7° N) **0.5564 *E***

46. **CRITICAL THINKING** Use the sum and difference formulas for sine and cosine to derive formulas for $\tan(\alpha + \beta)$ and $\tan(\alpha - \beta)$. See pp. 811A–811N.

www.algebra2.com/self_check_quiz

40.

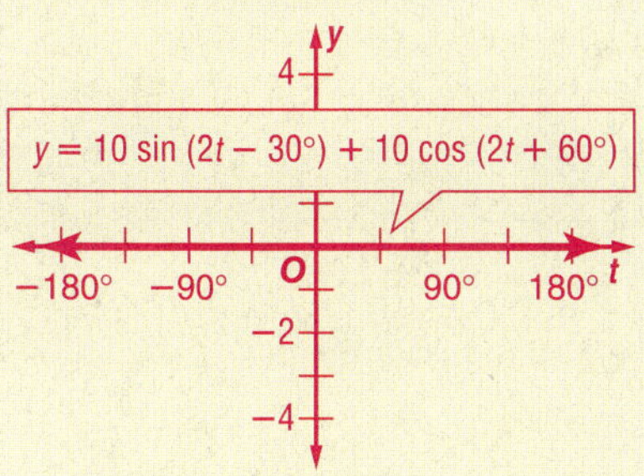

Study Guide and Intervention, p. 861 (shown) and p. 862

Sum and Difference Formulas The following formulas are useful for evaluating an expression like sin 15° from the known values of sine and cosine of 60° and 45°.

Sum and Difference of Angles	The following identities hold true for all values of α and β. $\cos(\alpha \pm \beta) = \cos\alpha \cdot \cos\beta \mp \sin\alpha \cdot \sin\beta$ $\sin(\alpha \pm \beta) = \sin\alpha \cdot \cos\beta \pm \cos\alpha \cdot \sin\beta$

Example **Find the exact value of each expression.**

a. cos 345°

$$\cos 345° = \cos(300° + 45°) = \cos 300° \cdot \cos 45° - \sin 300° \cdot \sin 45° = \frac{1}{2} \cdot \frac{\sqrt{2}}{2} - \left(-\frac{\sqrt{3}}{2}\right) \cdot \frac{\sqrt{2}}{2} = \frac{\sqrt{2}+\sqrt{6}}{4}$$

b. sin (−105°)

$$\sin(-105°) = \sin(45° - 150°) = \sin 45° \cdot \cos 150° - \cos 45° \cdot \sin 150° = \frac{\sqrt{2}}{2} \cdot \left(-\frac{\sqrt{3}}{2}\right) - \frac{\sqrt{2}}{2} \cdot \frac{1}{2} = -\frac{\sqrt{2}+\sqrt{6}}{4}$$

Exercises

Find the exact value of each expression.

1. $\sin 105°$ $\frac{\sqrt{2}+\sqrt{6}}{4}$
2. $\cos 285°$ $\frac{\sqrt{6}-\sqrt{2}}{4}$
3. $\cos(-75°)$ $\frac{\sqrt{6}-\sqrt{2}}{4}$
4. $\cos(-165°)$ $-\frac{\sqrt{2}+\sqrt{6}}{4}$
5. $\sin 195°$ $\frac{\sqrt{2}-\sqrt{6}}{4}$
6. $\cos 420°$ $\frac{1}{2}$
7. $\sin(-75°)$ $-\frac{\sqrt{2}+\sqrt{6}}{4}$
8. $\cos 135°$ $-\frac{\sqrt{2}}{2}$
9. $\cos(-15°)$ $\frac{\sqrt{2}+\sqrt{6}}{4}$
10. $\sin 345°$ $\frac{\sqrt{2}-\sqrt{6}}{4}$
11. $\cos(-105°)$ $\frac{\sqrt{2}-\sqrt{6}}{4}$
12. $\sin 495°$ $\frac{\sqrt{2}}{2}$

Skills Practice, p. 863 and Practice, p. 864 (shown)

Find the exact value of each expression.

1. $\cos 75°$ $\frac{\sqrt{6}-\sqrt{2}}{4}$
2. $\cos 375°$ $\frac{\sqrt{6}+\sqrt{2}}{4}$
3. $\sin(-165°)$ $\frac{\sqrt{2}-\sqrt{6}}{4}$
4. $\sin(-105°)$ $\frac{-\sqrt{2}-\sqrt{6}}{4}$
5. $\sin 150°$ $\frac{1}{2}$
6. $\cos 240°$ $-\frac{1}{2}$
7. $\sin 225°$ $-\frac{\sqrt{2}}{2}$
8. $\sin(-75°)$ $\frac{-\sqrt{2}-\sqrt{6}}{4}$
9. $\sin 195°$ $\frac{\sqrt{2}-\sqrt{6}}{4}$

Verify that each of the following is an identity.

10. $\cos(180° - \theta) = -\cos\theta$
$\cos(180° - \theta) \stackrel{?}{=} -\cos\theta$
$\cos 180° \cos\theta + \sin 180° \sin\theta \stackrel{?}{=} -\cos\theta$
$-1\cos\theta + 0\sin\theta \stackrel{?}{=} -\cos\theta$
$-\cos\theta = -\cos\theta$

11. $\sin(360° + \theta) = \sin\theta$
$\sin(360° + \theta) \stackrel{?}{=} \sin\theta$
$\sin 360° \cos\theta + \cos 360° \sin\theta \stackrel{?}{=} \sin\theta$
$0\cos\theta + 1\sin\theta \stackrel{?}{=} \sin\theta$
$\sin\theta = \sin\theta$

12. $\sin(45° + \theta) - \sin(45° - \theta) = \sqrt{2}\sin\theta$
$\sin(45° + \theta) - \sin(45° - \theta)$
$\stackrel{?}{=} \sin 45°\cos\theta + \cos 45°\sin\theta - (\sin 45°\cos\theta - \cos 45°\sin\theta)$
$\stackrel{?}{=} 2 \cdot \cos 45°\sin\theta$
$\stackrel{?}{=} 2 \cdot \frac{\sqrt{2}}{2} \cdot \sin\theta$
$= \sqrt{2}\sin\theta$

13. $\cos\left(x - \frac{\pi}{6}\right) + \sin\left(x - \frac{\pi}{3}\right) = \sin x$
$\cos\left(x - \frac{\pi}{6}\right) + \sin\left(x - \frac{\pi}{3}\right)$
$\stackrel{?}{=} \cos x\cos\frac{\pi}{6} + \sin x\sin\frac{\pi}{6} + \sin x\cos\frac{\pi}{3} - \cos x\sin\frac{\pi}{3}$
$\stackrel{?}{=} \frac{\sqrt{3}}{2}\cos x + \frac{1}{2}\sin x + \frac{1}{2}\sin x - \frac{\sqrt{3}}{2}\cos x$
$= \sin x$

14. **SOLAR ENERGY** On March 21, the maximum amount of solar energy that falls on a square foot of ground at a certain location is given by $E\sin(90° - \phi)$, where ϕ is the latitude of the location and E is a constant. Use the difference of angles formula to find the amount of solar energy, in terms of $\cos\phi$, for a location that has a latitude of ϕ. $E\cos\phi$

ELECTRICITY **In Exercises 15 and 16, use the following information.**
In a certain circuit carrying alternating current, the formula $i = 2\sin(120t)$ can be used to find the current i in amperes after t seconds.

15. Rewrite the formula using the sum of two angles. Sample answer: $i = 2\sin(90t + 30t)$

16. Use the sum of angles formula to find the exact current at $t = 1$ second. $\sqrt{3}$ amperes

Reading to Learn Mathematics, p. 865 ELL

Pre-Activity **How are the sum and difference formulas used to describe communication interference?**

Read the introduction to Lesson 14-5 at the top of page 786 in your textbook.

Consider the functions $y = \sin x$ and $y = 2\sin x$. Do the graphs of these two functions have *constructive* interference or *destructive* interference? constructive

Reading the Lesson

1. Match each expression from the list on the left with an expression from the list on the right that is equal to it for all values of the variables. (Some of the expressions from the list on the right may be used more than once or not at all.)

a. $\sin(\alpha - \beta)$ v	i. $\sin\beta$
b. $\cos(\alpha + \beta)$ vi	ii. $\sin\alpha\cos\beta + \cos\alpha\sin\beta$
c. $\sin(180° + \beta)$ vii	iii. $-\cos\beta$
d. $\sin(180° - \beta)$ i	iv. $\cos\alpha\cos\beta + \sin\alpha\sin\beta$
e. $\cos(180° + \beta)$ iii	v. $\sin\alpha\cos\beta - \cos\alpha\sin\beta$
f. $\sin(\alpha + \beta)$ ii	vi. $\cos\alpha\cos\beta - \sin\alpha\sin\beta$
g. $\cos(90° - \beta)$ i	vii. $-\sin\beta$
h. $\cos(\alpha - \beta)$ iv	viii. $\cos\beta$

2. Which expressions are equal to sin 15°? (There may be more than one correct choice.) B and C

A. $\sin 45°\cos 30° + \cos 45°\sin 30°$
B. $\sin 45°\cos 30° - \cos 45°\sin 30°$
C. $\sin 60°\cos 45° - \cos 60°\sin 45°$
D. $\cos 60°\cos 45° - \sin 60°\sin 45°$

Helping You Remember

3. Some students have trouble remembering which signs to use on the right-hand sides of the sum and difference of angle formulas. What is an easy way to remember this?

Sample answer: In the *sine* identities, the signs are the *same* on both sides. In the *cosine* identities, the signs are *opposite* on the two sides.

Enrichment, p. 866

Identities for the Products of Sines and Cosines

By adding the identities for the sines of the sum and difference of the measures of two angles, a new identity is obtained.

$$\sin(\alpha + \beta) = \sin\alpha\cos\beta + \cos\alpha\sin\beta$$
$$\sin(\alpha - \beta) = \sin\alpha\cos\beta - \cos\alpha\sin\beta$$
$$(i) \quad \sin(\alpha + \beta) + \sin(\alpha - \beta) = 2\sin\alpha\cos\beta$$

This new identity is useful for expressing certain products as sums.

Example **Write $\sin 3\theta \cos\theta$ as a sum.**

In the identity let $\alpha = 3\theta$ and $\beta = \theta$ so that
$2\sin 3\theta\cos\theta = \sin(3\theta + \theta) + \sin(3\theta - \theta)$. Thus,
$\sin 3\theta\cos\theta = \frac{1}{2}\sin 4\theta + \frac{1}{2}\sin 2\theta$.

By subtracting the identities for $\sin(\alpha + \beta)$ and $\sin(\alpha - \beta)$, a similar identity for expressing a product as a difference is obtained.

4 Assess

Open-Ended Assessment

Modeling Show students a graph of the function $y = \sin x$. Have students point out on the graph which values would be most useful to use with the sum and difference of angles formulas, and ask them to explain their reasoning.

Getting Ready for Lesson 14-6

PREREQUISITE SKILL Students will find values using half-angle formulas in Lesson 14-6. The half-angle formulas include expressions within square root symbols, so students must be comfortable evaluating square roots. Use Exercises 67–74 to determine your students' familiarity with the Square Root Property.

Answers

47. Sample answer: To determine communication interference, you need to determine the sine or cosine of the sum or difference of two angles. Answers should include the following information.
    - Interference occurs when waves pass through the same space at the same time. When the combined waves have a greater amplitude, constructive interference results and when the combined waves have a smaller amplitude, destructive interference results.

58. $\sin\theta = -\frac{3\sqrt{34}}{34}$, $\cos\theta = \frac{5\sqrt{34}}{34}$, $\tan\theta = -\frac{3}{5}$, $\csc\theta = -\frac{\sqrt{34}}{3}$, $\sec\theta = \frac{\sqrt{34}}{5}$, $\cot\theta = -\frac{5}{3}$

59. $\sin\theta = -\frac{4}{5}$, $\cos\theta = -\frac{3}{5}$, $\tan\theta = \frac{4}{3}$, $\csc\theta = -\frac{5}{4}$, $\sec\theta = -\frac{5}{3}$, $\cot\theta = \frac{3}{4}$

60. $\sin\theta = 1$, $\cos\theta = 0$, $\tan\theta =$ undefined, $\csc\theta = 1$, $\sec\theta =$ undefined, $\cot\theta = 0$

47. WRITING IN MATH Answer the question that was posed at the beginning of the lesson. **See margin.**

How are the sum and difference formulas used to describe communication interference?

Include the following in your answer:
- an explanation of the difference between constructive and destructive interference, and
- a description of how you would explain wave interference to a friend.

Standardized Test Practice

48. Find the exact value of $\sin\theta$. **A**

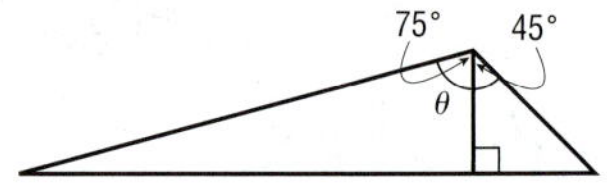

Ⓐ $\frac{\sqrt{3}}{2}$ Ⓑ $\frac{\sqrt{2}}{2}$

Ⓒ $\frac{1}{2}$ Ⓓ $\frac{\sqrt{3}}{3}$

49. Find the exact value of $\cos(-210°)$. **C**

Ⓐ $\frac{\sqrt{3}}{2}$ Ⓑ 0.5 Ⓒ $-\frac{\sqrt{3}}{2}$ Ⓓ -0.5

Maintain Your Skills

Mixed Review

Verify that each of the following is an identity. *(Lesson 14-4)* **50–53. See pp. 811A–811N.**

50. $\cot\theta + \sec\theta = \frac{\cos^2\theta + \sin\theta}{\sin\theta\cos\theta}$

51. $\sin^2\theta + \tan^2\theta = (1 - \cos^2\theta) + \frac{\sec^2\theta}{\csc^2\theta}$

52. $\sin\theta(\sin\theta + \csc\theta) = 2 - \cos^2\theta$

53. $\frac{\sec\theta}{\tan\theta} = \csc\theta$

Simplify each expression. *(Lesson 14-3)*

54. $\frac{\tan\theta\csc\theta}{\sec\theta}$ **1**

55. $4\left(\sec^2\theta - \frac{\sin^2\theta}{\cos^2\theta}\right)$ **4**

56. $(\cot\theta + \tan\theta)\sin\theta$ **sec θ**

57. $\csc\theta\tan\theta + \sec\theta$ **2 sec θ**

Find the exact values of the six trigonometric functions of θ if the terminal side of θ in standard position contains the given point. *(Lesson 13-3)* **58–60. See margin**

58. $(5, -3)$ 59. $(-3, -4)$ 60. $(0, 2)$

Evaluate each expression. *(Lesson 12-2)*

61. $P(6, 4)$ **360**

62. $P(12, 7)$ **3,991,680**

63. $C(8, 3)$ **56**

64. $C(10, 4)$ **210**

65. **AVIATION** A pilot is flying from Chicago to Columbus, a distance of 300 miles. In order to avoid an area of thunderstorms, she alters her initial course by 15° and flies on this course for 75 miles. How far is she from Columbus? *(Lesson 13-5)* **65. about 228 mi**

66. Write $6y^2 - 34x^2 = 204$ in standard form. *(Lesson 8-5)* $\frac{y^2}{34} - \frac{x^2}{6} = 1$

Getting Ready for the Next Lesson

PREREQUISITE SKILL Solve each equation.
*(To review **solving equations using the Square Root Property**, see Lesson 6-4.)*

67. $x^2 = \frac{20}{16}$ $\pm\frac{\sqrt{5}}{2}$

68. $x^2 = \frac{9}{25}$ $\pm\frac{3}{5}$

69. $x^2 = \frac{5}{25}$ $\pm\frac{\sqrt{5}}{5}$

70. $x^2 = \frac{18}{32}$ $\pm\frac{3}{4}$

71. $x^2 - 1 = \frac{1}{2}$ $\pm\frac{\sqrt{6}}{2}$

72. $x^2 - 1 = \frac{4}{5}$ **72.** $\pm\frac{3\sqrt{5}}{5}$

73. $x^2 = \frac{\sqrt{3}}{2} - \frac{1}{2}$ **73.** $\pm\frac{\sqrt{\sqrt{6} - \sqrt{2}}}{2}$

74. $x^2 = \frac{\sqrt{2}}{2} - 1$ **74.** $\pm\frac{\sqrt{2 - 2\sqrt{2}}}{2}$

790 Chapter 14 Trigonometric Graphs and Identities

14-6 Double-Angle and Half-Angle Formulas

What You'll Learn

- Find values of sine and cosine involving double-angle formulas.
- Find values of sine and cosine involving half-angle formulas.

How can trigonometric functions be used to describe music?

Stringed instruments such as a piano, guitar, or violin rely on waves to produce the tones we hear. When the strings are struck or plucked, they vibrate. If the motion of the strings were observed in slow motion, you could see that there are places on the string, called *nodes*, that do not move under the vibration. Halfway between each pair of consecutive nodes are *antinodes* that undergo the maximum vibration. The nodes and antinodes form *harmonics*. These harmonics can be represented using variations of the equations $y = \sin 2\theta$ and $y = \sin \frac{1}{2}\theta$.

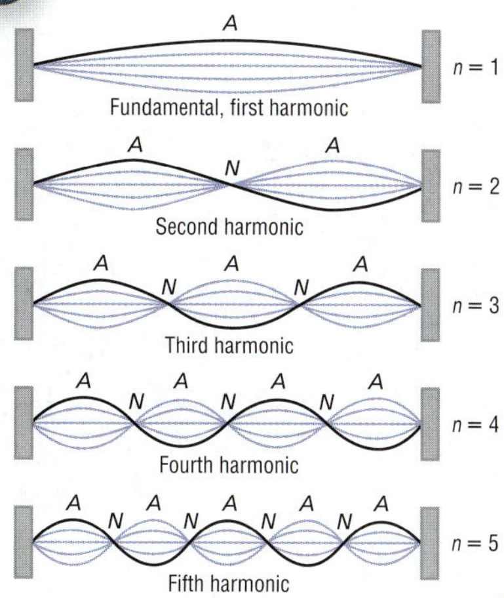

DOUBLE-ANGLE FORMULAS You can use the formula for $\sin(\alpha + \beta)$ to find the sine of twice an angle θ, $\sin 2\theta$, and the formula for $\cos(\alpha + \beta)$ to find the cosine of twice an angle θ, $\cos 2\theta$.

$$\begin{aligned}\sin 2\theta &= \sin(\theta + \theta)\\ &= \sin\theta\cos\theta + \cos\theta\sin\theta\\ &= 2\sin\theta\cos\theta\end{aligned}$$

$$\begin{aligned}\cos 2\theta &= \cos(\theta + \theta)\\ &= \cos\theta\cos\theta - \sin\theta\sin\theta\\ &= \cos^2\theta - \sin^2\theta\end{aligned}$$

You can find alternate forms for $\cos 2\theta$ by making substitutions into the expression $\cos^2\theta - \sin^2\theta$.

$\cos^2\theta - \sin^2\theta = (1 - \sin^2\theta) - \sin^2\theta$ Substitute $1 - \sin^2\theta$ for $\cos^2\theta$.

$= 1 - 2\sin^2\theta$ Simplify.

$\cos^2\theta - \sin^2\theta = \cos^2\theta - (1 - \cos^2\theta)$ Substitute $1 - \cos^2\theta$ for $\sin^2\theta$.

$= 2\cos^2\theta - 1$ Simplify.

These formulas are called the **double-angle formulas**.

Key Concept — Double-Angle Formulas

The following identities hold true for all values of θ.

$\sin 2\theta = 2\sin\theta\cos\theta$

$\cos 2\theta = \cos^2\theta - \sin^2\theta$

$\cos 2\theta = 1 - 2\sin^2\theta$

$\cos 2\theta = 2\cos^2\theta - 1$

14-6 Lesson Notes

1 Focus

5-Minute Check Transparency 14-6 Use as a quiz or review of Lesson 14-5.

Mathematical Background notes are available for this lesson on page 760D.

How can trigonometric functions be used to describe music?

Ask students:

- What do the values of n represent on the right side of the diagram? **the number of antinodes**
- What is occurring at an antinode? **maximum vibration**
- At what point do you think a guitar string would be more likely to break, a node or an antinode? **at an antinode**

2 Teach

DOUBLE-ANGLE FORMULAS

Teaching Tip Point out that the double-angle formulas are derived using the sum and difference of angles formulas presented in Lesson 14-5.

Resource Manager

Workbook and Reproducible Masters

Chapter 14 Resource Masters

- Study Guide and Intervention, pp. 867–868
- Skills Practice, p. 869
- Practice, p. 870
- Reading to Learn Mathematics, p. 871
- Enrichment, p. 872
- Assessment, p. 894

Transparencies

5-Minute Check Transparency 14-6
Answer Key Transparencies

Technology

Interactive Chalkboard

Teaching Tip Remind students of the identity $\sin^2\theta + \cos^2\theta = 1$, and its two variations that occur by subtracting either $\sin^2\theta$ or $\cos^2\theta$ from both sides. Explain that there are three versions of the formula for $\cos 2\theta$ because of the three variations of the identity $\sin^2\theta + \cos^2\theta = 1$.

In-Class Example PowerPoint®

1 Find the exact value of each expression if $\sin\theta = \frac{3}{4}$ and θ is between 0° and 90°.

a. $\sin 2\theta$ $\frac{3\sqrt{7}}{8}$

b. $\cos 2\theta$ $-\frac{1}{8}$

Example 1 Double-Angle Formulas

Find the exact value of each expression if $\sin\theta = \frac{4}{5}$ and θ is between 90° and 180°.

a. $\sin 2\theta$

Use the identity $\sin 2\theta = 2\sin\theta\cos\theta$.

First, find the value of $\cos\theta$.

$\cos^2\theta = 1 - \sin^2\theta$ $\cos^2\theta + \sin^2\theta = 1$

$\cos^2\theta = 1 - \left(\frac{4}{5}\right)^2$ $\sin\theta = \frac{4}{5}$

$\cos^2\theta = \frac{9}{25}$ Subtract.

$\cos\theta = \pm\frac{3}{5}$ Find the square root of each side.

Since θ is in the second quadrant, cosine is negative. Thus, $\cos\theta = -\frac{3}{5}$.

Now find $\sin 2\theta$.

$\sin 2\theta = 2\sin\theta\cos\theta$ Double-angle formula

$\sin 2\theta = 2\left(\frac{4}{5}\right)\left(-\frac{3}{5}\right)$ $\sin\theta = \frac{4}{5}$, $\cos\theta = -\frac{3}{5}$

$= -\frac{24}{25}$ The value of $\sin 2\theta$ is $-\frac{24}{25}$.

b. $\cos 2\theta$

Use the identity $\cos 2\theta = 1 - 2\sin^2\theta$.

$\cos 2\theta = 1 - 2\sin^2\theta$ Double-angle formula

$= 1 - 2\left(\frac{4}{5}\right)^2$ $\sin\theta = \frac{4}{5}$

$= -\frac{7}{25}$ The value of $\cos 2\theta$ is $-\frac{7}{25}$.

HALF-ANGLE FORMULAS You can derive formulas for the sine and cosine of half a given angle using the double-angle formulas.

Find $\sin\frac{\alpha}{2}$.

$1 - 2\sin^2\theta = \cos 2\theta$ Double-angle formula

$1 - 2\sin^2\frac{\alpha}{2} = \cos\alpha$ Substitute $\frac{\alpha}{2}$ for θ and α for 2θ.

$\sin^2\frac{\alpha}{2} = \frac{1 - \cos\alpha}{2}$ Solve for $\sin^2\frac{\alpha}{2}$.

$\sin\frac{\alpha}{2} = \pm\sqrt{\frac{1 - \cos\alpha}{2}}$ Take the square root of each side.

Find $\cos\frac{\alpha}{2}$.

$2\cos^2\theta - 1 = \cos 2\theta$ Double-angle formula

$2\cos^2\frac{\alpha}{2} - 1 = \cos\alpha$ Substitute $\frac{\alpha}{2}$ for θ and α for 2θ.

$\cos^2\frac{\alpha}{2} = \frac{1 + \cos\alpha}{2}$ Solve for $\cos^2\frac{\alpha}{2}$.

$\cos\frac{\alpha}{2} = \pm\sqrt{\frac{1 + \cos\alpha}{2}}$ Take the square root of each side.

These are called the **half-angle formulas**. The signs are determined by the function of $\frac{\alpha}{2}$.

Key Concept *Half-Angle Formulas*

The following identities hold true for all values of α.

$$\sin \frac{\alpha}{2} = \pm\sqrt{\frac{1 - \cos \alpha}{2}} \qquad \cos \frac{\alpha}{2} = \pm\sqrt{\frac{1 + \cos \alpha}{2}}$$

Example 2 Half-Angle Formulas

Find $\cos \frac{\alpha}{2}$ if $\sin \alpha = -\frac{3}{4}$ and α is in the third quadrant.

Since $\cos \frac{\alpha}{2} = \pm\sqrt{\frac{1 + \cos \alpha}{2}}$, we must find $\cos \alpha$ first.

$\cos^2 \alpha = 1 - \sin^2 \alpha$ — $\cos^2 \alpha + \sin^2 \alpha = 1$

$\cos^2 \alpha = 1 - \left(-\frac{3}{4}\right)^2$ — $\sin \alpha = -\frac{3}{4}$

$\cos^2 \alpha = \frac{7}{16}$ — Simplify.

$\cos \alpha = \pm\frac{\sqrt{7}}{4}$ — Take the square root of each side.

Since α is in the third quadrant, $\cos \alpha = -\frac{\sqrt{7}}{4}$.

$\cos \frac{\alpha}{2} = \pm\sqrt{\frac{1 + \cos \alpha}{2}}$ — Half-angle formula

$= \pm\sqrt{\frac{1 - \frac{\sqrt{7}}{4}}{2}}$ — $\cos \alpha = -\frac{\sqrt{7}}{4}$

$= \pm\sqrt{\frac{4 - \sqrt{7}}{8}}$ — Simplify the radicand.

$= \pm\frac{\sqrt{4 - \sqrt{7}}}{2\sqrt{2}} \cdot \frac{\sqrt{2}}{\sqrt{2}}$ — Rationalize.

$= \pm\frac{\sqrt{8 - 2\sqrt{7}}}{4}$ — Multiply.

Since α is between 180° and 270°, $\frac{\alpha}{2}$ is between 90° and 135°. Thus, $\cos \frac{\alpha}{2}$ is negative and equals $-\frac{\sqrt{8 - 2\sqrt{7}}}{4}$.

Study Tip

Choosing the Sign

You may want to determine the quadrant in which the terminal side of $\frac{\alpha}{2}$ will lie in the first step of the solution. Then you can use the correct sign from the beginning.

Example 3 Evaluate Using Half-Angle Formulas

Find the exact value of each expression by using the half-angle formulas.

a. sin 105°

$\sin 105° = \sin \frac{210°}{2}$

$= \sqrt{\frac{1 - \cos 210°}{2}}$ — $\sin \frac{\alpha}{2} = \pm\sqrt{\frac{1 - \cos \alpha}{2}}$

(continued on the next page)

HALF-ANGLE FORMULAS

Tips for New Teachers

Intervention Stress the Study Tip provided in the margin next to Example 2. Determining the proper sign for the answer at the beginning of the computation will help some students avoid forgetting this step at the end of their computations.

In-Class Example Power Point®

2 Find $\cos \frac{\alpha}{2}$ if $\sin \alpha = \frac{4}{5}$ and α is in the second quadrant. $\frac{\sqrt{5}}{5}$

DAILY INTERVENTION — **Differentiated Instruction**

Auditory/Musical If possible, ask a music teacher at your school to talk to students about harmonics. Students playing stringed instruments may also be willing to share what they have learned about harmonics and waves. If a music teacher is not available, a physics teacher may also be able to demonstrate harmonics or bring a device that creates standing waves to class.

In-Class Examples

Power Point®

3 Find the exact value of each expression by using the half-angle formulas.

a. $\sin 165°$ $\dfrac{\sqrt{2-\sqrt{3}}}{2}$

b. $\cos \dfrac{9\pi}{8}$ $-\dfrac{\sqrt{2+\sqrt{2}}}{2}$

4 Verify that $\sin\theta\,(\cos^2\theta - \cos 2\theta) = \sin^3\theta$ is an identity.

$\sin\theta\,(\cos^2\theta - \cos 2\theta)$
$\stackrel{?}{=} \sin\theta\,[\cos^2\theta - (\cos^2\theta - \sin^2\theta)]$
$\stackrel{?}{=} \sin\theta\,(\cos^2\theta - \cos^2\theta + \sin^2\theta)$
$\stackrel{?}{=} \sin\theta\,(\sin^2\theta)$
$= \sin^3\theta$

3 Practice/Apply

Study Notebook

Have students—
- add the definitions/examples of the vocabulary terms to their Vocabulary Builder worksheets for Chapter 14.
- record the double-angle and half-angle formulas.
- include any other item(s) that they find helpful in mastering the skills in this lesson.

$$= \sqrt{\frac{1-\left(-\frac{\sqrt{3}}{2}\right)}{2}} \qquad \cos 210° = -\frac{\sqrt{3}}{2}$$

$$= \sqrt{\frac{2+\sqrt{3}}{4}} \qquad \text{Simplify the radicand.}$$

$$= \frac{\sqrt{2+\sqrt{3}}}{2} \qquad \text{Simplify the denominator.}$$

b. $\cos \dfrac{\pi}{8}$

$$\cos\frac{\pi}{8} = \cos\frac{\frac{\pi}{4}}{2}$$

$$= \sqrt{\frac{1+\cos\frac{\pi}{4}}{2}} \qquad \cos\frac{\alpha}{2} = \pm\sqrt{\frac{1+\cos\alpha}{2}}$$

$$= \sqrt{\frac{1+\frac{\sqrt{2}}{2}}{2}} \qquad \cos\frac{\pi}{4} = \frac{\sqrt{2}}{2}$$

$$= \sqrt{\frac{2+\sqrt{2}}{4}} \qquad \text{Simplify the radicand.}$$

$$= \frac{\sqrt{2+\sqrt{2}}}{2} \qquad \text{Simplify the denominator.}$$

Recall that you can use the sum and difference formulas to verify identities. Double- and half-angle formulas can also be used to verify identities.

Example 4 Verify Identities

Verify that $(\sin\theta + \cos\theta)^2 = 1 + \sin 2\theta$ is an identity.

$$(\sin\theta+\cos\theta)^2 \stackrel{?}{=} 1+\sin 2\theta \qquad \text{Original equation}$$

$$\sin^2\theta + 2\sin\theta\cos\theta + \cos^2\theta \stackrel{?}{=} 1+\sin 2\theta \qquad \text{Multiply.}$$

$$1 + 2\sin\theta\cos\theta \stackrel{?}{=} 1+\sin 2\theta \qquad \sin^2\theta+\cos^2\theta = 1$$

$$1+\sin 2\theta = 1+\sin 2\theta \qquad \text{Double-angle formula}$$

Check for Understanding

Concept Check

1–3. See margin.

1. **Explain** how to find $\cos\frac{x}{2}$ if x is in the third quadrant.
2. **Find a counterexample** to show that $\cos 2\theta = 2\cos\theta$ is not an identity.
3. **OPEN ENDED** Describe the conditions under which you would use each of the three identities for $\cos 2\theta$.

Guided Practice

GUIDED PRACTICE KEY	
Exercises	Examples
4–7	1–2
8, 9, 12	3
10, 11	4

Find the exact values of $\sin 2\theta$, $\cos 2\theta$, $\sin\frac{\theta}{2}$, and $\cos\frac{\theta}{2}$ for each of the following.

4. $\cos\theta = \frac{3}{5};\ 0° < \theta < 90°$
5. $\cos\theta = -\frac{2}{3};\ 180° < \theta < 270°$
6. $\sin\theta = \frac{1}{2};\ 0° < \theta < 90°$
7. $\sin\theta = -\frac{3}{4};\ 270° < \theta < 360°$

4–7. See margin.

Find the exact value of each expression by using the half-angle formulas.

8. $\sin 195°$ $-\dfrac{\sqrt{2-\sqrt{3}}}{2}$
9. $\cos\dfrac{19\pi}{12}$ $\dfrac{\sqrt{2-\sqrt{3}}}{2}$

794 Chapter 14 Trigonometric Graphs and Identities

Answers

1. Sample answer: If x is in the third quadrant, then $\frac{x}{2}$ is between 90° and 135°. Use the half-angle formula for cosine knowing that the value is negative.
2. Sample answer: 45°; $\cos 2(45°) = \cos 90°$ or 0, $2\cos 45° = 2\cdot\frac{\sqrt{2}}{2}$ or $\sqrt{2}$
3. Sample answer: The identity used for $\cos 2\theta$ depends on whether you know the value of $\sin\theta$, $\cos\theta$, or both values.
4. $\frac{24}{25}, -\frac{7}{25}, \frac{\sqrt{5}}{5}, \frac{2\sqrt{5}}{5}$
5. $\frac{4\sqrt{5}}{9}, -\frac{1}{9}, \frac{\sqrt{30}}{6}, -\frac{\sqrt{6}}{6}$
6. $\frac{\sqrt{3}}{2}, \frac{1}{2}, \frac{\sqrt{2-\sqrt{3}}}{2}, \frac{\sqrt{2+\sqrt{3}}}{2}$
7. $-\frac{3\sqrt{7}}{8}, -\frac{1}{8}, \frac{\sqrt{8-2\sqrt{7}}}{4}, -\frac{\sqrt{8+2\sqrt{7}}}{4}$

Verify that each of the following is an identity. **10–11. See margin.**

10. $\cot x = \dfrac{\sin 2x}{1 - \cos 2x}$

11. $\cos^2 2x + 4\sin^2 x \cos^2 x = 1$

Application

12. **AVIATION** When a jet travels at speeds greater than the speed of sound, a sonic boom is created by the sound waves forming a cone behind the jet. If θ is the measure of the angle at the vertex of the cone, then the Mach number M can be determined using the formula $\sin\frac{\theta}{2} = \frac{1}{M}$. Find the Mach number of a jet if a sonic boom is created by a cone with a vertex angle of 75°. **1.64**

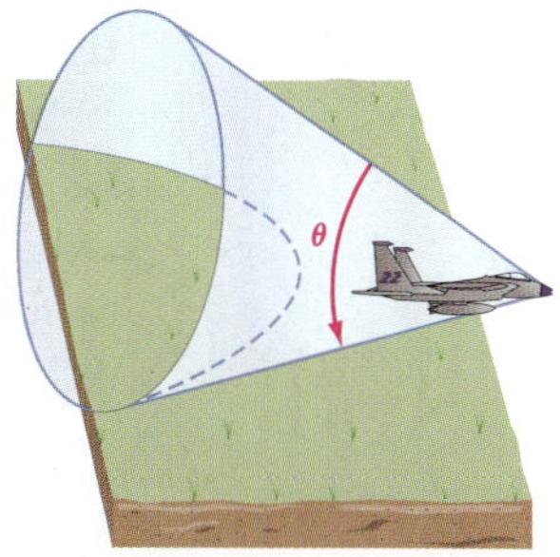

Practice and Apply

Homework Help

For Exercises	See Examples
13–24, 38, 39	1, 2
25–30, 37	3
31–36	4

Extra Practice See page 861.

13–24. See margin.

Find the exact values of $\sin 2\theta$, $\cos 2\theta$, $\sin\frac{\theta}{2}$, and $\cos\frac{\theta}{2}$ for each of the following.

13. $\sin\theta = \frac{5}{13}$; $90° < \theta < 180°$

14. $\cos\theta = \frac{1}{5}$; $270° < \theta < 360°$

15. $\cos\theta = -\frac{1}{3}$; $180° < \theta < 270°$

16. $\sin\theta = -\frac{3}{5}$; $180° < \theta < 270°$

17. $\sin\theta = -\frac{3}{8}$; $270° < \theta < 360°$

18. $\cos\theta = -\frac{1}{4}$; $90° < \theta < 180°$

19. $\cos\theta = \frac{1}{6}$; $0° < \theta < 90°$

20. $\cos\theta = -\frac{12}{13}$; $180° < \theta < 270°$

21. $\sin\theta = -\frac{1}{3}$; $270° < \theta < 360°$

22. $\sin\theta = -\frac{1}{4}$; $180° < \theta < 270°$

23. $\cos\theta = \frac{2}{3}$; $0° < \theta < 90°$

24. $\sin\theta = \frac{2}{5}$; $90° < \theta < 180°$

Find the exact value of each expression by using the half-angle formulas.

25. $\cos 165°$ $-\dfrac{\sqrt{2+\sqrt{3}}}{2}$

26. $\sin 22\frac{1}{2}°$ $\dfrac{\sqrt{2-\sqrt{2}}}{2}$

27. $\cos 157\frac{1}{2}°$ $-\dfrac{\sqrt{2+\sqrt{2}}}{2}$

28. $\sin 345°$ $-\dfrac{\sqrt{2-\sqrt{3}}}{2}$

29. $\sin\frac{7\pi}{8}$ $\dfrac{\sqrt{2-\sqrt{2}}}{2}$

30. $\cos\frac{7\pi}{12}$ $-\dfrac{\sqrt{2-\sqrt{3}}}{2}$

Verify that each of the following is an identity. **31–36. See pp. 811A–811N.**

31. $\sin 2x = 2\cot x \sin^2 x$

32. $2\cos^2\frac{x}{2} = 1 + \cos x$

33. $\sin^4 x - \cos^4 x = 2\sin^2 x - 1$

34. $\sin^2 x = \frac{1}{2}(1 - \cos 2x)$

35. $\tan^2\frac{x}{2} = \dfrac{1 - \cos x}{1 + \cos x}$

36. $\dfrac{1}{\sin x \cos x} - \dfrac{\cos x}{\sin x} = \tan x$

More About. . .

Optics A rainbow appears when the sun shines through water droplets that act as a prism.

37. **OPTICS** If a glass prism has an apex angle of measure α and an angle of deviation of measure β, then the index of refraction n of the prism is given by $n = \dfrac{\sin\left[\frac{1}{2}(\alpha + \beta)\right]}{\sin\frac{\alpha}{2}}$.

What is the angle of deviation of a prism with an apex angle of 40° and an index of refraction of 2? **46.3°**

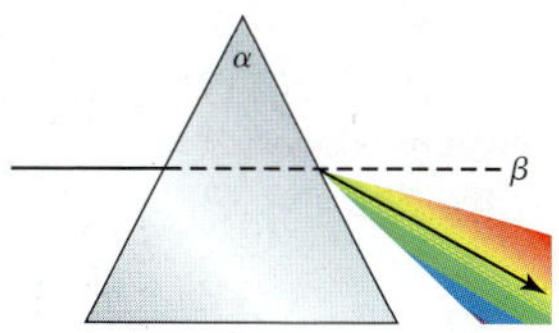

www.algebra2.com/self_check_quiz

About the Exercises...

Organization by Objective

- **Double-Angle Formulas:** 13–24, 31
- **Half-Angle Formulas:** 13–30, 32, 35

Odd/Even Assignments

Exercises 13–36 are structured so that students practice the same concepts whether they are assigned odd or even problems.

Assignment Guide

Basic: 13–37 odd, 38, 39, 41–65

Average: 13–37 odd, 38, 39, 41–65

Advanced: 14–38 even, 39–59 (optional: 60–65)

All: Practice Quiz 2 (1–10)

Answers

10. $\cot x \stackrel{?}{=} \dfrac{\sin 2x}{1 - \cos 2x}$
$\stackrel{?}{=} \dfrac{2\sin x \cos x}{1 - (1 - 2\sin^2 x)}$
$\stackrel{?}{=} \dfrac{2\sin x\cos x}{2\sin^2 x}$
$\stackrel{?}{=} \dfrac{\cos x}{\sin x}$
$= \cot x$

11. $\cos^2 2x + 4\sin^2 x\cos^2 x \stackrel{?}{=} 1$
$\cos^2 2x + \sin^2 2x \stackrel{?}{=} 1$
$1 = 1$

13. $-\frac{120}{169}, \frac{119}{169}, \frac{5\sqrt{26}}{26}, \frac{\sqrt{26}}{26}$

14. $-\frac{4\sqrt{6}}{25}, -\frac{23}{25}, \frac{\sqrt{10}}{5}, -\frac{\sqrt{15}}{5}$

15. $\frac{4\sqrt{2}}{9}, -\frac{7}{9}, \frac{\sqrt{6}}{3}, -\frac{\sqrt{3}}{3}$

16. $\frac{24}{25}, \frac{7}{25}, \frac{3\sqrt{10}}{10}, -\frac{\sqrt{10}}{10}$

17. $-\frac{3\sqrt{55}}{32}, \frac{23}{32}, \frac{\sqrt{8-\sqrt{55}}}{4}, -\frac{\sqrt{8+\sqrt{55}}}{4}$

18. $-\frac{\sqrt{15}}{8}, -\frac{7}{8}, \frac{\sqrt{10}}{4}, \frac{\sqrt{6}}{4}$

19. $\frac{\sqrt{35}}{18}, -\frac{17}{18}, \frac{\sqrt{15}}{6}, \frac{\sqrt{21}}{6}$

20. $\frac{120}{169}, \frac{119}{169}, \frac{5\sqrt{26}}{26}, -\frac{\sqrt{26}}{26}$

21. $-\frac{4\sqrt{2}}{9}, \frac{7}{9}, \frac{\sqrt{18-12\sqrt{2}}}{6}, -\frac{\sqrt{18-12\sqrt{2}}}{6}$

22. $\frac{\sqrt{15}}{8}, \frac{7}{8}, \frac{\sqrt{8+2\sqrt{15}}}{4}, -\frac{\sqrt{8-2\sqrt{15}}}{4}$

23. $\frac{4\sqrt{5}}{9}, -\frac{1}{9}, \frac{\sqrt{6}}{6}, \frac{\sqrt{30}}{6}$

24. $-\frac{4\sqrt{21}}{5}, \frac{17}{25}, \frac{\sqrt{5\sqrt{2}+10\sqrt{21}}}{10}, \frac{\sqrt{5\sqrt{2}-10\sqrt{21}}}{10}$

Study Guide and Intervention, p. 867 (shown) and p. 868

Double-Angle Formulas

Double-Angle Formulas	The following identities hold true for all values of θ. $\sin 2\theta = 2\sin\theta\cdot\cos\theta$ $\cos 2\theta = \cos^2\theta - \sin^2\theta$ $\cos 2\theta = 1 - 2\sin^2\theta$ $\cos 2\theta = 2\cos^2\theta - 1$

Example **Find the exact values of $\sin 2\theta$ and $\cos 2\theta$ if $\sin\theta = -\frac{9}{10}$ and $180° < \theta < 270°$.**

First, find the value of $\cos\theta$.

$\cos^2\theta = 1 - \sin^2\theta$ $\cos^2\theta + \sin^2\theta = 1$

$\cos^2\theta = 1 - \left(-\frac{9}{10}\right)^2$ $\sin\theta = -\frac{9}{10}$

$\cos^2\theta = \frac{19}{100}$

$\cos\theta = \pm\frac{\sqrt{19}}{10}$

Since θ is in the third quadrant, $\cos\theta$ is negative. Thus $\cos\theta = -\frac{\sqrt{19}}{10}$.

To find $\sin 2\theta$, use the identity $\sin 2\theta = 2\sin\theta\cdot\cos\theta$.

$\sin 2\theta = 2\sin\theta\cdot\cos\theta = 2\left(-\frac{9}{10}\right)\left(-\frac{\sqrt{19}}{10}\right) = \frac{9\sqrt{19}}{50}$

The value of $\sin 2\theta$ is $\frac{9\sqrt{19}}{50}$.

To find $\cos 2\theta$, use the identity $\cos 2\theta = 1 - 2\sin^2\theta$.

$\cos 2\theta = 1 - 2\sin^2\theta = 1 - 2\left(-\frac{9}{10}\right)^2 = -\frac{31}{50}$.

The value of $\cos 2\theta$ is $-\frac{31}{50}$.

Exercises

Find the exact values of $\sin 2\theta$ and $\cos 2\theta$ for each of the following.

1. $\sin\theta = \frac{1}{4}$, $0° < \theta < 90°$ $\frac{\sqrt{15}}{8}, \frac{7}{8}$
2. $\sin\theta = -\frac{1}{8}$, $270° < \theta < 360°$ $-\frac{3\sqrt{7}}{32}, \frac{31}{32}$
3. $\cos\theta = -\frac{3}{5}$, $180° < \theta < 270°$ $\frac{24}{25}, -\frac{7}{25}$
4. $\cos\theta = -\frac{4}{5}$, $90° < \theta < 180°$ $-\frac{24}{25}, \frac{7}{25}$
5. $\sin\theta = -\frac{3}{5}$, $270° < \theta < 360°$ $-\frac{24}{25}, \frac{7}{25}$
6. $\cos\theta = -\frac{2}{3}$, $90° < \theta < 180°$ $-\frac{4\sqrt{5}}{9}, -\frac{1}{9}$

Skills Practice, p. 869 and Practice, p. 870 (shown)

Find the exact values of $\sin 2\theta$, $\cos 2\theta$, $\sin\frac{\theta}{2}$, and $\cos\frac{\theta}{2}$ for each of the following.

1. $\cos\theta = \frac{5}{13}$, $0° < \theta < 90°$ $\frac{120}{169}, -\frac{119}{169}, \frac{2\sqrt{13}}{13}, \frac{3\sqrt{13}}{13}$
2. $\sin\theta = \frac{8}{17}$, $90° < \theta < 180°$ $-\frac{240}{289}, \frac{161}{289}, \frac{4\sqrt{17}}{17}, \frac{\sqrt{17}}{17}$
3. $\cos\theta = \frac{1}{4}$, $270° < \theta < 360°$ $-\frac{\sqrt{15}}{8}, -\frac{7}{8}, \frac{\sqrt{6}}{4}, \frac{\sqrt{10}}{4}$
4. $\sin\theta = -\frac{2}{3}$, $180° < \theta < 270°$ $\frac{4\sqrt{5}}{9}, \frac{1}{9}, \frac{\sqrt{18+6\sqrt{5}}}{6}, -\frac{\sqrt{18-6\sqrt{5}}}{6}$

Find the exact value of each expression by using the half-angle formulas.

5. $\tan 105°$ $-2-\sqrt{3}$
6. $\tan 15°$ $2-\sqrt{3}$
7. $\cos 67.5°$ $\frac{\sqrt{2-\sqrt{2}}}{2}$
8. $\sin\left(-\frac{\pi}{8}\right)$ $-\frac{\sqrt{2-\sqrt{2}}}{2}$

Verify that each of the following is an identity.

9. $\sin^2\frac{\theta}{2} = \frac{\tan\theta - \sin\theta}{2\tan\theta}$ $\left(\pm\sqrt{\frac{1-\cos\theta}{2}}\right)^2 \stackrel{?}{=} \frac{\tan\theta - \sin\theta}{2\tan\theta}$; $\frac{1-\cos\theta}{2} \stackrel{?}{=} \frac{\frac{\tan\theta}{\tan\theta} - \frac{\sin\theta}{\tan\theta}}{2\frac{\tan\theta}{\tan\theta}}$; $\frac{1-\cos\theta}{2} = \frac{1-\cos\theta}{2}$

10. $\sin 4\theta = 4\cos 2\theta\sin\theta\cos\theta$

$\sin 4\theta \stackrel{?}{=} 4\cos 2\theta\sin\theta\cos\theta$
$\sin 2(2\theta) \stackrel{?}{=} 4\cos 2\theta\sin\theta\cos\theta$
$2\sin 2\theta\cos 2\theta \stackrel{?}{=} 4\cos 2\theta\sin\theta\cos\theta$
$2(2\sin\theta\cos\theta)(\cos 2\theta) \stackrel{?}{=} 4\cos 2\theta\sin\theta\cos\theta$
$4\cos 2\theta\sin\theta\cos\theta = 4\cos 2\theta\sin\theta\cos\theta$

11. **AERIAL PHOTOGRAPHY** In aerial photography, there is a reduction in film exposure for any point X not directly below the camera. The reduction E_θ is given by $E_\theta = E_0\cos^4\theta$, where θ is the angle between the perpendicular line from the camera to the ground and the line from the camera to point X, and E_0 is the exposure for the point directly below the camera. Using the identity $2\sin^2\theta = 1 - \cos 2\theta$, verify that $E_0\cos^4\theta = E_0\left(\frac{1}{2} + \frac{\cos 2\theta}{2}\right)^2$.

$E_0\cos^4\theta = E_0(\cos^2\theta)^2 = E_0(1-\sin^2\theta)^2 = E_0\left(1 - \frac{2\sin^2\theta}{2}\right)^2 = E_0\left(1 - \frac{1-\cos 2\theta}{2}\right)^2 = E_0\left(\frac{1}{2} + \frac{\cos 2\theta}{2}\right)^2$

12. **IMAGING** A scanner takes thermal images from altitudes of 300 to 12,000 meters. The width W of the swath covered by the image is given by $W = 2H'\tan\theta$, where H' is the height and θ is half the scanner's field of view. Verify that $\frac{2H'\sin 2\theta}{1+\cos 2\theta} = 2H'\tan\theta$.

$\frac{2H'\sin 2\theta}{1+\cos 2\theta} = \frac{4H'\sin\theta\cos\theta}{1+(2\cos^2\theta-1)} = \frac{4H'\sin\theta\cos\theta}{2\cos^2\theta} = \frac{2H'\sin\theta}{\cos\theta} = 2H'\tan\theta$

Reading to Learn Mathematics, p. 871 ELL

Pre-Activity **How can trigonometric functions be used to describe music?**

Read the introduction to Lesson 14-6 at the top of page 791 in your textbook.

Suppose that the equation for the second harmonic is $y = \sin a\theta$. Then what would be the equations for the fundamental tone (first harmonic), third harmonic, fourth harmonic, and fifth harmonic?
$y = \sin 0.5a\theta$; $y = \sin 1.5a\theta$; $y = \sin 2a\theta$; $y = \sin 2.5a\theta$

Reading the Lesson

1. Match each expression from the list on the left with *all* expressions from the list on the right that are equal to it for all values of β.

a. $\sin\frac{\beta}{2}$ v   i. $2\sin\beta\cos\beta$
b. $\cos 2\beta$ ii and iii   ii. $1 - 2\sin^2\beta$
c. $\cos\frac{\beta}{2}$ iv   iii. $\cos^2\beta - \sin^2\beta$
d. $\sin 2\beta$ i   iv. $\pm\sqrt{\frac{1+\cos\beta}{2}}$
v. $\pm\sqrt{\frac{1-\cos\beta}{2}}$

2. Determine whether you would use the *positive* or *negative* square root in the half-angle identities for $\sin\frac{\alpha}{2}$ and $\cos\frac{\alpha}{2}$ in each of the following situations. (Do not actually calculate $\sin\frac{\alpha}{2}$ and $\cos\frac{\alpha}{2}$.)

a. $\sin\frac{\alpha}{2}$, if $\cos\alpha = \frac{2}{5}$ and α is in Quadrant I **positive**
b. $\cos\frac{\alpha}{2}$, if $\cos\alpha = -0.9$ and α is in Quadrant II **positive**
c. $\cos\frac{\alpha}{2}$, if $\sin\alpha = -0.75$ and α is in Quadrant III **negative**
d. $\sin\frac{\alpha}{2}$, if $\sin\alpha = -0.8$ and α is in Quadrant IV **positive**

Helping You Remember

3. Many students find it difficult to remember a large number of identities. How can you obtain all three of the identities for $\cos 2\theta$ by remembering only one of them and using a Pythagorean identity?
Sample answer: Just remember the identity $\cos 2\theta = \cos^2\theta - \sin^2\theta$. Using the Pythagorean identity $\cos^2\theta + \sin^2\theta = 1$, you can substitute either $1 - \sin^2\theta$ for $\cos^2\theta$ or $1 - \cos^2\theta$ for $\sin^2\theta$ to get the other two identities for $\cos 2\theta$.

38. $\dfrac{1 \pm \sqrt{\frac{1-\cos L}{1+\cos L}}}{1 \mp \sqrt{\frac{1-\cos L}{1+\cos L}}}$

GEOGRAPHY **For Exercises 38 and 39, use the following information.**
A Mercator projection map uses a flat projection of Earth in which the distance between the lines of latitude increases with their distance from the equator. The calculation of the location of a point on this projection uses the expression $\tan\left(45° + \frac{L}{2}\right)$, where L is the latitude of the point.

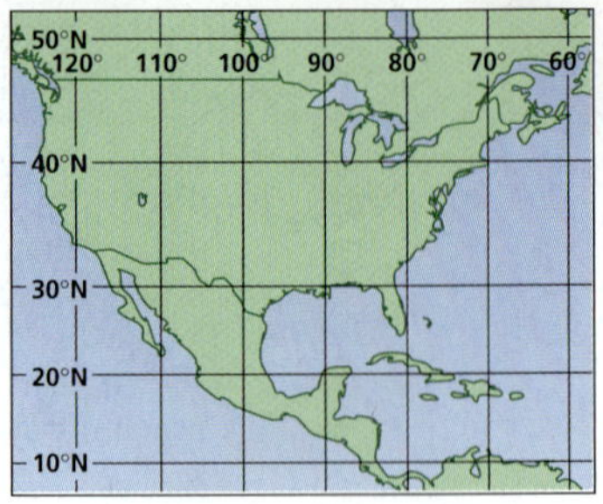

38. Write this expression in terms of a trigonometric function of L.

39. Find the exact value of the expression if $L = 60°$. $2 + \sqrt{3}$

PHYSICS **For Exercises 40 and 41, use the following information.**
An object is propelled from ground level with an initial velocity of v at an angle of elevation θ. **40. See pp. 811A–811N.**

40. The horizontal distance d it will travel can be determined using $d = \frac{v^2\sin 2\theta}{g}$, where g is acceleration due to gravity. Verify that this expression is the same as $\frac{2}{g}v^2(\tan\theta - \tan\theta\sin^2\theta)$.

41. The maximum height h the object will reach can be determined using the formula $h = \frac{v^2\sin^2\theta}{2g}$. Find the ratio of the maximum height attained to the horizontal distance traveled. $\frac{1}{4}\tan\theta$

CRITICAL THINKING **For Exercises 42–46, use the following information.**
Consider the functions $f(x) = \sin 2x$, $g(x) = \sin^2 x$, $h(x) = -\cos^2 x$, and $k(x) = -\frac{1}{2}\cos 2x$. **42–46. See pp. 811A–811N.**

42. Draw the graphs of $y = g(x)$, $y = h(x)$, and $y = k(x)$ on the same coordinate plane on the interval from $x = -2\pi$ to $x = 2\pi$. What do you notice about the graphs?

43. Where do the maxima and minima of g, h, and k occur?

44. Draw the graph of $y = f(x)$ on a separate coordinate plane.

45. What is the behavior of the graph of $f(x)$ at the locations found in Exercise 43?

46. Use what you know about transformations to determine c and d so that $g(x) = h(x) + c = k(x) + d$.

47. **WRITING IN MATH** Answer the question that was posed at the beginning of the lesson. **See margin.**

How can trigonometric functions be used to describe music?

Include the following in your answer:
- a description of what happens to the graph of the function of a vibrating string as it moves from one harmonic to the next, and
- an explanation of what happens to the period of the function as you move from the nth harmonic to the $(n + 1)$th harmonic.

Standardized Test Practice

48. Find the exact value of $\cos 2\theta$ if $\sin\theta = \frac{-\sqrt{5}}{3}$ and $180° < \theta < 270°$. **D**

(A) $\frac{-\sqrt{6}}{6}$ (B) $\frac{-\sqrt{30}}{6}$ (C) $\frac{-4\sqrt{5}}{9}$ (D) $\frac{-1}{9}$

49. Find the exact value of $\sin\frac{\theta}{2}$ if $\cos\theta = \frac{\sqrt{3}}{2}$ and $0° < \theta < 90°$. **B**

(A) $\frac{\sqrt{3}}{2}$ (B) $\frac{\sqrt{2-\sqrt{3}}}{2}$ (C) $\frac{\sqrt{2+\sqrt{3}}}{2}$ (D) $\frac{1}{2}$

Enrichment, p. 872

Alternating Current

The figure at the right represents an alternating current generator. A rectangular coil of wire is suspended between the poles of a magnet. As the coil of wire is rotated, it passes through the magnetic field and generates current.

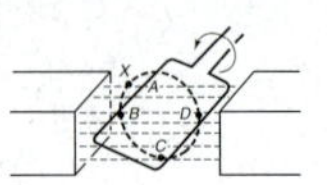

The maximum current may have a positive or negative value.

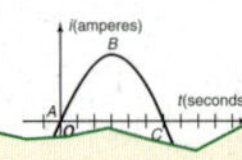

As point X on the coil passes through the points A and C, its motion is along the direction of the magnetic field between the poles. Therefore, no current is generated. However, through points $B$and D, the motion of X is perpendicular to the magnetic field. This induces maximum current in the coil. Between A and B, B and C, C and D, and D and A, the current in the coil will have an intermediate value. Thus, the graph of the current of an alternating current generator is closely related to the sine curve.

The actual current, i, in a household current is given

Maintain Your Skills

Mixed Review

52. $-\frac{\sqrt{2}}{2}$

53. $-\frac{\sqrt{3}}{2}$

Find the exact value of each expression. *(Lesson 14-5)*

50. $\cos 15°$ $\frac{\sqrt{6}+\sqrt{2}}{4}$
51. $\sin 15°$ $\frac{\sqrt{6}-\sqrt{2}}{4}$
52. $\sin(-135°)$
53. $\cos 150°$
54. $\sin 105°$ $\frac{\sqrt{6}+\sqrt{2}}{4}$
55. $\cos(-300°)$ $\frac{1}{2}$

Verify that each of the following is an identity. *(Lesson 14-4)*

56. $\cot^2 \theta - \sin^2 \theta = \frac{\cos^2 \theta \csc^2 \theta - \sin^2 \theta}{\sin^2 \theta \csc^2 \theta}$ **See pp. 811A–811N.**
57. $\cos \theta (\cos \theta + \cot \theta) = \cot \theta \cos \theta (\sin \theta + 1)$ **See pp. 811A–811N.**

EARTHQUAKE **For Exercises 58 and 59, use the following information.** The magnitude of an earthquake M measured on the Richter scale is given by $M = \log_{10}x$, where x represents the amplitude of the seismic wave causing ground motion. *(Lesson 10-2)* **58. 10^1 or 10**

58. How many times as great was the 1960 Chile earthquake as the 1938 Indonesia earthquake?
59. The largest aftershock of the 1964 Alaskan earthquake was 6.7 on the Richter scale. How many times as great was the main earthquake as this aftershock? **$10^{2.5}$ or about 316 times as great**

USA TODAY Snapshots®

Strongest quakes in 20th century

	Magnitude
Chile, 1960	9.5
Alaska, 1964	9.2
Russia, 1952	9.0
Ecuador, 1906	8.8
Alaska, 1957	8.8
Kuril Islands, 1958	8.7
Alaska, 1965	8.7
India, 1950	8.6
Chile, 1922	8.5
Indonesia, 1938	8.5

Source: U.S. Geological Survey

By William Risser and Marcy E. Mullins, USA TODAY

Getting Ready for the Next Lesson

PREREQUISITE SKILL **Solve each equation.**
*(To review **solving equations using the Zero Product Property**, see Lesson 6-3.)*

60. $(x + 6)(x - 5) = 0$ **−6, 5**
61. $(x - 1)(x + 1) = 0$ **1, −1**
62. $x(x + 2) = 0$ **0, −2**
63. $(2x - 5)(x + 2) = 0$ $\frac{5}{2}, -2$
64. $(2x + 1)(2x - 1) = 0$ $-\frac{1}{2}, \frac{1}{2}$
65. $x^2(2x + 1) = 0$ $0, -\frac{1}{2}$

Practice Quiz 2

Lessons 14-4 through 14-6

Verify that each of the following is an identity. *(Lessons 14-5)* **1–3. See pp. 811A–811N.**

1. $\sin \theta \sec \theta = \tan \theta$
2. $\sec \theta - \cos \theta = \sin \theta \tan \theta$
3. $\sin \theta + \tan \theta = \frac{\sin \theta (\cos \theta + 1)}{\cos \theta}$

Verify that each of the following is an identity. *(Lessons 14-4 and 14-5)* **4–6. See pp. 811A–811N.**

4. $\sin(90° + \theta) = \cos \theta$
5. $\cos\left(\frac{3\pi}{2} - \theta\right) = -\sin \theta$
6. $\sin(\theta + 30°) + \cos(\theta + 60°) = \cos \theta$

Find the exact value of each expression by using the double-angle or half-angle formulas. *(Lesson 14-6)*

7. $\sin 2\theta$ if $\cos \theta = -\frac{\sqrt{3}}{2}$; $180° < \theta < 270°$ $\frac{\sqrt{3}}{2}$
8. $\cos \frac{\theta}{2}$ if $\sin \theta = -\frac{9}{41}$; $270° \leq \theta < 360°$ $-\frac{9\sqrt{82}}{82}$
9. $\sin 165°$ $\frac{\sqrt{2-\sqrt{3}}}{2}$
10. $\cos \frac{5\pi}{8}$ $-\frac{\sqrt{2-\sqrt{2}}}{2}$

4 Assess

Open-Ended Assessment

Writing Have students write their own problems like Examples 2 and 3, and have them write an explanation of how to use a double-angle or half-angle formula to solve their examples.

Getting Ready for Lesson 14-7

PREREQUISITE SKILL In Lesson 14-7, students will solve trigonometric equations using the Zero Product Property. Use Exercises 60–65 to determine your students' familiarity with the Zero Product Property.

Assessment Options

Practice Quiz 2 The quiz provides students with a brief review of the concepts and skills in Lessons 14-4 through 14-6. Lesson numbers are given to the right of the exercises or instruction lines so students can review concepts not yet mastered.

Quiz (Lessons 14-5 and 14-6) is available on p. 894 of the *Chapter 14 Resource Masters.*

Answer (p. 796)

47. Sample answer: The sound waves associated with music can be modeled using trigonometric functions. Answers should include the following information.
    - In moving from one harmonic to the next, the number of vibrations that appear as sine waves increases by 1.
    - The period of the function as you move from the nth harmonic to the $(n + 1)$th harmonic decreases from $\frac{2\pi}{n}$ to $\frac{2\pi}{n+1}$.

Online Lesson Plans

USA TODAY Education

USA TODAY Education's Online site offers resources and interactive features connected to each day's newspaper. *Experience TODAY*, USA TODAY's daily lesson plan, is available on the site and delivered daily to subscribers. This plan provides instruction for integrating USA TODAY graphics and key editorial features into your mathematics classroom. Log on to www.education.usatoday.com.

A Preview of Lesson 14-7

Getting Started

Approximate Solutions In Example 1, approximate solutions can also be found by using the Trace feature. In most situations however, the Intersect feature will give more accurate solutions.

Teach

- You might wish to demonstrate the technique shown in Example 1 by first using a simple quadratic equation like $x^2 = 9$, whose solutions students will readily know. Graph $y = x^2$ and $y = 9$ to see that they intersect at two points, where $x = -3$ and $x = 3$, just as students will expect.
- Remind students that the solutions of the equation are the x values of the points of intersection, not the y values.
- If the expression on the right side of an equation is just 0 (as in Exercises 3 and 6), you can graph the function for the left side of the equation and then just use the Zero feature on the CALC menu to find approximate solutions.

Assess

In **Exercises 3–4**, check to see that students can explain why there are no real solutions.

In **Exercise 6**, make sure students found all four possible values. Students who found fewer than four are not using the correct domain for x.

Graphing Calculator Investigation

A Preview of Lesson 14-7

Solving Trigonometric Equations

The graph of a trigonometric function is made up of points that represent all values that satisfy the function. To solve a trigonometric equation, you need to find all values of the variable that satisfy the equation. You can use a TI-83 Plus to solve trigonometric equations by graphing each side of the equation as a function and then locating the points of intersection.

Example 1 Use a graphing calculator to solve $\sin x = 0.2$ if $0° \le x < 360°$.

Rewrite the equation as two functions, $y = \sin x$ and $y = 0.2$. Then graph the two functions. Look for the point of intersection.

Make sure that your calculator is in degree mode to get the correct viewing window.

KEYSTROKES: [MODE] [▼] [▼] [▶] [ENTER] [WINDOW] 0 [ENTER] 360 [ENTER] 90 [ENTER] −2 [ENTER] 1 [ENTER] 1 [ENTER] [Y=] [SIN] [X,T,θ,n] [ENTER] 0.2 [ENTER] [GRAPH]

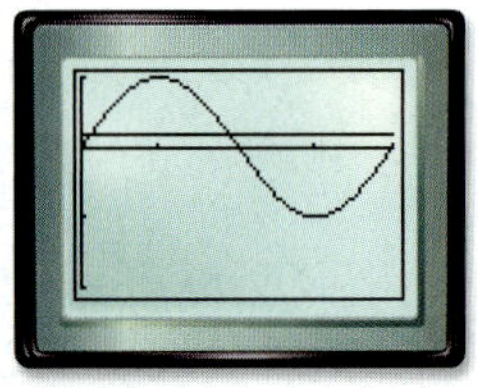

[0, 360] scl: 90 by [−2, 1] scl: 1

Based on the graph, you can see that there are two points of intersection in the interval $0° \le x < 360°$. Use [Zoom] or [2nd] [CALC] 5 to approximate the solutions. The approximate solutions are 168.5° and 11.5°.

Like other equations you have studied, some trigonometric equations have no real solutions. Carefully examine the graphs over their respective periods for points of intersection. If there are no points of intersection, then the trigonometric equation has no real solutions.

Example 2 Use a graphing calculator to solve $\tan^2 x \cos x + 5 \cos x = 0$ if $0° \le x < 360°$.

Because the tangent function is not continuous, place the calculator in Dot mode. The related functions to be graphed are $y = \tan^2 x \cos x + 5 \cos x$ and $y = 0$.

These two functions do not intersect. Therefore, the equation $\tan^2 x \cos x + 5 \cos x = 0$ has no real solutions.

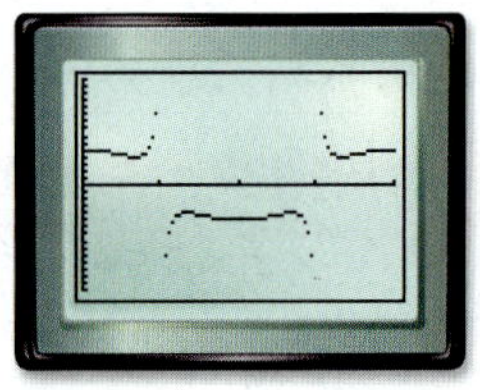

[0, 360] scl: 90 by [−15, 15] scl: 1

1–6. See pp. 811A–811N for graphs.

Exercises **1. 53.1°, 126.9° 3. no real solution 4. no real solution**

Use a graphing calculator to solve each equation for the values of x indicated.

1. $\sin x = 0.8$ if $0° \le x < 360°$
2. $\tan x = \sin x$ if $0° \le x < 360°$ **0°, 180°**
3. $2 \cos x + 3 = 0$ if $0° \le x < 360°$
4. $0.5 \cos x = 1.4$ if $-720° \le x < 720°$
5. $\sin 2x = \sin x$ if $0° \le x < 360°$ **60°, 180°, 300°**
6. $\sin 2x - 3 \sin x = 0$ if $-360° \le x < 360°$ **−360°, −180°, 0°, 180°**

www.algebra2.com/other_calculator_keystrokes

798 Chapter 14 Trigonometric Graphs and Identities

14-7 Solving Trigonometric Equations

What You'll Learn

- Solve trigonometric equations.
- Use trigonometric equations to solve real-world problems.

Vocabulary

- trigonometric equation

How can trigonometric equations be used to predict temperature?

The average daily high temperature for a region can be described by a trigonometric function. For example, the average daily high temperature for each month in Orlando, Florida, can be modeled by the function $T = 11.56 \sin(0.4516x - 1.641) + 80.89$, where T represents the average daily high temperature in degrees Fahrenheit and x represents the month of the year. This equation can be used to predict the months in which the average temperature in Orlando will be at or above a desired temperature.

SOLVE TRIGONOMETRIC EQUATIONS You have seen that trigonometric identities are true for *all* values of the variable for which the equation is defined. However, most **trigonometric equations** like some algebraic equations, are true for *some* but not *all* values of the variable.

Example 1 Solve Equations for a Given Interval

Find all solutions of each equation for the given interval.

a. $\cos^2\theta = 1$; $0° \le \theta < 360°$

$\cos^2\theta = 1$ — Original equation

$\cos^2\theta - 1 = 0$ — Solve for 0.

$(\cos\theta + 1)(\cos\theta - 1) = 0$ — Factor.

Now use the Zero Product Property.

$\cos\theta + 1 = 0$ or $\cos\theta - 1 = 0$

$\cos\theta = -1$ $\qquad$ $\cos\theta = 1$

$\theta = 180°$ $\qquad$ $\theta = 0°$

The solutions are 0° and 180°.

b. $\sin 2\theta = 2\cos\theta$; $0 \le \theta < 2\pi$

$\sin 2\theta = 2\cos\theta$ — Original equation

$2\sin\theta\cos\theta = 2\cos\theta$ — $\sin 2\theta = 2\sin\theta\cos\theta$

$2\sin\theta\cos\theta - 2\cos\theta = 0$ — Solve for 0.

$2\cos\theta(\sin\theta - 1) = 0$ — Factor.

(continued on the next page)

14-7 Lesson Notes

1 Focus

5-Minute Check Transparency 14-7 Use as a quiz or review of Lesson 14-6.

Mathematical Background notes are available for this lesson on page 760D.

How can trigonometric equations be used to predict temperature?

Ask students:

- Why would temperature not be modeled by a quadratic function? **Sample answer: Temperature varies periodically and does not continue upward or downward to infinity. So temperature should be modeled by a trigonometric, not a quadratic, function.**
- What can you tell about the range of temperatures by studying the function T? **The temperature varies 11.56 degrees (the amplitude of the function) above and below a temperature of 80.89° F (the midline of the function).**
- What can you tell about the maximum temperature by studying the function? **The maximum temperature is 11.56 + 80.89, or 92.45° F.**

Resource Manager

Workbook and Reproducible Masters

Chapter 14 Resource Masters

- Study Guide and Intervention, pp. 873–874
- Skills Practice, p. 875
- Practice, p. 876
- Reading to Learn Mathematics, p. 877
- Enrichment, p. 878
- Assessment, p. 894

School-to-Career Masters, p. 28

Science and Mathematics Lab Manual, pp. 145–148

Teaching Algebra With Manipulatives Masters, p. 304

Transparencies

5-Minute Check Transparency 14-7

Answer Key Transparencies

Technology

Interactive Chalkboard

2 Teach

Building on Prior Knowledge

In Lesson 6-3, students learned to use the Zero Product Property to solve equations. In this lesson, students will use the Zero Product Property to solve trigonometric equations.

SOLVE TRIGONOMETRIC EQUATIONS

In-Class Examples

Power Point®

1 Find all solutions of each equation for the given interval.

a. $2\cos^2\theta - 1 = \sin\theta$; $0° < \theta \leq 360°$ **30°, 150°, 270°**

b. $\sin\theta = \sin 2\theta$; $0 < \theta \leq 2\pi$ $\frac{\pi}{3}, \pi, \frac{5\pi}{3}, 2\pi$

Teaching Tip In Example 2, show students how to look for patterns in the solution of part a. Students should look for pairs of solutions that differ by exactly π or 2π.

2

a. Solve $2\sin\theta\cos\theta = \cos\theta$ for all values of θ if θ is measured in radians. $\frac{\pi}{6} + 2k\pi, \frac{\pi}{2} + 2k\pi, \frac{5\pi}{6} + 2k\pi, \frac{3\pi}{2} + 2k\pi$, **where k is any integer**

b. Solve $\cos\theta = -\cos 2\theta$ for all values of θ if θ is measured in degrees. **$60° + k \cdot 360°$, $180° + k \cdot 360°$, $300° + k \cdot 360°$, where k is any integer**

Use the Zero Product Property.

$2\cos\theta = 0$ or $\sin\theta - 1 = 0$

$\cos\theta = 0$ $\quad$ $\sin\theta = 1$

$\theta = \frac{\pi}{2}$ or $\frac{3\pi}{2}$ $\quad$ $\theta = \frac{\pi}{2}$

The solutions are $\frac{\pi}{2}$ and $\frac{3\pi}{2}$.

Trigonometric equations are usually solved for values of the variable between 0° and 360° or 0 radians and 2π radians. There are solutions outside that interval. These other solutions differ by integral multiples of the period of the function.

Example 2 Solve Trigonometric Equations

a. Solve $2\sin\theta = -1$ for all values of θ if θ is measured in radians.

$2\sin\theta = -1$ Original equation

$\sin\theta = -\frac{1}{2}$ Divide each side by 2.

Look at the graph of $y = \sin\theta$ to find solutions of $\sin\theta = -\frac{1}{2}$.

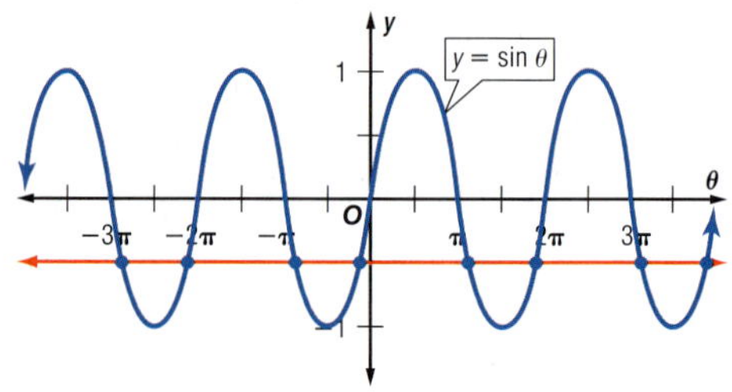

The solutions are $\frac{7\pi}{6}, \frac{11\pi}{6}, \frac{19\pi}{6}, \frac{23\pi}{6}$, and so on, and $-\frac{7\pi}{6}, -\frac{11\pi}{6}, -\frac{19\pi}{6}, -\frac{23\pi}{6}$, and so on. The only solutions in the interval 0 to 2π are $\frac{7\pi}{6}$ and $\frac{11\pi}{6}$. The period of the sine function is 2π radians. So the solutions can be written as $\frac{7\pi}{6} + 2k\pi$ and $\frac{11\pi}{6} + 2k\pi$, where k is any integer.

b. Solve $\cos 2\theta + \cos\theta + 1 = 0$ for all values of θ if θ is measured in degrees.

$\cos 2\theta + \cos\theta + 1 = 0$ Original equation

$2\cos^2\theta - 1 + \cos\theta + 1 = 0$ $\cos 2\theta = 2\cos^2\theta - 1$

$2\cos^2\theta + \cos\theta = 0$ Simplify.

$\cos\theta(2\cos\theta + 1) = 0$ Factor.

Solve for θ in the interval 0° to 360°.

$\cos\theta = 0$ or $2\cos\theta + 1 = 0$

$\theta = 90°$ or $270°$ $\quad$ $2\cos\theta = -1$

$\cos\theta = -\frac{1}{2}$

$\theta = 120°$ or $240°$

The solutions are $90° + k \cdot 180°$, $120° + k \cdot 360°$, and $240° + k \cdot 360°$.

Study Tip

Expressing Solutions as Multiples

The expression $90° + k \cdot 180°$ includes 270° and its multiples, so it is not necessary to list them separately.

If an equation cannot be solved easily by factoring, try rewriting the expression using trigonometric identities. However, using identities and some algebraic operations, such as squaring, may result in extraneous solutions. So, it is necessary to check your solutions using the original equation.

800 Chapter 14 Trigonometric Graphs and Identities

Example 3 Solve Trigonometric Equations Using Identities

Solve $\cos\theta\tan\theta - \sin^2\theta = 0$.

$\cos\theta\tan\theta - \sin^2\theta = 0$ Original equation

$\cos\theta\left(\frac{\sin\theta}{\cos\theta}\right) - \sin^2\theta = 0$ $\tan\theta = \frac{\sin\theta}{\cos\theta}$

$\sin\theta - \sin^2\theta = 0$ Multiply.

$\sin\theta(1 - \sin\theta) = 0$ Factor.

$\sin\theta = 0$ or $1 - \sin\theta = 0$

$\theta = 0°, 180°,$ or $360°$ $\sin\theta = 1$

$\theta = 90°$

CHECK

$\cos\theta\tan\theta - \sin^2\theta = 0$

$\cos 0°\tan 0° - \sin^2 0° \stackrel{?}{=} 0$ $\theta = 0°$

$1 \cdot 0 - 0 \stackrel{?}{=} 0$

$0 = 0$ ✓

$\cos\theta\tan\theta - \sin^2\theta = 0$

$\cos 180°\tan 180° - \sin^2 180° \stackrel{?}{=} 0$ $\theta = 180°$

$-1 \cdot 0 - 0 \stackrel{?}{=} 0$

$0 = 0$ ✓

$\cos\theta\tan\theta - \sin^2\theta = 0$

$\cos 360°\tan 360° - \sin^2 360° \stackrel{?}{=} 0$ $\theta = 360°$

$1 \cdot 0 - 0 \stackrel{?}{=} 0$

$0 = 0$ ✓

$\cos\theta\tan\theta - \sin^2\theta = 0$

$\cos 90°\tan 90° - \sin^2 90° \stackrel{?}{=} 0$ $\theta = 90°$

$\tan 90°$ is undefined.

Thus, 90° is not a solution.

The solution is $0° + k \cdot 180°$.

Some trigonometric equations have no solution. For example, the equation $\cos x = 4$ has no solution since all values of $\cos x$ are between -1 and 1, inclusive. Thus, the solution set for $\cos x = 4$ is empty.

Example 4 Determine Whether a Solution Exists

Solve $3\cos 2\theta - 5\cos\theta = 1$.

$3\cos 2\theta - 5\cos\theta = 1$ Original equation

$3(2\cos^2\theta - 1) - 5\cos\theta = 1$ $\cos 2\theta = 2\cos^2\theta - 1$

$6\cos^2\theta - 3 - 5\cos\theta = 1$ Multiply.

$6\cos^2\theta - 5\cos\theta - 4 = 0$ Subtract 1 from each side.

$(3\cos\theta - 4)(2\cos\theta + 1) = 0$ Factor.

$3\cos\theta - 4 = 0$ or $2\cos\theta + 1 = 0$

$3\cos\theta = 4$ $2\cos\theta = -1$

$\cos\theta = \frac{4}{3}$ $\cos\theta = -\frac{1}{2}$

Not possible since $\cos\theta$ cannot be greater than 1. $\theta = 120°$ or $240°$

Thus, the solutions are $120° + k \cdot 360°$ and $240° + k \cdot 360°$.

www.algebra2.com/extra_examples

In-Class Examples

PowerPoint®

3 Solve $\sin\theta\cot\theta = \cos^2\theta$. **$90° + k \cdot 180°$, where k is any integer**

4 Solve $\sin^2\theta = \frac{1}{4} + \cos\theta$. **$60° + k \cdot 360°$, $300° + k \cdot 360°$, where k is any integer**

DAILY INTERVENTION — Differentiated Instruction

Interpersonal As students work through this lesson, have them create a class list on the chalkboard that identifies common errors they made. Encourage students to add suggestions for how to avoid their errors. For example, one common error is having one's calculator set to degrees when it needs to be set to radians for a problem, and vice versa.

USE TRIGONOMETRIC EQUATIONS

In-Class Example

Power Point®

Teaching Tip As students work Example 5, some of them may calculate $\sin^{-1} 0.392$ as 23.079. Point out that this result occurs when their calculator is set to degrees instead of radians.

5 **GARDENING** Refer to Example 5 in the Student Edition. If Rhonda decides to wait until there are 15 hours of daylight, on what day can she plant her flowers? **around the 114th day of the year, or around April 24**

3 Practice/Apply

Study Notebook

Have students—

- *complete the definitions/examples for the remaining terms on their Vocabulary Builder worksheets for Chapter 14.*
- *summarize some techniques they used for solving trigonometric equations.*
- *include any other item(s) that they find helpful in mastering the skills in this lesson.*

About the Exercises...

Organization by Objective

- **Solve Trigonometric Equations:** 15–40
- **Use Trigonometric Equations:** 41–44

Odd/Even Assignments

Exercises 15–40 are structured so that students practice the same concepts whether they are assigned odd or even problems.

Assignment Guide

Basic: 15–39 odd, 41–55

Average: 15–39 odd, 41–55

Advanced: 16–40 even, 41–55

USE TRIGONOMETRIC EQUATIONS Trigonometric equations are often used to solve real-world situations.

Example 5 Use a Trigonometric Equation

GARDENING **Rhonda wants to wait to plant her flowers until there are at least 14 hours of daylight. The number of hours of daylight H in her town can be represented by $H = 11.45 + 6.5 \sin (0.0168d - 1.333)$, where d is the day of the year and angle measures are in radians. On what day is it safe for Rhonda to plant her flowers?**

$H = 11.45 + 6.5 \sin (0.0168d - 1.333)$	Original equation
$14 = 11.45 + 6.5 \sin (0.0168d - 1.333)$	$H = 14$
$2.55 = 6.5 \sin (0.0168d - 1.333)$	Subtract 11.45 from each side.
$0.392 = \sin (0.0168d - 1.333)$	Divide each side by 6.5.
$0.403 = 0.0168d - 1.333$	$\sin^{-1} 0.392 = 0.403$
$1.736 = 0.0168d$	Add 1.333 to each side.
$103.333 = d$	Divide each side by 0.0168.

Rhonda can safely plant her flowers around the 104th day of the year, or around April 14.

Check for Understanding

Concept Check **1–3. See margin.**

1. **Tell** why the equation $\sec \theta = 0$ has no solutions.
2. **Explain** why the number of solutions to the equation $\sin \theta = \frac{\sqrt{3}}{2}$ is infinite.
3. **OPEN ENDED** Write an example of a trigonometric equation that has no solution.

Guided Practice

GUIDED PRACTICE KEY	
Exercises	Examples
4–7	1
8–11	2
12, 13	3
14	5

Find all solutions of each equation for the given interval. **6.** $\frac{\pi}{6}, \frac{\pi}{2}, \frac{5\pi}{6}, \frac{3\pi}{2}$

4. $4 \cos^2 \theta = 1; 0° \leq \theta < 360°$ **60°, 120°, 240°, 300°**
5. $2 \sin^2 \theta - 1 = 0; 90° < \theta < 270°$ **135°, 225°**
6. $\sin 2\theta = \cos \theta; 0 \leq \theta < 2\pi$
7. $3 \sin^2 \theta - \cos^2 \theta = 0; 0 \leq \theta < \frac{\pi}{2}$ $\frac{\pi}{6}$

Solve each equation for all values of θ if θ is measured in radians.

8. $\cos 2\theta = \cos \theta$ $0 + \frac{2k\pi}{3}$
9. $\sin \theta + \sin \theta \cos \theta = 0$ $0 + k\pi$

Solve each equation for all values of θ if θ is measured in degrees.

10. $\sin \theta = 1 + \cos \theta$ **$90° + k \cdot 360°$, $180° + k \cdot 360°$**
11. $2 \cos^2 \theta + 2 = 5 \cos \theta$ **$60° + k \cdot 360°$, $300 + k \cdot 360°$**

Solve each equation for all values of θ.

12. $2 \sin^2 \theta - 3 \sin \theta - 2 = 0$
13. $2 \cos^2 \theta + 3 \sin \theta - 3 = 0$

12. $\frac{7\pi}{6} + 2k\pi, \frac{11\pi}{6} + 2k\pi$ or $210° + k \cdot 360°, 330° + k \cdot 360°$

13. $\frac{\pi}{6} + 2k\pi, \frac{5\pi}{6} + 2k\pi, \frac{\pi}{2} + 2k\pi$ or $30° + k \cdot 360°, 150° + k \cdot 360°, 90° + k \cdot 360°$

Application

14. **PHYSICS** According to Snell's law, the angle at which light enters water α is related to the angle at which light travels in water β by the equation $\sin \alpha = 1.33 \sin \beta$. At what angle does a beam of light enter the water if the beam travels at an angle of 23° through the water? **31.3°**

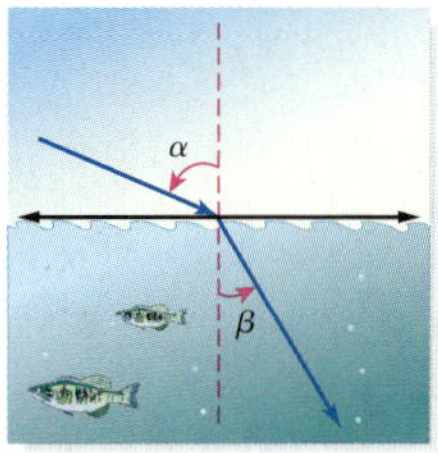

Answers

1. Sample answer: If $\sec \theta = 0$ then $\frac{1}{\cos \theta} = 0$. Since no value of θ makes $\frac{1}{\cos \theta} = 0$, there are no solutions.
2. Sample answer: The function is periodic with two solutions in each of its infinite number of periods.
3. Sample answer: $\sin \theta = 2$

23. $\frac{\pi}{3} + 2k\pi, \frac{5\pi}{3} + 2k\pi$

24. $\pi + 2k\pi, \frac{\pi}{3} + 2k\pi, \frac{5\pi}{3} + 2k\pi$

25. $\frac{2\pi}{3} + 2k\pi, \frac{4\pi}{3} + 2k\pi$

26. $0 + 2k\pi$

27. $\frac{\pi}{3} + 2k\pi, \frac{5\pi}{3} + 2k\pi$

28. $0 + k\pi, \frac{\pi}{6} + 2k\pi, \frac{5\pi}{6} + 2k\pi$

Practice and Apply

Homework Help

For Exercises	See Examples
15–22	1
23–34	2
35–40	3, 4
41–43	5

Extra Practice
See page 861.

Find all solutions of each equation for the given interval.

15. $2\cos\theta - 1 = 0;\ 0° \le \theta < 360°$
16. $2\sin\theta = -\sqrt{3};\ 180° < \theta < 360°$
17. $4\sin^2\theta = 1;\ 180° < \theta < 360°$
18. $4\cos^2\theta = 3;\ 0° \le \theta < 360°$
19. $2\cos^2\theta = \sin\theta + 1;\ 0 \le \theta < 2\pi$
20. $\sin^2\theta - 1 = \cos^2\theta;\ 0 \le \theta < \pi$
21. $2\sin^2\theta + \sin\theta = 0;\ \pi < \theta < 2\pi$
22. $2\cos^2\theta = -\cos\theta;\ 0 \le \theta < 2\pi$

15. 60°, 300°
16. 240°, 300°
17. 210°, 330°
18. 30°, 150°, 210°, 330°
19. $\frac{\pi}{6}, \frac{5\pi}{6}, \frac{3\pi}{2}$
20. $\frac{\pi}{2}$
21. $\frac{7\pi}{6}, \frac{11\pi}{6}$
22. $\frac{\pi}{2}, \frac{3\pi}{2}, \frac{2\pi}{3}, \frac{4\pi}{3}$

Solve each equation for all values of θ if θ is measured in radians.

23. $\cos 2\theta + 3\cos\theta - 1 = 0$
24. $2\sin^2\theta - \cos\theta - 1 = 0$
25. $\cos^2\theta - \frac{5}{2}\cos\theta - \frac{3}{2} = 0$
26. $\cos\theta = 3\cos\theta - 2$
27. $4\cos^2\theta - 4\cos\theta + 1 = 0$
28. $\cos 2\theta = 1 - \sin\theta$

23–28. See margin.

Solve each equation for all values of θ if θ is measured in degrees.

29. $\sin\theta = \cos\theta$
30. $\tan\theta = \sin\theta$
31. $\sin^2\theta - 2\sin\theta - 3 = 0$
32. $4\sin^2\theta - 4\sin\theta + 1 = 0$
33. $\tan^2\theta - \sqrt{3}\tan\theta = 0$
34. $\cos^2\theta - \frac{7}{2}\cos\theta - 2 = 0$

29–34. See margin.

Solve each equation for all values of θ. 35–40. See pp. 811A–811N.

35. $\sin^2\theta + \cos 2\theta - \cos\theta = 0$
36. $2\sin^2\theta - 3\sin\theta - 2 = 0$
37. $\sin^2\theta = \cos^2\theta - 1$
38. $2\cos^2\theta + \cos\theta = 0$
39. $\sin\frac{\theta}{2} + \cos\theta = 1$
40. $\sin\frac{\theta}{2} + \cos\frac{\theta}{2} = \sqrt{2}$

41. $S = \frac{352}{\tan\theta}$ or $S = 352\cot\theta$

LIGHT For Exercises 41 and 42, use the information shown.

41. The length of the shadow S of the International Peace Memorial at Put-In-Bay, Ohio, depends upon the angle of inclination of the Sun, θ. Express S as a function of θ.

42. Find the angle of inclination θ that will produce a shadow 560 feet long. about 32°

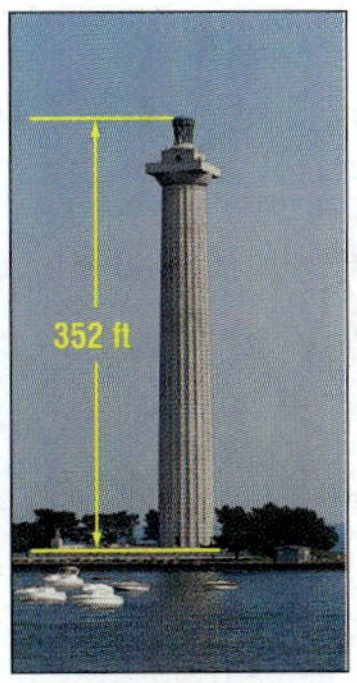

More About. . .

Waves
In the oceans, the height and period of water waves are determined by wind velocity, the duration of the wind, and the distance the wind has blown across the water.
Source: www.infoplease.com

WAVES For Exercises 43 and 44, use the following information.
For a short time after a wave is created by a boat, the height of the wave can be modeled using $y = \frac{1}{2}h + \frac{1}{2}h\sin\frac{2\pi t}{P}$, where h is the maximum height of the wave in feet, P is the period in seconds, and t is the propagation of the wave in seconds.

43. If $h = 3$ and $P = 2$ seconds, write the equation for the wave and draw its graph over a 10-second interval. See pp. 811A–811N.

44. How many times over the first 10 seconds does the graph predict the wave to be one foot high? 10

www.algebra2.com/self_check_quiz

Answers

29. $45° + k \cdot 180°$
30. $0° + k \cdot 180°$
31. $270° + k \cdot 360°$
32. $30° + k \cdot 360°,\ 150° + k \cdot 360°$
33. $0° + k \cdot 180°,\ 60° + k \cdot 180°$
34. $120° + k \cdot 360°,\ 240° + k \cdot 360°$

Study Guide and Intervention, p. 873 (shown) and p. 874

Solve Trigonometric Equations You can use trigonometric identities to solve trigonometric equations, which are true for only certain values of the variable.

Example 1 **Find all solutions of $4\sin^2\theta - 1 = 0$ for the interval $0° < \theta < 360°$.**

$4\sin^2\theta - 1 = 0$
$4\sin^2\theta = 1$
$\sin^2\theta = \frac{1}{4}$
$\sin\theta = \pm\frac{1}{2}$
$\theta = 30°, 150°, 210°, 330°$

Example 2 **Solve $\sin 2\theta + \cos\theta = 0$ for all values of θ. Give your answer in both radians and degrees.**

$\sin 2\theta + \cos\theta = 0$
$2\sin\theta\cos\theta + \cos\theta = 0$
$\cos\theta(2\sin\theta + 1) = 0$
$\cos\theta = 0$ or $2\sin\theta + 1 = 0$
$\sin\theta = -\frac{1}{2}$
$\theta = 90° + k \cdot 180°$; $\theta = \frac{\pi}{2} + k \cdot \pi$
$\theta = 210° + k \cdot 360°, 330° + k \cdot 360°$; $\theta = \frac{7\pi}{6} + k \cdot 2\pi, \frac{11\pi}{6} + k \cdot 2\pi$

Exercises

Find all solutions of each equation for the given interval.

1. $2\cos^2\theta + \cos\theta = 1, 0 \le \theta < 2\pi$ $\frac{\pi}{3}, \pi, \frac{5\pi}{3}$
2. $\sin^2\theta\cos^2\theta = 0, 0 \le \theta < 2\pi$ $0, \frac{\pi}{2}, \pi, \frac{3\pi}{2}$
3. $\cos 2\theta = \frac{\sqrt{3}}{2}, 0° \le \theta < 360°$ 15°, 165°, 195°, 345°
4. $2\sin\theta - \sqrt{3} = 0, 0 \le \theta < 2\pi$ $\frac{\pi}{3}, \frac{2\pi}{3}$

Solve each equation for all values of θ if θ is measured in radians.

5. $4\sin^2\theta - 3 = 0$ $\frac{\pi}{3} + k \cdot \pi, \frac{2\pi}{3} + k \cdot \pi$
6. $2\cos\theta\sin\theta + \cos\theta = 0$ $\frac{\pi}{2} + k \cdot 2\pi, \frac{3\pi}{2} + k \cdot 2\pi, \frac{7\pi}{6} + k \cdot 2\pi, \frac{11\pi}{6} + k \cdot 2\pi$

Solve each equation for all values of θ if θ is measured in degrees.

7. $\cos 2\theta + \sin^2\theta = \frac{1}{2}$ $45° + k \cdot 90°$
8. $\tan 2\theta = -1$ $67.5° + k \cdot 360°, 157.5° + k \cdot 360°$

Skills Practice, p. 875 and Practice, p. 876 (shown)

Find all solutions of each equation for the given interval.

1. $\sin 2\theta = \cos\theta, 90° \le \theta < 180°$ 90°, 150°
2. $\sqrt{2}\cos\theta = \sin 2\theta, 0° \le \theta < 360°$ 45°, 90°, 135°, 270°
3. $\cos 4\theta = \cos 2\theta, 180° \le \theta < 360°$ 180°, 240°, 300°
4. $\cos\theta + \cos(90 - \theta) = 0, 0 \le \theta < 2\pi$ $\frac{3\pi}{4}, \frac{7\pi}{4}$
5. $2 + \cos\theta = 2\sin^2\theta, \pi \le \theta \le \frac{3\pi}{2}$ $\frac{4\pi}{3}, \frac{3\pi}{2}$
6. $\tan^2\theta + \sec\theta = 1, \frac{\pi}{2} \le \theta < \pi$ $\frac{2\pi}{3}$

Solve each equation for all values of θ if θ is measured in radians.

7. $\cos^2\theta = \sin^2\theta$ $\frac{\pi}{4} + k\frac{\pi}{2}$
8. $\cot\theta = \cot^3\theta$ $\frac{\pi}{2} + k\pi$ and $\frac{\pi}{4} + k\frac{\pi}{2}$
9. $\sqrt{2}\sin^3\theta = \sin^2\theta$ $k\pi, \frac{\pi}{4} + 2k\pi$, and $\frac{3\pi}{4} + 2k\pi$
10. $\cos^2\theta\sin\theta = \sin\theta$ $k\pi$
11. $2\cos 2\theta = 1 - 2\sin^2\theta$ $\frac{\pi}{4} + k\frac{\pi}{2}$
12. $\sec^2\theta = 2$ $\frac{\pi}{4} + k\frac{\pi}{2}$

Solve each equation for all values of θ if θ is measured in degrees.

13. $\sin^2\theta\cos\theta = \cos\theta$ $90° + k \cdot 180°$
14. $\csc^2\theta - 3\csc\theta + 2 = 0$ $30° + k \cdot 360°, 90° + k \cdot 360°$, and $150° + k \cdot 360°$
15. $\frac{3}{1 + \cos\theta} = 4(1 - \cos\theta)$ $60° + k \cdot 180°$ and $120° + k \cdot 180°$
16. $\sqrt{2}\cos^2\theta = \cos^2\theta$ $90° + k \cdot 180°$ and $450° + k \cdot 360°$

Solve each equation for all values of θ.

17. $4\sin^2\theta = 3$ $\frac{\pi}{3} + k\pi$ and $\frac{2\pi}{3} + k\pi$, or $60° + k \cdot 180°$ and $120° + k \cdot 180°$
18. $4\sin^2\theta - 1 = 0$ $\frac{\pi}{6} + k\pi$ and $\frac{5\pi}{6} + k\pi$, or $30° + k \cdot 180°$ and $150° + k \cdot 180°$
19. $2\sin^2\theta - 3\sin\theta = -1$ $\frac{\pi}{6} + \frac{\pi k}{3}$, or $30° + k \cdot 60°$
20. $\cos 2\theta + \sin\theta - 1 = 0$ $k\pi$ and $\frac{\pi}{6} + 2k\pi$, or $k \cdot 180°$ and $30° + k \cdot 360°$
21. **WAVES** Waves are causing a buoy to float in a regular pattern in the water. The vertical position of the buoy can be described by the equation $h = 2\sin x$. Write an expression that describes the position of the buoy when its height is at its midline. $k\pi$ or $k \cdot 180°$
22. **ELECTRICITY** The electric current in a certain circuit with an alternating current can be described by the formula $i = 3\sin 240t$, where i is the current in amperes and t is the time in seconds. Write an expression that describes the times at which there is no current. $0.75kt$

Reading to Learn Mathematics, p. 877 ELL

Pre-Activity **How can trigonometric equations be used to predict temperature?**

Read the introduction to Lesson 14-7 at the top of page 799 in your textbook.

Describe how you could use a graphing calculator to determine the months in which the average daily high temperature is above 80°F. (Assume that $x = 1$ represents January.) Specify the graphing window that you would use. Sample answer: Graph the functions $y = 11.56\sin(0.4516x - 1.641) + 80.89$ (using radian mode) and $y = 80$ on the same screen. Use the window [1, 12] by [60, 100] with Xscl = 1 and Yscl = 4. Note the x values for which the curve is above the horizontal line.

Reading the Lesson

1. Identify which equations have no solution. C, E, and G

A. $\sin\theta = 1$	B. $\tan\theta = 0.001$	C. $\sec\theta = \frac{1}{2}$
D. $\csc\theta = -3$	E. $\cos\theta = 1.01$	F. $\cot\theta = -1000$
G. $\cos\theta + 2 = -1$	H. $\sec\theta - 1.5 = 0$	I. $\sin\theta - 0.009 = 0.99$

2. Use a trigonometric identity to write the first step in the solution of each trigonometric equation. (Do not complete the solution.)
   a. $\tan\theta = \cos^2\theta + \sin^2\theta, 0 \le \theta < 2\pi$ $\tan\theta = 1$
   b. $\sin^2\theta - 2\sin\theta + 1 = 0, 0° \le \theta < 360°$ $(\sin\theta - 1)^2 = 0$
   c. $\cos 2\theta = \sin\theta, 0° \le \theta < 360°$ $1 - 2\sin^2\theta = \sin\theta$
   d. $\sin 2\theta = \cos\theta, 0 \le \theta < 2\pi$ $2\sin\theta\cos\theta = \cos\theta$
   e. $2\cos 2\theta + 3\cos\theta = -1, 0° \le \theta < 360°$ $2(2\cos^2\theta - 1) + 3\cos\theta = -1$
   f. $3\tan^2\theta + 5\tan\theta - 2 = 0$ $(3\tan\theta - 1)(\tan\theta + 2) = 0$

Helping You Remember

3. A good way to remember something is to explain it to someone else. How would you explain to a friend the difference between verifying a trigonometric identity and solving a trigonometric equation. Sample answer: Verifying a trigonometric identity means showing that the two sides are equal for *all* values of the variable for which the functions involved are defined. This is done by transforming one or both sides until the same expression is obtained on both sides. Solving a trigonometric equation means finding the values of the variable for which both sides are equal. This process may require simplifying trigonometric expressions, but it also requires finding the angles for which a trigonometric function has a particular value.

Enrichment, p. 878

Families of Curves

Use these graphs for the problems below.

The Family $y = x^n$

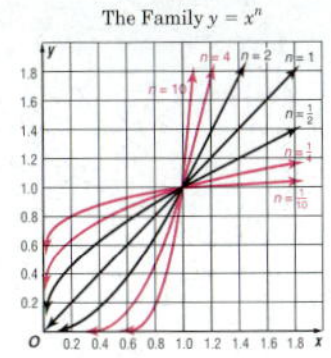

The Family $y = e^{mx}$

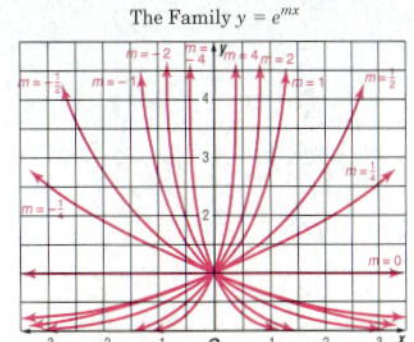

4 Assess

Open-Ended Assessment

Speaking Have students show you a trigonometric equation they solved and explain step by step how they performed each step of their computation.

Assessment Options

Quiz (Lesson 14-7) is available on p. 894 of the *Chapter 14 Resource Masters*.

Answer

46. Sample answer: Temperatures are cyclic and can be modeled by trigonometric functions. Answers should include the following information.

- **A temperature could occur twice in a given period such as when the temperature rises in the spring and falls in autumn.**

45. **CRITICAL THINKING** Computer games often use transformations to distort images on the screen. In one such transformation, an image is rotated counterclockwise using the equations $x' = x\cos\theta - y\sin\theta$ and $y' = x\sin\theta + y\cos\theta$. If the coordinates of an image point are (3, 4) after a 60° rotation, what are the coordinates of the preimage point? **(4.964, −0.598)**

46. **WRITING IN MATH** Answer the question that was posed at the beginning of the lesson. **See margin.**

How can trigonometric equations be used to predict temperature?

Include the following in your answer:

- an explanation of why the sine function can be used to model the average daily temperature, and
- an explanation of why, during one period, you might find a specific average temperature twice.

Standardized Test Practice

47. Which of the following is *not* a possible solution of $0 = \sin\theta + \cos\theta\tan^2\theta$? **D**

(A) $\frac{3\pi}{4}$ (B) $\frac{7\pi}{4}$ (C) 2π (D) $\frac{5\pi}{2}$

48. The graph of the equation $y = 2\cos\theta$ is shown. Which is a solution for $2\cos\theta = 1$? **B**

(A) $\frac{8\pi}{3}$ (B) $\frac{13\pi}{3}$

(C) $\frac{10\pi}{3}$ (D) $\frac{15\pi}{3}$

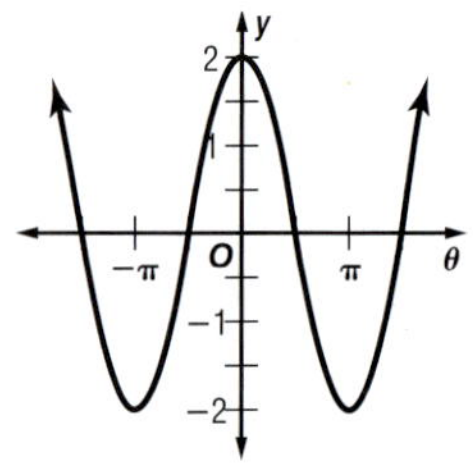

Maintain Your Skills

Mixed Review

Find the exact value of $\sin 2\theta$, $\cos 2\theta$, $\sin\frac{\theta}{2}$, and $\cos\frac{\theta}{2}$ for each of the following. *(Lesson 14-6)*

49. $\sin\theta = \frac{3}{5}$; $0° < \theta < 90°$

50. $\cos\theta = \frac{1}{2}$; $0° < \theta < 90°$

51. $\cos\theta = \frac{5}{6}$; $0° < \theta < 90°$

52. $\sin\theta = \frac{4}{5}$; $0° < \theta < 90°$

49. $\frac{24}{25}, \frac{7}{25}, \frac{\sqrt{10}}{10}, \frac{3\sqrt{10}}{10}$

50. $\frac{\sqrt{3}}{2}, -\frac{1}{2}, \frac{1}{2}, \frac{\sqrt{3}}{2}$

51. $\frac{5\sqrt{11}}{18}, \frac{7}{18}, \frac{\sqrt{3}}{6}, \frac{\sqrt{33}}{6}$

52. $\frac{24}{25}, -\frac{7}{25}, \frac{\sqrt{5}}{5}, \frac{2\sqrt{5}}{5}$

Find the exact value of each expression. *(Lesson 14-5)*

53. $\sin 240°$ **$-\frac{\sqrt{3}}{2}$**

54. $\cos 315°$ **$\frac{\sqrt{2}}{2}$**

55. Solve $\triangle ABC$. Round measures of sides and angles to the nearest tenth. *(Lesson 13-4)*
$b = 11.0$, $c = 12.2$, $m\angle C = 78$

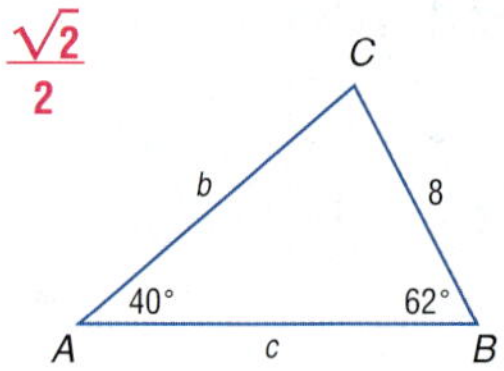

WebQuest Internet Project

Trig Class Angles for Lessons in Lit

It is time to complete your project. Use the information and data you have gathered about the applications of trigonometry to prepare a poster, report, or Web page. Be sure to include graphs, tables, or diagrams in the presentation.

www.algebra2.com/webquest

804 Chapter 14 Trigonometric Graphs and Identities

Chapter 14 Study Guide and Review

Vocabulary and Concept Check

amplitude (p. 763)
double-angle formula (p. 791)
half-angle formula (p. 793)
midline (p. 771)
phase shift (p. 769)
trigonometric equation (p. 799)
trigonometric identity (p. 777)
vertical shift (p. 771)

Choose the correct letter that best matches each phrase.

1. horizontal translation of a trigonometric function **h**
2. a reference line about which a graph oscillates **b**
3. vertical translation of a trigonometric function **d**
4. the formula used to find $\cos 22\frac{1}{2}°$ **f**
5. $\sin 2\theta = 2 \sin \theta \cos \theta$ **e**
6. a measure of how long it takes for a graph to repeat itself **c**
7. $\cos(\alpha - \beta) = \cos \alpha \cos \beta + \sin \alpha \sin \beta$ **g**
8. the absolute value of half the difference between the maximum and minimum values of a periodic function **a**

a. amplitude
b. midline
c. period
d. vertical shift
e. double-angle formula
f. half-angle formula
g. difference of angles formula
h. phase shift

Lesson-by-Lesson Review

14-1 Graphing Trigonometric Functions

See pages 762–768.

Concept Summary

- For trigonometric functions of the form $y = a \sin b\theta$ and $y = a \cos b\theta$, the amplitude is $|a|$, and the period is $\frac{360°}{|b|}$ or $\frac{2\pi}{|b|}$.
- The period of $y = a \tan b\theta$ is $\frac{180°}{|b|}$ or $\frac{\pi}{|b|}$.

Example **Find the amplitude and period of $y = 2 \cos 4\theta$. Then graph the function.**

The amplitude is $|2|$ or 2.

The period is $\frac{360°}{|4|}$ or 90°.

Use the amplitude and period to graph the function.

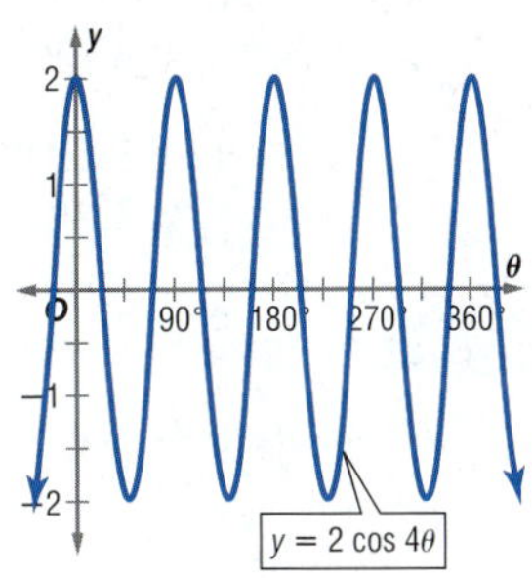

Exercises **Find the amplitude, if it exists, and period of each function. Then graph each function.** *See Example 1 on page 765.* **9–14. See pp. 811A–811N.**

9. $y = -\frac{1}{2} \cos \theta$
10. $y = 4 \sin 2\theta$
11. $y = \sin \frac{1}{2}\theta$
12. $y = 5 \sec \theta$
13. $y = \frac{1}{2} \csc \frac{2}{3}\theta$
14. $y = \tan 4\theta$

www.algebra2.com/vocabulary_review

FOLDABLES™ Study Organizer

For more information about Foldables, see *Teaching Mathematics with Foldables.*

Have students look through the chapter to make sure they have included notes and examples of graphs for each lesson in this chapter in their Foldable.

Encourage students to refer to their Foldables while completing the Study Guide and Review and to use them in preparing for the Chapter Test.

Chapter 14 Study Guide and Review

Vocabulary and Concept Check

- This alphabetical list of vocabulary terms in Chapter 14 includes a page reference where each term was introduced.
- **Assessment** A vocabulary test/review for Chapter 14 is available on p. 892 of the *Chapter 14 Resource Masters.*

Lesson-by-Lesson Review

For each lesson,
- the main ideas are summarized,
- additional examples review concepts, and
- practice exercises are provided.

Vocabulary PuzzleMaker

ELL The Vocabulary PuzzleMaker software improves students' mathematics vocabulary using four puzzle formats—crossword, scramble, word search using a word list, and word search using clues. Students can work on a computer screen or from a printed handout.

MindJogger Videoquizzes

ELL MindJogger Videoquizzes provide an alternative review of concepts presented in this chapter. Students work in teams in a game show format to gain points for correct answers. The questions are presented in three rounds.

Round 1 Concepts (5 questions)
Round 2 Skills (4 questions)
Round 3 Problem Solving (4 questions)

Answers

15. $-1, \frac{1}{2}, 180°, 60°$

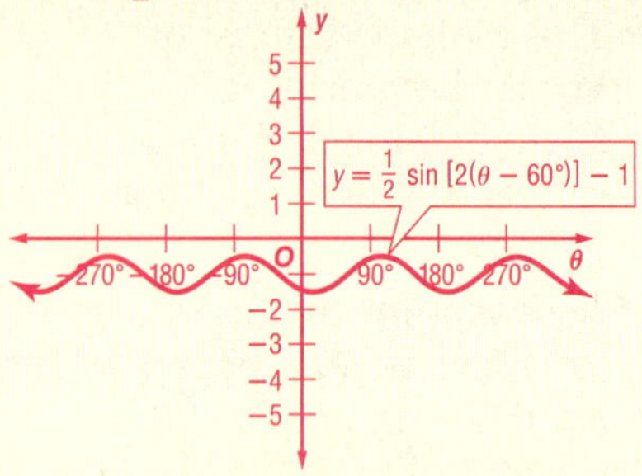

16. 3, does not exist, 720°, 90°

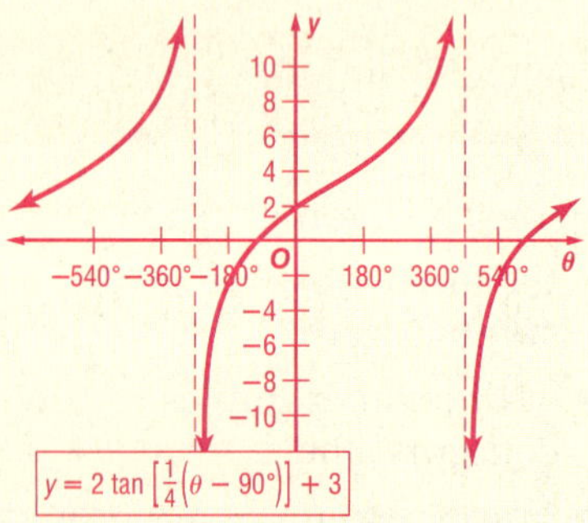

17. 1, does not exist, 4π, $-\frac{\pi}{4}$

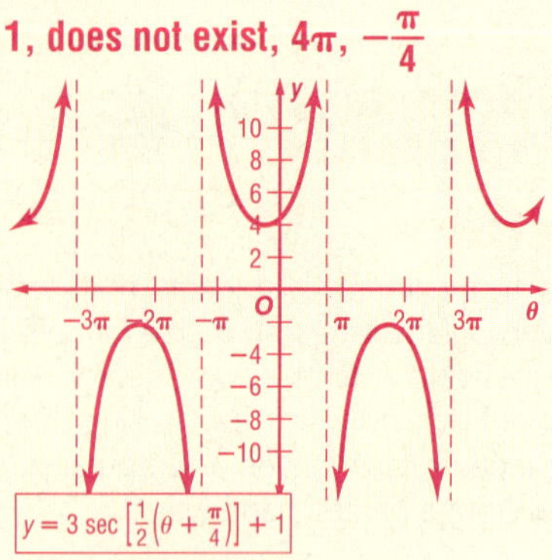

18. $-2, \frac{1}{3}, 6\pi, \frac{2\pi}{3}$

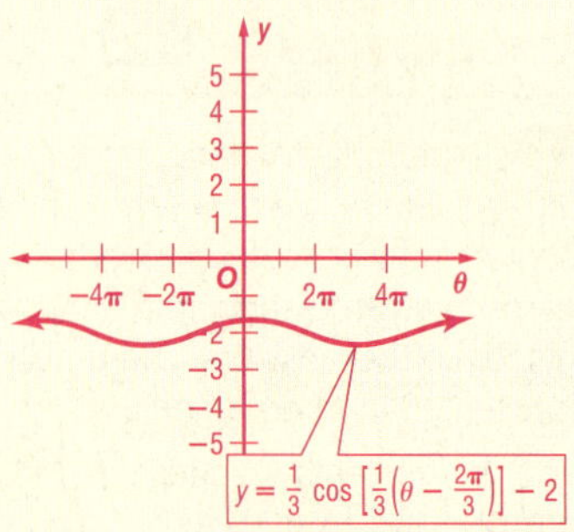

14-2 Translations of Trigonometric Graphs

See pages 769–776.

Concept Summary

- For trigonometric functions of the form $y = a \sin b(\theta - h)$, $y = a \cos(\theta - h)$, and $y = a \tan(\theta - h)$, the phase shift is to the right when $h > 0$ and to the left when $h < 0$.
- For trigonometric functions of the form $y = a \sin b(\theta - h) + k$, $y = a \cos(\theta - h) + k$, and $y = a \tan(\theta - h) + k$, the vertical shift is up when $k > 0$ and down when $k < 0$.

Example **State the vertical shift, amplitude, period, and phase shift of $y = 3\sin\left[2\left(\theta - \frac{\pi}{2}\right)\right] - 2$. Then graph the function.**

Identify the values of k, a, b, and h.

$k = -2$, so the vertical shift is –2.

$a = 3$, so the amplitude is 3.

$b = 2$, so the period is $\frac{2\pi}{|2|}$ or π.

$h = \frac{\pi}{2}$, so the phase shift is $\frac{\pi}{2}$ to the right.

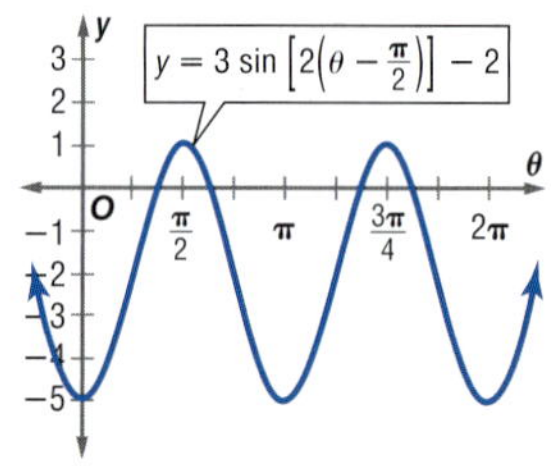

Exercises **State the vertical shift, amplitude, period, and phase shift of each function. Then graph the function.** *See Example 3 on page 772.* **15–18. See margin.**

15. $y = \frac{1}{2}\sin[2(\theta - 60°)] - 1$

16. $y = 2\tan\left[\frac{1}{4}(\theta - 90°)\right] + 3$

17. $y = 3\sec\left[\frac{1}{2}\left(\theta + \frac{\pi}{4}\right)\right] + 1$

18. $y = \frac{1}{3}\cos\left[\frac{1}{3}\left(\theta - \frac{2\pi}{3}\right)\right] - 2$

14-3 Trigonometric Identities

See pages 777–781.

Concept Summary

- Quotient Identities: $\tan\theta = \frac{\sin\theta}{\cos\theta}$, $\cot\theta = \frac{\cos\theta}{\sin\theta}$
- Reciprocal Identities: $\csc\theta = \frac{1}{\sin\theta}$, $\sec\theta = \frac{1}{\cos\theta}$, $\cot\theta = \frac{1}{\tan\theta}$
- Pythagorean Identities: $\cos^2\theta + \sin^2\theta = 1$, $\tan^2\theta + 1 = \sec^2\theta$, $\cot^2\theta + 1 = \csc^2\theta$

Example **Simplify $\sin\theta\cot\theta\cos\theta$.**

$$\sin\theta\cot\theta\cos\theta = \frac{\sin\theta}{1}\cdot\frac{\cos\theta}{\sin\theta}\cdot\frac{\cos\theta}{1} \quad \cot\theta = \frac{\cos\theta}{\sin\theta}$$

$$= \cos^2\theta \quad \text{Multiply.}$$

Exercises **Find the value of each expression.** *See Example 1 on page 778.*

19. $\cot\theta$, if $\csc\theta = -\frac{5}{3}$; $270° < \theta < 360°$ $-\frac{4}{3}$

20. $\sec\theta$, if $\sin\theta = \frac{1}{2}$; $0° \le \theta < 90°$ $\frac{2\sqrt{3}}{3}$

Simplify each expression. *See Example 2 on page 778.* **21. $\sin^2\theta$ 22. $\cot\theta$ 23. $\sec\theta$**

21. $\sin\theta\csc\theta - \cos^2\theta$

22. $\cos^2\theta\sec\theta\csc\theta$

23. $\cos\theta + \sin\theta\tan\theta$

14-4 Verifying Trigonometric Identities

See pages 782–785.

Concept Summary

- Use the basic trigonometric identities to transform one or both sides of a trigonometric equation into the same form.

Example Verify that $\tan\theta + \cot\theta = \sec\theta\csc\theta$.

$\tan\theta + \cot\theta \stackrel{?}{=} \sec\theta\csc\theta$	Original equation
$\frac{\sin\theta}{\cos\theta} + \frac{\cos\theta}{\sin\theta} \stackrel{?}{=} \sec\theta\csc\theta$	$\tan\theta = \frac{\sin\theta}{\cos\theta}$, $\cot\theta = \frac{\cos\theta}{\sin\theta}$
$\frac{\sin^2\theta + \cos^2\theta}{\cos\theta\sin\theta} \stackrel{?}{=} \sec\theta\csc\theta$	Rewrite using the LCD, $\cos\theta\sin\theta$.
$\frac{1}{\cos\theta\sin\theta} \stackrel{?}{=} \sec\theta\csc\theta$	$\sin^2\theta + \cos^2\theta = 1$
$\frac{1}{\cos\theta}\cdot\frac{1}{\sin\theta} \stackrel{?}{=} \sec\theta\csc\theta$	Rewrite as the product of two expressions.
$\sec\theta\csc\theta = \sec\theta\csc\theta$	$\frac{1}{\cos\theta} = \sec\theta$, $\frac{1}{\sin\theta} = \csc\theta$

Exercises Verify that each of the following is an identity.
See Examples 1–3 on pages 782–783. **24–27. See margin.**

24. $\frac{\sin\theta}{\tan\theta} + \frac{\cos\theta}{\cot\theta} = \cos\theta + \sin\theta$
25. $\frac{\sin\theta}{1-\cos\theta} = \csc\theta + \cot\theta$
26. $\cot^2\theta\sec^2\theta = 1 + \cot^2\theta$
27. $\sec\theta(\sec\theta - \cos\theta) = \tan^2\theta$

14-5 Sum and Difference of Angles Formulas

See pages 786–790.

Concept Summary

- For all values of α and β: $\cos(\alpha \pm \beta) = \cos\alpha\cos\beta \mp \sin\alpha\sin\beta$
  $\sin(\alpha \pm \beta) = \sin\alpha\cos\beta \pm \cos\alpha\sin\beta$

Example Find the exact value of $\sin 195°$.

$\sin 195° = \sin(150° + 45°)$	$195° = 150° + 45°$
$= \sin 150°\cos 45° + \cos 150°\sin 45°$	$\alpha = 150°, \beta = 45°$
$= \left(\frac{1}{2}\right)\left(\frac{\sqrt{2}}{2}\right) + \left(-\frac{\sqrt{3}}{2}\right)\left(\frac{\sqrt{2}}{2}\right)$	Evaluate each expression.
$= \frac{\sqrt{2} - \sqrt{6}}{4}$	Simplify.

Exercises Find the exact value of each expression. *See Example 1 on page 787.*

28. $\cos 15°$ $\frac{\sqrt{6}+\sqrt{2}}{4}$
29. $\cos 285°$ $\frac{\sqrt{6}-\sqrt{2}}{4}$
30. $\sin 150°$ $\frac{1}{2}$
31. $\sin 195°$
32. $\cos(-210°)$ $-\frac{\sqrt{3}}{2}$
33. $\sin(-105°)$ $\frac{-\sqrt{6}-\sqrt{2}}{4}$

31. $\frac{\sqrt{2}-\sqrt{6}}{4}$

Verify that each of the following is an identity. *See Example 3 on page 788.*

34–37. See pp. 811A–811N.

34. $\cos(90° + \theta) = -\sin\theta$
35. $\sin(30° - \theta) = \cos(60° + \theta)$
36. $\sin(\theta + \pi) = -\sin\theta$
37. $-\cos\theta = \cos(\pi + \theta)$

Answers

25. $\frac{\sin\theta}{1-\cos\theta} \stackrel{?}{=} \csc\theta + \cot\theta$
$\frac{\sin\theta}{1-\cos\theta} \stackrel{?}{=} \frac{1}{\sin\theta} + \frac{\cos\theta}{\sin\theta}$
$\frac{\sin\theta}{1-\cos\theta} \stackrel{?}{=} \frac{1+\cos\theta}{\sin\theta}$
$\frac{\sin\theta}{1-\cos\theta} \stackrel{?}{=} \frac{1+\cos\theta}{\sin\theta}\cdot\frac{1-\cos\theta}{1-\cos\theta}$
$\frac{\sin\theta}{1-\cos\theta} \stackrel{?}{=} \frac{1-\cos^2\theta}{\sin\theta(1-\cos\theta)}$
$\frac{\sin\theta}{1-\cos\theta} \stackrel{?}{=} \frac{\sin^2\theta}{\sin\theta(1-\cos\theta)}$
$\frac{\sin\theta}{1-\cos\theta} = \frac{\sin\theta}{1-\cos\theta}$

26. $\cot^2\theta\sec^2\theta \stackrel{?}{=} 1 + \cot^2\theta$
$\frac{\cos^2\theta}{\sin^2\theta}\cdot\frac{1}{\cos^2\theta} \stackrel{?}{=} 1 + \cot^2\theta$
$\frac{1}{\sin^2\theta} \stackrel{?}{=} 1 + \cot^2\theta$
$\csc^2\theta \stackrel{?}{=} 1 + \cot^2\theta$
$1 + \cot^2\theta = 1 + \cot^2\theta$

27. $\sec\theta(\sec\theta - \cos\theta) \stackrel{?}{=} \tan^2\theta$
$\frac{1}{\cos\theta}\left(\frac{1}{\cos\theta} - \cos\theta\right) \stackrel{?}{=} \tan^2\theta$
$\frac{1}{\cos^2\theta} - 1 \stackrel{?}{=} \tan^2\theta$
$\sec^2\theta - 1 \stackrel{?}{=} \tan^2\theta$
$\tan^2\theta = \tan^2\theta$

Answer

24. $\frac{\sin\theta}{\tan\theta} + \frac{\cos\theta}{\cot\theta} \stackrel{?}{=} \cos\theta + \sin\theta$
$\frac{\sin\theta}{\frac{\sin\theta}{\cos\theta}} + \frac{\cos\theta}{\frac{\cos\theta}{\sin\theta}} \stackrel{?}{=} \cos\theta + \sin\theta$
$\sin\theta\cdot\frac{\cos\theta}{\sin\theta} + \cos\theta\cdot\frac{\sin\theta}{\cos\theta} \stackrel{?}{=} \cos\theta + \sin\theta$
$\cos\theta + \sin\theta = \cos\theta + \sin\theta$

Chapter 14 For More ...
- Extra Practice, see pages 859–861.
- Mixed Problem Solving, see page 875.

Answers

38. $\frac{\sqrt{15}}{8}, \frac{7}{8}, \frac{\sqrt{8-2\sqrt{15}}}{4}, \frac{\sqrt{8+2\sqrt{15}}}{4}$

39. $\frac{120}{169}, \frac{119}{169}, \frac{5\sqrt{26}}{26}, -\frac{\sqrt{26}}{26}$

40. $-\frac{20\sqrt{66}}{289}, -\frac{239}{289}, \frac{\sqrt{187}}{17}, \frac{\sqrt{102}}{17}$

41. $-\frac{120}{169}, \frac{119}{169}, \frac{\sqrt{26}}{26}, -\frac{5\sqrt{26}}{26}$

Answers (p. 809)

5. 5, $\frac{2}{3}$, 180°, no phase shift

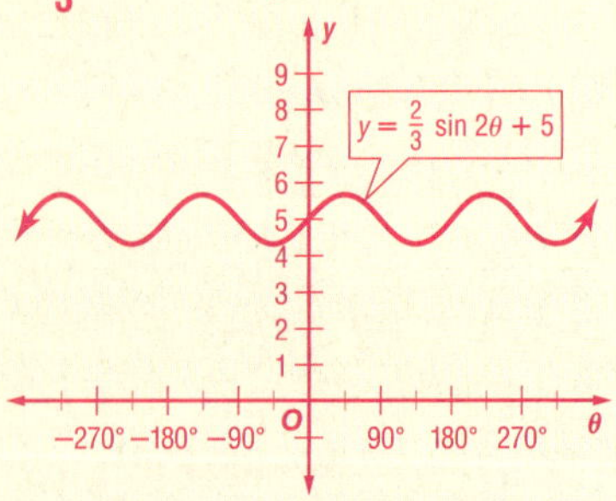

6. −1, 4, 720°, −30°

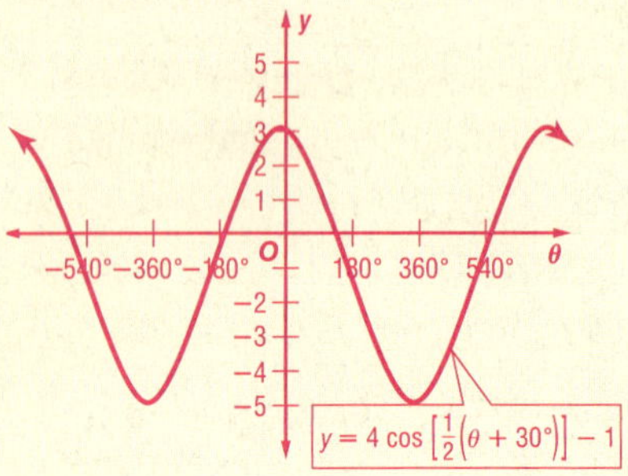

14-6 Double-Angle and Half-Angle Formulas

See pages 791–797.

Concept Summary

- Double-angle formulas: $\sin 2\theta = 2\sin\theta\cos\theta$, $\cos 2\theta = \cos^2\theta - \sin^2\theta$, $\cos 2\theta = 1 - 2\sin^2\theta$, $\cos 2\theta = 2\cos^2\theta - 1$
- Half-angle formulas: $\sin\frac{\alpha}{2} = \pm\sqrt{\frac{1-\cos\alpha}{2}}$, $\cos\frac{\alpha}{2} = \pm\sqrt{\frac{1+\cos\alpha}{2}}$

Example Verify that $\csc 2\theta = \frac{\sec\theta}{2\sin\theta}$ is an identity.

$\csc 2\theta \stackrel{?}{=} \frac{\sec\theta}{2\sin\theta}$ Original equation

$\frac{1}{\sin 2\theta} \stackrel{?}{=} \frac{\frac{1}{\cos\theta}}{2\sin\theta}$ $\csc\theta = \frac{1}{\sin\theta}$, $\sec\theta = \frac{1}{\cos\theta}$

$\frac{1}{\sin 2\theta} \stackrel{?}{=} \frac{1}{2\sin\theta\cos\theta}$ Simplify the complex fraction.

$\frac{1}{\sin 2\theta} = \frac{1}{\sin 2\theta}$ $2\sin\theta\cos\theta = \sin 2\theta$

Exercises Find the exact values of $\sin 2\theta$, $\cos 2\theta$, $\sin\frac{\theta}{2}$, and $\cos\frac{\theta}{2}$ for each of the following. *See Examples 1 and 2 on pages 792 and 793.* **38–41. See margin.**

38. $\sin\theta = \frac{1}{4}$; $0° < \theta < 90°$
39. $\sin\theta = -\frac{5}{13}$; $180° < \theta < 270°$
40. $\cos\theta = -\frac{5}{17}$; $90° < \theta < 180°$
41. $\cos\theta = \frac{12}{13}$; $270° < \theta < 360°$

14-7 Solving Trigonometric Equations

See pages 799–804.

Concept Summary

- Solve trigonometric equations by factoring or by using trigonometric identities.

Example Solve $\sin 2\theta + \sin\theta = 0$ if $0° \le \theta < 360°$.

$\sin 2\theta + \sin\theta = 0$ Original equation

$2\sin\theta\cos\theta + \sin\theta = 0$ $\sin 2\theta = 2\sin\theta\cos\theta$

$\sin\theta(2\cos\theta + 1) = 0$ Factor.

$\sin\theta = 0$ or $2\cos\theta + 1 = 0$

$\theta = 0°$ or $180°$ $\theta = 120°$ or $240°$

Exercises Find all solutions of each equation for the interval $0° \le \theta < 360°$. *See Example 1 on page 799.*

42. $2\sin 2\theta = 1$ **15°, 75°, 195°, 255°**
43. $2\cos^2\theta + \sin^2\theta = 2\cos\theta$ **0°**

Solve each equation for all values of θ if θ is measured in radians. *See Example 2 on page 800.* **44. $\frac{7\pi}{6} + 2k\pi, \frac{11\pi}{6} + 2k\pi$ 45. $\frac{\pi}{6} + 2k\pi, \frac{5\pi}{6} + 2k\pi$**

44. $6\sin^2\theta - 5\sin\theta - 4 = 0$
45. $2\cos^2\theta = 3\sin\theta$

808 Chapter 14 Trigonometric Graphs and Identities

Chapter 14 Practice Test

Vocabulary and Concepts

Choose the correct term to complete each sentence.

1. The (*period*, *phase shift*) of $y = 3 \sin 2(\theta - 60°) + 2$ is 180°.
2. A midline is used with a (*phase shift*, *vertical shift*) of a trigonometric function.
3. The amplitude of $y = \frac{1}{3} \cos [3(\theta + 4)] - 1$ is $\left(\frac{1}{3}, 3\right)$.
4. The (*cosine*, *cosecant*) has no amplitude.

Skills and Applications

State the vertical shift, amplitude, period, and phase shift of each function. Then graph the function. 5–6. See margin.

5. $y = \frac{2}{3} \sin 2\theta + 5$
6. $y = 4 \cos \left[\frac{1}{2}(\theta + 30°)\right] - 1$

Find the value of each expression.

7. $\tan \theta$, if $\sin \theta = \frac{1}{2}$; $90° < \theta < 180°$ $-\frac{\sqrt{3}}{3}$
8. $\sec \theta$, if $\cot \theta = \frac{3}{4}$; $180° < \theta < 270°$ $-\frac{5}{3}$

Verify that each of the following is an identity. 9–12. See pp. 811A–811N.

9. $(\sin \theta - \cos \theta)^2 = 1 - \sin 2\theta$
10. $\frac{\cos \theta}{1 - \sin^2 \theta} = \sec \theta$
11. $\frac{\sec \theta}{\sin \theta} - \frac{\sin \theta}{\cos \theta} = \cot \theta$
12. $\frac{1 + \tan^2 \theta}{\cos^2 \theta} = \sec^4 \theta$

Find the exact value of each expression. 13. $\frac{\sqrt{6} - \sqrt{2}}{4}$ 14. $\frac{\sqrt{2} - \sqrt{6}}{4}$ 15. $\frac{\sqrt{2}}{2}$

13. $\cos 285°$
14. $\sin 345°$
15. $\sin (-225°)$
16. $\cos 480°$ $-\frac{1}{2}$
17. $\cos 67.5°$ $\frac{\sqrt{2 - \sqrt{2}}}{2}$
18. $\sin 75°$ $\frac{\sqrt{6} + \sqrt{2}}{4}$

Solve each equation for all values of θ if θ is measured in degrees.

19. $\sec \theta = 1 + \tan \theta$ $0° + k \cdot 360°$
20. $\cos 2\theta = \cos \theta$ $0° + k \cdot 360°$, $120° + k \cdot 360°$, $240° + k \cdot 360°$
21. $\cos 2\theta + \sin \theta = 1$ $0° + k \cdot 180°$, $30° + k \cdot 360°$, $150° + k \cdot 360°$
22. $\sin \theta = \tan \theta$ $0° + k \cdot 180°$

GOLF **For Exercises 23 and 24, use the following information.**
A golf ball is hit with an initial velocity of 100 feet per second. The distance the ball travels is found by the formula $d = \frac{v_0^2}{g} \sin 2\theta$, where v_0 is the initial velocity, g is the acceleration due to gravity, 32 feet per second squared, and θ is the measurement of the angle that the path of the ball makes with the ground.

23. Find the distance that the ball travels if the angle between the path of the ball and the ground measures 60°. 270.6 ft
24. If a ball travels 312.5 feet, what was the angle the path of the ball made with the ground to the nearest degree? 45°
25. **STANDARDIZED TEST PRACTICE** Identify the equation of the graphed function. B

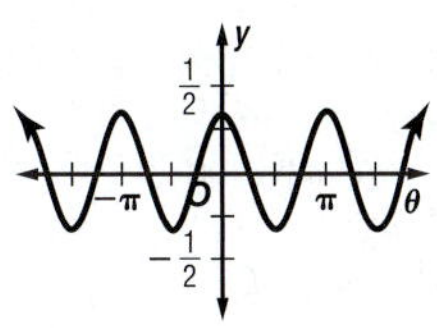

(A) $y = 3 \cos 2\theta$ (B) $y = \frac{1}{3} \cos 2\theta$

(C) $y = 3 \cos \frac{1}{2}\theta$ (D) $y = \frac{1}{3} \cos \frac{1}{2}\theta$

www.algebra2.com/chapter_test

Portfolio Suggestion

Introduction In mathematics, trigonometric functions can be used to model real-world problems.

Ask Students Find a real-world problem modeled in this chapter that interests you and show how you solved it. Explain how the function models the real-world problem and what could be gained by understanding the real-world problem better. Place your work in your portfolio.

Chapter 14 Practice Test

Assessment Options

Vocabulary Test A vocabulary test/review for Chapter 14 can be found on p. 892 of the *Chapter 14 Resource Masters*.

Chapter Tests There are six Chapter 14 Tests and an Open-Ended Assessment task available in the *Chapter 14 Resource Masters*.

Chapter 14 Tests			
Form	Type	Level	Pages
1	MC	basic	879–880
2A	MC	average	881–882
2B	MC	average	883–884
2C	FR	average	885–886
2D	FR	average	887–888
3	FR	advanced	889–890

MC = multiple-choice questions
FR = free-response questions

Open-Ended Assessment Performance tasks for Chapter 14 can be found on p. 891 of the *Chapter 14 Resource Masters*. A sample scoring rubric for these tasks appears on p. A28.

Unit 5 Test A unit test/review can be found on pp. 899–900 of the *Chapter 14 Resource Masters*.

End-of-Year Tests A Second Semester Test for Chapters 8–14 and a Final Test for Chapters 1–14 can be found on pp. 901–910 of the *Chapter 14 Resource Masters*.

ExamView® Pro

Use the networkable **ExamView® Pro** to:

- Create **multiple versions** of tests.
- Create **modified** tests for *Inclusion* students.
- **Edit** existing questions and **add** your own questions.
- Use built-in **state curriculum correlations** to create tests aligned with state standards.

These two pages contain practice questions in the various formats that can be found on the most frequently given standardized tests.

A practice answer sheet for these two pages can be found on p. A1 of the *Chapter 14 Resource Masters*.

Standardized Test Practice Student Recording Sheet, p. A1

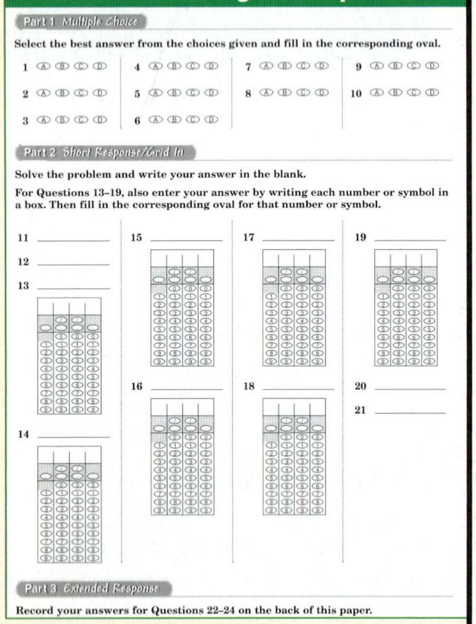
Part 1 Multiple Choice

Select the best answer from the choices given and fill in the corresponding oval.

Part 2 Short Response/Grid In

Solve the problem and write your answer in the blank.

For Questions 13–19, also enter your answer by writing each number or symbol in a box. Then fill in the corresponding oval for that number or symbol.

Part 3 Extended Response

Record your answers for Questions 22–24 on the back of this paper.

Additional Practice

See pp. 897–898 in the *Chapter 14 Resource Masters* for additional standardized test practice.

Chapter 14 Standardized Test Practice

Part 1 Multiple Choice

Record your answers on the answer sheet provided by your teacher or on a sheet of paper.

1. Which of the following is *not* equal to 3.5×10^{-2}? **D**

Ⓐ $\frac{35}{1000}$ Ⓑ 0.035 Ⓒ $\frac{7}{200}$ Ⓓ (0.5)(0.007)

2. The sum of five consecutive odd integers is 55. What is the sum of the greatest and least of these integers? **B**

Ⓐ 11 Ⓑ 22 Ⓒ 26 Ⓓ 30

3. If 8 bananas cost a cents and 6 oranges cost b cents, what is the cost of 2 bananas and 2 oranges in terms of a and b? **D**

Ⓐ $\frac{ab}{12}$ Ⓑ $3a + \frac{b}{3}$ Ⓒ $3a + 4b$ Ⓓ $\frac{3a + 4b}{12}$

4. A bag contains 16 peppermint candies, 10 butterscotch candies, and 8 cherry candies. Emma chooses one piece at random, puts it in her pocket, and then repeats the process. If she has chosen 3 peppermint candies, 2 butterscotch candies, and 1 cherry candy, what is the probability that the next piece of candy she chooses will be cherry? **C**

Ⓐ $\frac{7}{34}$ Ⓑ $\frac{8}{34}$ Ⓒ $\frac{1}{4}$ Ⓓ $\frac{3}{4}$

5. What is the value of $\frac{\sin \frac{\pi}{6}}{\cos \frac{2\pi}{3}}$? **B**

Ⓐ $-\sqrt{3}$ Ⓑ -1 Ⓒ $-\frac{\sqrt{3}}{3}$ Ⓓ 1

6. In right triangle QRS, what is the value of $\tan R$? **D**

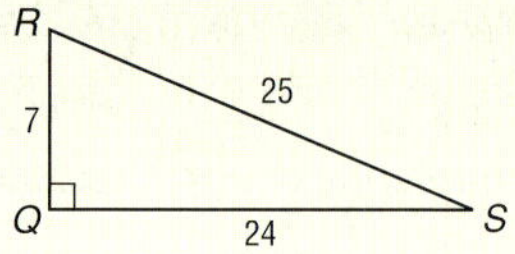

Ⓐ $\frac{7}{25}$ Ⓑ $\frac{7}{24}$ Ⓒ $\frac{25}{24}$ Ⓓ $\frac{24}{7}$

7. What is the value of $\sin\left(\cos^{-1}\frac{1}{3}\right)$ in Quadrant I? **B**

Ⓐ $\frac{2}{3}$ Ⓑ $\frac{2\sqrt{2}}{3}$ Ⓒ $\frac{\sqrt{2}}{3}$ Ⓓ $\frac{\sqrt{6}}{3}$

8. What is the least positive value for x where $y = \sin 2x$ reaches its minimum? **C**

Ⓐ $\frac{\pi}{2}$ Ⓑ π Ⓒ $\frac{3\pi}{4}$ Ⓓ $\frac{3\pi}{2}$

9. Which of the following is equivalent to $\frac{\sin^2 \theta + \cos^2 \theta}{\sec^2 \theta}$? **A**

Ⓐ $\cos^2 \theta$ Ⓑ $\sin^2 \theta$ Ⓒ $\tan^2 \theta$ Ⓓ $\sin^2 \theta + 1$

10. If $\cos \theta = -\frac{1}{2}$ and θ is in Quadrant II, what is the value of $\sin 2\theta$? **D**

Ⓐ $\frac{1}{2}$ Ⓑ $-\frac{1}{2}$ Ⓒ $\frac{\sqrt{3}}{2}$ Ⓓ $-\frac{\sqrt{3}}{2}$

ExamView® Pro

Special banks of standardized test questions similar to those on the SAT, ACT, TIMSS 12, NAEP 12, and state proficiency tests can be found on this CD-ROM.

810 Chapter 14 Trigonometric Graphs and Identities

Preparing for Standardized Tests
For test-taking strategies and more practice, see pages 877–892.

Part 2 Short Response/Grid In

Record your answers on the answer sheet provided by your teacher or on a sheet of paper.

11. If k is a positive integer, and $7k + 3$ equals a prime number that is less than 50, then what is one possible value of $7k + 3$? **17 or 31**

12. It costs \$8 to make a book. The selling price will include an additional 200%. What will be the selling price? **\$24**

13. The mean of seven numbers is 0. The sum of three of the numbers is −9. What is the sum of the remaining four numbers? **9**

14. If $4a - 6b = 0$ and $c = 9b$, what is the ratio of a to c? **1/6**

15. What is the value of x if $\frac{3^3 \cdot 3}{\sqrt{81}} = 3^x$? **2**

16. The ages of children at a party are 6, 7, 6, 6, 7, 7, 8, 6, 7, 8, 9, 7, and 7. Let N represent the median of their ages and m represent the mode. What is $N - m$? **0**

17. In the figure below, $CEFG$ is a square, ABD is a right triangle, D is the midpoint of side CE, H is the midpoint of side CG, and C is the midpoint of side BD. $BCDE$ is a line segment, and AHD is a line segment. If the measure of the area of square $CEFG$ is 16, what is the measure of the area of quadrilateral $ABCH$? **6**

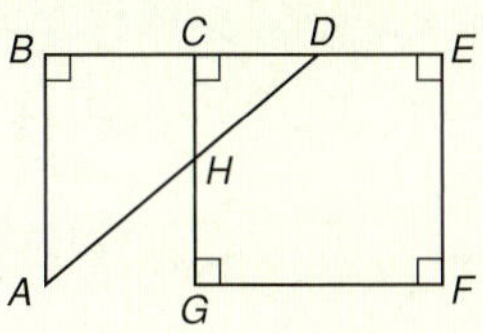

Test-Taking Tip

Always write down your calculations on scrap paper or in the test booklet, even if you think you can do the calculations in your head. Writing down your calculations will help you avoid making simple mistakes.

www.algebra2.com/standardized_test

18. A line with a slope of $\frac{3}{8}$ passes through $(6, 4n)$ and $(0, n)$. What is the value of n? **3/4**

19. If $\sin 60° = \frac{\sqrt{3}}{2}$, what is the value of $\sin^2 30° + \cos^2 30°$? **1**

20. A and B are both positive acute angles. If $\cos A = \frac{12}{13}$ and $\sin B = \frac{9}{41}$, find the value of $\sin(A + B)$. $\mathbf{\frac{308}{533}}$

21. The diagram below shows a portion of a graph of a trigonometric function. Write an equation that could represent the graph. $y = 3 \cos \frac{3}{2}x$

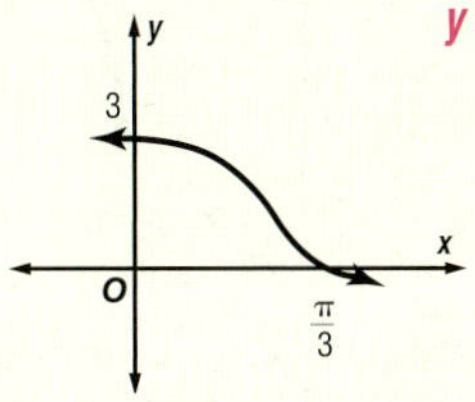

Part 3 Extended Response

Record your answers on a sheet of paper. Show your work.

For Exercises 22–24, use the following information.

The outdoor temperature T varies with the time of day. During the course of a certain day, the outdoor temperature can be modeled by the equation $T = 16.9 \sin 6.9t + 59.9$, where t is the number of hours since midnight.

22. Use the equation to find the maximum temperature of the day. At about what time was that maximum reached? **76.8°; 1:00 P.M.**

23. Use the equation to find the minimum temperature of the day. At about what time was that minimum reached? **59.9°; 12:00 A.M.**

24. The air conditioning in a home operates when the outside temperature is 75°F or above. Determine, to the nearest tenth of an hour, the amount of time in this day that the air conditioning is on in the home. **7.7 h**

Evaluating Extended Response Questions

Extended Response questions are graded by using a multilevel rubric that guides you in assessing a student's knowledge of a particular concept.

Goal: Determine the dimensions of a rectangular garden for a given perimeter.

Sample Scoring Rubric: The following rubric is a sample scoring device. You may wish to add more detail to this sample to meet your individual scoring needs.

Score	Criteria
4	A correct solution that is supported by well-developed, accurate explanations
3	A generally correct solution, but may contain minor flaws in reasoning or computation
2	A partially correct interpretation and/or solution to the problem
1	A correct solution with no supporting evidence or explanation
0	An incorrect solution indicating no mathematical understanding of the concept or task, or no solution is given

Pages 767–768, Lesson 14-1

4. amplitude: $\frac{1}{2}$; period 360° or 2π

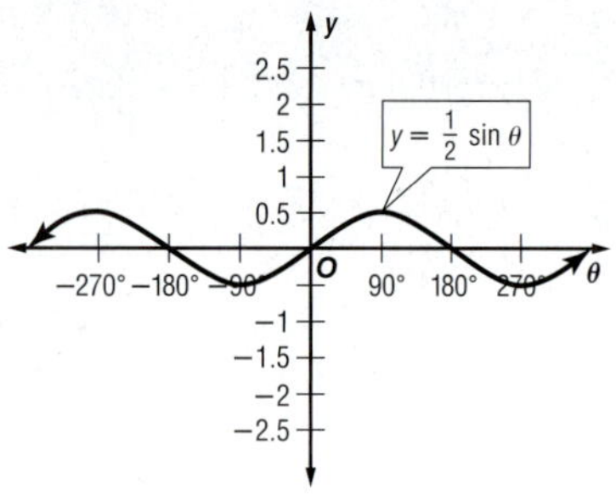

5. amplitude: 2; period: 360° or 2π

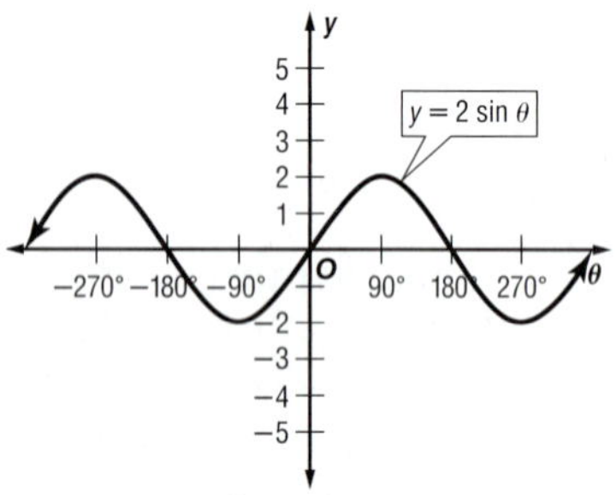

6. amplitude: $\frac{2}{3}$; period: 360° or 2π

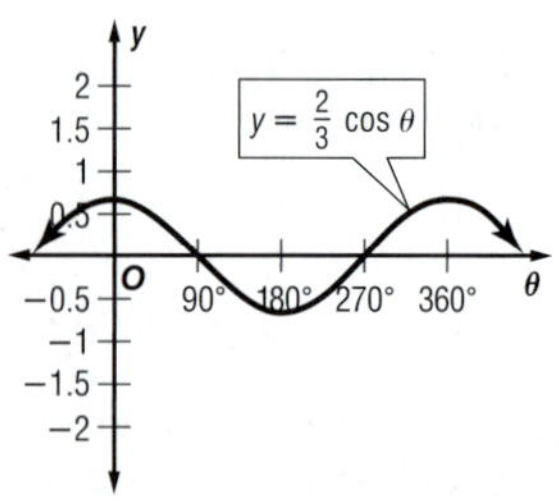

7. amplitude: does not exist; period: 180° or π

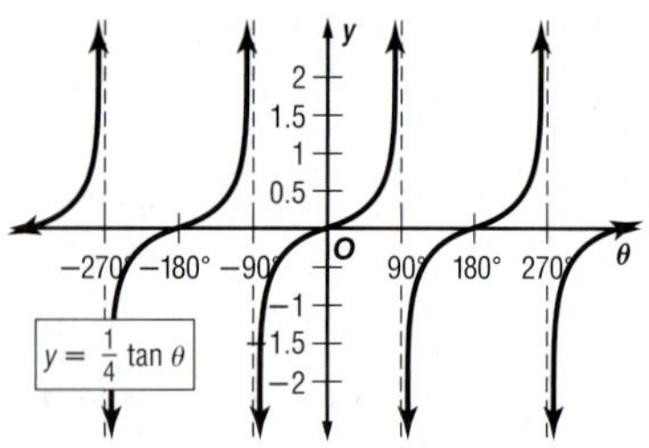

8. amplitude: does not exist; period: 180° or π

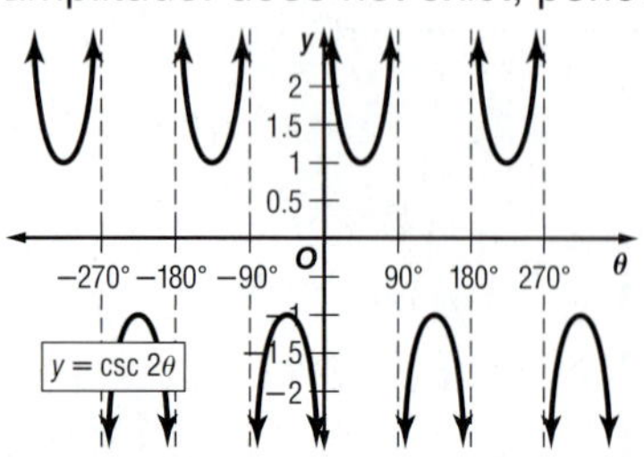

9. amplitude: 4; period: 180° or π

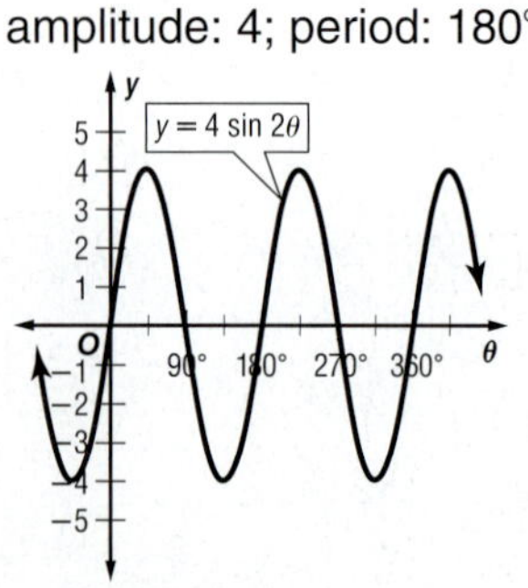

10. amplitude: 4; period: 480° or $\frac{8\pi}{3}$

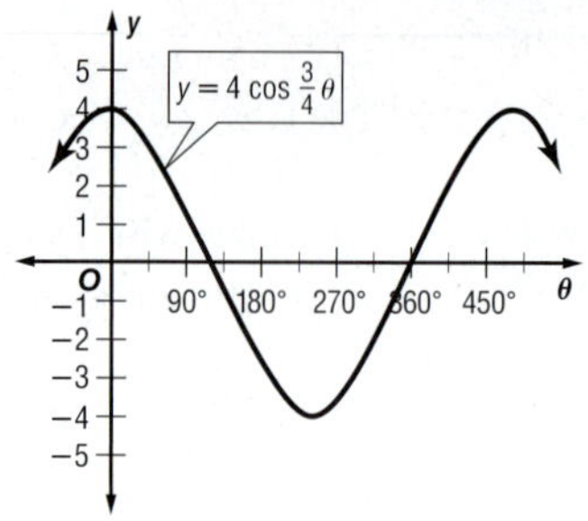

11. amplitude: does not exist; period: 120° or $\frac{2\pi}{3}$

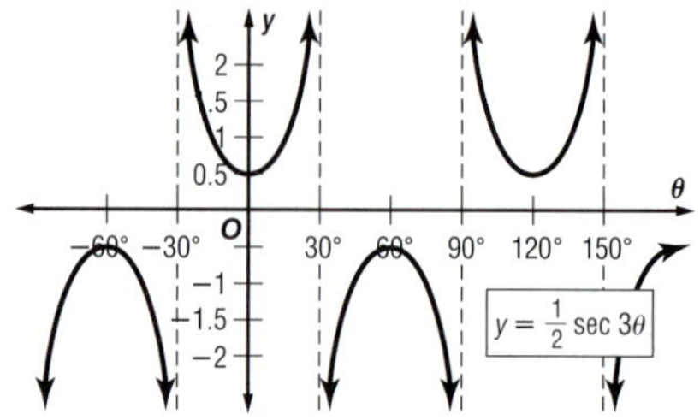

12. amplitude: $\frac{3}{4}$; period: 720° or 4π

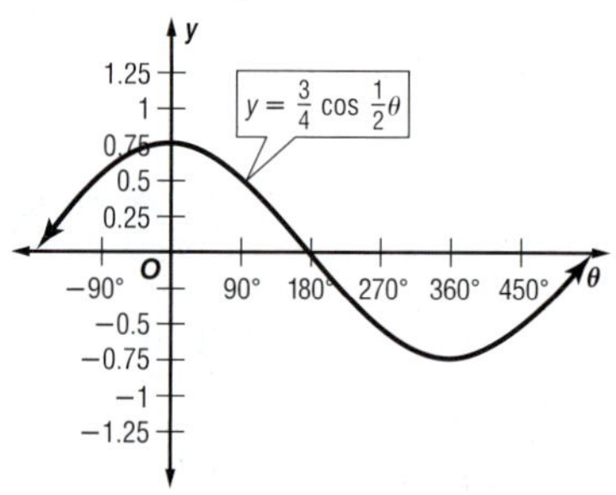

15. amplitude: 3; period: 360° or 2π

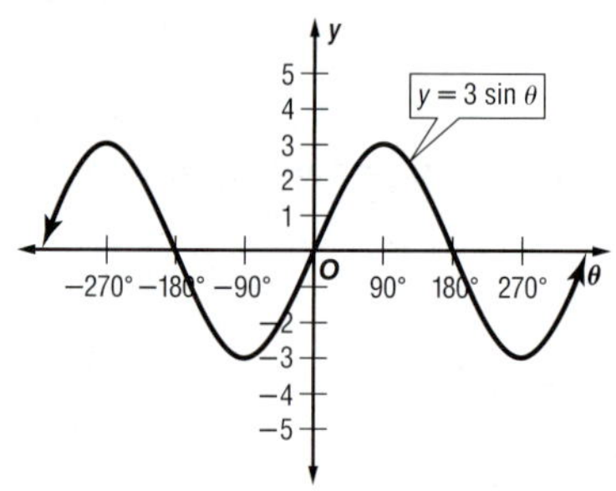

16. amplitude: 5; period: 360° or 2π

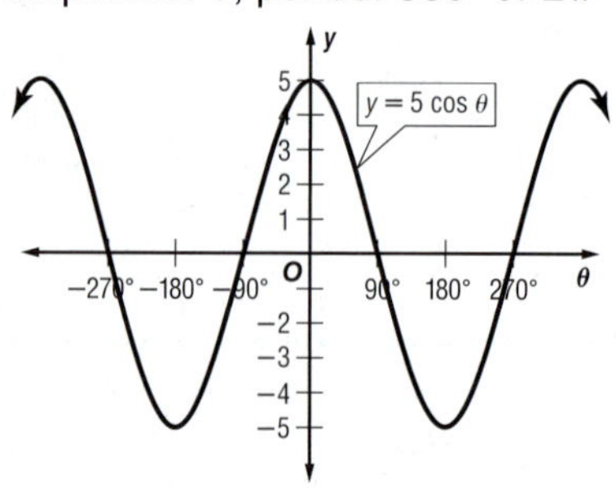

17. amplitude: does not exist; period: 360° or 2π

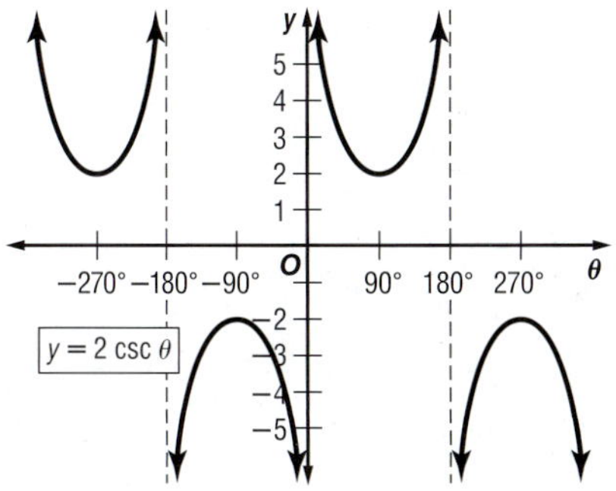

18. amplitude: does not exist; period: 180° or π

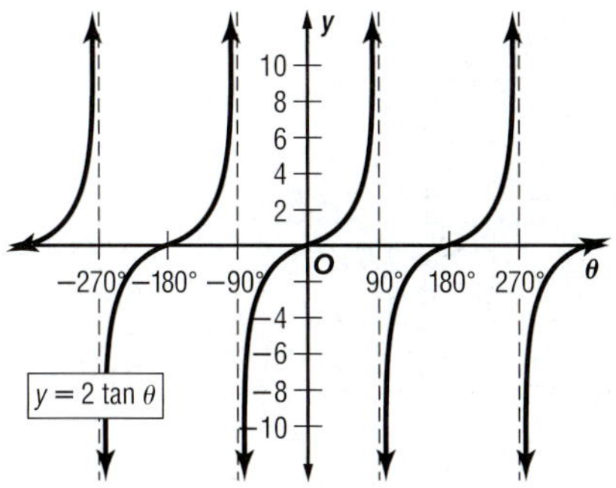

19. amplitude: $\frac{1}{5}$; period: 360° or 2π

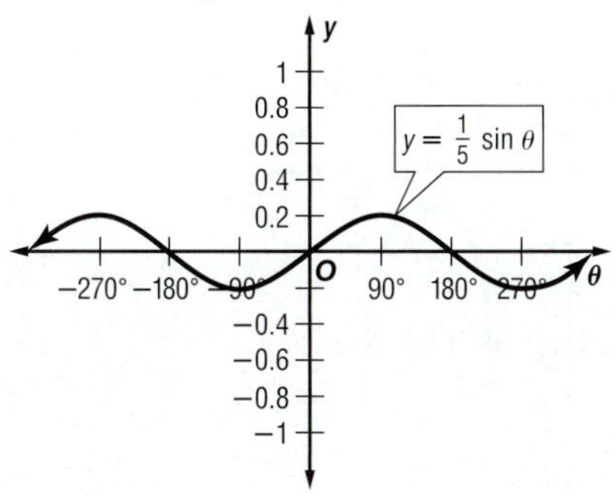

20. amplitude: does not exist; period: 360° or 2π

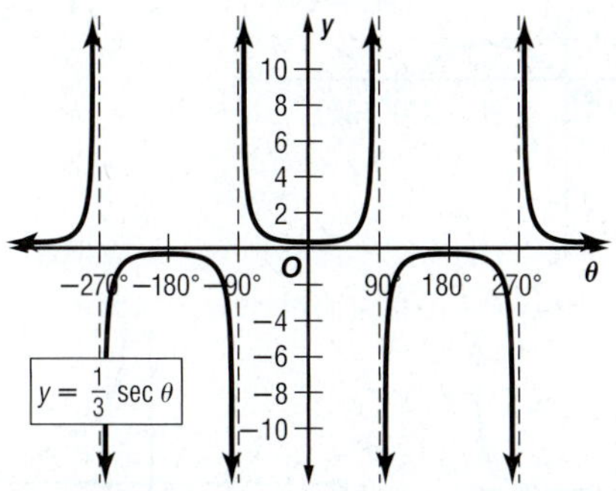

21. amplitude: 1; period 90° or $\frac{\pi}{2}$

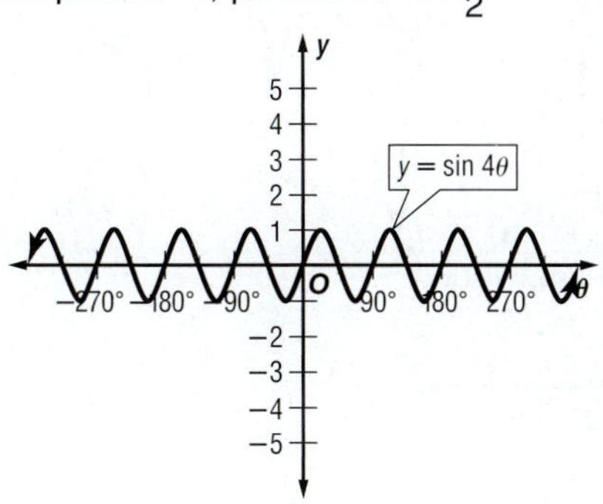

22. amplitude: 1; period: 180° or π

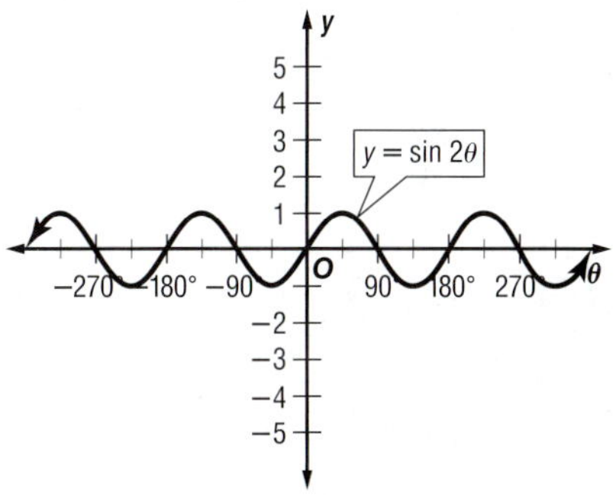

23. amplitude: does not exist; period: 120° or $\frac{2\pi}{3}$

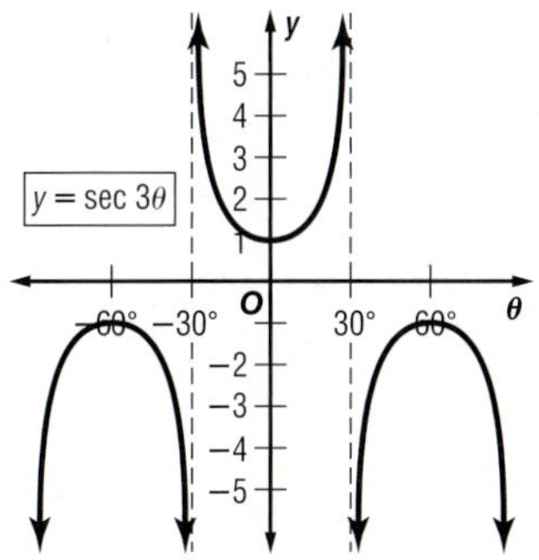

24. amplitude: does not exist; period: 36° or $\frac{\pi}{5}$

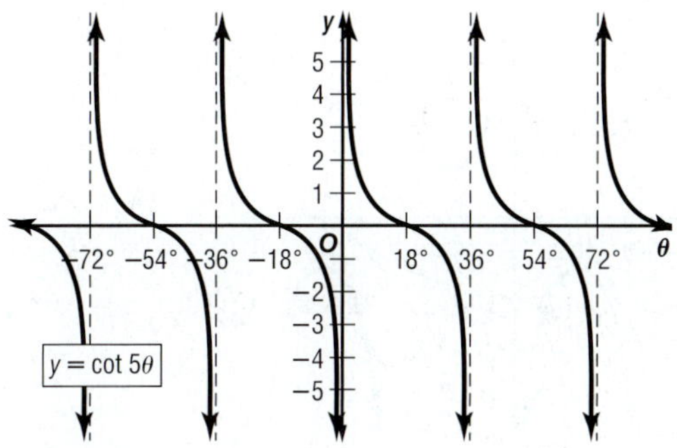

25. amplitude: does not exist; period: 540° or 3π

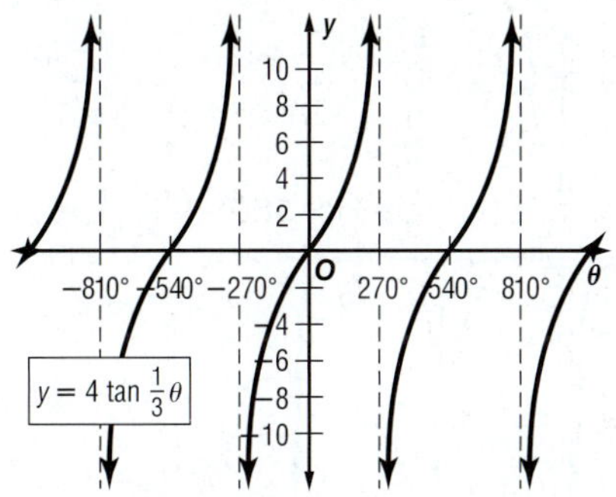

26. amplitude: does not exist; period: 360° or 2π

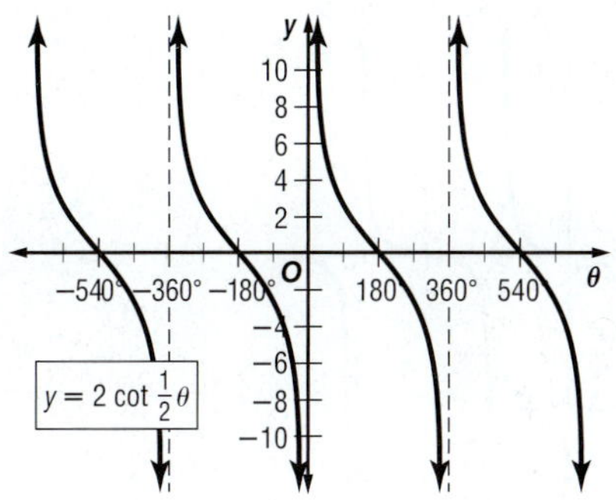

27. amplitude: 6; period: 540° or 3π

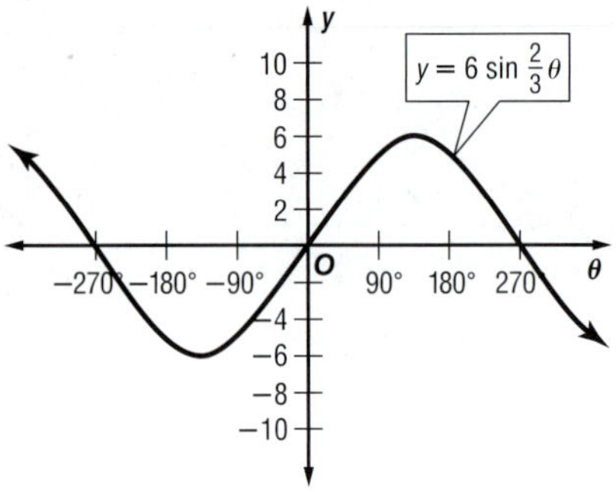

28. amplitude: 3; period: 720° or 4π

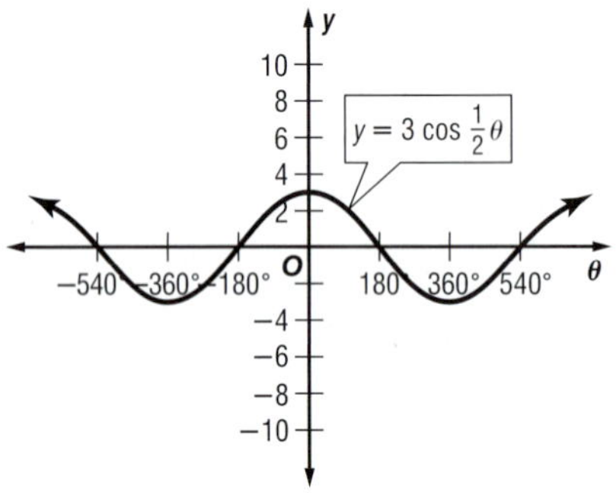

29. amplitude: does not exist; period: 720° or 4π

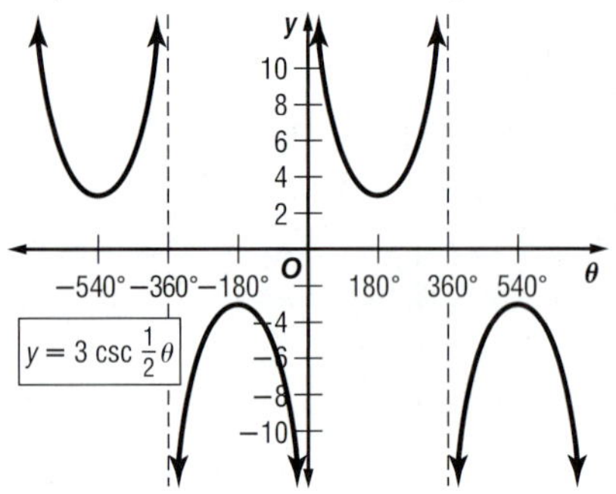

30. amplitude: does not exist; period: 90° or $\frac{\pi}{2}$

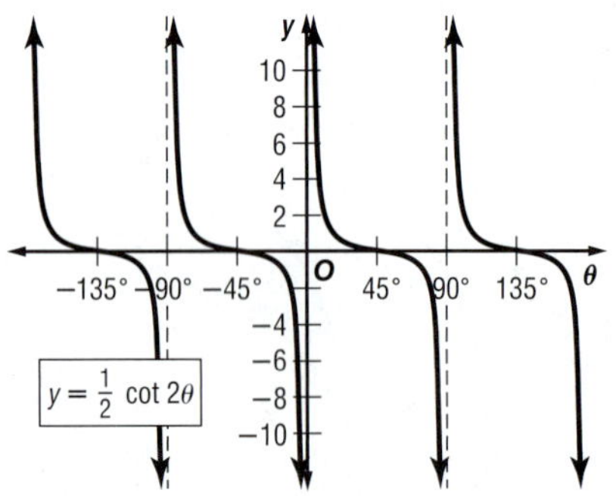

31. amplitude: does not exist; period: 180° or π

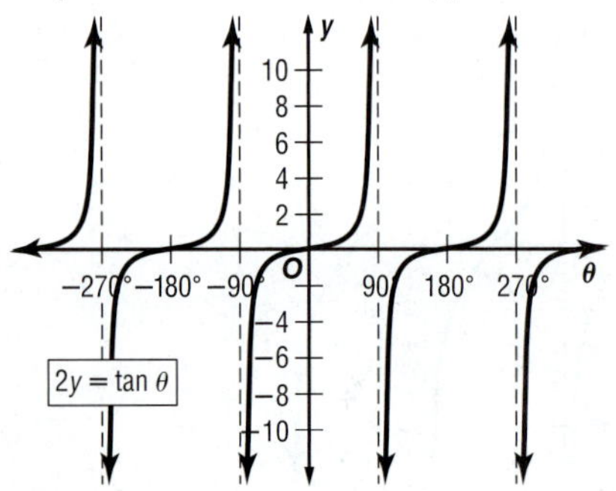

32. amplitude: $\frac{8}{9}$; period: 600° or $\frac{10\pi}{3}$

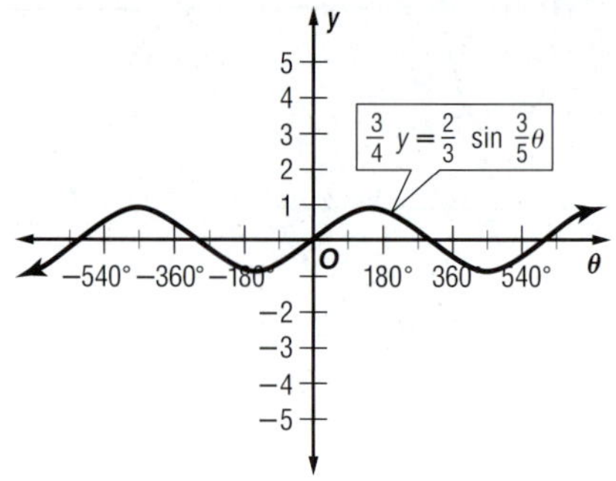

38.

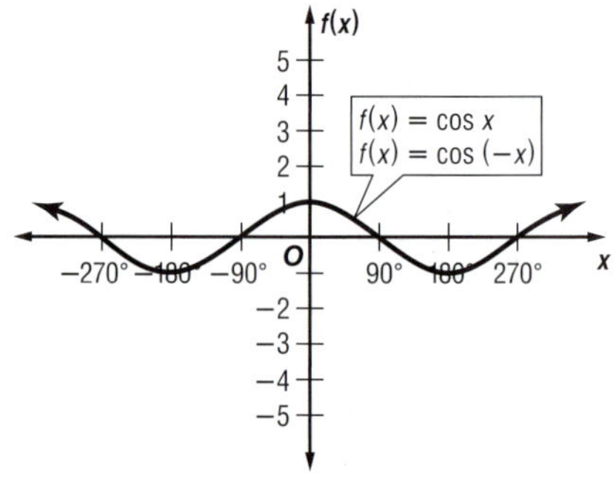

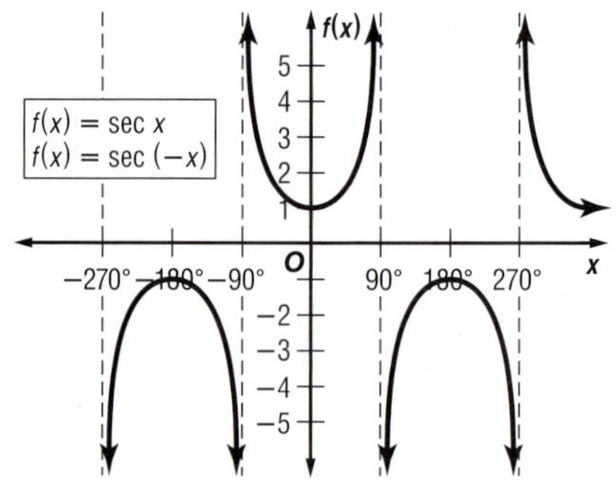

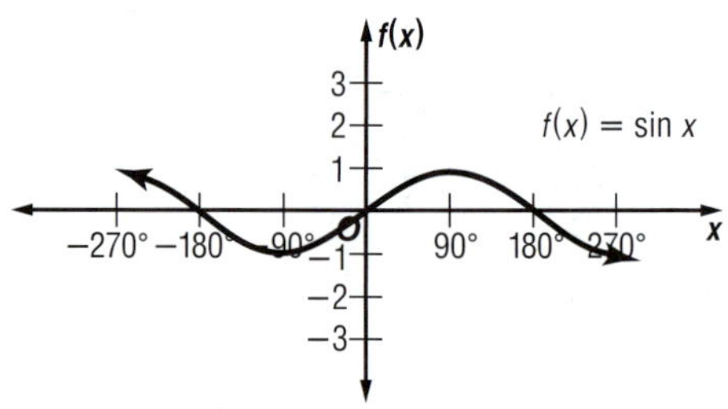

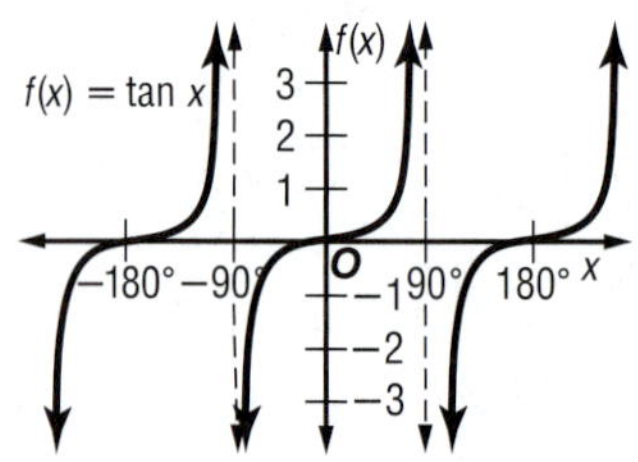

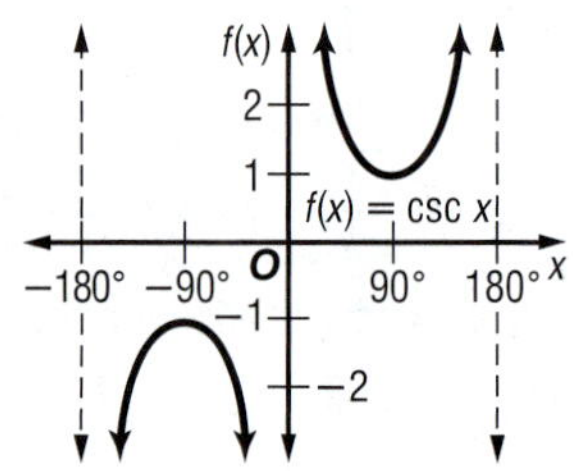

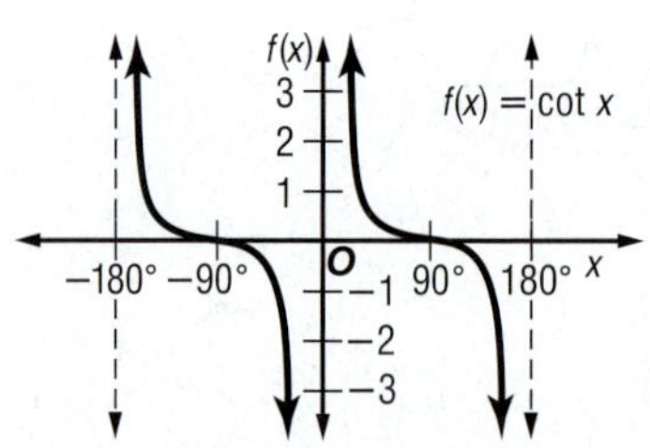

53.

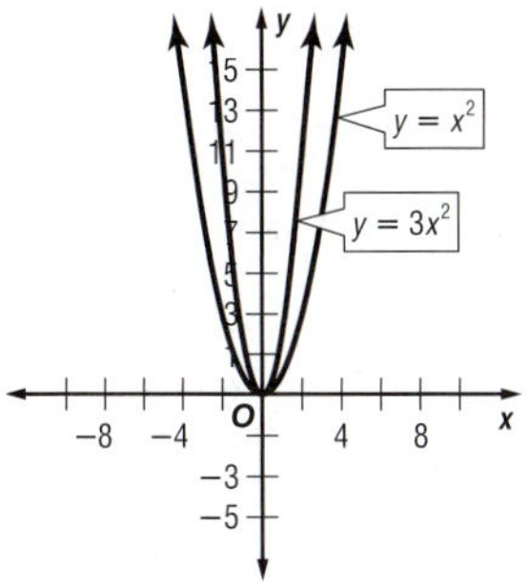

54.

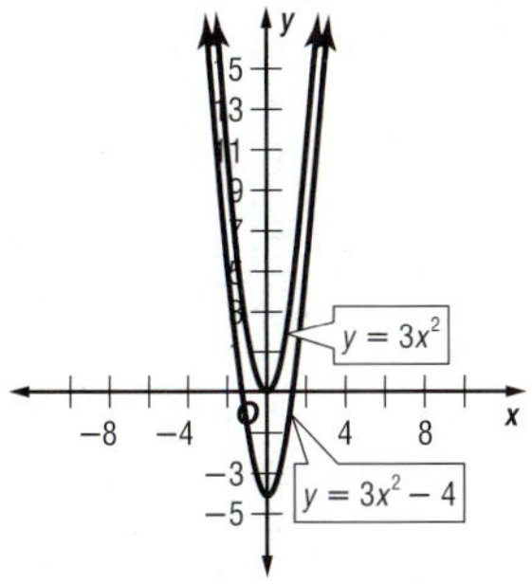

55.

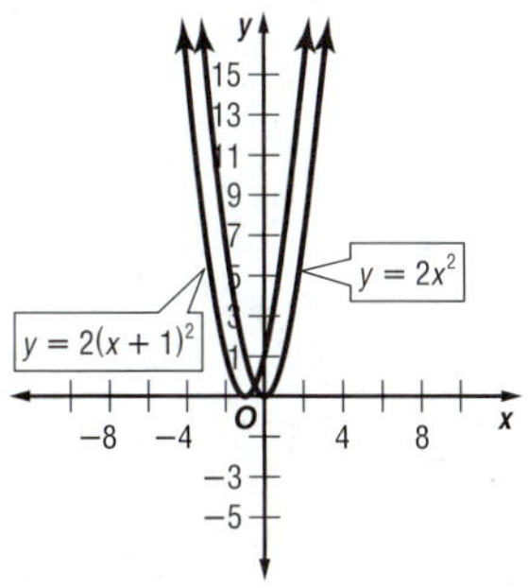

56.

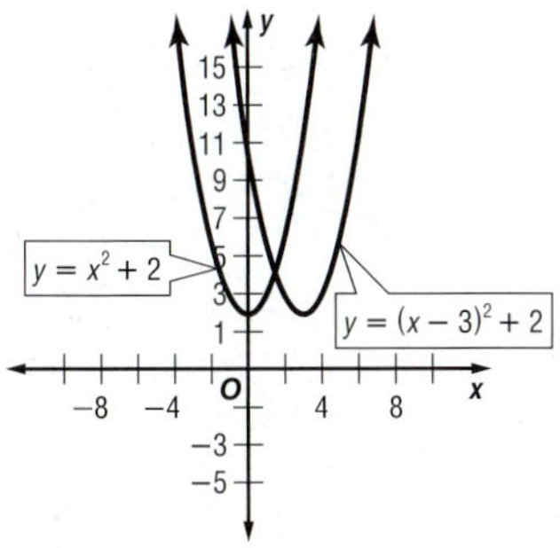

Page 769, Lesson 14-2 Graphing Calculator Investigation

4.

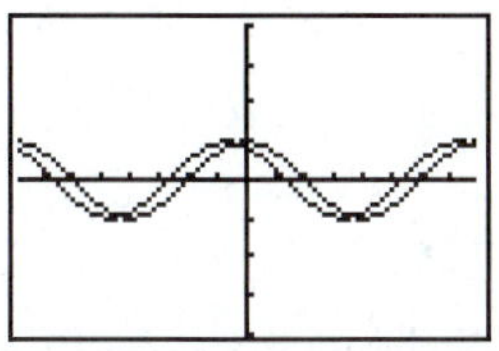

[−360, 360] scl: 90 by [−5, 5] scl: 1

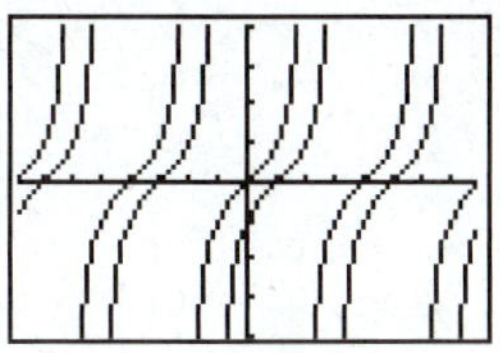

[−360, 360] scl: 90 by [−5, 5] scl: 1

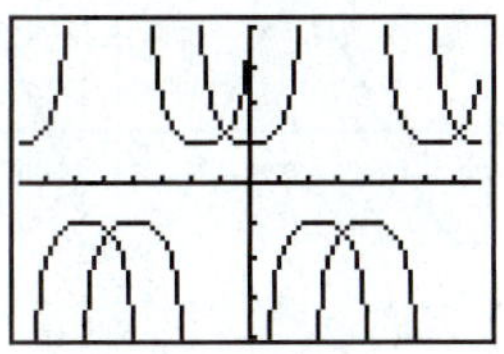

[−360, 360] scl: 90 by [−5, 5] scl: 1

Pages 774–776, Lesson 14-2

4.

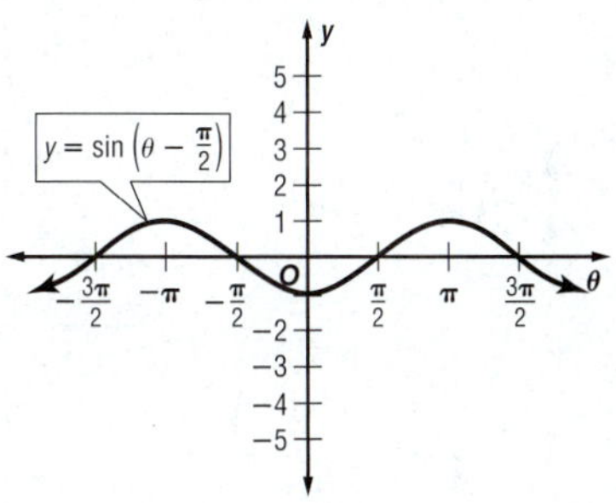

5.

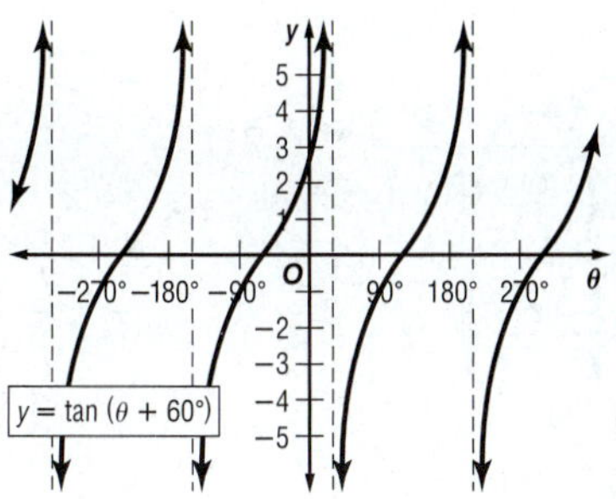

6.

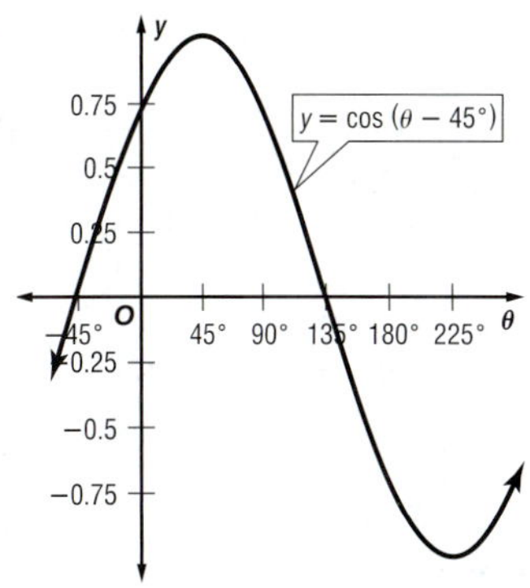

7.

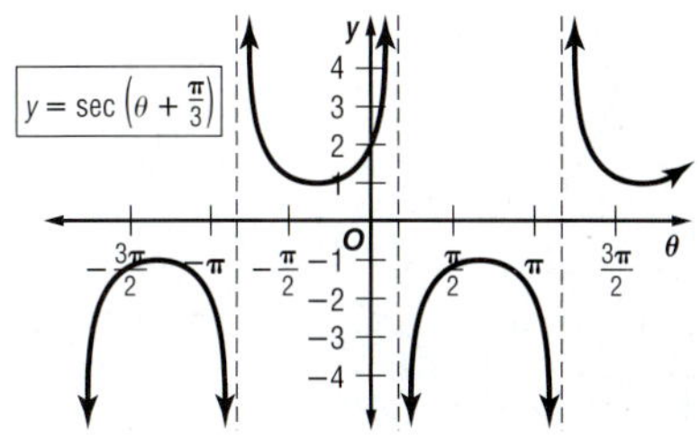

8.

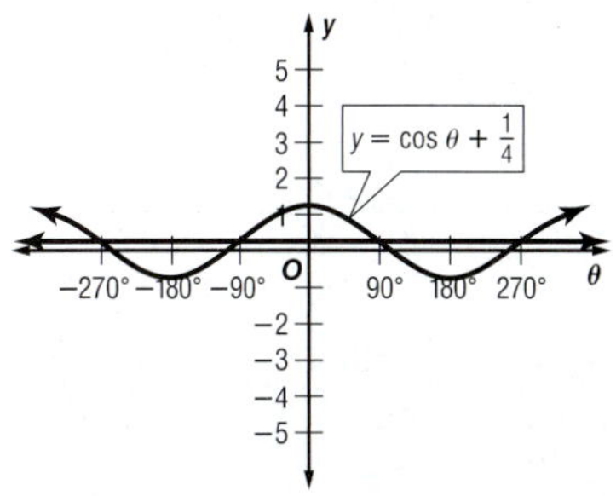

9.

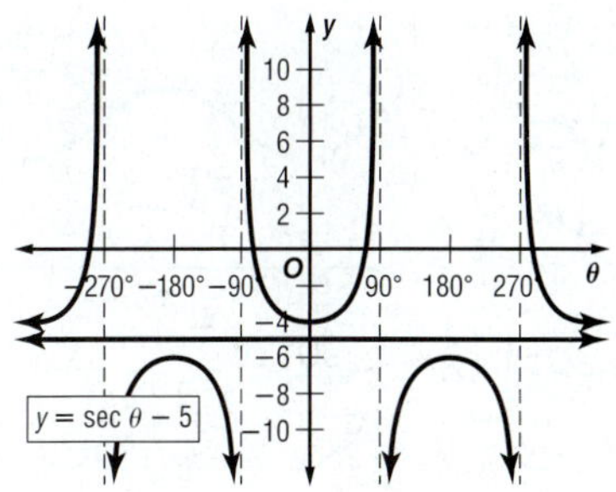

10.

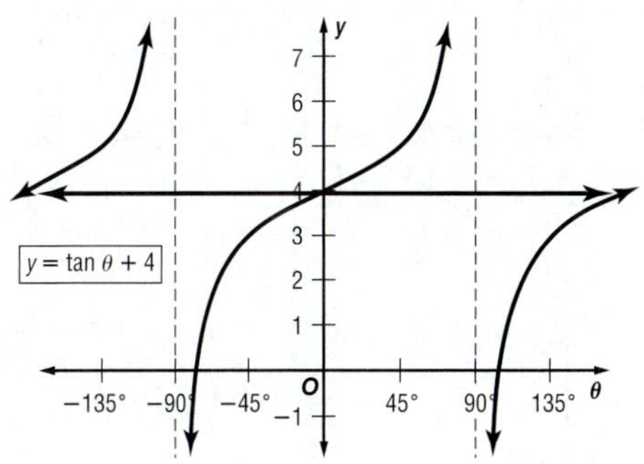

11.

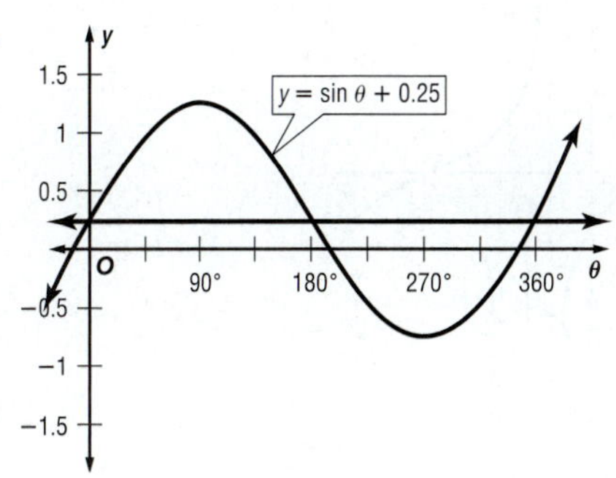

12.

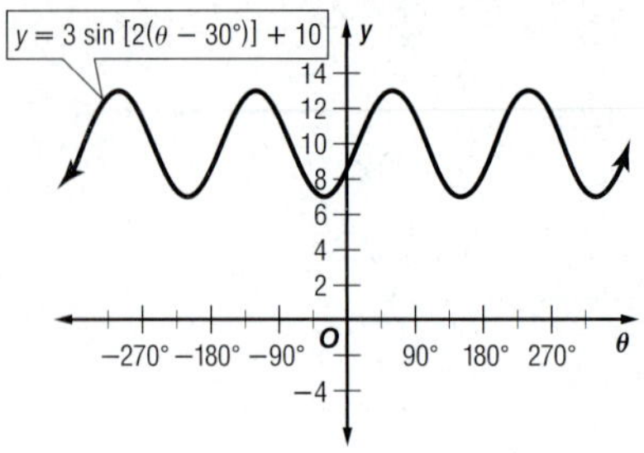

13.

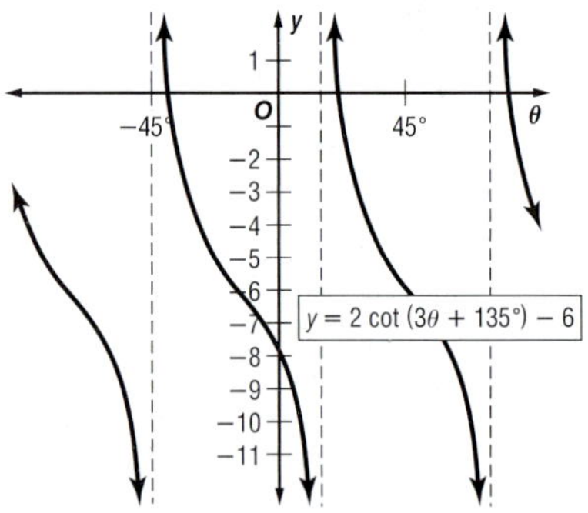

14.

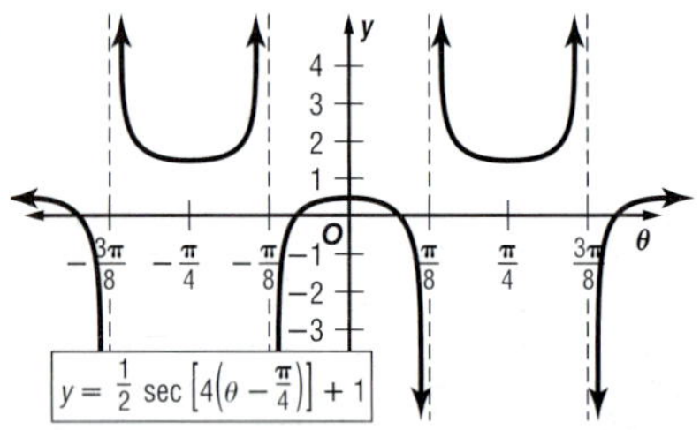

15.

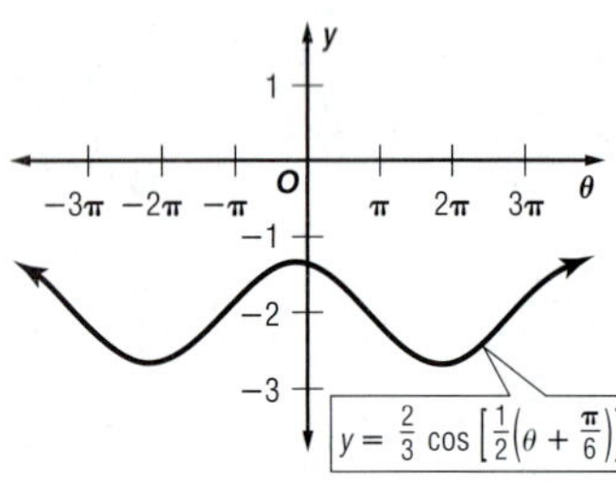

18.

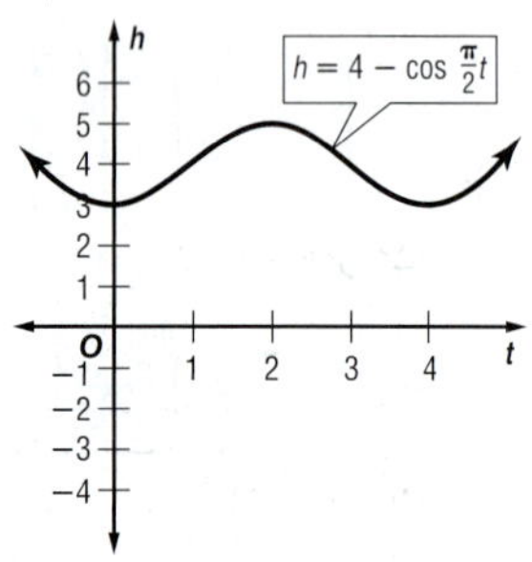

19.

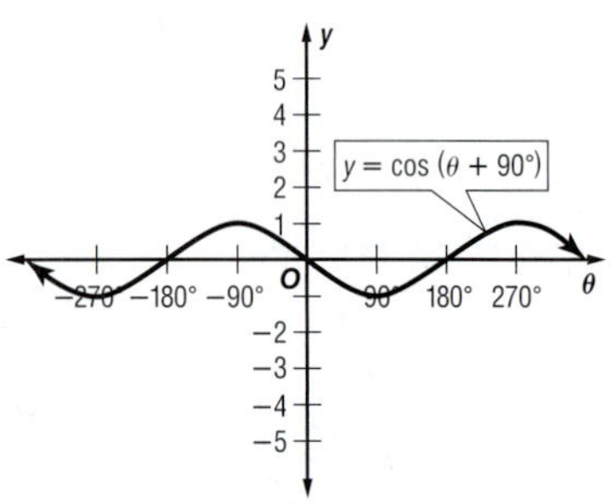

20.

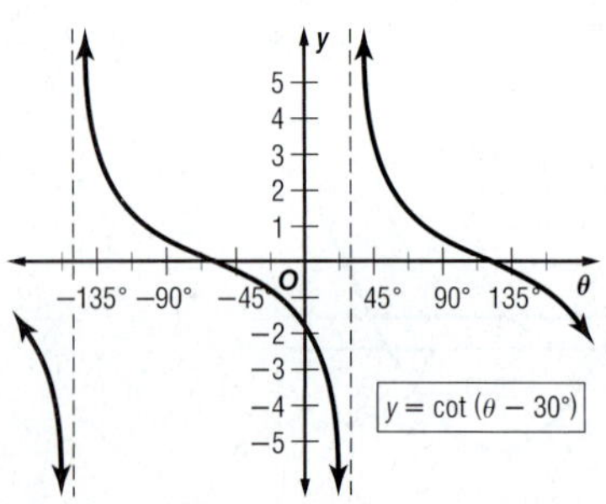

21.

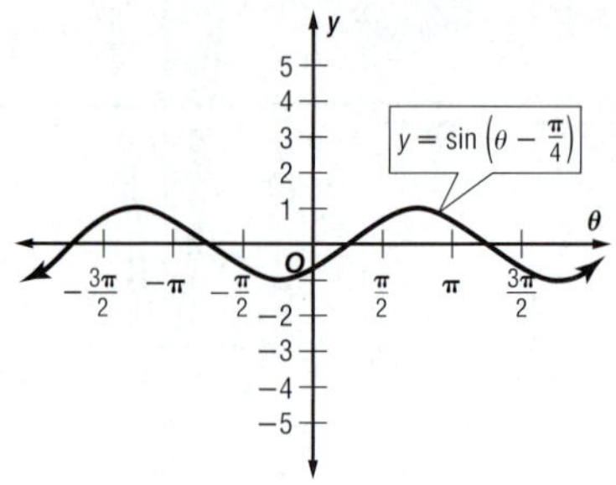

22.

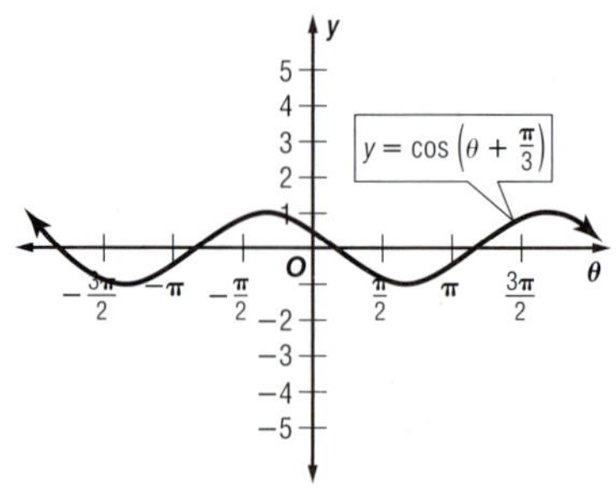

23.

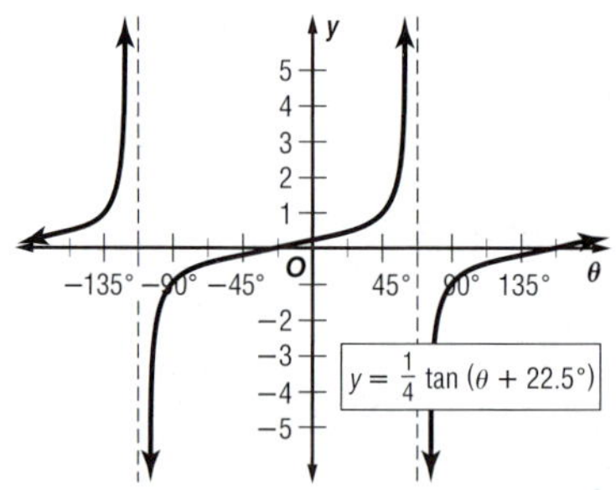

24.

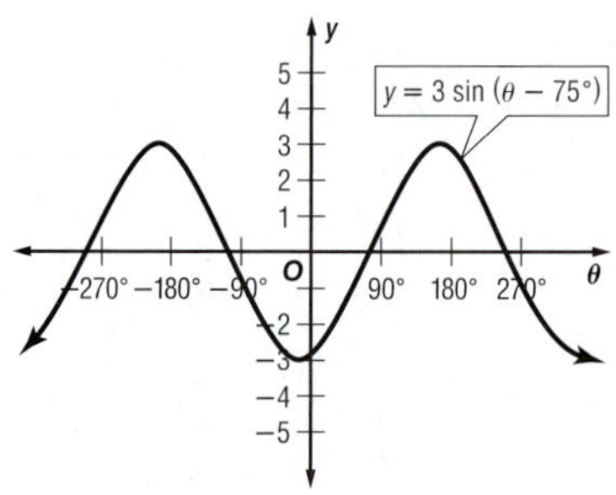

25.

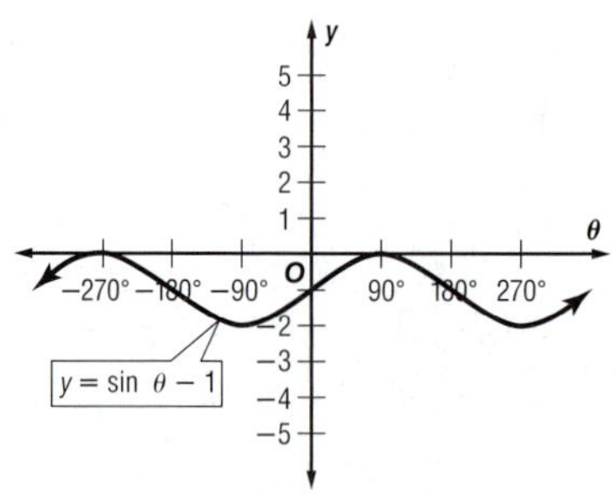

26.

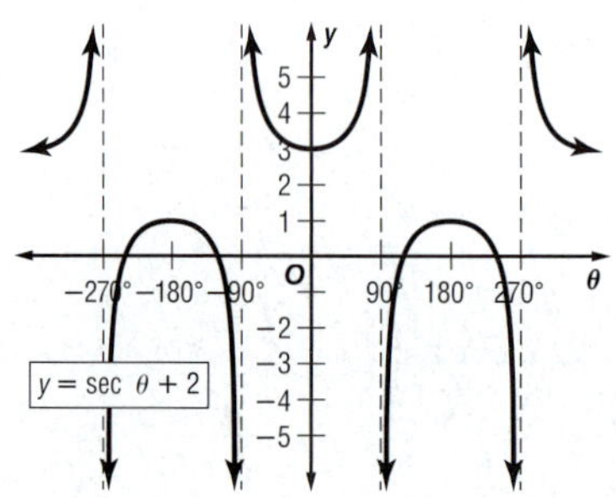

27.

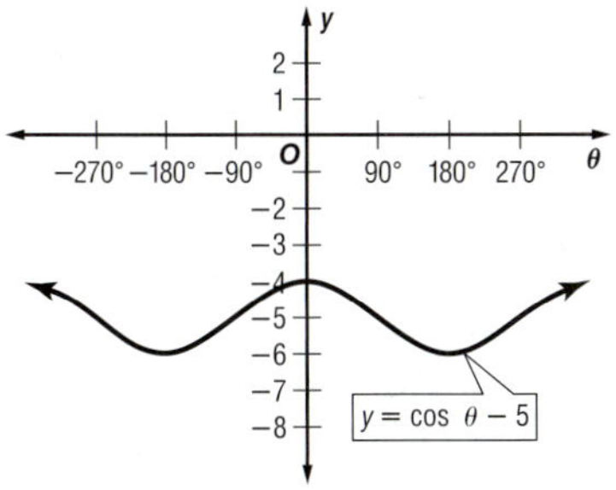

28.

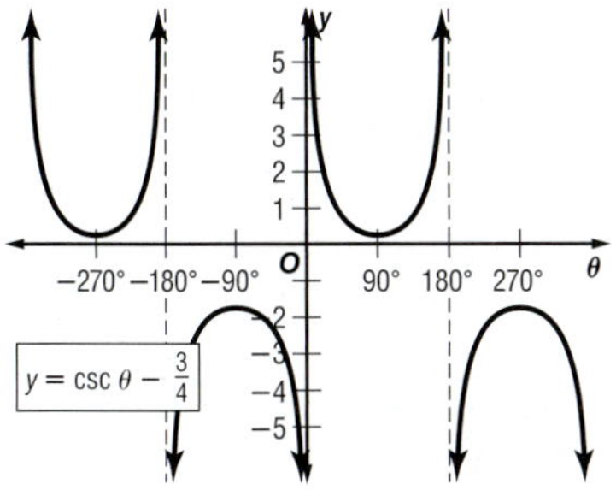

29.

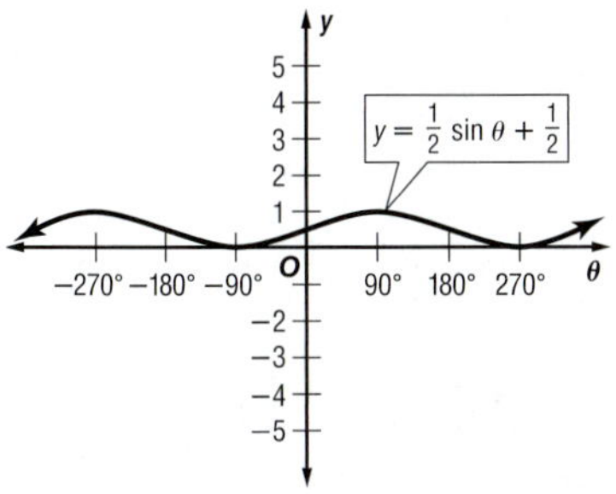

30.

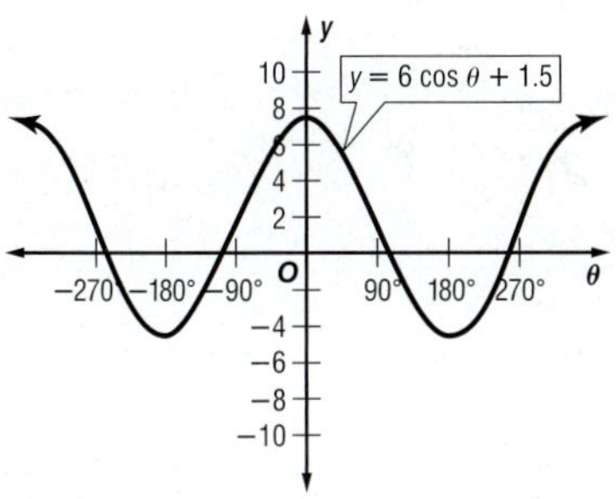

33. 1; 2; 120°; 45°

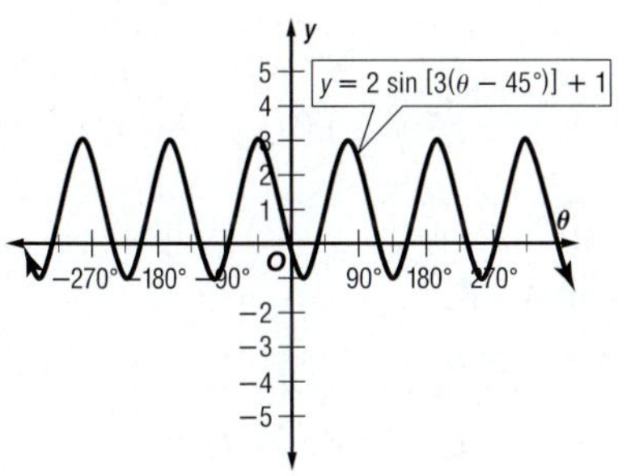

34. −5; 4; 180°; −30°

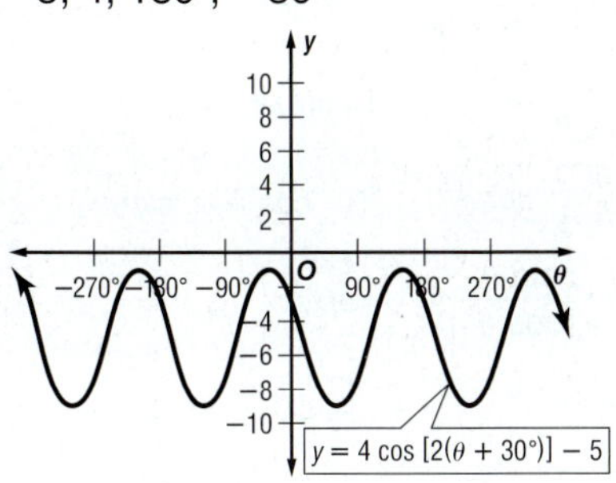

35. −3.5; does not exist; 720°; −60°

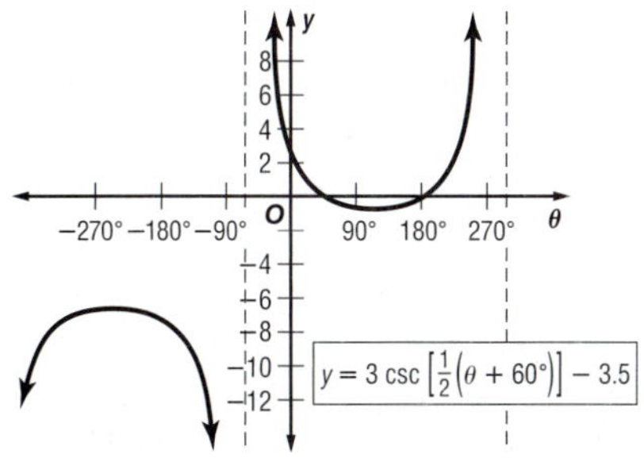

36. 0.75; does not exist; 270°; 90°

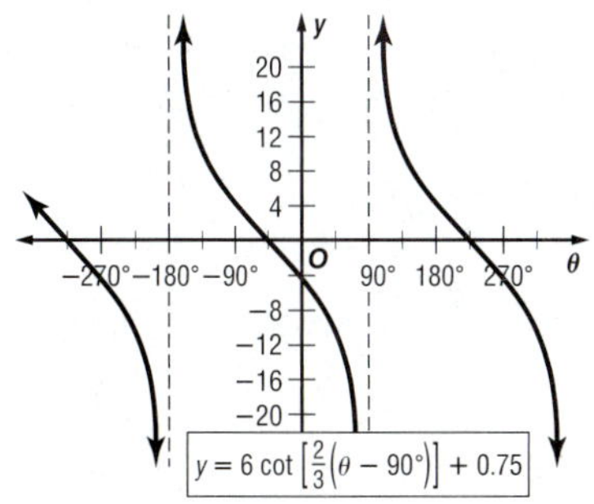

37. 1; $\frac{1}{4}$; 180°; 75°

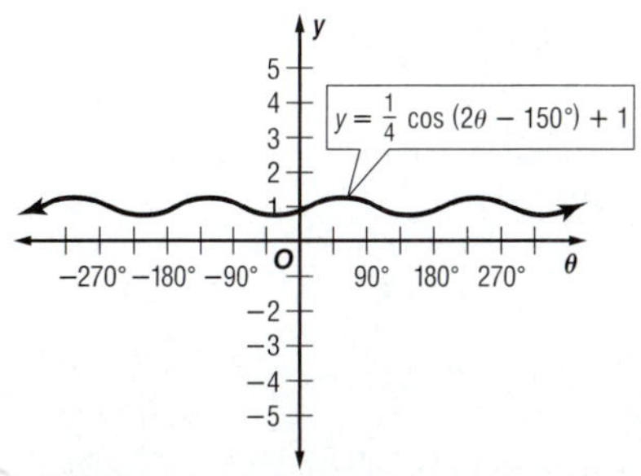

38. −4; does not exist; 30°; −22.5°

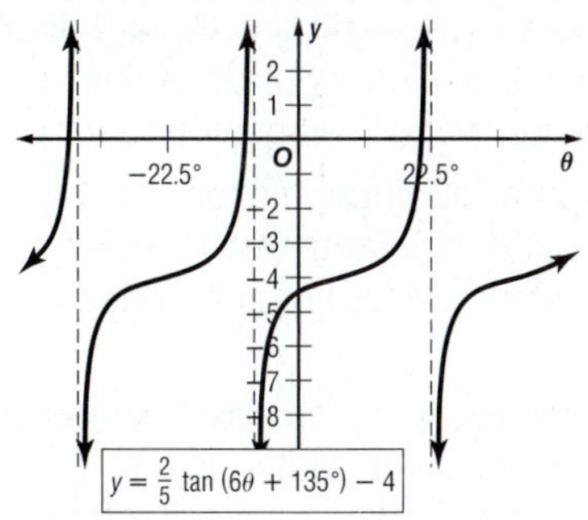

39. 3; 2; π; $-\frac{\pi}{4}$

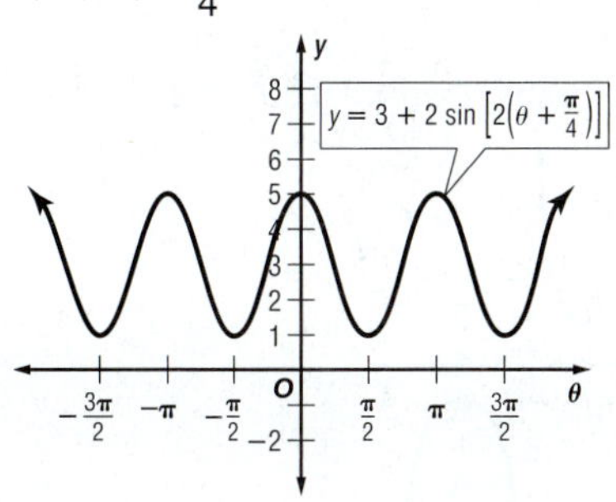

40. 4; does not exist; 6π; $-\frac{2\pi}{3}$

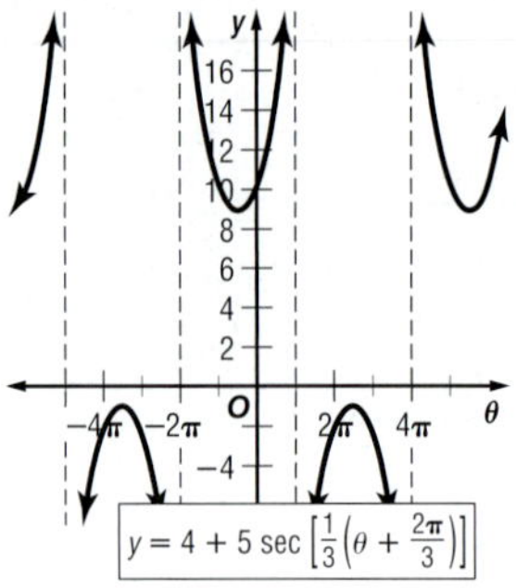

41.

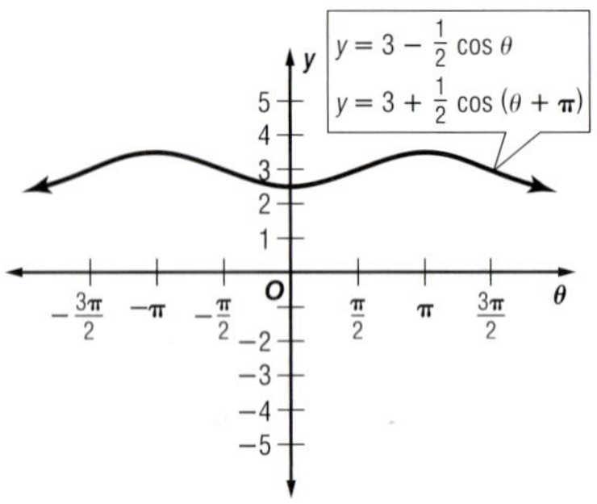

The graphs are identical.

42.

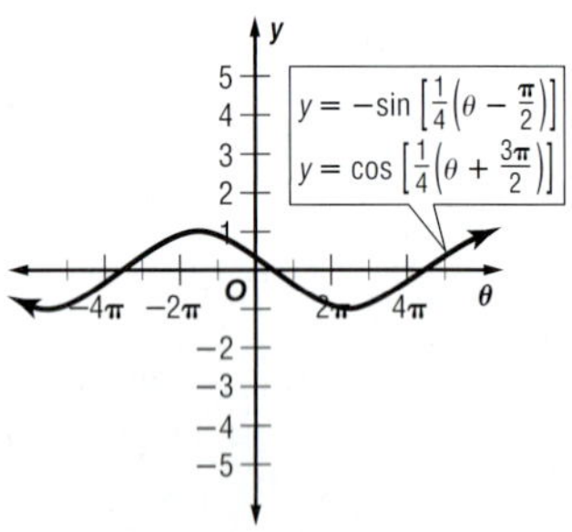

The graphs are identical.

49. Sample answer: You can use changes in amplitude and period along with vertical and horizontal shifts to show an animal population's starting point and display changes to that population over a period of time. Answers should include the following information.

- The equation shows a rabbit population that begins at 1200, increases to a maximum of 1450 then decreases to a minimum of 950 over a period of 4 years.
- Relative to $y = a\cos bx$, $y = a\cos bx + k$ would have a vertical shift of k units, while $y = a\cos[b(x - h)]$ has a horizontal shift of h units.

Page 781, Lesson 14-3

50. amplitude: does not exist; period: 180° or π

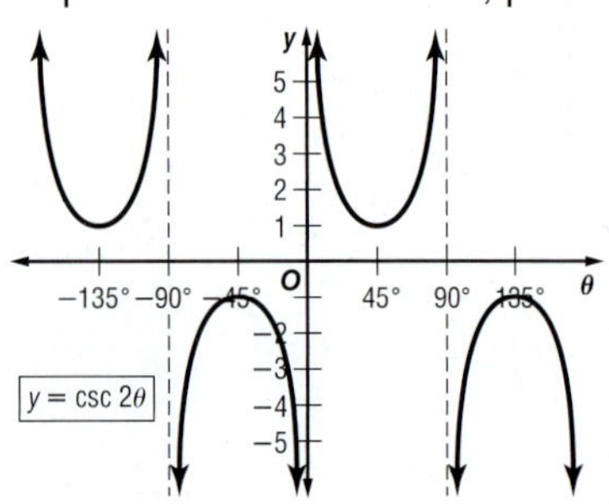

51. amplitude: 1; period: 120° or $\frac{2\pi}{3}$

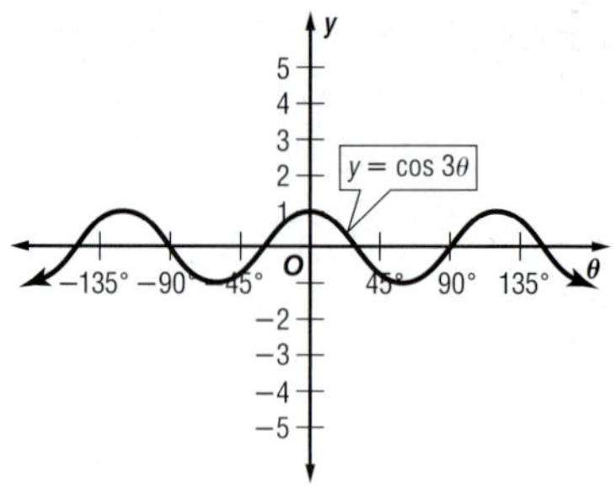

52. amplitude: does not exist; period: 36° or $\frac{\pi}{5}$

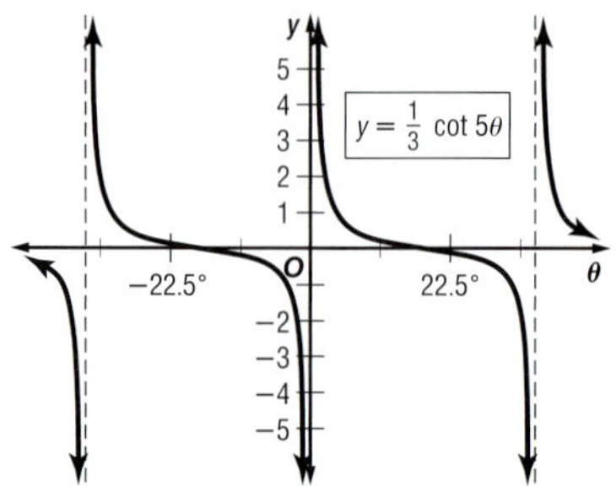

Page 781, Practice Quiz 1

1.

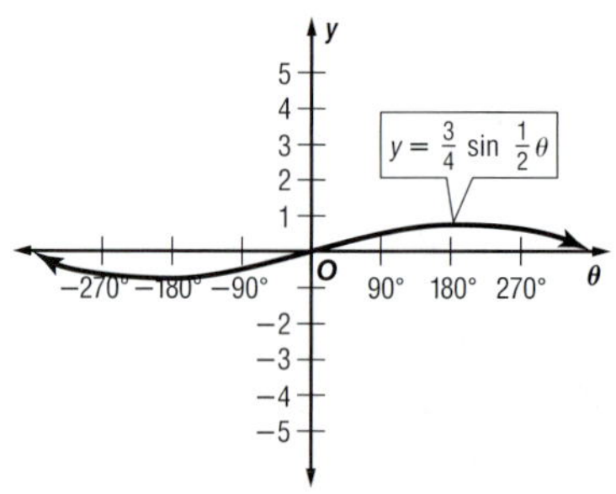

2.

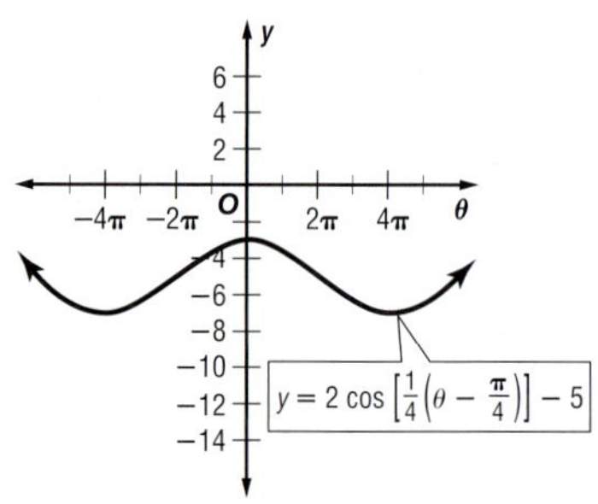

Pages 784–785, Lesson 14-4

1. $\sin\theta\tan\theta \stackrel{?}{=} \sec\theta - \cos\theta$

$\sin\theta\tan\theta \stackrel{?}{=} \frac{1}{\cos\theta} - \cos\theta$	$\sec\theta = \frac{1}{\cos\theta}$
$\sin\theta\tan\theta \stackrel{?}{=} \frac{1}{\cos\theta} - \frac{\cos^2\theta}{\cos\theta}$	Multiply by the LCD, $\cos\theta$.
$\sin\theta\tan\theta \stackrel{?}{=} \frac{1 - \cos^2\theta}{\cos\theta}$	Subtract.
$\sin\theta\tan\theta \stackrel{?}{=} \frac{\sin^2\theta}{\cos\theta}$	$1 - \cos^2\theta = \sin^2\theta$
$\sin\theta\tan\theta \stackrel{?}{=} \sin\theta \cdot \frac{\sin\theta}{\cos\theta}$	Factor.
$\sin\theta\tan\theta = \sin\theta\tan\theta$	$\frac{\sin\theta}{\cos\theta} = \tan\theta$

2. Sample answer: Use various identities, multiply or divide terms to form an equivalent expression, factor, and simplify rational expressions.

3. Sample answer: $\sin^2\theta = 1 + \cos^2\theta$; it is not an identity because $\sin^2\theta = 1 - \cos^2\theta$.

4.
$$\tan\theta\,(\cot\theta + \tan\theta) \stackrel{?}{=} \sec^2\theta$$
$$1 + \tan^2\theta \stackrel{?}{=} \sec^2\theta$$
$$\sec^2\theta = \sec^2\theta$$

5.
$$\tan^2\theta\cos^2\theta \stackrel{?}{=} 1 - \cos^2\theta$$
$$\frac{\sin^2\theta}{\cos^2\theta}\cdot\cos^2\theta \stackrel{?}{=} \sin^2\theta$$
$$\sin^2\theta = \sin^2\theta$$

6.
$$\frac{\cos^2\theta}{1-\sin\theta} \stackrel{?}{=} 1 + \sin\theta$$
$$\frac{1-\sin^2\theta}{1-\sin\theta} \stackrel{?}{=} 1 + \sin\theta$$
$$\frac{(1-\sin\theta)(1+\sin\theta)}{1-\sin\theta} \stackrel{?}{=} 1 + \sin\theta$$
$$1 + \sin\theta = 1 + \sin\theta$$

7.
$$\frac{1+\tan^2\theta}{\csc^2\theta} \stackrel{?}{=} \tan^2\theta$$
$$\frac{\sec^2\theta}{\csc^2\theta} \stackrel{?}{=} \tan^2\theta$$
$$\frac{\frac{1}{\cos^2\theta}}{\frac{1}{\sin^2\theta}} \stackrel{?}{=} \tan^2\theta$$
$$\frac{1}{\cos^2\theta}\cdot\sin^2\theta \stackrel{?}{=} \tan^2\theta$$
$$\tan^2\theta = \tan^2\theta$$

8.
$$\frac{\sin\theta}{\sec\theta} \stackrel{?}{=} \frac{1}{\tan\theta + \cot\theta}$$
$$\frac{\sin\theta}{\sec\theta} \stackrel{?}{=} \frac{1}{\frac{\sin\theta}{\cos\theta} + \frac{\cos\theta}{\sin\theta}}$$
$$\frac{\sin\theta}{\sec\theta} \stackrel{?}{=} \frac{1}{\frac{\sin^2\theta + \cos^2\theta}{\sin\theta\cos\theta}}$$
$$\frac{\sin\theta}{\sec\theta} \stackrel{?}{=} \frac{\sin\theta\cos\theta}{\sin^2\theta + \cos^2\theta}$$
$$\frac{\sin\theta}{\sec\theta} \stackrel{?}{=} \frac{\sin\theta\cos\theta}{1}$$
$$\frac{\sin\theta}{\sec\theta} = \frac{\sin\theta}{\sec\theta}$$

9.
$$\frac{\sec\theta + 1}{\tan\theta} \stackrel{?}{=} \frac{\tan\theta}{\sec\theta - 1}$$
$$\frac{\sec\theta + 1}{\tan\theta} \stackrel{?}{=} \frac{\tan\theta}{\sec\theta - 1}\cdot\frac{\sec\theta + 1}{\sec\theta + 1}$$
$$\frac{\sec\theta + 1}{\tan\theta} \stackrel{?}{=} \frac{\tan\theta\cdot(\sec\theta + 1)}{\sec^2\theta - 1}$$
$$\frac{\sec\theta + 1}{\tan\theta} \stackrel{?}{=} \frac{\tan\theta\cdot(\sec\theta + 1)}{\tan^2\theta}$$
$$\frac{\sec\theta + 1}{\tan\theta} = \frac{\sec\theta + 1}{\tan\theta}$$

11.
$$\cos^2\theta + \tan^2\theta\cos^2\theta \stackrel{?}{=} 1$$
$$\cos^2\theta + \frac{\sin^2\theta}{\cos^2\theta}\cdot\cos^2\theta \stackrel{?}{=} 1$$
$$\cos^2\theta + \sin^2\theta \stackrel{?}{=} 1$$
$$1 = 1$$

12.
$$\cot\theta\,(\cot\theta + \tan\theta) \stackrel{?}{=} \csc^2\theta$$
$$\cot^2\theta + \cot\theta\tan\theta \stackrel{?}{=} \csc^2\theta$$
$$\cot^2\theta + \frac{\sin\theta}{\cos\theta}\cdot\frac{\cos\theta}{\sin\theta} \stackrel{?}{=} \csc^2\theta$$
$$\cot^2\theta + 1 \stackrel{?}{=} \csc^2\theta$$
$$\csc^2\theta = \csc^2\theta$$

13.
$$1 + \sec^2\theta\sin^2\theta \stackrel{?}{=} \sec^2\theta$$
$$1 + \frac{1}{\cos^2\theta}\cdot\sin^2\theta \stackrel{?}{=} \sec^2\theta$$
$$1 + \tan^2\theta \stackrel{?}{=} \sec^2\theta$$
$$\sec^2\theta = \sec^2\theta$$

14.
$$\sin\theta\sec\theta\cot\theta \stackrel{?}{=} 1$$
$$\sin\theta\cdot\frac{1}{\cos\theta}\cdot\frac{\cos\theta}{\sin\theta} \stackrel{?}{=} 1$$
$$1 = 1$$

15.
$$\frac{1-\cos\theta}{1+\cos\theta} \stackrel{?}{=} (\csc\theta - \cot\theta)^2$$
$$\frac{1-\cos\theta}{1+\cos\theta} \stackrel{?}{=} \csc^2\theta - 2\cot\theta\csc\theta + \cot^2\theta$$
$$\frac{1-\cos\theta}{1+\cos\theta} \stackrel{?}{=} \frac{1}{\sin^2\theta} - 2\cdot\frac{\cos\theta}{\sin\theta}\cdot\frac{1}{\sin\theta} + \frac{\cos^2\theta}{\sin^2\theta}$$
$$\frac{1-\cos\theta}{1+\cos\theta} \stackrel{?}{=} \frac{1}{\sin^2\theta} - \frac{2\cos\theta}{\sin^2\theta} + \frac{\cos^2\theta}{\sin^2\theta}$$
$$\frac{1-\cos\theta}{1+\cos\theta} \stackrel{?}{=} \frac{1 - 2\cos\theta + \cos^2\theta}{\sin^2\theta}$$
$$\frac{1-\cos\theta}{1+\cos\theta} \stackrel{?}{=} \frac{(1-\cos\theta)(1-\cos\theta)}{1-\cos^2\theta}$$
$$\frac{1-\cos\theta}{1+\cos\theta} \stackrel{?}{=} \frac{(1-\cos\theta)(1-\cos\theta)}{(1-\cos\theta)(1+\cos\theta)}$$
$$\frac{1-\cos\theta}{1+\cos\theta} = \frac{1-\cos\theta}{1+\cos\theta}$$

16.
$$\frac{1-2\cos^2\theta}{\sin\theta\cos\theta} \stackrel{?}{=} \tan\theta - \cot\theta$$
$$\frac{(1-\cos^2\theta)-\cos^2\theta}{\sin\theta\cos\theta} \stackrel{?}{=} \tan\theta - \cot\theta$$
$$\frac{\sin^2\theta-\cos^2\theta}{\sin\theta\cos\theta} \stackrel{?}{=} \tan\theta - \cot\theta$$
$$\frac{\sin^2\theta}{\sin\theta\cos\theta} - \frac{\cos^2\theta}{\sin\theta\cos\theta} \stackrel{?}{=} \tan\theta - \cot\theta$$
$$\frac{\sin\theta}{\cos\theta} - \frac{\cos\theta}{\sin\theta} \stackrel{?}{=} \tan\theta - \cot\theta$$
$$\tan\theta - \cot\theta = \tan\theta - \cot\theta$$

17.
$$\cot\theta\csc\theta \stackrel{?}{=} \frac{\cot\theta + \csc\theta}{\sin\theta + \tan\theta}$$
$$\cot\theta\csc\theta \stackrel{?}{=} \frac{\frac{\cos\theta}{\sin\theta} + \frac{1}{\sin\theta}}{\sin\theta + \frac{\sin\theta}{\cos\theta}}$$
$$\cot\theta\csc\theta \stackrel{?}{=} \frac{\frac{\cos\theta + 1}{\sin\theta}}{\frac{\sin\theta\cos\theta + \sin\theta}{\cos\theta}}$$
$$\cot\theta\csc\theta \stackrel{?}{=} \frac{\frac{\cos\theta + 1}{\sin\theta}}{\frac{\sin\theta\,(\cos\theta + 1)}{\cos\theta}}$$
$$\cot\theta\csc\theta \stackrel{?}{=} \frac{\cos\theta + 1}{\sin\theta}\cdot\frac{\cos\theta}{\sin\theta\,(\cos\theta + 1)}$$
$$\cot\theta\csc\theta \stackrel{?}{=} \frac{\cos\theta}{\sin\theta}\cdot\frac{1}{\sin\theta}$$
$$\cot\theta\csc\theta = \cot\theta\csc\theta$$

18. $$\sin\theta + \cos\theta \stackrel{?}{=} \frac{1+\tan\theta}{\sec\theta}$$
$$\sin\theta + \cos\theta \stackrel{?}{=} \frac{1+\frac{\sin\theta}{\cos\theta}}{\frac{1}{\cos\theta}}$$
$$\sin\theta + \cos\theta \stackrel{?}{=} \frac{\frac{\sin\theta+\cos\theta}{\cos\theta}}{\frac{1}{\cos\theta}}$$
$$\sin\theta + \cos\theta \stackrel{?}{=} \frac{\sin\theta+\cos\theta}{\cos\theta}\cdot\cos\theta$$
$$\sin\theta + \cos\theta = \sin\theta + \cos\theta$$

19. $$\frac{\sec\theta}{\sin\theta} - \frac{\sin\theta}{\cos\theta} \stackrel{?}{=} \cot\theta$$
$$\frac{\frac{1}{\cos\theta}}{\sin\theta} - \frac{\sin\theta}{\cos\theta} \stackrel{?}{=} \cot\theta$$
$$\frac{1}{\sin\theta\cos\theta} - \frac{\sin^2\theta}{\sin\theta\cos\theta} \stackrel{?}{=} \cot\theta$$
$$\frac{1-\sin^2\theta}{\sin\theta\cos\theta} \stackrel{?}{=} \cot\theta$$
$$\frac{\cos^2\theta}{\sin\theta\cos\theta} \stackrel{?}{=} \cot\theta$$
$$\frac{\cos\theta}{\sin\theta} \stackrel{?}{=} \cot\theta$$
$$\cot\theta = \cot\theta$$

20. $$\frac{\sin\theta}{1-\cos\theta} + \frac{1-\cos\theta}{\sin\theta} \stackrel{?}{=} 2\csc\theta$$
$$\frac{\sin\theta}{\sin\theta}\cdot\frac{\sin\theta}{1-\cos\theta} + \frac{1-\cos\theta}{1-\cos\theta}\cdot\frac{1-\cos\theta}{\sin\theta} \stackrel{?}{=} 2\csc\theta$$
$$\frac{\sin^2\theta}{\sin\theta(1-\cos\theta)} + \frac{1-2\cos\theta+\cos^2\theta}{\sin\theta(1-\cos\theta)} \stackrel{?}{=} 2\csc\theta$$
$$\frac{\sin^2\theta+\cos^2\theta+1-2\cos\theta}{\sin\theta(1-\cos\theta)} \stackrel{?}{=} 2\csc\theta$$
$$\frac{2-2\cos\theta}{\sin\theta(1-\cos\theta)} \stackrel{?}{=} 2\csc\theta$$
$$\frac{2(1-\cos\theta)}{\sin\theta(1-\cos\theta)} \stackrel{?}{=} 2\csc\theta$$
$$\frac{2}{\sin\theta} \stackrel{?}{=} 2\csc\theta$$
$$2\csc\theta = 2\csc\theta$$

21. $$\frac{1+\sin\theta}{\sin\theta} \stackrel{?}{=} \frac{\cot^2\theta}{\csc\theta-1}$$
$$\frac{1+\sin\theta}{\sin\theta} \stackrel{?}{=} \frac{\cot^2\theta}{\csc\theta-1}\cdot\frac{\csc\theta+1}{\csc\theta+1}$$
$$\frac{1+\sin\theta}{\sin\theta} \stackrel{?}{=} \frac{\cot^2\theta(\csc\theta+1)}{\csc^2\theta-1}$$
$$\frac{1+\sin\theta}{\sin\theta} \stackrel{?}{=} \frac{\cot^2\theta(\csc\theta+1)}{\cot^2\theta}$$
$$\frac{1+\sin\theta}{\sin\theta} \stackrel{?}{=} \csc\theta+1$$
$$\frac{1+\sin\theta}{\sin\theta} \stackrel{?}{=} \frac{1}{\sin\theta}+\frac{\sin\theta}{\sin\theta}$$
$$\frac{1+\sin\theta}{\sin\theta} = \frac{1+\sin\theta}{\sin\theta}$$

22. $$\frac{1+\tan\theta}{1+\cot\theta} \stackrel{?}{=} \frac{\sin\theta}{\cos\theta}$$
$$\frac{1+\frac{\sin\theta}{\cos\theta}}{1+\frac{\cos\theta}{\sin\theta}} \stackrel{?}{=} \frac{\sin\theta}{\cos\theta}$$
$$\frac{\frac{\sin\theta+\cos\theta}{\cos\theta}}{\frac{\sin\theta+\cos\theta}{\sin\theta}} \stackrel{?}{=} \frac{\sin\theta}{\cos\theta}$$
$$\frac{\sin\theta+\cos\theta}{\cos\theta}\cdot\frac{\sin\theta}{\sin\theta+\cos\theta} \stackrel{?}{=} \frac{\sin\theta}{\cos\theta}$$
$$\frac{\sin\theta}{\cos\theta} = \frac{\sin\theta}{\cos\theta}$$

23. $$\frac{1}{\sec^2\theta} + \frac{1}{\csc^2\theta} \stackrel{?}{=} 1$$
$$\cos^2\theta + \sin^2\theta \stackrel{?}{=} 1$$
$$1 = 1$$

24. $$1 + \frac{1}{\cos\theta} \stackrel{?}{=} \frac{\tan^2\theta}{\sec\theta-1}$$
$$1 + \frac{1}{\cos\theta} \stackrel{?}{=} \frac{\tan^2\theta}{\sec\theta-1}\cdot\frac{\sec\theta+1}{\sec\theta+1}$$
$$1 + \frac{1}{\cos\theta} \stackrel{?}{=} \frac{\tan^2\theta(\sec\theta+1)}{\sec^2\theta-1}$$
$$1 + \frac{1}{\cos\theta} \stackrel{?}{=} \frac{\tan^2\theta(\sec\theta+1)}{\tan^2\theta}$$
$$1 + \frac{1}{\cos\theta} \stackrel{?}{=} \sec\theta+1$$
$$1 + \frac{1}{\cos\theta} = 1 + \frac{1}{\cos\theta}$$

25. $$1-\tan^4\theta \stackrel{?}{=} 2\sec^2\theta - \sec^4\theta$$
$$(1-\tan^2\theta)(1+\tan^2\theta) \stackrel{?}{=} \sec^2\theta(2-\sec^2\theta)$$
$$[(1-(\sec^2\theta-1)](\sec^2\theta) \stackrel{?}{=} (2-\sec^2\theta)(\sec^2\theta)$$
$$(2-\sec^2\theta)(\sec^2\theta) = (2-\sec^2\theta)(\sec^2\theta)$$

26. $$\cos^4\theta - \sin^4\theta \stackrel{?}{=} \cos^2\theta - \sin^2\theta$$
$$(\cos^2\theta-\sin^2\theta)(\cos^2\theta+\sin^2\theta) \stackrel{?}{=} \cos^2\theta - \sin^2\theta$$
$$(\cos^2\theta-\sin^2\theta)\cdot 1 \stackrel{?}{=} \cos^2\theta - \sin^2\theta$$
$$\cos^2\theta - \sin^2\theta = \cos^2\theta - \sin^2\theta$$

27. $$\frac{1-\cos\theta}{\sin\theta} \stackrel{?}{=} \frac{\sin\theta}{1+\cos\theta}$$
$$\frac{1-\cos\theta}{\sin\theta}\cdot\frac{1+\cos\theta}{1+\cos\theta} \stackrel{?}{=} \frac{\sin\theta}{1+\cos\theta}$$
$$\frac{1-\cos^2\theta}{\sin\theta(1+\cos\theta)} \stackrel{?}{=} \frac{\sin\theta}{1+\cos\theta}$$
$$\frac{\sin^2\theta}{\sin\theta(1+\cos\theta)} \stackrel{?}{=} \frac{\sin\theta}{1+\cos\theta}$$
$$\frac{\sin\theta}{1+\cos\theta} = \frac{\sin\theta}{1+\cos\theta}$$

28. $$\frac{\cos\theta}{1+\sin\theta} + \frac{\cos\theta}{1-\sin\theta} \stackrel{?}{=} 2\sec\theta$$
$$\frac{\cos\theta}{1+\sin\theta}\cdot\frac{1-\sin\theta}{1-\sin\theta} + \frac{\cos\theta}{1-\sin\theta}\cdot\frac{1+\sin\theta}{1+\sin\theta} \stackrel{?}{=} 2\sec\theta$$
$$\frac{\cos\theta(1-\sin\theta)+\cos\theta(1+\sin\theta)}{(1+\sin\theta)(1-\sin\theta)} \stackrel{?}{=} 2\sec\theta$$
$$\frac{\cos\theta - \sin\theta\cos\theta + \cos\theta + \sin\theta\cos\theta}{1-\sin^2\theta} \stackrel{?}{=} 2\sec\theta$$
$$\frac{2\cos\theta}{\cos^2\theta} \stackrel{?}{=} 2\sec\theta$$
$$\frac{2}{\cos\theta} \stackrel{?}{=} 2\sec\theta$$
$$2\sec\theta = 2\sec\theta$$

29. $\tan\theta \sin\theta \cos\theta \csc^2\theta \stackrel{?}{=} 1$

$$\frac{\sin\theta}{\cos\theta} \cdot \sin\theta \cdot \cos\theta \cdot \frac{1}{\sin^2\theta} \stackrel{?}{=} 1$$

$$1 = 1$$

30.

$$\frac{\sin^2\theta}{1 - \cos\theta} \stackrel{?}{=} 1 + \cos\theta$$

$$\frac{\sin^2\theta}{1 - \cos\theta} \cdot \frac{1 + \cos\theta}{1 + \cos\theta} \stackrel{?}{=} 1 + \cos\theta$$

$$\frac{\sin^2\theta\,(1 + \cos\theta)}{1 - \cos^2\theta} \stackrel{?}{=} 1 + \cos\theta$$

$$\frac{\sin^2\theta\,(1 + \cos\theta)}{\sin^2\theta} \stackrel{?}{=} 1 + \cos\theta$$

$$1 + \cos\theta = 1 + \cos\theta$$

31.

$$\frac{v^2 \tan^2\theta}{2g \sec^2\theta} \stackrel{?}{=} \frac{v^2 \frac{\sin^2\theta}{\cos^2\theta}}{2g \frac{1}{\cos^2\theta}}$$

$$\stackrel{?}{=} \frac{v^2}{2g} \cdot \frac{\sin^2\theta}{\cos^2\theta} \cdot \frac{\cos^2\theta}{1}$$

$$= \frac{v^2 \sin^2\theta}{2g}$$

34. Sample answer: Trigonometric identities are verified in a similar manner to proving theorems in geometry before using them. Answers should include the following.

- The expressions have not yet been shown to be equal, so you could not use the properties of equality on them.
- To show two expressions you must transform one, or both independently.
- Graphing two expressions could result in identical graphs for a set interval, that are different elsewhere.

37.

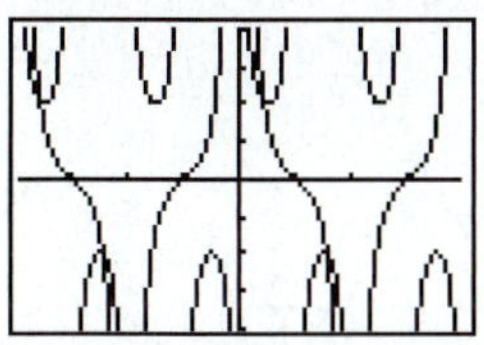

[−360, 360] scl: 90 by [−5, 5] scl: 1

38.

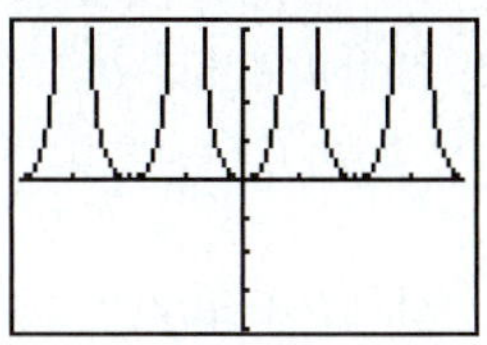

[−360, 360] scl: 90 by [−5, 5] scl: 1

39.

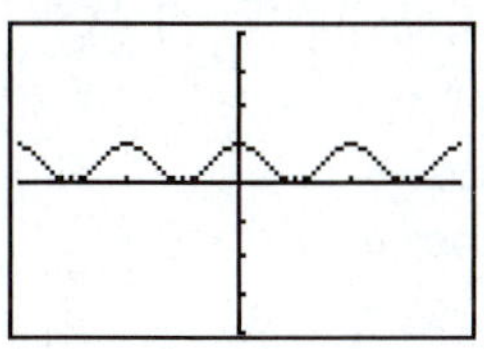

[−360, 360] scl: 90 by [−5, 5] scl: 1

40.

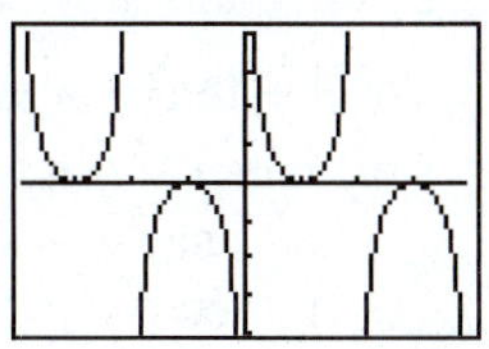

[−360, 360] scl: 90 by [−5, 5] scl: 1

41.

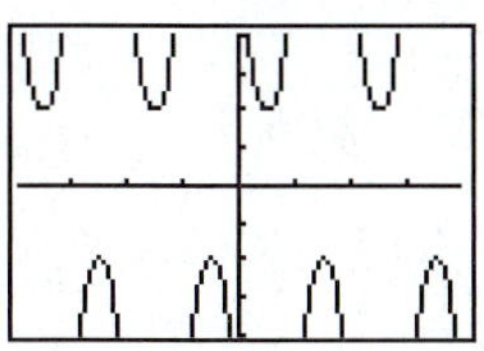

[−360, 360] scl: 90 by [−5, 5] scl: 1

42.

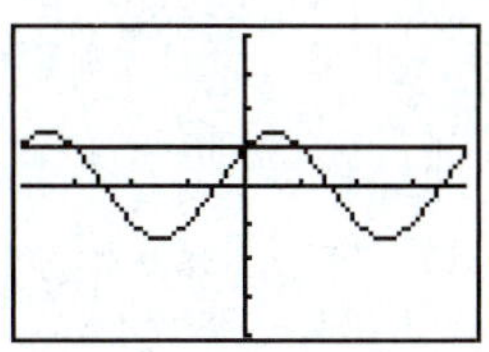

[−360, 360] scl: 90 by [−5, 5] scl: 1

47.

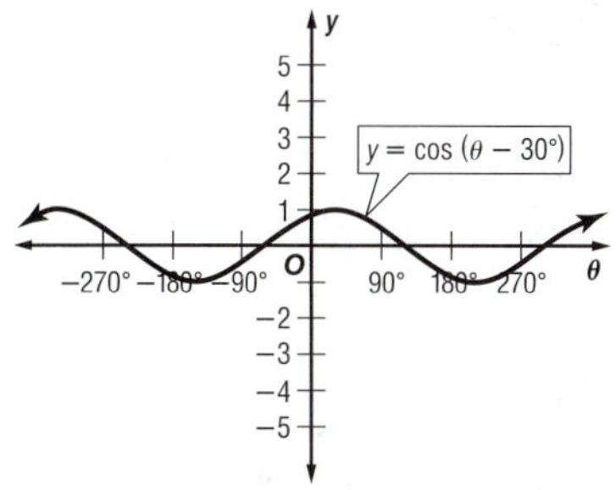

48.

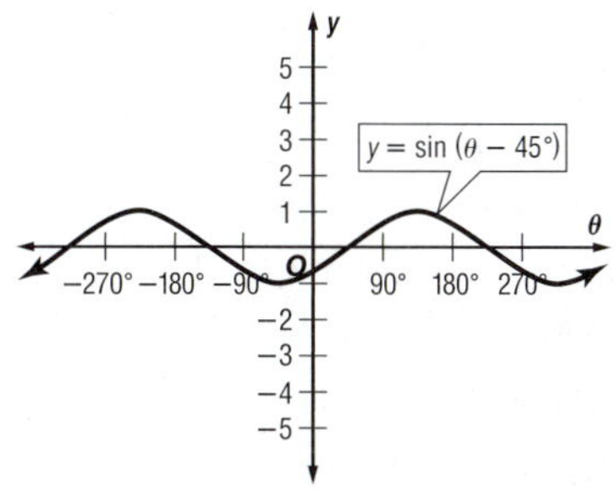

49.

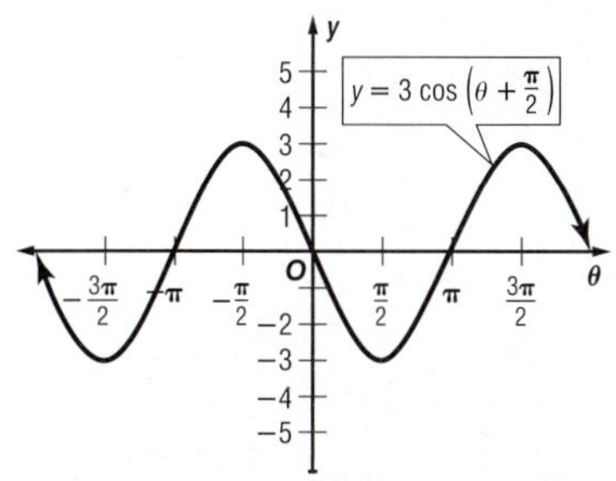

Pages 788–790, Lesson 14-5

10. $\cos(270° - \theta) \stackrel{?}{=} \cos 270° \cos\theta + \sin 270° \sin\theta$

$\stackrel{?}{=} 0 + (-1 \sin\theta)$

$= -\sin\theta$

11.

$$\sin\left(\theta + \frac{\pi}{2}\right) \stackrel{?}{=} \cos\theta$$

$$\sin\theta \cos\frac{\pi}{2} + \cos\theta \sin\frac{\pi}{2} \stackrel{?}{=} \cos\theta$$

$$\sin\theta \cdot 0 + \cos\theta \cdot 1 \stackrel{?}{=} \cos\theta$$

$$\cos\theta = \cos\theta$$

12. $\sin(\theta + 30°) + \cos(\theta + 60°)$

$\stackrel{?}{=} \sin\theta \cos 30° + \cos\theta \sin 30° + \cos\theta \cos 60° - \sin\theta \sin 60°$

$\stackrel{?}{=} \frac{\sqrt{3}}{2}\sin\theta + \frac{1}{2}\cos\theta + \frac{1}{2}\cos\theta - \frac{\sqrt{3}}{2}\sin\theta$

$\stackrel{?}{=} \frac{1}{2}\cos\theta + \frac{1}{2}\cos\theta$

$= \cos\theta$

28. $\sin(270° - \theta) \stackrel{?}{=} \sin 270° \cos\theta - \cos 270° \sin\theta$

$\stackrel{?}{=} -1 \cos\theta - 0$

$= -\cos\theta$

29. $\cos(90° + \theta) \stackrel{?}{=} \cos 90° \cos\theta - \sin 90° \sin\theta$

$\stackrel{?}{=} 0 - 1 \sin\theta$

$= -\sin\theta$

30. $\cos(90° - \theta) \stackrel{?}{=} \cos 90° \cos\theta + \sin 90° \sin\theta$

$\stackrel{?}{=} 0 \cdot \cos\theta + 1 \cdot \sin\theta$

$= \sin\theta$

31.
$$\begin{aligned}\sin(90^\circ - \theta) &\stackrel{?}{=} \cos\theta\\ \sin 90^\circ \cos\theta - \cos 90^\circ \sin\theta &\stackrel{?}{=} \cos\theta\\ 1\cdot\cos\theta - 0\cdot\sin\theta &\stackrel{?}{=} \cos\theta\\ \cos\theta - 0 &\stackrel{?}{=} \cos\theta\\ \cos\theta &= \cos\theta\end{aligned}$$

32.
$$\begin{aligned}\sin\left(\theta + \frac{3\pi}{2}\right) &\stackrel{?}{=} -\cos\theta\\ \sin\theta\cos\frac{3\pi}{2} + \cos\theta\sin\frac{3\pi}{2} &\stackrel{?}{=} -\cos\theta\\ \sin\theta\cdot 0 + \cos\theta\cdot(-1) &\stackrel{?}{=} -\cos\theta\\ 0 + (-\cos\theta) &\stackrel{?}{=} -\cos\theta\\ -\cos\theta &= -\cos\theta\end{aligned}$$

33.
$$\begin{aligned}\cos(\pi - \theta) &\stackrel{?}{=} -\cos\theta\\ \cos\pi\cos\theta + \sin\pi\sin\theta &\stackrel{?}{=} -\cos\theta\\ -1\cdot\cos\theta + 0\cdot\sin\theta &\stackrel{?}{=} -\cos\theta\\ -\cos\theta &= -\cos\theta\end{aligned}$$

34.
$$\begin{aligned}\cos(2\pi + \theta) &\stackrel{?}{=} \cos\theta\\ \cos 2\pi\cos\theta - [\sin 2\pi\sin\theta] &\stackrel{?}{=} \cos\theta\\ 1\cdot\cos\theta - [0\cdot\sin\theta] &\stackrel{?}{=} \cos\theta\\ 1\cdot\cos\theta - 0 &\stackrel{?}{=} \cos\theta\\ \cos\theta &= \cos\theta\end{aligned}$$

35.
$$\begin{aligned}\sin(\pi - \theta) &\stackrel{?}{=} \sin\theta\\ \sin\pi\cos\theta - [\cos\pi\sin\theta] &\stackrel{?}{=} \sin\theta\\ 0\cdot\cos\theta - [-1\cdot\sin\theta] &\stackrel{?}{=} \sin\theta\\ 0 - [-\sin\theta] &\stackrel{?}{=} \sin\theta\\ \sin\theta &= \sin\theta\end{aligned}$$

36. $\sin(60^\circ + \theta) + \sin(60^\circ - \theta)$
$$\begin{aligned}&\stackrel{?}{=} \sin 60^\circ\cos\theta + \cos 60^\circ\sin\theta + \sin 60^\circ\cos\theta - \cos 60^\circ\sin\theta\\ &\stackrel{?}{=} \frac{\sqrt{3}}{2}\cos\theta + \frac{1}{2}\sin\theta + \frac{\sqrt{3}}{2}\cos\theta - \frac{1}{2}\sin\theta\\ &= \sqrt{3}\cos\theta\end{aligned}$$

37. $\sin\left(\theta + \frac{\pi}{3}\right) - \cos\left(\theta + \frac{\pi}{6}\right)$
$$\begin{aligned}&\stackrel{?}{=} \sin\theta\cos\frac{\pi}{3} + \cos\theta\sin\frac{\pi}{3} - \cos\theta\cos\frac{\pi}{6} + \sin\theta\sin\frac{\pi}{6}\\ &\stackrel{?}{=} \frac{1}{2}\sin\theta + \frac{\sqrt{3}}{2}\cos\theta - \frac{\sqrt{3}}{2}\cos\theta + \frac{1}{2}\sin\theta\\ &\stackrel{?}{=} \frac{1}{2}\sin\theta + \frac{1}{2}\sin\theta\\ &= \sin\theta\end{aligned}$$

38. $\sin(\alpha + \beta)\sin(\alpha - \beta) \stackrel{?}{=} \sin^2\alpha - \sin^2\beta$
$$\begin{aligned}&\stackrel{?}{=} (\sin\alpha\cos\beta + \cos\alpha\sin\beta)(\sin\alpha\cos\beta - \cos\alpha\sin\beta)\\ &\stackrel{?}{=} \sin^2\alpha\cos^2\beta - \cos^2\alpha\sin^2\beta\\ &\stackrel{?}{=} \sin^2\alpha(1 - \sin^2\beta) - (1 - \sin^2\alpha)\sin^2\beta\\ &\stackrel{?}{=} \sin^2\alpha - \sin^2\alpha\sin^2\beta - \sin^2\beta + \sin^2\alpha\sin^2\beta\\ &= \sin^2\alpha - \sin^2\beta\end{aligned}$$

39.
$$\begin{aligned}\cos(\alpha + \beta) &\stackrel{?}{=} \frac{1 - \tan\alpha\tan\beta}{\sec\alpha\sec\beta}\\ \cos(\alpha + \beta) &\stackrel{?}{=} \frac{1 - \frac{\sin\alpha}{\cos\alpha}\cdot\frac{\sin\beta}{\cos\beta}}{\frac{1}{\cos\alpha}\cdot\frac{1}{\cos\beta}}\\ \cos(\alpha + \beta) &\stackrel{?}{=} \frac{1 - \frac{\sin\alpha}{\cos\alpha}\cdot\frac{\sin\beta}{\cos\beta}}{\frac{1}{\cos\alpha}\cdot\frac{1}{\cos\beta}}\cdot\frac{\cos\alpha\cos\beta}{\cos\alpha\cos\beta}\\ \cos(\alpha + \beta) &\stackrel{?}{=} \frac{\cos\alpha\cos\beta - \sin\alpha\sin\beta}{1}\\ \cos(\alpha + \beta) &= \cos(\alpha + \beta)\end{aligned}$$

46.
$$\begin{aligned}\tan(\alpha + \beta) &\stackrel{?}{=} \frac{\sin(\alpha + \beta)}{\cos(\alpha + \beta)}\\ &\stackrel{?}{=} \frac{\sin\alpha\cos\beta + \cos\alpha\sin\beta}{\cos\alpha\cos\beta - \sin\alpha\sin\beta}\\ &\stackrel{?}{=} \frac{\frac{\sin\alpha\cos\beta}{\cos\alpha\cos\beta} + \frac{\cos\alpha\sin\beta}{\cos\alpha\cos\beta}}{\frac{\cos\alpha\cos\beta}{\cos\alpha\cos\beta} - \frac{\sin\alpha\sin\beta}{\cos\alpha\cos\beta}}\\ &= \frac{\tan\alpha + \tan\beta}{1 - \tan\alpha\tan\beta}\end{aligned}$$

$$\begin{aligned}\tan(\alpha - \beta) &\stackrel{?}{=} \frac{\sin(\alpha - \beta)}{\cos(\alpha - \beta)}\\ &\stackrel{?}{=} \frac{\sin\alpha\cos\beta - \cos\alpha\sin\beta}{\cos\alpha\cos\beta + \sin\alpha\sin\beta}\\ &\stackrel{?}{=} \frac{\frac{\sin\alpha\cos\beta}{\cos\alpha\cos\beta} - \frac{\cos\alpha\sin\beta}{\cos\alpha\cos\beta}}{\frac{\cos\alpha\cos\beta}{\cos\alpha\cos\beta} + \frac{\sin\alpha\sin\beta}{\cos\alpha\cos\beta}}\\ &= \frac{\tan\alpha - \tan\beta}{1 + \tan\alpha\tan\beta}\end{aligned}$$

50.
$$\begin{aligned}\cot\theta + \sec\theta &\stackrel{?}{=} \frac{\cos^2\theta + \sin\theta}{\sin\theta\cos\theta}\\ \cot\theta + \sec\theta &\stackrel{?}{=} \frac{\cos^2\theta}{\sin\theta\cos\theta} + \frac{\sin\theta}{\sin\theta\cos\theta}\\ \cot\theta + \sec\theta &\stackrel{?}{=} \frac{\cos\theta}{\sin\theta} + \frac{1}{\cos\theta}\\ \cot\theta + \sec\theta &= \cot\theta + \sec\theta\end{aligned}$$

51.
$$\begin{aligned}\sin^2\theta + \tan^2\theta &\stackrel{?}{=} (1 - \cos^2\theta) + \frac{\sec^2\theta}{\csc^2\theta}\\ \sin^2\theta + \tan^2\theta &\stackrel{?}{=} \sin^2\theta + \frac{\sec^2\theta}{\csc^2\theta}\\ \sin^2\theta + \tan^2\theta &\stackrel{?}{=} \sin^2\theta + \frac{1}{\cos^2\theta} \div \frac{1}{\sin^2\theta}\\ \sin^2\theta + \tan^2\theta &\stackrel{?}{=} \sin^2\theta + \frac{\sin^2\theta}{\cos^2\theta}\\ \sin^2\theta + \tan^2\theta &= \sin^2\theta + \tan^2\theta\end{aligned}$$

52.
$$\begin{aligned}\sin\theta(\sin\theta + \csc\theta) &\stackrel{?}{=} 2 - \cos^2\theta\\ \sin^2\theta + 1 &\stackrel{?}{=} 2 - \cos^2\theta\\ 1 - \cos^2\theta + 1 &\stackrel{?}{=} 2 - \cos^2\theta\\ 2 - \cos^2\theta &= 2 - \cos^2\theta\end{aligned}$$

53.
$$\begin{aligned}\frac{\sec\theta}{\tan\theta} &\stackrel{?}{=} \csc\theta\\ \frac{1}{\cos\theta} \div \frac{\sin\theta}{\cos\theta} &\stackrel{?}{=} \csc\theta\\ \frac{1}{\cos\theta}\cdot\frac{\cos\theta}{\sin\theta} &\stackrel{?}{=} \csc\theta\\ \frac{1}{\sin\theta} &\stackrel{?}{=} \csc\theta\\ \csc\theta &= \csc\theta\end{aligned}$$

Pages 795–797, Lesson 14-6

31.
$$\begin{aligned}\sin 2x &\stackrel{?}{=} 2\cot x\sin^2 x\\ 2\sin x\cos x &\stackrel{?}{=} 2\cdot\frac{\cos x}{\sin x}\cdot\sin^2 x\\ 2\sin x\cos x &= 2\sin x\cos x\end{aligned}$$

32. $2\cos^2\frac{x}{2} \stackrel{?}{=} 1 + \cos x$

$2\left(\pm\sqrt{\frac{1+\cos x}{2}}\right)^2 \stackrel{?}{=} 1 + \cos x$

$2\left(\frac{1+\cos x}{2}\right) \stackrel{?}{=} 1 + \cos x$

$1 + \cos x = 1 + \cos x$

33. $\sin^4 x - \cos^4 x \stackrel{?}{=} 2\sin^2 x - 1$

$(\sin^2 x - \cos^2 x)(\sin^2 x + \cos^2 x) \stackrel{?}{=} 2\sin^2 x - 1$

$(\sin^2 x - \cos^2 x) \cdot 1 \stackrel{?}{=} 2\sin^2 x - 1$

$[\sin^2 x - (1 - \sin^2 x)] \cdot 1 \stackrel{?}{=} 2\sin^2 x - 1$

$\sin^2 x - 1 + \sin^2 x \stackrel{?}{=} 2\sin^2 x - 1$

$2\sin^2 x - 1 = 2\sin^2 x - 1$

34. $\sin^2 x \stackrel{?}{=} \frac{1}{2}(1 - \cos 2x)$

$\sin^2 x \stackrel{?}{=} \frac{1}{2}[1 - (1 - 2\sin^2 x)]$

$\sin^2 x \stackrel{?}{=} \frac{1}{2}(2\sin^2 x)$

$\sin^2 x = \sin^2 x$

35. $\tan^2\frac{x}{2} \stackrel{?}{=} \frac{1-\cos x}{1+\cos x}$

$\frac{\sin^2\frac{x}{2}}{\cos^2\frac{x}{2}} \stackrel{?}{=} \frac{1-\cos x}{1+\cos x}$

$\frac{\left(\pm\sqrt{\frac{1-\cos x}{2}}\right)^2}{\left(\pm\sqrt{\frac{1+\cos x}{2}}\right)^2} \stackrel{?}{=} \frac{1-\cos x}{1+\cos x}$

$\frac{1-\cos x}{1+\cos x} = \frac{1-\cos x}{1+\cos x}$

36. $\frac{1}{\sin x\cos x} - \frac{\cos x}{\sin x} \stackrel{?}{=} \tan x$

$\frac{1-\cos^2 x}{\sin x\cos x} \stackrel{?}{=} \tan x$

$\frac{\sin^2 x}{\sin x\cos x} \stackrel{?}{=} \tan x$

$\frac{\sin x}{\cos x} \stackrel{?}{=} \tan x$

$\tan x = \tan x$

40. $\frac{2}{g}v^2(\tan\theta - \tan\theta\sin^2\theta) \stackrel{?}{=} \frac{2}{g}v^2\tan\theta\,(1 - \sin^2\theta)$

$\stackrel{?}{=} \frac{2}{g}v^2\tan\theta\cos^2\theta$

$\stackrel{?}{=} \frac{2}{g}v^2\sin\theta\cos\theta$

$= \frac{v^2\sin 2\theta}{g}$

42.

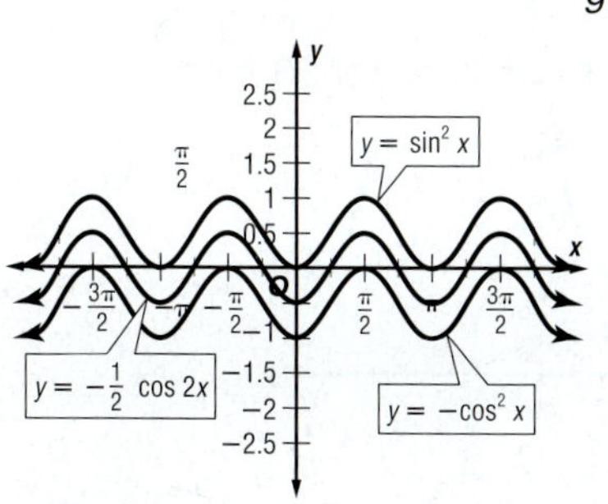

Sample answer: They all have the same shape and are vertical translations of each other.

43. The maxima occur at $\pm\frac{\pi}{2}$ and $\pm\frac{3\pi}{2}$. The minima occur at $x = 0$, $\pm\pi$, and $\pm 2\pi$.

44.

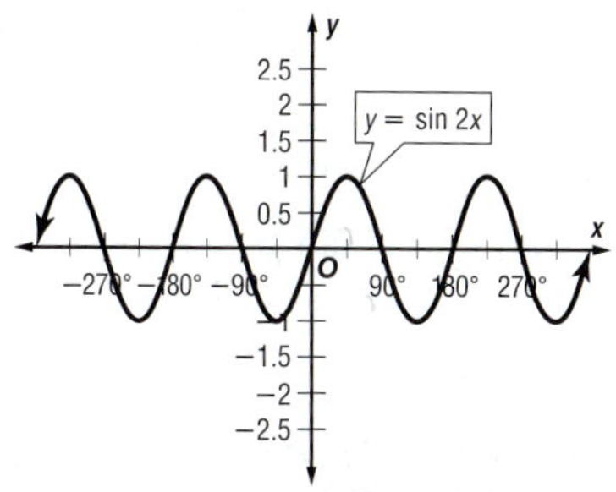

45. The graph of $f(x)$ crosses the x-axis at the points specified in Exercise 43.

46. $c = 1$ and $d = 0.5$

56. $\cot^2\theta - \sin^2\theta \stackrel{?}{=} \frac{\cos^2\theta\csc^2\theta - \sin^2\theta}{\sin^2\theta\csc^2\theta}$

$\cot^2\theta - \sin^2\theta \stackrel{?}{=} \frac{\cos^2\theta\,\frac{1}{\sin^2\theta} - \sin^2\theta}{\sin^2\theta\,\frac{1}{\sin^2\theta}}$

$\cot^2\theta - \sin^2\theta \stackrel{?}{=} \frac{\cot^2\theta - \sin^2\theta}{1}$

$\cot^2\theta - \sin^2\theta = \cot^2\theta - \sin^2\theta$

57. $\cos\theta\,(\cos\theta + \cot\theta) \stackrel{?}{=} \cot\theta\cos\theta\,(\sin\theta + 1)$

$\cos\theta\,(\cos\theta + \cot\theta) \stackrel{?}{=} \frac{\cos\theta}{\sin\theta}\cos\theta\sin\theta + \cot\theta\cos\theta$

$\cos\theta\,(\cos\theta + \cot\theta) \stackrel{?}{=} \cos^2\theta + \cot\theta\cos\theta$

$\cos\theta\,(\cos\theta + \cot\theta) = \cos\theta\,(\cos\theta + \cot\theta)$

Page 797, Practice Quiz 2

1. $\sin\theta\sec\theta \stackrel{?}{=} \tan\theta$

$\sin\theta \cdot \frac{1}{\cos\theta} \stackrel{?}{=} \tan\theta$

$\frac{\sin\theta}{\cos\theta} \stackrel{?}{=} \tan\theta$

$\tan\theta = \tan\theta$

2. $\sec\theta - \cos\theta \stackrel{?}{=} \sin\theta\tan\theta$

$\frac{1}{\cos\theta} - \cos\theta \cdot \frac{\cos\theta}{\cos\theta} \stackrel{?}{=} \sin\theta\tan\theta$

$\frac{1}{\cos\theta} - \frac{\cos^2\theta}{\cos\theta} \stackrel{?}{=} \sin\theta\tan\theta$

$\frac{1-\cos^2\theta}{\cos\theta} \stackrel{?}{=} \sin\theta\tan\theta$

$\frac{\sin^2\theta}{\cos\theta} \stackrel{?}{=} \sin\theta\tan\theta$

$\sin\theta\,\frac{\sin\theta}{\cos\theta} \stackrel{?}{=} \sin\theta\tan\theta$

$\sin\theta\tan\theta = \sin\theta\tan\theta$

3. $\sin\theta + \tan\theta \stackrel{?}{=} \frac{\sin\theta\,(\cos\theta + 1)}{\cos\theta}$

$\sin\theta + \tan\theta \stackrel{?}{=} \frac{\sin\theta\cos\theta + \sin\theta}{\cos\theta}$

$\sin\theta + \tan\theta \stackrel{?}{=} \frac{\sin\theta\cos\theta}{\cos\theta} + \frac{\sin\theta}{\cos\theta}$

$\sin\theta + \tan\theta = \sin\theta + \tan\theta$

4. $\sin(90° + \theta) \stackrel{?}{=} \cos\theta$

$\sin 90°\cos\theta + \cos 90°\sin\theta \stackrel{?}{=} \cos\theta$

$\cos\theta + 0 \stackrel{?}{=} \cos\theta$

$\cos\theta = \cos\theta$

5.
$$\cos\left(\frac{3\pi}{2} - \theta\right) \stackrel{?}{=} -\sin\theta$$
$$\cos\frac{3\pi}{2}\cos\theta + \sin\frac{3\pi}{2}\sin\theta \stackrel{?}{=} -\sin\theta$$
$$0 + (-1 \cdot \sin\theta) \stackrel{?}{=} -\sin\theta$$
$$-\sin\theta = -\sin\theta$$

6. $\sin(\theta + 30°) + \cos(\theta + 60°)$
$$\stackrel{?}{=} (\sin\theta\cos 30° + \cos\theta\sin 30°) + (\cos\theta\cos 60° - \sin\theta\sin 60°)$$
$$\stackrel{?}{=} \left(\frac{\sqrt{3}}{2}\sin\theta + \frac{1}{2}\cos\theta\right) + \left(\frac{1}{2}\cos\theta - \frac{\sqrt{3}}{2}\sin\theta\right)$$
$$\stackrel{?}{=} \frac{1}{2}\cos\theta + \frac{1}{2}\cos\theta$$
$$= \cos\theta$$

Page 798, Preview of Lesson 14-7 Graphing Calculator Investigation

1.

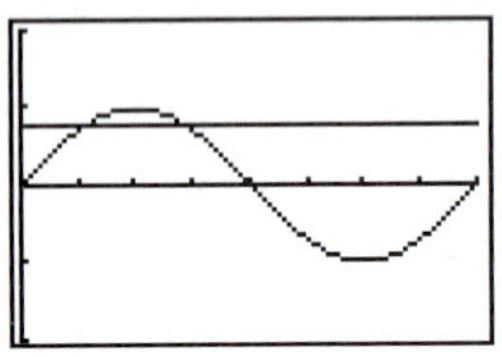

[0, 360] scl: 45 by [−2, 2] scl: 1

2.

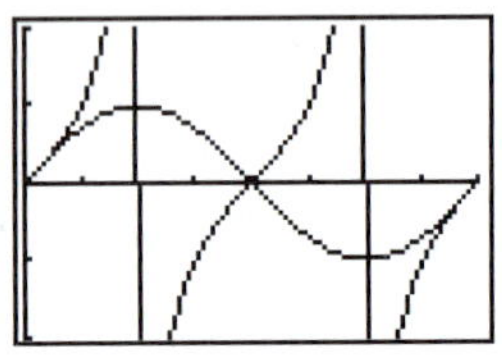

[0, 360] scl: 45 by [−2, 2] scl: 1

3.

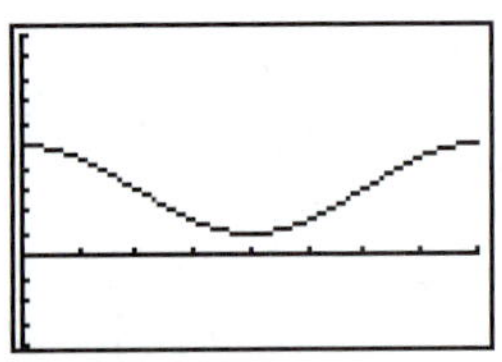

[0, 360] scl: 45 by [−5, 10] scl: 1

4.

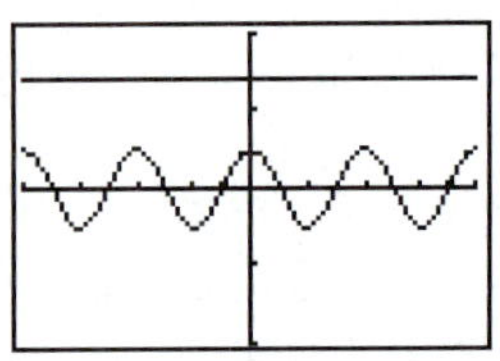

[−720, 720] scl: 90 by [−2, 2] scl: 1

5.

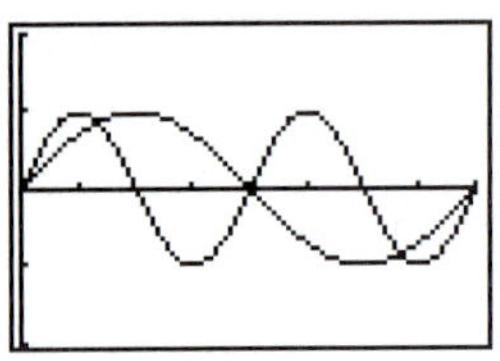

[0, 360] scl: 45 by [−2, 2] scl: 1

6.

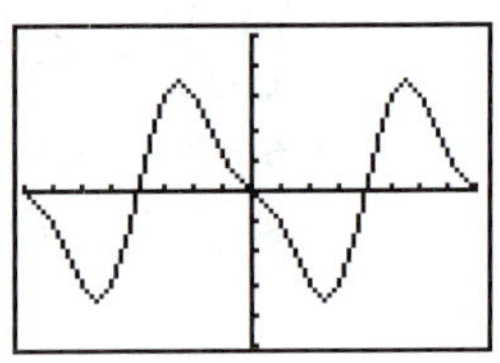

[−360, 360] scl: 45 by [−5, 5] scl: 1

Page 803, Lesson 14-7

35. $0 + 2k\pi, \frac{\pi}{2} + 2k\pi, \frac{3\pi}{2} + 2k\pi$ or $0° + k \cdot 360°$, $90° + k \cdot 360°, 270° + k \cdot 360°$

36. $\frac{7\pi}{6} + 2k\pi, \frac{11\pi}{6} + 2k\pi$ or $210° + k \cdot 360°$, $330° + k \cdot 360°$

37. $0 + k\pi$ or $0° + k \cdot 180°$

38. $\frac{\pi}{2} + k\pi, \frac{2\pi}{3} + 2k\pi, \frac{4\pi}{3} + 2k\pi$ or $90° + k \cdot 180°$, $120° + k \cdot 360°, 240° + k \cdot 360°$

39. $0 + 2k\pi, \frac{\pi}{3} + 2k\pi, \frac{5\pi}{3} + 2k\pi$ or $0° + k \cdot 360°$, $60° + k \cdot 360°, 300° + k \cdot 360°$

40. $\frac{\pi}{2} + 4k\pi$ or $90° + k \cdot 720°$

43. $y = \frac{3}{2} + \frac{3}{2}\sin(\pi t)$

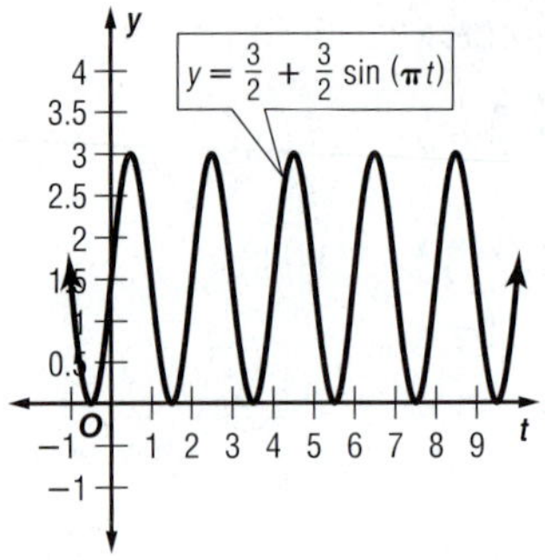

Pages 805–807, Chapter 14 Study Guide and Review

9. amplitude: $\frac{1}{2}$; period: 360° or 2π

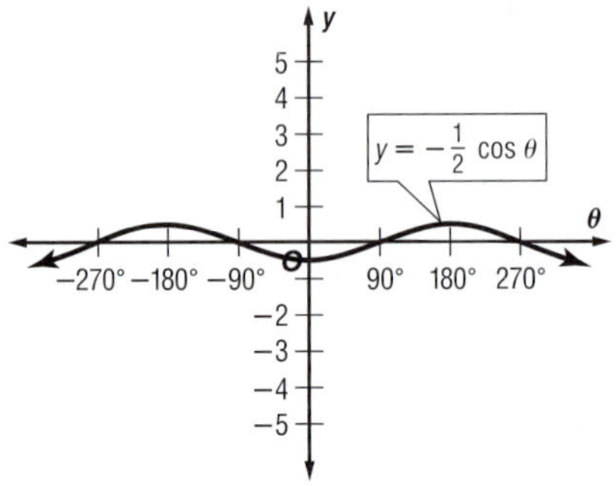

10. amplitude: 4; period: 180° or π

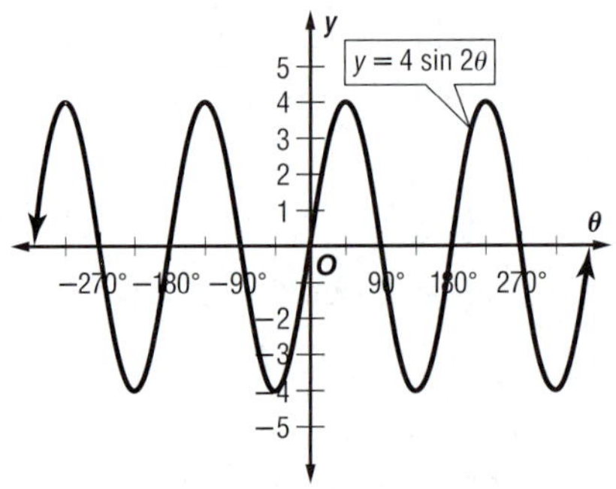

11. amplitude: 1; period: 720° or 4π

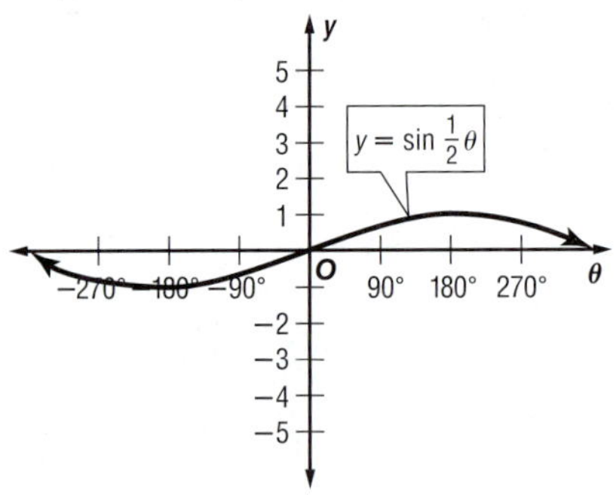

12. amplitude: does not exist; period: 360° or 2π

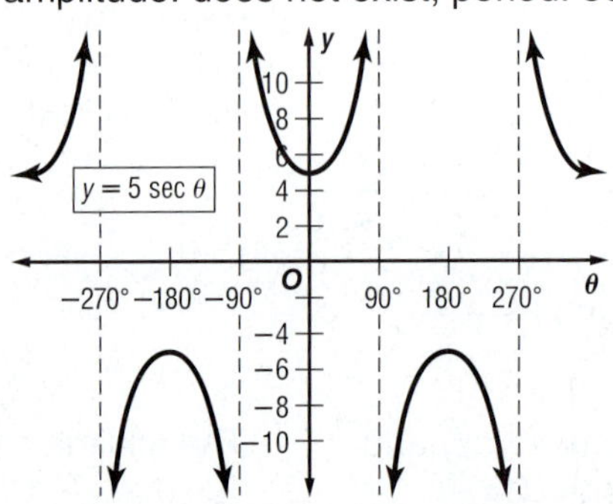

13. amplitude: does not exist; period: 540° or 3π

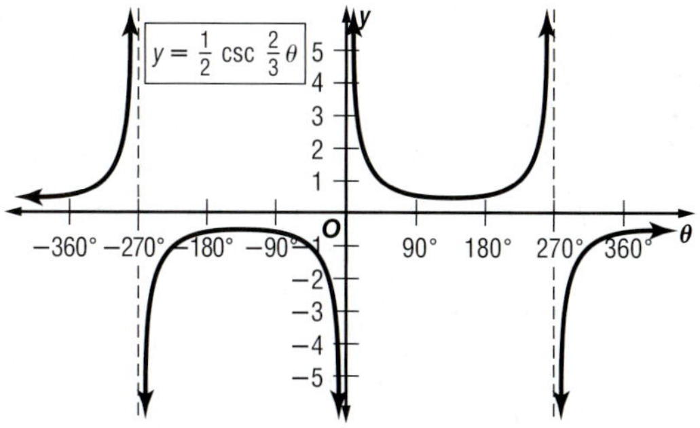

14. amplitude: does not exist; period: 45° or $\frac{\pi}{4}$

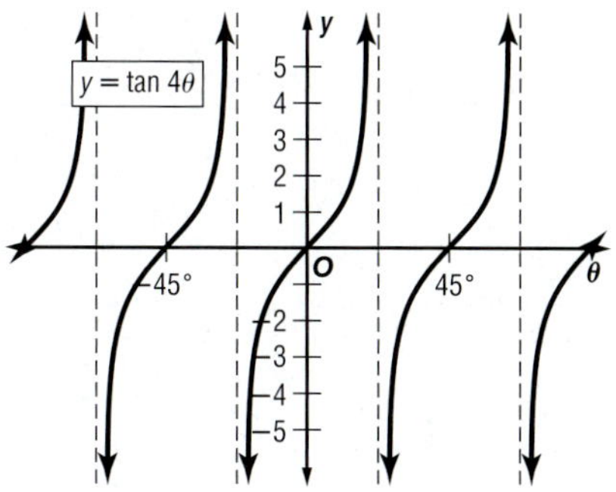

34.
$$\cos(90° + \theta) \stackrel{?}{=} -\sin\theta$$
$$\cos 90° \cos\theta - \sin 90° \sin\theta \stackrel{?}{=} -\sin\theta$$
$$0 \cdot \cos\theta - 1 \cdot \sin\theta \stackrel{?}{=} -\sin\theta$$
$$-\sin\theta = -\sin\theta$$

35.
$$\sin(30° - \theta) \stackrel{?}{=} \cos(60° + \theta)$$
$$\sin 30° \cos\theta - \cos 30° \sin\theta \stackrel{?}{=} \cos 60° \cos\theta - \sin 60° \sin\theta$$
$$\frac{1}{2}\cos\theta - \frac{\sqrt{3}}{2}\sin\theta = \frac{1}{2}\cos\theta - \frac{\sqrt{3}}{2}\sin\theta$$

36.
$$\sin(\theta + \pi) \stackrel{?}{=} -\sin\theta$$
$$\sin\theta\cos\pi + \cos\theta\sin\pi \stackrel{?}{=} -\sin\theta$$
$$(\sin\theta)(-1) + (\cos\theta)(0) \stackrel{?}{=} -\sin\theta$$
$$-\sin\theta = -\sin\theta$$

37.
$$-\cos\theta \stackrel{?}{=} \cos(\pi + \theta)$$
$$-\cos\theta \stackrel{?}{=} \cos\pi\cos\theta - \sin\pi\sin\theta$$
$$-\cos\theta \stackrel{?}{=} -1 \cdot \cos\theta - 0 \cdot \sin\theta$$
$$-\cos\theta = -\cos\theta$$

Page 809, Chapter 14 Practice Test

9.
$$(\sin\theta - \cos\theta)^2 \stackrel{?}{=} 1 - \sin 2\theta$$
$$\sin^2\theta - 2\sin\theta\cos\theta + \cos^2\theta \stackrel{?}{=} 1 - \sin 2\theta$$
$$(\sin^2\theta + \cos^2\theta) - 2\sin\theta\cos\theta \stackrel{?}{=} 1 - \sin 2\theta$$
$$1 - \sin 2\theta = 1 - \sin 2\theta$$

10.
$$\frac{\cos\theta}{1 - \sin^2\theta} \stackrel{?}{=} \sec\theta$$
$$\frac{\cos\theta}{\cos^2\theta} \stackrel{?}{=} \sec\theta$$
$$\frac{1}{\cos\theta} \stackrel{?}{=} \sec\theta$$
$$\sec\theta = \sec\theta$$

11.
$$\frac{\sec\theta}{\sin\theta} - \frac{\sin\theta}{\cos\theta} \stackrel{?}{=} \cot\theta$$
$$\frac{1}{\sin\theta\cos\theta} - \frac{\sin\theta}{\cos\theta} \stackrel{?}{=} \cot\theta$$
$$\frac{1}{\sin\theta\cos\theta} - \frac{\sin^2\theta}{\sin\theta\cos\theta} \stackrel{?}{=} \cot\theta$$
$$\frac{\cos^2\theta}{\sin\theta\cos\theta} \stackrel{?}{=} \cot\theta$$
$$\frac{\cos\theta}{\sin\theta} \stackrel{?}{=} \cot\theta$$
$$\cot\theta = \cot\theta$$

12.
$$\frac{1 + \tan^2\theta}{\cos^2\theta} \stackrel{?}{=} \sec^4\theta$$
$$\frac{\sec^2\theta}{\cos^2\theta} \stackrel{?}{=} \sec^4\theta$$
$$\sec^2\theta\sec^2\theta \stackrel{?}{=} \sec^4\theta$$
$$\sec^4\theta = \sec^4\theta$$

Student Handbook

Skills

Prerequisite Skills

1 Comparing and Ordering Real Numbers **814**
2 Factoring Polynomials **815**
3 Congruent and Similar Figures **817**
4 Pythagorean Theorem **820**
5 Mean, Median, and Mode **822**
6 Bar and Line Graphs **824**
7 Stem-and-Leaf Plots **825**
8 Box-and-Whisker Plots **826**

Extra Practice **828**

Mixed Problem Solving **862**

Preparing for Standardized Tests

Becoming a Better Test-Taker **877**
Multiple-Choice Questions and Practice **878**
Gridded-Response Questions and Practice **880**
Short-Response Questions and Practice **884**
Extended-Response Questions and Practice **888**

Reference

English-Spanish Glossary **R1**

Selected Answers **R20**

Photo Credits **R86**

Index **R87**

Symbols and Properties, Formulas .... **Inside Back Cover**

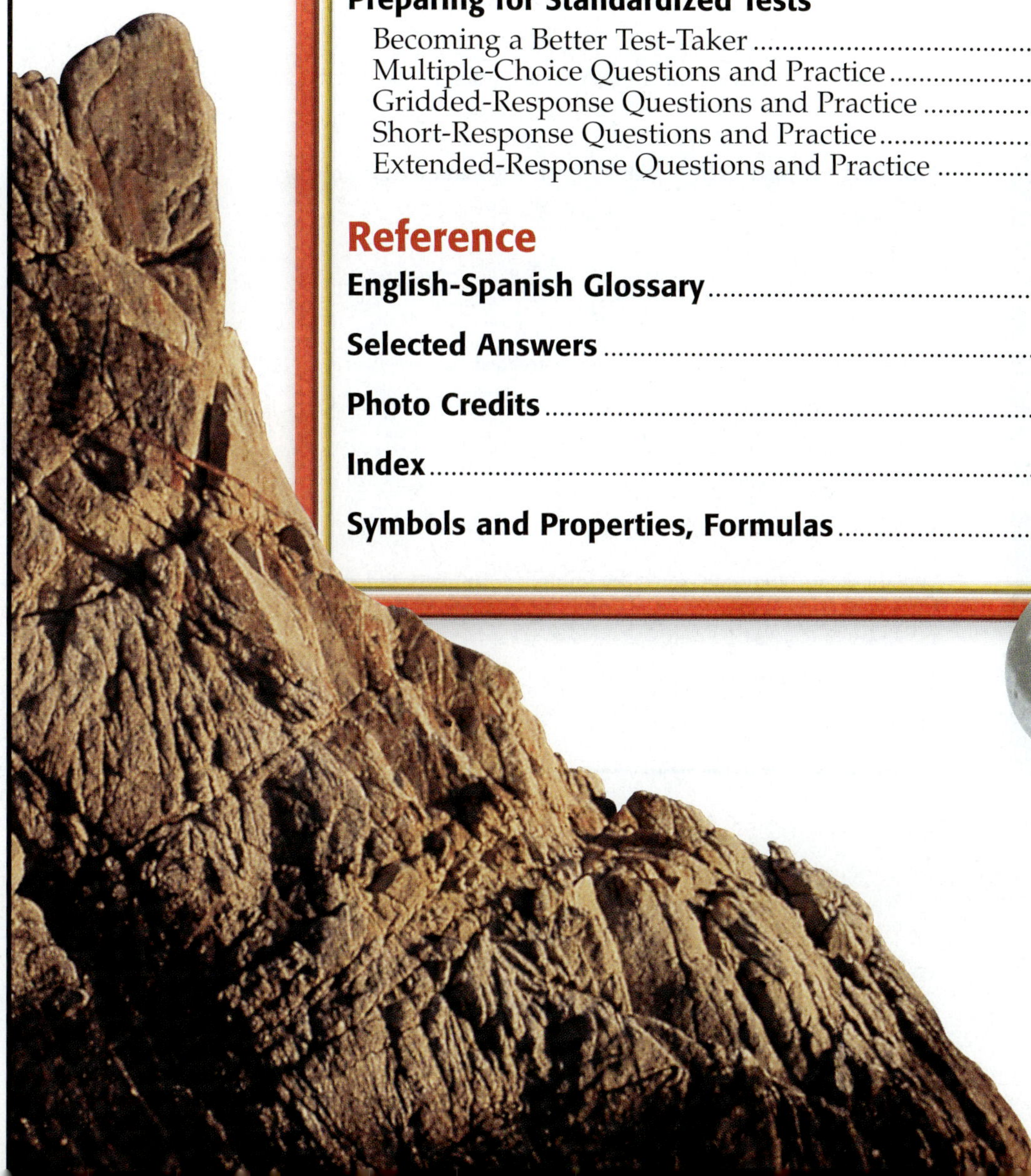

Prerequisite Skills

❶ Comparing and Ordering Real Numbers

- To determine which of two real numbers is greater, express each number as a decimal. Then compare the numbers.

Example 1 **Replace each ● with <, >, or = to make a true sentence.**

a. $\frac{4}{7}$ ● $0.\overline{5}$

$\frac{4}{7} \approx 0.57$
$0.\overline{5} \approx 0.56$

Round to the nearest hundredth.

Since $0.57 > 0.56$, $\frac{4}{7} > 0.\overline{5}$.

b. $\frac{1}{8}$ ● $\frac{1}{\sqrt{18}}$

$\frac{1}{8} = 0.125$
$\frac{1}{\sqrt{18}} \approx 0.236$

Use a calculator to find a rational approximation of $\frac{1}{\sqrt{18}}$.

Since $0.125 < 0.236$, $\frac{1}{8} < \frac{1}{\sqrt{18}}$.

- To order real numbers, first express each number as a decimal. Then write the decimals in order from least to greatest, and write the corresponding real numbers in the same order.

Example 2 **Order $2.\overline{54}$, $-\sqrt{6}$, $\frac{9}{4}$, $\frac{22}{-9}$ from least to greatest.**

$2.\overline{54} = 2.545454\ldots$ or about 2.55

$-\sqrt{6} = -2.44948974\ldots$ or about -2.45

$\frac{9}{4} = 2.25$

$\frac{22}{-9} = -2.4444\ldots$ or about -2.44

$-2.45 < -2.44 < 2.25 < 2.55$

Thus, the order from least to greatest is $-\sqrt{6}$, $\frac{22}{-9}$, $\frac{9}{4}$, $2.\overline{54}$.

Exercises **Replace each ● with <, >, or = to make a true sentence.**

1. $\frac{2}{3}$ ● 0.6 **>**
2. 0.35 ● $\frac{3}{8}$ **<**
3. $-\frac{3}{5}$ ● $-\frac{5}{7}$ **>**
4. $\frac{1}{39}$ ● $-\frac{2}{15}$ **>**
5. $0.\overline{4}$ ● $\frac{4}{9}$ **=**
6. $\frac{7}{11}$ ● $0.\overline{63}$ **=**
7. $\sqrt{5}$ ● $2\frac{1}{4}$ **<**
8. $\frac{1}{\sqrt{3}}$ ● $\frac{5}{9}$ **>**
9. $\frac{1}{\sqrt{2}}$ ● $\frac{11}{12}$ **<**

Order each set of numbers from least to greatest.

10. 0.1, 0.01, $\frac{3}{10}$, 0.2 **0.01, 0.1, 0.2, $\frac{3}{10}$**
11. $\frac{1}{3}$, 0.3, 0.4, $\frac{1}{4}$ **$\frac{1}{4}$, 0.3, $\frac{1}{3}$, 0.4**
12. 26.1, 26, 25.9, $\frac{181}{7}$ **$\frac{181}{7}$, 25.9, 26, 26.1**
13. $\frac{8}{9}$, 0.89, $0.8\overline{9}$, $\frac{8}{11}$ **$\frac{8}{11}$, $\frac{8}{9}$, 0.89, $0.8\overline{9}$**
14. 1.32, $-\sqrt{3}$, $\frac{4}{3}$, $\frac{-15}{11}$ **$-\sqrt{3}$, $\frac{-15}{11}$, 1.32, $\frac{4}{3}$**
15. $\frac{9}{2}$, $\frac{40}{-9}$, $-4.\overline{05}$, $\sqrt{18}$ **$\frac{40}{-9}$, $-4.\overline{05}$, $\sqrt{18}$, $\frac{9}{2}$**
16. $7.\overline{8}$, $8.\overline{7}$, $8.\overline{8}$, $8.\overline{78}$ **$7.\overline{8}$, $8.\overline{7}$, $8.\overline{78}$, $8.\overline{8}$**
17. 3.04, $4.0\overline{3}$, $3.0\overline{4}$, 4.03 **3.04, $3.0\overline{4}$, 4.03, $4.0\overline{3}$**

Explain the difference between the following numbers and arrange them in order from least to greatest. **18–21. See margin.**

18. $3.\overline{54}$, $3.5\overline{4}$, 3.5, 3.54, $3.\overline{5}$
19. $2.\overline{987}$, $2.98\overline{7}$, 2.987, $2.9\overline{87}$, $2.\overline{98}$
20. 8.6, $8.6\overline{7}$, $8\frac{2}{3}$, 8.7676…
21. $-4.\overline{10}$, $-4\frac{1}{9}$, $-4.121231234\ldots$, $-4.1\overline{2}$

18. $3.\overline{54} = 3.5454\ldots$, $3.5\overline{4} = 3.5444\ldots$, $3.5 = 3.5000\ldots$, $3.54 = 3.5400\ldots$, $3.\overline{5} = 3.5555\ldots$; The order from least to greatest is 3.5, 3.54, $3.5\overline{4}$, $3.\overline{54}$, $3.\overline{5}$.

19. $2.\overline{987} = 2.987987\ldots$, $2.98\overline{7} = 2.987777\ldots$, $2.987 = 2.987000\ldots$, $2.9\overline{87} = 2.987878\ldots$, $2.\overline{98} = 2.989898\ldots$; The order from least to greatest is 2.987, $2.98\overline{7}$, $2.9\overline{87}$, $2.\overline{987}$, $2.\overline{98}$.

20. $8.6 = 8.600$, $8.6\overline{7} = 8.6777\ldots$, $8\frac{2}{3} = 8.666\ldots$; The order from least to greatest is 8.6, $8\frac{2}{3}$, $8.6\overline{7}$, 8.7676….

21. $-4.\overline{10} = -4.1010\ldots$, $-4\frac{1}{9} = -4.1111\ldots$, $-4.1\overline{2} = -4.1222\ldots$; The order from least to greatest is $-4.1\overline{2}$, $-4.121231234\ldots$, $-4\frac{1}{9}$, $-4.\overline{10}$.

2 Factoring Polynomials

- Some polynomials can be factored using the Distributive Property.

Example 1 **Factor $4a^2 + 8a$.**

Find the GCF of $4a^2$ and $8a$.

$4a^2 = 2 \cdot 2 \cdot a \cdot a$
$8a = 2 \cdot 2 \cdot 2 \cdot a$
GCF: $2 \cdot 2 \cdot a$ or $4a$

$4a^2 + 8a = 4a(a) + 4a(2)$ Rewrite each term using the GCF.
$= 4a(a + 2)$ Distributive Property

Thus, the completely factored form of $4a^2 + 8a$ is $4a(a + 2)$.

- To factor quadratic trinomials of the form $x^2 + bx + c$, find two integers m and n whose product is equal to c and whose sum is equal to b. Then write $x^2 + bx + c$ using the pattern $(x + m)(x + n)$.

Example 2 **Factor each polynomial.**

a. $x^2 + 5x + 6$ ← Both b and c are positive.

In this trinomial, b is 5 and c is 6.
Find two numbers whose product is 6 and whose sum is 5.

Factors of 6	Sum of Factors
1, 6	7
2, 3	5

The correct factors are 2 and 3.

$x^2 + 5x + 6 = (x + m)(x + n)$ Write the pattern.
$= (x + 2)(x + 3)$ $m = 2$ and $n = 3$

CHECK Multiply the binomials to check the factorization.
$(x + 2)(x + 3) = x^2 + 3x + 2x + 2(3)$ FOIL
$= x^2 + 5x + 6$ ✓

b. $x^2 - 8x + 12$ ← b is negative and c is positive.

In this trinomial, $b = -8$ and $c = 12$. This means that $m + n$ is negative and mn is positive. So m and n must both be negative.

Factors of 12	Sum of Factors
−1, −12	−13
−2, −6	−8

The correct factors are −2 and −6.

$x^2 - 8x + 12 = (x + m)(x + n)$ Write the pattern.
$= [x + (-2)][x + (-6)]$ $m = -2$ and $n = -6$
$= (x - 2)(x - 6)$ Simplify.

c. $x^2 + 14x - 15$ ← b is positive and c is negative.

In this trinomial, $b = 14$ and $c = -15$. This means that $m + n$ is positive and mn is negative. So either m or n must be negative, but not both.

Factors of −15	Sum of Factors
1, −15	−14
−1, 15	14

The correct factors are −1 and 15.

$x^2 + 14x - 15 = (x + m)(x + n)$ Write the pattern.
$= [x + (-1)](x + 15)$ $m = -1$ and $n = 15$
$= (x - 1)(x + 15)$ Simplify.

Prerequisite Skills

Prerequisite Skills 815

Prerequisite Skills

- To factor quadratic trinomials of the form $ax^2 + bx + c$, find two integers m and n whose product is equal to ac and whose sum is equal to b. Write $ax^2 + bx + c$ using the pattern $ax^2 + mx + nx + c$. Then factor by grouping.

Example 3 **Factor $6x^2 + 7x - 3$.**

In this trinomial, $a = 6$, $b = 7$ and $c = -3$.
Find two numbers whose product is $6 \cdot (-3)$ or -18 and whose sum is 7.

Factors of −18	Sum of Factors
1, −18	−17
−1, 18	17
2, −9	−7
−2, 9	7

The correct factors are −2 and 9.

$6x^2 + 7x - 3 = 6x^2 + mx + nx - 3$ Write the pattern.
$= 6x^2 + (-2)x + 9x - 3$ $m = -2$ and $n = 9$
$= (6x^2 - 2x) + (9x - 3)$ Group terms with common factors.
$= 2x(3x - 1) + 3(3x - 1)$ Factor the GCF from each group.
$= (2x + 3)(3x - 1)$ Distributive Property

- Here are some special products.

Perfect Square Trinomials

$(a + b)^2 = (a + b)(a + b)$
$= a^2 + 2ab + b^2$

$(a - b)^2 = (a - b)(a - b)$
$= a^2 - 2ab + b^2$

Difference of Squares

$a^2 - b^2 = (a + b)(a - b)$

Example 4 **Factor each polynomial.**

a. $4x^2 + 20x + 25$

The first and last terms are perfect squares.
The middle term is equal to $2(2x)(5)$.
This is a perfect square trinomial of the form $(a + b)^2$.

$4x^2 + 20x + 25 = (2x)^2 + 2(2x)(5) + 5^2$ Write as $a^2 + 2ab + b^2$.
$= (2x + 5)^2$ Factor using the pattern.

b. $x^2 - 4$

This is a difference of squares.

$x^2 - 4 = x^2 - (2)^2$ Write in the form $a^2 - b^2$.
$= (x + 2)(x - 2)$ Factor the difference of squares.

Exercises **Factor the following polynomials.**

1. $12x^2 + 4x$ **$4x(3x + 1)$**
2. $6x^2y + 2x$ **$2x(3xy + 1)$**
3. $8ab^2 - 12ab$ **$4ab(2b - 3)$**
4. $x^2 + 5x + 4$ **$(x + 1)(x + 4)$**
5. $y^2 + 12y + 27$ **$(y + 3)(y + 9)$**
6. $x^2 + 6x + 8$ **$(x + 2)(x + 4)$**
7. $3y^2 + 13y + 4$ **$(3y + 1)(y + 4)$**
8. $7x^2 + 51x + 14$ **$(7x + 2)(x + 7)$**
9. $3x^2 + 28x + 32$ **$(3x + 4)(x + 8)$**
10. $x^2 - 5x + 6$ **$(x - 3)(x - 2)$**
11. $y^2 - 5y + 4$ **$(y - 4)(y - 1)$**
12. $6x^2 - 13x + 5$ **$(3x - 5)(2x - 1)$**
13. $6a^2 - 50ab + 16b^2$ **$2(3a - b)(a - 8b)$**
14. $11x^2 - 78x + 7$ **$(11x - 1)(x - 7)$**
15. $18x^2 - 31xy + 6y^2$ **$(2x - 3y)(9x - 2y)$**
16. $x^2 + 4xy + 4y^2$ **$(x + 2y)^2$**
17. $9x^2 - 24x + 16$ **$(3x - 4)^2$**
18. $4a^2 + 12ab + 9b^2$ **$(2a + 3b)^2$**
19. $x^2 - 144$ **$(x + 12)(x - 12)$**
20. $4c^2 - 9$ **$(2c + 3)(2c - 3)$**
21. $16y^2 - 1$ **$(4y + 1)(4y - 1)$**
22. $25x^2 - 4y^2$ **$(5x + 2y)(5x - 2y)$**
23. $36y^2 - 16$ **$4(3y + 2)(3y - 2)$**
24. $9a^2 - 49b^2$ **$(3a + 7b)(3a - 7b)$**

Prerequisite Skills

816 Prerequisite Skills

3 Congruent and Similar Figures

Congruent figures have the same size and the same shape.

- Two polygons are congruent if their corresponding sides are congruent and their corresponding angles are congruent.

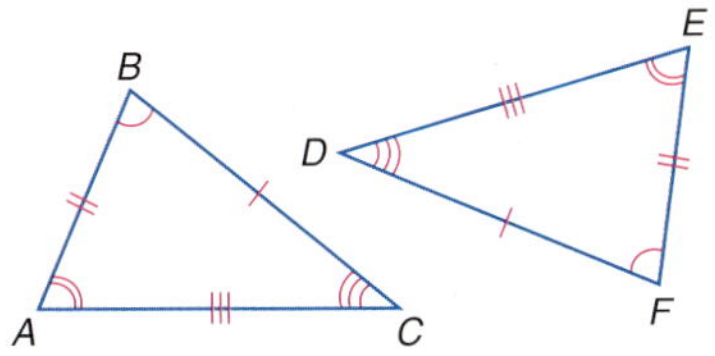

$\triangle ABC \cong \triangle EFD$

The order of the vertices indicates the corresponding parts.

Congruent Angles	Congruent Sides
$\angle A \cong \angle E$	$\overline{AB} \cong \overline{EF}$
$\angle B \cong \angle F$	$\overline{BC} \cong \overline{FD}$
$\angle C \cong \angle D$	$\overline{AC} \cong \overline{ED}$

Read the symbol $\cong$ as *is congruent to.*

Example 1 **If $\triangle XYZ \cong \triangle PQR$, name the congruent angles and sides.**

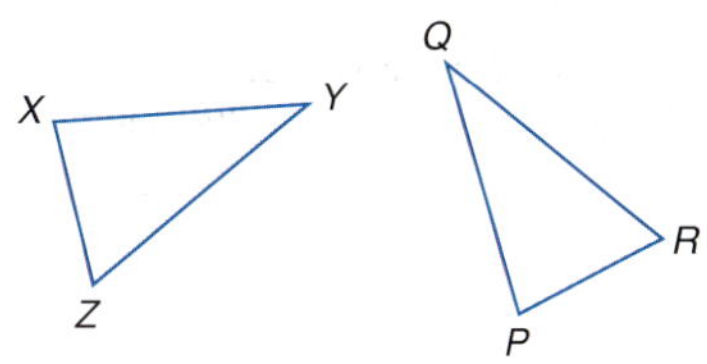

Name the pairs of congruent angles by looking at the order of the vertices in the statement $\triangle XYZ \cong \triangle PQR$.

So, $\angle X \cong \angle P$, $\angle Y \cong \angle Q$, and $\angle Z \cong \angle R$.

Since X corresponds to P, and Y corresponds to Q, $\overline{XY} \cong \overline{PQ}$.

Since Y corresponds to Q, and Z corresponds to R, $\overline{YZ} \cong \overline{QR}$.

Since Z corresponds to R, and X corresponds to P, $\overline{ZX} \cong \overline{RP}$.

Example 2 **The corresponding parts of two congruent triangles are marked on the figure. Write a congruence statement for the two triangles.**

List the congruent angles and sides.

$\angle A \cong \angle D$	$\overline{AB} \cong \overline{DE}$
$\angle B \cong \angle E$	$\overline{AC} \cong \overline{DC}$
$\angle ACB \cong \angle DCE$	$\overline{BC} \cong \overline{EC}$

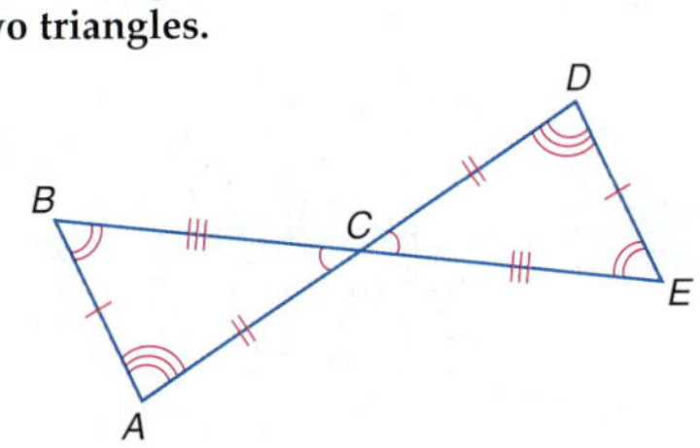

Match the vertices of the congruent angles. Therefore, $\triangle ABC \cong \triangle DEC$.

Similar figures have the same shape, but not necessarily the same size.

- In similar figures, corresponding angles are congruent, and the measures of corresponding sides are proportional. (They have equivalent ratios.)

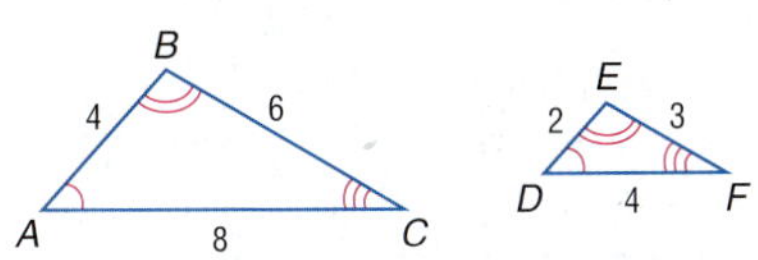

Congruent Angles

$\angle A \cong \angle D$, $\angle B \cong \angle E$, $\angle C \cong \angle F$

Proportional Sides

$\frac{AB}{DE} = \frac{BC}{EF} = \frac{AC}{DF}$

$\triangle ABC \sim \triangle DEF$

Read the symbol $\sim$ as *is similar to.*

Prerequisite Skills 817

Prerequisite Skills

Example 3 **Determine whether the polygons are similar. Justify your answer.**

a.

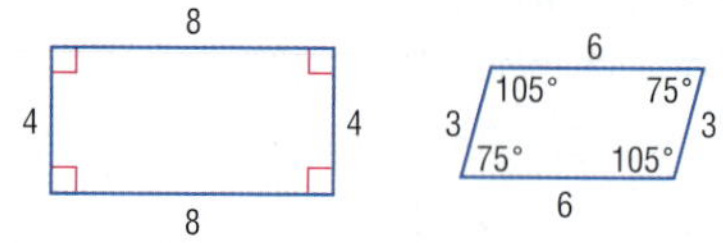

Since $\frac{4}{3} = \frac{8}{6} = \frac{4}{3} = \frac{8}{6}$, the measures of the sides of the polygons are proportional. However, the corresponding angles are not congruent. The polygons are not similar.

b.

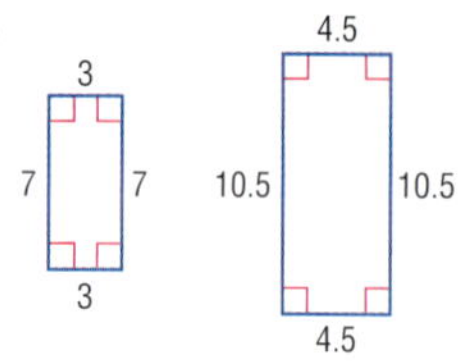

Since $\frac{7}{10.5} = \frac{3}{4.5} = \frac{7}{10.5} = \frac{3}{4.5}$, the measures of the sides of the polygons are proportional. The corresponding angles are congruent. Therefore, the polygons are similar.

Example 4 **The triangles are similar. Find the values of x and y.**

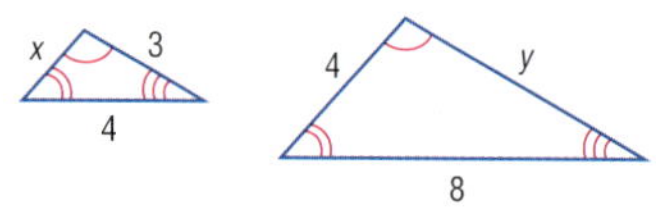

Write proportions using corresponding parts. Then solve to find the missing measures.

$\frac{x}{4} = \frac{4}{8}$	Definition of similar polygons	$\frac{3}{y} = \frac{4}{8}$	Definition of similar polygons
$x(8) = 4(4)$	Cross products	$3(8) = y(4)$	Cross products
$8x = 16$	Simplify.	$24 = 4y$	Simplify.
$\frac{8x}{8} = \frac{16}{8}$	Divide each side by 8.	$\frac{24}{4} = \frac{4y}{4}$	Divide each side by 4.
$x = 2$	Simplify.	$6 = y$	Simplify.

Example 5 **CIVIL ENGINEERING** **The city of Mansfield plans to build a bridge across Pine Lake. Use the information in the diagram to find the distance across Pine Lake.**

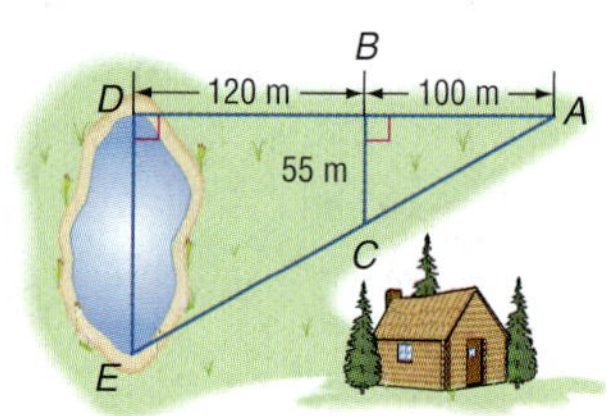

$\triangle ABC \sim \triangle ADE$

$\frac{AB}{AD} = \frac{BC}{DE}$	Definition of similar polygons
$\frac{100}{220} = \frac{55}{DE}$	$AB = 100$, $AD = 100 + 120 = 220$, $BC = 55$
$100DE = 220(55)$	Cross products
$100DE = 12{,}100$	Simplify.
$DE = 121$	Divide each side by 100.

The distance across the lake is 121 meters.

Exercises **If $\triangle GHI \cong \triangle JKL$, name the part that is congruent to each angle or segment.**

1. $\angle K$ **$\angle H$**
2. $\overline{HI}$ **$\overline{KL}$**
3. $\angle I$ **$\angle L$**
4. $\angle J$ **$\angle G$**
5. $\overline{JK}$ **$\overline{GH}$**
6. $\overline{GI}$ **$\overline{JL}$**

818 Prerequisite Skills

Complete each congruence statement.

7. 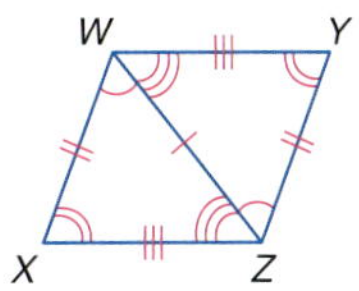

$\triangle XWZ \cong \triangle$ _?_ **YZW**

8. 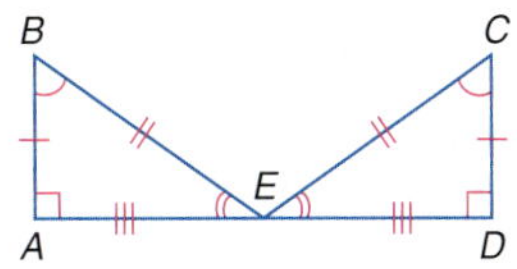

$\triangle ABE \cong \triangle$ _?_ **DCE**

9. 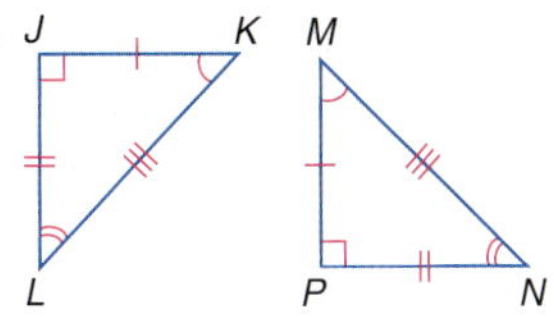

$\triangle JKL \cong \triangle$ _?_ **PMN**

Determine whether each pair of figures is *similar*, *congruent*, or *neither*.

10. **similar**

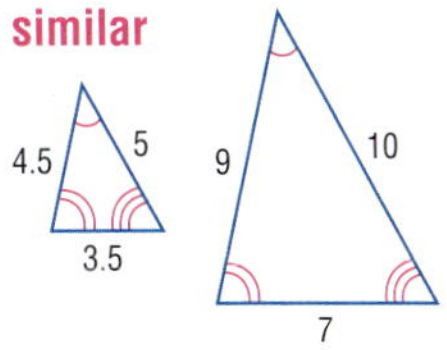

11. 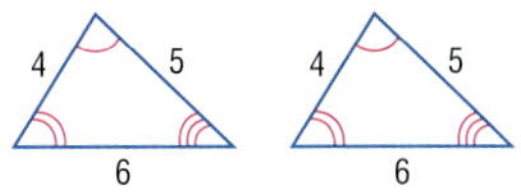

congruent

12. **neither**

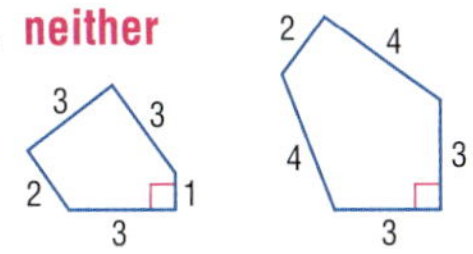

13. 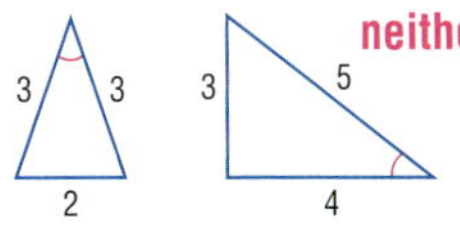

neither

14. 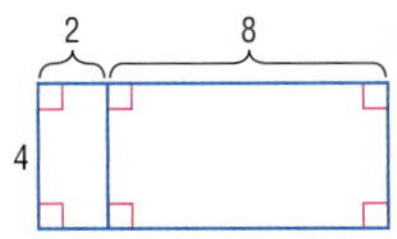

similar

15. 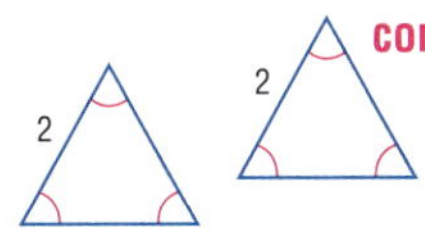

congruent

16. 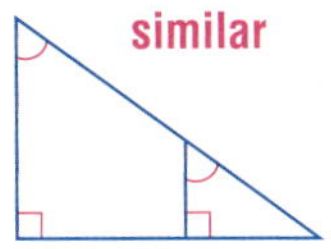

similar

17. 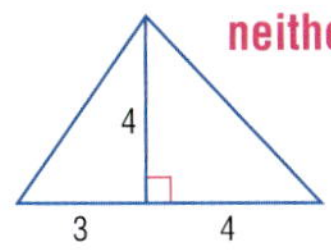

neither

18. 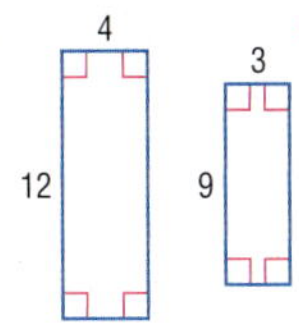

similar

Each pair of polygons is similar. Find the values of *x* and *y*.

19. 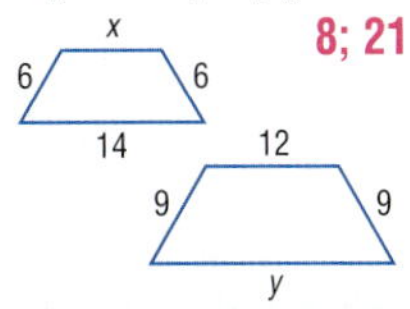

8; 21

20. 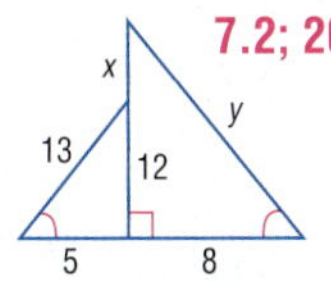

7.2; 20.8

21. 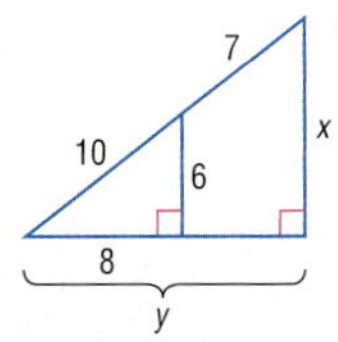

10.2; 13.6

22. What are the dimensions of a scale drawing of a room that measures 10 feet by 12 feet if $\frac{1}{2}$ inch represents 1 foot? **5 in. by 6 in.**

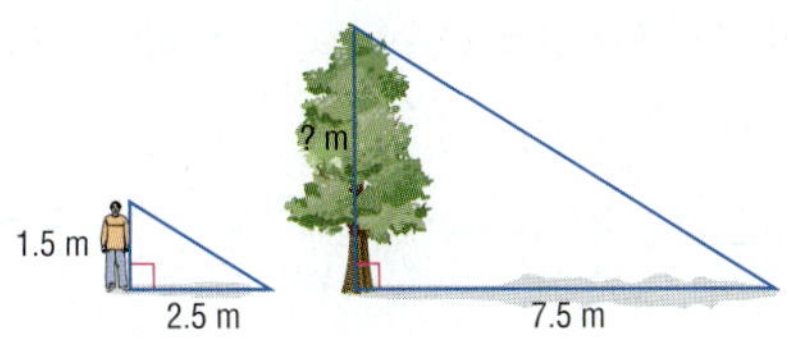

23. **SHADOWS** On a sunny day, Jason measures the length of his shadow and the length of a tree's shadow. Use the figures at the right to find the height of the tree. **4.5 m**

24. **PHOTOGRAPHY** A photo that is 4 inches wide by 6 inches long must be reduced to fit in a space 3 inches wide. How long will the reduced photo be? **$4\frac{1}{2}$ in.**

25. **SURVEYING** Surveyors use instruments to measure objects that are too large or too far away to measure by hand. They can use the shadows that objects cast to find the height of the objects without measuring them. A surveyor finds that a telephone pole that is 25 feet tall is casting a shadow 20 feet long. A nearby building is casting a shadow 52 feet long. What is the height of the building? **65 ft**

Prerequisite Skills

Prerequisite Skills

Prerequisite Skills 819

Prerequisite Skills

4 Pythagorean Theorem

The **Pythagorean Theorem** states that in a right triangle, the square of the length of the hypotenuse c is equal to the sum of the squares of the lengths of the legs a and b.

That is, in any right triangle, $c^2 = a^2 + b^2$.

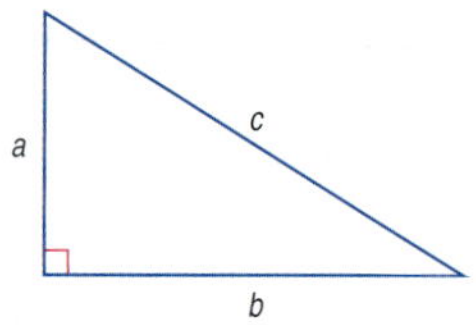

Example 1 **Find the length of the hypotenuse of each right triangle.**

a.

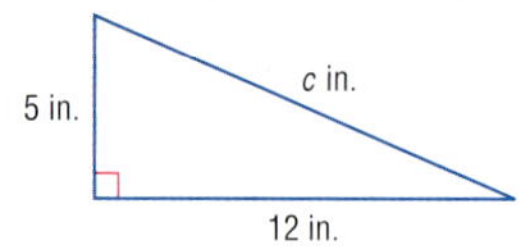

$c^2 = a^2 + b^2$	Pythagorean Theorem
$c^2 = 5^2 + 12^2$	Replace a with 5 and b with 12.
$c^2 = 25 + 144$	Simplify.
$c^2 = 169$	Add.
$c = \sqrt{169}$	Take the square root of each side.
$c = 13$	Simplify.

The length of the hypotenuse is 13 inches.

b.

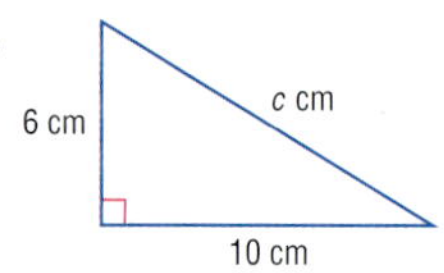

$c^2 = a^2 + b^2$	Pythagorean Theorem
$c^2 = 6^2 + 10^2$	Replace a with 6 and b with 10.
$c^2 = 36 + 100$	Simplify.
$c^2 = 136$	Add.
$c = \sqrt{136}$	Take the square root of each side.
$c \approx 11.7$	Use a calculator to find the square root of 136. Round to the nearest tenth.

To the nearest tenth, the length of the hypotenuse is 11.7 centimeters.

Example 2 **Find the length of the missing leg in each right triangle.**

a.

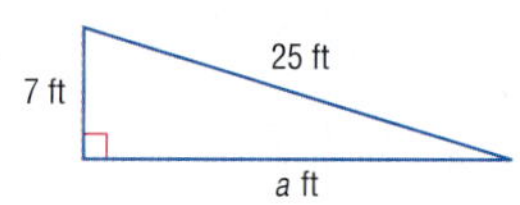

$c^2 = a^2 + b^2$	Pythagorean Theorem
$25^2 = a^2 + 7^2$	Replace c with 25 and b with 7.
$625 = a^2 + 49$	Simplify.
$625 - 49 = a^2 + 49 - 49$	Subtract 49 from each side.
$576 = a^2$	Simplify.
$\sqrt{576} = a$	Take the square root of each side.
$24 = a$	Simplify.

The length of the leg is 24 feet.

820 Prerequisite Skills

b.

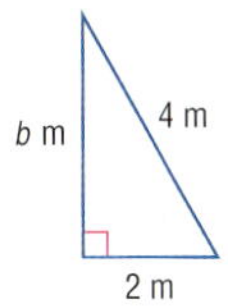

$c^2 = a^2 + b^2$	Pythagorean Theorem
$4^2 = 2^2 + b^2$	Replace *c* with 4 and *a* with 2.
$16 = 4 + b^2$	Simplify.
$12 = b^2$	Subtract 4 from each side.
$\sqrt{12} = b$	Take the square root of each side.
$3.5 \approx b$	Use a calculator to find the square root of 12. Round to the nearest tenth.

To the nearest tenth, the length of the leg is 3.5 meters.

Example 3 **The lengths of the three sides of a triangle are 5, 7, and 9 inches. Determine whether this triangle is a right triangle.**

Since the longest side is 9 inches, use 9 as *c*, the measure of the hypotenuse.

$c^2 = a^2 + b^2$	Pythagorean Theorem
$9^2 \stackrel{?}{=} 5^2 + 7^2$	Replace *c* with 9, *a* with 5, and *b* with 7.
$81 \stackrel{?}{=} 25 + 49$	Evaluate 9^2, 5^2, and 7^2.
$81 \neq 74$	Simplify.

Since $c^2 \neq a^2 + b^2$, the triangle is *not* a right triangle.

Exercises **Find each missing measure. Round to the nearest tenth, if necessary.**

1. **39**

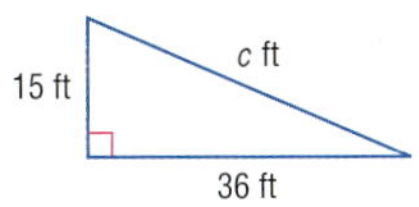

2. **24**

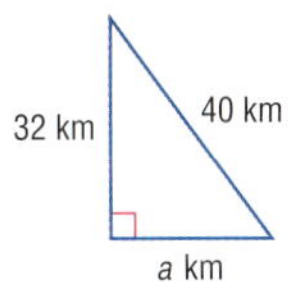

3. **8.3**

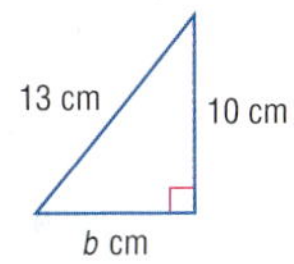

4. **12.5**

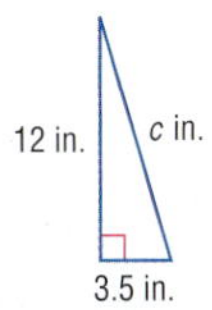

5. **12.4**

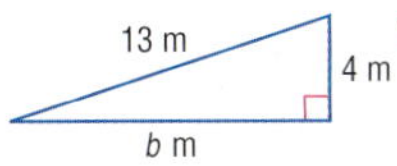

6. **6.4**

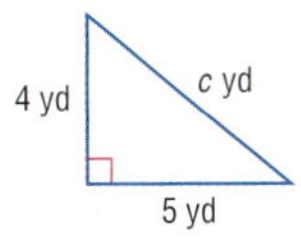

7. $a = 3, b = 4, c = ?$ **5**
8. $a = ?, b = 12, c = 13$ **5**
9. $a = 14, b = ?, c = 50$ **48**
10. $a = 2, b = 9, c = ?$ **9.2**
11. $a = 6, b = ?, c = 13$ **11.5**
12. $a = ?, b = 7, c = 11$ **8.5**

The lengths of three sides of a triangle are given. Determine whether each triangle is a right triangle.

13. 5 in., 7 in., 8 in. **no**
14. 9 m, 12 m, 15 m **yes**
15. 6 cm, 7 cm, 12 cm **no**
16. 11 ft, 12 ft, 16 ft **no**
17. 10 yd, 24 yd, 26 yd **yes**
18. 11 km, 60 km, 61 km **yes**

19. **FLAGPOLES** Mai-Lin wants to find the distance from her feet to the top of the flagpole. If the flagpole is 30 feet tall and Mai-Lin is standing a distance of 15 feet from the flagpole, what is the distance from her feet to the top of the flagpole? **about 33.5 ft**

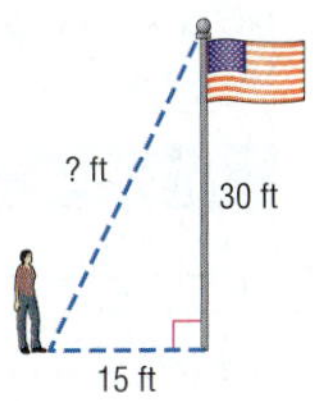

20. **CONSTRUCTION** The walls of the Downtown Recreation Center are being covered with paneling. The doorway into one room is 0.9 meters wide and 2.5 meters high. What is the width of the widest rectangular panel that can be taken through this doorway? **about 2.66 m**

Prerequisite Skills

Prerequisite Skills 821

5 Mean, Median, and Mode

Mean, median, and mode are measures of central tendency that are often used to represent a set of data.

- To find the **mean**, find the sum of the data and divide by the number of items in the data set. (The mean is often called the average.)
- To find the **median**, arrange the data in numerical order. The median is the middle number. If there is an even number of data, the median is the mean of the two middle numbers.
- The **mode** is the number (or numbers) that appears most often in a set of data. If no item appears most often, the set has no mode.

Example 1 **Michelle is saving to buy a car. She saved \$200 in June, \$300 in July, \$400 in August, and \$150 in September. What was her mean (or average) monthly savings?**

$$\text{mean} = \frac{\text{sum of monthly savings}}{\text{number of months}}$$
$$= \frac{\$200 + \$300 + \$400 + \$150}{4}$$
$$= \frac{\$1050}{4} \text{ or } \$262.50$$

Michelle's mean monthly savings was \$262.50.

Example 2 **Find the median of the data.**

Peter's Best Running Times	
Week	Minutes to Run a Mile
1	4.5
2	3.7
3	4.1
4	4.1
5	3.6
6	3.4

To find the median, order the numbers from least to greatest.
The median is in the middle.
3.4, 3.6, 3.7, 4.1, 4.1, 4.5

$$\frac{3.7 + 4.1}{2} = 3.9$$

There is an even number of data. Find the mean of the middle two.

Example 3 GOLF **Four players tied for first in the 2001 PGA Tour Championship. The scores for each player for each round are shown in the table below. What is the mode score?**

Player	Round 1	Round 2	Round 3	Round 4
Mike Weir	68	66	68	68
David Toms	73	66	64	67
Sergio Garcia	69	67	66	68
Ernie Els	69	68	65	68

Source: ESPN

The mode is the score that occurred most often. Since the score of 68 occurred 6 times, it is the mode of these data.

822 Prerequisite Skills

- The **range** of a set of data is the difference between the greatest and the least values of the set. It describes how a set of data varies.

Example 4 **Find the range of the data.**
{6, 11, 18, 4, 9, 15, 6, 3}

The greatest value is 18 and the least value is 3.
So, the range is 18 − 3 or 15.

Exercises **Find the mean, median, mode, and range for each set of data. Round to the nearest tenth if necessary.** **6. 502.5; 502.5; 502 and 503; 3**

1. {2, 8, 12, 13, 15} **10; 12; no mode; 13**
2. {66, 78, 78, 64, 34, 88} **68; 72; 78; 54**
3. {87, 95, 84, 89, 100, 82} **89.5; 88; no mode; 18**
4. {99, 100, 85, 96, 94, 99} **95.5, 97.5, 99, 15**
5. {9.9, 9.9, 10, 9.9, 8.8, 9.5, 9.5} **9.6; 9.9; 9.9; 1.2**
6. {501, 503, 502, 502, 502, 504, 503, 503}
7. {7, 19, 15, 13, 11, 17, 9} **13; 13; no mode; 12**
8. {6, 12, 21, 43, 1, 3, 13, 8} **13.4; 10; no mode; 42**
9. {0.8, 0.04, 0.9, 1.1, 0.25} **0.6; 0.8; no mode; 1.06**
10. $\left\{2\frac{1}{2}, 1\frac{7}{8}, 2\frac{5}{8}, 2\frac{3}{4}, 2\frac{1}{8}\right\}$ **$2\frac{3}{8}$; $2\frac{1}{2}$; no mode; $\frac{7}{8}$**

11. **CHARITY** The table shows the amounts collected by classes at Jackson High School. Find the mean, median, mode, and range of the data. **$110; $77.50; no mode; $290**

Amounts Collected for Charity	
Class	**Amount**
A	$150
B	$300
C	$55
D	$40
E	$10
F	$25
G	$200
H	$100

12. **SCHOOL** The table shows Pilar's grades in chemistry class for the semester. Find her mean, median, and mode scores, and the range of her scores. **93.3; 95; 95; 13**

Chemistry Grades	
Assignment	**Grade (out of 100)**
Homework	100
Electron Project	98
Test I	87
Atomic Mass Project	95
Test II	88
Phase Change Project	90
Test III	95

13. **WEATHER** The table shows the precipitation for the month of July in Cape Hatteras, North Carolina in various years. Find the mean, median, mode, and range of the data. **5.20; 4.585; no mode; 9.77**

July Precipitation in Cape Hatteras, North Carolina												
Year	1990	1991	1992	1993	1994	1995	1996	1997	1998	1999	2000	2001
Inches	4.23	8.58	5.28	2.03	3.93	1.08	9.54	4.94	10.85	2.66	6.04	3.26

Source: National Climatic Data Center

14. **SCHOOL** Kaitlyn's scores on her first five algebra tests are 88, 90, 91, 89, and 92. What test score must Kaitlyn earn on the sixth test so that her mean score will be at least 90? **at least 90**
15. **GOLF** Colin's average for three rounds of golf is 94. What is the highest score he can receive for the fourth round to have an average (mean) of 92? **86**
16. **SCHOOL** Mika has a mean score of 21 on his first four Spanish quizzes. If each quiz is worth 25 points, what is the highest possible mean score he can have after the fifth quiz? **21.8**
17. **SCHOOL** To earn a grade of B in math, Latisha must have an average (mean) score of at least 84 on five math tests. Her scores on the first three tests are 85, 89, and 82. What is the lowest total score that Latisha must have on the last two tests to earn a B test average? **164**

Prerequisite Skills

Prerequisite Skills 823

6 Bar and Line Graphs

A **bar graph** compares different categories of data by showing each as a bar whose length is related to the frequency. A **double bar graph** compares two sets of data. Another way to represent data is by using a **line graph**. A line graph usually shows how data changes over a period of time.

Example 1 MARRIAGE **The table shows the average age at which Americans marry for the first time. Make a double bar graph to display the data.**

Average Age to Marry		
Year	1990	1998
Men	26	27
Women	22	25

Source: U.S. Census Bureau

Step 1 Draw a horizontal and a vertical axis and label them as shown.

Step 2 Draw side-by-side bars to represent each category.

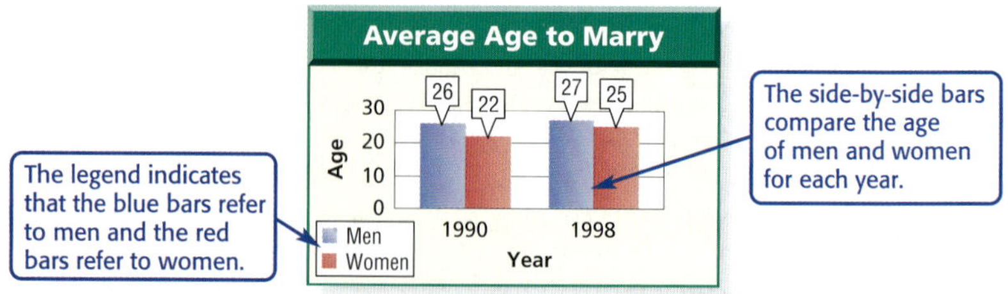

Example 2 HEALTH **The table shows Mark's height at 2-year intervals. Make a line graph to display the data.**

Age	2	4	6	8	10	12	14	16
Height (feet)	2.8	3.5	4.0	4.6	4.9	5.2	5.8	6

Step 1 Draw a horizontal and a vertical axis. Label them as shown.

Step 2 Plot the points.

Step 3 Draw a line connecting each pair of consecutive points.

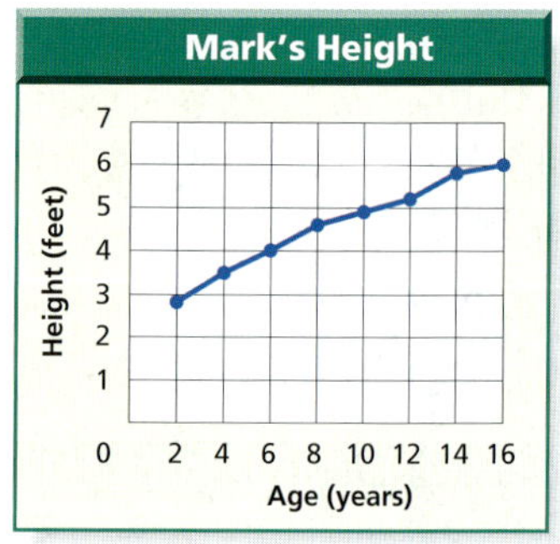

Exercises

1. **HEALTH** The table below shows the life expectancy for Americans born in each year listed. Make a double-bar graph to display the data. **See margin.**

Life Expectancy		
Year of Birth	Male	Female
1980	70.0	77.5
1985	71.2	78.2
1990	71.8	78.8
1995	72.5	78.9
1998	73.9	79.4

2. **MONEY** The amount of money in Becky's savings account from August through March is shown in the table below. Make a line graph to display the data. **See margin.**

Month	Amount	Month	Amount
August	$300	December	$780
September	$400	January	$800
October	$700	February	$950
November	$780	March	$900

1.

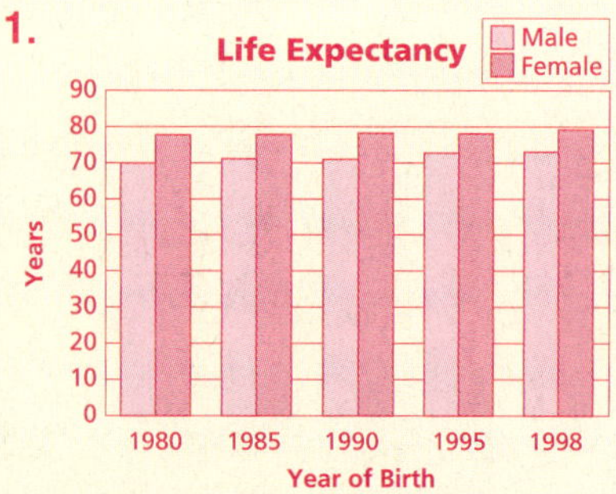

2.

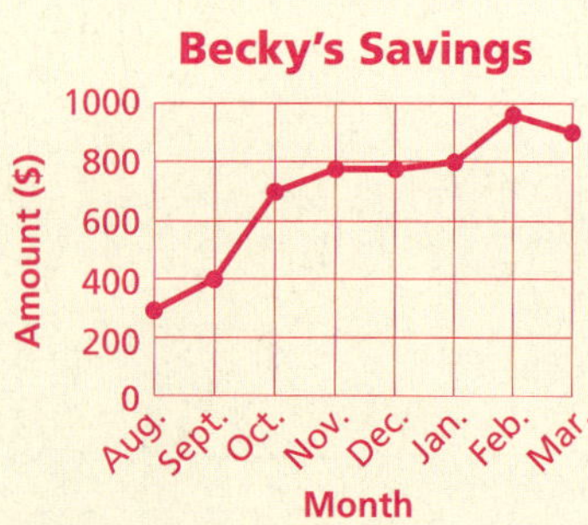

824 Prerequisite Skills

Prerequisite Skills

7 Stem-and-Leaf Plots

In a **stem-and-leaf plot**, data are organized in two columns. The greatest place value of the data is used for the stems. The next greatest place value forms the leaves. Stem-and-leaf plots are useful for organizing long lists of numbers.

Example

SCHOOL Isabella has collected data on the GPAs (grade point average) of the 16 students in the art club. Display the data in a stem-and-leaf plot.
{4.0, 3.9, 3.1, 3.9, 3.8, 3.7, 1.8, 2.6, 4.0, 3.9, 3.5, 3.3, 2.9, 2.5, 1.1, 3.5}

Step 1 Find the least and the greatest number. Then identify the greatest place-value digit in each number. In this case, ones.

least data: 1.1 — The least number has 1 in the ones place.

greatest data: 4.0 — The greatest number has 4 in the ones place.

Step 2 Draw a vertical line and write the stems from 1 to 4 to the left of the line.

Step 3 Write the leaves to the right of the line, with the corresponding stem. For example, write 0 to the right of 4 for 4.0.

Stem	Leaf
1	8 1
2	5 6 9
3	9 1 9 8 7 9 5 3 5
4	0 0

Step 4 Rearrange the leaves so they are ordered from least to greatest.

Step 5 Include a key or an explanation.

Stem	Leaf
1	1 8
2	5 6 9
3	1 3 5 5 7 8 9 9 9
4	0 0 *3|1 = 3.1*

Exercises

GAMES For Exercises 1–4, use the following information.
The stem-and-leaf plot at the right shows Charmaine's scores for her favorite computer game.

Stem	Leaf
9	0 0 0 1 3 4 5 5 7 8 8 8 9 9
10	0 3 4 4 5 6 9
11	0 3 9 9
12	1 2 6
13	0 *12|6 = 126*

1. What are Charmaine's highest and lowest scores? **130; 90**
2. Which score(s) occurred most frequently? **90, 98**
3. How many scores were above 115? **6**
4. Has Charmaine ever scored 123? **no**

5. **SCHOOL** The class scores on a 50-item test are shown in the table at the right. Make a stem-and-leaf plot of the data. **See margin.**

Test Scores					
45	15	30	40	28	35
39	29	38	18	43	49
46	44	48	35	36	30

6. **GEOGRAPHY** The table shows the land area of each county in Wyoming. Round each area to the nearest hundred square miles and organize the data in a stem-and-leaf plot. **See margin.**

County	Area (mi²)	County	Area (mi²)	County	Area (mi²)
Albany	4273	Hot Springs	2004	Sheridan	2523
Big Horn	3137	Johnson	4166	Sublette	4883
Campbell	4797	Laramie	2686	Sweetwater	10,425
Carbon	7896	Lincoln	4069	Teton	4008
Converse	4255	Natrona	5340	Unita	2082
Crook	2859	Niobrara	2626	Washakie	2240
Fremont	9182	Park	6942	Weston	2398
Goshen	2225	Platte	2085		

Source: *The World Almanac*

5.

Stem	Leaf
1	5 8
2	8 9
3	0 0 5 5 6 8 9
4	0 3 4 5 6 8 9 *2|8 = 28*

6.

Stem	Leaf
2	0 1 1 2 2 4 5 6 7 9
3	1
4	0 1 2 3 3 8 9
5	3
6	9
7	9
8	
9	2
10	4 *2|1 = 2100*

Prerequisite Skills

8 Box-and-Whisker Plots

In a set of data, **quartiles** are values that divide the data into four equal parts.

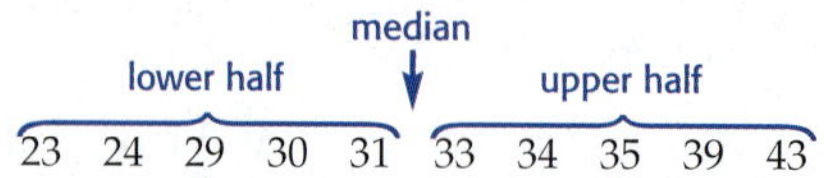

The median of the lower half of a set of data is the **lower quartile**, or LQ.

The median of the upper half of a set of data is the **upper quartile**, or UQ.

- To make a **box-and-whisker plot**, draw a box around the quartile values, and lines or *whiskers* to represent the values in the lower fourth of the data and the upper fourth of the data.

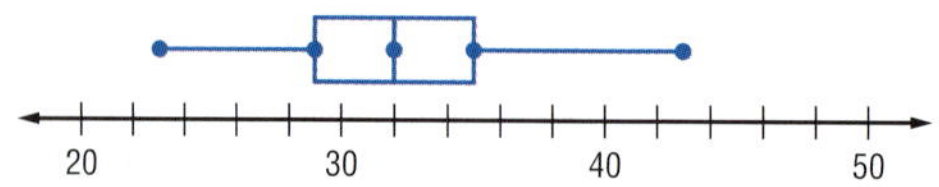

Example 1 MONEY **The amount spent in the cafeteria by 20 students is shown. Display the data in a box-and-whisker plot.**

Amount Spent			
\$2.00	\$2.00	\$1.00	\$4.00
\$1.00	\$2.50	\$2.50	\$2.00
\$2.50	\$1.00	\$4.00	\$2.50
\$3.50	\$2.00	\$3.00	\$2.50
\$4.00	\$4.00	\$5.50	\$1.50

Step 1 Find the least and greatest number. Then draw a number line that covers the range of the data. In this case, the least value is 1 and the greatest value is 5.5.

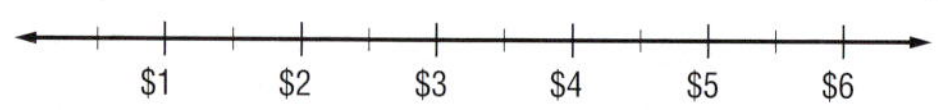

Step 2 Find the median, the extreme values, and the upper and lower quartiles. Mark these points above the number line.

1, 1, 1, 1.5, 2, 2, 2, 2, 2.5, 2.5, 2.5, 2.5, 2.5, 3, 3.5, 4, 4, 4, 4, 5.5

$LQ = \frac{2 + 2}{2}$ or 2 $M = \frac{2.5 + 2.5}{2}$ or 2.5 $UQ = \frac{3.5 + 4}{2}$ or 3.75

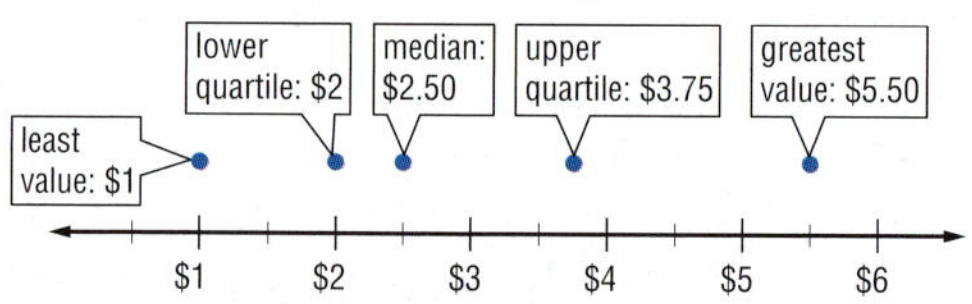

Step 3 Draw a box and the whiskers.

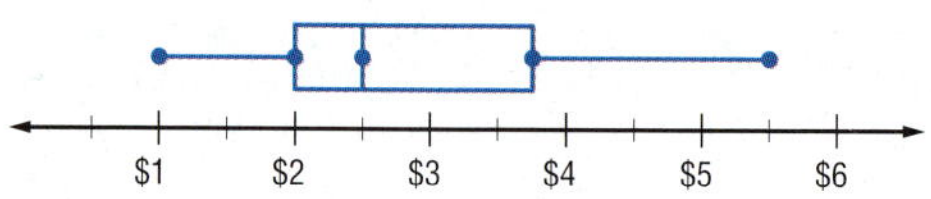

- The **interquartile range (IQR)** is the range of the middle half of the data and contains 50% of the data in the set.

$$\text{Interquartile range} = UQ - LQ$$

The interquartile range of the data in Example 1 is 3.75 − 2 or 1.75.

- An **outlier** is any element of a set that is at least 1.5 interquartile ranges less than the lower quartile or greater than the upper quartile. The whisker representing the data is drawn from the box to the least or greatest value that is not an outlier.

826 Prerequisite Skills

Example 2 **SCHOOL** **The number of hours José studied each day for the last month is shown in the box-and-whiskers plot below.**

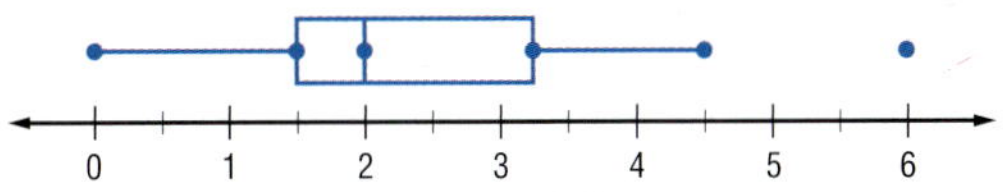

a. **What percent of the data lies between 1.5 and 3.25?**
The value 1.5 is the lower quartile and 3.25 is the upper quartile. The values between the lower and upper quartiles represent 50% of the data.

b. **What was the greatest amount of time José studied in a day?**
The greatest value in the plot is 6, so the greatest amount of time José studied in a day was 6 hours.

c. **What is the interquartile range of this box-and-whisker plot?**
The interquartile range is UQ − LQ. For this plot, the interquartile range is 3.25 − 1.5 or 1.75 hours.

d. **Identify any outliers in the data.**
An outlier is at least 1.5(1.75) less than the lower quartile or more than the upper quartile. Since 3.25 + (1.5)(1.75) = 5.875, and 6 > 5.875, the value 6 is an outlier, and was not included in the whisker.

Exercises

DRIVING **For Exercises 1–3, use the following information.**
Tyler surveyed 20 randomly chosen students at his school about how many miles they drive in an average day. The results are shown in the box-and-whisker plot.

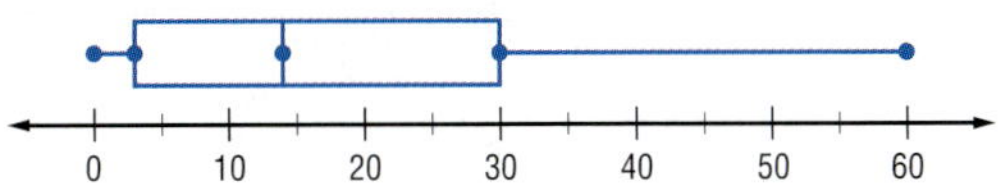

1. What percent of the students drive more than 30 miles in a day? **25%**
2. What is the interquartile range of the box-and-whisker plot? **27 mi**
3. Does a student at Tyler's school have a better chance to meet someone who drives the same mileage they do if they drive 50 miles in a day or 15 miles in a day? Why? **See margin.**
4. **SOFT DRINKS** Carlos surveyed his friends to find the number of cans of soft drink they drink in an average week. Make a box-and-whisker plot of the data.
{0, 0, 0, 1, 1, 1, 2, 2, 3, 4, 4, 5, 5, 7, 10, 10, 10, 11, 11} **See margin.**
5. **ANIMALS** The average life span of some animals commonly found in a zoo are given below. Make a box-and-whisker plot of the data.
{1, 7, 7, 10, 12, 12, 15, 15, 18, 20, 20, 20, 25, 40, 100} **See margin.**
6. **BASEBALL** The table shows the number of sacrifice hits made by teams in the National Baseball League in the 2001 season. Make a box-and-whisker plot of the data. **See margin.**

Team	Home Runs	Team	Home Runs
Arizona	71	Milwaukee	65
Atlanta	64	Montreal	64
Chicago	117	New York	52
Cincinnati	66	Philadelphia	67
Colorado	81	Pittsburgh	60
Florida	60	San Diego	29
Houston	71	San Francisco	67
Los Angeles	57	St. Louis	83

Source: ESPN

3. You have a better chance to find someone who drives the same mileage if you drive 15 miles per day because more of the data is located closer to 15 miles than to 50 miles.

4.

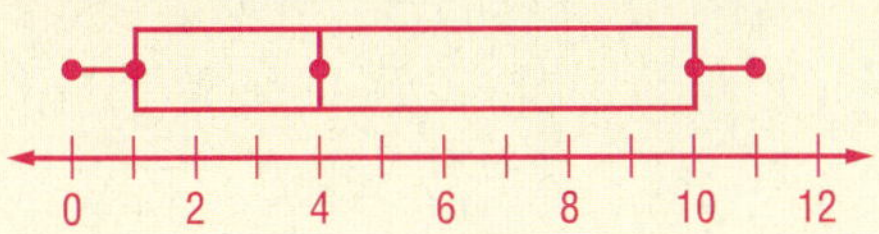

5.

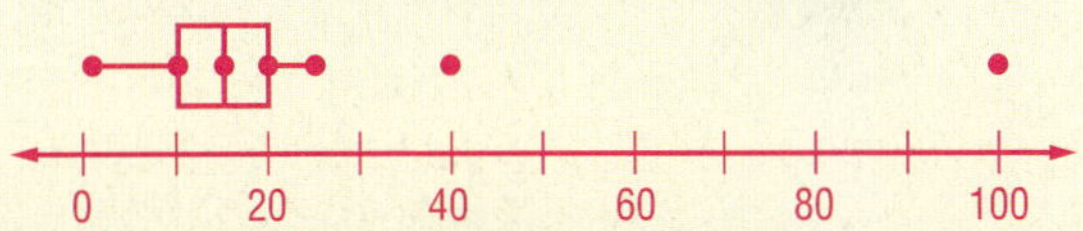

6.

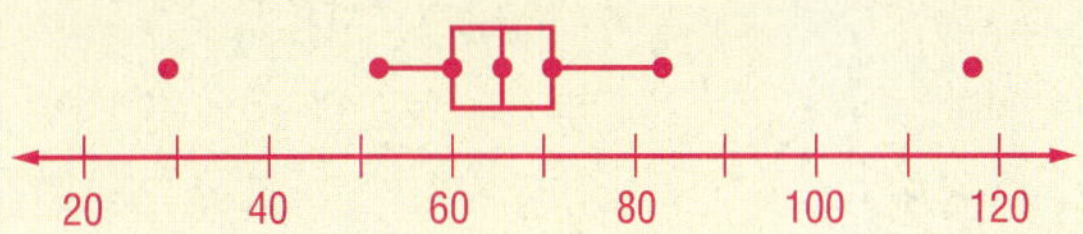

Prerequisite Skills 827

Extra Practice

Lesson 1-1 (pages 6–10)

Find the value of each expression.

1. $2(3 + 8) - 3$ **19**
2. $(5 + 3) - 16 \div 4$ **4**
3. $4 + 8(4) \div 2 - 10$ **10**
4. $15 \div 3 \cdot 5 + 1$ **26**
5. $3(2^2 + 3)$ **21**
6. $5 + 3^2 - 16 + 4$ **2**
7. $[(4 + 8)^2 \div 9] \cdot 5$ **80**
8. $5 + 8^2 \div 4 \cdot 3$ **53**
9. $5 \cdot 7 - 2(5 + 1) \div 3$ **31**
10. $3 + 7^2 - 16 \div 2$ **44**
11. $12 + 20 \div 4 - 5$ **12**
12. $0.5[7 - (8 - 6)^2] - 1$ **0.5**
13. $\frac{1}{2}(3^2 + 5 \cdot 7) - 8$ **14**
14. $\frac{3 \cdot 5 + 3^2}{2^3}$ **3**
15. $\frac{6^2 + 4(2^4)}{28 + 9 \cdot 8}$ **1**
16. $\frac{2^3 - 8(4^2)}{18 + 2^5}$ **−2.4**

Evaluate each expression if $a = -0.5$, $b = 4$, $c = 5$, and $d = -3$.

17. $3b + 4d$ **0**
18. $ab^2 + c$ **−3**
19. $bc + d \div a$ **26**
20. $7ab - 3d$ **−5**
21. $ad + b^2 - c$ **12.5**
22. $\frac{4a + 3c}{3b}$ **$\frac{13}{12}$**
23. $\frac{3ab^2 - d^3}{a}$ **−6**
24. $\frac{5a + ad}{bc}$ **−0.05**

Lesson 1-2 (pages 11–18)

Name the sets of numbers to which each number belongs. (Use N, W, Z, Q, I, and R.)

1. 8.2 **Q, R**
2. −9 **Z, Q, R**
3. $\sqrt{36}$ **N, W, Z, Q, R**
4. $-\frac{1}{3}$ **Q, R**
5. $\sqrt{2}$ **I, R**
6. $-0.\overline{24}$ **Q, R**

Name the property illustrated by each equation. **7. Comm. (×) 9. Distributive**

7. $(4 + 9a)2b = 2b(4 + 9a)$
8. $3\left(\frac{1}{3}\right) = 1$ **Mult. Inv.**
9. $a(3 - 2) = a \cdot 3 - a \cdot 2$
10. $(-3b) + 3b = 0$ **Add. Inv.**
11. $jk + 0 = jk$ **Add. Iden.**
12. $(2a)b = 2(ab)$ **Assoc. (×)**

Name the additive inverse and multiplicative inverse for each number.

13. 3 **$-3; \frac{1}{3}$**
14. $-\frac{1}{8}$ **$\frac{1}{8}; -8$**
15. 0.2 **−0.2; 5**
16. $-2\frac{2}{7}$ **$2\frac{2}{7}; -\frac{7}{16}$**

Simplify each expression. **18. $27a - 2b$ 19. $-4x - 28y$**

17. $7s + 9t + 2s - 7t$ **$9s + 2t$**
18. $6(2a + 3b) + 5(3a - 4b)$
19. $4(3x - 5y) - 8(2x + y)$
20. $0.2(5m - 8) + 0.3(6 - 2m)$ **$0.4m + 0.2$**
21. $\frac{1}{2}(7p + 3q) + \frac{3}{4}(6p - 4q)$ **$8p - \frac{3}{2}q$**
22. $\frac{4}{5}(3v - 2w) - \frac{1}{5}(7v - 2w)$ **$v - \frac{6}{5}w$**

Lesson 1-3 (pages 20–27)

Write an algebraic expression to represent each verbal expression. **1. $12 - n^2$ 2. $2[n + (-9)]$**

1. twelve decreased by the square of a number
2. twice the sum of a number and negative nine
3. the product of the square of a number and 6 **$6n^2$**
4. the square of the sum of a number and 11 **$(n + 11)^2$**

Name the property illustrated by each statement. **6. Substitution (=) 8. Transitive (=)**

5. If $a + 1 = 6$, then $3(a + 1) = 3(6)$ **Multiplication (=)**
6. If $x + (4 + 5) = 21$, then $x + 9 = 21$
7. If $7x = 42$, then $7x - 5 = 42 - 5$ **Subtraction (=)**
8. If $3 + 5 = 8$ and $8 = 2 \cdot 4$, then $3 + 5 = 2 \cdot 4$.

Solve each equation.

9. $5t + 8 = 88$ **16**
10. $27 - x = -4$ **31**
11. $\frac{3}{4}y = \frac{2}{3}y + 5$ **60**
12. $8s - 3 = 5(2s + 1)$ **−4**
13. $3(k - 2) = k + 4$ **5**
14. $0.5z + 10 = z + 4$ **12**
15. $8q - \frac{q}{3} = 46$ **6**
16. $-\frac{2}{7}r + \frac{3}{7} = 5$ **−16**
17. $d - 1 = \frac{1}{2}(d - 2)$ **0**

Solve each equation or formula for the specified variable.

18. $C = \pi r$; for r **$r = \frac{C}{\pi}$**
19. $I = Prt$, for t **$t = \frac{I}{Pr}$**
20. $m = \frac{n - 2}{n}$, for n **$n = \frac{-2}{m - 1}$**

828 Extra Practice

Lesson 1-4

(pages 28–32)

Evaluate each expression if $x = -5$, $y = 3$, and $z = -2.5$.

1. $|2x|$ **10**
2. $|-3y|$ **9**
3. $|2x + y|$ **7**
4. $|y + 5z|$ **9.5**
5. $-|x + z|$ **−7.5**
6. $8 - |5y - 3|$ **−4**
7. $2|x| - 4|2 + y|$ **−10**
8. $|x + y| - 6|z|$ **−13**

Solve each equation.

9. $|d + 1| = 7$ **{6, −8}**
10. $|a - 6| = 10$ **{−4, 16}**
11. $2|x - 5| = 22$ **{−6, 16}**
12. $|t + 9| - 8 = 5$ **{−22, 4}**
13. $|p + 1| + 10 = 5$ **∅**
14. $6|g - 3| = 42$ **{−4, 10}**
15. $2|y + 4| = 14$ **{−11, 3}**
16. $|3b - 10| = 2b$ **{2, 10}**
17. $|3x + 7| + 4 = 0$ **∅**
18. $|2c + 3| - 15 = 0$ **{−9, 6}**
19. $7 - |m - 1| = 3$ **{−3, 5}**
20. $3 + |z + 5| = 10$ **{−12, 2}**
21. $4|h + 1| = 32$ **{−9, 7}**
22. $2|2x + 3| = 34$ **{−10, 7}**
23. $3|a - 5| - 4 = 14$ **{−1, 11}**
24. $2|2d - 7| + 1 = 35$ **{−5, 12}**
25. $|3t + 6| + 9 = 30$ **{−9, 5}**
26. $|d - 3| = 2d + 9$ **{−2}**
27. $|4y - 5| + 4 = 7y + 8$ **$\left\{\frac{1}{11}\right\}$**
28. $|2b + 4| - 3 = 6b + 1$ **{0}**
29. $|5t| + 2 = 3t + 18$ **{−2, 8}**

Lesson 1-5

(pages 33–39)

Solve each inequality. Describe the solution set using set-builder or interval notation. Then, graph the solution set on a number line. **1–21. See margin for interval notation and graphs.**

1. $2z + 5 \le 7$ **$\{z \mid z \le 1\}$**
2. $3r - 8 > 7$ **$\{r \mid r > 5\}$**
3. $0.75b < 3$ **$\{b \mid b < 4\}$**
4. $-3x > 6$ **$\{x \mid x < -2\}$**
5. $2(3f + 5) \ge 28$ **$\{f \mid f \ge 3\}$**
6. $-33 > 5g + 7$ **$\{g \mid g < -8\}$**
7. $-3(y - 2) \ge -9$ **$\{y \mid y \le 5\}$**
8. $7a + 5 > 4a - 7$ **$\{a \mid a > -4\}$**
9. $5(b - 3) \le b - 7$ **$\{b \mid b \le 2\}$**
10. $3(2x - 5) < 5(x - 4)$ **$\{x \mid x < -5\}$**
11. $2(4m - 1) + 3(m + 4) \ge 6m$
12. $8(2c - 1) > 11c + 22$ **$\{c \mid c > 6\}$**
13. $5y - 4(2y + 1) \le 2(0.5 - 2y)$
14. $2(d + 4) - 5 \ge 5(d + 3)$
15. $8 - 3t < 4(3 - t)$ **$\{t \mid t < 4\}$**
16. $-x \ge \frac{x + 4}{7}$ **$\{x \mid x \le -0.5\}$**
17. $\frac{a + 8}{4} \le \frac{7 + a}{3}$ **$\{a \mid a \ge -4\}$**
18. $-y < \frac{y + 5}{2}$ **$\left\{y \mid y > -\frac{5}{3}\right\}$**
19. $2 + 4(d - 2) \le 3 - (d - 1)$
20. $5(x - 1) - 4x \ge 3(3 - x)$
21. $6s - (4s + 7) > 5 - s$ **$\{s \mid s > 4\}$**

11. $\{m \mid m \ge -2\}$
13. $\{y \mid y \le 5\}$
14. $\{d \mid d \le -4\}$
19. $\{d \mid d \le 2\}$
20. $\left\{x \mid x \ge \frac{7}{2}\right\}$

Define a variable and write an inequality for each problem. Then solve the resulting inequality. **22–25. Let n = the number.**

22. The product of 7 and a number is greater than 42. **$7n > 42$; $n > 6$**
23. The difference of twice a number and 3 is at most 11. **$2n - 3 \le 11$; $n \le 7$**
24. The product of −10 and a number is greater than or equal to 20. **$-10n \ge 20$; $n \le -2$**
25. Thirty increased by a number is less than twice the number plus three. **$30 + n < 2n + 3$; $n > 27$**

Lesson 1-6

(pages 40–46)

Write an absolute value inequality for each of the following. Then graph the solution set on a number line. **1–3. See margin for graphs.**

1. all numbers less than −9 and greater than 9 **$|n| > 9$**
2. all numbers between −5.5 and 5.5 **$|n| < 5.5$**
3. all numbers greater than or equal to −2 and less than or equal to 2 **$|n| \le 2$**

Solve each inequality. Graph the solution set on a number line. **4–24. See pp. 861A–861T.**

4. $3m - 2 < 7$ or $2m + 1 > 13$
5. $2 < n + 4 < 7$
6. $-3 \le s - 2 \le 5$
7. $5t + 3 \le -7$ or $5t - 2 \ge 8$
8. $7 \le 4x + 3 \le 19$
9. $4x + 7 < 5$ or $2x - 4 > 12$
10. $|7x| \ge 21$
11. $|8p| \le 16$
12. $|7d| \ge -42$
13. $|a + 3| < 1$
14. $|t - 4| > 1$
15. $|2y - 5| < 3$
16. $|3d + 6| \ge 3$
17. $|4x - 1| < 5$
18. $|6v + 12| > 18$
19. $|2r + 4| < 6$
20. $|5w - 3| \ge 9$
21. $|z + 2| \ge 0$
22. $12 + |2q| < 0$
23. $|3h| + 15 < 0$
24. $|5n| - 16 \ge 4$

10. $(-\infty, -5)$
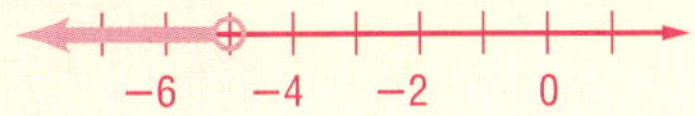

11. $[-2, +\infty)$
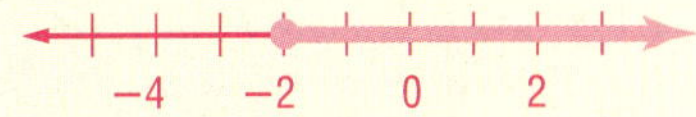

12. $(6, +\infty)$
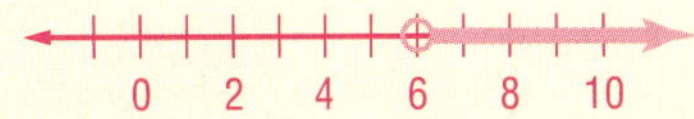

13. $(-\infty, 5]$
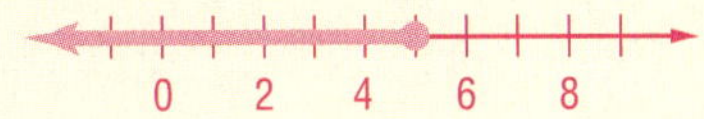

14. $(-\infty, -4]$
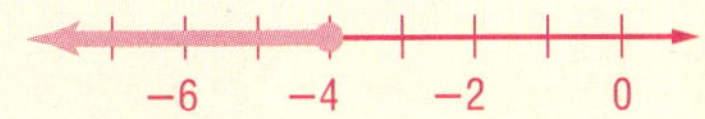

15. $(-\infty, 4)$
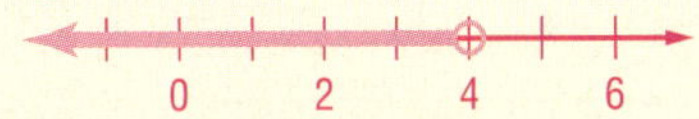

16. $(-\infty, -0.5]$
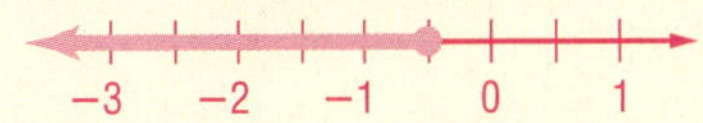

17. $[-4, +\infty)$
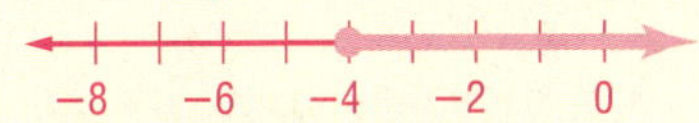

18. $\left(-\frac{5}{3}, +\infty\right)$

19. $(-\infty, 2]$
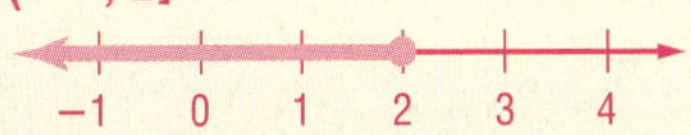

20. $\left[\frac{7}{2}, +\infty\right)$

21. $(4, +\infty)$
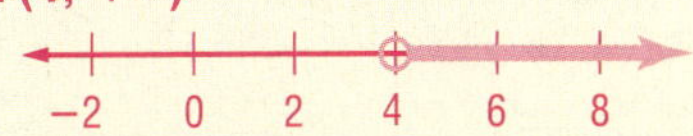

Lesson 1-6

1. 

2. 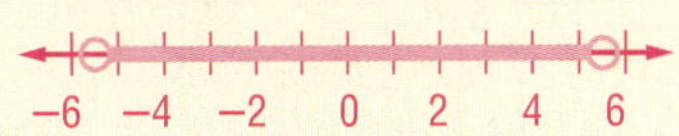

3. 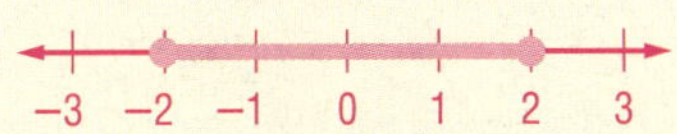

Lesson 1-5

1. $(-\infty, 1]$
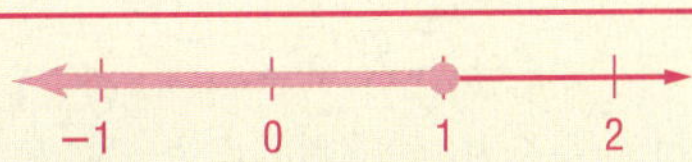

2. $(5, +\infty)$
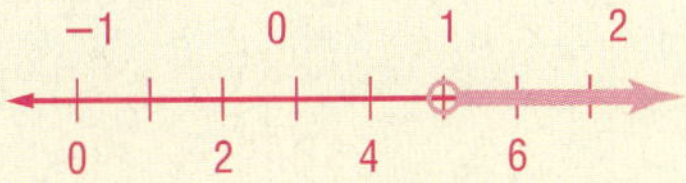

3. $(-\infty, 4)$
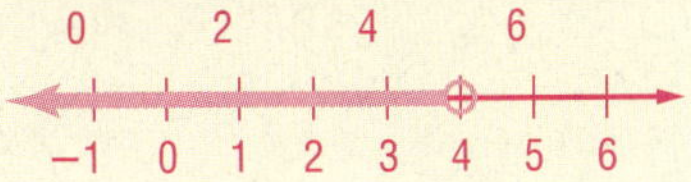

4. $(-\infty, -2)$
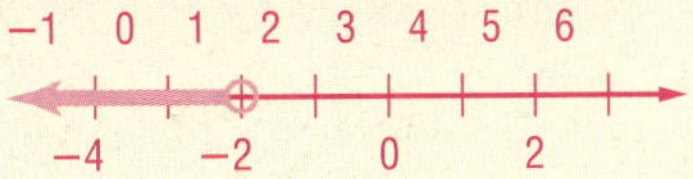

5. $[3, +\infty)$
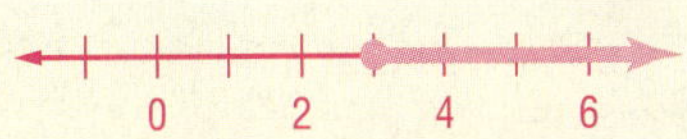

6. $(-\infty, -8)$

7. $(-\infty, 5]$
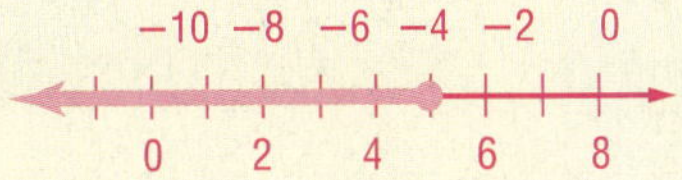

8. $(-4, +\infty)$
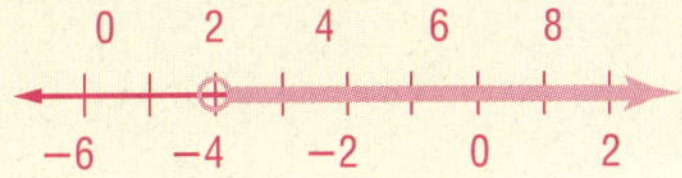

9. $(-\infty, 2]$
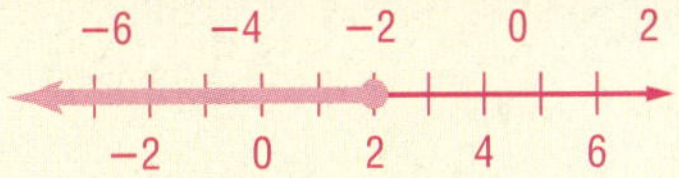

Extra Practice 829

Lesson 2-1

(pages 56–62)

Determine whether each relation is a function. Write *yes* or *no*.

1.

Year	Population
1970	11,605
1980	13,468
1990	15,630
2000	18,140

yes

2.

x	y
1	5
2	5
3	5
4	5

yes

3. 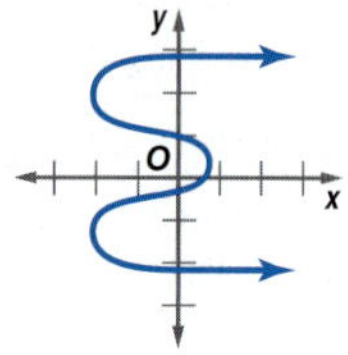

no

Graph each relation or equation and find the domain and range. Then determine whether the relation or equation is a function. **4–9. See margin.**

4. $\{(1, 2), (2, 3), (3, 4), (4, 5)\}$
5. $\{(0, 3), (0, 2), (0, 1), (0, 0)\}$
6. $y = -x$
7. $y = 2x - 1$
8. $y = 2x^2$
9. $y = -x^2$

Find each value if $f(x) = x + 7$ and $g(x) = (x + 1)^2$.

10. $f(2)$ **9**
11. $f(-4)$ **3**
12. $f(a + 2)$ **$a + 9$**
13. $g(4)$ **25**
14. $g(-2)$ **1**
15. $f(0.5)$ **7.5**
16. $g(b - 1)$ **b^2**
17. $g(3c)$ **$9c^2 + 6c + 1$**

Lesson 2-2

(pages 63–67)

State whether each equation or function is linear. Write *yes* or *no*. If no, explain. **2, 3, 6. See margin for explanations.**

1. $\frac{x}{2} - y = 7$ **yes**
2. $\sqrt{x} = y + 5$ **no**
3. $g(x) = \frac{2}{x - 3}$ **no**
4. $x = 3 + y$ **yes**
5. $f(x) = 7$ **yes**
6. $\frac{3}{x} - \frac{1}{4} = \frac{4}{3}$ **no**

Write each equation in standard form. Identify *A*, *B*, and *C*.

7. $x + 7 = y$ **$x - y = -7$; 1, −1, −7**
8. $x = -3y$ **$x + 3y = 0$; 1, 3, 0**
9. $5x = 7y + 3$ **$5x - 7y = 3$; 5, −7, 3**
10. $y = \frac{2}{3}x + 8$ **$2x - 3y = -24$; 2, −3, −24**
11. $-0.4x = 10$ **$x = -25$; 1, 0, −25**
12. $0.75y = -6$ **$y = -8$; 0, 1, −8**

Find the *x*-intercept and the *y*-intercept of the graph of each equation. Then graph the equation. **13–18. See pp. 861A–861T for graphs.**

13. $2x + y = 6$ **3; 6**
14. $3x - 2y = -12$ **−4; 6**
15. $y = -x$ **0; 0**
16. $x = 3y$ **0; 0**
17. $\frac{3}{4}y - x = 1$ **$-1; \frac{4}{3}$**
18. $y = -3$ **none; −3**

Lesson 2-3

(pages 68–74)

Find the slope of the line that passes through each pair of points.

1. (0, 3), (5, 0) **$-\frac{3}{5}$**
2. (2, 3), (5, 7) **$\frac{4}{3}$**
3. (2, 8), (2, −8) **undefined**
4. (1.5, −1), (3, 1.5) **$\frac{5}{3}$**
5. $\left(-\frac{1}{2}, \frac{3}{5}\right), \left(\frac{3}{10}, -\frac{1}{4}\right)$ **$-\frac{17}{16}$**
6. (−3, *c*), (4, *c*) **0**

Graph the line passing through the given point with the given slope. **7–9. See margin.**

7. (0, 3); 1
8. (2, 3); 0
9. (−1, 1); $-\frac{1}{3}$

Graph the line that satisfies each set of conditions. **10–13. See pp. 861A–861T.**

10. passes through (0, 1), parallel to a line whose slope is −2
11. passes through (2, −3), perpendicular to a line whose slope is 5
12. passes through (1, −1), parallel to the graph of $x + y = 3$
13. passes through (4, −5), perpendicular to the graph of $-2x + 5y = 1$

Lesson 2-1

4. D = {1, 2, 3, 4}; R = {2, 3, 4, 5}; yes

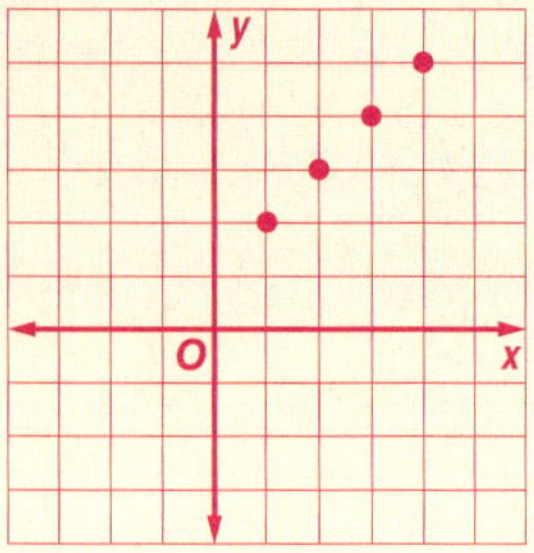

5. D = {0}; R = {0, 1, 2, 3}; no

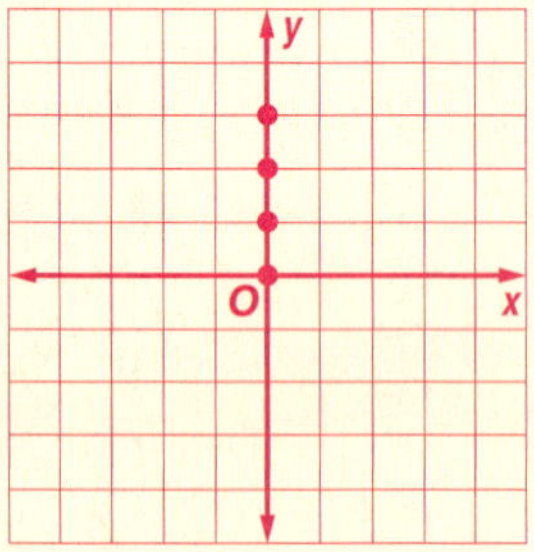

6. D = all real numbers; R = all real numbers; yes

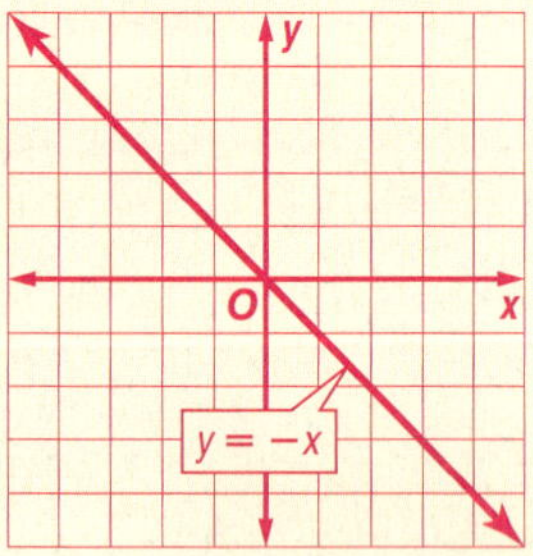

7. D = all real numbers; R = all real numbers; yes

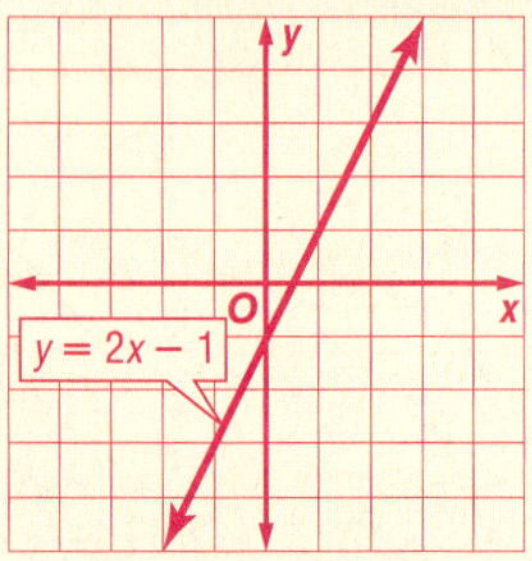

8. D = all real numbers; R = {y | y ≥ 0}; yes

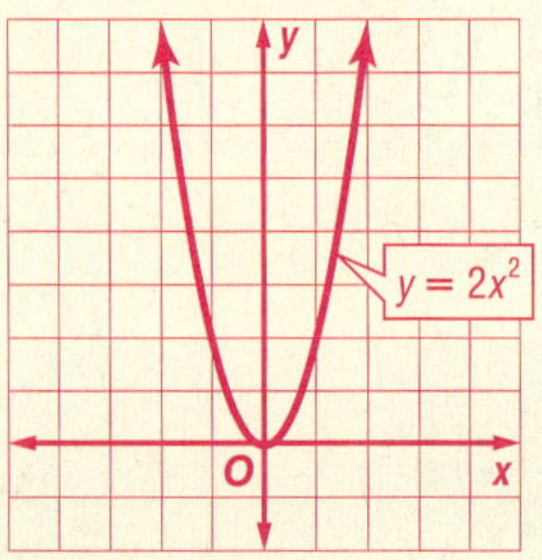

9. D = all real numbers; R = {y | y ≤ 0}; yes

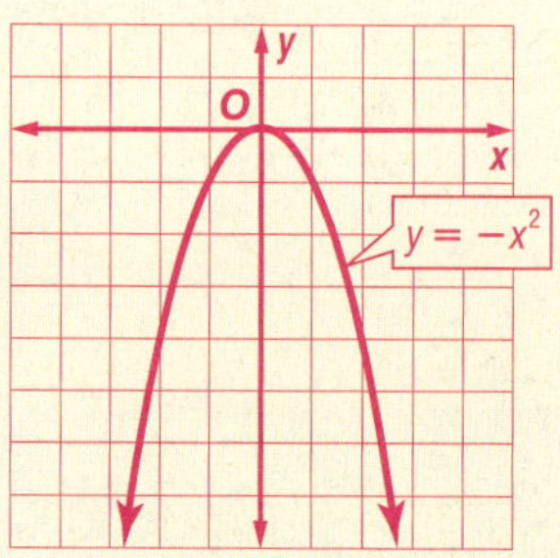

Lesson 2-2

2. *x* is inside a square root.

3. *x* appears in a denominator.

6. *x* appears in a denominator.

Lesson 2-4 *(pages 75–80)*

State the slope and y-intercept of the graph of each equation.

1. $y = -3x + 4$ **−3; 4**
2. $x - y = 5$ **1; −5**
3. $x = -4$ **undefined, none**

Write an equation in slope-intercept form for each graph.

4. 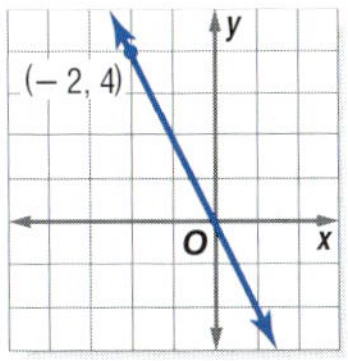

$y = -2x$

5. 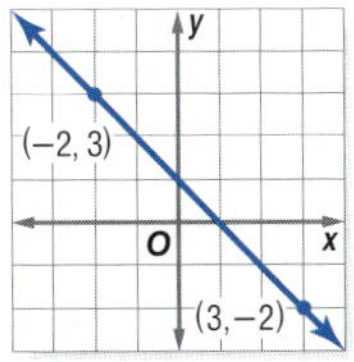

$y = -x + 1$

Write an equation in slope-intercept form for the line that satisfies each set of conditions. **6–12. See margin.**

6. slope −1, passes through (7, 2)
7. slope $\frac{3}{4}$, passes through the origin
8. passes through (1, −3) and (−1, 2)
9. x-intercept −5, y-intercept 2
10. passes through (1, 1), parallel to the graph of $2x + 3y = 5$
11. passes through (6, −2), perpendicular to the graph of $-x + 8y = 5$
12. passes through (0, 0), perpendicular to the graph of $2y + 3x = 4$

Lesson 2-5 *(pages 81–86)*

Complete parts a–c for each set of data in Exercises 1–3. **1–3. See margin.**

a. Draw a scatter plot.
b. Use two ordered pairs to write a prediction equation.
c. Use your prediction equation to predict the missing value.

1.

Telephone Costs	
Minutes	**Cost ($)**
1	0.20
3	0.52
4	0.68
6	1.00
9	1.48
15	?

2.

Washington	
Year	**Population**
1960	2,853,214
1970	3,413,244
1980	4,132,353
1990	4,866,669
2000	5,894,121
2010	?

Source: *The World Almanac*

3.

Federal Minimum Wage	
Year	**Wage**
1981	$3.35
1990	$3.80
1991	$4.25
1996	$4.75
1997	$5.15
2005	?

Source: *The World Almanac*

Lesson 2-6 *(pages 89–95)*

Identify each function as S for step, C for constant, A for absolute value, or P for piecewise.

1. 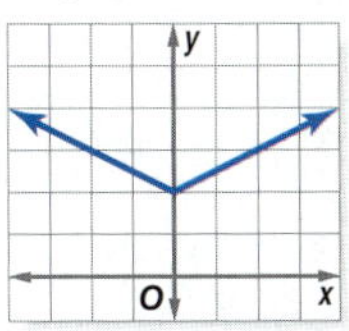

A

2. 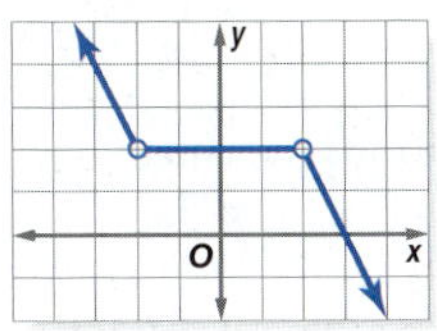

P

Graph each function. Identify the domain and range. **3–10. See pp. 861A–861T.**

3. $f(x) = [\![x + 5]\!]$
4. $g(x) = [\![x]\!] - 2$
5. $f(x) = -2[\![x]\!]$
6. $h(x) = |x| - 3$
7. $h(x) = |x - 1|$
8. $g(x) = |2x| + 2$
9. $h(x) = \begin{cases} x \text{ if } x < -2 \\ 4 \text{ if } x \geq -2 \end{cases}$
10. $f(x) = \begin{cases} -3 \text{ if } x \leq 1 \\ -x \text{ if } x > 1 \end{cases}$

Extra Practice

Lesson 2-4

6. $y = -x + 9$
7. $y = \frac{3}{4}x$
8. $y = -\frac{5}{2}x - \frac{1}{2}$
9. $y = \frac{2}{5}x + 2$
10. $y = -\frac{2}{3}x + \frac{5}{3}$
11. $y = -8x + 46$
12. $y = \frac{2}{3}x$

Lesson 2-5

1a.

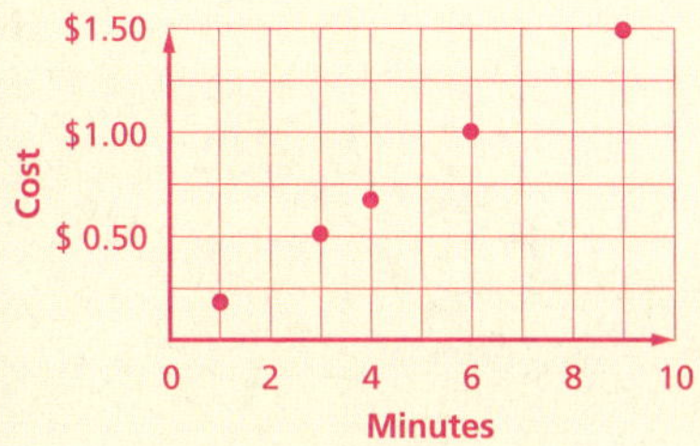

1b. $y = 0.16x + 0.04$

1c. $2.44

2a.

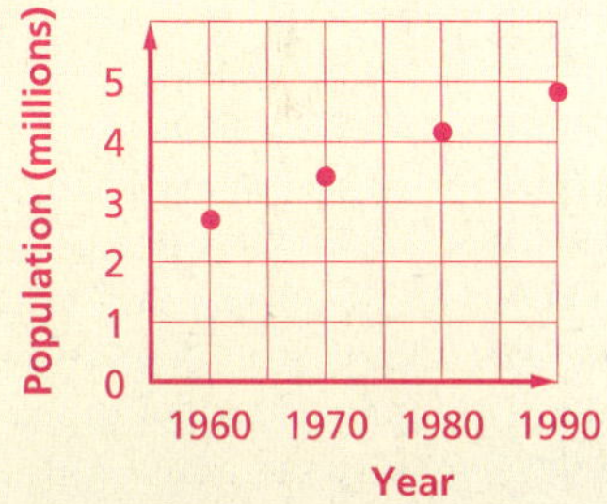

2b. Sample answer using (1960, 2853214) and (2000, 5894121): $y = 76{,}022.675x - 146{,}151{,}229$

2c. Sample answer: 6,654,348

3a.

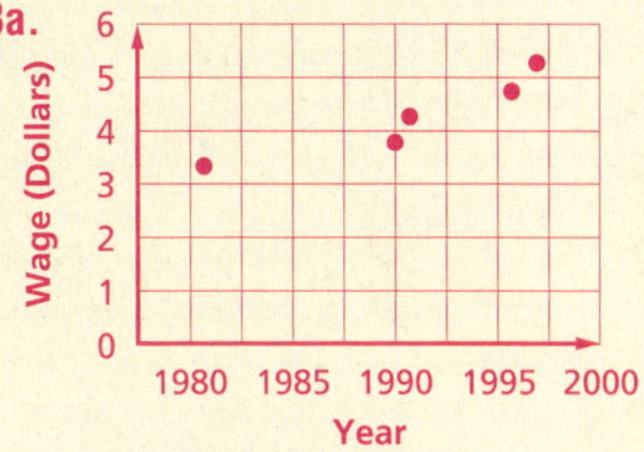

3b. Sample answer using (1991, 4.25) and (1996, 4.75): $y = 0.1x - 194.85$

3c. Sample answer: $5.65

Lesson 2-3

7. 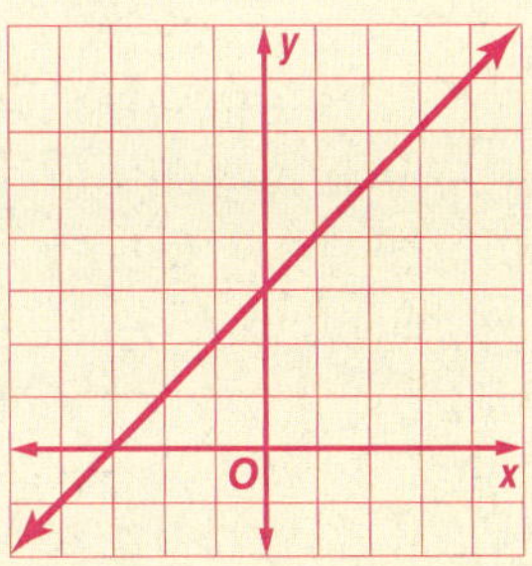

8. 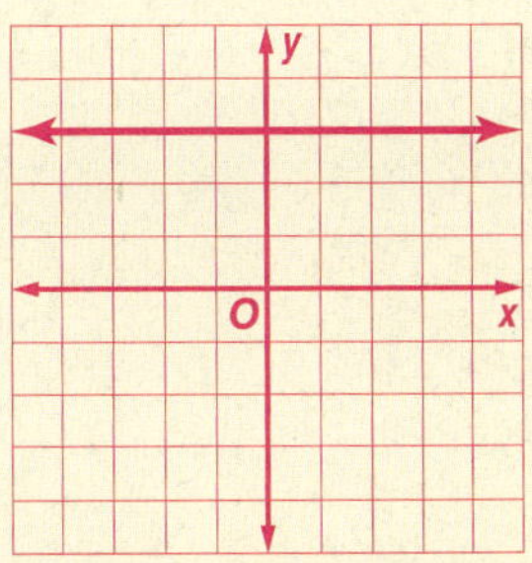

9. 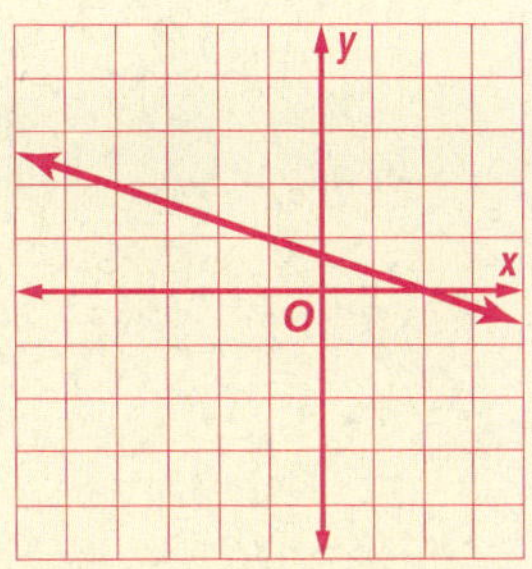

Lesson 3-1

1.
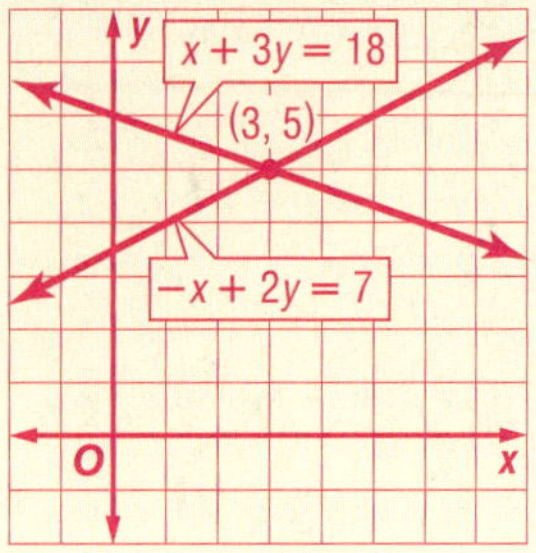

2.
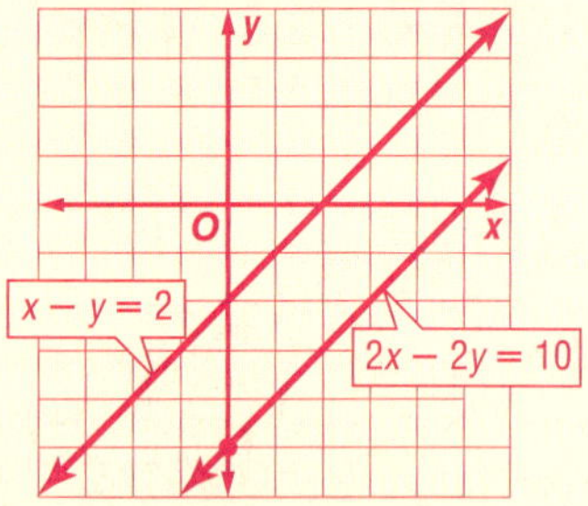

3.
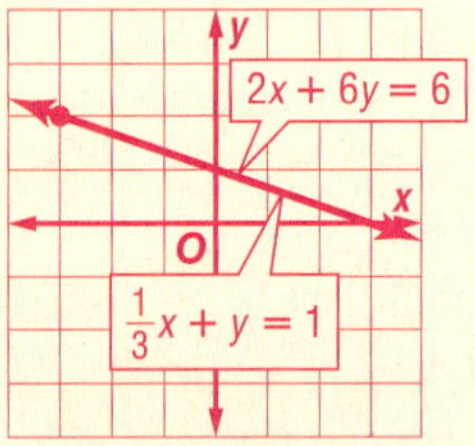

4.
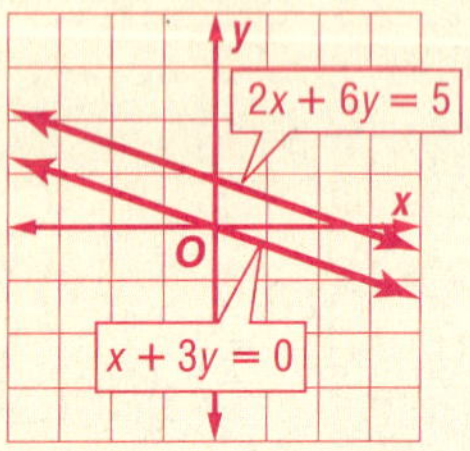

5.
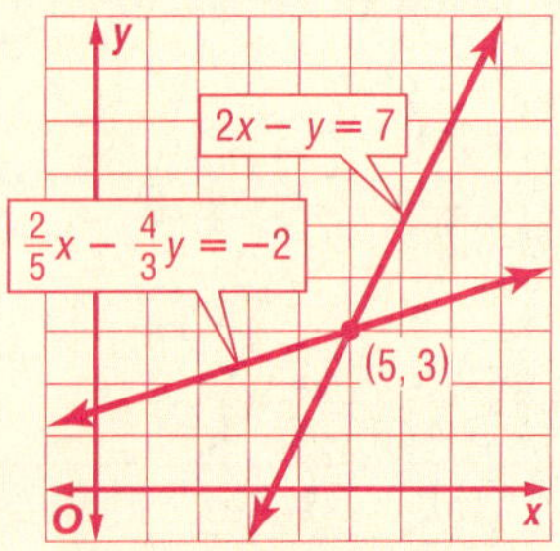

6.
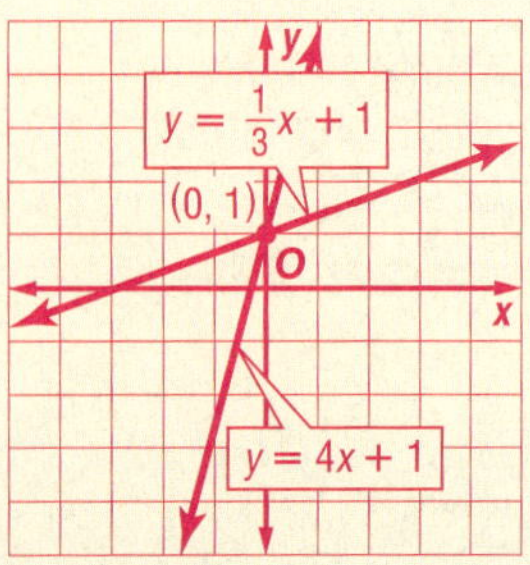

Extra Practice

Extra Practice

Lesson 2-7

(pages 96–99)

Graph each inequality. 1–21. See pp. 861A–861T.

1. $y \geq x - 2$
2. $y < -3x - 1$
3. $4y \leq -3x + 8$
4. $3x > y$
5. $x + 2 \geq y - 7$
6. $2x < 5 - y$
7. $y > \frac{1}{5}x - 8$
8. $2y - 5x \leq 8$
9. $-2x + 5 \leq \frac{2}{3}y$
10. $3x + 2y \geq 0$
11. $x \leq 2$
12. $\frac{y}{2} \leq x - 1$
13. $y - 3 < 5$
14. $y \geq -|x|$
15. $|x| \leq y + 3$
16. $y > |5x - 3|$
17. $y \leq |8 - x|$
18. $y < |x + 3| - 1$
19. $y + |2x| \geq 4$
20. $y \geq |2x - 1| + 5$
21. $y < \left|\frac{2x}{3}\right| - 1$

Lesson 3-1

(pages 110–115)

Solve each system of equations by graphing. 1–6. See margin for graphs.

1. $x + 3y = 18$
   $-x + 2y = 7$ **(3, 5)**
2. $x - y = 2$
   $2x - 2y = 10$ **no solution**
3. $2x + 6y = 6$
   $\frac{1}{3}x + y = 1$ **infinite solutions**
4. $x + 3y = 0$
   $2x + 6y = 5$ **no solution**
5. $2x - y = 7$
   $\frac{2}{5}x - \frac{4}{3}y = -2$ **(5, 3)**
6. $y = \frac{1}{3}x + 1$
   $y = 4x + 1$ **(0, 1)**

Graph each system of equations and describe it as *consistent and independent*, *consistent and dependent*, or *inconsistent*. 7–12. See pp. 861A–861T for graphs.

7. $2x + 3y = 5$
   $-6x - 9y = -15$ **consistent, dependent**
8. $x - 2y = 4$
   $y = x - 2$ **consistent, independent**
9. $y = 0.5x$
   $2y = x + 4$ **inconsistent**
10. $9x - 5 = 7y$
    $4.5x - 3.5y = 2.5$ **consistent, dependent**
11. $\frac{3}{4}x - y = 0$
    $\frac{1}{3}y + \frac{1}{2}x = 6$ **consistent, independent**
12. $\frac{2}{3}x = \frac{5}{3}y$
    $2x - 5y = 0$ **consistent, dependent**

Lesson 3-2

(pages 116–122)

Solve each system of equations by using substitution.

1. $2x + 3y = 10$
   $x + 6y = 32$ **(−4, 6)**
2. $x = 4y - 10$
   $5x + 3y = -4$ **(−2, 2)**
3. $3x - 4y = -27$
   $2x + y = -7$ **(−5, 3)**

Solve each system of equations by using elimination.

4. $7x + y = 9$
   $5x - y = 15$ **(2, −5)**
5. $r + 5s = -17$
   $2r - 6s = -2$ **(−7, −2)**
6. $6p + 8q = 20$
   $5p - 4q = -26$ **(−2, 4)**

Solve each system of equations by using either substitution or elimination.

7. $2x - 3y = 7$
   $3x + 6y = 42$ **(8, 3)**
8. $2a + 5b = -13$
   $3a - 4b = 38$ **(6, −5)**
9. $3c + 4d = -1$
   $6c - 2d = 3$ $\left(\frac{1}{3}, -\frac{1}{2}\right)$
10. $7x - y = 35$
    $y = 5x - 19$ **(8, 21)**
11. $3m + 4n = 28$
    $5m - 3n = -21$ **(0, 7)**
12. $x = 2y - 1$
    $4x - 3y = 21$ **(9, 5)**
13. $2.5x + 1.5y = -2$
    $3.5x - 0.5y = 18$ **(4, −8)**
14. $\frac{5}{2}x + \frac{1}{3}y = 13$
    $\frac{1}{2}x - y = -7$ **(4, 9)**
15. $\frac{2}{7}c - \frac{4}{3}d = 16$
    $\frac{4}{7}c + \frac{8}{3}d = -16$ **(14, −9)**

832 Extra Practice

Lesson 3-3 *(pages 123–127)*

Solve each system of inequalities by graphing. **1–12. See pp. 861A–861T.**

1. $x \le 5$ $y \ge -3$	**2.** $y < 3$ $y - x \ge -1$	**3.** $x + y < 5$ $x < 2$
4. $y + x < 2$ $y \ge x$	**5.** $x + y \le 2$ $y - x \le 4$	**6.** $y \le x + 4$ $y - x \ge 1$
7. $y < \frac{1}{3}x + 5$ $y > 2x + 1$	**8.** $y + x \ge 1$ $y - x \ge -1$	**9.** $\lvert x \rvert > 2$ $\lvert y \rvert \le 5$
10. $\lvert x - 3 \rvert \le 3$ $4y - 2x \le 6$	**11.** $4x + 3y \ge 12$ $2y - x \ge -1$	**12.** $y \le -1$ $3x - 2y \ge 6$

Find the coordinates of the vertices of the figure formed by each system of inequalities. **13–15. See margin for graphs.**

13. $y \le 3$ $x \le 2$ **(0, 3), (2, 3), (2, 0)** $y \ge -\frac{3}{2}x + 3$	**14.** $y \ge -1$ $y \le x$ $y \le -x + 4$ **(2, 2), (−1, −1), (5, −1)**	**15.** $y \le \frac{1}{3}x + \frac{7}{3}$ $4x - y \le 5$ $y \ge -\frac{3}{2}x + \frac{1}{2}$ **(2, 3), (1, −1), (−1, 2)**

Lesson 3-4 *(pages 129–135)*

A feasible region has vertices at (−3, 2), (1, 3), (6, 1), and (2, −2). Find the maximum and minimum values of each function.

1. $f(x, y) = 2x - y$ **11; −8**	**2.** $f(x, y) = x + 5y$ **16; −8**	**3.** $f(x, y) = y - 4x$ **14; −23**
4. $f(x, y) = -x + 3y$ **9; −8**	**5.** $f(x, y) = 3x - y$ **17; −11**	**6.** $f(x, y) = 2y - 2x$ **10; −10**

Graph each system of inequalities. Name the coordinates of the vertices of the feasible region. Find the maximum and minimum values of the given function for this region. **7–12. See margin for graphs.**

7. $4x - 5y \le -10$ $y \le 6$ **(0, 2), (5, 6), (−2, 6)** $2x + y \ge 2$ **11; 2** $f(x, y) = x + y$	**8.** $x \le 5$ $y \ge 2$ **(0, 2), (5, 2), (5, 4)** $2x - 5y \ge -10$ **19; 2** $f(x, y) = 3x + y$	**9.** $x - 2y \ge -7$ $x + y \le 8$ **(−1, 3);** $y \ge 5x + 8$ **no maximum;** $f(x, y) = 3x - 4y$ **no minimum**
10. $y \le 4x + 6$ **(−1, 2), (3, 1),** $x + 4y \ge 7$ $\left(\frac{1}{6}, 6\frac{2}{3}\right)$ $2x + y \le 7$ $f(x, y) = 2x - y$ **5; $-6\frac{1}{3}$**	**11.** $y \ge 0$ $y \le 5$ **(4, 0), (7, 0), (2, 5), (1, 5)** $y \le -x + 7$ **12; 4** $5x + 3y \ge 20$ $f(x, y) = x + 2y$	**12.** $y \ge 0$ **(0, 0), (2, 3), (5, 2), (8, 0)** $3x - 2y \ge 0$ **32; 0** $x + 3y \le 11$ $2x + 3y \le 16$ $f(x, y) = 4x + y$

Lesson 3-5 *(pages 138–144)*

For each system of equations, an ordered triple is given. Determine whether or not it is a solution of the system.

1. $4x + 2y - 6z = -38$ $5x - 4y + z = -18$ $x + 3y + 7z = 38$; $(-3, 2, 5)$ **yes**	**2.** $u + 3v + w = 14$ $2u - v + 3w = -9$ $4u - 5v - 2w = -2$; $(1, 5, -2)$ **no**	**3.** $x + y = -6$ $x + z = -2$ $y + z = 2$; $(-4, -2, 2)$ **no**

Solve each system of equations.

4. $5a = 5$ $6b - 3c = 15$ $2a + 7c = -5$ **(1, 2, −1)**	**5.** $s + 2t = 5$ $7r - 3s + t = 20$ $2t = 8$ **(1, −3, 4)**	**6.** $2u - 3v = 13$ $3v + w = -3$ $4u - w = 2$ **(2, −3, 6)**
7. $4a + 2b - c = 5$ $2a + b - 5c = -11$ $a - 2b + 3c = 6$ **(1, 2, 3)**	**8.** $x + 2y - z = 1$ $x + 3y + 2z = 7$ $2x + 6y + z = 8$ **(3, 0, 2)**	**9.** $2x + y - z = 7$ $3x - y + 2z = 15$ $x - 4y + z = 2$ **(4, 1, 2)**

Extra Practice

Lesson 3-4

7.
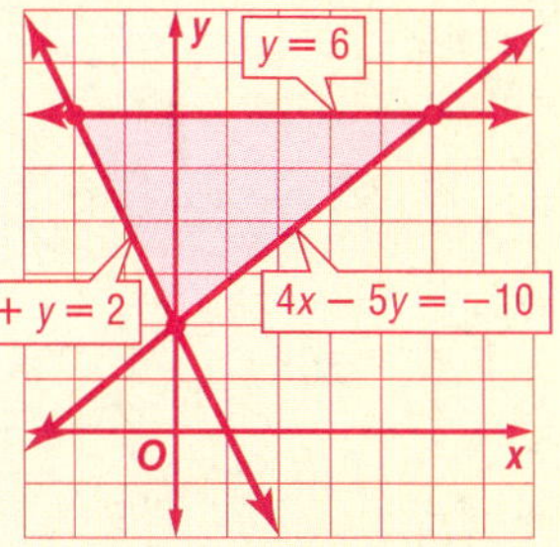

8.
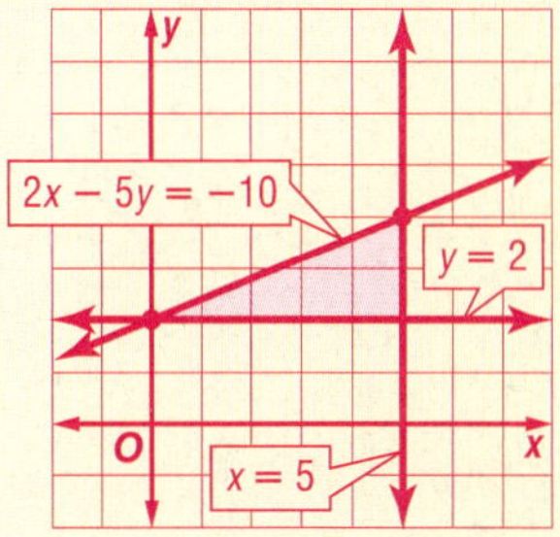

9.
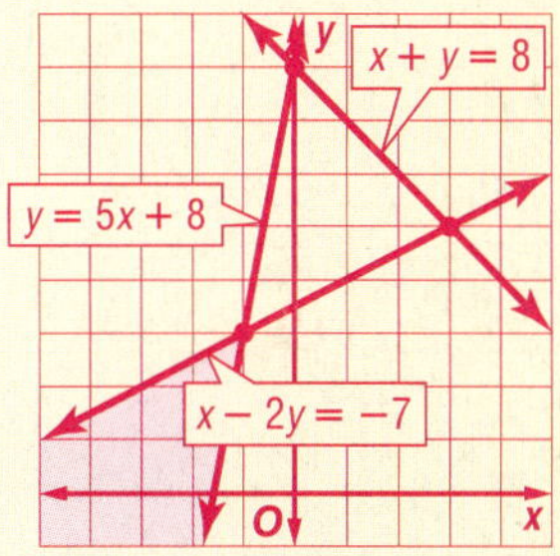

10.
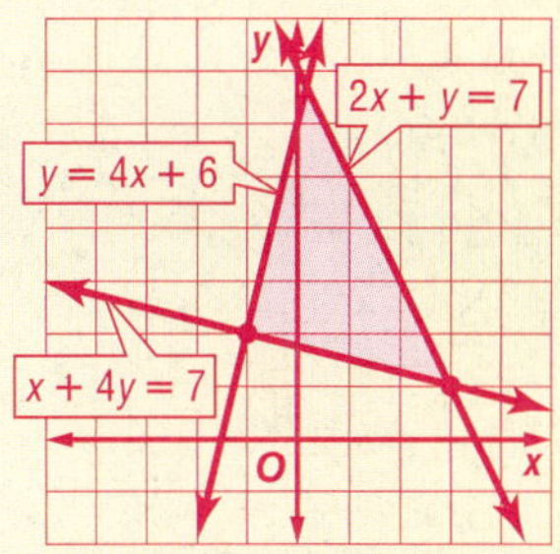

11.
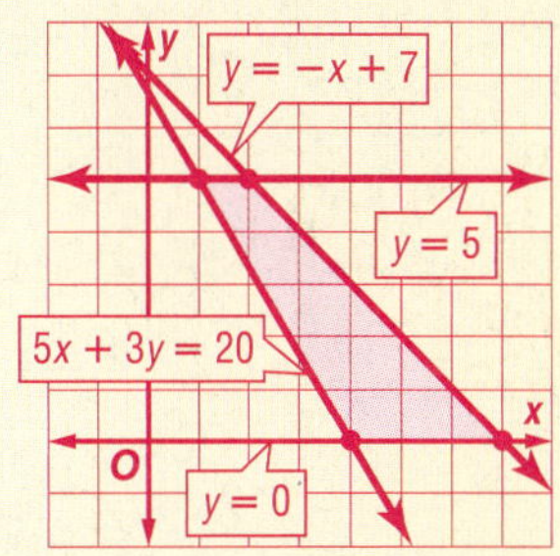

12.
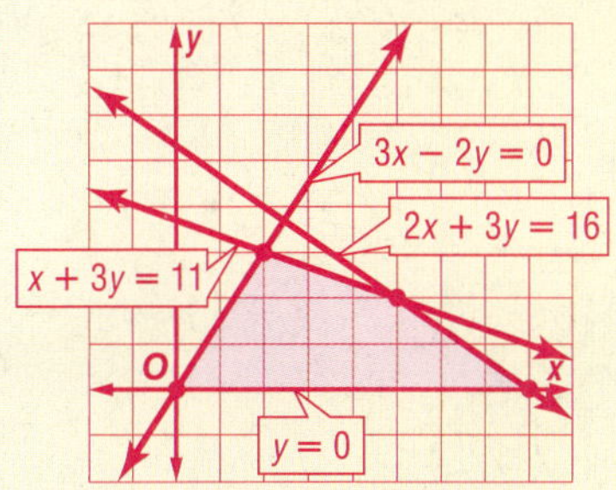

Lesson 3-3

13.
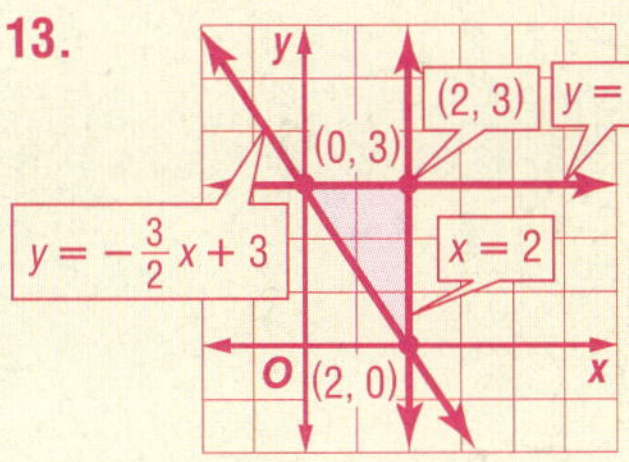

14.
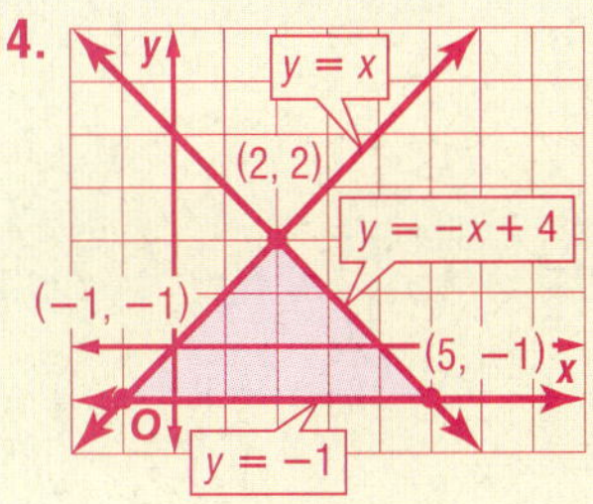

15.
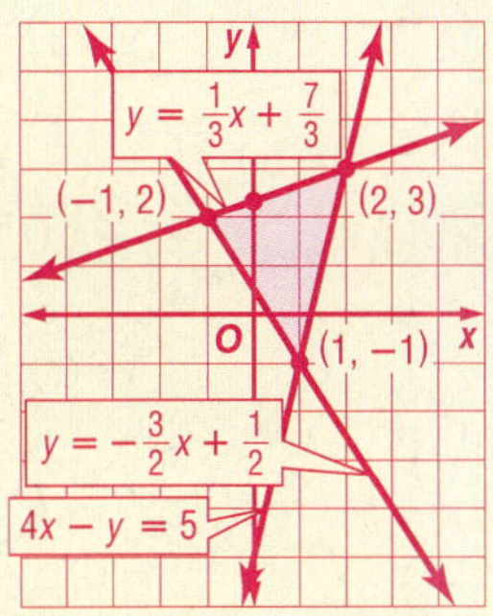

Lesson 4-1

(pages 154–158)

State the dimensions of each matrix.

1. $[8 \quad 4 \quad 3]$ **1 × 3**
2. $\begin{bmatrix} 2 & -1 \\ 0 & 6 \end{bmatrix}$ **2 × 2**
3. $\begin{bmatrix} 4 \\ -3 \end{bmatrix}$ **2 × 1**

Solve each matrix equation. **4. $x = 5, y = 5, z = -15$**

4. $[2x \quad 3y \quad -z] = [2y \quad -z \quad 15]$
5. $\begin{bmatrix} x + y \\ 4x - 3y \end{bmatrix} = \begin{bmatrix} 1 \\ 11 \end{bmatrix}$ **$x = 2, y = -1$**
6. $-2\begin{bmatrix} w + 5 & x - z \\ 3y & 8 \end{bmatrix} = \begin{bmatrix} -16 & -4 \\ 6 & 2x + 8z \end{bmatrix}$
7. $y\begin{bmatrix} 2 & x \\ 5 & 1 \end{bmatrix} = \begin{bmatrix} 4 & -10 \\ 10 & 2z \end{bmatrix}$ **$x = -5, y = 2, z = 1$**
8. $\begin{bmatrix} 2x \\ -y \\ 3z \end{bmatrix} = \begin{bmatrix} 16 \\ 18 \\ -21 \end{bmatrix}$ **$x = 8, y = -18, z = -7$**
9. $\begin{bmatrix} x - 3y \\ 4y - 3x \end{bmatrix} = -5\begin{bmatrix} 2 \\ x \end{bmatrix}$ **$x = -4, y = 2$**
10. $\begin{bmatrix} x^2 + 4 & y + 6 \\ x - y & 2 - y \end{bmatrix} = \begin{bmatrix} 5 & 7 \\ 0 & 1 \end{bmatrix}$ **$x = 1, y = 1$**
11. $\begin{bmatrix} x + y & 3 \\ y & 6 \end{bmatrix} = \begin{bmatrix} 0 & 2y - x \\ z & 4 - 2x \end{bmatrix}$ **$x = -1, y = 1, z = 1$**

6. $w = 3, x = 0, y = -1, z = -2$

Lesson 4-2

(pages 160–166)

Perform the indicated matrix operations. If the matrix does not exist, write impossible. **1–8. See margin.**

1. $\begin{bmatrix} 3 & 5 \\ -7 & 2 \end{bmatrix} + \begin{bmatrix} -2 & 6 \\ 8 & -1 \end{bmatrix}$
2. $[0 \quad -1 \quad 3] + \begin{bmatrix} 5 \\ -2 \\ -3 \end{bmatrix}$
3. $\begin{bmatrix} 45 & 36 & 18 \\ 63 & 29 & 5 \end{bmatrix} - \begin{bmatrix} 45 & -2 & 36 \\ 18 & 9 & -10 \end{bmatrix}$
4. $4[-8 \quad 2 \quad 9] - 3[2 \quad -7 \quad 6]$
5. $\begin{bmatrix} -3 & 6 & -9 \\ 4 & -3 & 0 \\ 8 & -2 & 3 \end{bmatrix} - \begin{bmatrix} 1 & 5 & 7 \\ 5 & 2 & -6 \\ 3 & 0 & -2 \end{bmatrix}$
6. $5\begin{bmatrix} 3 & 1 & 0 \\ 0 & 0 & 2 \\ 1 & -1 & -1 \end{bmatrix} + \begin{bmatrix} 2 & 0 & 3 \\ 1 & 1 & 2 \\ 2 & 1 & -1 \end{bmatrix}$
7. $5\begin{bmatrix} 6 & -2 \\ 5 & 4 \end{bmatrix} - 2\begin{bmatrix} 6 & -2 \\ 5 & 4 \end{bmatrix} + 4\begin{bmatrix} 7 & -6 \\ -4 & 2 \end{bmatrix}$
8. $1.3\begin{bmatrix} 3.7 \\ -5.4 \end{bmatrix} + 4.1\begin{bmatrix} 6.4 \\ -3.7 \end{bmatrix} - 6.2\begin{bmatrix} -0.8 \\ 7.4 \end{bmatrix}$

Use the matrices A, B, C, D, and E to find the following. **9–16. See margin.**

$A = \begin{bmatrix} 1 & 0 \\ 0 & 1 \end{bmatrix}$, $B = \begin{bmatrix} -1 & 0 \\ 0 & -1 \end{bmatrix}$, $C = \begin{bmatrix} 2 & -2 \\ -3 & 3 \end{bmatrix}$, $D = \begin{bmatrix} -2 & 2 \\ 3 & -3 \end{bmatrix}$, $E = \begin{bmatrix} 5 & -3 \\ -2 & 4 \end{bmatrix}$

9. $A + B$
10. $C + D$
11. $A - B$
12. $4B$
13. $D - C$
14. $E + 2A$
15. $D - 2B$
16. $2A + 3E - D$

Lesson 4-3

(pages 167–174)

Determine whether each matrix product is defined. If so, state the dimensions of the product. **3. undefined**

1. $A_{4 \times 3} \cdot B_{3 \times 4}$ **4 × 4**
2. $R_{2 \times 5} \cdot S_{2 \times 5}$ **undefined**
3. $G_{2 \times 5} \cdot H_{2 \times 3}$
4. $M_{3 \times 8} \cdot N_{8 \times 2}$ **3 × 2**
5. $X_{4 \times 3} \cdot Y_{2 \times 6}$ **undefined**
6. $J_{5 \times 3} \cdot K_{3 \times 6}$ **5 × 6**
7. $C_{m \times n} \cdot D_{n \times p}$ **$m \times p$**
8. $X_{2 \times 9} \cdot Y_{9 \times 7}$ **2 × 7**

Find each product, if possible.

9. $[-3 \quad 4] \cdot \begin{bmatrix} -1 \\ 2 \end{bmatrix}$ **[11]**
10. $\begin{bmatrix} 2 & -4 \\ 0 & 5 \end{bmatrix} \cdot \begin{bmatrix} 1 & 3 \\ -2 & -1 \end{bmatrix}$ $\begin{bmatrix} 10 & 10 \\ -10 & -5 \end{bmatrix}$
11. $\begin{bmatrix} 1 & 3 \\ -2 & -1 \end{bmatrix} \cdot \begin{bmatrix} 2 & -4 \\ 0 & 5 \end{bmatrix}$ $\begin{bmatrix} 2 & 11 \\ -4 & 3 \end{bmatrix}$
12. $\begin{bmatrix} 3 & 2 \\ 5 & 2 \end{bmatrix} \cdot \begin{bmatrix} -8 \\ 15 \end{bmatrix}$ $\begin{bmatrix} 6 \\ -10 \end{bmatrix}$
13. $\begin{bmatrix} -1 \\ 2 \\ 1 \end{bmatrix} \cdot \begin{bmatrix} 7 & 6 & 1 \\ 2 & -4 & 0 \end{bmatrix}$ **not possible**
14. $\begin{bmatrix} 0 & 1 & -2 \\ 5 & 3 & -4 \\ -1 & 0 & 0 \end{bmatrix} \cdot \begin{bmatrix} 1 & -3 & 0 \\ 2 & 0 & -1 \\ 0 & 1 & -2 \end{bmatrix}$ $\begin{bmatrix} 2 & -2 & 3 \\ 11 & -19 & 5 \\ -1 & 3 & 0 \end{bmatrix}$
15. $\begin{bmatrix} 3 & -2 \\ 4 & 5 \end{bmatrix} \cdot \begin{bmatrix} 1 & 0 \\ 0 & 1 \end{bmatrix}$ $\begin{bmatrix} 3 & -2 \\ 4 & 5 \end{bmatrix}$
16. $\begin{bmatrix} -1 & 0 & 2 \\ -6 & 5 & -3 \end{bmatrix} \cdot \begin{bmatrix} -2 \\ 1 \\ 7 \end{bmatrix}$ $\begin{bmatrix} 16 \\ -4 \end{bmatrix}$

Lesson 4-2

1. $\begin{bmatrix} 1 & 11 \\ 1 & 1 \end{bmatrix}$
2. impossible
3. $\begin{bmatrix} 0 & 38 & -18 \\ 45 & 20 & 15 \end{bmatrix}$
4. $[-38 \quad 29 \quad 18]$
5. $\begin{bmatrix} -4 & 1 & -16 \\ -1 & -5 & 6 \\ 5 & -2 & 5 \end{bmatrix}$
6. $\begin{bmatrix} 17 & 5 & 3 \\ 1 & 1 & 12 \\ 7 & -4 & -6 \end{bmatrix}$
7. $\begin{bmatrix} 46 & -30 \\ -1 & 20 \end{bmatrix}$
8. $\begin{bmatrix} 36.01 \\ -68.07 \end{bmatrix}$
9. $\begin{bmatrix} 0 & 0 \\ 0 & 0 \end{bmatrix}$
10. $\begin{bmatrix} 0 & 0 \\ 0 & 0 \end{bmatrix}$
11. $\begin{bmatrix} 2 & 0 \\ 0 & 2 \end{bmatrix}$
12. $\begin{bmatrix} -4 & 0 \\ 0 & -4 \end{bmatrix}$
13. $\begin{bmatrix} -4 & 4 \\ 6 & -6 \end{bmatrix}$
14. $\begin{bmatrix} 7 & -3 \\ -2 & 6 \end{bmatrix}$
15. $\begin{bmatrix} 0 & 2 \\ 3 & -1 \end{bmatrix}$
16. $\begin{bmatrix} 19 & -11 \\ -9 & 17 \end{bmatrix}$

Lesson 4-4

(pages 175–181)

For Exercises 1–3, use the following information.
Triangle XYZ with vertices $X(2, 5)$, $Y(-3, 1)$, and $Z(1, -4)$ is translated 3 units right and 2 units down.

1. Write the translation matrix. **See margin.**
2. Find the coordinates of $\triangle X'Y'Z'$. **$X'(5, 3)$, $Y'(0, -1)$, and $Z'(4, -6)$**
3. Graph the preimage and the image. **See margin.**

For Exercises 4–6, use the following information.
The vertices of quadrilateral $ABCD$ are $A(1, 1)$, $B(-2, 3)$, $C(-4, -1)$, and $D(2, -3)$. The quadrilateral is dilated so that its perimeter is 2 times the original perimeter.

4. Write the coordinates for $ABCD$ in a vertex matrix. **See margin.**
5. Find the coordinates of the image $A'B'C'D'$. **$A'(2, 2)$, $B'(-4, 6)$, $C'(-8, -2)$, $D'(4, -6)$**
6. Graph $ABCD$ and $A'B'C'D'$. **See margin.**

For Exercises 7–13, use the following information.
The vertices of $\triangle MQN$ are $M(2, 4)$, $Q(3, -5)$, and $N(1, -1)$.

7. Write the coordinates of $\triangle MQN$ in a vertex matrix. **See margin.**
8. Write the reflection matrix for reflecting over the line $y = x$. **See margin.**
9. Find the coordinates of $\triangle M'Q'N'$ after the reflection. **$M'(4, 2)$, $Q'(-5, 3)$, $N'(-1, 1)$**
10. Graph $\triangle MQN$ and $\triangle M'Q'N'$. **See margin.**
11. Write a rotation matrix for rotating $\triangle MQN$ 90° counterclockwise about the origin. **See margin.**
12. Find the coordinates of $\triangle M'Q'N'$ after the rotation. **$M'(-4, 2)$, $Q'(5, 3)$, $N'(1, 1)$**
13. Graph $\triangle MQN$ and $\triangle M'Q'N'$. **See margin.**

Lesson 4-5

(pages 182–188)

Find the value of each determinant.

1. $\begin{vmatrix} 1 & -5 \\ 3 & 4 \end{vmatrix}$ **19**
2. $\begin{vmatrix} 1 & 1 \\ 1 & 1 \end{vmatrix}$ **0**
3. $\begin{vmatrix} 2 & -2 \\ 3 & -3 \end{vmatrix}$ **0**
4. $\begin{vmatrix} 2 & -3 \\ -3 & -2 \end{vmatrix}$ **−13**

Evaluate each determinant using expansion by minors.

5. $\begin{vmatrix} 2 & -3 & 5 \\ 1 & -2 & -7 \\ -1 & 4 & -3 \end{vmatrix}$ **48**
6. $\begin{vmatrix} 0 & -1 & 2 \\ -2 & 1 & 0 \\ 2 & 0 & -1 \end{vmatrix}$ **−2**
7. $\begin{vmatrix} 4 & 3 & -2 \\ 2 & 5 & -8 \\ 6 & 4 & -1 \end{vmatrix}$ **14**
8. $\begin{vmatrix} -3 & 0 & 2 \\ 1 & -2 & -1 \\ 0 & 5 & 0 \end{vmatrix}$ **−5**

Evaluate each determinant using diagonals.

9. $\begin{vmatrix} 3 & 2 & -1 \\ 2 & 3 & 0 \\ -1 & 0 & 3 \end{vmatrix}$ **12**
10. $\begin{vmatrix} 1 & 0 & 0 \\ 0 & 1 & 0 \\ 0 & 0 & 1 \end{vmatrix}$ **1**
11. $\begin{vmatrix} 6 & 4 & -1 \\ 2 & 5 & -8 \\ 4 & 3 & -2 \end{vmatrix}$ **−14**
12. $\begin{vmatrix} 6 & 12 & 15 \\ 9 & 3 & 14 \\ 5 & 6 & 3 \end{vmatrix}$ **651**

Lesson 4-6

(pages 189–194)

Use Cramer's Rule to solve each system of equations.

1. $5x - 3y = 19$; $7x + 2y = 8$ **(2, −3)**
2. $4p - 3q = 22$; $2p + 8q = 30$ **(7, 2)**
3. $-x + y = 5$; $2x + 4y = 38$ **(3, 8)**
4. $\frac{1}{3}x - \frac{1}{2}y = -8$; $\frac{3}{5}x + \frac{5}{6}y = -4$ **(−15, 6)**
5. $\frac{1}{4}c + \frac{2}{3}d = 6$; $\frac{3}{4}c - \frac{5}{3}d = -4$ **(8, 6)**
6. $0.3a + 1.6b = 0.44$; $0.4a + 2.5b = 0.66$ **(0.4, 0.2)**
7. $x + y + z = 6$; $2x - y - z = -3$; $3x + y - 2z = -1$ **(1, 2, 3)**
8. $2a + b - c = -6$; $a - 2b + c = 8$; $-a - 3b + 2c = 14$ **(0, −2, 4)**
9. $r + 2s - t = 10$; $-2r + 3s + t = 6$; $3r - 2s + 2t = -19$ **(−1, 3, −5)**

Extra Practice

Lesson 4-4

1. $\begin{bmatrix} 3 & 3 & 3 \\ -2 & -2 & -2 \end{bmatrix}$

3.

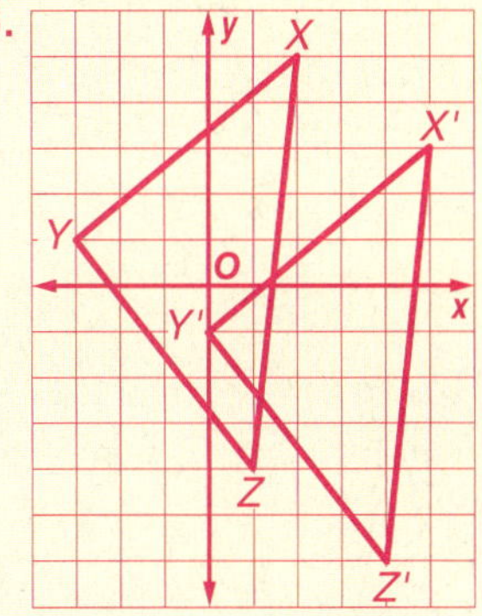

4. $\begin{bmatrix} 1 & -2 & -4 & 2 \\ 1 & 3 & -1 & -3 \end{bmatrix}$

6.

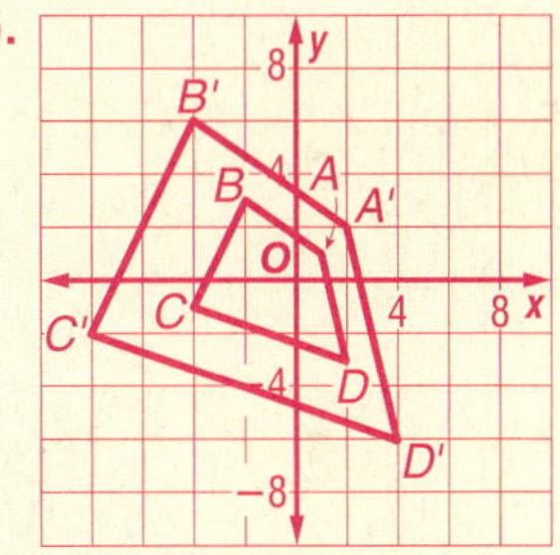

7. $\begin{bmatrix} 2 & 3 & 1 \\ 4 & -5 & -1 \end{bmatrix}$

8. $\begin{bmatrix} 0 & 1 \\ 1 & 0 \end{bmatrix}$

10.

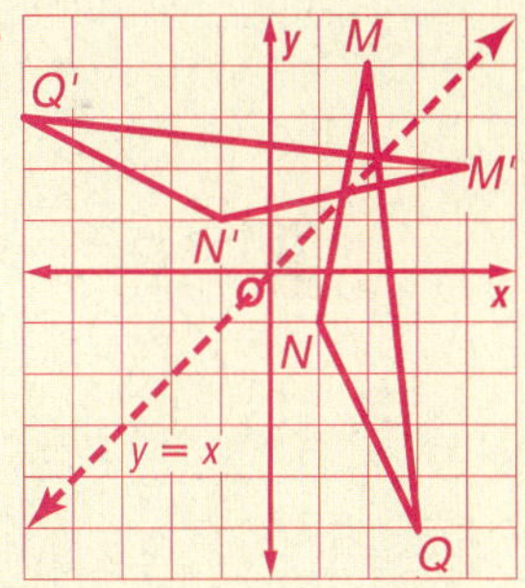

11. $\begin{bmatrix} 0 & -1 \\ 1 & 0 \end{bmatrix}$

13.

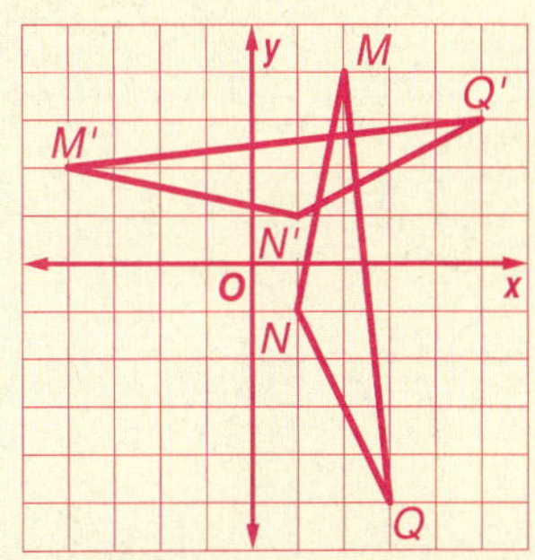

Lesson 4-7

5. $\begin{bmatrix} -1 & 3 \\ 1 & -2 \end{bmatrix}$

6. $-\frac{1}{12}\begin{bmatrix} -4 & -2 \\ 0 & 3 \end{bmatrix}$

7. $-\frac{1}{3}\begin{bmatrix} -1 & -8 \\ 0 & 3 \end{bmatrix}$

8. no inverse exists

9. $-\frac{1}{2}\begin{bmatrix} 3 & -4 \\ -2 & 2 \end{bmatrix}$

10. $\frac{1}{2}\begin{bmatrix} 4 & 5 \\ 6 & 8 \end{bmatrix}$

11. $-\frac{1}{35}\begin{bmatrix} -2 & -3 \\ -5 & 10 \end{bmatrix}$

12. $-\frac{1}{8}\begin{bmatrix} 8 & -4 \\ 4 & -3 \end{bmatrix}$

13. $-\frac{1}{2}\begin{bmatrix} -4 & 3 \\ 2 & -1 \end{bmatrix}$

14. $\begin{bmatrix} -1 & 0 \\ 0 & -1 \end{bmatrix}$

15. no inverse exists

16. $-\frac{120}{7}\begin{bmatrix} \frac{1}{5} & \frac{1}{2} \\ \frac{1}{4} & \frac{1}{3} \end{bmatrix}$

Lesson 4-8

1. $\begin{bmatrix} 5 & 3 \\ 2 & -1 \end{bmatrix} \cdot \begin{bmatrix} a \\ b \end{bmatrix} = \begin{bmatrix} 6 \\ 9 \end{bmatrix}$

2. $\begin{bmatrix} 3 & 4 \\ 2 & -3 \end{bmatrix} \cdot \begin{bmatrix} x \\ y \end{bmatrix} = \begin{bmatrix} -8 \\ 6 \end{bmatrix}$

3. $\begin{bmatrix} 1 & 3 \\ 4 & -1 \end{bmatrix} \cdot \begin{bmatrix} m \\ n \end{bmatrix} = \begin{bmatrix} 1 \\ -22 \end{bmatrix}$

4. $\begin{bmatrix} 4 & -3 \\ 5 & -2 \end{bmatrix} \cdot \begin{bmatrix} c \\ d \end{bmatrix} = \begin{bmatrix} -1 \\ 39 \end{bmatrix}$

5. $\begin{bmatrix} 1 & 2 & -1 \\ -2 & 3 & 1 \\ 1 & 1 & 3 \end{bmatrix} \cdot \begin{bmatrix} x \\ y \\ z \end{bmatrix} = \begin{bmatrix} 6 \\ 1 \\ 8 \end{bmatrix}$

6. $\begin{bmatrix} 2 & -3 & -1 \\ 4 & 1 & 1 \\ 1 & -1 & -1 \end{bmatrix} \cdot \begin{bmatrix} a \\ b \\ c \end{bmatrix} = \begin{bmatrix} 4 \\ 15 \\ -2 \end{bmatrix}$

Extra Practice

Lesson 4-7

(pages 195–201)

Determine whether each pair of matrices are inverses.

1. $A = \begin{bmatrix} -7 & -6 \\ 8 & 7 \end{bmatrix}, B = \begin{bmatrix} -7 & -6 \\ 8 & 7 \end{bmatrix}$ **yes**

2. $C = \begin{bmatrix} -3 & 4 \\ 2 & -2 \end{bmatrix}, D = \begin{bmatrix} -2 & -2 \\ -4 & -3 \end{bmatrix}$ **no**

3. $X = \begin{bmatrix} 1 & 0 \\ 0 & 1 \end{bmatrix}, Y = \begin{bmatrix} 1 & 0 \\ 0 & 1 \end{bmatrix}$ **yes**

4. $N = \begin{bmatrix} 1 & 0 \\ 0 & 1 \end{bmatrix}, M = \begin{bmatrix} 1 & 1 \\ 1 & 1 \end{bmatrix}$ **no**

Find the inverse of each matrix, if it exists. **5–16. See margin.**

5. $\begin{bmatrix} 2 & 3 \\ 1 & 1 \end{bmatrix}$

6. $\begin{bmatrix} 3 & 2 \\ 0 & -4 \end{bmatrix}$

7. $\begin{bmatrix} 3 & 8 \\ 0 & -1 \end{bmatrix}$

8. $\begin{bmatrix} 3 & -6 \\ 2 & -4 \end{bmatrix}$

9. $\begin{bmatrix} 2 & 4 \\ 2 & 3 \end{bmatrix}$

10. $\begin{bmatrix} 8 & -5 \\ -6 & 4 \end{bmatrix}$

11. $\begin{bmatrix} 10 & 3 \\ 5 & -2 \end{bmatrix}$

12. $\begin{bmatrix} -3 & 4 \\ -4 & 8 \end{bmatrix}$

13. $\begin{bmatrix} -1 & -3 \\ -2 & -4 \end{bmatrix}$

14. $\begin{bmatrix} -1 & 0 \\ 0 & -1 \end{bmatrix}$

15. $\begin{bmatrix} 3 & -3 \\ 3 & -3 \end{bmatrix}$

16. $\begin{bmatrix} \frac{1}{3} & -\frac{1}{2} \\ -\frac{1}{4} & \frac{1}{5} \end{bmatrix}$

Lesson 4-8

(pages 202–207)

Write a matrix equation for each system of equations. **1–6. See margin.**

1. $5a + 3b = 6$
   $2a - b = 9$

2. $3x + 4y = -8$
   $2x - 3y = 6$

3. $m + 3n = 1$
   $4m - n = -22$

4. $4c - 3d = -1$
   $5c - 2d = 39$

5. $x + 2y - z = 6$
   $-2x + 3y + z = 1$
   $x + y + 3z = 8$

6. $2a - 3b - c = 4$
   $4a + b + c = 15$
   $a - b - c = -2$

Solve each matrix equation or system of equations by using inverse matrices.

7. $\begin{bmatrix} 3 & 4 \\ 2 & -5 \end{bmatrix} \cdot \begin{bmatrix} x \\ y \end{bmatrix} = \begin{bmatrix} 33 \\ -1 \end{bmatrix}$ **(7, 3)**

8. $\begin{bmatrix} -1 & 1 \\ 7 & -6 \end{bmatrix} \cdot \begin{bmatrix} x \\ y \end{bmatrix} = \begin{bmatrix} 0 \\ 3 \end{bmatrix}$ **(3, 3)**

9. $\begin{bmatrix} 1 & 0 \\ 0 & 1 \end{bmatrix} \cdot \begin{bmatrix} x \\ y \end{bmatrix} = \begin{bmatrix} -29 \\ 52 \end{bmatrix}$ **(−29, 52)**

10. $5x - y = 7$; $8x + 2y = 4$ **(1, −2)**

11. $3m + n = 4$; $2m + 2n = 3$ **$\left(\frac{5}{4}, \frac{1}{4}\right)$**

12. $6c + 5d = 7$; $3c - 10d = -4$ **$\left(\frac{2}{3}, \frac{3}{5}\right)$**

13. $3a - 5b = 1$; $a + 3b = 5$ **(2, 1)**

14. $2r - 7s = 24$; $-r + 8s = -21$ **(5, −2)**

15. $x + y = -3$; $3x - 10y = 43$ **(1, −4)**

16. $2m - 3n = 3$; $-4m + 9n = -8$ **$\left(\frac{1}{2}, -\frac{2}{3}\right)$**

17. $x + y = 1$; $2x - 2y = -12$ **(−2.5, 3.5)**

Lesson 5-1

(pages 222–228)

Simplify. Assume that no variable equals 0. **11. $5.00315451 \times 10^{16}$**

1. $x^7 \cdot x^3 \cdot x$ **x^{11}**
2. $m^8 \cdot m \cdot m^{10}$ **m^{19}**
3. $7^5 \cdot 7^2$ **823,543**
4. $(-3)^4(-3)$ **−243**
5. $\frac{t^{12}}{t}$ **t^{11}**
6. $-\frac{16x^8}{8x^2}$ **$-2x^6$**
7. $\frac{6^5}{6^3}$ **36**
8. $\frac{p^5q^7}{p^2q^5}$ **p^3q^2**
9. $-(m^3)^8$ **$-m^{24}$**
10. $\frac{x}{x^7}$ **$\frac{1}{x^6}$**
11. $(3^5)^7$
12. -3^4 **−81**
13. $(abc)^3$ **$a^3b^3c^3$**
14. $(x^2)^5$ **x^{10}**
15. $-\left(\frac{x}{5}\right)^2$ **$-\frac{x^2}{25}$**
16. $(b^4)^6$ **b^{24}**
17. $(-2y^5)^2$ **$4y^{10}$**
18. $3x^0$ **3**
19. $(5x^4)^{-2}$ **$\frac{1}{25x^8}$**
20. $\left(\frac{5a^7}{2b^5c}\right)^3$ **$\frac{125a^{21}}{8b^{15}c^3}$**
21. $(-3)^{-2}$ **$\frac{1}{9}$**
22. -3^{-2} **$-\frac{1}{9}$**
23. $\frac{1}{x^{-3}}$ **x^3**
24. $\frac{5^6a^{x+y}}{5^4a^{x-y}}$ **$25a^{2y}$**

Evaluate. Express the result in scientific notation.

25. $(8.95 \times 10^9)(1.82 \times 10^7)$ **1.6289×10^{17}**

26. $(3.1 \times 10^5)(7.9 \times 10^{-8})$ **2.449×10^{-2}**

27. $\frac{(2.38 \times 10^{13})(7.56 \times 10^{-5})}{(4.2 \times 10^{18})}$ **4.284×10^{-10}**

Lesson 5-2 7. $2x^3 - 2x^2 + 5x - 7y^2$ 9. $-11x^2 + 2x + 36$ 10. $5x^2 + 12x + 10$ *(pages 229–232)*

Determine whether each expression is a polynomial. If it is a polynomial, state the degree of the polynomial.

1. $5x - 3x^2 + 7$ **yes, 2**
2. $\sqrt{7} - u^3$ **yes, 3**
3. $5r^3 + 7r^2s - 3rs^2 + 6s^3$ **yes, 3**
4. $\frac{xy}{3}$ **yes, 2**
5. $2 + \frac{7}{w}$ **no**
6. $\sqrt{a - 1}$ **no**

Simplify. 11. $7x^2 + 12x + 26$ 13. $28uw^2 - 35 + \frac{1}{u^2}$ 14. $12x^9 + 4x^8 - 4x^6 - 28x^5$

7. $(4x^3 + 5x - 7x^2) + (-2x^3 + 5x^2 - 7y^2)$
8. $(2x^2 - 3x + 11) + (7x^2 + 2x - 8)$ **$9x^2 - x + 3$**
9. $(-3x^2 + 7x + 23) + (-8x^2 - 5x + 13)$
10. $(-3x^2 + 7x + 23) - (-8x^2 - 5x + 13)$
11. $(5x^2 + 7x + 23) - (x^2 - 9x - 5) - (-3x^2 + 4x + 2)$
12. $5a^2b(4a - 3b)$ **$20a^3b - 15a^2b^2$**
13. $\frac{7}{uw}\left(4u^2w^3 - 5uw + \frac{w}{7u}\right)$
14. $-4x^5(-3x^4 - x^3 + x + 7)$
15. $(2x - 3)(4x + 7)$ **$8x^2 + 2x - 21$**
16. $(3x - 5)(-2x - 1)$
17. $(3x - 5)(2x - 1)$ **$6x^2 - 13x + 5$**
18. $(2x + 5)(2x - 5)$ **$4x^2 - 25$**
19. $(3x - 7)(3x + 7)$ **$9x^2 - 49$**
20. $(5 + 2w)(5 - 2w)$ **$25 - 4w^2$**
21. $(2a^2 + 8)(2a^2 - 8)$ **$4a^4 - 64$**
22. $(-5x + 10)(-5x - 10)$
23. $(4x - 3)^2$ **$16x^2 - 24x + 9$**
24. $(5x + 6)^2$ **$25x^2 + 60x + 36$**
25. $(-x + 1)^2$ **$x^2 - 2x + 1$**
26. $\frac{3}{4}x(x^2 + 4x + 14)$
27. $-\frac{1}{2}a^2(a^3 - 6a^2 + 5a)$

16. $-6x^2 + 7x + 5$ 22. $25x^2 - 100$ 26. $\frac{3}{4}x^3 + 3x^2 + \frac{21}{2}x$ 27. $-\frac{1}{2}a^5 + 3a^4 - \frac{5}{2}a^3$

Lesson 5-3 *(pages 233–238)*

Simplify. 5–11. See margin.

1. $\frac{18r^3s^2 + 36r^2s^3}{9r^2s^2}$ **$2r + 4s$**
2. $\frac{15v^3w^2 - 5v^4w^3}{-5v^4w^3}$ **$\frac{-3}{vw} + 1$**
3. $\frac{x^2 - x + 1}{x}$ **$x - 1 + \frac{1}{x}$**
4. $(5bh + 5ch) \div (b + c)$ **$5h$**
5. $(25c^4d + 10c^3d^2 - cd) \div 5cd$
6. $(16f^{18} + 20f^9 - 8f^6) \div 4f^3$
7. $(33m^5 + 55mn^5 - 11m^3)(11m)^{-1}$
8. $(8g^3 + 19g^2 - 12g + 9) \div (g + 3)$
9. $(p^{21} + 3p^{14} + p^7 - 2)(p^7 + 2)^{-1}$
10. $(15x^3 + 19x^2y - 7xy^2 + 5y^3) \div (3x + 5y)$
11. $(8k^2 - 56k + 98) \div (2k - 7)$
12. $(2r^2 + 5r - 3) \div (r + 3)$ **$2r - 1$**
13. $(n^3 + 125) \div (n + 5)$ **$n^2 - 5n + 25$**

Use synthetic division to find each quotient. 14–23. See margin.

14. $(10y^4 + 3y^2 - 7) \div (2y^2 - 1)$
15. $(q^4 + 8q^3 + 3q + 17) \div (q + 8)$
16. $(15v^3 + 8v^2 - 21v + 6) \div (5v - 4)$
17. $(-2x^3 + 15x^2 - 10x + 3) \div (x + 3)$
18. $(6a^4 - 22a^3 - 9a^2 + 9a - 17) \div (a - 4)$
19. $(5s^3 + s^2 - 7) \div (s + 1)$
20. $(t^4 - 2t^3 + t^2 - 3t + 2) \div (t - 2)$
21. $(z^4 - 3z^3 - z^2 - 11z - 4) \div (z - 4)$
22. $(3r^4 - 6r^3 - 2r^2 + r - 6) \div (r + 1)$
23. $(2b^3 - 11b^2 + 12b + 9) \div (b - 3)$

Lesson 5-4 *(pages 239–244)*

Factor completely. If the polynomial is not factorable, write *prime*. 1–26. See margin.

1. $14a^3b^3c - 21a^2b^4c + 7a^2b^3c$
2. $10ax - 2xy - 15ab + 3by$
3. $x^2 + x - 42$
4. $2x^2 + 5x + 3$
5. $6x^2 + 71x - 12$
6. $6x^4 - 12x^3 + 3x^2$
7. $x^2 - 6x + 2$
8. $x^2 - 2x - 15$
9. $6x^2 + 23x + 20$
10. $24x^2 - 76x + 40$
11. $6p^2 - 13pq - 28q^2$
12. $2x^2 - 6x + 3$
13. $x^2 + 49 - 14x$
14. $9x^2 - 64$
15. $36 - t^{10}$
16. $x^2 + 16$
17. $a^4 - 81b^4$
18. $3a^3 + 12a^2 - 63a$
19. $x^3 - 8x^2 + 15x$
20. $x^2 + 6x + 9$
21. $18x^3 - 8x$
22. $3x^2 - 42x + 40$
23. $2x^2 + 4x - 1$
24. $2x^3 + 6x^2 + x + 3$
25. $35ac - 3bd - 7ad + 15bc$
26. $5h^2 - 10hj + h - 2j$

Simplify. Assume that no denominator is equal to 0. 27–30. See margin.

27. $\frac{x^2 + 8x + 15}{x^2 + 4x + 3}$
28. $\frac{x^2 + x - 2}{x^2 - 6x + 5}$
29. $\frac{x^2 - 15x + 56}{x^2 - 4x - 21}$
30. $\frac{x^2 + x - 6}{x^3 + 9x^2 + 27x + 27}$

Extra Practice

Lesson 5-4

1. $7a^2b^3c(2a - 3b + 1)$
2. $(2x - 3b)(5a - y)$
3. $(x + 7)(x - 6)$
4. $(2x + 3)(x + 1)$
5. $(6x - 1)(x + 12)$
6. $3x^2(2x^2 - 4x + 1)$
7. prime
8. $(x - 5)(x + 3)$
9. $(2x + 5)(3x + 4)$
10. $4(2x - 5)(3x - 2)$
11. $(2p - 7q)(3p + 4q)$
12. prime
13. $(x - 7)^2$
14. $(3x - 8)(3x + 8)$
15. $(6 - t^5)(6 + t^5)$
16. prime
17. $(a + 3b)(a - 3b)(a^2 + 9b^2)$
18. $3a(a + 7)(a - 3)$
19. $x(x - 5)(x - 3)$
20. $(x + 3)^2$
21. $2x(3x - 2)(3x + 2)$
22. prime
23. prime
24. $(2x^2 + 1)(x + 3)$
25. $(7a + 3b)(5c - d)$
26. $(5h + 1)(h - 2j)$
27. $\frac{x + 5}{x + 1}$
28. $\frac{x + 2}{x - 5}$
29. $\frac{x - 8}{x + 3}$
30. $\frac{x - 2}{(x + 3)^2}$

Lesson 5-3

5. $5c^3 + 2c^2d - \frac{1}{5}$
6. $4f^{15} + 5f^6 - 2f^3$
7. $3m^4 + 5n^5 - m^2$
8. $8g^2 - 5g + 3$
9. $p^{14} + p^7 - 1$
10. $5x^2 - 2xy + y^2$
11. $4k - 14$
14. $5y^2 + 4 - \frac{3}{2y^2 - 1}$
15. $q^3 + 3 - \frac{7}{q + 8}$
16. $3v^2 + 4v - 1 + \frac{2}{5v - 4}$
17. $-2x^2 + 21x - 73 + \frac{222}{x + 3}$
18. $6a^3 + 2a^2 - a + 5 + \frac{3}{a - 4}$
19. $5s^2 - 4s + 4 - \frac{11}{s + 1}$
20. $t^3 + t - 1$
21. $z^3 + z^2 + 3z + 1$
22. $3r^3 - 9r^2 + 7r - 6$
23. $2b^2 - 5b - 3$

Lesson 5-6

1. $5\sqrt{3}$
2. $14\sqrt{3}$
3. $3\sqrt[3]{3}$
4. $r^2\sqrt{5r}$
5. $49|xy|\sqrt[4]{xy^2}$
6. $9\sqrt{5}$
7. $-2\sqrt{2}$
8. $13\sqrt[3]{4}$
9. 18
10. $26\sqrt{3}$
11. $3\sqrt[3]{2} - 2\sqrt[3]{3}$
12. $2\sqrt{10} - 5\sqrt{2}$
13. $-6\sqrt{2} + 3\sqrt{21}$
14. $15 + 3\sqrt{2} + 5\sqrt{3} + \sqrt{6}$
15. -1
16. $75 + 16\sqrt{11}$
17. -3
18. $21 + 4\sqrt{26}$
19. $-3 - 3\sqrt{7}$
20. $53 - 20\sqrt{7}$
21. $\dfrac{|m|\sqrt{2mn}}{4|n^3|}$
22. $\dfrac{3}{4}$
23. $\dfrac{r\sqrt[3]{4r^2st^2}}{st}$
24. $\dfrac{\sqrt[3]{196}}{7}$
25. $\dfrac{2\sqrt[5]{a}}{a}$
26. $\dfrac{\sqrt{6}}{12}$
27. $-15 - 5\sqrt{10}$
28. $\dfrac{-\sqrt{5} + \sqrt{15}}{2}$
29. $\dfrac{11 - 4\sqrt{7}}{3}$
30. $\dfrac{-1 + \sqrt{3} + 2\sqrt{2} - 2\sqrt{6}}{7}$
31. $-5 - 2\sqrt{6}$
32. $\dfrac{x^2 + 2x\sqrt{5} + 5}{x^2 - 5}$

Extra Practice

Lesson 5-5

(pages 245–249)

Use a calculator to approximate each value to three decimal places.

1. $\sqrt{289}$ **17**
2. $\sqrt{7832}$ **88.499**
3. $\sqrt[4]{0.0625}$ **0.5**
4. $\sqrt[3]{-343}$ **−7**
5. $\sqrt[10]{32^4}$ **4**
6. $\sqrt[3]{49}$ **3.659**
7. $\sqrt[5]{5}$ **1.380**
8. $-\sqrt[4]{25}$ **−2.236**

Simplify. **14. not a real number 19. $p^5q^3rs^4$ 20. $(2x^2 - y^8)^2$ 22. $-(2x - y)$ 24. $|3a + 1|$**

9. $\sqrt{9h^{22}}$ **$3|h^{11}|$**
10. $\sqrt[5]{0}$ **0**
11. $\sqrt{\dfrac{16}{9}}$ **$\frac{4}{3}$**
12. $\sqrt{\left(-\dfrac{2}{3}\right)^4}$ **$\frac{4}{9}$**
13. $\sqrt[5]{-32}$ **−2**
14. $-\sqrt{-144}$
15. $\sqrt[4]{a^{16}b^8}$ **a^4b^2**
16. $\pm\sqrt[4]{81x^4}$ **$\pm 3x$**
17. $\sqrt[5]{\dfrac{1}{100{,}000}}$ **$\frac{1}{10}$**
18. $\sqrt[3]{-d^6}$ **$-d^2$**
19. $\sqrt[5]{p^{25}q^{15}r^5s^{20}}$
20. $\sqrt[4]{(2x^2 - y^8)^8}$
21. $\pm\sqrt{16m^6n^2}$ **$\pm 4m^3|n|$**
22. $-\sqrt[3]{(2x - y)^3}$
23. $\sqrt[4]{(r + s)^4}$ **$|r + s|$**
24. $\sqrt{9a^2 + 6a + 1}$
25. $\sqrt{4y^2 + 12y + 9}$ **$|2y + 3|$**
26. $-\sqrt{x^2 - 2x + 1}$ **$-|x - 1|$**
27. $\pm\sqrt{x^2 + 2x + 1}$ **$\pm|x + 1|$**
28. $\sqrt[3]{a^3 + 6a^2 + 12a + 8}$ **$a + 2$**

Lesson 5-6

(pages 250–256)

Simplify. **1–32. See margin.**

1. $\sqrt{75}$
2. $7\sqrt{12}$
3. $\sqrt[3]{81}$
4. $\sqrt{5r^5}$
5. $\sqrt[4]{7^8x^5y^6}$
6. $3\sqrt{5} + 6\sqrt{5}$
7. $\sqrt{18} - \sqrt{50}$
8. $4\sqrt[3]{32} + \sqrt[3]{500}$
9. $\sqrt{12}\sqrt{27}$
10. $3\sqrt{12} + 2\sqrt{300}$
11. $\sqrt[3]{54} - \sqrt[3]{24}$
12. $\sqrt{10}(2 - \sqrt{5})$
13. $-\sqrt{3}(2\sqrt{6} - \sqrt{63})$
14. $(5 + \sqrt{2})(3 + \sqrt{3})$
15. $(2 + \sqrt{5})(2 - \sqrt{5})$
16. $(8 + \sqrt{11})^2$
17. $(\sqrt{3} + \sqrt{6})(\sqrt{3} - \sqrt{6})$
18. $(\sqrt{8} + \sqrt{13})^2$
19. $(1 - \sqrt{7})(4 + \sqrt{7})$
20. $(5 - 2\sqrt{7})^2$
21. $\sqrt{\dfrac{3m^3}{24n^5}}$
22. $\dfrac{\sqrt{18}}{\sqrt{32}}$
23. $2\sqrt[3]{\dfrac{r^5}{2s^2t}}$
24. $\sqrt[3]{\dfrac{4}{7}}$
25. $\sqrt[5]{\dfrac{32}{a^4}}$
26. $\sqrt{\dfrac{2}{3}} - \sqrt{\dfrac{3}{8}}$
27. $\dfrac{5}{3 - \sqrt{10}}$
28. $\dfrac{\sqrt{5}}{1 + \sqrt{3}}$
29. $\dfrac{-2 + \sqrt{7}}{2 + \sqrt{7}}$
30. $\dfrac{1 - \sqrt{3}}{1 + \sqrt{8}}$
31. $\dfrac{\sqrt{2} + \sqrt{3}}{\sqrt{2} - \sqrt{3}}$
32. $\dfrac{x + \sqrt{5}}{x - \sqrt{5}}$

Lesson 5-7

(pages 257–262)

Write each expression in radical form.

1. $10^{\frac{1}{3}}$ **$\sqrt[3]{10}$**
2. $8^{\frac{1}{4}}$ **$\sqrt[4]{8}$**
3. $a^{\frac{2}{3}}$ **$\sqrt[3]{a^2}$**
4. $(b^2)^{\frac{3}{4}}$ **$|b|\sqrt{b}$**

Write each radical using rational exponents.

5. $\sqrt{35}$ **$35^{\frac{1}{2}}$**
6. $\sqrt[4]{32}$ **$32^{\frac{1}{4}}$**
7. $\sqrt[3]{27a^2x}$ **$3a^{\frac{2}{3}}x^{\frac{1}{3}}$**
8. $\sqrt[5]{25ab^3c^4}$ **$5^{\frac{2}{5}}a^{\frac{1}{5}}b^{\frac{3}{5}}c^{\frac{4}{5}}$**

Evaluate each expression.

9. $2401^{\frac{1}{4}}$ **7**
10. $27^{\frac{4}{3}}$ **81**
11. $(-32)^{\frac{2}{5}}$ **4**
12. $-81^{\frac{3}{4}}$ **−27**
13. $(-125)^{-\frac{2}{3}}$ **$\frac{1}{25}$**
14. $16^{\frac{5}{2}} \cdot 16^{\frac{1}{2}}$ **4096**
15. $8^{-\frac{2}{3}} \cdot 64^{\frac{1}{6}}$ **0.5**
16. $\left(\dfrac{48}{1875}\right)^{-\frac{5}{4}}$ **$\frac{3125}{32}$**

Simplify each expression.

17. $7^{\frac{5}{9}} \cdot 7^{\frac{4}{9}}$ **7**
18. $32^{\frac{2}{3}} \cdot 32^{\frac{3}{5}}$ **$32^{\frac{19}{15}}$**
19. $\left(k^{\frac{8}{5}}\right)^5$ **k^8**
20. $x^{\frac{2}{5}} \cdot x^{\frac{8}{5}}$ **x^2**
21. $m^{\frac{2}{5}} \cdot m^{\frac{4}{5}}$ **$m^{\frac{6}{5}}$**
22. $\left(p^{\frac{5}{4}} \cdot q^{\frac{7}{2}}\right)^{\frac{8}{3}}$ **$p^{\frac{10}{3}}q^{\frac{28}{3}}$**
23. $\left(4^{\frac{9}{2}}c^{\frac{3}{2}}\right)^2$ **$262{,}144c^3$**
24. $\dfrac{7^{\frac{3}{4}}}{7^{\frac{5}{3}}}$ **$\dfrac{7^{\frac{1}{12}}}{7}$**
25. $\dfrac{1}{t^{\frac{9}{5}}}$ **$\dfrac{t^{\frac{1}{5}}}{t^2}$**
26. $a^{\frac{-8}{7}}$ **$\dfrac{a^{\frac{6}{7}}}{a^2}$**
27. $\dfrac{r}{r^{\frac{7}{5}}}$ **$\dfrac{r^{\frac{3}{5}}}{r}$**
28. $\sqrt[4]{36}$ **$\sqrt{6}$**
29. $\sqrt[4]{9a^2}$ **$\sqrt{3a}$**
30. $\sqrt[3]{\sqrt{81}}$ **$\sqrt[3]{9}$**
31. $\dfrac{v^{\frac{11}{7}} - v^{\frac{4}{7}}}{v^{\frac{4}{7}}}$ **$v - 1$**
32. $\dfrac{1}{5^{\frac{1}{2}} + 3^{\frac{1}{2}}}$ **$\dfrac{5^{\frac{1}{2}} - 3^{\frac{1}{2}}}{2}$**

Lesson 5-8 16. no solution 19. $-0.2 < y < 3$ (pages 263–267)

Solve each equation or inequality.

1. $\sqrt{x} = 16$ **256**
2. $\sqrt{z + 3} = 7$ **46**
3. $\sqrt[3]{a + 5} = 1$ **−4**
4. $5\sqrt{s} - 8 = 3$ $\frac{121}{25}$
5. $\sqrt[4]{m + 7} + 11 = 9$ **no solution**
6. $d + \sqrt{d^2 - 8} = 4$ **3**
7. $g\sqrt{5} + 4 = g + 4$ **0**
8. $\sqrt{x - 8} = \sqrt{13 + x}$ **no solution**
9. $\sqrt{3x + 10} = 1 + \sqrt{2x + 5}$ **−2, 2**
10. $\sqrt{3x + 9} > 2$ $x > -\frac{5}{3}$
11. $\sqrt{3n - 1} \le 5$ $\frac{1}{3} \le n \le \frac{26}{3}$
12. $2 - 4\sqrt{21 - 6c} < -6$ $c < \frac{17}{6}$
13. $\sqrt{5y + 4} > 8$ $y > 12$
14. $\sqrt{2w + 3} + 5 \ge 7$ $w \ge 0.5$
15. $\sqrt{x + 29} - 3 = \sqrt{x - 16}$ **52**
16. $\sqrt{3x + 25} + \sqrt{10 - 2x} = 0$
17. $\sqrt{2c + 3} - 7 > 0$ $c > 23$
18. $\sqrt{3z - 5} - 3 = 1$ **7**
19. $\sqrt{5y + 1} + 6 < 10$
20. $\sqrt{3n + 1} - 2 \le 6$
21. $\sqrt{y - 5} - \sqrt{y} \ge 1$ **no solution**
22. $(5n - 1)^{\frac{1}{2}} = 0$ $\frac{1}{5}$
23. $(7x - 6)^{\frac{1}{3}} + 1 = 3$ **2**
24. $(6a - 8)^{\frac{1}{4}} + 9 \ge 10$ $a \ge 1.5$

20. $-\frac{1}{3} \le n \le 21$

Lesson 5-9 (pages 270–275)

Simplify. **10. $6 + 14i$**

1. $\sqrt{-289}$ $17i$
2. $\sqrt{-\frac{25}{121}}$ $\frac{5}{11}i$
3. $\sqrt{-625b^8}$ $25b^4i$
4. $\sqrt{-\frac{28t^6}{27s^5}}$ $\frac{2t^3i\sqrt{21s}}{9s^3}$
5. $(7i)^2$ **−49**
6. $(6i)(-2i)(11i)$ $132i$
7. $(\sqrt{-8})(\sqrt{-12})$ $-4\sqrt{6}$
8. $-i^{22}$ **1**
9. $i^{17} \cdot i^{12} \cdot i^{26}$ $-i$
10. $(14 - 5i) + (-8 + 19i)$
11. $(7i) - (2 + 3i)$ $-2 + 4i$
12. $(2 + 2i) - (5 + i)$ $-3 + i$
13. $(7 + 3i)(7 - 3i)$ **58**
14. $(8 - 2i)(5 + i)$ $42 - 2i$
15. $(6 + 8i)^2$ $-28 + 96i$
16. $\frac{3}{6 - 2i}$ $\frac{9 + 3i}{20}$
17. $\frac{5i}{3 + 4i}$ $\frac{4 + 3i}{5}$
18. $\frac{3 - 7i}{5 + 4i}$ $\frac{-13 - 47i}{41}$

Solve each equation.

19. $x^2 + 8 = 3$ $\pm i\sqrt{5}$
20. $\frac{4x^2}{49} + 6 = 3$ $\pm\frac{7i\sqrt{3}}{2}$
21. $8x^2 + 5 = 1$ $\pm\frac{i\sqrt{2}}{2}$
22. $12 - 9x^2 = 38$ $\pm\frac{i\sqrt{26}}{3}$
23. $9x^2 + 7 = 4$ $\pm\frac{i\sqrt{3}}{3}$
24. $\frac{1}{2}x^2 + 1 = 0$ $\pm i\sqrt{2}$

Lesson 6-1 (pages 286–293)

For Exercises 1–12, complete parts a–c for each quadratic function.

a. Find the y-intercept, the equation of the axis of symmetry, and the x-coordinate of the vertex.

b. Make a table of values that includes the vertex.

c. Use this information to graph the function. **1–12. See pp. 861A–861T.**

1. $f(x) = 6x^2$
2. $f(x) = -x^2$
3. $f(x) = x^2 + 5$
4. $f(x) = -x^2 - 2$
5. $f(x) = 2x^2 + 1$
6. $f(x) = -3x^2 + 6x$
7. $f(x) = x^2 + 6x - 3$
8. $f(x) = x^2 - 2x - 8$
9. $f(x) = -3x^2 - 6x + 12$
10. $f(x) = x^2 + 5x - 6$
11. $f(x) = 2x^2 + 7x - 4$
12. $f(x) = -5x^2 + 10x + 1$

Determine whether each function has a maximum or a minimum value. Then find the maximum or minimum value of each function.

13. $f(x) = 9x^2$ **min.; 0**
14. $f(x) = 9 - x^2$ **max.; 9**
15. $f(x) = x^2 - 5x + 6$ **min.; −0.25**
16. $f(x) = 2 + 7x - 6x^2$ **max.;** $4\frac{1}{24}$
17. $f(x) = 4x^2 - 9$ **min.; −9**
18. $f(x) = x^2 + 2x + 1$ **min.; 0**
19. $f(x) = 8 - 3x - 4x^2$ **max.;** $8\frac{9}{16}$
20. $f(x) = x^2 - x + \frac{5}{4}$ **min.; 1**
21. $f(x) = -x^2 + \frac{14}{3}x + \frac{5}{3}$ **max.;** $7\frac{1}{9}$

Extra Practice

Extra Practice

Extra Practice 839

Lesson 6-4

1. $(x-2)^2$
2. $(x+10)^2$
3. $\left(x-\frac{11}{2}\right)^2$
4. $\left(x-\frac{1}{3}\right)^2$
5. $(x+15)^2$
6. $\left(x+\frac{3}{16}\right)^2$
7. $\left(x-\frac{1}{5}\right)^2$
8. $\left(x-\frac{3}{2}\right)^2$
9. −4, 1
10. −5, 0
11. −9, 7
12. $-\frac{5}{3}$, 7
13. $\frac{-7 \pm i\sqrt{3}}{2}$
14. $\frac{4 \pm \sqrt{6}}{5}$
15. $3 \pm i\sqrt{2}$
16. 6
17. $\frac{-13 \pm 3\sqrt{33}}{16}$
18. $\frac{-5 \pm i\sqrt{47}}{6}$
19. $-7 \pm 5\sqrt{2}$
20. 0.5, 7.5
21. $-\frac{1}{3}$, 4
22. −14, 6
23. $\frac{7 \pm \sqrt{29}}{2}$
24. $\frac{-3 \pm \sqrt{41}}{2}$
25. $\frac{5 \pm \sqrt{65}}{2}$
26. $\frac{6 \pm 2\sqrt{6}}{3}$
27. −15, −5
28. −3, 8
29. −3.5, 3

Extra Practice

Lesson 6-2

(pages 294–299)

Use the related graph of each equation to determine its solutions.

1. $x^2 + x - 6 = 0$ **−3, 2**

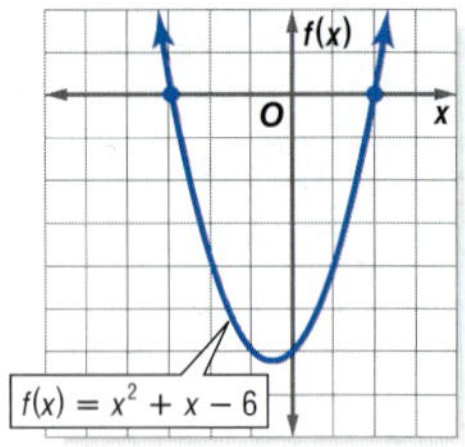

2. $-2x^2 = 0$ **0**

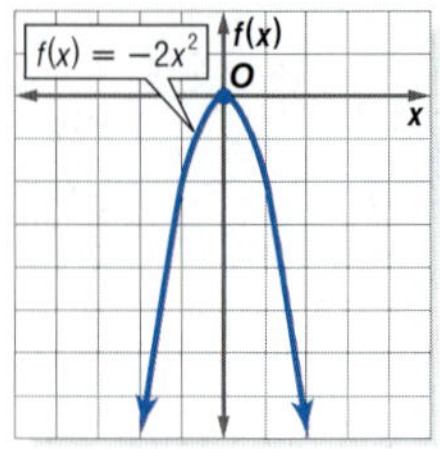

3. $x^2 - 4x - 5 = 0$ **−1, 5**

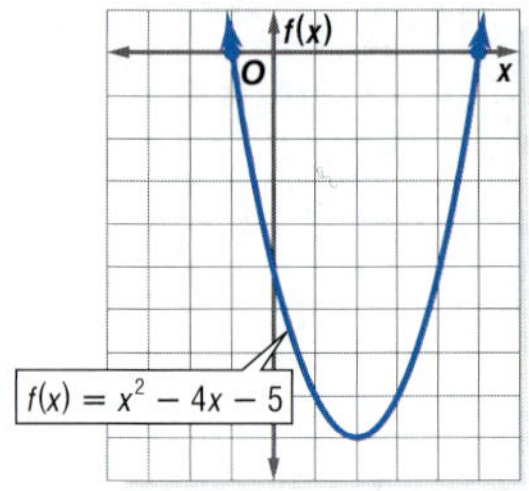

9. between −8 and −7; between 0 and 1 11. between −1 and 0; between 1 and 2

Solve each equation by graphing. If exact roots cannot be found, state the consecutive integers between which the roots are located. **6. between 0 and 1; between 4 and 5**

4. $x^2 - 2x = 0$ **0, 2**
5. $x^2 + 8x - 20 = 0$ **−10, 2**
6. $-2x^2 + 10x - 5 = 0$
7. $-5x + 2x^2 - 3 = 0$ **$-\frac{1}{2}$, 3**
8. $3x^2 - x + 8 = 0$ **no real solutions**
9. $-x^2 + 2 = 7x$
10. $4x^2 - 4x + 1 = 0$ **0.5**
11. $4x + 1 = 3x^2$
12. $x^2 = -9x$ **0, −9**
13. $x^2 + 6x - 27 = 0$ **−9, 3**
14. $0.4x^2 + 1 = 0$ **no real solutions**
15. $0.5x^2 + 3x - 2 = 0$ **between −7 and −6; between 0 and 1**

Lesson 6-3

(pages 301–305)

Solve each equation by factoring. **1. −2, −5 6. $-\frac{1}{12}$, 6 11. $-\frac{3}{2}$, −1 14. −1.5, 0.25**

1. $x^2 + 7x + 10 = 0$
2. $3x^2 = 75x$ **0, 25**
3. $2x^2 + 7x = 9$ **−4.5, 1**
4. $8x^2 = 48 - 40x$ **−6, 1**
5. $5x^2 = 20x$ **0, 4**
6. $12x^2 - 71x - 6 = 0$
7. $16x^2 - 64 = 0$ **±2**
8. $5x^2 - 45x + 90 = 0$ **3, 6**
9. $24x^2 - 15 = 2x$ **$-\frac{3}{4}$, $\frac{5}{6}$**
10. $x^2 = 72 - x$ **−9, 8**
11. $2x^2 + 5x + 3 = 0$
12. $4x^2 + 9 = 12x$ **$\frac{3}{2}$**
13. $2x^2 - 8x = 0$ **0, 4**
14. $8x^2 + 10x = 3$
15. $12x^2 - 5x = 3$
16. $x^2 + 8x + 12 = 0$
17. $x^2 + 9x + 14 = 0$ **−7, −2**
18. $9x^2 + 1 = 6x$ **$\frac{1}{3}$**
19. $6x^2 + 7x = 3$ **$\frac{1}{3}$, $-\frac{3}{2}$**
20. $x^2 - 4x = 21$ **−3, 7**

15. $\frac{3}{4}$, $-\frac{1}{3}$ 16. −6, −2

Write a quadratic equation with the given roots. Write the equation in the form $ax^2 + bx + c = 0$, where a, b, and c are integers.

21. 2, 1 **$x^2 - 3x + 2 = 0$**
22. −3, 4 **$x^2 - x - 12 = 0$**
23. −1, −7 **$x^2 + 8x + 7 = 0$**
24. $-1, \frac{1}{2}$ **$2x^2 + x - 1 = 0$**
25. $-5, \frac{1}{4}$ **$4x^2 + 19x - 5 = 0$**
26. $-\frac{1}{3}, -\frac{1}{2}$ **$6x^2 + 5x + 1 = 0$**

Lesson 6-4

(pages 306–312)

Find the value of c that makes each trinomial a perfect square. Then write the trinomial as a perfect square. **1–8. See margin for perfect square.**

1. $x^2 - 4x + c$ **4**
2. $x^2 + 20x + c$ **100**
3. $x^2 - 11x + c$ **$\frac{121}{4}$**
4. $x^2 - \frac{2}{3}x + c$ **$\frac{1}{9}$**
5. $x^2 + 30x + c$ **225**
6. $x^2 + \frac{3}{8}x + c$ **$\frac{9}{256}$**
7. $x^2 - \frac{2}{5}x + c$ **$\frac{1}{25}$**
8. $x^2 - 3x + c$ **$\frac{9}{4}$**

Solve each equation by completing the square. **9–29. See margin.**

9. $x^2 + 3x - 4 = 0$
10. $x^2 + 5x = 0$
11. $x^2 + 2x - 63 = 0$
12. $3x^2 - 16x - 35 = 0$
13. $x^2 + 7x + 13 = 0$
14. $5x^2 - 8x + 2 = 0$
15. $x^2 - 6x + 11 = 0$
16. $x^2 - 12x + 36 = 0$
17. $8x^2 + 13x - 4 = 0$
18. $3x^2 + 5x + 6 = 0$
19. $x^2 + 14x - 1 = 0$
20. $4x^2 - 32x + 15 = 0$
21. $3x^2 - 11x - 4 = 0$
22. $x^2 + 8x - 84 = 0$
23. $x^2 - 7x + 5 = 0$
24. $x^2 + 3x - 8 = 0$
25. $x^2 - 5x - 10 = 0$
26. $3x^2 - 12x + 4 = 0$
27. $x^2 + 20x + 75 = 0$
28. $x^2 - 5x - 24 = 0$
29. $2x^2 + x - 21 = 0$

Lesson 6-5

1a. −3
1b. 2 complex
1c. $\frac{-7 \pm i\sqrt{3}}{2}$
2a. 540
2b. 2 irrational
2c. $\frac{-1 \pm \sqrt{15}}{2}$
3a. −55
3b. 2 complex
3c. $\frac{5 \pm i\sqrt{55}}{10}$
4a. 0
4b. 1 rational
4c. $-2\frac{1}{3}$
5a. 208
5b. 2 irrational
5c. $\frac{4 \pm \sqrt{13}}{2}$
6a. 49
6b. 2 rational
6c. −0.5, 3

Lesson 6-5 *(pages 313–319)*

For Exercises 1–16, complete parts a–c for each quadratic equation.

a. Find the value of the discriminant.

b. Describe the number and type of roots.

c. Find the exact solutions by using the Quadratic Formula. **1–16. See margin.**

1. $x^2 + 7x + 13 = 0$	**2.** $6x^2 + 6x - 21 = 0$	**3.** $5x^2 - 5x + 4 = 0$	**4.** $9x^2 + 42x + 49 = 0$
5. $4x^2 - 16x + 3 = 0$	**6.** $2x^2 = 5x + 3$	**7.** $x^2 + 81 = 18x$	**8.** $3x^2 - 30x + 75 = 0$
9. $24x^2 + 10x = 43$	**10.** $9x^2 + 4 = 2x$	**11.** $7x = 8x^2$	**12.** $18x^2 = 9x + 45$
13. $x^2 - 4x + 4 = 0$	**14.** $4x^2 + 16x + 15 = 0$	**15.** $x^2 - 6x + 13 = 0$	**16.** $3x^2 = 108x$

Solve each equation by using the method of your choice. Find the exact solutions. **17–24. See margin.**

17. $x^2 + 4x + 29 = 0$	**18.** $4x^2 + 3x - 2 = 0$	**19.** $2x^2 + 5x = 9$	**20.** $x^2 = 8x - 16$
21. $7x^2 = 4x$	**22.** $2x^2 + 6x + 5 = 0$	**23.** $9x^2 - 30x + 25 = 0$	**24.** $3x^2 - 4x + 2 = 0$

Lesson 6-6 *(pages 322–328)*

Write each quadratic function in vertex form, if not already in that form. Then identify the vertex, axis of symmetry, and direction of opening. **1–8. See margin.**

1. $y = (x + 6)^2 - 1$	**2.** $y = 2(x - 8)^2 - 5$	**3.** $y = -(x + 1)^2 + 7$	**4.** $y = -9(x - 7)^2 + 3$
5. $y = -x^2 + 10x - 3$	**6.** $y = -2x^2 + 16x + 7$	**7.** $y = 3x^2 + 9x + 8$	**8.** $y = \frac{3}{4}x^2 - 6x - 5$

Graph each function. **9–20. See pp. 861A–861T.**

9. $y = x^2 - 2x + 4$	**10.** $y = -3x^2 + 18x$	**11.** $y = -2x^2 - 4x + 1$	**12.** $y = 2x^2 - 8x + 9$
13. $y = \frac{1}{3}x^2 + 2x + 7$	**14.** $y = x^2 + 6x + 9$	**15.** $y = x^2 + 3x + 6$	**16.** $y = 2x^2 + 8x + 9$
17. $y = x^2 - 8x + 9$	**18.** $y = -x^2 - x + 10$	**19.** $y = -0.5x^2 + 4x - 3$	**20.** $y = -2x^2 - 8x - 1$

Write an equation for the parabola with the given vertex that passes through the given point.

21. vertex: $(-1, 5)$ point: $(2, -4)$ $y = -1(x + 1)^2 + 5$

22. vertex: $(2, -1)$ point: $(-2, 7)$ $y = 0.5(x - 2)^2 - 1$

23. vertex: $(-5, -3)$ point: $(-1, 5)$ $y = 0.5(x + 5)^2 - 3$

24. vertex: $(0, -8)$ point: $(2, -2)$ $y = 1.5(x + 0)^2 - 8$

Lesson 6-7 *(pages 329–335)*

Graph each inequality. **1–8. See pp. 861A–861T.**

1. $y \le 5x^2 + 3x - 2$	**2.** $y > -3x^2 + 2$	**3.** $y \ge x^2 - 8x$	**4.** $y \ge -x^2 - x + 3$
5. $y \le 3x^2 + 4x - 8$	**6.** $y \le -5x^2 + 2x - 3$	**7.** $y > 4x^2 + x$	**8.** $y \ge -x^2 - 3$

Use the graph of its related function to write the solutions of each inequality.

9. $x^2 - 4 \le 0$ $-2 \le x \le 2$

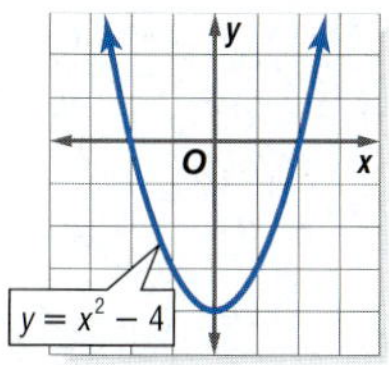

10. $-x^2 + 6x - 9 \ge 0$ 3

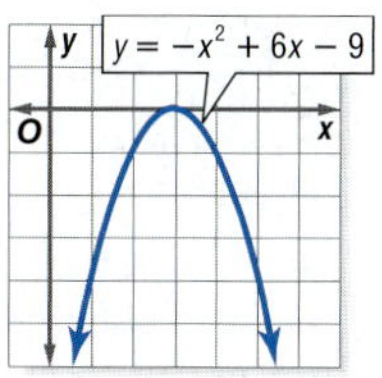

11. $x^2 + 4x - 5 < 0$ $-5 < x < 1$

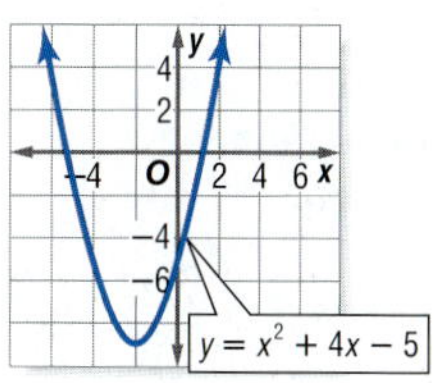

Solve each inequality algebraically. **12–19. See margin.**

12. $x^2 - 1 < 0$	**13.** $10x^2 - x - 2 \ge 0$	**14.** $-x^2 - 5x - 6 > 0$	**15.** $-3x^2 \ge 5$
16. $x^2 - 2x - 8 \le 0$	**17.** $2x^2 \ge 5x + 12$	**18.** $x^2 + 3x - 4 > 0$	**19.** $2x - x^2 \le -15$

Extra Practice

Extra Practice

7a. 0
7b. 1 rational
7c. 9
8a. 0
8b. 1 rational
8c. 5
9a. 4228
9b. 2 irrational
9c. $\frac{-5 \pm \sqrt{1057}}{24}$
10a. −140
10b. 2 complex
10c. $\frac{1 \pm i\sqrt{35}}{9}$
11a. 49
11b. 2 rational
11c. $0, \frac{7}{8}$
12a. 3321
12b. 2 irrational
12c. $\frac{1 \pm \sqrt{41}}{4}$
13a. 0
13b. 1 rational
13c. 2
14a. 16
14b. 2 rational
14c. −1.5, −2.5
15a. −16
15b. 2 complex
15c. $3 \pm 2i$
16a. 11,664
16b. 2 rational
16c. 0, 36
17. $-2 \pm 5i$
18. $\frac{-3 \pm \sqrt{41}}{8}$
19. $\frac{-5 \pm \sqrt{97}}{4}$
20. 4
21. $0, \frac{4}{7}$
22. $\frac{-3 \pm i}{2}$
23. $\frac{5}{3}$
24. $\frac{2 \pm i\sqrt{2}}{3}$

Lesson 6-6

1. $(-6, -1)$; $x = -6$; up
2. $(8, -5)$; $x = 8$; up
3. $(-1, 7)$; $x = -1$; down
4. $(7, 3)$; $x = 7$; down
5. $y = -(x - 5)^2 + 22$; $(5, 22)$; $x = 5$; down
6. $y = -2(x - 4)^2 + 39$; $(4, 39)$; $x = 4$; down
7. $y = 3(x + 1.5)^2 + 1.25$; $(-1.5, 1.25)$; $x = -1.5$; up
8. $y = \frac{3}{4}(x - 4)^2 - 17$; $(4, -17)$; $x = 4$; up

Lesson 6-7

12. $\{x \mid -1 < x < 1\}$
13. $\{x \mid x \le -0.4 \text{ or } x \ge 0.5\}$
14. $\{x \mid -3 < x < -2\}$
15. $\varnothing$
16. $\{x \mid -2 \le x \le 4\}$
17. $\{x \mid x \le -1.5 \text{ or } x \ge 4\}$
18. $\{x \mid x < -4 \text{ or } x > 1\}$
19. $\{x \mid x \le -3 \text{ or } x \ge 5\}$

Lesson 7-1

1. No; the polynomial contains two variables, n and m.
2. 2; −3
3. 5; 1
4. No; the term $\frac{2}{a}$ cannot be written in the form x^n, where n is a nonnegative integer.
5. 2; 1
6. 6; −7

16. $n^3 - 1$
17. $-8b^2 + 10b + 1$
18. $z^9 - 1$
19. $-18m^4 + 15m^2 + 1$
20. $x^3 + 3x^2 + 3x$
21. $-2x^2 + 7x - 2$
22. $a^6 - 6a^4 + 12a^2 - 9$
23. $3h^3 - 27h^2 + 81h - 84$
24. $-10c^2 + 105c - 255$
25. $n^6 + n^3 - 6n^2 + 12n - 10$
26. $98a^2 - 65a - 4$
27. $2d^6 + 6d^4 + 3d^3 + 6d^2 - 3$

Extra Practice

Lesson 7-1 *(pages 346–352)*

State the degree and leading coefficient of each polynomial in one variable. If it is not a polynomial in one variable, explain why. **1–6. See margin.**

1. $2n^2 + 2m^2$
2. $5 - 3a^2$
3. $(x^2 + 2)(x^3 - 5)$
4. $-3a^3 + 5a - \frac{2}{a}$
5. $5c + c^2 - \frac{3}{5}$
6. $-4m - 7m^6 + 6m^5 + 3m^2$

Find $p(5)$ and $p(-1)$ for each function. **10. −1513; 23**

7. $p(x) = 7x - 3$ **32; −10**
8. $p(x) = -3x^2 + 5x - 4$ **−54; −12**
9. $p(x) = 5x^4 + 2x^2 - 2x$ **3165; 9**
10. $p(x) = -13x^3 + 5x^2 - 3x + 2$
11. $p(x) = x^6 - 2$ **15,623; −1**
12. $p(x) = \frac{2}{3}x^2 + 5x$ **$41\frac{2}{3}$; $-4\frac{1}{3}$**
13. $p(x) = x^3 + x^2 - x + 1$ **146; 2**
14. $p(x) = x^4 - x^2 - 1$ **599; −1**
15. $p(x) = 1 - x^3$ **−124; 2**

If $p(x) = -2x^2 + 5x + 1$ and $q(x) = x^3 - 1$, find each value. **16–27. See margin.**

16. $q(n)$
17. $p(2b)$
18. $q(z^3)$
19. $p(3m^2)$
20. $q(x + 1)$
21. $p(3 - x)$
22. $q(a^2 - 2)$
23. $3q(h - 3)$
24. $5[p(c - 4)]$
25. $q(n - 2) + q(n^2)$
26. $-3p(4a) - p(a)$
27. $2[q(d^2 + 1)] + 3q(d)$

Lesson 7-2 *(pages 353–358)*

For Exercises 1–16, complete each of the following. **1–16. See pp. 861A–861T.**
a. Graph each function by making a table of values.
b. Determine the values of x between which the real zeros are located.
c. Estimate the x-coordinates at which the relative maxima and relative minima occur.

1. $f(x) = x^3 + x^2 - 3x$
2. $f(x) = -x^4 + x^3 + 5$
3. $f(x) = x^3 - 3x^2 + 8x - 7$
4. $f(x) = 2x^5 + 3x^4 - 8x^2 + x + 4$
5. $f(x) = x^4 - 5x^3 + 6x^2 - x - 2$
6. $f(x) = 2x^6 + 5x^4 - 3x^2 - 5$
7. $f(x) = -x^3 - 8x^2 + 3x - 7$
8. $f(x) = -x^4 - 3x^3 + 5x$
9. $f(x) = x^5 - 7x^4 - 3x^3 + 2x^2 - 4x + 9$
10. $f(x) = x^4 - 5x^3 + x^2 - x - 3$
11. $f(x) = x^4 - 128x^2 + 960$
12. $f(x) = -x^5 + x^4 - 208x^2 + 145x + 9$
13. $f(x) = x^5 - x^3 - x + 1$
14. $f(x) = x^3 - 2x^2 - x + 5$
15. $f(x) = 2x^4 - x^3 + x^2 - x + 1$
16. $f(x) = -x^3 - x^2 - x - 1$

Lesson 7-3 *(pages 360–364)*

Write each polynomial in quadratic form, if possible. **5. $\left(x^{\frac{1}{2}}\right)^2 - 10\left(x^{\frac{1}{2}}\right) + 25$**

1. $5x^{10} - 6x^5 - 3$ **$5(x^5)^2 - 6(x^5) - 3$**
2. $2y^6 + 3y^4 + 10$ **not possible**
3. $z^6 - 8z^3$ **$(z^3)^2 - 8(z^3)$**
4. $y^5 - 6y^3 + 4$ **not possible**
5. $x - 10x^{\frac{1}{2}} + 25$
6. $x^4 - 7x + 12$ **not possible**
7. $3r + 2r^{\frac{1}{2}} - 7$ **$3\left(r^{\frac{1}{2}}\right)^2 + 2\left(r^{\frac{1}{2}}\right) - 7$**
8. $r^{\frac{2}{3}} - 5r^{\frac{1}{3}} + 6$ **$\left(r^{\frac{1}{3}}\right)^2 - 5\left(r^{\frac{1}{3}}\right) + 6$**
9. $x^{\frac{1}{2}} + 9x^{\frac{1}{4}} + 18$ **$\left(x^{\frac{1}{4}}\right)^2 + 9\left(x^{\frac{1}{4}}\right) + 18$**

Solve each equation.

10. $8x^3 - 27 = 0$ **$\frac{3}{2}, \frac{-3 \pm 3i\sqrt{3}}{4}$**
11. $2m^4 = 3m^2 + 5$ **$\pm\frac{\sqrt{10}}{2}, \pm i$**
12. $3b^5 = 7b^3$ **$0, \pm\frac{\sqrt{21}}{3}$**
13. $x^3 + 10x^2 + 16x = 0$ **0, −8, −2**
14. $y^4 - 3y^2 + 2 = 0$ **$\pm 1, \pm\sqrt{2}$**
15. $a^3 = 125$ **$5, \frac{-5 \pm 5i\sqrt{3}}{2}$**
16. $5z^{\frac{2}{3}} - z^{\frac{1}{3}} = 4$ **$-\frac{64}{125}, 1$**
17. $x^{\frac{1}{2}} - 6x^{\frac{1}{4}} + 8 = 0$ **16, 256**
18. $3m + m^{\frac{1}{2}} - 2 = 0$ **$\frac{4}{9}$**
19. $m - 9\sqrt{m} + 8 = 0$ **1, 64**
20. $r^2 - 12r + 20 = 0$ **2, 10**
21. $\sqrt[3]{x^2} - 8\sqrt[3]{x} + 15 = 0$ **27, 125**
22. $m - 11\sqrt{m} + 30 = 0$ **25, 36**
23. $y^3 - 8\sqrt{y^3} + 16 = 0$ **$2\sqrt[3]{2}$**
24. $g^{\frac{2}{3}} - 2g^{\frac{1}{3}} - 8 = 0$ **−8, 64**

Extra Practice

842 Extra Practice

Lesson 7-4 (pages 365–370)

Use synthetic substitution to find $f(3)$ and $f(-4)$ for each function.

1. $f(x) = x^2 - 6x + 2$ **−7, 42**
2. $f(x) = x^3 + 5x - 6$ **36, −90**
3. $f(x) = x^3 - x^2 - 3x + 1$ **10, −67**
4. $f(x) = -3x^3 + 5x^2 + 7x - 3$ **−18, 241**
5. $f(x) = 3x^5 - 5x^3 + 2x - 8$ **592, −2768**
6. $f(x) = -2x^4 + 7x^3 + 8x^2 - 3x + 5$ **95, −815**
7. $f(x) = 10x^3 + 2$ **272, −638**
8. $f(x) = x^5 + x^4 + x^3 + x^2 + x + 1$ **364, −819**

Given a polynomial and one of its factors, find the remaining factors of the polynomial. Some factors may not be binomials. **9–22. See margin.**

9. $(x^3 - x^2 + x + 14); (x + 2)$
10. $(5x^3 - 17x^2 + 6x); (x - 3)$
11. $(2x^3 + x^2 - 41x + 20); (x - 4)$
12. $(x^3 - 8); (x - 2)$
13. $(x^2 + 6x + 5); (x + 1)$
14. $(x^4 + x^3 + x^2 + x); (x + 1)$
15. $(x^3 - 8x^2 + x + 42); (x - 7)$
16. $(6x^4 + 13x^3 - 36x^2 - 43x + 30); (x - 2)$
17. $(x^4 + 5x^3 - 27x - 135); (x - 3)$
18. $(2x^3 - 15x^2 - 2x + 120); (2x + 5)$
19. $(6x^3 - 17x^2 + 6x + 8); (3x - 4)$
20. $(30x^3 - 68x^2 + 10x + 12); (5x - 3)$
21. $(10x^3 + x^2 - 46x + 35); (5x - 7)$
22. $(x^3 + 9x^2 + 23x + 15); (x + 1)$

Lesson 7-5 **3. −2, 4, −i, i; 2 real, 2 imaginary** (pages 371–377)

Solve each equation and state the number and type of roots.

1. $-5x - 7 = 0$ **$-\frac{7}{5}$; 1 real**
2. $3x^2 + 10 = 0$ **$\pm\frac{i\sqrt{30}}{3}$; 2 imaginary**
3. $x^4 - 2x^3 - 7x^2 - 2x - 8 = 0$
4. $x^4 - 2x^3 = 23x^2 - 60x$ **−5, 0, 3, 4; 4 real**

State the number of positive real zeros, negative real zeros, and imaginary zeros for each function. **5–12. See margin.**

5. $f(x) = 5x^8 - x^6 + 7x^4 - 8x^2 - 3$
6. $f(x) = 6x^5 - 7x^2 + 5$
7. $f(x) = -2x^6 - 5x^5 + 8x^2 - 3x + 1$
8. $f(x) = 4x^3 + x^2 - 38x + 56$
9. $f(x) = 3x^8 - 15x^5 - 7x^4 - 8x^3 - 3$
10. $f(x) = -x^6 - 8x^5 - 5x^4 - 11x^3 - 2x^2 - 5x - 1$
11. $f(x) = 3x^4 - 5x^3 + 2x^2 - 7x + 5$
12. $f(x) = x^5 - x^4 + 7x^3 - 25x^2 + 8x - 13$

Find all of the zeros of the function. **13–20. See margin.**

13. $f(x) = x^3 - 7x^2 + 16x - 10$
14. $f(x) = 10x^3 + 7x^2 - 82x + 56$
15. $f(x) = x^3 - 16x^2 + 79x - 114$
16. $f(x) = -3x^3 + 6x^2 + 5x - 8$
17. $f(x) = 6x^4 + 13x^3 - 18x^2 - 7x + 6$
18. $f(x) = 4x^4 + 36x^3 + 57x^2 + 225x + 200$
19. $f(x) = 24x^3 + 64x^2 + 6x - 10$
20. $f(x) = 2x^3 + 2x^2 - 34x + 30$

Write a polynomial function of least degree with integral coefficients that has the given zeros. **21–24. See margin.**

21. −3, 1, 2
22. −5, −3, 3, 5
23. −6, 6, $-5i$, $5i$
24. 3, $\pm i\sqrt{7}$

Lesson 7-6 (pages 378–382)

List all of the possible rational zeros for each function. **1–3. See margin.**

1. $f(x) = 3x^5 - 7x^3 - 8x + 6$
2. $f(x) = 4x^3 + 2x^2 - 5x + 8$
3. $f(x) = 6x^9 - 7$

Find all of the rational zeros for each function. **4–11. See margin.**

4. $f(x) = x^4 + 3x^3 - 7x^2 - 27x - 18$
5. $f(x) = 6x^4 - 31x^3 - 119x^2 + 214x + 560$
6. $f(x) = 20x^4 - 16x^3 + 11x^2 - 12x - 3$
7. $f(x) = 2x^4 - 30x^3 + 117x^2 - 75x + 280$
8. $f(x) = 3x^4 + 8x^3 + 9x^2 + 32x - 12$
9. $f(x) = 2x^6 - 12x^5 + 17x^4 + 6x^3 - 10x^2 + 6x - 9$
10. $f(x) = 2x^5 - 6x^4 + 18x^3 - 54x^2 - 20x + 60$
11. $f(x) = x^5 - x^4 + x^3 + 3x^2 - x$

Find all of the zeros of each function. **12–15. See margin.**

12. $f(x) = x^4 + 8x^2 - 9$
13. $f(x) = 2x^4 - 6x^3 - 70x^2 - 30x - 400$
14. $f(x) = 3x^4 - 9x^2 - 12$
15. $f(x) = 4x^4 + 19x^2 - 63$

Lesson 7-4

9. $x^2 - 3x + 7$
10. $x(5x - 2)$
11. $(x + 5)(2x - 1)$
12. $x^2 + 2x + 4$
13. $x + 5$
14. $x(x^2 + 1)$
15. $(x - 3)(x + 2)$
16. $(2x + 3)(3x^2 + 8x - 5)$
17. $(x + 5)(x^2 + 3x + 9)$
18. $(x - 6)(x - 4)$
19. $(2x + 1)(x - 2)$
20. $2(3x + 1)(x - 2)$
21. $(2x + 5)(x - 1)$
22. $(x + 3)(x + 5)$

Lesson 7-5

5. 3 or 1; 3 or 1; 2, 4, or 6
6. 2 or 0; 1; 2 or 4
7. 3 or 1; 1; 2 or 4
8. 2 or 0; 1; 0 or 2
9. 1; 3 or 1; 4 or 6
10. 0; 0, 2, 4, or 6; 0, 2, 4, or 6
11. 4, 2, or 0; 0; 4, 2, or 0
12. 5, 3, 1; 0; 4, 2, or 0
13. 1, $3 \pm i$
14. $-\frac{7}{2}, \frac{4}{5}, 2$
15. 6, $5 \pm \sqrt{6}$
16. 1, $\frac{3 \pm \sqrt{105}}{6}$
17. $-3, -\frac{2}{3}, 1, \frac{1}{2}$
18. $-8, -1, \pm\frac{5i}{2}$
19. $-\frac{5}{2}, -\frac{1}{2}, \frac{1}{3}$
20. −5, 1, 3
21. $f(x) = x^3 - 7x + 6$
22. $f(x) = x^4 - 34x^2 + 225$
23. $f(x) = x^4 - 11x^2 - 900$
24. $f(x) = x^3 - 3x^2 + 7x - 21$

Lesson 7-6

1. $\pm1, \pm\frac{1}{3}, \pm\frac{2}{3}, \pm2, \pm3, \pm6$
2. $\pm1, \pm2, \pm4, \pm8, \pm\frac{1}{2}, \pm\frac{1}{4}$
3. $\pm1, \pm7, \pm\frac{1}{6}, \pm\frac{7}{6}, \pm\frac{1}{2}, \pm\frac{7}{2}, \pm\frac{1}{3}, \pm\frac{7}{3}$
4. ±3, −1, −2
5. $-\frac{5}{2}, -2, \frac{8}{3}, 7$
6. $1, -\frac{1}{5}$
7. 7, 8
8. $-3, \frac{1}{3}$
9. ±1, 3
10. ±1, 3
11. 0
12. $\pm1, \pm3i$
13. $8, -5, \pm i\sqrt{5}$
14. $\pm2, \pm i$
15. $\pm\frac{3}{2}, \pm i\sqrt{7}$

Lesson 7-7

1. $4x + 2$; $2x + 8$; $3x^2 - 4x - 15$; $\dfrac{3x + 5}{x - 3}$, $x \neq 3$
2. $x^2 + \sqrt{x}$; $\sqrt{x} - x^2$; $x^2\sqrt{x}$; $\dfrac{\sqrt{x}}{x^2}$; $x \neq 0$
3. $\dfrac{6x^2 - 14x + 16}{3}$; $\dfrac{6x^2 - 16x + 32}{3}$; $\dfrac{2x^3 - 21x^2 + 48x - 64}{3}$; $\dfrac{6x^2 - 15x + 24}{x - 8}$, $x \neq 8$
4. $2x^2$; -10; $x^4 - 25$; $\dfrac{x^2 - 5}{x^2 + 5}$, $x \neq \pm 5i$
5. $x^2 + 5x + 6$; $-x^2 - 3x - 2$; $x^3 + 6x^2 + 12x + 8$; $\dfrac{1}{x + 2}$, $x \neq -2$
6. $x^2 + x + 2$; $x^2 - x$; $x^3 + x^2 + x + 1$; $\dfrac{x^2 + 1}{x + 1}$, $x \neq -1$
7. {(1, 1), (−1, −1), (5, 5)}; {(−1, −1), (2, 2), (−3, −3)}
8. does not exist; {(0, 4), (5, 3), (−9, 1)}
9. {(2, 2), (5, 5), (4, 4), (0, 0)}; {(8, 8), (6, 6), (−3, −3), (1, 1)}
10. {(−4, 4), (−7, 6), (−2, 0)}; {(−1, −9)}

Lesson 7-8

1. {(7, −2), (0, 3), (−8, 5)}
2. {(9, −3), (4, −2), (9, 3), (1, −1)}
3. {(5, 1), (3, 2), (3, 4), (5, −1)}

Lesson 7-9

13.

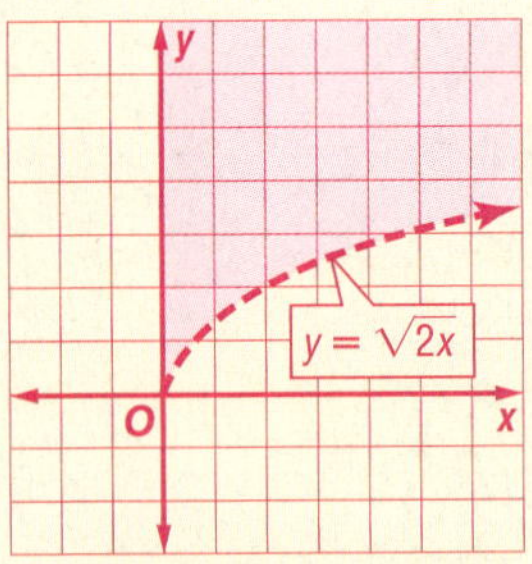

14.

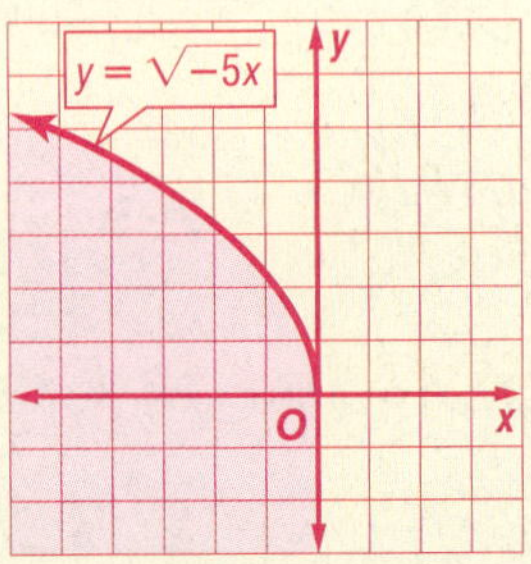

Extra Practice

Lesson 7-7 (pages 383–389)

Find $(f + g)(x)$, $(f - g)(x)$, $(f \cdot g)(x)$, and $\left(\frac{f}{g}\right)(x)$ for each $f(x)$ and $g(x)$. **1–6. See margin.**

1. $f(x) = 3x + 5$, $g(x) = x - 3$
2. $f(x) = \sqrt{x}$, $g(x) = x^2$
3. $f(x) = 2x^2 - 5x + 8$, $g(x) = \dfrac{x - 8}{3}$
4. $f(x) = x^2 - 5$, $g(x) = x^2 + 5$
5. $f(x) = x + 2$, $g(x) = x^2 + 4x + 4$
6. $f(x) = x^2 + 1$, $g(x) = x + 1$

For each set of ordered pairs, find $f \circ g$ and $g \circ f$ if they exist. **7–10. See margin.**

7. $f = \{(-1, 1), (2, -1), (-3, 5)\}$, $g = \{(1, -1), (-1, 2), (5, -3)\}$
8. $f = \{(0, 6), (5, -8), (-9, 2)\}$, $g = \{(-8, 3), (6, 4), (2, 1)\}$
9. $f = \{(8, 2), (6, 5), (-3, 4), (1, 0)\}$, $g = \{(2, 8), (5, 6), (4, -3), (0, 1)\}$
10. $f = \{(10, 4), (-1, 2), (5, 6), (-1, 0)\}$, $g = \{(-4, 10), (2, -9), (-7, 5), (-2, -1)\}$

Find $[g \circ h](x)$ and $[h \circ g](x)$. **12. $9x^2 + 12x - 3$; $3x^2 - 19$ 14. $-9x + 17$; $-9x - 1$**

11. $g(x) = 8 - 2x$, $h(x) = 3x$ **$8 - 6x$; $24 - 6x$**
12. $g(x) = x^2 - 7$, $h(x) = 3x + 2$
13. $g(x) = 2x + 7$, $h(x) = \dfrac{x - 7}{2}$ **x; x**
14. $g(x) = 3x + 2$, $h(x) = 5 - 3x$

If $f(x) = x^2 + 1$, $g(x) = 2x$, and $h(x) = x - 1$, find each value. **22. $4c^4 + 8c^2 + 5$**

15. $g[f(1)]$ **4**
16. $[f \circ h](3)$ **5**
17. $[h \circ f](3)$ **9**
18. $[g \circ f](-2)$ **10**
19. $g[h(-20)]$ **−42**
20. $f[h(-3)]$ **17**
21. $g[f(a)]$ **$2a^2 + 2$**
22. $[f \circ (g \circ f)](c)$

Lesson 7-8 (pages 390–394)

Find the inverse of each relation. **1–3. See margin.**

1. $\{(-2, 7), (3, 0), (5, -8)\}$
2. $\{(-3, 9), (-2, 4), (3, 9), (-1, 1)\}$
3. $\{(1, 5), (2, 3), (4, 3), (-1, 5)\}$

Find the inverse of each function. Then graph the function and its inverse. **4–15. See pp. 861A–861T.**

4. $f(x) = x - 7$
5. $y = 2x + 8$
6. $g(x) = 3x - 8$
7. $h(x) = \frac{x}{5} + 1$
8. $y = -2$
9. $g(x) = 5 - 2x$
10. $y = -5x - 6$
11. $h(x) = -\frac{2}{3}x$
12. $y = \dfrac{x - 5}{3}$
13. $y = \frac{1}{2}x - 1$
14. $f(x) = \dfrac{3x + 8}{4}$
15. $g(x) = \dfrac{2x - 1}{3}$

Determine whether each pair of functions are inverse functions.

16. $f(x) = \dfrac{2x - 3}{5}$, $g(x) = \dfrac{3x - 5}{3}$ **no**
17. $f(x) = 5x - 6$, $g(x) = \dfrac{x + 6}{5}$ **yes**
18. $f(x) = 6 - 3x$, $g(x) = 2 - \frac{1}{3}x$ **yes**
19. $f(x) = 3x - 7$, $g(x) = \frac{1}{3}x + 7$ **no**

Lesson 7-9 (pages 395–399)

Graph each function. State the domain and range of each function. **1–12. See pp. 861A–861T.**

1. $y = \sqrt{x - 4}$
2. $y = \sqrt{x + 3} - 1$
3. $y = \frac{1}{3}\sqrt{x + 2}$
4. $y = \sqrt{2x + 5}$
5. $y = -\sqrt{4x}$
6. $y = 2\sqrt{x}$
7. $y = -3\sqrt{x}$
8. $y = \sqrt{x} + 5$
9. $y = \sqrt{2x} - 1$
10. $y = 5\sqrt{x} + 1$
11. $y = \sqrt{x + 1} - 2$
12. $y = 6 - \sqrt{x + 3}$

Graph each inequality. **13–18. See margin.**

13. $y > \sqrt{2x}$
14. $y \leq \sqrt{-5x}$
15. $y \geq \sqrt{x + 6} + 6$
16. $y < \sqrt{3x + 1} + 2$
17. $y \geq \sqrt{8x - 3} + 1$
18. $y < \sqrt{5x - 1} + 3$

15.

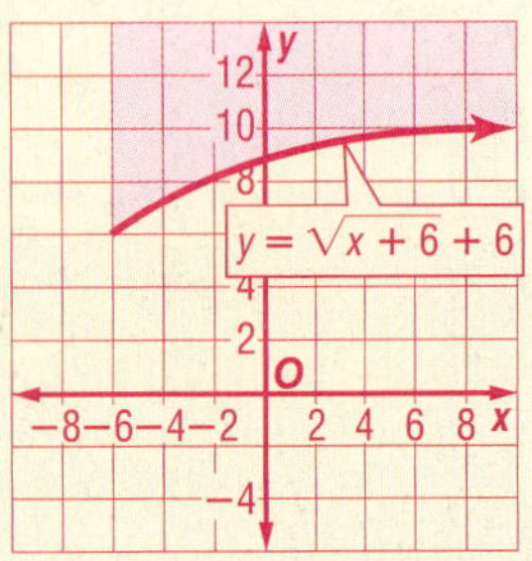

16.

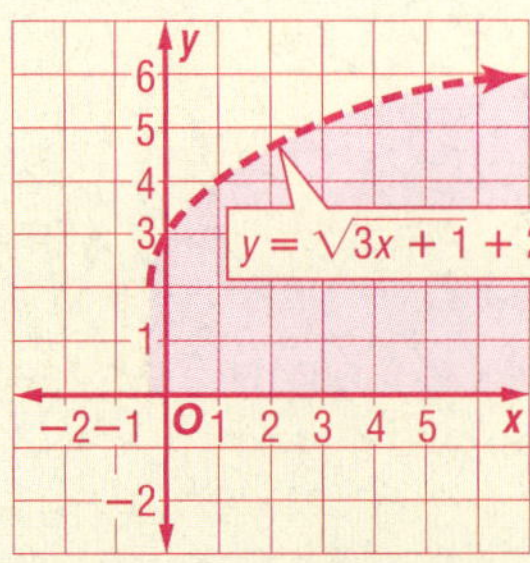

17.

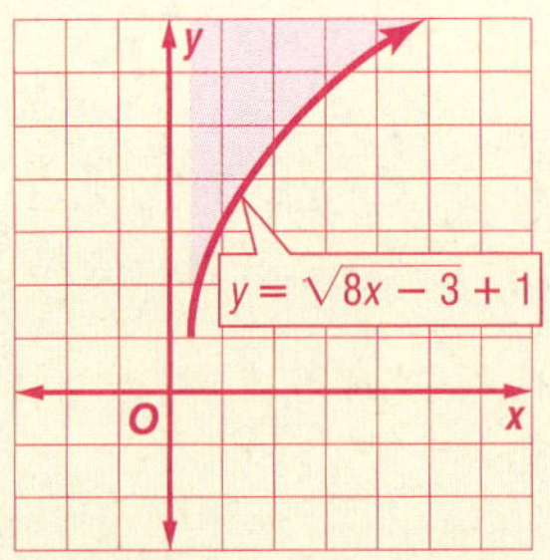

Lesson 8-1
(pages 412–416)

Find the midpoint of the line segment with endpoints at the given coordinates. **3. (−17.5, −25)**

1. (7, −3), (−11, 13) **(−2, 5)**
2. (16, 29), (−7, 2) **(4.5, 15.5)**
3. (43, −18), (−78, −32)
4. (−7.54, 3.42), (4.89, −9.28) **(−1.325, −2.93)**
5. $\left(\frac{1}{2}, \frac{1}{4}\right), \left(\frac{2}{3}, \frac{3}{5}\right)$ $\left(\frac{7}{12}, \frac{17}{40}\right)$
6. $\left(-\frac{1}{4}, \frac{2}{3}\right), \left(-\frac{1}{2}, -\frac{1}{2}\right)$ $\left(-\frac{3}{8}, \frac{1}{12}\right)$

Find the distance between each pair of points with the given coordinates.

7. (5, 7), (3, 19) **$2\sqrt{37}$ units**
8. (−2, −1), (5, 3) **$\sqrt{65}$ units**
9. (−3, 15), (7, −8) **$\sqrt{629}$ units**
10. (6, −3), (−4, −9) **$2\sqrt{34}$ units**
11. (3.89, −0.38), (4.04, −0.18) **0.25 unit**
12. $(5\sqrt{3}, 2\sqrt{2}), (-11\sqrt{3}, -4\sqrt{2})$ **$2\sqrt{210}$ units**
13. $\left(\frac{1}{4}, 0\right), \left(-\frac{2}{3}, \frac{1}{2}\right)$ **$\frac{\sqrt{157}}{12}$ units**
14. $\left(4, -\frac{5}{6}\right), \left(-2, \frac{1}{6}\right)$ **$\sqrt{37}$ units**
15. A circle has a radius with endpoints at (−3, 1) and (2, −5). Find the circumference and area of the circle. Write the answer in terms of π. **$2\pi\sqrt{61}$ units; 61π units²**
16. Triangle *ABC* has vertices *A*(0, 0), *B*(−3, 4), and *C*(2, 6). Find the perimeter of the triangle. **$5 + 2\sqrt{10} + \sqrt{29}$ units**

Lesson 8-2
(pages 419–425)

Write each equation in standard form. **1. $y = (x - 2)^2 + 3$ 2. $y = 2(x + 3)^2 - 1$ 3. $x = 3(y - 1)^2 + 2$**

1. $y = x^2 - 4x + 7$
2. $y = 2x^2 + 12x + 17$
3. $x = 3y^2 - 6y + 5$

Identify the coordinates of the vertex and focus, the equations of the axis of symmetry and directrix, and the direction of opening of the parabola with the given equation. Then find the length of the latus rectum and graph the parabola. **4–18. See pp. 861A–861T.**

4. $y + 4 = x^2$
5. $y = 5(x + 2)^2$
6. $4(y + 2) = 3(x - 1)^2$
7. $5x + 3y^2 = 15$
8. $y = 2x^2 - 8x + 7$
9. $x = 2y^2 - 8y + 7$
10. $3(x - 8)^2 = 5(y + 3)$
11. $x = 3(y + 4)^2 + 1$
12. $8y + 5x^2 + 30x + 101 = 0$
13. $x = -\frac{1}{5}y^2 + \frac{8}{5}y - 7$
14. $6x = y^2 - 6y + 39$
15. $-8y = x^2$
16. $y = 4x^2 + 24x + 38$
17. $y = x^2 - 6x + 3$
18. $y = x^2 + 4x + 1$

Write an equation for each parabola described below. Then draw the graph. **19–21. See margin for graphs.**

19. focus (1, 1), directrix $y = -1$ **$y = \frac{1}{4}(x - 1)^2 + 0$**
20. vertex (−1, 2), directrix $y = -4$ **$y = \frac{1}{24}(x + 1)^2 + 2$**
21. vertex (2, −3), focus (0, −3) **$x = -\frac{1}{8}(y + 3)^2 + 2$**

Lesson 8-3
(pages 426–431)

Write an equation for the circle that satisfies each set of conditions. **1–9. See margin.**

1. center (3, 2), $r = 5$ units
2. center (−5, 8), $r = 3$ units
3. center (1, −6), $r = \frac{2}{3}$ units
4. center (0, 7), tangent to *x*-axis
5. center (−2, −4), tangent to *y*-axis
6. endpoints of a diameter at (−9, 0) and (2, −5)
7. endpoints of a diameter at (4, 1) and (−3, 2)
8. center (6, −10), passes through origin
9. center (0.8, 0.5), passes through (2, 2)

Find the center and radius of the circle with the given equation. Then graph the circle. **10–21. See pp. 861A–861T.**

10. $x^2 + y^2 = 36$
11. $\left(x - \frac{3}{4}\right)^2 + \left(y + \frac{2}{3}\right)^2 = \frac{32}{49}$
12. $(x - 5)^2 + (y + 4)^2 = 1$
13. $x^2 + 3x + y^2 - 5y = 0.5$
14. $x^2 + y^2 = 14x - 24$
15. $x^2 + y^2 = 2(y - x)$
16. $x^2 + 10x + (y - \sqrt{3})^2 = 11$
17. $x^2 + y^2 = 4x + 9$
18. $x^2 + y^2 + 12x - 10y + 45 = 0$
19. $x^2 + y^2 - 6x + 4y = 156$
20. $x^2 + y^2 - 2x + 7y = 1$
21. $16(x^2 + y^2) - 8(3x + 5y) + 33 = 0$

18.

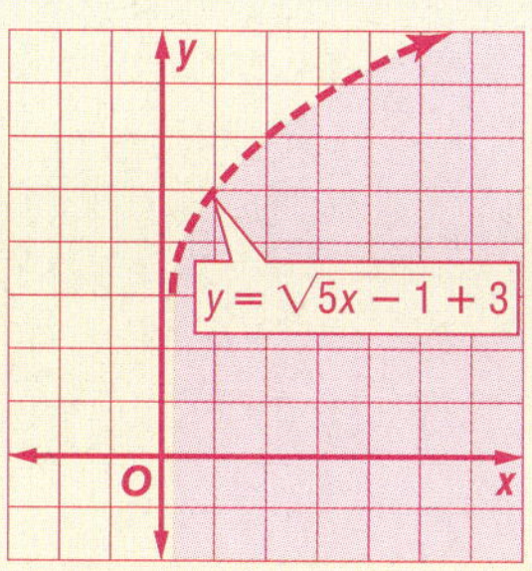

Extra Practice

Lesson 8-2

19.

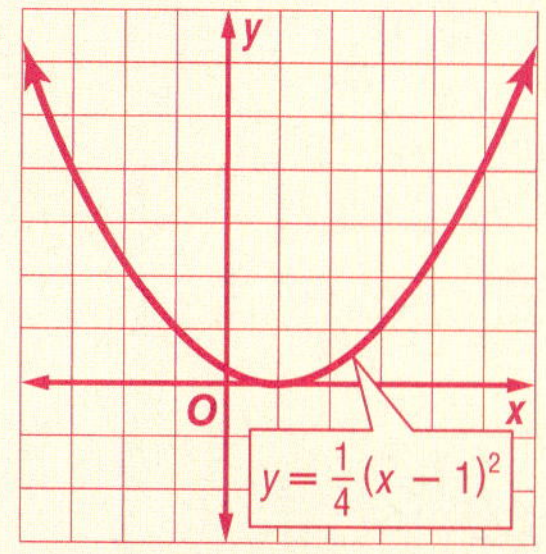

20.

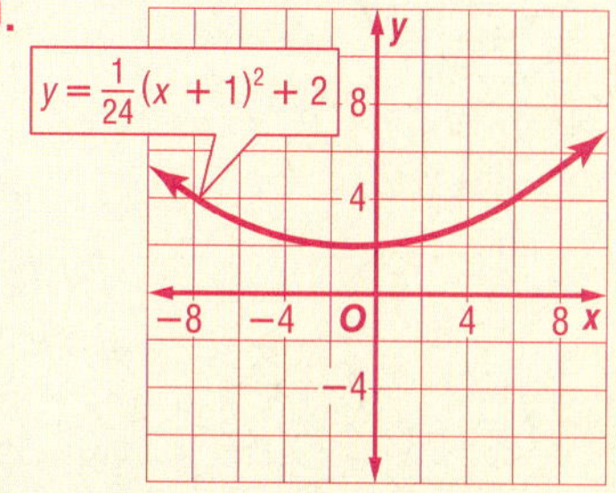

21.

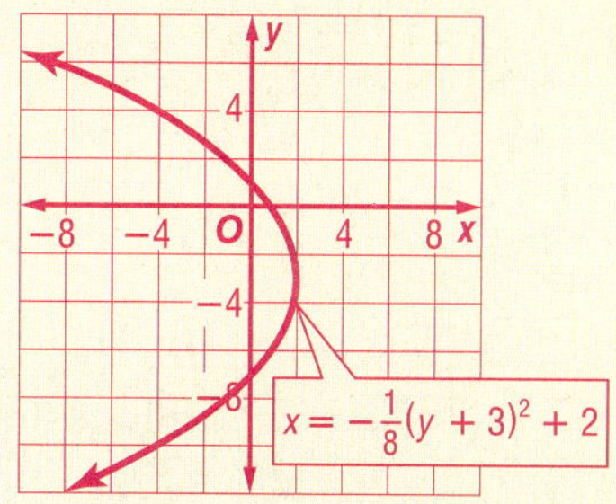

Lesson 8-3

1. $(x - 3)^2 + (y - 2)^2 = 25$
2. $(x + 5)^2 + (y - 8)^2 = 9$
3. $(x - 1)^2 + (y + 6)^2 = \frac{4}{9}$
4. $(x - 0)^2 + (y - 7)^2 = 49$
5. $(x + 2)^2 + (y + 4)^2 = 4$
6. $(x + 3.5)^2 + (y + 2.5)^2 = 36.5$
7. $(x - 0.5)^2 + (y - 1.5)^2 = 12.5$
8. $(x - 6)^2 + (y + 10)^2 = 136$
9. $(x - 0.8)^2 + (y - 0.5)^2 = 3.69$

Lesson 8-4

1. $\frac{(x-1)^2}{9}+\frac{(y-7)^2}{4}=1$
2. $\frac{(x-1)^2}{25}+\frac{(y-0.5)^2}{20.25}=1$
3. $\frac{(x-0)^2}{4}+\frac{(y-3)^2}{144}=1$

Lesson 8-5

17. $\frac{x^2}{9}-\frac{y^2}{16}=1$
18. $\frac{y^2}{49}-\frac{x^2}{156.25}=1$
19. $\frac{x^2}{36}-\frac{y^2}{25}=1$

Extra Practice

Lesson 8-4

(pages 433–440)

Write an equation for the ellipse that satisfies each set of conditions. 1–3. See margin.

1. endpoints of major axis at $(-2, 7)$ and $(4, 7)$, endpoints of minor axis at $(1, 5)$ and $(1, 9)$
2. endpoints of minor axis at $(1, -4)$ and $(1, 5)$, endpoints of major axis at $(-4, 0.5)$ and $(6, 0.5)$
3. major axis 24 units long and parallel to the y-axis, minor axis 4 units long, center at $(0, 3)$

Find the coordinates of the center and foci and the lengths of the major and minor axes for the ellipse with the given equation. Then graph the ellipse. 4–16. See pp. 861A–861T.

4. $\frac{x^2}{36}+\frac{y^2}{81}=1$
5. $\frac{x^2}{121}+\frac{(y-5)^2}{16}=1$
6. $\frac{(x+2)^2}{12}+\frac{(y+1)^2}{16}=1$
7. $\frac{(x+2)^2}{36}+\frac{(y-4)^2}{40}=1$
8. $\frac{(x+8)^2}{121}+\frac{(y-7)^2}{64}=1$
9. $\frac{(x-4)^2}{16}+\frac{(y+1)^2}{9}=1$
10. $8x^2+2y^2=32$
11. $7x^2+3y^2=84$
12. $9x^2+16y^2=144$
13. $169x^2-338x+169+25y^2=4225$
14. $x^2+4y^2+8x-64y=-128$
15. $4x^2+5y^2=6(6x+5y)+658$
16. $9x^2+16y^2-54x+64y+1=0$

Extra Practice

Lesson 8-5

(pages 441–448)

Find the coordinates of the vertices and foci and the equations of the asymptotes for the hyperbola with the given equation. Then graph the hyperbola. 1–16. See pp. 861A–861T.

1. $\frac{y^2}{25}-\frac{x^2}{9}=1$
2. $\frac{x^2}{4}-\frac{y^2}{9}=1$
3. $\frac{x^2}{81}-\frac{y^2}{36}=1$
4. $\frac{x^2}{9}-\frac{y^2}{16}=1$
5. $\frac{y^2}{100}-\frac{x^2}{144}=1$
6. $\frac{x^2}{16}-\frac{y^2}{4}=1$
7. $\frac{(x-4)^2}{64}-\frac{(y+1)^2}{16}=1$
8. $\frac{(y-7)^2}{2.25}-\frac{(x-3)^2}{4}=1$
9. $(x+5)^2-\frac{(y+3)^2}{48}=1$
10. $x^2-9y^2=36$
11. $4x^2-9y^2=72$
12. $49x^2-16y^2=784$
13. $144x^2+1152x-25y^2-100y=1396$
14. $576y^2=49x^2+490x+29449$
15. $23.04y^2-46.08y-1.96x^2-3.92x=24.0784$
16. $25(y+5)^2-20(x-1)^2=500$

Write an equation for the hyperbola that satisfies each set of conditions. 17–19. See margin.

17. vertices $(-3, 0)$ and $(3, 0)$; conjugate axis of length 8 units
18. vertices $(0, -7)$ and $(0, 7)$; conjugate axis of length 25 units
19. center $(0, 0)$; horizontal transverse axis of length 12 units and a conjugate axis of length 10 units

Lesson 8-6

(pages 449–452)

Write each equation in standard form. State whether the graph of the equation is a *parabola*, *circle*, *ellipse*, or *hyperbola*. Then graph the equation. 1–12. See pp. 861A–861T.

1. $9x^2-36x+36=4y^2+24y+72$
2. $x^2+4x+2y^2+16y+32=0$
3. $x^2+6x+y^2-6y+9=0$
4. $9y^2=25x^2+400x+1825$
5. $2y^2+12y-x+6=0$
6. $x^2+y^2=10x+2y+23$
7. $3x^2+y=12x-17$
8. $9x^2-18x+16y^2+160y=-265$
9. $x^2+10x+5=4y^2+16$
10. $\frac{(y-5)^2}{4}-(x+1)^2=4$
11. $9x^2+49y^2=441$
12. $4x^2-y^2=4$

Without writing the equation in standard form, state whether the graph of each equation is a *parabola*, *circle*, *ellipse*, or *hyperbola*.

13. $(x+3)^2=8(y+2)$ parabola
14. $x^2+4x+y^2-8y=2$ circle
15. $9x^2+9y^2=9$ circle
16. $y-x^2=x+3$ parabola
17. $2x^2-13y^2+5=0$ hyperbola
18. $16(x-3)^2+81(y+4)^2=1296$ ellipse
19. $x^2+5y^2=16$ ellipse
20. $4x^2-y^2=16$ hyperbola

846 Extra Practice

Lesson 8-7

(pages 455–460)

Solve each system of inequalities by graphing. **1–4. See margin.**

1. $\frac{x^2}{16} - \frac{y^2}{1} \geq 1$, $x^2 + y^2 \leq 49$
2. $\frac{x^2}{25} + \frac{y^2}{16} \leq 1$, $y \leq x - 2$
3. $y \geq x + 3$, $x^2 + y^2 < 25$
4. $4x^2 + (y - 3)^2 \leq 16$, $x + 2y \geq 4$

Find the exact solution(s) of each system of equations. **5–13. See margin.**

5. $\frac{x^2}{16} + \frac{y^2}{16} = 1$, $y = x + 3$
6. $x = y^2$, $(x + 3)^2 + y^2 = 53$
7. $\frac{x^2}{3} - \frac{(y + 2)^2}{4} = 1$, $x^2 = y^2 + 11$
8. $\frac{(x - 1)^2}{5} + \frac{y^2}{2} = 1$, $y = x + 1$
9. $x^2 + y^2 = 13$, $x^2 - y^2 = -5$
10. $\frac{x^2}{25} - \frac{y^2}{5} = 1$, $y = x - 4$
11. $x^2 + y = 0$, $x + y = -2$
12. $x^2 - 9y^2 = 36$, $x = y$
13. $4x^2 + 6y^2 = 360$, $y = x$

Lesson 9-1

(pages 472–478)

Simplify each expression.

1. $\frac{25xy^2}{15y}$ **$\frac{5xy}{3}$**
2. $\frac{-4a^2b^3}{28ab^4}$ **$-\frac{a}{7b}$**
3. $\frac{(-2cd^3)^2}{8c^2d^5}$ **$\frac{d}{2}$**
4. $\frac{3x^3}{-2} \cdot \frac{-4}{9x}$ **$\frac{2x^2}{3}$**
5. $\frac{21x^2}{-5} \cdot \frac{10}{7x^3}$ **$-\frac{6}{x}$**
6. $\frac{2u^2}{3} \div \frac{6u^3}{5}$ **$\frac{5}{9u}$**
7. $\frac{15x^3}{14} \div \frac{18x}{7}$ **$\frac{5x^2}{12}$**
8. $\frac{xy^2}{2} \cdot \frac{x^2}{2y} \cdot \frac{2}{x^2y}$ **$\frac{x}{2}$**
9. $axy \div \frac{ax}{y}$ **y^2**
10. $\frac{9u^2}{28v} \div \frac{27u^2}{8v^2}$ **$\frac{2v}{21}$**
11. $\frac{x^2 - 4}{4x^2 - 1} \cdot \frac{2x - 1}{x + 2}$ **$\frac{x - 2}{2x + 1}$**
12. $\frac{x^2 - 1}{2x^2 - x - 1} \div \frac{x^2 - 4}{2x^2 - 3x - 2}$ **$\frac{x + 1}{x + 2}$**
13. $\frac{2x^2 + x - 1}{2x^2 + 3x - 2} \div \frac{x^2 - 2x + 1}{x^2 + x - 2}$ **$\frac{x + 1}{x - 1}$**
14. $\frac{\frac{(ab)^2}{c}}{\frac{xa^3b}{cx^2}}$ **$\frac{bx}{a}$**
15. $\frac{\frac{x^4 - y^4}{x^3 + y^3}}{\frac{x^3 - y^3}{x + y}}$ **$\frac{(x^2 + y^2)(x + y)}{(x^2 - xy + y^2)(x^2 + xy + y^2)}$**

Lesson 9-2

(pages 479–484)

Find the LCM of each set of polynomials.

1. $2a^2b, 4ab^2, 20a$ **$20a^2b^2$**
2. $48c^2d, 72cd^2$ **$144c^2d^2$**
3. $x^2 - 4x - 12, x^2 + 7x + 10$ **$(x - 6)(x + 2)(x + 5)$**

Simplify each expression. **16–21. See margin.**

4. $\frac{12}{7d} - \frac{3}{14d}$ **$\frac{3}{2d}$**
5. $\frac{x + 1}{x} - \frac{x - 1}{x^2}$ **$\frac{x^2 + 1}{x^2}$**
6. $\frac{2x + 1}{4x^2} - \frac{x + 3}{6x}$ **$\frac{3 - 2x^2}{12x^2}$**
7. $\frac{7x}{13y^2} + \frac{4y}{6x^2}$ **$\frac{21x^3 + 26y^3}{39x^2y^2}$**
8. $\frac{x}{x - 1} + \frac{1}{1 - x}$ **1**
9. $\frac{1}{3v^2} + \frac{1}{uv} + \frac{3}{4u^2}$ **$\frac{4u^2 + 12uv + 9v^2}{12u^2v^2}$**
10. $\frac{1}{x^2 - x} + \frac{1}{x^2 + x}$ **$\frac{2}{(x - 1)(x + 1)}$**
11. $\frac{1}{x^2 - 1} - \frac{1}{(x - 1)^2}$ **$\frac{-2}{(x + 1)(x - 1)^2}$**
12. $\frac{5}{x} - \frac{3}{x + 5}$ **$\frac{2x + 25}{x(x + 5)}$**
13. $y - 1 + \frac{1}{y - 1}$ **$\frac{y^2 - 2y + 2}{y - 1}$**
14. $3m + 1 - \frac{2m}{3m + 1}$ **$\frac{9m^2 + 4m + 1}{3m + 1}$**
15. $\frac{3x}{x - y} + \frac{4x}{y - x}$ **$-\frac{x}{x - y}$**
16. $\frac{6}{4m^2 - 12mn + 9n^2} + \frac{2}{2mn - 3n^2}$
17. $\frac{3}{x^2 + 5ax + 6a^2} + \frac{2}{x^2 - 4a^2}$
18. $\frac{x}{x^2 + 5x + 6} - \frac{2}{x^2 + 4x + 4}$
19. $\frac{4}{a^2 - 4} - \frac{3}{a^2 + 4a + 4}$
20. $\frac{4}{3 - 3z^2} - \frac{2}{z^2 + 5z + 4}$
21. $\frac{2c}{c^2 - 9} - \frac{1}{c^2 + 6c + 9}$
22. $\frac{\frac{1}{x + y}}{\frac{1}{x} + \frac{1}{y}}$ **$\frac{xy}{(x + y)^2}$**
23. $\frac{1 - \frac{1}{x + 1}}{1 + \frac{1}{x - 1}}$ **$\frac{x - 1}{x + 1}$**
24. $\frac{4 + \frac{1}{x - 2}}{3 - \frac{1}{x - 2}}$ **$\frac{4x - 7}{3x - 7}$**

Lesson 9-2

16. $\frac{4m}{n(2m - 3n)^2}$
17. $\frac{5x}{(x + 3a)(x + 2a)(x - 2a)}$
18. $\frac{x^2 - 6}{(x + 2)^2(x + 3)}$
19. $\frac{a + 14}{(a - 2)(a + 2)^2}$
20. $-\frac{10}{3(z - 1)(z + 4)}$
21. $\frac{2c^2 + 5c + 3}{(c^2 - 9)(c + 3)}$

Lesson 8-7

1.

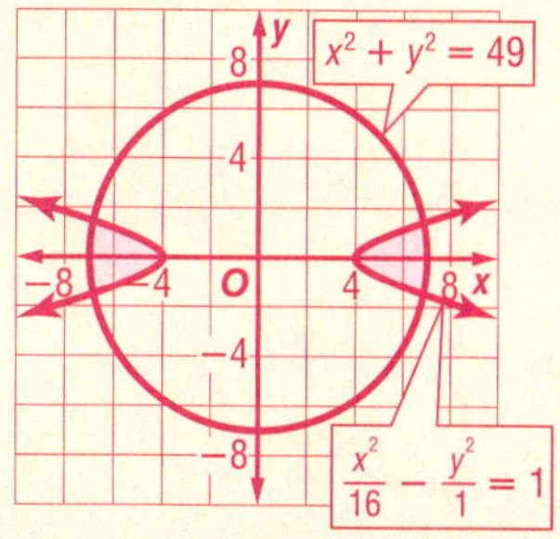

2.

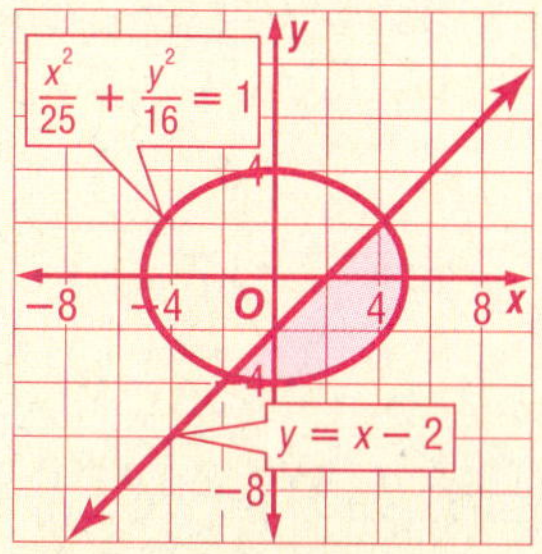

3.

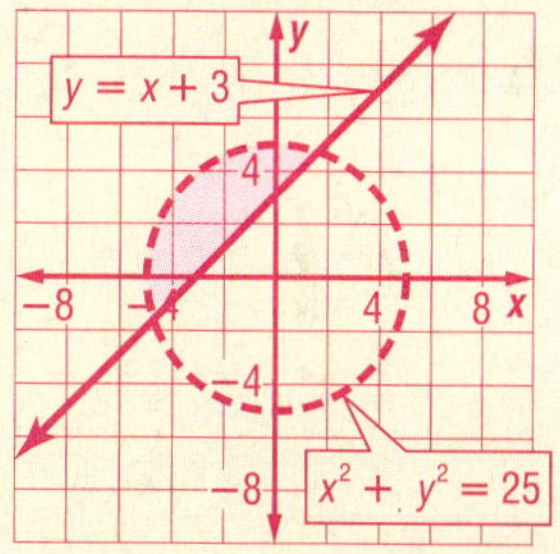

4.

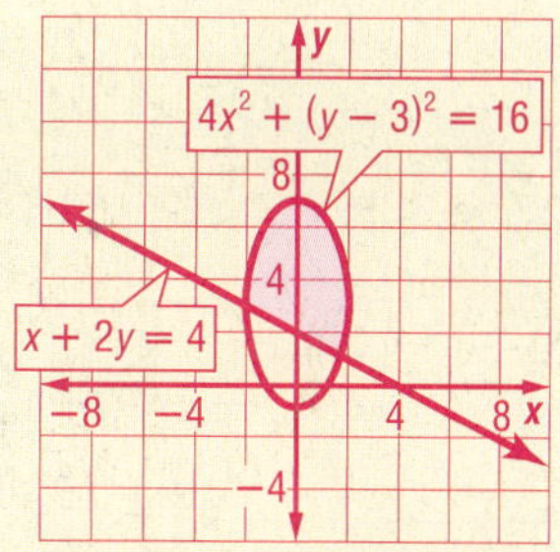

5. $\left(\frac{-3 + \sqrt{23}}{2}, \frac{3 + \sqrt{23}}{2}\right)$, $\left(\frac{-3 - \sqrt{23}}{2}, \frac{3 - \sqrt{23}}{2}\right)$
6. $(4, -2), (4, 2)$
7. $(\sqrt{111}, 10), (-\sqrt{111}, 10), (\sqrt{15}, 2), (-\sqrt{15}, 2)$
8. $\left(\frac{-3 + \sqrt{30}}{7}, \frac{4 + \sqrt{30}}{7}\right)$, $\left(\frac{-3 - \sqrt{30}}{7}, \frac{4 - \sqrt{30}}{7}\right)$
9. $(2, 3), (2, -3), (-2, 3), (-2, -3)$
10. no real solution
11. $(2, -4), (-1, -1)$
12. no real solution
13. $(6, 6), (-6, -6)$

Extra Practice 847

Lesson 9-3 3. asymptotes: $x = -1$, $x = 8$ *(pages 485–490)*

Determine the equations of any vertical asymptotes and the values of x for any holes in the graph of each rational function. 6. asymptote: $x = -5$, hole: $x = -3$

1. $f(x) = \frac{1}{x+4}$ asymptote: $x = -4$
2. $f(x) = \frac{x-2}{x+3}$ asymptote: $x = -3$
3. $f(x) = \frac{5}{(x+1)(x-8)}$
4. $f(x) = \frac{x}{x+2}$ asymptote: $x = -2$
5. $f(x) = \frac{x^2-4}{x+2}$ hole: $x = -2$
6. $f(x) = \frac{x^2+x-6}{x^2+8x+15}$

Graph each rational function. 7–15. See pp. 861A–861T.

7. $f(x) = \frac{1}{x-5}$
8. $f(x) = \frac{3x}{x+1}$
9. $f(x) = \frac{x^2-16}{x-4}$
10. $f(x) = \frac{x}{x-6}$
11. $f(x) = \frac{1}{(x-3)^2}$
12. $f(x) = \frac{2}{(x+3)(x-4)}$
13. $f(x) = \frac{x+4}{x^2-1}$
14. $f(x) = \frac{x+2}{x+3}$
15. $f(x) = \frac{x^2+5x-14}{x^2+9x+14}$

Lesson 9-4 *(pages 492–498)*

State whether each equation represents a *direct*, *joint*, or *inverse* variation. Then name the constant of variation.

1. $xy = 10$ inverse; 10
2. $x = 6y$ direct; $\frac{1}{6}$
3. $\frac{x}{7} = y$ direct; $\frac{1}{7}$
4. $\frac{x}{y} = -6$ direct; $-\frac{1}{6}$
5. $10x = y$ direct; 10
6. $x = \frac{2}{y}$ inverse; 2
7. $A = lw$ joint; 1
8. $\frac{1}{4}b = -\frac{3}{5}c$ direct; $-\frac{12}{5}$
9. $D = rt$ joint, 1

Find each value.

10. If y varies directly as x and $y = 16$ when $x = 4$, find y when $x = 12$. 48
11. If x varies inversely as y and $x = 12$ when $y = -3$, find x when $y = -18$. 2
12. If m varies directly as w and $m = -15$ when $w = 2.5$, find m when $w = 12.5$. -75
13. If y varies jointly as x and z and $y = 10$ when $z = 4$ and $x = 5$, find y when $x = 4$ and $z = 2$. 4
14. If y varies inversely as x and $y = \frac{1}{4}$ when $x = 24$, find y when $x = \frac{3}{4}$. 8
15. If y varies jointly as x and z and $y = 45$ when $x = 9$ and $z = 15$, find y when $x = 25$ and $z = 12$. 100

Lesson 9-5 4–12. See margin for graphs. *(pages 499–504)*

Identify the type of function represented by each graph.

1. 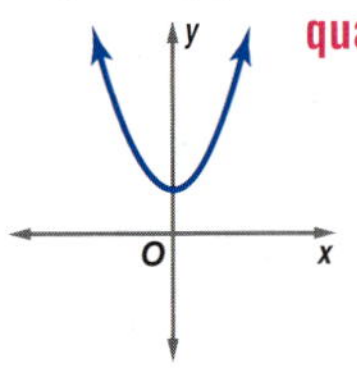 quadratic
2. 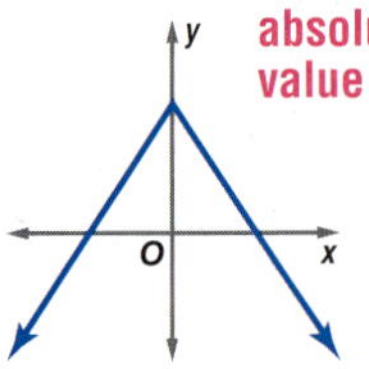 absolute value
3. 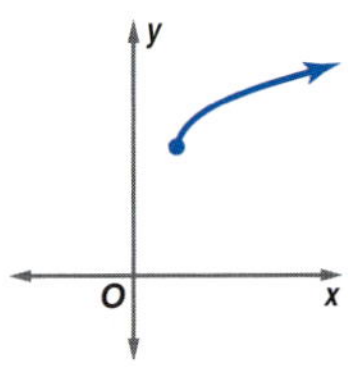 square root

Identify the function represented by each equation. Then graph the equation.

4. $y = \sqrt{5x}$ square root
5. $y = \frac{3}{4}x$ direct variation
6. $y = |x| + 3$ absolute value
7. $y = x^2 - 2$ quadratic
8. $y = \frac{2}{x}$ inverse variation
9. $y = 2[\![x]\!]$ greatest integer
10. $y = -2x^2 + 1$ quadratic
11. $y = \frac{x^2+2x-3}{x^2+7x+12}$ rational
12. $y = -3$ constant

Extra Practice

Lesson 9-5

4. 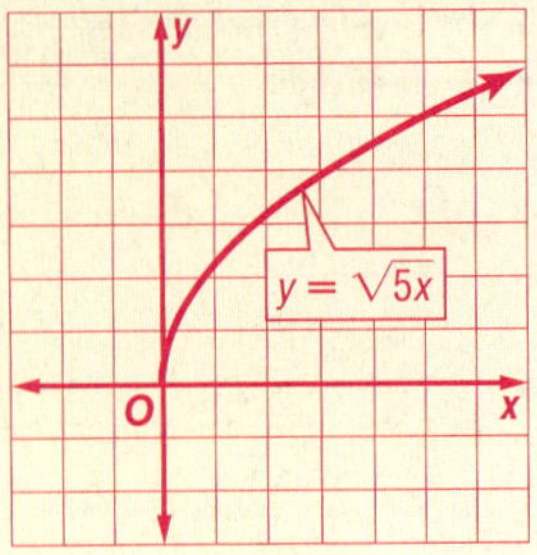

5. 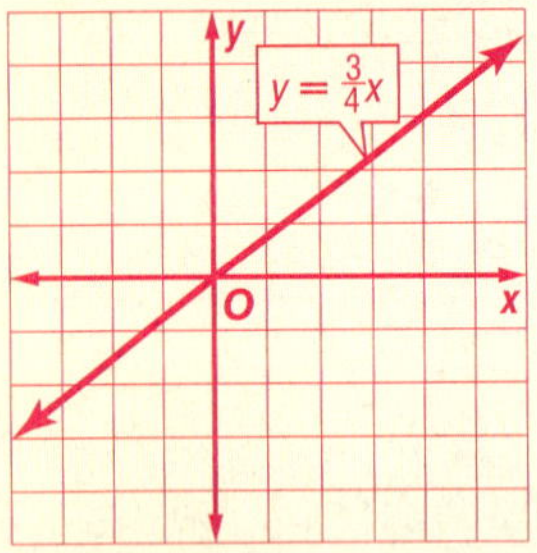

6. 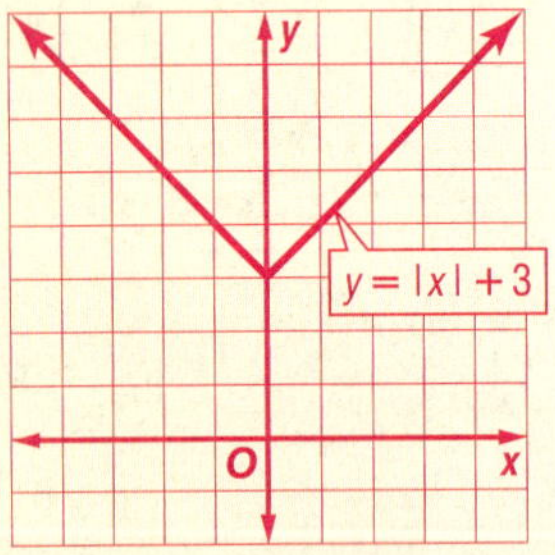

7. 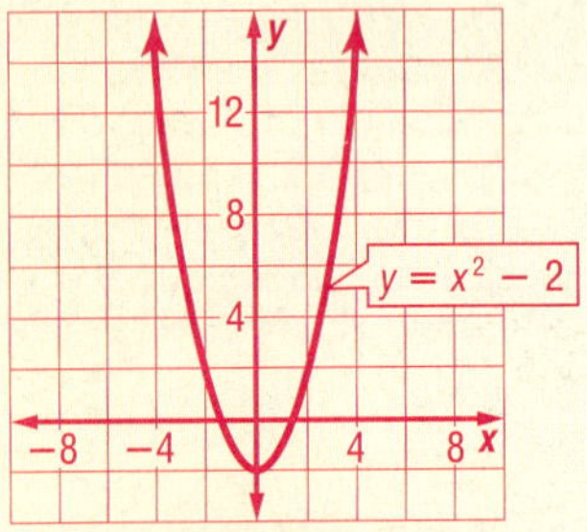

8. 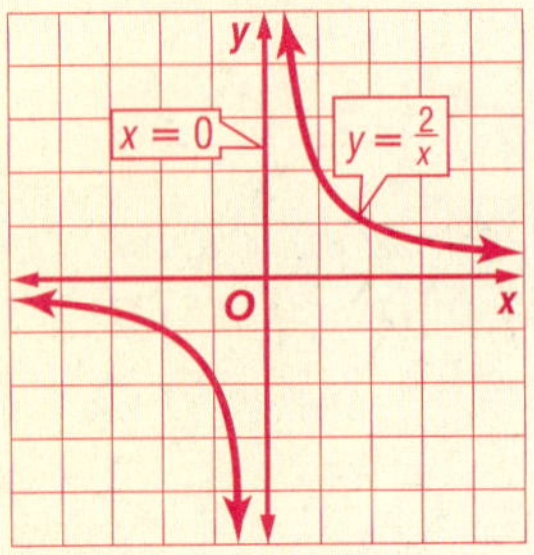

9. 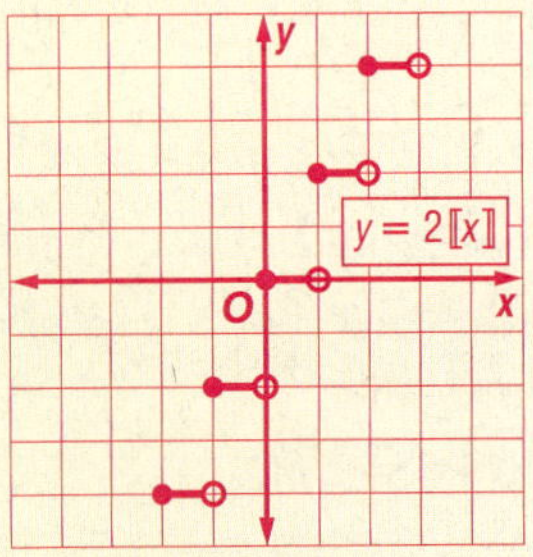

10. 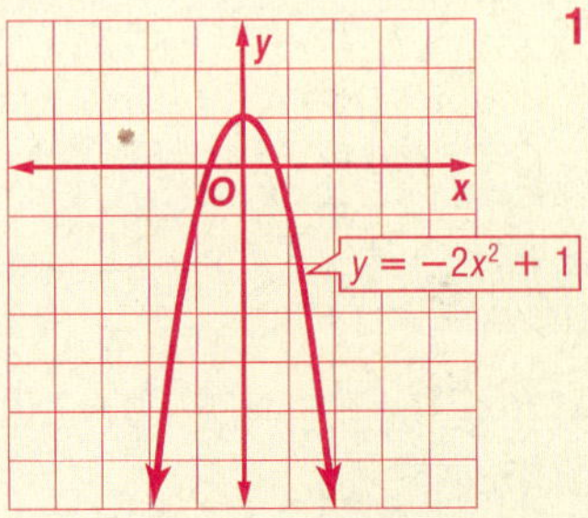

11. 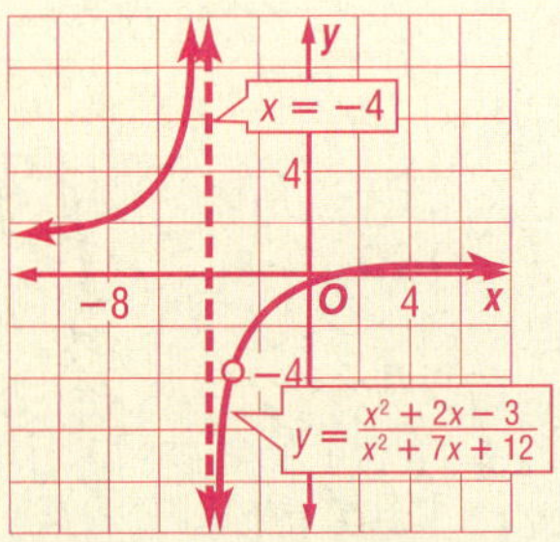

12. 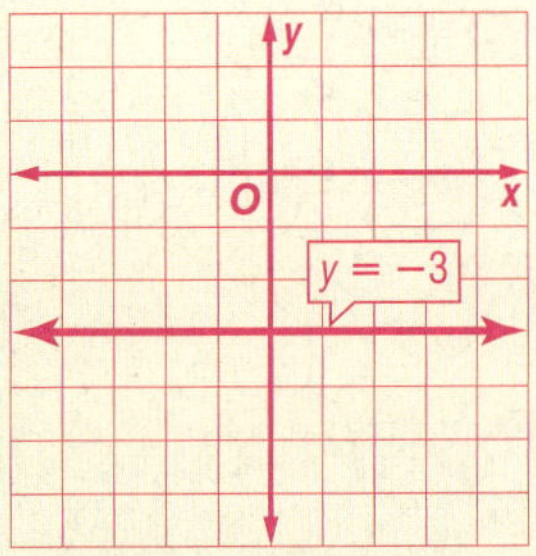

Lesson 9-6

(pages 505–511)

Solve each equation or inequality. Check your solutions.

1. $\frac{x}{x-3}=\frac{1}{4}$ **-1**
2. $\frac{5}{x}+\frac{3}{5}=\frac{2}{x}$ **-5**
3. $\frac{5}{b-2}<5$ **$b<2$ or $b>3$**
4. $\frac{4}{a+3}>2$ **$-3<a<-1$**
5. $\frac{x-2}{x}=\frac{x-4}{x-6}$ **3**
6. $-6-\frac{8}{n}<n$ **$-4<n<-2$ or $n>0$**
7. $\frac{2}{d}+\frac{1}{d-2}=1$ **$1, 4$**
8. $\frac{1}{2+3x}+\frac{2}{2-3x}=0$ **-2**
9. $\frac{1}{n+1}+\frac{1}{n-1}=\frac{2}{n^2-1}$ **$\varnothing$**
10. $\frac{1}{x-3}+\frac{1}{x+5}=\frac{x+1}{x-3}$ **$-3, -1$**
11. $\frac{4}{x^2-2x-3}=\frac{-x}{3-x}-\frac{1}{x+1}$ **1**
12. $\frac{p}{p+1}+\frac{3}{p-3}+1=0$ **$0, 1$**
13. $\frac{3x}{x^2+2x-8}=\frac{1}{x-2}+\frac{x}{x+4}$ **$\varnothing$**
14. $\frac{5z+2}{z^2-4}=\frac{-5z}{2-z}+\frac{2}{z+2}$ **$\frac{3}{5}$**
15. $\frac{1}{x-3}+\frac{2}{x^2-9}=\frac{5}{x+3}$ **5**
16. $\frac{1}{m^2-1}=\frac{2}{m^2+m-2}$ **0**
17. $\frac{12}{x^2-16}-\frac{24}{x-4}=3$ **$-6, -2$**
18. $n+\frac{1}{n+3}=\frac{n^2}{n-1}$ **-1**

Lesson 10-1

(pages 523–530)

Sketch the graph of each function. Then state the function's domain and range. **1–4. See margin.**

1. $y=3(5)^x$
2. $y=0.5(2)^x$
3. $y=3\left(\frac{1}{4}\right)^x$
4. $y=2(1.5)^x$

Determine whether each function represents exponential growth or decay.

5. $y=4(3)^x$ **growth**
6. $y=10^{-x}$ **decay**
7. $y=5\left(\frac{1}{2}\right)^x$ **decay**
8. $y=2\left(\frac{5}{4}\right)^x$ **growth**

Write an exponential function whose graph passes through the given points.

9. (0, 6) and (2, 54) **$y=6(3)^x$**
10. (0, −4) and (−4, −64) **$y=-4(0.5)^x$**
11. (0, 1.5) and (3, 40.5) **$y=1.5(3)^x$**
12. (0, −3.7) and (5, −118.4) **$y=-3.7(2)^x$**

Simplify each expression.

13. $4^{\sqrt{2}}\cdot 4^{\sqrt{8}}$ **$4^{3\sqrt{2}}$**
14. $(5^{\sqrt{5}})^{\sqrt{45}}$ **5^{15}**
15. $(w^{\sqrt{6}})^{\sqrt{3}}$ **$w^{3\sqrt{2}}$**
16. $27^{\sqrt{5}}\div 3^{\sqrt{5}}$ **$3^{2\sqrt{5}}$**
17. $8^{2\sqrt{3}}\cdot 4^{\sqrt{3}}$ **$2^{8\sqrt{3}}$**
18. $5^{\sqrt{2}}\cdot 5^{\sqrt{3}}$ **$5^{\sqrt{2}+\sqrt{3}}$**
19. $7^{\sqrt{3}}\cdot 7^{2\sqrt{3}}$ **$7^{3\sqrt{3}}$**
20. $(y^{\sqrt{3}})^{\sqrt{27}}$ **y^9**

Solve each equation or inequality. Check your solution.

21. $27^{2x-1}=3$ **$\frac{2}{3}$**
22. $8^{2+x}\geq 2$ **$x\geq -\frac{5}{3}$**
23. $4^{2x+5}<8^{x+1}$ **$x<-7$**
24. $6^{x+1}=36^{x-1}$ **3**
25. $10^{x-1}>100^{4-x}$ **$x>3$**
26. $\left(\frac{1}{5}\right)^{x-3}=125$ **0**
27. $2^{x^2+1}=32$ **± 2**
28. $36^x=6^{x^2-3}$ **$-1, 3$**

Lesson 10-2

(pages 531–538)

Write each equation in logarithmic form.

1. $3^5=243$ **$\log_3 243=5$**
2. $10^3=1000$ **$\log_{10} 1000=3$**
3. $4^{-3}=\frac{1}{64}$ **$\log_4 \frac{1}{64}=-3$**

Write each equation in exponential form.

4. $\log_2 \frac{1}{8}=-3$ **$2^{-3}=\frac{1}{8}$**
5. $\log_{25} 5=\frac{1}{2}$ **$25^{\frac{1}{2}}=5$**
6. $\log_7 \frac{1}{7}=-1$ **$7^{-1}=\frac{1}{7}$**

Evaluate each expression.

7. $\log_4 16$ **2**
8. $\log_{10} 10{,}000$ **4**
9. $\log_3 \frac{1}{9}$ **-2**
10. $\log_2 1024$ **10**
11. $\log_6 6^5$ **5**
12. $\log_{\frac{1}{2}} 8$ **-3**
13. $\log_{11} 121$ **2**
14. $5^{\log_5 10}$ **10**

Solve each equation or inequality. Check your solution.

15. $\log_8 b=2$ **64**
16. $\log_4 x<3$ **$0<x<64$**
17. $\log_{\frac{1}{9}} n=-\frac{1}{2}$ **3**
18. $\log_x 7=1$ **7**
19. $\log_{\frac{2}{3}} a<3$ **$0<a<\frac{8}{27}$**
20. $\log_2 (x^2-9)=4$ **± 5**

Extra Practice

Lesson 10-1

1. **D = {x | x is any real number.}; R = {y | y > 0}**

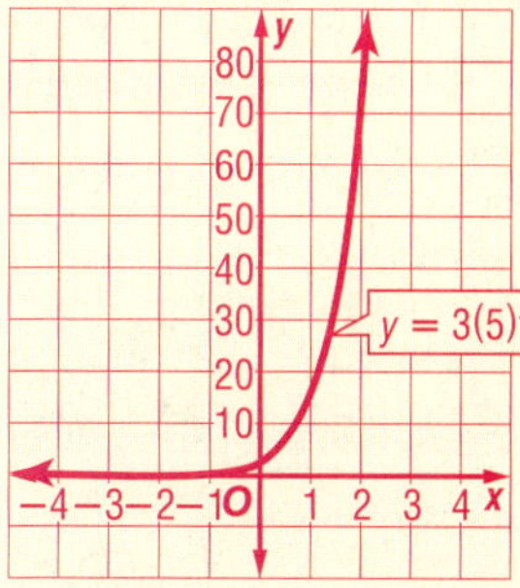

2. **D = {x | x is any real number.}; R = {y | y > 0}**

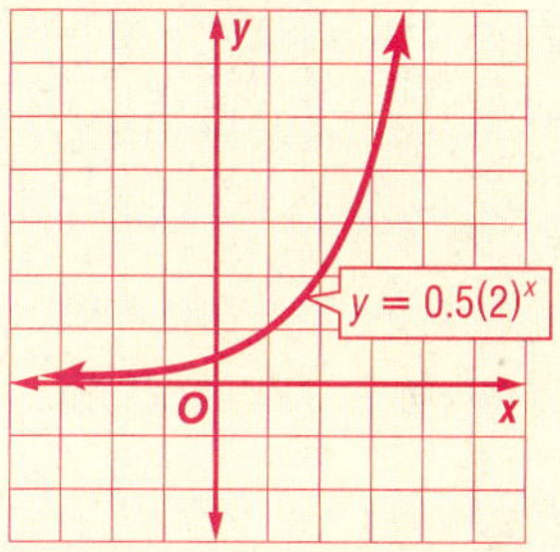

3. **D = {x | x is any real number.}; R = {y | y > 0}**

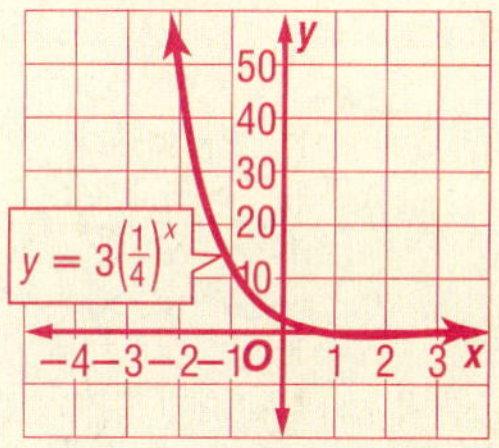

4. **D = {x | x is any real number.}; R = {y | y > 0}**

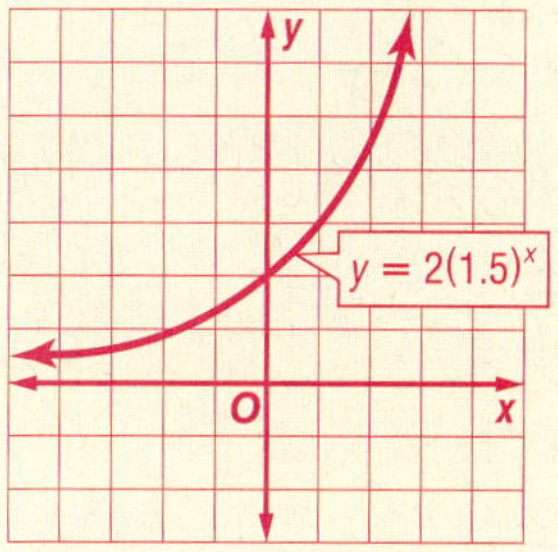

Extra Practice

Lesson 10-4

19. $\frac{\log 21}{\log 3} \approx 2.7712$

20. $\frac{\log 62}{\log 4} \approx 2.9771$

21. $\frac{\log 28}{\log 5} \approx 2.0704$

22. $\frac{\log 25}{\log 2} \approx 4.6439$

23. $\frac{\log 30}{\log 12} \approx 1.3687$

24. $\frac{\log 63}{\log 4} \approx 2.9886$

25. $\frac{\log 35}{\log 7} \approx 1.8271$

26. $\frac{\log 100}{\log 6} \approx 2.5702$

Extra Practice

Extra Practice

Lesson 10-3

(pages 541–546)

Use $\log_3 5 \approx 1.4651$ and $\log_3 7 \approx 1.7712$ to approximate the value of each expression.

1. $\log_3 \frac{7}{5}$ **0.3061**
2. $\log_3 245$ **5.0075**
3. $\log_3 35$ **3.2363**

Solve each equation. Check your solutions.

4. $\log_2 x + \log_2 (x - 2) = \log_2 3$ **3**
5. $\log_3 x = 2 \log_3 3 + \log_3 5$ **45**
6. $\log_5 (x^2 + 7) = \frac{2}{3} \log_5 64$ **±3**
7. $\log_7 (3x + 5) - \log_7 (x - 5) = \log_7 8$ **9**
8. $\log_2 (x^2 - 9) = 4$ **±5**
9. $\log_3 (x + 2) + \log_3 6 = 3$ **2.5**
10. $\log_6 x + \log_6 (x - 5) = 2$ **9**
11. $\log_5 (x + 3) = \log_5 8 - \log_5 2$ **1**
12. $2 \log_3 x - \log_3 (x - 2) = 2$ **3, 6**
13. $\log_6 x = \frac{3}{2} \log_6 9 + \log_6 2$ **54**
14. $\log_8 (x + 6) + \log_8 (x - 6) = 2$ **10**
15. $\frac{1}{2}\log_4 (x + 2) + \frac{1}{2}\log_4 (x + 2) = \frac{2}{3}\log_4 27$ **7**
16. $\log_3 14 + \log_3 x = \log_3 42$ **3**
17. $\log_{10} x = \frac{1}{2}\log_{10} 81$ **9**

Lesson 10-4

(pages 547–551)

Use a calculator to evaluate each expression to four decimal places.

1. $\log 55$ **1.7404**
2. $\log 6.7$ **0.8261**
3. $\log 3.3$ **0.5185**
4. $\log 0.08$ **−1.0969**
5. $\log 9.9$ **0.9956**
6. $\log 0.6$ **−0.2218**

Solve each equation or inequality. Round to four decimal places.

7. $2^x = 15$ **3.9069**
8. $4^{2a} > 45$ **$a > 1.3730$**
9. $7^{2x} = 35$ **0.9135**
10. $11^{x+4} > 57$ **$x > -2.3139$**
11. $1.5^{a-7} = 9.6$ **12.5782**
12. $3^{b^2} = 64$ **±1.9457**
13. $7^{3c} < 35^{2c-1}$ **$c < 2.7930$**
14. $5^{m^2+1} = 30$ **±1.0551**
15. $7^{3y-1} < 2^{2y+4}$ **$y < 1.0600$**
16. $9^{n-3} = 2^{n+3}$ **5.7651**
17. $11^{t+1} \leq 22^{t+3}$ **$t \leq -9.9189$**
18. $2^{3a-1} = 3^{a+2}$ **2.9469**

Express each logarithm in terms of common logarithms. Then approximate its value to four decimal places. **19–26. See margin.**

19. $\log_3 21$
20. $\log_4 62$
21. $\log_5 28$
22. $\log_2 25$
23. $\log_{12} 30$
24. $\log_4 63$
25. $\log_7 35$
26. $\log_6 100$

Lesson 10-5

(pages 554–559)

Use a calculator to evaluate each expression to four decimal places.

1. e^3 **20.0855**
2. $e^{0.75}$ **2.1170**
3. e^{-4} **0.0183**
4. $e^{-2.5}$ **0.0821**
5. $\ln 5$ **1.6094**
6. $\ln 8$ **2.0794**
7. $\ln 8.4$ **2.1282**
8. $\ln 0.6$ **−0.5108**

Write an equivalent exponential or logarithmic equation.

9. $e^x = 10$ **$\ln 10 = x$**
10. $\ln x \approx 2.3026$ **$x \approx e^{2.3026}$**
11. $e^3 = 9x$ **$\ln 9x = 3$**
12. $\ln 0.2 = x$ **$e^x = 0.2$**

Evaluate each expression.

13. $e^{\ln 5}$ **5**
14. $\ln e^{x-2}$ **$x - 2$**
15. $\ln e^{-3n}$ **$-3n$**
16. $e^{\ln 2.5}$ **2.5**

Solve each equation or inequality.

17. $25e^x = 1000$ **3.6889**
18. $e^{0.075x} > 25$ **$x > 42.9183$**
19. $e^x < 3.8$ **$x < 1.3350$**
20. $-2e^x + 5 = 1$ **0.6931**
21. $5 + 4e^{2x} = 17$ **0.5493**
22. $e^{-3x} \leq 15$ **$x \geq -0.9027$**
23. $\ln 7x = 10$ **3146.6380**
24. $\ln 4x = 8$ **745.2395**
25. $3 \ln 2x \geq 9$ **$x \geq 10.0428$**
26. $\ln (x + 2) = 4$ **52.5982**
27. $\ln (2x + 3) > 0$ **$x > -1$**
28. $\ln (3x - 1) = 5$ **49.8044**

Lesson 10-6
(pages 560–565)

1. Mr. Rogers purchased a combine for \$175,000 for his farming operation. It is expected to depreciate at a rate of 18% per year. What will be the value of the combine in 3 years? **\$96,489.40**
2. The Jacksons bought a house for \$65,000 in 1992. Houses in the neighborhood have appreciated at the rate of 4.5% a year. How much is the house worth in 2003? **\$105,485.45**
3. In 1950, the population of a city was 50,000. Since then, the population has increased by 2.25% per year. If it continues to grow at this rate, what will the population be in 2005? **about 170,000**
4. In a particular state, the population of black bears has been decreasing at the rate of 0.75% per year. In 1990, it was estimated that there were 400 black bears in the state. If the population continues to decline at the same rate, what will the population be in 2010? **344 bears**
5. Scott invests \$8500 in a certificate of deposit (CD). The interest is calculated on the money once per year. If the interest rate for the CD is 6.875% and he invests in a 4-year CD, how much money will he have at the end of the investment period? **\$11,089.79**
6. Mrs. Nguyen takes a job as a teacher with a starting salary of \$20,050. Over the past few years in this district, teachers have received a 4% increase in pay each year. If the pay continues to increase at the same rate, what will be her salary in 5 years? **\$24,393.89**

Lesson 11-1
(pages 578–582)

Find the next four terms of each arithmetic sequence.

1. 9, 7, 5, … **3, 1, −1, −3**
2. 3, 4.5, 6, … **7.5, 9, 10.5, 12**
3. 40, 35, 30, … **25, 20, 15, 10**
4. 2, 5, 8, … **11, 14, 17, 20**

Find the first five terms of each arithmetic sequence described.

5. $a_1 = 1, d = 7$ **1, 8, 15, 22, 29**
6. $a_1 = -5, d = 2$ **−5, −3, −1, 1, 3**
7. $a_1 = 1.2, d = 3.7$ **1.2, 4.9, 8.6, 12.3, 16**
8. $a_1 = -\frac{5}{4}, d = -\frac{1}{2}$ $-\frac{5}{4}, -\frac{7}{4}, -\frac{9}{4}, -\frac{11}{4}, -\frac{13}{4}$

Find the indicated term of each arithmetic sequence.

9. $a_1 = 4, d = 5, n = 10$ **49**
10. $a_1 = -30, d = -6, n = 5$ **−54**
11. $a_1 = -3, d = 32, n = 8$ **221**
12. $a_1 = \frac{3}{4}, d = -\frac{1}{4}, n = 72$ **−17**
13. $a_1 = -\frac{1}{5}, d = \frac{3}{5}, n = 17$ $\frac{47}{5}$
14. $a_1 = 20, d = -3, n = 16$ **−25**

Write an equation for the nth term of each arithmetic sequence.

15. 3, 5, 7, 9, … $a_n = 2n + 1$
16. 2, −1, −4, −7, … $a_n = -3n + 5$
17. 20, 28, 36, 44, … $a_n = 8n + 12$

Find the arithmetic means in each sequence.

18. 2, _?_, _?_, _?_, 34 **10, 18, 26**
19. 0, _?_, _?_, _?_, −28 **−7, −14, −21**
20. −10, _?_, _?_, _?_, 14 **−4, 2, 8**

Lesson 11-2
(pages 583–587)

Find S_n for each arithmetic series described.

1. $a_1 = 3, a_n = 20, n = 6$ **69**
2. $a_1 = 15, a_n = -12, n = 30$ **45**
3. $a_1 = 90, a_n = -4, n = 10$ **430**
4. $a_1 = 16, a_n = 14, n = 12$ **180**
5. $a_1 = -80, a_n = 120, n = 18$ **360**
6. $a_1 = -3, a_n = -72, n = 14$ **−525**
7. $a_1 = -1, d = 10, n = 30$ **4320**
8. $a_1 = 4, d = -5, n = 11$ **−231**
9. $a_1 = 5, d = -\frac{1}{2}, n = 17$ **17**

Find the sum of each arithmetic series.

10. 3 + 12 + 21 + 30 + … + 57 **210**
11. 1 + 4 + 7 + 10 + … + 31 **176**
12. 8 + 16 + 24 + … + 80 **440**
13. $\sum_{n=1}^{6} (n + 2)$ **33**
14. $\sum_{n=5}^{10} (2n - 5)$ **60**
15. $\sum_{k=1}^{5} (40 - 2k)$ **170**
16. $\sum_{k=8}^{12} (6 - 3k)$ **−120**
17. $\sum_{n=1}^{4} (10n + 2)$ **108**
18. $\sum_{n=6}^{10} (2 + 3n)$ **130**

Find the first three terms of each arithmetic series described.

19. $a_1 = 11, a_n = 38, S_n = 245$ **11, 14, 17**
20. $a_1 = -6, a_n = -34, S_n = -300$ **−6, −8, −10**
21. $a_1 = 0, a_n = 165, S_n = 990$ **0, 15, 30**
22. $n = 12, a_n = 13, S_n = -42$ **−20, −17, −14**
23. $n = 11, a_n = 5, S_n = 0$ **−5, −4, −3**
24. $a_1 = -25, a_n = 15, S_n = -45$ **−25, −20, −15**

Extra Practice 851

Lesson 11-3

7. −2, −12, −72, −432, −2592
8. 4, −20, 100, −500, 2500
9. 1, −3, 9, −27, 81
10. 1, 0.8, 0.64, 0.512, 0.4096
11. 0.8, 2, 5, 12.5, 31.25
12. 1.5, −4.5, 13.5, −40.5, 121.5
13. $\frac{2}{5}, -\frac{6}{5}, \frac{18}{5}, -\frac{54}{5}, \frac{162}{5}$
14. $-\frac{1}{3}, \frac{1}{5}, -\frac{3}{25}, \frac{9}{125}, -\frac{27}{625}$

Lesson 11-3 (pages 588–592)

Find the next two terms of each geometric sequence.

1. 5, 15, 45, … **135, 405**
2. 2, 10, 50, … **250, 1250**
3. 64, 16, 4, … **1, 0.25**
4. −9, 27, −81, … **243, −729**
5. 0.5, 0.75, 1.125, … **1.6875, 2.53125**
6. $\frac{1}{2}, -\frac{3}{8}, \frac{9}{32}, \ldots$ **$-\frac{27}{128}, \frac{81}{512}$**

Find the first five terms of each geometric sequence described. **7–14. See margin.**

7. $a_1 = -2, r = 6$
8. $a_1 = 4, r = -5$
9. $a_1 = 1, r = -3$
10. $a_1 = 1, r = 0.8$
11. $a_1 = 0.8, r = 2.5$
12. $a_1 = 1.5, r = -3$
13. $a_1 = \frac{2}{5}, r = -3$
14. $a_1 = -\frac{1}{3}, r = -\frac{3}{5}$

Find the indicated term of each geometric sequence.

15. $a_1 = 5, r = 7, n = 6$ **84,035**
16. $a_1 = 200, r = -\frac{1}{2}, n = 10$ **$-\frac{25}{64}$**
17. $a_1 = 60, r = -2, n = 4$ **−480**
18. $a_1 = 300, r = \frac{1}{4}, n = 6$ **$\frac{75}{256}$**
19. $a_1 = 8, r = -2, n = 8$ **−1024**
20. $a_1 = 1, r = -1, n = 30$ **−1**

Write an equation for the *n*th term of each geometric sequence.

21. 20, 40, 80, … **$a_n = 20(2)^{n-1}$**
22. $-\frac{1}{2}, -\frac{1}{8}, -\frac{1}{32}, \ldots$ **$a_n = -\frac{1}{2}\left(\frac{1}{4}\right)^{n-1}$**

Find the geometric means in each sequence.

23. 1, __?__, __?__, __?__, 81 **3, 9, 27 or −3, 9, −27**
24. 5, __?__, __?__, __?__, 6480 **30, 180, 1080 or −30, 180 −1080**

Lesson 11-4 (pages 594–598)

Find S_n for each geometric series described.

1. $a_1 = \frac{1}{81}, r = 3, n = 6$ **$\frac{364}{81}$**
2. $a_1 = 1, r = -2, n = 7$ **43**
3. $a_1 = 5, r = 4, n = 5$ **1705**
4. $a_1 = -27, r = -\frac{1}{3}, n = 6$ **$-\frac{182}{9}$**
5. $a_1 = 1000, r = \frac{1}{2}, n = 7$ **1984.375**
6. $a_1 = 125, r = -\frac{2}{5}, n = 5$ **$\frac{451}{5}$**
7. $a_1 = 10, r = 3, n = 6$ **3640**
8. $a_1 = 1250, r = -\frac{1}{5}, n = 5$ **1042**
9. $a_1 = 1215, r = \frac{1}{3}, n = 5$ **1815**
10. $a_1 = 16, r = \frac{3}{2}, n = 5$ **211**
11. $a_1 = 7, r = 2, n = 7$ **889**
12. $a_1 = -\frac{3}{2}, r = -\frac{1}{2}, n = 6$ **$-\frac{63}{64}$**

Find the sum of each geometric series.

13. $\sum_{k=1}^{5} 2^k$ **62**
14. $\sum_{n=0}^{3} 3^{-n}$ **$\frac{40}{27}$**
15. $\sum_{n=0}^{3} 2(5^n)$ **312**
16. $\sum_{k=2}^{5} -(-3)^{k-1}$ **−60**

Find the indicated term for each geometric series described.

17. $S_n = 300, a_n = 160, r = 2; a_1$ **20**
18. $S_n = -171, n = 9, r = -2; a_5$ **−16**
19. $S_n = -4372, a_n = -2916, r = 3; a_4$ **−108**

Lesson 11-5 (pages 599–604)

Find the sum of each infinite geometric series, if it exists. **2. does not exist** **3. $833\frac{1}{3}$**

1. $a_1 = 54, r = \frac{1}{3}$ **81**
2. $a_1 = 2, r = -1$
3. $a_1 = 1000, r = -0.2$
4. $a_1 = 7, r = \frac{3}{7}$ **$\frac{49}{4}$**
5. 49 + 14 + 4 + … **$\frac{343}{5}$**
6. $\frac{3}{4} + \frac{1}{2} + \frac{1}{3} + \ldots$ **$\frac{9}{4}$**
7. $12 - 4 + \frac{4}{3} - \ldots$ **9**
8. 3 − 9 + 27 − … **does not exist**
9. $3 - 2 + \frac{4}{3} - \ldots$ **$\frac{9}{5}$**
10. $\sum_{n=1}^{\infty} 3\left(\frac{1}{4}\right)^{n-1}$ **4**
11. $\sum_{n=1}^{\infty} 5\left(-\frac{1}{10}\right)^{n-1}$ **$\frac{50}{11}$**
12. $\sum_{n=1}^{\infty} -\frac{2}{3}\left(-\frac{3}{4}\right)^{n-1}$ **$-\frac{8}{21}$**

Write each repeating decimal as a fraction.

13. $0.\overline{4}$ **$\frac{4}{9}$**
14. $0.\overline{27}$ **$\frac{3}{11}$**
15. $0.\overline{123}$ **$\frac{41}{333}$**
16. $0.\overline{645}$ **$\frac{215}{333}$**
17. $0.6\overline{7}$ **$\frac{61}{90}$**
18. $0.8\overline{53}$ **$\frac{169}{198}$**

852 Extra Practice

Lesson 11-6

(pages 606–610)

Find the first five terms of each sequence. **6.** $\frac{1}{3}, \frac{1}{4}, \frac{7}{12}, \frac{5}{6}, \frac{17}{12}$

1. $a_1 = 4, a_{n+1} = 2a_n + 1$ **4, 9, 19, 39, 79**
2. $a_1 = 6, a_{n+1} = a_n + 7$ **6, 13, 20, 27, 34**
3. $a_1 = 16, a_{n+1} = a_n + (n + 4)$ **16, 21, 27, 34, 42**
4. $a_1 = 1, a_{n+1} = \frac{n}{n+2} \cdot a_n$ $1, \frac{1}{3}, \frac{1}{6}, \frac{1}{10}, \frac{1}{15}$
5. $a_1 = -\frac{1}{2}, a_{n+1} = 2a_n + \frac{1}{4}$ $-\frac{1}{2}, -\frac{3}{4}, -\frac{5}{4}, -\frac{9}{4}, -\frac{17}{4}$
6. $a_1 = \frac{1}{3}, a_2 = \frac{1}{4}, a_{n+1} = a_n + a_{n-1}$

Find the first three iterates of each function for the given initial value.

7. $f(x) = 3x - 1, x_0 = 3$ **8, 23, 68**
8. $f(x) = 2x^2 - 8, x_0 = -1$ **−6, 64, 8184**
9. $f(x) = 4x + 5, x_0 = 0$ **5, 25, 105**
10. $f(x) = 3x^2 + 1, x_0 = 1$ **4, 49, 7204**
11. $f(x) = x^2 + 4x + 4, x_0 = 1$ **9; 121; 15,129**
12. $f(x) = x^2 + 9, x_0 = 2$ **13; 178; 31,693**
13. $f(x) = 2x^2 + x + 1, x_0 = -\frac{1}{2}$ **1, 4, 37**
14. $f(x) = 3x^2 + 2x - 1, x_0 = \frac{2}{3}$ $\frac{5}{3}, \frac{32}{3}, \frac{1085}{3}$

Lesson 11-7

(pages 612–617)

Evaluate each expression.

1. $6!$ **720**
2. $4!$ **24**
3. $\frac{13!}{6!}$ **8,648,640**
4. $\frac{10!}{3!7!}$ **120**
5. $\frac{14!}{4!10!}$ **1001**
6. $\frac{7!}{2!5!}$ **21**
7. $\frac{12!}{5!}$ **3,991,680**
8. $\frac{9!}{8!}$ **9**
9. $\frac{10!}{10!0!}$ **1**

Expand each power. **10–21. See margin.**

10. $(z - 3)^5$
11. $(m + 1)^4$
12. $(x + 6)^4$
13. $(z - y)^2$
14. $(m + n)^5$
15. $(a - b)^4$
16. $(2n + 1)^4$
17. $(3n - 4)^3$
18. $(2n - m)^0$
19. $(4x - a)^4$
20. $(3r - 4s)^5$
21. $\left(\frac{b}{2} - 1\right)^4$

Find the indicated term of each expansion.

22. sixth term of $(x + 3)^8$ $13{,}608x^3$
23. fourth term of $(x - 2)^7$ $-280x^4$
24. fifth term of $(a + b)^6$ $15a^2b^4$
25. fourth term of $(x - y)^9$ $-84x^6y^3$
26. sixth term of $(x + 4y)^7$ $21{,}504x^2y^5$
27. fifth term of $(3x + 5y)^{10}$ $95{,}681{,}250x^6y^4$

Lesson 11-8

(pages 618–621)

Prove that each statement is true for all positive integers. **1–6. See pp. 861A–861T.**

1. $2 + 4 + 6 + \ldots + 2n = n^2 + n$
2. $1^3 + 3^3 + 5^3 + \ldots + (2n - 1)^3 = n^2(2n^2 - 1)$
3. $1 \cdot 3 + 3 \cdot 5 + 5 \cdot 7 + \ldots + (2n - 1)(2n + 1) = \frac{n(4n^2 + 6n - 1)}{3}$
4. $\frac{1}{1 \cdot 3} + \frac{1}{2 \cdot 4} + \frac{1}{3 \cdot 5} + \ldots + \frac{1}{n(n+2)} = \frac{n(3n+5)}{4(n+1)(n+2)}$
5. $1 \cdot 3 + 2 \cdot 4 + 3 \cdot 5 + \ldots + n(n + 2) = \frac{n(n+1)(2n+7)}{6}$
6. $\frac{5}{1 \cdot 2} \cdot \frac{1}{3} + \frac{7}{2 \cdot 3} \cdot \frac{1}{3^2} + \frac{9}{3 \cdot 4} \cdot \frac{1}{3^3} + \ldots + \frac{2n+3}{n(n+1)} \cdot \frac{1}{3^n} = 1 - \frac{1}{3^n(n+1)}$

Find a counterexample for each statement.

7. $n^2 + 2n - 1$ is divisible by 2. $n = 2$
8. $2^n + 3^n$ is prime. $n = 3$
9. $2^{n-1} + n = 2^n + 2 - n$ for all integers $n \geq 2$. $n = 4$
10. $3^n - 2n = 3^n - 2^n$ for all integers $n \geq 1$ $n = 3$

Lesson 11-7

10. $z^5 - 15z^4 + 90z^3 - 270z^2 + 405z - 243$
11. $m^4 + 4m^3 + 6m^2 + 4m + 1$
12. $x^4 + 24x^3 + 216x^2 + 864x + 1296$
13. $z^2 - 2zy + y^2$
14. $m^5 + 5m^4n + 10m^3n^2 + 10m^2n^3 + 5mn^4 + n^5$
15. $a^4 - 4a^3b + 6a^2b^2 - 4ab^3 + b^4$
16. $16n^4 + 32n^3 + 24n^2 + 8n + 1$
17. $27n^3 - 108n^2 + 144n - 64$
18. 1
19. $256x^4 - 256x^3a + 96x^2a^2 - 16xa^3 + a^4$
20. $243r^5 - 1620r^4s + 4320r^3s^2 - 5760r^2s^3 + 3840rs^4 - 1024s^5$
21. $\frac{b^4}{16} - \frac{b^3}{2} + \frac{3b^2}{2} - 2b + 1$

Lesson 12-1

1. H1, H2, H3, H4, H5, H6, T1, T2, T3, T4, T5, T6
2. dark blue with buttons, dark blue with snaps, stone washed with buttons, stone washed with snaps, black with buttons, black with snaps
3. thin pepperoni jack, thin pepperoni mozzarella, thick pepperoni jack, thick pepperoni mozzarella, thin sausage jack, thin sausage mozzarella, thick sausage jack, thick sausage mozzarella, thin vegetable jack, thin vegetable mozzarella, thick vegetable jack, thick vegetable mozzarella

Lesson 12-1 *(pages 632–637)*

List the possible outcomes for each situation. **1–3. See margin.**

1. tossing a penny and rolling a number cube
2. choosing a denim jacket that comes in dark blue, stone washed, or black that has buttons or snaps
3. ordering a large pizza with thin or thick crust, and one topping of either pepperoni, sausage, or vegetables, and either jack or mozzarella cheese

State whether the events are *independent* or *dependent*.

4. A comedy video and an action video are selected from the video store. **independent**
5. The numbers 1–10 are written on pieces of paper and are placed in a hat. Three of them are selected one after the other without replacing any of the pieces of paper. **dependent**

Solve each problem.

6. On a bookshelf there are 10 different algebra books, 6 different geometry books, and 4 different calculus books. In how many ways can you choose 3 books, one of each kind? **240**
7. In how many different ways can a 10-question true-false test be answered? **1024**
8. A student council has 6 seniors, 5 juniors, and 1 sophomore as members. In how many ways can a 3-member council committee be formed that includes one member of each class? **30**
9. How many license plates of 5 symbols can be made using a letter for the first symbol and digits for the remaining 4 symbols? **260,000**

Lesson 12-2 *(pages 638–643)*

Evaluate each expression.

1. $P(3, 2)$ **6**
2. $P(5, 2)$ **20**
3. $P(10, 6)$ **151,200**
4. $P(4, 3)$ **24**
5. $P(12, 2)$ **132**
6. $P(7, 2)$ **42**
7. $C(8, 6)$ **28**
8. $C(20, 17)$ **1140**
9. $C(9, 4) \cdot C(5, 3)$ **1260**
10. $C(6, 1) \cdot C(4, 1)$ **24**
11. $C(10, 5) \cdot C(8, 4)$ **17,640**
12. $C(7, 6) \cdot C(3, 1)$ **21**

Determine whether each situation involves a *permutation* or a *combination*. Then find the number of possibilities.

13. choosing a team of 9 players from a group of 20 **combination; 167,960**
14. selecting the batting order of 9 players in a baseball game **permutation; 362,880**
15. arranging the order of 8 songs on a CD **permutation; 40,320**
16. finding the number of 5-card hands that include 4 diamonds and 1 club **combination; 9295**

Lesson 12-3 *(pages 644–650)*

A jar contains 3 red, 4 green, and 5 orange marbles. If three marbles are drawn at random and not replaced, find each probability.

1. $P(\text{all green})$ $\frac{1}{55}$
2. $P(\text{1 red, then 2 not red})$ $\frac{9}{55}$
3. $P(\text{2 orange, then 1 not orange})$ $\frac{7}{66}$

Find the odds of an event occurring, given the probability of the event.

4. $\frac{5}{9}$ **5:4**
5. $\frac{4}{8}$ **1:1**
6. $\frac{3}{10}$ **3:7**

Find the probability of an event occurring, given the odds of the event.

7. $\frac{2}{7}$ $\frac{2}{9}$
8. $\frac{6}{13}$ $\frac{6}{19}$
9. $\frac{1}{19}$ $\frac{1}{20}$

The table shows the number of ways to achieve each product when two dice are tossed. Find each probability.

Product	1	2	3	4	5	6	8	9	10	12	15	16	18	20	24	25	30	36
Ways	1	2	2	3	2	4	2	1	2	4	2	1	2	2	2	1	2	1

10. $P(6)$ $\frac{1}{9}$
11. $P(12)$ $\frac{1}{9}$
12. $P(\text{not } 36)$ $\frac{35}{36}$
13. $P(\text{not } 12)$ $\frac{8}{9}$

854 Extra Practice

Lesson 12-4

(pages 651–657)

An octahedral die is rolled twice. The sides are numbered 1–8. Find each probability.

1. $P(1, \text{then } 8)$ $\frac{1}{64}$
2. $P(\text{two 7s})$ $\frac{1}{64}$
3. $P(8, \text{then any number})$ $\frac{1}{8}$
4. $P(\text{two of the same number})$ $\frac{1}{8}$
5. $P(\text{two different numbers})$ $\frac{7}{8}$
6. $P(\text{no 8s})$ $\frac{49}{64}$

Two cards are drawn from a standard deck of cards. Find each probability if no replacement occurs.

7. $P(\text{jack, jack})$ $\frac{1}{221}$
8. $P(\text{heart, club})$ $\frac{13}{204}$
9. $P(\text{two diamonds})$ $\frac{1}{17}$
10. $P(\text{2 of hearts, diamond})$ $\frac{1}{204}$
11. $P(\text{2 red cards})$ $\frac{25}{102}$
12. $P(\text{2 black aces})$ $\frac{1}{1326}$

Determine whether the events are *independent* or *dependent*. Then find the probability.

13. According to the weather reports, the probability of rain on a certain day is 70% in Yellow Falls and 50% in Copper Creek. What is the probability that it will rain in both cities? **independent; 0.35**
14. A contestant on a game show reaches into a container without looking and picks two paper bills. There are 2 \$100 bills, 4 \$50 bills, 10 \$20 bills, and 20 \$10 bills. What is the probability that the contestant draws 2 \$100 bills one after the other without replacement? **dependent;** $\frac{1}{630}$
15. The odds of winning a carnival game are 1 to 5. What is the probability that a player will win the game three consecutive times? **independent;** $\frac{1}{216}$

Lesson 12-5

(pages 658–663)

An octahedral die is rolled. The sides are numbered 1–8. Find each probability.

1. $P(7 \text{ or } 8)$ $\frac{1}{4}$
2. $P(\text{less than } 4)$ $\frac{3}{8}$
3. $P(\text{greater than } 6)$ $\frac{1}{4}$
4. $P(\text{not prime})$ $\frac{1}{2}$
5. $P(\text{odd or prime})$ $\frac{5}{8}$
6. $P(\text{multiple of 5 or odd})$ $\frac{1}{2}$

Ten slips of paper are placed in a container. Each is labeled with a number from 1 through 10. Determine whether the events are *mutually exclusive* or *inclusive*. Then find the probability.

7. $P(1 \text{ or } 10)$ **mutually exclusive;** $\frac{1}{5}$
8. $P(3 \text{ or odd})$ **inclusive;** $\frac{1}{2}$
9. $P(6 \text{ or less than } 7)$ **inclusive;** $\frac{3}{5}$

Find each probability. **10.** $\frac{3}{50}$

10. Two letters are chosen at random from the word GEESE and two are chosen at random from the word PLEASE. What is the probability that all four letters are Es or none of the letters is an E?
11. Three dice are thrown. What is the probability that all three dice show the same number? $\frac{1}{36}$
12. Two marbles are simultaneously drawn at random from a bag containing 3 red, 5 blue, and 6 green marbles.
    a. $P(\text{at least one red marble})$ $\frac{36}{91}$
    b. $P(\text{at least one green marble})$ $\frac{9}{13}$
    c. $P(\text{two marbles of the same color})$ $\frac{4}{13}$
    d. $P(\text{two marbles of different colors})$ $\frac{9}{13}$

Lesson 12-6

(pages 664–670)

Find the mean, median, mode, and standard deviation of each set of data. Round to the nearest hundredth, if necessary. **1–8. See margin.**

1. {4, 1, 2, 1, 1}
2. {86, 71, 74, 65, 45, 42, 76}
3. {16, 20, 15, 14, 24, 23, 25, 10, 19}
4. {18, 24, 16, 24, 22, 24, 22, 22, 24, 13, 17, 18, 16, 20, 16, 7, 22, 5, 4, 24}
5. {55, 50, 50, 55, 65, 50, 45, 35, 50, 40, 70, 40, 70, 50, 90, 30, 35, 55, 55, 40, 75, 35, 40, 45, 65, 50, 60}
6. {364, 305, 217, 331, 305, 311, 352, 319, 272, 238, 311, 226, 220, 226, 215, 160, 123, 4, 24, 238, 99}
7. {25.5, 26.7, 20.9, 23.4, 26.8, 24.0, 25.7}
8. The following high temperatures were recorded during a cold spell in Cleveland lasting thirty-eight days.
   29° 26° 17° 12° 5° 4° 25° 17° 23° 18° 13° 6° 25° 20° 27° 22° 26° 30° 31°
   2° 12° 27° 16° 27° 16° 30° 6° 16° 5° 0° 5° 29° 18° 16° 22° 29° 8° 23°

Extra Practice

Lesson 12-6

1. 1.8, 1, 1, 1.20
2. 65.57, 71, none, 15.15
3. 18.44, 19, none, 4.79
4. 17.9, 19, 24, 6.21
5. 51.85, 50, 50, 13.75
6. 231.43; 238; modes: 305, 311, 238, 226; 98.74
7. 24.71, 25.5, none, 1.95
8. 17.97°, 18°, 16°, 9.12°

Extra Practice

Lesson 12-7

(pages 671–675)

1. Determine whether the data in the table appear to be *positively skewed*, *negatively skewed*, or *normally distributed*. The average size of a farm in each U.S. state was determined. **positively skewed**

Acres	85–559	560–1034	1035–1509	1510–1984	1985–2459	2460–2934	2935–3409	3410–3884
States	37	4	3	1	2	1	0	2

Source: *The World Almanac*

2. The diameters of metal fittings made by a machine are normally distributed. The diameters have a mean of 7.5 centimeters and a standard deviation of 0.5 centimeters.
   a. What percent of the fittings have diameters between 7.0 and 8.0 centimeters? **68%**
   b. What percent of the fittings have diameters between 7.5 and 8.0 centimeters? **34%**
   c. What percent of the fittings have diameters greater than 6.5 centimeters? **97.5%**
   d. Of 100 fittings, how many will have a diameter between 6.0 and 8.5 centimeters? **97**
3. A college entrance exam was administered at a state university. The scores were normally distributed with a mean of 510, and a standard deviation of 80.
   a. What percent would you expect to score above 510? **50%**
   b. What percent would you expect to score between 430 and 590? **68%**
   c. What is the probability that a student chosen at random scored between 350 and 670? **95%**

Lesson 12-8

(pages 676–680)

Find each probability if a coin is tossed 5 times.

1. P(0 heads) $\frac{1}{32}$
2. P(exactly 4 heads) $\frac{5}{32}$
3. P(exactly 3 tails) $\frac{5}{16}$

Find each probability.

4. Ten percent of a batch of toothpaste is defective. Five tubes of toothpaste are selected at random from this batch.
   a. P(0 defective) **0.59045 or 59.045%**
   b. P(exactly one defective) **0.32805 or 32.805%**
   c. P(at least three defective) **0.00856 or 0.856%**
   d. P(less than three defective) **0.99144 or 99.144%**
5. On a 20-question true-false test, you guess at every question.
   a. P(all answers correct) $\frac{1}{1,048,576}$
   b. P(exactly 10 correct) $\frac{46,189}{262,144}$

Lesson 12-9

(pages 682–685)

Determine whether each situation would produce a random sample. Write *yes* or *no* and explain your answer. **1–3. See margin.**

1. finding the most often prescribed pain reliever by asking all of the doctors at a hospital
2. taking a poll of the most popular baby girl names this year by studying birth announcements in newspapers from different cities across the country
3. polling people who are leaving a pizza parlor about their favorite restaurant in the city

Find the margin of sampling error.

4. $p = 45\%$, $n = 125$ **about 8.90%**
5. $p = 62\%$, $n = 240$ **about 6.27%**
6. $p = 24\%$, $n = 600$ **about 3.49%**
7. $p = 67\%$, $n = 180$ **about 7.01%**
8. $p = 82\%$, $n = 1000$ **about 2.43%**
9. $p = 15\%$, $n = 2500$ **about 1.43%**
10. A poll conducted on the favorite breakfast choice of students in your school showed that 75% of the 2250 students asked indicated oatmeal as their favorite breakfast. **about 1.83%**
11. Of the 420 people polled at a supermarket, 56% felt that they were easily swayed by the sample items in the aisles and purchased those items, even though they were not intending to when they arrived at the store. **about 4.84%**
12. Of 3000 women between the ages of 25 and 35 polled, only 45% felt that they consume the recommended daily allowance of calcium by the National Institute of Health. **about 1.82%**

Extra Practice

Lesson 12-9

1. No; you would not be giving doctors who do not work at the hospital a chance to respond.
2. No; you would not see the names of babies whose parents do not place birth announcements in the paper.
3. No; people leaving a pizza parlor would probably favor pizza.

Lesson 13-1

1. $\sin\theta = \frac{19\sqrt{26}}{130}$; $\cos\theta = \frac{17\sqrt{26}}{130}$, $\tan\theta = \frac{19}{17}$; $\csc\theta = \frac{5\sqrt{26}}{19}$; $\sec\theta = \frac{5\sqrt{26}}{17}$; $\cot\theta = \frac{17}{19}$
2. $\sin\theta = \frac{1}{3}$; $\cos\theta = \frac{2\sqrt{2}}{3}$; $\tan\theta = \frac{\sqrt{2}}{4}$; $\csc\theta = 3$; $\sec\theta = \frac{3\sqrt{2}}{4}$; $\cot\theta = 2\sqrt{2}$
3. $\sin\theta = \frac{\sqrt{21}}{14}$; $\cos\theta = \frac{5\sqrt{7}}{14}$; $\tan\theta = \frac{\sqrt{3}}{5}$; $\csc\theta = \frac{2\sqrt{21}}{3}$; $\sec\theta = \frac{2\sqrt{7}}{5}$; $\cot\theta = \frac{5\sqrt{3}}{3}$
4. $A = 48°$, $a = 22.3$, $b = 20.1$
5. $B = 6°$, $b = 0.4$, $c = 4.0$
6. $A = 71°$, $a = 98.7$, $c = 104.4$
7. $B = 15°$, $a = 53.1$, $b = 14.2$
8. $A = 48°$, $B = 42°$, $a = 26.8$
9. $A = 26°$, $B = 64°$, $b = 103.1$
10. $A = 24°$, $B = 66°$, $b = 27.5$, $c = 30$
11. $A = 56°$, $B = 34°$, $a = 33$, $c = 39.7$

Lesson 13-2

1.

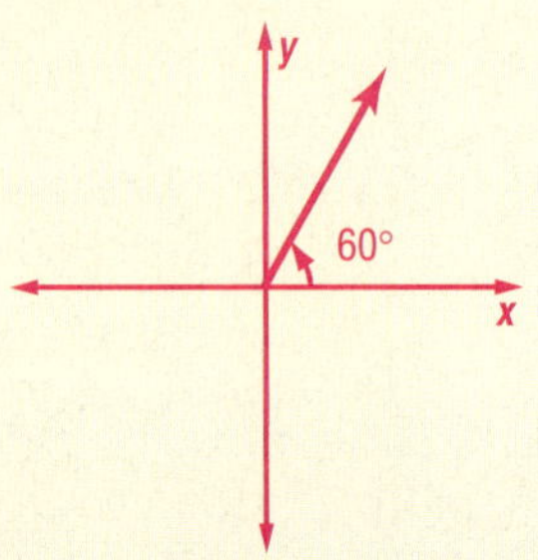

2.

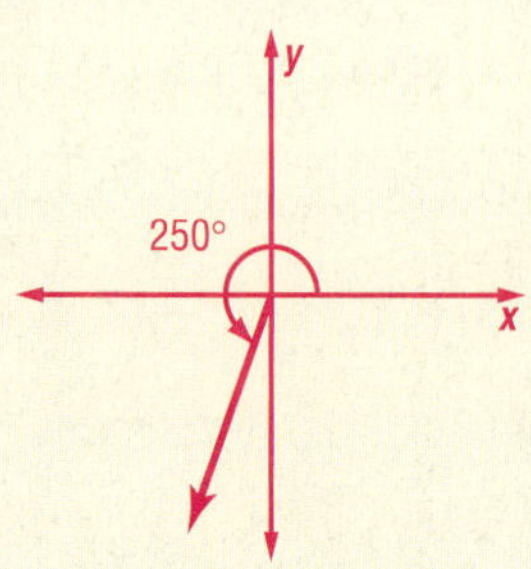

3.

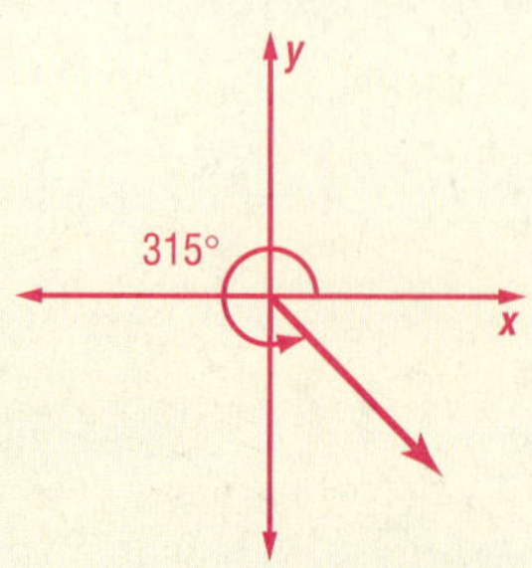

4.

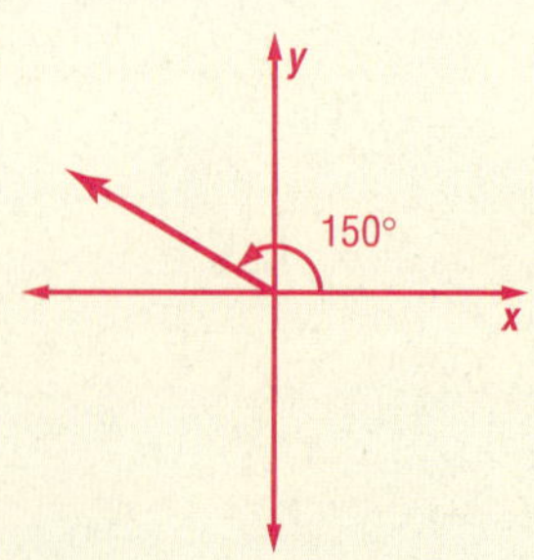

856 Extra Practice

Lesson 13-1

(pages 701–708)

Find the values of the six trigonometric functions for angle θ. **1–3. See margin.**

1. 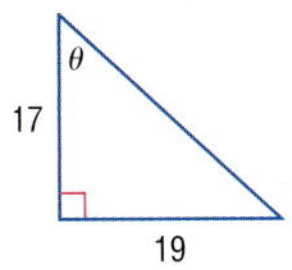

2. 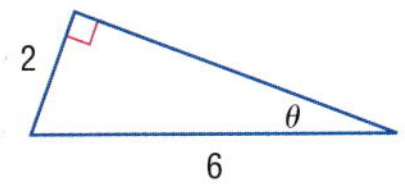

3. 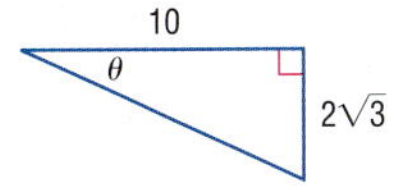

Solve $\triangle ABC$ using the diagram at the right and the given measurements. Round measures of sides to the nearest tenth and measures of angles to the nearest degree. **4–11. See margin.**

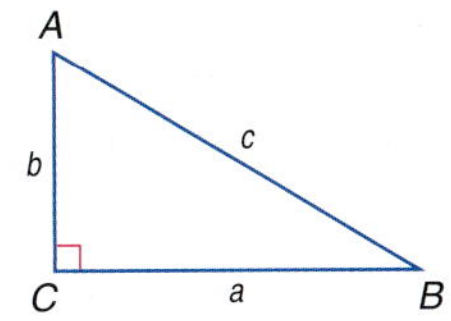

4. $B = 42°, c = 30$
5. $A = 84°, a = 4$
6. $B = 19°, b = 34$
7. $A = 75°, c = 55$
8. $b = 24, c = 36$
9. $a = 51, c = 115$
10. $\cos B = \frac{2}{5}, a = 12$
11. $\tan A = \frac{3}{2}, b = 22$

Lesson 13-2

(pages 709–715)

Draw an angle with the given measure in standard position. **1–4. See margin.**

1. 60°
2. 250°
3. 315°
4. 150°

Rewrite each degree measure in radians and each radian measure in degrees.

5. $-135°$ **$-\frac{3\pi}{4}$**
6. $-315°$ **$-\frac{7\pi}{4}$**
7. $45°$ **$\frac{\pi}{4}$**
8. $80°$ **$\frac{4\pi}{9}$**
9. $24°$ **$\frac{2\pi}{15}$**
10. $-54°$ **$-\frac{3\pi}{10}$**
11. $-\pi$ **$-180°$**
12. $\frac{9\pi}{4}$ **$405°$**
13. $\frac{3\pi}{2}$ **$270°$**
14. $-\frac{7\pi}{2}$ **$-630°$**
15. $\frac{9\pi}{10}$ **$162°$**
16. $\frac{17\pi}{30}$ **$102°$**
17. $\frac{7\pi}{12}$ **$105°$**
18. 1 **about 57.3°**
19. $-2\frac{1}{3}$ **about −133.7°**

Find one angle with positive measure and one angle with negative measure coterminal with each angle. **21. 285°, −435° 23. 320°, −40°**

20. 50° **410°, −310°**
21. −75°
22. 125° **485°, −235°**
23. −400°
24. 550° **190°, −170°**
25. 3π **$\pi, -\pi$**
26. -2π **$2\pi, -4\pi$**
27. $\frac{2\pi}{3}$ **$\frac{8\pi}{3}, -\frac{4\pi}{3}$**
28. $\frac{12\pi}{5}$ **$\frac{22\pi}{5}, -\frac{8\pi}{5}$**
29. 0 **$2\pi, -2\pi$**

Lesson 13-3

(pages 717–724)

Find the exact values of the six trigonometric functions of θ if the terminal side of θ in standard position contains the given point. **1–5. See margin.**

1. $P(3, -4)$
2. $P(1, \sqrt{3})$
3. $P(0, -4)$
4. $P(-5, -5)$
5. $P(-\sqrt{2}, -\sqrt{2})$

Find the exact value of each trigonometric function.

6. $\cos 225°$ **$-\frac{\sqrt{2}}{2}$**
7. $\sin\left(-\frac{5\pi}{3}\right)$ **$\frac{\sqrt{3}}{2}$**
8. $\tan \frac{7\pi}{6}$ **$\frac{\sqrt{3}}{3}$**
9. $\tan(-300°)$ **$\sqrt{3}$**
10. $\cos \frac{7\pi}{4}$ **$\frac{\sqrt{2}}{2}$**

Suppose θ is an angle in standard position whose terminal side is in the given quadrant. For each function, find the exact values of the remaining five trigonometric functions of θ. **11–19. See margin.**

11. $\cos\theta = -\frac{1}{3}$; Quadrant III
12. $\sec\theta = 2$; Quadrant IV
13. $\sin\theta = \frac{2}{3}$; Quadrant II
14. $\tan\theta = -4$; Quadrant IV
15. $\csc\theta = -5$; Quadrant III
16. $\cot\theta = -2$; Quadrant II
17. $\tan\theta = \frac{1}{3}$; Quadrant III
18. $\cos\theta = \frac{1}{4}$; Quadrant I
19. $\csc\theta = -\frac{5}{2}$; Quadrant IV

Extra Practice

Lesson 13-3

1. $\sin\theta = -\frac{4}{5}$, $\cos\theta = \frac{3}{5}$, $\tan\theta = -\frac{4}{3}$, $\csc\theta = -\frac{5}{4}$, $\sec\theta = \frac{5}{3}$, $\cot\theta = -\frac{3}{4}$

2. $\sin\theta = \frac{\sqrt{3}}{2}$, $\cos\theta = \frac{1}{2}$, $\tan\theta = \sqrt{3}$, $\csc\theta = \frac{2\sqrt{3}}{3}$, $\sec\theta = 2$, $\cot\theta = \frac{\sqrt{3}}{3}$

3. $\sin\theta = -1$, $\cos\theta = 0$, $\tan\theta$ undefined, $\csc\theta = -1$, $\sec\theta$ undefined, $\cot\theta = 0$

4. $\sin\theta = -\frac{\sqrt{2}}{2}$, $\cos\theta = -\frac{\sqrt{2}}{2}$, $\tan\theta = 1$, $\csc\theta = -\sqrt{2}$, $\sec\theta = -\sqrt{2}$, $\cot\theta = 1$

5. $\sin\theta = -\frac{\sqrt{2}}{2}$, $\cos\theta = -\frac{\sqrt{2}}{2}$, $\tan\theta = 1$, $\csc\theta = -\sqrt{2}$, $\sec\theta = -\sqrt{2}$, $\cot\theta = 1$

11. $\sin\theta = -\frac{2\sqrt{2}}{3}$, $\tan\theta = 2\sqrt{2}$, $\csc\theta = -\frac{3\sqrt{2}}{4}$, $\sec\theta = -3$, $\cot\theta = \frac{\sqrt{2}}{4}$

12. $\sin\theta = -\frac{\sqrt{3}}{2}$, $\cos\theta = \frac{1}{2}$, $\tan\theta = -\sqrt{3}$, $\csc\theta = -\frac{2\sqrt{3}}{3}$, $\cot\theta = -\frac{\sqrt{3}}{3}$

13. $\cos\theta = -\frac{\sqrt{5}}{3}$, $\tan\theta = -\frac{2\sqrt{5}}{5}$, $\csc\theta = \frac{3}{2}$, $\sec\theta = -\frac{3\sqrt{5}}{5}$, $\cot\theta = -\frac{\sqrt{5}}{2}$

14. $\sin\theta = -\frac{4\sqrt{17}}{17}$, $\cos\theta = \frac{\sqrt{17}}{17}$, $\csc\theta = -\frac{\sqrt{17}}{4}$, $\sec\theta = \sqrt{17}$, $\cot\theta = -\frac{1}{4}$

15. $\sin\theta = -\frac{1}{5}$, $\cos\theta = -\frac{2\sqrt{6}}{5}$, $\tan\theta = \frac{\sqrt{6}}{12}$, $\sec\theta = -\frac{5\sqrt{6}}{12}$, $\cot\theta = 2\sqrt{6}$

16. $\sin\theta = \frac{\sqrt{5}}{5}$, $\cos\theta = -\frac{2\sqrt{5}}{5}$, $\tan\theta = -\frac{1}{2}$, $\csc\theta = \sqrt{5}$, $\sec\theta = -\frac{\sqrt{5}}{2}$

17. $\sin\theta = -\frac{\sqrt{10}}{10}$, $\cos\theta = -\frac{3\sqrt{10}}{10}$, $\csc\theta = -\sqrt{10}$, $\sec\theta = -\frac{\sqrt{10}}{3}$, $\cot\theta = 3$

18. $\sin\theta = \frac{\sqrt{15}}{4}$, $\tan\theta = \sqrt{15}$, $\csc\theta = \frac{4\sqrt{15}}{15}$, $\sec\theta = 4$, $\cot\theta = \frac{\sqrt{15}}{15}$

19. $\sin\theta = -\frac{2}{5}$, $\cos\theta = \frac{\sqrt{21}}{5}$, $\tan\theta = -\frac{2\sqrt{21}}{21}$, $\sec\theta = \frac{5\sqrt{21}}{21}$, $\cot\theta = -\frac{\sqrt{21}}{2}$

Lesson 13-4 (pages 725–732)

Find the area of $\triangle ABC$. Round to the nearest tenth.

1. $a = 11$ m, $b = 13$ m, $C = 31°$ **36.8 m^2**
2. $a = 15$ ft, $b = 22$ ft, $C = 90°$ **165 ft^2**
3. $a = 12$ cm, $b = 12$ cm, $C = 50°$ **55.2 cm^2**

Solve each triangle. Round to the nearest tenth. **4–9. See margin.**

4. $A = 18°, B = 37°, a = 15$
5. $A = 60°, C = 25°, c = 3$
6. $B = 40°, C = 32°, b = 10$
7. $B = 10°, C = 23°, c = 8$
8. $A = 12°, B = 60°, b = 5$
9. $A = 35°, C = 45°, a = 30$

Determine whether each triangle has *no* solution, *one* solution, or *two* solutions. Then solve each triangle. Round to the nearest tenth. **10–21. See margin.**

10. $A = 40°, B = 60°, c = 20$
11. $B = 70°, C = 58°, a = 84$
12. $A = 40°, a = 5, b = 12$
13. $A = 58°, a = 26, b = 29$
14. $A = 38°, B = 63°, c = 15$
15. $A = 150°, a = 6, b = 8$
16. $A = 57°, a = 12, b = 19$
17. $A = 25°, a = 125, b = 150$
18. $C = 98°, a = 64, c = 90$
19. $A = 40°, B = 60°, c = 20$
20. $A = 132°, a = 33, b = 50$
21. $A = 45°, a = 83, b = 79$

Lesson 13-5 (pages 733–738)

Determine whether each triangle should be solved by beginning with the Law of Sines or Law of Cosines. Then solve each triangle. **1–18. See margin.**

1.
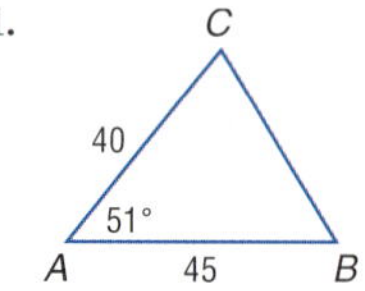

2.
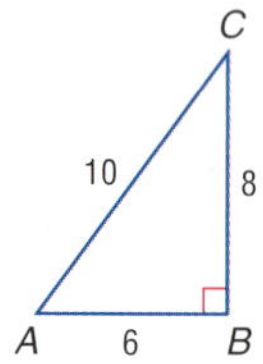

3.
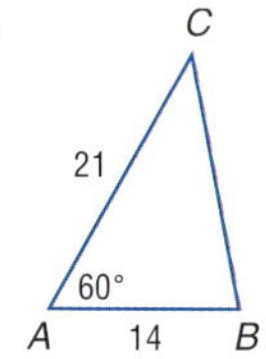

4. $a = 14, b = 15, c = 16$
5. $B = 41°, C = 52°, c = 27$
6. $a = 19, b = 24.3, c = 21.8$
7. $A = 112°, a = 32, c = 20$
8. $b = 8, c = 7, A = 28°$
9. $a = 5, b = 6, c = 7$
10. $C = 25°, a = 12, b = 9$
11. $a = 8, A = 49°, B = 58°$
12. $A = 42°, b = 120, c = 160$
13. $c = 10, A = 35°, C = 65°$
14. $a = 10, b = 16, c = 19$
15. $B = 45°, a = 40, c = 48$
16. $B = 100°, a = 10, c = 8$
17. $A = 40°, B = 45°, c = 4$
18. $A = 20°, b = 100, c = 84$

Lesson 13-6 (pages 739–745)

The given point P is located on the unit circle. Find $\sin\theta$ and $\cos\theta$. **1–5. See margin.**

1. $P\left(\frac{4}{5}, \frac{3}{5}\right)$
2. $P\left(\frac{12}{13}, -\frac{5}{13}\right)$
3. $P\left(-\frac{8}{17}, -\frac{15}{17}\right)$
4. $P\left(\frac{3}{7}, \frac{2\sqrt{10}}{7}\right)$
5. $P\left(-\frac{2}{3}, \frac{\sqrt{5}}{3}\right)$

Find the exact value of each function.

6. $\sin 210°$ **$-\frac{1}{2}$**
7. $\cos 150°$ **$-\frac{\sqrt{3}}{2}$**
8. $\cos(-135°)$ **$-\frac{\sqrt{2}}{2}$**
9. $\cos \frac{3\pi}{4}$ **$-\frac{\sqrt{2}}{2}$**
10. $\sin 570°$ **$-\frac{1}{2}$**
11. $\sin 390°$ **$\frac{1}{2}$**
12. $\sin \frac{4\pi}{3}$ **$-\frac{\sqrt{3}}{2}$**
13. $\cos\left(-\frac{7\pi}{3}\right)$ **$\frac{1}{2}$**
14. $\cos 30° + \cos 60°$ **14. $\frac{\sqrt{3}+1}{2}$**
15. $5(\sin 45°)(\cos 45°)$ **$\frac{5}{2}$**
16. $\frac{\sin 210° + \cos 240°}{2}$ **$-\frac{1}{2}$**
17. $\frac{6 \cos 120° + 4 \sin 150°}{5}$ **$-\frac{1}{5}$**

Determine the period of each function.

18. **270°**
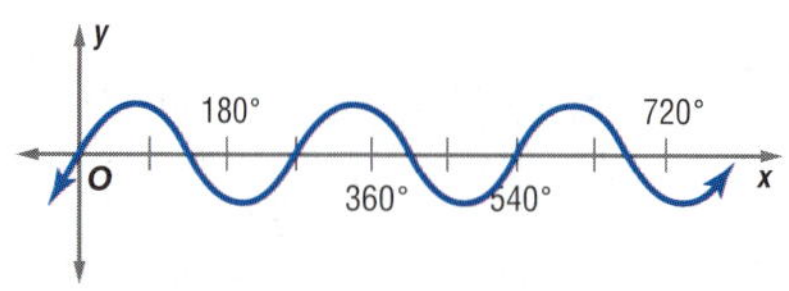

19. **3π**
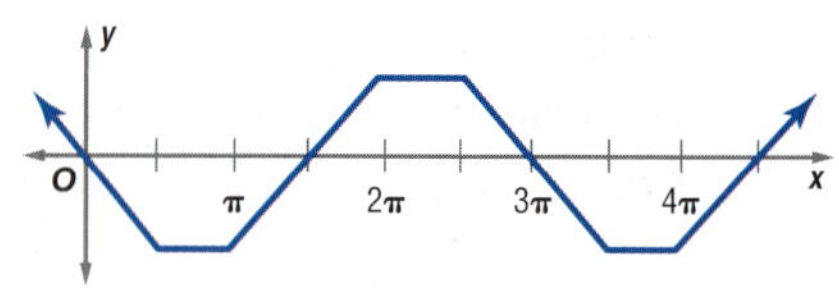

Extra Practice

Lesson 13-4

4. $C = 125°, b \approx 29.2, c \approx 39.8$
5. $B = 95°, a \approx 6.1, b \approx 7.1$
6. $A = 108°, a \approx 14.8, c \approx 8.2$
7. $A = 147°, a \approx 11.2, b \approx 3.6$
8. $C = 108°, a \approx 1.2, c \approx 5.5$
9. $B = 100°, b \approx 51.5, c \approx 37.0$
10. one solution; $C = 80°, a \approx 13.1, b \approx 17.6$
11. one solution; $A = 52°, b = 100.2, c = 90.4$
12. no solution
13. two solutions; $B \approx 71.1°, C \approx 50.9°, c = 23.8$; $B \approx 108.9°, C \approx 13.1°, c \approx 6.9$
14. one solution; $C = 79°, a \approx 9.4, b \approx 13.6$
15. no solution
16. no solution
17. two solutions; $B \approx 30.5°, C \approx 124.5°, c = 243.8$; $B \approx 149.5°, C \approx 5.5°, c \approx 28.3$
18. one solution; $A \approx 44.8°, B \approx 37.2°, b \approx 54.9$
19. one solution; $C = 80°, a \approx 13.1, b \approx 17.6$
20. no solution
21. one solution; $B \approx 42.3°, C \approx 92.7°, c \approx 117.2$

Lesson 13-5

1. cosines; $a \approx 36.9, B \approx 57.4°, C \approx 71.6°$
2. cosines; $A \approx 53.1°; B = 90°, C \approx 36.9°$
3. cosines; $a \approx 18.5, B \approx 79.1°, C \approx 40.9°$
4. cosines; $A \approx 53.6°; B \approx 59.6°, C \approx 66.8°$
5. sines; $A = 87°, a \approx 34.2, b \approx 22.5$
6. cosines; $A \approx 48.3°; B \approx 72.7°, C \approx 59°$
7. sines; $b \approx 18.6, B \approx 32.6°, C \approx 35.4°$
8. cosines; $a \approx 3.8, B \approx 91.0°, C \approx 61.0°$
9. cosines; $A \approx 44.4°; B \approx 57.1°, C \approx 78.5°$
10. cosines; $A \approx 110.1°; B \approx 44.9°, c \approx 5.4$
11. sines; $C = 73°, b \approx 9.0, c \approx 10.1$
12. cosines; $a \approx 107.1, B \approx 48.6°, C \approx 89.4°$
13. sines; $B = 80°, a \approx 5.8, b \approx 9.2$
14. cosines; $A \approx 31.8°; B \approx 57.5°, C \approx 90.7°$
15. cosines; $A \approx 55.1°; C \approx 79.9°, b \approx 34.5$
16. cosines; $A \approx 45.5°; C \approx 34.5°, b \approx 13.8$
17. sines; $C = 95°, a \approx 2.6, b \approx 2.8$
18. cosines; $B \approx 106.3°; C \approx 53.7°, a \approx 35.6$

Lesson 13-6

1. $\sin\theta = \frac{3}{5}, \cos\theta = \frac{4}{5}$
2. $\sin\theta = -\frac{5}{13}, \cos\theta = \frac{12}{13}$
3. $\sin\theta = -\frac{15}{17}, \cos\theta = -\frac{8}{17}$
4. $\sin\theta = \frac{2\sqrt{10}}{7}, \cos\theta = \frac{3}{7}$
5. $\sin\theta = \frac{\sqrt{5}}{3}, \cos\theta = -\frac{2}{3}$

Lesson 13-7

(pages 746–751)

Write each equation in the form of an inverse function.

1. $\text{Sin } m = n$ **$m = \text{Arcsin } n$**
2. $\text{Tan } 45° = 1$ **$45° = \text{Arctan } 1$**
3. $\text{Cos } x = \frac{1}{2}$ **$x = \text{Arccos } \frac{1}{2}$**
4. $\text{Sin } 65° = a$ **$\text{Arcsin } a = 65°$**
5. $\text{Tan } 60° = \sqrt{3}$ **$60° = \text{Arctan } \sqrt{3}$**
6. $\text{Sin } x = \frac{\sqrt{2}}{2}$ **$x = \text{Arcsin}\left(\frac{\sqrt{2}}{2}\right)$**

Solve each equation.

7. $y = \text{Sin}^{-1}\left(-\frac{\sqrt{2}}{2}\right)$ **$-45°$**
8. $\text{Tan}^{-1}(1) = x$ **$45°$**
9. $a = \text{Arccos}\left(\frac{\sqrt{3}}{2}\right)$ **$30°$**
10. $\text{Arcsin}(0) = x$ **0**
11. $y = \text{Cos}^{-1}\left(\frac{1}{2}\right)$ **$60°$**
12. $y = \text{Sin}^{-1}(1)$ **$90°$**

Find each value. Round to the nearest hundredth.

13. $\text{Arccos}\left(-\frac{\sqrt{2}}{2}\right)$ **$135°$**
14. $\text{Sin}^{-1}(-1)$ **$-90°$**
15. $\cos\left[\text{Arcsin}\left(\frac{\sqrt{2}}{2}\right)\right]$ **0.71**
16. $\tan\left[\text{Sin}^{-1}\left(\frac{5}{13}\right)\right]$ **0.42**
17. $\sin 2\left[\text{Arccos}\left(\frac{1}{2}\right)\right]$ **0.87**
18. $\sin\left[\text{Arccos}\left(\frac{5}{17}\right)\right]$ **0.96**
19. $\sin\left[\text{Tan}^{-1}\left(\frac{5}{12}\right)\right]$ **0.38**
20. $\tan\left[\text{Arccos}\left(-\frac{\sqrt{3}}{2}\right)\right]$ **-0.58**
21. $\sin^{-1}[\text{Cos}^{-1}(1) - 1]$ **$-90°$**
22. $\text{Cos}^{-1}\left[\tan\left(\frac{\pi}{4}\right)\right]$ **0**
23. $\cos\left[\text{Sin}^{-1}\left(\frac{1}{2}\right)\right]$ **0.87**
24. $\sin[\text{Cos}^{-1}(0)]$ **1**

Lesson 14-1

(pages 762–768)

Find the amplitude, if it exists, and period of each function. Then graph each function. **1–15. See pp. 861A–861T for graphs.**

1. $y = 2\cos\theta$ **$2; 360°$ or 2π**
2. $y = \frac{1}{3}\sin\theta$ **$\frac{1}{3}; 360°$ or 2π**
3. $y = \sin 3\theta$ **$1; 120°$ or $\frac{2\pi}{3}$**
4. $y = 3\sec\theta$ **none; $360°$ or 2π**
5. $y = \sec\frac{1}{3}\theta$ **none; $1080°$ or 6π**
6. $y = 2\csc\theta$ **none; $360°$ or 2π**
7. $y = 3\tan\theta$ **none; $180°$ or π**
8. $y = 3\sin\frac{2}{3}\theta$ **$3; 540°$ or 3π**
9. $y = 2\sin\frac{1}{5}\theta$ **$2; 1800°$ or 10π**
10. $y = 3\sin 2\theta$ **$3; 180°$ or π**
11. $y = \frac{1}{2}\cos\frac{3}{4}\theta$ **$\frac{1}{2}; 480°$ or $\frac{8\pi}{3}$**
12. $y = 5\csc 3\theta$ **none; $120°$ or $\frac{2\pi}{3}$**
13. $y = 2\cot 6\theta$ **none; $60°$ or $\frac{\pi}{3}$**
14. $y = 2\csc 6\theta$ **none; $60°$ or $\frac{\pi}{3}$**
15. $y = 3\tan\frac{1}{3}\theta$ **none; $540°$ or 3π**

Lesson 14-2

(pages 769–776)

State the phase shift for each function. Then graph the function. **1–4. See margin for graphs.**

1. $y = \sin(\theta + 60°)$ **$-60°$**
2. $y = \cos(\theta - 90°)$ **$90°$**
3. $y = \tan\left(\theta + \frac{\pi}{2}\right)$ **$-\frac{\pi}{2}$**
4. $y = \sin\left(\theta + \frac{\pi}{6}\right)$ **$-\frac{\pi}{6}$**

State the vertical shift and the equation of the midline for each function. Then graph the function. **5–10. See pp. 861A–861T for graphs.**

5. $y = \cos\theta + 3$ **$3; y = 3$**
6. $y = \sin\theta - 2$ **$-2; y = -2$**
7. $y = \sec\theta + 5$ **$5; y = 5$**
8. $y = \csc\theta - 6$ **$-6; y = -6$**
9. $y = 2\sin\theta - 4$ **$-4; y = -4$**
10. $y = \frac{1}{3}\sin\theta + 7$ **$7; y = 7$**

State the vertical shift, amplitude, period, and phase shift of each function. Then graph the function. **11–16. See pp. 861A–861T.**

11. $y = 3\cos[2(\theta + 30°)] + 4$
12. $y = 2\tan[3(\theta - 60°)] - 2$
13. $y = \frac{1}{2}\sin[4(\theta - 45°)] + 1$
14. $y = \frac{2}{5}\cos[6(\theta + 45°)] - 5$
15. $y = 6 + 2\sin\left[3\left(\theta + \frac{\pi}{2}\right)\right]$
16. $y = 3 + 3\cos\left[2\left(\theta - \frac{\pi}{3}\right)\right]$

Lesson 14-2

1.

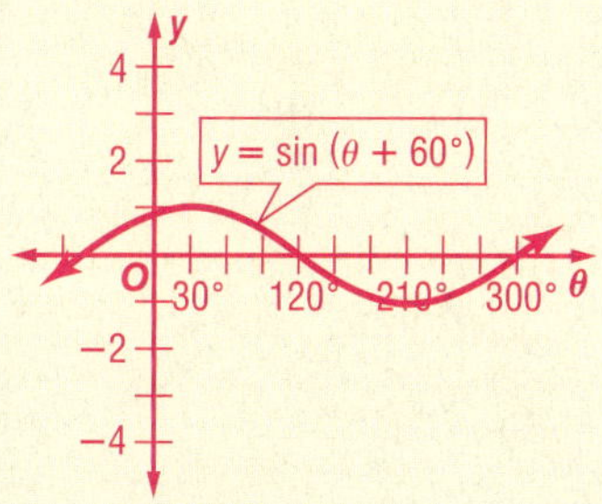

2.

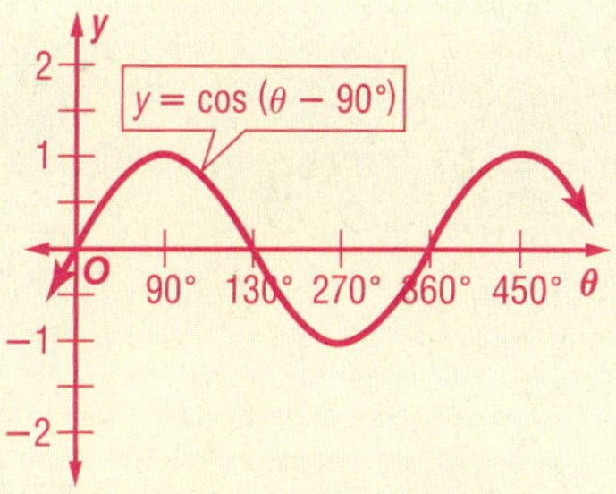

3.

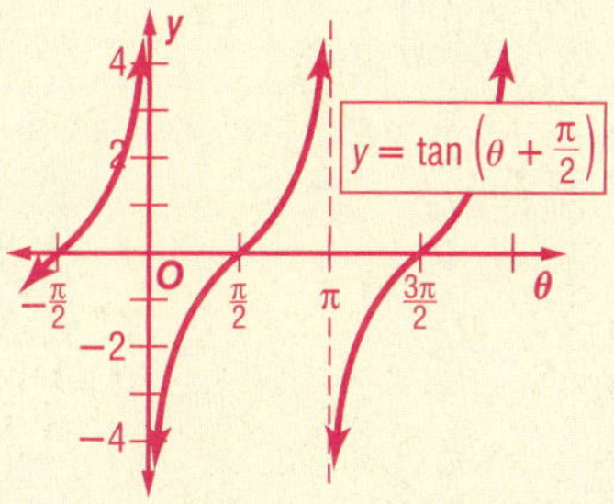

4.

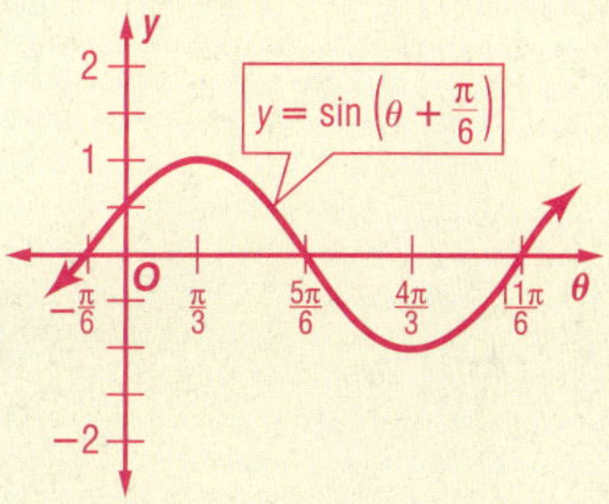

Extra Practice 859

Extra Practice

Lesson 14-5

1. $\frac{\sqrt{2}-\sqrt{6}}{4}$
2. $\frac{\sqrt{6}-\sqrt{2}}{4}$
3. $\frac{-\sqrt{2}-\sqrt{6}}{4}$
4. $\frac{\sqrt{6}+\sqrt{2}}{4}$
5. $\frac{\sqrt{6}+\sqrt{2}}{4}$
6. $\frac{\sqrt{6}-\sqrt{2}}{4}$
7. $\frac{\sqrt{6}+\sqrt{2}}{4}$
8. $\frac{\sqrt{6}-\sqrt{2}}{4}$
9. $\frac{\sqrt{2}}{2}$
10. $-\frac{\sqrt{3}}{2}$
11. $-\frac{\sqrt{2}}{2}$
12. $-\frac{1}{2}$
13. $\frac{\sqrt{3}}{2}$
14. $-\frac{\sqrt{2}}{2}$
15. $\frac{\sqrt{3}}{2}$

Extra Practice

Lesson 14-3

(pages 777–781)

Find the value of each expression.

1. $\sin\theta$, if $\cos\theta = \frac{4}{5}$; $0° \le \theta \le 90°$ $\frac{3}{5}$
2. $\tan\theta$, if $\sin\theta = \frac{1}{2}$; $0° \le \theta \le 90°$ $\frac{\sqrt{3}}{3}$
3. $\csc\theta$, if $\sin\theta = \frac{3}{4}$; $90° \le \theta \le 180°$ $\frac{4}{3}$
4. $\cos\theta$, if $\tan\theta = -4$; $90° \le \theta \le 180°$ $-\frac{\sqrt{17}}{17}$
5. $\sec\theta$, if $\tan\theta = -4$; $90° \le \theta \le 180°$ $-\sqrt{17}$
6. $\sin\theta$, if $\cot\theta = -\frac{1}{4}$; $270° \le \theta \le 360°$ $-\frac{4\sqrt{17}}{17}$
7. $\tan\theta$, if $\sec\theta = -3$; $90° \le \theta \le 180°$ $-2\sqrt{2}$
8. $\sin\theta$, if $\cos\theta = \frac{3}{5}$; $270° \le \theta \le 360°$ $-\frac{4}{5}$
9. $\cos\theta$, if $\sin\theta = -\frac{1}{2}$; $270° \le \theta \le 360°$ $\frac{\sqrt{3}}{2}$
10. $\csc\theta$, if $\cot\theta = -\frac{1}{4}$; $90° \le \theta \le 180°$ $\frac{\sqrt{17}}{4}$
11. $\csc\theta$, if $\sec\theta = -\frac{5}{3}$; $180° \le \theta \le 270°$ $-\frac{5}{4}$
12. $\cos\theta$, if $\tan\theta = 5$; $180° \le \theta \le 270°$ $-\frac{\sqrt{26}}{26}$

Simplify each expression.

13. $\csc^2\theta - \cot^2\theta$ **1**
14. $\sin\theta\tan\theta\csc\theta$ $\tan\theta$
15. $\tan\theta\csc\theta$ $\sec\theta$
16. $\sec\theta\cot\theta\cos\theta$ $\cot\theta$
17. $\cos\theta(1-\cos^2\theta)$ $\cos\theta \cdot \sin^2\theta$
18. $\frac{1-\sin^2\theta}{\cos^2\theta}$ **1**
19. $\frac{\sin^2\theta+\cos^2\theta}{\cos^2\theta}$ $\sec^2\theta$
20. $\frac{1+\tan^2\theta}{1+\cot^2\theta}$ $\tan^2\theta$
21. $\frac{1}{1+\sin\theta} + \frac{1}{1-\sin\theta}$ $2\sec^2\theta$

Lesson 14-4

(pages 782–785)

Verify that each of the following is an identity. **1–21. See pp. 861A–861T.**

1. $\sin^2\theta + \cos^2\theta + \tan^2\theta = \sec^2\theta$
2. $\frac{\tan\theta}{\sin\theta} = \sec\theta$
3. $\frac{\tan\theta}{\cot\theta} = \tan^2\theta$
4. $\csc^2\theta(1-\cos^2\theta) = 1$
5. $1 - \cot^4\theta = 2\csc^2\theta - \csc^4\theta$
6. $\sin^4\theta - \cos^4\theta = \sin^2\theta - \cos^2\theta$
7. $\sin^2\theta + \cot^2\theta\sin^2\theta = 1$
8. $\frac{\cos\theta}{\csc\theta} - \frac{\csc\theta}{\sec\theta} = -\frac{\cos^3\theta}{\sin\theta}$
9. $\frac{\cos\theta}{\sec\theta-1} + \frac{\cos\theta}{\sec\theta+1} = 2\cot^2\theta$
10. $\frac{1+\cos\theta}{\sin\theta} = \frac{\sin\theta}{1-\cos\theta}$
11. $\sec\theta + \tan\theta = \frac{\cos\theta}{1-\sin\theta}$
12. $\tan\theta + \cot\theta = \csc\theta\sec\theta$
13. $\frac{\cot^2\theta}{1+\cot^2\theta} = 1 - \sin^2\theta$
14. $\frac{\tan\theta - \sin\theta}{\sec\theta} = \frac{\sin^3\theta}{1+\cos\theta}$
15. $\sin^2\theta(1-\cos^2\theta) = \sin^4\theta$
16. $\sin^2\theta + \sin^2\theta\tan^2\theta = \tan^2\theta$
17. $\frac{\sec\theta-1}{\sec\theta+1} + \frac{\cos\theta-1}{\cos\theta+1} = 0$
18. $\tan^2\theta(1-\sin^2\theta) = \sin^2\theta$
19. $\tan\theta + \frac{\cos\theta}{1+\sin\theta} = \sec\theta$
20. $\frac{\tan\theta}{\sec\theta+1} = \frac{1-\cos\theta}{\sin\theta}$
21. $\csc\theta - \frac{\sin\theta}{1+\cos\theta} = \cot\theta$

Lesson 14-5

(pages 786–790)

Find the exact value of each expression. **1–15. See margin.**

1. $\sin 195°$
2. $\cos 285°$
3. $\sin 255°$
4. $\sin 105°$
5. $\cos 15°$
6. $\sin 15°$
7. $\cos 375°$
8. $\sin 165°$
9. $\sin(-225°)$
10. $\cos(-210°)$
11. $\cos(-225°)$
12. $\sin(-30°)$
13. $\sin 120°$
14. $\sin 225°$
15. $\cos(-30°)$

Verify that each of the following is an identity. **16–20. See pp. 861A–861T.**

16. $\sin(90° + \theta) = \cos\theta$
17. $\cos(180° - \theta) = -\cos\theta$
18. $\sin(\pi + \theta) = -\sin\theta$
19. $\sin(\theta + 30°) + \sin(\theta + 60°) = \frac{\sqrt{3}+1}{2}(\sin\theta + \cos\theta)$
20. $\cos(30° - \theta) + \cos(30° + \theta) = \sqrt{3}\cos\theta$

Lesson 14-6

(pages 791–797)

Find the exact value of $\sin 2\theta$, $\cos 2\theta$, $\sin \frac{\theta}{2}$, and $\cos \frac{\theta}{2}$ for each of the following.

1. $\cos \theta = \frac{7}{25}$; $0 < \theta < 90°$ $\frac{336}{625}$; $-\frac{527}{625}$; $\frac{3}{5}$; $\frac{4}{5}$
2. $\sin \theta = \frac{2}{7}$; $0 < \theta < 90°$ $\frac{12\sqrt{5}}{49}$; $\frac{41}{49}$; $\frac{\sqrt{98 - 42\sqrt{5}}}{14}$; $\frac{\sqrt{98 + 42\sqrt{5}}}{14}$
3. $\cos \theta = -\frac{1}{8}$; $180 < \theta < 270°$ $\frac{3\sqrt{7}}{32}$; $-\frac{31}{32}$; $\frac{3}{4}$; $-\frac{\sqrt{7}}{4}$
4. $\sin \theta = -\frac{5}{13}$; $270 < \theta < 360°$ $-\frac{120}{169}$; $\frac{119}{169}$; $\frac{\sqrt{26}}{26}$; $-\frac{5\sqrt{26}}{26}$
5. $\sin \theta = \frac{\sqrt{35}}{6}$; $0 < \theta < 90°$ $\frac{\sqrt{35}}{18}$; $-\frac{17}{18}$; $\frac{\sqrt{15}}{6}$; $\frac{\sqrt{21}}{6}$
6. $\cos \theta = -\frac{17}{18}$; $90 < \theta < 180°$ $-\frac{17\sqrt{35}}{162}$; $\frac{127}{162}$; $\frac{\sqrt{35}}{6}$; $\frac{1}{6}$

Find the exact value of each expression by using the half-angle formulas.

7. $\sin 75°$ 8. $\cos 75°$ 9. $\sin \frac{\pi}{8}$
10. $\cos \frac{13\pi}{12}$ 11. $\cos 22.5°$ 12. $\cos \frac{\pi}{4}$ $\frac{\sqrt{2}}{2}$

Verify that each of the following is an identity. **13–18. See pp. 861A–861T.**

13. $\frac{\sin 2\theta}{2 \sin^2 \theta} = \cot \theta$
14. $1 + \cos 2\theta = \frac{2}{1 + \tan^2 \theta}$
15. $\csc \theta \sec \theta = 2 \csc 2\theta$
16. $\sin 2\theta (\cot \theta + \tan \theta) = 2$
17. $\frac{1 - \tan^2 \theta}{1 + \tan^2 \theta} = \cos 2\theta$
18. $\frac{\cos \theta + \sin \theta}{\cos \theta - \sin \theta} = \frac{1 + \sin 2\theta}{\cos 2\theta}$

7. $\frac{\sqrt{2 + \sqrt{3}}}{2}$ 8. $\frac{\sqrt{2 - \sqrt{3}}}{2}$ 9. $\frac{\sqrt{2 - \sqrt{2}}}{2}$ 10. $-\frac{\sqrt{2 + \sqrt{3}}}{2}$ 11. $\frac{\sqrt{2 + \sqrt{2}}}{2}$

Lesson 14-7

(pages 799–804)

Find all the solutions for each equation for $0° \leq \theta < 360°$.

1. $\cos \theta = -\frac{\sqrt{3}}{2}$ 150°, 210°
2. $\sin 2\theta = -\frac{\sqrt{3}}{2}$ 120°, 150°, 300°, 330°
3. $\cos 2\theta = 8 - 15 \sin \theta$ 30°, 150°
4. $\sin \theta + \cos \theta = 1$ 0°, 90°
5. $2 \sin^2 \theta + \sin \theta = 0$ 0°, 180°, 210°, 330°
6. $\sin 2\theta = \cos \theta$ 30°, 90°, 150°, 270°

Solve each equation for all values of θ if θ is measured in radians.

7. $\cos 2\theta \sin \theta = 1$ $\frac{3\pi}{2} + 2k\pi$
8. $\sin \frac{\theta}{2} + \cos \frac{\theta}{2} = \sqrt{2}$ $\frac{\pi}{2} + 4k\pi$
9. $\cos 2\theta + 4 \cos \theta = -3$ $\pi + 2k\pi$
10. $\sin \frac{\theta}{2} + \cos \theta = 1$ $0 + 2k\pi$; $\frac{\pi}{3} + 4k\pi$; $\frac{5\pi}{3} + 4k\pi$
11. $3 \tan^2 \theta - \sqrt{3} \tan \theta = 0$ $0 + k\pi$; $\frac{\pi}{6} + k\pi$
12. $4 \sin \theta \cos \theta = -\sqrt{3}$ $\frac{2\pi}{3} + k\pi$; $\frac{5\pi}{6} + k\pi$

Solve each equation for all values of θ if θ is measured in degrees.

13. $2 \sin^2 \theta - 1 = 0$ $45° + k \cdot 90°$
14. $\cos \theta - 2 \cos \theta \sin \theta = 0$ $30° + k \cdot 120°$; $90° + k \cdot 180°$
15. $\cos 2\theta \sin \theta = 1$ $270° + k \cdot 360°$
16. $(\tan \theta - 1)(2 \cos \theta + 1) = 0$ $45° + k \cdot 180°$; $120° + k \cdot 360°$
17. $2 \cos^2 \theta = 0.5$ $60° + k \cdot 180°$; $120° + k \cdot 180°$
18. $\sin \theta \tan \theta - \tan \theta = 0$ $0° + k \cdot 180°$

Solve each equation for all values of θ. **21–24. See margin.**

19. $\tan \theta = 1$ $45° + k \cdot 180°$ or $\frac{\pi}{4} + k \cdot \pi$
20. $\cos 8\theta = 1$ $0° + k \cdot 45°$ or $0 + k \cdot \frac{\pi}{4}$
21. $\sin \theta + 1 = \cos 2\theta$
22. $8 \sin \theta \cos \theta = 2\sqrt{3}$
23. $\cos \theta = 1 + \sin \theta$
24. $2 \cos^2 \theta = \cos \theta$

Lesson 14-7

21. $0° + k \cdot 180°$, $210° + k \cdot 360°$, and $330° + k \cdot 360°$ or $0 + k\pi$, $\frac{7\pi}{6} + 2k\pi$, and $\frac{11\pi}{6} + 2k\pi$
22. $30° + k \cdot 180°$ and $60° + k \cdot 180°$ or $\frac{\pi}{6} + k\pi$ and $\frac{\pi}{3} + k\pi$
23. $0° + k \cdot 360°$ and $270° + k \cdot 360°$ or $0 + 2k\pi$ and $\frac{3\pi}{2} + 2k\pi$
24. $90° + k \cdot 180°$, $60° + k \cdot 360°$, and $300° + k \cdot 360°$, or $\frac{\pi}{2} + k\pi$, $\frac{\pi}{3} + 2\pi \cdot k$ and $\frac{5\pi}{6} + 2\pi \cdot k$

Extra Practice

Extra Practice 861

Extra Practice

Page 829, Lesson 1-6

4. $\{m \mid m < 3 \text{ or } m > 6\}$ 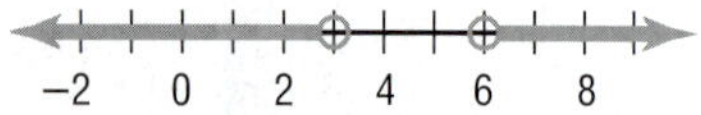

5. $\{n \mid -2 < n < 3\}$ 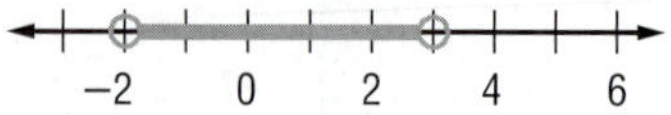

6. $\{s \mid -1 \le s \le 7\}$ 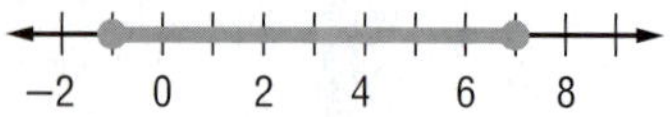

7. $\{t \mid t \le -2 \text{ or } t \ge 2\}$ 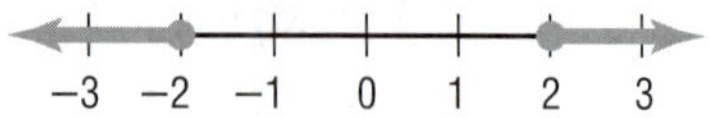

8. $\{x \mid 1 \le x \le 4\}$ 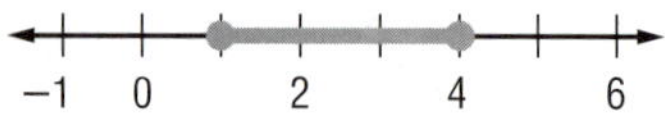

9. $\{x \mid x < -0.5 \text{ or } x > 8\}$ 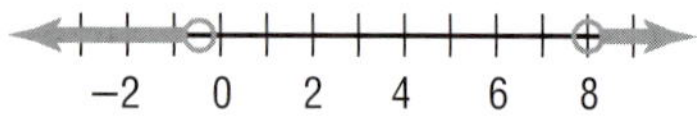

10. $\{x \mid x \le -3 \text{ or } x \ge 3\}$ 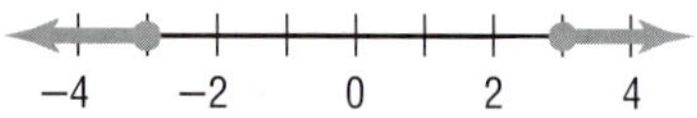

11. $\{p \mid -2 \le p \le 2\}$ 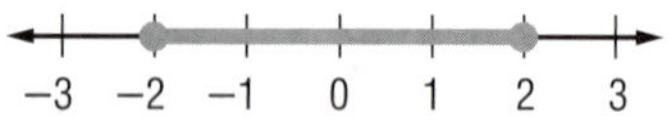

12. all real numbers 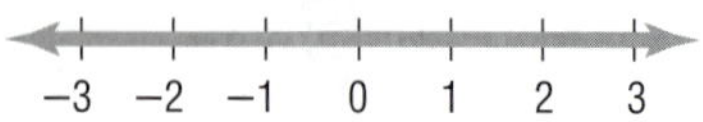

13. $\{a \mid -4 < a < -2\}$ 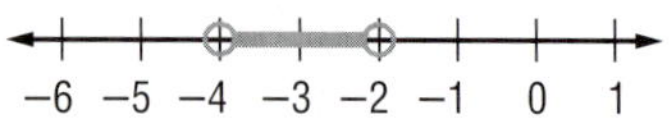

14. $\{t \mid t < 3 \text{ or } t > 5\}$ 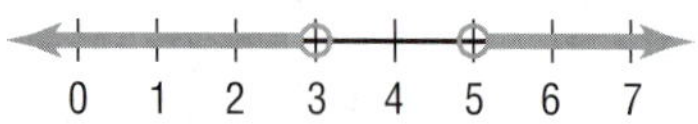

15. $\{y \mid 1 < y < 4\}$ 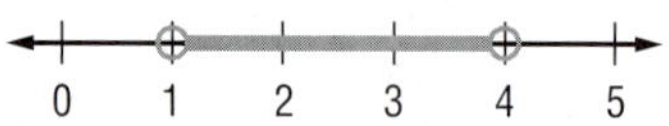

16. $\{d \mid d \le -3 \text{ or } d \ge -1\}$

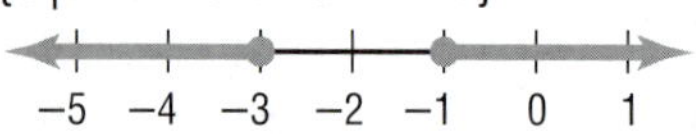

17. $\{x \mid -1 < x < 1.5\}$ 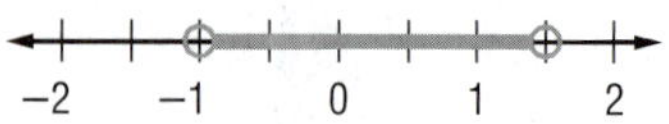

18. $\{v \mid v < -5 \text{ or } v > 1\}$ 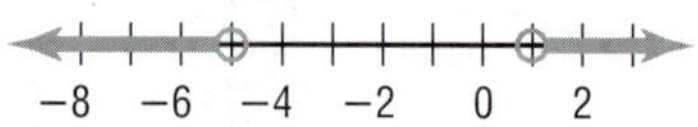

19. $\{r \mid -5 < r < 1\}$ 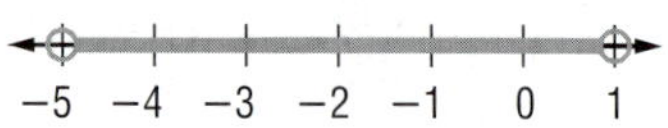

20. $\{w \mid w \le -1.2 \text{ or } w \ge 2.4\}$

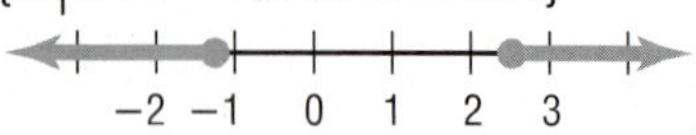

21. all real numbers 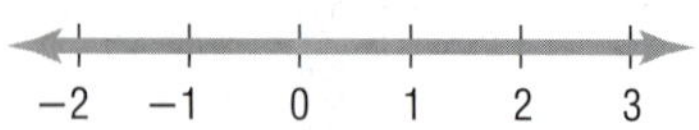

22. $\varnothing$ 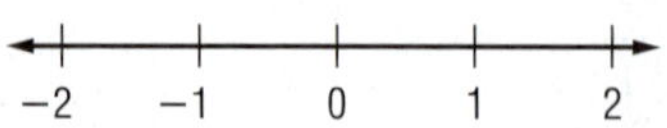

23. $\varnothing$ 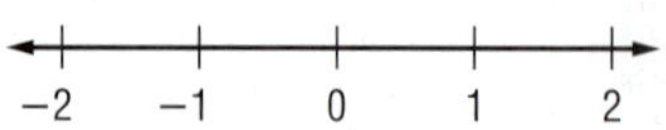

24. $\{n \mid n \le -4 \text{ or } n \ge 4\}$ 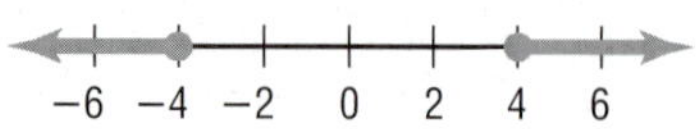

Page 830, Lesson 2-2

13. 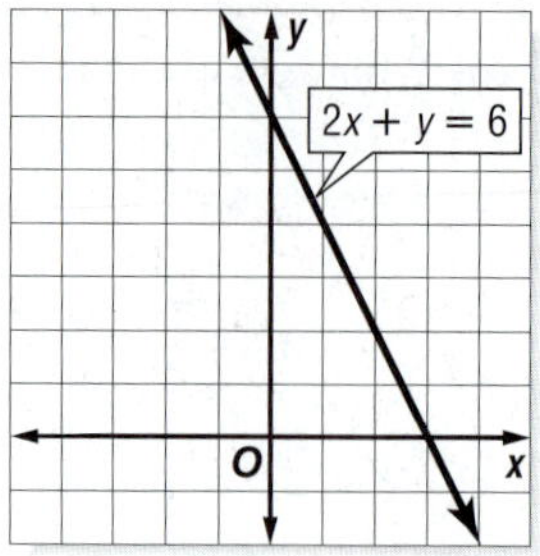

14. 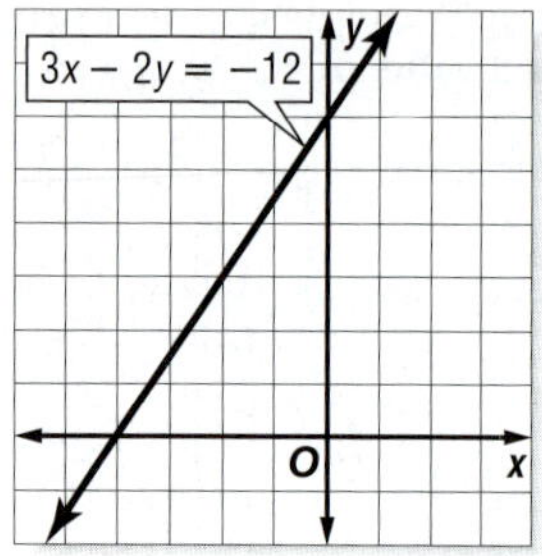

15. 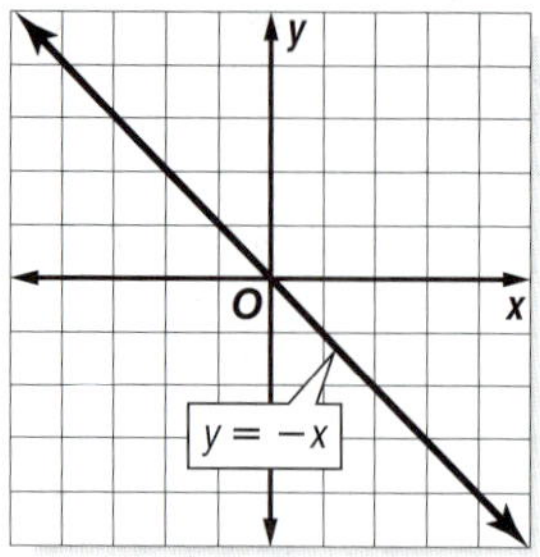

16. 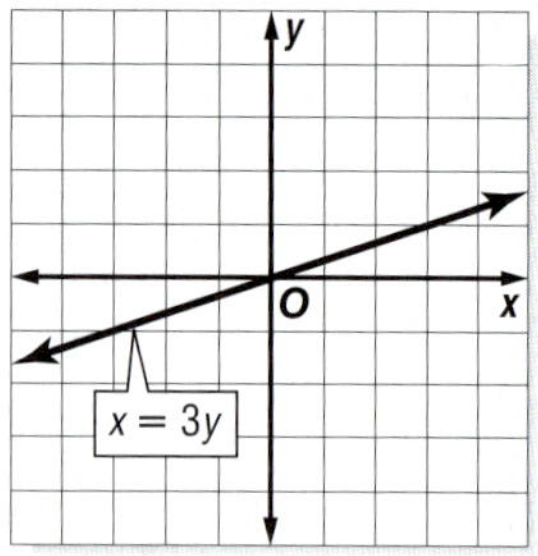

17. 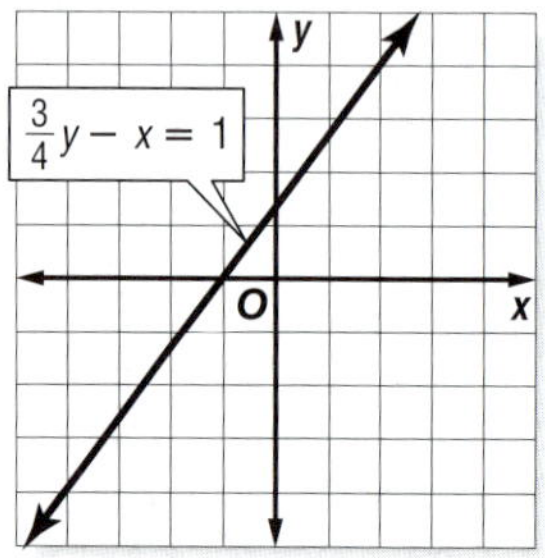

18. 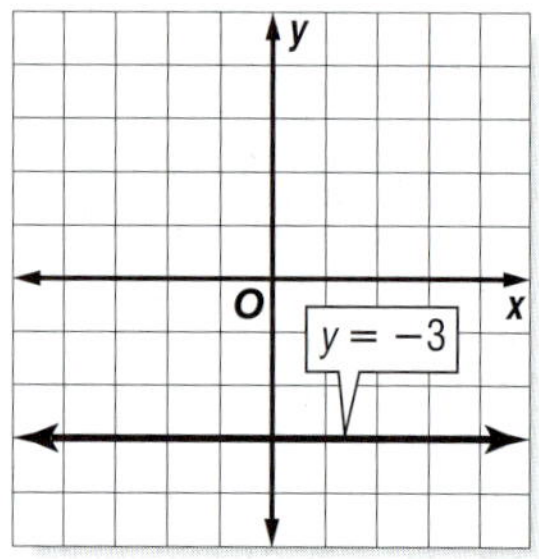

Page 830, Lesson 2-3

10. 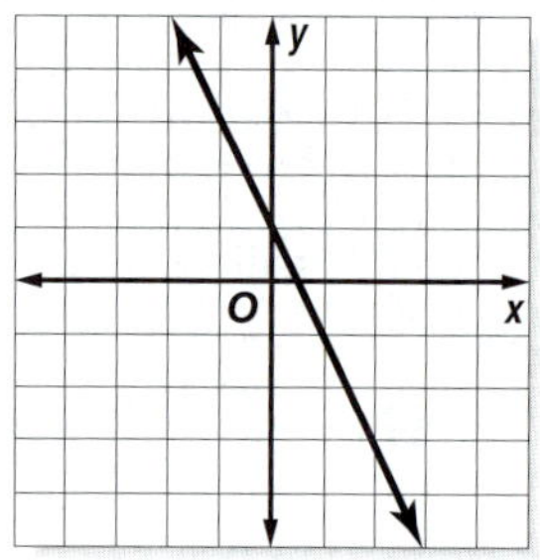

11. 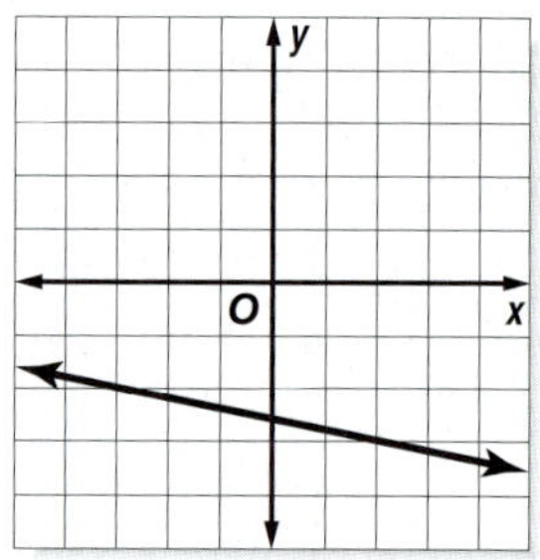

12. 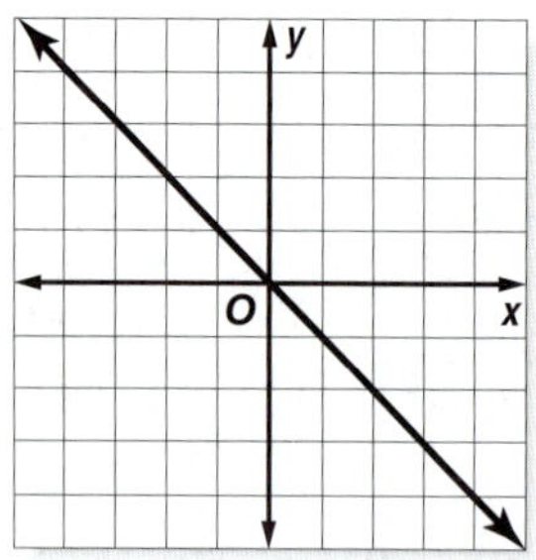

13. 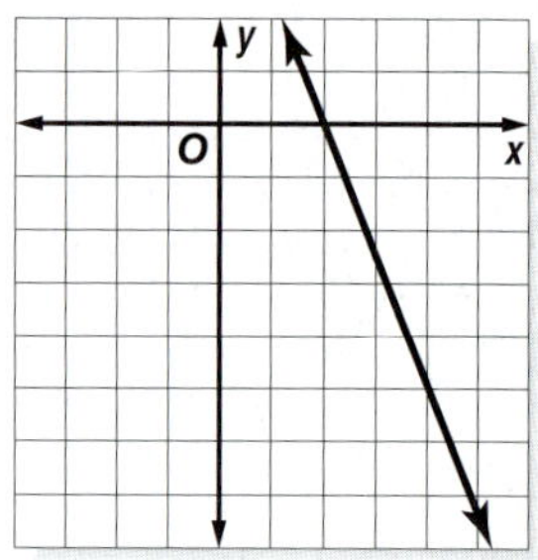

Page 831, Lesson 2-6

3. D = all real numbers; R = all integers

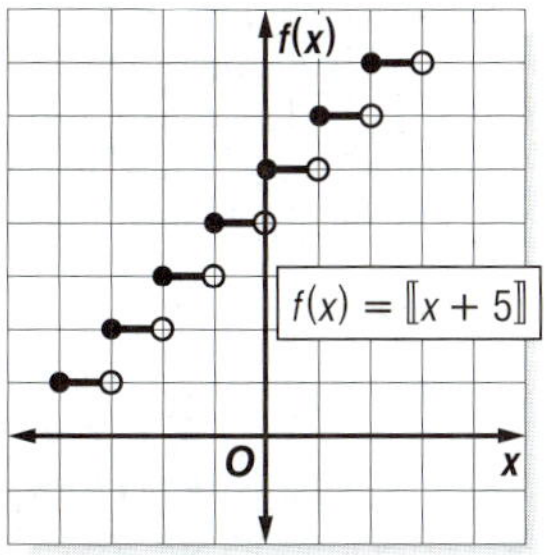

4. D = all real numbers; R = all integers

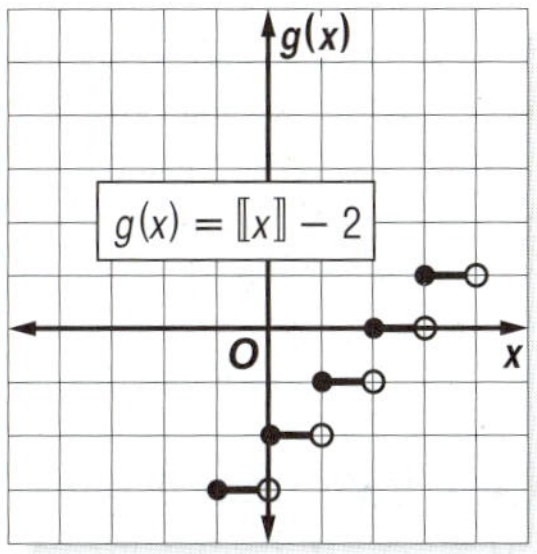

5. D = all real numbers; R = even integers

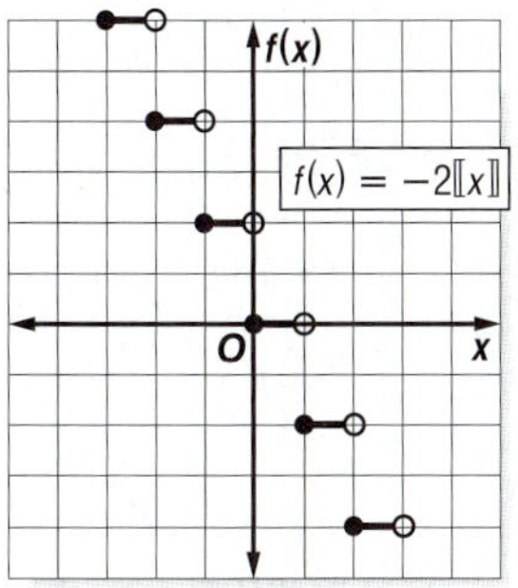

6. D = all real numbers; R = $\{y \mid y \geq -3\}$

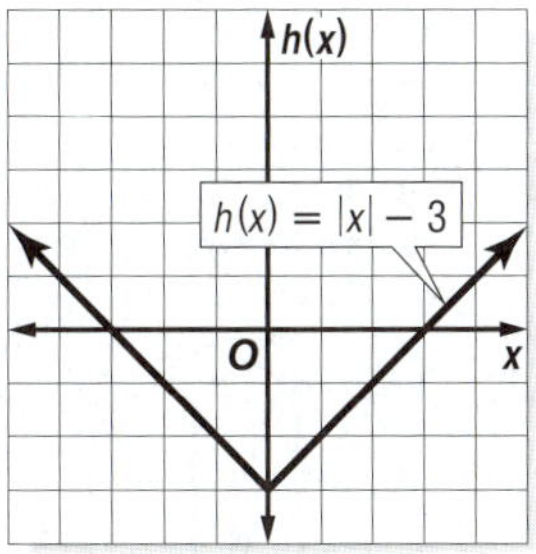

7. D = all real numbers; R = $\{y \mid y \geq 0\}$

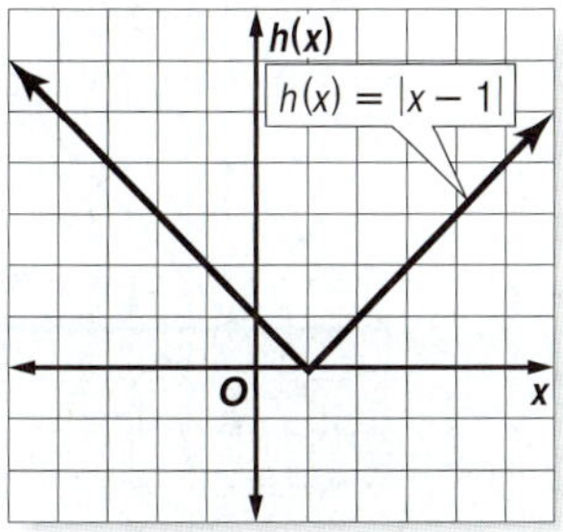

8. D = all real numbers; R = $\{y \mid y \geq 2\}$

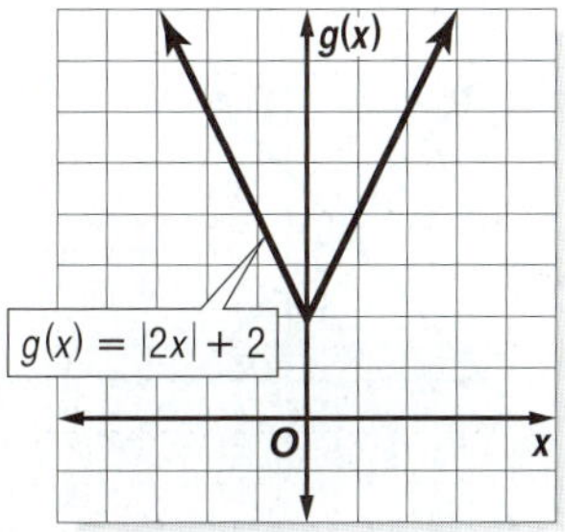

9. D = all real numbers; R = $\{y \mid y < -2 \text{ or } y = 4\}$

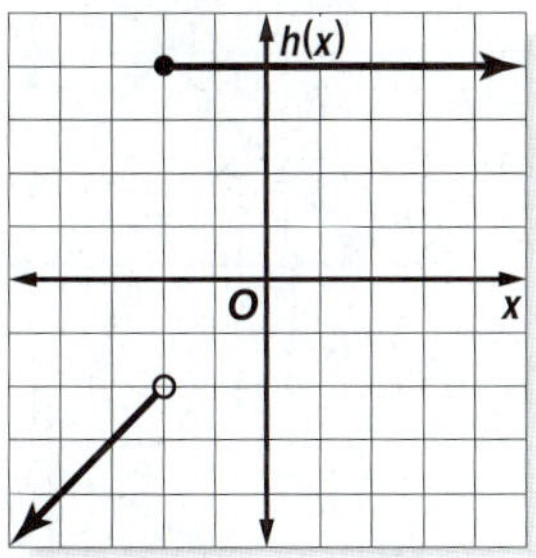

10. D = all real numbers; R = $\{y \mid y < -1\}$

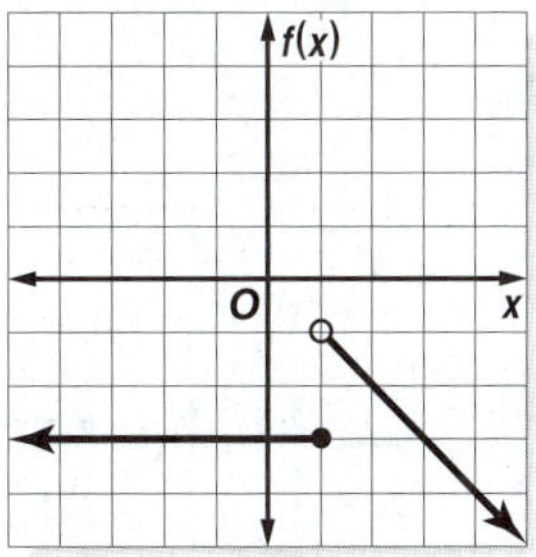

Page 832, Lesson 2-7

1.

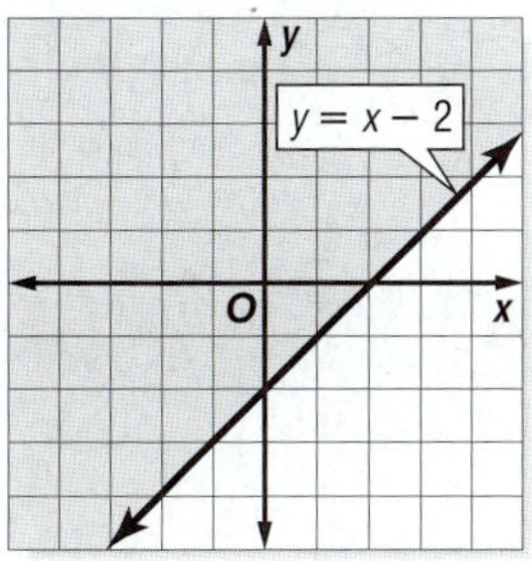

2.

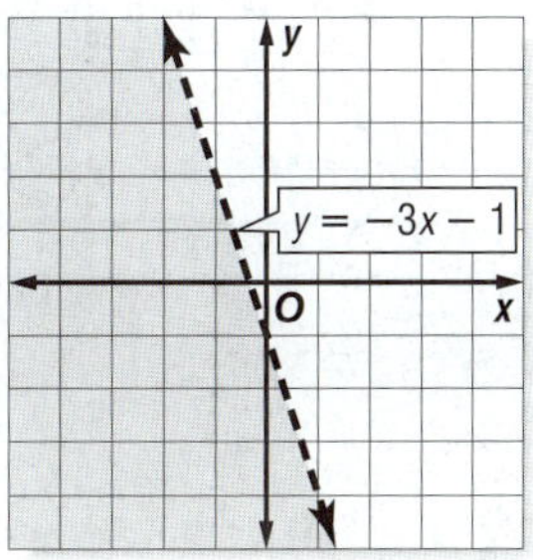

3.

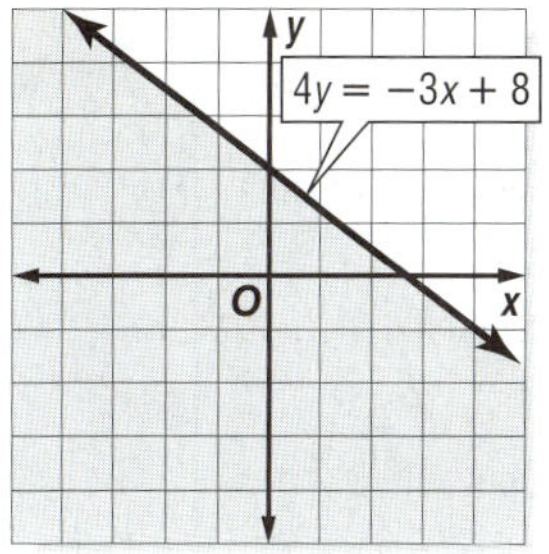

4.

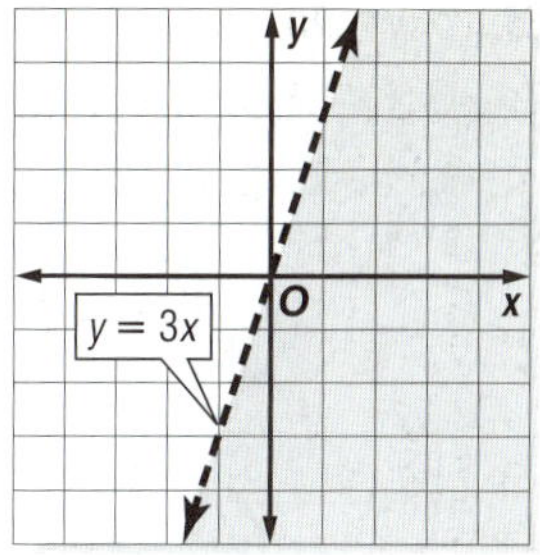

5.

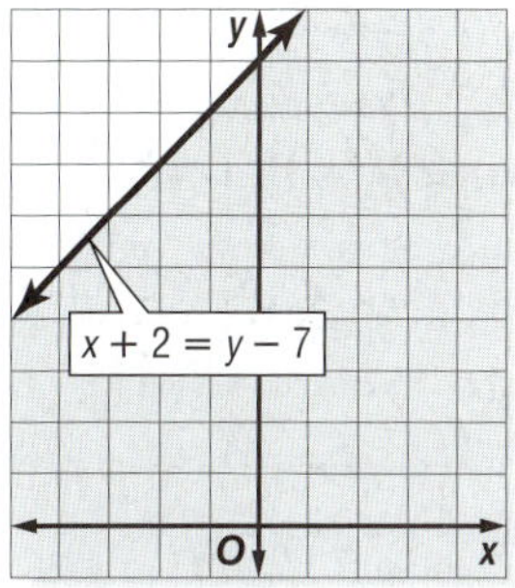

6.

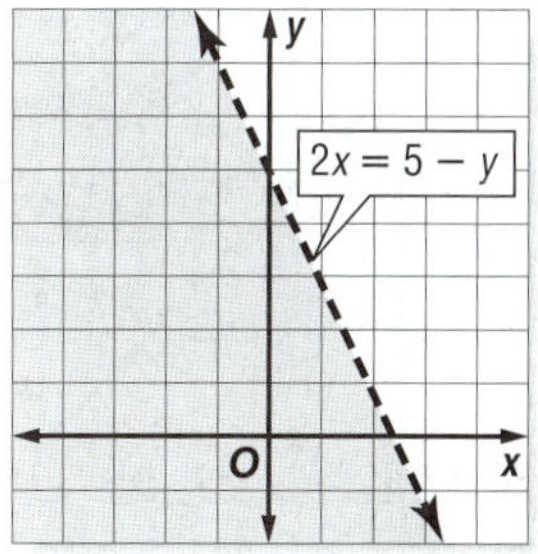

7.

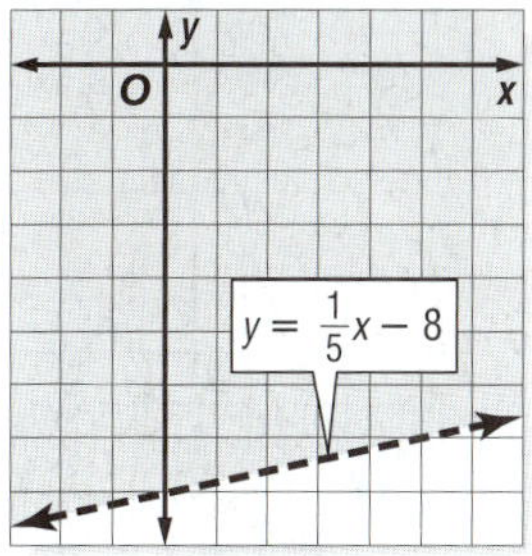

8.

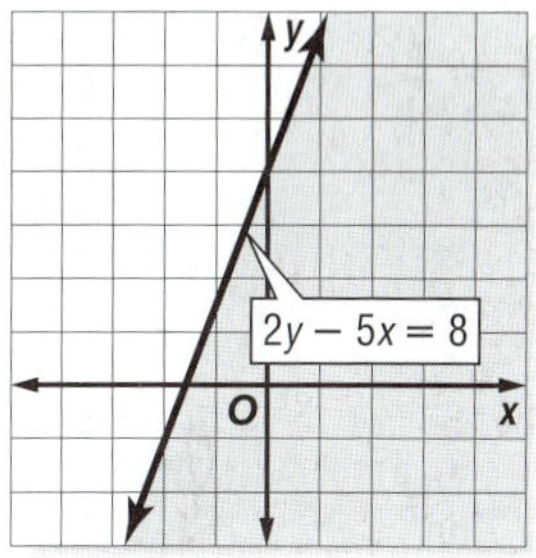

9.

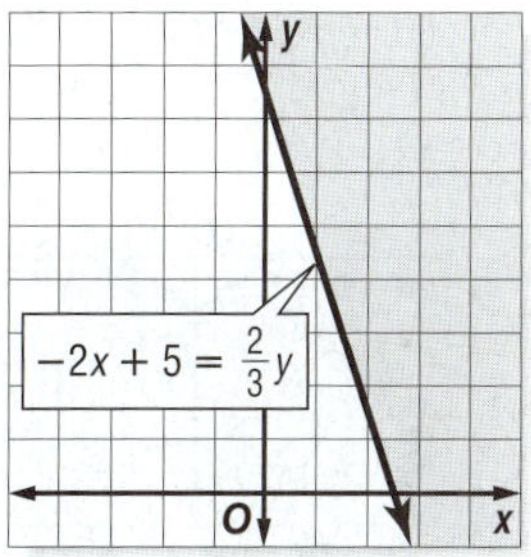

10.

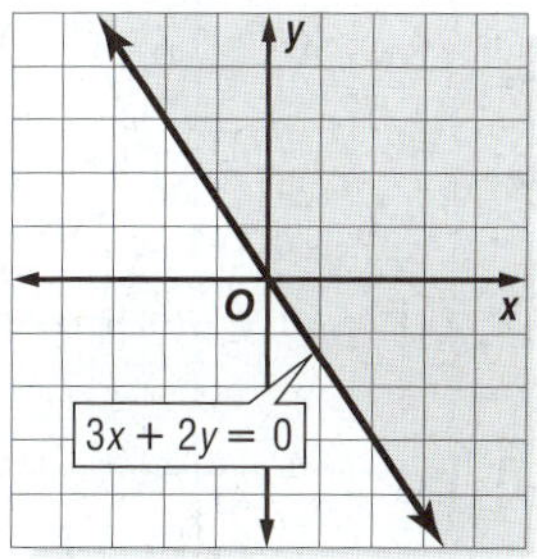

11.

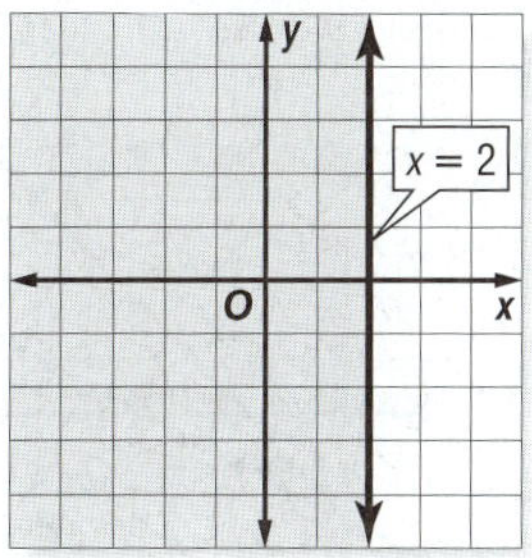

12.

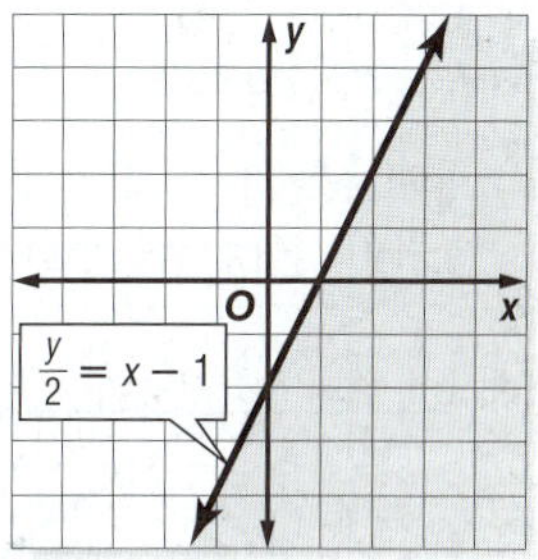

13.

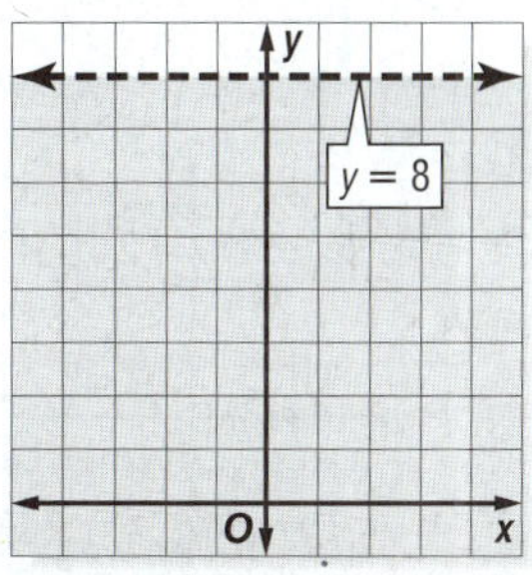

14.

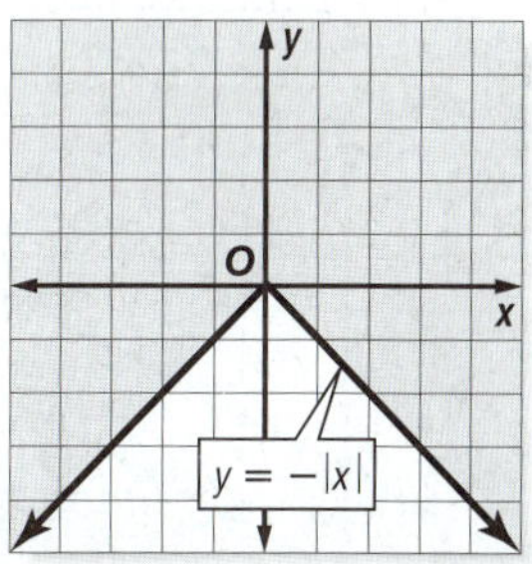

15.
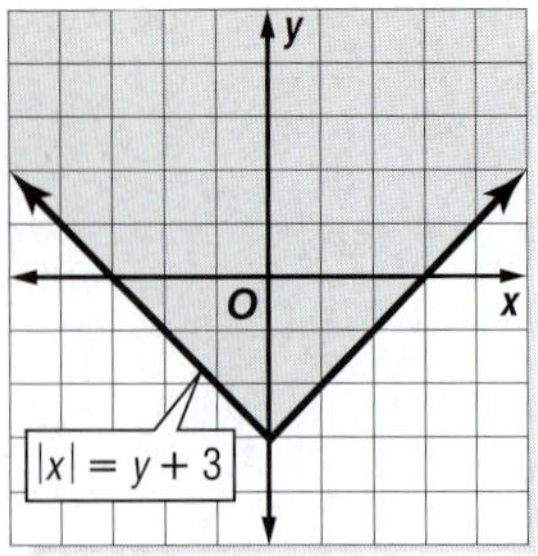

16.
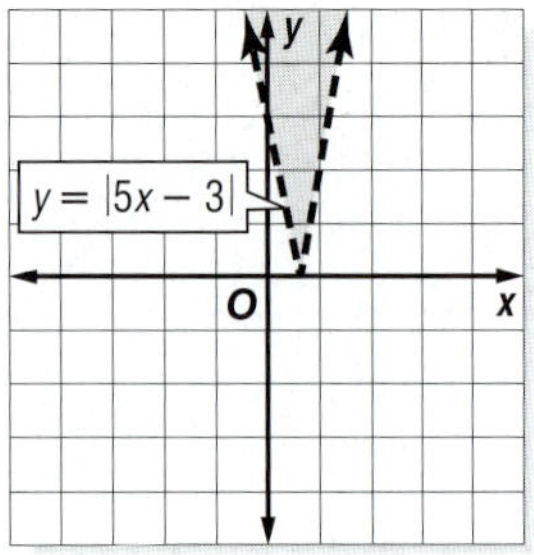

17.
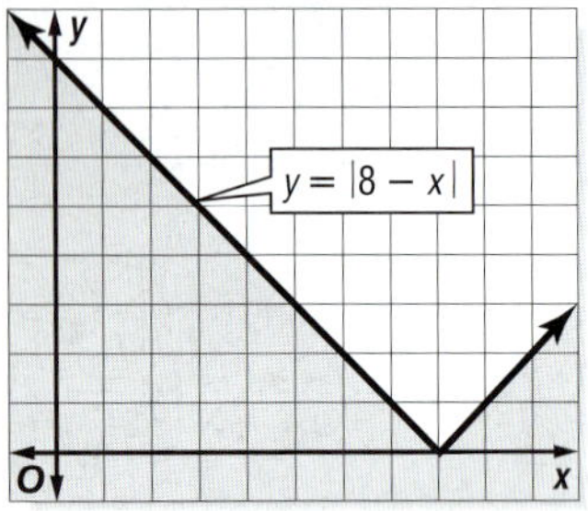

18.
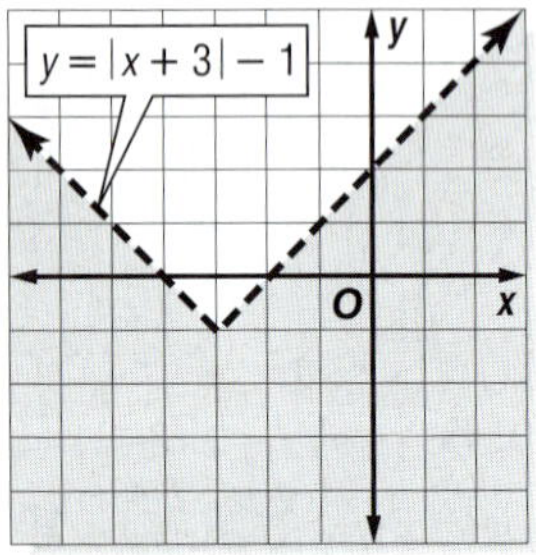

19.
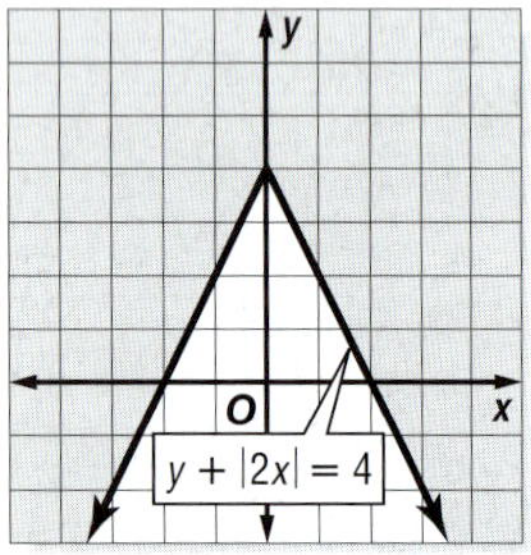

20.
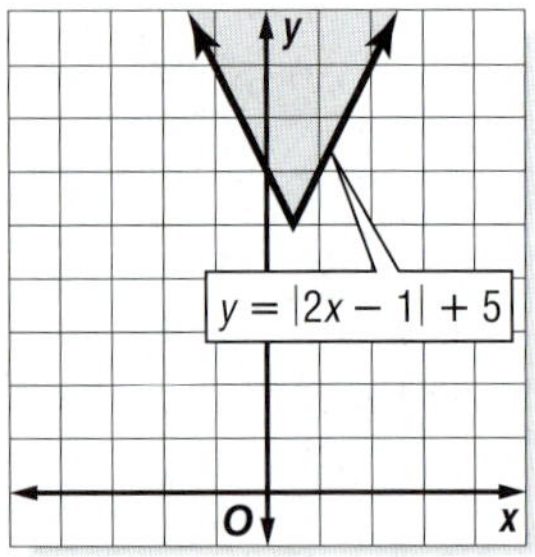

21.
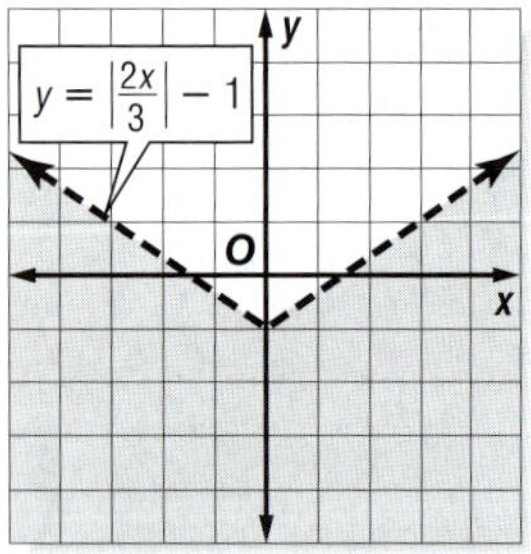

Page 832, Lesson 3-1

7.
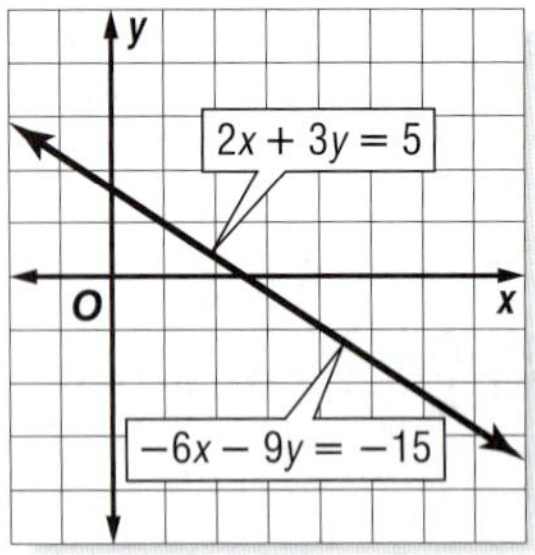

8.
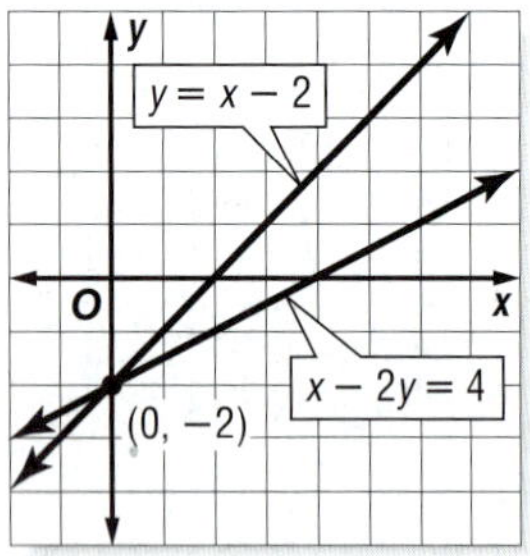

9.
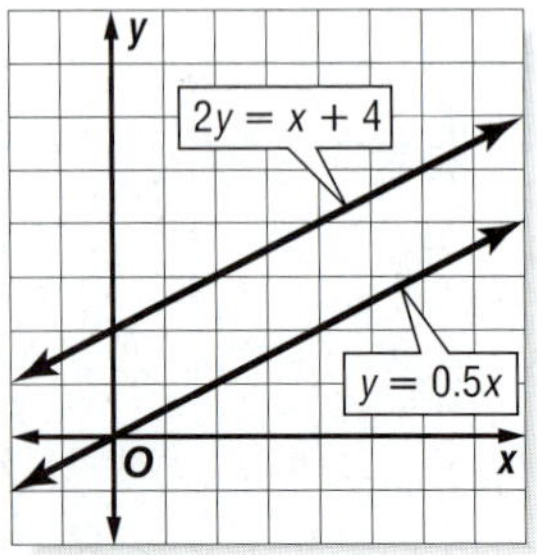

10.
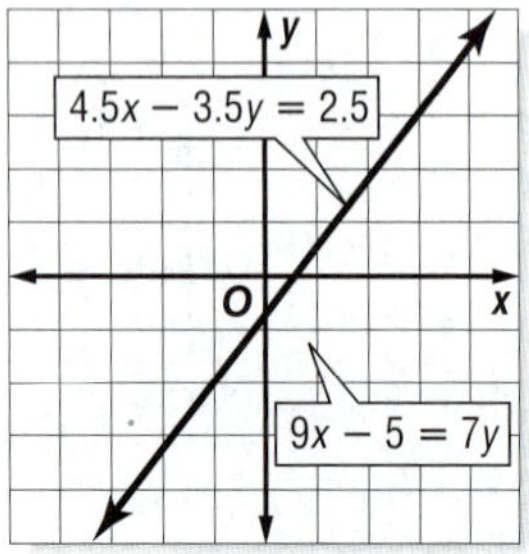

11.
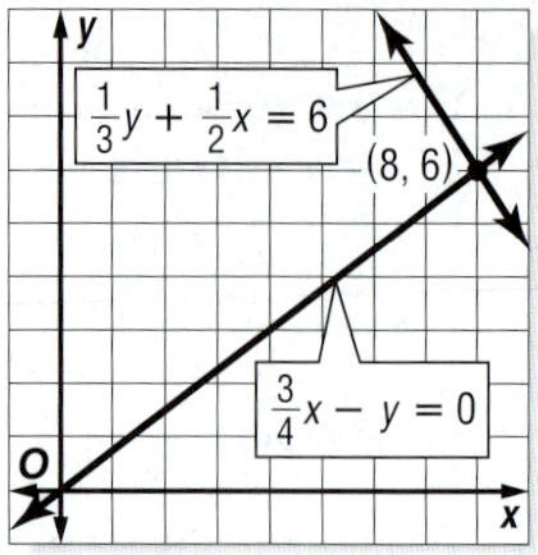

12.
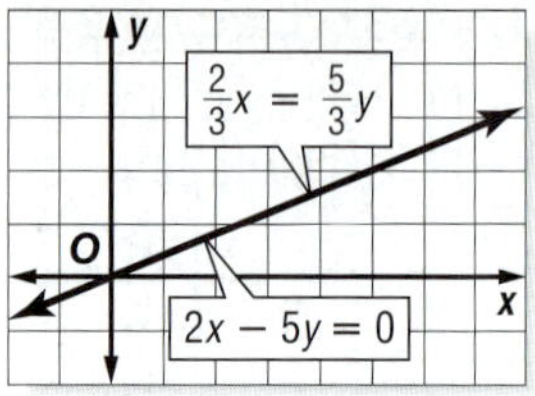

Page 833, Lesson 3-3

1.
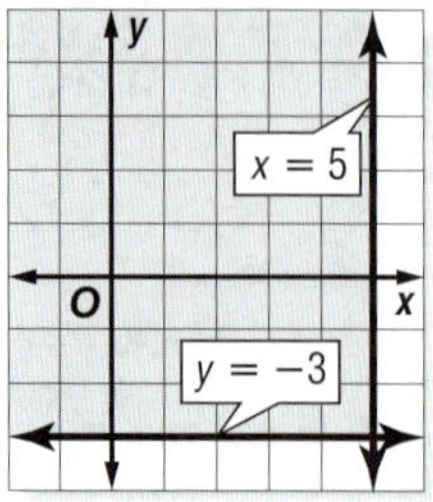

2.
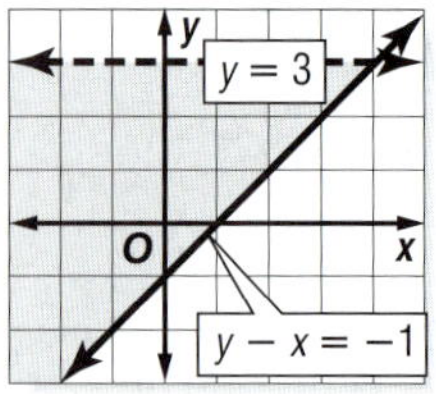

3.
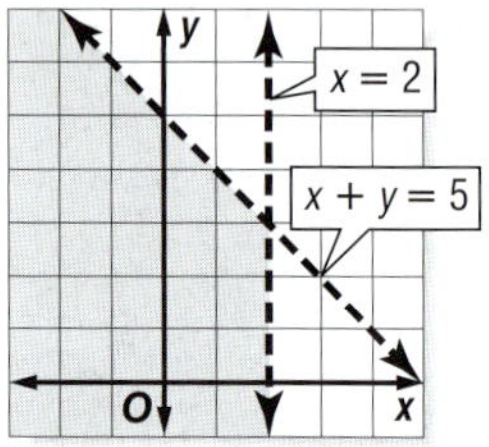

4.
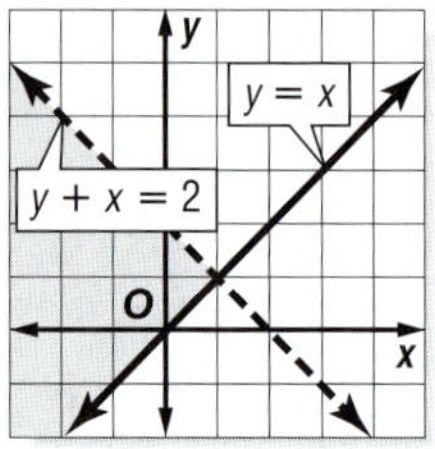

5.
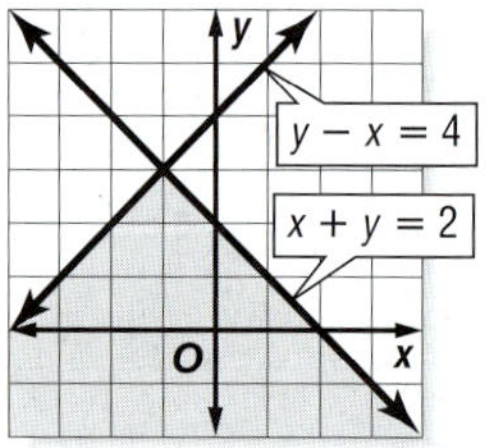

6.
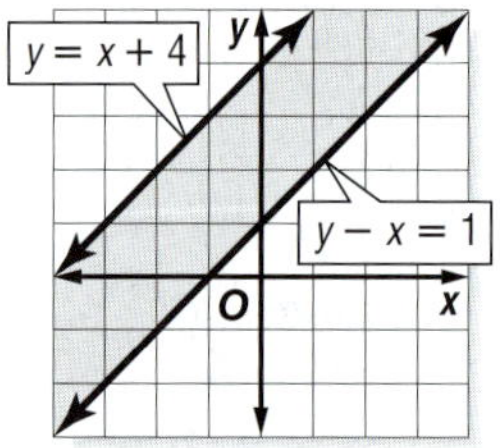

7.
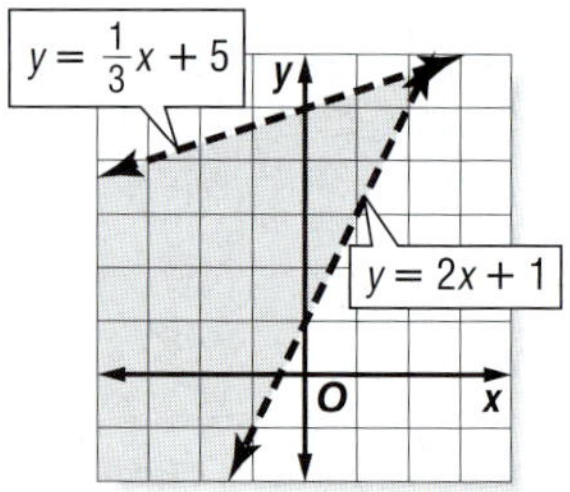

8.
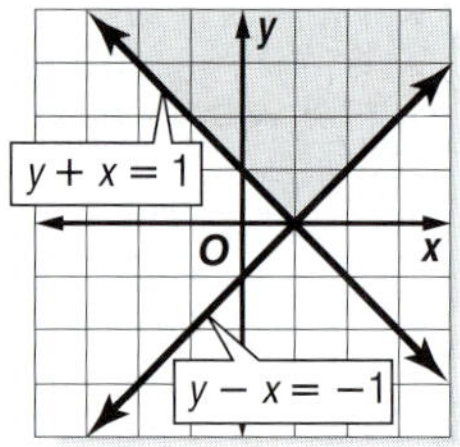

9.
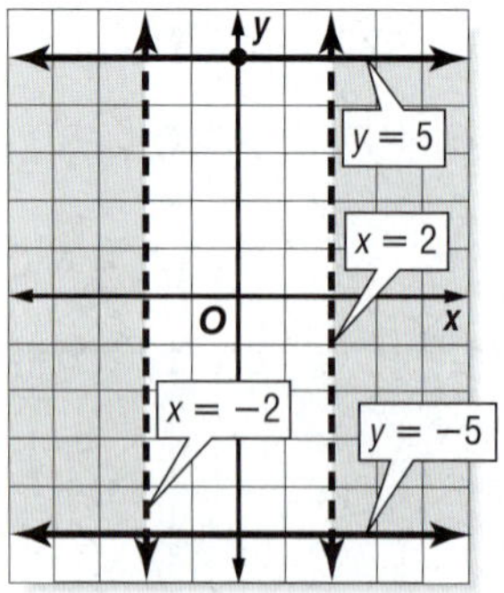

10.
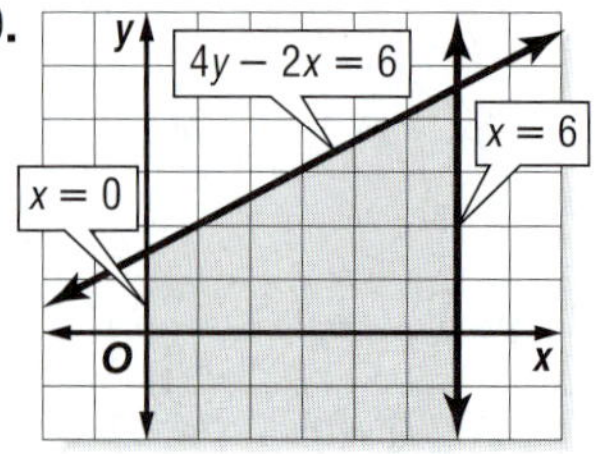

11.

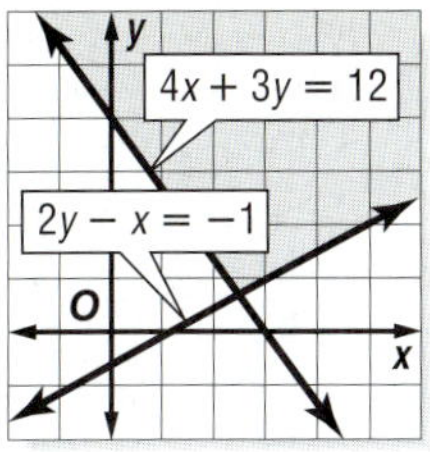

12.

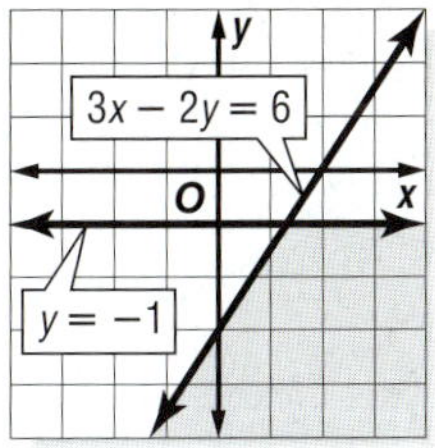

Page 839, Lesson 6-1

1a. y-intercept: 0; axis of symmetry: $x = 0$; x-coordinate of vertex: 0

1b.

x	$f(x)$
−2	24
−1	6
0	0
1	6
2	24

1c.

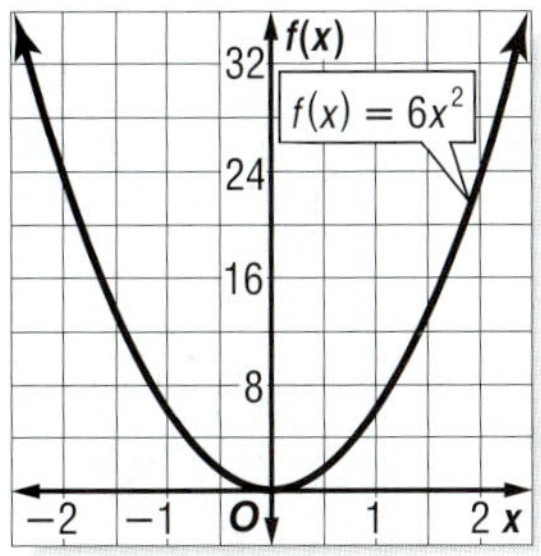

2a. y-intercept: 0; axis of symmetry: $x = 0$; x-coordinate of vertex: 0

2b.

x	$f(x)$
−2	−4
−1	−1
0	0
1	−1
2	−4

2c.

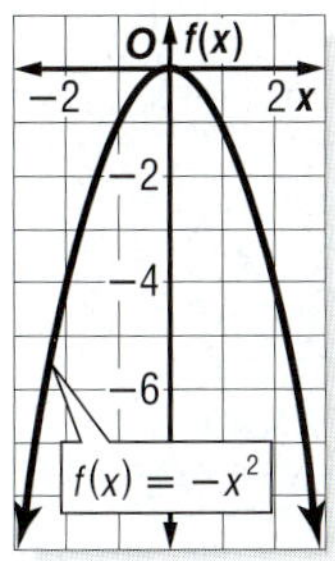

3a. y-intercept: 5; axis of symmetry: $x = 0$; x-coordinate of vertex: 0

3b.

x	$f(x)$
−2	9
−1	6
0	5
1	6
2	9

3c.

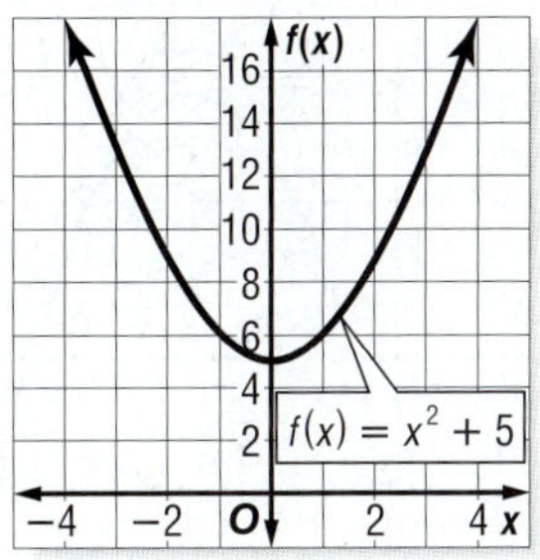

4a. y-intercept: −2; axis of symmetry: $x = 0$; x-coordinate of vertex: 0

4b.

x	$f(x)$
−2	−6
−1	−3
0	−2
1	−3
2	−6

4c.

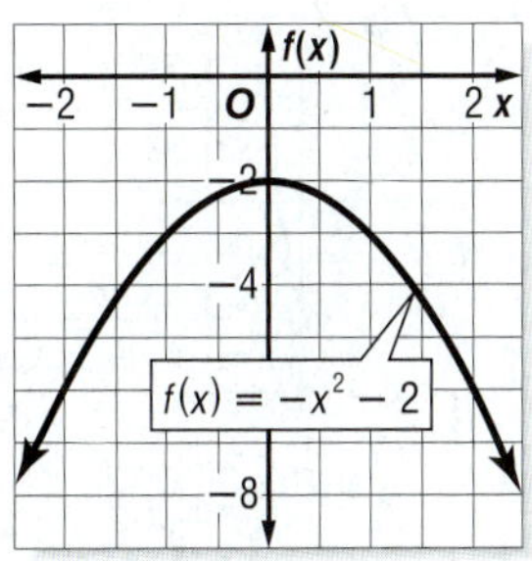

5a. y-intercept: 1; axis of symmetry: $x = 0$; x-coordinate of vertex: 0

5b.

x	$f(x)$
−2	9
−1	3
0	1
1	3
2	9

5c.

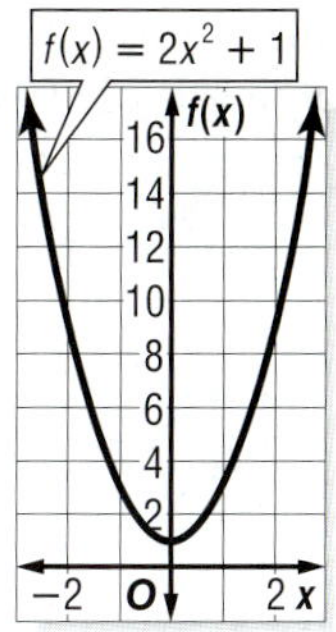

6a. y-intercept: 0; axis of symmetry: $x = 1$; x-coordinate of vertex: 1

6b.

x	$f(x)$
−1	−9
0	0
1	3
2	0
3	−9

6c.

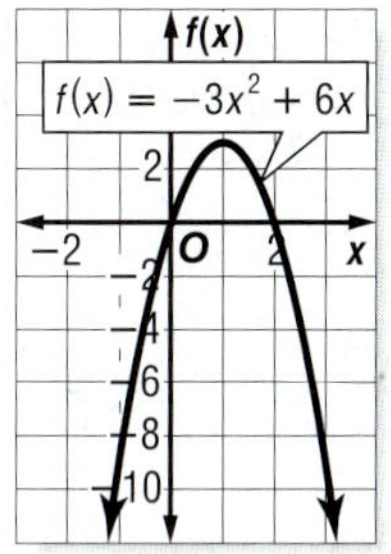

7a. y-intercept: −3; axis of symmetry: $x = -3$; x-coordinate of vertex: −3

7b.

x	$f(x)$
−7	4
−5	−8
−3	−12
−1	−8
1	4

7c.

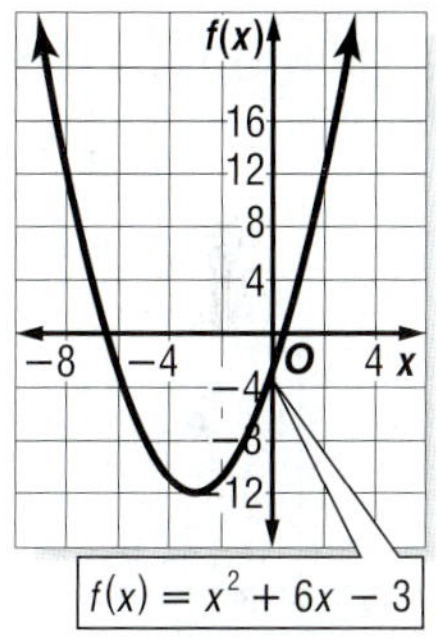

8a. y-intercept: −8; axis of symmetry: $x = 1$; x-coordinate of vertex: 1

8b.

x	$f(x)$
−2	0
0	−8
1	−9
2	−8
4	0

8c.

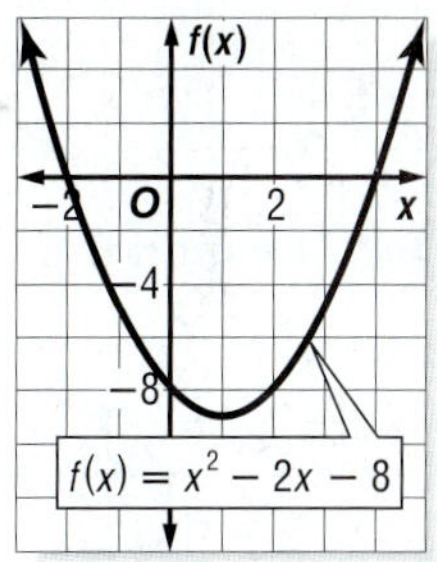

9a. y-intercept: 12; axis of symmetry: $x = -1$; x-coordinate of vertex: −1

9b.

x	$f(x)$
−3	3
−2	12
−1	15
0	12
1	3

9c.

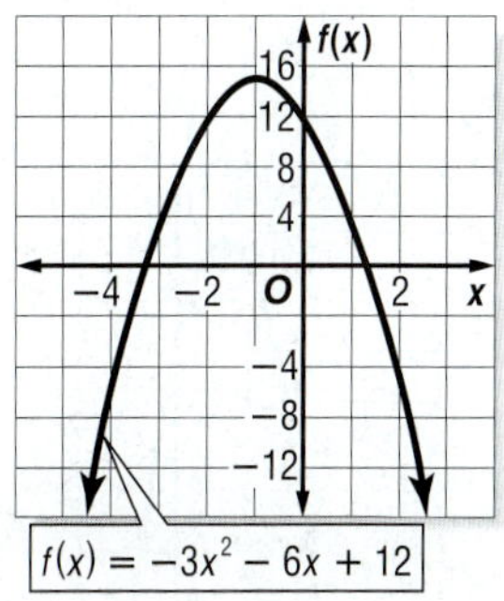

10a. *y*-intercept: −6; axis of symmetry: $x = -2.5$; *x*-coordinate of vertex: −2.5

10b.

x	*f*(*x*)
−6	0
−5	−6
−2.5	−12.25
0	−6
1	0

10c.

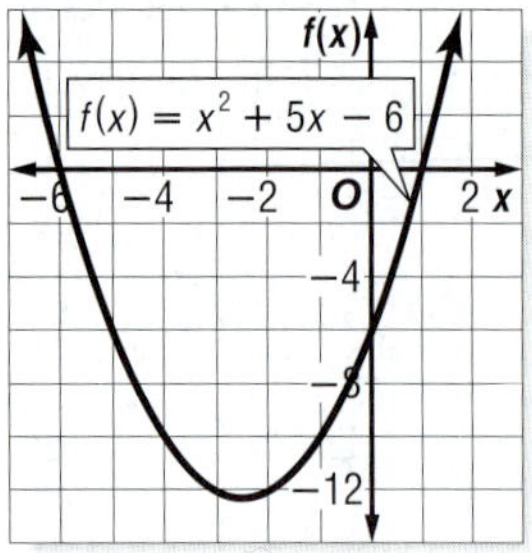

11a. *y*-intercept: −4; axis of symmetry: $x = -1.75$; *x*-coordinate of vertex: −1.75

11b.

x	*f*(*x*)
−4	0
−3.5	−4
−1.75	−10.125
0	−4
0.5	0

11c.

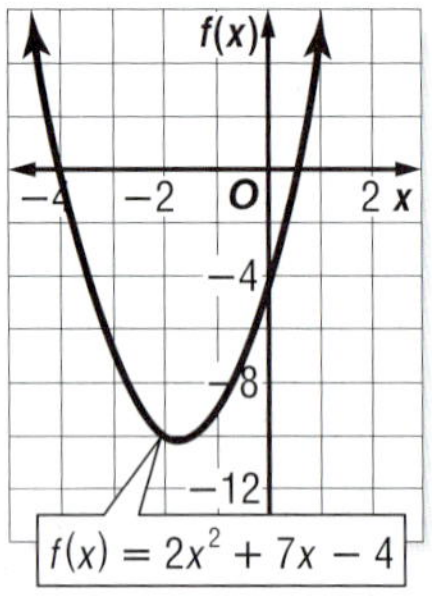

12a. *y*-intercept: 1; axis of symmetry: $x = 1$; *x*-coordinate of vertex: 1

12b.

x	*f*(*x*)
−1	−14
0	1
1	6
2	1
3	−14

12c.

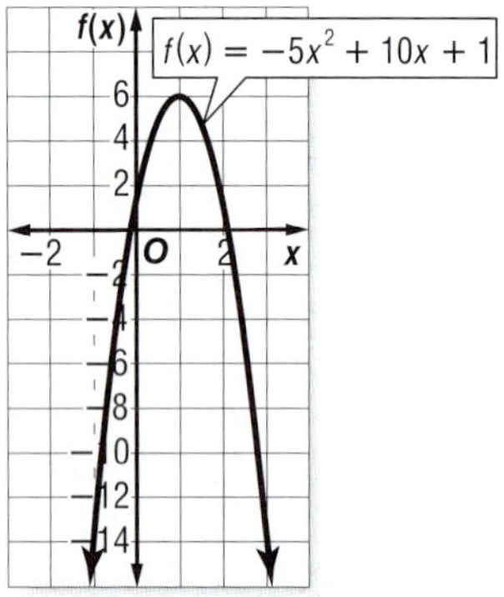

Page 841, Lesson 6-6

9.

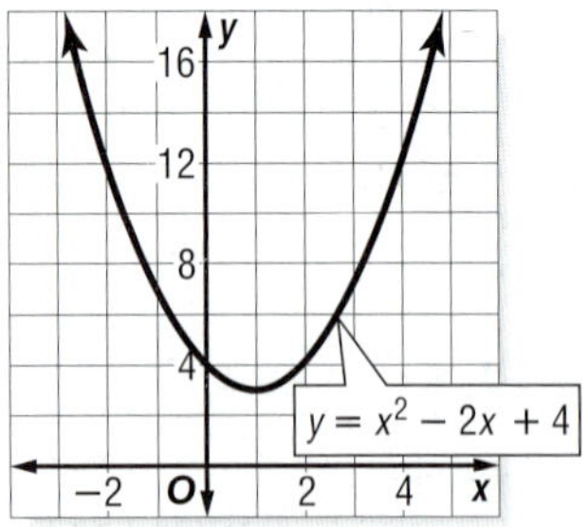

10.

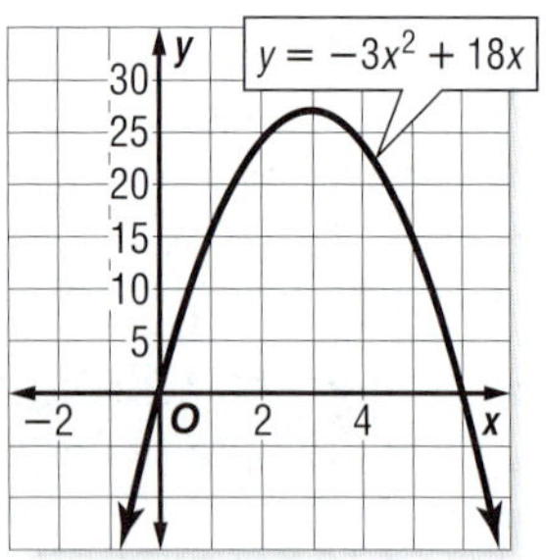

11.

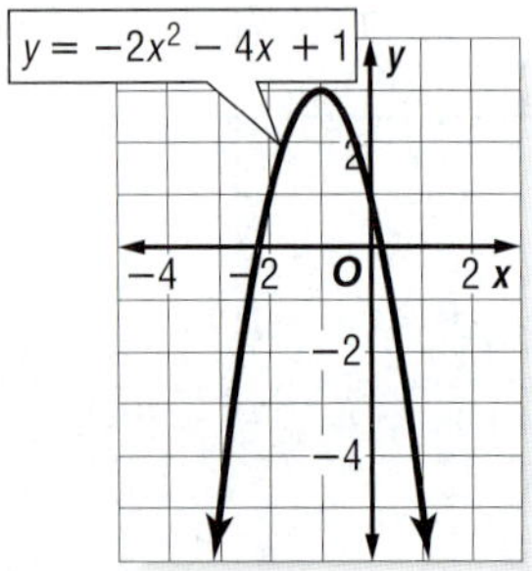

12.

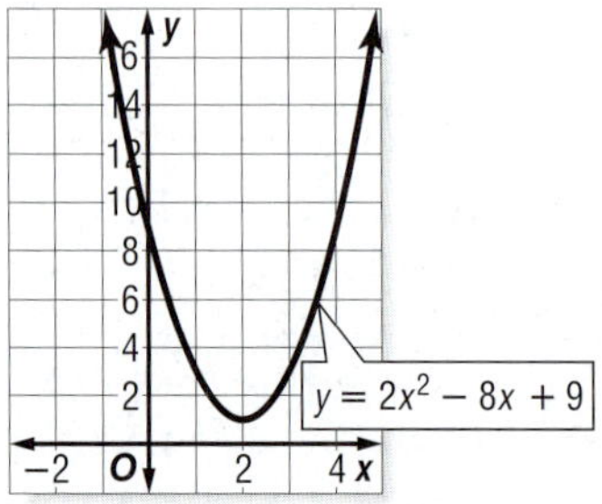

13.

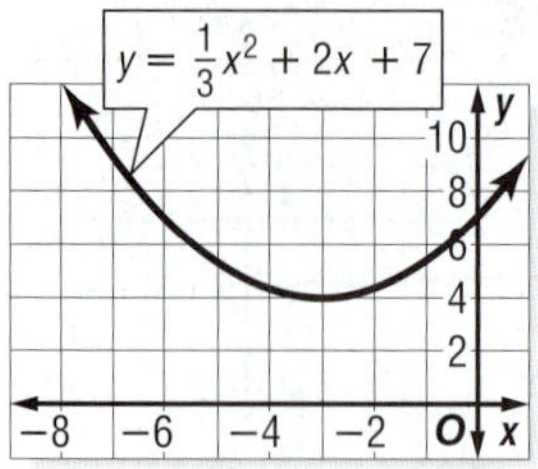

14.

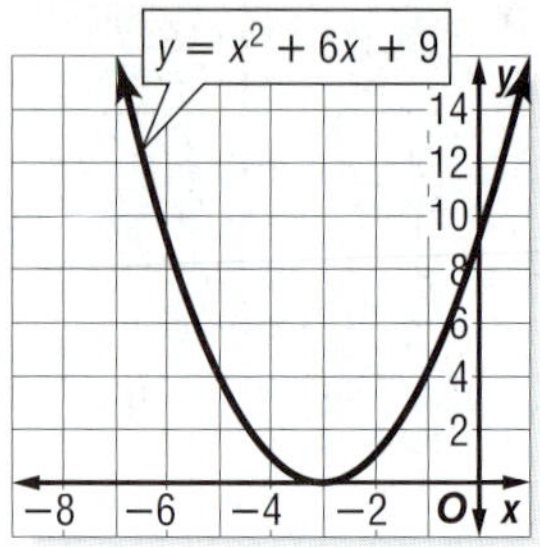

15.

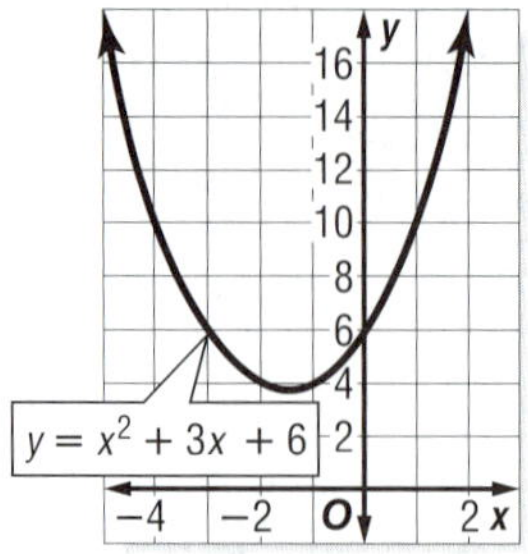

16.

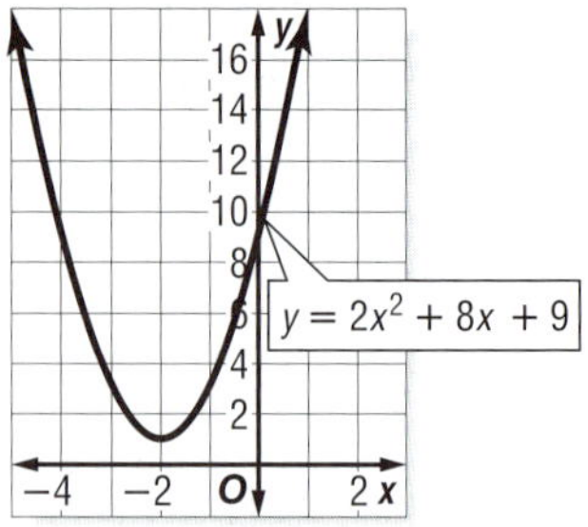

17.

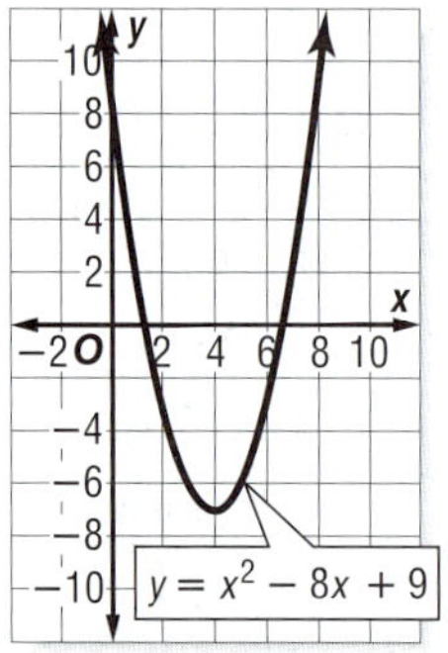

18.

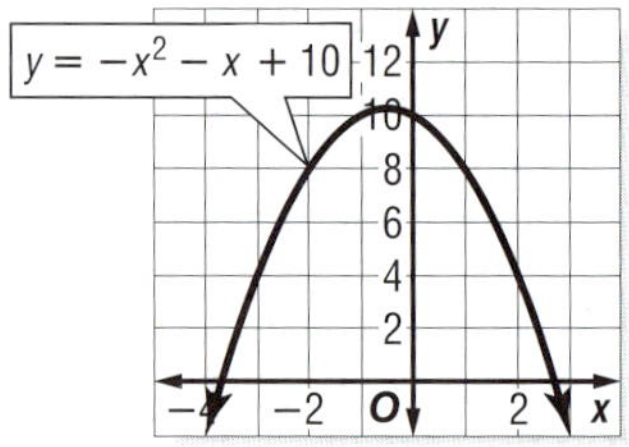

19.

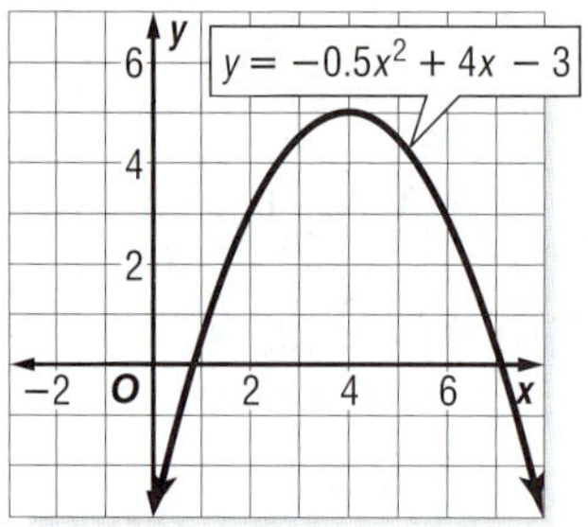

20.

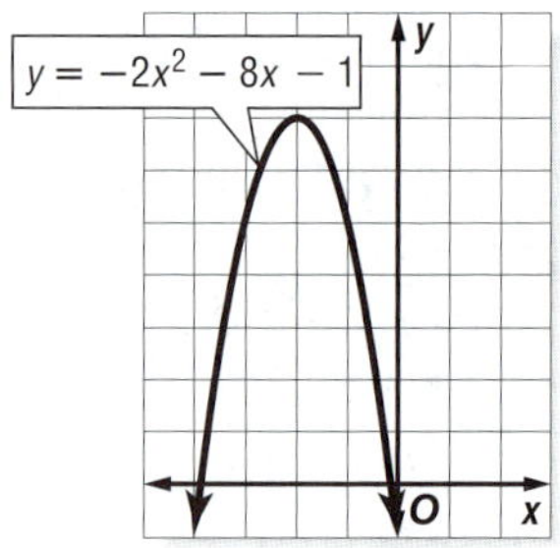

Page 841, Lesson 6-7

1.

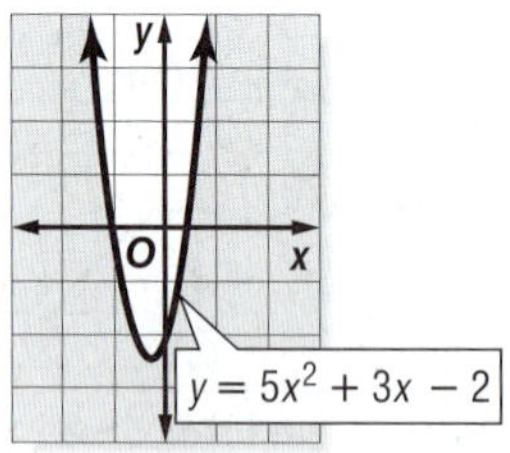

2.

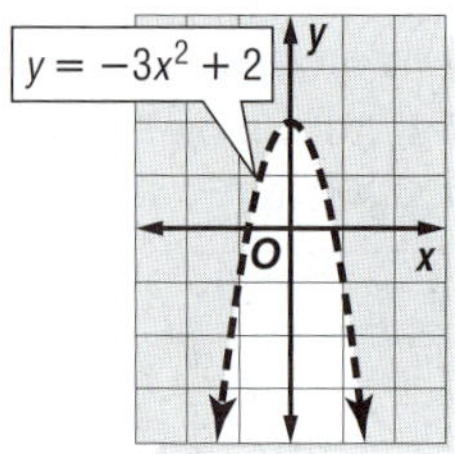

3.

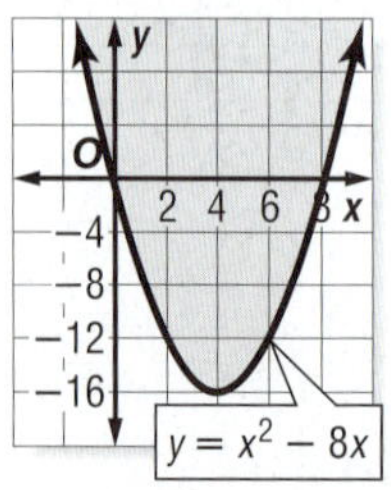

4.

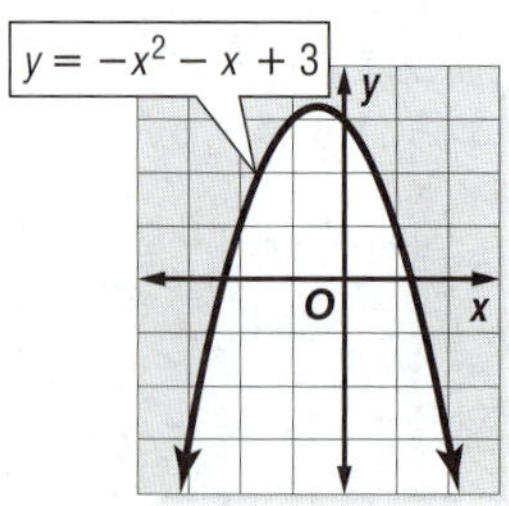

5.

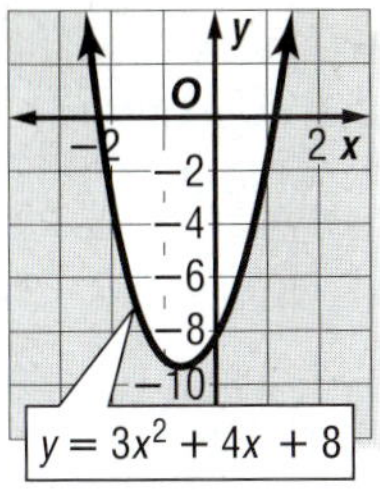

6.

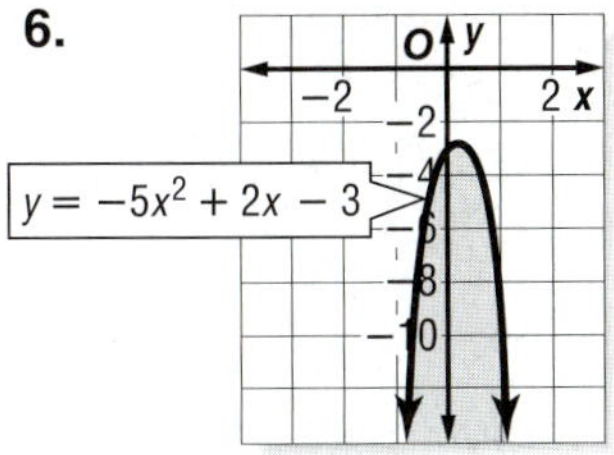

7.

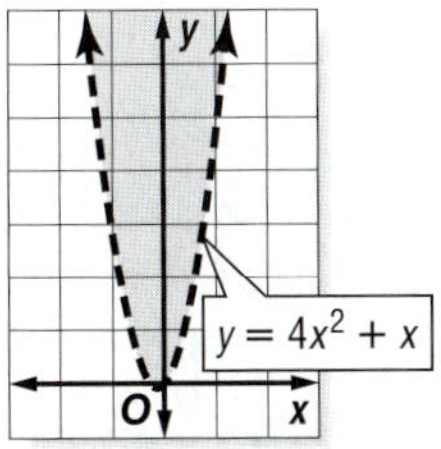

8.

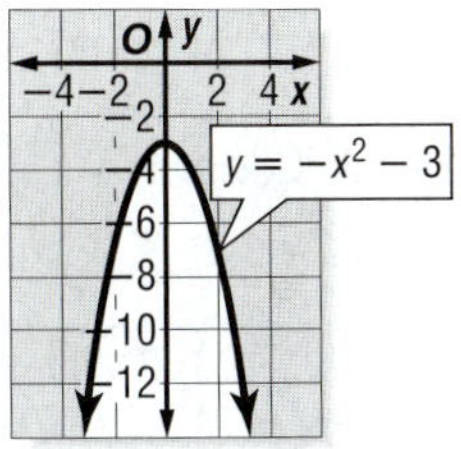

Page 842, Lesson 7-2

1a.

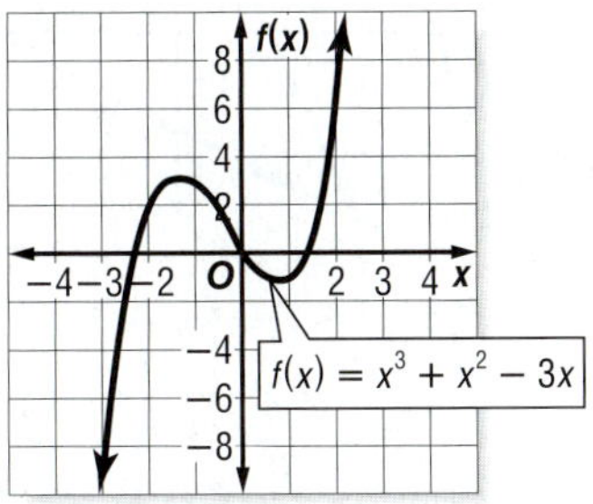

1b. between −3 and −2, at 0, between 1 and 2

1c. Sample answer: relative maximum at $x = -1.4$, relative minimum at $x = 0.7$

2a.

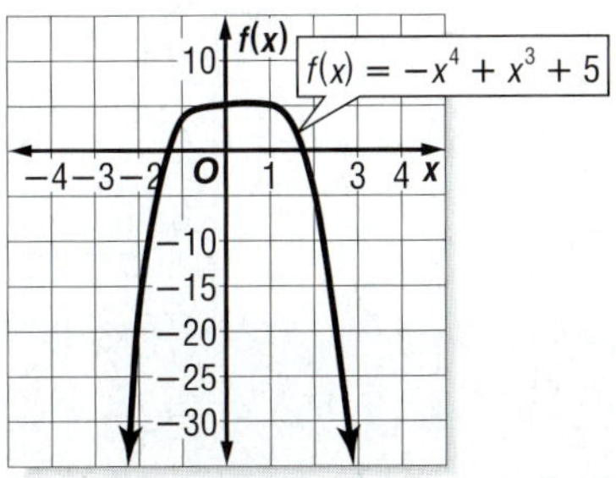

2b. between −2 and −1; between 1 and 2

2c. Sample answer: relative maximum at $x = 0.9$

3a.

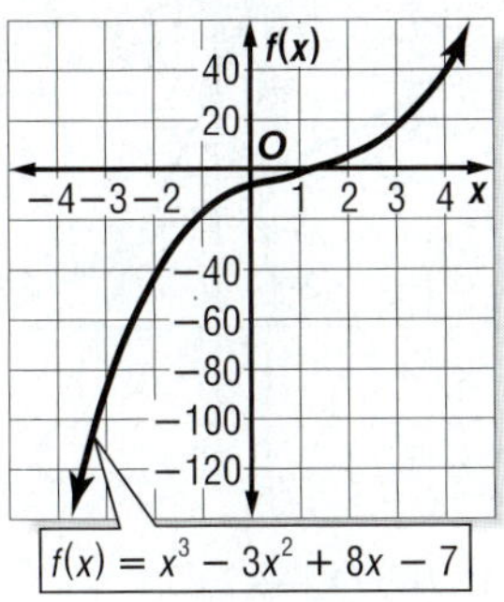

3b. between 1 and 2

3c. Sample answer: no relative maximum or minimum

4a.

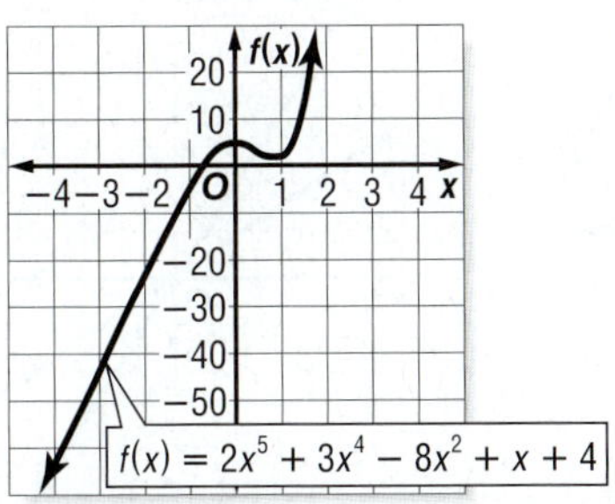

4b. between −1 and 0

4c. Sample answer: relative maximum at $x = 0.1$, relative minimum at $x = 0.9$

5a.

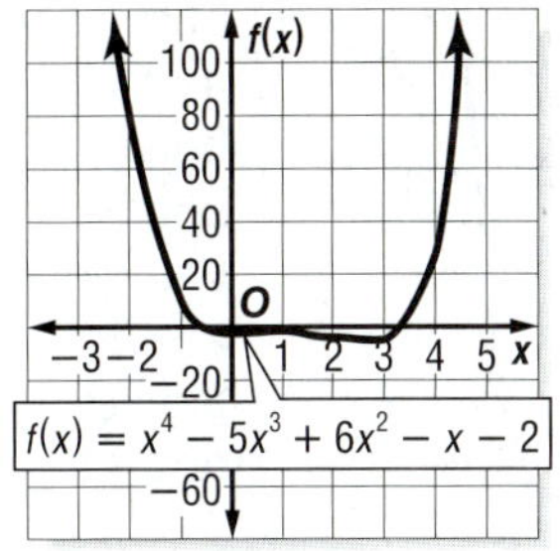

5b. between −1 and 0; between 3 and 4

5c. Sample answer: relative maximum at $x = 1.0$, relative minima at $x = 0.1$ and $x = 2.7$

6a.

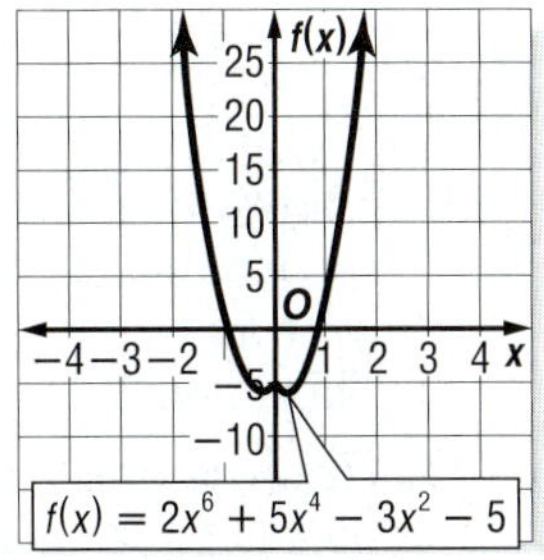

6b. between −2 and −1; between 1 and 2

6c. Sample answer: relative maximum at $x = 0$, relative minima at $x = -0.5$ and $x = 0.5$

7a.

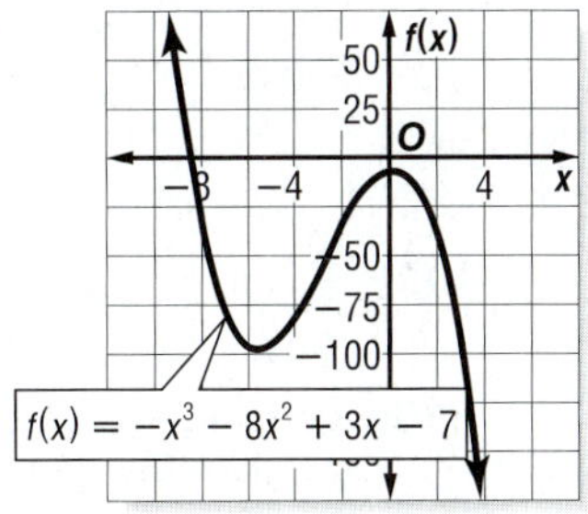

7b. between −9 and −8

7c. Sample answer: relative maximum at $x = 0.2$, relative minimum at $x = -5.5$

8a.

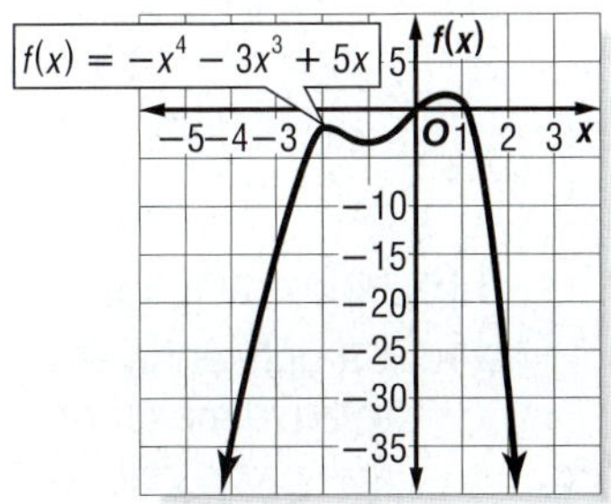

8b. at $x = 0$; between 1 and 2

8c. Sample answer: relative maxima at $x = -1.9$ and $x = 0.7$, relative minimum at $x = -1$

9a.

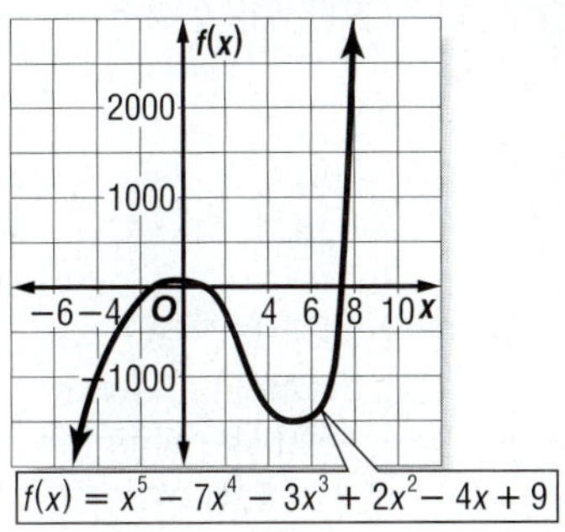

9b. between −2 and −1; 0 and 1; 7 and 8

9c. Sample answer: relative maximum at $x = -0.7$, relative minimum at $x = 5.9$

10a.

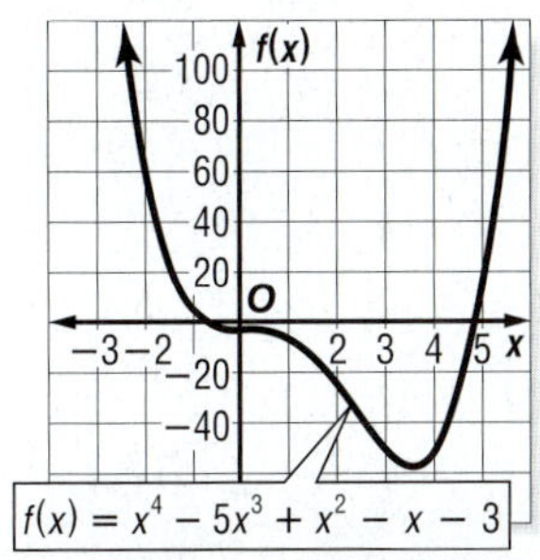

10b. between −1 and 0; between 4 and 5

10c. Sample answer: relative minimum at $x = 3.6$

11a.

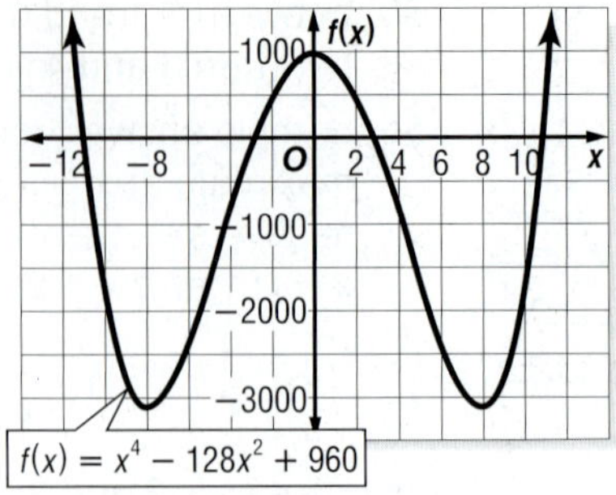

11b. between −11 and −10; between −3 and −2; between 2 and 3; between 10 and 11

11c. Sample answer: relative maximum at $x = 0$, relative minima at $x = -8$ and $x = 8$

12a.

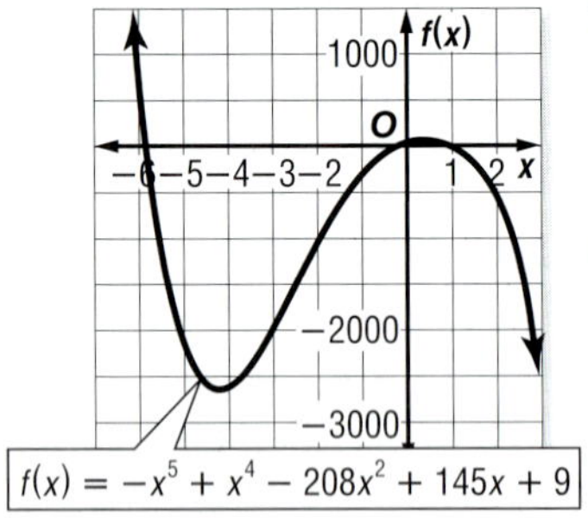

12b. between −6 and −5; between −1 and 0; between 0 and 1

12c. Sample answer: relative maximum at $x = 0.3$, relative minimum at $x = -4.2$

13a.

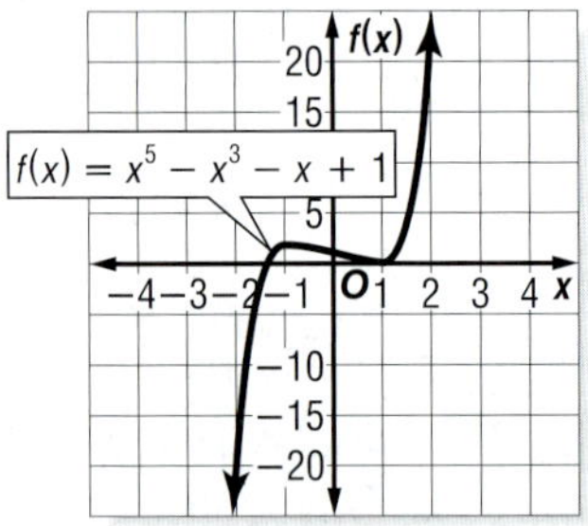

13b. $x = 1$; between −2 and −1

13c. Sample answer: relative maximum at $x = -0.9$, relative minimum at $x = 0.9$

14a.

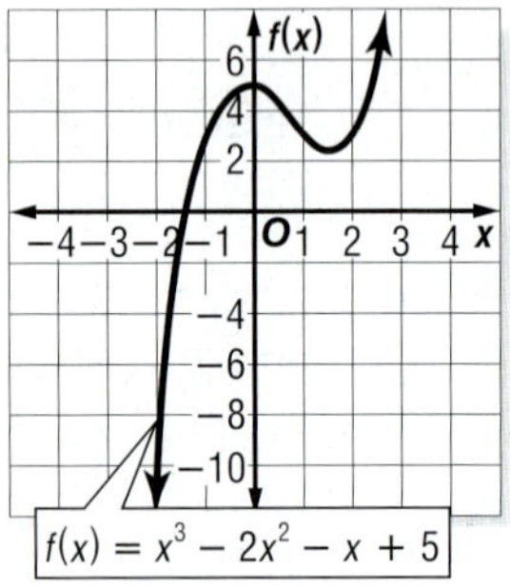

14b. between −2 and −1

14c. Sample answer: relative maximum at $x = -0.2$, relative minimum at $x = 1.5$

15a.

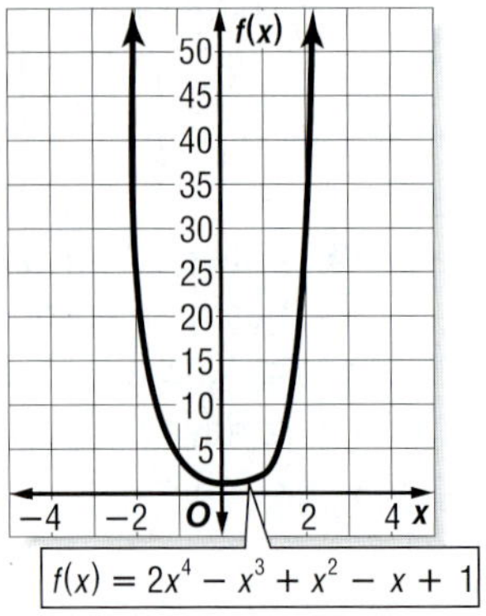

15b. no real roots

15c. Sample answer: relative minimum at $x = 0.4$

16a.

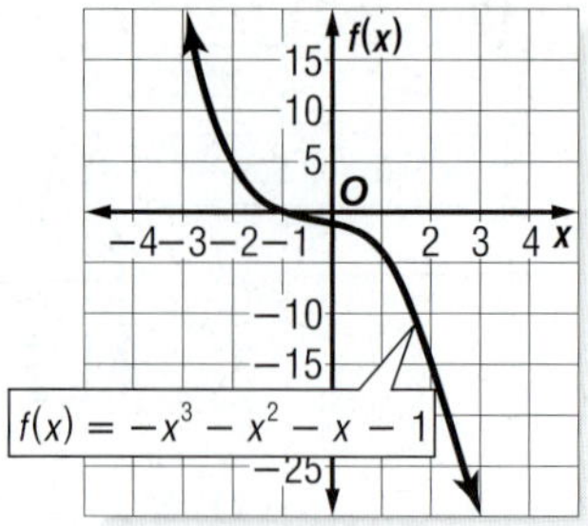

16b. $x = -1$

16c. no maxima or minima

Page 844, Lesson 7-8

4. $f^{-1}(x) = x + 7$

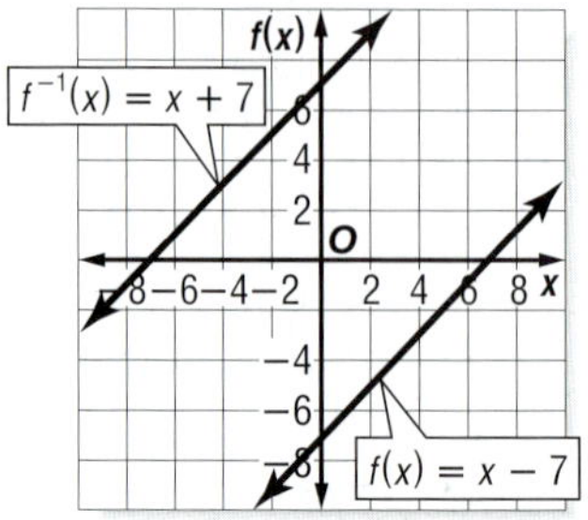

5. $y = \frac{x - 8}{2}$

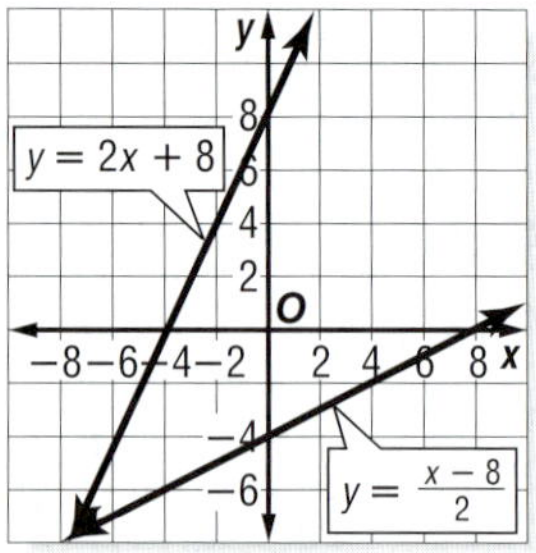

6. $g^{-1}(x) = \frac{x + 8}{3}$

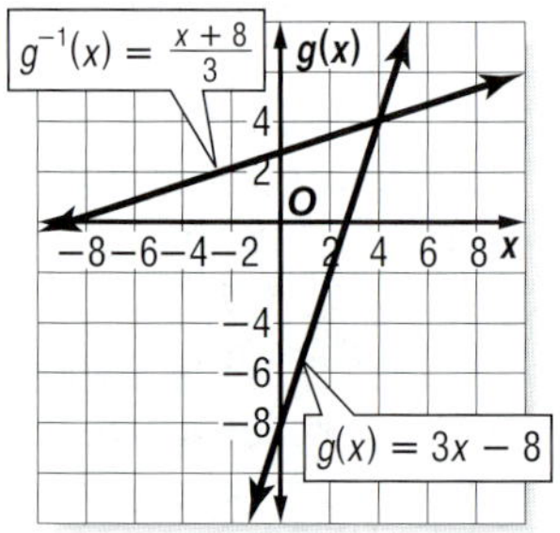

7. $h^{-1}(x) = 5x - 5$

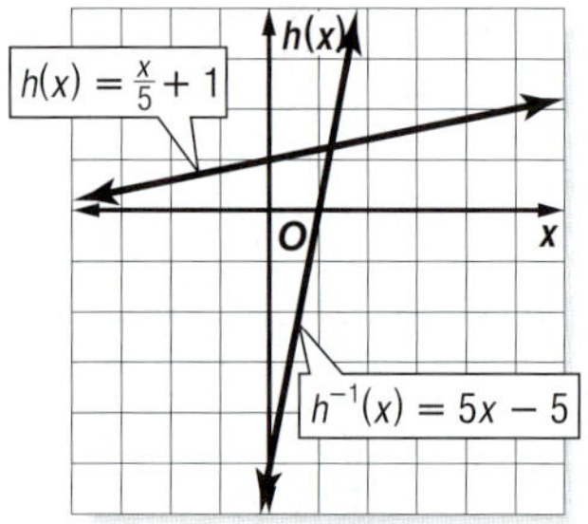

8. $x = -2$

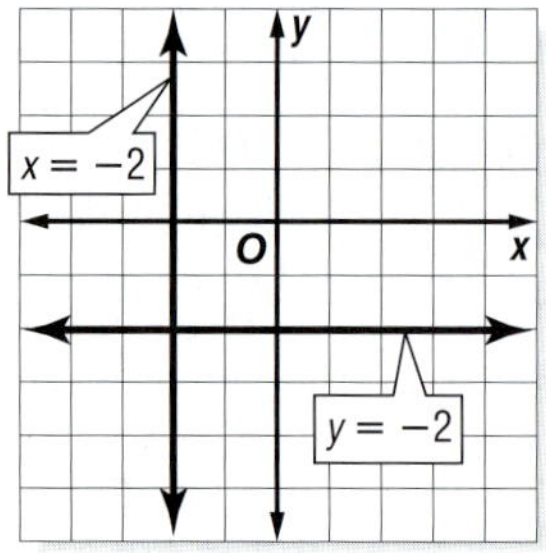

9. $g^{-1}(x) = \frac{x - 5}{-2}$

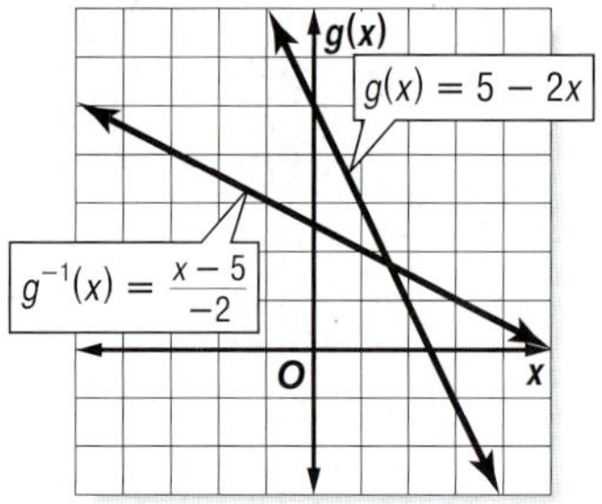

10. $y = \frac{x + 6}{-5}$

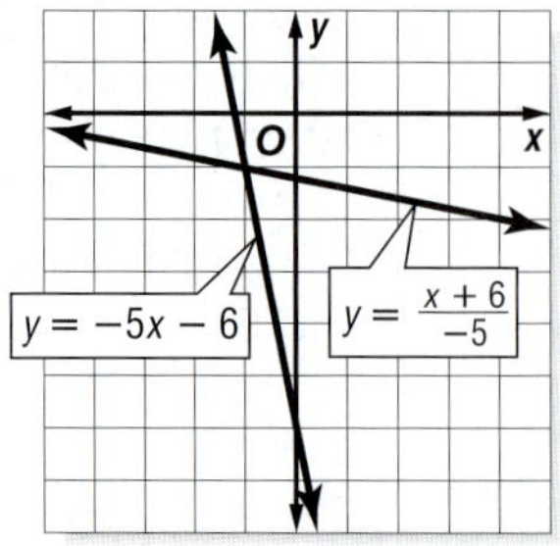

11. $h^{-1}(x) = -\frac{3}{2}x$

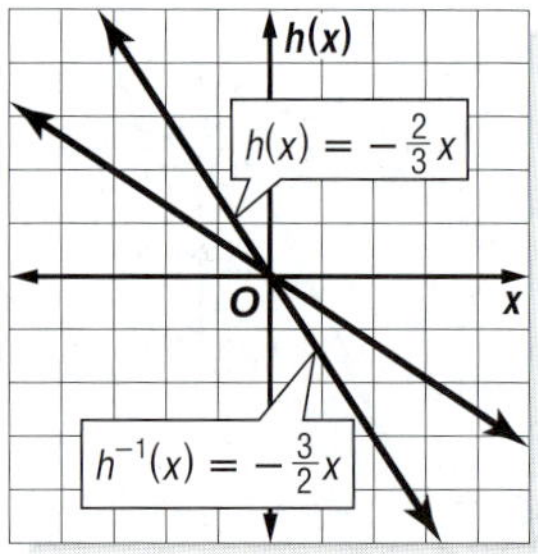

12. $y = 3x + 5$

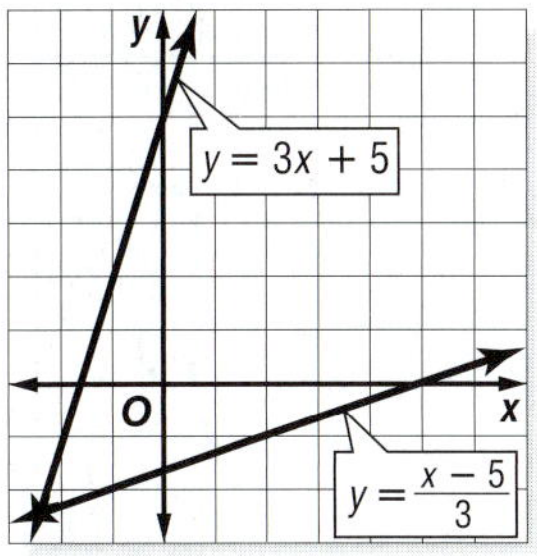

13. $y = 2x + 2$

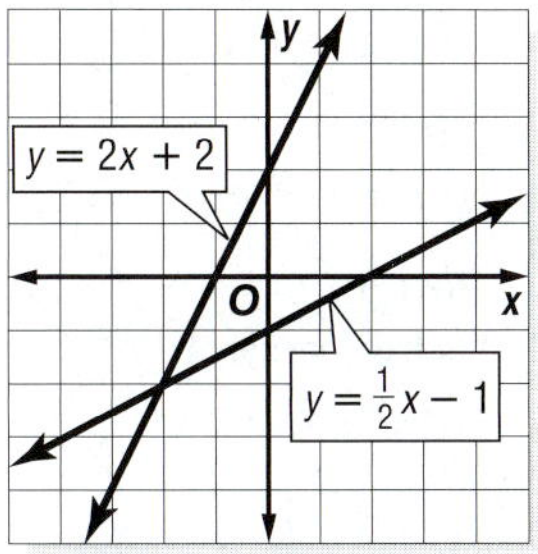

14. $f^{-1}(x) = \dfrac{4x - 8}{3}$

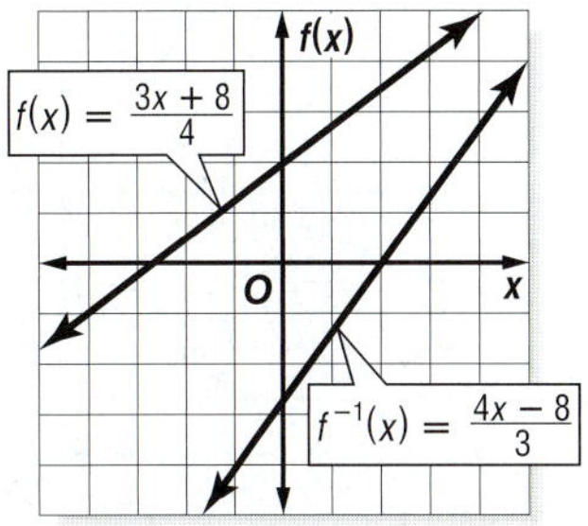

15. $g^{-1}(x) = \dfrac{3x + 1}{2}$

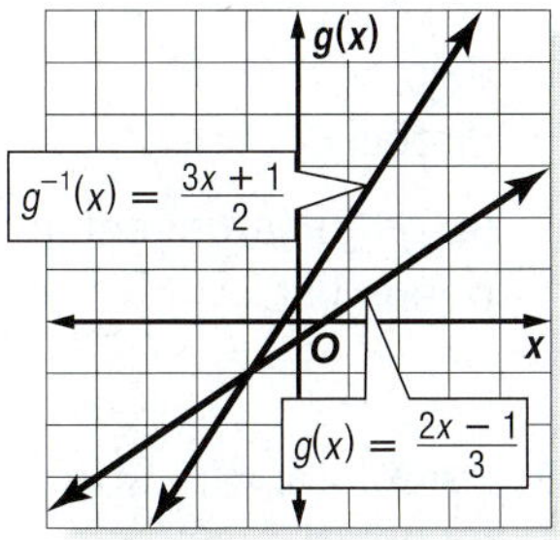

Page 844, Lesson 7-9

1. D: $x \geq 4$, R: $y \geq 0$

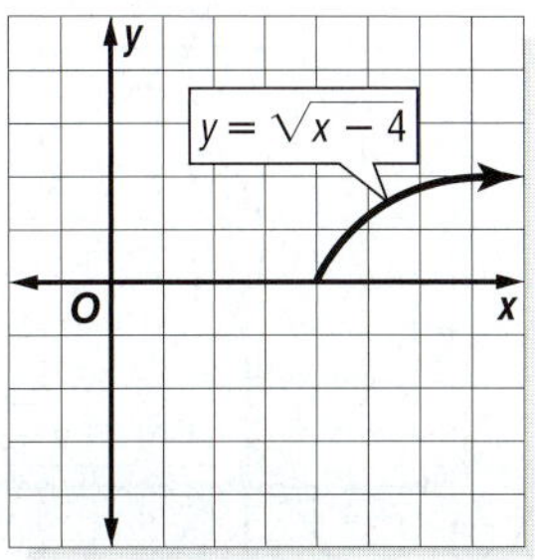

2. D: $x \geq -3$, R: $y \geq -1$

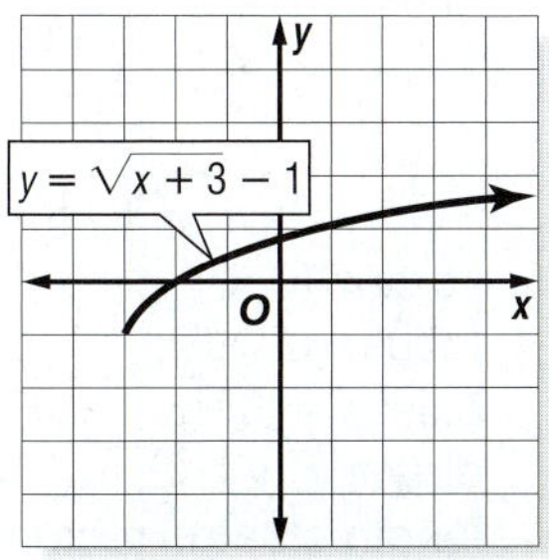

3. D: $x \geq -2$, R: $y \geq 0$

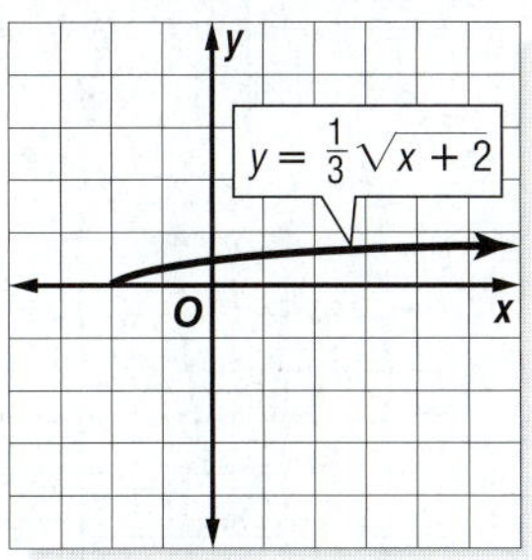

4. D: $x \geq -2.5$, R: $y \geq 0$

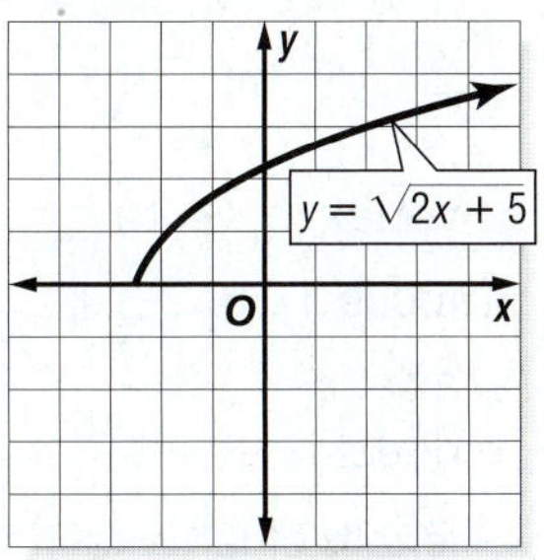

5. D: $x \geq 0$, R: $y \leq 0$

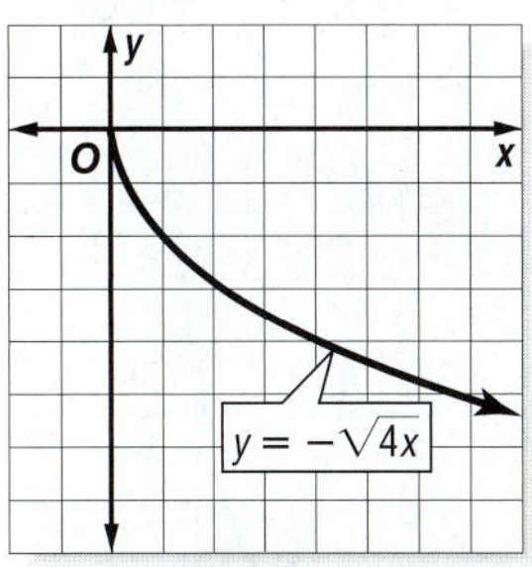

6. D: $x \geq 0$, R: $y \geq 0$

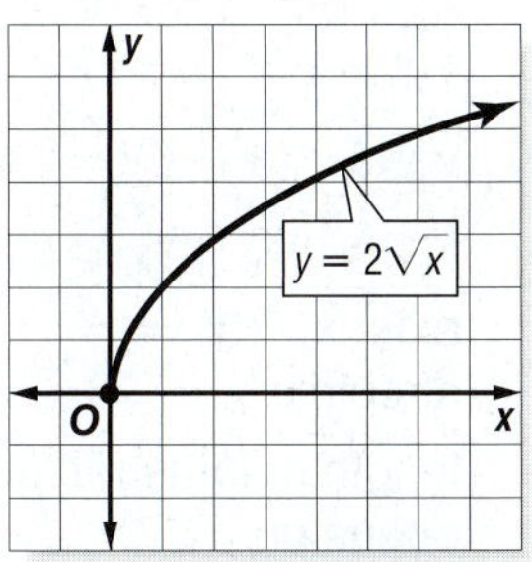

7. D: $x \geq 0$, R: $y \leq 0$

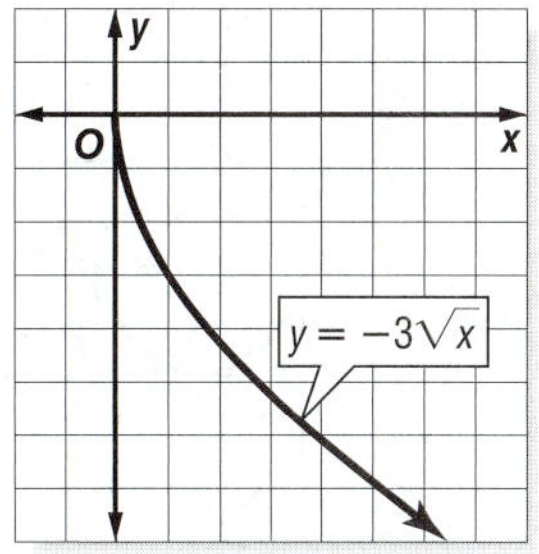

8. D: $x \geq 0$, R: $y \geq 5$

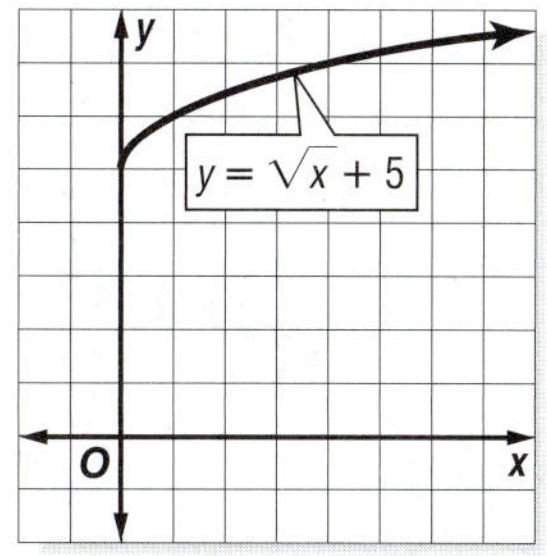

9. D: $x \geq 0$, R: $y \geq -1$

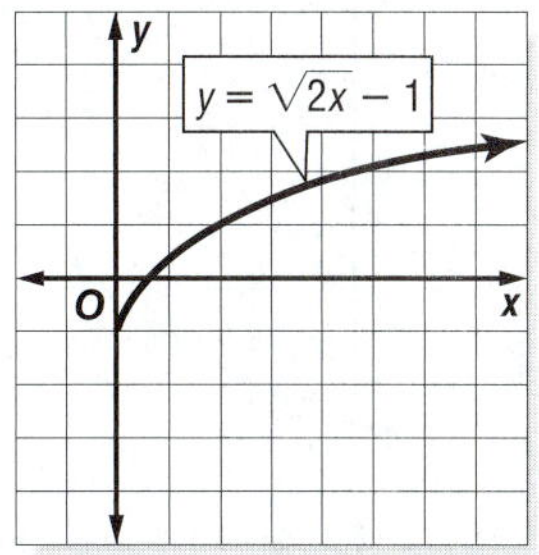

10. D: $x \geq 0$, R: $y \geq 1$

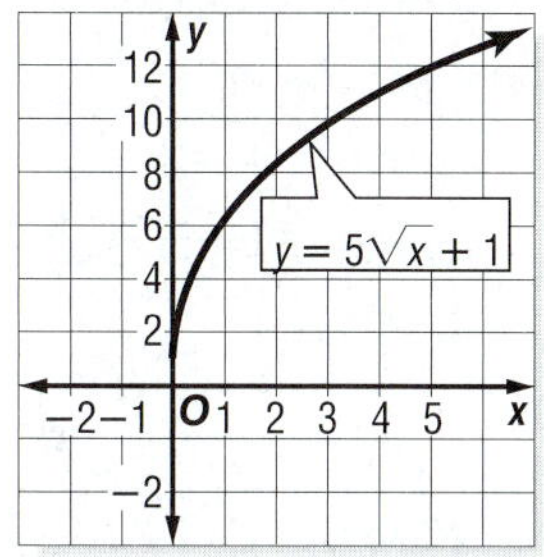

11. D: $x \geq -1$, R: $y \geq -2$

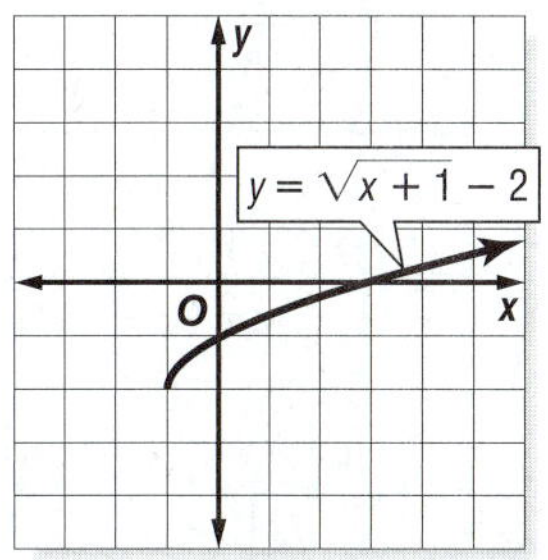

12. D: $x \geq -3$, R: $y \leq 6$

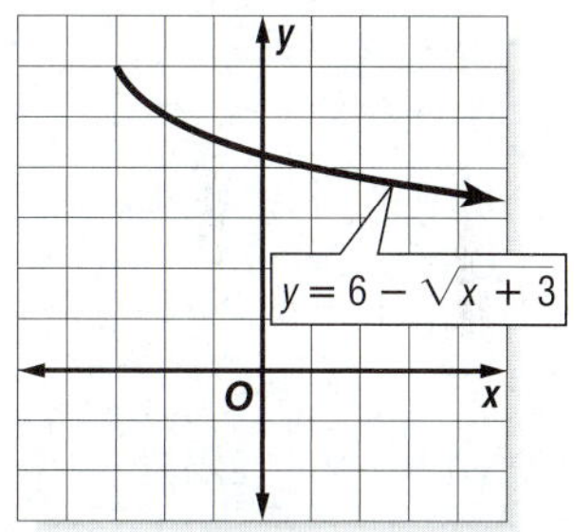

Page 845, Lesson 8-2

4. vertex: $(0, -4)$;
focus: $(0, -3.75)$;
axis: $x = 0$;
directrix: $y = -4.25$;
length of latus rectum: 1;
opens up

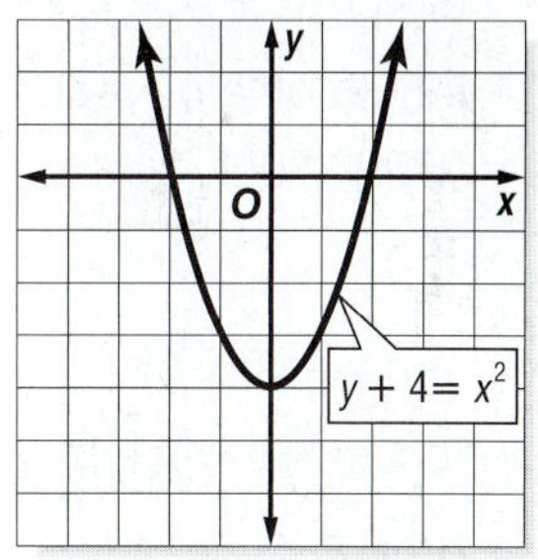

5. vertex: $(-2, 0)$;
focus: $\left(-2, \frac{1}{20}\right)$;
axis: $x = -2$;
directrix: $y = -\frac{1}{20}$;
length of latus rectum: $\frac{1}{5}$;
opens up

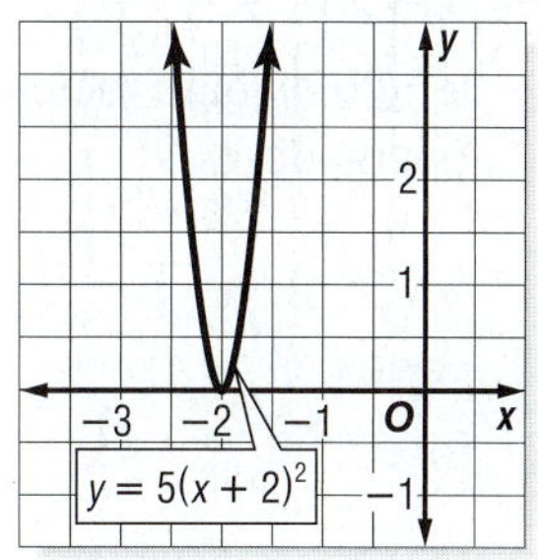

6. vertex: (1, −2);
focus: $\left(1, -1\frac{2}{3}\right)$;
axis: $x = 1$;
directrix: $y = -2\frac{1}{3}$;
length of latus rectum: $\frac{4}{3}$;
opens up

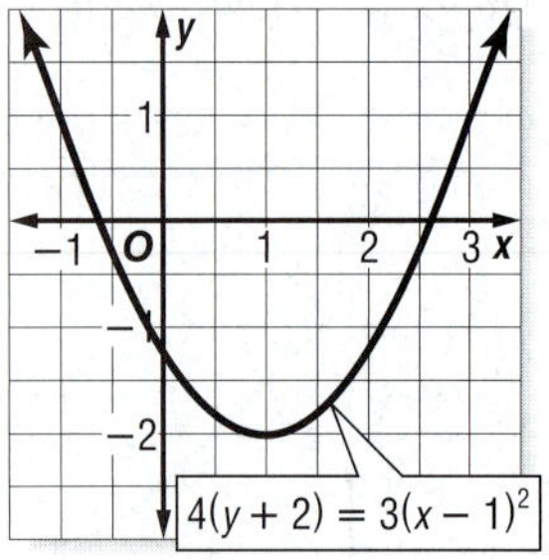

7. vertex: (3, 0);
focus: $\left(2\frac{7}{12}, 0\right)$;
axis: $y = 0$;
directrix: $x = 3\frac{5}{12}$;
length of latus rectum: $\frac{5}{3}$;
opens left

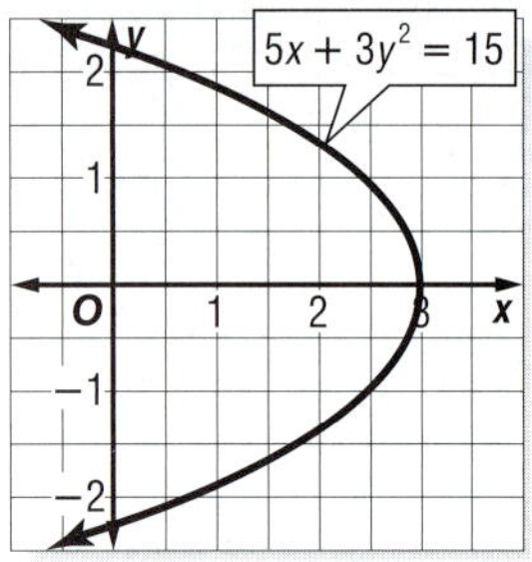

8. vertex: (2, −1);
focus: $\left(2, -\frac{7}{8}\right)$;
axis: $x = 2$;
directrix: $y = -1\frac{1}{8}$;
length of latus rectum: $\frac{1}{2}$;
opens up

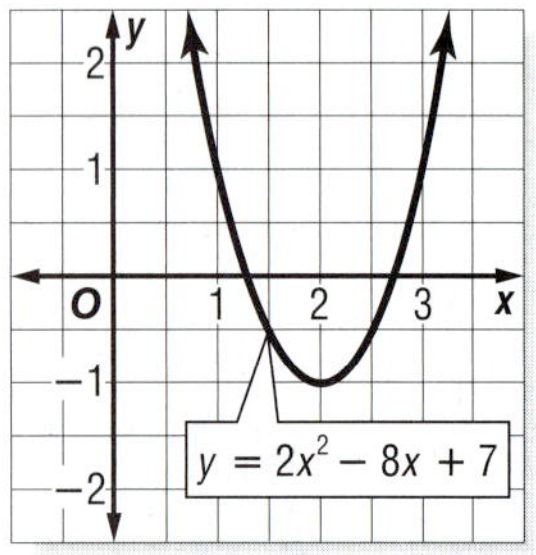

9. vertex: (−1, 2);
focus: $\left(-\frac{7}{8}, 2\right)$;
axis: $y = 2$;
directrix: $x = -1\frac{1}{8}$;
length of latus rectum: $\frac{1}{2}$;
opens right

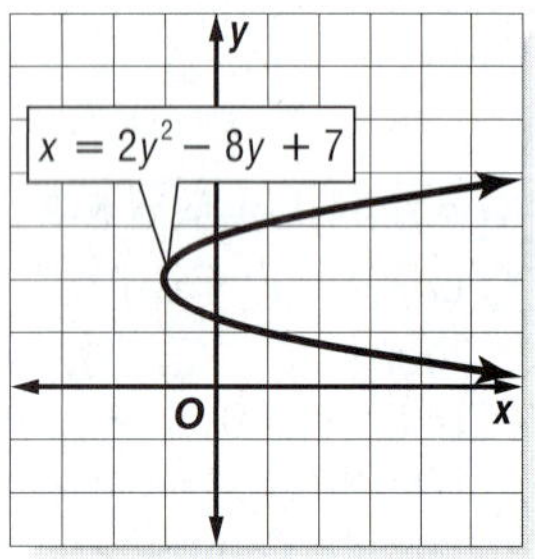

10. vertex: (8, −3);
focus: $\left(8, -2\frac{7}{12}\right)$;
axis: $x = 8$;
directrix: $y = -3\frac{5}{12}$;
length of latus rectum: $\frac{5}{3}$;
opens up

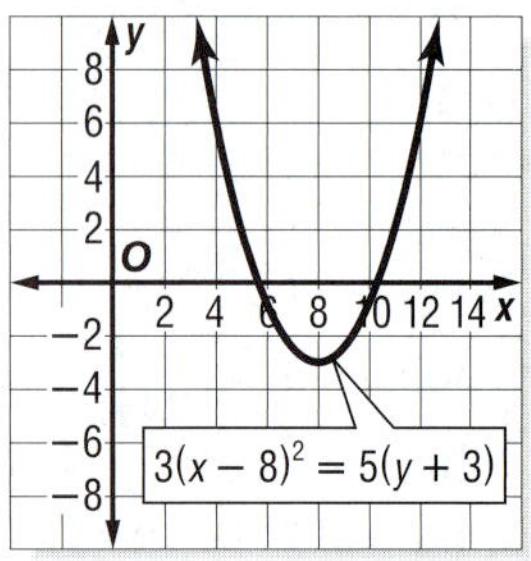

11. vertex: (1, −4);
focus: $\left(1\frac{1}{12}, -4\right)$;
axis: $y = -4$;
directrix: $x = \frac{11}{12}$;
length of latus rectum: $\frac{1}{3}$;
opens right

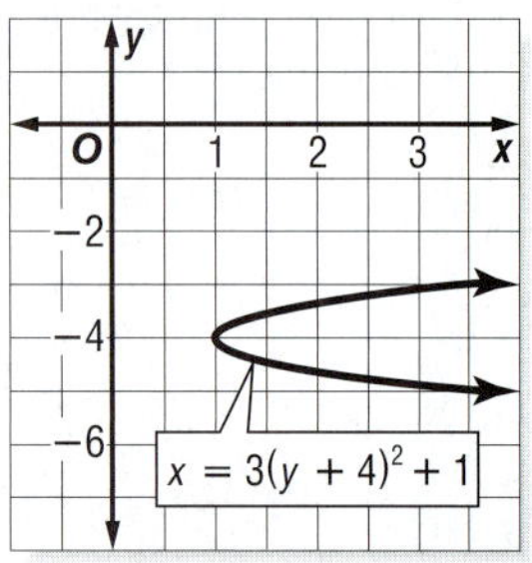

12. vertex: (−3, −7);
focus: $\left(-3, -7\frac{2}{5}\right)$;
axis: $x = -3$;
directrix: $y = -6\frac{3}{5}$;
length of latus rectum: $\frac{8}{3}$;
opens down

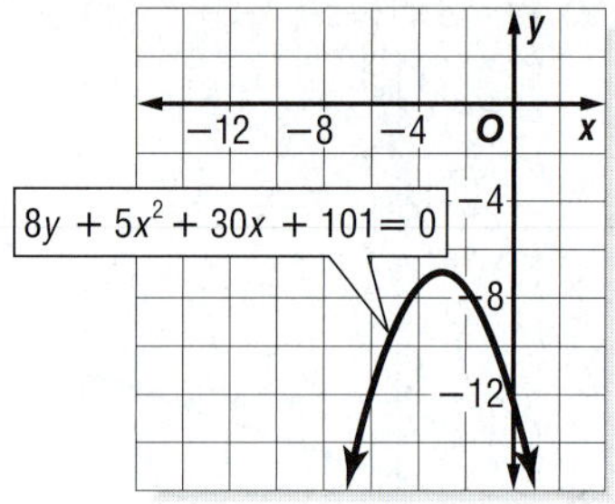

13. vertex: $\left(-3\frac{4}{5}, 4\right)$;
focus: $\left(-5\frac{1}{20}, 4\right)$;
axis: $y = 4$;
directrix: $x = -2\frac{11}{20}$;
length of latus rectum: 5;
opens left

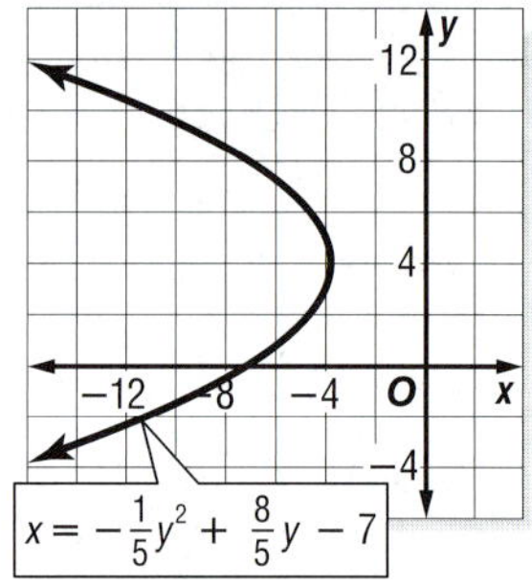

14. vertex: (5, 3);
focus: (6.5, 3);
axis: $y = 3$;
directrix: $x = 3.5$;
length of latus rectum: 6;
opens right

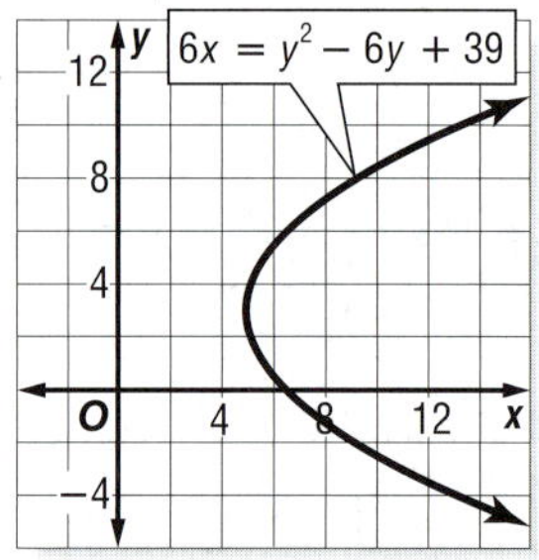

15. vertex: (0, 0);
focus: (0, −2);
axis: $x = 0$;
directrix: $y = 2$;
length of latus rectum: 8;
opens down

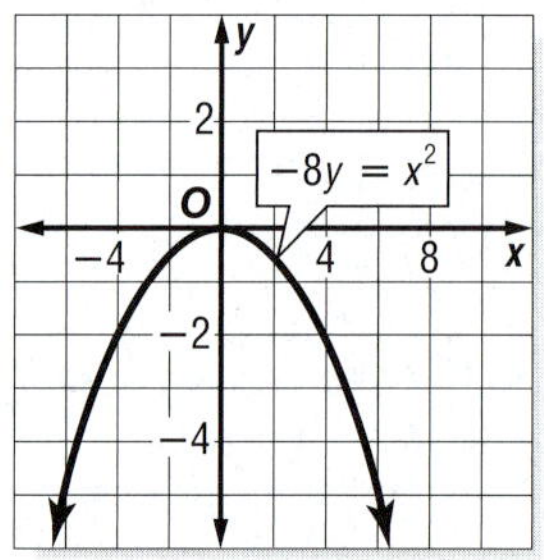

16. vertex: (−3, 2);
focus: $\left(-3, 2\frac{1}{16}\right)$;
axis: $x = -3$;
directrix: $y = 1\frac{15}{16}$;
length of latus rectum: $\frac{1}{4}$;
opens up

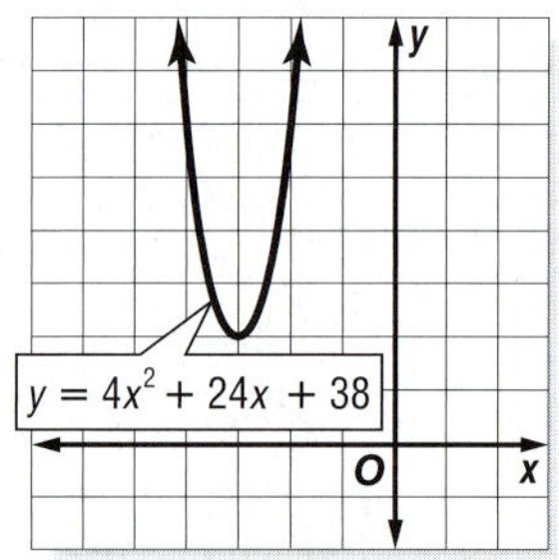

17. vertex: (3, −6);
focus: $\left(3, -5\frac{3}{4}\right)$;
axis: $x = 3$;
directrix: $y = -6\frac{1}{4}$;
length of latus rectum: 1;
opens up

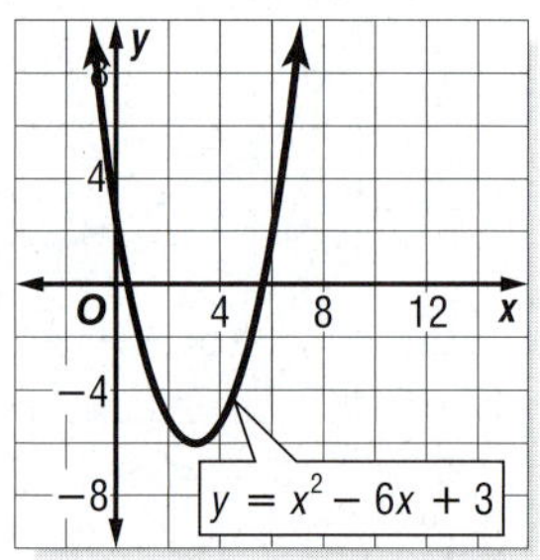

18. vertex: $(-2, -3)$;
focus: $\left(-2, -2\frac{3}{4}\right)$;
axis: $x = -2$;
directrix: $y = -3\frac{1}{4}$;
length of latus rectum: 1;
opens up

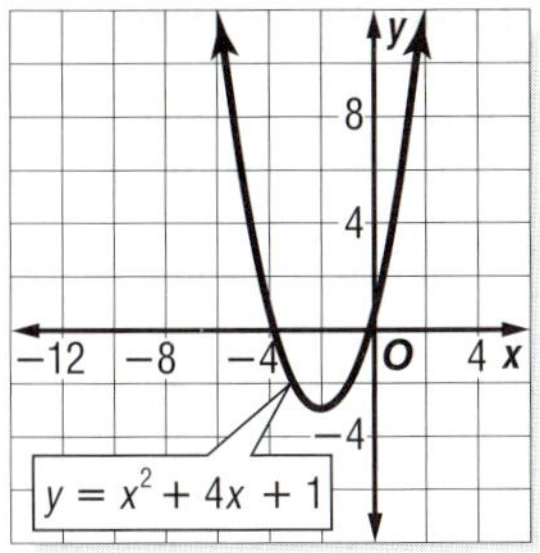

Page 845, Lesson 8-3

10. $(0, 0)$; 6

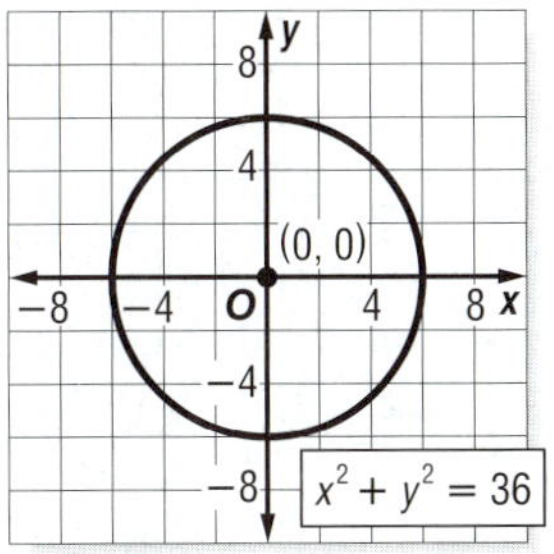

11. $\left(\frac{3}{4}, -\frac{2}{3}\right)$; $\frac{4\sqrt{2}}{7}$

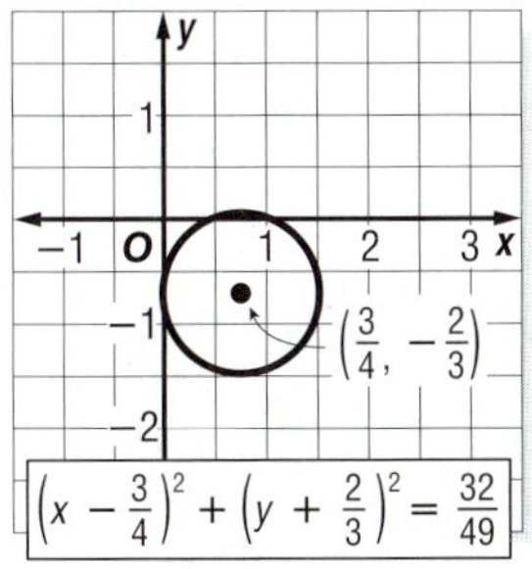

12. $(5, -4)$; 1

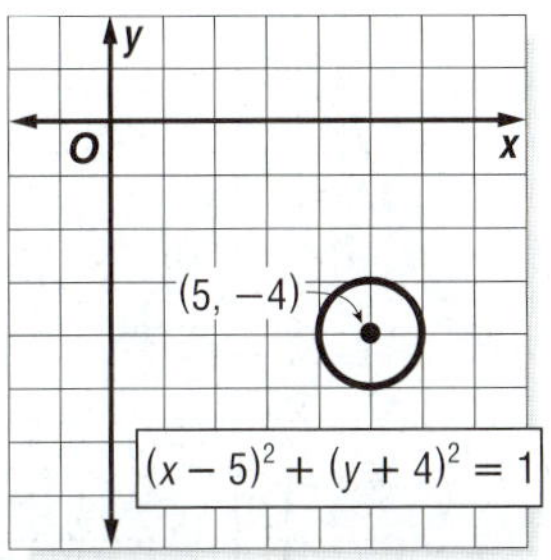

13. $(-1.5, 2.5)$; 3

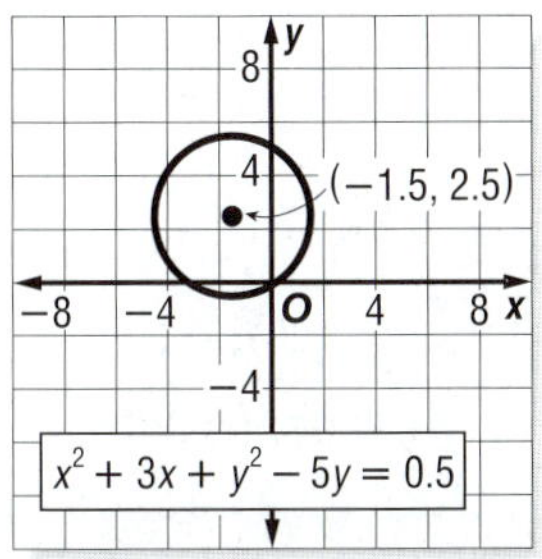

14. $(7, 0)$; 5

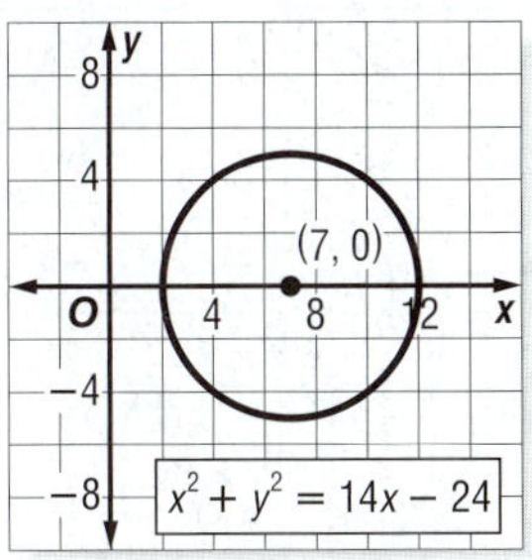

15. $(-1, 1)$; $\sqrt{2}$

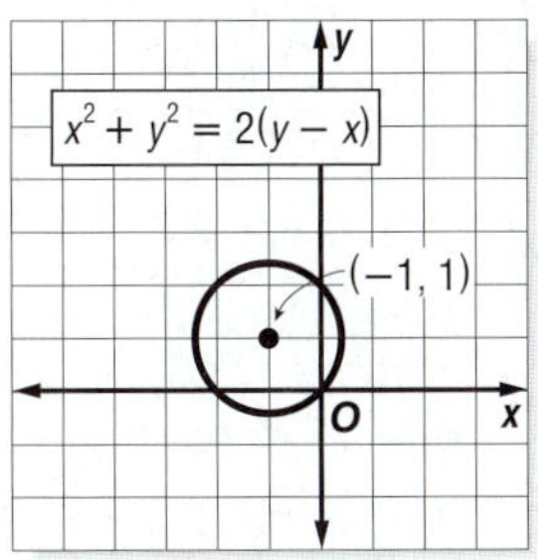

16. $(-5, \sqrt{3})$; 6

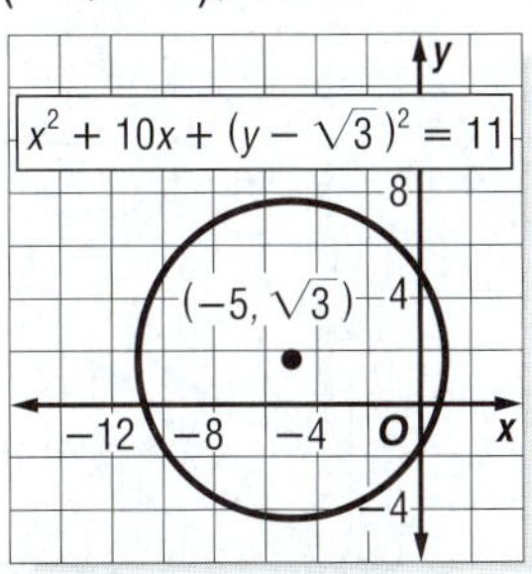

17. $(2, 0)$; $\sqrt{13}$

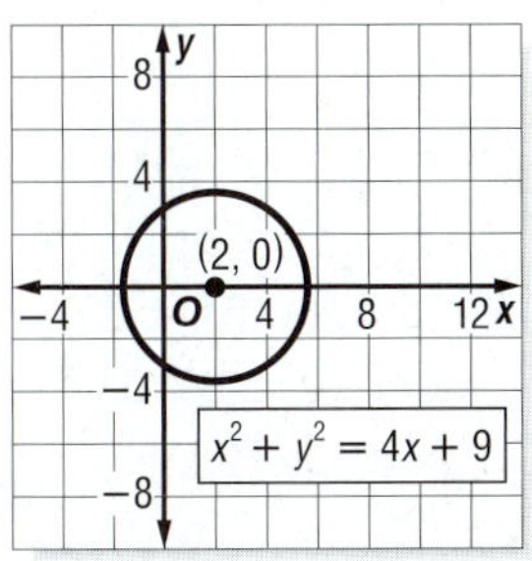

18. $(-6, 5)$; 4

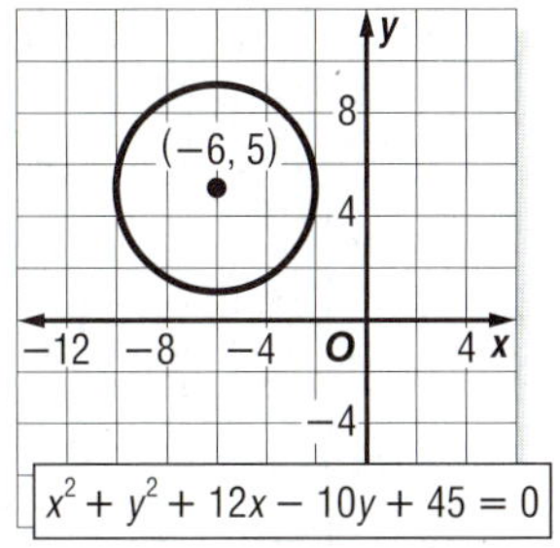

19. $(3, -2)$; 13

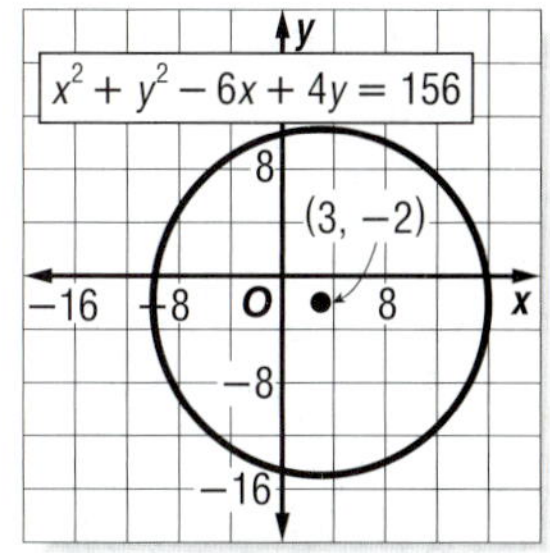

20. $(1, -3.5)$; $\sqrt{14.25}$

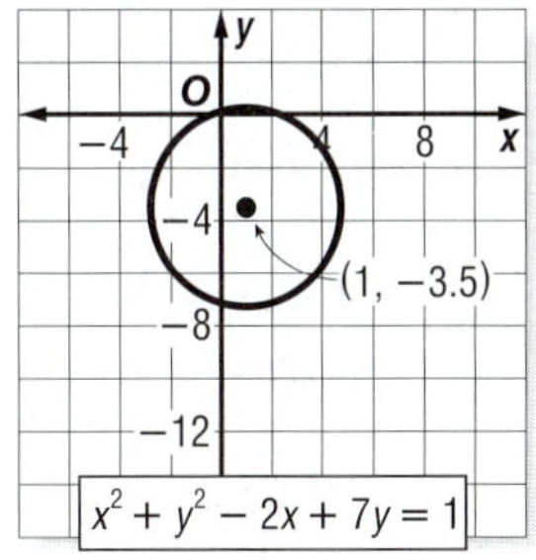

21. $(0.75, 1.25)$; 0.25

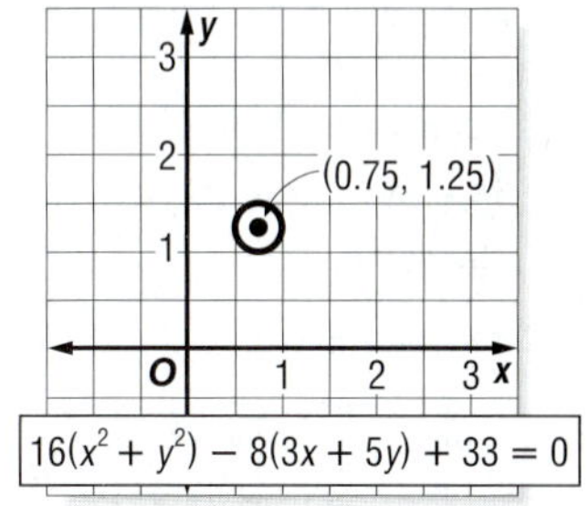

Page 846, Lesson 8-4

4. center: $(0, 0)$;
foci: $(0, \pm 3\sqrt{5})$;
major axis: 18;
minor axis: 12

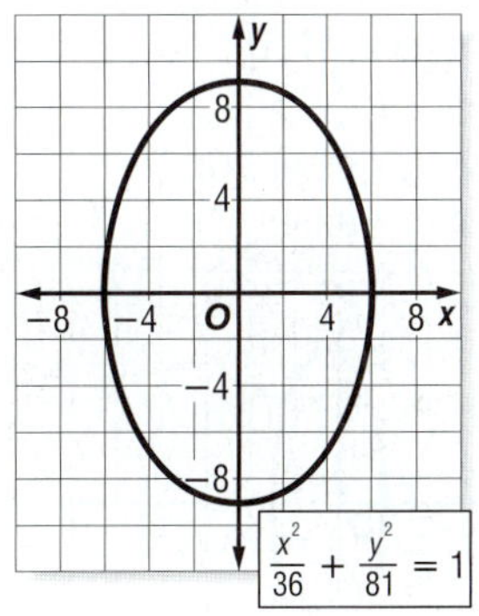

5. center: $(0, 5)$;
foci: $(\pm\sqrt{105}, 5)$;
major axis: 22;
minor axis: 8

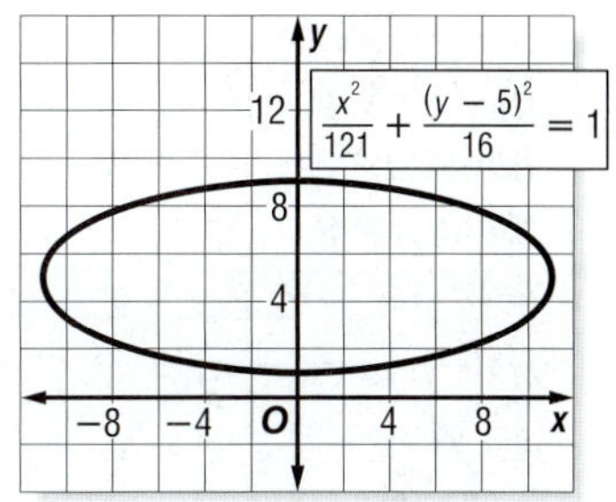

6. center: $(-2, -1)$;
foci: $(-2, -3)$, $(-2, 1)$;
major axis: 8;
minor axis: $4\sqrt{3}$

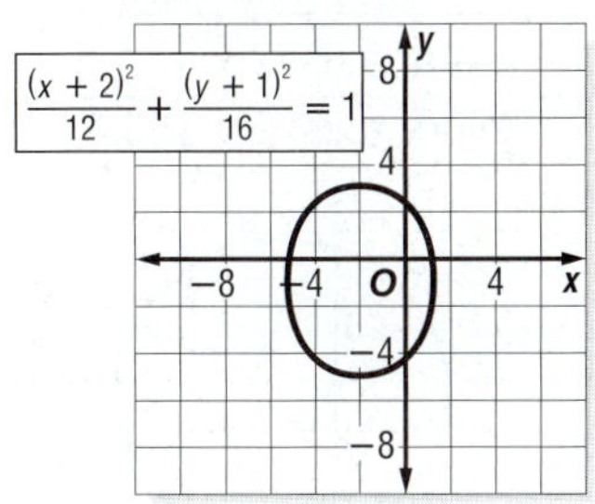

7. center: $(-2\ 4)$;
foci: $(-2, 6)$, $(-2, 2)$;
major axis: $4\sqrt{10}$;
minor axis: 12

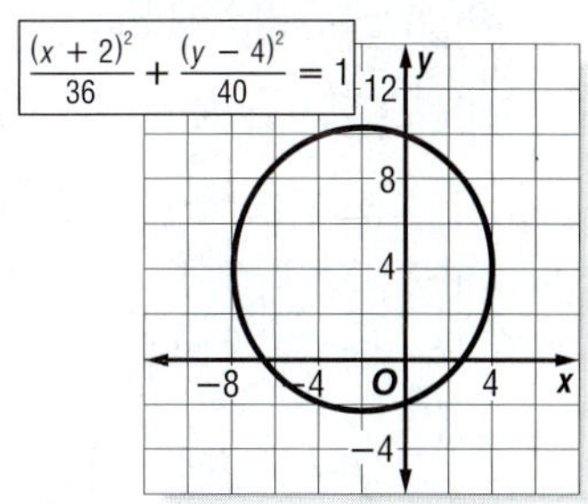

8. center: $(-8, 7)$;
foci: $\left(-8 \pm \sqrt{57}, 7\right)$;
major axis: 22;
minor axis: 16

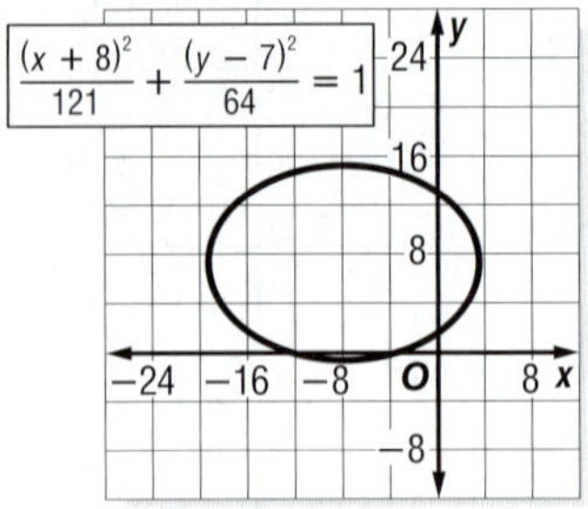

9. center: $(4, -1)$;
foci: $\left(4 \pm \sqrt{7}, -1\right)$;
major axis: 8;
minor axis: 6

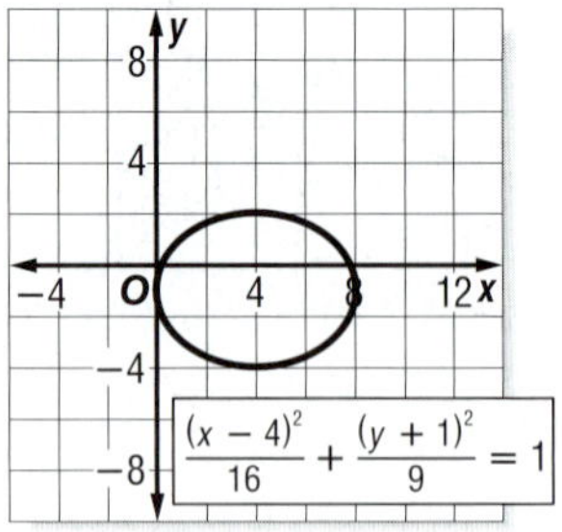

10. center: $(0, 0)$;
foci: $\left(0, \pm 2\sqrt{3}\right)$;
major axis: 8;
minor axis: 4

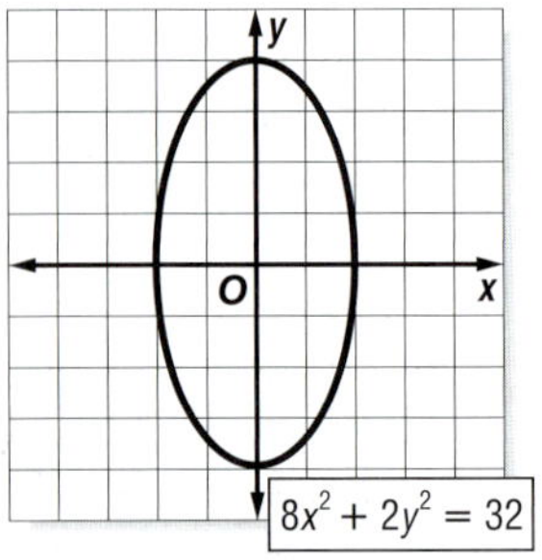

11. center: $(0, 0)$;
foci: $(0, \pm 4)$;
major axis: $4\sqrt{7}$;
minor axis: $4\sqrt{3}$

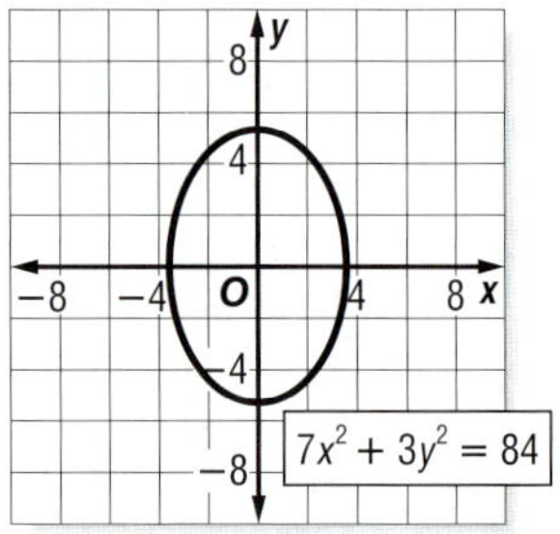

12. center: $(0, 0)$;
foci: $\left(\pm\sqrt{7}, 0\right)$;
major axis: 8;
minor axis: 6

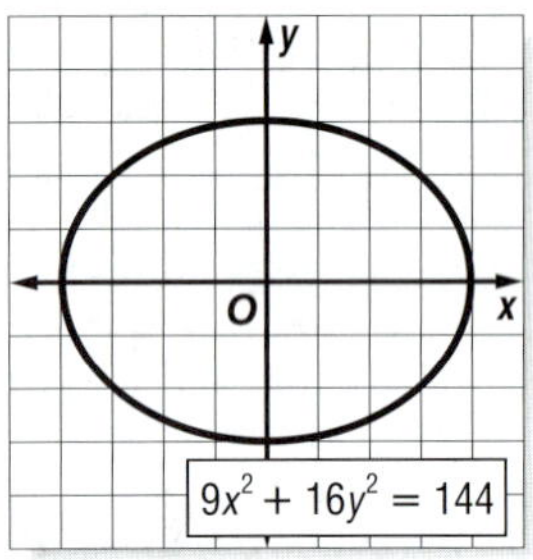

13. center: $(1, 0)$;
foci: $(1, \pm 12)$;
major axis: 26;
minor axis: 10

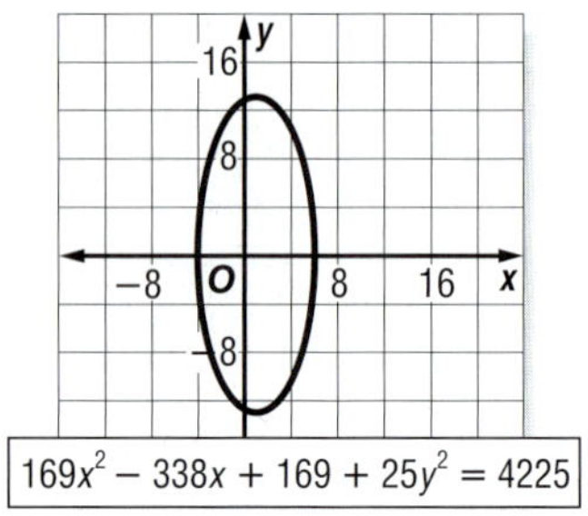

14. center: $(-4, 8)$;
foci: $\left(-4 \pm 6\sqrt{3}, 8\right)$;
major axis: 24;
minor axis: 12

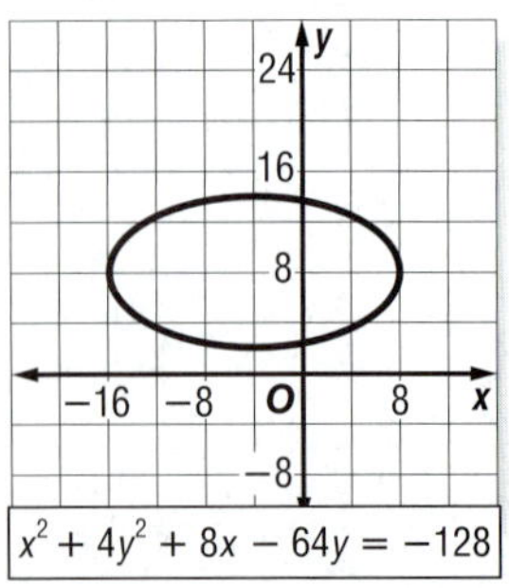

15. center: $(4.5, 3)$;
foci: $\left(4.5 \pm \sqrt{39.2}, 3\right)$;
major axis: 28;
minor axis: $2\sqrt{156.8}$

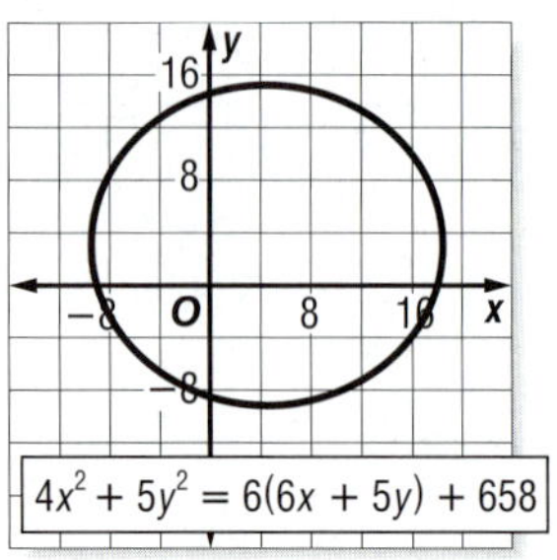

16. center: $(3, -2)$;
foci: $\left(3 \pm \sqrt{7}, -2\right)$;
major axis: 8;
minor axis: 6

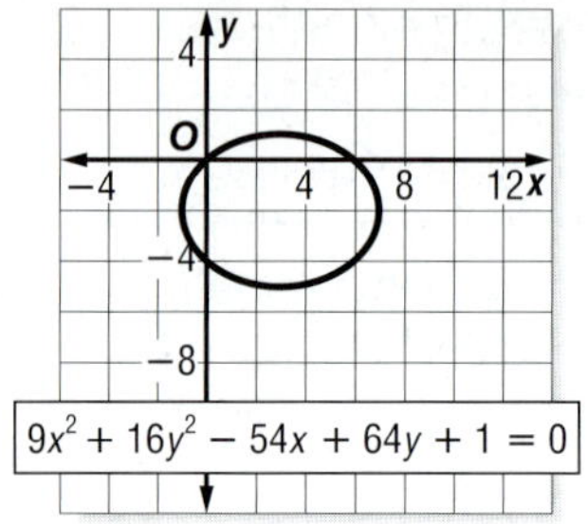

Page 846, Lesson 8-5

1. vertices: $(0, \pm 5)$;
foci: $\left(0, \pm\sqrt{34}\right)$;
asymptotes: $y = \pm\frac{5}{3}x$

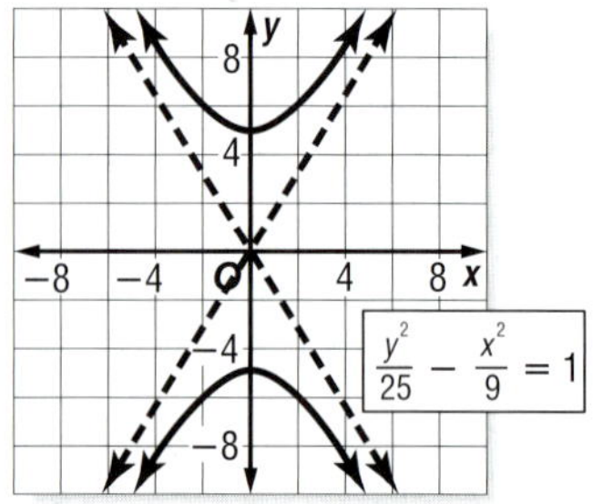

2. vertices: $(\pm 2, 0)$;
foci: $\left(\pm\sqrt{13}, 0\right)$;
asymptotes: $y = \pm\frac{3}{2}x$

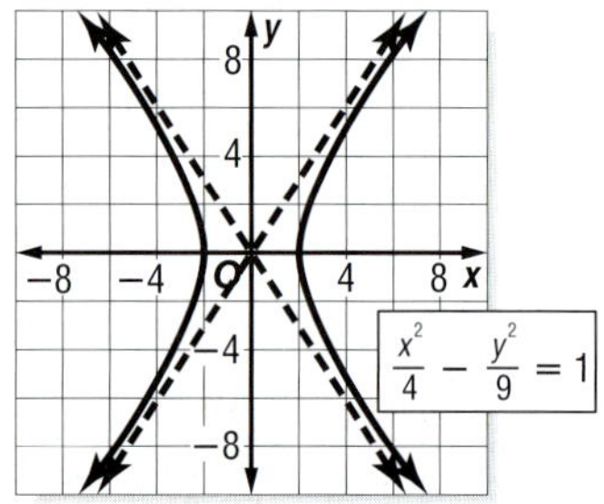

3. vertices: $(\pm 9, 0)$;
foci: $\left(\pm 3\sqrt{13}, 0\right)$;
asymptotes:
$y = \pm\frac{2}{3}x$

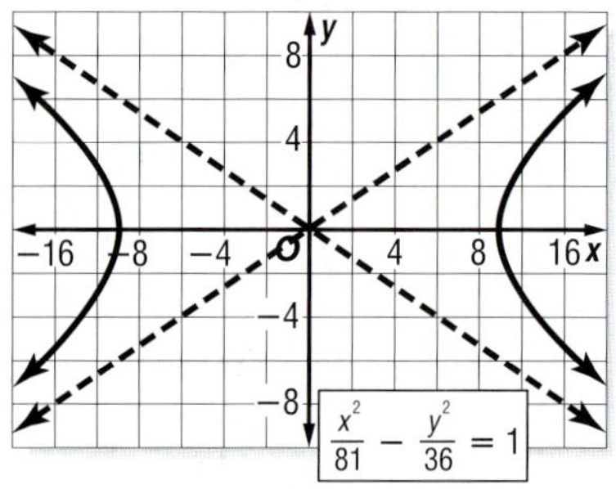

4. vertices: $(\pm 3, 0)$;
foci: $(\pm 5, 0)$;
asymptotes: $y = \pm\frac{4}{3}x$

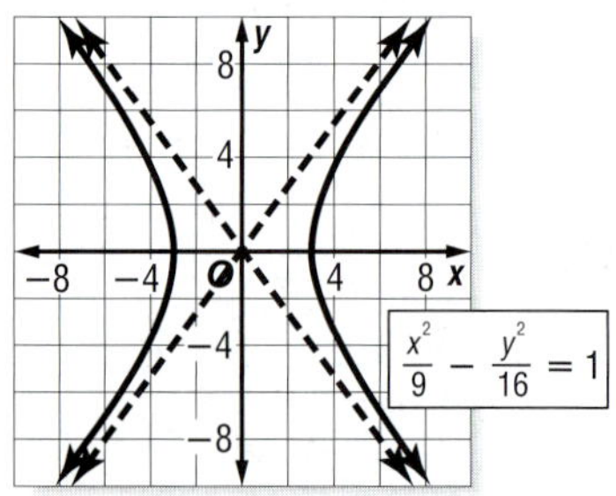

5. vertices: $(0, \pm 10)$;
foci: $\left(0, \pm 2\sqrt{61}\right)$;
asymptotes: $y = \pm\frac{5}{6}x$

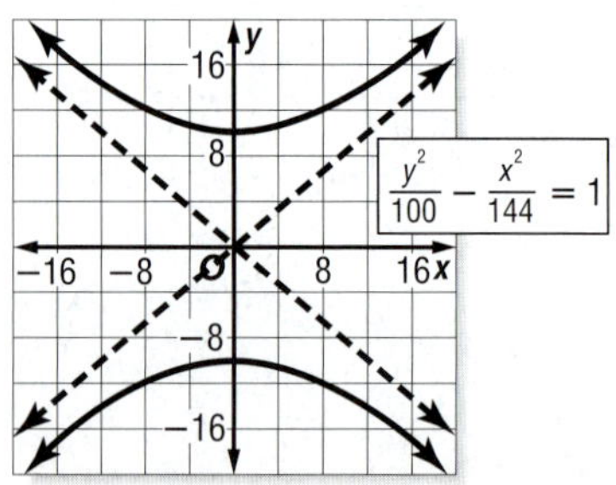

6. vertices: $(\pm 4, 0)$;
foci: $(\pm 2\sqrt{5}, 0)$;
asymptotes: $y = \pm\frac{1}{2}x$

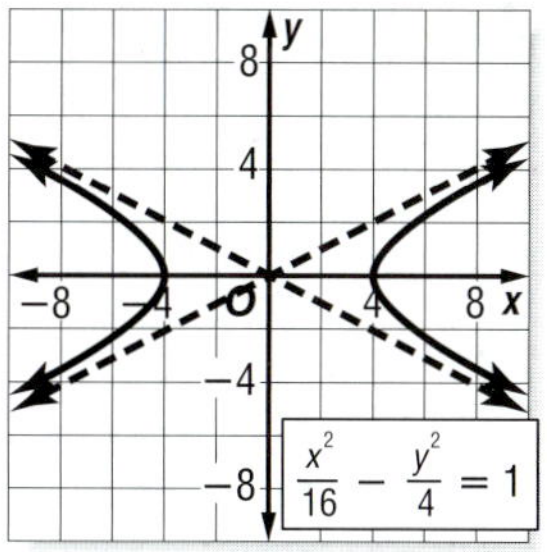

7. vertices: $(-4, -1)$, $(12, -1)$;
foci: $(4 + 4\sqrt{5}, -1)$, $(4 - 4\sqrt{5}, -1)$;
asymptotes:
$y + 1 = \pm\frac{1}{2}(x - 4)$

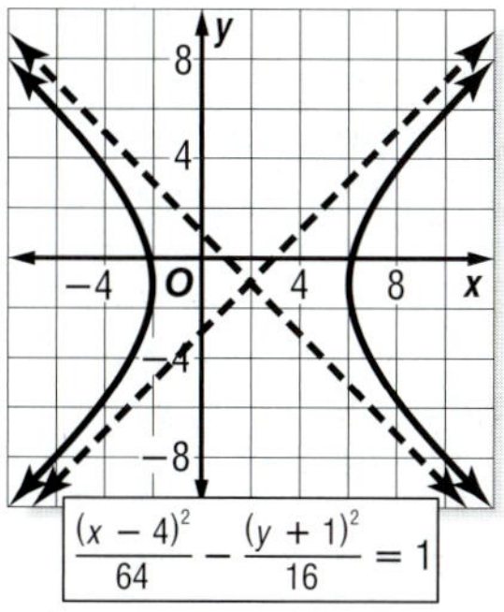

8. vertices: $(3, 5.5)$, $(3, 8.5)$;
foci: $(3, 4.5)$, $(3, 9.5)$;
asymptotes:
$y - 7 = \pm\frac{3}{4}(x - 3)$

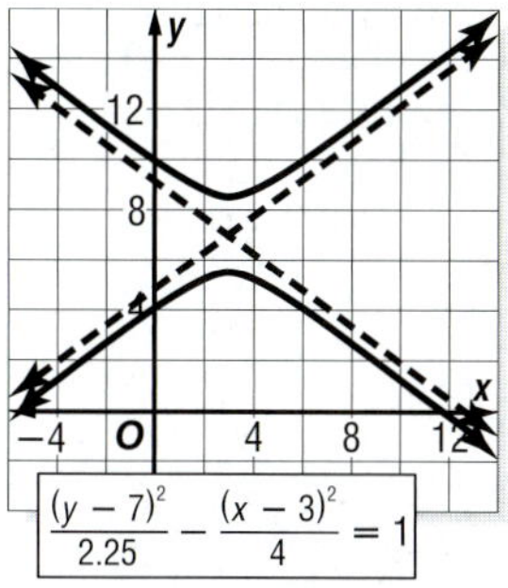

9. vertices: $(-4, -3)$, $(-6, -3)$;
foci: $(2, -3)$, $(-12, -3)$;
asymptotes:
$y + 3 = \pm 4\sqrt{3}(x + 5)$

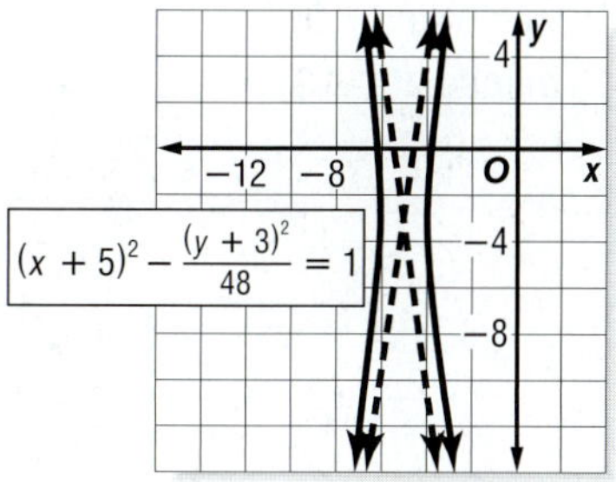

10. vertices: $(\pm 6, 0)$;
foci: $(\pm 2\sqrt{10}, 0)$;
asymptotes: $y = \pm\frac{1}{3}x$

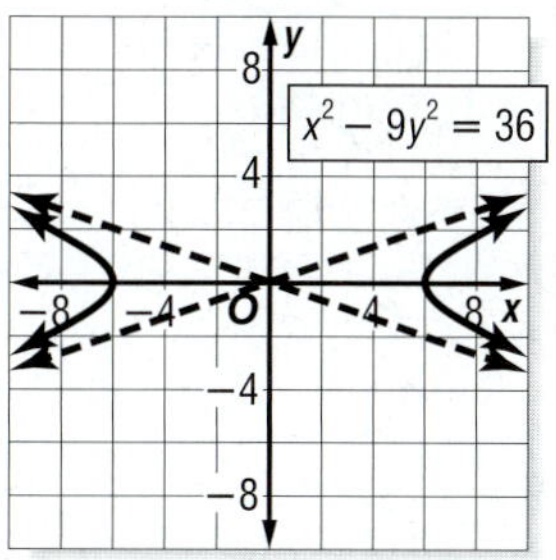

11. vertices: $(\pm 3\sqrt{2}, 0)$;
foci: $(\pm\sqrt{26}, 0)$;
asymptotes: $y = \pm\frac{2}{3}x$

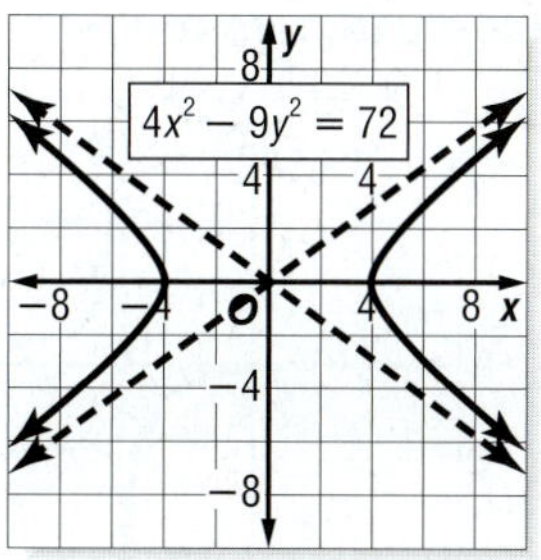

12. vertices: $(\pm 4, 0)$;
foci: $(\pm\sqrt{65}, 0)$;
asymptotes: $y = \pm\frac{7}{4}x$

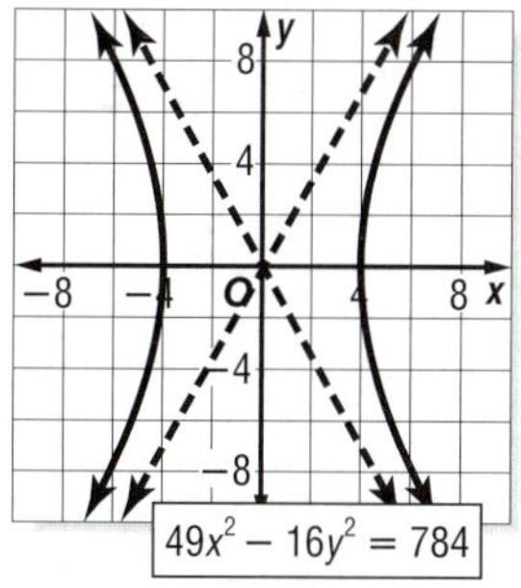

13. vertices: $(-9, -2)$, $(1, -2)$;
foci: $(9, -2)$, $(-17, -2)$;
asymptotes:
$y + 2 = \pm\frac{12}{5}(x + 4)$

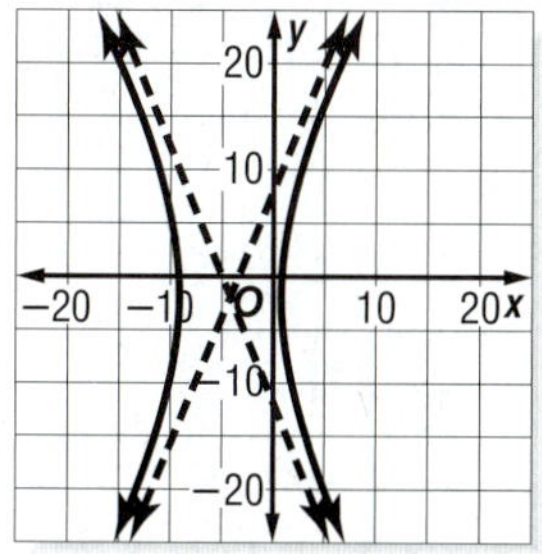

14. vertices: $(-5, \pm 7)$;
foci: $(-5, \pm 25)$;
asymptotes:
$y = \pm\frac{7}{24}(x + 5)$

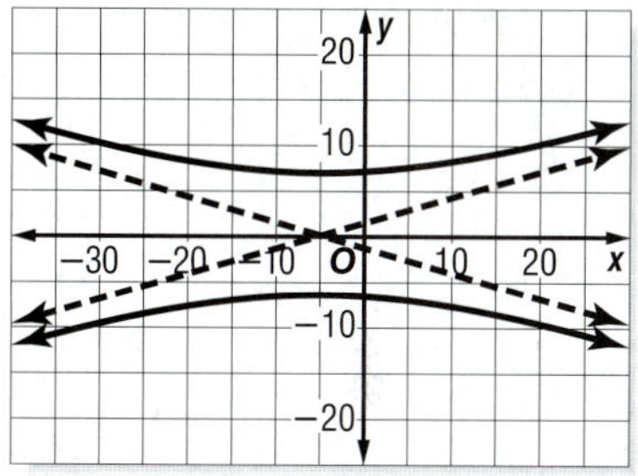

15. vertices: $(-1, 2.4)$, $(-1, -0.4)$;
foci: $(-1, 6)$, $(-1, -4)$;
asymptotes:
$y - 1 = \pm\frac{7}{24}(x + 1)$

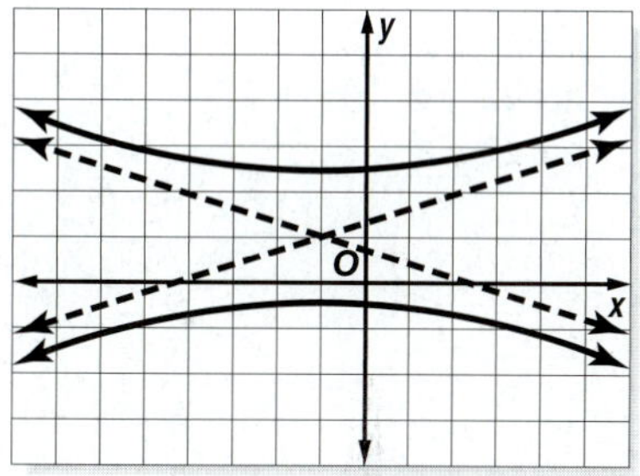

16. vertices: $(1, -5 \pm 2\sqrt{5})$;
foci: $(1, -5 \pm 3\sqrt{5})$;
asymptotes:
$y + 5 = \pm\frac{2\sqrt{5}}{5}(x - 1)$

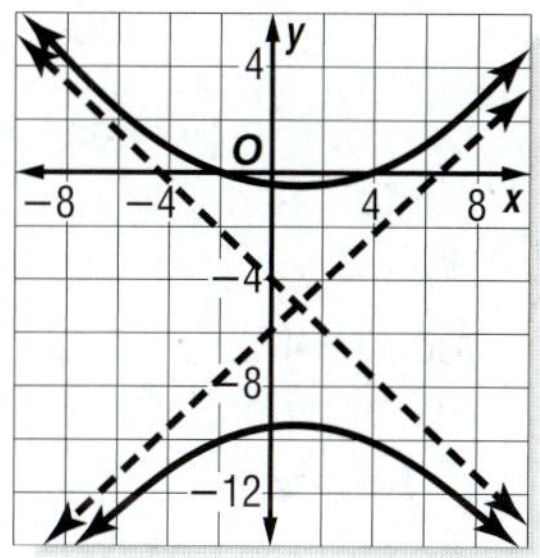

Page 846, Lesson 8-6

1. $\frac{(x-2)^2}{4} - \frac{(y+3)^2}{9} = 1$; hyperbola

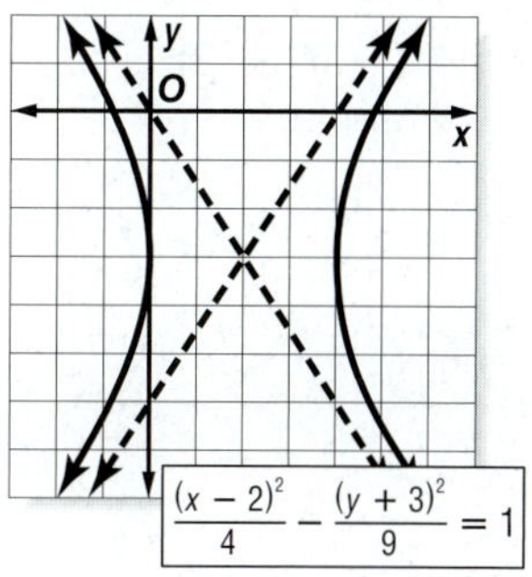

2. $\frac{(x+2)^2}{4} + \frac{(y+4)^2}{2} = 1$; ellipse

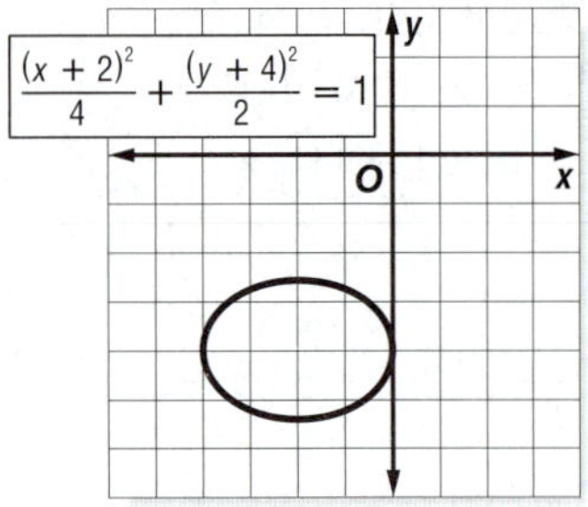

3. $(x+3)^2 + (y-3)^2 = 9$; circle

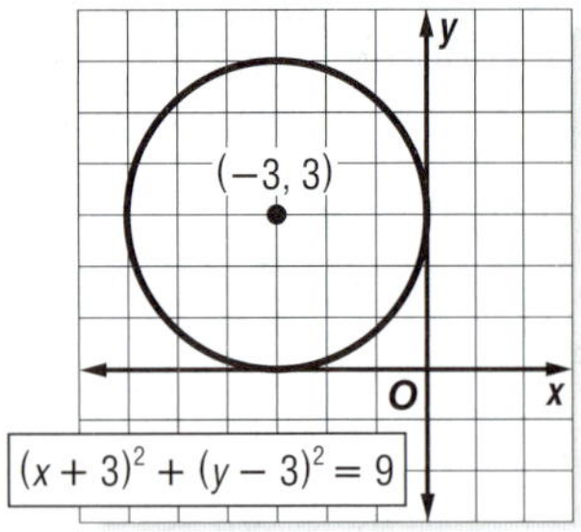

4. $\frac{(y-0)^2}{25} - \frac{(x+8)^2}{9} = 1$; hyperbola

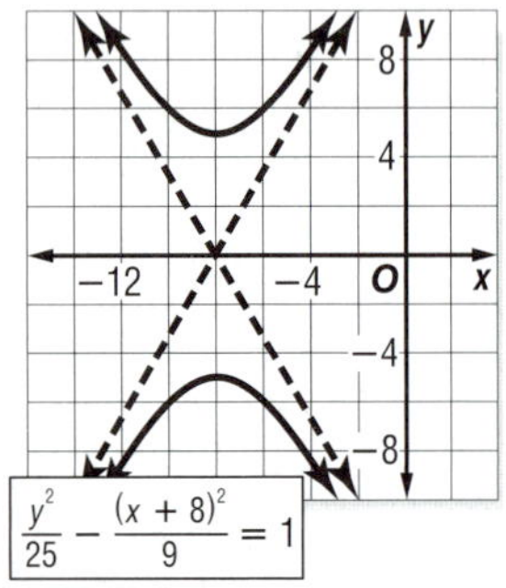

5. $x = 2(y+3)^2 - 12$; parabola

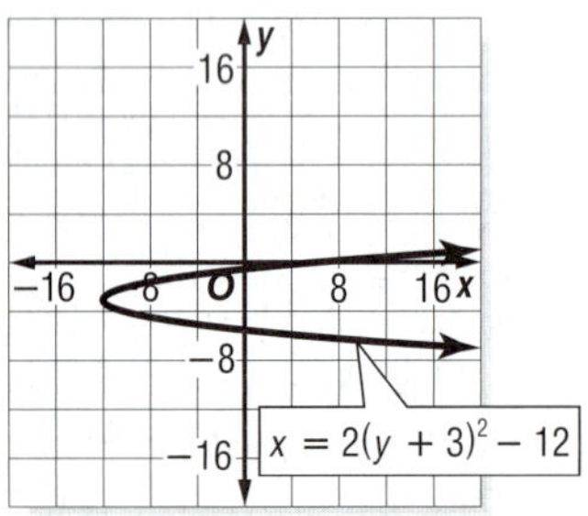

6. $(x-5)^2 + (y-1)^2 = 49$; circle

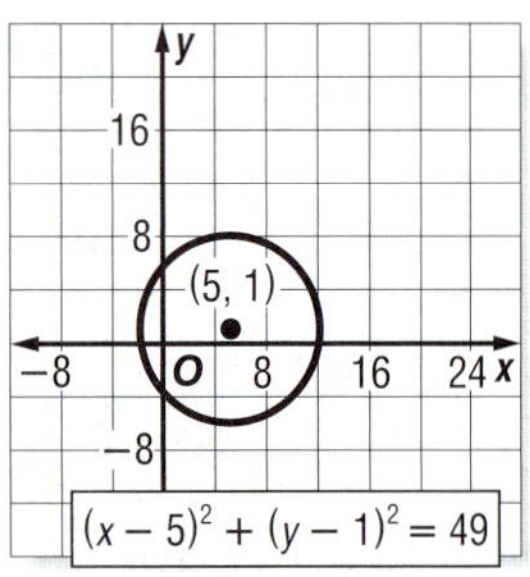

7. $y = -3(x-2)^2 - 5$; parabola

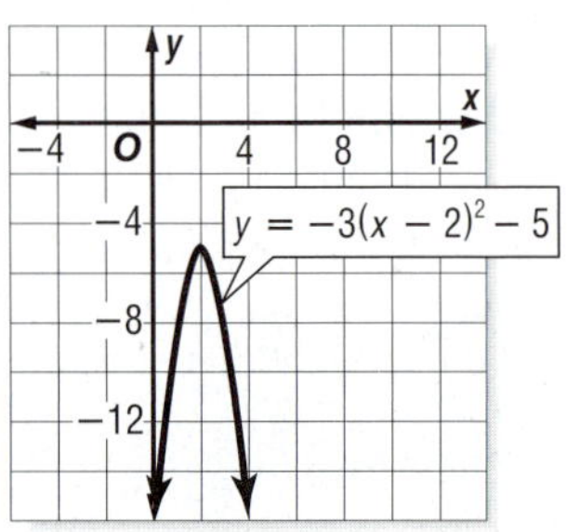

8. $\frac{(x-1)^2}{16} + \frac{(y+5)^2}{9} = 1$; ellipse

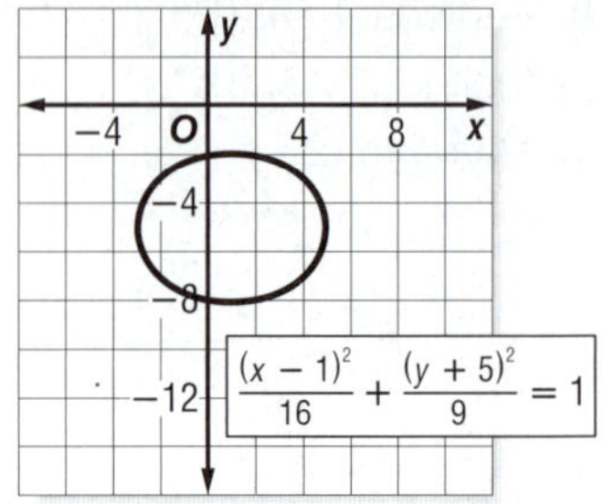

9. $\frac{(x+5)^2}{36} - \frac{(y-0)^2}{9} = 1$; hyperbola

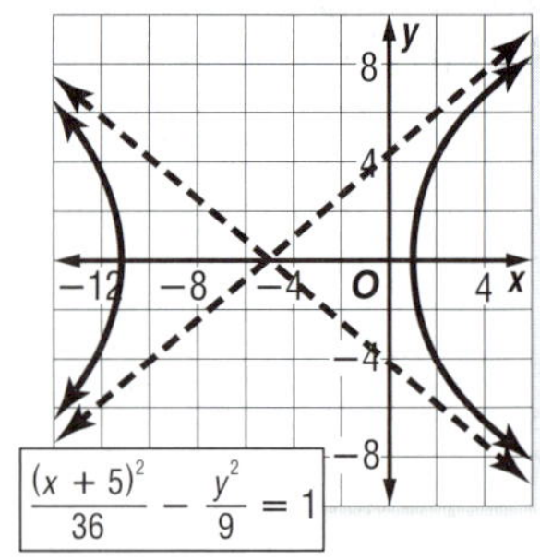

10. $\frac{(y-5)^2}{16} - \frac{(x+1)^2}{4} = 1$; hyperbola

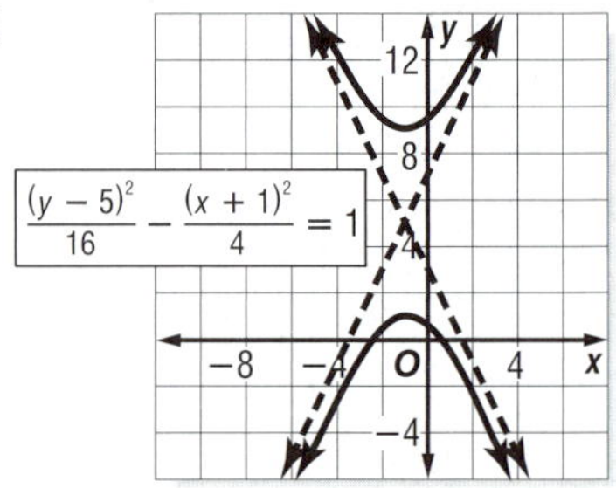

11. $\frac{x^2}{49} + \frac{y^2}{9} = 1$; ellipse

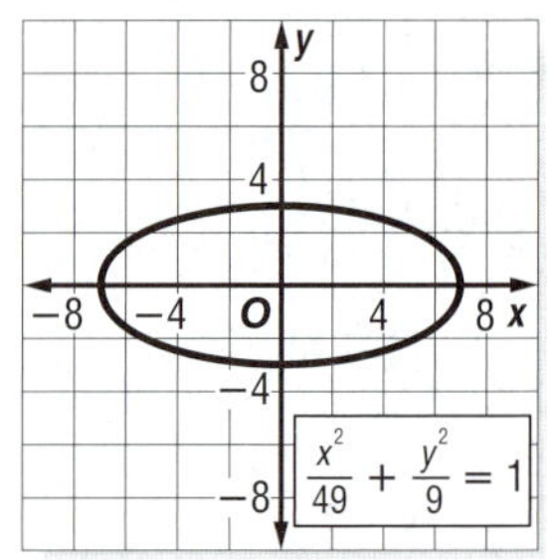

12. $\frac{x^2}{1} - \frac{y^2}{4} = 1$; hyperbola

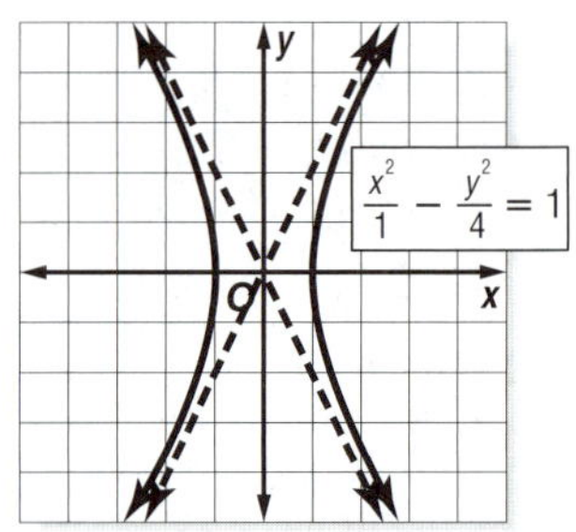

Page 848, Lesson 9-3

7.

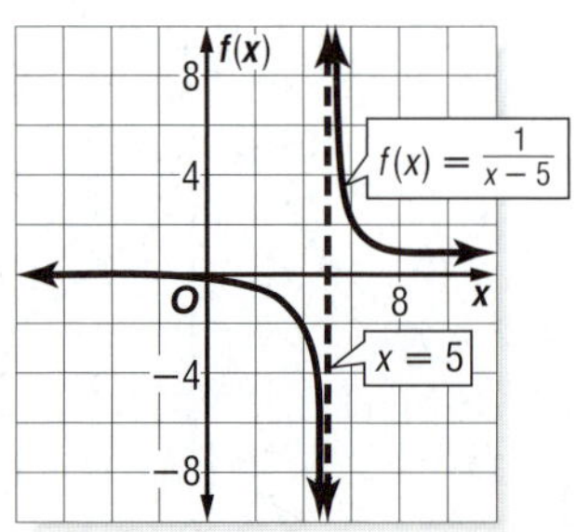

8.

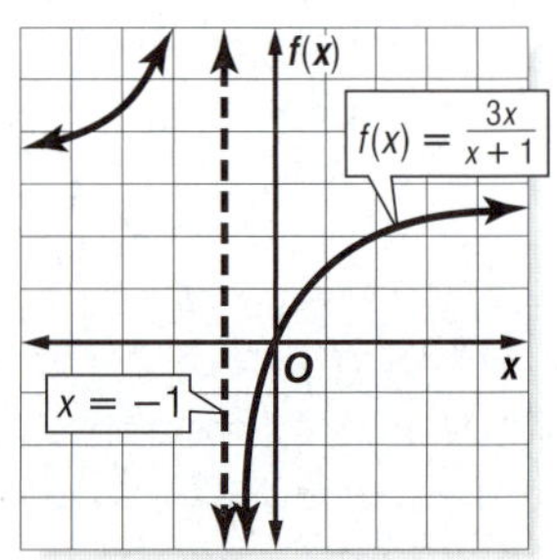

9.

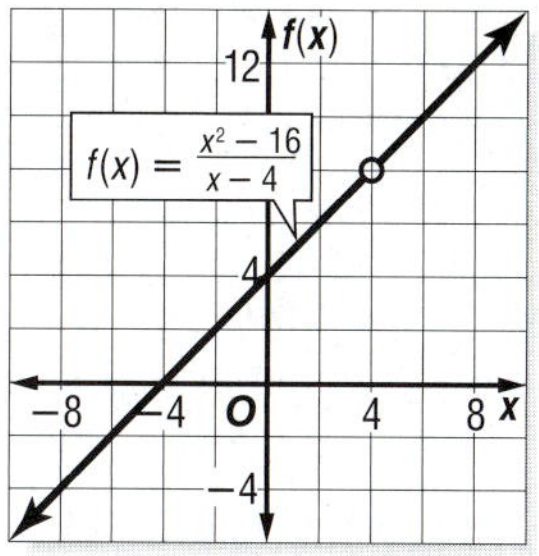

10.

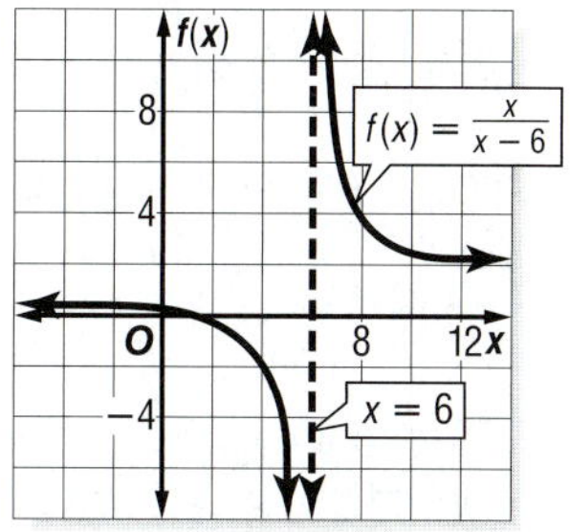

11.

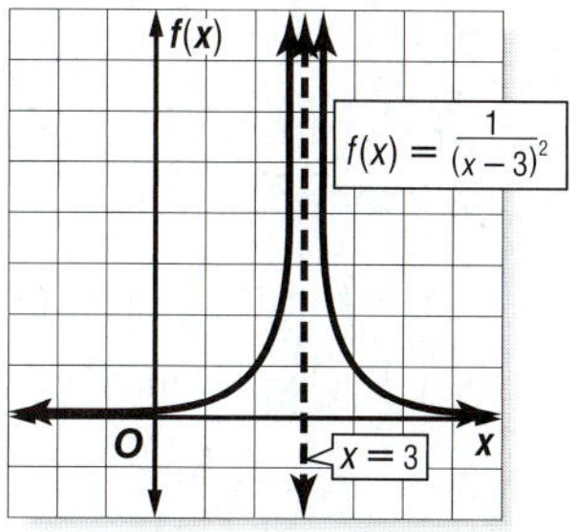

12.

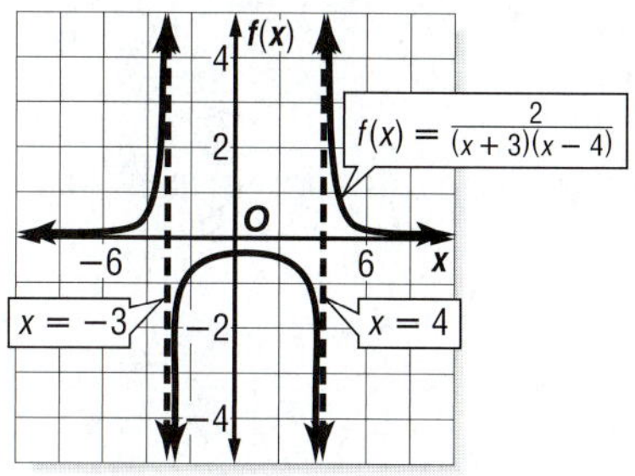

13.

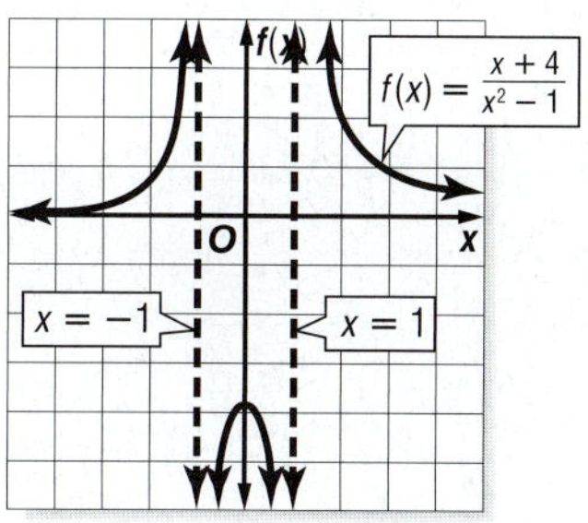

14.

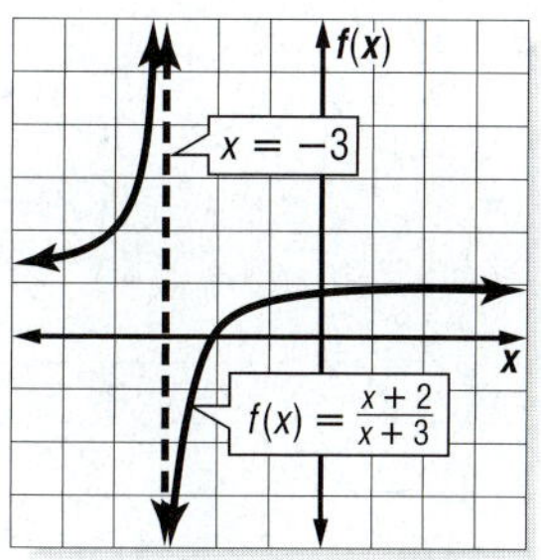

15.

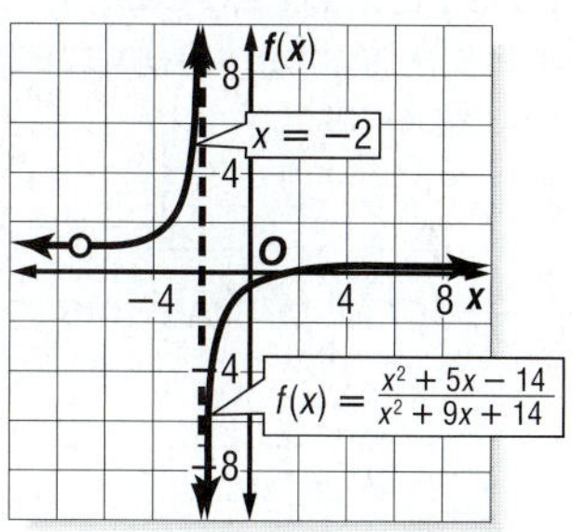

Page 853, Lesson 11-8

1. **Step 1:** When $n = 1$, the left side of the given equation is 2. The right side is $1^2 + 1$ or 2. Thus, the equation is true for $n = 1$.

   Step 2: Assume that $2 + 4 + 6 + \ldots + 2k = k^2 + k$ for some positive integer k.

   Step 3: Show that the given equation is true for $n = k + 1$.

   $$\begin{aligned} &2 + 4 + 6 + \ldots + 2k + 2(k + 1) \\ &\quad = k^2 + k + 2(k + 1) \\ &\quad = k^2 + k + 2k + 2 \\ &\quad = k^2 + 2k + 1 + k + 1 \\ &\quad = (k + 1)^2 + (k + 1) \end{aligned}$$

   The last expression above is the right side of the equation to be proved, where n has been replaced by $k + 1$. Thus, the equation is true for $n = k + 1$.

   Therefore, $2 + 4 + 6 + \ldots + 2n = n^2 + n$ for all positive integers n.

2. **Step 1:** When $n = 1$, the left side of the given equation is 1^3 or 1. The right side is $1^2[2(1)^2 - 1]$ or 1. Thus, the equation is true for $n = 1$.

   Step 2: Assume that $1^3 + 3^3 + 5^3 + \ldots + (2k - 1)^3 = k^2(2k^2 - 1)$ for some positive integer k.

   Step 3: Show that the given equation is true for $n = k + 1$.

   $$\begin{aligned} &1^3 + 3^3 + 5^3 + \ldots + (2k - 1)^3 + [2(k + 1) - 1]^3 \\ &\quad = k^2(2k^2 - 1) + [2(k + 1) - 1]^3 \\ &\quad = 2k^4 - k^2 + (2k + 2 - 1)^3 \\ &\quad = 2k^4 - k^2 + (2k + 1)^3 \\ &\quad = 2k^4 - k^2 + (8k^3 + 12k^2 + 6k + 1) \\ &\quad = 2k^4 + 8k^3 + 11k^2 + 6k + 1 \\ &\quad = (k^2 + 2k + 1)(2k^2 + 4k + 1) \\ &\quad = (k + 1)^2[2(k^2 + 2k + 1) - 1] \\ &\quad = (k + 1)^2[2(k + 1)^2 - 1] \end{aligned}$$

   The last expression above is the right side of the equation to be proved, where n has been replaced by $k + 1$. Thus, the equation is true for $n = k + 1$.

   Therefore, $1^3 + 3^3 + 5^3 + \ldots + (2n - 1)^3 = n^2(2n^2 - 1)$ for all positive integers n.

3. **Step 1:** When $n = 1$, the left side of the given equation is $1 \cdot 3$ or 3. The right side is $\frac{1(4 + 6 - 1)}{3}$ or 3. Thus, the equation is true for $n = 1$.

   Step 2: Assume that $1 \cdot 3 + 3 \cdot 5 + 5 \cdot 7 + \ldots + (2k - 1)(2k + 1) = \frac{k(4k^2 + 6k - 1)}{3}$ for some positive integer k.

   Step 3: Show that the given equation is true for $n = k + 1$.

   $$\begin{aligned} &1 \cdot 3 + 3 \cdot 5 + 5 \cdot 7 + \ldots + (2k - 1)(2k + 1) + \\ &\qquad [2(k + 1) - 1][2(k + 1) + 1] \\ &\quad = \frac{k(4k^2 + 6k - 1)}{3} + [2(k + 1) - 1][2(k + 1) + 1] \\ &\quad = \frac{k(4k^2 + 6k - 1)}{3} + (2k + 1)(2k + 3) \\ &\quad = \frac{4k^3 + 6k^2 - k}{3} + \frac{3(4k^2 + 8k + 3)}{3} \\ &\quad = \frac{4k^3 + 6k^2 - k + 12k^2 + 24k + 9}{3} \\ &\quad = \frac{4k^3 + 18k^2 + 23k + 9}{3} \\ &\quad = \frac{(k + 1)(4k^2 + 14k + 9)}{3} \\ &\quad = \frac{(k + 1)[(4k^2 + 8k + 4) + (6k + 6) - 1]}{3} \\ &\quad = \frac{(k + 1)[4(k + 1)^2 + 6(k + 1) - 1]}{3} \end{aligned}$$

The last expression above is the right side of the equation to be proved, where n has been replaced by $k + 1$. Thus, the equation is true for $n = k + 1$.

Therefore, $1 \cdot 3 + 3 \cdot 5 + 5 \cdot 7 + \ldots + (2n - 1)(2n + 1) = \frac{n(4n^2 + 6n - 1)}{3}$ for all positive integers n.

4. Step 1: When $n = 1$, the left side of the given equation is $\frac{1}{1 \cdot 3}$ or $\frac{1}{3}$. The right side is $\frac{1[3(1) + 5]}{4(1 + 1)(1 + 2)}$ or $\frac{1}{3}$. Thus, the equation is true for $n = 1$.

Step 2: Assume that $\frac{1}{1 \cdot 3} + \frac{1}{2 \cdot 4} + \frac{1}{3 \cdot 5} + \ldots + \frac{1}{k(k + 2)} = \frac{k(3k + 5)}{4(k + 1)(k + 2)}$ for some positive integer k.

Step 3: Show that the given equation is true for $n = k + 1$.

$$\frac{1}{1 \cdot 3} + \frac{1}{2 \cdot 4} + \frac{1}{3 \cdot 5} + \ldots + \frac{1}{k(k + 2)} + \frac{1}{(k + 1)[(k + 1) + 2]}$$
$$= \frac{k(3k + 5)}{4(k + 1)(k + 2)} + \frac{1}{(k + 1)[(k + 1) + 2]}$$
$$= \frac{3k^2 + 5k}{4(k + 1)(k + 2)} + \frac{1}{(k + 1)(k + 3)}$$
$$= \frac{(3k^2 + 5k)(k + 3) + 4(k + 2)}{4(k + 1)(k + 2)(k + 3)}$$
$$= \frac{3k^3 + 14k^2 + 15k + 4k + 8}{4(k + 1)(k + 2)(k + 3)}$$
$$= \frac{3k^3 + 14k^2 + 19k + 8}{4(k + 1)(k + 2)(k + 3)}$$
$$= \frac{3k^2 + 11k + 8}{4(k + 2)(k + 3)}$$
$$= \frac{(3k + 8)(k + 1)}{4(k + 2)(k + 3)}$$
$$= \frac{(k + 1)[3(k + 1) + 5]}{4[(k + 1) + 1][(k + 1) + 2]}$$

The last expression above is the right side of the equation to be proved, where n has been replaced by $k + 1$. Thus, the equation is true for $n = k + 1$.

Therefore, $\frac{1}{1 \cdot 3} + \frac{1}{2 \cdot 4} + \frac{1}{3 \cdot 5} + \ldots + \frac{1}{n(n + 2)} = \frac{n(3n + 5)}{4(n + 1)(n + 2)}$ for all positive integers n.

5. Step 1: When $n = 1$, the left side of the given equation is $1 \cdot 3$ or 3. The right side is $\frac{1(1 + 1)(2 + 7)}{6}$ or 3. Thus, the equation is true for $n = 1$.

Step 2: Assume that $1 \cdot 3 + 2 \cdot 4 + 3 \cdot 5 + \ldots + k(k + 2) = \frac{k(k + 1)(2k + 7)}{6}$ for some positive integer k.

Step 3: Show that the given equation is true for $n = k + 1$.

$$1 \cdot 3 + 2 \cdot 4 + 3 \cdot 5 + \ldots + k(k + 2) + (k + 1)[(k + 1) + 2]$$
$$= \frac{k(k + 1)(2k + 7)}{6} + (k + 1)[(k + 1) + 2]$$
$$= \frac{2k^3 + 9k^2 + 7k}{6} + k^2 + 4k + 3$$
$$= \frac{2k^3 + 9k^2 + 7k}{6} + \frac{6(k^2 + 4k + 3)}{6}$$
$$= \frac{2k^3 + 9k^2 + 7k + 6k^2 + 24k + 18}{6}$$
$$= \frac{2k^3 + 15k^2 + 31k + 18}{6}$$
$$= \frac{(k + 1)(2k^2 + 13k + 18)}{6}$$
$$= \frac{(k + 1)[(k + 2)(2k + 9)]}{6}$$
$$= \frac{(k + 1)[(k + 1) + 1][2(k + 1) + 7]}{6}$$

The last expression above is the right side of the equation to be proved, where n has been replaced by $k + 1$. Thus, the equation is true for $n = k + 1$.

Therefore, $1 \cdot 3 + 2 \cdot 4 + 3 \cdot 5 + \ldots + n(n + 2) = \frac{n(n + 1)(2n + 7)}{6}$ for all positive integers n.

6. Step 1: When $n = 1$, the left side of the given equation is $\frac{5}{2} \cdot \frac{1}{3}$ or $\frac{5}{6}$. The right side is $1 - \frac{1}{3(2)}$ or $\frac{5}{6}$. Thus, the equation is true for $n = 1$.

Step 2: Assume that $\frac{5}{1 \cdot 2} \cdot \frac{1}{3} + \frac{7}{2 \cdot 3} \cdot \frac{1}{3^2} + \frac{9}{3 \cdot 4} \cdot \frac{1}{3^3} + \ldots + \frac{2k + 3}{k(k + 1)} \cdot \frac{1}{3^k} = 1 - \frac{1}{3^k(k + 1)}$ for some positive integer k.

Step 3: Show that the given equation is true for $n = k + 1$.

$$\frac{5}{1 \cdot 2} \cdot \frac{1}{3} + \frac{7}{2 \cdot 3} \cdot \frac{1}{3^2} + \frac{9}{3 \cdot 4} \cdot \frac{1}{3^3} + \ldots + \frac{2k + 3}{k(k + 1)} \cdot \frac{1}{3^k} + \frac{2(k + 1) + 3}{(k + 1)[(k + 1) + 1]} \cdot \frac{1}{3^{k + 1}}$$
$$= 1 + \frac{-1}{3^k(k + 1)} + \frac{2(k + 1) + 3}{(k + 1)(k + 2)} \cdot \frac{1}{3^{k + 1}}$$
$$= 1 + \frac{(-3)(k + 2) + 2k + 5}{3^{k + 1}(k + 1)(k + 2)}$$
$$= 1 + \frac{(-k - 1)}{3^{k + 1}(k + 1)(k + 2)}$$
$$= 1 - \frac{k + 1}{3^{k + 1}(k + 1)(k + 2)}$$
$$= 1 - \frac{1}{3^{k + 1}[(k + 1) + 1]}$$

The last expression above is the right side of the equation to be proved, where n has been replaced by $k + 1$. Thus, the equation is true for $n = k + 1$.

Therefore, $\frac{5}{1 \cdot 2} \cdot \frac{1}{3} + \frac{7}{2 \cdot 3} \cdot \frac{1}{3^2} + \frac{9}{3 \cdot 4} \cdot \frac{1}{3^3} + \ldots + \frac{2n + 3}{n(n + 1)} \cdot \frac{1}{3^n} = 1 - \frac{1}{3^n(n + 1)}$ for all positive integers n.

Page 859, Lesson 14–1

1.

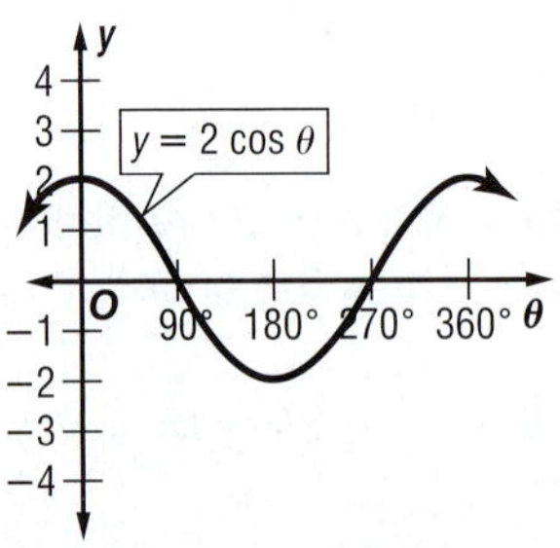

2.

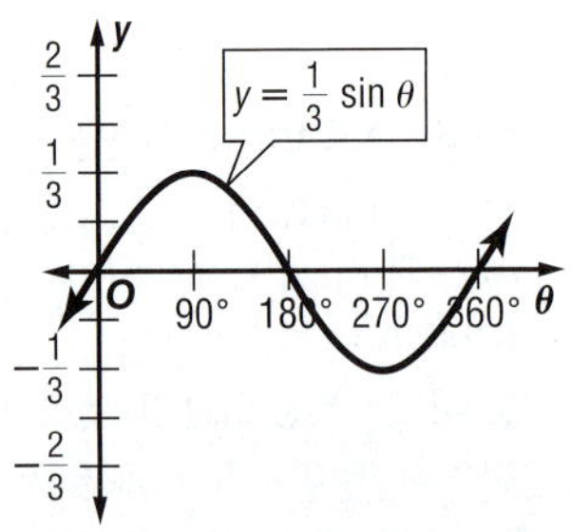

3.

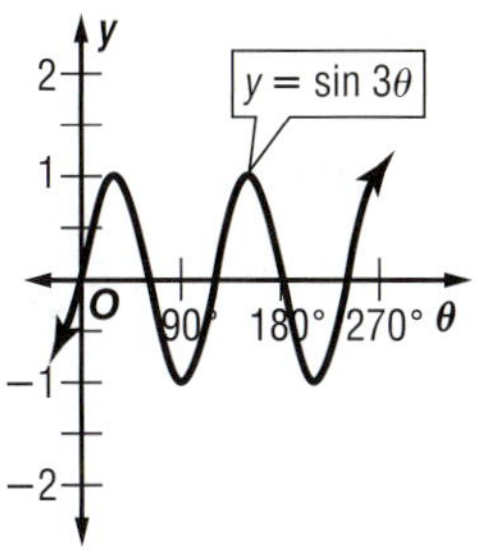

4.

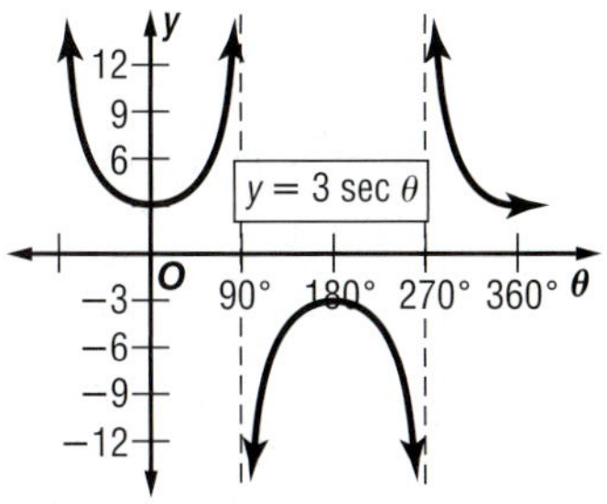

5.

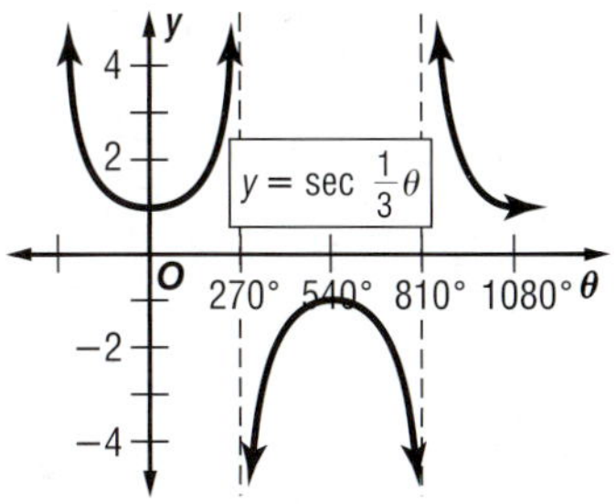

6.

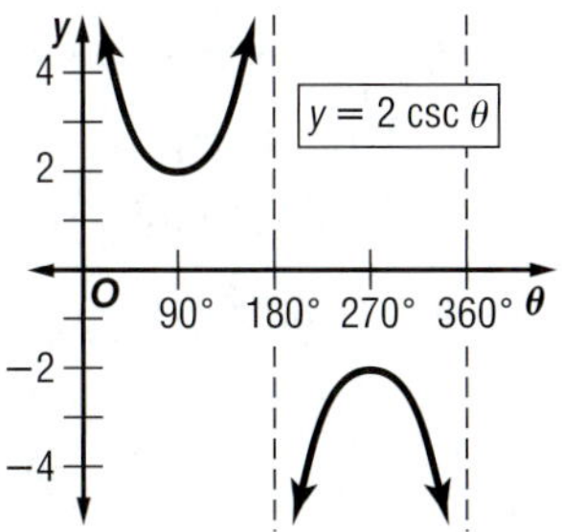

7.

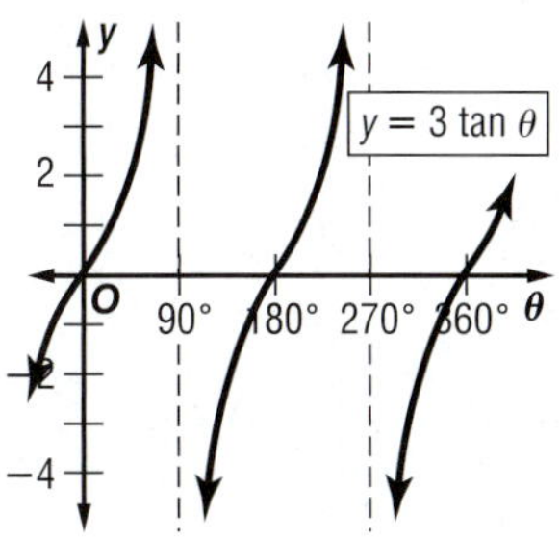

8.

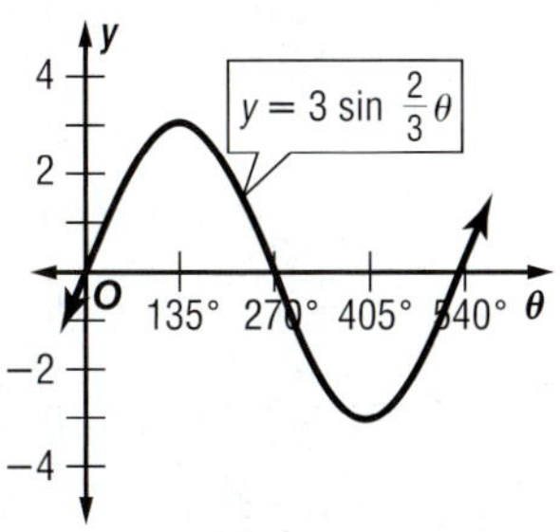

9.

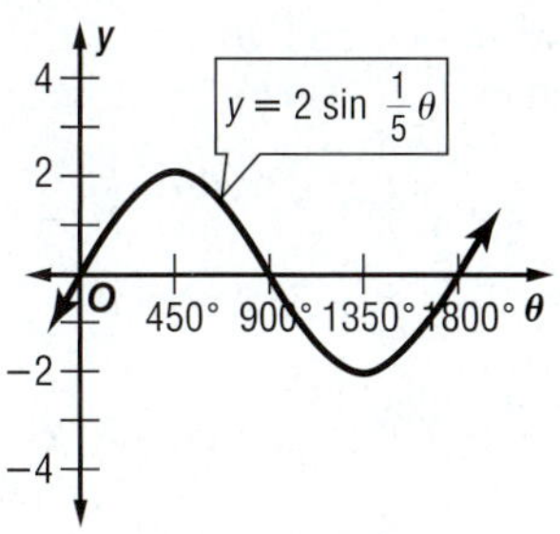

10.

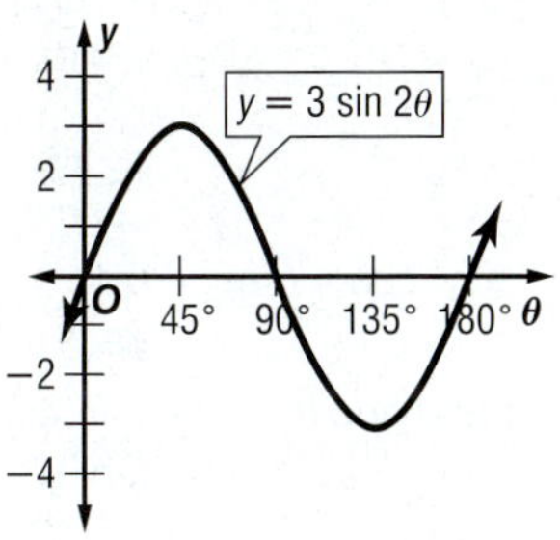

11.

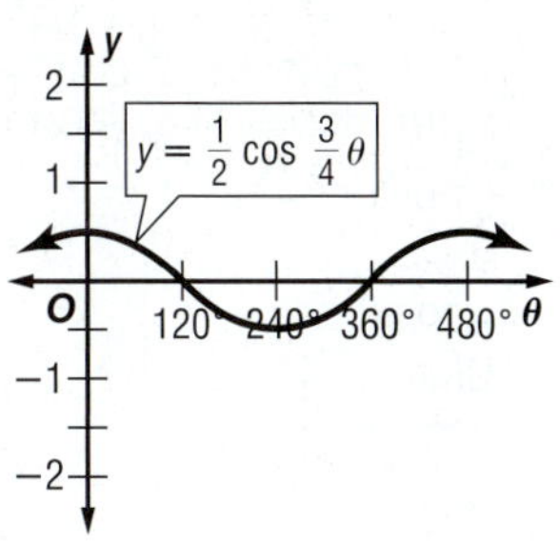

12.

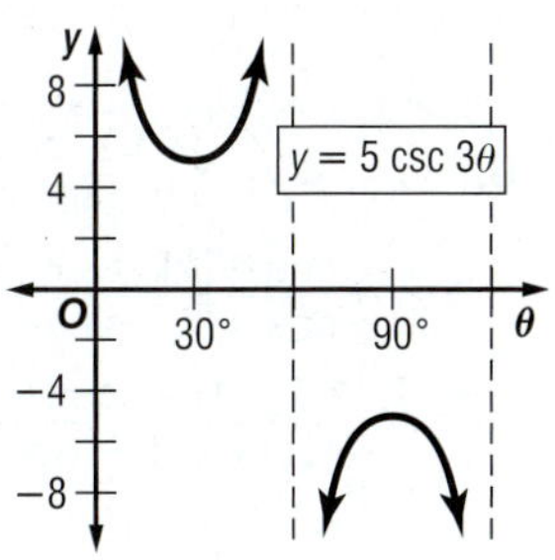

13.

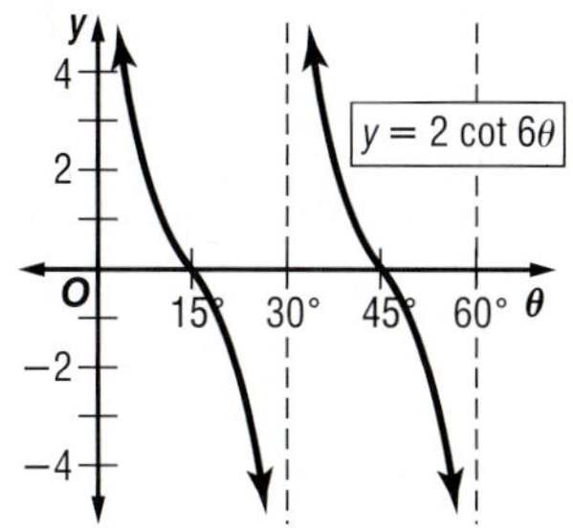

14.

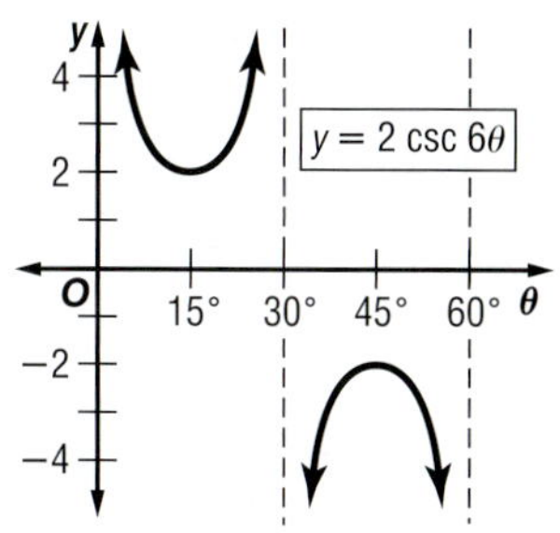

15.

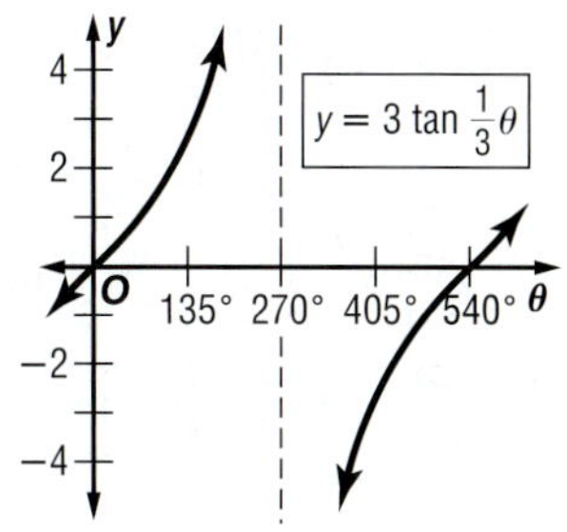

Page 859, Lesson 14-2

5.

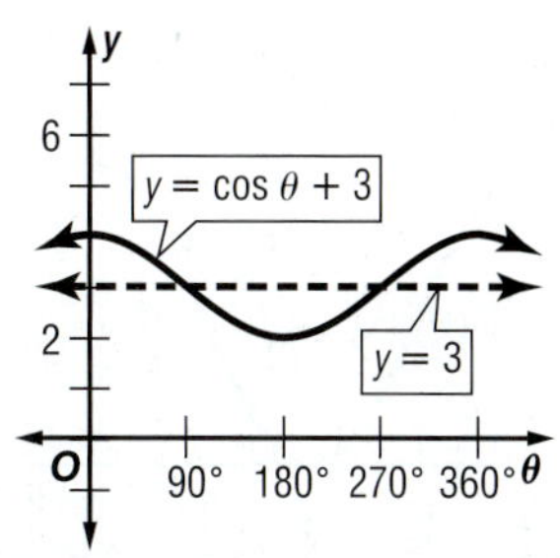

6.

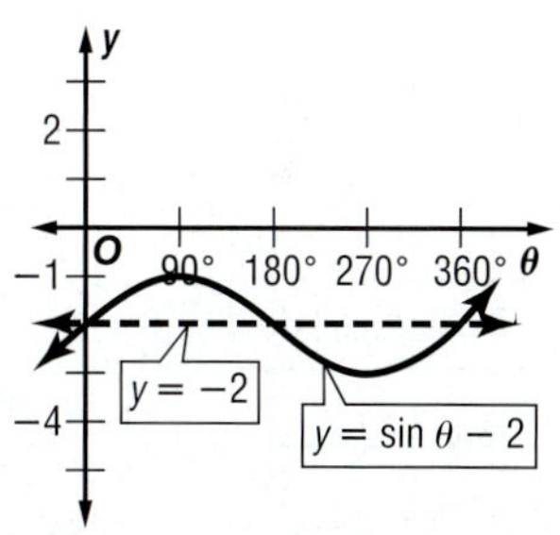

7.

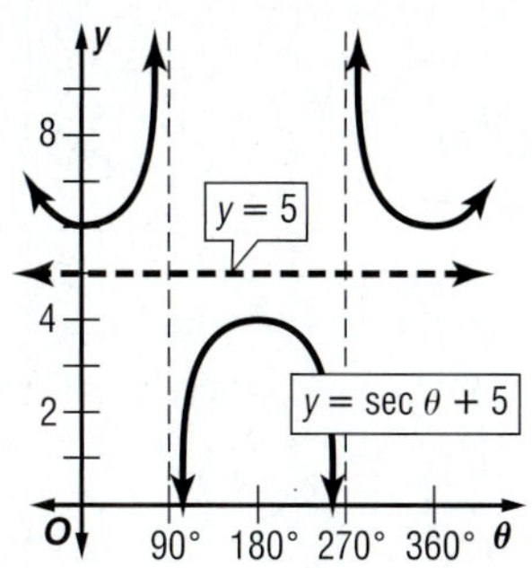

8.

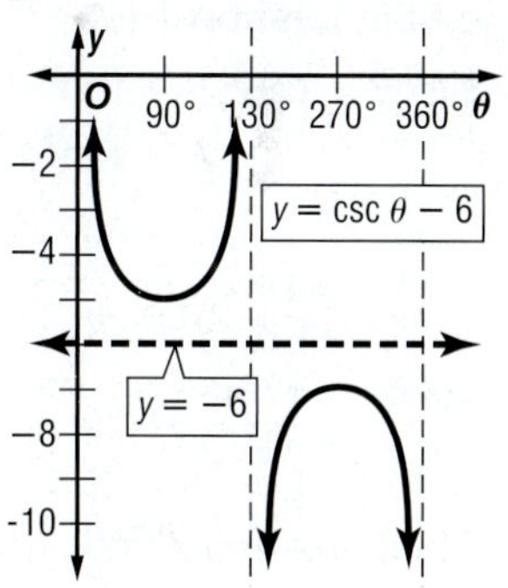

9.

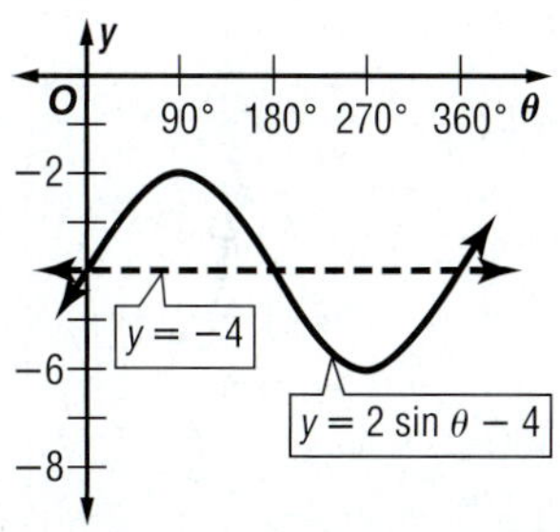

10.

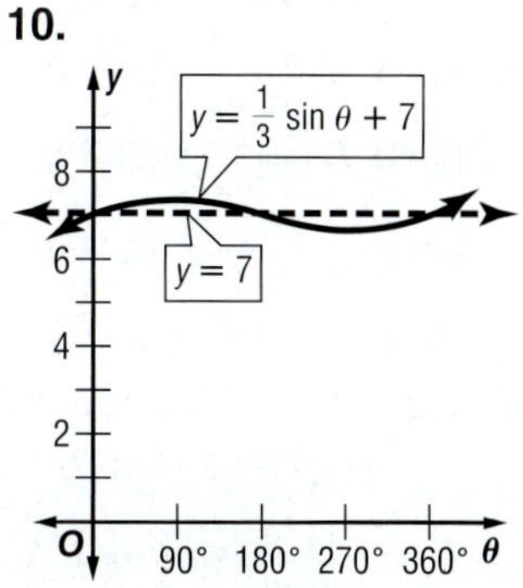

11. 4; 3; 180°; −30°

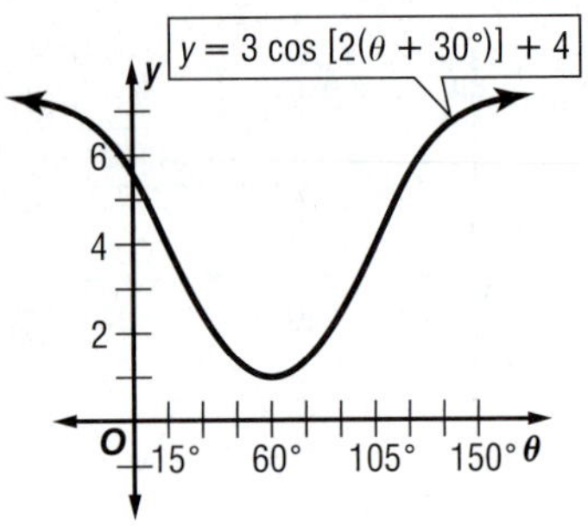

12. −2; none; 60°; 60°

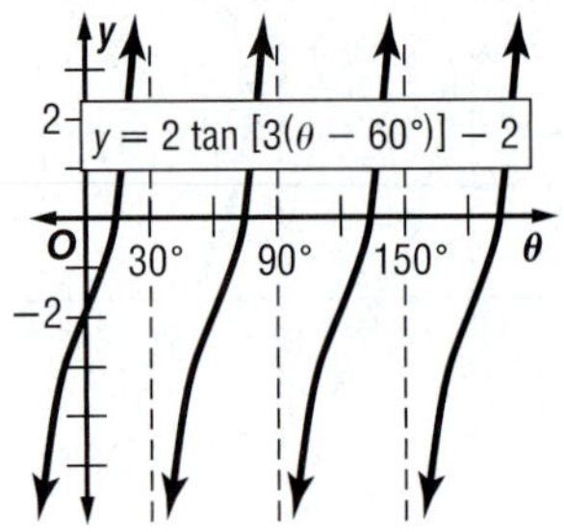

13. 1; $\frac{1}{2}$; 90°; 45°

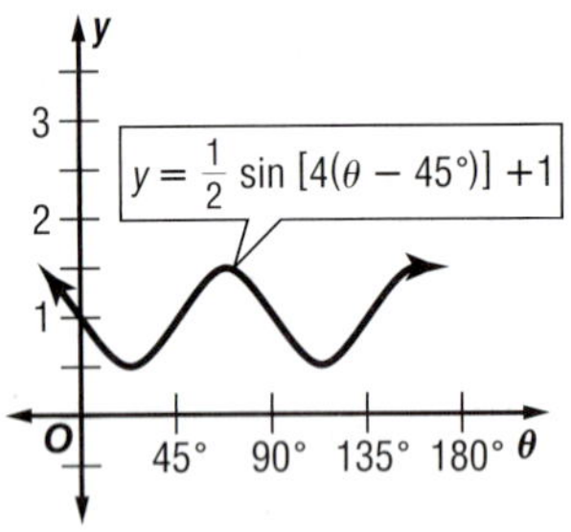

14. −5; $\frac{2}{5}$; 60°; −45°

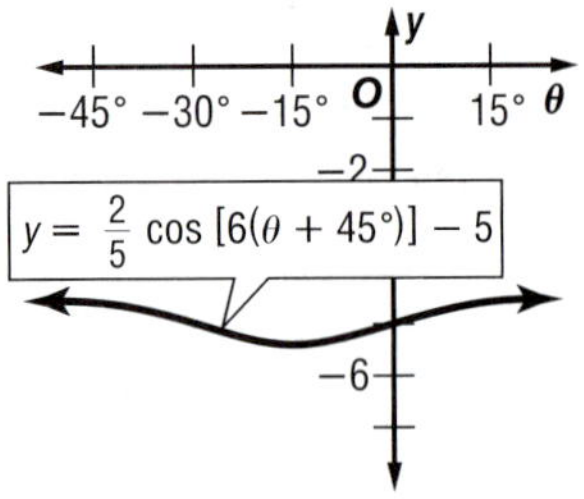

15. 6; 2; $\frac{2\pi}{3}$; $-\frac{\pi}{2}$

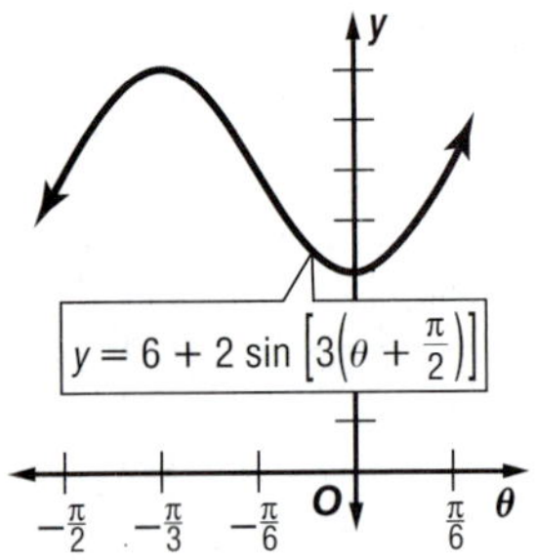

16. 3; 3; π; $\frac{\pi}{3}$

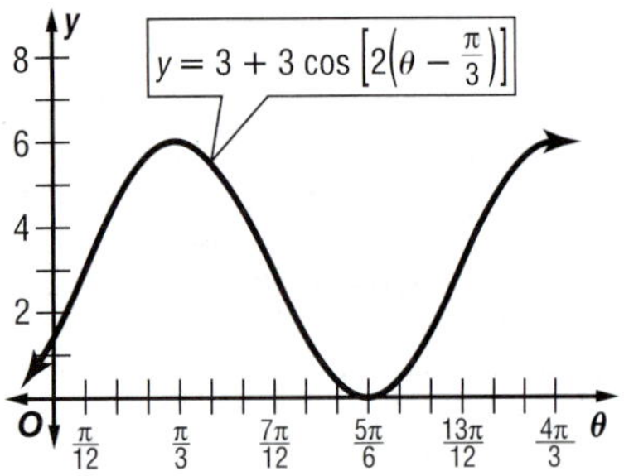

Page 860, Lesson 14-4

1.
$$\sin^2\theta + \cos^2\theta + \tan^2\theta \stackrel{?}{=} \sec^2\theta$$
$$1 + \tan^2\theta \stackrel{?}{=} \sec^2\theta$$
$$\sec^2\theta = \sec^2\theta$$

2.
$$\frac{\tan\theta}{\sin\theta} \stackrel{?}{=} \sec\theta$$
$$\frac{\sin\theta}{\cos\theta}\cdot\frac{1}{\sin\theta} \stackrel{?}{=} \sec\theta$$
$$\frac{1}{\cos\theta} \stackrel{?}{=} \sec\theta$$
$$\sec\theta = \sec\theta$$

3.
$$\frac{\tan\theta}{\cot\theta} \stackrel{?}{=} \tan^2\theta$$
$$\tan\theta \div \cot\theta \stackrel{?}{=} \tan^2\theta$$
$$\tan\theta \div \frac{1}{\tan\theta} \stackrel{?}{=} \tan^2\theta$$
$$\tan\theta\cdot\tan\theta = \tan^2\theta$$

4.
$$\csc^2\theta\,(1-\cos^2\theta) \stackrel{?}{=} 1$$
$$\csc^2\theta\,(\sin^2\theta) \stackrel{?}{=} 1$$
$$\frac{1}{\sin^2\theta}(\sin^2\theta) \stackrel{?}{=} 1$$
$$1 = 1$$

5.
$$1 - \cot^4\theta \stackrel{?}{=} 2\csc^2\theta - \csc^4\theta$$
$$(1-\cot^2\theta)(1+\cot^2\theta) \stackrel{?}{=} 2\csc^2\theta - \csc^4\theta$$
$$(1-\cot^2\theta)(\csc^2\theta) \stackrel{?}{=} 2\csc^2\theta - \csc^4\theta$$
$$[1-(\csc^2\theta - 1)](\csc^2\theta) \stackrel{?}{=} 2\csc^2\theta - \csc^4\theta$$
$$(2-\csc^2\theta)(\csc^2\theta) \stackrel{?}{=} 2\csc^2\theta - \csc^4\theta$$
$$2\csc^2\theta - \csc^4\theta = 2\csc^2\theta - \csc^4\theta$$

6.
$$\sin^4\theta - \cos^4\theta \stackrel{?}{=} \sin^2\theta - \cos^2\theta$$
$$(\sin^2\theta - \cos^2\theta)(\sin^2\theta + \cos^2\theta) \stackrel{?}{=} \sin^2\theta - \cos^2\theta$$
$$(\sin^2\theta - \cos^2\theta)(1) \stackrel{?}{=} \sin^2\theta - \cos^2\theta$$
$$\sin^2\theta - \cos^2\theta = \sin^2\theta - \cos^2\theta$$

7.
$$\sin^2\theta + \cot^2\theta\sin^2\theta \stackrel{?}{=} 1$$
$$\sin^2\theta\,(1+\cot^2\theta) \stackrel{?}{=} 1$$
$$\sin^2\theta\,(\csc^2\theta) \stackrel{?}{=} 1$$
$$\sin^2\theta\left(\frac{1}{\sin^2\theta}\right) \stackrel{?}{=} 1$$
$$1 = 1$$

8.
$$\frac{\cos\theta}{\csc\theta} - \frac{\csc\theta}{\sec\theta} \stackrel{?}{=} -\frac{\cos^3\theta}{\sin\theta}$$
$$\left(\cos\theta \div \frac{1}{\sin\theta}\right) - \left(\frac{1}{\sin\theta} \div \frac{1}{\cos\theta}\right) \stackrel{?}{=} -\frac{\cos^3\theta}{\sin\theta}$$
$$(\cos\theta\cdot\sin\theta) - \left(\frac{1}{\sin\theta}\cdot\cos\theta\right) \stackrel{?}{=} -\frac{\cos^3\theta}{\sin\theta}$$
$$\cos\theta\sin\theta - \frac{\cos\theta}{\sin\theta} \stackrel{?}{=} -\frac{\cos^3\theta}{\sin\theta}$$
$$\frac{\cos\theta\sin^2\theta - \cos\theta}{\sin\theta} \stackrel{?}{=} -\frac{\cos^3\theta}{\sin\theta}$$
$$\frac{-\cos\theta\,(1-\sin^2\theta)}{\sin\theta} \stackrel{?}{=} -\frac{\cos^3\theta}{\sin\theta}$$
$$\frac{-\cos\theta\,(\cos^2\theta)}{\sin\theta} \stackrel{?}{=} -\frac{\cos^3\theta}{\sin\theta}$$
$$-\frac{\cos^3\theta}{\sin\theta} = -\frac{\cos^3\theta}{\sin\theta}$$

9.
$$\frac{\cos\theta}{\sec\theta - 1} + \frac{\cos\theta}{\sec\theta + 1} \stackrel{?}{=} 2\cot^2\theta$$
$$\frac{\cos\theta\,(\sec\theta + 1) + \cos\theta\,(\sec\theta - 1)}{\sec^2\theta - 1} \stackrel{?}{=} 2\cot^2\theta$$
$$\frac{\cos\theta\sec\theta + \cos\theta + \cos\theta\sec\theta - \cos\theta}{\tan^2\theta} \stackrel{?}{=} 2\cot^2\theta$$
$$\frac{2}{\tan^2\theta} \stackrel{?}{=} 2\cot^2\theta$$
$$2\cot^2\theta = 2\cot^2\theta$$

10.
$$\frac{1+\cos\theta}{\sin\theta} \stackrel{?}{=} \frac{\sin\theta}{1-\cos\theta}$$
$$\left(\frac{1-\cos\theta}{1-\cos\theta}\right)\left(\frac{1+\cos\theta}{\sin\theta}\right) \stackrel{?}{=} \frac{\sin\theta}{1-\cos\theta}$$
$$\frac{1-\cos^2\theta}{\sin\theta\,(1-\cos\theta)} \stackrel{?}{=} \frac{\sin\theta}{1-\cos\theta}$$
$$\frac{\sin^2\theta}{\sin\theta\,(1-\cos\theta)} \stackrel{?}{=} \frac{\sin\theta}{1-\cos\theta}$$
$$\frac{\sin\theta}{1-\cos\theta} = \frac{\sin\theta}{1-\cos\theta}$$

11. $$\sec\theta + \tan\theta \stackrel{?}{=} \frac{\cos\theta}{1-\sin\theta}$$
$$\frac{1}{\cos\theta} + \frac{\sin\theta}{\cos\theta} \stackrel{?}{=} \frac{\cos\theta}{1-\sin\theta}$$
$$\frac{1+\sin\theta}{\cos\theta} \stackrel{?}{=} \frac{\cos\theta}{1-\sin\theta}$$
$$\left(\frac{1-\sin\theta}{1-\sin\theta}\right)\left(\frac{1+\sin\theta}{\cos\theta}\right) \stackrel{?}{=} \frac{\cos\theta}{1-\sin\theta}$$
$$\frac{1-\sin^2\theta}{\cos\theta\,(1-\sin\theta)} \stackrel{?}{=} \frac{\cos\theta}{1-\sin\theta}$$
$$\frac{\cos^2\theta}{\cos\theta\,(1-\sin\theta)} \stackrel{?}{=} \frac{\cos\theta}{1-\sin\theta}$$
$$\frac{\cos\theta}{1-\sin\theta} = \frac{\cos\theta}{1-\sin\theta}$$

12. $$\tan\theta + \cot\theta \stackrel{?}{=} \csc\theta\sec\theta$$
$$\frac{\sin\theta}{\cos\theta} + \frac{\cos\theta}{\sin\theta} \stackrel{?}{=} \csc\theta\sec\theta$$
$$\frac{\sin^2\theta+\cos^2\theta}{\cos\theta\sin\theta} \stackrel{?}{=} \csc\theta\sec\theta$$
$$\frac{1}{\cos\theta\sin\theta} \stackrel{?}{=} \csc\theta\sec\theta$$
$$\frac{1}{\cos\theta}\cdot\frac{1}{\sin\theta} \stackrel{?}{=} \csc\theta\sec\theta$$
$$\sec\theta\csc\theta \stackrel{?}{=} \csc\theta\sec\theta$$
$$\csc\theta\sec\theta = \csc\theta\sec\theta$$

13. $$\frac{\cot^2\theta}{1+\cot^2\theta} \stackrel{?}{=} 1-\sin^2\theta$$
$$\frac{\cot^2\theta}{\csc^2\theta} \stackrel{?}{=} 1-\sin^2\theta$$
$$\frac{\cos^2\theta}{\sin^2\theta} \div \frac{1}{\sin^2\theta} \stackrel{?}{=} 1-\sin^2\theta$$
$$\frac{\cos^2\theta}{\sin^2\theta}\cdot\sin^2\theta \stackrel{?}{=} 1-\sin^2\theta$$
$$1-\sin^2\theta = 1-\sin^2\theta$$

14. $$\frac{\tan\theta-\sin\theta}{\sec\theta} \stackrel{?}{=} \frac{\sin^3\theta}{1+\cos\theta}$$
$$\left(\frac{\sin\theta}{\cos\theta}-\sin\theta\right)\div\frac{1}{\cos\theta} \stackrel{?}{=} \frac{\sin^3\theta}{1+\cos\theta}$$
$$\left(\frac{\sin\theta-\sin\theta\cos\theta}{\cos\theta}\right)\cdot\cos\theta \stackrel{?}{=} \frac{\sin^3\theta}{1+\cos\theta}$$
$$\sin\theta-\sin\theta\cos\theta \stackrel{?}{=} \frac{\sin^3\theta}{1+\cos\theta}$$
$$\left(\frac{1+\cos\theta}{1+\cos\theta}\right)(\sin\theta-\sin\theta\cos\theta) \stackrel{?}{=} \frac{\sin^3\theta}{1+\cos\theta}$$
$$\frac{(1+\cos\theta)(1-\cos\theta)(\sin\theta)}{1+\cos\theta} \stackrel{?}{=} \frac{\sin^3\theta}{1+\cos\theta}$$
$$\frac{(1-\cos^2\theta)(\sin\theta)}{1+\cos\theta} \stackrel{?}{=} \frac{\sin^3\theta}{1+\cos\theta}$$
$$\frac{\sin^2\theta\,(\sin\theta)}{1+\cos\theta} \stackrel{?}{=} \frac{\sin^3\theta}{1+\cos\theta}$$
$$\frac{\sin^3\theta}{1+\cos\theta} = \frac{\sin^3\theta}{1+\cos\theta}$$

15. $$\sin^2\theta\,(1-\cos^2\theta) \stackrel{?}{=} \sin^4\theta$$
$$\sin^2\theta\,(\sin^2\theta) \stackrel{?}{=} \sin^4\theta$$
$$\sin^4\theta = \sin^4\theta$$

16. $$\sin^2\theta + \sin^2\theta\tan^2\theta \stackrel{?}{=} \tan^2\theta$$
$$\sin^2\theta\,(1+\tan^2\theta) \stackrel{?}{=} \tan^2\theta$$
$$\sin^2\theta\,(\sec^2\theta) \stackrel{?}{=} \tan^2\theta$$
$$\sin^2\theta\left(\frac{1}{\cos^2\theta}\right) \stackrel{?}{=} \tan^2\theta$$
$$\frac{\sin^2\theta}{\cos^2\theta} \stackrel{?}{=} \tan^2\theta$$
$$\tan^2\theta = \tan^2\theta$$

17. $$\frac{\sec\theta-1}{\sec\theta+1} + \frac{\cos\theta-1}{\cos\theta+1} \stackrel{?}{=} 0$$
$$\frac{(\sec\theta-1)(\cos\theta+1)+(\cos\theta-1)(\sec\theta+1)}{(\sec\theta+1)(\cos\theta+1)} \stackrel{?}{=} 0$$
$$\frac{1+\sec\theta-\cos\theta-1+1-\sec\theta+\cos\theta-1}{(\sec\theta+1)(\cos\theta+1)} \stackrel{?}{=} 0$$
$$0 = 0$$

18. $$\tan^2\theta\,(1-\sin^2\theta) \stackrel{?}{=} \sin^2\theta$$
$$\tan^2\theta\,(\cos^2\theta) \stackrel{?}{=} \sin^2\theta$$
$$\frac{\sin^2\theta}{\cos^2\theta}(\cos^2\theta) \stackrel{?}{=} \sin^2\theta$$
$$\sin^2\theta = \sin^2\theta$$

19. $$\tan\theta + \frac{\cos\theta}{1+\sin\theta} \stackrel{?}{=} \sec\theta$$
$$\frac{\tan\theta+\tan\theta\sin\theta+\cos\theta}{1+\sin\theta} \stackrel{?}{=} \sec\theta$$
$$\left(\frac{\sin\theta}{\cos\theta}+\frac{\sin\theta}{\cos\theta}\cdot\frac{\sin\theta}{1}+\cos\theta\right)\div(1+\sin\theta) \stackrel{?}{=} \sec\theta$$
$$\frac{\sin\theta+\sin^2\theta+\cos^2\theta}{\cos\theta}\cdot\frac{1}{1+\sin\theta} \stackrel{?}{=} \sec\theta$$
$$\frac{\sin\theta+1}{\cos\theta}\cdot\frac{1}{1+\sin\theta} \stackrel{?}{=} \sec\theta$$
$$\frac{1}{\cos\theta} \stackrel{?}{=} \sec\theta$$
$$\sec\theta = \sec\theta$$

20. $$\frac{\tan\theta}{\sec\theta+1} \stackrel{?}{=} \frac{1-\cos\theta}{\sin\theta}$$
$$\left(\frac{\sin\theta}{\cos\theta}\right)\div\left(\frac{1}{\cos\theta}+1\right) \stackrel{?}{=} \frac{1-\cos\theta}{\sin\theta}$$
$$\left(\frac{\sin\theta}{\cos\theta}\right)\div\left(\frac{1+\cos\theta}{\cos\theta}\right) \stackrel{?}{=} \frac{1-\cos\theta}{\sin\theta}$$
$$\left(\frac{\sin\theta}{\cos\theta}\right)\cdot\left(\frac{\cos\theta}{1+\cos\theta}\right) \stackrel{?}{=} \frac{1-\cos\theta}{\sin\theta}$$
$$\frac{\sin\theta}{1+\cos\theta} \stackrel{?}{=} \frac{1-\cos\theta}{\sin\theta}$$
$$\left(\frac{1-\cos\theta}{1-\cos\theta}\right)\left(\frac{\sin\theta}{1+\cos\theta}\right) \stackrel{?}{=} \frac{1-\cos\theta}{\sin\theta}$$
$$\frac{\sin\theta\,(1-\cos\theta)}{1-\cos^2\theta} \stackrel{?}{=} \frac{1-\cos\theta}{\sin\theta}$$
$$\frac{\sin\theta\,(1-\cos\theta)}{\sin^2\theta} \stackrel{?}{=} \frac{1-\cos\theta}{\sin\theta}$$
$$\frac{1-\cos\theta}{\sin\theta} = \frac{1-\cos\theta}{\sin\theta}$$

21. $\csc\theta - \dfrac{\sin\theta}{1+\cos\theta} \stackrel{?}{=} \cot\theta$

$$\frac{1}{\sin\theta} - \frac{\sin\theta}{1+\cos\theta} \stackrel{?}{=} \cot\theta$$

$$\frac{1+\cos\theta-\sin^2\theta}{\sin\theta\,(1+\cos\theta)} \stackrel{?}{=} \cot\theta$$

$$\frac{\cos\theta+\cos^2\theta}{\sin\theta\,(1+\cos\theta)} \stackrel{?}{=} \cot\theta$$

$$\frac{\cos\theta\,(1+\cos\theta)}{\sin\theta\,(1+\cos\theta)} \stackrel{?}{=} \cot\theta$$

$$\frac{\cos\theta}{\sin\theta} \stackrel{?}{=} \cot\theta$$

$$\cot\theta = \cot\theta$$

Page 860, Lesson 14-5

16. $\sin(90° + \theta) \stackrel{?}{=} \cos\theta$

$$\sin 90°\cos\theta + \cos 90°\sin\theta \stackrel{?}{=} \cos\theta$$

$$(1)\cos\theta + (0)\sin\theta \stackrel{?}{=} \cos\theta$$

$$\cos\theta = \cos\theta$$

17. $\cos(180° - \theta) \stackrel{?}{=} -\cos\theta$

$$\cos 180°\cos\theta + \sin 180°\sin\theta \stackrel{?}{=} -\cos\theta$$

$$(-1)\cos\theta + (0)\sin\theta \stackrel{?}{=} -\cos\theta$$

$$-\cos\theta = -\cos\theta$$

18. $\sin(\pi + \theta) \stackrel{?}{=} -\sin\theta$

$$\sin\pi\cos\theta + \cos\pi\sin\theta \stackrel{?}{=} -\sin\theta$$

$$(0)\cos\theta + (-1)\sin\theta \stackrel{?}{=} -\sin\theta$$

$$-\sin\theta = -\sin\theta$$

19. $\sin(\theta + 30°) + \sin(\theta + 60°) \stackrel{?}{=} \dfrac{\sqrt{3}+1}{2}(\sin\theta + \cos\theta)$

$$(\sin\theta\cos 30° + \cos\theta\sin 30°) + (\sin\theta\cos 60° + \cos\theta\sin 60°) \stackrel{?}{=} \frac{\sqrt{3}+1}{2}(\sin\theta + \cos\theta)$$

$$\left(\frac{\sqrt{3}}{2}\sin\theta + \frac{1}{2}\sin\theta\right) + \left(\frac{1}{2}\cos\theta + \frac{\sqrt{3}}{2}\cos\theta\right) \stackrel{?}{=} \frac{\sqrt{3}+1}{2}(\sin\theta + \cos\theta)$$

$$\frac{\sqrt{3}+1}{2}\sin\theta + \frac{\sqrt{3}+1}{2}\cos\theta \stackrel{?}{=} \frac{\sqrt{3}+1}{2}(\sin\theta + \cos\theta)$$

$$\frac{\sqrt{3}+1}{2}(\sin\theta + \cos\theta) = \frac{\sqrt{3}+1}{2}(\sin\theta + \cos\theta)$$

20. $\cos(30° - \theta) + \cos(30° + \theta) \stackrel{?}{=} \sqrt{3}\cos\theta$

$$(\cos 30°\cos\theta + \sin 30°\sin\theta) + (\cos 30°\cos\theta - \sin 30°\sin\theta) \stackrel{?}{=} \sqrt{3}\cos\theta$$

$$\left(\frac{\sqrt{3}}{2}\cos\theta + \frac{1}{2}\sin\theta\right) + \left(\frac{\sqrt{3}}{2}\cos\theta - \frac{1}{2}\sin\theta\right) \stackrel{?}{=} \sqrt{3}\cos\theta$$

$$\left(\frac{\sqrt{3}}{2} + \frac{\sqrt{3}}{2}\right)\cos\theta \stackrel{?}{=} \sqrt{3}\cos\theta$$

$$\sqrt{3}\cos\theta = \sqrt{3}\cos\theta$$

Page 861, Lesson 14-6

13. $\dfrac{\sin 2\theta}{2\sin^2\theta} \stackrel{?}{=} \cot\theta$

$$\frac{2\sin\theta\cos\theta}{2\sin^2\theta} \stackrel{?}{=} \cot\theta$$

$$\frac{\cos\theta}{\sin\theta} \stackrel{?}{=} \cot\theta$$

$$\cot\theta = \cot\theta$$

14. $1 + \cos 2\theta \stackrel{?}{=} \dfrac{2}{1+\tan^2\theta}$

$$1 + (\cos^2\theta - \sin^2\theta) \stackrel{?}{=} \frac{2}{1 + \frac{\sin^2\theta}{\cos^2\theta}}$$

$$\cos^2\theta + \cos^2\theta \stackrel{?}{=} \frac{2}{\frac{\cos^2\theta + \sin^2\theta}{\cos^2\theta}}$$

$$2\cos^2\theta \stackrel{?}{=} \frac{2}{\frac{1}{\cos^2\theta}}$$

$$2\cos^2\theta = 2\cos^2\theta$$

15. $\csc\theta\sec\theta \stackrel{?}{=} 2\csc 2\theta$

$$\frac{1}{\sin\theta}\cdot\frac{1}{\cos\theta} \stackrel{?}{=} \frac{2}{\sin 2\theta}$$

$$\frac{1}{\sin\theta\cos\theta} \stackrel{?}{=} \frac{2}{2\sin\theta\cos\theta}$$

$$\frac{1}{\sin\theta\cos\theta} = \frac{1}{\sin\theta\cos\theta}$$

16. $\sin 2\theta(\cot\theta + \tan\theta) \stackrel{?}{=} 2$

$$2\sin\theta\cos\theta\left(\frac{\cos\theta}{\sin\theta} + \frac{\sin\theta}{\cos\theta}\right) \stackrel{?}{=} 2$$

$$\frac{2\sin\theta\cos\theta\,(\cos\theta)}{\sin\theta} + \frac{2\sin\theta\cos\theta\,(\sin\theta)}{\cos\theta} \stackrel{?}{=} 2$$

$$2\cos^2\theta + 2\sin^2\theta \stackrel{?}{=} 2$$

$$2(\cos^2\theta + \sin^2\theta) \stackrel{?}{=} 2$$

$$2 = 2$$

17.

$$\frac{1 - \tan^2\theta}{1 + \tan^2\theta} \stackrel{?}{=} \cos 2\theta$$

$$\frac{1 - \dfrac{\sin^2\theta}{\cos^2\theta}}{1 + \dfrac{\sin^2\theta}{\cos^2\theta}} \stackrel{?}{=} \cos 2\theta$$

$$\frac{\cos^2\theta - \sin^2\theta}{\cos^2\theta} \cdot \frac{\cos^2\theta}{\cos^2\theta + \sin^2\theta} \stackrel{?}{=} \cos 2\theta$$

$$\frac{\cos^2\theta - \sin^2\theta}{\cos^2\theta + \sin^2\theta} \stackrel{?}{=} \cos 2\theta$$

$$\cos^2\theta - \sin^2\theta \stackrel{?}{=} \cos 2\theta$$

$$\cos 2\theta = \cos 2\theta$$

18.

$$\frac{\cos\theta + \sin\theta}{\cos\theta - \sin\theta} \stackrel{?}{=} \frac{1 + \sin 2\theta}{\cos 2\theta}$$

$$\left(\frac{\cos\theta + \sin\theta}{\cos\theta + \sin\theta}\right)\left(\frac{\cos\theta + \sin\theta}{\cos\theta - \sin\theta}\right) \stackrel{?}{=} \frac{1 + \sin 2\theta}{\cos 2\theta}$$

$$\frac{\cos^2\theta + 2\cos\theta\sin\theta + \sin^2\theta}{\cos^2\theta - \sin^2\theta} \stackrel{?}{=} \frac{1 + \sin 2\theta}{\cos 2\theta}$$

$$\frac{1 + 2\cos\theta\sin\theta}{\cos 2\theta} \stackrel{?}{=} \frac{1 + \sin 2\theta}{\cos 2\theta}$$

$$\frac{1 + \sin 2\theta}{\cos 2\theta} = \frac{1 + \sin 2\theta}{\cos 2\theta}$$

17. $3750 < x(1.5x) < 15{,}000$; $50 < x < 100$

Mixed Problem Solving

Chapter 1 Solving Equations and Inequalities (pages 4–53)

GEOMETRY **For Exercises 1 and 2, use the following information.**
The formula for the surface area of a sphere is $SA = 4\pi r^2$, and the formula for the volume of a sphere is $V = \frac{4}{3}\pi r^3$. *(Lesson 1-1)*

1. Find the volume and surface area of a sphere with radius 2 inches. Write your answer in terms of π.
2. Is it possible for a sphere to have the same numerical value for the surface area and volume? If so, find the radius of such a sphere.

1. $\frac{32}{3}\pi$ in^3; 16π in^2 2. Yes; 3 units

3. **CONSTRUCTION** The Birtic family is building a family room on their house. The dimensions of the room are 26 feet by 28 feet. Show how to use the Distributive Property to mentally calculate the area of the room. *(Lesson 1-2)* $26 \cdot 28 = 26(20 + 8) = 520 + 208 = 728$

GEOMETRY **For Exercises 4–6, use the following information.** 5. 78π cm^2
The formula for the surface area of a cylinder is $SA = 2\pi r^2 + 2\pi rh$. *(Lesson 1-2)*

4. Use the Distributive Property to rewrite the formula by factoring out the greatest common factor of the two terms. $SA = 2\pi r(r + h)$
5. Find the surface area for a cylinder with radius 3 centimeters and height 10 centimeters using both formulas. Leave the answer in terms of π.
6. Which formula do you prefer? Explain your reasoning. Sample answer: The formula in Exercise 4 is quicker.

POPULATION **For Exercises 7 and 8, use the following information.**
In 1990, the population of Mankato, Minnesota, was 31,460. For each of the next eight years, the population decreased by an average of 85 people per year. *(Lesson 1-3)*

7. What was the population in 1998? 30,780
8. If the population continues to decline at the same rate as from 1990 to 1998, what would you expect the population to be in 2005? 30,185

9. **WEATHER** The average yearly temperature for a particular coastal city in California is 64°F. The temperature seldom varies more than 7 degrees from the average. Write and solve an equation describing the maximum and minimum temperatures for this city. *(Lesson 1-4)* $|t - 64| = 7$; 71°; 57°

ASTRONOMY **For Exercises 10 and 11, use the following information.** 10. 3647.5 million miles
The planets in our solar system travel in orbits that are not circular. For example, Pluto's farthest distance from the Sun is 4539 million miles, and its closest distance is 2756 million miles. *(Lesson 1-4)*

10. What is the average of the two distances?
11. Write an equation that can be solved to find the minimum and maximum distances from the Sun to Pluto. $|t - 3647.5| = 891.5$

HEALTH **For Exercises 12 and 13, use the following information.**
The National Heart Association recommends that less than 30% of a person's total daily Caloric intake come from fat. One gram of fat yields nine Calories. Jason is a healthy 21-year old male whose average daily Caloric intake is between 2500 and 3300 Calories. *(Lesson 1-5)*

12. Write an inequality that represents the suggested fat intake for Jason. $750 \leq x \leq 990$
13. What is the greatest suggested fat intake for Jason? 110 grams

TRAVEL **For Exercises 14 and 15, use the following information.** 14. $5f + 375 + 5(85) \leq 1000$
Bonnie is planning a 5-day trip to a convention. She wants to spend no more than $1000. The plane ticket is $375, and the hotel is $85 per night. *(Lesson 1-5)*

14. Let f represent the cost of food for one day. Write an inequality to represent this situation.
15. Solve the inequality and interpret the solution. $f \leq 40$; Bonnie can spend no more than $40 per day on food.
16. **PAINTING** Phil owns and operates a home remodeling business. He estimates that he will need 12–15 gallons of paint for a particular project. If each gallon of paint costs $18.99, write and solve a compound inequality to determine what the cost c of the paint could be. *(Lesson 1-6)* $12 \leq \frac{c}{18.99} \leq 15$; between $227.88 and $284.85

CONSTRUCTION **For Exercises 17 and 18, use the following information.** 17. See margin.
A new playground is to be built in the shape of a rectangle. The length must be 1.5 times the width. The playground must be more than 3750 square feet, but less than 15,000 square feet. *(Lesson 1-6)*

17. Write and solve a compound inequality to determine possible dimensions for the playground.
18. Give three possible dimensions for the playground. Sample answer: 52 ft by 78 ft; 95 ft by 142.5 ft; 70 ft by 105 ft

862 Mixed Problem Solving

AGRICULTURE **For Exercises 1–3, use the following information.**
The table shows the average prices received by farmers for a bushel of corn from 1940–1999. *(Lesson 2-1)* **1–2. See margin.**

Year	Price	Year	Price
1940	$0.62	1980	$3.11
1950	$1.52	1990	$2.28
1960	$1.00	1999	$1.90
1970	$1.33		

Source: *The World Almanac*

1. Write a relation to represent the data.
2. Graph the relation.
3. Is the relation a function? Explain your reasoning. **Yes; each domain value is paired with only one range value.**

MEASUREMENT **For Exercises 4 and 5, use the following information.** **4. 39.37 in.**
The equation $y = 0.3937x$ can be used to convert any number of centimeters x to inches y. *(Lesson 2-2)*

4. Find the number of inches in 100 centimeters.
5. Find the number of centimeters in 12 inches. **about 30.5 cm**

POPULATION **For Exercises 6–8, use the following information.**
The table shows the population of Miami, Florida, for various years since 1950. *(Lesson 2-3)*

Year	Population	Year	Population
1950	249,276	1990	358,648
1970	334,859	1998	368,624
1980	346,681	2001	365,127

Source: *The World Almanac*

6. Graph the data in the table. **See margin.**
7. Find the average rate of change in population from 1950 to 2001. **about 2272**
8. Find the rate of change from 1998 to 2001. What does your answer mean? **About −1166 per year; the population has been decreasing since 1998.**

HEALTH **For Exercises 9–11, use the following information.**
In 1985, 39% of people in the United States age 12 and over reported using cigarettes. The percent of people using cigarettes has decreased about 1% per year following 1985. **Source:** *The World Almanac* *(Lesson 2-4)*

9. Write an equation that represents how many people use cigarettes in x years. $y = -x + 39$
10. If the percent of people using cigarettes continues to decrease at the same rate, what percent of people would you predict to be using cigarettes in 2005? **19%**
11. If the trend continues, when would you predict there to be no people using cigarettes in the U.S.? How accurate is your prediction? **See margin.**

EMPLOYMENT **For Exercises 12–16, use the following information.**
The table shows the number of unemployed people in the United States and the percent of the population unemployed for 1993 to 1999. *(Lesson 2-5)*

Year	Number Unemployed	Percent Unemployed
1993	8,940,000	6.9
1994	7,996,000	6.1
1995	7,404,000	5.6
1996	7,236,000	5.4
1997	6,739,000	4.9
1998	6,210,000	4.5
1999	5,880,000	4.2

Source: *The World Almanac*

12. Draw two scatter plots of the data: one with year and number unemployed and the other with year and percent unemployed. **See margin.**
13. Use two ordered pairs to write a prediction equation for each scatter plot. **See margin.**
14. Compare the two equations. **See margin.**
15. Predict the percent of people that will be unemployed in 2005. **1.8%**
16. In 1999, what was the total number of people in the United States? **140,000,000 people**
17. **EDUCATION** At Madison Elementary, each classroom of students can have a maximum of 25 students. Draw a graph of a step function that shows the relationship between the number of students x and the number of classrooms y that are needed. *(Lesson 2-6)* **See margin.**

CRAFTS **For Exercises 18–20, use the following information.** **18.** $10s + 15\ell \geq 350$
Priscilla makes stuffed animals and plans to sell them at a local craft show. She charges $10 for the small animals and $15 for the large animals. To cover expenses at the craft show, she needs to sell at least $350 worth of animals. *(Lesson 2-7)*

18. Write an inequality that describes this situation.
19. Graph the inequality. **See margin.**
20. If she sells 10 small and 15 large animals, will she cover her expenses? **no**

Mixed Problem Solving

11. 2024; Not completely accurate, because there will probably always be some people who use cigarettes as long as they are available.

12. **Number of People Unemployed**

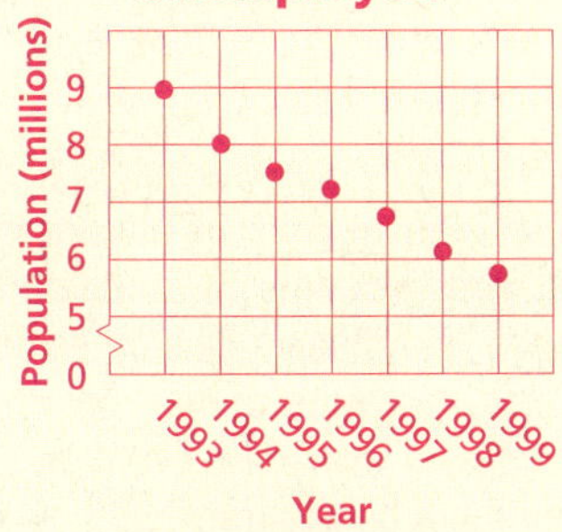

Percent of People Unemployed

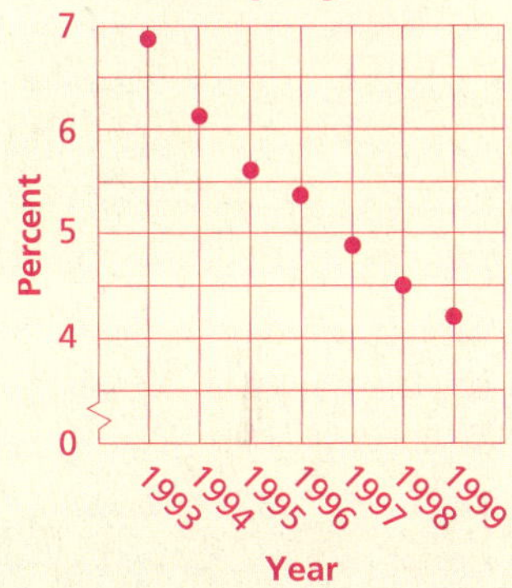

13. Sample answer using 1996 and 1999:
$y = -452{,}000x + 909{,}428{,}000$;
$y = -0.4x + 803.8$

14. The two equations both have negative slope, but the values of the slopes and the y-intercepts are very different.

17.

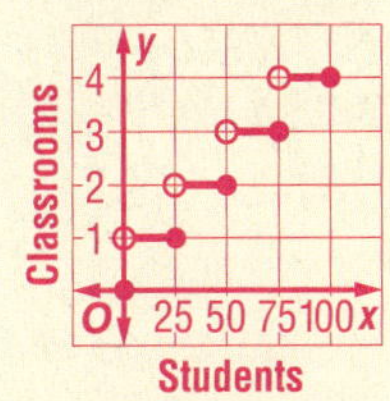

19.

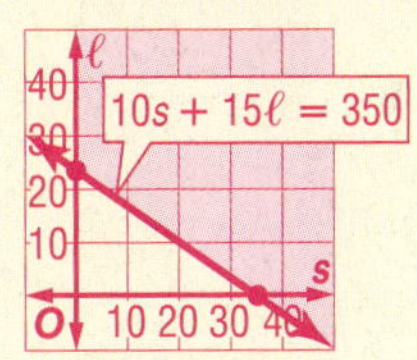

1. {(1940, 0.62), (1950, 1.52), (1960, 1.00), (1970, 1.33), (1980, 3.11), (1990, 2.28), (1999, 1.90)}

2. **Corn Price**

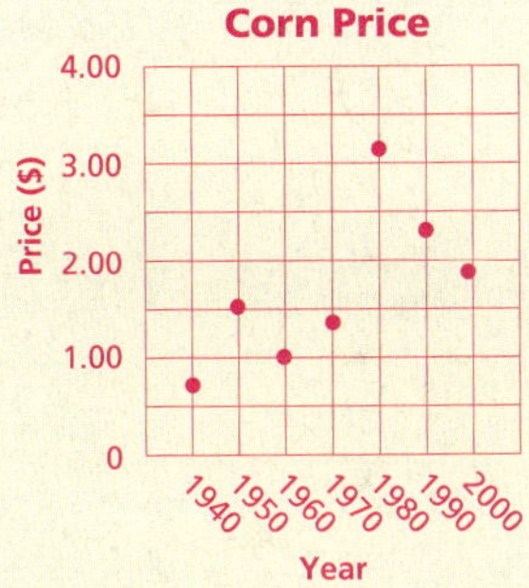

6. **Miami, FL Population**

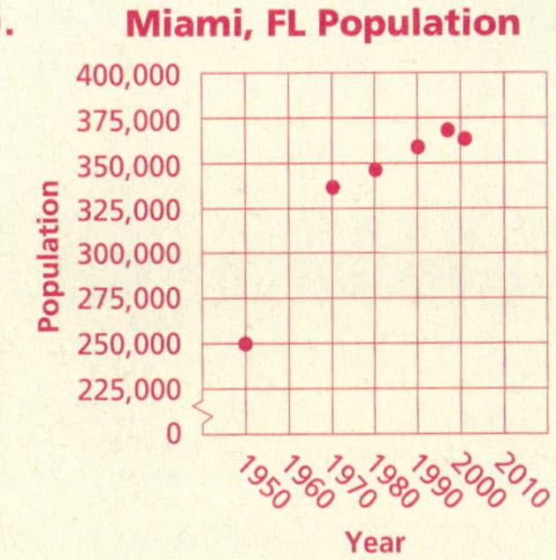

2. 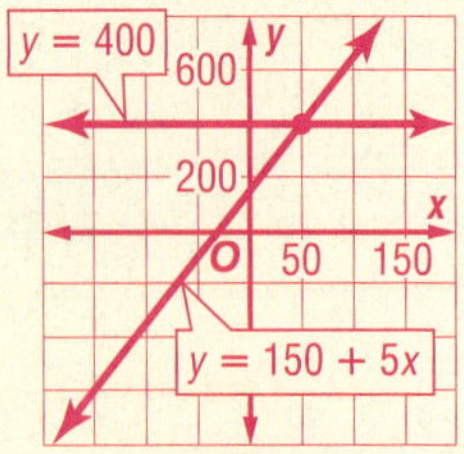

3. **It means that the options cost the same if you visit 50 times in a year.**

8. **Sample answer: If the situation remains the same, it would make sense, but if society changes and either more men or more women stay home with children, then it would not be accurate.**

11. 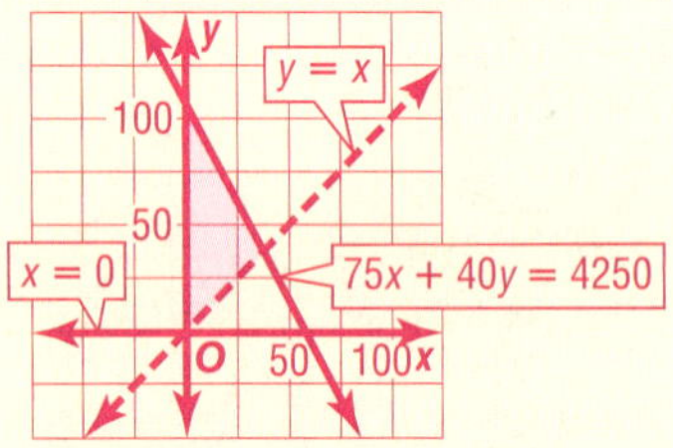

12. **Sample answer: 40 boots, 35 skis; 45 boots, 32 skis; 50 boots, 30 skis**

14. **$y \geq 0$, $x \geq 0$, $2x + y \leq 40$, $x + 3y \leq 60$**

15. 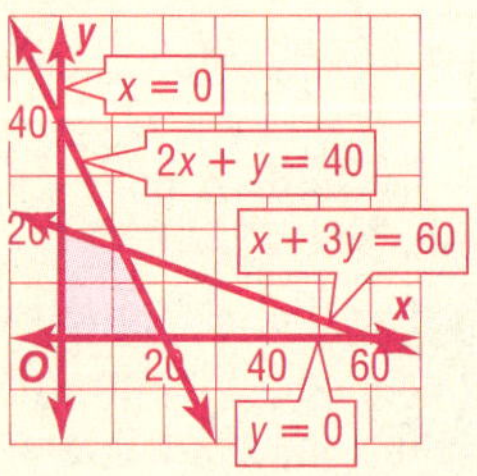

19. **$a + 2b + c = 58$, $\frac{a + c}{2} = 19$, $c = 2b + 7$**

20. **$a = 11$ in., $b = 10$ in., $c = 27$ in.**

Mixed Problem Solving

EXERCISE **For Exercises 1–4, use the following information.** **1. $y = 400$; $y = 150 + 5x$**
At Everybody's Gym, you have two options for becoming a member. You can pay \$400 per year or you can pay \$150 per year plus \$5 per visit. *(Lesson 3-1)* **2. See margin for graph; (50, 400).**

1. For each option, write an equation that represents the cost of belonging to the gym.
2. Graph the equations. Estimate the break-even point for the gym memberships. **3. See margin.**
3. Explain what the break-even point means.
4. If you plan to visit the gym at least once per week during the year, which option should you choose? **\$400 per year**

POPULATION **For Exercises 5–8, use the following information.**
In 1994, there were about 66.5 million men in the United States work force and 56.6 million women. Over the years, the number of men has increased by about 0.98 million per year, and the number of women has increased by about 1.08 million per year.
Source: *The World Almanac* *(Lesson 3-2)*

5. Write a system of equations that represents the number of men and women in the work force y for any number of years x.
6. Solve the system to determine the year in which the number of men and women in the work force will be the same. **2093**
7. What would be the number of men in the work force in the year in Exercise 6? **163.52 million**
8. Do you think the solution to this system makes sense for this situation? Explain. **See margin.**

5. men: $y = 66.5 + 0.98x$; women: $y = 56.6 + 1.08x$

9. **GEOMETRY** Find the coordinates of the vertices of the parallelogram whose sides are contained in the lines whose equations are $y = 3$, $y = 7$, $y = 2x$, and $y = 2x - 13$. *(Lesson 3-2)*
(1.5, 3), (3.5, 7), (8, 3), (10, 7)

EDUCATION **For Exercises 10–13, use the following information.**
Mr. Gunlikson needs to purchase equipment for his physical education classes. His budget for the year is \$4250. He decides to purchase cross-country ski equipment. He is able to find skis for \$75 per pair and boots for \$40 per pair. He knows that he should buy more boots than skis because the skis are adjustable to several sizes of boots. *(Lesson 3-3)*

10. Let y be the number of pairs of boots and x be the number of pairs of skis. Write a system of inequalities for this situation. (Remember that the number of pairs of boots and skis must be positive.) **$y \geq 0$, $x \geq 0$, $y > x$, $75x + 40y \leq 4250$**
11. Graph the region that shows how many pairs of boots and skis he can buy. **See margin.**
12. Give an example of three different purchases that Mr. Gunlikson can make. **See margin.**
13. Suppose Mr. Gunlikson wants to spend all of the money. What combination of skis and boots should he buy? Explain.
50 boots and 30 skis cost exactly \$4250.

MANUFACTURING **For Exercises 14–18, use the following information.**
A shoe manufacturer makes outdoor and indoor soccer shoes. There is a two-step process for both kinds of shoes. Each pair of outdoor shoes requires 2 hours in step one, 1 hour in step two, and produces a profit of \$20. Each pair of indoor shoes requires 1 hour in step one, 3 hours in step two, and produces a profit of \$15. The company has 40 hours of labor per day available for step one and 60 hours available for step 2. *(Lesson 3-4)* **17. $P(x) = 20x + 15y$**

14. Let x represent the number of pairs of outdoor shoes and let y represent the number of indoor shoes that can be produced per day. Write a system of inequalities to represent the number of pairs of outdoor and indoor soccer shoes that can be produced in one day. **See margin.**
15. Draw the graph showing the feasible region.
16. List the coordinates of the vertices of the feasible region. **(0, 0), (0, 20), (12, 16), (20, 0)**
17. Write a function for the total profit on the shoes.
18. What is the maximum profit? What is the combination of shoes for this profit?
\$480; 12 outdoor, 16 indoor **15. See margin.**

GEOMETRY **For Exercises 19–21, use the following information.**
An isosceles trapezoid has shorter base of measure a, longer base of measure c, and congruent legs of measure b. The perimeter of the trapezoid is 58 inches. The average of the bases is 19 inches and the longer base is twice the leg plus 7. *(Lesson 3-5)*

19. Write a system of three equations that represents this situation. **See margin.**
20. Find the lengths of the sides of the trapezoid.
21. Find the area of the trapezoid. **114 in^2**

20. See margin.

22. **EDUCATION** The three American universities with the greatest endowments in 2000 were Harvard, Yale, and Stanford. Their combined endowments are \$38.1 billion. Harvard had \$0.1 billion more in endowments than Yale and Stanford together. Stanford's endowments trailed Harvard's by \$10.2 billion. What were the endowments of each of these universities? *(Lesson 3-5)* **Harvard, \$19.1 billion; Yale, \$10.1 billion; Stanford, \$8.9 billion**

Mixed Problem Solving

864 Mixed Problem Solving

AGRICULTURE **For Exercises 1 and 2, use the following information.**
In 1999, the United States produced 62,662,000 metric tons of wheat, 9,546,000 metric tons of rice, and 239,719,000 metric tons of corn. In that same year, Russia produced 30,960,000 metric tons of wheat, 444,000 metric tons of rice, and 1,070,000 metric tons of corn. **Source:** *The World Almanac* *(Lesson 4-1)*

1. Organize the data in two different matrices.
2. What are the dimensions of the matrices in Exercise 1? **2 × 3; 3 × 2**

1. See margin.

LIFE EXPECTANCY **For Exercises 3–5, use the following information.**
The table shows the life expectancy for males and females for the given years. **Source:** *The World Almanac* *(Lesson 4-2)* **3–5. See margin.**

Year	1910	1930	1950	1970	1990
Male	48.4	58.1	65.6	67.1	71.8
Female	51.8	61.6	71.1	74.7	78.8

3. Organize all the data in a matrix.
4. Show how to organize the data in two matrices in such a way that you can find the difference between the life expectancies of males and females for the given years. Find the difference.
5. Does addition of any two matrices you can write for the data make sense? Explain.

CRAFTS **For Exercises 6 and 7, use the following information.**
Mrs. Long is selling crocheted items at a craft fair. She sells large afghans for \$60, baby blankets for \$40, doilies for \$25, and pot holders for \$5. She takes the following number of items to the fair: 12 afghans, 25 baby blankets, 45 doilies, and 50 pot holders. *(Lesson 4-3)* **6–7. See margin.**

6. Write an inventory matrix for the number of each item and a cost matrix for the price of each item.
7. Suppose Mrs. Long sells all of the items. Find her total income expressed as a matrix.

GEOMETRY **For Exercises 8–11, use the following information.**
A trapezoid has vertices $T(3, 3)$, $R(-1, 3)$, $A(-2, -4)$, and $P(5, -4)$. *(Lesson 4-4)* **8–11. See margin.**

8. Show how to use a reflection matrix to find the vertices of *TRAP* after a reflection over the *x*-axis.
9. The area of a trapezoid is found by multiplying one-half the sum of the bases by the height. Find the areas of *TRAP* and *T′R′A′P′*. How do they compare?
10. Show how to use a matrix and scalar multiplication to find the vertices of *TRAP* after a dilation that triples the perimeter of *TRAP*.
11. Find the areas of *TRAP* and *T′R′A′P′* in Exercise 10. How do they compare?

AGRICULTURE **For Exercises 12 and 13, use the following information.**
A farm has a triangular plot planted with alfalfa defined by the coordinates $\left(-\frac{1}{2}, -\frac{1}{4}\right)$, $\left(\frac{1}{3}, \frac{1}{2}\right)$, and $\left(\frac{2}{3}, -\frac{1}{2}\right)$, where units are in square miles. *(Lesson 4-5)*

12. Find the area of the region in square miles.
13. One square miles equals 640 acres. To the nearest acre, how many acres are in the triangular plot? **347 acres**

12. $\frac{13}{24}$ **mi²**

ART **For Exercises 14 and 15, use the following information.**
Alberda sells beads to use in making Native American jewelry. Small beads sell for \$5.80 per pound, and large beads sell for \$4.60 per pound. Bernadette bought a bag of beads for \$33.00 that contained 3 times as many pounds of the small beads as the large beads. *(Lesson 4-6)*

14. Write a system of equations using the information given.
15. How many pounds of small and large beads did Bernadette buy? **4.5 of small, 1.5 of large**

14. $x - 3y = 0$, $5.8x + 4.6y = 33.00$

MATRICES **For Exercises 16 and 17, use the following information.** **16–17. See margin.**
Two 2 × 2 inverse matrices have a sum of $\begin{bmatrix} -2 & 0 \\ 0 & -2 \end{bmatrix}$. The value of each entry is no less than −3 and no greater than 2. *(Lesson 4-7)*

16. Find the two matrices that satisfy the conditions.
17. Explain your method for finding the matrices.

18. **CONSTRUCTION** Alan enjoys building decks during the summer. For labor he charges \$750 to build a small deck and \$1250 to build a large deck. During the spring and summer of 2001, he built 5 more small decks than large decks. If he earned \$11,750, how many of each type of deck did he build? *(Lesson 4-8)* **9 small, 4 large**

6. $[12 \quad 25 \quad 45 \quad 50]$, $\begin{bmatrix} 60 \\ 40 \\ 25 \\ 5 \end{bmatrix}$

7. $[3095]$

8. $\begin{bmatrix} 1 & 0 \\ 0 & -1 \end{bmatrix} \cdot \begin{bmatrix} 3 & -1 & -2 & 5 \\ 3 & 3 & -4 & -4 \end{bmatrix} = \begin{bmatrix} 3 & -1 & -2 & 5 \\ -3 & -3 & 4 & 4 \end{bmatrix}$

9. **38.5 units², 38.5 units²; The areas are the same.**

10. $3\begin{bmatrix} 3 & -1 & -2 & 5 \\ 3 & 3 & -4 & -4 \end{bmatrix} = \begin{bmatrix} 9 & -3 & -6 & 15 \\ 9 & 9 & -12 & -12 \end{bmatrix}$

11. **38.5 units², 346.5 units²; The area has been multiplied by 9.**

16. **Sample answer:** $\begin{bmatrix} -1 & 0 \\ 0 & -1 \end{bmatrix}$ **and** $\begin{bmatrix} -1 & 0 \\ 0 & -1 \end{bmatrix}$ **or** $\begin{bmatrix} 1 & -2 \\ 2 & -3 \end{bmatrix}$ **and** $\begin{bmatrix} -3 & 2 \\ -2 & 1 \end{bmatrix}$ **or any two matrices that multiply to** $\begin{bmatrix} 1 & 0 \\ 0 & 1 \end{bmatrix}$ **and add to** $\begin{bmatrix} -2 & 0 \\ 0 & -2 \end{bmatrix}$.

17. **Sample answer: guess and check**

1. $\begin{array}{c} \\ \text{U.S.} \\ \text{Russia} \end{array} \begin{array}{c} \begin{array}{ccc} \text{wheat} & \text{rice} & \text{corn} \end{array} \\ \begin{bmatrix} 62{,}662{,}000 & 9{,}546{,}000 & 239{,}719{,}000 \\ 30{,}960{,}000 & 444{,}000 & 1{,}070{,}000 \end{bmatrix} \end{array}$; $\begin{array}{c} \\ \text{wheat} \\ \text{rice} \\ \text{corn} \end{array} \begin{array}{c} \begin{array}{cc} \text{U.S.} & \text{Russia} \end{array} \\ \begin{bmatrix} 62{,}662{,}000 & 30{,}960{,}000 \\ 9{,}546{,}000 & 444{,}000 \\ 239{,}719{,}000 & 1{,}070{,}000 \end{bmatrix} \end{array}$

3. $\begin{array}{c} \\ \text{Male} \\ \text{Female} \end{array} \begin{array}{c} \begin{array}{ccccc} 1910 & 1930 & 1950 & 1970 & 1990 \end{array} \\ \begin{bmatrix} 48.4 & 58.1 & 65.6 & 67.1 & 71.8 \\ 51.8 & 61.6 & 71.1 & 74.7 & 78.8 \end{bmatrix} \end{array}$

4. Sample answer: $[51.8 \quad 61.6 \quad 71.1 \quad 74.7 \quad 78.8] - [48.4 \quad 58.1 \quad 65.6 \quad 67.1 \quad 71.8] = [3.4 \quad 3.5 \quad 5.5 \quad 7.6 \quad 7]$

5. No; adding life expectancies would make no sense.

1. 2.826146×10^6; 4.6534687×10^7

3. $\$3.10014084794 \times 10^{11}$; \$310,014,084,794

4. Mexico City: $18{,}131{,}000(1 + r)^2 = 18{,}131{,}000 + 36{,}262{,}000r + 18{,}131{,}000r^2$; Bombay: $18{,}066{,}000(1 + r)^2 = 18{,}066{,}000 + 36{,}132{,}000r + 18{,}066{,}000r^2$

5. Mexico City: 19,249,854; Bombay: 28,146,239

8. No; for example, if $x = 1$, the ratio is 3:5:8, but if $x = 2$, the ratio is 4:7:11. The ratios are not equivalent.

13. With the new formula, the answer to Exercise 12 is about 42 mph. $2\sqrt{5} \neq 5.5$.

16. 1 AU; an astronomical unit is the distance of a planet from the Sun where one unit is the distance of the Earth from the Sun. The calculations were based upon the Earth year which is also one unit. So the answer is 1.

20. Sample: Write the equation $(a + bi) + (a - bi) = 12$. Solving results in $a = 6$. Then write the equation $(a + bi)(a - bi) = 40$. Solving results in $a^2 - b^2i^2 = 40$, $36 - b^2i^2 = 40$, $-b^2i^2 = 4$, $b^2 = 4$, and $b = \pm 2$.

Mixed Problem Solving

Chapter 5 Polynomials

(pages 220–283)

EDUCATION For Exercises 1–3, use the following information.

In 1998 in the United States, there were 2,826,146 classroom teachers and 46,534,687 students. An average of \$6662 was spent per student.
Source: *The World Almanac* (Lesson 5-1)

1. Write the numbers of teachers and students in scientific notation. **See margin.**
2. Find the number of students per teacher. Write the answer in standard notation rounded to the nearest whole number. **16**
3. Find the total amount of money spent for students in 1998. Write the answer in both scientific and standard notation. **See margin.**

POPULATION For Exercises 4–6, use the following information.

In 2000, the population of Mexico City, Mexico, was 18,131,000, and the population of Bombay, India, was 18,066,000. It is projected that, until the year 2015, the population of Mexico City will increase at the rate of 0.4% per year and the population of Bombay will increase at the rate of 3% per year. **Source:** *The World Almanac* (Lesson 5-2) **4–5. See margin.**

4. Let r represent the rate of increase in population for each city. Write a polynomial to represent the population of each city in 2002.
5. Predict the population of each city in 2015.
6. If the projected rates are accurate, in what year will the two cities have approximately the same population? **before the end of 2001**

GEOMETRY For Exercises 7 and 8, use the following information.

A rectangular box for a new product is designed in such a way that the three dimensions always have a particular relationship defined by the variable x. The volume of the box can be written as $6x^3 + 31x^2 + 53x + 30$, and the height is always $x + 2$. (Lesson 5-3)

7. What are the width and length of the box in terms of x? **$2x + 3$ and $3x + 5$**
8. Will the ratio of the dimensions of the box always be the same regardless of the value of x? Explain. **See margin.**

GEOMETRY For Exercises 9 and 10, use the following information.

Hero's formula for the area of a triangle is given by $A = \sqrt{s(s - a)(s - b)(s - c)}$, where a, b, and c are the lengths of the sides of the triangle and $s = 0.5(a + b + c)$. (Lesson 5-4)

9. Find the lengths of the sides of the triangle given in this application of Hero's formula: $A = \sqrt{s^4 - 12s^3 + 47s^2 - 60s}$. **3, 4, 5**
10. What type of triangle is this? **right**

11. **PHYSICS** The speed of sound in a liquid is $s = \sqrt{\frac{B}{d}}$, where B is known as the bulk modulus of the liquid and d is the density of the liquid. For water $B = 2.1 \cdot 10^9$ N/m^2 and $d = 10^3$ kg/m^3. Find the speed of sound in water to the nearest meter per second. (Lesson 5-5) **1449 m/s**

LAW ENFORCEMENT For Exercises 12 and 13, use the following information.

The approximate speed s in miles per hour that a car was traveling if it skidded d feet is given by the formula $s = 5.5\sqrt{kd}$, where k is the coefficient of friction. (Lesson 5-6) **13. See margin.**

12. For a dry concrete road, $k = 0.8$. If a car skids 110 feet on a dry concrete road, find its speed in miles per hour to the nearest whole number. **about 52 mph**
13. Another formula using the same variables is $s = 2\sqrt{5kd}$. Compare the results using the two formulas. Explain any variations in the answers.

PHYSICS For Exercises 14–16, use the following information. **15. about 39.44 AU**

Kepler's Third Law of planetary motion states that the square of the orbital period of any planet, in Earth years, is equal to the cube of the planet's distance from the Sun in astronomical units (AU).
Source: *The World Almanac* (Lesson 5-7) **14. about 0.387 AU**

14. The orbital period of Mercury is 87.97 Earth days. What is Mercury's distance from the Sun in AU?
15. Pluto's period of revolution is 247.66 Earth years. What is Pluto's distance from the Sun?
16. What is Earth's distance from the Sun in AU? Explain your result. **See margin.**

PHYSICS For Exercises 17 and 18, use the following information. **17. about 9.5 s**

The time T in seconds that it takes a pendulum to make a complete swing back and forth is given by the formula $T = 2\pi\sqrt{\frac{L}{g}}$, where L is the length of the pendulum in feet and g is the acceleration due to gravity, 32 feet per second squared. (Lesson 5-8)

17. In Tokyo, Japan, a huge pendulum in the Shinjuku building measures 73 feet 9.75 inches. How long does it take for the pendulum to make a complete swing? **Source:** *The Guinness Book of Records*
18. A clockmaker wants to build a pendulum that takes 20 seconds to swing back and forth. How long should the pendulum be? **about 324 ft**

NUMBER THEORY For Exercises 19 and 20, use the following information.

Two complex conjugate numbers have a sum of 12 and a product of 40. (Lesson 5-9)

19. Find the two numbers. **$6 + 2i$, $6 - 2i$**
20. Explain the method you used to find the numbers. **See margin.**

PHYSICS For Exercises 1–3, use the following information.
A model rocket is shot straight up from the top of a 100-foot building at a velocity of 800 feet per second. *(Lesson 6-1)* **1. $h(t) = -16t^2 + 800t + 100$**

1. The height $h(t)$ of the model rocket t seconds after firing is given by $h(t) = -16t^2 + at + b$ where a is the initial velocity in feet per second and b is the initial height of the rocket above the ground. Write an equation for the model rocket.
2. Find the maximum height reached by the rocket and the time that the height is reached.
3. Suppose a rocket is fired from the ground (initial height is 0). Find values for a, initial velocity, and t, time, such that the rocket reaches a height of 32,000 feet at time t. **Sample answer: $a \approx 1547$, $t = 30$**

2. 10,100 ft; 25 s

RIDES For Exercises 4 and 5, use the following information.
An amusement park ride carries riders to the top of a 225-foot tower. The riders then free-fall in their seats until they reach 30 feet above the ground. *(Lesson 6-2)*

4. Use the formula $h(t) = -16t^2 + h_0$, where the time t is in seconds and the initial height h_0 is in feet to find how long the riders are in free-fall. **about 3.5 s**
5. Suppose the designer of the ride wants the riders to experience free-fall for 5 seconds before stopping 30 feet above the ground. What should be the height of the tower? **430 ft**

CONSTRUCTION For Exercises 6 and 7, use the following information.
Nicole's new house has a small deck that measures 6 feet by 12 feet. She would like to build a larger deck. *(Lesson 6-3)* **6. 6 ft 7. 12 ft by 18 ft**

6. By what amount must each dimension be increased to triple the area of the original deck?
7. What are the new dimensions of the deck?

CONSTRUCTION For Exercises 8 and 9, use the following information.
A contractor wants to construct a rectangular pool with a length that is twice the width. He plans to build a two-meter-wide walkway around the pool. He wants the area of the walkway to equal the surface area of the pool. *(Lesson 6-4)*

8. Find the dimensions of the pool to the nearest tenth of a meter. **7.1 m by 14.2 m**
9. What is the surface area of the pool to the nearest square meter? **101 m²**

PHYSICS For Exercises 10–12, use the following information.
A ball is thrown vertically into the air with a velocity of 112 feet per second. The ball was released 6 feet above the ground. The height above the ground t seconds after release is modeled by the equation $h(t) = -16t^2 + 112t + 6$. *(Lesson 6-5)*

10. When will the ball reach a height of 130 feet?
11. Will the ball ever reach 250 feet? Explain your reasoning.
12. In how many seconds after its release will the ball hit the ground? **in about 7 s**

10. at about 1.4 s and 5.6 s
11. No; if you graph the function, the vertex is at about $y = 201$, so the height will never be 250.

WEATHER For Exercises 13–15, use the following information.
The table shows the normal high temperatures for Albany, New York.
Source: *The World Almanac* *(Lesson 6-6)*

Month	Temperature (°F)
January	21
February	24
March	34
April	46
May	58
June	67
July	72
August	70
September	61
October	50
November	40
December	27

13. Suppose the months are numbered such that January = 1, February = 2, and so on. A graphing calculator gave the following function as a model for the data: $y = -1.5x^2 + 21.2x - 8.5$. Graph the points in the table and the function on the same coordinate plane. **See margin.**
14. Identify the vertex, axis of symmetry, and direction of opening for the calculator function.
15. Discuss how well you think the function models the actual temperature data. **See margin.**

14. vertex: about (7.1, 66.4); axis: $x = 7.1$; opens down

16. **MODELS** John is building a display table for model cars. He wants the perimeter of the table to be 26 feet, but he wants the area of the table to be no more than 30 square feet. What could the width of the table be? *(Lesson 6-7)*
0 to 3 ft or 10 to 13 ft

Mixed Problem Solving

13.

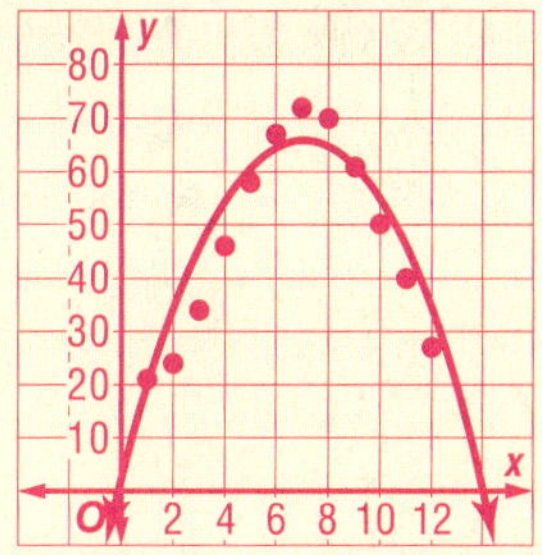

15. The function does not fit the first two months very well and the temperature represented by its vertex is not quite as high as the highest normal temperature. However, the data does appear to need a curved model and this function fits quite a few of the points well.

Mixed Problem Solving

2. the high and low temperatures

3.

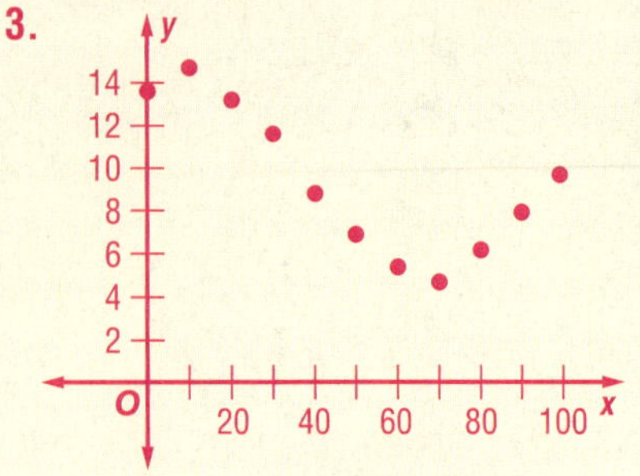

4. There is a relative maximum at $x = 10$ and a relative minimum at $x = 70$ and the graph is increasing as $x \to +\infty$.

8. \$1161.7 billion; \$1042 billion

9. Sample answer: No; the graph of the function has a relative maximum at about $x = 8$ or the year 2004 and then the values for sales decrease. It is not likely that the sales will decrease in the future.

10. $(x + 12)(x + 16)(x + 18) = 5985$

12. bases: 25 ft, 83 ft; height: 76 ft

14. The function represents the difference in the number of men and women employed in the U.S.

16–17.

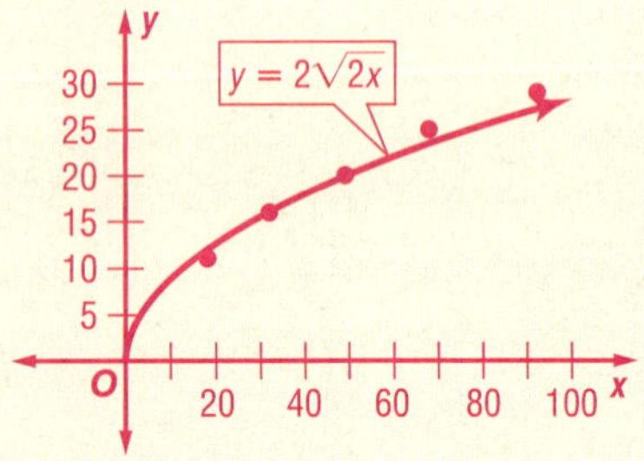

17. Sample answer: The graph of $y = 2\sqrt{2x}$ fits the data quite well.

WEATHER For Exercises 1 and 2, use the following information.

The graph models the average monthly temperature of Boise, Idaho. The values of x are $x = 1$ for January, $x = 2$ for February, and so on. The temperatures are in °F. Source: *The World Almanac* (Lesson 7-1) 2. See margin.

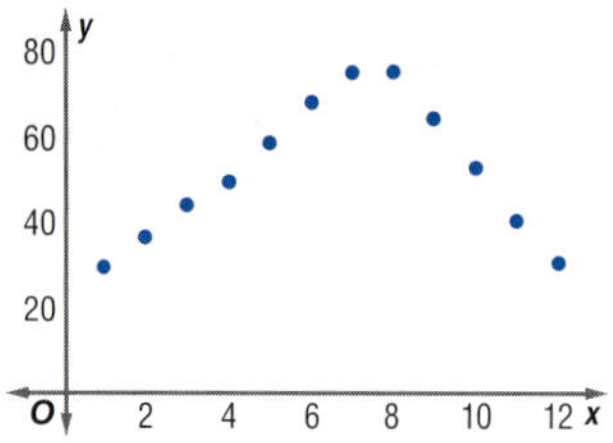

1. Is the function odd or even? even
2. What do the relative maxima and minima represent?

POPULATION For Exercises 3–5, use the following information.

The table shows the percent of the U.S. population that was foreign-born during various years. The x values are years since 1900 and the y values are the percent of the population. Source: *The World Almanac* (Lesson 7-2)

U.S. Foreign-Born Population

x	y	x	y
0	13.6	60	5.4
10	14.7	70	4.7
20	13.2	80	6.2
30	11.6	90	8.0
40	8.8	99	9.7
50	6.9		

3. Graph the function. 3–4. See margin.
4. Describe the turning points of the graph and its end behavior.
5. If this graph was modeled by a polynomial equation, what is the least degree the equation could have? 3

DESIGN For Exercises 6 and 7, use the following information.

A cylindrical can has a volume of approximately 628 cubic inches and a height of 9 inches. The volume of the can is represented by $V = \pi(a + 3)^2(9)$. (Lesson 7-3)

6. Use $\pi = 3.14$ to find a to the nearest tenth. 1.7
7. What is the radius of the can? 4.7 in.

SALES For Exercises 8 and 9, use the following information. 8–9. See margin.

The sales of items related to information technology can be modeled by $S(x) = -1.7x^3 + 18x^2 + 26.4x + 678$, where x is the number of years since 1996 and y is billions of dollars. Source: *Wall Street Almanac* (Lesson 7-4)

8. Use synthetic substitution to estimate the sales for 2003 and 2006.
9. Do you think this model is useful in estimating the sales of this industry in the future? Explain.

MANUFACTURING For Exercises 10 and 11, use the following information. 11. 3 in.

A box measures 12 inches by 16 inches by 18 inches. The manufacturer will increase each dimension of the box by the same number of inches and have a new volume of 5985 cubic inches. (Lesson 7-5)

10. Write a polynomial equation to model this situation. See margin.
11. How much should be added to each dimension?
12. **CONSTRUCTION** A picnic area has the shape of a trapezoid. The longer base is 8 more than 3 times the length of the shorter base and the height is 1 more than 3 times the shorter base. What are the dimensions of the trapezoid if the area is 4104 square feet? (Lesson 7-6) See margin.

EMPLOYMENT For Exercises 13 and 14, use the following information. 13. $y = 2085.6x + 123{,}060$

From 1994 to 1999, the number of employed women and men in the United States, age 16 and over, can be modeled by the following equations where x is the number of years since 1994 and y is the number of people in thousands. Source: *The World Almanac* (Lesson 7-7)

women: $y = 1086.4x + 56{,}610$
men: $y = 999.2x + 66{,}450$

13. Write a function that models the total number of men and women employed in the United States.
14. If f is the function for the number of men and g is the function for the number of women, what does $(f - g)(x)$ represent? See margin.
15. **HEALTH** The average weight of a baby born at a certain hospital is $7\frac{1}{2}$ pounds, and the average length is 19.5 inches. One kilogram is about 2.2 pounds, and 1 centimeter is about 0.3937 inches. Find the average weight in kilograms and the length in centimeters. (Lesson 7-8) about 3.4 kg, about 49.5 cm

SAFETY For Exercises 16 and 17, use the following information.

The table shows the total stopping distance x, in meters, of a vehicle and the speed y, in meters per second. (Lesson 7-9) 16–17. See margin.

Distance	92	68	49	32	18
Speed	29	25	20	16	11

16. Graph the data in the table.
17. Graph the function $y = 2\sqrt{2x}$ on the same coordinate plane. How well do you think this function models the given data? Explain.

868 Mixed Problem Solving

GEOMETRY **For Exercises 1–4, use the following information.**
Triangle ABC has vertices $A(2, 1)$, $B(-6, 5)$, and $C(-2, -3)$. *(Lesson 8-1)* **3–4. See margin.**

1. An isosceles triangle has two sides with equal length. Is triangle ABC isosceles? Explain your reasoning. **Yes; $AB = BC = 4\sqrt{5}$.**
2. An equilateral triangle has three sides of equal length. Is triangle ABC equilateral? Explain your reasoning. **No; $AC = 4\sqrt{2}$.**
3. Triangle EFG is formed by joining the midpoints of the sides of triangle ABC. What type of triangle is $\triangle EFG$? Explain your reasoning.
4. Describe any relationship between the lengths of the sides of the two triangles.

ENERGY **For Exercises 5–8, use the following information.**
Solar energy plays an important role in space satellites. As they become more efficient, solar batteries are being used for more purposes on Earth. A parabolic mirror can be used to collect solar energy. The mirrors reflect the rays from the Sun to the focus of the parabola. The latus rectum of a particular mirror is 40 feet long. *(Lesson 8-2)* **5–8. See margin.**

5. Write an equation for the parabola formed by the mirror if the vertex of the mirror is 9.75 feet below the origin.
6. One foot is exactly 0.3048 meter. Rewrite the equation for the mirror in terms of meters.
7. Graph one of the equations for the mirror.
8. Which equation did you choose to graph? Explain.

COMMUNICATION **For Exercises 9–11, use the following information.**
Radio waves carry information from the transmitter to the receivers. The radio tower for KCGM, Voice of the Prairies, has a circular radius for broadcasting of 65 miles. The radio tower for KVCK has a circular radius for broadcasting of 85 miles. *(Lesson 8-3)*

9. Let the radio tower for KCGM be located at the origin of a coordinate system. Write an equation for the set of points at the maximum broadcast distance from the tower. **$x^2 + y^2 = 4225$**
10. The radio tower for KVCK is 50 miles south and 15 miles west of the KCGM tower. Let each mile represent one unit on the coordinate system. Write an equation for the set of points at the maximum broadcast distance from the KVCK tower. **$(x + 15)^2 + (y + 50)^2 = 7225$**
11. Graph the two equations and show the area where the radio signals overlap. **See margin.**

ASTRONOMY **For Exercises 12–14, use the following information.**
The table shows the closest and farthest distances of Venus and Jupiter from the Sun in millions of miles. **Source:** *The World Almanac* *(Lesson 8-4)*

Planet	Closest	Farthest
Venus	66.8	67.7
Jupiter	460.1	507.4

12. Write an equation for the orbit of each planet, assuming that the center of the orbit is the origin, the center of the Sun is a focus of the ellipse, and the Sun lies on the x-axis. **See margin.**
13. Find the eccentricity e, or the ratio $\frac{c}{a}$, for each planet. **Venus: 0.0067; Jupiter: 0.0489**
14. Which planet has an orbit that is closer to looking like a circle? Explain your reasoning. **See margin.**
15. A comet follows a path that is one branch of a hyperbola. Suppose Earth is the center of the hyperbolic curve and has coordinates (0, 0). Write an equation for the path of the comet if $c = 5{,}225{,}000$ miles and $a = 2{,}500{,}000$ miles. Let the x-axis be the transverse axis. *(Lesson 8-5)*
$$\frac{x^2}{6.25 \times 10^{12}} - \frac{y^2}{2.1050625 \times 10^{13}} = 1$$

AVIATION **For Exercises 16–18, use the following information.**
A military jet performs for an air show. The path of the plane during one trick can be modeled by a conic section with equation $24x^2 + 1000y - 31{,}680x - 45{,}600 = 0$, where distances are represented in feet.

16. Identify the shape of the curved path of the jet. Write the equation in standard form.
17. If the jet begins its path upward or ascent at (0, 0), what is the horizontal distance traveled by the jet from the beginning of the ascent to the end of the descent? **about 1321 ft**
18. What is the maximum height of the jet? **10,500 ft**

16. parabola: $y = -0.024(x - 660)^2 + 10{,}500$

SATELLITES **For Exercises 19–21, use the following information.**
Two satellites are placed in orbit about Earth. The equations of the two orbits are $\frac{x^2}{(300)^2} + \frac{y^2}{(900)^2} = 1$ and $\frac{x^2}{(600)^2} + \frac{y^2}{(690)^2} = 1$, where distances are in km and Earth is the center of each curve. *(Lesson 8-7)*

19. Solve each equation for y. **19–21. See margin.**
20. Use a graphing calculator to estimate the intersection points of the two orbits.
21. Compare the orbits of the two satellites.

3. Triangle EFG is also isosceles with two sides of measure $2\sqrt{5}$.

4. The side lengths are one-half the side lengths of the sides of the other triangle.

5. $y = \frac{1}{40}x^2 - 9.75$

6. $y = \frac{1}{12.192}x^2 - 2.9718$

7.

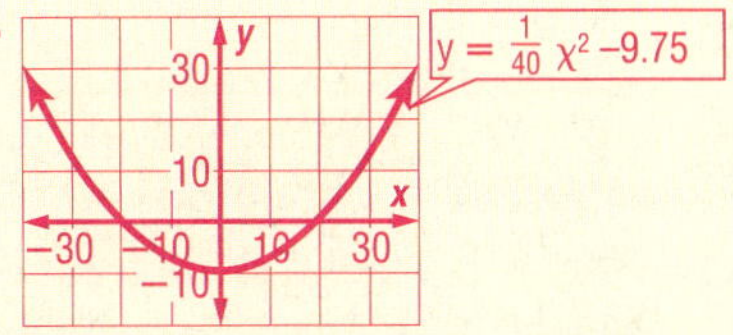

8. Sample: The equation in feet is easier to graph since the numbers have fewer decimal places.

11.

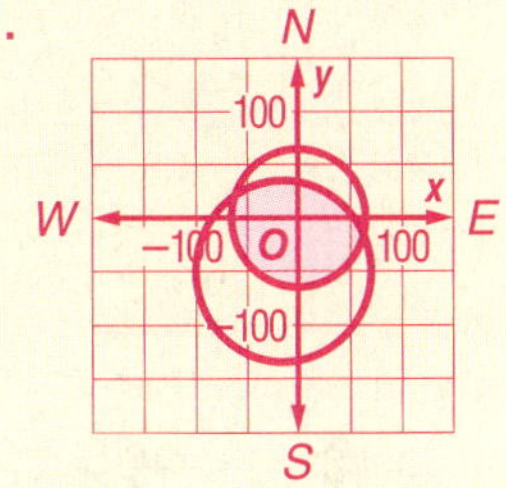

12. Venus:
$$\frac{x^2}{4522.5625} + \frac{y^2}{4522.36} = 1;$$
Jupiter:
$$\frac{x^2}{234{,}014.06} + \frac{y^2}{233{,}454.74} = 1$$

14. Venus; the value of e is closer to 0 than for Jupiter.

19. $y = \pm 900\sqrt{1 - \frac{x^2}{(300)^2}}$;
$y = \pm 690\sqrt{1 - \frac{x^2}{(600)^2}}$

20. Sample answer: (209, 647), (−209, 647), (−209, −647), (209, −647)

21. Sample answer: The orbit of the satellite modeled by the second equation is closer to a circle than the other orbit. The distance on the x-axis is twice as great for one satellite than the other.

Mixed Problem Solving 869

4. $\frac{1}{p} = \frac{1}{f} - \frac{1}{q}$

5. $\frac{q - f}{fq}$

7.

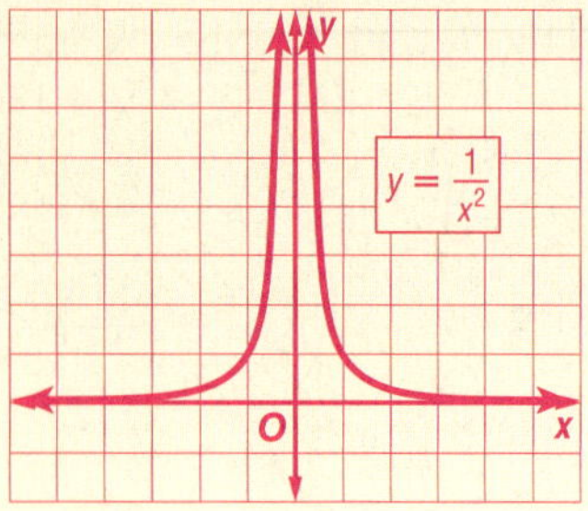

12.

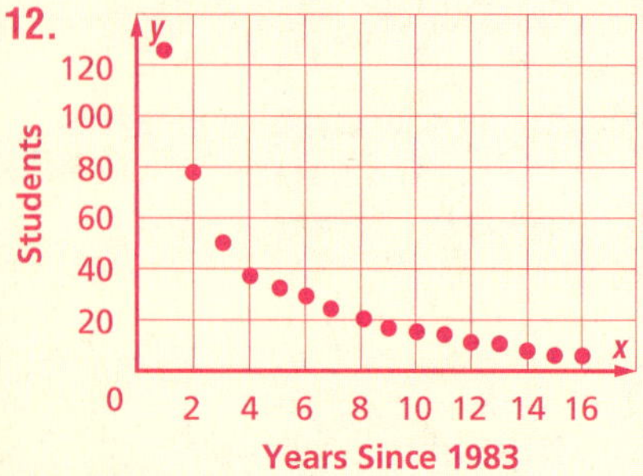

13. Sample answer: inverse variation or exponential

14. Sample answer: $y = 162x^{-1}$ or $y = \frac{162}{x}$; or $y = 93.3(0.83)^x$

MANUFACTURING **For Exercises 1–3, use the following information.**
A shipping container in the shape of a rectangular prism is designed using any value for x such that the volume can be represented by the polynomial $6x^3 + 11x^2 + 4x$, where the height is x. *(Lesson 9-1)*

1. Find the length and width of the container in terms of x. **$2x + 1$, $3x + 4$**
2. Find the ratio of the three dimensions of the container when $x = 2$. **2:5:10**
3. Will the ratio of the three dimensions be the same for all values of x? **no**

PHOTOGRAPHY **For Exercises 4–6, use the following information.**
The formula $\frac{1}{q} = \frac{1}{f} - \frac{1}{p}$ can be used to determine how far the film should be placed from the lens of a camera to create a perfect photograph. The variable q represents the distance from the lens to the film, f represents the focal length of the lens, and p represents the distance from the object to the lens. *(Lesson 9-2)*

4. Solve the formula for $\frac{1}{p}$. **4–5. See margin.**
5. Write the expression containing f and q as a single rational expression.
6. If a camera has a focal length of 8 centimeters and the lens is 10 centimeters from the film, how far should an object be from the lens so that the picture will be in focus? **40 cm**

PHYSICS **For Exercises 7–9, use the following information.**
Isaac Newton's law of universal gravitation depends upon the Inverse Square Law. It states that the relationship between two variables is related to the equation $y = \frac{1}{x^2}$. *(Lesson 9-3)*

7. Graph $y = \frac{1}{x^2}$. **See margin.**
8. Give the equations of any asymptotes of the graph. **$x = 0$, $y = 0$**
9. If x represents distance in an application of the Inverse Square Law, what values of x make sense for the situation? **positive values of x**

PHYSICS **For Exercises 10 and 11, use the following information.**
In 1798, Henry Cavendish found a value for the universal gravitational constant G that Newton used in his formula $F = G\frac{m_A m_B}{d^2}$ for finding the gravitational force between two objects. The variables in the formula are defined as follows: F is the gravitational force between the objects, G is the universal constant, m_A is the mass of the first object, m_B is the mass of the second object, and d is the distance between the centers of the objects. *(Lesson 9-4)*

10. If the mass of object A is constant, does Newton's formula represent a *direct* or *inverse* variation between the mass of object B and the distance?
11. Cavendish found the value of G to be 6.67×10^{-11} N · m^2/kg^2. If two objects each weighing 5 kilograms are placed so that their centers are 0.5 meter apart, what is the gravitational force between the two objects? **6.67×10^{-9} N**

10. inverse

EDUCATION **For Exercises 12–14, use the following information.**
The table shows the average number of students per computer in United States public schools for various years. *(Lesson 9-5)* **12–14. See margin.**

Year	Students	Year	Students
1984	125	1992	18
1985	75	1993	16
1986	50	1994	14
1987	37	1995	10.5
1988	32	1996	10
1989	25	1997	7.8
1990	22	1998	6.1
1991	20	1999	5.7

Source: *The World Almanac*

12. Let x represent years where 1984 = 1, 1985 = 2, and so on. Let y represent the number of students. Graph the data.
13. What type of function does the graph most closely resemble?
14. Use a graphing calculator to find an equation that models the data.

TRAVEL **For Exercises 15–17, use the following information.**
A trip between two towns in Montana takes 4 hours under ideal conditions. The first 150 miles of the trip is on an interstate, and the second 130 miles is on a highway with a speed limit that is 10 miles per hour less than on the interstate. *(Lesson 9-6)*

15. If x represents the speed limit on the interstate, write an expression that represents the time spent at that speed. **$\frac{150}{x}$**
16. Write an expression for the time spent on the other highway. **$\frac{130}{x - 10}$**
17. Write and solve an equation to find the speed limit on the interstate and on the highway. **$\frac{150}{x} + \frac{130}{x - 10} = 4$; 75 mph, 65 mph**

870 Mixed Problem Solving

POPULATION **For Exercises 1–4, use the following information.**
The population of the world has been growing rapidly. In 1950, the world population was about 2.556 billion. By 1980, it had increased to about 4.458 billion. **Source:** *The World Almanac* *(Lesson 10-1)*

1. Write an exponential function of the form $y = ab^x$ that could be used to model the world population y in billions for 1950 to 1980. Write the equation in terms of x, the number of years since 1950. (Round the value of b to the nearest ten thousandth.) **$y = 2.556(1.0187)^x$**
2. Suppose the world population continued to grow at that rate. Estimate the population for the year 2000. **about 6.455 billion**
3. In 2000, the population of the world was about 6.08 billion. Compare your estimate to the actual population. **See margin.**
4. Use the equation you wrote in Exercise 1 to estimate the world population in the year 2020. How accurate do you think the estimate is? Explain your reasoning. **See margin.**

EARTHQUAKES **For Exercises 5–8, use the following information.**
The table shows the Richter scale that measures earthquake intensity. Column 2 shows the increase in intensity between each number. For example, an earthquake that measures 7 is 10 times more intense than one measuring 6. *(Lesson 10-2)*

Richter Number	Increase in Magnitude
1	1
2	10
3	100
4	1000
5	10,000
6	100,000
7	1,000,000
8	10,000,000

Source: *The New York Public Library* **5–8. See margin.**

5. Graph this function where x is the Richter number and y is the increase in magnitude.
6. Write an equation of the form $y = b^{x - c}$ for the function in Exercise 5. (*Hint*: Write the values in the second column as powers of 10 to see a pattern and find the value of c.)
7. Graph the inverse of the function in Exercise 6.
8. Write an equation of the form $y = \log_{10}x + c$ for the function in Exercise 7.

EARTHQUAKES **For Exercises 9–11, use the following information.**
The table shows the magnitude on the Richter scale of some major earthquakes. *(Lesson 10-3)*
9–10. See margin.

Year/Location	Magnitude
1939/Turkey	8.0
1963/Yugoslavia	6.0
1970/Peru	7.8
1988/Armenia	7.0
1995/Japan	6.9

Source: *The World Almanac*

9. Name two earthquakes such that the intensity of one was 10 times the intensity of the other.
10. Name two earthquakes such that the intensity of one was 100 times the intensity of the other.
11. What would be the magnitude of an earthquake that is 1000 times as intense as the 1963 earthquake in Yugoslavia? **9.0** **12. See margin.**
12. Suppose you know that $\log_7 2 \approx 0.3562$ and $\log_7 3 \approx 0.5646$. Describe two different methods that you could use to approximate $\log_7 2.5$. (You can use a calculator, of course.) Then describe how you can check your result. *(Lesson 10-4)*

WEATHER **For Exercises 13–15, use the following information.** **13–14. See margin.**
The atmospheric pressure P, in bars, of a given height on Earth can be found by using the formula $P = s \cdot e^{-\frac{h}{H}}$. In the formula, s is the surface pressure on Earth, which is approximately 1 bar, h is the altitude for which you want to find the pressure in kilometers, and H is always 7 kilometers. *(Lesson 10-5)*

13. Find the pressure for 2 kilometers, 4 kilometers, and 7 kilometers.
14. What do you notice about the pressure as altitude increases?
15. What is the pressure on top of Mount Everest at an altitude of 8700 meters? **0.29 bars**

AGRICULTURE **For Exercises 16–19, use the following information.**
An equation that models the decline in the number of U.S. farms is $y = 3{,}962{,}520(0.98)^x$, where x is years since 1960 and y is the number of farms.
"**Source:** *Wall Street Journal* *(Lesson 10-6)* **16. $b < 1$**

16. By examining the equation, how can you determine that the number of farms is declining?
17. By what rate per year is the number of farms declining? **2%**
18. How many farms were there in 1960? **3,962,520**
19. Predict when the number of farms will be less than 1.5 million. **in about 2008**

Mixed Problem Solving

3. The prediction was about 375 million greater than the actual.

4. About 9.3498 billion; since the prediction for the year 2000 was greater than the actual population, this prediction is probably even higher than the actual population at that time.

5.

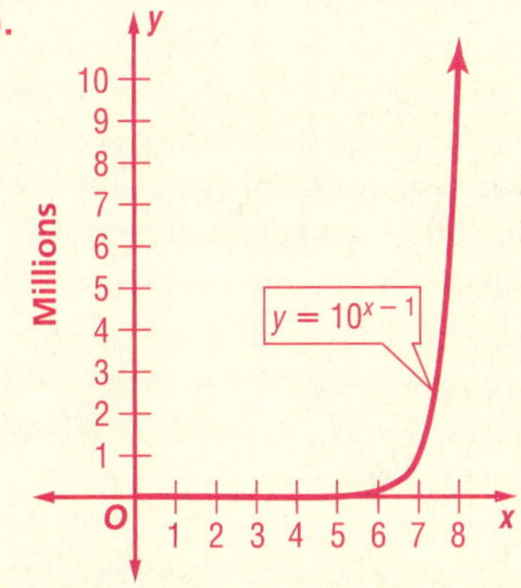

6. $y = 10^{x - 1}$

7.

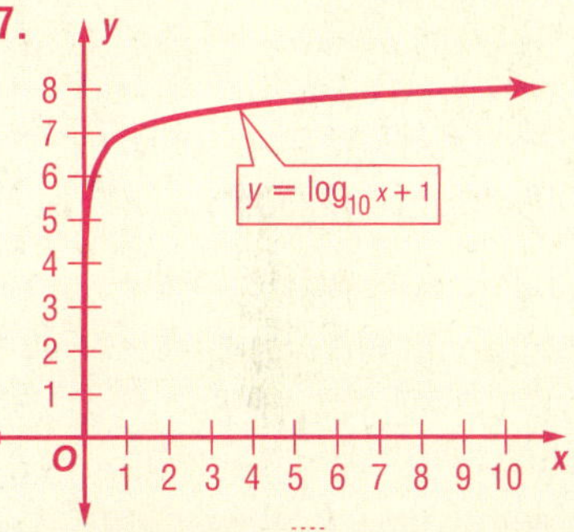

8. $y = \log_{10} x + 1$

9. Armenia and Yugoslavia or Turkey and Armenia

10. Turkey and Yugoslavia

12. Sample answer: Method 1: Use the change of base formula and find that the value is about 0.4709. Method 2: Use the values $\log_7 2 \approx 0.3562$ and $\log_7 3 \approx 0.5646$. First, average the values and then guess and check by raising 7 to the various powers. Continue until you get the desired closeness to 2.5. To check, find $7^{0.4709}$ which is very close to 2.5.

13. 0.75 bars; 0.56 bars; 0.37 bars

14. The pressure decreases as altitude increases.

CLUBS **For Exercises 1 and 2, use the following information.**
Kim and her mother are in a quilting club consisting of 9 members. Each week, the club meets, and each member must bring one completed quilt square. *(Lesson 11-1)* **1. 9, 18, 27, 36, 45, 54, 63, 72**

1. Find the first eight terms of the sequence that describes the total number of squares that have been made for the quilt after each meeting.
2. One particular quilt measures 72 inches by 84 inches and is being designed with 4-inch squares. After how many meetings will the quilt be complete? **42 meetings**

ART **For Exercises 3 and 4, use the following information.**
Alberta is making a Native American beadwork design consisting of rows of colored beads. The first row consists of 10 beads, and each consecutive row will have 15 more beads than the previous row. *(Lesson 11-2)*

3. Write an equation for the number of beads in the nth row. $a_n = 15n - 5$
4. Find the number of beads in the design if it contains 25 rows. **4750 beads**

GAMES **For Exercises 5 and 6, use the following information. 5–6. See margin.**
An audition is held for a TV game show. At the end of each round, one-half of the prospective contestants are eliminated from the competition. On a particular day, 524 contestants begin the audition. *(Lesson 11-3)*

5. Write an equation for finding the number of contestants that are left after n rounds.
6. Using this method, will the number of contestants that are to be eliminated always be a whole number? Explain.

SPORTS **For Exercises 7–9, use the following information.**
Caitlin is training for a marathon (about 26 miles). Her trainer advises her to begin by running 2 miles. Then she is to run every other day. During each session, she is to multiply the distance she ran the previous session by one and a half. *(Lesson 11-4)*

7. Write the first five terms of a sequence describing the number of miles she is to run during consecutive training sessions. **See margin.**
8. When will she exceed 26 miles in one training session? **the eighth session**
9. When will she have run at least 100 total miles? **during the ninth session**

GEOMETRY **For Exercises 10–12, use the following information.**
You can illustrate the sum of an infinite geometric series by using a square of paper. *(Lesson 11-5)*

10. Cut a square of paper at least 8 inches on a side. Let the square be one unit. Cut away one-half of the square. Call this piece Term 1. Next, cut away one-half of the remaining sheet of paper. Call this piece Term 2. Continue cutting the remaining paper in half and labeling the pieces with a term number as long as possible. List the fractions represented by the pieces. **See margin.**
11. If you could cut squares indefinitely, you would have an infinite series. Find the sum of the series. **1**
12. How does the sum of the series relate to the original square of paper? **See margin.**

BIOLOGY **For Exercises 13–15, use the following information.**
In a particular forest, scientists are interested in how the population of wolves will change over the next two years. One mathematical model for animal population is the Verhulst population model. The formula for this model is $p_{n+1} = p_n + rp_n(1 - p_n)$, where n represents the number of time periods that have passed, p_n represents the percent of the maximum sustainable population that exists at time n, and r is the growth factor. *(Lesson 11-6)*

13. To find the population of the wolves after one year, you must evaluate the expression $p_1 = 0.45 + 1.5(0.45)(1 - 0.45)$. What is the value of the expression? **0.82125**
14. Explain what each number in the expression in Exercise 13 represents. **See margin.**
15. The current population of wolves is 165. Find the new population by multiplying 165 by the value in Exercise 13. **about 136 wolves**

16. **PASCAL'S TRIANGLE** Study the first eight rows of Pascal's triangle. Write the sum of the terms in each row as a list. Make a conjecture about the sums of the rows of Pascal's triangle. *(Lesson 11-7)* **See margin.**

17. **NUMBER THEORY** Two statements that can be proved using mathematical induction are $\frac{1}{3} + \frac{1}{3^2} + \frac{1}{3^3} + \cdots + \frac{1}{3^n} = \frac{1}{2}\left(1 - \frac{1}{3^n}\right)$ and $\frac{1}{4} + \frac{1}{4^2} + \frac{1}{4^3} + \cdots + \frac{1}{4^n} = \frac{1}{3}\left(1 - \frac{1}{4^n}\right)$.
Write and prove a conjecture involving $\frac{1}{5}$ that is similar to the two given statements. *(Lesson 11-8)* **See margin.**

5. $a_n = 524\left(\frac{1}{2}\right)^n$
6. No; at the beginning of the third round there will be 131 contestants and one-half of that is 65.5.
7. 2, 3, 4.5, 6.75, 10.125
10. $\frac{1}{2}, \frac{1}{4}, \frac{1}{8}, \frac{1}{16}, \ldots$
12. The original square has area 1 unit and the area of all the pieces cannot exceed 1.
14. 0.45 represents the maximum sustainable population of 45% and 1.5 is the growth factor.
16. 1, 2, 4, 8, 16, 32, 64, 128; The sum of the nth row is 2^n.
17. $\frac{1}{5} + \frac{1}{5^2} + \frac{1}{5^3} + \ldots + \frac{1}{5^n} = \frac{1}{4}\left(1 - \frac{1}{5^n}\right)$

Step 1: When $n = 1$, the left side of the given equation is $\frac{1}{5}$. The right side is $\frac{1}{4}\left(1 - \frac{1}{5}\right)$ or $\frac{1}{5}$. Thus, the equation is true for $n = 1$.

Step 2: Assume that $\frac{1}{5} + \frac{1}{5^2} + \frac{1}{5^3} + \ldots + \frac{1}{5^k} = \frac{1}{4}\left(1 - \frac{1}{5^k}\right)$ for some positive integer k.

Step 3: Show that the given equation is true for $n = k + 1$.

$$\frac{1}{5} + \frac{1}{5^2} + \frac{1}{5^3} + \ldots + \frac{1}{5^k} + \frac{1}{5^{k+1}}$$
$$= \frac{1}{4}\left(1 - \frac{1}{5^k}\right) + \frac{1}{5^{k+1}}$$
$$= \frac{1}{4} - \frac{1}{4(5^k)} + \frac{1}{5^{k+1}}$$
$$= \frac{5^{k+1} - 5 + 4}{4(5^{k+1})}$$
$$= \frac{5^{k+1} - 1}{4(5^{k+1})}$$
$$= \frac{5^{k+1}}{4(5^{k+1})} - \frac{1}{4(5^{k+1})}$$
$$= \frac{1}{4}\left(1 - \frac{1}{5^{k+1}}\right)$$

The last expression above is the right side of the equation to be proved, where n has been replaced by $k + 1$. Thus, the equation is true for $n = k + 1$.

Therefore, $\frac{1}{5} + \frac{1}{5^2} + \frac{1}{5^3} + \ldots + \frac{1}{5^n} = \frac{1}{4}\left(1 - \frac{1}{5^n}\right)$ for all positive integers n.

872 Mixed Problem Solving

NUMBER THEORY **For Exercises 1–3, use the following information.**
According to the Rational Zero Theorem, if $\frac{p}{q}$ is a rational root, then p is a factor of the constant of the polynomial, and q is a factor of the leading coefficient. *(Lesson 12-1)*

1. What is the maximum number of possible rational roots that you may need to check for the polynomial $3x^4 - 5x^3 + 2x^2 - 7x + 10$? Explain your answer. **See margin.**
2. Why may you not need to check the maximum number of possible roots? **See margin.**
3. Are choosing the numerator and the denominator for a possible rational root independent or dependent events? **independent**

4. **GARDENING** A gardener is selecting plants for a special display. There are 15 varieties of pansies from which to choose. The gardener can only use 9 varieties in the display. How many ways can 9 varieties be chosen from the 15 varieties? *(Lesson 12-2)* **5005**

SPEED LIMITS **For Exercises 5–7, use the following information.**
In 1995, states were allowed to set their own highway speed limits. The table shows the number of states having each speed limit for their rural interstates. *(Lesson 12-3)*

Speed Limit	Number of States
55	1
65	20
70	18
75	11

Source: *The World Almanac*

5. If a state is randomly selected, what is the probability that its speed limit is 75? **0.22 or 22%**
6. If a state is randomly selected, what is the probability that its speed limit is 55? **0.02 or 2%**
7. If a state is randomly selected, what is the probability that its speed limit is 55 or greater? **1 or 100%**

8. **LOTTERIES** A lottery number for a particular lottery has seven digits which can be any digit from 0 to 9. It is advertised that the odds of winning the lottery are 1 to 10,000,000. Is this statement about the odds correct? Explain your reasoning. *(Lesson 12-4)* **See margin.**

For Exercises 9 and 10, use the following information. **10. about 110 cars**
The table shows the results of a survey of the most popular colors for luxury cars in 1999. *(Lesson 12-5)*

Color	% of cars	Color	% of cars
white	16.1	gold	7.0
silver	14.8	green	6.1
lt. brown	12.9	red	6.0
black	9.4	blue	4.9
gray	8.3	other	14.5

Source: *The World Almanac*

9. If a car sold in 1999 is randomly selected, what is the probability that it is white or silver? **30.9%**
10. In a parking lot of 1000 cars sold in 1999, how many cars would you expect to be blue or green?

EDUCATION **For Exercises 11–13, use the following information.** **11–13. See margin.**
The list of numbers given are the average scores for each state for the ACT for 1999–2000. *(Lesson 12-6)*
20.2, 21.3, 21.5, 20.3, 21.4, 21.5, 21.3, 20.6, 20.6, 19.9, 21.6, 21.4, 21.5, 21.4, 22.0, 21.6, 20.1, 19.6, 21.9, 20.7, 21.9, 21.3, 22.0, 18.7, 21.6, 21.8, 21.7, 21.5, 22.5, 20.7, 20.1, 22.2, 19.5, 21.4, 21.4, 20.8, 22.7, 21.4, 21.1, 19.3, 21.5, 20.0, 20.3, 21.5, 22.2, 20.5, 22.4, 20.2, 22.2, 21.6

11. Compare the mean and median of the data.
12. Find the standard deviation of the data. Round to the nearest hundredth.
13. Suppose the state with an average score of 20.0 incorrectly reported the results. The score for the state is actually 22.5. How are the mean and median of the data affected by this data change?
14. **HEALTH** The heights of students at Madison High School are normally distributed with a mean of 66 inches and a standard deviation of 2 inches. Of the 1080 students in the school, how many would you expect to be less than 62 inches tall? *(Lesson 12-7)* **27 students**

EDUCATION **For Exercises 15 and 16, use the following information.**
A mathematics contest consists of four tests. Each test has ten multiple-choice questions with four answer options. *(Lesson 12-8)* **15–16. See margin.**

15. What is the probability that a student who randomly answers all questions on a test will get a score of 5 correct?
16. What is the probability that a student who randomly answers all questions on a test will get every question incorrect?
17. **SURVEY** A poll of 1750 people shows that 78% enjoy travel. Find the margin of the sampling error for the survey. *(Lesson 12-9)* **about 1.98%**

Mixed Problem Solving

1. 32; There are 4 factors of 3 and 8 factors of 10, so there are $4 \cdot 8$ or 32 possible permutations of choices for p and q.
2. Some of the potential roots may be repeated when you form the fractions $\frac{p}{q}$.
8. No; there are 10^7 different numbers possible. If there is only one winner, then the probability is 1 in 10^7 or 1 in 10,000,000. The odds cannot be 1 to 10,000,000.
11. The mean is 21.128 and the median is 21.4. They are very close.
12. 0.88
13. The new mean is 21.178 and the median is still 21.4. The mean was slightly raised, but the median was not affected.
15. $\frac{61{,}236}{1{,}048{,}576}$
16. $\frac{59{,}049}{1{,}048{,}576}$

1.

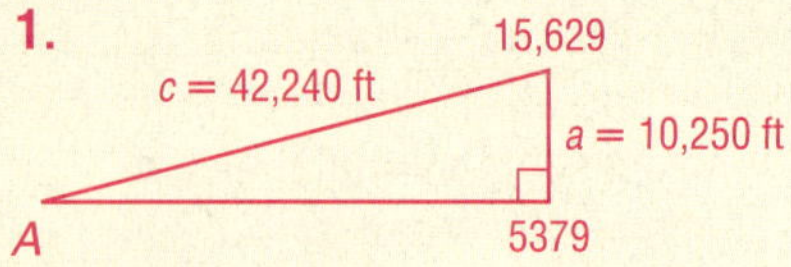

7. Sample answer: $V_0 = 27.8$ ft/s and $\theta = 75°$

12.

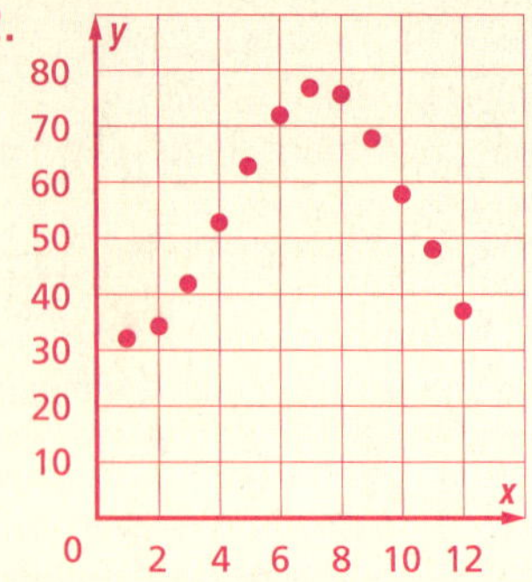

13. Sample answer: The trigonometric model fits the data better than the quadratic model if you graph the points and both functions on a calculator screen. The trigonometric model comes closer to more of the temperatures.

14. $\sin \theta_1 = \frac{n_2 \sin \theta_2}{n_1}$

15. $\sin^{-1} \theta_1 = \frac{n_2 \sin \theta_2}{n_1}$ or $\theta_1 = \text{Arcsin}\left(\frac{n_2 \sin \theta_2}{n_1}\right)$

Mixed Problem Solving

CABLE CARS For Exercises 1 and 2, use the following information. **1. See margin.**
The longest cable car route in the world is located in Venezuela. It begins at an altitude of 5379 feet and ends at an altitude of 15,629 feet. The ride is 8 miles long. **Source:** *The Guinness Book of Records* *(Lesson 13-1)*

1. Draw a diagram to represent this situation.
2. To the nearest degree, what is the average angle of elevation of the cable car ride? **14°**

RIDES For Exercises 3–5, use the following information.
In 2000, a gigantic Ferris wheel, the London Eye, opened in England. The wheel has 32 cars evenly spaced around the circumference. *(Lesson 13-2)*

3. What is the measure, in degrees, of the angle between any two consecutive cars? **11.25°**
4. What is the measure, in radians, of the angle between any two consecutive cars? $\frac{\pi}{16}$
5. If a car is located such that the measure in standard position is $-60°$, what are the measures of one angle with positive measure and one angle with negative measure coterminal with the angle of this car? **300°, −420°**

BASKETBALL For Exercises 6 and 7, use the following information.
During the halftime of a basketball game, a person is selected to try to make a shot at a distance of 12 feet from the basket. The formula $R = \frac{V_0^2 \sin 2\theta}{32}$ gives the distance of a basketball shot with an initial velocity of V_0 feet per second at an angle of θ with the ground. *(Lesson 13-3)*

6. If the basketball was shot with an initial velocity of 24 feet per second at an angle of 75°, how far will the basketball travel? **9 ft**
7. Find an initial velocity and angle of release that would result in the ball traveling approximately 12 feet to the basket. **See margin.**
8. **COMMUNICATIONS** A telecommunications tower needs to be supported by two wires. The angle between the ground and the tower on one side must be 35° and the angle between the ground and the second tower must be 72°. The distance between the two wires is 110 feet.

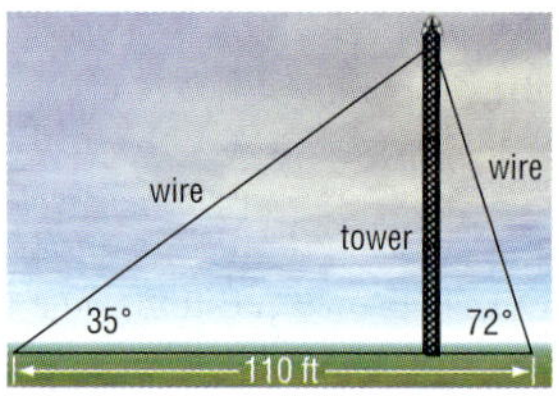

To the nearest foot, what should be the lengths of the two wires? *(Lesson 13-4)* **66 ft, 109 ft**

SURVEYING For Exercises 9–11, use the following information.
A triangular plot of farm land measures 0.9 by 0.5 by 1.25 miles. *(Lesson 13-5)* **9. 19°, 37°, 124°**

9. If the plot of land is fenced on the border, what will be the angles at which the fences of the three sides meet? Round to the nearest degree.
10. What is the area of the plot of land? (*Hint*: Use the area formula in Lesson 13-4.) **about 0.19 m^2**
11. One square mile is 640 acres. How many acres are in the plot of land? **about 122**

WEATHER For Exercises 12 and 13, use the following information.
The monthly normal temperatures, in degrees Fahrenheit, for New York City are given in the table. January is assigned a value of 1, February a value of 2, and so on. *(Lesson 13-6)* **12–13. See margin.**

Month	Temperature	Month	Temperature
1	32	7	77
2	34	8	76
3	42	9	68
4	53	10	58
5	63	11	48
6	72	12	37

Source: *The World Almanac*

12. Graph the data in a scatter plot.
13. A trigonometric model for the temperature T in degrees Fahrenheit of New York City at t months is given by $T = 22.5 \sin\left(\frac{\pi}{6}x - 2.25\right) + 54.3$. A quadratic model for the same situation is $T = -1.34x^2 + 18.84x + 5$. Which model do you think best fits the data? Explain your reasoning.

PHYSICS For Exercises 14–16, use the following information.
When light passes from one substance to another, it may be reflected and refracted. Snell's law can be used to find the angle of refraction as a beam of light passes from one substance to another. One form of the formula for Snell's law is $n_1 \sin \theta_1 = n_2 \sin \theta_2$, where n_1 and n_2 are the indices of refraction for the two substances and θ_1 and θ_2 are the angles of the light rays passing through the two substances. *(Lesson 13-7)* **14–15. See margin.**

14. Solve the equation for $\sin \theta_1$.
15. Write an equation in the form of an inverse function that allows you to find θ_1.
16. If a light beam in air with index of refraction of 1.00 hits a diamond with index of 2.42 at an angle of 30°, find the angle of refraction. **11.9°**

Mixed Problem Solving

874 Mixed Problem Solving

TIDES For Exercises 1–3, use the following information.
The world's record for the highest tide is held by the Minas Basin in Nova Scotia, Canada, with a tidal range of 54.6 feet. A tide is at equilibrium when it is at its normal level halfway between its highest and lowest points. *(Lesson 14-1)* **1–3. See margin.**

1. Write an equation to represent the height h of the tide. Assume that the tide is at equilibrium at $t = 0$, that the high tide is beginning, and that the tide completes one cycle in 12 hours.
2. Mobile, Alabama, has a very small tidal range at only one foot six inches. Write an equation to represent the height h of the tide in Mobile.
3. Graph the functions for the tides in Minas Basin and Mobile on the same axis system. How do the graphs compare?

RIDES For Exercises 4–7, use the following information.
The Cosmoclock 21 is a huge Ferris wheel in Yokohama City, Japan. The diameter is 328 feet. Suppose that a rider enters the ride at 0 feet and then rotates in 90° increments counterclockwise. The table shows the angle measures of rotation and the height above the ground of the rider. *(Lesson 14-2)*

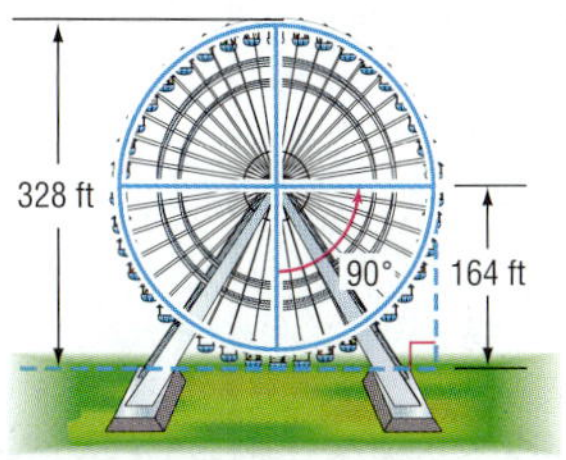

Angle	Height	Angle	Height
0°	0	450°	**164**
90°	164	540°	**328**
180°	328	630°	**164**
270°	**164**	720°	**0**
360°	**0**		

4. Copy and complete the table. Then graph the points (angle, height). **See margin for graph.**
5. A function that models the data is $y = 164 \cdot (\sin (x - 90°)) + 164$. Identify the vertical shift, amplitude, period, and phase shift of the graph. **164; 164; 360°, 90°**
6. Write an equation using the sine that models the position of a rider on the Vienna Giant Ferris Wheel in Vienna, Austria, with a diameter of 200 feet. Check your equation by plotting the points and the equation with a graphing calculator. **$y = 100 \cdot (\sin (x - 90°)) + 100$**
7. Write an equation using the cosine that models the position of a rider on the Vienna Giant Ferris Wheel. **$y = 100 \cdot (\cos (x - 180°)) + 100$**
8. **TRIGONOMETRY** Using the exact values for the sine and cosine functions, show that the identity $\cos^2 \theta + \sin^2 \theta = 1$ is true for angles of measure 30°, 45°, 60°, 90°, and 180°. *(Lesson 14-3)* **See margin.**
9. **ROCKETS** In the formula $h = \frac{v^2 \sin^2 \theta}{2g}$, h is the maximum height reached by a rocket, θ is the angle between the ground and the initial path of the object, v is the rocket's initial velocity, and g is the acceleration due to gravity. Verify the identity $\frac{v^2 \sin^2 \theta}{2g} = \frac{v^2 \cos^2 \theta}{2g \cot^2 \theta}$. *(Lesson 14-4)* **See margin.**

WEATHER For Exercises 10–12, use the following information.
The monthly high temperatures for Minneapolis, Minnesota, can be modeled by the equation $y = 31.65 \sin \left(\frac{\pi}{6}x - 2.09\right) + 52.35$, where the months x are January = 1, February = 2, and so on. The monthly low temperatures for Minneapolis can be modeled by the equation $y = 30.15 \sin \left(\frac{\pi}{6}x - 2.09\right) + 32.95$. *(Lesson 14-5)* **10–12. See margin.**

10. Write a new function by adding the expressions on the right side of each equation and dividing the result by 2.
11. Graph the two original functions and the new function on the same calculator screen. What do you notice?
12. What is the meaning of the function you wrote in Exercise 10? **13. See students' work.**
13. Begin with one of the Pythagorean Identities. Perform equivalent operations on each side to create a new trigonometric identity. Then show that the identity is true. *(Lesson 14-6)*
14. **TELEVISION** The tallest structure in the world is a television transmitting tower located near Fargo, North Dakota, with a height of 2064 feet.

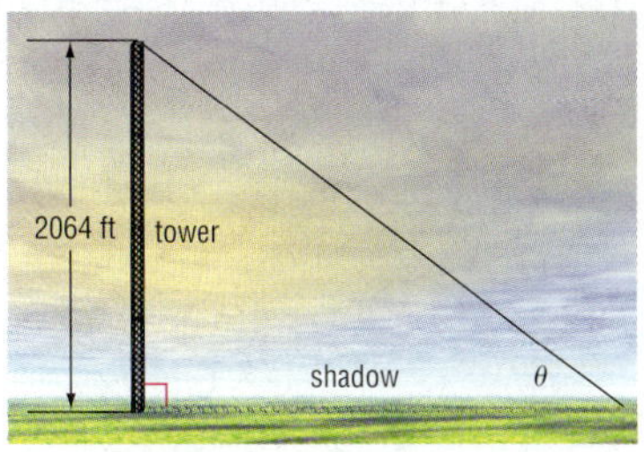

What is the measure of θ if the length of the shadow is 1 mile? **Source:** *The Guinness Book of Records* *(Lesson 14-7)* **about 21°**

Mixed Problem Solving

1. $y = 27.3 \sin \frac{\pi}{6}t$
2. $0.75 \sin \frac{\pi}{6}t$
3. The graphs of the two functions have the same period, but the amplitudes are very different.

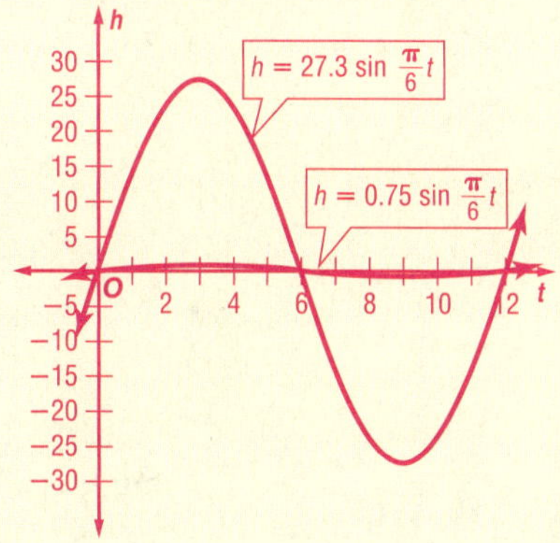

4.

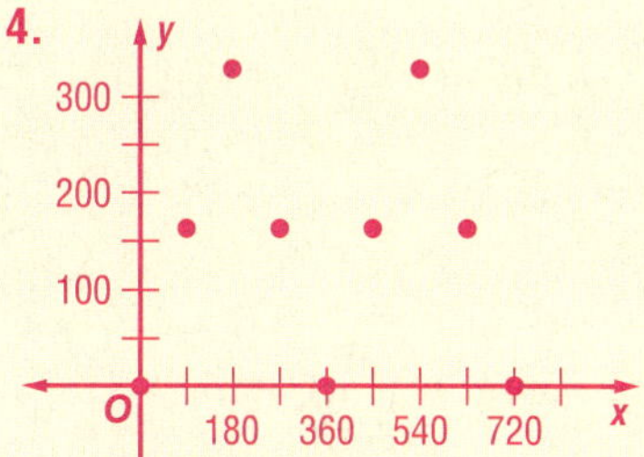

8. 30°: $\cos^2(30°) + \sin^2(30°) = \left(\frac{\sqrt{3}}{2}\right)^2 + \left(\frac{1}{2}\right)^2 = \frac{3}{4} + \frac{1}{4} = 1$;
45°: $\cos^2(45°) + \sin^2(45°) = \left(\frac{\sqrt{2}}{2}\right)^2 + \left(\frac{\sqrt{2}}{2}\right)^2 = \frac{1}{2} + \frac{1}{2} = 1$;
60°: $\cos^2(60°) + \sin^2(60°) = \left(\frac{1}{2}\right)^2 + \left(\frac{\sqrt{3}}{2}\right)^2 = \frac{1}{4} + \frac{3}{4} = 1$;
90°: $\cos^2(90°) + \sin^2(90°) = (0)^2 + (1)^2 = 0 + 1 = 1$;
180°: $\cos^2(180°) + \sin^2(180°) = (-1)^2 + (0)^2 = 1 + 0 = 1$
9. $$\frac{v^2 \sin^2 \theta}{2g} \stackrel{?}{=} \frac{v^2 \cos^2 \theta}{2g \cot^2 \theta}$$
$$\frac{v^2 \sin^2 \theta}{2g} \stackrel{?}{=} \frac{v^2 \cos^2 \theta}{\frac{2g \cos^2 \theta}{\sin^2 \theta}}$$
$$\frac{v^2 \sin^2 \theta}{2g} \stackrel{?}{=} \frac{v^2 \cos^2 \theta}{1} \cdot \frac{\sin^2 \theta}{2g \cos^2 \theta}$$
$$\frac{v^2 \sin^2 \theta}{2g} = \frac{v^2 \sin^2 \theta}{2g}$$
10. $y = 30.9 \sin \left(\frac{\pi}{6}x - 2.09\right) + 42.65$
11. The graph of the function in Exercise 10 lies between the graphs of the other two functions.
12. The new function represents the average of the high and low temperatures for each month.

Preparing For Standardized Tests

Becoming a Better Test-Taker

At some time in your life, you will probably have to take a standardized test. Sometimes this test may determine if you go on to the next grade or course, or even if you will graduate from high school. This section of your textbook is dedicated to making you a better test-taker.

TYPES OF TEST QUESTIONS In the following pages, you will see examples of four types of questions commonly seen on standardized tests. A description of each type of question is shown in the table below.

Type of Question	Description	See Pages
multiple choice	Four or five possible answer choices are given from which you choose the best answer.	878–879
gridded response	You solve the problem. Then you enter the answer in a special grid and shade in the corresponding circles.	880–883
short response	You solve the problem, showing your work and/or explaining your reasoning.	884–887
extended response	You solve a multi-part problem, showing yourwork and/or explaining your reasoning	888–892

PRACTICE After being introduced to each type of question, you can practice that type of question. Each set of practice questions is divided into five sections that represent the concepts most commonly assessed on standardized tests.

- Number and Operations
- Algebra
- Geometry
- Measurement
- Data Analysis and Probability

USING A CALCULATOR On some tests, you are permitted to use a calculator. You should check with your teacher to determine if calculator use is permitted on the test you will be taking, and if so, what type of calculator can be used.

TEST-TAKING TIPS In addition to Test-Taking Tips like the one shown at the right, here are some additional thoughts that might help you.

- Get a good night's rest before the test. Cramming the night before does not improve your results.
- Budget your time when taking a test. Don't dwell on problems that you cannot solve. Just make sure to leave that question blank on your answer sheet.
- Watch for key words like NOT and EXCEPT. Also look for order words like LEAST, GREATEST, FIRST, and LAST.

Test-Taking Tip

If you are allowed to use a calculator, make sure you are familiar with how it works so that you won't waste time trying to figure out the calculator when taking the test.

Preparing for Standardized Tests

Preparing for Standardized Tests 877

Multiple-Choice Questions

Multiple-choice questions are the most common type of questions on standardized tests. These questions are sometimes called *selected-response questions*. You are asked to choose the best answer from four or five possible answers. To record a multiple-choice answer, you may be asked to shade in a bubble that is a circle or an oval or to just write the letter of your choice. Always make sure that your shading is dark enough and completely covers the bubble.

The answer to a multiple-choice question is usually not immediately obvious from the choices, but you may be able to eliminate some of the possibilities by using your knowledge of mathematics. Another answer choice might be that the correct answer is not given.

Incomplete Shading
Ⓐ Ⓑ Ⓒ Ⓓ
Too light shading
Ⓐ Ⓑ Ⓒ Ⓓ
Correct shading
Ⓐ Ⓑ Ⓒ Ⓓ

Preparing for Standardized Tests

Example 1

Strategy

Reasonableness
Check to see that your answer is reasonable with the given information.

White chocolate pieces sell for \$3.25 per pound while dark chocolate pieces sell for \$2.50 per pound. How many pounds of white chocolate are needed to produce a 10-pound mixture of both kinds that sells for \$2.80 per pound?

Ⓐ 2 lb Ⓑ 4 lb Ⓒ 6 lb Ⓓ 10 lb

The question asks you to find the number of pounds of the white chocolate. Let w be the number of pounds of white chocolate and let d be the number of pounds of dark chocolate. Write a system of equations.

$w + d = 10$ — There is a total of 10 pounds of chocolate.

$3.25w + 2.50d = 2.80(10)$ — The price is \$2.80 × 10 for the mixed chocolate.

Use substitution to solve.

$3.25w + 2.50d = 2.80(10)$	Original equation
$3.25w + 2.50(10 - w) = 28$	Solve the first equation for *d* and substitute.
$3.25w + 25 - 2.5w = 28$	Distributive Property
$0.75w = 3$	Simplify.
$w = 4$	Divide each side by 0.75.

The answer is B.

Example 2

Strategy

Diagrams
Drawing a diagram for a situation may help you to answer the question.

Josh throws a baseball upward at a velocity of 105 feet per second, releasing the baseball when it is 5 feet above the ground. The height of the baseball *t* seconds after being thrown is given by the formula $h(t) = -16t^2 + 105t + 5$. Find the time at which the baseball reaches its maximum height. Round to the nearest tenth of a second.

Ⓕ 1.0 s Ⓖ 3.3 s Ⓗ 6.6 s Ⓙ 177.3 s

Graph the equation. The graph is a parabola. Make sure to label the horizontal axis as *t* (time in seconds) and the vertical axis as *h* for height in feet. The ball is at its maximum height at the vertex of the graph.

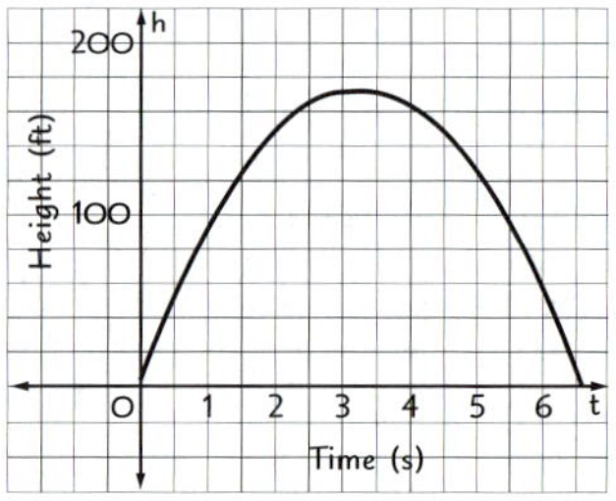

The graph indicates that the maximum height is achieved between 3 and 4 seconds after launch.

The answer is G.

Preparing for Standardized Tests

878 Preparing for Standardized Tests

Multiple-Choice Practice

Choose the best answer.

Number and Operations

1. In 2002, 1.8123×10^8 people in the United States and Canada used the Internet while 5.442×10^8 people worldwide used the Internet. What percent of users were from the United States and Canada? **A**

Ⓐ 33% Ⓑ 35% Ⓒ 37% Ⓓ 50%

2. Serena has 6 plants to put in her garden. How many different ways can she arrange the plants? **D**

Ⓐ 21 Ⓑ 30 Ⓒ 360 Ⓓ 720

Algebra

3. The sum of Kevin's, Anna's, and Tia's ages is 40. Anna is 1 year more than twice as old as Tia. Kevin is 3 years older than Anna. How old is Anna? **C**

Ⓐ 7 Ⓑ 14 Ⓒ 15 Ⓓ 18

4. Rafael's Theatre Company sells tickets for \$10. At this price, they sell 400 tickets. Rafael estimates that they would sell 40 less tickets for each \$2 price increase. What charge would give the most income? **C**

Ⓐ 10 Ⓑ 13 Ⓒ 15 Ⓓ 20

Geometry

5. Hai stands 75 feet from the base of a building and sights the top at a 35° angle. What is the height of the building to the nearest tenth of a foot? **C**

Ⓐ 0.0 ft Ⓑ 43.0 ft

Ⓒ 52.5 ft Ⓓ 61.4 ft

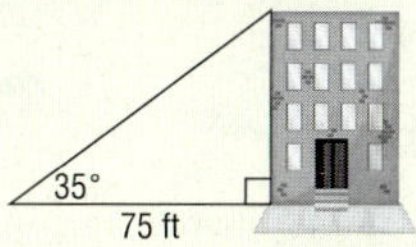

6. Samone draws $\triangle QRS$ on grid paper to use for a design in her art class. She needs to rotate the triangle 180° counterclockwise. What will be the y-coordinate of the image of S? **B**

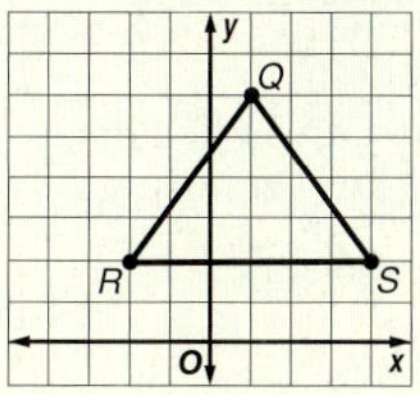

Ⓐ −6 Ⓑ −2 Ⓒ −1 Ⓓ 2

Measurement

7. Lakeisha is teaching a summer art class for children. For one project, she estimates that she will need $\frac{2}{3}$ yard of string for each 3 students. How many yards will she need for 16 students? **B**

Ⓐ 3 yd Ⓑ $3\frac{5}{9}$ yd

Ⓒ $10\frac{2}{3}$ yd Ⓓ 16 yd

8. Kari works at night so she needs to make her room as dark as possible during the day to sleep. How much black paper will she need to cover the window in her room, which is shaped as shown. Use 3.14 for π. Round to the nearest tenth of a square foot. **C**

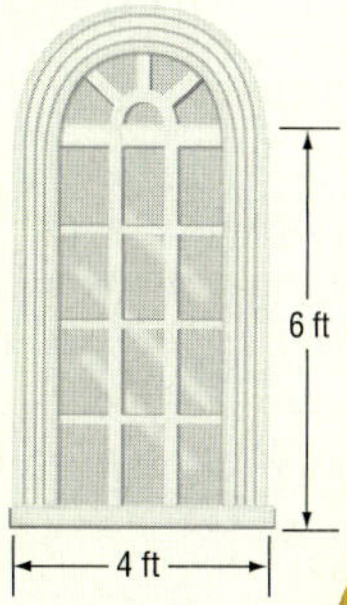

Ⓐ 24.0 ft² Ⓑ 24.6 ft²

Ⓒ 30.3 ft² Ⓓ 36.6 ft²

Data Analysis and Probability

9. A card is drawn from a standard deck of 52 cards. If one card is drawn, what is the probability that it is a heart or a 2? **D**

Ⓐ $\frac{1}{52}$ Ⓑ $\frac{1}{13}$ Ⓒ $\frac{1}{4}$ Ⓓ $\frac{4}{13}$

10. The weight of candy in boxes is normally distributed with a mean of 12 ounces and a standard deviation of 0.5 ounce. About what percent of the time will you get a box that weighs over 12.5 ounces? **B**

Ⓐ 13.5% Ⓑ 16% Ⓒ 50% Ⓓ 68%

Test-Taking Tip

Question 8

Many standardized tests include a reference sheet with common formulas that you may use during the test. If it is available before the test, familiarize yourself with the reference sheet for quick reference during the test.

Preparing for Standardized Tests

Preparing for Standardized Tests

Preparing for Standardized Tests 879

Preparing for Standardized Tests

Gridded-Response Questions

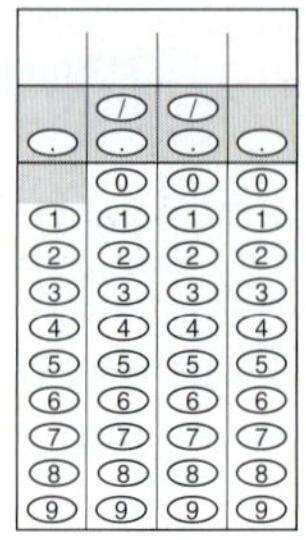

Gridded-response questions are another type of question on standardized tests. These questions are sometimes called *student-produced response* or *grid-in*, because you must create the answer yourself, not just choose from four or five possible answers.

For gridded response, you must mark your answer on a grid printed on an answer sheet. The grid contains a row of four or five boxes at the top, two rows of ovals or circles with decimal and fraction symbols, and four or five columns of ovals, numbered 0–9. At the right is an example of a grid from an answer sheet.

Example 1 **Find the x-coordinate for the solution of the given system of equations.**

$$4x - y = 14$$
$$-3x + y = -11$$

What value do you need to find?

You need to find only the x-coordinate of the point where the graphs of the two equations intersect. You could graph the system, but that takes time. The easiest method is probably the substitution method since the second equation can be solved easily for y.

$-3x + y = -11$	Second equation
$y = -11 + 3x$	Solve the second equation for y.

$4x - y = 14$	First equation
$4x - (-11 + 3x) = 14$	Substitute for y.
$4x + 11 - 3x = 14$	Distributive Property
$x = 3$	Simplify.

The answer to be filled in on the grid is 3.

How do you fill in the grid for the answer?

- Write your answer in the answer boxes.
- Write only one digit or symbol in each answer box.
- Do not write any digits or symbols outside the answer boxes.
- You may write your answer with the first digit in the left answer box, or with the last digit in the right answer box. You may leave blank any boxes you do not need on the right or the left side of your answer.

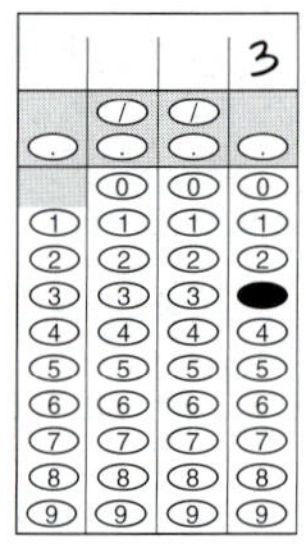

- Fill in only one bubble for every answer box that you have written in. Be sure not to fill in a bubble under a blank answer box.

Many gridded response questions result in an answer that is a fraction or a decimal. These values can also be filled in on the grid.

880 Preparing for Standardized Tests

Example 2 **Zuri is solving a problem about the area of a room. The equation she needs to solve is $4x^2 + 11x - 3 = 0$. Since the answer will be a length, she only needs to find the positive root. What is the solution?**

Since you can see the equation is not easily factorable, use the Quadratic Formula.

$$x = \frac{-b \pm \sqrt{b^2 - 4ac}}{2a}$$

$$= \frac{-11 \pm \sqrt{11^2 - 4(4)(-3)}}{2(4)}$$

$$= \frac{-11 + 13}{8}$$

There are two roots, but you only need the positive one.

$$= \frac{2}{8} \text{ or } \frac{1}{4}$$

How do you grid the answer?

You can either grid the fraction $\frac{1}{4}$, or rewrite it as 0.25 and grid the decimal. Be sure to write the decimal point or fraction bar in the answer box. The following are acceptable answer responses that represent $\frac{1}{4}$ and 0.25.

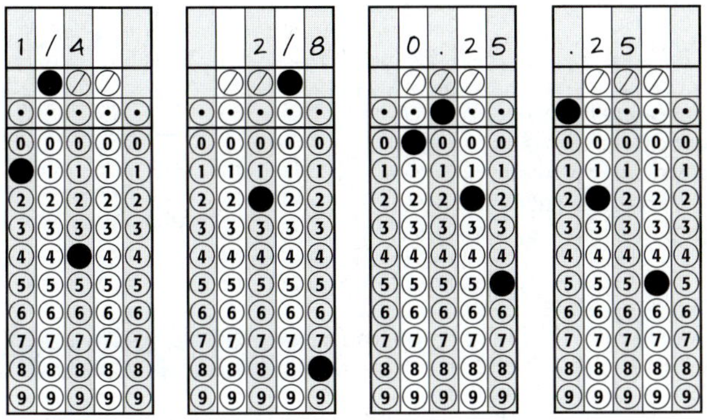

Do not leave a blank answer box in the middle of an answer.

Some problems may result in an answer that is a mixed number. Before filling in the grid, change the mixed number to an equivalent improper fraction or decimal. For example, if the answer is $1\frac{1}{2}$, do not enter 1 1/2 as this will be interpreted as $\frac{11}{2}$. Instead, either enter 3/2 or 1.5.

Example 3 **José is using this figure for a computer graphics design. He wants to dilate the figure by a scale factor of $\frac{7}{4}$. What will be the y-coordinate of the image of D?**

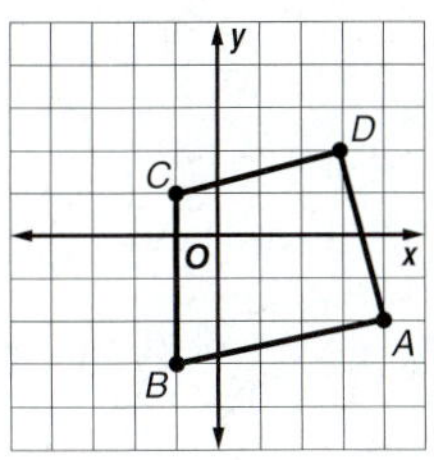

To find the y-coordinate of the image of D, multiply the x-coordinate by the scale factor of $\frac{7}{4}$.

$$2 \cdot \frac{7}{4} = \frac{7}{2}$$

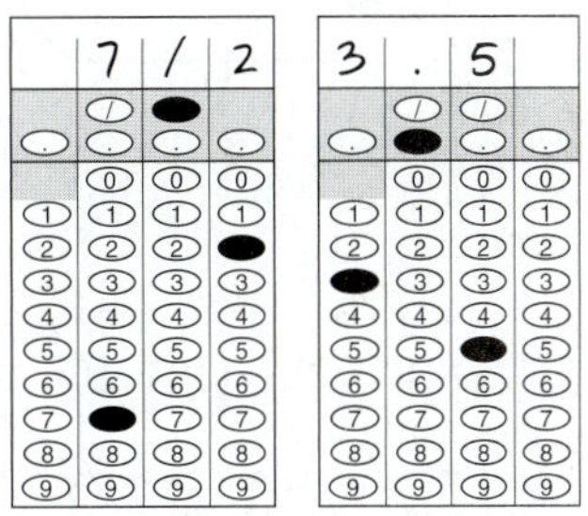

Grid in 7/2 or 3.5. Do not grid 3 1/2.

Preparing for Standardized Tests

Preparing for Standardized Tests

Preparing for Standardized Tests 881

Gridded-Response Practice

Solve each problem. Then copy and complete a grid.

Number and Operations

1. Rewrite 16^3 as a power of 2. What is the value of the exponent for 2? **12**

2. Wolf 359 is the fourth closest star to Earth. It is 45,531,250 million miles from Earth. A light-year, the distance light travels in a year, is 5.88×10^{12} miles. What is the distance from Earth to Wolf 359 in light-years? Round to the nearest tenth. **7.7**

3. A store received a shipment of coats. The coats were marked up 50% to sell to the customers. At the end of the season, the coats were discounted 60%. Find the ratio of the discounted price of a coat to the original cost of the coat to the store. **0.9**

4. Find the value of the determinant $\begin{vmatrix} -1 & 4 \\ -3 & 0 \end{vmatrix}$. **12**

5. Kendra is displaying eight sweaters in a store window. There are four identical red sweaters, three identical brown sweaters, and one white sweater. How many different arrangements of the eight sweaters are possible? **280**

6. An electronics store reduced the price of a DVD player by 10% because it was used as a display model. If the reduced price was $107.10, what was the cost in dollars before it was reduced? Round to the nearest cent if necessary. **119**

Algebra

7. The box shown can be purchased to ship merchandise at the Pack 'n Ship Store. The volume of the box is 945 cubic inches. What is the measure of the greatest dimension of the box in inches? **15**

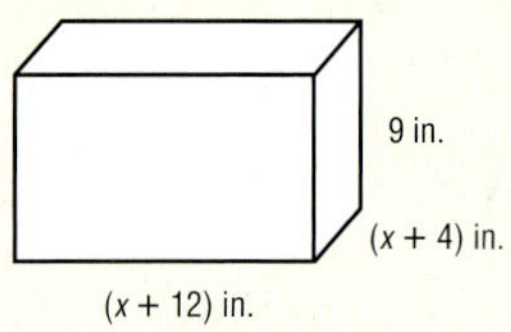

8. If $f(x) = 2x^2 - 3x + 10$, find $f(-1)$. **15**

Test-Taking Tip

Question 22
Fractions do not have to be written in lowest terms. Any equivalent fraction that fits the grid is correct.

9. Find the positive root of $\frac{10}{a^2 - 16} - \frac{6}{a + 4} = \frac{4}{9}$. **5**

10. The graph shows the retail price per gallon of unleaded gasoline in the U.S. from 1980 to 2000. What is the slope if a line is drawn through the points for 15 and 20 years since 1980? **0.072**

Retail Price of Unleaded Gasoline

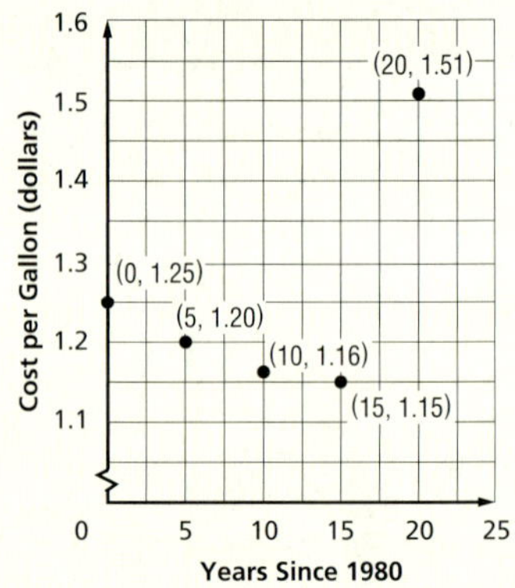

Source: U.S. Dept. of Energy

11. Solve $\sqrt{x + 11} - 9 = \sqrt{x} - 8$. **25**

Geometry

12. Polygon *DABC* is rotated 90° counterclockwise and then reflected over the line $y = x$. What is the x-coordinate of the final image of *A*? **3**

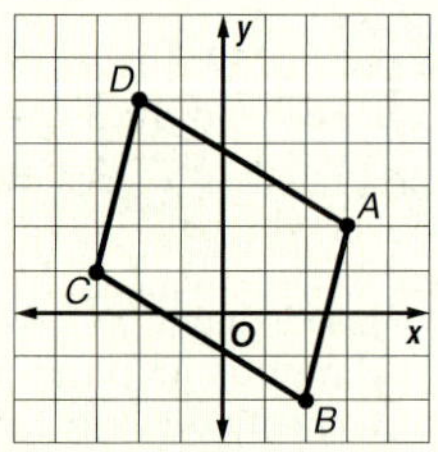

13. A garden is shaped as shown below. What is the measure of $\angle A$ to the nearest degree? **25**

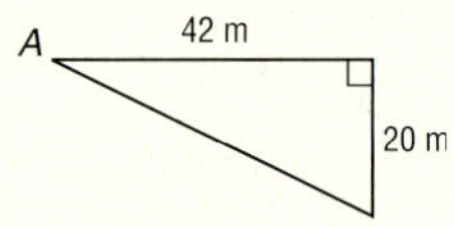

14. A circle of radius r is circumscribed about a square. What is the ratio of the area of the circle to the area of the square? Express the ratio as a decimal rounded to the nearest hundredth. **1.57**

882 Preparing for Standardized Tests

15. Find the degree measure of $\angle 1$. **34**

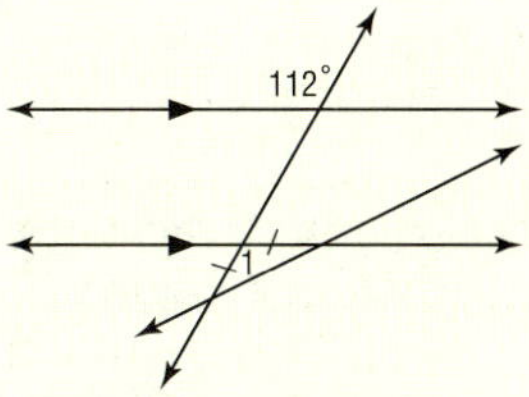

16. Each angle of a regular polygon measures 150°. How many sides does the polygon have? **12**

Measurement

17. The Pasacal (P) is a measure of pressure that is equivalent to 1 Newton per square meter. The typical pressure in an automobile tire is 2×10^5 P while typical blood pressure is 1.6×10^4 P. How many times greater is the pressure in a tire than typical blood pressure? **12.5**

18. A circular ride at an amusement park rotated $\frac{7\pi}{4}$ radians while loading riders. What is the degree measure of the rotation? **315**

19. Find the value of x in the triangle. Round to the nearest tenth of a foot. **21.2**

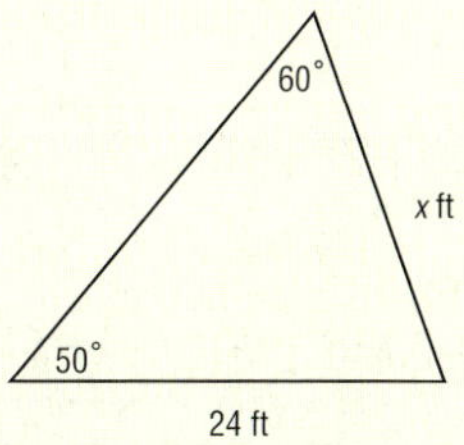

20. Four equal-sized cylindrical juice cans are packed tightly in the box shown. What is the volume of space in the box that is not occupied by the cans in cubic inches? Use 3.14 for π and round to the nearest cubic inch. **660**

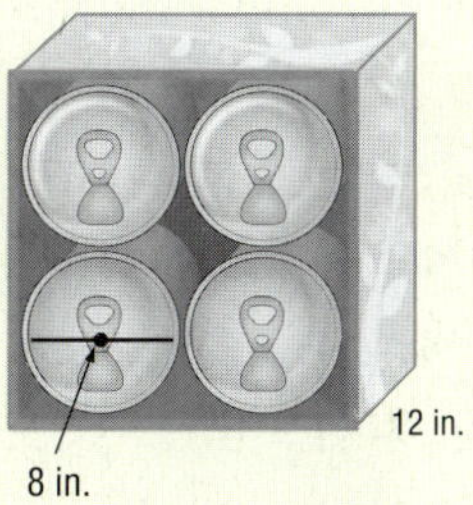

21. Caroline is making a quilt. The diagram shows a piece of cloth she will cut for a portion of the pattern. Find the area of the entire hexagonal piece to the nearest tenth of a square inch. **53.0**

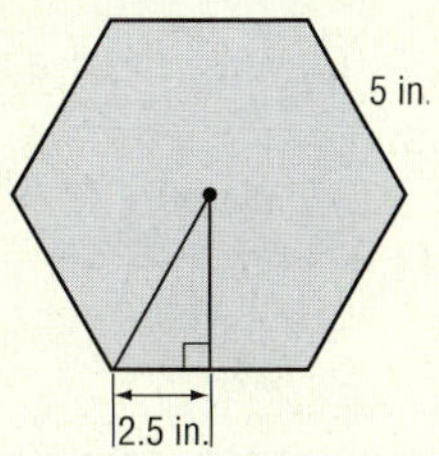

Data Analysis and Probability

22. Of ten girls on a team, three have blue eyes. If two girls are chosen at random, what is the probability that neither has blue eyes? **7/15**

23. In order to win a game, Miguel needs to advance his game piece 4 spaces. What is the probability that the sum of the numbers on the two dice he rolls will be 4? **1/12**

24. The table shows the number of televisions owned per 1000 people in each country. What is the absolute value of the difference between the mean and the median of the data? **7**

Country	Televisions
United States	844
Latvia	741
Japan	719
Canada	715
Australia	706
United Kingdom	652
Norway	648
Finland	643
France	623

Source: International Telecommunication Union

25. The table shows the amount of breakfast cereal eaten per person each year by the ten countries that eat the most. Find the standard deviation of the data set. Round to the nearest tenth of a pound. **4.2**

Country	Cereal (lb)
Sweden	23
Canada	17
Australia	16
United Kingdom	15
Nauru	14
New Zealand	14
Ireland	12
United States	11
Finland	10
Denmark	7

Source: Euromonitor

26. Two number cubes are rolled. If the two numbers appearing on the faces of the number cubes are different, find the probability that the sum is 6. Round to the nearest hundredth. **0.13**

Preparing for Standardized Tests

Preparing for Standardized Tests 883

Short-Response Questions

Short-response questions require you to provide a solution to the problem, as well as any method, explanation, and/or justification you used to arrive at the solution. These are sometimes called *constructed-response, open-response, open-ended, free-response,* or *student-produced questions.* The following is a sample rubric, or scoring guide, for scoring short-response questions.

Criteria	Score
Full credit: The answer is correct and a full explanation is provided that shows each step in arriving at the final answer.	2
Partial credit: There are two different ways to receive partial credit. • The answer is correct, but the explanation provided is incomplete or incorrect. • The answer is incorrect, but the explanation and method of solving the problem is correct.	1
No credit: Either an answer is not provided or the answer does not make sense.	0

On some standardized tests, no credit is given for a correct answer if your work is not shown.

Preparing for Standardized Tests

Preparing for Standardized Tests

Example

Mr. Youngblood has a fish pond in his backyard. It is circular with a diameter of 10 feet. He wants to build a walkway of equal width around the pond. He wants the total area of the pond and walkway to be about 201 square feet. To the nearest foot, what should be the width of the walkway?

Full Credit Solution

Strategy

Diagrams

Draw a diagram of the pond and the walkway. Label important information.

First draw a diagram to represent the situation.

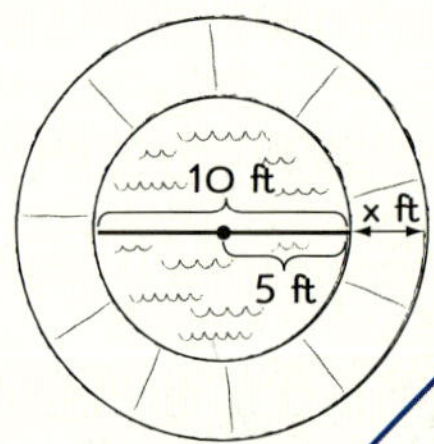

Since the diameter of the pond is 10 feet, the radius is 5 ft. Let the width of the walkway be x feet.

$A = \pi r^2$

$201 = \pi(x + 5)^2$ $r = x + 5, A = 201$

$201 = \pi(x^2 + 10x + 25)$ Multiply.

$\frac{201}{\pi} = \frac{\pi(x^2 + 10x + 25)}{\pi}$ Divide by π, using 3.14 for π.

$64 = x^2 + 10x + 25$ or $x^2 + 10x - 39 = 0$

Use the Quadratic Formula.

$$x = \frac{-b \pm \sqrt{b^2 - 4ac}}{2a}$$

$$= \frac{-10 \pm \sqrt{(-10)^2 - 4(1)(-39)}}{2(1)} \qquad a = 1, b = 10, c = -39$$

$$= \frac{-10 + \sqrt{256}}{2} \text{ or } 3$$

The width of the walkway should be 3 feet.

The steps, calculations, and reasoning are clearly stated.

Before taking a standardized test, memorize common formulas, like the Quadratic Formula, to save time.

Since length must be positive, eliminate the negative solution.

884 Preparing for Standardized Tests

Partial Credit Solution

In this sample solution, the equation that can be used to solve the problem is correct. However, there is no justification for any of the calculations.

There is no explanation of how the quadratic equation was found.

$x^2 + 10x - 39 = 0$

$x = \dfrac{-10 + \sqrt{256}}{2}$

$= -13 \text{ or } 3$

The walkway should be 3 feet wide.

Partial Credit Solution

In this sample solution, the answer is incorrect because the wrong root was chosen.

Since the diameter of the pond is 10 feet, the radius is 5 ft. Let the width of the walkway be x feet. Use the formula for the area of a circle.

$A = \pi r^2$

$201 = \pi(x + 5)^2$

$201 = \pi(x^2 + 10x + 25)$

$\dfrac{201}{\pi} = \dfrac{\pi(x^2 + 10x + 25)}{\pi}$

$64 = x^2 \text{ a } 10x \text{ a } 25 \text{ or } x^2 \text{ a } 10x - 39 = 0$

Use the Quadratic Formula.

$x = \dfrac{-b \pm \sqrt{b^2 - 4ac}}{2a}$

$= -13 \text{ or } 3$

The walkway should be 13 feet wide.

The negative root was chosen as the solution.

No Credit Solution

Use the formula for the area of a circle.

$A = \pi r^2 x$

$201 = \pi(5)^2 x$

$201 = 3.14(25)x$

$201 = 78.5x$

$x = 2.56$

Build the walkway 3 feet wide.

The width of the walkway x is used incorrectly in the area formula for a circle. However, when the student rounds the value for the width of the walkway, the answer is correct. No credit is given for an answer achieved using faulty reasoning.

Preparing for Standardized Tests

Preparing for Standardized Tests 885

3. Let a matrix $A = \begin{bmatrix} a & b \\ c & d \end{bmatrix}$, then.

$\begin{bmatrix} 1 & 0 \\ 0 & 1 \end{bmatrix} \cdot \begin{bmatrix} a & b \\ c & d \end{bmatrix} = \begin{bmatrix} a & b \\ c & d \end{bmatrix}$

and $\begin{bmatrix} a & b \\ c & d \end{bmatrix} \cdot \begin{bmatrix} 1 & 0 \\ 0 & 1 \end{bmatrix} = \begin{bmatrix} a & b \\ c & d \end{bmatrix}$.

8. 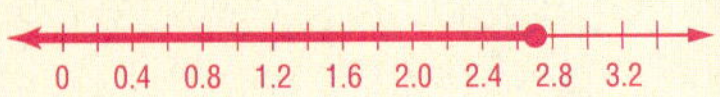

10. $y = 17.50 + 15.50x$; the slope is the charge per hour and the y-intercept is the initial charge for traveling to the customer's yard

12. 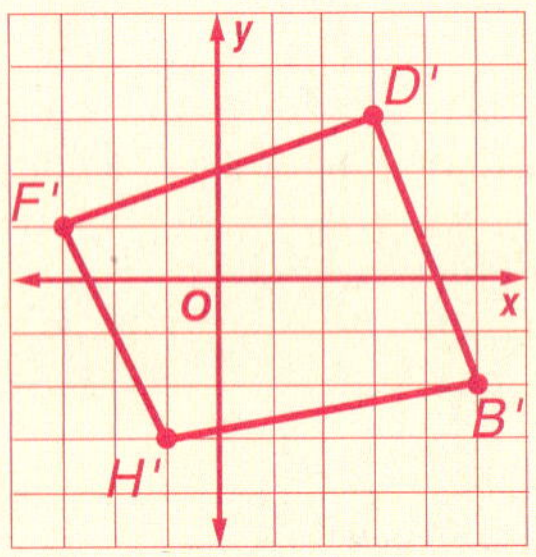

Short-Response Practice

Solve each problem. Show all your work.

Number and Operations

1. An earthquake that measures a value of 1 on the Richter scale releases the same amount of energy as 170 grams of TNT, while one that measures 4 on the scale releases the energy of 5 metric tons of TNT. One metric ton is 1000 kilograms and 1000 grams is 1 kilograms. How many times more energy is released by an earthquake measuring 4 than one measuring 1? **about 29,412 times**

2. In 2000, Cook County, Illinois was the second largest county in the U.S with a population of about 5,377,000. This was about 43.3% of the population of Illinois. What was the approximate population of Illinois in 2000? **about 12,418,000**

3. Show why $\begin{bmatrix} 1 & 0 \\ 0 & 1 \end{bmatrix}$ is the identity matrix for multiplication for 2×2 matrices. **See margin.**

4. The total volume of the oceans on Earth is 3.24×10^8 cubic miles. The total surface area of the water of the oceans is 139.8 million square miles. What is the average depth of the oceans? **about 2.3 mi**

5. At the Blaine County Fair, there are 12 finalists in the technology project competition. How many ways can 1st, 2nd, 3rd, and 4th place be awarded? **11,880 ways**

Algebra

6. Factor $3x^2a^2 - 3x^2b^2$. Explain each step. $\mathbf{3x^2(a + b)(a - b)}$

7. Solve and graph $7 - 2a > \frac{15 - 2a}{6}$. **$a < 2.7$; See margin for graph.**

8. Solve the system of equations.
$x^2 + 9y^2 = 25 \quad y - x = -5$ **(5, 0), (4, −1)**

9. The table shows what Miranda Richards charges for landscaping services for various numbers of hours. Write an equation to find the charge for any amount of time, where y is the total charge in dollars and x is the amount of time in hours. Explain the meaning of the slope and y-intercept of the graph of the equation. **See margin.**

Hours	Charge (dollars)	Hours	Charge (dollars)
0	17.50	3	64.00
1	33.00	4	79.50
2	48.50	5	95.00

10. Write an equation that fits the data in the table. $\mathbf{y = x^2 + 3}$

x	y
−3	12
−1	4
0	3
2	7
4	19

Geometry

11. A sledding hill at the local park has an angle of elevation of 15°. Its vertical drop is 400 feet. What is the length of the sledding path? **about 1545.5 ft**

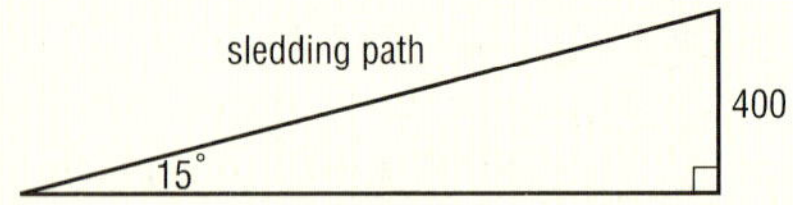

12. Polygon $BDFH$ is transformed using the matrix $\begin{bmatrix} 0 & 1 \\ 1 & 0 \end{bmatrix}$. Graph $B'D'F'H'$ and identify the type of transformation. **See margin for graph; reflection over line $y = x$.**

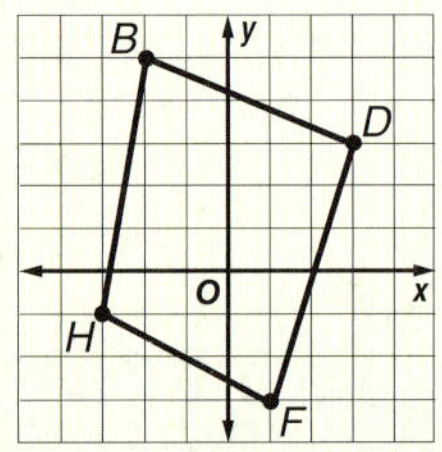

13. The map shows the trails that connect three hiking destinations. If Amparo hikes from Deer Ridge to Egg Mountain to Lookout Point and back to Deer Ridge, what is the distance she will have traveled? **about 35.9 mi**

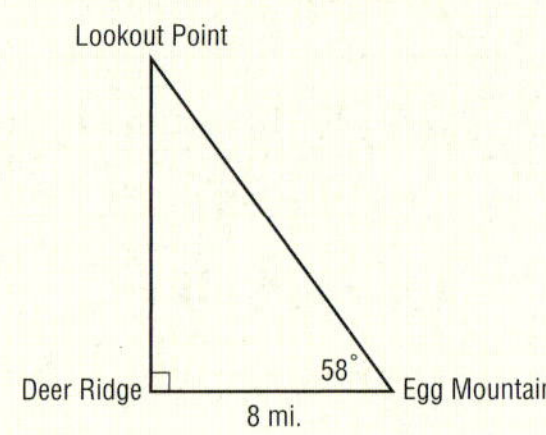

14. Mr Washington is making a cement table for his backyard. The tabletop will be circular with a diameter of 6 feet and a depth of 6 inches. How much cement will Mr. Washington need to make the top of the table? Use 3.14 for π and round to the nearest cubic foot. **14.1 ft³**

15. A triangular garden is plotted on grid paper, where each unit is 1 meter. Its sides are segments that are parts of the lines with equations $y = -\frac{5}{4}x + 2$, $2y - 5x = 4$, and $y = -3$. Graph the triangle and find its area. **15 m²; See margin for graph.**

16. Dylan is flying a kite. He wants to know how high above the ground it is. He knows that he has let out 75 feet of string and that it is flying directly over a nearby fence post. If he is 50 feet from the fence post, how high is the kite? Round to the nearest tenth of a foot. **55.9 ft**

Measurement

17. The temperature of the Sun can reach 27,000,000°F. The relationship between Fahrenheit F and Celsius C temperatures is given by the equation $F = 1.8C + 32$. Find the temperature of the Sun in degrees Celsius. **about 15,000,000°C**

18. In 2003, Monaco was the most densely populated country in the world. There were about 32,130 people occupying the country at the rate of 16,477 people per square kilometer. What is the area of Monaco? **about 1.95 km²**

19. A box containing laundry soap is a cylinder with a diameter of 10.5 inches and a height of 16 inches. What is the surface area of the box? **about 701 in²**

20. Light travels at 186,291 miles per second or 299,792 kilometers per second. What is the relationship between miles and kilometers? **1 mi ≈ 1.609 km**

21. Home Place Hardware sells storage buildings for your backyard. The front of the building is a trapezoid as shown. The store manager wants to advertise the total volume of the building. Find the volume in cubic feet. **2520 ft³**

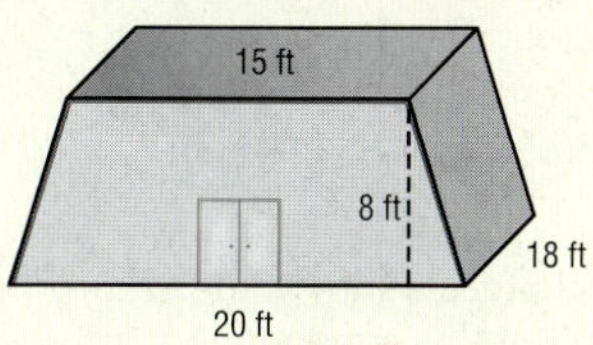

Test-Taking Tip

Questions 15, 16, and 22

Be sure to read the instructions of each problem carefully. Some questions ask for more than one solution, specify how to round answers, or require an explanation.

Data Analysis and Probability

22. The table shows the 2000 populations of the six largest cities in Tennessee. Which measure, mean or median, do you think best represents the data? Explain your answer. **See margin.**

City	Population
Chattanooga	155,404
Clarksville	105,898
Jackson	60,635
Knoxville	173,661
Memphis	648,882
Nashville-Davidson	545,915

Source: International Telecommunication Union

23. The scatter plot shows the number of hours worked per week for U.S. production workers from 1970 through 2000. Let y be the hours worked per week and x be the years since 1970. Write an equation that you think best models the data.

Average Hours Worked per Week for Production Workers

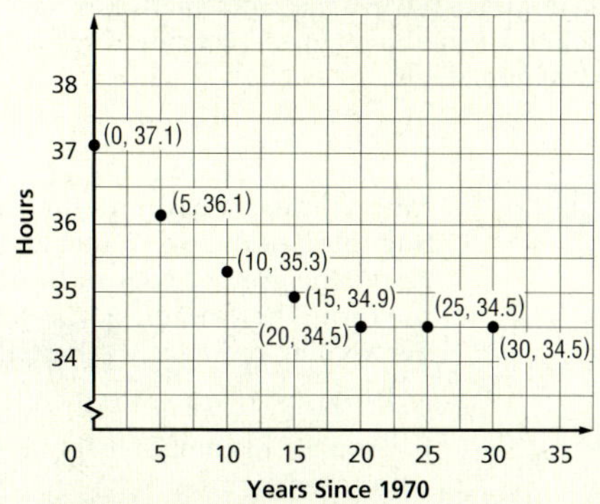

Source: Bureau of Labor Statistics

23. See margin.

24. A day camp has 240 participants. Children can sign up for various activities. Suppose 135 children take swimming, 160 take soccer, and 75 take both swimming and soccer. What is the probability that a child selected at random takes swimming or soccer? $\frac{11}{12}$

25. In how many different ways can seven members of a student government committee sit around a circular table? **720 ways**

26. Illinois residents can choose to buy an environmental license plate to support Illinois parks. Each environmental license plate displays 3 or 4 letters followed by a number 1 thru 99. How many different environmental license plates can be issued? **46,980,648**

15.

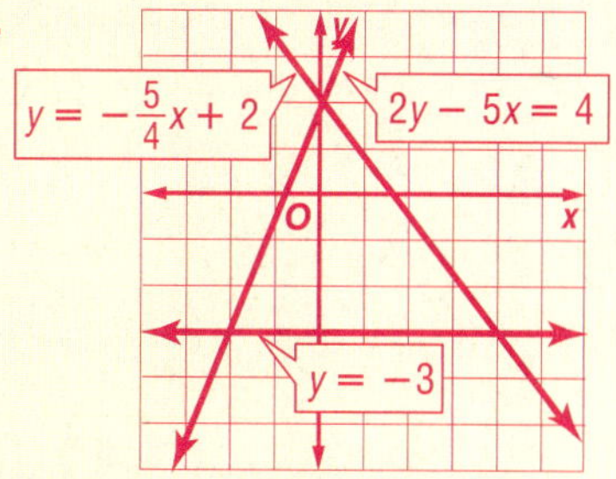

22. Sample answer: The mean is 281,733 and the median is 164,532. The median is better because it is lower and the populations are in general much lower than the highest two populations.

23. Sample answer: A linear model using a graphing calculator is $y = -0.084x + 36.536$.

Extended-Response Questions

Extended-response questions are often called *open-ended* or *constructed-response questions*. Most extended-response questions have multiple parts. You must answer all parts to receive full credit.

Extended-response questions are similar to short-response questions in that you must show all of your work in solving the problem and a rubric is used to determine whether you receive full, partial, or no credit. The following is a sample rubric for scoring extended-response questions.

Criteria	Credit
Full credit: The answer is correct and a full explanation is provided that shows each step in arriving at the final answer.	2
Partial credit: There are two different ways to receive partial credit. • The answer is correct, but the explanation provided is incomplete or incorrect. • The answer is incorrect, but the explanation and method of solving the problem is correct.	1
No credit: Either an answer is not provided or the answer does not make sense.	0

On some standardized tests, no credit is given for a correct answer if your work is not shown.

Make sure that when the problem says to *Show your work*, you show every aspect of your solution including figures, sketches of graphing calculator screens, or reasoning behind computations.

Preparing for Standardized Tests

Example **Libby throws a ball into the air with a velocity of 64 feet per second. She releases the ball 5 feet above the ground. The height of the ball above the ground t seconds after release is modeled by an equation of the form $h(t) = -16t^2 + v_0t + h_0$ where v_0 is the initial velocity in feet per second and h_0 is the height at which the ball is released.**

a. Write an equation for the flight of the ball. Sketch the graph of the equation.

b. Find the maximum height that the ball reaches and the time that this height is reached.

c. Change only the speed of the release of the ball such that the ball will reach a maximum height greater than 100 feet. Write an equation for the flight of the ball.

Full Credit Solution

Part a A complete graph includes appropriate scales and labels for the axes, and points that are correctly graphed.

- A complete graph also shows the basic characteristics of the graph. The student should realize that the graph of this equation is a parabola opening downward with a maximum point reached at the vertex.
- The student should choose appropriate points to show the important characteristics of the graph.
- Students should realize that t and x and $h(t)$ and y are interchangeable on a graph on the coordinate plane.

Preparing for Standardized Tests

888 Preparing for Standardized Tests

To write the equation for the ball, I substituted $v_0 = 64$ and $h_0 = 5$ into the equation $h(t) = -16t^2 + v_0t + h_0$, so the equation is $h(t) = -16t^2 + 64t + 5$. To graph the equation, I found the equation of the axis of symmetry and the vertex.

$x = -\frac{b}{2a}$

$= -\frac{64}{2(-16)}$ or 2

The equation of the axis of symmetry is $x = 2$, so the x–coordinate of the vertex is 2. You let $t = x$ and then $h(t) = -16t^2 + 64t + 5 = -16(2)^2 + 64(2) + 5$ or 69. The vertex is (2, 69). I found some other points and sketched the graph. I graphed points $(t, h(t))$ as (x, y).

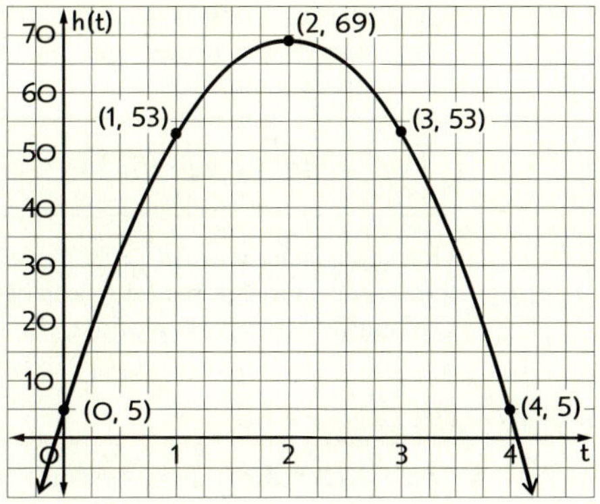

Part b

The maximum height of the ball is reached at the vertex of the parabola. So, the maximum height is 69 feet and the time it takes to reach the maximum height is 2 seconds.

Part c

In part c, any equation whose graph has a vertex with y-coordinate greater than 100 would be a correct answer earning full credit.

Since I have a graphing calculator, I changed the value of v_0 until I found a graph in which the y or $h(t)$ coordinate was greater than 100. The equation I used was $h(t) = -16t^2 + 80t + 5$.

Partial Credit Solution

Part a This sample answer does not earn full credit because it includes no explanation of how the equation was written or the vertex was found.

$h(t) = -16t^2 + 64t + 5$; (2, 69)

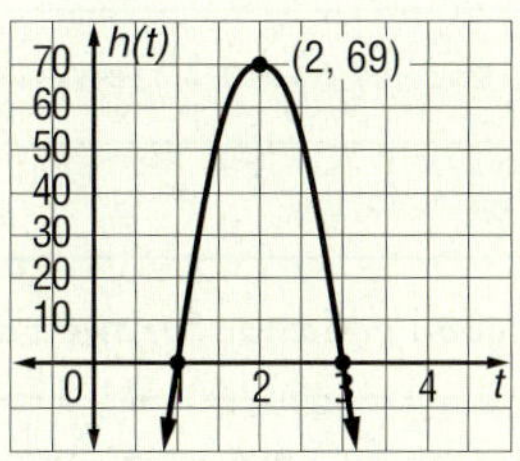

Preparing for Standardized Tests

Preparing for Standardized Tests 889

Part b Full credit is given because the vertex is correct and is interpreted correctly.

> *The vertex shows the maximum height of the ball. The time it takes to reach the maximum height of 69 feet is 2 seconds.*

Part c Partial credit is given for part c since no explanation is given for using this equation. The student did not mention that the vertex would have a y-coordinate greater than 100.

> *I will write the equation $h(t) = -16t^2 + 100t + 5$.*

This sample answer might have received a score of 2 or 1, depending on the judgment of the scorer. Had the student sketched a more accurate graph and given more complete explanations for Parts a and c, the score would probably have been a 3.

No Credit Solution

Part a No credit is given because the equation is incorrect with no explanation and the sketch of the graph has no labels, making it impossible to determine whether the student understands the relationship between the equation for a parabola and the graph.

> *$h(t) = -16t^2 + 5t + 64$*

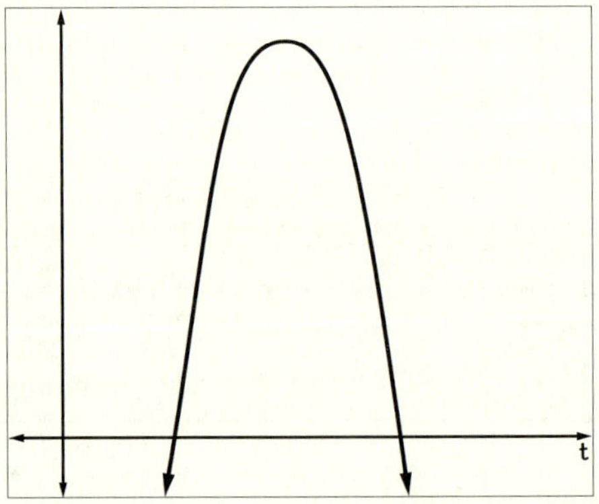

Part b

> *It reaches about 10 feet.*

Part c

> *A good equation for the ball is $h(t) = -16t^2 + 5t + 100$.*

In this sample answer, the student does not understand how to substitute the given information into the equation, graph a parabola, or interpret the vertex of a parabola.

Preparing for Standardized Tests

890 Preparing for Standardized Tests

Extended Response Practice

Solve each problem. Show all your work.

Number and Operations

1. Mrs. Ebbrect is assigning identification (ID) numbers to freshman students. She plans to use only the digits 2, 3, 5, 6, 7, and 9. The ID numbers will consist of three digits with no repetitions.

a. How many 3-digit ID numbers can be formed? **120**

b. How many more ID numbers can Mrs. Ebbrect make if she allows repetitions? **96**

c. What type of system could Mrs. Ebbrect use to choose the numbers if there are at least 400 students who need ID numbers? **See margin.**

2. Use these four matrices. **See margin.**

$A = \begin{bmatrix} -1 & 0 \\ 3 & -2 \end{bmatrix}$ $\quad B = \begin{bmatrix} 4 & -6 & 5 \\ 0 & 1 & -3 \end{bmatrix}$

$C = \begin{bmatrix} -7 & 3 \\ -6 & 2 \end{bmatrix}$ $\quad D = [-3 \ \ 1]$

a. Find $A + C$.

b. Compare the dimensions of AB and DB.

c. Compare the matrices BC and CB.

Algebra

3. Roger is using the graph showing the gross cash income for all farms in the U.S. from 1930 through 2000 to make some predictions for the future.

Source: U.S. Department of Agriculture

a. Write an equation in slope-intercept form for the line passing through the point for 1930 and the point for 1970. $y = 1150x - 2{,}210{,}500$

b. Write an equation in slope-intercept form for the line passing through the point for 1980 and the point for 2000. Compare the slope of this line to the slope of the line in part a. **See margin.**

c. Which equation, if any, do you think Roger should use to model the data? Explain. **See margin.**

d. Suggest an equation that is not linear for Roger to use. **Sample answer:** $y = 53.8x^2 - 207{,}929x + 201{,}076{,}251$

4. Brad is coaching the bantam age division (8 years old and younger) swim team. On the first day of practice, he has the team swim 4 laps of the 25-meter pool. For each of the next practices, he increases the laps by 3. In other words, the children swim 4 laps the first day, 7 laps the second day, 10 laps the third day, and so on.

a. Write a formula for the nth term of the sequence of the number of laps each day. Explain how you found the formula.

b. How many laps will the children swim on the 10th day? **31 laps**

c. Brad's goal is to have the children swim at least one mile during practice on the 20th day. If one mile is approximately 1.6 kilometers, will Brad reach his goal? **4a, c. See margin.**

Geometry

5. Alejandra is planning to use a star shape in a galaxy-themed mural on her wall. The pentagon in the center is regular, and the triangles forming the points are isosceles.

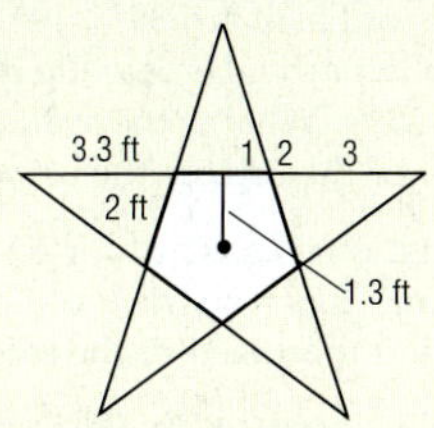

a. Find the measures of $\angle 1$, $\angle 2$, and $\angle 3$. Explain your method. **108°, 172°, 36°**

b. The approximate dimensions of the design are given. The segment of length 1.3 feet is the apothem of the pentagon. Find the approximate area of the design. **about 13.75 ft²**

c. If Alejandra circumscribes a circle about the star, what is the area of the circle? **about 55.42 ft²**

1c. Sample answer: Use 4-digit numbers using those digits and repetitions. That allows for 6^4 or 1296 different ID numbers.

2a. $\begin{bmatrix} -8 & 3 \\ -3 & 0 \end{bmatrix}$

2b. AB is a 2×3 and DB is a 1×3.

2c. BC does not exist, while CB is a 2×3 matrix.

3b. $y = 4350x - 8{,}470{,}000$; the slope of the line in part b is almost 4 times steeper than the slope of the line in part a.

3c. Sample answer: Neither equation; the equation in part a will give a low prediction and the equation in part b will give a prediction that may be too high.

4a. $a_n = 3n + 1$

4c. No; on the 20th day, they will swim 61 laps, which is 61×25 or 1525 m or 1.525 km. 1.525 km is approximately 0.95 mile.

Preparing for Standardized Tests 891

6a.

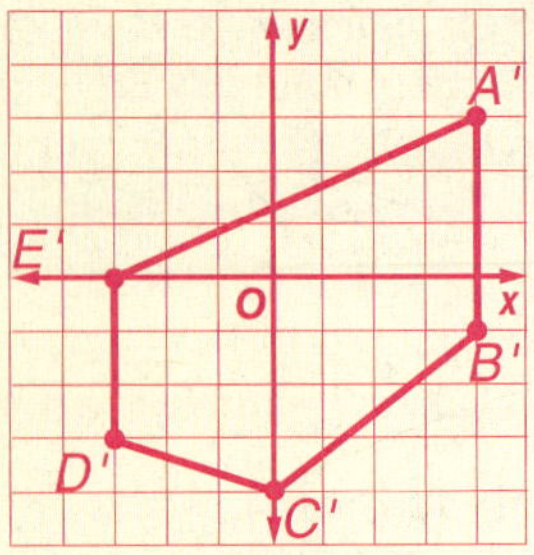

$A'(4, 3)$, $B'(4, -1)$, $C'(0, -4)$, $D'(-3, -3)$, $E'(-3, 0)$; The image of a point with coordinates (a, b) has coordinates $(b, -a)$.

6b.

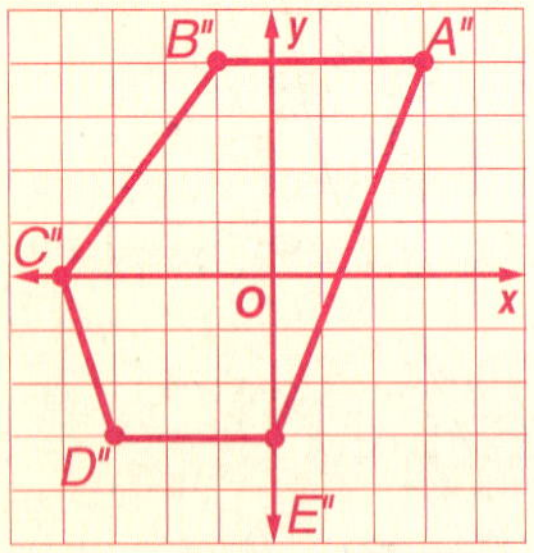

$A''(3, 4)$, $B''(-1, 4)$, $C''(-4, 0)$, $D''(-3, -3)$, $E''(0, -3)$; The image of a point with coordinates (a, b) has coordinates (b, a).

6c. Reflecting *ABCDE* directly over the *y*-axis will produce *A″B″C″D″E″* in one transformation.

7c. Its speed decreases. If you increase the height of the satellite, thus increasing *r*, and try it in the formula, the value of *v* decreases.

9b. Sample graph:

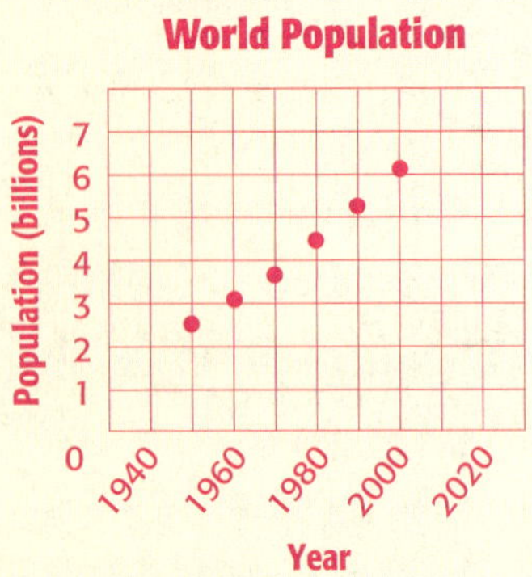

9c. Sample using population in billions, such as 2.6 for 1950 and years as they are such as 1950: $y = 0.072x - 138$

6. Kareem is using polygon *ABCDE*, shown on a coordinate plane, as a basis for a computer graphics design. He plans to perform various transformations on the polygon to produce a variety of interesting designs.

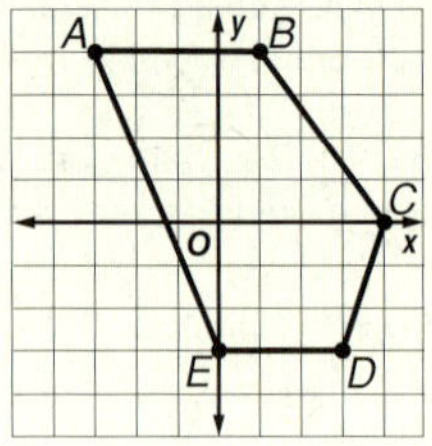

a. First, Kareem creates polygon *A′B′C′D′E′* by rotating *ABCDE* counterclockwise about the origin 270°. Graph polygon *A′B′C′D′E′* and describe the relationship between the coordinates of *ABCDE* and *A′B′C′D′E′*.

b. Next, Kareem reflects polygon *A′B′C′D′E′* in the line $y = x$ to produce polygon *A″B″C″D″E″*. Graph *A″B″C″D″E″* and describe the relationship between the coordinates of *A′B′C′D′E′* and *A″B″C″D″E″*.

c. Describe how Kareem could transform polygon *ABCDE* to polygon *A″B″C″D″E″* with only one transformation. **6a–6c. See margin.**

Measurement

7. The speed of a satellite orbiting Earth can be found using the equation $v = \sqrt{\frac{\frac{Gm_E}{r}}{m_E}}$. *G* is the gravitational constant for Earth, m_E is the mass of Earth, and *r* is the radius of the orbit which includes the radius of Earth and the height of the satellite.

a. The radius of Earth is 6.38×10^6 meters. The distance of a particular satellite above Earth is 350 kilometers. What is the value of *r*? (*Hint:* The center of the orbit is the center of Earth.) **6.73×10^6 m**

b. The gravitational constant for Earth is 6.67×10^{-11} N • m²/kg². The mass of Earth is 5.97×10^{24} kg. Find the speed of the satellite in part a. **about 7.7×10^3 m/s**

c. As a satellite increases in distance from the Earth, what is the effect on the speed of the orbit? Explain your reasoning. **See margin.**

Test-Taking Tip

Question 6

When questions require graphing, make sure your graph is accurate to receive full credit for your correct solution.

8. A cylindrical cooler has a diameter of 9 inches and a height of 11 inches. Scott plans to use it for soda cans that have a diameter of 2.5 inches and a height of 4.75 inches.

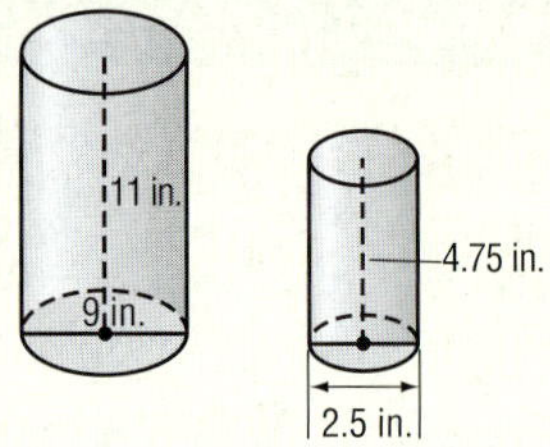

a. Scott plans to place two layers consisting of 9 cans each into the cooler. What is the volume of the space that will not be filled with cans? **280.1 in³**

b. Find the ratio of the volume of the cooler to the volume of the cans in part b. **about 1.7**

Data Analysis and Probability

9. The table shows the total world population from 1950 through 2000. **9a. 1980 to 1990**

Year	Population	Year	Population
1950	2,566,000,053	1980	4,453,831,714
1960	3,039,451,023	1990	5,278,639,789
1970	3,706,618,163	2000	6,082,966,429

a. Between which two decades was the percent increase in population the greatest?

b. Make a scatter plot of the data. **See margin.**

c. Find a function that models the data. **See margin.**

d. Predict the world population for 2030. **Using equation in part c, about 8.2 billion**

10. Each year, a university sponsors a conference for women. The Venn diagram shows the number of participants in three activities for the 680 women that attended. Suppose women who attended are selected at random for a survey.

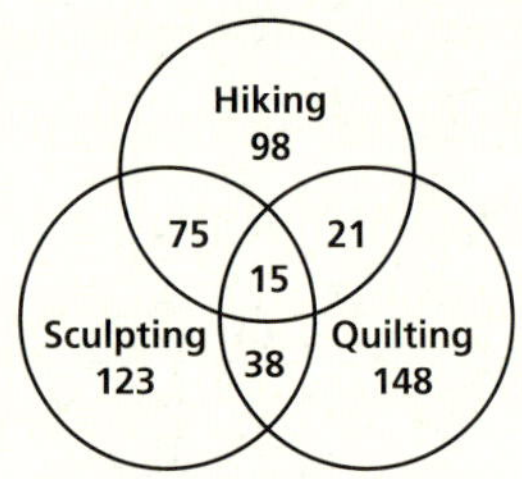

a. What is the probability that a woman selected participated in hiking or sculpting? **about 0.54**

b. Describe a set of women such that the probability of their being selected is about 0.39. **Sample answer: Women who participated in hiking or quilting but not sculpting**

Glossary/Glosario

Cómo usar el glosario en español:

1. Busca el término en inglés que desees encontrar.
2. El término en español, junto con la definición, se encuentran en la columna de la derecha.

English

A

absolute value (28) A number's distance from zero on the number line, represented by $|x|$.

absolute value function (90) A function written as $f(x) = |x|$, where $f(x) \geq 0$ for all values of x.

absolute value inequalities (42) For all real numbers a and b, $b > 0$, the following statements are true.
1. If $|a| < b$, then $-b < a < b$
2. If $|a| > b$, then $a > b$ or $a < -b$.

algebraic expression (7) An expression that contains at least one variable.

amplitude (763) For functions in the form $y = a \sin b\theta$ or $y = a \cos b\theta$, the amplitude is $|a|$.

angle of depression (705) The angle between a horizontal line and the line of sight from the observer to an object at a lower level.

angle of elevation (705) The angle between a horizontal line and the line of sight from the observer to an object at a higher level.

arccosine (747) The inverse of $y = \cos x$, written as $x = \arccos y$.

arcsine (747) The inverse of $y = \sin x$, written as $x = \arcsin y$.

arctangent (747) The inverse of $y = \tan x$ written as $x = \arctan y$.

arithmetic mean (580) The terms between any two nonconsecutive terms of an arithmetic sequence.

arithmetic sequence (578) A sequence in which each term after the first is found by adding a constant, the common difference d, to the previous term.

arithmetic series (583) The indicated sum of the terms of an arithmetic sequence.

asymptote (442, 485) A line that a graph approaches but never crosses.

augmented matrix (208) A coefficient matrix with an extra column containing the constant terms.

axis of symmetry (287) A line about which a figure is symmetric.

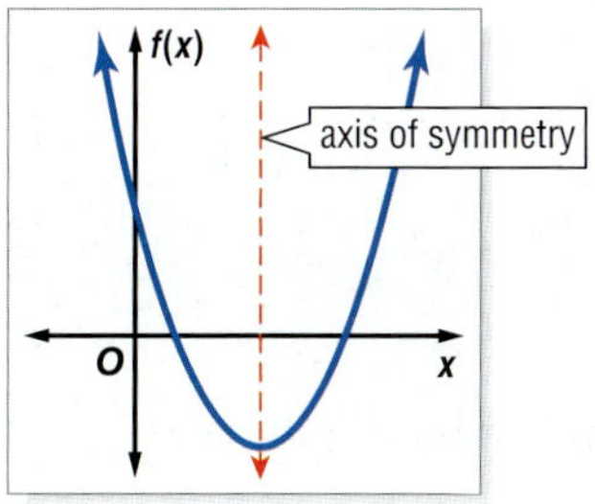

Español

valor absoluto Distancia entre un número y cero en una recta numérica; se denota con $|x|$.

función del valor absoluto Una función que se escribe $f(x) = |x|$, donde $f(x) \geq 0$, para todos los valores de x.

desigualdades con valor absoluto Para todo número real a y b, $b > 0$, se cumple lo siguiente.
1. Si $|a| < b$, entonces $-b < a < b$
2. Si $|a| > b$, entonces $a > b$ o $a < -b$.

expresión algebraica Expresión que contiene al menos una variable.

amplitud Para funciones de la forma $y = a$ sen $b\theta$ o $y = a \cos b\theta$, la amplitud es $|a|$.

ángulo de depresión Ángulo entre una recta horizontal y la línea visual de un observador a una figura en un nivel inferior.

ángulo de elevación Ángulo entre una recta horizontal y la línea visual de un observador a una figura en un nivel superior.

arcocoseno La inversa de $y = \cos x$, que se escribe como $x = \arccos y$.

arcoseno La inversa de $y =$ sen x, que se escribe como $x =$ arcsen y.

arcotangente La inversa de $y = \tan x$ que se escribe como $x = \arctan y$.

media aritmética Cualquier término entre dos términos no consecutivos de una sucesión aritmética.

sucesión aritmética Sucesión en que cualquier término después del primero puede hallarse sumando una constante, la diferencia común d, al término anterior.

serie aritmética Suma específica de los términos de una sucesión aritmética.

asíntota Recta a la que se aproxima una gráfica, sin jamás cruzarla.

matriz ampliada Matriz coeficiente con una columna extra que contiene los términos constantes.

eje de simetría Recta respecto a la cual una figura es simétrica.

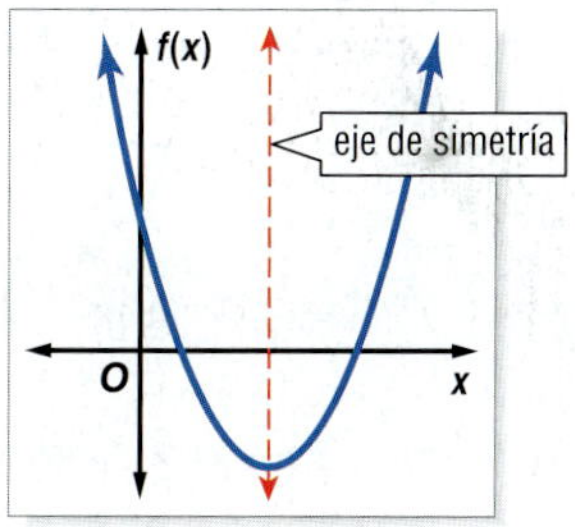

B

$b^{\frac{1}{n}}$ (257) For any real number b and for any positive integer n, $b^{\frac{1}{n}} = \sqrt[n]{b}$, except when $b < 0$ and n is even.

binomial (229) A polynomial that has two unlike terms.

binomial experiment (677) An experiment in which there are exactly two possible outcomes for each trial, a fixed number of independent trials, and the probabilities for each trial are the same.

Binomial Theorem (613) If n is a nonnegative integer, then $(a + b)^n =$
$1a^nb^0 + \frac{n}{1}a^{n-1}b^1 + \frac{n(n-1)}{1 \cdot 2}a^{n-2}b^2 + \ldots + 1a^0b^n$.

boundary (96) A line or curve that separates the coordinate plane into two regions.

bounded (129) A region is bounded when the graph of a system of constraints is a polygonal region.

$b^{\frac{1}{n}}$ Para cualquier número real b y para cualquier entero positivo n, $b^{\frac{1}{n}} = \sqrt[n]{b}$, excepto cuando $b < 0$ y n es par.

binomio Polinomio con dos términos diferentes.

experimento binomial Experimento con exactamente dos resultados posibles para cada prueba, un número fijo de pruebas independientes y en el cual cada prueba tiene igual probabilidad.

Teorema del binomio Si n es un entero no negativo, entonces $(a + b)^n =$
$1a^nb^0 + \frac{n}{1}a^{n-1}b^1 + \frac{n(n-1)}{1 \cdot 2}a^{n-2}b^2 + \ldots + 1a^0b^n$.

frontera Recta o curva que divide un plano de coordenadas en dos regiones.

acotada Una región está acotada cuando la gráfica de un sistema de restricciones es una región poligonal.

C

Cartesian coordinate plane (56) A plane divided into four quadrants by the intersection of the x-axis and the y-axis at the origin.

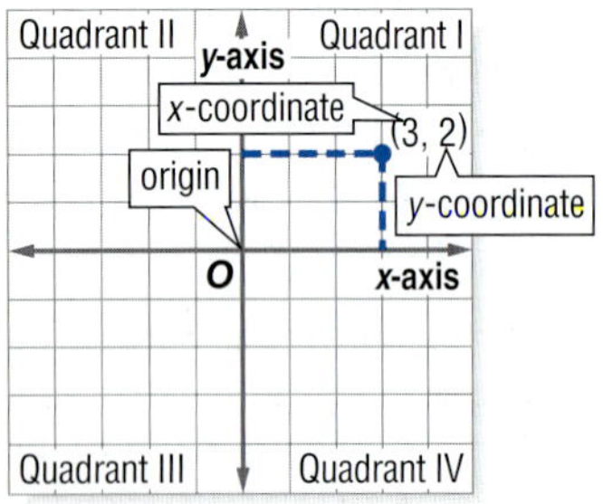

center of a circle (426) The point from which all points on a circle are equidistant.

center of an ellipse (434) The point at which the major axis and minor axis of an ellipse intersect.

center of a hyperbola (442) The midpoint of the segment whose endpoints are the foci.

change of base formula (548) For all positive numbers a, b, and n, where $a \neq 1$ and $b \neq 1$, $\log_a n = \frac{\log_b n}{\log_b a}$.

circle (426) The set of all points in a plane that are equidistant from a given point in the plane, called the center.

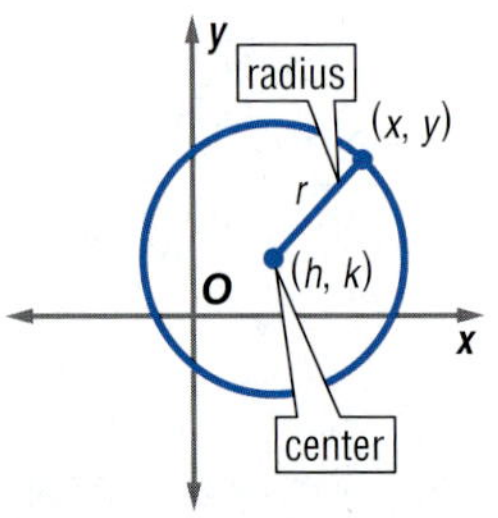

circular functions (740) Functions defined using a unit circle.

plano de coordenadas cartesiano Plano dividido en cuatro cuadrantes mediante la intersección en el origen de los ejes x y y.

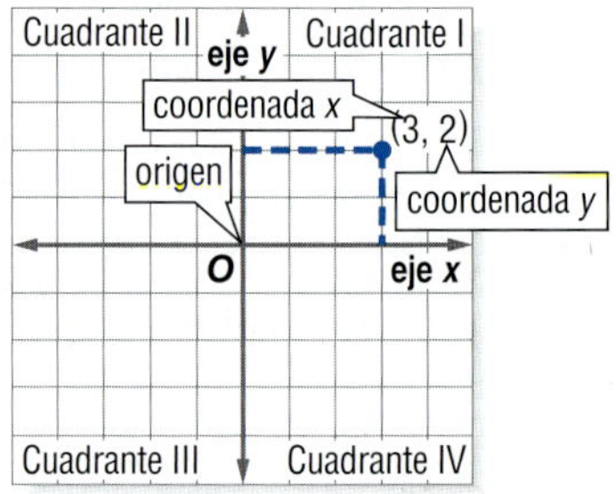

centro de un círculo El punto desde el cual todos los puntos de un círculo están equidistantes.

centro de una elipse Punto de intersección de los ejes mayor y menor de una elipse.

centro de una hipérbola Punto medio del segmento cuyos extremos son los focos.

fórmula del cambio de base Para todo número positivo a, b y n, donde $a \neq 1$ y $b \neq 1$, $\log_b n = \frac{\log_b n}{\log_b a}$.

círculo Conjunto de todos los puntos en un plano que equidistan de un punto dado del plano llamado centro.

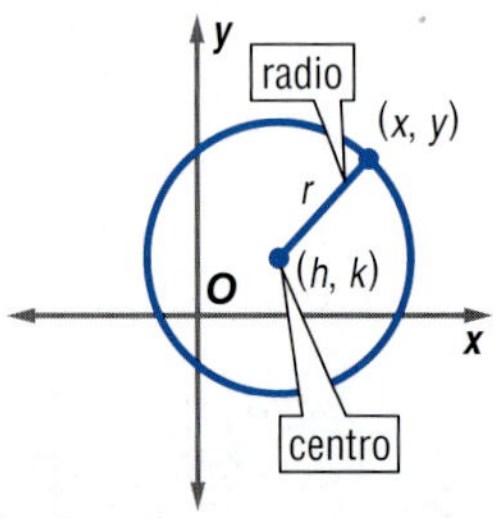

funciones circulares Funciones definidas en un círculo unitario.

coefficient (222) The numerical factor of a monomial.

column matrix (155) A matrix that has only one column.

combination (640) An arrangement of objects in which order is not important.

common difference (578) The difference between the successive terms of an arithmetic sequence.

common logarithms (547) Logarithms that use 10 as the base.

common ratio (588) The ratio of successive terms of a geometric sequence.

Commutative Property of Addition (12) For any real numbers a and b, $a + b = b + a$.

Commutative Property of Multiplication (12) For any real numbers a and b, $a \cdot b = b \cdot a$.

completing the square (307) A process used to make a quadratic expression into a perfect square trinomial.

complex conjugates (273) Two complex numbers of the form $a + bi$ and $a - bi$.

complex fraction (475) A rational expression whose numerator and/or denominator contains a rational expression.

complex number (271) Any number that can be written in the form $a + bi$, where a and b are real numbers and $\boldsymbol{i}$ is the imaginary unit.

composition of functions (384) A function is performed, and then a second function is performed on the result of the first function. The composition of f and g is denoted by $f \circ g$, and $[f \circ g](x) = f[g(x)]$.

compound event (658) Two or more simple events.

compound inequality (40) Two inequalities joined by the word *and* or *or*.

conic section (419) Any figure that can be obtained by slicing a double cone.

conjugate axis (442) The segment of length $2b$ units that is perpendicular to the transverse axis at the center.

conjugates (253) Binomials of the form $a\sqrt{b} + c\sqrt{d}$ and $a\sqrt{b} - c\sqrt{d}$, where a, b, c, and d are rational numbers.

consistent (111) A system of equations that has at least one solution.

constant (222) Monomials that contain no variables.

constant function (90) A linear function of the form $f(x) = b$.

constant of variation (492) The constant k used with direct or inverse variation.

coeficiente Factor numérico de un monomio.

matriz columna Matriz que sólo tiene una columna.

combinación Arreglo de elementos en que el orden no es importante.

diferencia común Diferencia entre términos consecutivos de una sucesión aritmética.

logaritmos comunes El logaritmo de base 10.

razón común Razón entre términos consecutivos de una sucesión geométrica.

Propiedad conmutativa de la adición Para cualquier número real a y b, $a + b = b + a$.

Propiedad conmutativa de la multiplicación Para cualquier número real a y b, $a \cdot b = b \cdot a$.

completar el cuadrado Proceso mediante el cual una expresión cuadrática se transforma en un trinomio cuadrado perfecto.

conjugados complejos Dos números complejos de la forma $a + bi$ y $a - bi$.

fracción compleja Expresión racional cuyo numerador o denominador contiene una expresión racional.

número complejo Cualquier número que puede escribirse de la forma $a + bi$, donde a y b son números reales e $\boldsymbol{i}$ es la unidad imaginaria.

composición de funciones Se evalúa una función y luego se evalúa una segunda función en el resultado de la primera función. La composición de f y g se define con $f \circ g$ y $[f \circ g](x) = f[g(x)]$.

evento compuesto Dos o más eventos simples.

desigualdad compuesta Dos desigualdades unidas por las palabras *y* u *o*.

sección cónica Cualquier figura obtenida mediante el corte de un cono doble.

eje conjugado El segmento de $2b$ unidades de longitud que es perpendicular al eje transversal en el centro.

conjugados Binomios de la forma $a\sqrt{b} + c\sqrt{d}$ y $a\sqrt{b} - c\sqrt{d}$, donde a, b, c y d son números racionales.

consistente Sistema de ecuaciones que posee por lo menos una solución.

constante Monomios que carecen de variables.

función constante Función lineal de la forma $f(x) = b$.

constante de variación La constante k que se usa en variación directa o inversa.

constant term (286) In $f(x) = ax^2 + bx + c$, c is the constant term.

término constante En $f(x) = ax^2 + bx + c$, c es el término constante.

constraints (129) Conditions given to variables, often expressed as linear inequalities.

restricciones Condiciones a que están sujetas las variables, a menudo escritas como desigualdades lineales.

continuity (485) A graph of a function that can be traced with a pencil that never leaves the paper.

continuidad La gráfica de una función que se puede calcar sin levantar nunca el lápiz del papel.

continuous probability distribution (671) The outcome can be any value in an interval of real numbers, represented by curves.

distribución de probabilidad continua El resultado puede ser cualquier valor de un intervalo de números reales, representados por curvas.

cosecant (701) For any angle, with measure α, a point $P(x, y)$ on its terminal side, $r = \sqrt{x^2 + y^2}$, $\csc \alpha = \frac{r}{y}$.

cosecante Para cualquier ángulo de medida α, un punto $P(x, y)$ en su lado terminal, $r = \sqrt{x^2 + y^2}$, $\csc \alpha = \frac{r}{y}$.

cosine (701) For any angle, with measure α, a point $P(x, y)$ on its terminal side, $r = \sqrt{x^2 + y^2}$, $\cos \alpha = \frac{x}{r}$.

coseno Para cualquier ángulo de medida α, un punto $P(x, y)$ en su lado terminal, $r = \sqrt{x^2 + y^2}$, $\cos \alpha = \frac{x}{r}$.

cotangent (701) For any angle, with measure α, a point $P(x, y)$ on its terminal side, $r = \sqrt{x^2 + y^2}$, $\cot \alpha = \frac{x}{y}$.

cotangente Para cualquier ángulo de medida α, un punto $P(x, y)$ en su lado terminal, $r = \sqrt{x^2 + y^2}$, $\cot \alpha = \frac{x}{y}$.

coterminal angles (711) Two angles in standard position that have the same terminal side.

ángulos coterminales Dos ángulos en posición estándar que tienen el mismo lado terminal.

Cramer's Rule (189) A method that uses determinants to solve a system of linear equations.

Regla de Crámer Método que usa determinantes para resolver un sistema de ecuaciones lineales.

D

degree (222) The sum of the exponents of the variables of a monomial.

grado Suma de los exponentes de las variables de un monomio.

degree of a polynomial in one variable (346) The greatest exponent of the variable of the polynomial.

grado de un polinomio de una variable El exponente máximo de la variable del polinomio.

dependent events (633) The outcome of one event does affect the outcome of another event.

eventos dependientes El resultado de un evento afecta el resultado de otro evento.

dependent system (111) A consistent system of equations that has an infinite number of solutions.

sistema dependiente Sistema de ecuaciones que posee un número infinito de soluciones.

dependent variable (59) The other variable in a function, usually y, whose values depend on x.

variable dependiente La otra variable de una función, por lo general y, cuyo valor depende de x.

depressed polynomial (366) The quotient when a polynomial is divided by one of its binomial factors.

polinomio reducido El cociente cuando se divide un polinomio entre uno de sus factores binomiales.

determinant (182) A square array of numbers or variables enclosed between two parallel lines.

determinante Arreglo cuadrado de números o variables encerrados entre dos rectas paralelas.

dilation (176) A transformation in which a geometric figure is enlarged or reduced.

dilatación Transformación en que se amplía o reduce una figura geométrica.

dimensional analysis (225) Performing operations with units.

análisis dimensional Realizar operaciones con unidades.

dimensions of a matrix (155) The number of rows, m, and the number of columns, n, of the matrix written as $m \times n$.

tamaño de una matriz El número de filas, m, y columnas, n, de una matriz, lo que se escribe $m \times n$.

directrix (419) See parabola.

direct variation (492) y varies directly as x if there is some nonzero constant k such that $y = kx$. k is called the constant of variation.

discrete probability distributions (671) Probabilities that have a finite number of possible values.

discriminant (316) In the Quadratic Formula, the expression $b^2 - 4ac$.

Distance Formula (413) The distance between two points with coordinates (x_1, y_1) and (x_2, y_2) is given by $d = \sqrt{(x_2 - x_1)^2 + (y_2 - y_1)^2}$.

domain (56) The set of all x-coordinates of the ordered pairs of a relation.

directriz Véase parábola.

variación directa y varía directamente con x si hay una constante no nula k tal que $y = kx$. k se llama la constante de variación.

distribución de probabilidad discreta Probabilidades que tienen un número finito de valores posibles.

discriminante En la fórmula cuadrática, la expresión $b^2 - 4ac$.

Fórmula de la distancia La distancia entre dos puntos (x_1, y_1) y (x_2, y_2) viene dada por $d = \sqrt{(x_2 - x_1)^2 + (y_2 - y_1)^2}$.

dominio El conjunto de todas las coordenadas x de los pares ordenados de una relación.

E

e (554) The irrational number 2.71828.... e is the base of the natural logarithms.

element (155) Each value in a matrix.

elimination method (118) Eliminate one of the variables in a system of equations by adding or subtracting the equations.

ellipse (433) The set of all points in a plane such that the sum of the distances from two given points in the plane, called foci, is constant.

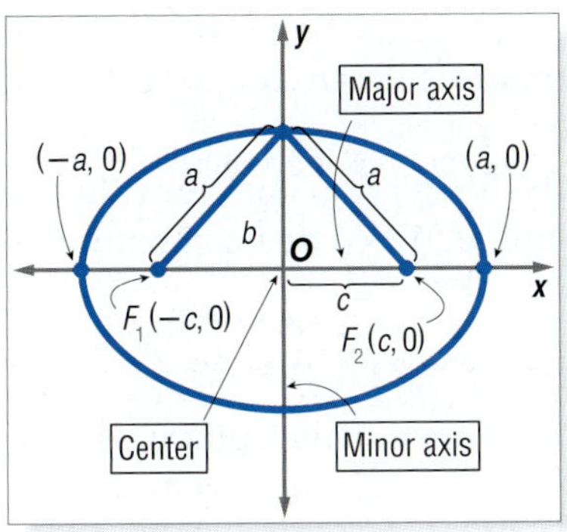

empty set (29) The solution set for an equation that has no solution, symbolized by { } or ∅.

end behavior (349) The behavior of the graph as x approaches positive infinity $(+\infty)$ or negative infinity $(-\infty)$.

equal matrices (155) Two matrices that have the same dimensions and each element of one matrix is equal to the corresponding element of the other matrix.

equation (20) A mathematical sentence stating that two mathematical expressions are equal.

expansion by minors (183) A method of evaluating a third or high order determinant by using determinants of lower order.

e El número irracional 2.71828.... e es la base de los logaritmos naturales.

elemento Cada valor de una matriz.

método de eliminación Eliminar una de las variables de un sistema de ecuaciones sumando o restando las ecuaciones.

elipse Conjunto de todos los puntos de un plano en los que la suma de sus distancias a dos puntos dados del plano, llamados focos, es constante.

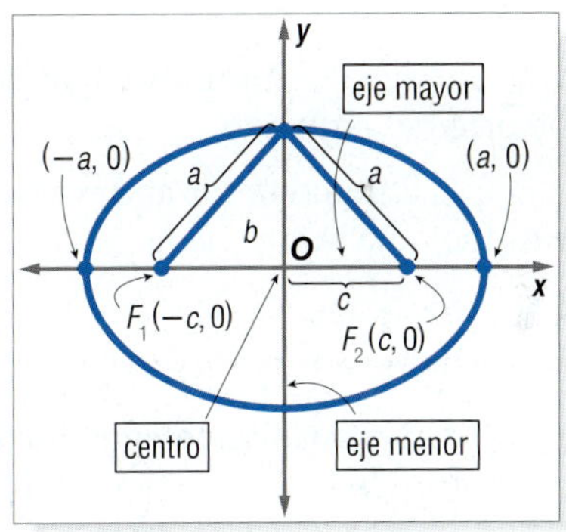

conjunto vacío Conjunto solución de una ecuación que no tiene solución, denotado por { } o ∅.

comportamiento final El comportamiento de una gráfica a medida que x tiende a más infinito $(+\infty)$ o menos infinito $(-\infty)$.

matrices iguales Dos matrices que tienen las mismas dimensiones y en las que cada elemento de una de ellas es igual al elemento correspondiente en la otra matriz.

ecuación Enunciado matemático que afirma la igualdad de dos expresiones matemáticas.

expansión por determinantes menores Un método de calcular el determinante de tercer orden o mayor mediante el uso de determinantes de orden más bajo.

exponential decay (524) Exponential decay occurs when a quantity decreases exponentially over time.

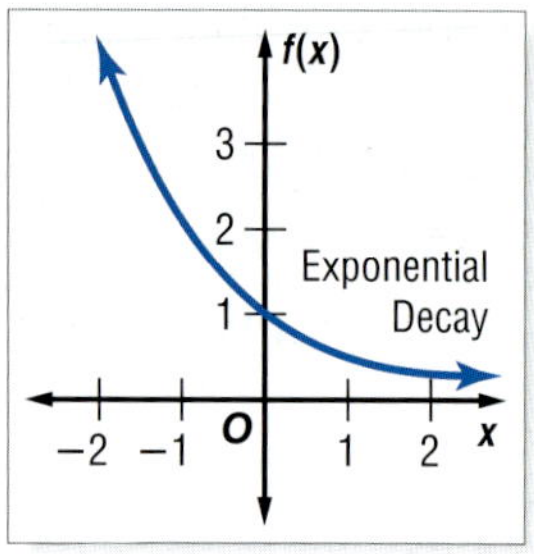

exponential equation (526) An equation in which the variables occur as exponents.

exponential function (524) A function of the form $y = ab^x$, where $a \neq 0$, $b > 0$, and $b \neq 1$.

exponential growth (524) Exponential growth occurs when a quantity increases exponentially over time.

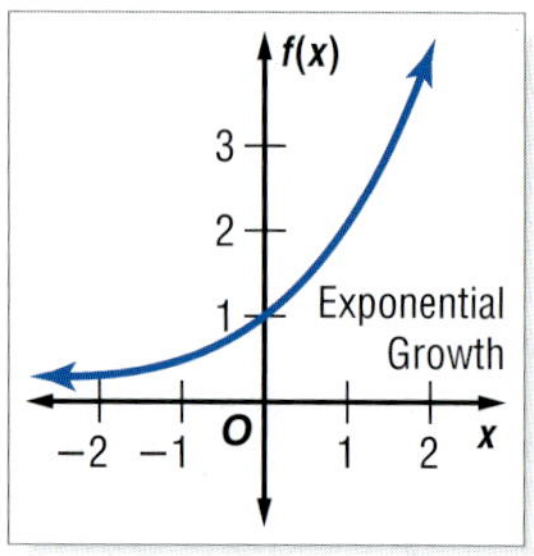

extraneous solution (263) A number that does not satisfy the original equation.

extrapolation (82) Predicting for an x-value greater than any in the data set.

F

factorial (613) If n is a positive integer, then $n! = n(n - 1)(n - 2) \ldots 2 \cdot 1$.

failure (644) Any outcome other than the desired outcome.

family of graphs (70) A group of graphs that displays one or more similar characteristics.

feasible region (129) The intersection of the graphs in a system of constraints.

Fibonacci sequence (606) A sequence in which the first two terms are 1 and each of the additional terms is the sum of the two previous terms.

focus (419, 433, 441) See parabola, ellipse, hyperbola.

FOIL method (230) The product of two binomials is the sum of the products of **F** the *first* terms, **O** the *outer* terms, **I** the *inner* terms, and **L** the *last* terms.

formula (8) A mathematical sentence that expresses the relationship between certain quantities.

desintegración exponencial Ocurre cuando una cantidad disminuye exponencialmente con el tiempo.

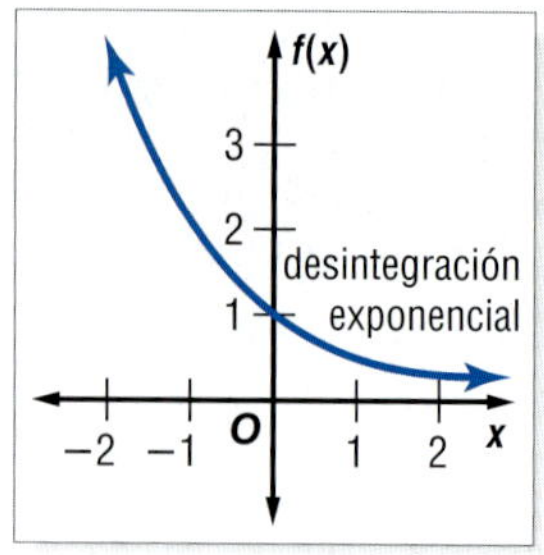

ecuación exponencial Ecuación en que las variables aparecen en los exponentes.

función exponencial Una función de la forma $y = ab^x$, donde $a \neq 0$, $b > 0$, y $b \neq 1$.

crecimiento exponencial El que ocurre cuando una cantidad aumenta exponencialmente con el tiempo.

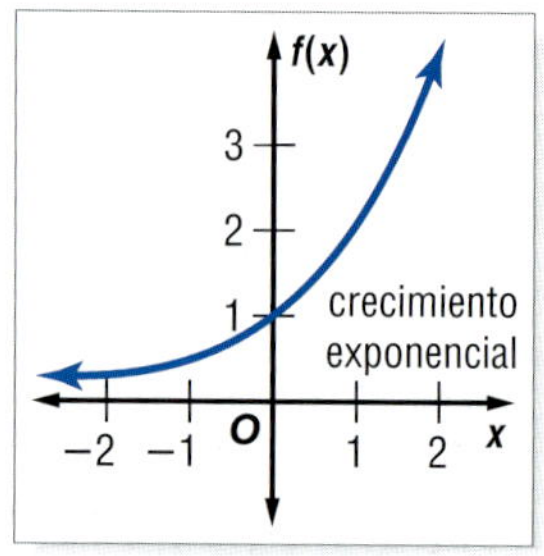

solución extraña Número que no satisface la ecuación original.

extrapolación Predicción para un valor de x mayor que cualquiera de los de un conjunto de datos.

factorial Si n es un entero positivo, entonces $n! = n(n - 1)(n - 2) \ldots 2 \cdot 1$.

fracaso Cualquier resultado distinto del deseado.

familia de gráficas Grupo de gráficas que presentan una o más características similares.

región viable Intersección de las gráficas de un sistema de restricciones.

sucesión de Fibonacci Sucesión en que los dos primeros términos son iguales a 1 y cada término que sigue es igual a la suma de los dos anteriores.

foco Véase parábola, elipse, hipérbola.

método FOIL El producto de dos binomios es la suma de los productos de los primeros (*First*) términos, los términos exteriores (*Outer*), los términos interiores (*Inner*) y los últimos (*Last*) términos.

fórmula Enunciado matemático que describe la relación entre ciertas cantidades.

function (57) A relation in which each element of the domain is paired with exactly one element in the range.

function notation (59) An equation of y in terms of x can be rewritten so that $y = f(x)$. For example, $y = 2x + 1$ can be written as $f(x) = 2x + 1$.

función Relación en que a cada elemento del dominio le corresponde un solo elemento del rango.

notación funcional Una ecuación de y en términos de x puede escribirse en la forma $y = f(x)$. Por ejemplo, $y = 2x + 1$ puede escribirse como $f(x) = 2x + 1$.

G

geometric mean (590) The terms between any two nonsuccessive terms of a geometric sequence.

geometric sequence (588) A sequence in which each term after the first is found by multiplying the previous term by a constant r, called the common ratio.

geometric series (594) The sum of the terms of a geometric sequence.

greatest integer function (89) A step function, written as $f(x) = [\![x]\!]$, where $f(x)$ is the greatest integer less than or equal to x.

media geométrica Cualquier término entre dos términos no consecutivos de una sucesión geométrica.

sucesión geométrica Sucesión en que cualquier término después del primero puede hallarse multiplicando el término anterior por una constante r, llamada razón común .

serie geométrica La suma de los términos de una sucesión geométrica.

función del máximo entero Una función etapa que se escribe $f(x) = [x]$, donde $f(x)$ es el meaximo entero que es menor que o igual a x.

H

hyperbola (441) The set of all points in the plane such that the absolute value of the difference of the distances from two given points in the plane, called foci, is constant.

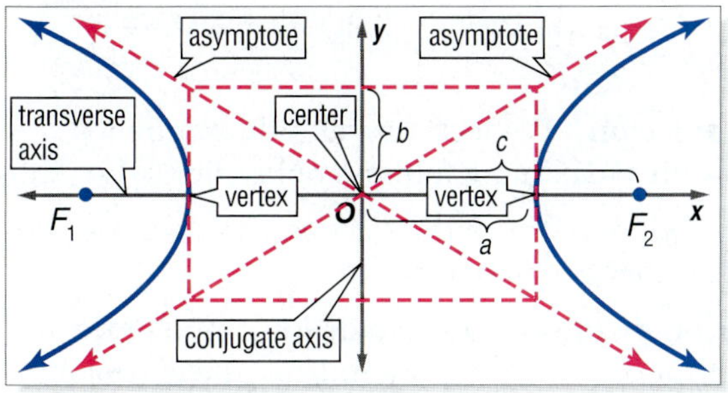

hypothesis (686) A statement to be tested.

hipérbola Conjunto de todos los puntos de un plano en los que el valor absoluto de la diferencia de sus distancias a dos puntos dados del plano, llamados focos, es constante.

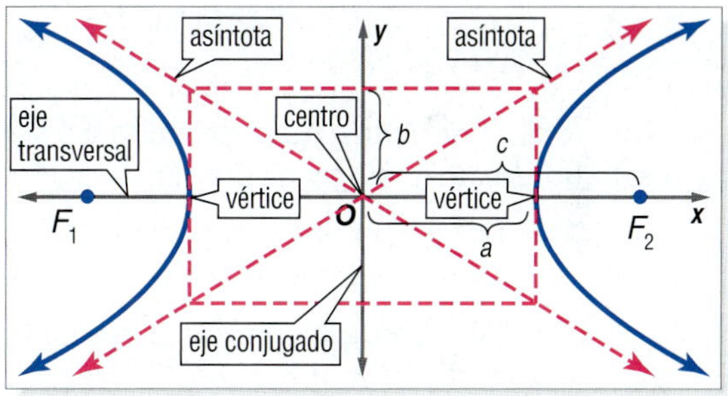

hipótesis Proposición que debe ser verificada.

I

identity function (90, 391) The function $I(x) = x$.

identity matrix (195) A square matrix that, when multiplied by another matrix, equals that same matrix. If A is any $n \times n$ matrix and I is the $n \times n$ identity matrix, then $A \cdot I = A$ and $I \cdot A = A$.

image (175) The graph of an object after a transformation.

imaginary unit (270) i, or the principal square root of -1.

inclusive (659) Two events whose outcomes may be the same.

función identidad La función $I(x) = x$.

matriz identidad Matriz cuadrada que al multiplicarse por otra matriz, es igual a la misma matriz. Si A es una matriz de $n \times n$ e I es la matriz identidad de $n \times n$, entonces $A \cdot I = A$ y $I \cdot A = A$.

imagen Gráfica de una figura después de una transformación.

unidad imaginaria i, o la raíz cuadrada principal de -1.

inclusivo Dos eventos que pueden tener los mismos resultados.

inconsistent (111) A system of equations that has no solutions.

independent (111) A system of equations that has exactly one solution.

independent events (632) Events that do not affect each other.

independent variable (59) In a function, the variable, usually x, whose values make up the domain.

index of summation (585) The variable used with the summation symbol. In the expression below, the index of summation is n.

$$\sum_{n=1}^{3} 4n$$

inductive hypothesis (618) The assumption that a statement is true for some positive integer k, where $k \geq n$.

infinite geometric series (599) A geometric series with an infinite number of terms.

initial side of an angle (709) The fixed ray of an angle.

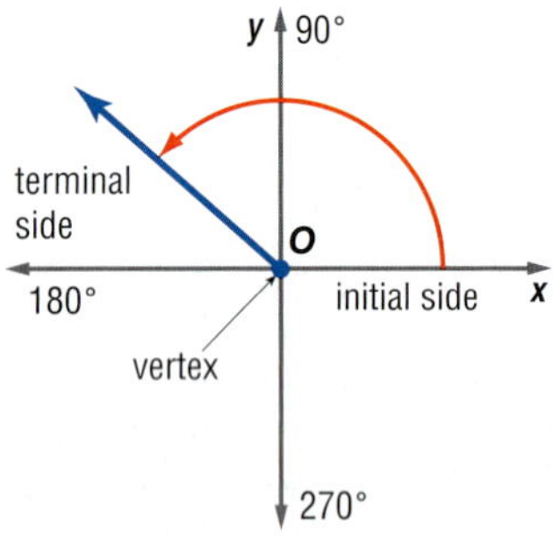

interpolation (82) Predicting for an x-value between the least and greatest values of the set.

intersection (40) The graph of a compound inequality containing *and*.

interval notation (35) Using the infinity symbols, $+\infty$ and $-\infty$, to indicate that the solution set of an inequality is unbounded in the positive or negative direction, respectively.

inverse (195) Two $n \times n$ matrices are inverses of each other if their product is the identity matrix.

inverse function (391) Two functions f and g are inverse functions if and only if both of their compositions are the identity function.

inverse of a trigonometric function (746) The arccosine, arcsine, and arctangent relations.

inverse relations (390) Two relations are inverse relations if and only if whenever one relation contains the element (a, b) the other relation contains the element (b, a).

inverse variation (493) y varies inversely as x if there is some nonzero constant k such that $xy = k$ or $y = \frac{k}{x}$.

inconsistente Sistema de ecuaciones que no tiene solución alguna.

independiente Sistema de ecuaciones que sólo tiene una solución.

eventos independientes Eventos que no se afectan mutuamente.

variable independiente En una función, la variable, por lo general x, cuyos valores forman el dominio.

índice de suma Variable que se usa con el símbolo de suma. En la siguiente expresión, el índice de suma es n.

$$\sum_{n=1}^{3} 4n$$

hipótesis inductiva El suponer que un enunciado es verdadero para algún entero positivo k, donde $k \geq n$.

serie geométrica infinita Serie geométrica con un número infinito de términos.

lado inicial de un ángulo El rayo fijo de un ángulo.

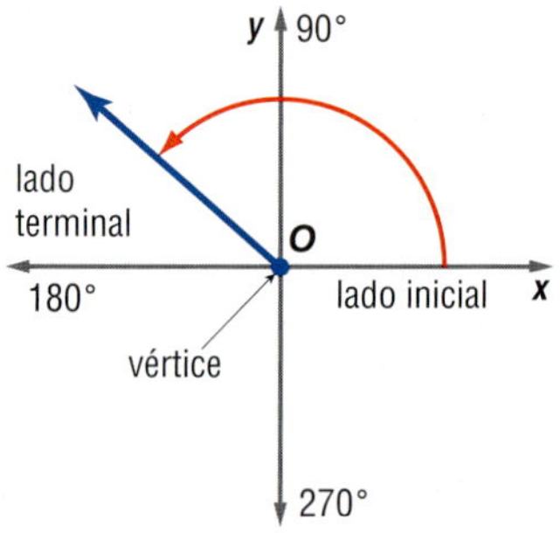

interpolación Predecir un valor de x entre los valores máximo y mínimo del conjunto de datos.

intersección Gráfica de una desigualdad compuesta que contiene la palabra *y*.

notación de intervalo Uso de los símbolos de infinito, $+\infty$ y $-\infty$, para indicar que el conjunto solución de una desigualdad no es acotado en la dirección positiva o negativa, respectivamente.

inversa Dos matrices de $n \times n$ son inversas mutuas si su producto es la matriz identidad.

función inversa Dos funciones f y g son inversas mutuas si y sólo si las composiciones de ambas son la función identidad.

inversa de una función trigonométrica Las relaciones arcocoseno, arcoseno y arcotangente.

relaciones inversas Dos relaciones son relaciones inversas mutuas si y sólo si cada vez que una de las relaciones contiene el elemento (a, b), la otra contiene el elemento (b, a).

variación inversa y varía inversamente con x *si* hay una constante no nula k tal que $xy = k$ o $y = \frac{k}{x}$.

irrational number (11) A real number that is not rational. The decimal form neither terminates nor repeats.

número irracional Número que no es racional. Su expansión decimal no es ni terminal ni periódica.

isometry (175) A transformation in which the image and preimage are congruent figures.

isometría Transformación en que la imagen y la preimagen son figuras congruentes.

iteration (608) The process of composing a function with itself repeatedly.

iteración Proceso de componer una función consigo misma repetidamente.

J

joint variation (493) y varies jointly as x and z if there is some nonzero constant k such that $y = kxz$, where $x \neq 0$ and $z \neq 0$.

variación conjunta y varía conjuntamente con x y z si hay una constante no nula k tal que $y = kxz$, donde $x \neq 0$ y $z \neq 0$.

L

latus rectum (421) The line segment through the focus of a parabola and perpendicular to the axis of symmetry.

latus rectum El segmento de recta que pasa por el foco de una parábola y que es perpendicular a su eje de simetría.

Law of Cosines (733–734) Let $\triangle ABC$ be any triangle with a, b, and c representing the measures of sides, and opposite angles with measures A, B, and C, respectively. Then the following equations are true.

$a^2 = b^2 + c^2 - 2bc \cos A$
$b^2 = a^2 + c^2 - 2ac \cos B$
$c^2 = a^2 + b^2 - 2ab \cos C$

Ley de los cosenos Sea $\triangle ABC$ un triángulo cualquiera, con a, b y c las longitudes de los lados y con ángulos opuestos de medidas A, B y C, respectivamente. Entonces se cumplen las siguientes ecuaciones.

$a^2 = b^2 + c^2 - 2bc \cos A$
$b^2 = a^2 + c^2 - 2ac \cos B$
$c^2 = a^2 + b^2 - 2ab \cos C$

Law of Sines (726) Let $\triangle ABC$ be any triangle with a, b, and c representing the measures of sides opposite angles with measurements A, B, and C, respectively. Then $\frac{\sin A}{a} = \frac{\sin B}{b} = \frac{\sin C}{c}$.

Ley de los senos Sea $\triangle ABC$ cualquier triángulo con a, b y c las longitudes de los lados y con ángulos opuestos de medidas A, B y C, respectivamente. Entonces $\frac{\sin A}{a} = \frac{\sin B}{b} = \frac{\sin C}{c}$.

leading coefficient (346) The coefficient of the term with the highest degree.

coeficiente líder Coeficiente del término de mayor grado.

like radical expressions (252) Two radical expressions in which both the radicands and indices are alike.

expresiones radicales semejantes Dos expresiones radicales en que tanto los radicandos como los índices son semejantes.

like terms (229) Monomials that can be combined.

términos semejantes Monomios que pueden combinarse.

limit (593) The value that the terms of a sequence approach.

límite El valor al que tienden los términos de una sucesión.

linear equation (63) An equation that has no operations other than addition, subtraction, and multiplication of a variable by a constant.

ecuación lineal Ecuación sin otras operaciones que las de adición, sustracción y multiplicación de una variable por una constante.

linear function (63) A function whose ordered pairs satisfy a linear equation.

función lineal Función cuyos pares ordenados satisfacen una ecuación lineal.

linear permutation (638) The arrangement of objects or people in a line.

permutación lineal Arreglo de personas o figuras en una línea.

linear programming (130) The process of finding the maximum or minimum values of a function for a region defined by inequalities.

programación lineal Proceso de hallar los valores máximo o mínimo de una función lineal en una región definida por las desigualdades.

linear term (286) In the equation $f(x) = ax^2 + bx + c$, bx is the linear term.

término lineal En la ecuación $f(x) = ax^2 + bx + c$, el término lineal es bx.

line of fit (81) A line that closely approximates a set of data.

recta de ajuste Recta que se aproxima estrechamente a un conjunto de datos.

logarithm (531) In the function $x = b^y$, y is called the logarithm, base b, of x. Usually written as $y = \log_b x$ and is read "y equals log base b of x."

logaritmo En la función $x = b^y$, y es el logaritmo en base b, de x. Generalmente escrito como $y = \log_b x$ y se lee "y es igual al logaritmo en base b de x."

logarithmic equation (533) An equation that contains one or more logarithms.

ecuación logarítmica Ecuación que contiene uno o más logaritmos.

logarithmic function (532) The function $y = \log_b x$, where $b > 0$ and $b \neq 1$, which is the inverse of the exponential function $y = b^x$.

función logarítmica La función $y = \log_b x$, donde $b > 0$ y $b \neq 1$, inversa de la función exponencial $y = b^x$.

M

$m \times n$ matrix (155) A matrix with m rows and n columns.

matriz de $m \times n$ Matriz de m filas y n columnas.

major axis (434) The longer of the two line segments that form the axes of symmetry of an ellipse.

eje mayor El más largo de dos segmentos de recta que forman los ejes de simetría de una elipse.

mapping (57) How each member of the domain is paired with each member of the range.

transformaciones La correspondencia entre cada miembro del dominio con cada miembro del rango.

margin of sampling error (ME) (682) The limit on the difference between how a sample responds and how the total population would respond.

margen de error muestral (EM) Límite en la diferencia entre las respuestas obtenidas con una muestra y cómo pudiera responder la población entera.

mathematical induction (618) A method of proof used to prove statements about positive integers.

inducción matemática Método de demostrar enunciados sobre los enteros positivos.

matrix (154) Any rectangular array of variables or constants in horizontal rows and vertical columns.

matriz Arreglo rectangular de variables o constantes en filas horizontales y columnas verticales.

maximum value (288) The y-coordinate of the vertex of the quadratic function $f(x) = ax^2 + bx + c$, where $a < 0$.

valor máximo La coordenada y del vértice de la función cuadrática $f(x) = ax^2 + bx + c$, donde $a < 0$.

measure of central tendency (665) A number that represents the center or middle of a set of data.

medida de tendencia central Número que representa el centro o medio de un conjunto de datos.

measure of variation (664) A representation of how spread out or scattered a set of data is.

medida de variación Número que representa la dispersión de un conjunto de datos.

midline (771) A horizontal axis used as the reference line about which the graph of a periodic function oscillates.

recta central Eje horizontal que se usa como recta de referencia alrededor de la cual oscila la gráfica de una función periódica.

minimum value (288) The y-coordinate of the vertex of the quadratic function $f(x) = ax^2 + bx + c$, where $a > 0$.

valor mínimo La coordenada y del vértice de la función cuadrática $f(x) = ax^2 + bx + c$, donde $a > 0$.

minor (183) The determinant formed when the row and column containing that element are deleted.

determinante menor El que se forma cuando se descartan la fila y columna que contienen dicho elemento.

minor axis (434) The shorter of the two line segments that form the axes of symmetry of an ellipse.

eje menor El más corto de los dos segmentos de recta de los ejes de simetría de una elipse.

monomial (222) An expression that is a number, a variable, or the product of a number and one or more variables.

monomio Expresión que es un número, una variable o el producto de un número por una o más variables.

mutually exclusive (658) Two events that cannot occur at the same time.

mutuamente exclusivos Dos eventos que no pueden ocurrir simultáneamente.

N

n^{th} root (245) For any real numbers a and b, and any positive integer n, if $a^n = b$, then a is an nth root of b.

raíz *enésima* Para cualquier número real a y b y cualquier entero positivo n, si $a^n = b$, entonces a se llama una raíz *enésima* de b.

natural base exponential function (554) An exponential function with base e, $y = e^x$.

función exponencial natural La función exponencial de base e, $y = e^x$.

natural logarithm (554) Logarithms with base e, written $\ln x$.

logaritmo natural Logaritmo de base e, el que se escribe $\ln x$.

natural logarithmic function (554) $y = \ln x$, the inverse of the natural base exponential function $y = e^x$.

función logarítmica natural $y = \ln x$, la inversa de la función exponencial natural $y = e^x$.

negative exponent (222) For any real number $a \neq 0$ and any integer n, $a^{-n} = \frac{1}{a^n}$ and $\frac{1}{a^{-n}} = a^n$.

exponente negativo Para cualquier número real $a \neq 0$ cualquier entero positivo n, $a^{-n} = \frac{1}{a^n}$ y $\frac{1}{a^{-n}} = a^n$.

normal distribution (671) A frequency distribution that often occurs when there is a large number of values in a set of data: about 68% of the values are within one standard deviation of the mean, 95% of the values are within two standard deviations from the mean, and 99% of the values are within three standard deviations.

Normal Distribution

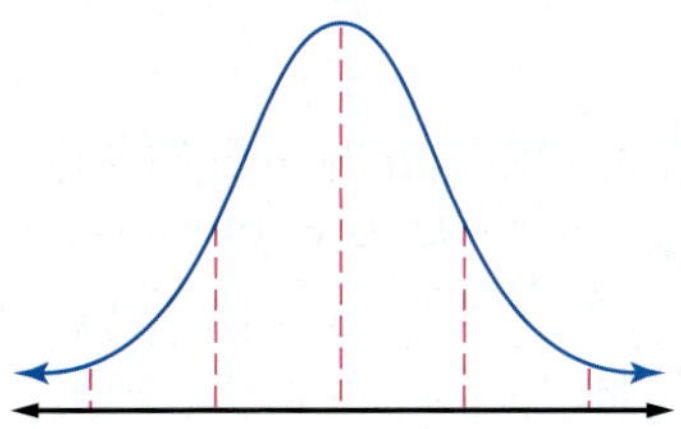

distribución normal Distribución de frecuencia que aparece a menudo cuando hay un número grande de datos: cerca del 68% de los datos están dentro de una desviación estándar de la media, 95% están dentro de dos desviaciones estándar de la media y 99% están dentro de tres desviaciones estándar de la media.

Distribución normal

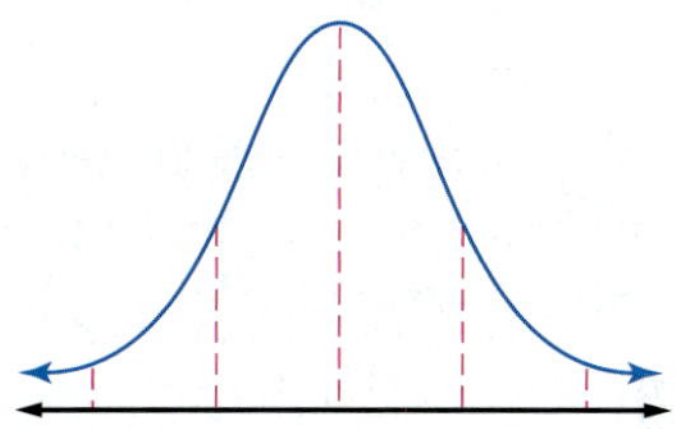

O

octants (136) The eight regions of three-dimensional space.

octantes Las ocho regiones del espacio tridimensional.

odds (645) The ratio of the number of the successes of an event to the number of failures.

posibilidades Razón del número de éxitos de un evento a su número de fracasos.

one-to-one function (57, 392) **1.** A function where each element of the range is paired with exactly one element of the domain **2.** A function whose inverse is a function.

función biunívoca **1.** Función en la que a cada elemento del rango le corresponde sólo un elemento del dominio. **2.** Función cuya inversa es una función.

open sentence (20) A mathematical sentence containing one or more variables.

enunciado abierto Enunciado matemático que contiene una o más variables.

ordered pair (56) A pair of coordinates, written in the form (x, y), used to locate any point on a coordinate plane.

par ordenado Un par de números, escrito en la forma (x, y), que se usa para ubicar cualquier punto en un plano de coordenadas.

ordered triple (136, 139) **1.** The coordinates of a point in space **2.** The solution of a system of equations in three variables x, y, and z.

triple ordenado **1.** Las coordenadas de un punto en el espacio **2.** Solución de un sistema de ecuaciones en tres variables x, y y z.

Order of Operations (6)
- **Step 1** Evaluate expressions inside grouping symbols.
- **Step 2** Evaluate all powers.
- **Step 3** Do all multiplications and/or divisions from left to right.
- **Step 4** Do all additions and subtractions from left to right.

Orden de las operaciones
- **Paso 1** Evalúa las expresiones dentro de símbolos de agrupamiento.
- **Paso 2** Evalúa todas las potencias.
- **Paso 3** Ejecuta todas las multiplicaciones y divisiones de izquierda a derecha.
- **Paso 4** Ejecuta todas las adiciones y sustracciones de izquierda a derecha.

outcomes (632) The results of a probability experiment/an event.

resultados Lo que produce un experimento o evento probabilístico.

outlier (826) A data point that does not appear to belong to the rest of the set.

valor atípico Dato que no parece pertenecer al resto el conjunto.

P

parabola (286, 419) The set of all points in a plane that are the same distance from a given point, called the focus, and a given line, called the directrix.

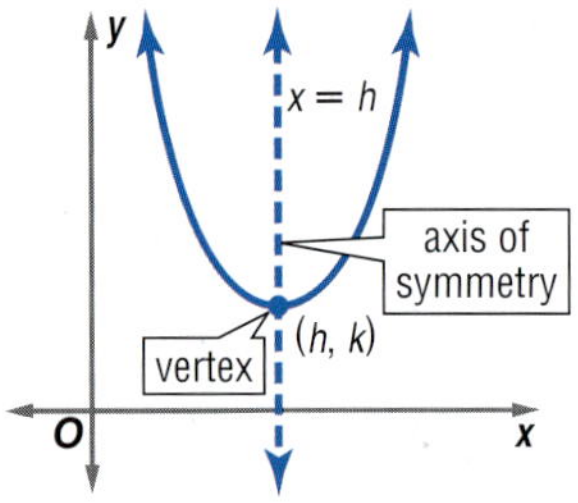

parábola Conjunto de todos los puntos de un plano que están a la misma distancia de un punto dado, llamado foco, y de una recta dada, llamada directriz.

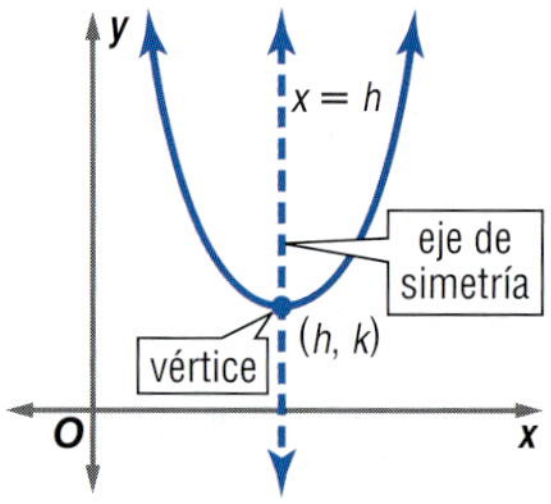

parallel lines (70) Nonvertical coplanar lines with the same slope.

rectas paralelas Rectas coplanares no verticales con la misma pendiente.

parent graph (70) The simplest of graphs in a family.

gráfica madre La gráfica más sencilla en una familia de gráficas.

partial sum (599) The sum of the first n terms of a series.

suma parcial La suma de los primeros n términos de una serie.

Pascal's triangle (612) A triangular array of numbers such that the $(n + 1)^{th}$ row is the coefficient of the terms of the expansion $(x + y)^n$ for $n = 0, 1, 2$ ...

Triángulo de Pascal Arreglo triangular de números en el que la fila $(n + 1)^n$ proporciona los coeficientes de los términos de la expansión de $(x + y)^n$ para $n = 0, 1, 2$ ...

period (741) The least possible value of a for which $f(x) = f(x + a)$.

período El menor valor positivo posible para a, para el cual $f(x) = f(x + a)$.

periodic function (741) A function is called periodic if there is a number a such that $f(x) = f(x + a)$ for all x in the domain of the function.

función periódica Función para la cual hay un número a tal que $f(x) = f(x + a)$ para todo x en el dominio de la función .

permutation (638) An arrangement of objects in which order is important.

permutación Arreglo de elementos en que el orden es importante.

perpendicular lines (71) In a plane, any two oblique lines the product of whose slopes is -1.

rectas perpendiculares En un plano, dos rectas oblicuas cualesquiera cuyas pendientes tienen un producto igual a -1.

phase shift (769) A horizontal translation of a trigonometric function.

desvío de fase Traslación horizontal de una función trigonométrica.

piecewise function (91) A function that is written using two or more expressions.

función a intervalos Función que se escribe usando dos o más expresiones.

point discontinuity (485) If the original function is undefined for $x = a$ but the related rational expression of the function in simplest form is defined for $x = a$, then there is a hole in the graph at $x = a$.

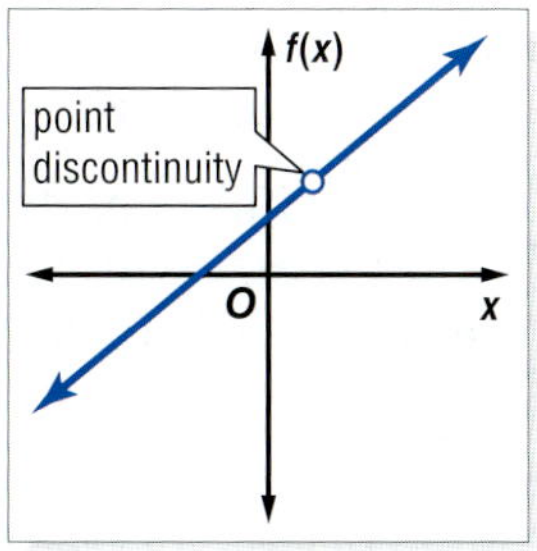

discontinuidad evitable Si la función original no está definida en $x = a$ pero la expresión racional reducida correspondiente de la función está definida en $x = a$, entonces la gráfica tiene una ruptura o corte en $x = a$.

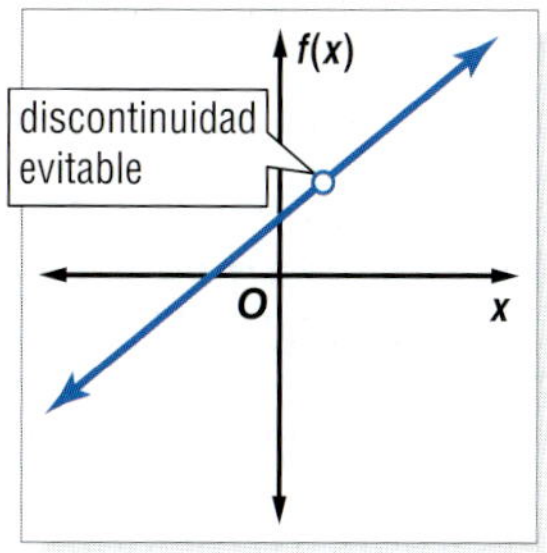

point-slope form (76) An equation in the form $y - y_1 = m(x - x_1)$ where (x_1, y_1) are the coordinates of a point on the line and m is the slope of the line.

forma punto-pendiente Ecuación de la forma $y - y_1 = m(x - x_1)$ donde (x_1, y_1) es un punto en la recta y m es la pendiente de la recta.

polynomial (229) A monomial or a sum of monomials.

polinomio Monomio o suma de monomios.

polynomial function (347) A function that is represented by a polynomial equation.

función polinomial Función representada por una ecuación polinomial.

polynomial in one variable (346) $a_0x^n + a_1x^{n-1} + \ldots + a_{n-2}x^2 + a_{n-1}x + a_n$, where the coefficients $a_0, a_1, \ldots, a_n$ represent real numbers, and a_0 is not zero and n is a nonnegative integer.

polinomio de una variable $a_0x^n + a_1x^{n-1} + \ldots + a_{n-2}x^2 + a_{n-1}x + a_n$, donde los coeficientes $a_0, a_1, \ldots, a_n$ son números reales, a_0 no es nulo y n es un entero no negativo.

power (222) An expression of the form x^n.

potencia Expresión de la forma x^n.

power function (704) An equation in the form $f(x) = ax^b$, where a and b are real numbers.

función potencia Ecuación de la forma $f(x) = ax^b$, donde a y b son números reales.

prediction equation (81) An equation suggested by the points of a scatter plot that is used to predict other points.

ecuación de predicción Ecuación sugerida por los puntos de una gráfica de dispersión y que se usa para predecir otros puntos.

preimage (175) The graph of an object before a transformation.

preimagen Gráfica de una figura antes de una transformación.

principal root (246) The nonnegative root.

raíz principal La raíz no negativa.

principal values (746) The values in the restricted domains of trigonometric functions.

valores principales Valores en los dominios restringidos de las funciones trigonométricas.

probability (644) A ratio that measures the chances of an event occurring.

probabilidad Razón que mide la posibilidad de que ocurra un evento.

probability distribution (646) A function that maps the sample space to the probabilities of the outcomes in the sample space for a particular random variable.

distribución de probabilidad Función que aplica el espacio muestral a las probabilidades de los resultados en el espacio muestral obtenidos para una variable aleatoria particular.

pure imaginary number (270) The square roots of negative real numbers. For any positive real number b, $\sqrt{-b^2} = \sqrt{b^2} \cdot \sqrt{-1}$, or bi.

número imaginario puro Raíz cuadrada de un número real negativo. Para cualquier número real positivo b, $\sqrt{-b^2} = \sqrt{b^2} \cdot \sqrt{-1}$ ó bi.

Q

quadrantal angle (718) An angle in standard position whose terminal side coincides with one of the axes.

ángulo de cuadrante Ángulo en posición estándar cuyo lado terminal coincide con uno de los ejes.

quadrants (56) The four areas of a Cartesian coordinate plane.

cuadrantes Las cuatro regiones de un plano de coordenadas cartesiano.

quadratic equation (294) A quadratic function set equal to a value, in the form $ax^2 + bx + c$, where $a \neq 0$.

ecuación cuadrática Función cuadrática igual a un valor, de la forma $ax^2 + bx + c$, donde $a \neq 0$.

quadratic form (360) For any numbers a, b, and c, except for $a = 0$, an equation that can be written in the form $a[f(x)^2] + b[f(x)] + c = 0$, where $f(x)$ is some expression in x.

forma de ecuación cuadrática Para cualquier número a, b y c, excepto $a = 0$, una ecuación que puede escribirse de la forma $a[f(x)^2] + b[f(x)] + c = 0$, donde $f(x)$ es una expresión en x.

Quadratic Formula (313) The solutions of a quadratic equation of the form $ax^2 + bx + c = 0$, where $a \neq 0$, are given by the Quadratic Formula, which is $x = \frac{-b \pm \sqrt{b^2 - 4ac}}{2a}$.

Fórmula cuadrática Las soluciones de una ecuación cuadrática de la forma $ax^2 + bx + c = 0$, donde $a \neq 0$, se dan por la fórmula cuadrática, que es $x = \frac{-b \pm \sqrt{b^2 - 4ac}}{2a}$.

quadratic function (286) A function described by the equation $f(x) = ax^2 + bx + c$, where $a \neq 0$.

función cuadrática Función descrita por la ecuación $f(x) = ax^2 + bx + c$, donde $a \neq 0$.

quadratic term (286) In the equation $f(x) = ax^2 + bx + c$, ax^2 is the quadratic term.

término cuadrático En la ecuación $f(x) = ax^2 + bx + c$, el término cuadrático es ax^2.

R

radian (710) The measure of an angle θ in standard position whose rays intercept an arc of length 1 unit on the unit circle.

radián Medida de un ángulo θ en posición normal cuyos rayos intersecan un arco de 1 unidad de longitud en el círculo unitario.

radical equation (263) An equation with radicals that have variables in the radicands.

ecuación radical Ecuación con radicales que tienen variables en el radicando.

radical inequality (264) An inequality that has a variable in the radicand.

desigualdad radical Desigualdad que tiene una variable en el radicando.

random (645) All outcomes have an equally likely chance of happening.

aleatorio Todos los resultados son equiprobables.

random variable (646) The outcome of a random process that has a numerical value.

variable aleatoria El resultado de un proceso aleatorio que tiene un valor numérico.

range (56) The set of all y-coordinates of a relation.

rango Conjunto de todas las coordenadas y de una relación.

rate of change (69) How much a quantity changes on average, relative to the change in another quantity, often time.

tasa de cambio Lo que cambia una cantidad en promedio, respecto al cambio en otra cantidad, por lo general el tiempo.

rate of decay (560) The percent decrease r in the equation $y = a(1 - r)^t$.

tasa de desintegración Disminución porcentual r en la ecuación $y = a(1 - r)^t$.

rate of growth (562) The percent increase r in the equation $y = a(1 + r)^t$.

tasa de crecimiento Aumento porcentual r en la ecuación $y = a(1 + r)^t$.

rational equation (505) Any equation that contains one or more rational expressions.

ecuación racional Cualquier ecuación que contiene una o más expresiones racionales.

rational exponent (258) For any nonzero real number b, and any integers m and n, with $n > 1$, $b^{\frac{m}{n}} = \sqrt[n]{b^m} = (\sqrt[n]{b})^m$, except when $b < 0$ and n is even.

exponente racional Para cualquier número real no nulo b y cualquier entero m y n, con $n > 1$, $b^{\frac{m}{n}} = \sqrt[n]{b^m} = (\sqrt[n]{b})^m$, excepto cuando $b < 0$ y n es par.

rational expression (472) A ratio of two polynomial expressions.

expresión racional Razón de dos expresiones polinomiales.

rational function (472) An equation of the form $f(x) = \frac{p(x)}{q(x)}$, where $p(x)$ and $q(x)$ are polynomial functions, and $q(x) \neq 0$.

función racional Ecuación de la forma $f(x) = \frac{p(x)}{q(x)}$, donde $p(x)$ y $q(x)$ son funciones polinomiales y $q(x) \neq 0$.

rational inequality (508) Any inequality that contains one or more rational expressions.

rationalizing the denominator (251) To eliminate radicals from a denominator or fractions from a radicand.

rational number (11) Any number $\frac{m}{n}$, where m and n are integers and n is not zero. The decimal form is either a terminating or repeating decimal.

real numbers (11) All numbers used in everyday life; the set of all rational and irrational numbers.

recursive formula (606) Each term is formulated from one or more previous terms.

reference angle (718) The acute angle formed by the terminal side of an angle in standard position and the x-axis.

reflection (177) A transformation in which every point of a figure is mapped to a corresponding image across a line of symmetry.

reflection matrix (177) A matrix used to reflect an object over a line or plane.

regression line (87) A line of best fit.

relation (56) A set of ordered pairs.

relative frequency histogram (646) A table of probabilities or a graph to help visualize a probability distribution.

relative maximum (354) A point on the graph of a function where no other nearby points have a greater y-coordinate.

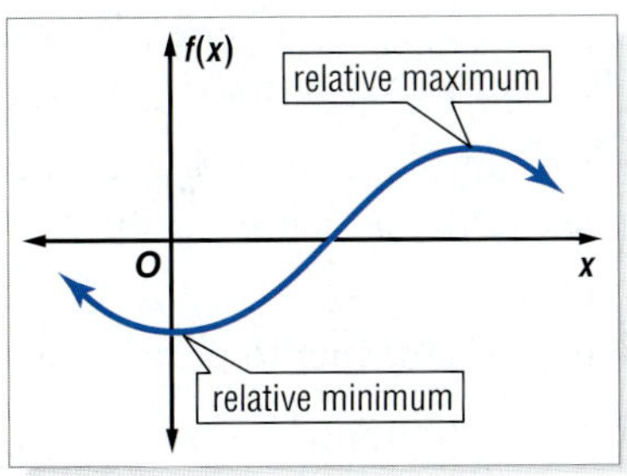

relative minimum (354) A point on the graph of a function where no other nearby points have a lesser y-coordinate.

root (294) The solutions of a quadratic equation.

rotation (178) A transformation in which an object is moved around a center point, usually the origin.

rotation matrix (178) A matrix used to rotate an object.

row matrix (155) A matrix that has only one row.

desigualdad racional Cualquier desigualdad que contiene una o más expresiones racionales.

racionalizar el denominador La eliminación de radicales de un denominador o de fracciones de un radicando.

número racional Cualquier número $\frac{m}{n}$, donde m y n son enteros y n no es cero. Su expansión decimal es o terminal o periódica.

números reales Todos los números que se usan en la vida cotidiana; el conjunto de los todos los números racionales e irracionales.

fórmula recursiva Cada término proviene de uno o más términos anteriores.

ángulo de referencia El ángulo agudo formado por el lado terminal de un ángulo en posición estándar y el eje x.

reflexión Transformación en que cada punto de una figura se aplica a través de una recta de simetría a su imagen correspondiente.

matriz de reflexión Matriz que se usa para reflejar una figura sobre una recta o plano.

recta de regresión Una recta de óptimo ajuste.

relación Conjunto de pares ordenados.

histograma de frecuencia relativa Tabla de probabilidades o gráfica para asistir en la visualización de una distribución de probabilidad.

máximo relativo Punto en la gráfica de una función en donde ningún otro punto cercano tiene una coordenada y mayor.

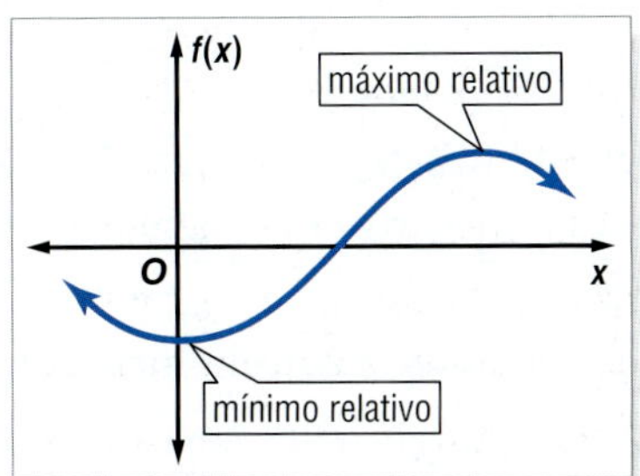

mínimo relativo Punto en la gráfica de una función en donde ningún otro punto cercano tiene una coordenada y menor.

raíz Las soluciones de una ecuación cuadrática.

rotación Transformación en que una figura se hace girar alrededor de un punto central, generalmente el origen.

matriz de rotación Matriz que se usa para hacer girar un objeto.

matriz fila Matriz que sólo tiene una fila.

sample space (632) The set of all possible outcomes of an experiment.

scalar (162) A constant.

scalar multiplication (162) Multiplying any matrix by a constant called a scalar; the product of a scalar k and an $m \times n$ matrix.

scatter plot (81) A set of data graphed as ordered pairs in a coordinate plane.

scientific notation (225) The expression of a number in the form $a \times 10^n$, where $1 \leq a < 10$ and n is an integer.

secant (701) For any angle, with measure α, a point $P(x, y)$ on its terminal side, $r = \sqrt{x^2 + y^2}$, $\sec \alpha = \frac{r}{x}$.

second-order determinant (182) The determinant of a 2×2 matrix.

sequence (578) A list of numbers in a particular order.

series (583) The sum of the terms of a sequence.

set-builder notation (34) The expression of the solution set of an inequality, for example $\{x \mid x > 9\}$.

sigma notation (585) For any sequence $a_1, a_2, a_3, \ldots,$ the sum of the first k terms may be written $\sum_{n=1}^{k} a_n$, which is read "the summation from $n = 1$ to k of a_n." Thus, $\sum_{n=1}^{k} a_n = a_1 + a_2 + a_3 + \ldots + a_k$, where k is an integer value.

simple event (658) One event.

simplify (222) To rewrite an expression without parentheses or negative exponents.

simulation (681) The use of a probability experiment to mimic a real-life situation.

sine (701) For any angle, with measure α, a point $P(x, y)$ on its terminal side, $r = \sqrt{x^2 + y^2}$, $\sin \alpha = \frac{y}{r}$.

skewed distribution (671) A curve or histogram that is not symmetric.

Positively Skewed

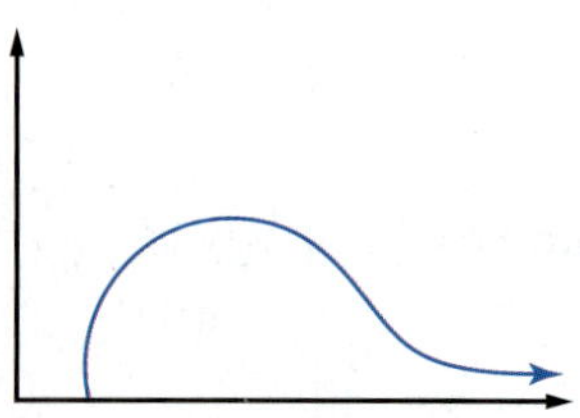

Negatively Skewed

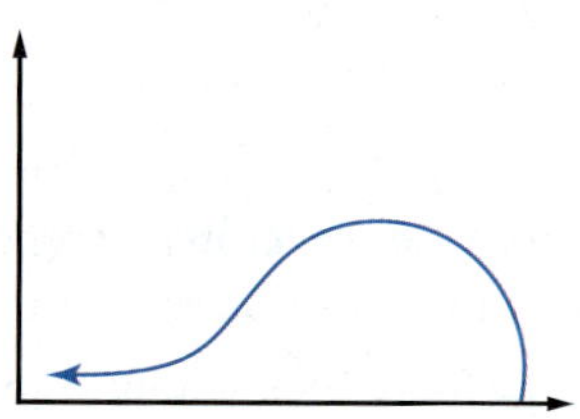

espacio muestral Conjunto de todos los resultados posibles de un experimento probabilístico.

escalar Una constante.

multiplicación por escalares Multiplicación de una matriz por una constante llamada escalar; producto de un escalar k y una matriz de $m \times n$.

gráfica de dispersión Conjuntos de datos graficados como pares ordenados en un plano de coordenadas.

notación científica Escritura de un número en la forma $a \times 10^n$, donde $1 \leq a < 10$ y n es un entero.

secante Para cualquier ángulo de medida α, un punto $P(x, y)$ en su lado terminal, $r = \sqrt{x^2 + y^2}$, $\sec \alpha = \frac{r}{x}$.

determinante de segundo orden El determinante de una matriz de 2×2.

sucesión Lista de números en un orden particular.

serie Suma específica de los términos de una sucesión.

notación de construcción de conjuntos Escritura del conjunto solución de una desigualdad, por ejemplo, $\{x \mid x > 9\}$.

notación de suma Para cualquier sucesión $a_1, a_2, a_3, \ldots,$ la suma de los k primeros términos puede escribirse $\sum_{n=1}^{k} a_n$, lo que se lee "la suma de $n = 1$ a k de los a_n." Así, $\sum_{n=1}^{k} a_n = a_1 + a_2 + a_3 + \ldots + a_k$, donde k es un valor entero.

evento simple Un solo evento.

reducir Escribir una expresión sin paréntesis o exponentes negativos.

simulación Uso de un experimento probabilístico para imitar una situación de la vida real.

seno Para cualquier ángulo, de medida α, un punto $P(x, y)$ en su lado terminal, $r = \sqrt{x^2 + y^2}$, sin $\alpha = \frac{y}{r}$.

distribución asimétrica Curva o histograma que no es simétrico.

Positivamente Alabeada

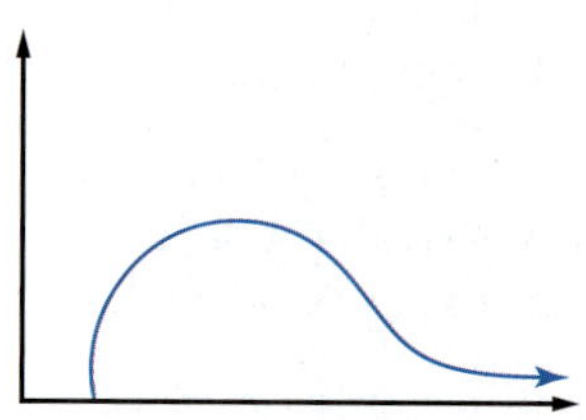

Negativamente Alabeada

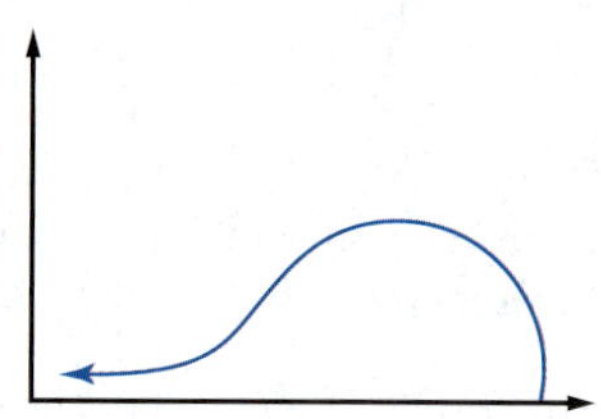

slope (68) The ratio of the change in y-coordinates to the change in x-coordinates.

pendiente La razón del cambio en coordenadas y al cambio en coordenadas x.

slope-intercept form (75) The equation of a line in the form $y = mx + b$, where m is the slope and b is the y-intercept.

forma pendiente-intersección Ecuación de una recta de la forma $y = mx + b$, donde m es la pendiente y b la intersección.

solution (20) A replacement for the variable in an open sentence that results in a true sentence.

solución Sustitución de la variable de un enunciado abierto que resulta en un enunciado verdadero.

solving a right triangle (704) The process of finding the measures of all of the sides and angles of a right triangle.

resolver un triángulo rectángulo Proceso de hallar las medidas de todos los lados y ángulos de un triángulo rectángulo.

square matrix (155) A matrix with the same number of rows and columns.

matriz cuadrada Matriz con el mismo número de filas y columnas.

square root (245) For any real numbers a and b, if $a^2 = b$, then a is a square root of b.

raíz cuadrada Para cualquier número real a y b, si $a^2 = b$, entonces a es una raíz cuadrada de b.

square root function (395) A function that contains a square root of a variable.

función radical Función que contiene la raíz cuadrada de una variable.

Square Root Property (306) For any real number n, if $x^2 = n$, then $x = \pm\sqrt{n}$.

Propiedad de la raíz cuadrada Para cualquier número real n, si $x^2 = n$, entonces $x = \pm\sqrt{n}$.

standard deviation (665) The square root of the variance, represented by α.

desviación estándar La raíz cuadrada de la varianza, la que se escribe α.

standard form (64) A linear equation written in the form $Ax + By = C$, where A, B, and C are real numbers and A and B are not both zero.

forma estándar Ecuación lineal escrita de la forma $Ax + By = C$, donde A, B, y C son números reales y A y B no son cero simultáneamente.

standard position (709) An angle positioned so that its vertex is at the origin and its initial side is along the positive x-axis.

posición estándar Ángulo en posición tal que su vértice está en el origen y su lado inicial está a lo largo del eje x positivo.

step function (89) A function whose graph is a series of line segments.

función etapa Función cuya gráfica es una serie de segmentos de recta.

substitution method (116) A method of solving a system of equations in which one equation is solved for one variable in terms of the other.

método de sustitución Método para resolver un sistema de ecuaciones en que una de las ecuaciones se resuelve en una de las variables en términos de la otra.

success (644) The desired outcome of an event.

éxito El resultado deseado de un evento.

synthetic division (234) A method used to divide a polynomial by a binomial.

división sintética Método que se usa para dividir un polinomio entre un binomio.

synthetic substitution (365) The use of synthetic division to evaluate a function.

sustitución sintética Uso de la división sintética para evaluar una función polinomial.

system of equations (110) A set of equations with the same variables.

sistema de ecuaciones Conjunto de ecuaciones con las mismas variables.

system of inequalities (123) A set of inequalities with the same variables.

sistema de desigualdades Conjunto de desigualdades con las mismas variables.

T

tangent (427, 701) **1.** A line that intersects a circle at exactly one point. **2.** For any angle, with measure α, a point $P(x, y)$ on its terminal side, $r = \sqrt{x^2 + y^2}$, $\tan \alpha = \frac{y}{x}$.

tangente **1.** Recta que interseca un círculo en un solo punto. **2.** Para cualquier ángulo, de medida α, un punto $P(x, y)$ en su lado terminal, $r = \sqrt{x^2 + y^2}$, $\tan \alpha = \frac{y}{x}$.

term (229, 578) **1.** The monomials that make up a polynomial. **2.** Each number in a sequence or series.

terminal side of an angle (709) A ray of an angle that rotates about the center.

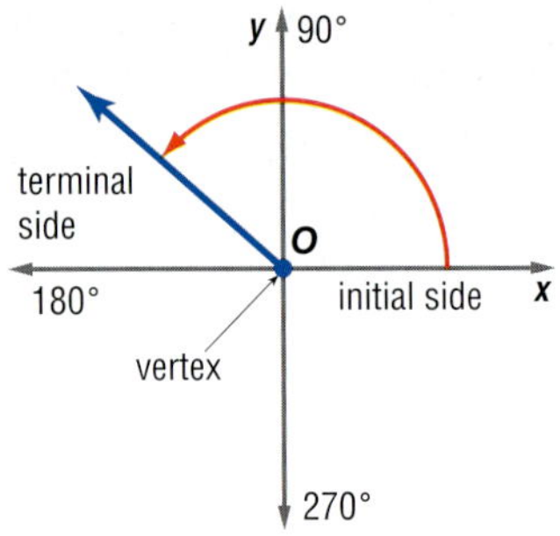

third-order determinant (183) Determinants of a 3×3 matrix.

transformation (175) Functions that map points of a pre-image onto its image.

translation (175) A figure is moved from one location to another on the coordinate plane without changing its size, shape, or orientation.

translation matrix (175) A matrix that represents a translated figure.

transverse axis (442) The segment of length $2a$ whose endpoints are the vertices of a hyperbola.

trigonometric equation (799) An equation containing at least one trigonometric function that is true for some but not all values of the variable.

trigonometric functions (701, 717) For any angle, with measure α, a point $P(x, y)$ on its terminal side, $r = \sqrt{x^2 + y^2}$, the trigonometric functions of α are as follows.

$$\sin \alpha = \frac{y}{r} \quad \cos \alpha = \frac{x}{r} \quad \tan \alpha = \frac{y}{x}$$

$$\csc \alpha = \frac{r}{y} \quad \sec \alpha = \frac{r}{x} \quad \cot \alpha = \frac{x}{y}$$

trigonometric identity (777) An equation involving a trigonometric function that is true for all values of the variable.

trigonometry (701) The study of the relationships between the angles and sides of a right triangle.

trinomial (229) A polynomial with three unlike terms.

término **1.** Los monomios que constituyen un polinomio. **2.** Cada número de una sucesión o serie.

lado terminal de un ángulo Rayo de un ángulo que gira alrededor de un centro.

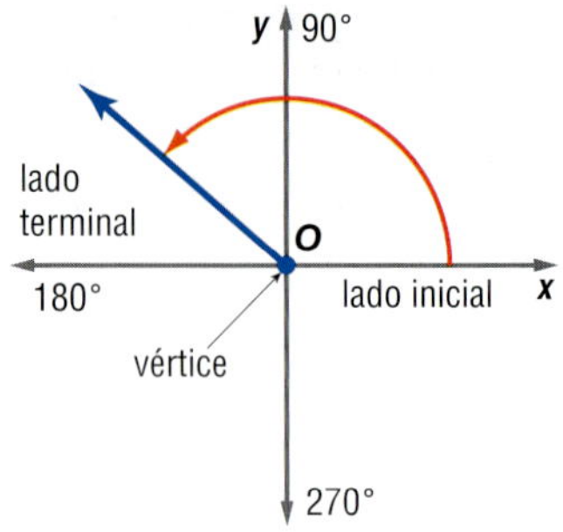

determinante de tercer orden Determinante de una matriz de 3×3.

transformación Funciones que aplican puntos de una preimagen en su imagen.

traslación Se mueve una figura de un lugar a otro en un plano de coordenadas sin cambiar su tamaño, forma u orientación.

matriz de traslación Matriz que representa una figura trasladada.

eje transversal El segmento de longitud $2a$ cuyos extremos son los vértices de una hipérbola.

ecuación trigonométrica Ecuación que contiene por lo menos una función trigonométrica y que sólo se cumple para algunos valores de la variable.

funciones trigonométricas Para cualquier ángulo, de medida α, un punto $P(x, y)$ en su lado terminal, $r = \sqrt{x^2 + y^2}$, las funciones trigonométricas de α son las siguientes.

$$\text{sen } \alpha = \frac{y}{r} \quad \cos \alpha = \frac{x}{r} \quad \tan \alpha = \frac{y}{x}$$

$$\csc \alpha = \frac{r}{y} \quad \sec \alpha = \frac{r}{x} \quad \cot \alpha = \frac{x}{y}$$

identidad trigonométrica Ecuación que involucra una o más funciones trigonométricas y que se cumple para todos los valores de la variable.

trigonometría Estudio de las relaciones entre los lados y ángulos de un triángulo rectángulo.

trinomio Polinomio con tres términos diferentes.

U

unbiased sample (682) A sample in which every possible sample has an equal chance of being selected.

unbounded (130) A system of inequalities that forms a region that is open.

union (41) The graph of a compound inequality containing *or*.

muestra no sesgada Muestra en que cualquier muestra posible tiene la misma posibilidad de seleccionarse.

no acotado Sistema de desigualdades que forma una región abierta.

unión Gráfica de una desigualdad compuesta que contiene la palabra *o*.

unit circle (710) A circle of radius 1 unit whose center is at the origin of a coordinate system.

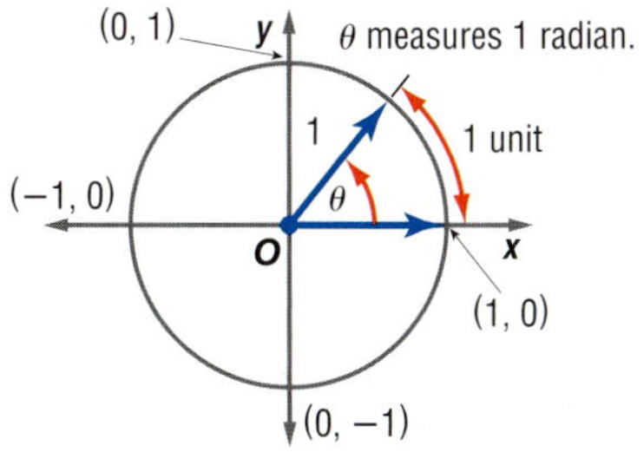

círculo unitario Círculo de radio 1 cuyo centro es el origen de un sistema de coordenadas.

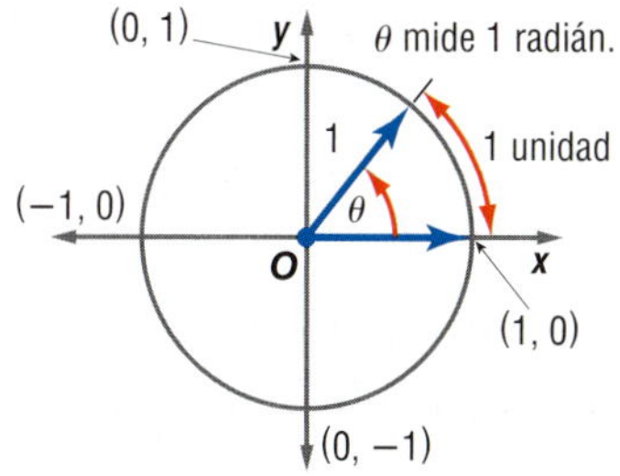

V

variables (7) Symbols, usually letters, used to represent unknown quantities.

variables Símbolos, por lo general letras, que se usan para representar cantidades desconocidas.

variance (665) The mean of the squares of the deviations from the arithmetic mean.

varianza Media de los cuadrados de las desviaciones de la media aritmética.

vertex (287, 442) **1.** The point at which the axis of symmetry intersects a parabola. **2.** The point on each branch nearest the center of a hyperbola.

vértice **1.** Punto en el que el eje de simetría interseca una parábola. **2.** El punto en cada rama más cercano al centro de una hipérbola.

vertex form (322) A quadratic function in the form $y = a(x - h)^2 + k$, where (h, k) is the vertex of the parabola and $x = h$ is its axis of symmetry.

forma de vértice Función cuadrática de la forma $y = a(x - h)^2 + k$, donde (h, k) es el vértice de la parábola y $x = h$ es su eje de simetría.

vertex matrix (175) A matrix used to represent the coordinates of the vertices of a polygon.

matriz de vértice Matriz que se usa para escribir las coordenadas de los vértices de un polígono.

vertical asymptote (485) If the related rational expression of a function is written in simplest form and is undefined for $x = a$, then $x = a$ is a vertical asymptote.

asíntota vertical Si la expresión racional que corresponde a una función racional se reduce y está no definida en $x = a$, entonces $x = a$ es una asíntota vertical.

vertical line test (57) If no vertical line intersects a graph in more than one point, then the graph represents a function.

prueba de la recta vertical Si ninguna recta vertical interseca una gráfica en más de un punto, entonces la gráfica representa una función.

vertices (129) The maximum or minimum value that a linear function has for the points in a feasible region.

vértices El valor máximo o mínimo que una función lineal tiene para los puntos en una región viable.

X

x-intercept (65) The x-coordinate of the point at which a graph crosses the x-axis.

intersección x La coordenada x del punto o puntos en que una gráfica interseca o cruza el eje x.

Y

y-intercept (65) The y-coordinate of the point at which a graph crosses the y-axis.

intersección y La coordenada y del punto o puntos en que una gráfica interseca o cruza el eje y.

Z

zeros (294) The x-intercepts of the graph of a quadratic equation; the points for which $f(x) = 0$.

ceros Las intersecciones x de la gráfica de una ecuación cuadrática; los puntos x para los que $f(x) = 0$.

zero matrix (155) A matrix in which every element is zero.

matriz nula Matriz cuyos elementos son todos igual a cero.

Selected Answers

Chapter 1 Solving Equations and Inequalities

Page 5 Chapter 1 Getting Started

1. 19.84 **3.** -17.51 **5.** $-\frac{5}{12}$ **7.** $-2\frac{1}{6}$ **9.** 0.48 **11.** 1.1 **13.** $-2\frac{2}{3}$ **15.** $8\frac{4}{5}$ **17.** 8 **19.** 49 **21.** 0.64 **23.** $\frac{4}{9}$ **25.** false **27.** true **29.** false **31.** true

Pages 8–10 Lesson 1-1

1. First, find the sum of c and d. Divide this sum by e. Multiply the quotient by b. Finally, add a. **3.** b; The sum of the cost of adult and children tickets should be subtracted from 50. Therefore parentheses need to be inserted around this sum to insure that this addition is done before subtraction. **5.** 6 **7.** 1 **9.** 119 **11.** -23 **13.** \$432 **15.** \$1162.50 **17.** 3 **19.** 25 **21.** -34 **23.** 5 **25.** -31 **27.** 14 **29.** -3 **31.** 162 **33.** 2.56 **35.** $25\frac{1}{3}$ **37.** 31.25 drops per min **39.** 2 **41.** -4.2 **43.** -4 **45.** 1.4 **47.** -8 **49.** $2\frac{1}{6}$ **51.** -16 **53.** \$14,439.83 **55.** Sample answer: $4 - 4 + 4 \div 4 = 1$; $4 \div 4 + 4 \div 4 = 2$; $(4 + 4 + 4) \div 4 = 3$; $4 \times (4 - 4) + 4 = 4$; $(4 \times 4 + 4) \div 4 = 5$; $(4 + 4) \div 4 + 4 = 6$; $44 \div 4 - 4 = 7$; $(4 + 4) \times (4 \div 4) = 8$; $4 + 4 + 4 \div 4 = 9$; $(44 - 4) \div 4 = 10$ **57.** C **59.** 3 **61.** 10 **63.** -2 **65.** $\frac{2}{3}$

Pages 14–17 Lesson 1-2

1a. Sample answer: 2 **1b.** Sample answer: 5 **1c.** Sample answer: -11 **1d.** Sample answer: 1.3 **1e.** Sample answer: $\sqrt{2}$ **1f.** Sample answer: -1.3 **3.** 0; Zero does not have a multiplicative inverse since $\frac{1}{0}$ is undefined. **5.** N, W, Z, Q, R **7.** Multiplicative Inverse **9.** Additive Identity **11.** $-\frac{1}{3}, 3$ **13.** $-2x + 4y$ **15.** $3c + 18d$ **17.** $1.5(10 + 15 + 12 + 8 + 19 + 22 + 31)$ or $1.5(10) + 1.5(15) + 1.5(12) + 1.5(8) + 1.5(19) + 1.5(22) + 1.5(31)$ **19.** W, Z, Q, R **21.** N, W, Z, Q, R **23.** I, R **25.** N, W, Z, Q, R **27.** Q, R; $2.4, 2.49, 2.\overline{49}, 2.4\overline{9}, 2.\overline{9}$ **29.** Associative ($\times$) **31.** Associative (+) **33.** Multiplicative Inverse **35.** Multiplicative Identity **37.** $-m$; Additive Inverse **39.** 1 **41.** $\sqrt{2}$ units **43.** 10; $-\frac{1}{10}$ **45.** 0.125; -8 **47.** $-\frac{4}{3}, \frac{3}{4}$ **49.** $3a - 2b$ **51.** $40x - 7y$ **53.** $-12r + 4t$ **55.** $-3.4m + 1.8n$ **57.** $-8 + 9y$ **59.** true **61.** false; 6 **63.** $6.5(4.5 + 4.25 + 5.25 + 6.5 + 5)$ or $6.5(4.5) + 6.5(4.25) + (6.5)5.25 + 6.5(6.5) + 6.5(5)$

65.

$3\left(2\frac{1}{4}\right) + 2\left(1\frac{1}{8}\right)$	
$= 3\left(2 + \frac{1}{4}\right) + 2\left(1 + \frac{1}{8}\right)$	*Definition of a mixed number*
$= 3(2) + 3\left(\frac{1}{4}\right) + 2(1) + 2\left(\frac{1}{8}\right)$	*Distributive Property*
$= 6 + \frac{3}{4} + 2 + \frac{1}{4}$	*Multiply.*
$= 6 + 2 + \frac{3}{4} + \frac{1}{4}$	*Commutative Property (+)*
$= 8 + \frac{3}{4} + \frac{1}{4}$	*Add.*
$= 8 + \left(\frac{3}{4} + \frac{1}{4}\right)$	*Associative Property (+)*
$= 8 + 1$ or 9	*Add.*

67. 4700 ft^2 **69.** \$62.15 **71.** Answers should include the following.

- Instead of doubling each coupon value and then adding these values together, the Distributive Property could be applied allowing you to add the coupon values first and then double the sum.
- If a store had a 25% off sale on all merchandise, the Distributive Property could be used to calculate these savings. For example, the savings on a \$15 shirt, \$40 pair of jeans, and \$25 pair of slacks could be calculated as $0.25(15) + 0.25(40) + 0.25(25)$ or as $0.25(15 + 40 + 25)$ using the Distributive Property.

73. C **75.** False; $0 - 1 = -1$, which is not a whole number. **77.** False; $2 \div 3 = \frac{2}{3}$, which is not a whole number. **79.** 6 **81.** -2.75 **83.** -11 **85.** -4.3

Page 17 Practice Quiz 1

1. 14 **3.** 6 **5.** 2 amperes **7.** N, W, Z, Q, R **9.** $-\frac{6}{7}, \frac{7}{6}$

Pages 24–27 Lesson 1-3

1. Sample answer: $2x = -14$ **3.** Jamal; his method can be confirmed by solving the equation using an alternative method.

$$C = \frac{5}{9}(F - 32)$$
$$C = \frac{5}{9}F - \frac{5}{9}(32)$$
$$C + \frac{5}{9}(32) = \frac{5}{9}F$$
$$\frac{9}{5}\left[C + \frac{5}{9}(32)\right] = F$$
$$\frac{9}{5}C + 32 = F$$

5. $2n - n^3$ **7.** Sample answer: 5 plus 3 times the square of a number is twice that number. **9.** Addition (=) **11.** 14 **13.** -4.8 **15.** 16 **17.** $p = \frac{I}{rt}$ **19.** $5 + 3n$ **21.** $n^2 - 4$ **23.** $5(9 + n)$ **25.** $\left(\frac{n}{4}\right)^2$ **27.** $2\pi rh + 2\pi r^2$ **29.** Sample answer: 5 less than a number is 12. **31.** Sample answer: A number squared is equal to 4 times the number. **33.** Sample answer: A number divided by 4 is equal to twice the sum of that number and 1. **35.** Substitution (=) **37.** Transitive (=) **39.** Symmetric (=) **41.** 7 **43.** 3.2 **45.** $\frac{1}{12}$ **47.** -8 **49.** -7 **51.** 1 **53.** $\frac{1}{4}$ **55.** $-\frac{55}{2}$ **57.** $\frac{d}{t} = r$ **59.** $\frac{3V}{\pi r^2} = h$ **61.** $b = \frac{x(c - 3)}{a} + 2$ **63.** n = number of games; $2(1.50) + n(2.50) = 16.75$; 5 **65.** x = cost of gasoline per mile; $972 + 114 + 105 + 7600x = 1837$; 8.5¢/mi **67.** a = Chun-Wei's age; $a + (2a + 8) + (2a + 8 + 3) = 94$; Chun-Wei: 15 yrs old, mother: 38 yrs old, father: 41 yrs old **69.** n = number of lamps broken; $12(125) - 45n = 1365$; 3 lamps **71.** 15.1 mi/month **73.** The Central Pacific had to lay their track through the Rocky Mountains, while the Union Pacific mainly built track over flat prairie. **75.** the product of 3 and the difference of a number and 5 added to the product of four times the number and the sum of the number and 1 **77.** B **79.** $-6x + 8y + 4z$ **81.** 6.6 **83.** 105 cm^2 **85.** 3 **87.** $-\frac{1}{4}$ **89.** $-5 + 6y$

Pages 30–32 Lesson 1-4

1. $|a| = -a$ when a is a negative number and the negative of a negative number is positive. **3.** Always; since the opposite of 0 is still 0, this equation has only one case, $ax + b = 0$. The solution is $\frac{-b}{a}$. **5.** 8 **7.** −17 **9.** $\{-18, -12\}$ **11.** $\{-32, 36\}$ **13.** $\{8\}$ **15.** least: 158°F; greatest: 162°F **17.** 15 **19.** 0 **21.** 3 **23.** −4 **25.** −9.4 **27.** 55 **29.** $\{8, 42\}$ **31.** $\{-45, 21\}$ **33.** $\{-2, 16\}$ **35.** $\left\{\frac{3}{2}\right\}$ **37.** $\left\{2, \frac{9}{2}\right\}$ **39.** $\varnothing$ **41.** $\{-5, 11\}$ **43.** $\left\{-\frac{11}{3}, -3\right\}$ **45.** $\{8\}$ **47.** $|x - 200| = 5$; maximum: 205°F; minimum: 195°F **49.** $|x - 13| = 5$; maximum: 18 km, minimum: 8 km **51.** sometimes; true only if $c \geq 0$ **53.** B **55.** $|x + 1| + 2 = x + 4$; $|x + 1| + 2 = -(x + 4)$ **57.** $\{-1.5\}$ **59.** $2(n - 11)$ **61.** $\frac{16}{3}$ **63.** 14 **65.** Distributive **67.** Additive Identity **69.** true **71.** false; 1.2 **73.** 364 ft^2 **75.** 8 **77.** $\frac{2}{3}$ **79.** $-\frac{3}{4}$

Pages 37–39 Lesson 1-5

1. Dividing by a number is the same as multiplying by its inverse. **3.** Sample answer: $x + 2 < x + 1$

5. $\left\{x \mid x \leq \frac{5}{3}\right\}$ or $\left(-\infty, \frac{5}{3}\right]$

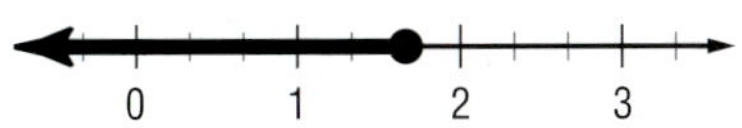

7. $\{y \mid y > 6\}$ or $(6, +\infty)$

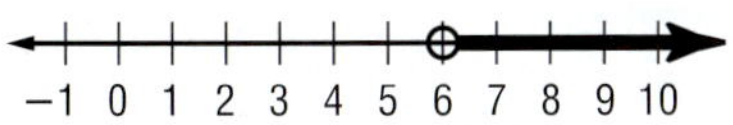

9. $\{p \mid p > 15\}$ or $(15, +\infty)$

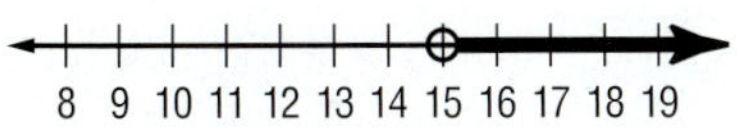

11. all real numbers or $(-\infty, +\infty)$

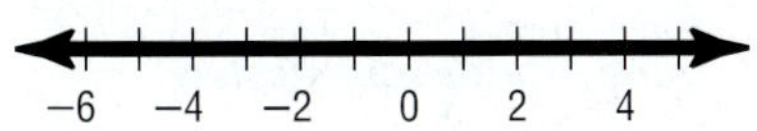

13. $2n - 3 \leq 5$; $n \leq 4$

15. $\{n \mid n \geq -11\}$ or $[-11, +\infty)$

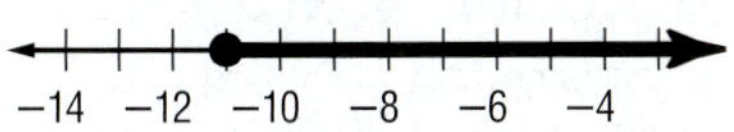

17. $\{x \mid x < 7\}$ or $(-\infty, 7)$

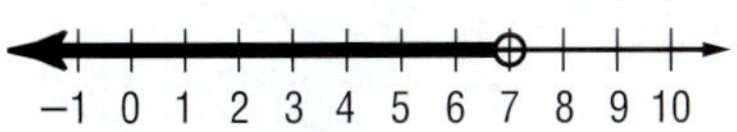

19. $\{g \mid g \leq 27\}$ or $(-\infty, 27]$

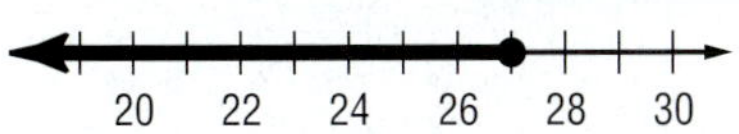

21. $\{k \mid k \geq -3.5\}$ or $[-3.5, +\infty)$

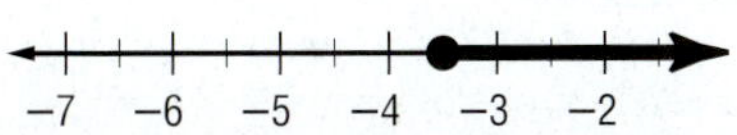

23. $\{m \mid m > -4\}$ or $(-4, +\infty)$

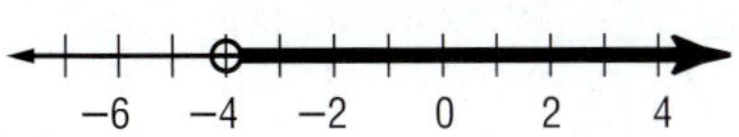

25. $\{t \mid t \leq 0\}$ or $(-\infty, 0]$

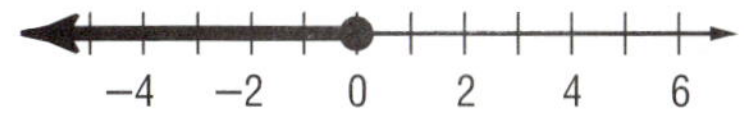

27. $\{n \mid n \geq 1.75\}$ or $[1.75, +\infty)$

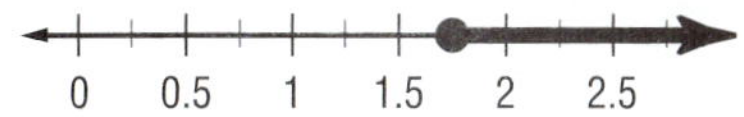

29. $\{x \mid x < -279\}$ or $(-\infty, -279)$

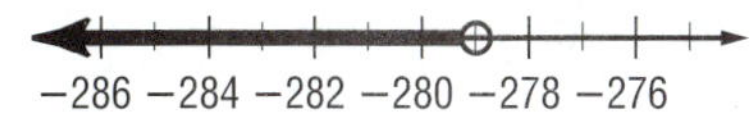

31. $\{d \mid d \geq -5\}$ or $[-5, +\infty)$

33. $\{g \mid g < 2\}$ or $(-\infty, 2)$

35. $\left\{y \mid y < \frac{1}{5}\right\}$ or $\left(-\infty, \frac{1}{5}\right)$

37. $\varnothing$

39. at least 25 h **41.** $n + 8 > 2$; $n > -6$ **43.** $\frac{1}{2}n - 7 \geq 5$; $n \geq 24$ **45.** $2(n + 5) \leq 3n + 11$; $n \geq -1$ **47.** $2(7m) \geq 17$; $m \geq \frac{17}{14}$; at least 2 child care staff members **49.** $n \geq 34.97$; She must sell at least 35 cars. **51.** $s \geq 91$; Ahmik must score at least 91 on her next test to have an A test average. **53.** Answers should include the following.

- $150 < 400$
- Let n equal the number of minutes used. Write an expression representing the cost of Plan 1 and for Plan 2 for n minutes. The cost for Plan 1 would include a monthly access fee of \$35 plus 40¢ for each minute over 150 minutes or $35 + 0.4(n - 150)$. The cost for Plan 2 for 400 minutes or less would be \$55. To find where Plan 2 would cost less than Plan 1 solve $55 < 35 + 0.4(n - 150)$ for n. The solution set is $\{n \mid n > 200\}$, which means that for more than 200 minutes of calls, Plan 2 is cheaper.

55. D **57.** $x \geq -2$ **59.** $\{-14, 20\}$ **61.** $\varnothing$ **63.** N, W, Z, Q, R **65.** I, R **67.** $\{-7, 7\}$ **69.** $\left\{4, -\frac{4}{5}\right\}$ **71.** $\{-11, -1\}$

Page 39 Practice Quiz 2

1. 0.5 **3.** 14 **5.** $\left\{m \mid m > \frac{4}{9}\right\}$ or $\left(\frac{4}{9}, +\infty\right)$

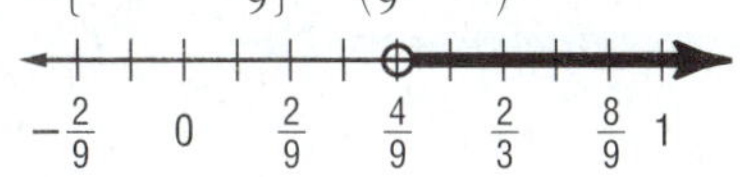

Pages 43–46 Lesson 1-6

1. $5 \leq c \leq 15$ **3.** Sabrina; an absolute value inequality of the form $|a| > b$ should be rewritten as an *or* compound inequality, $a > b$ or $a < -b$.

5. $|n| > 3$

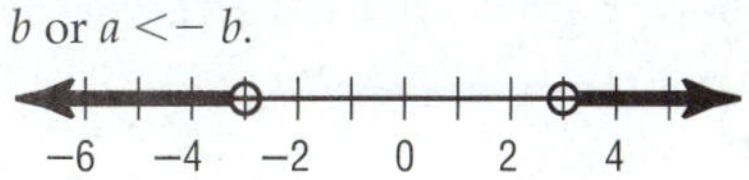

Selected Answers

7. $|n| < 2$

9. $\{d \mid -2 < d < 3\}$ 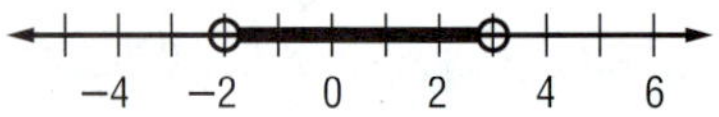

11. $\{g \mid -13 \leq g \leq 5\}$

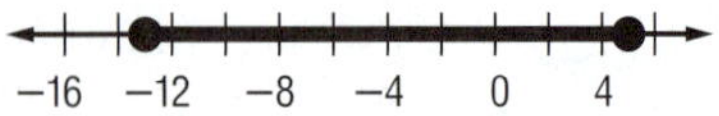

13. all real numbers

15. $|n| \geq 5$ 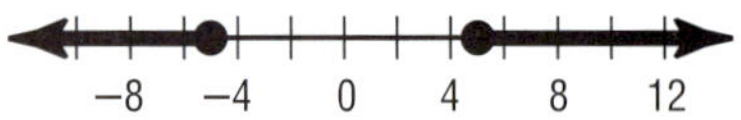

17. $|n| < 4$ 

19. $|n| > 8$ 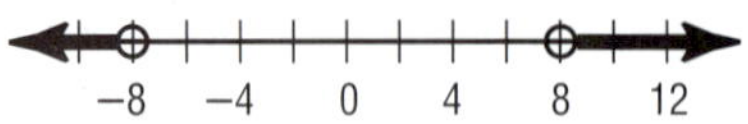

21. $|n| > 1$ 23. $|n| \geq 1.5$ 25. $|n + 1| > 1$

27. $\{p \mid p \leq 2 \text{ or } p \geq 8\}$

29. $\{x \mid -2 < x < 4\}$

31. $\{f \mid -7 < f < -5\}$

33. $\{g \mid -9 \leq g \leq 9\}$

35. $\varnothing$ 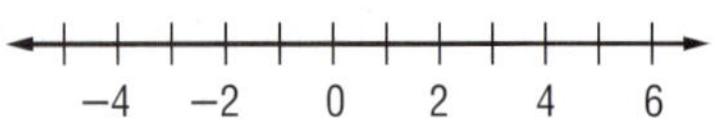

37. $\{b \mid b > 10 \text{ or } b < -2\}$

39. $\left\{w \mid -\frac{7}{3} \leq w \leq 1\right\}$

41. all real numbers

43. $\left\{n \mid n = \frac{7}{2}\right\}$ 

45. $6.8 < x < 7.4$ 47. $45 \leq s \leq 55$

49. 108 in. $< L + D \leq$ 130 in.

51. $a + b > c, a + c > b, b + c > a$

53a. 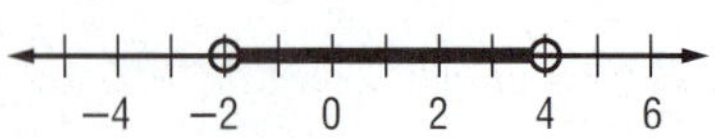

53b. 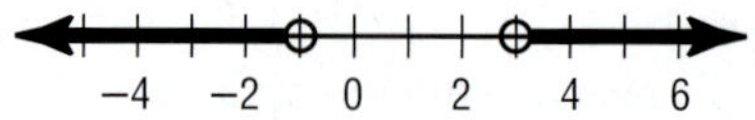

53c. 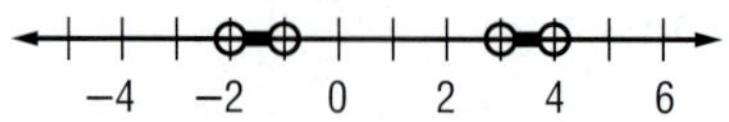

53d. $3 < |x + 2| \leq 8$ can be rewritten as $|x + 2| > 3$ and $|x + 2| \leq 8$. The solution of $|x + 2| > 3$ is $x > 1$ or $x < -5$. The solution of $|x + 2| \leq 8$ is $-10 \leq x \leq 6$. Therefore, the union of these two sets is ($x > 1$ or $x < -5$) and ($-10 \leq x \leq 6$). The union of the graph of $x > 1$ or $x < -5$ and the graph of $-10 \leq x \leq 6$ is shown below. From this we can see that solution can be rewritten as ($-10 \leq x < -5$) or ($1 < x \leq 6$).

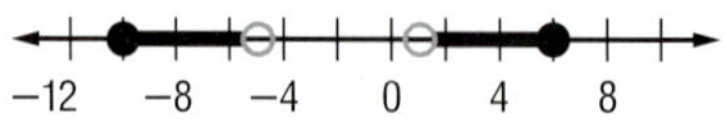

55. $x > -5$ or $x < -6$

57. 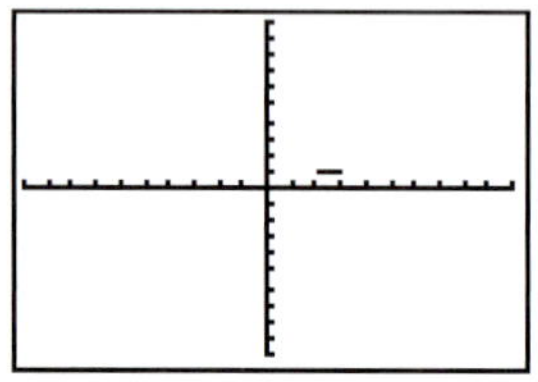

59. $(5x + 2 \geq 3)$ or $(5x + 2 \leq -3)$; $\{x \mid x \geq 0.2 \text{ or } x \leq -1\}$

61. $\{d \mid d \geq -6\}$ or $[-6, +\infty)$

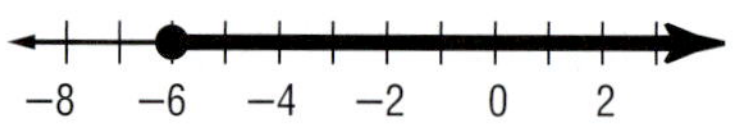

63. $\{n \mid n < -1\}$ or $(-\infty, -1)$

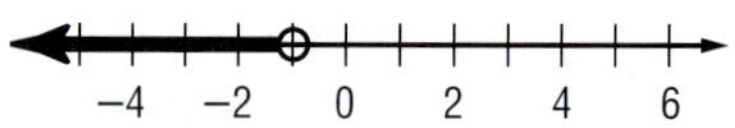

65. $\{-10, 16\}$ 67. $\varnothing$ 69. Symmetric (=) 71. $3a + 7b$ 73. 2 75. -7

Pages 47–50 Chapter 1 Study Guide and Review

1. compound inequality 3. Commutative (×) 5. Reflexive (=) 7. Multiplicative Inverse 9. absolute value 11. 22 13. -49 15. -23 17. 37.5 19. Q, R 21. I, R 23. $-5a + 24b$ 25. -14 27. -13 29. -4 31. $x = \frac{C - By}{A}$ 33. $p = \frac{A}{1 + rt}$ 35. $\{6, -18\}$ 37. $\{6\}$ 39. $\left\{-\frac{3}{2}, -1\right\}$

41. $\{x \mid x \geq 5\}$ or $[5, +\infty)$

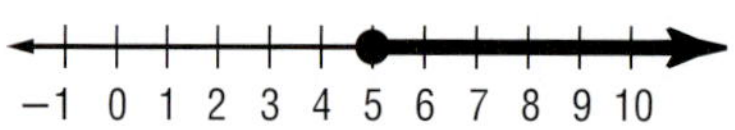

43. $\{a \mid a > 2\}$ or $(2, +\infty)$

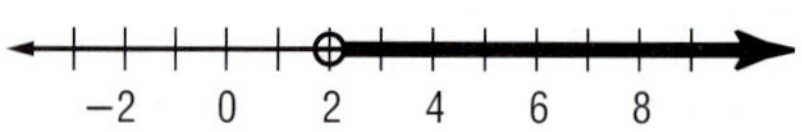

45. $\{x \mid x > -1.8\}$ or $(-1.8, +\infty)$

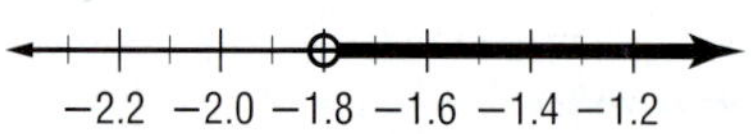

47. $\left\{y \mid \frac{5}{3} < y \leq 5\right\}$

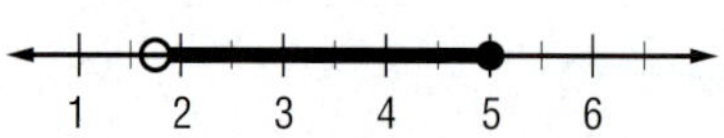

49. $\{y \mid -9 \le y \le 18\}$

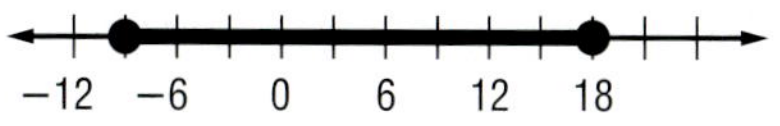

51. $\left\{b \mid b < -4 \text{ or } b > -\frac{10}{3}\right\}$

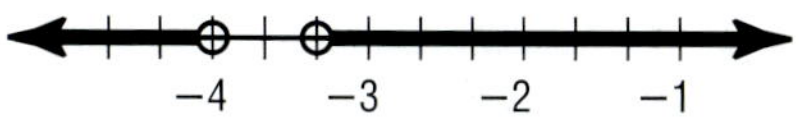

Chapter 2 Linear Relations and Functions

Page 55 Chapter 2 Getting Started

1. $(-3, 3)$ **3.** $(-3, -1)$ **5.** $(0, -4)$ **7.** -2 **9.** 9 **11.** 2 **13.** $x + 1$ **15.** $2x + 6$ **17.** $\frac{1}{2}x + 2$ **19.** 3 **21.** 15 **23.** 2.5

Pages 60–62 Lesson 2-1

1. Sample answer: $\{(-4, 3), (-2, 3), (1, 5), (-2, 1)\}$

3. Molly; to find $g(2a)$, replace x with $2a$. Teisha found $2g(a)$, not $g(2a)$. **5.** yes

7. D = {7}, R = {−1, 2, 5, 8}, no

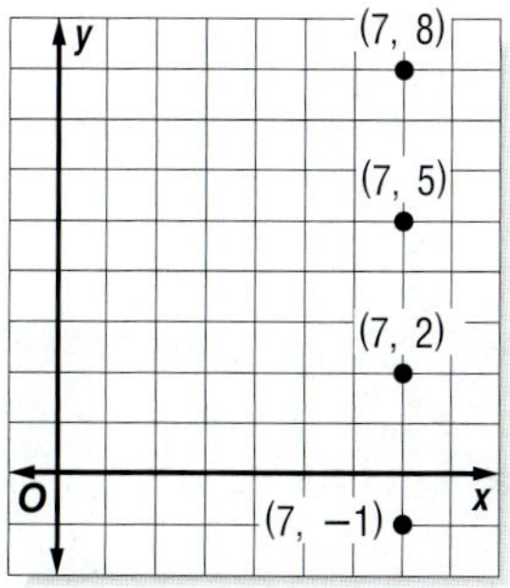

9. D = all reals, R = all reals, yes

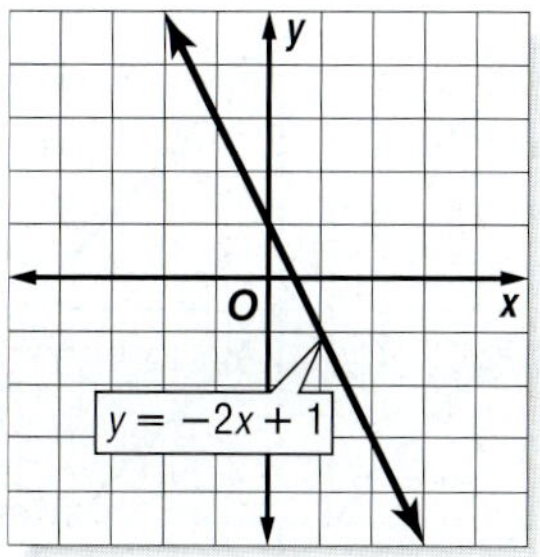

11. 10 **13.** D = {70, 72, 88}, R = {95, 97, 105, 114}

15.

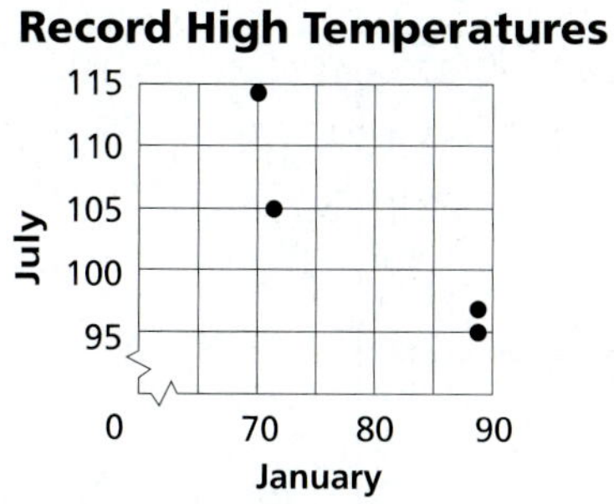

17. yes **19.** no **21.** yes

23. D = {−3, 1, 2}, R = {0, 1, 5}; yes

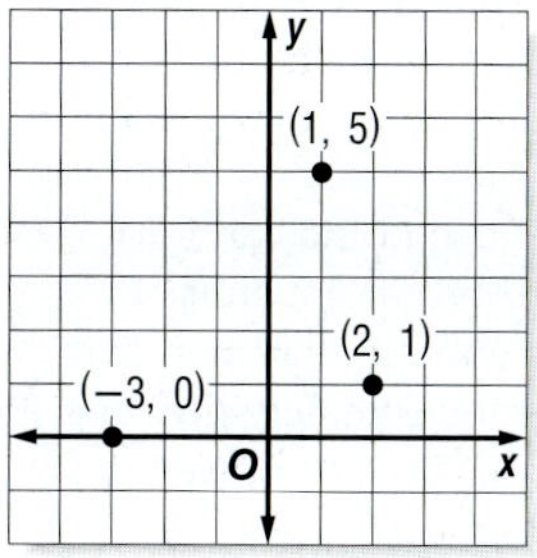

25. D = {−2, 3}, R = {5, 7, 8}; no

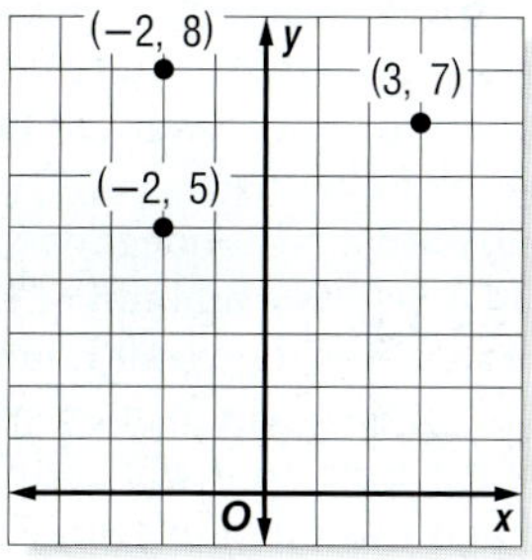

27. D = {−3.6, 0, 1.4, 2}, R = {−3, −1.1, 2, 8}; yes

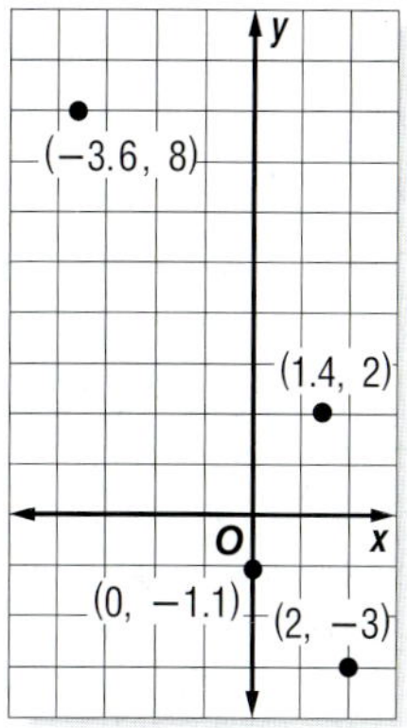

29. D = all reals, R = all reals; yes

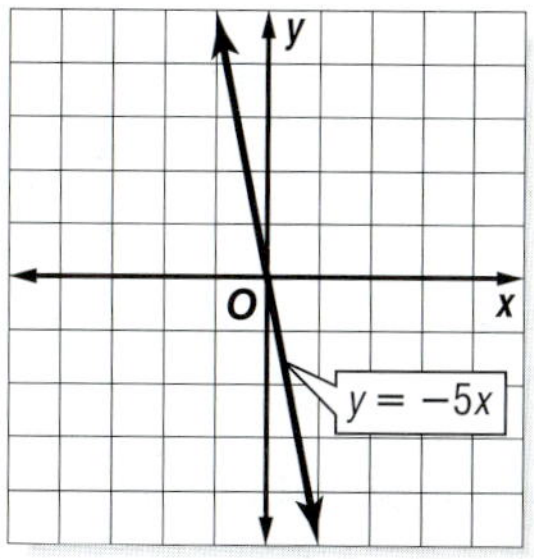

31. D = all reals, R = all reals; yes

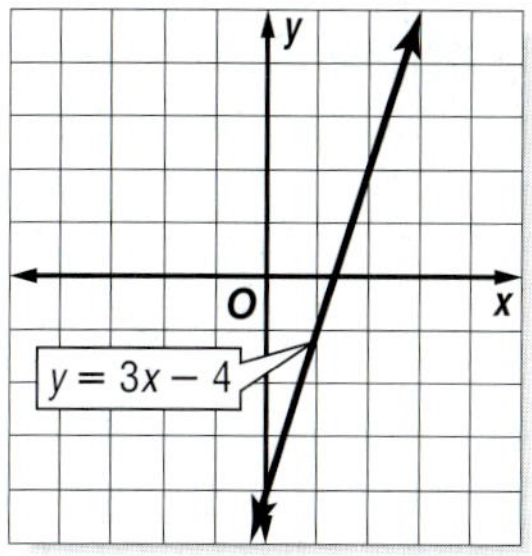

33. D = all reals, R = $\{y \mid y \ge 0\}$; yes

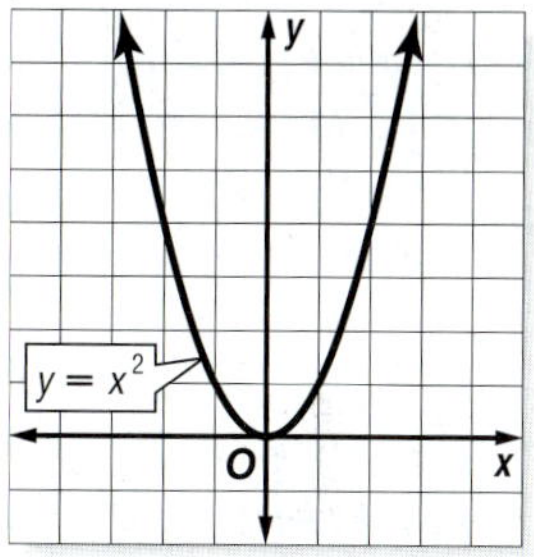

35.

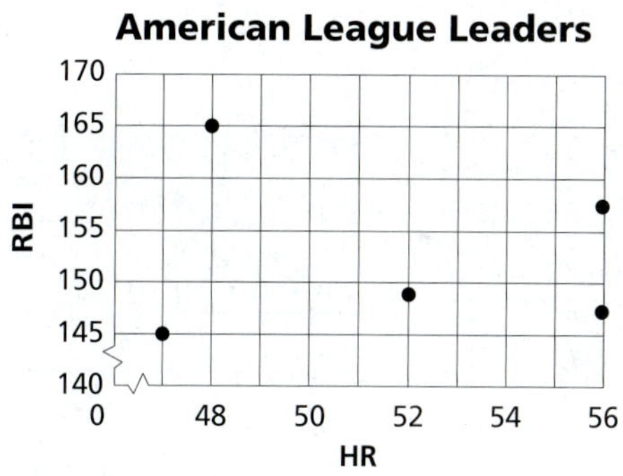

37. No; the domain value 56 is paired with two different range values.

39.

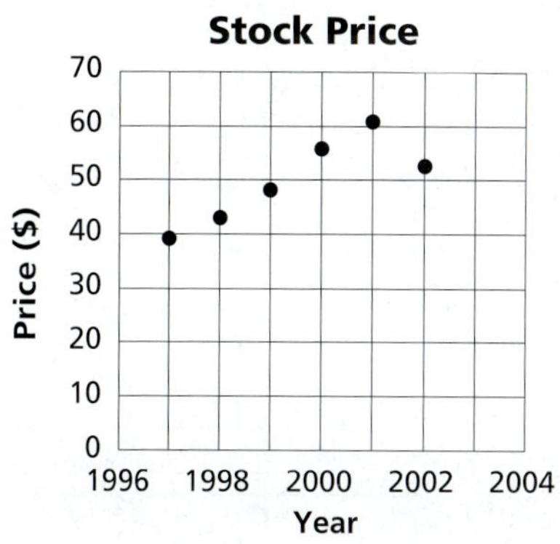

41. Yes; each domain value is paired with only one range value.

43.

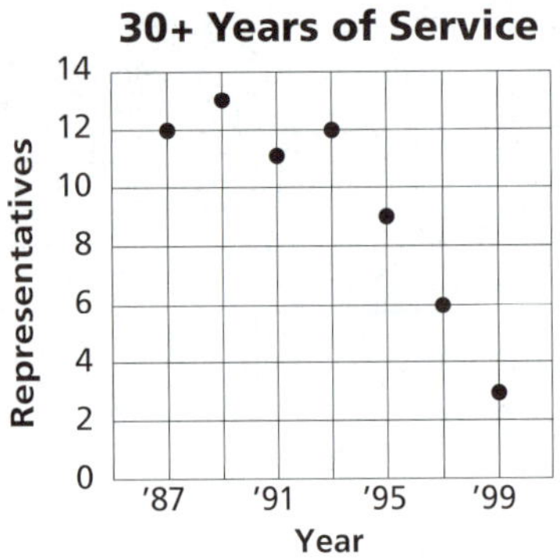

45. Yes; no; each domain value is paired with only one range value so the relation is a function, but the range value 12 is paired with two domain values so the function is not one-to-one. **47.** 6 **49.** −3 **51.** $25n^2 - 5n$ **53.** 11 **55.** $f(x) = 4x - 3$ **57.** B **59.** discrete **61.** discrete **63.** $\{y \mid -8 < y < 6\}$ **65.** $\{x \mid x < 5.1\}$ **67.** \$29.82 **69.** $31a + 10b$ **71.** 2 **73.** 15

Pages 65–67 Lesson 2-2

1. The function can be written as $f(x) = \frac{1}{2}x + 1$, so it is of the form $f(x) = mx + b$, where $m = \frac{1}{2}$ and $b = 1$. **3.** Sample answer: $x + y = 2$ **5.** yes **7.** $2x - 5y = 3$; 2, −5, 3

9. $-\frac{5}{3}$, −5

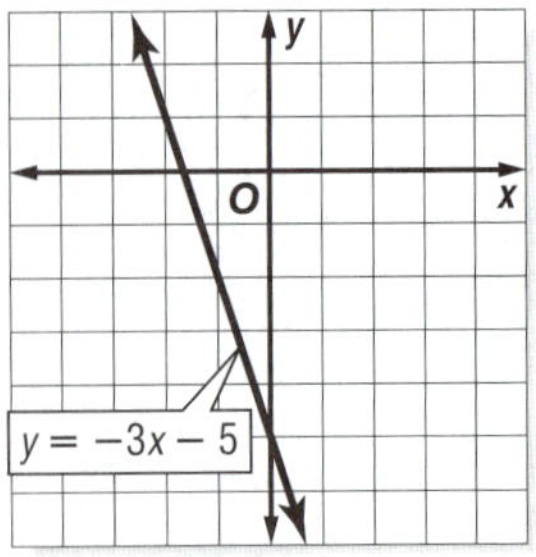

11. 2, 3

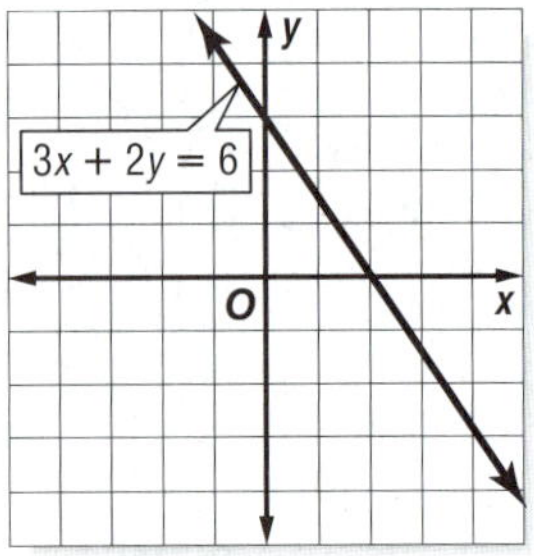

13. \$177.62 **15.** yes **17.** No; y is inside a square root. **19.** No; x appears in a denominator. **21.** No; x has an exponent other than 1. **23.** $x^2 + 5y = 0$ **25.** 7200 m **27.** $3x + y = 4$; 3, 1, 4 **29.** $x - 4y = -5$; 1, −4, −5 **31.** $2x - y = 5$; 2, −1, 5 **33.** $x + y = 12$; 1, 1, 12 **35.** $x = 6$; 1, 0, 6 **37.** $25x + 2y = 9$; 25, 2, 9

39. 3, 5

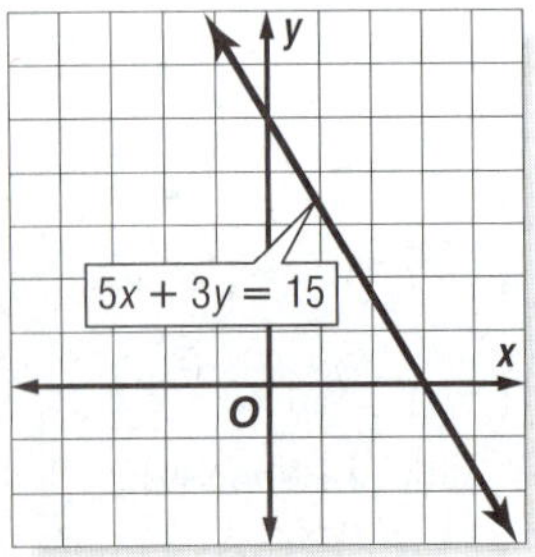

41. $\frac{10}{3}$, $-\frac{5}{2}$

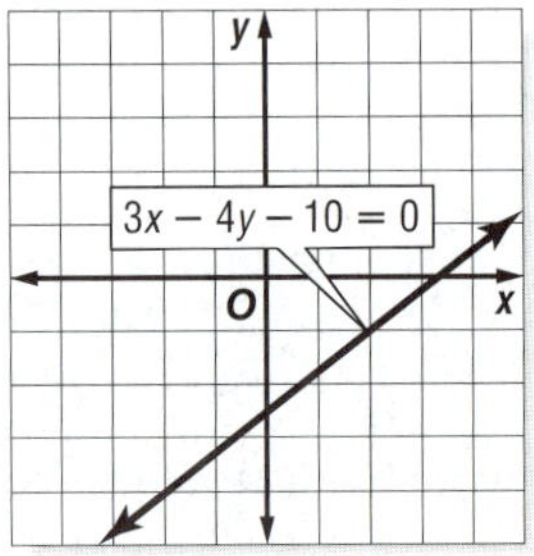

43. 0, 0

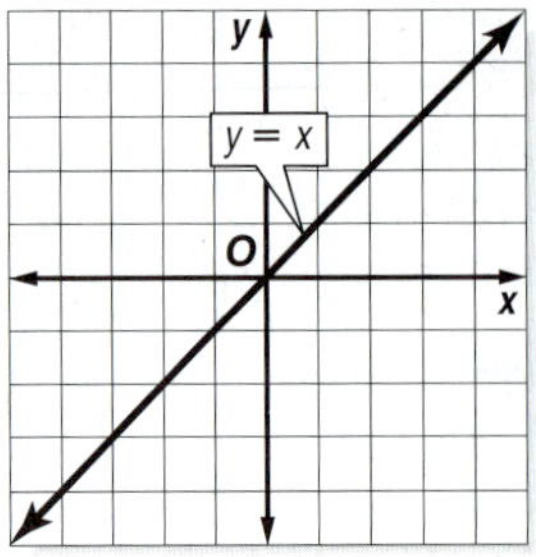

45. none, −2

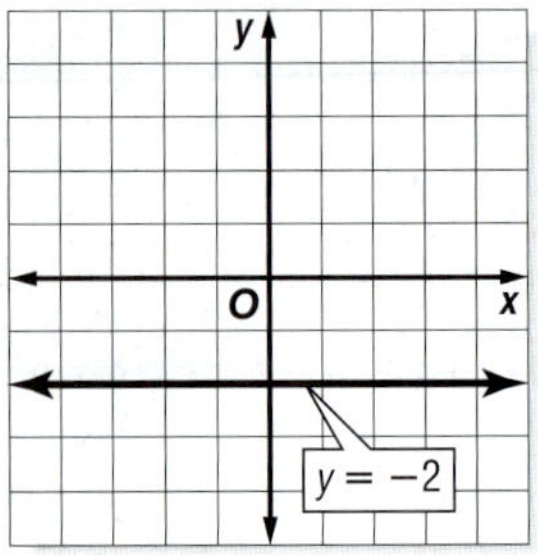

47. 8, none

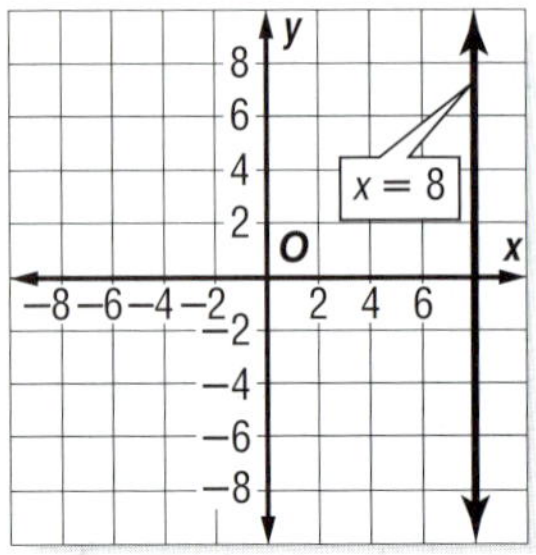

49. $\frac{1}{4}$, −1

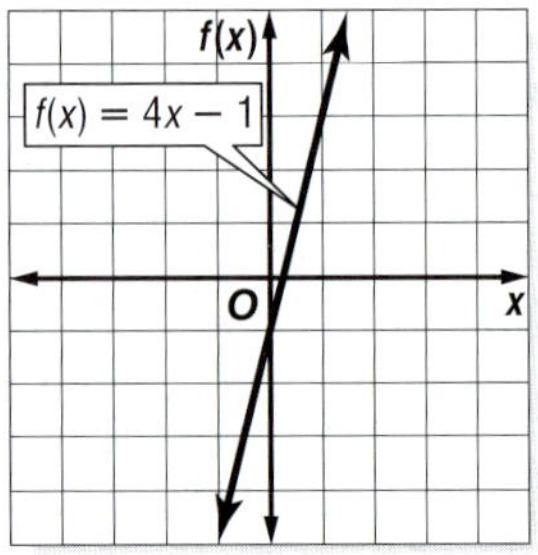

51.

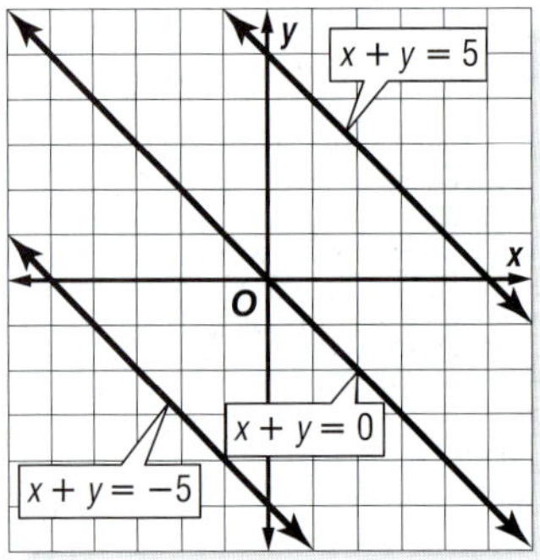

The lines are parallel but have different y-intercepts. **53.** 90°C

55.

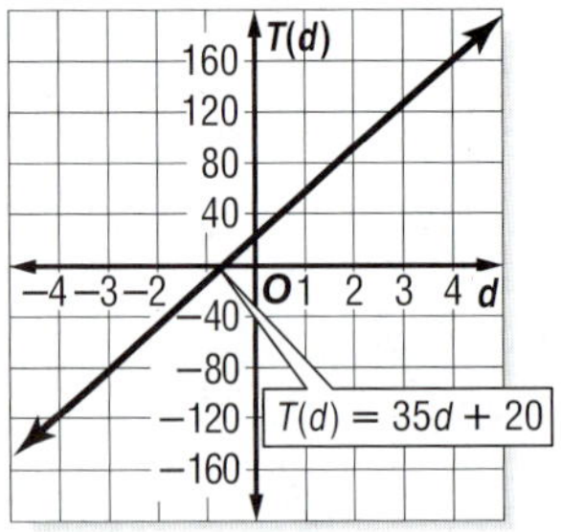

57.

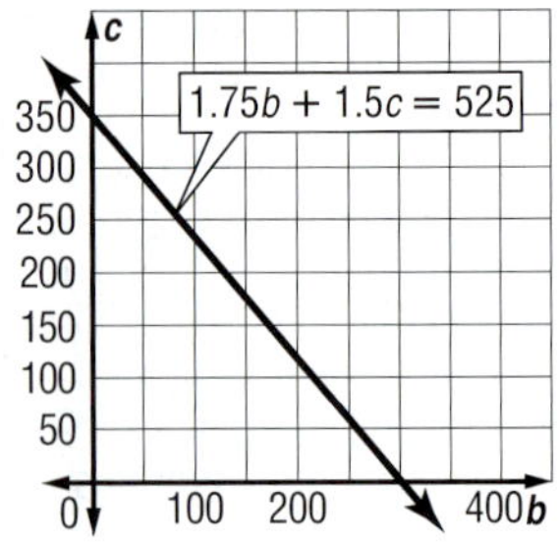

59. no **61.** A linear equation can be used to relate the amounts of time that a student spends on each of two subjects if the total amount of time is fixed. Answers should include the following.

- x and y must be nonnegative because Lolita cannot spend a negative amount of time studying a subject.
- The intercepts represent Lolita spending all of her time on one subject. The x-intercept represents her spending all of her time on math, and the y-intercept represents her spending all of her time on chemistry.

63. B **65.** D = {0, 1, 2}, R = {−1, 0, 2, 3}; no

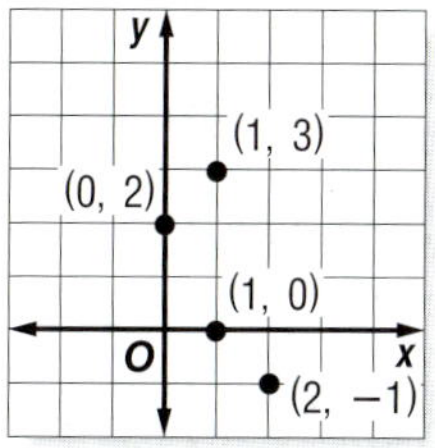

67. $\{x \mid x < -6 \text{ or } x > -2\}$
69. $3s + 14$ **71.** $\frac{1}{3}$ **73.** 2
75. −5 **77.** 0.4

Pages 71–74 Lesson 2-3

1. Sample answer: $y = 1$ **3.** Luisa; Mark did not subtract in a consistent manner when using the slope formula. If $y_2 = 5$ and $y_1 = 4$, then x_2 must be −1 and x_1 must be 2, not vice versa. **5.** $-\frac{1}{2}$

7.

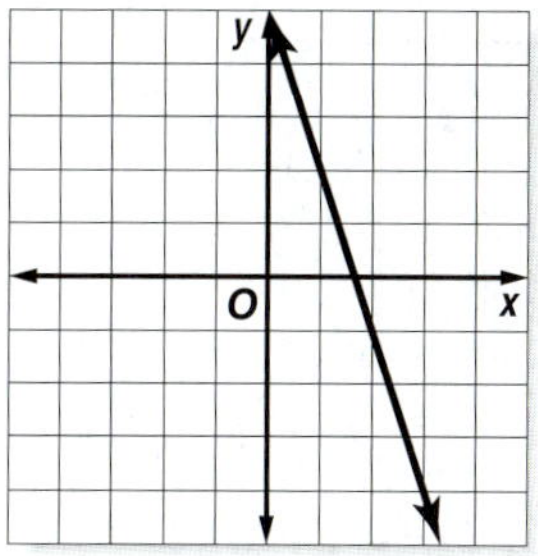

9.

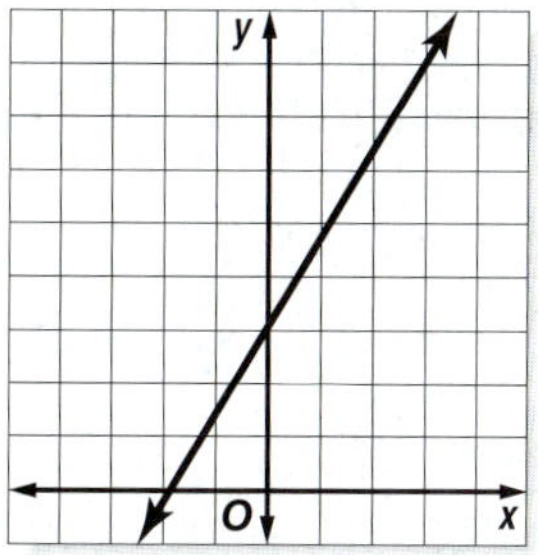

11.

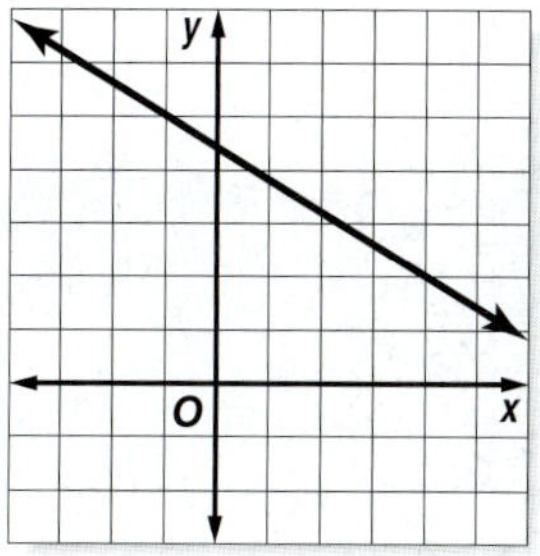

13. 1.25°/hr **15.** $-\frac{5}{2}$ **17.** $\frac{3}{5}$
19. 0 **21.** 8 **23.** −4
25. undefined **27.** 1
29. about 0.6

31.

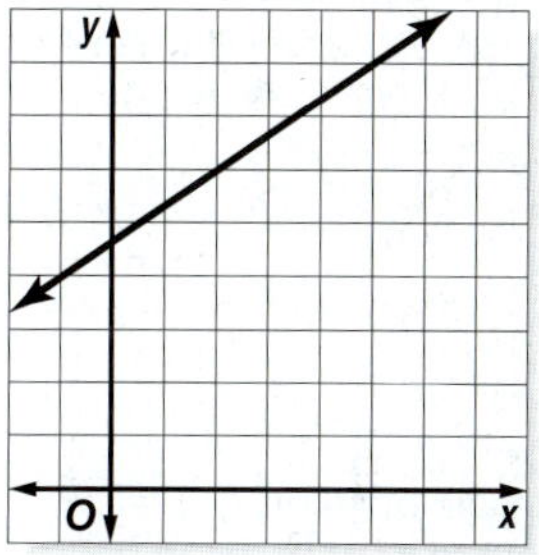

33.

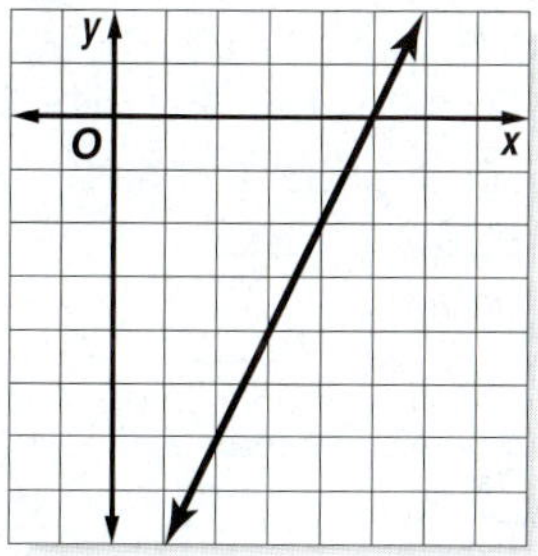

35.

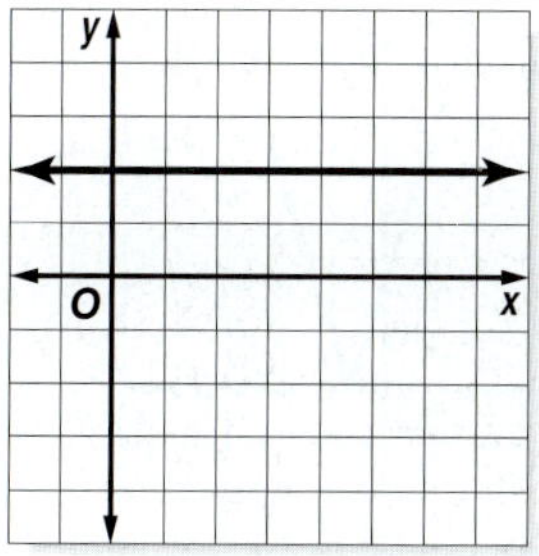

37. about 68 million per year
39. The number of cassette tapes shipped has been decreasing. **41.** 45 mph

43.

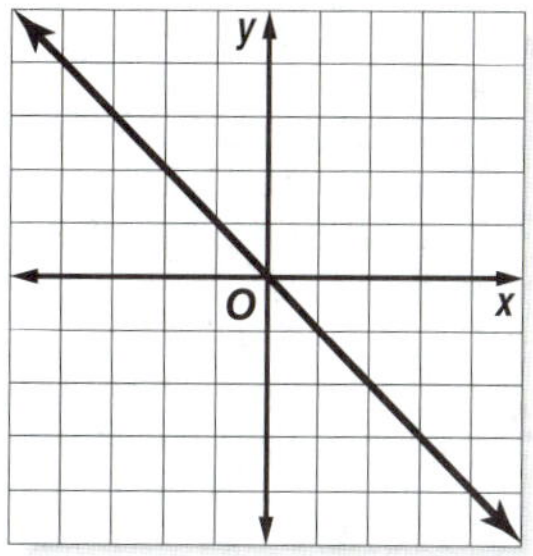

45.

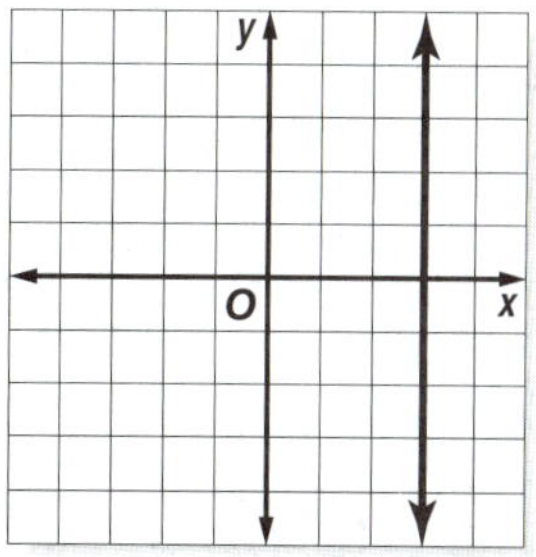

47.

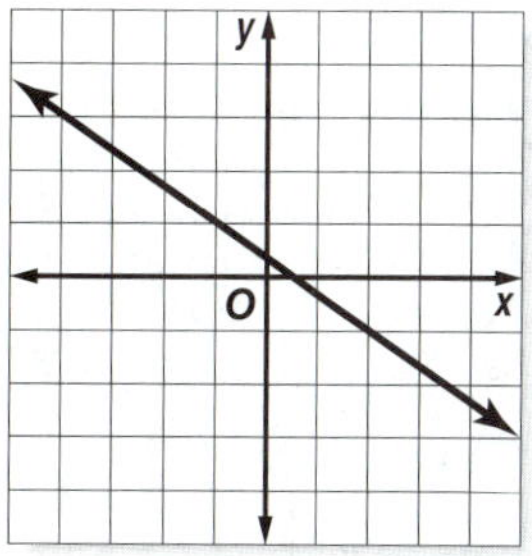

49.

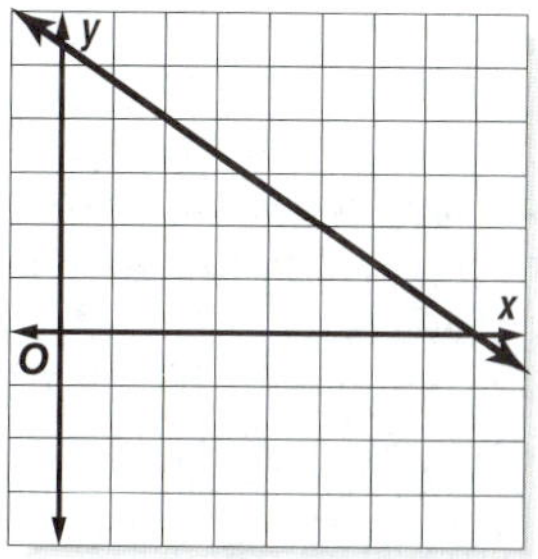

51. Yes; slopes show that adjacent sides are perpendicular.
53. The grade or steepness of a road can be interpreted mathematically as a slope. Answers should include the following.

- Think of the diagram at the beginning of the lesson as being in a coordinate plane. Then the rise is a change in y-coordinates and the horizontal distance is a change in x-coordinates. Thus, the grade is a slope expressed as a percent.
- 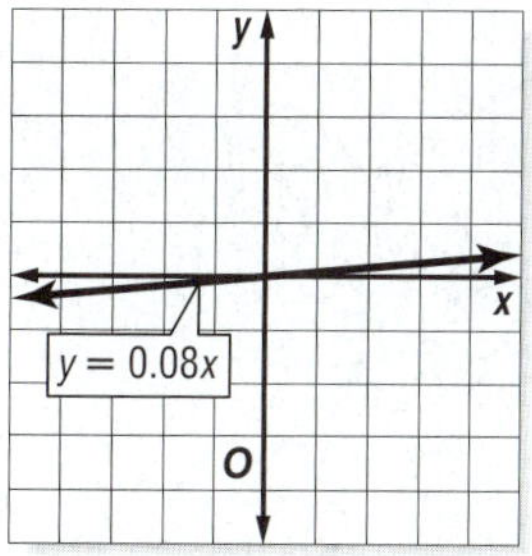

55. D **57.** The graphs have the same y-intercept. As the slopes become more negative, the lines get steeper.

59. $-2, \frac{8}{3}$

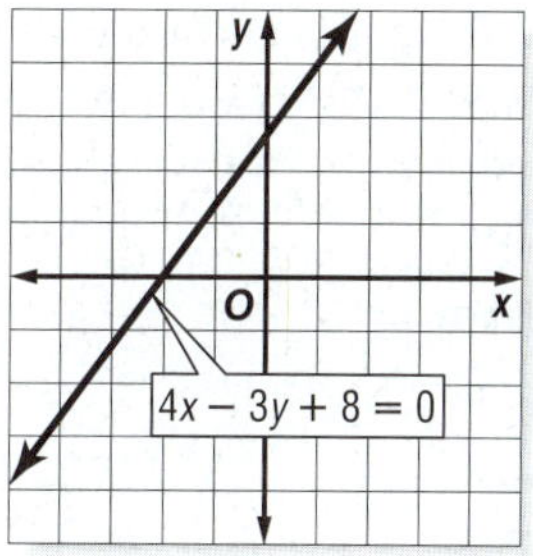

61. −7 **63.** $-\frac{5}{2}$ **65.** $\{x \mid -1 < x < 3\}$ **67.** at least 8 **69.** 9

71. $y = -4x + 2$ **73.** $y = \frac{5}{2}x - \frac{1}{2}$ **75.** $y = -\frac{2}{3}x + \frac{11}{3}$

Page 74 Practice Quiz 1

1. D = {−7, −3, 0, 2}, R = {−2, 1, 2, 4, 5} **3.** $6x + y = 4$

5. 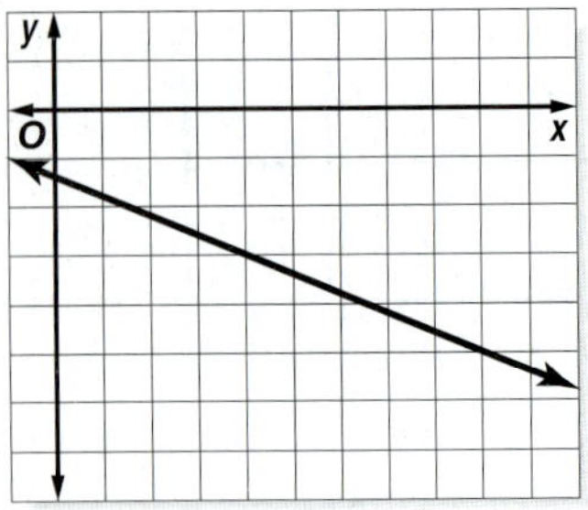

Pages 78–80   Lesson 2-4

1. Sample answer: $y = 3x + 2$ **3.** Solve the equation for y to get $y = \frac{3}{5}x - \frac{2}{5}$. The slope of this line is $\frac{3}{5}$. The slope of a parallel line is the same. **5.** $-\frac{3}{2}, 5$ **7.** $y = -\frac{3}{4}x + 2$ **9.** $y = -\frac{3}{5}x + \frac{16}{5}$ **11.** $y = \frac{5}{4}x + 7$ **13.** $-\frac{2}{3}, -4$ **15.** $\frac{1}{2}, -\frac{5}{2}$ **17.** undefined, none **19.** $y = 0.8x$ **21.** $y = -4$ **23.** $y = 3x - 6$ **25.** $y = -\frac{1}{2}x + \frac{7}{2}$ **27.** $y = -0.5x - 2$ **29.** $y = -\frac{4}{5}x + \frac{17}{5}$ **31.** $y = 0$ **33.** $y = x + 4$ **35.** $y = \frac{2}{3}x + \frac{10}{3}$ **37.** $y = -\frac{1}{15}x - \frac{23}{5}$ **39.** $y = 3x - 2$ **41.** $d = 180c - 360$ **43.** 540° **45.** 10 mi **47.** 68°F **49.** $y = 0.35x + 1.25$ **51.** $y = 2x + 4$ **53.** C **55.** $\frac{x}{\frac{5}{2}} - \frac{y}{5} = 1$ **57.** -2 **59.** 0 **61.** ∅ **63.** $\{r \mid r \geq 6\}$ **65.** 6.5 **67.** 5.85

Pages 83–86   Lesson 2-5

1. d **3.** Sample answer using (4, 130.0) and (6, 140.0): $y = 5x + 110$

5a. **Cable Television**

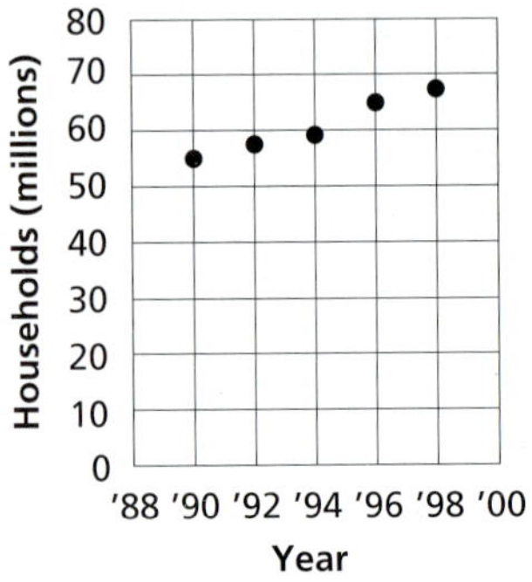

5b. Sample answer using (1992, 57) and (1998, 67): $y = 1.67x - 3269.64$ **5c.** Sample answer: about 87 million

7a. **2000–2001 Detroit Red Wings**

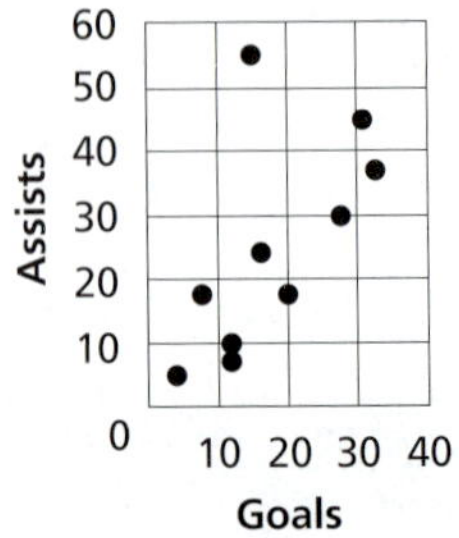

7b. Sample answer using (4, 5) and (32, 37): $y = 1.14x + 0.44$
7c. Sample answer: about 13

9a. **Broadway Play Revenue**

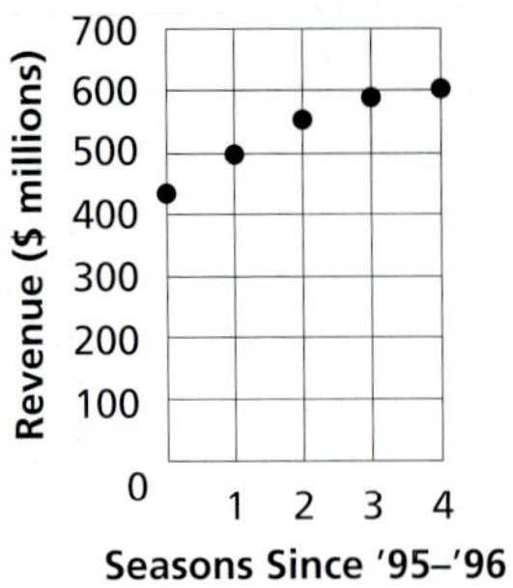

9b. Sample answer using (1, 499) and (3, 588): $y = 44.5x + 454.5$, where x is the number of seasons since 1995–1996 **9c.** Sample answer: about $1078 million or $1.1 billion **11.** Sample answer: $1091 **13.** Sample answer: Using the data for August and November, a prediction equation for Company 1 is $y = -0.86x + 25.13$, where x is the number of months since August. The negative slope suggests that the value of Company 1's stock is going down. Using the data for October and November, a prediction equation for Company 2 is $y = 0.38x + 31.3$, where x is the number of months since August. The positive slope suggests that the value of Company 2's stock is going up. Since the value of Company 1's stock appears to be going down, and the value of Company 2's stock appears to be going up, Della should buy Company 2.

15. **World Cities**

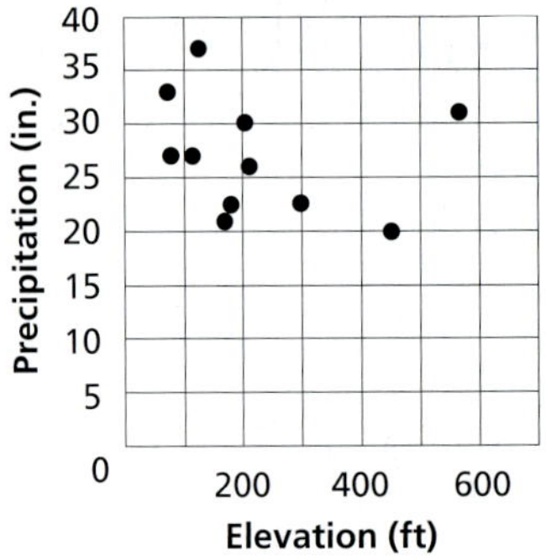

17. Sample answer: about 23 in. **19.** Sample answer: Using (1975, 62.5) and (1995, 81.7): 96.1% **23.** D **25.** 1988, 1993, 1998; 247, 360.5, 461 **27.** 354 **29.** $y = 21.4x - 42{,}294.03$ **31.** $y = 4x + 6$ **33.** 3 **35.** $\frac{29}{3}$ **37.** $\{x \mid x < -7 \text{ or } x > -1\}$ **39.** 11 **41.** $\frac{2}{3}$

Pages 92–95   Lesson 2-6

1. Sample answer: $[\![1.9]\!] = 1$ **3.** Sample answer: $f(x) = |x - 1|$ **5.** S

7. D = all reals, R = all integers

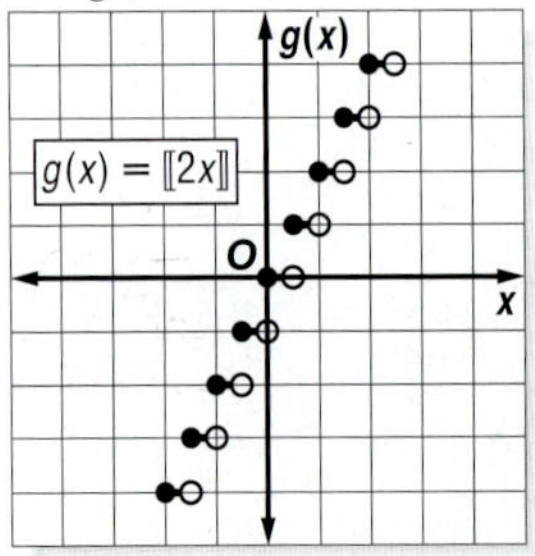

9. D = all reals, R = all nonnegative reals

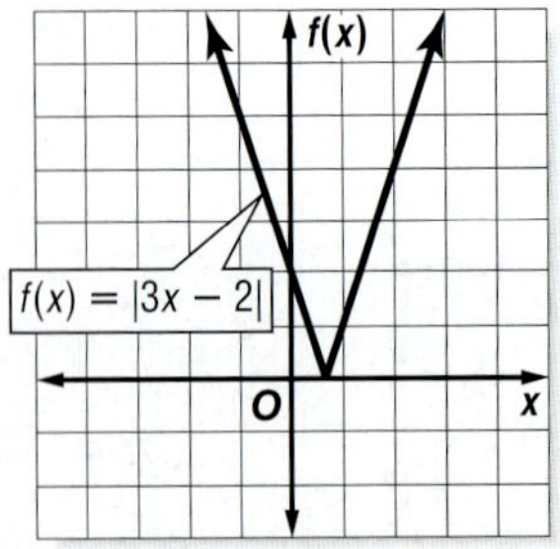

11. D = all reals, R = all reals

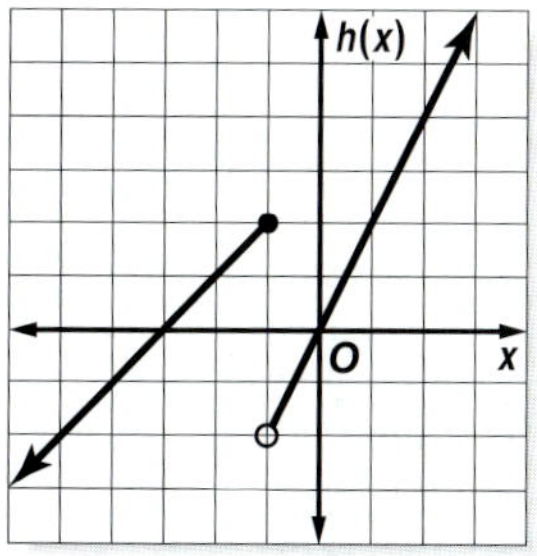

13.

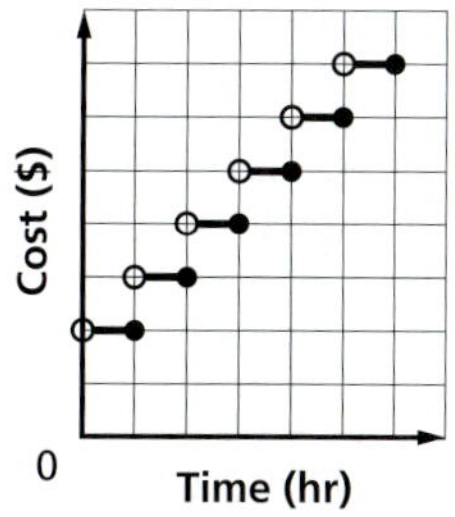

15. C **17.** S **19.** A **21.**

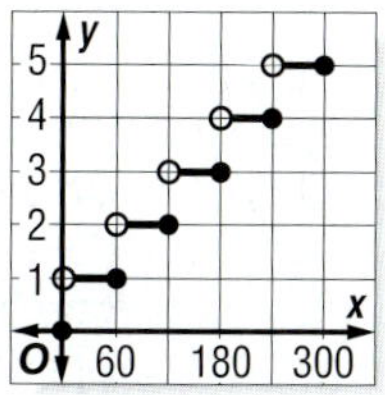

23. $1.00

25. D = all reals, R = all integers

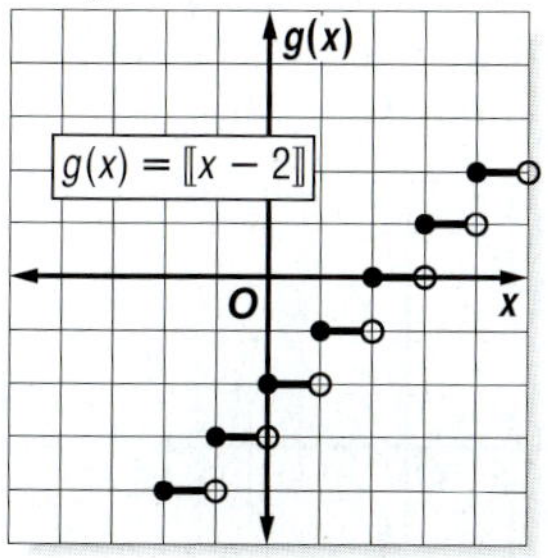

27. D = all reals, R = {3*a* | *a* is an integer.}

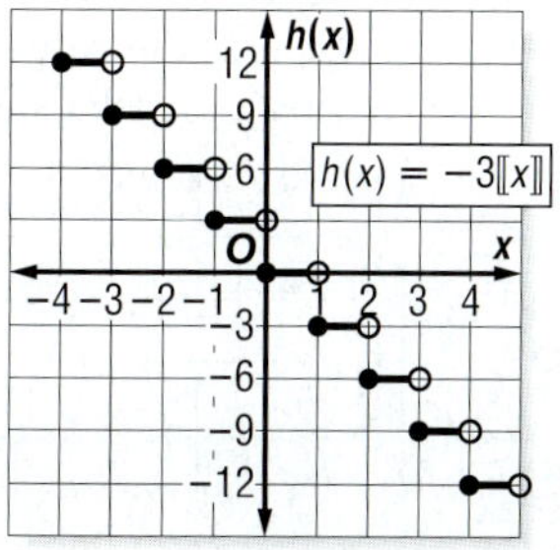

29. D = all reals, R = all integers

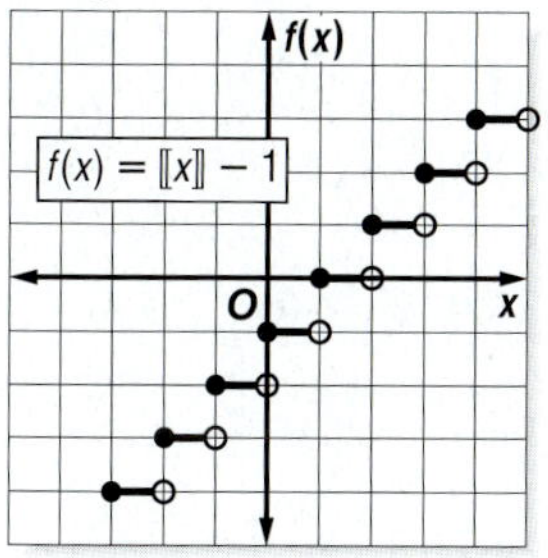

31. D = all reals, R = all nonnegative reals

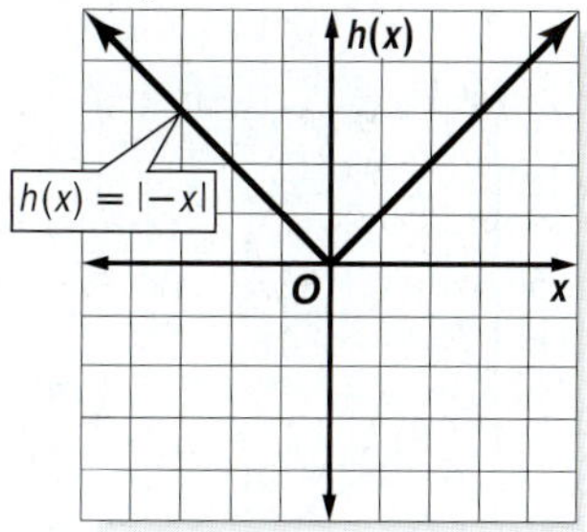

33. D = all reals, R = $\{y \mid y \geq -4\}$

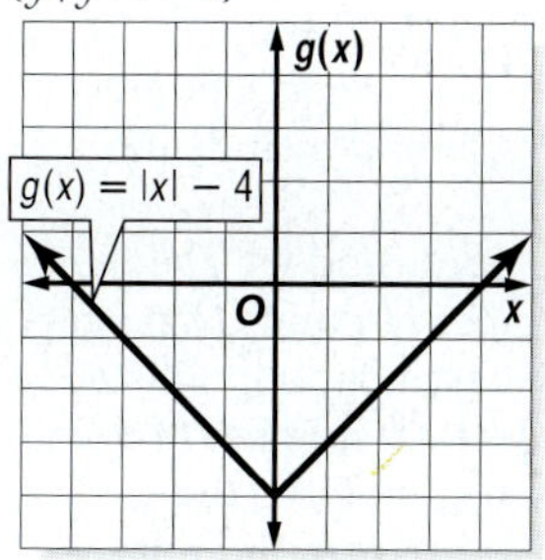

35. D = all reals, R = all nonnegative reals

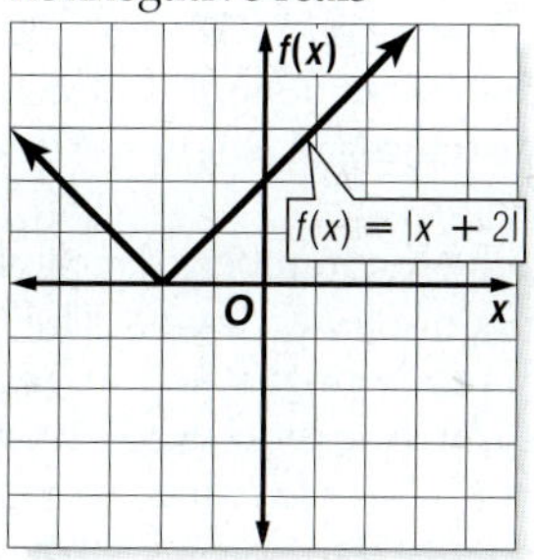

37. D = all reals, R = all nonnegative reals

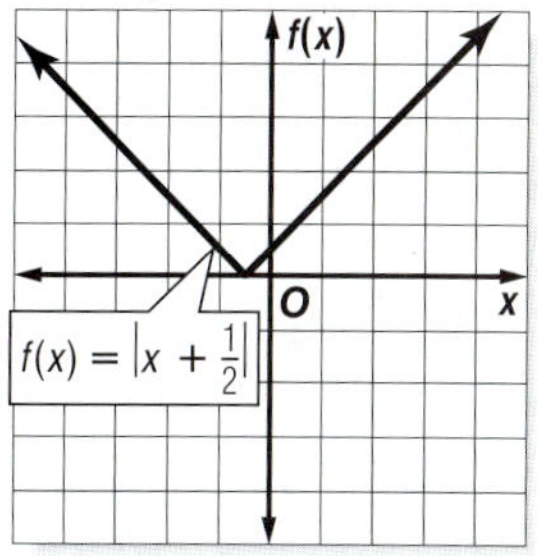

39. D = $\{x \mid x < -2 \text{ or } x > 2\}$, R = $\{-1, 1\}$

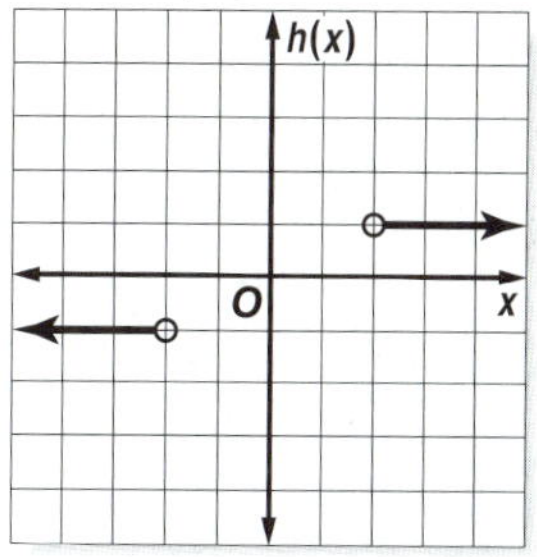

41. D = all reals, R = $\{y \mid y < 2\}$

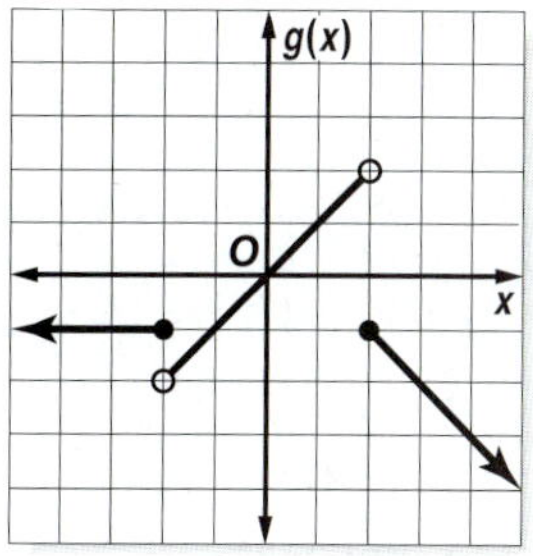

43. D = all reals, R = all whole numbers

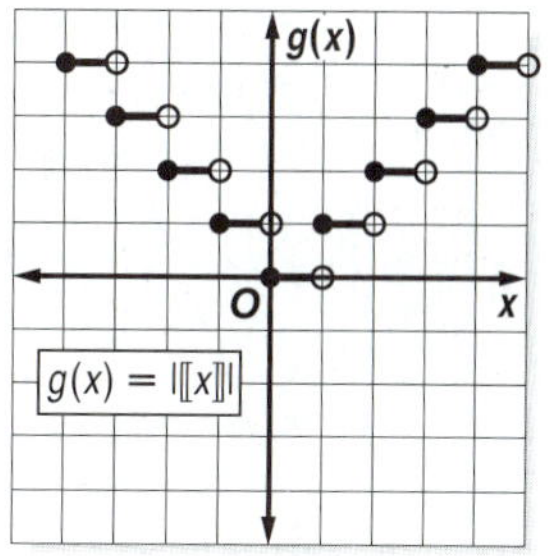

45. $f(x) = |x - 2|$

47.

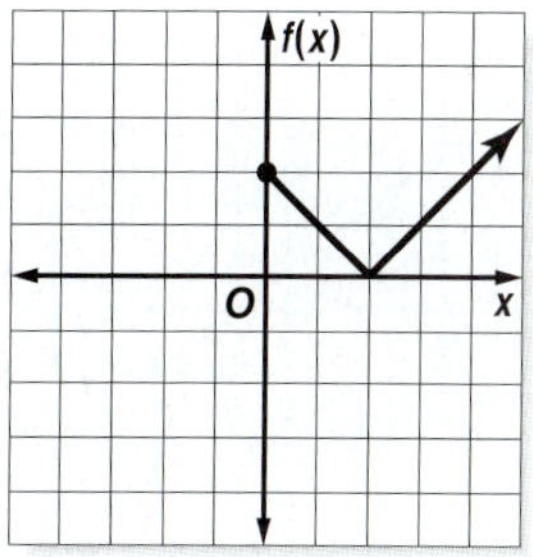

49.

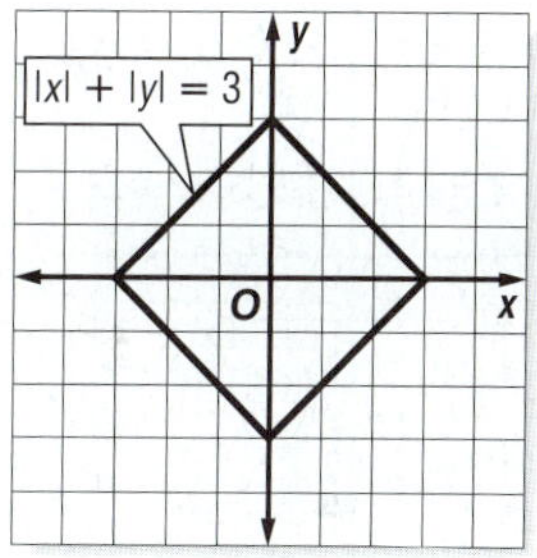

51. B **53.**

Life Expectancy

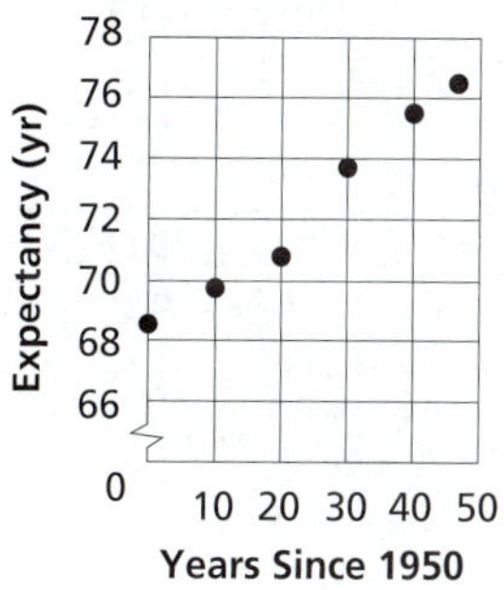

55. Sample answer: 78.7 yr **57.** $y = x - 2$

59. $\left\{y \mid y > \frac{5}{6}\right\}$

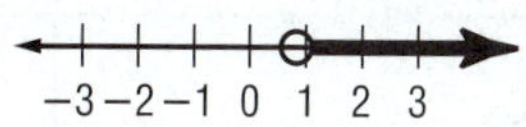

61. no **63.** yes **65.** yes

Page 95 Practice Quiz 2

1. $y = -\frac{2}{3}x + \frac{11}{3}$ **3.** Sample answer using (66, 138) and (74, 178): $y = 5x - 192$ **5.** D = all reals, R = nonnegative reals

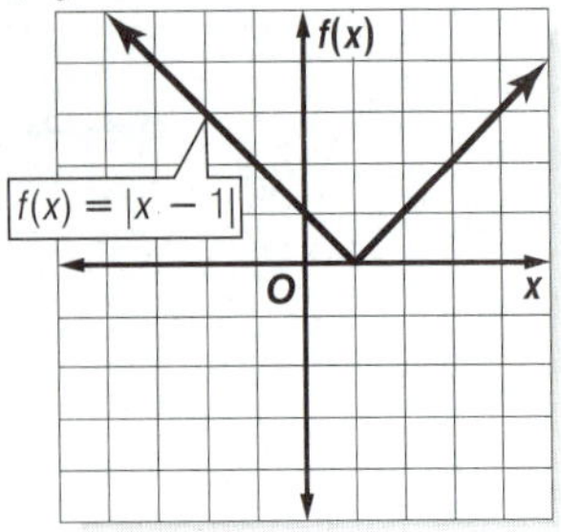

Pages 98–99 Lesson 2-7

1. $y \leq -3x + 4$ **3.** Sample answer: $y \geq |x|$

5.

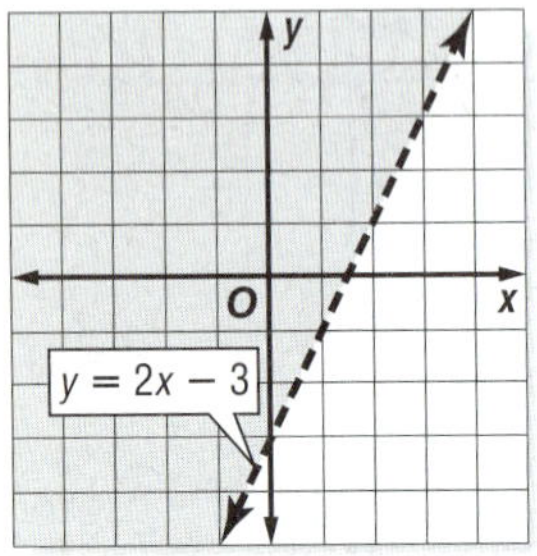

7.

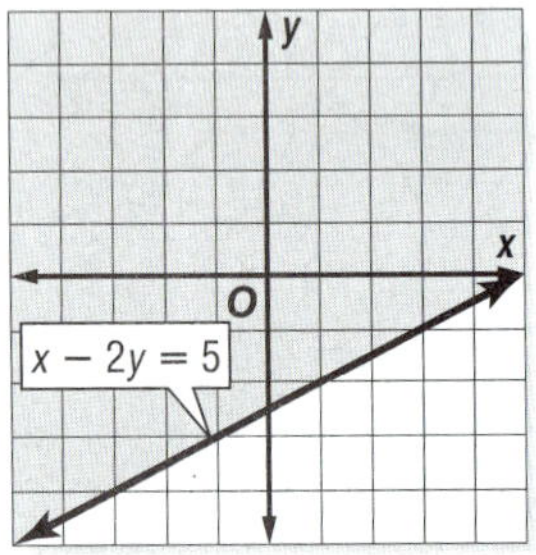

9.

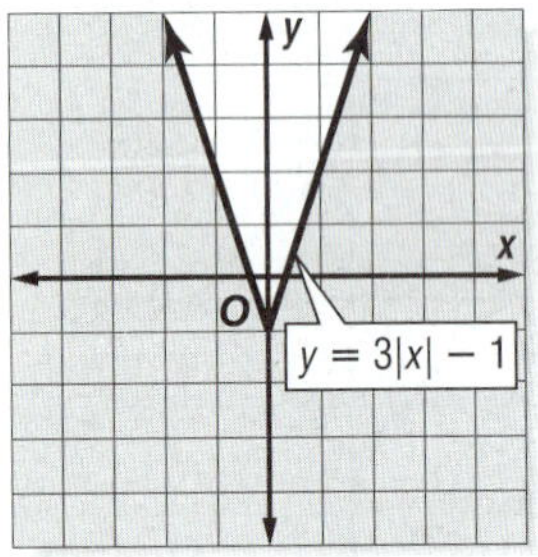

11.

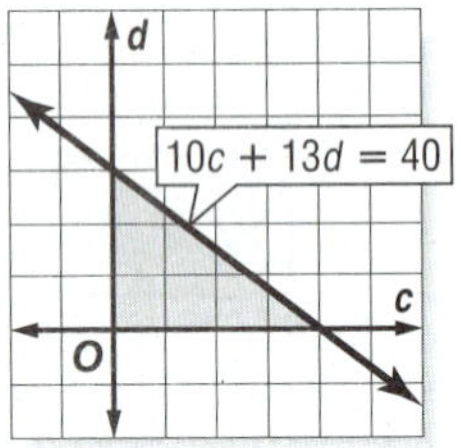

13.

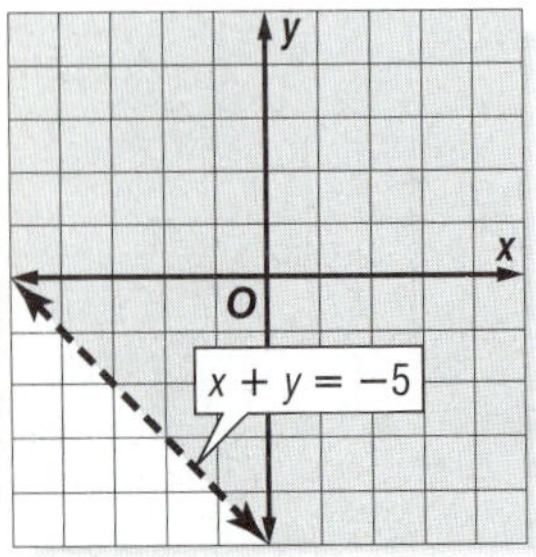

15.

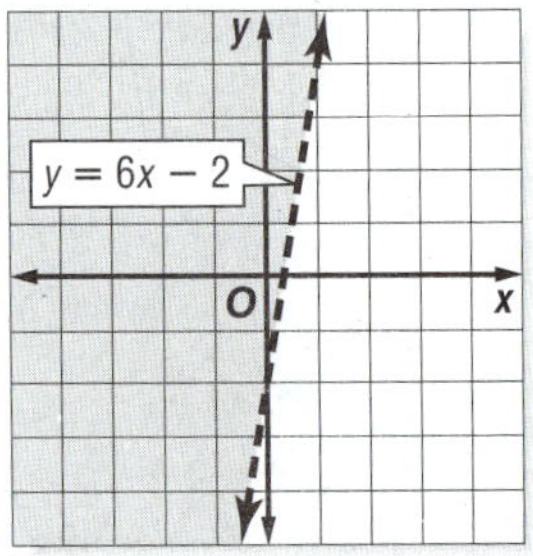

17.

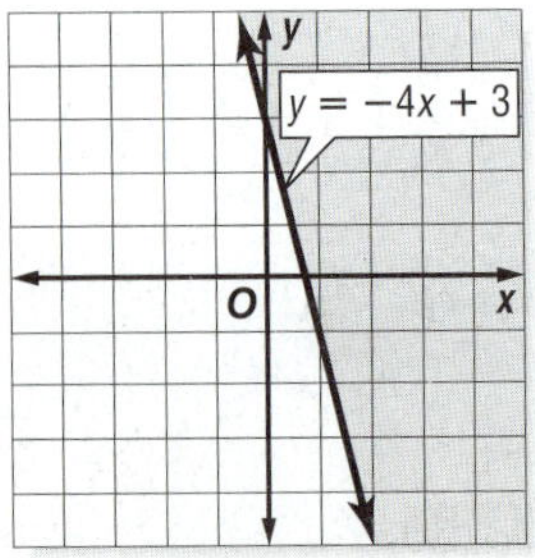

19.

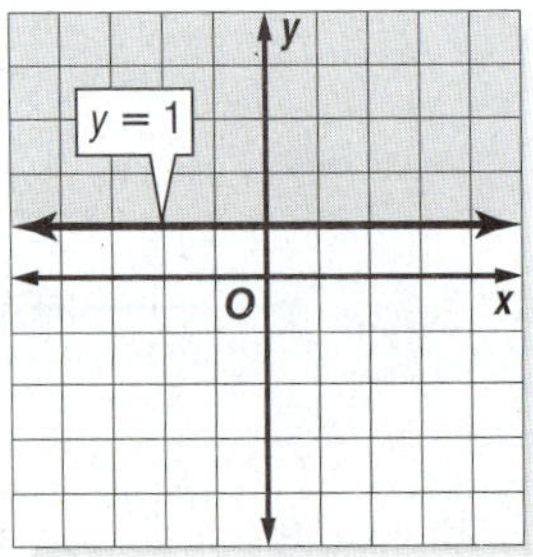

21.

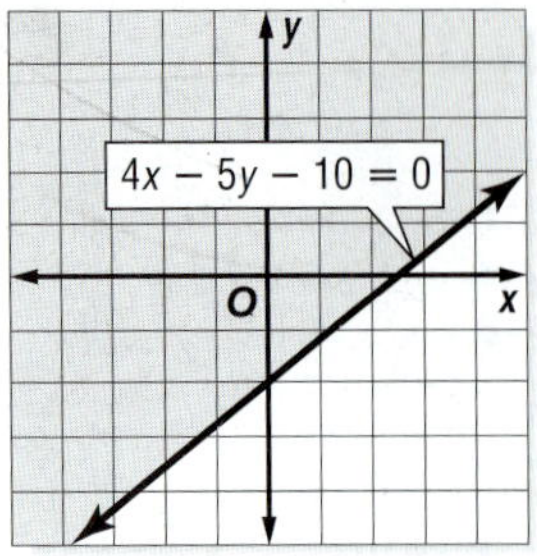

23.

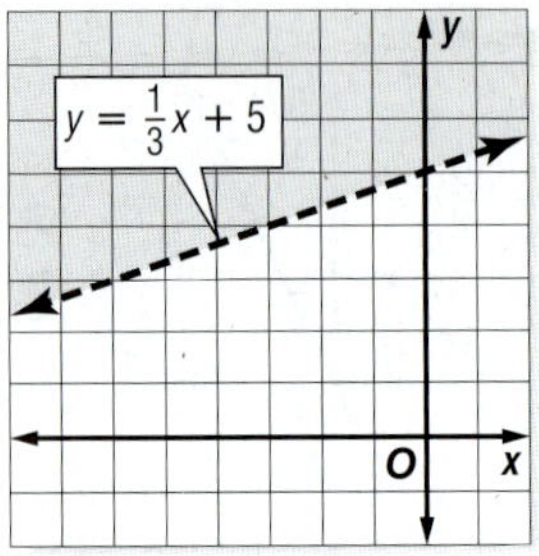

25.

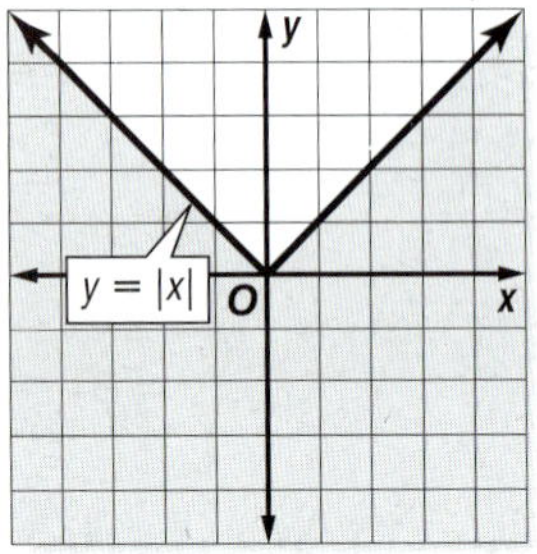

27.

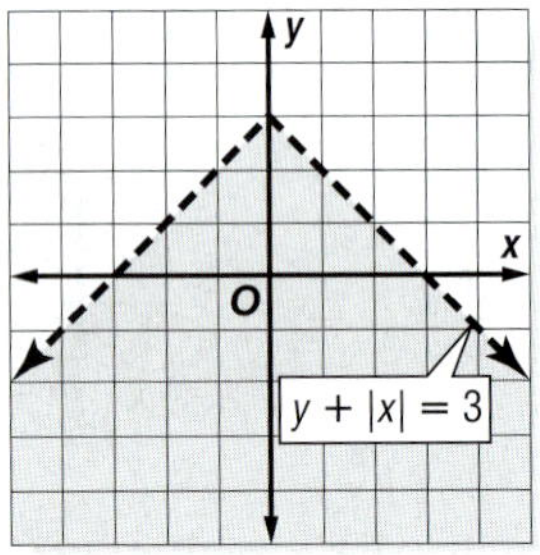

29.

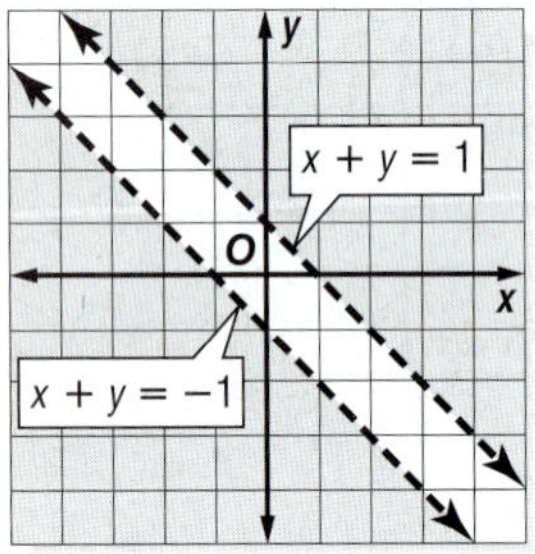

31. $x < -2$

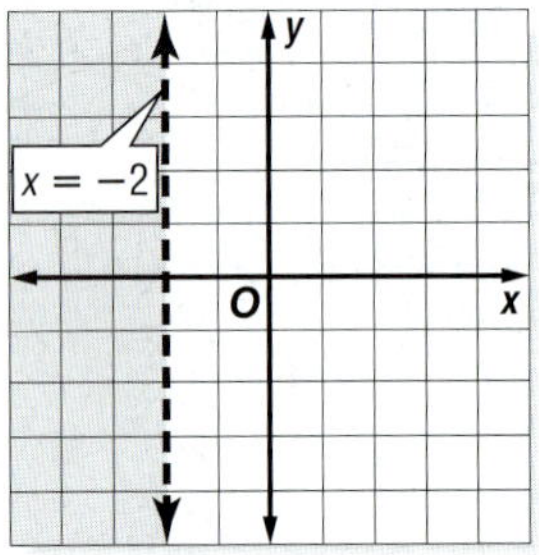

33.

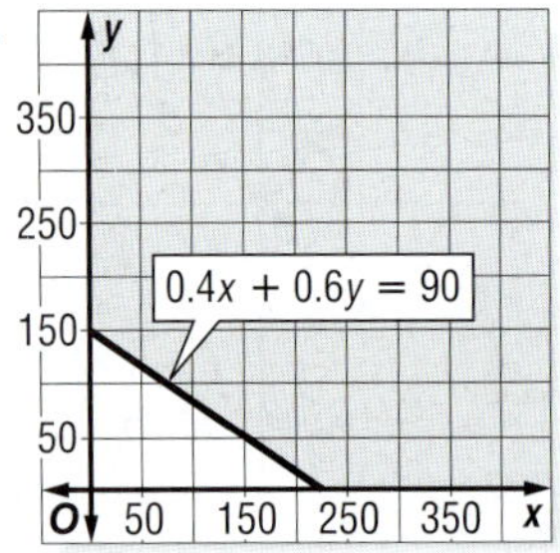

35. $4a + 3s \geq 2000$ **37.** yes **39.** yes **41.** Linear inequalities can be used to track the performance of players in fantasy football leagues. Answers should include the following.

- Let x be the number of receiving yards and let y be the number of touchdowns. The number of points Dana gets from receiving yards is $5x$ and the number of points he gets from touchdowns is $100y$. His total number of points is $5x + 100y$. He wants at least 1000 points, so the inequality $5x + 100y \geq 1000$ represents the situation.

- 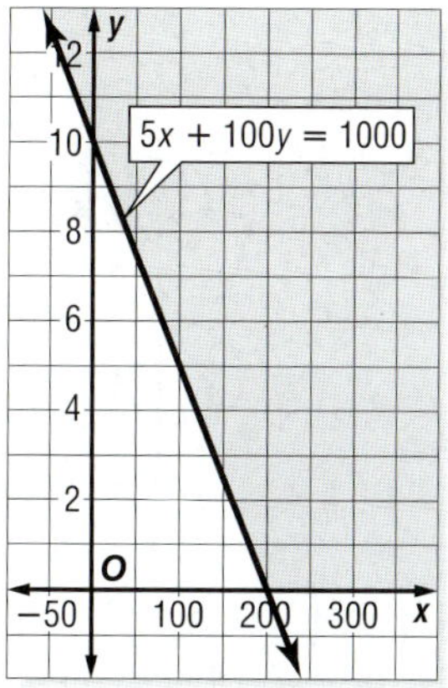

- the first one

43. B 45. 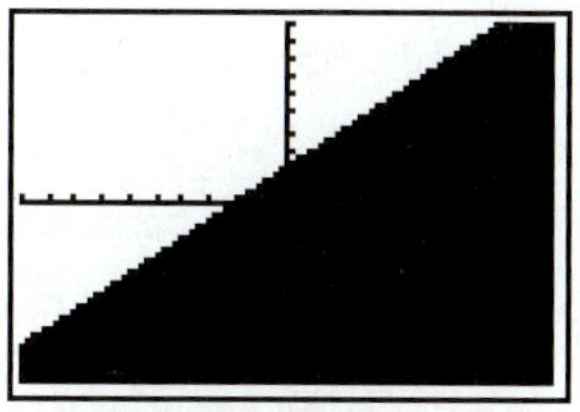

[−10, 10] scl: 1 by [−10, 10] scl: 1

47. 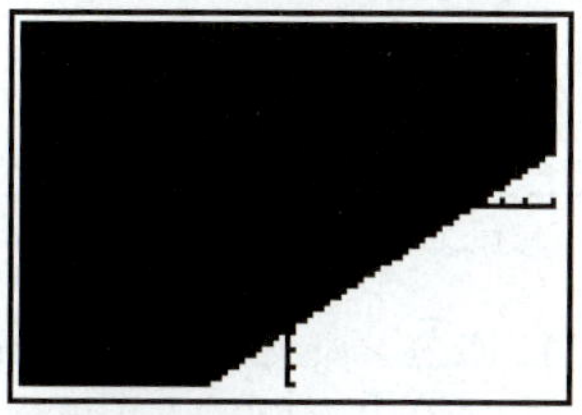

[−10, 10] scl: 1 by [−10, 10] scl: 1

49. D = all reals, R = $\{y \mid y \geq -1\}$

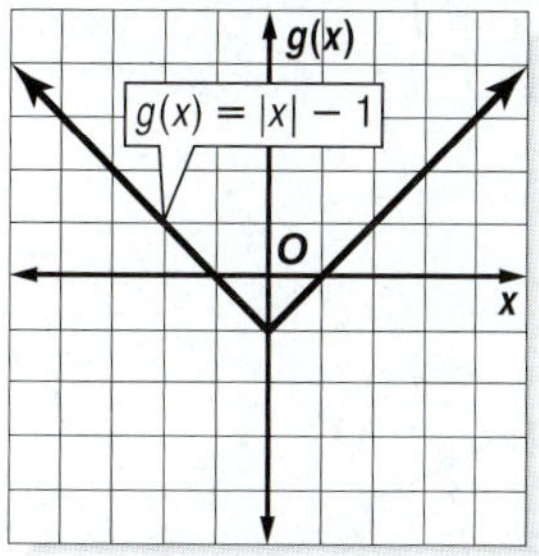

51. **Sales vs. Experience**

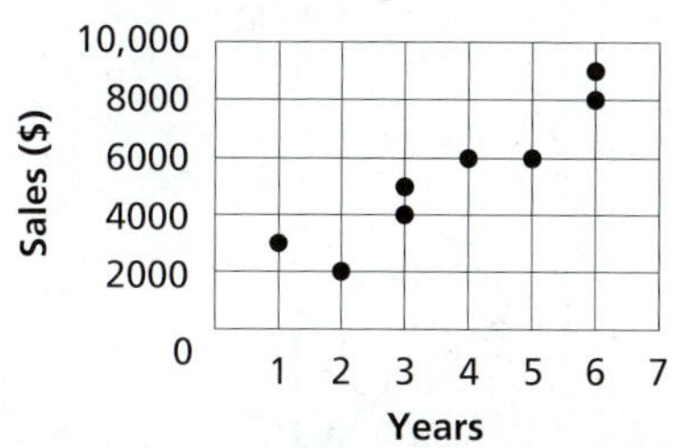

53. Sample answer: $10,000 55. 3

Pages 100–104 Chapter 2 Study Guide and Review

1. identity 3. standard 5. domain 7. slope

9. D = $\{-2, 2, 6\}$, R = $\{1, 3\}$; yes

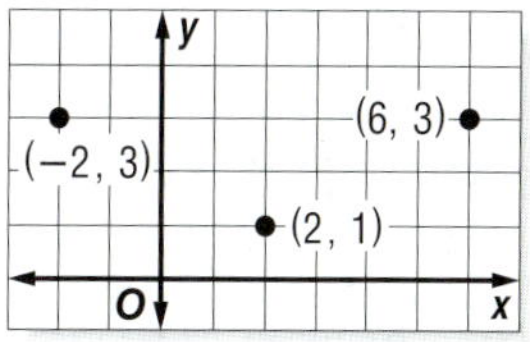

11. D = all reals, R = all reals; yes

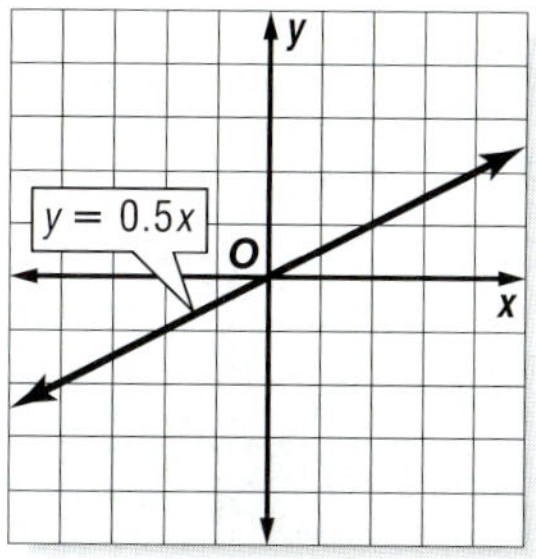

13. 21 15. $5y - 9$ 17. No; x has an exponent other than 1. 19. No; x is inside a square root. 21. $5x + 2y = -4$; 5, 2, −4

23. −4, −20

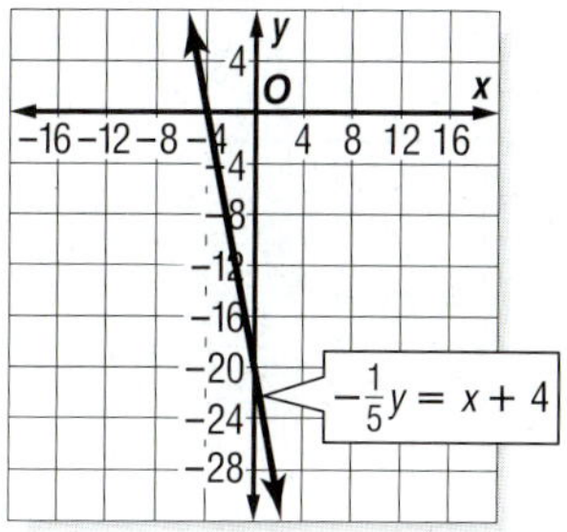

25. 9, −9

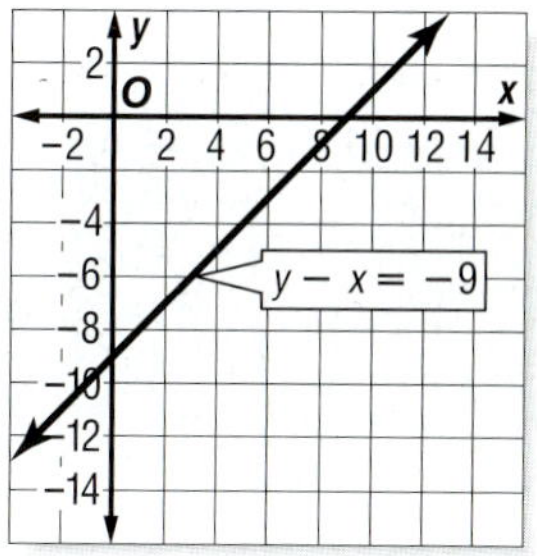

27. $-\frac{3}{11}$

29. 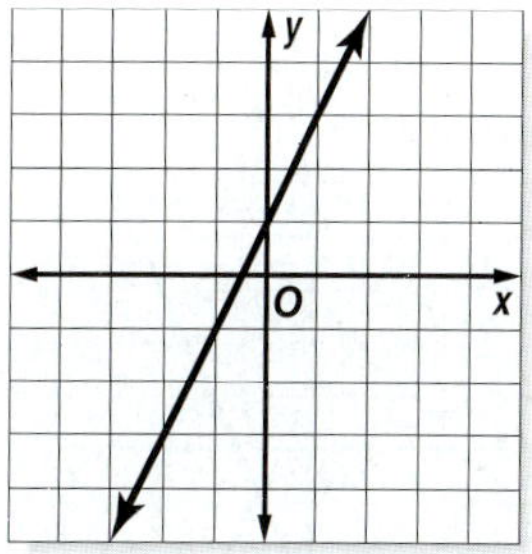

31. 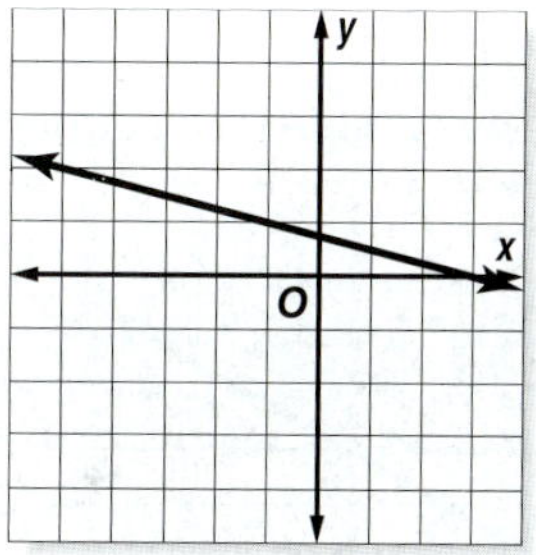

33. 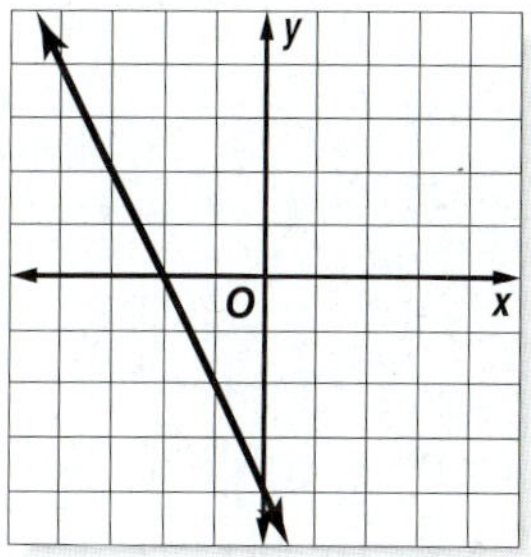

35. 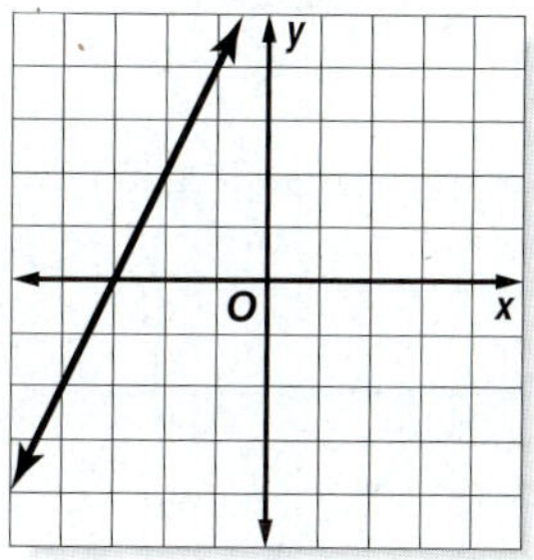

37. $y = -\frac{5}{3}x - 3$ 39. $y = -\frac{3}{4}x + \frac{17}{4}$ 41. Sample answer using (1980, 29.3) and (1990, 33.6): $y = 0.43x - 822.1$

43. D = all reals, R = all integers

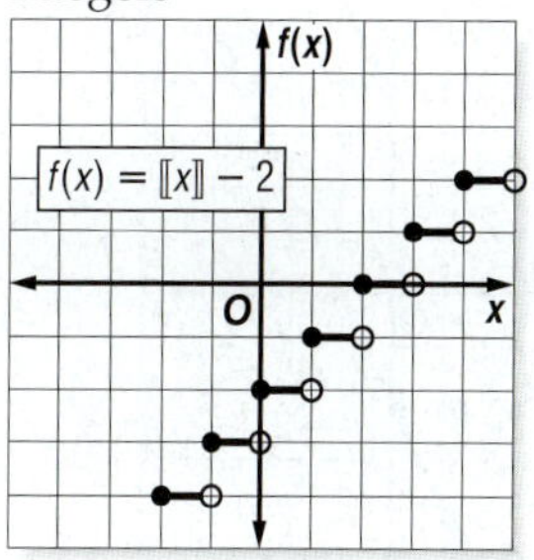

45. D = all reals, R = $\{y \mid y \geq 4\}$

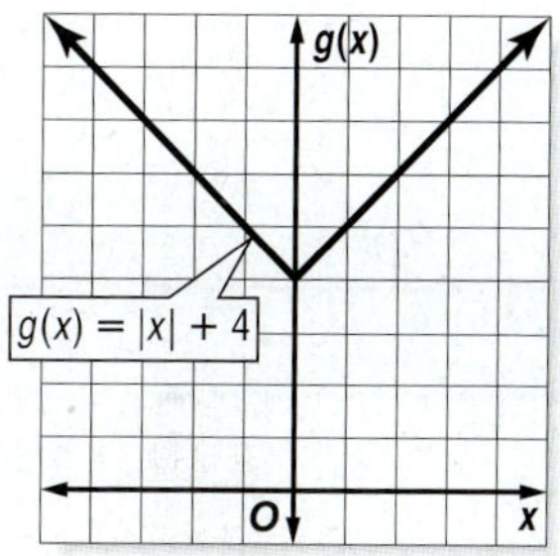

47. D = all reals, R = $\{y \mid y \leq 0 \text{ or } y = 2\}$

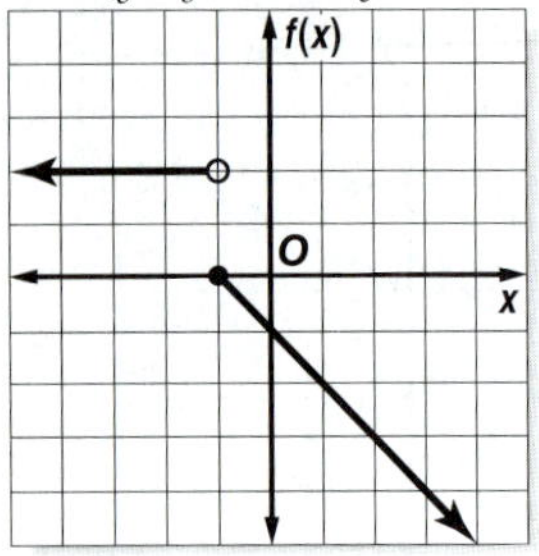

49.

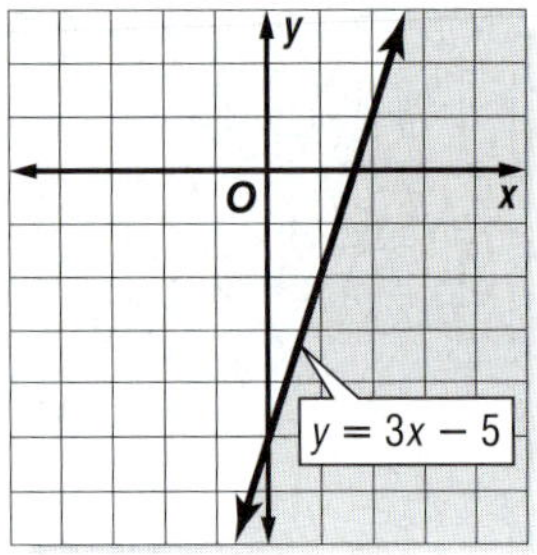

51.

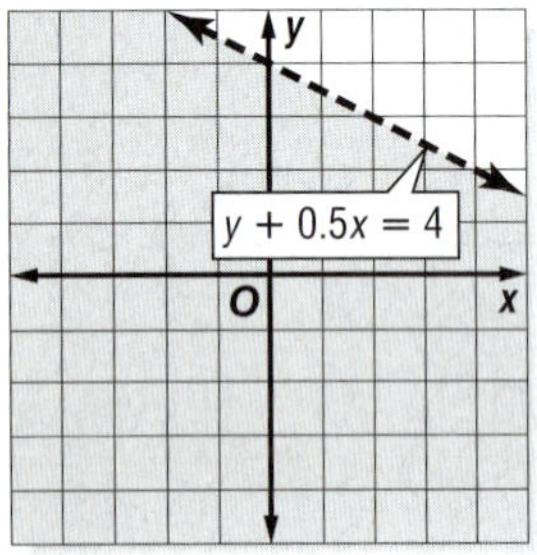

53.

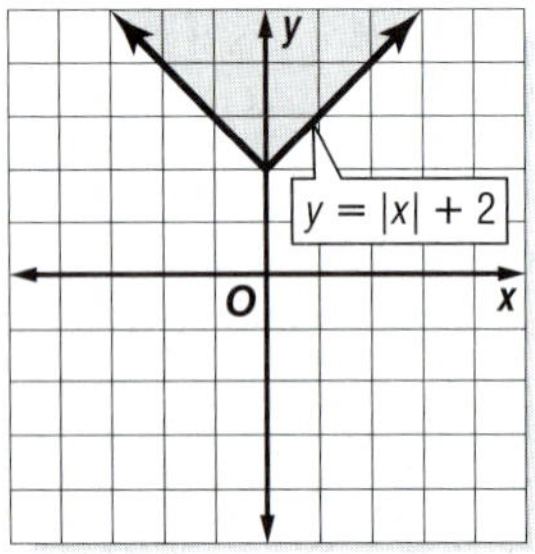

Chapter 3 Systems of Equations and Inequalities

Page 109 Chapter 3 Getting Started

1.

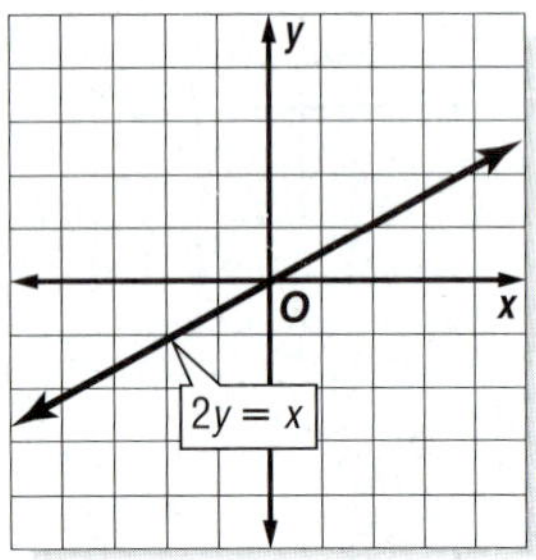

3.

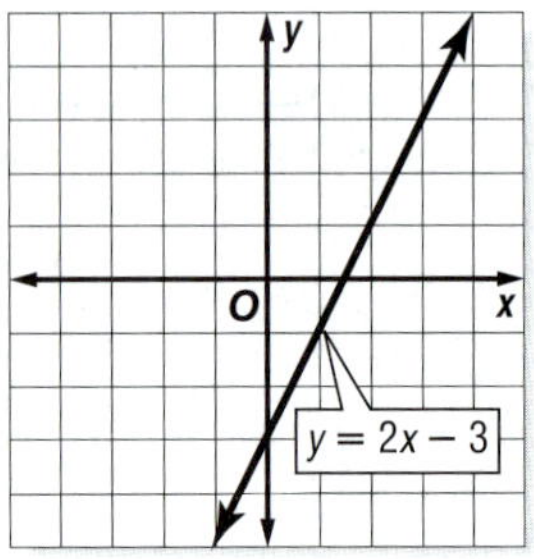

5.

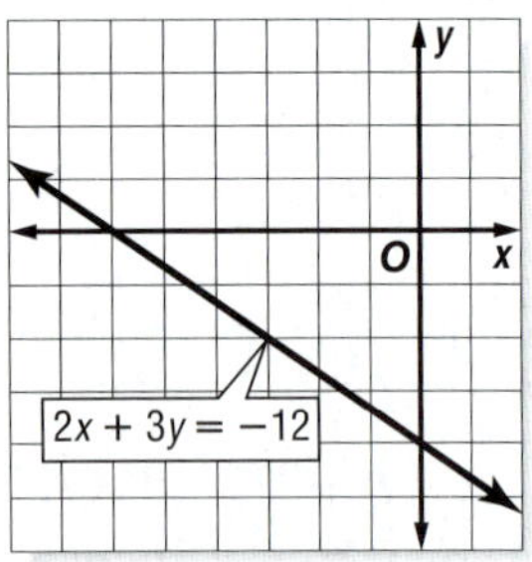

7. $y = -2x$ 9. $y = 6 - 3x$
11. $y = 2 - 6x$

13.

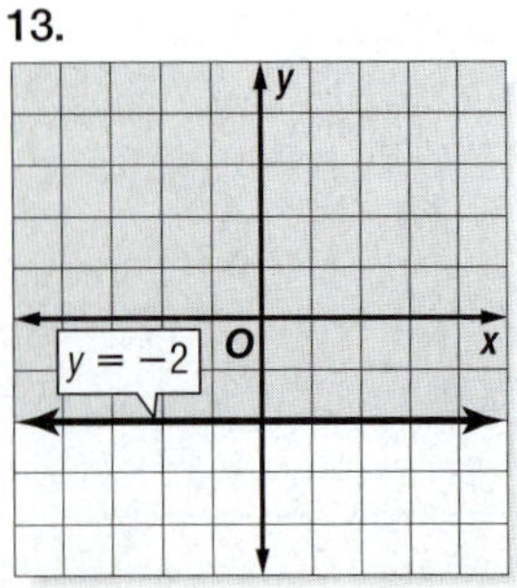

15.

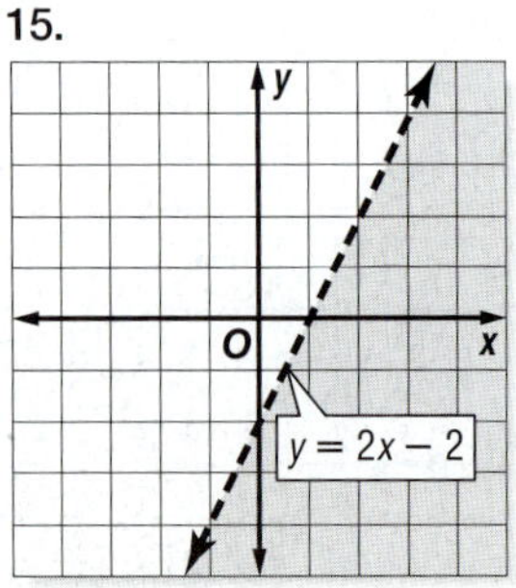

17.

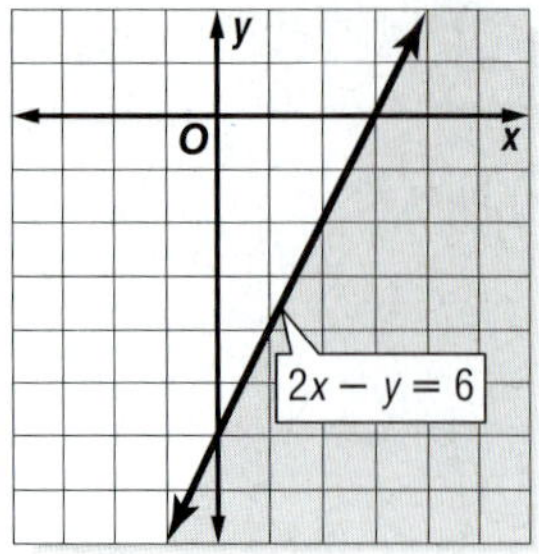

19. −9 21. 0 23. −22

Pages 112–115 Lesson 3-1

1. Two lines cannot intersect in exactly two points.
3. A graph is used to estimate the solution. To determine that the point lies on both lines, you must check that it satisfies both equations.

5.

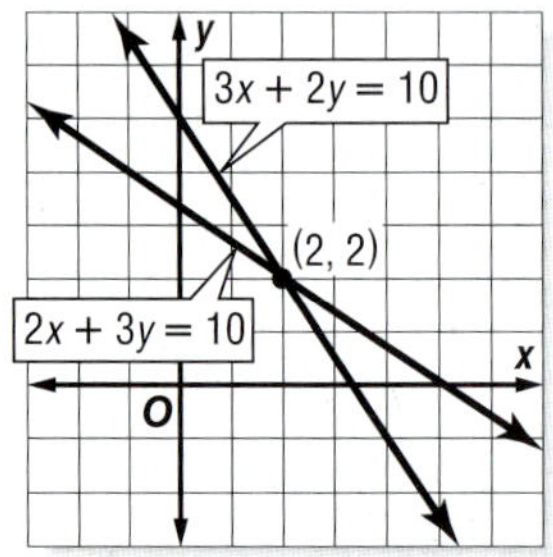

7. consistent and independent

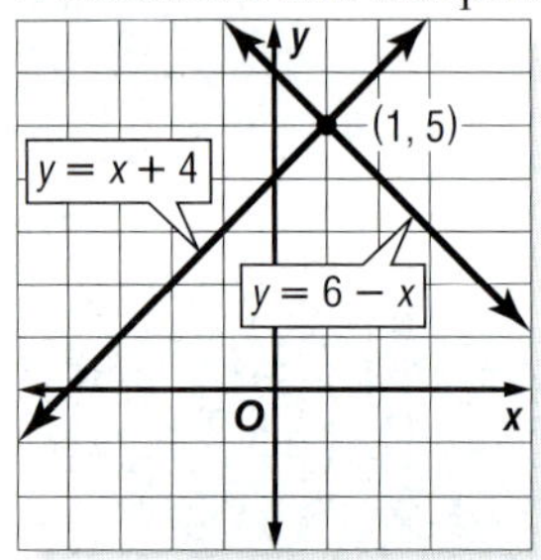

9. consistent and dependent

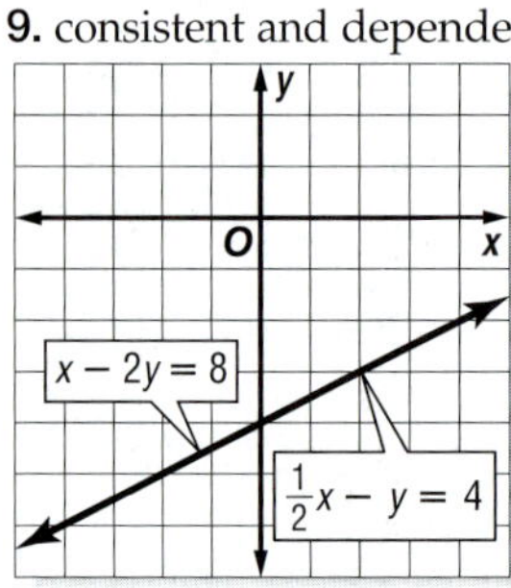

11. The cost is \$5.60 for both stores to develop 30 prints.

13.

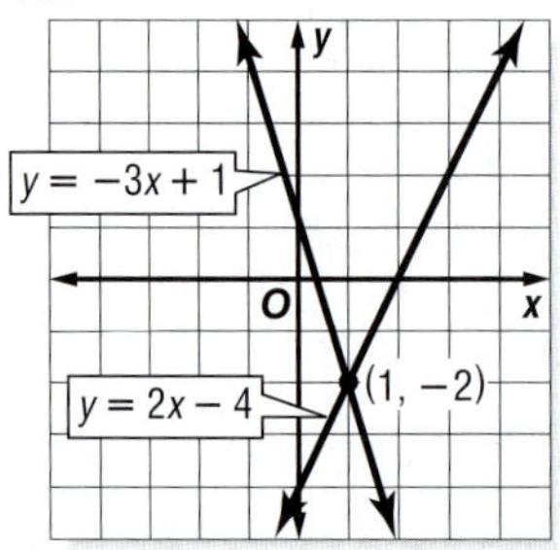

15.

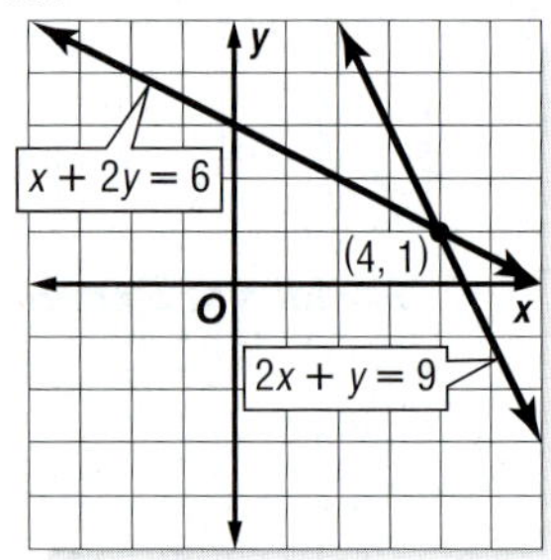

17.

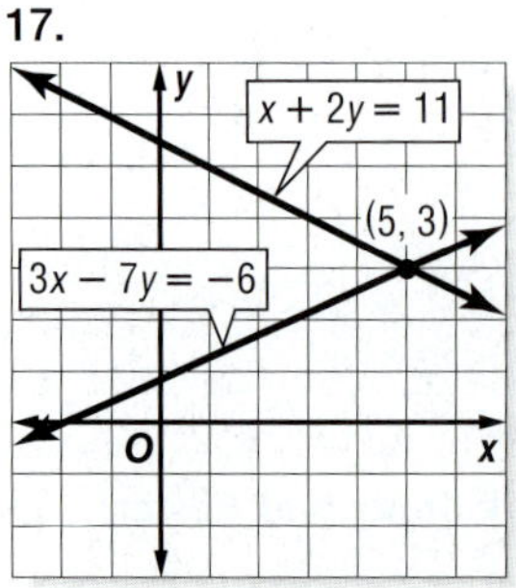

19.

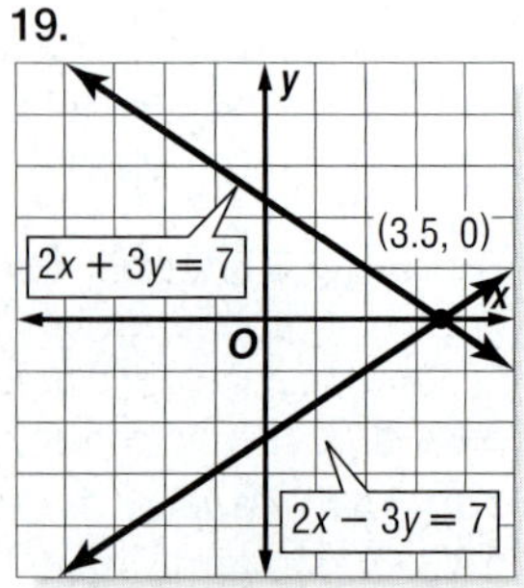

21.

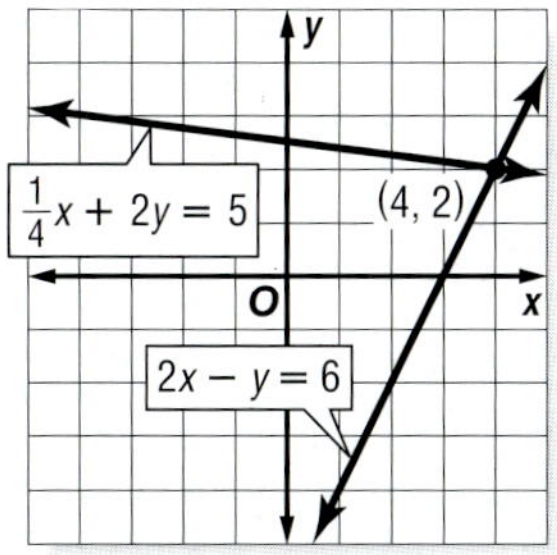

23.

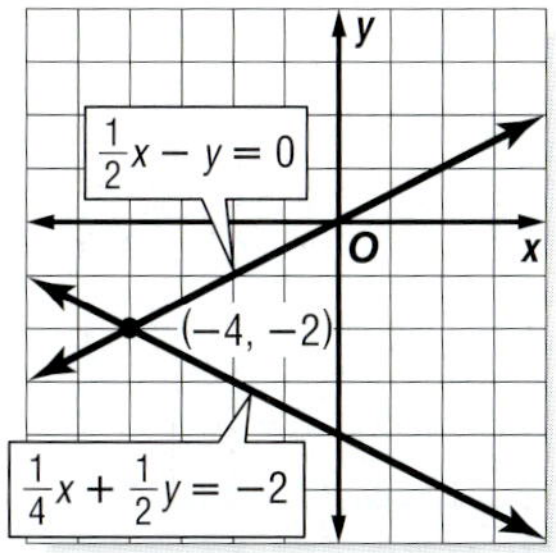

25. inconsistent

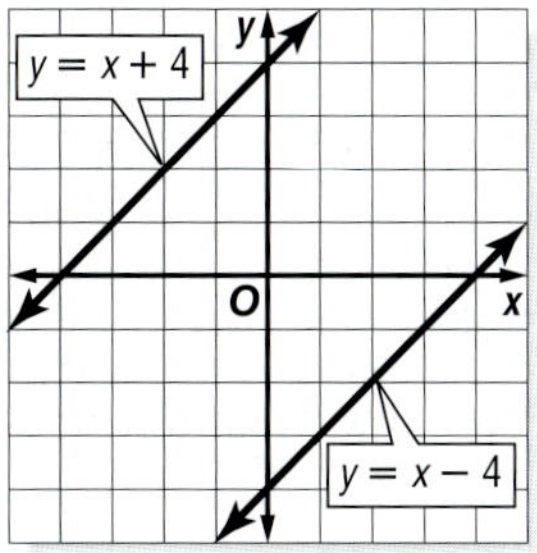

27. consistent and independent

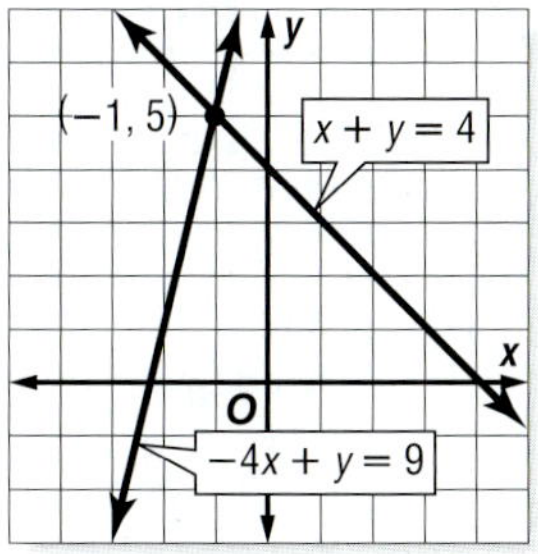

29. inconsistent

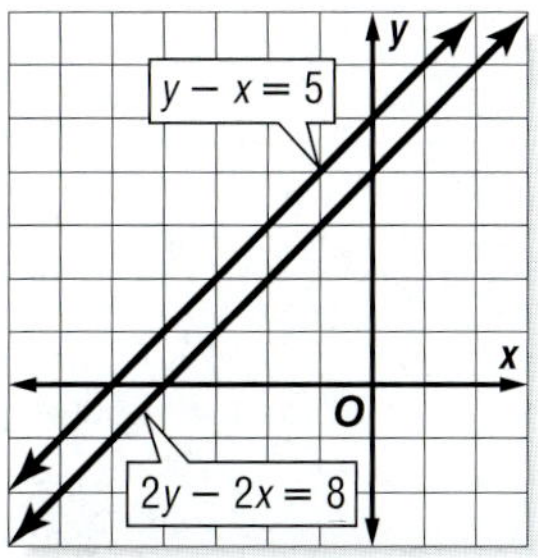

31. consistent and independent

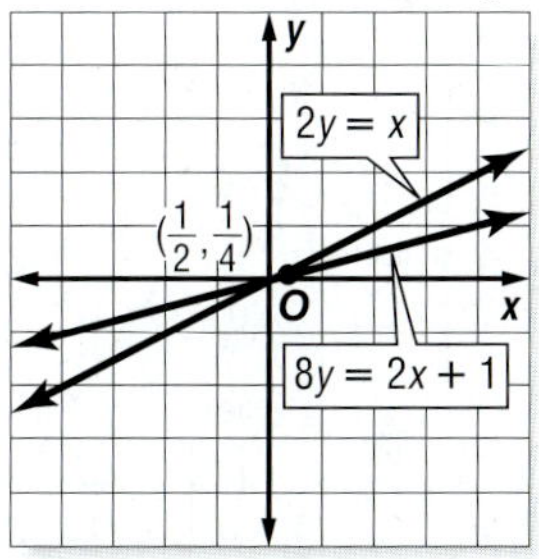

33. consistent and independent

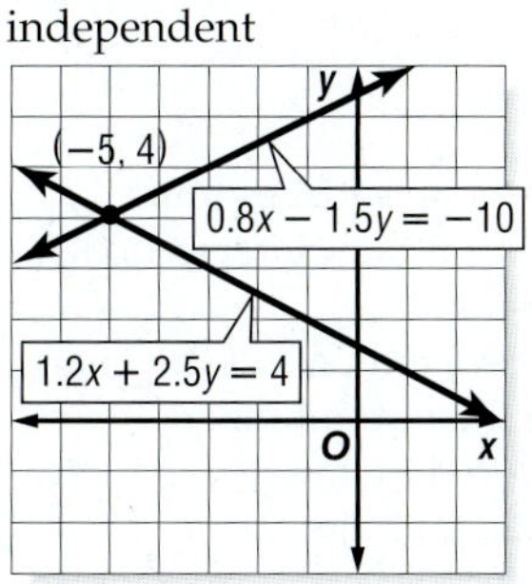

35. inconsistent

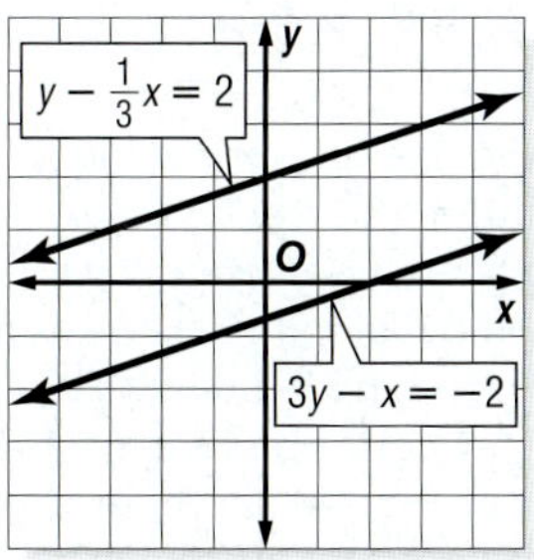

37. (−3, 1) **39.** $y = 52 + 0.23x$, $y = 80$ **41.** Deluxe Plan **43.** Supply, 300,000; demand, 200,000; prices will tend to fall. **45.** $y = 304x + 15{,}982$, $y = 98.6x + 18{,}976$ **47.** FL will probably be ranked third by 2020. The graphs intersect in the year 2015, so NY will still have a higher population in 2010, but FL will have a higher population in 2020.

49. You can use a system of equations to track sales and make predictions about future growth based on past performance and trends in the graphs. Answers should include the following.

- The coordinates (6, 54) represent that 6 years after 1999 both the in-store sales and online sales will be $54,000.
- The in-store sales and the online sales will never be equal and in-store sales will continue to be higher than online sales.

51. C **53.** (−5.56, 12) **55.** no solution **57.** (2.64, 42.43)

59.

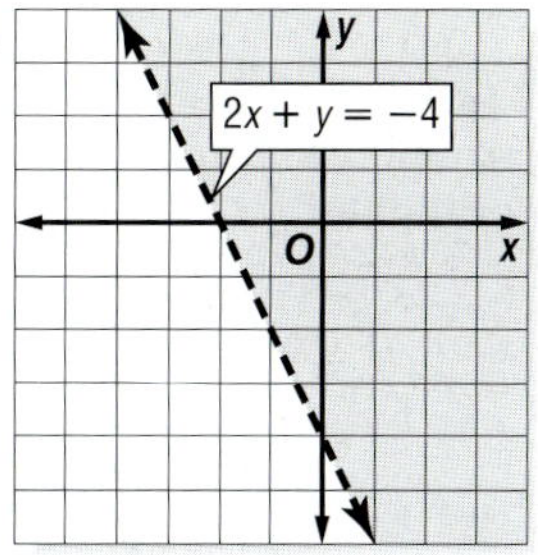

61. A **63.** P **65.** {−15, 9} **67.** {−2, 3} **69.** {9} **71.** $x^2 - 6$ **73.** $\frac{z}{3} + 1$ **75.** $9y + 1$ **77.** $12x + 18y - 6$ **79.** $x + 4y$

Pages 119–122 Lesson 3-2

3. Vincent; Juanita subtracted the two equations incorrectly; $-y - y = -2y$, not 0. **5.** (1, 3) **7.** (5, 2) **9.** (6, −20) **11.** $\left(3\frac{1}{3}, 2\frac{2}{3}\right)$ **13.** (9, 5) **15.** (3, −2) **17.** no solution **19.** (4, 3) **21.** (2, 0) **23.** (10, −1) **25.** (4, −3) **27.** (−8, −3) **29.** no solution **31.** $\left(-\frac{1}{2}, \frac{3}{2}\right)$ **33.** (−6, 11) **35.** (1.5, 0.5) **37.** 8, 6 **39.** $x + y = 28$, $16x + 19y = 478$ **41.** 4 2-bedroom, 2 3-bedroom **43.** $x + y = 30$, $700x + 200y = 15{,}000$ **45.** $2x + 4y = 100$, $y = 2x$ **47.** Yes; they should finish the test within 40 minutes. **49.** 25 min of step aerobics, 15 min of stretching **51.** You can use a system of equations to find the monthly fee and rate per minute charged during the months of January and February. Answers should include the following.

- The coordinates of the point of intersection are (0.08, 3.5).
- Currently, Yolanda is paying a monthly fee of $3.50 and an additional 8¢ per minute. If she graphs y = 0.08x + 3.5 (to represent what she is paying currently) and y = 0.10x + 3 (to represent the other long-distance plan) and finds the intersection, she can identify which plan would be better for a person with her level of usage.

53. A

55. consistent and dependent

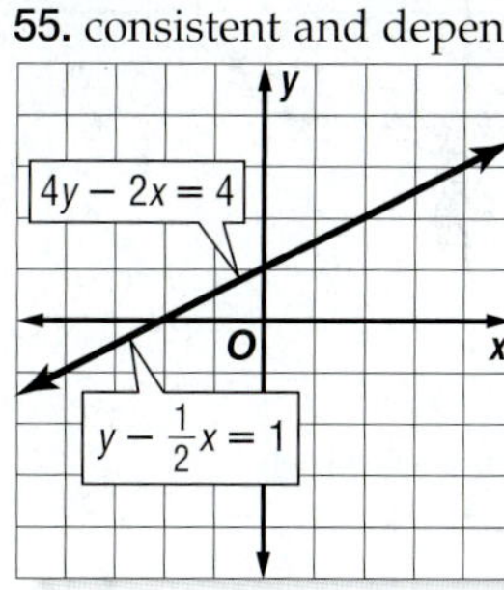

57.

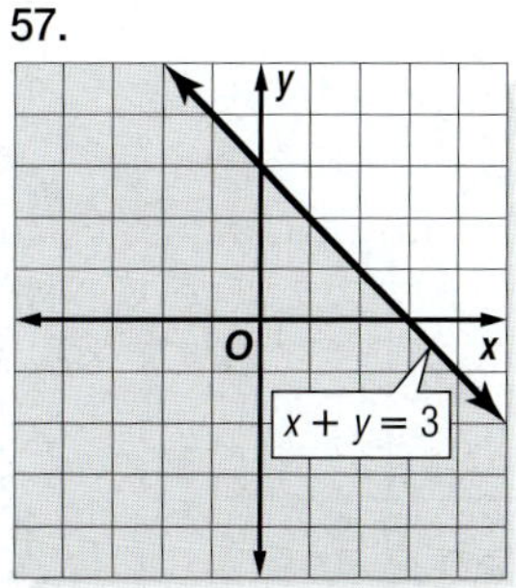

59.

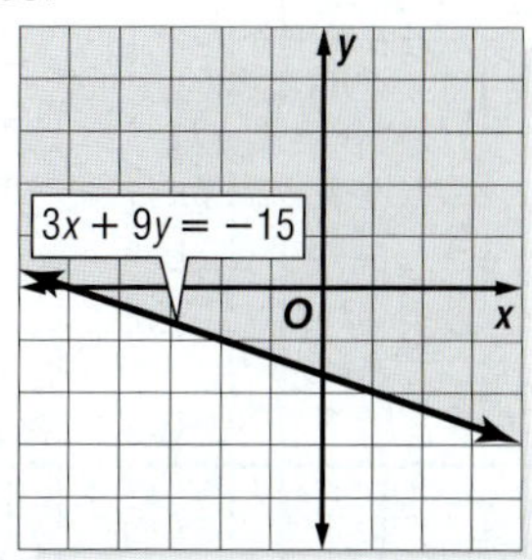

61. $x - y = 0$; 1, −1, 0 **63.** $2x - y = -3$; 2, −1, −3 **65.** $3x + 2y = 21$; 3, 2, 21 **67.** yes **69.** no

Selected Answers

Page 122 Practice Quiz 1

1.

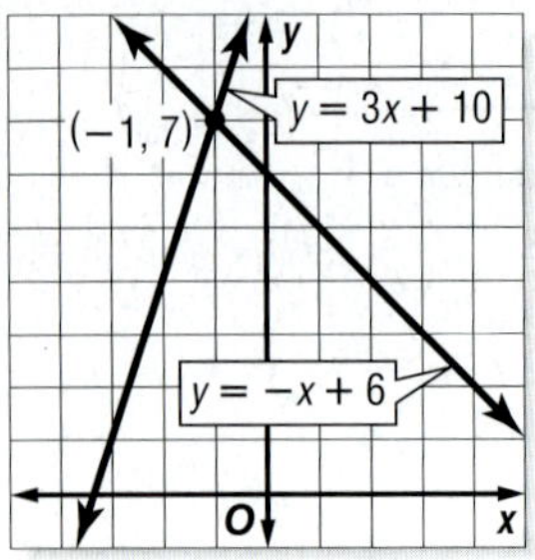

3. (2, 7) **5.** Hartsfield, 78 million; O'Hare, 72.5 million

Pages 125–127 Lesson 3-3

1. Sample answer: $y > x + 3$, $y < x - 2$ **3a.** 4 **3b.** 2 **3c.** 1 **3d.** 3

5.

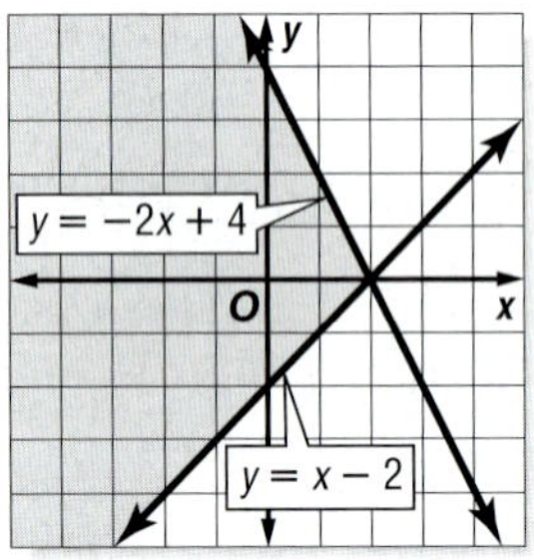

7.

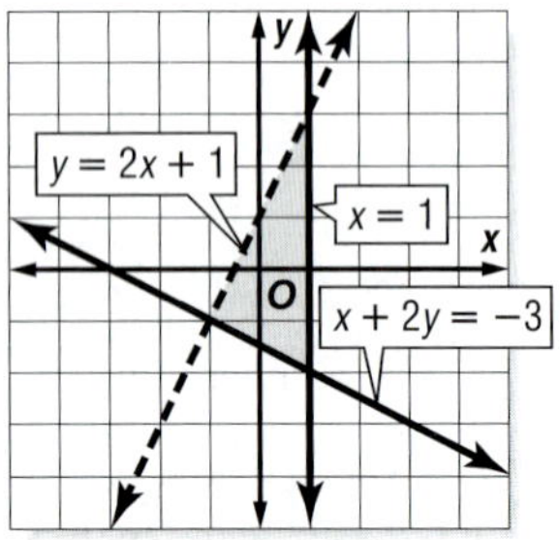

9. (−4, 3), (1, −2), (2, 9), (7, 4) **11.** Sample answer: 3 packages of bagels, 4 packages of muffins; 4 packages of bagels, 4 packages of muffins; 3 packages of bagels, 5 packages of muffins

13.

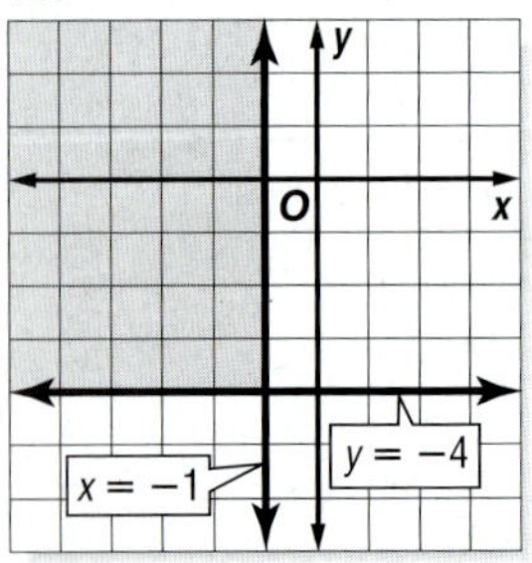

15.

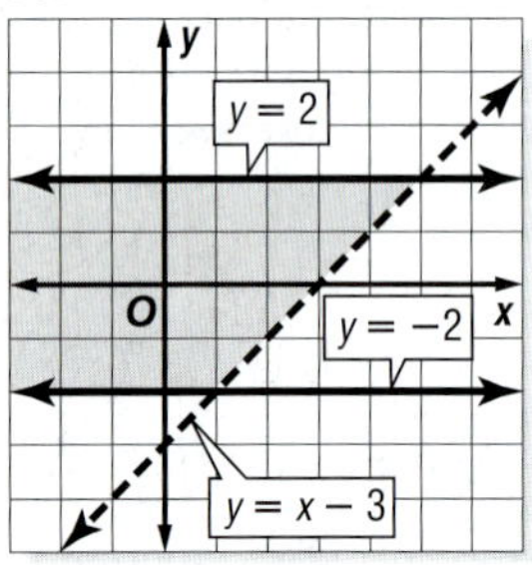

17.

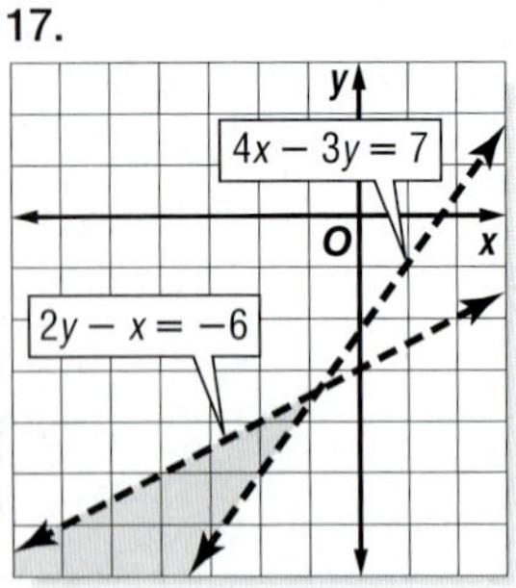

19. no solution

21.

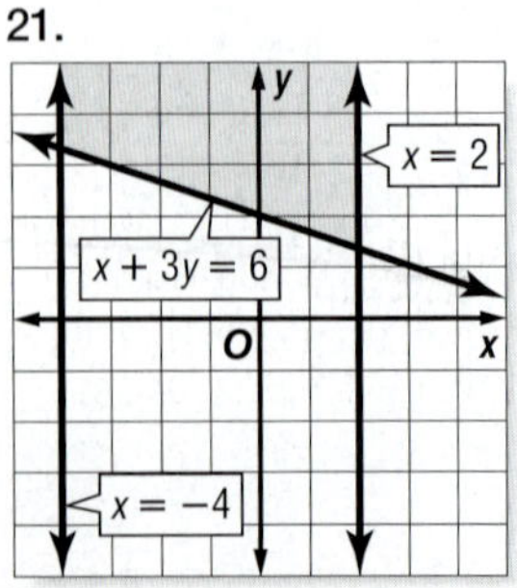

23.

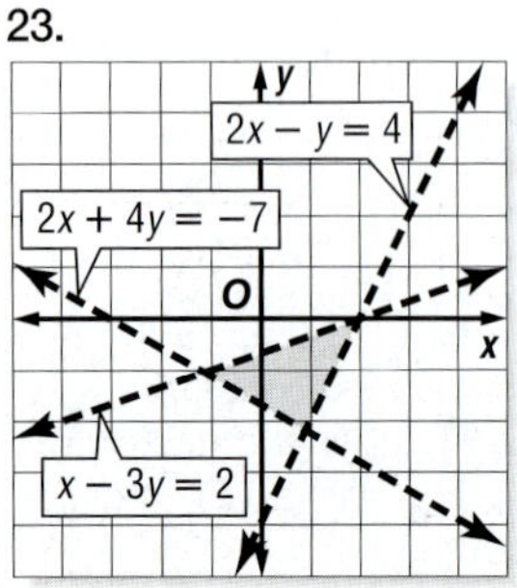

25. (−3, −4), (5, −4), (1, 4) **27.** (−6, −9), (2, 7), (10, −1)

29. (−4, 3), (−2, 7), (4, −1), $\left(7\frac{1}{3}, 2\frac{1}{3}\right)$ **31.** 64 units^2

33. $s \geq 111$, $s \leq 130$, $h \geq 9$, $h \leq 12$

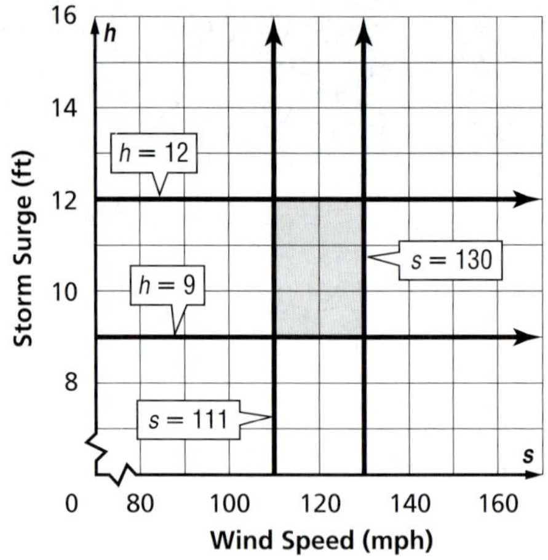

35.

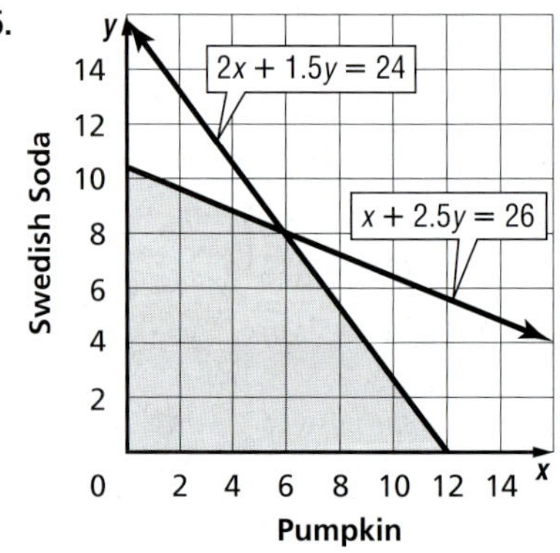

37. 6 pumpkin, 8 soda

39. The range for normal blood pressure satisfies four inequalities that can be graphed to find their intersection. Answers should include the following.
- Graph the blood pressure as an ordered pair; if the point lies in the shaded region, it is in the normal range.
- High systolic pressure is represented by the region to the right of $x = 140$ and high diastolic pressure is represented by the region above $y = 90$.

41. Sample answer: $y \leq 6$, $y \geq 2$, $x \leq 5$, $x \geq 1$ **43.** (6, 5)

45.

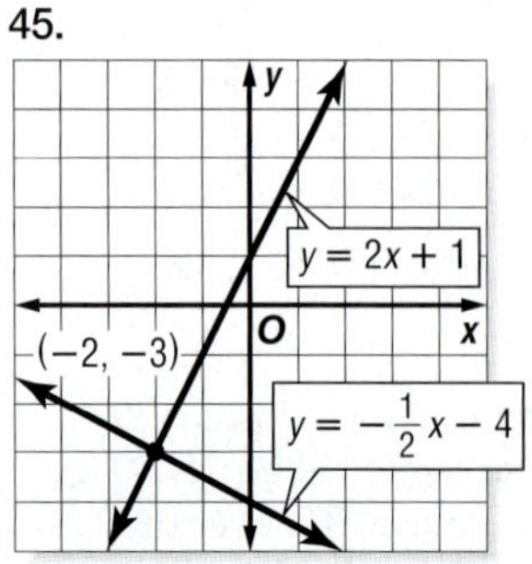

47.

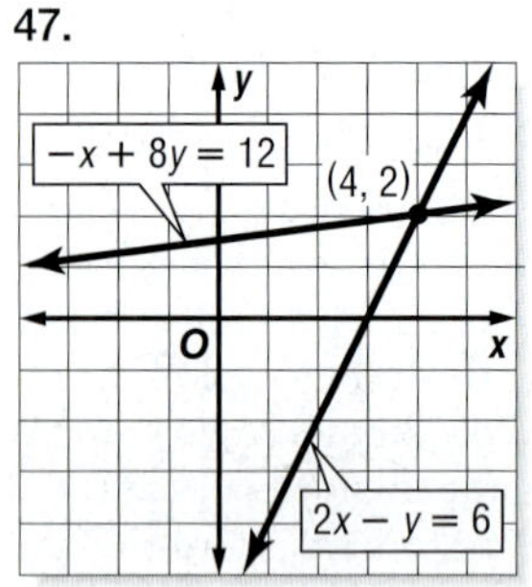

49. −5 **51.** 8 **53.** 5

Pages 132–135 Lesson 3-4

1. sometimes

3.

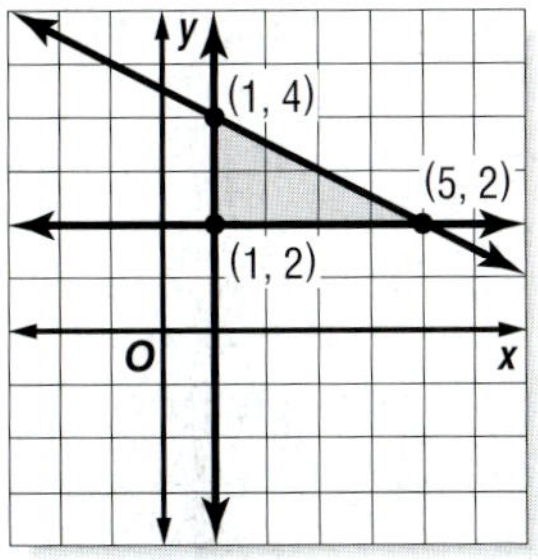

vertices: (1, 2), (1, 4), (5, 2);
max: $f(5, 2) = 4$,
min: $f(1, 4) = -10$

5.

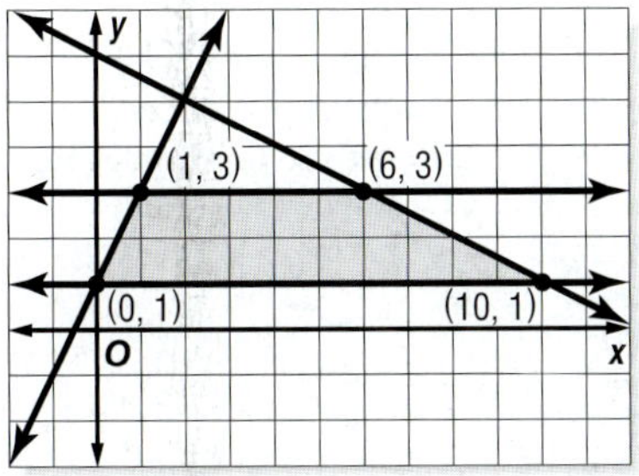

vertices: (0, 1), (1, 3), (6, 3), (10, 1); max: $f(10, 1) = 31$, min: $f(0, 1) = 1$

7.

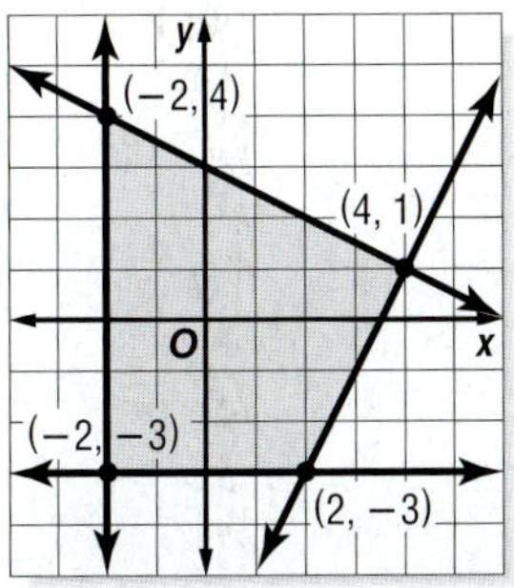

vertices: (−2, 4), (−2, −3), (2, −3), (4, 1);
max: $f(2, -3) = 5$;
min: $f(-2, 4) = -6$

9. $c \geq 0$, $\ell \geq 0$, $c + 3\ell \leq 56$, $4c + 2\ell \leq 104$ **11.** (0, 0), (26, 0), (20, 12), $\left(0, 18\frac{2}{3}\right)$

13. 20 canvas tote bags and 12 leather tote bags

15.

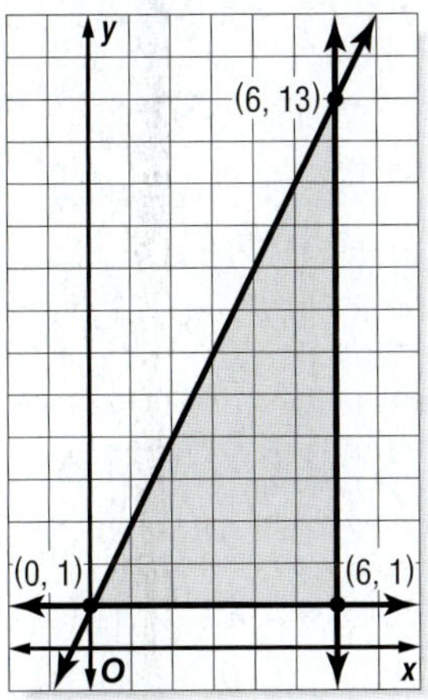

vertices: (0, 1), (6, 1), (6, 13);
max: $f(6, 13) = 19$;
min: $f(0, 1) = 1$

17.

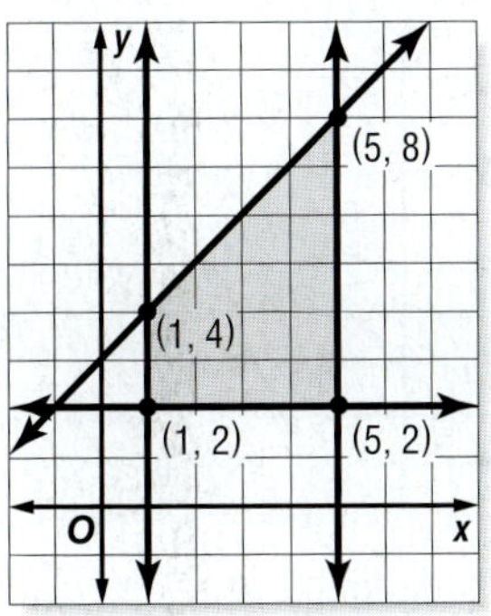

vertices: (1, 4), (5, 8), (5, 2), (1, 2); max: $f(5, 2) = 11$, min: $f(1, 4) = -5$

19.

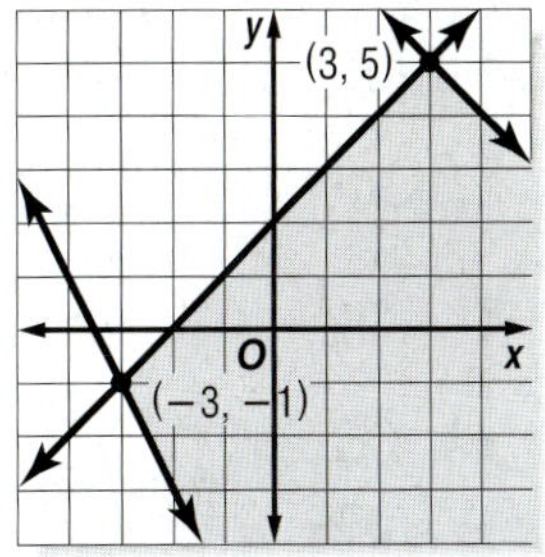

vertices: (−3, −1), (3, 5);
min: $f(-3, -1) = -9$;
no maximum

21.

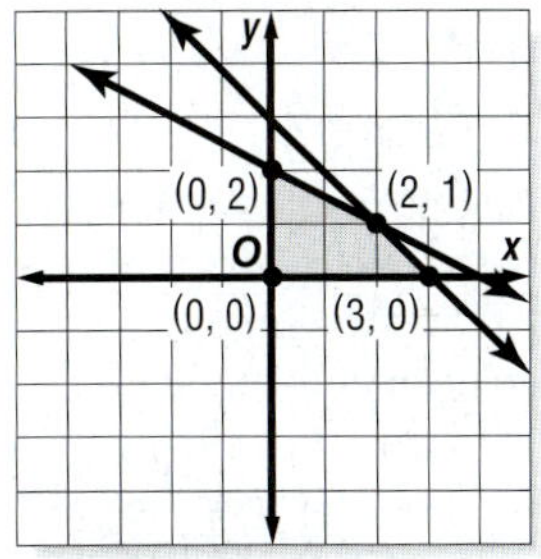

vertices: (0, 0), (0, 2), (2, 1), (3, 0); max: $f(0, 2) = 6$;
min: $f(3, 0) = -12$

23.

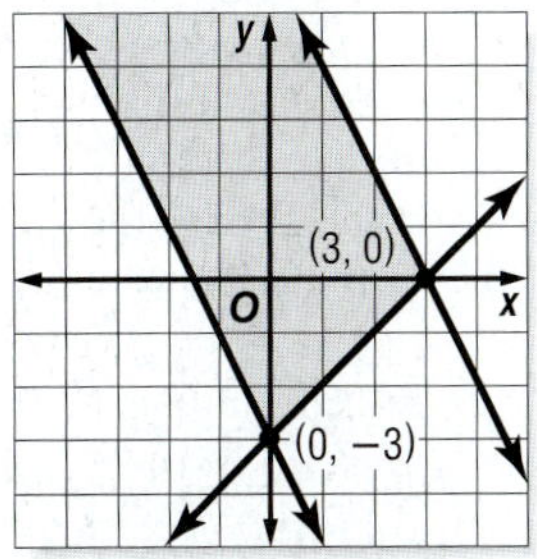

vertices: (3, 0), (0, −3);
min: $f(0, -3) = -12$;
no maximum

25.

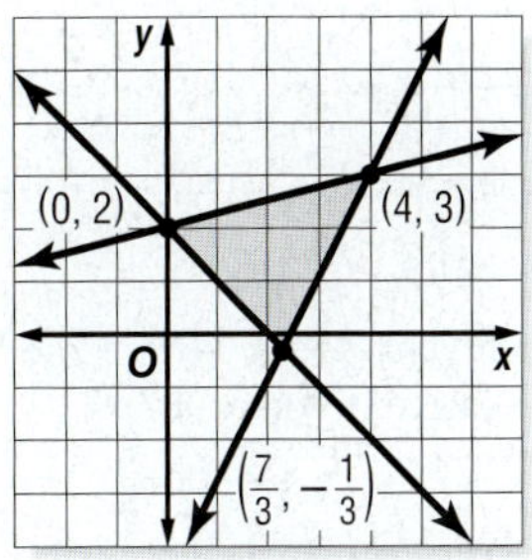

vertices: (0, 2), (4, 3), $\left(\frac{7}{3}, -\frac{1}{3}\right)$; max: $f(4, 3) = 25$,
min: $f(0, 2) = 6$

27.

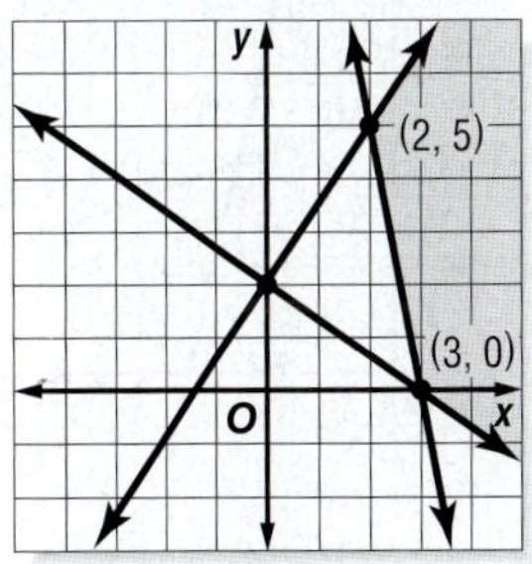

vertices: (2, 5), (3, 0);
no maximum;
no minimum

29.

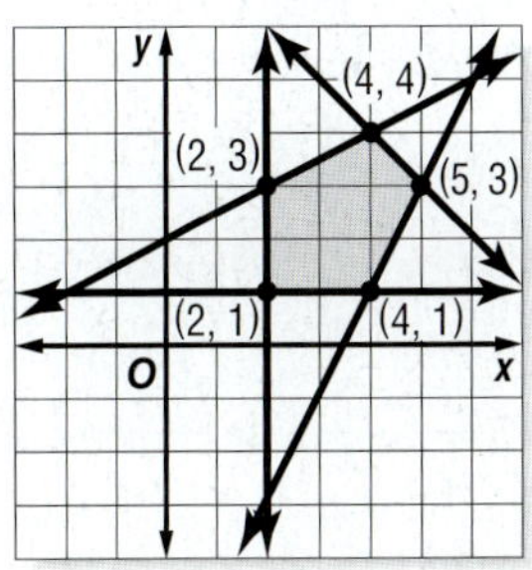

vertices: (2, 1), (2, 3), (4, 1), (4, 4), (5, 3);
max: $f(4, 1) = 0$,
min: $f(4, 4) = -12$

31. $g \geq 0, c \geq 0, 1.5g + c \leq 85, 2g + 0.5c \leq 40$ **33.** (0, 0), (0, 20), (80, 0) **35.** 0 graphing calculators, 80 CAS calculators

39.
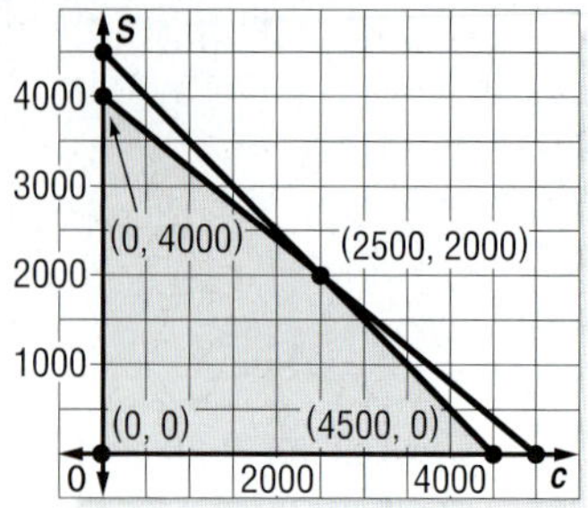

(0, 0), (0, 4000), (2500, 2000), (4500, 0)

41. 4500 acres corn, 0 acres soybeans; \$130,500

43. There are many variables in scheduling tasks. Linear programming can help make sure that all the requirements are met. Answers should include the following.

- Let x = the number of buoy replacements and let y = the number of buoy repairs. Then, $x \geq 0, y \geq 0, x \leq 8$ and $2.5x + y \leq 24$.
- The captain would want to maximize the number of buoys that a crew could repair and replace so $f(x, y) = x + y$.
- Graph the inequalities and find the vertices of the intersection of the graphs. The coordinate (0, 24) maximizes the function. So the crew can service the maximum number of buoys if they replace 0 and repair 24 buoys.

45. C **47.**
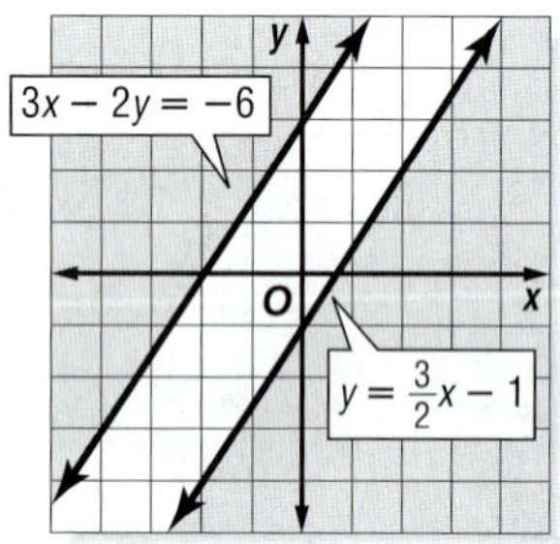

∅

49. (2, 3) **51.** c = average cost each year; $15c + 3479 = 7489$ **53.** Additive Inverse **55.** Multiplicative Inverse **57.** 9 **59.** 16 **61.** 8

Page 135 Practice Quiz 2

1.
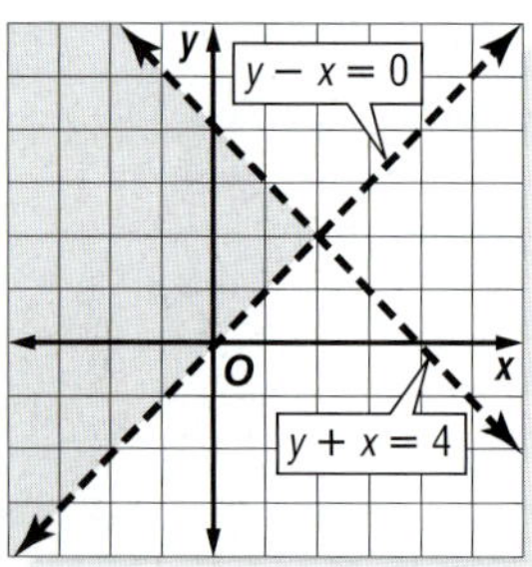

3.
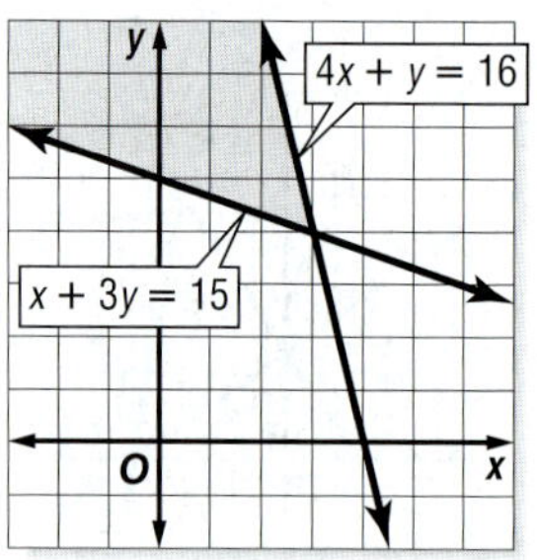

5.
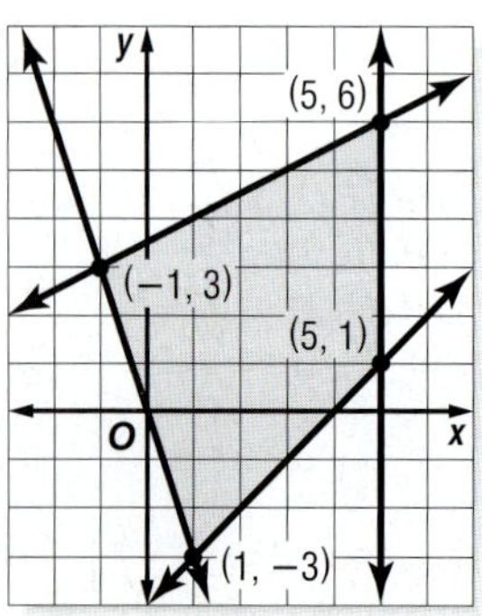

vertices: (1, −3), (−1, 3), (5, 6), (5, 1); max: $f(5, 1) = 17$, min: $f(-1, 3) = -13$

Pages 142–144 Lesson 3-5

1. You can use elimination or substitution to eliminate one of the variables. Then you can solve two equations in two variables.

3. Sample answer: $x + y + z = 4$, $2x - y + z = -9$, $x + 2y - z = 5$; $-3 + 5 + 2 = 4$, $2(-3) - 5 + 2 = -9$, $-3 + 2(5) - 2 = 5$ **5.** (−1, −3, 7) **7.** (5, 2, −1) **9.** (4, 0, 8) **11.** $4\frac{1}{2}$ lb chicken, 3 lb sausage, 6 lb rice **13.** (−2, 1, 5) **15.** (4, 0, −1) **17.** (1, 5, 7) **19.** infinitely many **21.** $\left(\frac{1}{3}, -\frac{1}{2}, \frac{1}{4}\right)$ **23.** (−5, 9, 4) **25.** 8, 1, 3 **27.** enchilada, \$2.50; taco, \$1.95; burrito, \$2.65 **29.** $x + y + z = 355$, $x + 2y + 3z = 646$, $y = z + 27$ **31.** $a = \frac{3}{2}, b = 0, c = 3$; $y = \frac{3}{2}x^2 + 0x + 3$ or $y = \frac{3}{2}x^2 + 3$ **33.** D **35.** 120 units of notebook paper and 80 units of newsprint

37.
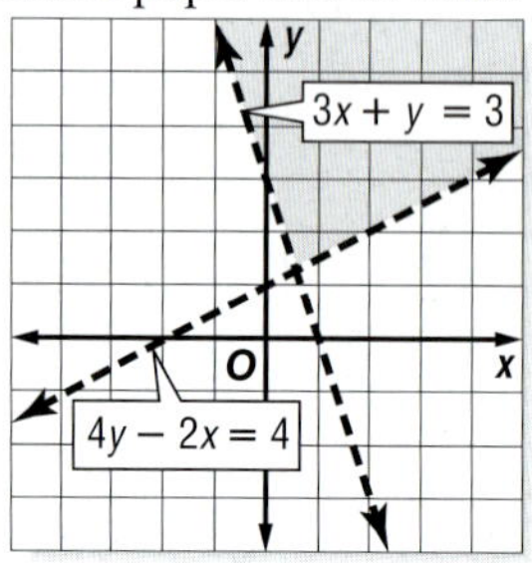

39. Sample answer using (7, 15) and (14, 22): $y = x + 8$ **41.** $x + 3y$ **43.** $9s + 4t$

Pages 145–148 Study Guide and Review

1. c **3.** f **5.** a **7.** h **9.** d

11.
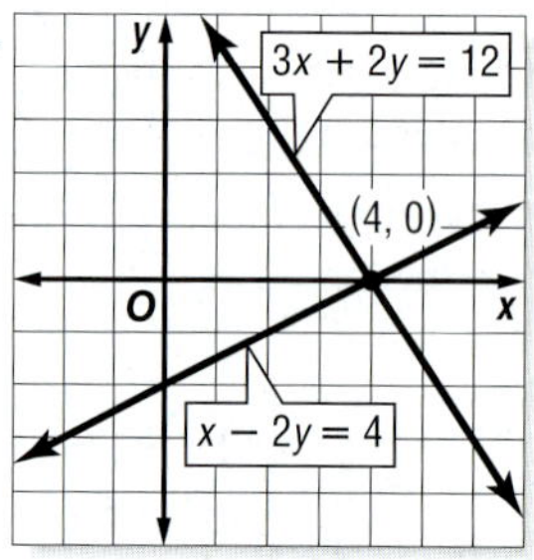

13.
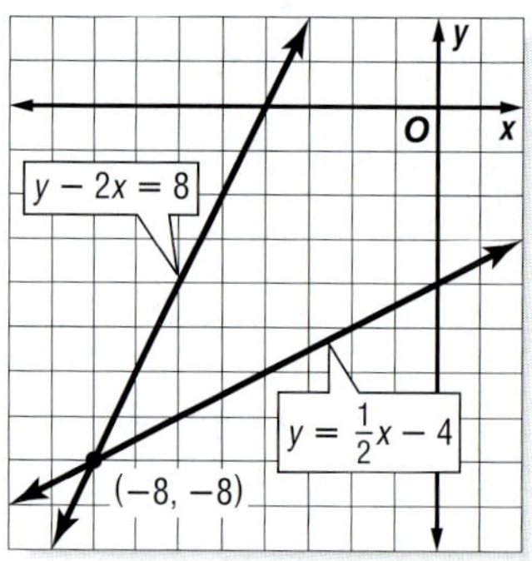

15. (3, 2) **17.** (9, 4) **19.** (−1, 2)

21.
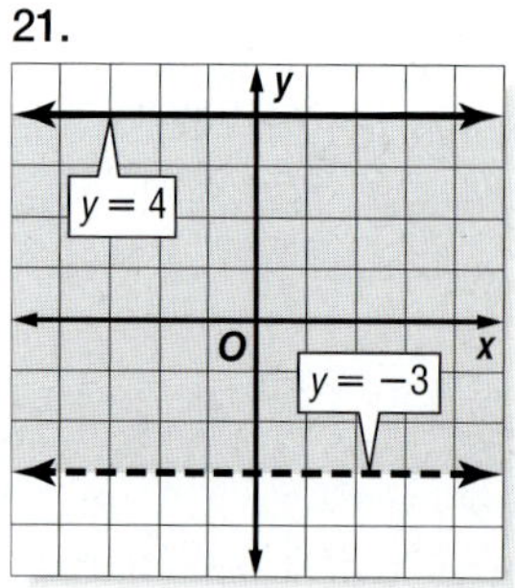

23.
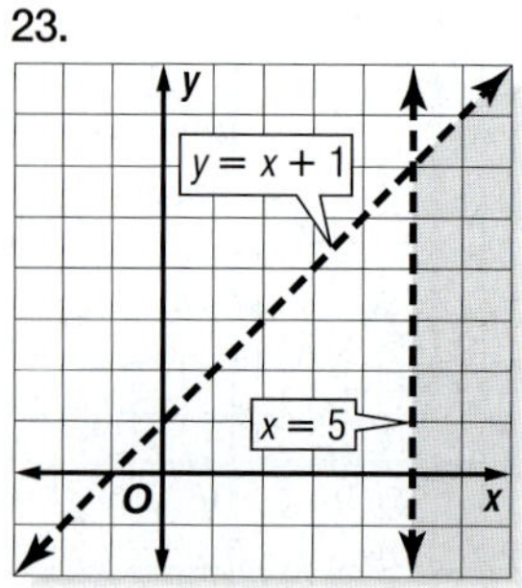

25. 160 My Real Babies, 320 My First Babies **27.** (4, −2, 1)

Selected Answers

Chapter 4 Matrices

Page 153 Chapter 4 Getting Started

1. 6 **3.** $4\frac{3}{4}$ **5.** −13 **7.** $-3; \frac{1}{3}$ **9.** $-8; \frac{1}{8}$ **11.** −1.25; 0.8

13. $\frac{8}{3}; -\frac{3}{8}$

15.

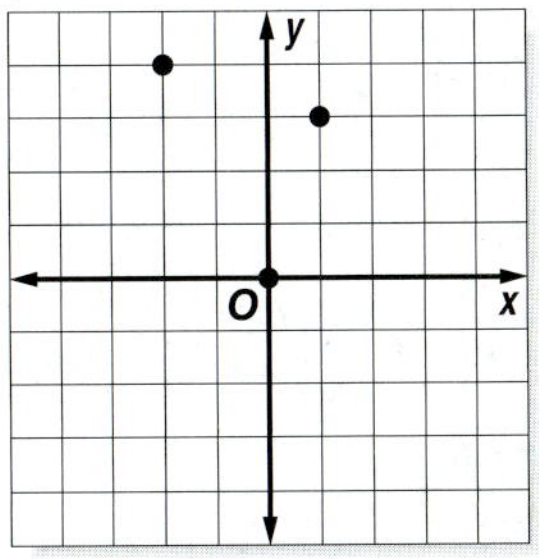

17.

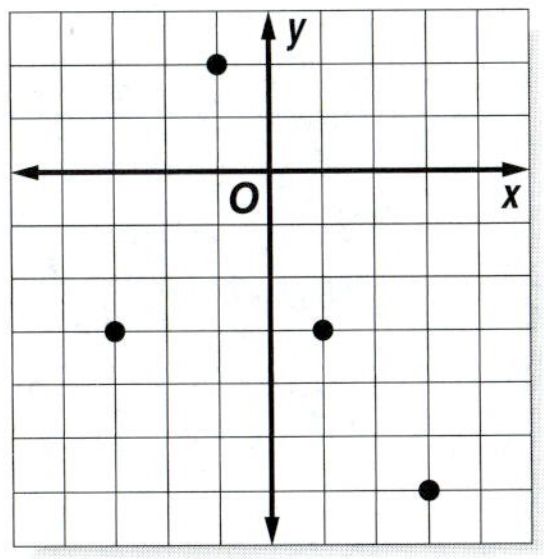

19. (6, 1) **21.** (8, −5) **23.** (2, −2)

Pages 156–158 Lesson 4-1

1. The matrices must have the same dimensions and each element of one matrix must be equal to the corresponding element of the other matrix. **3.** Corresponding elements are elements in the same row and column positions.
5. 3 × 4 **7.** (3, 3) **9.** 2 × 5 **11.** 3 × 1 **13.** 3 × 3
15. 3 × 2 **17.** $\left(3, -\frac{1}{3}\right)$ **19.** (3, −5, 6) **21.** (4, −3)
23. (14, 15) **25.** (5, 3, 2) **27.** 3 × 3 **29.** Sample answer: Mason's Steakhouse; it was given the highest rating possible for service and atmosphere, location was given one of the highest ratings, and it is moderately priced.

31.

	Single	Double	Suite
Weekday	60	70	75
Weekend	79	89	95

33. row 6, column 9 **35.** B **37.** (7, 5, 4) **39.** $\left(-\frac{4}{5}, \frac{3}{5}, -11\right)$

41.

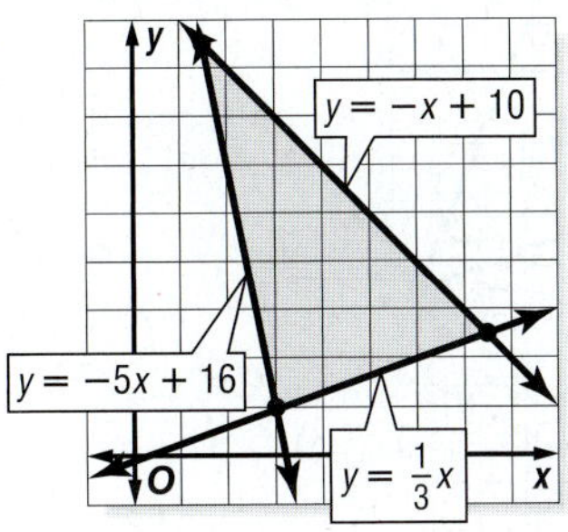

vertices: (3, 1), $\left(\frac{15}{2}, \frac{5}{2}\right)$, $\left(\frac{3}{2}, \frac{17}{2}\right)$; max: $f\left(\frac{15}{2}, \frac{5}{2}\right) =$ 35, min: $f\left(\frac{3}{2}, \frac{17}{2}\right) = -1$

43.

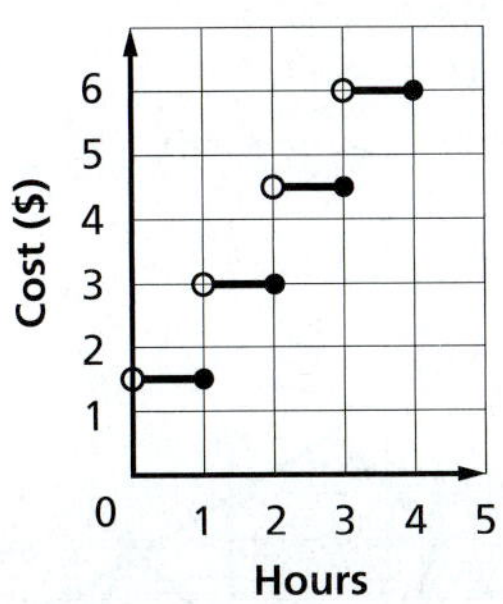

45. \$4.50 **47.** 2 **49.** 20 **51.** −10 **53.** −18 **55.** −3 **57.** $\frac{3}{2}$

Pages 163–166 Lesson 4-2

1. They must have the same dimensions.

3. $\begin{bmatrix} 4 & 4 \\ 4 & 4 \\ 4 & 4 \end{bmatrix}$ **5.** $\begin{bmatrix} 1 & 10 \\ -7 & 5 \end{bmatrix}$ **7.** $\begin{bmatrix} -22 & 8 \\ 3 & 24 \end{bmatrix}$ **9.** $\begin{bmatrix} -21 & 29 \\ 12 & -22 \end{bmatrix}$

11. Males = $\begin{bmatrix} 16{,}763 & 549{,}499 \\ 14{,}620 & 477{,}960 \\ 14{,}486 & 455{,}305 \\ 9041 & 321{,}416 \\ 5234 & 83{,}411 \end{bmatrix}$, Females = $\begin{bmatrix} 16{,}439 & 456{,}873 \\ 14{,}545 & 405{,}163 \\ 12{,}679 & 340{,}480 \\ 7931 & 257{,}586 \\ 5450 & 133{,}235 \end{bmatrix}$

13. No; many schools offer the same sport for males and females, so those schools would be counted twice.

15. impossible **17.** $\begin{bmatrix} -4 & 8 & -2 \\ 6 & -10 & -16 \\ -14 & -12 & 4 \end{bmatrix}$ **19.** $\begin{bmatrix} -13 \\ -3 \\ 23 \end{bmatrix}$

21. $\begin{bmatrix} 1.5 & 3 \\ 4.5 & 9 \end{bmatrix}$ **23.** $\begin{bmatrix} -5\frac{1}{2} & 3 & 9 \\ 10\frac{2}{3} & 1\frac{2}{3} & -2\frac{1}{2} \end{bmatrix}$ **25.** $\begin{bmatrix} -2 & -1 \\ 4 & -1 \\ -7 & -4 \end{bmatrix}$

27. $\begin{bmatrix} 38 & 4 \\ 32 & -6 \\ 18 & 42 \end{bmatrix}$ **29.** $\begin{bmatrix} 2 & 4\frac{2}{3} \\ 1 & 5 \\ 6 & -1 \end{bmatrix}$ **31.** $\begin{bmatrix} 232 & 184 & 120 & 149 \\ 164 & 124 & 75 & 130 \\ 160 & 182 & 72 & 108 \end{bmatrix}$

33. $\begin{bmatrix} 245 \\ 228 \\ 319 \\ 227 \\ 117 \end{bmatrix}$ **35.** 1996, floods; 1997, floods; 1998, floods; 1999, tornadoes; 2000, lightning

37. $\begin{bmatrix} 1.50 & 2.25 \\ 1.00 & 1.75 \end{bmatrix}$ **39.** $\begin{bmatrix} 1.00 & 1.00 \\ 1.50 & 1.50 \end{bmatrix}$

41. You can use matrices to track dietary requirements and add them to find the total each day or each week. Answers should include the following.

- Breakfast = $\begin{bmatrix} 566 & 18 & 7 \\ 482 & 12 & 17 \\ 530 & 10 & 11 \end{bmatrix}$, Lunch = $\begin{bmatrix} 785 & 22 & 19 \\ 622 & 23 & 20 \\ 710 & 26 & 12 \end{bmatrix}$,
  Dinner = $\begin{bmatrix} 1257 & 40 & 26 \\ 987 & 32 & 45 \\ 1380 & 29 & 38 \end{bmatrix}$
- Add the three matrices: $\begin{bmatrix} 2608 & 80 & 52 \\ 2091 & 67 & 82 \\ 2620 & 65 & 61 \end{bmatrix}$.

43. A **45.** 1 × 4 **47.** 3 × 3 **49.** 4 × 3 **51.** (5, 3, 7)
53. (2, 5) **55.** (6, −1)

57.

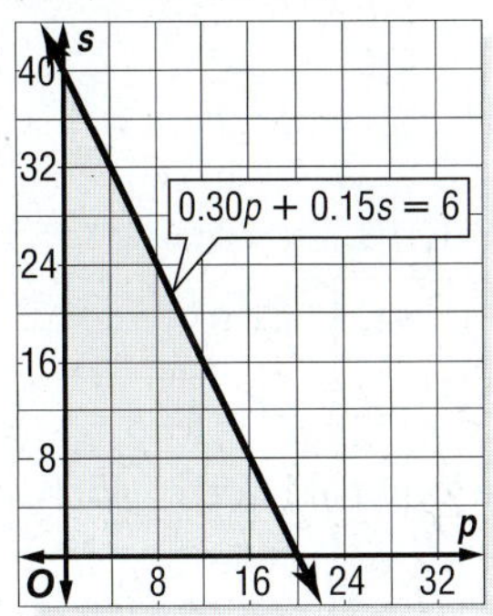

59. Multiplicative Inverse
61. Distributive

Pages 171–174 Lesson 4-3

1. Sample answer: $\begin{bmatrix} 1 & 2 \\ 3 & 4 \\ 5 & 6 \end{bmatrix} \cdot \begin{bmatrix} 7 & 8 \\ 9 & 10 \end{bmatrix}$ **3.** The Right Distributive Property says that $(A + B)C = AC + BC$, but $AC + BC \neq CA + CB$ since the Commutative Property does not hold for matrix multiplication in most cases. **5.** undefined

7. $\begin{bmatrix} 15 & -5 & 20 \\ 24 & -8 & 32 \end{bmatrix}$ **9.** $\begin{bmatrix} 24 \\ 41 \end{bmatrix}$ **11.** $[45 \quad 55 \quad 65]$, $\begin{bmatrix} 350 & 280 \\ 320 & 165 \\ 180 & 120 \end{bmatrix}$

Selected Answers

13. 4×2 **15.** undefined **17.** undefined

19. [6] **21.** not possible **23.** $\begin{bmatrix} 1 & -25 & 2 \\ 29 & 1 & -30 \end{bmatrix}$

25. $\begin{bmatrix} 24 & 16 \\ -32 & -5 \\ -48 & -11 \end{bmatrix}$

27. yes

$$AC + BC = \begin{bmatrix} 1 & -2 \\ 4 & 3 \end{bmatrix} \cdot \begin{bmatrix} 5 & 1 \\ 2 & -4 \end{bmatrix} + \begin{bmatrix} -5 & 2 \\ 4 & 3 \end{bmatrix} \cdot \begin{bmatrix} 5 & 1 \\ 2 & -4 \end{bmatrix}$$
$$= \begin{bmatrix} 1 & 9 \\ 26 & -8 \end{bmatrix} + \begin{bmatrix} -21 & -13 \\ 26 & -8 \end{bmatrix}$$
$$= \begin{bmatrix} -20 & -4 \\ 52 & -16 \end{bmatrix}$$

$$(A + B)C = \left(\begin{bmatrix} 1 & -2 \\ 4 & 3 \end{bmatrix} + \begin{bmatrix} -5 & 2 \\ 4 & 3 \end{bmatrix}\right) \cdot \begin{bmatrix} 5 & 1 \\ 2 & -4 \end{bmatrix}$$
$$= \begin{bmatrix} -4 & 0 \\ 8 & 6 \end{bmatrix} \cdot \begin{bmatrix} 5 & 1 \\ 2 & -4 \end{bmatrix}$$
$$= \begin{bmatrix} -20 & -4 \\ 52 & -16 \end{bmatrix}$$

29. no

$$C(A + B) = \begin{bmatrix} 5 & 1 \\ 2 & -4 \end{bmatrix} \cdot \left(\begin{bmatrix} 1 & -2 \\ 4 & 3 \end{bmatrix} + \begin{bmatrix} -5 & 2 \\ 4 & 3 \end{bmatrix}\right)$$
$$= \begin{bmatrix} 5 & 1 \\ 2 & -4 \end{bmatrix} \cdot \begin{bmatrix} -4 & 0 \\ 8 & 6 \end{bmatrix}$$
$$= \begin{bmatrix} -12 & 6 \\ -40 & -24 \end{bmatrix}$$

$$AC + BC = \begin{bmatrix} 1 & -2 \\ 4 & 3 \end{bmatrix} \cdot \begin{bmatrix} 5 & 1 \\ 2 & -4 \end{bmatrix} + \begin{bmatrix} -5 & 2 \\ 4 & 3 \end{bmatrix} \cdot \begin{bmatrix} 5 & 1 \\ 2 & -4 \end{bmatrix}$$
$$= \begin{bmatrix} 1 & 9 \\ 26 & -8 \end{bmatrix} + \begin{bmatrix} -21 & -13 \\ 26 & -8 \end{bmatrix}$$
$$= \begin{bmatrix} -20 & -4 \\ 52 & -16 \end{bmatrix}$$

31. $\begin{bmatrix} 290 & 165 & 210 \\ 175 & 240 & 190 \\ 110 & 75 & 0 \end{bmatrix}$ **33.** $\begin{bmatrix} 14{,}285 \\ 13{,}270 \\ 4295 \end{bmatrix}$

35. any two matrices $\begin{bmatrix} a & b \\ c & d \end{bmatrix}$ and $\begin{bmatrix} e & f \\ g & h \end{bmatrix}$ where $bg = cf$, $a = d$, and $e = h$ **37.** $\begin{bmatrix} 96.50 \\ 99.50 \\ 118 \\ 117 \end{bmatrix}$ **39.** \$431 **41.** \$26,360

43. $a = 1, b = 0, c = 0, d = 1$; the original matrix **45.** B

47. $\begin{bmatrix} 12 & -6 \\ -3 & 21 \end{bmatrix}$ **49.** $\begin{bmatrix} -20 & 2 \\ -28 & 12 \end{bmatrix}$ **51.** (5, −9) **53.** \$2.50; \$1.50

55. 8; −16

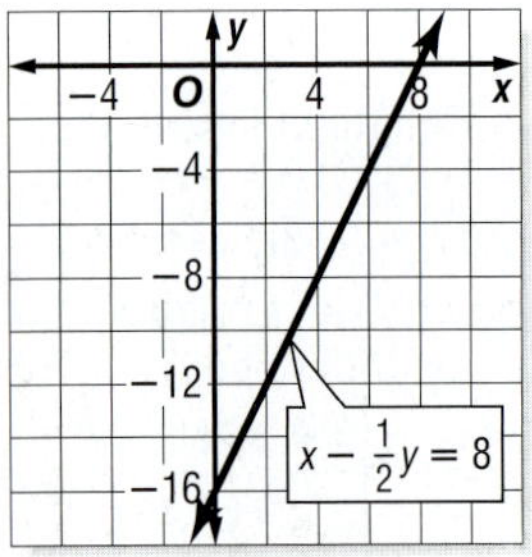

57.

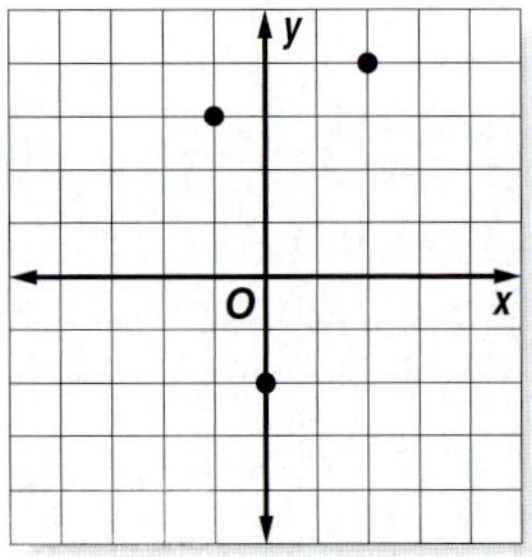

59.

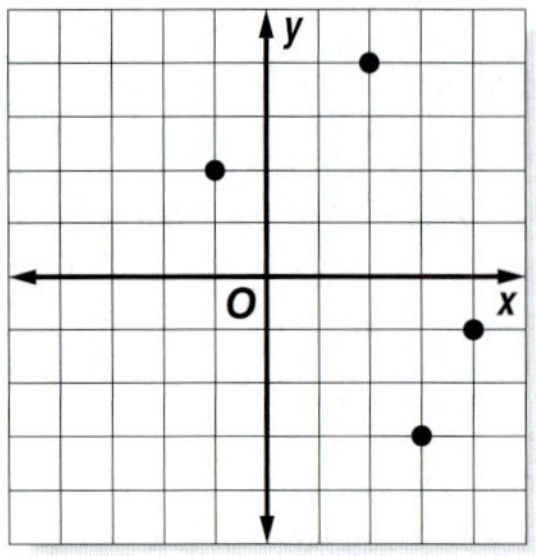

Page 174 Practice Quiz 1

1. (6, 3) **3.** (1, 3, 5) **5.** $\begin{bmatrix} 232 & 159 & 120 & 149 \\ 134 & 200 & 159 & 103 \end{bmatrix}$ **7.** $\begin{bmatrix} 4 & 3 \\ 1 & 3 \end{bmatrix}$

9. not possible

Pages 178–181 Lesson 4-4

1.

Transformation	Size	Shape	Isometry
reflection	same	same	yes
rotation	same	same	yes
translation	same	same	yes
dilation	changes	same	no

3. Sample answer: $\begin{bmatrix} -4 & -4 & -4 \\ 1 & 1 & 1 \end{bmatrix}$ **5.** $A'(4, 3)$, $B'(5, -6)$, $C'(-3, -7)$ **7.** $\begin{bmatrix} 0 & 5 & 5 & 0 \\ 4 & 4 & 0 & 0 \end{bmatrix}$ **9.** $A'(0, -4)$, $B'(5, -4)$, $C'(5, 0)$, $D'(0, 0)$ **11.** B **13.** $D'(-3, 6)$, $E'(-2, -3)$, $F'(-10, -4)$

15. $\begin{bmatrix} 0 & 1.5 & -2.5 \\ 2 & -1.5 & 0 \end{bmatrix}$

17.

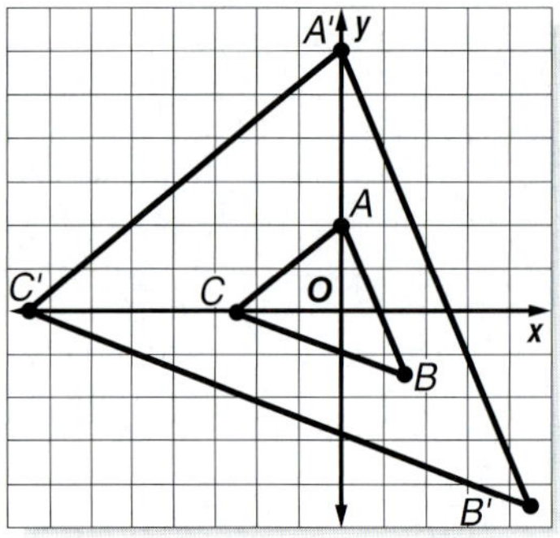

19. $X'(-1, 1)$, $Y'(-4, 2)$, $Z'(-1, 7)$

21. $\begin{bmatrix} 2 & 5 & 4 & 1 \\ 4 & 4 & 1 & 1 \end{bmatrix}$

23.

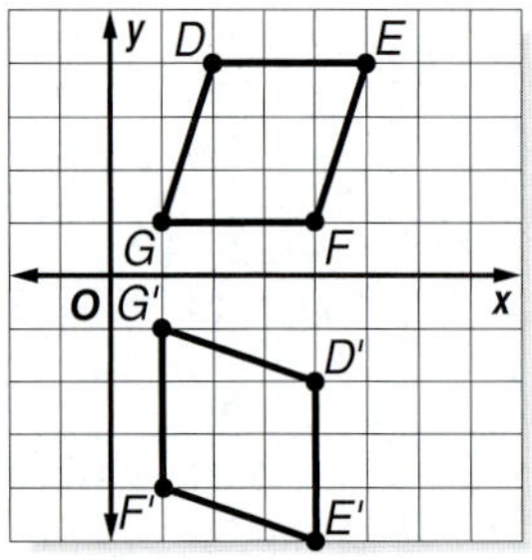

25. $J(-5, 3)$, $K(7, 2)$, $L(4, -1)$ **27.**

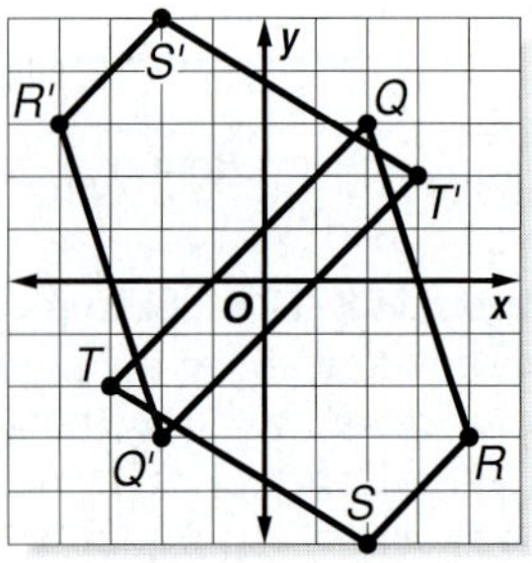

29. $\begin{bmatrix} 4 & -4 & -4 & 4 \\ -4 & -4 & 4 & 4 \end{bmatrix}$ **31.** $\begin{bmatrix} 4 & 4 & -4 & -4 \\ -4 & 4 & 4 & -4 \end{bmatrix}$

33. (−1.5, −1.5), (−4.5, −1.5), (−6, −3.75), (−3, −3.75)

35. $\begin{bmatrix} 3 \\ 4 \end{bmatrix}$ **37.** (−8, 7), (−7, −8), and (8, −7) **39.** Multiply the coordinates by $\begin{bmatrix} 1 & 0 \\ 0 & -1 \end{bmatrix}$, then add the result to $\begin{bmatrix} 6 \\ 0 \end{bmatrix}$.

41. (17, −2), (23, 2)

43. Transformations are used in computer graphics to create special effects. You can simulate the movement of an object, like in space, which you wouldn't be able to recreate otherwise. Answers should include the following.

- A figure with points (a, b), (c, d), (e, f), (g, h), and (i, j) could be written in a 2 × 5 matrix $\begin{bmatrix} a & c & e & g & i \\ b & d & f & h & j \end{bmatrix}$ and multiplied on the left by the 2 × 2 rotation matrix.
- The object would get smaller and appear to be moving away from you.

45. A **47.** undefined **49.** $\begin{bmatrix} 11 & 24 & -7 \\ 18 & -13 & 8 \\ 33 & -8 & 21 \end{bmatrix}$

51.

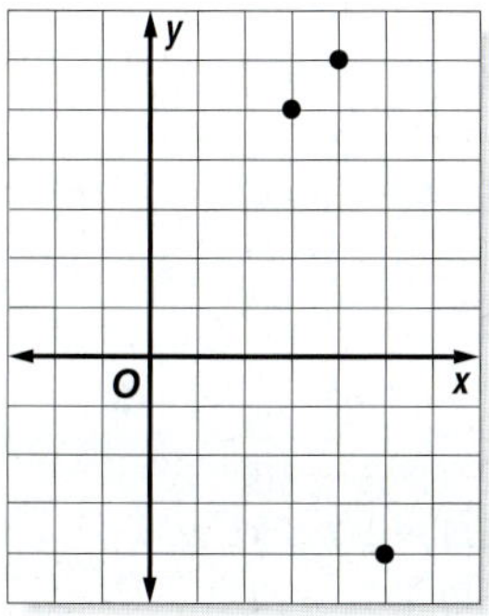

D = {3, 4, 5},
R = {−4, 5, 6}; yes

53.

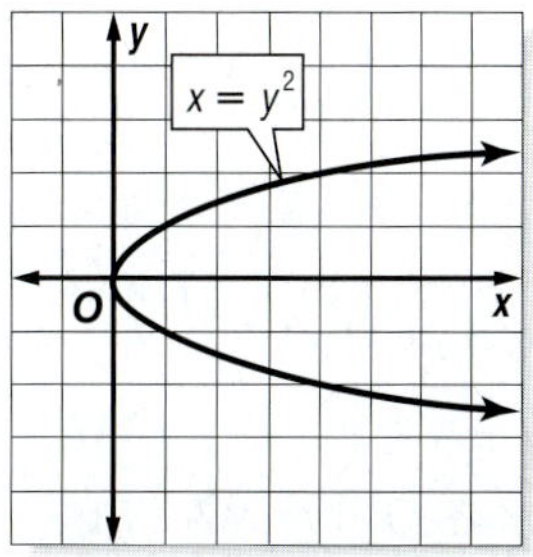

D = $\{x \mid x \geq 0\}$,
R = {all real numbers}; no

55. $|x| < 2.8$ **57.** $|x - 1| < 1$ **59.** 6 **61.** 28 **63.** $\frac{9}{4}$

Pages 185–188 Lesson 4-5

1. Sample answer: $\begin{bmatrix} 2 & 1 \\ 8 & 4 \end{bmatrix}$ **3.** It is not a square matrix.

5. Cross out the column and row that contains 6. The minor is the remaining 2 × 2 matrix. **7.** −38 **9.** −40 **11.** −43 **13.** 45 **15.** 20 **17.** −22 **19.** −29 **21.** 63 **23.** 32 **25.** 32 **27.** −58 **29.** 62 **31.** 172 **33.** −22 **35.** −5 **37.** −141 **39.** −6 **41.** 14.5 units^2 **43.** about 26 ft^2

45. Sample answer: $\begin{vmatrix} 1 & 1 & 1 \\ 1 & 1 & 1 \\ 1 & 1 & 1 \end{vmatrix}$

47. If you know the coordinates of the vertices of a triangle, you can use a determinant to find the area. This is convenient since you don't need to know any additional information such as the measure of the angles. Answers should include the following.

- You could place a coordinate grid over a map of the Bermuda Triangle with one vertex at the origin. By using the scale of the map, you could determine coordinates to represent the other two vertices and use a determinant to estimate the area.
- The determinant method is advantageous since you don't need to physically measure the lengths of each side or the measure of the angles between the vertices.

49. C **51.** −36.9 **53.** −493 **55.** −3252 **57.** $A'(-5, 2.5)$, $B'(2.5, 5)$, $C'(5, -7.5)$ **59.** [−4] **61.** undefined **63.** [14 −8] **65.** 138,435 ft **67.** $y = -\frac{4}{3}x$ **69.** $y = \frac{1}{2}x + 5$ **71.** (1, 9) **73.** (−1, 1) **75.** (4, 7)

Pages 192–194 Lesson 4-6

1. The determinant of the coefficient matrix cannot be zero. **3.** $3x + 5y = -6$, $4x - 2y = 30$ **5.** (0.75, 0.5) **7.** no solution **9.** $\left(6, -\frac{1}{2}, 2\right)$ **11.** savings account, \$1500; certificate of deposit, \$2500 **13.** (−12, 4) **15.** (6, 3) **17.** (−0.75, 3) **19.** (−8.5625, −19.0625) **21.** (4, −8) **23.** $\left(\frac{2}{3}, \frac{5}{6}\right)$ **25.** (3, −4) **27.** (2, −1, 3) **29.** $\left(\frac{141}{29}, -\frac{102}{29}, \frac{244}{29}\right)$ **31.** $\left(-\frac{155}{28}, \frac{143}{70}, \frac{673}{140}\right)$ **33.** race car, 5 plays; snowboard, 3 plays **35.** silk, \$34.99; cotton, \$24.99 **37.** peanuts, 2 lb; raisins, 1 lb; pretzels, 2 lb **39.** Cramer's Rule is a formula for the variables x and y where (x, y) is a solution for a system of equations. Answers should include the following.

- Cramer's Rule uses determinants composed of the coefficients and constants in a system of linear equations to solve the system.
- Cramer's Rule is convenient when coefficients are large or involve fractions or decimals. Finding the value of the determinant is sometimes easier than trying to find a greatest common factor if you are solving by using elimination or substituting complicated numbers.

41. 111°, 69° **43.** 40 **45.** $\begin{bmatrix} 1 & 1 & 1 \\ 3 & 3 & 3 \end{bmatrix}$

47.

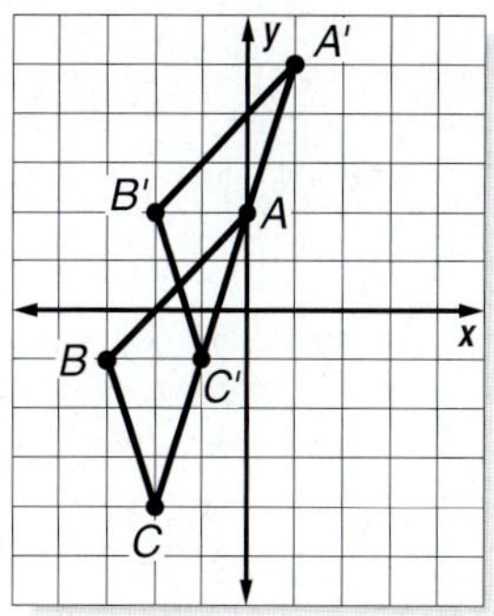

49.

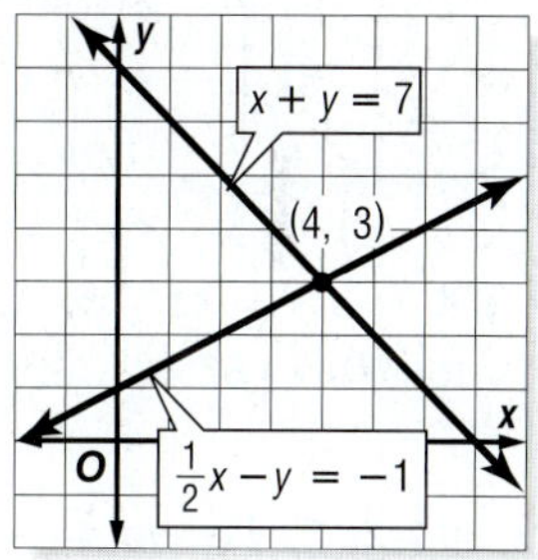

(4, 3)

51. $c = 10h + 35$ **53.** $\begin{bmatrix} 72 & 9 \\ 66 & -23 \end{bmatrix}$

Page 194 Practice Quiz 2

1. $\begin{bmatrix} 1 & 4 & 1 & -2 \\ 2 & -1 & -4 & -1 \end{bmatrix}$

3.

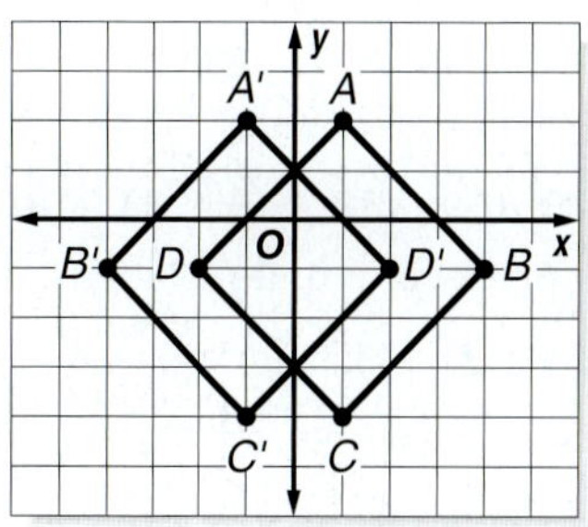

5. −58 **7.** 26 **9.** (4, −5)

Pages 198–201 Lesson 4-7

1. $\begin{bmatrix} 1 & 0 & 0 & 0 \\ 0 & 1 & 0 & 0 \\ 0 & 0 & 1 & 0 \\ 0 & 0 & 0 & 1 \end{bmatrix}$ **3.** Sample answer: $\begin{bmatrix} 3 & 3 \\ 3 & 3 \end{bmatrix}$ **5.** yes

7. no inverse exists **11.** yes **13.** no **15.** yes **17.** true **19.** false **21.** no inverse exists **23.** $\frac{1}{7}\begin{bmatrix} 1 & -1 \\ 4 & 3 \end{bmatrix}$

25. $\frac{1}{4}\begin{bmatrix} -6 & -7 \\ -2 & -3 \end{bmatrix}$ **27.** $-\frac{1}{12}\begin{bmatrix} 6 & 0 \\ -5 & -2 \end{bmatrix}$ **29.** $\frac{1}{32}\begin{bmatrix} 1 & 5 \\ -6 & 2 \end{bmatrix}$

31. $10\begin{bmatrix} \frac{3}{4} & -\frac{5}{8} \\ -\frac{1}{5} & \frac{3}{10} \end{bmatrix}$ **33a.** yes

33b. Sample answer:

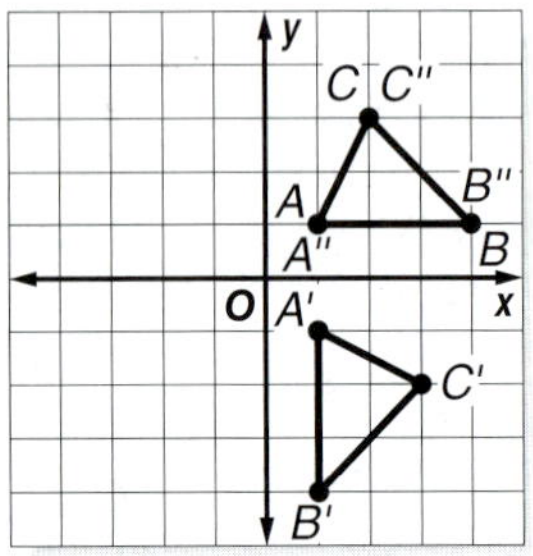

35. $\begin{bmatrix} 0 & -4 & 4 & 8 \\ 0 & 4 & 12 & 8 \end{bmatrix}$ **37.** dilation by a scale factor of $\frac{1}{2}$

39. MEET_IN_THE_LIBRARY **41.** BRING_YOUR_BOOK

43. $a = \pm 1, d = \pm 1, b = c = 0$ **45.** A **47.** $\begin{bmatrix} -5 & -9 \\ -6 & -11 \end{bmatrix}$

49. $\begin{bmatrix} \frac{3}{5} & -\frac{1}{5} \\ \frac{1}{5} & -\frac{2}{5} \end{bmatrix}$ **51.** $\begin{bmatrix} -1 & 1 & \frac{1}{3} \\ -\frac{1}{3} & \frac{2}{3} & 0 \\ \frac{7}{3} & -\frac{8}{3} & -\frac{1}{3} \end{bmatrix}$ **53.** $(2, -4)$

55. $(-5, 4, 1)$ **57.** -14 **59.** 1 **61.** -5 **63.** $\frac{5}{2}$ **65.** 7.82 tons/in^2 **67.** $5\frac{1}{2}$ **69.** 3 **71.** 300 **73.** -2 **75.** 4 **77.** -34

Pages 205–207 Lesson 4-8

1. $2r - 3s = 4, r + 4s = -2$ **3.** Tommy; a 2×1 matrix cannot be multiplied by a 2×2 matrix.

5. $\begin{bmatrix} 2 & 3 \\ -4 & -7 \end{bmatrix} \cdot \begin{bmatrix} g \\ h \end{bmatrix} = \begin{bmatrix} 8 \\ -5 \end{bmatrix}$ **7.** $(5, -2)$ **9.** $(-3, 5)$

11. $h = 1, c = 12$ **13.** $\begin{bmatrix} 4 & -7 \\ 3 & 5 \end{bmatrix} \cdot \begin{bmatrix} x \\ y \end{bmatrix} = \begin{bmatrix} 2 \\ 9 \end{bmatrix}$

15. $\begin{bmatrix} 3 & -7 \\ 6 & 5 \end{bmatrix} \cdot \begin{bmatrix} m \\ n \end{bmatrix} = \begin{bmatrix} -43 \\ -10 \end{bmatrix}$ **17.** $\begin{bmatrix} 3 & -5 & 2 \\ 1 & -7 & 3 \\ 4 & 0 & -3 \end{bmatrix} \cdot \begin{bmatrix} x \\ y \\ z \end{bmatrix} = \begin{bmatrix} 9 \\ 11 \\ -1 \end{bmatrix}$

19. $\begin{bmatrix} 3 & -5 & 6 \\ 11 & -12 & 16 \\ -5 & 8 & -3 \end{bmatrix} \cdot \begin{bmatrix} r \\ s \\ t \end{bmatrix} = \begin{bmatrix} 21 \\ 15 \\ -7 \end{bmatrix}$ **21.** $(3, 4)$ **23.** $(6, 1)$

25. $\left(-\frac{1}{3}, 4\right)$ **27.** $(-2, -2)$ **29.** $(0, 9)$ **31.** $\left(\frac{3}{2}, \frac{1}{3}\right)$ **33.** 2010 **35.** The solution set is the empty set or infinite solutions.

37. D **39.** $(-6, 2, 5)$ **41.** $(0, -1, 3)$ **43.** $\begin{bmatrix} 4 & -5 \\ -7 & 9 \end{bmatrix}$

45. $(4, -2)$ **47.** $(-6, -8)$ **49.** $\{-4, 10\}$ **51.** $\{2, 7\}$

Pages 209–214 Chapter 4 Study Guide and Review

1. identity matrix **3.** Scalar multiplication **5.** determinant **7.** dimensions **9.** equal matrices **11.** $(-5, -1)$ **13.** $(-1, 0)$

15. $\begin{bmatrix} -3 & 0 \\ -2 & -6 \end{bmatrix}$ **17.** $\begin{bmatrix} 1 & -2 \\ -14 & 9 \end{bmatrix}$ **19.** $[-18]$ **21.** not possible

23. $A'(1, 0), B'(8, -2), C'(3, -7)$ **25.** $A'(3, 5), B'(-4, 3), C'(1, -2)$ **27.** 109 **29.** 0 **31.** -52 **33.** $\left(\frac{2}{3}, 5\right)$ **35.** $(-1, -3)$

37. $(1, 2, -1)$ **39.** $-\frac{1}{14}\begin{bmatrix} -2 & -2 \\ -4 & 3 \end{bmatrix}$ **41.** $\frac{1}{24}\begin{bmatrix} 6 & -4 \\ 3 & 2 \end{bmatrix}$

43. $-\frac{1}{10}\begin{bmatrix} -4 & -2 \\ -5 & 0 \end{bmatrix}$ **45.** $(4, 2)$ **47.** $(-3, 1)$

Chapter 5 Polynomials

Page 221 Chapter 5 Getting Started

1. $2 + (-7)$ **3.** $x + (-y)$ **5.** $2xy + (-6yz)$

7. $-8x^3 - 2x + 6$ **9.** $-x + 3$ **11.** $-\frac{3}{2}a - 1$ **13.** 6.3; reals, rationals **15.** 17; reals, rationals, integers, whole numbers, natural numbers **17.** 4; reals, rationals, integers, whole numbers, natural numbers

Pages 226–228 Lesson 5-1

1. Sample answer: $(2x^2)^3 = 8x^6$ since $(2x^2)^3 = (2x^2)^3 \cdot (2x^2)^3 \cdot (2x^2)^3 = 2x^2 \cdot 2x^2 \cdot 2x^2 = 2x \cdot x \cdot 2x \cdot x \cdot 2x \cdot x = 8x^6$

3. Alejandra; when Kyle used the Power of a Product property in his first step, he forgot to put an exponent of -2 on a. Also, in his second step, $(-2)^{-2}$ should be $\frac{1}{4}$, not 4.

5. $16b^4$ **7.** $-6y^2$ **9.** $9p^2q^3$ **11.** $\frac{9}{c^2d^2}$ **13.** 4.21×10^5

15. 3.762×10^3 **17.** about 1.28 s **19.** b^4 **21.** z^{10} **23.** $-8c^3$

25. $-y^3z^2$ **27.** $-21b^5c^3$ **29.** $-24r^7s^5$ **31.** $90a^4b^4$

33. $\frac{a^2c^2}{3b^4}$ **35.** $-\frac{m^4n^9}{3}$ **37.** $\frac{8y^3}{x^6}$ **39.** $\frac{1}{v^3w^6}$ **41.** $\frac{2x^3y^2}{5z^7}$ **43.** 7

45. 4.32×10^4 **47.** 6.81×10^{-3} **49.** 6.754×10^8

51. 6.02×10^{-5} **53.** 6.2×10^{10} **55.** 1.681×10^{-7}

57. 2×10^{-7} m **59.** about 330,000 times

61. Definition of an exponent

63. Economics often involves large amounts of money. Answers should include the following.

- The national debt in 2000 was five trillion, six hundred seventy-four billion, two hundred million or 5.6742×10^{12} dollars. The population was two hundred eighty-one million or 2.81×10^8.
- Divide the national debt by the population. $\frac{5.6742 \times 10^{12}}{2.81 \times 10^8} \approx \$2.0193 \times 10^4$ or about \$20,193 per person.

65. B **67.** $(-3, 3)$ **69.** $\begin{bmatrix} -\frac{1}{2} & \frac{3}{2} \\ 1 & -2 \end{bmatrix}$ **71.** 7 **73.** $(2, 0, 4)$

75. Sample answer using (0, 4.9) and (28, 8.3): $y = 0.12x + 4.9$ **77.** 7 **79.** $2x + 2y$ **81.** $4x + 8$ **83.** $-5x + 10y$

Pages 231–232 Lesson 5-2

1. Sample answer: $x^5 + x^4 + x^3$

3.

	x	x	x
x	x^2	x^2	x^2
2	x	x	x
	x	x	x

5. yes, 3 **7.** $10a - 2b$ **9.** $6xy + 18x$ **11.** $y^2 - 3y - 70$ **13.** $4z^2 - 1$ **15.** $7.5x^2 + 12.5x$ ft^2 **17.** yes, 3 **19.** no **21.** yes, 7 **23.** $-3y - 3y^2$ **25.** $10m^2 + 5m - 15$

27. $7x^2 - 8xy + 4y^2$ **29.** $12a^3 + 4ab$
31. $6x^2y^4 - 8x^2y^2 + 4xy^5$ **33.** $2a^4 - 3a^3b + 4a^4b^4$
35. $-0.001x^2 + 5x - 500$ **37.** $p^2 + 2p - 24$ **39.** $b^2 - 25$
41. $6x^2 + 34x + 48$ **43.** $a^6 - b^2$ **45.** $x^2 - 6xy + 9y^2$
47. $d^2 - 2 + \frac{1}{d^4}$ **49.** $27b^3 - 27b^2c + 9bc^2 - c^3$
51. $9c^2 - 12cd + 7d^2$ **53.** $R^2 + 2RW + W^2$
55. The expression for how much an amount of money will grow to is a polynomial in terms of the interest rate. Answers should include the following.

- If an amount A grows by r percent for n years, the amount will be $A(1 + r)^n$ after n years. When this expression is expanded, a polynomial results.
- $8820(1 + r)^3$, $8820r^3 + 26{,}460r^2 + 26{,}460r + 8820$
- Evaluate one of the expressions when $r = 0.04$. For example, $8820(1 + r)^3 = 8820(1.04)^3$ or \$9921.30 to the nearest cent. The value given in the table is \$9921 rounded to the nearest dollar.

57. B **59.** $20r^3t^4$ **61.** $\frac{b^2}{4a^2}$

63.

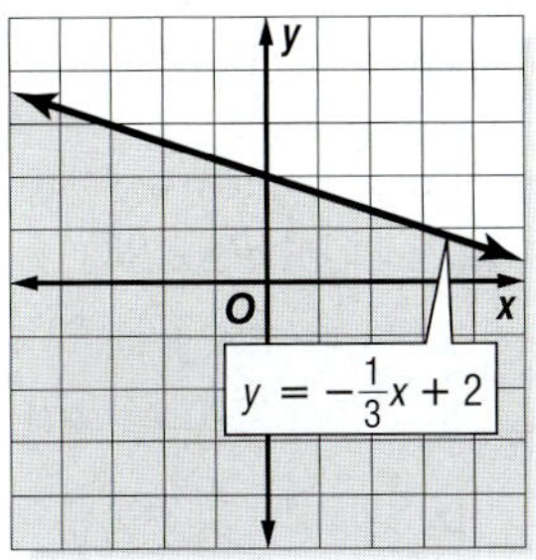

65.

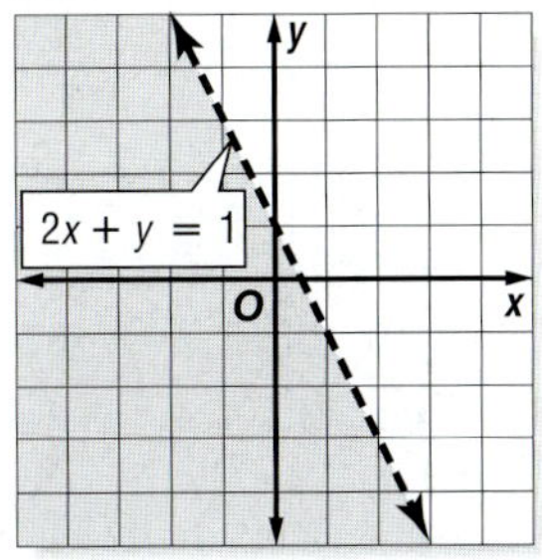

67. $2y^3$ **69.** $3a^2$

Pages 236–238 Lesson 5-3

1. Sample answer: $(x^2 + x + 5) \div (x + 1)$ **3.** Jorge; Shelly is subtracting in the columns instead of adding.
5. $5b - 4 + 7a$ **7.** $3a^3 - 9a^2 + 7a - 6$ **9.** $x^2 - xy + y^2$
11. $b^3 + b - 1$ **13.** $3b + 5$ **15.** $3ab - 6b^2$ **17.** $2c^2 - 3d + 4d^2$ **19.** $2y^2 + 4yz - 8y^3z^4$ **21.** $b^2 + 10b$ **23.** $n^2 - 2n + 3$
25. $x^3 - 5x^2 + 11x - 22 + \frac{39}{x + 2}$ **27.** x^2 **29.** $y^2 - y - 1$
31. $a^3 - 6a^2 - 7a + 7 + \frac{3}{a + 1}$
33. $x^4 - 3x^3 + 2x^2 - 6x + 19 - \frac{56}{x + 3}$ **35.** $g + 5$
37. $t^4 + 2t^3 + 4t^2 + 5t + 10$ **39.** $3t^2 - 2t + 3$
41. $3d^2 + 2d + 3 - \frac{2}{3d - 2}$ **43.** $x^3 - x - \frac{6}{2x + 3}$ **45.** $x - 3$
47. $x + 2$ **49.** $x^2 - x + 3$ **51.** $\$0.03x + 4 + \frac{1000}{x}$
53. $170 - \frac{170}{t^2 + 1}$ **55.** $x^3 + x^2 + 6x - 24$ ft
57. $x^2 + 3x + 12$ ft/s **59.** Division of polynomials can be used to solve for unknown quantities in geometric formulas that apply to manufacturing situations. Answers should include the following.

- $8x$ in. by $4x + s$ in.
- The area of a rectangle is equal to the length times the width. That is, $A = \ell w$.
- Substitute $32x^2 + x$ for A, $8x$ for ℓ, and $4x + s$ for w. Solving for s involves dividing $32x^2 + x$ by $8x$.

$$A = \ell w$$
$$32x^2 + x = 8x(4x + s)$$
$$\frac{32x^2 + x}{8x} = 4x + s$$
$$4x + \frac{1}{8} = 4x + s$$
$$\frac{1}{8} = s$$

The seam is $\frac{1}{8}$ inch.

61. D **63.** $y^4z^4 - y^3z^3 + 3y^2z$ **65.** $a^2 - 2ab + b^2$
67. $y = -x + 2$ **69.** 9 **71.** 4 **73.** 6

Page 238 Practice Quiz 1

1. 6.53×10^8 **3.** $-108x^8y^3$ **5.** $\frac{x^2}{z^6}$ **7.** $3t^2 + 2t - 8$
9. $m^2 - 3 - \frac{19}{m - 4}$

Pages 242–244 Lesson 5-4

1. Sample answer: $x^2 + 2x + 1$ **3.** sometimes
5. $a(a + 5 + b)$ **7.** $(y - 2)(y - 4)$ **9.** $3(b - 4)(b + 4)$
11. $(h + 20)(h^2 - 20h + 400)$ **13.** $\frac{2y}{y - 4}$ **15.** $2x(y^3 - 5)$
17. $2cd^2(6d - 4c + 5c^4d)$ **19.** $(2z - 3)(4y - 3)$
21. $(x + 1)(x + 6)$ **23.** $(2a + 1)(a + 1)$ **25.** $(2c + 3)(3c + 2)$
27. $3(n + 8)(n - 1)$ **29.** $(x + 6)^2$ **31.** prime
33. $(y^2 + z)(y^2 - z)$ **35.** $(z + 5)(z^2 - 5z + 25)$
37. $(p^2 + 1)(p + 1)(p - 1)$ **39.** $(7a + 2b)(c + d)(c - d)$
41. $(a - b)(5ax + 4by + 3cz)$ **43.** $(3x - 2)(x + 1)$
45. 30 ft by 40 ft **47.** $\frac{x + 5}{x - 6}$ **49.** $\frac{x - 4}{x^2 + 2x + 4}$ **51.** $x + 2$
53. $16x + 16$ ft/s **55.** $(8p^n + 1)^2$ **57.** B **59.** yes **61.** no; $(2x + 1)(x - 3)$ **63.** $t^2 - 2t + 1$ **65.** $x^2 + 2$
67. $4x^2 + 3xy - 3y^2$ **69.** $[-2]$ **71.** 15 in. by 28 in. **73.** no
75. Associative Property (+) **77.** irrational **79.** rational
81. rational

Pages 247–249 Lesson 5-5

1. Sample answer: 64 **3.** Sometimes; it is true when $x > 0$.
5. -2.668 **7.** 4 **9.** -3 **11.** x **13.** $6|a|b^2$ **15.** about 3.01 mi
17. -12.124 **19.** 2.066 **21.** -7.830 **23.** 3.890 **25.** 4.647
27. 59.161 **29.** ± 13 **31.** 18 **33.** -2 **35.** $\frac{1}{5}$ **37.** -0.4
39. $-|x|$ **41.** $8a^4$ **43.** $-c^2$ **45.** $4z^2$ **47.** $6x^2z^2$
49. $3p^6|q^3|$ **51.** $-3c^3d^4$ **53.** $p + q$ **55.** $|z + 4|$ **57.** not a real number **59.** -5 **61.** about 1.35 m **63.** $x = 0$ and $y \geq 0$, or $y = 0$ and $x \geq 0$ **65.** B **67.** $7xy^2(y - 2xy^3 + 4x^2)$
69. $(2x + 5)(x + 5)$ **71.** $4x^2 + x + 5 + \frac{8}{x - 2}$
73. $\begin{bmatrix} 810 & 2320 \\ 1418 & 2504 \end{bmatrix}$ **75.** $(1, -3)$ **77.** $x^2 + 11x + 24$
79. $a^2 - 7a - 18$ **81.** $x^2 - 9y^2$

Pages 254–256 Lesson 5-6

1. Sometimes; $\frac{1}{\sqrt[n]{a}} = \sqrt[n]{a}$ only when $a = 1$. **3.** The product of two conjugates yields a difference of two squares. Each square produces a rational number and the difference of two rational numbers is a rational number. **5.** $2x|y|\sqrt[4]{x}$
7. $-24\sqrt{35}$ **9.** $2a^2b^2\sqrt{3}$ **11.** $22\sqrt[3]{2}$ **13.** $2 + \sqrt{5}$
15. $9\sqrt{3}$ **17.** $3\sqrt[3]{2}$ **19.** $5x^2\sqrt{2}$ **21.** $3|x|y\sqrt{2y}$
23. $6y^2z\sqrt[3]{7}$ **25.** $\frac{1}{3}c|d|\sqrt[4]{c}$ **27.** $\frac{\sqrt[3]{6}}{2}$ **29.** $\frac{a^2\sqrt{b}}{b^2}$ **31.** $36\sqrt{7}$
33. $\frac{\sqrt{6}}{2}$ **35.** $3\sqrt{3}$ **37.** $7\sqrt{3} - 2\sqrt{2}$
39. $25 - 5\sqrt{2} + 5\sqrt{6} - 2\sqrt{3}$ **41.** $13 - 2\sqrt{22}$
43. $\frac{28 + 7\sqrt{3}}{13}$ **45.** $\frac{-1 - \sqrt{3}}{2}$ **47.** $\frac{\sqrt{x^2 - 1}}{x - 1}$ **49.** $6 + 16\sqrt{2}$ yd, $24 + 6\sqrt{2}$ yd² **51.** 0 ft/s **53.** about 18.18 m **55.** x and y are nonnegative. **57.** B **59.** $12z^4$ **61.** $|y + 2|$
63. $\frac{x + 1}{x + 4}$ **65.** $\begin{bmatrix} 1 & 4 \\ -5 & -4 \end{bmatrix}$ **67.** consistent and independent
69. -5 **71.** $-2, 4$ **73.** $\{x \mid x > 6\}$ **75.** $\frac{1}{4}$ **77.** $\frac{5}{6}$ **79.** $\frac{13}{24}$
81. $\frac{3}{8}$

Selected Answers

Page 256 Practice Quiz 2

1. $x^2y(3x + y + 1)$ **3.** $a(x + 3)^2$ **5.** $6|x||y^3|$ **7.** $|2n + 3|$ **9.** $-1 - \sqrt{7}$

Pages 260–262 Lesson 5-7

1. Sample answer: 64 **3.** In exponential form $\sqrt[n]{b^m}$ is equal to $(b^m)^{\frac{1}{n}}$. By the Power of a Power Property, $(b^m)^{\frac{1}{n}} = b^{\frac{m}{n}}$. But, $b^{\frac{m}{n}}$ is also equal to $\left(b^{\frac{1}{n}}\right)^m$ by the Power of a Power Property. This last expression is equal to $\left(\sqrt[n]{b}\right)^m$. Thus, $\sqrt[n]{b^m} = \left(\sqrt[n]{b}\right)^m$. **5.** $\sqrt[3]{x^2}$ or $\left(\sqrt[3]{x}\right)^2$ **7.** $6^{\frac{1}{3}}x^{\frac{5}{3}}y^{\frac{7}{3}}$ **9.** $\frac{1}{3}$ **11.** 2 **13.** $x^{\frac{2}{3}}$ **15.** $a^{\frac{3}{2}}b^{\frac{2}{3}}$ **17.** $\frac{z(x - 2y)^{\frac{1}{2}}}{x - 2y}$ **19.** $\sqrt{3}$ **21.** $\sqrt[5]{6}$ **23.** $\sqrt[5]{c^2}$ or $\left(\sqrt[5]{c}\right)^2$ **25.** $23^{\frac{1}{2}}$ **27.** $2z^{\frac{1}{2}}$ **29.** 2 **31.** $\frac{1}{5}$ **33.** $\frac{1}{9}$ **35.** 81 **37.** $\frac{2}{3}$ **39.** $\frac{4}{3}$ **41.** y^4 **43.** $b^{\frac{1}{5}}$ **45.** $\frac{w^{\frac{1}{5}}}{w}$ **47.** $t^{\frac{1}{4}}$ **49.** $\frac{a^{\frac{5}{12}}}{6a}$ **51.** $\frac{y^2 - 2y^{\frac{3}{2}}}{y - 4}$ **53.** $\sqrt{5}$ **55.** $17\sqrt[6]{17}$ **57.** $\sqrt[4]{5x^2y^2}$ **59.** $\frac{xy\sqrt{z}}{z}$ **61.** $\sqrt{2}$ **63.** $2\sqrt{6} - 5$ **65.** $2^{\frac{3}{2}} + 3^{\frac{1}{2}}$

67. 880 vibrations per second **69.** about 336

71. The equation that determines the size of the region around a planet where the planet's gravity is stronger than the Sun's can be written in terms of a fractional exponent. Answers should include the following.

- The radical form of the equation is

$$r = D\sqrt[5]{\left(\frac{M_p}{M_s}\right)^2} \text{ or } r = D\sqrt[5]{\frac{M_p^2}{M_s^2}}.$$ Multiply the fraction under the radical by $\frac{M_s^3}{M_s^3}$.

$$r = D\sqrt[5]{\frac{M_p^2}{M_s^2} \cdot \frac{M_s^3}{M_s^3}}$$

$$= D\sqrt[5]{\frac{M_p^2M_s^3}{M_s^5}}$$

$$= D\frac{\sqrt[5]{M_p^2M_s^3}}{\sqrt[5]{M_s^5}}$$

$$= \frac{D\sqrt[5]{M_p^2M_s^3}}{M_s}$$

The simplified radical form is $r = \frac{D\sqrt[5]{M_p^2M_s^3}}{M_s}$.

- If M_p and M_s are constant, then r increases as D increases because r is a linear function of D with positive slope.

73. C **75.** $36\sqrt{2}$ **77.** 8 **79.** $\frac{1}{2}x^2$ **81.** $x - 2$ **83.** $x + 2\sqrt{x} + 1$

Pages 265–267 Lesson 5-8

1. Since x is not under the radical, the equation is a linear equation, not a radical equation. The solution is $x = \frac{\sqrt{3} - 1}{2}$. **3.** Sample answer: $\sqrt{x} + \sqrt{x + 3} = 3$ **5.** -9 **7.** 15 **9.** 31 **11.** $0 \le b < 4$ **13.** 16 **15.** no solution **17.** 9 **19.** -1 **21.** -20 **23.** no solution **25.** $x > 1$ **27.** $x \le -11$ **29.** no solution **31.** 3 **33.** $0 \le x \le 2$ **35.** $b \ge 5$ **37.** 3 **39.** 1152 lb **41.** 34 ft **43.** Since $\sqrt{x + 2} \ge 0$ and $\sqrt{2x - 3} \ge 0$, the left side of the equation is nonnegative. Therefore, the left side of the equation cannot equal -1. Thus, the equation has no solution. **45.** D **47.** $5^{\frac{3}{7}}$

49. $(x^2 + 1)^{\frac{2}{3}}$ **51.** $\frac{\sqrt[3]{100}}{10}$

53. $x + y = 7, 30x + 20y = 160; (2, 5)$

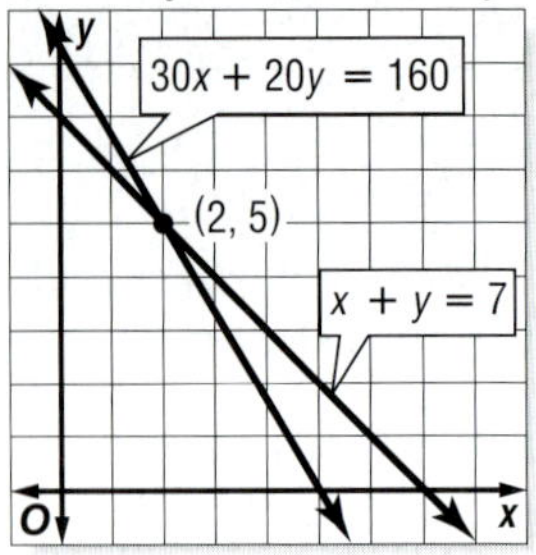

55. $1 - y$ **57.** -11 **59.** $-3 - 10x - 8x^2$

Pages 273–275 Lesson 5-9

1a. true **1b.** true **3.** Sample answer: $1 + 3i$ and $1 - 3i$ **5.** $5i|xy|\sqrt{2}$ **7.** $-180\sqrt{3}$ **9.** $6 + 3i$ **11.** $\frac{7}{17} - \frac{11}{17}i$ **13.** $\pm 2i\sqrt{2}$ **15.** $3, -3$ **17.** $10 + 3j$ amps **19.** $9i$ **21.** $10a^2|b|i$ **23.** -12 **25.** $-75i$ **27.** 1 **29.** $-i$ **31.** 6 **33.** $4 - 5i$ **35.** $6 - 7i$ **37.** $-8 + 4i$ **39.** $\frac{10}{17} - \frac{6}{17}i$ **41.** $\frac{2}{5} + \frac{1}{5}i$ **43.** $20 + 15i$ **45.** $-\frac{1}{3} - \frac{2\sqrt{2}}{3}i$ **47.** $(5 - 2i)x^2 + (-1 + i)x + 7 + i$ **49.** $\pm 4i$ **51.** $\pm 2i\sqrt{3}$ **53.** $\pm 2i\sqrt{10}$ **55.** $\pm\frac{\sqrt{5}}{2}i$ **57.** $4, -3$ **59.** $\frac{5}{3}, 4$ **61.** $\frac{67}{11}, \frac{19}{11}$ **63.** $13 + 18j$ volts

65. Case 1: $i > 0$

Multiply each side by i to get $i^2 > 0 \cdot i$ or $-1 > 0$. This is a contradiction.

Case 2: $i < 0$

Since you are assuming i is negative in this case, you must change the inequality symbol when you multiply each side by i. The result is again $i^2 > 0 \cdot i$ or $-1 > 0$, a contradiction.

Since both possible cases result in contradictions, the order relation "$<$" cannot be applied to the complex numbers.

67. C **69.** $-1, -i, 1, i, -1, -i, 1, i, -1$ **71.** 12 **73.** 4 **75.** $y^{\frac{1}{3}}$ **77.** $\begin{bmatrix} 2 & 1 & -2 \\ 3 & -2 & 1 \end{bmatrix}$ **79.** $\begin{bmatrix} 2 & 1 & -2 \\ -3 & 2 & -1 \end{bmatrix}$

81. sofa: \$1200, love seat: \$600, coffee table: \$250

83.

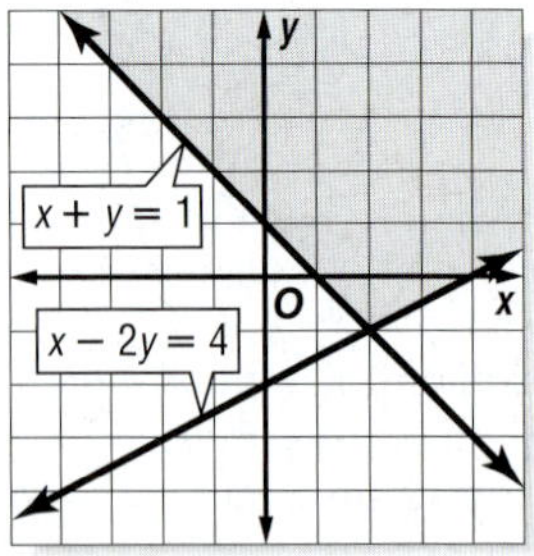

85. 0

Pages 276–280 Chapter 5 Study Guide and Review

1. scientific notation **3.** FOIL method **5.** extraneous solution **7.** square root **9.** principal root **11.** $\frac{1}{f^3}$ **13.** $8xy^4$ **15.** 1.7×10^8 **17.** 9×10^2 **19.** $4x^2 + 22x - 34$ **21.** $x^3y + x^2y^4$ **23.** $4a^4 + 24a^2 + 36$ **25.** $2x^3 + x - \frac{3}{x - 3}$ **27.** $x - 4$ **29.** $50(2x + 1)(2x - 1)$ **31.** $(5w^2 + 3)(w - 4)$ **33.** $(s + 8)(s^2 - 8s + 64)$ **35.** ± 16 **37.** 8 **39.** $|x^4 - 3|$ **41.** $2m^2$ **43.** $2\sqrt[6]{2}$ **45.** $-5\sqrt{3}$ **47.** $20 + 8\sqrt{6}$ **49.** 9

51. $\frac{2\sqrt{10}-\sqrt{5}}{7}$ **53.** 81 **55.** $\frac{y^{\frac{3}{5}}}{y}$ **57.** $3x^{\frac{5}{3}}+4x^{\frac{8}{3}}$ **59.** 343
61. 4 **63.** 5 **65.** 8 **67.** $8m^6i$ **69.** 72 **71.** $23+14i$ **73.** i
75. $\frac{-3-21i}{10}$

Chapter 6 Quadratic Functions and Inequalities

Page 284 Chapter 6 Getting Started

1.

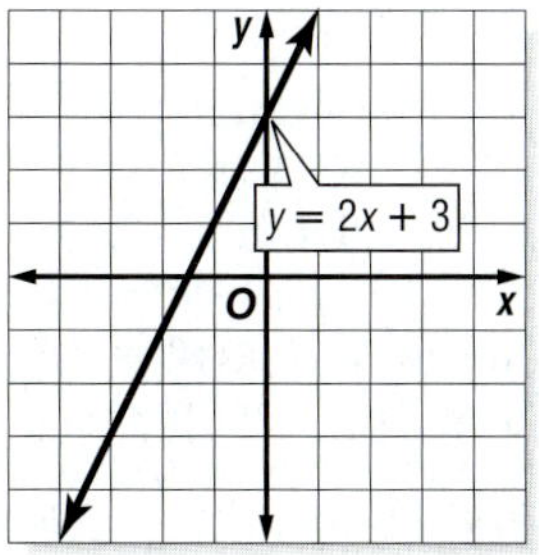

3.

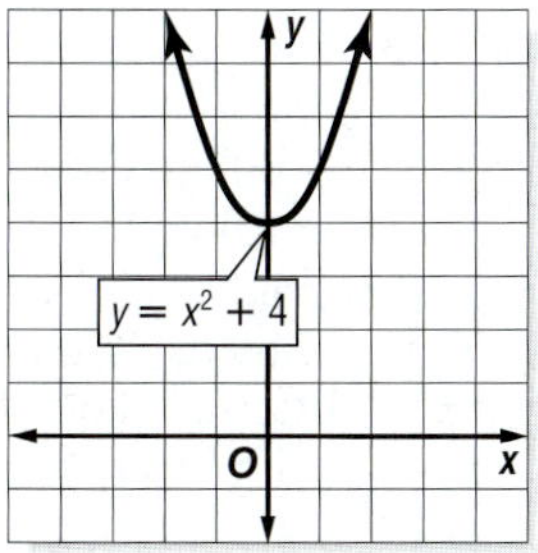

5. $7x^2-16x-48$ **7.** $9x^2-6x+1$ **9.** $(x+6)(x+5)$
11. $(x-8)(x+7)$ **13.** prime **15.** $(x-11)^2$ **17.** 15
19. $6\sqrt{5}$ **21.** $5i$ **23.** $3i\sqrt{30}$

Pages 290–293 Lesson 6-1

1. Sample answer: $f(x)=3x^2+5x-6$; $3x^2$, $5x$, -6
3a. up; min. **3b.** down; max. **3c.** down; max.
3d. up; min. **5a.** 0; $x=-1$; -1

5b.

x	f(x)
−3	3
−2	0
−1	−1
0	0
1	3

5c.

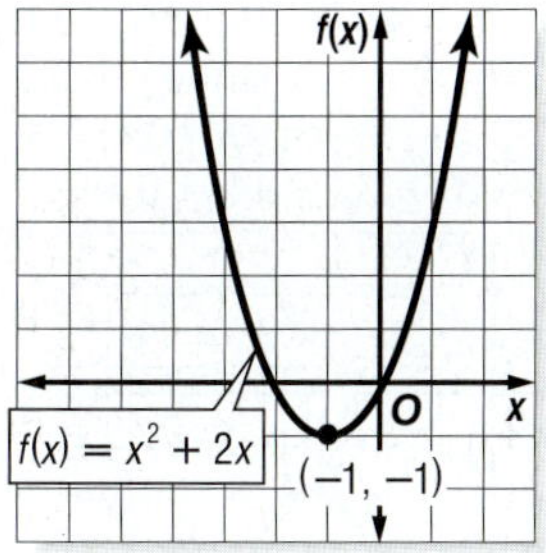

7a. 3; $x=-4$; -4

7b.

x	f(x)
−6	−9
−5	−12
−4	−13
−3	−12
−2	−9

7c.

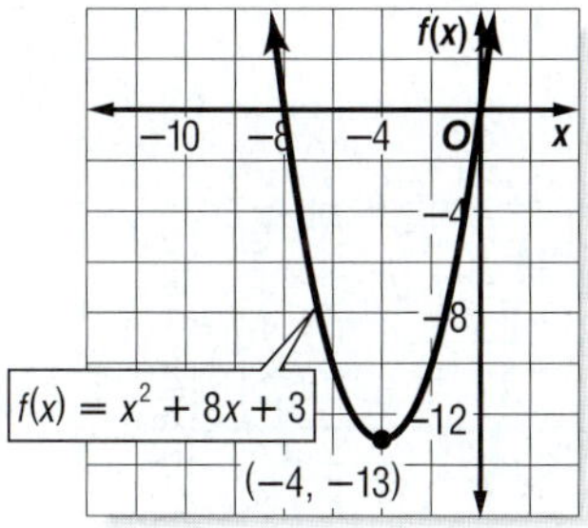

9a. 0; $x=-\frac{5}{3}$; $-\frac{5}{3}$

9b.

x	f(x)
−3	−3
−2	−8
$-\frac{5}{3}$	$-\frac{25}{3}$
−1	−7
0	0

9c.

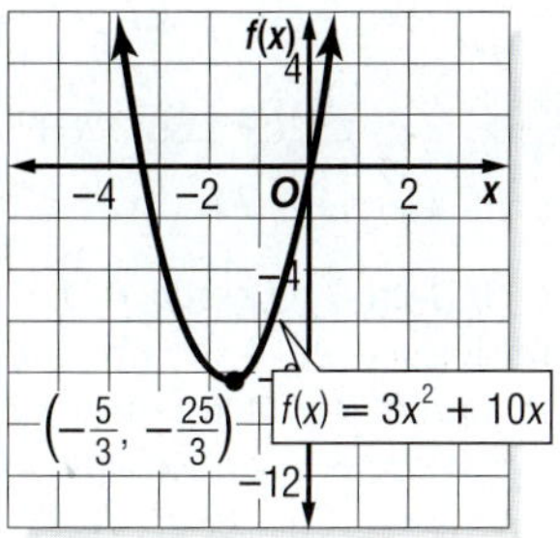

11. min.; $-\frac{25}{4}$ **13.** \$8.75 **15a.** 0; $x=0$; 0

15b.

x	f(x)
−2	−20
−1	−5
0	0
1	−5
2	−20

15c.

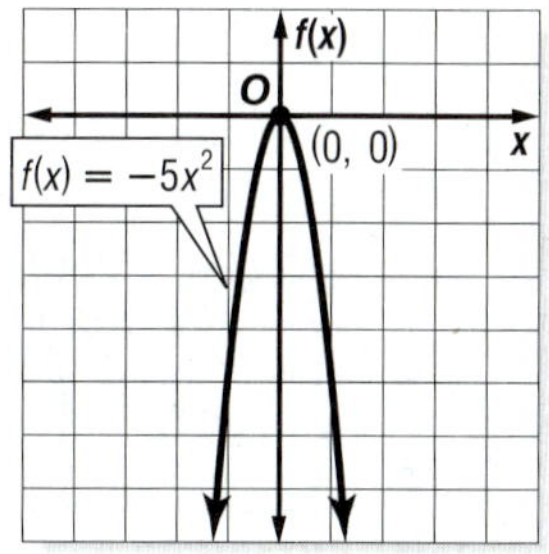

17a. -9; $x=0$; 0

17b.

x	f(x)
−2	−5
−1	−8
0	−9
1	−8
2	−5

17c.

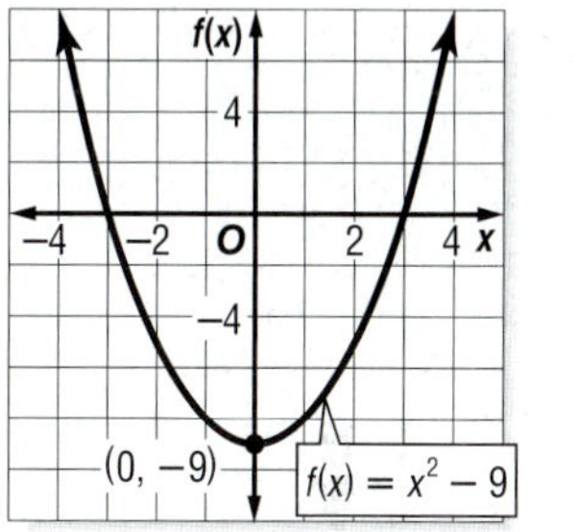

19a. 1; $x=0$; 0

19b.

x	f(x)
−2	13
−1	4
0	1
1	4
2	13

19c.

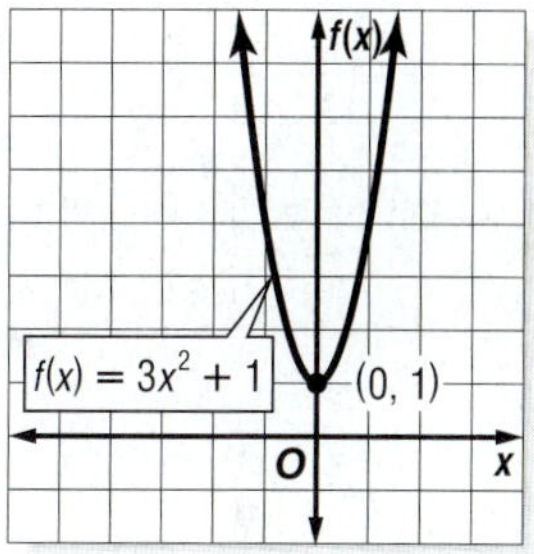

21a. 9; $x=4.5$; 4.5

21b.

x	f(x)
3	−9
4	−11
4.5	−11.25
5	−11
6	−9

21c.

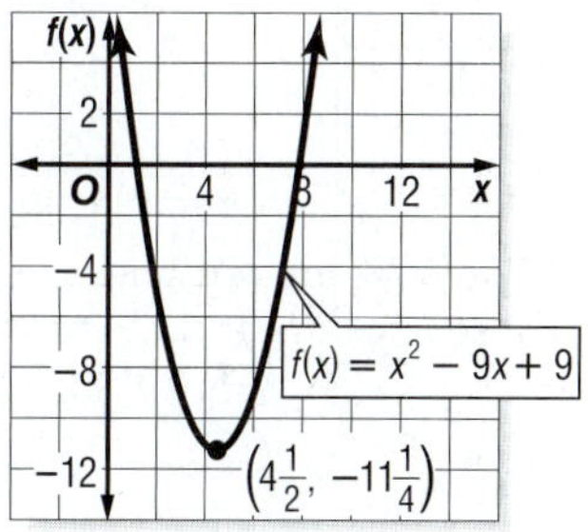

23a. 36; $x=-6$; -6

23b.

x	f(x)
−8	4
−7	1
−6	0
−5	1
−4	4

23c.

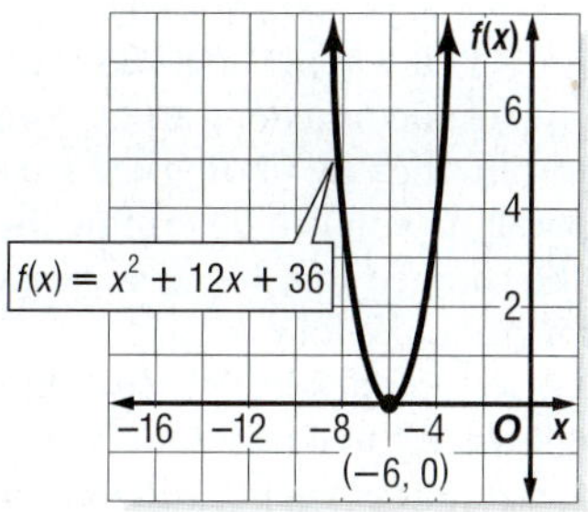

25a. $-3; x = 2, 2$

25b.

x	f(x)
0	−3
1	3
2	5
3	3
4	−3

25c.

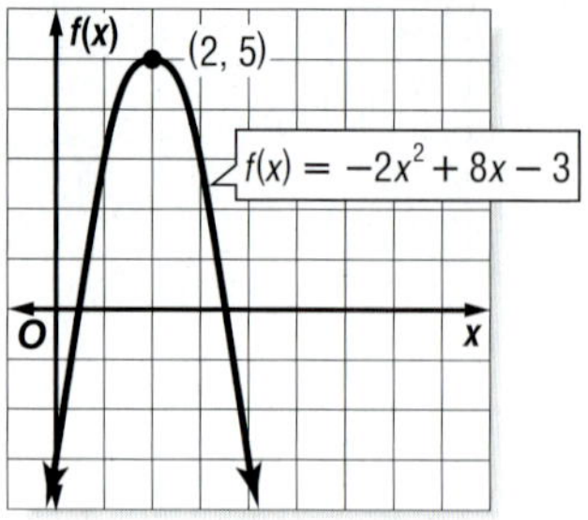

27a. $0; x = -\frac{5}{4}; -\frac{5}{4}$

27b.

x	f(x)
−3	−3
−2	−2
$-\frac{5}{4}$	$-\frac{25}{8}$
−1	−3
0	0

27c.

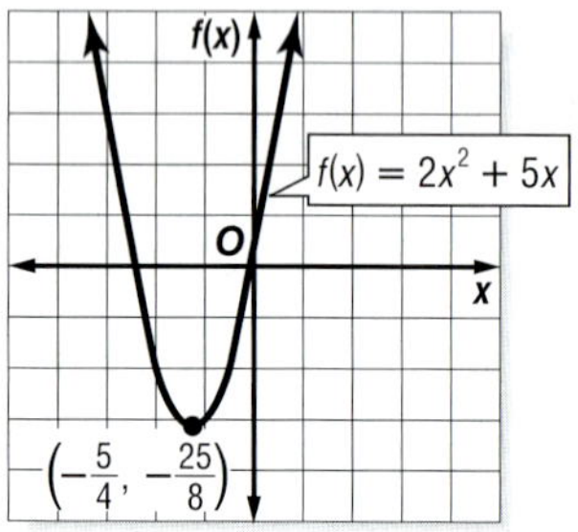

29a. $0; x = -6; -6$

29b.

x	f(x)
−8	8
−7	8.75
−6	9
−5	8.75
−4	8

29c.

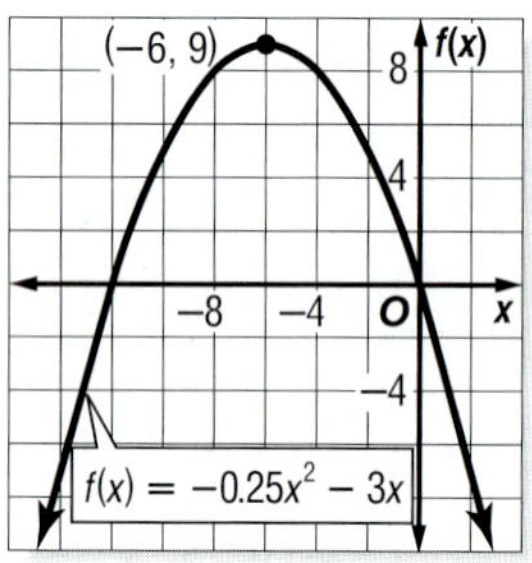

31a. $-\frac{8}{9}; x = \frac{1}{3}; \frac{1}{3}$

31b.

x	f(x)
−1	$\frac{7}{9}$
0	$-\frac{8}{9}$
$\frac{1}{3}$	−1
1	$-\frac{5}{9}$
2	$1\frac{7}{9}$

31c.

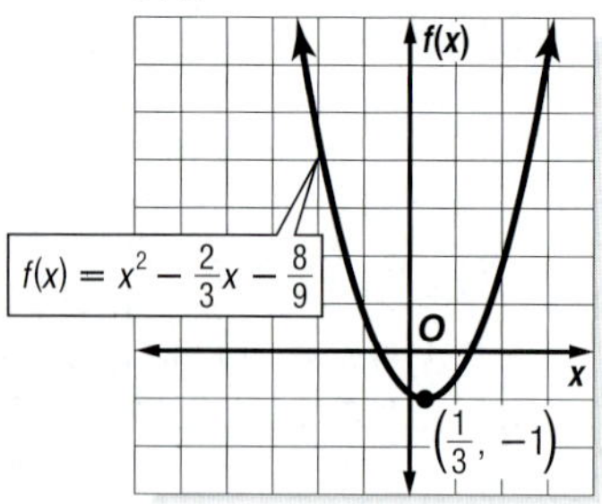

33. max.; −9 **35.** min.; −11 **37.** max.; 12 **39.** max.; $-\frac{7}{8}$ **41.** min.; −11 **43.** min.; $-10\frac{1}{3}$ **45.** 40 m
47. The y-intercept is the initial height of the object.
49. 60 ft by 30 ft **51.** \$11.50 **53.** 5 in. by 4 in.
55. If a quadratic function can be used to model ticket price versus profit, then by finding the x-coordinate of the vertex of the parabola you can determine the price per ticket that should be charged to achieve maximum profit. Answers should include the following.

- If the price of a ticket is too low, then you won't make enough money to cover your costs, but if the ticket price is too high fewer people will buy them.
- You can locate the vertex of the parabola on the graph of the function. It occurs when $x = 40$. Algebraically, this is found by calculating $x = -\frac{b}{2a}$ which, for this case, is $x = \frac{-4000}{2(-50)}$ or 40. Thus the ticket price should be set at \$40 each to achieve maximum profit.

57. C **59.** 3.20 **61.** 3.38 **63.** 1.56 **65.** $-1 + 3i$ **67.** 23

69. 4 **71.** $[5 \quad -13 \quad 8]$ **73.** $\begin{bmatrix} 6 & 0 & -24 \\ 14 & -\frac{2}{3} & -8 \end{bmatrix}$ **75.** 5 **77.** −2

Pages 297–299 Lesson 6-2

1a. The solution is the value that satisfies an equation. **1b.** A root is a solution of an equation. **1c.** A zero is the x value of a function that makes the function equal to 0. **1d.** An x-intercept is the point at which a graph crosses the x-axis. The solutions, or roots, of a quadratic equation are the zeros of the related quadratic function. You can find the zeros of a quadratic function by finding the x-intercepts of its graph. **3.** The x-intercepts of the related function are the solutions to the equation. You can estimate the solutions by stating the consecutive integers between which the x-intercepts are located. **5.** −2, 1 **7.** −7, 0 **9.** −7, 4 **11.** between −2 and −1, 3 **13.** −2, 7 **15.** 3 **17.** 0 **19.** no real solutions **21.** 0, 4 **23.** between −1 and 0; between 2 and 3 **25.** 3, 6 **27.** 6 **29.** $-\frac{1}{2}, 2\frac{1}{2}$ **31.** $-2\frac{1}{2}, 3$ **33.** between 0 and 1; between 3 and 4 **35.** between −3 and −2; between 2 and 3 **37.** no real solutions
39. Let x be the first number.
Then, $7 - x$ is the other number.

$$x(7 - x) = 14$$
$$-x^2 + 7x - 14 = 0$$

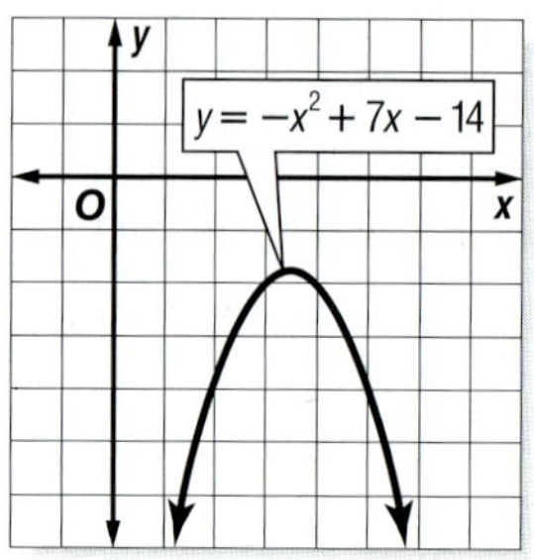

Since the graph of the related function does not intersect the x-axis, this equation has no real solutions. Therefore no such numbers exist. **41.** −2, 14 **43.** 3 s **45.** about 35 mph **47.** −4 and −2; The value of the function changes from negative to positive, therefore the value of the function is zero between these two numbers. **49.** A **51.** −1 **53.** 3, 5 **55.** ±1.33

57. $4, x = 3; 3$

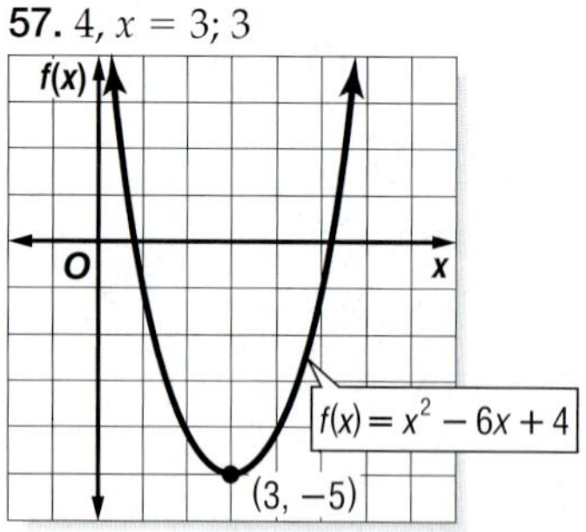

59. $4; x = -6; -6$

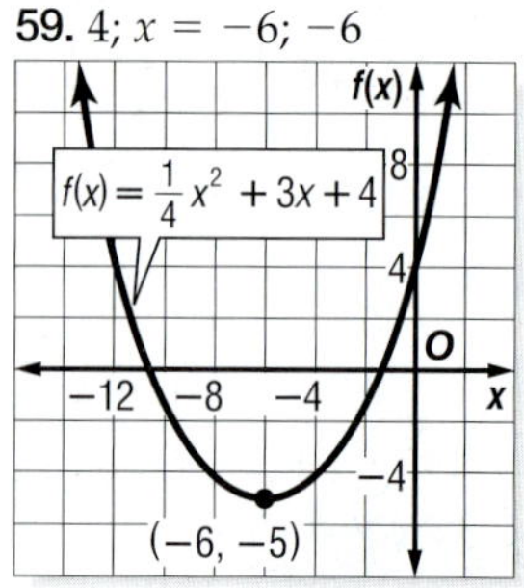

61. $\frac{10}{13} + \frac{2}{13}i$ **63.** 24 **65.** −60 **67.** $x(x + 5)$
69. $(x - 7)(x - 4)$ **71.** $(3x + 2)(x + 2)$

Pages 303–305 Lesson 6-3

1. Sample answer: If the product of two factors is zero, then at least one of the factors must be zero. **3.** Kristin; the Zero Product Property applies only when one side of the equation is 0. **5.** {−8, 2} **7.** {3} **9.** {−3, 4}

11. $6x^2 - 11x + 4 = 0$ **13.** D **15.** $\{-4, 7\}$ **17.** $\{-9, 9\}$ **19.** $\{-3, 7\}$ **21.** $\left\{0, -\frac{3}{4}\right\}$ **23.** $\{8\}$ **25.** $\left\{\frac{1}{4}, 4\right\}$ **27.** $\left\{-\frac{2}{3}, -\frac{3}{2}\right\}$ **29.** $\left\{\frac{3}{4}, \frac{9}{4}\right\}$ **31.** $\{-3, 1\}$ **33.** 0, −3, 3 **35.** $x^2 - 5x - 14 = 0$ **37.** $x^2 + 14x + 48 = 0$ **39.** $3x^2 - 16x + 5 = 0$ **41.** $10x^2 + 23x + 12 = 0$ **43.** 14, 16 or −14, −16 **45.** $B = D^2 - 8D + 16$

47. $y = (x - p)(x - q)$

$y = x^2 - px - qx + pq$

$y = x^2 - (p + q)x + pq$

$a = 1, b = -(p + q), c = +pq$

axis of symmetry: $x = -\frac{b}{2a}$

$x = -\frac{-(p + q)}{2(1)}$

$x = \frac{p + q}{2}$

The axis of symmetry is the average of the x-intercepts. Therefore the axis of symmetry is located halfway between the x-intercepts. **49.** −6 **51.** D **53.** −5, 1 **55.** between −1 and 0; between 3 and 4 **57.** $3\sqrt{2} - 2\sqrt{3}$ **59.** $33 + 20\sqrt{2}$ **61.** (3, −5) **63.** $2\sqrt{2}$ **65.** $3\sqrt{3}$ **67.** $2i\sqrt{3}$

Page 305 Practice Quiz 1

1. 4; $x = 2$; 2

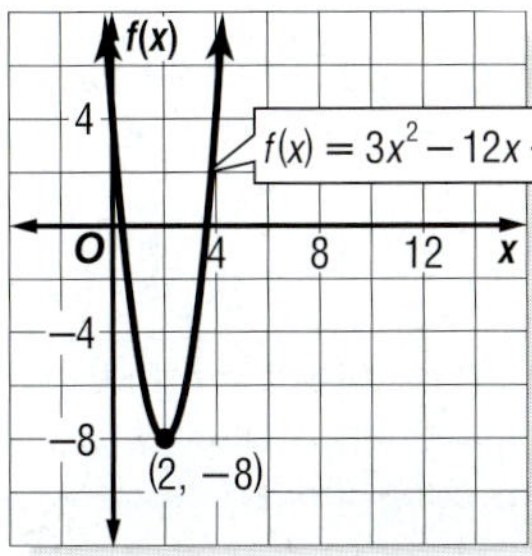

3. $1\frac{1}{2}$, 4

5. $3x^2 + 11x - 4 = 0$

Pages 310–312 Lesson 6-4

1. Completing the square allows you to rewrite one side of a quadratic equation in the form of a perfect square. Once in this form, the equation is solved by using the Square Root Property. **3.** Tia; before completing the square, you must first check to see that the coefficient of the quadratic term is 1. If it is not, you must first divide the equation by that coefficient. **5.** $\left\{\frac{4 \pm \sqrt{2}}{3}\right\}$ **7.** $\frac{9}{4}; \left(x - \frac{3}{2}\right)^2$ **9.** $\{4 \pm \sqrt{5}\}$ **11.** $\left\{\frac{3 \pm \sqrt{33}}{4}\right\}$ **13.** Earth: 4.5 s, Jupiter: 2.9 s **15.** $\{-2, 12\}$ **17.** $\{3 \pm 2\sqrt{2}\}$ **19.** $\left\{\frac{-5 \pm \sqrt{11}}{3}\right\}$ **21.** $\{-1.6, 0.2\}$ **23.** about 8.56 s **25.** 81; $(x - 9)^2$ **27.** $\frac{49}{4}; \left(x + \frac{7}{2}\right)^2$ **29.** 1.44; $(x - 1.2)^2$ **31.** $\frac{25}{16}; \left(x + \frac{5}{4}\right)^2$ **33.** $\{-12, 10\}$ **35.** $\{2 \pm \sqrt{3}\}$ **37.** $\{-3 \pm 2i\}$ **39.** $\left\{\frac{1}{2}, 1\right\}$ **41.** $\left\{\frac{2 \pm \sqrt{10}}{3}\right\}$ **43.** $\left\{\frac{-5 \pm i\sqrt{23}}{6}\right\}$ **45.** $\{0.7, 4\}$ **47.** $\left\{\frac{3}{4} \pm \sqrt{2}\right\}$ **49.** $\frac{x}{1}, \frac{1}{x - 1}$

51. Sample answers: The golden rectangle is found in much of ancient Greek architecture, such as the Parthenon, as well as in modern architecture, such as in the windows of the United Nations building. Many songs have their climax at a point occurring 61.8% of the way through the piece, with 0.618 being about the reciprocal of the golden ratio. The reciprocal of the golden ratio is also used in the design of some violins. **53.** 18 ft by 32 ft or 64 ft by 9 ft

55. D **57.** $x^2 - 3x + 2 = 0$ **59.** $3x^2 - 19x + 6 = 0$ **61.** between −4 and −3; between 0 and 1 **63.** $-4, -1\frac{1}{2}$ **65.** (2, −5) **67.** $|x - (-257)| = 2$ **69.** 37 **71.** 121

Pages 317–319 Lesson 6-5

1a. Sample answer:

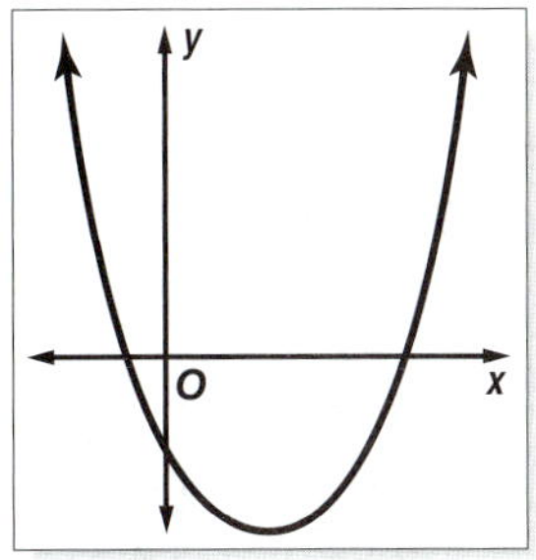

1b. Sample answer:

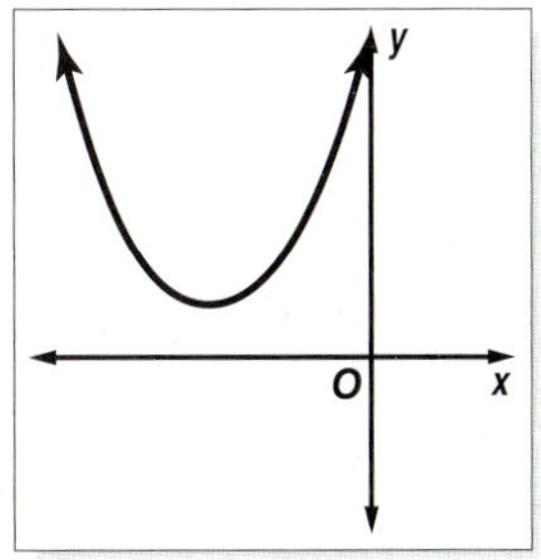

1c. Sample answer:

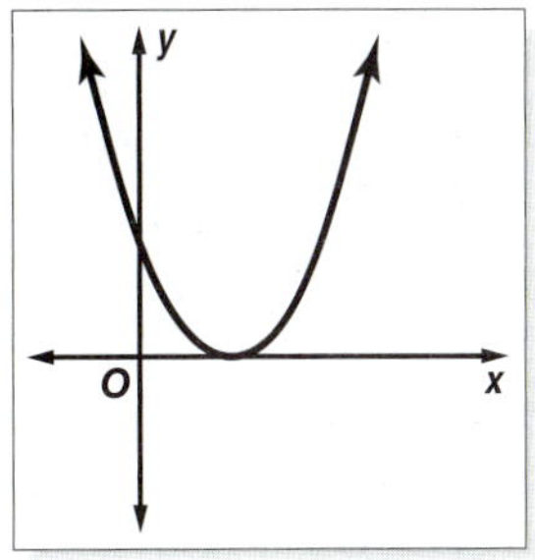

3. $b^2 - 4ac$ must equal 0. **5a.** 8 **5b.** 2 irrational **5c.** $\frac{2 \pm \sqrt{2}}{2}$ **7a.** −3 **7b.** two complex **7c.** $\frac{-3 \pm i\sqrt{3}}{2}$ **9.** −3, −2 **11.** $\frac{-5 \pm i\sqrt{2}}{2}$ **13.** No; the discriminant of $-16t^2 + 85t = 120$ is −455, indicating that the equation has no real solutions. **15a.** 240 **15b.** 2 irrational **15c.** $8 \pm 2\sqrt{15}$ **17a.** −23 **17b.** 2 complex **17c.** $\frac{1 \pm i\sqrt{23}}{2}$ **19a.** 49 **19b.** 2 rational **19c.** $-2, \frac{1}{3}$ **21a.** 24 **21b.** 2 irrational **21c.** $-1 \pm \sqrt{6}$ **23a.** 0 **23b.** one rational **23c.** $-\frac{5}{2}$ **25a.** −135 **25b.** 2 complex **25c.** $\frac{-1 \pm i\sqrt{15}}{4}$ **27a.** 1.48 **27b.** 2 irrational **27c.** $\frac{-1 \pm 2\sqrt{0.37}}{0.8}$ **29.** $\pm i\frac{\sqrt{21}}{7}$ **31.** $\frac{-3 \pm \sqrt{15}}{2}$ **33.** $\frac{9}{2}$ **35.** $\frac{5 \pm \sqrt{46}}{3}$ **37.** $0, -\frac{3}{10}$ **39.** −2, 6 **41.** This means that the cables do not touch the floor of the bridge, since the graph does not intersect the x-axis and the roots are imaginary. **43.** 1998 **45a.** $k = \pm 6$ **45b.** $k < -6$ or $k > 6$ **45c.** $-6 < k < 6$ **47.** D **49.** −14, −4 **51.** $\frac{1 \pm 2\sqrt{2}}{2}$ **53.** −2, 7 **55.** a^4b^{10} **57.** $4b^2c^2$

59.

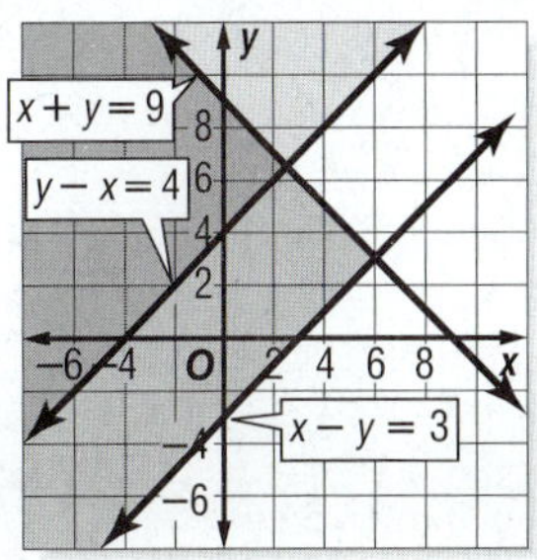

61. no **63.** yes; $(2x + 3)^2$ **65.** no

Selected Answers

Pages 325–328 Lesson 6-6

1a. $y = 2(x + 1)^2 + 5$ **1b.** $y = 2(x + 1)^2$ **1c.** $y = 2(x + 3)^2 + 3$ **1d.** $y = 2(x - 2)^2 + 3$ **1e.** Sample answer: $y = 4(x + 1)^2 + 3$ **1f.** Sample answer: $y = (x + 1)^2 + 3$ **1g.** $y = -2(x + 1)^2 + 3$ **3.** Sample answer: $y = 2(x - 2)^2 - 1$ **5.** $(-3, -1)$; $x = -3$; up **7.** $y = -3(x + 3)^2 + 38$; $(-3, 38)$; $x = -3$; down

9.

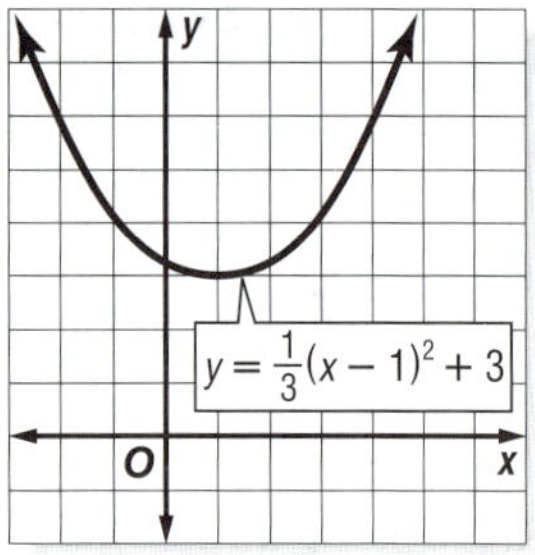

11. $y = 4(x - 2)^2$ **13.** $y = -\frac{1}{2}(x + 2)^2 - 3$ **15.** $(-3, 0)$; $x = -3$; down **17.** $(0, -6)$; $x = 0$; up **19.** $y = -(x + 2)^2 + 12$; $(-2, 12)$; $x = -2$; down **21.** $y = -3(x - 2)^2 + 12$; $(2, 12)$; $x = 2$; down **23.** $y = 4(x + 1)^2 - 7$; $(-1, -7)$; $x = -1$; up **25.** $y = 3\left(x + \frac{1}{2}\right)^2 - \frac{7}{4}$; $\left(-\frac{1}{2}, -\frac{7}{4}\right)$; $x = -\frac{1}{2}$; up

27.

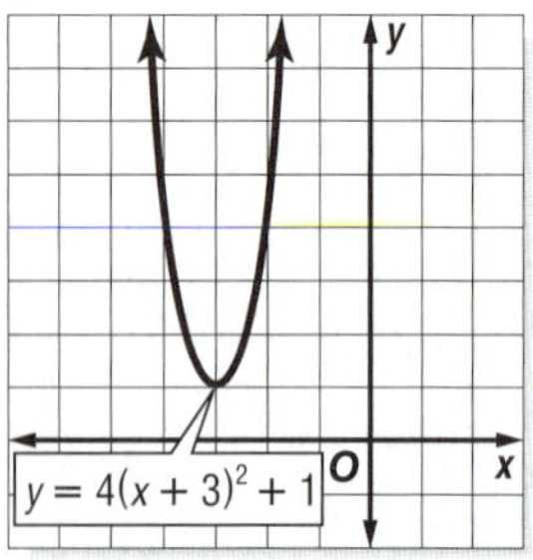

29.

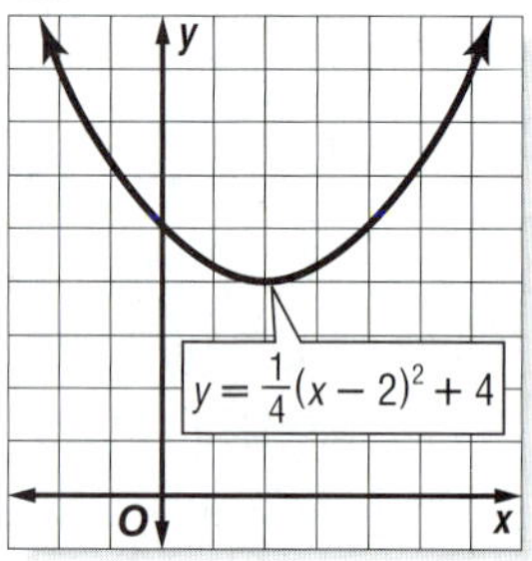

31.

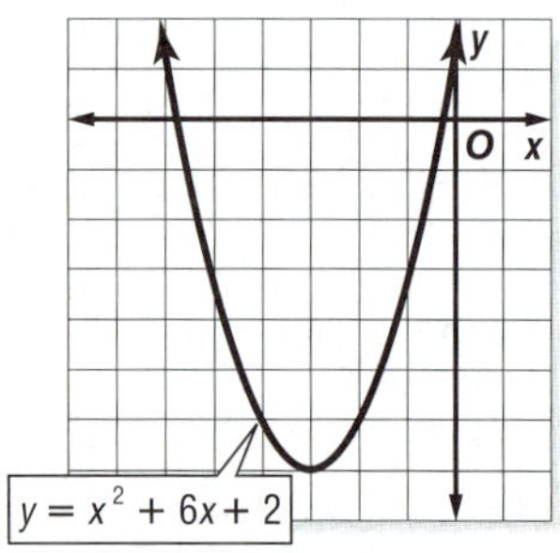

33.

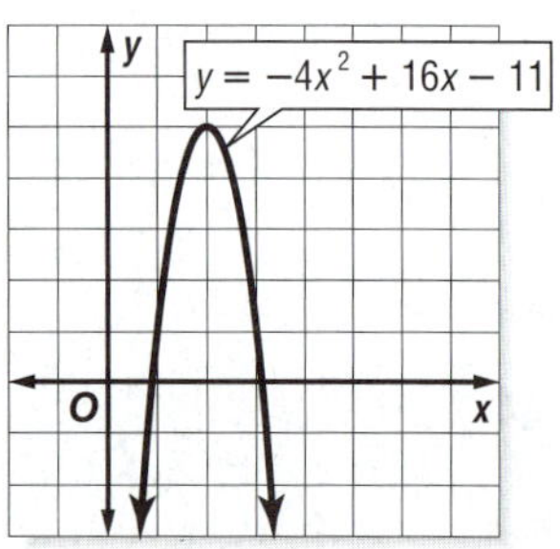

35.

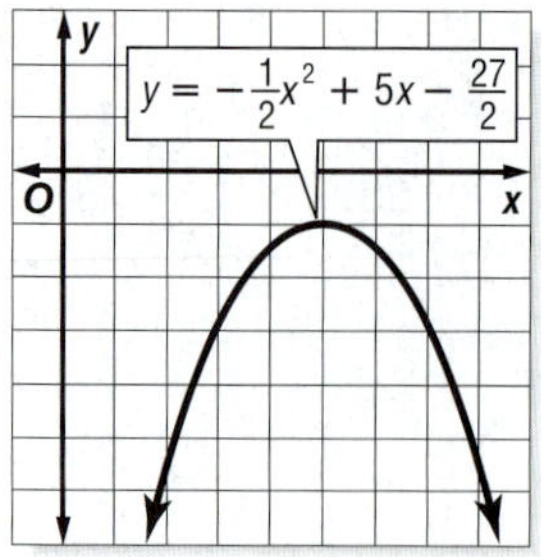

37. Sample answer: the graph of $y = 0.4(x + 3)^2 + 1$ is narrower than the graph of $y = 0.2(x + 3)^2 + 1$. **39.** $y = 9(x - 6)^2 + 1$ **41.** $y = -\frac{2}{3}(x - 3)^2$ **43.** $y = \frac{1}{3}x^2 + 5$ **45.** $y = -2x^2$ **47.** 34,000 feet; 32.5 s after the aircraft begins its parabolic flight **49.** $d(t) = -16t^2 + 8t + 50$ **51.** Angle A; the graph of the equation for angle A is higher than the other two since 3.27 is greater than 2.39 or 1.53. **53.** $y = ax^2 + bx + c$

$y = a\left(x^2 + \frac{b}{a}x\right) + c$

$y = a\left[x^2 + \frac{b}{a}x + \left(\frac{b}{2a}\right)^2\right] + c - a\left(\frac{b}{2a}\right)^2$

$y = a\left(x + \frac{b}{2a}\right)^2 + c - \frac{b^2}{4a}$

The axis of symmetry is $x = h$ or $-\frac{b}{2a}$. **55.** D **57.** 12; 2 irrational **59.** -23; 2 complex **61.** $\{3 \pm 3i\}$ **63.** $2t^2 + 2t - \frac{3}{t - 1}$ **65.** $n^3 - 3n^2 - 15n - 21$ **67a.** Sample answer using (1994, 76,302) and (1997, 99,448): $y = 7715x - 15{,}307{,}408$ **67b.** 161,167 **69.** no **71.** no

Page 328 Practice Quiz 2

1. $\{-7 \pm 2\sqrt{3}\}$ **3.** -11; 2 complex **5.** $\left\{\frac{-9 \pm 5\sqrt{5}}{2}\right\}$ **7.** $y = \frac{2}{3}(x - 2)^2 - 5$ **9.** $y = -(x - 6)^2$; $(6, 0)$, $x = 6$; down

Pages 332–335 Lesson 6-7

1. $y \geq (x - 3)^2 - 1$ **3a.** $x = -1, 5$ **3b.** $x \leq -1$ or $x \geq 5$ **3c.** $-1 \leq x \leq 5$

5.

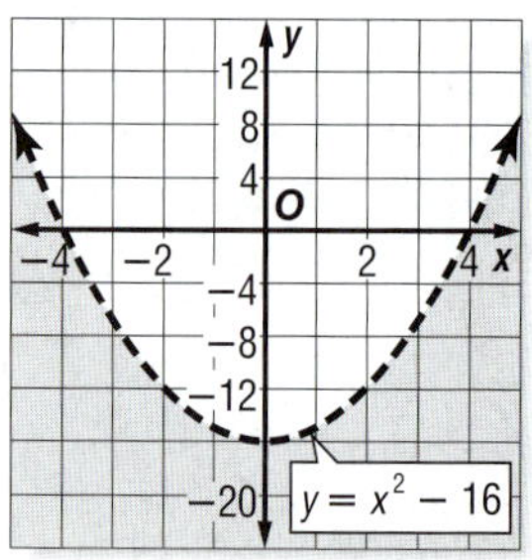

7.

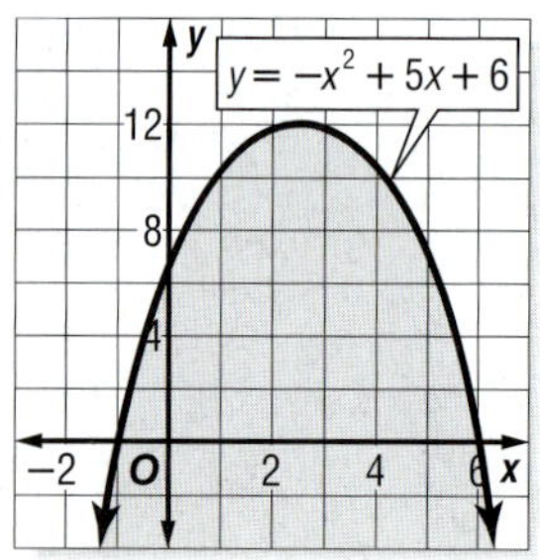

9. $\{x \mid -1 < x < 7\}$ **11.** $\varnothing$ **13.** about 6.1 s

15.

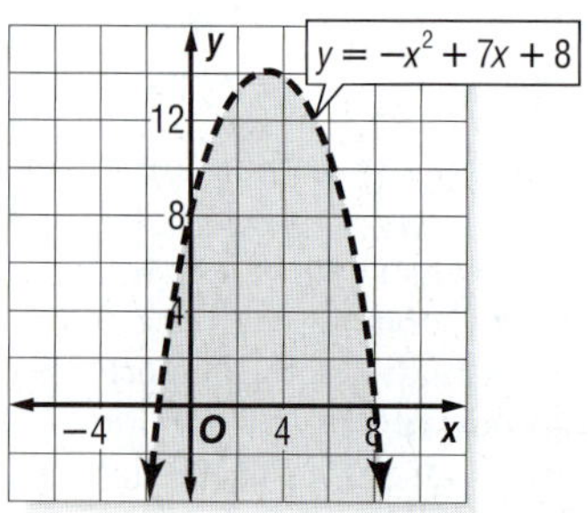

17.

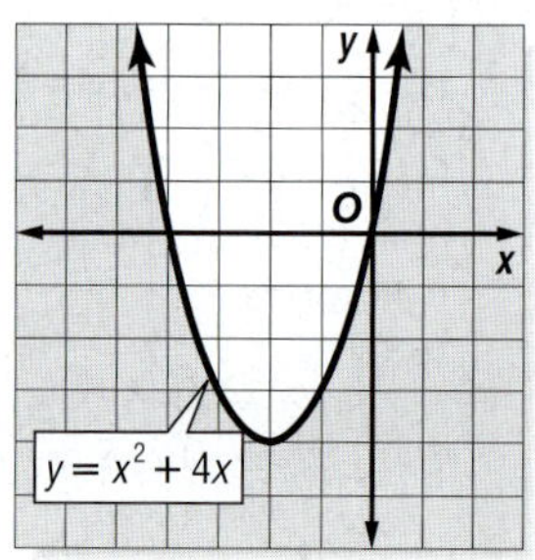

19.

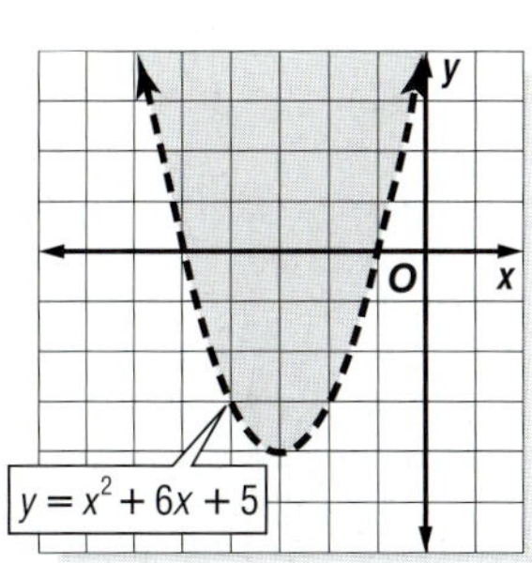

21.

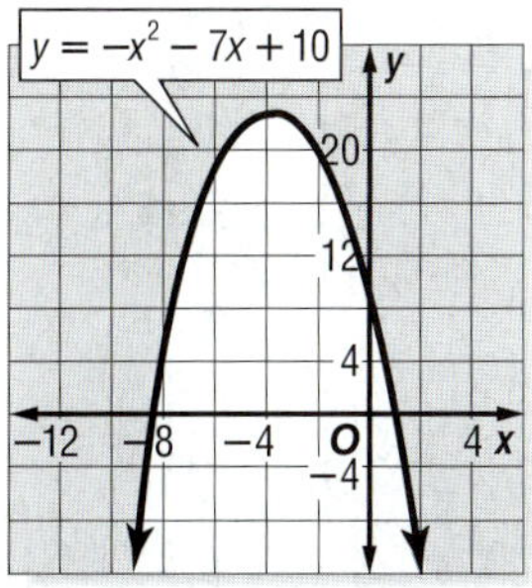

23.

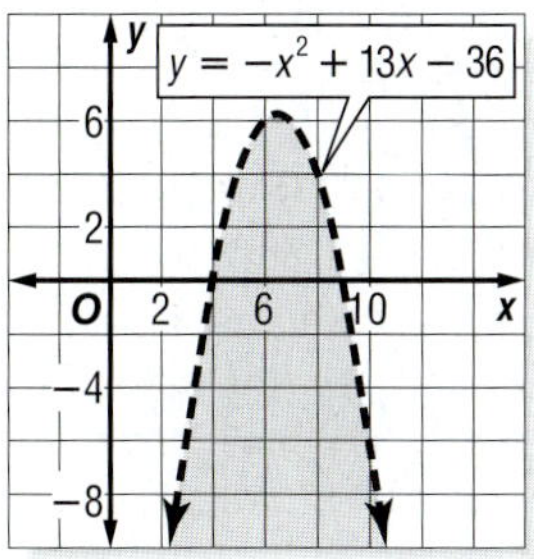

25.

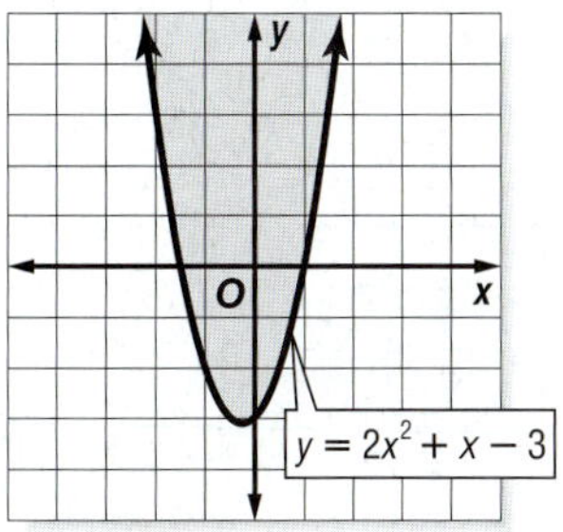

27. $-2 \leq x \leq 6$ **29.** $x < -7$ or $x > -3$ **31.** $\{x \mid -7 < x < 4\}$ **33.** $\{x \mid x \leq -6$ or $x \geq 4\}$ **35.** $\{x \mid x \leq -7$ or $x \geq 1\}$ **37.** all reals **39.** $\{x \mid x = 7\}$ **41.** $\varnothing$ **43.** 0 to 10 ft or 24 to 34 ft **45.** The width should be greater than 12 cm and the length should be greater than 18 cm **47.** 6

49.

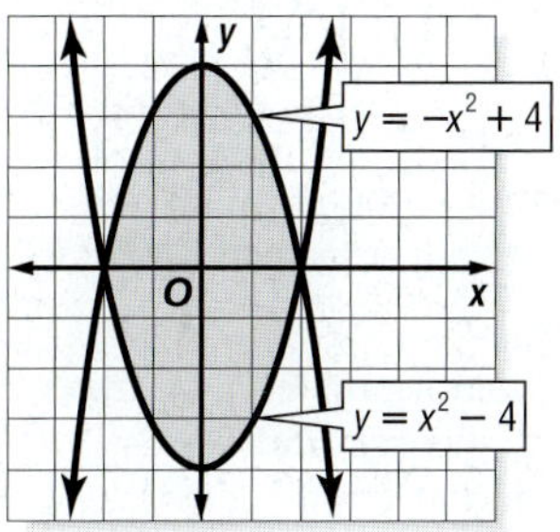

51. C **53.** $\{x \mid$ all reals, $x \neq 2\}$ **55.** $\{x \mid x < -9$ or $x > 3\}$ **57.** $\{x \mid -1.2 \leq x \leq -0.4\}$ **59.** $y = (x - 1)^2 + 8$; $(1, 8)$, $x = 1$; up **61.** $y = \frac{1}{2}(x + 6)^2$; $(-6, 0)$, $x = -6$; up **63.** $\frac{-5 \pm i\sqrt{3}}{2}$ **65.** $4a^2b^2 + 2a^2b + 4ab^2 + 12a - 7b$ **67.** $xy^3 + y + \frac{1}{x}$ **69.** $\begin{bmatrix} -21 & 48 \\ -13 & 22 \end{bmatrix}$ **71.** $|x - 0.08| \leq 0.002$; $0.078 \leq x \leq 0.082$

Pages 336–340 Chapter 6 Study Guide and Review

1. f **3.** a **5.** i **7.** c **9a.** 20; $x = -3$; -3

9b.

x	$f(x)$
−5	15
−4	12
−3	11
−2	12
−1	15

9c.

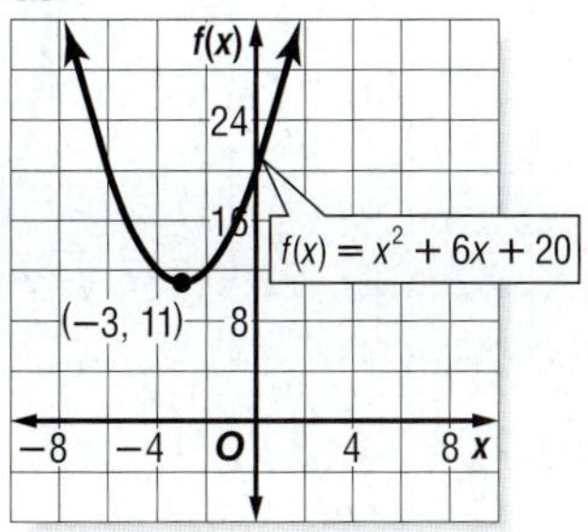

11a. 7; $x = 4$; 4

11b.

x	$f(x)$
2	−5
3	−8
4	−9
5	−8
6	−5

11c.

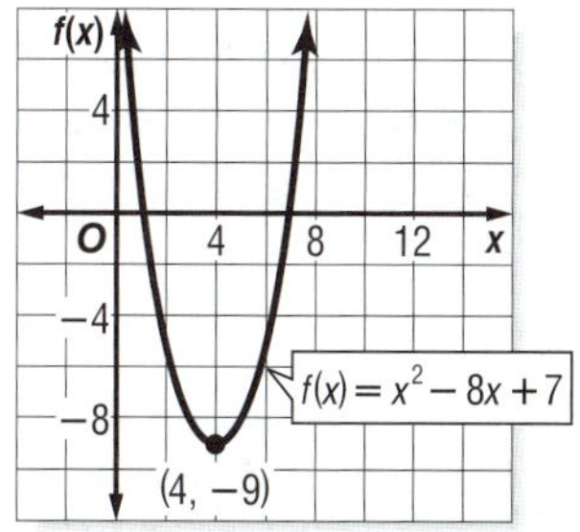

13a. −3; $x = -2$; −2

13b.

x	$f(x)$
−4	−3
−3	0
−2	1
−1	0
0	−3

13c.

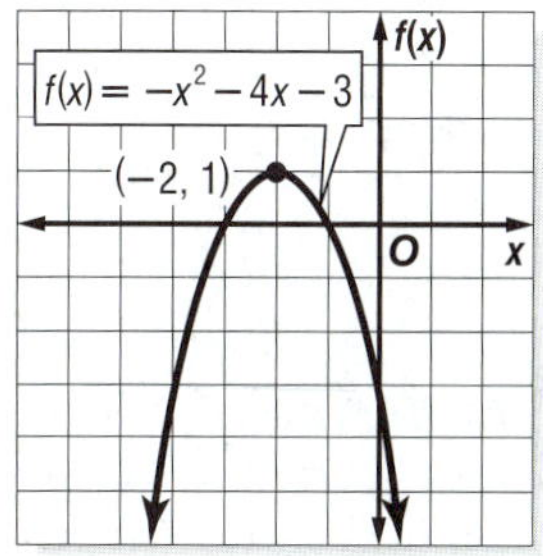

15. min.; $-\frac{89}{16}$ **17.** max.; 7 **19.** 2, −5 **21.** between −3 and −2; between −38 and −37 **23.** 2, −8 **25.** $\{-1\}$ **27.** $\{-11, 2\}$ **29.** $\left\{\frac{1}{3}, -\frac{3}{2}\right\}$ **31.** $x^2 - 3x - 70 = 0$ **33.** 289; $(x + 17)^2$ **35.** $\frac{49}{16}$; $\left(x + \frac{7}{4}\right)^2$ **37.** $3 \pm 2\sqrt{5}$ **39a.** −24 **39b.** 2 complex **39c.** $-1 \pm \sqrt{6}i$ **41a.** 73 **41b.** 2 irrational **41c.** $\frac{-7 + \sqrt{73}}{6}$ **43.** $y = 5\left(x + \frac{7}{2}\right)^2 - \frac{13}{4}$; $\left(-\frac{7}{2}, -\frac{13}{4}\right)$; $x = -\frac{7}{2}$; up

45.

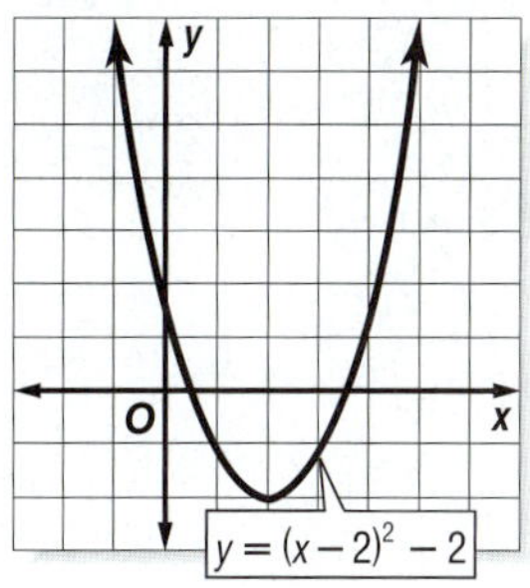

47.

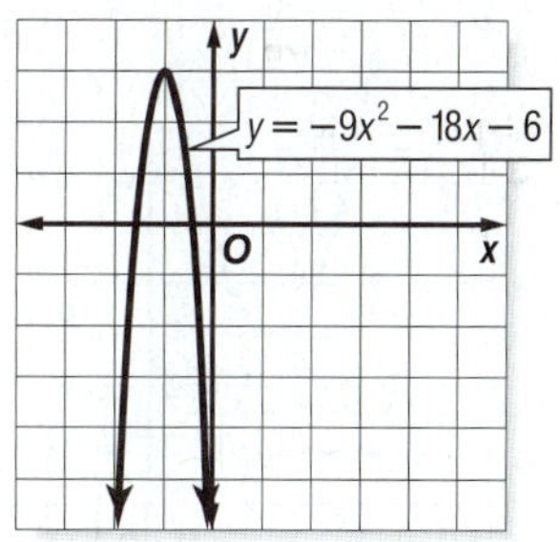

49. $y = \frac{1}{2}(x + 2)^2 + 3$

51.

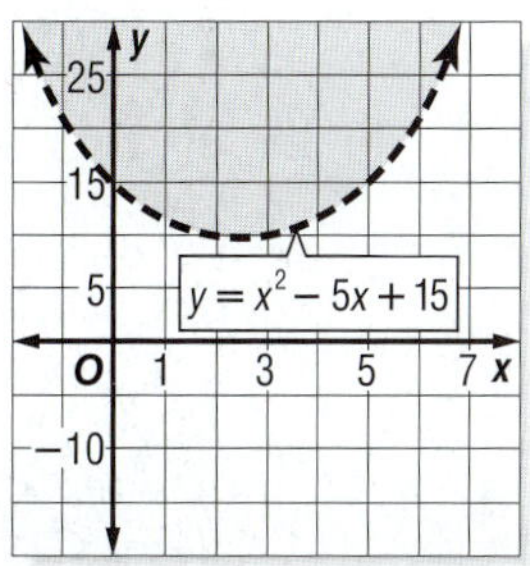

53.

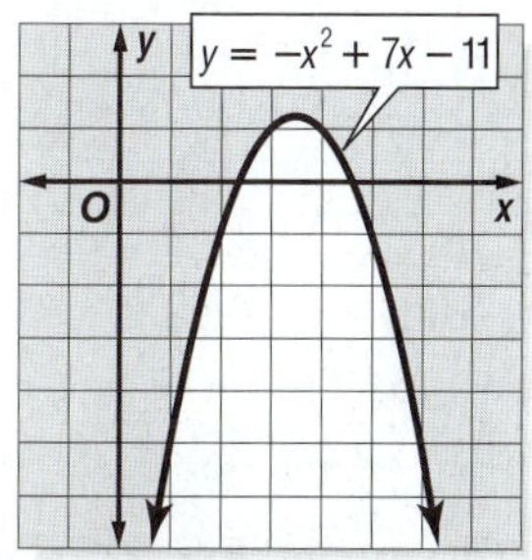

55. all reals **57.** $\left\{x \mid x < -\frac{1}{2} \text{ or } x > 3\right\}$

59. $\left\{x \mid x < \frac{3 - 2\sqrt{6}}{3} \text{ or } x > \frac{3 + 2\sqrt{6}}{3}\right\}$

Selected Answers

Chapter 7 Polynomial Functions

Page 345 Chapter 7 Getting Started

1. between 0 and 1, between 4 and 5 **3.** between −5 and −4, between 0 and 1 **5.** $-\frac{3}{2}, -\frac{1}{7}$ **7.** $3x + 4$ **9.** −19 **11.** $18b^2 - 3b - 6$

Pages 350–352 Lesson 7-1

1. $4 = 4x^0$; $x = x^1$ **3.** Sample answer given.

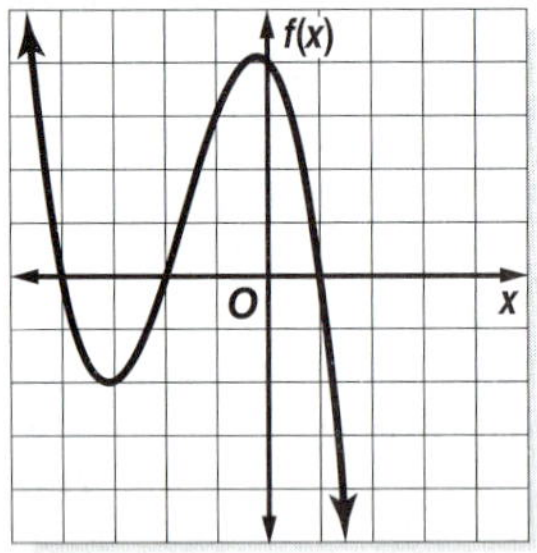

5. 6; 5 **7.** −21; 3 **9.** $2a^9 + 6a^3 - 12$ **11.** $6a^3 - 5a^2 + 8a - 45$ **13a.** $f(x) \to +\infty$ as $x \to +\infty$, $f(x) \to +\infty$ as $x \to -\infty$ **13b.** even **13c.** 0 **15.** 109 lumens **17.** 3; 1 **19.** 4; 6 **21.** No, this is not a polynomial because the term $\frac{1}{c}$ cannot be written in the form x^n, where n is a nonnegative integer. **23.** 12; 18 **25.** 1008; −36 **27.** 86; 56 **29.** 7; 4 **31.** $12a^2 - 8a + 20$ **33.** $12a^6 - 4a^3 + 5$ **35.** $3x^4 + 16x^2 + 26$ **37.** $-x^6 + x^3 + 2x^2 + 4x + 2$ **39a.** $f(x) \to +\infty$ as $x \to +\infty$, $f(x) \to -\infty$ as $x \to -\infty$ **39b.** odd **39c.** 3 **41a.** $f(x) \to -\infty$ as $x \to +\infty$, $f(x) \to -\infty$ as $x \to -\infty$ **41b.** even **41c.** 0 **43a.** $f(x) \to +\infty$ as $x \to +\infty$, $f(x) \to -\infty$ as $x \to -\infty$ **43b.** odd **43c.** 1 **45.** 5.832 units **47.** $f(x) \to -\infty$ as $x \to +\infty$; $f(x) \to -\infty$ as $x \to -\infty$ **49.** $\frac{1}{2}$ **51.** $f(x)\frac{1}{2}x^3 - \frac{3}{2}x^2 - 2x$ **53.** 4 **55.** 8 points **57.** C **59.** $\{x \mid 2 < x < 6\}$ **61.** $\left\{x \mid -1 \le x \le \frac{4}{5}\right\}$

63.

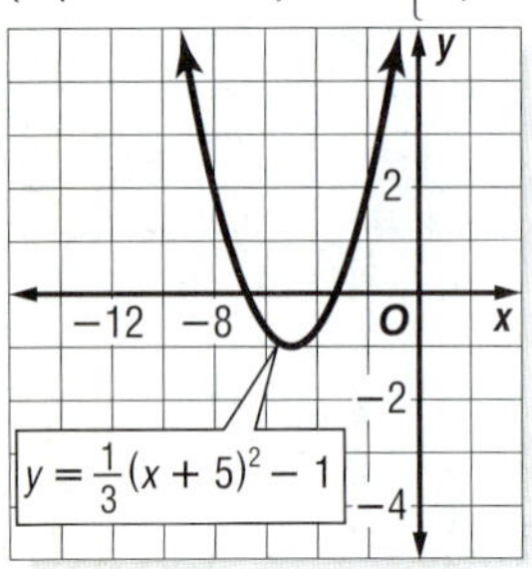

65. $\{4 \pm 3\sqrt{2}\}$ **67.** $23{,}450(1 + p)$; $23{,}450(1 + p)^3$

69.

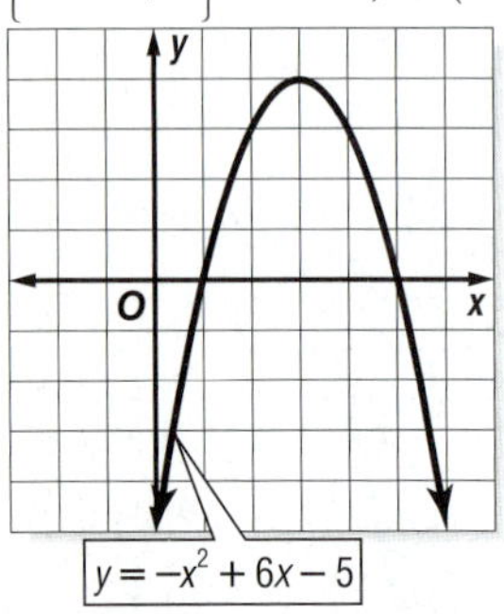

Pages 356–358 Lesson 7-2

1. There must be at least one real zero between two points on a graph when one of the points lies below the x-axis and the other point lies above the x-axis.

3.

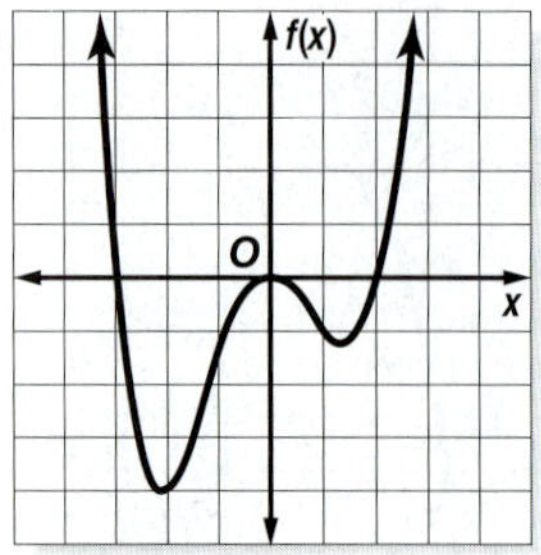

5.

x	$f(x)$
−3	20
−2	−9
−1	−2
0	5
1	0
2	−5
3	26

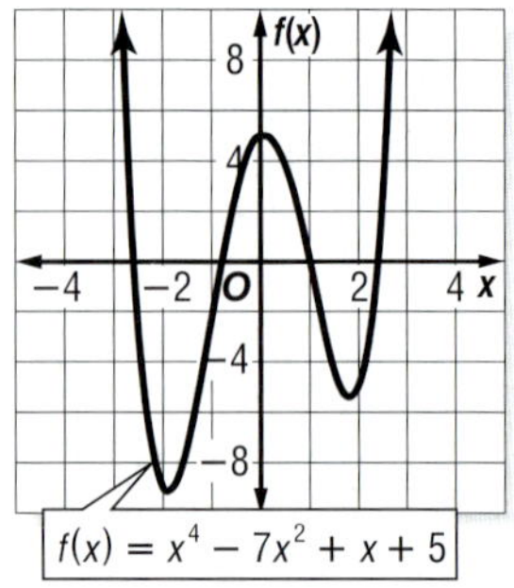

7. between −2 and −1, between −1 and 0, between 0 and 1, and between 1 and 2

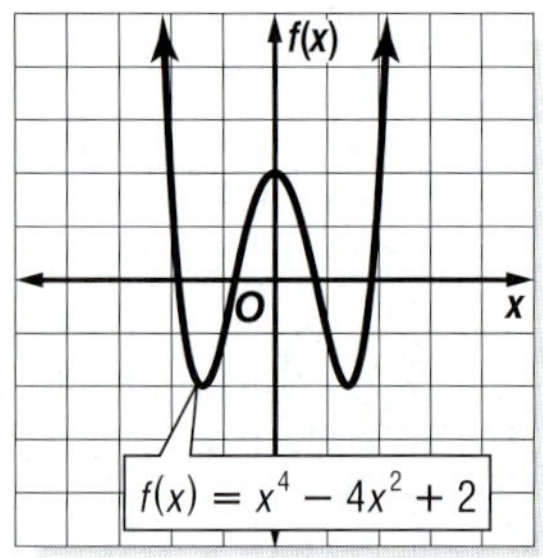

9.

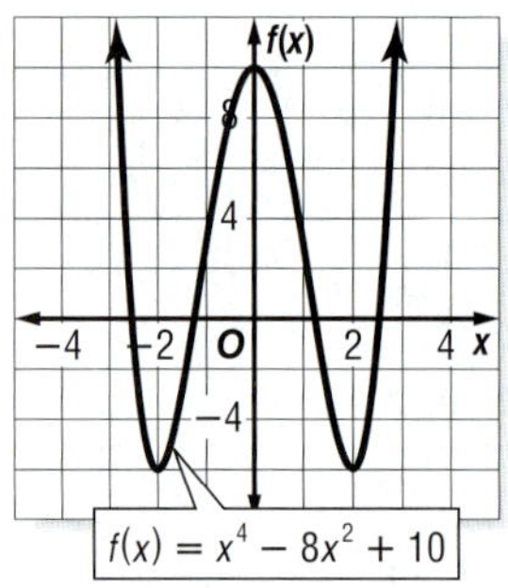

Sample answer: rel. max. at $x = 0$, rel. min. at $x = -2$ and at $x = 2$

11. rel. max. between $x = 15$ and $x = 16$, and no rel. min.; $f(x) \to -\infty$ as $x \to -\infty$, $f(x) \to -\infty$ as $x \to +\infty$.

13a.

x	$f(x)$
−5	25
−4	0
−3	−9
−2	−8
−1	−3
0	0
1	−5
2	−24

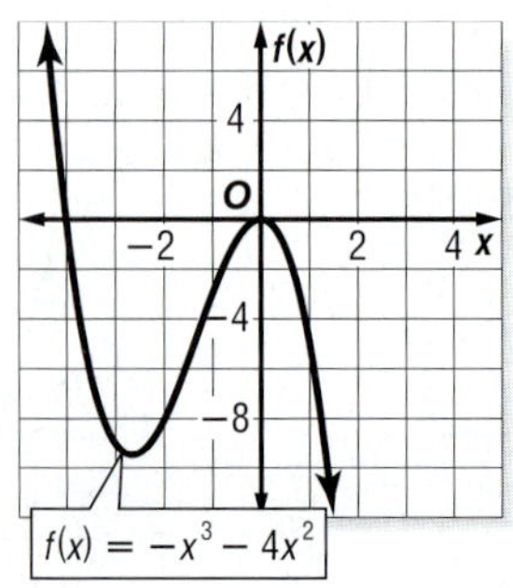

13b. at $x = -4$ and $x = 0$ **13c.** Sample answer: rel. max. at $x = 0$, rel. min. at $x = -3$

15a.

x	$f(x)$
−2	−18
−1	−2
0	2
1	0
2	−2
3	2
4	18

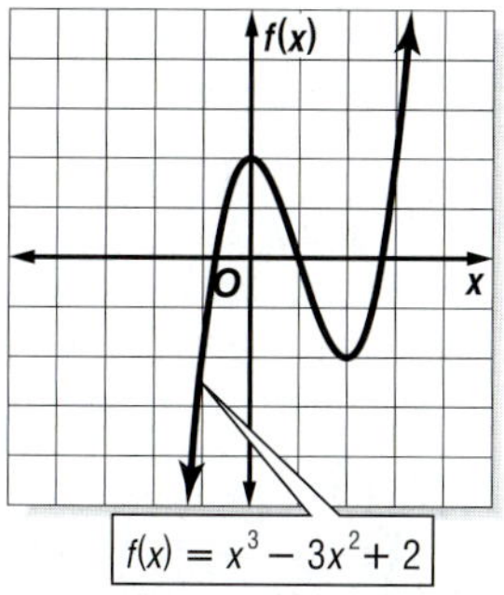

15b. at $x = 1$, between -1 and 0, and between 2 and 3
15c. Sample answer: rel. max. at $x = 0$, rel. min. at $x = 2$

17a.

x	$f(x)$
−1	75
0	16
1	−3
2	0
3	7
4	0
5	−39

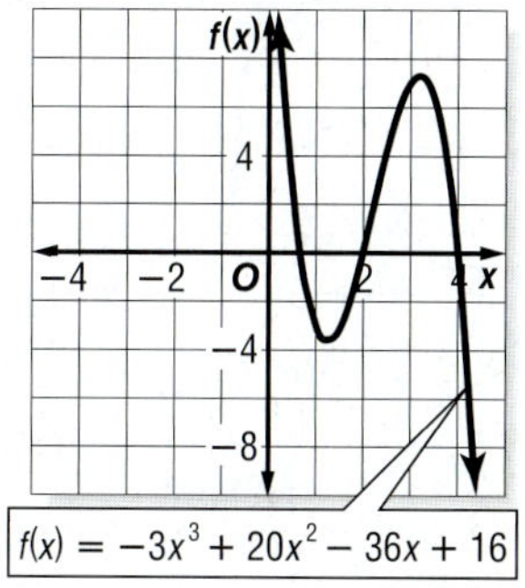

17b. between 0 and 1, at $x = 2$, and at $x = 4$
17c. Sample answer: rel. max. at $x = 3$, rel. min. at $x = 1$

19a.

x	$f(x)$
−3	73
−2	8
−1	−7
0	−8
1	−7
2	8
3	73

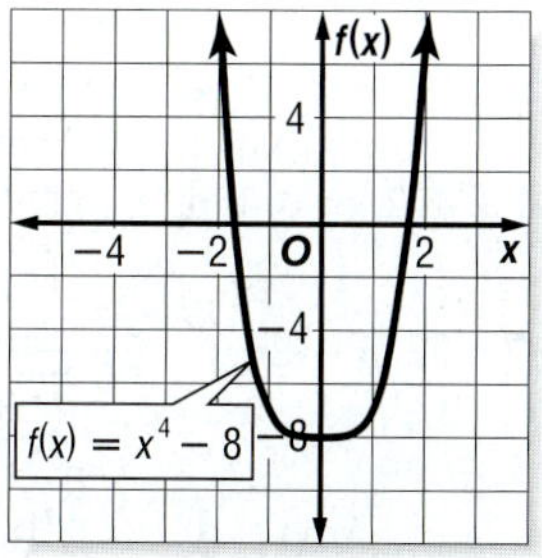

19b. between -2 and -1 and between 1 and 2
19c. Sample answer: no rel. max., rel. min. at $x = 0$

21a.

x	$f(x)$
−4	−169
−3	−31
−2	7
−1	5
0	−1
1	1
2	−1
3	−43

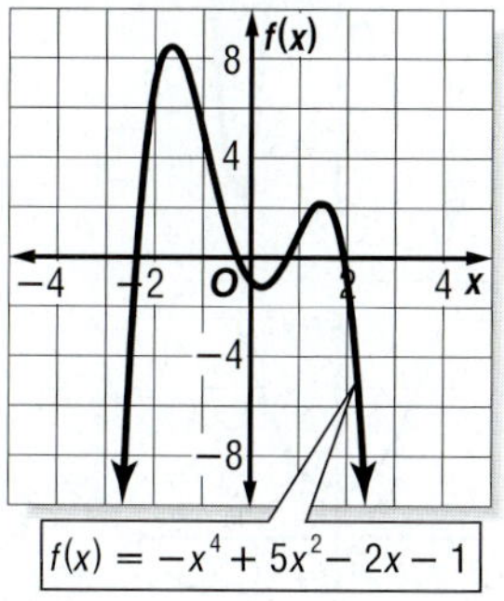

21b. between -3 and -2, between -1 and 0, between 0 and 1, and between 1 and 2 **21c.** Sample answer: rel. max. at $x = -2$ and at $x = 1.5$, rel. min. at $x = 0$

23a.

x	$f(x)$
−1	65
0	6
1	−1
2	2
3	−3
4	−10
5	11

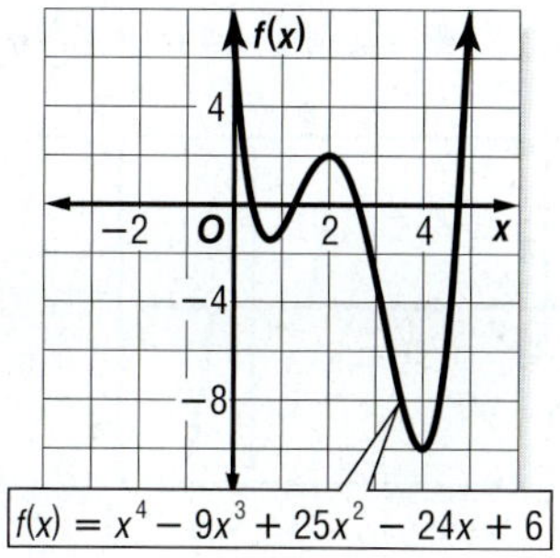

23b. between 0 and 1, between 1 and 2, between 2 and 3, and between 4 and 5 **23c.** Sample answer: rel. max. at $x = 2$, rel. min. at $x = 0.5$ and at $x = 4$

25a.

x	$f(x)$
−4	−77
−3	30
−2	7
−1	−2
0	3
1	−2
2	55

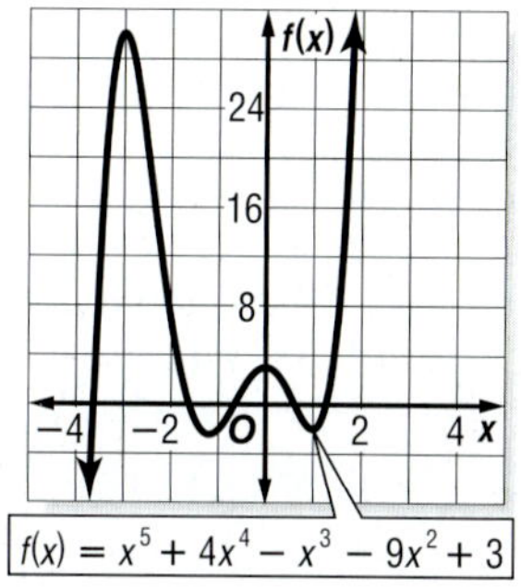

25b. between -4 and -3, between -2 and -1, between -1 and 0, between 0 and 1, and between 1 and 2 **25c.** Sample answer: rel. max. at $x = -3$ and at $x = 0$, rel. min. at $x = -1$ and at $x = 1$ **27.** highest: 1982; lowest: 2000 **29.** 5

31.

x	0	2	4	6	8	10	12	14	16	18	20
$B(x)$	25	34	40	45	50	54	59	64	68	71	71
$G(x)$	26	33	39	44	49	53	56	59	61	61	60

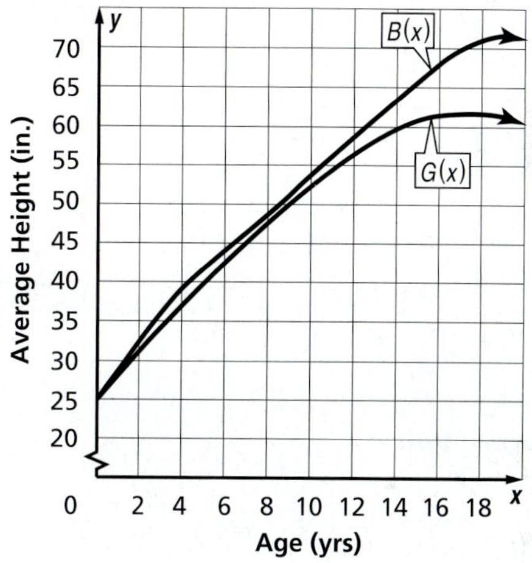

33. 0 and between 5 and 6 **35.** 3.4 s

37.

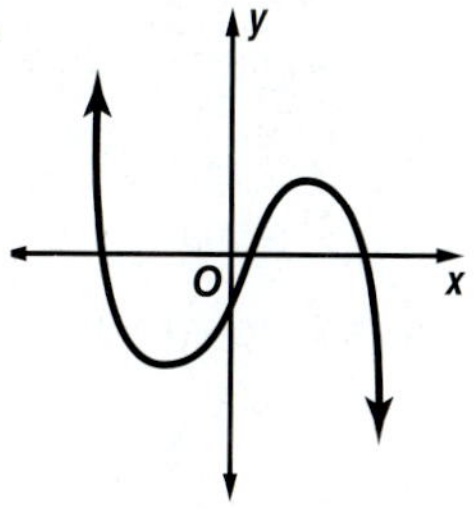

39.

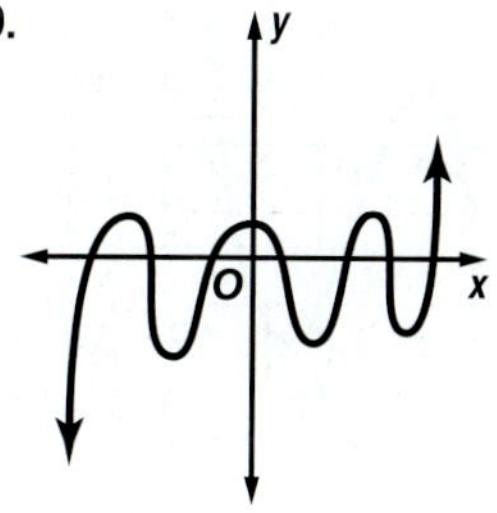

41. D **43.** $-1.90; 1.23$ **45.** $0; -1.22, 1.22$ **47.** $24a^3 - 4a^2 - 2$
49. $8a^4 - 10a^2 + 4$ **51.** $2x^4 + 11x^2 + 16$

53.

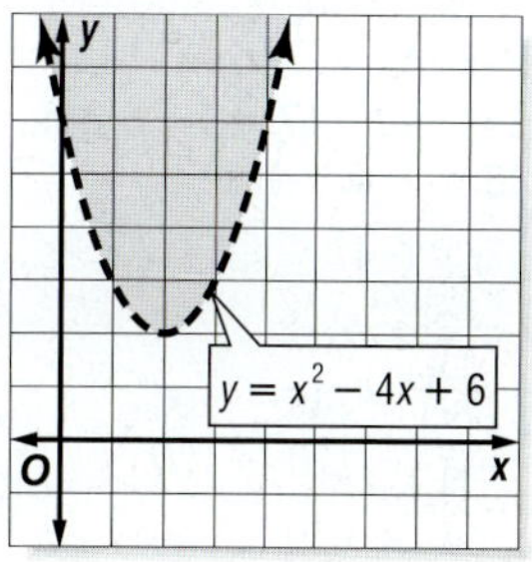

55.

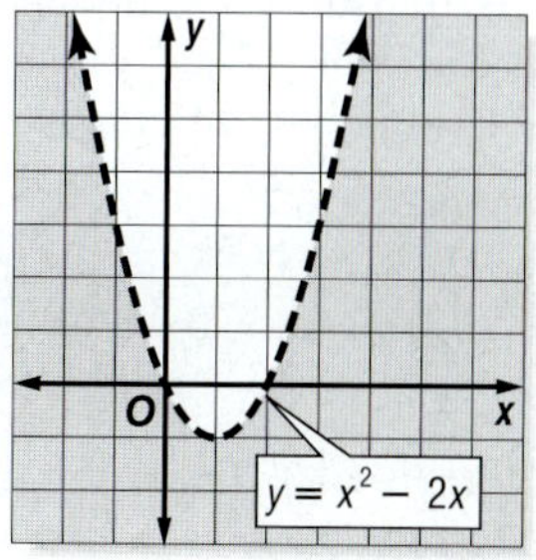

57. $(-3, -2)$ **59.** $(1, 3)$ **61.** $(x + 5)(x - 6)$ **63.** $(3a + 1)(2a + 5)$ **65.** $(t - 3)(t^2 + 3t + 9)$

Pages 362–364 Lesson 7-3

1. Sample answer: $16x^4 - 12x^2 = 0$; $4[4(x^2)^2 - 3x^2] = 0$ **3.** Factor out an x and write the equation in quadratic form so you have $x[(x^2)^2 - 2(x^2) + 1] = 0$. Factor the trinomial and solve for x using the Zero Product Property. The solutions are -1, 0, and 1. **5.** $84(n^2)^2 - 62(n^2)$ **7.** $-4, -1, 4, 1$ **9.** 64 **11.** $2(x^2)^2 + 6(x^2) - 10$ **13.** $11(n^3)^2 + 44(n^3)$ **15.** not possible **17.** $0, -4, -3$ **19.** $-\sqrt{3}, \sqrt{3}, -i\sqrt{3}, i\sqrt{3}$ **21.** $2, -2, 2\sqrt{2}, -2\sqrt{2}$ **23.** $-9, \frac{9 + 9i\sqrt{3}}{2}, \frac{9 - 9i\sqrt{3}}{2}$ **25.** 81, 625 **27.** 225, 16 **29.** $1, -1, 4$ **31.** $w = 4$ cm, $\ell = 8$ cm, $h = 2$ cm **33.** 3×3 in. **35.** $h^2 + 4, 3h + 2, h + 3$ **37.** Write the equation in quadratic form, $u^2 - 9x - 8 = 0$, where $u = |a - 3|$. Then factor and use the Zero Product Property to solve for a; 11, 4, 2, and -5. **39.** D

41.

x	$f(x)$
−2	−21
−1	−1
0	5
1	3
2	−1
3	−1
4	9
5	35

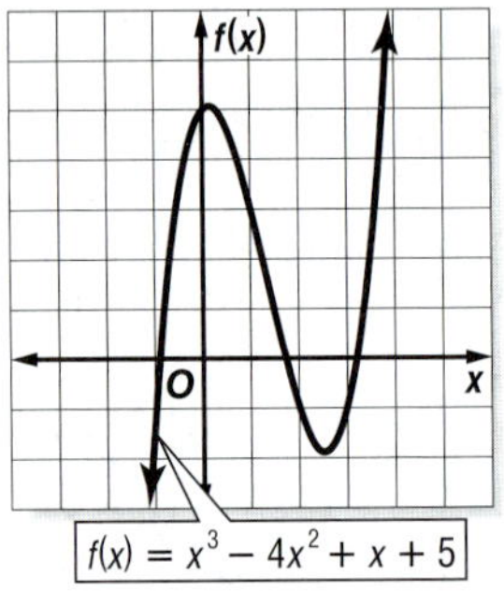

43. 17; 27 **45.** $\frac{1715}{3}$; 135 **47.** $A'(-1, -2), B'(3, -3), C'(1, 3)$ **49.** $x^2 + 5x - 4$ **51.** $x^3 - 6x - 20 - \frac{54}{x - 3}$

Page 364 Practice Quiz 1

1. $2a^3 - 6a^2 + 5a - 1$

3. Sample answer: maximum at $x = -2$, minimum at $x = 0.5$

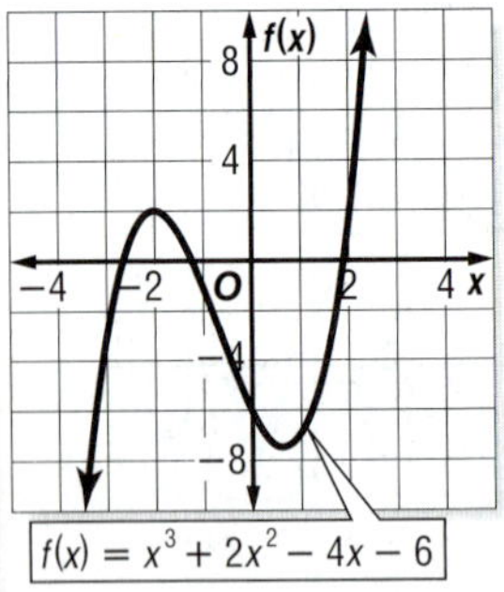

5. $-3, 3, -i\sqrt{3}, i\sqrt{3}$

Pages 368–370 Lesson 7-4

1. Sample answer: $f(x) = x^2 - 2x - 3$ **3.** dividend: $x^3 + 6x + 32$; divisor: $x + 2$; quotient: $x^2 - 2x + 10$; remainder: 12 **5.** 353, 1186 **7.** $x - 1, x + 2$ **9.** $x - 2, x^2 + 2x + 4$ **11.** \$2.894 billion **13.** $-9, 54$ **15.** 14, -42 **17.** $-19, -243$ **19.** 450, -1559 **21.** $x + 1, x + 2$ **23.** $x - 4, x + 1$ **25.** $x + 3, x - \frac{1}{2}$ or $2x - 1$ **27.** $x + 7, x - 4$ **29.** $x - 1, x^2 + 2x + 3$ **31.** $x - 2, x + 2, x^2 + 1$ **33.** 3 **35.** 1, 4 **37.**

5	1	−14	69	−140	100
		5	−45	120	−100
	1	−9	24	−20	0

39. 7.5 ft/s, 8 ft/s, 7.5 ft/s **41.** By the Remainder Theorem, the remainder when $f(x)$ is divided by $x - 1$ is equivalent to $f(1)$, or $a + b + c + d + e$. Since $a + b + c + d + e = 0$, the remainder when $f(x)$ is divided by $x - 1$ is 0. Therefore, $x - 1$ is a factor of $f(x)$. **43.** \$16.70 **45.** No, he will still owe \$4.40. **47.** D **49.** $(x^2)^2 - 8(x^2) + 4$ **51.** not possible **53.** Sample answer: rel. max. and $x = -1$ and $x = 1.5$, rel. min. at $x = 1$

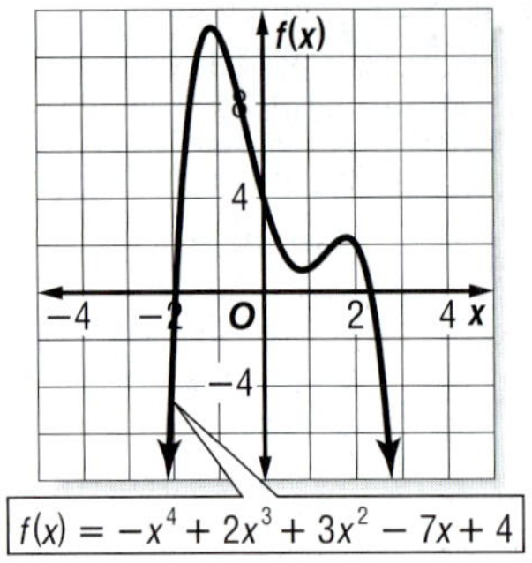

55. $(4, -2)$ **57.** A **59.** S **61.** $\frac{9 \pm \sqrt{57}}{6}$

Pages 375–377 Lesson 7-5

1. Sample answer: $p(x) = x^3 - 6x^2 + x + 1$; $p(x)$ has either 2 or 0 positive real zeros, 1 negative real zero, and 2 or 0 imaginary zeros. **3.** 6 **5.** -7, 0, and 3; 3 real **7.** 2 or 0; 1; 2 or 4 **9.** $2, 1 + i, 1 - i$ **11.** $2 + 3i, 2 - 3i, -1$ **13.** $-\frac{8}{3}$; 1 real **15.** $0, 3i, -3i$; 1 real, 2 imaginary **17.** $2, -2, 2i$, and $-2i$; 2 real, 2 imaginary **19.** 2 or 0; 1; 2 or 0 **21.** 3 or 1; 0; 2 or 0 **23.** 4, 2, or 0; 1; 4, 2, or 0 **25.** $-2, -2 + 3i, -2 - 3i$ **27.** $2i, -2i, \frac{i}{2}, -\frac{i}{2}$ **29.** $-\frac{3}{2}, 1 + 4i, 1 - 4i$ **31.** $4 - i, 4 + i, -3$ **33.** $3 - 2i, 3 + 2i, -1, 1$ **35.** $f(x) = x^3 - 2x^2 - 19x + 20$ **37.** $f(x) = x^4 + 7x^2 - 144$ **39.** $f(x) = x^3 - 11x^2 + 23x - 45$

41a.

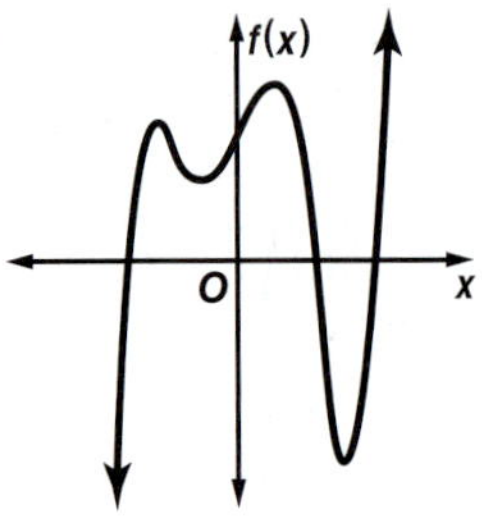

41b.

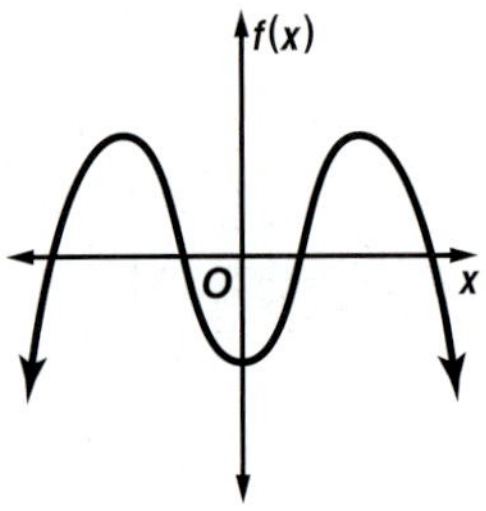

41c.

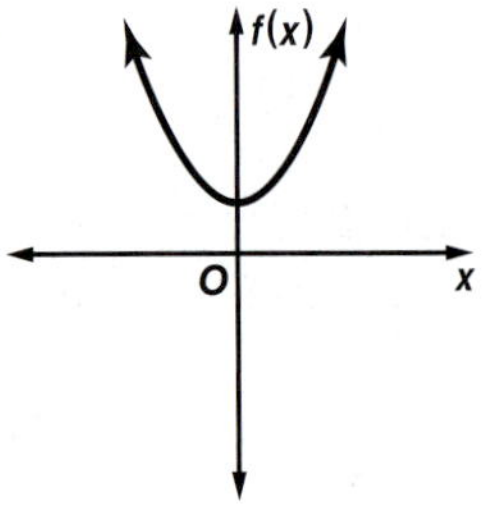

43. 1 ft **45.** radius = 4 m, height = 21 m **47.** $-24.1, -4.0$, 0, and 3.1

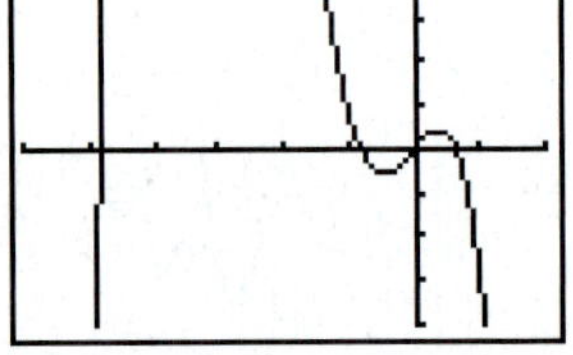

[−30, 10] scl: 5 by [−20, 20] scl: 5

49. Sample answer: $f(x) = x^3 - 6x^2 + 5x + 12$ and $g(x) = 2x^3 - 12x^2 + 10x + 24$; each have zeros at $x = 4$, $x = -2$, and $x = 3$.
51. If the equation models the level of a medication in a patient's bloodstream, a doctor can use the roots of the equation to determine how often the patient should take the medication to maintain the necessary concentration in the body. Answers should include the following.

- A graph of this equation reveals that only the first positive real root of the equation, 5, has meaning for this situation, since the next positive real root occurs after the medication level in the bloodstream has dropped below 0 mg. Thus according to this model, after 5 hours there is no significant amount of medicine left in the bloodstream.
- The patient should not go more than 5 hours before taking their next dose of medication.

53. C **55.** −254, 915 **57.** min.; −13 **59.** min.; −7
61. $(6p - 5)(2p - 9)$ **63.** $\begin{bmatrix} -3 & 2 \\ 3 & -4 \\ -2 & 9 \end{bmatrix}$ **65.** $\begin{bmatrix} 29 & -8 \\ 8 & 9 \\ 16 & -16 \end{bmatrix}$
67. $\pm\frac{1}{2}, \pm1, \pm\frac{5}{2}, \pm5$ **69.** $\pm\frac{1}{9}, \pm\frac{1}{3}, \pm1, \pm3$

Pages 380–382 Lesson 7-6

1. Sample answer: You limit the number of possible solutions. **3.** Luis; Lauren found numbers in the form $\frac{q}{p}$, not $\frac{p}{q}$ as Luis did according to the Rational Zero Theorem. **5.** $\pm1, \pm2, \pm\frac{1}{2}, \pm\frac{1}{3}, \pm\frac{1}{6}, \pm\frac{2}{3}$ **7.** −2, −4, 7 **9.** $-2, 2, \frac{7}{2}$ **11.** 10 cm × 11 cm × 13 cm **13.** ±1, ±2, ±3, ±6 **15.** ±1, ±2, ±3, ±6, ±9, ±18 **17.** $\pm1, \pm\frac{1}{3}, \pm\frac{1}{9}, \pm3, \pm9, \pm27$ **19.** −1, −1, 2 **21.** 0, 9 **23.** 0, 2, −2 **25.** −2, −4 **27.** $\frac{1}{2}, -\frac{1}{3}, -2$ **29.** $-\frac{1}{2}, \frac{1}{3}, \frac{1}{2}, \frac{3}{4}$ **31.** $\frac{4}{5}, 0, \frac{5 \pm i\sqrt{3}}{2}$ **33.** $-1, -2, 5, i, -i$ **35.** $2, -3 \pm i\sqrt{3}; 2$ **37.** $V = 2h^3 - 8h^2 - 64h$ **39.** $V = \frac{1}{3}\ell^3 - 3\ell^2$ **41.** $\ell = 30$ in., $w = 30$ in., $h = 21$ in.
43. The Rational Zero Theorem helps factor large numbers by eliminating some possible zeros because it is not practical to test all of them using synthetic substitution. Answers should include the following.

- The polynomial equation that represents the volume of the compartment is $V = w^3 + 3w^2 - 40w$.
- Reasonable measures of the width of the compartment are, in inches, 1, 2, 3, 4, 6, 7, 9, 12, 14, 18, 21, 22, 28, 33, 36, 42, 44, 63, 66, 77, and 84. The solution shows that $w = 14$ in., $\ell = 22$ in., and $d = 9$ in.

45. Sample answer $x^5 - x^4 - 27x^3 + 41x^2 + 106x - 120$ **47.** $-4, 2 + i, 2 - i$ **49.** $-7, 5 + 2i, 5 - 2i$ **51.** $x - 4, 3x^2 + 2$ **53.** $\pm3xy\sqrt{2x}$ **55.** 6 cm, 8 cm, 10 cm **57.** $4x^2 - 8x + 3$ **59.** $x^5 - 7x^4 - 8x^3 + 106x^2 - 85x + 25$ **61.** $x^2 + x - 4 + \frac{5}{x+1}$

Page 382 Practice Quiz 2

1. −930, −145 **3.** $x^4 - 4x^3 - 7x^2 + 22x + 24 = 0$ **5.** $-\frac{3}{2}$

Pages 386–389 Lesson 7-7

1. Sometimes; sample answer: If $f(x) = x - 2$, $g(x) = x + 8$, then $f \circ g = x + 6$ and $g \circ f = x + 6$.
3. Danette; $[g \circ f](x) = g[f(x)]$ means to evaluate the f function first and then the g function. Marquan evaluated the functions in the wrong order. **5.** $x^2 + x - 1; x^2 - x + 7; x^3 - 4x^2 + 3x - 12; \frac{x^2+3}{x-4}, x \neq 4$ **7.** {(2, −7)}; {(1, 0), (2, 10)} **9.** $x^2 + 11; x^2 + 10x + 31$ **11.** 11 **13.** $p(x) = \frac{3}{4}x; c(x) = x - 5$ **15.** $33.74; price of CD when coupon is subtracted and then 25% discount is taken **17.** $2x; 18; x^2 - 81; \frac{x+9}{x-9}, x \neq 9$ **19.** $2x^2 - x + 8; 2x^2 + x - 8; -2x^3 + 16x^2; \frac{2x^2}{8-x}, x \neq 8$ **21.** $\frac{x^3+x^2-1}{x+1}, x \neq -1; \frac{x^3+x^2-2x-1}{x+1}, x \neq -1; x^2 - x, x \neq -1; \frac{x^3+x^2-x-1}{x}, x \neq 0$ **23.** {(1, −3), (−3, 1), (2, 1)}; {(1, 0), (0, 1)} **25.** {(0, 0), (8, 3), (3, 3)}; {(3, 6), (4, 4), (6, 6), (7, 8)} **27.** {(5, 1), (8, 9)}; {(2, −4)} **29.** $8x - 4; 8x - 1$ **31.** $x^2 + 2; x^2 + 4x + 4$ **33.** $2x^3 + 2x^2 + 2x + 2; 8x^3 + 4x^2 + 2x + 1$ **35.** −12 **37.** 39 **39.** 25 **41.** 2 **43.** 79 **45.** 226 **47.** $P(x) = -50x + 1939$ **49.** $p(x) = 0.70x; s(x) = 1.0575x$ **51.** $110.30 **53.** 373 K; 273 K **55.** $700, $661.20, $621.78, $581.73, $541.04 **57.** Answers should include the following.

- Using the revenue and cost functions, a new function that represents the profit is $p(x) = r(c(x))$.
- The benefit of combining two functions into one function is that there are fewer steps to compute and it is less confusing to the general population of people reading the formulas.

59. C **61.** $\pm1, \pm\frac{1}{2}, \pm\frac{1}{4}, \pm2, \pm3, \pm\frac{3}{2}, \pm\frac{3}{4}, \pm6$ **63.** $x^3 - 4x^2 - 17x + 60$ **65.** $6x^3 - 13x^2 + 9x - 2$ **67.** $x^3 - 9x^2 + 31x - 39$ **69.** $10 + 2j$ **71.** $\begin{bmatrix} 3 & -2 \\ -1 & 1 \end{bmatrix}$ **73.** $-\frac{1}{2}\begin{bmatrix} -1 & -2 \\ -3 & -4 \end{bmatrix}$ **75.** $-\frac{1}{16}\begin{bmatrix} -5 & -2 \\ -3 & 2 \end{bmatrix}$ **77.** $y = \frac{1-4x^2}{-5x}$ **79.** $t = \frac{I}{pr}$ **81.** $m = \frac{Fr^2}{GM}$

Pages 393–394 Lesson 7-8

1. no **3.** Sample answer: $f(x) = 2x, f^{-1}(x) = 0.5x; f[f^{-1}(x)] = f^{-1}[f(x)] = x$ **5.** {(4, 2), (1, −3), (8, 2)}
7. $f^{-1}(x) = -x$

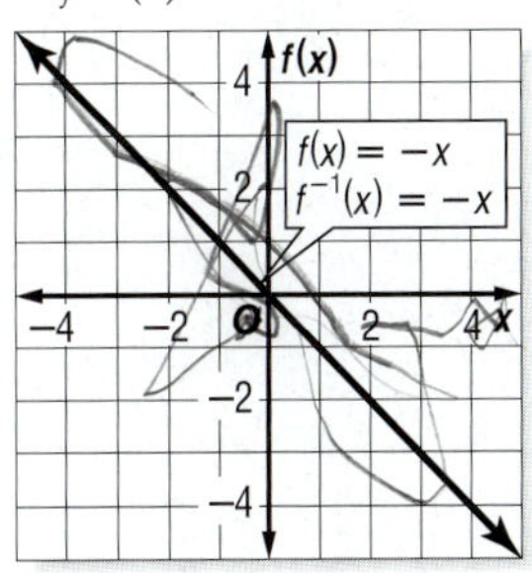

9. $y = 2x - 10$

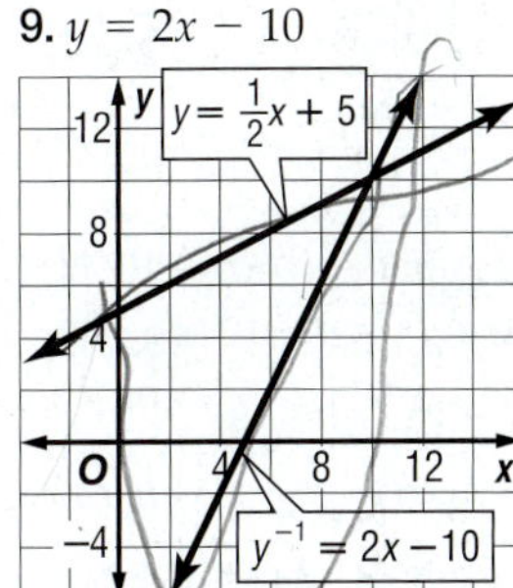

11. no **13.** 15.24 m/s² **15.** {(8, 3), (−2, 4), (−3, 5)} **17.** {(−2, −1), (−2, −3), (−4, −1), (6, 0)} **19.** {(8, 2), (5, −6), (2, 8), (−6, 5)}
21. $g^{-1}(x) = -\frac{1}{2}x$

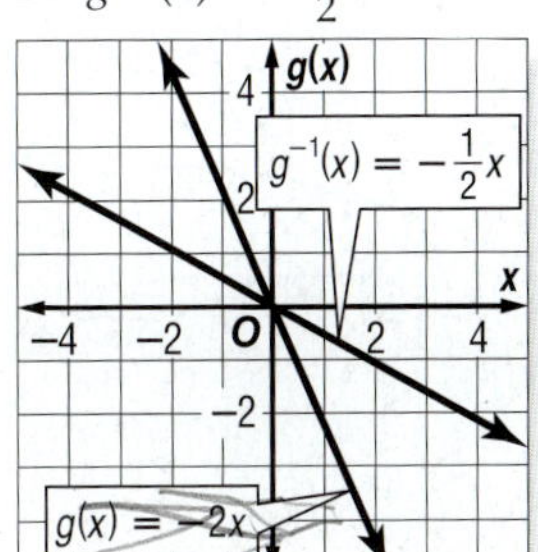

23. $g^{-1}(x) = x - 4$

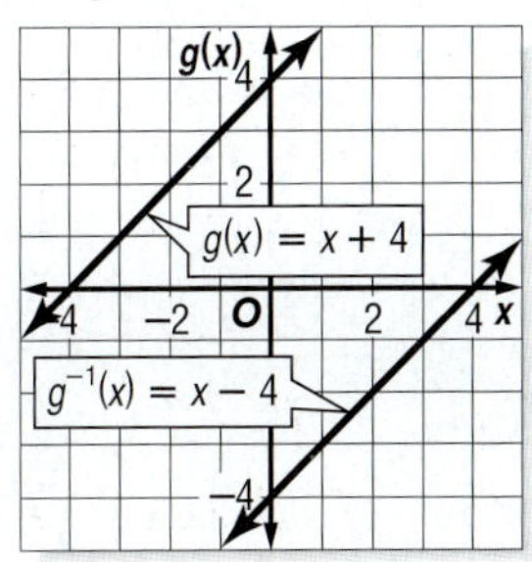

25. $y = -\frac{1}{2}x - \frac{1}{2}$

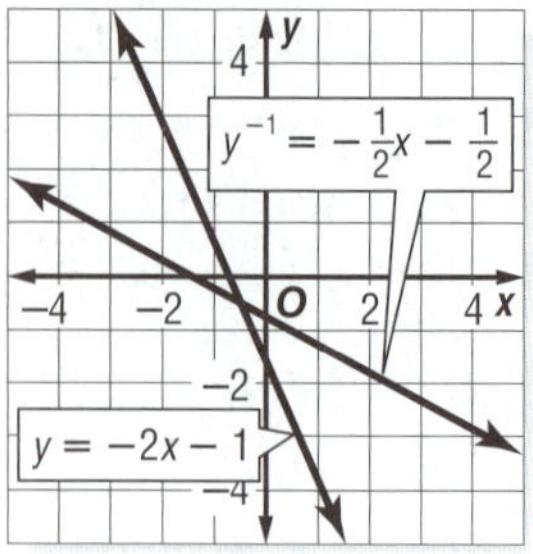

27. $f^{-1}(x) = \frac{8}{5}x$

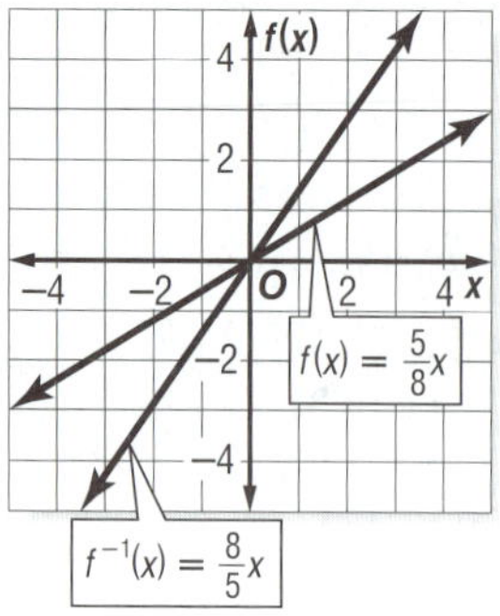

29. $f^{-1}(x) = \frac{5}{4}x + \frac{35}{4}$

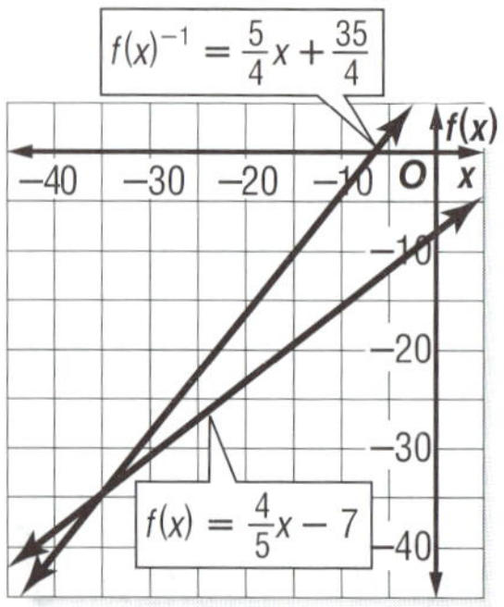

31. $f^{-1}(x) = \frac{8}{7}x + \frac{4}{7}$

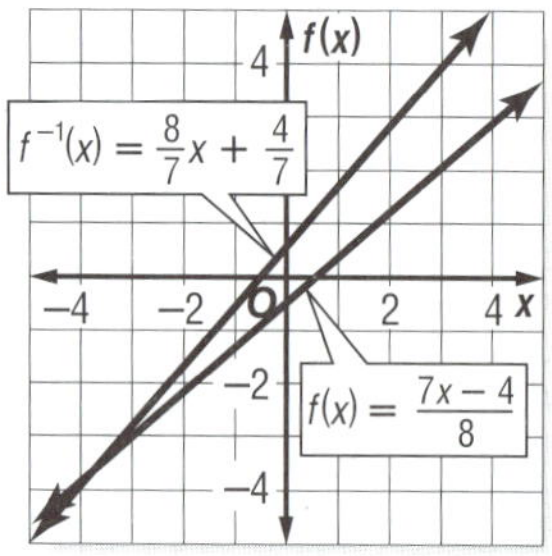

33. no 35. yes 37. yes 39. $y = \frac{1}{2}x - \frac{11}{2}$ 41. $I(m) = 320 + 0.04m$; \$4500 43. It can be used to convert Celsius to Fahrenheit.

45. Inverses are used to convert between two units of measurement. Answers should include the following.

- Even if it is not necessary, it is helpful to know the imperial units when given the metric units because most measurements in the U.S. are given in imperial units so it is easier to understand the quantities using our system.
- To convert the speed of light from meters per second to miles per hour,

$$f(x) \approx \frac{3.0 \times 10^8 \text{ meters}}{1 \text{ second}} \cdot \frac{3600 \text{ seconds}}{1 \text{ hour}} \cdot \frac{1 \text{ mile}}{1600 \text{ meters}}$$
$$\approx 675{,}000{,}000 \text{ mi/hr}$$

47. B 49. $g[h(x)] = 6x - 10$; $h[g(x)] = 6x$ 51. −7, −2, 3 53. 64 55. 3 57. 117 59. −7 61. $\frac{25}{4}$

Pages 397–399 Lesson 7-9

1. In order for it to be a square root function, only the nonnegative range can be considered. 3. Sample answer: $y = \sqrt{2x - 4}$

5.

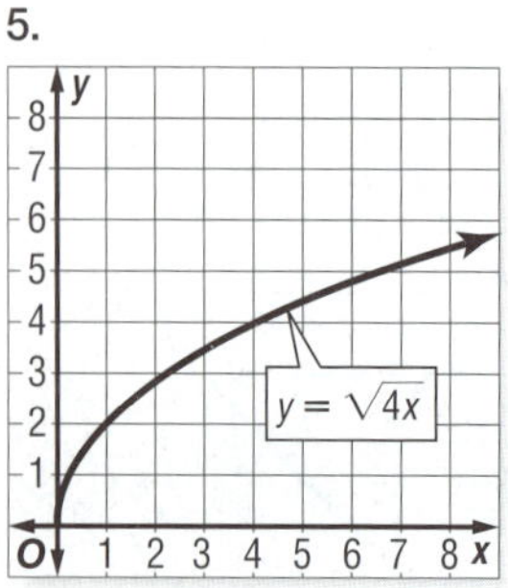

D: $x \geq 0$; R: $y \geq 0$

7.

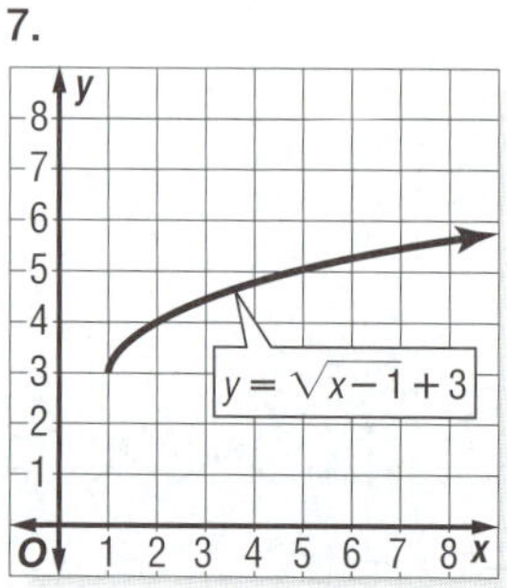

D: $x \geq 1$; R: $y \geq 3$

9.

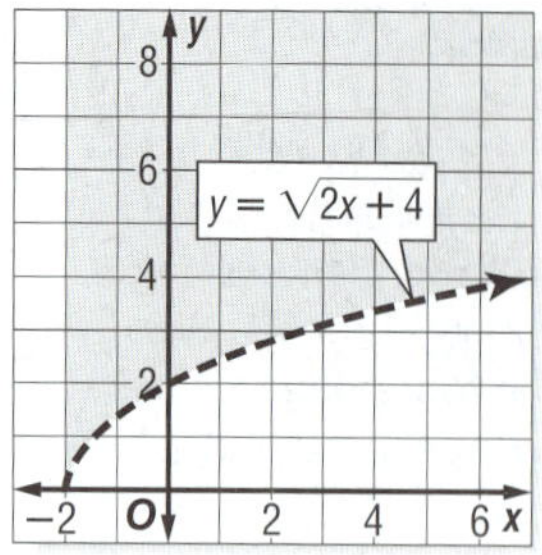

11.

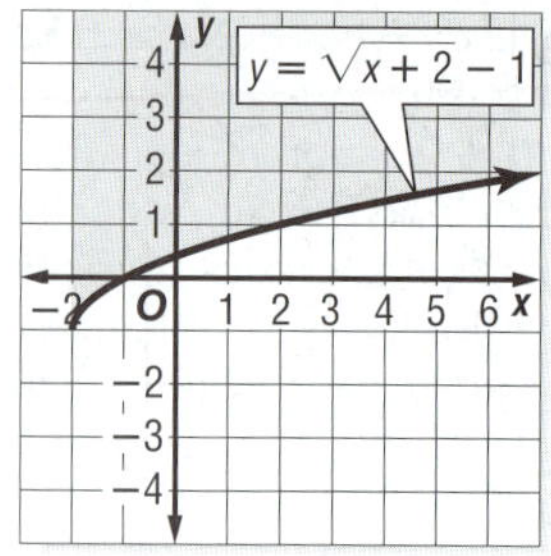

13. Yes; sample answer: The advertised pump will reach a maximum height of 87.9 ft.

15.

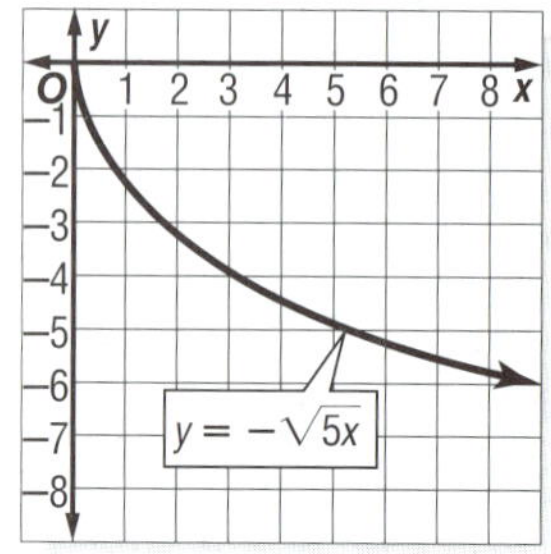

D: $x \geq 0$, R: $y \leq 0$

17.

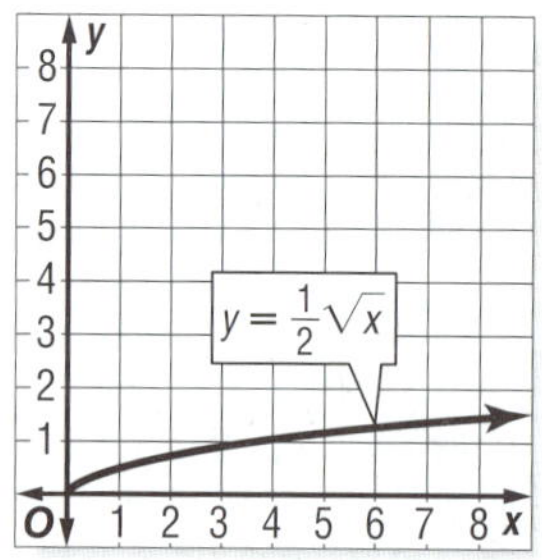

D: $x \geq 0$, R: $y \geq 0$

19.

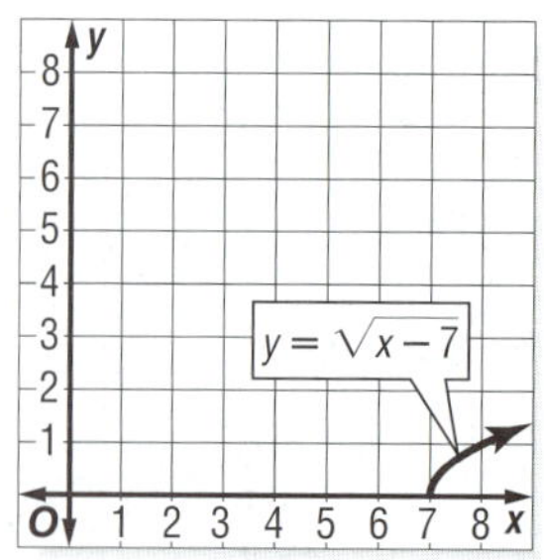

D: $x \geq 7$, R: $y \geq 0$

21.

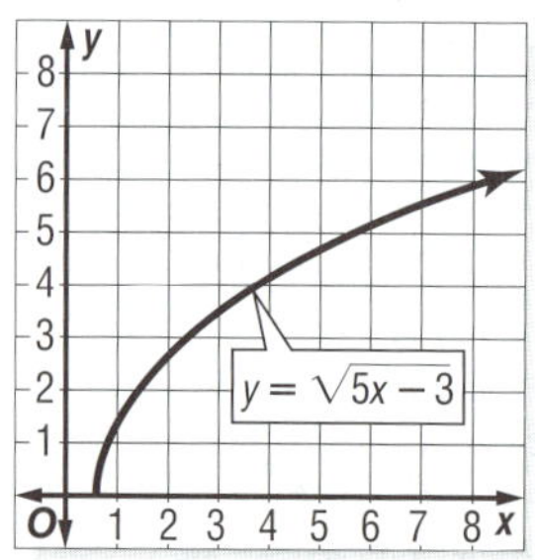

D: $x \geq 0.6$, R: $y \geq 0$

23.

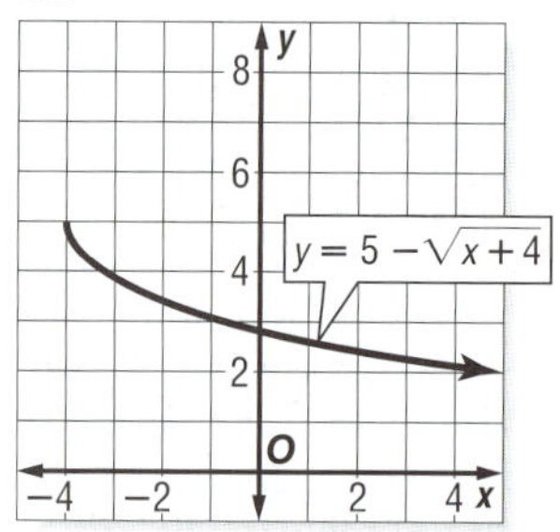

D: $x \geq -4$, R: $y \leq 5$

25.

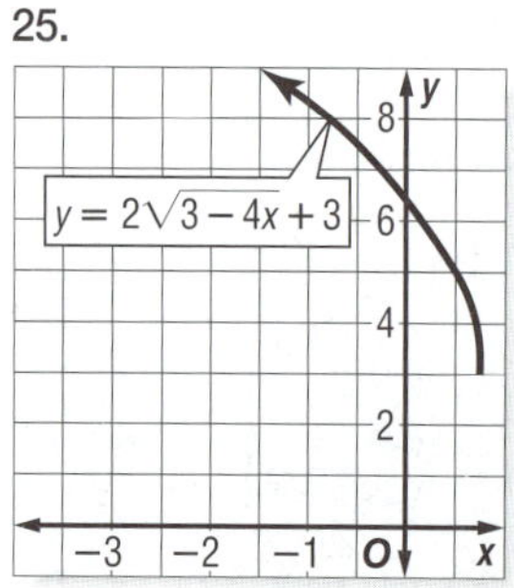

D: $x \leq 0.75$, R: $y \geq 3$

27.

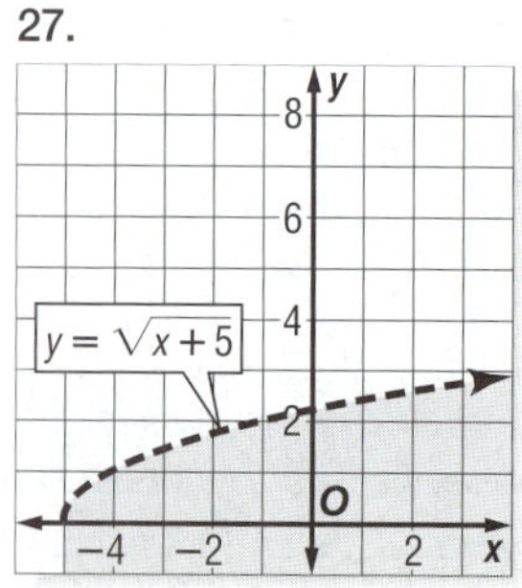

29.

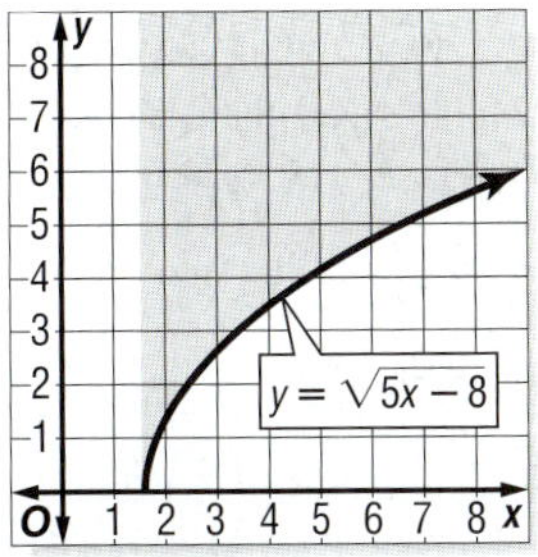

31.

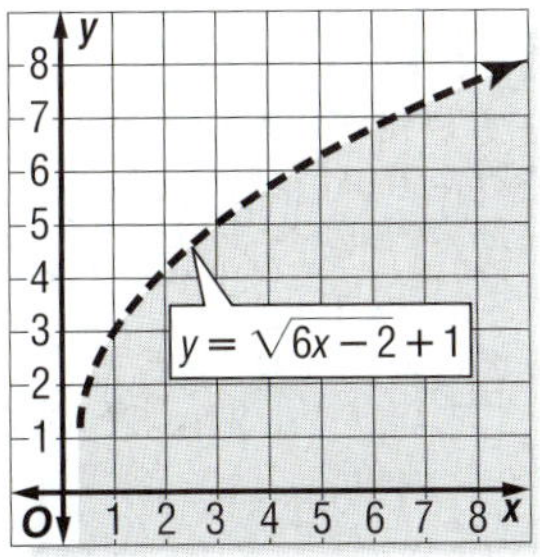

33. 317.29 mi

37. Square root functions are used in bridge design because the engineers must determine what diameter of steel cable needs to be used to support a bridge based on its weight. Answers should include the following.

- Sample answer: When the weight to be supported is less than 8 tons.
- 13,608 tons

39. D **41.** no **43.** $2x + 2$; 8; $x^2 + 2x - 15$; $\frac{x+5}{x-3}$, $x \neq 3$

45. $\frac{8x^3 + 12x^2 - 18x - 26}{2x + 3}$, $x \neq -\frac{3}{2}$; $\frac{8x^3 + 12x^2 - 18x - 28}{2x + 3}$, $x \neq -\frac{3}{2}$; $2x - 3$, $x \neq -\frac{3}{2}$; $8x^3 + 12x^2 - 18x - 27$, $x \neq -\frac{3}{2}$

47. $2x^2 - 4x - 16$ **49.** $a^3 - 1$

Pages 400–404 Chapter 7 Study Guide and Review

1. f **3.** a **5.** e **7.** -6; $x + h - 2$ **9.** -21; $6x + 6h + 3$
11. 20; $x^2 + 2xh + h^2 - x - h$

13a.

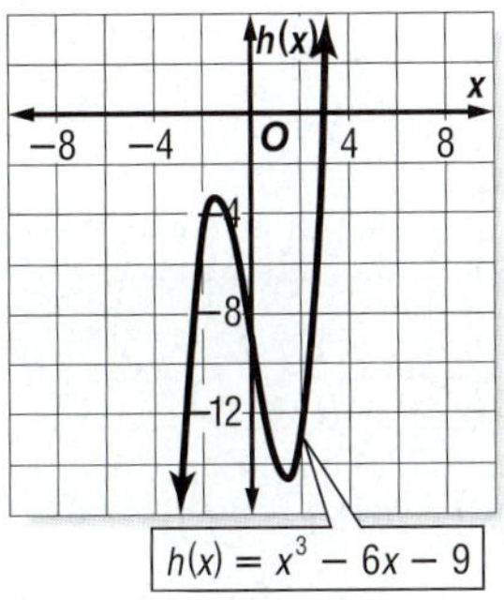

13b. at $x = 3$
13c. Sample answer: rel. max. at $x = -1.4$, rel. min. at $x = 1.4$

15a.

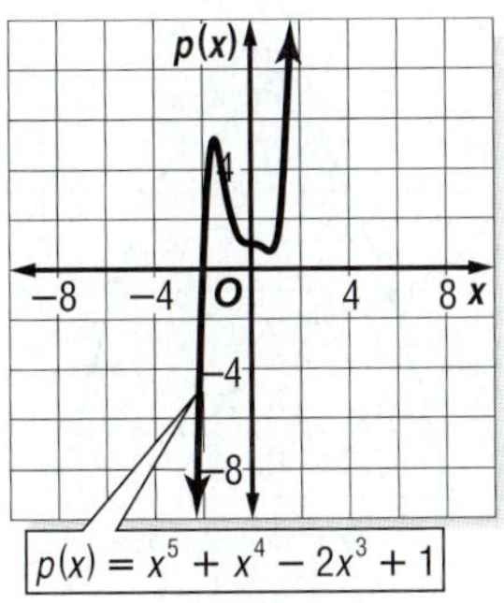

15b. between -2 and -3
15c. Sample answer: rel. max. at $x = -1.6$, rel. min. at $x = 0.8$

17a.

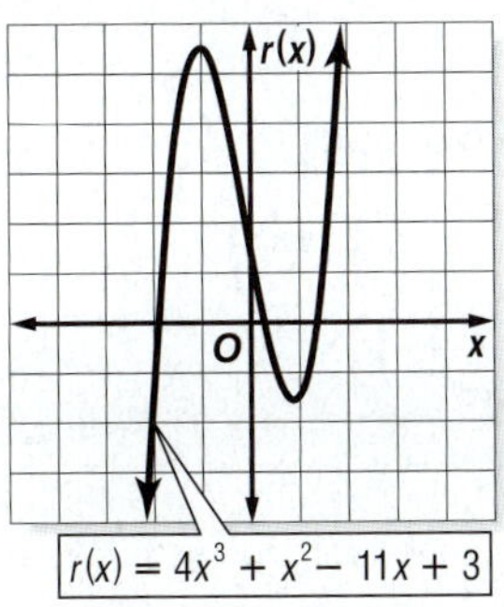

17b. between -2 and -1, between 0 and 1, and between 1 and 2
17c. Sample answer: rel. max. at $x = -1$, rel. min. at $x = 0.9$

19. $\frac{5}{3}$, -3, 0 **21.** 4, $-2 \pm 2i\sqrt{3}$ **23.** 2, -2 **25.** 4, -1

27. 20, -20 **29.** $x^2 + 2x + 3$ **31.** 1; 0; 2 **33.** 3 or 1; 1; 0 or 2 **35.** 2 or 0; 2 or 0; 4, 2, or 0 **37.** -1, -1
39. 1, 2, 4, -3 **41.** $\frac{1}{2}$, 2 **43.** $x^2 - 1$; $x^2 - 6x + 11$
45. $-15x - 5$; $-15x + 25$

47. $|x + 4|$; $|x| + 4$

49. $f^{-1}(x) = \frac{-x-3}{2}$

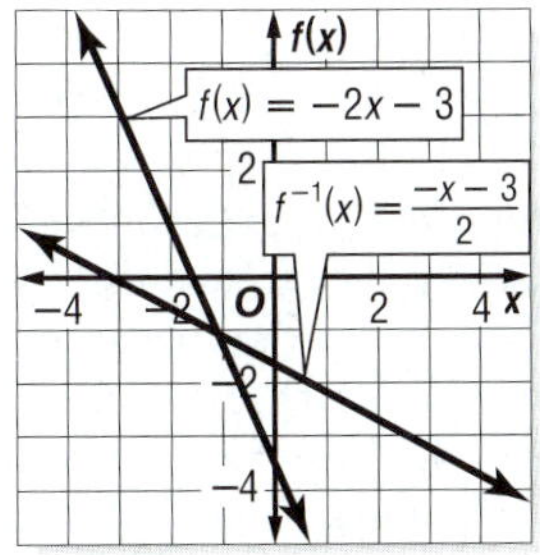

51. $f^{-1}(x) = \frac{2x-1}{-3}$

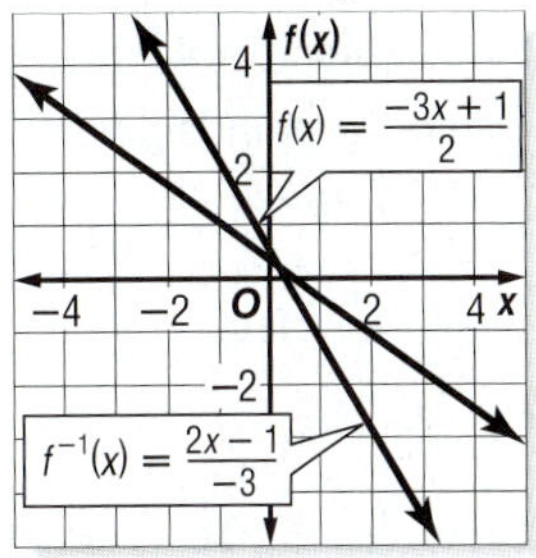

53. $y^{-1} = \pm\frac{1}{2}\sqrt{x} - \frac{3}{2}$

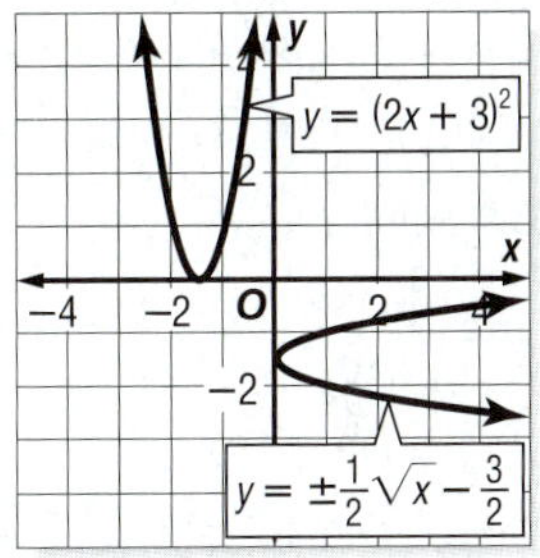

55. D: $x \geq \frac{3}{5}$, R: $y \geq 0$

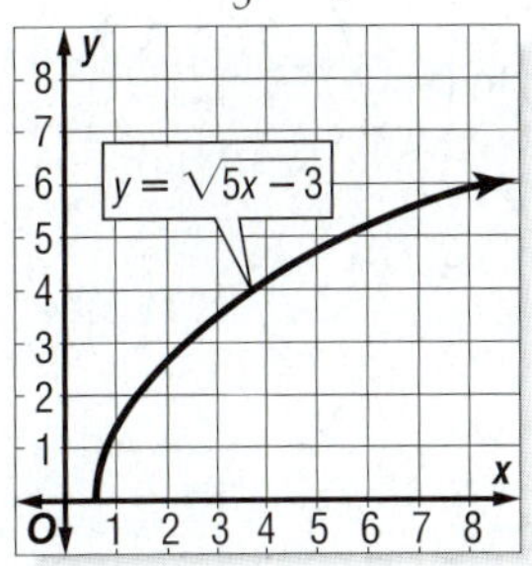

57.

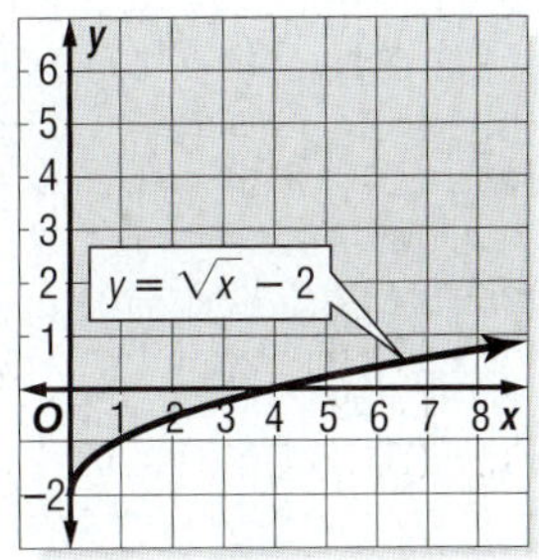

Chapter 8 Conic Sections

Page 411 Chapter 8 Getting Started

1. $\{-4, -6\}$ **3.** $\left\{\frac{3}{2}, -4\right\}$ **5a.** $\begin{bmatrix} -2 & 4 & -1 \\ 2 & 0 & -2 \end{bmatrix}$

5b. $\begin{bmatrix} 5 & 5 & 5 \\ -3 & -3 & -3 \end{bmatrix}$ **5c.** $\begin{bmatrix} 3 & 9 & 4 \\ -1 & -3 & -5 \end{bmatrix}$

7.

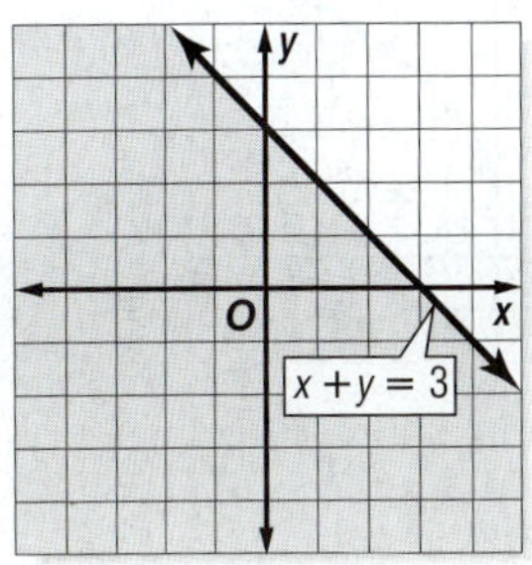

Selected Answers

Pages 414–416 Lesson 8-1

1. Since the sum of the x-coordinates of the given points is negative, the x-coordinate of the midpoint is negative. Since the sum of the y-coordinates of the given points is positive, the y-coordinate of the midpoint is positive. Therefore, the midpoint is in Quadrant II. **3.** Sample answer: (0, 0) and (5, 2) **5.** (2.5, 2.25) **7.** $\sqrt{122}$ units **9.** D **11.** $(-4, -2)$

13. $\left(\frac{17}{2}, \frac{27}{2}\right)$ **15.** (3.1, 2.7) **17.** $\left(\frac{1}{24}, \frac{5}{8}\right)$ **19.** (7, 11)

21. Sample answer: Draw several line segments across the U.S. One should go from the northeast corner to the southwest corner; another should go from the southeast corner to the northwest corner; another should go across the middle of the U.S. from east to west; and so on. Find the midpoints of these segments. Locate a point to represent all of these midpoints.

25. 25 units **27.** $3\sqrt{17}$ units **29.** $\sqrt{70.25}$ units **31.** 1 unit **33.** $\frac{\sqrt{813}}{12}$ units **35.** $7\sqrt{2} + \sqrt{58}$ units, 10 units^2

37. $\sqrt{130}$ units **39.** about 0.9 h **41.** The slope of the line through (x_1, y_1) and (x_2, y_2) is $\frac{y_2 - y_1}{x_2 - x_1}$ and the point-slope form of the equation of the line is $y - y_1 = \frac{y_2 - y_1}{x_2 - x_1}(x - x_1)$. Substitute $\left(\frac{x_1 + x_2}{2}, \frac{y_1 + y_2}{2}\right)$ into this equation. The left side is $\frac{y_1 + y_2}{2} - y_1$ or $\frac{y_2 - y_1}{2}$. The right side is $\frac{y_2 - y_1}{x_2 - x_1}\left(\frac{x_1 + x_2}{2} - x_1\right) = \frac{y_2 - y_1}{x_2 - x_1}\left(\frac{x_2 - x_1}{2}\right)$ or $\frac{y_2 - y_1}{2}$. Therefore, the point with coordinates $\left(\frac{x_1 + x_2}{2}, \frac{y_1 + y_2}{2}\right)$ lies on the line through (x_1, y_1) and (x_2, y_2).

The distance from $\left(\frac{x_1 + x_2}{2}, \frac{y_1 + y_2}{2}\right)$ to (x_1, y_1) is $\sqrt{\left(x_1 - \frac{x_1 + x_2}{2}\right)^2 + \left(y_1 - \frac{y_1 + y_2}{2}\right)^2}$ or $\sqrt{\left(\frac{x_1 - x_2}{2}\right)^2 + \left(\frac{y_1 - y_2}{2}\right)^2}$. The distance from $\left(\frac{x_1 + x_2}{2}, \frac{y_1 + y_2}{2}\right)$ to (x_2, y_2) is

$$\sqrt{\left(x_2 - \frac{x_1 + x_2}{2}\right)^2 + \left(y_2 - \frac{y_1 + y_2}{2}\right)^2} = \sqrt{\left(\frac{x_2 - x_1}{2}\right)^2 + \left(\frac{y_2 - y_1}{2}\right)^2}$$

or $\sqrt{\left(\frac{x_1 - x_2}{2}\right)^2 + \left(\frac{y_1 - y_2}{2}\right)^2}$. Therefore the point with coordinates $\left(\frac{x_1 + x_2}{2}, \frac{y_1 + y_2}{2}\right)$ is equidistant from (x_1, y_1) and (x_2, y_2). **43.** C **45.** on the line with equation $y = x$

47. $D = \{x \mid x \geq 2\}$, $R = \{y \mid y \geq 0\}$

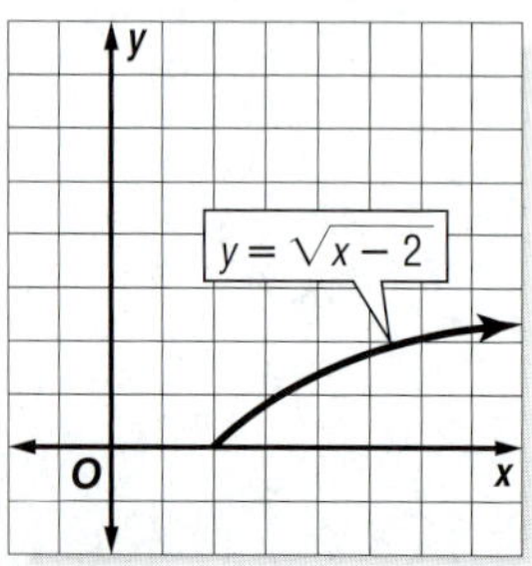

49. $D = \{x \mid x \geq 0\}$, $R = \{y \mid y \geq 1\}$

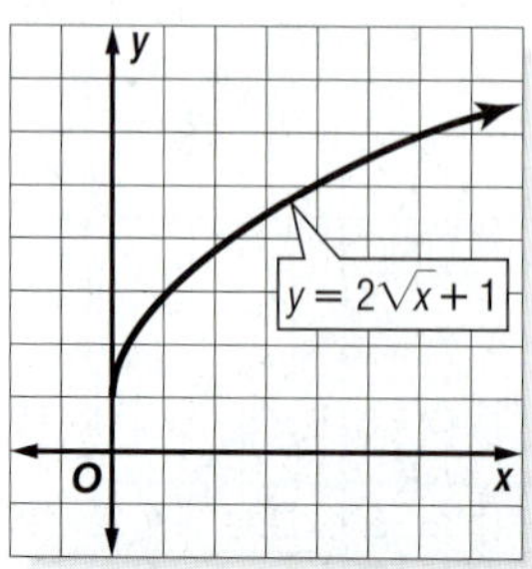

51. $-1 + 13i$ **53.** $4 - 3i$ **55.** $y = (x - 2)^2 - 3$ **57.** $y = 3(x - 1)^2 + 2$ **59.** $y = -3(x + 3)^2 + 17$

Pages 423–425 Lesson 8-2

1. $(3, -7)$, $\left(3, -6\frac{15}{16}\right)$, $x = 3$, $y = -7\frac{1}{16}$ **3.** When she added 9 to complete the square, she forgot to also subtract 9. The standard form is $y = (x + 3)^2 - 9 + 4$ or $y = (x + 3)^2 - 5$.

5. $(3, -4)$, $\left(3, -3\frac{3}{4}\right)$, $x = 3$, $y = -4\frac{1}{4}$, upward, 1 unit

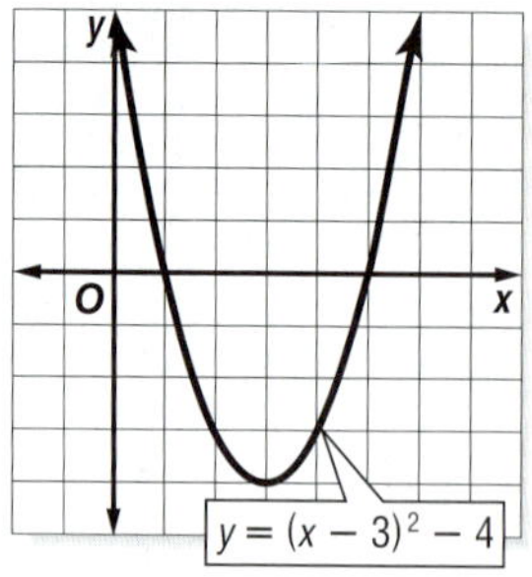

7. $\left(-\frac{4}{3}, -\frac{2}{3}\right)$, $\left(-\frac{4}{3}, -\frac{3}{4}\right)$, $x = -\frac{4}{3}$, $y = -\frac{7}{12}$, downward, $\frac{1}{3}$ unit

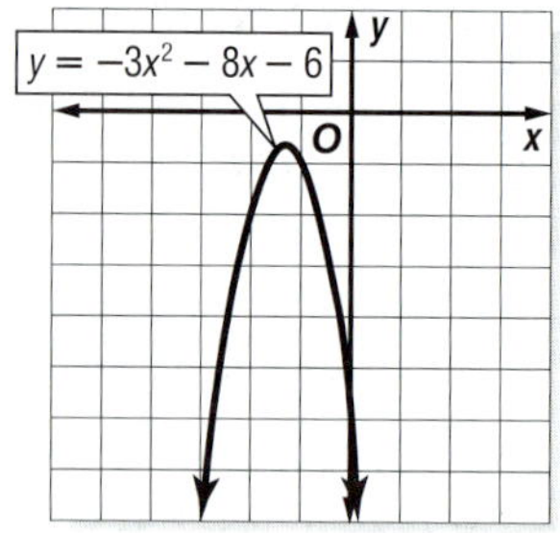

9. $y = \frac{1}{8}(x - 3)^2 + 6$

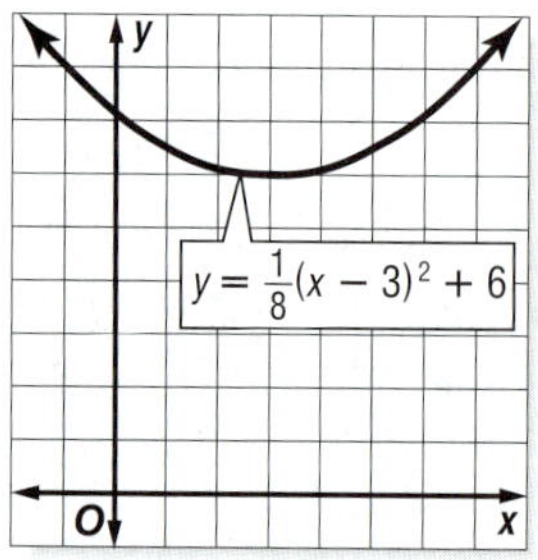

11. $x = \frac{1}{24}y^2 - 6$
13. $x = (y + 7)^2 - 29$
15. $x = 3\left(y + \frac{5}{6}\right)^2 - 11\frac{1}{12}$

17. $(0, 0)$, $\left(\frac{1}{2}, 0\right)$, $y = 0$, $x = -\frac{1}{2}$, right, 2 units

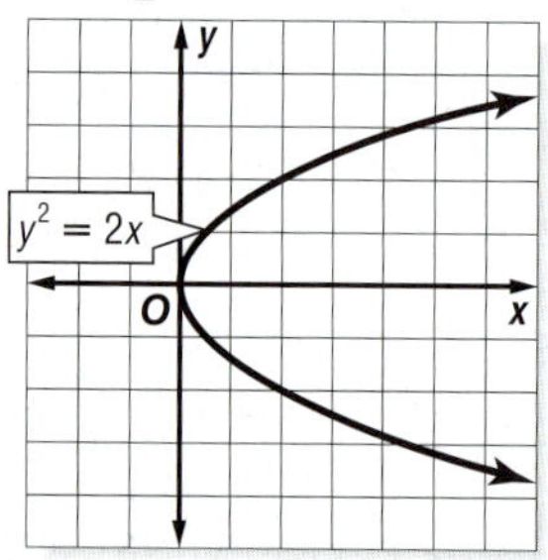

19. $(1, 4)$, $\left(1, 3\frac{1}{2}\right)$, $x = 1$, $y = 4\frac{1}{2}$, downward, 2 units

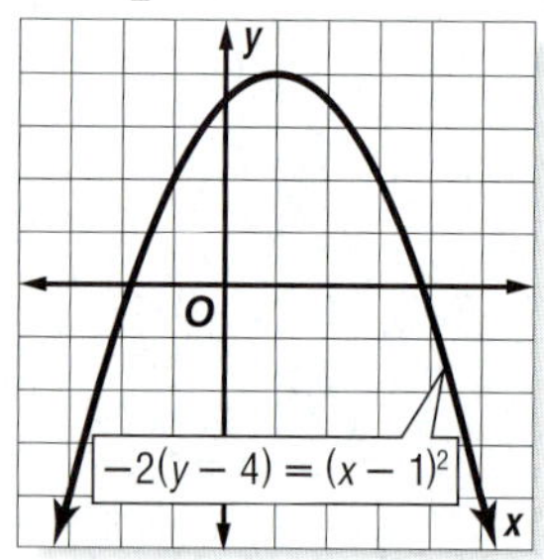

21. $(4, 8)$, $(3, 8)$, $y = 8$, $x = 5$, left, 4 units

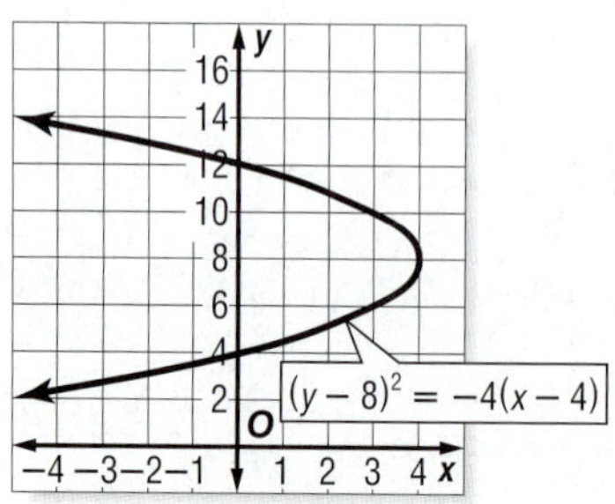

23. $(-24, 7), \left(-23\frac{3}{4}, 7\right), y = 7, x = -24\frac{1}{4}$, right, 1 unit

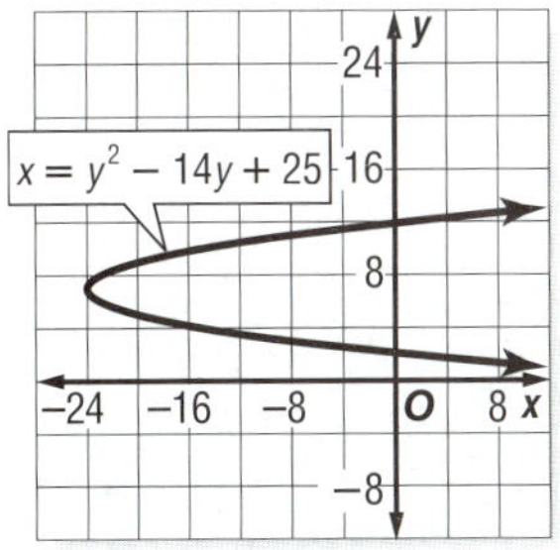

25. $(4, 2), \left(4, 2\frac{1}{12}\right), x = 4, y = 1\frac{11}{12}$, upward, $\frac{1}{3}$ unit

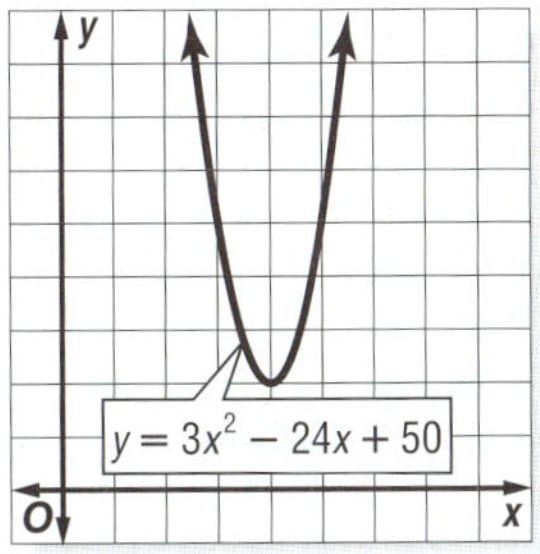

27. $\left(\frac{17}{4}, \frac{3}{4}\right), \left(\frac{67}{16}, \frac{3}{4}\right), y = \frac{3}{4}, x = \frac{69}{16}$, left, $\frac{1}{4}$ unit

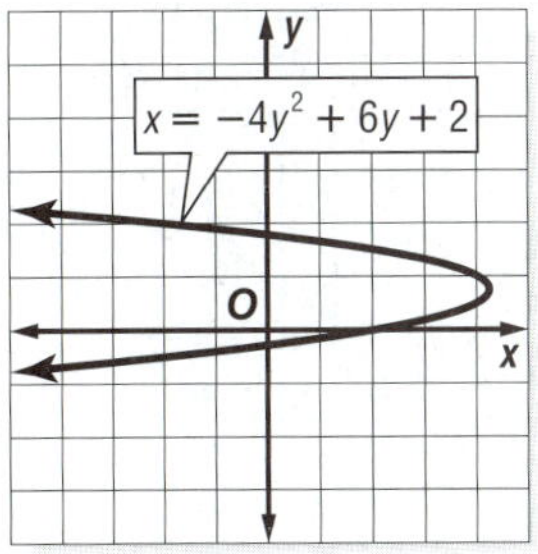

29. $(123, -18), \left(122\frac{1}{4}, -18\right), y = -18, x = 123\frac{3}{4}$, left, 3 units

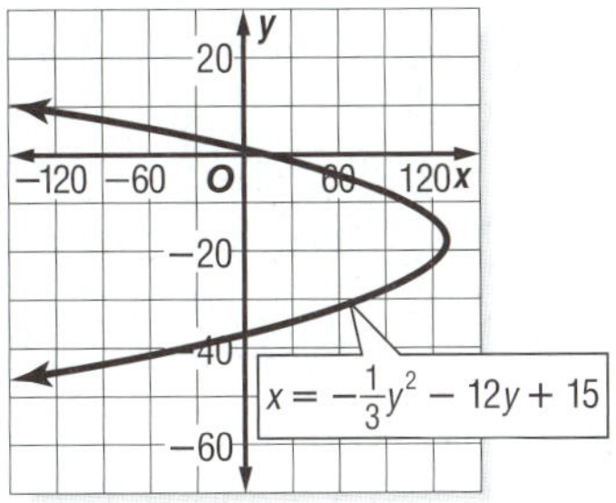

31. 1 **33.** $y = -\frac{2}{3}$ **35.** 0.75 cm

37. $x = -\frac{1}{24}(y - 6)^2 + 8$

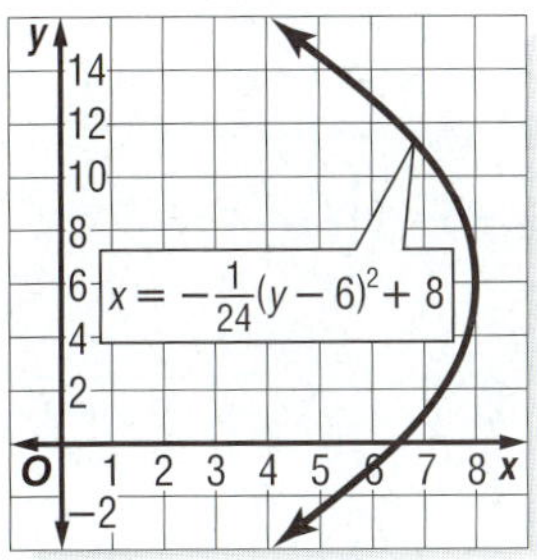

39. $y = \frac{1}{16}(x - 1)^2 + 7$

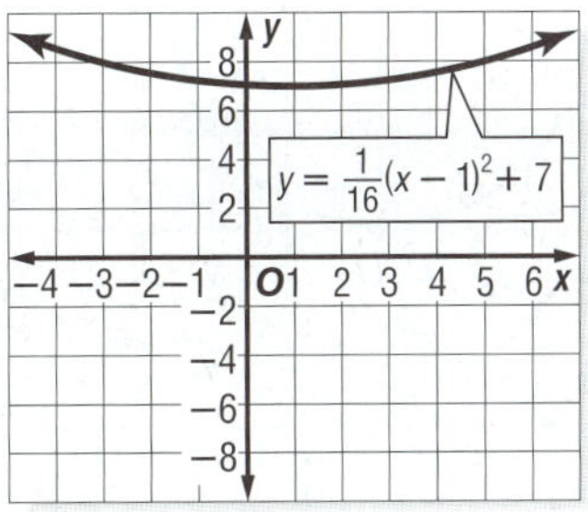

41. $x = \frac{1}{4}(y - 3)^2 + 4$

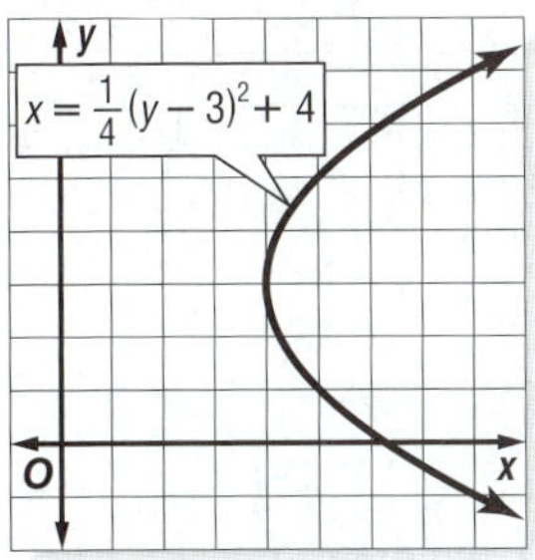

43. about $y = -0.00046x^2 + 325$ **45.** $y = -\frac{1}{26{,}200}x^2 + 6550$

47. A parabolic reflector can be used to make a car headlight more effective. Answers should include the following.

- Reflected rays are *focused* at that point.
- The light from an unreflected bulb would shine in all directions. With a parabolic reflector, most of the light can be directed forward toward the road.

49. A **51.** 10 units

53.

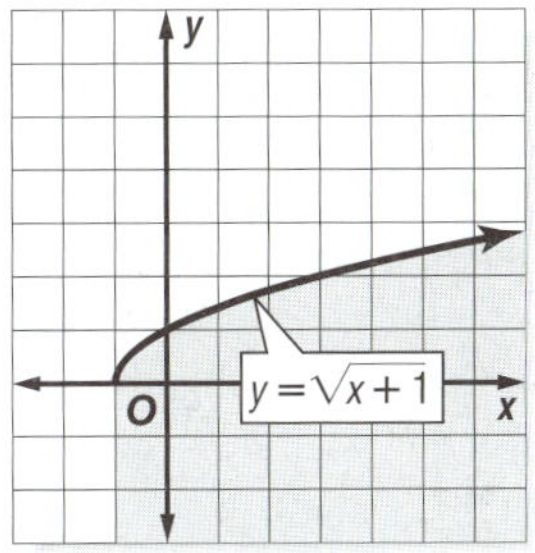

55. 4 **57.** 9 **59.** $2\sqrt{3}$ **61.** $4\sqrt{3}$

Pages 428–431 Lesson 8-3

1. Sample answer: $(x - 6)^2 + (y + 2)^2 = 16$ **3.** Lucy; 36 is the *square* of the radius, so the radius is 6 units.
5. $(x + 1)^2 + (y + 5)^2 = 4$ **7.** $(x - 3)^2 + (y + 7)^2 = 9$

9. $(0, 14)$, $\sqrt{34}$ units

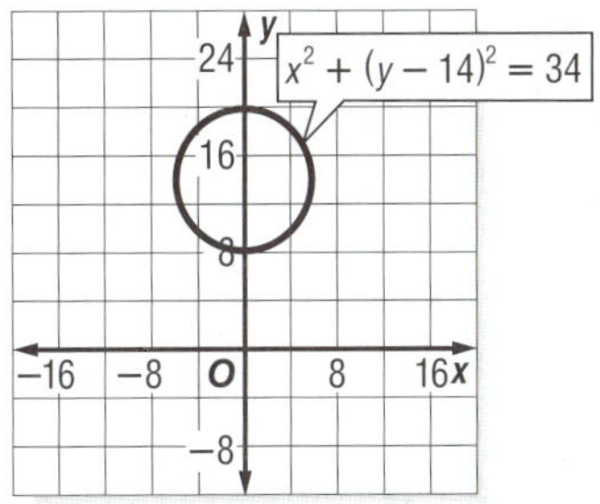

11. $\left(-\frac{2}{3}, \frac{1}{2}\right), \frac{2\sqrt{2}}{3}$ unit

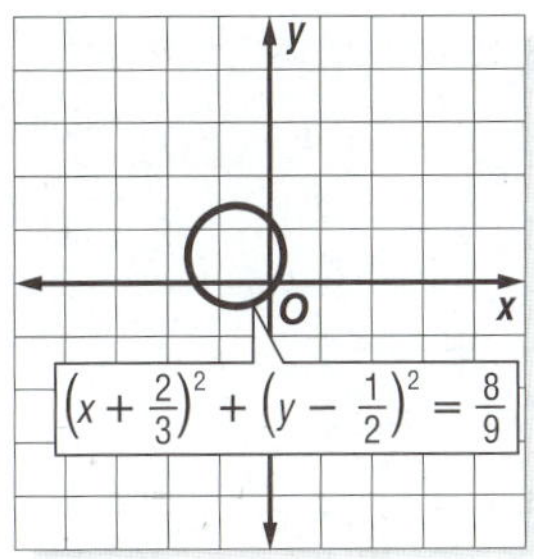

13. $(-2, 0)$, $2\sqrt{3}$ units

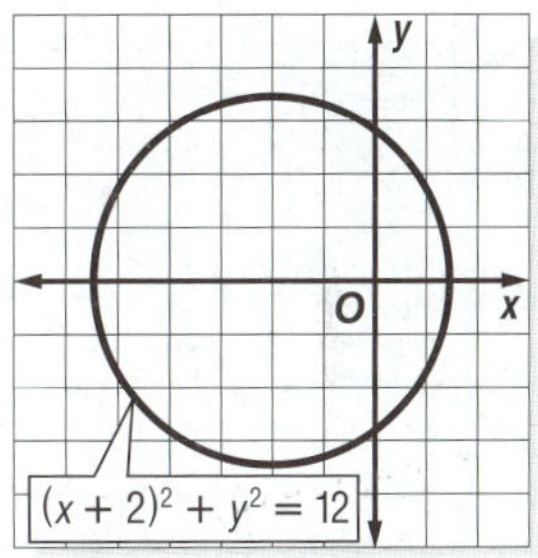

15.

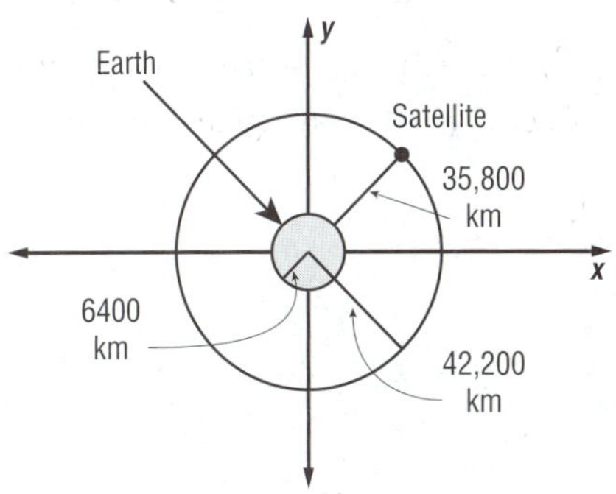

17. $(x - 2)^2 + (y + 1)^2 = 4$ **19.** $(x + 8)^2 + (y - 7)^2 = \frac{1}{4}$

21. $(x + 1)^2 + \left(y + \frac{1}{2}\right)^2 = \frac{1945}{4}$

23. $\left(x + \sqrt{13}\right)^2 + (y - 42)^2 = 1777$
25. $(x - 4)^2 + (y - 2)^2 = 4$ **27.** $(x + 5)^2 + (y - 4)^2 = 25$
29. $(x + 2.5)^2 + (y + 2.8)^2 = 1600$

31. (0, 0), 12 units

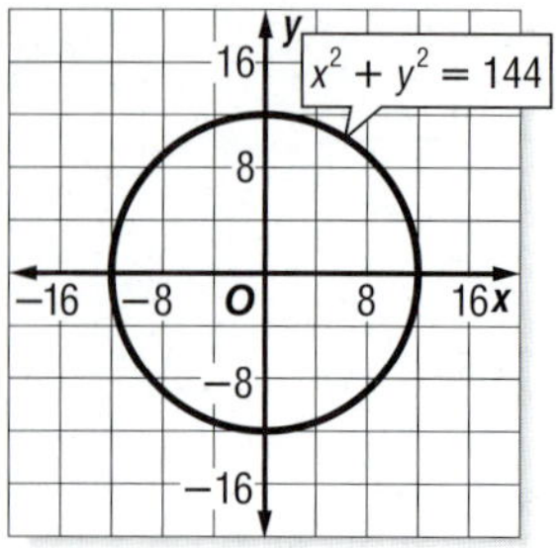

33. (−3, −7), 9 units

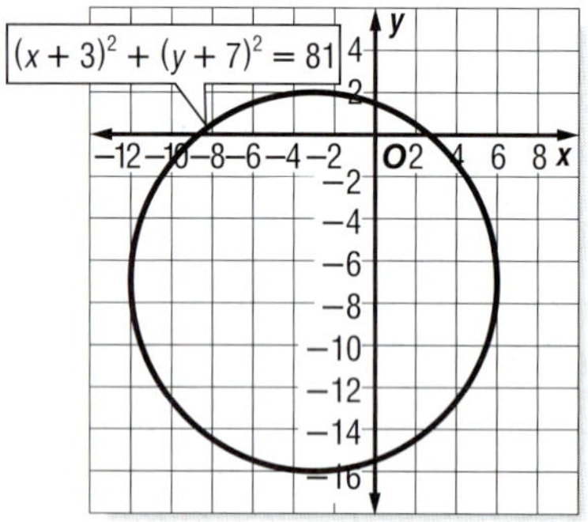

35. (3, −7), $5\sqrt{2}$ units

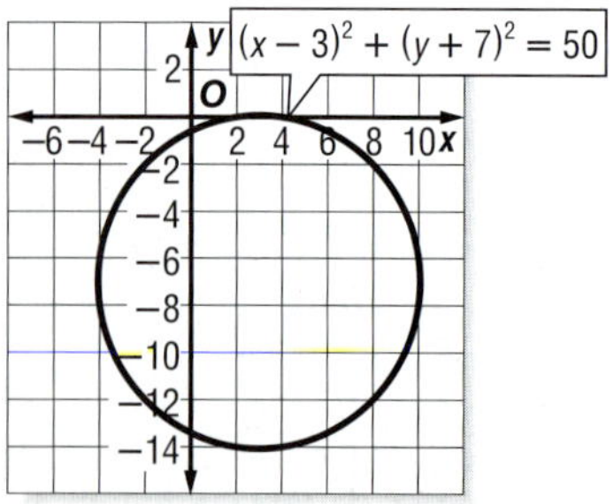

37. $\left(-2, \sqrt{3}\right)$, $\sqrt{29}$ units

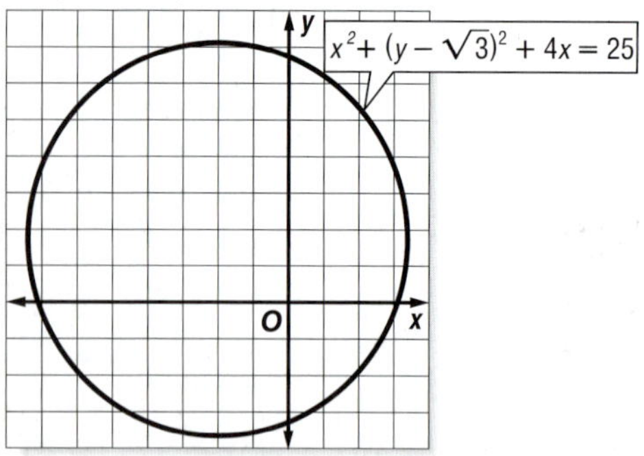

39. (0, 3), 5 units

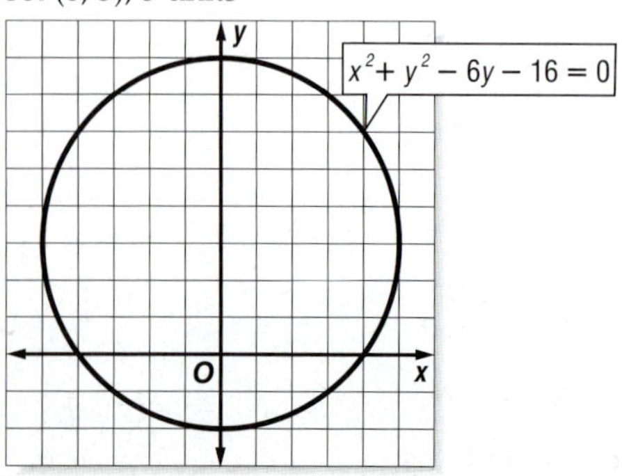

41. (9, 9), $\sqrt{109}$ units

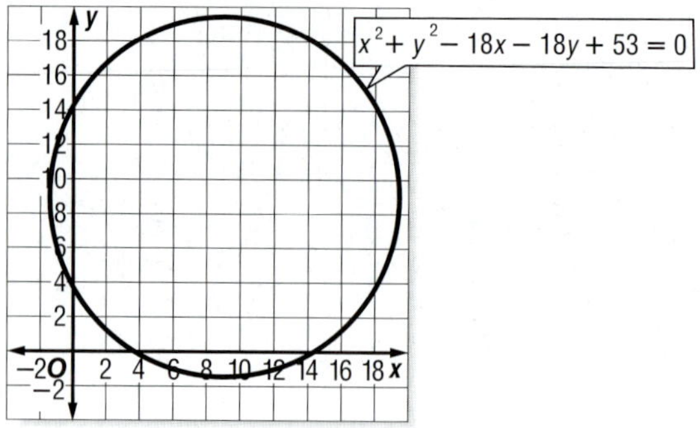

43. $\left(\frac{3}{2}, -4\right)$, $\frac{3\sqrt{17}}{2}$ units

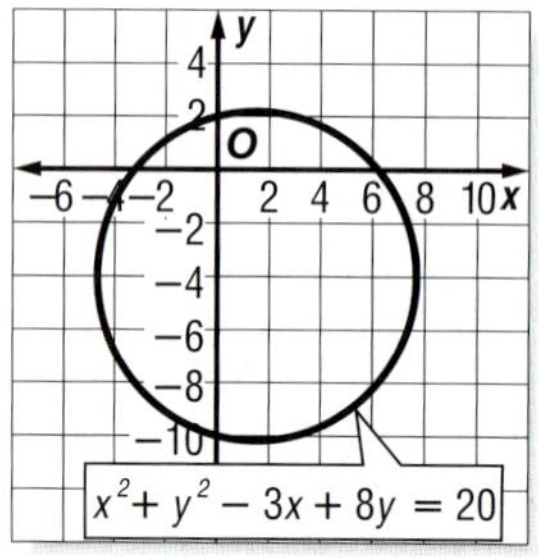

45. (−1, −2), $\sqrt{14}$ units

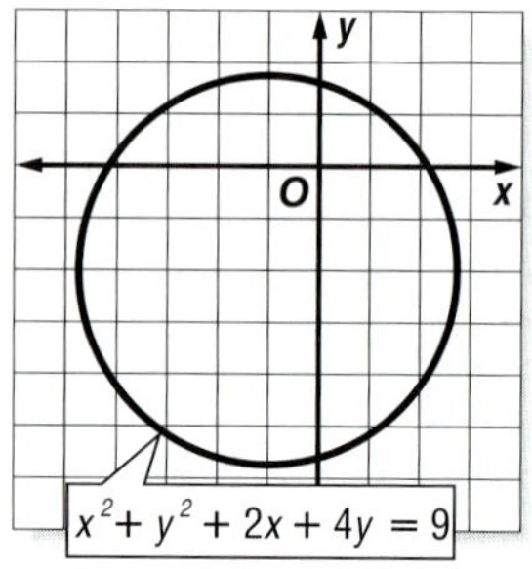

47. $\left(0, -\frac{9}{2}\right)$, $\sqrt{19}$ units

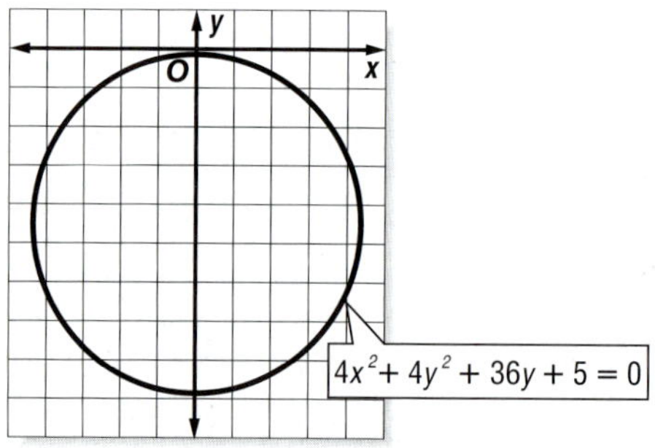

49. $(x + 1)^2 + (y + 2)^2 = 5$ **51.** A **53.** $y = \pm\sqrt{16 - (x + 3)^2}$

55.

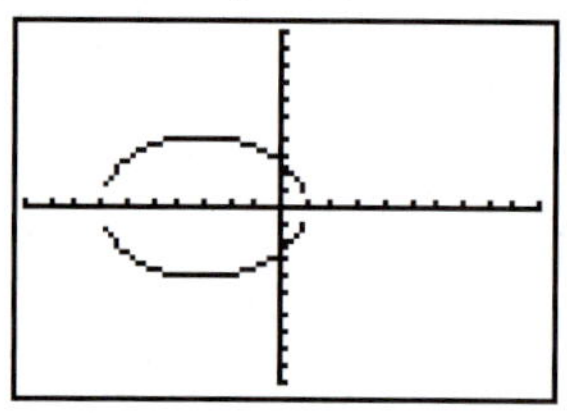

[−10, 10] scl:1 by [−10, 10] scl:1

57. (1, 0), $\left(\frac{11}{12}, 0\right)$, $y = 0$, $x = 1\frac{1}{12}$, left, $\frac{1}{3}$ unit

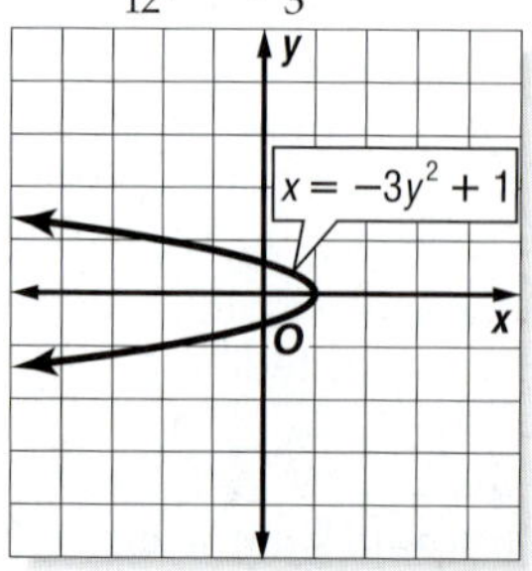

59. (−2, −4), $\left(-2, -3\frac{3}{4}\right)$, $x = -2$, $y = -4\frac{1}{4}$, upward, 1 unit

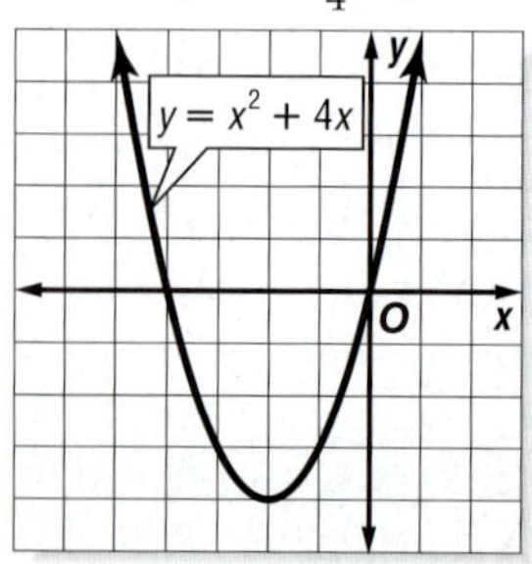

61. (−1, −2) **63.** −4, −2, 1 **65.** 28 in. by 15 in. **67.** 6 **69.** 25 **71.** $2\sqrt{2}$

Page 431 Practice Quiz 1

1. 13 units

3. (0, 0), $\left(1\frac{1}{2}, 0\right)$, $y = 0$, $x = -1\frac{1}{2}$, right, 6 units

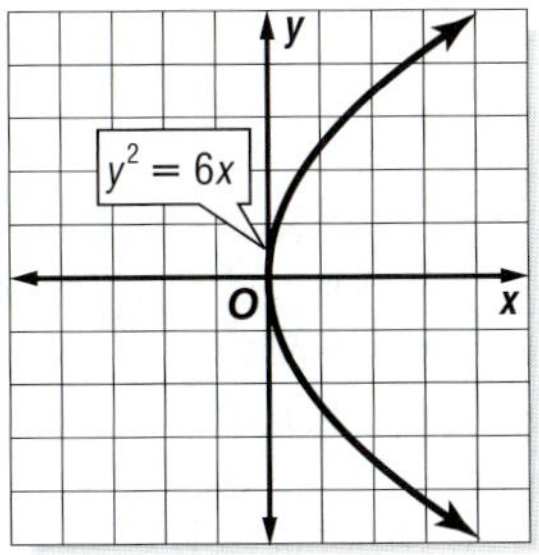

5. (0, 4), 7 units

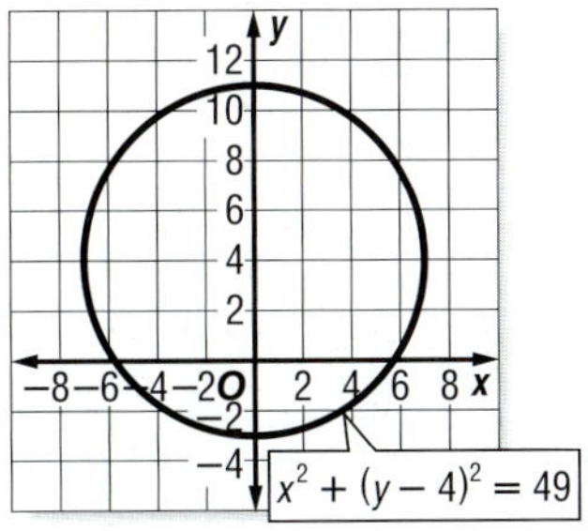

Pages 437–440 Lesson 8-4

1. $x = -1, y = 2$ **3.** Sample answer: $\frac{(x-2)^2}{4} + \frac{(y+5)^2}{1} = 1$

5. $\frac{(y+4)^2}{36} + \frac{(x-2)^2}{4} = 1$

7. (0, 0): (0, ±3); $6\sqrt{2}$; 6

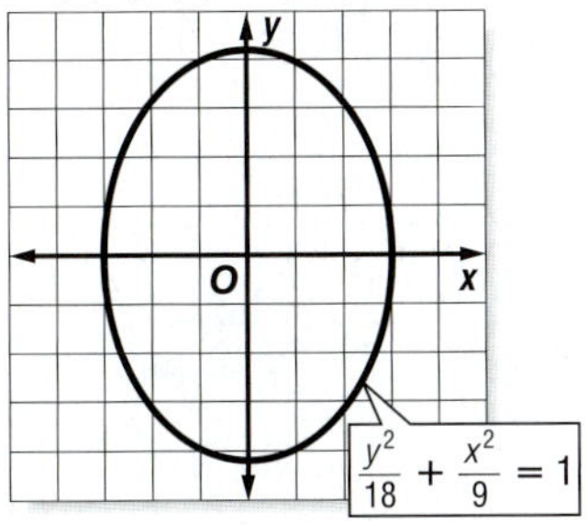

9. (0, 0); (±2, 0); $4\sqrt{2}$; 4

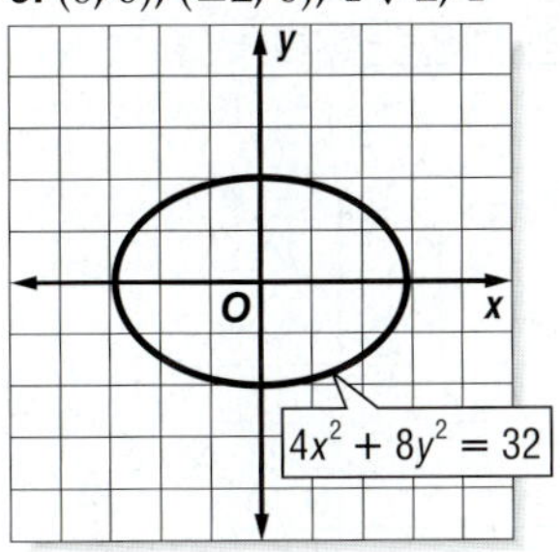

11. about $\frac{x^2}{1.32 \times 10^{15}} + \frac{y^2}{1.27 \times 10^{15}} = 1$ **13.** $\frac{x^2}{16} + \frac{y^2}{7} = 1$

15. $\frac{y^2}{16} + \frac{(x+2)^2}{4} = 1$ **17.** $\frac{(y-4)^2}{64} + \frac{(x-2)^2}{4} = 1$

19. $\frac{(x-5)^2}{64} + \frac{(y-4)^2}{\frac{81}{4}} = 1$ **21.** $\frac{x^2}{169} + \frac{y^2}{25} = 1$

23. about $\frac{x^2}{2.02 \times 10^{16}} + \frac{y^2}{2.00 \times 10^{16}} = 1$ **25.** $\frac{y^2}{20} + \frac{x^2}{4} = 1$

27. (0, 0); $\left(0, \pm\sqrt{5}\right)$; $2\sqrt{10}$; $2\sqrt{5}$

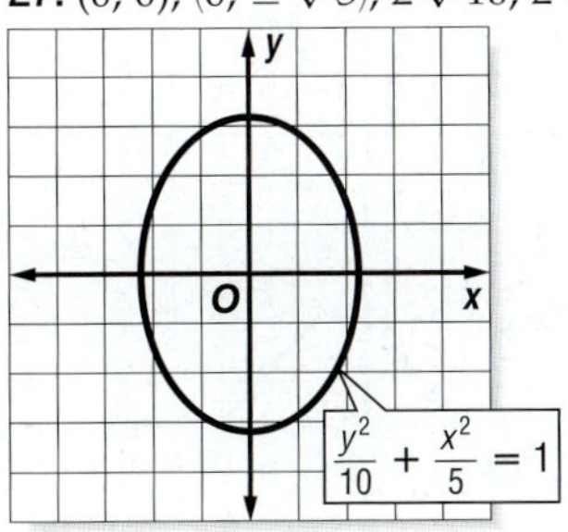

29. (−8, 2); $\left(-8 \pm 3\sqrt{7}, 2\right)$; 24; 18

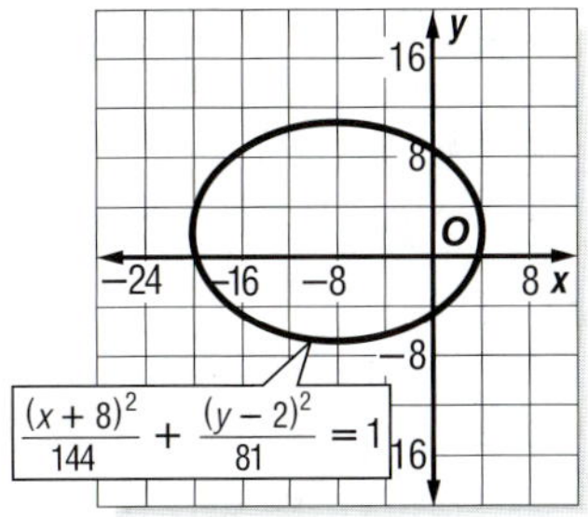

31. (0, 0); $\left(\pm\sqrt{6}, 0\right)$; 6; $2\sqrt{3}$

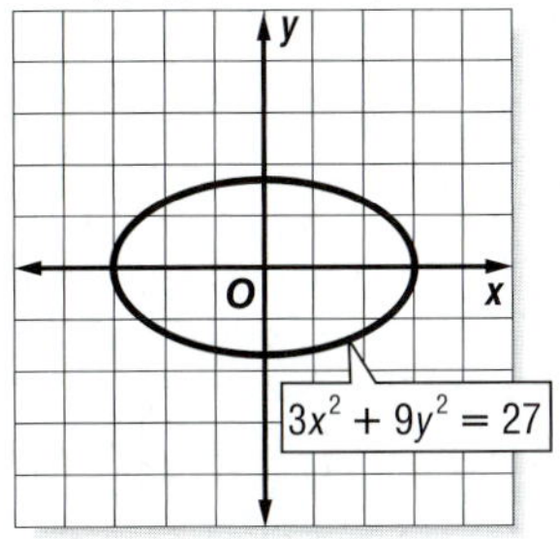

33. (0, 0); $\left(0, \pm\sqrt{7}\right)$; 8; 6

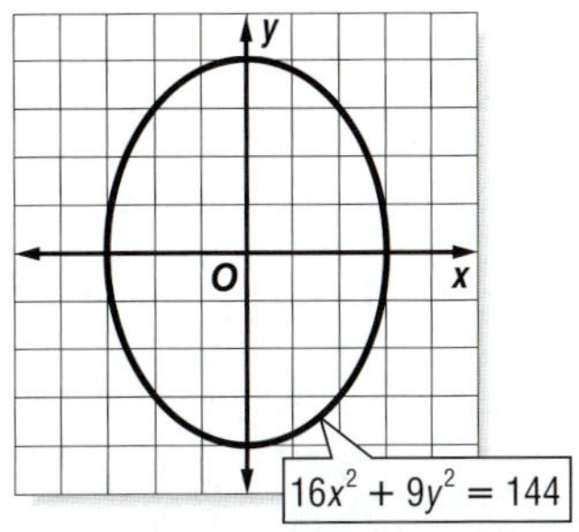

35. (−3, 1); (−3, 5), (−3, −3); $4\sqrt{6}$; $4\sqrt{2}$

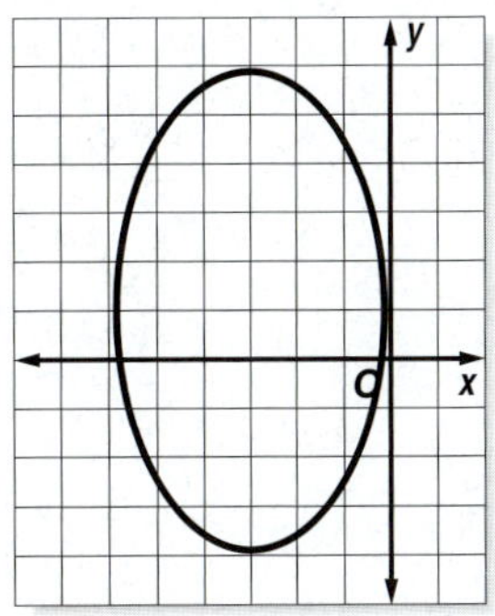

37. (2, 2); (2, 4), (2, 0); $2\sqrt{7}$; $2\sqrt{3}$

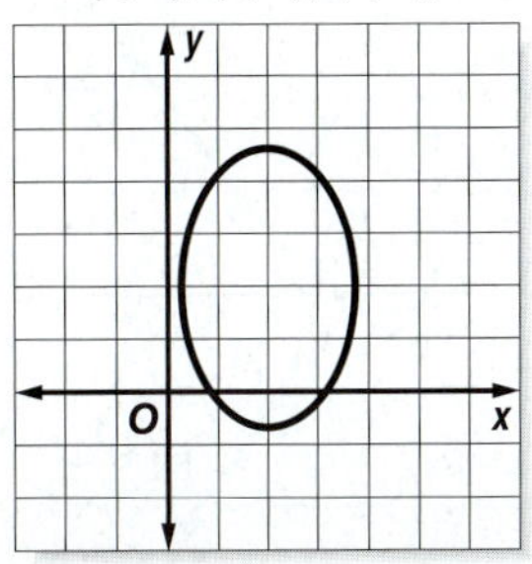

39. $\frac{x^2}{12} + \frac{y^2}{9} = 1$ **41.** C **43.** about $\frac{x^2}{1.35 \times 10^{19}} + \frac{y^2}{1.26 \times 10^{19}} = 1$

45. $(x-4)^2 + (y-1)^2 = 101$ **47.** $(x-4)^2 + (y+1)^2 = 16$

49. **Married Americans**

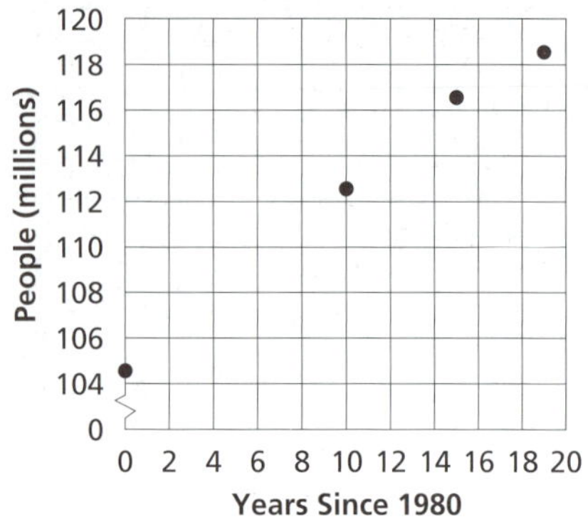

51. Sample answer: 128,600,000

53.

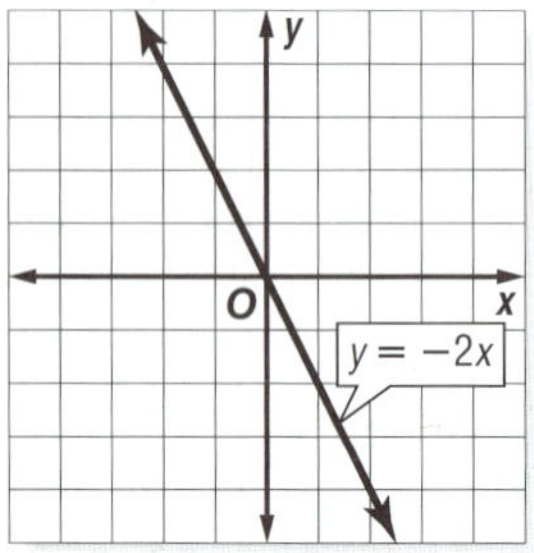

55.

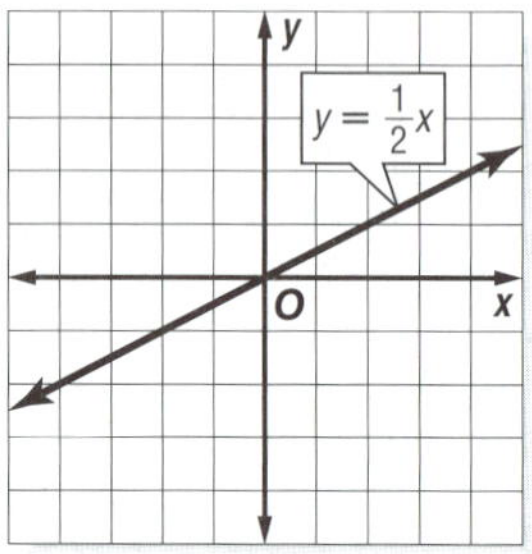

57.

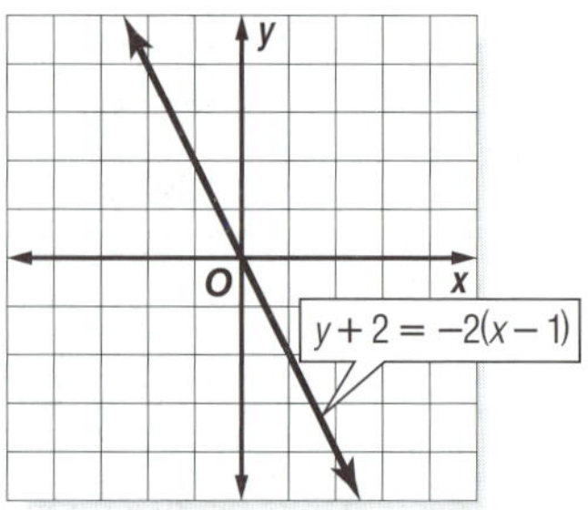

Pages 445–448 Lesson 8-5

1. sometimes 3. Sample answer: $\frac{x^2}{4} - \frac{y^2}{9} = 1$

5. $\frac{x^2}{1} - \frac{y^2}{15} = 1$

7. $\left(1, -6 \pm 2\sqrt{5}\right)$; $\left(1, -6 \pm 3\sqrt{5}\right)$; $y + 6 = \pm\frac{2\sqrt{5}}{5}(x - 1)$

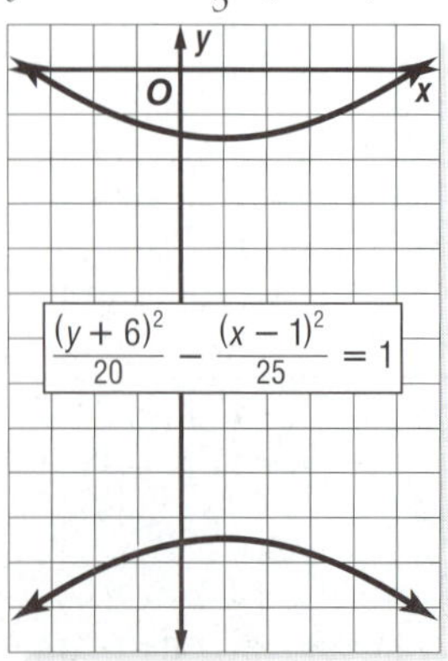

9. $\left(4 \pm 2\sqrt{5}, -2\right)$; $\left(4 \pm 3\sqrt{5}, -2\right)$; $y + 2 = \pm\frac{\sqrt{5}}{2}(x - 4)$

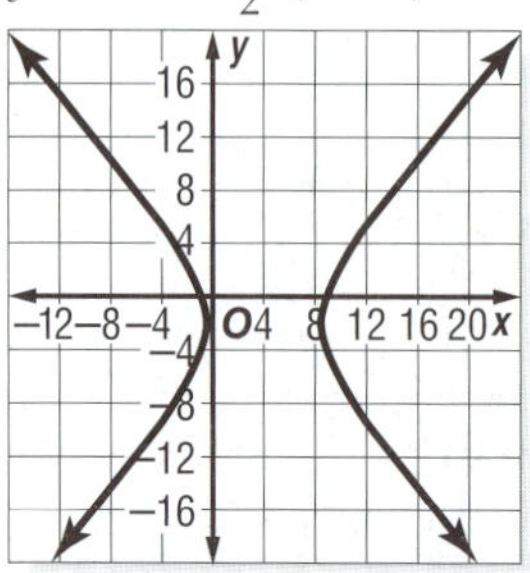

11. $\frac{x^2}{4} - \frac{y^2}{12} = 1$ 13. $\frac{\left(y + \frac{11}{2}\right)^2}{\frac{25}{4}} - \frac{x^2}{6} = 1$

15. $\frac{x^2}{25} - \frac{y^2}{36} = 1$ 17. $\frac{(x-2)^2}{49} - \frac{(y+3)^2}{4} = 1$

19. $\frac{x^2}{16} - \frac{y^2}{9} = 1$

21. $(\pm 9, 0)$; $\left(\pm\sqrt{130}, 0\right)$; $y = \pm\frac{7}{9}x$

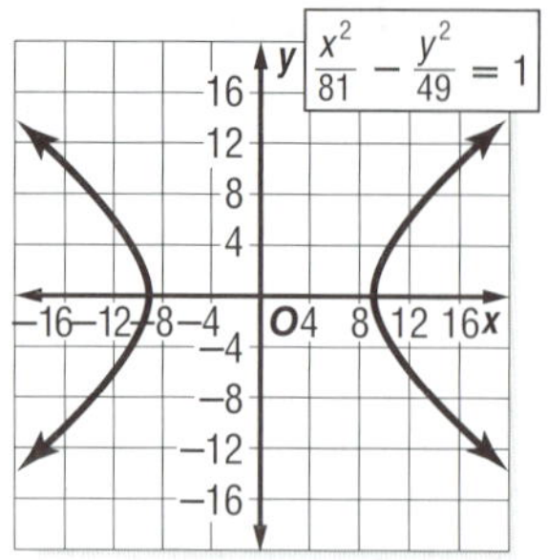

23. $(0, \pm 4)$; $\left(0, \pm\sqrt{41}\right)$; $y = \pm\frac{4}{5}x$

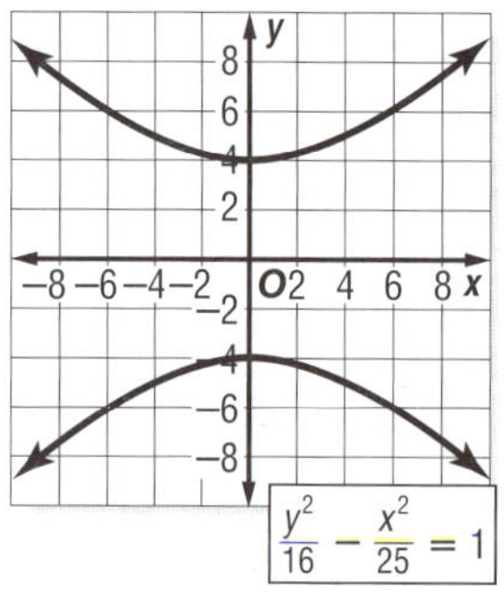

25. $\left(\pm\sqrt{2}, 0\right)$; $\left(\pm\sqrt{3}, 0\right)$; $y = \pm\frac{\sqrt{2}}{2}x$

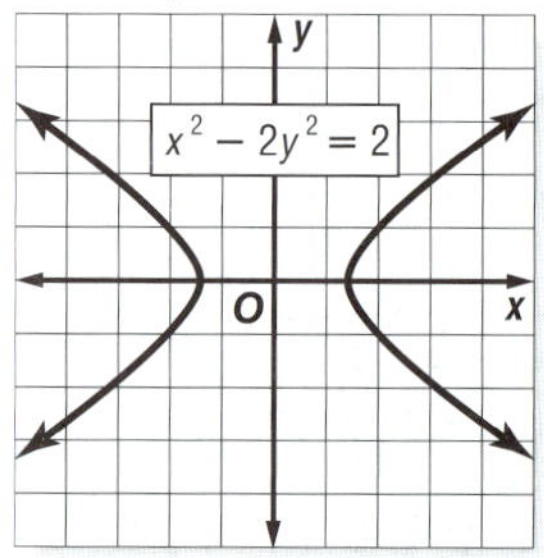

27. $(0, \pm 6)$; $\left(0, \pm 3\sqrt{5}\right)$; $y = \pm 2x$

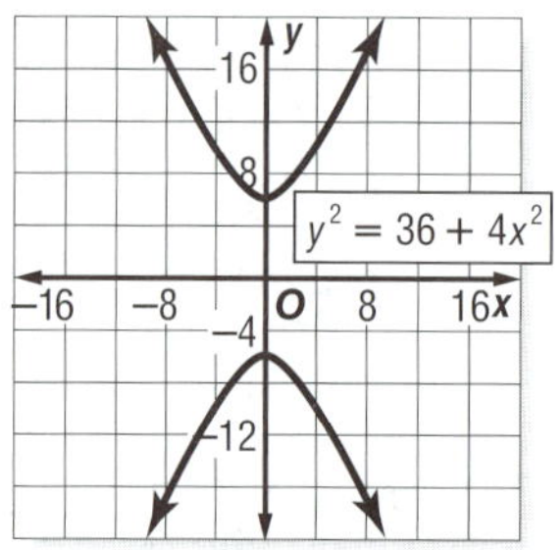

29. $(-2, 0)$, $(-2, 8)$; $(-2, -1)$, $(-2, 9)$; $y - 4 = \pm\frac{4}{3}(x + 2)$

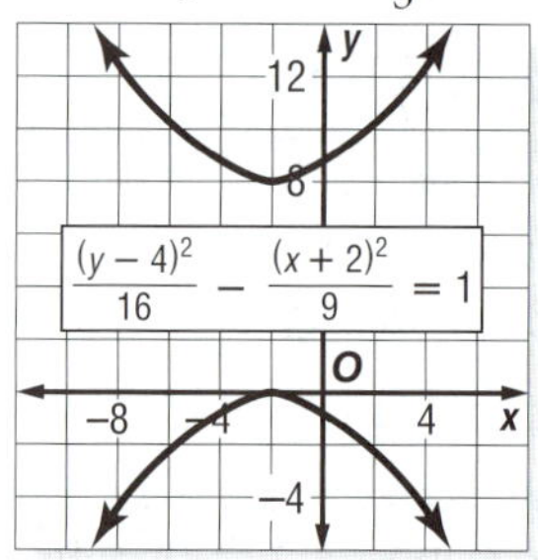

31. $(-3, -3)$, $(1, -3)$; $\left(-1 \pm \sqrt{13}, -3\right)$; $y + 3 = \pm\frac{3}{2}(x + 1)$

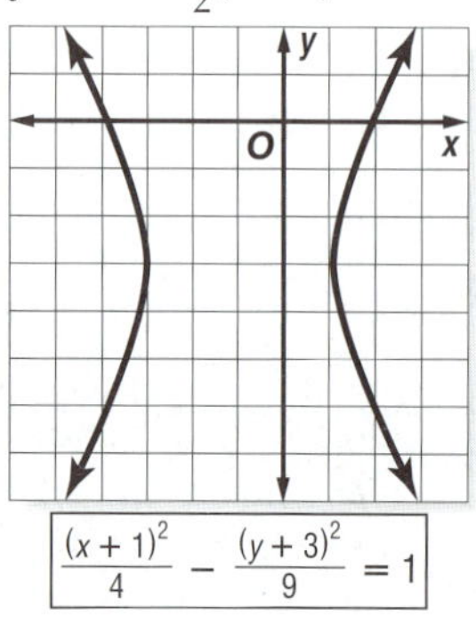

33. $\left(1, -3 \pm 2\sqrt{6}\right)$; $\left(1, -3 \pm 4\sqrt{2}\right)$; $y + 3 = \pm\sqrt{3}(x - 1)$

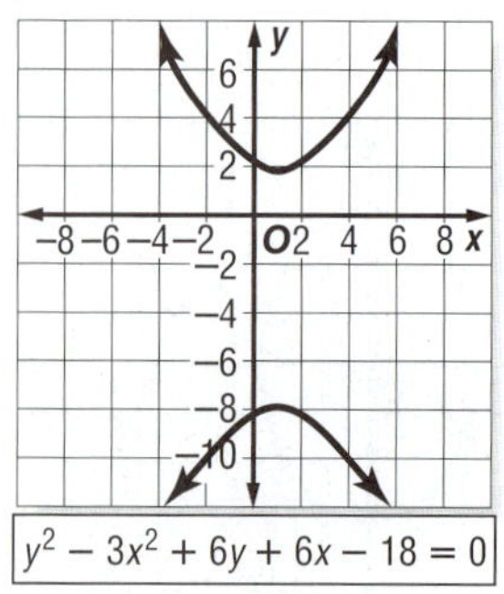

35. $\frac{x^2}{1.1025} - \frac{y^2}{7.8975} = 1$ 37. 120 cm, 100 cm
39. about 47.32 ft 41. C

43.

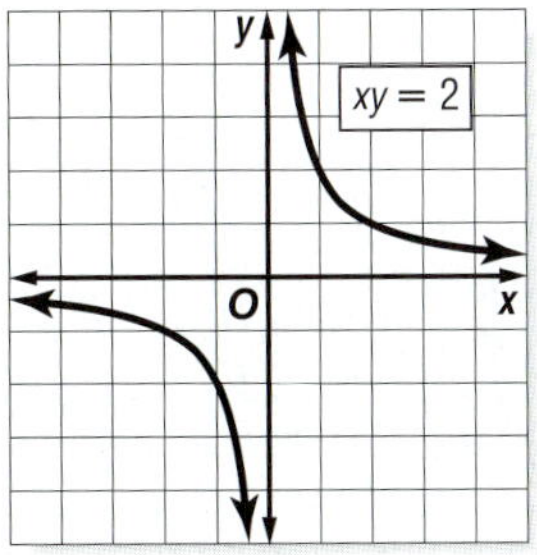

45.

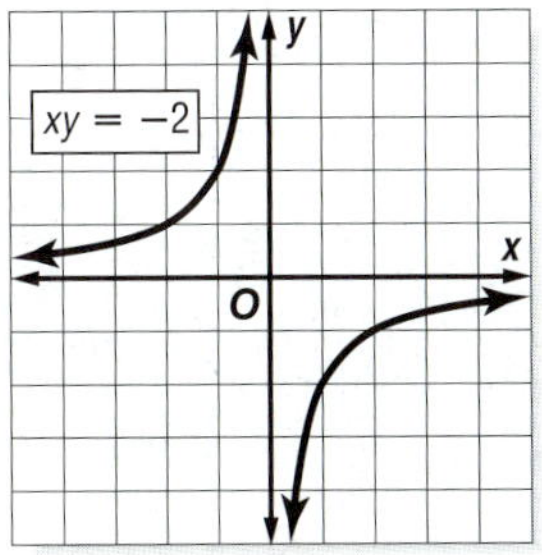

47. $\frac{(x-5)^2}{16} + \frac{(y-2)^2}{1} = 1$ 49. $\frac{(x-1)^2}{25} + \frac{(y-4)^2}{9} = 1$

51. −4, −2 53. $\begin{bmatrix} -7 & 0 \\ 5 & 20 \end{bmatrix}$ 55. about 5,330,000 subscribers per year 57. $2x + 17y$ 59. 1, −2, 9 61. 5, 0, −2
63. 0, 1, 0

Page 448 Practice Quiz 2

1. $\frac{(y-1)^2}{81} + \frac{(x-3)^2}{32} = 1$

3. $(-1, 1)$; $(-1, 1 \pm \sqrt{11})$; 8; $2\sqrt{5}$

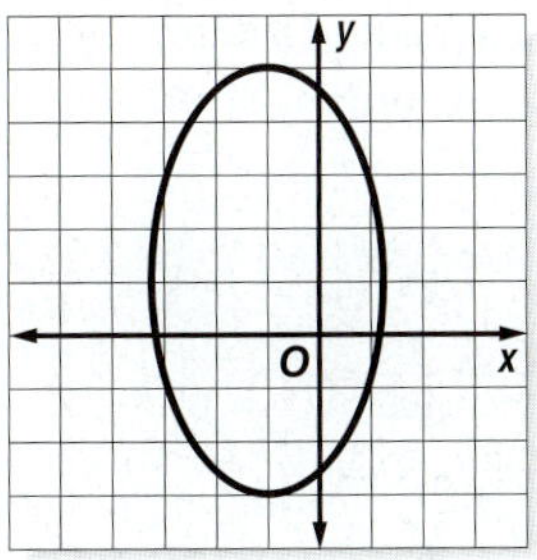

5. $\frac{(x-2)^2}{16} - \frac{(y-2)^2}{5} = 1$

Pages 450–452 Lesson 8-6

1. Sample answer: $2x^2 + 2y^2 - 1 = 0$ 3. The standard form of the equation is $(x-2)^2 + (y+1)^2 = 0$. This is an equation of a circle centered at $(2, -1)$ with radius 0. In other words, $(2, -1)$ is the only point that satisfies the equation.

5. $\frac{y^2}{16} - \frac{x^2}{8} = 1$, hyperbola

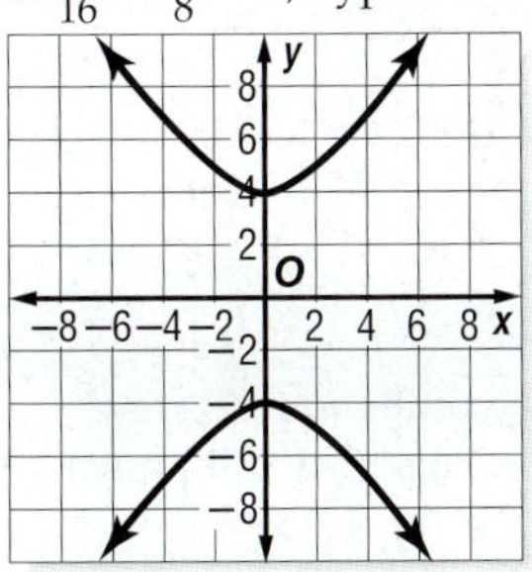

7. $\frac{(x+1)^2}{4} + \frac{(y-3)^2}{1} = 1$, ellipse

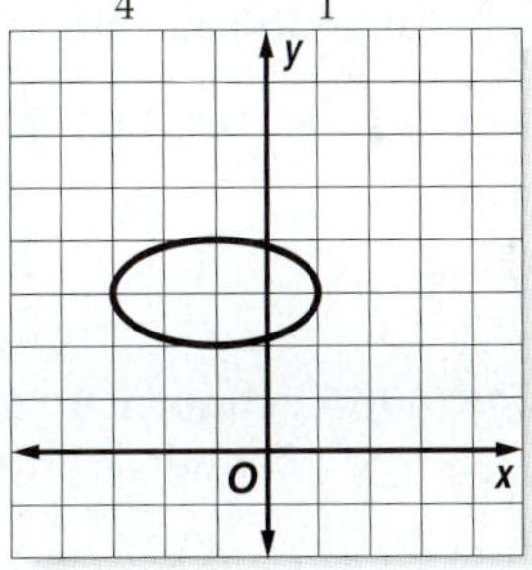

9. ellipse

11.

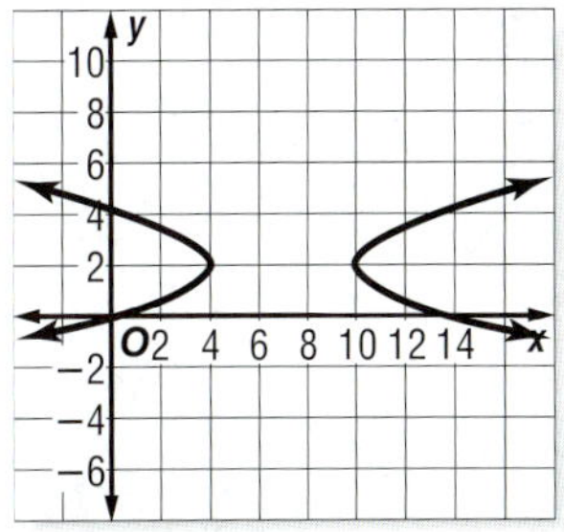

13. $\frac{y^2}{4} + \frac{x^2}{2} = 1$, ellipse

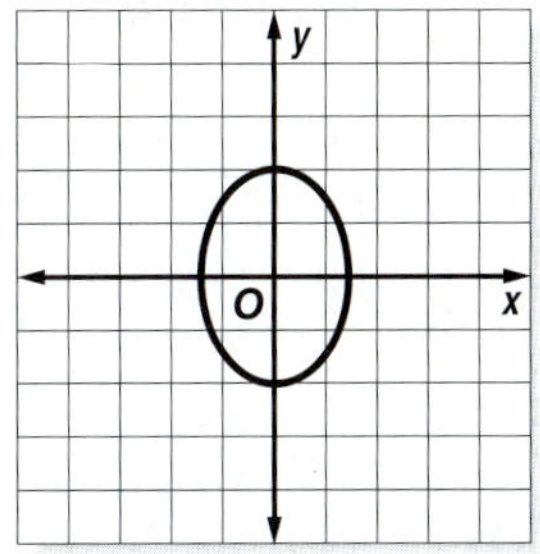

15. $\frac{x^2}{4} - \frac{y^2}{1} = 1$, hyperbola

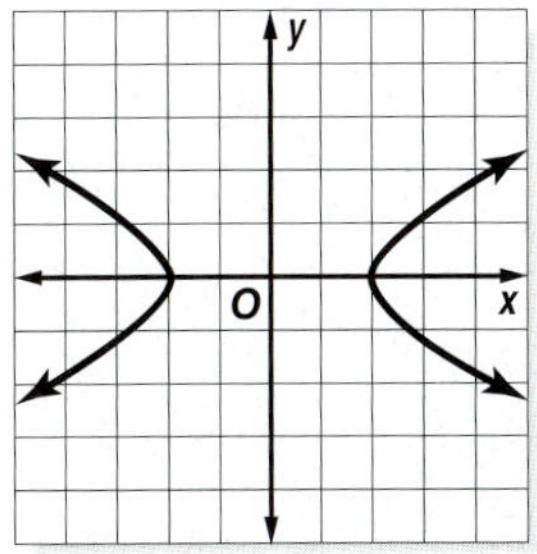

17. $y = (x-2)^2 - 4$, parabola

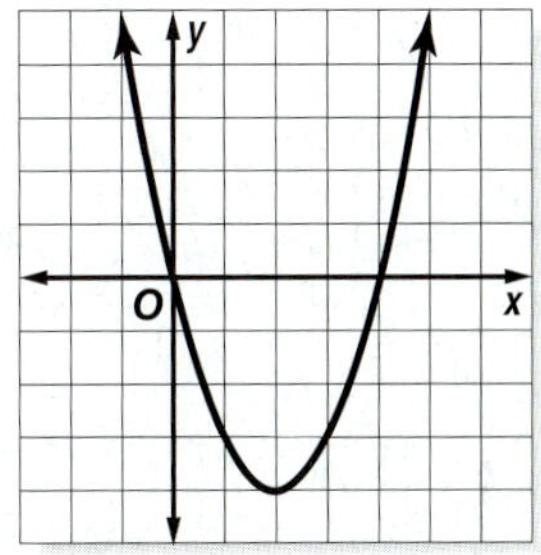

19. $(x+2)^2 + (y-3)^2 = 9$, circle

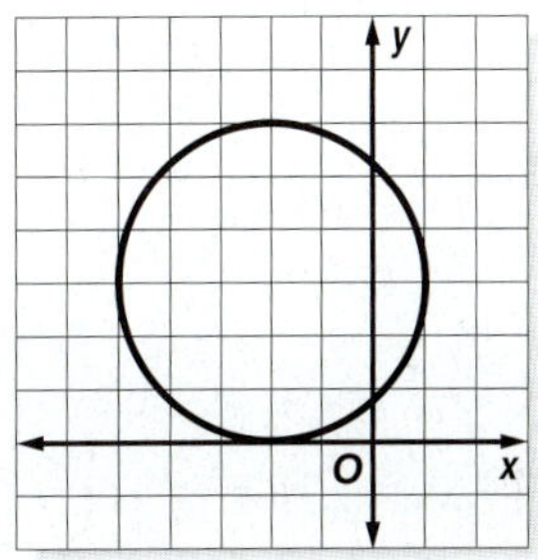

21. $\frac{(x+4)^2}{32} - \frac{y^2}{32} = 1$, hyperbola

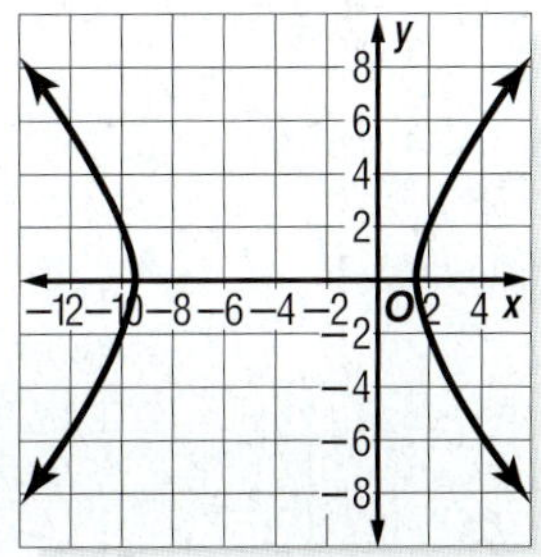

23. $x^2 + (y-4)^2 = 5$, circle

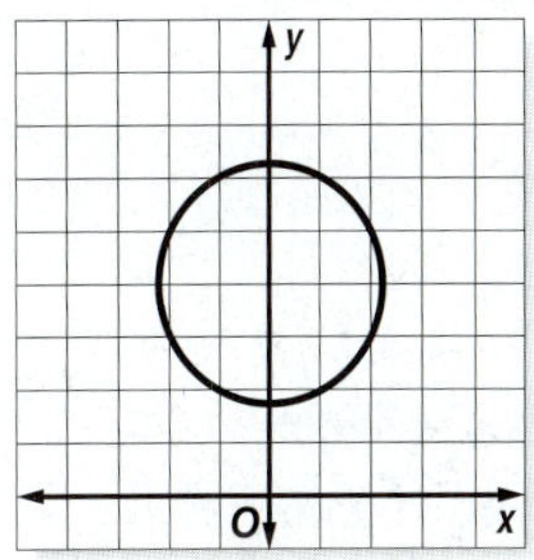

25. $\frac{x^2}{4} + \frac{(y+1)^2}{3} = 1$, ellipse

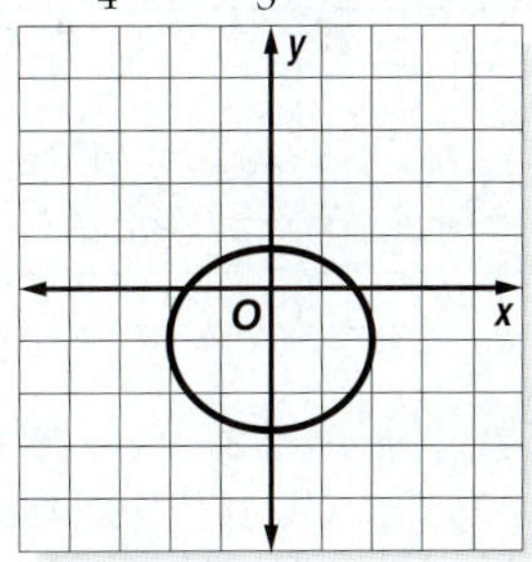

27. $y = -(x+4)^2 - 7$, parabola

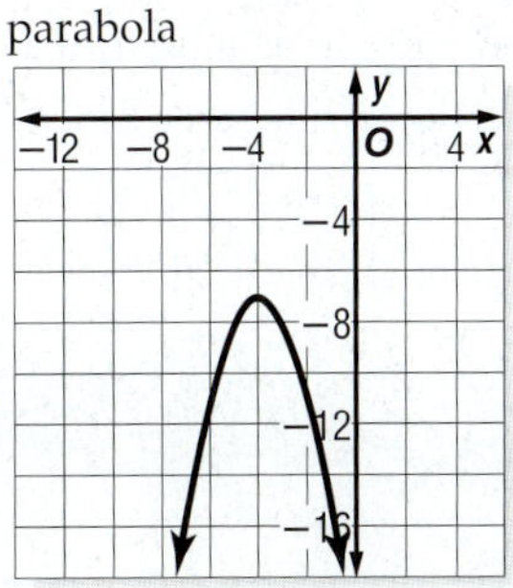

29. $\frac{(x-3)^2}{25} + \frac{(y-1)^2}{9} = 1$, ellipse

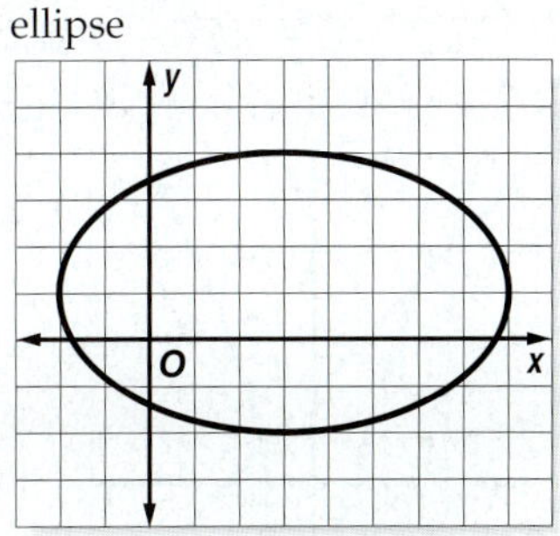

31. hyperbola 33. circle 35. parabola 37. ellipse

39. parabola **41.** b **43.** c **45.** The plane should be vertical and contain the axis of the double cone.

47. D **49.** $0 < e < 1, e > 1$ **51.** $\frac{(x-3)^2}{9} - \frac{(y+6)^2}{4} = 1$

53. x^{12} **55.** $\frac{x^7}{y^4}$ **57.** (2, 6) **59.** (0, 2)

Pages 458–460 Lesson 8-7

1a. $(-3, -4), (3, 4)$

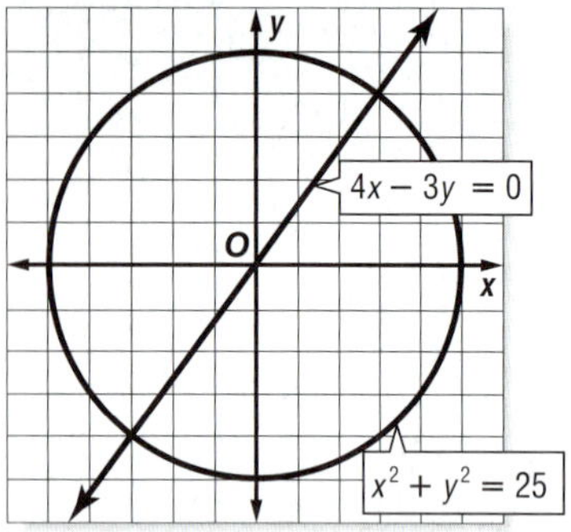

1b. $(\pm 1, 4)$

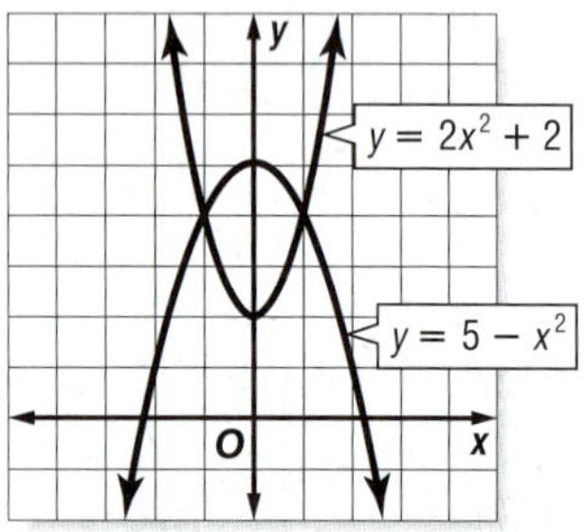

3. Sample answer: $x^2 + y^2 = 40, y = x^2 + x$

5. $(-4, -3), (3, 4)$ **7.** $(1, \pm 5), (-1, \pm 5)$

9.

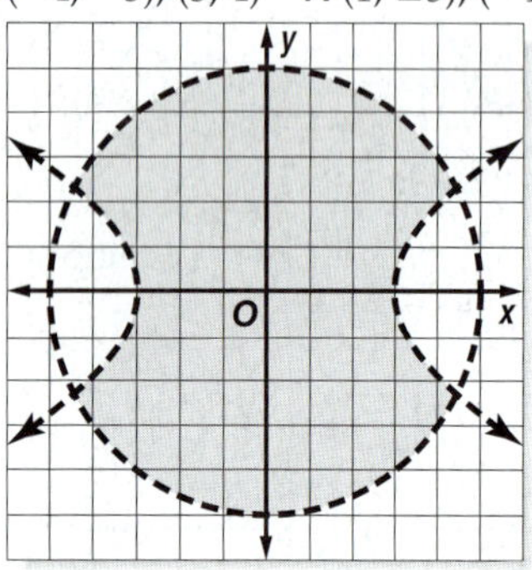

11. $(2, 4), (-1, 1)$ **13.** $\left(-1 + \sqrt{17}, 1 + \sqrt{17}\right), \left(-1 - \sqrt{17}, 1 - \sqrt{17}\right)$ **15.** $\left(\sqrt{5}, \sqrt{5}\right), \left(-\sqrt{5}, -\sqrt{5}\right)$

17. $(5, 0), (-4, \pm 6)$ **19.** $(\pm 8, 0)$ **21.** no solution

23. $(-5, 5), (-5, 1), (3, 3)$ **25.** $\left(-\frac{5}{3}, -\frac{7}{3}\right), (1, 3)$ **27.** 0.5 s

29. $\left(\frac{40 - 24\sqrt{5}}{5}, \frac{45 - 12\sqrt{5}}{5}\right)$ **31.** No; the comet and Pluto may not be at either point of intersection at the same time.

33.

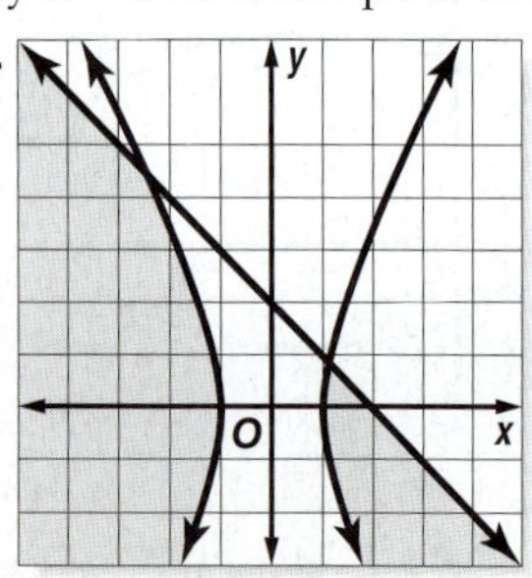

35.

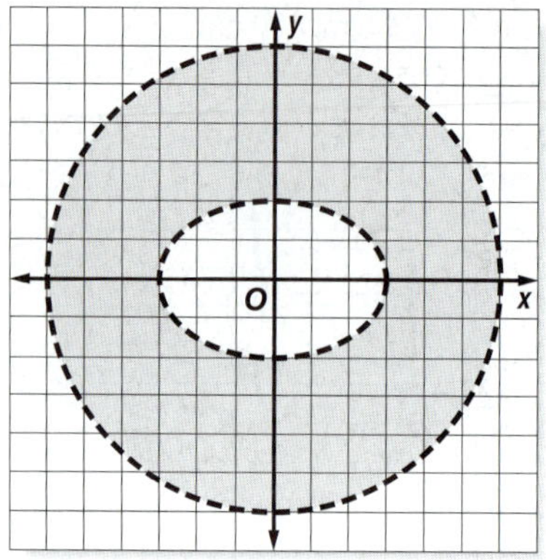

37.

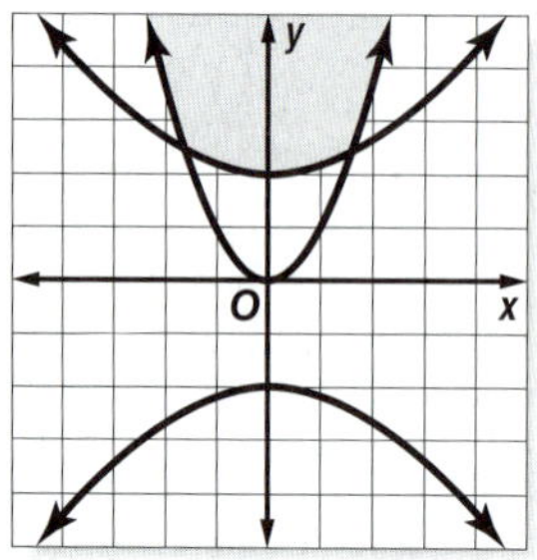

39. none **41.** none

43. Systems of equations can be used to represent the locations and/or paths of objects on the screen. Answers should include the following.

- $y = 3x, x^2 + y^2 = 2500$
- The y-intercept of the graph of the equation $y = 3x$ is 0, so the path of the spaceship contains the origin.
- $\left(-5\sqrt{10}, -15\sqrt{10}\right)$ or about $(-15.81, -47.43)$

45. B **47.** Sample answer: $x^2 + y^2 = 36, \frac{(x+2)^2}{16} - \frac{y^2}{4} = 1$

49. Sample answer: $x^2 + y^2 = 81, \frac{x^2}{4} + \frac{y^2}{100} = 1$

51. impossible

53. $\frac{(y-3)^2}{9} + \frac{x^2}{4} = 1$, ellipse

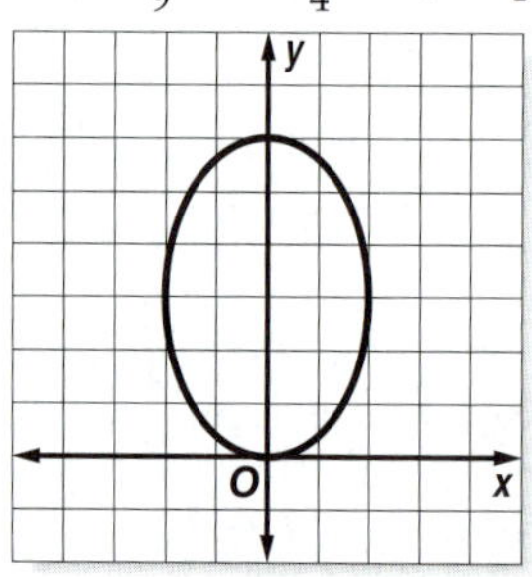

55. $-7, 0$ **57.** $-7, 3$ **59.** $-\frac{4}{3}$ **61a.** 40

61b. two real, irrational **61c.** $\pm\frac{\sqrt{10}}{5}$ **63.** $2 + 9i$

65. $\frac{8}{5} - \frac{1}{5}i$ **67.** 6 **69.** -51 **71.** $y = 3x - 2$

Pages 461–466 Chapter 8 Study Guide and Review

1. true **3.** true **5.** true **7.** true **9.** False; the midpoint formula is given by $\left(\frac{x_1 + x_2}{2}, \frac{y_1 + y_2}{2}\right)$. **11.** $\left(\frac{5}{2}, 4\right)$

13. $\left(\frac{17}{40}, -\frac{43}{40}\right)$ **15.** $\sqrt{290}$ units

17. (1, 1); (1, 4); $x = 1$; $y = -2$; upward; 12 units

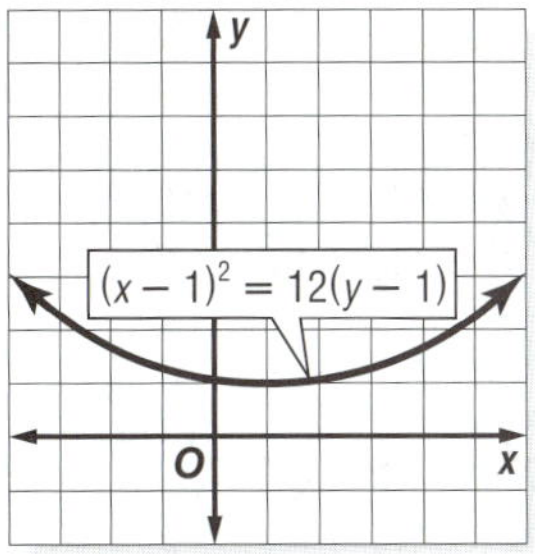

19. (4, −2); (4, −4); $x = 4$; $y = 0$; downward; 8 units

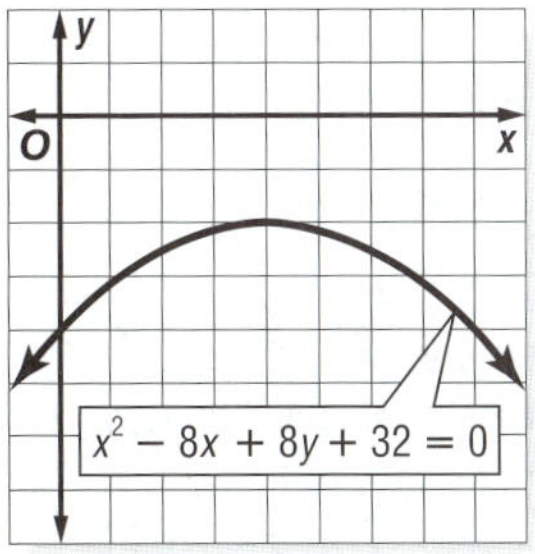

21. $y = -\frac{1}{8}x^2 + 1$

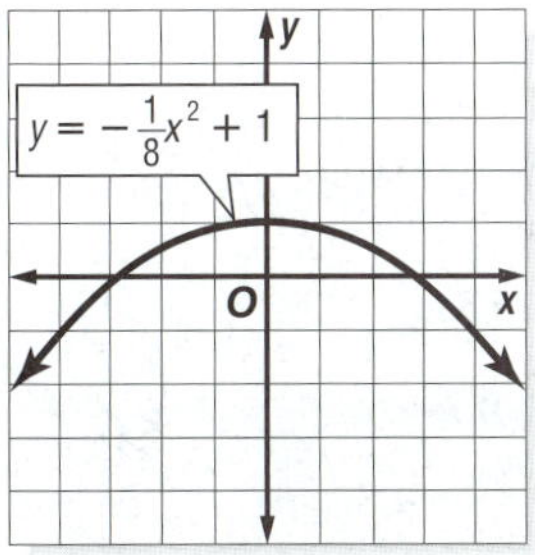

23. $(x + 4)^2 + y^2 = \frac{9}{16}$ **25.** $(x + 1)^2 + (y - 2)^2 = 4$

27. (−5, 11); 7 units

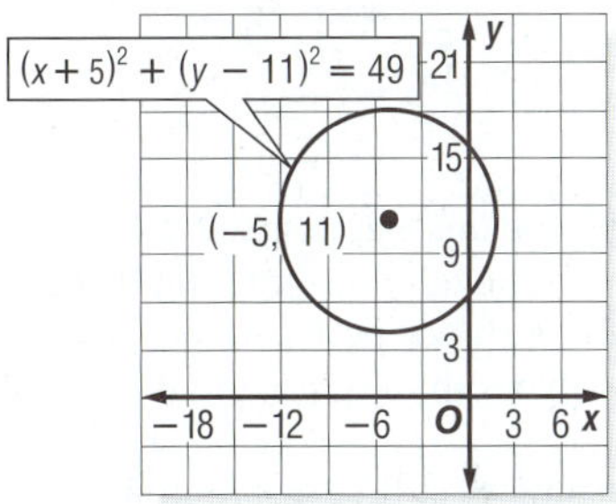

29. (−3, 1); 5 units

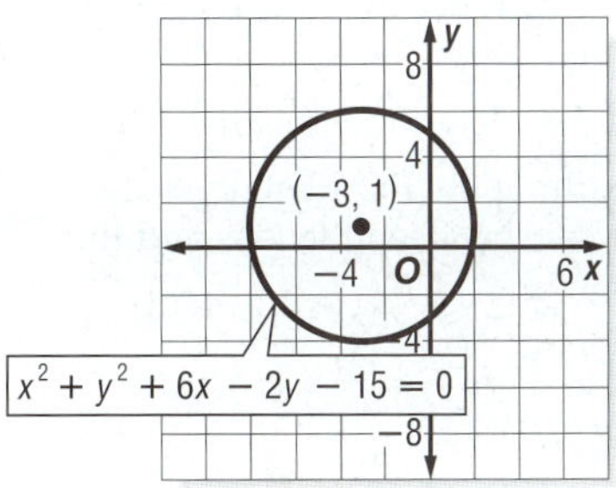

31. (0, 0); (0, ±3); 10; 8

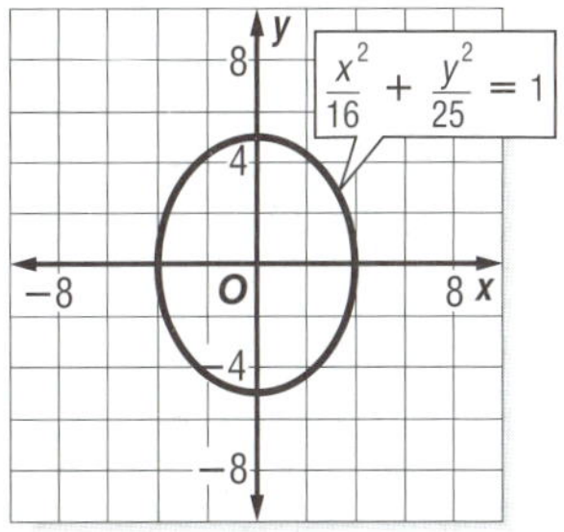

33. (1, −2); $(1 \pm \sqrt{3}, -2)$; 4; 2

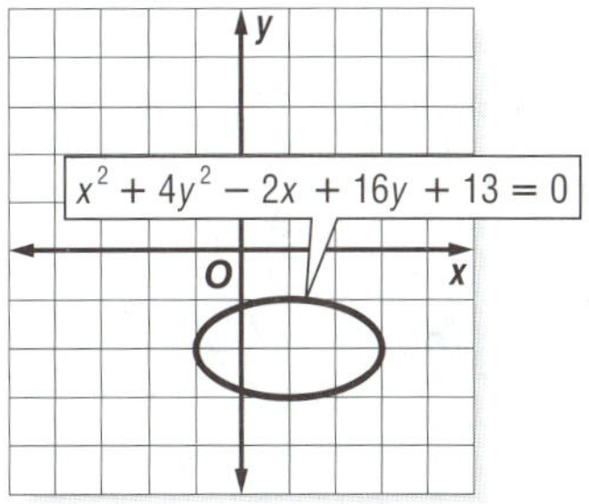

35. (0, ±2); $(0, \pm\sqrt{13})$; $y = \pm\frac{2}{3}x$

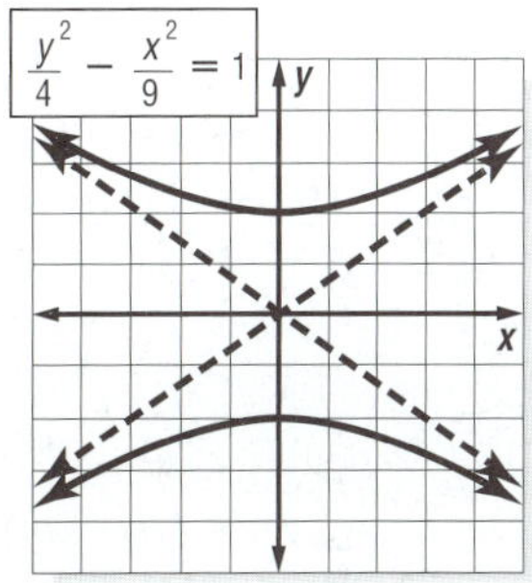

37. (0, ±4); (0, ±5); $y = \pm\frac{4}{3}x$

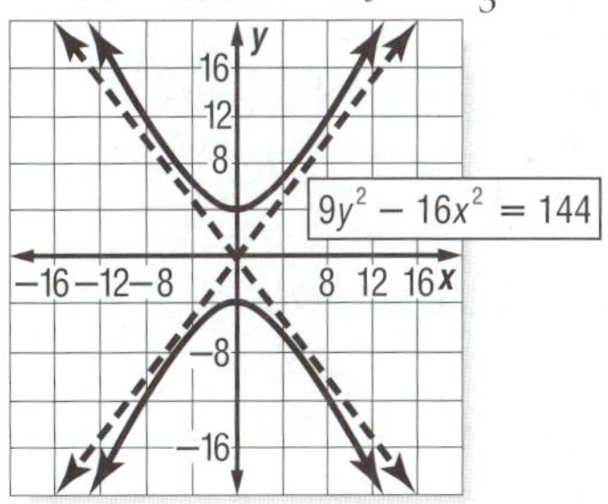

39. $y = (x + 2)^2 - 4$; parabola

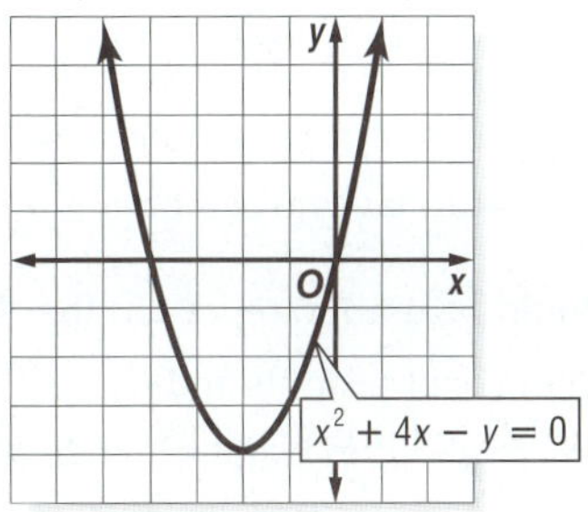

41. $\frac{y^2}{4} - \frac{(x-1)^2}{1} = 1$; hyperbola

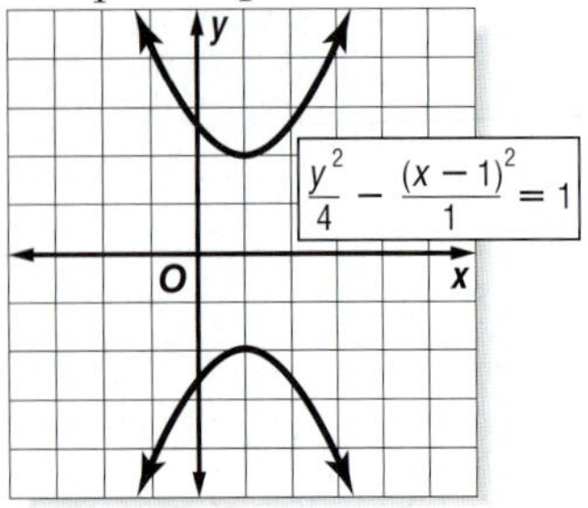

43. ellipse 45. circle 47. (6, −8), (12, −16)

49.

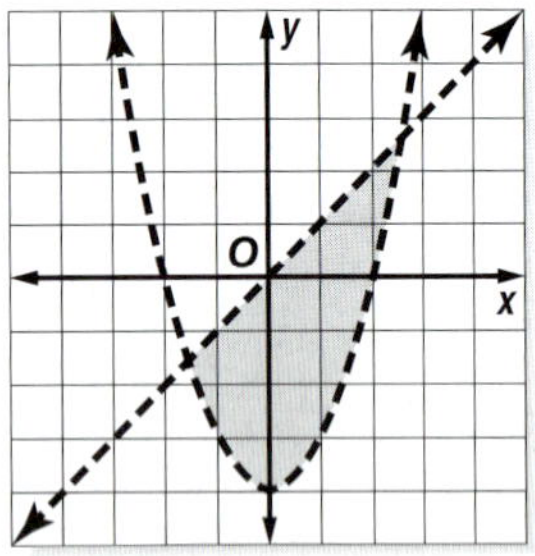

55. $\frac{(x-7)^2}{9} - \frac{(y-2)^2}{1} = 1$; hyperbola

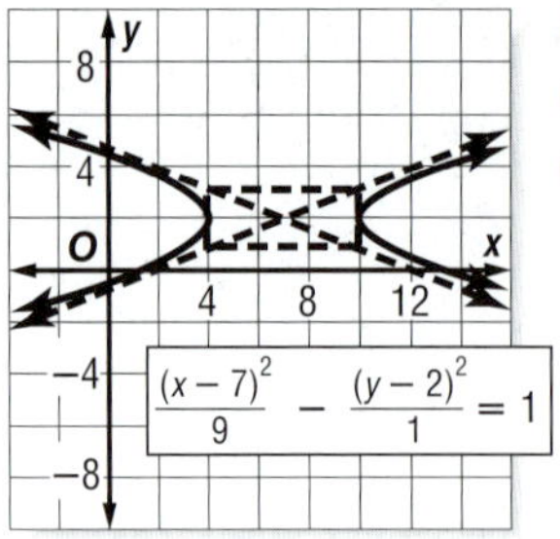

57. odd; 3 59. {−1, 4} 61. {0, 5} 63. ∅ 65. $-1\frac{1}{9}$ 67. $1\frac{4}{15}$ 69. $-\frac{11}{18}$

Chapter 9 Rational Expressions and Equations

Page 471 Chapter 9 Getting Started

1. $\frac{1}{6}$ 3. $\frac{5}{8}$ 5. 16 7. $2\frac{1}{2}$ 9. $1\frac{1}{24}$

11.

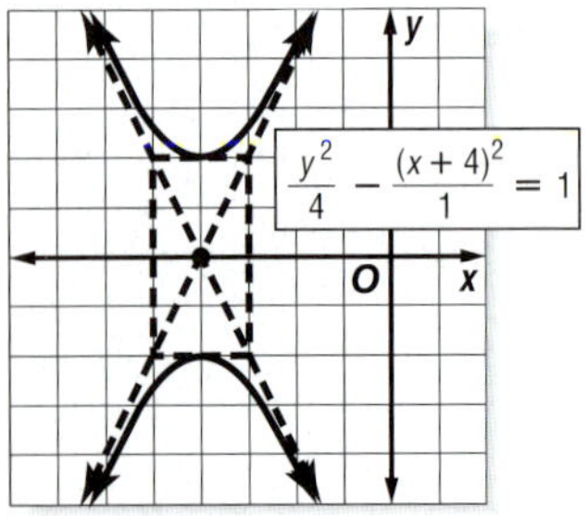

13. 12 15. 15 17. 15 19. 6 21. $7\frac{1}{2}$

Pages 476–478 Lesson 9-1

1. Sample answer: $\frac{4}{6}, \frac{4(x+2)}{6(x+2)}$ 3. Never; solving the equation using cross products leads to 15 = 10, which is never true. 5. $\frac{1}{a-b}$ 7. $\frac{3c}{20b}$ 9. $\frac{6}{5}$ 11. cd^2x 13. D 15. $-\frac{n^2}{7m}$ 17. $\frac{s}{3}$ 19. $\frac{1}{2}$ 21. $\frac{a+1}{2a+1}$ 23. $-\frac{4bc}{27a}$ 25. $-2p^2$ 27. $\frac{b^3}{x^2y^2}$ 29. $\frac{4}{3}$ 31. 1 33. $\frac{w-3}{w-4}$ 35. $\frac{2(a+5)}{(a-2)(a+2)}$ 37. $-2p$ 39. $\frac{2x+y}{2x-y}$ 41. $\frac{4}{3}$ 43. $a = -b$ or b 45. $\frac{6827+m}{13{,}129+a}$ 47. $(2x^2 + x - 15)$ m^2

49. A rational expression can be used to express the fraction of a nut mixture that is peanuts. Answers should include the following.

- The rational expression $\frac{8+x}{13+x}$ is in simplest form because the numerator and the denominator have no common factors.
- Sample answer: $\frac{8+x}{13+x+y}$ could be used to represent the fraction that is peanuts if x pounds of peanuts and y pounds of cashews were added to the original mixture.

51. A 53. $(\pm\sqrt{17}, \pm 2\sqrt{2})$

Pages 481–484 Lesson 9-2

1. Catalina: you need a common denominator, not a common numerator, to subtract two rational expressions. 3a. Always; since a, b, and c are factors of abc, abc is always a common denominator of $\frac{1}{a} + \frac{1}{b} + \frac{1}{c}$. 3b. Sometimes; if a, b, and c have no common factors, then abc is the LCD of $\frac{1}{a} + \frac{1}{b} + \frac{1}{c}$. 3c. Sometimes; if a and b have no common factors and c is a factor of ab, then ab is the LCD of $\frac{1}{a} + \frac{1}{b} + \frac{1}{c}$. 3d. Sometimes; if a and c are factors of b, then b is the LCD of $\frac{1}{a} + \frac{1}{b} + \frac{1}{c}$. 3e. Always; since $\frac{1}{a} + \frac{1}{b} + \frac{1}{c} = \frac{bc}{abc} + \frac{ac}{abc} + \frac{ab}{abc}$, the sum is always $\frac{bc+ac+ab}{abc}$. 5. $80a^2b^3c$ 7. $\frac{2-x^3}{x^2y}$ 9. $\frac{37}{42m}$ 11. $\frac{3a-10}{(a-5)(a+4)}$ 13. $\frac{13x^2+4x-9}{2x(x-1)(x+1)}$ units 15. $180x^2yz$ 17. $36p^3q^4$ 19. $x^2(x-y)(x+y)$ 21. $(n-4)(n-3)(n+2)$ 23. $\frac{31}{12v}$ 25. $\frac{2x+15y}{3y}$ 27. $\frac{25b-7a^3}{5a^2b^2}$ 29. $\frac{110w-423}{90w}$ 31. $\frac{a+3}{a-4}$ 33. $\frac{y(y-9)}{(y+3)(y-3)}$ 35. $\frac{-8d+20}{(d-4)(d+4)(d-2)}$ 37. $\frac{x^2-6}{(x+2)^2(x+3)}$ 39. $\frac{2y^2+y-4}{(y-1)(y-2)}$ 41. −1 43. $\frac{a+7}{a+2}$ 45. 12 ohms 47. $\frac{24}{x-4}$ h 49. $\frac{2md}{(d-L)^2(d+L)^2}$ or $\frac{2md}{(d^2-L^2)^2}$

51. Subtraction of rational expressions can be used to determine the distance between the lens and the film if the focal length of the lens and the distance between the lens and the object are known. Answers should include the following.

- To subtract rational expressions, first find a common denominator. Then, write each fraction as an equivalent fraction with the common denominator. Subtract the numerators and place the difference over the common denominator. If possible, reduce the answer.
- $\frac{1}{q} = \frac{1}{10} - \frac{1}{60}$ could be used to determine the distance between the lens and the film if the focal length of the lens is 10 cm and the distance between the lens and the object is 60 cm.

53. C 55. $\frac{a(a+2)}{a+1}$

Selected Answers

57.

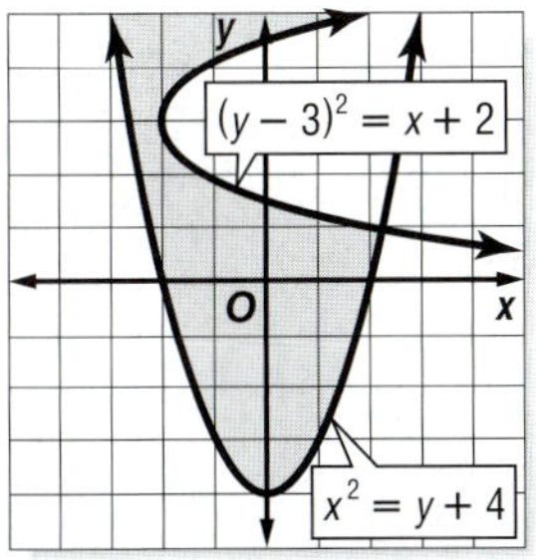

59.

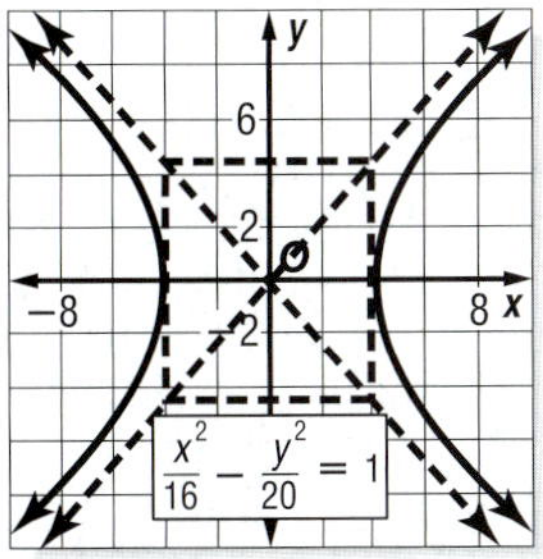

61.

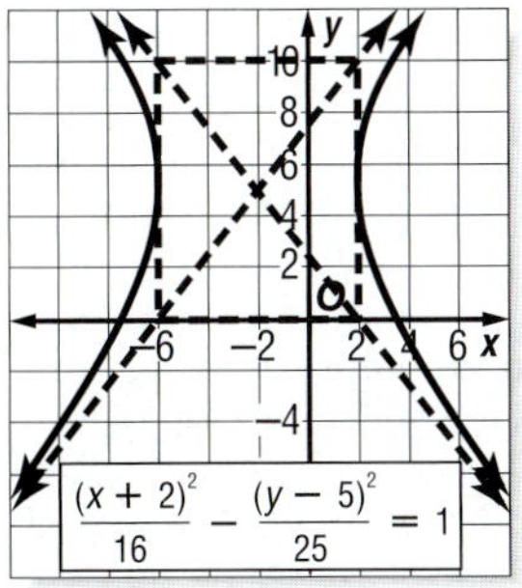

Page 484 Practice Quiz 1

1. $\frac{t+2}{t-3}$ **3.** $-\frac{y^2}{32}$ **5.** $(w+4)(3w+4)$ **7.** $\frac{4a+1}{a+b}$
9. $\frac{n-29}{(n+6)(n-1)}$

Pages 488–490 Lesson 9-3

1. Sample answer: $f(x) = \frac{1}{(x+5)(x-2)}$ **3.** $x = 2$ and $y = 0$ are asymptotes of the graph. The y-intercept is 0.5 and there is no x-intercept because $y = 0$ is an asymptote.
5. asymptote: $x = -5$; hole: $x = 1$

7.

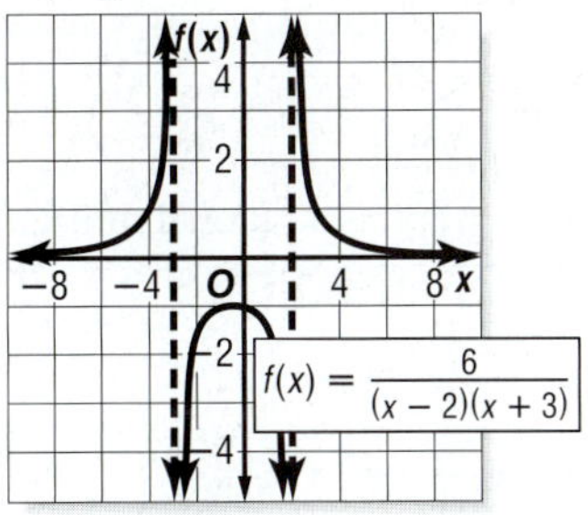

9.

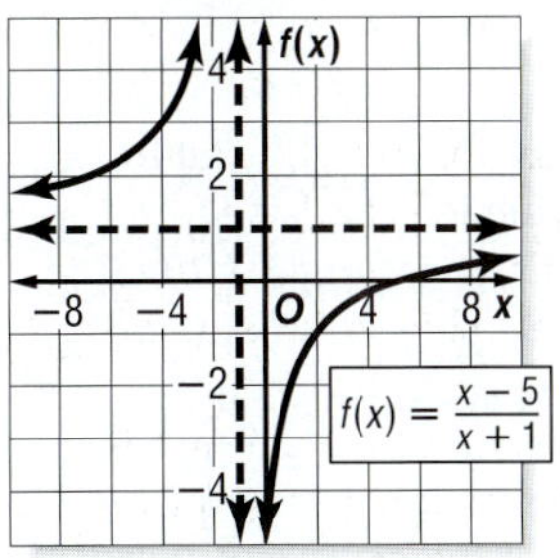

11.

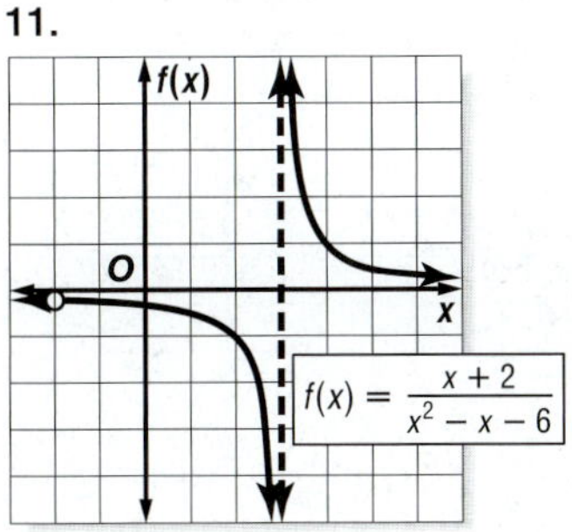

13.

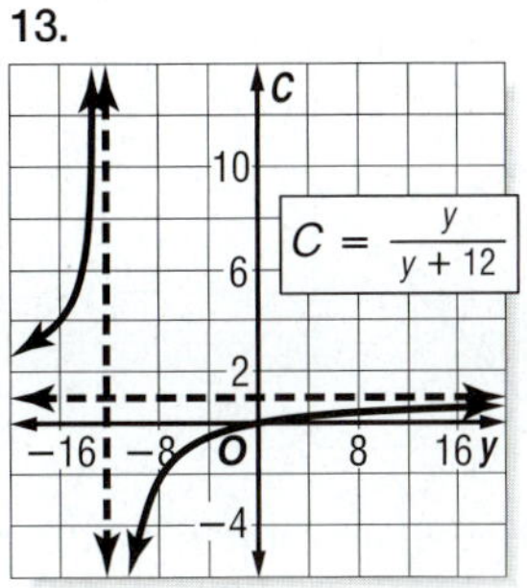

15. $y > 0$ and $0 < C < 1$ **17.** asymptotes: $x = -4$, $x = 2$
19. asymptotes: $x = -1$, hole: $x = 5$ **21.** hole: $x = 1$

23.

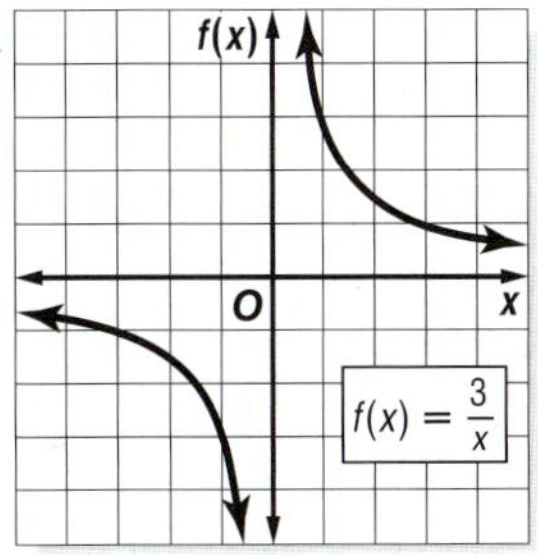

25.

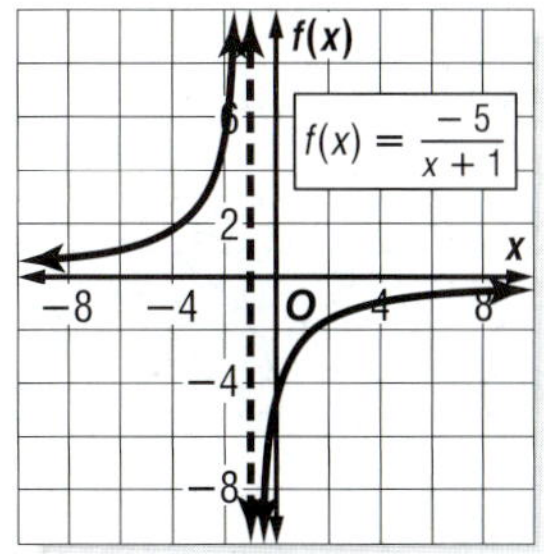

27.

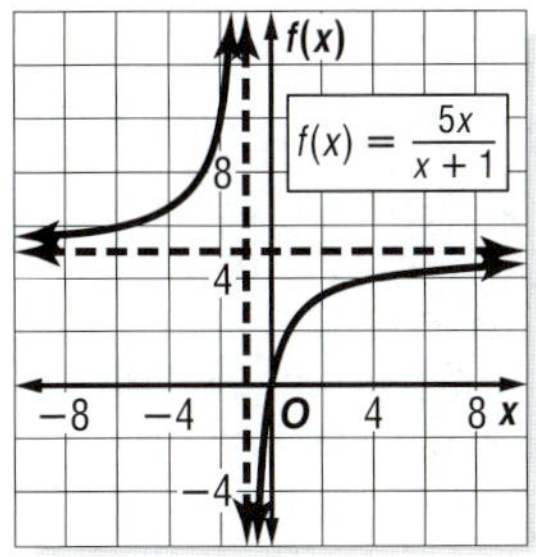

29.

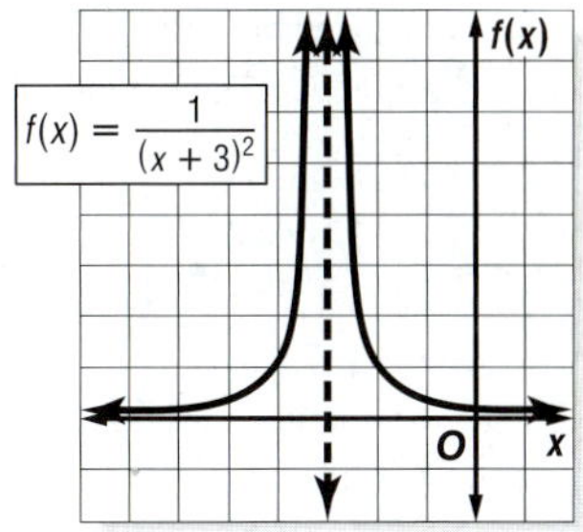

31.

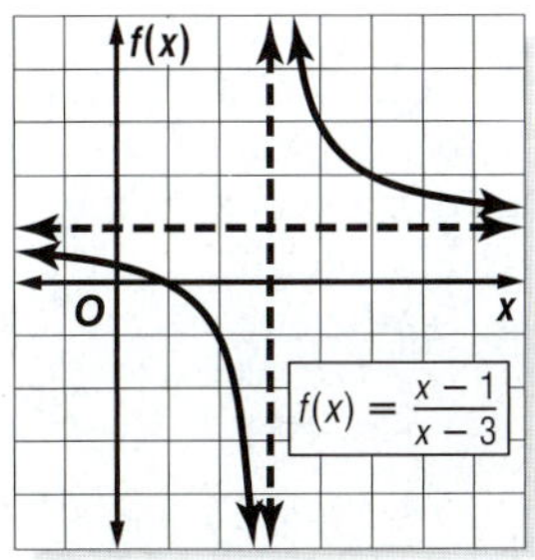

33.

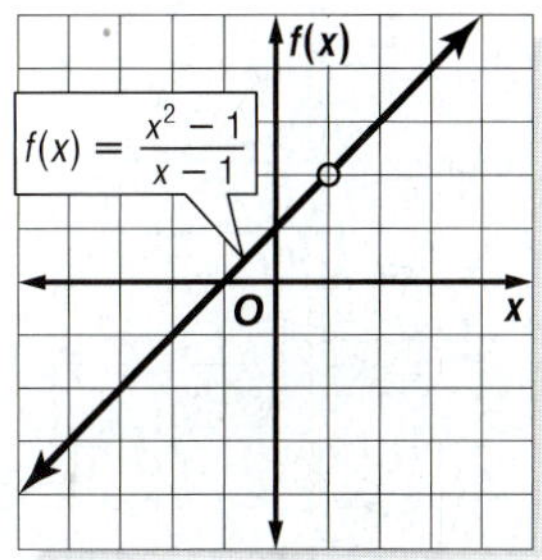

35.

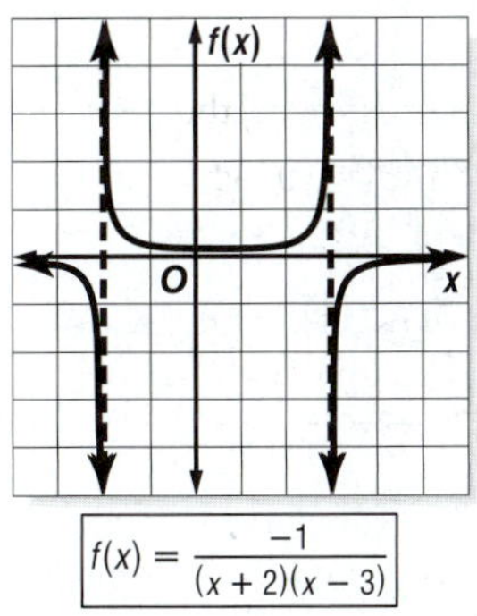

37.

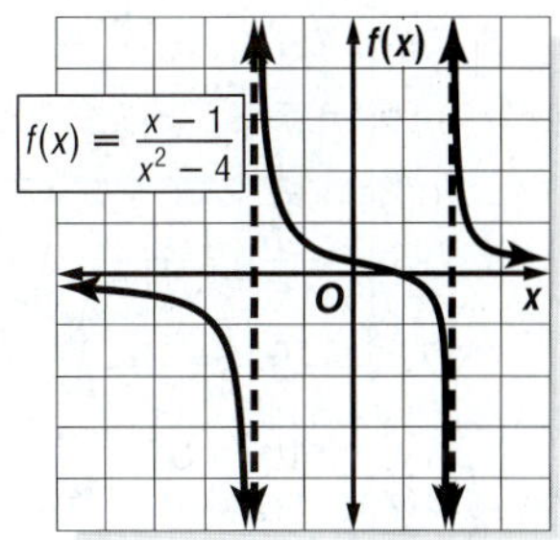

39.

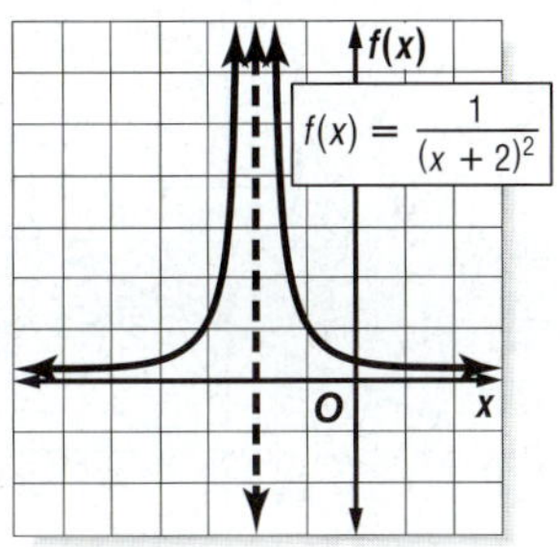

Selected Answers

41. The graph is bell-shaped with a horizontal asymptote at $f(x) = 0$.

43.

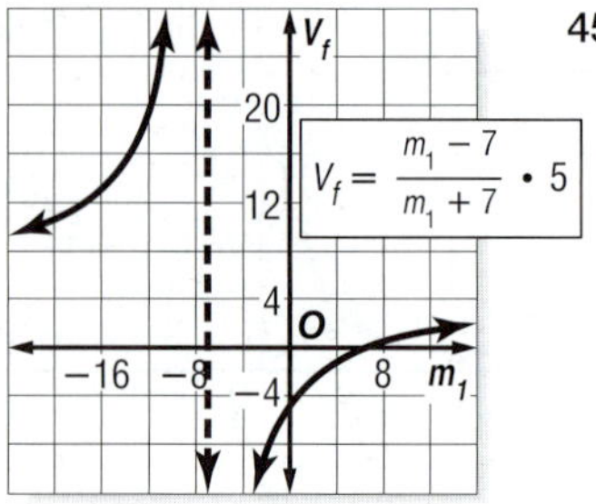

45. about −0.83 m/s

47.

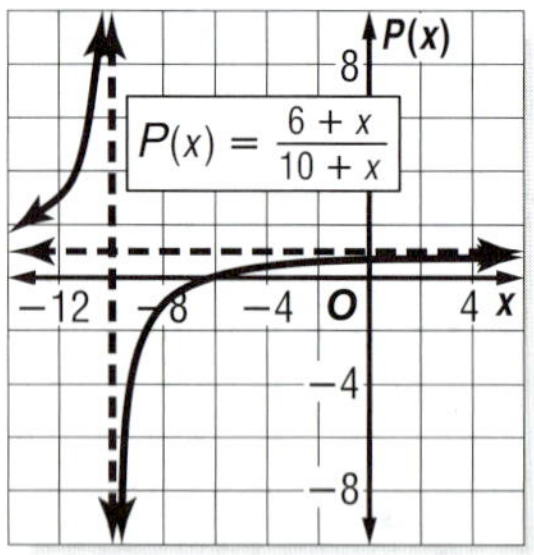

49. It represents her original free-throw percentage of 60%.

51. A rational function can be used to determine how much each person owes if the cost of the gift is known and the number of people sharing the cost is s. Answers should include the following.

- 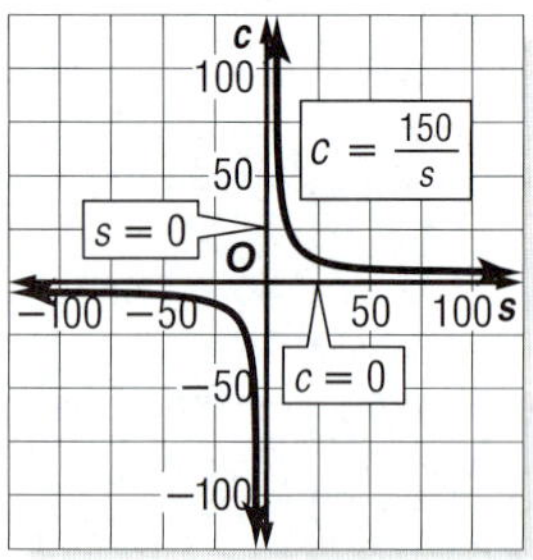

- Only the portion in the first quadrant is significant in the real world because there cannot be a negative number of people nor a negative amount of money owed for the gift.

53. B **55.** $\frac{3x - 16}{(x + 3)(x - 2)}$

57. (6, 2); 5

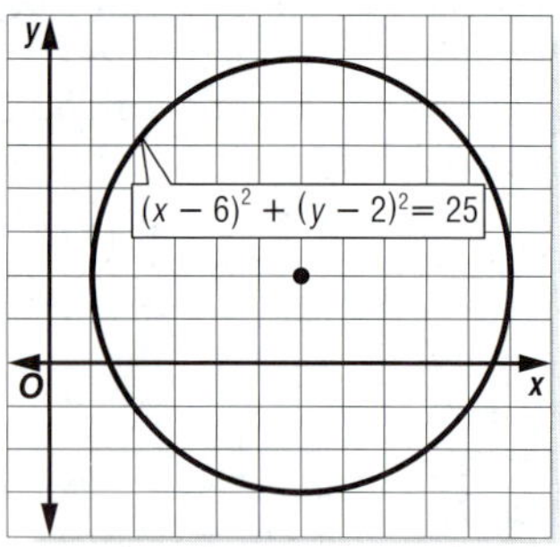

59. $65,892

61. {−12, 10} **63.** 4.5 **65.** 20

Pages 495–498 Lesson 9-4

1a. inverse **1b.** direct **3.** Sample answers: wages and hours worked, total cost and number of pounds of apples; distances traveled and amount of gas remaining in the tank, distance of an object and the size it appears **5.** direct; −0.5 **7.** 24 **9.** −8 **11.** 25.8 psi

13.

Depth (ft)	Pressure (psi)
0	0
1	0.43
2	0.86
3	1.29
4	1.72

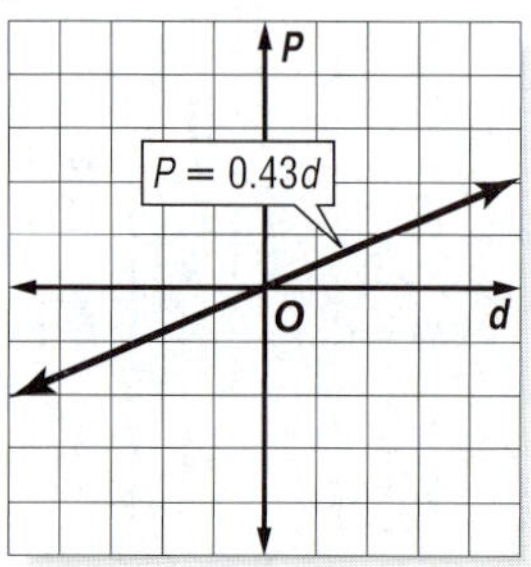

15. joint; 5 **17.** direct; 3 **19.** direct; −7 **21.** inverse; 2.5 **23.** $V = kt$ **25.** 118.5 km **27.** 20 **29.** 64 **31.** 4 **33.** 9.6 **35.** 0.83 **37.** $\frac{1}{6}$ **39.** 100.8 cm³ **41.** $m = 20sd$ **43.** 1860 lb

45. joint **47.** $I = \frac{k}{d^2}$ **49.** The sound will be heard $\frac{1}{4}$ as intensely. **51.** about 127,572 calls **53.** no; $d \neq 0$

55. A direct variation can be used to determine the total cost when the cost per unit is known. Answers should include the following.

- Since the total cost T is the cost per unit u times the number of units n or $T = un$, the relationship is a direct variation. In this equation u is the constant of variation.
- Sample answer: The school store sells pencils for 20¢ each. John wants to buy 5 pencils. What is the total cost of the pencils? ($1.00)

57. C **59.** asymptotes: $x = -4$, $x = 3$ **61.** $\frac{x}{y - x}$

63. $\frac{m(m + 1)}{m + 5}$ **65.** 0.4; 1.2 **67.** $-\frac{3}{5}$; 3 **69.** A

71. P **73.** C

Page 498 Practice Quiz 2

1.

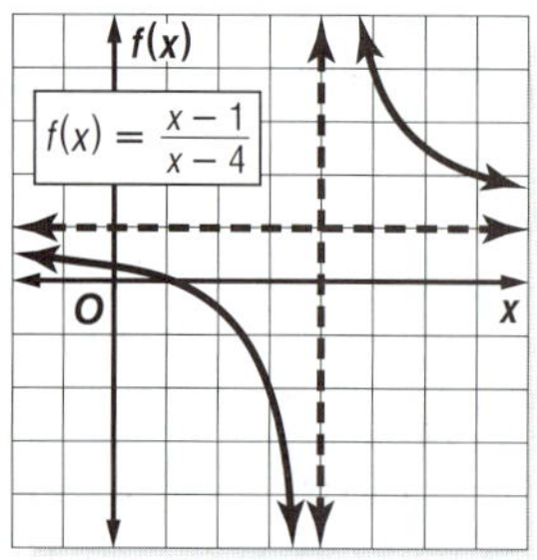

3. 49 **5.** 112

Pages 501–504 Lesson 9-5

1. Sample answer:

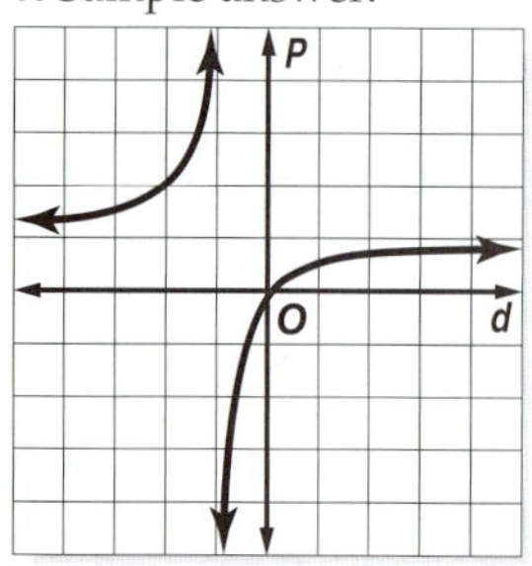

This graph is a rational function. It has an asymptote at $x = -1$.

3. The equation is a greatest integer function. The graph looks like a series of steps. **5.** inverse variation or rational

7. c

9. identity or direct variation

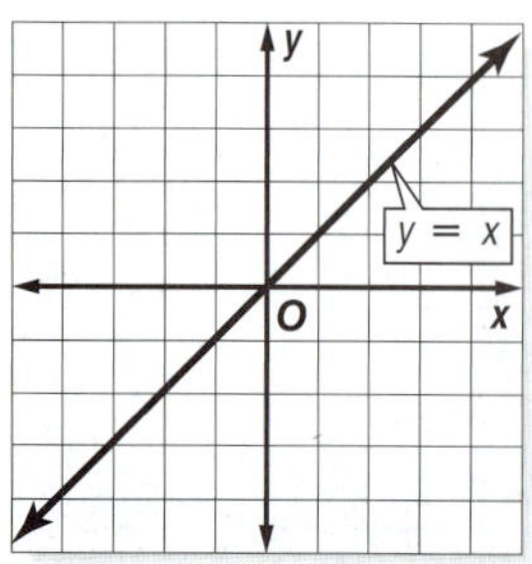

11. absolute value

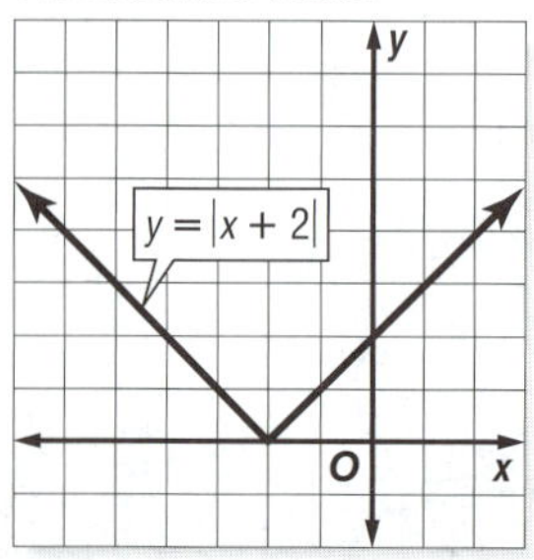

13. absolute value **15.** rational **17.** quadratic **19.** b **21.** g

23. constant

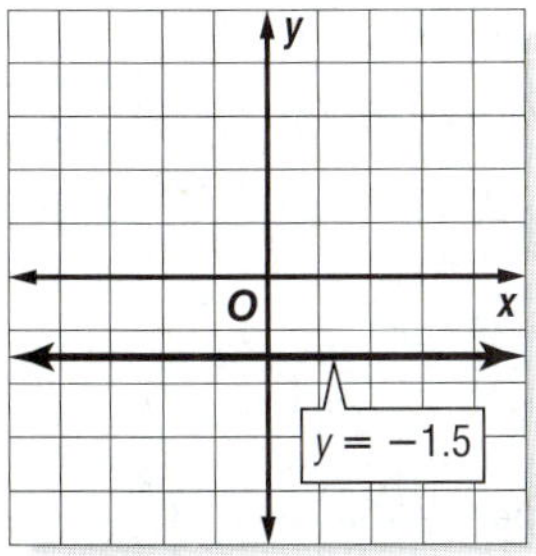

25. square root

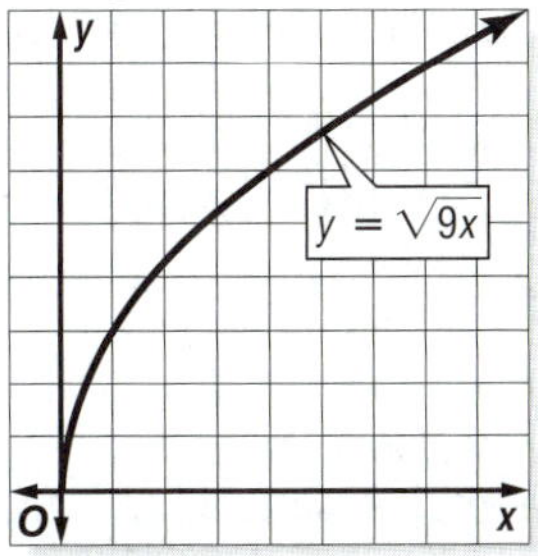

27. rational

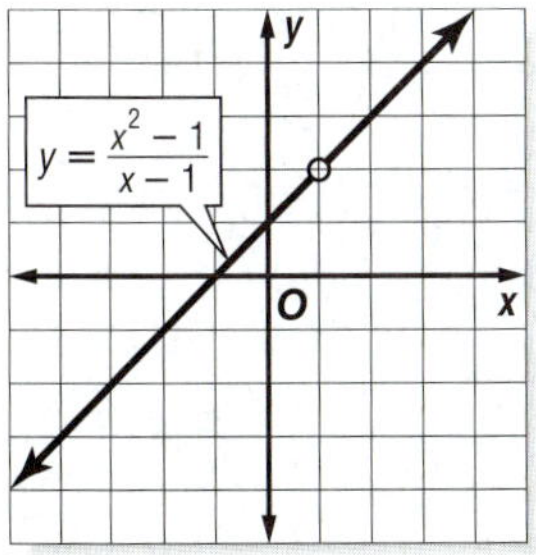

29. absolute value

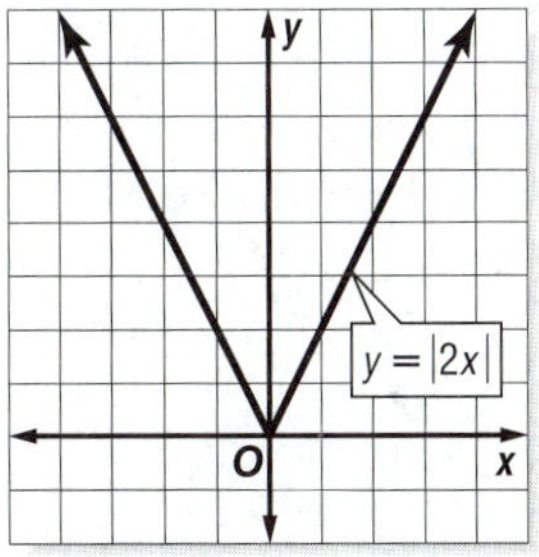

31. $C = 4.5m$ **33.** a line slanting to the right and passing through the origin

35.

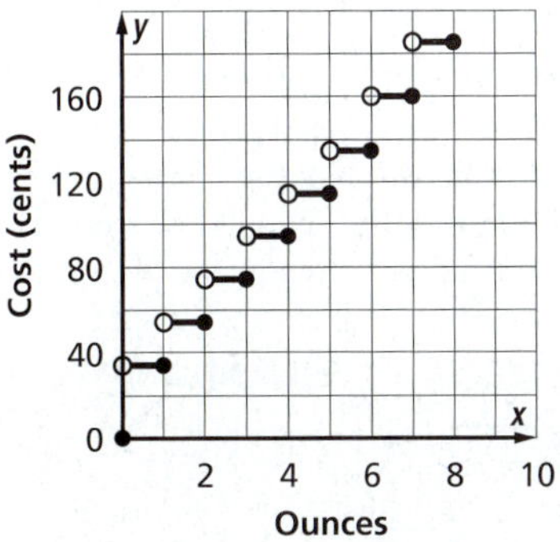

37a. absolute value **37b.** quadratic **37c.** greatest integer **37d.** square root **39.** C **41.** 22

43.

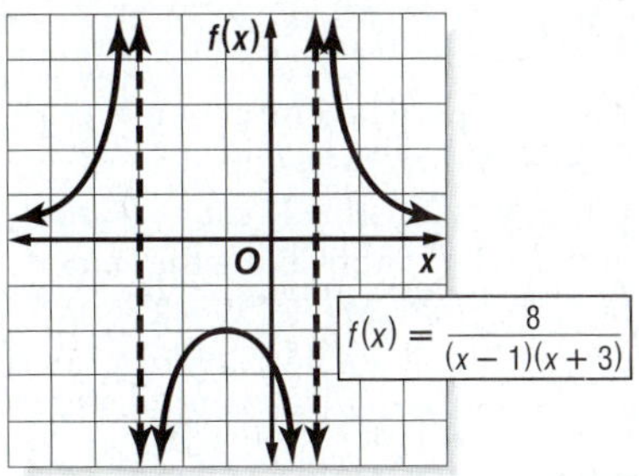

45. $(8, -1)$; $\left(8, -\frac{7}{8}\right)$; $x = 8$; $y = -1\frac{1}{8}$; up; $\frac{1}{2}$ unit

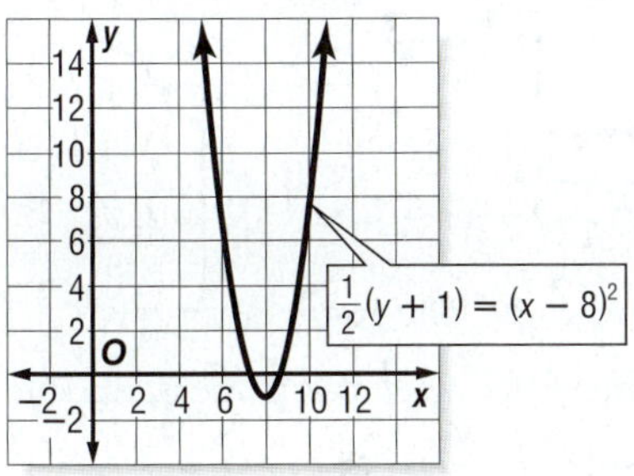

47. $(5, -4)$; $\left(5\frac{3}{4}, -4\right)$; $y = -4$; $x = 4\frac{1}{4}$; right; 3 units

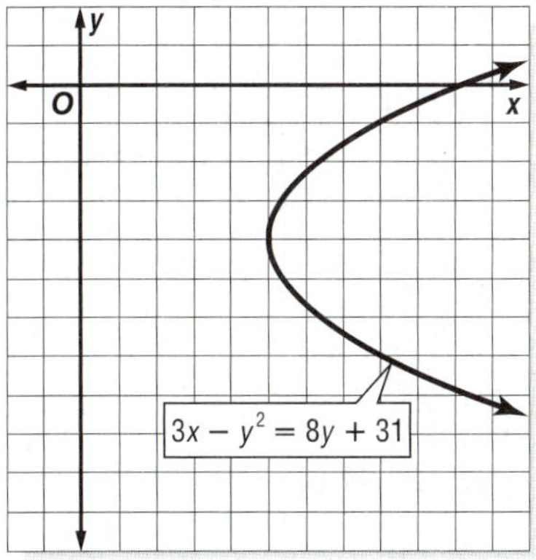

49. impossible **51.** $\left(\frac{1}{3}, 2\right)$ **53.** 1 **55.** $-\frac{17}{6}$ **57.** $45x^3y^3$
59. $3(x - y)(x + y)$ **61.** $(t - 5)(t + 6)(2t + 1)$

Pages 509–511 Lesson 9-6

1. Sample answer: $\frac{1}{5} + \frac{2}{a+2} = 1$ **3.** Jeff; when Dustin multiplied by $3a$, he forgot to multiply the 2 by $3a$. **5.** 2, 6 **7.** −6, −2 **9.** $v < 0$ or $v > 1\frac{1}{6}$ **11.** 2 **13.** −6, 1 **15.** $-1 < a < 0$ **17.** 11 **19.** $t < 0$ or $t > 3$ **21.** $0 < y < 2$ **23.** 14 **25.** ∅ **27.** 7 **29.** $\frac{-3 \pm 3\sqrt{2}}{2}$ **31.** 32 **33.** band, 80 members; chorale, 50 members **35.** 24 cm **37.** 5 mL **39.** 6.15

41. If something has a general fee and cost per unit, rational equations can be used to determine how many units a person must buy in order for the actual unit price to be a given number. Answers should include the following.

- To solve $\frac{500 + 5x}{x} = 6$, multiply each side of the equation by x to eliminate the rational expression. Then subtract $5x$ from each side. Therefore, $500 = x$. A person would need to make 500 minutes of long distance minutes to make the actual unit price 6¢.
- Since the cost is 5¢ per minute plus $5.00 per month, the actual cost per minute could never be 5¢ or less.

43. C **45.** square root

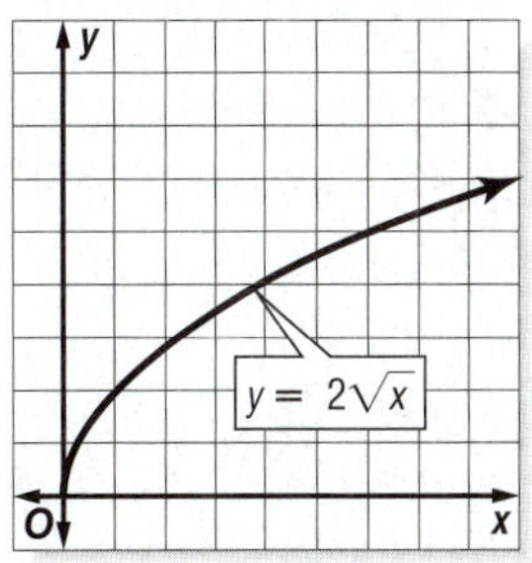

47. 36 **49.** $2\sqrt{130}$ **51.** $\sqrt{137}$ **53.** $\{x \mid 0 \le x \le 4\}$

Pages 513–516 Chapter 9 Study Guide and Review

1. false; point discontinuity **3.** false; rational **5.** true
7. $\frac{-4bc}{33a}$ **9.** $(y + 3)(y - 6)$ **11.** $\frac{2}{n-3}$ **13.** $\frac{7(x-4)}{x-5}$ **15.** $\frac{19}{3y}$
17. $-\frac{3}{20b}$

19.

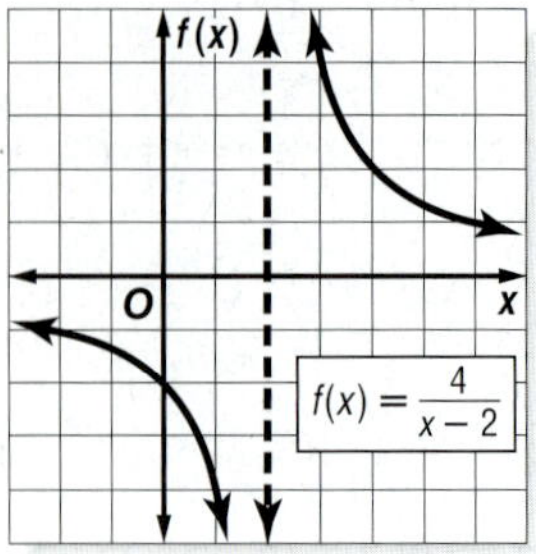

21.

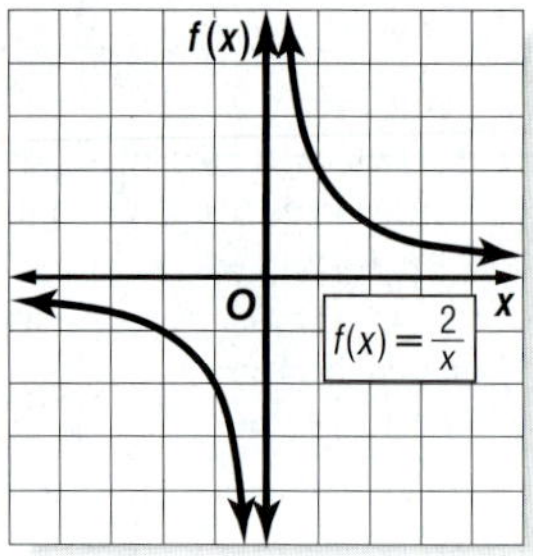

23.

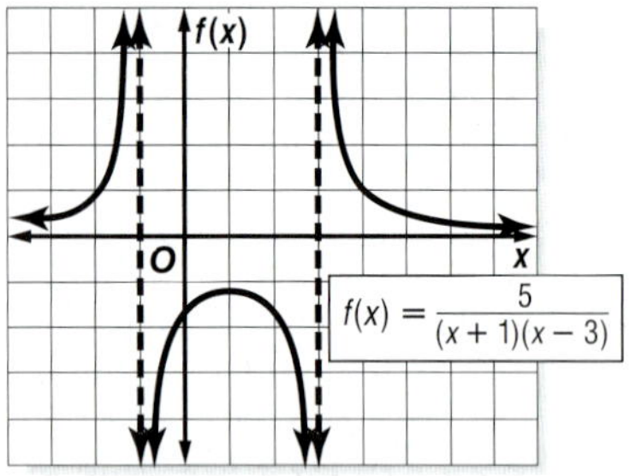

25. $-1\frac{2}{3}$ **27.** 8 **29.** 80 **31.** absolute value **33.** $1\frac{1}{9}$ **35.** 3 **37.** $1\frac{1}{2}$

Chapter 10 Exponential and Logarithmic Relations

Page 521 Chapter 10 Getting Started

1. x^{12} **3.** $-\frac{12x^3}{7y^5z}$ **5.** $a < -14$ **7.** $y \geq -2$

9. $f^{-1}(x) = -\frac{1}{2}x$

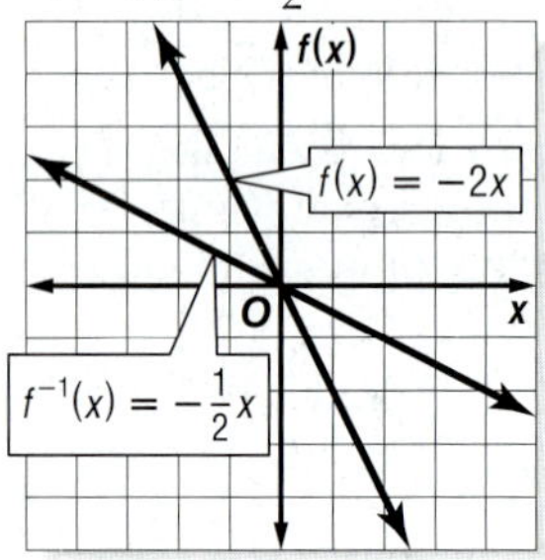

11. $f^{-1}(x) = -x + 1$

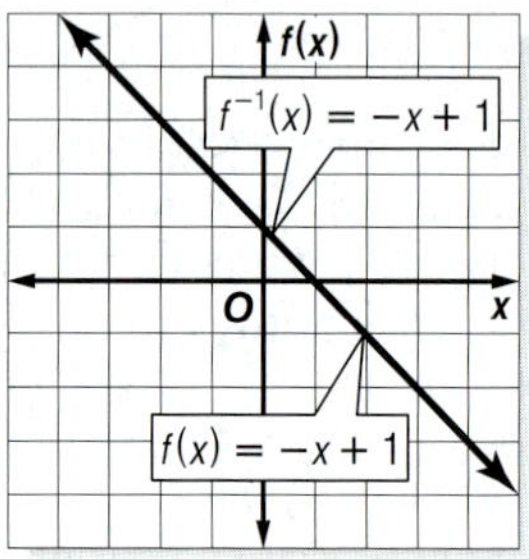

13. $g[h(x)] = 3x + 2$; $h[g(x)] = 3x - 2$
15. $g[h(x)] = x^2 - 8x + 16$; $h[g(x)] = x^2 - 4$

Pages 527–530 Lesson 10-1

1. Sample answer: 0.8 **3.** c **5.** b
7. D = $\{x \mid x$ is all real numbers.$\}$, R = $\{y \mid y > 0\}$

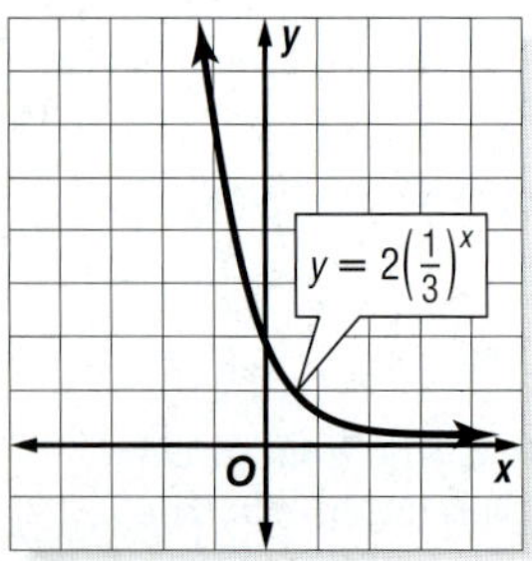

9. decay **11.** $y = 3\left(\frac{1}{2}\right)^x$ **13.** $2^{2\sqrt{7}}$ or $4^{\sqrt{7}}$
15. $3^{3\sqrt{2}}$ or $27^{\sqrt{2}}$ **17.** $x \leq 0$ **19.** $y = 65{,}000(6.20)^x$

21. D = $\{x \mid x$ is all real numbers.$\}$, R = $\{y \mid y > 0\}$

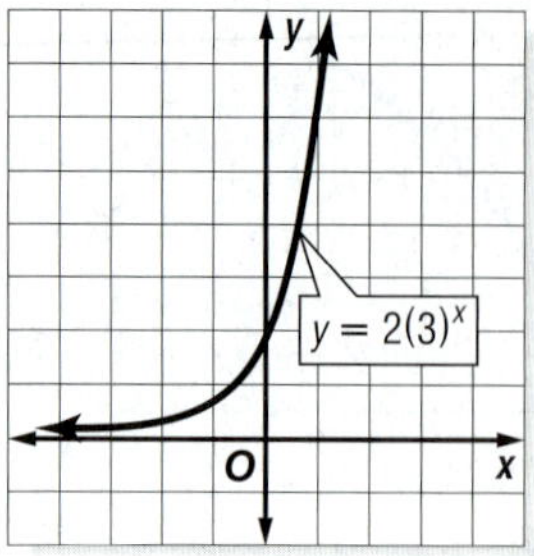

23. D = $\{x \mid x$ is all real numbers.$\}$, R = $\{y \mid y > 0\}$

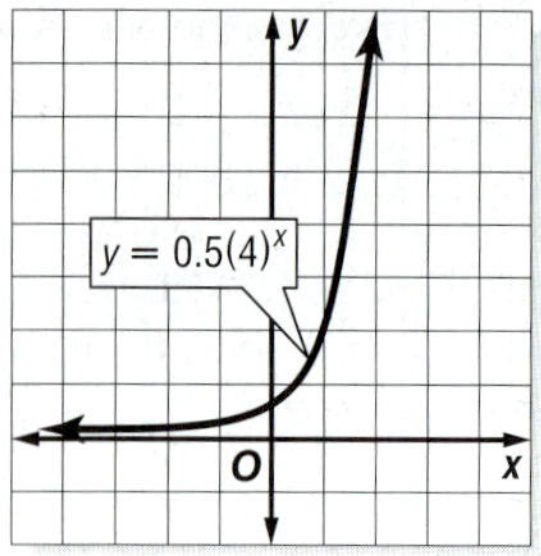

25. D = $\{x \mid x$ is all real numbers.$\}$, R = $\{y \mid y < 0\}$

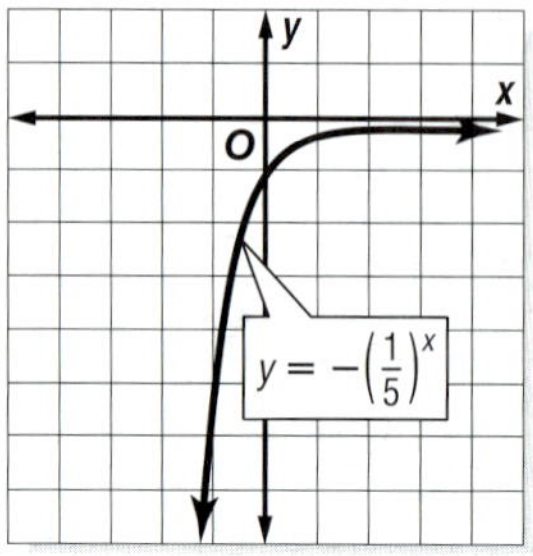

27. growth **29.** decay **31.** decay **33.** $y = -2\left(\frac{1}{4}\right)^x$
35. $y = 7(3)^x$ **37.** $y = 0.2(4)^x$ **39.** 5^4 or 625 **41.** $7^{4\sqrt{2}}$
43. $n^{2+\pi}$ **45.** $n = 5$ **47.** 1 **49.** $-\frac{8}{3}$ **51.** $n < 3$ **53.** -3
55. 10 **57.** $y = 100(6.32)^x$ **59.** $y = 3.93(1.35)^x$
61. 2144.97 million; 281.42 million; No, the growth rate has slowed considerably. The population in 2000 was much smaller than the equation predicts it would be.
63. $A(t) = 1000(1.01)^{4t}$ **65.** $s \cdot 4^x$ **67.** Sometimes; true when $b > 1$, but false when $b < 1$. **69.** A

71.

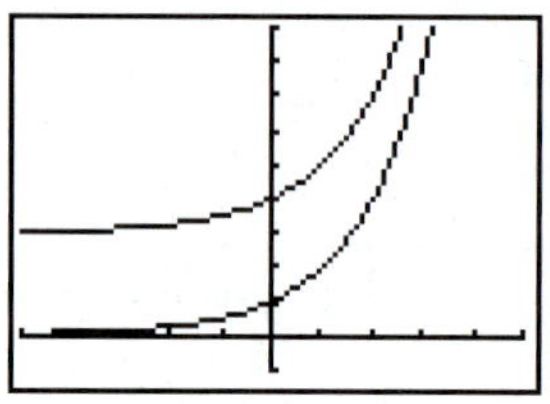

[−5, 5] scl: 1 by [−1, 9] scl: 1

The graphs have the same shape. The graph of $y = 2^x + 3$ is the graph of $y = 2^x$ translated three units up. The asymptote for the graph of $y = 2^x$ is the line $y = 0$ and for $y = 2^x + 3$ is the line $y = 3$. The graphs have the same domain, all real numbers, but the range of $y = 2^x$ is $y > 0$ and the range of $y = 2^x + 3$ is $y > 3$. The y-intercept of the graph of $y = 2^x$ is 1 and for the graph of $y = 2^x + 3$ is 4.

73.

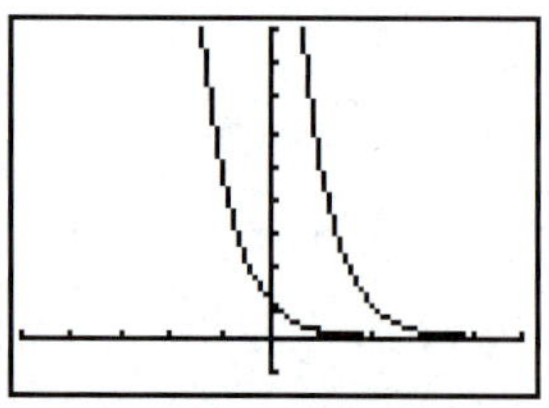

[−5, 5] scl: 1 by [−1, 9] scl: 1

The graphs have the same shape. The graph of $y = \left(\frac{1}{5}\right)^{x-2}$ is the graph of $y = \left(\frac{1}{5}\right)^x$ translated two units to the right. The asymptote for the graph of $y = \left(\frac{1}{5}\right)^x$ and for $y = \left(\frac{1}{5}\right)^{x-2}$ is

the line $y = 0$. The graphs have the same domain, all real numbers, and range, $y > 0$. The y-intercept of the graph of $y = \left(\frac{1}{5}\right)^x$ is 1 and for the graph of $y = \left(\frac{1}{5}\right)^{x-2}$ is 25. **75.** For $h > 0$, the graph of $y = 2^x$ is translated $|h|$ units to the right. For $h < 0$, the graph of $y = 2^x$ is translated $|h|$ units to the left. For $k > 0$, the graph of $y = 2^x$ is translated $|k|$ units up. For $k < 0$, the graph of $y = 2^x$ is translated $|k|$ units down. **77.** 1, 6 **79.** $0 < x < 3$ or $x > 6$
81. greatest integer

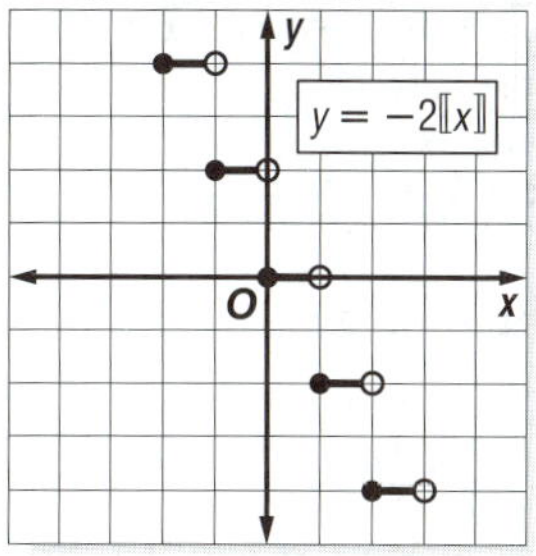

83. $\begin{bmatrix} 1 & 0 \\ 0 & 1 \end{bmatrix}$ **85.** $\frac{1}{51}\begin{bmatrix} 3 & -6 \\ 11 & -5 \end{bmatrix}$ **87.** $g[h(x)] = 2x - 6$; $h[g(x)] = 2x - 11$ **89.** $g[h(x)] = -2x - 2$; $h[g(x)] = -2x + 11$

Pages 535–538 Lesson 10-2

1. Sample answer: $x = 5^y$ and $y = \log_5 x$ **3.** Scott; the value of a logarithmic equation, 9, is the exponent of the equivalent exponential equation, and the base of the logarithmic expression, 3, is the base of the exponential equation. Thus, $x = 3^9$ or 19,683. **5.** $\log_7 \frac{1}{49} = -2$
7. $36^{\frac{1}{2}} = 6$ **9.** -3 **11.** -1 **13.** 1000 **15.** $\frac{1}{2}$, 1 **17.** 3
19. $10^{7.5}$ **21.** $\log_8 512 = 3$ **23.** $\log_5 \frac{1}{125} = -3$
25. $\log_{100} 10 = \frac{1}{2}$ **27.** $5^3 = 125$ **29.** $4^{-1} = \frac{1}{4}$ **31.** $8^{\frac{2}{3}} = 4$
33. 4 **35.** $\frac{1}{2}$ **37.** -5 **39.** 7 **41.** $n - 5$ **43.** -3 **45.** $10^{18.8}$
47. 81 **49.** $0 < y \leq 8$ **51.** 7 **53.** $x \geq 24$ **55.** 4 **57.** 2
59. 5 **61.** $a > 3$

63.

$\log_5 25 \stackrel{?}{=} 2 \log_5 5$	*Original equation*
$\log_5 5^2 \stackrel{?}{=} 2 \log_5 5^1$	*$25 = 5^2$ and $5 = 5^1$*
$2 \stackrel{?}{=} 2(1)$	*Inverse Property of Exponents and Logarithms*
$2 = 2\ \checkmark$	*Simplify.*

65.

$\log_7 [\log_3 (\log_2 8)] \stackrel{?}{=} 0$	*Original equation*
$\log_7 [\log_3 (\log_2 2^3)] \stackrel{?}{=} 0$	*$8 = 2^3$*
$\log_7 (\log_3 3) \stackrel{?}{=} 0$	*Inverse Property of Exponents and Logarithms*
$\log_7 (\log_3 3^1) \stackrel{?}{=} 0$	*$3 = 3^1$*
$\log_7 1 \stackrel{?}{=} 0$	*Inverse Property of Exponents and Logarithms*
$\log_7 7^0 \stackrel{?}{=} 0$	*$1 = 7^0$*
$0 = 0\ \checkmark$	*Inverse Property of Exponents and Logarithms*

67a.

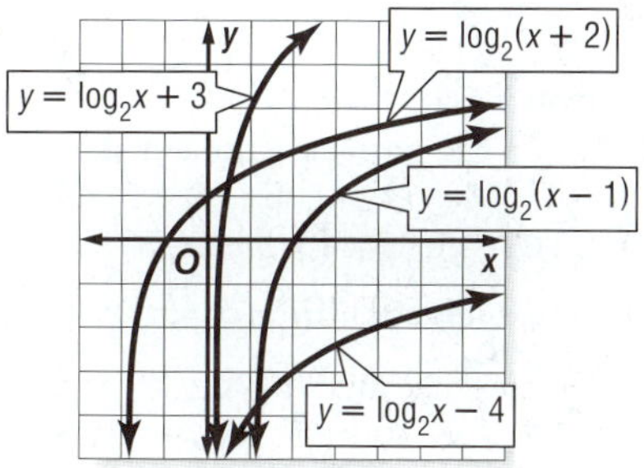

67b. The graph of $y = \log_2 x + 3$ is the graph of $y = \log_2 x$ translated 3 units up. The graph of $y = \log_2 x - 4$ is the graph of $y = \log_2 x$ translated 4 units down. The graph of $\log_2 (x - 1)$ is the graph of $y = \log_2 x$ translated 1 unit to the right. The graph of $\log_2 (x + 2)$ is the graph of $y = \log_2 x$ translated 2 units to the left. **69.** $10^{1.4}$ or about 25 times as great **71.** 2 and 3; Sample answer: 5 is between 2^2 and 2^3. **73.** A logarithmic scale illustrates that values next to each other vary by a factor of 10. Answers should include the following.

- Pin drop: 1×10^0; Whisper: 1×10^2; Normal conversation: 1×10^6; Kitchen noise: 1×10^{10}; Jet engine: 1×10^{12}
- 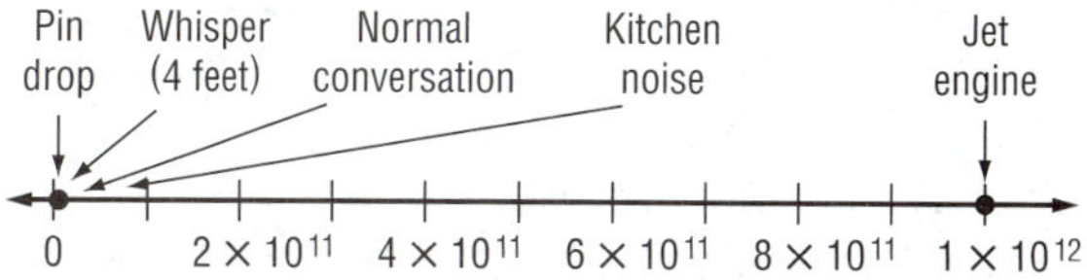

- On the scale shown above, the sound of a pin drop and the sound of normal conversation appear not to differ by much at all, when in fact they do differ in terms of the loudness we perceive. The first scale shows this difference more clearly.

75. D **77.** b^{12} **79.** $-3, \frac{14}{5}$ **81.** $\frac{5 \pm \sqrt{73}}{4}$
83. $\frac{6x - 58}{(x - 3)(x + 3)(x + 7)}$ **85.** x^{10} **87.** $8a^6b^3$ **89.** $\frac{x^3}{y^2z^3}$

Page 538 Practice Quiz 1

1. growth **3.** $\log_4 4096 = 6$ **5.** $\frac{4}{3}$ **7.** $\frac{3}{5}$ **9.** $x > 26$

Page 544–546 Lesson 10-3

1. properties of exponents **3.** Umeko; Clemente incorrectly applied the product and quotient properties of logarithms.
$\log_7 6 + \log_7 3 = \log_7 (6 \cdot 3)$ or $\log_7 18$
Product Property of Logarithms
$\log_7 18 - \log_7 2 = \log_7 (18 \div 2)$ or $\log_7 9$
Quotient Property of Logarithms
5. 2.6310 **7.** 6 **9.** 3 **11.** $\text{pH} = 6.1 + \log_{10} \frac{B}{C}$ **13.** 1.3652
15. -0.2519 **17.** 2.4307 **19.** -0.4307 **21.** 2 **23.** 4 **25.** 14
27. 2 **29.** $\varnothing$ **31.** 10 **33.** $\frac{x^3}{4}$ **35.** False; $\log_2 (2^2 + 2^3) = \log_2 12$, $\log_2 2^2 + \log_2 2^3 = 2 + 3$ or 5, and $\log_2 12 \neq 5$, since $2^5 \neq 12$. **37.** 2 **39.** about 0.4214 kilocalories per gram
41. 3 **43.** About 95 decibels; $L = 10 \log_{10} R$, where L is the loudness of the sound in decibels and R is the relative intensity of the sound. Since the crowd increased by a factor of 3, we assume that the intensity also increases by a factor of 3. Thus, we need to find the loudness of $3R$.
$L = 10 \log_{10} 3R$
$L = 10 (\log_{10} 3 + \log_{10} R)$
$L = 10 \log_{10} 3 + 10 \log_{10} R$
$L \approx 10(0.4771) + 90$
$L \approx 4.771 + 90$ or about 95
45. 7.5
47. Let $b^x = m$ and $b^y = n$. Then $\log_b m = x$ and $\log_b n = y$.

$\frac{b^x}{b^y} = \frac{m}{n}$	
$b^{x-y} = \frac{m}{n}$	*Quotient Property*
$\log_b b^{x-y} = \log_b \frac{m}{n}$	*Property of Equality for Logarithmic Equations*
$x - y = \log_b \frac{m}{n}$	*Inverse Property of Exponents and Logarithms*
$\log_b m - \log_b n = \log_b \frac{m}{n}$	*Replace x with $\log_b m$ and y with $\log_b n$.*

49. A **51.** 4 **53.** $2x$ **55.** -8

57. odd; 3 59. $\frac{3b}{a}$ 61. $\frac{5}{3x}$ 63. 1 65. $x > \frac{5}{3}$

Pages 549–551 Lesson 10-4

1. 10; common logarithms 3. A calculator is not programmed to find base 2 logarithms. 5. 1.3617
7. 1.7325 9. 4.9824 11. 11.5665 13. $\frac{\log 5}{\log 7}$; 0.8271
15. $\frac{\log 9}{\log 2}$; 3.1699 17. 0.6990 19. 0.8573 21. −0.0969
23. 11 25. 2.1 27. $\{x \mid x \geq 2.0860\}$ 29. $\{a \mid a < 1.1590\}$
31. 0.4341 33. 4.7820 35. ±1.1909 37. $\{n \mid n < -1.0178\}$
39. 3.7162 41. 0.5873 43. −7.6377 45. $\frac{\log 13}{\log 2} \approx 3.7004$
47. $\frac{\log 3}{\log 7} \approx 0.5646$ 49. $\frac{2 \log 1.6}{\log 4} \approx 0.6781$ 51. between 0.000000001 and 0.000001 mole per liter 53. Sirius
55. Vega 57. about 3.75 yr or 3 yr 9 mo
59. Comparisons between substances of different acidities are more easily distinguished on a logarithmic scale. Answers should include the following.
Sample Answer:
- Tomatoes: 6.3×10^{-5} mole per liter
  Milk: 3.98×10^{-7} mole per liter
  Eggs: 1.58×10^{-8} mole per liter
- Those measurements correspond to pH measurements of 5 and 4, indicating a weak acid and a stronger acid. On the logarithmic scale we can see the difference in these acids, whereas on a normal scale, these hydrogen ion concentrations would appear nearly the same. For someone who has to watch the acidity of the foods they eat, this could be the difference between an enjoyable meal and heartburn.

61. C 63. 1.6938 65. 64 67. 62 69. $(d + 2)(3d - 4)$
71. prime 73. $3^2 = x$ 75. $\log_5 45 = x$ 77. $\log_b x = y$

Pages 557–559 Lesson 10-5

1. the number e 3. Elsu; Colby tried to write each side as a power of 10. Since the base of the natural logarithmic function is e, he should have written each side as a power of e; $10^{\ln 4x} \neq 4x$. 5. 0.0334 7. −2.3026 9. $e^0 = 1$ 11. $5x$
13. 1.0986 15. $0 < x < 403.4288$ 17. ±90.0171
19. about 15,066 ft 21. 148.4132 23. 1.6487 25. 2.3026
27. −3.5066 29. about 49.5 cm 31. $2 = \ln 6x$ 33. $e^x = 5.2$
35. y 37. 45 39. −0.6931 41. $x > 0.4700$ 43. 0.5973
45. $x \geq -0.9730$ 47. 49.4711 49. 14.3891 51. 45.0086
53. 1 55. $t = \frac{100 \ln 2}{r}$ 57. $t = \frac{110}{r}$ 59. about 55 yr
61. about 21 min
63. The number e is used in the formula for continuously compounded interest, $A = Pe^{rt}$. Although no banks actually pay interest compounded continually, the equation is so accurate in computing the amount of money for quarterly compounding, or daily compounding, that it is often used for this purpose. Answers should include the following.
- If you know the annual interest rate r and the principal P, the value of the account after t years is calculated by multiplying P times e raised to the r times t power. Use a calculator to find the value of e^{rt}.
- If you know the value A you wish the account to achieve, the principal P, and the annual interest rate r, the time t needed to achieve this value is found by first taking the natural logarithm of A minus the natural logarithm of P. Then, divide this quantity by r.

65. 1946, 1981, 2015; It takes between 34 and 35 years for the population to double.

67. $\frac{\log 0.047}{\log 6} = -1.7065$ 69. 5 71. inverse; 4 73. direct; −7
75. 3.32 77. 1.43 79. 13.43

Page 559 Practice Quiz 2

1. $\frac{\log 5}{\log 4}$; 1.1610 3. 3 5. 1.3863

Pages 563–565 Lesson 10-6

1. $y = a(1 + r)^t$, where $r > 0$ represents exponential growth and $r < 0$ represents exponential decay 3. Sample answer: money in a bank 5. about 33.5 watts 7. $y = 212{,}000e^{0.025t}$
9. C 11. at most $108,484.93 13. No; the bone is only about 21,000 years old, and dinosaurs died out 63,000,000 years ago. 15. about 0.0347 17. $12,565 billion
19. after the year 2182 21. Never; theoretically, the amount left will always be half of the previous amount.
23. about 19.5 yr 25. $\ln y = 3$ 27. $4x^2 = e^8$ 29. $p > 3.3219$
31. $\frac{0.5(0.08p)}{6} + \frac{0.5(0.08p)}{4}$ 33. $\frac{p}{150}$ 35. ellipse 37. circle
39. 8×10^7

Pages 566–570 Chapter 10 Study Guide and Review

1. true 3. false; common logarithm 5. true
7. false; logarithmic function 9. false; exponential function
11. growth 13. $y = 7\left(\frac{1}{5}\right)^x$ 15. −1 17. $x \leq -\sqrt{6}$ or $x \geq \sqrt{6}$ 19. $\log_5 \frac{1}{25} = -2$ 21. $4^3 = 64$ 23. $6^{-2} = \frac{1}{36}$
25. −5 27. 2 29. $\frac{3}{2}$ 31. $\frac{1}{3} < y < 3$ 33. −4, 3 35. 1.7712
37. 3 39. 6 41. 15 43. 5.7279 45. $x < 7.3059$
47. $x \geq 5.8983$ 49. $\frac{\log 11}{\log 4}$; 1.7297 51. $\frac{\log 1000}{\log 20}$; 2.3059
53. $e^x = 7.4$ 55. $7x$ 57. $x > 1.1632$ 59. $0 < x \leq 49.4711$
61. 74.2066 63. 5.05 days 65. about 3.6%

Chapter 11 Sequences and Series

Page 577 Chapter 11 Getting Started

1. 6 3. −5 5. $\frac{1}{2}$

7.

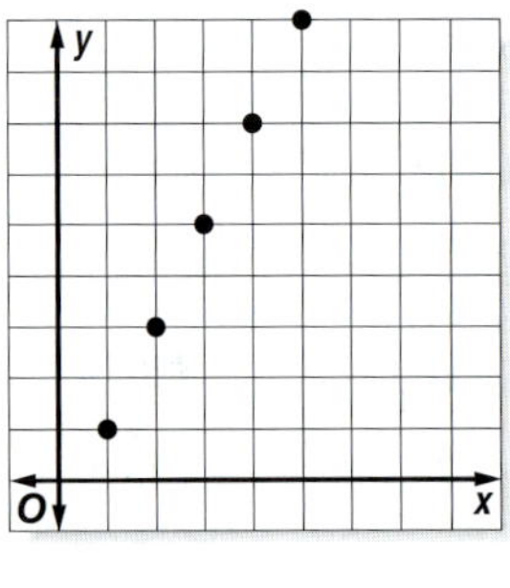

9.

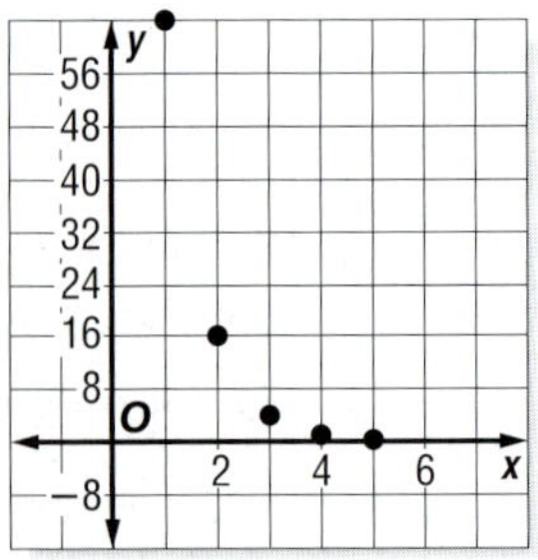

11. 17 13. $\frac{1}{32}$ 15. $\frac{3}{5}$

Pages 580–582 Lesson 11-1

1. The differences between the terms are not constant.
3. Sample answer: 1, −4, −9, −14, … 5. −3, −5, −7, −9
7. 14, 12, 10, 8, 6 9. −112 11. 15 13. 56, 68, 80
15. 30, 37, 44, 51 17. 6, 10, 14, 18 19. $\frac{7}{3}, 3, \frac{11}{3}, \frac{13}{3}$
21. 5.5, 5.1, 4.7, 4.3 23. 2, 15, 28, 41, 54
25. 6, 2, −2, −6, −10 27. $\frac{4}{3}, 1, \frac{2}{3}, \frac{1}{3}, 0$ 29. 28 31. 94

33. 335 **35.** $\frac{26}{3}$ **37.** 27 **39.** 61 **41.** 37.5 in. **43.** 30th
45. 82nd **47.** $a_n = -7n + 25$
49. 13, 17, 21

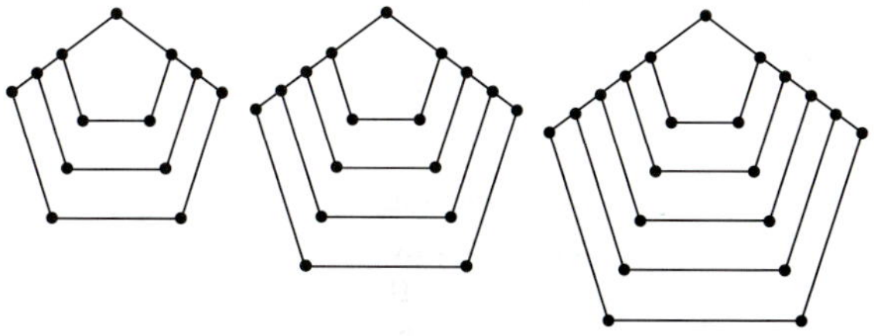

51. Yes; it corresponds to $n = 100$. **53.** 4, −2
55. 7, 11, 15, 19, 23
57. Arithmetic sequences can be used to model the numbers of shingles in the rows on a section of roof. Answers should include the following.

- One additional shingle is needed in each successive row.
- One method is to successively add 1 to the terms of the sequence: $a_8 = 9 + 1$ or 10, $a_9 = 10 + 1$ or 11, $a_{10} = 11 + 1$ or 12, $a_{11} = 12 + 1$ or 13, $a_{12} = 13 + 1$ or 14, $a_{13} = 14 + 1$ or 15, $a_{14} = 15 + 1$ or 16, $a_{15} = 16 + 1$ or 17. Another method is to use the formula for the nth term: $a_{15} = 3 + (15 - 1)1$ or 17.

59. B **61.** −0.4055 **63.** 146.4132 **65.** 2, 5, 8, 11
67. 11, 15, 19, 23, 27

Pages 586–587 Lesson 11-2

1. In a series, the terms are added. In a sequence, they are not. **3.** Sample answer: $\sum_{n=1}^{4} (3n + 4)$ **5.** 230 **7.** 552
9. 260 **11.** 95 **13.** −6, 0, 6 **15.** 344 **17.** 1501 **19.** −9 **21.** 104 **23.** −714 **25.** 14 **27.** 10 rows **29.** 721 **31.** 162 **33.** 108 **35.** −195 **37.** 315,150 **39.** 1,001,000 **41.** 17, 26, 35 **43.** −12, −9, −6 **45.** 265 ft **47.** False; for example, $7 + 10 + 13 + 16 = 46$, but $7 + 10 + 13 + 16 + 19 + 22 + 25 + 28 = 140$. **49.** C **51.** 5555 **53.** 6683 **55.** −135
57. $-\frac{9}{2}$ **59.** $\frac{3 \pm \sqrt{89}}{2}$ **61.** $26\sqrt{21}$ **63.** 16 **65.** $\frac{2}{27}$

Pages 590–592 Lesson 11-3

1a. Geometric; the terms have a common ratio of −2.
1b. Arithmetic; the terms have a common difference of −3.
3. Marika; Lori divided in the wrong order when finding r.
5. 2, −4 **7.** $\frac{15}{64}$ **9.** −4 **11.** 3, 9 **13.** 15, 5 **15.** 54, 81
17. $\frac{20}{27}, \frac{40}{81}$ **19.** −2.16, 2.592 **21.** 2, −6, 18, −54, 162
23. 243, 81, 27, 9, 3 **25.** $\frac{3}{16}$ **27.** 729 **29.** 243 **31.** 1
33. 78,125 **35.** −8748 **37.** 655.36 lb **39.** $a_n = 36\left(\frac{1}{3}\right)^{n-1}$
41. $a_n = -2(-5)^{n-1}$ **43.** ±18, 36, ±72 **45.** 16, 8, 4, 2
47. 8 days **49.** False; the sequence 1, 4, 9, 16, ..., for example, is neither arithmetic nor geometric.
51. The heights of the bounces of a ball and the heights from which a bouncing ball falls each form geometric sequences. Answers should include the following.

- 3, 1.8, 1.08, 0.648, 0.3888
- The common ratios are the same, but the first terms are different. The sequence of heights from which the ball falls is the sequence of heights of the bounces with the term 3 inserted at the beginning.

53. C **55.** 203 **57.** −12, −16, −20 **59.** 127 **61.** $\frac{61}{81}$

Page 592 Practice Quiz 1

1. 46 **3.** 187 **5.** 1

Pages 596–598 Lesson 11-4

1. Sample answer: $4 + 2 + 1 + \frac{1}{2}$
3. Sample answer: The first term is $a_1 = 2$. Divide the second term by the first to find that the common ratio is $r = 6$. Therefore, the nth term of the series is given by $2 \cdot 6^{n-1}$. There are five terms, so the series can be written as $\sum_{n=1}^{5} 2 \cdot 6^{n-1}$. **5.** 39,063 **7.** 165 **9.** 129 **11.** $\frac{1093}{9}$
13. 3 **15.** 728 **17.** 1111 **19.** 244 **21.** 2101 **23.** $\frac{728}{3}$
25. 1040.984 **27.** 6564 **29.** 1,747,625 **31.** 3641 **33.** $\frac{5461}{16}$
35. 2555 **37.** $\frac{387}{4}$ **39.** 3,145,725 **41.** 243 **43.** 2 **45.** 80
47. about 7.13 in. **49.** If the number of people that each person sends the joke to is constant, then the total number of people who have seen the joke is the sum of a geometric series. Answers should include the following.

- The common ratio would change from 3 to 4.
- Increase the number of days that the joke circulates so that it is inconvenient to find and add all the terms of the series.

51. C **53.** 3.99987793 **55.** $\pm\frac{1}{4}, \frac{3}{2}, \pm 9$ **57.** 232
59. **Drive-In Movie Screens**

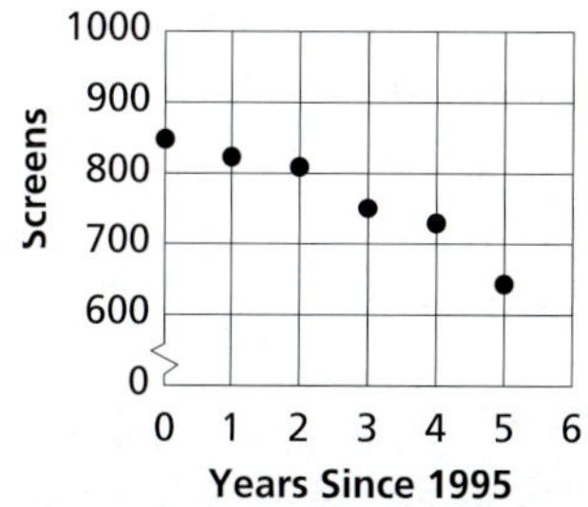

61. Sample answer: 294 **63.** 2 **65.** $\frac{2}{3}$ **67.** 0.6

Pages 602–604 Lesson 11-5

1. Sample answer: $\sum_{n=1}^{\infty} \left(\frac{1}{2}\right)^n$ **3.** Beth; the common ratio for the infinite geometric series is $-\frac{4}{3}$. Since $\left|-\frac{4}{3}\right| \geq 1$, the series does not have a sum and the formula $S = \frac{a_1}{1 - r}$ does not apply. **5.** does not exist **7.** $\frac{3}{4}$ **9.** 100 **11.** $\frac{73}{99}$
13. 96 cm **15.** does not exist **17.** 45 **19.** −16 **21.** $\frac{54}{5}$
23. does not exist **25.** 1 **27.** $\frac{2}{3}$ **29.** $\frac{3}{2}$ **31.** 2
33. $40 + 20\sqrt{2} + 20 + \ldots$ **35.** 900 ft **37.** 75, 30, 12
39. $-8, -3\frac{1}{5}, -1\frac{7}{25}, -\frac{64}{125}$ **41.** $\frac{1}{9}$ **43.** $\frac{82}{99}$ **45.** $\frac{427}{999}$ **47.** $\frac{229}{990}$
49. The total distance that a ball bounces, both up and down, can be found by adding the sums of two infinite geometric series. Answers should include the following.

- $a_n = a_1 \cdot r^{n-1}$, $S_n = \frac{a_1(1 - r^n)}{1 - r}$, or $S = \frac{a_1}{1 - r}$
- The total distance the ball falls is given by the infinite geometric series $3 + 3(0.6) + 3(0.6)^2 + \ldots$. The sum of this series is $\frac{3}{1 - 0.6}$ or 7.5. The total distance the ball bounces up is given by the infinite geometric series $1.8(0.6) + 1.8(0.6)^2 + 1.8(0.6)^3 + \ldots$. The sum of this

Selected Answers

series is $\frac{1.8(0.6)}{1-0.6}$ or 2.7. Thus, the total distance the ball travels is 7.5 + 2.7 or 10.2 feet.

51. C **53.** $\frac{8744}{81}$ **55.** 3 **57.** $x \geq 5$ **59.** $\frac{-x+7}{(x-3)(x+1)}$

61. $(x-2)^2 + (y-4)^2 = 36$ **63.** $-\frac{1}{2}, \frac{3}{2}, \frac{7}{2}$ **65.** $x^2 - 36 = 0$

67. $x^2 - 10x + 24 = 0$ **69.** The number of visitors was decreasing. **71.** 3 **73.** $\frac{1}{2}$ **75.** -4

Pages 608–610 Lesson 11-6

1. $a_n = a_{n-1} + d$; $a_n = r \cdot a_{n-1}$ **3.** Sometimes; if $f(x) = x^2$ and $x_1 = 2$, then $x_2 = 2^2$ or 4, so $x_2 \neq x_1$. But, if $x_1 = 1$, then $x_2 = 1$, so $x_2 = x_1$. **5.** $-3, -2, 0, 3, 7$ **7.** 1, 2, 5, 14, 41
9. $1, 3, -1$ **11.** $b_n = 1.05b_{n-1} - 10$ **13.** $-6, -3, 0, 3, 6$
15. $2, 1, -1, -4, -8$ **17.** 9, 14, 24, 44, 84 **19.** $-1, 5, 4, 9, 13$

21. $\frac{7}{2}, \frac{7}{4}, \frac{7}{6}, \frac{7}{8}, \frac{7}{10}$ **23.** 67 **25.** 1, 1, 2, 3, 5, …

27. \$99,921.21, \$99,841.95, \$99,762.21, \$99,681.99, \$99,601.29, \$99,520.11, \$99,438.44, \$99,356.28 **29.** $t_n = t_{n-1} + n$
31. 16, 142, 1276 **33.** $-7, -16, -43$ **35.** $-3, 13, 333$

37. $\frac{5}{2}, \frac{37}{2}, \frac{1445}{2}$ **39.** \$75.78

41. Under certain conditions, the Fibonacci sequence can be used to model the number of shoots on a plant. Answers should include the following.
- The 13th term of the sequence is 233, so there are 233 shoots on the plant during the 13th month.
- The Fibonacci sequence is not arithmetic because the differences $(0, 1, 1, 2, \ldots)$ of the terms are not constant. The Fibonacci sequence is not geometric because the ratios $\left(1, 2, \frac{3}{2}, \ldots\right)$ of the terms are not constant.

43. C **45.** $\frac{1}{6}$ **47.** -5208 **49.** $3x + 7$ units **51.** 5040

53. 20 **55.** 210

Pages 615–617 Lesson 11-7

1. 1, 8, 28, 56, 70, 56, 28, 8, 1 **3.** Sample answer: $(5x + y)^4$
5. 17,160 **7.** $p^5 + 5p^4q + 10p^3q^2 + 10p^2q^3 + 5pq^4 + q^5$
9. $x^4 - 12x^3y + 54x^2y^2 - 108xy^3 + 81y^4$ **11.** $1{,}088{,}640a^6b^4$
13. 362,880 **15.** 72 **17.** 495 **19.** $a^3 - 3a^2b + 3ab^2 - b^3$
21. $r^8 + 8r^7s + 28r^6s^2 + 56r^5s^3 + 70r^4s^4 + 56r^3s^5 + 28r^2s^6 + 8rs^7 + s^8$ **23.** $x^5 + 15x^4 + 90x^3 + 270x^2 + 405x + 243$
25. $16b^4 - 32b^3x + 24b^2x^2 - 8bx^3 + x^4$ **27.** $243x^5 - 810x^4y + 1080x^3y^2 - 720x^2y^3 + 240xy^4 - 32y^5$ **29.** $\frac{a^5}{32} + \frac{5a^4}{8} + 5a^3 + 20a^2 + 40a + 32$ **31.** $27x^3 + 54x^2 + 36x + 8$ cm^3 **33.** 45

35. $924x^6y^6$ **37.** $5670a^4$ **39.** $145{,}152x^6y^3$ **41.** $-\frac{63}{8}x^5$

43. The coefficients in a binomial expansion give the numbers of sequences of births resulting in given numbers of boys and girls. Answers should include the following.
- $(b+g)^5 = b^5 + 5b^4g + 10b^3g^2 + 10b^2g^3 + 5bg^4 + g^5$; There is one sequence of births with all five boys, five sequences with four boys and one girl, ten sequences with three boys and two girls, ten sequences with two boys and three girls, five sequences with one boy and four girls, and one sequence with all five girls.
- The number of sequences of births that have exactly k girls in a family of n children is the coefficient of $b^{n-k}g^k$ in the expansion of $(b+g)^n$. According to the Binomial Theorem, this coefficient is $\frac{n!}{(n-k)!k!}$.

45. C **47.** 3, 5, 9, 17, 33 **49.** $\frac{\log 5}{\log 2}$; 2.3219 **51.** $\frac{\log 8}{\log 5}$; 1.2920 **53.** asymptotes: $x = -4$, $x = 1$ **55.** hyperbola

57. yes **59.** True; $\frac{1(1+1)}{2} = \frac{1(2)}{2}$ or 1. **61.** True; $\frac{1^2(1+1)^2}{4} = \frac{1(4)}{4}$ or 1.

Page 617 Practice Quiz 2

1. 1,328,600 **3.** 24 **5.** 1, 5, 13, 29, 61 **7.** $5, -13, 41$
9. $a^6 + 12a^5 + 60a^4 + 160a^3 + 240a^2 + 192a + 64$

Pages 619–621 Lesson 11-8

1. Sample answers: formulas for the sums of powers of the first n positive integers and statements that expressions involving exponents of n are divisible by certain numbers
3. Sample answer: $3^n - 1$
5. Step 1: When $n = 1$, the left side of the given equation is $\frac{1}{2}$. The right side is $1 - \frac{1}{2}$ or $\frac{1}{2}$, so the equation is true for $n = 1$.

Step 2: Assume $\frac{1}{2} + \frac{1}{2^2} + \frac{1}{2^3} + \ldots + \frac{1}{2^k} = 1 - \frac{1}{2^k}$ for some positive integer k.

Step 3:
$$\begin{aligned}\frac{1}{2} + \frac{1}{2^2} + \frac{1}{2^3} + \ldots + \frac{1}{2^k} + \frac{1}{2^{k+1}} &= 1 - \frac{1}{2^k} + \frac{1}{2^{k+1}}\\ &= 1 - \frac{2}{2^{k+1}} + \frac{1}{2^{k+1}}\\ &= 1 - \frac{1}{2^{k+1}}\end{aligned}$$

The last expression is the right side of the equation to be proved, where $n = k + 1$. Thus, the equation is true for $n = k + 1$.

Therefore, $\frac{1}{2} + \frac{1}{2^2} + \frac{1}{2^3} + \ldots + \frac{1}{2^n} = 1 - \frac{1}{2^n}$ for all positive integers n.

7. Step 1: $5^1 + 3 = 8$, which is divisible by 4. The statement is true for $n = 1$.
Step 2: Assume that $5^k + 3$ is divisible by 4 for some positive integer k. This means that $5^k + 3 = 4r$ for some positive integer r.
Step 3:
$$\begin{aligned}5^k + 3 &= 4r\\ 5^k &= 4r - 3\\ 5^{k+1} &= 20r - 15\\ 5^{k+1} + 3 &= 20r - 12\\ 5^{k+1} + 3 &= 4(5r - 3)\end{aligned}$$
Since r is a positive integer, $5r - 3$ is a positive integer. Thus, $5^{k+1} + 3$ is divisible by 4, so the statement is true for $n = k + 1$.
Therefore, $5^n + 3$ is divisible by 4 for all positive integers n.
9. Sample answer: $n = 3$
11. Step 1: When $n = 1$, the left side of the given equation is 1. The right side is $1[2(1) - 1]$ or 1, so the equation is true for $n = 1$.
Step 2: Assume $1 + 5 + 9 + \ldots + (4k - 3) = k(2k - 1)$ for some positive integer k.
Step 3:
$$\begin{aligned}&1 + 5 + 9 + \ldots + (4k - 3) + [4(k+1) - 3]\\ &= k(2k-1) + [4(k+1) - 3]\\ &= 2k^2 - k + 4k + 4 - 3\\ &= 2k^2 + 3k + 1\\ &= (k+1)(2k+1)\\ &= (k+1)[2(k+1) - 1]\end{aligned}$$
The last expression is the right side of the equation to be proved, where $n = k + 1$. Thus, the equation is true for $n = k + 1$.
Therefore, $1 + 5 + 9 + \ldots + (4n - 3) = n(2n - 1)$ for all positive integers n.
13. Step 1: When $n = 1$, the left side of the given equation is 1^3 or 1. The right side is $\frac{1^2(1+1)^2}{4}$ or 1, so the equation is true for $n = 1$.
Step 2: Assume $1^3 + 2^3 + 3^3 + \ldots + k^3 = \frac{k^2(k+1)^2}{4}$ for some

Selected Answers

positive integer k.

Step 3: $1^3 + 2^3 + 3^3 + \ldots + k^3 + (k+1)^3$

$$= \frac{k^2(k+1)^2}{4} + (k+1)^3$$
$$= \frac{k^2(k+1)^2 + 4(k+1)^3}{4}$$
$$= \frac{(k+1)^2[k^2 + 4(k+1)]}{4}$$
$$= \frac{(k+1)^2(k^2 + 4k + 4)}{4}$$
$$= \frac{(k+1)^2(k+2)^2}{4}$$
$$= \frac{(k+1)^2[(k+1)+1]^2}{4}$$

The last expression is the right side of the equation to be proved, where $n = k + 1$. Thus, the equation is true for $n = k + 1$.

Therefore, $1^3 + 2^3 + 3^3 + \ldots + n^3 = \frac{n^2(n+1)^2}{4}$ for all positive integers n.

15. Step 1: When $n = 1$, the left side of the given equation is $\frac{1}{3}$. The right side is $\frac{1}{2}\left(1 - \frac{1}{3}\right)$ or $\frac{1}{3}$, so the equation is true for $n = 1$.

Step 2: Assume $\frac{1}{3} + \frac{1}{3^2} + \frac{1}{3^3} + \ldots + \frac{1}{3^k} = \frac{1}{2}\left(1 - \frac{1}{3^k}\right)$ for some positive integer k.

Step 3: $\frac{1}{3} + \frac{1}{3^2} + \frac{1}{3^3} + \ldots + \frac{1}{3^k} + \frac{1}{3^{k+1}} = \frac{1}{2}\left(1 - \frac{1}{3^k}\right) + \frac{1}{3^{k+1}}$

$$= \frac{1}{2} - \frac{1}{2 \cdot 3^k} + \frac{1}{3^{k+1}}$$
$$= \frac{3^{k+1} - 3 + 2}{2 \cdot 3^{k+1}}$$
$$= \frac{3^{k+1} - 1}{2 \cdot 3^{k+1}}$$
$$= \frac{1}{2}\left(\frac{3^{k+1} - 1}{3^{k+1}}\right)$$
$$= \frac{1}{2}\left(1 - \frac{1}{3^{k+1}}\right)$$

The last expression is the right side of the equation to be proved, where $n = k + 1$. Thus, the equation is true for $n = k + 1$.

Therefore, $\frac{1}{3} + \frac{1}{3^2} + \frac{1}{3^3} + \ldots + \frac{1}{3^n} = \frac{1}{2}\left(1 - \frac{1}{3^n}\right)$ for all positive integers n.

17. Step 1: $8^1 - 1 = 7$, which is divisible by 7. The statement is true for $n = 1$.

Step 2: Assume that $8^k - 1$ is divisible by 7 for some positive integer k. This means that $8^k - 1 = 7r$ for some whole number r.

Step 3:
$$8^k - 1 = 7r$$
$$8^k = 7r + 1$$
$$8^{k+1} = 56r + 8$$
$$8^{k+1} - 1 = 56r + 7$$
$$8^{k+1} - 1 = 7(8r + 1)$$

Since r is a whole number, $8r + 1$ is a whole number. Thus, $8^{k+1} - 1$ is divisible by 7, so the statement is true for $n = k + 1$.

Therefore, $8^n - 1$ is divisible by 7 for all positive integers n.

19. Step 1: $12^1 + 10 = 22$, which is divisible by 11. The statement is true for $n = 1$.

Step 2: Assume that $12^k + 10$ is divisible by 11 for some positive integer k. This means that $12^k + 10 = 11r$ for some positive integer r.

Step 3:
$$12^k + 10 = 11r$$
$$12^k = 11r - 10$$
$$12^{k+1} = 132r - 120$$
$$12^{k+1} + 10 = 132r - 110$$
$$12^{k+1} + 10 = 11(12r - 10)$$

Since r is a positive integer, $12r - 10$ is a positive integer. Thus, $12^{k+1} + 10$ is divisible by 11, so the statement is true for $n = k + 1$.

Therefore, $12^n + 10$ is divisible by 11 for all positive integers n.

21. Step 1: There are 6 bricks in the top row, and $1^2 + 5(1) = 6$, so the formula is true for $n = 1$.

Step 2: Assume that there are $k^2 + 5k$ bricks in the top k rows for some positive integer k.

Step 3: Since each row has 2 more bricks than the one above, the numbers of bricks in the rows form an arithmetic sequence. The number of bricks in the $(k + 1)$st row is $6 + [(k + 1) - 1](2)$ or $2k + 6$. Then the number of bricks in the top $k + 1$ rows is $k^2 + 5k + (2k + 6)$ or $k^2 + 7k + 6$. $k^2 + 7k + 6 = (k + 1)^2 + 5(k + 1)$, which is the formula to be proved, where $n = k + 1$. Thus, the formula is true for $n = k + 1$.

Therefore, the number of bricks in the top n rows is $n^2 + 5n$ for all positive integers n.

23. Step 1: When $n = 1$, the left side of the given equation is a_1. The right side is $\frac{1}{2}[2a_1 + (1 - 1)d]$ or a_1, so the equation is true for $n = 1$.

Step 2: Assume $a_1 + (a_1 + d) + (a_1 + 2d) + \ldots + [a_1 + (k - 1)d] = \frac{k}{2}[2a_1 + (k - 1)d]$ for some positive integer k.

Step 3: $a_1 + (a_1 + d) + (a_1 + 2d) + \ldots + [a_1 + (k - 1)d] + [a_1 + (k + 1 - 1)d]$

$$= \frac{k}{2}[2a_1 + (k-1)d] + [a_1 + (k + 1 - 1)d]$$
$$= \frac{k}{2}[2a_1 + (k-1)d] + a_1 + kd$$
$$= \frac{k[2a_1 + (k-1)d] + 2(a_1 + kd)}{2}$$
$$= \frac{k \cdot 2a_1 + (k^2 - k)d + 2a_1 + 2kd}{2}$$
$$= \frac{(k+1)2a_1 + (k^2 - k + 2k)d}{2}$$
$$= \frac{(k+1)2a_1 + k(k+1)d}{2}$$
$$= \frac{k+1}{2}(2a_1 + kd)$$
$$= \frac{k+1}{2}[2a_1 + (k + 1 - 1)d]$$

The last expression is the right side of the formula to be proved, where $n = k + 1$. Thus, the formula is true for $n = k + 1$.

Therefore, $a_1 + (a_1 + d) + (a_1 + 2d) + \ldots + [a_1 + (n - 1)d] = \frac{n}{2}[2a_1 + (n - 1)d]$ for all positive integers n.

25. Sample answer: $n = 3$ **27.** Sample answer: $n = 2$ **29.** Sample answer: $n = 11$ **31.** Write 7^n as $(6 + 1)^n$. Then use the Binomial Theorem.

$$7^n - 1 = (6+1)^n - 1$$
$$= 6^n + n \cdot 6^{n-1} + \frac{n(n-1)}{2}6^{n-2} + \ldots + n \cdot 6 + 1 - 1$$
$$= 6^n + n \cdot 6^{n-1} + \frac{n(n-1)}{2}6^{n-2} + \ldots + n \cdot 6$$

Since each term in the last expression is divisible by 6, the whole expression is divisible by 6. Thus, $7^n - 1$ is divisible by 6. **33.** C **35.** $x^6 + 6x^5y + 15x^4y^2 + 20x^3y^3 + 15x^2y^4 + 6xy^5 + y^6$ **37.** $256x^8 + 1024x^7y + 1792x^6y^2 + 1792x^5y^3 + 1120x^4y^4 + 448x^3y^5 + 112x^2y^6 + 16xy^7 + y^8$ **39.** 2, 14, 782 **41.** 0, 1

Selected Answers

Pages 622–626 Chapter 11 Study Guide and Review

1. partial sum **3.** sigma notation **5.** Binomial Theorem **7.** arithmetic series **9.** 38 **11.** −11 **13.** −3, 1, 5 **15.** 6, 3, 0, −3 **17.** 2322 **19.** −220 **21.** 32 **23.** 3 **25.** 6, 12 **27.** 4, 2, 1, $\frac{1}{2}$ **29.** 1452 **31.** $\frac{14,197}{16}$ **33.** 72 **35.** $-\frac{16}{13}$ **37.** 3, 2, −2, −18, −82 **39.** 1, 3, 4, 7, 11 **41.** 10, 66, 458 **43.** −1, 4, −31 **45.** $x^4 - 8x^3 + 24x^2 - 32x + 16$ **47.** $160x^3y^3$

49. Step 1: When $n = 1$, the left side of the given equation is 1. The right side is $2^1 - 1$ or 1, so the equation is true for $n = 1$.

Step 2: Assume $1 + 2 + 4 + \ldots + 2^{k-1} = 2^k - 1$ for some positive integer k.

Step 3: $1 + 2 + 4 + \ldots + 2^{k-1} + 2^{(k+1)-1} = 2^k - 1 + 2k$
$= 2 \cdot 2^k - 1$
$= 2^{k+1} - 1$

The last expression is the right side of the equation to be proved, where $n = k + 1$. Thus, the equation is true for $n = k + 1$.

Therefore, $1 + 2 + 4 + \ldots + 2^{n-1} = 2^n - 1$ for all positive integers n.

Chapter 12 Probability and Statistics

Page 631 Chapter 12 Getting Started

1. $\frac{1}{6}$ **3.** $\frac{1}{2}$ **5.** $\frac{2}{3}$

7.

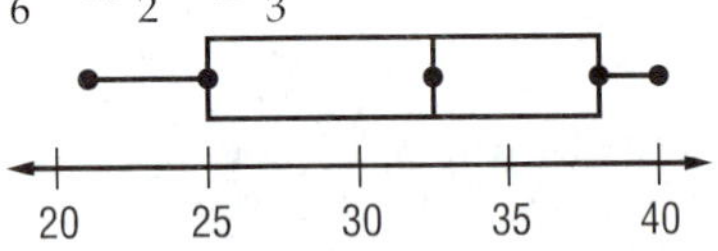

9.

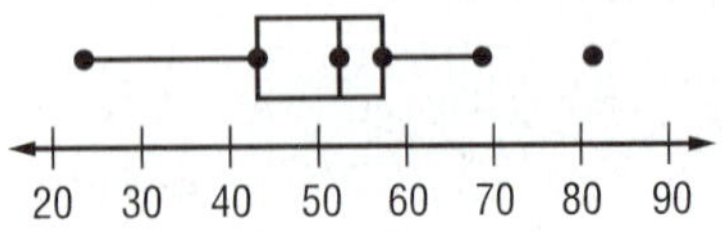

11. 3 **13.** $\sqrt{13}$ **15.** $a^3 + 3a^2b + 3ab^2 + b^3$ **17.** $m^5 - 5m^4n + 10m^3n^2 - 10m^2n^3 + 5mn^4 - n^5$

Pages 634–637 Lesson 12-1

1. HHH, HHT, HTH, HTT, THH, THT, TTH, TTT **3.** The available colors for the car could be different from those for the truck. **5.** dependent **7.** 256 **9.** D **11.** independent **13.** dependent **15.** 16 **17.** 30 **19.** 1024 **21.** 10,080 **23.** 362,880 **25.** 27,216 **27.** 800

29. The maximum number of license plates is a product with factors of 26s and 10s, depending on how many letters are used and how many digits are used. Answers should include the following.

- There are 26 choices for the first letter, 26 for the second, and 26 for the third. There are 10 choices for the first number, 10 for the second, and 10 for the third. By the Fundamental Counting Principle, there are $26^3 \cdot 10^3$ or 17,576,000 possible license plates.
- Replace positions containing numbers with letters.

31. C **33.** 20 mi **35.** $28x^6y^2$ **37.** 7 **39.** $\frac{1}{2}$ **41.** $-\frac{x}{x+5y}$ **43.** ±1, ±2 **45.** $y = (x - 3)^2 + 2$ **47.** $y = -\frac{1}{2}x^2 + 8$ **49.** 3 **51.** $\frac{1}{7}\begin{bmatrix} 1 & -1 \\ 4 & 3 \end{bmatrix}$ **53.** no inverse exists **55.** $y = \frac{2}{3}x + \frac{1}{3}$ **57.** 30 **59.** 720 **61.** 15 **63.** 1

Pages 641–643 Lesson 12-2

1. Sample answer: There are six people in a contest. How many ways can the first, second, and third prizes be awarded? **3.** Sometimes; the statement is only true when $r = 1$. **5.** 120 **7.** 6 **9.** permutation; 5040 **11.** 84 **13.** 9 **15.** 665,280 **17.** 70 **19.** 210 **21.** 1260 **23.** combination; 28 **25.** permutation; 120 **27.** permutation; 3360 **29.** combination; 455 **31.** 60 **33.** 111,540 **35.** 80,089,128

37. $C(n-1, r) + C(n-1, r-1)$

$= \frac{(n-1)!}{(n-1-r)!r!} + \frac{(n-1)!}{[n-1-(r-1)]!(r-1)!}$

$= \frac{(n-1)!}{(n-r-1)!r!} + \frac{(n-1)!}{(n-r)!(r-1)!}$

$= \frac{(n-1)!}{(n-r-1)!r!} \cdot \frac{n-r}{n-r} + \frac{(n-1)!}{(n-r)!(r-1)!} \cdot \frac{r}{r}$

$= \frac{(n-1)!(n-r)}{(n-r)!r!} + \frac{(n-1)!r}{(n-r)!r!}$

$= \frac{(n-1)!(n-r+r)}{(n-r)!r!}$

$= \frac{(n-1)!n}{(n-r)!r!}$

$= \frac{n!}{(n-r)!r!}$

$= C(n, r)$

39. D **41.** 24 **43.** 120 **45.** 80 **47.** Sample answer: $n = 2$ **49.** $x > 0.8047$ **51.** 20 days **53.** $\frac{(y-4)^2}{9} + \frac{(x-4)^2}{4} = 1$ **55.** −4; 128 **57.** {−2, 5} **59.** $8\sqrt{2}$ **61.** $4\sqrt{5}$ **63.** (0, 2) **65.** $-\frac{6}{7}$ **67.** {−7, 15} **69.** $\frac{3}{5}$ **71.** $\frac{1}{5}$

Pages 647–650 Lesson 12-3

1. Sample answer: The event *July comes before June* has a probability of 0. The event *June comes before July* has a probability of 1. **3.** There are 6 · 6 or 36 possible outcomes for the two dice. Only 1 outcome, 1 and 1, results in a sum of 2, so $P(2) = \frac{1}{36}$. There are 2 outcomes, 1 and 2 as well as 2 and 1, that result in a sum of 3, so $P(3) = \frac{2}{36}$ or $\frac{1}{18}$. **5.** $\frac{2}{7}$ **7.** 8:1 **9.** 2:7 **11.** $\frac{10}{11}$ **13.** $\frac{1}{8}$ **15.** $\frac{1}{10}$ **17.** $\frac{2}{25}$ **19.** $\frac{6}{55}$ **21.** $\frac{28}{55}$ **23.** $\frac{11}{115}$ **25.** $\frac{6}{115}$ **27.** $\frac{24}{115}$ **29.** 0 **31.** 0.007 **33.** 0.109 **35.** 3:5 **37.** 5:3 **39.** 1:4 **41.** 3:1 **43.** $\frac{3}{10}$ **45.** $\frac{4}{9}$ **47.** $\frac{1}{9}$ **49.** $\frac{3}{5}$ **51.** 2:23 **53.** 1:4 **55.** $\frac{1}{20}$ **57.** $\frac{9}{20}$ **59.** $\frac{9}{20}$ **61.** $\frac{1}{120}$ **63.** Probability and odds are good tools for assessing risk. Answers should include the following.

- $P(\text{struck by lightning}) = \frac{s}{s+f} = \frac{1}{750,000}$, so Odds = 1:(750,000 − 1) or 1:749,999. $P(\text{surviving a lightning strike}) = \frac{s}{s+f} = \frac{3}{4}$, so Odds = 3:(4 − 3) or 3:1.
- In this case, success is being struck by lightning or surviving the lightning strike. Failure is not being struck by lightning or not surviving the lightning strike.

65. D **67.** experimental; about 0.307 **69.** theoretical; $\frac{1}{17}$ **71.** permutation; 1260 **73.** 16 **75.** direct variation **77.** (4, 4) **79.** $\frac{6}{35}$ **81.** $\frac{1}{4}$ **83.** $\frac{9}{20}$

Page 650 Practice Quiz 1

1. 24 **3.** 18,720 **5.** 56 **7.** combination; 20,358,520 **9.** $\frac{13}{102}$

Pages 654–657 Lesson 12-4

1. Sample answer: putting on your socks, and then your shoes **3.** Mario; the probabilities of rolling a 4 and rolling a 2 are both $\frac{1}{6}$. **5.** $\frac{1}{4}$ **7.** $\frac{4}{663}$ **9.** $\frac{1}{4}$ **11.** dependent; $\frac{21}{220}$ **13.** $\frac{1}{12}$ **15.** $\frac{25}{36}$ **17.** $\frac{1}{6}$ **19.** $\frac{5}{6}$ **21.** $\frac{1}{49}$ **23.** $\frac{10}{21}$ **25.** 0 **27.** $\frac{2}{15}$ **29.** $\frac{2}{15}$ **31.** independent; $\frac{25}{81}$ **33.** dependent; $\frac{1}{21}$ **35.** dependent; $\frac{81}{2401}$

37.

		First Spin		
		Blue $\frac{1}{3}$	Yellow $\frac{1}{3}$	Red $\frac{1}{3}$
Second Spin	Blue $\frac{1}{3}$	BB $\frac{1}{9}$	BY $\frac{1}{9}$	BR $\frac{1}{9}$
	Yellow $\frac{1}{3}$	YB $\frac{1}{9}$	YY $\frac{1}{9}$	YR $\frac{1}{9}$
	Red $\frac{1}{3}$	RB $\frac{1}{9}$	RY $\frac{1}{9}$	RR $\frac{1}{9}$

39. $\frac{1}{3}$ **41.** $\frac{19}{1,160,054}$ **43.** $\frac{6327}{20,825}$ **45.** about 4.87% **47.** no

49. Sample answer: As the number of trials increases, the results become more reliable. However, you cannot be absolutely certain that there are no black marbles in the bag without looking at all of the marbles. **51.** Probability can be used to analyze the chances of a player making 0, 1, or 2 free throws when he or she goes to the foul line to shoot 2 free throws. Answers should include the following.

- One of the decimals in the table could be used as the value of p, the probability that a player makes a given free throw. The probability that a player misses both free throws is $(1 - p)(1 - p)$ or $(1 - p)^2$. The probability that a player makes both free throws is $p \cdot p$ or p^2. Since the sum of the probabilities of all the possible outcomes is 1, the probability that a player makes exactly 1 of the 2 free throws is $1 - (1 - p)^2 - p^2$ or $2p(1 - p)$.
- The result of the first free throw could affect the player's confidence on the second free throw. For example, if the player makes the first free throw, the probability of he or she making the second free throw might increase. Or, if the player misses the first free throw, the probability that he or she makes the second free throw might decrease.

53. C **55.** $\frac{3}{340}$ **57.** 1440 ways **59.** 36 **61.** $x, x - 4$

63.

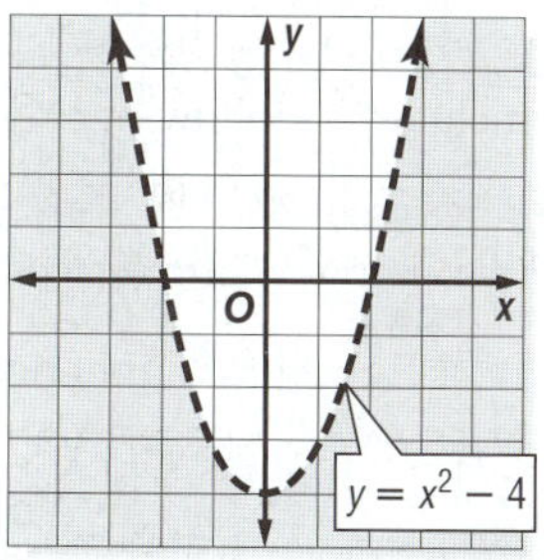

65. 153 **67.** $|b|$ **69.** (1, 2) **71.** (−2, 4) **73.** $\frac{5}{6}$ **75.** $\frac{11}{12}$ **77.** $1\frac{5}{12}$

Pages 660–663 Lesson 12-5

1. Sample answer: mutually exclusive events: tossing a coin and rolling a die; inclusive events: drawing a 7 and a diamond from a standard deck of cards **3.** The events are not mutually exclusive, so the chance of rain is less than 100%. **5.** $\frac{1}{3}$ **7.** $\frac{1}{2}$ **9.** $\frac{2}{3}$ **11.** inclusive; $\frac{4}{13}$ **13.** $\frac{5}{6}$ **15.** $\frac{25}{42}$ **17.** $\frac{35}{143}$ **19.** $\frac{3}{143}$ **21.** $\frac{38}{143}$ **23.** mutually exclusive; $\frac{7}{9}$ **25.** inclusive; $\frac{21}{34}$ **27.** $\frac{4}{13}$ **29.** $\frac{55}{221}$ **31.** $\frac{188}{663}$ **33.** $\frac{1}{8}$ **35.** $\frac{1}{4}$ **37.** $\frac{1}{780}$ **39.** $\frac{9}{130}$ **41.** $\frac{11}{780}$ **43.** $\frac{3}{5}$ **45.** $\frac{17}{27}$

47. Subtracting $P(A \text{ and } B)$ from each side and adding $P(A \text{ or } B)$ to each side results in the equation $P(A \text{ or } B) = P(A) + P(B) - P(A \text{ and } B)$. This is the equation for the probability of inclusive events. If A and B are mutually exclusive, then $P(A \text{ and } B) = 0$, so the equation simplifies to $P(A \text{ or } B) = P(A) + P(B)$, which is the equation for the probability of mutually exclusive events. Therefore, the equation is correct in either case. **49.** C **51.** $\frac{1}{216}$ **53.** $\frac{1}{216}$

55. 4:1 **57.** 2:5 **59.** 254 **61.** (±8, −10) **63.** $(x + 1)^2(x - 1)(x^2 + 1)$ **65.** min: (−0.42, 0.62); max: (−1.58, 1.38)

67.

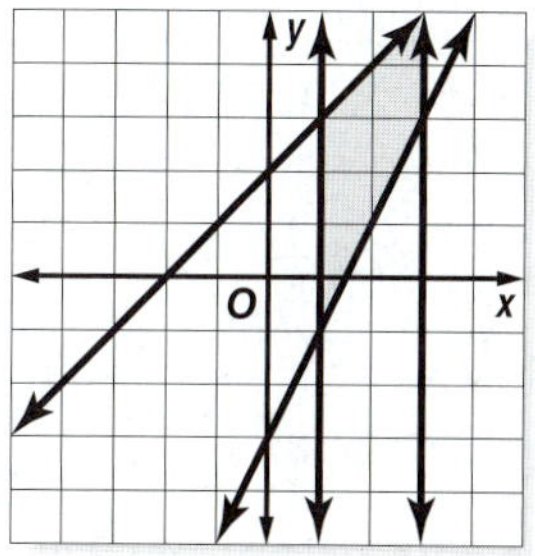

(1, 3), (1, −1), (3, 3), (3, 5); max: $f(3, 5) = 23$; min: $f(1, -1) = -3$
69. direct variation
71. 35.4, 34, no mode, 72
73. 63.75, 65, 50 and 65, 30
75. 12.98, 12.9, no mode, 4.7

Pages 666–670 Lesson 12-6

1. Sample answer: {10, 10, 10, 10, 10, 10}

3. $\sigma = \sqrt{\frac{1}{n}\sum_{i=1}^{n}(x_i - \bar{x})^2}$ **5.** 8.3, 2.9 **7.** \$7300.50, \$5335.25 **9.** 2500, 50 **11.** 3.1, 1.7 **13.** 37,691.2, 194.1 **15.** 82.9, 9.1 **17.** 77.7; 32; 19 **19.** Mean; it is highest. **21.** \$1047.88, \$1049.50, \$695 **23.** Mean or median; they are nearly equal and are more representative of the prices than the mode. **25.** Mode; it is lowest. **27.** 19.3 **29.** 19.5 **31.** 59.8, 7.7 **33.** 100% **35.** Sample answer: The first graph might be used by a sales manager to show a salesperson that he or she does not deserve a big raise. It appears that sales are steady but not increasing fast enough to warrant a big raise.
37. A: 2.5, 2.5, 0.7, 0.8; B: 2.5, 2.5, 1.1, 1.0
39. The statistic(s) that best represent a set of test scores depends on the distribution of the particular set of scores. Answers should include the following.

- mean, 73.9; median, 76.5; mode, 94
- The mode is not representative at all because it is the highest score. The median is more representative than the mean because it is influenced less than the mean by the two very low scores of 34 and 19.
- Each measure is increased by 5.

41. D **43.** 1.9 **45.** inclusive; $\frac{4}{13}$ **47.** $\frac{1}{169}$ **49.** $\frac{13}{204}$ **51.** (0, ±9); $(0, \pm\sqrt{106})$; $\pm\frac{9}{5}$ **53.** 17 **55.** 12 cm^3 **57.** (1, 5) **59.** 136 **61.** 380 **63.** 396

Page 670 Practice Quiz 2

1. $\frac{3}{20}$ **3.** $\frac{2}{9}$ **5.** $\frac{1}{6}$ **7.** $\frac{3}{4}$ **9.** 23.6, 4.9

Pages 673–675 Lesson 12-7

1. Sample answer:

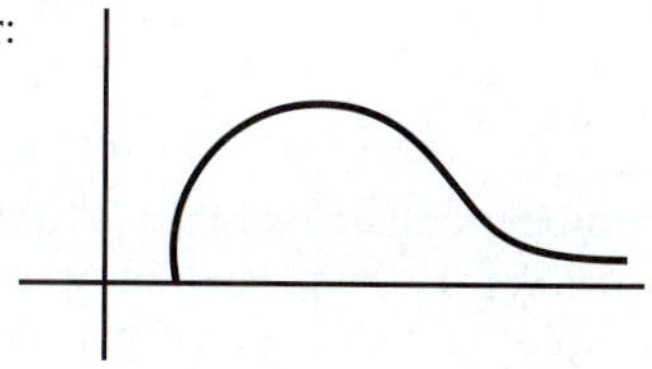

The use of cassettes since CDs were introduced.

3. Since 99% of the data is within 3 standard deviations of the mean, 1% of the data is more than 3 standard deviations from the mean. By symmetry, half of this, or 0.5%, is more than 3 standard deviations above the mean. **5.** 68% **7.** 95% **9.** 250 **11.** 81.5% **13.** normally distributed **15.** 68% **17.** 0.5% **19.** 50% **21.** 95% **23.** 815 **25.** 16% **27.** The mean would increase by 25; the standard deviation would not change; and the graph would be translated 25 units to the right. **29.** A **31.** 17.5, 4.2 **33.** $\frac{2}{13}$ **35.** $\frac{4}{13}$ **37.** $-3, 2, 4$ **39.** $\frac{1}{4}, 1$ **41.** 0.76 h **43.** $56c^5d^3$

Pages 678–680 Lesson 12-8

1. Sample answer: In a 5-card hand, what is the probability that at least 2 cards are hearts? **3a.** Each trial has more than two possible outcomes. **3b.** The number of trials is not fixed. **3c.** The trials are not independent. **5.** $\frac{1}{8}$ **7.** $\frac{1}{28{,}561}$ **9.** $\frac{27{,}648}{28{,}561}$ **11.** about 0.37 **13.** $\frac{1}{16}$ **15.** $\frac{1}{4}$ **17.** $\frac{11}{16}$ **19.** $\frac{125}{3888}$ **21.** $\frac{23}{648}$ **23.** $\frac{1}{1024}$ **25.** $\frac{135}{512}$ **27.** $\frac{53}{512}$ **29.** $\frac{105}{512}$ **31.** $\frac{319}{512}$ **33.** about 0.44 **35.** about 0.32 **37.** $\frac{7}{32}$

39. Getting a right answer and a wrong answer are the outcomes of a binomial experiment. The probability is far greater that guessing will result in a low grade than in a high grade. Answers should include the following.

- Use $(r + w)^5 = r^5 + 5r^4w + 10r^3w^2 + 10r^2w^3 + 5rw^4 + w^5$ and the chart on page 48 to determine the probabilities of each combination of right and wrong.
- $P(\text{5 right})$: $r^5 = \left(\frac{1}{4}\right)^5 = \frac{1}{1024}$ or about 0.098%; $P(\text{4 right, 1 wrong})$: $\frac{15}{1024}$ or about 1.5%; $P(\text{3 right, 2 wrong})$: $10r^3w^2 = 10\left(\frac{1}{4}\right)^3\left(\frac{3}{4}\right)^2 = \frac{45}{512}$ or about 8.8%; $P(\text{3 wrong, 2 right})$: $10r^2w^3 = 10\left(\frac{1}{4}\right)^2\left(\frac{3}{4}\right)^3 = \frac{135}{512}$ or about 26.4%; $P(\text{4 wrong, 1 right})$: $5rw^4 = 5\left(\frac{1}{4}\right)\left(\frac{3}{4}\right)^4 = \frac{405}{1024}$ or about 39.6%; $P(\text{5 wrong})$: $w^5 = \left(\frac{3}{4}\right)^5 = \frac{243}{1024}$ or about 23.7%.

41. B **43.** normal distribution **45.** 10 **47.** Mean; it is highest.

49.

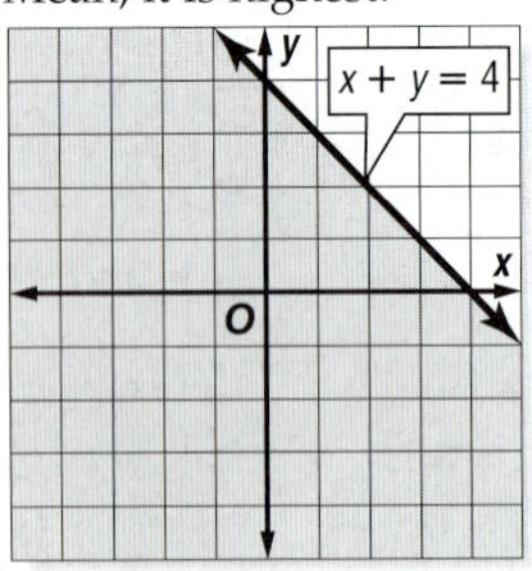

51. 0.1 **53.** 0.039 **55.** 0.041

Pages 684–685 Lesson 12-9

1. Sample answer: If a sample is not random, the results of a survey may not be valid. **3.** The margin of sampling error decreases when the size of the sample n increases. As n increases, $\frac{p(1-p)}{n}$ decreases. **5.** No; these students probably study more than average. **7.** about 4% **9.** The probability is 0.95 that the percent of Americans ages 12 and older who listen to the radio every day is between 72% and 82% **11.** No; you would tend to point toward the middle of the page. **13.** Yes; a wide variety of people would be called since almost everyone has a phone. **15.** about 8% **17.** about 4% **19.** about 3% **21.** about 4% **23.** about 3% **25.** about 2% **27.** about 983 **29.** A political candidate can use the statistics from an opinion poll to analyze his or her standing and to help plan the rest of the campaign. Answers should include the following.

- The candidate could decide to skip areas where he or she is way ahead or way behind, and concentrate on areas where the polls indicate the race is close.
- about 3.5%
- The margin of error indicates that with a probability of 0.95 the percent of the Florida population that favored Bush was between 43.5% and 50.5%. The margin of error for Gore was also about 3.5%, so with probability 0.95 the percent that favored Gore was between 40.5% and 47.5%. Therefore, it was possible that the percent of the Florida population that favored Bush was less than the percent that favored Gore.

31. C **33.** $\frac{5}{32}$ **35.** 95% **37.** 97.5%

Pages 687–692 Chapter 12 Study Guide and Review

1. c **3.** a **5.** d **7.** f **9.** 5040 codes **11.** 4 **13.** 1:3 **15.** 7:5 **17.** 2:3 **19.** independent; $\frac{1}{36}$ **21.** dependent; $\frac{1}{7}$ **23.** mutually exclusive; $\frac{2}{3}$ **25.** inclusive; $\frac{7}{13}$ **27.** 341.0, 18.5 **29.** 3400 **31.** 800 **33.** $\frac{1}{32}$ **35.** $\frac{1}{2{,}176{,}782{,}336}$ **37.** $\frac{14{,}437{,}500}{2{,}176{,}782{,}336}$ **39.** 460 mothers

Chapter 13 Trigonometric Functions

Page 699 Chapter 13 Getting Started

1. 10 **3.** 16.7 **5.** $x = 7, y = 7\sqrt{2}$ **7.** $x = 4\sqrt{3}, y = 8$

9. $f^{-1}(x) = x - 3$

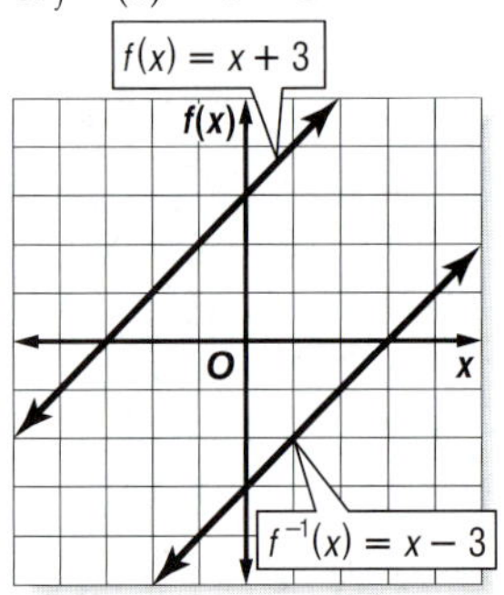

11. $f^{-1}(x) = \pm\sqrt{x + 4}$

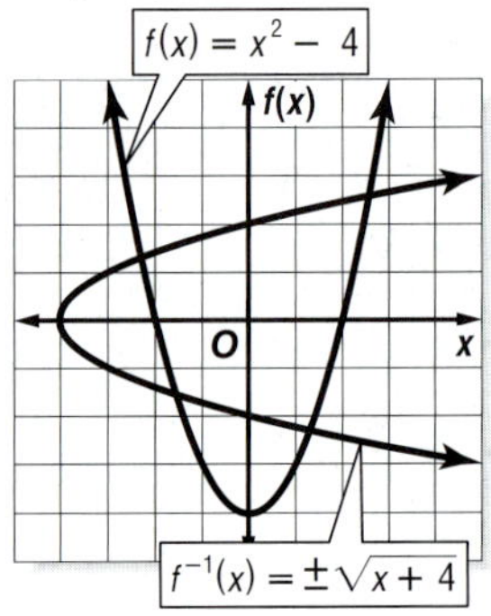

Pages 706–708 Lesson 13-1

1. Trigonometry is the study of the relationships between the angles and sides of a right triangle. **3.** Given only the measures of the angles of a right triangle, you cannot find the measures of its sides. **5.** $\sin\theta = \frac{\sqrt{85}}{11}$; $\cos\theta = \frac{6}{11}$; $\tan\theta = \frac{\sqrt{85}}{6}$; $\csc\theta = \frac{11\sqrt{85}}{85}$; $\sec\theta = \frac{11}{6}$; $\cot\theta = \frac{6\sqrt{85}}{85}$ **7.** $\cos 23° = \frac{32}{x}$; $x \approx 34.8$ **9.** $B = 45°, a = 6, c \approx 8.5$ **11.** $a \approx 16.6, A \approx 67°, B \approx 23°$ **13.** 1660 ft **15.** $\sin\theta = \frac{4}{11}$; $\cos\theta = \frac{\sqrt{105}}{11}$; $\tan\theta = \frac{4\sqrt{105}}{105}$; $\csc\theta = \frac{11}{4}$; $\sec\theta = \frac{11\sqrt{105}}{105}$; $\cot\theta = \frac{\sqrt{105}}{4}$ **17.** $\sin\theta = \frac{\sqrt{7}}{4}$; $\cos\theta = \frac{3}{4}$; $\tan\theta = \frac{\sqrt{7}}{3}$; $\csc\theta = \frac{4\sqrt{7}}{7}$; $\sec\theta = \frac{4}{3}$; $\cot\theta = \frac{3\sqrt{7}}{7}$ **19.** $\sin\theta = \frac{\sqrt{5}}{5}$;

$\cos\theta = \frac{2\sqrt{5}}{5}$; $\tan\theta = \frac{1}{2}$; $\csc\theta = \sqrt{5}$; $\sec\theta = \frac{\sqrt{5}}{2}$; $\cot\theta = 2$ **21.** $\tan 30° = \frac{x}{10}$, $x \approx 5.8$ **23.** $\sin 54° = \frac{17.8}{x}$, $x \approx 22.0$ **25.** $\cos x° = \frac{15}{36}$, $x \approx 65$

27a. $\sin 30° = \frac{\text{opp}}{\text{hyp}}$ *sine ratio*

$\sin 30° = \frac{x}{2x}$ *Replace opp with x and hyp with 2x.*

$\sin 30° = \frac{1}{2}$ *Simplify.*

27b. $\cos 30° = \frac{\text{adj}}{\text{hyp}}$ *cosine ratio*

$\cos 30° = \frac{\sqrt{3}x}{2x}$ *Replace adj with $\sqrt{3}x$ and hyp with 2x.*

$\cos 30° = \frac{\sqrt{3}}{2}$ *Simplify.*

27c. $\sin 60° = \frac{\text{opp}}{\text{hyp}}$ *sine ratio*

$\sin 60° = \frac{\sqrt{3}x}{2x}$ *Replace opp with $\sqrt{3}x$ and hyp with 2x.*

$\sin 60° = \frac{\sqrt{3}}{2}$ *Simplify.*

29. $B = 74°, a \approx 3.9, b \approx 13.5$ **31.** $B = 56°, b \approx 14.8, c \approx 17.9$ **33.** $A = 60°, a \approx 19.1, c = 22$ **35.** $A = 72°, b \approx 1.3, c \approx 4.1$ **37.** $A \approx 63°, B \approx 27°, a \approx 11.5$ **39.** $A \approx 49°, B \approx 41°, a = 8, c \approx 10.6$ **41.** about 300 ft **43.** about 6° **45.** 93.53 units^2 **47.** The sine and cosine ratios of acute angles of right triangles each have the longest measure of the triangle, the hypotenuse, as their denominator. A fraction whose denominator is greater than its numerator is less than 1. The tangent ratio of an acute angle of a right triangle does not involve the measure of the hypotenuse, $\frac{\text{opp}}{\text{adj}}$. If the measure of the opposite side is greater than the measure of the adjacent side, the tangent ratio is greater than 1. If the measure of the opposite side is less than the measure of the adjacent side, the tangent ratio is less than 1. **49.** C **51.** No; band members may be more likely to like the same kinds of music. **53.** $\frac{3}{8}$ **55.** $\frac{15}{16}$ **57.** $\{-2, -1, 0, 1, 2\}$ **59.** 20 qt **61.** 12 m^2

Pages 712–715 Lesson 13-2

1. reals

3.

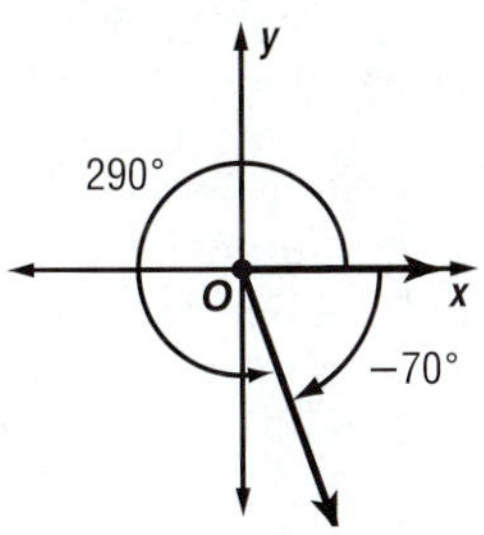

5.

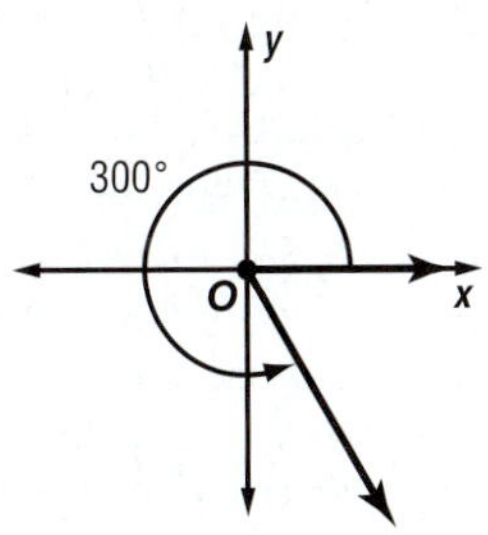

7.

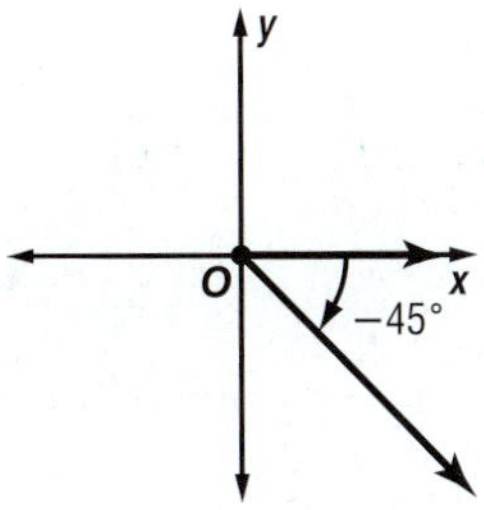

9. $-\frac{\pi}{18}$ **11.** 135° **13.** 1140° **15.** 785°, −295° **17.** 21 h

19.

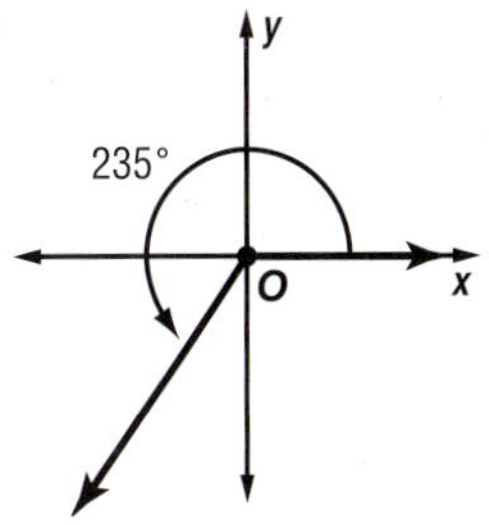

21.

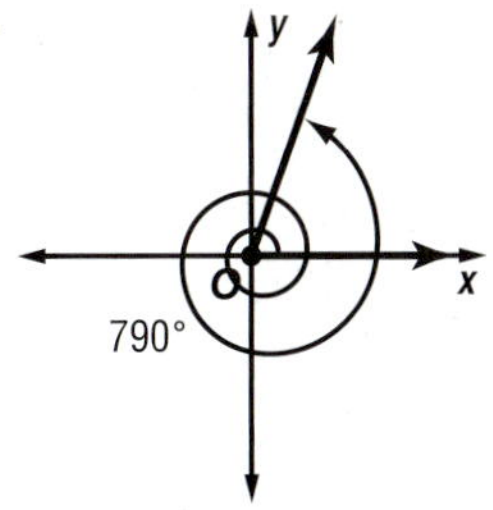

23.

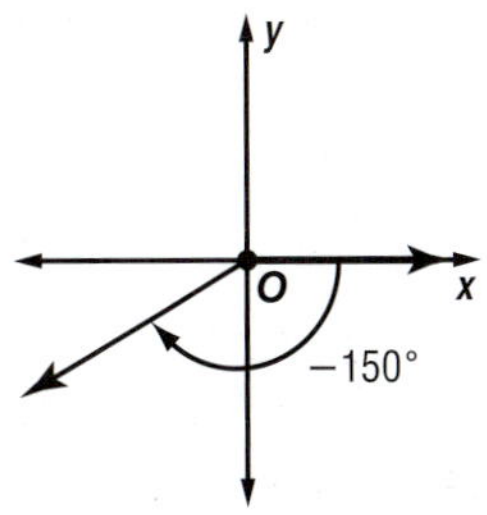

25.

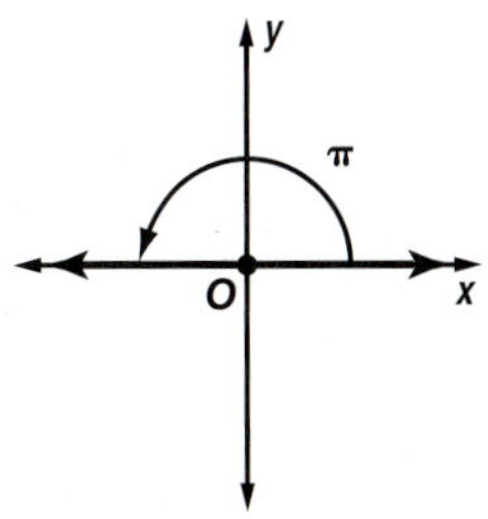

27. $\frac{2\pi}{3}$ **29.** $-\frac{\pi}{12}$ **31.** $\frac{11\pi}{3}$ **33.** $\frac{79\pi}{90}$ **35.** 150° **37.** −45° **39.** 1305° **41.** $\frac{1620}{\pi} \approx 515.7°$ **43.** Sample answer: 585°, −135° **45.** Sample answer: 345°, −375° **47.** Sample answer: 8°, −352° **49.** Sample answer: $\frac{11\pi}{4}, -\frac{5\pi}{4}$ **51.** Sample answer: $\frac{3\pi}{4}, -\frac{13\pi}{4}$ **53.** Sample answer: $\frac{13\pi}{2}, -\frac{3\pi}{2}$ **55.** 2689° per second; 47 radians per second **57.** about 188.5 m^2 **59.** about 640.88 in^2 **61.** Student answers should include the following.

- An angle with a measure of more than 180° gives an indication of motion in a circular path that ended at a point more than halfway around the circle from where it started.
- Negative angles convey the same meaning as positive angles, but in an opposite direction. The standard convention is that negative angles represent rotations in a clockwise direction.
- Rates over 360° per minute indicate that an object is rotating or revolving more than one revolution per minute.

63. D **65.** $A = 22°, a \approx 5.9, c \approx 15.9$ **67.** $c = 0.8, A = 30°, B = 60°$ **69.** about 7.07% **71.** combination, 35 **73.** $[g \circ h](x) = 4x^2 - 6x + 23$, $[h \circ g](x) = 8x^2 + 34x + 44$ **75.** 1418.2 or about 1418; the number of sports radio stations in 2008 **77.** $\frac{3\sqrt{5}}{5}$ **79.** $\frac{\sqrt{10}}{2}$ **81.** $\frac{\sqrt{10}}{4}$

Page 715 Practice Quiz 1

1. $B = 42°, a \approx 13.3, c \approx 17.9$

3.

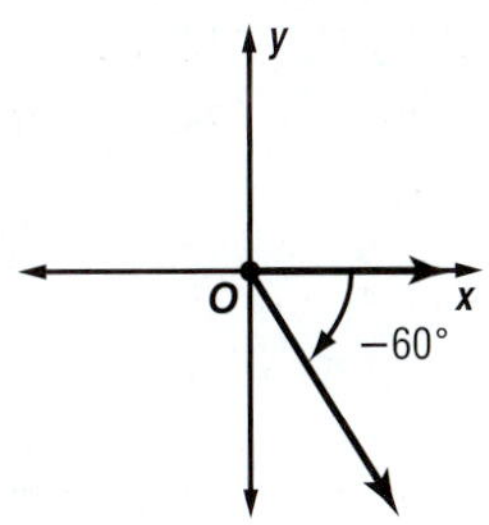

5. $\frac{19\pi}{18}$ **7.** 210° **9.** 305°; −415°

Pages 722–724 Lesson 13-3

1. False; $\sec 0° = \frac{r}{r}$ or 1 and $\tan 0° = \frac{0}{r}$ or 0.

3. To find the value of a trigonometric function of θ, where θ is greater than 90°, find the value of the trigonometric function for θ', then use the quadrant in which the terminal

Selected Answers

side of θ lies to determine the sign of the trigonometric function value of θ. **5.** $\sin\theta = 0$, $\cos\theta = -1$, $\tan\theta = 0$, $\csc\theta =$ undefined, $\sec\theta = -1$, $\cot\theta =$ undefined

7. 55° **9.** 60º

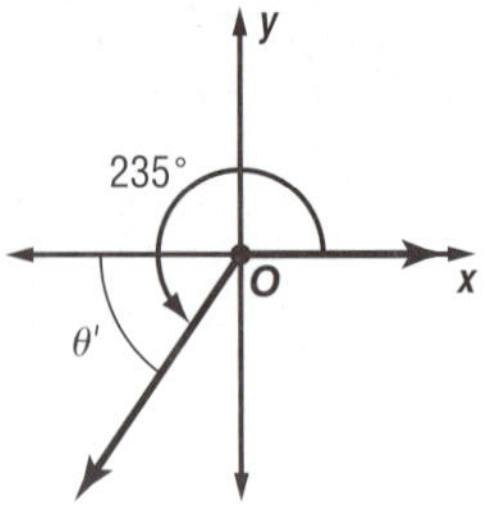

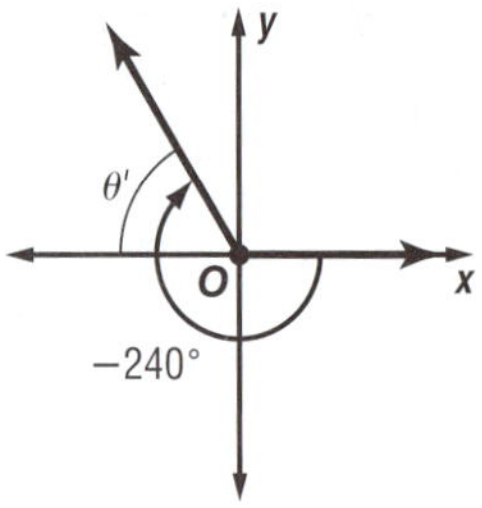

11. −1 **13.** $-\frac{2\sqrt{3}}{3}$ **15.** $\sin\theta = -\frac{\sqrt{6}}{3}$, $\cos\theta = \frac{\sqrt{3}}{3}$, $\tan\theta = -\sqrt{2}$, $\csc\theta = -\frac{\sqrt{6}}{2}$, $\sec\theta = \sqrt{3}$ **17.** $\sin\theta = \frac{24}{25}$, $\cos\theta = \frac{7}{25}$, $\tan\theta = \frac{24}{7}$, $\csc\theta = \frac{25}{24}$, $\sec\theta = \frac{25}{7}$, $\cot\theta = \frac{7}{24}$ **19.** $\sin\theta = -\frac{8\sqrt{89}}{89}$, $\cos\theta = \frac{5\sqrt{89}}{89}$, $\tan\theta = -\frac{8}{5}$, $\csc\theta = -\frac{\sqrt{89}}{8}$, $\sec\theta = \frac{\sqrt{89}}{5}$, $\cot\theta = -\frac{5}{8}$ **21.** $\sin\theta = -1$, $\cos\theta = 0$, $\tan\theta =$ undefined, $\csc\theta = -1$, $\sec\theta =$ undefined, $\cot\theta = 0$ **23.** $\sin\theta = -\frac{\sqrt{2}}{2}$, $\cos\theta = \frac{\sqrt{2}}{2}$, $\tan\theta = -1$, $\csc\theta = -\sqrt{2}$, $\sec\theta = \sqrt{2}$, $\cot\theta = -1$

25. 45° **27.** 30°

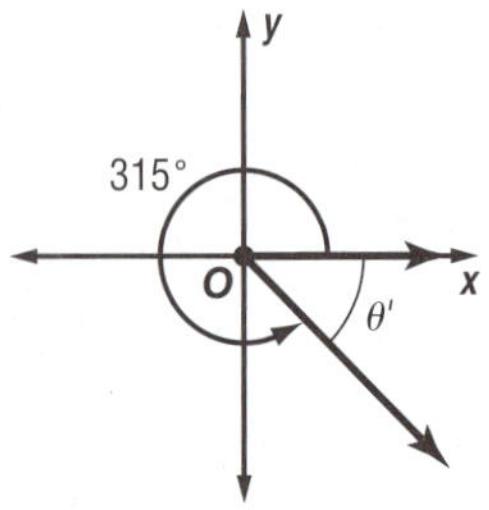

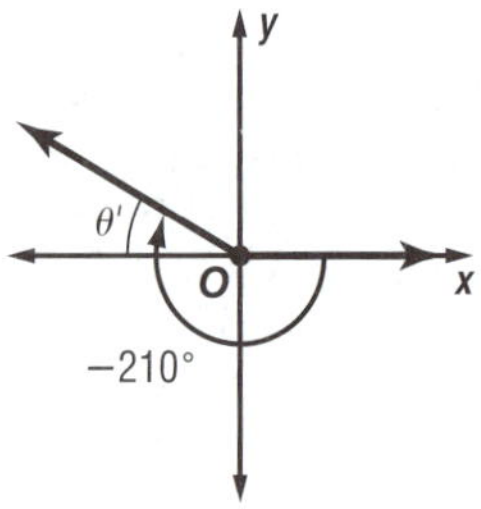

29. $\frac{\pi}{4}$ **31.** $\frac{\pi}{7}$

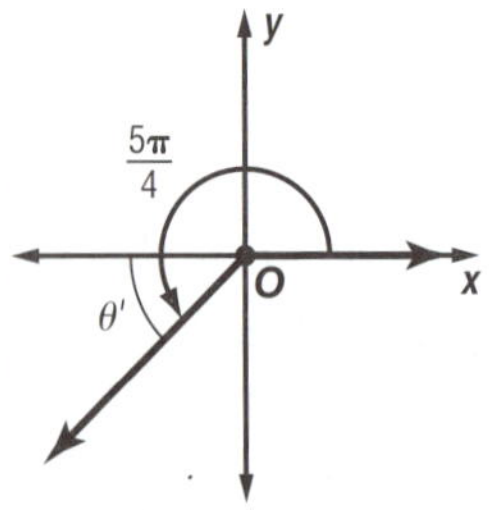

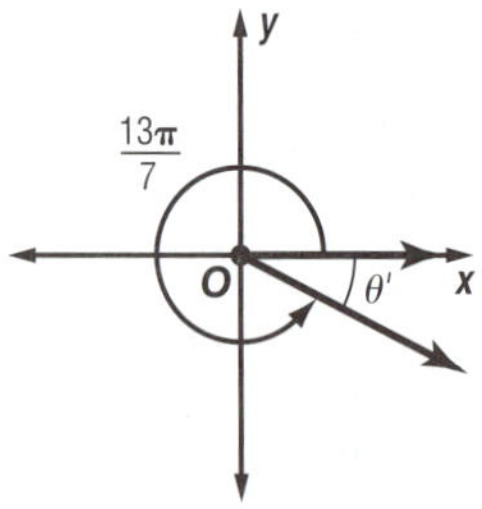

33. $-\frac{\sqrt{3}}{2}$ **35.** $-\sqrt{3}$ **37.** undefined **39.** $\sqrt{3}$ **41.** undefined **43.** $\frac{\sqrt{3}}{2}$ **45.** 0.2, 0, −0.2, 0, 0.2, 0, and −0.2; or about 11.5°, 0°, −11.5°, 0°, 11.5°, 0°, and −11.5° **47.** $\sin\theta = -\frac{4}{5}$, $\tan\theta = -\frac{4}{3}$, $\csc\theta = -\frac{5}{4}$, $\sec\theta = \frac{5}{3}$, $\cot\theta = -\frac{3}{4}$ **49.** $\cos\theta = -\frac{2\sqrt{2}}{3}$, $\tan\theta = -\frac{\sqrt{2}}{4}$, $\csc\theta = 3$, $\sec\theta = -\frac{3\sqrt{2}}{4}$, $\cot\theta = -2\sqrt{2}$ **51.** $\sin\theta = -\frac{3\sqrt{10}}{10}$, $\cos\theta = -\frac{\sqrt{10}}{10}$, $\tan\theta = 3$, $\csc\theta = -\frac{\sqrt{10}}{3}$, $\cot\theta = \frac{1}{3}$

53. about 173.2 ft **55.** 9 meters **57.** II

59. Answers should include the following.

- The cosine of any angle is defined as $\frac{x}{r}$, where x is the x-coordinate of any point on the terminal ray of the angle and r is the distance from the origin to that point. This means that for angles with terminal sides to the left of the y-axis, the cosine is negative, and those with terminal sides to the right of the y-axis, the cosine is positive. Therefore the cosine function can be used to model real-world data that oscillate between being positive and negative.
- If we knew the length of the cable we could find the vertical distance from the top of the tower to the rider. Then if we knew the height of the tower we could subtract from it the vertical distance calculated previously. This will leave the height of the rider from the ground.

61. $\left(\frac{5}{2}, -\frac{5\sqrt{3}}{2}\right)$ **63.** 300° **65.** $\sin 28° = \frac{x}{12}$, 5.6 **67.** $\sin x° = \frac{5}{13}$, 23 **69.** (7, 2) **71.** (5, −4) **73.** 15.1 **75.** 32.9° **77.** 39.6°

Pages 729–732 Lesson 13-4

1. Sometimes; only when A is acute, $a = b\sin A$ or $a > b$ and when A is obtuse, $a > b$.

3. Gabe;

$$\frac{\sin 64°}{15} = \frac{\sin B}{8}$$
$$\sin B = \frac{8\sin 64°}{15}$$
$$B \approx 28.6°$$

$$m\angle C \approx 180° - (64° + 28.6°) \approx 87.4°$$
$$\text{Area} = \frac{1}{2}ab\sin C \approx \frac{1}{2}(15)(8)\sin 87.4° \approx 59.9 \text{ m}^2$$

5. 6.4 cm^2 **7.** $B = 80°$, $a \approx 32.0$, $b \approx 32.6$ **9.** no solution **11.** one; $B \approx 24°$, $C \approx 101°$, $c \approx 12.0$ **13.** 5.5 m **15.** 19.5 yd^2 **17.** 62.4 cm^2 **19.** 14.6 mi^2 **21.** $C = 73°$, $a \approx 55.6$, $b \approx 48.2$ **23.** $B \approx 46°$, $C \approx 69°$, $c \approx 5.1$ **25.** $A \approx 40°$, $B \approx 65°$, $b \approx 2.8$ **27.** $A = 20°$, $a \approx 22.1$, $c \approx 39.8$ **29.** one; $B \approx 36°$, $C \approx 45°$, $c \approx 1.8$ **31.** no **33.** one; $B \approx 18°$, $C \approx 101°$, $c \approx 25.8$ **35.** two; $B \approx 85°$, $C \approx 15°$, $c \approx 2.4$; $B \approx 95°$, $C \approx 5°$, $c \approx 0.8$ **37.** two; $B \approx 65°$, $C \approx 68°$, $c \approx 84.9$; $B \approx 115°$, $C \approx 18°$, $c \approx 28.3$ **39.** 7.5 mi from Ranger B, 10.9 mi from Ranger A **41.** 107 mph **43.** Answers should include the following.

- If the height of the triangle is not given, but the measure of two sides and their included angle are given, then the formula for the area of a triangle using the sine function should be used.
- You might use this formula to find the area of a triangular piece of land, since it might be easier to measure two sides and use surveying equipment to measure the included angle than to measure the perpendicular distance from one vertex to its opposite side.
- The area of $\triangle ABC$ is $\frac{1}{2}ah$.

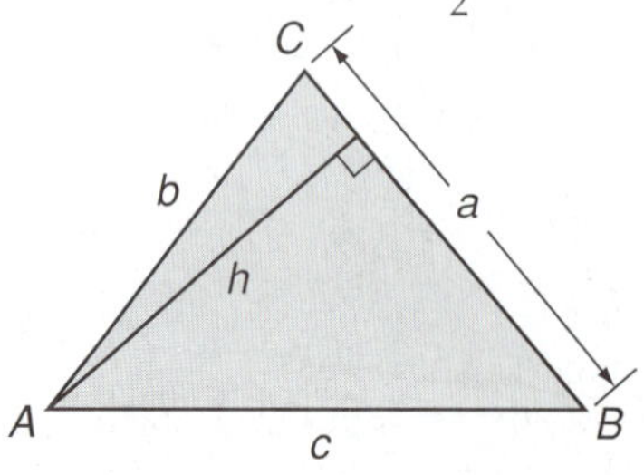

$\sin B = \frac{h}{c}$ or $h = c\sin B$

• Area $= \frac{1}{2}ah$ or Area $= \frac{1}{2}a(c \sin B)$
45. $B = 78°, a \approx 50.1, c \approx 56.1$ **47.** $\frac{\sqrt{3}}{3}$ **49.** 660°, −60°
51. $\frac{17\pi}{6}, -\frac{7\pi}{6}$ **53.** $\frac{55}{221}$ **55.** 5.6 **57.** 39.4°

Pages 735–738 Lesson 13-5

1. Mateo; the angle given is not between the two sides; therefore the Law of Sines should be used.
3. Sample answer:

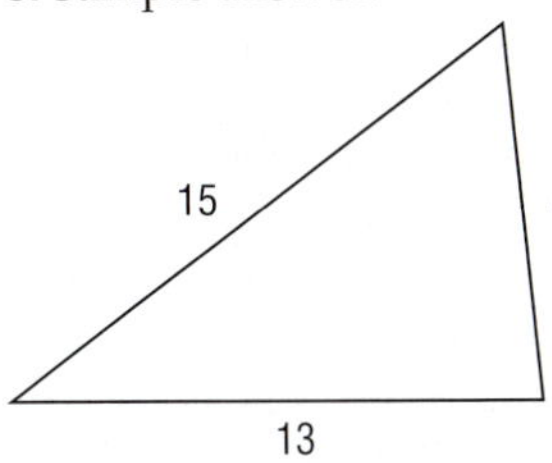

5. sines; $B = 70°, a \approx 9.6, b = 14$ **7.** cosines; $A \approx 23°$, $B \approx 67°, C = 90°$ **9.** 94.3° **11.** cosines; $A \approx 48°, B \approx 63°$, $C \approx 70°$ **13.** sines; $B \approx 102°, C \approx 44°, b \approx 21.0$
15. $A = 80°, a \approx 10.9, c \approx 5.4$ **17.** cosines; $A \approx 30°$, $B \approx 110°, C \approx 40°$ **19.** sines; $C \approx 77°, b \approx 31.7, c \approx 31.6$
21. no **23.** cosines; $A \approx 52°, C \approx 109°, b \approx 21.0$
25. cosines; $A \approx 24°, B \approx 125°, C \approx 31°$ **27.** sines; $B \approx 49°, C \approx 91°, c \approx 9.3$ **29.** about 100.1° **31.** 4.4 cm, 9.0 cm **33.** 91.6°
35. Answers should include the following.
• The Law of Cosines can be used when you know all three sides of a triangle or when you know two sides and the included angle. It can even be used with two sides and the nonincluded angle. This set of conditions leaves a quadratic equation to be solved. It may have one, two, or no solution just like the SSA case with the Law of Sines.
• Given the latitude of a point on the surface of Earth, you can use the radius of the Earth and the orbiting height of a satellite in geosynchronous orbit to create a triangle. This triangle will have two known sides and the measure of the included angle. Find the third side using the Law of Cosines and then use the Law of Sines to determine the angles of the triangle. Subtract 90 degrees from the angle with its vertex on Earth's surface to find the angle at which to aim the receiver dish.
37. A **39.** Sample answer: 100.2° **41.** one; $B \approx 46°$, $C \approx 79°, c \approx 9.6$ **43.** $\sin\theta = \frac{12}{13}, \cos\theta = \frac{5}{13}, \tan\theta = \frac{12}{5}$, $\csc\theta = \frac{13}{12}, \sec\theta = \frac{13}{5}, \cot\theta = \frac{5}{12}$ **45.** $\sin\theta = \frac{\sqrt{6}}{4}$, $\cos\theta = \frac{\sqrt{10}}{4}, \tan\theta = \frac{\sqrt{15}}{5}, \csc\theta = \frac{2\sqrt{6}}{3}, \sec\theta = \frac{2\sqrt{10}}{5}$, $\cot\theta = \frac{\sqrt{15}}{3}$ **47.** $\{x \mid x > -0.6931\}$ **49.** 405°, −315°
51. 540°, −180° **53.** $\frac{19\pi}{6}, -\frac{5\pi}{6}$

Page 738 Practice Quiz 2

1. $\sin\theta = \frac{3\sqrt{13}}{13}; \cos\theta = -\frac{2\sqrt{13}}{13}; \tan\theta = -\frac{3}{2}$; $\csc\theta = \frac{\sqrt{13}}{3}; \sec\theta = -\frac{\sqrt{13}}{2}; \cot\theta = -\frac{2}{3}$ **3.** 27.7 m²

5. cosines; $c \approx 15.9, C \approx 59°, B \approx 43°$

Pages 742–745 Lesson 13-6

1. The terminal side of the angle θ in standard position must intersect the unit circle at $P(x, y)$. **3.** Sample answer: The graphs have the same shape, but cross the x-axis at different points. **5.** $\sin\theta = \frac{\sqrt{2}}{2}; \cos\theta = \frac{\sqrt{2}}{2}$ **7.** $-\frac{1}{2}$
9. 2 s **11.** $\sin\theta = \frac{4}{5}; \cos\theta = -\frac{3}{5}$ **13.** $\sin\theta = \frac{15}{17}; \cos\theta = \frac{8}{17}$
15. $\sin\theta = \frac{\sqrt{3}}{2}; \cos\theta = -\frac{1}{2}$ **17.** $-\frac{1}{2}$ **19.** −1 **21.** 1
23. $\frac{1}{4}$ **25.** $\frac{1-\sqrt{3}}{2}$ **27.** $-3\sqrt{3}$ **29.** 6 **31.** 2π **33.** $\frac{1}{440}$ s
35. $\left(\frac{1}{2}, \frac{\sqrt{3}}{2}\right), \left(-\frac{1}{2}, \frac{\sqrt{3}}{2}\right), (-1, 0), \left(-\frac{1}{2}, -\frac{\sqrt{3}}{2}\right), \left(\frac{1}{2}, -\frac{\sqrt{3}}{2}\right)$
37. $\frac{y}{x}$ **39.** $-\frac{x}{y}$ **41.** $\sqrt{3}$ **43.** sine: D = {all reals}, R = $\{-1 \le y \le 1\}$; cosine: D = {all reals}, R = $\{-1 \le y \le 1\}$ **45.** A
47. cosines; $c \approx 12.4, B \approx 59°, A \approx 76°$ **49.** 27.0 in²
51. 6800 **53.** 5000 **55.** 250 **57.** does not exist **59.** 8
61. $2x + 9$ **63.** $2y + 7 + \frac{5}{y-3}$ **65.** 110° **67.** 80° **69.** 89°

Pages 749–751 Lesson 13-7

1. Restricted domains are denoted with a capital letter.
3. They are inverses of each other. **5.** α = Arccos 0.5 **7.** 0°
9. $\pi \approx 3.14$ **11.** 0.75 **13.** 0.58 **15.** β = Arcsin α
17. y = Arccos x **19.** Arccos y = 45° **21.** 60° **23.** 45°
25. 45° **27.** 2.09 **29.** 0.52 **31.** 0.5 **33.** 0.60 **35.** 0.8
37. 0.5 **39.** −0.5 **41.** 0.71 **43.** 0.96 **45.** 60° south of west
47. No; with this point on the terminal side of the throwing angle θ, the measure of θ is found by solving the equation $\tan\theta = \frac{17}{18}$. Thus $\theta = \tan^{-1}\frac{17}{18}$ or about 43.4°, which is greater than the 40° requirement. **49.** 31° **51.** Suppose $P(x_1, y_1)$ and $Q(x_2, y_2)$ lie on the line $y = mx + b$. Then $m = \frac{y_2 - y_1}{x_2 - x_1}$. The tangent of the angle θ the line makes with the positive x-axis is equal to the ratio $\frac{\text{opp}}{\text{adj}}$ or $\frac{y_2 - y_1}{x_2 - x_1}$. Thus $\tan\theta = m$.

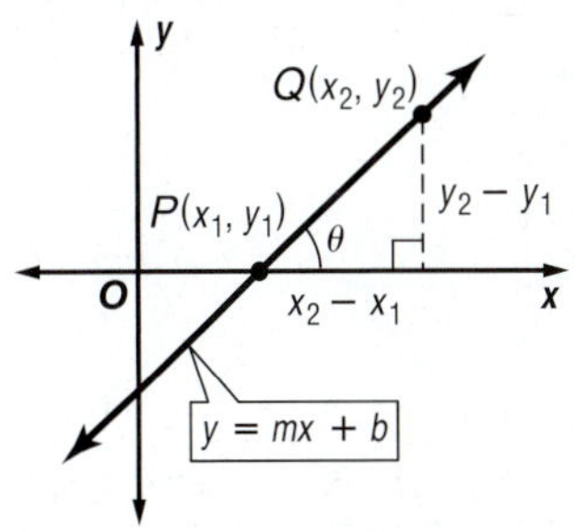

53. 37°
55.

x	0	$\frac{1}{2}$	$\frac{\sqrt{2}}{2}$	$\frac{\sqrt{3}}{2}$	1	$-\frac{1}{2}$	$-\frac{\sqrt{2}}{2}$	$-\frac{\sqrt{3}}{2}$	−1
y	$\frac{\pi}{2}$	$\frac{\pi}{2}$	$\frac{\pi}{2}$	$\frac{\pi}{2}$	$\frac{\pi}{2}$	$\frac{\pi}{2}$	$\frac{\pi}{2}$	$\frac{\pi}{2}$	$\frac{\pi}{2}$

57. From a right triangle perspective, if an acute angle θ has a given sine, say x, then the complementary angle $\frac{\pi}{2} - \theta$ has that same value as its cosine. This can be verified by looking at a right triangle. Therefore, the sum of the angle whose sine is x and the angle whose cosine is x should be $\frac{\pi}{2}$.
59. −1 **61.** sines; $B \approx 69°, C \approx 81°, c \approx 6.1$ or $B \approx 111°$, $C \approx 39°, c \approx 3.9$ **63.** 46, 39 **65.** 11, 109

Pages 752–756 Chapter 13 Study Guide and Review

1. false, coterminal **3.** true **5.** true **7.** false, an angle that has its terminal side on an axis where x or y is equal to zero **9.** false, terminal **11.** $B = 65°, a \approx 2.5, b \approx 5.4$ **13.** $A = 7°, a \approx 0.7, c \approx 5.6$ **15.** $A \approx 76°, B \approx 14°, b \approx 1.0, c \approx 4.1$ **17.** $-\frac{7\pi}{6}$ **19.** $-720°$ **21.** $320°, -400°$ **23.** $\frac{\pi}{4}; -\frac{15\pi}{4}$ **25.** $\sin\theta = -\frac{8}{17}, \cos\theta = \frac{15}{17}, \tan\theta = -\frac{8}{15}, \csc\theta = -\frac{17}{8}, \sec\theta = \frac{17}{15}, \cot\theta = -\frac{15}{8}$ **27.** $-\sqrt{3}$ **29.** $\frac{2\sqrt{3}}{3}$ **31.** two; $B \approx 53°, C \approx 87°, c \approx 12.4$; $B \approx 127°, C \approx 13°, c \approx 3.0$ **33.** no **35.** one; $A = 51°, a = 70.2, c = 89.7$ **37.** sines; $C = 105°, a \approx 28.3, c \approx 38.6$ **39.** cosines; $A \approx 34°, B \approx 81°, c \approx 6.4$ **41.** cosines; $B \approx 26°, C \approx 125°, a \approx 8.3$ **43.** $\frac{1}{2}$ **45.** $-\frac{\sqrt{2}}{2}$ **47.** $-\sqrt{3}$ **49.** 1.05 **51.** 0

Chapter 14 Trigonometric Graphs and Identities

Page 761 Chapter 14 Getting Started

1. $\frac{\sqrt{2}}{2}$ **3.** 0 **5.** $-\frac{\sqrt{2}}{2}$ **7.** $-\frac{1}{2}$ **9.** $-\frac{\sqrt{3}}{2}$ **11.** 1 **13.** not defined **15.** $\frac{1}{2}$ **17.** $-5x(3x + 1)$ **19.** prime **21.** $(2x + 1)(x - 2)$ **23.** 8, −3 **25.** −8, 5 **27.** $-4, -\frac{3}{2}$

Pages 766–768 Lesson 14-1

1. Sample answer: Amplitude is half the difference between the maximum and minimum values of a graph; $y = \tan\theta$ has no maximum or minimum value. **3.** Jamile; The amplitude is 3 and the period is 3π. **5.** amplitude: 2; period: 360° or 2π

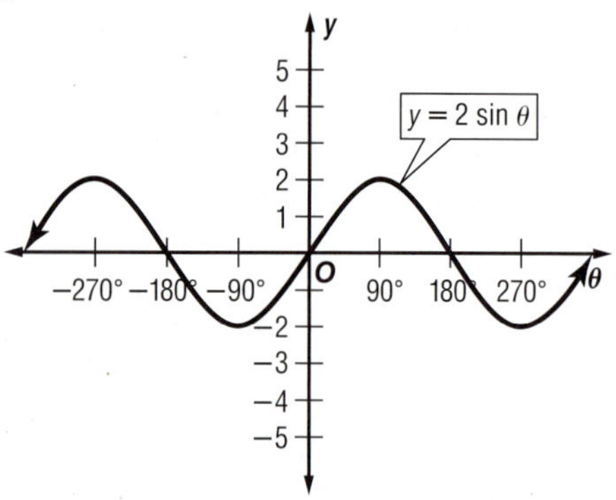

7. amplitude : does not exist; period: 180° or π

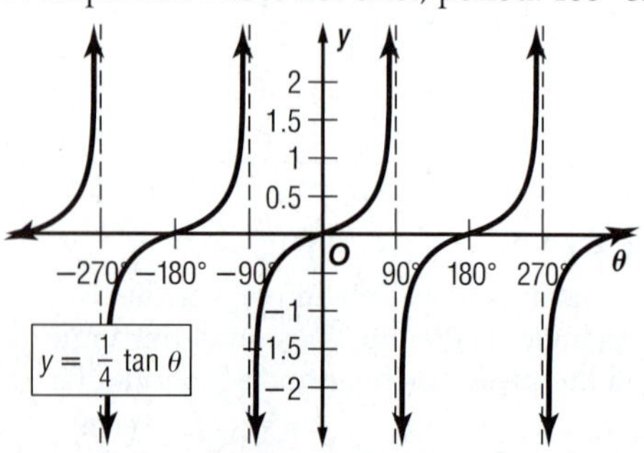

9. amplitude: 4; period: 180° or π

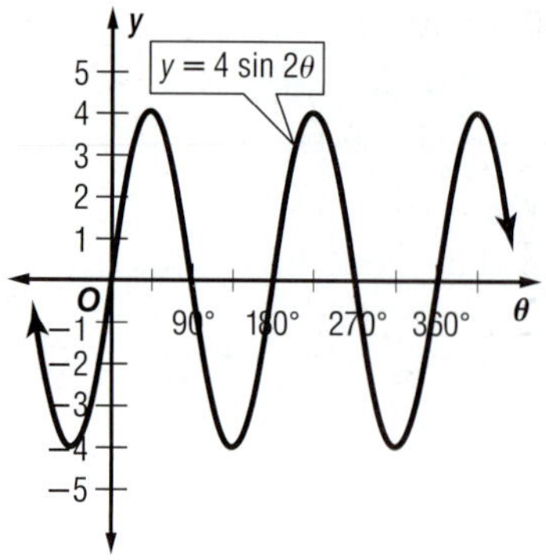

11. amplitude: does not exist; period: 120° or $\frac{2\pi}{3}$

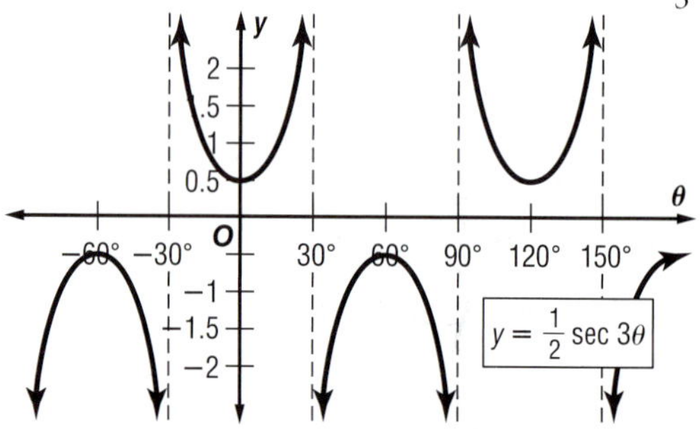

13. 12 months; Sample answer: The pattern in the population will repeat itself every 12 months.

15. amplitude: 3; period: 360° or 2π

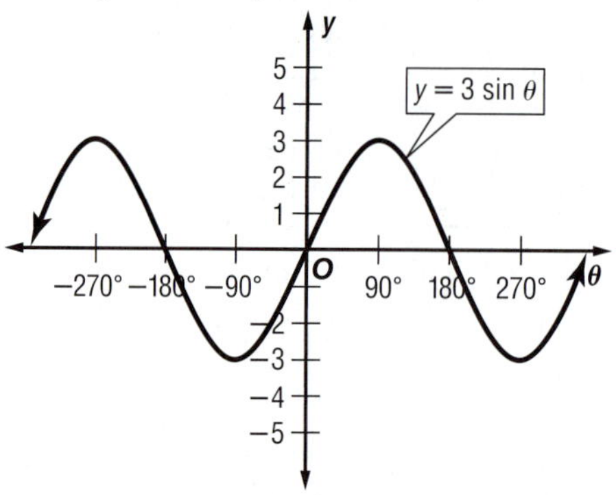

17. amplitude: does not exist; period: 360° or 2π

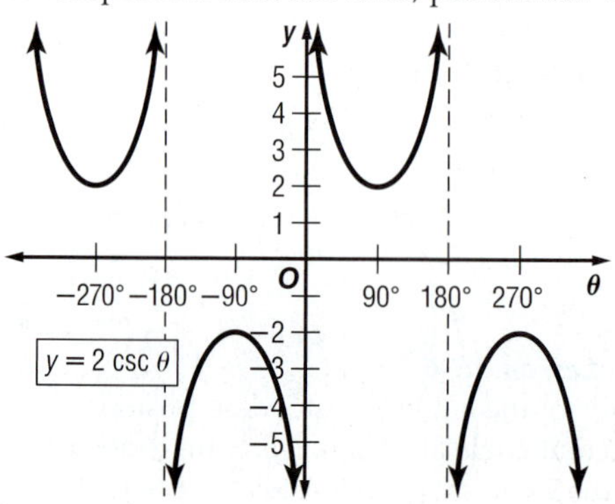

19. amplitude: $\frac{1}{5}$; period: 360° or 2π

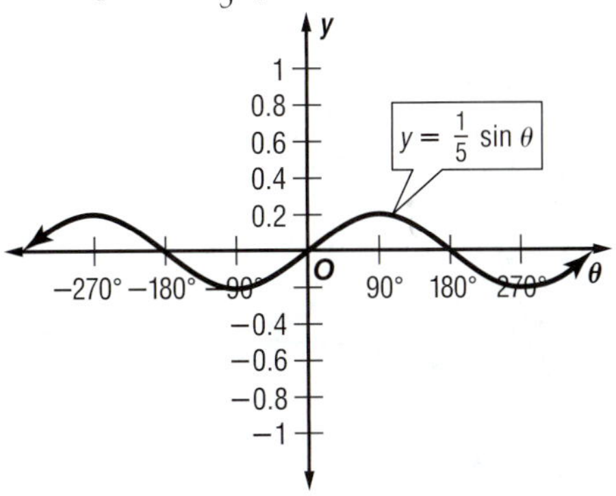

21. amplitude: 1; period 90° or $\frac{\pi}{2}$

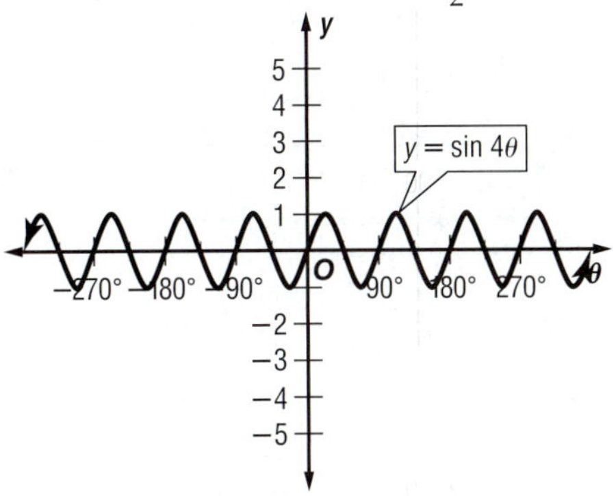

23. amplitude: does not exist; period: 120° or $\frac{2\pi}{3}$

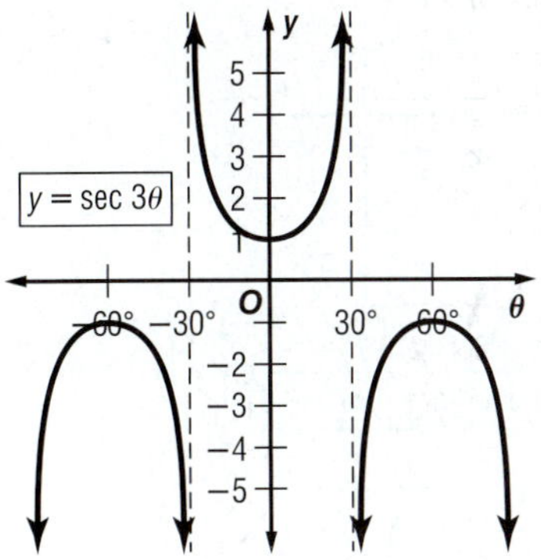

25. amplitude: does not exist; period: 540° or 3π

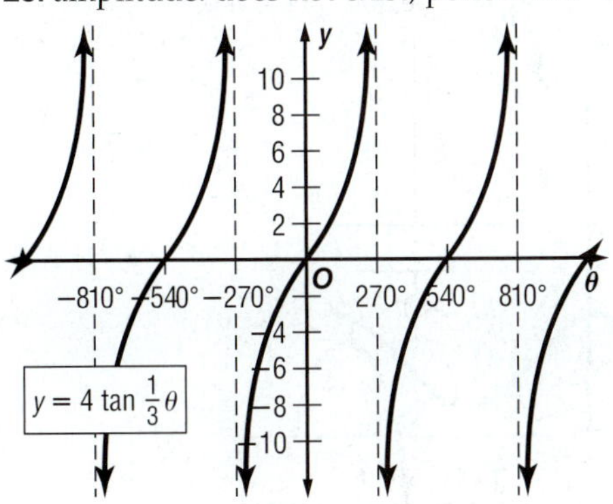

27. amplitude: 6; period: 540° or 3π

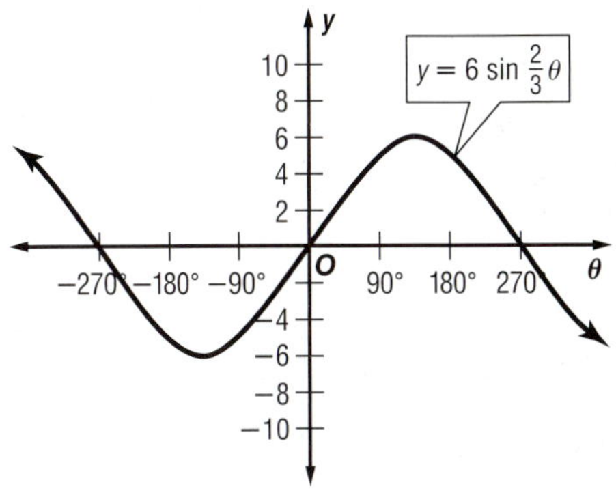

29. amplitude: does not exist; period: 720° or 4π

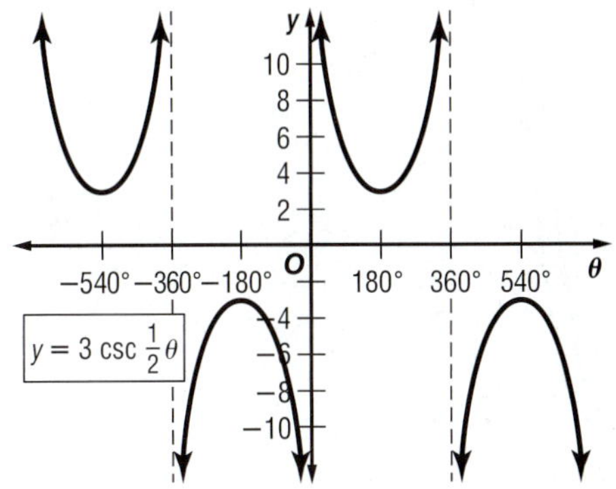

31. amplitude: does not exist; period: 180° or π

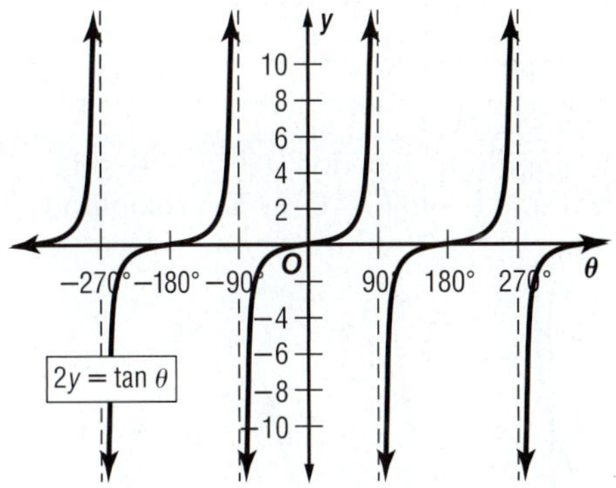

33.

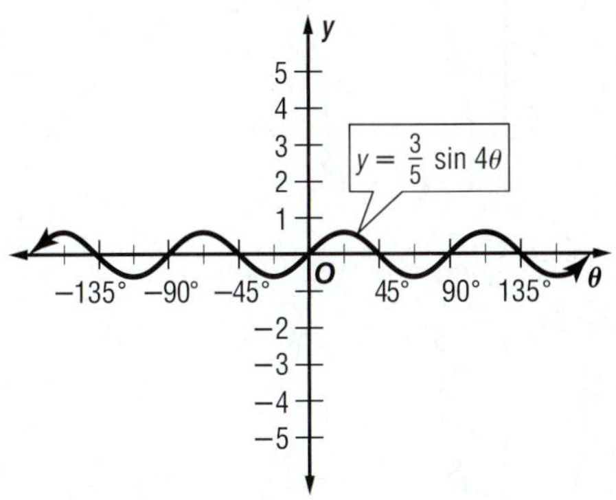

$y = \frac{3}{5}\sin 4\theta$ **35.** $\frac{1}{10^7}$ **37.** Sample answer: The amplitudes are the same. As the frequency increases, the period decreases.

39. $y = 2\sin\frac{\pi}{5}t$

41. about 1.9 ft **43.** A **45.** 90° **47.** 45° **49.** $\frac{\sqrt{2}}{2}$ **51.** $\frac{13}{16}$

53.

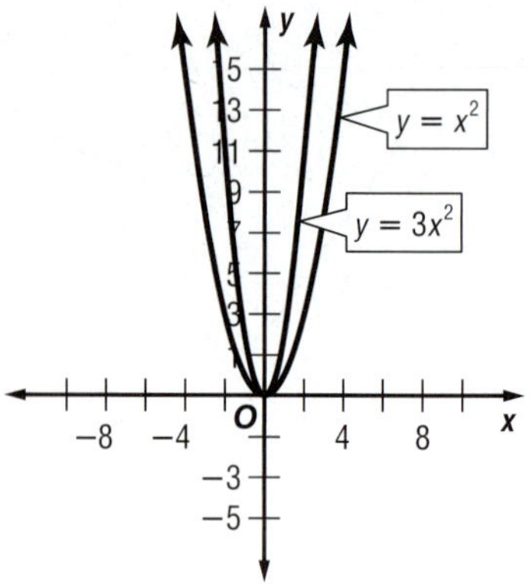

55.

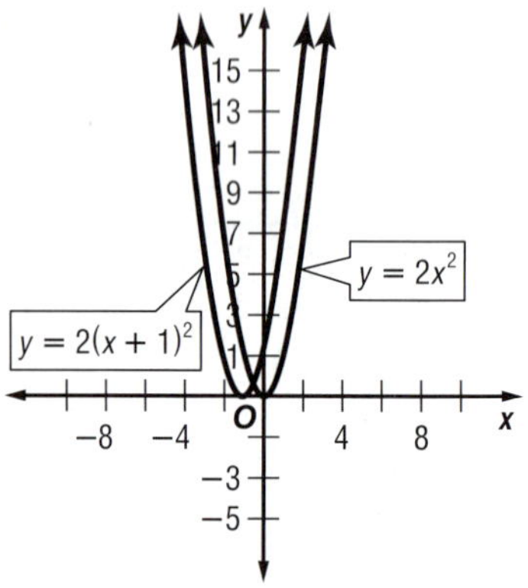

Pages 774–776 Lesson 14-2

1. vertical shift: 15; amplitude: 3; period: 180°; phase shift: 45° **3.** Sample answer: $y = \sin(\theta + 45°)$ **5.** no amplitude; 180°; −60°

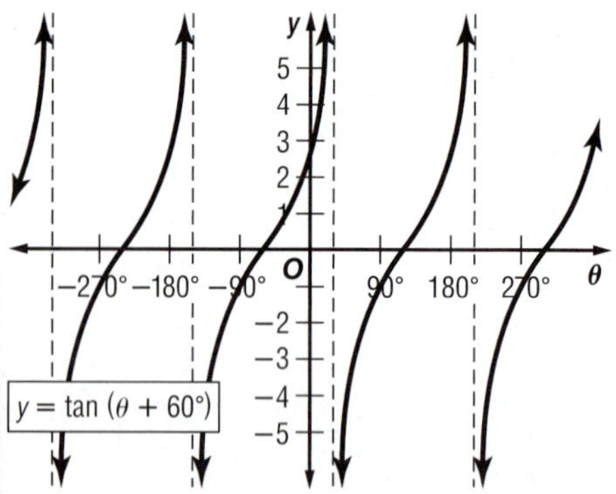

7. no amplitude; 2π; $-\frac{\pi}{3}$

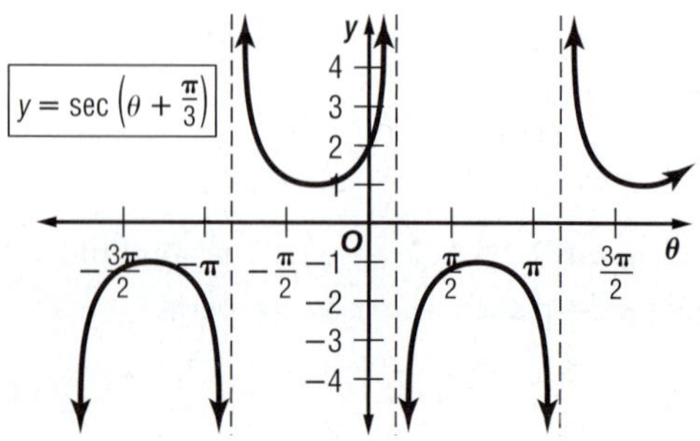

9. −5; $y = -5$; no amplitude; 360°

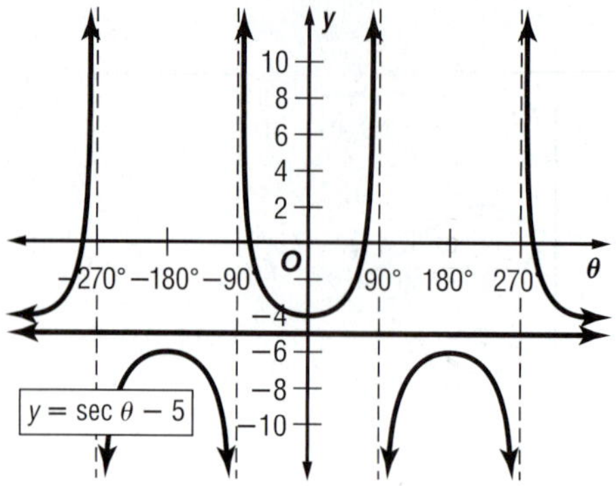

11. 0.25; $y = 0.25$; 1; 360°

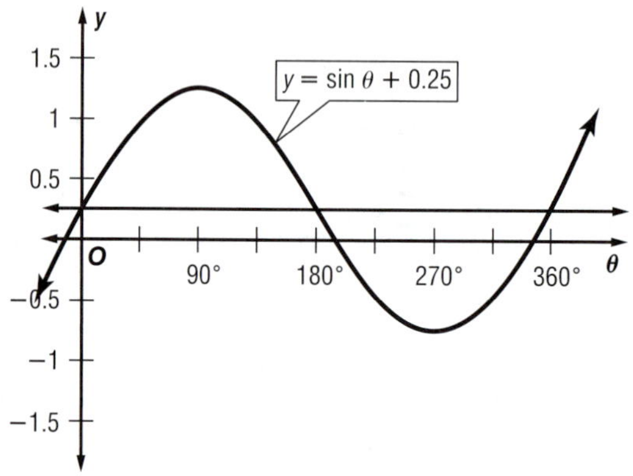

13. −6; no amplitude; 60°; −45°

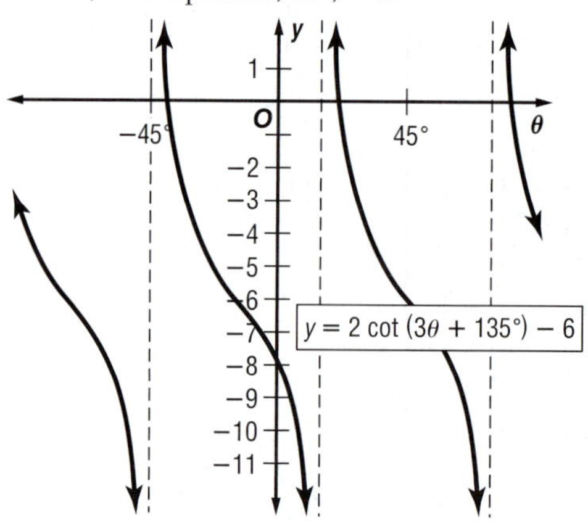

15. −2; $\frac{2}{3}$; 4π; $-\frac{\pi}{6}$

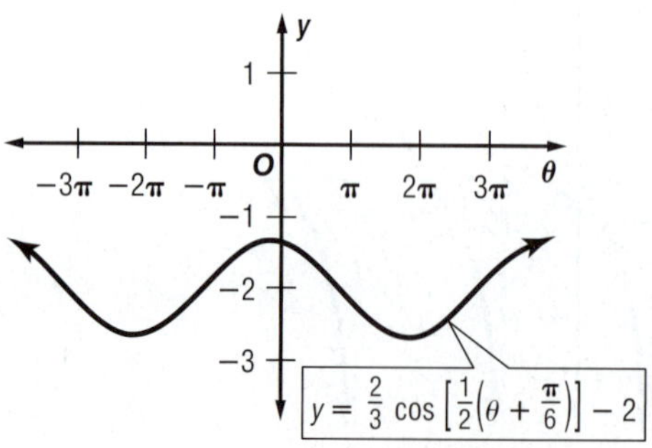

17. $h = 4 - \cos \frac{\pi}{2}t$ or $h = 4 - \cos 90°t$

19. 1; 360°; −90°

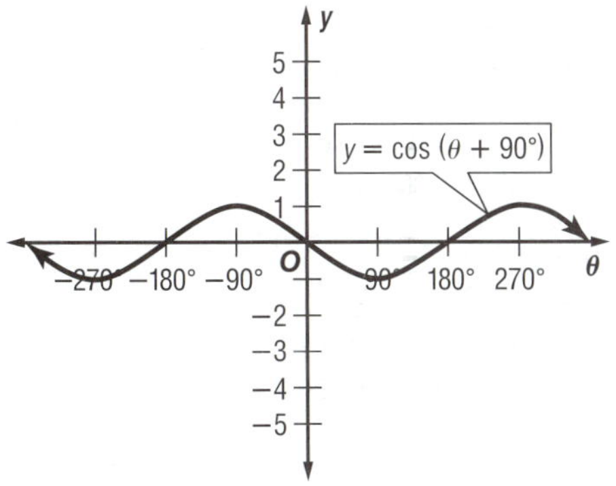

21. 1; 2π; $\frac{\pi}{4}$

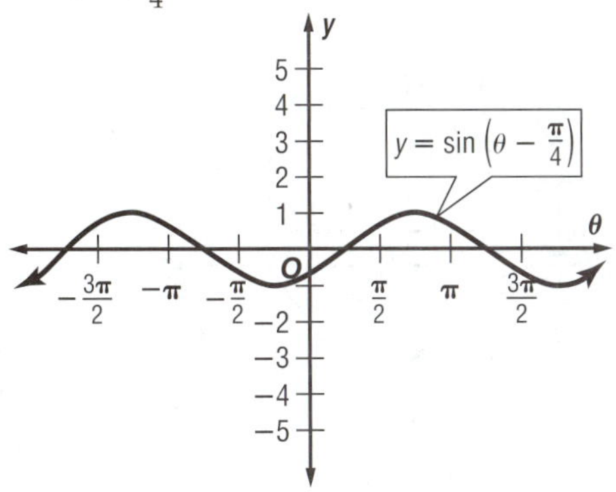

23. no amplitude; 180°; −22.5°

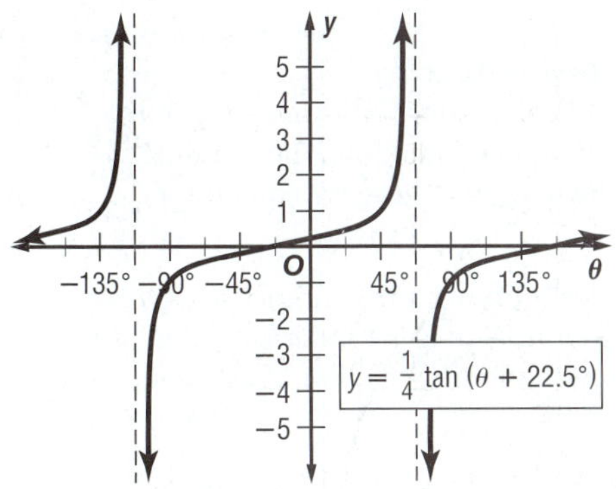

25. −1; $y = -1$; 1; 360°

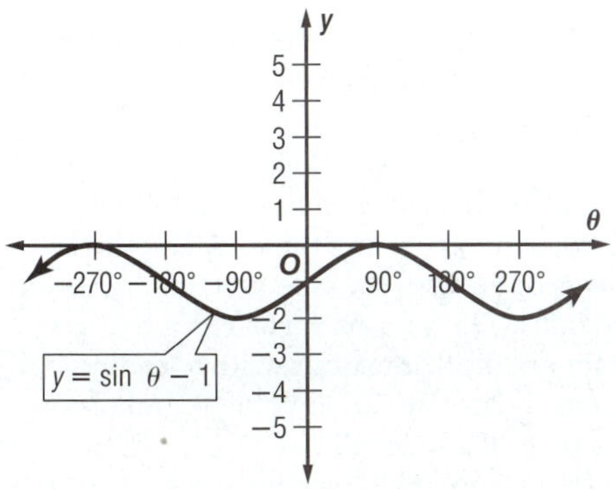

27. −5; $y = -5$; 1; 360°

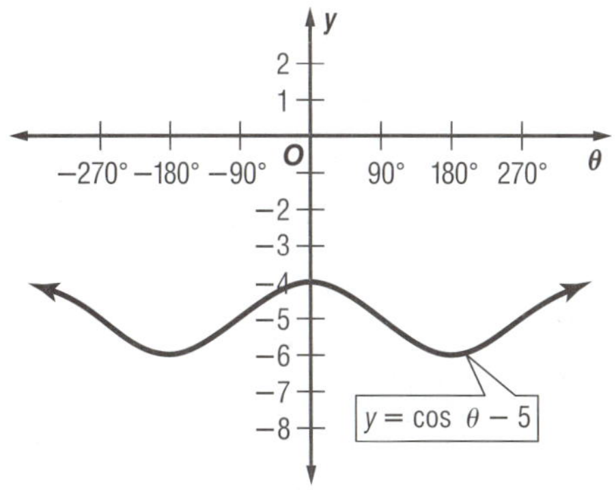

29. $\frac{1}{2}$; $y = \frac{1}{2}$; $\frac{1}{2}$; 360°

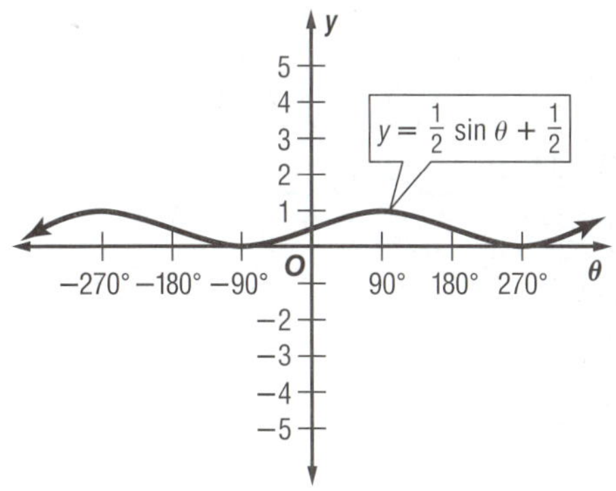

31.

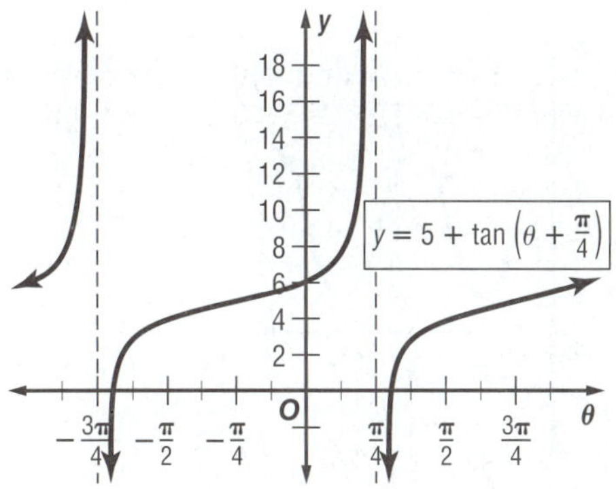

translation $\frac{\pi}{4}$ units left and 5 units up

33. 1; 2; 120°; 45°

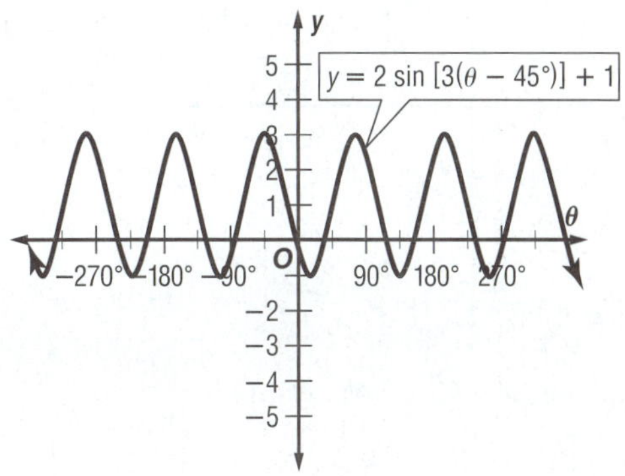

35. −3.5; does not exist; 720°; −60°

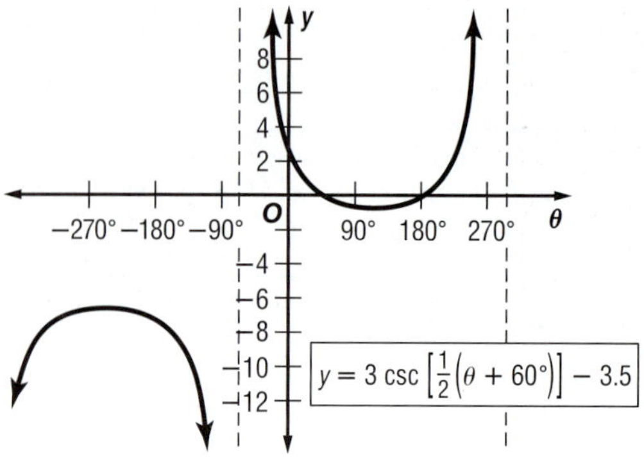

37. $1; \frac{1}{4}$; 180°; 75°

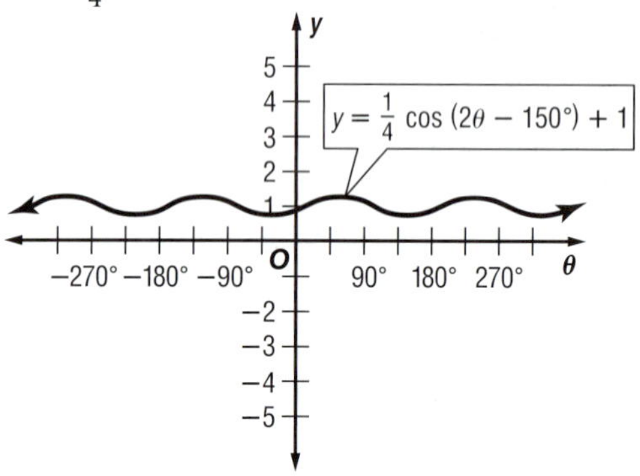

39. 3; 2; π; $-\frac{\pi}{4}$

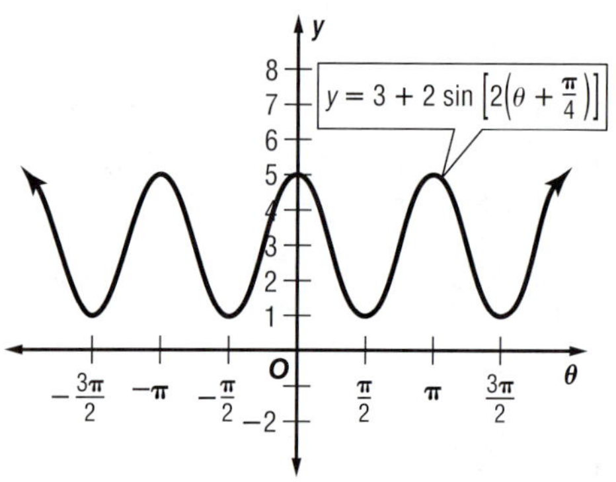

41.

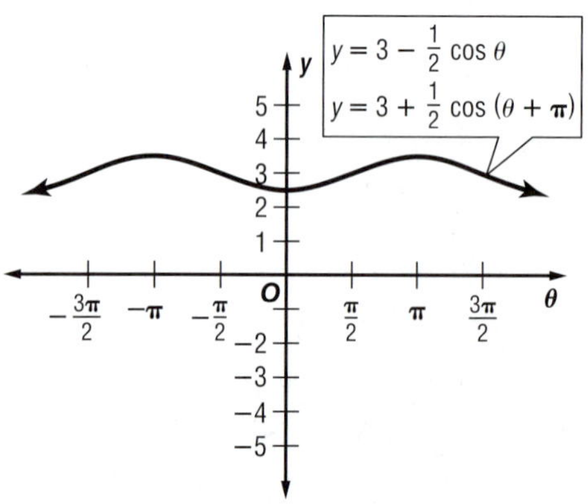

The graphs are identical. 43. c 45. 300; 14.5 yr

47. $h = 9 + 6\sin\left[\frac{\pi}{9}(t - 1.5)\right]$

49. Sample answer: You can use changes in amplitude and period along with vertical and horizontal shifts to show an animal population's starting point and display changes to that population over a period of time. Answers should include the following information.

- The equation shows a rabbit population that begins at 1200, increases to a maximum of 1450 then decreases to a minimum of 950 over a period of 4 years.
- Relative to $y = a\cos bx$, $y = a\cos bx + k$ would have a vertical shift of k units, while $y = a\cos[b(x - h)]$ has a horizontal shift of h units.

51. D 53. amplitude: 1; period: 720° or 4π

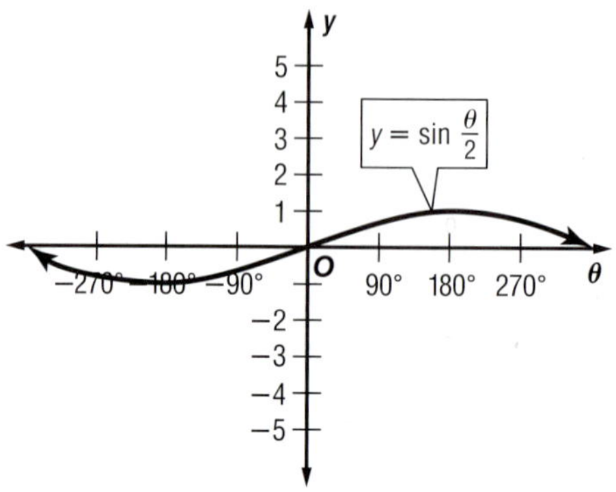

55. 0.75 57. 0.83 59. 35 61. 0.66 63. $\frac{5a - 13}{(a - 2)(a - 3)}$
65. $\frac{3y^2 + 10y + 5}{2(y - 5)(y + 3)}$ 67. −1 69. $\frac{1}{2}$ 71. $\frac{\sqrt{3}}{3}$ 73. 1

Pages 779–781 Lesson 14-3

1. Sample answer: The sine function is negative in the third and fourth quadrants. Therefore, the terminal side of the angle must lie in one of those two quadrants.

3. Sample answer: Simplifying a trigonometric expression means writing the expression as a numerical value or in terms of a single trigonometric function, if possible.

5. $-\frac{5}{4}$ 7. $\sqrt{2}$ 9. $\tan^2\theta$ 11. $\csc\theta$ 13. $\frac{1}{2}$ 15. $-\sqrt{5}$
17. $\frac{5}{4}$ 19. $\frac{\sqrt{3}}{2}$ 21. $\frac{3}{4}$ 23. $-\frac{4\sqrt{7}}{7}$ 25. $\cot\theta$ 27. $\cos\theta$
29. 2 31. $\cot^2\theta$ 33. 1 35. $\csc^2\theta$ 37. about 11.5°
39. about 9.4° 41. No; $R^2 = \frac{I\tan\theta\cos\theta}{E}$ simplifies to $E = \frac{I\sin\theta}{R^2}$. 43. $P = I^2R - \frac{I^2R}{1 + \tan^2 2\pi ft}$.

45. Sample answer: You can use equations to find the height and the horizontal distance of a baseball after it has been hit. The equations involve using the initial angle the ball makes with the ground with the sine function. Answers should include the following information.

- Both equations are quadratic in nature with a leading negative coefficient. Thus, both are inverted parabolas which model the path of a baseball.
- model rockets, hitting a golf ball, kicking a rock

47. A **49.** 12; $y = 12$; no amplitude; 180°

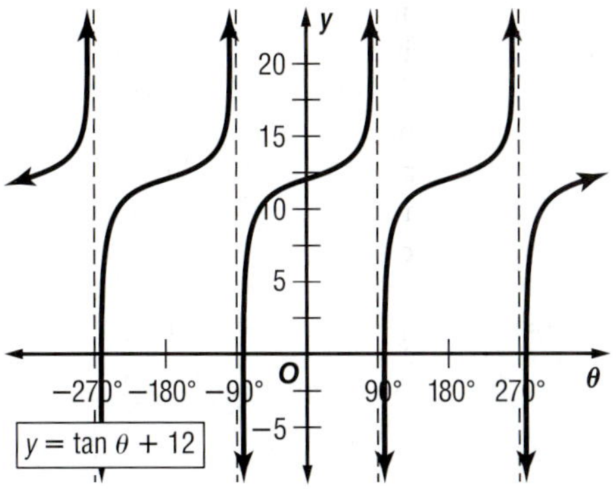

51. amplitude: 1; period: 120° or $\frac{2\pi}{3}$

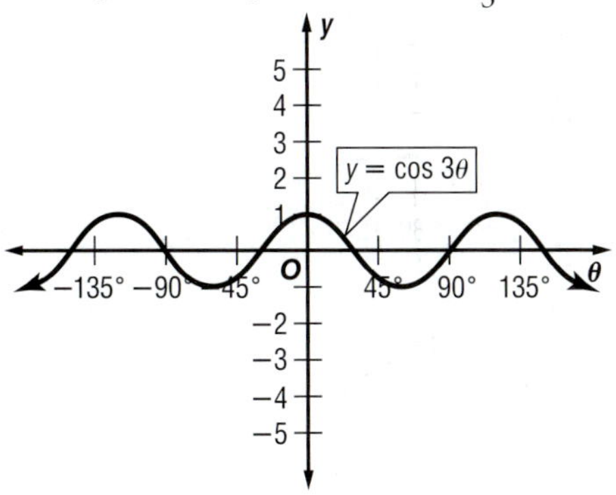

53. 93 **55.** Symmetric (=) **57.** Multiplication (=)

Page 781 Practice Quiz 1

1. $\frac{3}{4}$, 720° or 4π

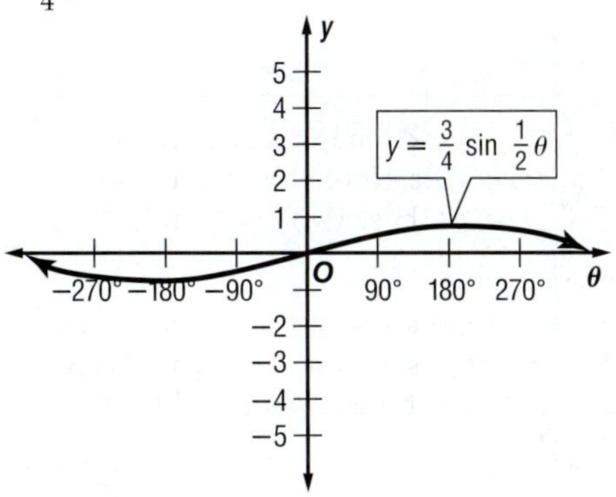

3. $-\frac{3}{5}$ **5.** $\frac{\sqrt{5}}{2}$

Pages 784–785 Lesson 14-4

1. $\sin\theta\tan\theta \stackrel{?}{=} \sec\theta - \cos\theta$

$\sin\theta\tan\theta \stackrel{?}{=} \frac{1}{\cos\theta} - \cos\theta$ *$\sec\theta = \frac{1}{\cos\theta}$*

$\sin\theta\tan\theta \stackrel{?}{=} \frac{1}{\cos\theta} - \frac{\cos^2\theta}{\cos\theta}$ *Multiply by the LCD, $\cos\theta$.*

$\sin\theta\tan\theta \stackrel{?}{=} \frac{1-\cos^2\theta}{\cos\theta}$ *Subtract.*

$\sin\theta\tan\theta \stackrel{?}{=} \frac{\sin^2\theta}{\cos\theta}$ *$1 - \cos^2\theta = \sin^2\theta$*

$\sin\theta\tan\theta \stackrel{?}{=} \sin\theta \cdot \frac{\sin\theta}{\cos\theta}$ *Factor.*

$\sin\theta\tan\theta = \sin\theta\tan\theta$ *$\frac{\sin\theta}{\cos\theta} = \tan\theta$*

3. Sample answer: $\sin^2\theta = 1 + \cos^2\theta$; it is not an identity because $\sin^2\theta = 1 - \cos^2\theta$.

5. $\tan^2\theta\cos^2\theta \stackrel{?}{=} 1 - \cos^2\theta$

$\frac{\sin^2\theta}{\cos^2\theta} \cdot \cos^2\theta \stackrel{?}{=} \sin^2\theta$

$\sin^2\theta = \sin^2\theta$

7. $\frac{1+\tan^2\theta}{\csc\theta} \stackrel{?}{=} \tan^2\theta$

$\frac{\sec^2\theta}{\csc^2\theta} \stackrel{?}{=} \tan^2\theta$

$\frac{\frac{1}{\cos^2\theta}}{\frac{1}{\sin^2\theta}} \stackrel{?}{=} \tan^2\theta$

$\frac{1}{\cos^2\theta} \cdot \sin^2\theta \stackrel{?}{=} \tan^2\theta$

$\tan^2\theta = \tan^2\theta$

9. $\frac{\sec\theta+1}{\tan\theta} \stackrel{?}{=} \frac{\tan\theta}{\sec\theta-1}$

$\frac{\sec\theta+1}{\tan\theta} \stackrel{?}{=} \frac{\tan\theta}{\sec\theta-1} \cdot \frac{\sec\theta+1}{\sec\theta+1}$

$\frac{\sec\theta+1}{\tan\theta} \stackrel{?}{=} \frac{\tan\theta\cdot(\sec\theta+1)}{\sec^2\theta-1}$

$\frac{\sec\theta+1}{\tan\theta} \stackrel{?}{=} \frac{\tan\theta\cdot(\sec\theta+1)}{\tan^2\theta}$

$\frac{\sec\theta+1}{\tan\theta} = \frac{\sec\theta+1}{\tan\theta}$

11. $\cos^2\theta + \tan^2\theta\cos^2\theta \stackrel{?}{=} 1$

$\cos^2\theta + \frac{\sin^2\theta}{\cos^2\theta} \cdot \cos^2\theta \stackrel{?}{=} 1$

$\cos^2\theta + \sin^2\theta \stackrel{?}{=} 1$

$1 = 1$

13. $1 + \sec^2\theta\sin^2\theta \stackrel{?}{=} \sec^2\theta$

$1 + \frac{1}{\cos^2\theta} \cdot \sin^2\theta \stackrel{?}{=} \sec^2\theta$

$1 + \tan^2\theta \stackrel{?}{=} \sec^2\theta$

$\sec^2\theta = \sec^2\theta$

15. $\frac{1-\cos\theta}{1+\cos\theta} \stackrel{?}{=} (\csc\theta - \cot\theta)^2$

$\frac{1-\cos\theta}{1+\cos\theta} \stackrel{?}{=} \csc^2\theta - 2\cot\theta\csc\theta + \cot^2\theta$

$\frac{1-\cos\theta}{1+\cos\theta} \stackrel{?}{=} \frac{1}{\sin^2\theta} - 2 \cdot \frac{\cos\theta}{\sin\theta} \cdot \frac{1}{\sin\theta} + \frac{\cos^2\theta}{\sin^2\theta}$

$\frac{1-\cos\theta}{1+\cos\theta} \stackrel{?}{=} \frac{1}{\sin^2\theta} - \frac{2\cos\theta}{\sin^2\theta} + \frac{\cos^2\theta}{\sin^2\theta}$

$\frac{1-\cos\theta}{1+\cos\theta} \stackrel{?}{=} \frac{1-2\cos\theta+\cos^2\theta}{\sin^2\theta}$

$\frac{1-\cos\theta}{1+\cos\theta} \stackrel{?}{=} \frac{(1-\cos\theta)(1-\cos\theta)}{1-\cos^2\theta}$

$\frac{1-\cos\theta}{1+\cos\theta} \stackrel{?}{=} \frac{(1-\cos\theta)(1-\cos\theta)}{(1-\cos\theta)(1+\cos\theta)}$

$\frac{1-\cos\theta}{1+\cos\theta} = \frac{1-\cos\theta}{1+\cos\theta}$

17. $\cot\theta\csc\theta \stackrel{?}{=} \frac{\cot\theta+\csc\theta}{\sin\theta+\tan\theta}$

$\cot\theta\csc\theta \stackrel{?}{=} \frac{\frac{\cos\theta}{\sin\theta} + \frac{1}{\sin\theta}}{\sin\theta + \frac{\sin\theta}{\cos\theta}}$

$\cot\theta\csc\theta \stackrel{?}{=} \frac{\frac{\cos\theta+1}{\sin\theta}}{\frac{\sin\theta\cos\theta+\sin\theta}{\cos\theta}}$

$\cot\theta\csc\theta \stackrel{?}{=} \frac{\frac{\cos\theta+1}{\sin\theta}}{\frac{\sin\theta(\cos\theta+1)}{\cos\theta}}$

$\cot\theta\csc\theta \stackrel{?}{=} \frac{\cos\theta+1}{\sin\theta} \cdot \frac{\cos\theta}{\sin\theta(\cos\theta+1)}$

$\cot\theta\csc\theta \stackrel{?}{=} \frac{\cos\theta}{\sin\theta} \cdot \frac{1}{\sin\theta}$

$\cot\theta\csc\theta = \cot\theta\csc\theta$

19.
$$\frac{\sec\theta}{\sin\theta} - \frac{\sin\theta}{\cos\theta} \stackrel{?}{=} \cot\theta$$
$$\frac{\frac{1}{\cos\theta}}{\sin\theta} - \frac{\sin\theta}{\cos\theta} \stackrel{?}{=} \cot\theta$$
$$\frac{1}{\sin\theta\cos\theta} - \frac{\sin^2\theta}{\sin\theta\cos\theta} \stackrel{?}{=} \cot\theta$$
$$\frac{1-\sin^2\theta}{\sin\theta\cos\theta} \stackrel{?}{=} \cot\theta$$
$$\frac{\cos^2\theta}{\sin\theta\cos\theta} \stackrel{?}{=} \cot\theta$$
$$\frac{\cos\theta}{\sin\theta} \stackrel{?}{=} \cot\theta$$
$$\cot\theta = \cot\theta$$

21.
$$\frac{1+\sin\theta}{\sin\theta} \stackrel{?}{=} \frac{\cot^2\theta}{\csc\theta - 1}$$
$$\frac{1+\sin\theta}{\sin\theta} \stackrel{?}{=} \frac{\cot^2\theta}{\csc\theta - 1}\cdot\frac{\csc\theta+1}{\csc\theta+1}$$
$$\frac{1+\sin\theta}{\sin\theta} \stackrel{?}{=} \frac{\cot^2\theta(\csc\theta+1)}{\csc^2\theta - 1}$$
$$\frac{1+\sin\theta}{\sin\theta} \stackrel{?}{=} \frac{\cot^2\theta(\csc\theta+1)}{\cot^2\theta}$$
$$\frac{1+\sin\theta}{\sin\theta} \stackrel{?}{=} \csc\theta + 1$$
$$\frac{1+\sin\theta}{\sin\theta} \stackrel{?}{=} \frac{1}{\sin\theta} + \frac{\sin\theta}{\sin\theta}$$
$$\frac{1+\sin\theta}{\sin\theta} = \frac{1+\sin\theta}{\sin\theta}$$

23.
$$\frac{1}{\sec^2\theta} + \frac{1}{\csc^2\theta} \stackrel{?}{=} 1$$
$$\cos^2\theta + \sin^2\theta \stackrel{?}{=} 1$$
$$1 = 1$$

25.
$$1 - \tan^4\theta \stackrel{?}{=} 2\sec^2\theta - \sec^4\theta$$
$$(1-\tan^2\theta)(1+\tan^2\theta) \stackrel{?}{=} \sec^2\theta\,(2-\sec^2\theta)$$
$$[1-(\sec^2\theta - 1)](\sec^2\theta) \stackrel{?}{=} (2-\sec^2\theta)(\sec^2\theta)$$
$$(2-\sec^2\theta)(\sec^2\theta) = (2-\sec^2\theta)(\sec^2\theta)$$

27.
$$\frac{1-\cos\theta}{\sin\theta} \stackrel{?}{=} \frac{\sin\theta}{1+\cos\theta}$$
$$\frac{1-\cos\theta}{\sin\theta}\cdot\frac{1+\cos\theta}{1+\cos\theta} \stackrel{?}{=} \frac{\sin\theta}{1+\cos\theta}$$
$$\frac{1-\cos^2\theta}{\sin\theta(1+\cos\theta)} \stackrel{?}{=} \frac{\sin\theta}{1+\cos\theta}$$
$$\frac{\sin^2\theta}{\sin\theta(1+\cos\theta)} \stackrel{?}{=} \frac{\sin\theta}{1+\cos\theta}$$
$$\frac{\sin\theta}{1+\cos\theta} = \frac{\sin\theta}{1+\cos\theta}$$

29.
$$\tan\theta\sin\theta\cos\theta\csc^2\theta \stackrel{?}{=} 1$$
$$\frac{\sin\theta}{\cos\theta}\cdot\sin\theta\cdot\cos\theta\cdot\frac{1}{\sin^2\theta} \stackrel{?}{=} 1$$
$$1 = 1$$

31.
$$\frac{v^2\tan^2\theta}{2\sec^2\theta} = \frac{v^2\,\frac{\sin^2\theta}{\cos^2\theta}}{2g\,\frac{1}{\cos^2\theta}}$$
$$= \frac{v^2}{2g}\cdot\frac{\sin^2\theta}{\cos^2\theta}\cdot\frac{\cos^2\theta}{1}$$
$$= \frac{v^2\sin^2\theta}{2g}$$

33. Sample answer: Consider a right triangle *ABC* with right angle at *C*. If an angle, say *A*, has a sine of *x*, then angle *B* must have a cosine of *x*. Since *A* and *B* are both in a right triangle and neither is the right angle, their sum must be $\frac{\pi}{2}$. **35.** D

37.
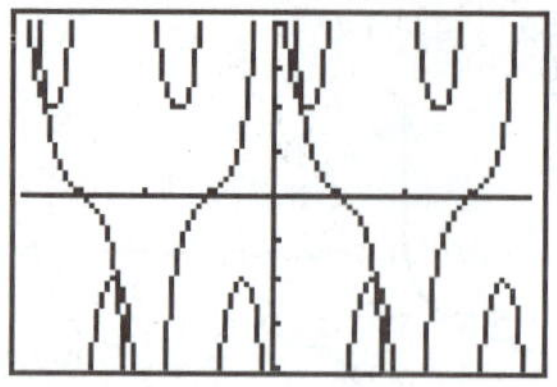

[−360, 360] scl: 90 by [−5, 5] scl: 1

is not

39.
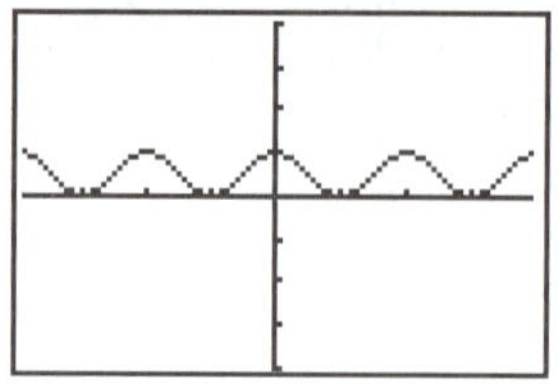

[−360, 360] scl: 90 by [−5, 5] scl: 1

may be

41.
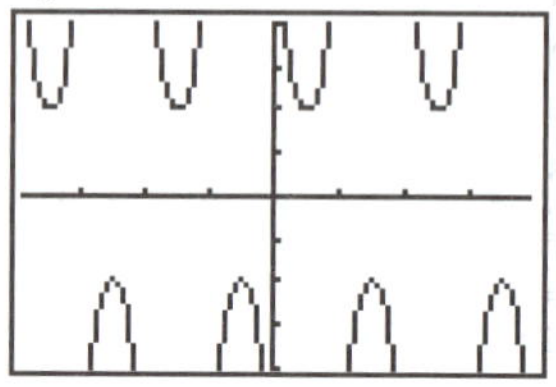

[−360, 360] scl: 90 by [−5, 5] scl: 1

may be

43. $\frac{\sqrt{5}}{2}$ **45.** $\frac{\sqrt{193}}{12}$ **47.** 1: 360°; 30°

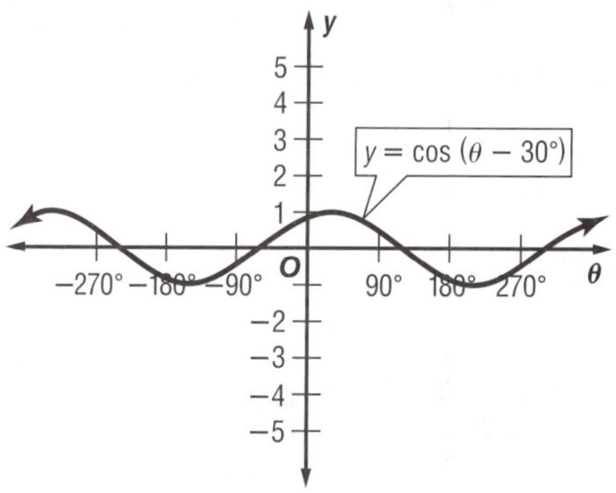

49. 3; 2π; $-\frac{\pi}{2}$

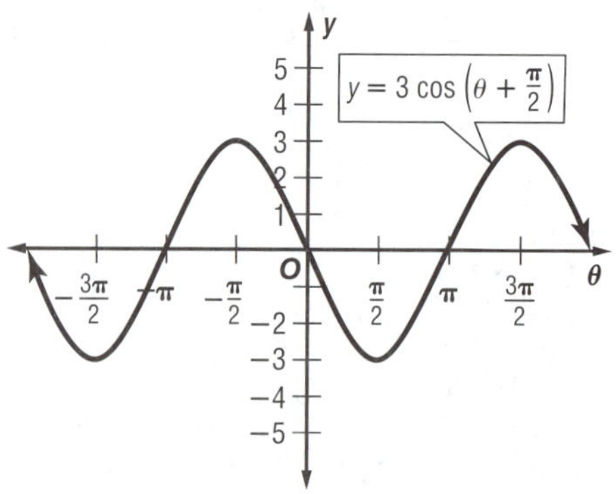

51. $\frac{\sqrt{6}}{4}$ **53.** $\frac{\sqrt{6}+2\sqrt{2}}{4}$

Pages 788–790 Lesson 14-5

1. $\sin(\alpha+\beta) - \sin\alpha + \sin\beta$

$\sin\alpha\cos\beta + \cos\alpha\sin\beta \neq \sin\alpha + \sin\beta$

3. Sometimes; sample answer: The cosine function can equal 1. **5.** $\frac{\sqrt{6}-\sqrt{2}}{4}$ **7.** $\frac{\sqrt{3}}{2}$

9. $-\frac{1}{2}$

11.
$$\begin{aligned}\sin\left(\theta+\tfrac{\pi}{2}\right) &\stackrel{?}{=} \cos\theta\\ \sin\theta\cos\tfrac{\pi}{2}+\cos\theta\sin\tfrac{\pi}{2} &\stackrel{?}{=} \cos\theta\\ \sin\theta\cdot 0+\cos\theta\cdot 1 &\stackrel{?}{=} \cos\theta\\ \cos\theta &= \cos\theta\end{aligned}$$

13. $\frac{5-\sqrt{3}}{1+5\sqrt{3}}$ **15.** $\frac{\sqrt{2}-\sqrt{6}}{4}$ **17.** $\frac{-\sqrt{6}-\sqrt{2}}{4}$

19. $\frac{-\sqrt{6}-\sqrt{2}}{4}$ **21.** $-\frac{\sqrt{2}}{2}$ **23.** $\frac{\sqrt{2}}{2}$ **25.** $\frac{\sqrt{2}-\sqrt{6}}{4}$

27. $\frac{-\sqrt{6}-\sqrt{2}}{4}$

29.
$$\begin{aligned}\cos(90^\circ+\theta) &\stackrel{?}{=} \cos 90^\circ\cos\theta-\sin 90^\circ\sin\theta\\ &\stackrel{?}{=} 0-1\sin\theta\\ &= -\sin\theta\end{aligned}$$

31.
$$\begin{aligned}\sin(90^\circ-\theta) &\stackrel{?}{=} \cos\theta\\ \sin 90^\circ\cos\theta-\cos 90^\circ\sin\theta &\stackrel{?}{=} \cos\theta\\ 1\cdot\cos\theta-0\cdot\sin\theta &\stackrel{?}{=} \cos\theta\\ \cos\theta-0 &\stackrel{?}{=} \cos\theta\\ \cos\theta &= \cos\theta\end{aligned}$$

33.
$$\begin{aligned}\cos(\pi-\theta) &\stackrel{?}{=} -\cos\theta\\ \cos\pi\cos\theta+\sin\pi\sin\theta &\stackrel{?}{=} -\cos\theta\\ -1\cdot\cos\theta+0\cdot\sin\theta &\stackrel{?}{=} -\cos\theta\\ -\cos\theta &= -\cos\theta\end{aligned}$$

35.
$$\begin{aligned}\sin(\pi-\theta) &\stackrel{?}{=} \sin\theta\\ \sin\pi\cos\theta-[\cos\pi\sin\theta] &\stackrel{?}{=} \sin\theta\\ 0\cdot\cos\theta-[-1\cdot\sin\theta] &\stackrel{?}{=} \sin\theta\\ 0-[-\sin\theta] &\stackrel{?}{=} \sin\theta\\ \sin\theta &= \sin\theta\end{aligned}$$

37.
$$\begin{aligned}&\sin\left(\theta+\tfrac{\pi}{3}\right)-\cos\left(\theta+\tfrac{\pi}{6}\right)\\ &\stackrel{?}{=} \sin\theta\cos\tfrac{\pi}{3}+\cos\theta\sin\tfrac{\pi}{3}-\cos\theta\cos\tfrac{\pi}{6}+\sin\theta\sin\tfrac{\pi}{6}\\ &\stackrel{?}{=} \tfrac{1}{2}\sin\theta+\tfrac{\sqrt{3}}{2}\cos\theta-\tfrac{\sqrt{3}}{2}\cos\theta+\tfrac{1}{2}\sin\theta\\ &\stackrel{?}{=} \tfrac{1}{2}\sin\theta+\tfrac{1}{2}\sin\theta\\ &= \sin\theta\end{aligned}$$

39.
$$\begin{aligned}\cos(\alpha+\beta) &\stackrel{?}{=} \frac{1-\tan\alpha\tan\beta}{\sec\alpha\sec\beta}\\ \cos(\alpha+\beta) &\stackrel{?}{=} \frac{1-\frac{\sin\alpha}{\cos\alpha}\cdot\frac{\sin\beta}{\cos\beta}}{\frac{1}{\cos\alpha}\cdot\frac{1}{\cos\beta}}\\ \cos(\alpha+\beta) &\stackrel{?}{=} \frac{1-\frac{\sin\alpha}{\cos\alpha}\cdot\frac{\sin\beta}{\cos\beta}}{\frac{1}{\cos\alpha}\cdot\frac{1}{\cos\beta}}\cdot\frac{\cos\alpha\cos\beta}{\cos\alpha\cos\beta}\\ \cos(\alpha+\beta) &\stackrel{?}{=} \frac{\cos\alpha\cos\beta-\sin\alpha\sin\beta}{1}\\ \cos(\alpha+\beta) &= \cos(\alpha+\beta)\end{aligned}$$

41. Destructive; the resulting graph has a smaller amplitude than the two initial graphs. **43.** 0.4179 *E* **45.** 0.5564 *E*

47. Sample answer: To determine communication interference, you need to determine the sine or cosine of the sum or difference of two angles. Answers should include the following information.

- Interference occurs when waves pass through the same space at the same time. When the combined waves have a greater amplitude, constructive interference results and when the combined waves have a smaller amplitude, destructive interference results.

49. C

51.
$$\begin{aligned}\sin^2\theta+\tan^2\theta &\stackrel{?}{=} (1-\cos^2\theta)+\frac{\sec^2\theta}{\csc^2\theta}\\ \sin^2\theta+\tan^2\theta &\stackrel{?}{=} \sin^2\theta+\frac{\sec^2\theta}{\csc^2\theta}\\ \sin^2\theta+\tan^2\theta &\stackrel{?}{=} \sin^2\theta+\frac{1}{\cos^2\theta}\div\frac{1}{\sin^2\theta}\\ \sin^2\theta+\tan^2\theta &\stackrel{?}{=} \sin^2\theta+\frac{\sin^2\theta}{\cos^2\theta}\\ \sin^2\theta+\tan^2\theta &= \sin^2\theta+\tan^2\theta\end{aligned}$$

53.
$$\begin{aligned}\frac{\sec\theta}{\tan\theta} &\stackrel{?}{=} \csc\theta\\ \frac{1}{\cos\theta}\div\frac{\sin\theta}{\cos\theta} &\stackrel{?}{=} \csc\theta\\ \frac{1}{\cos\theta}\cdot\frac{\cos\theta}{\sin\theta} &\stackrel{?}{=} \csc\theta\\ \frac{1}{\sin\theta} &\stackrel{?}{=} \csc\theta\\ \csc\theta &= \csc\theta\end{aligned}$$

55. 4 **57.** 2 sec θ **59.** $\sin\theta=-\frac{4}{5}$, $\cos\theta=-\frac{3}{5}$, $\tan\theta=\frac{4}{3}$, $\csc\theta=-\frac{5}{4}$, $\sec\theta=-\frac{5}{3}$, $\cot\theta=\frac{3}{4}$ **61.** 360 **63.** 56

65. about 228 mi **67.** $\pm\frac{\sqrt{5}}{2}$ **69.** $\pm\frac{\sqrt{5}}{5}$ **71.** $\pm\frac{\sqrt{6}}{2}$

73. $\frac{\sqrt{\sqrt{6}-\sqrt{2}}}{2}$

Pages 794–797 Lesson 14-6

1. Sample answer: If x is in the third quadrant, then $\frac{x}{2}$ is between 90° and 135°. Use the half-angle formula for cosine knowing that the value is negative. **3.** Sample answer: The identity used for cos 2θ depends on whether you know the value of sin θ, cos θ or both values.

5. $\frac{4\sqrt{5}}{9}$, $-\frac{1}{9}$, $\frac{\sqrt{30}}{6}$, $-\frac{\sqrt{6}}{6}$ **7.** $-\frac{3\sqrt{7}}{8}$, $-\frac{1}{8}$, $\frac{\sqrt{8-2\sqrt{7}}}{4}$, $-\frac{\sqrt{8+2\sqrt{7}}}{4}$ **9.** $\frac{\sqrt{2-\sqrt{3}}}{2}$

11.
$$\begin{aligned}\cos^2 2x+4\sin^2 x\cos^2 x &\stackrel{?}{=} 1\\ \cos^2 2x+\sin^2 2x &\stackrel{?}{=} 1\\ 1 &= 1\end{aligned}$$

13. $-\frac{120}{169}$, $\frac{119}{169}$, $\frac{5\sqrt{26}}{26}$, $\frac{\sqrt{26}}{26}$ **15.** $\frac{4\sqrt{2}}{9}$, $-\frac{7}{9}$, $\frac{\sqrt{6}}{3}$, $-\frac{\sqrt{3}}{3}$

17. $-\frac{3\sqrt{55}}{32}$, $\frac{23}{32}$, $\frac{\sqrt{8-\sqrt{55}}}{4}$, $-\frac{\sqrt{8+\sqrt{55}}}{4}$

19. $\frac{\sqrt{35}}{18}$, $-\frac{17}{18}$, $\frac{\sqrt{15}}{6}$, $\frac{\sqrt{21}}{6}$ **21.** $-\frac{4\sqrt{2}}{9}$, $\frac{7}{9}$, $\frac{\sqrt{18-12\sqrt{2}}}{6}$, $-\frac{\sqrt{18-12\sqrt{2}}}{6}$ **23.** $\frac{4\sqrt{5}}{9}$, $-\frac{1}{9}$, $\frac{\sqrt{6}}{6}$, $\frac{\sqrt{30}}{6}$ **25.** $-\frac{\sqrt{2+\sqrt{3}}}{2}$

27. $-\frac{\sqrt{2+\sqrt{2}}}{2}$ **29.** $\frac{\sqrt{2-\sqrt{2}}}{2}$

31.
$$\begin{aligned}\sin 2x &\stackrel{?}{=} 2\cot x\sin^2 x\\ 2\sin x\cos x &\stackrel{?}{=} 2\,\frac{\cos x}{\sin x}\cdot\sin^2 x\\ 2\sin x\cos x &= 2\sin x\cos x\end{aligned}$$

33.
$$\begin{aligned}\sin^4 x-\cos^4 x &\stackrel{?}{=} 2\sin^2 x-1\\ (\sin^2 x-\cos^2 x)(\sin^2 x+\cos^2 x) &\stackrel{?}{=} 2\sin^2 x-1\\ (\sin^2 x-\cos^2 x)\cdot 1 &\stackrel{?}{=} 2\sin^2 x-1\\ [\sin^2 x-(1-\sin^2 x)]\cdot 1 &\stackrel{?}{=} 2\sin^2 x-1\\ \sin^2 x-1+\sin^2 x &\stackrel{?}{=} 2\sin^2 x-1\\ 2\sin^2 x-1 &= 2\sin^2 x-1\end{aligned}$$

35.
$$\begin{aligned}\tan^2\frac{x}{2} &\stackrel{?}{=} \frac{1-\cos x}{1+\cos x}\\ \frac{\sin^2\frac{x}{2}}{\cos^2\frac{x}{2}} &\stackrel{?}{=} \frac{1-\cos x}{1+\cos x}\\ \frac{\left(\pm\sqrt{\frac{1-\cos x}{2}}\right)^2}{\left(\pm\sqrt{\frac{1+\cos x}{2}}\right)^2} &\stackrel{?}{=} \frac{1-\cos x}{1+\cos x}\\ \frac{1-\cos x}{1+\cos x} &\stackrel{?}{=} \frac{1-\cos x}{1+\cos x}\end{aligned}$$

37. 46.3° **39.** $2 + \sqrt{3}$ **41.** $\frac{1}{4}\tan\theta$ **43.** The maxima occur at $x = \pm\frac{\pi}{2}$ and $\pm\frac{3\pi}{2}$. The minima occur at $x = 0, \pm\pi$ and $\pm 2\pi$. **45.** The graph of $f(x)$ crosses the x-axis at the points specified in Exercise 43. **47.** Sample answer: The sound waves associated with music can be modeled using trigonometric functions. Answers should include the following information.

- In moving from one harmonic to the next, the number of vibrations that appear as sine waves increase by 1.
- The period of the function as you move from the nth harmonic to the $(n + 1)$th harmonic decreases from $\frac{2\pi}{n}$ to $\frac{2\pi}{n+1}$.

49. B **51.** $\frac{\sqrt{6}-\sqrt{2}}{4}$ **53.** $-\frac{\sqrt{3}}{2}$ **55.** $\frac{1}{2}$

57. $\cos\theta(\cos\theta + \cot\theta) \stackrel{?}{=} \cot\theta\cos\theta(\sin\theta + 1)$

$\cos\theta(\cos\theta + \cot\theta) \stackrel{?}{=} \frac{\cos\theta}{\sin\theta}\cos\theta\sin\theta + \cot\theta\cos\theta$

$\cos\theta(\cos\theta + \cot\theta) \stackrel{?}{=} \cos^2\theta + \cot\theta\cos\theta$

$\cos\theta(\cos\theta + \cot\theta) = \cos\theta(\cos\theta + \cot\theta)$

59. $10^{2.5}$ or about 316 times greater **61.** 1, −1 **63.** $\frac{5}{2}, -2$ **65.** $0, -\frac{1}{2}$

Page 797 Practice Quiz 2

1. $\sin\theta\sec\theta \stackrel{?}{=} \tan\theta$

$\sin\theta \cdot \frac{1}{\cos\theta} \stackrel{?}{=} \tan\theta$

$\frac{\sin\theta}{\cos\theta} \stackrel{?}{=} \tan\theta$

$\tan\theta = \tan\theta$

3. $\sin\theta + \tan\theta \stackrel{?}{=} \frac{\sin\theta(\cos\theta + 1)}{\cos\theta}$

$\sin\theta + \tan\theta \stackrel{?}{=} \frac{\sin\theta\cos\theta + \sin\theta}{\cos\theta}$

$\sin\theta + \tan\theta \stackrel{?}{=} \frac{\sin\theta\cos\theta}{\cos\theta} + \frac{\sin\theta}{\cos\theta}$

$\sin\theta + \tan\theta = \sin\theta + \tan\theta$

5. $\cos\left(\frac{3\pi}{2} - \theta\right) \stackrel{?}{=} -\sin\theta$

$\cos\frac{3\pi}{2}\cos\theta + \sin\frac{3\pi}{2}\sin\theta \stackrel{?}{=} -\sin\theta$

$0 + (-1 \cdot \sin\theta) \stackrel{?}{=} -\sin\theta$

$-\sin\theta = -\sin\theta$

7. $\frac{\sqrt{3}}{2}$ **9.** $\frac{\sqrt{2-\sqrt{3}}}{2}$

Pages 802–804 Lesson 14-7

1. Sample answer: If $\sec\theta = 0$ then $\frac{1}{\cos\theta} = 0$. Since no value of θ makes $\frac{1}{\cos\theta} = 0$, there are no solutions. **3.** Sample answer: $\sin\theta = 2$ **5.** 135°, 225° **7.** $\frac{\pi}{6}$ **9.** $0 + k\pi$ **11.** $60° + k \cdot 360°, 300° + k \cdot 360°$ **13.** $\frac{\pi}{6} + 2k\pi, \frac{5\pi}{6} + 2k\pi, \frac{\pi}{2} + 2k\pi$ or $30° + k \cdot 360°, 150° + k \cdot 360°, 90° + k \cdot 360°$ **15.** 60°, 300° **17.** 210°, 330° **19.** $\frac{\pi}{6}, \frac{5\pi}{6}, \frac{3\pi}{2}$ **21.** $\frac{7\pi}{6}, \frac{11\pi}{6}$ **23.** $\frac{\pi}{3} + 2k\pi, \frac{5\pi}{3} + 2k\pi$ **25.** $\frac{2\pi}{3} + 2k\pi, \frac{4\pi}{3} + 2k\pi$ **27.** $\frac{\pi}{3} + 2k\pi, \frac{5\pi}{3} + 2k\pi$ **29.** $45° + k \cdot 180°$ **31.** $270° + k \cdot 360°$ **33.** $0° + k \cdot 180°, 60° + k \cdot 180°$ **35.** $0 + 2k\pi, \frac{\pi}{2} + 2k\pi, \frac{3\pi}{2} + 2k\pi$ or $0° + k \cdot 360°, 90° + k \cdot 360°, 270° + k \cdot 360°$ **37.** $0 + k\pi$ or $0° + k \cdot 180°$ **39.** $0 + 2k\pi, \frac{\pi}{3} + 2k\pi, \frac{5\pi}{3} + 2k\pi$, or $0° + k \cdot 360°, 60° + k \cdot 360°, 300° + k \cdot 360°$ **41.** $S = \frac{352}{\tan\theta}$ or $S = 352\cot\theta$

43. $y = \frac{3}{2} + \frac{3}{2}\sin(\pi t)$

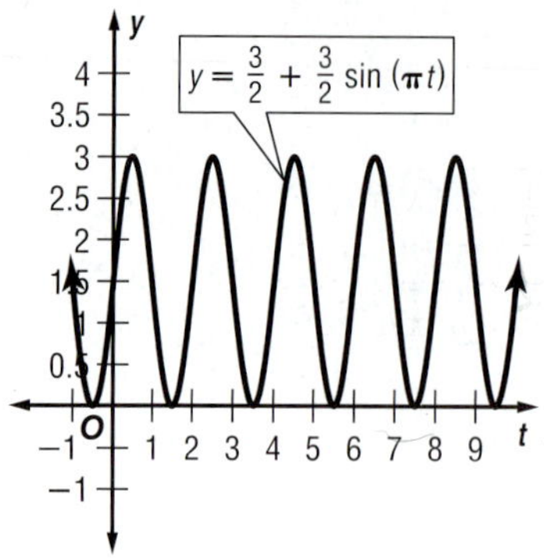

45. (4.964, −0.598) **47.** D **49.** $\frac{24}{25}, \frac{7}{25}, \frac{\sqrt{10}}{10}, \frac{3\sqrt{10}}{10}$ **51.** $\frac{5\sqrt{11}}{18}, \frac{7}{18}, \frac{\sqrt{3}}{6}, \frac{\sqrt{33}}{6}$ **53.** $-\frac{\sqrt{3}}{2}$ **55.** $b = 11.0, c = 12.2, m\angle C = 78$

Pages 805–808 Chapter 14 Study Guide and Review

1. h **3.** d **5.** e **7.** g **9.** amplitude: $\frac{1}{2}$; period: 360° or 2π

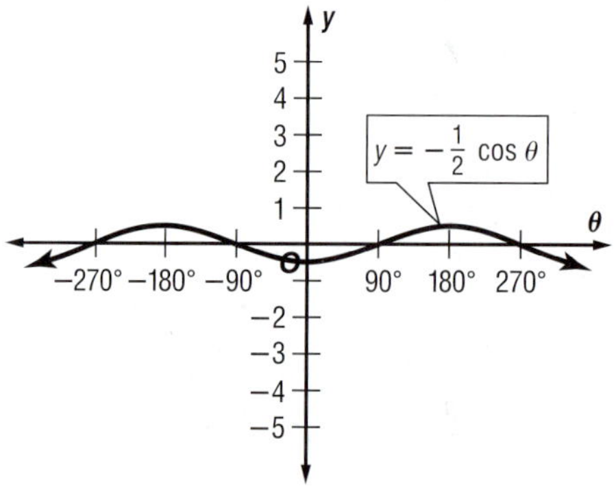

11. amplitude: 1; period: 720° or 4π

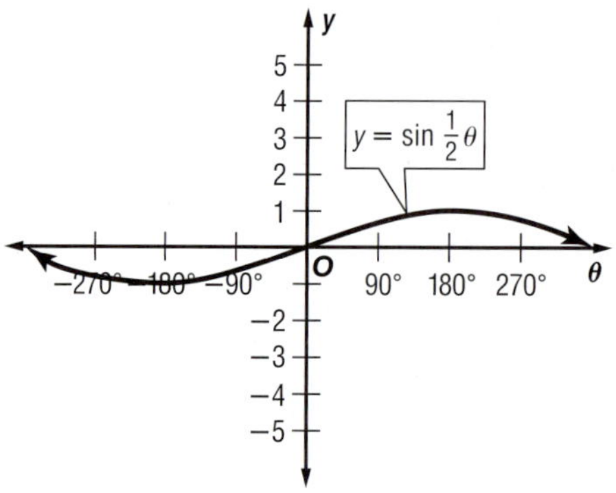

13. amplitude: does not exist; period: 540° or 3π

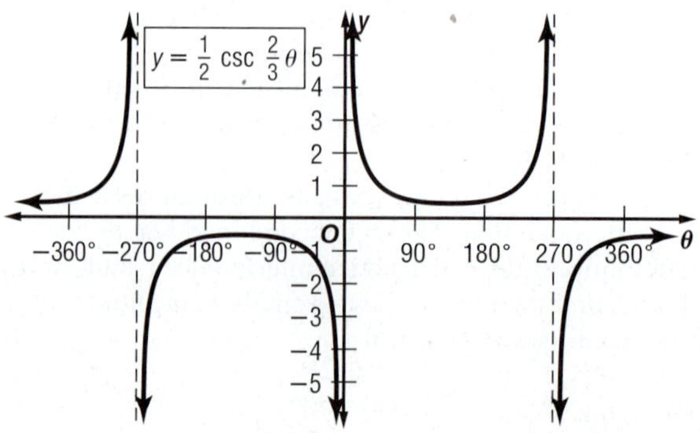

15. $-1, \frac{1}{2}, 180°, 60°$

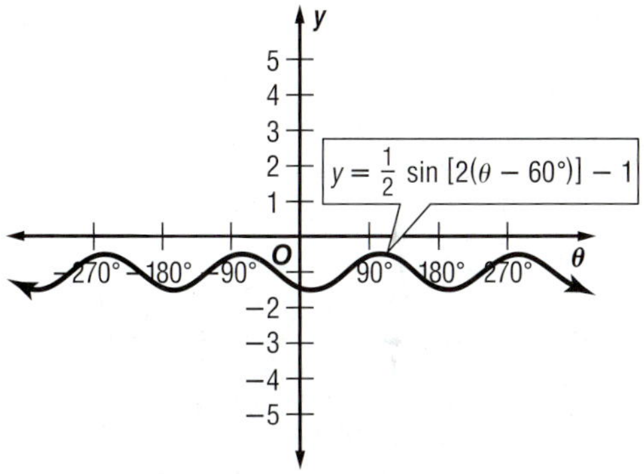

17. 1, does not exist, $4\pi, -\frac{\pi}{4}$

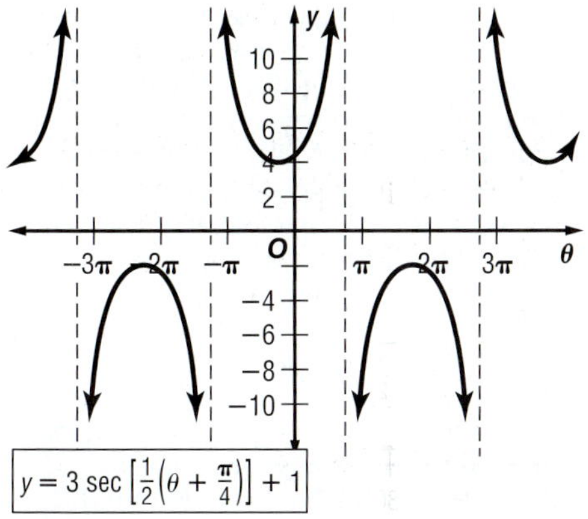

19. $-\frac{4}{3}$ 21. $\sin^2\theta$ 23. $\sec\theta$

25.
$$\frac{\sin\theta}{1-\cos\theta} \stackrel{?}{=} \csc\theta + \cot\theta$$
$$\frac{\sin\theta}{1-\cos\theta} \stackrel{?}{=} \frac{1}{\sin\theta} + \frac{\cos\theta}{\sin\theta}$$
$$\frac{\sin\theta}{1-\cos\theta} \stackrel{?}{=} \frac{1+\csc\theta}{\sin\theta}$$
$$\frac{\sin\theta}{1-\cos\theta} \stackrel{?}{=} \frac{1+\cos\theta}{\sin\theta} \cdot \frac{1-\cos\theta}{1-\cos\theta}$$
$$\frac{\sin\theta}{1-\cos\theta} \stackrel{?}{=} \frac{1-\cos^2\theta}{\sin\theta\,(1-\cos\theta)}$$
$$\frac{\sin\theta}{1-\cos\theta} \stackrel{?}{=} \frac{\sin^2\theta}{\sin\theta\,(1-\cos\theta)}$$
$$\frac{\sin\theta}{1-\cos\theta} = \frac{\sin\theta}{1-\cos\theta}$$

27.
$$\sec\theta(\sec\theta - \cos\theta) \stackrel{?}{=} \tan^2\theta$$
$$\frac{1}{\cos\theta}\left(\frac{1}{\cos\theta} - \cos\theta\right) \stackrel{?}{=} \tan^2\theta$$
$$\frac{1}{\cos^2\theta} - 1 \stackrel{?}{=} \tan^2\theta$$
$$\sec^2\theta - 1 \stackrel{?}{=} \tan^2\theta$$
$$\tan^2\theta = \tan^2\theta$$

29. $\frac{\sqrt{6}-\sqrt{2}}{4}$ 31. $\frac{\sqrt{2}-\sqrt{6}}{4}$ 33. $\frac{-\sqrt{6}-\sqrt{2}}{4}$

35.
$$\sin(30 - \theta) \stackrel{?}{=} \cos(60 + \theta)$$
$$\sin 30°\cos\theta - \cos 30°\sin\theta \stackrel{?}{=} \cos 60°\cos\theta - \sin 60°\sin\theta$$
$$\frac{1}{2}\cos\theta - \frac{\sqrt{3}}{2}\sin\theta = \frac{1}{2}\cos\theta - \frac{\sqrt{3}}{2}\sin\theta$$

37.
$$-\cos\theta \stackrel{?}{=} \cos(\pi + \theta)$$
$$-\cos\theta \stackrel{?}{=} \cos\pi\cos\theta - \sin\pi\sin\theta$$
$$-\cos\theta \stackrel{?}{=} -1\cdot\cos\theta - 0\cdot\sin\theta$$
$$-\cos\theta = -\cos\theta$$

39. $\frac{120}{169}, \frac{119}{169}, \frac{5\sqrt{26}}{26}, -\frac{\sqrt{26}}{26}$ 41. $-\frac{120}{169}, \frac{119}{169}, \frac{\sqrt{26}}{26}, -\frac{5\sqrt{26}}{26}$

43. 0° 45. $\frac{\pi}{6} + 2k\pi, \frac{5\pi}{6} + 2k\pi$

Selected Answers

Photo Credits

About the Cover: Alexander Calder (1898–1976) was one of America's most acclaimed sculptors. Renowned for his invention of the *mobile*, or movable sculpture, Calder also created sculptures called *stabiles*, or immovable sculptures. The cover photograph illustrates his *Grand Stabile Rouge*, located in Paris. One of Calder's last great public works, this sculpture is reminiscent of another of his stabiles, *Flamingo*, in Chicago. Both stabiles feature large red arches that resemble parabolas.

Cover Vanni Archive/CORBIS; **x** D & K Tapparel/Getty Images; **xi** Telegraph Colour Library/Getty Images; **xii** DEX Images Inc./CORBIS Stock Market; **xiii** AFP/CORBIS; **xiv** CORBIS; **xix** Jenny Hager/ImageState; **xv** Brownie Harris/CORBIS Stock Market; **xvi** Ray F. Hillstrom Jr.; **xvii** Kunio Owaki/CORBIS Stock Market; **xviii** Jane Burton/Bruce Coleman; **xx** Food & Drug Administration/SPL/Photo Researchers; **xxi** R. Ian Lloyd/Masterfile; **xxii** Getty Images; **2** David De Lossy/Getty Images; **2–3** Bryan Peterson/Getty Images; **4** Johnny Stockshooter/International Stock; **4–5** Orion/International Stock; **6** Mark Harmel/Getty Images; **14** Amy C. Etra/PhotoEdit; **16** Archivo Iconografico, S.A./CORBIS; **19** Aaron Haupt; **20** SuperStock; **23** Michael Newman/PhotoEdit; **26** Pictor; **28** Robert Yager/Getty Images; **31** E.L. Shay; **38** Lawrence Migdale; **40** Index Stock/Ewing Galloway; **43** PhotoDisc; **44** Rudi Von Briel/PhotoEdit; **54–55** Jack Dykinga/Getty Images; **56** William J. Weber; **61** Bettmann/CORBIS; **64** D & K Tapparel/Getty Images; **67** Lynn M. Stone; **72** (l)SuperStock, (r)Richard T. Nowitz/CORBIS; **80** VCG/Getty Images; **82** John Evans; **85** Matt Meadows; **94** David Ball/CORBIS Stock Market; **99** Getty Images; **108** PhotoDisc; **108–109** NASA/TSADO/Tom Stack & Assotes; **111** Dave Starrett/Masterfile; **114** Telegraph Colour Library/Getty Images; **121** Will Hart/PhotoEdit; **124** NASA; **126** Doug Martin; **129** AFP/CORBIS; **131** Caroline Penn/CORBIS; **138** S. Carmona/CORBIS; **140** M. Angelo/CORBIS; **143** Andy Lyons/Allsport; **152** CORBIS; **152–153** William Sallaz/DUOMO; **157** Bettman/CORBIS; **161** Tui De Roy/Bruce Coleman, Inc.; **165** PhotoDisc; **169** Andy Lyons STF/Allsport; **172** Mark Tomalty/Masterfile; **175** Mark Richards/PhotoEdit; **180** Michael Denora/Getty Images; **187** Jonathan Blair/CORBIS; **190** FDR Library; **193** DEX Images Inc./CORBIS Stock Market; **195** Jose Luis Pelaez Inc./CORBIS Stock Market; **197** Volker Steger/SPL/Photo Researchers; **203** Ken Eward/Science Source/Photo Researchers; **218** www.comstock.com; **218–219** Rafael Marcia/Photo Researchers; **220–221** Carl Purcell/Words and Pictures/PictureQuest; **225** AFP/CORBIS; **227** K.G. Murti/Visuals Unlimited; **229** David Umberger/Purdue University Photo; **243** Steve Rayer/CORBIS; **249** Roger Ressmeyer/CORBIS; **255** Roy Ooms/Masterfile; **259** AFP/CORBIS; **262** Victoria & Albert Museum, London/Art Resource, NY; **267** Lori Adamski Peek/Getty Images; **274** Kaluzny/Thatcher/Getty Images; **284** Allsport Concepts/Getty Images; **284–285** Ed Pritchard/Getty Images; **291** Aidan O'Rourke; **292** Getty Images; **298** SuperStock; **304** Matthew McVay/Stock Boston; **306** DUOMO/CORBIS; **311** CORBIS; **313** Jeff Kaufman/Getty Images; **318** Bruce Hands/Getty Images; **327** NASA; **329** Nick Wilson/Allsport; **331** Todd Rosenberg/Allsport; **334** Aaron Haupt; **344–345** Guy Grenier/Masterfile; **346** Brownie Harris/CORBIS Stock Market; **351** Martha Swope/Timepix; **355** VCG/Getty Images; **357** Michael Newman/PhotoEdit; **363** Gregg Mancuso/Stock Boston; **365** Boden/Ledingham/Masterfile; **372** National Library of Medicine/Mark Marten/Photo Researchers; **376** VCG/Getty Images; **381** Bob Krist/CORBIS; **383** Ed Bock/CORBIS Stock Market; **388** SuperStock; **392** Matt Meadows; **394** SuperStock; **395** Raymond Gehman/CORBIS; **396** Frank Rossotto/Stocktreck/CORBIS Stock Market; **398** Getty Images; **408** Michael S. Yamashita/CORBIS; **408–409** Jose Fuste Raga/eStock Photo; **410** Photographers Library LTD/eStock Photo; **410–411** James Hackett/eStock Photo; **424** James Rooney; **426** SuperStock; **432** Matt Meadows; **435** Ray F. Hillstrom Jr.; **439** James P. Blair/CORBIS; **443** CORBIS; **446** Bob Krist/Getty Images; **459** (l)Space Telescope Science Institute/NASA/SPL/Photo Researchers, (r)Michael Newman/PhotoEdit; **470–471** David Fleetham/Getty Images; **477** AFP/CORBIS; **483** Pascal Rondeau/Allsport; **487** Aaron Haupt; **489** Bettmann/CORBIS; **494** JPL/TSADO/Tom Stack & Associates; **496** Geoff Butler; **497** Lynn M. Stone/Bruce Coleman, Inc.; **499** Picture Press/CORBIS; **503** Kunio Owaki/CORBIS Stock Market; **505** (b)Phil Schermeister/CORBIS, (t)Bruce Ayres/Getty Images; **507** Reuters NewMedia Inc./CORBIS; **511** Keith Wood/Getty Images; **520–521** Michael S. Yamashita/CORBIS; **522** Aaron Haupt; **525** Ariel Skelley/CORBIS Stock Market; **529** Jeff Zaruba/CORBIS Stock Market; **536** (l)Mark Jones/Minden Pictures, (r)Jane Burton/Bruce Coleman, Inc.; **537** David Weintraub/Photo Researchers; **542** SuperStock; **545** Bettman/CORBIS; **558** Jim Craigmyle/Masterfile; **561** Richard T. Nowitz/Photo Researchers; **564** Karl Weatherly/CORBIS; **574** Amanda Kaye; **574–575** Bryan Barr; **576–577** Christine Osborne/CORBIS; **579** SuperStock; **583** Michele Wigginton; **584** Michelle Bridwell/PhotoEdit; **595** Hank Morgan/Photo Researchers; **599** IN THE BLEACHERS ©1997 Steve Moore. Reprinted with permission of Universal Press Syndicate. All rights reserved; **603** Jenny Hager/ImageState; **609** Jeff Greenberg/Visuals Unlimited; **612** SPL/Photo Researchers; **630** Steve Liss/Timepix; **630–631** Greg Mathieson/Timepix; **632** D.F. Harris; **635** ©1979 United Feature Syndicate, Inc.; **636** (l)Mitch Kezar/Getty Images, (r)Jim Erickson/CORBIS Stock Market; **638** Mark C. Burnett/Photo Researchers; **642** Matt Meadows; **644** CORBIS; **648** Food & Drug Administration/SPL/Photo Researchers; **651** Chris Trotman/DUOMO; **656** Charles Gupton/CORBIS Stock Market; **660** *The Born Loser* reprinted by permission of Newspaper Enterprise Association, Inc.; **662** Bob Daemmrich/Stock Boston; **667** Greg Fiume/New Sport/CORBIS; **668** SuperStock; **671** AFP/CORBIS; **675** Will & Deni McIntyre/Photo Researchers; **677** Gregg Forwerck/SportsChrome USA; **679** Steve Chenn/CORBIS; **683** Aaron Haupt; **685** HMS Images/Getty Images; **686** Aaron Haupt; **696** Photofest; **696–697** Ed and Chris Kumler; **698–699** Bill Ross/CORBIS; **705** John P. Kelly/Getty Images; **707** SuperStock; **709** L. Clarke/CORBIS; **713** Ray Juno/CORBIS Stock Market; **716** Aaron Haupt; **717** courtesy Skycoaster of Florida; **721** Reuters NewMedia Inc./CORBIS; **723** Otto Greule/Allsport; **729** Peter Miller/Photo Researchers; **731** SuperStock; **735** Roy Ooms/Masterfile; **737** John T. Carbone/Photonica; **744** R. Ian Lloyd/Masterfile; **746** Doug Plummer/Photonica; **748** SuperStock; **750** Steven E. Sutton/DUOMO; **760–761** Boden/Ledingham/Masterfile; **766** Larry Hamill; **773** Ben Edwards/Getty Images; **780** James Schot/Martha's Vineyard Preservation Trust; **789** Cosmo Condina/Getty Images; **795** SuperStock; **799** SuperStock; **803** (l)Getty Images, (r)Frank Wiewandt/Image Finders.

T1 T5 PhotoDisc; **T9** Corbis Image 100; **T14** file photo; **T16** Corbis Image 100.

Index

Red type denotes items only in the Teacher's Wraparound Edition.

A

Absolute value equations, 28, 53
graphing, 299
solving, 28–32, 39, 49

Absolute value functions, 90, 91, 92, 104, 115, 247, 272, 370, 499, 502, 503, 515, 599, 831, 848

Absolute value inequalities, 40–46, 86, 829
graphing, 97, 335
multi-step, 42
solving, 50

Addition
Associative Property, 15, 162, 166, 828
Commutative Property, 15, 162
complex numbers, 270, 272
Distributive Property, 221
functions, 383, 403
matrices, 160
polynomials, 229, 277
probabilities, 658–663, 689–690
properties, 25
radicals, 252, 253
rational expressions, 480, 514
signs for, 46
solving inequality, 34

Addition Property of Equality, 21

Addition Property of Inequality, 33

Additive identity, 15, 32, 162, 828

Additive inverses, 13, 15, 16, 18, 153, 828

Addressing Individual Needs, T15, *See also* Prerequisite Skills and Differentiated Instruction

Algebra Activity
Adding Radicals, 252
Area Diagrams, 651
Arithmetic Sequences, 580
Completing the Square, 308
Conic Sections, 453–454
Distributive Property, 13
Factoring Trinomials, 240
Fractals, 611
Graphing Equations in Three Variables, 136–137
Head versus Height, 83
Inverses of Functions, 392
Investigating Ellipses, 432
Investigating Exponential Functions, 522
Investigating Polygons and Patterns, 19
Investigating Regular Polygons using Trigonometry, 716
Locating Foci, 437
Midpoint and Distance Formulas in Three Dimensions, 417–418
Multiplying Binomials, 230
Parabolas, 421
Rational Functions, 487
Simulations, 681
Special Sequences, 607
Testing Hypotheses, 686

Algebraic expressions, 828
evaluating, 7, 8, 9, 18, 27, 30, 53, 109
fraction bar, 7
simplifying, 14, 15, 16, 27, 48, 53, 62
verbal expressions, 20, 24, 115

Algebra tiles. *See also* Modeling
binomials, 230
complete the square, 308
modeling binomials, 230
polynomials, 240

Alge2PASS, *See* Tutorials

Algorithms, division, 233–234

Alternative hypothesis, 686

Alternative method, 580, 590, 652

Alternative representations, 726

Amortization, 605
schedule, 605

Amplitude, 763, 764, 765, 766, 767, 771, 774, 775, 776, 781, 785, 805, 806, 859

"And" compound inequalities, 40

Angles, 709–716, 734, 753
coterminal, 711, 738
depression, 705
elevation, 705
finding, 721
general, 754
inclination, 779
measurement, 709, 711, 712, 745, 748, 753
quadrantal, 718
reference, 718–719, 722, 776
trigonometric function of general, 717–724
vertex, 192
vertices, 113, 192

Angles formulas
differences, 786–790, 807
sum, 786–790, 807

Angular velocity, 714

Antinodes, 791

Apothem, 716

Applications, T3, T5, *See also* Cross-Curriculum Connections; More About
acidity, 550
activities, 510
advertising, 459, 668
aeronautics, 732
aerospace, 425, 429, 587
aerospace engineering, 266
agriculture, 565, 863, 865, 871
airports, 122
altitude, 557
amusement parks, 380
ancient cultures, 72
animal control, 528
animals, 319, 827
archery, 298
architecture, 497, 503, 749
art, 490, 865, 872
astronomy, 226, 238, 262, 310, 438, 439, 440, 445–447, 459, 478, 498, 550, 712, 862, 869
auto maintenance, 517
automobiles, 380
automotive engineering, 255
auto safety, 489
aviation, 450, 706, 737, 790, 795, 869
babysitting, 39
baking, 16, 127
ballooning, 341
band boosters, 15
banking, 9, 538, 608
baseball, 88, 333, 722, 779, 827
basketball, 17, 95, 143, 490, 874
boating, 298, 768
bowling, 25
bridge construction, 705
bridges, 424
building design, 550
bulbs, 745
business, 26, 79, 80, 89, 97, 158, 165, 174, 181, 194, 237, 256, 334, 352, 565, 570, 670
cable cars, 874
cable TV, 356
caffeine, 560
camera supplies, 174
card games, 642, 649
car expenses, 26
carousels, 723
car rental, 51
cars, 679, 713
car sales, 38
cartography, 637

charity, 823
child's play, 602
clocks, 602, 617
clubs, 872
coffee, 31
coins, 571
communications, 423, 559, 767, 789, 869, 874
community service, 299
computers, 563, 582
construction, 266, 292, 586, 643, 821, 862, 865, 867
contest, 46
cooking, 142
crafts, 863, 865
cryptography, 199, 200
cycling, 510
deliveries, 36
dentistry, 227
design, 363, 627, 868
diet, 549
dining, 143
dining out, 157
diving, 304, 327
drama, 98
driving, 827
earthquake, 797
earthquakes, 458, 545, 547, 871
ecology, 79, 207
economics, 66, 114, 261, 564, 610, 684
education, 328, 648, 660, 667, 863, 864, 866, 870, 873
elections, 655
electricity, 18, 122, 273, 274, 483, 517
electronics, 389, 780
employment, 357, 863, 868
energy, 869
engineering, 369
entertainment, 73, 237, 399, 581, 598, 635, 713
e-sales, 231
exercise, 121, 707, 864
extreme spots, 296
family, 26
farming, 134
figure skating, 638
finance, 61, 85, 173, 388
financial planning, 405
firefighting, 398
fish, 248
flagpoles, 821
flooring, 44
food, 30, 632, 674
football, 318, 668
footprints, 180
forestry, 731
fountains, 326, 750
framing, 311
fund-raising, 67, 173, 334
furniture, 275
games, 193, 616, 825, 872
gardening, 607, 802
gardens, 484
genealogy, 648
genetics, 648
golf, 809, 822, 823
government, 61, 88, 641
gymnastics, 180
health, 45, 84, 95, 267, 425, 452, 503, 597, 824, 863, 873
highway safety, 319
hobbies, 62
hockey, 84
home security, 636
hotels, 157
housing, 81, 82, 121
hurricanes, 126
insurance, 94
interior design, 193, 438
Internet, 80
intramurals, 616
inventory, 121
investing, 192
investments, 140
kennel, 312
landscaping, 180, 243, 334, 412, 430, 597
laughter, 497
law enforcement, 254, 298, 335, 866
lawn care, 327
life expectancy, 865
light, 803
lighting, 780
loans, 609
lotteries, 642, 648
mail, 45, 503
manufacturing, 31, 132, 147, 149, 424, 674, 868, 870
marriage, 440, 824
measurement, 863
media, 684
medicine, 10, 84, 237, 376, 488, 544, 563, 592
meteorology, 675
mirrors, 459
models, 867
money, 10, 27, 529, 551, 824, 826
movies, 641
movie screens, 310
music, 775
navigation, 507, 723, 732
newspapers, 291
noise ordinance, 537
nuclear power, 426
number games, 394
nursing, 9
nutrition, 94
oceanography, 201, 249
optics, 248, 750
ownership, 564
packaging, 26, 134, 363
pagers, 448
paleontology, 561, 563
paper, 144
parking, 93
parties, 620
part-time jobs, 37, 126
passwords, 635
personal finance, 231, 369
pets, 341
photography, 244, 304, 431, 447, 451, 657, 819, 870
photos, 113
physics, 767
pilot training, 206
pool, 363
population, 227, 529, 558, 562, 570, 862, 863, 864, 866, 868, 871
population growth, 388, 563
pricing, 193
produce, 172
production, 133
puzzles, 620
quality control, 673
radio, 430, 731
radioactivity, 587
real estate, 415, 563
recreation, 88, 105, 165, 604, 627
retail, 377
retail sales, 215
rides, 867, 874, 875
rockets, 458, 875
roller coasters, 398
rumors, 558
running, 188
safety, 84, 868
salaries, 591
sales, 77, 99, 868
satellites, 869
savings, 556, 557
school, 634
schools, 26, 37, 51, 74, 98, 135, 206, 615, 641, 661, 662, 668, 674, 823, 825
school shopping, 17
school trip, 26
scrapbooks, 166
sculpting, 376
shadows, 819
shopping, 39, 62, 98, 125, 149, 387, 685
skiing, 120
skycoasting, 723
skydiving, 281
slope, 744
soft drinks, 827
sound, 535, 542, 545
space, 563
space exploration, 124
speed limits, 45, 873
speed skating, 663
sports, 164, 171, 248, 255, 358, 425, 451, 678, 872
stamps, 144
state fair, 37
statistics, 511
structural design, 446
surveying, 707, 737, 819, 874
surveys, 873
swimming, 495

taxes, 67, 386
telecommunications, 497
telephones, 80
television, 83, 875
temperature, 388
tennis, 298
test grades, 38
theatre, 84
thinking, 564
tides, 775, 875
tourism, 292
toys, 586
transportation, 88, 93, 228, 487, 636
travel, 69, 73, 113, 143, 249, 415, 707, 750, 862, 870
tunnels, 467
utilities, 656
vending, 674
water, 451
water supply, 496
weather, 60, 72, 156, 597, 823, 862, 867, 868, 871, 874, 875
weekly pay, 103
White House, 439
woodworking, 416, 730
work, 16, 496, 509
world cultures, 586
world records, 536
writing, 649

Arcsine function, 747

Area
circles, 9, 415, 502
diagrams, 651
hexagons, 707
parallelograms, 477
polygons, 187
rectangles, 255, 334
trapezoids, 8, 67, 865
triangles, 32, 184, 185, 186, 187, 231, 281, 866

Area tiles. *See* Algebra tiles; Modeling

Arithmetic expressions, simplifying, 6

Arithmetic means, 580, 582, 590, 592, 622, 623, 851

Arithmetic operations, 383–384

Arithmetic sequences, 578–582, 579, 583, 622–623, 768, 851
modeling, 580
*n*th term, 579, 591, 851

Arithmetic series, 583–587, 620, 623, 851
sum, 583, 584, 586, 592

Assessment, T4, T10–T11, 4E, 4F, 54E, 54F, 108E, 108F, 152E, 152F, 220E, 220F, 284E, 284F, 344E, 344F, 410E, 410F, 470E, 470F, 520E, 520F, 576E, 576F, 630E, 630F, 698E, 698F, 760E, 760F
Open-Ended Assessment
Modeling, 18, 32, 67, 74, 95, 115, 127, 174, 188, 244, 249, 299, 319, 352, 389, 394, 416, 452, 498, 504, 530, 565, 592, 650, 663, 708, 732, 745, 768, 790
Speaking, 10, 27, 62, 98, 122, 144, 166, 194, 207, 228, 256, 262, 275, 293, 305, 335, 364, 370, 382, 431, 440, 460, 478, 546, 559, 582, 587, 610, 643, 657, 680, 684, 715, 751, 781, 784, 803
Writing, 39, 46, 80, 86, 135, 158, 181, 201, 232, 238, 267, 312, 328, 358, 377, 399, 425, 448, 484, 490, 511, 538, 551, 598, 604, 617, 637, 670, 675, 724, 738, 776, 797
Practice Chapter Test, T10, 51, 105, 149, 215, 281, 341, 405, 467, 517, 571, 627, 693, 757, 809
Practice Quiz, T10, 18, 74, 95, 122, 135, 174, 194, 238, 256, 328, 364, 382, 431, 448, 484, 498, 538, 559, 592, 617, 650, 670, 715, 738, 781, 797
Prerequisite Skills, 5, 10, 18, 27, 32, 39, 55, 62, 67, 74, 80, 86, 95, 109, 115, 122, 127, 135, 153, 158, 166, 174, 181, 188, 194, 201, 221, 228, 232, 238, 244, 249, 256, 262, 267, 285, 293, 299, 305, 312, 319, 328, 345, 352, 358, 364, 370, 377, 382, 389, 394, 411, 416, 425, 431, 440, 448, 452, 471, 478, 484, 490, 498, 504, 521, 530, 538, 546, 551, 559, 577, 582, 587, 592, 598, 604, 610, 617, 631, 637, 643, 657, 663, 670, 675, 680, 699, 708, 715, 724, 732, 738, 745, 761, 768, 776, 781, 785, 790, 797, 814–827
Standardized Test Practice, T3, 10, 17, 23, 24, 27, 31, 39, 46, 51, 67, 74, 76, 78, 80, 86, 95, 99, 105, 115, 117, 120, 122, 127, 134, 144, 149, 158, 166, 173, 176, 179, 181, 187, 194, 201, 207, 215, 228, 232, 236, 238, 243, 244, 249, 255, 267, 281, 292, 299, 302, 303, 305, 312, 319, 327, 335, 341, 352, 358, 364, 370, 374, 375, 377, 382, 389, 394, 399, 405, 413, 414, 416, 425, 431, 439, 447, 452, 459, 467, 473, 476, 478, 484, 490, 498, 503, 511, 517, 530, 537, 546, 559, 562, 563, 564, 582, 587, 588, 591, 592, 598, 603, 610, 616, 621, 627, 634, 636, 642, 649, 657, 662, 669, 675, 685, 693, 706, 708, 724, 732, 737, 745, 757, 768, 776, 781, 784, 785, 790, 796, 804, 809
Extended Response, 53, 107, 151, 217, 343, 407, 469, 519, 551, 573, 621, 629, 695, 714, 759, 811
Grid In, 680, 751
Multiple Choice, 52, 106, 150, 216, 342, 406, 468, 518, 572, 628, 633, 694, 702, 758, 783, 810
Open Ended, *See* Extended Response
Short Response/Grid In, 53, 107, 151, 217, 343, 407, 469, 519, 573, 629, 695, 759, 811
Test-Taking Tips, 23, 52, 76, 106, 117, 151, 176, 217, 234, 282, 302, 342, 407, 468, 473, 519, 562, 572, 588, 628, 633, 695, 702, 758, 783, 811
See also Preparing for Standardized Tests

Associative Property
Addition, 15, 162, 166, 828
Multiplication, 15, 171

Asymptotes, 491, 530
determining, 471
hyperbola, 846, 848
vertical, 617, 763

Auditory/Musical, *See* Differentiated Instruction

Augmented matrices, 208

Axis
conjugate, 442
minor, 434
symmetry, 287–288, 290, 291, 299, 339, 839
transverse, 442

B

Bar graphs, 824

Base *e* equations, 555, 569

Base *e* inequalities, 556

Base *e* logarithms, 554–559
inverse property, 555

Base formula, 548, 549

Bias, 682

Biased sample, 682

Binomials, 229, 366, 368, 382
expansions, 631, 676
experiments, 676–681, 677, 691–692

Binomial Theorem, 612–617, 625–626
factorial form, 614

Bivariate data, 81

Boundary, 96

Bounded region, 129–130

Box-and-whisker plots, 631, 826–827

Break-even point analysis, 110, 111

Building on Prior Knowledge, 6, 11, 20, 28, 33, 40, 56, 63, 68, 75, 81, 89, 110, 129, 138, 154, 162, 170, 239, 286, 301, 306, 313, 329, 353, 360, 366, 371, 412, 419, 449, 485, 499, 505, 523, 599, 676, 701, 762, 770, 800

C

Calculator. *See* Graphing calculator; Graphing Calculator Investigation

Career Choices, T5
- archaeologist, 187
- atmospheric scientist, 126
- chemist, 511
- cost analyst, 237
- designer, 363
- electrical engineering, 274
- finance, 85
- forester, 446
- interior design, 193
- landscape architect, 334
- paleontologist, 561
- physician, 685
- real estate agent, 609
- sound technician, 542
- surveyor, 707
- travel agent, 496
- veterinary medicine, 131

Cartesian coordinate plane, 56

Center
- circles, 426, 845
- ellipses, 434

Central tendency measures of, 664, 822–823

Change of Base Formula, 548, 569

Checking solutions, 13, 22, 24, 25, 29, 30, 31, 34, 39, 46, 49, 51, 62, 110, 115, 117, 197, 207, 263, 264, 294, 302, 309, 314, 315, 325, 361, 362, 367, 379, 481, 506, 509, 516, 526, 527, 528, 530, 533, 534, 535, 536, 538, 542, 543, 544, 546, 548, 551, 555, 559, 580, 604, 621, 643, 657, 801, 849, 850

Circles, 426–431, 450, 451, 460, 463, 467, 565, 617
- area, 9, 415, 502
- center, 426, 845
- circumference, 496, 710
- connecting points, 352
- eccentricity, 440
- equations, 426, 846
- graphing, 428, 429
- radius, 845
- sectors, 713
- unit, 710, 740, 742, 743

Circular functions, 739–745, 756, 761

Circular permutation, 642

Circumference, 496, 710

Closure Property, 18

Coefficients, 222, 448
- integral, 376
- leading, 379
- least, 389

Column matrix, 155, 156

Combinations, 638–643, 640, 641, 650, 688, 715

Combined variation, 497

Common difference, 578

Common logarithms, 547–553, 559, 617

Common Misconception, 7, 12, 29, 118, 130, 289, 308, 523, 659, 703, 782 *See also* Find the Error

Common ratio, 588, 603

Communication, 633
- compare and contrast, 178, 673, 742
- copy, 60
- decide, 71, 242, 273, 590
- define, 297, 706, 712, 774
- describe, 8, 156, 163, 185, 192, 317, 350, 362, 397, 445, 535, 619, 683, 749, 779, 784, 788
- determine, 8, 24, 171, 226, 247, 254, 273, 310, 332, 386, 393, 445, 476, 495, 641, 722, 729, 788
- disprove, 14
- draw, 660
- evaluate, 92
- examine, 332
- explain, 14, 30, 37, 65, 78, 98, 112, 142, 156, 171, 185, 198, 236, 247, 260, 265, 297, 310, 317, 325, 350, 356, 362, 375, 380, 393, 397, 414, 450, 476, 549, 563, 580, 586, 596, 602, 619, 634, 673, 678, 722, 736, 749, 766, 779, 784, 794, 802
- graph, 458
- identify, 78, 231–232, 290, 414, 423, 437, 527, 615, 774
- list, 615, 634
- make, 119
- name, 65, 544, 549, 557, 712
- show, 185, 265, 641
- sketch, 458
- state, 78, 125, 290, 356, 368, 375, 509, 608, 742
- tell, 125, 350, 802
- verify, 647
- write, 30, 37, 43, 98, 163, 178, 192, 198, 205, 303, 325, 368, 428, 450, 563, 608, 647, 654

Commutative Property
- Addition, 15, 162
- Multiplication, 15, 32, 166, 170, 828

Comparison
- quantitative, 117, 120
- real numbers, 5, 814

Completing the square, 306–312, 328, 338, 352, 411, 490, 587, 840

Complex conjugates, 273, 374–375

Complex fractions, 475, 481

Complex numbers, 270–275, 280, 370
- addition, 270, 272
- division, 272–273
- multiplication, 272–273
- subtraction, 270, 272

Complex roots, 315

Composition, functions, 384–386, 530, 532

Compound event, 658

Compound inequalities, 40–46, 50
- and, 40
- or, 41

Concept Check, 13, 29, 35, 112, 118, 224, 241, 250, 315, 764

Concept Summary, 47, 48, 49, 57, 69, 92, 100, 101, 102, 103, 104, 112, 146, 162, 171, 177, 178, 209, 210, 211, 212, 213, 214, 239, 246, 251, 260, 265, 276, 277, 278, 279, 280, 317, 323, 336, 337, 338, 339, 340, 349, 371, 400, 401, 402, 403, 404, 422, 449, 450, 461, 462, 463, 464, 465, 466, 499, 513, 514, 515, 516, 566, 567, 568, 569, 570, 622, 623, 624, 625, 626, 634, 664, 687, 688, 689, 690, 691, 692, 735, 747, 752, 753, 754, 755, 756, 772, 805, 806, 807, 808

Conditional probability, 653

Cones, surface areas, 22, 266

Congruent angles, 817

Conic sections, 419, 449–452, 453–454, 465–466, 869

Conjectures, 19, 32, 83, 119, 240, 252, 432, 437, 489, 522, 558, 585, 607, 681, 686, 716

Conjugate axis, 442

Conjugates, 253

Conjunctions, 42

Constant functions, 90, 92, 115, 370, 499, 502, 515, 831

Constants, 104, 222, 530
- variation, 492

Constraints, 129

Constructed Response. *See* Preparing for Standardized Tests

Continuous functions, 62, 524

Continuously compounded interest, 556

Continuous probability distribution, 671

Convergent series, 599, 622

Coordinate matrix, 175

Coordinate plane, 110

Coordinates, finding, 721

Coordinate system, 56

Corollary, 372

Corresponding elements, 156

Cosecant function, 701

Cosine function, 701, 706, 707, 740, 747, 767, 770, 771
definition, 739
value, 747

Cotangent function, 701

Coterminal angles, 711, 712, 738

Counterexamples, 14, 16, 32, 92, 185, 242, 580, 619, 620, 621, 643, 666, 706, 794, 853

Counting Principle, 632–637, 644, 687–688

Course Planning Guide, *See* Pacing

Cramer's Rule, 189–194, 207, 213, 724, 835
solving systems of equations, 670
three variables, 191
two variables, 189

Critical Thinking, T3, 10, 17, 27, 31, 38, 45, 62, 66, 73, 80, 85, 94, 99, 114, 121, 127, 133, 143, 157, 166, 172, 173, 181, 187, 193, 200, 207, 227, 232, 237, 243, 249, 255, 262, 267, 275, 292, 298, 304, 311, 319, 327, 334, 357, 364, 369, 376, 377, 380, 389, 394, 398, 416, 425, 430, 439, 446, 452, 459, 477, 483, 489, 497, 503, 511, 529, 537, 545, 546, 550, 558, 582, 587, 592, 598, 603, 610, 616, 621, 635, 642, 649, 656, 662, 669, 675, 679, 685, 708, 714, 723, 732, 737, 744, 750, 767, 776, 780, 785, 789, 796, 804

Cross-Curriculum Connections. *See also* Applications; More About
anthropology, 563
biology, 62, 227, 262, 350, 497, 529, 545, 564, 570, 621, 744, 767, 872
chemistry, 203, 205, 206, 312, 460, 496, 511, 570
geography, 58, 85, 187, 415, 451, 647, 796, 825
geology, 67, 581, 708, 757
history, 489
literature, 656, 724
physical science, 779
physics, 66, 237, 267, 292, 318, 370, 393, 510, 546, 557, 604, 743, 751, 774, 784, 788, 789, 796, 802, 866, 867, 870, 874
physiology, 357, 672
science, 80, 83
spelling, 656
zoology, 775

Cross products, 181

Cube root equation, 264

Cubes, volumes, 615

Curve fitting, 300, 359, 539

Cylinders, surface areas, 25, 862

D

Dashed boundary, 96–97

Data
analyzing, 522, 681, 716
box-and-whisker plots, 631, 826–827
collecting, 522, 681, 716
distribution, 672
graphs of polynomial functions, 353, 357
modeling real-world, 359
organizing, 154, 159
scatter plots, 81–86, 87, 95, 99, 103, 598, 831
skew, 856
stem-and-leaf plots, 667, 825

Decay
exponential, 524, 525, 528, 560–565, 561, 567, 570, 849
rate of, 560

Decimals, 838, 850
approximations for irrational numbers, 247
repeating, 601, 602, 603, 852

Degrees, 222, 724, 753, 757, 802, 803
converting radian measures between, 711
measurement, 711
polynomials, 229, 346, 350, 400, 837, 842

Denominators
monomials, 480
polynomials, 475, 480

Dependent events, 633–634, 635, 653, 654, 655, 687, 689, 854, 855

Dependent variable, 59

Depressed polynomial, 366

Depression, angle of, 705

Descartes, René, 372

Descartes' Rule of Signs, 372–373, 379

Determinants, 182–188, 212
evaluating
using diagonals, 835
using expansion by minors, 835
finding value, 186, 835
second-order, 182
third-order, 182, 183
3×3 matrices, 182, 183
2×2 matrices, 182
value, 185, 194

Deviation, mean, 669

Diagonals, 19, 182, 183, 184, 201, 642
in decagon, 776
evaluating determinants, 186, 835

Differences
rewriting as sums, 221
squares, 816

Differentiated Instruction, T6, T15
Auditory/Musical, 59, 170, 242, 259, 289, 397, 434, 495, 527, 587, 720, 793
Interpersonal, 24, 97, 111, 184, 225, 236, 349, 481, 501, 543, 590, 634, 667, 784, 801
Intrapersonal, 35, 78, 191, 251, 331, 367, 450, 476, 660, 728
Kinesthetic, 14, 42, 155, 309, 373, 422, 556, 582, 608, 672, 677, 712, 773
Logical, 119, 198, 204, 232, 265, 316, 379, 391, 443, 456, 509, 561, 601, 778
Naturalist, 71, 82, 324, 385, 428, 549, 596, 646, 654, 742, 787
Verbal/Linguistic, 29, 92, 125, 141, 162, 271, 296, 356, 615, 683, 735
Visual/Spatial, 8, 65, 131, 177, 247, 303, 362, 414, 488, 534, 619, 640, 704, 748, 763

Dilations, 175, 176, 177

Dimensional analysis, 225, 708

Dimensions, 155

Directrix, 419

Direct substitution, 366, 368

Direct variation, 492–493, 495, 496, 499, 502, 515, 559, 650, 848

Discrete function, 62

Discrete Mathematics
See Matrices
arithmetic sequences, 578–582, 583, 622–623, 768, 851
arithmetic series, 583–587, 623, 851
Binomial Theorem, 612–617, 625–626
combinations, 638–643, 650, 688, 715
Fibonacci sequences, 606, 609, 610
finite graphs, 636
geometric sequences, 588–593, 623–624, 852
geometric series, 594–598, 624, 781
induction, 618–621, 626
number sets, 11–12, 18, 828
number theory, 15, 295, 297, 298, 304, 510, 866, 872, 873
permulations, 638–643, 650, 688, 715
recursion, 606–611, 625
recursive formula, 606, 607, 608

Discrete probability distributions, 671

Discriminant, 328
quadratic formula, 313–319, 339

Disjunctions, 42

Index

Distance Formulas, 413–414, 415, 416, 417–418, 425, 441, 461–462, 467

Distributions
continuous probability, 671
discrete probability, 671
normal, 671–675, 680
probability, 646
skewed, 671

Distributive Property, 12, 13, 14, 15, 17, 32, 162, 166
Addition, 221
Multiplication, 170, 171, 228, 828

Divisibility, 619

Division
algorithm, 233–234
complex numbers, 272–273
functions, 384, 403
polynomials, 233, 277, 364, 365–366
properties of equality, 21
rational expressions, 474, 513
simplifying expressions, 223
solving inequality, 35
synthetic, 345, 745, 837

Division Property of Equality, 21

Division Property of Inequality, 34

Domain, 56, 57, 58, 61, 93, 94, 95, 99, 100, 101, 104, 181, 397, 398, 416, 523, 527, 528, 530, 830, 831, 844, 849
range, 67

Double-angle formulas, 791–798, 808

Double root, 302

Doubling time, 558

E

Eccentricity, 440, 452

Edges, 636

Elements, 155
corresponding, 156

Elevation, angle of, 705

Elimination, 146, 149, 153, 504
simplifying rational expressions, 473
solving systems of equations, 118–119, 120, 122, 135, 166, 832

Ellipses, 432, 433–440, 450, 451, 452, 460, 464, 467, 565, 617
center, 846
equations, 433–435, 643
graphing, 435–437
major axes, 846
minor axes, 846
writing equations, 846

Empty set, 29

End behavior, 349

Endpoints, 418

Energy, 530

English-Language Learner (ELL), T6, T15, 4, 9, 17, 26, 29, 31, 38, 45, 47, 54F, 54, 61, 66, 73, 79, 85, 92, 94, 99, 100, 108, 114, 121, 125, 126, 134, 141, 143, 145, 152F, 152, 157, 162, 165, 173, 180, 187, 193, 200, 206, 209, 220, 227, 231, 237, 243, 248, 255, 261, 266, 271, 274, 276, 284F, 284, 292, 296, 298, 304, 311, 318, 327, 334, 336, 344, 351, 356, 357, 363, 369, 376, 381, 388, 393, 398, 400, 410F, 410, 415, 424, 430, 439, 447, 451, 459, 461, 470, 477, 483, 489, 496, 503, 510, 513, 520F, 520, 529, 537, 545, 550, 558, 564, 566, 576, 581, 586, 591, 597, 603, 609, 615, 616, 621, 622, 630F, 630, 636, 642, 649, 656, 662, 669, 674, 679, 683, 685, 687, 698, 707, 714, 723, 731, 735, 737, 744, 750, 752, 760F, 760, 767, 775, 780, 785, 789, 796, 803, 805

Enrichment, T4, *See* Critical Thinking; Extending the Lesson

Equal matrices, 209

Equate complex numbers, 271

Equations, 23. *See also* Quadratic equations; Systems of equations
absolute value, 28–32, 39, 49, 299
circles, 426
complex solutions, 309
cube root, 264
ellipses, 433–435
equivalent exponential, 565, 850
exponential, 526, 548, 570
solving, with logarithms, 548
writing equivalent, 570
forms, 75–80
graphing, 471
hyperbolas, 441–443
imaginary solutions, 271
irrational roots, 307
linear, 63–67, 75–80, 86, 101, 102, 109, 189, 191, 452, 830
logarithmic, 533, 543, 546, 551, 565, 570, 850
matrix, 202–203, 358, 370, 834, 836
solving, 205, 834, 836
writing, 202–203, 836
midline, 771, 774, 775, 781
multi-step, 22, 201
for *n*th term, 579, 589
one-step, 21
parabolas, 419–420
polynomial, 360–364, 401, 837
prediction, 81–82, 83, 84, 95, 99, 598
quadratic, 604
radical, 263–269, 280, 362
rational roots, 306, 505–509, 516
regression, 87
rounding, 776
solving, 20–27, 25, 48–49, 153, 157, 174, 535, 536, 538, 544, 546, 549, 550, 557, 558, 559, 565, 568, 569, 570, 577, 582, 604, 621, 637, 643, 657, 708, 747, 768, 828, 829, 839, 849, 850, 862
involving matrices, 155–156, 202
with inverses, 746
using Properties of Logarithms, 543
with rational numbers, 471
trigonometric, 799–804, 808
two-variable matrix, 202

Equilateral triangles, 869

Equivalent exponential equations, 565, 850

Equivalent expressions, 555

Error, 692
measurement, 704, 738
sampling, 682–686, 714

Error Analysis. *See* Find The Error; Common Misconceptions

Estimating, 225, 296

Evaluation, *See also* Assessment

Events, 632
compound, 658
dependent, 633–634, 634, 635, 653, 654, 655, 687, 689, 854, 855
inclusive, 659, 660, 661, 670, 689, 690, 855
independent, 632–633, 634, 651, 652, 654, 687, 689, 854, 855
multiple, 640
mutually exclusive, 658–659, 661, 670, 689, 690, 855
odds, 854

Excluded values, 472

Exclusive events, mutually, 658–659, 661, 670, 689, 690, 855

Expansion by minors, 182, 183, 186, 201
evaluating determinants, 186, 835

Expansions, binomials, 631, 676

Expected value, 681

Experimental probability, 649

Exponential decay, 524, 525, 528, 560–565, 561, 567, 570, 849

Exponential equations, 526
solving, with logarithms, 548
writing equivalent, 570

Exponential form, 257, 532, 535, 536, 568, 849

Exponential functions, 520, 566–567
graphing, 523
property of equality, 526

property of inequality, 527
solving, 526
writing, 525, 528

Exponential growth, 524, 525, 528, 560–565, 562, 567, 570, 849

Exponential inequalities, solving, 527
with logarithms, 548

Exponential relations, 871

Exponents
inverse property, 533
irrational, 526
negative, 222
radical, 279
rational, 257–262, 361–362, 838

Expressions, 47–48, 53, 779. *See also* Algebraic expressions; Arithmetic expressions; Radical expressions; Rational expressions; Verbal expressions
evaluating, 158, 201, 394, 535, 536, 546, 557, 558, 568, 570, 577, 582, 610, 615, 617, 631, 637, 641, 643, 650, 779, 780, 790, 828, 829, 838, 853, 854
simplifying, 223–224, 528, 538, 546, 604, 637, 776, 778, 779, 780, 790, 828, 838, 847

Extended Response, 364. *See also* Preparing for Standardized Tests

Extending the Lesson, 18, 32, 62, 80, 86, 275, 299, 335, 416, 440, 447, 452, 636, 642, 649, 669, 738

Extraneous solutions, 263–264, 534

Extra Practice, 828–861

F

Factorial, 613, 614, 637

Factoring, 367, 460
polynomials, 239–241, 358, 377, 815–816, 837
solving quadratic equations by, 301–305, 338, 840
solving system of equations, 643

Factors, polynomials, 366

Factor Theorem, 365–370, 402

Failure, 644
probability, 644

Families of graphs, 70
absolute value graphs, 91
parabolas, 320–321

Feasible region, 129, 134, 833

Fibonacci sequence, 606, 609, 610

Field, 12

Figures
congruent, 817–819
similar, 817–819
translating, 175

Find the Error, 24, 43, 60, 71, 119, 142, 185, 205, 226, 236, 303, 310, 325, 380, 386, 423, 428, 481, 509, 535, 544, 557, 590, 602, 654, 660, 730, 735, 766. *See also* Common Misconceptions

Finite graph, 636

Focus
ellipse, 432
parabola, 419

FOIL Method, 230, 240

Foldables™ Study Organizers, T5–T6, T14–T15, 5, 47, 55, 100, 109, 145, 153, 209, 221, 276, 285, 336, 345, 400, 411, 461, 471, 513, 521, 566, 577, 622, 631, 687, 699, 752, 761, 805

Forms of equations, 75–80

Formulas, 6–10, 25, 47–48, 122
angles, 786–790
area, 184
base, 548, 549
change of base, 548, 569
differences, 786–787, 790
distance, 413–414, 415, 416, 417–418, 425, 441, 461–462, 467
double-angle, 791, 808
half-angle, 791–798, 792, 793, 794, 795, 797, 808, 861
midpoint, 412, 414, 416, 417–418, 461–462, 467
quadratic, 313–319, 339, 345, 370, 460, 841
recursive, 606, 607, 608
summation, 618
sums, 596, 600, 786–787, 790

45°-45°-90° triangles, 699, 703, 707

Fourth term, 589

Fractals, 611

Fraction bar, 7

Fractions
complex, 475, 481
repeating decimals, 601, 602, 603, 852

Free Response. *See* Preparing for Standardized Tests

Function notation, 59

Function values, 348, 604

Functions, 57, 100–102, 830
absolute value, 91, 92, 115, 370, 499, 502, 503, 515, 831, 848
addition, 383, 403
circular, 739–745,756, 761
classes, 499–504, 515
composition, 384–386, 521
constant, 370, 831
division, 384, 403
equations, 58–62
exponential, 523–530
graphing, 577, 768, 863
inverse, 390–394, 404, 405, 521, 617, 699, 749
inverse trigonometric, 746–751, 756
iterating, 608
multiplication, 384, 403
operations, 383–389, 403
periodic, 741
piecewise, 89–95, 104, 370, 831
step, 89–95, 370, 831
subtraction, 383, 403
zero, 376

Fundamental Counting Principle, 633, 644, 687

Fundamental Theorem of Algebra, 344, 371–372

G

General angles, 717, 754

Geometric means, 590, 591, 598, 623, 852

Geometric sequences, 588–593, 594, 623–624, 852
limits, 593
*n*th term, 589, 852
sums, 852
terms, 594

Geometric series, 594–598, 617, 620, 624, 781
infinite, 599–605, 624–625, 745, 852
sum, 595, 597, 610

Geometry, 186
areas
circles, 9, 415, 502
hexagons, 707
parallelograms, 477
polygons, 187
rectangles, 255, 334
trapezoids, 8, 67, 865
triangles, 32, 185, 186, 187, 231, 281, 866
arrays of numbers, 582
circumferences of circles, 496
degrees in convex polygon, 79
diagonals in decagons, 776
dimensions of inscribed rectangle, 292
equilateral triangles, 869
exact coordinates, 744
factoring, 243
height of parallelogram, 477
isosceles triangles, 869
leg of right triangle, 243
matrix multiplication, 200
measures of diagonals, 737

midpoint, 414
ordered pairs, 390
perimeters
octagons, 26
quadrilaterals, 415, 482
rectangles, 255
right triangles, 382
squares, 603
triangles, 592
perpendicular lines, 73
slope of a line, 481
squares, 609
surface areas
cones, 22, 266
cylinders, 25, 862
pyramids, 27
rectangular prisms, 18
spheres, 862
triangular numbers, 609
vertices
angles, 113, 192
parallelograms, 121, 192
triangles, 113, 415
volumes
cubes, 615
rectangles, 866
rectangular prism, 367
rectangular solid, 379, 380
width
rectangle, 242
rectangular prism, 363

Golden ratio, 311

Golden rectangle, 311

Graph functions, 285

Graphing
absolute value equations, 299
absolute value inequalities, 335
circles, 428, 429
ellipses, 435–437
exponential function, 523
horizontal translations, 770
hyperbola, 846
inequalities, 657, 680, 832, 841, 844
linear inequalities, 329
linear relations and functions, 863
parabolas, 420–423
polynomial functions, 348–349, 353–358, 401
polynomial model, 355
quadratic equations, 294–299, 337
quadratic functions, 286–293, 336–337
quadratic inequalities, 329–333, 340
rational functions, 514, 848
square root functions, 395–396, 404
square root inequalities, 397–399, 404
systems of equations, 110–115, 194, 832
systems of inequalities, 123–127, 135, 484, 833, 847
table of values, 352, 356, 364
transformations, 772
trigonometric functions, 762–768, 772
vertical translation, 771

Graphing calculator, T4, 39, 431, 444, 455, 456, 460, 585, 613
addition of trigonometric inverses, 751
approximating value, 247, 248
binomial distribution, 680
check factoring, 244
family of graphs, 74
families of graphs, 530
intersect feature, 115
inverse functions, 201
inverse matrices, 207
logic menu, 46
matrix function, 188
maxima, 293, 358
minima, 293, 358
shade command, 99
sum of each arithmetic series, 587
sum of geometric series, 598
verifying trigonometric identities, 785
Zero function, 296, 307

Graphing Calculator Investigation
augmented matrices, 208
factoring polynomials, 241
families of absolute value graphs, 91
families of exponential functions, 524
families of parabolas, 320–321
graphing rational functions, 491
horizontal translations, 769
limits, 593
lines of regression, 87–88
lines with same slope, 70
matrix operations, 163
maximum and minimum points, 355–356
modeling real-world data, 300, 359, 539–540
one-variable statistics, 666
order of operations, 7
point discontinuity, 491
quadratic systems, 457
sine and cosine on unit circle, 740
solving exponential and logarithmic equations and inequalities, 552–553
solving inequalities, 36
solving radical equations and inequalities, 268–269
solving rational equations by graphing, 512
solving trigonometric equations, 798
square root functions, 396
sums of series, 585
systems of linear inequalities, 128
systems of three equations in three variables, 205

Graphing functions, 577

Graph relations, 56–62

Graphs
bar, 824
finite, 636
line, 824

Greatest common factor, 239, 302

Greatest integer function, 89, 104, 499, 503, 515, 517, 530

Gridded Response. *See* Preparing for Standardized Tests

Grid In, 530, 708. *See also* Assessment

Grouping, 240
symbols, 6

Growth, exponential, 524, 525, 528, 560–565, 562, 567, 570, 849
rate of, 562

H

Half-angle formulas, 791–798, 792, 793, 794, 795, 797, 808, 861

Harmonics, 791

Hexagons, area, 707

Histogram, 669, 671
relative-frequency, 646, 647

Homework Help, 15, 24, 31, 37, 44, 60, 66, 72, 78, 84, 93, 98, 113, 120, 126, 132, 142, 156, 164, 172, 179, 186, 192, 199, 206, 226, 231, 237, 243, 248, 254, 261, 266, 274, 291, 304, 310, 318, 326, 333, 350, 356, 368, 375, 380, 387, 393, 398, 414, 424, 429, 458, 476, 482, 489, 496, 502, 510, 528, 536, 544, 550, 557, 563, 581, 586, 591, 597, 602, 609, 615, 620, 635, 641, 648, 655, 661, 667, 674, 678, 684, 706, 713, 722, 730, 736, 743, 749, 767, 775, 780, 784, 789, 795, 803

Horizontal lines, 65, 70

Horizontal line test, 392

Horizontal translations, 769–770
graphing, 770

Hyperbolas, 441–448, 450, 451, 452, 460, 464–465, 467, 565, 617, 670
equations, 441–443, 846
graphing, 443–444, 846

Hypothesis, 686

I

Identify functions, 92

Identify matrices, 213

Identify properties, real numbers, 13

Identities, 12, 861
additive, 15, 32, 162, 828

multiplicative, 15, 199
Pythagorean, 777
quotient, 777
reciprocal, 777
trigonometric, 777, 785, 806
verifying, 784, 788, 794

Identity function, 90, 391–392, 499, 515

Identity matrices, 195

Image, 175

Imaginary unit, 270

Imaginary zeros, 375, 402, 843

Inclination, angle of, 779

Included angle, 734

Inclusive events, 659, 660, 661, 670, 689, 690, 855

Independent events, 632–633, 634, 651, 652, 654, 687, 689, 854, 855

Independent variable, 59

Index of summation, 585

Indicated sum, 583

Indicated terms of expansion, 853

Indirect measurement, 705

Induction, mathematical, 618–621, 626

Inductive hypothesis, 618

Inequalities, 95, 122
absolute value, 829
graphing, 96–99, 104, 109, 115, 657, 680, 832, 841, 844
logarithmic functions property, 534
solving, 33–39, 39, 49–50, 62, 67, 74, 80, 352, 358, 521, 533, 534, 535, 536, 538, 546, 549, 550, 557, 558, 559, 565, 568, 569, 570, 604, 643, 829, 839, 849, 850, 862
writing, 36

Infinite geometric series, 599–605, 624–625, 745, 852
sigma notation, 601
sum, 600, 610

Infinity symbol, 601

Initial side, 709

Instructional Design, T2–T3

Integers, 11, 32, 48
positive, 620, 853

Integral coefficients, 376, 389

Integral Zero Theorem, 378, 403

Interactive Chalkboard, 7, 57, 117, 162, 223, 288, 348, 413, 473, 525, 579, 633, 702, 771

Intercept form, 80

Internal notation, 829

Internet Connections
T5, T7, T9, T11–T13, T15, T17, 4D, 4E, 54D, 54E, 108D, 108E, 152D, 152E, 220D, 220E, 284D, 284E, 344D, 344E, 410D, 410E, 470D, 470E, 520D, 520E, 576D, 576E, 630D, 630E, 698D, 698E, 760D, 760E
www.algebra2.com/careers, 26, 85, 121, 126, 187, 193, 237, 274, 334, 363, 446, 496, 511, 542, 561, 609, 685, 707
www.algebra2.com/chapter_test, 51, 105, 149, 215, 281, 341, 405, 467, 517, 571, 627, 693, 757, 809
www.algebra2.com/data_update, 10, 66, 143, 165, 255, 318, 357, 440, 477, 558, 598, 667, 723, 775
www.algebra2.com/extra_examples, 7, 13, 21, 23, 29, 35, 41, 59, 65, 69, 77, 83, 91, 97, 111, 117, 125, 131, 139, 155, 161, 169, 177, 182, 183, 191, 197, 203, 223, 229, 235, 241, 247, 251, 259, 265, 271, 289, 295, 303, 307, 315, 323, 331, 347, 355, 361, 379, 385, 391, 397, 413, 421, 427, 435, 443, 449, 457, 473, 481, 487, 493, 501, 507, 525, 533, 543, 549, 555, 561, 579, 585, 589, 595, 601, 607, 613, 619, 633, 639, 645, 653, 659, 665, 673, 677, 683, 685, 703, 711, 719, 727, 735, 741, 747, 765, 771, 779, 783, 787, 793, 801
www.algebra2.com/other_calculator_keystrokes, 86, 128, 208, 268, 320, 359, 491, 512, 539, 552, 593, 798
www.algebra2.com/self_check_quiz, 9, 15, 17, 31, 37, 45, 61, 73, 79, 85, 93, 99, 113, 121, 133, 143, 157, 165, 173, 179, 187, 193, 199, 207, 227, 231, 243, 249, 255, 261, 267, 275, 291, 297, 305, 311, 319, 327, 333, 351, 357, 363, 369, 375, 379, 381, 387, 393, 399, 415, 425, 429, 439, 445, 451, 459, 477, 483, 489, 497, 503, 511, 529, 537, 545, 551, 557, 563, 581, 587, 591, 597, 603, 609, 615, 621, 635, 641, 649, 655, 661, 667, 675, 679, 707, 713, 723, 731, 737, 743, 749, 767, 775, 781, 785, 789, 795, 803
www.algebra2.com/standardized_test, 53, 107, 151, 217, 283, 343, 407, 469, 519, 573, 629, 695, 759, 811
www.algebra2.com/usa_today, 16, 39, 69, 84, 135, 207, 228, 328, 368, 448, 493, 535, 565, 604, 659, 715, 797
www.algebra2.com/vocabulary_review, 47–50, 145, 209, 276, 400, 461, 513, 566, 622, 687, 752, 805
www.algebra2.com/webquest, 3, 27, 84, 120, 192, 207, 219, 227, 328, 369, 399, 409, 429, 504, 529, 565, 575, 616, 635, 697, 708, 775, 804

Interpersonal, *See* Differentiated Instruction

Interquartile range, 827

Intersecting lines, 111

Intervals, 803, 808
notation, 35, 37, 40, 41, 51

Intervention, T4–T5, T8–T9, 4E, 4F, 54E, 54F, 108E, 108F, 152E, 152F, 220E, 220F, 284E, 284F, 344E, 344F, 410E, 410F, 470E, 470F, 520E, 520F, 576E, 576F, 630E, 630F, 698E, 698F, 760E, 760F
Daily Intervention, 8, 14, 15, 18, 22, 24, 29, 35, 42, 44, 58, 59, 60, 65, 71, 78, 82, 92, 97, 111, 119, 125, 131, 140, 141, 142, 155, 161, 162, 170, 177, 183, 184, 185, 191, 198, 204, 205, 223, 225, 226, 232, 235, 236, 242, 244, 246, 247, 251, 253, 258, 259, 265, 271, 288, 289, 295, 296, 303, 309, 310, 316, 324, 325, 331, 349, 354, 356, 361, 362, 367, 373, 375, 379, 380, 385, 386, 391, 397, 414, 420, 422, 423, 428, 429, 434, 435, 442, 443, 450, 456, 474, 476, 481, 486, 488, 494, 495, 501, 509, 527, 534, 535, 542, 543, 544, 548, 549, 556, 557, 561, 579, 582, 587, 590, 596, 600, 601, 602, 608, 615, 619, 634, 639, 640, 646, 654, 660, 667, 672, 677, 683, 704, 712, 718, 720, 726, 728, 730, 735, 742, 748, 763, 766, 773, 778, 784, 787, 793, 801
Tips for New Teachers, 10, 27, 62, 74, 90, 144, 159, 166, 228, 238, 256, 262, 267, 275, 288, 305, 312, 323, 384, 416, 440, 448, 478, 484, 487, 498, 504, 511, 534, 582, 587, 592, 598, 604, 610, 617, 620, 648, 668, 703, 711, 793

Intrapersonal, *See* Differentiated Instruction

Inverse Cosine, 747

Inverse functions, 390–394, 399, 404, 405, 521, 531, 617, 699, 749, 844, 859

Inverse matrices, 195, 196, 201, 205, 206, 207, 213, 214, 228, 312, 358, 637

Inverse property
exponents, 533
logarithms, 533

Inverse relations, 390–394, 399, 404, 405, 844

Inverses, 195, 836

Index

additive, 13, 15, 16, 18, 153, 828
multiplicative, 13, 14, 15, 16, 32, 153, 199, 828
verifying, 196

Inverse Sine, 747

Inverse Tangent, 747

Inverse trigonometric functions, 746–751, 756

Inverse variation, 493–495, 496, 500, 515, 517, 559, 848

Irrational numbers, 11, 32

Irrational roots, 315

Isometry, 175

Isosceles triangles, 869

Iteration, 608, 853

J

Joint variation, 492–493, 496, 515, 559, 848

K

Key Concept, T2, T6, 6, 11, 12, 21, 28, 33, 34, 40, 41, 42, 57, 64, 68, 70, 75, 76, 130, 138, 160, 161, 162, 168, 182, 183, 184, 189, 191, 195, 196, 222, 223, 224, 230, 245, 250, 251, 257, 258, 271, 287, 288, 295, 301, 306, 307, 313, 316, 346, 347, 354, 360, 365, 374, 378, 383, 384, 390, 391, 412, 413, 420, 426, 434, 435, 442, 443, 474, 485, 492, 493, 494, 524, 526, 532, 533, 534, 541, 542, 543, 548, 579, 583, 589, 595, 600, 613, 614, 618, 633, 638, 639, 640, 644, 645, 652, 653, 658, 660, 665, 672, 677, 682, 701, 703, 711, 717, 718, 719, 725, 726, 727, 733, 739, 741, 747, 764, 770, 771, 777, 787, 791, 793

Keystrokes. *See* Graphing Calculator; Graphing Calculator Investigations; Internet Connections

Kinesthetic, *See* Differentiated Instruction

L

Law of Cosines, 733–738, 755, 858

Law of Large Numbers, 682

Law of Sines, 725–732, 726, 736, 754–755, 858

Leading coefficient, 346, 350, 379

Leaf, 667

Least common denominator (LCD), 505–506, 516

Least common multiple (LCM)
monomials, 479
polynomials, 479, 480, 482, 504, 847

Lesson Objectives, 4A, 54A, 108A, 152A, 220A, 284A, 344A, 410A, 470A, 520A, 576A, 630A, 698A, 760A

Like radical expressions, 252

Like terms, 229

Limits, 593

Linear correlation coefficient, 87

Linear equations, 63–67, 86, 101, 830
graphing, 109
identifying, 63
solving systems, 452
standard form, 64
systems of three, 191
systems of two, 189
writing, 75–80, 102

Linear function, 64, 830

Linear inequalities, graphing, 96, 329, 411

Linear permutations, 638

Linear programming, 129–135, 147

Linear-quadratic system, 455–456

Linear relations, graphing, 863

Line graphs, 824

Line of best fit, 87

Line of fit, 81–86

Lines
horizontal, 65
intersecting, 111
parallel, 70, 77–78, 101, 112
perpendicular, 70–71, 77–78, 101
slope, 68–74, 80, 82, 101–102, 201, 643, 830, 831
vertical, 65

Line segment, midpoint, 845

Loans, amortization, 605

Location Principle, 353, 354

Logarithmic equations, 551, 850
solving, 533, 534, 543, 546
writing, 565, 570

Logarithmic expressions, evaluating, 532

Logarithmic form, 532, 535, 536, 568, 849

Logarithmic functions, 531–540, 532

Logarithmic inequalities, solving, 546

Logarithmic relations, 871

Logarithmic to exponential form, 532

Logarithmic to exponential inequality, 533

Logarithms, 520, 531–540
base *b*, 532
base *e*, 554–559
common, 547–553, 569, 617
functions, 567
inverse property, 533
natural, 554–559, 569
power property, 543
properties, 541–546, 568
using, 548

Logical, *See* Differentiated Instruction

Logical reasoning. *See* Critical Thinking

Lower quartile, 826

M

Major axis, 434

Manipulatives, T5, 4B, 54B, 108B, 152B, 220B, 284B, 344B, 410B, 470B, 520B, 576B, 630B, 698B, 760B

Mapping, 57

Margin of error, 683

Margin of sampling error, 682, 684

Mathematical Background, 4C, 4D, 6, 11, 20, 28, 33, 40, 54C, 54D, 56, 63, 68, 75, 81, 89, 108C, 108D, 110, 129, 138, 152C, 152D, 154, 162, 170, 220C, 220D, 239, 284C, 284D, 286, 301, 306, 313, 329, 344C, 344D, 353, 360, 366, 371, 410C, 410D, 412, 419, 449, 470C, 470D, 485, 499, 505, 520C, 520D, 523, 576C, 576D, 599, 630C, 630D, 676, 698C, 698D, 701, 760C, 760D, 762, 770, 800

Mathematical induction, 618–621, 620, 626

Matrices, 152–217, 865
addition, 160
column, 155, 156
coordinate, 175
determinants of 3 × 3, 183
dimensions, 155, 156, 166, 834
equal, 209
identity, 195–201, 213
inverse, 195–201, 205, 206, 207, 213, 214, 358, 637, 836
modeling real-world data, 161
multiplication, 167–174, 210, 211
different dimensions, 169
scalar, 162
square, 168
operations, 160–166, 210, 834
organizing data, 154

reflection, 177
rotation, 178
row, 155, 156
solving systems of equations, 155–156, 202–208, 214
square, 155, 156, 198
subtraction, 161
transformations, 175–181, 211
translation, 175, 176
zero, 155, 156

Matrix multiplication, Associative Property, 171

Matrix operations, 163
combination, 163
properties, 162

Matrix products, 167

Maximum points, 354–356, 358, 364

Maximum values, 129, 158, 288–289, 290, 291, 293, 337, 377, 663, 839

Mean deviation, 669

Mean, 663, 664, 667, 668, 669, 822–823, 855
arithmetic, 580, 582, 590, 592, 622, 623, 851
geometric, 590, 591, 598, 623, 852

Measurement
angles, 709, 711, 712, 713, 745, 748, 753
conversions, 390, 394
tendency, 664
variation, 665

Measures of central tendency, 664, 822–823

Median, 82, 663, 664, 667, 668, 669, 822–823, 855

Median-fit line, 86

Midline, 771

Midpoint, 414
formula, 412, 414, 416, 417–418, 461–462, 467
line segments, 845

Minimum points, 354–356, 358, 364

Minimum values, 129, 158, 288–289, 290, 291, 293, 337, 377, 663, 839

Minor axis, 434

Misconceptions, 15, 18, 22, 58, 140, 161, 183, 223, 235, 244, 246, 253, 258, 288, 295, 354, 361, 375, 420, 435, 442, 474, 486, 494, 542, 548, 579, 600, 639, 718, 726

Mixed Problem Solving, 862–875

Mixed Review. *See* Review

Mode, 663, 664, 667, 668, 822–823, 855

Modeling, T6, *See also* Open-Ended Assessment
absolute value, 28
algebra tiles, 308
area diagrams, 651
arithmetic sequences, 580
circular functions, 739
complex numbers, 272
conic sections, 453–454
data, 159
distance formula, 413
distributive property, 13
ellipses, 432
fractals, 611
irrational numbers, 252
location principle, 354
midpoint formula, 412
parabolas, 421
parallel lines, 70
perpendicular line, 71
point discontinuity, 485–487
polynomials, 240
quadratic equations, 295
quadratic functions, 287
radicals, 252
Real-World Data, 103
real-world data, 81–86, 300, 359, 539–540
slope-intercept form, 75
solving inequalities, 36
special sequences, 607
vertical asymptotes, 485–487
vertical line test, 57

Monomials, 222–228, 276–277
denominators, 480
division, 233, 521, 538
least common multiple, 479
multiplication, 521, 538

More About
aerospace, 327, 398
amusement parks, 255, 780
animals, 161
architecture, 291, 503
area codes, 636
astronomy, 225, 459
aviation, 603
ballooning, 731
baseball, 723
basketball, 143, 477, 667
betta fish, 44
bicycling, 483
bridges, 318
building, 243
card games, 642
child care, 38
child development, 357
computers, 529
construction, 579
cryptography, 197
dinosaurs, 737
drawbridges, 748
driving, 713
earthquakes, 537
elections, 190
emergency medicine, 735
Empire State Building, 298
energy, 355
engineering, 311
entrance tests, 648
farming, 525
finance, 99
fireworks, 10
food, 380
food service, 14
football, 331
forestry, 304
genealogy, 595
genetics, 232
guitar, 744
health, 267, 675, 683, 773
home improvement, 23
Internet, 679
investments, 140
job hunting, 43
lighthouses, 729
magnets, 483
math history, 16
meteorology, 31
military, 64
money, 558
movies, 157
museums, 435
music, 111, 262
navigation, 443
nutrition, 94
oceanography, 766
Olympics, 564
optics, 795
Pascal's triangle, 612
population, 114
radio, 584
railroads, 26
recycling, 662
René Descartes, 372
robotics, 721
satellite TV, 422
shopping, 388, 668
skiing, 705
space, 494
space exploration, 124, 376
space science, 249
spelling, 656
sports, 61, 677
star light, 545
submarines, 396
technology, 180
temperature, 394
theater, 351
tourism, 292
track and field, 169, 750
tunnels, 507
veterinary medicine, 131
waves, 803
weather, 165

Index

weight lifting, 259
White House, 439
world cultures, 661

Multiple Choice. *See* Assessment

Multiple events, 640

Multiple Representations, 11, 12, 21, 28, 40, 42, 57, 68, 71, 75, 160, 161, 162, 168, 182, 195, 223, 245, 250, 251, 257, 258, 271, 287, 295, 301, 307, 346, 347, 378, 390, 391, 412, 413, 474, 485, 526, 527, 532, 533, 534, 541, 543, 548, 633, 634, 658, 660, 725, 764

Multiplication, 781, 828
Associative Property, 15, 171
Commutative Property, 15, 32, 166, 170, 828
complex numbers, 272–273
Distributive Property, 170, 171, 228, 828
functions, 384, 403
matrices, 167–174, 168, 210, 211
polynomials, 230, 277
probabilities, 651–657, 689
pure imaginary numbers, 270, 272
radicals, 252
rational expressions, 474, 513
scalar, 162, 163, 211
scientific notation, 225
simplifying expressions, 222–223

Multiplication Property of Equality, 21

Multiplication Property of Inequality, 34, 35

Multiplicative identities, 15, 199

Multiplicative inverses, 13, 14, 15, 16, 32, 153, 199, 828

Multi-step equations, solving, 22, 201

Multi-step inequality, solving, 35

Mutually exclusive events, 658–659, 661, 670, 689, 690, 855

N

Natural base, *e*, 554

Natural base exponential function, 554

Natural base expressions, evaluating, 554

Natural logarithmic equations, solving, 556

Natural logarithmic expressions, evaluating, 555

Natural logarithmic function, 554

Natural logarithmic inequalities, solving, 556

Natural logarithms, 554–559, 569
inverse property, 555

Natural numbers, 11, 17, 32, 48

Naturalist, *See* Differentiated Instruction

NCTM Principles and Standards, T22–T24, 4, 54, 108, 152, 220, 284, 344, 410, 470, 520, 576, 630, 698, 760

Negative angle, 709, 712

Negative base, 258

Negative exponents, 222

Negative measure, 713, 732, 754

Negative numbers, square roots of, 270

Negative zeros, 373, 375, 402, 843

Nodes, 791

Normal distribution, 671–675, 672, 680, 685, 691

Notation
function, 59
internal, 829
intervals, 35, 37, 40, 41, 51
scientific, 225, 226, 227, 836
set-builder, 34, 37, 51, 829
sigma, 585, 595, 601, 602
standard, 225

***n*th root,** 245, 246

***n*th term**
arithmetic sequences, 579, 591, 851
geometric sequences, 852

Null hypothesis, 686

Number line, 44, 46

Numbers
classification, 221
complex, 270–275, 280, 370
irrational, 11, 32
natural, 11, 17, 32, 48
pure imaginary, 270, 272
rational, 5, 11, 32, 48, 471
real, 5, 11–18, 32, 48, 245–249, 278, 814
triangular, 609
whole, 11, 18, 48

Number theory, 15, 295, 297, 298, 304, 510, 866, 872, 873

Numerators, polynomials, 475

O

Oblique triangle, 735

Octagons, perimeter, 26

Octants, 136

Odds, 644, 645–646, 647, 648, 663, 854

One-step equations, solving, 21

One-to-one functions, 57, 392, 524

Online Research, *See also* Internet Connections
career choices, 121, 187, 193, 237, 274, 334, 363, 446, 496, 511, 542, 561, 609, 685, 707
data update, 10, 66, 143, 165, 255, 318, 357, 440, 477, 558, 598, 667, 723, 775

Open Ended, 8, 14, 24, 30, 37, 43, 60, 65, 71, 78, 83, 92, 98, 112, 119, 125, 132, 142, 156, 171, 178, 185, 192, 198, 205, 226, 231, 236, 242, 247, 254, 260, 265, 273, 290, 297, 303, 317, 325, 332, 350, 356, 362, 368, 375, 380, 382, 386, 393, 397, 414, 423, 428, 437, 445, 450, 458, 476, 481, 488, 495, 501, 509, 527, 535, 544, 549, 557, 563, 580, 586, 590, 596, 602, 608, 615, 634, 647, 654, 660, 666, 673, 678, 683, 706, 712, 722, 729, 736, 742, 749, 766, 774, 779, 784, 788, 794, 802

Open-Ended Assessment
Modeling, 18, 32, 67, 74, 95, 115, 127, 174, 188, 244, 249, 299, 319, 352, 389, 394, 416, 452, 498, 504, 530, 565, 592, 650, 663, 708, 732, 745, 768, 790
Speaking, 10, 27, 62, 98, 122, 144, 166, 194, 207, 228, 256, 262, 275, 293, 305, 335, 364, 370, 382, 431, 440, 460, 478, 546, 559, 582, 587, 610, 643, 657, 680, 684, 715, 751, 781, 784, 803
Writing, 39, 46, 80, 86, 135, 158, 181, 201, 232, 238, 267, 312, 328, 358, 377, 399, 425, 448, 484, 490, 511, 538, 551, 598, 604, 617, 637, 670, 675, 724, 738, 776, 797

Open Response. *See* Preparing for Standardized Tests

Open sentences, 20

Operations
arithmetic, 383–384
functions, 383–389, 403
radicals, 252

Or compound inequalities, 41

Ordered arrays, 154

Ordered pairs, 56, 78, 83, 84, 153, 387–388, 522, 831, 844

Ordered triples, 136, 139, 833

Ordering real numbers, 814

Order of operations, 6–7

Outcomes, 632, 854

Outliers, 83, 827

P

Pacing, T20–21, 4A, 54A, 108A, 152A, 220A, 284A, 344A, 410A, 470A, 520A, 576A, 630A, 698A, 760A

Parabolas, 419–425, 450, 451, 460, 462–463, 467, 565, 617, 637
equations, 419–420, 841, 845, 846
graphing, 420–423

Parallel lines, 70, 77–78, 101, 112

Parallelograms, 192
area, 477
vertices, 121, 192

Parent graph, 70

Partial sum, 599

Pascal's triangle, 612, 613, 625–626, 872

Patterns, 352

Perfect square trinomials, 310, 816, 840

Perimeter
octagons, 26
quadrilaterals, 415, 482
rectangles, 255
right triangles, 382
squares, 603
triangles, 592

Period, 741, 762, 764, 765, 767, 771, 774, 775, 781, 785, 805, 806, 859

Periodic functions, 741, 742, 743, 762

Permutations, 638–643, 650, 688, 715
circular, 642
linear, 638
repetition, 639

Perpendicular lines, 70–71, 77–78, 101

Phase shift, 769, 770, 774, 785, 806, 859

Piecewise functions, 90–91, 92, 104, 115, 370, 831

Planning Guide, *See* Pacing

Plots
box-and-whisker, 631, 826–827
stem-and-leaf, 667, 825

Point discontinuity, 485–487

Point-slope form, 76, 78, 102

Polygonal region, vertices, 124–125, 126

Polygons
area, 187
finding areas, 187

Polynomial equations
simplifying, 837
solving using quadratic techniques, 360–364, 401

Polynomial functions, 344–407, 400, 868
end behavior, 349
evaluating, 347
even-degree, 349, 357
graphing, 348–349, 353–358, 401
odd-degree, 349, 357
zero, 371

Polynomials, 229–232, 866
addition, 229, 277
degrees, 229, 346, 350, 400, 837, 842
denominators, 475, 480
depressed, 366
division, 233, 277, 364, 365–366
factoring, 239–241, 278, 358, 366, 377, 761, 815–816, 837
least common multiple, 479, 480, 482, 504, 847
multiplication, 230, 277, 285
numerator, 475
one variable, 346, 350
operations, 382
simplifying, 244
subtraction, 229, 277

Portfolios, 51, 105, 149, 215, 281, 341, 405, 467, 517, 571, 627, 693, 757, 809

Positive angle, 709, 712

Positive integers, 620, 853

Positive measure, 713, 732, 754

Positive zeros, 373, 375, 402, 843

Power function, 347, 853

Power Property of Logarithms, 543

Powers, 5, 222
expanding, 615, 617, 621
simplifying expressions, 224

Practice Chapter Test. *See* Assessment

Practice Quiz. *See* Assessment

Prediction equations, 81–82, 83, 84, 95, 99, 598

Preimage, 175

Preparing for Standardized Tests, 877–892
Constructed Response, 884
Free Response, 884
Grid In, 880
Gridded Response, 880–883
Multiple Choice, 878, 879
Open Response, 884–887
Selected Response, 884–887
Student-Produced Questions, 884
Student-Produced Response, 880
Test-Taking Tips, 877, 879, 883, 887, 891

Prerequisite Skills. *See also* Assessment
T3, T5, T8, T15
bar and line graphs, 824
box-and-whisker plots, 826–827
comparing and ordering real numbers, 814
congruent and similar figures, 817–819
factoring polynomials, 815–816
Getting Ready for the Next Lesson, 10, 18, 27, 32, 39, 62, 67, 74, 80, 86, 95, 115, 122, 127, 135, 158, 166, 174, 181, 188, 194, 201, 228, 232, 238, 244, 249, 256, 262, 267, 293, 299, 305, 312, 319, 328, 352, 358, 364, 370, 377, 382, 389, 394, 416, 425, 431, 440, 448, 452, 478, 484, 490, 498, 504, 530, 538, 546, 551, 559, 582, 587, 592, 598, 604, 610, 617, 637, 643, 657, 663, 670, 675, 680, 708, 715, 724, 732, 738, 745, 768, 776, 781, 785, 790, 797
Getting Started, 5, 53, 55, 109, 153, 221, 285, 345, 411, 471, 521, 577, 631, 699, 761
mean, median, and mode, 822–823
Pythagorean Theorem, 820–821
stem-and-leaf plots, 825

Prime, 239, 242

Principal root, 246

Principal values, 746

Probability, 644–650, 655, 660, 663, 670, 688–689, 708, 732, 768, 785, 854, 855, 856, 873
addition, 658–663, 689–690
combinations, 645
conditional, 653
distribution, 646
events, 647, 648
dependent, 633–634, 634, 635, 653, 654, 655, 687, 689, 854, 855
inclusive, 659, 660, 661, 670, 689, 690, 855
independent, 632–633, 634, 651, 652, 654, 687, 689, 854, 855
mutually exclusive, 658–659, 661, 670, 689, 690, 855
experimental, 649
failure, 644
multiplication, 651–657, 689
odds, 644, 645–646, 647, 648, 663, 854
simple, 631
success, 644
theoretical, 649

Problem-based learning. *See* WebQuest, Applications.

Problem solving, 854
distributive property, 14
inverses, 197

Index

matrix equation, 203
mixed, 862–875
right triangles, 703
translations, 773

Product of powers, 223

Product Property, 542
Logarithms, 541–542
Radicals, 250

Professional Development, *See* Staff Development

Program Validation, T18–T19

Project CRISS, T15, 4F, 108F, 220F, 344F, 470F, 576F, 698F

Proof, 618–621, 626

Properties of Equality, 21, 23, 566, 781
Logarithmic Functions, 567

Properties of Inequality, 566
Logarithmic Functions, 567

Properties of Logarithms, solving equations using, 543

Properties of Matrix Multiplication, 171

Properties of Order, 33

Properties of Powers, 224, 226, 526

Proportional sides, 817

Proportions, 181
solving, 471, 490

Pure imaginary numbers, 270
multiplication, 270, 272

Pyramid, surface area, 27

Pythagoras, 16

Pythagorean identities, 777, 779

Pythagorean Theorem, 699, 720, 820–821

Q

Quadrantal angle, 718

Quadrants, 56, 720

Quadratic equations, 328, 604, 841
solving, 761
by completing the square, 306–312, 328, 338, 352, 411, 490, 587, 840
by factoring, 301–305, 338, 840
by graphing, 294–299, 337, 345, 352
for variables, 389

Quadratic form, 360, 363, 370, 842

Quadratic Formula, 345, 370, 460, 841
discriminant, 313–319, 339

Quadratic functions, 286, 499, 502, 503, 515, 839, 848, 867
graphing, 286–293, 322–328, 336–337, 339–340

Quadratic identities, 375

Quadratic inequalities, 839, 867
graphing, 329–333, 340
solving, 329–333, 340

Quadratic-quadratic system, 456–457

Quadratic solutions, 271

Quadratic systems, solving, 455–460, 466

Quadratic techniques, 401
solving polynomial equations using, 360–364

Quadrilaterals, perimeter, 415, 482

Quartile, 826
lower, 826
upper, 826

Quotient identities, 777

Quotient of Powers, 223

Quotient Property
logarithms, 542
radicals, 251

Quotients, 328, 364
simplifying, 242, 251
trinomials, 242

R

Radian measure, 710, 711, 713
conversion, 711

Radians, 710, 713, 724, 749, 753, 757, 802, 803, 808, 857
measuring, 711, 712

Radical equations, 263–269, 280
solving, 263, 362

Radical exponents, 279

Radical expressions, 250–256, 255, 279, 285

Radical form, 257, 838

Radical inequalities, solving, 264–265

Radicals
addition, 252, 253
approximating, 247
multiplication, 252
simplifying, 245
subtraction, 253

Radius, 426

Random, 645

Random sample, 682, 856

Random variable, 646

Range, 56, 57, 58, 61, 93, 94, 95, 99, 101, 104, 181, 397, 398, 416, 523, 527, 528, 530, 663, 823, 830, 831, 844, 849

Rate of change, 69, 560

Rate of decay, 560

Rate of growth, 562

Rate problem, 507

Ratio
common, 588, 603
finding term given, 589

Rational equations, 505–509
solving, 505–509, 516

Rational exponents, 257–262, 838
solving equations, 361–362

Rational expressions, 472, 870
addition, 480, 514
division, 474, 513
multiplication, 474, 513
simplifying, 472–475
subtraction, 480, 514

Rational functions, 500, 502, 504, 515
graphing, 485–490, 514, 848

Rational inequalities, solving, 505–509, 508–509, 516

Rationalizing denominators, 251, 253, 715

Rational numbers, 11, 32, 48
operations, 5
solving equations, 471

Rational zeros, 379, 381, 394, 403, 675, 843

Rational Zero Theorem, 378–382, 403

Reading and Writing, 5, 53, 109, 153, 221, 285, 345, 411, 471, 521, 577, 631, 699, 761

Reading and Writing Mathematics, T4, T6–T7, T14, 4F, 4, 9, 17, 26, 29, 31, 38, 39, 45, 46, 47, 54F, 54, 61, 66, 73, 79, 80, 85, 86, 92, 94, 99, 100, 108F, 108, 114, 121, 125, 126, 134, 135, 141, 143, 145, 152F, 152, 157, 158, 162, 165, 173, 180, 181, 187, 193, 200, 201, 206, 209, 220F, 220, 227, 231, 232, 237, 238, 243, 248, 255, 261, 266, 267, 271, 274, 276, 284F, 284, 292, 296, 298, 304, 311, 312, 318, 327, 328, 334, 336, 344F, 344, 351, 356, 357, 358, 363, 369, 376, 377, 381, 388, 393, 398, 399, 400, 410F, 410, 415, 424, 425, 430, 439, 447, 448, 451, 459, 461, 470F, 470, 477, 483, 484, 489, 490, 496, 503, 510, 511, 513, 520F, 520, 529, 537, 538, 545, 550, 551, 558, 564, 566, 576F, 576, 581, 586, 591, 597, 598, 603, 604, 609, 615, 616, 617, 621, 622, 630F, 630, 636, 637, 642, 649, 656, 662, 669, 670, 674, 675, 679, 683, 685, 687, 698F,

698, 707, 714, 723, 724, 731, 735, 737, 738, 744, 750, 752, 760F, 760, 767, 775, 776, 780, 785, 789, 796, 797, 803, 805

Reading Math, 11, 12, 56, 59, 71, 82, 154, 175, 182, 229, 252, 270, 271, 272, 273, 306, 313, 316, 323, 442, 449, 606, 619, 633, 638, 644, 646, 665, 669, 709, 711, 718, 740, 786, 788

Real numbers, 11–18, 32
comparing and ordering, 5, 814
Identify Properties, 13
properties, 48
roots, 245–249, 278

Real-world applications. *See* Applications; More About

Real-world data, modeling, 81–86, 103

Reciprocal identities, 777

Rectangles
area, 255, 334
golden, 311
perimeter, 255
volumes, 866
width, 242

Rectangular prisms
surface areas, 18
volumes, 367
width, 363

Rectangular solid, volumes, 379, 380

Recursion, 606–611, 625

Recursive formula, 606, 607, 608

Reference angles, 718–719, 722, 776
finding trigonometric value, 720

Reflection, 177

Reflection matrices, 177

Reflexive Property of Equality, 21

Regression equation, 87

Regression line, 87

Relations, 56, 100–102

Relative-frequency histogram, 646, 647

Relative maximum, 354, 356, 842

Relative minimum, 354, 356, 842

Remainder Theorem, 365–370, 402

Repeating decimals, as fractions, 601, 602, 603, 852

Repetition, permutation, 639

Replacement sets, 377

Research, 85, 133, 200, 227, 311, 398, 415, 497, 529, 545, 592, 636. *See also* Online Research

Research on Program, T18–T19

Residuals, 540

Resources for Teachers, T4–T11, T13–T17, 4B, 4E, 4F, 54B, 54E, 54F, 108B, 108E, 108F, 152B, 152E, 152F, 220B, 220E, 220F, 284B, 284E, 284F, 344B, 344E, 344F, 410B, 410E, 410F, 470B, 470E, 470F, 520B, 520E, 520F, 576B, 576E, 576F, 630B, 630E, 630F, 698B, 698E, 698F, 760B, 760E, 760F

Review, T10
Lesson-by-Lesson, 47–50, 100–104, 145–148, 209–214, 276–280, 336–340, 400–404, 461–466, 513, 566–570, 622–626, 687–692, 752–756, 805–808
Mixed Review, T3, 18, 27, 32, 46, 62, 67, 74, 80, 86, 95, 99, 115, 122, 127, 135, 144, 158, 166, 174, 181, 188, 194, 201, 207, 228, 232, 238, 244, 249, 256, 262, 267, 275, 293, 299, 305, 312, 319, 328, 335, 352, 358, 364, 370, 377, 382, 389, 394, 399, 416, 425, 431, 440, 447, 452, 460, 478, 484, 490, 498, 504, 511, 530, 538, 546, 551, 559, 565, 582, 587, 592, 598, 604, 610, 617, 621, 637, 643, 650, 657, 663, 670, 675, 680, 685, 708, 714, 724, 732, 738, 745, 751, 768, 776, 781, 785, 790, 797, 804

Right triangles, 700, 704
perimeter, 382

Right triangle trigonometry, 701–708, 752

Roots, 296, 371–377, 376, 840, 843
complex, 315
double, 302
irrational, 315
*n*th, 245, 246
principal, 246
real number, 245–249, 278
square, 245, 249, 362, 530, 650

Rotation matrices, 178

Rotations, 177, 178

Rounding, 358, 549, 550, 565, 569, 663, 704, 706, 714, 724, 730, 731, 732, 736, 738, 745, 748, 749, 751, 753, 756, 821, 823, 855, 858, 859

Row matrix, 155, 156

Rubric, 53, 107, 151, 217, 283, 343, 407, 469, 519, 573, 629, 695, 759, 811, 884, 888

S

Sample
bias, 682
random, 682, 856
unbiased, 682

Sample space, 632

Sampling, 692

Sampling error, 682–686, 714
margin, 682, 684

Scalar multiplication, 162, 163, 211
Associative Property, 171

Scatter plots, 81–86, 87, 95, 99, 103, 598, 831

Scientific notation, 225, 226, 227, 836

Secant, 701, 708

Second-order determinant, 182

Sector, 713

Selected Response. *See* Preparing for Standardized Tests

Sequences, 578, 872
arithmetic, 578–582, 583, 622–623, 768, 851
Fibonacci, 606, 609, 610
geometric, 588–593, 623–624, 852

Series, 583, 872
arithmetic, 583–587, 623, 851
geometric, 594–598, 624, 781
infinite geometric, 599–605, 624–625

Set-builder notation, 34, 37, 51, 829

Sets, 18, 828
empty, 29
replacement, 377
solution, 37, 41, 44, 46, 95, 829

Short Response, 546, 559, 564, 724, 732, 745. *See also* Assessment *and* Preparing for Standardized Tests

Sides, 734
initial, 709
proportional, 817
terminal, 709

Sigma notation, 585, 595, 601, 602
evaluating sum, 585, 595
infinite series, 601

Similar figures, 817–819

Simple event, 658

Simple probability, 631

Simplify Powers of *i*, 270, 272

Simulations, 681

Sin^{-1}, 747

Sine function, 701, 706, 707, 747, 767, 770, 771
definition, 739
finding, 740
value, 747

Skewed distributions, 671

Slope-intercept form, 75, 78, 79, 86, 102, 188, 637, 831

Slope of line, 68–74, 80, 82, 101–102, 201, 643, 830, 831

Solid boundary, 97

Solution, 20, 801

Solution set, 37, 41, 44, 46, 95, 829

Speaking, See Open-Ended Assessment

Special angles, 703

Special functions, 89–95, 104

Special sequences, 606–611

Special values, 533

Spheres, surface areas, 862

Spreadsheet Investigation
amortizing loans, 605
organizing data, 159
special right triangles, 700

Square matrix, 155, 156, 198

Square root, 245, 249, 362, 530, 650
approximate, 247
negative numbers, 270

Square root functions, 395–396, 398, 399, 404, 500, 502, 503, 515, 848

Square root inequalities, 397–399, 404
graphing, 404

Square Root Property, 250–251, 306, 310, 313, 790

Squares, perimeter, 603

Staff Development, T2–T24, 4C, 4D, 54C, 54D, 108C, 108D, 152C, 152D, 220C, 220D, 284C, 284D, 344C, 344D, 410C, 410D, 470C, 470D, 520C, 520D, 576C, 576D, 630C, 630D, 698C, 698D, 760C, 760D

Standard deviation, 665, 666, 667, 669, 670, 675, 685, 690, 855

Standard form, 64, 101, 122, 423, 424, 428, 449, 460, 478, 830, 845, 846

Standardized Test Practice. See Assessment

Standard notation, 225

Standard position, 709

Statistics, 664–670, 690, 873

Stem, 667

Stem-and-leaf plots, 667, 825

Step functions, 89–90, 92, 115, 158, 370, 831

Student-Produced Questions. See Preparing for Standardized Tests

Student-Produced Response. See Preparing for Standardized Tests

Student Works, T7, 4A, 54A, 108A, 152A, 220A, 284A, 344A, 410A, 470A, 520A, 576A, 630A, 698A, 760A

Study Notebook, T6, 8, 15, 19, 24, 30, 37, 43, 60, 65, 71, 78, 83, 93, 97, 113, 119, 125, 132, 136, 142, 156, 163, 171, 178, 185, 192, 198, 205, 226, 230, 236, 242, 247, 254, 260, 265, 273, 290, 297, 303, 310, 317, 325, 332, 350, 356, 362, 368, 375, 380, 386, 392, 397, 414, 418, 423, 429, 432, 437, 445, 450, 454, 458, 476, 481, 488, 495, 501, 509, 522, 527, 535, 544, 549, 557, 563, 580, 585, 590, 596, 602, 605, 608, 611, 615, 635, 641, 647, 654, 660, 667, 673, 678, 681, 684, 686, 706, 713, 716, 722, 730, 735, 743, 749, 766, 774, 779, 783, 788, 794, 802

Study Organizer. See Foldables™ Study Organizers

Study Tips, T6
absolute value, 90, 599
absolute value inequalities, 42
additive identity, 162
A is acute, 728
algebra tiles, 240
alternative method, 77, 264, 474, 580, 590, 652, 728, 734
alternative representations, 726
amplitude and period, 764
angle measure, 748
area formula, 184
checking solutions, 110, 265, 481, 543
choosing a committee, 659
choosing the independent variable, 81
choosing the sign, 793
coefficient, 116
combinations, 640
combining functions, 386
common factors, 480
common misconception, 7, 12, 29, 118, 130, 289, 308, 523, 659, 703, 782
conditional probability, 653
continuously compounded interest, 556
coterminal angles, 712
deck of cards, 640
depressed polynomial, 366
Descartes' Rule of Signs, 379
element, 155
elimination, 139
equations with ln, 556
equations with roots, 303
error in measurement, 704
exponential growth and decay, 524
expressing solutions as multiples, 800
extraneous solutions, 506, 534
factor first, 475
factoring, 367
finding zeros, 374
focus of parabola, 419
formula for sum, 600
graphing calculators, 225, 247, 436, 444, 456, 525, 585, 613
graphing polynomial functions, 353
graphing quadratic inequalities, 457
graphing rational functions, 486
graphs of piecewise functions, 92
greatest integer function, 89
horizontal lines, 70
identity matrix, 204
indicated sum, 583
inequality phrases, 36
interval notation, 40, 41
inverse functions, 392
Law of Large Numbers, 682
location of roots, 296
look back, 91, 97, 123, 189, 204, 273, 329, 361, 365, 371, 420, 485, 508, 524, 526, 531, 532, 608, 634, 664, 676, 720, 747, 762, 771, 772
matrix operations, 163
memorize trigonometric ratios, 702
message, 198
midpoints, 412
missing steps, 614
multiplication and division properties of equality, 22
multiplying matrices, 168
negative base, 258
normal distribution, 672
number of zeros, 349
outliers, 83
one real solution, 295
parallel lines, 112
permutations, 640
power function, 347
properties of equality, 21
properties of Inequality, 33
quadratic formula, 314
quadratic solutions, 271
radian measure, 710
rate of change, 560
rationalizing denominator, 251
reading math, 11, 12, 34, 35, 56, 59, 71, 82, 124, 129, 154, 175, 182, 229, 246, 252, 270, 271, 272, 273, 294, 306, 313, 316, 323, 354, 372, 384, 391, 442, 449, 606, 619, 638, 644, 646, 665, 669, 701, 709, 711, 718, 740, 786, 788
sequences, 578
sides and angles, 734
sigma notation, 585
simplified expressions, 224
skewed distributions, 671
slope, 68
slope-intercept form, 75
solutions to inequalities, 35

solving quadratic inequalities algebraically, 332
solving quadratic inequalities by graphing, 330
special values, 533
step 1, 618
substitution, 361
symmetry, 288
technology, 547
terms of geometric sequences, 594
using the discriminant, 316
using logarithms, 548
using quadratic formula, 315
verifying a graph, 770
verifying inverses, 196
vertical and horizontal lines, 65
vertical line test, 58
vertical method, 230
vertices of ellipses, 434
zero at origin, 372
zero product property, 302

Substitution, 21, 146, 149, 153, 504, 781, 828
direct, 366, 368
solving systems of equations, 116, 119, 120, 122, 135, 166, 832
synthetic, 365–366, 368, 369, 377, 402, 715, 751, 843

Substitution Property of Equality, 25

Subtraction
complex numbers, 270, 272
functions, 383, 403
matrices, 161
polynomials, 229, 277
radicals, 253
rational expressions, 480, 514
solving inequality, 34

Subtraction Property of Equality, 21

Subtraction Property of Inequality, 33

Success, 644
probability, 644

Sum and difference formulas, 786–787

Summation formula, 618

Sums, 657, 787
arithmetic series, 583, 584, 586, 587, 592, 598
geometric series, 595, 596, 597, 610
infinite geometric series, 602, 610
partial, 599
rewriting differences, 221
series, 585, 663
sigma notation, 585, 595
two cubes, 361

Surface area
cones, 22, 266
cylinders, 25, 862
pyramids, 27
rectangular prisms, 18
spheres, 862

Symbols, infinity, 601

Symmetric Property of Equality, 21, 25, 46, 781

Symmetry, 288, 767

Synthetic division, 234–236, 345, 745, 837

Synthetic substitution, 365–366, 368, 369, 372–373, 377, 402, 551, 715, 751, 843

Systems of equations, 110, 158, 864
consistent, 111, 112, 113, 122, 293
Cramer's Rule in solving, 835
dependent, 111, 112, 113, 122, 293
inconsistent, 111, 112, 113, 122, 293
independent, 111, 112, 113, 122, 293
solving, 166, 188, 203, 657, 724
algebraically, 116–122, 146
elimination, 153, 832
graphing, 110–115, 122, 145, 146, 147, 148, 194, 832
matrices, 205, 206, 214
substitution, 832
three variables, 138–144, 148

Systems of inequalities, 663
solving, graphing, 123–127, 135, 144, 147, 158, 484, 833, 847

Systems of quadratic inequalities, 457

Systems of three linear equations, 191

Systems of two linear equations, 189

T

Table of values, 286, 288, 290, 291, 299, 352, 356, 364, 839

Tangent, 706, 707

Tangent function, 427, 701, 747, 770, 771

Tangent ratio, 708

Teacher to Teacher, 41, 69, 86, 118, 196, 234, 314, 372, 427, 452, 497, 500, 525, 589, 641, 719, 765

TeacherWorks, T5

Teaching Tips, 12, 13, 21, 29, 34, 35, 36, 41, 43, 57, 59, 77, 90, 91, 112, 118, 130, 139, 140, 155, 161, 177, 183, 191, 197, 216, 222, 223, 224, 234, 235, 240, 246, 247, 251, 252, 258, 259, 264, 265, 271, 287, 288, 295, 302, 307, 308, 309, 315, 323, 330, 331, 332, 347, 348, 366, 367, 373, 379, 392, 395, 413, 420, 421, 422, 427, 434, 435, 442, 444, 456, 457, 473, 474, 480, 481, 486, 487, 493, 494, 506, 507, 508, 524, 532, 542, 547, 548, 555, 561, 579, 580, 584, 585, 589, 595, 600, 601, 607, 613, 614, 633, 652, 665, 666, 677, 694, 702, 703, 710, 718, 726, 747, 748, 758, 763, 771, 772, 778, 787, 791, 792, 800, 802

Technology, T5, T7, T9, T11–T13, T15–T17

Terminal side, 709

Terms, 229, 578, 615
finding, 578, 579, 588, 589
like, 229
series, 596

Testing hypotheses, 686

Test preparation. *See* Assessment

Test-taking tips. *See* Assessment

Theoretical probability, 649

Third-order determinant, 182, 183

30°-60°-90° triangles, 699, 703, 707

3 × 3 matrices, determinants, 183

Tips for New Teachers
Intervention, 10, 27, 62, 74, 90, 144, 159, 166, 228, 238, 256, 262, 267, 275, 288, 305, 312, 323, 384, 416, 440, 448, 478, 484, 487, 498, 504, 511, 534, 582, 587, 592, 598, 604, 610, 617, 620, 648, 668, 703, 711, 793
Reading Tip, 249

***Towers of Hanoi* game,** 607

Transformations, 175
graphing, 772
matrices, 175–181, 211
verifying, 783

Transitive, 21, 46, 828

Transitive Property of Equality, 25

Translation matrix, 175, 176

Translations, 175
horizontal, 769–770
trigonometric graphs, 769–776
vertical, 771–772

Transverse axis, 442

Trapezoid, area, 8, 67, 865

Triangle Inequality Theorem, 45

Triangles
area, 32, 184, 185, 186, 187, 231, 281, 725, 866
equilateral, 869
45°-45°-90°, 699, 703, 707
isosceles, 869
Pascal's, 612, 625–626, 872
perimeter, 592
right, 382, 700, 704
30°-60°-90°, 699, 703, 707
vertices, 113, 415

Trichotomy Property, 33

Trigonometric identities, 875

Trigonometric equations, 799, 802
solving, 799–804, 800, 801, 808

Trigonometric functions, 698–759, 701, 717, 722, 723, 732, 738, 754, 761, 790, 796, 857, 874
evaluating, 717, 741, 742, 778
general angles, 717–724
graphing, 762–768, 765, 772, 805
inverse, 746–751
solving equations, 724, 732
using, 766
variations, 764

Trigonometric graphs, 875
translations, 769–776, 806

Trigonometric identities, 777, 785, 806
basic, 777
verifying, 782–785, 785, 807

Trigonometric inverses, addition, 751

Trigonometric values, 703, 720, 761
finding, 702, 748, 777

Trigonometry, 701, 875
right triangle, 701–708, 752

Trinomials, 229, 310
perfect square, 816, 840
quotient of two, 242

Tutorials
Alge2PASS, T9, T15, 4F, 54F, 108F, 152F, 220F, 284F, 344F, 410F, 470F, 520F, 576F, 630F, 698F, 760F

2 × 2 matrices, determinants, 182

Two-variable matrix equation, 202

U

Unbiased sample, 682

Unbounded region, 130

Uniform distribution, 646

Union, 41

Unit circle, 710, 739, 742, 743

Univariate data, 664

Unlocking Misconceptions, See Misconceptions

Upper quartile, 826

USA TODAY, T12, Snapshots, 3, 69, 84, 135, 206, 219, 228, 328, 368, 409, 448, 492, 535, 565, 575, 604, 697, 715, 797

V

Values
maximum, 158, 663
minimum, 158, 663

Variables, 7, 25
dependent, 59
functional values, 348
independent, 59
random, 646
solving for, 22, 109, 389
systems of equations, 138–144, 148

Variance, 665, 666, 667, 669, 670, 675, 690

Variations
direct, 496, 559, 650, 848
inverse, 496, 559, 848
joint, 496, 559, 848

Velocity, angular, 714

Venn diagram, 12, 271

Verbal expressions, 828
algebraic expressions, 20, 24, 115

Verbal/Linguistic, See Differentiated Instruction

Vertex form, 322–328, 335

Vertex matrix, 175

Vertical asymptotes, 485–487, 617, 763

Vertical lines, 65

Vertical Line Test, 57, 58

Vertical shift, 771, 774, 775, 781, 806, 859

Vertical translations, 771–772

Vertices, 129, 287–288, 290, 291, 299, 339, 636
angles, 113, 192
coordinates, 846
exact coordinates, 744
parallelograms, 121, 192
polygonal region, 124–125, 126
triangles, 113, 415

Visual/Spatial, See Differentiated Instruction

Vocabulary Builder, T4, 4, 54, 108, 152, 220, 284, 344, 410, 470, 520, 576, 630, 698, 760

Volumes
cubes, 615
rectangular prism, 367, 866
rectangular solid, 379, 380

Von Koch snowflake, 611

W

WebQuest, T4, T6, T13, 3, 27, 120, 192, 207, 219, 227, 328, 369, 399, 409, 430, 502, 529, 565, 575, 616, 635, 685, 697, 708, 775, 804

Whole numbers, 11, 18, 48

Work problem, 507

Writing in Math, 10, 17, 27, 31, 38, 45, 62, 67, 73, 80, 86, 94, 99, 114, 121, 127, 134, 144, 158, 166, 173, 181, 187, 193, 200, 207, 227, 232, 238, 243, 255, 262, 267, 275, 292, 299, 305, 312, 319, 327, 334, 352, 357, 364, 370, 377, 382, 389, 394, 399, 416, 425, 430, 439, 447, 452, 459, 477, 484, 490, 498, 503, 530, 537, 546, 551, 559, 564, 582, 587, 592, 598, 603, 610, 616, 621, 636, 642, 649, 657, 662, 675, 679, 685, 708, 714, 724, 732, 737, 744, 751, 768, 776, 781, 785, 790, 796, 804

Writing in Mathematics
See Reading and Writing Mathematics
See Open-Ended Assessment

X

x-coordinate, 68, 290, 299, 348, 354, 356, 401, 839, 842

x-intercept, 65, 66, 70, 74, 101, 174, 330, 830

Y

y-coordinate, 68

y-intercept, 65, 66, 70, 74, 78, 82, 101, 174, 287–288, 291, 299, 530, 830, 831

Z

Zero matrix, 155, 156

Zero Product Property, 301, 302, 305, 361, 362
solving equations, 797

Zeros, 294, 371–377, 604
function, 294, 348, 349, 354
imaginary, 375, 402, 843
negative, 373, 375, 402
negative real, 843
origin, 372
positive, 373, 375, 402
positive real, 843
rational, 379, 381, 394, 403, 675, 843
synthetic substitution, 373–374